McGraw-Hill
Dictionary of Scientific and Technical Terms

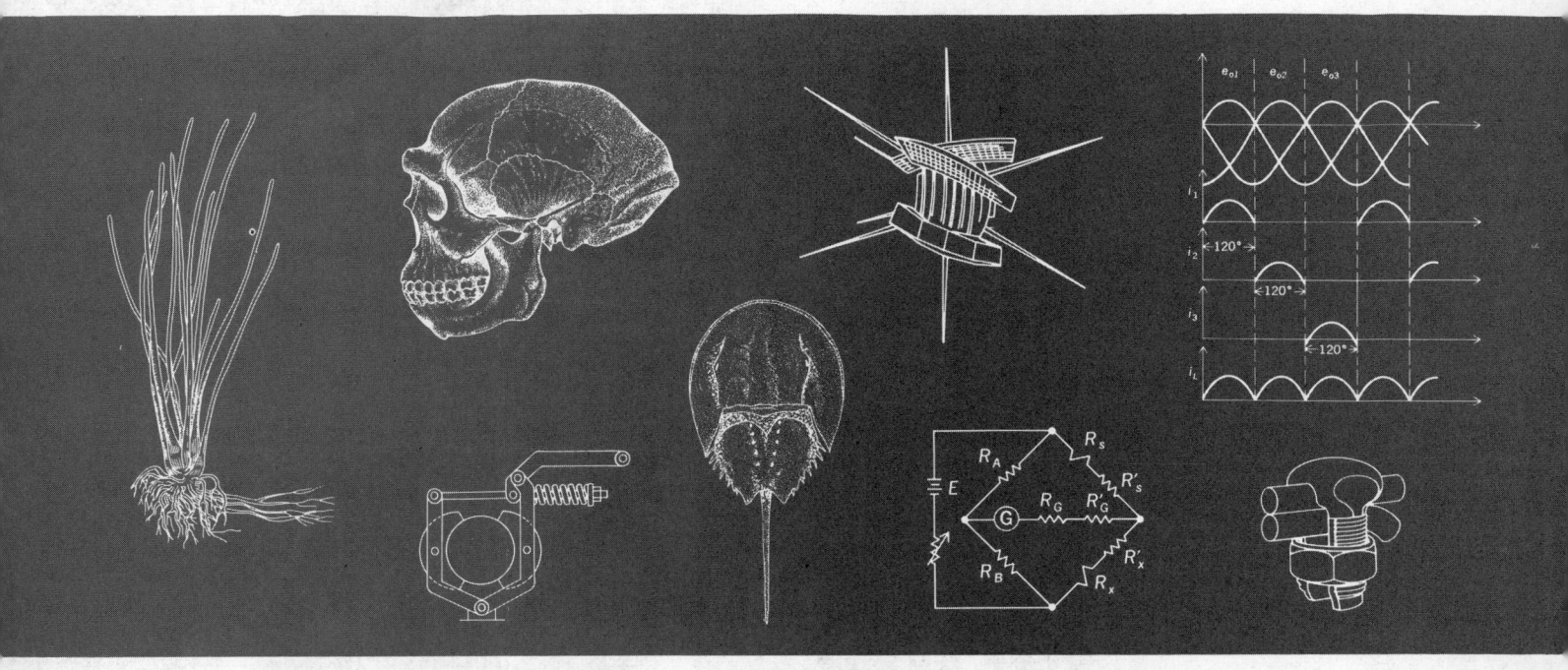

McGraw-Hill
Dictionary of Scientific and Technical Terms

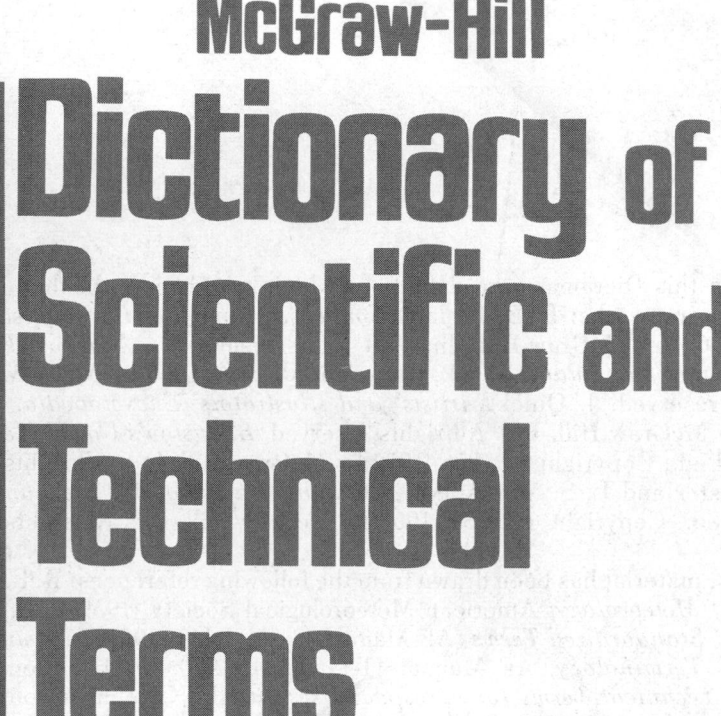

rocker
spring
valve rod
valve
injection nozzle

DANIEL N. LAPEDES Editor in Chief

McGRAW-HILL BOOK COMPANY
New York   St. Louis   San Francisco

Auckland
Düsseldorf
Johannesburg
Kuala Lumpur
London
Mexico
Montreal
New Delhi
Panama
Paris
São Paulo
Singapore
Sydney
Tokyo
Toronto

Included in this Dictionary are definitions which have been published previously in the following works: P. B. Jordain, *Condensed Computer Encyclopedia,* Copyright © 1969 by McGraw-Hill, Inc. All rights reserved. J. Markus, *Electronics and Nucleonics Dictionary,* 3d ed., Copyright © 1960, 1966 by McGraw-Hill, Inc. All rights reserved. J. Quick, *Artists' and Illustrators' Encyclopedia,* Copyright © 1969 by McGraw-Hill, Inc. All rights reserved. *Blakiston's Gould Medical Dictionary,* 3d ed., Copyright © 1956, 1972 by McGraw-Hill, Inc. All rights reserved. T. Baumeister and L. S. Marks, eds., *Standard Handbook for Mechanical Engineers,* 7th ed., Copyright © 1958, 1967 by McGraw-Hill, Inc. All rights reserved.

In addition, material has been drawn from the following references: R. E. Huschke, *Glossary of Meteorology,* American Meteorological Society, 1959; *U.S. Air Force Glossary of Standardized Terms,* AF Manual 11–1, vol. 1, 1972; *Communications-Electronics Terminology,* AF Manual 11-1, vol. 3, 1970; W. H. Allen, ed., *Dictionary of Technical Terms for Aerospace Use,* 1st ed., National Aeronautics and Space Administration, 1965; J. M. Gilliland, *Solar-Terrestrial Physics: A Glossary of Terms and Abbreviations,* Royal Aircraft Establishment Technical Report 67158, 1967; *Glossary of Air Traffic Control Terms,* Federal Aviation Agency; *A Glossary of Range Terminology, White Sands Missile Range, New Mexico,* National Bureau of Standards, AD 467–424; *A DOD Glossary of Mapping, Charting and Geodetic Terms,* 1st ed., Department of Defense, 1967; P. W. Thrush, comp. and ed., *A Dictionary of Mining, Mineral, and Related Terms,* Bureau of Mines, 1968; *Nuclear Terms: A Glossary,* 2d ed., Atomic Energy Commission; F. Casey, ed., *Compilation of Terms in Information Sciences Technology,* Federal Council for Science and Technology, 1970; *Glossary of Stinfo Terminology,* Office of Aerospace Research, U.S. Air Force, 1963; *Naval Dictionary of Electronic, Technical, and Imperative Terms,* Bureau of Naval Personnel, 1962; *ADP Glossary,* Department of the Navy, NAVSO P-3097.

10 9 8 7 6 5 4

Library of Congress Cataloging in Publication Data

McGraw-Hill dictionary of scientific and technical terms.

   1. Science—Dictionaries. 2. Technology—Dictionaries. I. Lapedes, Daniel N., ed. II. Title: Dictionary of scientific and technical terms.
Q123.M15   503   74-16193
ISBN 0-07-045257-1

Contents

Preface

The vocabulary of science and technology has never been adequately represented in the many general dictionaries of the English language. The *McGraw-Hill Dictionary of Scientific and Technical Terms,* by concentrating solely on this vocabulary, aims to fill this gap and to serve scientists and technologists as well as the general public.

Certainly the publication of such a comprehensive scientific dictionary seems long overdue. The growth of science and technology in recent decades has been of astronomical proportions. The fields of nuclear engineering, molecular biology, integrated circuitry, and aerospace are but some of the new disciplines that have evolved. With each new development and as each new discipline has come into being, there has been an expansion of the vocabulary used by workers in these areas to describe the problems in which they have been involved, to communicate inside and outside their fields.

The *McGraw-Hill Dictionary of Scientific and Technical Terms,* with almost 100,000 definitions, is a major compendium of the vocabulary of science and technology. The definitions are written in a clear and simple style that is understandable and yet consistent with the level of complexity of the term being defined. The definitions are amplified by approximately 2800 illustrations.

During the preparation of the *McGraw-Hill Dictionary of Scientific and Technical Terms* the vocabulary of science and technology was surveyed by a select group of consulting editors. They considered all terms and all definitions a term could have. The terms and the definitions included in the Dictionary are those considered important in each consultant's particular area of specialization.

The definitions were written by the staff of the *McGraw-Hill Encyclopedia of Science and Technology* and the contributing editors. Their efforts were supplemented by the existing resources of the McGraw-Hill Book Company, including such published works as the *Condensed Computer Encyclopedia* by Philip Jordain, *Artists' and Illustrators' Encyclopedia* by John Quick, *Electronics and Nucleonics Dictionary* by John Markus, *Blakiston's Gould Medical Dictionary,* and *Standard Handbook for Mechanical Engineers* edited by Theodore Baumeister and Lionel Marks. Other works from which material was drawn are credited on the copyright page. Each definition was reviewed by the consulting editor who selected the term, and by the library consultant.

As each definition was written, it was identified with the field of science or technology in which it is primarily used. An alphabetical list of the abbreviations for these fields with an explanation of the scope of each begins on page xi. The fields on the list include some that are quite specialized, such as crystallography [CRYSTAL], atomic physics [ATOM PHYS], and nuclear physics [NUC PHYS], and others that are less specialized, such as physics [PHYS] and science and technology [SCI TECH]. A definition that is identified with more than one specialized field is placed in a more general field. For example, a definition that is used in both invertebrate zoology [INV ZOO] and vertebrate zoology [VERT ZOO] is assigned to the field of zoology [ZOO].

The Dictionary was typeset by the latest technique in graphic arts—computer composition—to provide a method for updating it in future printings. The contents of the Dictionary are stored on magnetic tape masters, which can be readily modified and used to generate the pages of type for subsequent editions. Instead of being "frozen" in content, therefore, this Dictionary will accommodate new terms and definitions as they evolve in the various disciplines. The computer was also an aid in ensuring that technical and scientific terms used in the text of a definition were defined elsewhere in the Dictionary.

The emphasis in the Dictionary has been placed on providing definitions for terms, rather than on pronunciation, word derivation, and syllabication. Economies derived from this approach have permitted the Dictionary to be published at a more moderate price.

In preparation for the worldwide use of the International System (SI) of units, metric equivalents for U.S. Customary measurement units presently in use are provided in each definition in which units appear. The Appendix has a full explanation of the International System with conversion tables for units. It also has lists of the chemical elements and elementary particles, abbreviations for scientific and technical organizations, and symbols used in science and technology, including those used in circuit diagrams.

The reexamination that the taxonomy of microorganisms—the bacteria and the rickettsiae in particular—has undergone since 1958 was published in the eighth edition of *Bergey's Manual of Determinative Bacteriology* when this Dictionary was on press. Because the editors have not been able to incorporate the new information in definitions relating to these microorganisms, the editors have included in the Appendix the taxonomy from the seventh edition of *Bergey's Manual,* used in this Dictionary, and have also included the taxonomy from the eighth edition for comparison.

An explanation of how to use the Dictionary, describing alphabetization, cross-referencing, etc., is on page x.

The *McGraw-Hill Dictionary of Scientific and Technical Terms* is the culmination of the ideas and efforts of many people—the editorial staff, the contributing editors, and the consulting editors. In producing this reference tool, they hope that it will facilitate communication of ideas throughout science and technology.

DANIEL N. LAPEDES
Editor in Chief

Editorial Staff

Daniel N. Lapedes, Editor in Chief

Sybil Parker, Senior Editor
Marvin Yelles, Senior Editor
Jonathan Weil, Editor

Edward J. Fox, Art Director
Richard A. Roth, Art Editor
Ann D. Bonardi, Art Coordinator

Joe Faulk, Editing Manager
Catherine Engel, Copy Editor
Patricia Walsh, Editing Assistant
William Blaszczak, Editing Assistant

Consulting Editors

Brig. Gen. Peter C. Sandretto — Formerly, Director, Engineering Management, International Telephone and Telegraph Corporation. NAVIGATION.

Prof. Frederick Schwab — Department of Geology, Washington and Lee University. GEOLOGY; PHYSICAL GEOGRAPHY.

Dr. Raymond Siever — Department of Geological Sciences, Harvard University. GEOLOGY; PHYSICAL GEOGRAPHY.

Dr. W. R. Sistrom — Department of Biology, University of Oregon. MICROBIOLOGY.

Dr. Leonard Spero — Walter Reed Hospital Unit, Fort Dietrick, MD. CHEMISTRY.

Dr. C. N. Touart — Senior Scientist, Air Force Cambridge Research Laboratory. GEOCHEMISTRY; GEOPHYSICS; METEOROLOGY.

Prof. H. H. Uhlig — Department of Metallurgy and Materials Science, Massachusetts Institute of Technology. METALLURGICAL ENGINEERING.

Dr. Joachim Weindling — Professor of System Engineering and Operations Research, Polytechnic Institute of Brooklyn. INDUSTRIAL AND PRODUCTION ENGINEERING.

Prof. George S. Bonn — Graduate School of Library Science, University of Illinois. LIBRARY CONSULTANT.

Contributing Editors

Prof. Theodore Baumeister — Consulting Engineer; Stevens Professor of Mechanical Engineering, Emeritus, Columbia University. MECHANICAL POWER ENGINEERING.

Robert L. Davidson — Editor, "Chemical Engineering." CHEMICAL ENGINEERING; CHEMISTRY; PETROLEUM ENGINEERING.

Philip B. Jordain — Senior Research Officer, First National City Bank of New York. COMPUTERS.

John Markus — Author and Consultant. ELECTRONICS.

Dr. Nathaniel Martin — Department of Mathematics, University of Virginia. MATHEMATICS.

Dr. John Quick — Arthur D. Little, Inc., Cambridge, MA. ARMAMENTS: GRAPHIC ARTS.

Dr. Aaron Strauss — Department of Mathematics, University of Maryland. MATHEMATICS.

How to Use the Dictionary

I. ALPHABETIZATION

The terms in the *McGraw-Hill Dictionary of Scientific and Technical Terms* are alphabetized on a letter-by-letter basis; word spacing, hyphen, comma, solidus, and apostrophe in a term are ignored in the sequencing. For example, an ordering of terms would be:

air-earth current
air ejector
air field
air filter
AKF diagram

Also ignored in the sequencing of terms (usually, chemistry terms) are italic elements, numbers, small capitals, and Greek letters. For example, the following terms appear within alphabet letter "A":

D-**aminobenzyl penicillin**
2-**aminobutane**
α-**amino-hydrocinnamic acid**
ortho-**aminophenol**

II. FORMAT

The basic format for a defining entry provides the term in boldface, the field in small capitals, and the single definition in lightface:

term [FIELD] Definition.

A field may be followed by multiple definitions, each introduced by a boldface number:

term [FIELD] **1.** Definition. **2.** Definition. **3.** Definition.

A term may have definitions in two or more fields:

term [BOT] Definition. [GEOL] Definition.

A simple cross-reference entry appears as:

term *See* another term.

A cross-reference may also appear in combination with definitions:

term [BOT] Definition. [GEOL] *See* another term.

III. CROSS-REFERENCING

A cross-reference entry directs the user to the defining entry. For example, the user looking up "avalanche diode" finds:

avalanche diode *See* Zener diode.

The user then turns to the "Z" terms for the definition.

Cross-references are also made from variant spellings, acronyms, abbreviations, and symbols.

aesthacyte *See* esthacyte.
ASROC *See* antisubmarine rocket.
at wt *See* atomic weight.
Au *See* gold.

The user should observe that an element ignored in alphabetizing may appear in a cross-reference entry. For example, the following directs the user to a term in alphabet letter "V," not in "N":

amyl aldehyde *See n*-valeraldehyde.

IV. ALSO KNOWN AS . . . , etc.

A definition may conclude with a mention of a synonym of the term, a variant spelling, an abbreviation for the term, or other such information, introduced by "Also known as . . . ," "Also spelled . . . ," "Abbreviated . . . ," "Symbolized . . . ," "Derived from . . . ." When a term has more than one definition, the positioning of any of these phrases conveys the extent of applicability. For example:

term [BOT] **1.** Definition. Also known as synonym. **2.** Definition. Symbolized T.

In the above arrangement, "Also known as . . ." applies only to the first definition: "Symbolized . . ." applies only to the second definition.

term [BOT] **1.** Definition. **2.** Definition. [GEOL] Definition. Also known as synonym.

In the above arrangement "Also known as . . ." applies only to the second field.

term [BOT] Also known as synonym. **1.** Definition. **2.** Definition. [GEOL] Definition.

In the above arrangement, "Also known as . . ." applies to both definitions in the first field.

term Also known as synonym. [BOT] **1.** Definition. **2.** Definition. [GEOL] Definition.

In the above arrangement, "Also known as . . ." applies to all definitions in both fields.

V. CHEMICAL FORMULAS

Chemistry definitions may include either an empirical formula (say, for acetaldehyde, C_2H_4O) or a line formula (for acrylic acid, $CH_2CHCOOH$), whichever is appropriate.

Field Abbreviations

ACOUS	acoustics	INV ZOO	invertebrate zoology
ADP	automatic data processing	LAP	lapidary
AERO ENG	aerospace engineering	MAP	mapping
AGR	agriculture	MATER	materials
ANALY CHEM	analytical chemistry	MATH	mathematics
ANAT	anatomy	MECH	mechanics
ANTHRO	anthropology	MECH ENG	mechanical engineering
ARCH	architecture	MED	medicine
ARCHEO	archeology	MET	metallurgy
ASTRON	astronomy	METEOROL	meteorology
ASTROPHYS	astrophysics	MICROBIO	microbiology
ATOM PHYS	atomic physics	MIN ENG	mining engineering
BIOCHEM	biochemistry	MINERAL	mineralogy
BIOL	biology	MOL BIO	molecular biology
BIOPHYS	biophysics	MYCOL	mycology
BOT	botany	NAV	navigation
BUILD	building construction	NAV ARCH	naval architecture
CHEM	chemistry	NUCLEO	nucleonics
CHEM ENG	chemical engineering	NUC PHYS	nuclear physics
CIV ENG	civil engineering	OCEANOGR	oceanography
CLIMATOL	climatology	OPTICS	optics
COMMUN	communications	ORD	ordnance
CONT SYS	control systems	ORG CHEM	organic chemistry
CRYO	cryogenics	PALEOBOT	paleobotany
CRYSTAL	crystallography	PALEON	paleontology
CYTOL	cytology	PARTIC PHYS	particle physics
DES ENG	design engineering	PATH	pathology
ECOL	ecology	PETR	petrology
ELEC	electricity	PETRO ENG	petroleum engineering
ELECTR	electronics	PHARM	pharmacology
ELECTROMAG	electromagnetism	PHYS	physics
EMBRYO	embryology	PHYS CHEM	physical chemistry
ENG	engineering	PHYSIO	physiology
ENG ACOUS	engineering acoustics	PL PATH	plant pathology
EVOL	evolution	PL PHYS	plasma physics
FL MECH	fluid mechanics	PSYCH	psychology
FOOD ENG	food engineering	QUANT MECH	quantum mechanics
FOR	forestry	RELAT	relativity
GEN	genetics	SCI TECH	science and technology
GEOCHEM	geochemistry	SOLID STATE	solid-state physics
GEOD	geodesy	SPECT	spectroscopy
GEOGR	geography	STAT	statistics
GEOL	geology	STAT MECH	statistical mechanics
GEOPHYS	geophysics	SYS ENG	systems engineering
GRAPHICS	graphic arts	SYST	systematics
HISTOL	histology	TEXT	textiles
HOROL	horology	THERMO	thermodynamics
HYD	hydrology	VERT ZOO	vertebrate zoology
IMMUNOL	immunology	VET MED	veterinary medicine
IND ENG	industrial engineering	VIROL	virology
INORG CHEM	inorganic chemistry	ZOO	zoology

Scope of Fields

acoustics—The science of the production, transmission, and effects of sound.

aerospace engineering—Engineering pertaining to the design and construction of aircraft and space vehicles and of power units, and dealing with the special problems of flight in both the earth's atmosphere and space, such as in the flight of air vehicles and the launching, guidance, and control of missiles, earth satellites, and space vehicles and probes.

agriculture—The production of plants and animals useful to humans, involving soil cultivation and the breeding and management of crops and livestock.

analytical chemistry—Science and art of determining composition of materials in terms of elements and compounds which they contain.

anatomy—The branch of morphology concerned with the gross and microscopic structure of animals, especially humans.

anthropology—The study of the interrelations of biological, cultural, geographical, and historical aspects of the human race.

archeology—The scientific study of the material remains of the cultures of historical and prehistorical peoples.

architecture—The art or practice of designing structures, especially habitable structures in accordance with principles determined by esthetic and practical or material considerations.

astronomy—The science concerned with celestial bodies and with the observation and interpretation of radiation received in the vicinity of earth from the component parts of the universe.

astrophysics—A branch of astronomy that treats of the physical properties of celestial bodies, such as luminosity, size, mass, density, temperature, and chemical composition, and their origin and evolution.

atomic physics—A branch of physics concerned with the structures of the atom, the characteristics of the electrons and other elementary particles of which the atom is composed, the arrangement of the atom's energy states, and the processes involved in the radiation of light and x-rays.

automatic data processing—The machine performance, with little or no human assistance, of any of a variety of tasks involving informational data; examples include automatic and responsive reading, computation, writing, speaking, directing artillery, and running of an entire factory.

biochemistry—The study of the chemical substances that occur in living organisms, the processes by which these substances enter into or are formed in the organisms and react with each other and the environment, and the methods by which the substances and processes are identified, characterized, and measured.

biology—The science of living organisms, concerned with the study of embryology, anatomy, physiology, cytology, morphology, taxonomy, genetics, evolution, and ecology.

biophysics—The hybrid science involving the methods and ideas of physics and chemistry to study and explain the structures of living organisms and the mechanics of life processes.

botany—That branch of biological science which embraces the study of plants and plant life, including algae; deals with taxonomy, morphology, physiology, and other aspects.

building construction—The art of business of assembling materials into a structure, especially one designated for occupancy.

chemical engineering—A branch of engineering that deals with the development and application of manufacturing processes, such as refinery processes, which chemically convert raw materials into a variety of products, and that deals with the design and operation of plants and equipment to perform such work.

chemistry—The scientific study of the properties, composition, and structure of matter, the changes in structure and composition of matter, and accompanying energy changes.

civil engineering—The planning, design, construction, and maintenance of fixed structures and ground facilities for industry, for transportation, for use and control of water, for occupancy, and for harbor facilities.

climatology—That branch of meteorology concerned with the mean physical state of the atmosphere together with its statistical variations in both space and time as reflected in the weather behavior over a period of many years.

communications—The science and technology by which information is collected from an originating source, transformed into electric currents or fields, transmitted over electrical networks or space to another point, and reconverted into a form suitable for interpretation by a receiver.

control systems—The study of those systems in which one or more outputs are forced to change in a desired manner as time progresses.

cryogenics—The science of producing and maintaining very low temperatures, of phenomena at those temperatures, and of technical operations performed at very low temperatures.

crystallography—The branch of science that deals with the geometric description of crystals, their internal arrangement, and their properties.

cytology—The branch of biological science which deals with the structure, behavior, growth, and reproduction of cells and the function and chemistry of cells and cell components.

design engineering—A branch of engineering concerned with the design of a product or facility according to generally accepted uniform standards and procedures, such as the specification of a linear dimension, or a manufacturing practice, such as the consistent use of a particular size of screw to fasten covers.

ecology—The study of the interrelationships between organisms and their environment.

electricity—The science of physical phenomena involving electric charges and their effects when at rest and when in motion.

electromagnetism—The branch of physics dealing with the observations and laws relating electricity to magnetism, and with magnetism produced by an electric current.

electronics—The branch of science and technology relating to the conduction of electricity through gases or vacuum or through semiconducting materials; concerned with the design, manufacture, and application of electron tubes.

embryology—The study of the development of the organism from the zygote, or fertilized egg.

engineering—The science by which the properties of matter and the sources of power in nature are made useful to man in structures, machines, and products.

engineering acoustics—A field of acoustics that deals with the production, detection, and control of sound by electrical devices, including the study, design, and construction of such things as microphones, loudspeakers, sound recorders and reproducers, and public address systems.

evolution—The processes of biological and organic change in organisms by which descendants come to differ from their ancestors, and a history of the sequence of such change.

fluid mechanics—The science concerned with fluids, either at rest or in motion, and dealing with pressures, velocities, and accelerations in the fluid, including fluid deformation and compression or expansion.

food engineering—Technical discipline involved in food manufacturing and processing.

forestry—The science of developing, cultivating, and managing forest lands for wood, forage, water, wildlife, and recreation; the management of growing timber.

genetics—The science concerned with biological inheritance, that is, with the causes of the resemblances and differences among related individuals.

geochemistry—The study of the chemical composition of the various phases of the earth and the physical and chemical processes which have produced the observed distribution of the elements and nuclides in these phases.

geodesy—A subdivision of geophysics which includes determinations of the size and shape of the earth, the earth's gravitational field, and the location of points fixed to the earth's crust in an earth-referred coordinate system.

geography—The science that deals with the description of land, sea, and air and the distribution of plant and animal life, including humans.

geology—The study or science of earth, its history, and its life as recorded in the rocks; includes the study of the geologic features of an area, such as the geometry of rock formations, weathering and erosion, and sedimentation.

geophysics—A branch of geology in which the principles and practices of physics are used to study the earth and its environment, that is, earth, air, and (by extension) space.

graphic arts—The fine and applied arts of representation, decoration, and writing or printing on flat surfaces together with the techniques and crafts associated with each; includes painting, drawing, engraving, etching, lithography, photography, and printing arts.

histology—The study of the structure and chemical composition of animal tissues as related to their function.

horology—Science of time measurement and the principles and technology of constructing time-measuring instruments.

hydrology—The science that treats of the surface and ground waters of the earth; their occurrence, circulation, and distribution; their chemical and physical properties; and their reaction with their environment.

immunology—The division of biological science concerned with the native or acquired resistance of higher animal forms and humans to infection with microorganisms.

industrial engineering—The application of engineering principles and training and the techniques of scientific management to the maintenance of a high level of productivity at optimum cost in industrial enterprises, as by analytical study, improvement, and installation of methods and systems, operating procedures, quantity and quality measurements and controls, safety measures, and personnel administration.

inorganic chemistry—A branch of chemistry that deals with reactions and properties of all chemical elements and their compounds, excluding hydrocarbons but usually including carbides and other simple carbon compounds (such as CO_2, CO, and HCN).

invertebrate zoology—A branch of zoology concerned with the taxonomy, behavior, and morphology of invertebrate animals.

lapidary—The study relating to precious stones or the art of cutting them.

mapping—The art and practice of making a drawing or other representation, usually on a flat surface, of the whole or part of an area (as the surface of the earth or some other planet), indicating relative position and size according to a specified scale or projection of selected features, as countries, cities, rock formations, or bodies of water.

materials—The study of admixtures of matter or the basic matter from which products are made; includes adhesives, building materials, fuels, paints, leathers, and so on.

mathematics—The deductive study of shape, quantity, and dependence; the two main areas are applied mathematics and pure mathematics, the former arising from the study of physical phenomena, the latter involving the intrinsic study of mathematical structures.

mechanical engineering—The branch of engineering concerned with the generation, transmission, and utilization of heat and mechanical power, and with the production and operation of tools, machinery, and their products.

mechanics—The branch of physics which seeks to formulate general rules for predicting the behavior of a physical system under the influence of any type of interaction with its environment.

medicine—The study of cause and treatment of human disease, including the healing arts dealing with diseases which are treated by a physician or a surgeon.

metallurgy—The branch of engineering concerned with the production of metals and alloys, their adaptation to use, and their performance in service; and the study of chemical reactions involved in the processes by which metals are produced, and of the laws governing the physical, chemical, and mechanical behavior of metallic materials.

meteorology—The science concerned primarily with the observation of the atmosphere and its phenomena, including temperature, density, winds, clouds, and precipitation.

microbiology — The science and study of microorganisms, especially bacteria and rickettsiae, and of antibiotic substances.

mineralogy — The science concerning the study of natural inorganic substances called minerals, including origin, description, and classification.

mining engineering — A branch of engineering concerned with the location and evaluation of coal and mineral deposits, the survey of mining areas, the layout and equipment of mines, the supervision of mining operations, and the cleaning, sizing, and dressing of the product.

molecular biology — That branch of biology which attempts to interpret biological events in terms of the molecules in the cell.

mycology — A branch of biological science concerned with the study of fungi.

naval architecture — The study of the physical characteristics and the design and construction of buoyant structures which operate in water, and of the construction and operation of the power plant and other mechanical equipment of these structures.

navigation — The science or art of conducting ships or aircraft from one place to another, especially the method of determining position, course, and distance traveled over the surface of the earth by the principles of geometry and astronomy and by reference to devices (radar, beacons, and instruments) designed as aids.

nuclear physics — The study of the characteristics, behavior, and internal structure of the atomic nucleus.

nucleonics — The technology based on phenomena of the atomic nucleus such as radioactivity, fission, and fusion; includes nuclear reactors, various applications of radioisotopes and radiation, particle accelerators, and radiation detection devices.

oceanography — The scientific study and exploration of the oceans and seas in all their aspects.

optics — The study of phenomena associated with the generation, transmission, and detection of electromagnetic radiation in the spectral range extending from the long-wave edge of the x-ray region to the short-wave edge of the radio region; and the science of light.

ordnance — That military area concerned with supplies, including weapons, ammunition, combat vehicles, and the necessary repair equipment; and with heavy firearms discharged from mounts, including cannons and artillery.

organic chemistry — The study of the composition, reactions, and properties of carbon compounds except CO_2, CO, and certain ionic compounds such as Na_2CO_3 and $NaCN$.

paleobotany — The study of fossil plants and vegetation of the geologic past.

paleontology — The study of life in the geologic past as recorded by fossil remains.

pathology — The branch of biological science which deals with the nature of disease, through study of its causes, its processes, and its effects, together with the associated alterations of structure and function; and the laboratory findings of disease, as distinguished from clinical signs and symptoms.

particle physics — The branch of physics concerned with understanding the properties, behavior, and structure of elementary particles, especially through study of collisions or decays involving energies of hundreds of MeV or more.

petroleum engineering — A branch of engineering concerned with the search for and extraction of oil, gas, and liquefiable hydrocarbons.

petrology — The branch of geology dealing with the origin, occurrence, structure, and history of rocks, especially igneous and metamorphic rocks.

pharmacology — The science of detection and measurement of the effects of drugs or other chemicals on biological systems; includes all chemicals used as drugs.

physical chemistry — The description and prediction of chemical behavior by means of physical theory, with extensive use of graphs and mathematical formulas; main subject areas are structure, thermodynamics, and kinetics.

physics — The science concerned with those aspects of nature which can be understood in terms of elementary principles and laws.

physiology — The branch of biological science concerned with the basic activities that occur in cells and tissues of living organisms and involving physical and chemical studies of these organisms.

plant pathology — The branch of botany concerned with diseases of plants.

plasma physics — The study of highly ionized gases.

psychology — The science of the function of the mind and the behavior of an organism, both animal and human, in relation to its environment.

quantum mechanics — The modern theory of matter, of electromagnetic radiation, and of the interaction between matter and radiation; it differs from classical physics, which it generalizes and supersedes, mainly in the realm of atomic and subatomic phenomena.

relativity — The study of physics theory which recognizes the universal character of the propagation speed of light and the consequent dependence of space, time, and other mechanical measurements on the motion of the observer performing the measurements; the two main divisions are special theory and general theory.

science and technology — The study of the natural sciences and the application of this knowledge for practical purposes.

solid-state physics — The branch of physics centering on the physical properties of solid materials; it is usually concerned with the properties of crystalline materials only, but it is sometimes extended to include the properties of glasses or polymers.

spectroscopy — The branch of physics concerned with the production, measurement, and interpretation of electromagnetic spectra arising from either emission or absorption of radiant energy by various substances.

statistical mechanics — That branch of physics which endeavors to explain and predict the macroscopic properties and behavior of a system on the basis of the known characteristics and interactions of the microscopic constituents of the system, usually when the number of such constituents is very large.

statistics — The science dealing with the collection, analysis, interpretation, and presentation of masses of numerical data.

systematics — The science of animal and plant classification.

systems engineering — The branch of engineering dealing with the design of a complex interconnection of many elements (a system) to maximize an agreed-upon measure of system performance.

textiles — Area of industry involving the production of fibers, filaments, or yarn, and the cloth made from these materials.

thermodynamics — The branch of physics which seeks to derive, from a few basic postulates, relations between properties of substances, especially those which are affected by changes in temperature, and a description of the conversion of energy from one form to another.

vertebrate zoology — A branch of zoology concerned with the taxonomy, behavior, and morphology of vertebrate animals.

veterinary medicine — That branch of medical practice which treats of the diseases and injuries of animals.

virology — The science that deals with the study of viruses.

zoology — The science that deals with the taxonomy, behavior, and morphology of animal life.

a *See* ampere; atto-.

aΩ *See* abohm.

(aΩ)⁻¹ *See* abmho.

A *See* ampere; angstrom.

A+ *See* A positive.

A-1 *See* Skyraider.

A-3 *See* Skywarrior.

A-4 *See* V-2.

A-5 *See* Vigilante.

A1 time [ASTRON] A particular atomic time scale, established by the U.S. Naval Observatory, with the origin on Jan. 1, 1958, at zero hours Universal Time and with the unit (second) equal to 9,192,631,770 cycles of cesium at zero field.

aA *See* abampere.

AA *See* antiaircraft.

AAA *See* antiaircraft artillery.

aaa disease [MED] An endemic hookworm disease accompanied by anemia that occurred in ancient Egypt and is designated in the Ebers Papyrus.

aa channel [GEOL] A narrow, sinuous channel in which a lava river moves down and away from a central vent to feed an aa flow.

aAcm² *See* abampere centimeter squared.

aA/cm² *See* abampere per square centimeter.

aa lava [GEOL] Type of lava with a rough, fragmental surface; consists of clinkers and scoria.

Aalenian [GEOL] Lowermost Middle or uppermost Lower Jurassic geologic time.

AAM *See* air-to-air missile; antiaircraft missile.

A AND NOT B gate *See* AND NOT gate.

aapamoor [ECOL] A moor with elevated areas or mounds supporting dwarf shrubs and sphagnum, interspersed with low areas containing sedges and sphagnum, thus forming a mosaic.

aardvark [VERT ZOO] A nocturnal, burrowing, insectivorous mammal of the genus *Orycteropus* in the order Tubulidentata. Also known as earth pig.

aardwolf [VERT ZOO] *Proteles cristatus.* A hyenalike African mammal of the family Hyaenidae.

Aaron's rod [ARCH] A decorative rounded molding on which are entwined a single serpent and sometimes vines and leaves.

a axis [CRYSTAL] One of the crystallographic axes used as reference in crystal description, usually oriented horizontally, front to back. [GEOL] The direction of movement or transport in a tectonite.

ab- [ELECTROMAG] A prefix used to identify centimeter-gram-second electromagnetic units, as in abampere, abcoulomb, abfarad, abhenry, abmho, abohm, and abvolt.

abac *See* nomograph.

abaca [BOT] *Musa textilis.* A plant of the banana family native to Borneo and the Philippines, valuable for its hard fiber. Also known as Manila hemp.

abactinal [INV ZOO] In radially symmetrical animals, pertaining to the surface opposite the side where the mouth is located.

abacus [ARCH] A slab forming the topmost division of the capital of a column. [MATH] An instrument for performing arithmetical calculations manually by sliding markers on rods or in grooves.

abaft [NAV ARCH] In a direction farther aft in a ship than a specified reference position, such as abaft the mast.

abalienation [PSYCH] Mental deterioration or derangement.

abalone [INV ZOO] A gastropod mollusk composing the single genus, *Haliotis,* of the family Haliotidae. Also known as ear shell; ormer; paua.

abampere [ELEC] The unit of electric current in the electromagnetic centimeter-gram-second system; 1 abampere equals 10 amperes in the absolute meter-kilogram-second-ampere system. Abbreviated aA. Also known as Bi; biat.

abampere centimeter squared [ELECTROMAG] The unit of magnetic moment in the electromagnetic centimeter-gram-second system. Abbreviated aAcm².

abampere per square centimeter [ELEC] The unit of current density in the electromagnetic centimeter-gram-second system. Abbreviated aA/cm².

abamurus [ARCH] A masonry block, in the form of a buttress, used to support a structure.

A band [HISTOL] The region between two adjacent I bands in a sarcomere; characterized by partial overlapping of actin and myosin filaments.

abandon [ENG] To stop drilling and remove the drill rig from the site of a borehole before the intended depth or target is reached.

abandoned channel *See* oxbow.

abandoned mine *See* abandoned workings.

abandoned workings [MIN ENG] Deserted excavations, either caved or sealed, in which further mining is not intended, and opening workings which are not ventilated and inspected regularly. Also known as abandoned mine.

abandonment [MIN ENG] Failure to perform work, by conveyance, by absence, and by lapse of time, on a mining claim. [PETRO ENG] *See* abandonment contour.

abandonment contour [PETRO ENG] A graph of actual cumulative yield of an oil well compared with its estimated ultimate yield; useful in determining the most economic time to abandon an oil well. Also known as abandonment.

abapertural [INV ZOO] Away from the shell aperture, referring to mollusks.

abapical [BIOL] On the opposite side to, or directed away from, the apex.

abarognosis [MED] Lack of ability to estimate the weight of an object one is holding.

abasia [MED] Lack of muscular coordination in walking.

abatement [ENG] **1.** The waste produced in cutting a timber, stone, or metal piece to a desired size and shape. **2.** A decrease in the amount of a substance or other quantity, such as atmospheric pollution.

abat-jour [BUILD] A device that is used to deflect daylight downward as it streams through a window.

A battery [ELECTR] The battery that supplies power for filaments or heaters of electron tubes in battery-operated equipment.

abattoir [IND ENG] A building in which cattle or other animals are slaughtered.

abat-vent [BUILD] A series of sloping boards or metal strips, or some similar contrivance, to break the force of wind without being an obstruction to the passage of air or sound, as in a louver or chimney cowl.

abaxial [BIOL] On the opposite side to, or facing away from, the axis of an organ or organism.

abb [TEXT] Yarn made of abb wool. [VERT ZOO] A coarse wool from the fleece areas of lesser quality.

abbazzo [GRAPHICS] A rough sketch, draft, or model.

Abbe condenser [OPTICS] A variable large-aperture lens system arranged substage to image a light source into the focal plane of a microscope objective.

Abbe number [OPTICS] A number which expresses the deviating effect of an optical glass on light of different wavelengths.

Abbe prism [OPTICS] A system used for image erection which is composed of two double right-angle prisms and involves four reflections.

Abbe refractometer [OPTICS] An optical instrument for the measurement of the refractive index of liquids.

Abbe's sine condition [OPTICS] A relationship which must hold to prevent aberration of a mirror or lens from producing a coma.

AARDVARK

The aardvark *(Orycteropus afer),* a nocturnal, burrowing animal ranging from Ethiopia to southern Africa.

ABACUS

Drawing of an abacus.

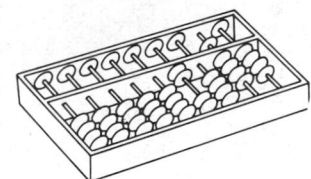

ABALONE

Typical abalone ear-shaped shell perforated by pores.

ABERRATION

The aberration of light as seen in astronomy. Starlight arriving along AB and seen in this direction by a stationary observer (left) appears to the observer in transverse motion AA' (right) to come from the direction AB' (or A'B). *(From G. de Vaucouleurs, Discovery of the Universe, Faber, 1957)*

Abbe's theory [OPTICS] The theory that for a lens to produce a true image, it must be large enough to transmit the entire diffraction pattern of the object.

abbreviated dialing [COMMUN] A method comprising special circuits which require less than the usual number of dialing operations to connect two or more subscribers.

ABC *See* automatic brightness control.

abcoulomb [ELEC] The unit of electric charge in the electromagnetic centimeter-gram-second system, equal to 10 coulombs. Abbreviated aC.

abcoulomb centimeter [ELEC] The electromagnetic centimeter-gram-second unit of electric dipole moment. Abbreviated aCcm.

abcoulomb per cubic centimeter [ELEC] The electromagnetic centimeter-gram-second unit of volume density of charge. Abbreviated aC/cm^3.

abcoulomb per square centimeter [ELEC] The electromagnetic centimeter-gram-second unit of surface density of charge, electric polarization, and displacement. Abbreviated aC/cm^2.

ABC system [GEOPHYS] A procedure in seismic surveying to determine the effect of irregular weathering thickness. [MAP] *See* airborne control system. [ORD] An atomic, biological, or chemical weapons system.

Abderhalden reaction [PATH] A chemical blood test for the identification of certain enzymes associated with pregnancy and a few diseases.

abdomen [ANAT] **1.** The portion of the vertebrate body between the thorax and the pelvis. **2.** The cavity of this part of the body. [INV ZOO] The elongate region posterior to the thorax in arthropods.

abdominal apoplexy [MED] Vascular occlusion and hemorrhage in an abdominal organ, usually the small intestine, or in the peritoneal cavity.

abdominal depth [ANTHRO] Maximum horizontal contact dimension, measured front to back.

abdominal gestation [MED] Development of a fetus outside the uterus in the abdominal cavity.

abdominal hernia *See* ventral hernia.

abdominal hysterectomy [MED] Surgical removal of all or part of the uterus through an incision in the abdomen.

abdominal regions [ANAT] Nine theoretical areas delineated on the abdomen by two horizontal and two parasagittal lines: above, the right hypochondriac, epigastric, and left hypochondriac; in the middle, the right lateral, umbilical, and left lateral; and below, the right inguinal, hypogastric, and left inguinal.

abducens [ANAT] The sixth cranial nerve in vertebrates; a paired, somatic motor nerve arising from the floor of the fourth ventricle of the brain and supplying the lateral rectus eye muscles.

abduction [PHYSIO] Movement of an extremity or other body part away from the axis of the body.

abductor [PHYSIO] Any muscle that draws a part of the body or an extremity away from the body axis.

abeam *See* on the beam.

Abegg's rule [CHEM] An empirical rule, holding for a large number of elements, that the sum of the maximum positive and negative valencies of an element equals eight.

Abelian domain *See* Abelian field.

Abelian field [MATH] A set of elements $a, b, c, \ldots$ forming Abelian groups with addition and multiplication as group operations where $a(b + c) = ab + ac$. Also known as Abelian domain; domain.

Abelian group [MATH] A group whose binary operation is commutative; that is, $ab = ba$ for each a and b in the group.

abelite [MATER] A substance made of ammonium nitrate and a nitrated aromatic hydrocarbon and used as an explosive.

Abel's inequality [MATH] An inequality which states that the absolute value of the sum of n terms, each in the form ab, where the bs are positive numbers, is not greater than the product of the largest b with the largest absolute value of a partial sum of the as.

Abel's integral equation [MATH] The equation

$$f(x) = \int_a^x u(z)(x - z)^{-a}dz \ (0 < a < 1, \ x \geq a),$$

where $f(x)$ is a known function and $u(z)$ is the function to be determined; when $a = \frac{1}{2}$, this equation has application to Abel's problem.

Abel's problem [MATH] The problem which asks what path a particle will follow if it moves under the influence of gravity alone and its altitude-time function is to follow a specific law.

Abel tester [PHYS CHEM] A laboratory instrument used in testing the flash point of kerosine and other volatile oils having flash points below 120°F (48.9°C); the oil is contained in a closed cup which is heated by a fixed flame below and a movable flame above.

Abel theorem [MATH] **1.** A theorem stating that if a power series in z converges for $z = a$, it converges absolutely for $|z| < |a|$. **2.** A theorem stating that if a power series in z converges to $f(z)$ for $|z| < 1$ and to a for $z = 1$, then the limit of $f(z)$ as z approaches $1 = a$. **3.** A theorem stating that if the three series with nth term a_n, b_n, and $c_n = a_0b_n + a_1b_{n-1} + \cdots + a_nb_0$, respectively, converge, then the third series equals the product of the first two series.

abenteric [MED] Involving abdominal organs and structures outside the intestine.

aberrant [BIOL] An atypical group, individual, or structure, especially one with an aberrant chromosome number.

aberration [ASTRON] The apparent angular displacement of the position of a celestial body in the direction of motion of the observer, caused by the combination of the velocity of the observer and the velocity of light. [OPTICS] Any deviation from perfect reproduction so that a point is not imaged as a point, a straight line as straight, or an angle as an equal angle.

abfarad [ELEC] A unit of capacitance in the electromagnetic centimeter-gram-second system equal to 10^9 farads. Abbreviated aF.

abhenry [ELEC] A unit of inductance in the electromagnetic centimeter-gram-second system of units equal to 10^{-9} henry. Abbreviated aH.

Abies [BOT] The firs, a genus of trees in the pine family characterized by erect cones, absence of resin canals in the wood, and flattened needlelike leaves.

abietic acid [ORG CHEM] $C_{20}H_{30}O_2$ A tricyclic, crystalline acid obtained from rosin; used in making esters for plasticizers.

abiocoen [ECOL] A nonbiotic habitat.

abiogenesis [BIOL] The obsolete concept that plant and animal life arise from nonliving organic matter. Also known as autogenesis; spontaneous generation.

abiotic [BIOL] Referring to the absence of living organisms.

abiotic environment [ECOL] All physical and nonliving chemical factors, such as soil, water, and atmosphere, which influence living organisms.

abiotic substance [ECOL] Any fundamental chemical element or compound in the environment.

abiotrophy [MED] Disordered functioning of an organ or system, as in Huntington's chorea, due to an inherited pathologic trait, which trait, however, may remain latent in the individual rather than becoming apparent; this mechanism is still conceptual.

abjection [MYCOL] The discharge or casting off of spores by the spore-bearing structure of a fungus.

ablastin [IMMUNOL] An antibodylike substance elicited by *Trypanosoma lewisi* in the blood serum of infected rats that inhibits reproduction of the parasite.

ablation [AERO ENG] The carrying away of heat, generated by aerodynamic heating, from a vital part by arranging for its absorption in a nonvital part, which may melt or vaporize and then pass away, taking the heat with it. Also known as ablative cooling. [GEOL] The wearing away of rocks, as by erosion or weathering. [HYD] The reduction in volume of a glacier due to melting and evaporation. [MED] The removal of tissue or a part of the body by surgery, such as by excision or amputation.

ablation area [HYD] The section in a glacier or snowfield where ablation exceeds accumulation.

ablation cone [HYD] A debris-covered cone of ice, firn, or snow formed by differential ablation.

ablation factor [HYD] The rate at which a snow or ice surface wastes away.

ablation form [HYD] A feature on a snow or ice surface caused by melting or evaporation.

ablation moraine [GEOL] **1.** A layer of rock particles overlying ice in the ablation of a glacier. **2.** Drift deposited from a superglacial position through the melting of underlying stagnant ice.

ablative agent [MATER] A material from which the surface layer is to be removed, often for the purpose of dissipating extreme heat energy, as in space vehicles reentering the earth's atmosphere. Also known as ablative material.

ablative cooling *See* ablation.

ablative material *See* ablative agent.

ablative shielding [AERO ENG] A covering of material designed to reduce heat transfer to the internal structure through sublimation and loss of mass.

ablatograph [ENG] An instrument that records ablation by measuring the distance a snow or ice surface falls during the observation period.

ABM *See* antiballistic missile.

abmho [ELEC] A unit of conductance in the electromagnetic centimeter-gram-second system of units equal to 10^9 mhos. Abbreviated $(a\Omega)^{-1}$. Also known as absiemens (aS).

Abney effect [OPTICS] A shift in the apparent hue of a light which occurs as colored light is desaturated by the addition of white light.

Abney level [ENG] A hand level with a vertical circle, used to measure vertical angles, especially in ascertaining tree heights by trigonometric relationships.

abnormal anticlinorium [GEOL] An anticlinorium with axial planes of subsidiary folds diverging upward.

abnormal behavior [PSYCH] Personality functioning that is socially undesirable or that renders the individual unable to cope with day-to-day living. Also known as behavior disorder.

abnormality [SCI TECH] Any deviation from normal characteristics.

abnormal magnetic variation [GEOPHYS] The anomalous value in magnetic compass readings made in some local areas containing unknown sources that deflect the compass needle from the magnetic meridian.

abnormal place [MIN ENG] An area in a coal mine where geological conditions render mining uneconomical.

abnormal propagation [COMMUN] Phenomena of unstable or changing atmospheric or ionospheric conditions acting upon transmitted radio waves, preventing such waves from following their normal path through space, and causing difficulties and disruptions of communications.

abnormal psychology [PSYCH] A branch of psychology that deals with behavior disorders and internal psychic conflict in addition to certain normal phenomena such as dreams, motivations, and anxiety.

abnormal reflections [ELECTROMAG] Sharply defined reflections of substantial intensity at frequencies greater than the critical frequency of the ionized layer of the ionosphere.

abnormal statement [ADP] An element of a FORTRAN V (UNIVAC) program which specifies that certain function subroutines must be called every time they are referred to.

abnormal synclinorium [GEOL] A synclinorium with axial planes of subsidiary folds converging downward.

ABO blood group [IMMUNOL] An immunologically distinct, genetically determined group of human erythrocyte antigens represented by two blood factors (A and B) and four blood types (A, B, AB, and O).

abohm [ELEC] The unit of electrical resistance in the centimeter-gram-second system; 1 abohm equals 10^{-9} ohm in the meter-kilogram-second system. Abbreviated aΩ.

abohm centimeter [ELEC] The centimeter-gram-second unit of resistivity. Abbreviated aΩcm.

abomasitis [VET MED] Inflammation of the abomasum in ruminants.

abomasum [VERT ZOO] The final chamber of the complex stomach of ruminants; has a glandular wall and corresponds to a true stomach.

A bomb *See* atomic bomb.

aboral [INV ZOO] Opposite to the mouth.

abort [AERO ENG] **1.** To cut short or break off an action, operation, or procedure with an aircraft, space vehicle, or the like, especially because of equipment failure. **2.** An aircraft, space vehicle, or the like which aborts. **3.** An act or instance of aborting.

aborted firing [ORD] A firing of a gun or launching of a missile which is cut off either manually or automatically after the firing command has been given but before ignition has been initiated.

abortifacient [MED] Any agent that induces abortion.

abortion [MED] The spontaneous or induced expulsion of the fetus prior to the time of viability, most often during the first 20 weeks of the human gestation period.

abortive [BIOL] Imperfectly formed or developed.

abortive transduction [MICROBIO] Failure of exogenous fragments that were introduced into a bacterial cell by viruses to become inserted into the bacterial chromosome.

abortus [MED] An aborted fetus.

abort zone [AERO ENG] The area surrounding the launch within which malperforming missiles will be contained with known and acceptable probability.

about-sledge [MET] A large hammer used in blacksmithing.

a-b plane [GEOL] The surface along which differential movement takes place.

AB power pack [ELEC] **1.** Assembly in a single unit of the A battery and B battery for a battery-operated vacuum-tube circuit. **2.** Unit that supplies the necessary A and B direct-current voltages from an alternating-current source of power.

abrachiocephalia [MED] Congenital lack of arms and head. Also known as acephalobrachia.

abrade [GEOL] To wear away by abrasion or friction.

Abraham's tree [METEOROL] The popular name given to a form of cirrus radiatus clouds, consisting of an assemblage of long feathers and plumes of cirrus that seems to radiate from a single point on the horizon.

abranchiate [ZOO] Without gills.

abrasion [ENG] The removal of surface material from any solid through the frictional action of another solid, a liquid, or a gas or combination thereof. [GEOL] Wearing away of sedimentary rock chiefly by currents of water laden with sand and other rock debris and by glaciers. [MED] A spot denuded of skin, mucous membrane, or superficial epithelium by rubbing or scraping.

abrasion platform [GEOL] An uplifted marine peneplain or plain, according to the smoothness of the surface produced by wave erosion, which is of large area.

abrasion resistance [MATER] The ability of a surface to resist wearing due to contact with another surface moving with respect to it.

abrasion test [MECH ENG] The measurement of abrasion resistance, usually by the weighing of a material sample before and after subjecting it to a known abrasive stress throughout a known time period, or by reflectance or surface finish comparisons, or by dimensional comparisons.

abrasive [GEOL] A small, hard, sharp-cornered rock fragment, used by natural agents in abrading rock material or land surfaces. Also known as abrasive ground. [MATER] **1.** A material used, usually as a grit sieved by a specified mesh but also as a solid shape or as a paste or slurry or air suspension, for grinding, honing, lapping, superfinishing, polishing, pressure blasting, or barrel tumbling. **2.** A material sintered or formed into a solid mass such as a hone or a wheel disk, cone, or burr for grinding or polishing other materials. **3.** Having qualities conducive to or derived from abrasion.

abrasive belt [MECH ENG] A cloth, leather, or paper band impregnated with grit and rotated as an endless loop to abrade materials through continuous friction.

abrasive blasting [MECH ENG] The cleaning or finishing of surfaces by the use of an abrasive entrained in a blast of air.

abrasive cloth [MECH ENG] Tough cloth to whose surface an abrasive such as sand or emery has been bonded for use in grinding or polishing.

abrasive cone [MECH ENG] An abrasive sintered or shaped into a solid cone to be rotated by an arbor for abrasive machining.

abrasive disk [MECH ENG] An abrasive sintered or shaped into a disk to be rotated by an arbor for abrasive machining.

abrasive drilling [MIN ENG] A rotary drilling method in which drilling is effected by the abrasive action of the drill steel or drilling medium which rotates while being pressed against the rock.

abrasive ground   *See* abrasive.

abrasive jet cleaning   [ENG] The removal of dirt from a solid by a gas or liquid jet carrying abrasives to ablate the surface.

abrasive machining   [MECH ENG] Grinding, drilling, shaping, or polishing by abrasion.

abrasiveness   [MATER] **1.** The property of a material causing wear of a surface by friction. **2.** The quality or characteristic of being able to scratch, abrade, or wear away another material.

abrasive paper   [MATER] Tough paper to whose surface an abrasive, such as sand or emery, has been bonded for use in grinding or polishing.

abrasive sand   [MATER] Grit used as abrasive, usually graded as to which sieve mesh it will pass through.

abreast milling   [MECH ENG] A milling method in which parts are placed in a row parallel to the axis of the cutting tool and are milled simultaneously.

abreuvoir   [CIV ENG] A space between stones in masonry to be filled with mortar.

Abridged Nautical Almanac   *See* Nautical Almanac.

abrupt   [BOT] Ending suddenly, as though broken off.

abs   [ADP] A special function in ALGOL, which yields the absolute value, or modulus, of its argument. [METEOROL] *See* absolute.

ABS   *See* acrylonitrile butadiene styrene.

absarokite   [PETR] An alkalic basalt of about equal portions of olivine, augite, labradorite, and sanidine with accessory biotite, apatite, and opaque oxides; leucite is occasionally present in small amounts.

abscess   [MED] A localized collection of pus surrounded by inflamed tissue.

abscisic acid   [BIOCHEM] $C_{16}H_{20}O_4$ A plant hormone produced by fruits and leaves that promotes abscission and dormancy and retards vegetative growth. Formerly known as abscisin.

abscisin   *See* abscisic acid.

abscissa   [MATH] One of the coordinates of a two-dimensional coordinate system, usually the horizontal coordinate, denoted by x.

abscission   [BOT] A physiological process promoted by abscisic acid whereby plants shed a part, such as a leaf, flower, seed, or fruit.

absence-of-ground searching selector   [COMMUN] In dial telephones, an automatic switch which rotates, or rises vertically and rotates, in search of an ungrounded contact.

absiemens   *See* abmho.

absinthe   [FOOD ENG] A green liqueur having a bitter licorice flavor and a high alcohol content.

absolute   [METEOROL] Referring to the highest or lowest recorded value of a meteorological element, whether at a single station or over an area, during a given period. Abbreviated abs.

absolute address   [ADP] The numerical identification of each storage location which is wired permanently into a computer by the manufacturer.

absolute age   [GEOL] The geologic age of a fossil, or a geologic event or structure expressed in units of time, usually years. Also known as actual age.

absolute alcohol   [ORG CHEM] Ethyl alcohol that contains no more than 1% water. Also known as anhydrous alcohol.

absolute altimeter   [ENG] An instrument which employs radio, sonic, or capacitive technology to produce on its indicator the measurement of distance from the aircraft to the terrain below. Also known as terrain clearance indicator.

absolute altitude   [ENG] Altitude above the actual surface, either land or water, of a planet or natural satellite.

absolute angle of attack   [AERO ENG] The acute angle between the chord of an airfoil at any instant in flight and the chord of that airfoil at zero lift.

absolute blocking   [CIV ENG] A control arrangement for rail traffic in which a track is divided into sections or blocks upon which a train may not enter until the preceding train has left.

absolute boiling point   [CHEM] The boiling point of a substance expressed in the unit of an absolute temperature scale.

absolute ceiling   [AERO ENG] The greatest altitude at which an aircraft can maintain level flight in a standard atmosphere and under specified conditions.

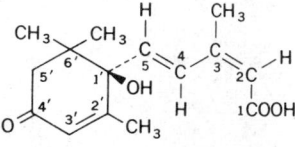

ABSCISIC ACID

Structural formula for *S*-abscisic acid, the naturally occurring form.

absolute code   [ADP] A code used when the addresses in a program are to be written in machine language exactly as they will appear when the instructions are executed by the control circuits.

absolute convergence   [MATH] That property of an infinite series (or infinite product) of real or complex numbers if the series (product) of absolute values converges; absolute convergence implies convergence.

absolute coordinate system   [NAV] The inertial coordinate system which has its origin on the axis of the earth and is fixed with respect to the stars. Also known as absolute reference frame.

absolute delay   [NAV] In loran, the time interval between transmission of a signal from the A station and transmission of the next signal from the B station.

absolute density   *See* absolute gravity.

absolute deviation   [ORD] The shortest distance between the center of the target and the point where a projectile hits or bursts. [STAT] The difference, without regard to sign, between a variate value and a given value.

absolute drought   [METEOROL] In Britain, a period of at least 15 consecutive days during which no measurable daily precipitation has fallen.

absolute efficiency   [ENG ACOUS] The ratio of the power output of an electroacoustic transducer, under specified conditions, to the power output of an ideal electroacoustic transducer.

absolute electrometer   [ELEC] A very precise type of attracted disk electrometer in which the attraction between two disks is balanced against the force of gravity.

absolute error   [MATH] In an approximate number, the numerical difference between the number and a number considered exact. [ORD] **1.** Shortest distance between the center of impact or the center of burst of a group of shots and the point of impact or burst of a single shot within the group. **2.** Error of a sight consisting of its error in relation to a master service sight with which it is tested and of the known error of the master service sight.

absolute gain of an antenna   [ELECTROMAG] Gain in a given direction when the reference antenna is an isotropic antenna isolated in space. Also known as isotropic gain of an antenna.

absolute geopotential topography   *See* geopotential topography.

absolute gravity   [CHEM] Density or specific gravity of a fluid reduced to standard conditions; for example, with gases, to 760 mm Hg pressure and 0°C temperature. Also known as absolute density.

absolute humidity   [METEOROL] The ratio of the mass of water vapor in a sample of air to the volume of the sample.

absolute inequality   *See* unconditional inequality.

absolute instability   [METEOROL] The state of a column of air in the atmosphere when it has a superadiabatic lapse rate of temperature, that is, greater than the dry-adiabatic lapse rate. Also known as autoconvective instability; mechanical instability.

absolute instrument   [ENG] An instrument which measures a quantity (such as pressure or temperature) in absolute units by means of simple physical measurements on the instrument.

absolute isohypse   [METEOROL] A line that has the properties of both constant pressure and constant height above mean sea level.

absolute linear momentum   *See* absolute momentum.

absolute luminosity   [OPTICS] The luminosity of an object expressed in units of fundamental quantities.

absolute magnetometer   [ENG] An instrument used to measure the intensity of a magnetic field without reference to other magnetic instruments.

absolute magnitude   [ASTROPHYS] **1.** A measure of the brightness of a star equal to the magnitude the star would have at a distance of 10 parsecs from the observer. **2.** The stellar magnitude any meteor would have if placed in the observer's zenith at a height of 100 kilometers. [MATH] The absolute value of a number or quantity.

absolute manometer   [ENG] **1.** A gas manometer whose calibration, which is the same for all ideal gases, can be calculated from the measurable physical constants of the instrument. **2.** A manometer that measures absolute pressure.

absolute momentum [METEOROL] The sum of the (vector) momentum of a particle relative to the earth and the (vector) momentum of the particle due to the earth's rotation. Also known as absolute linear momentum.

absolute motion [NAV] Motion relative to a point fixed on the earth or to an apparently fixed celestial point. [PHYS] The motion of an object described by its measurement in a frame of reference that is preferred over all other frames.

absolute orientation [NAV] The adjusting to proper scale, orientating of the model datum parallel to sea level or another given vertical datum, and positioning of the model with reference to the horizontal datum of a stereoscopic model or group of models.

absolute parallax *See* absolute stereoscopic parallax.

absolute permeability [ELECTROMAG] The ratio of the magnetic flux density to the intensity of the magnetic field in a medium; measurement is in webers per square meter in the meter-kilogram-second system. Also known as induced capacity.

absolute pitch [ACOUS] The pitch of a musical tone expressed as the frequency of the sound wave of that tone.

absolute potential vorticity *See* potential vorticity.

absolute pressure [PHYS] The pressure above the absolute zero value of pressure that theoretically obtains in empty space or at the absolute zero of temperature, as distinguished from gage pressure.

absolute pressure gage [ENG] A device that measures the pressure exerted by a fluid relative to a perfect vacuum; used to measure pressures very close to a perfect vacuum.

absolute pressure transducer [ENG] A device that responds to absolute pressure as the input and provides a measurable output of a nature different than but proportional to absolute pressure.

absolute programming [ADP] Programming with the use of absolute code.

absolute reaction rate [PHYS CHEM] The rate of a chemical reaction as calculated by means of the (statistical-mechanics) theory of absolute reaction rates.

absolute reference frame *See* absolute coordinate system.

absolute roof [MIN ENG] The entire mass of strata overlying a subsurface point of reference.

absolute scale *See* absolute temperature scale.

absolute space-time [PHYS] A concept underlying Newtonian mechanics which postulates the existence of a preferred reference system of time and spatial coordinates; replaced in relativistic mechanics by Einstein's equivalency principle. Also known as absolute time.

absolute specific gravity [MECH] The ratio of the weight of a given volume of a substance in a vacuum at a given temperature to the weight of an equal volume of water in a vacuum at a given temperature.

absolute stability [METEOROL] The state of a column of air in the atmosphere when its lapse rate of temperature is less than the saturation-adiabatic lapse rate.

absolute standard [PHYS] A particle or object designated as a standard by assigning to it a mass of one unit; used in defining quantities in Newton's second law of motion.

absolute stereoscopic parallax [GRAPHICS] Considering a pair of aerial photographs of equal principal distance, the absolute stereoscopic parallax of a point is the algebraic difference of the distances of the two images from their respective photograph nadirs, measured in a horizontal plane and parallel to the air base. Also known as absolute parallax; horizontal parallax; linear parallax; parallax; stereoscopic parallax; x-parallax.

absolute stop [CIV ENG] A railway signal which indicates that the train must make a full stop and not proceed until there is a change in the signal. Also known as stop and stay.

absolute system of units [PHYS] A set of units for measuring physical quantities, defined by interrelated equations in terms of arbitrary fundamental quantities of length, mass, time, and charge or current.

absolute temperature [THERMO] **1.** The temperature measurable in theory on the thermodynamic temperature scale. **2.** The temperature in Celsius degrees relative to the absolute zero at −273.16°C (the Kelvin scale) or in Fahrenheit degrees relative to the absolute zero at −459.69°F (the Rankine scale).

absolute temperature scale [THERMO] A scale with which temperatures are measured relative to absolute zero. Also known as absolute scale.

absolute threshold [PHYSIO] The minimum stimulus energy that an organism can detect.

absolute time [GEOL] Geologic time measured in years, as determined by radioactive decay of elements. [PHYS] *See* absolute space-time.

absolute unit [PHYS] A unit defined in terms of units of fundamental quantities such as length, time, mass, and charge or current.

absolute vacuum [PHYS] A void completely empty of matter. Also known as perfect vacuum.

absolute-value computer [ADP] A computer that processes the values of the variables rather than their increments.

absolute value of a complex number [MATH] The modulus of a complex number; the square root of the sum of the squares of its real and imaginary part. Also known as magnitude of a complex number.

absolute value of a real number [MATH] The number if it is nonnegative, and the negative of the number if it is negative. Also known as magnitude of a real number; numerical value of a real number.

absolute value of a vector [MATH] The length of a vector, disregarding its direction; the square root of the sum of the squares of its orthogonal components. Also known as magnitude of a vector.

absolute velocity [PHYS] The vector sum of the velocity of a fluid parcel relative to the earth and the velocity of the parcel due to the earth's rotation; the east-west component is the only one affected.

absolute viscosity [FL MECH] The tangential force per unit area of two parallel planes at unit distance apart when the space between them is filled with a fluid and one plane moves with unit velocity in its own plane relative to the other.

absolute vorticity [FL MECH] The vorticity of a fluid relative to an absolute coordinate system; especially, the vorticity of the atmosphere relative to axes not rotating with the earth.

absolute wavemeter [ELECTROMAG] A type of wavemeter in which the frequency of an injected radio-frequency voltage is determined by measuring the length of a resonant line.

absolute zero [THERMO] The temperature of −273.16°C, or −459.69°F, or 0 K, thought to be the temperature at which molecular motion vanishes and a body would have no heat energy.

absorb [CHEM] To take up matter in bulk. [ELECTROMAG] To take up energy from radiation. [PHYS] To take up matter or radiation.

absorbance [PHYS CHEM] The common logarithm of the reciprocal of the transmittance of a pure solvent. Also known as absorbancy; extinction.

absorbancy *See* absorbance.

absorbed dose [NUCLEO] The amount of energy imparted by ionizing particles to a unit mass of irradiated material at a place of interest. Also known as dosage; dose.

absorbency [CHEM] Penetration of one substance into another.

absorbency index *See* absorptivity.

absorbent cotton [MATER] A cotton fiber that absorbs water because its natural waxes have been removed.

absorbent paper [MATER] Paper capable of absorbing and holding liquids by the capillarity of the pores between or within the closely matted cellulosic fibers.

absorber [CHEM ENG] Equipment in which a gas is absorbed by contact with a liquid. [ELECTR] A material or device that takes up and dissipates radiated energy; may be used to shield an object from the energy, prevent reflection of the energy, determine the nature of the radiation, or selectively transmit one or more components of the radiation. [NUCLEO] A material that absorbs neutrons or other ionizing radiation.

absorber control *See* absorption control.

absorber oil *See* absorption oil.

absorbing boom [CIV ENG] A device that floats on the water and is used to stop the spread of an oil spill and aid in its removal.

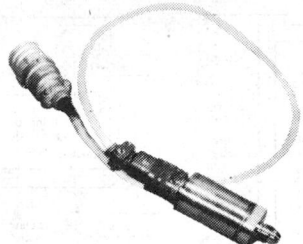

ABSOLUTE PRESSURE TRANSDUCER

A miniaturized absolute pressure transducer used in flight testing.

ABSOLUTE TEMPERATURE

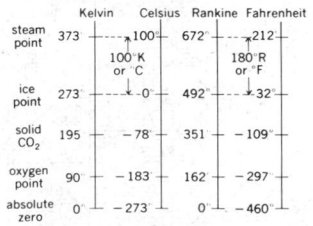

	Kelvin	Celsius	Rankine	Fahrenheit
steam point	373	100	672	212
		100°K or °C		180°R or °F
ice point	273	0	492	32
solid CO_2	195	−78	351	−109
oxygen point	90	−183	162	−297
absolute zero	0	−273	0	−460

Comparisons of Kelvin, Celsius, Rankine, and Fahrenheit temperature scales. Temperatures are rounded off to nearest degree. *(From M. W. Zemansky, Temperatures Very Low and Very High, Van Nostrand, 1964)*

absorbing rod   See control rod.

absorbing state   [MATH] A special case of recurrent state in a Markov process in which the transition probability, P_{ii}, equals 1; a process will never leave an absorbing state once it enters.

absorbing well   [CIV ENG] A shaft that permits water to drain through an impermeable stratum to a permeable stratum.

absorptance   [PHYS] The ratio of the total unabsorbed radiation to the total incident radiation; equal to one (unity) minus the transmittance.

absorptiometer   [ANALY CHEM] 1. An instrument equipped with a filter system or other simple dispersing system to measure the absorption of nearly monochromatic radiation in the visible range by a gas or a liquid, and so determine the concentration of the absorbing constituents in the gas or liquid. 2. A device for regulating the thickness of a liquid in spectrophotometry.

absorptiometric analysis   [ANALY CHEM] Chemical analysis of a gas or a liquid by measurement of the peak electromagnetic absorption wavelengths that are unique to a specific material or element.

absorption   [CHEM] The taking up of matter in bulk by other matter, as in dissolving of a gas by a liquid. [ELEC] The property of a dielectric in a capacitor which causes a small charging current to flow after the plates have been brought up to the final potential, and a small discharging current to flow after the plates have been short-circuited, allowed to stand for a few minutes, and short-circuited again. Also known as dielectric soak. [ELECTROMAG] The taking up of energy from radiation by the medium through which the radiation is passing. [HYD] The entrance of surface water into the lithosphere. [PHYSIO] Passage of a chemical substance through a body membrane.

absorption atelectasis   See obstructive atelectasis.

absorption band   [PHYS] A range of wavelengths or frequencies in the electromagnetic spectrum within which radiant energy is absorbed by a substance.

absorption circuit   [ELECTR] A series-resonant circuit used to absorb power at an unwanted signal frequency by providing a low impedance to ground at this frequency.

absorption coefficient   Also known as absorption factor; absorption ratio; coefficient of absorption. [ACOUS] The ratio of the sound energy absorbed by a surface of a medium or material to the sound energy incident on the surface. [PHYS] If a flux through a material decreases with distance x in proportion to e^{-ax}, then a is called the absorption coefficient.

absorption column   See absorption tower.

absorption constant   See absorptivity.

absorption control   [ELECTR] See absorption modulation. [NUCLEO] Control of a nuclear reactor by a material that absorbs neutrons, such as cadmium or boron steel. Also known as absorber control.

absorption cross section   [ELECTROMAG] In radar, the ratio of the amount of power removed from a beam by absorption of radio energy by a target to the power in the beam incident upon the target.

absorption current   [ELEC] The component of a dielectric current that is proportional to the rate of accumulation of electric charges within the dielectric.

absorption curve   [PHYS] A graph showing the curvilinear relationship of the variation in absorbed radiation as a function of wavelength.

absorption cycle   [MECH ENG] In refrigeration, the process whereby a circulating refrigerant, for example, ammonia, is evaporated by heat from an aqueous solution at elevated pressure and subsequently reabsorbed at low pressure, displacing the need for a compressor.

absorption dynamometer   [ENG] A device for measuring mechanical forces or power in which the mechanical energy input is absorbed by friction or electrical resistance.

absorption edge   [SPECT] The wavelength corresponding to a discontinuity in the variation of the absorption coefficient of a substance with the wavelength of the radiation. Also known as absorption limit.

absorption-emission pyrometer   [ENG] A thermometer for determining gas temperature from measurement of the radiation emitted by a calibrated reference source before and after

this radiation has passed through and been partially absorbed by the gas.

absorption factor   See absorption coefficient.

absorption fading   [COMMUN] Slow type of fading, primarily caused by variations in the absorption rate along the radio path.

absorption gasoline   [MATER] A gasoline obtained by using an oil to absorb the natural or refinery gas containing the gasoline and then distilling it from the oil.

absorption hygrometer   [ENG] An instrument with which the water vapor content of the atmosphere is measured by means of the absorption of vapor by a hygroscopic chemical.

absorption index   [OPTICS] The complex index of refraction may be written as $n(1 + ik)$; the coefficient k is the absorption index. Also known as index of absorption.

absorption lens   [OPTICS] Glass which prevents selected wavelengths from passing through it; used in eyeglasses.

absorption limit   See absorption edge.

absorption line   [SPECT] A minute range of wavelength or frequency in the electromagnetic spectrum within which radiant energy is absorbed by the medium through which it is passing.

absorption loss   [COMMUN] That part of the transmission loss due to the dissipation or conversion of either sound energy or electromagnetic energy into other forms of energy, either within the medium or attendant upon a reflection.

absorption meter   [ENG] An instrument designed to measure the amount of light transmitted through a transparent substance, using a photocell or other light detector.

absorption modulation   [ELECTR] A system of amplitude modulation in which a variable-impedance device is inserted in or coupled to the output circuit of the transmitter. Also known as absorption control; loss modulation.

absorption number   [ENG] A dimensionless group used in the field of gas absorption in a wetted-wall column; represents the liquid side mass-transfer coefficient.

absorption oil   [MATER] A petroleum or coal tar oil that is contacted with a vapor or gas mixture to remove heavy components, as in the recovery of natural gasoline from wet natural gas. Also known as absorber oil; scrubbing oil; wash oil.

absorption peak   [SPECT] A wavelength of maximum electromagnetic absorption by a chemical sample; used to identify specific elements, radicals, or compounds.

absorption plant   [CHEM ENG] A facility to recover the condensable portion of natural or refinery gas.

absorption ratio   See absorption coefficient.

absorption refrigeration   [MECH ENG] Refrigeration in which cooling is effected by the expansion of liquid ammonia into gas and absorption of the gas by water; the ammonia is reused after the water evaporates.

absorption spectrophotometer   [SPECT] An instrument used to measure the relative intensity of absorption spectral lines and bands. Also known as difference spectrophotometer.

absorption spectroscopy   [SPECT] The study of spectra obtained by the passage of radiant energy from a continuous source through a cooler, selectively absorbing medium.

absorption spectrum   [SPECT] The array of absorption lines and absorption bands which results from the passage of radiant energy from a continuous source through a cooler, selectively absorbing medium.

absorption test   [IMMUNOL] Analysis of the antigenic components of bacterial cells and large macromolecules by a series of precipitation or agglutination reactions with specific antibodies.

absorption tower   [ENG] A vertical tube in which a rising gas is partially absorbed by a liquid in the form of falling droplets. Also known as absorption column.

absorption tube   [CHEM] A tube filled with a solid absorbent and used to absorb gases and vapors.

absorption unit   See sabin.

absorption wavemeter   [ELECTR] A frequency- or wavelength-measuring instrument consisting of a calibrated tunable circuit and a resonance indicator.

absorptive power   See absorptivity.

absorptivity   [ANALY CHEM] The constant a in the Beer's law relation $A = abc$, where A is the absorbance, b is the path

ABSORPTION CYCLE

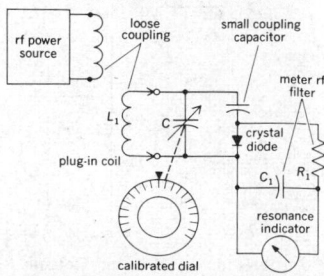

high-pressure refrigeration vapor

Basic absorption cycle for an air-conditioning system.

ABSORPTION WAVEMETER

Schematic diagram of inductance-capacitance type of absorption wavemeter (for frequencies between approximately 50 kilohertz and 1000 megahertz).

length, and *c* the concentration of solution. Also known as absorptive power. Formerly known as absorbency index; absorption constant; extinction coefficient.

absorptivity-emissivity ratio [ASTROPHYS] In space applications, the ratio of absorptivity for solar radiation of a material to its infrared emissivity. Also known as A/E ratio.

ABS resin *See* acrylonitrile butadiene styrene resin.

abstinence syndrome [MED] A disturbance of metabolic equilibrium that occurs when a narcotic drug is withdrawn from the user.

abstract algebra [MATH] The study of mathematical systems consisting of a set of elements, one or more binary operations by which two elements may be combined to yield a third, and several rules (axioms) for the interaction of the elements and the operations; includes group theory, ring theory, and number theory.

abstract automata theory [ADP] The mathematical theory which characterizes automata by three sets: input signals, internal states, and output signals; and two functions: input functions and output functions.

abstraction [HYD] The draining of water from a stream by another having more rapid corroding action.

abstract theory [SCI TECH] A theory in which a system is described without specifying a structure.

abstriction [MYCOL] In fungi, the cutting off of spores in hyphae by formation of septa followed by abscission of the spores, especially by constriction.

abT *See* gauss.

abterminal [BIOL] Referring to movement from the end toward the middle; specifically, describing the mode of electric current flow in a muscle.

abtesla *See* gauss.

Abt track [CIV ENG] One of the cogged rails used for railroad tracking in mountains and so arranged that the cogs are not opposite one another on any pair of rails.

Abukuma-type facies [PETR] A type of dynathermal regional metamorphism characterized by low pressure.

abulia [PSYCH] Loss of ability to make decisions.

abundance ratio [NUCLEO] The ratio of the number of atoms of one isotope to the total number of atoms in a mixture of isotopes.

aburton [NAV ARCH] Of an object, having its length directed across a ship from side to side.

abutment [CIV ENG] A surface or mass provided to withstand thrust; for example, end supports of an arch or a bridge.

abvolt [ELEC] The unit of electromotive force in the electromagnetic centimeter-gram-second system; 1 abvolt equals 10^{-8} volt in the absolute meter-kilogram-second system. Abbreviated aV.

abvolt per centimeter [ELEC] The electromagnetic centimeter-gram-second unit of electric field strength. Abbreviated aV/cm.

abwatt [ELEC] The unit of electrical power in the centimeter-gram-second system; 1 abwatt equals 1 watt in the absolute meter-kilogram-second system.

abWb *See* maxwell.

abweber *See* maxwell.

abyssal [GEOL] *See* plutonic. [OCEANOGR] Pertaining to the abyssal zone.

abyssal-benthic [OCEANOGR] Pertaining to the bottom of the abyssal zone.

abyssal cave *See* submarine fan.

abyssal fan *See* submarine fan.

abyssal floor [GEOL] The ocean floor, or bottom of the abyssal zone.

abyssal gap [GEOL] A gap in a sill, ridge, or rise that lies between two abyssal plains.

abyssal hill [GEOL] A hill 2000 to 3000 feet (600 to 900 meters) high and a few miles wide within the deep ocean.

abyssal injection [GEOL] The process of driving magmas, originating at considerable depths, up through deep-seated contraction fissures in the earth's crust.

abyssal plain [GEOL] A flat, almost level area occupying the deepest parts of many of the ocean basins.

abyssal rock [GEOL] Plutonic, or deep-seated, igneous rocks.

abyssal theory [GEOL] A theory of the origin of ores involving the separation of ore silicates from the liquid stage during the cooling of the earth.

abyssal zone [OCEANOGR] The biogeographic realm of the great depths of the ocean beyond the limits of the continental shelf, generally below 1000 meters.

abyssolith [GEOL] A molten mass of eruptive material passing up without a break from the zone of permanently molten rock within the earth.

abyssopelagic [OCEANOGR] Pertaining to the open waters of the abyssal zone.

ac *See* alternating current.

aC *See* abcoulomb.

Ac *See* actinium; altocumulus cloud.

Ac_0 [MET] The temperature at which a magnetic change occurs in cementite; the Curie point of cementite.

Ac_1 [MET] The temperature at which austenite begins to be formed upon heating a steel.

Ac_2 [MET] The Curie point of ferrite.

Ac_3 [MET] The temperature at which the transformation of ferrite to austenite is completed upon heating a steel.

Ac_4 [MET] The temperature at which delta iron is formed from gamma iron upon heating a steel.

acacia gum *See* gum arabic.

Acadian orogeny [GEOL] The period of formation accompanied by igneous intrusion that took place during the Middle and Late Devonian in the Appalachian Mountains.

Acala [BOT] A type of cotton indigenous to Mexico and cultivated in Texas, Oklahoma, and Arkansas.

acalyculate [BOT] Lacking a calyx.

Acalyptratae [INV ZOO] A large group of small, two-winged flies in the suborder Cyclorrhapha characterized by small or rudimentary calypters.

acantha [BIOL] A sharp spine; a spiny process, as on vertebrae.

Acanthaceae [BOT] A family of dicotyledonous plants in the order Scrophulariales distinguished by their usually herbaceous habit, irregular flowers, axile placentation, and dry, dehiscent fruits.

acanthaceous [BOT] Having sharp points or prickles; prickly.

Acantharia [INV ZOO] A subclass of essentially pelagic protozoans in the class Actinopodea characterized by skeletal rods constructed of strontium sulfate (celestite).

Acanthaster [INV ZOO] A genus of Indo-Pacific starfishes, including the crown-of-thorns, of the family Asteriidae; economically important as a destroyer of oysters in fisheries.

acanthella [INV ZOO] A transitional larva of the phylum Acanthocephala in which rudiments of reproductive organs, lemnisci, a proboscis, and a proboscis receptacle are formed.

acanthite [MINERAL] Ag_2S A blackish to lead-gray silver sulfide mineral, crystallizing in the orthorhombic system.

acanthocarpous [BOT] Having spiny fruit.

Acanthocephala [INV ZOO] The spiny-headed worms, a phylum of helminths; adults are parasitic in the alimentary canal of vertebrates.

Acanthocheilonema perstans [INV ZOO] A tropical filarial worm, parasitic in man.

acanthocheilonemiasis [MED] A parasitic infection of man caused by the filarial nematode *Acanthocheilonema perstans.*

acanthocladous [BOT] Having spiny branches.

acanthocytosis [MED] A disorder of erythrocytes in which spiny projections appear on the blood cells.

Acanthodes [PALEON] A genus of Carboniferous and Lower Permian eellike acanthodian fishes of the family Acanthodidae.

Acanthodidae [PALEON] A family of extinct acanthodian fishes in the order Acanthodiformes.

Acanthodiformes [PALEON] An order of extinct fishes in the class Acanthodii having scales of acellular bone and dentine, one dorsal fin, and no teeth.

Acanthodii [PALEON] A class of extinct fusiform fishes, the first jaw-bearing vertebrates in the fossil record.

acanthoid [BIOL] Shaped like a spine.

Acanthometrida [INV ZOO] An order of marine protozoans in the subclass Acantharia with 20 or less skeletal rods.

Acanthophis antarcticus [VERT ZOO] The death adder, a venomous snake found in Australia and New Guinea; venom is neurotoxic.

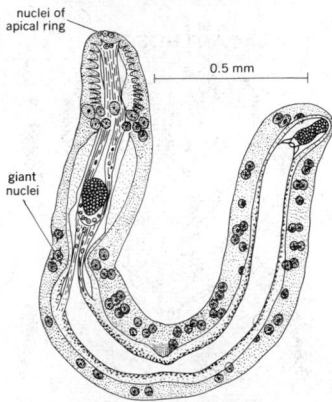

ACANTHELLA

nuclei of apical ring

0.5 mm

giant nuclei

A stage in the life history of *Moniliformis dubius*, a helminth, with the acanthella dissected from its enveloping sheath.

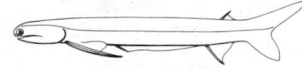

ACANTHODES

A lateral view of *Acanthodes* species, about 30 centimeters long. (*After D. M. S. Watson*)

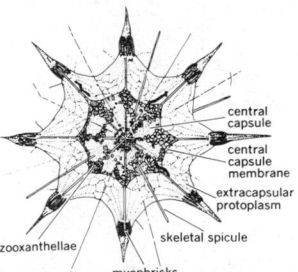

ACANTHOMETRIDA

central capsule

central capsule membrane

extracapsular protoplasm

skeletal spicule

zooxanthellae

myophrisks

A drawing of *Acanthometra* showing the characteristic pattern of the radially arranged rods (skeletal spicules). (*From L. H. Hyman, The Invertebrates, vol. 1, McGraw-Hill, 1940*)

Acanthophractida [INV ZOO] An order of marine protozoans in the subclass Acantharia; skeleton includes a latticework shell and skeletal rods.

acanthopodous [BOT] Having a spiny or prickly petiole or peduncle.

acanthopore [PALEON] A tubular spine in some fossil bryozoans.

Acanthopteri [VERT ZOO] An equivalent name for the Perciformes.

Acanthopterygii [VERT ZOO] An equivalent name for the Perciformes.

acanthosis [MED] Any thickening of the prickle-cell layer of the epidermis; associated with many skin diseases.

acanthosoma [INV ZOO] The last primitive larval stage, the mysis, in the family Sergestidae.

Acanthosomatidae [INV ZOO] A small family of insects in the order Hemiptera.

acanthosphere [BOT] A specialized ciliated body in *Nitella* cells.

acanthostegous [INV ZOO] Being overlaid with two series of spines, as the ovicell or ooecium of certain bryozoans.

acanthozooid [INV ZOO] A specialized individual in a bryozoan colony that secretes tubules which project as spines above the colony's outer surface.

Acanthuridae [VERT ZOO] The surgeonfishes, a family of perciform fishes in the suborder Acanthuroidei.

Acanthuroidei [VERT ZOO] A suborder of chiefly herbivorous fishes in the order Perciformes.

acanthus [ARCH] A sculptured ornamentation representing the leaves of an *Acanthus,* a Mediterranean prickly herb.

acapnia [MED] Absence of carbon dioxide in the blood and tissues.

Acari [INV ZOO] The equivalent name for Acarina.

acariasis [MED] Any skin disease resulting from infestation with acarids or mites.

acaricide [MATER] A pesticide used to destroy mites on domestic animals, crops, and man. Also known as miticide.

Acaridiae [INV ZOO] A group of pale, weakly sclerotized mites in the suborder Sarcoptiformes, including serious pests of stored food products and skin parasites of warm-blooded vertebrates.

Acarina [INV ZOO] The ticks and mites, a large order of the class Arachnida, characterized by lack of body demarcation into cephalothorax and abdomen.

acaroid resin [ORG CHEM] A gum resin from certain aloelike trees of the genus *Xanthorrhoea* in Australia and Tasmania; used in varnishes and inks. Also known as gum accroides; yacca gum.

acarology [INV ZOO] A branch of zoology dealing with the mites and ticks.

acarophily [ECOL] A symbiotic relationship between plants and mites.

acarophobia [PSYCH] Abnormal fear of mites.

acarpellous [BOT] Lacking carpels.

acarpous [BOT] Not producing fruit.

acatalasemia [MED] Lack of catalase in the blood.

acatalasia [MED] Congenital absence of the enzyme catalase.

acatamathesia [MED] **1.** Inability to understand conversation. **2.** Morbid blunting or deterioration of the senses, as in mental deafness and blindness.

acaulous [BOT] **1.** Lacking a stem. **2.** Being apparently stemless but having a short underground stem.

acaustobiolith [PETR] A noncombustible organic rock, or one formed by organic accumulation of minerals.

acaustophytolith [PETR] An acaustobiolith resulting from plant activity, such as a pelagic ooze that contains diatoms.

accelerated erosion [GEOL] The process of weathering at a rate greater than normal for the site, brought about by man, usually through reduction of the vegetation.

accelerated hypertension *See* malignant hypertension.

accelerated test [ELEC] A test of the serviceability of an electric cable in use for some time by applying twice the voltage normally carried.

accelerating agent [MATER] **1.** A substance which increases the speed of a chemical reaction. **2.** A compound which hastens and improves the curing of rubber.

ACANTHOSOMA

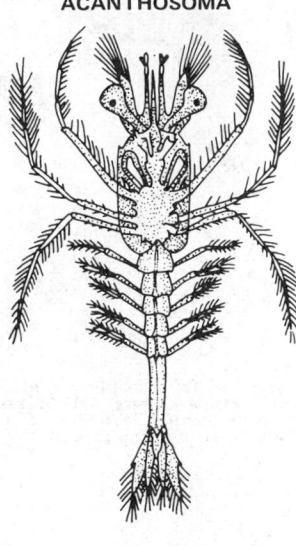

1 mm

The mysis (acanthosoma) larva of sergestid shrimp.
(Smithsonian Institution)

accelerating electrode [ELECTR] An electrode used in cathode-ray tubes and other electron tubes to increase the velocity of the electrons that contribute the space current or form a beam.

accelerating incentive *See* differential piece-rate system.

acceleration [MECH] The rate of change of velocity with respect to time.

acceleration analysis [MECH ENG] A mathematical technique, often done graphically, by which accelerations of parts of a mechanism are determined.

acceleration error [NAV] That error resulting from change in a craft's velocity vector: specifically, either the deviation of an aircraft magnetic compass caused by the action of the vertical component of the earth's magnetic field on the compass magnets when the compass card is thrown off level by accelerations of the aircraft; or the deflection of the apparent vertical, as indicated by an artificial horizon, due to acceleration.

acceleration error constant [CONT SYS] The ratio of the acceleration of a controlled variable of a servomechanism to the actuating error when the actuating error is constant.

acceleration feedback [AERO ENG] The use of accelerometers strategically located within the body of a missile so that they sense body accelerations during flight and interact with another device on board the missile or with a control center on the ground or in an airplane to keep the missile's speed within design limits.

acceleration globulin [BIOCHEM] A globulin that acts to accelerate the conversion of prothrombin to thrombin in blood clotting; found in blood plasma in an inactive form.

acceleration measurement [MECH] The technique of determining the magnitude and direction of acceleration, including translational and angular acceleration. [NAV] A fundamental measurement required for the operation of the inertial navigator.

acceleration mechanisms [ASTROPHYS] The ways in which cosmic-ray and solar-flare particles may have acquired their high energies.

acceleration of gravity [MECH] The acceleration imparted to bodies by the attractive force of the earth; has an international standard value of 980.665 cm/sec² but varies with latitude and elevation. Also known as apparent gravity.

acceleration stress [MED] The effect of an increase in gravitational force upon man's physiology and behavior, particularly during takeoff and reentry in space flight.

acceleration time [ADP] The time required for a magnetic tape transport or any other mechanical device to attain its operating speed.

acceleration tolerance [ENG] The degree to which personnel or equipment withstands acceleration. [PHYSIO] The maximum *g* forces an individual can withstand without losing control or consciousness.

acceleration voltage [ELECTR] The voltage between a cathode and accelerating electrode of an electron tube.

accelerator [MATER] Any substance added to stucco, plaster, mortar, concrete, cement, and so on to hasten the set. [MECH ENG] A device for varying the speed of an automotive vehicle by varying the supply of fuel. [PHYS] *See* particle accelerator.

accelerator catalyst [MATER] A catalyst that increases the rate of a chemical reaction.

accelerator jet [MECH ENG] The jet through which the fuel is injected into the incoming air in the carburetor of an automotive vehicle with rapid demand for increased power output.

accelerator linkage [MECH ENG] The linkage connecting the accelerator pedal of an automotive vehicle to the carburetor throttle valve or fuel injection control.

accelerator pedal [MECH ENG] A pedal that operates the carburetor throttle valve or fuel injection control of an automotive vehicle.

accelerator pump [MECH ENG] A small cylinder and piston controlled by the throttle of an automotive vehicle so as to provide an enriched air-fuel mixture during acceleration.

acceleratory reflex [PHYSIO] Any reflex originating in the labyrinth of the inner ear in response to a change in the rate of movement of the head.

accelerogram [ENG] A record made by an accelerograph.

accelerograph [ENG] An accelerometer having provisions for recording the acceleration of a point on the earth during an earthquake or for recording any other type of acceleration.

accelerometer [ENG] An instrument which measures acceleration or gravitational force capable of imparting acceleration.

accelofilter [CHEM] A filtration device that uses a vacuum or pressure to draw or force the liquid through the filter to increase the rate of filtration.

accentuation [ELECTR] The enhancement of signal amplitudes in selected frequency bands with respect to other signals.

accentuator [ELECTR] A circuit that provides for the first part of a process for increasing the strength of certain audio frequencies with respect to others, to help these frequencies override noise or to reduce distortion. Also known as accentuator circuit.

accentuator circuit *See* accentuator.

accept [ADP] A data transmission statement which is used in FORTRAN when the computer is in conversational mode, and which enables the programmer to input, through the teletypewriter, data the programmer wishes stored in memory.

acceptability [ENG] State or condition of meeting minimum standards for use, as applied to methods, equipment, or consumable products.

acceptable quality level [IND ENG] The maximum percentage of defects that has been determined tolerable as a process average for a sampling plan during inspection or test of a product with respect to economic and functional requirements of the item. Abbreviated AQL.

acceptable reliability level [IND ENG] The required level of reliability for a part, system, device, and so forth; may be expressed in a variety of terms, for example, number of failures allowable in 1000 hours of operating life. Abbreviated ARL.

acceptance criteria [IND ENG] Standards of judging the acceptability of manufactured items.

acceptance number [IND ENG] The maximum allowable number of defective pieces in a sample of specified size.

acceptance test [IND ENG] A test used to determine conformance of a product to design specifications, as a basis for its acceptance.

accepted indicator [NAV] An airborne indicator which has been proven capable of accurate and reliable measurement.

acceptor [CHEM] 1. A chemical whose reaction rate with another chemical increases because the other substance undergoes another reaction. 2. A species that accepts electrons, protons, electron pairs, or molecules such as dyes. [SOLID STATE] An impurity element that increases the number of holes in a semiconductor crystal such as germanium or silicon; aluminum, gallium, and indium are examples. Also known as acceptor impurity; acceptor material.

acceptor atom [SOLID STATE] An atom of a substance added to a semiconductor crystal to increase the number of holes in the conduction band.

acceptor circuit [ELECTR] A series-resonant circuit that has a low impedance at the frequency to which it is tuned and a higher impedance at all other frequencies.

acceptor impurity *See* acceptor.

acceptor material *See* acceptor.

access [ADP] The reading of data from storage or the writing of data into storage. [CIV ENG] Freedom, ability, or the legal right to pass without obstruction from a given point on earth to some other objective, such as the sea or a public highway.

access arm [ADP] The mechanical device which positions the read/write head on a magnetic storage unit.

access code [COMMUN] 1. Numeric identification for internetwork or facility switching. 2. The preliminary digits that a user must dial to be connected through an automatic PBX to the serving switching center, as in AUTOVON.

access-control register [ADP] A storage device which controls the word-by-word transmission over a given channel.

access-control words [ADP] Permanently wired instructions channeling transmitted words into reserved locations.

access door [BUILD] A provision for access to concealed plumbing or other equipment without disturbing the wall or fixtures.

access line [COMMUN] Four-wire circuit between a subscriber or a local PBX to the serving switching center.

access method [ADP] A set of programming routines which links programs and the data that these programs transfer into and out of memory.

access mode [ADP] A programming clause in COBOL which is required when using a random-access device so that a specific record may be read out of or written into a mass storage bin.

accessorius [ANAT] Any muscle that reinforces the action of another.

accessory [MECH ENG] A part, subassembly, or assembly that contributes to the effectiveness of a piece of equipment without changing its basic function; may be used for testing, adjusting, calibrating, recording, or other purposes.

accessory bud [BOT] An embryonic shoot occurring above or to the side of an axillary bud. Also known as supernumerary bud.

accessory cell [BOT] A morphologically distinct epidermal cell adjacent to, and apparently functionally associated with, guard cells on the leaves of many plants.

accessory chromosome *See* supernumerary chromosome.

accessory cloud [METEOROL] A cloud form that is dependent, for its formation and continuation, upon the existence of one of the major cloud genera; may be an appendage of the parent cloud or an immediately adjacent cloudy mass.

accessory ejecta [GEOL] Pyroclastic material formed from solidified volcanic rocks that are from the same volcano as the ejecta.

accessory element *See* trace element.

accessory gland [ANAT] A mass of glandular tissue separate from the main body of a gland. [INV ZOO] A gland associated with the male reproductive organs in insects.

accessory mineral [MINERAL] A minor mineral in an igneous rock that does not affect its general character.

accessory movement *See* synkinesia.

accessory nerve [ANAT] The eleventh cranial nerve in tetrapods, a paired visceral motor nerve; the bulbar part innervates the larynx and pharynx, and the spinal part innervates the trapezius and sternocleidomastoid muscles.

accessory plate [OPTICS] Thin plate of quartz, gypsum, or mica used with a petrological microscope to modify the effects of polarized light and intensify qualities in translucent minerals.

access road [CIV ENG] A route, usually paved, that enables vehicles to reach a designated facility expeditiously.

access time [ADP] The time period required for reading out of or writing into the computer memory. Also known as read time.

access tunnel [CIV ENG] A tunnel provided for an access road.

accident [HYD] An interruption in a river that interferes with, or sometimes stops, the normal development of the river system.

accidental ejecta [GEOL] Pyroclastic rock formed from preexisting nonvolcanic rocks or from volcanic rocks unrelated to the erupting volcano.

accidental error [SCI TECH] In experimental observations, an error which does not always recur when an observation is repeated under the same conditions.

accidental inclusion *See* xenolith.

accidental point [GRAPHICS] The vanishing point of a group of lines in a perspective drawing that are parallel neither to the line of sight nor to the horizon.

accidental whorl [ANAT] A type of whorl fingerprint pattern which is a combination of two different types of pattern, with the exception of the plain arch, with two or more deltas; or a pattern which possesses some of the requirements for two or more different types; or a pattern which conforms to none of the definitions; in accidental whorl tracing three types appear: an outer (O), inner (I), or meeting (M).

accident block [GEOL] A solid chip of rock broken off from the subvolcanic basement and ejected from a volcano.

accident-cause code [IND ENG] Sponsored by the American Standards Association, the code that classifies accidents

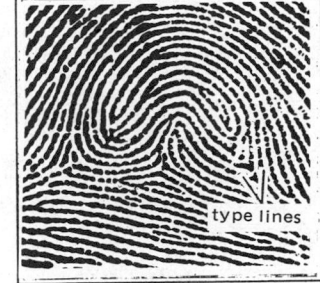

ACCIDENTAL WHORL

A reproduction of an I tracing of an accidental whorl. *(Federal Bureau of Investigation)*

under eight defective working conditions and nine improper working practices.

accident frequency rate [IND ENG] The number of all disabling injuries per million man-hours of exposure.

accident prone [MED] Predisposed to sustain more accidents than others exposed to the same hazard.

accident severity rate [IND ENG] The number of man-days lost as a result of disabling injuries per thousand man-hours of exposure.

Accipitridae [VERT ZOO] The diurnal birds of prey, the largest and most diverse family of the order Falconiformes, including hawks, eagles, and kites.

acclimated microorganism [ECOL] Any microorganism that is able to adapt to environmental changes such as a change in temperature, or a change in the quantity of oxygen or other gases.

acclimation *See* acclimatization.

acclimatization [EVOL] Adaptation of a species or population to a changed environment over several generations. Also known as acclimation.

acclivity [GEOL] A slope that is ascending from a reference point.

aCcm *See* abcoulomb centimeter.

Ac$_{cm}$ [MET] The temperature at which the solution of cementite in austenite is completed upon heating a hypereutectoid steel.

aC/cm² *See* abcoulomb per square centimeter.

aC/cm³ *See* abcoulomb per cubic centimeter.

accolade [ARCH] Decorative molding in which two ogee curves meet centrally over the top of a window or door.

accommodation [MAP] The limits or range within which a stereo-plotting instrument is capable of operating. [PHYSIO] A process in most vertebrates whereby the focal length of the eye is changed by automatic adjustment of the lens to bring images of objects from various distances into focus on the retina.

accommodation coefficient [STAT MECH] The ratio of the average energy actually transferred between a surface and impinging gas molecules scattered by the surface, to the average energy which would theoretically be transferred if the impinging molecules reached complete thermal equilibrium with the surface.

accommodation ladder [NAV ARCH] A light ladder or similar structure, usually portable, hung over a ship's side at the gangway to permit access to small boats.

accommodation reflex [PHYSIO] Changes occurring in the eyes when vision is focused from a distant to a near object; involves pupil contraction, increased lens convexity, and convergence of the eyes.

accordant fold [GEOL] One of several folds that are similarly oriented.

accordant summit level [GEOL] A hypothetical horizontal plane that can be drawn over a broad region connecting mountain summits of similar elevation.

accordion door [BUILD] A door that folds and unfolds like an accordion when it is opened and closed.

accordion roller conveyor [MECH ENG] A conveyor with a flexible latticed frame which permits variation in length.

accounting machine [ADP] A machine that produces tabulations or accounting records of a specified unvarying format.

accounting package [ADP] A set of special routines that allow collection of information about the usage level of various components of a computer system by each production program.

accouplement [ARCH] A pair of elements of a structure that are very close or touching, such as two columns.

accrescent [BOT] Growing continuously with age, especially after flowering.

accretion [CIV ENG] Artificial buildup of land due to the construction of a groin, breakwater, dam, or beach fill. [GEOL] **1.** Gradual buildup of land on a shore due to wave action, tides, currents, airborne material, or alluvial deposits. **2.** The process whereby stones or other inorganic masses add to their bulk by adding particles to their surfaces. Also known as aggradation. [METEOROL] The growth of a precipitation particle by the collision of a frozen particle (ice crystal or snowflake) with a supercooled liquid droplet which freezes upon contact.

accretionary lapilli *See* mud ball.

accretionary lava ball [GEOL] A rounded ball of lava that occurs on the surface of an aa lava flow.

accretionary limestone [PETR] A type of limestone formed by the slow accumulation of organic remains.

accretionary ridge [GEOL] A beach ridge located inland from the modern beach, indicating that the coast has been built seaward.

accretion hypothesis [ASTRON] Any hypothesis which assumes that the earth originated by the gradual addition of solid bodies, such as meteorites, that were formerly revolving about the sun but were drawn by gravitation to the earth.

accretion line [HISTOL] A microscopic line on a tooth, marking the addition of a layer of enamel or dentin.

accretion theory [ASTRON] A theory that the solar system originated from vortices in a disk-shaped mass.

accretion topography [GEOL] Topographic features built by accumulation of sediment.

accretion vein [GEOL] A type of vein formed by the repeated filling of channels followed by their opening because of the development of fractures in the zone undergoing mineralization.

accretion zone [GEOL] Any beach area undergoing accretion.

accumbent [BOT] Describing an organ that leans against another; specifically referring to cotyledons having their edges folded against the hypocotyl.

accumulated discrepancy [ENG] The sum of the separate discrepancies which occur in the various steps of making a survey.

accumulated divergence [MAP] In making a map, the algebraic sum of the divergences for the sections of a line of levels, from the beginning of the line to any section end at which it is desired to compute the total divergence.

accumulated total punching [ADP] A checking procedure to ensure that no punch-card item has been dropped from a file.

accumulating reproducer [ADP] An electromechanical device which reads a sorted deck of cards and creates a set of subtotals on additional cards according to some preset criterion.

accumulation [HYD] The quantity of snow or other solid form of water added to a glacier or snowfield by alimentation. [MIN ENG] **1.** In coal mining, firedamp that collects in higher parts of mine workings and at the edge of wastes. **2.** Oil or gas in some form of trap.

accumulation area [HYD] The portion of a glacier above the firn line, where the accumulation exceeds ablation. Also known as firn field; zone of accumulation.

accumulation factor [MATH] The quantity $(1+r)$ in the formula for compound interest, where r is the rate of interest; measures the rate at which the principal grows.

accumulation point *See* cluster point.

accumulation zone [GEOL] The area where the bulk of the snow contributing to an avalanche was originally deposited.

accumulative error *See* cumulative error.

accumulative timing [IND ENG] A time-study method that allows direct reading of the time for each element of an operation by the use of two stopwatches which operate alternately.

accumulator [ADP] A specific register, in the arithmetic unit of a computer, in which the result of an arithmetic or logical operation is formed; here numbers are added or subtracted, and certain operations such as sensing, shifting, and complementing are performed. Also known as accumulator register; counter. [AERO ENG] A device sometimes incorporated in the fuel system of a gas-turbine engine to store fuel and release it under pressure as an aid in starting. [ELEC] *See* storage battery. [MECH ENG] **1.** A device, such as a bag containing pressurized gas, which acts upon hydraulic fluid in a vessel, discharging it rapidly to give high hydraulic power, after which the fluid is returned to the vessel with the use of low hydraulic power. **2.** A device connected to a steam boiler to enable a uniform boiler output to meet an irregular steam demand.

accumulator battery *See* storage battery.

accumulator jump instruction [ADP] An instruction which programs a computer to ignore the previously established program sequence depending on the status of the accumulator. Also known as accumulator transfer instruction.

accumulator plant [BOT] A plant or tree that grows in a metal-bearing soil and accumulates an abnormal content of the metal.

accumulator register *See* accumulator.

accumulator shift instruction [ADP] A computer instruction which causes the word in a register to be displaced a specified number of bit positions to the left or right.

accumulator transfer instruction *See* accumulator jump instruction.

accuracy [SCI TECH] The extent to which the results of a calculation or the readings of an instrument approach the true values of the calculated or measured quantities, and are free from error.

accuracy checking [MAP] The procurement of presumptive evidence of a map's compliance with specified accuracy standards; indicates the relative (rather than the absolute) accuracy of map features.

accuracy control system [ADP] Any method which attempts error detection and control, such as random sampling and squaring.

accuracy life [ORD] The estimated average number of rounds that a particular weapon can fire before its tube becomes so worn that its accuracy tolerance is exceeded.

accuracy of fire [ORD] The measurement of the precision of fire expressed as the distance of the center of impact from the center of the target.

accuracy testing [MAP] The procurement of confirmed evidence, on a sampling basis, of a map's compliance with specified accuracy standards; indicates both the relative and absolute accuracy of map features.

accurate contour [MAP] A contour line whose accuracy lies within one-half of the basic vertical interval. Also known as normal contour.

accustomization [ENG] The process of learning the techniques of living with a minimum of discomfort in an extreme or new environment.

ac/dc motor *See* universal motor.

ac/dc receiver [ELECTR] A radio receiver designed to operate from either an alternating- or direct-current power line. Also known as universal receiver.

Acele [TEXT] Trade name of an acetate fiber made by Du-Pont.

acellular [BIOL] Not composed of cells.

acellular gland [PHYSIO] A gland, such as intestinal glands, the pancreas, and the parotid gland, that secretes a noncellular product.

acellular slime mold [MYCOL] The common name for members of the Myxomycetes.

Ac-Em *See* actinon.

acenaphthene [ORG CHEM] $C_{12}H_8$ An unsaturated hydrocarbon whose colorless crystals melt at 92°C; insoluble in water; used as a dye intermediate and as an agent for inducing polyploidy.

acentric [BIOL] Not oriented around a middle point. [GEN] A chromosome or chromosome fragment lacking a centromere.

acentrous [VERT ZOO] Lacking vertebral centra and having the notochord persistent throughout life, as in certain primitive fishes.

Acephalina [INV ZOO] A suborder of invertebrate parasites in the protozoan order Eugregarinida characterized by nonseptate trophozoites.

acephalobrachia *See* abrachiocephalia.

acephalocardia [MED] Congenital lack of a head and a heart.

acephalochiria [MED] Congenital lack of a head and hands.

acephalocyst [INV ZOO] An abnormal cyst of the *Echinococcus granulosus* larva, lacking a head and brood capsules, found in human organs.

acephalopodia [MED] Congenital lack of a head and feet.

acephalorrhachia [MED] Congenital lack of a head and vertebral column.

acephalostomia [MED] Congenital lack of a head, with a mouthlike orifice in the neck or chest.

acephalothoracica [MED] Congenital lack of a head and thorax.

acephalous [BOT] Having the style originate at the base instead of at the apex of the ovary. [ZOO] Lacking a head.

Acer [BOT] A genus of broad-leaved, deciduous trees of the order Sapindales, commonly known as the maples; the sugar or rock maple (*A. saccharum*) is the most important commercial species.

acerate [BOT] Needle-shaped, specifically referring to leaves.

acerbophobia [PSYCH] Abnormal fear of sour taste sensations. Also known as acerophobia.

Acerentomidae [INV ZOO] A family of wingless insects belonging to the order Protura; the body lacks tracheae and spiracles.

acerophobia *See* acerbophobia.

acervate [BIOL] Growing in heaps or dense clusters.

acervulus [MYCOL] A cushion- or disk-shaped mass of hyphae, peculiar to the Melanconiales, on which there are dense aggregates of conidiophores.

acetabulum [ANAT] A cup-shaped socket on the hipbone that receives the head of the femur. [INV ZOO] **1.** A cavity on an insect body into which a leg inserts for articulation. **2.** The sucker of certain invertebrates such as trematodes and tapeworms.

acetal [ORG CHEM] **1.** $CH_3CH(OC_2H_5)_2$ A colorless, flammable, volatile liquid used as a solvent and in manufacture of perfumes. Also known as 1,1-diethoxyethane. **2.** Any of a class of compounds, stable ethers, formed from aldehydes and 1,1-dihydroxy alcohols.

acetaldehydase [BIOCHEM] An enzyme that catalyzes the oxidation of acetaldehyde to acetic acid.

acetaldehyde [ORG CHEM] C_2H_4O A colorless, flammable liquid used chiefly to manufacture acetic acid. Also known as ethanal.

para-**acetaldehyde** *See* paraldehyde.

acetaldehyde cyanohydrin *See* lactonitrile.

acetal resins [ORG CHEM] Linear, synthetic resins produced by the polymerization of formaldehyde (acetal homopolymers) or of formaldehyde with trioxane (acetal copolymers); hard, tough plastics used as substitutes for metals. Also known as polyacetals.

acetamide [ORG CHEM] CH_3CONH_2 The crystalline, colorless amide of acetic acid, used in organic synthesis and as a solvent. Also known as ethanamide.

acetaminophen [PHARM] $C_8H_9O_2N$ A drug used as an analgesic and antipyretic.

acetanilide [PHARM] $C_6H_5NHCOCH_3$ A white, crystalline compound used medicinally to relieve pain and reduce fever.

acetanisidine *See* methacetin.

acetate [ORG CHEM] One of two compounds derived from acetic acid, HC_2H_3O; one type is the negative acetate ion, $C_2H_3O_2^-$; the second type is an ester such as ethyl acetate. [TEXT] The official name for the textile fiber produced from partially hydrolyzed cellulose acetate. Formerly known as acetate rayon.

acetate film [MATER] A cellulose acetate resin sheet that is transparent, airproof, hygienic, and resistant to grease, oil, and dust; used for photographic film, magnetic tapes, and packaging.

acetate process [CHEM ENG] Acetylation of cellulose (wood pulp or cotton linters) with acetic acid or acetic anhydride and sulfuric acid catalyst to make cellulose acetate resin or fiber.

acetate rayon *See* acetate.

acetenyl *See* ethinyl.

acetic acid [ORG CHEM] CH_3COOH **1.** A clear, colorless liquid or crystalline mass with a pungent odor, miscible with water or alcohol; crystallizes in deliquescent needles; a component of vinegar. Also known as ethanoic acid. **2.** A mixture of the normal and acetic salts; used as a mordant in the dyeing of wool.

acetic acid bacteria *See* Acetobacter.

acetic anhydride [ORG CHEM] $(CH_3CO)_2O$ A liquid with a pungent odor that combines with water to form acetic acid; used as an acetylating agent.

acetic ester *See* ethyl acetate.

acetic ether *See* ethyl acetate.

acetic fermentation [MICROBIO] Oxidation of alcohol to produce acetic acid by the action of bacteria of the genus *Acetobacter*.

acetic thiokinase [BIOCHEM] An enzyme that catalyzes the formation of acetyl coenzyme A from acetate and adenosine-triphosphate.

acetidin *See* ethyl acetate.

acetin [ORG CHEM] An ester of acetic acid made by heating glycerin and acetic acid together; used as a solvent and in explosives. Also spelled acetine.

acetine *See* acetin.

acetoacetate [ORG CHEM] A salt containing the CH_3COCH_2COO radical; derived from acetoacetic acid.

acetoacetic acid [ORG CHEM] CH_3COCH_2COOH A colorless liquid miscible with water; derived from β-hydroxybutyric acid in the body.

acetoacetic ester *See* ethyl acetoacetate.

acetoacetyl coenzyme A [BIOCHEM] $C_{25}H_{41}O_{18}N_7P_3S$ An intermediate product in the oxidation of fatty acids.

Acetobacter [MICROBIO] A genus of aerobic, peritrichously flagellated bacteria in the family Achromobacteraceae. Also known as acetic acid bacteria; vinegar bacteria.

Acetobacter aceti [MICROBIO] An aerobic, rod-shaped bacterium capable of efficient oxidation of glucose, ethyl alcohol, and acetic acid; found in vinegar, beer, and souring fruits and vegetables.

Acetobacter suboxydans [MICROBIO] A short, nonmotile vinegar bacterium that can oxidize ethanol to acetic acid; useful for industrial production of ascorbic and tartaric acids.

acetoin [ORG CHEM] $CH_3COCHOHCH_3$ A colorless liquid; a condensation product of two molecules of acetic acid. Also known as acetylmethylcarbinol; 3-hydroxy-2-butanone.

acetol [ORG CHEM] CH_3COCH_2OH A colorless liquid soluble in water; a reducing agent. Also known as 1-hydroxy-2-propanone.

acetolactic acid [BIOCHEM] $C_5H_8O_4$ A monocarboxylic acid formed as an intermediate in the synthesis of the amino acid valine.

acetolysis [ORG CHEM] Decomposition of an organic molecule through the action of acetic acid or acetic anhydride.

Acetomonas [MICROBIO] A genus of aerobic, polarly flagellated vinegar bacteria in the family Pseudomonadaceae; used industrially to produce vinegar, gluconic acid, and L-sorbose.

acetone [ORG CHEM] CH_3COCH_3 A colorless, volatile, extremely flammable liquid, miscible with water; used as a solvent and reagent. Also known as 2-propanone.

acetone-benzol process [CHEM ENG] A dewaxing process in petroleum refining, with acetone and benzol used as solvents.

acetone body *See* ketone body.

acetone cyanohydrin [ORG CHEM] $(CH_3)_2COHCN$ A colorless liquid obtained from condensation of acetone with hydrocyanic acid; used as an insecticide or as an organic chemical intermediate.

acetone fermentation [MICROBIO] Formation of acetone by the metabolic action of certain anaerobic bacteria on carbohydrates.

acetone glucose *See* acetone sugar.

acetonemia [MED] A condition characterized by large amounts of acetone bodies in the blood. Also known as ketonemia.

acetone number [CHEM] A ratio used to estimate the degree of polymerization of materials such as drying oils; it is the weight in grams of acetone added to 100 grams of a drying oil to cause an insoluble phase to form.

acetone pyrolysis [ORG CHEM] Thermal decomposition of acetone into ketene.

acetone sugar [ORG CHEM] Any reducing sugar that contains acetone; examples are 1,2-monoacetone-D-glucofuranose and 1,2-5,6-diacetone-D-glucofuranose. Also known as acetone glucose.

acetonitrile [ORG CHEM] CH_3CN A colorless liquid soluble in water; used in organic synthesis. Also known as ethane nitrile; methyl cyanide.

acetophenone [ORG CHEM] $C_6H_5COCH_3$ Colorless crystals with a melting point of 19.6°C and a specific gravity of 1.028; used as a chemical intermediate. Also known as phenyl methyl ketone.

acetyl [ORG CHEM] CH_3CO- A two-carbon organic radical containing a methyl group and a carbonyl group.

acetylacetone [ORG CHEM] $CH_3COCH_2COCH_3$ A colorless liquid. Also known as 2,4-pentanedione.

acetylase [BIOCHEM] Any enzyme that catalyzes the formation of acetyl esters.

acetylating agent [ORG CHEM] A reagent, such as acetic anhydride, capable of bonding an acetyl group onto an organic molecule.

acetylation [ORG CHEM] The process of bonding an acetyl group onto an organic molecule.

acetylcholine [BIOCHEM] $C_7H_{17}O_3N$ A compound released from certain autonomic nerve endings which acts in the transmission of nerve impulses to excitable membranes.

acetylcholinesterase [BIOCHEM] An enzyme found in excitable membranes that inactivates acetylcholine.

acetyl coenzyme A [BIOCHEM] $C_{23}H_{39}O_{17}N_7P_3S$ A coenzyme, derived principally from the metabolism of glucose and fatty acids, that takes part in many biological acetylation reactions; oxidized in the Krebs cycle.

acetylene [ORG CHEM] C_2H_2 A colorless, highly flammable gas that is explosive when compressed; the simplest compound containing a triple bond; used in organic synthesis and as a welding fuel. Also known as ethyne.

acetylene black [ORG CHEM] A form of carbon with high electrical conductivity; made by decomposing acetylene by heat.

acetylene cutting *See* oxyacetylene cutting.

acetylene dichloride *See* sym-dichloroethylene.

acetylene generator [ENG] A steel cylinder or tank that provides for controlled mixing of calcium carbide and water to generate acetylene.

acetylene series [ORG CHEM] A series of unsaturated aliphatic hydrocarbons, each containing at least one triple bond and having the general formula C_nH_{2n-2}.

acetylene tetrachloride *See* sym-tetrachloroethane.

acetylene torch *See* oxyacetylene torch.

acetylene welding *See* oxyacetylene welding.

acetylenic [ORG CHEM] Pertaining to acetylene or being like acetylene, such as having a triple bond.

acetylenyl *See* ethinyl.

acetylide [ORG CHEM] A compound formed from acetylene with the H atoms replaced by metals, as in cuprous acetylide (Cu_2C_2).

acetyl ketene *See* diketene.

acetylmethylcarbinol *See* acetoin.

acetyl number [ANALY CHEM] A measure of free hydroxyl groups in fats or oils determined by the amount of potassium hydroxide used to neutralize the acetic acid formed by saponification of acetylated fat or oil.

acetyl phosphate [BIOCHEM] $C_2H_5O_5P$ The anhydride of acetic and phosphoric acids occurring in the metabolism of pyruvic acid by some bacteria; phosphate is used by some microorganisms, in place of ATP, for the phosphorylation of hexose sugars.

acetylpropionic acid *See* levulinic acid.

acetylsalicylic acid [ORG CHEM] $CH_3COOC_6H_4COOH$ A white, crystalline, weakly acidic substance, with melting point 137°C; slightly soluble in water; used medicinally as an antipyretic. Also known by trade name aspirin.

acetylurea [ORG CHEM] $CH_3CONHCONH_2$ Crystals that are colorless and are slightly soluble in water.

ACF diagram [PETR] A triangular diagram showing the chemical character of a metamorphic rock; the three components plotted are A = $Al_2O_3 + Fe_2O_3 - (Na_2O + K_2O)$, C = CaO, F = FeO + MgO + MnO.

acfm *See* actual cubic feet per minute.

a-c fracture [CRYSTAL] A type of tension fracture lying parallel to the a-c fabric plane and normal to plane b in a crystal.

a-c girdle [GEOL] A girdle of points in a petrofabric diagram that have a tread parallel with the plane of the a and c fabric axes.

Achaenodontidae [PALEON] A family of Eocene dichobunoids, piglike mammals belonging to the suborder Palaeodonta.

achalasia [MED] Inability of a hollow muscular organ or ring of muscle (sphincter) to relax.

achalasia of the cardia [MED] Enlargement of the esophagus as a result of cardiospasm.

ache [MED] A constant dull or throbbing pain.

acheb [ECOL] Short-lived vegetation regions of the Sahara composed principally of mustards (Cruciferae) and grasses (Gramineae).

achene [BOT] A small, dry, indehiscent fruit formed from a simple ovary bearing a single seed.

Acheulean [ARCHEO] Lower Paleolithic archeological time, characterized by biface tools having cutting edges all around.

achievement age [PSYCH] Accomplishment, or actual level of scholastic performance, expressed as equivalent to the age in years of the average child showing similar attainments.

achievement quotient [PSYCH] The ratio, usually multiplied by 100, between the achievement age, or actual scholastic level, and the mental age.

achilary [BOT] In flowers, having the lip (labellum) undeveloped or lacking.

Achilles [ASTRON] An asteroid; member of the group known as the Trojan planets.

Achilles jerk [PHYSIO] A reflex action seen as plantar flection in response to a blow to the Achilles tendon. Also known as Achilles tendon reflex.

Achilles tendon [ANAT] The tendon formed by union of the tendons of the calf muscles, the soleus and gastrocnemius, and inserted into the heel bone.

Achilles tendon reflex *See* Achilles jerk.

achlamydeous [BOT] Lacking a perianth.

achlorhydria [MED] Absence of hydrochloric acid in the stomach.

achluophobia [PSYCH] Abnormal fear of darkness.

acholia [MED] Suppression or absence of bile secretion into the small intestine.

achondrite [GEOL] A stony meteorite that contains no chondrules.

achondroplasia [MED] A hereditary deforming disease of the skeletal system, inherited in man as an autosomal dominant trait and characterized by insufficient growth of the long bones, resulting in reduced length. Also known as chondrodystrophy fetalis.

achondroplastic dwarf [MED] A human with short legs and arms due to achondroplasia.

achordate [VERT ZOO] Lacking a notochord.

achreocythemia [MED] An anemia characterized by pale erythrocytes due to hemoglobin deficiency.

achroglobin [BIOCHEM] A colorless respiratory pigment present in some mollusks and urochordates.

Achromatiaceae [MICROBIO] A family of large, spherical to ovoid bacteria in the order Beggiatoales.

achromatic [OPTICS] Capable of transmitting light without decomposing it into its constituent colors.

achromatic lens [OPTICS] A combination of two or more lenses having a focal length that is the same for two quite different wavelengths, thereby removing a major portion of chromatic aberration. Also known as achromat.

achromatic prism [OPTICS] A prism consisting of two or more prisms with different refractive indices combined so that light passing through the device is deviated but not dispersed.

achromatin [CYTOL] The portion of the cell nucleus which does not stain easily with basic dyes.

achromatophilia [BIOL] The property of not staining readily.

achromic [BIOL] Colorless; lacking normal pigmentation.

Achromobacter [MICROBIO] A genus of motile and nonmotile, gram-negative, rod-shaped bacteria in the family Achromobacteraceae.

Achromobacteraceae [MICROBIO] A family of true bacteria, order Eubacteriales, characterized by aerobic metabolism.

Achromycin [MICROBIO] Trade name for the antibiotic tetracycline.

achylia [MED] Absence of chyle.

achylia gastrica [MED] Lack of secretion of hydrochloric acid and proteolytic enzymes by the stomach.

ACI *See* acoustic comfort index.

acicular [SCI TECH] Needlelike; slender and pointed.

acicular ice [HYD] Fresh-water ice composed of many long crystals and layered hollow tubes of varying shape containing air bubbles. Also known as fibrous ice; satin ice.

acicular powder [MET] A metal powder whose grains are needle-shaped.

aciculignosa [ECOL] Narrow sclerophyll or coniferous vegetation that is mostly subalpine, subarctic, or continental.

acid [CHEM] **1.** Any of a class of chemical compounds whose aqueous solutions turn blue litmus paper red, react with and dissolve certain metals to form salts, and react with bases to form salts. **2.** A compound capable of transferring a hydrogen ion in solution. **3.** By extension of the term, a substance that ionizes in solution to yield the positive ion of the solvent. **4.** A molecule or ion that combines with another molecule or ion by forming a covalent bond with two electrons from the other species.

π-acid [ORG CHEM] An acid that readily forms stable complexes with aromatic systems.

acid acceptor [ORG CHEM] A stabilizer compound added to plastic and resin polymers to combine with trace amounts of acids formed by decomposition of the polymers.

acid alcohol [ORG CHEM] A compound containing both a carboxyl group ($-COOH$) and an alcohol group ($-CH_2OH$, $=CHOH$, or 6 COH).

acid ammonium tartrate *See* ammonium bitartrate.

acid anhydride [CHEM] An acid with one or more molecules of water removed; for example, SO_3 is the acid anhydride of H_2SO_4, sulfuric acid.

acid-base balance [PHYSIO] Physiologically maintained equilibrium of acids and bases in the body.

acid-base catalysis [CHEM] The increase in speed of certain chemical reactions due to the presence of acids and bases.

acid-base equilibrium [CHEM] The condition when acidic and basic ions in a solution exactly neutralize each other; that is, the pH is 7.

acid-base indicator [ANALY CHEM] A substance that reveals, through characteristic color changes, the degree of acidity or basicity of solutions.

acid-base pair [CHEM] A concept in the Brönsted theory of acids and bases; the pair consists of the source of the proton (acid) and the base generated by the transfer of the proton.

acid-base titration [ANALY CHEM] A titration in which an acid of known concentration is added to a solution of base of unknown concentration, or the converse.

acid blowcase *See* blowcase.

acid bottom and lining [MET] A melting furnace's inner bottom and lining composed of materials that at operating temperatures of the furnace react with the melt and slag to give an acid reaction; examples of materials are sand, siliceous rock, and silica brick.

acid brittleness [MET] Low ductility of a metal due to its absorption of hydrogen gas, which may occur during an electrolytic process or during cleaning. Also known as hydrogen embrittlement.

acid bronze [MET] A copper-tin alloy containing lead and nickel; used in pumping equipment.

acid calcium phosphate *See* calcium phosphate.

acid cell [HISTOL] A parietal cell of the stomach. [PHYS CHEM] An electrolytic cell whose electrolyte is an acid.

acid chloride [ORG CHEM] A compound containing the radical $-COCl$; an example is benzoyl chloride.

acid clay [GEOL] A type of clay that gives off hydrogen ions when it dissolves in water.

acid dilution [PETRO ENG] Dilution of concentrated hydrochloric acid with water prior to oil-well acidizing.

acid dye [ORG CHEM] Any of a group of sodium salts of sulfonic and carboxylic acids used to dye natural and synthetic fibers, leather, and paper.

acid egg *See* blowcase.

acid electrolyte [INORG CHEM] A compound, such as sulfuric acid, that dissociates into ions when dissolved, forming an acidic solution that conducts an electric current.

acidemia [MED] A condition in which the pH of the blood falls below normal.

acid-fast bacteria [MICROBIO] Bacteria, especially mycobacteria, that stain with basic dyes and fluorochromes and resist decoloration by acid solutions.

acid-fast stain [MICROBIO] A differential stain used in identifying species of *Mycobacterium* and one species of *Nocardia*.

acid gases [CHEM ENG] The hydrogen sulfide and carbon dioxide found in natural and refinery gases which, when combined with moisture, form corrosive acids; known as sour gases when hydrogen sulfide and mercaptans are present.

acid halide [ORG CHEM] A compound of the type RCOX, where R is an alkyl or aryl radical and X is a halogen.

acid heat test [ANALY CHEM] The determination of degree of unsaturation of organic compounds by reacting with sulfuric acid and measuring the heat of reaction.

acidic [CHEM] 1. Pertaining to an acid or to its properties. 2. Forming an acid during a chemical process.

acidic group [ORG CHEM] The radical COOH, present in organic acids.

acidic oxide [INORG CHEM] An oxygen compound of a non-metal, for example, SO_2 or P_2O_5, which yields an oxyacid with water.

acidic rock [PETR] Igneous rock containing more than 66% SiO_2, silicic.

acidic titrant [ANALY CHEM] An acid solution of known concentration used to determine the basicity of another solution by titration.

acidification [CHEM] Addition of an acid to a solution until the pH falls below 7.

acidimeter [ANALY CHEM] An apparatus or a standard solution used to determine the amount of acid in a sample.

acidimetry [ANALY CHEM] The titration of an acid with a standard solution of base.

acidity [CHEM] The state of being acid.

acidity coefficient [GEOCHEM] The ratio of the oxygen content of the bases in a rock to the oxygen content in the silica. Also known as oxygen ratio.

acidizing [PETRO ENG] Well-stimulation method to increase oil production by injecting hydrochloric acid into the oil-bearing formation; the acid dissolves rock to enlarge the porous passages through which the oil must flow.

acid jetting [PETRO ENG] The jetting, from a device lowered through oil-well tubing, of an acid spray onto bottom-hole rock to clean away mud and scale interfering with oil flow.

acid magnesium citrate *See* dibasic magnesium citrate.

acid manganous phosphate [INORG CHEM] $MnHPO_4 \cdot 3H_2O$ Acid-soluble pink powder. Also known as manganese hydrogen phosphate; secondary manganous phosphate.

acid mine drainage [MIN ENG] Drainage from bituminous coal mines containing a large concentration of acidic sulfates, especially ferrous sulfate.

acid mine water [MIN ENG] Mine water with free sulfuric acid, due to the weathering of iron pyrites.

acid number [CHEM] *See* acid value. [ENG] A number derived from a standard test indicating the acid or base composition of lubricating oils; it in no way indicates the corrosive attack of the used oil in service. Also known as corrosion number.

acidolysis [ORG CHEM] A chemical reaction involving the decomposition of a molecule, with the addition of the elements of an acid to the molecule; the reaction is comparable to hydrolysis or alcoholysis, in which water or alcohol, respectively, is used in place of the acid. Also known as acyl exchange.

acid open-hearth process [MET] A steelmaking process employing an open-hearth furnace lined with siliceous-type refractories.

acidophil [BIOL] 1. Any substance, tissue, or organism having an affinity for acid stains. 2. An organism having a preference for an acid environment. [HISTOL] 1. An alpha cell of the adenohypophysis. 2. *See* eosinophil.

acidophilia *See* eosinophilia.

acidophilic erythroblast *See* normoblast.

acidosis [MED] A condition of decreased alkali reserve of the blood and other body fluids.

acidotrophic [BIOL] Having an acid nutrient requirement.

acid phosphatase [BIOCHEM] Any nonspecific phosphatase requiring an acid medium for optimum activity.

acid phosphate [INORG CHEM] A mono- or dihydric phosphate; for example, M_2HPO_4 or MH_2PO_4, where M represents a metal atom.

acid phthalic anhydride *See* phthalic anhydride.

ACINOUS GLAND

An example of a typical compound type of acinous gland.

acid pickle [MATER] Industrial waste water that is the spent liquor from a chemical process used to clean metal surfaces.

acid potassium sulfate *See* potassium bisulfate.

acid process [MET] A melting process carried out in a furnace lined with acidic materials which combine readily with the oxides in the ore.

acid-proof coating [MATER] Material in liquid form suitable for application, by spraying, to the wall of projectile or bomb cavities to protect the metal from attack by explosives or other shell fillers.

acid reaction [CHEM] A chemical reaction produced by an acid.

acid refractory [MATER] A refractory that is composed principally of silica and reacts at high temperatures with bases such as lime, alkalies, and basic oxides.

acid-resistant [MATER] Able to withstand chemical reaction with or degeneration by acids.

acid salt [CHEM] A compound derived from an acid and base in which only a part of the hydrogen is replaced by a basic radical; for example, the acid sulfate $NaHSO_4$.

acid slag [MET] Furnace slag in which there is more silica and silicates than lime and magnesia.

acid sludge [CHEM ENG] The residue left after treating petroleum oil with sulfuric acid for the removal of impurities.

acid sodium tartrate *See* sodium bitartrate.

acid soil [GEOL] A soil with pH less than 7; results from presence of exchangeable hydrogen and aluminum ions.

acid solution [CHEM] An aqueous solution containing more hydrogen ions than hydroxyl ions.

acid soot [ENG] Carbon particles that have absorbed acid fumes as a by-product of combustion; hydrochloric acid absorbed on carbon particulates is frequently the cause of metal corrosion in incineration.

acid steel [MET] Steel produced in a melting furnace employing siliceous-type refractories.

acid tartrate *See* bitartrate.

acid tide [MED] A period of increased acidity of urine and body fluids.

acid treating [CHEM ENG] A refining process in which unfinished petroleum products, such as gasoline, kerosine, and diesel oil, are contacted with sulfuric acid to improve their color, odor, and other properties.

acidulant [FOOD ENG] One of a class of chemicals added to food to increase either tartness or acidity, such as malic or citric acids for tartness and phosphoric acid for acidity.

acidulous water [HYD] Mineral water either with dissolved carbonic acid or dissolved sulfur compounds such as sulfates.

acid value [CHEM] The acidity of a solution expressed in terms of normality. Also known as acid number.

acid wash [MATER] A solution of phosphoric acid applied to steel parts that removes and neutralizes the alkaline solutions used for grease removal after machining; also leaves a metallic phosphate coating which accepts paint well and, of itself, provides a degree of protection against rust.

acid water pollution [ENG] Industrial waste waters that are acidic; usually appear in effluent from the manufacture of chemicals, batteries, artificial and natural fiber, fermentation processes (beer), and mining.

acieration [MET] Electrolytic coating of a thin metal plate with iron; the iron hardens to steellike strength.

acinar [ANAT] Pertaining to an acinus.

acinar cell [ANAT] Any of the cells lining an acinous gland.

Acinetobacter [MICROBIO] A term suggested as the genus name for some species of *Achromobacter*.

aciniform [ZOO] Shaped like a berry or a bunch of grapes.

acinotubular gland *See* tubuloalveolar gland.

acinous [BIOL] Of or pertaining to acini.

acinous gland [ANAT] A multicellular gland with sac-shaped secreting units. Also known as alveolar gland.

acinus [ANAT] The small terminal sac of an acinous gland, lined with secreting cells. [BOT] An individual drupelet of a multiple fruit.

Acipenser [VERT ZOO] A genus of actinopterygian fishes in the sturgeon family, Acipenseridae.

Acipenseridae [VERT ZOO] The sturgeons, a family of actinopterygian fishes in the order Acipenseriformes.

Acipenseriformes [VERT ZOO] An order of the subclass Actinopterygii represented by the sturgeons and paddlefishes.

Ackerman linkage *See* Ackerman steering gear.

Ackerman steering gear [MECH ENG] Differential gear or linkage that turns the two steered road wheels of a self-propelled vehicle so that all wheels roll on circles with a common center. Also known as Ackerman linkage.

aclastic [OPTICS] Having the property of not refracting light.

aclinal [GEOL] Without dip; horizontal.

aclinic [GEOPHYS] Referring to a situation where a freely suspended magnetic needle remains in a horizontal position.

aclinic line *See* magnetic equator.

aΩcm *See* abohm centimeter.

Acmaeidae [INV ZOO] A family of gastropod mollusks in the order Archaeogastropoda; includes many limpets.

acme [PALEON] The time of largest abundance or variety of a fossil taxon; the taxon may be either general or local.

acme harrow [AGR] A type of harrow having a transverse horizontal frame with stiff curved blades. Also known as blade harrow; curved knife-tooth harrow; pulverizer.

acme screw thread [DES ENG] A standard thread having a profile angle of 29° and a flat crest; used on power screws in such devices as automobile jacks, presses, and lead screws on lathes. Also known as acme thread.

acme thread *See* acme screw thread.

acmite [MINERAL] $NaFeSi_2O_6$ A brown or green silicate mineral of the pyroxene group, often in long, pointed prismatic crystals; hardness is 6–6.5 on Mohs scale, and specific gravity is 3.50–3.55; found in igneous and metamorphic rocks.

acne [MED] A pleomorphic, inflammatory skin disease involving sebaceous follicles of the face, back, and chest and characterized by blackheads, whiteheads, papules, pustules, and nodules.

acne rosacea [MED] A form of acne occurring in older persons and seen as reddened inflamed areas on the forehead, nose, and cheeks.

Acnidosporidia [INV ZOO] An equivalent name for the Haplosporea.

acnode *See* isolated point.

Acoela [INV ZOO] An order of marine flatworms in the class Turbellaria characterized by the lack of a digestive tract and coelomic cavity.

Acoelea [INV ZOO] An order of gastropod mollusks in the subclass Opistobranchia; includes many sea slugs.

Acoelomata [INV ZOO] A subdivision of the animal kingdom; individuals are characterized by lack of a true body cavity.

acoelous [ZOO] **1.** Lacking a true body cavity or coelom. **2.** Lacking a true stomach or digestive tract.

acolpate [BOT] Of pollen grains, lacking furrows or grooves.

Aconchulinida [INV ZOO] An order of protozoans in the subclass Filosia comprising a small group of naked amebas having filopodia.

aconitase [BIOCHEM] An enzyme involved in the Krebs cycle in muscles that catalyzes the breakdown of citric acid to *cis*-aconitic and isocitric acids.

aconite [BOT] Any plant of the genus *Aconitum*. Also known as friar's cowl; monkshood; mousebane; wolfsbane. [PHARM] A toxic drug obtained from the dried tuberous root of *Aconitum napellus;* the principal alkaloid is aconitine.

aconitic acid [ORG CHEM] $C_6H_6O_2$ A white, crystalline organic acid found in sugarcane and sugarbeet; obtained during manufacture of sugar.

aconitine [PHARM] $C_{34}H_{47}O_{11}N$ A poisonous, white, crystalline alkaloid compound obtained from aconites such as monkshood (*Aconitum napellus*).

acorn [BOT] The nut of the oak tree, usually surrounded at the base by a woody involucre.

acorn barnacle [INV ZOO] Any of the sessile barnacles that are enclosed in conical, flat-bottomed shells and attach to ships and near-shore rocks and piles.

acorn disease [PL PATH] A virus disease of citrus plants characterized by malformation of the fruit, which is somewhat acorn-shaped.

acorn tube [ELECTR] An ultra-high-frequency electron tube resembling an acorn in shape and size.

acorn worm [INV ZOO] Any member of the class Enteropneusta, free-living animals that usually burrow in sand or mud. Also known as tongue worm.

acotyledon [BOT] A plant without cotyledons.

acouchi [VERT ZOO] A hystricomorph rodent represented by two species in the family Dasyproctidae; believed to be a dwarf variety of the agouti.

acoustic [ACOUS] Relating to, containing, producing, arising from, actuated by, or carrying sound.

acoustic absorption *See* sound absorption.

acoustic absorption coefficient *See* sound absorption coefficient.

acoustic absorptivity *See* sound absorption coefficient.

acoustical [ACOUS] Having a characteristic concerning sound, of an object or quantity that in and of itself does not have properties associated with sound, such as a device, measurement, or symbol.

acoustical Doppler effect [ACOUS] The change in pitch of a sound observed when there is relative motion between source and observer.

acoustical material [MATER] Any natural or synthetic material that absorbs sound; acoustical tile is an example.

acoustical scintillation [COMMUN] Irregular fluctuation in the received intensity of sounds propagated through the atmosphere from a source of uniform output; produced by nonhomogeneous structure of the atmosphere along the path of the sound.

acoustical stiffness [ACOUS] The acoustic reactance associated with the potential energy of a medium multiplied by 2π times the sound frequency.

acoustical treatment [CIV ENG] That part of building planning that is designed to provide a proper acoustical environment; includes the use of acoustical material.

acoustic amplifier [ELECTR] A device that amplifies mechanical vibrations directly at audio and ultrasonic frequencies.

acoustic array [ENG ACOUS] A sound-transmitting or -receiving system whose elements are arranged to give desired directional characteristics.

acoustic bearing *See* sonic bearing.

acoustic branch [SOLID STATE] One of the parts of the dispersion relation, frequency as a function of wave number, for crystal lattice vibrations, representing vibration at low (acoustic) frequencies.

acoustic bridge [ELECTR] A device, based on the principle of the electrical Wheatstone bridge, used for analysis of deafness.

acoustic capacitance [ACOUS] A measure of volume displacement in a sound medium, per dyne per square centimeter.

acoustic capacitance unit [ACOUS] The centimeter-to-the-fifth-power per dyne.

acoustic center [ENG ACOUS] The center of the spherical sound waves radiating outward from an acoustic transducer.

acoustic clarifier [ENG ACOUS] System of cones loosely attached to the baffle of a loudspeaker and designed to vibrate and absorb energy during sudden loud sounds to suppress these sounds.

acoustic comfort index [ACOUS] An arbitrarily designed scale to indicate the noise inside the passenger cabin of an aircraft; on this scale +100 represents ideal conditions or zero noise, 0 represents barely tolerable conditions, and −100 represents intolerable conditions. Abbreviated ACI.

acoustic compliance [ACOUS] The reciprocal of acoustic stiffness.

acoustic delay [ENG ACOUS] A delay which is deliberately introduced in sound reproduction by having the sound travel a certain distance along a pipe before conversion into electric signals.

acoustic delay line [ELECTR] A device in which acoustic signals are propagated in a medium to make use of the sonic propagation time to obtain a time delay for the signals. Also known as sonic delay line.

acoustic detection [ENG] Determination of the profile of a geologic formation, an ocean layer, or some object in the ocean by measuring the reflection of sound waves off the object.

acoustic detector [ELECTR] The stage in a receiver at which demodulation of a modulated radio wave into its audio component takes place.

ACKERMAN STEERING GEAR

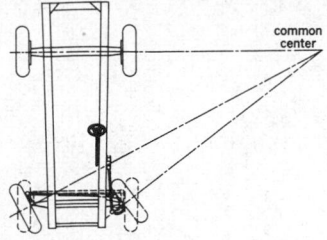

All wheels turn about a common center.

ACONCHULINIDA

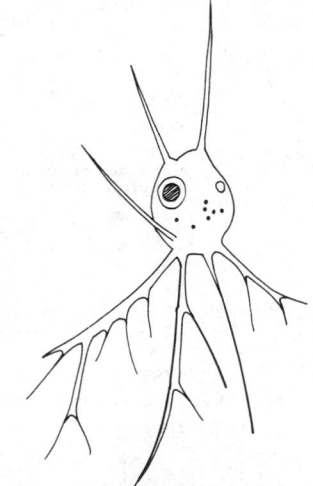

A drawing of *Penardia cometa* showing the filamentous pseudopodia (filopodia). *(After De Saedeleer)*

ACORN WORM

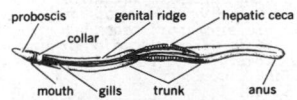

A drawing of the acorn worm, *Saccoglossus*, showing anatomical features. *(From T. I. Storer and R. L. Usinger, General Zoology, 4th ed., McGraw-Hill, 1965)*

acoustic dispersion [ACOUS] A complex sound wave's separation into its frequency components as it passes through a medium; usually measured by the rate of change of velocity with frequency.

acoustic energy *See* sound energy.

acoustic fatigue [MECH] The tendency of a material, such as a metal, to lose strength after acoustic stress.

acoustic feedback [ENG ACOUS] The reverberation of sound waves from a loudspeaker to a preceding part of an audio system, such as to the microphone, in such a manner as to reinforce, and distort, the original input. Also known as acoustic regeneration.

acoustic filter *See* filter.

acoustic fix *See* sonic fix.

acoustic generator [ENG ACOUS] A transducer which converts electrical, mechanical, or other forms of energy into sound.

acoustic homing [NAV] The following of a path of acoustic energy to or toward its source or point of reflection.

acoustic horn *See* horn.

acoustic image [ACOUS] The geometric space figure that is made up of the acoustic foci of an acoustic lens, mirror, or other acoustic optical system and is the acoustic counterpart of an extended source of sound. Also known as image.

acoustic impedance [ACOUS] The complex ratio of the sound pressure on a given surface to the sound flux through that surface, expressed in acoustic ohms.

acoustic inertance *See* acoustic mass.

acoustic insulation [MATER] A material used to diminish sound energy that passes through it or strikes its surface.

acoustic interferometer [ACOUS] A device for measuring the velocity and attenuation of sound waves in a gas or liquid by an interference method.

acoustic jamming [ENG ACOUS] The deliberate radiation or reradiation of mechanical or electroacoustic signals with the objectives of obliterating or obscuring signals which the enemy is attempting to receive and of deterring enemy weapons systems.

acoustic labyrinth [ENG ACOUS] Special baffle arrangement used with a loudspeaker to prevent cavity resonance and to reinforce bass response.

acoustic lens [MATER] Selected materials shaped to refract sound waves in accordance with the principles of geometrical optics, as is done for light. Also known as lens.

acoustic line [ENG ACOUS] The acoustic equivalent of an electrical transmission line, involving baffles, labyrinths, or resonators placed at the rear of a loudspeaker and arranged to help reproduce the very low audio frequencies.

acoustic line of position *See* sonic line of position.

acoustic Mach meter [AERO ENG] A device which registers data on sound propagation for the calculation of Mach number.

acoustic mass [ACOUS] The quantity which, after multiplication by twice the frequency, results in the acoustic reactance associated with the kinetic energy of the sound medium. Also known as acoustic inertance.

acoustic mass unit [ACOUS] Usually, the gram per centimeter-to-the-fourth-power.

acoustic measurement [ACOUS] The process of quantitatively determining one or more properties of sound.

acoustic memory [ADP] A computer memory that uses an acoustic delay line, in which a train of pulses travels through a medium such as mercury or quartz.

acoustic mine [ORD] A naval mine which is activated by acoustic means.

acoustic mode [SOLID STATE] The type of crystal lattice vibrations which for long wavelengths act like an acoustic wave in a continuous medium, but which for shorter wavelengths approach the Debye frequency, showing a dispersive decrease in phase velocity.

acoustic navigation *See* sonic navigation.

acoustic nerve *See* auditory nerve.

acoustic noise [ACOUS] Noise in the acoustic spectrum; usually measured in decibels.

acoustic ohm [ACOUS] The unit of acoustic impedance. Also known as acoustic reactance unit; acoustic resistance unit.

acousticophobia [PSYCH] Abnormal fear of sounds.

acoustic phonon [SOLID STATE] A quantum of excitation of an acoustic mode of vibration.

acoustic plaster [MATER] Plaster having good acoustic absorbing properties; it contains metal which, upon contact with water, evolves gas to aerate the mass.

acoustic power *See* sound power.

acoustic radiation [ACOUS] Infrasonic, sonic, or ultrasonic waves propagating through a solid, liquid, or gaseous medium.

acoustic radiation pressure [ACOUS] A unidirectional, steady-state pressure exerted upon a surface exposed to a sound wave.

acoustic radiator [ENG ACOUS] A vibrating surface that produces sound waves, such as a loudspeaker cone or a headphone diaphragm.

acoustic radiometer [ENG] An instrument for measuring sound intensity by determining the unidirectional steady-state pressure caused by the reflection or absorption of a sound wave at a boundary.

acoustic ratio [ENG ACOUS] The ratio of the intensity of sound radiated directly from a source to the intensity of sound reverberating from the walls of an enclosure, at a given point in the enclosure.

acoustic reactance [ACOUS] The imaginary component of the acoustic impedance.

acoustic reactance unit *See* acoustic ohm.

acoustic receiver [ELECTR] The complete equipment required for receiving modulated radio waves and converting them into sound.

acoustic reciprocity theorem [ACOUS] A theorem which states that in the acoustic field due to a sound source at point A, the sound pressure received at any other point B is the same as that which would be produced at A if the source were placed at B, and that this can be generalized for multiple sources and receivers.

acoustic reflection coefficient *See* acoustic reflectivity.

acoustic reflectivity [ACOUS] Ratio of the rate of flow of sound energy reflected from a surface, on the side of incidence, to the incident rate of flow. Also known as acoustic reflection coefficient; sound reflection coefficient.

acoustic reflex [PHYSIO] Brief, involuntary closure of the eyes due to stimulation of the acoustic nerve by a sudden sound.

acoustic reflex enclosure [ENG ACOUS] A loudspeaker cabinet designed with a port to allow a low-frequency contribution from the rear of the speaker cone to be radiated forward.

acoustic refraction [ACOUS] Variation of the direction of sound transmission due to spatial variation of the wave velocity in the medium.

acoustic regeneration *See* acoustic feedback.

acoustic resistance [ACOUS] The real component of the acoustic impedance.

acoustic resistance unit *See* acoustic ohm.

acoustic resonance [ACOUS] A phenomenon exhibited by an acoustic system, such as an organ pipe or Helmholtz resonator, in which the response of the system to sound waves becomes very large when the frequency of the sound approaches a natural vibration frequency of the air in the system.

acoustic resonator [ACOUS] An enclosure that produces sound-wave resonance at a particular frequency.

acoustics [PHYS] **1.** The science of the production, transmission, and effects of sound. **2.** The characteristics of a room that determine the qualities of sound in it relevant to hearing.

acoustic scattering [ACOUS] The irregular reflection, refraction, and diffraction of sound in many directions.

acoustic seal [ENG ACOUS] A joint between two parts to provide acoustical coupling with low losses of energy, such as between an earphone and the human ear.

acoustic shielding [ACOUS] A sound barrier that prevents the transmission of acoustic energy.

acoustic signature [ENG] In acoustic detection, the profile characteristic of a particular object or class of objects, such as a school of fish or a specific ocean-bottom formation.

acoustic spectrometer [ENG ACOUS] An instrument that measures the intensities of the various frequency components of a complex sound wave. Also known as audio spectrometer.

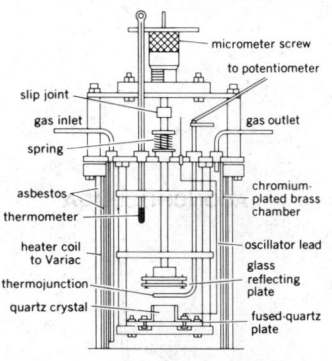

ACOUSTIC INTERFEROMETER

micrometer screw
to potentiometer
slip joint
gas inlet
gas outlet
spring
asbestos
chromium-plated brass chamber
thermometer
heater coil to Variac
oscillator lead
glass reflecting plate
thermojunction
quartz crystal
fused-quartz plate

The interferometric chamber in which measurements of change in velocity and attenuation of sound waves are made.

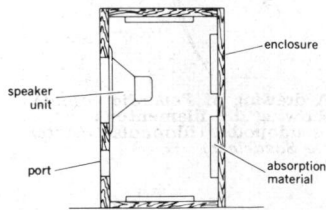

ACOUSTIC REFLEX ENCLOSURE

enclosure
speaker unit
port
absorption material

Sectional view of one type of reflex enclosure; large port area near diaphragm obtains maximum aid from port radiation; phase shift of backside radiation is obtained by choice of enclosure volume and port mass. (*From K. Henney, ed., Radio Engineering Handbook, 5th ed., McGraw-Hill, 1959*)

acoustic spectrum [ACOUS] The range of acoustic frequencies, extending from subsonic to ultrasonic frequencies, that is, approximately from zero to at least 1 megahertz.

acoustic stiffness *See* stiffness.

acoustic streaming [FL MECH] Unidirectional flow currents in a fluid that are due to the presence of sound waves.

acoustic theory [AERO ENG] The linearized small-disturbance theory used to predict the approximate airflow past an airfoil when the disturbance velocities caused by the flow are small compared to the flight speed and to the speed of sound.

acoustic tile [MATER] A thin, often decorative tile with sound-absorbing properties, used to cover ceilings and walls.

acoustic torpedo [ORD] A naval torpedo which is directed toward its target either by the noise emitted by the target or by sonar.

acoustic transducer [ENG ACOUS] A device that converts acoustic energy to electrical or mechanical energy, such as a microphone or phonograph pickup.

acoustic transformer [ENG ACOUS] A device, such as a horn or megaphone, for increasing the efficiency of sound radiation.

acoustic transmission coefficient *See* sound transmission coefficient.

acoustic transmissivity *See* sound transmission coefficient.

acoustic transponder [NAV] A device used in underwater navigation which, on being interrogated by coded acoustic signals, emits acoustic reply.

acoustic treatment [BUILD] The use of sound-absorbing materials to give a room a desired degree of freedom from echo and reverberation.

acoustic velocity *See* speed of sound.

acoustic vibration [ACOUS] Any vibration transmitted through a gas: subsonic, sonic, or ultrasonic.

acoustic wave *See* sound.

acoustic well logging [ENG] A ground exploration method that uses a high-energy sound source and a receiver, both underground.

acoustooptic interaction [OPTICS] A way to influence the propagation characteristics of an optical wave by applying a low-frequency acoustical field to the medium through which the wave passes.

acoustooptic modulator [OPTICS] A device utilizing acoustooptic interaction ultrasonically to vary the amplitude or the phase of a light beam.

ACP *See* acyl carrier protein.

a-c plane [CRYSTAL] A plane at right angles to the surface of movement in a crystal.

acquire [ELECTR] 1. Of acquisition radars, the process of detecting the presence and location of a target in sufficient detail to permit identification. 2. Of tracking radars, the process of positioning a radar beam so that a target is in that beam to permit the effective employment of weapons. Also known as target acquisition.

acquired [BIOL] Not present at birth, but developed by an individual in response to the environment and not subject to hereditary transmission.

acquired drive [PSYCH] Any aroused state of behavior that originates from and is nurtured by experience rather than inheritance.

acquired immunity [IMMUNOL] Resistance to a microbial or other antigenic substance taken on by a naturally susceptible individual; may be either active or passive.

acquired immunological tolerance [IMMUNOL] Failure of immunological responsiveness, that is, inability of antigen-sensitive cells to synthesize antibodies; induced by exposure to large amounts of an antigen. Also known as immunological paralysis.

acquisition [ENG] The process of pointing an antenna or a telescope so that it is properly oriented to allow gathering of tracking or telemetry data from a satellite or space probe.

acquisition and tracking radar [ENG] A radar set capable of locking onto a received signal and tracking the object emitting the signal; the radar may be airborne or on the ground.

acrania [MED] Partial or complete absence of the cranium at birth.

Acrania [ZOO] A group of lower chordates with no cranium, jaws, vertebrae, or paired appendages; includes the Tunicata and Cephalochordata.

acrasia [PSYCH] Lack of self-control.

Acrasiales [BIOL] A group of microorganisms that have plant and animal characteristics; included in the phylum Myxomycophyta by botanists and Mycetozoia by zoologists.

Acrasida [MYCOL] An order of Mycetozoia containing cellular slime molds.

Acrasieae [BIOL] An equivalent name for the Acrasiales.

acrasin [BIOCHEM] The chemotactic substance thought to be secreted by, and to effect aggregation of, myxamebas during their fruiting phase.

acraspedote [INV ZOO] Describing tapeworm segments which are not overlapping.

acre [MECH] A unit of area, equal to 43,560 square feet, or to 4046.8564224 square meters.

Acree's reaction [ANALY CHEM] A test for protein in which a violet ring appears when concentrated sulfuric acid is introduced below a mixture of the unknown solution and a formaldehyde solution containing a trace of ferric chloride.

acre-foot [HYD] The volume of water required to cover 1 acre to a depth of 1 foot, hence 43,560 cubic feet; a convenient unit for measuring irrigation water, runoff volume, and reservoir capacity.

acre foot per day [HYD] The United States unit of volume rate of water flow. Abbreviated acre ft/d.

acre ft/d *See* acre foot per day.

acre-in. *See* acre-inch.

acre-inch [HYD] A unit of volume used in the United States for water flow, equal to 3630 cubic feet. Abbreviated acre-in.

acre-yield [GEOL] The average amount of oil, gas, or water taken from one acre of a reservoir.

acridine [ORG CHEM] $(C_6H_4)_2NCH$ A typical member of a group of organic heterocyclic compounds containing benzene rings fused to the 2,3 and 5,6 positions of pyridene; derivatives include dyes and medicines.

acridine orange [ORG CHEM] A dye with an affinity for nucleic acids; the complexes of nucleic acid and dye fluoresce orange with RNA and green with DNA when observed in the fluorescence microscope.

acriflavine [ORG CHEM] $C_{14}H_{14}N_3Cl$ A yellow acridine dye obtained from proflavine by methylation in the form of red crystals; used as an antiseptic in solution.

Acrilan [TEXT] Trade name of a man-made fiber that is an acrylonitrile-vinyl acetate copolymer; manufactured by the Monsanto Company.

acroagnosis [MED] Loss or absence of sense perception in a limb.

acrobatholithic [GEOL] A stage in batholithic erosion where summits of cupolas and stocks are exposed without any exposure of the surface separating the barren interior of the batholith from the mineralized upper part.

acroblast [CYTOL] A vesicular structure in the spermatid formed from Golgi material.

acrocarpous [BOT] In some mosses of the subclass Eubrya, having the sporophyte at the end of a stem and therefore exhibiting the erect habit.

acrocentric chromosome [CYTOL] A chromosome having the centromere close to one end.

acrocephalosyndactylism [MED] A congenital malformation consisting of an enlarged, pointed skull and defective separation of fingers and toes. Also known as Apert's syndrome.

acrocephaly *See* oxycephaly.

Acroceridae [INV ZOO] The humpbacked flies, a family of orthorrhaphous dipteran insects in the series Brachycera.

acrodomatia [ECOL] Specialized structures on certain plants adapted to shelter mites; relationship is presumably symbiotic.

acrodont [ANAT] Having teeth fused to the edge of the supporting bone.

acrodynia [MED] A childhood syndrome associated with mercury ingestion and characterized by periods of irritability alternating with apathy, anorexia, pink itching hands and feet, photophobia, sweating, tachycardia, hypertension, and hypotonia.

acrolein [ORG CHEM] CH_2CHCHO The simplest member of a class of unsaturated aldehydes; a yellow, irritant, pungent liquid. Also known as acrylaldehyde; propenal.

acrolein dimer [ORG CHEM] $C_6H_8O_2$ A flammable, water-soluble liquid used as an intermediate for resins, dyestuffs,

ACOUSTIC TORPEDO

The Mark 32 acoustic torpedo shown leaving its launcher. *(Official U. S. Navy photograph)*

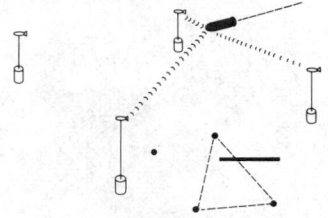

ACOUSTIC TRANSPONDER

pilot's dead reckoning tracer (DRT) presentation

Submersible navigating from underwater acoustic transponders.

ACRIDINE

Structural formula for acridine showing the carbon positions by number.

and pharmaceuticals. Also known as 2-formyl-3,4-dihydro-2*H*-pyran.

acrolein test [ANALY CHEM] A test for the presence of glycerin or fats; a sample is heated with potassium bisulfate, and acrolein is released if the test is positive.

acromegaly [MED] A chronic condition in adults caused by hypersecretion of the growth hormone and marked by enlarged jaws, extremities, and viscera, accompanied by certain physiological changes.

acromelalgia *See* erythromelalgia.

acromere [HISTOL] The distal portion of a rod or cone in the retina.

acrometer [ENG] An instrument to measure the density of oils.

acromion [ANAT] The flat process on the outer end of the scapular spine that articulates with the clavicle and forms the outer angle of the shoulder.

acromorph [GEOL] A salt dome.

acron [EVOL] Unsegmented head of the ancestral arthropod. [INV ZOO] **1.** The preoral, nonsegmented portion of an arthropod embryo. **2.** The prostomial region of the trochophore larva of some mollusks.

acroparesthesia [MED] A chronic self-limited symptom complex associated with a variety of systemic diseases, characterized by tingling, pins-and-needles sensations, numbness or stiffness, and occasionally pains in the hands and feet.

acropetal [BOT] From the base toward the apex, as seen in the formation of certain organs or the spread of a pathogen.

acrophobia [PSYCH] Abnormal fear of great heights.

Acrosaleniidae [PALEON] A family of Jurassic and Cretaceous echinoderms in the order Salenoida.

acroscopic [BOT] Facing, or on the side toward, the apex.

acrosome [CYTOL] The anterior, crescent-shaped body of spermatozoon, formed from Golgi material of the spermatid. Also known as perforatorium.

acrospore [MYCOL] In fungi, a spore formed at the outer tip of a hypha.

acrotarsium [ANAT] Instep of the foot.

acroterion [ARCH] **1.** A pedestal on a pediment to support an ornamental, such as a statue. **2.** An ornamental placed on such a pedestal.

Acrothoracica [INV ZOO] A small order of burrowing barnacles in the subclass Cirripedia that inhabit corals and the shells of mollusks and barnacles.

Acrotretacea [PALEON] A family of Cambrian and Ordovician inarticulate brachipods of the suborder Acrotretidina.

Acrotretida [INV ZOO] An order of brachiopods in the class Inarticulata; representatives are known from Lower Cambrian to the present.

Acrotretidina [INV ZOO] A suborder of inarticulate brachiopods of the order Acrotretida; includes only species with shells composed of calcium phosphate.

acrozone *See* range zone.

acrylamide [ORG CHEM] $CH_2CHCONH_2$ A crystalline amide of acrylic acid.

acrylamide copolymer [ORG CHEM] A thermosetting resin formed of acrylamide with other resins, such as the acrylic resins.

acrylate [ORG CHEM] **1.** A salt or ester of acrylic acid. **2.** *See* acrylate resin.

acrylate resin [ORG CHEM] Acrylic acid or ester polymer with a $-CH_2-CH(COOR)-$ structure; used in paints, sizings and finishes for paper and textiles, adhesives, and plastics. Also known as acrylate.

acrylic acid [ORG CHEM] $CH_2CHCOOH$ An easily polymerized, colorless, corrosive liquid used as a monomer for acrylate resins.

acrylic ester [ORG CHEM] An ester of acrylic acid.

acrylic fiber [TEXT] Any of numerous synthetic textile fibers made by polymerization of acrylonitrile.

acrylic resin [ORG CHEM] A thermoplastic synthetic organic polymer made by the polymerization of acrylic derivatives such as acrylic acid, methacrylic acid, ethyl acrylate, and methyl acrylate; used for adhesives, protective coatings, and finishes.

acrylic rubber [ORG CHEM] Synthetic rubber containing acrylonitrile; for example, nitrile rubber.

acrylonitrile [ORG CHEM] CH_2CHCN A colorless liquid compound used in the manufacture of acrylic rubber and fibers. Also known as vinylcyanide.

acrylonitrile-butadiene rubber *See* nitrile rubber.

acrylonitrile butadiene styrene resin [ORG CHEM] A polymer made by blending acrylonitrile-styrene copolymer with a butadiene-acrylonitrile rubber or by interpolymerizing polybutadiene with styrene and acrylonitrile; combines the advantages of hardness and strength of the vinyl resin component with the toughness and impact resistance of the rubbery component. Abbreviated ABS resin.

acrylonitrile copolymer [ORG CHEM] Oil-resistant synthetic rubber made by polymerization of acrylonitrile with compounds such as butadiene or acrylic acid.

acrylonitrile rubber *See* nitrile rubber.

ACSR *See* aluminum cable steel-reinforced.

Actaeonidae [INV ZOO] A family of gastropod mollusks in the order Tectibranchia.

Actaletidae [INV ZOO] A family of insects belonging to the order Collembola characterized by simple tracheal systems.

ACTH *See* adrenocorticotropic hormone.

Actidione [MICROBIO] Trade name for the antibiotic cyclohexamide.

actin [BIOCHEM] A muscle protein that is the chief constituent of the Z-band myofilaments of each sarcomere.

actinal [INV ZOO] In radially symmetrical animals, referring to the part from which the tentacles or arms radiate or to the side where the mouth is located.

Actiniaria [INV ZOO] The sea anemones, an order of coelenterates in the subclass Zoantharia.

actinic [PHYS] Pertaining to electromagnetic radiation capable of initiating photochemical reactions, as in photography or the fading of pigments.

actinic focus [OPTICS] The point in an optical system at which the chemically most effective rays (usually those in the ultraviolet) converge. Also known as chemical focus.

actinic glass [OPTICS] Glass that transmits more of the visible components of incident radiation and less of the infrared and ultraviolet components.

actinide series [CHEM] The group of elements of atomic number 89 through 103. Also known as actinoid elements.

actinism [CHEM] The production of chemical changes in a substance upon which electromagnetic radiation is incident.

actinium [CHEM] A radioactive element, symbol Ac, atomic number 89; its longest-lived isotope is Ac^{227} with a half-life of 21.7 years; the element is trivalent; chief use is, in equilibrium with its decay products, as a source of alpha rays.

actinium decay series [NUCLEO] A series of radioactive disintegration products derived from uranium-235.

actinium emanation *See* actinon.

actinobacillosis [VET MED] A bacterial disease of domestic animals caused by *Actinobacillus lignieresii*.

Actinobacillus [MICROBIO] A genus of aerobic, gram-negative bacteria in the family Brucellaceae; species are pathogenic for animals, occasionally for man.

Actinobacillus mallei [MICROBIO] A gram-negative, nonmotile, rod-shaped bacterium, identified as the causative agent of glanders.

actinocarpous [BOT] Having flowers and fruit radiating from one point.

actinochemistry [CHEM] A branch of chemistry concerned with chemical reactions produced by light or other radiation.

actinochitin [BIOCHEM] A form of birefringent or anisotropic chitin found in the seta of certain mites.

Actinochitinosi [INV ZOO] A group name for two closely related suborders of mites, Trombidiformes and Sarcoptiformes.

actinodielectric [ELEC] Of a substance, exhibiting an increase in electrical conductivity when electromagnetic radiation is incident upon it.

actinoelectricity [ELEC] The electromotive force produced in a substance by electromagnetic radiation incident upon it.

actinogram [ENG] The record of heat from a source, such as the sun, as detected by a recording actinometer.

actinograph [ENG] A recording actinometer.

actinoid elements *See* actinide series.

actinolite [MINERAL] $Ca_2(Mg,Fe)_5Si_8O_{22}(OH)_2$ A green,

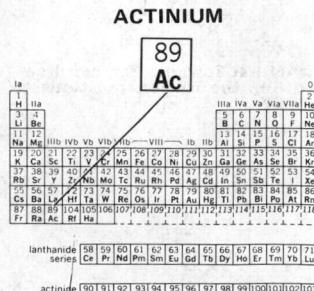

ACTINIUM

89
Ac

Periodic table of the chemical elements showing the position of actinium.

monoclinic rock-forming amphibole; a variety of asbestos occurring in needlelike crystals and in fibrous or columnar forms; specific gravity 3–3.2.

actinology [PHYS] The branch of physics dealing with electromagnetic radiation and its chemical effects.

actinomere [INV ZOO] One of the segments composing the body of a radially symmetrical animal.

actinometer [ENG] Any instrument used to measure the intensity of radiant energy, particularly that of the sun.

actinometry [ASTROPHYS] The science of measurement of radiant energy, particularly that of the sun, in its thermal, chemical, and luminous aspects.

actinomorphic [BIOL] Descriptive of an organism, organ, or part that is radially symmetrical.

Actinomyces [MICROBIO] A genus of anaerobic and microaerophilic pathogenic bacteria belonging to the family Actinomycetaceae.

Actinomyces bovis [MICROBIO] A nonmotile, gram-positive, anaerobic bacterium that causes actinomycosis.

Actinomycetaceae [MICROBIO] A family of bacteria belonging to the order Actinomycetales characterized by the formation of a true mycelium.

Actinomycetales [MICROBIO] An order of bacteria in the class Schizomycetes; individuals produce filamentous cells or hyphae.

actinomycete [MICROBIO] Any member of the bacterial family Actinomycetaceae.

actinomycin [MICROBIO] The collective name for a large number of red chromoprotein antibiotics elaborated by various strains of *Streptomyces*.

actinomycosis [MED] An infectious bacterial disease caused by *Actinomyces bovis* in cattle, hogs, and occasionally in man. Also known as lumpy jaw.

Actinomyxida [INV ZOO] An order of protozoan invertebrate parasites of the class Myxosporidea characterized by trivalved spores with three polar capsules.

actinon [NUC PHYS] A radioactive isotope of radon, symbol An, atomic number 86, atomic weight 219, belonging to the actinium series. Also known as actinium emanation (Ac-Em).

actinophage [MICROBIO] A bacteriophage that infects and lyses members of the order Actinomycetales.

Actinophryida [INV ZOO] An order of protozoans in the subclass Heliozoia; individuals lack an organized test, a centroplast, and a capsule.

Actinoplanaceae [MICROBIO] A family of bacteria in the order Actinomycetales with well-developed mycelia and spores formed on sporangia.

Actinoplanes [MICROBIO] A genus of bacteria in the family Actinoplanaceae having aerial mycelia and spherical to subspherical sporangiospores with a tuft of flagella.

Actinopodea [INV ZOO] A class of protozoans belonging to the superclass Sarcodina; most are free-floating, with highly specialized pseudopodia.

Actinopteri [VERT ZOO] An equivalent name for the Actinopterygii.

Actinopterygii [VERT ZOO] The ray-fin fishes, a subclass of the Osteichthyes distinguished by the structure of the paired fins, which are supported by dermal rays.

actinostele [BOT] A protostele characterized by xylem that is either star-shaped in cross section or has ribs radiating from the center.

actinostome [BIOL] **1.** The mouth of a radiate animal. **2.** The peristome of an echinoderm.

Actinostromariidae [PALEON] A sphaeractinoid family of extinct marine hydrozoans.

actinotherapy *See* radiation therapy.

actinotrocha [INV ZOO] The free-swimming larva of *Phoronis*, a genus of small, marine, tubiculous worms.

actinouranium [NUC PHYS] A naturally occurring radioactive isotope of the actinium series, emitting only alpha decay; symbol AcU; atomic number 92; mass number 235; half-life 7.1×10^8 years; isotopic symbol U^{235}.

actinula [INV ZOO] A larval stage of some hydrozoans that has tentacles and a mouth; attaches and develops into a hydroid in some species, or metamorphoses into a medusa.

action [MECH] An integral associated with the trajectory of a system in configuration space, equal to the sum of the integrals of the generalized momenta of the system over their canonically conjugate coordinates. Also known as phase integral. [ORD] The mechanism of a gun, usually breech-loading, by which it is loaded, fired, and unloaded.

action at a distance theory [PHYS] A theory of the interaction of two bodies separated in space, without concern for a detailed mechanism of the propagation of effects between bodies.

action current [PHYSIO] The electric current accompanying membrane depolarization and repolarization in an excitable cell.

action potential [PHYSIO] A transient change in electric potential at the surface of a nerve or muscle cell occurring at the moment of excitation.

action-reaction law [PHYS] The law that when one body exerts force on another, the second body exerts a collinear force on the first equal in magnitude but oppositely directed.

action spectrum [PHYSIO] Graphic representation of the comparative effects of different wavelengths of light on living systems or their components.

action variable [PHYS] The integral $\int p \, dq$ over a cycle of a dynamical system; q is some coordinate, and p the conjugate momentum.

actium [ECOL] A rocky seashore community.

activate [ELEC] To make a cell or battery operative by addition of a liquid. [ELECTR] To treat the filament, cathode, or target of a vacuum tube to increase electron emission. [ENG] To set up conditions so that the object will function as designed or required. [NUCLEO] To induce radioactivity through bombardment by neutrons or by other types of radiation. [ORD] **1.** To bring into existence by official order a unit, post, camp, station, base, or shore activity which has previously been constituted and designated by name or number, or both, so that it can be organized to function in its assigned capacity. **2.** To prepare for active service a naval ship or craft which has been in an inactive or reserve status. [PHYS] To start activity or motion in a device or material.

activated alumina [MATER] Highly porous, granular aluminum oxide that preferentially absorbs liquids from gases and vapors, and moisture from some liquids; also used as a catalyst or catalyst carrier, as an absorbent to remove fluorides from drinking water, and in chromatography.

activated carbon [MATER] A powdered, granular, or pelleted form of amorphous carbon characterized by very large surface area per unit volume because of an enormous number of fine pores. Also known as activated charcoal.

activated cathode [ELECTR] A thermionic cathode consisting of a tungsten filament to which thorium has been added, and then brought to the surface, by a process such as heating in the absence of an electric field in order to increase thermionic emission.

activated charcoal *See* activated carbon.

activated clay [MATER] Bentonite, or other clay, treated with acid to enhance its ability to absorb or bleach.

activated coal plough [MIN ENG] A type of power-operated cutting blade used for coal seams too hard to be sheared by a normal blade.

activated complex [PHYS CHEM] An energetically excited state which is intermediate between reactants and products in a chemical reaction.

activated sintering [MET] Sintering of a metal powder compact in contact with a gaseous atmosphere which reacts with the metal surfaces and enhances the joining of metal particles.

activated sludge [CIV ENG] A semiliquid mass removed from the liquid flow of sewage and subjected to aeration and aerobic microbial action; the end product is dark to golden brown, partially decomposed, granular, and flocculent, and has an earthy odor when fresh.

activated-sludge effluent [CIV ENG] The liquid from the activated-sludge treatment that is further processed by chlorination or by oxidation.

activated-sludge process [CIV ENG] A sewage treatment process in which the sludge in the secondary stage is put into aeration tanks to facilitate aerobic decomposition by microorganisms; the sludge and supernatant liquor are separated in

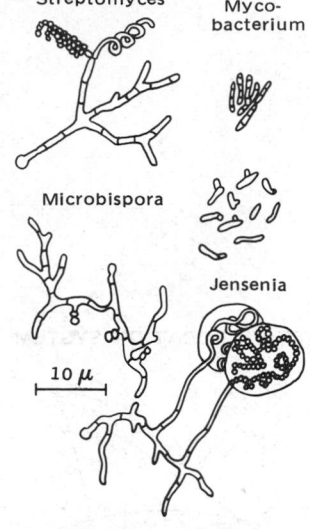

ACTINOMYCETALES

Streptomyces

Mycobacterium

Microbispora

Jensenia

10 μ

Streptosporangium

Micromonospora

Nocardia

Some genera of Actinomycetales.

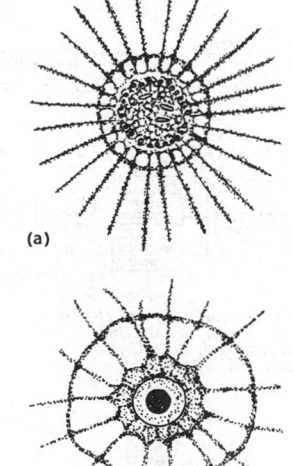

ACTINOPHRYIDA

(a)

(b)

Examples of Actinophryida. *(a)* Single specimen of *Actinosphaerium eichorni* (after Pernard). *(b)* Single specimen of *Actinophrys pontica*. (From R. P. Hall, Protozoology, Prentice-Hall, 1953)

a settling tank; the supernatant liquor or effluent is further treated by chlorination or oxidation.

activating reagent [MATER] Material added to another material or mixture so that a physical or chemical change will take place more rapidly or completely.

activation [CHEM] Treatment of a substance by heat, radiation, or activating reagent to produce a more complete or rapid chemical or physical change. [ELEC] The process of adding liquid to a manufactured cell or battery to make it operative. [ELECTR] The process of treating the cathode or target of an electron tube to increase its emission. Also known as sensitization. [MET] 1. A process of facilitating the separation and collection of ore powders by the use of substances which change the response of the particle surfaces to a flotation fluid. 2. A process that increases the rate of pressing and heating a metal powder into cohesion. [NUCLEO] The process of inducing radioactivity by bombardment with neutrons or other types of radiation. [PHYSIO] The designation for all changes in the ovum during fertilization, from sperm contact to the dissolution of nuclear membranes.

activation analysis [NUCLEO] A method of chemical analysis based on the detection of characteristic radionuclides following a nuclear bombardment. Also known as radioactivity analysis.

activation energy [PHYS CHEM] The energy, in excess over the ground state, which must be added to an atomic or molecular system to allow a particular process to take place.

activator [CHEM] 1. A substance that increases the effectiveness of a rubber vulcanization accelerator; for example, zinc oxide or litharge. 2. A trace quantity of a substance that imparts luminescence to crystals; for example, silver or copper in zinc sulfide or cadmium sulfide pigments.

activator RNA [GEN] Ribonucleic acid molecules which form a sequence-specific complex with receptor genes linked to producer genes.

active anaphylaxis [IMMUNOL] The allergic response following reintroduction of an antigen into a hypersensitive individual.

active antiroll system [NAV ARCH] A system of antiroll tanks in a ship in which pumps are used to transfer the liquid, through a connecting channel, from one tank in a pair to the other.

active area [ELECTR] The area of a metallic rectifier that acts as the rectifying junction and conducts current in the forward direction.

active balance [COMMUN] Summation of all return currents, in telephone repeater operation, at a terminal network balanced against the impedance of the local circuit or drop.

active center [ASTRON] A localized, transient region of the solar atmosphere in which sunspots, faculae, plages, prominences, solar flares, and so forth are observed.

active communications satellite [AERO ENG] Satellite which receives, regenerates, and retransmits signals between stations.

active component *See* active current; active element.

active computer [ADP] When two or more computers are installed, the one that is on-line and processing data.

active current [ELEC] The component of an electric current in a branch of an alternating-current circuit that is in phase with the voltage. Also known as active component; watt current.

active detection system [ENG] A guidance system which emits energy as a means of detection; for example, sonar and radar.

active electric network [ELEC] Electric network containing one or more sources of energy.

active electronic countermeasures [ELECTR] The major subdivision of electronic countermeasures concerning electronic jamming and electronic deceptions.

active element [ELECTR] Any generator of voltage or current in an impedance network. Also known as active component. [NUC PHYS] A chemical element which has one or more radioactive isotopes.

active entry [MIN ENG] An entry in which coal is being mined from a portion or from connected sections.

active front [METEOROL] A front, or portion thereof, which produces appreciable cloudiness and, usually, precipitation.

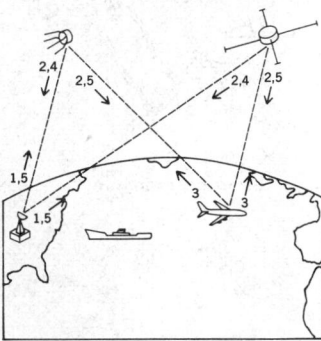

ACTIVE LOCATION SYSTEM

Sequence of events in active location procedure for traffic control showing the relay of questions and answers between the plane, the satellites, and the ground station.

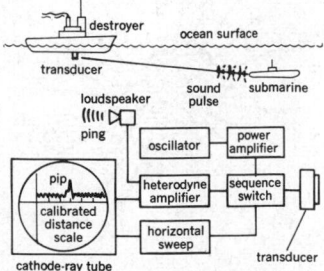

ACTIVE SONAR

Active sonar system. (From H. F. Olson, Acoustical Engineering, 3d ed., Van Nostrand, 1957)

active glacier [HYD] A glacier in which some of the ice is flowing.

active homing [NAV] 1. The homing of an aerodynamic missile by radar, in which radio signals are transmitted from the missile to the target and reflected to the missile to direct it toward the target. 2. Homing in which the homing device on the missile reveals the presence of the missile to the target.

active immunity [IMMUNOL] Disease resistance in an individual due to antibody production after exposure to a microbial antigen following disease, inapparent infection, or inoculation.

active infrared detection system [ENG] An infrared detection system in which a beam of infrared rays is transmitted toward possible targets, and rays reflected from a target are detected.

active jamming *See* jamming.

active layer [GEOL] That part of the soil which is within the suprapermafrost layer and which usually freezes in winter and thaws in summer. Also known as frost zone.

active leg [ELECTR] An electrical element within a transducer which changes its electrical characteristics as a function of the application of a stimulus.

active location system [NAV] A navigation system in which the navigation satellite interrogates the craft, and the craft responds; useful for surveillance by a ground station, or for automated navigation if the satellite subsequently transmits data.

active master file [ADP] A relatively active computer master file, as determined by usage data.

active master item [ADP] A relatively active item in a computer master file, as determined by usage data.

active material [ELEC] 1. A fluorescent material used in screens for cathode-ray tubes. 2. An energy-storing material, such as lead oxide, used in the plates of a storage battery. 3. A material, such as the iron of a core or the copper of a winding, that is involved in energy conversion in a circuit. [ELECTR] The material of the cathode of an electron tube that emits electrons when heated. [NUCLEO] A material capable of releasing substantial quantities of nuclear energy during fission.

active permafrost [GEOL] Permanently frozen ground (permafrost) which, after thawing by artificial or unusual natural means, reverts to permafrost under normal climatic conditions.

active power [ELEC] The product of the voltage across a branch of an alternating-current circuit and the component of the electric current that is in phase with the voltage.

active prominence [ASTRON] A classification of prominences of the sun; such a prominence is rapidly moving, and is the most frequent type.

active prominence region [ASTRON] Portions of the solar limb that display active prominences, characterized by downflowing knots and streamers, sprays, frequent surges, and curved loops. Abbreviated APR.

active region [ASTRON] A localized, transient, nonuniform region on the sun's surface, penetrating well down into the lower chromosphere.

active satellite [AERO ENG] A satellite which transmits a signal.

active sonar [ENG] A system consisting of one or more transducers to send and receive sound, equipment for the generation and detection of the electrical impulses to and from the transducer, and a display or recorder system for the observation of the received signals.

active system [ENG] In radio and radar, a system that requires transmitting equipment, such as a beacon or transponder.

active tracking system [NAV] A system with a transponder or transmitter on board the vehicle to repeat, transmit, or retransmit information to the tracking equipment; for example, Dovap, Secor, Azusa, Miran, and Minitrack.

active transducer [ELECTR] A transducer whose output is dependent upon sources of power, apart from that supplied by any of the actuating signals, which power is controlled by one or more of these signals.

active transport [PHYSIO] The pumping of ions or other

substances across a cell membrane against an osmotic gradient, that is, from a lower to a higher concentration.

active volcano [GEOL] A volcano capable of venting lava, pyroclastic material, or gases.

active voltage [ELEC] In an alternating-current circuit, the component of voltage which is in phase with the current.

active workings [MIN ENG] All places in a mine that are ventilated and inspected regularly.

activity [ADP] The use or modification of information contained in a file. [NUC PHYS] The intensity of a radioactive source. Also known as radioactivity. [PHYS CHEM] A thermodynamic function that correlates changes in the chemical potential with changes in experimentally measurable quantities, such as concentrations or partial pressures, through relations formally equivalent to those for ideal systems.

activity chart [IND ENG] A tabular presentation of a series of operations of a process plotted against a time scale.

activity coefficient [PHYS CHEM] A characteristic of a quantity expressing the deviation of a solution from ideal thermodynamic behavior; often used in connection with electrolytes.

activity level [ADP] 1. The value assumed by a structural variable during the solution of a programming problem. 2. A measure of the number of times that use or modification is made of the information contained in a file.

activity ratio [ADP] The ratio between used or modified records and the total number of records in a file. [GEOL] The ratio of plasticity index to percentage of clay-sized minerals in sediment.

activity sampling *See* work sampling.

actol *See* silver lactate.

actomyosin [BIOCHEM] A protein complex consisting of myosin and actin; the major constituent of a contracting muscle fibril.

actophilous [ECOL] Having a seashore growing habit.

actual age *See* absolute age.

actual cost [IND ENG] Cost determined by an allocation of cost factors recorded during production.

actual cubic feet per minute [CHEM ENG] A measure of the volume of gas at operating temperature and pressure, as distinct from volume of gas at standard temperature and pressure. Abbreviated acfm.

actual decimal point [ADP] The period appearing on a printed report as opposed to the virtual point defined only by the data structure within the computer.

actual elevation [METEOROL] The vertical distance above mean sea level of the ground at the meteorological station.

actual exhaust velocity [AERO ENG] 1. The real velocity of the exhaust gas leaving a nozzle as determined by accurately measuring at a specified point in the nozzle exit plane. 2. The velocity obtained when the kinetic energy of the gas flow produces actual thrust.

actual height [ELECTROMAG] Highest altitude at which refraction of radio waves actually occurs.

actual horsepower *See* actual power.

actual key [ADP] A data item in COBOL computer language which can be used as an address.

actual motion [NAV] Motion of a craft relative to the earth.

actual power [MECH ENG] The power delivered at the output shaft of a source of power. Also known as actual horsepower.

actual pressure [METEOROL] The atmospheric pressure at the level of the barometer (elevation of ivory point), as obtained from the observed reading after applying the necessary corrections for temperature, gravity, and instrumental errors.

actual relative movement *See* slip.

actual time [IND ENG] Time taken by a worker to perform a given task.

actual time of arrival [NAV] The time at which a craft arrives at a specified point at a destination.

actual time of departure [NAV] 1. The time of leaving a specified point at a place. 2. The actual time an aircraft becomes airborne.

actual time of interception [NAV] The time of intercepting a craft by another craft.

actuate [MECH ENG] To put into motion or mechanical action, as by an actuator.

actuated roller switch [MECH ENG] A centrifugal sequence-control switch that is placed in contact with a belt conveyor, immediately preceding the conveyor it controls.

actuating system [CONT SYS] An electric, hydraulic, or other system that supplies and transmits energy for the operation of other mechanisms or systems.

actuator [CONT SYS] A mechanism to activate process control equipment by use of pneumatic, hydraulic, or electronic signals; for example, a valve actuator for opening or closing a valve to control the rate of fluid flow. [ENG ACOUS] An auxiliary external electrode used to apply a known electrostatic force to the diaphragm of a microphone for calibration purposes. Also known as electrostatic actuator. [ORD] Part of the receiver mechanism in certain types of automatic weapons.

acuate [BIOL] 1. Having a sharp point. 2. Needle-shaped.

acuity [BIOL] Sharpness of sense perception, as of vision or hearing.

Aculeata [INV ZOO] A group of seven superfamilies that constitute the stinging forms of hymenopteran insects in the suborder Apocrita.

aculeus [INV ZOO] 1. A sharp, hairlike spine, as on the wings of certain lepidopterans. 2. An insect stinger modified from an ovipositor.

Aculognathidae [INV ZOO] The ant-sucking beetles, a family of coleopteran insects in the superfamily Cucujoidea.

acuminate [BOT] Tapered to a slender point, especially referring to leaves.

acupuncture [MED] The ancient Chinese art of puncturing the body with long, fine gold or silver needles to relieve pain and cure disease.

acutance [OPTICS] An objective measure of the ability of a photographic system to show a sharp edge between contiguous areas of low and high illuminance.

acute [BIOL] Ending in a sharp point. [MED] Referring to a disease or disorder of rapid onset, short duration, and pronounced symptoms.

acute alcoholism [MED] Drunkenness accompanied by an acute, transient disturbance of physiological and mental functions.

acute angle [MATH] An angle of less than 90°.

acute angle block [GEOL] A fault block in which the strike of strata on the down-dip side meets a diagonal fault at an acute angle.

acute appendicitis [MED] A sudden, severe attack of appendicitis characterized by abdominal pain, usually localized in the lower right quadrant, with nausea, vomiting, and constipation.

acute arthritis [MED] A severe joint inflammation with a short course.

acute ascending myelitis [MED] Severe inflammation of the spinal cord beginning in the lower segments and progressing toward the head.

acute bacterial endocarditis [MED] Fulminant, rapidly progressive endocarditis, usually associated with a significant systemic infection.

acute benign lymphoblastosis *See* infectious mononucleosis.

acute berylliosis [MED] Severe chemical pneumoconiosis caused by inhalation of beryllium salts.

acute bisectrix [MINERAL] A bisecting line of the acute angle of the optic axes of biaxial minerals.

acute cerebellar ataxia [MED] A severe childhood syndrome of sudden onset characterized by muscular incoordination, impaired articulation, oscillations of the eyeballs, and decreased intraocular pressure.

acute dermatitis [MED] Any severe inflammation of the skin.

acute glomerulonephritis [MED] Severe kidney inflammation, usually following infection with group A hemolytic streptococci, particularly type 12.

acute granulocytic leukemia [MED] A severe blood disorder in which the abnormal white cells are immature forms of granulocytes. Also known as myeloblastic leukemia.

acute infective encephalomyelitis *See* epidemic neuromyasthenia.

acute inflammation [MED] Severe inflammation with rapid progress and pronounced symptoms.

acute leukemia [MED] A severe blood disorder characterized

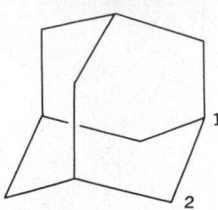

ADAMANTANE

Structure of adamantane showing bridgehead carbon labeled 1 and alternate position for substitution at carbon 2.

by rapid onset and progress, with anemia and hemorrhagic manifestations; immature forms of leukocytes are predominant.

acute lymphocytic leukemia [MED] A severe blood disorder in which abnormal leukocytes are identified as immature forms of lymphocytes. Also known as lymphoblastic leukemia.

acute monocytic leukemia [MED] A severe blood disorder in which abnormal leukocytes are identified as immature forms of monocytes. Also known as monoblastic leukemia.

acute necrotizing hemorrhagic encephalomyelitis [MED] A sudden, severe central nervous system disease with variable symptoms; pathology includes hemorrhages and necrosis of the white matter.

acute nonsuppurative hepatitis *See* interstitial hepatitis.

acute radiation syndrome [MED] A complex of symptoms involving the intestinal tract, blood-forming organs, and skin following whole-body irradiation.

acute respiratory disease [MED] Severe adenovirus infection of the respiratory tract characterized by fever, sore throat, and cough.

acute rheumatic fever [MED] A severe infectious process caused by beta hemolytic streptococci; characterized by fever and frequently accompanied by painful inflamed joints, endocarditis, chorea, or glomerulonephritis.

acute rhinitis [MED] Inflammation of the nasal mucous membrane due to either infection or allergy.

acute toxic encephalopathy [MED] A severe childhood syndrome characterized by sudden onset of coma or stupor, fever, convulsions, and impaired respiratory and cardiovascular functioning.

acute triangle [MATH] A triangle each of whose angles is less than 90°.

acute tubular necrosis *See* lower nephron nephrosis.

acute yellow atrophy [MED] Rapid liver destruction following viral hepatitis, toxic chemicals, or other agents.

acutifoliate [BOT] Having sharply pointed leaves.

acutilobate [BOT] Having sharply pointed lobes.

acyclic [BOT] Having flowers arranged in a spiral instead of a whorl. [MATH] **1.** A transformation on a set to itself for which no nonzero power leaves an element fixed. **2.** A chain complex all of whose homology groups are trivial. [PHYS] Continually varying without a regularly repeated pattern.

acyclic compound [ORG CHEM] A chemical compound with an open-chain molecular structure rather than a ring-shaped structure; for example, the alkane series.

acyclic feeding [ADP] A method employed by alphanumeric readers in which the trailing edge or some other document characteristic is used to activate the feeding of the succeeding document.

acyclic machine *See* homopolar generator.

acyclic motion *See* irrotational flow.

acyl [ORG CHEM] A radical formed from an organic acid by removal of a hydroxyl group; the general formula is RCO, where R may be aliphatic, alicyclic, or aromatic.

acylation [ORG CHEM] Any process (with the exception of the Friedel-Crafts method) whereby the acyl group is incorporated into a molecule by substitution.

acyl carrier protein [BIOCHEM] A protein in fatty acid synthesis that picks up aceytl and malonyl groups from acetyl coenzyme A and malonyl coenzyme A and links them by condensation to form β-keto acid acyl carrier protein, releasing carbon dioxide and the sulfhydryl form of acyl carrier protein. Abbreviated ACP.

acyl exchange *See* acidolysis.

acyl halide [ORG CHEM] One of a large group of organic substances containing the halocarbonyl group; for example, acyl fluoride.

acyloin [ORG CHEM] An organic compound that may be synthesized by condensation of aldehydes; an example is benzoin, $C_6H_5COCHOHC_6H_5$.

acyloin condensation [ORG CHEM] The reaction of an aliphatic ester with metallic sodium to form intermediates converted by hydrolysis into aliphatic α-hydroxyketones called acyloins.

AD *See* average deviation.

ADA *See* air defense artillery.

adalert [GEOPHYS] An advance alert issued by a regional warning center to give prompt warning of a change in solar activity.

adamantane [ORG CHEM] A $C_{10}H_{16}$ alicyclic hydrocarbon whose structure has the same arrangement of carbon atoms as does the basic unit of the diamond lattice.

adamantine drill [MECH ENG] A core drill with hardened steel shot pellets that revolve under the rim of the rotating tube; employed in rotary drilling in very hard ground.

adamantine spar [MINERAL] A silky brown variety of corundum.

adamantinoma *See* ameloblastoma.

adambulacral [INV ZOO] Lying adjacent to the ambulacrum.

adamellite *See* quartz monzonite.

adamite [MINERAL] $Zn_2(AsO_4)(OH)$ A colorless, white, or yellow mineral consisting of basic zinc arsenate, crystallizing in the orthorhombic system; hardness is 3.5 on Mohs scale, and specific gravity is 4.34–4.35.

Adams-Bashforth process [MATH] A method of numerically integrating a differential equation of the form $(dy/dx) = f(x,y)$ that uses one of Gregory's interpolation formulas to expand f.

adamsite [ORG CHEM] $C_6H_4 \cdot NH \cdot C_6H_4 \cdot AsCl$ A yellow crystalline arsenical; used in leather tanning and in warfare and riot control to produce skin and eye irritation, chest distress, and nausea. U.S. army code is DM. Also known as diphenylaminechloroarsine; phenarsazine chloride. [MINERAL] Greenish-black mica.

ada mud [ENG] A conditioning material added to drilling mud to obtain satisfactory cores and samples of formations.

adapertural [INV ZOO] Near the aperture, specifically of a conch.

adapical [BOT] Near or toward the apex or tip.

adaptability test [ORD] A test to ascertain the adaptability of a standardized item or equipment to a particular unit or organization.

adaptation [GEN] The occurrence of genetic changes in a population or species as the result of natural selection so that it adjusts to new or altered environmental conditions. [PHYSIO] The occurrence of physiological changes in an individual exposed to changed conditions; for example, tanning of the skin in sunshine, or increased red blood cell counts at high altitudes.

adaptation brightness *See* adaptation luminance.

adaptation illuminance *See* adaptation luminance.

adaptation level *See* adaptation luminance.

adaptation luminance [OPTICS] The average luminance, or brightness, of objects and surfaces in the immediate vicinity of an observer estimating the visual range. Also known as adaptation brightness; adaptation illuminance; adaptation level; brightness level; field brightness; field luminance.

adaptation syndrome [MED] Endocrine-mediated stress reaction of the body in response to systemic injury; involves an initial stage of shock, followed by resistance or adaptation and then healing or exhaustion.

adapter [ADP] A device which converts bits of information received serially into parallel bit form for use in the inquiry buffer unit. [ENG] A device used to make electrical or mechanical connections between items not originally intended for use together. [MET] A connecting piece, usually made of fireclay, between a horizontal zinc retort and the condenser in which the molten zinc collects. [OPTICS] An attachment to a camera that permits its use in a manner for which it was not designed.

adapter skirt [AERO ENG] A flange or extension of a space vehicle that provides a ready means for fitting some object to a stage or section.

adapter transformer [ELEC] A transformer designed to supply a single electric lamp; its primary terminals are designed to fit into an ordinary lampholder, its secondary terminals into a lampholder of a low-voltage lamp.

adaptive behavior [PSYCH] Any behavior that helps the organism adjust to its environment.

adaptive colitis *See* irritable colon.

adaptive control [CONT SYS] A control method in which one or more parameters are sensed and used to vary the feedback control signals in order to satisfy the performance criteria.

adaptive disease [PHYSIO] The physiologic changes impairing an organism's health as the result of exposure to an unfamiliar environment.

adaptive divergence [EVOL] Divergence of new forms from a common ancestral form due to adaptation to different environmental conditions.

adaptive enzyme [MICROBIO] Any bacterial enzyme formed in response to the presence of a substrate specific for that enzyme.

adaptive radiation [EVOL] Diversification of a dominant evolutionary group into a large number of subsidiary types adapted to more restrictive modes of life (different adaptive zones) within the range of the larger group.

adaptive system [ADP] A system that can change itself to meet new requirements.

adaptive system theory [ADP] The branch of automata theory dealing with adaptive, or self-organizing, systems.

adatom [PHYS CHEM] An atom adsorbed on a surface so that it will migrate over the surface.

adaxial [BIOL] On the same side as or facing toward the axis of an organ or organism.

Adcock antenna [ELECTROMAG] A pair of vertical antennas separated by a distance of one-half wavelength or less and connected in phase opposition to produce a radiation pattern having the shape of a figure eight.

ADCON *See* address constant.

add [ORD] A fire correction term used by observers in adjusting fire to indicate that an increase in range (of so many yards) will follow and is desired.

addend [MATH] One of a collection of numbers to be added.

addendum [DES ENG] The radial distance between two concentric circles on a gear, one being that whose radius extends to the top of a gear tooth (addendum circle) and the other being that which will roll without slipping on a circle on a mating gear (pitch line).

addendum circle [DES ENG] The circle on a gear passing through the tops of the teeth.

adder [ADP] A computer device that can form the sum of two or more numbers or quantities. [ELECTR] A circuit in which two or more signals are combined to give an output-signal amplitude that is proportional to the sum of the input-signal amplitudes. Also known as adder circuit. [VERT ZOO] Any of the venomous viperine snakes included in the family Viperidae.

adder circuit *See* adder.

addiction [MED] Habituation to a specific practice, such as drinking alcoholic beverages or using drugs.

adding circuit [ELECTR] A circuit that performs the mathematical operation of addition.

adding machine [ADP] A device which performs the arithmetical operation of addition and subtraction.

adding tape [ENG] A surveyor's tape that is calibrated from 0 to 100 by full feet (or meters) in one direction, and has 1 additional foot (or meter) beyond the zero end which is subdivided in tenths or hundredths.

Addis count [PATH] A renal function test which estimates the blood cell count in a 12-hour urine specimen.

Addison's disease [MED] A primary failure or insufficiency of the adrenal cortex to secrete hormones.

addition [MATH] An operation by which two elements of a set are combined to yield a third; denoted + ; usually reserved for the operation in an Abelian group or the group operation in a ring or vector space.

addition agent [PHYS CHEM] A substance added to a plating solution to change characteristics of the deposited substances.

addition item [ADP] An item which is to be filed in its proper place in a computer.

addition of complex quantities [MATH] The combining of complex quantities in which the individual real parts and the individual imaginary parts are separately added.

addition of vectors [MATH] The combining of vectors in a prescribed way; for example, by algebraically adding corresponding components of vectors or by forming the third side of the triangle whose other sides each represent a vector.

addition polymer [ORG CHEM] A polymer formed by the chain addition of unsaturated monomer molecules, such as olefins, with one another without the formation of a by-product, as water; examples are polyethylene, polypropylene, and polystyrene. Also known as addition resin.

addition reaction [ORG CHEM] A type of reaction of unsaturated hydrocarbons with hydrogen halogens, halogen acids, and other reagents, so that no change in valency is observed and the organic compound forms a more complex one.

addition record [ADP] A new record inserted into an updated master file.

addition resin *See* addition polymer.

addition solid solution [CRYSTAL] Random addition of atoms or ions in the interstices within a crystal structure.

additive [MATER] **1.** A substance added to another to improve, strengthen, or otherwise alter it; for example, tetraethyllead added to gasoline to prevent engine knock. **2.** *See* admixture. [MATH] Pertaining to addition. [STAT] That property of a process in which increments of the dependent variable are independent for nonoverlapping intervals of the independent variable.

additive function [MATH] Any function which preserves addition, such as $f(x+y) = f(x) + f(y)$.

additive set function [MATH] A set function with the property that the value of the function at a finite union of disjoint sets is equal to the sum of the values at each set in the union.

address [ADP] The number or name that uniquely identifies a register, memory location, or storage device in a computer.

address computation [ADP] The modification by a computer of an address within an instruction, or of an instruction based on results obtained so far. Also known as address modification.

address constant [ADP] A value, or its expression, used in the calculation of storage addresses from relative addresses for computers. Abbreviated ADCON. Also known as base address; presumptive address; reference address.

address conversion [ADP] The use of an assembly program to translate symbolic or relative computer addresses.

address counter [ADP] A counter which increments an initial memory address as a block of data is being transferred into the memory locations indicated by the counter.

address format [ADP] A description of the number of addresses included in a computer instruction.

address generation [ADP] An addressing technique which facilitates addressing large storages and implementing dynamic program relocation; the effective main storage address is obtained by adding together the contents of the base register of the index register and of the displacement field.

addressing mode [ADP] The specific technique by means of which a memory reference instruction will be spelled out if the computer word is too small to contain the memory address.

addressing system [ADP] A labeling technique used to identify storage locations within a computer system.

address modification *See* address computation.

address part [ADP] That part of a computer instruction which contains the address of the operand, of the result, or of the next instruction.

address register [ADP] A register wherein the address part of an instruction is stored by a computer.

address sort routine [ADP] A debugging routine which scans all instructions of the program being checked for a given address.

add-subtract time [ADP] The time required to perform an addition or subtraction, exclusive of the time required to obtain the quantities from storage and put the sum or difference back into storage.

add-to-memory technique [ADP] In direct-memory-access systems, a technique which adds a data word to a memory location; permits linear operations such as data averaging on process data.

adduct [CHEM] **1.** A chemical compound that forms from chemical addition of two species; for example, reaction of butadiene with styrene forms an adduct, 4-phenyl-1-cyclo-hexene. **2.** The complex compound formed by association of an inclusion complex.

adduction [PHYSIO] Movement of one part of the body toward another or toward the median axis of the body.

adductor [ANAT] Any muscle that draws a part of the body toward the median axis.

Adeleina [INV ZOO] A suborder of protozoan invertebrate parasites in the order Eucoccida in which the sexual and asexual stages are in different hosts.

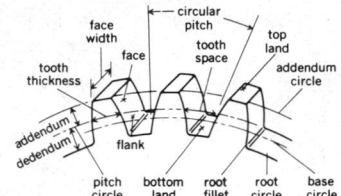

ADDENDUM

Drawing of the principal features of a gear tooth showing the addendum in relation to the other parts.

ADENINE

Structural formula of adenine; the carbon atoms are numbered.

ADENOSINE

Structural formula of adenosine.

adelphous [BOT] Having stamens fused together by their filaments.

adenase [BIOCHEM] An enzyme that catalyzes the hydrolysis of adenine to hypoxanthine and ammonia.

adenine [BIOCHEM] $C_5H_5N_5$ A purine base, 6-aminopurine; one of the fundamental components of all nucleic acids.

adenitis [MED] Inflammation of a gland or lymph node.

adenoacanthoma [MED] An adenocarcinoma, common in the endometrium, in which squamous cells replace the cylindrical epithelium.

adeno-associated satellite virus [VIROL] A defective virus that is unable to reproduce without the help of an adenovirus.

adenocarcinoma [MED] A malignant tumor originating in glandular or ductal epithelium and tending to produce acinic structures.

adenohypophysis [ANAT] The glandular part of the pituitary gland, composing the anterior and intermediate lobes.

adenoid [ANAT] **1.** A mass of lymphoid tissue. **2.** Lymphoid tissue of the nasopharynx. Also known as pharyngeal tonsil.

adenoma [MED] A benign tumor of glandular origin and structure.

adenomatoid tumor [MED] A benign genital-tract tumor composed of stroma whose spaces are lined by cells that resemble epithelium, endothelium, and mesothelium.

adenomatosis [MED] A condition characterized by multiple adenomas within an organ or in several related organs.

adenomatous goiter [MED] An asymmetric goiter due to isolated nodular masses of thyroid tissue. Also known as multiple colloid goiter; nodular goiter.

adenomere [EMBRYO] The embryonic structure which will become the functional portion of a gland.

adenomyoma [MED] A benign tumor of glandular and muscular elements occurring principally in the uterus and rectum.

adenomyosis [MED] **1.** The invasion of muscular tissue, such as of the uterine wall or Fallopian tubes, by endometrial tissue. **2.** Any abnormal growth of muscle or glandular tissues.

adenopathy [MED] Any glandular disease; common usage limits the term to any abnormal swelling or enlargement of lymph nodes.

Adenophorea [INV ZOO] A class of unsegmented worms in the phylum Nematoda.

adenophyllous [BOT] Having leaves with glands.

adenosine [BIOCHEM] $C_{10}H_{13}N_5O_4$ A nucleoside composed of adenine and D-ribose.

adenosinediphosphatase [BIOCHEM] An enzyme that catalyzes the hydrolysis of adenosinediphosphate. Abbreviated ADPase.

adenosinediphosphate [BIOCHEM] $C_{10}H_{15}N_5O_{10}P_2$ A coenzyme composed of adenosine and two molecules of phosphoric acid that is important in intermediate cellular metabolism. Abbreviated ADP.

adenosinemonophosphate *See* adenylic acid.

adenosinetriphosphatase [BIOCHEM] An enzyme that catalyzes the hydrolysis of adenosinetriphosphate. Abbreviated ATPase.

adenosinetriphosphate [BIOCHEM] $C_{10}H_{16}N_5O_{12}P_3$ A coenzyme composed of adenosinediphosphate with an additional phosphate group; an important energy compound in metabolism. Abbreviated ATP.

adenosis [MED] Any nonneoplastic glandular disease, especially one involving the lymph nodes.

adeno-SV40 hybrid virus [VIROL] A defective virus particle in which part of the genetic material of papovavirus SV40 is encased in an adenovirus protein coat.

adenovirus [VIROL] A group of animal viruses which cause febrile catarrhs and other respiratory diseases.

adenylic acid [BIOCHEM] **1.** A generic term for a group of isomeric nucleotides. **2.** The phosphoric acid ester of adenosine. Also known as adenosinemonophosphate (AMP).

adeoniform [INV ZOO] **1.** A lobate, bilamellar zooarium. **2.** Resembling the fossil bryozoan *Adeona*.

Adephaga [INV ZOO] A suborder of insects in the order Coleoptera characterized by fused hind coxae that are immovable.

adequate contact [MED] The degree of contact required between an infectious and a susceptible individual to cause infection of the latter.

adequate stimulus [PHYSIO] The energy of any specific mode that is sufficient to elicit a response in an excitable tissue.

ader wax *See* ozocerite.

ADF *See* automatic direction finder.

ADF bearing indicator [NAV] An instrument used with an airborne radio direction finder to indicate automatically the relative, magnetic, or true bearing (or reciprocal) of a transmitter.

adfreezing [HYD] The process by which one object adheres to another by the binding action of ice; applied to permafrost studies.

ADF reversal [NAV] The swinging of the needle on the direction indicator of an airborne automatic direction finder (ADF) through 180°, indicating that the station to which the direction finder is tuned has been passed.

ADH *See* vasopressin.

Adhara [ASTRON] A star of spectral type B2II. Also known as ε Canis Majoris.

adhesion [BOT] Growing together of members of different and distinct whorls. [ELECTROMAG] Any mutually attractive force holding together two magnetic bodies, or two oppositely charged nonconducting bodies. [ENG] Intimate sticking together of metal surfaces under compressive stresses by formation of metallic bonds. [MECH] The force of static friction between two bodies, or the effects of this force. [MED] The abnormal union of an organ or part with some other part by formation of fibrous tissue. [PHYS] The tendency, due to intermolecular forces, for matter to cling to other matter.

adhesive [MATER] A substance used to bond two or more solids so that they act or can be used as a single piece; examples are resins, formaldehydes, glue, paste, cement, putty, and polyvinyl resin emulsions.

adhesive bonding [ENG] The fastening together of two or more solids by the use of glue, cement, or other adhesive.

adhesive cell [INV ZOO] Any of various glandular cells in ctenophores, turbellarians, and hydras used for adhesion to a substrate and for capture of prey. Also known as colloblast; glue cell; lasso cell.

adhesive strength [ENG] The strength of an adhesive bond, usually measured as a force required to separate two objects of standard bonded area, by either shear or tensile stress.

adhesive tape [MATER] Tape coated with a substance that binds or sticks to a surface.

adiabatic [THERMO] Referring to any change in which there is no gain or loss of heat.

adiabatic atmosphere [METEOROL] A model atmosphere characterized by a dry-adiabatic lapse rate throughout its vertical extent.

adiabatic calorimeter [PHYS CHEM] An instrument used to study chemical reactions which have a minimum loss of heat.

adiabatic chart *See* Stuve chart.

adiabatic compression [THERMO] A reduction in volume of a substance without heat flow, in or out.

adiabatic condensation pressure *See* condensation pressure.

adiabatic condensation temperature *See* condensation temperature.

adiabatic cooling [THERMO] A process in which the temperature of a system is reduced without any heat being exchanged between the system and its surroundings.

adiabatic demagnetization [CRYO] A method of cooling paramagnetic salts to temperatures of 10^{-3} K; the sample is cooled to boiling point of helium in a strong magnetic field, thermally isolated, and then removed from the field to demagnetize it. Also known as Giaque-Debye method; magnetic cooling; paramagnetic cooling.

adiabatic engine [MECH ENG] A heat engine or thermodynamic system which is adiabatic.

adiabatic equilibrium [METEOROL] A vertical distribution of temperature and pressure in an atmosphere in hydrostatic equilibrium such that an air parcel displaced adiabatically will continue to possess the same temperature and pressure as its surroundings, so that no restoring force acts on a parcel displaced vertically. Also known as convective equilibrium.

adiabatic equivalent temperature *See* equivalent temperature.

adiabatic expansion [THERMO] Increase in volume without heat flow, in or out.

adiabatic flame temperature [PHYS CHEM] The highest possible temperature of combustion obtained under the conditions that the burning occurs in an adiabatic vessel, that it is complete, and that dissociation does not occur.

adiabatic flow [FL MECH] Movement of a fluid without heat transfer.

adiabatic invariant [PHYS] A physical quantity which may be quantized and which, to a certain degree of approximation, remains unchanged under the slow variation of any parameter.

adiabatic lapse rate *See* dry-adiabatic lapse rate.

adiabatic law [PHYS] The relationship which states that, for adiabatic expansion of gases, $P\rho^{-\gamma}$ = constant, where P = pressure, ρ = density, and γ = ratio of specific heats C_P/C_V.

Adiabatic Low-energy Injection and Capture Experiment [NUCLEO] An experimental apparatus for research on controlled fusion that uses a beam of neutral deuterium atoms. Also known as Alice.

adiabatic process [THERMO] Any thermodynamic procedure which takes place in a system without the exchange of heat with the surroundings.

adiabatic recovery temperature [FL MECH] **1.** The temperature reached by a moving fluid when brought to rest through an adiabatic process. Also known as recovery temperature; stagnation temperature. **2.** The final and initial temperature in an adiabatic, Carnot cycle.

adiabatic saturation pressure *See* condensation pressure.

adiabatic saturation temperature *See* condensation temperature.

adiabatic system [SCI TECH] A body or system whose condition is altered without gaining heat from or losing heat to the surroundings.

adiabatic wall temperature [FL MECH] The temperature assumed by a wall in a moving fluid stream when there is no heat transfer between the wall and the stream.

adiadochokinesis [MED] A type of motor incoordination associated with cerebellar damage in which repetitive movements controlled by antagonistic muscles cannot be performed without severe muscular incoordination.

adiagnostic [PETR] Pertaining to a rock texture in which identification of individual components is not possible macroscopically or microscopically; applied especially to igneous rock.

Adie's syndrome [MED] Impaired pupillary reaction to light and absent tendon reflexes.

Adimeridae [INV ZOO] An equivalent name for the Colydiidae.

adinole [GEOL] An argillaceous sediment that has undergone albitization at the margin of a basic intrusion.

adipate [ORG CHEM] Salt produced by reaction of adipic acid with a basic compound.

adiphenine [PHARM] $C_{20}H_{25}NO_2 \cdot HCl$ The compound 2-diethylaminoethyl diphenylacetate hydrochloride; a cholinergic blocking agent.

adipic acid [ORG CHEM] $HOOC(CH_2)_4COOH$ A colorless crystalline dicarboxylic acid, sparingly soluble in water; used in nylon manufacture. Also known as butane dicarboxylic acid; hexanedioic acid.

adipinketone *See* cyclopentanone.

adipocellulose [BIOCHEM] A type of cellulose found in the cell walls of cork tissue.

adipocere [MED] A light-colored, waxy material formed by postmortem conversion of body fats to higher fatty acids.

adipocerite *See* hatchettite.

adipogenesis [PHYSIO] The formation of fat or fatty tissue.

adiponecrosis neonatorum [MED] Localized fatty-tissue necrosis occurring in large, healthy infants born after difficult labor.

adiponitrile [ORG CHEM] $NC(CH_2)_4CN$ The high-boiling liquid dinitrile of adipic acid; used to make nylon intermediates. Also known as hexanedinitrile.

adipose [BIOL] Fatty; of or relating to fat.

adipose fin [VERT ZOO] A modified posterior dorsal fin that is fleshy and lacks rays; found in salmon and typical catfishes.

adipose tissue [HISTOL] A type of connective tissue specialized for lipid storage.

adiposis dolorosa [MED] An uncommon type of obesity in which the excess fat deposits are tender and painful.

adiposogenital dystrophy [MED] A syndrome involving obesity, retarded gonad development, and sometimes diabetes insipidus resulting from impaired functioning of the pituitary and hypothalamus. Also known as Froehlich's syndrome.

A display [ELECTR] A radar oscilloscope display in cartesian coordinates; the targets appear as vertical deflection lines; their Y coordinates are proportional to signal intensity; their X coordinates are proportional to distance to targets.

adit [CIV ENG] An access tunnel used for excavation of the main tunnel. [MIN ENG] A nearly horizontal tunnel for access, drainage, or ventilation of a mine. Also known as side drift.

adjacency [ADP] A condition in character recognition in which two consecutive graphic characters are separated by less than a specified distance.

adjacent angle [MATH] One of a pair of angles with common side formed by two intersecting straight lines.

adjacent-channel interference [COMMUN] Interference that is caused by a transmitter operating in an adjacent channel when the side bands of the adjacent-channel transmitter beat with the carrier signal of the desired station. Also known as monkey chatter; side-band interference; side-band splash.

adjacent-channel selectivity [ELECTR] The ability of a radio receiver to respond to the desired signal and to reject signals in adjacent frequency channels.

adjacent sea [GEOGR] A sea connected with the oceans but semienclosed by land; examples are the Caribbean Sea and North Polar Sea.

adjective dye [CHEM] Any dye that needs a mordant.

adjoining sheets [MAP] Those maps that are contiguous to one or more sides and corners of a map series.

adjoint of a matrix *See* adjugate of a matrix; Hermitian conjugate of a matrix.

adjoint operator [MATH] An operator B such that the inner products (Ax,y) and (x,By) are equal for a given operator A and for all elements x and y of a Hilbert space. Also known as associate operator; Hermitian conjugate operator.

adjoint variable [PHYS] In classical dynamics, the canonically conjugate p_i interpreted as generalized momenta.

adjoint vector space [MATH] The complete normed vector space constituted by a class of bounded, linear, homogeneous scalar functions defined on a normed vector space.

adjoint wave functions [QUANT MECH] Functions in the Dirac electron theory which are formed by applying the Dirac matrix B to the Hermitian conjugates of the original wave functions.

adjugate of a matrix [MATH] The matrix obtained by replacing each element with the cofactor of the transposed element. Also known as adjoint of a matrix.

adjustable propeller [NAV ARCH] A screw propeller with blades that can be rotated around their axes to change the pitch. Also known as controllable-pitch propeller.

adjustable resistor [ELEC] A resistor having one or more sliding contacts whose position may be changed.

adjustable transformer *See* variable transformer.

adjusted decibel [ELECTR] A unit used to show the relationship between the interfering effect of a noise frequency, or band of noise frequencies, and a reference noise power level of -85 dBm.

adjusted elevation [GEOD] **1.** The elevation resulting from the application of an adjustment correction to an orthometric elevation. **2.** The elevation resulting from the application of both an orthometric correction and an adjustment correction to a preliminary elevation.

adjusted position [MAP] An adjusted value of the coordinate position of a point.

adjusted stream [HYD] A stream which flows mostly parallel to the strike and as little as necessary in other courses.

adjusted value [SCI TECH] A value of a quantity derived from observed data by some orderly process which eliminates discrepancies arising from errors in those data.

adjusting point [ORD] A distinctive terrain feature or some

A DISPLAY

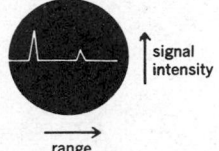

Drawing of a frequently employed display for presentation of radar outputs.

portion of the target at or near the center of the area upon which the observer wishes to place fire.

adjusting ring [ORD] Part of a fuse setter which is used to adjust the setting element on the fuse of an explosive projectile.

adjustment [GEOD] **1.** The determination and application of corrections to orthometric differences of elevation or to orthometric elevations to make the elevation of all bench marks consistent and independent of the circuit closures. **2.** The placing of detail or control stations in their positions relative to other detail or control stations. [PSYCH] The process of mental change to allow an individual to function harmoniously with his environment.

adjustment correction See arbitrary correction.

adjustment of fire [ORD] The determining and applying of corrections to firing data to bring the center of impact or of burst, or the cone of fire of automatic weapons, to the adjusting point and to keep it there.

adjustment reaction [PSYCH] A transient, situational personality disorder occurring in reaction to some significant person, immediate event, or internal emotional conflict.

adjustor neuron [ANAT] Any of the interconnecting nerve cells between sensory and motor neurons of the central nervous system.

adjutage [ENG] A tube attached to a container of liquid at an orifice to facilitate or regulate outflow.

adjuvant [PHARM] A material that enhances the action of a drug or antigen.

ad lib [BIOL] Shortened form for ad libitum; without limit or restraint.

adlittoral [OCEANOGR] Of, pertaining to, or occurring in shallow waters adjacent to a shore.

administrative map [MAP] **1.** A map with graphically recorded information pertaining to administrative matters, such as supply and evacuation installations, medical facilities, service and maintenance areas, main supply roads, boundaries, and other details necessary to show the administrative situation in relation to the tactical situation. **2.** Any map on which are delineated political subdivisions and boundaries of a country or countries.

Administrative Terminal System [ADP] A system developed by the International Business Machine Corporation to enable the handling by computer of texts that would otherwise require copying by a typist. Abbreviated ATS.

admiralty brass [MET] An alloy containing 71% copper, 28% zinc, and 0.75–1.0% tin for additional corrosion resistance.

admiralty coal [MIN ENG] A high-quality, smokeless steam coal.

Admiralty constant [NAV ARCH] Any of the numbers that give the horsepower of a ship required for a given speed at a given displacement and midship section.

admittance [ELEC] A measure of how readily alternating current will flow in a circuit; the reciprocal of impedance, it is expressed in mhos.

admixture [GEOL] One of the lesser or subordinate grades of sediment. [MATER] A material (other than aggregate, cement, or water) added in small quantities to concrete to produce some desired change in properties. Also known as additive.

adnate [BIOL] United through growth; used especially for unlike parts. [BOT] Pertaining to growth with one side adherent to a stem.

adnexa [BIOL] Subordinate or accessory parts, such as eyelids, Fallopian tubes, and extraembryonic membranes.

adobe [GEOL] Heavy-textured clay soil found in the southwestern United States and in Mexico.

adobe brick [MATER] An earth or clay, straw brick of large but varying dimensions, roughly molded and sun-dried.

adobe flats [GEOL] Broad flats that are floored with sandy clay and have been formed from sheet floods.

adolescence [GEOL] Stage in the cycle of erosion following youth and preceding maturity. [PSYCH] The period of life from puberty to maturity.

adolescent coast [GEOL] A type of shoreline characterized by low but nearly continuous sea cliffs.

adolescent river [HYD] A river with a graded bed and a well-cut channel that reaches base level at its mouth, its waterfalls

and lakes of the youthful stage having been destroyed.

adolescent stream [HYD] A stream characterized by a well-cut, smoothly graded channel that may reach base level at its mouth.

adont hinge [INV ZOO] A type of ostracod hinge articulation which either lacks teeth and has overlapping valves or has a ridge and groove.

adoral [ZOO] Near the mouth.

ADP See adenosinediphosphate; automatic data processing.

ADPase See adenosinediphosphatase.

ADPE See automatic data-processing equipment.

adrenal cortex [ANAT] The cortical moitie of the suprarenal glands which secretes glucocorticoids, mineralocorticoids, androgens, estrogens, and progestagens.

adrenal cortex hormone [BIOCHEM] Any of the steroids produced by the adrenal cortex. Also known as adrenocortical hormone; corticoid.

adrenal cortical insufficiency [MED] Failure of the adrenal cortex to secrete adequate hormones.

adrenalectomy [MED] Surgical removal of an adrenal gland.

adrenal gland [ANAT] An endocrine organ located close to the kidneys of vertebrates and consisting of two morphologically distinct components, the cortex and medulla. Also known as suprarenal gland.

adrenaline See epinephrine.

adrenal medulla [ANAT] The hormone-secreting chromaffin cells of the adrenal gland that produce epinephrine and norepinephrine.

adrenal virilism [MED] **1.** The development of male characteristics in the female resulting from excessive production of adrenal hormones with androgenic activity. **2.** A rare form of pseudohermaphroditism.

adrenergic [PHYSIO] Describing the chemical activity of epinephrine or epinephrine-like substances.

adrenergic blocking agent [BIOCHEM] Any substance that blocks the action of epinephrine or an epinephrine-like substance.

adrenochrome [BIOCHEM] $C_9H_9O_3N$ A brick-red oxidation product of epinephrine which can convert hemoglobin into methemoglobin.

adrenocortical hormone See adrenal cortex hormone.

adrenocorticotropic hormone [BIOCHEM] The chemical secretion of the adenohypophysis that stimulates the adrenal cortex. Abbreviated ACTH. Also known as adrenotropic hormone.

adrenogenital syndrome [MED] A group of symptoms associated with hypersecretion of adrenal cortex hormones; effects vary with sex and time of development.

adrenotropic [PHYSIO] Of or pertaining to an effect on the adrenal cortex.

adrenotropic hormone See adrenocorticotropic hormone.

adsorbate [CHEM] A solid, liquid, or gas which is adsorbed as molecules, atoms, or ions by such substances as charcoal, silica, metals, water, and mercury.

adsorbent [CHEM] A solid or liquid that adsorbs other substances; for example, charcoal, silica, metals, water, and mercury.

adsorption [CHEM] The surface retention of solid, liquid, or gas molecules, atoms, or ions by a solid or liquid, as opposed to absorbtion, the penetration of substances into the bulk of the solid or liquid.

adsorption chromatography [ANALY CHEM] Separation of a chemical mixture (gas or liquid) by passing it over an adsorbent bed which adsorbs different compounds at different rates.

adsorption gasoline [MATER] Gasoline extracted from natural gas or refinery gas.

adsorption indicator [ANALY CHEM] An indicator used in solutions to detect slight excess of a substance or ion; precipitate becomes colored when the indicator is adsorbed. An example is fluorescein.

adsorption isobar [PHYS CHEM] A graph showing how adsorption varies with some parameter, such as temperature, while holding pressure constant.

adsorption isotherm [PHYS CHEM] The relationship between the gas pressure p and the amount w, in grams, of a gas or vapor taken up per gram of solid at a constant temperature.

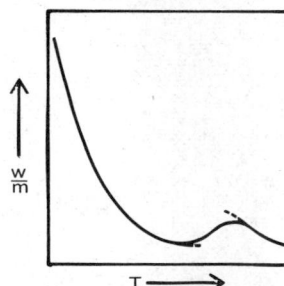

ADSORPTION ISOBAR

A typical adsorption isobar; w/m is weight of material adsorbed per unit weight of adsorbent, and T is absolute temperature.

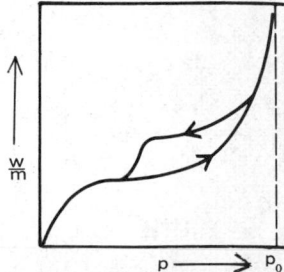

ADSORPTION ISOTHERM

A typical adsorption isotherm; w/m is weight of material adsorbed per weight of adsorbent, and p is pressure.

adsorption system [MECH ENG] A device that dehumidifies air by bringing it into contact with a solid adsorbing substance.

adularia [MINERAL] A weakly triclinic form of the mineral orthoclase occurring in transparent, colorless to milky-white pseudo-orthorhombic crystals.

adularization [GEOL] Replacement by or introduction of the mineral adularia.

adult rickets *See* osteomalacia.

ad valorem tax [PETRO ENG] Property tax for oil-producing properties, assessed at a flat rate for each net barrel of oil produced.

advance [GEOL] **1.** A continuing movement of a shoreline toward the sea. **2.** A net movement over a specified period of time of a shoreline toward the sea. [HYD] The forward movement of a glacier. [NAV] **1.** In making a turn, the distance a vessel moves in its initial direction from the point where the rudder is started over until the heading has changed 90°. **2.** The distance a vessel moves in the initial direction for heading changes of less than 90°.

advanced gallery [MIN ENG] A small heading driven in advance of the main tunnel in tunnel excavation.

advanced gas-cooled reactor [NUCLEO] A power-generating nuclear reactor which has steel-clad uranium dioxide fuel elements and is cooled by carbon dioxide gas.

advanced line of position [NAV] A line of position which has been moved forward along the course line to allow for the run since the line was established; the opposite is a retired line of position.

advanced potential [ELECTROMAG] Any electromagnetic potential arising as a solution of the classical Maxwell field equations, analogous to a retarded potential solution, but lying on the future light cone of space-time; the potential appears, at present, to have no physical interpretation.

advance feed tape [ADP] Computer tape punched so that the leading edges of its feed holes will line up with the leading edges of the data holes in the tape usage device.

advance overburden [MIN ENG] Overburden in excess of the average overburden-to-ore ratio that must be removed in opencut mining.

advance signal [CIV ENG] A signal in a block system up to which a train may proceed within a block that is not completely cleared.

advance stripping [MIN ENG] The removal of barren or sub-ore-grade earthy or rock materials to expose the minable grade of ore.

advance wave [MIN ENG] The air pressure wave preceding the flame in a coal dust explosion.

advancing [MIN ENG] Mining outward from the shaft toward the boundary.

advancing fire *See* assault fire.

advancing longwall [MIN ENG] Mining coal outward from the shaft pillar, with roadways maintained through the worked-out portion of the mine.

advection [METEOROL] The process of transport of an atmospheric property solely by the mass motion of the atmosphere. [OCEANOGR] The process of transport of water, or of an aqueous property, solely by the mass motion of the oceans, most typically via horizontal currents.

advection fog [METEOROL] A type of fog caused by the horizontal movement of moist air over a cold surface and the consequent cooling of that air to below its dew point.

advective hypothesis [METEOROL] The assumption that local temperature changes are the result only of horizontal or isobaric advection.

advective model [METEOROL] A mathematical or dynamic model of fluid flow which is characterized by the advective hypothesis.

advective thunderstorm [METEOROL] A thunderstorm resulting from static instability produced by advection of relatively colder air at high levels or relatively warmer air at low levels or by a combination of both conditions.

adventitia [ANAT] The external, connective-tissue covering of an organ or blood vessel. Also known as tunica adventitia.

adventitious [BIOL] Also known as adventive. **1.** Acquired spontaneously or accidentally, not by heredity. **2.** Arising, as a tissue or organ, in an unusual or abnormal place.

adventitious bud [BOT] A bud that arises at points on the plant other than at the stem apex or a leaf axil.

adventitious embryo [MED] An embryo developing outside the uterus.

adventitious root [BOT] A root that arises from any plant part other than the primary root (radicle) or its branches.

adventitious vein [INV ZOO] The vessel between the intercalary and accessory veins on certain insect wings.

adventive [BIOL] **1.** An organism that is introduced accidentally and is imperfectly naturalized; not native. **2.** *See* adventitious.

adventive cone [GEOL] A volcanic cone that is on the flank of and subsidiary to a larger volcano. Also known as lateral cone; parasitic cone.

adventive crater [GEOL] A crater opened on the flank of a large volcanic cone.

advisory area [NAV] A designated area within a flight information region to which air-traffic advisory notices apply.

advolution [BIOL] Development or growth with increasing similarities; growth toward; the opposite of evolution.

adz [DES ENG] A cutting tool with a thin arched blade, sharpened on the concave side, at right angles on the handle; used for rough dressing of timber.

adz block [MECH ENG] The part of a machine for wood planing that carries the cutters.

Ae [MET] The temperature of equilibrium for corresponding phase change Ac; Ae_{cm}, Ae_3, and Ae_4 are similarly related to Ac_{cm}, Ac_3, and Ac_4.

Ae₁ [MET] The temperature, attained without thermal lag, at which cementite-austenite conversion takes place in a hypoeutectoid steel or ferrite-austenite conversion takes place in a hypereutectoid steel.

aebi [BIOL] A unit for the standardization of a phosphatase.

Aechminidae [PALEON] A family of extinct ostracods in the order Paleocopa in which the hollow central spine is larger than the valve.

aeciospore [MYCOL] A spore produced by an aecium.

aecium [MYCOL] The fruiting body or sporocarp of rust fungi.

aedeagus [INV ZOO] The copulatory organ of a male insect.

Aedes [INV ZOO] A genus of the dipterous subfamily Culicinae in the family Culicidae, with species that are vectors for many diseases of man.

Aedes aegypti [INV ZOO] A cosmopolitan mosquito that transmits the etiological agents of yellow fever, dengue, equine encephalomyelitis, and Bancroft's filariasis.

Aeduellidae [PALEON] A family of Lower Permian palaeoniscoid fishes in the order Palaeonisciformes.

Aegeriidae [INV ZOO] The clearwing moths, a family of lepidopteran insects in the suborder Heteroneura characterized by the lack of wing scales.

Aegialitidae [INV ZOO] An equivalent name for the Salpingidae.

Aegidae [INV ZOO] A family of isopod crustaceans in the suborder Flabellifera whose members are economically important as fish parasites.

aegirine [MINERAL] $NaFe(SiO_3)_2$ A brown or green clinopyroxene occurring in alkali-rich igneous rocks. Also known as aegirite.

aegirite *See* aegirine.

aegithognathous [VERT ZOO] Referring to a bird palate in which the vomers are completely fused and truncate in appearance.

Aegothelidae [VERT ZOO] A family of small Australo-Papuan owlet-nightjars in the avian order Caprimulgiformes.

Aegypiinae [VERT ZOO] The Old World vultures, a subfamily of diurnal carrion feeders of the family Accipitridae.

Aegyptopithecus [PALEON] A primitive primate that is thought to represent the common ancestor of both the family of man and that of the apes.

aelophilous [BOT] Describing a plant whose disseminules are dispersed by wind.

Aelosomatidae [INV ZOO] A family of microscopic freshwater annelid worms in the class Oligochaeta characterized by a ventrally ciliated prostomium.

aenigmatite *See* enigmatite.

aeolian *See* eolian.

aeolotropic See anisotropic.

aeolotropy See anisotropy.

aeon [ASTRON] A billion (10^9) years.

Aepophilidae [INV ZOO] A family of bugs in the hemipteran superfamily Saldoidea.

Aepyornis [PALEON] A genus of extinct ratite birds representing the family Aepyornithidae.

Aepyornithidae [PALEON] The single family of the extinct avian order Aepyornithiformes.

Aepyornithiformes [PALEON] The elephant birds, an extinct order of ratite birds in the superorder Neognathae.

aerated concrete [MATER] Concrete made by adding substances which will liberate gases by chemical reaction; entrapped gases reduce its density and enhance insulation properties.

aerated flow [ENG] Flowing liquid in which gas is dispersed as fine bubbles throughout the liquid.

aerating agent [FOOD ENG] A gas such as carbon dioxide, nitrous oxide, or food-grade Freon used to flavor beverages or to dispense liquids from charged containers.

A/E ratio See absorptivity-emissivity ratio.

aeration [ENG] **1.** Exposing to the action of air. **2.** Causing air to bubble through. **3.** Introducing air into a solution by spraying, stirring, or similar method. **4.** Supplying or infusing with air, as in sand or soil. [FOOD ENG] Charging a liquid with some gas, such as water with carbon dioxide (soda water). [MIN ENG] The introduction of air into the pulp in a flotation cell to form air bubbles.

aeration cell [PHYS CHEM] An electrolytic cell whose electromotive force is due to electrodes of the same material located in different concentrations of dissolved air. Also known as oxygen cell.

aeration tank [ENG] A fluid-holding tank with provisions to aerate its contents by bubbling air or another gas through the liquid or by spraying the liquid into the air.

aerator [DES ENG] A tool having a roller equipped with hollow fins; used to remove cores of soil from turf. [ENG] **1.** One who aerates. **2.** Equipment used for aeration. **3.** Any device for supplying air or gas under pressure, as for fumigating, welding, or ventilating. [MECH ENG] Equipment used to inject compressed air into sewage in the treatment process. [MET] A device which decreases the density of sand by mixing it with air, thus facilitating the movement of sand particles in packing.

aerenchyma [BOT] A specialized tissue in some water plants characterized by thin-walled cells and large intercellular air spaces.

aerial [BIOL] Of, in, or belonging to the air or atmosphere. [ELECTR] See antenna. [ORD] **1.** Of or pertaining to operations in the air or to aircraft. **2.** Of weapons or missiles used in aircraft or launched, dropped, or shot from aircraft.

aerial archeology [ARCHEO] Location of ancient earthworks, walls, village sites, ditches, and other features through the use of aerial surveys.

aerial bomb [ORD] A bomb designed to be dropped from an aircraft, carrying either a high explosive or another agent, and normally detonated on contact or by a timing device.

aerial burst See airburst.

aerial cableway See aerial tramway.

aerial camera [OPTICS] A camera designed for use in aircraft and containing a mechanism to expose the film in continuous sequence at a steady rate. Also known as aerocamera.

aerial cannon See aircraft cannon.

aerial cartographic photography See mapping photography.

aerial dart [ORD] A metal dart designed to be dropped as a missile from an aircraft. Also known as flechette.

aerial film [GRAPHICS] Specially designed roll film supplied in many lengths and widths, with various emulsion types, for use in aerial cameras.

aerial mapping [MAP] The making of planimetric and contoured maps and charts on the basis of photographs of the ground surface from an aircraft, spacecraft, or rocket. Also known as aerocartography.

aerial mine [ORD] **1.** A mine designed to be dropped from an aircraft, especially into water. **2.** An early World War II light case bomb, the predecessor of the blockbuster, that was normally dropped by parachute.

aerial mycelium [MYCOL] A mass of hyphae that occurs above the surface of a substrate.

aerial observation [ORD] Observation from aircraft of artillery fire.

aerial perspective [OPTICS] The effect produced by diffusion of light in the atmosphere whereby more distant objects have less clarity of outline and are lighter in tone.

aerial photogrammetry [ENG] Use of aerial photographs to make accurate measurements in surveying and mapmaking.

aerial photographic reconnaissance See aerial photoreconnaissance.

aerial photography [ENG] The making of photographs of the ground surface from an aircraft, spacecraft, or rocket. Also known as aerophotography.

aerial photoreconnaissance [ENG] The obtaining of information by air photography; the three types are strategic, tactical, and survey-cartographic photoreconnaissance. Also known as aerial photographic reconnaissance.

aerial reconnaissance [ENG] The collection of information by visual, electronic, or photographic means while aloft.

aerial root [BOT] A root exposed to the air, usually anchoring the plant to a tree, and often functioning in photosynthesis.

aerial ropeway See aerial tramway.

aerial sound ranging [AERO ENG] The process of locating an aircraft by means of the sounds it emits.

aerial spud [MECH ENG] A cable for moving and anchoring a dredge.

aerial stem [BOT] A stem with an erect or vertical growth habit above the ground.

aerial survey [ENG] A survey utilizing photographic, electronic, or other data obtained from an airborne station. Also known as aerosurvey; air survey.

aerial torpedo [ORD] **1.** A torpedo designed or adapted to be launched from a low-flying aircraft into water. **2.** Formerly, the explosive projectile thrown by a trench mortar and designed so as to fall point down.

aerial tramway [MECH ENG] A system for transporting bulk materials that consists of one or more cables supported by steel towers and is capable of carrying a traveling carriage from which loaded buckets can be lowered or raised. Also known as aerial cableway; aerial ropeway.

aeriform [PHYS] Having the form or nature of air.

aeroallergen [MED] Any airborne particulate matter that can induce allergic responses in sensitive persons.

Aerobacter [MICROBIO] A genus of nonpathogenic, rod-shaped, gram-negative bacteria of the family Enterobacteriaceae.

Aerobacter aerogenes [MICROBIO] A widely distributed species of coliform bacteria commonly associated with urinary tract infections in man.

aeroballistics [MECH] The study of the interaction of projectiles or high-speed vehicles with the atmosphere.

aerobe [BIOL] An organism that requires air or free oxygen to maintain its life processes.

aerobic-anaerobic interface [CIV ENG] That point in bacterial action in the body of a sewage sludge or compost heap where both aerobic and anaerobic microorganisms participate, and the decomposition of the material goes no further.

aerobic-anaerobic lagoon [CIV ENG] A pond in which the solids from a sewage plant are placed in the lower layer; the solids are partially decomposed by anaerobic bacteria, while air or oxygen is bubbled through the upper layer to create an aerobic condition.

aerobic bacteria [MICROBIO] Any bacteria requiring free oxygen for the metabolic breakdown of materials.

aerobic lagoon [CIV ENG] An aerated pond in which sewage solids are placed, and are decomposed by aerobic bacteria.

aerobic process [BIOL] A process requiring the presence of oxygen.

aerobiology [BIOL] The study of the atmospheric dispersal of airborne fungus spores, pollen grains, and microorganisms; and, more broadly, of airborne propagules of algae and protozoans, minute insects such as aphids, and pollution gases and particles which exert specific biologic effects.

aerobioscope [MICROBIO] An apparatus for collecting and determining the bacterial content of a sample of air.

aerobiosis [BIOL] Life existing in air or oxygen.

AERIAL PHOTOGRAPHY

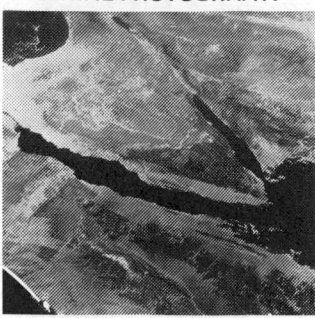

Aerial photograph of view over Sinai Peninsula from *Gemini 2* looking northeast; eastern desert of Egypt in foreground, Israel and Jordan in background. *(NASA photograph)*

aerocamera   *See* aerial camera.

aerocartography   *See* aerial mapping.

aerochlorination   [CIV ENG] Treatment of sewage with compressed air and chlorine gas to remove fatty substances.

AERO code   [METEOROL] An international code used to encode for transmission, in words five numerical digits long, synoptic weather observations of particular interest to aviation operations.

aerocyst   [BOT] An air vesicle in certain species of algae.

aerodiscone antenna   [ELECTROMAG] Electrically small antenna for airborne applications in the very-high-frequency and ultra-high-frequency bands; it is derived from, and preserves, the desirable electrical characteristics of the discone antenna and can be designed in various physical shapes.

aerodontalgia   [MED] A toothache brought on by atmospheric decompression.

aerodrome   *See* airport.

aeroduct   [AERO ENG] A ramjet type of engine designed to scoop up ions and electrons freely available in the outer reaches of the atmosphere or in the atmospheres of other spatial bodies and, by a metachemical process within the engine duct, to expel particles derived from the ions and electrons as a propulsive jetstream.

aerodynamic   [FL MECH] Pertaining to forces acting upon any solid or liquid body moving relative to a gas (especially air).

aerodynamically rough surface   [FL MECH] A surface whose irregularities are sufficiently high that the turbulent boundary layer reaches right down to the surface.

aerodynamically smooth surface   [FL MECH] A surface whose irregularities are sufficiently small to be entirely embedded in the laminar sublayer.

aerodynamic balance   [ENG] A balance used for the measurement of the forces exerted on the surfaces of instruments exposed to flowing air; frequently used in tests made on models in wind tunnels.

aerodynamic center   [AERO ENG] A point on a cross section of a wing or rotor blade through which the forces of drag and lift are acting and about which the pitching moment coefficient is practically constant.

aerodynamic characteristics   [AERO ENG] The performance of a given airfoil profile as related to lift and drag, to angle of attack, and to velocity, density, viscosity, compressibility, and so on.

aerodynamic chord   [AERO ENG] A straight line intersecting or touching an airfoil profile at two points; specifically, that part of such a line between two points of intersection.

aerodynamic coefficient   [FL MECH] Any nondimensional coefficient relating to aerodynamic forces or moments, such as a coefficient of drag or a coefficient of lift.

aerodynamic configuration   [AERO ENG] The form of an aircraft, incorporating desirable aerodynamic qualities.

aerodynamic control   [AERO ENG] A control surface whose use causes local aerodynamic forces.

aerodynamic drag   [FL MECH] A retarding force that acts upon a body moving through a gaseous fluid and that is parallel to the direction of motion of the body; it is a component of the total fluid forces acting on the body. Also known as aerodynamic resistance.

aerodynamic force   [FL MECH] The force between a body and a gaseous fluid caused by their relative motion. Also known as aerodynamic load.

aerodynamic heating   [FL MECH] The heating of a body produced by passage of air or other gases over its surface; caused by friction and by compression processes and significant chiefly at high speeds.

aerodynamic instability   [AERO ENG] An unstable state caused by oscillations of a structure that are generated by spontaneous and more or less periodic fluctuations in the flow, particularly in the wake of the structure.

aerodynamic lift   [FL MECH] That component of the total aerodynamic force acting on a body perpendicular to the undisturbed airflow relative to the body. Also known as lift.

aerodynamic load   *See* aerodynamic force.

aerodynamic missile   [AERO ENG] A missile with surfaces which produce lift during flight.

aerodynamic moment   [AERO ENG] The torque about the center of gravity of a missile or projectile moving through the

atmosphere, produced by any aerodynamic force which does not act through the center of gravity.

aerodynamic noise   [ACOUS] Acoustic noise caused by turbulent airflow over the surface of a body.

aerodynamic phenomena   [FL MECH] Acoustic, thermal, electrical, and mechanical effects, among others, that result from the flow of air over a body.

aerodynamic resistance   *See* aerodynamic drag.

aerodynamics   [FL MECH] The science that deals with the motion of air and other gaseous fluids and with the forces acting on bodies when they move through such fluids or when such fluids move against or around the bodies.

aerodynamic size   [PHYS] Particle size determined from inertia or settling velocity, assuming Stokes' law for the resistance to a sphere moving through a fluid. Also known as inertial size.

aerodynamic stability   [AERO ENG] The property of a body in the air, such as an aircraft or rocket, to maintain its attitude, or to resist displacement, and if displaced, to develop aerodynamic forces and moments tending to restore the original condition.

aerodynamic trail   [FL MECH] A condensation trail formed by adiabatic cooling to saturation (or slight supersaturation) of air passing over the surfaces of high-speed aircraft.

aerodynamic trajectory   [MECH] A trajectory or part of a trajectory in which the missile or vehicle encounters sufficient air resistance to stabilize its flight or to modify its course significantly.

aerodynamic turbulence   [FL MECH] A state of fluid flow in which the instantaneous velocities exhibit irregular and apparently random fluctuations.

aerodynamic vehicle   [AERO ENG] A device, such as an airplane or glider, capable of flight only within a sensible atmosphere and relying on aerodynamic forces to maintain flight.

aerodynamic wave drag   [FL MECH] The force retarding an airplane, especially in supersonic flight, as a consequence of the formation of shock waves ahead of it.

aerodyne   [AERO ENG] Any heavier-than-air craft that derives its lift in flight chiefly from aerodynamic forces, such as the conventional airplane, glider, or helicopter.

aeroelasticity   [MECH] The deformation of structurally elastic bodies in response to aerodynamic loads.

aeroembolism   [MED] A condition marked by the presence of nitrogen bubbles in the blood and other body tissues resulting from a sudden fall in atmospheric pressure. Also known as air embolism.

aerofall mill   [MECH ENG] A grinding mill of large diameter with either lumps of ore, pebbles, or steel balls as crushing bodies; the dry load is airswept to remove mesh material.

aerofilter   [CIV ENG] A filter bed for sewage treatment consisting of coarse material and operated at high speed, often with recirculation.

aerofoil   *See* airfoil.

aerogel   [CHEM] A porous solid formed from a gel by replacing the liquid with a gas with little change in volume so that the solid is highly porous.

aerogenerator   [ELEC] A generator that is driven by the wind, designed to utilize wind power on a commercial scale.

aerogeography   [GEOGR] The geographic study of earth features by means of aerial observations and aerial photography.

aerogeology   [GEOL] The geologic study of earth features by means of aerial observations and aerial photography.

aerogram   [COMMUN] A message transmitted by radio or carried by an aircraft.

aerograph   [ENG] Any self-recording instrument carried aloft by any means to obtain meteorological data.

aerography   [METEOROL] **1.** The study of the air or atmosphere. **2.** The practice of weather observation, map plotting, and maintaining records. **3.** *See* descriptive meteorology.

aerohydrous mineral   [MINERAL] A mineral containing water in small cavities.

aerolite   *See* stony meteorite.

aerologation   [NAV] The science of long-range navigation by altimetry. Also known as minimal flight path; pressure pattern flying; single heading flight.

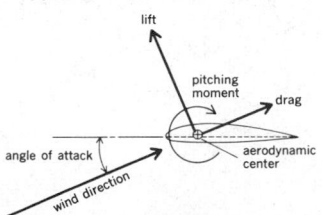

AERODYNAMIC CENTER

Aerodynamic forces on an airfoil section, showing aerodynamic center. *(From C. D. Perkins and R. E. Hage, Airplane Performance Stability and Control, Wiley, 1949)*

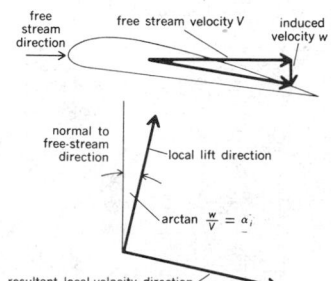

AERODYNAMIC FORCE

Effect of the induced velocity on the direction of local velocity and lift.

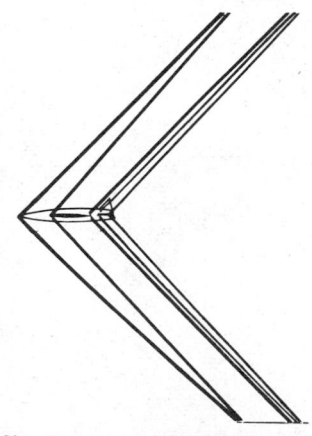

AERODYNAMIC WAVE DRAG

Shock waves generated by an airplane at supersonic speeds act to retard the plane.

aerological days [METEOROL] Specified days on which additional upper-air observations are made; an outgrowth of the International Polar Year.

aerology [METEOROL] **1.** Synonym for meteorology, according to official usage in the U.S. Navy until 1957. **2.** The study of the free atmosphere throughout its vertical extent, as distinguished from studies confined to the layer of the atmosphere near the earth's surface.

aeromagnetic surveying [GEOPHYS] The mapping of the magnetic field of the earth through the use of electronic magnetometers suspended from aircraft.

aeromechanics [FL MECH] The science of air and other gases in motion or equilibrium; has two branches, aerostatics and aerodynamics.

aeromedicine *See* aerospace medicine.

aerometeorograph [ENG] A self-recording instrument used on aircraft for the simultaneous recording of atmospheric pressure, temperature, and humidity.

aerometer [ENG] An instrument to ascertain the weight or density of air or other gases.

Aeromonas [MICROBIO] A genus of short, gram-negative, rod-shaped, motile, heterotrophic bacteria in the family Pseudomonadaceae; the majority are pathogenic to marine and fresh-water animals.

aeromotor [AERO ENG] An engine designed to provide motive power for an aircraft.

aeronaut [AERO ENG] A person who operates or travels in an airship or balloon.

aeronautical beacon [NAV] A visual aid to navigation, displaying flashes of white or colored light or both, used to indicate the location of airports, landmarks, and certain points of the Federal airways in mountainous terrain and to mark hazards.

aeronautical chart [MAP] A basic map of countries or lands made primarily on Lambert conformal ionic projection and layer-tinted; air navigational data are overprinted.

aeronautical climatology [METEOROL] The application of the data and techniques of climatology to aviation meteorological problems.

aeronautical engineering [AERO ENG] The branch of engineering concerned primarily with the design and construction of aircraft structures and power units, and with the special problems of flight in the atmosphere.

aeronautical flutter [FL MECH] An aeroelastic, self-excited vibration in which the external source of energy is the airstream and which depends on the elastic, inertial, and dissipative forces of the system in addition to the aerodynamic forces.

aeronautical information overprint [NAV] Additional information which is printed or stamped on a map or chart for the specific purpose of air navigation.

aeronautical light [NAV] A luminous or lighted aid to navigation intended primarily for air navigation.

aeronautical meteorology [METEOROL] The study of the effects of weather upon aviation.

aeronautical mile *See* air mile.

aeronautical pilotage chart [NAV] An aeronautical chart designed primarily for air navigation. Also known as contact chart.

aeronautical planning chart [NAV] An aeronautical chart of small scale designed to satisfy long-range air navigation and mission-planning requirements.

aeronautics [FL MECH] The science that deals with flight through the air.

aeronomy [GEOPHYS] The study of the atmosphere of the earth or other bodies, particularly in relation to composition, properties, relative motion, and radiation from outer space or other bodies.

aerootitis *See* barotitis.

aerophare *See* radio beacon.

aerophotography *See* aerial photography.

aerophysics [AERO ENG] The physics dealing with the design, construction, and operation of aerodynamic devices.

aerophyte *See* epiphyte.

aeropulse engine *See* pulsejet engine.

aerosiderite [GEOL] A meteorite composed principally of iron.

aerosinusitis *See* barosinusitis.

aerosol [CHEM] A gaseous suspension of ultramicroscopic particles of a liquid or a solid.

aerosol generator [MECH ENG] A mechanical means of producing a system of dispersed phase and dispersing medium, that is, an aerosol.

aerosol propellant [MATER] Compressed gas or vapor in a container which, upon release of pressure and expansion through a valve, carries another substance from the container; used for cosmetics, household cleaners, and so on; examples are butanes, propane, nitrogen, fluorocarbons, and carbon dioxide.

aerospace *See* airspace.

aerospace electronics [ELECTR] The field of electronics as applied to aircraft and spacecraft.

aerospace engineering [ENG] Engineering pertaining to the design and construction of aircraft and space vehicles and of power units, and to the special problems of flight in both the earth's atmosphere and space, as in the flight of air vehicles and in the launching, guidance, and control of missiles, earth satellites, and space vehicles and probes.

aerospace environment [GEOPHYS] **1.** The conditions, influences, and forces that are encountered by vehicles, missiles, and so on in the earth's atmosphere or in space. **2.** External conditions which resemble those of atmosphere and space, and in which a piece of equipment, a living organism, or a system operates.

aerospace ground equipment [AERO ENG] Support equipment for air and space vehicles. Abbreviated AGE.

aerospace industry [ENG] Industry concerned with the use of vehicles in both the earth's atmosphere and space.

aerospace medicine [MED] The branch of medicine dealing with the effects of flight in the atmosphere or space upon the human body and with the prevention or cure of physiological or psychological malfunctions arising from these effects. Also known as aeromedicine; aviation medicine.

aerospace vehicle [AERO ENG] A vehicle capable of flight both within and outside the sensible atmosphere.

Aerosporin [MICROBIO] Trade name for the antibiotic polymyxin B.

aerostat [AERO ENG] Any aircraft that derives its buoyancy or lift from a lighter-than-air gas contained within its envelope or one of its compartments; for example, ships and balloons.

aerostatic balance [ENG] An instrument for weighing air.

aerostatics [FL MECH] The science of the equilibrium of gases and of solid bodies immersed in them when under the influence only of natural gravitational forces.

aerosurvey *See* aerial survey.

aerotaxis [BIOL] The movement of an organism, especially aerobic and anaerobic bacteria, with reference to the direction of oxygen or air.

aerothermochemistry [FL MECH] The study of gases which takes into account the effect of motion, heat, and chemical changes.

aerothermodynamic border [GEOPHYS] An altitude of about 100 miles (161 km), above which the atmosphere is so rarefied that the skin of an object moving through it at high speeds generates no significant heat.

aerothermodynamics [FL MECH] The study of aerodynamic phenomena at sufficiently high gas velocities that thermodynamic properties of the gas are important.

aerothermoelasticity [FL MECH] The study of the response of elastic structures to the combined effects of aerodynamic heating and loading.

aerotropism [BOT] A response in which the growth direction of a plant component changes due to modifications in oxygen tension.

aerozine [MATER] Hydrazine mixed 50:50 with dimethylhydrazine; the most-used bipropellant rocket fuel.

AES *See* Auger electron spectroscopy.

aeschynomenous [BOT] Having sensitive leaves that droop when touched, such as members of the Leguminosae.

Aesculus [BOT] A genus of deciduous trees or shrubs belonging to the order Sapindales. Commonly known as buckeye.

Aeshnidae [INV ZOO] A family of odonatan insects in the

suborder Anisoptera distinguished by partially fused eyes.

aesthacyte   *See* esthacyte.

aesthesia   *See* esthesia.

aesthesiometer   *See* esthesiometer.

aesthete   [BOT] A plant organ with the capacity to respond to definite physical stimuli.

aestidurilignosa   [ECOL] A mixed woodland of evergreen and deciduous hardwoods.

aestilignosa   [ECOL] A woodland of tropophytic vegetation in temperate regions.

aestival   *See* estival.

aestivation   [BOT] The arrangement of floral parts in a bud. [PHYSIO] The condition of dormancy or torpidity.

Aetosauria   [PALEON] A suborder of Triassic archosaurian quadrupedal reptiles in the order Thecodontia armored by rings of thick, bony plates.

af   *See* audio frequency.

aF   *See* abfarad.

af amplifier   *See* audio-frequency amplifier.

AFC   *See* automatic frequency control.

affect   [PSYCH] Conscious awareness of feelings; mood.

affection   [MED] Any pathology or diseased state of the body. [PSYCH] The feeling aspect of consciousness.

affective psychosis   [PSYCH] A severe disturbance marked by extremes of mood, such as depression or manic elation.

affectivity   [PSYCH] The state of being susceptible to emotional stimuli.

afferent   [PHYSIO] Conducting or conveying inward or toward the center, specifically in reference to nerves and blood vessels.

afferent neuron   [ANAT] A nerve cell that conducts impulses toward a nerve center, such as the central nervous system.

affination   [FOOD ENG] Removing the film of adhering molasses from raw sugar crystals by treating with a heavy sugar syrup.

affine deformation   [GEOL] A type of deformation in which very thin layers slip against each other so that each moves equally with respect to its neighbors; generally does not result in folding. [MAP] A deformation in which the scale along one axis, or reference plane, is different from the scale along the other axis, or plane, normal to the first.

affine geometry   [MATH] The study of geometry using the methods of linear algebra.

affine strain   [GEOPHYS] A strain in the earth that does not differ from place to place.

affine transformation   [MATH] A function on a linear space to itself, which is the sum of a linear transformation and a fixed vector.

affinity   [BIOL] A relationship between species or higher taxonomic groups that depends on anatomical similarities and indicates a common origin. [CHEM] The selective tendency of an element to combine with one rather than another element when conditions are equal.

affinity chromatography   [ANALY CHEM] A chromatographic technique that utilizes the ability of biological molecules to bend to certain ligands specifically and reversibly; used in protein biochemistry.

afforestation   [FOR] Establishment of a new forest by seeding or planting on nonforested land.

affreightment   [IND ENG] The lease of a vessel for the transportation of goods.

afibrinogenemia   [MED] Complete absence of fibrinogen in the blood.

aflatoxicosis   [MED] Aflatoxin poisoning.

aflatoxin   [BIOCHEM] The toxin produced by some strains of the fungus *Aspergillus flavus,* the most potent carcinogen yet discovered.

afocal lens   [OPTICS] A lens of zero convergent power, whose focal points are infinitely distant.

afocal system   [OPTICS] An optical system of zero convergent power, for example, a telescope.

a format   [ADP] A nonexecutable statement in FORTRAN which permits alphanumeric characters to be transmitted in a manner similar to numeric data.

A frame   [BUILD] A dwelling whose main frames are in the shape of the letter A. [ENG] Two poles supported in an upright position by braces or guys and used for lifting equipment. Also known as double mast. [OCEANOGR] An A-shaped frame used for outboard suspension of oceanographic gear on a research vessel.

Africa   [GEOGR] The second largest continent, with an area of 11,700,000 square miles; bisected midway by the Equator, above and below which it shows symmetry of climate and vegetation zones.

African horse sickness   [VET MED] An infectious, mosquito-borne virus disease of equines characterized by fever and edematous swelling.

African sleeping sickness   [MED] A disease of man confined to tropical Africa, caused by the protozoans *Trypanosoma gambiense* or *T. rhodesiense;* symptoms include local reaction at the site of the bite, fever, enlargement of adjacent lymph nodes, skin rash, edema, and during the late phase, somnolence and emaciation. Also known as African trypanosomiasis; maladie du sommeil; sleeping sickness.

African swine fever   *See* hog cholera.

African trypanosomiasis   *See* African sleeping sickness.

African violet   [BOT] *Saintpaulia ionantha.* A flowering plant typical of the family Gesneriaceae.

aft   [NAV ARCH] Near, toward, or at the stern of a ship.

afterbirth   [EMBRYO] The placenta and fetal membranes expelled from the uterus following birth of offspring in viviparous mammals.

afterblow   [MET] A blow in a Bessemer process, occurring after the flame for the removal of carbon has dropped, to remove phosphorus.

afterbody   [AERO ENG] **1.** A companion body that trails a satellite. **2.** A section or piece of a rocket or spacecraft that enters the atmosphere unprotected behind the nose cone or other body that is protected for entry. **3.** The afterpart of a vehicle.

afterbreak   [MIN ENG] A phenomenon occurring during mine subsidence; material slides inward after the main break, assumed at right angles to the plane of the seam.

afterburner   [AERO ENG] A device for augmenting the thrust of a jet engine by burning additional fuel in the uncombined oxygen in the gases from the turbine.

afterburning   [AERO ENG] The function of an afterburner. [MECH ENG] Combustion in an internal combustion engine following the maximum pressure of explosion.

afterburnt   [AERO ENG] Descriptive of the condition following the complete transformation of the solid propellant to gaseous form.

afterburst   [MIN ENG] **1.** A tremor that sometimes follows a rock blast as the ground adjusts to the new stress distribution. **2.** A sudden collapse of rock in an underground mine subsequent to a rock burst.

afterchromed dye   [MATER] A dye that is improved in color quality or fastness after the textile is dyed by treatment with sodium dichromate, copper sulfate, or similar materials.

aftercondenser   [MECH ENG] A condenser in the second stage of a two-stage ejector; used in steam power plants, refrigeration systems, and air conditioning systems.

aftercooling   [MECH ENG] The cooling of a gas after its compression. [NUCLEO] The cooling of a reactor after it has been shut down.

afterdamp   [MIN ENG] The mixture of gases which remains in a mine after a mine fire or an explosion of firedamp.

afterflaming   [AERO ENG] With liquid- or solid-propellant rocket thrust chambers, a characteristic low-grade combustion that takes place in the thrust chamber assembly and around its nozzle exit after the main propellant flow has been stopped.

aftergases   [MIN ENG] Gases produced by mine explosions or mine fires.

afterglow   [ATOM PHYS] *See* phosphorescence. [METEOROL] A broad, high arch of radiance or glow seen occasionally in the western sky above the highest clouds in deepening twilight, caused by the scattering effect of very fine particles of dust suspended in the upper atmosphere. [PL PHYS] The transient decay of a plasma after the power has been turned off.

afterheat   [NUCLEO] Heat derived from residual radioactivity after a reactor has been shut down.

AETOSAURIA

Restoration of *Desmatosuchus,* a quadrupedal thecodont, length to 3 meters. *(From E. H. Colbert, Evolution of the Vertebrates, Wiley, 1956)*

AFTERBURNER

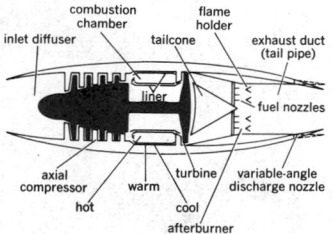

Diagram of turbojet engine showing afterburner.

afterimage [PHYSIO] A visual sensation occurring after the stimulus to which it is a response has been removed.

afterpeak [NAV ARCH] The compartment closest to the stern in the hold of a ship.

after-perpendicular [NAV ARCH] A vertical line through the intersection of the design waterline with the after-side of the straight portion of the rudder post of a ship. Abbreviated AP.

afterpotential [PHYSIO] A small positive or negative wave that follows and is dependent on the main spike potential, seen in the oscillograph tracing of an action potential passing along a nerve.

afterripening [BOT] A period of dormancy after a seed is shed during which the synthetic machinery of the seed is prepared for germination and growth.

aftershaft [VERT ZOO] An accessory, plumelike feather near the upper umbilicus on the feathers of some birds.

aftershock [GEOPHYS] A small earthquake following a larger earthquake and originating at or near the larger earthquake's epicenter.

afterwind [NUCLEO] A wind produced by the updraft accompanying the rise of the fireball of a nuclear explosion and directed toward the burst center.

Aftonian interglacial [GEOL] Post-Nebraska interglacial geologic time.

afwillite [MINERAL] Ca$_3$Si$_2$O$_4$(OH)$_6$ A colorless mineral consisting of a hydrous calcium silicate and occurring in monoclinic crystals; specific gravity is 2.6.

Ag *See* silver.

agalactia [MED] Nonsecretion or imperfect secretion of milk after childbirth.

agalmatolite [GEOL] A soft, waxy, gray, green, yellow, or brown mineral or stone, such as pinite and steatite; used by the Chinese for carving images. Also known as figure stone; lardite; pagodite.

Agamemnon [ASTRON] An asteroid, one of a group of Trojan planets whose periods of revolution are approximately equal to that of Jupiter, or about 12 years.

agameon [BIOL] An organism which reproduces only by asexual means. Also known as agamospecies.

agamete [BIOL] An asexual reproductive cell that develops into an adult individual.

Agamidae [VERT ZOO] A family of Old World lizards in the suborder Sauria that have acrodont dentition.

agammaglobulinemia [MED] The condition characterized by lack of or extremely low levels of gamma globulin in the blood, together with defective antibody production and frequent infections; primary agammaglobulinemia occurs in three clinical forms: congenital, acquired, and transient.

agamogony [BIOL] Asexual reproduction, specifically schizogony.

agamospecies *See* agameon.

agamospermy [BOT] Apogamy in which sexual union is incomplete because of abnormal development of the pollen and the embryo sac.

Agaontidae [INV ZOO] A family of small hymenopteran insects in the superfamily Chalcidoidea; commonly called fig insects for their role in cross-pollination of figs.

agar [MATER] A gelatinous product extracted from certain red algae and used chiefly as a gelling agent in culture media.

Agarbacterium [MICROBIO] A genus of motile and nonmotile, rod-shaped, gram-negative bacteria in the family Achromobacteraceae that digest agar.

agar-gel reaction [IMMUNOL] A precipitin type of antigen-antibody reaction in which the reactants are introduced into different regions of an agar gel and allowed to diffuse toward each other.

agaric acid [ORG CHEM] C$_{19}$H$_{36}$(OH)(COOH)$_3$ An acid with melting point 141°C; soluble in water, insoluble in benzene; used as an irritant. Also known as agaricin.

Agaricales [MYCOL] An order of fungi in the class Basidiomycetes containing all forms of fleshy, gilled mushrooms.

agaricin *See* agaric acid.

agaric mineral *See* rock milk.

agarophyte [BOT] Any seaweed that yields agar.

Agassiz orogeny [GEOL] A phase of diastrophism confined to North America Cordillera occurring at the boundary between the Middle and Late Jurassic.

Agassiz trawl [OCEANOGR] A dredge consisting of a net attached to an iron frame with a hoop at each end that is used to collect organisms, particularly invertebrates, living on the ocean bottom.

Agassiz Valleys [GEOL] Undersea valleys in the Gulf of Mexico between Cuba and Key West.

agate [GRAPHICS] A type size in printing of about 5½ points, where 72 points equals 1 inch. [MINERAL] SiO$_2$ A fine-grained, fibrous variety of chalcedony with color banding or irregular clouding.

agate glass [MATER] Multicolor glass made by blending glasses of two or more colors or by rolling transparent glass into powdered glass of various colors.

agate jasper [MINERAL] An impure variety of quartz consisting of jasper and agate. Also known as jaspagate.

agatized wood *See* silicified wood.

Agavaceae [BOT] A family of flowering plants in the order Liliales characterized by parallel, narrow-veined leaves, a more or less corolloid perianth, and an agavaceous habit.

AGC *See* automatic gain control.

age [BIOL] Period of time from origin or birth to a later time designated or understood; length of existence. [GEOL] **1.** Any one of the named epochs in the history of the earth marked by specific phases of physical conditions or organic evolution, such as the Age of Mammals. **2.** One of the smaller subdivisions of the epoch as geologic time, corresponding to the stage or the formation, such as the Lockport Age in the Niagara Epoch. [NUCLEO] *See* Fermi age.

AGE *See* aerospace ground equipment.

age coating [ELEC] The black deposit that is formed on the inner surface of an electric lamp by material evaporated from the filament.

aged [GEOL] Of a ground configuration, having been reduced to base level.

age determination [GEOL] Identification of the geologic age of a biological or geological specimen by using the methods of dendrochronology or radiometric dating.

aged shore [GEOL] A shore long established at a constant level and adjusted to the waves and currents of the sea.

age-hardening [MET] Increasing the hardness of an alloy by a relatively low-temperature heat treatment that causes precipitation of components or phases of the alloy from the supersaturated solid solution.

Agena rocket [AERO ENG] A liquid-fuel, upper-stage rocket usually used with a first-stage Atlas or Thor booster in certain space satellite projects.

agenda [ADP] The sequence of control statements required to carry out the solution of a computer problem.

agenesis [BIOL] Absence of a tissue or organ due to lack of development.

age of diurnal inequality [GEOPHYS] The time interval between the maximum semimonthly north or south declination of the moon and the time that the maximum effect of the declination upon the range of tide or speed of the tidal current occurs. Also known as age of diurnal tide; diurnal age.

age of diurnal tide *See* age of diurnal inequality.

Age of Fishes [GEOL] An informal designation of the Silurian and Devonian periods of geologic time.

Age of Mammals [GEOL] An informal designation of the Cenozoic era of geologic time.

Age of Man [GEOL] An informal designation of the Quaternary period of geologic time.

age of parallax inequality [GEOPHYS] The time interval between the perigee of the moon and the maximum effect of the parallax (distance of the moon) upon the range of tide or speed of tidal current. Also known as parallax age.

age of phase inequality [GEOPHYS] The time interval between the new or full moon and the maximum effect of these phases upon the range of tide or speed of tidal current. Also known as age of tide; phase age.

age of the moon [ASTRON] The elapsed time, usually expressed in days, since the last new moon.

age of tide *See* age of phase inequality.

ageostrophic wind *See* geostrophic deviation.

age ratio [GEOL] The ratio of the amount of daughter to parent isotope in a mineral being dated radiometrically.

age structure [ANTHRO] Categorization of the population of communities or countries by age groups, allowing demographers to make projections of the growth or decline of the particular population.

agglomeration [FOOD ENG] A technique that combines powdered material to form larger, more soluble particles by intermingling in a humid atmosphere. [MET] Conversion of small pieces of low-grade iron ore into larger lumps by application of heat. [METEOROL] The process in which particles grow by collision with and assimilation of cloud particles or other precipitation particles. Also known as coagulation. [SCI TECH] An indiscriminately formed cluster of particles.

agglomeration test [MIN ENG] A test of a button of coke whose results are used as a measure of the binding qualities of the coal.

agglutinate cone *See* spatter cone.

agglutination reaction [IMMUNOL] Clumping of a particulate suspension of antigen by a reagent, usually an antibody.

agglutinin [IMMUNOL] An antibody from normal or immune serum that causes clumping of its complementary particulate antigen, such as bacteria or erythrocytes.

agglutinogen [IMMUNOL] An antigen that stimulates production of a specific antibody (agglutinin) when introduced into an animal body.

agglutinoid [IMMUNOL] An agglutin that lacks the power to agglutinate but has the ability to unite with its agglutinogen.

aggradation [GEOL] *See* accretion. [HYD] A process of shifting equilibrium of stream deposition, with upbuilding approximately at grade.

aggraded valley floor [GEOL] The surface of a flat deposit of alluvium which is thicker than the stream channel's depth and is formed where a stream has aggraded its valley.

aggregate [GEOL] A collection of soil grains or particles gathered into a mass. [MATER] The natural sands, gravels, and crushed stone used for mixing with cementing material in making mortars and concretes.

aggregate recoil [NUC PHYS] The ejection of atoms from the surface of a sample as a result of their being attached to one atom that is recoiling as the result of α-particle emission.

aggregate structure [GEOL] A mass composed of separate small crystals, scales, and grains that, under a microscope, extinguish at different intervals during the rotation of the stage.

aggression [PSYCH] Feelings and behavior of anger and hostility usually manifested by punitive or destructive actions; often associated with frustration.

aggressive carbon dioxide [CHEM ENG] The carbon dioxide dissolved in water in excess of the amount required to precipitate a specified concentration of calcium ions as calcium carbonate; used as a measure of the corrosivity and scaling properties of water.

aggressive magma [GEOL] A magma that forces itself into place.

aggressive water [HYD] Any of the waters which force their way into place.

Agilon [TEXT] Trade name of a stretch nylon yarn made by Deering Milliken, Inc.

aging [BIOL] Growing older. [ELEC] Allowing a permanent magnet, capacitor, meter, or other device to remain in storage for a period of time, sometimes with a voltage applied, until the characteristics of the device become essentially constant. [ELECTROMAG] Change in the magnetic properties of iron with passage of time, for example, increase in the hysteresis. [ENG] 1. The changing of the characteristics of a device due to its use. 2. Operation of a product before shipment to stabilize characteristics or detect early failures. [MATER] 1. Change in the properties of any substance with time. 2. Change occurring in powders or slips with the passage of time. 3. Curing of ceramic materials, such as clays and glazes, by a definite period of time under controlled storage conditions. [MET] 1. Change in properties of an alloy or metal which generally proceeds slowly at room temperatures and faster at elevated temperatures. 2. Strain relief, occurring through long storage outdoors under varying temperatures, of iron castings intended for use as toolroom plates or lathe-bed supports. 3. A second heat treatment of an alloy at a lower temperature, causing precipitation of the unstable phase and

increasing hardness, strength, and electrical conductivity. [NUCLEO] The slowing down of neutrons.

aging-lung emphysema [MED] An asymptomatic pulmonary disease associated with aging, characterized by alveolar dilation due to loss of tissue elasticity.

agitator [MECH ENG] A device for keeping liquids and solids in liquids in motion by mixing, stirring, or shaking.

Aglaspida [PALEON] An order of Cambrian and Ordovician merostome arthropods in the subclass Xiphosurida characterized by a phosphatic exoskeleton and vaguely trilobed body form.

aglomerular [HISTOL] Lacking glomeruli.

Aglossa [VERT ZOO] A suborder of anuran amphibians represented by the single family Pipidea and characterized by the absence of a tongue.

aglycon [BIOCHEM] The nonsugar compound resulting from the hydrolysis of glycosides; an example is 3,5,7,3',4'-pentahydroxyflavylium, or cyanidin.

aglyphous [VERT ZOO] Having solid teeth.

agmatite [PETR] 1. A migmatite that contains xenoliths. 2. Fragmental plutonic rock with granitic cement.

agnate [BIOL] Related exclusively through male descent.

Agnatha [VERT ZOO] The most primitive class of vertebrates, characterized by the lack of true jaws.

agnathia [MED] Lack of the jaws.

agnosia [MED] Loss of the ability to recognize persons or objects and their meaning.

agonic line [GEOPHYS] The imaginary line through all points on the earth's surface at which the magnetic declination is zero; that is, the locus of all points at which magnetic north and true north coincide.

Agonidae [VERT ZOO] The poachers, a small family of marine perciform fishes in the suborder Cottoidei.

agonistic behavior [PSYCH] In social animals, fighting and escape behavior common in males during the rutting season.

agoraphobia [PSYCH] Abnormal fear of open places.

agouti [VERT ZOO] A hystricomorph rodent, *Dasyprocta,* in the family Dasyproctidae, represented by 13 species.

agpaite [PETR] A group of igneous rocks containing feldspathoids; includes naujaite, lujavrite, and kakortokite.

agranular leukocyte [HISTOL] A type of white blood cell, including lymphocytes and monocytes, characterized by the absence of cytoplasmic granules and by a relatively large spherical or indented nucleus.

agranulocytosis [MED] An acute febrile illness, usually resulting from drug hypersensitivity, manifested as severe leukopenia, often with complete disappearance of granulocytes.

agraphia [MED] Loss of the ability to write.

agravic [GEOPHYS] Of or pertaining to a condition of no gravitation.

agravic illusion [MED] An apparent movement of a target in the visual field due to otolith response in zero gravity. Also known as oculoagravic illusion.

agrestal [ECOL] Growing wild in the fields.

agricere [GEOL] A waxy or resinous organic coating on soil particles.

agricolite *See* eulytite.

agricultural chemicals [MATER] Fertilizers, soil conditioners, fungicides, insecticides, weed killers, and other chemicals used to increase farm crop productivity and quality.

agricultural chemistry [AGR] The science of chemical compositions and changes involved in the production, protection, and use of crops and livestock; includes all the life processes through which food and fiber are obtained for man and his animals, and control of these processes to increase yields, improve quality, and reduce costs.

agricultural climatology [AGR] In general, the study of climate as to its effect on crops; it includes, for example, the relation of growth rate and crop yields to the various climatic factors and hence the optimum and limiting climates for any given crop. Also known as agroclimatology.

agricultural engineering [AGR] A discipline concerned with developing and improving the means for providing food and fiber for mankind's needs.

agricultural geology [GEOL] A branch of geology that deals with the nature and distribution of soils, the occurrence of

AGGLUTINATION REACTION

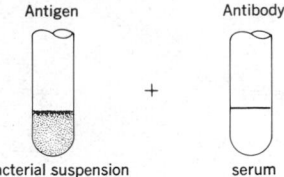
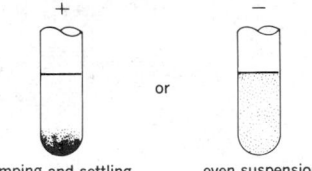
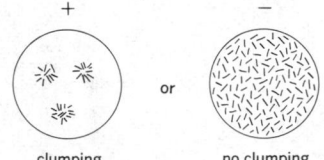

Macroscopic or microscopic results of agglutination reaction. *(From C. J. Witton, Microbiology with Application to Nursing, 2d ed., McGraw-Hill, 1956)*

AGLYCON

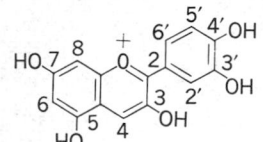

Cyanidin, the commonest aglycon.

AGOUTI

The agouti *(Dasyprocta aguti),* a rodent found in Mexico, South America, and the West Indies.

mineral fertilizers, and the behavior of underground water.

agricultural machinery [AGR] Machines utilized for tillage, planting, cultivation, and harvesting of crops.

agricultural meteorology [AGR] The study and application of relationships between meteorology and agriculture, involving problems such as timing the planting of crops. Also known as agrometeorology.

agricultural science [AGR] A discipline dealing with the selection, breeding, and management of crops and domestic animals for more economical production.

agricultural wastes [AGR] Those liquid or solid wastes that result from agricultural practices, such as cattle manure, crop residue (for example, corn stalks), pesticides, and fertilizers.

agriculture [BIOL] The production of plants and animals useful to man, involving soil cultivation and the breeding and management of crops and livestock.

Agriochoeridae [PALEON] A family of extinct tylopod ruminants in the superfamily Merycoidodontoidea.

agrioecology [ECOL] The ecology of cultivated plants.

Agrionidae [INV ZOO] A family of odonatan insects in the suborder Zygoptera characterized by black or red markings on the wings.

Agrobacterium [MICROBIO] A genus of rod-shaped, gram-negative soil bacteria composed of four species, three of which are plant pathogens.

Agrobacterium tumefaciens [MICROBIO] A pathogenic soil bacterium that causes root and stem galls on plants in over 40 families of angiosperms.

Agromyzidae [INV ZOO] A family of myodarian cyclorrhaphous dipteran insects of the subsection Acalypteratae; commonly called leaf-miner flies because the larvae cut channels in leaves.

agronomy [AGR] The principles and procedures of soil management and of field crop and special-purpose plant improvement, management, and production.

agrophilous [ECOL] Having a natural habitat in grain fields.

agrostology [BOT] A division of systematic botany concerned with the study of grasses.

Agulhas Current [OCEANOGR] A fast current flowing in a southwestward direction along the southeastern coast of Africa.

agyiophobia [PSYCH] Abnormal fear of crossing streets.

aH *See* abhenry.

Ah *See* ampere-hour.

ahaptoglobinemia [MED] An inherited lack of haptoglobin, a blood serum protein.

ahermatypic [INV ZOO] Non-reef-building, applied to corals.

aichmophobia [PSYCH] Abnormal fear of sharp or pointed objects.

aided matching [ORD] Mechanical system for transferring firing or other data from a data-transmission line to a gun data computer or other device.

aided tracking [ENG] A system of radar-tracking a target signal in bearing, elevation, or range, or any combination of these variables, in which the rate of motion of the tracking equipment is machine-controlled in collaboration with an operator so as to minimize tracking error.

aided-tracking mechanism [ENG] A device consisting of a motor and variable-speed drive which provides a means of setting a desired tracking rate into a director or other fire-control instrument, so that the process of tracking is carried out automatically at the set rate until it is changed manually.

aided-tracking ratio [ENG] The ratio between the constant velocity of the aided-tracking mechanism and the velocity of the moving target.

aid to navigation [NAV] A device external to a craft, designed to assist in determination of position of the craft, or of a safe course, or to warn of dangers or obstructions.

aiguille [GEOL] The needle-top of the summit of certain glaciated mountains, such as near Mont Blanc.

aikinite [MINERAL] PbCuBiS₃ A mineral crystallizing in the orthorhombic system and occurring massive and in gray needle-shaped crystals; hardness is 2 on Mohs scale, and specific gravity is 7.07. Also known as needle ore.

aileron [AERO ENG] The hinged rear portion of an aircraft wing moved differentially on each side of the aircraft to obtain lateral or roll control moments.

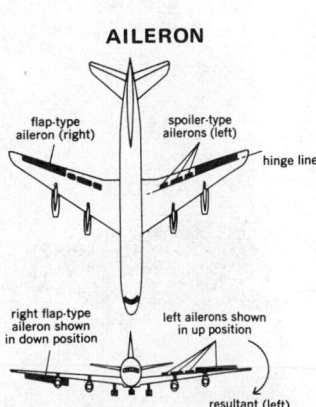

AILERON

flap-type aileron (right)

spoiler-type ailerons (left)

hinge line

right flap-type aileron shown in down position

left ailerons shown in up position

resultant (left) rolling moment

Flap- and spoiler-type ailerons on jet transport airplane.

ailsyte [PETR] An alkalic microgranite containing a considerable amount of riebeckite. Also known as paisanite.

ailurophobia [PSYCH] Abnormal fear of cats.

AIM *See* Airmen's Information Manual.

aiming circle [ENG] An instrument for measuring angles in azimuth and elevation in connection with artillery firing and general topographic work; equipped with fine and coarse azimuth micrometers and a magnetic needle.

aiming point [ORD] **1.** An object or point on which the sight of a weapon is laid for direction, or on which an observer orients his observing instrument. **2.** The point used by an aerial bombardier or pilot to determine where to release bombs, rockets, mines, torpedoes, and so forth.

aiming post [ORD] In mortar firing, a wooden or metal post having contrasting painted transverse bands and a metal point or stake for driving into the ground; used as a sighting point in direct fire.

aimless drainage [HYD] Drainage without a well-developed system, as in areas of glacial drift or karst topography.

A/in.² *See* ampere per square inch.

A indicator *See* A scope.

ainhum [MED] A tropical disease of unknown etiology that is peculiar to male Negroes, in which a toe is slowly and spontaneously amputated by a fibrous ring.

aiophyllous *See* evergreen.

air [METEOROL] A predominantly mechanical mixture of a variety of individual gases forming the earth's enveloping atmosphere.

air-acetylene welding [MET] A gas-welding process in which the heat is obtained from the combustion of acetylene and air.

AIRAD [NAV] Airmen advisory on NOTAM which is given only local dissemination during preflight or inflight briefings.

air adit [MIN ENG] An adit driven for the purpose of ventilating a mine.

Air Almanac [NAV] **1.** A periodical publication of astronomical data useful to and designed primarily for air navigation. **2.** A joint publication of the U.S. Naval Observatory and the Royal Greenwich Observatory, listing the Greenwich hour angle and declination of various celestial bodies at 10-minute intervals, the time of sunrise, sunset, moonrise, and moonset, and other astronomical information arranged in a form convenient for navigators; each publication covers 4 months.

air-arc furnace [ENG] An arc furnace designed to power wind tunnels, the air being superheated to 20,000 K and expanded to emerge at supersonic speeds.

air armament [ORD] All equipment through which a combat aircraft can release destructive power on a target.

air barrage [MIN ENG] An airtight wall dividing a ventilation gallery in a mine into two parts, so that air is led in through one part and out through the other. [ORD] **1.** Bombs dropped in a predetermined pattern from a fleet of aircraft to reduce the effectiveness of ground military personnel and equipment. **2.** The shells fired by an antiaircraft battery of guns at attacking airplanes.

air base [AERO ENG] **1.** In the U.S. Air Force, an establishment, comprising an airfield, its installations, facilities, personnel, and activities, for the flight operation, maintenance, and supply of aircraft and air organizations. **2.** A similar establishment belonging to any other air force. **3.** In a restricted sense, only the physical installation. [MAP] **1.** The line joining two air stations, or the length of this line. **2.** The distance, at the scale of the stereoscopic model, between adjacent perspective centers as reconstructed in the plotting instrument.

air battery [ELEC] A connected group of two or more air cells; also, a single air cell.

air-bend die [MET] A device for forming metals in which only the two edges of the lower section are in contact with the metal.

airblast [MIN ENG] A disturbance in underground workings accompanied by a strong rush of air.

airblast circuit breaker [ELEC] An electric switch which, on opening, utilizes a high-pressure gas blast (air or sulfur hexafluoride) to break the arc.

airblasting [ENG] A blasting technique in which air at very high pressure is piped to a steel shell in a shot hole and discharged. Also known as air breaking.

air-blower noise [ACOUS] Noise in blowers in heater and air-conditioning systems due to air turbulence.

air-blown asphalt [MATER] A bituminous product made by reacting the residual oil of petroleum distillation with air at 400–600°F (204–316°C).

airboat [AERO ENG] *See* seaplane. [NAV ARCH] A shallow boat that is propelled by an airplane propeller and steered by an airplane rudder.

airborne [AERO ENG] Of equipment and material, carried or transported by aircraft. [ORD] **1.** Of a force or organization, transported, or designed to be transported, by aircraft. **2.** Of an action or operation, carried out with transport aircraft.

airborne alert [ORD] A state of aircraft readiness wherein combat-equipped aircraft are rapidly airborne and ready for immediate action; it is designed to reduce reaction time and to increase the survival factor.

airborne assault weapon [ORD] An unarmored, mobile, full-tracked gun providing a mobile antitank capability for airborne troops; it can be air-dropped.

airborne collision warning system [ENG] A system such as a radar set or radio receiver carried by an aircraft to warn of the danger of possible collision.

airborne command post [ORD] A suitably equipped aircraft used by the commander for the control of his forces.

airborne control system [GEOD] A survey system for fourth-order horizontal and vertical control surveys involving electromagnetic distance measurements and horizontal and vertical measurements from two or more known positions to a helicopter hovering over the unknown position. Also known as ABC system.

airborne detector [ENG] A device, transported by an aircraft, whose function is to locate or identify an air or surface object.

airborne early warning [ORD] The detection of enemy air or surface units by radar or other equipment carried in an airborne vehicle, and the transmitting of a warning to friendly units.

airborne early warning and control [ORD] Air surveillance and control provided by airborne early-warning vehicles which are equipped with search and height-finding radar and communications equipment for controlling weapons.

airborne early-warning station [ORD] An airborne radar station for detecting aircraft, which are incapable of being detected by a ground radar station because of line-of-sight propagation limitations.

airborne electronic survey control [ENG] The airborne portion of very accurate positioning systems used in controlling surveys from aircraft.

airborne intercept radar [ENG] Airborne radar used to track and "lock on" to another aircraft to be intercepted or followed.

airborne magnetometer [ENG] An airborne instrument used to measure the magnetic field of the earth.

airborne operation [ORD] An operation involving the movement and delivery by air, into an objective area, of combat forces and their logistic support for execution of a tactical or a strategic mission; the means may be any combination of airborne units, air-transportable units, and types of transport aircraft, depending on the mission and the overall situation.

airborne profile [GEOD] Continuous terrain-profile data produced by an absolute altimeter in an aircraft which is making an altimeter-controlled flight along a prescribed course.

airborne profile recorder [ENG] An electronic instrument that emits a pulsed-type radar signal from an aircraft to measure vertical distances between the aircraft and the earth's surface. Abbreviated APR. Also known as terrain profile recorder (TPR).

airborne radar [ENG] Radar equipment carried by aircraft to assist in navigation by pilotage, to determine drift, and to locate weather disturbances; a very important use is locating other aircraft either for avoidance or attack.

airborne troops [ORD] Ground units whose primary mission is to make assault landings from the air; may refer specifically to troops landed by aircraft as distinguished from parachute troops, or may refer only to those troops landed by parachute or glider as distinguished from those landed in powered aircraft.

airborne unit [ORD] A ground unit organized, trained, and equipped for airborne assault.

airborne waste [ENG] Vapors, gases, or particulates introduced into the atmosphere by evaporation, chemical, or combustion processes; a frequent cause of smog and an irritant to eyes and breathing passages.

air brake [MECH ENG] An energy-conversion mechanism activated by air pressure and used to retard, stop, or hold a vehicle or, generally, any moving element.

air breaking *See* airblasting.

air-break switch *See* air switch.

air breakup [AERO ENG] The breakup of a test reentry body within the atmosphere.

air-breathing [MECH ENG] Of an engine or aerodynamic vehicle, required to take in air for the purpose of combustion.

air-breathing missile [ORD] A missile having an engine that requires air intake for combustion of its fuel, such as a ramjet or turbojet; it cannot operate beyond the atmosphere.

air brick [MATER] A brick or brick-sized metal box that is hollow or perforated and used for ventilation.

airbrush [GRAPHICS] A pencil-shaped air gun that fine-sprays paint; used in retouching photographs and in shading drawings.

airburst [ORD] Any burst in the air, but usually the bursting of a projectile or bomb above the ground with resulting spray of fragments. Also known as aerial burst.

air capacitor [ELEC] A capacitor having only air as the dielectric material between its plates. Also known as air condenser.

air casing [ENG] A metal casing surrounding a pipe or reservoir and having a space between to prevent heat transmission.

air cell [ELECTR] A cell in which depolarization at the positive electrode is accomplished chemically by reduction of the oxygen in the air. [ZOO] A cavity or receptacle for air such as an alveolus, an air sac in birds, or a dilation of the trachea in insects.

air chamber [MECH ENG] A pressure vessel, partially filled with air, for converting pulsating flow to steady flow of water in a pipeline, as with a reciprocating pump.

air check [ENG ACOUS] A recording made of a live radio broadcast for filing purposes at the broadcasting facility.

air classifier [MECH ENG] A device to separate particles by size through the action of a stream of air.

air cleaner [ENG] Any of various devices designed to remove particles and aerosols of specific sizes from air; examples are screens, settling chambers, filters, wet collectors, and electrostatic precipitators.

Airco-Hoover sweetening [CHEM ENG] Removal of mercaptans from gasoline by caustic and water washes, then heating the dried gasoline and passing it with some oxygen through a reactor containing a slurry of diatomaceous earth impregnated with copper chloride; the oxygen regenerates the catalyst.

air composition [METEOROL] The kinds and amounts of the constituent substances of air, the amounts being expressed as percentages of the total volume or mass.

air compression [PHYS] The decrease of volume of a quantity of air as a result of an increase in pressure, as is accomplished by a piston moving in a cylinder.

air compressor [MECH ENG] A machine that increases the pressure of air by increasing its density and delivering the fluid against the connected system resistance on the discharge side.

air-compressor unloader [MECH ENG] A device for control of air volume flowing through an air compressor.

air-compressor valve [MECH ENG] A device for controlling the flow into or out of the cylinder of a compressor.

air condenser [ELEC] *See* air capacitor. [MECH ENG] **1.** A steam condenser in which the heat exchange occurs through metal walls separating the steam from cooling air. Also known as air-cooled condenser. **2.** A device that removes vapors, such as of oil or water, from the airstream in a compressed-air line.

air conditioner [MECH ENG] A mechanism primarily for comfort cooling that lowers the temperature and reduces the humidity of air in buildings.

air conditioning [MECH ENG] The maintenance of certain aspects of the environment within a defined space to facilitate the function of that space; aspects controlled include air

AIRBORNE MAGNETOMETER

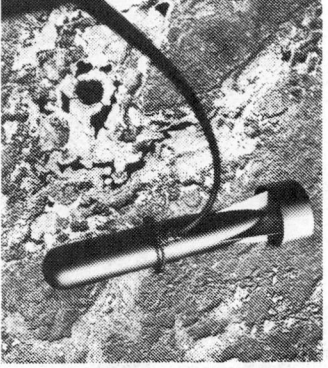

View from aircraft of suspended magnetometer in use on a magnetic survey in an inaccessible region of moderate to low relief. *(Aero Service Corporation)*

AIR CAPACITOR

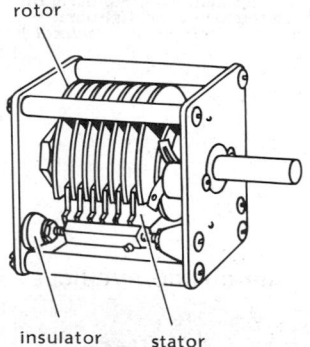

A drawing of a variable air capacitor showing the flat, parallel metallic plates separated by air.

AIR CONDITIONER

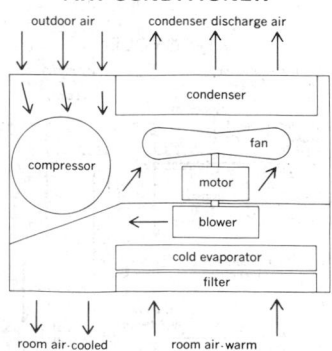

Schematic of room air conditioner. *(From American Society of Heating, Refrigerating and Air-conditioning Engineers, Inc., Guide and Data Book, 1967)*

temperature and motion, radiant heat level, moisture, and concentration of pollutants such as dust, microorganisms, and gases. Also known as climate control. [TEXT] A chemical process by which small fibers are sealed into yarn.

air-control center [COMM] An area set aside in a submarine for the control of aircraft; it is the equivalent of a combat information center on an aircraft or a ship.

air conveyor *See* pneumatic conveyor.

air-cooled condenser *See* air condenser.

air-cooled engine [MECH ENG] An engine cooled directly by a stream of air without the interposition of a liquid medium.

air-cooled heat exchanger [MECH ENG] A finned-tube (extended-surface) heat exchanger with hot fluids inside the tubes, and cooling air that is fan-blown (forced draft) or fan-pulled (induced draft) across the tube bank.

air cooling [MECH ENG] Lowering of air temperature for comfort, process control, or food preservation.

air coordinates [ORD] A system of coordinates used in determining the forces acting on a missile.

air-core coil [ELECTR] An inductor without a magnetic core.

air-core transformer [ELECTROMAG] Transformer (usually radio-frequency) having a nonmetallic core.

air course *See* airway.

air cover [ORD] **1.** The protection against attack, especially air attack, given by airplanes to surface or airborne forces. **2.** The airplanes giving, or designated to give, this protection.

aircraft [AERO ENG] Any structure, machine, or contrivance, especially a vehicle, designed to be supported by the air, either by the dynamic action of the air upon the surfaces of the structure or object or by its own buoyancy. Also known as air vehicle.

aircraft ammunition [ORD] Ammunition designed to be shot, launched, or dropped from aircraft.

aircraft antenna [ELECTR] An airborne device used to detect or radiate electromagnetic waves.

aircraft axes *See* axes of an aircraft.

aircraft bonding [AERO ENG] Electrically connecting together all of the metal structure of the aircraft, including the engine and metal covering on the wiring.

aircraft cannon [ORD] A cannon designed or modified for use in an aircraft. Also known as aerial cannon.

aircraft carrier [NAV ARCH] A ship that carries aircraft, has a takeoff and landing deck, and is otherwise designed and equipped to serve as a base of operations for the aircraft. Also known as carrier.

aircraft ceiling [METEOROL] After United States weather observing practice, the ceiling classification applied when the reported ceiling value has been determined by a pilot while in flight within 1.5 nautical miles of any runway of the airport.

aircraft compass system [NAV] A wholly airborne direction-giving system consisting of means for sensing the earth's magnetic field, stabilizing means, and usually remote indicating equipment.

aircraft control and warning system [ORD] A system established to control and report the movement of aircraft; consists of observation facilities (radar, passive electronic, visual, or other means), control centers, and necessary communications.

aircraft decibel rating [ELECTROMAG] The ratio of the radar reflectivity of a specific type of aircraft to that of a selected reference aircraft, measured in decibels.

aircraft detection [ENG] The sensing and discovery of the presence of aircraft; major techniques include radar, acoustical, and optical methods.

aircraft dispersal area [ORD] An area on a military installation designed primarily for the dispersal of parked aircraft, to make such aircraft less vulnerable in the event of an enemy air raid.

aircraft early-warning station [NAV] A station with long-range radar for detecting the approach of aircraft at a distance.

aircraft electrification [METEOROL] **1.** The accumulation of a net electric charge on the surface of an aircraft. **2.** The separation of electric charge into two concentrations of opposite sign on distinct portions of an aircraft surface.

aircraft engine [AERO ENG] A component of an aircraft that develops either shaft horsepower or thrust and incorporates design features advantageous for aircraft propulsion.

aircraft fuel [MATER] The material used as a source of energy by an aircraft engine to provide propulsion.

aircraft gun [ORD] A gun of any type mounted in a combat aircraft, such as a machine gun or cannon.

aircraft icing [METEOROL] The accumulation of ice on the exposed surfaces of aircraft when flying through supercooled water drops (cloud or precipitation).

aircraft impactor [ENG] An instrument carried by an aircraft for the purpose of obtaining samples of airborne particles.

aircraft instrumentation [AERO ENG] Electronic, gyroscopic, and other instruments for detecting, measuring, recording, telemetering, processing, or analyzing different values or quantities in the flight of an aircraft.

aircraft instrument panel [AERO ENG] A coordinated instrument display arranged to provide the pilot and flight crew with information about the aircraft's speed, altitude, attitude, heading, and condition; also advises the pilot of the aircraft's response to his control efforts.

aircraft low-approach system [NAV] Any of the various means for furnishing guidance in the vertical and horizontal planes to aircraft during descent from an initial approach altitude to a point near the ground.

aircraft machine-gun turret [ORD] An armored enclosure installed in an aircraft for housing the armament and related accessories; it is designed to rotate about one or more axes, thus permitting positioning and firing of a machine gun in a number of directions or angles.

aircraft noise [ACOUS] Effective sound output of the various sources of noise associated with aircraft operation, such as propeller and engine exhaust, jet noise, and sonic boom.

aircraft propeller [AERO ENG] A hub-and-multiblade device for transforming the rotational power of an aircraft engine into thrust power for the purpose of moving an aircraft through the air.

aircraft propulsion [AERO ENG] The means, other than gliding, whereby an aircraft moves through the air; effected by the rearward acceleration of matter through the use of a jet engine or by the reactive thrust of air on a propeller.

aircraft pylon [AERO ENG] A suspension device externally installed under the wing or fuselage of an aircraft; it is aerodynamically designed to fit the configuration of specific aircraft so as to create the least amount of drag; it provides a means of attaching fuel tanks, bombs, rockets, torpedoes, rocket motors, or machine-gun pods.

aircraft report *See* pilot report.

aircraft rocket [ORD] A rocket missile designed to be launched from an aircraft and employing warheads that are armor-piercing, incendiary, or explosive; it is carried beneath the wings of fighter or ground-attack airplanes.

aircraft testing [AERO ENG] The subjecting of an aircraft or its components to simulated or actual flight conditions while measuring and recording pertinent physical phenomena that indicate operating characteristics.

aircraft thermometry [METEOROL] The science of temperature measurement from aircraft.

aircraft vectoring [NAV] The directional control of in-flight aircraft through transmission of bearing and altitude instructions from the ground.

aircraft weather reconnaissance [METEOROL] The making of detailed weather observations or investigations from aircraft in flight.

air crossing [MIN ENG] A mine passage in which two airways cross each other.

air current [FL MECH] **1.** Very generally, any moving stream of air. **2.** *See* air-earth conduction current. [MIN ENG] The flow of air ventilating the workings of a mine. Also known as airflow.

air cushion [MECH ENG] A mechanical device using trapped air to arrest motion without shock.

air-cushion vehicle [MECH ENG] A transportation device supported by low-pressure, low-velocity air capable of traveling equally well over water, ice, marsh, or relatively level land. Also known as ground-effect machine (GEM); hovercraft.

air cycle [MECH ENG] A refrigeration cycle characterized by the working fluid, air, remaining as a gas throughout the cycle

AIRCRAFT INSTRUMENT PANEL

Integrated flight instrument panel designed by Wright Air Development Center, used on high-performance fighters. (*Aviat. Week Space Technol.*)

AIR-CUSHION VEHICLE

Princeton University experimental 20-foot air-cushion vehicle.

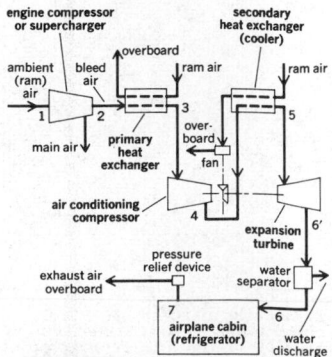

AIR CYCLE

Open air-cycle bootstrap system for airplanes.

rather than being condensed to a liquid; used primarily in airplane air conditioning.

air cylinder [MECH ENG] A cylinder in which air is compressed by a piston, compressed air is stored, or air drives a piston.

air defense [ORD] All measures designed to destroy, nullify, or reduce the effectiveness of enemy attack by aircraft or missiles after they are airborne.

air defense artillery [ORD] Weapons and equipment for actively combating air targets from the ground. Abbreviated ADA.

air delivery [ORD] A method of air movement wherein personnel, supplies, and equipment are unloaded from aircraft in flight. Also known as air drop.

air density [MECH] The mass per unit volume of air.

air depolarized battery [ELEC] A primary battery which is kept depolarized by atmospheric oxygen rather than chemical compounds. Also known as metal-air battery.

air discharge [GEOPHYS] 1. A form of lightning discharge, intermediate in character between a cloud discharge and a cloud-to-ground discharge, in which the multibranching lightning channel descending from a cloud base does not reach the ground, but succeeds only in neutralizing the space charge distributed in the subcloud layer. 2. A type of diffuse electrical discharge occasionally reported as occurring in the region above an active thunderstorm.

air door [MIN ENG] A door placed in a mine roadway to prevent the passage of air.

air drainage [METEOROL] General term for gravity-induced, downslope flow of relatively cold air.

airdraulic [MECH ENG] Combining pneumatic and hydraulic action for operation.

air drift [MIN ENG] A roadway, generally inclined, driven in stone for ventilation purposes.

air drill [MECH ENG] A drill powered by compressed air.

air drop See air delivery.

air drying [ENG] Removing moisture from a material by exposure to air to the extent that no further moisture is released on contact with air; important in lumber manufacture.

air duct See airflow duct.

air-earth conduction current [GEOPHYS] That part of the air-earth current contributed by the electrical conduction of the atmosphere itself; represented as a downward movement of positive space charge in storm-free regions all over the world. Also known as air current.

air-earth current [GEOPHYS] The transfer of electric charge from the positively charged atmosphere to the negatively charged earth; made up of the air-earth conduction current, a precipitation current, a convection current, and miscellaneous smaller contributions.

air ejector [MECH ENG] A device that uses a fluid jet to remove air or other gases, as from a steam condenser.

air embolism See aeroembolism.

air endway [MIN ENG] A narrow roadway driven in a coal seam parallel and close to a winning headway for ventilation.

air engine [MECH ENG] An engine in which compressed air is the actuating fluid.

air-entrained cement [MATER] Cement with improved qualities due to the introduction of air bubbles in its preparation.

air-entraining agent [MATER] An admixture, usually a resin, soap, or grease, for portland cement or concrete to effect air entrainment and, thus, superior properties.

air entrainment [ENG] The inclusion of minute bubbles of air in cement or concrete through the addition of some material during grinding or mixing to reduce the surface tension of the water, giving improved properties for the end product.

air environment [COMMUN] The aggregate of all airborne equipment which is a part of a communications-electronics system.

airfield [CIV ENG] The area of an airport for the takeoff and landing of airplanes.

air filter [ENG] A device that reduces the concentration of solid particles in an airstream to a level that can be tolerated in a process or space occupancy; a component of most systems in which air is used for industrial processes, ventilation, or comfort air conditioning.

airfloat clay [MIN ENG] Fine particles of clay obtained by air separation from coarser particles following a grinding operation.

airflow [FL MECH] 1. A flow or stream of air which may take place in a wind tunnel or, as a relative airflow, past the wing or other parts of a moving craft. Also known as airstream. 2. A rate of flow, measured by mass or volume per unit of time. [MIN ENG] See air current.

airflow duct [ENG] A pipe, tube, or channel through which air moves into or out of an enclosed space. Also known as air duct.

airflow orifice [ENG] An opening through which air moves out of an enclosed space.

airflow pipe [ENG] A tube through which air is conveyed from one location to another.

airflow stack effect [FL MECH] The variation of pressure with height in air flowing in a vertical duct due to a difference in temperature between the flowing air and the air outside the duct.

airfoil [AERO ENG] A body of such shape that the force exerted on it by its motion through a fluid has a larger component normal to the direction of motion than along the direction of motion; examples are the wing of an airplane and the blade of a propeller. Also known as aerofoil.

airfoil profile [AERO ENG] The closed curve defining the external shape of the cross section of an airfoil. Also known as airfoil section; airfoil shape; wing section.

airfoil section See airfoil profile.

airfoil shape See airfoil profile.

airfoil-vane fan [MIN ENG] A centrifugal-type mine fan; the vanes are curved backward from the direction of rotation.

airframe [AERO ENG] The basic assembled structure of any aircraft or rocket vehicle, except lighter-than-air craft, necessary to support aerodynamic forces and inertia loads imposed by the weight of the vehicle and contents.

air furnace [MET] 1. A furnace using a natural air draft. 2. A furnace in which the metal is melted by a flame originating from fuel burned at one end, passing over the hearth in the middle, and exiting at the other end.

air gage [ENG] 1. A device that measures air pressure. 2. A device that compares the shape of a machined surface to that of a reference surface by measuring the rate of passage of air between the surfaces.

air gap [ELECTR] 1. A gap or an equivalent filler of nonmagnetic material across the core of a choke, transformer, or other magnetic device. 2. A spark gap consisting of two electrodes separated by air. [GEOL] See wind gap.

air gas [MATER] A gaseous fuel made by blowing air through a coal or coke bed so that CO_2 is reduced to CO.

airglow [GEOPHYS] The quasi-steady radiant emission from the upper atmosphere over middle and low latitudes, as distinguished from the sporadic emission of auroras which occur over high latitudes. Also known as light-of-the-night-sky; night-sky light; night-sky luminescence; permanent aurora.

air-ground communication [COMMUN] Two-way communication between aircraft and stations on the ground.

air gun See air rifle.

air-hardening steel [MET] A steel whose content of carbon and other alloying elements is sufficient for the steel to harden fully by cooling in air or any other atmosphere from a temperature above its transformation range. Also known as self-hardening steel.

airhead [ORD] A designated geographical area in an area of operations used as a base of supply and evacuation by air.

air heater See air preheater.

air-heating system See air preheater.

air heave [GEOL] Deformation of plastic sediments on a tidal flat as a result of the growth of air pockets in them; the growth occurs by accretion of smaller air bubbles oozing through the sediment.

air hoar [HYD] A hoar growing on objects above the ground or snow.

air hoist [MECH ENG] A lifting tackle or tugger constructed with cylinders and pistons for reciprocating motion and air motors for rotary motion, all powered by compressed air. Also known as pneumatic hoist.

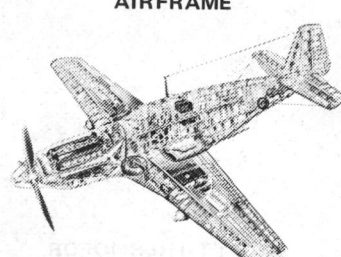

AIRFRAME

Airframe of piston-engine P-51A pursuit airplane. *(North American Aviation, Inc.)*

airhole [MIN ENG] A small excavation or hole made to improve ventilation by communication with other workings or with the surface.

air horsepower [MECH ENG] The theoretical (minimum) power required to deliver the specified quantity of air under the specified pressure conditions in a fan, blower, compressor, or vacuum pump. Abbreviated air hp.

air hp *See* air horsepower.

air hunger [MED] The deep, gasping respiration characteristic of severe diabetic acidosis and coma.

air-injection system [MECH ENG] A device that uses compressed air to inject the fuel into the cylinder of an internal combustion engine.

air intake [AERO ENG] An open end of an air duct or similar projecting structure so that the motion of the aircraft is utilized in capturing air to be conducted to an engine or ventilator. [MIN ENG] A device for supplying a compressor with clean air at the lowest possible temperature.

air knife [ENG] A device that uses a thin, flat jet of air to remove the excess coating from freshly coated paper.

air-knife coating [ENG] An even film of coating left on paper after treatment with an air knife.

air-lance [ENG] To direct a pressurized-air stream to remove unwanted accumulations, as in boiler-wall cleaning.

air launch [AERO ENG] Launching from an aircraft in the air.

air-launched ballistic missile [ORD] A ballistic missile launched from an airborne vehicle.

air layering [BOT] A method of vegetative propagation, usually of a wounded part, in which the branch or shoot is enclosed in a moist medium until roots develop, and then it is severed and cultivated as an independent plant.

air lift *See* air-lift pump.

airlift [AERO ENG] **1.** To transport passengers and cargo by the use of aircraft. **2.** The total weight of personnel or cargo carried by air.

air-lift pump [MECH ENG] A device composed of two pipes, one inside the other, used to extract water from a well; the lower end of the pipes is submerged, and air is delivered through the inner pipe to form a mixture of air and water which rises in the outer pipe above the water in the well; also used to move corrosive liquids, mill tailings, and sand. Also known as air lift.

Airlift Thermofor catalytic cracking [CHEM ENG] A continuous catalytic process for the conversion of heavy gas oils into lighter products; utilizes a moving bed of small pellets or beads of catalyst traveling continuously through the oil-cracking vessel and then through a regeneration kiln.

airlight [METEOROL] In determinations of visual range, light from sun and sky which is scattered into the eyes of an observer by atmospheric suspensoids (and, to slight extent, by air molecules) lying in the observer's cone of vision.

airlight formula [OPTICS] A fundamental equation of visual-range theory, relating the apparent luminance of a distant black object, the apparent luminance of the background sky above the horizon, and the extinction coefficient of the air layer near the ground.

air line [ENG] A fault, in the form of an elongated bubble, in glass tubing. Also known as hairline. [SPECT] Lines in a spectrum due to the excitation of air molecules by spark discharges, and not ordinarily present in arc discharges.

air-line lubricator *See* line oiler.

air-line main [MIN ENG] The pipe column supplying air from the compressors to the quarry face.

air lock [ENG] **1.** A chamber capable of being hermetically sealed that provides for passage between two places of different pressure, such as between an altitude chamber and the outside atmosphere, or between the outside atmosphere and the work area in a tunnel or shaft being excavated through soil subjected to water pressure higher than atmospheric. **2.** An air bubble in a pipeline which impedes liquid flow. [MIN ENG] A casing atop an upcast mine shaft to minimize surface air leakage into the fan.

air log [AERO ENG] A distance-measuring device used especially in certain guided missiles to control range.

airman [MIN ENG] A man who constructs brattices. [ORD] An enlisted man in the military air force.

air mass [METEOROL] An extensive body of the atmosphere which approximates horizontal homogeneity in its weather characteristics, particularly with reference to temperature and moisture distribution.

air-mass analysis [METEOROL] In general, the theory and practice of synoptic surface-chart analysis by the so-called Norwegian methods, which involve the concepts of the polar front and of the broad-scale air masses which it separates.

air-mass climatology [CLIMATOL] The representation of the climate of a region by the frequency and characteristics of the air masses under which it lies; basically, a type of synoptic climatology.

air-mass precipitation [METEOROL] Any precipitation that can be attributed only to moisture and temperature distribution within an air mass when that air mass is not, at that location, being influenced by a front or by orographic lifting.

air-mass shower [METEOROL] A shower that is produced by local convection within an unstable air mass; the most common type of air-mass precipitation.

air-mass source region [METEOROL] An extensive area of the earth's surface over which bodies of air frequently remain for a sufficient time to acquire characteristic temperature and moisture properties imparted by that surface.

air-measuring station [MIN ENG] A place in a mine airway where the volume of air passing is measured periodically.

Airmen's Information Manual [NAV] Notices to airmen regarding the status of components of the national airspace system. Abbreviated AIM.

air meter [ENG] A device that measures the flow of air, or gas, expressed in volumetric or weight units per unit time. Also known as airometer.

air mile [NAV] A unit of length used in air navigation and equal, since 1954, to 1 international nautical mile. Also known as aeronautical mile.

air mileage [NAV] The number of miles flown relative to the air; true air speed multiplied by time.

air mileage indicator [ENG] An instrument on an airplane which continuously indicates mileage through the air.

air mileage unit [ENG] A device which derives continuously and automatically the air distance flown, and feeds this information into other units, such as an air mileage indicator.

air monitoring [CIV ENG] A practice of continuous air sampling by various levels of government or particular industries.

air motor [MECH ENG] A device in which the pressure of confined air causes the rotation of a rotor or the movement of a piston.

air mover [MIN ENG] A portable compressed-air appliance, used as a blower or exhauster.

air navigation [NAV] The process of directing and monitoring the progress of an aircraft between selected geographic points or with respect to some predetermined plan. Also known as avigation.

airometer [ENG] **1.** An apparatus for both holding air and measuring the quantity of air admitted into it. **2.** *See* air meter.

air parcel [METEOROL] An imaginary body of air to which may be assigned any or all of the basic dynamic and thermodynamic properties of atmospheric air.

air permeability [TEXT] The quality of porosity of a fabric; indicates the wind resistance, and therefore warmth or coolness, of the fabric.

air pickets [ORD] Airborne early-warning aircraft disposed around a position, area, or formation primarily to detect, report, and track approaching enemy aircraft and to control intercepts.

airplane [AERO ENG] A heavier-than-air vehicle designed to use the pressures created by its motion through the air to lift and transport useful loads.

airplane flare [ENG] A flare, often magnesium, that is dropped from an airplane to illuminate a ground area; a small parachute decreases the rate of descent.

air plot [NAV] A continuous plot of the position of an airborne object represented graphically to show true headings steered and distances flown.

air-plot wind velocity [NAV] The wind velocity calculated from a knowledge of an air position and a fix.

air pocket [ENG] An air-filled space that is normally occupied by a liquid. Also known as air trap. [METEOROL] An expression used in the early days of aviation for a downdraft;

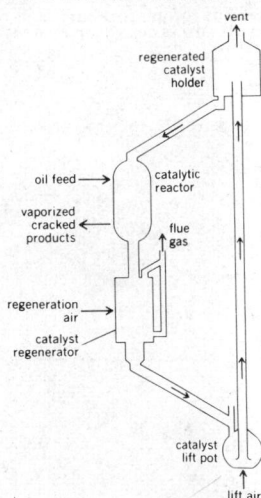

AIRLIFT THERMOFOR CATALYTIC CRACKING

Components of the Airlift Thermofor catalytic cracking process.

such downdrafts were thought to be pockets in which there was insufficient air to support the plane.

air pollution [ECOL] The presence in the outdoor atmosphere of one or more contaminants such as dust, fumes, gas, mist, odor, smoke, or vapor in quantities and of characteristics and duration such as to be injurious to human, plant, or animal life or to property, or to interfere unreasonably with the comfortable enjoyment of life and property.

air-pollution control [ENG] A practical means of treating polluting sources to maintain a desired degree of air cleanliness.

airport [CIV ENG] A terminal facility used for aircraft takeoff and landing and including facilities for handling passengers and cargo and for servicing aircraft. Also known as aerodrome.

airport engineering [CIV ENG] The planning, design, construction, and operation and maintenance of facilities providing for the landing and takeoff, loading and unloading, servicing, maintenance, and storage of aircraft.

airport light beacon [NAV] A visual navigational aid, usually a rotating beacon displaying flashes of white or colored light or both, located at or near an airport to indicate its specific or general location.

airport surface detection equipment [NAV] A short-range radar using millimeter waves and giving a panoramic presentation of all aircraft and vehicles, moving or stationary, on the surface of an aerodrome; used by air traffic controllers for expeditious movement of surface aircraft on the ramp, taxiway, and runway.

airport traffic [NAV] Aircraft both in the air and on the active runways in the airport zone.

airport traffic control tower [NAV] The terminal control point at an airport for all takeoff and landing operations, departure and approach operations, and ground movements of aircraft and airport vehicles. Abbreviated ATCT.

air position [NAV] The theoretical position of an aircraft or missile at a given moment, assuming it to have been unaffected in flight by wind. Also known as no-wind position.

air-position indicator [NAV] An airborne computing system which presents a continuous indication of the aircraft position on the basis of aircraft heading, airspeed, and elapsed time. Abbreviated API.

air preheater [MECH ENG] A device used in steam boilers to transfer heat from the flue gases to the combustion air before the latter enters the furnace. Also known as air heater; air-heating system.

air pressure [PHYS] The force per unit area that the air exerts on any surface in contact with it, arising from the collisions of the air molecules with the surface.

air-pressure drop [FL MECH] The pressure lost in overcoming friction along an airway.

airproof *See* airtight.

air propeller [AERO ENG] A hub-and-multiblade device for changing rotational power of an aircraft engine into thrust power for the purpose of propelling an aircraft through the air. [MECH ENG] A rotating fan for moving air.

air properties [PHYS] Characteristics of air as a gas, such as density, molecular weight, specific heats, boiling point, critical temperature, and critical pressure.

air pump [MECH ENG] A device for removing air from an enclosed space or for adding air to an enclosed space.

Air Pump *See* Antlia.

air puncher [ENG] A machine consisting essentially of a reciprocating chisel or pick, driven by air.

air raid [ORD] An attack of a place by aircraft; may consist of a bombing or strafing attack against a position or a city.

air raid shelter [CIV ENG] A chamber, often underground, provided with living facilities and food, for sheltering people against air attacks.

air register [ENG] A device attached to an air-distributing duct for the purpose of controlling the discharge of air into the space to be heated, cooled, or ventilated.

air regulator [MIN ENG] An adjustable door installed in permanent air stoppings to control ventilating current.

air resistance [MECH] Wind drag giving rise to forces and wear on buildings and other structures.

air rifle [ORD] A low-powered rifle that shoots small metal pellets (BBs) by the action of compressed air. Also known as air gun; BB gun.

air route [NAV] The navigable airspace between two points, identified to the extent necessary for the application of flight rules.

air-route traffic control [NAV] Traffic control service provided to aircraft which are operating under instrument flight rules within controlled airspace, and principally en route.

air-route traffic control center [NAV] A place from which traffic is directed along a controlled air route.

air sac [GEOL] *See* vesicle. [INV ZOO] One of large, thin-walled structures associated with the tracheal system of some insects. [VERT ZOO] In birds, any of the small vesicles that are connected with the respiratory system and located in bones and muscles to increase buoyancy.

air sampling [ENG] The collection and analysis of samples of air to measure the amounts of various pollutants or other substances in the air, or the air's radioactivity.

air scoop [DES ENG] An air-duct cowl projecting from the outer surface of an aircraft or automobile, which is designed to utilize the dynamic pressure of the airstream to maintain a flow of air.

air screw [MECH ENG] A screw propeller that operates in air.

air-seasoned [ENG] Treated by exposure to air to give a desired quality.

air separator [MECH ENG] A device that uses an air current to separate a material from another of greater density or particles from others of greater size.

air-setting mortar [MATER] Mortar that sets in air at atmospheric pressure and ordinary temperatures.

air sextant [NAV] A sextant designed primarily for air navigation; generally provided with some mechanical devices for furnishing horizon indications.

air shaft [BUILD] An open space surrounded by the walls of a building or buildings to provide ventilation for windows. Also known as air well. [MIN ENG] A usually vertical earth bore or shaft to supply surface air to an underground facility such as a mine.

airship [AERO ENG] A propelled and steered aerial vehicle, dependent on gases for flotation.

air shot [ENG] A shot prepared by loading (charging) so that an air space is left in contact with the explosive for the purpose of lessening its shattering effect.

air shower [GEOPHYS] A grouping of cosmic-ray particles observed in the atmosphere; a cascade shower in the atmosphere. Also known as shower.

airsickness [MED] Motion sickness associated with flying due to the effects of acceleration.

air-slaked [CHEM] Having the property of a substance, such as lime, that has been at least partially converted to a carbonate by exposure to air.

Airslide conveyor [MECH ENG] An air-activated gravity-type conveyor, of the Fuller Company, using low-pressure air to aerate or fluidize pulverized material to a degree which will permit it to flow on a slight incline by the force of gravity.

air sounding [METEOROL] The act of measuring atmospheric phenomena or determining atmospheric conditions at altitude, especially by means of apparatus carried by balloons or rockets.

air space [ENG] An enclosed space containing air in a wall for thermal insulation. [ORD] The space in a firearm between the powder charge and projectile.

airspace [AERO ENG] **1.** The space occupied by an aircraft formation or used in a maneuver. **2.** The area around an airplane in flight that is considered an integral part of the plane in order to prevent collision with another plane; the space depends on the speed of the plane. [METEOROL] **1.** Of or pertaining to both the earth's atmosphere and space. Also known as aerospace. **2.** The portion of the atmosphere above a particular land area, especially a nation or other political subdivision.

air-spaced coax [ELECTROMAG] Coaxial cable in which air is basically the dielectric material; the conductor may be centered by means of a spirally wound synthetic filament, beads, or braided filaments.

airspace reservation [NAV] An airspace designated by government authority in which flight is prohibited or restricted.

airspace warning area [NAV] An airspace which is outside

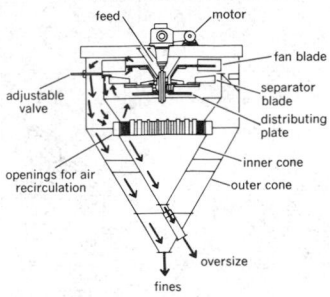

AIR SEPARATOR

The double-cone air separator, a type of air classifier. (*From W. L. McCabe and J. C. Smith, Unit Operations of Chemical Engineering, McGraw-Hill, 1956*)

AIRSHIP

(a)

(b)

Two types of airships, (a) rigid and (b) nonrigid.

the territorial limits of any country and in which exists an actual or potential hazard to aircraft in flight; such an area has no legal status.

airspeed [AERO ENG] The speed of an airborne object relative to the atmosphere; in a calm atmosphere, airspeed equals ground speed.

airspeed head [ENG] Any instrument or device, usually a pitot tube, mounted on an aircraft for receiving the static and dynamic pressures of the air used by the airspeed indicator.

airspeed indicator [ENG] A device that computes and displays the speed of an aircraft relative to the air mass in which the aircraft is flying.

air spora [BIOL] Airborne fungus spores, pollen grains, and microorganisms.

air spring [MECH ENG] A spring in which the energy storage element is air confined in a container that includes an elastomeric bellows or diaphragm.

air stack [AERO ENG] A group of planes flying at prescribed heights while waiting to land at an airport. [MIN ENG] A chimney for ventilating a coal mine.

air-standard cycle [THERMO] A thermodynamic cycle in which the working fluid is considered to be a perfect gas with such properties of air as a volume of 12.4 ft³/lb at 14.7 psi and 492°R and a ratio of specific heats of 1:4.

air-standard engine [MECH ENG] A heat engine operated in an air-standard cycle.

airstart [AERO ENG] An act or instance of starting an aircraft's engine while in flight, especially a jet engine after flameout.

air station *See* camera station.

airstream *See* airflow.

air strike [ORD] An attack on specific objectives by fighter, bomber, or attack air craft on an offensive mission; may consist of several air organizations under a single command in the air.

air strip *See* landing strip.

air-supply mask *See* air-tube breathing apparatus.

air surveillance [ENG] Systematic observation of the airspace by visual, electronic, or other means, primarily for identifying all aircraft in that airspace, and determining their movements.

air surveillance radar [ENG] Radar of moderate range providing position of aircraft by azimuth and range data without elevation data; used for air-traffic control.

air survey *See* aerial survey.

air-suspension system [MECH ENG] Parts of an automotive vehicle that are intermediate between the wheels and the frame, and support the car body and frame by means of a cushion of air to absorb road shock caused by passage of the wheels over irregularities.

air sweetening [CHEM ENG] A process in which air or oxygen is used to oxidize lead mercaptides to disulfides instead of using elemental sulfur.

air switch [ELEC] A switch in which the breaking of the electric circuit takes place in air. Also known as air-break switch.

air system [MECH ENG] A mechanical refrigeration system in which air serves as the refrigerant in a cycle comprising compressor, heat exchanger, expander, and refrigerating core.

air taxi [AERO ENG] A carrier of passengers and cargo engaged in charter flights, feeder air services to large airline facilities, or contract airmail transportation.

air temperature [METEOROL] 1. The temperature of the atmosphere which represents the average kinetic energy of the molecular motion in a small region and is defined in terms of a standard or calibrated thermometer in thermal equilibrium with the air. 2. The temperature that the air outside of the aircraft is assumed to have as indicated on a cockpit instrument.

air temperature correction [NAV] 1. A correction to a sextant altitude reading, required because of the air temperature. 2. A correction required in the airspeed as read by a pitot tube to arrive at true airspeed.

air terminal [CIV ENG] A facility providing a place of assembly and amenities for airline passengers and space for administrative functions. [ELEC] A structure, such as a tower, that serves as a lightning arrester.

air thermometer [ENG] A device that measures the tempera-

ture of an enclosed space by means of variations in the pressure or volume of air contained in a bulb placed in the space.

airtight [ENG] Not permitting the passage of air. Also known as airproof.

air-to-air missile [ORD] A missile launched from an aircraft at an airborne target. Abbreviated AAM.

air-to-ground missile [ORD] A missile launched from an aircraft at a ground target. Abbreviated AGM.

air-to-space missile [ORD] A missile launched from an aircraft at a space target, such as an earth satellite.

air-to-surface missile [ORD] A missile launched from aircraft at a ground or sea target. Abbreviated ASM.

air-to-underwater missile [ORD] A missile launched from an aircraft at an underwater target.

air-track drill [MIN ENG] A heavy drilling machine for quarry or opencast blasting, equipped with caterpillar tracks and operated by independent air motors.

air traffic [NAV] Aircraft operating in the air or on an airport surface.

air-traffic area [NAV] The airspace within a circular limit defined by a 5-statute-mile horizontal radius from the geographical center of an airport at which an operative airport traffic control tower is located, and an upward extent to 2000 feet above the surface.

air-traffic control [NAV] 1. A service which promotes the safe and fast movement of aircraft operating in the air or on an airport surface by providing rules, procedures, and information and advisory services for pilots. Abbreviated ATC. 2. A system comprising enabling legislation, operating procedures, and navigation and communication equipment which is intended to make for the safe and expeditious movement of aircraft from the time that they leave the departure gates to arrival at the terminal gates.

air-traffic control center [NAV] A place for receiving information regarding aircraft movement, interpreting this information, and issuing instructions to aircraft to promote safe, orderly, expeditious flow of traffic; it directs traffic along controlled airspace.

air-traffic controller [NAV] A person responsible for controlling air traffic. Also known as air traffic control officer (British usage).

air-traffic control radar beacon system [NAV] A system adopted by the Federal Aviation Agency for use in controlling air traffic over the United States; the aircraft carry identification transponders designed to transmit an airplane identity code, altitude, and additional message when interrogated by an air-traffic controller's equipment. Abbreviated ATCRBS.

air-traffic control zone *See* control zone.

air transport [MIN ENG] Movement from one place to another of the filling material in a mine through pneumatic pipelines. Also known as air transportation.

air transportation [AERO ENG] The use of aircraft, predominantly airplanes, to move passengers and cargo. [MIN ENG] *See* air transport.

air trap [CIV ENG] A device that prevents the escape of foul air or gas from such systems as drains and sewers. [ENG] *See* air pocket.

air-tube breathing apparatus [ENG] A device consisting of a smoke helmet, mask, or mouthpiece supplied with fresh air by means of a flexible tube. Also known as air-supply mask.

air turbulence [METEOROL] Highly irregular atmospheric motion characterized by rapid changes in wind speed and direction and by the presence, usually, of up and down currents.

air valve [MECH ENG] A valve that automatically lets air out of or into a liquid-carrying pipe when the internal pressure drops below atmospheric.

air vane [AERO ENG] A vane that acts in the air, as contrasted to a jet vane which acts within a jetstream.

air vehicle *See* aircraft.

air-velocity measurement [FL MECH] The measurement of the rate of displacement of air or gas at a specific location, as when ascertaining wind speed or airspeed of an aircraft.

air volcano [GEOL] An eruptive opening in the earth from

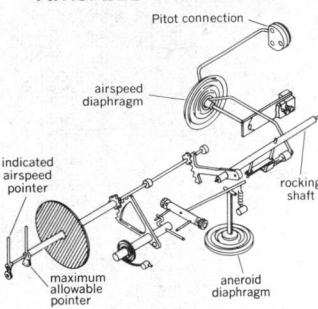

AIRSPEED INDICATOR

Drawing of moving parts of indicated airspeed meter.

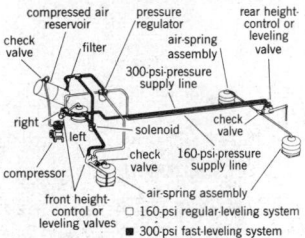

AIR-SUSPENSION SYSTEM

Basic components of an air suspension system in an automobile.

AIR-TO-GROUND MISSILE

Air-to-ground missile, Hound Dog, launched from B-52 strategic bomber. *(Official U.S. Air Force photograph)*

which large volumes of gas emanate, in addition to mud and stones; a variety of mud volcano.

air washer [MECH ENG] **1.** A device for cooling and cleaning air in which the entering warm, moist air is cooled below its dew point by refrigerated water so that although the air leaves close to saturation with water, it has less moisture per unit volume than when it entered. **2.** Apparatus to wash particulates and soluble impurities from air by passing the airstream through a liquid bath or spray.

air-water vapor mixture [PHYS] A mixture of dry air and water vapor, such as the atmosphere.

airwave [ELECTR] A radio wave used in radio and television broadcasting. [METEOROL] A wavelike oscillation in the pattern of wind flow aloft, usually with reference to the stronger portion of the westerly current.

airway [MIN ENG] A passage for air in a mine. Also known as air course. [NAV] A designated route of passage for aircraft.

airway beacon [NAV] A revolving light indicating the course of an airway.

airways code See U.S. airways code.

airways forecast See aviation weather forecast.

airways observation See aviation weather observation.

air well See air shaft.

Airy differential equation [MATH] The differential equation $(d^2f/dz^2) - zf = 0$, where z is the independent variable and f is the value of the function; used in studying the diffraction of light near caustic surface.

Airy disk [OPTICS] The bright, diffuse central spot of light formed by an optical system imaging a point source of light.

Airy function [MATH] Either of the solutions of the Airy differential equation.

Airy isostasy [GEOPHYS] A theory of hydrostatic equilibrium of the earth's surface which contends that mountains are floating on a fluid lava of higher density, and that higher mountains have a greater mass and deeper roots.

Airy phase [ACOUS] An acoustic wave formed by an explosion in shallow water over a flat bottom.

Airy points [ENG] The points at which a horizontal rod is optionally supported to avoid its bending.

aisle [ARCH] **1.** A passageway between or alongside blocks of seats, as in an auditorium. **2.** One of the parts of a basilica which are located at the sides of the nave, with each aisle separated from it by a row of columns.

Aistopoda [PALEON] An order of Upper Carboniferous amphibians in the subclass Lepospondyli characterized by reduced or absent limbs and an elongate, snakelike body.

Aitken dust counter [ENG] An instrument for determining the dust content of the atmosphere. Also known as Aitken nucleus counter.

Aitken nuclei [METEOROL] The microscopic particles in the atmosphere which serve as condensation nuclei for droplet growth during the rapid adiabatic expansion produced by an Aitken dust counter.

Aitken nucleus counter See Aitken dust counter.

Aitken's formula [ASTRON] The expression used to determine the separation limit for true binary stars: $\log p'' = 2.5 - 0.2m$, where p'' = limit, m = magnitude.

Aitoff equal-area map projection [MAP] A Lambert equal-area azimuthal projection of a hemisphere converted into a map projection of the entire sphere by a manipulation suggested by Aitoff.

Aizoaceae [BOT] A family of flowering plants in the order Caryophyllales; members are unarmed leaf-succulents, chiefly of Africa.

Ajax metal [MET] A bronze of the Ajax Metal Company, used for bearings.

Ajax powder [MATER] A high-strength, high-density, gelatinous permitted explosive having good water resistance; contains nitroglycerine, potassium perchlorate, ammonium oxalate, and wood flour.

ajmaline [ORG CHEM] $C_{20}H_{26}N_2O_2$ An amber, crystalline alkaloid obtained from *Rauwolfia* plants, especially *R. serpentina*.

AJO breathing apparatus [MIN ENG] A breathing device consisting of a Siebe-Gorman mining gas mask with a small oxygen cylinder and a canister which neutralizes mining gases, such as carbon monoxide, sulfureted hydrogen, and nitrous fumes.

ajowan oil [MATER] A yellow essential oil distilled from seeds of the herbaceous plant *Carum copticum* (*Ptychotis ajowan*) and used in pharmaceuticals. Also known as ptychotis oil.

akaganeite [MINERAL] β-FeO(OH) A mineral found in meteorites and considered to be formed in flight or by alteration.

akaryote [CYTOL] A cell that lacks a nucleus.

akenobeite [PETR] A form of aplite composed of orthoclase and oligoclase with quartz in the interstices.

akerite [PETR] A rock composed of quartz syenite containing soda microcline, oligoclase, and augite.

akermanite [MINERAL] $Ca_2MgSi_2O_7$ Anhydrous calcium-magnesium silicate found in igneous rocks; a melilite.

AKF diagram [PETR] A triangular diagram showing the chemical character of a metamorphic rock in which the three components plotted are $A = Al_2O_3 + Fe_2O_3 + (CaO + Na_2O)$, $K = K_2O$, and $F = FeO + MgO + MnO$.

akimbo span [ANTHRO] The distance measured between elbow points when the subject stands with arms flexed in a horizontal plane, with the wrists straight, palms down, fingers straight and together, and the thumbs touching the chest; or the distance measured with upper arms horizontal, forearms vertical.

akinesia [MED] **1.** Loss of or impaired motor function. **2.** Immobility from any cause.

akinete [BOT] A thick-walled resting cell of unicellular and filamentous green algae.

Akins' classifier [MIN ENG] A device for separating fine-size solids from coarser solids in a wet pulp; consists of an interrupted-flight screw conveyor, operating in an inclined trough.

akrochordite [MINERAL] $Mn_4Mg(AsO_4)_2(OH)_4\cdot4H_2O$ Mineral consisting of a hydrous basic manganese magnesium arsenate and occurring in reddish-brown rounded aggregates; hardness is 3 on Mohs scale, and specific gravity is 3.2.

aktological [GEOL] Nearshore shallow-water areas, conditions, sediments, or life.

akureyri disease See epidemic neuromyasthenia.

Al See aluminum.

ala [BIOL] A wing or winglike structure.

alabandite [MINERAL] MnS A complex sulfide mineral that is a component of meteorites and usually occurs in iron-black massive or granular form. Also known as manganblende.

alabaster [MINERAL] **1.** $CaSO_4\cdot2H_2O$ A fine-grained, colorless gypsum. **2.** See onyx marble.

alabaster glass [MATER] A glass that contains small inclusions of different diffractive indexes and shows no color reaction to light.

alalia [MED] Loss of speech.

alamosite [MINERAL] $PbSiO_3$ A white or colorless monoclinic mineral consisting of lead silicate and occurring in radiating fibers; hardness is 4.5 on Mohs scale, and specific gravity is 6.5.

alang-alang See cogon.

alanine [BIOCHEM] $C_3H_7NO_2$ A white, crystalline, nonessential amino acid of the pyruvic acid family.

alanyl [ORG CHEM] The radical CH_3CHNH_2CO-; occurs in, for example, alanyl alanine, a dipeptide.

alar [BIOL] Winglike or pertaining to a wing.

alarm gage [ENG] A device that actuates a signal either when the steam pressure in a boiler is too high or when the water level in a boiler is too low.

alarm reaction [BIOL] The sum of all nonspecific phenomena which are elicited by sudden exposure to stimuli, which affect large portions of the body, and to which the organism is quantitatively or qualitatively not adapted.

alarm signal [ELECTR] The international radiotelegraph alarm signal transmitted to actuate automatic devices that sound an alarm indicating that a distress message is about to be broadcast.

alarm song [INV ZOO] A stress signal occurring in many families of beetles.

alarm system [ENG] A system which operates a warning device after the occurrence of a dangerous or undesirable condition.

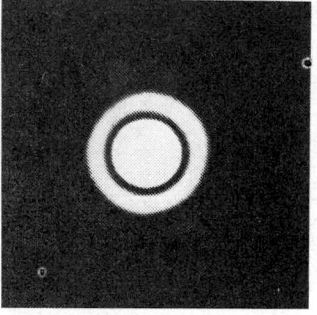

AIR WASHER

chilled water supply / droplet eliminator / entering / leaving / warm, moist air / cool, dry air / return water to chiller

Schematic of air washer.

AIRY DISK

Picture of an Airy disk, one type of diffraction pattern. (R. W. Ditchburn)

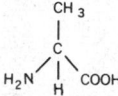

ALANINE

$$\begin{array}{c} CH_3 \\ | \\ C \\ / \ \backslash \\ H_2N \ \ \ \ COOH \\ H \end{array}$$

Structural formula of the amino acid alanine.

alarm valve [ENG] A device that sounds an alarm when water flows in an automatic sprinkler system.

Alaska Current [OCEANOGR] A current that flows northwestward and westward along the coasts of Canada and Alaska to the Aleutian Islands.

Alaska Integrated Communications Exchange [ELECTR] A network of radio stations, generally using scatter-propagation equipment, that links early-warning radar stations. Also known as Alice; White Alice.

alaskite *See* aplite.

Alaudidae [VERT ZOO] The larks, a family of Oscine birds in the order Passeriformes.

Albada finder [OPTICS] A viewfinder used with a camera held at eye level; the field of view is enclosed by a white frame that is made to appear very distant by reflection from the rear surface of the objective lens.

albafite [MINERAL] Greenish to brownish bitumen which becomes white when exposed to air; contains up to 15% oxygen; fusible; insoluble in organic solvents; varies from soft to hard, porous to compact; atomic ratio H/C 1.75–2.25.

Albamycin [MICROBIO] A trade name for the antibiotic novobiocin.

albanite [PETR] A melanocratic leucitite found near Rome, Italy.

albatross [VERT ZOO] Any of the large, long-winged oceanic birds composing the family Diomedeidae of the order Procellariformes.

albedo [NUCLEO] The reflection factor a surface, such as paraffin, has for neutrons. [OPTICS] That fraction of the total light incident on a reflecting surface, especially a celestial body, which is reflected back in all directions.

albedometer [ENG] An instrument used for the measurement of the reflecting power, that is, the albedo, of a surface.

albedo neutrons *See* albedo particles.

albedo particles [GEOPHYS] Neutrons or other particles, such as electrons or protons, which leave the earth's atmosphere, having been produced by nuclear interactions of energetic particles within the atmosphere. Also known as albedo neutrons.

Alberger process [CHEM ENG] A method of manufacturing salt by heating brine at high pressure and passing it to a graveler which removes calcium sulfate; the salt crystallizes as the pressure is reduced and thus is separated from the brine.

Albers projection [MAP] An equal-area projection of the conical type, on which the meridians are straight lines that meet in a common point beyond the limits of the map, and the parallels are concentric circles whose center is at the point of intersection of the meridians.

Alberta low [METEOROL] A low centered on the eastern slope of the Canadian Rockies in the province of Alberta, Canada.

albertite [MINERAL] Jet-black, brittle natural hydrocarbon with conchoidal fracture, hardness of 1–2, and specific gravity of approximately 1.1. Also known as asphaltite coal.

albertype *See* photogelatin printing plate.

Albian [GEOL] Uppermost Lower Cretaceous geologic time.

albinism [MED] A hereditary, metabolic disorder transmitted as an autosomal recessive and characterized by the inability to form melanin in the skin, hair, and eyes due to tyrosinase deficiency.

Albionian [GEOL] Lower Silurian geologic time.

albite [MINERAL] NaAlSi₃O₈ $NaAlSi_3O_8$ A colorless or milky-white variety of plagioclase of the feldspar group found in granite and various igneous and metamorphic rocks. Also known as sodaclase; sodium feldspar; white feldspar; white schorl.

albite-epidote-amphibolite facies [PETR] Rocks of metamorphic type formed under intermediate temperature and pressure conditions by regional metamorphism or in the outer contact metamorphic zone.

albite law [CRYSTAL] A rule specifying the orientation of alternating lamellae in multiple twin feldspar crystals; the twinning plane is brachypinacoid and is common in albite.

albitite [PETR] A porphyritic dike rock that is coarse-grained and composed almost wholly of albite; common accessory minerals are muscovite, garnet, apatite, quartz, and opaque oxides.

albitization [PETR] The formation of albite in a rock as a secondary mineral.

albitophyre [PETR] A porphyritic rock that contains albite phenocrysts in a groundmass composed mostly of albite.

albolite [MATER] A plastic cement composed principally of magnesia and silica.

albomycin [MICROBIO] An antibiotic produced by *Actinomyces subtropicus*; effective against penicillin-resistant pneumococci and staphylococci.

alboranite [PETR] Olivine-free hypersthene basalt.

albuginea [HISTOL] A layer of white, fibrous connective tissue investing an organ or other body part.

albumen [CYTOL] The white of an egg, composed principally of albumin.

albumin [BIOCHEM] Any of a group of plant and animal proteins which are soluble in water, dilute salt solutions, and 50% saturated ammonium sulfate.

albumin-globulin ratio [BIOCHEM] The ratio of the concentrations of albumin to globulin in blood serum.

albumin glue [MATER] A bonding agent composed of soluble dried blood with minor additives and giving strong, durable bonds when coagulated in plywood joints at temperatures of 160–180° F (71–82°C).

albuminoid [BIOL] Having the characteristics of albumin.

albuminuria [MED] The presence of albumin in the urine; usually indicating renal disease.

albumose [BIOCHEM] A protein derivative formed by the action of a hydrolytic enzyme, such as pepsin.

alburnum *See* sapwood.

Alcaligenes [MICROBIO] A genus of motile and nonmotile, gram-negative, rod-shaped bacteria of the family Achromobacteraceae, found generally in the intestinal tract of vertebrates.

Alcaligenes bookeri [MICROBIO] A motile, gram-negative, rod-shaped bacterium associated with diarrhea in human infants.

Alcaligenes faecalis [MICROBIO] A gram-negative, rod-shaped, peritrichously flagellated bacterium, the type species of the genus, in the family Achromobacteraceae.

alcaptonuria *See* alkaptonuria.

Alcedinidae [VERT ZOO] The kingfishers, a worldwide family of colorful birds in the order Coraciiformes; characterized by large heads, short necks, and heavy, pointed bills.

alchemy [CHEM] A speculative chemical system having as its central aims the transmutation of base metals to gold and the discovery of the philosopher's stone.

Alcidae [VERT ZOO] A family of shorebirds, predominantly of northern coasts, in the order Charadriiformes, including auks, puffins, murres, and guillemots.

Alciopidae [INV ZOO] A pelagic family of errantian annelid worms in the class Polychaeta.

Alclad [MET] A composite metal in sheet form, of the Aluminum Company of America, having a face usually of relatively pure aluminum and a base of higher-strength aluminum alloy. Also known as clad.

alcogel [CHEM] A gel formed by an alcosol.

alcohol [ORG CHEM] **1.** C₂H₅OH C_2H_5OH A colorless, volatile liquid; boiling point of pure liquid is 78.3°C; it is soluble in water, chloroform, and methyl alcohol; used as solvent and in manufacture of many chemicals and medicines. Also known as ethanol; ethyl alcohol. **2.** Any of a class of organic compounds containing the hydroxyl group, OH.

alcoholate [ORG CHEM] A compound formed by the reaction of an alcohol with an alkali metal. Also known as alkoxide.

alcohol C-9 *See* n-nonyl alcohol.

alcohol C-12 *See* lauryl alcohol.

alcohol dehydrogenase [BIOCHEM] The enzyme that catalyzes the oxidation of ethanol to acetaldehyde.

alcohol fuel [MATER] A motor fuel of gasoline blended with 5–25% of anhydrous ethyl alcohol; used particularly in Europe.

alcoholic [MED] An individual who consumes excess amounts of alcoholic beverages to the extent of being addicted, habituated, or dependent.

alcoholic beverage [FOOD ENG] A potable preparation containing ethyl alcohol.

alcoholic fermentation [MICROBIO] The process by which

ALBATROSS

The laysan albatross, with the characteristic hooked bill and long, tubular nostrils of oceanic birds.

ALBITE

Crystals of albite from Amelia Court House, Virginia.
(Specimen from Department of Geology, Bryn Mawr College)

certain yeasts decompose sugars in the absence of oxygen to form alcohol and carbon dioxide; method for production of ethanol, wine, and beer.

alcoholimeter *See* alcoholometer.

alcoholism [MED] Compulsive consumption of and dependence on alcoholic beverages, usually leading to pathology of the digestive and nervous systems.

alcoholmeter *See* alcoholometer.

alcoholometer [ENG] A device, such as a form of hydrometer, that measures the quantity of an alcohol contained in a liquid. Also known as alcoholimeter; alcoholmeter.

alcoholysis [ORG CHEM] The breaking of a carbon-to-carbon bond by addition of an alcohol.

Alcor [ASTRON] The star 80 Ursae Majoris.

alcosol [CHEM] Mixture of an alcohol and a colloid.

alcove [ARCH] **1.** A recessed part of a room. **2.** A small room that opens into a larger one. **3.** An arched opening in a wall. [GEOL] A large niche formed by a stream in a face of horizontal strata.

alcove lands [GEOL] Terrain where the mud rocks or sandy clays and shales that compose the hills (badlands) are interstratified by occasional harder beds; the slopes are terraced.

Alcumite [MET] A corrosion-resistant copper alloy, of the Duriron Company, containing aluminum, iron, and either manganese or nickel.

Alcyonacea [INV ZOO] The soft corals, an order of littoral anthozoans of the subclass Alcyonaria.

Alcyonaria [INV ZOO] A subclass of the Anthozoa; members are colonial coelenterates, most of which are sedentary and littoral.

Aldebaran [ASTRON] A red giant star of visual magnitude 1.06, spectral classification K5-III, in the constellation Taurus; the star α Tauri.

aldehyde [ORG CHEM] One of a class of organic compounds containing the CHO radical.

aldehyde C-12 lauric *See* lauryl aldehyde.

aldehyde lyase [BIOCHEM] Any enzyme that catalyzes the nonhydrolytic cleavage of an aldehyde.

aldehyde polymer [ORG CHEM] Any of the plastics based on aldehydes, such as formaldehyde, acetaldehyde, butyraldehyde, or acrylic aldehyde (acrolein).

alder [BOT] The common name for several trees of the genus Alnus.

Aldis signaling lamp [COMMUN] A hand-carried signaling lamp used to transmit messages from ships and aircraft.

aldohexose [ORG CHEM] A hexose, such as glucose or mannose, containing the aldehyde group.

aldol [ORG CHEM] $CH_3CHOHCH_2CHO$ A colorless liquid, a condensation product of acetaldehyde; one of a class of condensation products of aldehydes. Also known as 3-hydroxybutanal.

aldolase [BIOCHEM] An enzyme in anaerobic glycolysis that catalyzes the cleavage of fructose 1,6-diphosphate to glyceraldehyde 3-phosphate; used also in the reverse reaction.

aldol condensation [ORG CHEM] Aldol formation by polymerization of an aldehyde in the presence of a dilute acid or alkali.

aldose [ORG CHEM] A class of monosaccharide sugars; the molecule contains an aldehyde group.

aldosterone [BIOCHEM] $C_{21}H_{28}O_5$ A steroid hormone extracted from the adrenal cortex that functions chiefly in regulating sodium and potassium metabolism.

aldosteronism [MED] Hypertension induced by excessive secretion of aldosterone.

Aldrich syndrome [MED] A recessive, sex-linked disease characterized by a complex of symptoms, including eczematoid dermatitis, thrombocytopenia, black stool, and a deficiency of immune globulins.

aldrin [ORG CHEM] $C_{12}H_8Cl_6$ A water-insoluble, white crystalline compound, consisting mainly of chlorinated dimethanonaphthalene; used as a pesticide.

ale [FOOD ENG] A fermented malt beverage, differing from beer in containing up to 8% alcohol by volume and being hopped more heavily.

alecithal [CYTOL] Referring to an egg without yolk, such as the eggs of placental mammals.

alee basin [GEOL] A basin formed in the deep sea by turbidi-ty currents aggrading courses where the currents were deflected around a submarine ridge.

aleishtite [GEOL] A bluish or greenish mixture of dickite and other clay minerals.

Alembert *See* d'Alembert entries.

aleph null [MATH] The cardinal number of any set which can be put in one-to-one correspondence with the set of positive integers.

Alepocephaloidei [VERT ZOO] The slickheads, a suborder of deap-sea teleostean fishes of the order Salmoniformes.

alert [ORD] A state of readiness against impending danger or for going into action.

alerting signal [COMMUN] Specific signal that is applied to subscriber access lines to indicate an incoming call.

aletophyte [ECOL] A weedy plant growing on the roadside or in fields where natural vegetation has been disrupted by man.

aleukemia [MED] Leukemia in which the white blood cell count is normal or low.

aleuriospore [MYCOL] A simple terminal or lateral, thick-walled, nondeciduous spore produced by some fungi of the order Moniliales.

aleuron [BOT] Protein in the form of grains stored in the embryo, endosperm, or perisperm of many seeds.

Aleutian Current [OCEANOGR] A current setting southwestward along the southern coasts of the Aleutian Islands.

Aleutian disease [VET MED] A disease of mink characterized by accumulations of plasma cells ih several organs, hyaline changes in the walls of small arteries, and interstitial fibrosis of the kidneys.

Aleutian low [METEOROL] The low-pressure center located near the Aleutian Islands on mean charts of sea-level pressure; represents one of the main centers of action in the atmospheric circulation of the Northern Hemisphere.

alewife [VERT ZOO] *Pomolobus pseudoharengus.* A food fish of the herring family that is very abundant on the Atlantic coast.

Alexandrian [GEOL] Lower Silurian geologic time.

alexandrite [MINERAL] A gem variety of chrysoberyl; emerald green in natural light but red in transmitted or artificial light.

alexia [MED] Loss of the ability to read.

Alexinic unit [BIOL] A unit for the standardization of blood serum.

Aleyrodidae [INV ZOO] The whiteflies, a family of homopteran insects included in the series Sternorrhyncha; economically important as plant pests.

alfalfa [BOT] *Medicago sativa.* A herbaceous perennial legume in the order Rosales, characterized by a deep taproot. Also known as lucerne.

alfa process [FOOD ENG] A German process for butter manufacture in which high-fat cream (78%) is run through a three-cylinder cooler, inside of which cooled spiral-ribbed rollers turn.

alfin catalyst [ORG CHEM] A catalyst derived from reaction of an alkali alcoholate with an olefin halide; used to convert olefins (for example, ethylene, propylene, or butylenes) into polyolefin polymers.

Alfisol [GEOL] An order of soils with gray to brown surface horizons, a medium-to-high base supply, and horizons of clay accumulation.

Alford loop [ELECTROMAG] An antenna utilizing multielements which usually are contained in the same horizontal plane and adjusted so that the antenna has approximately equal and in-phase currents uniformly distributed along each of its peripheral elements and produces a substantially circular radiation pattern in the plane of polarization; it is known for its purity of polarization.

Alfvén number [PHYS] The ratio of the speed of the Alfvén wave to the speed of the fluid at a point in the fluid.

Alfvén speed [PHYS] The speed of motion of the Alfvén wave, which is $v_a = B_0/\sqrt{\rho\mu}$, where B_0 is the magnetic field strength, ρ the fluid density, and μ the magnetic permeability (in meter-kilogram-second units).

Alfvén wave [PHYS] A hydromagnetic shear wave which moves along magnetic field lines; a major accelerative mechanism of charged particles in plasma physics and astrophysics.

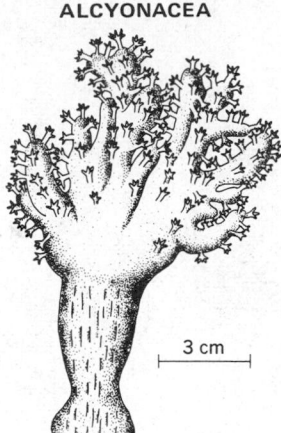

ALCYONACEA

3 cm

Representative alcyonacean *Alcyonium palmatum (after Y. Delage).*

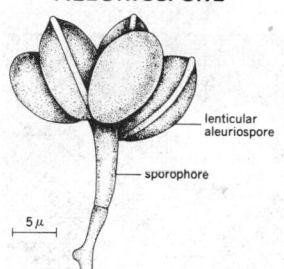

ALEURIOSPORE

lenticular aleuriospore

sporophore

5 μ

Papularia sphaerosperma (Coniosporium arundinis) with dark, aleuriospores. *(After G. Goidanich, 1938)*

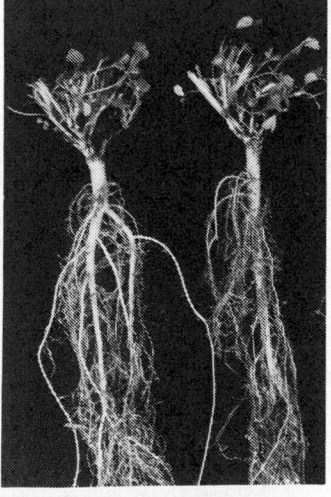

ALFALFA

Buffalo alfalfa, showing crown and root in 8-month-old plants. *(Iowa State University Photo Service)*

algae [BOT] General name for the chlorophyll-bearing organisms in the plant subkingdom Thallobionta.

algae bloom [ECOL] A heavy growth of algae in and on a body of water as a result of high phosphate concentration from farm fertilizers and detergents.

algal [BOT] Of or pertaining to algae. [GEOL] Formed from or by algae.

algal biscuit [GEOL] A disk-shaped or spherical mass, up to 20 centimeters in diameter, made up of carbonate that is probably the result of precipitation by algae.

algal limestone [PETR] A type of limestone either formed from the remains of calcium-secreting algae or formed when algae bind together the fragments of other lime-secreting organisms.

algal pit [GEOL] An ablation depression that is small and contains algae.

algal reef [GEOL] An organic reef which has been formed largely of algal remains and in which algae are or were the main lime-secreting organisms.

algal ridge [GEOL] Elevated margin of a windward coral reef built by actively growing calcareous algae.

algal rim [GEOL] Low rim built by actively growing calcareous algae on the lagoonal side of a leeward reef or on the windward side of a patch reef in a lagoon.

algal structure [GEOL] A deposit, most frequently calcareous, with banding, irregular concentric structures, crusts, and pseudopisolites or pseudoconcretionary forms resulting from organic, colonial secretion and precipitation.

algebra [MATH] 1. A method of solving practical problems by using symbols, usually letters, for unknown quantities. 2. The study of the formal manipulations of equations involving symbols and numbers. 3. An abstract mathematical system consisting of a vector space together with a multiplication by which two vectors may be combined to yield a third, and some axioms relating this multiplication to vector addition and scalar multiplication. Also known as hypercomplex system.

algebraic addition [MATH] The addition of algebraic quantities in the sense that adding a negative quantity is the same as subtracting a positive one.

algebraic closure of a field [MATH] An algebraic extension field which has no algebraic extensions but itself.

algebraic curve [MATH] 1. The set of points in the plane satisfying a polynomial equation in two variables. 2. More generally, the set of points in n-space satisfying a polynomial equation in n variables.

algebraic equation [MATH] An equation in which zero is set equal to an algebraic expression.

algebraic expression [MATH] An expression which is obtained by performing a finite number of the following operations on symbols representing numbers: addition, subtraction, multiplication, division, raising to a power.

algebraic extension of a field [MATH] A field which contains both the given field and all roots of polynomials with coefficients in the given field.

algebraic function [MATH] A function whose value is obtained by performing only the following operations to its argument: addition, subtraction, multiplication, division, raising to a rational power.

algebraic geometry [MATH] The study of geometric properties of figures using methods of abstract algebra.

algebraic number [MATH] Any root of a polynomial with rational coefficients.

algebraic number theory [MATH] The study of properties of real numbers, especially integers, using the methods of abstract algebra.

algebraic sum [MATH] The result of the addition of two or more quantities, with the addition of a negative quantity equivalent to subtraction of the corresponding positive quantity.

algebraic topology [MATH] The study of topological properties of figures using the methods of abstract algebra; includes homotopy theory, homology theory, and cohomology theory.

alged malaria See falciparum malaria.

Algenib [ASTRON] A star in the constellation Pegasus.

Algerian onyx See onyx marble.

algesia [PHYSIO] Sensitivity to pain.

algesimeter [PHYSIO] A device used to determine pain thresholds.

algesiroreceptor [PHYSIO] A pain-sensitive cutaneous sense organ.

algicide [MATER] A chemical used to kill algae.

algin See alginic acid.

alginate [ORG CHEM] One of a class of salts of algin, such as sodium alginate.

algin fibers [ORG CHEM] A polymer of d-mannuronic acid.

alginic acid [ORG CHEM] $(C_6H_8O_6)_n$ An insoluble colloidal acid obtained from brown marine algae; it is hard when dry and absorbent when moist. Also known as algin.

alginite See algite.

Alginobacter [MICROBIO] A genus of gram-negative, peritrichously flagellated, rod-shaped soil bacteria in the family Enterobacteriaceae that ferment alginic acid and glucose, with the production of acid and gas.

Alginomonas [MICROBIO] A genus of gram-negative, nonsporeforming, rod-shaped bacteria in the family Pseudomonadaceae; species decompose alginic acid.

algite [PETR] The petrological unit that constitutes algal material present in considerable amounts in algal or boghead coal. Also known as alginite.

Algol [ASTRON] An eclipsing variable star of spectral classification B8 in the constellation Perseus; the star β Persei. Also known as Demon Star.

ALGOL [ADP] An algorithmic and procedure-oriented computer language used principally in the programming of scientific problems.

algology [BOT] The study of algae.

Algoman orogeny [GEOL] Orogenic episode affecting Archean rocks of Canada about 2.4 billion years ago. Also known as Kenoran orogeny.

algometer [MED] An instrument for measuring pressure stimuli which produce pain.

Algonkian [GEOL] Geologic time between the Archean and Paleozoic. Also known as Proterozoic.

algophobia [PSYCH] Abnormal fear of pain.

algorithm [MATH] A set of well-defined rules for the solution of a problem in a finite number of steps.

algorithmic language [ADP] A language in which a procedure or scheme of calculations can be expressed accurately.

algorithm translation [ADP] A step-by-step computerized method of translating one programming language into another programming language.

algor mortis [PATH] Postmortem cooling of the body.

alias [ADP] An alternative entry point in a computer subroutine at which its execution may begin, if so instructed by another routine.

aliasing [MATH] Introduction of error into the computed amplitudes of the lower frequencies in a Fourier analysis of a function carried out using discrete time samplings whose interval does not allow the proper analysis of the higher frequencies present in the analyzed function.

Alice See Adiabatic Low-energy Injection and Capture Experiment; Alaska Integrated Communications Exchange.

alicyclic [ORG CHEM] 1. Having the properties of both aliphatic and cyclic substances. 2. Referring to a class of organic compounds containing only carbon and hydrogen atoms joined to form one or more rings.

alidade [ENG] 1. An instrument for topographic surveying and mapping by the plane-table method. 2. Any sighting device employed for angular measurement.

alignment [ELECTR] The process of adjusting components of a system for proper interrelationship, including the adjustment of tuned circuits for proper frequency response and the time synchronization of the components of a system. [ENG] Placing of surveying points along a straight line. [MAP] Representing of the correct direction, character, and relationships of a line or feature on a map. [MIN ENG] The act of laying out a tunnel or regulating by line; adjusting to a line. [NUC PHYS] A population p^m of the $2I + 1$ orientational substates of a nucleus; $m = -I$ to $+I$, such that $p(m) = p(-m)$.

alignment chart See nomograph.

alignment correction [ENG] A correction applied to the measured length of a line to allow for not holding the tape exactly in a vertical plane of the line.

ALIDADE

A surveying alidade, having a telescope with attached graduated vertical circle, mounted on a flat base that can be moved about the table. *(Kern Instruments, Inc.)*

alignment pin [DES ENG] Pin in the center of the base of an octal, loctal, or other tube having a single vertical projecting rib that aids in correctly inserting the tube in its socket.

alignment wire *See* ground wire.

alimentary [BIOL] Of or relating to food, nutrition, or diet.

alimentary canal [ANAT] The tube through which food passes; in man, includes the mouth, pharynx, esophagus, stomach, and intestine.

alimentation [BIOL] Providing nourishment by feeding.

Alioth [ASTRON] Traditional name for a second-magnitude star in the Big Dipper; the star ε Ursae Majoris.

aliphatic [ORG CHEM] Of or pertaining to any organic compound of hydrogen and carbon characterized by a straight chain of the carbon atoms; three subgroups of such compounds are alkanes, alkenes, and alkynes.

aliphatic acid [ORG CHEM] Any organic acid derived from aliphatic hydrocarbons.

aliphatic acid ester [ORG CHEM] Any organic ester derived from aliphatic acids.

aliphatic polycyclic hydrocarbon [ORG CHEM] A hydrocarbon compound in which at least two of the aliphatic structures are cyclic or closed.

aliphatic polyene compound [ORG CHEM] Any unsaturated aliphatic or alicyclic compound with more than four carbons in the chain and with at least two double bonds; for example, hexadiene.

aliphatic series [ORG CHEM] A series of open-chained carbon-hydrogen compounds; the two major classes are the series with saturated bonds and with the unsaturated.

Alismataceae [BOT] A family of flowering plants belonging to the order Alismatales characterized by schizogenous secretory cells, a horseshoe-shaped embryo, and one of two ovules.

Alismatales [BOT] A small order of flowering plants in the subclass Alismatidae, including aquatic and semiaquatic herbs.

Alismatidae [BOT] A relatively primitive subclass of aquatic or semiaquatic herbaceous flowering plants in the class Liliopsida, generally having apocarpous flowers, and trinucleate pollen and lacking endosperm.

alisphenoid [ANAT] **1.** The bone forming the greater wing of the sphenoid in adults. **2.** Of or pertaining to the sphenoid wing.

alite [MATER] A constituent of portland cement clinker consisting mostly of calcium silicate.

Aliva concrete sprayer [MIN ENG] A compressed-air machine for spraying concrete on the roof and sides of mine roadways.

alive [ELEC] *See* energized. [MIN ENG] That portion of a lode that is productive.

alivincular [INV ZOO] In some bivalves, having the long axis of the short ligament transverse to the hinge line.

alizarin [ORG CHEM] $C_{14}H_6O_2(OH)_2$ An orange crystalline compound, insoluble in cold water; made synthetically from anthraquinone; used in the manufacture of dyes and red pigments. Also known as 1,2-dihydroxyanthraquinone.

alizarin dye [ORG CHEM] Sodium salts of sulfonic acids derived from alizarin.

alizarin red [ORG CHEM] Any of several red dyes derived from anthraquinone.

alizarin yellow [MATER] A dye useful as an acid-base indicator; solutions change color from yellow (acid) to purple (basic) in the pH range 10.1 to 12.0. Also known as *para*-nitrophenylazosalicylate sodium.

alizarin yellow C *See* gallacetophenone.

alkalemia [MED] An increase in blood pH above normal levels.

alkalescence [CHEM] The property of a substance that is alkaline, that is, having a pH greater than 7.

alkali [CHEM] Any compound having highly basic qualities. [PETR] *See* alkalic.

alkali-aggregate reaction [CHEM] The chemical reaction of an aggregate with the alkali in a cement, resulting in a weakening of the concrete.

alkali alcoholate [ORG CHEM] A compound formed from an alcohol and an alkali metal base; the alkali metal replaces the hydrogen in the hydroxyl group.

alkali blue [ORG CHEM] The sodium salt of triphenylrosanilinesulfonic acid; used as an indicator.

alkalic Also known as alkali. [PETR] **1.** Of igneous rock, containing more than average alkali (K_2O and Na_2O) for that clan in which they are found. **2.** Of igneous rock, having feldspathoids or other minerals, such as acmite, so that the molecular ratio of alkali to silica is greater than 1:6. **3.** Of igneous rock, having a low alkali-lime index (51 or less).

alkali-calcic series [PETR] The series of igneous rocks with weight percentage of silica in the range 51–55, and weight percentages of CaO and $K_2O + Na_2O$ equal.

alkali cellulose [MATER] Product of wood pulp steeped with sodium hydroxide; first step in manufacture of viscose rayon and other cellulosics.

alkali chlorosis [PL PATH] Yellowing of plant foliage due to excess amounts of soluble salts in the soil.

alkali disease [MED] Selenium poisoning. [VET MED] **1.** Botulism of ducks. **2.** Trembles of cattle.

alkali emission [GEOPHYS] Light from free lithium, potassium, and especially sodium in the atmosphere, observed in twilight.

alkali feldspar [MINERAL] A feldspar composed of potassium feldspar and sodium feldspar, such as orthoclase, microcline, albite, and anorthoclase; all are considered alkali-rich.

alkali flat [GEOL] A level lakelike plain formed by the evaporation of water in a depression and deposition of its fine sediment and dissolved minerals.

alkali lake [HYD] A lake with large quantities of dissolved sodium and potassium carbonates as well as sodium chloride.

alkali-lime index [PETR] The percentage by weight of silica in a sequence of igneous rocks on a variation diagram where the weight percentages of CaO and of K_2O and Na_2O are equal.

alkali metal [CHEM] Any of the elements of group Ia in the periodic table: lithium, sodium, potassium, rubidium, cesium, and francium.

alkalimeter [ANALY CHEM] **1.** An apparatus for measuring the quantity of alkali in a solid or liquid. **2.** An apparatus for measuring the quantity of carbon dioxide formed in a reaction.

alkalimetry [ANALY CHEM] Quantitative measurement of the concentration of bases or the quantity of one free base in a solution; techniques include titration and other analytical methods.

alkaline [CHEM] **1.** Having properties of an alkali. **2.** Having a pH greater than 7.

alkaline cleaner [MET] An aqueous solution of an alkali used for metal cleaning.

alkaline earth [INORG CHEM] An oxide of an element of group IIa in the periodic table, such as barium, calcium, and strontium. Also known as alkaline-earth oxide.

alkaline-earth metals [CHEM] The heaviest members of group IIa in the periodic table; usually calcium, strontium, magnesium, and barium.

alkaline-earth oxide *See* alkaline earth.

alkaline phosphatase [BIOCHEM] A phosphatase active in alkaline media.

alkaline storage battery [ELEC] A storage battery in which the electrolyte consists of an alkaline solution, usually potassium hydroxide.

alkaline tide [PHYSIO] The temporary decrease in acidity of urine and body fluids after eating, attributed by some to the withdrawal of acid from the body due to gastric digestion.

alkalinity [CHEM] The property of having excess hydroxide ions in solution.

alkali-resisting paint [MATER] A paint, such as one made with a synthetic resin, that does not undergo saponification when used in such places as bathrooms or on such materials as new concretes.

alkaloid [ORG CHEM] One of a group of nitrogenous bases of plant origin, such as nicotine, cocaine, and morphine.

alkalometry [ANALY CHEM] The measurement of the quantity of alkaloids present in a substance.

alkalosis [MED] A condition of high blood alkalinity caused either by high intake of sodium bicarbonate or by loss of hydrochloric acid or blood carbon dioxide.

alkamine [ORG CHEM] A compound that has both the alcohol and amino groups. Also known as amino alcohol.

alkane [ORG CHEM] A member of a series of saturated aliphatic hydrocarbons having the empirical formula C_nH_{2n+2}.

alkannin [ORG CHEM] $C_{15}H_{14}O$ A red powder used as a color for fats and oils; insoluble in water; softens about 100°C.

Alkanol [ORG CHEM] A trademark for a group of fatty alcohols, ethylene oxide condensation products; they are used as surface-active agents in the wetting of paper, textiles, and leather.

alkanolamine [ORG CHEM] One of a group of viscous, water-soluble amino alcohols of the aliphatic series.

alkaptonuria [MED] A hereditary metabolic disorder transmitted as an autosomal recessive in man in which large amounts of homogentisic acid (alkapton) are excreted in the urine due to a deficiency of homogentisic acid oxidase. Also spelled alcaptonuria.

Alkar process [CHEM ENG] Catalytic alkylation of aromatic hydrocarbons with olefins to produce alkylaromatics; for example, production of ethylbenzene from benzene and ethylene.

alkarsine See cacodyl oxide.

alkene [ORG CHEM] One of a class of unsaturated aliphatic hydrocarbons containing one or more carbon-to-carbon double bonds.

alkoxide See alcoholate.

alkoxy [ORG CHEM] An alkyl radical attached to a molecule by oxygen, such as the ethoxy radical.

alkyd resin [ORG CHEM] A class of adhesive resins made from unsaturated acids and glycerol.

alkyl [ORG CHEM] A monovalent radical, C_nH_{2n+1}, which may be considered to be formed by loss of a hydrogen atom from an alkane; usually designated by R.

alkylamine [ORG CHEM] A compound consisting of an alkyl group attached to the nitrogen of an amine; an example is ethylamine, $C_2H_5NH_2$.

alkylaryl sulfonates [ORG CHEM] General name for alkylbenzene sulfonates.

alkylate [ORG CHEM] A product of the alkylation process in petroleum refining.

alkylate bottom [CHEM ENG] Residue from fractionation of total alkylate which boils at a higher temperature than aviation gasolines.

alkylation [CHEM ENG] A refinery process for chemically combining isoparaffin with olefin hydrocarbons. [ORG CHEM] A chemical process in which an alkyl radical is introduced into an organic compound by substitution or addition.

alkylbenzene sulfonates [ORG CHEM] Widely used nonbiodegradable detergents, commonly dodecylbenzene or tridecylbenzene sulfonates.

alkylene [ORG CHEM] An organic radical formed from an unsaturated aliphatic hydrocarbon; for example, the ethylene radical C_2H_3-.

alkyl halide [ORG CHEM] A compound consisting of an alkyl group and a halogen; an example is ethylbromide.

alkyne [ORG CHEM] One of a group of organic compounds containing a carbon-to-carbon triple bond.

allachesthesia [MED] A tactile sensation experienced remote from the point of stimulation but on the same side of the body.

allactite [MINERAL] $Mn_7(AsO_4)_2(OH)_8$ Brownish-red mineral consisting of a basic manganese arsenate.

allalinite [PETR] An altered gabbro with original texture and euhedral pseudomorphs.

allanite [MINERAL] $(Ca,Ce,La,Y)_2(Al,Fe)_3Si_3O_{12}(OH)$ Monoclinic mineral distinguished from all other members of the epidote group of silicates by a relatively high content of rare earths. Also known as bucklandite; cerine; orthite; treanorite.

Allan's metal [MET] A high-lead bronze, of A. Allan and Son, containing 40% lead, 55% copper, and 5% tin; used for bearings.

allantoic acid [BIOCHEM] $C_4H_8N_4O_4$ A crystalline acid obtained by hydrolysis of allantoin; intermediate product in nucleic acid metabolism.

allantoin [BIOCHEM] $C_4H_6N_4O_3$ A crystallizable oxidation product of uric acid found in allantoic and amniotic fluids and in fetal urine.

allantoinase [BIOCHEM] An enzyme, occurring in nonmammalian vertebrates, that catalyzes the hydrolysis of allantoin.

allantois [EMBRYO] A fluid-filled, saclike, extraembryonic membrane lying between the chorion and amnion of reptilian, bird, and mammalian embryos.

allantoxanic acid [BIOCHEM] $C_4H_3N_3O_4$ An acid formed by oxidation of uric acid or allantoin.

allanturic acid [BIOCHEM] $C_3H_4N_2O_3$ An acid formed principally by the oxidation of allantoin.

Allard's law [OPTICS] A mathematical formula defining the relationship between the intensity of a light, atmospheric conditions, and the amount of light received at any given distance.

all-around traverse [ORD] A turn in a complete circle in a horizontal plane; a weapon has this capability when it can be turned 360° by its traversing mechanism.

all-burnt time [AERO ENG] The point in time at which a rocket has consumed its propellants.

all-burnt velocity See burnout velocity.

all-diffused monolithic integrated circuit [ELECTR] Microcircuit consisting of a silicon substrate into which all of the circuit parts (both active and passive elements) are fabricated by diffusion and related processes.

Alleculidae [INV ZOO] The comb claw beetles, a family of mostly tropical coleopteran insects in the superfamily Tenebrionoidea.

Allee's principle [GEN] The concept of an intermediate optimal population density by which groups of organisms often flourish best if neither too few nor too many individuals are present.

alleghanyite [MINERAL] $Mn_5(SiO_4)_2(OH)_2$ A pink mineral consisting of basic manganese silicate.

Alleghenian [GEOL] Lower Middle Pennsylvanian geologic time.

Alleghenian life zone [ECOL] A biome that includes the eastern mixed coniferous and deciduous forests of New England.

Alleghenian orogeny [GEOL] Pennsylvanian and Early Permian orogenic episode which deformed the rocks of the Appalachian Valley and the Ridge and Plateau provinces.

allele [GEN] One of a pair of genes, or of multiple forms of a gene, located at the same locus of homologous chromosomes. Also known as allelomorph.

allelic mutant [GEN] A cell or organism with characters different from those of the parent due to alterations of one or more alleles.

allelomimetic behavior [PSYCH] Behavior in social animals in which each animal does the same thing as those nearby.

allelomorph See allele.

allelopathy [BOT] The harmful influence on a plant by another living plant that secretes a toxic substance.

allelotropism [BIOL] A mutual attraction between two cells or organisms.

allemontite [MINERAL] AsSb Rhombohedric, gray or reddish, native antimony aresenide occurring in reniform masses. Also known as arsenical antimony.

Allen-Doisy unit [BIOL] A unit for the standardization of estrogens.

allene [ORG CHEM] C_3H_4 An unsaturated aliphatic hydrocarbon with two double bonds. Also known as propadiene.

Allen-Moore diaphragm cell [CHEM ENG] A rectangular electrolyte diaphragm used early in the history of chlorine manufacture.

Allen screw [DES ENG] A screw or bolt which has an axial hexagonal socket in its head.

Allen wrench [DES ENG] A wrench made from a straight or bent hexagonal rod, used to turn an Allen screw.

allergen [IMMUNOL] Any antigen, such as pollen, a drug, or food, that induces an allergic state in humans or animals.

allergic arteritis [MED] Inflammation of the arterial walls resulting from an allergic state.

allergic dermatitis [MED] Inflammation of the skin following contact of an allergen with sensitized tissue.

allergic reaction See allergy.

allergic rhinitis See hay fever.

ALLANITE

2.5 cm

Allanite from Goiaz, Brazil.
(Specimen from Department of Geology, Bryn Mawr College)

allergic vasculitis syndrome [MED] A skin disease, possibly immunologic, characterized by ulcers which result from destructive inflammation of underlying blood vessels.

allergy [MED] A type of antigen-antibody reaction marked by an exaggerated physiologic response to a substance that causes no symptoms in nonsensitive individuals. Also known as allergic reaction.

allethrin [ORG CHEM] An insecticide, a synthetic pyrethroid, more effective than pyrethrin.

alliaceous [SCI TECH] Having a garlic- or onionlike smell.

alliance [SYST] A group of related families ranking between an order and a class.

allicin [ORG CHEM] $C_6H_{10}OS_2$ A liquid that decomposes when heated; soluble in alcohol and benzene, slightly soluble in water; found in onion oil; has some antibacterial activity. Also known as S-oxodiallyl sulfide.

alligator [VERT ZOO] Either of two species of archosaurian reptiles in the genus *Alligator* of the family Alligatoridae.

alligator clip [ELEC] A long, narrow spring clip with meshing jaws; used with test leads to make temporary connections quickly. Also known as crocodile clip.

alligator effect *See* orange peel.

Alligatorinae [VERT ZOO] A subgroup of the crocodilian family Crocodylidae that includes alligators, caimans, *Melanosuchus*, and *Paleosuchus*.

alligatoring [MATER] Cracking of a film of paint or varnish, with broad, deep cracks through one or more coats. Also known as crocodiling. [MET] **1.** A splitting of an end of a rolled steel slab in which the plane of the split is parallel to the rolled surface. Also known as fishmouthing. **2.** The roughening of a sheet-metal surface during forming due to the coarse grain of the metal used.

alligator shears *See* lever shears.

alligator squeezer [MET] A tool with a fixed upper jaw and a movable lower jaw used to squeeze a ball of iron produced by the paddling process into a bloom or billet.

alligator wrench [DES ENG] A wrench having fixed jaws forming a V, with teeth on one or both jaws.

Allihn condenser [CHEM ENG] A condenser whose condensing surface is a series of interconnected glass bulbs.

all-inertial guidance [NAV] **1.** The guidance of a vehicle entirely by use of inertial devices. **2.** A self-contained system which can automatically determine the position, velocity, and attitude of a moving vehicle for the purpose of directing its future course.

Allium [BOT] A genus of bulbous herbs in the family Liliaceae including leeks, onions, and chives.

allivalite [PETR] A form of gabbro composed of anorthite and olivine; accessories are augite, apatite, and opaque iron oxides.

allo- [CHEM] Prefix applied to the stabler form of two isomers.

allobar [NUC PHYS] A form of an element differing in its atomic weight from the naturally occurring form and hence being of different isotopic composition. [PHYS] A barometric pressure change.

allocation [IND ENG] The assigning of a portion of a resource to an activity.

allocheiria [MED] A form of allachesthesia in which the tactile sensation is experienced on the side opposite the one to which the stimulus was applied; seen in tabes dorsalis and other conditions.

allochem [GEOL] Sediment formed by chemical or biochemical precipitation within a depositional basin; includes intraclasts, oolites, fossils, and pellets.

allochemical metamorphism [PETR] Metamorphism accompanied by addition or removal of material so that the bulk chemical composition of the rock is changed.

allochetite [PETR] A porphyritic igneous rock composed of phenocrysts of labradorite, orthoclase, titanaugite, nepheline, magnetite, and apatite in a groundmass of augite, biotite, magnetite, hornblende, nepheline, and orthoclase.

allochoric [BOT] Describing a species that inhabits two or more closely related communities, such as forest and grassland, in the same region.

allochromatic crystal [CRYSTAL] A crystal having photoconductive properties due to the presence of small particles within it.

allochthon [GEOL] A rock that was transported a great distance from its original deposition by some tectonic process, generally related to overthrusting, recumbent folding, or gravity sliding.

allochthonous [PETR] Of rocks whose primary constituents have not been formed in situ.

allochthonous coal [GEOL] A type of coal arising from accumulations of plant debris moved from their place of growth and deposited elsewhere.

allochthonous stream [HYD] A stream flowing in a channel that it did not form.

Alloeocoela [INV ZOO] An order of platyhelminthic worms of the class Turbellaria distinguished by a simple pharynx and a diverticulated intestine.

allogene [GEOL] A mineral or rock that has been moved to the site of deposition. Also known as allothigene.

allogenic [ECOL] Caused by external factors, as in reference to the change in habitat of a natural community resulting from draught. [GEOL] *See* allothogenic.

allograft *See* homograft.

Allogromiidae [INV ZOO] A little-known family of protozoans in the order Foraminiferida; adults are characterized by a chitinous test.

Allogromiina [INV ZOO] A suborder of marine and freshwater protozoans in the order Foraminiferida characterized by an organic test of protein and acid mucopolysaccharide.

allogyric birefringence [OPTICS] The phenomenon in active optical media whereby circularly polarized light is transmitted unchanged but the velocity of right-handed circularly polarized light is different from that of left-handed.

allomerism [CRYSTAL] A constancy in crystal form in spite of a variation in chemical composition.

allometry [BIOL] **1.** The quantitative relation between a part and the whole or another part as the organism increases in size. Also known as heterauxesis; heterogony. **2.** The quantitative relation between the size of a part and the whole or another part, in a series of related organisms that differ in size.

allomorphism *See* paramorphism.

allomorphite [MINERAL] A mineral consisting of barite that is pseudomorphous after anhydrite.

allomorphosis [EVOL] Allometry in phylogenetic development.

Allomyces [MYCOL] A genus of aquatic phycomycetous fungi in the order Blastocladiales characterized by basal rhizoids, terminal hyphae, and zoospores with a single posterior flagellum.

allopathy [MED] A system of medicine that employs remedies whose effects are unlike those of the disease, in contrast to homeopathy.

allopelagic [ECOL] Relating to organisms living at various depths in the sea in response to influences other than temperature.

allophanamide *See* biuret.

allophane [GEOL] $Al_2O_3 \cdot SiO_2 \cdot nH_2O$ A clay mineral composed of hydrated aluminosilicate gel of variable composition; P_2O_5 may be present in appreciable quantity.

allophore [HISTOL] A chromatophore which contains a red pigment that is soluble in alcohol; found in the skin of fishes, amphibians, and reptiles.

allopolyploid [GEN] An organism or strain arising from a combination of genetically distinct chromosome sets from two diploid organisms.

allopurinol [PHARM] $C_5H_4ON_4$ The compound 4-hydroxypyrazolo-3,4-d-pyrimidine; inhibits xanthine oxidase, an enzyme required for uric acid formation.

all-or-none law [PHYSIO] The principle that transmission of a nerve impulse is either at full strength or not at all.

allosome [GEN] **1.** Sex chromosome. **2.** Any atypical chromosome.

allosteric enzyme [BIOCHEM] Any of the regulatory bacterial enzymes, such as those involved in end-product inhibition.

allostery [BIOCHEM] The property of an enzyme able to shift reversibly between an active and an inactive configuration.

Allotheria [PALEON] A subclass of Mammalia that appeared in the Upper Jurassic and became extinct in the Cenozoic.

allothigene *See* allogene.

ALLIGATOR

The American alligator (*Alligator mississipiensis*).

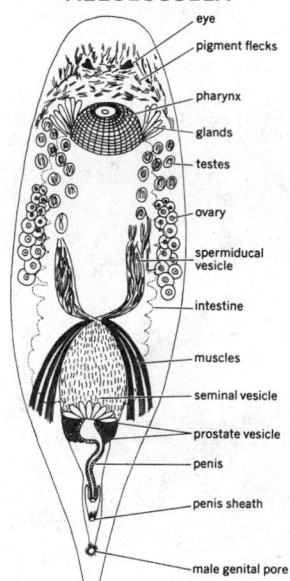

ALLOEOCOELA

eye
pigment flecks
pharynx
glands
testes
ovary
spermiducal vesicle
intestine
muscles
seminal vesicle
prostate vesicle
penis
penis sheath
male genital pore

Plagiostomum morgani, one of the two largest and best-known genera of the Alloeocoela. (*After L. Graff, 1911*)

allothimorph [GEOL] A metamorphic rock constituent which retains its original crystal outlines in the new rock.

allothogenic [GEOL] Formed from preexisting rocks which have been transported from another location. Also known as allogenic.

allotrioblast *See* xenoblast.

Allotriognathi [VERT ZOO] An equivalent name for the Lampridiformes.

allotriomorphic [MINERAL] Of minerals in igneous rock not bounded by their own crystal faces but having their outlines impressed on them by the adjacent minerals. Also known as anhedral; xenomorphic.

allotriomorphism *See* allotropy.

allotrope [CHEM] A form of an element or other substance showing allotropy.

allotropism *See* allotropy.

allotropy [CHEM] The assumption by an element or other substance of two or more different forms or structures which are most frequently stable in different temperature ranges, such as different crystalline forms of carbon as charcoal, graphite, or diamond. Also known as allotriomorphism; allotropism.

allotter [COMMUN] A telephone term referring to a distributor, associated with the finder control group relay assembly, which allots an idle line-finder in preparation for an additional call.

allotter relay [COMMUN] A telephone term referring to a relay of the line-finder circuit whose function is to preallot an idle line-finder to the next incoming call from the line, and to guard relays.

allotype [SYST] A paratype of the opposite sex to the holotype.

allowable load [MECH] The maximum force that may be safely applied to a solid, or is permitted by applicable regulators.

allowable oil [PETRO ENG] The oil an operator is permitted by law to remove from a well in one day.

allowable stress [MECH] The maximum force per unit area that may be safely applied to a solid.

allowance [DES ENG] An intentional difference in sizes of two mating parts, allowing clearance usually for a film of oil, for running or sliding fits.

allowed energy bands [SOLID STATE] The restricted regions of possible electron energy levels in a solid.

allowed hours *See* standard hour.

allowed time [IND ENG] Amount of time allowed each employee for personal needs during a work cycle.

allowed transition [QUANT MECH] A transition between two states which is permitted by the selection rules and which consequently has a relatively high priority.

alloxan [BIOCHEM] $C_4H_2N_2O_4$ Crystalline oxidation product of uric acid; induces diabetes experimentally by selective destruction of pancreatic beta cells. Also known as mesoxalyurea.

alloy [MET] Any of a large number of substances having metallic properties and consisting of two or more elements; with few exceptions, the components are usually metallic elements.

Alloy 600 *See* Inconel.

alloying [MET] The addition of a metal or alloy to another metal or alloy.

alloy junction [ELECTR] A junction produced by alloying one or more impurity metals to a semiconductor to form a *p* or *n* region, depending on the impurity used. Also known as fused junction.

alloy junction diode [ELECTR] A junction diode made by placing a pill of doped alloying material on a semiconductor material and heating until the molten alloy melts a portion of the semiconductor, resulting in a *pn* junction when the dissolved semiconductor recrystallizes. Also known as fused-junction diode.

alloy-junction transistor [ELECTR] A junction transistor made by placing pellets of a *p*-type impurity such as indium above and below an *n*-type wafer of germanium, then heating until the impurity alloys with the germanium to give a *pnp* transistor. Also known as fused-junction transistor.

alloy nuclear fuel [NUCLEO] A material used in nuclear reactors that is an alloy of a fissionable substance and a nonfissionable metal or metals.

alloy plating [MET] The codeposition of two or more metals on an electrode by electrolysis.

alloy steel [MET] A steel whose distinctive properties are due to the presence of one or more elements other than carbon.

all-pass network [ELECTR] A network designed to introduce a phase shift in a signal without introducing an appreciable reduction in energy of the signal at any frequency.

allspice [BOT] The dried, unripe berries of a small, tropical evergreen tree, *Pimenta officinalis*, of the myrtle family; yields a pungent, aromatic spice.

allspice oil *See* pimenta oil.

allulose [ORG CHEM] $CH_2OHCO(CHOH)_3CH_2OH$ A constituent of cane sugar molasses; it is nonfermentable. Also known as D-piscose; D-ribo-2-ketohexose.

alluvial [GEOL] **1.** Of a placer, or its associated valuable mineral, formed by the action of running water. **2.** Pertaining to or consisting of alluvium, or deposited by running water.

alluvial cone [GEOL] An alluvial fan with steep slopes formed of loose material washed down the slopes of mountains by ephemeral streams and deposited as a conical mass of low slope at the mouth of a gorge. Also known as cone delta; cone of dejection; cone of detritus; debris cone; dry delta; hemicone; wash.

alluvial dam [GEOL] A sedimentary deposit which is built by an overloaded stream and dams its channel; especially characteristic of distributaries on alluvial fans.

alluvial deposit *See* alluvium.

alluvial fan [GEOL] A fan-shaped deposit formed by a stream either where it issues from a narrow moutain valley onto a plain or broad valley, or where a tributary stream joins a main stream.

alluvial flat [GEOL] A small alluvial plain having a slope of about 5 to 20 feet per mile and built of fine sandy clay or adobe deposited during flood.

alluvial mining [MIN ENG] The exploitation of alluvial deposits by dredging, hydraulicking, or drift mining.

alluvial ore deposit [GEOL] A deposit in which the valuable mineral particles have been transported and left by a stream.

alluvial plain [GEOL] A plain formed from the deposition of alluvium usually adjacent to a river that periodically overflows. Also known as aggraded valley plain; river plain; wash plain; waste plain.

alluvial slope [GEOL] A surface of alluvium which slopes down from mountainsides and merges with the plain or broad valley floor.

alluvial soil [GEOL] A soil deposit developed on floodplain and delta deposits.

alluvial terrace [GEOL] A terraced embankment of loose material adjacent to the sides of a river valley. Also known as built terrace; drift terrace; fill terrace; stream-built terrace.

alluvial valley [GEOL] A valley filled with a stream deposit.

alluviation [GEOL] Deposition of sediment by a river.

alluvion *See* alluvium.

alluvium [GEOL] The detrital materials eroded, transported, and deposited by streams; an important constituent of shelf deposits. Also known as alluvial deposit; alluvion.

all-wave receiver [ELECTR] A radio receiver capable of being tuned from about 535 kilohertz to at least 20 megahertz; some go above 100 megahertz and thus cover the FM band also.

all-weather aircraft [AERO ENG] Aircraft that are designed or equipped to perform by day or night under any weather conditions.

all-weather airport [CIV ENG] An airport with facilities to permit the landing of qualified aircraft and aircrewmen without regard to operational weather limits.

all-weather fighter [AERO ENG] A fighter aircraft equipped with radar and other special devices which enable it to intercept its target in the dark, or in daylight weather conditions that do not permit visual interception; it is usually a multiplace (pilot plus navigator-observer) airplane.

all-weather landing system [NAV] An instrument landing system having optimum operational capability in low and zero visibility.

allyl [ORG CHEM] C_3H_5- An unsaturated radical found in compounds such as allylbromide (3-bromopropene).

allylalcohol [ORG CHEM] CH_2CHCH_2OH Colorless, pun-

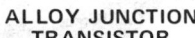
ALLOY JUNCTION TRANSISTOR

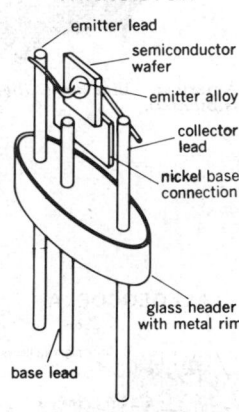

emitter lead
semiconductor wafer
emitter alloy
collector lead
nickel base connection
glass header with metal rim
base lead

View of an alloy junction transistor.

ALLSPICE

A branch of *Pimenta officinalis* with the berry fruit.

gent liquid, boiling at 96°C; soluble in water; made from allyl chloride by hydrolysis. Also known as propenol.

allylamine [ORG CHEM] $CH_2CHCH_2NH_2$ A yellow oil that is miscible with water; boils at 58°C; prepared from mustard oil. Also known as 3-aminopropene.

5-allyl-5-sec-butylbarbituric acid See talbutal.

allyl chloride [ORG CHEM] CH_2CHCH_2Cl A volatile, pungent, toxic, flammable, colorless liquid, boiling at 46°C; insoluble in water; made by chlorination of propylene at high temperatures. Also known as 3-chloropropene.

allylene [ORG CHEM] $CH_3C{:}CH$ An acetylenic, three-carbon hydrocarbon; a colorless gas boiling at $-24°C$; soluble in ether. Also known as methyl acetylene; propine; propyne.

4-allyl guaiacol See eugenol.

1-allyl-4-hydroxybenzene See chavicol.

allylic rearrangement [ORG CHEM] In a three-carbon molecule, the shifting of a double bond from the 1,2 carbon position to the 2,3 position, with the accompanying migration of an entering substituent or substituent group from the third carbon to the first.

allyl isosulfocyanate See allyl isothiocyanate.

allyl isothiocyanate [ORG CHEM] $CH_2CH{:}CH_2NCS$ A pungent, colorless to pale-yellow liquid; soluble in alcohol, slightly soluble in water; irritating odor; boiling point 152°C; used as a fumigant and as a poison gas. Also known as allyl isosulfocyanate; mustard oil.

para-allyl phenol See chavicol.

allyl plastic See allyl resin.

allyl resin [ORG CHEM] Any of a class of thermosetting synthetic resins derived from esters of allyl alcohol or allyl chloride; used in making cast and laminated products. Also known as allyl plastic.

allylsulfocarbamide See allylthiourea.

allyl sulfourea See allylthiourea.

allylthiourea [ORG CHEM] $C_3H_5NHCSNH_2$ A white, crystalline solid that melts at 78°C; soluble in water; used as a corrosion inhibitor. Also known as allylsulfocarbamide; allyl sulfourea; thiosinamine.

alm [ECOL] A meadow in alpine or subalpine mountain regions.

almandine [MINERAL] $Fe_3Al_2(SiO_4)_3$ A variety of garnet, deep red to brownish red, found in igneous and metamorphic rocks in many parts of world; used as a gemstone and an abrasive. Also known as almandite.

almandite See almandine.

almerite See natroalunite.

almond [BOT] *Prunus amygdalus.* A small deciduous tree of the order Rosales; it produces a drupaceous edible fruit with an ellipsoidal, slightly compressed nutlike seed.

almost periodic function [MATH] A continuous function $f(x)$ such that for any positive number ε there is a number M so that for any real number x, any interval of length M contains a nonzero number t such that $|f(x+t) - f(x)| < \varepsilon$.

Almquist unit [BIOL] A unit for the standardization of vitamin K.

almucantar See parallel of altitude.

alnico [MET] One of a series of ferrous alloys containing aluminum, nickel, and cobalt, valued because of their highly retentive magnetic properties; usually designated with a roman-numeral number, such as alnico VII. Also known as aluminum-nickel-cobalt alloy.

alnico magnet [ELECTROMAG] A permanent magnet made of alnico.

Alnilam [ASTRON] A star in the constellation Orion.

alnoite [PETR] A variety of biotite lamprophyres characterized by lepidomelane phenocrysts; it is feldspar-free but contains melitite, perovskite, olivine, and carbonate in the matrix.

aloisite [MINERAL] A brown to violet mineral consisting of a hydrous subsilicate of calcium, iron, magnesium, and sodium, and occurring in amorphous masses.

alongshore current See littoral current.

alopecia [MED] Loss of hair; baldness.

Alopiidae [VERT ZOO] A family of pelagic isurid elasmobranchs commonly known as thresher sharks because of their long, whiplike tail.

Aloxite [MATER] Aluminum oxide, of the Carborundum Company, used as an abrasive; adapted to the grinding of high-tensile-strength materials such as steel and annealed malleable iron.

alpaca [VERT ZOO] *Lama pacos.* An artiodactyl of the camel family (Camelidae); economically important for its long, fine wool.

alpage [ECOL] A summer grazing area composed of natural plant pasturage in upland or mountainous regions.

Alpax [MET] A strong, light alloy, of Light Alloys, Ltd., composed of 87% aluminum and 13% silicon and well suited for casting.

alpenglow [METEOROL] A reappearance of sunset colors on a mountain summit after the original mountain colors have faded into shadow; also, a similar phenomenon preceding the regular coloration at sunrise.

alpha [SCI TECH] The first letter in the Greek alphabet; α, A.

alphabet [SCI TECH] Any ordered set of unique graphics called characters, such as the 26 letters of the Roman alphabet.

alphabetic coding [ADP] 1. Abbreviation of words for computer input. 2. A system of coding with a number system of base 26, the letters of the alphabet being used instead of the cardinal numbers.

alphabetic shift [ADP] The status of a card punch when the program control is off (star wheels are raised).

alpha brass [MET] An alloy of copper and zinc containing up to 36% zinc dissolved, rather than chemically combined, with the copper; ductile, easily cold-worked, and corrosion resistant; used for hot-water pipes.

alpha cell [HISTOL] Any of the acidophilic chromophiles in the anterior lobe of the adenohypophysis.

alpha cross section [NUCLEO] Total cross section for interaction with α-particles.

alpha cutoff frequency [ELECTR] The frequency at the high end of a transistor's range at which current amplification drops 3 decibels below its low-frequency value.

alpha decay [NUC PHYS] A radioactive transformation in which an alpha particle is emitted by a nuclide.

alpha emission [NUC PHYS] Ejection of alpha particles from the atom's nucleus.

alpha globulin [BIOCHEM] A heterogeneous fraction of serum globulins containing the proteins of greatest electrophoretic mobility.

alpha helix [MOL BIO] A spatial configuration of the polypeptide chains of proteins in which the chain assumes a helical form, 0.54 nanometer in pitch, 3.6 amino acids per turn, presenting the appearance of a hollow cylinder with radiating side groups.

alpha hemolysis [MICROBIO] Partial hemolysis of red blood cells with green discoloration in a blood agar medium by certain hemolytic streptococci.

alpha iron [MET] Iron with a body-centered cubic structure which is stable below 1670°F (910°C).

alpha irradiation [NUCLEO] Subjection of a substance to a flux of alpha particles.

alphameric characters See alphanumeric characters.

alphameric typebar [ADP] A metal bar containing the alphabet, the ten numerical characters, and the ampersand, in use in electromechanical accounting machines.

alphanumeric characters [ADP] All characters used by a computer, including letters, numerals, punctuation marks, and such signs as \$, @, and #. Also known as alphameric characters.

alphanumeric display device [ELECTR] A device which visibly represents alphanumeric output information from some signal source.

alphanumeric grid See atlas grid.

alphanumeric instruction [ADP] The name given to instructions which can be read equally well with alphabetic or numeric kinds of fields of data.

alphanumeric reader [ELECTR] A device capable of reading alphabetic, numeric, and special characters and punctuation marks.

alpha olefin [ORG CHEM] An olefin where the unsaturation (double bond) is at the alpha position, that is, between the two end carbons of the carbon chain.

alpha particle [ATOM PHYS] A positively charged particle

ALMOND

Almond twig with leaves and fruit.

ALPACA

The alpaca (*Lama pacos*), sheared every 2 years, to yield 18–24 pounds of wool in its lifetime.

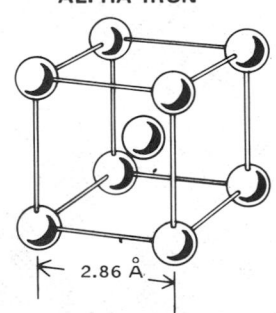

ALPHA IRON

2.86 Å

The body-centered cubic crystal structure of alpha iron.

consisting of two protons and two neutrons, identical with the nucleus of the helium atom; emitted by several radioactive substances.

alpha-particle detector [NUCLEO] A device used to indicate the presence of alpha particles.

alpha-particle scattering [ATOM PHYS] Deviation at various angles of a stream of alpha particles passing through a foil of material.

alpha-particle spectroscopy [SPECT] The science of alpha-particle spectra.

alpha position [ORG CHEM] In chemical nomenclature, the position of a substituting group of atoms in the main group of a molecule; for example, in a straight-chain compound such as α-hydroxypropionic acid (CH₃CHOHCOOH), the hydroxyl radical is in the alpha position.

alpha ray [NUCLEO] A stream of alpha particles.

alpha-ray vacuum gage [ENG] An ionization gage in which the ionization is produced by alpha particles emitted by a radioactive source, instead of by electrons emitted from a hot filament; used chiefly for pressures from 10^{-3} to 10 torrs. Also known as alphatron.

alpha rhythm [PHYSIO] An electric current from the occipital region of the brain cortex having a pulse frequency of 8 to 13 per second; associated with a relaxed state in normal human adults.

alpha system [COMMUN] A signaling system in which alphabetic characters designate the signaling code to be used.

alpha taxonomy [SYST] The initial or descriptive phase in the history of animal systematics.

alphatopic [NUCLEO] Pertaining to the relationship between two nuclides that differ in composition or in mass by an alpha particle.

alphatron *See* alpha-ray vacuum gage.

Alpheidae [INV ZOO] The snapping shrimp, a family of decapod crustaceans included in the section Caridea.

Alphonsus [ASTRON] A moon crater.

Alpides [GEOL] Great east-west structural belt including the Alps of Europe and the Himalayas and related mountains of Asia; mostly folded in Tertiary times.

alpine [ECOL] Any plant native to mountain peaks or boreal regions.

alpine glacier [HYD] A glacier lying on or occupying a depression in mountainous terrain. Also known as mountain glacier.

Alpine orogeny [GEOL] Jurassic through Tertiary orogeny which affected the Alpide orogenic belt.

alpine tundra [ECOL] Large, flat or gently sloping, treeless tracts of land above the timberline.

alpine-type facies [PETR] High-pressure, low-temperature (150–400°C) dynamothermal metamorphism characterized by the presence of the pumpellyite and glaucophane schist facies.

alpinotype tectonics [GEOL] Tectonics of the alpine-type geosynclinal mountain belts characterized by deep-seated plastic folding, plutonism, and lateral thrusting.

alsbachite [PETR] A plutonic rock of sodic plagioclase, quartz, and subordinate orthoclase and accessory garnet, biotite, and muscovite; a variety of porphyritic granodiorite.

alsifilm [MATER] A bentonite gel in the form of sheets used primarily for electrical insulation, because of its properties of heat and oil resistance.

alstonite *See* bromlite.

alt *See* altitude.

Altair [ASTRON] A star that is 16.5 light-years from the sun; spectral type A7IV-V. Also known as α Aquilae.

altaite [MINERAL] PbTe A tin-white lead-tellurium mineral occurring as isometric crystals with tin ores in central Asia.

Altar *See* Ara.

altazimuth [ENG] An instrument equipped with both horizontal and vertical graduated circles, for the simultaneous observation of horizontal and vertical directions or angles. Also known as astronomical theodolite; universal instrument.

alteration [PETR] A change in a rock's mineral composition.

alteration switch [ADP] A hand-operated switch mounted on the console of a computer, used to feed a single bit of information into a program. Also known as sense switch.

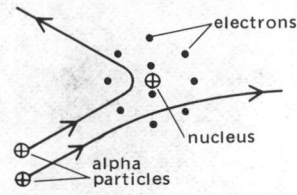

ALPHA–PARTICLE SCATTERING

electrons
nucleus
alpha particles

Scattering by an atom of alpha particles.

alter ego [PSYCH] A close friend who represents a second self to an individual.

alternate [BOT] **1.** Of the arrangement of leaves on opposite sides of the stem at different levels. **2.** Of the arrangement of the parts of one whorl between members of another whorl.

alternate airport [NAV] An airport designated in a pilot's flight plan at which an aircraft will have the capability to land if a landing at the intended airport becomes inadvisable.

alternate angles [MATH] A pair of nonadjacent angles that a transversal forms with each of two lines; they lie on opposite sides of the transversal, and are both interior, or both exterior, to the two lines.

alternate-channel interference [COMMUN] Interference that is caused in one communications channel by a transmitter operating in the next channel beyond an adjacent channel. Also known as second-channel interference.

alternate immersion test [MET] A corrosion test in which a specimen is repeatedly carried through a cycle of immersion in and removal from a corrosive medium over definite time intervals.

alternate routing [COMMUN] The operation of a switching center when all circuits are found busy in a programmed route to the destination, and the call is offered to another programmed route.

alternate track [ADP] The disk track used if, after a disk volume is initialized, a defective track is sensed by the system.

alternate traversing fire [ORD] Method of covering a target that has both width and depth by firing a succession of traversing groups whose normal range dispersion will provide for distribution in depth.

alternating copolymer [ORG CHEM] A polymer formed of two different monomer molecules that alternate in sequence in the polymer chain.

alternating current [ELEC] Electric current that reverses direction periodically, usually many times per second. Abbreviated ac.

alternating-current circuit theory [ELEC] The mathematical description of conditions in an electric circuit driven by an alternating source or sources.

alternating-current coupling [ELECTR] A coupling which passes alternating-current signals but blocks direct-current signals.

alternating-current/direct-current [ELECTR] Pertaining to electronic equipment capable of operation from either an alternating-current or direct-current primary power source.

alternating-current dump [ELECTR] The removal of all alternating-current power from a computer intentionally, accidentally, or conditionally.

alternating-current erase [ELECTR] The use of an alternating current to energize a tape recorder erase head in order to remove previously recorded signals from a tape.

alternating-current erasing head [ELECTR] In magnetic recording, an erasing head which uses alternating current to produce the magnetic field necessary for erasing.

alternating-current generator [ELEC] A machine, usually rotary, which converts mechanical power into alternating-current electric power.

alternating-current magnetic biasing [ELECTR] Biasing with alternating current, usually well above the signal frequency range, in magnetic tape recording.

alternating-current motor [ELEC] A machine that converts alternating-current electrical energy into mechanical energy by utilizing forces exerted by magnetic fields produced by the current flow through conductors.

alternating-current power supply [ELEC] A power supply that provides one or more alternating-current output voltages, such as an ac generator, dynamotor, inverter, or transformer.

alternating-current resistance *See* high-frequency resistance.

alternating-current transmission [ELECTR] In television, that form of transmission in which a fixed setting of the controls makes any instantaneous value of signal correspond to the same value of brightness for only a short time.

alternating-current welder [ENG] A welding machine utilizing alternating current for welding purposes.

alternating flashing light [NAV] A light showing one or more flashes with color variations at regular intervals, the duration of light being less than that of darkness.

alternating function [MATH] A function in which the interchange of two independent variables causes the dependent variable to change sign.

alternating gradient [ELECTROMAG] A magnetic field in which successive magnets have gradients of opposite sign, so that the field increases with radius in one magnet and decreases with radius in the next; used in synchrotrons and cyclotrons.

alternating-gradient synchrotron [NUCLEO] A proton synchrotron using an alternating magnetic-field gradient for focusing; beams of protons having extremely high energy (above 25 GeV) are produced.

alternating group [MATH] A group made up of all the even permutations of n objects.

alternating group flashing light [NAV] A navigation light having groups of total eclipses at regular intervals and having color variations, the duration of light being equal to or greater than that of darkness. Also known as alternating group occulting light.

alternating group occulting light See alternating group flashing light.

alternating occulting light [NAV] A navigation light having one or more total eclipses at regular intervals and having color variations, the duration of light being equal to or greater than that of darkness.

alternating series [MATH] Any series of real numbers in which consecutive terms have opposite signs.

alternating stress [MECH] A stress produced in a material by forces which are such that each force alternately acts in opposite directions.

alternating voltage [ELEC] Periodic voltage, the average value of which over a period is zero.

alternation [PHYS] Variation, either positive or negative, of a waveform from zero to maximum and back to zero, equaling one-half of a cycle.

alternation of generations See metagenesis.

alternation of multiplicities law [CHEM] The law that the periodic table arranges the elements in such a sequence that their number of orbital electrons, and hence their multiplicities, alternates between even and odd numbers.

alternator [ELEC] A mechanical, electrical, or electromechanical device which supplies alternating current.

alterne [ECOL] A community exhibiting alternating dominance with other communities in the same area.

altherbosa [ECOL] Communities of tall herbs, usually succeeding where forests have been destroyed.

altigraph [ENG] A pressure altimeter that has a recording mechanism to show the changes in altitude.

altimeter [ENG] An instrument which determines the altitude of an object with respect to a fixed level, such as sea level; there are two common types: the aneroid altimeter and the radio altimeter.

altimeter corrections [ENG] Corrections which must be made to the readings of a pressure altimeter to obtain true altitudes; involve horizontal pressure gradient error and air temperature error.

altimeter setting [ENG] The value of atmospheric pressure to which the scale of an aneroid altimeter is set; after United States practice, the pressure that will indicate airport elevation when the altimeter is 10 feet above the runway (approximately cockpit height).

altimeter-setting indicator [ENG] A precision aneroid barometer calibrated to indicate directly the local altimeter setting.

altimetry [ENG] The measurement of heights in the atmosphere (altitude), generally by means of an altimeter.

altiplanation [GEOL] A phase of solifluction that may be seen as terracelike forms, flattened summits, and passes that are mainly accumulations of loose rock.

altiplanation surface [GEOL] A flat area fronted by scarps a few to hundreds of feet in height; the area ranges from several square rods to hundreds of acres. Also known as altiplanation terrace.

altiplanation terrace See altiplanation surface.

altithermal [GEOPHYS] Period of high temperature, particularly the postglacial thermal optimum.

Altithermal [GEOL] A dry postglacial interval centered about 5500 years ago during which temperatures were warmer than at present.

altithermal soil [GEOL] Soil recording a period of rising or high temperature.

altitude Abbreviated alt. [ENG] **1.** Height, measured as distance along the extended earth's radius above a given datum, such as average sea level. **2.** Angular displacement above the horizon measured by an altitude curve. [MATH] The perpendicular distance from the base to the top (a vertex or parallel line) of a geometric figure such as a triangle or parallelogram.

altitude acclimatization [PHYSIO] A physiological adaptation to reduced atmospheric and oxygen pressure.

altitude azimuth [ENG] An azimuth determined by solution of the navigational triangle with altitude, declination, and latitude given.

altitude chamber [ENG] A chamber within which the air pressure, temperature, and so on can be adjusted to simulate conditions at different altitudes; used for experimentation and testing.

altitude circle [ASTRON] See parallel of altitude. [ELECTROMAG] A bright circle which surrounds the central dark portion of a plan position indicator display or photograph, and which results from ground clutter.

altitude-contour See C factor.

altitude curve [ENG] The arc of a vertical circle between the horizon and a point on the celestial sphere, measured upward from the horizon. [NAV] A graphical representation of the altitude of a celestial body as it would appear from a single assumed position or a series of assumed positions over a period of time; such curves are precomputed.

altitude datum [ENG] The arbitrary level from which heights are reckoned.

altitude delay [ELECTR] Synchronization delay introduced between the time of transmission of the radar pulse and the start of the trace on the indicator to eliminate the altitude/height hole on the plan position indicator-type display.

altitude difference [ENG] The difference between computed and observed altitudes, or between precomputed and sextant altitudes. Also known as altitude intercept; intercept.

altitude hole [ELECTR] The blank area in the center of a plan position indicator-type radarscope display caused by the time interval between transmission of a pulse and the receipt of the first ground return.

altitude intercept See altitude difference.

altitude reservation [NAV] **1.** The prior approval by the appropriate air traffic control agencies of flight plans, requesting use of certain airspace for the purpose of expediting mass movement of aircraft or other special air operations. **2.** A flight altitude assigned to an aircraft by an air-traffic control agency.

altitude sickness [MED] In general, any sickness brought on by exposure to reduced oxygen tension and barometric pressure.

altitude signal [ELECTR] The radio signals returned to an airborne electronics device by the ground or sea surface directly beneath the aircraft.

altitude tints See hypsometric tinting.

altitude valve [AERO ENG] A valve that adjusts the composition of the air-fuel mixture admitted into an airplane carburetor as the air density varies with altitude.

altitude wind tunnel [AERO ENG] A wind tunnel in which the air pressure, temperature, and humidity can be varied to simulate conditions at different altitudes.

altitudinal vegetation zone [ECOL] A geographical band of physiognomically similar vegetation correlated with vertical and horizontal gradients of environmental conditions.

alto [GRAPHICS] The negative which is electrodeposited during the preparation of plates for printing currency.

altocumulus cloud [METEOROL] A principal cloud type, white or gray or both white and gray in color; occurs as a layer or patch with a waved aspect, the elements of which appear as laminae, rounded masses, or rolls; frequently appears at different levels in a given sky. Abbreviated Ac.

ALTIMETER

Aneroid altimeter.

ALTOCUMULUS CLOUD

Altocumulus, which occurs at intermediate levels. (G. A. Lott, U.S. Weather Bureau)

ALTOSTRATUS CLOUD

Altostratus, a middle-level layer cloud. Thick layers of such cloud, with bases extending down to low levels, produce prolonged rain or snow, and are then called nimbostratus. (C. F. Brooks, U.S. Weather Bureau)

ALUMINUM

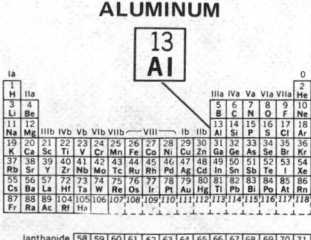

Periodic table of the chemical elements showing the position of aluminum.

altostratus cloud [METEOROL] A principal cloud type in the form of a gray or bluish (never white) sheet or layer of striated, fibrous, or uniform appearance; very often totally covers the sky and may cover an area of several thousand square miles; vertical extent may be from several hundred to thousands of meters. Abbreviated As.

altricial [VERT ZOO] Pertaining to young that are born or hatched immature and helpless, thus requiring extended development and parental care.

ALU See arithmetical unit.

Aludur [MET] The trade name for an alloy of aluminum with less than 1% each of magnesium, silicon, and perhaps iron and with high tensile strength; used for conductors in overhead transmission lines.

alula [ZOO] **1.** Digit of a bird wing homologous to the thumb. **2.** See calypter.

alum [INORG CHEM] **1.** Any of a group of double sulfates of trivalent metals such as aluminum, chromium, or iron and a univalent metal such as potassium or sodium. **2.** See aluminum sulfate; ammonium aluminum sulfate; potassium aluminum sulfate. [MINERAL] $KAl(SO_4)_2 \cdot 12H_2O$ A colorless, white, astringent-tasting evaporite mineral.

alum cake [MATER] A material composed of silica and aluminum sulfate produced by the action of sulfuric acid on clay.

alum coal [GEOL] Argillaceous brown coal rich in pyrite in which alum is formed on weathering.

Alumel [MET] An alloy, of the Hoskins Manufacturing Company, containing 94% nickel, 2% aluminum, 2% manganese, and 1% silicon; used in chromel-alumel thermocouples and less frequently in electrical resistance wire in heaters and other electrical appliances at elevated temperatures.

alumina [INORG CHEM] Al_2O_3 The native form of aluminum oxide occurring as corundum or in hydrated forms, as a powder or crystalline substance.

alumina brick [MATER] A group of fireclay bricks containing 50, 60, or 70% alumina, used in high temperature applications.

alumina cement [MATER] A cement made with bauxite and containing a high percentage of aluminate, having the property of setting to high strength in 24 hours.

alumina porcelain [MATER] Porcelain composed principally of alumina; used to make spark plugs.

aluminate [INORG CHEM] A negative ion usually assigned the formula AlO_2^- and derived from aluminum hydroxide.

alumina trihydrate [INORG CHEM] $Al_2O_3 \cdot 3H_2O$, or $Al(OH)_3$ A white powder; insoluble in water, soluble in hydrochloric or sulfuric acid or sodium hydroxide; used in the manufacture of ceramic glasses and in paper coating. Also known as aluminum hydrate; aluminum hydroxide; hydrated alumina; hydrated aluminum oxide.

aluminite [MINERAL] $Al_2(SO_4)(OH)_4 \cdot 7H_2O$ Native monoclinic hydrous aluminum sulfate; used in tanning, papermaking, and water purification. Also known as websterite.

aluminite powder [MATER] A powdered aluminum ingredient in blasting agents and high explosives.

aluminium See aluminum.

aluminize [ENG] To apply a film of aluminum to a material, such as glass. [MET] To form a protective surface alloy on a metal by treatment at elevated temperature with aluminum or an aluminum compound.

aluminized explosive [MATER] An explosive to which aluminum has been added.

aluminosilicate [INORG CHEM] $3Al_2O_3 \cdot 2SiO_2$ A colorless, crystalline combination of silicate and aluminate in the form of rhombic crystals.

aluminosis [MED] A lung disorder caused by inhalation of alumina dust.

aluminothermy [MET] The process of reducing a metallic oxide to the metal and producing great heat by mixing finely divided aluminum with the oxide, which is reduced as the aluminum is oxidized.

aluminotype [GRAPHICS] A printing plate of the relief type; the raised surface is made of aluminum.

aluminous cement [MATER] A cement with a higher alumina content than portland cement; gets very hot during hardening and is acid-resistant and refractory. Also known as high-alumina cement.

aluminum [CHEM] A chemical element, symbol Al, atomic number 13, and atomic weight 26.9815. Also spelled aluminium.

aluminum acetate [ORG CHEM] $Al(CH_3COO)_3$ A white, amorphous powder that is soluble in water; used in aqueous solution as an antiseptic.

aluminum alloy [MET] An alloy of aluminum and relatively small amounts of other metals, such as copper, magnesium, or manganese.

aluminum ammonium sulfate See ammonium aluminum sulfate.

aluminum arrester See aluminum-cell arrester.

aluminum-base grease See aluminum-soap grease.

aluminum brass [MET] **1.** A casting brass to which aluminum has been added as a flux to improve the casting qualities and, with the addition of lead, the machining qualities. **2.** A wrought brass to which aluminum has been added to improve the extruding and forging qualities and the oxidation resistance.

aluminum bronze [MET] A copper-aluminum alloy which may also contain iron, manganese, nickel, or zinc.

aluminum cable steel-reinforced [ELEC] A type of power transmission line made of an aluminum conductor provided with a core of steel. Abbreviated ACSR.

aluminum-cell arrester [ELEC] A lightning arrester consisting of a number of electrolytic cells in series formed from aluminum trays containing electrolyte. Also known as aluminum arrester; electrolytic arrester.

aluminum chloride [INORG CHEM] $AlCl_3$ or Al_2Cl_6 A deliquescent compound in the form of white to colorless hexagonal crystals; fumes in air and reacts explosively with water; used as a catalyst.

aluminum coating [MET] A film of aluminum applied to a metallic surface by, for example, spraying, electrolysis, or hot dipping.

aluminum conductor [ELEC] Any of several aluminum alloys employed for conducting electric current; because its weight is one-half that of copper for the same conductance, it is used in high-voltage transmission lines.

aluminum fluoride [INORG CHEM] $AlF_3 \cdot 3\frac{1}{2}H_2O$ A white, crystalline powder, insoluble in cold water.

aluminum foil [MET] Aluminum in the form of a sheet of thickness not exceeding 0.005 inch (0.127 mm).

aluminum halide [INORG CHEM] A compound of aluminum with a halogen element, such as aluminum chloride.

aluminum hydrate See alumina trihydrate.

aluminum hydroxide See alumina trihydrate.

aluminum-magnesium alloy [MET] An alloy of aluminum, 5–10% magnesium, and sometimes small amounts of other metals, characterized by high resistance to corrosion and high machinability.

aluminum-nickel-cobalt alloy See alnico.

aluminum nitrate [INORG CHEM] $Al(NO_3)_3 \cdot 9H_2O$ A colorless, rhombic, deliquescent compound.

aluminum oleate [ORG CHEM] A soaplike compound of aluminum and oleic acid, used in lubricating oils and greases to improve their viscosity.

aluminum ore [GEOL] A natural material from which aluminum may be economically extracted.

aluminum orthophosphate [INORG CHEM] $AlPO_4$ White crystals, melting above 1500°C; insoluble in water, soluble in acids and bases; useful in ceramics, paints, pulp, and paper. Also known as aluminum phosphate.

aluminum oxide [INORG CHEM] Al_2O_3 A compound in the form of a white powder or colorless hexagonal crystals; melts at 2020°C; insoluble in water.

aluminum paint [MATER] A mixture of oil varnish and aluminum pigment in the form of thin flakes which overlap in the paint film; reflects the sun's radiation well and retains well the heat in hot-air or hot-water pipes or tanks.

aluminum palmitate [ORG CHEM] $Al(C_{16}H_{31}O_2) \cdot H_2O$ An aluminum soap used in waterproofing fabrics, paper, and leather and as a drier in paints.

aluminum paste [MATER] Aluminum powder finely ground in oil; used in aluminum paints.

aluminum phosphate See aluminum orthophosphate.

aluminum plate [GRAPHICS] A polymer-coated, anodized aluminum printing plate used in offset-lithographic printing.

aluminum potassium sulfate *See* potassium aluminum sulfate.

aluminum silicate [INORG CHEM] $Al_2(SiO_3)_3$ A white solid that is insoluble in water; used as a refractory in glassmaking.

aluminum-silicon alloy [MET] An alloy of aluminum, 5–22% silicon, and sometimes small amounts of other metals, characterized by ease of casting and welding, light weight, and high resistance to corrosion.

aluminum soap [ORG CHEM] Any of various salts of higher carboxylic acids and aluminum that are insoluble in water and soluble in oils; used in lubricating greases, paints, varnishes, and waterproofing substances.

aluminum-soap grease [MATER] A lubricating grease consisting of a petroleum oil thickened with aluminum soap. Also known as aluminum-base grease.

aluminum solder [MET] A solder containing up to 15% aluminum, having a melting point above that of the tin-lead solders, and applied with a brazing torch.

aluminum stearate [ORG CHEM] $Al(C_{17}H_{35}COO)_3$ An aluminum soap in the form of a white powder that is insoluble in water and soluble in oils; used for waterproofing fabrics and concrete and as a drier in paints and varnishes.

aluminum sulfate [INORG CHEM] $Al_2(SO_4)_3 \cdot 18H_2O$ A colorless salt in the form of monoclinic crystals that decompose in heat and are soluble in water; used in papermaking, water purification, and tanning, and as a mordant in dyeing. Also known as alum.

aluminum therapy [MED] Therapy intended mainly for prevention rather than treatment of silicosis; provides for inhalation of powdered aluminum and alumina (Al_2O_3) dust by miners in the change house.

aluminum triacetate [ORG CHEM] $Al(C_2H_3O_2)_3$ A white solid, very slightly soluble in cold water.

alumite *See* alunite.

alum rock *See* alunite.

alum schist *See* alum shale.

alum shale [PETR] A shale containing pyrite that is decomposed by weathering to form sulfuric acid, which acts on potash and alumina constituents to form alum. Also known as alum schist; alum slate.

alum slate *See* alum shale.

alumstone *See* alunite.

Alundum [MATER] Aluminum oxide, of the Norton Company, used as an abrasive; adapted to the grinding of high-tensile-strength materials such as steel and annealed malleable iron.

alunite [MINERAL] $KAl_3(SO_4)_2(OH)_6$ A mineral composed of a basic potassium aluminum sulfate; it occurs as a hydrothermal-alteration product in feldspathic igneous rocks and is used in the manufacture of alum. Also known as alumite; alum rock; alumstone.

alunitization [GEOL] Introduction of or replacement by alunite.

alunogen [MINERAL] $Al_2(SO_4)_3 \cdot 18H_2O$ A white mineral occurring as a fibrous incrustation of hydrated aluminum sulfate by volcanic action or decomposition of pyrite. Also known as feather alum; hair salt.

alurgite [MINERAL] A purple manganiferous variety of muscovite mica.

alvar [ECOL] Dwarfed vegetation characteristic of certain Scandinavian steppelike communities with a limestone base.

alveator [INV ZOO] A type of pedicellaria in echinoderms.

alveolar [BIOL] Of or relating to an alveolus.

alveolar-capillary block syndrome [MED] Arterial oxygen deficiency due to improper functioning of the membranes between the alveoli and capillaries.

alveolar gland *See* acinous gland.

alveolar oxygen pressure [PHYSIO] The oxygen pressure in the alveoli; the value is about 105 mm Hg.

alveolated cell *See* epithelioid cell.

alveolitoid [INV ZOO] A type of tabulate coral having a vaulted upper wall and a lower wall parallel to the surface of attachment.

alveolus [ANAT] **1.** A tiny air sac of the lung. **2.** A tooth socket. **3.** A sac of a compound gland.

Alydidae [INV ZOO] A family of hemipteran insects in the superfamily Coreoidea.

alymphocytosis [MED] Absence or deficiency of blood lymphocytes.

alyphite [GEOL] Bitumen that yields a high percentage of open-chain aliphatic hydrocarbons upon distillation.

Alzheimer's disease [MED] A type of presenile dementia associated with sclerosis of the cerebral cortex.

Am *See* americium; ammonium.

AM *See* amplitude modulation.

A/m *See* ampere per meter.

Am² *See* ampere meter squared.

A/m² *See* ampere per square meter.

Am²/Js *See* ampere square meter per joule second.

amacratic lens *See* amasthenic lens.

Amagat density unit [PHYS] A unit of density in the Amagat system, used in the study of the behavior of gases under pressure; it is equal to the density of a gas at a pressure of 1 atmosphere and a temperature of 0°C; for an ideal gas this is 44.616 ± 0.006 moles per cubic meter.

Amagat diagram [PHYS] A diagram that plots a series of isothermal curves for a gas pressure versus the gas pressure-volume product.

Amagat law *See* Amagat-Leduc rule.

Amagat-Leduc rule [PHYS] The rule which states that the volume taken up by a gas mixture equals the sum of the volumes each gas would occupy at the temperature and pressure of the mixture. Also known as Amagat law; Leduc law.

Amagat system [PHYS] A system of units in which the unit of pressure is the atmosphere and the unit of volume is the gram-molecular volume (22.4 liters at standard conditions).

Amagat volume unit [PHYS] A unit of volume in the Amagat system, used in the study of the behavior of gases under pressure; it is equal to the volume occupied by 1 mole of a gas at a pressure of 1 atmosphere and a temperature of 0°C; for an ideal gas this is 0.022413 ± 0.000003 cubic meter.

amalgam [MET] An alloy of mercury.

amalgamate [MET] **1.** To unite a metal in an alloy with mercury. **2.** To unite two dissimilar metals. **3.** To cover the zinc elements of a galvanic battery with mercury.

amalgamating table [MET] A sloping wooden table covered with a copper plate on which mercury is spread to amalgamate with precious-metal particles.

amalgamation [MET] Also known as amalgam treatment. **1.** The process of separating metal from ore by alloying the metal with mercury; formerly used for gold and silver recovery, where it has been superseded by the cyanide process. **2.** The formation of an alloy of a metal with mercury.

amalgamation pan [MET] A circular cast-iron pan in which gold or silver ore is ground and the precious metal particles are amalgamated with mercury added to the pan.

amalgamator [MET] A device for bringing pulverized ore into contact with mercury to form an amalgam from which the metal is subsequently recovered.

amalgam barrel [MET] A small batching mill used to grind auriferous concentrates with mercury.

amalgam retort [MET] A retort in which mercury is distilled off from gold, or silver amalgam is obtained in amalgamation.

amalgam treatment *See* amalgamation.

amantadine [PHARM] $C_{10}H_{17}N$ A symmetrical amine used as a viral chemoprophylactic because it selectively inhibits certain myxoviruses; also of value in the treatment of parkinsonism. Also known as 1-aminoadamantane.

amanthophilous [BOT] Of plants having a habitat in sandy plains or hills.

Amaranthaceae [BOT] The characteristic family of flowering plants in the order Caryophyllales; they have a syncarpous gynoecium, a monochlamydeous perianth that is more or less scarious, and mostly perfect flowers.

amarillite [MINERAL] $NaFe(SO_4)2 \cdot 6H_2O$ A pale greenish-yellow mineral consisting of a hydrous sodium ferric sulfate.

Amaryllidaceae [BOT] The former designation for a family of plants now included in the Liliaceae.

amasthenic lens [OPTICS] A lens that refracts the rays of light into one focus. Also known as amacratic lens.

amateur bands [COMMUN] Bands of frequencies assigned exclusively to licensed radio amateurs.

amateur radio [ELECTR] A radio used for two-way radio

communications by private individuals as leisure-time activity. Also known as ham radio.

amathophobia [PSYCH] Abnormal fear of dust.

amatol [MATER] An explosive mixture composed of ammonium nitrate and trinitrotoluene; mixtures with 50% and 80% ammonium nitrate are used for small and large shells, respectively.

amaurosis [MED] Total or partial blindness.

amaurotic familial idiocy [MED] A hereditary condition, transmitted as an autosomal recessive, predominantly in Jewish children, characterized by blindness, muscular weakness, and subnormal mental development; when onset is in infancy, the disease is known commonly as Tay-Sachs disease.

amazonite [MINERAL] An apple-green, bright-green, or blue-green variety of microcline found in the United States and Soviet Union; sometimes used as a gemstone. Also known as amazon stone.

amazon stone *See* amazonite.

ambatoarinite [MINERAL] A mineral consisting of a carbonate of cerium metals and strontium.

amber [MINERAL] A transparent yellow, orange, or reddish-brown fossil resin derived from a coniferous tree; used for ornamental purposes; it is amorphous, has a specific gravity of 1.05–1.10, and a hardness of 2–2.5 on Mohs scale.

amber glass [MATER] A tinted glass made by using different mixtures of sulfur and iron oxide; the color can vary from pale yellow to ruby amber.

ambergris [PHYSIO] A fatty substance formed in the intestinal tract of the sperm whale; used in the manufacture of perfume.

amberite [MATER] A smokeless powder composed of guncotton, barium nitrate, and paraffin.

amber mutation [GEN] Alteration of another triplet to *UAG,* a terminator codon controlling termination of polypepide chain synthesis in bacteria.

amber oil [MATER] **1.** A yellowish to brown essential oil made by destructive distillation of amber; has an acrid taste. **2.** A light essential oil prepared by destructive distillation of rosin.

ambidextrous [PHYSIO] Capable of using both hands with equal skill.

ambient [ENG] Surrounding; especially, of or pertaining to the environment about a flying aircraft or other body but undisturbed or unaffected by it, as in ambient air or ambient temperature.

ambient light [OPTICS] The surrounding light, such as that reaching a television picture-tube screen from light sources in a room.

ambient noise [ACOUS] The pervasive noise associated with a given environment, being usually a composite of sounds from sources both near and distant.

ambient pressure [FL MECH] The pressure of the surrounding medium, such as a gas or liquid, which comes into contact with an apparatus or with a reaction.

ambient stress field [GEOPHYS] The distribution and numerical value of the stresses present in a rock environment prior to its disturbance by man. Also known as in-place stress field; primary stress field; residual stress field.

ambient temperature [PHYS] The temperature of the surrounding medium, such as gas or liquid, which comes into contact with the apparatus.

ambigenous [BOT] Of a perianth whose outer leaves resemble the calyx while the inner leaves resemble the corolla.

ambiguity [ELECTR] The condition in which a synchro system or servosystem seeks more than one null position. [NAV] The condition in which navigation coordinates derived from a navigational instrument define more than one point, direction, line of position, or surface of position.

ambipolar [SCI TECH] Simultaneously operating in two opposite directions; for example, an electric current arising from the movement of positive and negative ions.

ambipolar diffusion [PHYS] The diffusion in a plasma of charged particles, such as electrons or ions, as a result of the almost exact local charge neutrality required.

ambisexual [MED] An individual having undifferentiated

primordia of both sexes. [PSYCH] Having feelings and exhibiting behavior common to both sexes.

ambitus [BIOL] The periphery or external edge, as of a mollusk shell or leaf.

ambivalence [PSYCH] The coexistence of conflicting reactions toward a person or object.

amblygonite [MINERAL] (Li,Na)AlPO$_4$(F,OH) A mineral occurring in white or greenish cleavable masses and found in the United States and Europe; important ore of lithium.

amblyopia [MED] Dimness of vision, especially that not due to refractive errors or organic disease of the eye; may be congenital or acquired.

Amblyopsidae [VERT ZOO] The cave fishes, a family of actinopterygian fishes in the order Percopsiformes.

Amblyopsiformes [VERT ZOO] An equivalent name for the Percopsiformes.

Amblypygi [INV ZOO] An order of chelicerate arthropods in the class Arachnida, commonly known as the tailless whip scorpions.

amboceptor [IMMUNOL] According to P. Ehrlich, an antibody present in the blood of immunized animals which contains two specialized elements: a cytophil group that unites with a cellular antigen, and a complementophil group that joins with the complement.

amboceptor unit [BIOL] A unit for the standardization of blood serum.

ambonite [PETR] Any of a group of hornblende-biotite andesites and dacites containing cordierite.

ambrite [MINERAL] A yellow-gray, semitransparent fossil resin resembling amber; found in large masses in New Zealand coal fields and regarded as a semiprecious stone.

ambrosine [MINERAL] A yellowish to clove-brown variety of amber rich in succinic acid; occurs as rounded masses in phosphate beds near Charleston, S.C.

ambrotype [GRAPHICS] An obsolete method of photography in which a negative is formed in a collodion emulsion on glass; when backed with black velvet or black varnish, the collodion surface reflects positive highlights and the resulting effect is that of a positive.

ambulacrum [INV ZOO] In echinoderms, any of the radial series of plates along which the tube feet are arranged.

ambulatorial [ZOO] **1.** Capable of walking. **2.** In reference to a forest animal, having adapted to walking, as opposed to running, crawling, or leaping.

ambulatory schizophrenia [PSYCH] A condition in which a person exhibits symptoms of both manic-depressive and schizophrenic psychosis but is not considered to require institutionalization.

Ambystoma [VERT ZOO] A genus of common salamanders; the type genus of the family Ambystomatidae.

Ambystomatidae [VERT ZOO] A family of urodele amphibians in the suborder Salamandroidea; neoteny occurs frequently in this group.

Ambystomoidea [VERT ZOO] A suborder to which the family Ambystomatidae is sometimes elevated.

ameba [INV ZOO] The common name for a number of species of naked unicellular protozoans of the order Amoebida. An example is a member of the genus *Amoeba.*

Amebelodontinae [PALEON] A subfamily of extinct elephantoid proboscideans in the family Gomphotheriidae.

amebiasis [MED] A parasitic disease of man caused by the ameba *Entamoeba histolytica,* characterized by clinical-pathological intestinal manifestations, including an acute dysentery phase. Also known as amebic dysentery.

amebic abscess [MED] Liquefactive necrosis of the brain and liver, without suppuration, caused by amebas, usually *Entamoeba histolytica.*

amebic dysentery *See* amebiasis.

amebicide [MATER] A chemical used to kill amebas, especially parasitic species.

amebocyte [INV ZOO] One of the wandering ameboid cells in the tissues and fluids of many invertebrates that function in assimilation and excretion.

ameboid movement [CYTOL] A type of cellular locomotion involving the formation of pseudopodia.

ameiosis [GEN] Nonreduction of chromosome number due

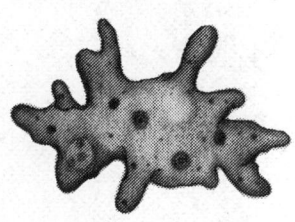

AMEBA

Typical species of ameba showing numerous pseudopodia, which are used for locomotion.

to suppression of one of the miotic divisions, as in parthenogenesis.

Ameiuridae [VERT ZOO] A family of North American catfishes belonging to the suborder Siluroidei.

ameloblast [EMBRYO] One of the columnar cells of the enamel organ that form dental enamel in developing teeth.

ameloblastic odontoma [MED] A neoplasm of epithelial and mesenchymal odontogenic tissue. Also known as odontoblastoma.

ameloblastoma [MED] An epithelial tumor associated with the enamel organ; cells of basal layers resemble the ameloblast. Also known as adamantinoma.

amemolite [GEOL] A stalactite with one or more changes in its axis of growth.

amenorrhea [MED] Absence of menstruation due to either normal or abnormal conditions.

ament [BOT] A catkin. [MED] A person with congenital mental deficiency; an idiot.

amentia [MED] Congenital subnormal intellectual development.

Amera [INV ZOO] One of the three divisions of the phylum Vermes proposed by O. Bütschli in 1910 and given the rank of a subphylum.

American basement [BUILD] A basement located above ground level and containing the building's main entrance.

American bond [CIV ENG] A bond in which every fifth, sixth, or seventh course of a wall consists of headers and the other courses consist of stretchers. Also known as common bond; Scotch bond.

American boreal faunal region [ECOL] A zoogeographic region comprising marine littoral animal communities of the coastal waters off east-central North America.

American boring system [MIN ENG] A rope system of percussive boring, with a derrick which enables the complete set of boring tools to be raised clear of the hole. Also known as American system.

American caisson *See* box caisson.

American dillweed oil *See* dill oil.

American-Egyptian cotton [BOT] A type of cotton developed by hybridization of Egyptian and American plants.

American Ephemeris and Nautical Almanac [ASTRON] An annual publication of the U.S. Naval Observatory containing tables of the predicted positions of various celestial bodies and other data of use to astronomers and navigators.

American explosive [MATER] One of many explosives that have passed U.S. Bureau of Mines tests and are used under certain conditions.

American filter *See* disk filter.

American lion *See* puma.

American melting point [CHEM ENG] A temperature 3°F (1.7°C) higher than the ASTM Method D87 paraffin-wax melting point.

American mucocutaneous leishmaniasis [MED] A form of leishmaniasis caused by *Leishmania braziliensis*, transmitted by sandflies of the genus *Phlebotomus*, and characterized by skin ulcers and ulceration and necrosis of the mucosa of the mouth and nose. Also known as South American leishmaniasis.

American run [TEXT] A system of numbering woolen yarns of 100 yards each in 1 ounce; thus if 1400 yards weigh 1 ounce, it would be a 14-run yarn.

American spotted fever *See* Rocky Mountain spotted fever.

American Standard Code for Information Interchange [COMMUN] Coded character set to be used for the general interchange of information among information-processing systems, communications systems, and associated equipment. Abbreviated ASCII.

American standard pipe thread [DES ENG] Taper, straight, or dryseal pipe thread whose dimensions conform to those of a particular series of specified sizes established as a standard in the United States. Also known as Briggs pipe thread.

American standard screw thread [DES ENG] Screw thread whose dimensions conform to those of a particular series of specified sizes established as a standard in the United States; used for bolts, nuts, and machine screws.

American system *See* American boring system.

American system drill *See* churn drill.

American Table of Distances [ENG] Published data concerning the safe storage of explosives and ammunition.

American wire gage [MET] A particular series of specified diameters and thicknesses established as a standard in the United States and used for nonferrous sheets, rods, and wires. Abbreviated AWG. Also known as Brown and Sharp gage (B and S gage).

American wormseed oil *See* chenopodium oil.

americium [CHEM] A chemical element, symbol Am, atomic number 95; the mass number of the isotope with the longest half-life is 243.

Amerosporae [MYCOL] A spore group of the Fungi Imperfecti characterized by one-celled or threadlike spores.

amesite [MINERAL] $(Mg,Fe)_4Al_4Si_2O_{10}(OH)_8$ Apple-green phyllosilicate mineral occurring in foliated hexagonal plates.

ametabolous metamorphosis [INV ZOO] A growth stage of certain insects characterized by an increase in size without distinct external changes.

amethocaine hydrochloride *See* tetracaine hydrochloride.

amethopterin [PHARM] $C_{20}H_{22}N_8O_5 \cdot H_2O$ An antimetabolite effective as a folic acid antagonist and used for treatment of acute and subacute leukemia. Also known as methotrexate.

amethyst [MINERAL] The transparent purple to violet variety of the mineral quartz; used as a jeweler's stone.

ametoecious [ECOL] Of a parasite that remains with the same host.

ametropia [MED] Any deficiency in the refractive ability of the eye that causes an unfocused image to fall on the retina.

AM field signature [ELECTR] The characteristic pattern of an alternating magnetic field, as displayed by detection and classification equipment.

amherstite [PETR] A syenodiorite containing andesine and antiperthite.

amianthus [MINERAL] A fine, silky variety of asbestos, such as chrysotile.

amicable numbers [MATH] Two numbers such that the exact divisors of each number (except the number itself) add up to the other number.

Amici prism [OPTICS] A compound prism, used in direct-vision spectroscopes, that disperses a beam of light into a spectrum without causing the beam as a whole to undergo any net deviation; it is made up of alternate crown and flint glass components, refracting in opposite directions.

amicron [PHYS CHEM] A particle having a size of 10^{-7} centimeter or less, which is a size in a system of classification of particle sizes in colloid chemistry.

amictic [INV ZOO] 1. In rotifers, producing diploid eggs that are incapable of being fertilized. 2. Pertaining to the egg produced by the amictic female.

amidase [BIOCHEM] Any enzyme that catalyzes the hydrolysis of nonpeptide $C=N$ linkages.

amidation [ORG CHEM] The process of forming an amide; for example, in the laboratory benzyl reacts with methyl amine to form N-methylbenzamide.

amide [ORG CHEM] One of a class of organic compounds containing the $CONH_2$ radical.

amidine [ORG CHEM] A compound which contains the radical $CNHNH_2$.

amido [ORG CHEM] Indicating the NH_2 radical when it is present in a molecule with the CO radical.

amidocarbonic acid *See* carbamic acid.

amidohydrolase [BIOCHEM] An enzyme that catalyzes deamination.

Amidol [ORG CHEM] $C_6H_3(NH_2)_2OH \cdot HCl$ A grayish-white crystalline salt; soluble in water, slightly soluble in alcohol; used as a developer in photography and as an analytical reagent. Also known as 2,4-diaminophenol hydrochloride.

amidourea hydrochloride *See* semicarbazide hydrochloride.

amidships [NAV ARCH] At or toward the middle of a ship.

Amiidae [VERT ZOO] A family of actinopterygian fishes in the order Amiiformes represented by a single living species, the bowfin (*Amia calva*).

Amiiformes [VERT ZOO] An order of actinopterygian fishes characterized by an abbreviate heterocercal tail, fusiform body, and median fin rays.

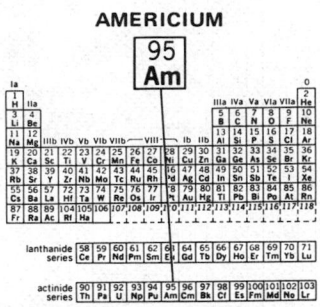

AMERICIUM

Periodic table of the chemical elements showing the position of americium.

A min See ampere-minute.

amin- See amino-.

amination [ORG CHEM] The preparation of amines.

amine [ORG CHEM] One of a class of organic compounds which can be considered to be derived from ammonia by replacement of one or more hydrogens by organic radicals.

amine oxidase [BIOCHEM] An enzyme that catalyzes the oxidation of tyramine and tryptamine to aldehyde.

aminiform See cystamine.

amino- [CHEM] Having the property of a compound in which the group NH_2 is attached to a radical other than an acid radical. Also spelled amin-.

aminoacetic acid See glycine.

amino acid [BIOCHEM] Any of the organic compounds that contain one or more basic amino groups and one or more acidic carboxyl groups and that are polymerized to form peptides and proteins; only 20 of the more than 80 amino acids found in nature serve as building blocks for proteins; examples are tyrosine and lysine.

amino aciduria [MED] A group of disorders in which excess amounts of amino acids are excreted in the urine; caused by abnormal protein metabolism.

1-aminoadamantane See amantadine.

amino alcohol See alkamine.

para-aminobenzenesulfonamide See sulfanilamide.

meta-aminobenzenesulfonic acid See metanillic acid.

para-aminobenzenesulfonic acid See sulfanilic acid.

para-aminobenzoic acid [BIOCHEM] $C_7H_7O_2N$ A yellow-red, crystalline compound that is part of the folic acid molecule; essential in metabolism of certain bacteria. Abbreviated PABA.

para-aminobenzoyldiethylaminoethanol base See procaine base.

D-aminobenzyl penicillin See Ampicillin.

1-aminobutane See n-butyl amine.

2-aminobutane See sec-butyl amine.

amino diabetes [MED] A congenital disorder characterized by excessive quantities of amino acids, glucose, and phosphate in the urine, resulting from deficient resorption in the proximal convoluted tubules of the kidney. Also known as Fanconi's syndrome.

aminodimethylbenzene See xylidine.

2-amino-4,6-dinitrophenol See picramic acid.

aminodithioformic acid See dithiocarbamic acid.

aminoethane See ethyl amine.

2-aminoethanesulfonic acid See taurine.

2-aminoethanol See ethanolamine.

4-aminofolic acid See aminopterin.

aminoformic acid See carbamic acid.

aminohexane See hexylamine.

α-aminohydrocinnamic acid See phenylalanine.

2-aminoisovaleric acid See valine.

α-aminoisovaleric acid See valine.

aminomercuric chloride See ammoniated mercury.

aminomethane See methylamine.

2-amino-3-methylbutyric acid See valine.

2-amino-2-methyl-1-propanol-8-bromotheophyllinate See pamabrom.

1-aminonaphthalene-4-sulfonic acid See naphthionic acid.

4-amino-1-naphthalene sulfonic acid See naphthionic acid.

1-amino-8-naphthol-4,6-disulfonic acid See K acid.

8-amino-1-naphthol-3,5-disulfonic acid See K acid.

8-amino-1-naphthol-3,6-disulfonic acid See H acid.

2-amino-5-naphthol-7-sulfonic acid [ORG CHEM] $C_{10}H_5NH_2$ $OHSO_3H$ Gray or white needles that are soluble in hot water; used as a dye intermediate. Also known as 6-amino-1-naphthol-3-sulfonic acid; J acid.

5-amino-1-naphthol-3-sulfonic acid See M acid.

6-amino-1-naphthol-3-sulfonic acid See 2-amino-5-naphthol-7-sulfonic acid.

amino nitrogen [CHEM] Nitrogen combined with hydrogen in the amino group. Also known as ammonia nitrogen.

aminopeptidase [BIOCHEM] An enzyme which catalyzes the liberation of an amino acid from the end of a peptide having a free amino group.

4-aminophenetole See phenetidine.

aminophenol [ORG CHEM] A type of compound containing the NH_2 and OH groups joined to the benzene ring; examples are para-aminophenol and ortho-hydroxylaniline.

ortho-aminophenol See ortho-hydroxylaniline.

para-aminophenol [ORG CHEM] $p\text{-}HOC_6H_4NO_2$ A phenol in which an amino ($-NH_2$) group is located on the benzene ring of carbon atoms para (p) to the hydroxyl ($-OH$) group; used as a photographic developer and as an intermediate in dye manufacture.

α-amino-β-phenylpropionic acid See phenylalanine.

3-aminophthalic acid cyclic hydrazide See luminol.

aminophylline [PHARM] $C_{16}H_{24}N_{10}O_4$ A drug in the form of white or slightly yellowish, water-soluble granules or powder, used as a smooth-muscle relaxant, myocardial stimulant, and diuretic.

aminopolypeptidase [BIOCHEM] A proteolytic enzyme that cleaves polypeptides containing either a free amino group or a basic nitrogen atom having at least one hydrogen atom.

3-aminopropene See allylamine.

aminoprotease [BIOCHEM] An enzyme that hydrolyzes a protein and unites with its free amino group.

aminopterin [PHARM] $C_{19}H_{20}N_8O_5 \cdot 2H_2O$ A yellow crystalline acid which is similar to folic acid and is used clinically as an antagonist of folic acid. Trade name for 4-aminofolic acid; 4-aminopteroylglutamic acid.

4-aminopteroylglutamic acid See aminopterin.

aminopyrine [PHARM] $C_{13}H_{17}N_3O$ White, crystalline compound, used as an antipyretic and analgesic.

amino resin [ORG CHEM] A type of resin prepared by condensation polymerization, with an aldehyde, of a compound containing an amino group.

para-aminosalicylic acid [PHARM] $C_7H_7NO_3$ White, crystalline drug used with other drugs in the treatment of tuberculosis. Abbreviated PAS.

amino sugar [BIOCHEM] A monosaccharide in which a nonglycosidic hydroxyl group is replaced by an amino or substituted amino group; an example is D-glucosamine.

amino terminal [BIOCHEM] The end part of a polypeptide chain which contains a free α-amino group.

2-aminothiazole [ORG CHEM] $C_3H_4N_2S$ Pale-yellow crystals that melt at 92°C; soluble in cold water, slightly soluble in ethyl alcohol; used as an intermediate in the synthesis of sulfathiazole.

aminothiourea See thiosemicarbazide.

meta-aminotoluene See meta-toluidine.

ortho-aminotoluene See ortho-toluidine.

para-aminotoluene See para-toluidine.

aminotransferase See transaminase.

aminourea hydrochloride See semicarbazide hydrochloride.

aminoxylene See xylidine.

amitosis [CYTOL] Cell division by simple fission of the nucleus and cytoplasm without chromosome differentiation.

Ammanian [GEOL] Middle Upper Cretaceous geologic time.

ammeter [ENG] An instrument for measuring the magnitude of electric current flow.

ammine [INORG CHEM] One of a group of complex compounds formed by coordination of ammonia molecules with metal ions.

ammocoete [ZOO] A protracted larval stage of lampreys.

ammocolous [ECOL] Describing plants having a habitat in dry sand.

Ammodiscacea [INV ZOO] A superfamily of foraminiferal protozoans in the suborder Textulariina, characterized by a simple to labyrinthic test wall.

Ammodytoidei [VERT ZOO] The sand lances, a suborder of marine actinopterygian fishes in the order Perciformes, characterized by slender, eel-shaped bodies.

ammonal [MATER] A high-explosive mixture, made of ammonium nitrate, trinitrotoluene (TNT), and flaked or powdered aluminum.

ammonation [INORG CHEM] A reaction in which ammonia is added to other molecules or ions by covalent bond formation utilizing the unshared pair of electrons on the nitrogen atom, or through ion-dipole electrostatic interactions.

ammonia [INORG CHEM] NH_3 A colorless gaseous alkaline compound that is very soluble in water, has a characteristic pungent odor, is lighter than air, and is formed as a result of the decomposition of most nitrogenous organic material; used as a fertilizer and as a chemical intermediate.

PARA-AMINOPHENOL

Structural formula for para-aminophenol.

PARA-AMINOSALICYLIC ACID

Structural formula for para-aminosalicylic acid.

AMINO SUGAR

The Haworth formula of D-glucosamine.

ammonia alum *See* ammonium aluminum sulfate.

ammonia-beam maser [PHYS] A gas maser using ammonia as the paramagnetic material.

ammoniac [INORG CHEM] *See* ammoniacal. [MATER] Gum resin obtained from the stems of the ammoniac plant; used in medicine, perfume, plaster, concrete, and adhesive. Also known as ammoniac gum; ammoniacum; gum ammoniac; Persian ammoniac.

ammoniacal [INORG CHEM] Pertaining to ammonia or its properties. Also known as ammoniac.

ammoniac gum *See* ammoniac.

ammonia clock [HOROL] A time-measuring device dependent on the pyramidal ammonia molecule's property of turning inside out readily and oscillating between the two extreme positions at the precise frequency of 2.387013×10^{10} hertz.

ammonia compressor [MECH ENG] A device that decreases the volume of a quantity of gaseous ammonia by the amplification of pressure; used in refrigeration systems.

ammonia condenser [MECH ENG] A device in an ammonia refrigerating system that raises the pressure of the ammonia gas in the evaporating coil, conditions the ammonia, and delivers it to the condensing system.

ammoniacum *See* ammoniac.

ammonia dynamite [CHEM] Dynamite with part of the nitroglycerin replaced by ammonium nitrate.

ammonia gelatin [MATER] An explosive of the gelatin dynamite class containing ammonium nitrate.

ammonia liquor [CHEM ENG] Water solution of ammonia, ammonium compounds, and impurities, obtained from destructive distillation of bituminous coal.

ammonia maser clock [HOROL] A gas maser that utilizes the transition of high-energy ammonia molecules to generate a stable microwave output signal for use as a time standard.

ammonia nitrogen *See* amino nitrogen; imino nitrogen.

ammonia permissible [MATER] A permissible explosive that is an ammonia dynamite.

ammonia synthesis [CHEM ENG] Chemical combination of nitrogen and hydrogen gases at high temperature and pressure in the presence of a catalyst to form ammonia.

ammoniated mercuric chloride *See* ammoniated mercury.

ammoniated mercury [INORG CHEM] $HgNH_2Cl$ A white powder that darkens on light exposure; insoluble in water and alcohol, soluble in ammonium carbonate solutions and in warm acids; used in pharmaceuticals and as a local anti-infective in medicine. Also known as aminomercuric chloride; ammoniated mercuric chloride; ammoniated mercury chloride; ammonobasic mercuric chloride; mercury cosmetic.

ammoniated mercury chloride *See* ammoniated mercury.

ammoniated ruthenium oxychloride *See* ruthenium red.

ammoniated superphosphate [INORG CHEM] A fertilizer containing 5 parts of ammonia to 100 parts of superphosphate.

ammoniation [CHEM] Treating or combining with ammonia.

ammonia valve [ENG] A valve that is resistant to corrosion by ammonia.

ammonia water [CHEM] A water solution of ammonia; a clear colorless liquid that is basic because of dissociation of NH_4OH to produce hydroxide ions; used as a reagent, solvent, and neutralizing agent.

ammonification [CHEM] Addition of ammonia or ammonia compounds, especially to the soil.

ammonifiers [ECOL] Fungi, or actinomycetous bacteria, that participate in the ammonification part of the nitrogen cycle and release ammonia (NH_3) by decomposition of organic matter.

ammonioborite [MINERAL] $(NH_4)_2B_{10}O_{16} \cdot 5H_2O$ A white mineral consisting of a hydrous ammonium borite and occurring as aggregates of minute plates.

ammoniojarosite [MINERAL] $(NH_4)Fe_3(SO_4)_2(OH)_6$ Pale-yellow mineral consisting of basic ferric ammonium sulfate.

ammonite [MATER] An explosive containing 70–95% ammonium nitrate. [PALEON] A fossil shell of the cephalopod order Ammonoidea.

ammonium [CHEM] The radical NH_4^+. Abbreviated Am.

ammonium acetate [ORG CHEM] **1.** CH_3COONH_4 A normal salt formed by the neutralization of acetic acid with ammonium hydroxide; a white, crystalline, deliquescent material used in solution for the standardization of electrodes for hydrogen ions. **2.** $CH_3COONH_4 \cdot CH_3COOH$ An acid salt resulting from the distillation of the neutral salt or from its solution in hot acetic acid; crystallizes in deliquescent needles. **3.** A mixture of the normal and acetic salts; used as a mordant in the dyeing of wool.

ammonium acid carbonate *See* ammonium bicarbonate.

ammonium acid fluoride *See* ammonium bifluoride.

ammonium acid tartrate *See* ammonium bitartrate.

ammonium alginate [ORG CHEM] $(C_6H_7O_6 \cdot NH_4)_n$ A high-molecular-weight, hydrophilic colloid; used as a thickening agent/stabilizer in ice cream, cheese, canned fruits, and other food products.

ammonium alum *See* ammonium aluminum sulfate.

ammonium aluminum sulfate [INORG CHEM] $NH_4Al(SO_4)_2 \cdot 12H_2O$ Colorless, odorless crystals that are soluble in water; used in manufacturing medicines and baking powder, dyeing, papermaking, and tanning. Also known as alum; aluminum ammonium sulfate; ammonia alum; ammonium alum.

ammonium benzoate [ORG CHEM] $NH_4C_7H_5O_2$ A salt of benzoic acid prepared as a coarse, white powder; used as a preservative in certain adhesives and rubber latex.

ammonium bicarbonate [INORG CHEM] NH_4HCO_3 White, crystalline, water-soluble salt; used in baking powders and in fire-extinguishing mixtures. Also known as ammonium acid carbonate; ammonium hydrogen carbonate.

ammonium bichromate *See* ammonium dichromate.

ammonium bifluoride [INORG CHEM] $NH_4F \cdot HF$ A salt that crystallizes in the orthorhombic system and is soluble in water; prepared in the form of white flakes from ammonia treated with hydrogen fluoride; used in solution as a fungicide and wood preservative. Also known as ammonium acid fluoride; ammonium hydrogen fluoride.

ammonium bitartrate [ORG CHEM] $NH_4HC_4H_4O_6$ Colorless crystals that are soluble in water; used to make baking powder and to detect calcium. Also known as acid ammonium tartrate; ammonium acid tartrate; ammonium hydrogen tartrate; monoammonium tartrate.

ammonium bithiolicum *See* ichthammol.

ammonium borate [INORG CHEM] NH_4BO_3 A white, crystalline, water-soluble salt which decomposes at 198°C; used as a fire retardant on fabrics.

ammonium bromide [INORG CHEM] NH_4Br An ammonium halide that crystallizes in the cubic system; made by the reaction of ammonia with hydrobromic acid or bromine; used in photography and for pharmaceutical preparations (sedatives).

ammonium carbamate [INORG CHEM] $NH_4NH_2CO_2$ A salt that forms colorless, rhombic crystals, which are very soluble in cold water; an important, unstable intermediate in the manufacture of urea; found in commercial ammonium carbonate.

ammonium carbonate [INORG CHEM] **1.** $(NH_4)_2CO_3$ The normal ammonium salt of carbonic acid, prepared by passing gaseous carbon dioxide into an aqueous solution of ammonia and allowing the vapors (ammonia, carbon dioxide, water) to crystallize. **2.** $NH_4HCO_3 \cdot NH_2COONH_4$ A white, crystalline double salt of ammonium bicarbonate and ammonium carbamate obtained commercially; the principal ingredient of smelling salts.

ammonium chloride [INORG CHEM] NH_4Cl A white crystalline salt that occurs naturally as a sublimation product of volcanic action or is manufactured; used as an electrolyte in dry cells, as a flux for soldering, tinning, and galvanizing, and as an expectorant.

ammonium chromate [INORG CHEM] $(NH_4)_2CrO_4$ A salt that forms yellow, monoclinic crystals; made from ammonium hydroxide and ammonium dichromate; used in photography as a sensitizer for gelatin coatings.

ammonium citrate [ORG CHEM] $(NH_4)_2HC_6H_5O_7$ White, granular material; used as a reagent.

ammonium dichromate [INORG CHEM] $(NH_4)_2Cr_2O_7$ A salt that forms orange, monoclinic crystals; made from ammonium sulfate and sodium dichromate; soluble in water and alcohol; ignites readily; used in photography, lithography, pyrotechnics, and dyeing. Also known as ammonium bichromate.

ammonium fluoride [INORG CHEM] NH_4F A white, unstable, crystalline salt with a strong odor of ammonia.

ammonium halide [INORG CHEM] A compound with the ammonium ion bonded to an ion formed from one of the halogen elements.

ammonium hydrogen carbonate *See* ammonium bicarbonate.

ammonium hydrogen fluoride *See* ammonium bifluoride.

ammonium hydrogen tartrate *See* ammonium bitartrate.

ammonium hydroxide [INORG CHEM] NH_4OH A hydrate of ammonia, crystalline below $-79°C$; it is a weak base known only in solution as ammonia water. Also known as aqua ammonia.

ammonium ichthosulfonate *See* ichthammol.

ammonium iodide [INORG CHEM] NH_4I A salt prepared from ammonia and hydrogen iodide or iodine; it forms colorless, regular crystals which sublime when heated; used in photography and for pharmaceutical preparations.

ammonium molybdate [INORG CHEM] $(NH_4)_2MoO_4$ White, crystalline salt used as an analytic reagent, as a precipitant of phosphoric acid, and in pigments.

ammonium nitrate [INORG CHEM] NH_4NO_3 A colorless crystalline salt; very insensitive and stable high explosive; also used as a fertilizer.

ammonium oxalate [ORG CHEM] $(NH_4)_2C_2O_4 \cdot H_2O$ A salt in the form of colorless, rhombic crystals.

ammonium perchlorate [INORG CHEM] NH_4ClO_4 A salt that forms colorless or white rhombic and regular crystals, which are soluble in water; it decomposes at $150°C$, and the reaction is explosive at higher temperatures.

ammonium perchlorate explosive [MATER] Any of several compositions consisting of ammonium perchlorate in combination with a high explosive and powdered metal; produces powerful blast.

ammonium phosphate [INORG CHEM] $(NH_4)_2HPO_4$ A salt of ammonia and phosphoric acid that forms white monoclinic crystals, which are soluble in water; used as a fertilizer and fire retardant.

ammonium picrate [ORG CHEM] $NH_4C_6H_2O(NO_2)_3$ Compound with stable yellow and metastable red forms of orthorhombic crystals; used as a military explosive for armor-piercing shells. Also known as ammonium trinitrophenolate; explosive D.

ammonium salt [INORG CHEM] A product of a reaction between ammonia and various acids; examples are ammonium chloride and ammonium nitrate.

ammonium soap [ORG CHEM] A product from reaction of a fatty acid with ammonium hydroxide; used in toiletry preparations such as soaps and in emulsions.

ammonium sulfate [INORG CHEM] $(NH_4)_2SO_4$ Colorless, rhombic crystals which melt at $140°C$ and are soluble in water.

ammonium sulfoichthyolate *See* ichthammol.

ammonium tartrate [ORG CHEM] $(NH_4)_2C_2H_4O_6$ Colorless, monoclinic crystals; used in textiles and in medicine.

ammonium trinitrophenolate *See* ammonium picrate.

ammonobasic mercuric chloride *See* ammoniated mercury.

ammonoid [PALEON] A cephalopod of the order Ammonoidea.

Ammonoidea [PALEON] An order of extinct cephalopod mollusks in the subclass Tetrabranchia; important as index fossils.

ammonolysis [CHEM] **1.** A dissociation reaction of the ammonia molecule producing H^+ and NH_2^- species. **2.** Breaking of a bond by addition of ammonia.

Ammon's law [ANTHRO] The law stating that cephalic index and stature vary inversely.

Ammotheidae [INV ZOO] A family of marine arthropods in the subphylum Pycnogonida.

ammunition [ORD] **1.** All kinds of missiles to be thrown against an enemy. **2.** Missiles not for direct use against an enemy, with such purposes as illumination, signaling, and decelerating. **3.** A complete round and all its components, that is, the material required for firing a weapon such as a pistol.

ammunition belt [ORD] A fabric or metal band with loops for carrying cartridges that are fed from it into a machine gun or other automatic weapon; a feed belt.

ammunition box [ORD] A box in which linked ammunition is folded and from which it can be fed into a machine gun.

ammunition carrier [ORD] **1.** A vehicle that accompanies guns and carries ammunition for them. **2.** A member of a gun or mortar squad who carries ammunition and helps load in actual firing.

ammunition chest [ORD] A receptacle, as on a caisson, limber, or gun carriage, in which ammunition is kept.

ammunition day of supply [ORD] The estimated quantity of ammunition required per day to sustain operations in an active theater.

ammunition depot [ORD] A storage and supply depot for projectiles, bombs, or other ammunition.

ammunition dump [ORD] An ammunition storage point, usually established in the field for temporary use.

amnesia [MED] The pathological loss or impairment of memory brought about by psychogenic or physiological disturbances.

amnesic aphasia [MED] Loss of memory for the appropriate names of objects, conditions, or relations, accompanied by fragmented or hesitant speech.

amnicolous [ECOL] Describing plants having a habitat on sandy riverbanks.

amniocentesis [MED] A procedure during pregnancy by which the abdominal wall and fetal membranes are punctured with a cannula to withdraw amniotic fluid.

amnion [EMBRYO] A thin extraembryonic membrane forming a closed sac around the embryo in birds, reptiles, and mammals.

Amniota [VERT ZOO] A collective term for the Reptilia, Aves, and Mammalia, all of which have an amnion during development.

amniotic fluid [PHYSIO] A substance that fills the amnion to protect the embryo from desiccation and shock.

amodiaquine [PHARM] $C_{20}H_{22}ClN_3O$ A crystalline compound that melts at $208°C$; used as an antimalarial.

Amoeba [INV ZOO] A genus of naked, rhizopod protozoans in the order Amoebida characterized by a thin pellicle and thick, irregular pseudopodia.

Amoebida [INV ZOO] An order of rhizopod protozoans in the subclass Lobosia characterized by the absence of a protective covering (test).

Amoebobacter [MICROBIO] A sulfur purple bacteria, a genus of spherical to rod-shaped photosynthetic bacteria in the family Thiorhodaceae.

amoeboid fold [GEOL] A fold or structure, such as an anticline, having no prevailing trend or definite shape.

amoeboid glacier [HYD] A glacier connected with its snowfield for a portion of the year only.

amorphous [PHYS] Pertaining to a solid which is noncrystalline, having neither definite form nor structure.

amorphous frost [HYD] Hoar frost which possesses no apparent simple crystalline structure; opposite of crystalline frost.

amorphous memory array [ADP] An array of memory switches made of amorphous material.

amorphous mineral [MINERAL] A mineral without definite crystalline structure.

amorphous peat [GEOL] Peat composed of fine grains of organic matter; is plastic like wet, heavy soil, with all original plant structures destroyed by decomposition of cellulosic matter.

amorphous semiconductor [SOLID STATE] A semiconductor material which is not entirely crystalline, having only short-range order in its structure.

amorphous sky [METEOROL] A sky characterized by an abundance of fractus clouds, usually accompanied by precipitation falling from a higher, overcast cloud layer.

amorphous snow [HYD] A type of snow with irregular crystalline structure.

amortisseur winding *See* damper winding.

amortize [IND ENG] To reduce gradually an obligation, such as a mortgage, by periodically paying a part of the principal as well as the interest.

amosite [MINERAL] A monoclinic amphibole form of asbestos having long fibers and a high iron content; used in insulation.

amount limit [IND ENG] In a test for a fixed quantity of work, the time required to complete the work or the total amount of work that can be completed in an unlimited time.

para-**amoxyphenol** See *para*-pentyloxyphenol.

amp See amperage; ampere.

AMP See adenylic acid.

ampacity [ELEC] Current-carrying capacity in amperes; used as a rating for power cables.

ampangabeite See samarskite.

Ampeliscidae [INV ZOO] A family of tube-dwelling amphipod crustaceans in the suborder Gammaridea.

ampelite [PETR] A graphite schist containing silica, alumina, and sulfur; used as a refractory.

amperage [ELEC] The amount of electric current in amperes. Abbreviated amp.

ampere [ELEC] The unit of electric current in the rationalized meter-kilogram-second system of units; defined in terms of the force of attraction between two parallel current-carrying conductors. Abbreviated a; A; amp.

Ampère currents [ELECTROMAG] Postulated "molecular-ring" currents to explain the phenomena of magnetism as well as the apparent nonexistence of isolated magnetic poles.

ampere-hour [ELEC] A unit for the quantity of electricity, obtained by integrating current flow in amperes over the time in hours for its flow; used as a measure of battery capacity. Abbreviated Ah; amp-hr.

ampere-hour capacity [ELEC] The charge, measured in ampere-hours, that can be delivered by a storage battery up to the limit to which the battery may be safely discharged.

Ampère law [ELECTROMAG] **1.** A law giving the magnetic induction at a point due to given currents in terms of the current elements and their positions relative to the point. Also known as Laplace law. **2.** A law giving the line integral over a closed path of the magnetic induction due to given currents in terms of the total current linking the path.

ampere meter squared [ELECTROMAG] The SI unit of electromagnetic moment. Abbreviated Am^2.

ampere-minute [ELEC] A unit of electrical charge, equal to the charge transported in 1 minute by a current of 1 ampere, or to 60 coulombs. Abbreviated A min.

ampere per meter [ELECTROMAG] The SI unit of magnetic field strength and magnetization. Abbreviated A/m.

ampere per square inch [ELEC] A unit of current density, equal to the uniform current density of a current of 1 ampere flowing through an area of 1 square inch. Abbreviated A/in^2.

ampere per square meter [ELEC] The SI unit of current density. Abbreviated A/m^2.

Ampère rule [ELECTROMAG] The rule which states that the direction of the magnetic field surrounding a conductor will be clockwise when viewed from the conductor if the direction of current flow is away from the observer.

ampere square meter per joule second [ELECTROMAG] The SI unit of gyromagnetic ratio. Abbreviated Am^2/Js.

Ampère theorem [ELECTROMAG] The theorem which states that an electric current flowing in a circuit produces a magnetic field at external points equivalent to that due to a magnetic shell whose bounding edge is the conductor and whose strength is equal to the strength of the current.

ampere-turn [ELECTROMAG] A unit of magnetomotive force in the meter-kilogram-second system defined as the force of a closed loop of one turn when there is a current of 1 ampere flowing in the loop. Abbreviated amp-turn.

amperometric titration [PHYS CHEM] A titration that involves measuring an electric current or changes in current during the course of the titration.

amperometry [PHYS CHEM] Chemical analysis by techniques which involve measuring electric currents.

Ampharetidae [INV ZOO] A large, deep-water family of polychaete annelids belonging to the Sedentaria.

Ampharetinae [INV ZOO] A subfamily of annelids belonging to the family Ampharetidae.

amphetamine [PHARM] $C_6H_5CH_2CHNHCH_3$ A volatile, colorless liquid used as a central nervous system stimulant. Also known as racemic 1-phenyl-2-aminopropane and by the trade name Benzedrine.

amphiarthrosis [ANAT] An articulation of limited movement in which bones are connected by fibrocartilage, such as that between vertebrae or that at the tibiofibular junction.

amphiaster [INV ZOO] Type of spicule found in some sponges.

Amphibia [VERT ZOO] A class of vertebrate animals in the superclass Tetrapoda characterized by a moist, glandular skin, gills at some stage of development, and no amnion during the embryonic stage.

Amphibicorisae [INV ZOO] A subdivision of the insect order Hemiptera containing surface water bugs with exposed antennae.

Amphibioidei [INV ZOO] A family of tapeworms in the order Cyclophyllidea.

amphibiotic [ZOO] Being aquatic during the larval stage and terrestrial in the adult stage.

amphibious [BIOL] Capable of living both on dry or moist land and in water. [MECH ENG] Of vehicles or equipment designed to be operated or used on either land or water. [ORD] A military operation conducted by coordinated action of land, sea, and air forces.

amphibious-assault landing model See assault-landing model.

amphibious assault ship [NAV ARCH] A ship designed to transport and land troops, equipment, and supplies by means of embarked helicopters.

amphibious command ship [NAV ARCH] A naval ship from which a commander exercises control in amphibious operations.

amphibious mine [ORD] A mine designed especially to hinder beach landing and river-crossing operations by damaging or destroying landing craft, small boats, water-fording vehicles, and floating bridges; it may be of various types, such as contact, controlled, or drifting.

amphibious tank [ORD] A vehicle mounting a howitzer or cannon, capable of delivering direct fire from the water as well as from land; used in providing early artillery support in amphibious operations.

amphibious tractor [ORD] A vehicle used for the movement of troops and cargo from ship to shore in the assault phase of amphibious operations, or for limited movement of troops and cargo over land or water. Abbreviated amtrac.

amphiblastic cleavage [EMBRYO] The unequal but complete cleavage of telolecithal eggs.

amphiblastula [EMBRYO] A blastula resulting from amphiblastic cleavage. [INV ZOO] The free-swimming flagellated larva of many sponges.

amphibole [MINERAL] Any of a group of rock-forming, ferromagnesian silicate minerals commonly found in igneous and metamorphic rocks; includes hornblende, anthophyllite, tremolite, and actinolite (asbestos minerals).

amphibolic [MED] Uncertain; wavering; refers to the stage of a disease when prognosis is uncertain. [ZOO] Possessing the ability to turn either backward or forward, as the outer toe of certain birds.

amphibolic pathway [BIOCHEM] A microbial biosynthetic and energy-producing pathway, such as the glycolytic pathway.

Amphibolidae [INV ZOO] A family of gastropod mollusks in the order Basommatophora.

amphibolite [PETR] A crystalloblastic metamorphic rock composed mainly of amphibole and plagioclase; quartz may be present in small quantities.

amphibolite facies [PETR] Rocks produced by medium- to high-grade regional metamorphism.

amphibolization [PETR] Formation of amphibole in a rock as a secondary mineral.

amphicarpic [BOT] Having two types of fruit, differing either in form or ripening time.

Amphichelydia [PALEON] A suborder of Triassic to Eocene anapsid reptiles in the order Chelonia; these turtles did not have a retractable neck.

amphichrome [BOT] A plant that produces flowers of different colors on the same stalk.

Amphicoela [VERT ZOO] A small suborder of amphibians in the order Anura characterized by amphicoelous vertebrae.

AMPERE LAW

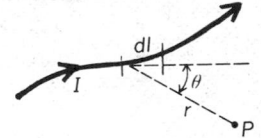

A graphic representation of the Ampère law (def. 1). Contribution of current element to magnetic induction of P is proportional to $I\,dl\,\sin\,\theta/r^2$. I = current, dl = length of current element, r = distance of point P from current elements, angle θ is between current element and the line forming the element to point P.

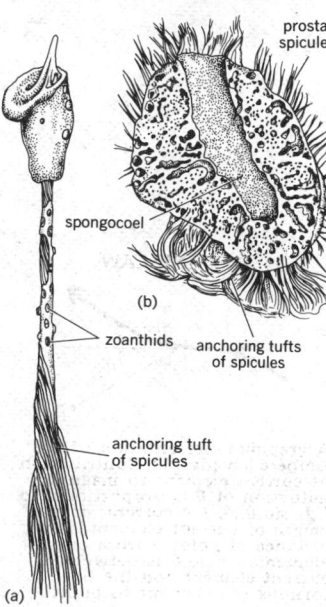

AMPHIDISCOPHORA

prostal
spicules

spongocoel

zoanthids

anchoring tufts
of spicules

anchoring tuft
of spicules

(b)

(a)

Representative
amphidiscophorans.
*(a) Hyalonema, with zoanthids
encrusting root tuft of spicules
(after Hyman, 1940);
(b) Pheronema, sectioned
longitudinally (from Hyman,
1940, after Schulze, 1887)*

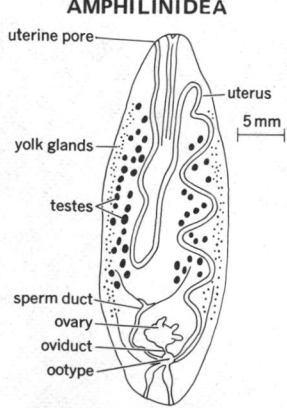

AMPHILINIDEA

uterine pore

uterus

5 mm

yolk glands

testes

sperm duct

ovary

oviduct

ootype

Amphilina, the only tapeworm in
the order Amphilinidea whose
life history is completely known.

amphicoelous [VERT ZOO] Describing vertebrae that have biconcave centra.

amphicribral [BOT] Having the phloem surrounded by the xylem, as seen in certain vascular bundles.

amphicryptophyte [BOT] A marsh plant with amphibious vegetative organs.

Amphicyonidae [PALEON] A family of extinct giant predatory carnivores placed in the infraorder Miacoidea by some authorities.

amphicytula [EMBRYO] A zygote that is capable of holoblastic unequal cleavage.

amphid [INV ZOO] Either of a pair of sensory receptors in nematodes, believed to be chemoreceptors and situated laterally on the anterior end of the body.

amphidetic [INV ZOO] Of a bivalve ligament, extending both before and behind the beak.

amphidiploid [GEN] An organism having a diploid set of chromosomes from each parent.

Amphidiscophora [INV ZOO] A subclass of sponges in the class Hexactinellida characterized by an anchoring tuft of spicules and no hexasters.

Amphidiscosa [INV ZOO] An order of hexactinellid sponges in the subclass Amphidiscophora characterized by amphidisc spicules, that is, spicules having a stellate disk at each end.

amphidromic [OCEANOGR] Of or pertaining to progression of a tide wave or bulge around a point or center of little or no tide.

amphidromic point [MAP] On a chart of cotidal lines, a no-tide or nodal point from which the cotidal lines radiate.

amphidromic region [MAP] An area surrounding an amphidromic point in which the cotidal lines radiate from the no-tide point and progress through all hours of the tide cycle.

amphigastrium [BOT] Any of the small appendages located ventrally on the stem of some liverworts.

amphigean [ECOL] An organism that is native to both Old and New Worlds.

amphigene *See* leucite.

Amphilestidae [PALEON] A family of Jurassic triconodont mammals whose subclass is uncertain.

Amphilinidea [INV ZOO] An order of tapeworms in the subclass Cestodaria characterized by a protrusible proboscis, anterior frontal glands, and no holdfast organ; they inhabit the coelom of sturgeon and other fishes.

Amphimerycidae [PALEON] A family of late Eocene to early Oligocene tylopod ruminants in the superfamily Amphimerycoidea.

Amphimerycoidea [PALEON] A superfamily of extinct ruminant artiodactyls in the infraorder Tylopoda.

amphimixis [PHYSIO] The union of egg and sperm in sexual reproduction.

Amphimonadidae [INV ZOO] A family of zoomastigophorean protozoans in the order Kinetoplastida.

amphimorphic [GEOL] A rock or mineral formed by two geologic processes.

Amphineura [INV ZOO] A class of the phylum Mollusca; members are bilaterally symmetrical, elongate marine animals, such as the chitons.

Amphinomidae [INV ZOO] The stinging or fire worms, a family of amphinomorphan polychaetes belonging to the Errantia.

Amphinomorpha [INV ZOO] Group name for three families of errantian polychaetes: Amphenomidae, Euphrosinidae, and Spintheridae.

amphioxus [ZOO] Former designation for the lancelet, *Branchiostoma.*

amphiphloic [BOT] Pertaining to the central vascular cylinder of stems having phloem on both sides of the xylem.

amphiphyte [ECOL] A plant growing on the boundary zone of wet land.

amphiplatyan [ANAT] Describing vertebrae having centra that are flat both anteriorly and posteriorly.

amphipneustic [VERT ZOO] Having both gills and lungs through all life stages, as in some amphibians.

Amphipoda [INV ZOO] An order of crustaceans in the subclass Malacostraca; individuals lack a carapace, bear unstalked eyes, and respire through thoracic branchiae or gills.

amphipodous [INV ZOO] Having both walking and swimming legs.

amphiprotic *See* amphoteric.

amphisapropel [GEOL] Cellulosic ooze containing coarse plant debris.

amphisarca [BOT] An indehiscent fruit characterized by many cells and seeds, pulpy flesh, and a hard rind; melon is an example.

Amphisbaenidae [VERT ZOO] A family of tropical snakelike lizards in the suborder Sauria.

Amphisopidae [INV ZOO] A family of isopod crustaceans in the suborder Phreactoicoidea.

amphispore [MYCOL] A specialized urediospore with a thick, colorful wall; a resting spore.

Amphissitidae [PALEON] A family of extinct ostracods in the suborder Beyrichicopina.

Amphistaenidae [VERT ZOO] The worm lizards, a family of reptiles in the suborder Sauria; structural features are greatly reduced, particularly the limbs.

amphistome [INV ZOO] An adult type of digenetic trematode having a well-developed ventral sucker (acetabulum) on the posterior end.

amphistylar [ARCH] Having free columns in porticoes either at both ends or at both sides and across the full ends of sides.

amphistylic [VERT ZOO] Having the jaw suspended from the brain case and the hyomandibular cartilage, as in some sharks.

amphitene *See* zygotene.

amphitheater [ARCH] A structure or large room containing oval, circular, or semicircular tiers of seats facing an open space. [GEOGR] A valley or gulch having an oval or circular floor surrounded by glacial action.

amphithecium [BOT] The external cell layer during development of the sporangium in mosses.

Amphitheriidae [PALEON] A family of Jurassic therian mammals in the infraclass Pantotheria.

amphitriaene [INV ZOO] A poriferan spicule having three divergent rays at each end.

amphitrichous [BIOL] Having flagella at both ends, as in certain bacteria.

Amphitritinae [INV ZOO] A subfamily of sedentary polychaete worms in the family Terebellidae.

amphitropous [BOT] Having a half-inverted ovule with the funiculus attached near the middle.

Amphiumidae [VERT ZOO] A small family of urodele amphibians in the suborder Salamandroidea composed of three species of large, eellike salamanders with tiny limbs.

amphivasal [BOT] Having the xylem surrounding the phloem, as seen in certain vascular bundles.

Amphizoidae [INV ZOO] The trout stream beetles, a small family of coleopteran insects in the suborder Adephaga.

ampholyte [CHEM] An amphoteric electrolyte.

ampholytic detergent [CHEM] A detergent that is cationic in acidic solutions and anionic in basic solutions.

Amphoriscidae [INV ZOO] A family of calcareous sponges in the order Sycettida.

amphoteric [CHEM] Having both acidic and basic characteristics. Also known as amphiprotic.

amphotericin [MICROBIO] An amphoteric antifungal antibiotic produced by *Streptomyces nodosus* and having of two components, A and B.

amphotericin A [MICROBIO] The relatively inactive component of amphotericin.

amphotericin B [MICROBIO] $C_{46}H_{73}O_{20}N$ The active component of amphotericin, suitable for systemic therapy of deep or superficial mycotic infections.

amphoterism [CHEM] The property of being able to react either as an acid or a base.

amphoterite [GEOL] A stony meteorite containing bronzite and olivine with some oligoclase and nickel-rich iron.

amp-hr *See* ampere-hour.

Ampicillin [MICROBIO] $C_{16}H_{19}N_3O_4S$ Semisynthetic broad-spectrum penicillin produced initially by *Penicillium chrysogenum* and then chemically modified; used as an antibiotic. Trade name for D-aminobenzyl penicillin.

amplexicaul [BOT] Pertaining to a sessile leaf with the base or stipules embracing the stem.

amplexus [BOT] Having the edges of a leaf overlap the edges of a leaf above it in vernation. [VERT ZOO] The copulatory embrace of frogs and toads.

ampliate [BIOL] Widened or enlarged.

amplidyne [ELEC] A rotating magnetic amplifier having special windings and brush connections so that small changes in power input to the field coils produce large changes in power output.

amplification [SCI TECH] The production of an output of greater magnitude than the input.

amplification factor [ELECTR] In a vacuum tube, the ratio of the incremental change in plate voltage to a given small change in grid voltage, under the conditions that the plate current and all other electrode voltages are held constant.

amplified back bias [ELECTR] Degenerative voltage developed across a fast time-constant circuit within a stage of an amplifier and fed back into a preceding stage.

amplifier [ENG] A device capable of increasing the magnitude or power level of a physical quantity, such as an electric current or a hydraulic mechanical force, that is varying with time, without distorting the wave shape of the quantity.

amplifier-type meter [ENG] An electric meter whose characteristics have been enhanced by the use of preamplification for the signal input eventually used to actuate the meter.

amplifying delay line [ELECTR] Delay line used in pulse-compression systems to amplify delayed signals in the super-high-frequency region.

amplitron [ELECTR] Crossed-field continuous cathode reentrant beam backward-wave amplifier for microwave frequencies.

amplitude [GEOPHYS] Linear index of geomagnetic activity; for example, the amplitude in gammas of the variation in the most disturbed field component during a 3-hour interval. [MATH] The angle between a vector representing a specified complex number on an Argand diagram and the positive real axis. Also known as argument. [NAV] Angular distance north or south of the prime vertical; the arc of the horizon, or the angle at the zenith between the prime vertical and a vertical circle, measured north or south from the prime vertical to the vertical circle.

amplitude discriminator See pulse-height discriminator.

amplitude distortion See frequency distortion.

amplitude factor See crest factor.

amplitude fading [COMMUN] Fading in which the amplitudes of all frequency components of a modulated carrier wave are uniformly attenuated.

amplitude-frequency distortion See frequency distortion.

amplitude-frequency response See frequency response.

amplitude gate [ELECTR] A circuit which transmits only those portions of an input signal which lie between two amplitude boundary level values. Also known as slicer; slicer amplifier.

amplitude level [PHYS] The natural logarithm of the ratio of two amplitudes, each measured in the same units.

amplitude limiter See limiter.

amplitude-limiting circuit See limiter.

amplitude-modulated indicator [ENG] A general class of radar indicators, in which the sweep of the electron beam is deflected vertically or horizontally from a base line to indicate the existence of an echo from a target. Also known as deflection-modulated indicator; intensity-modulated indicator.

amplitude modulation [ELECTR] Abbreviated AM. **1.** Modulation in which the aplitude of a wave is the characteristic varied in accordance with the intelligence to be transmitted. **2.** In telemetry, those systems of modulation in which each component frequency f of the transmitted intelligence produces a pair of sideband frequencies at carrier frequency plus f and carrier minus f.

amplitude-modulation noise [COMMUN] Noise produced by undesirable amplitude variations of a radio-frequency signal.

amplitude-modulation radio [COMMUN] Also known as AM radio. **1.** The system of radio communication employing amplitude modulation of a radio-frequency carrier to convey the intelligence. **2.** A receiver used in such a system.

amplitude modulator [PHYS] Any device which imposes amplitude modulation upon a carrier wave in accordance with a desired program.

amplitude noise [ELECTROMAG] Effect on radar accuracy of the fluctuations in the amplitude of the signal returned by the target; these fluctuations are caused by any change in aspect if the target is not a point source.

amplitude resonance [PHYS] The frequency at which a given sinusoidal excitation produces the maximum amplitude of oscillation in a resonant system.

amplitude response [ELECTR] The maximum output amplitude obtainable at various points over the frequency range of an instrument operating under rated conditions.

amplitude selector See pulse-height selector.

amplitude separator [ELECTR] A circuit used to isolate the portion of a waveform with amplitudes above or below a given value or between two given values.

amplitude splitting [OPTICS] A technique in which light falls on a partially reflecting surface; part of the light is transmitted, part reflected, and after further manipulation, these parts are recombined to give interference.

amplitude suppression ratio [ELECTR] Ratio, in frequency modulation, of the undesired output to the desired output of a frequency-modulated receiver when the applied signal has simultaneous amplitude and frequency modulation.

amplitude versus frequency distortion [ELECTR] Distortion caused by the nonuniform attenuation or gain of the system, with respect to frequency under specified terminal conditions.

amp-turn See ampere-turn.

Ampulicidae [INV ZOO] A small family of hymenopteran insects in the superfamily Sphecoidea.

ampulla [ANAT] A dilated segment of a gland or tubule. [BOT] A small air bladder in some aquatic plants. [INV ZOO] The sac at the base of a tube foot in certain echinoderms.

ampulla of Lorenzini [VERT ZOO] Any of the cutaneous receptors in the head region of elasmobranchs; thought to have a thermosensory function.

ampulla of Vater [ANAT] Dilation at the junction of the bile and pancreatic ducts and the duodenum in man. Also known as papilla of Vater.

amputation [MED] The surgical, congenital, or spontaneous removal of a limb or projecting body part.

AM radio See amplitude-modulation radio.

AM signature [COMMUN] A graphic representation of the significant identifying characteristics of an amplitude-modulated signal.

amtrac See amphibious tractor.

amu See atomic mass unit.

amychophobia [PSYCH] Abnormal fear of being scratched or clawed.

amygdalic acid See mandelic acid.

amygdalin [BIOCHEM] $C_6H_5CH(CN)OC_{12}H_{21}O_{10}$ A glucoside occurring in the kernels of certain plants of the genus *Prunus.*

amygdaloid [GEOL] Lava rock containing amygdules. Also known as amygdaloidal lava.

amygdaloidal lava See amygdaloid.

amygdule [GEOL] **1.** A mineral filling formed in vesicles (cavities) of lava flows; it may be chalcedony, opal, calcite, chlorite, or prehnite. **2.** An agate pebble.

amyl [ORG CHEM] Any of the eight isomeric arrangements of the radical C_5H_{11} or a mixture of them. Also known as pentyl.

amyl acetate [ORG CHEM] $CH_3COO(CH_2)_2CH(CH_3)_2$ A colorless liquid, boiling at 142°C; soluble in alcohol and ether, slightly soluble in water; used in flavors and perfumes. Also known as banana oil; isoamyl acetate.

amyl alcohol [ORG CHEM] **1.** A colorless liquid that is a mixture of isomeric alcohols. **2.** An optically active liquid composed of isopentyl alcohol and active amyl alcohol.

amyl aldehyde See n-valeraldehyde.

amylase [BIOCHEM] An enzyme that hydrolyzes reserve carbohydrates, starch in plants and glycogen in animals.

amyl benzoate See isoamyl benzoate.

amyl carbinol See hexyl alcohol.

amyl ether [ORG CHEM] **1.** Either of two isomeric compounds, *n*-amyl ether or isoamyl ether; both may be represented by the formula $(C_5H_{11})_2O$. **2.** A mixture mainly of isoamyl ether and *n*-amyl ether formed in preparation of amyl alcohols from amyl chloride; very slightly soluble in water; used mainly as a solvent.

amyl nitrate [ORG CHEM] $C_5H_{11}ONO_2$ An ester of amyl alcohol to diesel fuel to raise the cetane number.

amyl nitrite [ORG CHEM] $(CH_3)_2CH(CH_2)_2NO_2$ A yellow liquid; soluble in alcohol, very slightly soluble in water; fruity odor; it is flammable and the vapor is explosive; used in medicine and perfumes. Also known as isoamyl nitrite.

amyloid [PATH] An abnormal protein deposited in tissues, formed from the infiltration of an unknown substance, probably a carbohydrate.

amyloid body [PATH] Any of the microscopic hyaline bodies that stain like amyloid with metachromatic aniline dyes.

amyloidosis [MED] Deposition of amyloid in one or more organs of the body.

amylolysis [BIOCHEM] The enzyme-catalyzed hydrolysis of starch to soluble products.

amylolytic enzyme [BIOCHEM] A type of enzyme capable of denaturing starch molecules; used in textile manufacture to remove starch added to slash sizing agents.

amylopectin [BIOCHEM] A highly branched, high-molecular-weight carbohydrate polymer composed of about 80% corn starch.

amyloplast [BOT] A colorless cell plastid packed with starch grains and occurring in cells of plant storage tissue.

amylopsin [BIOCHEM] An enzyme in pancreatic juice that acts to hydrolyze starch into maltose.

amylose [BIOCHEM] A linear starch polymer.

amyl salicylate [ORG CHEM] $C_6H_4OHCOOC_5H_{11}$ A clear liquid that occasionally has a yellow tinge; boils at 280°C; soluble in alcohol, insoluble in water; used in soap and perfumes. Also known as isoamyl salicylate.

amyl xanthate [ORG CHEM] A salt formed by replacing the hydrogen attached to the sulfur in amylxanthic acid by a metal; used as collector agent in the flotation of certain minerals.

Amynodontidae [PALEON] A family of extinct hippopotamuslike perissodactyl mammals in the superfamily Rhinoceratoidea.

amyotonia [MED] Absence of muscle tone.

amyotonia congenita [MED] A congenital disease of the central nervous system characterized by absence of voluntary muscle tone and reflexes. Also known as Oppenheim's disease.

amyotrophic lateral sclerosis [MED] A degenerative disease of the pyramidal tracts and lower motor neurons characterized by motor weakness and a spastic condition of the limbs associated with muscular atrophy, fibrillary twitching, and final involvement of nuclei in the medulla. Also known as lateral sclerosis.

An *See* actinon.

Anabaena [BOT] A genus of blue-green algae in the class Cyanophycaea; members fix atmospheric nitrogen.

Anabantidae [VERT ZOO] A fresh-water family of actinopterygian fishes in the order Perciformes, including climbing perches and gourami.

Anabantoidei [VERT ZOO] A suborder of fresh-water labyrinth fishes in the order Perciformes.

anabatic wind [METEOROL] An upslope wind; usually applied only when the wind is blowing up a hill or mountain as the result of a local surface heating, and apart from the effects of the larger-scale circulation.

anabiosis [BIOL] State of suspended animation induced by desiccation and reversed by addition of moisture; can be achieved in rotifers.

anabohitsite [PETR] A variety of olivine-pyroxenite containing hornblende and hypersthene and a high proportion (about 30%) of magnetite and ilmenite.

anabolic [BIOCHEM] Pertaining to anabolism. [EVOL] Pertaining to anaboly.

anabolism [BIOCHEM] A part of metabolism involving the union of smaller molecules into larger molecules; the method of synthesis of tissue structure.

anaboly [EVOL] The addition, through evolutionary differentiation, of a new terminal stage to the morphogenetic pattern.

anabranch [HYD] A diverging branch of a stream or river that loses itself in sandy soil or rejoins the main flow downstream.

Anacanthini [VERT ZOO] An equivalent name for the Gadiformes.

Anacardiaceae [BOT] A family of flowering plants, the sumacs, in the order Sapindales; many species are allergenic to man.

anaclinal [GEOL] Having a downward inclination opposite to that of a stratum.

anaclitic depression [PSYCH] Seriously impaired physical, social, and intellectual development in some infants following sudden separation from a loving mother or mother figure.

anaclitic object choice [PSYCH] The choosing of a love object resembling the person upon whom the individual was emotionally dependent during infancy.

anaconda [VERT ZOO] *Eunectes murinus.* The largest living snake, an arboreal-aquatic member of the boa family (Boidae).

anacoustic zone [GEOPHYS] The zone of silence in space, starting at about 100 miles altitude, where the distance between air molecules is greater than the wavelength of sound, and sound waves can no longer be propagated.

Anactinochitinosi [INV ZOO] A group name for three closely related suborders of mites and ticks: Onychopalpida, Mesostigmata, and Ixodides.

Anacystis [BOT] A genus of blue-green algae in the class Cyanophycea.

anadromous [VERT ZOO] Said of a fish, such as the salmon and shad, that ascends fresh-water streams from the sea to spawn.

Anadyomenaceae [BOT] A family of green marine algae in the order Siphonocladales characterized by the expanded blades of the thallus.

anaerobe [BIOL] An organism that does not require air or free oxygen to maintain its life processes.

anaerobic bacteria [MICROBIO] Any bacteria that can survive in the partial or complete absence of air; two types are facultative and obligate.

anaerobic condition [BIOL] The absence of oxygen, preventing normal life for organisms that depend on oxygen.

anaerobic glycolysis [BIOCHEM] A metabolic pathway in plants by which, in the absence of oxygen, hexose is broken down to lactic acid and ethanol with some adenosinetriphosphate synthesis.

anaerobic petri dish [MICROBIO] A glass laboratory dish for plate cultures of anaerobic bacteria; a thioglycollate agar medium and restricted air space give proper conditions.

anaerobic process [SCI TECH] A process from which air or oxygen not in chemical combination is excluded.

anaerobiosis [BIOL] A mode of life carried on in the absence of molecular oxygen.

anaerophyte [ECOL] A plant that does not need free oxygen for respiration.

anafront [METEOROL] A front at which the warm air is ascending the frontal surface up to high altitudes.

anaglyph [GRAPHICS] **1.** A stereogram in which the two views are printed or projected superimposed in complementary colors, usually red and blue; by viewing through filter spectacles of corresponding complementary colors, a stereoscopic image is formed. **2.** A surface worked in low relief.

anakinesis [BIOCHEM] A process in living organisms by which energy-rich molecules, such as adenosinetriphosphate, are formed.

anal [ANAT] Relating to or located near the anus.

analbite [MINERAL] A triclinic albite which is not stable and becomes monoclinic at about 700°C.

analbuminemia [MED] A disorder transmitted as an autosomal recessive, characterized by drastic reduction or absence of serum albumin.

anal character [PSYCH] A type of personality in which anal erotic traits dominate beyond the period of childhood.

analcime [MINERAL] $NaAlSi_2O_6 \cdot H_2O$ A white or slightly colored isometric zeolite found in diabase and in alkali-rich basalts. Also known as analcite.

ANADYOMENACEAE

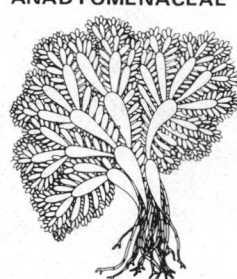

Anadyomene, a genus in Anadyomenaceae, with expanded blades.

ANAEROBIC PETRI DISH

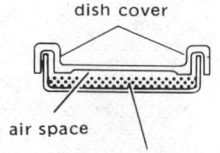

dish cover

air space

anaerobic agar

Brewer anaerobic petri dish. *(Courtesy BioQuest, Division of Becton, Dickinson and Co.)*

analcimite [PETR] An extrusive or hypabyssal rock that consists primarily of pyroxene and analcime.

analcimization [GEOL] The replacement in igneous rock of feldspars or feldspathoids by analcime.

analcite See analcime.

analemma [ASTRON] A figure-eight-shaped diagram on a globe showing the declination of the sun throughout the year and also the equation of time.

analeptic [PHARM] Any drug used to restore respiration and a wakeful state.

anal fin [VERT ZOO] An unpaired fin located medially on the posterior ventral part of the fish body.

analgesia [PHYSIO] Insensibility to pain with no loss of consciousness.

analgesic [PHARM] Any drug, such as salicylates, morphine, or opiates, used primarily for the relief of pain.

anal gland [INV ZOO] A gland in certain mollusks that secretes a purple substance. [VERT ZOO] A gland located near the anus or opening into the rectum in many vertebrates.

anallobaric center See pressure-rise center.

analog [ELECTR] A physical variable which remains similar to another variable insofar as the proportional relationships are the same over some specified range; for example, a temperature may be represented by a voltage which is its analog. [METEOROL] A past large-scale synoptic weather pattern which resembles a given (usually current) situation in its essential characteristics.

analog channel [ELECTR] A channel on which the information transmitted can have any value between the channel limits, such as a voice channel.

analog communications [COMMUN] System of telecommunications employing a nominally continuous electric signal that varies in frequency, amplitude, and so on, in some direct correlation to nonelectrical information (sound, light, and so on) impressed on a transducer.

analog comparator [ELECTR] **1.** A comparator that checks digital values to determine whether they are within predetermined upper and lower limits. **2.** A comparator that produces high and low digital output signals when the sum of two analog voltages is positive and negative, respectively.

analog computer [ADP] A computer is which quantities are represented by physical variables; problem parameters are translated into equivalent mechanical or electrical circuits as an analog for the physical phenomenon being investigated.

analog data [ADP] Data represented in a continuous form, as contrasted with digital data having discrete values.

analog-digital computer See hybrid computer.

analog indicator [ELECTR] A device in which the result of a measurement is indicated by a pointer deflection or other visual quantity.

analog multiplexer [ELECTR] A multiplexer that provides switching of analog input signals to allow use of a common analog-to-digital converter.

analogous [BIOL] Referring to structures that are similar in function and general appearance but not in origin, such as the wing of an insect and the wing of a bird.

analogous pole [SOLID STATE] The pole of a crystal that acquires a positive charge when the crystal is heated.

analog output [CONT SYS] Transducer output in which the amplitude is continuously proportional to a function of the stimulus.

analog simulation [ADP] The representation of physical systems and phenomena by variables such as translation, rotation, resistance, and voltage.

analog-to-digital converter [ELECTR] A device which translates continuous analog signals into proportional discrete digital signals.

anal plate [EMBRYO] An embryonic plate formed of endoderm and ectoderm through which the anus later ruptures. [VERT ZOO] **1.** One of the plates on the posterior portion of the plastron in turtles. **2.** A large scale anterior to the anus of most snakes.

anal sphincter [ANAT] Either of two muscles, one voluntary and the other involuntary, controlling closing of the anus in vertebrates.

analysis [ANALY CHEM] The determination of the composition of a substance. [MATH] The branch of mathematics most explicitly concerned with the limit process or the concept of convergence; includes the theories of differentiation, integration and measure, infinite series, and analytic functions. Also known as mathematical analysis. [METEOROL] A detailed study in synoptic meteorology of the state of the atmosphere based on actual observations, usually including a separation of the entity into its component patterns and involving the drawing of families of isopleths for various elements.

analysis of fire [ORD] The study of previous antiaircraft preparatory firings for the purpose of determining corrections which will improve fire for effect.

analytical aerotriangulation [ENG] Analytical phototriangulation, performed with aerial photographs.

analytical balance [ENG] A balance with a sensitivity of 0.1–0.01 milligram.

analytical centrifugation [ENG] Centrifugation following precipitation to separate solids from solid-liquid suspensions; faster than filtration.

analytical chemistry [CHEM] The branch of chemistry dealing with techniques which yield any type of information about chemical systems.

analytical distillation [ANALY CHEM] Precise resolution of a volatile liquid mixture into its components; the mixture is vaporized by heat or vacuum, and the vaporized components are recondensed into liquids at their respective boiling points.

analytical engine [ADP] An early-19th-century form of mechanically operated digital computer.

analytical extraction [ANALY CHEM] Precise transfer of one or more components of a mixture (liquid to liquid, gas to liquid, solid to liquid) by contacting the mixture with a solvent in which the component of interest is preferentially soluble.

analytical nadir-point triangulation [ENG] Radial triangulation performed by computational routines in which nadir points are utilized as radial centers.

analytical orientation [ENG] The computational steps required to determine tilt, direction of principal line, flight height, angular elements, and linear elements in preparing aerial photographs for rectification.

analytical photogrammetry [ENG] A method of photogrammetry in which solutions are obtained by mathematical methods.

analytical photography [ENG] Photography, either motion picture or still, accomplished to determine (by qualitative, quantitative, or any other means) whether a particular phenomenon does or does not occur.

analytical phototriangulation [ENG] A phototriangulation procedure in which the spatial solution is obtained by computational routines.

analytical radar prediction [ENG] Prediction based on proven formulas, power tables, or graphs; considers surface height, structural and terrain information, and criteria for radar reflectivity together with the aspect angle and range to the target.

analytical radial triangulation [ENG] Radial triangulation performed by computational routines.

analytical three-point resection radial triangulation [ENG] A method of computing the coordinates of the ground principal points of overlapping aerial photographs by resecting on three horizontal control points appearing in the overlap area.

analytical ultracentrifuge [ENG] An ultracentrifuge that uses one of three optical systems (schlieren, Rayleigh, or absorption) for the accurate determination of sedimentation velocity or equilibrium.

analytic continuation [MATH] The process of extending an analytic function to a domain larger than the one on which it was originally defined.

analytic curve [MATH] A curve whose parametric equations are real analytic functions of the same real variable.

analytic function [MATH] A function which can be represented by a convergent Taylor series. Also known as holomorphic function.

analytic geometry [MATH] The study of geometric figures and curves using a coordinate system and the methods of algebra. Also known as cartesian geometry.

analytic mechanics [MECH] The application of differential

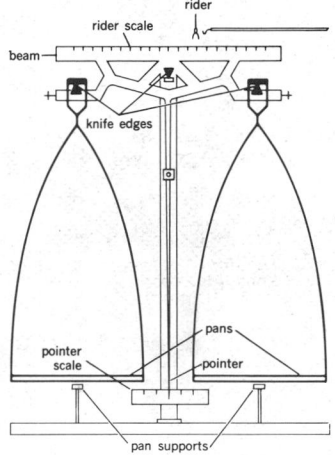

ANALYTICAL BALANCE

The rider analytical balance; simplest type of analytical balance.

and integral calculus to classical (nonquantum) mechanics.

analytic number theory [MATH] The study of problems concerning the discrete domain of integers by means of the mathematics of continuity.

analytic psychology [PSYCH] The school of psychology that regards the libido not as an expression of the sex instinct, but of the will to live; the unconscious mind is thought to express certain archaic memories of race. Also known as Jungian psychology.

analytic trigonometry [MATH] The study of the properties and relations of the trigonometric functions.

analyzer [ADP] 1. A routine for the checking of a program. 2. One of several types of computers used to solve differential equations. [ENG] A multifunction test meter, measuring volts, ohms, and amperes. Also known as set analyzer. [MECH ENG] The component of an absorption refrigeration system where the mixture of water vapor and ammonia vapor leaving the generator meets the relatively cool solution of ammonia in water entering the generator and loses some of its vapor content. [OPTICS] A device, such as a Nicol prism, which passes only plane polarized light; used in the eyepiece of instruments such as the polariscope.

anamigmatism [GEOL] A process of high-temperature, high-pressure remelting of sediment to yield magma.

anamnesis [MED] Information gained from the patient and others regarding his medical history. [PSYCH] The faculty of memory.

anamnestic response [IMMUNOL] A rapidly increased antibody level following renewed contact with a specific antigen, even after several years. Also known as booster response.

Anamnia [VERT ZOO] Vertebrate animals which lack an amnion in development, including Agnatha, Chondrichthyes, Osteichthyes, and Amphibia.

Anamniota [VERT ZOO] The equivalent name for Anamnia.

anamorphic lens [OPTICS] A lens that produces different magnifications along lines in different directions in the image plane.

anamorphic system [OPTICS] An optical system incorporating a cylindrical surface in which the image is distorted so that the angle of coverage in a direction perpendicular to the cylinder is different for the image than for the object.

anamorphic zone [GEOL] The zone of rock flow, as indicated by reactions that may involve decarbonation, dehydration, and deoxidation; silicates are built up, and the formation of denser minerals and of compact crystalline structure takes place.

anamorphism [EVOL] *See* anamorphosis. [GEOL] A kind of metamorphism at considerable depths in the earth's crust and under great pressure, resulting in the formation of complex minerals from simple ones.

anamorphoscope [OPTICS] An optical instrument, usually consisting of a cylindrical lens or mirror, that restores an image distorted by anamorphosis to its normal proportions.

anamorphosis [EVOL] Gradual increase in complexity of form and function during evolution of a group of animals or plants. Also known as anamorphism. [OPTICS] The production of a distorted image by an optical system.

anamorphote lens [OPTICS] A lens designed to produce anamorphosis.

Anancinae [PALEON] A subfamily of extinct proboscidean placental mammals in the family Gomphotheriidae.

anapaite [MINERAL] $Ca_2Fe(PO_4)_2 \cdot 4H_2O$ A pale-green or greenish-white triclinic mineral consisting of a ferrous iron hydrous phosphate and occurring in crystals and massive forms; hardness is 3–4 on Mohs scale, and specific gravity is 3.81.

anapeirean *See* Pacific suite.

anaphase [CYTOL] 1. The stage in mitosis and in the second meiotic division when the centromere splits and the chromatids separate and move to opposite poles. 2. The stage of the first meiotic division when the two halves of a bivalent chromosome separate and move to opposite poles.

anaphoresis [MED] Deficient functioning of sweat glands. [PHYSIO] Movement of positively charged ions into tissues under the influence of an electric current.

anaphylactic shock [MED] A syndrome seen as one of the clinical manifestations of anaphylaxis.

anaphylactoid reaction [MED] A nonallergic reaction resembling anaphylaxis and depending on the toxicity of the inductant.

anaphylaxis [MED] Hypersensitivity following parenteral injection of an antigen; local or systemic allergic reaction occurs when the antigen is reintroduced after a time lapse.

anaphylotoxin [IMMUNOL] The vasodilator principle, a toxic substance released by tissues of sensitized animals when antigen and antibody react.

anaplasia [MED] Reversion of cells to an embryonic, immature, or undifferentiated state; degree usually corresponds to malignancy of a tumor.

Anaplasma [INV ZOO] A genus of parasites that infect the erythrocytes of cattle, causing anaplasmosis.

Anaplasmataceae [MICROBIO] A family of small bacteria in the order Rickettsiales, parasitic in red blood cells of domestic cloven-hoofed animals.

Anaplotheriidae [PALEON] A family of extinct tylopod ruminants in the superfamily Anaplotherioidea.

Anaplotherioidea [PALEON] A superfamily of extinct ruminant artiodactyls in the infraorder Tylopoda.

anapolysis [INV ZOO] Lifetime retention of ripe proglottids in some tapeworms.

anapophysis [ANAT] An accessory process on the dorsal side of the transverse process of the lumbar vertebrae in man and other mammals.

Anapsida [VERT ZOO] A subclass of reptiles characterized by a roofed temporal region in which there are no temporal openings.

anasarca [MED] Generalized edema of the subcutaneous connective tissue and serous body cavities.

Anasca [PALEON] A suborder of extinct bryozoans in the order Cheilostomata.

Anaspida [PALEON] An order of extinct fresh- or brackish-water vertebrates in the class Agnatha.

Anaspidacea [INV ZOO] An order of the crustacean superorder Syncarida.

Anaspididae [INV ZOO] A family of crustaceans in the order Anaspidacea.

anastatic process [GRAPHICS] Reproduction of a printed page, either type or pictures, by moistening it with dilute acid and pressing it against a zinc plate; the acid etches the zinc wherever in contact with unprinted portions; the plate can then be inked and printed.

anastatic water [HYD] That part of the subterranean water in the capillary fringe between the zone of aeration and the zone of saturation in the soil.

anastigmatic lens [OPTICS] A compound lens corrected for astigmatism and curvature of field.

anastomosis [SCI TECH] The union or intercommunication of branched systems in either two or three dimensions. Also known as inosculation.

anastral [CYTOL] Lacking asters.

anatase [MINERAL] The brown, dark-blue, or black tetragonal crystalline form of titanium dioxide, TiO_2; used to make a white pigment. Also known as octahedrite.

anatexis [GEOL] A high-temperature process of metamorphosis by which plutonic rock in the lowest levels of the crust is melted and regenerated as a magma.

anathermal [GEOL] A period of time between the age of other strata or units of reference in which the temperature is increasing.

Anatidae [VERT ZOO] A family of waterfowl, including ducks, geese, mergansers, pochards, and swans, in the order Anseriformes.

anatomical dead space *See* dead space.

anatomy [BIOL] A branch of morphology dealing with the structure of animals and plants.

anatropous [BOT] Having the ovule fully inverted so that the micropyle adjoins the funiculus.

anauxite [MINERAL] $Al_2(SiO_7)(OH)_4$ A clay mineral that is a mixture of kaolinite and quartz. Also known as ionite.

Anavor [TEXT] Trade name for a filament polyester manufactured by the Dow Badische Company.

anaxial [BIOL] Lacking an axis, therefore being irregular in form.

ANASPIDA

Pharyngolepis oblongus Kiaer of the Anaspida, reconstruction. *(From A. Ritchie)*

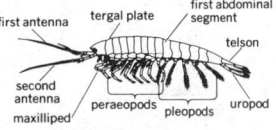

ANASPIDIDAE

A male *Anaspides tasmaniae* Thompson about 5 centimeters long, found in freshwater habitats of Tasmania. *(From R. E. Snodgrass, A Textbook of Arthropod Anatomy, Comstock, 1952)*

Anbauhobel [MIN ENG] A rapid plough, traveling at a speed of 75 feet per minute, for use on longwall faces.

ancestroecium [INV ZOO] The tube that encloses an ancestrula.

ancestrula [INV ZOO] The first polyp of a bryozoan colony.

anchieutectic [GEOL] A type of magma which is incapable of undergoing further notable main-stage differentiation because its mineral composition is practically in eutectic proportions.

anchimonomineralic [PETR] Of rock composed mostly of one kind of mineral.

anchor [CIV ENG] A device connecting a structure to a heavy masonry or concrete object to a metal plate or to the ground to hold the structure in place. [MECH ENG] A vehicle used in steam plowing and located on the side of the field opposite that of the engine while maintaining the tension on the endless wire by means of a pulley. [MET] A device that prevents the movement of sand cores in molds. [NAV ARCH] A device attached by cable to a ship and dropped overboard so that its hooks or flukes engage the bottom and hold the ship at that location.

anchorage [CIV ENG] **1.** An area where a vessel anchors or may anchor because of either suitability or designation. Also known as anchor station. **2.** See deadman.

anchorage chart [MAP] A nautical chart showing prescribed or recommended anchorages.

anchor and collar [DES ENG] A door or gate hinge whose socket is attached to an anchor embedded in the masonry.

anchor ball [NAV ARCH] **1.** A projectile with grappling hooks which is fired into the rigging of a wrecked vessel for lifesaving purposes. **2.** A black, circular shape hoisted between the bow and foremast of a vessel to indicate that it is anchored in or near a channel.

anchor block See deadman.

anchor bolster See hawse bolster.

anchor bolt [DES ENG] A bolt used with its head embedded in masonry or concrete and its threaded part protruding to hold a structure or machinery in place. Also known as anchor rod.

anchor buoy [NAV] A buoy connected to an anchor or indicating the location of an anchor.

anchor charge [ENG] A procedure that allows several charges to be preloaded in a seismic shot hole; the bottom charges are fired first, and the upper charges are held down by anchors.

anchored bulkhead [CIV ENG] A bulkhead secured to anchor piles.

anchored dune [GEOL] A sand dune stabilized by growth of vegetation.

anchor escapement [HOROL] A clock escapement in which pallets of an anchor-shaped component cause the escape wheel to recoil slightly as the wheel is arrested. Also known as recoil escapement.

anchor gear [NAV ARCH] Shipboard apparatus consisting of anchor windlass, chain stoppers, and hawsepipes; more generally includes anchor and chain.

anchor ice [HYD] Ice formed beneath the surface of water, as in a lake or stream, and attached to the bottom or to submerged objects. Also known as bottom ice; ground ice.

anchorite [PETR] A variety of diorite having nodules of mafic minerals and veins of felsic minerals.

anchor light [NAV ARCH] A mast light shown by a vessel at anchor during the night.

anchor log [CIV ENG] A log, beam, or concrete block buried in the earth and used to hold a guy rope firmly. Also known as deadman.

anchor nut [DES ENG] A nut in the form of a tapped insert forced under steady pressure into a hole in sheet metal.

anchor packer [PETRO ENG] A device used in oil wells to seal the annular space between the tubing and its surrounding casing to help control the oil-producing gas lift.

anchor pile [CIV ENG] A pile that is located on the land side of a bulkhead or pier and anchors it through such devices as rods, cables, and chains.

anchor plate [CIV ENG] A metal or wooden plate fastened to or embedded in a support, such as a floor, and used to hold a supporting cable firmly.

anchor rod See anchor bolt.

anchor shackle [NAV ENG] A shackle by which a chain is joined to the ring of an anchor. Also known as bending shackle.

anchor station [CIV ENG] See anchorage. [OCEANOGR] An anchoring site by a research vessel for the purpose of making a set of scientific observations.

anchor stone [GEOL] A rock or pebble that has marine plants attached to it.

anchor tower [CIV ENG] **1.** A tower which is a part of a crane staging or stiffleg derrick and serves as an anchor. **2.** A tower that supports and anchors an overhead transmission line.

anchor wall See deadman.

anchor well [NAV ARCH] A well in a ship's forward overhang for holding anchors.

anchor windlass [NAV ARCH] A machine, generally located on the forecastle head of a ship, designed to raise or lower an anchor; it consists of a horizontal barrel that is fitted with gearlike projections that engage the links of the anchor chain, and is turned by steam or electrical power. Also known as windlass.

anchovy [VERT ZOO] Any member of the Engraulidae, a family of herringlike fishes harvested commercially for human consumption.

anchylosis See ankylosis.

ancipital [BOT] Having two edges, specifically referring to flattened stems, as of certain grasses.

ancon [ARCH] A bracket, elbow, or console at the top of a wall or window jamb to support a cornice.

ancora [INV ZOO] The initial, anchor-shaped growth stage of graptolithinids.

ancylite [MINERAL] $SrCe(CO_3)_2(OH) \cdot H_2O$ A mineral consisting of hydrous basic carbonate of cerium and strontium.

ancyloid [INV ZOO] A limpet-shaped or patelliform shell with the apex directed anteriorly.

ancylopoda [PALEON] A suborder of extinct herbivorous mammals in the order Perissodactyla.

Ancylostoma [INV ZOO] A genus of roundworms, commonly known as hookworms, in the order Ancylostomidae; parasites of man, dogs, and cats.

Ancylostoma duodenale [INV ZOO] The Old World hookworm, a human intestinal parasite that causes microcytic hypochromic anemia.

Ancylostomidae [INV ZOO] A family of nematodes belonging to the group Strongyloidea.

And See Andromeda.

andalusite [MINERAL] Al_2SiO_5 A brown, yellow, green, red, or gray neosilicate mineral crystallizing in the orthorhombic system, usually found in metamorphic rocks.

AND circuit See AND gate.

Andean-type continental margin [GEOL] A continental margin, as along the Pacific coast of South America, where oceanic lithosphere descends beneath an adjacent continent producing andesitic continental margin volcanism.

Anderson bridge [ELECTR] A six-branch modification of the Maxwell-Wien bridge, used to measure self-inductance in terms of capacitance and resistance; bridge balance is independent of frequency.

andersonite [MINERAL] $Na_2Ca(UO_2)(CO_3)_3 \cdot 6H_2O$ Bright yellow-green secondary mineral consisting of a hydrous sodium calcium uranium carbonate.

Andes glow See Andes lightning.

andesine [MINERAL] A plagioclase feldspar with a composition ranging from $Ab_{70}An_{30}$ to $Ab_{50}An_{50}$, where $Ab = NaAlSi_3O_8$ and $An = CaAl_2Si_2O_8$; it is a primary constituent of intermediate igneous rocks, such as andesites.

andesite [PETR] Very finely crystalline extrusive rock of volcanic origin composed largely of plagioclase feldspar (oligoclase or andesine) with smaller amounts of dark-colored mineral (hornblende, biotite, or pyroxene), the extrusive equivalent of diorite.

andesite line [GEOL] The postulated geographic and petrographic boundary between the andesite-dacite-rhyolite rock association of the margin of the Pacific Ocean and the olivine-basalt-trachyte rock association of the Pacific Ocean basin.

andesitic glass [GEOL] A natural glass that is chemically equivalent to andesite.

Andes lightning [GEOPHYS] Electrical coronal discharges ob-

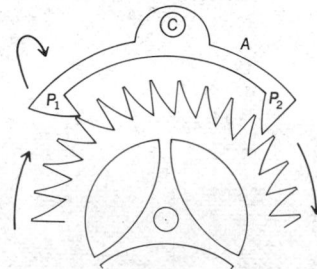

ANCHOR ESCAPEMENT

Common in domestic pendulum clocks, this escapement tends to compensate for changes in amplitude of the swing because of irregularly cut gears and varying lubrication; A = anchor, P_1, P_2 = pallets, C = point at which anchor swings about.

ANDALUSITE

Andalusite, variety chiastolite. Prismatic crystal specimens from Worchester County, Mass. (*American Museum of Natural History specimens*)

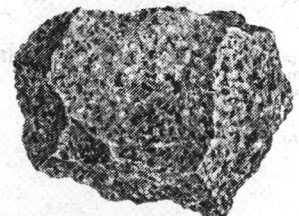

ANDESINE

Andesine grains with biotite, in a specimen from Zoutpansberg, Transvaal, South Africa. Andesine is rarely found except as grains in igneous rocks.

servable often as far as several hundred miles away, generally over any of the mountainous areas of the world when under disturbed electrical conditions. Also known as Andes glow.

AND function [ADP] An operation in logical algebra on statements *P, Q, R,* such that the operation is true if all the statements *P, Q, R, . . .* are true, and the operation is false if at least one statement is false.

AND gate [ELECTR] A circuit which has two or more input-signal ports and which delivers an output only if and when every input signal port is simultaneously energized. Also known as AND circuit; passive AND gate.

AND NOT gate [ELECTR] A coincidence circuit that performs the logic operation AND NOT, under which a result is true only if statement A is true and statement B is not. Also known as A AND NOT B gate.

AND-OR circuit [ELECTR] Gating circuit that produces a prescribed output condition when several possible combined input signals are applied; exhibits the characteristics of the AND gate and the OR gate.

andorite [MINERAL] $AgPbSb_3S_6$ A dark-gray or black orthorhombic mineral. Also known as sundtite.

Andr *See* Andromeda.

Andrade's creep law [MECH] A law which states that creep exhibits a transient state in which strain is proportional to the cube root of time and then a steady state in which strain is proportional to time.

andradite [MINERAL] The calcium-iron end member of the garnet group.

Andreaeales [BOT] The single order of mosses of the subclass Andreaeobrya.

Andreaeaceae [BOT] The single family of the Andreaeales, an order of mosses.

Andreaeobrya [BOT] The granite mosses, a subclass of the class Bryopsida.

Andrenidae [INV ZOO] The mining or burrower bees, a family of hymenopteran insects in the superfamily Apoidea.

andrewsite [MINERAL] $(Cu,Fe^{2+})Fe_3^{3+}(PO_4)_3(OH)_2$ A bluish-green mineral consisting of a basic phosphate of iron and copper.

andrite [GEOL] A meteorite composed principally of augite with some olivine and troilite.

androecium [BOT] The aggregate of stamens in a flower.

androgen [BIOCHEM] A class of steroid hormones produced in the testis and adrenal cortex which act to regulate masculine secondary sexual characteristics.

androgenesis [EMBRYO] Development of an embryo from a fertilized irradiated egg, involving only the male nucleus.

androgenetic merogony [EMBRYO] The fertilization of egg fragments that lack a nucleus.

androgenic gland [INV ZOO] Any of the accessory glands associated with the sperm duct in male crustaceans and required for differentiation of a functional male.

androgen unit [BIOL] A unit for the standardization of male sex hormones.

androgyny [MED] A form of pseudohermaphroditism in humans in which the individual has female external sexual characteristics, but has undescended testes. Also known as male pseudohermaphroditism.

android pelvis *See* masculine pelvis.

Andromeda [ASTRON] A constellation with a right ascension of 1 hour and a declination of 40°N. Abbreviated And; Andr.

Andromeda Galaxy [ASTRON] The spiral galaxy of type Sb nearest to the Milky Way. Also known as Andromeda Nebula.

Andromeda Nebula *See* Andromeda Galaxy.

andromerogony [EMBRYO] Development of an egg fragment following cutting, shaking, or centrifugation of a fertilized or unfertilized egg.

androphile [ECOL] An organism, such as a mosquito, showing a preference for humans as opposed to animals.

androphobia [PSYCH] Abnormal fear of men or of the male sex.

androphore [BOT] A stalk that supports stamens or antheridia. [INV ZOO] A gonophore in coelenterates in which only male elements develop.

androsin [BIOCHEM] $C_{15}H_{20}O_8$ A glucoside found in the herb *Apocynum androsaemifolium;* yields glucose and acetovanillone on hydrolysis.

androstane [BIOCHEM] $C_{19}H_{32}$ The parent steroid hydrocarbon for all androgen hormones. Also known as etioallocholane.

androstenedione [BIOCHEM] $C_{19}H_{26}O_2$ Any one of three isomeric androgens produced by the adrenal cortex.

androsterone [BIOCHEM] $C_{19}H_{30}O_2$ An androgenic hormone occurring as a hydroxy ketone in the urine of men and women.

anechoic chamber [ENG] **1.** A test room in which all surfaces are lined with a sound-absorbing material to reduce reflections of sound to a minimum. Also known as dead room; free-field room. **2.** A room completely lined with a material that absorbs radio waves at a particular frequency or over a range of frequencies; used principally at microwave frequencies, such as for measuring radar beam cross sections.

anelasticity [MECH] Deviation from a proportional relationship between stress and strain.

anelectric [PHYS] Not becoming charged by friction.

Anelytropsidae [VERT ZOO] A family of lizards represented by a single Mexican species.

anemia [MED] A condition marked by significant decreases in hemoglobin concentration and in the number of circulating red blood cells. Also known as oligochromemia.

anemic necrosis [MED] Tissue death following a critical decrease in blood flow or oxygen levels.

anemobiagraph [ENG] A recording pressure-tube anemometer in which the wind scale of the float manometer is linear through the use of springs; an example is the Dines anemometer.

anemochory [ECOL] Wind dispersal of plant and animal disseminules.

anemoclast [GEOL] A clastic rock that was fragmented and rounded by wind.

anemoclastic [GEOL] Referring to rock that was broken by wind erosion and rounded by wind action.

anemoclinometer [ENG] A type of instrument which measures the inclination of the wind to the horizontal plane.

anemogenic [MED] Causing anemia.

anemogram [ENG] A record made by an anemograph.

anemograph [ENG] **1.** An instrument which records wind velocities. **2.** A recording anemometer.

anemology [METEOROL] Scientific investigation of winds.

anemometer [ENG] A device which measures air speed.

anemometry [METEOROL] The study of measuring and recording the direction and speed (or force) of the wind, including its vertical component.

anemophilous [BOT] Pollinated by wind-carried pollen.

anemophobia [PSYCH] Abnormal fear of drafts or winds.

anemoscope [ENG] An instrument for indicating the direction of the wind.

anemotaxis [BIOL] Orientation movement of a free-living organism in response to wind.

anemotropism [BIOL] Orientation response of a sessile organism to air currents and wind.

anemovane [ENG] A combined contact anemometer and wind vane used in the Canadian Meteorological Service.

anencephalia [MED] A congenital malformation in which all or most of the brain and flat skull bones are absent.

anenterous [ZOO] Having no intestine, as a tapeworm.

Anepitheliocystidia [INV ZOO] A superorder of digenetic trematodes proposed by G. La Rue.

anergy [IMMUNOL] The condition of exhibiting no response to an antigen or antibody. [MED] The condition of exhibiting a lack of energy.

aneroid [ENG] **1.** Containing no liquid or using no liquid. **2.** *See* aneroid barometer.

aneroid altimeter [ENG] An altimeter containing an aneroid barometer that actuates the indicator.

aneroid barograph [ENG] An aneroid barometer arranged so that the deflection of the aneroid capsule actuates a pen which graphs a record on a rotating drum. Also known as aneroidograph; barograph; barometrograph.

aneroid barometer [ENG] A barometer which utilizes an aneroid capsule. Also known as aneroid.

ANDREAECEAE

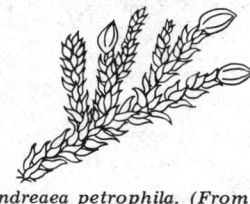

Andreaea petrophila. (From H. E. Jaques, Plant Families, How to Know Them, 2d ed., Brown, 1949)

ANEMOMETER

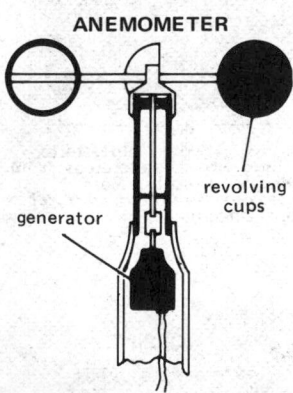

revolving cups

generator

Cutaway diagram of one type of anemometer: revolving-cup electric anemometer. (From D. M. Considine, ed., Process Instruments and Controls Handbook, McGraw-Hill, 1957)

aneroid calorimeter [ENG] A calorimeter that uses a metal of high thermal conductivity as a heat reservoir.

aneroid capsule [ENG] A thin, disk-shaped box or capsule, usually metallic, partially evacuated and sealed, held extended by a spring, which expands and contracts with changes in atmospheric or gas pressure. Also known as bellows.

aneroid diaphragm [ENG] A thin plate, usually metal, covering the end of an aneroid capsule and moving axially as the ambient gas pressure increases or decreases.

aneroid flowmeter [ENG] A mechanism to measure fluid flow rate by pressure of the fluid against a bellows counterbalanced by a calibrated spring.

aneroid liquid-level meter [ENG] A mechanism to measure fluid depth by pressure of the fluid against a bellows which in turn acts on a manometer or signal transmitter.

aneroidograph *See* aneroid barograph.

aneroid valve [MECH ENG] A valve actuated or controlled by an aneroid capsule.

anesthesia [PHYSIO] 1. Insensibility, general or local, induced by anesthetic agents. 2. Loss of sensation, of neurogenic or psychogenic origin.

anesthesiology [MED] A branch of medicine dealing with the administration of anesthetics.

anesthetic [PHARM] A drug, such as ether, that produces loss of sensibility.

anestrus [VERT ZOO] A prolonged period of inactivity between two periods of heat in cyclically breeding female mammals.

anethole [ORG CHEM] $C_{10}H_{12}O$ White crystals that melt at 22.5°C; very slightly soluble in water; affected by light; odor resembles oil of anise; used in perfumes and flavors, and as a sensitizer in color-bleaching processes in color photography. Also known as anise camphor; 1-methoxy-4-propenyl benzene; *para*-methoxypropenylbenzene; *para*-propenylanisole.

anethum oil *See* dill oil.

aneuploidy [GEN] Deviation from a normal haploid, diploid, or polyploid chromosome complement by the presence in excess of, or in defect of, one or more individual chromosomes.

aneurine *See* thiamine.

aneurysm [MED] Localized abnormal dilation of an artery due to weakening of the vessel wall.

angaralite [MINERAL] $Mg_2(Al,Fe)_{10}Si_6O_{29}$ A mineral of the chlorite group, occurring in thin black plates.

Angara Shield [GEOL] A shield area of crystalline rock in Siberia.

angel echo [ENG] A radar echo from a region where there are no visible targets; may be caused by insects, birds, or refractive index variations in the atmosphere.

angelica [FOOD ENG] 1. A spice from the perennial herb *Angelica archangelica* of the ginger family. 2. An amber or a yellow sweet wine without muscat flavor.

angelica oil [MATER] An essential oil with an odor that is strongly aromatic; soluble in alcohol; main ingredients are phellandrene and valeric acid; distilled from the seeds and roots of *Anglica archangelica;* used in medicines, liqueurs, and perfumes.

angiectasis [MED] Abnormal blood vessel dilation.

angina [MED] 1. A sore throat. 2. Any tense, constricting pain.

angina pectoris [MED] Constricting chest pain which may be accompanied by pain radiating down the arms, up into the jaw, or to other sites.

angioblast [EMBRYO] A mesenchyme cell derived from extraembryonic endoderm that differentiates into embryonic blood cells and endothelium.

angiocardiography [MED] Roentgenographic visualization of the heart chambers and thoracic vessels following injection of a radiopaque material.

Angiococcus [MICROBIO] A genus of bacteria in the family Myxococcaceae characterized by a fruiting body consisting of numerous round, thin-walled cysts.

angiogenesis [EMBRYO] The origin and development of blood vessels.

angiogram [MED] An x-ray photograph of blood vessels following injection of a radiopaque material.

angiography [MED] Roentgenographic visualization of blood vessels following injection of a radiopaque material.

angiology [MED] The branch of medicine concerned with the blood vessels and the lymphatic system.

angioma [MED] A tumor composed of lymphatic vessels or blood.

angioneurotic edema [MED] Acute, localized accumulations of tissue fluid causing swellings around the face; condition may be due either to heredity or a food allergy.

angiosarcoma [MED] A malignant soft-tissue tumor arising from vascular elements.

angioscotoma [MED] A visual-field disturbance caused by dilated blood-vessels in the retina.

angiosperm [BOT] The common name for members of the plant division Magnoliophyta.

Angiospermae [BOT] An equivalent name for Magnoliophyta.

angiotensin [BIOCHEM] A decapeptide hormone that influences blood vessel constriction and aldosterone secretion by the adrenal cortex. Also known as hypertensin.

angle [MATH] The geometric figure, arithmetic quantity, or algebraic signed quantity determined by two rays emanating from a common point or by two planes emanating from a common line.

angle back-pressure valve [MECH ENG] A back-pressure valve with its outlet opening at right angles to its inlet opening.

angle bisection [MATH] The division of an angle by a line or plane into two equal angles.

angle brace [ENG] A brace across the interior angle of two members that meet at an angle. Also known as angle tie.

angle bracket [ARCH] A bracket used in an angle or corner of a molded cornice. [GRAPHICS] Either of a pair of marks $< >$ enclosing a mutilated passage or the explanation of an abbreviation in a text, or to enclose quotations or illustrations in a reference work. Also known as broken bracket; pointed bracket.

angle cut [MIN ENG] A drilling pattern in which drill holes converge, so that a core can be blasted out and an open or relieved cavity or free face be left for the following shots, which are timed to ensue with a fractional delay.

Angledozer [MECH ENG] A power-operated machine fitted with a blade, adjustable in height and angle, for pushing, sidecasting, and spreading loose excavated material as for opencast pits, clearing land, or leveling runways. Also known as angling dozer.

angle equation [ENG] A condition equation which expresses the relationship between the sum of the measured angles of a closed figure and the theoretical value of that sum, the unknowns being the corrections to the observed directions or angles, depending on which are used in the adjustment. Also known as triangle equation.

angle fishplates [CIV ENG] Plates which join the rails and prevent the rail joint from sagging where heavy cars and locomotives are used. Also known as angle; angle bar.

angle gear *See* angular gear.

angle iron [CIV ENG] 1. An L-shaped cleat or brace. 2. A length of steel having a cross section resembling the letter L.

angle jamming [ELECTR] An electronic countermeasure in which azimuth and elevation information, from a scanning fire control radar present in the modulation components on the returning echo pulse, is jammed by transmitting a pulse similar to the radar pulse but with modulation information out of phase with the returning target angle modulation information.

angle method of adjustment [ENG] A method of adjustment of observations which determines corrections to observed angles.

angle modulation [ELECTR] The variation in the angle of a sine-wave carrier; particular forms are phase modulation and frequency modulation. Also known as sinusoidal angular modulation.

angle noise [ELECTROMAG] Tracking error introduced into radar by variations in the apparent angle of arrival of the echo from a target, because of finite target size.

angle of action [MECH ENG] The angle of revolution of either of two wheels in gear during which any particular tooth remains in contact.

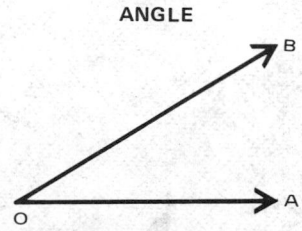

ANGLE

Angle formed by rays *OA* and *OB* emanating from common point *O. O* is vertex of angle; *OA* and *OB* are sides of angle.

angle of advance See angular advance.

angle of approach [CIV ENG] The maximum angle of an incline onto which a vehicle can move from a horizontal plane without interference. [MECH ENG] The angle that is turned through by either of paired wheels in gear from the first contact between a pair of teeth until the pitch points of these teeth fall together. [ORD] Angle between the line along which a moving target is traveling and the line along which the gun is pointed.

angle of arrival [ELECTROMAG] A measure of the direction of propagation of electromagnetic radiation upon arrival at a receiver (the term is most commonly used in radio); it is the angle between the plane of the phase front and some plane of reference, usually the horizontal, at the receiving antenna.

angle of attack [AERO ENG] The angle between a reference line fixed with respect to an airframe (usually the longitudinal axis) and the direction of movement of the body. [MIN ENG] The angle in a mine fan made by the direction of air approach and the chord of the aerofoil section.

angle of bite See angle of nip.

angle of cant [AERO ENG] In a spin-stabilized rocket, the angle formed by the axis of a venturi tube and the longitudinal axis of the rocket.

angle of clearance [ORD] The angle between the line along which a gun or launcher is pointed at the target and the line along which the weapon must be pointed for a projectile or missile fired from it to clear any obstruction between the weapon and the target.

angle of climb [AERO ENG] The angle between the flight path of a climbing vehicle and the local horizontal.

angle of contact [FL MECH] The angle between the surface of a liquid and the surface of a partially submerged object or of the container at the line of contact. Also known as contact angle.

angle of current [HYD] In stream gaging, the angular difference between 90° and the angle made by the current with a measuring section.

angle of cut [NAV] The smaller angular difference of two bearings or lines of position.

angle of deflection [ORD] The horizontal clockwise angle between the axis of the bore of a gun and the line of sighting when a gun is laid for direction.

angle of departure [AERO ENG] The vertical angle, at the origin, between the line of site and the line of departure. [CIV ENG] The maximum angle of an incline from which a vehicle can move onto a horizontal plane without interference, such as from rear bumpers.

angle of depression [ENG] The angle in a vertical plane between the horizontal and a descending line. Also known as depression angle; descending vertical angle; minus angle.

angle of descent [AERO ENG] The angle between the flight path of a descending vehicle and the local horizontal.

angle of dip See dip.

angle of divergence [NAV] In terminal instrument procedures, the smaller of the angles formed by the intersection of two courses, radials, bearings, or combinations thereof.

angle of elevation [ENG] The angle in a vertical plane between the local horizontal and an ascending line, as from an observer to an object; used in astronomy, surveying, and so on. Also known as ascending vertical angle; elevation angle. [ORD] 1. The vertical angle above the line of sight through which the axis of the gun bore must be raised so that the bullet or projectile will carry to the target. 2. In aerial gunnery, an acute angle between the bore axis of a gun and the horizontal; called quadrant elevation in ground gunnery.

angle of entry [ORD] The acute angle between the tangent to the trajectory at the point of impact of a bomb or projectile and the perpendicular to the surface of the ground or target at the point of impact.

angle of external friction [ENG] The angle between the abscissa and the tangent of the curve representing the relationship of shearing resistance to normal stress acting between soil and the surface of another material. Also known as angle of wall friction.

angle of fall [MECH] The vertical angle at the level point, between the line of fall and the base of the trajectory.

angle of glide [AERO ENG] Angle of descent for an airplane or missile in a glide.

angle of impact [MECH] The acute angle between the tangent to the trajectory at the point of impact of a projectile and the plane tangent to the surface of the ground or target at the point of impact.

angle of incidence [OPTICS] The angle formed by a ray arriving at a surface and the perpendicular to that surface at the point of arrival. Also known as incidence angle.

angle of lag See lag angle.

angle of lead See lead angle.

angle of nip [MECH ENG] The largest angle that will just grip a lump between the jaws, rolls, or mantle and ring of a crusher. Also known as angle of bite; nip. [MIN ENG] In a rock-crushing machine, the maximum angle subtended by its approaching jaws or roll surfaces at which a specified piece of ore can be gripped.

angle of obliquity See angle of pressure.

angle of orientation [MECH] Of a projectile in flight, the angle between the plane determined by the axis of the projectile and the tangent to the trajectory (direction of motion), and the vertical plane including the tangent to the trajectory.

angle of pitch [AERO ENG] The angle, as seen from the side, between the longitudinal body axis of an aircraft or similar body and a chosen reference line or plane, usually the horizontal plane.

angle of pressure [DES ENG] The angle between the profile of a gear tooth and a radial line at its pitch point. Also known as angle of obliquity.

angle of radiation [ELECTROMAG] Angle between the surface of the earth and the center of the beam of energy radiated upward into the sky from a transmitting antenna.

angle of recess [MECH ENG] The angle that is turned through by either of two wheels in gear, from the coincidence of the pitch points of a pair of teeth until the last point of contact of the teeth.

angle of reflection [PHYS] The angle between the direction of propagation of a wave reflected by a surface and the line perpendicular to the surface at the point of reflection. Also known as reflection angle.

angle of refraction [PHYS] The angle between the direction of propagation of a wave that is refracted by a surface and the line that is perpendicular to the surface at the point of refraction.

angle of repose [ENG] See angle of rest. [MECH] The angle between the horizontal and the plane of contact between two bodies when the upper body is just about to slide over the lower.

angle of rest [ENG] The maximum slope at which a heap of any loose or fragmented solid material will stand without sliding, or will come to rest when poured or dumped in a pile or on a slope. Also known as angle of repose.

angle of roll [AERO ENG] The angle that the lateral body axis of an aircraft or similar body makes with a chosen reference plane in rolling; usually, the angle between the lateral axis and a horizontal plane.

angle of safety [ORD] Minimum permissible angular clearance, at the weapon, of the path of a projectile or missile above friendly troops.

angle of shear [GEOL] The angle between the planes of maximum shear which is bisected by the axis of greatest compression.

angle of site [ORD] The vertical angle that is formed by the line of site and the horizontal.

angle of slide [MIN ENG] The slope, measured in degrees of deviation from the horizontal, on which loose or fragmented solid materials will start to slide.

angle of stall [AERO ENG] The angle of attack at which the flow of air begins to break away from the airfoil, the lift begins to decrease, and the drag begins to increase. Also known as stalling angle.

angle of thread [DES ENG] The angle occurring between the sides of a screw thread, measured in an axial plane.

angle of torsion [MECH] The angle through which a part of an object such as a shaft or wire is rotated from its normal position when a torque is applied. Also known as angle of twist.

angle of train [ORD] The azimuth element of firing data furnished by a remote-control system.

angle of traverse [ORD] 1. Horizontal angle through which a

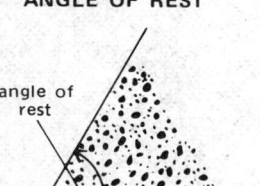

ANGLE OF REST

angle of rest

cement: 40°
round gravel: 30°

Angle of rest, one of the critical angles for bulk material.

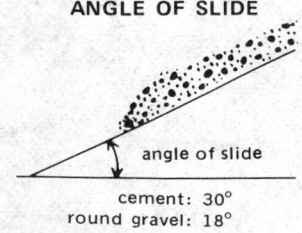

ANGLE OF SLIDE

angle of slide

cement: 30°
round gravel: 18°

Angle of slide, one of the critical angles for bulk material.

gun or launcher can be turned on its mount. **2.** Angle between the lines from a gun or launcher to the right and left limits of the front which is covered by its fire, that is, the angle through which the weapon is traversed.

angle of twist *See* angle of torsion.

angle of view [OPTICS] The angle subtended by an image at the second nodal point of a lens.

angle of wall friction *See* angle of external friction.

angle of yaw [AERO ENG] The angle, as seen from above, between the longitudinal body axis of an aircraft, a rocket, or the like and a chosen reference direction. Also known as yaw angle.

angle plate [DES ENG] An L-shaped plate or a plate having an angular section.

angle post [BUILD] A railing support used at a landing or other break in the stairs.

angle press [MECH ENG] A hydraulic plastics-molding press with both horizontal and vertical rams; used to produce complex moldings with deep undercuts.

angle rafter [BUILD] A rafter, such as a hip rafter, at the angle of the roof.

anglerfish [VERT ZOO] Any of several species of the order Lophiiformes characterized by remnants of a dorsal fin seen as a few rays on top of the head that are modified to bear a terminal bulb.

angle rib [ARCH] **1.** One of the diagonal ribs dividing each rectangle of a gothic vaulting; it is a structural member. **2.** Molding ornamenting an angle in decorative work.

angle set [MIN ENG] **1.** A timber set using an angle brace. **2.** One of a series of sets placed at angles to each other.

angle shot [GRAPHICS] **1.** A photograph taken with the camera tilted at an angle from the horizontal. **2.** In cinematography, a motion picture that is taken from a different angle; it may repeat the action in a previous shot or continue the action.

anglesite [MINERAL] $PbSO_4$ A mineral occurring in white or gray, tabular or prismatic orthorhombic crystals or compact masses. Also known as lead spar; lead vitriol.

angle-stem thermometer [ENG] A device used to measure temperatures in oil-custody tanks; the angle of the calibrated stem may be 90° or greater to the sensitive portion of the thermometer, as needed to fit the tank shell contour.

angle tracking noise [ELECTR] Any deviation of the tracking axis from the center of reflectivity of a radar target; it is the resultant of servo noise, receiver noise, angle noise, and amplitude noise.

angle valve [DES ENG] A manually operated valve with its outlet opening oriented at right angles to its inlet opening; used for regulating the flow of a fluid in a pipe.

angle variable [MECH] The dynamical variable w conjugate to the action variable J, defined only for periodic motion.

angling dozer *See* Angledozer.

Angoumian [GEOL] Upper middle Upper Cretaceous (Upper Turonian) geologic time.

angrite [GEOL] An achondrite stony meteorite composed principally of augite with a little olivine and troilite.

angstrom [MECH] A unit of length, 10^{-10} meter, used primarily to express wavelengths of optical spectra. Abbreviated A. Also known as tenthmeter.

Ångström coefficient [PHYS] The multiplying amplitude parameter inserted in Ångström's formula for the scattering of electromagnetic radiation by atmospheric dust.

Ångström compensation pyrheliometer [ASTROPHYS] An instrument for the measurement of direct solar radiation.

Ångström's formula [PHYS] A formula stating that the scattering coefficient for dust in the atmosphere is inversely proportional to a positive power of the wavelength of the radiation, with the power depending on the size of the dust particles.

anguclast [GEOL] An angular phenoclast.

Anguidae [VERT ZOO] A family of limbless, snakelike lizards in the suborder Sauria, commonly known as slowworms or glass snakes.

Anguilliformes [VERT ZOO] A large order of actinopterygian fishes containing the true eels.

Anguilloidei [VERT ZOO] The typical eels, a suborder of actinopterygian fishes in the order Anguilliformes.

angular [SCI TECH] **1.** Having sharp corners. **2.** Having or forming angles.

angular acceleration [MECH] The time rate of change of angular velocity.

angular advance [MECH ENG] The amount by which the angle between the crank of a steam engine and the virtual crank radius of the eccentric exceeds a right angle. Also known as angle of advance; angular lead.

angular aperture [OPTICS] The angle subtended at an axial object point of an optical instrument by the radius of the entrance pupil.

angular bitstalk *See* angular bitstock.

angular bitstock [MECH ENG] A bitstock whose handles are positioned to permit its use in corners and other cramped areas. Also known as angular bitstalk.

angular clearance [DES ENG] The relieved space located below the straight of a die, to permit passage of blanks or slugs.

angular-contact bearing [MECH ENG] A rolling-contact antifriction bearing designed to carry heavy thrust loads and also radial loads.

angular cutter [MECH ENG] A tool-steel cutter used for finishing surfaces at angles greater or less than 90° with its axis of rotation.

angular displacement [PHYS] A vector measure of the rotation of an object about an axis; the vector points along the axis according to the right-hand rule; the length of the vector is the rotation angle, in degrees or radians.

angular distance [NAV] **1.** The angular difference between two directions, numerically equal to the angle between two lines extending in the given directions. **2.** The arc of the great circle joining two points, expressed in angular units. **3.** The distance between two points, expressed in wavelengths at a specified frequency.

angular distortion [MAP] Distortion in a map projection because of nonconformality.

angular distribution [NUCLEO] The distribution in angle, relative to an experimentally specified direction, of the intensity of photons or particles resulting from a nuclear or extranuclear process.

angular error of closure *See* error of closure.

angular frequency [PHYS] For any oscillation, the number of vibrations per unit time, multiplied by 2π. Also known as angular velocity; radian frequency.

angular gear [MECH ENG] A gear that transmits motion between two rotating shafts that are not parallel. Also known as angle gear.

angular height [ORD] The vertical angle between the line of site and the horizontal.

angular impulse [MECH] The integral of the torque applied to a body over time.

angularity [SCI TECH] The sharpness of corners and edges.

angular lead *See* angular advance.

angular leaf spot [PL PATH] A bacterial disease of plants characterized by angular spots on leaves and caused by *Pseudomonas lachrymans* in cucumbers and *Xanthomonas malvacearum* in cotton.

angular length [MECH] A length expressed in the unit of the length per radian or degree of a specified wave.

angular magnification [OPTICS] For an optical system, the ratio of the angle subtended by the image at the eye to the angle subtended by the object at the eye.

angular milling [MECH ENG] Milling surfaces that are flat and at an angle to the axis of the spindle of the milling machine.

angular momentum [MECH] **1.** The cross product of a vector from a specified reference point to a particle, with the particle's linear momentum. Also known as moment of momentum. **2.** For a system of particles, the vector sum of the angular momenta (first definition) of the particles.

angular momentum operator [QUANT MECH] Any vector operator satisfying communication rules of the type $[J_x, J_y] = iJ_z$.

angular parallax [NAV] The angle between grid north and true north; the angle increases the farther the grid departs from the standard meridian.

angular perspective [GRAPHICS] A form of plane linear perspective in which some of the principal lines of the picture are

ANGLERFISH

Cryptopsaras couesi, the anglerfish, length up to 18 inches (45 centimeters). *(After G. B. Goode and T. H. Bean, Oceanic Ichthyology, U.S. Nat. Mus. Spec. Bull. no. 2, 1895)*

ANGLESITE

⊢ 2.5 cm ⊣

Crystals of anglesite on galena. *(Specimen from Department of Geology, Bryn Mawr College)*

ANGULAR-CONTACT BEARING

Standard angular-contact ball bearing. *(Marlin-Rockwell)*

either parallel or perpendicular to the picture plane and some are oblique.

angular pitch [DES ENG] The angle determined by the length along the pitch circle of a gear between successive teeth.

angular rate *See* angular speed; angular velocity.

angular resolution [ELECTROMAG] A measure of the ability of a radar to distinguish between two targets solely by the measurement of angles.

angular resolver *See* resolver.

angular shear [MECH ENG] A shear effected by two cutting edges inclined to each other to reduce the force needed for shearing.

angular speed [MECH] **1.** Change of direction per unit time, as of a target on a radar screen. Also known as angular rate. **2.** *See* angular velocity.

angular travel [ORD] Angular distance covered by a moving target in a given time.

angular travel error [MECH] The error which is introduced into a predicted angle obtained by multiplying an instantaneous angular velocity by a time of flight.

angular travel method [ORD] Method of calculating firing data based on the rate of angular travel of the target in direction and elevation.

angular unconformity [GEOL] An unconformity in which the older strata dip at a different angle (usually steeper) than the younger strata.

angular unit method [ORD] A method of adjusting antiaircraft artillery gunfire in which range deviations in mils obtained by a distant observer are converted into altitude corrections in yards for application at the data computer.

angular velocity [MECH] The time rate of change of angular displacement. Also known as angular rate; angular speed. [PHYS] *See* angular frequency.

angular wave number [METEOROL] The number of waves of a given wavelength required to encircle the earth at the latitude of the disturbance. Also known as hemispheric wave number.

angulator [ENG] An instrument for converting angles measured on an oblique plane to their corresponding projections on a horizontal plane; the rectoblique plotter and the photoangulator are types.

anhalonium alkaloid [PHARM] Any of the alkaloids found in mescal buttons, including hordenine and mescaline; used as a cerebral stimulant and motor depressant. Also known as cactus alkaloid.

anharmonicity [PHYS] **1.** Mechanical vibration where the restoring force acting on a system does not vary linearly with displacement from equilibrium position. **2.** Variation from a linear relationship of dipole moment with internuclear distance in the infrared portion of the electromagnetic spectrum.

anharmonic oscillator [PHYS] An oscillating system in which the restoring force opposing a displacement from the position of equilibrium is a nonlinear function of the displacement.

anhedral *See* allotriomorphic.

anhedron [PETR] Rock that has the organized internal structure of a crystal without the external geometric form of a crystal.

anhidrosis [MED] Absent or deficient secretion of sweat.

Anhimidae [VERT ZOO] The screamers, a family of birds in the order Anseriformes characterized by stout bills, webbed feet, and spurred wings.

Anhingidae [VERT ZOO] The anhingas or snakebirds, a family of swimming birds in the order Pelecaniformes.

anhydrase [BIOCHEM] Any enzyme that catalyzes the removal of water from a substrate.

anhydremia [MED] A decreased amount of water in the plasma.

anhydride [CHEM] A compound formed from another by removal of water.

anhydrite [MINERAL] $CaSO_4$ A mineral that represents gypsum without its water of crystallization, occurring commonly in white and grayish granular to compact masses; the hardness is 3–3.5 on Mohs scale, and specific gravity is 2.90–2.99. Also known as cube spar.

anhydrite evaporite [PETR] $CuSO_4$ A sedimentary rock composed chiefly of copper sulfate in compact granular form deposited by evaporation of water; resembles marble and

differs from gypsum in lack of water of hydration and hardness.

anhydrock [PETR] A sedimentary rock chiefly made of anhydrite.

anhydrous [CHEM] Being without water, especially water of crystallization.

anhydrous alcohol *See* absolute alcohol.

anhydrous ammonia [INORG CHEM] Liquid ammonia, a colorless liquid boiling at −33.3°C.

anhydrous ferric chloride *See* ferric chloride.

anhydrous hydrogen chloride [INORG CHEM] HCl Hazardous, toxic, colorless gas used in polymerization, isomerization, alkylation, nitration, and chlorination reactions; becomes hydrochloric acid in aqueous solutions.

anhydrous phosphoric acid *See* phosphoric anhydride.

anhydrous plumbic acid *See* lead dioxide.

anhydrous sodium carbonate *See* soda ash.

anhydrous sodium sulfate [INORG CHEM] Na_2SO_4 Water-soluble, white crystals with bitter, salty taste; melts at 888°C; used in the manufacture of glass, paper, pharmaceuticals, and textiles, and as an analytical reagent.

anilide [ORG CHEM] A compound that has the $C_6H_5NH_2-$ group; an example is benzanilide, $C_6H_5NHCOC_6H_5$.

Aniliidae [VERT ZOO] A small family of nonvenomous, burrowing snakes in the order Squamata.

aniline [ORG CHEM] $C_6H_5NH_2$ An aromatic amine compound that is a pale brown liquid at room temperature; used in the dye, pharmaceutical, and rubber industries.

aniline black [ORG CHEM] A black dye produced on certain textiles, such as cotton, by oxidizing aniline or aniline hydrochloride.

aniline chloride *See* aniline hydrochloride.

aniline N,N-dimethyl *See* N,N-dimethyl aniline.

aniline dye [ORG CHEM] A dye derived from aniline or benzene.

aniline-formaldehyde resin [CHEM] A thermoplastic resin made by polymerizing aniline and formaldehyde.

aniline hydrochloride [ORG CHEM] $C_6H_5NH_2 \cdot HCl$ White crystals, although sometimes the commercial variety has a greenish tinge; melting point 198°C; soluble in water and ethanol; used in dye manufacture, dyeing, and printing. Also known as aniline chloride; aniline salt.

aniline ink [MATER] A fast-drying printing ink that is a solution of a coal-tar dye in an organic solvent or a solution of a pigment in an organic solvent or water.

aniline leather [MATER] Leather whose grain pattern has been accentuated by impregnation with aniline dye.

aniline point [CHEM ENG] The minimum temperature for a complete mixing of aniline and materials such as gasoline; used in some specifications to indicate the aromatic content of oils and to calculate approximate heat of combustion.

aniline printing *See* flexography.

aniline process *See* flexography.

aniline salt *See* aniline hydrochloride.

***para*-anilinesulfonic acid** *See* sulfanilic acid.

anilinoacetic acid *See* N-phenylglycine.

animal [ZOO] Any living organism distinguished from plants by the lack of chlorophyll, the requirement for complex organic nutrients, the lack of a cell wall, limited growth, mobility, and greater irritability.

animal black [CHEM] Finely divided carbon made by calcination of animal bones or ivory; used for pigments, decolorizers, and purifying agents; varieties include bone black and ivory black.

animal charcoal [CHEM] Charcoal obtained by the destructive distillation of animal matter at high temperatures; used to adsorb organic coloring matter.

animal communication [PSYCH] The discipline within the field of animal behavior that deals with the receipt and use of signals by animals.

animal community [ECOL] An aggregation of animal species held together in a continuous or discontinuous geographic area by ties to the same physical environment, mainly vegetation.

animal ecology [ECOL] A study of the relationships of animals to their environment.

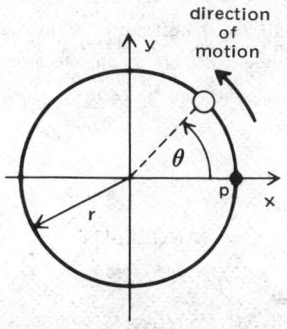

ANGULAR VELOCITY

Illustration of angular velocity. Particle *p* moves on circular path with radius *r*. *θ* is angular displacement of *p*. Angular velocity is rate of change of *θ* with respect to time.

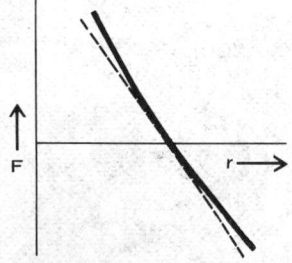

ANHARMONIC OSCILLATOR

Interatomic force *F* as a function of the atomic separation *r*. Anharmonicity is produced by the departure of the actual force (solid curve) from the dashed line.

ANHYDRITE

Anhydrite from Montanzas, Cuba. (*Specimen from Department of Geology, Bryn Mawr College*)

animal fiber [TEXT] A natural textile fiber of animal origin; wool and silk are the most important.

animal husbandry [AGR] A branch of agriculture concerned with the breeding and feeding of domestic animals.

Animalia [SYST] The animal kingdom.

animal kingdom [SYST] One of the two generally accepted major divisions of living organisms which live or have lived on earth (the other division being the plant kingdom).

animal locomotion [ZOO] Progressive movement of an animal body from one point to another.

animal oil *See* bone oil.

animal pole [CYTOL] The region of an ovum which contains the least yolk and where the nucleus gives off polar bodies during meiosis.

animal power [MECH ENG] The time rate at which muscular work is done by a work animal, such as a horse, bullock, or elephant.

animal virus [VIROL] A small infectious agent able to propagate only within living animal cells.

Animikean [GEOL] The middle subdivision of Proterozoic geologic time. Also known as Penokean; Upper Huronian.

animikite [GEOL] An ore of silver, composed of a mixture of sulfides, arsenides, and antimonides, and containing nickel and lead; occurs in white or gray granular masses.

anion [CHEM] An ion that is negatively charged.

anion exchange [CHEM] A type of ion exchange in which the immobilized functional groups on the solid resin are positive.

anionic detergent [MATER] A class of detergents having a negatively charged surface-active ion, such as sodium alkylbenzene sulfonate.

anionic polymerization [ORG CHEM] A type of polymerization in which Lewis bases, such as alkali metals and metallic alkyls, act as catalysts.

anionotropy [CHEM] The breaking off of an ion such as hydroxyl or bromide from a molecule so that a positive ion remains in a state of dynamic equilibrium.

Anisakidae [INV ZOO] A family of parasitic roundworms in the superfamily Ascaridoidea.

anisaldehyde [ORG CHEM] $C_6H_4(OCH_3)CHO$ A compound with melting point 2.5°C, boiling point 249.5°C; insoluble in water, soluble in alcohol and ether; used in perfumery and flavoring, and as an intermediate in production of antihistamines. Also known as *para*-anisic aldehyde; aubepine; 4-methoxybenzaldehyde; *para*-methoxybenzaldehyde.

anise [BOT] The small fruit of the annual herb *Pimpinella anisum* in the family Umbelliferae; fruit is used for food flavoring, and oil is used in medicines, soaps, and cosmetics.

anise alcohol *See* anisic alcohol.

Anisean [GEOL] Lower Middle Triassic geologic time.

anise camphor *See* anethole.

anisic alcohol [ORG CHEM] $C_8H_{10}O_2$ A colorless liquid that boils in the range 255–265°C; it is obtained by reduction of anisic aldehyde; used in perfumery, and as an intermediate in the manufacture of pharmaceuticals. Also known as anise alcohol; anisyl alcohol; *para*-methoxybenzyl alcohol.

para-anisic aldehyde *See* anisaldehyde.

anisocarpous [BOT] Referring to a flower whose number of carpels is different from the number of stamens, petals, and sepals.

anisochela [INV ZOO] A chelate sponge spicule with dissimilar ends.

anisocytosis [MED] A condition in which the erythrocytes show a considerable variation in size due to excessive quantities of hemoglobin.

anisodactylous [VERT ZOO] Having unequal digits, especially referring to birds with three toes forward and one backward.

anisodesmic [MINERAL] Pertaining to crystals or compounds in which the ionic bonds are unequal in strength.

anisogamete *See* heterogamete.

anisogamy *See* heterogamy.

anisole [ORG CHEM] $C_6H_5OCH_3$ A colorless liquid that is soluble in ether and alcohol, insoluble in water; boiling point is 155°C; vapors are highly toxic; used as a solvent and in perfumery. Also known as methoxybenzene; methylphenyl ether.

anisomerous [BOT] Referring to flowers that do not have the same number of parts in each whorl.

anisometric particle [VIROL] Any unsymmetrical, rod-shaped plant virus.

Anisomyaria [INV ZOO] An order of mollusks in the class Bivalvia containing the oysters, scallops, and mussels.

anisophyllous [BOT] Having leaves of two or more shapes and sizes.

Anisoptera [INV ZOO] The true dragonflies, a suborder of insects in the order Odonata.

anisostemonous [BOT] Referring to a flower whose number of stamens is different from the number of carpels, petals, and sepals.

Anisotomidae [INV ZOO] An equivalent name for Leiodidae.

anisotropic [PHYS] Showing different properties as to velocity of light transmission, conductivity of heat or electricity, compressibility, and so on, in different directions. Also known as aeolotropic.

anisotropy [BOT] The property of a plant that assumes a certain position in response to an external stimulus. [PHYS] The characteristic of a substance for which a physical property, such as index of refraction, varies in value with the direction in or along which the measurement is made. Also known as aeolotropy; eolotropy. [ZOO] The property of an egg that has a definite axis or axes.

anisotropy constant [ELECTROMAG] In a ferromagnetic material, temperature-dependent parameters relating the magnetization in various directions to the anisotropy energy.

anisotropy energy [ELECTROMAG] Energy stored in a ferromagnetic crystal by virtue of the work done in rotating the magnetization of a domain away from the direction of easy magnetization.

anisotropy factor *See* dissymmetry factor.

anisyl alcohol *See* anisic alcohol.

ankaramite [PETR] A mafic olivine basalt primarily composed of pyroxene with smaller amounts of olivine and plagioclase and accessory biotite, apatite, and opaque oxides.

ankaratrite *See* olivine nephelinite.

anker [MECH] A unit of capacity equal to 10 U.S. gallons; used to measure liquids, especially honey, oil, vinegar, spirits, and wine.

ankerite [MINERAL] $Ca(Fe,Mg,Mn)(CO_3)_2$ A white, red, or gray iron-rich carbonate mineral associated with iron ores and found in thin veins in coal seams; specific gravity is 2.95–3.1. Also known as cleat spear; ferroan dolomite.

ankle [ANAT] The joint formed by the articulation of the leg bones with the talus, one of the tarsal bones.

ankle breadth [ANTHRO] The distance measured between projections at lower ends of the tibia and fibula.

ankle thickness [ANTHRO] Distance measured perpendicular to ankle breadth.

Ankylosauria [PALEON] A suborder of Cretaceous dinosaurs in the reptilian order Ornithischia characterized by short legs and flattened, heavily armored bodies.

ankylosis Also spelled anchylosis. [MED] Stiffness or immobilization of a joint due to a surgical or pathologic process. [PHYS] The loss by a system of one or more degrees of freedom through development of one or more frictional constraints.

anlage [EMBRYO] Any group of embryonic cells when first identifiable as a future organ or body part. Also known as blastema; primordium.

annabergite [MINERAL] $(Ni,Co)_3(AsO_4)_2 \cdot 8H_2O$ A monoclinic mineral usually found as apple-green incrustations as an alteration product of nickel arsenides; it is isomorphous with erythrite. Also known as nickel bloom; nickel ocher.

annatto [BOT] *Bixa orellana.* A tree found in tropical America, characterized by cordate leaves and spinose, seed-filled capsules; a yellowish-red dye obtained from the pulp around the seeds is used as a food coloring.

anneal [ENG] To treat a metal, alloy, or glass with heat and then cool to remove internal stresses and to make the material less brittle.

annealing furnace [ENG] A furnace for annealing metals or glass. Also known as annealing oven.

annealing oven *See* annealing furnace.

annealing point [THERMO] The temperature at which the

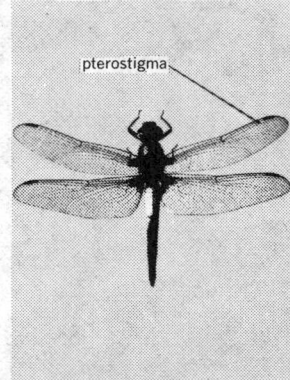

ANISOPTERA

pterostigma

An adult dragonfly, showing the thickened spot, pterostigma, on the costal margin of the wing.

ANKERITE

A specimen of ankerite. (*Specimen from Department of Geology, Bryn Mawr College*)

ANKYLOSAURIA

Restoration of the armored Cretaceous dinosaur *Ankylosaurus* (about 20 feet or 6 meters long).

viscosity of a glass is $10^{13.0}$ poises. Also known as annealing temperature; 13.0 temperature.

annealing temperature *See* annealing point.

Annedidae [VERT ZOO] A small family of limbless, snakelike, burrowing lizards of the suborder Sauria.

Annelida [INV ZOO] A diverse phylum comprising the multisegmented wormlike animals.

annex point [MAP] A point used to assist in the relative orientation of vertical and oblique photographs; selected in the overlap area between the vertical and its corresponding oblique photograph, about midway between the pass points.

Anniellidae [VERT ZOO] A family of limbless, snakelike lizards in the order Squamata.

annihilation [PARTIC PHYS] A process in which an antiparticle and a particle combine and release their rest energies in other particles.

annihilation operator [QUANT MECH] An operator which reduces the occupation number of a single state by unity; for example, an annihilation operator applied to a state of one particle yields the vacuum.

annihilation radiation [PARTIC PHYS] Electromagnetic radiation arising from the collision, and resulting annihilation, of an electron and a positron, or of any particle and its antiparticle.

anniversary clock [HOROL] A clock that can run as long as 400 days on a single winding because of a slow torsion pendulum. Also known as four-hundred-day clock.

Annonaceae [BOT] A large family of woody flowering plants in the order Magnoliales, characterized by hypogynous flowers, exstipulate leaves, a trimerous perianth, and distinct stamens with a short, thick filament.

annotated photograph [MAP] A photograph on which planimetric, geologic, cultural, hydrographic, or vegetation information has been added to identify, classify, outline, clarify, or describe features that would not otherwise be apparent in examination of an unmarked photograph.

annotation [ADP] Any comment or note included in a program or flow chart in order to clarify some point at issue.

annotation overprint [ORD] 1. The outline delimiting a target or installation. 2. A symbol which locates the position of a target together with an identifying reference number as depicted on a target graphic.

annotation text [ORD] A descriptive text containing the identification, function, location, physical characteristics, and other information concerning a target or installation.

annual aberration [ASTRON] Aberration caused by the velocity of the earth's revolution about the sun.

annual cost comparison [IND ENG] A method of selecting from among several alternative projects or courses of action on the basis of their annual costs, including depreciation.

annual flood [HYD] The highest flow at a point on a stream during any particular calendar year or water year.

annual inequality [OCEANOGR] Seasonal variation in water level or tidal current speed, more or less periodic, due chiefly to meteorological causes.

annual labor *See* assessment drilling.

annual layer [GEOL] 1. A sedimentary layer deposited, or presumed to have been deposited, during the course of a year; for example, a glacial varve. 2. A dark layer in a stratified salt deposit containing disseminated anhydrite.

annual magnetic change *See* magnetic annual change.

annual magnetic variation *See* magnetic annual variation.

annual parallax [ASTRON] The apparent displacement of a celestial body viewed from two separated observation points whose base line is the radius of the earth's orbit.

annual plant [BOT] A plant that completes its growth in one growing season and therefore must be planted annually.

annual ring [BOT] A line appearing on tree cross sections marking the end of a growing season and showing the volume of wood added during the year.

annual storage [HYD] The capacity of a reservoir that can handle a watershed's annual runoff but cannot carry over any portion of the water for longer than the year.

annuation [ECOL] The annual variation in the presence, absence, or abundance of a member of a plant community.

annular [MIN ENG] The space between the casing and wall of a hole or between a drill pipe and casing.

annular auger [DES ENG] A ring-shaped boring tool which cuts an annular channel, leaving the core intact.

annular budding [BOT] Budding by replacement of a ring of bark on a stock with a ring bearing a bud from a selected species or variety.

annular conductor [ELEC] A number of wires stranded in three reversed concentric layers around a saturated hemp core.

annular drainage pattern [HYD] A ringlike pattern subsequent in origin and associated with maturely dissected dome or basin structures.

annular eclipse [ASTRON] An eclipse in which a thin ring of the source of light appears around the obscuring body.

annular gear [DES ENG] A gear having a cylindrical form.

annular hernia *See* umbilical hernia.

annular nozzle [DES ENG] A nozzle with a ring-shaped orifice.

annular section [ENG] The open space between two concentric tubes, pipes, or vessels.

annular solid [MATH] A solid generated by rotating a closed plane curve about a line which lies in the plane of the curve and does not intersect the curve.

annular transistor [ELECTR] Mesa transistor in which the semiconductor regions are arranged in concentric circles about the emitter.

annular vault [ARCH] A vault arising from circular walls.

annulated shaft [ARCH] A column consisting of a cluster of shafts which seem to be held together at intervals by an annular band.

annulus [ANAT] Any ringlike anatomical part. [BOT] 1. An elastic ring of cells between the operculum and the mouth of the capsule in mosses. 2. A line of cells, partly or entirely surrounding the sporangium in ferns, which constricts, thus causing rupture of the sporangium to release spores. 3. A whorl resembling a calyx at the base of the strobilus in certain horsetails. [MATH] The ringlike figure that lies between two concentric circles. [MYCOL] A ring of tissue representing the remnant of the veil around the stipe of some agarics.

annulus conjecture [MATH] For dimension n, the assertion that if f and g are locally flat embeddings of the $(n-1)$ sphere, S^{n-1}, in real n space, R^n, with $f(S^{n-1})$ in the bounded component of $R^n - g(S^{n-1})$, then the closed region in R^n bounded by $f(S^{n-1})$ and $g(S^{n-1})$ is homeomorphic to the direct product of S^{n-1} and the closed interval [0,1]; it is established for $n \neq 4$.

annunciator [ENG] A signaling apparatus which operates electromagnetically and serves to indicate visually, or visually and audibly, whether a current is flowing, has flowed, or has changed direction of flow in one or more circuits.

Anobiidae [INV ZOO] The deathwatch beetles, a family of coleopteran insects of the superfamily Bostrichoidea.

anode [ELEC] The positive terminal of a primary cell or of a storage battery. [ELECTR] The collector of electrons in an electron tube. Also known as plate; positive electrode. [PHYS CHEM] The positive terminal of an electrolytic cell.

anode balancing coil [ELEC] Set of mutually coupled windings used to maintain approximately equal currents in anodes operating in parallel from the same transformer terminal.

anode circuit [ELECTR] Complete external electrical circuit connected between the anode and the cathode of an electron tube. Also known as plate circuit.

anode-circuit detector [ELECTR] Detector functioning by virtue of a nonlinearity in its anode-circuit characteristic. Also known as plate-circuit detector.

anode copper [MET] Slabs of refined blister copper used as anodes in the electrolytic refining of copper.

anode corrosion [MET] The disintegration of a metal acting as an anode.

anode current [ELECTR] The electron current flowing through an electron tube from the cathode to the anode. Also known as plate current.

anode detector [ELECTR] A detector in which rectification of radio-frequency signals takes place in the anode circuit of an electron tube. Also known as plate detector.

anode dissipation [ELECTR] Power dissipated as heat in the anode of an electron tube because of bombardment by electrons and ions.

anode effect [PHYS CHEM] A condition produced by polarization of the anode in the electrolysis of fused salts and characterized by a sudden increase in voltage and a corresponding decrease in amperage.

anode efficiency [ELECTR] The ratio of the ac load circuit power to the dc anode power input for an electron tube. Also known as plate efficiency.

anode fall [ELECTR] A very thin space-charge region in front of an anode surface, characterized by a steep potential gradient through the region.

anode film [CHEM] The portion of solution in immediate contact with the anode.

anode furnace [MET] A furnace in which blister copper or impure nickel is refined.

anode impedance [ELECTR] Total impedance between anode and cathode exclusive of the electron stream. Also known as plate impedance; plate-load impedance.

anode input power [ELECTR] Direct-current power delivered to the plate (anode) of a vacuum tube by the source of supply. Also known as plate input power.

anode metal [MET] The metal used as anode in an electroplating process.

anode modulation [ELECTR] Modulation produced by introducing the modulating signal into the anode circuit of any tube in which the carrier is present. Also known as plate modulation.

anode mud [MET] An insoluble substance or mixture that collects at the anode in an electrolytic refining or plating process. Also known as anode slime.

anode neutralization [ELECTR] Method of neutralizing an amplifier in which the necessary 180° phase shift is obtained by an inverting network in the plate circuit. Also known as plate neutralization.

anode pulse modulation [ELECTR] Modulation produced in an amplifier or oscillator by application of externally generated pulses to the plate circuit. Also known as plate-pulse modulation.

anode rays [ELECTR] Positive ions coming from the anode of an electron tube; generally due to impurities in the metal of the anode.

anode resistance [ELECTR] The resistance value obtained when a small change in the anode voltage of an electron tube is divided by the resulting small change in anode current. Also known as plate resistance.

anode saturation [ELECTR] The condition in which the anode current of an electron tube cannot be further increased by increasing the anode voltage; the electrons are then being drawn to the anode at the same rate as they are emitted from the cathode. Also known as current saturation; plate saturation; saturation; voltage saturation.

anode scrap [MET] Portions of anode copper retrieved from electrolytic refining of the metal.

anode sheath [ELECTR] The electron boundary which exists in a gas-discharge tube between the plasma and the anode when the current demanded by the anode circuit exceeds the random electron current at the anode surface.

anode slime *See* anode mud.

anodic [PHYS] Pertaining to the anode.

anodic cleaning [MET] The removal of a foreign substance from a metallic surface by electrolysis with the metal as the anode. Also known as anodic pickling; reverse-current cleaning.

anodic coating [MET] A film of oxide produced on a metal by electrolysis with the metal as the anode.

anodic pickling *See* anodic cleaning.

anodic polarization [PHYS CHEM] The change in potential of an anode caused by current flow.

anodic reaction [MET] The reaction in the mechanism of electrochemical corrosion in which the metal forming the anode dissolves in the electrolyte in the form of positively charged ions.

anodize [MET] The formation of a decorative or protective passive film on a metal part by making it the anode of a cell and applying electric current.

anodized aluminum [MET] Aluminum coated with a layer of aluminum oxide by an anodic process in a suitable electrolyte such as chromic acid or sulfuric acid solution.

anodized magnesium [MET] An anodic coating on magnesium produced in one of various electrolytes, mainly of fluorides, phosphates, or chromates.

anole [VERT ZOO] Any arboreal lizard of the genus *Anolis*, characterized by flattened adhesive digits and a prehensile outer toe.

anolyte [CHEM] The part of the electrolyte at or near the anode that is changed in composition by the reactions at the anode.

Anomalinacea [INV ZOO] A superfamily of marine and benthic sarcodinian protozoans in the order Foraminiferida.

anomalistic month [ASTRON] The average period of revolution of the moon from perigee to perigee, a period of 27 days 13 hours 18 minutes 33.2 seconds.

anomalistic period [ASTRON] The interval between two successive perigee passages of a satellite in orbit about a primary. Also known as perigee-to-perigee period.

anomalistic year [ASTRON] The period of one revolution of the earth about the sun from perihelion to perihelion; 365 days 6 hours 13 minutes 53.0 seconds in 1900 and increasing at the rate of 0.26 second per century.

anomaloscope [OPTICS] An optical instrument for testing color vision, in which a yellow light whose intensity may be varied is matched against red and green lights whose intensity is fixed.

anomalous [SCI TECH] Deviating from the normal; irregular.

anomalous dispersion [OPTICS] Extraordinary behavior in the curve of refractive index versus wavelength which occurs in the vicinity of absorption lines or bands in the absorption spectrum of a medium.

anomalous magma [GEOL] Magma formed or obviously changed by assimilation.

anomalous magnetic moment [PARTIC PHYS] The difference between the observed magnetic moment and the value predicted by Dirac's theory.

anomalous water *See* polywater.

anomalous Zeeman effect [SPECT] A type of splitting of spectral lines of a light source in a magnetic field which occurs for any line arising from a combination of terms of multiplicity greater than one; due to a nonclassical magnetic behavior of the electron spin.

Anomaluridae [VERT ZOO] The African flying squirrels, a small family in the order Rodentia characterized by the climbing organ, a series of scales at the root of the tail.

anomaly [ASTRON] In celestial mechanics, the angle between the radius vector to an orbiting body from its primary (the focus of the orbital ellipse) and the line of apsides of the orbit, measured in the direction of travel, from the point of closest approach to the primary (perifocus). Also known as true anomaly. [BIOL] An abnormal deviation from the characteristic form of a group. [GEOL] A local deviation from the general geological properties of a region. [MED] Any part of the body that is abnormal in position, form, or structure. [METEOROL] The deviation of the value of an element (especially temperature) from its mean value over some specified interval. [OCEANOGR] The difference between conditions actually observed at a serial station and those that would have existed had the water all been of a given arbitrary temperature and salinity.

anomaly of geopotential difference *See* dynamic-height anomaly.

anomer [ORG CHEM] One of a pair of isomers of cyclic carbohydrates; resulting from creation of a new point of symmetry when a rearrangement of the atoms occurs at the aldehyde or ketone position.

anomite [MINERAL] A variety of biotite different only in optical orientation.

Anomocoela [VERT ZOO] A suborder of toadlike amphibians in the order Anura characterized by a lack of free ribs.

anomocoelous [ANAT] Describing a vertebra with a centrum that is concave anteriorly and flat or convex posteriorly.

Anomphalacea [PALEON] A superfamily of extinct gastropod mollusks in the order Aspidobranchia.

Anomura [VERT ZOO] A section of the crustacean order Decapoda that includes lobsterlike and crablike forms.

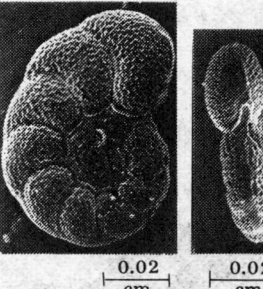

ANOMALINACEA

0.02 cm   0.02 cm

(a)   (b)

Scanning electron micrograph of the foraminiferan *Holmanella*, from the Miocene of California. (*a*) Spiral view and (*b*) edge view of bievolute planispiral test, with a coarsely perforate granular margin, and a slitlike aperture extending up the terminal face. (*R. B. MacAdam, Chevron Oil Field Research Co.*)

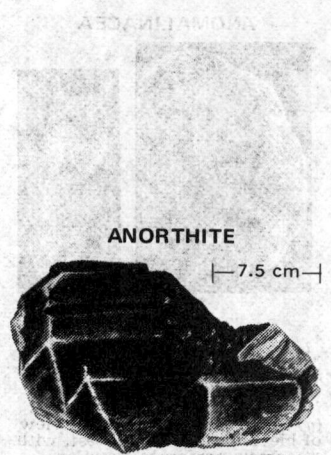

ANORTHITE

⊢—7.5 cm—⊣

Anorthite crystals from Miakejima, Japan. *(Specimen from Department of Geology, Bryn Mawr College)*

A-N RADIO RANGE

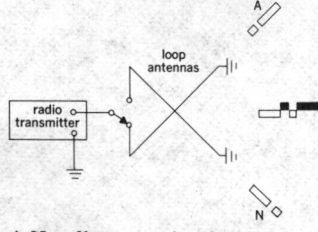

A-N radio range signals.

ANT

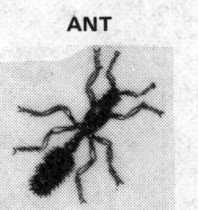

The female black ant *(Monororium minimum)*, about ¹/₁₆ inch (1.6 millimeters).

anonymous dimensionless group 1–4 [CHEM ENG] Four of the dimensionless groups, used to solve problems in transfer processes, gas absorption in wetted-wall columns, and laminar boundary-layer flow.

anoopsia [MED] Strabismus in which the eye is turned upward.

Anopheles [INV ZOO] A genus of mosquitoes in the family Culicidae; members are vectors of malaria, dengue, and filariasis.

anopheline [INV ZOO] Pertaining to mosquitoes of the genus *Anopheles* or a closely related genus.

Anopla [INV ZOO] A class or subclass of the phylum Rhynchocoela characterized by a simple tubular proboscis and by having the mouth opening posterior to the brain.

Anoplocephalidae [INV ZOO] A family of tapeworms in the order Cyclophyllidea.

Anoplura [INV ZOO] The sucking lice, a small group of mammalian ectoparasites usually considered to constitute an order in the class Insecta.

anorexia [MED] Loss of appetite.

anorgasmy [MED] Inability, usually psychic, to reach a climax during coitus.

anorogenic [GEOL] Of a feature, forming during tectonic quiescence between orogenic periods, that is, lacking in tectonic disturbance.

anorogenic time [GEOL] Geologic time when no significant deformation of the crust occurred.

anorthic crystal *See* triclinic crystal.

anorthite [MINERAL] The white, grayish, or reddish calcium-rich end member of the plagioclase feldspar series; composition ranges from $Ab_{10}An_{90}$ to Ab_0An_{100}, where $Ab = NaAlSi_3O_8$ and $An = CaAl_2Si_2O_8$. Also known as calciclase; calcium feldspar.

anorthite-basalt [PETR] A rock composed of a basic variety of basalt with anorthite instead of labradorite.

anorthoclase [MINERAL] A triclinic alkali feldspar having a chemical composition ranging from $Or_{40}Ab_{60}$ to $Or_{10}Ab_{90}$ to about 20 mole % An, where $Or = KAlSi_3O_8$, $Ab = NaAlSi_3O_8$, and $An = CaAl_2Si_2O_8$. Also known as anorthose; soda microcline.

anorthopia [MED] A defect of vision in which straight lines do not seem straight, and parallelism or symmetry is not properly perceived.

anorthosite [PETR] A visibly crystalline plutonic rock composed almost entirely of plagioclase feldspar (andesine to anorthite) with minor amounts of pyroxene and olivine.

anorthositization [GEOL] A process of anorthosite formation by replacement or metasomatism.

anoscope [MED] An instrument for examining the lower rectum and anal canal.

anosmia [MED] Absence of the sense of smell.

Anostraca [INV ZOO] An order of shrimplike crustaceans generally referred to the subclass Branchiopoda.

anotron [ELECTR] A cold-cathode glow-discharge diode having a copper anode and a large cathode of sodium or other material.

anoxia [MED] The failure of oxygen to gain access to, or to be utilized by, the body tissues.

A-N radio range [NAV] A type of radio beacon station whose signals provide definite track guidance for aircraft by establishing four radial lines of position which can be identified by a continuous-tone signal made up of keyed pulses of equal amplitude representing the Morse code letters A and N.

Ansbacher unit [BIOL] A unit for the standardization of vitamin K.

Anser [VERT ZOO] A genus of birds in the family Anatidae comprising the typical geese.

Anseranatini [VERT ZOO] A subfamily of aquatic birds in the family Anatidae represented by a single species, the magpie goose.

Anseriformes [VERT ZOO] An order of birds, including ducks, geese, swans, and screamers, characterized by a broad, flat bill and webbed feet.

Anson unit [BIOL] A unit for the standardization of trypsin and proteinases.

answering cord [ELEC] Cord nearest the face of the switch-

board which is used for answering subscribers' calls and incoming trunks.

answering jack [ELEC] Jack on which a station calls in and is answered by an operator.

answer lamp [ELEC] Telephone switchboard lamp that lights when an answer cord is plugged into a line jack; the lamp goes out when the call is completed.

ant [INV ZOO] The common name for insects in the hymenopteran family Formicidae; all are social, and colonies exhibit a highly complex organization.

Ant *See* Antlia.

antacid [CHEM] Any substance that counteracts or neutralizes acidity.

antagonism [BIOL] **1.** Mutual opposition as seen between organisms, muscles, physiologic actions, and drugs. **2.** Opposing action between drugs and disease or drugs and functions.

Antarctica [GEOGR] A continent roughly centered on the South Pole and surrounded by an ocean consisting of the southern parts of the Atlantic, Pacific, and Indian oceans.

antarctic air [METEOROL] A type of air whose characteristics are developed in an antarctic region.

antarctic anticyclone [METEOROL] The glacial anticyclone which has been said to overlie the continent of Antarctica; analogous to the Greenland anticyclone.

Antarctic Circle [GEOD] The parallel of latitude approximately 66°32′ south of the Equator.

Antarctic Circumpolar Current [OCEANOGR] The ocean current flowing from west to east through all the oceans around the Antarctic Continent. Also known as West Wind Drift.

Antarctic Convergence [OCEANOGR] The oceanic polar front indicating the boundary between the subantarctic and subtropical waters. Also known as Southern Polar Front.

antarctic faunal region [ECOL] A zoogeographic region describing both the marine littoral and terrestrial animal communities on and around Antarctica.

antarctic front [METEOROL] The semipermanent, semicontinuous front between the antarctic air of the Antarctic continent and the polar air of the southern oceans; generally comparable to the arctic front of the Northern Hemisphere.

Antarctic Intermediate Water [OCEANOGR] A water mass in the Southern Hemisphere, formed at the surface near the Antarctic Convergence between 45° and 55°S; it can be traced in the North Atlantic to about 25°N.

Antarctic Ocean [GEOGR] A circumpolar ocean belt including those portions of the Atlantic, Pacific, and Indian oceans which reach the Antarctic continent and are bounded on the north by the Subtropical Convergence; not recognized as a separate ocean.

Antarctic Zone [GEOGR] The region between the Antarctic Circle (66°32′S) and the South Pole.

Antares [ASTRON] A red supergiant variable binary star of stellar magnitude 0.9, 170 light-years from the sun, spectral classification M1-Ib, in the constellation Scorpius; the star α Scorpii.

anteater [VERT ZOO] Any of several mammals, in five orders, which live on a diet of ants and termites.

antebrachium *See* forearm.

antecedent platform [GEOL] A submarine platform 50 meters or more below sea level from which barrier reefs and atolls are postulated to grow toward the water's surface.

antecedent precipitation index [METEOROL] A weighted summation of daily precipitation amounts; used as an index of soil moisture.

antecedent stream [HYD] A stream that has retained its early course in spite of geologic changes since its course was assumed.

antecedent valley [GEOL] A stream valley that existed before uplift, faulting, or folding occurred and which has maintained itself during and after these events.

anteconsequent stream [HYD] A stream consequent to the form assumed by the earth's surface as the result of early movement of the earth but antecedent to later movement.

antecosta [INV ZOO] The internal, anterior ridge of the tergum or sternum of many insects that provides a surface for attachment of the longitudinal muscles.

antediluvial [GEOL] Formerly referred to time or deposits antedating Noah's flood.

anteflexion [MED] Forward bending of an organ on itself.

antelope [VERT ZOO] Any of the hollow-horned, hoofed ruminants assigned to the artiodactyl subfamily Antilopinae; confined to Africa and Asia.

ante meridian [ASTRON] **1.** A section of the celestial meridian; it lies below the horizon, and the nadir is included. **2.** Before noon, or the period of time between midnight (0000) and noon (1200).

antemortem [MED] Before death.

antenatal [MED] Occurring or existing before birth.

antenna [ELECTROMAG] A device used for radiating or receiving radio waves. Also known as aerial; radio antenna. [INV ZOO] Any one of the paired, segmented, and movable sensory appendages on the heads of many arthropods.

antenna amplifier [ELECTROMAG] One or more stages of wide-band electronic amplification placed within or physically close to a receiving antenna to improve signal-to-noise ratio and mutually isolate various devices receiving their feed from the antenna.

antenna array [ELECTROMAG] An arrangement of several individual antennas so spaced and phased that their individual contributions add in the preferred direction and cancel in other directions. Also known as array antennas; beam antennas.

antenna bearing [NAV] The generated bearing of the antenna of a radar set, as delivered to the indicator.

antenna circuit [ELECTR] A complete electric circuit which includes an antenna.

antenna coil [ELECTROMAG] Coil through which antenna current flows.

antenna coincidence [ELECTROMAG] That instance when two rotating, highly directional antennas are pointed toward each other.

antenna counterpoise See counterpoise.

antenna coupler [ELECTROMAG] A radio-frequency transformer, tuned line, or other device used to transfer energy efficiently from a transmitter to a transmission line or from a transmission line to a receiver.

antenna crosstalk [ELECTROMAG] The ratio or the logarithm of the ratio of the undesired power received by one antenna from another to the power transmitted by the other.

antenna detector [ELECTROMAG] Device consisting of an antenna and electronic equipment to warn aircraft crew members that they are being observed by radar sets.

antenna directivity diagram [ELECTROMAG] Curve representing, in polar or cartesian coordinates, a quantity proportional to the gain of an antenna in the various directions in a particular plane or cone.

antenna effect [ELECTROMAG] A distortion of the directional properties of a loop antenna caused by an input to the direction-finding receiver which is generated between the loop and ground, in contrast to that which is generated between the two terminals of the loop. Also known as electrostatic error; vertical component effect.

antenna effective area [ELECTROMAG] In any specified direction, the square of the wavelength multiplied by the power gain (or directive gain) in that direction, and divided by 4π.

antenna field [ELECTROMAG] A group of antennas placed in a geometric configuration.

antennafier [COMMUN] Integrated low-profile antenna and amplifier for use with compact, portable communications systems.

antenna gain [ELECTROMAG] A measure of the effectiveness of a directional antenna as compared to a standard nondirectional antenna. Also known as gain.

antenna ground system [ELECTROMAG] That portion of an antenna closely associated with the earth and including an extensive conducting surface, which may be the earth itself.

antennal gland [INV ZOO] An excretory organ in the cephalon of adult crustaceans and best developed in the Malacostraca. Also known as green gland.

antenna loading [ELECTR] **1.** The amount of inductance or capacitance in series with an antenna, which determines the antenna's electrical length. **2.** The practice of loading an antenna in order to increase its electrical length.

antenna matching [ELECTROMAG] Process of adjusting impedances so that the impedance of an antenna equals the characteristic impedance of its transmission line.

antennamitter [COMMUN] Antenna combined with an oscillator to function as a low-power transmitter.

antenna pair [ELECTROMAG] Two antennas located on a base line of accurately surveyed length. [NAV] Two antennas located on a common baseline to produce a directional pattern, often arranged so that the array may be rotated around an axis which is at their center.

antenna pattern See radiation pattern.

antenna power [ELECTROMAG] Radio-frequency power delivered to an antenna.

antenna power gain [ELECTROMAG] The power gain of an antenna in a given direction is 4π times the ratio of the radiation intensity in that direction to the total power delivered to the antenna.

antenna resistance [ELECTROMAG] The power supplied to an entire antenna divided by the square of the effective antenna current measured at the point where power is supplied to the antenna.

antenna scanner [ELECTROMAG] A microwave feed horn which moves in such a way as to illuminate sequentially different reflecting elements of an antenna array and thus produce the desired field pattern.

Antennata [INV ZOO] An equivalent name for the Mandibulata.

antenna tilt error [ENG] Angular difference between the tilt angle of a radar antenna shown on a mechanical indicator, and the electrical center of the radar beam.

antennaverter [COMMUN] Receiving antenna combined with a converter as a single unit, feeding directly into the intermediate-frequency amplifier of the receiver.

antenodal [INV ZOO] Before or in front of the nodus, a cross vein near the middle of the costal border of the wing of dragonflies.

antepartum [MED] Pertaining to the period before delivery or birth.

anter [INV ZOO] Part of a bryozoan operculum which serves to close off a portion of the operculum.

anteriad [ZOO] Toward the anterior portion of the body.

anterior [ZOO] Situated near or toward the front or head of an animal body.

anterior arm reach [ANTHRO] The distance measured from the wall to the tip of the right middle finger when the subject makes the maximum forward reach with both arms, standing with heels, buttocks, middle of back (in lateral sense), and occiput against the wall.

anterograde amnesia [MED] Loss of memory for the period subsequent to a sudden trauma or a seizure.

anteromedial [ZOO] Anterior and toward the middle of the body.

antetheca [PALEON] The last or exposed septum at any stage of fusulinid growth.

anthelate [BOT] An open, paniculate cyme.

anthelic arc [ASTRON] A rare type of halo phenomenon appearing in an area 180° from the sun's azimuth and at the sun's elevation.

anthelion [ASTRON] A luminous white spot which occasionally appears on the parhelic circle 180° in azimuth away from the sun. Also known as counter sun.

anthelminthic [PHARM] A chemical substance used to destroy tapeworms in domestic animals.

anther [BOT] The pollen-producing structure of a flower.

antheraxanthin [BIOCHEM] A neutral yellow plant pigment unique to the Euglenophyta.

antheridiophore [BOT] A specialized stemlike structure that supports an antheridium in some mosses and liverworts.

antheridium [BOT] **1.** The sex organ that produces male gametes in cryptogams. **2.** A minute structure within the pollen grain of seed plants.

antheriferous [BOT] Anther-bearing.

antherozoid [BOT] A motile male gamete produced by plants.

anther smut [MYCOL] *Ustilago violacea.* A smut fungus that attacks certain plants and forms spores in the anthers.

anthesis [BOT] The flowering period in plants.

ANTELOPE

The black buck or Indian antelope (*Antilope cervicapra*). The male is black with long, spiral horns, while the female is beige and hornless.

ANTHOPHYLLITE

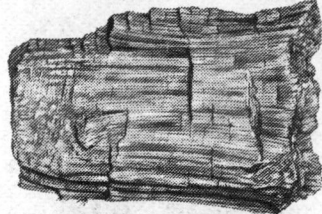

Anthophyllite from East Goshen, Pa. (*Specimen from Department of Geology, Bryn Mawr College*)

ANTHRAQUINONE

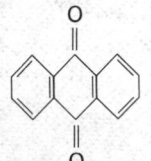

Structural formula of anthraquinone.

ANTHURIDEA

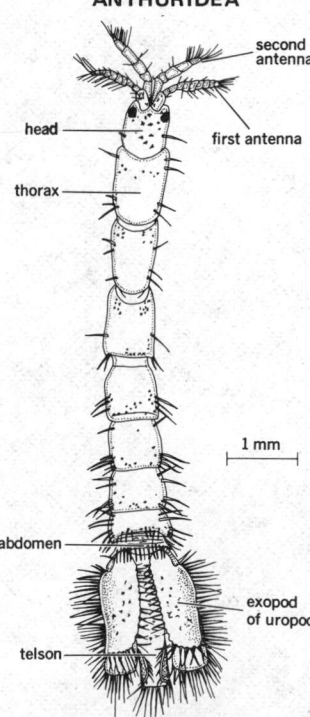

A drawing of *Paranthura infundibulata* Richardson, showing some of the characteristics of the Anthuridea.

Anthicidae [INV ZOO] The antlike flower beetles, a family of coleopteran insects in the superfamily Tenebrionoidea.

anthill [INV ZOO] A mound of earth deposited around the entrance to ant and termite nests in the ground. [MIN ENG] The cuttings around the hole collar in blast-hole drilling.

anthoblast [INV ZOO] A developmental stage of some corals; produced by budding.

anthocarpous [BOT] Describing fruits having accessory parts.

anthocaulis [INV ZOO] The stemlike basal portion of some solitary corals; the oral portion becomes a new zooid.

Anthocerotae [BOT] A small class of the plant division Bryophyta, commonly known as hornworts or horned liverworts.

anthocodium [INV ZOO] The free, oral end of an anthozoan polyp.

Anthocoridae [INV ZOO] The flower bugs, a family of hemipteran insects in the superfamily Cimicimorpha.

anthocyanidin [BIOCHEM] Any of the colored aglycone plant pigments obtained by hydrolysis of anthocyanins.

anthocyanin [BIOCHEM] Any of the intensely colored, sap-soluble glycoside plant pigments responsible for most scarlet, purple, mauve, and blue coloring in higher plants.

Anthocyathea [PALEON] A class of extinct marine organisms in the phylum Archaeocyatha characterized by skeletal tissue in the central cavity.

anthocyathus [INV ZOO] The oral portion of the anthocaulus of some solitary corals that becomes a new zooid.

anthodite [GEOL] Gypsum or aragonite growing in clumps of long needle- or hairlike crystals on the roof or wall of a cave.

anthoinite [MINERAL] $Al_2W_2O_9 \cdot 3H_2O$ A white mineral consisting of a hydrous basic aluminum tungstate.

Anthomedusae [INV ZOO] A suborder of hydrozoan coelenterates in the order Hydroida characterized by athecate polyps.

Anthomyzidae [INV ZOO] A family of cyclorrhaphous myodarian dipteran insects belonging to the subsection Acalypteratae.

anthophagous [ZOO] Feeding on flowers.

anthophobia [PSYCH] Abnormal fear of flowers.

anthophore [BOT] A stalklike extension of the receptacle bearing the pistil and corolla in certain plants.

anthophyllite [MINERAL] A clove-brown orthorhombic mineral of the amphibole group. A variety of asbestos occurring as lamellae, radiations, fibers, or massive in metamorphic rocks. Also known as bidalotite.

Anthosomidae [INV ZOO] A family of fish ectoparasites in the crustacean suborder Caligoida.

anthostele [INV ZOO] A thick-walled, nonretractile aboral region of certain coelenterates.

Anthozoa [INV ZOO] A class of marine organisms in the phylum Coelenterata including the soft, horny, stony, and black corals, the sea pens, and the sea anemones.

anthozooid [INV ZOO] Any of the individual zooids of a compound anthozoan.

anthracene [ORG CHEM] $C_{14}H_{10}$ A colorless, crystalline tricyclic hydrocarbon melting at 216.2°C and boiling at 340°C; obtained in the distillation of coal tar and used in coatings and as a luminescent material.

anthracene oil [MATER] A heavy, green oil that is a coal-tar fraction boiling above 270°C; used as a source of anthracene, phenanthrene, and carbazole.

anthracene violet *See* gallein.

anthraciferous coal [ORG CHEM] Anthracite-hard coal containing or yielding anthracene.

anthracite [MINERAL] A high-grade metamorphic coal having a semimetallic luster, high content of fixed carbon, and high density, and burning with a short blue flame and little smoke or odor. Also known as hard coal; Kilkenny coal; stone coal.

anthracite duff [MATER] In Wales, fine anthracite screenings used in briquets or mixed with bituminous coal to fuel cement kilns on chain grate stokers.

anthracite fines [MIN ENG] Small pieces of material from an anthracite coal preparation plant, usually below one-eighth-inch diameter.

anthracnose [PL PATH] A fungus disease of plants caused by

members of the Melanconiales and characterized by dark or black limited stem lesions.

Anthracosauria [PALEON] An order of Carboniferous and Permian labyrinthodont amphibians that includes the ancestors of living reptiles.

anthracosilicosis [MED] Chronic lung inflammation caused by inhalation of carbon and silicon particles.

anthracosis [MED] The accumulation of inhaled black coal dust particles in the lung accompanied by chronic inflammation. Also known as blacklung.

Anthracotheriidae [PALEON] A family of middle Eocene and early Pleistocene artiodactyl mammals in the superfamily Anthracotherioidea.

Anthracotherioidea [PALEON] A superfamily of extinct artiodactyl mammals in the suborder Paleodonta.

anthracoxene [GEOL] A brownish resin that occurs in brown coal; in ether it dissolves into an insoluble portion, anthracoxenite, and a soluble portion, schlanite.

anthranilic acid [ORG CHEM] $o\text{-}NH_2C_6H_4COOH$ A white, or pale yellow crystalline acid melting at 146°C; used as an intermediate in the manufacture of dyes, pharmaceuticals, and perfumes.

anthraquinone [ORG CHEM] $C_6H_4(CO)_2C_6H_4$ Yellow crystalline diketone that is insoluble in water; used in the manufacture of dyes. Also known as dihydrodiketoanthracene.

anthraquinone pigments [BIOCHEM] Coloring materials which occur in plants, fungi, lichens, and insects; consists of about 50 derivatives of the parent compound, anthraquinone.

anthrax [VET MED] An acute, infectious bacterial disease of sheep and cattle caused by *Bacillus anthracis*; transmissible to man. Also known as splenic fever; wool-sorter's disease.

anthraxolite [GEOL] Anthracite-like asphaltic material occurring in veins in Precambrian slate of Sudbury District, Ontario.

anthraxylon [GEOL] The vitreous-appearing components of coal derived from the woody tissues of plants.

Anthribidae [INV ZOO] The fungus weevils, a family of coleopteran insects in the superfamily Curculionoidea.

anthropic [ANTHRO] Pertaining to man or the period of his existence on earth.

anthropocentric [PSYCH] **1.** Regarding man as the most important factor in the universe. **2.** Evaluating all occurrences solely by human values.

anthropochory [ECOL] Dispersal of plant and animal disseminules by man.

anthropography [ANTHRO] A branch of anthropology that deals with the geographic distribution of divisions of humans based on physical character, language, customs, and institutions.

anthropoid [VERT ZOO] Pertaining to or resembling the Anthropoidea.

Anthropoidea [VERT ZOO] A suborder of mammals in the order Primates including New and Old World monkeys.

anthropology [BIOL] The study of the interrelations of biological, cultural, geographical, and historical aspects of man.

anthropometry [ANTHRO] Description of the physical variation in man by measurement; a basic technique of physical anthropology.

anthropophobia [PSYCH] Abnormal fear of people.

anthroposcopy [ANTHRO] The description of physical variation in man by visual inspection; a basic technique of physical anthropology.

anthroposphere [ECOL] A term sometimes used to refer to the biosphere of the great geological activities of man. Also known as noosphere.

Anthuridea [INV ZOO] A suborder of crustaceans in the order Isopoda characterized by slender, elongate, subcylindrical bodies, and by the fact that the outer branch of the paired tail appendage (uropod) arches over the base of the terminal abdominal segment, the telson.

antiagglutinin [IMMUNOL] A substance that neutralizes a corresponding agglutinin.

antiaggressin [IMMUNOL] An antibody that neutralizes aggressin, a substance produced by pathogenic microorganisms to enhance virulence.

antiaircraft [ORD] Used, or designed to be used, against airborne aircraft or missiles. Abbreviated AA.

antiaircraft artillery [ORD] Projectile weapons with related equipment, such as searchlights or radar, employed on the ground or on ships to strike at airborne aircraft or missiles. Abbreviated AAA.

antiaircraft barrage [ORD] **1.** A concentration of antiaircraft fire covering an area through which enemy aircraft are likely to fly. Also known as predicted barrage. **2.** Less precisely, any concentration of antiaircraft fire.

antiaircraft gun [ORD] A gun designed for use against aircraft, easily shifted in direction and elevation, having great range, and capable of firing at high angles of elevation.

antiaircraft missile [ORD] A guided missile intended to be launched from the surface against an airborne target. Abbreviated AAM.

antianaphylaxis [IMMUNOL] A condition in which a sensitized animal resists anaphylaxis.

Antiarchi [PALEON] A division of highly specialized placoderms restricted to fresh-water sediments of the Middle and Upper Devonian.

antiatom [ATOM PHYS] An atom made up of antiprotons, antineutrons, and positrons in the same way that an ordinary atom is made up of protons, neutrons, and electrons.

antibacterial agent [MICROBIO] A synthetic or natural compound which inhibits the growth and division of bacteria.

antiballistic missile [ORD] Any object thrown, dropped, fired, or otherwise projected with the purpose of intercepting a ballistic missile. Abbreviated ABM.

antibaryon [ATOM PHYS] One of a class of antiparticles, including the antinucleons and the antihyperons, with strong interactions, baryon number -1, and hypercharge and charge opposite to those for the particles.

antibiosis [ECOL] Antagonistic association between two organisms in which one is adversely affected.

antibiotic [MICROBIO] A chemical substance, produced by microorganisms and synthetically, that has the capacity in dilute solutions to inhibit the growth of, and even to destroy, bacteria and other microorganisms.

antibiotic assay [MICROBIO] A method for quantitatively determining the concentration of an antibiotic by its effect in inhibiting the growth of a susceptible microorganism.

antibody [IMMUNOL] A protein, found principally in blood serum, originating either normally or in response to an antigen and characterized by a specific reactivity with its complementary antigen. Also known as immune body.

antibody-deficiency syndrome [MED] Any of the human defects of antibody production, such as hypogammaglobulinemia, agammaglobulinemia, and dysgammaglobulinemia, usually associated with reduced serum concentrations of immunoglobulins.

antibonding orbital [PHYS] An atomic or molecular orbital whose energy increases as atoms are brought closer together, indicating a net repulsion rather than a net attraction and chemical bonding.

antiboreal faunal region [ECOL] A zoogeographic region including marine littoral faunal communities at the southern end of South America.

anticapacitance switch [ELECTR] A switch designed to have low capacitance between its terminals when open.

anticarcinogen [PHARM] Any substance which is antagonistic to the action of a carcinogen.

anticatalyst [CHEM] A material that slows down the action of a catalyst; an example is lead, which inhibits the action of platinum.

anticathode [ELECTR] The anode or target of an x-ray tube, on which the stream of electrons from the cathode is focused and from which x-rays are emitted.

anticenter [GEOL] The point on the surface of the earth that is diametrically opposite the epicenter of an earthquake. Also known as antiepicenter.

antichlor [CHEM ENG] A chemical used in the manufacture of paper or textiles to remove excess chlorine or bleaching solution.

anticholinesterase [BIOCHEM] Any agent, such as a nerve gas, that inhibits the action of cholinesterase and thereby destroys or interferes with nerve conduction.

anticlastic [MATH] Having the property of a surface or portion of a surface whose two principal curvatures at each point have opposite signs, so that one normal section is concave and the other convex.

anticlinal [BOT] Perpendicular to the surface or periphery of an organ. [GEOL] Folded as in an anticline.

anticlinal axis [GEOL] The median line of a folded structure from which the strata dip on either side.

anticlinal bend [GEOL] An upwardly convex flexure of rock strata in which one limb dips gently toward the apex of the strata and the other dips steeply away from it.

anticlinal mountain [GEOL] Ridges formed by a convex flexure of the strata.

anticlinal theory [GEOL] A theory relating trapped underground oil accumulation to anticlinal structures.

anticlinal valley [GEOL] A valley that follows an anticlinal axis.

anticline [GEOL] A fold in which layered strata are inclined down and away from the axes.

anticlinorium [GEOL] A series of anticlines and synclines that form a general arch or anticline.

anticlutter gain control [ELECTR] Device which automatically and smoothly increases the gain of a radar receiver from a low level to the maximum, within a specified period after each transmitter pulse, so that short-range echoes producing clutter are amplified less than long-range echoes.

anticoagulant [PHARM] An agent, such as sodium citrate, that prevents coagulation of a colloid, especially blood.

anticodon [GEN] A three-nucleotide sequence of transfer RNA that complements the codon in messenger RNA.

anticoincidence [NUC PHYS] The occurrence of an event at one place without a simultaneous event at another place.

anticoincidence circuit [ELECTR] Circuit that produces a specified output pulse when one (frequently predesignated) of two inputs receives a pulse and the other receives no pulse within an assigned time interval.

anticollision radar [ENG] A radar set designed to give warning of possible collisions during movements of ships or aircraft.

anticommutator [MATH] The anticommutator of two operators, A and B, is the operator $AB + BA$.

anticommute [MATH] Two operators anticommute if their anticommutator is equal to zero.

anticonvulsant [PHARM] An agent, such as Dilantin, that prevents or arrests a convulsion.

anticorona [OPTICS] A diffraction phenomenon appearing at a point before an observer with the sun or moon directly behind him; consists of rings of colored lights complementary to the coronal rings. Also known as Brocken bow.

anticreeper [CIV ENG] A device attached to a railroad rail to prevent it from moving in the direction of its length.

anticrepuscular rays [ASTRON] Extensions of crepuscular rays, converging toward a point 180° from the sun.

anticryptic [ECOL] Pertaining to protective coloration that makes an animal resemble its surroundings so that it is inconspicuous to its prey.

anticurl coating [GRAPHICS] A thin layer generally applied to or on the back of photographic material to prevent front warping.

anticusp [INV ZOO] An anterior, downward projection in conodonts.

anticyclogenesis [METEOROL] The process which creates an anticyclone or intensifies an existing one.

anticyclolysis [METEOROL] Any weakening of anticyclonic circulation in the atmosphere.

anticyclone [METEOROL] High-pressure atmospheric closed circulation whose relative direction of rotation is clockwise in the Northern Hemisphere, counterclockwise in the Southern Hemisphere, and undefined at the Equator. Also known as high; high-pressure area.

anticyclonic [METEOROL] Referring to a rotation about the local vertical that is clockwise in the Northern Hemisphere, counterclockwise in the Southern Hemisphere, undefined at the Equator.

anticyclonic shear [METEOROL] Horizontal wind shear of

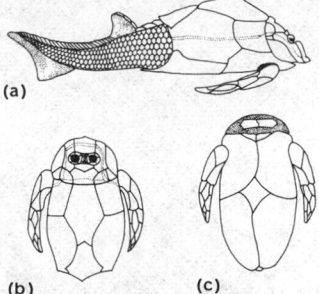

ANTIARCHI

Pterichthyodes (Pterichthys), a Middle Devonian antiarch. *(a)* Lateral view showing scale-covered tail; about 15 centimeters in length. *(b)* Dorsal view and *(c)* ventral view of the armor. *(After Traquair)*

ANTICLINE

Diagram relating anticlinal structure to topography.

such a nature that it tends to produce anticyclonic rotation of the individual air particles along the line of flow.

anticyclonic winds [METEOROL] The winds associated with a high pressure area and constituting part of an anticyclone.

antidepressant [PHARM] A drug, such as imipramine and tranylcypromine, that relieves depression by increasing central sympathetic activity.

antiderivative *See* indefinite integral.

antidetonant *See* antiknock.

antideuteron [ATOM PHYS] The antiparticle to the deuteron, composed of an antineutron and an antiproton.

antidiabetic [PHARM] An agent, such as insulin, that is effective in controlling diabetes.

antidiarrheal [PHARM] An agent, such as Kaopectate, that prevents or arrests diarrhea.

antidiuretic [PHARM] An agent, such as vasopressin, that prevents the excretion of urine.

antidiuretic hormone *See* vasopressin.

antidote [PHARM] An agent that relieves or counteracts the action of a poison.

antidrag [AERO ENG] **1.** Of structural members in an aircraft or missile that are designed or built to resist the effects of drag. **2.** Of a force acting against the force of drag.

antidune [GEOL] A temporary form of ripple on a stream bed analogous to a sand dune but migrating upcurrent.

antienzyme [BIOCHEM] An agent that selectively inhibits the action of an enzyme.

antiepicenter *See* anticenter.

antierythrite *See* erythritol.

antifading antenna [ELECTR] An antenna designed to confine radiation mainly to small angles of elevation to minimize the fading of radiation directed at larger angles of elevation.

antiferroelectric crystal [SOLID STATE] A crystalline substance characterized by a state of lower symmetry consisting of two interpenetrating sublattices with equal but opposite electric polarization, and a state of higher symmetry in which the sublattices are unpolarized and indistinguishable.

antiferromagnetic domain [SOLID STATE] A region in a solid within which equal groups of elementary atomic or molecular magnetic moments are aligned antiparallel.

antiferromagnetic resonance [ELECTROMAG] Magnetic resonance in antiferromagnetic materials which may be observed by rotating magnetic fields in either of two opposite directions.

antiferromagnetic substance [ELECTROMAG] A substance composed of antiferromagnetic domains.

antiferromagnetic susceptibility [ELECTROMAG] The magnetic response to an applied magnetic field of a substance whose atomic magnetic moments are aligned in antiparallel fashion.

antiferromagnetism [SOLID STATE] A property possessed by some metals, alloys, and salts of transition elements by which the atomic magnetic moments form an ordered array which alternates or spirals so as to give no net total moment in zero applied magnetic field.

antifertilizin [BIOCHEM] An immunologically specific substance produced by animal sperm to implement attraction by the egg before fertilization.

antifibrinolysin [BIOCHEM] Any substance that inhibits the proteolytic action of fibrinolysin.

antifix [ARCH] In classical architecture, either an ornament at the eaves to hide the ends of the joint tiles of the roof or an ornament of the cymatium of a classic cornice, sometimes pierced so that water can flow through.

antifoaming agent [ORG CHEM] A substance, such as silicones, organic phosphates, and alcohols, that inhibits the formation of bubbles in a liquid during its agitation by reducing its surface tension.

antifogging compound [MATER] A compound of one or more basic chemicals with filler or extenders for preventing condensation of moisture on glass and other transparent material, such as lenses or windshields.

antiform [GEOL] An anticlinelike structure whose stratigraphic sequence is not known.

antifouling coating [MATER] A special paint containing copper used on ships' bottoms to prevent marine organisms from attaching themselves.

antifreeze [CHEM] A substance added to a liquid to lower its freezing point; the principal automotive antifreeze component is ethylene glycol.

antifriction [MECH] Making friction smaller in magnitude. [MECH ENG] Employing a rolling contact instead of a sliding contact.

antifriction alloy [MET] An alloy generally having more than 50% tin as a base and cast as facings of machinery bearings, domestic equipment, and small parts in rolling contact.

antifriction device [ORD] A mechanism which reduces friction in pointing a gun; operates by lifting the gun off the main bearings and supporting it, while it is being moved, on light, free-turning bearings.

antifriction material [ENG] A machine element made of Babbitt metal, lignum vitae, rubber, or a combination of a soft, easily deformable metal overlaid on a hard, resistant one.

antifungal agent [MATER] A chemical compound that either destroys or inhibits the growth of fungi.

antigen [IMMUNOL] A substance which reacts with the products of specific humoral or cellular immunity, even those induced by related heterologous immunogens.

antigen-antibody reaction [IMMUNOL] The combination of an antigen with its antibody.

antigenic determinant [IMMUNOL] The portion of an antigen molecule that determines the specificity of the antigen-antibody reaction.

antigorite [MINERAL] $Mg_3Si_2O_5(OH)_4$ Brownish-green variety of the mineral serpentine. Also known as baltimorite; picrolite.

antigravity [PHYS] The repulsion of one body by another by means of a gravitational type of force; this has never been observed.

anti-g suit *See* g suit.

antihalation backing [GRAPHICS] A coating on the back of film to minimize reflection of light from the base into the emulsion.

antihemophilic factor [BIOCHEM] A soluble protein clotting factor in mammalian blood. Also known as factor VIII; thromboplastinogen.

antihemorrhagic vitamin *See* vitamin K.

antihistamine [PHARM] A drug that prevents or diminishes the effect of histamine; used in treating allergic reactions and common-cold symptoms.

antihunt circuit [ELECTR] A stabilizing circuit used in a closed-loop feedback system to prevent self-oscillations.

antihyperon [PARTIC PHYS] An antiparticle to a hyperon, having the same mass, lifetime, and spin as the hyperon, but with charge and magnetic moment reversed in sign.

antihypertensive agent [PHARM] A substance, such as reserpine, that reduces hypertension.

anti-icing [AERO ENG] The prevention of the formation of ice upon any object, especially aircraft, by means of windshield sprayer, addition of antifreeze materials to the carburetor, or heating the wings and tail.

anti-infective vitamin *See* vitamin A.

anti-inflammatory agent [PHARM] A substance, such as cortisone, that counteracts inflammation.

antijamming [ELECTR] Any system or technique used to counteract the jamming of communications or of radar operation.

antiknock [MATER] **1.** Resisting detonation or pinging in spark-ignited engines. **2.** A substance, such as tetraethyllead, added to motor and aviation gasolines to increase the resistance of the fuel to knock in spark-ignited engines. Also known as antidetonant.

antiknock blending value [ENG] The numerical improvement by an antiknock additive to gasoline octane, often a greater amount than the additive's own octane value.

antiknock gasoline [MATER] Gasoline containing, for example, tetraethyllead, which in small amounts prevents or lessens knocking in a spark-ignited engine.

antiknock rating [ENG] Measurement of the ability of an automotive gasoline to resist detonation or pinging in spark-ignited engines.

Antilles Current [OCEANOGR] A current formed by part of the North Equatorial Current that flows along the northern side of the Greater Antilles.

Antilocapridae [VERT ZOO] A family of artiodactyl mammals in the superfamily Bovoidea; the pronghorn is the single living species.

antilog *See* antilogarithm of a number.

antilogarithm of a number [MATH] A second number whose logarithm is the first number. Abbreviated antilog. Also known as inverse logarithm of a number.

antilogous pole [SOLID STATE] That crystal pole which becomes electrically negative when the crystal is heated or is expanded by decompression.

Antilopinae [VERT ZOO] The antelopes, a subfamily of artiodactyl mammals in the family Bovidae.

antilymphocyte serum [IMMUNOL] An immunosuppressive agent effective in prolonging the lives of homografts in experimental animals by reducing the circulating lymphocytes.

antilysin [IMMUNOL] A substance antagonistic to the action of a lysin.

antimagnetic [ENG] Constructed so as to avoid the influence of magnetic fields, usually by the use of nonmagnetic materials and by magnetic shielding.

antimalarial [PHARM] **1.** A drug, such as quinacrine, that prevents or suppresses malaria. **2.** Acting against malaria.

antimatter [PHYS] Material consisting of atoms which are composed of positrons, antiprotons, and antineutrons.

antimere [INV ZOO] Any one of the equivalent parts into which a radially symmetrical animal may be divided.

antimetabolite [PHARM] A substance, such as sulfanilamide or amethopterin, that inhibits utilization of an essential metabolite because it is an analog of the metabolite.

antimicrobial agent [MICROBIO] A chemical compound that either destroys or inhibits the growth of microscopic and submicroscopic organisms.

antimissile missile [ORD] An explosive missile designed to intercept and destroy another missile in flight. Also known as auntie.

antimitotic drug [PHARM] A substance, such as colchicine, vincristine, or vinblastine, that interferes with mitotic cellular division; used in the chemotherapy of leukemia.

antimolecule [ATOM PHYS] A molecule made up of antiprotons, antineutrons, and positrons in the same way that an ordinary molecule is made up of protons, neutrons, and electrons.

antimonate [CHEM] The negative radical $[Sh(OH)_6]^-$ in salts derived from antimony pentoxide, Sb_4O_{10}, and bases.

antimonial lead [MET] A lead alloy containing up to 25% antimony and possessing greater hardness and tensile strength than lead; used for storage-battery plates, pipes, cable coverings, and roofing.

antimonic [CHEM] Derived from or pertaining to pentavalent antimony.

antimonide [INORG CHEM] A binary compound of antimony with a more positive compound, for example, H_5Sb. Also known as stibide.

antimonite [MINERAL] Sb_2S_3 A lead-gray antimony sulfide mineral, the primary source of antimony; sometimes contains gold or silver; has a brilliant metallic luster, and occurs as prismatic orthorhombic crystals in massive forms. Also known as antimony glance; gray antimony; stibium; stibnite.

antimonous [CHEM] Pertaining to antimony, especially trivalent antimony.

antimonous chloride *See* antimony chloride.

antimonous sulfide *See* antimony trisulfide.

antimony [CHEM] A chemical element, symbol Sb, atomic number 51, atomic weight 121.75.

antimony-124 [NUC PHYS] Radioactive antimony with mass number of 124; 60-day half-life; used as tracer in solid-state and pipeline flow studies.

antimony black *See* antimony trisulfide.

antimony blende *See* kermesite.

antimony chloride [INORG CHEM] Either antimony trichloride or antimony pentachloride; the trichloride, $SbCl_3$, is a hygroscopic colorless crystalline mass, fumes slightly in air, is soluble in alcohol acetone, forms antimony oxychloride in water, and is used as a mordant, in antimony salts, as a chlorinating agent, and in the fireproofing of textiles; also known as antimonous chloride, caustic antimony; the penta-

chloride, $SbCl_5$ is an oily liquid with reddish-yellow color, is hygroscopic and solidifies after moisture is absorbed, decomposes in excess water, is soluble in hydrochloric acid and chloroform, and is used in analytical testing for cesium and alkaloids, for dyeing, and as an intermediary in synthesis; also known as antimony perchloride.

antimony(III) oxide [INORG CHEM] Sb_2O_3 Colorless, rhombic crystals, melting at 656°C; insoluble in water; powerful reducing agent.

antimonyl [CHEM] The inorganic radical SbO^-.

antimony needles *See* antimony trisulfide.

antimony orange *See* antimony trisulfide.

antimony pentasulfide [INORG CHEM] Sb_2S_5 An orange-yellow powder; soluble in alkali, soluble in concentrated hydrochloric acid, with hydrogen sulfide as a by-product, and insoluble in water; used as a red pigment. Also known as antimony persulfide; antimony red; antimony sulfide; golden antimony sulfide.

antimony perchloride *See* antimony chloride.

antimony persulfide *See* antimony pentasulfide.

antimony red *See* antimony pentasulfide.

antimony sulfate [INORG CHEM] $Sb_2(SO_4)_3$ Antimony(III) sulfate, a white, deliquescent powder; soluble in acids.

antimony sulfide *See* antimony pentasulfide; antimony trisulfide.

antimony trisulfide [INORG CHEM] Sb_2S_3 Black and orange-red rhombic crystals; soluble in concentrated hydrochloric acid and sulfide solutions, insoluble in water; melting point 546°C; used as a pigment, and in matches and pyrotechnics. Also known as antimonous sulfide; antimony black; antimony needles; antimony orange; antimony sulfide; black antimony.

antimony yellow *See* lead antimonite.

antimycin A [MICROBIO] $C_{28}H_{40}O_9N_2$ An antifungal antibiotic produced by *Streptomyces kitazawaensis;* the principal fraction of antimycin.

antineoplastic drug [PHARM] An agent, such as mercaptopurine compounds, that is antagonistic to the growth of a neoplasm.

antineutrino [PARTIC PHYS] The antiparticle to the neutrino; it has zero mass, spin ½, and positive helicity; there are two antineutrinos, one associated with electrons and one with muons.

antineutron [PARTIC PHYS] The antiparticle to the neutron; a strongly interacting baryon which has no charge, mass of 939.6 MeV, spin ½, and mean life of about 10^3 seconds.

antinode [ASTRON] Either of the two points on an orbit where a line in the orbit plane, perpendicular to the line of nodes and passing through the focus, intersects the orbit. [PHYS] A point, line, or surface in a standing-wave system at which some characteristic of the wave has maximum amplitude. Also known as loop.

antinoise microphone [ENG ACOUS] Microphone with characteristics which discriminate against acoustic noise.

antinucleon [PARTIC PHYS] An antineutron or antiproton, that is, particles having the same mass as their nucleon counterparts but opposite charge or opposite magnetic moment.

antinucleus [NUC PHYS] A nucleus made up of antineutrons and antiprotons in the same way that an ordinary nucleus is made up of neutrons and protons.

Antioch process [MET] A method of plaster molding in which a plaster-water mixture is poured over a pattern, after which the mold is steam-treated, allowed to set in air, dried in an oven, and cooled for use in casting certain alloys.

antioxidant [CHEM] An inhibitor, such as ascorbic acid, effective in preventing oxidation by molecular oxygen.

antiparasitic agent [PHARM] An agent, such as emetine or quinine, that destroys or suppresses human and animal parasites.

antiparticle [PARTIC PHYS] A counterpart to a particle having mass lifetime and spin identical to the particle but with charge and magnetic moment reversed in sign.

Antipatharia [INV ZOO] The black or horny corals, an order of tropical and subtropical coelenterates in the subclass Zoantharia.

antipersonnel [ORD] Applied to bombs, mines, and missiles

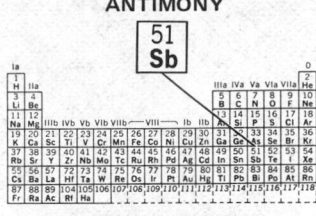

ANTIMONY

Periodic table of the chemical elements showing the position of antimony.

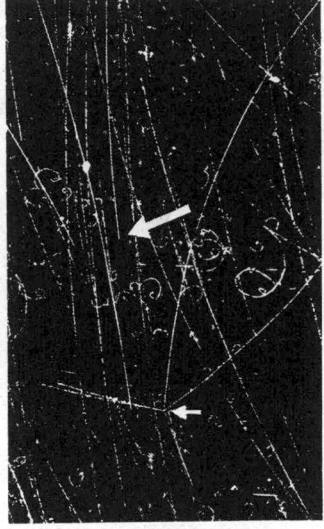

ANTINEUTRON

Propane bubble-chamber photograph taken in proton and meson beam of the bevatron. Large arrow indicates where antineutron was formed. Five-pronged star indicated by small arrow is annihilation star of antineutron. Actual distance between formation and annihilation is about 12 centimeters. Density of propane is 0.42 g/cm³. *(University of California Lawrence Berkeley Laboratory)*

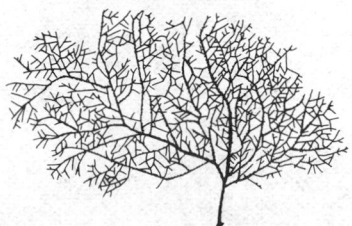

ANTIPATHARIA

Irregularly branching plantlike colony, *Antipathes rhipidion.* *(After F. Pax)*

designed to kill, wound, or obstruct people. Abbreviated apers.

antipersonnel weapon [ORD] All instruments of combat, either offensive or defensive, used to destroy, wound, obstruct, or threaten enemy personnel.

antiperthite [GEOL] Natural intergrowth of feldspars formed by separation of sodium feldspar (albite) and potassium feldspar (orthoclase) during slow cooling of molten mixtures; the potassium-rich phase is exolved in a plagioclase host, exactly the inverse of perthite.

antipetalous [BOT] Having stamens positioned opposite to, rather than alternating with, the petals.

antipitching fin [NAV ARCH] A control surface protruding underwater at the bow or stern to generate lift opposed to the pitching of the ship.

antipodal [BOT] Any of three cells grouped at the base of the embryo sac, that is, at the end farthest from the micropyle, in most angiosperms.

antipodes [GEOD] Diametrically opposite points on the earth.

antiproton [PARTIC PHYS] The antiparticle to the proton; a strongly interacting baryon which is stable, carries unit negative charge, has the same mass as the proton (983.3 MeV), and has spin ½.

antiprotonic atom [ATOM PHYS] An atom consisting of an ordinary nucleus with an orbiting antiproton.

antipruritic [PHARM] An agent, such as camphor, that relieves itching.

antipyretic [PHARM] Any agent, such as aspirin, that reduces or prevents fever.

antipyrine [PHARM] $C_{11}H_{12}ON_2$ A compound used as an antipyretic, analgesic, and antirheumatic.

antiquark [PARTIC PHYS] The hypothetical antiparticle of a quark, having electric charge, baryon number, and strangeness opposite in sign to that of the corresponding quark.

antique finish [MATER] A paper finish, somewhat rougher than eggshell, obtained by operating wet presses and calender stacks at reduced pressures.

antiquing [ENG] Producing a rich glow on the surface of a leather by applying stain, wax, or oil, allowing it to set, and rubbing or brushing the leather.

antirachitic vitamin *See* vitamin D.

antiradar coating [ENG] A surface treatment used to reduce the reflection of electromagnetic waves so as to avoid radar detection.

antireflection coating [ENG] The application of a thin film of dielectric material to a surface to reduce its reflection and to increase its transmission of light or other electromagnetic radiation.

antiresonance [ELEC] *See* parallel resonance. [ENG] The condition for which the impedance of a given electric, acoustic, or dynamic system is very high, approaching infinity.

antiresonant circuit *See* parallel resonant circuit.

anti-Rh agglutinin [IMMUNOL] An antibody against any Rh antigen; it must be acquired and is never natural.

anti-Rh immunoglobulin [IMMUNOL] A serum protein that destroys Rh-positive fetal erythrocytes in an Rh-negative mother when administered after delivery.

anti-Rh serum [IMMUNOL] A blood serum containing anti-Rh antibodies.

antiricochet device [ORD] Usually a parachute unit, fuse adapter, and fuse attached to the tail of a bomb to prevent ricochet, with consequent loss of effectiveness and possible danger to the dropping plane.

antiroll fin [NAV ARCH] A control surface protruding from either of the lower sides of a ship to generate lift opposed to the rolling of the ship.

antirolling gyroscope [NAV ARCH] A very large gyroscope having a vertical axis and mounted in a ship so that the axis can be tipped fore and aft by an electric motor, resulting in the gyroscope's exerting a large torque on the ship about the fore-and-aft axis.

antiroll tank [NAV ARCH] One of paired tanks at opposite sides within a ship which are partially filled with liquid whose flow from one tank to the other helps to offset ship rolls.

antirostrum [INV ZOO] The terminal segment of the appendages of certain mites.

antisatellite missile [ORD] A missile whose target is an orbiting satellite.

antiseptic [PHARM] A substance used to destroy or prevent the growth of infectious microorganisms on or in the body.

antiserum [IMMUNOL] Any immune serum that contains antibodies active chiefly in destroying a specific infecting virus or bacterium.

antisetoff powder [GRAPHICS] Fine-powdered starch or limestone applied to a freshly printed sheet of paper to reduce unintentional transfer of ink to neighboring sheets in a stack.

antishock agent [PHARM] A substance, such as a cesium salt, that relieves a state of shock.

antisideric [PHARM] A pharmaceutical that counteracts the physiological effects of iron.

anti-sidetone circuit [ELEC] Telephone circuit which prevents sound, introduced in the local transmitter, from being reproduced in the local receiver.

antiskid plate [ENG] A sheet of metal roughed on both sides and placed between piled objects, such as boxes in a freight car, to prevent sliding.

antismallpox vaccine *See* smallpox vaccine.

antisolar point [ASTRON] The point on the celestial sphere which lies directly opposite the sun from the observer, that is, on the line from the sun through the observer.

antispasmodic [PHARM] An agent, such as benzyl benzoate, that relieves convulsions and the pain of muscular spasms.

antistat *See* antistatic agent.

antistatic agent [MATER] A material used with textiles, plastics, paper products, or wax polishes to reduce static-electrical charges by allowing the charge to leak off. Also known as antistat.

antistatic brush [GRAPHICS] A brush whose bristles do not retain electrostatic charge.

anti-Stokes lines [SPECT] Lines of radiated frequencies which are higher than the frequency of the exciting incident light.

antistreptolysin [IMMUNOL] The antibody that neutralizes the streptolysin of group A hemolytic streptococci.

antistress mineral [MINERAL] Minerals such as leucite, nepheline, alkalic feldspar, andalusite, and cordierite which cannot form or are unstable in an environment of high shearing stress, and hence are not found in highly deformed rocks.

antistripping agent [MATER] An additive used in an asphaltic binder to overcome the affinity of an aggregate for water instead of asphalt; assists the asphalt to adhere to wet surfaces.

antisubmarine [ORD] Descriptive of equipment, mines, or missiles designed to attack or destroy submarines.

antisubmarine missile [ORD] A missile whose target is a submarine.

antisubmarine rocket [ORD] Solid-propellant-rocket-driven torpedo that is surface-launched, with final track underwater. Abbreviated ASROC.

antisubmarine weapon [ORD] Any naval device designed to destroy submarines.

antisymmetric dyadic [MATH] A dyadic equal to the negative of its conjugate.

antisymmetric matrix [MATH] A matrix which is equal to the negative of its transpose. Also known as skew-symmetric matrix.

antisymmetric tensor [MATH] A tensor in which interchanging two indices of an element changes the sign of the element.

antisymmetric wave function [PHYS] A many-particle wave function which changes its sign when the coordinates of two of the particles are interchanged.

antitail [ASTRON] A structure occasionally observed in comets that appears to extend from the coma toward the sun, and usually has the appearance of a spike.

antitank ditch [ORD] A deep ditch prepared as an obstacle to enemy tanks. Also known as tank ditch.

antitank grenade [ORD] A rifle grenade designed to be used against tanks or other armored vehicles.

antitank guided missile [ORD] An antitank missile whose flight path is controlled by a combination of optical sighting and command signals from an automatic computer through multiple wire command links; it may also contain an infrared homing device for final range correction; it is designed to be launched from any type of vehicle or ground emplacement; this definition excludes items whose trajectory cannot be altered in flight.

antitank gun [ORD] A gun suitable for use against tanks or other armored vehicles.

antitank mine activator [ORD] A nonmetallic item designed to adapt a firing device to an antitank mine; may be empty, inert-filled, or explosive-filled.

antitank mine field [ORD] **1.** Area in which antitank mines are planted to stop or slow down enemy tanks or other armored vehicles. **2.** Pattern of antitank mines.

antitank missile [ORD] A missile designed to destroy or immobilize enemy tanks.

antitank obstacle [ORD] Natural or manmade obstruction or barrier which will stop, slow down, or cause tanks or other armored vehicles to maneuver or change direction. Also known as tank obstacle.

antitank weapon [ORD] Any weapon suitable for use against tanks or other armored vehicles.

antitermination factor [BIOCHEM] Protein that interferes with normal termination of ribonucleic acid synthesis.

antithetic variable [STAT] One of two random variables having high negative correlation, used in the antithetic variate method of estimating the mean of a series of observations.

antithrombin [BIOCHEM] A substance in blood plasma that inactivates thrombin.

antitorpedo [ORD] Descriptive of equipment, missiles, and the like designed to combat or destroy torpedoes.

antitoxin [IMMUNOL] An antibody elaborated by the body in response to a bacterial toxin that will combine with and generally neutralize the toxin.

antitrades [METEOROL] A deep layer of westerly winds in the troposphere above the surface trade winds of the tropics.

anti-transmit-receive tube [ELECTR] A switching tube that prevents the received echo signal from being dissipated in the transmitter.

antitriptic wind [METEOROL] A wind for which the pressure force exactly balances the viscous force, in which the vertical transfers of momentum predominate.

antitubercular agent [PHARM] A substance, such as streptomycin or isoniazid, used in the treatment of tuberculosis.

antitumor antibiotic [MICROBIO] A substance, such as actinomycin, luteomycin, or mitomycin C, which is produced by microorganisms and is effective against some forms of cancer.

antitussive [PHARM] An agent, such as benylin expectorants, that relieves coughing.

antitwilight arch [METEOROL] The pink or purplish band of about 3° vertical angular width which lies just above the antisolar point at twilight; it rises with the antisolar point at sunset and sets with the antisolar point at sunrise.

antivenin [IMMUNOL] An immune serum that neutralizes the venoms of certain poisonous snakes and black widow spiders.

antivernalization [BOT] Delayed flowering in plants due to treatment with heat.

antiviral agent [PHARM] A substance, such as interferon or amantadine, that decreases virus multiplication in the body.

antivitamin [BIOCHEM] Any substance that prevents a vitamin from normal metabolic functioning.

anti-voice-operated transmission [COMMUN] Radio communications in which a voice-activated circuit prevents the operation of the transmitter while messages are being received on an associated receiver.

antixerophthalmic vitamin *See* vitamin A.

antler [VERT ZOO] One of a pair of solid bony, usually branched outgrowths on the head of members of the deer family (Cervidae); shed annually.

antlerite [MINERAL] $Cu_3SO_4(OH)_4$ Emerald- to blackish-green mineral occurring in aggregates of needlelike crystals; an ore of copper. Also known as vernadskite.

Antler orogeny [GEOL] Late Devonian and Early Mississippian orogeny in Nevada, resulting in the structural emplacement of eugeosynclinal rocks over microgeosynclinal rocks.

Antlia [ASTRON] A constellation with a right ascension of 10 hours and declination of 35°S. Abbreviated Ant. Also known as Air Pump.

ant lion [INV ZOO] The common name for insects of the family Myrmeleontidae in the order Neuroptera; larvae are commonly called doodlebugs.

antlophobia [PSYCH] Abnormal fear of floods.

AN-TNT slurry [MATER] A mixture of ammonium nitrate and trinitrotoluene; used as an explosive.

Antoine equation [PHYS] The empirical relationship between temperature and vapor pressure of liquids; $\log P = B - A/(C + T)$, where A, B, C are experimental constants, T is absolute temperature, and P is vapor pressure.

Antonoff's rule [PHYS] The rule which states that the surface tension at the interface between two saturated liquid layers in equilibrium is equal to the difference between the individual surface tensions of similar layers when exposed to air.

Antron [TEXT] Trade name for nylon having a trilobed cross section; made by DuPont.

antrorse [BIOL] Turned or directed forward or upward.

antrum [ANAT] A cavity of a hollow organ or a sinus.

Antrypol [PHARM] Trade name for suramin sodium, an antitrypanosomal and antifilarial agent.

Anura [VERT ZOO] An order of the class Amphibia comprising the frogs and toads.

anuresis [MED] Retention of urine in the urinary bladder due to inability to void.

anuria [MED] Complete absence of urinary output.

anus [ANAT] The posterior orifice of the alimentary canal.

anvil [ANAT] *See* incus. [ENG] **1.** The part of a machine that absorbs the energy delivered by a sharp force or blow. **2.** The stationary end of a micrometer caliper. [MET] **1.** A heavy wrought-iron, cast-iron, or steel block upon which metal is hammered in smith forging. **2.** The base of the hammer, holding the die bed and lower die part in drop forging.

anvil cloud [METEOROL] The popular name given to a cumulonimbus capillatus cloud, a thunderhead whose upper portion spreads in the form of an anvil with a fibrous or smooth aspect; it also refers to such an upper portion alone when it persists beyond the parent cloud.

anxiety [PSYCH] A physiological and mental state of apprehension and fear of something unknown to the conscious.

anxiety neurosis [PSYCH] A psychoneurotic disorder characterized by diffuse anxious expectation not restricted to definite situations, persons, or objects, and by emotional instability, irritability, apprehensiveness, and a sense of fatigue; caused by incomplete resolution of repressed emotional problems, and frequently associated with somatic symptoms.

Ao horizon [GEOL] That portion of the A horizon of a soil profile which is composed of pure humus.

Aoo horizon [GEOL] Uppermost portion of the A horizon of a soil profile which consists of undecomposed vegetable litter.

AOQL *See* average outgoing quality limit.

aorta [ANAT] The main vessel of systemic arterial circulation arising from the heart in vertebrates. [INV ZOO] The large dorsal or anterior vessel in many invertebrates.

aortic aneurysm [MED] Dilation of the wall of the aorta, usually the ascending portion.

aortic arch [ANAT] The portion of the aorta extending from the heart to the third thoracic vertebra; single in warm-blooded vertebrates and paired in fishes, amphibians, and reptiles.

aortic body *See* aortic paraganglion.

aortic paraganglion [ANAT] A structure in vertebrates belonging to the chromaffin system and found on the front of the abdominal aorta near the mesenteric arteries. Also known as aortic body; organs of Zuckerkandl.

aortic stenosis [MED] Abnormal narrowing of the aortic valve orifice; may be either congenital or acquired.

aortitis [MED] Inflammation of the aorta.

AP *See* after-perpendicular.

apachite [PETR] A phonolite consisting of enigmatite and hornblende in about the same quantity as the pyroxene, but of a later crystallization phase.

apandrous [BOT] Lacking male organs or having nonfunctional male organs.

ANT LION

The ant lion
(*Myrmeleon immaculatus*).

AORTIC ARCH

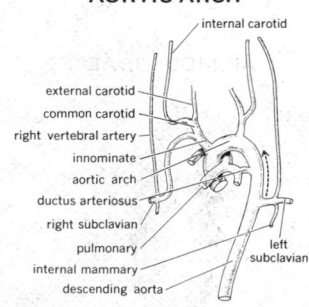

internal carotid
external carotid
common carotid
right vertebral artery
innominate
aortic arch
ductus arteriosus
right subclavian
pulmonary
left subclavian
internal mammary
descending aorta

Ventral view of the aortic arch in the human, showing the blood vessels that arise from it.
(*Modified from B. M. Patten*)

APATITE

├── 3.1 cm ──┤

Specimen of apatite from Eganville, Ontario, Canada. *(American Museum of Natural History specimen)*

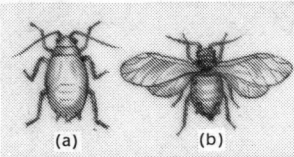

APHID

(a)   (b)

Two forms during the life cycle of an aphid: *(a)* wingless female, *(b)* winged female.

APHRODITIDAE

The sea mouse, *Aphrodita*, of the Aphroditidae.

apareon [ASTRON] The point on a Mars-centered orbit where a satellite is at its greatest distance from Mars.

apastron [ASTRON] That point of the orbit of one member of a binary star system at which the stars are farthest apart.

Apatemyidae [PALEON] A family of extinct rodentlike insectivorous mammals belonging to the Proteutheria.

apatetic [ECOL] Pertaining to the imitative protective coloration of an animal subject to being preyed upon.

Apathornithidae [PALEON] A family of Cretaceous birds, with two species, belonging to the order Ichthyornithiformes.

apatite [MINERAL] A group of phosphate minerals that includes 10 mineral species and has the general formula $X_5(YO_4)_3Z$, where X is usually Ca^{2+} or Pb^{3+}, Y is P^{5+} or As^{5+}, and Z is F^-, Cl^-, or OH^-.

ape [VERT ZOO] Any of the tailless primates of the families Hylobatidae and Pongidae in the same superfamily as man.

apeirophobia [PSYCH] Abnormal fear of infinity.

aperiodic [PHYS] Of irregular occurrence; not periodic; not displaying resonant response.

aperiodic antenna [ELECTROMAG] Antenna designed to have constant impedance over a wide range of frequencies because of the suppression of reflections within the antenna system; includes terminated wave and rhombic antennas.

aperiodic compass [NAV] Literally, a compass without a period, that is, a compass that, after being deflected, returns by one direct movement to its proper reading, without oscillation. Also known as deadbeat compass.

aperiodic damping [PHYS] Condition of a system in which the amount of damping is so large that, when the system is subjected to a single disturbance, either constant or instantaneous, the system comes to a position of rest without passing through that position; while an aperiodically damped system is not strictly an oscillating system, it has such properties that it should become an oscillating system if the damping were sufficiently reduced.

aperiodic waves [ELEC] The transient current wave in a series circuit with resistance R, inductance L, and capacitance C when $R^2C = 4L$. [PHYS] Waves without a definite repetitive pattern; for example, transient waves.

apers *See* antipersonnel.

apertometer [OPTICS] An instrument designed to measure the numerical aperture of microscope objectives.

Apert's syndrome *See* acrocephalosyndactylism.

aperture [ELECTR] An opening through which electrons, light, radio waves, or other radiation can pass. [OPTICS] The diameter of the objective of a telescope or other optical instrument, usually expressed in inches, but sometimes as the angle between lines from the principal focus to opposite ends of a diameter of the objective.

aperture aberration [OPTICS] Errors in optical imaging which occur because rays of different distances from the axis do not come to the same focus.

aperture antenna [ELECTROMAG] Antenna in which the beam width is determined by the dimensions of a horn, lens, or reflector.

aperture conductivity [ACOUS] The ratio of the density of a medium to the acoustic mass at an aperture.

aperture grill picture tube [ELECTR] An in-line gun-type picture tube in which the shadow mask is perforated by long, vertical stripes and the screen is painted with vertical phosphor stripes.

aperture illumination [ELECTROMAG] Field distribution in amplitude and phase over an aperture.

aperture mask *See* shadow mask.

aperture ratio [OPTICS] The ratio of the effective diameter of a lens to its focal length.

aperture sight [ORD] An irregularly shaped adjustable mechanical item usually integral to a rear sight; it functions as a peephole through which the sight at the opposite end of a gun is brought into view in aiming at a target or object.

aperture stop [OPTICS] That opening in an optical system that determines the size of the bundle of rays which traverse the system from a given point of the object to the corresponding point of the image.

apetalous [BOT] Lacking petals.

apex [ANAT] **1.** The upper portion of a lung extending into the root. **2.** The pointed end of the heart. **3.** The tip of the root of a tooth. [BOT] The pointed tip of a leaf. [GEOL] The part of a mineral vein nearest the surface of the earth. [MATH] **1.** The vertex of a triangle opposite the side which is regarded as the base. **2.** The vertex of a cone or pyramid.

apex impulse [PHYSIO] The point of maximum outward movement of the left ventricle of the heart during systole, normally localized in the fifth left intercostal space in the midclavicular line. Also known as left ventricular thrust.

aphagia [MED] Inability to swallow; may be organic or psychic in origin.

aphakia [MED] Absence of the lens of the eye.

aphaniphyric [PETR] Denoting a texture of porphyritic rocks with microaphanitic groundmasses. Also known as felsophyric.

aphanite [PETR] **1.** A general term applied to dense, homogeneous rocks whose constituents are too small to be distinguished by the unaided eye. **2.** A rock having aphanitic texture.

Aphanomyces [MYCOL] A genus of fungi in the phycomycetous order Saprolegniales; species cause root rot in plants.

aphasia [MED] Impairment in the use or comprehension of language caused by lesions of the cerebral cortex.

aphasic seizure [MED] A transient inability to speak due to an abnormal electrical discharge from the speech areas of the brain.

Aphasmidea [INV ZOO] An equivalent name for the Adenophorea.

Aphelenchoidea [INV ZOO] A superfamily of plant and insect-associated nematodes in the order Tylenchida.

aphelion [ASTRON] The point on a planetary orbit farthest from the sun.

Aphelocheiridae [INV ZOO] A family of hemipteran insects belonging to the superfamily Naucoroidea.

aphesperian [ASTRON] The farthest point of a satellite in its orbit about Venus.

aphid [INV ZOO] The common name applied to the soft-bodied insects of the family Aphididae; they are phytophagous plant pests and vectors for plant viruses and fungal parasites.

Aphididae [INV ZOO] The true aphids, a family of homopteran insects in the superfamily Aphidoidea.

Aphidoidea [INV ZOO] A superfamily of sternorrhynchan insects in the order Homoptera.

Aphis [INV ZOO] A genus of aphid, the type genus of the family Aphididae.

aphonia [MED] Loss of voice and power of speech.

aphotic zone [OCEANOGR] The deeper part of the ocean where sunlight is absent.

Aphredoderidae [VERT ZOO] A family of actinopterygian fishes in the order Percopsiformes containing one species, the pirate perch.

aphrodisiac [PHYSIO] Any chemical agent or odor that stimulates sexual desires.

Aphroditidae [INV ZOO] A family of scale-bearing polychaete worms belonging to the Errantia.

Aphrosalpingoidea [PALEON] A group of middle Paleozoic invertebrates classified with the calcareous sponges.

aphrosiderite *See* ripiddite.

aphtha [MED] A white, painful oral ulcer of unknown cause.

aphthitalite [MINERAL] $(K,Na)_3Na(SO_4)_2$ A white mineral crystallizing in the rhombohedral system and occurring massively or in crystals.

aphylactic map projection [MAP] A map projection which is neither conformal nor equal-area.

Aphylidae [INV ZOO] An Australian family of hemipteran insects composed of two species; not placed in any higher toxonomic group.

aphyllous [BOT] Lacking foliage leaves.

aphyric [PETR] Of the texture of fine-grained igneous rocks, showing two generations of the same mineral but without phenocrysts.

API *See* air-position indicator; armor-piercing incendiary.

apiary [AGR] A place where bees are kept, especially for breeding and honey making.

apical [BOT] Relating to the apex or tip.

apical angle [MECH] The angle between the tangents to the curve outlining the contour of a projectile at its tip.

apical bud *See* terminal bud.

apical dominance [BOT] Inhibition of lateral bud growth by the apical bud of a shoot, believed to be a response to auxins produced by the apical bud.

apical meristem [BOT] A region of embryonic tissue occurring at the tips of roots and stems.

apical plate [INV ZOO] A group of cells at the anterior end of certain trochophore larvae; believed to have nervous and sensory functions.

apiculate [BOT] Ending abruptly in a short, sharp point.

apiculture [AGR] Large-scale commercial beekeeping.

Apidae [INV ZOO] A family of hymenopteran insects in the superfamily Apoidea including the honeybees, bumblebees, and carpenter bees.

Apioceridae [INV ZOO] A family of orthorrhaphous dipteran insects in the series Brachycera.

apioid [PHYS] A pear-shaped form taken by a rapidly revolving mass of liquid due to the force of gravity.

apiology [INV ZOO] The scientific study of bees, particularly honeybees.

Apis [INV ZOO] A genus of bees, the type genus of the Apidae.

API scale [CHEM ENG] The American Petroleum Institute hydrometer scale for the measurement of the specific gravity of liquids; used primarily in the American petroleum industry.

Apis mellifera [INV ZOO] The western honeybee, an economically important species for its honey.

Apistobranchidae [INV ZOO] A family of spioniform annelid worms belonging to the Sedentaria.

apjohnite [MINERAL] $MnAl_2(SO_4)_4 \cdot 22H_2O$ A white, rose-green, or yellow mineral containing water and occurring in crusts, fibrous masses, or efflorescences.

Aplacophora [INV ZOO] A subclass of vermiform mollusks in the class Amphineura characterized by no shell and calcareous integumentary spicules.

aplanatic lens [OPTICS] A lens corrected for spherical aberration.

aplanogamete [BIOL] A gamete that lacks motility.

aplanospore [MYCOL] A nonmotile, asexual spore, usually a sporangiospore, common in the Phycomycetes.

aplasia [MED] Defective development resulting in the virtual absence of a tissue or organ; only a remnant appears.

aplastic anemia [MED] A blood disorder in which lymphocytes predominate while there is a deficiency of erythrocytes, hemoglobin, and granulocytes.

aplite [PETR] Fine-grained granitic dike rock made up of light-colored mineral constituents, mostly quartz and feldspar; used to manufacture glass and enamel. Also known as alaskite.

apnea [MED] A transient cessation of respiration.

Apneumonomorphae [INV ZOO] A suborder of arachnid arthropods in the order Araneida characterized by the lack of book lungs.

apneusis [PHYSIO] In certain lower vertebrates, sustained tonic contraction of the respiratory muscles to allow prolonged inspiration.

apo- [CHEM] A prefix that denotes formation from or relationship to another chemical compound.

apoapsis [ASTRON] The point in an orbit farthest from the center of attraction.

apoatropine [ORG CHEM] $C_{17}H_{21}NO_2$ An alkaloid melting at 61°C with decomposition of the compound; highly toxic; obtained by dehydrating atropine.

apob [METEOROL] An observation of pressure, temperature, and relative humidity taken aloft by means of an aerometeorograph; a type of aircraft sounding.

apocarpous [BOT] Having carpels separate from each other.

apocenter *See* apofocus.

apochromat *See* apochromatic lens.

apochromatic lens [OPTICS] A lens with corrections for chromatic and spherical aberration.

apochromatic system [OPTICS] An optical system which is free from both spherical and chromatic aberration for two or more colors.

apocrine gland [PHYSIO] A multicellular gland, such as a mammary gland or an axillary sweat gland, that extrudes part of the cytoplasm with the secretory product.

apocronus [ASTRON] The farthest point of a satellite in its orbit about Saturn.

Apocynaceae [BOT] A family of tropical and subtropical flowering trees, shrubs, and vines in the order Gentianales, characterized by a well-developed latex system, granular pollen, a poorly developed corona, and the carpels often united by the style and stigma; well-known members are oleander and periwinkle.

Apoda [VERT ZOO] The caecilians, a small order of wormlike, legless animals in the class Amphibia.

Apodacea [INV ZOO] A subclass of echinoderms in the class Holothuroidea characterized by simple or pinnate tentacles and reduced or absent tube feet.

apodeme [INV ZOO] An internal ridge or process on an arthropod exoskeleton to which organs and muscles attach.

Apodes [VERT ZOO] An equivalent name for the Anguilliformes.

Apodi [VERT ZOO] The swifts, a suborder of birds in the order Apodiformes.

Apodida [INV ZOO] An order of worm-shaped holothurian echinoderms in the subclass Apodacea.

Apodidae [VERT ZOO] The true swifts, a family of apodiform birds belonging to the suborder Apodi.

Apodiformes [VERT ZOO] An order of birds containing the hummingbirds and swifts.

apodization [ELECTR] A technique for modifying the response of a surface acoustic wave filter by varying the overlap between adjacent electrodes of the interdigital transducer. [OPTICS] The modification of the amplitude transmittance of the aperture of an optical system so as to reduce or suppress the energy of the diffraction rings relative to that of the central Airy disk.

apoenzyme [BIOCHEM] The protein moiety of an enzyme; determines the specificity of the enzyme reaction.

apofocus [ASTRON] The point on an elliptic orbit at the greatest distance from the principal focus. Also known as apocenter.

apogalacteum [ASTRON] The point at which a celestial body is farthest from the center of the Milky Way.

apogamy [BIOL] Asexual, parthenogenetic development of diploid cells, such as the development of a sporophyte from a gametophyte without fertilization.

apogee [ASTRON] That point in an orbit at which the moon or an artificial satellite is most distant from the earth; the term is sometimes loosely applied to positions of satellites of other planets.

apogeny [BOT] Loss of the function of reproduction.

apogeotropism [BOT] Negative geotropism; growth up or away from the soil.

Apogonidae [VERT ZOO] The cardinal fishes, a family of tropical marine fishes in the order Perciformes; males incubate eggs in the mouth.

Apoidea [INV ZOO] The bees, a superfamily of hymenopteran insects in the suborder Apocrita.

apojove [ASTRON] The farthest point of a satellite in its orbit about Jupiter.

Apollo [ASTRON] To the Greeks, the planet Mercury when it was a morning star.

Apollo program [AERO ENG] The scientific and technical program of the United States that involved placing men on the moon and returning them safely to earth.

apolune [ASTRON] Farthest point of a satellite in an elliptic orbit about the moon. Also known as aposelene.

apolysis [INV ZOO] In most tapeworms, the shedding of ripe proglottids.

apomeiosis [CYTOL] Meiosis that is either suppressed or imperfect.

apomercurian [ASTRON] The farthest point of a satellite in its orbit about Mercury.

apomixis [EMBRYO] Parthenogenetic development of sex cells without fertilization.

apomorph [SYST] Any derived character occurring at a branching point and carried through one descending group in a phyletic lineage.

apomorphine [PHARM] $C_{17}H_{17}NO_2$ A crystalline alkaloid obtained by dehydration of morphine; acts as a powerful emetic.

APLANOSPORE

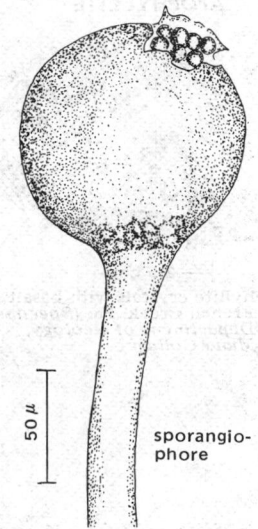

aplanospores

sporangium

sporangio-
phore

Several aplanospores and the sporangium that contains them.

APODIDA

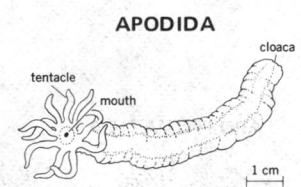

Typical appearance of an apodous holothurian.

aponeurosis [ANAT] A broad sheet of regularly arranged connective tissue that covers a muscle or serves to connect a flat muscle to a bone.

apophorometer [ENG] An apparatus used to identify minerals by sublimation.

apophyge [ARCH] A small curvature applied to the top or bottom of a column shaft where it expands to meet the fillet.

apophyllite [MINERAL] A hydrous calcium potassium silicate containing fluorine and occurring as a secondary mineral with zeolites with geodes and other igneous rocks; the composition is variable but approximates $KFCa_4(Si_2O_5)_4 \cdot 8H_2O$. Also known as fish-eye stone.

apophyllous [BOT] Having the parts of the perianth distinct.

apophysis [ANAT] An outgrowth or process on an organ or bone. [MYCOL] A swollen filament in fungi.

apoplexy [MED] **1.** A symptom complex caused by an acute vascular lesion of the brain and characterized by unconsciousness with various degrees of paralysis and sensory impairment. **2.** Sudden, severe hemorrhage into any organ.

apoplutonian [ASTRON] The farthest point of a satellite in its orbit about Pluto.

apoposeidon [ASTRON] The farthest point of a satellite in its orbit about Neptune.

apopyle [INV ZOO] Any one of the large pores in a sponge by which water leaves a flagellated chamber to enter the exhalant system.

Aporidea [INV ZOO] An order of tapeworms of uncertain composition and affinities; parasites of anseriform birds.

aporogamy [BOT] Entry of the pollen tube into the embryo sac through an opening other than the micropyle.

aposelene *See* apolune.

A positive [ELEC] Also known as A+. **1.** Positive terminal of an A battery or positive polarity of other sources of filament voltage. **2.** Denoting the terminal to which the positive side of the filament voltage source should be connected.

apospory [MYCOL] Suppression of spore formation with development of the haploid (sexual) generation directly from the diploid (asexual) generation.

a posteriori probability *See* empirical probability.

apostilb [OPTICS] A luminance unit equal to one ten-thousandth of a lambert. Also known as blandel.

Apostomatida [INV ZOO] An order of ciliated protozoans in the subclass Holotrichia; majority are commensals on marine crustaceans.

apotele [ANAT] A scalloped ridge around the edge of an otolith.

apothecaries' measure [PHARM] A system of units of volume, usually of liquid drugs, in which 16 fluid ounces equals 1 pint.

apothecaries' ounce *See* ounce.

apothecaries' pound *See* pound.

apothecaries' weight [PHARM] A system of units of mass, usually of drugs, in which 1 pound equals 5760 grains or 1 troy pound.

apothecium [MYCOL] A spore-bearing structure in some Ascomycetes and lichens in which the fruiting surface or hymenium is exposed during spore maturation.

apothem [MATH] The perpendicular distance from the center of a regular polygon to one of its sides.

apouranian [ASTRON] The farthest point of a satellite in its orbit about Uranus.

A power supply *See* A supply.

apozymase [BIOCHEM] The protein component of a zymase.

Appalachia [GEOL] Proposed borderland along the southeastern side of North America, seaward of the Appalachian geosyncline in Paleozoic time.

Appalachian orogeny [GEOL] An obsolete term referring to Late Paleozoic diastrophism beginning perhaps in the Late Devonian and continuing until the end of the Permian; now replaced by Allegheny orogeny.

apparatus [SCI TECH] A compound instrument designed to carry out a specific function.

apparent [ASTRON] A term used to designate certain measured or measurable astronomic quantities to refer them to real or visible objects, such as the sun or a star.

apparent additional mass [FL MECH] A fictitious mass of fluid added to the mass of the body to represent the force required to accelerate the body through the fluid.

apparent candlepower [OPTICS] For an extended source of light, at a specified distance, the candlepower of a point source that would produce the same illumination as the extended source at the same distance.

apparent density [MET] The weight per unit volume of a metal powder, in contrast to the weight per unit volume of the individual particles.

apparent dip [GEOL] Dip of a rock layer as it is exposed in any section not at a right angle to the strike.

apparent expansion [THERMO] The expansion of a liquid with temperature, as measured in a graduated container without taking into account the container's expansion.

apparent force [MECH] A force introduced in a relative coordinate system in order that Newton's laws be satisfied in the system; examples are the Coriolis force and the centrifugal force incorporated in gravity.

apparent gravity *See* acceleration of gravity.

apparent horizon *See* horizon.

apparent libration in longitude *See* lunar libration.

apparent luminance [OPTICS] Luminance, created by air light, of that portion of the visual field subtended by a dark, distant object; that is, the light scattered into the eye by particles, including air molecules, lying along the optic path from eye to object.

apparent magnitude [ASTRON] An index of a star's brightness relative to that of the other stars; it does not take into account the difference in distance between the stars and is not an indication of the star's true luminosity.

apparent motion [MECH] Movement relative to a specified or implied reference point which may itself be in motion. Also known as relative motion.

apparent movement of faults [GEOL] The apparent motion observed to have occurred in any chance section across a fault.

apparent noon [ASTRON] Twelve o'clock apparent time, or the instant the apparent sun is over the upper branch of the meridian.

apparent plunge [GEOL] Inclination of a normal projection of lineation in the plane of a vertical cross section.

apparent position [ASTRON] The position on the celestial sphere at which a heavenly body (or a space vehicle) would be seen from the center of the earth at a particular time.

apparent power [ELEC] The product of the root-mean-square voltage and the root-mean-square current delivered in an alternating-current circuit, no account being taken of the phase difference between voltage and current.

apparent precession *See* apparent wander.

apparent shoreline [MAP] The outer edge of marine vegetation (marsh, mangrove, cypress) delineated on photogrammetric surveys where the actual shoreline is obscured.

apparent slope [GRAPHICS] A true slope that appears vertically distorted or exaggerated in air photographs viewed with a stereoscope.

apparent solar day [ASTRON] The duration of one rotation of the earth on its axis with respect to the apparent sun. Also known as true solar day.

apparent solar time [ASTRON] Time measured by the apparent diurnal motion of the sun. Also known as apparent time; true solar time.

apparent sun [ASTRON] The sun as it appears to an observer. Also known as true sun.

apparent time *See* apparent solar time.

apparent vertical [GEOPHYS] The direction of the resultant of gravitational and all other accelerations. Also known as dynamic vertical.

apparent volume [PHYS] The difference between the volume of a binary solution and the volume of the pure solvent at the same temperature.

apparent wander [GEOPHYS] Apparent change in the direction of the axis of rotation of a spinning body, such as a gyroscope, due to rotation of the earth. Also known as apparent precession; wander.

apparent water table *See* perched water table.

apparent weight [MECH] For a body immersed in a fluid (such as air), the resultant of the gravitational force and the buoyant force of the fluid acting on the body; equal in

APOPHYLLITE

├── 2.5 cm ──┤

Apophyllite crystals with basalt from French Creek, Pa. *(Specimen from Department of Geology, Bryn Mawr College)*

APOSTOMATIDA

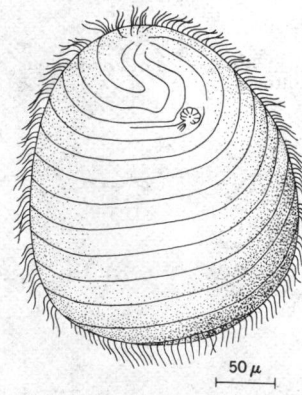

├── 50 μ ──┤

Foettingeria, an example of an apostomatid.

magnitude to the true weight minus the weight of the displaced fluid.

apparent wind *See* relative wind.

appearance potential [PHYS] The minimal potential which the electron beam in the ion source of a mass spectrometer must traverse in order to acquire enough energy to produce ions of a specified nuclide or molecular fragment.

appearance ratio *See* hyperstereoscopy.

appendage [BIOL] Any subordinate or nonessential structure associated with a major body part. [NAV ARCH] Any fitting installed outside the underwater hull proper, such as a rudder, shaft, shaft strut, and bilge keel. [ZOO] Any jointed, peripheral extension, especially limbs, of arthropod and vertebrate bodies.

appendectomy [MED] Surgical removal of the vermiform appendix.

appendicitis [MED] Inflammation of the vermiform appendix.

appendicular skeleton [ANAT] The bones of the pectoral and pelvic girdles and the paired appendages in vertebrates.

appendiculate [BIOL] Having or forming appendages.

appendix [ANAT] **1.** Any appendage. **2.** *See* vermiform appendix.

appendix testis [MED] A remnant of the cranial part of the paramesonephric or Müllers duct, attached to the testis. Also known as hydatid of Morgagni.

appersonification [PSYCH] Unconscious identification with and acquisition of another person's characteristics.

appestat [PHYSIO] The center for appetite regulation in the hypothalamus.

appinite [PETR] Hornblende-rich plutonic rock with high feldspar content.

apple [BOT] *Malus domestica.* A deciduous tree in the order Rosales which produces an edible, simple, fleshy, pome-type fruit.

apple coal [GEOL] Easily mined soft coal that breaks into small pieces the size of apples.

apple essence *See* isoamyl valerate.

Applegate diagram [ELECTR] A graph of the electron paths in a two-cavity klystron tube, showing how electron bunching occurs.

apple of Peru *See* jimsonweed.

apple oil *See* isoamyl valerate.

apple pox *See* blister canker.

Appleton layer *See* F_2 layer.

Apple Tube *See* beam-indexing picture tube.

appliance [ENG] A piece of equipment that draws electric or other energy and produces a desired work-saving or other result, such as an electric heater, a radio, or an electronic range.

applications technology satellite [AERO ENG] Any artificial satellite in the National Aeronautics and Space Administration program for the evaluation of advanced techniques and equipment for communications, meteorological, and navigation satellites. Abbreviated ATS.

application study [ADP] The detailed process of determining a system or set of procedures for using a computer for definite functions of operations, and establishing specifications to be used as a base for the selection of equipment suitable to the specific needs.

applied climatology [CLIMATOL] The scientific analysis of climatic data in the light of a useful application for an operational purpose.

applied ecology [ECOL] Activities involved in the management of natural resources.

applied epistemology [ADP] The use of machines or other models to simulate processes such as perception, recognition, learning, and selective recall, or the application of principles assumed to hold for human categorization, perception, storage, search, and so on, to the design of machines, machine programs, scanning, storage, and retrieval systems.

applied meteorology [METEOROL] The application of current weather data, analyses, or forecasts to specific practical problems.

applied research [ENG] Research directed toward using knowledge gained by basic research to make things or to create situations that will serve a practical or utilitarian purpose.

applied strategic research [ENG] Research done to provide a basic understanding of a current applied project.

appliqué [GRAPHICS] A decoration or ornament made by cutting out and attaching one piece of material to the surface of another. [OPTICS] A combination of lenses that provides for the same focal length at three or more wavelengths.

appliqué armor [ORD] Material or attachment which can be installed on a tank to give it additional protection against kinetic- or nonkinetic-energy projectiles.

applique circuit [ELEC] Special circuit which is provided to modify existing equipment to allow for special usage; for example, some carrier telephone equipment designed for ringdown manual operation can be modified through the use of an applique circuit to allow for use between points having dial equipment.

apposition beach [GEOL] One of a series of parallel beaches formed on the seaward side of an older beach.

apposition eye [INV ZOO] A compound eye in diurnal insects and crustaceans in which each ommatidium focuses on a small part of the whole field of light, producing a mosaic image.

apposition fabric [PETR] A primary orientation of the elements of a sedimentary rock that is developed or formed at time of deposition of the material; fabrics of most sedimentary rocks belong to this type. Also known as primary fabric.

appressed [BIOL] Pressed close to or lying flat against something.

approach [MECH ENG] The difference between the temperature of the water leaving a cooling tower and the wet-bulb temperature of the surrounding air. [NAV] In air operations, a maneuver executed by an aircraft in making its transit from high-altitude enroute flight to the point where it begins the landing approach; includes maneuvers (such as flying race-track pattern) required for traffic control.

approach and landing area [NAV] An airspace of defined dimensions together with its runways and water channels, used by aircraft arriving at or departing from or operating within the vicinity of the airport.

approach chart [NAV] An aeronautical chart which provides information about navigational facilities, flight patterns, radio aids and their frequencies, and so on, for use in making an approach to an airfield under either visual or instrument flight conditions.

approach course [NAV] A flight track to runway or approach fix defined by a visual or electronic aid. Also known as approach lane.

approach fix [NAV] A geographical position on a defined approach course.

approach gate [NAV] That point on the final approach course which is 1 mile from the approach fix on the side away from the airport or 5 miles from the landing threshold, whichever is farthest from the landing threshold.

approach lane *See* approach course.

approach lights [NAV] A group of aeronautical lights indicating a desirable line of approach to a landing area.

approach navigation [NAV] The navigation operation which must be performed in the proximity of an airport or harbor by an aircraft or ship when proceeding to terminal or dock.

approach path [NAV] That portion of the flight path which extends from the point where a descent is started to the point where the aircraft touches down on the runway.

approach segment [NAV] The basic functional division of an instrument approach procedure. Also known as approachway.

approach sequence [NAV] The position of an aircraft while awaiting approach clearance or while on approach.

approach signal [CIV ENG] A railway signal warning an engineer of a signal ahead that displays a restrictive indication.

approach surface base line [NAV] In terminal instrument procedures, an imaginary horizontal line at threshold elevation.

approach visibility [NAV] The distance from which a pilot on

APPLE

Apple *(Malus domestica);* mature fruit.

the instrument approach glide path can see landing aids at the runway threshold. Also known as slant visibility.

approachway   See approach segment.

approved flame safety lamp   [MIN ENG] A flame safety lamp which has been approved for use in gaseous coal mines.

approx   See approximate.

approximate   [MATH] **1.** To obtain a result that is not exact but is near enough to the correct result for some specified purpose. **2.** To obtain a series of results approaching the correct result. [SCI TECH] **1.** Close to the correct value. Abbreviated approx. **2.** To be close to.

approximate absolute temperature   [PHYS] A temperature scale with the ice point at 273° and boiling point of water at 373°; it is intended to approximate the Kelvin temperature scale with sufficient accuracy for many sciences, notably meteorology, and is widely used in the meteorological literature. Also known as tercentesimal thermometric scale.

approximate altitude   [NAV] Of a star, an altitude determined by estimation, by a star finder or star chart.

approximate contour   [MAP] A contour substituted for a normal contour whenever there is a question as to its reliability; reliability usually is evaluated as exceeding one-half the contour interval.

approximation   [MATH] **1.** A result that is not exact but is near enough to the correct result for some specified purpose. **2.** A procedure for obtaining such a result.

appulse   [ASTRON] **1.** The near approach of one celestial body to another on the celestial sphere, as in occultation or conjunction. **2.** A penumbral eclipse of the moon.

APR   See active prominence region; airborne profile recorder.

apraxia   [MED] The inability to perform purposeful acts as a result of brain lesions; characteristically, paralysis is absent and kinesthesia is unimpaired.

apricot   [BOT] *Prunus armeniaca*. A deciduous tree in the order Rosales which produces a simple fleshy stone fruit.

apricot kernel oil   See persic oil.

a priori   [MATH] Pertaining to deductive reasoning from assumed axioms or supposedly self-evident principles, supposedly without reference to experience.

a priori probability   See mathematical probability.

aproctous   [MED] Having an imperforate anus. [ZOO] Lacking an anus.

apron   [AERO ENG] A protective device specially designed to cover an area surrounding the fuel inlet on a rocket or spacecraft. [BUILD] **1.** A board on an interior wall beneath a windowsill. **2.** The vertical rear panel of a sink attached to a wall. [CIV ENG] A covering of a material such as concrete or timber over soil to prevent erosion by flowing water, as at the bottom of a dam. [GEOL] See outwash plain. [MECH ENG] A plate serving to protect or cover a machine. [MIN ENG] A canvas-covered frame set at such an angle in the miner's rocker that the gravel and water passing over it are carried to the head of the machine. [ORD] **1.** That portion of the superior slope of a parapet or the interior slope of a pit designed to protect the slopes against blast. **2.** The hinged portion of a shield. **3.** A removal screen of camouflage material placed over or in front of artillery guns. **4.** A hard-surfaced area, usually paved, adjacent to a ship or the like, used to park, load, unload, or service vehicles.

apron conveyor   [MECH ENG] A conveyor used for carrying granular or lumpy material and consisting of two strands of roller chain separated by overlapping plates, forming the carrying surface, with sides 2-6 inches high.

apron feeder   [MECH ENG] A limited-length version of apron conveyor used for controlled-rate feeding of pulverized materials to a process or packaging unit. Also known as plate-belt feeder; plate feeder.

apron piece   [BUILD] A beam that supports a landing or a series of winders in a staircase.

aprotic solvent   [CHEM] A solvent that does not yield or accept a proton.

Aps   See Apus.

APS   See armor-piercing sabot.

apse   See apsis.

Apsidospondyli   [VERT ZOO] A term used to include, as a subclass, amphibians in which the vertebral centra are formed from cartilaginous arches.

apsis   [ASTRON] In celestial mechanics, either of the two

orbital points nearest or farthest from the center of attraction. Also known as apse.

APT   See armor-piercing tracer; Automatic Programming Tool.

apterium   [VERT ZOO] A bare space between feathers on a bird's skin.

apterous   [BIOL] Lacking wings, as in certain insects, or winglike expansions, as in certain seeds.

Apterygidae   [VERT ZOO] The kiwis, a family of nocturnal ratite birds in the order Apterygiformes.

Apterygiformes   [VERT ZOO] An order of ratite birds containing three living species, the kiwis, characterized by small eyes, limited eyesight, and nostrils at the tip of the bill.

Apterygota   [INV ZOO] A subclass of the Insecta characterized by being primitively wingless.

Aptian   [GEOL] Lower Cretaceous geologic time, between Barremian and Albian. Also known as Vectian.

aptitude   [PSYCH] The natural inclination or capacity for skillful performance of an as yet unlearned task.

aptitude test   [PSYCH] Any standardized examination used to evaluate a person'a ability to learn a particular skill.

aptyalism   [MED] Deficiency or absence of saliva.

APU   See auxiliary power unit.

Apus   [ASTRON] A constellation with a right ascension of 16 hours and declination of 75°S. Abbreviated Aps. [VERT ZOO] A genus of birds comprising the Old World swifts.

apyrase   [BIOCHEM] Any enzyme that hydrolyzes adenosine-triphosphate, with liberation of phosphate and energy, and that is believed to be associated with actomyosin activity.

apyrexia   [MED] Absence of fever.

Aqil   See Aquila.

Aql   See Aquila.

AQL   See acceptable quality level.

Aqr   See Aquarius.

aqua   [CHEM] Latin for water.

aqua ammonia   See ammonium hydroxide.

aquaboard   See aquaplane.

aquaculture   See aquiculture.

aquadag   [ELECTR] Graphite coating on the inside of certain cathode-ray tubes for collecting secondary electrons emitted by the face of the tube.

aquafortis   See nitric acid.

Aquagel   [MATER] A hydrous silicate of alumina, of the Silica Products Company, used for waterproofing concrete.

aquagene tuff   See hyaloclastite.

aqualung   [ENG] A self-contained underwater breathing apparatus (scuba) of the demand or open-circuit type developed by J.Y. Cousteau.

aquamarine   [MINERAL] A pale-blue or greenish-blue transparent gem variety of the mineral beryl.

aquametry   [ANALY CHEM] Analytical processes to measure the water present in materials; methods include Karl Fischer titration, reactions with acid chlorides and anhydrides, oven drying, distillation, and chromatography.

aquaplane   [NAV ARCH] A board on which a person rides while being towed by a motorboat at such speed that the front part of the board rises out of the water. Also known as aquaboard.

aqua regia   [INORG CHEM] A mixture of three parts by volume of concentrated nitric acid and one part concentrated hydrochloric acid.

Aquarius   [ASTRON] A constellation with a right ascension of 23 hours and declination of 15°S. Abbreviated Aqr. Also known as Water Bearer.

aquatic   [BIOL] Living or growing in, on, or near water; having a water habitat.

aquatic weed cutter   [NAV ARCH] A device attached to the bow of a boat that can cut and clear a wide path through a weed-choked lake.

aquatint   [GRAPHICS] An etching process that produces several tones by varying the etching time of different areas of a copper plate; the resulting print resembles an ink or wash drawing.

aquation   [CHEM] Formation of a complex that contains water by replacement of other coordinated groups in the complex.

aquatone   [GRAPHICS] An offset printing process utilizing a

APRICOT

Apricot (*Prunus armeniaca*) branch showing fruit cluster.

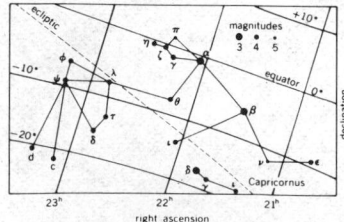

AQUARIUS

Line pattern of the constellation Aquarius. The grid lines represent the coordinates of the sky. The apparent brightness, or magnitude, of the stars is shown by the sizes of the dots, which are graded by appropriate numbers as indicated.

zinc plate that is gelatin-coated and hardened and sensitized to print type, line drawings, and fine-screen halftones.

aqueduct [CIV ENG] An artificial tube or channel for conveying water.

aqueous desert [ECOL] A marine bottom environment with little or no macroscopic invertebrate shelled life.

aqueous humor [PHYSIO] The transparent fluid filling the anterior chamber of the eye.

aqueous lava [GEOL] Mud lava produced by the mixing of volcanic ash with condensing volcanic vapor or other water.

aqueous rock [PETR] A sedimentary rock deposited by or in water. Also known as hydrogenic rock.

aqueous solution [CHEM] A solution with the solvent as water.

aquiclude [GEOL] A porous formation that absorbs water slowly but will not transmit it fast enough to furnish an appreciable supply for a well or spring.

aquiculture [BIOL] Cultivation of natural faunal resources of water. Also spelled aquaculture.

aquifer [HYD] A subsurface zone that yields economically important amounts of water to wells.

Aquifoliaceae [BOT] A family of woody flowering plants in the order Celastrales characterized by pendulous ovules, alternate leaves, imbricate petals, and drupaceous fruit; common members include various species of holly (*Ilex*).

aquifuge [GEOL] An impermeable body of rock which contains no interconnected openings or interstices and therefore neither absorbs nor transmits water.

aquiherbosa [ECOL] Herbaceous plant communities in wet areas, such as swamps and ponds.

Aquila [ASTRON] A constellation with a right ascension of 20 hours and declination of 5°N. Abbreviated Aqil; Aql.

α Aquilae *See* Altair.

aquiprata [ECOL] Communities of plants which are found in areas such as wet meadows where groundwater is a factor.

Aquitanian [GEOL] Lower lower Miocene or uppermost Oligocene geologic time.

aquo ion [CHEM] Any ion containing one or more water molecules.

Ar *See* argon.

Ar$_1$ [MET] The temperature at which conversion of austenite to ferrite or to ferrite plus cementite is completed upon cooling a steel.

Ar$_3$ [MET] The temperature at which austenite begins to convert to ferrite upon cooling a steel.

Ar$_4$ [MET] The temperature at which delta ferrite is converted to gamma iron (austenite) upon cooling a steel.

Ara [ASTRON] A constellation with a right ascension of 17 hours and declination of 55°S. Also known as Altar.

Arabellidae [INV ZOO] A family of polychaete worms belonging to the Errantia.

arabic numerals [MATH] The numerals 0, 1, 2, 3, 4, 5, 6, 7, 8, and 9.

arabinose [BIOCHEM] $C_5H_{10}O_5$ A pentose sugar obtained in crystalline form from plant polysaccharides such as gums, hemicelluloses, and some glycosides.

arabite *See* arabitol.

arabitol [ORG CHEM] $CH_2OH(CHOH)_3CH_2OH$ An alcohol that is derived from arabinose; a sweet, colorless crystalline material present in D and L forms; soluble in water; melts at 103°C. Also known as arabite.

Araceae [BOT] A family of herbaceous flowering plants in the order Arales; plants have stems, roots, and leaves, the inflorescence is a spadix, and the growth habit is terrestrial or sometimes more or less aquatic; well-known members include dumb cane (*Dieffenbachia*), jack-in-the-pulpit (*Arisaema*), and *Philodendron*.

arachic acid *See* eicosanoic acid.

arachidic acid *See* eicosanoic acid.

arachis oil *See* peanut oil.

arachnephobia [PSYCH] Abnormal fear of spiders.

Arachnida [INV ZOO] A class of arthropods in the subphylum Chelicerata characterized by four pairs of thoracic appendages.

arachnodactyly [MED] A rare congenital defect of the skeletal system marked by abnormally long hand and foot bones.

arachnoid [ANAT] A membrane that covers the brain and spinal cord and lies between the pia mater and dura mater.

[BOT] Of cobweblike appearance, caused by fine white hairs. Also known as araneose. [INV ZOO] Any invertebrate related to or resembling the Arachnida.

arachnoidal granulations [ANAT] Projections of the arachnoid layer of the cerebral meninges through the dura mater. Also known as arachnoid villi; Pacchionian bodies.

Arachnoidea [INV ZOO] The name used in some classification schemes to describe a class of primitive arthropods.

arachnoid villi *See* arachnoidal granulations.

arachnology [INV ZOO] The study of arachnids.

Aradidae [INV ZOO] The flat bugs, a family of hemipteran insects in the superfamily Aradoidea.

Aradoidea [INV ZOO] A small superfamily of hemipteran insects belonging to the subdivision Geocorisae.

Araeoscelidia [PALEON] A provisional order of extinct reptiles in the subclass Euryapsida.

Arago distance [ASTRON] The angular distance from the antisolar point to the Arago point.

aragonite [MINERAL] $CaCO_3$ A white, yellowish, or gray orthorhombic mineral species of calcium carbonate but with a crystal structure different from those of vaterite and calcite, the other two polymorphs of the same composition. Also known as Aragon spar.

Arago point [ASTRON] One of the three commonly detectable points along the vertical circle through the sun at which the degree of polarization of diffuse sky radiation goes to zero; a neutral point.

Aralac [TEXT] A fiber synthesized from casein.

Arales [BOT] An order of monocotyledonous plants in the subclass Arecidae.

Araliaceae [BOT] A family of dicotyledonous trees and shrubs in the order Umbellales; there are typically five carpels and the fruit, usually a berry, is fleshy or dry; well-known members are ginseng (*Panax*) and English ivy (*Hedera helix*).

aralkyl [ORG CHEM] A radical in which an aryl group is substituted for an alkyl H atom. Derived from arylated alkyl.

aramayoite [MINERAL] $Ag(Sb,Bi)S_2$ An iron-black mineral consisting of silver antimony bismuth sulfide.

Aramidae [VERT ZOO] The limpkins, a family of birds in the order Gruiformes.

Aramite [MATER] A miticide; used for the control of phytophagous mites.

Aran-Duchenne atrophy [MED] A muscular system disorder of adults involving progressive spinal muscular atrophy.

Araneae [INV ZOO] An equivalent name for Araneida.

Araneida [INV ZOO] The spiders, an order of arthropods in the class Arachnida.

araneology [INV ZOO] The study of spiders.

araneose *See* arachnoid.

arapahite [PETR] A dark-colored, porous, fine-grained basic basalt consisting of magnetite, bytownite, and augite.

Arbacioida [INV ZOO] An order of echinoderms in the superorder Echinacea.

arbitrary correction [ORD] The correction of firing data or sound-locator data, applied to correct for observed errors after allowance has been made for all known causes of deviation. Also known as adjustment correction.

arbitrary course computer *See* course-line computer.

arbitrary grid [MAP] Any reference system developed for use where no grid is available or practical, or where military security for the reference is desired.

arbitration [IND ENG] A semijudicial means of settling labor-management disputes in which both sides agree to be bound by the decision of one or more neutral persons selected by some method mutually agreed upon.

arbitration bar [MET] A cast-iron specimen, in the form of a standard-sized bar, to be tested for conformity to specifications of the American Society for Testing and Materials.

arbor [HOROL] The axle of a wheel in a watch or clock. [MECH ENG] **1.** A cylindrical device positioned between the spindle and outer bearing of a milling machine and designed to hold a milling cutter. **2.** A shaft or spindle used to hold a revolving cutting tool or the work to be cut. [MET] A device which supports sand cores in molds.

arbor collar [ENG] A cylindrical spacer that positions and secures a revolving cutter on an arbor.

ARABINOSE

The structural formula of β-L-arabinose, obtained from the cherry gum plant.

ARAGONITE

Specimen of aragonite from Girgenti, Sicily. (*American Museum of Natural History specimen*)

├── 5 cm ───┤

ARBORVITAE

Eastern aborvitae
(Thuja occidentalis).

ARCHAEOPTERYX

Archaeopteryx lithographica.
(From W. E. Swinton, Fossil Birds,
British Museum-Natural History,
1958)

ARCH DAM

East Canyon Dam, a thin-arch
concrete structure on the East
Canyon River, Utah. *(U.S.*
Bureau of Reclamation)

arboreal Also known as arboreous. [BOT] Relating to or resembling a tree. [ZOO] Living in trees.

arboreous [BOT] 1. Wooded. 2. *See* arboreal.

arborescence [BIOL] The state of being treelike in form and appearance.

arborescent powder *See* dendritic powder.

arboretum [BOT] An area where trees and shrubs are cultivated for educational and scientific purposes.

arbor hole [DES ENG] A hole in a revolving cutter or grinding wheel for mounting it on an arbor.

arboriculture [BOT] The cultivation of ornamental trees and shrubs.

arborization [BIOL] A treelike arrangement, such as a branched dendrite or axon.

arborization block *See* intraventricular heart block.

arbor press [MECH ENG] A machine used for forcing an arbor or a mandrel into drilled or bored parts preparatory to turning or grinding. Also known as mandrel press.

arbor support [ENG] A device to support the outer end or intermediate point of an arbor.

arbor vitae [ANAT] The treelike arrangement of white nerve tissue seen in a median section of the cerebellum.

arborvitae [BOT] Any of the ornamental trees, sometimes called the tree of life, in the genus *Thuja* of the order Pinales.

arbor vitae oil *See* thuja oil.

arboviral encephalitides [MED] Diseases which are caused by arthropod-borne viruses (arboviruses), such as the encephalitis infections.

arbovirus [VIROL] Small, arthropod-borne animal viruses that are unstable at room temperature and inactivated by sodium deoxycholate; cause several types of encephalitis. Also known as arthropod-borne virus.

Arbuckle orogeny [GEOL] Mid-Pennsylvanian episode of diastrophism in the Wichita and Arbuckle Mountains of Oklahoma.

arbutin [ORG CHEM] $C_{12}H_{16}O_7$ A bitter glycoside from the bearberry and certain other plants; sometimes used as a urinary antiseptic.

arc [ELEC] *See* electric arc. [ENG] The graduated scale of an instrument for measuring angles, as a marine sextant; readings obtained on that part of the arc beginning at zero and extending in the direction usually considered positive are popularly said to be on the arc, and those beginning at zero and extending in the opposite direction are said to be off the arc. [GEOL] A geologic or topographic feature that is repeated along a curved line on the surface of the earth. [MATH] A continuous piece of the circumference of a circle.

arcade [ARCH] 1. An arched building. 2. An arched passageway. [INV ZOO] A type of cell associated with the pharyngeal region of nematodes and united with like cells by an arch.

arcanite [MINERAL] K_2SO_4 A colorless, vitreous orthorhombic sulfate mineral. Also known as glaserite.

arcature [ARCH] 1. A small arcade. 2. A blind, usually decorative arcade.

arcback [ELECTR] The flow of a principal electron stream in the reverse direction in a mercury-vapor rectifier tube because of formation of a cathode spot on an anode; this results in failure of the rectifying action. Also known as backfire.

arc blow [MET] The shifting of the arc in various directions in electric-arc welding due to the magnetic fields at the arc.

arc brazing [MET] Brazing with the use of an electric arc.

arc chute [ELEC] A collection of insulating barriers in a circuit breaker for confining the arc and preventing it from causing damage.

arc converter [ELECTR] A form of oscillator using an electric arc as the generator of alternating or pulsating current.

arc cosecant of a number [MATH] 1. Any angle whose cosecant is that number. 2. The angle between $-\pi/2$ radians and $\pi/2$ radians whose cosecant is that number; it is the value of the inverse of the restriction of the cosecant function to the interval between $-\pi/2$ and $\pi/2$, at this number.

arc cosine of a number [MATH] 1. Any angle whose cosine is that number. 2. The angle between 0 radians and π radians whose sine is that number; it is the value of the inverse of the restriction of the cosine function to the interval between 0 and π, at this number.

arc cotangent of a number [MATH] 1. Any angle whose cotangent is that number. 2. The angle between 0 radians and

π radians whose cotangent is that number; it is the value of the inverse of the restriction of the cotangent function to the interval between 0 and π, at this number.

arc cutting [MET] A type of thermal cutting of metal using the temperature generated by an electric arc.

arc discharge [ELEC] A direct-current electrical current between electrodes in a gas or vapor, having high current density and relatively low voltage drop.

Arcellinida [INV ZOO] An order of rhizopodous protozoans in the subclass Lobosia characterized by lobopodia and a well-defined aperture in the test.

arc excitation [ATOM PHYS] Use of electric-arc energy to move electrons into higher energy orbits.

arc furnace [MET] A furnace used to heat materials by the energy from an electric arc. Also known as electric-arc furnace.

arch [CIV ENG] A structure curved and so designed that when it is subjected to vertical loads, its two end supports exert reaction forces with inwardly directed horizontal components; common uses for the arch are as a bridge, support for a roadway or railroad track, or part of a building.

Archaeoceti [PALEON] The zeuglodonts, a suborder of aquatic Eocene mammals in the order Cetacea; the oldest known cetaceans.

Archaeocidaridae [PALEON] A family of Carboniferous echinoderms in the order Cidaroida characterized by a flexible test and more than two columns of interambulacral plates.

Archaeocopida [PALEON] An order of Cambrian crustaceans in the subclass Ostracoda characterized by only slight calcification of the carapace.

Archaeogastropoda [INV ZOO] An order of gastropod mollusks that includes the most primitive snails.

Archaeopteridales [PALEOBOT] An order of Upper Devonian sporebearing plants in the class Polypodiopsida characterized by woody trunks and simple leaves.

Archaeopteris [PALEOBOT] A genus of fossil plants in the order Archaeopteridales; used sometimes as an index fossil of the Upper Devonian.

Archaeopterygiformes [PALEON] The single order of the extinct avian subclass Archaeornithes.

Archaeopteryx [PALEON] The earliest known bird; a genus of fossil birds in the order Archaeopterygiformes characterized by flight feathers like those of modern birds.

Archaeornithes [PALEON] A subclass of Upper Jurassic birds comprising the oldest fossil birds.

archaic [PSYCH] Designating elements, largely unconscious, in the psyche which are remnants of man's prehistoric past and which reappear in dreams and other symbolic manifestations.

archallaxis [BIOL] Deviation from an ancestral pattern early in development, eliminating duplication of the phylogenetic history.

Archangiaceae [MICROBIO] A family of slime bacteria in the order Myxobacterales, characterized by fruiting bodies and nonspherical microcysts surrounded by a wall or found in cysts.

Archanthropinae [PALEON] A subfamily of the Hominidae, set up by F. Weidenreich, which is no longer used.

arch beam [CIV ENG] A curved beam, used in construction, with a longitudinal section bounded by two arcs having different radii and centers of curvature so that the beam cross section is larger at either end than at the center.

arch blocks [MIN ENG] Blocks applied to the wooden voussoirs used in framing a timber support for the roof when driving a tunnel.

arch brick [MATER] 1. A wedge-shaped brick used in arches. 2. An overburned brick, resulting from contact with the fire in the arch of a kiln.

arch bridge [CIV ENG] A bridge having arches as the main supports.

arch center [CIV ENG] A temporary structure for support of the parts of a masonry or concrete arch during its construction.

arch dam [CIV ENG] A dam having a curved face on the downstream side, the curve being roughly a portion of a cylinder whose axis is vertical.

Archean [GEOL] A term, meaning ancient, which has been applied to the oldest rocks of the Precambrian; as more physical measurements of geologic time are made, the usage is changing; the term Early Precambrian is preferred.

arc heating [MET] The heating of a material by the heat energy from an electric arc, which has a very high temperature and very high concentration of heat energy. Also known as electric-arc heating.

archegoniophore [BOT] The stalk supporting the archegonium in liverworts and ferns.

archegonium [BOT] The multicellular female sex organ in all plants of the Embryobionta except the Pinophyta and Magnoliophyta.

archen *See* emodin.

archencephalon [EMBRYO] The primitive embryonic forebrain from which the forebrain and midbrain develop.

archenteron [EMBRYO] The cavity of the gastrula formed by ingrowth of cells in vertebrate embryos. Also known as gastrocoele; primordial gut.

archeocyte [INV ZOO] A type of ovoid amebocyte in sponges, characterized by large nucleolate nuclei and blunt pseudopodia; gives rise to germ cells.

archeological chemistry [ARCHEO] The application of chemical theories and experimental procedures, especially of analytical chemistry, to the solution of problems in archeology.

archeology [SCI TECH] The scientific study of the material remains of the cultures of historical or prehistorical peoples.

Archeozoic [GEOL] **1.** The era during which, or during the latter part of which, the oldest system of rocks was made. **2.** The last of three subdivisions of Archean time, when the lowest forms of life probably existed; as more physical measurements of geologic time are made, the usage is changing; it is now considered part of the Early Precambrian.

Archer *See* Sagittarius.

archerfish [VERT ZOO] The common name for any member of the fresh-water family Toxotidae in the order Perciformes; individuals eject a stream of water from the mouth to capture insects.

Archeria [PALEONTOL] Genus of amphibians, order Embolomeri, in early Permian in Texas; fish eaters.

archespore [BOT] A cell from which the spore mother cell develops in either the pollen sac or the ovule of an angiosperm.

archetype [EVOL] The original forerunner of a group of animals or plants.

arch girder [CIV ENG] A normal H-section steel girder bent to a circular shape.

arch-gravity dam [CIV ENG] An arch dam stabilized by gravity due to great mass and breadth of the base.

Archiacanthocephala [INV ZOO] An order of worms in the phylum Acanthocephala; adults are endoparasites of terrestrial vertebrates.

Archiannelida [INV ZOO] A group name applied to three families of unrelated annelid worms: Nerillidae, Protodrilidae, and Dinophilidae.

archibenthic zone [OCEANOGR] The biogeographic realm of the ocean extending from a depth of about 200 meters to 800–1100 meters (665 feet to 2625–3610 feet).

archibole *See* positive element.

Archichlamydeae [BOT] An artificial group of flowering plants, in the Englerian system of classification, consisting of those families of dicotyledons that lack petals or have petals separate from each other.

archicoel [ZOO] The segmentation cavity persisting between the ectoderm and endoderm as a body cavity in certain lower forms.

archigastrula [EMBRYO] A gastrula formed by invagination, as opposed to ingrowth of cells.

Archigregarinida [INV ZOO] An order of telosporean protozoans in the subclass Gregarinia; endoparasites of invertebrates and lower chordates.

Archimedean principle [PHYS] The principle that a body immersed in a fluid undergoes an apparent loss in weight equal to the weight of the fluid it displaces.

Archimedean solid [MATH] One of 13 possible solids whose faces are all regular polygons, though not necessarily all of the same type, and whose polyhedral angles are all equal.

Archimedes number [FL MECH] One of a dimensionless group of numbers denoting the ratio of gravitational force to viscous force.

Archimedes' problem [MATH] The problem of dividing a hemisphere into two parts of equal volume with a plane parallel to the base of the hemisphere; it cannot be solved by Euclidean methods.

Archimedes' screw [MECH ENG] A device for raising water by means of a rotating broad-threaded screw or spirally bent tube within an inclined hollow cylinder.

Archimedes' spiral *See* spiral of Archimedes.

archinephridium [INV ZOO] One of a pair of primitive nephridia found in each segment of some annelid larvae.

archinephros [VERT ZOO] The paired excretory organ of primitive vertebrates and the larvae of hagfishes and caecilians.

arching [GEOL] The folding of schists, gneisses, or sediments into anticlines. [MIN ENG] Curved support for roofs of openings in mines.

archipallium [PHYSIO] The olfactory pallium or the olfactory cerebral cortex; phylogenetically, the oldest part of the cerebral cortex.

archipelago [GEOGR] **1.** A large group of islands. **2.** A sea that has a large group of islands within it.

architect [ARCH] A person who is skilled and knowledgeable in the design of buildings and whose qualifications are recognized by a college degree and licensing by an appropriate professional organization.

architectonic [ARCH] **1.** Pertaining to or in accordance with the principles of architecture. **2.** Having the qualities of architecture in terms of structure and concept. [GEOL] Of forces that determine structure.

architect's scale [GRAPHICS] A rule with a scale on it so chosen that by placing the rule's edge on a reduced-scale drawing the scale of the drawing (say, in inches) may be converted directly into the dimensions of the object (say, in feet).

architectural acoustics [CIV ENG] The science of planning and building a structure to ensure the most advantageous flow of sound to all listeners.

architectural concrete [MATER] Concrete used for ornamentation or finish on exterior or interior surfaces of a building or other structure.

architectural engineering [CIV ENG] The branch of engineering dealing primarily with building materials and components and with the design of structural systems for buildings, in contrast to heavy construction such as bridges.

architecture [ENG] **1.** The art and science of designing buildings. **2.** The product of this art and science.

architrave [ARCH] **1.** The lowest division of an entablature that rests on the column capital. **2.** The molded band, group of moldings, or other architectural member around an opening, such as a door, especially if rectangular.

archivolt [ARCH] **1.** The band or molding surrounding an arch. **2.** The architectural members of the inner surface of an arch.

Archosauria [VERT ZOO] A subclass of reptiles composed of five orders: Thecodontia, Saurischia, Ornithschia, Pterosauria, and Crocodilia.

Archostemata [INV ZOO] A suborder of insects in the order Coleoptera.

arch pattern [ANAT] A fingerprint pattern in which ridges enter on one side of the impression, form a wave or angular upthrust, and flow out the other side.

arch press [MECH ENG] A punch press having an arch-shaped frame to permit operations on wide work.

arch rib [CIV ENG] One of a set of projecting molded members subdividing the undersurface of an arch.

arch ring [CIV ENG] A curved member that provides the main support of an arched structure.

arch truss [CIV ENG] A truss having the form of an arch or arches.

archway [ARCH] A way or passage over which an arch extends.

arc hyperbolic function [MATH] An inverse function of one of the hyperbolic functions.

arcing contacts [ELEC] Special contacts on which the arc is drawn after the main contacts of a switch or circuit breaker have opened.

ARCHERFISH

The archerfish *(Toxotes jaculator)*, maximum length 7 inches (18 centimeters).

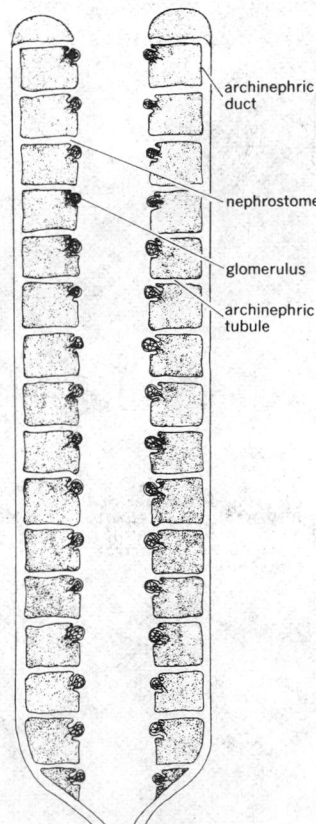

ARCHINEPHROS

archinephric duct

nephrostome

glomerulus

archinephric tubule

Diagram showing hypothetical structure of archinephros. *(From C. K. Weichert, Elements of Chordate Anatomy, 3d ed., McGraw-Hill, 1967)*

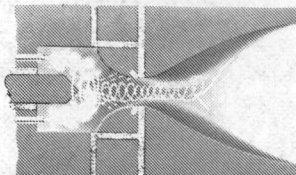

ARC JET ENGINE

Diagram of an arc jet engine.

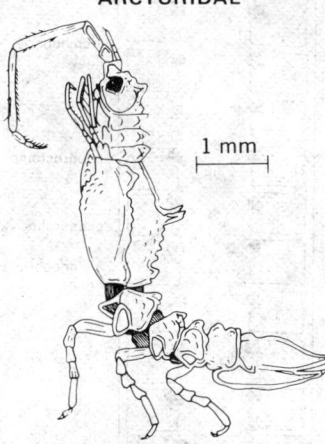

ARCTURIDAE

1 mm

Arcturella, a member of the family Arcturidae. (*From G. O. Sars, An Account of the Crustacea of Norway, vol. 2, 1899*)

arcing time [ELEC] **1.** Interval between the parting, in a switch or circuit breaker, of the arcing contacts and the extension of the arc. **2.** Time elapsing, in a fuse, from the severance of the fuse link to the final interruption of the circuit under a specified condition.

arc jet engine [AERO ENG] An electromagnetic propulsion engine used to supply motive power for flight; hydrogen and ammonia are used as the propellant, and some plasma is formed as the result of electric-arc heating.

arc lamp [ELEC] An electric lamp in which the light is produced by an arc made when current flows through ionized gas between two electrodes. Also known as electric-arc lamp.

Ar$_{cm}$ [MET] The temperature at which austenite is converted to cementite upon cooling a hypereutectoid steel.

arc measurement [GEOD] A survey method used to determine the size of the earth.

arc melting [MET] Melting and purification of metal in an electric-arc furnace.

arcmin *See* minute.

arc navigation [NAV] A navigation system in which the position of an airplane or ship is maintained along an arc of a circle which has a radius measured from a control station by means of electronic distance-measuring equipment, such as shoran or oboe.

arcocentrum [ANAT] A centrum formed of modified, fused mesial parts of the neural or hemal arches.

arc of action *See* arc of contact.

arc of approach [DES ENG] In toothed gearing, the part of the arc of contact along which the flank of the driving wheel contacts the face of the driven wheel.

arc of contact [MECH ENG] **1.** The angular distance over which a gear tooth travels while it is in contact with its mating tooth. Also known as arc of action. **2.** The angular distance a pulley travels while in contact with a belt or rope.

arc of parallel [GEOD] A part of an astronomic or a geodetic parallel.

arc of recess [DES ENG] In toothed gearing, the part of the arc of contact wherein the face of the driving wheel touches the flank of the driven wheel.

arc of visibility [NAV] The arc of a light sector, designated by its limiting bearings as observed from seaward.

arc-over [ELEC] An unwanted arc resulting from the opening of a switch or the breakdown of insulation.

arc process [CHEM ENG] A former process that used electric arcs for fixation (oxidation) of atmospheric nitrogen to manufacture nitric acid.

arc-seam weld [MET] A linear weld or overlapping spot welds made by an arc welding process.

arc secant of a number [MATH] **1.** Any angle whose secant is that number. **2.** The angle between 0 radians and π radians whose secant is that number; it is the value of the inverse of the restriction of the secant function to the interval between 0 and π, at this number.

arc sine of a number [MATH] **1.** Any angle whose sine is that number. **2.** The angle between $-\pi/2$ radians and $\pi/2$ radians whose sine is that number; it is the value of the inverse of the restriction of the sine function to the interval between $-\pi/2$ and $\pi/2$, at this number.

arc spectrum [SPECT] The spectrum of a neutral atom, as opposed to that of a molecule or an ion; it is usually produced by vaporizing the substance in an electric arc; designated by the roman numeral I following the symbol for the element, for example, HeI.

arc-spot weld [MET] A weld that covers a very small area of the surface in contact and was made by an arc welding process.

arc spraying [MET] Depositing on a surface a metal melted by an electric arc and blown at high speed in an atomized state.

arc tangent of a number [MATH] **1.** Any angle whose tangent is that number. **2.** The angle between $-\pi/2$ radians and $\pi/2$ radians whose tangent is that number; it is the value of the inverse of the restriction of the tangent function to the interval between $-\pi/2$ and $\pi/2$, at this number.

arc-through [ELECTR] Of a gas tube, a loss of control resulting in the flow of a principal electron stream in the normal direction during a scheduled nonconducting period.

arctic air [METEOROL] An air mass whose characteristics are developed mostly in winter over arctic surfaces of ice and snow.

arctic-alpine [ECOL] Of or pertaining to areas above the timberline in mountainous regions.

arctic anticyclone *See* arctic high.

Arctic Circle [GEOD] The parallel of latitude 66°32′N (often taken as 66½°N).

arctic climate *See* polar climate.

arctic front [METEOROL] The semipermanent, semicontinuous front between the deep, cold arctic air and the shallower, basically less cold polar air of northern latitudes.

arctic haze [METEOROL] A condition of reduced horizontal and slant visibility (but unimpeded vertical visibility) encountered by aircraft in flight (up to more than 30,000 feet, or 9140 meters) over arctic regions.

arctic high [METEOROL] A weak high that appears on mean charts of sea-level pressure over the Arctic Basin during late spring, summer, and early autumn. Also known as arctic anticyclone; polar anticyclone; polar high.

arcticization [ENG] The preparation of equipment for operation in an environment of extremely low temperatures.

arctic mist [METEOROL] A mist of ice crystals; a very light ice fog.

Arctic Ocean [GEOGR] The north polar ocean lying between North America, Greenland, and Asia.

arctic sea smoke [METEOROL] Steam fog; but often specifically applied to steam fog rising from small areas of open water within sea ice.

Arctic suite [PETR] A group of basic igneous rocks intermediate in composition between Atlantic and Pacific suites.

arctic tree line [ECOL] The northern limit of tree growth; the sinuous boundary between tundra and boreal forest.

Arctic Zone [GEOGR] The area north of the Arctic Circle (66°32′N).

Arctiidae [INV ZOO] The tiger moths, a family of lepidopteran insects in the suborder Heteroneura.

arc time [MET] The time interval during which an arc is maintained in the making of an arc weld.

arc-to-chord correction *See* conversion angle.

Arctocyonidae [PALEON] A family of extinct carnivorelike mammals in the order Condylarthra.

Arctolepiformes [PALEON] A group of the extinct joint-necked fishes belonging to the Arthrodira.

arc triangulation [ENG] A system of triangulation in which an arc of a great circle on the surface of the earth is followed in order to tie in two distant points.

Arcturidae [INV ZOO] A family of isopod crustaceans in the suborder Valvifera characterized by an almost cylindrical body and extremely long antennae.

Arcturus [ASTRON] A star that is 36 light-years from the sun; spectral classification K2IIIp. Also known as α Boötes.

arcuale [EMBRYO] Any of the four pairs of primitive cartilages from which a vertebra is formed.

arcuate [ANAT] Arched or curved; bow-shaped.

arcuate delta [GEOL] A bowed or curved delta with the convex margin facing the body of water. Also known as fan-shaped delta.

arcuation [GEOL] Production of an arc, as in rock flowage where movement proceeded in a fanlike manner.

arcus [METEOROL] A dense and horizontal roll-shaped accessory cloud, with more or less tattered edges, situated on the lower front part of the main cloud.

arcwall coal cutter [MIN ENG] A type of electric or compressed-air cutter for under- or overcutting a coal seam in narrow work.

arc-welder's disease *See* siderosis.

arc welding *See* electric-arc welding.

Arcyzonidae [PALEON] A family of Devonian paleocopan ostracods in the superfamily Kirkbyacea characterized by valves with a large central pit.

ARDC model atmosphere *See* standard atmosphere.

ardealite [MINERAL] $Ca_2(HPO_4)(SO_4)\cdot 4H_2O$ A white or light-yellow mineral consisting of a hydrous acid calcium phosphate-sulfate.

Ardeidae [VERT ZOO] The herons, a family of wading birds in the order Ciconiiformes.

Ardennian orogeny [GEOL] A short-lived orogeny during the Ludlovian stage of the Silurian period of geologic time.

ardennite [MINERAL] $Mn_5Al_5(VO_4)(SiO_4)_5(OH)_2 \cdot 2H_2O$ A yellow to yellowish-brown mineral consisting of a hydrous silicate vanadate and arsenate of manganese and aluminum.

arduinite *See* mordenite.

are [MECH] A unit of area, used mainly in agriculture, equal to 100 square meters.

area [MATH] A measure of the size of a two-dimensional surface, or of a region on such a surface.

area amniotica [EMBRYO] The transparent part of the blastodisc in mammals.

area bombing [ORD] Dropping of bombs on a general locality rather than on a specific target or in a pattern.

area control [NAV] A form of air-traffic control in which aircraft flying in designated areas are furnished flight advisory service.

area coverage [ENG] Complete coverage of an area by aerial photography having parallel overlapping flight lines and stereoscopic overlap between exposures in the line of flight.

area defense [ORD] A defense against air attack organized to protect an area, as distinguished from a point defense or line defense.

area delimiting line [MAP] A line fixing the boundary of an area.

area fire [ORD] Fire delivered on a prescribed area.

area forecast [METEOROL] A weather forecast for a specified geographic area; usually applied to a form of aviation weather forecast. Also known as regional forecast.

area landfill [CIV ENG] A sanitary landfill operation that takes care of the solid waste of more than one municipality in a region.

areal eruption [GEOL] Volcanic eruption resulting from collapse of the roof of a batholith; the volcanic rocks grade into parent plutonic rocks.

areal geology [GEOL] Distribution and form of rocks or geologic units of any relatively large area of the earth's surface.

areal pattern [PETRO ENG] Distribution pattern of oil-production wells and water- or gas-injection wells over a given oil reservoir.

areal sweep efficiency [PETRO ENG] Percentage of the total oil reservoir or pore volume which is within the area being swept of oil by a displacing fluid, as in a natural or artificial gas drive or water injection.

areal velocity [ASTROPHYS] In celestial mechanics, the area swept out by the radius vector per unit time.

area meter [ENG] A mechanism to measure fluid flow rate through a fixed-area conduit by the movement of a weighted piston or float supported by the flowing fluid; includes rotameters and piston-type meters.

area of error [NAV] An area (elliptical, circular, parallelogram-shaped, and so on) containing the true position, within which, for a stated level of probability, the true position is considered to lie.

area opaca [EMBRYO] The opaque peripheral area of the blastoderm of birds and reptiles, continuous with the yolk.

area pellucida [EMBRYO] The central transparent area of the blastoderm of birds and reptiles, overlying the subgerminal cavity.

area placentalis [EMBRYO] The part of the trophoblast in immediate contact with the uterine mucosa in the embryos of early placental vertebrates.

area rule [AERO ENG] A prescribed method of design for obtaining minimum zero-lift drag for a given aerodynamic configuration, such as a wing-body configuration, at a given speed.

area survey [ENG] A survey of areas large enough to require loops of control.

area target [ORD] A target attacked by area bombing, consisting of an area such as an entire munitions factory, rather than a single building or similar point target.

area triangulation [ENG] A system of triangulation designed to progress in every direction from a control point.

area vitellina [EMBRYO] The outer nonvascular zone of the area opaca; consists of ectoderm and endoderm.

areaway [CIV ENG] An open space at subsurface level adjacent to a building, providing access to and utilities for a basement.

Arecaceae [BOT] The palms, the single family of the order Arecales.

arecaidine methyl ester *See* arecoline.

Arecales [BOT] An order of flowering plants in the subclass Arecidae composed of the palms.

Arecidae [BOT] A subclass of flowering plants in the class Liliopsida characterized by numerous, small flowers subtended by a prominent spathe and often aggregated into a spadix, and broad, petiolate leaves without typical parallel venation.

arecoline [ORG CHEM] $C_8H_{13}O_2N$ An alkaloid from the betel nut; an oily, colorless liquid with a boiling point of 209°C; soluble in water, ethanol, and ether; combustible; used as a medicine. Also known as arecaidine methyl ester; methyl-1,2,5,6-tetrahydro-1-methylnicotinate.

areg [ECOL] A sand desert.

A register *See* arithmetic register.

arenaceous [GEOL] Of sediment or sedimentary rocks that have been derived from sand or that contain sand. Also known as psammitic; sabulous.

arendalite [MINERAL] A dark-green variety of epidote found in Arendal, Norway.

arene [ORG CHEM] A hydrocarbon containing at least one aromatic ring.

Arenicolidae [INV ZOO] The lugworms, a family of mud-swallowing worms belonging to the Sedentaria.

arenicolite [GEOL] A hole, groove, or other mark in a sedimentary rock, generally sandstone, interpreted as a burrow made by an arenicolous marine worm or a trail of a mollusk or crustacean.

arenicolous [ZOO] Living or burrowing in sand.

Arenigian [GEOL] A European stage including Lower Ordovician geologic time (above Tremadocian, below Llanvirnian). Also known as Skiddavian.

arenite [PETR] Consolidated sand-texture sedimentary rock of any composition. Also known as arenyte; psammite.

arenyte *See* arenite.

areocentric [ASTRON] With Mars as a center.

areodesy [ASTRON] Determination, by observation and measurement, of the exact positions of points on, and the figures and areas of large portions of, the surface of the planet Mars, or the shape and size of the planet Mars.

areographic [ASTRON] Referring to positions on Mars measured in latitude from the planet's equator and in longitude from a reference meridian.

areography [ASTRON] The study of the surface features of Mars, or its geography. [ECOL] Descriptive biogeography.

areola [ANAT] **1.** The portion of the iris bordering the pupil of the eye. **2.** A pigmented ring surrounding a nipple, vesicle, or pustule. **3.** A small space, interval, or pore in a tissue.

areola mammae [ANAT] The circular pigmented area surrounding the nipple of the breast. Also known as areola papillaris; mammary areola.

areola papillaris *See* areola mammae.

areology [ASTRON] The scientific study related to the properties of Mars.

Ares [ASTRON] The planet Mars.

arête [GEOL] Narrow, jagged ridge produced by the merging of glacial cirques. Also known as arris; crib; serrate ridge.

ARFOR [METEOROL] A code word used internationally to indicate an area forecast; usually applied to an aviation weather forecast.

ARFOT [METEOROL] A code word used internationally to indicate an area forecast with units in the English system; usually applied to an aviation weather forecast.

arfvedsonite [MINERAL] A black monoclinic amphibole, containing sodium and silicon trioxide with occluded water and some calcium. Also known as soda hornblende.

Arg *See* Argo.

Argand diagram [MATH] A two-dimensional cartesian coordinate system for representing the complex numbers, the number $x + iy$ being represented by the point whose coordinates are x and y.

Argand lamp [ENG] A gas lamp having a tube-shaped wick, allowing a current of air inside as well as outside the flame.

Argasidae [INV ZOO] The soft ticks, a family of arachnids in the suborder Ixodides; several species are important as ectoparasites and disease vectors for man and domestic animals.

ARGASIDAE

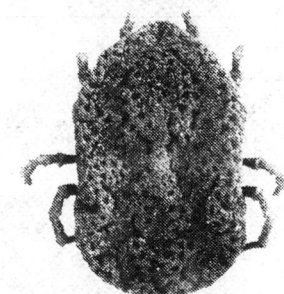

Argasid tick, *Ornithodoros coriaceus*, enlarged to about 4 times natural size.

ARGENTITE

|— 2.5 cm —|

Argentite with quartz crystals from Sarrabus, Sardinia. (*Specimen from Department of Geology, Bryn Maw College*)

ARGON

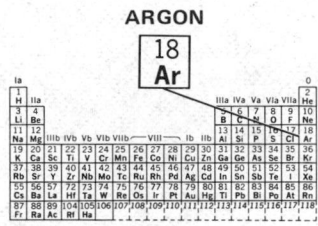

Periodic table of the chemical elements showing the position of argon.

ARIES

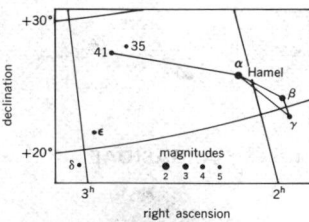

Line pattern of constellation Aries. Grid lines represent the coordinates of the sky. Apparent brightness, or magnitude, of stars is shown by sizes of the dots, which are graded by appropriate numbers as indicated.

Argelander method [ASTRON] A technique to estimate the brightness of variable stars; it involves estimating the difference in magnitude between the variable stars as compared to one or more stars that are invariable.

argentaffin cell [HISTOL] Any of the cells of the gastrointestinal tract that are thought to secrete serotonin.

argentaffin fiber *See* reticular fiber.

argentic [CHEM] Relating to or containing silver.

argentic oxide *See* silver suboxide.

Argentinoidei [VERT ZOO] A family of marine deepwater teleostean fishes, including deep-sea smelts, in the order Salmoniformes.

argentite [MINERAL] Ag_2S A lustrous, lead-gray ore of silver; it is a monoclinic mineral and is dimorphous with acanthite. Also known as argyrite; silver glance; vitreous silver.

argentocyanides [INORG CHEM] Complexes formed, for example, in the cyanidation of silver ores and in electroplating, when silver cyanide reacts with solutions of soluble metal cyanides. Also known as dicyanoargentates.

argentojarosite [MINERAL] $AgFe_3(SO_4)_2(OH)_6$ A yellow or brownish mineral consisting of basic silver ferric sulfate.

argentometer [ENG] A hydrometer used to find the amount of silver salt in a solution.

argentometry [ANALY CHEM] A volumetric analysis that employs precipitation of insoluble silver salts; the salts may be chromates or chlorides.

argentophil [BIOL] Of cells, tissues, or other structures, having an affinity for silver.

argentum [CHEM] Latin for silver.

Argidae [INV ZOO] A small family of hymenopteran insects in the superfamily Tenthredinoidea.

argillaceous [GEOL] Of rocks or sediments made of or largely composed of clay-size particles or clay minerals.

argillation [GEOL] Development of clay minerals by weathering of aluminum silicates.

argillic alteration [GEOL] A rock alteration in which certain minerals are converted to minerals of the clay group.

argilliferous [GEOL] Abounding in or producing clay.

argillite [PETR] A compact rock formed from siltstone, shale, or claystone but intermediate in degree of induration and structure between them and slate; argillite is more indurated than mudstone but lacks the fissility of shale.

arginase [BIOCHEM] An enzyme that catalyzes the splitting of urea from the amino acid arginine.

arginine [BIOCHEM] $C_6H_{14}N_4O_2$ A colorless, crystalline, water-soluble, essential amino acid of the α-ketoglutaric acid family.

Argo [ASTRON] The large Ptolemy constellation; a southern constellation, now divided into four groups (Carina, Pupis, Vela, and Pyxis Nautica). Abbreviated Arg. Also known as Ship.

argol [FOOD ENG] A deposit formed in casks during aging of wine. [MATER] Any of several manures used as fuel in parts of Asia.

argon [CHEM] A chemical element, symbol Ar, atomic number 18, atomic weight 39.998.

argon ionization detector [NUCLEO] An ionization chamber that is filled with argon gas.

argon laser [OPTICS] A gas laser using ionized argon; emits a 4880-angstrom line as well as infrared radiation.

Argovian [GEOL] Upper Jurassic (lower Lusitanian), a substage of geologic time in Great Britain.

Arguloida [INV ZOO] A group of crustaceans known as the fish lice; taxonomic status is uncertain.

argument [ASTRON] An angle or arc, as in argument of perigee. [MATH] *See* amplitude; independent variable.

argument of latitude [ASTRON] The angular distance measured in the orbit plane from the ascending node to the orbiting object; the sum of the argument of perigee and the true anomaly.

argument of perigee [ASTRON] The angle or arc, as seen from a focus of an elliptical orbit, from the ascending node to the closest approach of the orbiting body to the focus; the angle is measured in the orbital plane in the direction of motion of the orbiting body.

ε Argus [ASTRON] A star of visual magnitude 1.74, spectral type K0.

argyria [MED] A dusky-gray or bluish discoloration of the skin and mucous membranes produced by the prolonged administration or application of silver preparations.

argyrodite [MINERAL] Ag_8GeS_6 A steel-gray mineral, one of two germanium minerals and a source for germanium; crystallizes in the isometric system and is isomorphous with canfieldite.

argyrophil lattice fiber *See* reticular fiber.

Arhynchobdellae [INV ZOO] An order of annelids in the class Hirudinea characterized by the lack of an eversible proboscis; includes most of the important leech parasites of man and warm-blooded animals.

Arhynchodina [INV ZOO] A suborder of ciliophoran protozoans in the order Thigmotrichida.

arhythmicity [BIOL] A condition characterized by the absence of an expected behavioral or physiologic rhythm.

Ari *See* Aries.

ariboflavinosis [MED] Dietary deficiency of riboflavin, associated with the syndrome of angular cheilosis and stomatitis, corneal vascularity, nasolabial seborrhea, and genitorectal dermatitis.

arid biogeographic zone [ECOL] Any region of the world that supports relatively little vegetation due to lack of water.

arid climate [CLIMATOL] Any extremely dry climate.

arid erosion [GEOL] Erosion or wearing away of rock that occurs in arid regions, due largely to the wind.

Aridisol [GEOL] A soil suborder and a major soil of deserts characterized by an ochric epipedon and other pedogenic horizons.

aridity [CLIMATOL] The degree to which a climate lacks effective, life-promoting moisture.

aridity coefficient [CLIMATOL] A function of precipitation and temperature designed by W. Gorczynski to represent the relative lack of effective moisture (the aridity) of a place.

aridity index [CLIMATOL] An index of the degree of water deficiency below water need at any given station; a measure of aridity.

Arid Transition life zone [ECOL] The zone of climate and biotic communities occurring in the chaparrals and steppes from the Rocky Mountain forest margin to California.

arid zone *See* equatorial dry zone.

ariegite [PETR] A group of pyroxenites composed principally of clinopyroxene, orthopyroxene, and spinel.

Ariel [AERO ENG] A series of artificial satellites launched for Britain by the United States. [ASTRON] **1.** A satellite of the planet Uranus orbiting at a mean distance of 192,000 kilometers (119,000 miles). **2.** A constellation in the northern celestial hemisphere, seen in proximity to Taurus and Pisces.

Aries [ASTRON] A constellation with a right ascension of 3 hours and declination of 20°N. Abbreviated Ari. Also known as Ram.

arietiform [VERT ZOO] Shaped like a ram's horns; specifically, describing the dark facial marking that extends across the nose of kangaroo rats.

Ariidae [VERT ZOO] A family of tropical salt-water catfishes in the order Siluriformes.

Arikareean [GEOL] Lower Miocene geologic time.

aril [BOT] An outgrowth of the funiculus in certain seeds that either remains as an appendage or envelops the seed.

arilode [BOT] An aril originating from tissues in the micropyle region; a false aril.

Arionidae [INV ZOO] A family of mollusks in the order Stylommatophora, including some of the pulmonate slugs.

arista [INV ZOO] The bristlelike or hairlike structure in many organisms, especially at or near the tip of the antenna of many Diptera.

Aristarchus [ASTRON] A crater on the moon.

Aristolochiaceae [BOT] The single family of the plant order Aristolochiales.

Aristolochiales [BOT] An order of dicotyledonous plants in the subclass Magnoliidae; species are herbaceous to woody, often climbing, with perigynous to epigynous, apetalous flowers, uniaperturate or nonaperturate pollen, and seeds with a small embryo and copious endosperm.

aristopedia [INV ZOO] Replacement of the arista by a nearly perfect leg.

Aristotle's lantern [INV ZOO] A five-sided feeding and locomotor apparatus surrounding the esophagus of most sea urchins.

arithlog paper [MATH] Graph paper marked with a semilogarithmic coordinate system.

arithmetic [MATH] Addition, subtraction, multiplication, and division, usually of integers, rational numbers, real numbers, or complex numbers.

arithmetical addition [MATH] The addition of positive numbers or of the absolute values of signed numbers.

arithmetical element *See* arithmetical unit.

arithmetical operation [ADP] A digital computer operation in which numerical quantities are added, subtracted, multiplied, divided, or compared.

arithmetical unit [ADP] The section of the computer which carries out all arithmetic and logic operations. Also known as arithmetical element; arithmetic-logic unit (ALU); arithmetic section; logic-arithmetic unit; logic section.

arithmetic circuitry [ADP] The section of the computer circuitry which carries out the arithmetic operations.

arithmetic-logic unit *See* arithmetical unit.

arithmetic mean [MATH] The average of a collection of numbers obtained by dividing the sum of the numbers by the quantity of numbers. Also known as average (av).

arithmetic progression [MATH] A sequence of numbers for which there is a constant d such that the difference between any two successive terms is equal to d.

arithmetic register [ADP] A specific memory location reserved for intermediate results of arithmetic operations. Also known as A register.

arithmetic section *See* arithmetical unit.

arithmetic shift [ADP] A shift of the digits of a number, expressed in a positional notation system, in the register without changing the sign of the number.

arithmetic sum [MATH] **1.** The result of the addition of two or more positive quantities. **2.** The result of the addition of the absolute values of two or more quantities.

arithmetization [MATH] The study of various branches of higher mathematics by methods that make use of only the basic concepts and operations of arithmetic.

Arizona ruby [MINERAL] A ruby-red pyrope garnet of igneous origin found in the southwestern United States.

arizonite [MINERAL] $Fe_2Ti_3O_9$ A steel-gray mineral containing iron and titanium and found in irregular masses in pegmatite. [PETR] A dike rock composed of mostly quartz, some orthoclase, and accessory mica and apatite.

Arkansas stone [ENG] A whetstone made of Arkansas stone, for sharpening edged tools. [PETR] A variety of novaculite quarried in Arkansas.

arkite [PETR] A feldspathoid-rich rock consisting largely of pseudoleucite and nepheline, subordinate melanite and pyroxene, and accessory orthoclase, apatite, and sphene.

arkose [PETR] A sedimentary rock composed of sand-size fragments that contain a high proportion of feldspar in addition to quartz and other detrital minerals. Also known as feldspathic sandstone.

arkose quartzite *See* arkosite.

arkosic [PETR] Having wholly or partly the character of arkose.

arkosic bentonite [PETR] Bentonite derived from volcanic ash which contains 25-75% sandy impurities and whose detrital crystalline grains remain essentially unaltered. Also known as sandy bentonite.

arkosic limestone [PETR] An impure clastic limestone composed of a relatively high proportion of grains or crystals of feldspar.

arkosic sandstone [PETR] A sandstone in which much feldspar is present, ranging from unassorted products of granular disintegration of granite to partly sorted river-laid or even marine deposits.

arkosic wacke [PETR] Graywacke containing more feldspar than rock fragments. Also known as feldspathic graywacke.

arkosite [PETR] A quartzite with a high proportion of feldspar. Also known as arkose quartzite.

ARL *See* acceptable reliability level.

arm [ANAT] The upper or superior limb in man, consisting of the upper arm with one bone and the forearm with two bones. [ELEC] *See* branch. [ENG ACOUS] *See* tone arm. [MATH] A side of an angle. [NAV ARCH] The part of an anchor extend-

ing from the crown to one of the flukes. [ORD] **1.** A combat branch of a military force; specifically, a branch of the U.S. Army, such as the Infantry Armored Cavalry, the primary function of which is combat. **2.** *Often plural.* Weapons for use in war. **3.** To supply with arms. **4.** To ready ammunition for detonation, as by removal of safety devices or alignment of the explosive elements in the explosive train of the fuse. [PHYS] The perpendicular distance from the line along which a force is applied to a reference point.

armadillo [VERT ZOO] Any of 21 species of edentate mammals in the family Dasypodidae.

armament [ORD] **1.** The weapons of an airplane, tank, ship, or the like or of a unit or organized force. **2.** *Often plural.* War equipment, weapons, and supplies.

armament error [ORD] **1.** The dispersion of shots from a particular gun. **2.** The deviation of any shot from the center of impact of a series of shots from a gun after all errors of personnel and adjustment have been accounted for.

armangite [MINERAL] $Mn_3(AsO_3)_2$ A black mineral crystallizing in the rhombohedral system and consisting of manganese arsenite.

armature [ELECTROMAG] **1.** That part of an electric rotating machine that includes the main current-carrying winding in which the electromotive force produced by magnetic flux rotation is induced; it may be rotating or stationary. **2.** The movable part of an electromagnetic device, such as the movable iron part of a relay, or the spring-mounted iron part of a vibrator or buzzer.

armature chatter [ELECTROMAG] Vibration of the armature of a relay caused by pulsating coil current or by marginally low coil current.

armature contact *See* movable contact.

armature reactance [ELECTROMAG] The inductive reactance due to the flux produced by the armature current and enclosed by the conductors in the armature slots and the end connections.

armature reaction [ELECTROMAG] Interaction between the magnetic flux produced by armature current and that of the main magnetic field in an electric motor or generator.

armature resistance [ELEC] The ohmic resistance in the main current-carrying windings of an electric generator or motor.

Armco iron [MET] A relatively pure ingot iron which is made by the Armco process in a basic open-hearth furnace in a manner similar to that used for steel.

Armco process [MET] The basic open-hearth process used by Armco Steel Corporation to make nearly chemically pure iron for use as a construction material, in electromagnetic cores, and in special steels.

armed merchantman [NAV ARCH] An armed merchant ship of a neutral state; under international law, a vessel may be armed without prejudice to its status as a merchant ship.

arm elevator [MECH ENG] A chain elevator with protruding arms to cradle fixed-shape objects, such as drums or barrels, as they are moved upward.

armenite [MINERAL] $BaCa_2Al_6Si_8O_{28} \cdot 2H_2O$ Mineral composed of a hydrous calcium barium aluminosilicate.

ARMET [METEOROL] An international code word used to indicate an area forecast with units in the metric system.

Armillaria root rot [PL PATH] A fungus disease of forest and orchard trees initiated by invasion of the root system, then of the lower trunk, by *Armillaria mellea.* Also known as bark-splitting disease.

Armilliferidae [INV ZOO] A family of pentastomid arthropods belonging to the suborder Porocephaloidea.

arming [ORD] The changing of a fuse from a safe condition to a state of readiness for functioning.

arming device [ORD] Device for arming of a fuse under controlled conditions.

arming distance *See* arming range.

arming range [ORD] The distance from a weapon or launching point at which a fuse is expected to become armed. Also known as arming distance.

arming resistance [ORD] The resistance to the displacement of certain fuse components; must be overcome in order to arm a fuse.

arm length [ANTHRO] The length of the arm measured from

top of the clavicle to the tip of the middle finger, with the arm straight down at the side of the body.

armor [ELEC] Metal sheath enclosing a cable, primarily for mechanical protection. [ORD] **1.** Any physical protective covering, such as metal, used on vehicles or persons against projectiles or fragments. **2.** Armored units or forces. **3.** The component of a weapon system that gives protection to the vehicle or weapon on its way to the target.

armor castings [MET] A type of armor made of high-alloy steel frequently used when complicated shapes are involved.

armored artillery [ORD] Self-propelled artillery, with some armor, organic to the armored division or organized as separate battalions.

armored cable [ELEC] An electrical cable provided with a sheath of metal primarily for mechanical protection.

armored cavalry [ORD] A unit organized and equipped to perform missions requiring great mobility, firepower, and shock action.

armored infantry [ORD] A field army unit designed to close and destroy the enemy by fire and maneuver, to repel hostile assault in close combat, and to provide support for tanks.

armored mud ball [GEOL] A large (1–50 centimeters in diameter) subspherical mass of silt or clay coated with coarse sand and fine gravel. Also known as mud ball; pudding ball.

armored personnel carrier [ORD] An armored vehicle which provides protection from small-arms fire and shell fragments; used to transport personnel both on and off the battlefield.

armored vehicle [ORD] A wheeled or track-laying vehicle protected by armor plate that is used for combat security or cargo; for example, tanks, personnel carriers, armored cars, and self-propelled artillery.

armorer [ORD] One who services and repairs small arms, fills ammunition belts, and performs similar duties necessary to keep small arms ready for use.

Armorican orogeny [GEOL] Little-used term, now replaced by Hercynian or Variscan orogeny.

armor-piercing [ORD] Of ammunition, bombs, bullets, and projectiles, designed to penetrate armor and other resistant targets.

armor-piercing bomb [ORD] A missile, designed for dropping from aircraft, which is capable of penetrating the heaviest deck armor; also effective against reinforced-concrete structures; usually contains charge of explosive D, making up about 15% of the total weight.

armor-piercing bullet [ORD] A bullet having a hard metal core, a soft metal envelope, and a jacket; when the bullet strikes armor, the envelope and jacket are stopped, but the armor-piercing core continues forward and penetrates the armor.

armor-piercing incendiary [ORD] Armor-piercing projectile designed to set fires after piercing armor. Abbreviated API.

armor-piercing incendiary tracer [ORD] An armor-piercing projectile designed to set fires after piercing armor, and also fitted with a tracer for spotting.

armor-piercing sabot [ORD] A type of projectile which is armor-piercing and which incorporates a sabot. Abbreviated APS.

armor-piercing tracer [ORD] An armor-piercing projectile fitted with a tracer for spotting. Abbreviated APT.

armor plate [MET] Heavy, flat steel, either surface-hardened or hardened throughout, used as a sheathing for warships, tanks, and so forth to resist penetration and deformation from heavy gunfire.

Armour unit [BIOL] A unit for the standardization of adrenal cortical hormones and trypsin.

armrack [ORD] A frame with shelves, niches, hooks, or similar devices used to store small arms, to protect them, and to prevent unauthorized handling.

arms locker [ORD] A chest, cupboard, or the like used for the safekeeping of small arms.

Armstrong oscillator [ELECTR] Inductive feedback oscillator that consists of a tuned-grid circuit and an untuned-tickler coil in the plate circuit; control of feedback is accomplished by varying the coupling between the tickler and the grid circuit.

Armstrong's acid See naphthalene-1,5-disulfonic acid.

army [ORD] **1.** The land military forces of a nation. **2.** A unit of the U.S. Army made up of two or more army corps.

army artillery [ORD] Artillery assigned or attached to an army and retained under direct army command.

army depot [ORD] A depot located within the area of an army and designated by the army commander where supplies from the communications zone or from local sources are received, classified, stored, and distributed.

armyworm [INV ZOO] Any of the larvae of certain species of noctuid moths composing the family Phalaenidae; economically important pests of corn and other grasses.

Arndt-Eistert synthesis [ORG CHEM] A method of increasing the length of an aliphatic acid by one carbon by reacting diazomethane with acid chloride.

Arnel [TEXT] Trade name of a triacetate fiber manufactured by the Celanese Corporation.

Arneth's classification See Arneth's index.

Arneth's count See Arneth's index.

Arneth's formula See Arneth's index.

Arneth's index [HISTOL] A system for dividing peripheral blood granulocytes into five classes according to the number of nuclear lobes, the least mature cells being tabulated on the left, giving rise to the terms "shift to left" and "shift to right" as an indication of granulocytic immaturity or hypermaturity respectively. Also known as Arneth's classification; Arneth's count; Arneth's formula.

arnimite [MINERAL] $Cu_5(SO_4)_2(OH)_6 \cdot 3H_2O$ Mineral consisting of a hydrous copper sulfate.

Arnold sterilizer [MICROBIO] An apparatus that employs steam under pressure at 212°F (100°C) for fractional sterilization of specialized bacteriological culture media.

Arodoidea [INV ZOO] A superfamily of hemipteran insects belonging to the subdivision Geocorisae.

arolium [INV ZOO] A pad projecting between the tarsal claws of some insects and arachnids.

aromatic [ORG CHEM] Pertaining to or characterized by the presence of at least one benzene ring.

aromatic alcohol [ORG CHEM] Any of the compounds containing the hydroxyl group in a side chain to a benzene ring, such as benzyl alcohol.

aromatic aldehyde [ORG CHEM] An aromatic compound containing the CHO radical, such as benzaldehyde.

aromatic amine [ORG CHEM] An organic compound that contains one or more amino groups joined to an aromatic structure.

aromatic amino acid [BIOCHEM] An organic acid containing at least one amino group and one or more aromatic groups; for example, phenylalanine, one of the essential amino acids.

aromatic ketone [ORG CHEM] An aromatic compound containing the $-CO$ radical, such as acetophenone.

aromatic nucleus [ORG CHEM] The six-carbon ring characteristic of benzene and related series, or condensed six-carbon rings of naphthalene, anthracene, and so forth.

aromatic spirit of ammonia [PHARM] A flavored, hydroalcoholic solution of ammonia and ammonium carbonate having an aromatic, pungent odor; used as a reflex stimulant.

aromatic sulfuric acid [PHARM] A preparation consisting of sulfuric acid, tincture of ginger, oil of cinnamon, and alcohol; formerly used as a tonic and astringent.

aromatic vinegar [PHARM] A flavored solution of acetic acid used as smelling salts.

aromatization [CHEM ENG] Conversion of any nonaromatic hydrocarbon structure to aromatic hydrocarbon, particularly petroleum.

Arosorb process [CHEM ENG] Separation of aromatics from nonaromatics in refinery streams by adsorption on a gel from which aromatics are recovered by desorption.

aroyl [ORG CHEM] The radical RCO, where R is an aromatic (benzoyl, napthoyl) group.

aroylation [ORG CHEM] An acylation process by which an aromatic radical is incorporated in a molecule by substitution.

arquerite [MINERAL] A mineral consisting of a soft, malleable, silver-rich variety of amalgam, containing about 87% silver and 13% mercury.

arrastra See arrastre.

arrastre [MIN ENG] A mill comprising a circular, rock-lined pit in which broken ore is pulverized by stones, attached to horizontal poles fastened in a central pillar, which stones are dragged around the pit. Also spelled arrastra.

array [ELECTR] A group of components such as antennas, reflectors, or directors arranged to provide a desired variation of radiation transmission or reception with direction. [STAT] The arrangement of a sequence of items in statistics according to their values, such as from largest to smallest.

array antennas *See* antenna array.

array radar [ENG] A radar incorporating a multiplicity of phased antenna elements.

array sonar [ENG] A sonar system incorporating a phased array of radiating and receiving transducers.

arrested decay [GEOL] A stage in coal formation where biochemical action ceases.

arrested evolution [EVOL] Evolution that was extremely slow in comparison with that characteristic of most organic lineages.

arrester *See* lightning arrester.

arrester hook [AERO ENG] A hook in the tail section of an airplane; used to engage the arrester wires on an aircraft carrier's deck.

arrester wires [NAV ARCH] Cables that extend across the stern of an aircraft carrier's deck; they are raised above the deck on supports so that they may be engaged by the arrester hook of an aircraft when it lands on a carrier; the wires play out along the deck with increasing resistance until the aircraft stops moving. Also known as arresting gear.

arresting gear *See* arrester wires.

arrhenite [MINERAL] A variety of fergusonite.

Arrhenius equation [PHYS CHEM] The relationship that the specific reaction rate constant k equals the frequency factor constant s times exp $(-\Delta H_{act}/RT)$, where ΔH_{act} is the heat of activation, R the gas constant, and T the absolute temperature.

Arrhenius-Guzman equation [PHYS] The relation between the viscosity η of a liquid and the Kelvin temperature T at constant pressure: $\eta = A \exp (B/RT)$, where A and B are constants and R is the gas constant.

Arrhenius viscosity formulas [PHYS] A series of three equations which relate the viscosity of a liquid to the temperature, the viscosity of a solution to its concentration and to the viscosity of the solvent, and the viscosity of a sol to the viscosity of the medium.

arrhenoblastoma [MED] A solid, sometimes malignant, tumor of the ovary that usually produces male sex hormones, inducing virilism.

arrhenotoky [BIOL] Production of only male offspring by a parthenogenetic female.

arrhinencephalia [MED] A congenital malformation in which part or all of the rhinencephalon is absent and the nose is malformed.

arrhythmia [MED] Absence of rhythm, especially of heart beat or respiration.

Arridae [VERT ZOO] A family of catfishes in the suborder Siluroidei found from Cape Cod to Panama.

arris [ARCH] A short edge or angle at the junction of two surfaces, especially moldings and raised edges. [GEOL] *See* arête.

arris fillet [BUILD] A triangular wooden piece that raises the slates of a roof against a chimney or wall so that rain runs off.

arrival rate [IND ENG] The mean number of new calling units arriving at a service facility per unit time.

arrojadite [MINERAL] $Na_2(Fe,Mn)_5(PO_4)_4$ Dark-green mineral crystallizing in the monoclinic system, being isostructural with dickinsonite and occurring in masses.

Arrow *See* Sagitta.

arrowroot [BOT] Any of the tropical American plants belonging to the genus *Maranta* in the family Marantaceae.

arrowroot starch [FOOD ENG] A nutritive carbohydrate obtained from the underground stems of arrowroot plants and used as a food for infants and invalids.

arrow wing [AERO ENG] An aircraft wing of V-shaped planform, either tapering or of constant chord, suggesting a stylized arrowhead.

arrowworm [INV ZOO] Any member of the phylum Chaetognatha; useful indicator organism for identifying displaced masses of water.

arroyo [GEOL] Small, deep gully produced by flash flooding in arid and semiarid regions of the southwestern United States.

arsenal [ORD] **1.** An installation whose primary mission is research, development, and manufacture pertaining to assigned items or components. **2.** An installation having co-equal missions of maintenance and supply for assigned items or components.

arsenate [INORG CHEM] **1.** AsO_4^{-3} A negative ion derived from orthoarsenic acid, $H_3AsO_4 \cdot \frac{1}{2}H_2O$. **2.** A salt or ester of arsenic acid.

arsenic [CHEM] A chemical element, symbol As, atomic number 33, atomic weight 74.9216. [MINERAL] A brittle, steel-gray hexagonal mineral, the native form of the element.

arsenic acid [INORG CHEM] $H_3AsO_4 \cdot \frac{1}{2}H_2O$ White, poisonous crystals, soluble in water and alcohol; used in manufacturing insecticides, glass, and arsenates and as a defoliant. Also known as orthoarsenic acid.

arsenical [CHEM] **1.** Pertaining to arsenic. **2.** A compound that contains arsenic.

arsenical nickel *See* niccolite.

arsenic bloom *See* pharmacolite.

arsenic disulfide [INORG CHEM] As_2S_2 Red, orange, or black monoclinic crystals, insoluble in water; used in fireworks; occurs naturally as realgar.

arsenic oxide [INORG CHEM] **1.** An oxide of arsenic. **2.** *See* arsenic pentoxide. **3.** *See* arsenic trioxide.

arsenic pentasulfide [INORG CHEM] As_2S_5 Yellow crystals that are insoluble in water and readily decompose to the trisulfide and sulfur; used as a pigment.

arsenic pentoxide [INORG CHEM] As_2O_5 A white, deliquescent compound that decomposes by heat and is soluble in water. Also known as arsenic oxide.

arsenic trichloride [INORG CHEM] $AsCl_3$ An oily, colorless liquid that dissolves in water; used in ceramics, organic chemical syntheses, and in the preparation of pharmaceuticals.

arsenic trioxide [INORG CHEM] As_2O_3 A toxic compound, slightly soluble in water; octahedral crystals change to the monoclinic form by heating at 200°C; occurs naturally as arsenolite and claudetite; used in small quantities in some medicinal preparations. Also known as arsenic oxide; arsenious acid.

arsenic trisulfide [INORG CHEM] As_2S_3 An acidic compound in the form of yellow or red monoclinic crystals with a melting point at 300°C; occurs as the mineral orpiment; used as a pigment.

arsenide [CHEM] A binary compound of negative, trivalent arsenic; for example, H_3As or $GaAs$.

arseniopleite [MINERAL] A reddish-brown mineral consisting of a basic arsenate of manganese, calcium, iron, lead, and magnesium and occurring in cleavable masses.

arseniosiderite [MINERAL] $Ca_3Fe_4(AsO_4)_4(OH)_4 \cdot 4H_2O$ A yellowish-brown mineral consisting of a basic iron calcium arsenate and occurring as concretions.

arsenious acid *See* arsenic trioxide.

arsenite [INORG CHEM] **1.** AsO_3^{-3} A negative ion derived from aqueous solutions of As_4O_6. **2.** A salt or ester of arsenious acid.

arsenobenzene [ORG CHEM] $C_6H_5As{:}AsC_6H_5$ White needles that melt at 212°C; insoluble in cold water, soluble in benzenes; derivatives have some use in medicine.

arsenobismite [MINERAL] $Bi_2(AsO_4)(OH)_3$ A yellowish-green mineral consisting of a basic bismuth arsenate and occurring in aggregates.

arsenoclasite [MINERAL] $Mn_5(AsO_4)_2(OH)_4$ A red mineral consisting of a basic manganese arsenate. Also spelled arsenoklasite.

arsenoklasite *See* arsenoclasite.

arsenolite [MINERAL] As_2O_3 A mineral crystallizing in the isometric system and usually occurring as a white bloom or crust. Also known as arsenic bloom.

arsenopyrite [MINERAL] $FeAsS$ A white to steel-gray mineral crystallizing in the monoclinic system with pseudo-orthorhombic symmetry because of twinning; occurs in crystalline rock and is the principal ore of arsenic. Also known as mispickel.

arsenotherapy [MED] Treatment of disease by means of arsenical drugs.

arsenous oxide *See* arsenic trioxide.

arsine [INORG CHEM] H_3As A colorless, highly poisonous gas with an unpleasant odor.

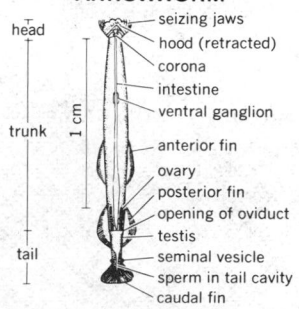

ARROWWORM

Dorsal view of *Sagitta enflata*, an arrowworm.

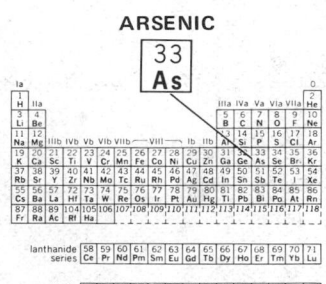

ARSENIC

Periodic table of the chemical elements showing the position of arsenic.

ARSENOPYRITE

A specimen of arsenopyrite from Tavistock, Devon, England. (*American Museum of Natural History*)

ARTACAMINAE

Dorsal view of *Artacamella*, a genus of the Artacaminae.

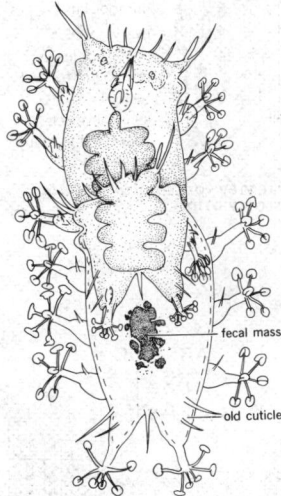

ARTHROTARDIGRADA

fecal mass

old cuticle

Batillipes, a genus of the Arthrotardigrada; defecation during molt.

ARTICHOKE

Edible artichoke flower heads. The bottom one is shown in cross section.

arsinic acid [INORG CHEM] An acid of general formula R_2AsO_2H; derived from trivalent arsenic; an example is cacodylic acid, or dimethylarsinic acid, $(CH_3)_2AsO_2H$.

arsoite [PETR] An olivine-bearing diopside trachyte.

arsonic acid [INORG CHEM] An acid derived from orthoarsenic acid, $OAs(OH)_3$; the type formula is generally considered to be $RAsO(OH)_2$; an example is *para*-aminobenzenearsonic acid, $NH_2C_6H_4AsO(OH)_2$.

arsonium [INORG CHEM] $-AsH_4$ A radical; it may be considered analogous to the ammonium radical in that a compound such as AsH_4OH may form.

Arsonval *See* d'Arsonval entries.

arsphenamine [PHARM] $C_{12}H_{12}As_2N_2O_2 \cdot 2HCl \cdot 2H_2O$ The antisyphilitic diaminodihydroxyarsenobenzene dihydrochloride, effective also on protozoan infections, first prepared by P. Ehrlich in 1909. Also known as Ehrlich's 606.

Artacaminae [INV ZOO] A subfamily of polychaete annelids in the family Terebellidae of the Sedentaria.

arteriogram [MED] A roentgenogram of an artery after injection with radiopaque material.

arteriography [MED] **1.** Graphic presentation of the pulse. **2.** Roentgenography of the arteries after the intravascular injection of a radiopaque substance.

arteriole [ANAT] An artery of small diameter that terminates in capillaries.

arteriolopathy [MED] Disease of the arterioles.

arteriolosclerosis [MED] Thickening of the lining of arterioles, usually due to hyalinization or fibromuscular hyperplasia.

arteriosclerosis [MED] A degenerative arterial disease marked by hardening and thickening of the vessel walls.

arteriosclerosis obliterans [MED] Hardening of the artery walls with obstruction of the lumen due to proliferation of the innermost vessel layer.

arteriotomy [MED] Incision or opening of an artery.

arteriovenous anastomosis [ANAT] A blood vessel that connects an arteriole directly to a venule without capillary intervention.

arteriovenous aneurysm [MED] **1.** Dilation of the walls of an artery and a vein via an abnormal canal (fistula) between the vessels. **2.** Dilation of an arteriovenous fistula.

arterite [PETR] **1.** A migmatite produced as a result of regional contact metamorphism during which residual magmas were injected into the host rock. **2.** Gneisses characterized by veins formed from the solution given off by deep-seated intrusions of molten granite. **3.** A veined gneiss in which the vein material was injected from a magma.

arteritic migmatite [GEOL] Injection gneiss supposedly produced by introduction of pegmatite, granite, or aplite into schist parallel to the foliation.

arteritis [MED] Inflammation of an artery.

artery [ANAT] A vascular tube that carries blood away from the heart.

artesian aquifer [HYD] An aquifer that is bounded above and below by impermeable beds and that contains artesian water. Also known as confined aquifer.

artesian basin [HYD] A geologic structural feature or combination of such features in which water is confined under artesian pressure.

artesian leakage [HYD] The slow percolation of water from artesian formations into the confining materials of a less permeable, but not strictly impermeable, character.

artesian spring [HYD] A spring whose water issues under artesian pressure, generally through some fissure or other opening in the confining bed that overlies the aquifer.

artesian water [HYD] Groundwater that is under sufficient pressure to rise above the level at which it encounters a well, but which does not necessarily rise to or above the surface of the ground.

artesian well [HYD] A well in which the water rises above the top of the water-bearing bed.

arthochromatic erythroblast *See* normoblast.

Arthoniaceae [BOT] A family of lichens in the order Hysteriales.

arthritis [MED] Any inflammatory process affecting joints or their component tissues.

arthritis deformans [MED] A chronic rheumatoid arthritis marked by deformation of affected joints.

arthritis urethritica *See* Reiter's syndrome.

Arthrobacter [MICROBIO] A genus of aerobic, cellulolytic rod-shaped soil bacteria.

arthrobranch [INV ZOO] In Malacostraca, the gill attached to the joint between the body and the first leg segment.

arthrodesis [MED] Fusion of a joint by removing the articular surfaces and securing bony union. Also known as operative ankylosis.

arthrodia [ANAT] A diarthrosis permitting only restricted motion between a concave and a convex surface, as in some wrist and ankle articulations. Also known as gliding joint.

Arthrodira [PALEON] The joint-necked fishes, an Upper Silurian and Devonian order of the Placodermi.

Arthrodonteae [BOT] A family of mosses in the subclass Eubrya characterized by thin, membranous peristome teeth composed of cell walls.

arthrogram [MED] A roentgenogram of a joint space after injection of radiopaque material.

arthrography [MED] Roentgenography of a joint space after the injection of radiopaque material.

arthrogryposis [MED] Permanent fixation of a joint in a flexed position.

Arthromitaceae [MICROBIO] A family of nonmotile bacteria in the order Caryophanales found in the intestine of millipedes, cockroaches, and toads.

arthropathy [MED] **1.** Any joint disease. **2.** A neurotrophic disorder of a joint, usually due to lack of pain sensation, found in association with tabes dorsalis, leprosy, syringomyelia, diabetic polyneuropathy, and occasionally multiple sclerosis and myelodysplasias.

arthroplasty [MED] **1.** The making of an artificial joint. **2.** Reconstruction of a new and functioning joint from an ankylosed one; a plastic operation upon a joint.

Arthropoda [INV ZOO] The largest phylum in the animal kingdom; adults typically have a segmented body, a sclerotized integument, and many-jointed segmental limbs.

arthropod-borne virus *See* arbovirus.

arthropodin [BIOCHEM] A water-soluble protein which forms part of the endocuticle of insects.

arthrosis [ANAT] An articulation or suture uniting two bones. [MED] Any degenerative joint disease.

arthrospore [BOT] A jointed, vegetative resting spore resulting from filament segmentation in some blue-green algae and hypha segmentation in many Basidiomycetes.

Arthrotardigrada [INV ZOO] A suborder of microscopic invertebrates in the order Heterotardigrada characterized by toelike terminations on the legs.

Arthur unit [BIOL] A unit for the standardization of splenin A.

Arthus reaction [IMMUNOL] An allergic reaction of the immediate hypersensitive type that results from the union of antigen and antibody, with complement present, in blood vessel walls.

artichoke [BOT] *Cynara scolymus.* A herbaceous perennial plant belonging to the order Asterales; the flower head is edible.

articulamentum [INV ZOO] The innermost layer of a calcareous plate in a chiton.

articular cartilage [ANAT] Cartilage that covers the articular surfaces of bones.

articular disk [ANAT] A disk of fibrocartilage, dividing the cavity of certain joints.

articular membrane [INV ZOO] A flexible region of the cuticle between sclerotized areas of the exoskeleton of an arthropod; functions as a joint.

Articulata [INV ZOO] **1.** A class of the Brachiopoda having hinged valves that usually bear teeth. **2.** The only surviving subclass of the echinoderm class Crinoidea.

articulated leader [MECH ENG] A wheel-mounted transport unit with a pivotal loading element used in earth moving.

articulated structure [CIV ENG] A structure in which relative motion is allowed to occur between parts, usually by means of a hinged or sliding joint or joints.

articulated train [ENG] A railroad train whose cars are permanently or semipermanently connected.

articulation [ANAT] *See* joint. [BOT] A joint between two parts of a plant that can separate spontaneously. [COMMUN] The percentage of speech units understood correctly by a listener in a communications system; it generally applies to unrelated words, as in code messages, in distinction to intelligibility. [INV ZOO] A joint between rigid parts of an animal body, as the segments of an appendage in insects. [PHYSIO] The act of enunciating speech.

articulation equivalent [COMMUN] Of a complete telephone connection, a measure of the articulation of speech reproduced over it, expressed numerically in terms of the trunk loss of a working reference system when the latter is adjusted to give equal articulation.

articulation index [PSYCH] An indication of the percentage of speech sounds that would be heard correctly when heard without a context.

articulite *See* itacolumite.

artifact [ARCHEO] Any man-made object of common use that reflects the skills of man in past cultures. [HISTOL] A structure in a fixed cell or tissue formed by manipulation or by the reagent.

artificial aging [MET] The heat treatment of an alloy at moderately elevated temperatures to accelerate precipitation of a component from the supersaturated solid solution.

artificial antenna *See* dummy antenna.

artificial asteroid [AERO ENG] A man-made object placed in orbit about the sun.

artificial camphor *See* terpene hydrochloride.

artificial delay line *See* delay line.

artificial echo [ELECTROMAG] **1.** Received reflections of a transmitted pulse from an artificial target, such as an echo box, corner reflector, or other metallic reflecting surface. **2.** Delayed signal from a pulsed radio-frequency signal generator.

artificial feel [AERO ENG] A type of force feedback incorporated in the control system of an aircraft or spacecraft whereby a portion of the forces acting on the control surfaces are transmitted to the cockpit controls.

artificial fiber [TEXT] A filament made from materials such as glass, rayon, or nylon. Also known as synthetic fiber.

artificial gold *See* stannic sulfide.

artificial gravity [AERO ENG] A simulated gravity established within a space vehicle by rotation or acceleration.

artificial harbor [CIV ENG] **1.** A harbor protected by breakwaters. **2.** A harbor formed by dredging.

artificial heart [MED] An endoprosthetic device used to replace or assist the heart.

artificial horizon [NAV] **1.** A gyro-operated flight instrument that shows the pitching and banking attitudes of an aircraft or spacecraft with respect to a reference line horizon, within limited degrees of movement, by means of the relative position of lines or marks on the face of the instrument representing the aircraft and the horizon. Also known as automatic horizon. **2.** A device, such as a spirit level or pendulum, that establishes a horizontal reference in a navigation instrument.

artificial insemination [MED] A process by which spermatozoa are collected from males and deposited in female genitalia by instruments rather than by natural service.

artificial intelligence [ADP] The property of a machine capable of reason by which it can learn functions normally associated with human intelligence.

artificial ionization [COMMUN] Introduction of an artificial reflecting or scattering layer into the atmosphere to permit beyond-the-horizon communications.

artificial kidney [MED] An apparatus that performs the work of the kidney in purifying blood; used only in cases of renal failure or shutdown.

artificial lift [PETRO ENG] Any method of lifting oil out of underground reservoirs, usually by injecting gas into the rock or sand formation to force fluids from wells; an example is a gas lift.

artificial line [ELEC] Circuit made up of lumped constants, which is used to simulate various characteristics of a transmission line.

artificial line duct [ELEC] Balancing network simulating the impedance of the real line and distant terminal apparatus, which is employed in a duplex circuit to make the receiving device unresponsive to outgoing signal currents.

artificial load [ELEC] Dissipative but essentially nonradiating device having the impedance characteristics of an antenna, transmission line, or other practical utilization circuit.

artificial monument [ENG] A relatively permanent man-made object, such as an abutment or stone marker, used to identify the location of a survey station or corner.

artificial neroli oil *See* methyl anthranilate.

artificial parthenogenesis [PHYSIO] Activation of an egg by chemical and physical stimuli in the absence of sperm.

artificial radiation belt [GEOPHYS] High-energy electrons trapped in the earth's geomagnetic field as a result of high-altitude nuclear explosions.

artificial radioactivity *See* induced radioactivity.

artificial recharge [CIV ENG] The recharge of an aquifer depleted by abnormally large withdrawals, by the use of injection wells and other techniques.

artificial respiration [MED] The maintenance of breathing by artificial ventilation, in the absence of normal spontaneous respiration; effective methods include mouth-to-mouth breathing and the use of a respirator.

artificial satellite [AERO ENG] Any man-made object placed in a near-periodic orbit in which it moves mainly under the gravitational influence of one celestial body, such as the earth, sun, another planet, or a planet's moon.

artificial scheelite *See* calcium tungstate.

artificial sweetener [FOOD ENG] A sugar substitute, such as saccharin.

artificial variable [IND ENG] One type of variable introduced in a linear program model in order to find an initial basic feasible solution; an artificial variable is used for equality constraints and for greater-than or equal inequality constraints.

artificial ventilation [MIN ENG] The inducing of a flow of air through a mine or part of a mine by mechanical or other means.

artificial voice [ENG ACOUS] **1.** Small loudspeaker mounted in a shaped baffle which is proportioned to simulate the acoustical constants of the human head; used for calibrating and testing close-talking microphones. **2.** Synthetic speech produced by a multiple tone generator; used to produce a voice reply in some real-time computer applications.

artillery [ORD] A gun or a rocket launcher with mounting too large or too heavy to be classed as a small arm. Abbreviated arty.

artillery bogie [ORD] The portion of an artillery weapon, consisting of wheels, axles, and various supporting appurtenances, which is the principal weight-bearing unit when the weapon is being transported.

artillery cart [ORD] A trailer that carries equipment used by artillery units for fire control, communications, and mapping; it is attached to a field gun for traveling.

artillery cartridge extractor [ORD] An extractor designed for pulling an empty cartridge case or unfired cartridge out of the chamber of an artillery weapon.

artillery cleaning staff [ORD] A round wooden or metal staff with or without a metal handle and with a head unit designed for holding a piece of fabric or cotton for swabbing the bore of a mortar or a short-barrel cannon.

artillery sled [ORD] A flat-bottomed steel item usually curved up at one end; it usually has wheel welds and attaching facilities for fastening the wheels of artillery mounts; used primarily to transport weapons over snow, ice, swamps, or rough terrain.

artillery survey [ORD] The process of determining, with sufficient exactness, the relative horizontal and vertical locations of the pieces and targets so that they may be plotted on the firing chart, and of providing accurate data for the pieces.

artillery train [ORD] A number of pieces of ordnance mounted on traveling carriages.

artinite [MINERAL] $Mg_2CO_3(OH)_2 \cdot 3H_2O$ A snow-white mineral crystallizing in the orthorhombic system and occurring in crystals or fibrous aggregates.

Artinskian [GEOL] A European stage of geologic time including Lower Permian (above Sakmarian, below Kungurian).

ARTIFICIAL HORIZON

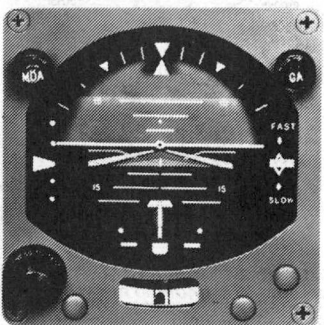

Artificial horizon indicator.
(Kollsman Instrument Co.)

Artiodactyla [VERT ZOO] An order of terrestrial, herbivorous mammals characterized by having an even number of toes and by having the main limb axes pass between the third and fourth toes.

artotype *See* photogelatin printing plate.

ARTS *See* automated radar terminal system.

artwork [GRAPHICS] **1.** Illustrative and decorative matter as distinguished from text. **2.** Illustrative matter and type proofs arranged on a mechanical.

arty *See* artillery.

Arundel method [MAP] A combination of graphical and analytical methods, based on radial triangulation, for point-by-point topographic mapping from aerial photographs.

arviculture [AGR] The cultivation of field crops.

aryl [ORG CHEM] An organic radical derived from an aromatic hydrocarbon by removal of one hydrogen.

aryl acid [ORG CHEM] An organic acid that has an aryl radical.

arylamine [ORG CHEM] An organic compound formed from an aromatic hydrocarbon that has at least one amine group joined to it, such as aniline.

arylated alkyl *See* aralkyl.

aryl compound [ORG CHEM] Molecules with the six-carbon aromatic ring structure characteristic of benzene or compounds derived from aromatics.

aryl diazo compound [ORG CHEM] A diazo compound bonded to the ring structure characteristic of benzene or any other aromatic derivative.

arylene [ORG CHEM] A radical that is bivalent and formed by removal of hydrogen from two carbon sites on an aromatic nucleus.

arylide [ORG CHEM] A compound formed from a metal and an aryl radical, for example, PbR_4, where R is the aryl radical.

aryloxy compound [ORG CHEM] One of a group of compounds useful as organic weed killers, such as 2,4-dichlorophenoxyacetic acid (2,4-D).

arytenoid [ANAT] Relating to either of the paired, pyramid-shaped, pivoting cartilages on the dorsal aspect of the larynx, in man and most other mammals, to which the vocal cords and arytenoid muscles are attached.

arzrunite [MINERAL] A bluish-green mineral consisting of a basic copper sulfate with copper chloride and lead, and occurring as incrustations.

aS *See* abmho.

As *See* altostratus cloud; arsenic.

asar *See* esker.

asarone [ORG CHEM] $C_{12}H_{16}O_3$ A crystalline substance with melting point 67°C; insoluble in water, soluble in alcohol; found in plants of the genus *Asarum*; used as a constituent in essential oils such as calamus oil. Also known as 2,4,5,-trimethyl-1-propenyl benzene.

asbestine [MATER] A material with the properties of asbestos.

asbestos [MINERAL] A general name for the useful, fibrous varieties of a number of rock-forming silicate minerals that are heat-resistant and chemically inert; two varieties exist: amphibole asbestos, the best grade of which approaches the composition $Ca_2Mg_5(OH)_2Si_8O_{22}$ (tremolite), and serpentine asbestos, usually chrysotile, $Mg_3Si_2(OH)_4O_5$.

asbestos cement [MATER] A building material composed of a mixture of asbestos fiber, portland cement, and water made into plain sheets, corrugated sheets, tiles, and piping.

asbestos-cement pipe [MATER] A concrete pipe made of a mixture of portland cement and asbestos fiber and highly resistant to corrosion; used in drainage systems, waterworks systems, and gas lines. Also known by the trade name Transite pipe.

asbestos insulation [MATER] A material composed of asbestos fibers bonded with mixtures of clay or sodium silicate; used as thermal insulation for temperatures above 1500°F (816°C).

asbestosis [MED] A chronic lung inflammation caused by inhalation of asbestos dust.

asbestos shingle [MATER] A shingle composed of asbestos cement formed under pressure; used on houses for roofing and siding that resist the destructive effects of time, weather, and fire.

asbolane *See* asbolite.

asbolite [MINERAL] A black, earthy mineral aggregate containing hydrated oxides of manganese and cobalt. Also known as asbolane; black cobalt; cobalt ocher; earthy cobalt.

A scale [ACOUS] A system used to filter out sound below 55 decibels; its characteristics are equal to those of the human ear.

A scan *See* A scope.

Ascaphidae [VERT ZOO] A family of amphicoelous frogs in the order Anura, represented by four living species.

ascariasis [MED] Any parasitic infection of man and other domestic mammals caused by species of *Ascaris*.

ascarid [INV ZOO] The common name for any roundworm belonging to the superfamily Ascaridoidea.

Ascaridata [INV ZOO] An equivalent name for the Ascaridina.

Ascaridida [INV ZOO] An order of parasitic nematodes in the subclass Phasmidia.

Ascarididae [INV ZOO] A family of parasitic nematodes in the superfamily Ascaridoidea.

Ascaridina [INV ZOO] A suborder of parasitic nematodes in the order Ascaridida.

Ascaridoidea [INV ZOO] A large superfamily of parasitic nematodes of the suborder Ascaridina.

ascaridole [ORG CHEM] $C_{10}H_{16}O_2$ A terpene peroxide, explosive when heated; used as an initiator in polymerization.

Ascaris [INV ZOO] A genus of roundworms that are intestinal parasites in mammals, including man.

Ascaris lumbricoides [INV ZOO] A large parasitic roundworm in the small intestine of man.

Ascaroidea [INV ZOO] An equivalent name for Ascaridoidea.

ascending aorta [ANAT] The first part of the aorta, extending from its origin in the heart to the aortic arch.

ascending branch [MECH] The portion of the trajectory between the origin and the summit on which a projectile climbs and its altitude constantly increases.

ascending chromatography [ANALY CHEM] A technique for the analysis of mixtures of two or more compounds in which the mobile phase (sample and carrier) rises through the fixed phase.

ascending node [ASTRON] Also known as northbound node. **1.** The point at which a planet, planetoid, or comet crosses to the north side of the ecliptic. **2.** The point at which a satellite crosses to the north side of the equatorial plane of its primary.

ascending series [MATH] **1.** A series each of whose terms is greater than the preceding term. **2.** *See* power series.

ascending vertical angle *See* angle of elevation.

ascent [AERO ENG] Motion of a craft in which the path is inclined upward with respect to the horizontal.

Ascheim-Zondek test [PATH] A human pregnancy test that uses the reaction of ovaries in immature white mice to an injection of urine from a woman.

Aschelmintha [INV ZOO] A theoretical grouping erected by B. G. Chitwood as a series that includes the phylum Nematoda.

Aschelminthes [INV ZOO] A heterogeneous phylum of small to microscopic wormlike animals; individuals are pseudo-coelomate and mostly unsegmented and are covered with a cuticle.

aschistic [GEOL] Pertaining to rocks of minor igneous intrusions that have not been differentiated into light and dark portions but that have essentially the same composition as the larger intrusions with which they are associated.

Aschoff body [MED] The lesion of rheumatic fever found around blood vessels in the myocardium.

Ascidiacea [INV ZOO] A large class of the phylum Tunicata; adults are sessile and may be solitary or colonial.

ascidiform [BOT] Pitcher-shaped, as certain leaves.

ascidium [BOT] A pitcher-shaped plant organ or part.

ASCII *See* American Standard Code for Information Interchange.

ascites [MED] An abnormal accumulation of serous fluid in the abdominal cavity.

Asclepiadaceae [BOT] A family of tropical and subtropical flowering plants in the order Gentianales characterized by a well-developed latex system; milkweed (*Asclepias*) is a well-known member.

ASCIDIACEA

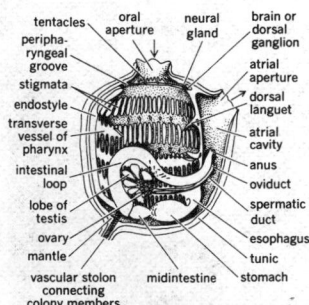

tentacles
peripha, ryngeal groove
stigmata
endostyle
transverse vessel of pharynx
intestinal loop
lobe of testis
ovary
mantle
vascular stolon connecting colony members

oral aperture
neural gland

brain or dorsal ganglion
atrial aperture
dorsal languet
atrial cavity
anus
oviduct
spermatic duct
esophagus
tunic

midintestine
stomach

Left side of a zooid of the colonial ascidian *Perophora*. The tunic, mantle, and anterior wall of the pharynx have been removed from the left. The two arrows indicate direction of water intake and expulsion.

ascocarp [MYCOL] The mature fruiting body bearing asci with ascospores in higher Ascomycetes.

ascogenous [MYCOL] Pertaining to or producing asci.

ascogonium [MYCOL] The specialized female sexual organ in higher Ascomycetes.

Ascolichenes [BOT] A class of the lichens characterized by the production of asci similar to those produced by Ascomycetes.

Ascomycetes [MYCOL] A class of fungi in the subdivision Eumycetes, distinguished by the ascus.

ascon [INV ZOO] A sponge or sponge larva having incurrent canals leading directly to the paragaster.

A scope [ELECTR] A radarscope on which the trace appears as a horizontal or vertical range scale and the signals appear as vertical or horizontal deflections. Also known as A indicator; A scan.

ascorbic acid [BIOCHEM] $C_6H_8O_6$ A white, crystalline, water-soluble vitamin found in many plant materials, especially citrus fruit. Also known as vitamin C.

ascospore [MYCOL] An asexual spore representing the final product of the sexual process, borne on an ascus in Ascomycetes.

Ascothoracica [INV ZOO] An order of marine crustaceans in the subclass Cirripedia occurring as endo- and ectoparasites of echinoderms and coelenterates.

ascus [MYCOL] An oval or tubular spore sac bearing ascospores in members of the Ascomycetes.

asdic [ELECTR] Acronym for Anti-Submarine Detection Investigation Committee; British term for sonar and underwater listening devices.

Aselloidea [INV ZOO] A group of free-living, fresh-water isopod crustaceans in the suborder Asellota.

Asellota [INV ZOO] A suborder of morphologically and ecologically diverse aquatic crustaceans in the order Isopoda.

asepsis [MED] The state of being free from pathogenic microorganisms.

asexual [BIOL] **1.** Not involving sex. **2.** Exhibiting absence of sex or of functional sex organs.

asexual reproduction [BIOL] Formation of new individuals from a single individual without the involvement of gametes.

ash [BOT] **1.** A tree of the genus *Fraxinus*, deciduous trees of the olive family (Oleaceae) characterized by opposite, pinnate leaflets. **2.** Any of various Australian trees having wood of great toughness and strength; used for tool handles and in work requiring flexibility. [CHEM] The incombustible matter remaining after a substance has been incinerated. [GEOL] Volcanic dust and particles less than 4 millimeters in diameter. [ENG] An undesirable constituent of diesel fuel whose quantitative measurement indicates degree of fuel cleanliness and freedom from abrasive material.

ash collector *See* dust chamber.

ash cone [GEOL] A volcanic cone built primarily of unconsolidated ash and generally shaped somewhat like a saucer, with a rim in the form of a wide circle and a broad central depression often nearly at the same elevation as the surrounding country.

ash conveyor [MECH ENG] A device that transports refuse from a furnace by fluid or mechanical means.

ash fall [GEOL] **1.** A fall of airborne volcanic ash from an eruption cloud; characteristic of Vulcanian eruptions. Also known as ash shower. **2.** Volcanic ash resulting from an ash fall and lying on the ground surface.

ash field [GEOL] A thick, extensive deposit of volcanic ash. Also known as ash plain.

ash flow [GEOL] **1.** An avalanche of volcanic ash, generally a highly heated mixture of volcanic gases and ash, traveling down the flanks of a volcano or along the surface of the ground. Also known as glowing avalanche; incandescent tuff flow; pyroclastic flow. **2.** A deposit of volcanic ash and other debris resulting from such a flow and lying on the surface of the ground.

ash furnace [ENG] A furnace in which materials are fritted for glassmaking.

ash fusibility [GEOL] The gradual softening and melting of coal ash that takes place with increase in temperature as a result of the melting of the constituents and chemical reactions.

Ashgillian [GEOL] A European stage of geologic time in the Upper Orodovician (above Upper Caradocian, below Llandoverian of Silurian).

ashing [ANALY CHEM] An analytical process in which the chemical material being analyzed is oven-heated to leave only noncombustible ash.

ashlar [CIV ENG] Masonry with an exposed side of square or rectangular stones.

ashlar brick [MATER] A brick having rough-hackled faces resembling stone.

ashlar line [BUILD] The outer line of a wall above any projecting base.

ash plain *See* ash field.

ash rock [GEOL] The material of arenaceous texture produced by volcanic explosions.

ash shower *See* ash fall.

ashstone [PETR] A rock composed of fine volcanic ash; particles are less than 0.06 millimeter in diameter.

ashtonite *See* mordenite.

ash viscosity [GEOL] The ratio of shearing stress to velocity gradient of molten ash; indicates the suitability of a coal ash for use in a slag-tap-type boiler furnace.

ashy grit [GEOL] **1.** Pyroclastic material of sand and smaller size. **2.** Mixture of ordinary sand and volcanic ash.

Asia [GEOGR] The largest continent, comprising the major portion of the broad east-west extent of the Northern Hemisphere land masses.

Asian flu [MED] An acute viral respiratory infection of man caused by influenza A-2 virus.

asiderite *See* stony meteorite.

Asilidae [INV ZOO] The robber flies, a family of predatory, orthorrhaphous, dipteran insects in the series Brachycera.

A size [ENG] One of a series of sizes to which trimmed paper and board are manufactured; for size A*N*, with *N* equal to any integer from 0 to 10, the length of the longer side is $2^{-(2N-1)/4}$ meters, while the length of the shorter side is $2^{-(2N+1)/4}$ meters, with both lengths rounded off to the nearest millimeter.

ASM *See* air-to-surface missile.

aso lava [GEOL] A type of indurated pyroclastic deposit produced during the explosive eruptions that formed the Aso Caldera of Hyushu, Japan.

asomatognosia [MED] Lacking awareness of paralysis because the brain is damaged.

Asopinae [INV ZOO] A family of hemipteran insects in the superfamily Pentatomoidea including some predators of caterpillars.

asparaginase [BIOCHEM] An enzyme that catalyzes the hydrolysis of asparagine to asparaginic acid and ammonia.

asparagine [BIOCHEM] $C_4H_8N_2O_3$ A white, crystalline amino acid found in many plant seeds.

asparagolite *See* asparagus stone.

asparagus [BOT] *Asparagus officinalis.* A dioecious, perennial monocot belonging to the order Liliales; the shoot of the plant is edible.

asparagus stone [MINERAL] A yellow-green variety of apatite occurring in crystals. Also known as asparagolite.

aspartase [BIOCHEM] A bacterial enzyme that catalyzes the deamination of aspartic acid to fumaric acid and ammonia.

aspartate [BIOCHEM] A compound that is an ester or salt of aspartic acid.

aspartic acid [BIOCHEM] $C_4H_7NO_4$ A nonessential, crystalline dicarboxylic amino acid found in plants and animals, especially in molasses from young sugarcane and sugarbeet.

aspartoyl [BIOCHEM] $- COCH_2CH(NH_2)CO -$ A bivalent radical derived from aspartic acid.

aspect [ASTRON] The apparent position of a celestial body relative to another; particularly, the apparent position of the moon or a planet relative to the sun. [CIV ENG] Of railway signals, what the engineer sees when viewing the blades or lights in their relative positions or colors. [ECOL] Seasonal appearance.

aspect angle [ENG] The angle formed between the longitudinal axis of a projectile in flight and the axis of a radar beam.

aspect card [ADP] A card on which is entered the accession numbers of documents in an information retrieval system; the

ASCOCARP

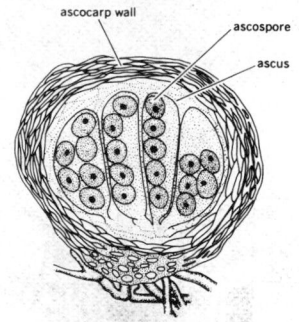

ascocarp wall
ascospore
ascus

Longitudinal section showing several asci in ascocarp of *Erysiphe aggregata.*

ASCORBIC ACID

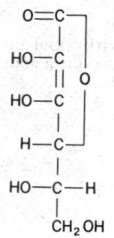

$$
\begin{array}{c}
O = C \\
| \\
HO - C \\
\quad\quad \| \; O \\
HO - C \\
| \\
H - C \\
| \\
HO - C - H \\
| \\
CH_2OH
\end{array}
$$

The structural formula of ascorbic acid.

ASCOTHORACICA

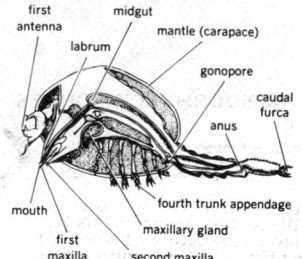

first antenna
midgut
labrum
mantle (carapace)
gonopore
caudal furca
anus
mouth
fourth trunk appendage
first maxilla
maxillary gland
second maxilla

Ascothorax ophioctenis, a parasite in the bursae of brittle stars. *(After Wagin from A Kaestner; from R. D. Barnes, Invertebrate Zoology, 2d ed., Saunders, 1968)*

ASPEN

A leaf scar with axial bud, leaf, and twig of the quaking aspen *(Populus tremuloides)*.

ASPIDORHYNCHIFORMES

Aspidorhynchus acutirostris (Blainville), Upper Jurassic, Bavaria, length to 3 feet (91 centimeters); a typical fish of the order Aspidorhynchiformes. *(After Assmann)*

documents are judged to be related importantly to the concept for which the card is established.

aspection [ECOL] Seasonal change in appearance or constitution of a plant community.

aspect ratio [AERO ENG] The ratio of the square of the span of an airfoil to the total airfoil area, or the ratio of its span to its mean chord. [DES ENG] The ratio of frame width to frame height in television; it is 4:3 in the United States and Britain.

aspen [BOT] Any of several species of poplars *(Populus)* characterized by weak, flattened leaf stalks which cause the leaves to flutter in the slightest breeze.

Aspergillaceae [MYCOL] Former name for the Eurotiaceae.

Aspergillales [MYCOL] Former name for the Eurotiales.

aspergillic acid [BIOCHEM] $C_{12}H_{20}O_2N_2$ A diketopiperazine-like antifungal antibiotic produced by certain strains of *Aspergillus flavus.*

aspergillin [MYCOL] **1.** A black pigment found in spores of some molds of the genus *Aspergillus.* **2.** A broad-spectrum antibacterial antibiotic produced by the molds *Aspergillus flavus* and *A. fumigatus.*

aspergillosis [MED] A rare fungus infection of man and animals caused by several species of *Aspergillus.*

Aspergillus [MYCOL] A genus of fungi including several species of common molds and some human and plant pathogens.

Aspergillus fumigatus [MYCOL] A pathogenic fungus causing aspergillosis in man and animals; source of the antibiotic fumagillin.

Aspergillus niger [MYCOL] A pathogenic fungus causing black mold in many plants, wet stem rot in sisal, and otomycosis in man; used industrially for fermentative production of enzymes and citric acid.

asperifoliate [BOT] Rough-leaved.

aspermatism [MED] **1.** Failure to ejaculate or secrete semen. **2.** Defective or absent sperm formation.

asperulate [BOT] Delicately roughened.

asphalt [MATER] A brown to black, hard, brittle, or plastic bituminous material composed principally of hydrocarbons; formed in oil-bearing rocks near the Dead Sea, and in Trinidad; prepared by pyrolysis from coal tar, certain petroleums, and lignite tar; melts on heating, insoluble in water but soluble in gasoline; used for paving and roofing and in paints and varnishes.

asphalt cement [MATER] Fluxed or unfluxed asphalt especially prepared for direct use in making bituminous pavements.

asphalt emulsion [MATER] Asphalt cement in water containing a small amount of emulsifying agent.

asphaltene [MATER] Any of the dark, solid constituents of crude oils and other bitumens which are soluble in carbon disulfide but insoluble in paraffin naphthas; they hold most of the organic constituents of bitumens.

asphalt flux [MATER] An oil used to reduce the consistency or viscosity of hard asphalt to the point required for use.

asphaltic road oil [MATER] A thick, fluid solution of asphalt.

asphaltic sand [GEOL] Deposits of sand grains cemented together with soft, natural asphalt.

asphaltite [GEOL] Any of the dark-colored, solid, naturally occurring bitumens that are insoluble in water, but more or less completely soluble in carbon disulfide, benzol, and so on, with melting points between 250 and 600°F (121–316°C); examples are gilsonite and grahamite.

asphaltite coal *See* albertite.

asphalt macadam [MATER] Pavement made with an asphalt binder rather than tar.

asphalt mastic [MATER] A mixture of asphalt with sand, asbestos, crushed rock, or similar material; used like cement.

asphalt paint [MATER] Asphaltic material dissolved in volatile solvent with or without pigments, drying oils, resins, and so on.

asphalt paper [MATER] A paper that is coated or impregnated with asphalt.

asphalt primer [MATER] Low-viscosity, liquid asphaltic material applied to and absorbed by nonbituminous surfaces, as in waterproofing.

asphalt rock [GEOL] Natural rock asphalt or asphalt-containing rock, such as porous sandstones and dolomites. Also known as asphalt stone; bituminous rock; rock asphalt.

asphalt roofing [MATER] A roofing material made by impregnating a dry roofing felt with a hot asphalt saturant, applying asphalt coatings to the weather and reverse sides, and embedding a mineral surfacing in the coating on the weather side.

asphalt tile [MATER] Floor tile composed of asbestos fibers, mineral coloring pigments, and inert fillers bound together; used on rigid subfloors or hardwood floors.

asphaltum [MATER] Bituminous material in oil of turpentine; used in photomechanical work because of its ability to be rendered insoluble in light.

aspheric surface [OPTICS] A lens or mirror surface which is altered slightly from a spherical surface in order to reduce aberrations.

asphradium [INV ZOO] An organ, believed to be a chemoreceptor, in mollusks. Also spelled osphradium.

asphyxia [MED] Suffocation due to oxygen deprivation, resulting in anoxia and carbon dioxide accumulation in the body.

aspiculate [INV ZOO] Lacking spicules, referring to Porifera.

Aspidiotinae [INV ZOO] A subfamily of homopteran insects in the superfamily Coccoidea.

Aspidiphoridae [INV ZOO] An equivalent name for the Sphindidae.

Aspidobothria [INV ZOO] An equivalent name for the Aspidogastrea.

Aspidobothroidea [INV ZOO] A group of trematodes accorded class rank by W. J. Hargis.

Aspidobranchia [INV ZOO] An equivalent name for the Archaeogastropoda.

Aspidochirotacea [INV ZOO] A subclass of bilaterally symmetrical echinoderms in the class Holothuroidea characterized by tube feet and 10–30 shield-shaped tentacles.

Aspidochirotida [INV ZOO] An order of holothurioid echinoderms in the subclass Aspidochirotacea characterized by respiratory trees and dorsal tube feet converted into tactile warts.

Aspidocotylea [INV ZOO] An equivalent name for the Aspidogastrea.

Aspidodiadematidae [INV ZOO] A small family of deep-sea echinoderms in the order Diadematoida.

Aspidogastrea [INV ZOO] An order of endoparasitic worms in the class Trematoda having strongly developed ventral holdfasts.

Aspidogastridae [INV ZOO] A family of trematode worms in the order Aspidogastrea occurring as endoparasites of mollusks.

Aspidorhynchidae [PALEON] The single family of the Aspidorhynchiformes, an extinct order of holostean fishes.

Aspidorhynchiformes [PALEON] A small, extinct order of specialized holostean fishes.

Aspinothoracida [PALEON] The equivalent name for Brachythoraci.

aspirating *See* dedusting.

aspirating screen [MIN ENG] A vibrating screen from which light, liberated particles are removed by suction.

aspiration [SCI TECH] Act or the result of removing, carrying along, or drawing by suction. [MED] The removal of fluids from a cavity by suction. [MICROBIOL] The use of suction to draw up a sample in a pipette.

aspiration condenser [NUCLEO] An ion-counter collecting element consisting of a cylindrical condenser which when charged produces a radial field that collects ions from the aspirated air.

aspiration meteorograph [ENG] An instrument for the continuous recording of two or more meteorological parameters, with the ventilation being provided by a suction fan.

aspiration psychrometer [ENG] A psychrometer in which the ventilation is provided by a suction fan.

aspiration thermograph [ENG] A thermograph in which ventilation is provided by a suction fan.

aspirator [ENG] Any instrument or apparatus that utilizes a vacuum to draw up gases or granular materials. [MIN ENG] A device made of wire gauze, of cloth, or of a fibrous mass held between pieces of meshed material and used to cover the mouth and nose to keep dusts from entering the lungs.

aspirin *See* acetylsalicylic acid.

aspite [GEOL] A cratered volcano with base wide in relation to height; for example, Mauna Loa.

asporogenic mutant [MICROBIO] A bacillus that is unable to form spores due to alterations at any of several gene loci.

asporogenous [BOT] Not producing spores, especially of certain yeasts.

Aspredinidae [VERT ZOO] A family of salt-water catfishes in the order Siluriformes found off the coast of South America.

ASROC See antisubmarine rocket.

ass [VERT ZOO] Any of several perissodactyl mammals in the family Equidae belonging to the genus *Equus*, especially *E. hemionus* and *E. asinus*.

assault [ORD] 1. Final phase of an attack; closing with the enemy in hand-to-hand fighting. 2. The landing of troops for attack on the enemy's beach defenses. 3. The landing of parachute and glider elements on unsecured and unprepared drop zones and landing zones to attack and seize an airhead. 4. A short, violent, but well-ordered attack against a local objective, such as a gun emplacement, fort, or machine gun nest.

assault aircraft [AERO ENG] Powered aircraft, including helicopters, which move assault troops and cargo into an objective area and which provide for their resupply.

assault boat [NAV ARCH] A small boat that can easily be transported on land; used for amphibious military attacks or to cross lakes and rivers in land warfare.

assault fire [ORD] 1. Fire delivered by attacking troops as they close with an enemy to engage him at close range or in hand-to-hand fighting, usually delivered from the hip or the standing position at a sustained rate. Also known as advancing fire. 2. In artillery, extremely accurate, short-range destruction fire at point targets.

assault gun [ORD] Any of various sizes and types of guns that are self-propelled or mounted on tanks and are used for direct fire from close range against point targets.

assault-landing model [ORD] A special form of assault model designed specifically for planning amphibious landings. Also known as amphibious-assault landing model.

assault model [ORD] Vehicle designed to provide direct fire in combat.

assay [ANALY CHEM] Qualitative or quantitative determination of the components of a material, as an ore or a drug.

assay balance [ENG] A sensitive balance used in the assaying of gold, silver, and other precious metals.

assay bar [MET] A bar of pure or nearly pure gold and silver; used by a government as a standard.

assay plan [MIN ENG] A mine map showing the assay, stope, width, and so forth of samples taken from positions marked.

assay pound [MIN ENG] A weight which varies from time to time but is sometimes 0.5 gram, and is used by assayers to proportionately represent a pound.

assay ton [MIN ENG] A unit of weight of ore equal to 29,167 milligrams; the number of milligrams of precious metal in this measure equals the number of troy ounces in a short ton.

assay value [MIN ENG] The amount of gold or silver as shown by assay of any given sample and represented by ounces per ton of ore.

assay walls [MIN ENG] The planes to which an ore body can be profitably mined, the limiting factor being the metal content of the country rock as determined from assays.

assemblage [ARCHEO] All related cultural traits and artifacts associated with one archeological manifestation. [ECOL] A group of organisms sharing a common habitant by chance. [ORD] A collection of items designed to accomplish one general function and identified and issued as a single item.

assemblage zone [PALEON] A biostratigraphic unit defined and identified by a group of associated fossils rather than by a single index fossil.

assembler [ADP] A program designed to convert symbolic instruction into a form suitable for execution on a computer. Also known as assembly program; assembly routine.

assembly [ADP] The automatic translation into machine language of a computer program written in symbolic language.

assembly drawing [GRAPHICS] A working-type engineering drawing depicting a complete unit, usually included with detail drawings of all parts in a set of working drawings.

assembly language [ADP] A low-level computer language one step above the binary machine language.

assembly line [IND ENG] A mass-production arrangement whereby the work in process is progressively transferred from one operation to the next until the product is assembled.

assembly-line balancing [IND ENG] Assigning numbers of operators or machines to each operation of an assembly line so as to meet the required production rate with a minimum of idle time.

assembly list [ADP] A printed list which is the by-product of an assembly procedure; it lists in logical instruction sequence all details of a routine, showing the coded and symbolic notation next to the actual notations established by the assembly procedure; this listing is highly useful in the debugging of a routine.

assembly method [IND ENG] The technique used to assemble a manufactured product, such as hand assembly, progressive line assembly, and automatic assembly.

assembly program See assembler.

assembly routine See assembler.

assembly unit [ADP] 1. A device which performs the function of associating and joining several parts or piecing together a program. 2. A portion of a program which is capable of being assembled into a larger program.

assessment drilling [MINE ENG] Drilling to fulfill the requirement that a prescribed amount of work be done annually on an unpatented mining claim to retain title. Also known as annual labor.

assessment work [MIN ENG] Annual work at an unpatented mining claim in the public domain performed under law to maintain the claim title.

assets [IND ENG] All the resources, rights, and property owned by a person or a company; the book value of these items as shown on the balance sheet.

assign [ADP] A control statement in FORTRAN which assigns a computed value i to a variable k, the latter representing the number of the statement to which control is then transferred.

assignable cause [IND ENG] Any identifiable factor which causes variation in a process outside the predicted limits, thereby altering quality.

assignment problem [ADP] A special case of the transportation problem in a linear program, in which the number of sources (assignees) equals the number of designations (assignments) and each supply and each demand equals 1.

assili cotton [TEXT] A long-staple Egyptian cotton characterized by high tensile strength.

assimilation [GEOL] Incorporation of solid or fluid material that was originally in the rock wall into a magma. [PHYSIO] Conversion of nutritive materials into protoplasm.

assimilative nitrate reduction [MICROBIO] The reduction of nitrates by some aerobic bacteria for purposes of assimilation.

assimilative sulfate reduction [MICROBIO] The reduction of sulfates by certain obligate anaerobic bacteria for purposes of assimilation.

assisted takeoff [AERO ENG] A takeoff of an aircraft or a missile by using a supplementary source of power, usually rockets.

Assmann psychrometer [ENG] A special form of the aspiration psychrometer in which the thermometric elements are well shielded from radiation.

associate [PSYCH] An item or event that is linked to another in the mind of an individual.

associate curve See Bertrand curve.

associated automatic movement See synkinesia.

associated corpuscular emission [GEOPHYS] The full complement of secondary charged particles associated with the passage of an x-ray or gamma-ray beam through air.

associated gas [PETRO ENG] Gaseous hydrocarbons occurring as a free-gas phase under original oil-reservoir conditions of temperature and pressure.

associated production [PARTIC PHYS] Production of strange particles invariably in twos, never one particle alone.

associate matrix See Hermitian conjugate of a matrix.

associate operator See adjoint operator.

association [CHEM] Combination or correlation of substances or functions. [ECOL] Major segment of a biome formed by a climax community, such as an oak-hickory forest

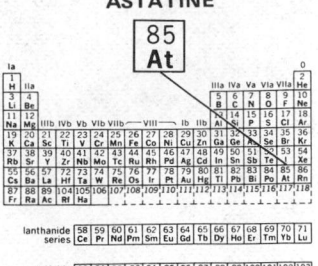

ASTATINE

Periodic table of the chemical elements showing the position of astatine.

ASTER

New England aster (*Aster novae-angliae*), showing ray and disk flowers. (*Courtesy of Alvin E. Staffan, from National Audubon Society*)

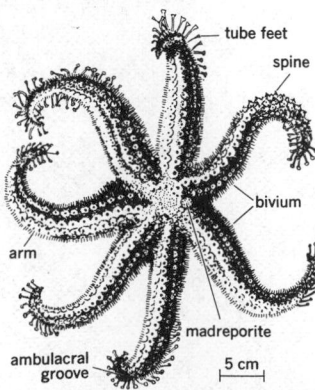

ASTEROIDEA

A representative asteroid, *Astrostole scabra*. Although five arms are common, species of this group may have 6–12 arms.

of the deciduous forest biome. [PSYCH] A connection formed through learning.

association area [PHYSIO] An area of the cerebral cortex that is thought to link and coordinate activities of the projection areas.

association center [INV ZOO] In invertebrates, a nervous center coordinating and distributing stimuli from sensory receptors.

association fiber [ANAT] One of the white nerve fibers situated just beneath the cortical substance and connecting the adjacent cerebral gyri.

association neuron [ANAT] A neuron, usually within the central nervous system, between sensory and motor neurons.

association test [PSYCH] Any test designed to determine the nature of the mental or emotional link between a stimulus and a response.

association trail [ADP] A linkage between two or more documents or items of information, discerned during the process of their examination and recorded with the aid of an information retrieval system.

associative algebra [MATH] An algebra in which the vector multiplication obeys the associative law.

associative facilitation [PSYCH] Ease in establishing a new association because of previous associations.

associative inhibition [PSYCH] Difficulty in establishing a new association because of previous associations.

associative law [MATH] For a binary operation designated ∘, the relationship expressed by $a \circ (b \circ c) = (a \circ b) \circ c$.

associative learning [PSYCH] The principle that items experienced together are mentally linked so that they tend to reinforce one another.

associative memory [ADP] A data-storage device in which a location is identified by its informational content rather than by names, addresses, or relative positions, and from which the data may be retrieved. Also known as associative storage. [PSYCH] Recalling a previously experienced item by thinking of something that is linked with it, thus invoking the association.

associative storage See associative memory.

associative thinking [PSYCH] **1.** The mental process of making associations between a given subject and all pertinent present factors without drawing on past experience. **2.** Free association.

associator [ADP] A device for bringing like entities into conjunction or juxtaposition.

assortative mating [GEN] Nonrandom mating with respect to phenotypes.

assumed plane coordinates [ENG] A local plane-coordinate system set up at the convenience of the surveyor.

assumed position [NAV] In celestial navigation, a point on the surface of the earth at which a craft is assumed to be located and for which the computed altitude is determined in the solution of a celestial observation. Also known as chosen position.

assured mineral See developed reserves.

assyntite [PETR] A plutonic rock consisting largely of orthoclase and pyroxene, lesser amounts of sodalite and nepheline, and accessory biotite, sphene, apatite, and opaque oxides.

astable circuit [ELECTR] A circuit that alternates automatically and continuously between two unstable states at a frequency dependent on circuit constants; for example, a blocking oscillator.

astable multivibrator [ELECTR] A multivibrator in which each active device alternately conducts and is cut off for intervals of time determined by circuit constants, without use of external triggers. Also known as free-running multivibrator.

Astacidae [INV ZOO] A family of fresh-water crayfishes belonging to the section Macrura in the order Decapoda, occurring in the temperate regions of the Northern Hemisphere.

astacin [BIOCHEM] $C_{40}H_{48}O_4$ A red carotenoid ketone pigment found in crustaceans, as in the shell of a boiled lobster.

Astacinae [INV ZOO] A subfamily of crayfishes in the family Astacidae including all North American species west of the Rocky Mountains.

A-stage resin [ORG CHEM] A resin formed by condensation of a phenol and an aldehyde in the presence of a catalyst and a base.

Astartian See Sequanian.

astasia [MED] Lack of muscular coordination in standing.

astatic [PHYS] Without orientation or directional characteristics; having no tendency to change position.

astatic galvanometer [ENG] A sensitive galvanometer designed to be independent of the earth's magnetic field.

astatic governor See isochronous governor.

astatic pair [ELECTROMAG] A pair of parallel magnets, equal in strength and having polarities in opposite directions, and perpendicular to an axis which bisects both of them; there is no net force or torque on the pair in a uniform field.

astatic wattmeter [ENG] An electrodynamic wattmeter designed to be insensitive to uniform external magnetic fields.

astatine [CHEM] A radioactive chemical element, symbol At, atomic number 85, the heaviest of the halogen elements.

A station [NAV] In loran, the designation applied to one transmitting station of a pair, the signal of which always occurs less than half a repetition period after the preceding signal and more than half a repetition period before the succeeding signal of the other station, designated a B station.

astatized gravimeter [ENG] A gravimeter, sometimes referred to as unstable, where the force of gravity is maintained in an unstable equilibrium with the restoring force.

astaxanthin [BIOCHEM] $C_{40}H_{52}O_4$ A violet carotenoid pigment found in combined form in certain crustacean shells and bird feathers.

Asteidae [INV ZOO] A small, obscure family of cyclorrhaphous myodarian dipteran insects in the subsection Acalypteratae.

astel [MIN ENG] An overhead boarding or arching in a mine gallery.

astelic [BOT] Lacking a stele or having a discontinuous arrangement of vascular bundles.

aster [BOT] Any of the herbaceous ornamental plants of the genus *Aster* belonging to the family Compositae.

Asteraceae [BOT] An equivalent name for the Compositae.

Asterales [BOT] An order of dicotyledonous plants in the subclass Asteridae, including aster, sunflower, zinnia, lettuce, artichoke, and dandelion.

astereognosis [MED] Loss of recognition of objects by touch, although recognition occurs through another sense, usually vision. Also known as tactile agnosia.

Asteridae [BOT] A large subclass of dicotyledonous plants in the class Magnoliopsida; plants are sympetalous, with unitegmic, tenuinucellate ovules and with the stamens usually as many as, or fewer than, the corolla lobes and alternate with them.

Asteriidae [INV ZOO] A large family of echinoderms in the order Forcipulatida, including many predatory sea stars.

Asterinidae [INV ZOO] The starlets, a family of echinoderms in the order Spinulosida.

asterism [ASTRON] A constellation or small group of stars. [OPTICS] A starlike optical phenomenon seen in gemstones called star stones; due to reflection of light by lustrous inclusions reduced to sharp lines of light by a domed cabochon style of cutting. [SPECT] A star-shaped pattern sometimes seen in x-ray spectrophotographs.

astern [ENG] To the rear of an aircraft, vehicle, or vessel; behind; from the back.

asternal [ANAT] **1.** Not attached to the sternum. **2.** Without a sternum.

asteroid [ASTRON] One of the many small celestial bodies revolving around the sun, most of the orbits being between those of Mars and Jupiter. Also known as minor planet; planetoid.

asteroid belt [ASTRON] The region between 2.1 and 3.5 astronomical units from the sun where most of the asteroids are found.

Asteroidea [INV ZOO] The starfishes, a subclass of echinoderms in the subphylum Asterozoa characterized by five radial arms.

Asteroschematidae [INV ZOO] A family of ophiuroid echinoderms in the order Phrynophiurida with individuals having a small disk and stout arms.

Asterozoa [INV ZOO] A subphylum of echinoderms characterized by a star-shaped body and radially divergent axes of symmetry.

aster wilt [PL PATH] A fungus disease of asters caused by *Fusarium oxysporum* f. *callistephi*.

aster yellows [PL PATH] A widespread virus disease of asters, other ornamental plants, and many vegetables, characterized by yellowing and dwarfing; leafhoppers are vectors.

asthenia [MED] Loss or lack of strength.

asthenolith [GEOL] A body of magma locally melted anywhere at any time within any solid portion of the earth.

asthenopia [MED] Weakness of the eye muscles or of visual acuity, sometimes accompanied by pain and headache.

asthenosphere *See* upper mantle.

asthma [MED] A pulmonary disease marked by labored breathing, wheezing, and coughing; cause may be emotional stress, chemical irritation, or exposure to an allergen.

Astian [GEOL] A European stage of geologic time: upper Pliocene, above Plaisancian, below the Pleistocene stage known as Villafranchian, Calabrian, or Günz.

astichous [BOT] Not arranged in rows.

astigmatism [ELECTR] In an electron-beam tube, a focus defect in which electrons in different axial planes come to focus at different points. [MED] A defect of vision due to irregular curvatures of the refractive surfaces of the eye so that focal points of light are distorted. [OPTICS] The failure of an optical system, such as a lens or a mirror, to image a point as a single point; the system images a point as a line segment.

astigmatizer [OPTICS] A device, as attached to a rangefinder, for drawing out a point of light into a line or band.

astigmometer [OPTICS] An instrument which measures the amount of astigmatism in an optical system.

ASTM-CFR engine [MECH ENG] A special engine developed by the Coordinating Fuel and Equipment Research Committee of the Coordinating Research Council, Inc., to determine the knock tendency of gasolines.

ASTM gum test [CHEM ENG] **1.** Determination of the amount of existing gum in a gasoline by evaporating a sample from a glass dish or an elevated temperature bath with the aid of circulating air. **2.** Any gum test carried out in accordance with an ASTM gum test procedure.

ASTM melting point [CHEM ENG] The temperature at which wax first shows a minimum rate of temperature change.

astogeny [INV ZOO] Morphological and size changes associated with zooids of aging colonial animals.

astomatal [BOT] Lacking stomata. Also known as astomous.

Astomatida [INV ZOO] An order of mouthless protozoans in the subclass Holotrichia; all species are invertebrate parasites, typically in oligochaete annelids.

astomatous [INV ZOO] Lacking a mouth, especially a cytostome, as in certain ciliates.

astomocnidae nematocyst [INV ZOO] A stinging cell whose thread has a closed end and either is adhesive or acts as a lasso to entangle prey.

astomous [BOT] **1.** Having a capsule that bursts irregularly and is not dehiscent by an operculum. **2.** *See* astomatal.

Aston dark space [ELECTR] A dark region in a glow-discharge tube which extends for a few millimeters from the cathode up to the cathode glow.

Aston process [MET] A process for making controlled-quality wrought iron synthetically.

Aston whole-number rule [PHYS] The rule which states that when expressed in atomic weight units, the atomic weights of isotopes are very nearly whole numbers, and the deviations found in samples of elements are due to the presence of several isotopes with different weights.

astraeid [INV ZOO] Of a group of corals that are imperforate.

astragalus [ANAT] The bone of the ankle which articulates with the bones of the leg. Also known as talus.

astrakanite *See* bloedite.

astral [ASTRON] Characteristic of a specific star or stars; stellar is the accepted term.

astral dome *See* astrodome.

astral lamp [ENG] An Argand lamp designed so that its light is not prevented from reaching a table beneath it by the flattened annular reservoir holding the oil.

astraphobia [PSYCH] Abnormal fear of lightning and thunderstorms. Also known as astrapophobia.

astrapophobia *See* astraphobia.

Astrapotheria [PALEON] A relatively small order of large, extinct South American mammals in the infraclass Eutheria.

Astrapotheroidea [PALEON] A suborder of extinct mammals in the order Astrapotheria, ranging from early Eocene to late Miocene.

Astrea [INV ZOO] A genus of mollusks in the class Gastropoda.

astre fictif [ASTRON] Any of several fictitious stars assumed to move along the celestial equator at uniform rates corresponding to the speeds of the several harmonic constituents of the tide-producing force.

astrionics [ELECTR] The science of adapting electronics to aerospace flight.

astro- [ASTRON] A prefix meaning star or stars and, by extension, sometimes used as the equivalent of celestial, as in astronautics.

astroballistics [MECH] The study of phenomena arising out of the motion of a solid through a gas at speeds high enough to cause ablation; for example, the interaction of a meteoroid with the atmosphere.

astrobiology [BIOL] The study of living organisms on celestial bodies other than the earth.

astrobleme [GEOL] A circular-shaped depression on the earth's surface produced by the impact of a cosmic body.

astrochanite *See* bloedite.

astrochronology [ASTRON] The use of stellar phenomena in chronology.

astrocompass [NAV] A direction-determining instrument into which can be set the coordinates of any celestial body and the latitude of the observer and which will then give an indication of azimuth, true north, and heading.

astrocyte [HISTOL] A star-shaped cell; specifically, a neuroglial cell.

astrocytoma [MED] A slow-growing glial tumor made up of cells resembling astrocytes; often it will undergo malignant change and assume the appearance and growth characteristics of a glioblastoma.

astrodome [AERO ENG] A transparent dome in the fuselage or body of an aircraft or spacecraft intended primarily to permit taking celestial observations in navigating. Also known as astral dome; navigation dome.

astrodynamics [AERO ENG] The practical application of celestial mechanics, astroballistics, propulsion theory, and allied fields to the problem of planning and directing the trajectories of space vehicles. [ASTROPHYS] The dynamics of celestial objects.

astrogation *See* astronavigation.

astrogeodetic datum orientation [GEOD] Adjustment of the ellipsoid of reference for a particular datum so that the sum of the squares of deflections of the vertical at selected points throughout the geodetic network is made as small as possible.

astrogeodetic deflection [GEOD] The angle at a point between the normal to the geoid and the normal to the ellipsoid of an astrogeodetically oriented datum. Also known as relative deflection.

astrogeodetic leveling [GEOD] A concept whereby the astrogeodetic deflections of the vertical are used to determine the separation of the ellipsoid and the geoid in studying the figure of the earth. Also known as astronomical leveling.

astrogeodetic undulations [GEOD] Separations between the geoid and astrogeodetic ellipsoid.

astrogeology [ASTRON] The science that applies the principles of geology, geochemistry, and geophysics to the moon and planets other than the earth.

astroglia [HISTOL] Neuroglia composed of astrocytes.

astrograph [MAP] A device for projecting a set of precomputed altitude curves onto a chart or plotting sheet, the curves moving with time such that if they are properly adjusted, they will remain in the correct position on the chart or plotting sheet; used in mapping the heavens.

astrographic position *See* astrometric position.

astrograph mean time [ASTRON] A form of mean time, used

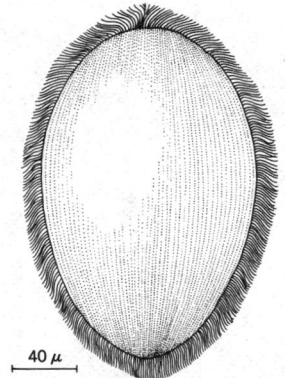

ASTOMATIDA

Anoplophyra, an example of an astomatid.

ASTRAPOTHERIA

Astrapotherium magnum skeleton. Genus is known from late Oligocene to late Miocene. *(From Riggs)*

in setting an astrograph; mean-time setting of 1200 occurs when the local hour angle of Aries is 0°.

astrogravimetric leveling [GEOD] A concept whereby a gravimetric map is used for the interpolation of the astrogeodetic deflections of the vertical to determine the separation of the ellipsoid and the geoid in studying the figure of the earth.

astrogravimetric points [GEOD] Astronomical positions corrected for the deflection of the vertical by gravimetric methods.

astrolabe [ENG] An instrument designed to observe the positions and measure the altitudes of celestial bodies.

astrometric position [ASTRON] The position of a heavenly body or space vehicle on the celestial sphere corrected for aberration but not for planetary aberration. Also known as astrographic position.

astrometry [ASTRON] The branch of astronomy dealing with the geometrical relations of the celestial bodies and their real and apparent motions.

astron [NUC PHYS] A proposed thermonuclear device in which a deuterium plasma is confined by an axial magnetic field produced by a shell of relativistic electrons.

astronaut [AERO ENG] In United States terminology, a person who rides in a space vehicle.

astronautical engineering [AERO ENG] The engineering aspects of flight in space.

astronautics [AERO ENG] 1. The art, skill, or activity of operating spacecraft. 2. The science of space flight.

astronavigation [NAV] The plotting and directing of the movement of a vehicle from within by means of observations on celestial bodies. Also known as astrogation; celestial navigation.

astronomic *See* astronomical.

astronomical [ASTRON] Of or pertaining to astronomy or to observations of the celestial bodies. Also known as astronomic.

astronomical almanac [ASTRON] A publication giving the tables of coordinates of a number of celestial bodies at a number of specific times during a given period.

astronomical azimuth [GEOD] The angle between the astronomical meridian plane of the observer and the plane containing the observed point and the true normal (vertical) of the observer, measured in the plane of the horizon, preferably clockwise from north.

astronomical bearing *See* true bearing.

astronomical camera [OPTICS] A camera designed to record either point sources (stars), extended sources (nebulae, galaxies, planets, or the sun and moon), or the spectra of celestial bodies.

astronomical clock [HOROL] A clock that shows sidereal time.

astronomical constants [ASTROPHYS] The elements of the orbits of the bodies of the solar system, their masses relative to the sun, their size, shape, orientation, rotation, and inner constitution, and the velocity of light.

astronomical coordinate system [ASTRON] Any system of spherical coordinates serving to locate astronomical objects on the celestial sphere.

astronomical date [ASTRON] Designation of epoch by year, month, day, and decimal fraction.

astronomical day [ASTRON] A mean solar day beginning at mean noon, 12 hours later than the beginning of the civil day of the same date; astronomers now generally use the civil day.

astronomical distance [ASTRON] The distance of a celestial body expressed in units such as the light-year, astronomical unit, and parsec.

astronomical eclipse *See* eclipse.

astronomical ephemeris *See* ephemeris.

astronomical equator [GEOD] An imaginary line on the surface of the earth connecting points having 0° astronomical latitude. Also known as terrestrial equator.

astronomical instruments [ENG] Specific kinds of telescopes and ancillary equipment used by astronomers to study the positions, motions, and composition of stars and members of the solar system.

astronomical latitude [GEOD] Angular distance between the direction of gravity (plumb line) and the plane of the celestial equator; applies only to positions on the earth and is reckoned from the astronomical equator.

astronomical leveling *See* astrogeodetic leveling.

astronomical longitude [GEOD] The angle between the plane of the reference meridian and the plane of the local celestial meridian.

astronomical meridian [GEOD] A line on the surface of the earth connecting points having the same astronomical longitude. Also known as terrestrial meridian.

astronomical meridian plane [GEOD] A plane that contains the vertical of the observer and is parallel to the instantaneous rotation axis of the earth.

astronomical nutation [ASTRON] A small periodic motion of the celestial pole of celestial bodies, including the earth, with respect to the pole of the ecliptic.

astronomical observatory [ASTRON] A building designed and equipped for making observations of astronomical phenomena.

astronomical parallel [GEOD] A line connecting points having the same astronomical latitude.

astronomical photography [OPTICS] The use of the photographic process to record surface features of celestial objects, their positions and motions (for measurement), and their radiation (photometry) and spectra (spectroscopy).

astronomical position [GEOD] 1. A point on the earth whose coordinates have been determined as a result of observation of celestial bodies. Also known as astronomical station. 2. A point on the earth defined in terms of astronomical latitude and longitude.

astronomical refraction [GEOPHYS] The bending of a ray of celestial radiation as it passes through atmospheric layers of increasing density.

astronomical scintillation [ASTROPHYS] Any scintillation phenomena, such as irregular oscillatory motion, variation of intensity, and color fluctuation, observed in the light emanating from an extraterrestrial source. Also known as stellar scintillation.

astronomical seeing *See* seeing.

astronomical spectrograph [SPECT] An instrument used to photograph spectra of stars.

astronomical spectroscopy [SPECT] The use of spectrographs in conjunction with telescopes to obtain observational data on the velocities and physical conditions of astronomical objects.

astronomical station *See* astronomical position.

astronomical surveying [GEOD] The celestial determination of latitude and longitude; separations are calculated by computing distances corresponding to measured angular displacements along the reference spheroid.

astronomical telescope [OPTICS] An optical instrument that increases the angle under which distant astronomical objects are seen.

astronomical theodolite *See* altazimuth.

astronomical tide [OCEANOGR] An equilibrium tide due to attractions of the sun and moon. Also known as astronomic tide.

astronomical time [ASTRON] Solar time in an astronomical day that begins at noon.

astronomical traverse [ENG] A survey traverse in which the geographic positions of the stations are obtained from astronomical observations, and lengths and azimuths of lines are obtained by computation.

astronomical twilight [ASTRON] The period of incomplete darkness when the center of the sun is more than 6° but more than 18° below the celestial horizon.

astronomical unit [ASTRON] Abbreviated AU. 1. A measure for distance within the solar system equal to the mean distance between earth and sun, that is, 149,599,000 kilometers. 2. The semimajor axis of the elliptical orbit of earth.

astronomical year *See* tropical year.

astronomic tide *See* astronomical tide.

astronomy [SCI TECH] The science concerned with celestial bodies and the observation and interpretation of the radiation received in the vicinity of the earth from the component parts of the universe.

Astropectinidae [INV ZOO] A family of echinoderms in the suborder Paxillosina occurring in all seas from tidal level downward.

astrophobia [PSYCH] Abnormal fear of stars and celestial space.

ASTRONOMICAL COORDINATE SYSTEM

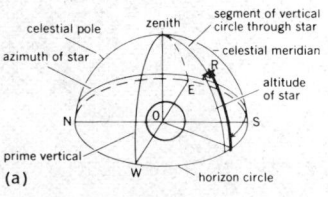

(a)

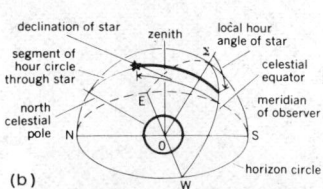

(b)

Two types of astronomical coordinate systems. (a) Horizon system. (b) Equatorial system.

astrophyllite [MINERAL] $(K,Na)_3 (Fe,Mn)_7 Ti_2Si_8O_{24}(O,OH)_7$ A mineral composed of a basic silicate of potassium or sodium, iron or manganese, and titanium.

astrophysics [ASTRON] A branch of astronomy that treats of the physical properties of celestial bodies, such as luminosity, size, mass, density, temperature, and chemical composition, and with their origin and evaluation.

astropyle [INV ZOO] A small, rounded projection from the central capsule of some radiolarians.

astrosphere [CYTOL] The center of the aster exclusive of the rays.

astrotracker [NAV] An automatic sextant which has the ability to sight on and continuously to track selected stars throughout the day and night, providing continuous heading and position data with no intervention on the part of the airman. Also known as star tracker.

Asturian orogeny [GEOL] Mid-Upper Carboniferous diastrophism.

asty [INV ZOO] A bryozoan colony.

astylar [ARCH] Not having columns or pilasters.

asulcal [BIOL] Without a sulcus.

A supply [ELECTR] Battery, transformer filament winding, or other voltage source that supplies power for heating filaments of vacuum tubes. Also known as A power supply.

asymbolia [MED] An aphasia in which there is an inability to understand or use acquired symbols, such as speech, writing, or gestures, as a means of communication.

asymmetrical bedding [GEOL] An order in which lithologic types or facies follow one another in a circuitous arrangement so that, for example, the sequence of types 1-2-3-1-2-3-1-2-3 indicates asymmetry (while the sequence 1-2-3-2-1-2-3-2-1 indicates symmetrical bedding).

asymmetrical cell [ELECTR] A cell, such as a photoelectric cell, in which the impedance to the flow of current in one direction is greater than in the other direction.

asymmetrical fold [GEOL] A fold in which one limb dips more steeply than the other.

asymmetrical laccolith [GEOL] A laccolith in which the beds dip at conspicuously different angles in different sectors.

asymmetrical ripple mark [GEOL] The normal form of ripple mark, with short downstream slopes and comparatively long, gentle upstream slopes.

asymmetrical-side-band transmission *See* vestigial-side-band transmission.

asymmetrical vein [GEOL] A crustified vein of geologic material with unlike layers on each side.

asymmetric carbon atom [ORG CHEM] A carbon atom with four different radicals or atoms bonded to it.

asymmetric rotor [MECH ENG] A rotating element for which the axis (center of rotation) is not centered in the element.

asymmetric synthesis [ORG CHEM] Chemical synthesis of a pure enantiomer, or of an enantiomorphic mixture in which one enantiomer predominates, without the use of resolution.

asymmetric top [MECH] A system that has all three principal moments of inertia different.

asymmetry [PHYS CHEM] The geometrical design of a molecule, atom, or ion that cannot be divided into like portions by one or more hypothetical planes. Also known as molecular asymmetry.

asymptote [MATH] **1.** A line approached by a curve in the limit as the curve approaches infinity. **2.** The limit of the tangents to a curve as the point of contact approaches infinity.

asymptote of convergence *See* convergence line.

asymptotic cone of acceptance [GEOPHYS] The solid angle in the celestial sphere from which particles have to come in order to contribute significantly to the counting rate of a given neutron monitor on the surface of the earth.

asymptotic curve [MATH] A curve on a surface whose osculating plane at each point is the same as the tangent plane to the surface.

asymptotic direction of arrival [GEOPHYS] The direction at infinity of a positively charged particle, with given rigidity, which impinges in a given direction at a given point on the surface of the earth, after passing through the geomagnetic field.

asymptotic expansion [MATH] A series of the form $a_0 + (a_1/x) + (a_2/x^2) + \ldots + (a_n/x^n) + \ldots$ is an asymptotic expansion of the function $f(x)$ if there exists a number N such that for all $n > N$ the quantity $x^n [f(x) - S_n(x)]$ approaches zero as x approaches infinity, where $S_n(x)$ is the sum of the first n terms in the series. Also known as asymptotic series.

asymptotic formula [MATH] A statement of equality between two functions which is not a true equality but which means the ratio of the two functions approaches 1 as the variable approaches some value, usually infinity.

asymptotic series *See* asymptotic expansion.

asynchronous [ADP] Operating at a speed determined by the circuit functions rather than by timing signals. [PHYS] Not synchronous.

asynchronous computer [ADP] A computer in which the performance of any operation starts as a result of a signal that the previous operation has been completed, rather than on a signal from a master clock.

asynchronous control [CONT SYS] A method of control in which the time allotted for performing an operation depends on the time actually required for the operation, rather than on a predetermined fraction of a fixed machine cycle.

asynchronous device [CONT SYS] A device in which the speed of operation is not related to any frequency in the system to which it is connected.

asynchronous machine [ELEC] An ac machine whose speed is not proportional to the frequency of the power line.

asynchronous timing [IND ENG] A simulation method for queues in which the system model is updated at each arrival or departure, resulting in the master clock being increased by a variable amount.

asyndetic [ADP] **1.** Omitting conjunctions or connectives. **2.** Pertaining to a catalog without cross references.

asynergia [MED] Faulty coordination of groups of organs or muscles normally acting in unison; particularly, the abnormal state of muscle antagonism in cerebellar disease.

asystole [MED] The absence of cardiac contraction; cardiac arrest.

at *See* technical atmosphere.

At *See* astatine.

ata [MECH] A unit of absolute pressure in the metric technical system equal to 1 technical atmosphere.

Atabrine [PHARM] A trade name for quinacrine.

atacamite [MINERAL] $Cu_2Cl(OH)_3$ A native, green hydrous copper oxychloride crystallizing in the orthorhombic system.

atactic [ORG CHEM] Of the configuration for a polymer, having the opposite steric configurations for the carbon atoms of the polymer chain occur in equal frequency and more or less at random.

atactostele [BOT] A type of monocotyledonous siphonostele in which the vascular bundles are dispersed irregularly throughout the center of the stem.

atavism [EVOL] Appearance of a distant ancestral form of an organism or one of its parts due to reactivation of ancestral genes.

ataxia [MED] Lack of muscular coordination due to any of several nervous system diseases.

ataxic [GEOL] Pertaining to unstratified ore deposits.

ataxiophobia [PSYCH] Abnormal fear of disorder.

ataxite [GEOL] An iron meteorite that lacks the structure of either hexahedrite or octahedrite and contains more than 10% nickel. [PETR] A taxitic rock whose components are arranged in a breccialike manner, that is, there is no specific arrangement.

ATC *See* air-traffic control.

ATCRBS *See* air-traffic control radar beacon system.

atectonic [GEOL] Of an event that occurs when orogeny is not taking place.

atectonic pluton [GEOL] A pluton that is emplaced when orogeny is not occurring.

atelectasis [MED] **1.** Total or partial collapsed state of the lung. **2.** Failure of the lung to expand at birth.

Ateleopoidei [VERT ZOO] A family of oceanic fishes in the order Cetomimiformes characterized by an elongate body, lack of a dorsal fin, and an anal fin continuous with the caudal fin.

ASTROTRACKER

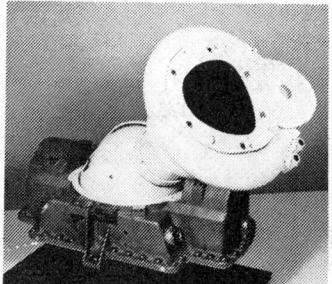

One of six course-alignment star trackers used on the orbiting astronomical observatory. (Kollsman Instrument Co.)

ASYMPTOTE

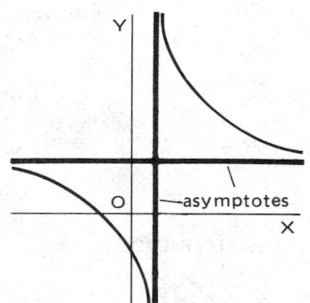

Asymptotes of a hyperbola.

ATACTIC

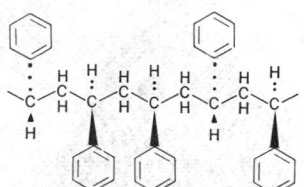

Atactic type of spatially oriented polymers.

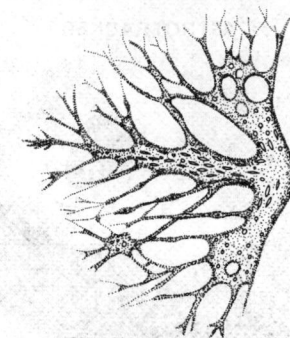

ATHALAMIDA

A representative athalamid, *Biomyxa vagans (after Leidy)*. *(From R. P. Hall, Protozoology, Prentice-Hall, 1953)*

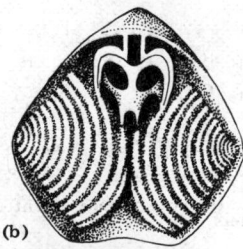

ATHYRIDIDINA

(a)

(b)

Composita, a genus in Athyrididina: *(a)* dorsal view, *(b)* ventral view with pedicle cut away to show spire. *(From R. C. Moore, ed., Treatise on Invertebrate Palentology, pt. H, Geological Society of America, Inc., and University of Kansas Press, 1965)*

atelestite [MINERAL] $Bi_8(AsO_4)_3O_5(OH)_5$ A yellow mineral consisting of basic bismuth arsenate and occurring in minute crystals; specific gravity is 6.82.

ateliosis [MED] Infantilism or dwarfism characterized by general, but proportional, underdevelopment and normal intelligence; associated with anterior pituitary deficiencies.

Atelopodidae [VERT ZOO] A family of small, brilliantly colored South and Central American frogs in the suborder Procoela.

Atelostomata [INV ZOO] A superorder of echinoderms in the subclass Euechinoidea characterized by a rigid, exocylic test and lacking a lantern, or jaw, apparatus.

Athalamida [INV ZOO] An order of naked amebas of the subclass Granuloreticulosia in which pseudopodia are branched and threadlike (reticulopodia).

Athecanephria [INV ZOO] An order of tube-dwelling, tentaculate animals in the class Pogonophora characterized by a saclike anterior coelom.

athecate [INV ZOO] Lacking a theca.

Atherinidae [VERT ZOO] The silversides, a family of actinopterygian fishes of the order Atheriniformes.

Atheriniformes [VERT ZOO] An order of actinopterygian fishes in the infraclass Teleostei, including flyingfishes, needlefishes, killifishes, silversides, and allied species.

athermalize [ENG] To make independent of temperature or of thermal effects.

athermal transformation [PHYS] A chemical or physical change not requiring a change in the temperature of the substance, as in the formation of martensite.

athermancy [ELECTROMAG] Property of a substance which cannot transmit infrared radiation.

atheroma [MED] **1.** Fatty degeneration of the inner arterial walls. **2.** A fatty cyst.

atherosclerosis [MED] Deposition of lipid with proliferation of fibrous connective tissue cells in the inner walls of the arteries.

athetosis [MED] Slow, recurrent, involuntary wormlike movements of various parts of the body associated with lesions of the basal ganglia.

athetotic speech [MED] Disorder of articulation rhythm involving a general jerkiness in speech production that interferes with the normal rate of speech; associated with athetosis.

Athey wheel [MECH ENG] A crawler wheel assembly used on tractors for moving over soft terrain.

Athiorhodaceae [MICROBIO] The nonsulfur photosynthetic bacteria, a family of small, gram-negative, nonsporeforming, motile bacteria in the suborder Rhodobacteriineae.

athlete's foot [MED] Dermatophytosis of the feet, usually affecting the skin between the toes.

athodyd [AERO ENG] A type of jet engine, consisting essentially of a duct or tube of varying diameter and open at both ends, which admits air at one end, compresses it by the forward motion of the engine, adds heat to it by the combustion of fuel, and discharges the resulting gases at the other end to produce thrust.

athrocyte [HISTOL] A cell that engulfs extraneous material and stores it as granules in the cytoplasm.

athrogenic [PETR] Of or pertaining to pyroclastics.

athwartship [NAV ARCH] Perpendicular to the fore and aft centerline of a ship.

Athyrididina [PALEON] A suborder of fossil articulate brachiopods in the order Spiriferida characterized by laterally or, more rarely, ventrally directed spires.

Atlantacea [INV ZOO] A superfamily of mollusks in the subclass Prosobranchia.

Atlantic Ocean [GEOGR] The large body of water separating the continents of North and South America from Europe and Africa and extending from the Arctic Ocean to the continent of Antarctica.

Atlantic series [PETR] A great group of igneous rocks, based on tectonic setting, found in nonorogenic areas, often associated with block sinking and great crustal instability, and erupted along faults and fissures or through explosion vents. Also known as Atlantic suite.

Atlantic standard time See Atlantic time.

Atlantic suite See Atlantic series.

Atlantic time [ASTRON] A time zone; the fourth zone west of Greenwich. Also known as Atlantic standard time.

Atlantic-type continental margin [GEOL] A continental margin typified by that of the Atlantic which is aseismic because oceanic and continental lithospheres are coupled.

atlantite [PETR] An olivine-bearing nepheline tephrite.

atlas [ANAT] The first cervical vertebra. [MAP] A collection of charts or maps kept loose or bound in a volume.

Atlas [ORD] A U.S. Air Force surface-to-surface intercontinental ballistic missile having a range of about 6000 miles and capable of carrying nuclear warheads.

Atlas-Centaur launch vehicle [AERO ENG] A two-stage rocket consisting of an Atlas first stage and a Centaur second stage; used for launching unmanned spacecraft.

atlas grid [MAP] A reference system that permits the designation of the location of a point or an area on a map, photo, or other graphic in terms of numbers and letters. Also known as alphanumeric grid.

Atlas-Johnson tubing joint [PETRO ENG] A tapered, screw-on joint for connecting lengths of tubing for oil-well casing strings.

atm See atmosphere.

atmidometer See atmometer.

atmoclast [GEOL] A fragment of rock broken off in place by atmospheric weathering.

atmoclastic [PETR] Of a clastic rock, composed of atmoclasts that have been recemented without rearrangement.

atmogenic [GEOL] Of rocks, minerals, and other deposits derived directly from the atmosphere by condensation, wind action, or deposition from volcanic vapors; for example, snow.

atmolith [GEOL] A rock precipitated from the atmosphere, that is, an atmogenic rock.

atmolysis [FL MECH] The separation of gas mixtures by using their relative diffusibility through a porous partition.

atmo-meter See meter-atmosphere.

atmometer [ENG] The general name for an instrument which measures the evaporation rate of water into the atmosphere. Also known as atmidometer; evaporation gage; evaporimeter.

atmophile element [METEOROL] **1.** Any of the most typical elements of the atmosphere (hydrogen, carbon, nitrogen, oxygen, iodine, mercury, and inert gases). **2.** Any of the elements which either occur in the uncombined state or, as volatile compounds, concentrate in the gaseous primordial atmosphere.

atmosphere [MECH] A unit of pressure equal to 1.013250×10^6 dynes/cm^2, which is the air pressure at mean sea level. Abbreviated atm. Also known as standard atmosphere. [METEOROL] The gaseous envelope surrounding a planet or celestial body.

atmospheric absorption [GEOPHYS] The reduction in energy of microwaves by gases and moisture in the atmosphere.

atmospheric acoustics [ACOUS] The science of sound waves in the open air.

atmospheric attenuation [GEOPHYS] A process in which the flux density of a parallel beam of energy decreases with increasing distance from the source as a result of absorption or scattering by the atmosphere.

atmospheric boil See terrestrial scintillation.

atmospheric boundary layer See surface boundary layer.

atmospheric braking [AERO ENG] **1.** Slowing down an object entering the atmosphere of the earth or other planet from space by using the drag exerted by air or other gas particles in the atmosphere. **2.** The action of the drag so exerted.

atmospheric chemistry [METEOROL] The study of the production, transport, modification, and removal of atmospheric constituents in the troposphere and stratosphere.

atmospheric composition [METEOROL] The chemical abundance in the earth's atmosphere of its constituents, including nitrogen, oxygen, argon, carbon dioxide, water vapor, ozone, neon, helium, krypton, methane, hydrogen, and nitrous oxide.

atmospheric condensation [METEOROL] The transformation of water in the air from a vapor phase to dew, fog, or cloud.

atmospheric control [ORD] **1.** Any device or system designed to operate movable aerodynamic control surfaces to direct a guided missile in an atmosphere dense enough for such controls to be effective. **2.** The control provided by such devices.

atmospheric cooler [MECH ENG] A fluids cooler that utilizes the cooling effect of ambient air surrounding the hot, fluids-filled tubes.

atmospheric corrosion [MET] The gradual destruction or alteration of a metal or alloy by contact with substances present in the atmosphere, such as oxygen, carbon dioxide, water vapor, and sulfur and chlorine compounds.

atmospheric density [METEOROL] The ratio of the mass of a portion of the atmosphere to the volume it occupies.

atmospheric diffusion [METEOROL] The exchange of fluid parcels between regions in the atmosphere in the apparently random motions of a scale too small to be treated by equations of motion.

atmospheric distillation [CHEM ENG] Distillation operation conducted at atmospheric pressure, in contrast to vacuum distillation or pressure distillation.

atmospheric disturbance [METEOROL] Any agitation or disruption of the atmospheric steady state.

atmospheric drag [FL MECH] A major perturbation of close artificial satellite orbits caused by the resistance of the atmosphere; the secular effects are decreasing eccentricity, semidiameter, and period.

atmospheric duct [GEOPHYS] A stratum of the troposphere within which the refractive index varies so as to confine within the limits of the stratum the propagation of an abnormally large proportion of any radiation of sufficiently high frequency, as in a mirage.

atmospheric electric field [GEOPHYS] The atmosphere's electric field strength in volts per meter at any specified point in time and space; near the earth's surface, in fair-weather areas, a typical datum is about 100 and the field is directed vertically in such a way as to drive positive charges downward.

atmospheric electricity [GEOPHYS] The electrical processes occurring in the lower atmosphere, including both the intense local electrification accompanying storms and the much weaker fair-weather electrical activity over the entire globe produced by the electrified storms continuously in progress.

atmospheric entry [AERO ENG] The penetration of any planetary atmosphere by any object from outer space; specifically, the penetration of the earth's atmosphere by a manned or unmanned capsule or spacecraft.

atmospheric evaporation [HYD] The exchange of water between the earth's oceans, lakes, rivers, ice, snow, and soil and the atmosphere.

atmospheric gas [METEOROL] One of the constituents of air, which is a gaseous mixture primarily of nitrogen, oxygen, argon, carbon dioxide, water vapor, ozone, neon, helium, krypton, methane, hydrogen, and nitrous oxide.

atmospheric general circulation [METEOROL] The statistical mean global flow pattern of the atmosphere.

atmospheric impurity [ENG] An extraneous substance mixed as a contaminant with the air of the atmosphere.

atmospheric interference [GEOPHYS] Electromagnetic radiation, caused by natural electrical disturbances in the atmosphere, which interferes with radio systems. Also known as atmospherics; sferics; strays.

atmospheric ionization [GEOPHYS] The process by which neutral atmospheric molecules are rendered electrically charged chiefly by collisions with high-energy particles.

atmospheric lapse rate *See* environmental lapse rate.

atmospheric layer *See* atmospheric shell.

atmospheric noise [ELECTR] Noise heard during radio reception due to atmospheric interference.

atmospheric optics *See* meteorological optics.

atmospheric physics [GEOPHYS] The study of the physical phenomena of the atmosphere.

atmospheric pressure [METEOROL] The pressure at any point in an atmosphere due solely to the weight of the atmospheric gases above the point concerned. Also known as barometric pressure.

atmospheric radiation [GEOPHYS] Infrared radiation emitted by or being propagated through the atmosphere.

atmospheric radio wave [ELECTROMAG] Radio wave that is propagated by reflection in the atmosphere; may include

either the ionospheric wave or the tropospheric wave, or both.

atmospheric radio window [ELECTROMAG] That portion of the frequency spectrum that will allow radio-frequency waves to pass through the earth's atmosphere (approximately 10 to 10,000 megahertz).

atmospheric refraction [GEOPHYS] **1.** The angular difference between the apparent zenith distance of a celestial body and its true zenith distance, produced by refraction effects as the light from the body penetrates the atmosphere. **2.** Any refraction caused by the atmosphere's normal decrease in density with height.

atmospheric region *See* atmospheric shell.

atmospherics *See* atmospheric interference.

atmospheric scattering [GEOPHYS] A change in the direction of propagation, frequency, or polarization of electromagnetic radiation caused by interaction with the atoms of the atmosphere.

atmospheric shell [METEOROL] Any one of a number of strata or layers of the earth's atmosphere; temperature distribution is the most common criterion used for denoting the various shells. Also known as atmospheric layer; atmospheric region.

atmospheric shimmer *See* terrestrial scintillation.

atmospheric sounding [METEOROL] A measurement of atmospheric conditions aloft, above the effective range of surface weather observations.

atmospheric structure [METEOROL] Atmospheric characteristics, including wind direction and velocity, altitude, air density, and velocity of sound.

atmospheric suspensoids [METEOROL] Moderately finely divided particles suspended in the atmosphere; dust is an example.

atmospheric tide [GEOPHYS] Periodic global motions of the earth's atmosphere, produced by gravitational action of the sun and moon; amplitudes are minute except in the upper atmosphere.

atmospheric turbulence [METEOROL] Apparently random fluctuations of the atmosphere that often constitute major deformations of its state of fluid flow.

Atokan [GEOL] A North American provincial series in lower Middle Pennsylvanian geologic time, above Morrowan, below Desmoinesian.

atoll [GEOGR] A ring-shaped coral reef that surrounds a lagoon without projecting land area and that is surrounded by open sea.

atoll texture [GEOL] The surrounding of a ring of one mineral with another mineral, or minerals, within and without the ring. Also known as core texture.

atom [CHEM] The individual structure which constitutes the basic unit of any chemical element. [MATH] An element, A, of a measure algebra, other than the zero element, which has the property that any element which is equal to or less than A is either equal to A or equal to the zero element.

atomic absorption coefficient [PHYS] The linear absorption coefficient divided by the number of atoms per unit volume.

atomic airburst [ORD] The explosion of an atomic weapon in the air at a height greater than the maximum radius of the fireball.

atomic ammunition [ORD] Formerly, ammunition deriving its explosive force from nuclear fission; now, all ammunition which derives its explosive force from a chain reaction of active nuclear material.

atomic battery *See* nuclear battery.

atomic beam [PHYS] A stream of atoms, which may or may not be ionized.

atomic-beam frequency standard [PHYS] A source of precisely timed signals which are derived from an atomic-beam resonance, such as a cesium-beam cell or a hydrogen maser.

atomic beam resonance [PHYS] Phenomenon in which an oscillating magnetic field, superimposed on a uniform magnetic field at right angles to it, causes transitions between states with different magnetic quantum numbers of the nuclei of atoms in a beam passing through the field; the transitions occur only when the frequency of the oscillating field assumes certain characteristic values.

atomic bomb [ORD] Also known as A bomb. **1.** A device for

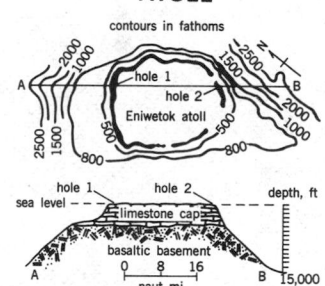

ATOLL

The Eniwetok atoll in the Marshall Islands.

ATOMIC ROCKET

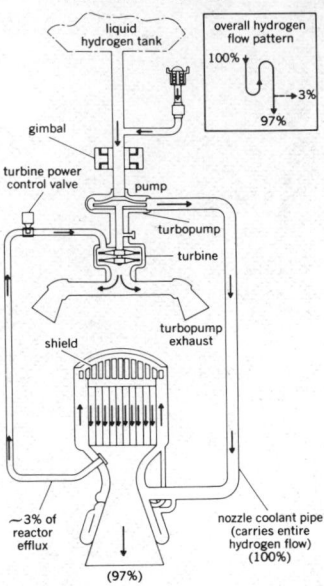

Schematic diagram of the hot-bleed cycle used for the Nerva nuclear rocket engine. The Nerva engine can start and restart itself using a "bootstrap" process, wherein start-up energy is derived from the pressurized propellant and sensible heat of the reactor. *(From W. R. Corliss, Nuclear Propulsion for Space, USAEC, 1967)*

suddenly producing an explosively rapid neutron chain reaction in a fissile material such as uranium-235 or plutonium-239. Also known as fission bomb. **2.** Any explosive device which derives its energy from nuclear reactions, including a fusion bomb. Also known as nuclear bomb.

atomic charge [ATOM PHYS] The electric charge of an ion, equal to the number of electrons the atom has gained or lost in its ionization multiplied by the charge on one electron.

atomic clock [HOROL] An electronic clock whose frequency is supplied or governed by the natural resonance frequencies of atoms or molecules of suitable substances.

atomic cloud [NUCLEO] The cloud of hot gases, smoke, dust, and other matter that is carried aloft after the explosion of a nuclear weapon in the air or near the surface; frequently has a mushroom shape.

atomic constants [PHYS] The physical constants which play a fundamental role in atomic physics, including the electronic charge, electronic mass, speed of light, Avogadro number, and Planck's constant.

atomic device *See* atomic weapon.

atomic diamagnetism [ATOM PHYS] Diamagnetic ionic susceptibility, important in providing correction factors for measured magnetic susceptibilities; calculated theoretically by considering electron density distributions summed for each electron shell.

atomic energy *See* nuclear energy.

atomic energy level [ATOM PHYS] A definite value of energy possible for an atom, either in the ground state or an excited condition.

atomic fission *See* fission.

atomic form factor *See* atomic scattering factor.

atomic fusion *See* fusion.

atomic ground state [ATOM PHYS] The state of lowest energy in which an atom can exist. Also known as atomic unexcited state.

atomic heat capacity [PHYS CHEM] The heat capacity of a gram-atomic weight of an element.

atomic hydrogen [CHEM] Gaseous hydrogen whose molecules are dissociated into atoms.

atomic hydrogen maser [PHYS] An active clock oscillator using a hydrogen maser to produce a very stable signal at 1420 megahertz.

atomic hydrogen welding [MET] An arc welding process in which hydrogen gas dissociated by the arc recombines outside the arc to provide intense heat and protection against oxidation for the weld.

atomic magnet [ATOM PHYS] An atom which possesses a magnetic moment either in the ground state or in an excited state.

atomic magnetic moment [ATOM PHYS] A magnetic moment, permanent or temporary, associated with an atom, measured in magnetons.

atomic mass [PHYS] The mass of a neutral atom usually expressed in atomic mass units.

atomic mass unit [PHYS] An arbitrarily defined unit in terms of which the masses of individual atoms are expressed; the standard is the unit of mass equal to $1/12$ the mass of the carbon atom, having as nucleus the isotope with mass number 12. Abbreviated amu. Also known as dalton.

atomic nucleus *See* nucleus.

atomic number [NUC PHYS] The number of protons in an atomic nucleus.

atomic orbital [ATOM PHYS] The space-dependent part of a wave function describing an electron in an atom.

atomic paramagnetism [ELECTROMAG] The result of a permanent magnetic moment in an atom.

atomic particle [ATOM PHYS] One of the particles of which an atom is constituted, as an electron, neutron, or proton.

atomic percent [CHEM] The number of atoms of an element in 100 atoms representative of a substance.

atomic photoelectric effect *See* photoionization.

atomic physics [PHYS] The science concerned with the structure of the atom, the characteristics of the elementary particles of which the atom is composed, and the processes involved in the interactions of radiant energy with matter.

atomic pile *See* nuclear reactor.

atomic polarization [PHYS CHEM] Polarization of a material

arising from the change in dipole moment accompanying the stretching of chemical bonds between unlike atoms in molecules.

atomic power plant [NUCLEO] A means for converting stored nuclear energy into work, such as a nuclear electric power generating station.

atomic radius [PHYS CHEM] **1.** Half the distance of closest approach of two like atoms not united by a bond. **2.** The experimentally determined radius of an atom in a covalently bonded compound.

atomic reactor *See* nuclear reactor.

atomic rocket [AERO ENG] A rocket propelled by an engine in which the energy for the jetstream is to be generated by nuclear fission or fusion. Also known as nuclear rocket.

atomic scattering factor [PHYS] A quantity which expresses the efficiency with which x-rays of a stated wavelength are scattered into a given direction by a particular atom, measured in terms of the corresponding scattering by a point electron. Also known as atomic form factor.

atomic second [PHYS] As defined in 1967, the duration of 9,192,631,770 periods of the radiation corresponding to the two hyperfine levels of the fundamental state of the atom of cesium-133.

atomic spectroscopy [SPECT] The branch of physics concerned with the production, measurement, and interpretation of spectra arising from either emission or absorption of electromagnetic radiation by atoms.

atomic spectrum [SPECT] The spectrum of radiations due to transitions between energy levels in an atom, either absorption or emission.

atomic standard [PHYS] Any supposedly immutable property of an atom, such as the wavelength or frequency of a characteristic spectral line, in terms of which a unit of a physical quantity is defined.

atomic stopping power [NUCLEO] For an ionizing particle passing through an element, the particle's energy loss per atom within a unit area normal to the particle's path; equal to the linear energy transfer (energy loss per unit path length) divided by the number of atoms per unit volume.

atomic structure [ATOM PHYS] The arrangement of the parts of an atom, which consists of a massive, positively charged nucleus surrounded by a cloud of electrons arranged in orbits describable in terms of quantum mechanics.

atomic surface burst [ORD] An atomic missile burst at an elevation such that the fireball touches the ground.

atomic susceptibility [ELECTROMAG] The magnetization of a material per atom per unit of applied field; measured in ergs/oersted/atom.

atomic theory [CHEM] The assumption that matter is composed of particles called atoms and that these are the limit to which matter can be subdivided.

atomic time [HOROL] Any time system standardized with reference to an atomic resonance, such as the international standard cesium-133 transition.

atomic underground burst [ORD] The explosion of an atomic weapon with its center beneath the surface of the ground.

atomic underwater burst [ORD] The explosion of an atomic weapon with its center beneath the surface of the water.

atomic unexcited state *See* atomic ground state.

atomic units *See* Hartree units.

atomic vibration [ATOM PHYS] Periodic, nearly harmonic changes in position of the atoms in a molecule giving rise to many properties of matter, including molecular spectra, heat capacity, and heat conduction.

atomic volume [PHYS CHEM] The volume occupied by 1 gram-atom of an element in the solid state.

atomic weapon [ORD] Any bomb, warhead, or projectile using active nuclear material to cause a chain reaction upon detonation. Also known as atomic device; nuclear weapon.

atomic weight [CHEM] The relative mass of an atom based on a scale in which a specific carbon atom (carbon-12) is assigned a mass value of 12. Abbreviated at. wt.

atomization [MECH ENG] The mechanical subdivision of a bulk liquid or meltable solid, such as certain metals, to produce drops, which vary in diameter depending on the process from under 10 to over 1000 micrometers.

atomizer [MECH ENG] A device that produces a mechanical subdivision of a bulk liquid, as by spraying, sprinkling, misting, or nebulizing.

atomizer burner [MECH ENG] A liquid-fuel burner that atomizes the unignited fuel into a fine spray as it enters the combustion zone.

atomizer mill [MECH ENG] A solids grinder, the product from which is a fine powder.

atom probe [ENG] An instrument for identifying a single atom or molecule on a metal surface; it consists of a field ion microscope with a probe hole in its screen opening into a mass spectrometer; atoms that are removed from the specimen by pulsed field evaporation fly through the probe hole and are detected in the mass spectrometer.

atom smasher *See* particle accelerator.

atony [MED] Absence or extremely low degree of tonus.

atopic allergy [MED] A type of immediate hypersensitivity in humans resulting from spontaneous sensitization, usually by inhaled or ingested antigens; for example, asthma, hay fever, or hives.

atopite [MINERAL] A yellow or brown variety of romeite that contains fluorine.

atopy [MED] Clinically evident hypersensitivity.

ATP *See* adenosinetriphosphate.

ATPase *See* adenosinetriphosphatase.

ATR *See* attenuated total reflectance.

A trace [ELECTR] The first trace of an oscilloscope, such as the upper trace of a loran indicator.

Atractidae [INV ZOO] A family of parasitic nematodes in the superfamily Oxyuroidea.

atran [NAV] An acronym for automatic terrain recognition and navigation, a system which depends upon the correlation of terrain images appearing on a radar cathode-ray tube with previously prepared maps or simulated radar images of the terrain.

atrazine [MATER] An organic herbicide that acts through the soil.

atresia [MED] Imperforation or closure of a natural orifice or passage of the body.

atrial flutter [MED] A cardiac arrhythmia characterized by rapid, irregular atrial impulses and ineffective atrial contractions; the heartbeat varies from 60 to 180 per minute and is grossly irregular in intensity and rhythm. Also known as auricular flutter.

atrial septum [ANAT] The muscular wall between the atria of the heart. Also known as interatrial septum.

atrichia [MED] Any congenital or acquired condition in which hair is essentially absent.

Atrichornithidae [VERT ZOO] The scrubbirds, a family of suboscine perching birds in the suborder Menurae.

atrichous [CYTOL] Lacking flagella.

atriopore [ZOO] The opening of an atrium as seen in lancelets and tunicates.

atrioventricular canal [EMBRYO] The common passage between the atria and ventricles in the heart of the mammalian embryo before division of the organ into right and left sides.

atrioventricularis communis [MED] A congenital malformation of the heart in which partitioning has not occurred.

atrioventricular valve [ANAT] A structure located at the orifice between the atrium and ventricle which maintains a unidirectional blood flow through the heart.

atrium [ANAT] **1.** The heart chamber that receives blood from the veins. **2.** The main part of the tympanic cavity, below the malleus. **3.** The external chamber to receive water from the gills in lancelets and tunicates.

atrophic arthritis *See* rheumatoid arthritis.

atrophic gastritis [MED] Chronic inflammation of the stomach with atrophy of the mucosa.

atrophy [MED] Diminution in the size of a cell, tissue, or organ that was once fully developed and of normal size.

atropine [PHARM] $C_{17}H_{23}O_3N$ An alkaloid extracted from *Atropa belladonna* and related plants of the family Solanaceae; used to relieve muscle spasms and pain, and to dilate the pupil of the eye.

atropous *See* orthotropous.

ATR tube *See* anti-transmit-receive tube.

Atrypidina [PALEON] A suborder of fossil articulate brachiopods in the order Spiriferida.

ATS *See* Administrative Terminal System; applications technology satellite.

attached groundwater [HYD] The portion of subsurface water adhering to pore walls.

attached shock *See* attached shock wave.

attached shock wave [FL MECH] An oblique or conical shock wave that appears to be in contact with the leading edge of an airfoil or the nose of a body in a supersonic flow field. Also known as attached shock.

attached thermometer [ENG] A thermometer which is attached to an instrument to determine its operating temperature.

attachment plug [ELEC] A device having an attached flexible cord containing conductors, and capable of being inserted in a receptacle so as to form an electrical connection between the conductors in the cord and conductors permanently connected to the receptacle.

attack director [ADP] An electromechanical analog computer which is designed for surface antisubmarine use and which computes continuous solution of several lines of submarine attack; it is part of several antisubmarine fire control systems.

attack heading [NAV] **1.** The interceptor heading during the attack phase which will achieve the desired track-crossing angle. **2.** The assigned magnetic-compass heading to be flown by aircraft during the delivery phase of an air strike.

attack plane [AERO ENG] A multiweapon carrier aircraft which can carry bombs, torpedoes, and rockets.

attapulgite [MINERAL] $(Mg,Al)_2Si_4O_{10}(OH)\cdot4H_2O$ A clay mineral with a needlelike shape from Georgia and Florida; active ingredient in most fuller's earth, and used as a suspending agent, as an oil well drilling fluid, as a thickener in latex paint. Also known as palygorskite.

attar of roses *See* rose oil.

attemperation [ENG] The regulation of the temperature of a substance.

attemperation of steam [MECH ENG] The controlled cooling, in a steam boiler, of steam at the superheater outlet or between the primary and secondary stages of the superheater to regulate the final steam temperature.

attendant's switchboard [COMMUN] Switchboard of one or more positions in a central-office location which permits the central-office operator to receive, transmit, or cut in on a call to or from one of the lines which the office services.

attention hypothesis [PSYCH] The theory that a person's attention to objects and other persons is selectively drawn toward areas of great personal interest.

attention span [PSYCH] The period of time a person is able to concentrate his attentions on a given item, usually with respect to learning.

attenuated total reflectance [SPECT] A method of spectrophotometric analysis based on the reflection of energy at the interface of two media which have different refractive indices and are in optical contact with each other. Abbreviated ATR. Also known as frustrated internal reflectance; internal reflectance spectroscopy.

attenuated vaccine [IMMUNOL] A suspension of weakened bacteria, viruses, or fractions thereof used to produce active immunity.

attenuation [BOT] Tapering, sometimes to a long point. [MICROBIO] Weakening or reduction of the virulence of a microorganism. [PHYS] The reduction in level of a quantity, such as the intensity of a wave, over an interval of a variable, such as the distance from a source.

attenuation coefficient [ELECTROMAG] The space rate of attenuation of any transmitted electromagnetic radiation.

attenuation constant [PHYS] A wave with space-time dependence $\exp[i(kx - wt)]$, where $k = k_r + ik_i$, is attenuated as it propagates according to the factor $\exp(=k_ix)$ and k_i is called the attenuation constant. Also known as attenuation factor.

attenuation distortion [COMMUN] **1.** In a circuit or system, departure from uniform amplification or attenuation over the frequency range required for transmission. **2.** The effect of such departure on a transmitted signal.

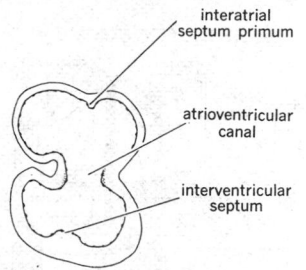

ATRIOVENTRICULAR CANAL

The atrioventricular canal in a 4–5 millimeter human embryo. *(Slightly modified from B. M Patten, Developmental defects at the foramen ovale, Amer. J. Pathol., 14(2):135–161, 1938)*

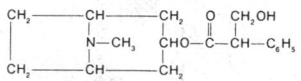

ATROPINE

Structural formula of atropine.

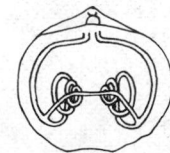

ATRYPIDINA

Spiralium of *Zygospira*.

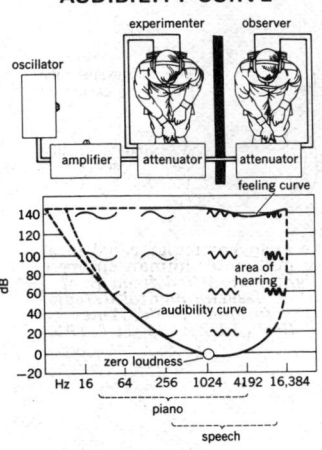

AUDIBILITY CURVE

The audibility curve plotted by measuring range of sound intensities in decibels (dB) against the sound frequencies in hertz (Hz).

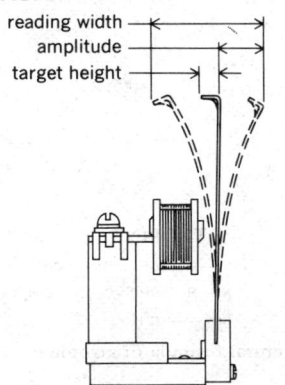

AUDIO-FREQUENCY METER

Reed-type frequency meter.

attenuation equalizer [ELECTR] Corrective network which is designed to make the absolute value of the transfer impedance, with respect to two chosen pairs of terminals, substantially constant for all frequencies within a desired range.

attenuation length [PHYS] The reciprocal of the attenuation coefficient.

attenuation network [ELECTR] Arrangement of circuit elements, usually impedance elements, inserted in circuitry to introduce a known loss or to reduce the impedance level without reflections.

attenuation ratio [PHYS] The magnitude of the propagation ratio.

attenuator [ELECTR] An adjustable or fixed transducer for reducing the amplitude of a wave without introducing appreciable distortion.

Atterberg scale [GEOL] A geometric and decimal grade scale for classification of particles in sediments based on the unit value of 2mm and involving a fixed ratio of 10 for each successive grade; subdivisions are geometric means of the limits of each grade.

attic [BUILD] The part of a building immediately below the roof and entirely or partly within the roof framing.

Attican orogeny [GEOL] Late Miocene diastrophism.

attitude [AERO ENG] The position or orientation of an aircraft, spacecraft, and so on, either in motion or at rest, as determined by the relationship between its axes and some reference line or plane or some fixed system of reference axes.

attitude control [AERO ENG] **1.** The regulation of the attitude of an aircraft, spacecraft, and so on. **2.** A device or system that automatically regulates and corrects attitude, especially of a pilotless vehicle.

attitude gyro [AERO ENG] Also known as attitude indicator. **1.** A gyro-operated flight instrument that indicates the attitude of an aircraft or spacecraft with respect to a reference coordinate system. **2.** Any gyro-operated instrument that indicates attitude.

attitude indicator *See* attitude gyro.

attitude jet [AERO ENG] **1.** A jetstream used to correct or alter the attitude of a flying body either in the atmosphere or in space. **2.** The nozzle that directs this jetstream.

atto- [PHYS] A prefix representing 10^{-18}, which is 0.000 000 000 000 000 001, or one-millionth of a millionth of a millionth. Abbreviated a.

attracted-disk electrometer [ELEC] A type of electrometer in which the attraction between two oppositely charged disks is measured.

attribute sampling [IND ENG] A quality-control inspection method in which the sampled articles are classified only as defective or nondefective.

attributes testing [ENG] A reliability test procedure in which the items under test are classified according to qualitative characteristics.

attrital coal [GEOL] A bright coal composed of anthraxylon and of attritus in which the translucent cell-wall degradation matter or translucent humic matter predominates, with the ratio of anthraxylon to attritus being less than 1:3.

attrition [GEOL] The act of wearing and smoothing of rock surfaces by the flow of water charged with sand and gravel, by the passage of sand drifts, or by the movement of galciers. [MATER] Wear caused by rubbing or friction; for metal surfaces, also known as scoring; scouring.

attrition mill [MECH ENG] A machine in which materials are pulverized between two toothed metal disks rotating in opposite directions.

attrition minesweeping [ORD] Minesweeping designed to minimize the number of live mines at any one time in a channel or area being subjected to heavy, continuous mining attack when clearance sweeping is impossible and port closure unacceptable.

attrition rate [ORD] A factor, normally expressed as a percentage, reflecting the degree of losses of personnel or nonconsumable supplies due to various causes within a specified period of time.

attritus [GEOL] **1.** Visible-to-ultramicroscopic particles of vegetable matter produced by microscopic and other organisms in vegetable deposits, particularly in swamps and bogs. **2.** The dull gray to nearly black, frequently striped portion of

material that makes up the bulk of some coals and alternate bands of bright anthraxylon in well-banded coals.

atu [PHYS] A unit of underpressure or pressure below atmospheric pressure in the metric technical system; equal to 1 technical atmosphere.

atü [PHYS] A unit of overpressure or gage pressure in the metric technical system; equal to 1 technical atmosphere.

Atwood machine [MECH ENG] A device comprising a pulley over which is passed a stretch-free cord with a weight hanging on each end.

at wt *See* atomic weight.

Atyidae [INV ZOO] A family of decapod crustaceans belonging to the section Caridea.

A-type star [ASTRON] In star classification based on spectral characteristics, the type of star in whose spectrum the hydrogen absorption lines are at a maximum.

AU *See* astronomical unit.

Au *See* gold.

aubepine *See* anisaldehyde.

Auberger blood group system [IMMUNOL] An immunologically distinct, genetically determined human erythrocyte antigen, demonstrated by reaction with anti-Au^a (anti-Auberger) antibody.

aubrite [GEOL] An enstatite achondrite (meteorite) consisting almost wholly of crystalline-granular enstatite (and clinoenstatite) poor in lime and practically free from ferrous oxide, with accessory oligoclase.

Auchenorrhyncha [INV ZOO] A group of homopteran families and one superfamily, in which the beak arises at the anteroventral extremity of the face and is not sheathed by the propleura.

audibility [ACOUS] **1.** The state or quality of being heard. **2.** The intensity of a received audio signal, usually expressed in decibels above or below 1 milliwatt using a stated single-frequency sine wave.

audibility curve [ACOUS] **1.** The limits of hearing represented graphically as an area by plotting the minimum audible intensity of a sine wave sound versus frequency. **2.** *See* equal loudness contour.

audibility threshold [ACOUS] The sound intensity at a given frequency which is the minimum perceptible by a normal human ear under specified standard conditions.

audio [ACOUS] **1.** Of or pertaining to sound in the range of frequencies considered audible at reasonable listening intensities to the average young adult listener, approximately 15 to 20,000 hertz. **2.** Pertaining to equipment for the recording, transmission, reproduction, or amplification of such sound.

audio amplifier *See* audio-frequency amplifier.

audio frequency [ACOUS] A frequency that can be detected as a sound by the average young adult, approximately 15 to 20,000 hertz. Abbreviated af. Also known as sonic frequency; sound frequency.

audio-frequency amplifier [ELECTR] An electronic circuit for amplification of signals within, and in some cases above, the audible range of frequencies in equipment used to record and reproduce sound. Also known as af amplifier; audio amplifier.

audio-frequency choke [ELECTROMAG] Choke used to impede the flow of audio-frequency currents; generally a coil wound on an iron core.

audio-frequency meter [ENG] One of a number of types of frequency meters usable in the audio range; for example, a resonant-reed frequency meter.

audio-frequency oscillator [ELECTR] An oscillator circuit using an electron tube, transistor, or other nonrotating device to produce an audio-frequency alternating current. Also known as audio oscillator.

audio-frequency peak limiter [ELEC] A circuit used in an audio-frequency system to cut off signal peaks that exceed a predetermined value. Also known as audio peak limiter.

audio-frequency range [ACOUS] The range of frequencies to which the human ear is sensitive, approximately 15 to 20,000 hertz. Also known as audio range.

audio-frequency shift modulation [COMMUN] System of facsimile transmission over radio, in which the frequency shift required is applied through an 800-hertz shift of an audio signal, rather than shifting the radio transmitter frequency;

the radio signal is modulated by the shifting audio signal, usually at 1500 to 2300 hertz.

audio-frequency transformer [ELEC] An iron-core transformer used for coupling audio-frequency circuits. Also known as audio transformer.

audiogenic seizure [MED] A transient episode of muscular, sensory, or psychic dysfunction induced by sound.

audiogram [ACOUS] A graph showing hearing loss, percent hearing loss, or percent hearing as a function of frequency.

audioimpedance measurement [ACOUS] The measurement of acoustic impedance, as in the direct assessment of the dynamic motor control of sound feedback of different parts of the ear.

audiology [ACOUS] The science of hearing.

audio masking *See* masking.

audiometer [ENG] An instrument composed of an oscillator, amplifier, and attenuator and used to measure hearing acuity for pure tones, speech, and bone conduction.

audiometry [ACOUS] The study of hearing ability by means of audiometers.

audio-modulated radiosonde [ENG] A radiosonde with a carrier wave modulated by audio-frequency signals whose frequency is controlled by the sensing elements of the instrument.

audio oscillator *See* audio-frequency oscillator.

audio patch bay [ENG ACOUS] Specific patch panels provided to terminate all audio circuits and equipment used in a channel and technical control facility; this equipment can also be found in transmitting and receiving stations.

audio peak limiter *See* audio-frequency peak limiter.

audio range *See* audio-frequency range.

audio signal [ACOUS] An electric signal having the frequency of a mechanical wave that can be detected as a sound by the human ear.

audio spectrometer *See* acoustic spectrometer.

audio system *See* sound-reproducing system.

audio transformer *See* audio-frequency transformer.

audiovisual [COMMUN] Pertaining to methods of education and training that make use of both hearing and sight.

audiphone [ENG ACOUS] A device that enables persons with certain types of deafness to hear, consisting of a plate or diaphragm that is placed against the teeth and transmits sound vibrations to the inner ear.

audit [ADP] The operations developed to corroborate the evidence as regards authenticity and validity of the data that are introduced into the data-processing problem or system.

audition [PHYSIO] Ability to hear.

auditory [PHYSIO] Pertaining to the act or the organs of hearing.

auditory association area [PHYSIO] The cortical association area in the brain just inferior to the auditory projection area, related to it anatomically and functionally by association fibers.

auditory impedance [PHYSIO] The acoustic impedance of the ear.

auditory nerve [ANAT] The eighth cranial nerve in vertebrates; either of a pair of sensory nerves composed of two sets of nerve fibers, the cochlear nerve and the vestibular nerve. Also known as acoustic nerve; vestibulocochlear nerve.

auditory perspective [ACOUS] Three-dimensional realism of sound, as produced by an actual orchestra or by a stereophonic sound system.

auditory placode [EMBRYO] An ectodermal thickening from which the inner ear develops in vertebrates.

audit trail [ADP] A system that provides a means for tracing items of data from processing step to step, particularly from a machine-produced report or other machine output back to the original source data.

Auerbach's plexus *See* myenteric plexus.

aufwuch [ECOL] A plant or animal organism which is attached or clings to surfaces of leaves or stems of rooted plants above the bottom stratum.

auganite [PETR] An olivine-free basalt (calcic plagioclase and augite are the essential mineral components) or an augite-bearing andesite.

augelite [MINERAL] Natural, basic aluminum phosphate.

augen [PETR] Large, lenticular eye-shaped mineral grain or mineral aggregate visible in some metamorphic rocks.

augend [MATH] A quantity to which another quantity is added.

augen kohle [GEOL] Coal characterized by the occurrence of small circular or elliptic structural disks or eyes which are arranged in parallel planes in, or normal to, the bedding and which have shiny surfaces that reflect light. Also known as eye coal.

augen schist [PETR] A mylonitic rock characterized by the presence of recrystallized minerals in schistose streaks and lenticles.

augen structure [PETR] A structure found in some gneisses and granites in which certain of the constituents are squeezed into elliptic or lens-shaped forms and, especially if surrounded by parallel flakes of mica, resemble eyes.

auger [DES ENG] **1.** A wood-boring tool that consists of a shank with spiral channels ending in two spurs, a central tapered feed screw, and a pair of cutting lips. **2.** A large augerlike tool for boring into soil.

auger bit [DES ENG] A bit shaped like an auger but without a handle; used for wood boring and earth drilling. [MIN ENG] Hard steel or tungsten carbide–tipped cutting teeth used in an auger running on a torque bar or in an auger-drill head running on a continuous-flight auger.

auger boring [MIN ENG] **1.** The hole drilled by the use of auger equipment. **2.** *See* auger drilling.

Auger coefficient [ATOM PHYS] The ratio of the number of Auger electrons to the number of ejected x-ray photons.

auger conveyor *See* screw conveyor.

auger drilling [ENG] A method of drilling in which penetration is accomplished by the cutting or gouging action of chisel-type cutting edges forced into the substance by rotation of the auger bit. Also known as auger boring.

Auger effect [ATOM PHYS] The radiationless transition of an electron in an atom from a discrete electronic level to an ionized continuous level of the same energy. Also known as autoionization.

Auger electron [ATOM PHYS] An electron that is expelled from an atom in the Auger effect.

Auger electron spectroscopy [SPECT] The energy analysis of Auger electrons produced when an excited atom relaxes by a radiationless process after ionization by a high-energy electron, ion, or x-ray beam. Abbreviated AES.

auger packer [MECH ENG] A feed mechanism that uses a continuous auger or screw inside a cylindrical sleeve to feed hard-to-flow granulated solids into shipping containers, such as bags or drums.

Auger recombination [ATOM PHYS] Recombination of an electron and a hole in which no electromagnetic radiation is emitted, and the excess energy and momentum of the recombining electron and hole are given up to another electron or hole.

Auger shower [ASTRON] A very large cosmic-ray shower. Also known as extensive air shower.

auget [ENG] A priming tube, used in blasting. Also spelled augette.

augette *See* auget.

augite [MINERAL] $(Ca,Mg,Fe)(Mg,Fe,Al)(Al,Si)_2O_6$ A general name for the monoclinic pyroxenes; occurs as dark green to black, short, stubby, prismatic crystals, often of octagonal outline.

augitite [PETR] A volcanic rock consisting of abundant phenocrysts of augite in a glassy groundmass containing microlites of nepheline and plagioclase, with accessory biotite, apatite, and opaque oxides.

augitophyre [PETR] A porphyritic rock in which the phenocrysts are augite and the groundmass is potash feldspar.

augmentation [ASTRON] The apparent increase in the semidiameter of a celestial body, as observed from the earth, as the body's altitude (angular distance above the horizon) increases, due to the reduced distance from the observer. The term is used principally in reference to the moon.

augmentation correction [NAV] A correction necessitated by augmentation, particularly that sextant altitude correction due to the apparent increase in the semidiameter of a celestial body as its altitude increases.

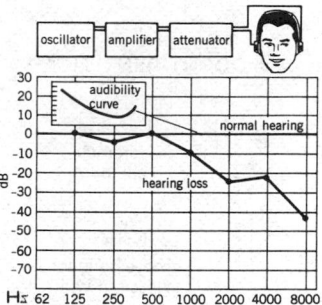

AUDIOGRAM

Audiogram for determining the audibility curve for pure-tone hearing loss at various frequency levels.

AUGITE

A typical specimen of augite from Latium, Italy. *(American Museum of Natural History specimen)*

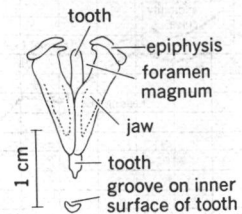

AULODONT DENTITION

Lantern structure in aulodont dentition; a single interradial jaw is shown and below it a cross section of the tooth.

AURICULARIA LARVA

Auricularia larva, showing the ciliated band.

AURORA

Photograph of aurora borealis taken at College, Alaska. (*V. P. Hessler, University of Alaska*)

augmentation distance [NUC PHYS] The extrapolation distance, which is the distance between the time boundary of a nuclear reactor and its boundary calculated by extrapolation.

augmented matrix [MATH] The matrix of the coefficients, together with the constant terms, in a system of linear equations.

augmented operation code [ADP] An operation code which is further defined by information from another portion of an instruction.

augmenter tube [AERO ENG] A tube or pipe, usually one of several, through which the exhaust gases from an aircraft reciprocating engine are directed to provide additional thrust.

Augustin process [MET] A silver extraction process by roasting to form a chloride, leaching with a salt solution, and precipitating by metallic copper.

auk [VERT ZOO] Any of several large, short-necked diving birds (*Alca*) of the family Alcidae found along North Atlantic coasts.

Aulodonta [INV ZOO] An order of echinoderms proposed by R. Jackson in 1921.

aulodont dentition [INV ZOO] In echinoderms, grooved teeth with epiphyses that do not meet so that the foramen magnum of the jaw is open.

Aulolepidae [PALEON] A family of marine fossil teleostean fishes in the order Ctenothrissiformes.

aulophobia [PSYCH] Abnormal fear of flutes.

aulophyte [ECOL] A nonparasitic plant that lives in the cavity of another plant for shelter.

Auloporidae [PALEON] A family of Paleozoic corals in the order Tabulata.

A* unit [PHYS] An atomic standard unit of length, based on the tungsten $K\alpha_1$ line, approximately 10^{-11} centimeter; used for measurements of x-ray wavelengths and of crystal dimensions.

auntie *See* antimissile missile.

Aur *See* Auriga.

aural [BIOL] Pertaining to the ear or the sense of hearing.

aural masking *See* masking.

aural null [NAV] A signal generated by the operator of a direction finder by the back-and-forth rotation of a loop antenna or the goniometer of an Adcock or other directional antenna system; the position of the rotating mechanism at which the null (minimum signal) is determined to be located indicates the azimuth or bearing of the radio station from which the emission is known to have originated.

aural radio range [ELECTR] A radio-range station providing lines of position by virtue of aural identification or comparison of signals at the output of a receiver.

aural signal [ACOUS] 1. A signal that can be heard. 2. The sound portion of a television signal.

aural transmitter [COMMUN] Radio equipment used for transmitting aural (sound) signals from a television broadcast station.

auramine hydrochloride [ORG CHEM] $C_{17}H_{22}ClN_3 \cdot H_2O$ A compound melting at 267°C; very soluble in water, soluble in ethanol; used as a dye and an antiseptic. Also known as yellow pyoktannin.

aurantia [ORG CHEM] $C_{12}H_8N_8O_{12}$ An orange aniline dye, used in stains in biology and in some photographic filters.

aurelia [INV ZOO] A morphological grouping of paramecia, including the elongate, cigar-shaped species which appear to be nearly circular in cross section.

Aurelia [INV ZOO] A genus of scyphozoans.

aureofacin [MICROBIC] An antifungal antibiotic produced by a strain of *Streptomyces aureofaciens*.

aureole [GEOL] A ring-shaped contact zone surrounding an igneous intrusion. Also known as contact aureole; contact zone; exomorphic zone; metamorphic aureole; metamorphic zone; thermal aureole. [METEOROL] A poorly developed corona in the atmosphere characterized by a bluish-white disk immediately around the luminous celestial body, as around the sun or moon in the fog.

Aureomycin [MICROBIO] The trade name for the antibiotic chlortetracycline.

aureothricin [MICROBIO] $C_9H_{10}O_2N_2S_2$ An antibacterial antibiotic produced by a strain of *Actinomyces*.

aureusidin [BIOCHEM] $C_{15}H_{11}O_5$ A yellow flavonoid pigment found typically in the yellow snapdragon. Also known as 4,6,3′,4′-tetrahydroxyaurone.

Auri *See* Auriga.

aurichalcite [MINERAL] $(Zn,Cu)_5(CO_3)_2(OH)_6$ Pale-green or pale-blue mineral consisting of a basic copper zinc carbonate and occurring in crystalline incrustations. Also known as brass ore.

auricle [ANAT] 1. An ear-shaped appendage to an atrium of the heart. 2. Any ear-shaped structure. 3. *See* pinna.

auric oxide *See* gold oxide.

auricular fibrillation [MED] Arrhythmic contractions of the auricles.

auricular flutter *See* atrial flutter.

auricular height [ANTHRO] Height of the cranium, measured from the auditory point to the vertex.

auricularia larva [INV ZOO] A barrel-shaped, food-gathering larval form with a winding ciliated band, common to holothurians and asteroids.

auricularis [ANAT] Any of the three muscles attached to the cartilage of the external ear.

auriferous [GEOL] Of a substance, especially a mineral deposit, bearing gold.

Auriga [ASTRON] A constellation with a right ascension of 6 hours and declination of 40°N. Abbreviated Aur; Auri.

aurin [ORG CHEM] $C_{19}H_{14}O_3$ A derivative of triphenylmethane; solid with red-brown color with green luster; melting point about 220°C; insoluble in water; used as a dye intermediate. Also known as pararosolic acid; rosolic acid.

aurophore [INV ZOO] A bell-shaped structure which is part of the float of certain coelenterates.

aurora [ELEC] *See* corona. [GEOPHYS] The most intense of the several lights emitted by the earth's upper atmosphere, seen most often along the outer realms of the Arctic and Antarctic, where it is called the aurora borealis and aurora australis, respectively; excited by charged particles from space.

aurora australis [GEOPHYS] The aurora of southern latitudes. Also known as southern lights.

aurora borealis [GEOPHYS] The aurora of northern latitudes. Also known as aurora polaris; northern lights.

aurora gating [ELECTR] Operator-controlled gating to eliminate undesirable radar returns from aurora.

auroral [ASTRON] The period of dusk before sunrise.

auroral absorption event [GEOPHYS] A large increase in D-region electron density and associated radio-signal absorption, caused by electron-bombardment of the atmosphere during an aurora or a geomagnetic storm.

auroral caps [GEOPHYS] The regions surrounding the auroral poles, lying between the poles and the auroral zones.

auroral electrojet [GEOPHYS] An intense electric current in the magnetosphere, flowing along the auroral zones during a polar substorm.

auroral forms [GEOPHYS] Auroral display types, of which two are basic: ribbonlike bands and cloudlike surfaces.

auroral frequency [GEOPHYS] The percentage of nights on which an aurora is seen at a particular place, or on which one would be seen if clouds did not interfere.

auroral isochasm [GEOPHYS] A line connecting places of equal auroral frequency, averaged over a number of years.

auroral line [SPECT] A prominent green line in the spectrum of the aurora at a wavelength of 5577 angstroms, resulting from a certain forbidden transition of oxygen.

auroral oval [GEOPHYS] An oval-shaped region centered on the earth's magnetic pole in which auroral emissions occur.

auroral poles [GEOPHYS] The points on the earth's surface on which the auroral isochasms are centered; coincide approximately with the magnetic-axis poles of the geomagnetic field.

auroral region [GEOPHYS] The region within 30° geomagnetic latitude of each auroral pole.

auroral storm [GEOPHYS] A rapid succession of auroral substorms, occurring in a short period, of the order of a day, during a geomagnetic storm.

auroral substorm [GEOPHYS] A characteristic sequence of auroral intensifications and movements occurring around midnight, in which a rapid poleward movement of auroral arcs produces a bulge in the auroral oval.

auroral zone [GEOPHYS] A roughly circular band around either geomagnetic pole within which there is a maximum of auroral activity; lies about 10–15° geomagnetic latitude from the geomagnetic poles.

auroral zone blackout [GEOPHYS] An increase of ionization in the lower ionosphere near the auroral zone.

aurora polaris [GEOPHYS] A high-altitude aurora borealis or aurora australis.

aurothioglucose [PHARM] $C_6H_{11}AuO_5S$ A compound of gold with thioglucose, used for the treatment of rheumatoid arthritis and nondisseminated lupus erythematosus; administered in oil suspension.

auscultation [MED] The act of listening to sounds from internal organs, especially the heart and lungs, to aid in diagnosing their physical state.

ausforging [MET] The forming of austenitic steel into required shapes by hammering or pressing after cooling.

ausrolling [MET] The working of austenitic steel by passing it, after cooling, between oppositely revolving rollers.

austausch *See* austausch coefficient.

austausch coefficient [FL MECH] A measure for the degree of turbulent flow of a fluid; it is the product of mass and transverse distance traveled in a unit of time by the fluid in turbulent motion as it passes through a unit area that lies parallel to the general direction of flow. Also known as austausch; eddy conductivity.

austempering [MET] A process for the heat treatment of austenitic steel.

austenite [MET] Gamma iron with carbon in solution.

austenitic [MET] Composed mainly of austenite.

austenitic cast iron [MET] The product resulting from changing the basic crystalline structure of gray or ductile iron by alloying it with substantial amounts of nickel, manganese, silicon, or other elements.

austenitic manganese steel *See* Hadfield manganese steel.

austenitic stainless steel [MET] Stainless steel composed principally of austenite made stable by alloying with nickel.

austenitic steel [MET] An alloy whose structure is typically that of austenite at room temperature.

austenitize [MET] To heat steel to the temperature range in which the crystalline form of the iron is austenite.

auster *See* ostria.

austinite [MINERAL] $CaZnAsO_4(OH)$ A colorless or yellowish mineral crystallizing in the orthorhombic system; consists of a basic calcium zinc arsenate; hardness is 4.5 on Mohs scale, and specific gravity is 4.13.

Austin Red-D-Gel [MATER] A gelatinous explosive used in mining.

austral [GEOD] Of or pertaining to south.

austral axis pole [GEOPHYS] The southern intersection of the geomagnetic axis with the earth's surface.

Australia [GEOGR] An island continent of 2,941,526 square miles, with low elevation and moderate relief, situated in the southern Pacific.

Australia antigen [IMMUNOL] An infectious agent that causes hepatitis in some people; similar to an inherited serum protein in being polymorphic.

Australian faunal region [ECOL] A zoogeographic region that includes the terrestrial animal communities of Australia and all surrounding islands except those of Asia.

Australian X disease *See* Murray Valley encephalitis.

australite [GEOL] A tektite found in southern Australia, occurring as glass balls and spheroidal dumbbell forms of green and black, similar to obsidian and probably of cosmic origin.

Australopithecinae [PALEON] The near-men, a subfamily of the family Hominidae composed of the single genus *Australopithecus*.

Australopithecus [PALEON] A genus of near-men in the subfamily Australopithecinae representing a side branch of human evolution.

austral region [ECOL] A North American biogeographic region including the region between transitional and tropical zones.

Austrian orogeny [GEOL] A short-lived orogeny during the end of the Early Cretaceous.

Austroastacidae [INV ZOO] A family of crayfish in the order Decapoda found in temperate regions of the Southern Hemisphere.

Austrodecidae [INV ZOO] A monogeneric family of marine arthropods in the subphylum Pycnogonida.

Austroriparian life zone [ECOL] The zone in which occurs the climate and biotic communities of the southeastern coniferous forests of North America.

autallotriomorphic [PETR] Pertaining to an aplitic texture in which all mineral constituents crystallized simultaneously, preventing the development of euhedral crystals.

autecology [ECOL] The ecological relations between a plant species and its environment.

authalic latitude [MAP] A latitude based on a sphere having the same area as the spheroid, and such that areas between successive parallels of latitude are exactly equal to the corresponding areas on the spheroid. Also known as equal-area latitude.

authalic map projection *See* equal-area map projection.

authentication [COMMUN] Security measure designed to protect a communications system against fraudulent transmissions and establish the authenticity of a message by an authenticator within the transmission derived from certain predetermined elements of the message itself.

authenticator [COMMUN] Letter, numeral, or groups of letters or numerals attesting to the authenticity of a message or transmission.

authigene [MINERAL] A mineral which has not been transported but has been formed in place. Also known as authigenic mineral.

authigenic [GEOL] Of constituents that came into existence with or after the formation of the rock of which they constitute a part; for example, the primary and secondary minerals of igneous rocks. [MINERAL] Of a mineral formed by sedimentary processes as a crystallographic unit at the place of its occurrence.

authigenic mineral *See* authigene.

authigenic sediment [GEOL] Sediment occurring in the place where it was originally formed.

authoritarian character [PSYCH] A personality that asks unquestioning subordination and obedience but is intolerant of weakness in others, rejects members of groups other than his own, is very rigid, and requires all issues to be decided in black and white, yet may accept superior authority in a servile way.

authorized carrier frequency [COMMUN] A specific carrier frequency authorized for use, from which the actual carrier frequency is permitted to deviate, solely because of frequency instability, by an amount not to exceed the frequency tolerance.

autism [PSYCH] A schizophrenic symptom characterized by absorption in fantasy to the exclusion of perceptual reality.

autistic [PSYCH] **1.** Pertaining to or characterized by autism. **2.** Behavior, most commonly in children, characterized by aphonia, disregard for reality, self-manipulation, and sometimes a preoccupation with pointed or shiny objects.

auto-abstract [ADP] **1.** To select key words from a document, commonly by an automatic or machine method, for the purpose of forming an abstract of the document. **2.** The material abstracted from a document by machine methods.

autoagglutination [IMMUNOL] Agglutination of an individual's erythrocytes by his own serum. Also known as autohemagglutination.

autoagglutinin [IMMUNOL] An antibody in an individual's blood serum that causes agglutination of his own erythrocytes.

autoalarm *See* automatic alarm receiver.

autoanalysis [PSYCH] Self-analysis, a technique of psychotherapy.

autoantibody [IMMUNOL] An antibody formed by an individual against his own tissues; common in hemolytic anemias.

autoantigen [IMMUNOL] A tissue within the body which acquires the ability to incite the formation of complementary antibodies.

autoasphyxiation [PHYSIO] Asphyxiation by the products of metabolic activity.

autobarotropy [METEOROL] The state of a fluid which is characterized by both barotropy and piezotropy when the coefficients of barotropy and piezotropy are equal.

autobasidium [MYCOL] An undivided basidium typically found in higher Basidiomycetes.

autoboat *See* motorboat.

autobrecciation [GEOL] The process whereby portions of the first consolidated crust of a lava flow are incorporated into the still-fluid portion.

autocarp [BOT] **1.** A fruit formed as the result of self-fertilization. **2.** A fruit consisting of the ripened pericarp without adnate parts.

autocarpy [BOT] Production of fruit by self-fertilization.

autocatalysis [CHEM] A catalytic reaction started by the products of a reaction that was itself catalytic.

autochory [ECOL] Active self-dispersal of individuals or their disseminules.

autochrome plate [GRAPHICS] A photographic plate once used for direct color photography on coated glass; the glass must be viewed as a transparency, with the image formed in the film of dyed starch grains coating the glass.

autochthon [GEOL] A succession of rock beds that have been moved comparatively little from their original site of formation, although they may be folded and faulted extensively. [PALEON] A fossil occurring where the organism once lived.

autochthonous coal [GEOL] Coal believed to have originated from accumulations of plant debris at the place where the plants grew. Also known as indigenous coal.

autochthonous microorganism [MICROBIO] An indigenous form of soil microorganisms, responsible for chemical processes that occur in the soil under normal conditions.

autochthonous stream [HYD] A stream flowing in its original channel.

autoclastic [GEOL] Of rock, fragmented in place by folding due to orogenic forces when the rock is not so heavily loaded as to render it plastic.

autoclastic schist [GEOL] Schist formed in place from massive rocks by crushing and squeezing.

autoclave [ENG] An airtight vessel for heating and sometimes agitating its contents under high steam pressure; used for industrial processing, sterilizing, and cooking with moist or dry heat at high temperatures.

autoclave molding [ENG] A method of curing reinforced plastics that uses an autoclave with 50–100 psi steam pressure to set the resin.

autoclaving [SCI TECH] Heating of liquids or sterilizing of equipment at high steam pressure.

autocode [ADP] The process of using a computer to convert automatically a symbolic code into a machine code. Also known as automatic code.

autocoder [ADP] A person or machine producing or using autocode as a part or the whole of some task.

autocollimation [OPTICS] A procedure for collimating a telescope or other optical instrument with objective and crosshairs, in which the instrument is directed toward a plane mirror and the crosshairs and lens are adjusted so that the crosshairs coincide with their reflected image.

autocollimator [OPTICS] **1.** A device by which a single lens collimates diverging light from a slit, and then focuses the light on an exit slit after it has passed through a prism to a mirror and been reflected back through the prism. **2.** A telescope which has a graduated reticle, enabling an observer to read off the angles subtended by distant objects. **3.** A convex mirror at the focus of the principal mirror of a reflecting telescope, which causes light to leave the telescope in a parallel beam. **4.** A telescope equipped with an eyepiece designed for autocollimation.

autoconsequent falls [HYD] Waterfalls in streams carrying a heavy load of calcium carbonate in solution which develop at particular sites along the stream course where warming, evaporation, and other factors cause part of the solution load to be precipitated.

autoconsequent stream [HYD] A stream in the process of building a fan or an alluvial plain, the course of which is guided by the slopes of the alluvium the stream itself has deposited.

autoconvection [METEOROL] The phenomenon of the spontaneous initiation of convection in an atmospheric layer in which the lapse rate is equal to or greater than the autoconvective lapse rate.

autoconvection gradient *See* autoconvective lapse rate.

autoconvective instability *See* absolute instability.

autoconvective lapse rate [METEOROL] The largely hypothetical environmental lapse rate of temperature in an atmosphere in which the density is constant with height (homogeneous atmosphere); 3°C per 100 meters in dry air. Also known as autoconvection gradient.

autocopulation [INV ZOO] Self-copulation; sometimes occurs in certain hermaphroditic worms.

autocorrelation [ELECTR] A technique used to detect cyclic activity in a complex signal.

autocorrelation function [MATH] For a function $f(t)$, the limit as T approaches infinity of $1/2T$ times the integral from $-T$ to T of $f(t)f(t-\tau)dt$, where τ is a time-delay parameter.

autocorrelator [ELECTR] A correlator in which the input signal is delayed and multiplied by the undelayed signal, the product of which is then smoothed in a low-pass filter to give an approximate computation of the autocorrelation function; used to detect a nonperiodic signal or a weak periodic signal hidden in noise.

autodecrement addressing [ADP] An addressing mode of minicomputers in which the register is first decremented and then used as a pointer.

autodyne circuit [ELECTR] A circuit in which the same tube elements serve as oscillator and detector simultaneously.

autodyne reception [COMMUN] System of heterodyne reception through the use of a device which is both an oscillator and a detector.

autoecious *See* autoicous.

Autofining process [CHEM ENG] A fixed-bed catalytic process for desulfurizing distillates; the pelleted catalyst is cobalt molybdate on alumina and may be regenerated in place.

autofocus rectifier [OPTICS] A precise, vertical photoenlarger which permits the correction of distortion in an aerial negative caused by tilt.

autofrettage [ENG] A process for manufacturing gun barrels; prestressing the metal increases the load at which its permanent deformation occurs.

autogamy [BIOL] A process of self-fertilization that results in homozygosis; occurs in some flowering plants, fungi, and protozoans.

autogenesis *See* abiogenesis.

autogenetic drainage [HYD] A self-established drainage system developed solely by headwater erosion.

autogenetic topography [GEOL] Conformation of land due to the physical action of rain and streams.

autogenous [SCI TECH] Self-generated; produced without external influence.

autogenous electrification [PHYS] The process by which net charge is built up on an object, such as an airplane, moving relative to air containing dust or ice crystals; produced by frictional effects (triboelectrification) accompanying contact between the object and the particulate matter.

autogenous grinding [MECH ENG] The secondary grinding of material by tumbling the material in a revolving cylinder, without balls or bars taking part in the operation.

autogenous ignition temperature *See* ignition temperature.

autogenous mill *See* autogenous tumbling mill.

autogenous tumbling mill [MECH ENG] A type of ball-mill grinder utilizing as the grinding medium the coarse feed (incoming) material. Also known as autogenous mill.

autogenous vaccine [IMMUNOL] A vaccine prepared from a culture of microorganisms taken directly from the infected person.

autogenous welding [MET] A fusion welding process using heat without the addition of filler metal to join two pieces of the same metal.

autogeosyncline [GEOL] A parageosyncline that subsides as an elliptical basin or trough nearly without associated highlands. Also known as intracratonic basin.

autogiro [AERO ENG] A type of aircraft which utilizes a rotating wing (rotor) to provide lift and a conventional engine-propeller combination to pull the vehicle through the air.

autograft [BIOL] A tissue transplanted from one part to another part of an individual's body.

autohemagglutination *See* autoagglutination.

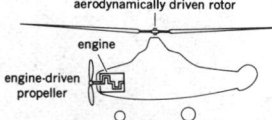

AUTOGIRO

aerodynamically driven rotor

engine

engine-driven propeller

The in-flight dynamic components of a typical autogiro.

autohemorrhage [INV ZOO] Voluntary exudation or ejection of nauseous or poisonous blood by certain insects as a defense against predators.

autohemotherapy [MED] Treatment of disease with the patient's own blood, withdrawn by venipuncture and then injected intramuscularly.

autoicous [BOT] Having male and female organs on the same plant but on different branches. Also spelled autoecious.

autoignition [MECH ENG] Spontaneous ignition of some or all of the fuel-air mixture in the combustion chamber of an internal combustion engine. [PHYS] *See* spontaneous combustion.

autoimmune disease [IMMUNOL] An illness involving the formation of autoantibodies which appear to cause pathological damage to the host.

autoimmunity [IMMUNOL] An immune state in which antibodies are formed against the person's own body tissues.

autoimmunization [IMMUNOL] Immunization obtained by natural processes within the body.

autoincrement addressing [ADP] An addressing mode of minicomputers in which the operand address is gotten from the specified register which is then incremented.

autoindexing *See* automatic indexing.

autoinfection [MED] Reinfection by an organism existing within the body or transferred from one part of the body to another.

autoinjection *See* autointrusion.

autoinoculation [MED] **1.** Spread of a disease from one part of the body to another. **2.** Injection of an autovaccine.

autointoxication [MED] Poisoning by metabolic products elaborated within the body; generally, toxemia of pathologic states.

autointrusion [GEOL] A process wherein the residual liquid of a differentiating magma is drawn into rifts formed in the crystal mesh at a late stage by deformation of unspecified origin. Also known as autoinjection.

autoionization *See* Auger effect.

autolith [PETR] **1.** A fragment of igneous rock enclosed in another igneous rock of later consolidation, each being regarded as a derivative from a common parent magma. **2.** A round, oval, or elongated accumulation of iron-magnesium minerals of uncertain origin in granitoid rock.

autolithography [GRAPHICS] A lithographic process in which the artist makes a drawing onto the printing surface directly.

autologous [BIOL] Derived from or produced by the individual in question, such as an autologous protein or an autologous graft.

autoluminescence [ATOM PHYS] Luminescence of a material (such as a radioactive substance) resulting from energy originating within the material itself.

autolysis [GEOCHEM] Return of a substance to solution, as of phosphate removed from seawater by plankton and returned when these organisms die and decay. [PATH] Self-digestion by body cells following somatic or organ death or ischemic injury.

autolysosome *See* autophagic vacuole.

autolytic enzyme [BIOCHEM] A bacterial enzyme, located in the cell wall, that causes disintegration of the cell following injury or death.

Autolytinae [INV ZOO] A subfamily of errantian polychaetes in the family Syllidae.

automanual system [CIV ENG] A railroad signal system in which signals are set manually but are activated automatically to return to the danger position by a passing train.

automata theory [ENG] The theory concerning the operation principles, application, and behavior characteristics of automatic devices.

automated radar terminal system [NAV] A system for carrying out air-traffic control in the vicinity of airports which uses both airport surveillance radar and the air-traffic radar beacon system; radar video, representing aircraft targets, is presented on the air-traffic controllers' displays, and the automation system automatically tracks controlled aircraft and presents alphanumeric information adjacent to their targets. Abbreviated ARTS.

automatic [ENG] Having a self-acting mechanism that performs a required act at a predetermined time or in response to certain conditions. [ORD] *See* automatic weapon.

automatic abstracting [ADP] Techniques whereby, on the basis of statistical properties, a subset of the sentences in a document is selected as representative of the general content of that document.

automatic alarm receiver [ELECTR] A complete receiving, selecting, and warning device capable of being actuated automatically by intercepted radio-frequency waves forming the international automatic alarm signal. Also known as autoalarm.

automatic-alarm-signal keying device [COMMUN] Device capable of automatically keying the radiotelegraph transmitter on board a vessel to transmit the international automatic-alarm signal, or to respond to receipt of an internationally agreed-upon distress signal and wake up the radio operator on ships not having a 24-hour radio watch.

automatic back bias [ELECTR] Radar technique which consists of one or more automatic gain control loops to prevent overloading of a receiver by large signals, whether jamming or actual radar echoes.

automatic background control *See* automatic brightness control.

automatic bass compensation [ELECTR] A circuit related to the volume control in some radio receivers and audio amplifiers to make bass notes sound properly balanced, in the audio spectrum, at low volume-control settings.

automatic breech mechanism [ORD] A device that utilizes the energy of recoil, or the pressure of the powder gases, to open the breech, withdraw the fired cartridge case, insert a new cartridge, and close the breech.

automatic brightness control [ELECTR] A circuit used in a television receiver to keep the average brightness of the reproduced image essentially constant. Abbreviated ABC. Also known as automatic background control.

automatic carriage [ADP] Any mechanism designed to feed continuous paper or plastic forms through a printing or writing device, often using sprockets to engage holes in the paper.

automatic casing hanger [PETRO ENG] Unitized hanger-seal assembly latched at the lower end of an oil-well casing string to support the next smaller string and make a seal between the two strings.

automatic C bias *See* self-bias.

automatic celestial navigation *See* celestial-inertial guidance.

automatic character recognition [ADP] The technology of using special machine systems to identify human-readable symbols, most often alphanumeric, and then to utilize this data.

automatic check [ADP] An error-detecting procedure performed by a computer as an integral part of the normal operation of a device, with no human attention required unless an error is actually detected.

automatic check-out system [CONT SYS] A system utilizing test equipment capable of automatically and simultaneously providing actions and information which will ultimately result in the efficient operation of tested equipment while keeping time to a minimum.

automatic code *See* autocode.

automatic coding [ADP] Any technique in which a computer is used to help bridge the gap between some intellectual and manual form of describing the steps to be followed in solving a given problem, and some final coding of the same problem for a given computer.

automatic computer [ADP] A computer which can carry out a special set of operations without human intervention.

automatic connection [ELECTR] Ability of electronic switching equipment to make a connection between users without human intervention.

automatic contrast control [ELECTR] A circuit that varies the gain of the radio-frequency and video intermediate-frequency

AUTOLYTINAE

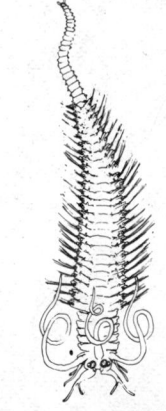

Procerastea of the Syllidae (Autolytinae); epitokous male, dorsal view.

amplifiers in such a way that the contrast of the television picture is maintained at a constant average level.

automatic control [CONT SYS] Control in which regulating and switching operations are performed automatically in response to predetermined conditions. Also known as automatic regulation.

automatic-control block diagram [CONT SYS] A diagrammatic representation of the mathematical relationships defining the flow of information and energy through the automatic control system, in which the components of the control system are represented as functional blocks in series and parallel arrangements according to their position in the actual control system.

automatic-control error coefficient [CONT SYS] Three numerical quantities that are used as a measure of the steady-state errors of an automatic control system when the system is subjected to constant, ramp, or parabolic inputs.

automatic-control frequency response [CONT SYS] The steady-state output of an automatic control system for sinusoidal inputs of varying frequency.

automatic controller [CONT SYS] An instrument that continuously measures the value of a variable quantity or condition and then automatically acts on the controlled equipment to correct any deviation from a desired preset value. Also known as automatic regulator; controller.

automatic-control servo valve [CONT SYS] A mechanically or electrically actuated servo valve controlling the direction and volume of fluid flow in a hydraulic automatic control system.

automatic-control stability [CONT SYS] The property of an automatic control system whose performance is such that the amplitude of transient oscillations decreases with time and the system reaches a steady state.

automatic control system [CONT SYS] A control system having one or more automatic controllers connected in closed loops with one or more processes. Also known as regulating system.

automatic-control transient analysis [CONT SYS] The analysis of the behavior of the output variable of an automatic control system as the system changes from one steady-state condition to another in terms of such quantities as maximum overshoot, rise time, and response time.

automatic coupling [MECH ENG] A device which couples rail cars when they are bumped together.

automatic cutout [ELEC] A device, usually operated by centrifugal force or by an electromagnet, that automatically shorts part of a circuit at a particular time.

automatic dam [MIN ENG] In placer mining, a dam with a gate that automatically discharges the water when it reaches a certain height behind the dam.

automatic data processing [ENG] The machine performance, with little or no human assistance, of any of a variety of tasks involving informational data; examples include automatic and responsive reading, computation, writing, speaking, directing artillery, and the running of an entire factory. Abbreviated ADP.

automatic data-processing auxiliary equipment [ADP] Equipment which is related in function to automatic data-processing equipment, other than peripheral equipment, and whose use is not exclusively and directly used with an automatic data-processing system; when it is so used, it supports the system in off-line operations such as card-punching equipment and paper-tape-preparing equipment; for example, a flexowriter.

automatic data-processing equipment [ADP] Electronic data-processing equipment and punched-card accounting machines, irrespective of use, application, or source of funding. Abbreviated ADPE.

automatic data-processing system [ADP] The equipment, personnel program, and application operations involved in the utilization of electronic data-processing equipment, along with associated electric accounting machines, to solve business and logistics data-processing problems, with a minimum of human intervention.

automatic data-processing system specifications [ADP] A description (devoid of any orientation to the specific equip-

ment of particular suppliers) of a requirement for, and operations to be performed by, automatic data-processing equipment, and generally including a workload description in terms of representative programs (benchmarks) and extension factors.

automatic degausser [ELECTR] An arrangement of degaussing coils mounted around a color television picture tube, combined with a special circuit that energizes these coils only while the set is warming up; demagnetizes any parts of the receiver that have been affected by the magnetic field of the earth or of any nearby home appliance.

automatic dialer [ELECTR] A device in which a telephone number up to a maximum of 14 digits can be stored in a memory and then activated, directly into the line, by the caller's pressing a button. Also known as mechanical dialer.

automatic dictionary [ADP] Any table within a computer memory which establishes a one-to-one correspondence between two sets of characters.

automatic direction finder [ELECTR] A direction finder that without manual manipulation indicates the direction of arrival of a radio signal. Abbreviated ADF. Also known as radio compass.

automatic drill [DES ENG] A straight brace for bits whose shank comprises a coarse-pitch screw sliding in a threaded tube with a handle at the end; the device is operated by pushing the handle.

automatic error correction [COMMUN] A technique, usually requiring the use of special codes or automatic retransmission, which detects and corrects errors occurring in transmission; the degree of correction depends upon coding and equipment configuration.

automatic exchange [ELECTR] A telephone, teletypewriter, or data-transmission exchange in which communication between subscribers is effected, without the intervention of an operator, by devices set in operation by the originating subscriber's instrument. Also known as automatic switching system; machine switching system.

automatic-feed punch [ADP] A card punch having a hopper, a card track, and a stacker; movement of cards through the punch is automatic.

automatic fire [ORD] Continuous fire from an automatic gun until the pressure on the trigger is released.

automatic focus [OPTICS] A device in a camera or enlarger which automatically keeps the objective lens in focus through a range of magnification.

automatic frequency control [ELECTR] Abbreviated AFC. **1.** A circuit used to maintain the frequency of an oscillator within specified limits, as in a transmitter. **2.** A circuit used to keep a superheterodyne receiver tuned accurately to a given frequency by controlling its local oscillator, as in an FM receiver. **3.** A circuit used in radar superheterodyne receivers to vary the local oscillator frequency so as to compensate for changes in the frequency of the received echo signal. **4.** A circuit used in television receivers to make the frequency of a sweep oscillator correspond to the frequency of the synchronizing pulses in the received signal.

automatic gain control [ELECTR] A control circuit that automatically changes the gain (amplification) of a receiver or other piece of equipment so that the desired output signal remains essentially constant despite variations in input signal strength. Abbreviated AGC.

automatic grid bias *See* self-bias.

automatic gun *See* automatic weapon.

automatic gun charger [ORD] A gun charger that includes a mechanism for the clearance of gun stoppages and for the retention of the breech mechanism to the rear of the gun receiver.

automatic horizon *See* artificial horizon.

automatic indexing [ADP] Selection of key words from a document by computer for use as index entries. Also known as autoindexing.

automatic intercept [COMMUN] Telephone service that automatically records messages a caller may leave when the called party is away from his telephone.

automatic landing system [NAV] The means for automatically guiding and controlling aircraft from an initial approach

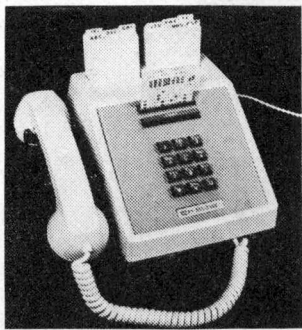

AUTOMATIC DIALER

A card-dialer-type telephone set used in automatic dialing. (*American Telephone and Telegraph Co.*)

altitude to a point where safe contact is made with the landing surface.

automatic level compensation [COMMUN] System which automatically compensates for variations in a circuit.

automatic light control [ELECTR] Automatic adjustment of illumination reaching a film, television camera, or other imaging device as a function of scene brightness.

automatic mathematical translator [ADP] Automatic-programming computer capable of receiving a mathematical equation from a remote input and returning an immediate solution.

automatic message accounting *See* automatic toll ticketing.

automatic message-switching center [COMMUN] A center in which messages are automatically routed according to information in them.

automatic microfiller [ENG] A device used to place microfilm in jackets at relatively high speeds.

automatic mold [ENG] A mold used in injection or compression molding of plastic objects so that repeated molding cycles are possible, including ejection, without manual assistance.

automatic peak limiter *See* limiter.

automatic pilot [NAV] Also known as autopilot. **1.** Equipment which automatically stabilizes the attitude of an aircraft about its pitch, roll, and yaw axes and keeps it on a predetermined heading. **2.** Equipment for automatically steering ships.

automatic pistol [ORD] A pistol able to operate as an automatic weapon.

automatic press [MECH ENG] A press in which mechanical feeding of the work is synchronized with the press action.

automatic programming [ADP] The preparation of machine-language instructions by use of a computer.

Automatic Programming Tool [ADP] A computer language used to program numerically controlled machine tools. Abbreviated APT.

automatic record changer [ENG ACOUS] An electric phonograph that automatically plays a number of records one after another.

automatic regulation *See* automatic control.

automatic regulator *See* automatic controller.

automatic relay [COMMUN] Means of selective switching which causes automatic equipment to record and retransmit communications.

automatic rifle [ORD] A rifle capable of operating as an automatic weapon.

automatic routine [ADP] A routine that is executed independently of manual operations, but only if certain conditions occur within a program or record, or during some other process.

automatic sampler [MECH ENG] A mechanical device to sample process streams (gas, liquid, or solid) either continuously or at preset time intervals.

automatic scanning receiver [ELECTR] A receiver which can automatically and continuously sweep across a preselected frequency, either to stop when a signal is found or to plot signal occupancy within the frequency spectrum being swept.

automatic screw machine [MECH ENG] A machine designed to automatically produce finished parts from bar stock at high production rates; the term is not an exact, specific machine-tool classification.

automatic sensitivity control [ELECTR] Circuit used for automatically maintaining receiver sensitivity at a predetermined level; it is similar to automatic gain control, but it affects the receiver constantly rather than during the brief interval selected by the range gate.

automatic sequences [ADP] The characteristic of a computer that can perform successive operations without human intervention.

automatic short-circuiter [ELEC] Device designed to automatically short-circuit the commutator bars in some forms of single-phase commutator motors.

automatic spider [MIN ENG] A foot or hydraulically actuated drill-rod clamping device similar to a Wommer safety clamp.

automatic stability [AERO ENG] Stability achieved with the controls operated by automatic devices, as by an automatic pilot.

automatic stabilization equipment [AERO ENG] Apparatus which automatically operates control devices to maintain an aircraft in a stable condition.

automatic stoker [MECH ENG] A device that supplies fuel to a boiler furnace by mechanical means. Also known as mechanical stoker.

automatic stop [ADP] An automatic halting of a computer processing operation as the result of an error detected by built-in checking devices.

automatic switchboard [COMMUN] Telephone switchboard in which the connections are made by using remotely controlled switches.

automatic switching system *See* automatic exchange.

automatic tank battery [PETRO ENG] Interconnected system of storage tanks with automatic controls to direct incoming oil to empty tanks in a desired sequence.

automatic telegraph transmission [COMMUN] Form of telegraphy in which telegraph signals are transmitted mechanically from a perforated tape.

automatic threshold variation [ELECTR] Constant false-alarm rate scheme that is an open-loop of automatic gain control in which the decision threshold is varied continuously in proportion to the incoming intermediate frequency and video noise level.

automatic time switch [ENG] Combination of a switch with an electric or spring-wound clock, arranged to turn an apparatus on and off at predetermined times.

automatic titrator [ANALY CHEM] **1.** Titration with quantitative reaction and measured flow of reactant. **2.** Electrically generated reactant with potentiometric, ampherometric, or colorimetric end-point or null-point determination.

automatic toll ticketing [COMMUN] System whereby toll calls are automatically recorded and timed, and toll tickets printed, under control of the calling telephone's dial pulses and without the intervention of an operator. Also known as automatic message accounting.

automatic tracking [NAV] **1.** Tracking in which a servomechanism automatically follows some characteristic of the signal; specifically, a process by which tracking or data-acquisition systems are enabled to keep their antennas continuously directed at a moving target without manual operation. **2.** An instrument which displays the actual course made good through the use of navigation derived from several sources.

automatic transfer equipment [ELEC] Equipment which automatically transfers a load so that a source of power may be selected from one of several incoming lines.

automatic traverse computer [NAV] An instrument which, by mechanical linkages and gears, converts courses into latitudes and departures; also totals the algebraic sum of the latitudes and departures.

automatic tuning system [CONT SYS] An electrical, mechanical, or electromechanical system that tunes a radio receiver or transmitter automatically to a predetermined frequency when a button or lever is pressed, a knob turned, or a telephone-type dial operated.

automatic-type belt tensioning device [MECH ENG] Any device which maintains a predetermined tension in a conveyor belt.

automatic video noise leveling [ELECTR] Constant false-alarm rate scheme in which the video noise level at the output of the receiver is sampled at the end of each range sweep and the receiver gain is readjusted accordingly to maintain a constant video noise level at the output.

automatic voltage regulator *See* voltage regulator.

automatic volume compressor *See* volume compressor.

automatic volume control [ELECTR] An automatic gain control that keeps the output volume of a radio receiver essentially constant despite variations in input-signal strength during fading or when tuning from station to station. Abbreviated AVC.

automatic volume expander *See* volume expander.

automatic wagon control [MIN ENG] A mechanism to keep the speed of wagons within certain designed limits; may consist of small hydraulic units fixed at intervals along the inside of the track.

automatic weapon [ORD] A gun that fires, extracts, ejects,

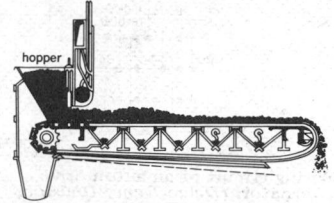

AUTOMATIC STOKER

hopper

A chain grate, one type of automatic stoker.

AUTOMOTIVE ALTERNATOR

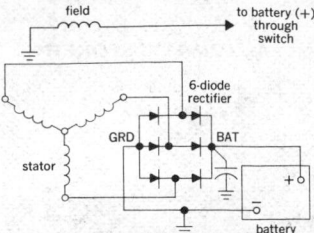

Wiring circuit of an automotive alternator. (*Delco-Remy Division, General Motors Corp.*)

AUTOMOTIVE BODY

Cutaway view showing conventional construction of all-steel, welded body for a four-door sedan. The body is bolted to a frame which carries engine and power train. (*Chevrolet Division, General Motors Corp.*)

AUTOMOTIVE STEERING

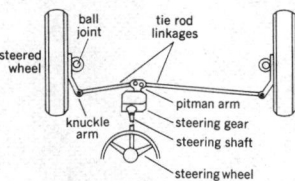

Components of a typical steering system.

AUTORADIOGRAPHY

A reproduction of a photograph produced by autoradiography of metatarsus of a calf. The dense areas indicate highest concentration of the radioisotope. (*From C. L. Comar, Radioisotopes in Biology and Agriculture, McGraw-Hill, 1955*)

and reloads without application of power from an outside source, repeating the cycle as long as the firing mechanism is held in the proper position. Also known as automatic; automatic gun.

automatic weather station [METEOROL] A weather station at which the services of an observer are not required; usually equipped with telemetric apparatus.

automatic welding [MET] Electric-arc welding with automatic control of the arc movement along the welding line, the electrode feed, and the arc-gap length.

automation [ENG] **1.** The replacement of human or animal labor by machines. **2.** The operation of a machine or device automatically or by remote control.

automation source data [ADP] The many methods of recording information in coded forms on paper tapes, punched cards, or tags that can be used over and over again to produce many other records without rewriting. Also known as source data automation (SDA).

automatism [BIOL] Spontaneous activity of tissues or cells. [MED] An act performed with no apparent exercise of will, as in sleepwalking and certain hysterical and epileptic states.

automaton [ADP] A robot which functions without step-by-step guidance by a human operator.

automechanism [CONT SYS] A machine or other deivce that operates automatically or under control of a servomechanism.

autometamorphism [PETR] Metamorphism of an igneous rock by the action of its own volatile fluids. Also known as autometasomatism.

autometasomatism *See* autometamorphism.

automobile [MECH ENG] A four-wheeled, trackless, self-propelled vehicle for land transportation of as many as eight people. Also known as car.

automobile chassis [MECH ENG] The automobile frame, together with the wheels, power train, brakes, engine, and steering system.

automorphic [PETR] Of minerals in igneous rock bounded by their own crystal faces. Also known as euhedral; idiomorphic.

automorphism [MATH] An isomorphism of an algebraic structure with itself.

autometamorphosis [PETR] Metamorphosis of solidified igneous rock by solutions from its heated interior.

automotive air conditioning [MECH ENG] A system for maintaining comfort of occupants of automobiles, buses, and trucks, limited to air cooling, air heating, ventilation, and occasionally dehumidification.

automotive alternator [ELEC] An ac generator used in an automotive vehicle to provide current for the vehicle's electrical systems.

automotive body [ENG] An enclosure mounted on and attached to the frame of an automotive vehicle, to contain passengers and luggage, or in the case of commercial vehicles the commodities being carried.

automotive brake [MECH ENG] A friction mechanism that slows or stops the rotation of the wheels of an automotive vehicle, so that tire traction slows or stops the vehicle.

automotive engine [MECH ENG] The fuel-consuming machine that provides the motive power for automobiles, airplanes, tractors, buses, and motorcycles and is carried in the vehicle.

automotive engineering [MECH ENG] The branch of mechanical engineering concerned primarily with the special problems of land transportation by a four-wheeled, trackless, automotive vehicle.

automotive frame [ENG] The basic structure of all automotive vehicles, except tractors, which is supported by the suspension and upon which or attached to which are the power plant, transmission, clutch, and body or seat for the driver.

automotive fuel [MATER] A material, generally a liquid fuel, gasoline, or distillate, whose combustion is used to supply chemical energy to provide the power for an automotive vehicle.

automotive ignition system [MECH ENG] A device in an automotive vehicle which initiates the chemical reaction between fuel and air in the cylinder charge.

automotive steering [MECH ENG] Mechanical means by which a driver controls the course of a moving automobile, bus, truck, or tractor.

automotive suspension [MECH ENG] The springs and related parts intermediate between the wheels and frame of an automotive vehicle that support the frame on the wheels and absorb road shock caused by passage of the wheels over irregularities.

automotive transmission [MECH ENG] A device for providing different gear or drive ratios between the engine and drive wheels of an automotive vehicle, a principal function being to enable the vehicle to accelerate from rest through a wide speed range while the engine operates within its most effective range.

automotive vehicle [MECH ENG] A self-propelled vehicle or machine for land transportation of people or commodities or for moving materials, such as a passenger car, bus, truck, motorcycle, tractor, airplane, motorboat, or earthmover.

automotive voltage regulator [ELEC] A device in the automotive electrical system to prevent generator or alternator overvoltage.

autonavigator [NAV] A navigation system that automatically measures absolute vehicle motions and computes distance and direction from the departure point.

autonomic agent [MED] A compound that reduces or enhances nerve-impulse transmission across synaptic junctions, especially in the autonomic nervous system.

autonomic nervous system [ANAT] The visceral or involuntary division of the nervous system in vertebrates, which enervates glands, viscera, and smooth, cardiac, and some striated muscles.

autonomic reflex system [PHYSIO] An involuntary biological control system characterized by the uncontrolled functioning of smooth muscles and glands to maintain an acceptable internal environment.

autooxidation [CHEM] **1.** Oxidation caused by the atmosphere. **2.** An oxidation reaction that is self-catalyzed and spontaneous. **3.** An oxidation reaction begun only by an inductor.

autopatrol [MECH ENG] A self-powered blade grader. Also known as motor grader.

autophagic vacuole [CYTOL] A membrane-bound cellular organelle that engulfs pieces of the substance of the cell itself. Also known as autolysosome.

autophagocytosis [CYTOL] The cellular process of phagocytizing a portion of protoplasm by a vacuole within the cell.

autophagy [CYTOL] The cellular process of self-digestion.

autophobia [PSYCH] Abnormal fear of one's self or of being alone.

autophytograph [GEOL] An imprint on a rock surface made by chemical activity of a plant or plant part.

autopilot *See* automatic pilot.

autoplotter [ADP] A machine which automatically draws a graph from input data.

autopneumatolysis [GEOL] The occurrence of metamorphic changes at the pneumatolytic stage of a cooling magma when temperatures are approximately 400–600°C.

autopolyploid [GEN] A cell or organism having three or more sets of chromosomes all derived from the same species.

autopositive [GRAPHICS] A film or paper which produces a positive image when exposed to a positive (or a negative image from a negative) and which is processed in a single development stage.

autoprotolysis [CHEM] Transfer of a proton from one molecule to another of the same substance.

autoprotolysis constant [CHEM] A constant denoting the equilibrium condition for the autoprotolysis reaction.

autopsy [PATH] A postmortem examination of the body to determine cause of death.

autoradar plot *See* chart comparison unit.

autoradiography [ENG] A technique for detecting radioactivity in a specimen by producing an image on a photographic film or plate. Also known as radioautography.

autorail [MECH ENG] A self-propelled vehicle having both flange wheels and pneumatic tires to permit operation on both rails and roadways.

autoreducing tachymeter [ENG] A class of tachymeter by

which horizontal and height distances are read simultaneously.

autoregressive series [MATH] A function of the form $f(t) = a_1 f(t-1) + a_2 f(t-2) + \ldots + a_m f(t-m) + k$, where k is any constant.

autorotation [MECH] **1.** Rotation about any axis of a body that is symmetrical and exposed to a uniform airstream and maintained only by aerodynamic moments. **2.** Rotation of a stalled symmetrical airfoil parallel to the direction of the wind.

autoserum [IMMUNOL] A serum obtained from and used for treatment of the patient himself.

autosexing [BIOL] Displaying differential sex characters at birth, noted particularly in fowl bred for sex-specific colors and patterns.

autoskeleton [INV ZOO] The endoskeleton of a sponge.

autosled [MECH ENG] A propeller-driven machine equipped with runners and wheels and adaptable to use on snow, ice, or bare roads.

autosome [GEN] Any chromosome other than a sex chromosome.

autostability [CONT SYS] The ability of a device (such as a servomechanism) to hold a steady position, either by virtue of its shape and proportions, or by control by a servomechanism.

autostarter [ELEC] **1.** Automatic starting and switchover generating system consisting of a standby generator coupled to the station load through an automatic power transfer control unit. **2.** See autotransformer starter.

autostoper [MIN ENG] A stoper, or light compressed-air rock drill, mounted on an air-leg support which not only supports the drill but also exerts pressure on the drill bit.

autostylic [VERT ZOO] Having the jaws attached directly to the cranium, as in chimeras, amphibians, and higher vertebrates.

Autosyn [ELECTR] Trademark of Bendix Aviation Corporation's line of synchros.

autosyndesis [CYTOL] The act of pairing of homologous chromosomes from the same parent during meiosis in polyploids.

autotomy [MED] Surgical removal of a part of one's own body. [ZOO] The process of self-amputation of appendages in crabs and other crustaceans and tails in some salamanders and lizards under stress.

autotransformer [ELEC] A power transformer having one continuous winding that is tapped; part of the winding serves as the primary and all of it serves as the secondary, or vice versa; small autotransformers are used to start motors.

autotransformer starter [ELEC] Motor starter having an autotransformer to furnish a reduced voltage for starting; includes the necessary switching mechanism. Also known as autostarter; compensator.

autotroph [BIOL] An organism capable of synthesizing organic nutrients directly from simple inorganic substances, such as carbon dioxide and inorganic nitrogen.

autozooecium [INV ZOO] The tube enclosing an autozooid.

autozooid [INV ZOO] An unspecialized feeding individual in a bryozoan colony, possessing fully developed organs and exoskeleton.

autumn [ASTRON] The season of the year which is the transition period from summer to winter, occurring as the sun approaches the winter solstice; beginning is marked by the autumnal equinox. Also known as fall.

autumnal [ASTRON] Pertaining to the season autumn.

autumnal equinox [ASTRON] The point on the celestial sphere at which the sun's rays at noon are 90° above the horizon at the Equator, or at an angle of 90° with the earth's axis, and neither North nor South Pole is inclined to the sun; occurs in the Northern Hemisphere on approximately September 23 and marks the beginning of autumn.

autumn ice [OCEANOGR] Sea ice in early stage of formation; comparatively salty, and crystalline in appearance.

Autunian [GEOL] A European stage of geologic time: Lower Permian (above Stephanian of Carboniferous, below Saxonian).

autunite [MINERAL] $Ca(UO_2)_2(PO_4)_2 \cdot 10H_2O$ A common fluorescent mineral that occurs as yellow tetragonal plates in uranium deposits; minor ore of uranium.

Auversian See Ledian.

auxanogram [MICROBIO] A plate culture provided with variable growth conditions to determine the effects of specific environmental factors.

auxanography [MICROBIO] The study of growth-inhibiting or growth-promoting agents by means of auxanograms.

auxanometer [ENG] An instrument used to detect and measure plant growth rate.

auxesis [PHYSIO] Growth resulting from increase in cell size.

auxiliary anode [MET] A supplementary anode that alters the current distribution in electroplating to give a more uniform plating thickness.

auxiliary circle [ASTRON] In celestial mechanics, a circumscribing circle to an orbital ellipse with radius a, the semimajor axis.

auxiliary contacts [ELEC] Contacts, in a switching device, in addition to the main circuit contacts, which function with the movement of the latter.

auxiliary fan [MIN ENG] A small fan installed underground for ventilating narrow coal drivages or hard headings which are not ventilated by the normal air current.

auxiliary fault [GEOL] A branch fault; a minor fault ending against a major one.

auxiliary fluid ignition [AERO ENG] A method of ignition of a liquid-propellant rocket engine in which a liquid that is hypergolic with either the fuel or the oxidizer is injected into the combustion chamber to initiate combustion.

auxiliary instruction buffer [ADP] A section of storage in the instruction unit, 16 bytes in length, used to hold prefetched instructions.

auxiliary landing gear [AERO ENG] The part or parts of a landing gear, such as an outboard wheel, which is intended to stabilize the craft on the surface but which bears no significant part of the weight.

auxiliary mineral [MINERAL] A light-colored, relatively rare or unimportant mineral in an igneous rock; examples are apatite, muscovite, corundrum, fluorite, and topaz.

auxiliary operation [ADP] An operation performed by equipment not under continuous control of the central processing unit of a computer.

auxiliary plane [GEOL] A plane at right angles to the net slip on a fault plane as determined from analysis of seismic data for an earthquake.

auxiliary power plant [MECH ENG] Ancillary equipment, such as pumps, fans, and soot blowers, used with the main boiler, turbine, engine, waterwheel, or generator of a power-generating station.

auxiliary power unit [AERO ENG] A power unit carried on an aircraft or spacecraft which can be used in addition to the main sources of power. Abbreviated APU.

auxiliary processor [ADP] Any equipment which performs an auxiliary operation in a computer.

auxiliary rafter [BUILD] A member strengthening the principal rafter in a truss.

auxiliary relay [ELEC] Relay that operates in response to the opening or closing of its operating circuit to assist another relay or device in performing a function.

auxiliary routine [ADP] A routine designed to assist in the operation of the computer and in debugging other routines.

auxiliary ship [NAV ARCH] A naval vessel other than a combat ship, such as a troop ship, repair ship, and cargo ship.

auxiliary storage [ADP] Storage device in addition to the main storage of a computer; for example, magnetic tape, disk, or magnetic drum.

auxiliary switch [ELEC] A switch actuated by the main device (such as a circuit breaker) for signaling, interlocking, or other purposes.

auxiliary thermometer [ENG] A mercury-in-glass thermometer attached to the stem of a reversing thermometer and read at the same time as the reversing thermometer so that the correction to the reading of the latter, resulting from change in temperature since reversal, can be computed.

auximone [BIOCHEM] Any of certain growth-promoting substances occurring principally in sphagnum peat decomposed by nitrogen bacteria.

auxin [BIOCHEM] Any organic compound which promotes plant growth along the longitudinal axis when applied to shoots free from indigenous growth-promoting substances.

auxoautotrophic [BIOL] Requiring no exogenous growth factors.

auxochrome [CHEM] Any substituent group such as $-NH_2$ and $-OH$ which, by affecting the spectral regions of strong absorption in chromophores, enhance the ability of the chromogen to act as a dye.

auxocyte [BIOL] A gamete-forming cell, such as an oocyte or spermatocyte, or a sporocyte during its growth period.

auxograph [ENG] An automatic device that records changes in the volume of a body.

auxoheterotrophic [BIOL] Requiring exogenous growth factors.

auxometer [ENG] An instrument that measures the magnification of a lens system.

auxospore [INV ZOO] A reproductive cell in diatoms formed in association with rejuvenescence by the union of two cells that have diminished in size through repeated divisions.

auxotonic [BOT] Induced by growth rather than by exogenous stimuli.

auxotrophic mutant [GEN] An organism that requires a specific growth factor, such as an amino acid, for its growth.

av *See* arithmetic mean.

aV *See* abvolt.

availability [ADP] Of data, data channels, and input-output devices in computers, the condition of being ready for use and not immediately committed to other tasks. [PHYS] The difference between the enthalpy per unit mass of substance and the product of entropy per unit mass multiplied by the lowest temperature available to the substance for heat discard; used in determining the ratio of actual work performed during a process by a working substance to that which theoretically should have been performed.

available chlorine [CHEM] The quantity of chlorine released by a bleaching powder when treated with acid.

available-chlorine method [MICROBIO] A technique for the standardization of chlorine disinfectants intended for use as germicidal rinses on cleaned surfaces; increments of bacterial inoculum are added to different disinfectant concentrations, and after incubation the results indicate the capacity of the disinfectant to handle an increasing bacterial load before exhaustion of available chlorine, the germicidal principle.

available energy [MECH ENG] Energy which can in principle be converted to mechanical work.

available heat [MECH ENG] The heat per unit mass of a working substance that could be transformed into work in an engine under ideal conditions for a given amount of heat per unit mass furnished to the working substance.

available line [ELECTR] Portion of the length of the scanning line which can be used specifically for picture signals in a facsimile system.

available machine time [ADP] Time during which a computer has the power turned on, is not under maintenance, and is known or believed to be operating correctly.

available moisture [HYD] Moisture in soil that is available for use by plants.

available power gain [ELECTR] Ratio, in an electronic transducer, of the available power from the output terminals of the transducer, under specified input termination conditions, to the available power from the driving generator.

available relief [GEOL] The vertical distance after uplift between the altitude of the original surface and the level at which grade is first attained.

avalanche [ELECTR] **1.** The cumulative process in which an electron or other charged particle accelerated by a strong electric field collides with and ionizes gas molecules, thereby releasing new electrons which in turn have more collisions, so that the discharge is thus self-maintained. Also known as avalanche effect; cascade; cumulative ionization; Townsend avalanche; Townsend ionization. **2.** Cumulative multiplication of carriers in a semiconductor as a result of avalanche breakdown. Also known as avalanche effect. [HYD] A mass of snow or ice moving rapidly down a mountain slope or cliff.

avalanche breakdown [ELECTR] Nondestructive breakdown in a semiconductor diode when the electric field across the barrier region is strong enough so that current carriers collide with valence electrons to produce ionization and cumulative multiplication of carriers.

avalanche conduction [PHYSIO] Conduction of a nerve impulse through several neurons which converge, increasing the discharge intensity by summation.

avalanche diode *See* Zener diode.

avalanche effect *See* avalanche.

avalanche wind [METEOROL] The rush of air produced in front of an avalanche of dry snow or in front of a landslide.

AVC *See* automatic volume control.

aV/cm *See* abvolt per centimeter.

aven [GEOL] *See* pothole. [MIN ENG] A vertical shaft leading upward from a cave passage, sometimes connecting with passages above.

Avena [BOT] A genus of grasses (family Gramineae), including oats, characterized by an inflorescence that is loosely paniculate, two-toothed lemmas, and deeply furrowed grains.

Avena unit [BIOL] A unit for the standardization of auxins.

avenin [BIOCHEM] The glutelin of oats.

aventurine [MINERAL] **1.** A glass or mineral containing sparkling gold-colored particles, usually copper or chromic oxide. **2.** A shiny red or green translucent quartz having small, but microscopically visible, exsolved hematite or included mica particles.

average *See* arithmetic mean.

average assay value [MIN ENG] The weighted result of assays obtained from a number of samples by multiplying the assay value of each sample by the width or thickness of the ore face over which it is taken and dividing the sum of these products by the total width of cross section sampled.

average bisector [NAV] A line extending through a four-course radio range station into opposing quadrants and midway between the lines (and their extensions) bisecting these two quadrants; used when the courses are not symmetrical.

average-calculating operation [ADP] A common or typical calculating operation longer than an addition and shorter than a multiplication; often taken as the mean of nine additions and one multiplication.

average deviation [MATH] In statistics, the average or arithmetic mean of the deviation, taken without regard to sign, from some fixed value, usually the arithmetic mean of the data. Abbreviated AD. Also known as mean deviation.

average discount factor *See* discount factor.

average-edge line [ADP] The imaginary line which traces or smooths the shape of any written or printed character to be recognized by a computer through optical, magnetic, or other means.

average effectiveness level *See* effectiveness level.

average heading [NAV] The average heading flown for a given period; it should be the same value as desired heading if the drift was predicted accurately.

average igneous rock [PETR] A hypothetical rock whose composition is thought to be similar to the average chemical composition of the outermost 10-mile shell of the earth.

average information content [COMMUN] The average of the information content per symbol emitted from a source.

average life *See* mean life.

average molecular weight [ORG CHEM] The calculated number to average the molecular weights of the varying-length polymer chains present in a polymer mixture.

average noise figure [ELECTR] Ratio in a transducer of total output noise power to the portion thereof attributable to thermal noise in the input termination, the total noise being summed over frequencies from zero to infinity, and the noise temperature of the input termination being standard (290K).

average outgoing quality limit [IND ENG] The average quality of all lots that pass quality inspection, expressed in terms of percent defective. Abbreviated AOQL.

average power output [ELECTR] Radio-frequency power, in an audio-modulation transmitter, delivered to the transmitter output terminals, averaged over a modulation cycle.

average wind [NAV] In air navigation, the resultant wind which would produce, or has produced, the same wind effect during a given period as the summation of the actual winds which will affect, or have affected, the flight of an aircraft.

averaging device [ENG] A device for obtaining the arithmetic mean of a number of readings, as on a bubble sextant.

aversive behavior [PSYCH] Avoidance behavior.

Aves [VERT ZOO] A class of animals composed of the birds, which are warm-blooded, egg-laying vertebrates primarily adapted for flying.

avgas *See* aviation gasoline.

avianize [VIROL] To attenuate a virus by repeated culture on chick embryos.

avian leukosis [VET MED] A disease complex in fowl probably caused by viruses and characterized by autonomous proliferation of blood-forming cells.

avian pneumoencephalitis *See* Newcastle disease.

avian pseudoplague *See* Newcastle disease.

avian tuberculosis [VET MED] A tuberculosis-like mycobacterial disease of fowl caused by *Mycobacterium avium*.

aviation [AERO ENG] **1.** The science and technology of flight through the air. **2.** The world of airplane business and its allied industries.

aviation gasoline [MATER] Stable fuel with high volatility and high octane, especially suited for use in aircraft reciprocating engines. Abbreviated avgas.

aviation medicine *See* aerospace medicine.

aviation method [ENG] Determination of knock-limiting power, under lean-mixture conditions, of fuels used in spark-ignition aircraft engines.

aviation mix [MATER] Antiknock fluid containing tetraethyllead, ethylene dibromide, and dye; used in aviation gasoline.

aviation weather forecast [METEOROL] A forecast of weather elements of particular interest to aviation, such as the ceiling, visibility, upper winds, icing, turbulence, and types of precipitation or storms. Also known as airways forecast.

aviation weather observation [METEOROL] An evaluation, according to set procedure, of those weather elements which are most important for aircraft operations. Also known as airways observation.

avicolous [ECOL] Living on birds, as of certain insects.

avicularium [INV ZOO] A specialized individual in a bryozoan colony with a beak that keeps other animals from settling on the colony.

aviculture [VERT ZOO] Care and breeding of birds, especially wild birds, in captivity.

avifauna [VERT ZOO] **1.** Birds, collectively. **2.** Birds characterizing a period, region, or environment.

avigation *See* air navigation.

aviolite [PETR] A mica-cordierite-hornfels.

avionics [ENG] The design and production of airborne electrical and electronic devices; term is derived from aviation electronics.

Avisco [TEXT] Trade name for rayon fibers manufactured by the FMC Corporation, American Viscose Division.

avitaminosis [MED] Any vitamin-deficiency disease.

Avitene [ORG CHEM] The trade name for a microcrystalline polymer form of collagen produced by American Viscose Corporation.

avocado [BOT] *Persea americana*. A subtropical evergreen tree of the order Magnoliales that bears a pulpy pear-shaped edible fruit.

avogadrite [MINERAL] $(K,Cs)BF_4$ An orthorhombic fluoborate mineral occurring in small crystals on Vesuvian lava.

Avogadro's number [PHYS] The number (6.02×10^{23}) of molecules in a gram-molecular weight of a substance.

Avogadro's hypothesis *See* Avogadro's law.

Avogadro's law [PHYS] The law which states that under the same conditions of pressure and temperature, equal volumes of all gases contain equal numbers of molecules; for example, 359 cubic feet at 32°F and 1 atmosphere for a perfect gas. Also known as Avogadro's hypothesis.

avogram [MECH] A unit of mass, equal to 1 gram divided by the Avogadro number.

avoidable delay [IND ENG] An interruption under the control of the operator during the normal operating time.

avoidance-avoidance conflict [PSYCH] Conflict within a person caught between two negative goals; as one goal is avoided, the other is brought closer.

avoirdupois pound *See* pound.

avoirdupois weight [MECH] The system of units commonly used in English-speaking countries for measurement of the mass of any substance except precious stones, precious metals, and drugs.

Avonian *See* Dinantian.

Avril [TEXT] Trade name for rayon fibers manufactured by the FMC Corporation, American Viscose Division.

avulsion [HYD] A sudden change in the course of a stream by which a portion of land is cut off, as where a stream cuts across and forms an oxbow. [MED] Tearing one part away from the other, either by trauma or surgery.

awaruite [MINERAL] Native nickel-iron alloy containing 57.7% nickel.

AWG *See* American wire gage.

awl [DES ENG] A point tool with a short wooden handle used to mark surfaces and to make small holes, as in leather or wood.

awn [BOT] Any of the bristles at the ends of glumes or bracts on the spikelets of oats, barley, and some wheat and grasses. Also known as beard.

awning deck [NAV ARCH] A light deck over the main deck, running the full length of a ship. Also known as hurricane deck.

awning window [BUILD] A window consisting of a series of vertically arranged, top-hinged rectangular sections; designed to admit air while excluding rain.

ax [DES ENG] An implement consisting of a heavy metal wedge-shaped head with one or two cutting edges and a relatively long wooden handle; used for chopping wood and felling trees.

axenic culture [BIOL] The growth of organisms of a single species in the absence of cells or living organisms of any other species.

axes of an aircraft [AERO ENG] Three fixed lines of reference, usually centroidal and mutually perpendicular: the longitudinal axis, the normal or yaw axis, and the lateral or pitch axis. Also known as aircraft axes.

axes of inertia [PHYS] The three principal axes of inertia, namely, one about which the moment of inertia is a maximum, one about which the moment of inertia is a minimum, and one perpendicular to both.

axhammer [DES ENG] An ax having one cutting edge and one hammer face.

axial [SCI TECH] Of, pertaining to, or along an axis.

axial angle [CRYSTAL] **1.** The acute angle between the two optic axes of a biaxial crystal. Also known as optic angle; optic-axial angle. **2.** In air, the larger angle between the optic axes after refraction on leaving the crystal.

axial compression [GEOL] A compression applied parallel with the cylinder axis in experimental work with rock cylinders.

axial culmination [GEOL] Distortion of the fold axis upward in a form similar to an anticline.

axial dipole field [GEOPHYS] A postulated magnetic field for the earth, consisting of a dipolar field centered at the earth's center, with its axis coincident with the earth's rotational axis.

axial element [CRYSTAL] The lengths, length ratios, and angles which define a crystal's unit cell.

axial fan [MECH ENG] A fan whose housing confines the gas flow to the direction along the rotating shaft at both the inlet and outlet.

axial filament [CYTOL] The central microtubule elements of a cilium or flagellum.

axial flow [FL MECH] Flow of fluid through an axially symmetric device such that the direction of the flow is along the axis of symmetry. Also known as axisymmetric flow.

axial-flow compressor [MECH ENG] A fluid compressor that accelerates the fluid in a direction generally parallel to the rotating shaft.

axial-flow jet engine [AERO ENG] **1.** A jet engine in which the general flow of air is along the longitudinal axis of the engine. **2.** A turbojet engine that utilizes an axial-flow compressor and turbine.

axial-flow pump [MECH ENG] A pump having an axial-flow or propeller-type impeller; used when maximum capacity and minimum head are desired. Also known as propeller pump.

AVOCADO

Avocado foliage and fruit.

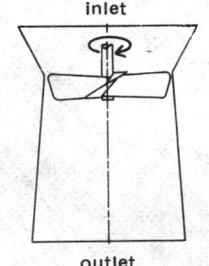

AXIAL FAN

inlet

outlet

Simplified diagram of an axial fan

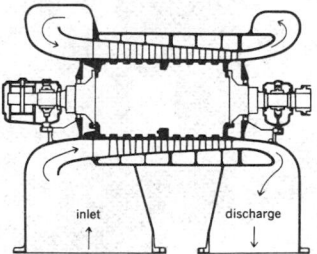

AXIAL-FLOW COMPRESSOR

inlet   discharge

Section drawing of an axial-flow compressor.

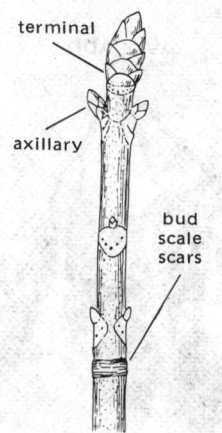

AXILLARY BUD

terminal

axillary

bud
scale
scars

Position of axillary bud in the buckeye.

AXINELLINA

Axinella polycapella, an axinelline sponge. *(After Hyman, 1940)*

axial gland [INV ZOO] A structure enclosing the stone canal in certain echinoderms; its function is uncertain.

axial gradient [EMBRYO] In some invertebrates, a graded difference in metobolic activity along the anterior-posterior, dorsal-ventral, and medial-lateral embryonic axes.

axial hydraulic thrust [MECH ENG] In single-stage and multi-stage pumps, the summation of unbalanced impeller forces acting in the axial direction.

axial jet [FL MECH] A flowing, turbulent stream which mixes with standing water in three dimensions.

axial lead [ELEC] A wire lead extending from the end along the axis of a resistor, capacitor, or other component.

axial load [MECH] A force with its resultant passing through the centroid of a particular section and being perpendicular to the plane of the section.

axial mining [ORD] Continuous or intermittent nuisance mining in great depth along the axis of enemy advance.

axial moment of inertia [MECH] One of the factors determining the stability of a spinning projectile.

axial musculature [ANAT] The muscles that lie along the longitudinal axis of the vertebrate body.

axial plane [CRYSTAL] 1. A plane that includes two of the crystallographic axes. 2. The plane of the optic axis of an optically biaxial crystal. [GEOL] A plane that intersects the crest or trough in such a manner that the limbs or sides of the fold are more or less symmetrically arranged with reference to it. Also known as axial surface.

axial plane cleavage [GEOL] Rock cleavage essentially parallel to the axial plane of a fold.

axial plane foliation [GEOL] Foliation developed in rocks parallel to the axial plane of a fold and perpendicular to the chief deformational pressure.

axial plane schistosity [GEOL] Schistosity developed parallel to the axial planes of folds.

axial plane separation [GEOL] The distance between axial planes of adjacent anticline and syncline.

axial rake [MECH ENG] The angle between the face of a blade of a milling cutter or reamer and a line parallel to its axis of rotation.

axial ratio [CRYSTAL] The ratio obtained by comparing the length of a crystallographic axis with one of the lateral axes taken as unity. [ELECTR] The ratio of the major axis to the minor axis of the polarization ellipse of a waveguide. Also known as ellipticity.

axial relief [MECH ENG] The relief behind the end cutting edge of a milling cutter.

axial runout [MECH ENG] The total amount, along the axis of rotation, by which the rotation of a cutting tool deviates from a plane.

axial skeleton [ANAT] The bones composing the skull, vertebral column, and associated structures of the vertebrate body.

axial stream [HYD] 1. The chief stream of an intermontane valley, the course of which is along the deepest part of the valley and is parallel to its longer dimension. 2. A stream whose course is along the axis of an anticlinal or a synclinal fold.

axial surface *See* axial plane.

axial symmetry [MATH] Property of a geometric configuration which is unchanged when rotated about a given line.

axial trace [GEOL] The intersection of the axial plane of a fold with the surface of the earth or any other specified surface; sometimes such a line is loosely and incorrectly called the axis.

axial trough [GEOL] Distortion of a fold axis downward into a form similar to a syncline.

axial vector *See* pseudovector.

axiation [EMBRYO] The formation or development of axial structures, such as the neural tube.

Axiidae [INV ZOO] A family of decapod crustaceans, including the hermit crabs, in the suborder Reptantia.

axil [BOT] The angle between a structure and the axis from which it arises, especially for branches and leaves.

axilla [ANAT] The depression between the arm and the thoracic wall; the armpit. [BOT] An axil.

axillary [ANAT] Of, pertaining to, or near the axilla or armpit. [BOT] Placed or growing in the axis of a branch or leaf.

axillary bud [BOT] A lateral bud borne in the axil of a leaf.

axillary sweat gland [ANAT] An apocrine gland located in the axilla.

Axinellina [INV ZOO] A suborder of sponges in the order Clavaxinellida.

axinite [MINERAL] $H_2(Ca, Fe, Mn)_4(BO)Al_2(SiO_4)_5$ Brown, blue, green, gray, or purplish gem mineral that commonly forms glassy triclinic crystals. Also known as glass schorl.

axinitization [GEOL] The replacement of rocks by axinite, as in the border zones of some granites.

axiolite [MINERAL] A variety of elongated spherulite in which there is an aggregation of minute acicular crystals arranged at right angles to a central axis.

axiom [MATH] Any of the assumptions upon which a mathematical theory (such as geometry, ring theory, and the real numbers) is based. Also known as postulate.

axiomatic S-matrix theory [PARTIC PHYS] An approach to the study of elementary particles that seeks to formulate S-matrix theory in a rigorous manner based on a few fundamental axioms that include Lorentz invariance, unitarity, analyticity near the physical values of the energy and momentum variables, and singularities in the physical region that correspond to known particles and scattering thresholds.

axis [ANAT] 1. The second cervical vertebra in higher vertebrates; the first vertebra of amphibians. 2. The center line of an organism, organ, or other body part. [GEOL] 1. A line where a folded bed has maximum curvature. 2. The central portion of a mountain chain. [GRAPHICS] The locus of intersection of two pencils of lines in perspective position. [MATH] 1. In a coordinate system, the line determining one of the coordinates, obtained by setting all other coordinates to zero. 2. A line of symmetry for a geometric figure. [MECH] A line about which a body rotates.

axis cylinder [CYTOL] 1. The central mass of a nerve fiber. 2. The core of protoplasm in a medullated nerve fiber.

axis of abscissas [MATH] The horizontal or x axis of a two-dimensional cartesian coordinate system, parallel to which abscissas are measured.

axis of freedom [DES ENG] An axis in a gyro about which a gimbal provides a degree of freedom.

axis of homology [MAP] The intersection of the plane of the photograph with the horizontal plane of the map or the plane of reference of the ground. Also known as axis of perspective; map parallel; perspective axis.

axis of ordinates [MATH] The vertical or y axis of a two-dimensional cartesian coordinate system, parallel to which ordinates are measured.

axis of perspective *See* axis of homology.

axis of rotation [MECH] A straight line passing through the points of a rotating rigid body that remain stationary, while the other points of the body move in circles about the axis.

axis of sighting [ENG] A line taken through the sights of a gun, or through the optical center and centers of curvature of lenses in any telescopic instrument.

axis of symmetry [MECH] An imaginary line about which a geometrical figure is symmetric. Also known as symmetry axis.

axis of the bore [ORD] The imaginary central line of the bore of a gun.

axis of thrust *See* thrust axis.

axis of tilt [GRAPHICS] A line through the perspective center perpendicular to the principal plane.

axis of trunnions [ORD] The axis about which a gun is rotated in elevation to increase or decrease the range of fire.

axis of weld [MET] A line along a weld used to describe the positions of the localized welds.

axisymmetric flow *See* axial flow.

axle [MECH ENG] A supporting member that carries a wheel and either rotates with the wheel to transmit mechanical power to or from it, or allows the wheel to rotate freely on it.

axle box [ENG] A bushing through which an axle passes in the hub of a wheel.

axle grease [MATER] A lubricating grease containing suspended lime particles and thickened with rosin soap.

axoblast [INV ZOO] 1. The germ cell in mesozoans; cells are linearly arranged in the longitudinal axis and produce the primary nematogens. 2. The individual scleroblast of the axis epithelium which produces spicules in octocorals.

axocoel [INV ZOO] The anterior pair of coelomic sacs in the dipleurula larval ancestral stage of echinoderms.

axogamy [BOT] Having sex organs on a leafy stem.

axolemma [CYTOL] The plasma membrane of an axon.

axolotl [VERT ZOO] The neotenous larva of some salamanders in the family Ambystomidae.

axolotl unit [BIOL] A unit for the standardization of thyroid extracts.

axometer [ENG] An instrument that locates the optical axis of a lens, particularly a lens used in eyeglasses.

axon [ANAT] The process or nerve fiber of a neuron that carries the unidirectional nerve impulse away from the cell body. Also known as neuraxon; neurite.

Axonolaimoidea [INV ZOO] A superfamily of free-living nematodes with species inhabiting marine and brackish-water environments.

axonometric projection [GRAPHICS] A drawing that shows an object's inclined position with respect to the planes of projection. Also known as isometric projection.

axoplasm [CYTOL] The protoplasm of an axon.

axopodium [INV ZOO] A semipermanent pseudopodium composed of axial filaments surrounded by a cytoplasmic envelope.

aye-aye [VERT ZOO] *Daubentonia madagascariensis.* A rare prosimian primate indigenous to eastern Madagascar; the single species of the family Daubentoniidae.

Ayre method [NAV ARCH] A method of calculating the effective horsepower of ships based on a review of model tests.

Ayrton-Jones balance [ELEC] A type of balance with which force between current-carrying conductors is measured; uses single-layer solenoids as the fixed and movable coils.

Ayrton shunt [ELEC] A shunt used to increase the range of a galvanometer without changing the damping. Also known as universal shunt.

9-azafluorene See carbazole.

azaserine [PHARM] $C_4H_7O_4N_3$ An antibiotic produced by a species of *Streptomyces* or by synthesis; used in treatment of acute leukemia.

azelaic acid [ORG CHEM] $HOOC(CH_2)_7CCOH$ Colorless leaflets; melting point 106.5°C; a dicarboxylic acid useful in lacquers, alkyd salts, organic synthesis, and formation of polyamides. Also known as 1,7-heptanedicarboxylic acid; nonanedioic acid.

azelate [ORG CHEM] A salt of azelaic acid, for example, sodium azelate.

azel display [ELECTR] Modified type of plan position indicator presentation showing two separate radar displays on one cathode-ray screen; one display presents bearing information and the other shows elevation.

azeotrope See azeotropic mixture.

azeotropic mixture [CHEM] A solution of two or more liquids, the composition of which does not change upon distillation. Also known as azeotrope.

azide [ORG CHEM] One of several types of compounds containing the $-N_3$ group and derived from hydrazoic acid, HN_3.

azimethane See diazomethane.

azimethylene See diazomethane.

azimidobenzene See 1,2,3-benzotriazole.

azimino See diazoamine.

azimuth [ASTRON] The horizontal direction of a celestial point from a terrestrial point, expressed as the angular distance from a reference direction, usually measured from 0° at the reference direction clockwise through 360°. [GEOD] Horizontal direction on the earth's surface.

azimuth-adjustment slide rule [ENG] A circular slide rule by which a known angular correction for fire at one elevation can be changed to the proper correction for any other elevation.

azimuthal chart [MAP] A chart on an azimuthal projection. Also known as zenithal chart.

azimuthal equidistant chart [MAP] A chart on the azimuthal equidistant map projection.

azimuthal map projection [MAP] The transformation of a spherical representation of the earth's surface to a tangent or intersecting plane established perpendicular to a right line passing through the center of the spherical representation.

azimuthal orthomorphic projection See stereographic projection.

azimuthal quantum number [ATOM PHYS] The orbital angular momentum quantum number l, such that the eigenvalue of L^2 is $l(l + 1)$.

azimuth angle [ENG] An angle in triangulation or in traverse through which the computation of azimuth is carried.

azimuth bar See azimuth instrument.

azimuth blanking [ELECTR] Blanking of the radar receiver as the scan traverses a selected azimuth region.

azimuth circle [DES ENG] A ring calibrated from 0 to 360 degrees over a compass, compass repeater, radar plan position indicator, direction finder, and so on, which provides means for observing compass bearings and azimuths.

azimuth compass [NAV] A compass with vertical sights so that the magnetic azimuths of celestial bodies may be read.

azimuth deviation [ORD] The angular difference in azimuth between the lines from the gun to the target and from the gun to the point where a projectile strikes or bursts.

azimuth dial [ENG] Any horizontal circle dial that reads azimuth.

azimuth equation [GEOD] A condition equation which expresses the relationship between the fixed azimuths of two lines that are connected by triangulation or traverse.

azimuth error [ENG] An error in the indicated azimuth of a target detected by radar.

azimuth gain reduction [ELECTR] Technique which allows control of the radar receiver system throughout any two azimuth sectors.

azimuth gating [ELECTR] The practice of selectively brightening and enhancing the gain-desired sectors of a radar plan position indicator display, usually by applying a step waveform to the automatic gain control circuit.

azimuth indicator [ENG] An approach-radar scope which displays azimuth information.

azimuth instrument [ENG] An instrument for measuring azimuths, particularly a device which fits over a central pivot in the glass cover of a magnetic compass. Also known as azimuth bar; bearing bar.

azimuth line [ENG] A radial line from the principal point, isocenter, or nadir point of a photograph, representing the direction to a similar point of an adjacent photograph in the same flight line; used extensively in radial triangulation.

azimuth marker [ENG] **1.** A scale encircling the plan position indicator scope of a radar on which the azimuth of a target from the radar may be measured. **2.** Any of the reference limits inserted electronically at 10 or 15° intervals which extend radially from the relative position of the radar on an off-center plan position indicator scope. [NAV] See electronic azimuth marker.

azimuth rate [ORD] In gunnery, the rate of change in azimuth measured in mils or degrees per second.

azimuth resolution [ELECTROMAG] Angle or distance by which two targets must be separated in azimuth to be distinguished by a radar set, when the targets are at the same range.

azimuth scale [ENG] A graduated angle-measuring device on instruments, gun carriages, and so forth that indicates azimuth.

azimuth-stabilized plan position indicator [ENG] A north-upward plan position indicator (PPI), a radarscope, which is stabilized by a gyrocompass so that either true or magnetic north is always at the top of the scope regardless of vehicle orientation.

azimuth tables [ASTRON] Publications providing tabulated azimuths or azimuth angles of celestial bodies for various combinations of declination, altitude, and hour angle; great-circle course angles can also be obtained by substitution of values.

azimuth transfer [ENG] Connecting, with a straight line, the nadir points of two vertical photographs selected from overlapping flights.

azimuth traverse [ENG] A survey traverse in which the direction of the measured course is determined by azimuth and verified by back azimuth.

azimuth versus amplitude [ELECTR] Electronic counter-countermeasures receiver with a plan position indicator type

AYE–AYE

Aye-aye *(Daubentonia madagascariensis)*, nocturnal, arboreal primate found only in eastern Madagascar.

AZOLE

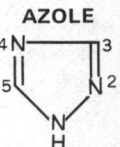

Structural formula for 1,2,4-triazole.

AZOTOBACTER

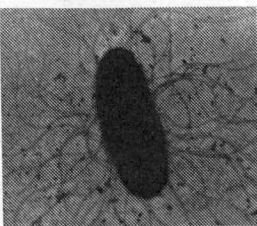

Azotobacter vinelandii with peritrichous flagella. *(From A. W. Hofer, Flagellation of Azotobacter, J. Bacteriol., 48:697-701, 1944)*

AZURITE

├── 2.5 cm ──┤

Crystals of azurite from Tsumeb, Southwest Africa. *(Specimen from Department of Geology, Bryn Mawr College)*

of display attached to the main antenna and used to display strobes due to jamming aircraft; it is useful in making passive fixes when two or more radar sites can operate together.

azine [ORG CHEM] A compound of six atoms in a ring; at least one of the atoms is nitrogen, and the ring structure resembles benzene; an example is pyridine.

azine dyes [ORG CHEM] Benzene-type dyes derived from phenazine; members of the group, such as nigrosines and safranines, are quite varied in application.

aziridine *See* ethyleneimine.

azlactone [ORG CHEM] A compound that is an anhydride of α-acylamino acid; the basic ring structure is the 5-oxazolone type.

azlon [TEXT] Any textile fiber derived from protein, such as casein.

azo- [ORG CHEM] A prefix indicating the radical $-N=N-$.

azobenzene [ORG CHEM] $C_6H_5N_2C_6H_5$ A compound existing in cis and trans geometric isomers; the cis form melts at 71°C; the trans form comprises orange-red leaflets, melting at 68.5°C; used in manufacture of dyes and accelerators for rubbers. Also known as azobenzidine; diphenyldiimide; phenylazobenzene.

azobenzidine *See* azobenzene.

azo dyes [ORG CHEM] Widely used commercial dyestuffs derived from amino compounds, with the $-N=N-$ chromophore group; can be made as acid, basic, direct, or mordant dyes.

Azoic [GEOL] That portion of the earlier Precambrian time in which there is no trace of life.

azoic dye [ORG CHEM] A water-insoluble azo dye that is formed by coupling of the components on a fiber. Also known as ice color; ingrain color.

azoic printing [GRAPHICS] The method whereby azoic compositions, that is, mixtures of napthols and diazotized products, temporarily inhibited from color development are printed on cloth; when the printed material is passed through steam containing formic acid vapor, the coupling reaction occurs and development takes place.

azoimide *See* hydrazoic acid.

azole [ORG CHEM] One of a class of organic compounds with a five-membered *N*-heterocycle containing two double bonds; an example is 1,2,4 triazole.

azomycin [MICROBIO] $C_3H_3O_2N_3$ An antimicrobial antibiotic produced by a strain of *Nocardia mesenterica.*

azon [ORD] A glide bomb used in World War II; its control surfaces in the trail were adjusted by radio signals to control the bomb in azimuth only (hence, azon).

azonal soil [GEOL] Any group of soils without well-developed profile characteristics, owing to their youth, conditions

of parent material, or relief that prevents development of normal soil-profile characteristics.

azophenylene *See* phenazine.

Azores high [METEOROL] The semipermanent subtropical high over the North Atlantic Ocean, especially when it is located over the eastern part of the ocean; when in the western part of the Atlantic, it becomes the Bermuda high.

azotemia [MED] The presence of excessive amounts of nitrogenous compounds in the blood.

Azotobacter [MICROBIO] A genus of large, usually motile, rod-shaped, oval, or spherical bacteria of the family Azotobacteraceae, found mostly in soils; the fixation of elemental nitrogen is their most important property.

Azotobacteraceae [MICROBIO] A family of aerobic bacteria in the order Eubacteriales capable of vigorous assimilation of elemental nitrogen.

azotometer *See* nitrometer.

Azotomonas [MICROBIO] A genus of soil bacteria in the family Pseudomonadaceae able to fix atmospheric nitrogen.

azoturia [MED] A condition characterized by excess amounts of urea or other nitrogenous substances in the urine. Also known as lumbago.

azoxybenzene [ORG CHEM] $C_6H_5NO=N-C_6H_5$ A compound existing in cis and trans forms; the cis form melts at 87°C; the trans form comprises yellow crystals, melting at 36°C, insoluble in water, soluble in ethanol.

azulene [ORG CHEM] $C_{16}H_{26}O$ The blue coloring matter of wormwood and other essential oils; an oily, blue liquid, boiling at 170°C; insoluble in water; used in cosmetics.

azulite [MINERAL] A translucent pale-blue variety of smithsonite found in large masses in Arizona and Greece.

azurite [MINERAL] $Cu_3(OH)_2CO_3$ A blue monoclinic mineral consisting of a basic carbonate of copper; an ore of copper. Also known as blue copper ore; blue malachite; chessylite.

azurmalachite [MINERAL] A mixture of azurite and malachite, usually occurring massive with concentric banding; used as an ornamental stone.

Azusa [ENG] A continuous-wave, high-accuracy, phase-comparison, single-station tracking system operating at C-band and giving two direction cosines and slant range which can be used to determine space position and velocity of a vehicle (usually a rocket or a missile).

azygospore [MYCOL] A spore which is morphologically similar to a zygospore but is formed parthenogenetically.

azygos vein [ANAT] A branch of the right precava which drains the intercostal muscles and empties into the superior vena cava.

azygote [BIOL] An individual produced by haploid parthenogenesis.

b *See* barn.

B *See* boron; brewster.

B-47 *See* Stratojet.

B-52 *See* Stratofortress.

Ba *See* barium.

babbitt metal [MET] Any of the white alloys composed primarily of tin or lead and of lesser amounts of antimony, copper, and perhaps other metals, and used for bearings.

babble [COMMUN] 1. Aggregate crosstalk from a large number of channels. 2. Unwanted disturbing sounds in a carrier or other multiple-channel system which result from the aggregate crosstalk or mutual interference from other channels.

Babcock coefficient of friction [FL MECH] An approximation to the coefficient of friction for steam flowing in a circular pipe of diameter d inches, given by $0.0027(1 + 3.6/d)$.

Babcock magnetograph [ASTRON] An instrument used to measure weak magnetic fields on the sun.

Babcock test [FOOD ENG] A test involving volumetric measurement of the fat content of a sample of milk; sulfuric acid is added, and the flask is heated to liquefy the fat, which is brought to the calibrated neck of the flask by centrifugal force.

Babesia [INV ZOO] The type genus of the Babesiidae, a protozoan family containing red blood cell parasites.

babesiasis [VET MED] A tick-borne protozoan disease of mammals other than man caused by species of *Babesia*.

Babesiidae [INV ZOO] A family of protozoans in the suborder Haemosporina containing parasites of vertebrate red blood cells.

Babinet compensator [OPTICS] A device for working with polarized light, made of two quartz prisms, assembled in a rhomb, to enable the optical retardation to be adjusted to positive or negative values.

Babinet point [GEOPHYS] One of the three commonly detectable points of zero polarization of skylight, the neutral points, lying along the vertical circle through the sun; the other two are the Arago point and Brewster point.

Babinet's principle [OPTICS] The principle that the diffraction patterns produced by complementary screens are identical; two screens are said to be complementary when the opaque parts of one correspond to the transparent parts of the other.

Babinski reflex [MED] An abnormal reflex after infancy associated with a disturbance of the pyramidal tract, characterized by extension of the great toe with fanning of the other toes on sharply stroking the lateral aspect of the sole.

baboon [VERT ZOO] Any of five species of large African and Asian terrestrial primates of the genus *Papio*, distinguished by a doglike muzzle, a short tail, and naked callosities on the buttocks.

Babo's law [PHYS CHEM] A law stating that the relative lowering of a solvent's vapor pressure by a solute is the same at all temperatures.

babs *See* blind approach beacon system.

babuina [VERT ZOO] A female baboon.

baby-pig disease [VET MED] Acute hypoglycemia of newborn pigs; usually fatal if untreated.

baby spot [ELEC] A small spotlight, usually equipped with a hood, used (as in the theater) to concentrate light on an area or an object a small distance from the spotlight.

baccate [BOT] 1. Bearing berries. 2. Having pulp like a berry.

bacciferous [BOT] Bearing berries.

Bacillaceae [MICROBIO] A family of rod-shaped, sporeforming bacteria in the order Eubacteriales.

Bacillariophyceae [BOT] The diatoms, a class of algae in the division Chrysophyta.

Bacillariophyta [BOT] An equivalent name for Bacillariophyceae.

bacillary [MICROBIO] 1. Rod-shaped. 2. Produced by, pertaining to, or resembling bacilli.

bacillary dysentery [MED] A highly infectious bacterial disease of man, localized in the bowels; caused by *Shigella*.

bacillary white diarrhea *See* pullorum disease.

bacillophobia [PSYCH] An abnormal fear of bacilli.

bacilluria [MED] The presence of bacilli in the urine.

bacillus [MICROBIO] Any rod-shaped bacterium.

Bacillus [MICROBIO] The type genus of rod-shaped, aerobic and sometimes anaerobic bacteria in the family Bacillaceae.

Bacillus anthracis [MICROBIO] A nonmotile pathogenic species of bacteria that causes anthrax.

Bacillus brevis [MICROBIO] A gram-variable, saprophytic species of bacteria important for production of the antibiotic tyrothricin.

Bacillus Calmette-Guerin vaccine [IMMUNOL] A vaccine prepared from attenuated human tubercle bacilli and used to immunize humans against tuberculosis. Abbreviated BCG vaccine.

Bacillus diphtheriae *See* Corynebacterium diphtheriae.

Bacillus licheniformis [MICROBIO] A gram-positive species of bacteria, recognized as a strain of *Bacillus subtilis*; important for production of the antibiotic bacitracin.

Bacillus megatherium [MICROBIO] A gram-positive species of bacteria that can produce levan from sucrose.

Bacillus polymyxa [MICROBIO] A gram-variable species of bacteria that produces gas from carbohydrates; important for production of the antibiotic polymyxin B.

Bacillus subtilis [MICROBIO] A gram-positive species of bacteria important for production of the antibiotic subtilin; the type species of the genus.

bacitracin [MICROBIO] A group of polypeptide antibiotics produced by *Bacillus licheniformis*.

back [ANAT] The part of the human body extending from the neck to the base of the spine. [GRAPHICS] The part of a book where the binding and pages are stitched together. [MIN ENG] 1. The upper part of any mining cavity. 2. A joint, usually a strike joint, perpendicular to the direction of working.

backacter *See* backhoe.

back azimuth [NAV] An azimuth 180° from a given azimuth.

back balance [MIN ENG] 1. A kind of self-acting incline in a mine. 2. The means of maintaining tension on a rope transmission or haulage system, consisting of the tension carriage, attached weight, and supporting structure.

back beach *See* backshore.

back bearing [NAV] A bearing along the reverse direction of a line. Also known as reciprocal bearing.

back-bent occlusion *See* bent-back occlusion.

back bias [ELECTR] 1. Degenerative or regenerative voltage which is fed back to circuits before its originating point; usually applied to a control anode of a tube or other device. 2. Voltage applied to a grid of a tube (or tubes) or electrode of another device to reduce a condition which has been upset by some external cause.

backblast area [ORD] Cone-shaped area to the rear of a recoilless weapon, rocket launcher, or rocket-assisted takeoff unit which is dangerous to personnel because of the expulsion of powder gases.

back bond [SOLID STATE] A chemical bond between an atom in the surface layer of a solid and an atom in the second layer.

backbone *See* spine.

backbreak *See* overbreak.

back bulb [BOT] A pseudobulb on certain orchid plants that remains on the plant after removal of the terminal growth, and that is used for propagation.

backcast stripping [MIN ENG] A stripping method using two

BABOON

A baboon, a representative of Old World cercopithecoid monkeys.

BACILLARIOPHYCEAE

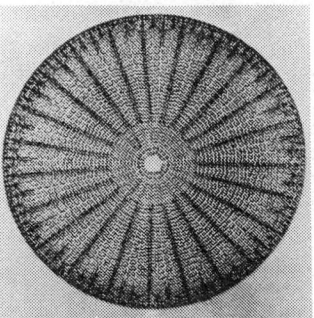

Arachnoidiscus ehrenbergii, a concentric diatom with radial symmetry. (*From H. J. Fuller and O. Tippo, College Botany, rev. ed., Holt, 1954*)

draglines; one strips and casts the overburden, and the other recasts a portion of the overburden.

back-coated mirror [OPTICS] Glass with a reflective coating applied against the rear surface.

back contact [ELEC] Normally closed stationary contact on a relay that is opened when the relay is energized.

back course [NAV] In the instrument low-approach system, the course which extends from the back of a localizer antenna system to furnish guidance in the horizontal plane at the rear of the localizer.

backcross [GEN] A cross between an F_1 heterozygote and an individual of P_1 genotype.

backdeep [GEOL] An epieugeosynclinal basin; a nonvolcanic postorogenic geosynclinal basin whose sediments are derived from an uplifted eugeosyncline.

backdigger *See* backhoe.

back-door cold front [METEOROL] A front which leads a cold air mass toward the south and southwest along the Atlantic seaboard of the United States.

back draft [MET] A reversed taper given to a casting model or pattern to prevent its withdrawal from the mold.

back echo [ELECTROMAG] Echo due to the back lobe of an antenna.

back echo reflection [ELECTR] A radar echo produced by radiation reflected to the target by a large, fixed obstruction; that is, the ray path is from antenna to obstruction to target to antenna, instead of antenna to target to antenna.

backed cloth [TEXT] Cloth made by weaving or knitting an extra weft or warp to the back to increase thickness or to obtain different color effects.

back-emission electron radiography [ELECTR] A technique used in microradiography to visualize, among other things, the presence of material of different atomic numbers in the surface of the specimen being observed; the polished side of the specimen is facing and in close contact with the emulsion side of a fine-grain photographic plate; a light-tight cover holds the specimen and plate in place to be subjected to hardened x-rays.

back end *See* thrust yoke.

backfill [CIV ENG] Earth refilling a trench or an excavation around a building, bridge abutment, and the like. [MIN ENG] Waste sand or rock used to support the mine roof after removal of ore.

backfire [ENG] Momentary backward burning of flame into the tip of a torch. Also known as flashback. [ORD] Rearward escapement of gases or cartridge fragments upon firing a gun.

backfire antenna [ELECTROMAG] An antenna which exhibits significant gain in a direction 180° from its principal lobe.

backflash [CHEM] Rapid combustion of a material occurring in an area that the reaction was not intended for.

backflooding [HYD] A reversal of flow of water at the water table resulting from changes in precipitation.

backflow preventer *See* vacuum breaker.

back focal length [OPTICS] The distance from the rear surface of a lens to its focal plane when the lens is focused on a distant object.

backfolding [GEOL] Process in mountain forming in which the folds are overturned toward the interior of an orogenic belt. Also known as backward folding.

back furrow *See* esker.

back gearing [MECH ENG] The technique of using gears on machine tools to obtain an increase in the number of speed changes that can be gotten with cone belt drives.

background [COMMUN] **1.** Picture white of the facsimile copy being scanned when the picture is black and white only. **2.** Undesired printing in the recorded facsimile copy of the picture being transmitted, resulting in shading of the background area. **3.** Noise heard during radio reception caused by atmospheric interference or the operation of the receiver at such high gain that inherent tube and circuit noises become noticeable.

background count [PHYS] Responses of the radiation counting system to radiation coming from sources other than the source to be measured.

background discrimination [ENG] The ability of a measuring instrument, circuit, or other device to distinguish signal from background noise.

background luminance [OPTICS] In visual-range theory, the brightness of the background against which a target is viewed.

background noise [ACOUS] The unwanted residual sound that is present whether or not the sound source being studied is in operation. [ENG] The undesired signals that are always present in an electronic or other system, independent of whether or not the desired signal is present.

background processing [ADP] The execution of lower-priority programs when higher-priority programs are not being handled by a data-processing system.

background radiation [NUCLEO] The radiation in man's natural environment, including cosmic rays and radiation from the naturally radioactive elements. Also known as natural radiation. [PHYS] Radiation which is due to sources other than the source of interest in a measurement of radiation and which is detected by the measuring apparatus.

background reflectance [ADP] The reflectance, relative to a standard, of the surface on which a printed or handwritten character has been inscribed in optical character recognition.

background returns [ENG] **1.** Signals on a radar screen from objects which are of no interest. **2.** *See* clutter.

backhand welding [MET] A welding technique in which the flame is directed back against the completed weld. Also known as backward welding.

back-haul [COMMUN] Use of excess circuit mileage by routing via switching centers that are not in a direct facility path from an originating office to a terminating office.

backhoe [MECH ENG] An excavator fitted with a hinged arm to which is rigidly attached a bucket that is drawn toward the machine in operation. Also known as backacter; backdigger; dragshovel; pullshovel.

back holes [MIN ENG] The holes which are shot last in mine shaft sinking.

backing [MET] *See* backing strip. [METEOROL] **1.** Internationally, a change in wind direction in a counterclockwise sense (for example, south to east) in either hemisphere of the earth. **2.** In United States usage, a change in wind direction in a counterclockwise sense in the Northern Hemisphere, clockwise in the Southern Hemisphere. [MIN ENG] **1.** Timbers across the top of a level, supported in notches cut in the rock. **2.** Rough masonry of a wall faced with finer work. **3.** Earth placed behind a retaining wall.

backing cloth [GRAPHICS] Fiber material on the reverse side of photographic paper for strength.

backing deals [MIN ENG] Boards, 1–4 inches thick, of sufficient length to bridge the space between timber or steel sets or between rings in skeleton tubbing.

backing off [ENG] Removing excessive body metal from badly worn bits.

backing plate [ENG] A plate used to support the hardware for the cavity used in plastics injection molding.

backing strip [MET] A piece of metal, asbestos, or other nonflammable material placed behind a joint to facilitate welding. Also known as backing.

backlands [GEOL] A section of a river floodplain lying behind a natural levee.

backlash [DES ENG] The amount by which the tooth space of a gear exceeds the tooth thickness of the mating gear along the pitch circles. [ELECTR] A small reverse current in a rectifier tube caused by the motion of positive ions produced in the gas by the impact of thermoelectrons. [ENG] **1.** Relative motion of mechanical parts caused by looseness. **2.** The difference between the actual values of a quantity when a dial controlling this quantity is brought to a given position by a clockwise rotation and when it is brought to the same position by a counterclockwise rotation.

backlight [GRAPHICS] A spotlight that illuminates from behind so that the subject is separated from the background; used in photography.

backlimb [GEOL] Of the two limbs of an asymmetrical anticline, the one that is more gently dipping.

backlining [GRAPHICS] Paper strip that is cemented to a book's backbone to bind the signatures and permit space between the backbone and the cover.

back lobe [ELECTROMAG] The three-dimensional portion of the radiation pattern of a directional antenna that is directed away from the intended direction.

backlog [IND ENG] **1.** An accumulation of orders promising

BACKHOE

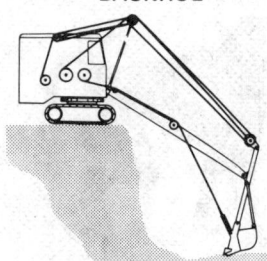

Reeving diagram for power crane fitted with backhoe. *(United States Steel Corp.)*

future work and profit. **2.** An accumulation of unprocessed materials or unperformed tasks.

back mixing [CHEM ENG] The tendency of reacted chemicals to intermingle with unreacted feed in reactors, such as stirred tanks, packed towers, and baffled tanks.

back off [ENG] **1.** To unscrew or disconnect. **2.** To withdraw the drill bit from a borehole. **3.** To withdraw a cutting tool or grinding wheel from contact with the workpiece.

back order [IND ENG] **1.** An order held for future completion. **2.** A new order placed for previously unavailable materials of an old order.

backout [AERO ENG] An undoing of previous steps during a countdown, usually in reverse order.

back porch [ELECTR] The period of time in a television circuit immediately following a synchronizing pulse during which the signal is held at the instantaneous amplitude corresponding to a black area in the received picture.

back pressure [MECH] Pressure due to a force that is operating in a direction opposite to that being considered, such as that of a fluid flow. [MECH ENG] Resistance transferred from rock into the drill stem when the bit is being fed at a faster rate than the bit can cut.

back-pressure curve [PETRO ENG] A graph used to arrive at the capacity of a natural-gas well to deliver gas into a pipeline at a sustained rate; uses data from back-pressure testing.

back-pressure testing [PETRO ENG] Method of estimating open-flow capacity of natural-gas wells by relating a series of gas-flow rates and their corresponding stabilized pressures at the bottom of the well bore.

back-pressure valve [PETRO ENG] A check valve installed in a natural-gas well bore to shut off gas flow while replacing the blowout preventer (used during drilling) with a christmas tree piping arrangement, which controls gas flow out of the completed well.

back radiation *See* backscattering; counterradiation.

back rake [DES ENG] An angle on a single-point turning tool measured between the plane of the tool face and the reference plane.

back range [NAV] A range (distance) measured astern, particularly one used as guidance for a craft moving away from the objects from which the distance information was deduced, forming the range.

back reef [GEOGR] The area between a reef and the land.

back resistance [ELECTR] The resistance between the contacts opposing the inverse current of a metallic rectifier.

backrope [NAV ARCH] **1.** Either of two ropes or chains on a sailing ship, extending aft from the lower end of the dolphin striker to each side of the bows. **2.** *See* cat back.

back-run process [CHEM ENG] A process for manufacturing water gas in which part of the run is made down, by passing steam through the superheater, thence up through the carburetor, down through the generator, and direct to the scrubbers.

back rush [OCEANOGR] Return of water seaward after the uprush of the waves.

backs [MIN ENG] Ore height available above a given working level.

backsaw [DES ENG] A fine-tooth saw with its upper edge stiffened by a metal rib to ensure straight cuts.

backsawing [FOR] A method of converting timber so that the growth rings meet the face in any part of an angle of less than 45°. Also known as bastard-sawing; crown-cut; plain-sawing; slash-sawing.

backscatter gage [ENG] A radar instrument used to measure the radiation scattered at 180° to the direction of the incident wave.

backscattering Also known as back radiation; backward scattering. [COMMUN] Propagation of extraneous signals by F- or E-region reflection in addition to the desired ionospheric scatter mode; the undesired signal enters the antenna through the back lobes. [ELECTROMAG] **1.** Radar echos from a target. **2.** Undesired radiation of energy to the rear by a directional antenna. [PHYS] The deflection of radiation or nuclear particles by scattering processes through angles greater than 90° with respect to the original direction of travel.

backscattering thickness gage [ENG] A device that uses a radioactive source for measuring the thickness of materials, such as coatings, in which the source and the instrument measuring the radiation are mounted on the same side of the material, the backscattered radiation thus being measured.

back-set bed [GEOL] Cross bedding that dips in a direction against the flow of a depositing current.

backshore [GEOL] The upper shore zone that is beyond the advance of the usual waves and tides. Also known as back beach; backshore beach.

backshore beach *See* backshore.

backshore terrace *See* berm.

back shot [MIN ENG] A shot used for widening an entry, placed at some distance from the head of an entry.

backsight [ENG] **1.** A sight on a previously established survey point or line. **2.** Reading a leveling rod in its unchanged position after moving the leveling instrument to a different location. [NAV] A marine sextant observation of a celestial body made by facing 180° from the azimuth of the body and using the visible horizon in the direction in which the observer is facing.

backsight method [ENG] **1.** A plane-table traversing method in which the table orientation produces the alignment of the alidade on an established map line, the table being rotated until the line of sight is coincident with the corresponding ground line. **2.** Sighting two pieces of equipment directly at each other in order to orient and synchronize one with the other in azimuth and elevation.

back slope *See* dip slope.

backspace [ADP] To move a recording medium one unit in the reverse or background direction. [MECH ENG] To move a typewriter carriage back one space by depressing a back space key.

backspring [NAV ARCH] A heavy line extending forward at an acute angle with a ship from the stern or midships to a wharf.

backstay [ENG] **1.** A supporting cable that prevents a more or less vertical object from falling forward. **2.** A spring used to keep together the cutting edges of purchase shears. **3.** A rod that runs from either end of a carriage's rear axle to the reach. **4.** A leather strip that covers and strengthens a shoe's back seam. [GRAPHICS] A rope or strap that keeps the carriage of a hand printing press from moving too far forward. [NAV ARCH] A rope, wire, or cable that runs from the top of a mast to the side of a ship and slants a little aft. [TEXT] A bar having a glass rod on its top, that runs across a loom beneath the lowest motion of the warp yarns.

backstep sequence [MET] Sequential deposition of weld beads in the direction opposite to the direction of welding.

back stoping *See* shrinkage stoping.

backtalk [ADP] Passage of information from a standby computer to the active computer.

backthrusting [GEOL] The thrusting in the direction of the interior of an orogenic belt, opposite the general structural trend.

back titration [CHEM] A titration to return to the end point which was passed.

back-to-front ratio [ELECTROMAG] Ratio used in connection with an antenna, metal rectifier, or any device in which signal strength or resistance in one direction is compared with that in the opposite direction.

backup [ENG] An item under development intended to perform the same general functions that another item also under development performs. [GRAPHICS] **1.** An image printed on the reverse side of a printed sheet. **2.** The printing of such an image.

backup arrangement *See* cascade.

backup relay [ELEC] A relay designed to protect a power system in case a primary relay fails to operate as desired.

backup system [SYS ENG] A system, normally redundant but kept available to replace a system which may fail in operation.

backward-acting regulator [ELECTR] Transmission regulator in which the adjustment made by the regulator affects the quantity which caused the adjustment.

backward-bladed aerodynamic fan [MECH ENG] A fan that consists of several streamlined blades mounted in a revolving casing.

backward diode [ELECTR] A semiconductor diode similar to

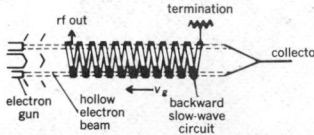

O-type backward-wave oscillator (or O-carcinotron), which uses a helix as the slow-wave circuit and has a hollow cylindrical electron beam; v_g is the velocity at which microwave energy travels along the helix toward the gun.

BACTERIAL CAPSULE

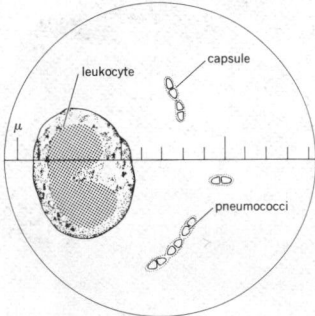

Drawing of pneumococci in sputum showing capsule surrounding the bacteria. *(From A. B. Sabin, J. Amer. Med. Assoc., 100(20):1585, 1933)*

BACTERIOPHAGE

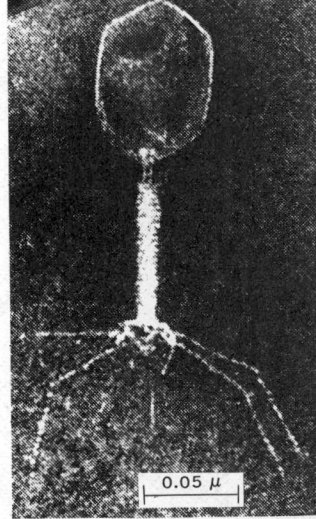

Electron micrograph of negatively stained T2 bacteriophage showing head and tail components. *(Courtesy of H. Fernandez-Moran)*

a tunnel diode except that it has no forward tunnel current; used as a low-voltage rectifier.

backward folding *See* backfolding.

backward scattering *See* backscattering.

backward wave [ELECTROMAG] An electromagnetic wave traveling opposite to the direction of motion of some other physical quantity in an electronic device such as a traveling-wave tube or mismatched transmission line.

backward-wave magnetron [ELECTR] A magnetron in which the electron beam travels in a direction opposite to the flow of the radio-frequency energy.

backward-wave oscillator [ELECTR] An electronic device which amplifies microwave signals simultaneously over a wide band of frequencies and in which the traveling wave produced is reflected backward so as to sustain the wave oscillations. Abbreviated BWO. Also known as carcinotron.

backward-wave tube [ELECTR] A type of microwave traveling-wave electron tube in which electromagnetic energy on a slow-wave circuit flows opposite in direction to the travel of electrons in a beam.

backward welding *See* backhand welding.

backwash [CHEM ENG] *See* blowback. [OCEANOGR] Water or waves thrown back by an obstruction such as a ship or breakwater.

backwash mark [GEOL] A crisscross ridge pattern in beach sand, caused by backwash.

backwash ripple mark [GEOL] Ripple marks that are broad and flat and parallel to the shoreline, with narrow, shallow troughs and crests about 30 centimeters apart; formed by backwash above the maximum wave retreat level.

backwater [HYD] **1.** A series of connected lagoons, or a creek parallel to a coast, narrowly separated from the sea and connected to it by barred outlets. **2.** Accumulation of water resulting from and held back by an obstruction. **3.** Water reversed in its course by an obstruction.

back wave [COMMUN] Signal emitted from a radio telegraph transmitter during spacing portions of the code characters. Also known as spacing wave.

backweld [MET] A weld placed behind a single groove weld.

back work [MIN ENG] Any kind of operation in a mine not immediately concerned with production or transport; literally, work behind the face, such as repairs to roads.

Bacon fuel cell [ELEC] A hydrogen-oxygen fuel cell, made by the Bacon Company, using porous nickel electrodes and operating at a relatively high temperature and pressure; used in the Apollo program.

bacteremia [MED] Presence of bacteria in the blood.

bacteremic shock [MED] A state of shock occurring during the course of bacteremia, especially if caused by gram-negative bacteria.

bacteria [MICROBIO] Extremely small, relatively simple prokaryotic microorganisms traditionally classified with the fungi as Schizomycetes.

Bacteriaceae [MICROBIO] Former designation for Brevibacteriaceae.

bacterial blight [PL PATH] Any blight disease of plants caused by bacteria, including common bacterial blight, halo blight, and fuscous blight.

bacterial bronchopneumonia [MED] Bacterial infection of the lung which has spread from infected bronchi.

bacterial brown spot [PL PATH] A bacterial blight disease of plants caused by *Pseudomonas syringae;* marked by water-soaked reddish-brown spots or cankers. Also known as bacterial canker.

bacterial canker *See* bacterial brown spot.

bacterial capsule [MICROBIO] A thick, mucous envelope, composed of polypeptides or carbohydrates, surrounding some bacteria.

bacterial coenzyme [MICROBIO] Organic molecules that participate directly in a bacterial enzymatic reaction and may be chemically altered during the reaction.

bacterial competence [MICROBIO] The ability of cells in a bacterial culture to accept and be transformed by a molecule of transforming deoxyribonucleic acid.

bacterial encephalitis [MED] Inflammation of the brain caused by primary or secondary bacterial infection.

bacterial endocarditis [MED] Inflammation of the endocardi-

um due to bacterial invasion. Also known as subacute bacterial endocarditis.

bacterial endoenzyme [MICROBIO] An enzyme produced and active within the bacterial cell.

bacterial endospore [MICROBIO] A body, resistant to extremes of temperature and to dehydration, produced within the cells of gram-positive, sporeforming rods of *Bacillus* and *Clostridium* and by the coccus *Sporosarcina*.

bacterial enzyme [MICROBIO] Any of the metabolic catalysts produced by bacteria.

bacterial genetics [GEN] The study of inheritance and variation patterns in bacteria.

bacterial infection [MED] Establishment of an infective bacterial agent in or on the body of a host.

bacterial leaf spot [PL PATH] A bacterial disease of plants characterized by spotty discolorations on the leaves; examples are angular leaf spot and leaf blotch.

bacterial luminescence [MICROBIO] A light-producing phenomenon exhibited by certain bacteria.

bacterial metabolism [MICROBIO] Total chemical changes carried out by living bacteria.

bacterial motility [MICROBIO] Self-propulsion in bacteria, either by gliding on a solid surface or by moving the flagella.

bacterial photosynthesis [MICROBIO] Use of light energy to synthesize organic compounds in green and purple bacteria.

bacterial pigmentation [MICROBIO] The organic compounds produced by certain bacteria which give color to both liquid cultures and colonies.

bacterial pneumonia [MED] Consolidation of the lung caused by inflammatory exudation due to bacterial infection.

bacterial pustule [PL PATH] A bacterial blight of plants caused by *Xanthomonas phaseoli;* characterized by blisters on the leaves.

bacterial soft rot [PL PATH] A bacterial disease of plants marked by disintegration of tissues.

bacterial speck [PL PATH] A bacterial disease of plants characterized by small lesions on plant parts.

bacterial spot [PL PATH] Any bacterial disease of plants marked by spotting of the infected part.

bacterial vaccine [IMMUNOL] A preparation of living, attenuated, or killed bacteria used to enhance the immune reaction in an individual already infected with the same bacteria.

bacterial wilt disease [PL PATH] A common bacterial disease of cucumber and muskmelon, caused by *Erwinia tracheiphila*, characterized by wilting and shriveling of the leaves and stems.

bactericide [MATER] An agent that destroys bacteria.

bactericidin [IMMUNOL] An antibody that kills bacteria in the presence of complement.

bacteriochlorophyll [BIOCHEM] $C_{52}H_{70}O_6N_4Mg$ A tetrahydroporphyrin chlorophyll compound occurring in the forms *a* and *b* in photosynthetic bacteria; there is no evidence that *b* has the empirical formula given.

bacteriocyte [INV ZOO] A modified fat cell found in certain insects that contains bacterium-shaped rods believed to be symbiotic bacteria.

bacteriogenic [MICROBIO] Caused by bacteria.

bacteriological warfare [ORD] Warfare conducted with pathogenic microorganisms as offensive weapons; a type of biological warfare.

bacteriologist [MICROBIO] A specialist in the study of bacteria.

bacteriology [MICROBIO] The science and study of bacteria; a specialized branch of microbiology.

bacteriolysin [MICROBIO] An antibody that is active against and causes lysis of specific bacterial cells.

bacteriolysis [MICROBIO] Dissolution of bacterial cells.

bacteriophage [VIROL] Any of the viruses that infect bacterial cells; each has a narrow host range. Also known as phage.

bacteriophobia [PSYCH] An abnormal fear of bacteria and other microorganisms.

bacteriosis [PL PATH] Any bacterial disease of plants.

bacteriostasis [MICROBIO] Inhibition of bacterial growth and metabolism.

bacteriotoxin [MICROBIO] **1.** Any toxin that destroys or inhibits growth of bacteria. **2.** A toxin produced by bacteria.

bacteriotropin [IMMUNOL] An antibody that is increased in amount during specific immunization and that renders the corresponding bacterium more susceptible to phagocytosis.

bacterioviridin See chlorobium chlorophyll.

Bacterium enteritidis See Salmonella enteritidis.

Bacterium paratyphosum B See Salmonella schottmülleri.

Bacterium paratyphosum C See Salmonella hirschfeldii.

Bacterium typhosum See Salmonella typhosum.

bacteriuria [MED] The occurrence of bacteria in the urine.

Bacteroides [MICROBIO] A genus of gram-negative, anaerobic, rod-shaped bacteria in the family Bacteroidaceae, found in the alimentary and urogenital tracts of man and animals.

Bacteroides fragilis [MICROBIO] The type species of the genus *Bacteroides;* simple, nonmotile, rod-shaped bacteria.

Bactoscilla [MICROBIO] A genus of gram-negative, slender trichomes, composed of rod-shaped cells, in the family Vitreoscillaceae.

Bactrian camel [VERT ZOO] *Camelus bactrianus.* The two-humped camel.

baculite [GEOL] A crystallite that looks like a dark rod.

baculum [VERT ZOO] The penis bone, or os priapi, in lower mammals.

baddeleyite [MINERAL] ZrO_2 A colorless, yellow, brown, or black monoclinic zirconium oxide mineral found in Brazil and Ceylon; used as heat- and corrosion-resistant linings for furnaces and muffles.

badge meter See film badge.

badger [VERT ZOO] Any of eight species of carnivorous mammals in six genera comprising the subfamily Melinae of the weasel family (Mustelidae).

Badger's rule [PHYS CHEM] An empirical relationship between the stretching force constant for a molecular bond and the bond length.

badlands [GEOGR] An erosive physiographic feature in semiarid regions characterized by sharp-edged, sinuous ridges separated by steep-sided, narrow, winding gullies.

bad top [MIN ENG] A weak roof in a coal mine; sometimes develops following a blast.

baeckeol [ORG CHEM] $C_{13}H_{18}O_4$ A phenolic ketone that is crystalline and pale yellow; found in oils from plants of species of the myrtle family.

Baeyer strain theory [ORG CHEM] The theory that the relative stability of penta- and hexamethylene ring compounds is caused by a propitious bond angle between carbons and a lack of bond strain.

baffle [ELEC] Device for deflecting oil or gas in a circuit breaker. [ENG ACOUS] A cabinet or partition used with a loudspeaker to reduce interaction between sound waves produced simultaneously by the two surfaces of the diaphragm. [ELECTR] An auxiliary member in a gas tube used, for example, to control the flow of mercury particles or deionize the mercury following conduction. [ENG] A plate that regulates the flow of a fluid, as in a steam-boiler flue or a gasoline muffler.

baffle plate [ELECTROMAG] Metal plate inserted in a waveguide to reduce the cross-sectional area for wave conversion purposes.

baffling wind [METEOROL] A wind that is shifting so that nautical movement by sailing vessels is impeded.

bag [ENG] **1.** A flexible cover used in bag molding. **2.** A container made of paper, plastic, or cloth without rigid walls to transport or store material.

bagasse [FOOD ENG] Remains of sugarcane after the juice has been extracted by pressure between the rolls of a mill; used to supply the fuel requirements of raw-sugar mills.

bagasse disease See bagassosis.

bagassosis [MED] A pneumoconiosis caused by the inhalation of bagasse dust, a dry sugarcane residue. Also known as bagasse disease.

bag filter [ENG] Filtering apparatus with porous cloth or felt bags through which dust-laden gases are sent, leaving the dust on the inner surfaces of the bags.

baghouse [ENG] The large chamber or room for holding bag filters used to filter gas streams from a furnace.

bag molding [ENG] A method of molding plastic or plywood-plastic combinations into curved shapes, in which fluid

pressure acting through a flexible cover, or bag, presses the material to be molded against a rigid die.

Bagnold number [ENG] A dimensionless number used in saltation studies.

bag powder [MATER] An explosive loaded in bags.

Bagridae [VERT ZOO] A family of semitropical catfishes in the suborder Siluroidei.

bahada See bajada.

bahamite [PETR] A consolidated limestone formed of sediment similar to a type currently found accumulating in the Bahamas.

bahiaite [PETR] Holocrystalline igneous rock formed mainly of hypersthene with subordinate hornblende and sometimes minor amounts of other minerals.

bai [METEOROL] A yellow mist prevalent in China and Japan in spring and fall, when the loose surface of the interior of China is churned up by the wind, and clouds of sand rise to a great height and are carried eastward, where they collect moisture and fall as a yellow mist.

baikerite [MINERAL] A waxlike mineral from the vicinity of Lake Baikal, Siberia; apparently about 60% ozocerite with other tarry, waxy, and resinous hydrocarbons.

bail [DES ENG] A supporting yolk or half-hoop.

bailer [ENG] A long, cylindrical vessel fitted with a bail at the upper end and a flap or tongue valve at the lower extremity; used to remove water, sand, and mud- or cuttings-laden fluids from a borehole. Also known as bailing bucket.

Bailey bridge [CIV ENG] A lattice bridge built of interchangeable panels connected at the corners with steel pins, permitting rapid construction; developed in Britain about 1942 as a military bridge.

Bailey meter [ENG] A flowmeter consisting of a helical quarter-turn vane which operates a counter to record the total weight of granular material flowing through vertical or near-vertical ducts, spouts, or pipes.

bailing [ENG] Removal of the cuttings from a well during cable-tool drilling, or of the liquid from a well, by means of a bailer.

bailing bucket See bailer.

bailout [AERO ENG] The exiting from a flying aircraft and descending by parachute in an emergency.

bailout bottle [AERO ENG] A personal supply of oxygen usually contained in a cylinder under pressure and utilized when the individual has left the central oxygen system, as in a parachute jump.

Baily's beads [ASTRON] Bright points of sunlight appearing around the edge of the moon just before and after the central phase of a total solar eclipse.

Bainbridge reflex [PHYSIO] A poorly understood reflex acceleration of the heart rate due to rise of pressure in the right atrium and vena cavae, possibly mediated through afferent vagal fibers.

bainite [MET] Steel formed by austempering, having an acicular structure of ferrite and carbides, exhibiting considerable toughness, and combining high strength with high ductility.

Bairdiacea [INV ZOO] A superfamily of ostracod crustaceans in the suborder Podocopa.

Baire function [MATH] The smallest class of functions on a topological space which contains the continuous functions and is closed under pointwise limits.

Bairstow number [FL MECH] A term previously used for Mach number.

bajada [GEOL] An alluvial plain formed as a result of lateral growth of adjacent alluvial fans until they finally coalesce to form a continuous inclined deposit along a mountain front. Also spelled bahada.

bajada breccia [PETR] An imperfectly stratified accumulation of coarse, angular rock fragments mixed with mud that formed in arid climates and results from a mudflow containing considerable water.

Bajocian [GEOL] A European stage: the middle Middle or lower Middle Jurassic geologic time; above Toarcian, below Bathonian.

Bakanae disease [PL PATH] A fungus disease of rice in Japan, caused by *Gibberella fujikurae;* a foot rot disease.

Bakelite [MATER] A trademark for a thermosetting plastic

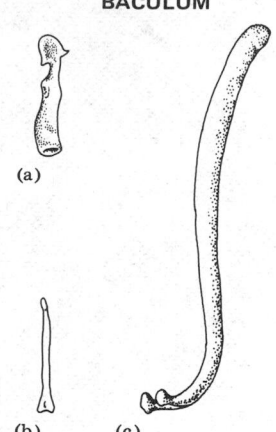

BACULUM

Bacula of *(a)* squirrel, *(b)* cotton mouse, *(c)* otter.

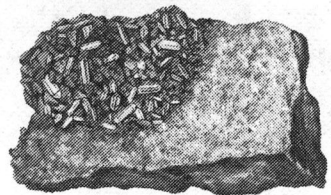

BADDELEYITE

⊢— 2.5 cm —⊣

Baddeleyite crystals with jacupirangite rock from Brazil. *(Specimen from Department of Geology, Bryn Mawr College)*

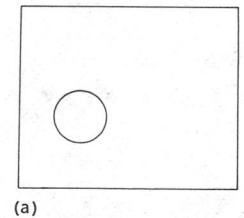

BAFFLE

(a)

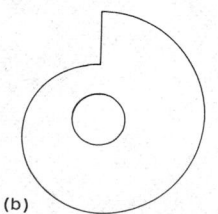

(b)

(a) Baffle with speaker mounted eccentrically and *(b)* irregular-shaped baffle, to broaden the frequency band of destructive interference between speaker front and back waves at listener's position.

used in a variety of articles; has high chemical and electrical resistance.

bakeout [ENG] The degassing of surfaces of a vacuum system by heating during the pumping process.

baker bell dolphin [CIV ENG] A dolphin consisting of a heavy bell-shaped cap pivoted on a group of piles; a blow from a ship will tilt the bell, thus absorbing energy.

bakerite [MINERAL] $8CaO \cdot 5B_2O_3 \cdot 6SiO_2 \cdot 6H_2O$ White mineral, occurring in fine-grained, nodular masses, resembling marble and unglazed porcelain, and consisting of hydrous calcium borosilicate.

Baker-Nunn camera [OPTICS] A large camera with a Schmidt-type lens system used to track earth satellites.

bakers' yeast [FOOD ENG] An industrial yeast used for baking purposes because of maximum growth and low alcohol production; composed of dry cells of one or more strains of *Saccharomyces cerevisiae*.

baking [FOOD ENG] The use of heat in an oven to convert flour, water, yeast, and such, into baked goods. [MET] Heating metal at low temperatures to remove gases.

baking finish [MATER] Varnish or paint that must be baked at temperatures greater than 150°F (66°C) to develop desired final properties of strength and hardness.

baking powder [FOOD ENG] A yeast substitute of sodium bicarbonate plus potassium tartrate, tartaric acid, anhydrous sodium aluminum sulfate, monocalcium phosphate, or any combination of these acids so formulated as to release carbon dioxide from the sodium bicarbonate (baking soda) when moistened.

baking soda *See* sodium bicarbonate.

baking varnish [MATER] A chemical-resistant varnish made of synthetic resins that requires baking to be dried.

BAL *See* dimercaprol.

Balaenicipitidae [VERT ZOO] A family of wading birds composed of a single species, the shoebill stork (*Balaeniceps rex*), in the order Ciconiiformes.

Balaenidae [VERT ZOO] The right whales, a family of cetacean mammals composed of five species in the suborder Mysticeti.

balance [ACOUS] The condition in a stereo system wherein both speakers produce the same average sound levels. [AERO ENG] **1.** The equilibrium attained by an aircraft, rocket, or the like when forces and moments are acting upon it so as to produce steady flight, especially without rotation about its axes. **2.** The equilibrium about any specified axis that counterbalances something, especially on an aircraft control surface, such as a weight installed forward of the hinge axis to counterbalance the surface aft of the hinge axis. [CHEM] To bring a chemical equation into balance so that reaction substances and reaction products obey the laws of conservation of mass and charge. [ELEC] The state of an electrical network when it is adjusted so that voltage in one branch induces or causes no current in another branch. [ENG] An instrument for measuring mass or weight. [MIN ENG] The counterpoise or weight attached by cable to the drum of a winding engine to balance the weight of the cage and hoisting cable and thus assist the engine in lifting the load out of the shaft.

Balance *See* Libra.

balance car [MIN ENG] In quarrying, a car loaded with iron or stone and connected by means of a steel cable with a channeling machine operating on an inclined track; used to counteract the force of gravity and thus enable the channeling machine to operate with equal ease uphill and downhill.

balance coil [ELEC] An iron-core solenoid with adjustable taps near the center; used to convert a two-wire circuit to a three-wire circuit, the taps furnishing a neutral terminal for the latter.

balance control [ELECTR] A control used in a stereo sound system to vary the volume of one loudspeaker system relative to the other while maintaining their combined volume essentially constant.

balanced amplifier [ELECTR] An electronic amplifier in which there are two identical signal branches connected so as to operate with the inputs in phase opposition and with the output connections in phase, each balanced to ground.

balanced anesthesia [MED] Anesthesia produced by safe doses of two or more agents or methods of anesthesia, each of which contributes to the total desired effect.

balanced armature unit [ENG ACOUS] Driving unit used in magnetic loudspeakers, consisting of an iron armature pivoted between the poles of a permanent magnet and surrounded by coils carrying the audio-frequency current; variations in audio-frequency current cause corresponding changes in armature magnetism and corresponding movements of the armature with respect to the poles of the permanent magnet.

balanced bridge [ELEC] Wheatstone bridge circuit which, when in a quiescent state, has an output voltage of zero.

balanced circuit [ELEC] A circuit whose two sides are electrically alike and symmetrical with respect to a common reference point, usually ground.

balanced cloth [TEXT] A fabric made up of equal numbers and sizes of warp and filling yarns.

balanced converter *See* balun.

balanced currents [ELEC] Currents flowing in the two conductors of a balanced line which, at every point along the line, are equal in magnitude and opposite in direction. Also known as push-pull currents.

balanced detector [ELECTR] A detector used in frequency-modulation receivers; in one form the audio output is the rectified difference between voltages produced across two resonant circuits, one being tuned slightly above the carrier frequency and one slightly below.

balanced fertilizer [MATER] A material of varying composition added to soil so as to provide essential mineral elements at required levels, improve soil structure, or enhance microbial activity.

balanced gasoline [MATER] An automotive gasoline blended from petroleum hydrocarbons of varying volatilities to provide desired performance for engine starting, warm-up, acceleration, and mileage.

balanced line [ELEC] A transmission line consisting of two conductors capable of being operated so that the voltages of the two conductors at any transverse plane are equal in magnitude and opposite in polarity with respect to ground. [IND ENG] A production line for which the time cycles of the operators are made approximately equal so that the work flows at a desired steady rate from one operator to the next.

balanced method [ENG] Method of measurement in which the reading is taken at zero; it may be a visual or audible reading, and in the latter case the null is the no-sound setting.

balanced modulator [ELECTR] A modulator in which the carrier and modulating signal are introduced in such a way that the output contains the two sidebands without the carrier.

balanced network [ELEC] Hybrid network in which the impedances of the opposite branches are equal.

balanced oscillator [ELECTR] Any oscillator in which, at the oscillator frequency, the impedance centers of the tank circuits are at ground potential, and the voltages between either end and their centers are equal in magnitude and opposite in phase.

balanced polymorphism [GEN] Maintenance in a population of two or more alleles in equilibrium at frequencies too high to be explained, particularly for the rarer of them, by mutation balanced by selection; for example, the selective advantage of heterozygotes over both homozygotes.

balanced range of error [STAT] A range of error in which the maximum and minimum possible errors are opposite in sign and equal in magnitude.

balanced rudder [NAV ARCH] A rudder, part of whose area is forward of the vertical axis, to counterbalance water pressure abaft of the axis.

balanced stock [ORD] **1.** That condition in supply when availability and requirements are in equilibrium for specific items. **2.** Accumulation of supplies in quantities determined necessary to meet requirements for a fixed period of time.

balanced surface [AERO ENG] A control surface that extends on both sides of the axis of the hinge or pivot, or that has auxiliary devices or extensions connected with it, in such a manner as to effect a small or zero resultant moment of the air forces about the hinge axis.

balanced translocation [GEN] Positional change of one or more chromosome segments in cells or gametes without alteration of the normal diploid or haploid complement of genetic material.

balanced voltages [ELEC] Voltages that are equal in magnitude and opposite in polarity with respect to ground. Also known as push-pull voltages.

balanced wire circuit [ELEC] Circuit wherein the two sides are electrically alike and symmetrical with respect to ground and other conductors.

balance equation [MATH] An equation expressing a balance of quantities in the sense that the local or individual rates of change are zero. [METEOROL] A diagnostic equation expressing a balance between the pressure field and the horizontal field of motion of the atmosphere.

balance method *See* null method.

balancer [ELEC] A mechanism for equalizing the loads on the outer lines of a three-wire system for electric power distribution, consisting of two similar shunt or compound machines coupled together with the armatures connected in series across the outer lines. [INV ZOO] *See* haltere. [VERT ZOO] Either of a pair of rodlike lateral appendages on the heads of some larval salamanders.

balance shot [MIN ENG] In coal mining, a shot for which the drill hole is parallel to the face of the coal that is to be broken by it.

balance spring [HOROL] An oscillating spring of spiral or (in a chronometer) cylindrical shape which governs the movement of a balance wheel in a timepiece. Also known as hairspring.

balance-to-unbalance transformer [ELEC] Device for matching a pair of lines, balanced with respect to earth, to a pair of lines not balanced with respect to earth.

balance wheel [MECH ENG] 1. A wheel which governs or stabilizes the movement of a mechanism. 2. *See* flywheel.

balancing a survey [ENG] Distributing corrections through any traverse to eliminate the error of closure and to obtain an adjusted position for each traverse station. Also known as traverse adjustment.

balancing band [NAV ARCH] A band at the balancing point of an anchor on either side of the shank, fitted with a ring.

balancing capacitor [ELECTR] A variable capacitor used to improve the accuracy of a radio direction finder. Also known as compensating capacitor.

balancing delay [IND ENG] In motion study, idleness of one hand while the other is active to catch up.

balancing unit [ELEC] 1. Antenna-matching device used to permit efficient coupling of a transmitter or receiver having an unbalanced output circuit to an antenna having a balanced transmission line. 2. Device for converting balanced to unbalanced transmission lines, and vice versa, by placing suitable discontinuities at the junction between the lines instead of using lumped components.

Balanidae [INV ZOO] A family of littoral, sessile barnacles in the suborder Balanomorpha.

balanitis [MED] Inflammation of the glans of the penis or of the clitoris.

Balanoglossus [INV ZOO] A cosmopolitan genus of tongue worms belonging to the class Enteropneusta.

Balanomorpha [INV ZOO] The symmetrical barnacles, a suborder of sessile crustaceans in the order Thoracica.

Balanopaceae [BOT] A small family of dioecious dicotyledonous plants in the order Fagales characterized by exstipulate leaves, seeds with endosperm, and the pistillate flower solitary in a multibracteate involucre.

Balanopales [BOT] An ordinal name suggested for the Balanopaceae in some classifications.

Balanophoraceae [BOT] A family of dicotyledonous terrestrial plants in the order Santalales characterized by dry nutlike fruit, one to five ovules, unisexual flowers, attachment to the stem of the host, and the lack of chlorophyll.

balanoposthitis [MED] Inflammation of the glans penis and of the prepuce.

Balanopsidales [BOT] An order in some systems of classification which includes only the Balanopaceae of the Fagales.

balantidiasis [MED] An intestinal infection of man caused by the protozoan *Balantidium coli*.

Balantidium [INV ZOO] A genus of protozoans in the order Trichostomatida containing the only ciliated protozoan species parasitic in man, *Balantidium coli*.

Balanus [INV ZOO] A genus of barnacles composed of sessile acorn barnacles; the type genus of the family Balanidae.

balata [MATER] A hard substance, similar to gutta-percha, used mainly in golf balls and belting, which is made by drying the milky juice of the bully tree (*Manilkara bidentata*). Also known as gutta-balata.

Balbiani rings [CYTOL] Localized swellings of a polytene chromosome.

balbriggan [TEXT] Knitted yarn made of cotton, wool, rayon, or other fiber and which is characteristically lightweight and unbleached.

bald [GEOGR] An elevated grassy, treeless area, as on the top of a mountain. [MIN ENG] 1. Without framing. 2. A mine timber which has a flat end.

baldheaded anticline [GEOL] An upfold with a crest that has been deeply eroded before later deposition.

baldness [MED] Loss or absence of hair.

Baldwin spot *See* bitter pit.

bale [IND ENG] 1. A large package of material, pressed tightly together, tied with rope, wire, or hoops and usually covered with wrapping. 2. The amount of material in a bale; sometimes used as a unit of measure, as 500 pounds of cotton in the United States.

baleen [VERT ZOO] A horny substance, growing as fringed filter plates suspended from the upper jaws of whalebone whales. Also known as whalebone.

baler [MECH ENG] A machine which takes large quantities of raw or finished materials and binds them with rope or metal straps or wires into a large package.

Balfour's law [EMBRYO] The law that the speed with which any part of the ovum segments is roughly proportional to the protoplasm's concentration in that area; the segment's size is inversely proportional to the protoplasm's concentration.

baling [CIV ENG] A technique used to convert loose refuse into heavy blocks by compaction; the blocks are then burned and are buried in sanitary landfill.

Bali wind [METEOROL] A strong east wind at the eastern end of Java.

balk [MIN ENG] A sudden thinning out, for a certain distance, of a bed of coal.

balking [IND ENG] The refusal of a customer to enter a queue for some reason, such as insufficient waiting room.

ball [GEOL] 1. A low sand ridge, underwater by high tide, which extends generally parallel with the shoreline; usually separated by an intervening trough from the beach. Also known as longshore bar. 2. A spheroidal mass of sedimentary material. 3. Common name for a nodule, especially of ironstone. [MECH ENG] In fine grinding, one of the crushing bodies used in a ball mill. [ORD] 1. A bullet for general use, as distinguished from bullets for special uses such as armor-piercing, incendiary, or high explosive. 2. A small-arms solid propellant which is oblate spheroidal in shape, generally a double-base propellant.

ball-and-race-type pulverizer [MECH ENG] A grinding machine in which balls rotate under an applied force between two races to crush materials, such as coal, to fine consistency. Also known as ball-bearing pulverizer.

ball-and-socket joint [ANAT] *See* enarthrosis. [MECH ENG] A joint in which a member ending in a ball is joined to a member ending in a socket so that relative movement is permitted within a certain angle in all planes passing through a line. Also known as ball joint.

ballas [MINERAL] A spherical aggregate of small diamond crystals; used in diamond drill bits and other diamond tools.

ballast [AERO ENG] A relatively dense substance that is placed in the car of a vehicle and can be thrown out to reduce the load or can be shifted to change the center of gravity. [CIV ENG] Crushed stone used in a railroad bed to support the ties, hold the track in line, and help drainage. [ELEC] A circuit element that serves to limit an electric current or to provide a starting voltage, as in certain types of lamps, such

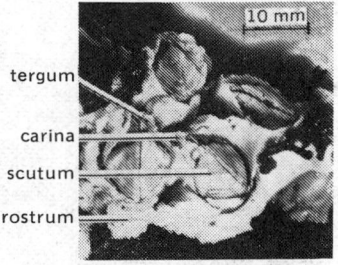

BALANUS

Balanus balanoides, apical view. (From D. P. Henry, *Studies on the sessile Cirripedia of the Pacific coast of North America, Univ. Wash. Publ. Oceanogr.*, 4(3):99–131, 1942)

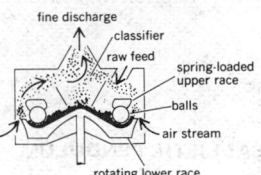

BALL-AND-RACE-TYPE PULVERIZER

Coarse raw material is ground by crushing and attrition between balls and races and is then withdrawn from the pulverizer by an air stream.

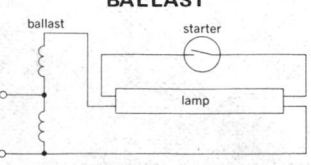

BALLAST

Position of the ballast in preheat lamp circuit.

as in fluorescent ceiling fixtures. [MATER] Coarse gravel used as an ingredient in concrete. [NAV ARCH] **1.** A relatively heavy material such as lead, iron, or water placed in a ship to ensure stability or to maintain the proper draft or trim. **2.** To pump seawater into empty fuel tanks of a ship to ensure its stability or suitable draft and trim for seaworthiness.

ballast lamp [ELEC] A light-producing electrical resistance device which maintains nearly constant current by increasing in resistance as the current increases.

ballast line [NAV ARCH] A ship's water line when the ship is ballasted.

ballast pump [NAV ARCH] A pump designed to discharge water ballast.

ballast resistor [ELEC] A resistor that increases in resistance as current through it increases, and decreases in resistance as current decreases. Also known as barretter.

ballast tank [NAV ARCH] **1.** One of several tanks in the hold of a ship which may be pumped full of water as ballast. **2.** One of the tanks in a submarine that are filled with water or air to submerge or surface.

ballast tube [ELEC] A ballast resistor mounted in an evacuated glass or metal envelope, like that of a vacuum tube, to reduce radiation of heat from the resistance element and thereby improve the voltage-regulating action.

ball bearing [MECH ENG] An antifriction bearing permitting free motion between moving and fixed parts by means of balls confined between outer and inner rings.

ball-bearing pulverizer *See* ball-and-race-type pulverizer.

ball bonding [ENG] The making of electrical connections in which a flame is used to cut a wire, the molten end of which solidifies as a ball, which is pressed against the bonding pad on an integrated circuit.

ball breaker [ENG] A steel or iron ball that is hoisted by a derrick and allowed to fall on blocks of waste stone to break them or to swing against old buildings to demolish them. Also known as skull cracker; wrecking ball.

ball burnishing [MET] A method of giving small stainless steel parts a lustrous finish by rotating them in a wood-lined barrel with water, burnishing soap, and hardened steel balls.

ball bushing [MECH ENG] A type of ball bearing that allows motion of the shaft in its axial direction.

ball clay [MATER] A clay used in ceramics that is characterized by strong binding properties, a tendency to ball, and excellent plasticity.

ball coal [GEOL] A variety of coal occurring in spheroidal masses. Also known as pebble coal.

ball-float liquid-level meter [ENG] A float which rises and falls with liquid level, actuating a pointer adjacent to a calibrated scale in order to measure the level of a liquid in a tank or other container.

ball grinder *See* ball mill.

ballhead [MECH ENG] That part of the governor which contains flyweights whose force is balanced, at least in part, by the force of compression of a speeder spring.

ballistic area [ORD] Space lying between the centers of impact of two groups of shots, one consisting entirely of shots over the target and the other entirely of shots short of the target.

ballistic body [ENG] A body free to move, behave, and be modified in appearance, contour, or texture by ambient conditions, substances, or forces, such as by the pressure of gases in a gun, by rifling in a barrel, by gravity, by temperature, or by air particles.

ballistic camera [OPTICS] A ground-based camera using multiple exposures on the same plate to record the trajectory of a rocket.

ballistic case [ORD] Any shell or casing with efficient ballistic characteristics, used to enclose elements for delivery on a target; any bomb, projectile, or rocket case.

ballistic coefficient [MECH] The numerical measure of the ability of a missile to overcome air resistance; dependent upon the mass, diameter, and form factor.

ballistic conditions [MECH] Conditions which affect the motion of a projectile in the bore and through the atmosphere, including muzzle velocity, weight of projectile, size and shape of projectile, rotation of the earth, density of the air, temperature or elasticity of the air, and the wind.

ballistic correction [ORD] An adjustment in firing data that is based on ballistic conditions, affecting the flight of a projectile; does not include adjustment based on observation of fire.

ballistic curve [MECH] The curve described by the path of a bullet, a bomb, or other projectile as determined by the ballistic conditions, by the propulsive force, and by gravity.

ballistic deflection [MECH] The deflection of a missile due to its ballistic characteristics.

ballistic density [MECH] A representation of the atmospheric density encountered by a projectile in flight, expressed as a percentage of the density according to the standard artillery atmosphere.

ballistic efficiency [MECH] **1.** The ability of a projectile to overcome the resistance of the air; depends chiefly on the weight, diameter, and shape of the projectile. **2.** The external efficiency of a rocket or other jet engine of a missile.

ballistic entry [MECH] Movement of a ballistic body from without to within a planetary atmosphere.

ballistic galvanometer [ELEC] A galvanometer having a long period of swing so that the deflection may measure the electric charge in a current pulse or the time integral of a voltage pulse.

ballistic lead [ORD] The correction or allowance for wind effects and gravity made when computing a lead angle.

ballistic limit [MECH] The minimum velocity at which a particular armor-piercing projectile is expected to consistently and completely penetrate armor plate of given thickness and physical properties at a specified angle of obliquity.

ballistic measurement [MECH] Any measurement in which an impulse is applied to a device such as the bob of a ballistic pendulum, or the moving part of a ballistic galvanometer, and the subsequent motion of the device is used to determine the magnitude of the impulse, and, from this magnitude, the quantity to be measured.

ballistic missile [ORD] A missile capable of guiding and propelling itself in a direction and to a velocity such that it will follow a ballistic trajectory to a desired point.

ballistic mortar [ORD] A heavy, short-barreled mortar, pendulum-mounted, for determining the relative power of explosives; a small sample of a test explosive is placed in the detonation chamber and a projectile is located forward of the charge; upon detonation the projectile is driven into a sand bank and the mortar swings through an arc; a marker records the maximum weight of the test explosive required to produce the same rise as 10 grams of TNT.

ballistic pendulum [ENG] A device which uses the deflection of a suspended weight to determine the momentum of a projectile.

ballistic range [ORD] A long, instrumented enclosure wherein tests of gun-launched projectiles are conducted.

ballistics [MECH] Branch of applied mechanics which deals with the motion and behavior characteristics of missiles, that is, projectiles, bombs, rockets, guided missiles, and so forth, and of accompanying phenomena.

ballistic separator [CIV ENG] A device that takes out noncompostable material like stones, glass, metal, and rubber, from solid waste by passing the waste over a rotor that has impellers to fling the material in the air; the lighter organic (compostable) material travels a shorter distance than the heavier (noncompostable) material.

ballistics of penetration [MECH] That part of terminal ballistics which treats of the motion of a projectile as it forces its way into targets of solid or semisolid substances, such as earth, concrete, or steel.

ballistic table [MECH] Compilation of ballistic data from which trajectory elements such as angle of fall, range to summit, time of flight, and ordinate at any time, can be obtained.

ballistic temperature [MECH] That temperature (in °F) which, when regarded as a surface temperature and used in conjunction with the lapse rate of the standard artillery atmosphere, would produce the same effect on a projectile as the actual temperature distribution encountered by the projectile in flight.

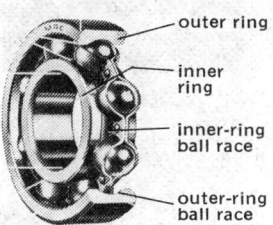

BALL BEARING

— outer ring
— inner ring
— inner-ring ball race
— outer-ring ball race

Deep-groove ball bearing. *(Marlin-Rockwell)*

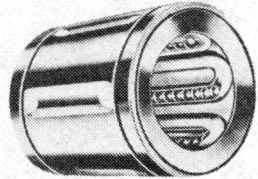

BALL BUSHING

Ball bushing that permits unlimited travel and linear motion. *(Thomson Industries)*

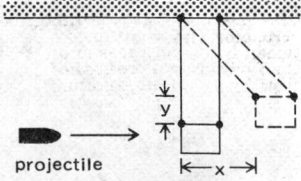

BALLISTIC PENDULUM

projectile

Ballistic pendulum before (solid lines) and after (broken lines) projectile impact. Measurements of *x* and *y* and knowledge of mass and initial velocity of projectile give its momentum.

ballistic test [ORD] Proof test of weapons or ammunition to determine suitability.

ballistic trajectory [MECH] The trajectory followed by a body being acted upon only by gravitational forces and resistance of the medium through which it passes.

ballistic tube [ORD] Gun tube used for ballistic tests; usually selected because of certain demonstrated characteristics.

ballistic uniformity [MECH] The capability of a propellant, when fired under identical conditions from round to round, to impart uniform muzzle velocity and produce similar interior ballistic results.

ballistic vehicle [ENG] A nonlifting vehicle; a vehicle that follows a ballistic trajectory.

ballistic wave [MECH] An audible disturbance caused by compression of air ahead of a missile in flight.

ballistic weapon [ORD] Any missile weapon, as a bomb, rocket, projectile, or bullet, affected by ballistic conditions.

ballistic wind [MECH] That constant wind which would produce the same effect upon the trajectory of a projectile as the actual wind encountered in flight.

ballistite [MATER] A smokeless propellant containing nitrocellulose and nitroglycerin; used in some rocket, mortar, and small-arms ammunition.

ballistocardiogram [MED] The recording made by a ballistocardiograph.

ballistocardiograph [MED] A device to measure the volume of blood passing through the heart in a given period of time.

ballistophobia [PSYCH] An abnormal fear of projectiles or missiles.

ballistospore [MYCOL] A type of fungal spore that is forcibly discharged at maturity.

ball joint *See* ball-and-socket joint.

ball lightning [GEOPHYS] A relatively rare form of lightning, consisting of a reddish, luminous ball, of the order of 1 foot in diameter, which may move rapidly along solid objects or remain floating in midair. Also known as globe lightning.

ball mill [MECH ENG] A pulverizer that consists of a horizontal rotating cylinder, up to three diameters in length, containing a charge of tumbling or cascading steel balls, pebbles, or rods. Also known as ball grinder.

balloon [AERO ENG] A nonporous, flexible spherical bag, inflated with a gas such as helium that is lighter than air, so that it will rise and float in the atmosphere; a large-capacity balloon can be used to lift a payload suspended from it.

balloon apron [ORD] An antiaircraft device consisting of cables hanging perpendicularly from other cables extending between two or more balloons.

balloon barrage [ORD] An antiaircraft defense consisting of a number of balloons, usually together and equipped with balloon aprons, held captive by steel cables and strategically moored near vital areas or installations.

balloon ceiling [METEOROL] The ceiling classification applied when the ceiling height is determined by timing the ascent and disappearance of a ceiling balloon or pilot balloon in United States weather observing practice.

balloon cover [AERO ENG] A cover which fits over a large, inflated balloon to facilitate handling in high or gusty winds. Also known as balloon shroud.

balloon drag [METEOROL] A small balloon, loaded with ballast and inflated so that it will explode at a predetermined altitude, which is attached to a larger balloon; frequently used to retard the ascent of a radiosonde during the early part of the flight so that more detailed measurements may be obtained.

balloonet [AERO ENG] One of the air cells in a blimp, fastened to the bottom or sides of the envelope, which are used to maintain the required pressure in the envelope without adding or valving gas as the ship ascends or descends.

balloon shroud *See* balloon cover.

balloon-type rocket [AERO ENG] A liquid-fuel rocket, such as the Atlas, that requires the pressure of its propellants (or other gases) within it to give it structural integrity.

balloting [MECH] A tossing or bounding movement of a projectile, within the limits of the bore diameter, while moving through the bore under the influence of the propellant gases.

ball pendulum test [ENG] A test for measuring the strength of explosives; consists of measuring the swing of a pendulum produced by the explosion of a weighed charge of material.

ball sealers [PETRO ENG] Balls of rubber, plastic, or metal that are dropped down the well bore to aid the acidizing of impermeable zones of an oil reservoir; they wedge into and plug the bottomhole tubing perforations that are adjacent to the more permeable reservoir zones.

ball sizing [MET] Finishing a hole to a precise diameter and burnishing the surface by forcing a steel ball through it.

ballstone [GEOL] 1. Large mass or concretion of fine, unstratified limestone resulting from growth of coral colonies. 2. A nodule of rock, especially ironstone, in a stratified unit.

balluster [GRAPHICS] One of the sections of wood used to make the frames on which silk, or some other material, is stretched in the silk-screen process.

ballute [AERO ENG] A cross between a balloon and a parachute, used to brake the free fall of sounding rockets.

ball valve [MECH ENG] A valve in which the fluid flow is regulated by a ball moving relative to a spherical socket as a result of fluid pressure and the weight of the ball.

Balmer lines [SPECT] Lines in the hydrogen spectrum, produced by transitions between $n = 2$ and $n > 2$ levels either in emission or absorption; here n is the principal quantum number.

Balmer series [SPECT] The set of Balmer lines.

balsa [BOT] *Ochroma lagopus*. A tropical American tree in the order Malvales; its wood is strong and lighter than cork.

balsam [MATER] An exudate of the balsam tree; a mixture of resins, essential oils, cinnamic acid, and benzoic acid.

Balsaminaceae [BOT] A family of flowering plants in the order Geraniales, including touch-me-not (*Impatiens*); flowers are irregular with five stamens and five carpels, leaves are simple with pinnate venation, and the fruit is an elastically dehiscent capsule.

balsam of Peru *See* Peru balsam.

Baltic Sea [GEOGR] An intracontinental, Mediterranean-type sea, connected with the North Sea and surrounded by Sweden, Denmark, Germany, Poland, the Baltic States, and Finland.

Baltimore oil *See* wormseed oil.

balun [ELEC] A device used for matching an unbalanced coaxial transmission line or system to a balanced two-wire line or system. Also known as balanced converter; bazooka; line-balance converter.

Bamberger's formula [ORG CHEM] A structural formula for naphthalene that shows the valencies of the benzene rings pointing toward the centers.

bamboo [BOT] The common name of various tropical and subtropical, perennial, ornamental grasses in five genera of the family Gramineae characterized by hollow woody stems up to 6 inches in diameter.

Bambusoideae [BOT] A subfamily of grasses, composed of bamboo species, in the family Gramineae.

Banach algebra [MATH] An algebra which is a Banach space satisfying the property that for every pair of vectors, the norm of the product of those vectors does not exceed the product of their norms.

Banach space [MATH] A real or complex vector space in which each vector has a non-negative length, or norm, and in which every Cauchy sequence converges to a point of the space. Also known as complete normed linear space.

Banach-Steinhaus theorem [MATH] If a sequence of bounded linear transformations of a Banach space is point-wise bounded, then it is uniformly bounded.

banakite [PETR] An alkalic basalt made up of plagioclase, sanidine, and biotite, with small quantities of analcime, augite, and olivine; quartz or leucite may be present.

banana [BOT] Any of the treelike, perennial plants of the genus *Musa* in the family Musaceae characterized by soft, pulpy flesh and a thin rind.

banana freckle [PL PATH] A fungus disease of the banana caused by *Macrophoma musae*, producing brown or black spots on the fruit and leaves.

banana jack [ELEC] A jack that fits a banana plug; generally designed for panel mounting.

banana oil *See* amyl acetate.

banana plug [ELEC] A plug having a spring-metal tip shaped

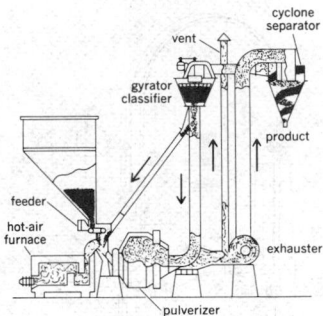

BALL MILL

Conical ball mill pulverizer in closed circuit with classifier. (Hardinge Co., Inc.)

BANANA

Pseudostem of commercial banana plant *(Musa sapientum)*, showing characteristic foliage and single stem of bananas. Plant grows to height of 30 feet (9 meters) or more.

BAND BRAKE

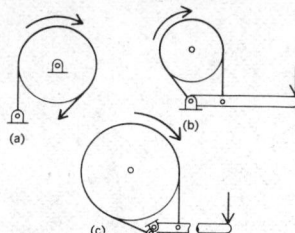

Band brakes. *(a)* Basic structure. *(b)* With direct-action lever. *(c)* Differential self-assisting feature.

BAND SAW

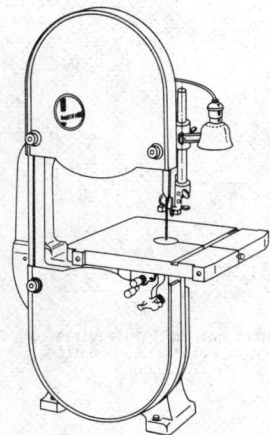

The narrow band saw, a flexible band of steel, can make curved as well as straight cuts even in thick pieces. *(Delta)*

BAND-STOP FILTER

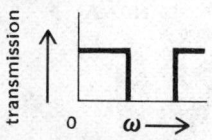

band-stop

Transmission function of a band-stop filter. Frequency (ω) components are largely attenuated at the stop band.

like a banana and used on test leads or as terminals for plug-in components.

Banbury mixer [MECH ENG] Heavy-duty batch mixer, with two counterrotating rotors, for doughy material; used mainly in the plastics and rubber industries.

banco [HYD] A meander or oxbow lake separated from a river by a change in its course.

Bancroft's filariasis *See* wuchereriasis.

band [ADP] A set of circular or cyclic recording tracks on a storage device such as a magnetic drum, disk, or tape loop. Also known as cylinder. [COMMUN] A range of electromagnetic-wave frequencies between definite limits, such as that assigned to a particular type of radio service. [DES ENG] A strip or cord crossing the back of a book to which the sections are sewn. [GEOD] Any latitudinal strip, designated by accepted units of linear or angular measurement, which circumscribes the earth. [GEOL] A thin layer or stratum of rock that is noticeable because its color is different from the colors of adjacent layers. [ORD] A metal sleeve joining together the barrel and stock of a gun. [SOLID STATE] A restricted range in which the energies of electrons in solids lie, or from which they are excluded, as understood in quantum-mechanical terms. [SPECT] *See* band spectrum.

bandage [ELEC] Rubber ribbon about 4 inches (10 centimeters) wide for temporarily protecting a telephone or coaxial splice from moisture. [MED] A strip of gauze, muslin, flannel, or other material, usually in the form of a roll, but sometimes triangular or tailed, used to hold dressing in place, to apply pressure, to immobilize a part, to support a dependent or injured part, to obliterate tissue cavities, or to check hemorrhage.

bandaite [PETR] A dacite type of extrusive rock composed of hypersthene and labradorite.

band brake [MECH ENG] A brake in which the frictional force is applied by increasing the tension in a flexible band to tighten it around the drum.

band chain [ENG] A steel or Invar tape, graduated in feet and at least 100 feet long, used for accurate surveying.

band clutch [MECH ENG] A friction clutch in which a steel band, lined with fabric, contracts onto the clutch rim.

banded [PETR] Pertaining to the appearance of rocks that have thin and nearly parallel bands of different textures, colors, and minerals.

banded anteater *See* marsupial anteater.

banded coal [GEOL] A variety of bituminous and subbituminous coal made up of a sequence of thin lenses of highly lustrous coalified wood or bark interspersed with layers of more or less striated bright or dull coal.

banded differentiate [PETR] A type of igneous rock made up of bands of different composition, frequently alternating between two varieties as in a layered intrusion.

banded ore [GEOL] Ore made up of layered bands composed either of the same minerals that differ from band to band in color or textures or proportion, or of different minerals.

banded peat [GEOL] Peat formed of alternate layers of vegetable debris.

banded structure [MET] The appearance of a metal showing light and dark parallel bands in the direction of rolling or working. [METEOROL] The appearance of precipitation echoes in the form of long bands as presented on radar plan position indicator (PPI) scopes.

banded vein [GEOL] A vein composed of layers of different minerals that lie parallel to the walls. Also known as ribbon vein.

band-elimination filter *See* band-stop filter.

band gap [SOLID STATE] An energy difference between two allowed bands of electron energy in a metal.

band groove [ORD] One of the channels cut into the rotating band of a projectile during the process of engraving.

band head [SPECT] A location on the spectrogram of a molecule a̲ ̲hich the lines of a band pile up.

bandicu̲ ̲ ̲ ̲1. Any of several large Indian rats of the genus *Ne*̲ ̲ ̲ ̲ ̲ ̲ ̲ ̲ ̲ insectivo̲ ̲ ̲ family Peramelidae and found in Tas̲ ̲ ̲ ̲ ̲ ̲ ̲ ̲astralia̲ ̲ ̲d New Guinea.

banding [DES ENG] A strip of fabric used for bands. [SCI TECH] A pattern of bands.

band land [ORD] The raised portion of the rotating band of a projectile after engraving has taken place, produced by the rifling groove of the gun tube.

band lightning *See* ribbon lightning.

band of position [NAV] An area which extends to either side of a line of position of imperfect accuracy and within which a craft is considered to be located.

bandoleer [ORD] A closed loop of fabric, provided with pockets designed to accommodate small-arms ammunition; used by the individual soldier for carrying ammunition, by suspending one or more bandoleers over the shoulders.

band-pass amplifier [ELECTR] An amplifier designed to pass a definite band of frequencies with essentially uniform response.

band-pass filter [ELECTR] An electric filter which transmits more or less uniformly in a certain band, outside of which frequency components are attenuated. [OPTICS] *See* Christiansen filter.

band-pass response [ELECTR] Response characteristics in which a definite band of frequencies is transmitted uniformly. Also known as flat top response.

band-rejection filter *See* band-stop filter.

band saw [MECH ENG] A power-operated woodworking saw consisting basically of a flexible band of steel having teeth on one edge, running over two vertical pulleys, and operated under tension.

band scheme [SOLID STATE] The identification of energy bands of a solid with the levels of independent atoms from which they arise as the atoms are brought together to form the solid, together with the width and spacing of the bands.

band selector [ELECTR] A switch that selects any of the bands in which a receiver, signal generator, or transmitter is designed to operate and usually has two or more sections to make the required changes in all tuning circuits simultaneously. Also known as band switch.

B and S gage *See* American wire gage.

band spectrum [SPECT] A spectrum consisting of groups or bands of closely spaced lines in emission or absorption, characteristic of molecular gases and chemical compounds. Also known as band.

band spreading [COMMUN] Method of double-sideband transmission in which the frequency band of the modulating wave is shifted upward in frequency so that the sidebands produced by modulation are separated in frequency from the carrier by an amount at least equal to the bandwidth of the original modulating wave, and second order distortion products may be filtered from the demodulator output.

bandspread tuning control [ELECTR] A tuning control provided on some shortwave receivers to spread the stations in a single band of frequencies over an entire tuning dial.

band-stop filter [ELECTR] An electric filter which transmits more or less uniformly at all frequencies of interest except for a band within which frequency components are largely attenuated. Also known as band-elimination filter; band-rejection filter.

band switch *See* band selector.

band theory of solids [SOLID STATE] A quantum-mechanical theory of the motion of electrons in solids that predicts certain restricted ranges or bands for the energies of these electrons.

bandwidth [COMMUN] The difference between the frequency limits of a band containing the useful frequency components of a signal. Abbreviated BW.

Bangalore torpedo [ORD] A metal tube or pipe packed with a high-explosive charge; chiefly used to clear a path through barbed wire or minefields.

bang-bang circuit [ELECTR] An operational amplifier with double feedback limiters that drive a high-speed relay (1–2 milliseconds) in an analog computer; involved in signal-controlled programming.

bang-bang control [ADP] Control of programming in an analog computer through a bang-bang circuit.

Bangiophyceae [BOT] A class of red algae in the plant division Rhodophyta.

Bang's disease *See* contagious abortion.

bank [AERO ENG] The lateral inward inclination of an airplane when it rounds a curve. [CIV ENG] *See* embankment. [ELEC] **1.** A number of similar electrical devices, such as resistors, connected together for use as a single device. **2.** An assemblage of fixed contacts over which one or more wipers or brushes move in order to establish electrical connections in automatic switching. [GEOL] **1.** The edge of a waterway. **2.** The rising ground bordering a body of water. [IND ENG] The amount of material allowed to accumulate at a point on a production line where it is not employed or worked upon, to permit reasonable fluctuations in line speed before and after the point. Also known as float. [MIN ENG] **1.** The top of the shaft. **2.** The surface around the mouth of a shaft. **3.** The whole, or sometimes only one side or one end, of a working place underground. **4.** To manipulate materials such as coal, gravel, or sand on a bank. **5.** A terrace-like bench in open-pit mining.

bank and turn indicator [AERO ENG] A device used to advise the pilot that the aircraft is turning at a certain rate, and that the wings are properly banked to preclude slipping or sliding of the aircraft as it continues in flight. Also known as bank indicator.

bank-and-wiper switch [ELEC] Switch in which electromagnetic ratchets or other mechanisms are used, first, to move the wipers to a desired group of terminals, and second, to move the wipers over the terminals of the group to the desired bank contacts.

bank cushion [NAV] In nautical navigation, a force acting on the bow of a ship in a manner which forces the ship away from the bank in a restricted channel, especially where the banks are steep; it is a force which opposes bank suction.

bank deposit [GEOL] Mounds, ridges, and terraces of sediment rising above and about the surrounding sea bottom.

banked winding [ELECTR] A radio-frequency coil winding which proceeds from one end of the coil to the other without return by having, side by side, many flat spirals formed by winding single turns one over the other, thereby reducing the distributed capacitance of the coil.

banket [GEOL] A conglomerate containing valuable metal to be exploited.

bank height [MIN ENG] The vertical height of a bank as measured between its highest point or crest and its toe at the digging level or bench. Also known as bench height; digging height.

bank indicator *See* bank and turn indicator.

banking pin [HOROL] One of the erect pins in the bottom plate of a watch that restrict the movement of the lever.

bank-inset reef [GEOL] A coral reef situated on island or continental shelves well inside the outer edges.

bank reef [GEOL] A reef which rises at a distance back from the outer margin of rimless shoals.

bank-run gravel [GEOL] A natural deposit of gravel or sand.

bank slope [MIN ENG] The angle, measured in degrees of deviation from the horizontal, at which the earthy or rock material will stand in an excavated, terracelike cut in an open-pit mine or quarry. Also known as bench slope.

banksman *See* lander.

banks oil *See* cod-liver oil.

bank storage [HYD] Water absorbed in the permeable bed and banks of a lake, reservoir, or stream.

bank suction [NAV] The bodily movement of a ship toward the near bank due to a decrease in pressure as a result of increased velocity of flow of water past the hull in a restricted channel.

Ban-Lon [TEXT] Trade name of Joseph Bancroft & Sons Company for fabrics and garments made from Textralized yarn of nylon, polyester, or fiber blends.

banner [BOT] The fifth or posterior petal of a butterfly-shaped (papilionaceous) flower.

banner cloud [METEOROL] A cloud plume often observed to extend downwind from isolated mountain peaks, even on otherwise cloud-free days. Also known as cloud banner.

bantam tube [ELECTR] Vacuum tube having a standard octal base, but a considerably smaller glass tube than a standard glass tube.

Banti's disease [MED] Portal hypertension, congestive splenomegaly, and hypersplenism due to an obstructive lesion in the splenic vein, portal vein, or intrahepatic veins.

bar [MECH] A unit of pressure equal to 10^5 pascals, or 10^5 newtons per square meter, or 10^6 dynes per square centimeter. [MET] An elongated piece of metal of simple uniform cross-section dimensions, usually rectangular, circular, or hexagonal, produced by forging or hot rolling. Also known as barstock. [MIN ENG] *See* bar drill.

BAR *See* Browning automatic rifle.

baraboo [GEOL] A monadnock buried by a series of strata and then reexposed by the partial erosion of these younger strata.

Bárány chair [ENG] A chair in which a person is revolved to test his susceptibility to vertigo.

barat [METEOROL] A heavy northwest squall in Manado Bay on the north coast of the island of Celebes, prevalent from December to February.

barathea [TEXT] Material, such as silk, cotton, or rayon, made with a pebble weave.

barb [METEOROL] A means of representing wind speed in the plotting of a synoptic chart, being a short, straight line drawn obliquely toward lower pressure from the end of a wind-direction shaft. Also known as feather. [VERT ZOO] A side branch on the shaft of a bird's feather.

Barbados earth [GEOL] A deposit of fossil radiolarians.

bar beach [GEOL] A straight beach of offshore bars that are separated by shallow bodies of water from the mainland.

barbed tributary [HYD] A tributary that enters the main stream in an upstream direction instead of pointing downstream.

barbel [VERT ZOO] **1.** A slender, tactile process near the mouth in certain fishes, such as catfishes. **2.** Any European fresh-water fish in the genus *Barbus*.

barbellate [BIOL] Having short, stiff, hooked bristles.

barber [METEOROL] A severe storm at sea during which spray and precipitation freeze onto the decks and rigging of ships.

barberite [MET] A nonferrous alloy with good resistance to sulfuric acid, sea water, and mine waters; 88.5% copper, 5% nickel, 5% tin, 1.5% silicon.

barbette [NAV ARCH] A nonrotating cylinder of armor that protects the rotating part of the turret of a warship below the gunhouse. [ORD] A mound of earth or a platform upon which guns are mounted to fire over a wall or parapet.

barbicel [VERT ZOO] One of the small, hook-bearing processes on a barbule of the distal side of a barb on a feather.

barbierite [MINERAL] $NaAlSi_3O_8$ A hypothetical soda feldspar thought to be isomorphous with orthoclase.

barbital [ORG CHEM] $C_8H_{12}N_2O_3$ A compound crystallizing in needlelike form from water; has a faintly bitter taste; melting point 188–192°C; used to make sodium barbital, a long-duration hypnotic and sedative. Also known as diethylbarbituric acid; diethylmalonylurea.

barbiturate [PHARM] Any of a group of ureides, such as phenobarbital, Amytal, or Seconal, that act as central nervous system depressants.

barbituric acid [ORG CHEM] $C_4H_4O_3N_2$ 2,4,6-Trioxypyrimidine, the parent compound of the barbiturates; colorless crystals melting at 245°C, slightly soluble in water. Also known as malonyl urea.

barbiturism [MED] Intoxication following an overdose of barbiturates; characterized by delirium, coma, and sometimes death.

bar buoy [NAV] A buoy marking the location of a bar at the mouth of a river or approach to a harbor.

barchan [GEOL] A crescent-shaped dune or drift of wind-blown sand or snow, the arms of which point downwind; formed by winds of almost constant direction and of moderate speeds. Also spelled barchane; barkhan; crescentic dune.

barchane *See* barchan.

bar chart *See* bar graph.

bar-code scanner [ADP] An optical scanning device that reads texts which have been converted into a special bar code.

Bardeen-Cooper-Schrieffer theory [SOLID STATE] A theory of superconductivity that describes quantum-mechanically those states of the system in which conduction electrons cooperate in their motion so as to reduce the total energy appreciably below that of other states by exploiting their

BAR DRAWING

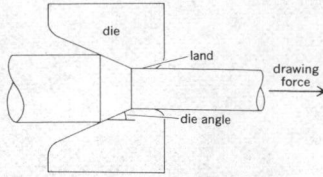

A cross section of a drawing die in operation showing the reduction in cross-sectional area of the bar being pulled through the die.

BARITE

—— 5 cm ——

A barite specimen from Dufton, Westmoreland, England. (*American Museum of Natural History*)

BARIUM

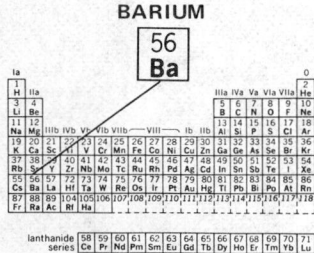

Periodic table of the chemical elements showing the position of barium.

effective mutual attraction; these states predominate in a superconducting material. Abbreviated BCS theory.

bar drawing [MET] An operation in which a metallic bar is pulled through a die so that the cross-sectional area of the bar is reduced.

bar drill [MIN ENG] A small diamond type or other type of rock drill mounted on a bar and used in an underground workplace. Also known as bar.

bareboat charter [IND ENG] An agreement to charter a ship without its crew or stores; the fee for its use for a predetermined period of time is based on the price per ton of cargo handled.

bare charge [ORD] An explosive charge without casing prepared for use in determining explosive blast characteristics.

bare electrode [MET] An uncoated electrode used in submerged arc automatic welding with a gas-shielded arc or a granular flux deposited in an elongated mound over a joint.

bareface fabric [TEXT] Any perfectly smooth fabric that has no nap.

baren [GRAPHICS] A pad for pressing paper against an inked wooden block to produce a print.

bare rock [NAV] In U.S. Coast and Geodetic Survey terminology, a rock extending above the plane of mean high water.

bar finger sand [GEOL] An elongated lenticular sand body that lies beneath a distributory in a birdfoot delta.

Barfoed's test [ANALY CHEM] A test for monosaccharides conducted in an acid solution; cupric acetate is reduced to cuprous oxide, a red precipitate.

bar folder [MET] A machine used to bend a metal sheet into a sharp, narrow, and accurate fold, or a rounded fold, along the edge.

barge [NAV ARCH] A large cargo-carrying craft which is towed or pushed by a tug on both seagoing and inland waters.

bar generator [ELECTR] Generator of pulses or repeating waves which are equally separated in time; these pulses are synchronized by the synchronizing pulses of a television system, so that they can produce a stationary bar pattern on a television screen.

barge spike *See* boat spike.

bar graph [STAT] A diagram of frequency-table data in which a rectangle with height proportional to the frequency is located at each value of a variate that takes only certain discrete values. Also known as bar chart.

baric topography *See* height pattern.

baric wind law *See* Buys-Ballot's law.

barines [METEOROL] Westerly winds in eastern Venezuela.

baring *See* overburden.

bar iron [MET] Wrought iron formed into bars.

Bari-Sol process [CHEM ENG] Removal of waxes from liquid hydrocarbons by extraction of the wax with a mixed ethylene dichloride–benzene solvent, followed by separation from the hydrocarbon in a centrifuge.

barite [MINERAL] $BaSO_4$ A white, yellow, or colorless orthorhombic mineral occurring in tabular crystals, granules, or compact masses; specific gravity is 4.5; used in paints and drilling muds and as a source of barium chemicals; the principal ore of barium. Also known as baryte; barytine; cawk; heavy spar.

barite dollar [MINERAL] Barite in the form of rounded disk-shaped masses; formed in a sandstone or sandy shale.

barium [CHEM] A chemical element, symbol Ba, with atomic number 56 and atomic weight of 137.34.

barium-140 [NUC PHYS] A radioactive isotope of barium with atomic mass 140; the half-life is 12.8 days, and the decay is by negative beta-particle emission.

barium acetate [INORG CHEM] $Ba(C_2H_3O_2)_2 \cdot H_2O$ A barium salt made by treating barium sulfide or barium carbonate with acetic acids; it forms colorless, triclinic crystals that decompose upon heating; used as a reagent for sulfates and chromates.

barium-base grease [MATER] A lubricating material made from lubricating oil and barium soap.

barium binoxide *See* barium peroxide.

barium bromate [INORG CHEM] $Ba(BrO_3)_2 \cdot H_2O$ A poisonous compound that forms colorless, monoclinic crystals, decomposing at 260°C; used for preparing other bromates.

barium chlorate [INORG CHEM] $Ba(ClO_3)_2 \cdot H_2O$ A salt pre-

pared by the reaction of barium chloride and sodium chlorate; it forms colorless, monoclinic crystals, soluble in water; used in pyrotechnics.

barium chloride [INORG CHEM] $BaCl_2$ A toxic salt obtained as colorless, water-soluble cubic crystals, melting at 963°C; used as a rat poison, in metal surface treatment, and as a laboratory reagent.

barium chromate [INORG CHEM] $BaCrO_4$ A toxic salt that forms yellow, rhombic crystals, insoluble in water; used as a pigment in overglazes.

barium dioxide *See* barium peroxide.

barium enema [MED] A suspension of barium sulfate administered as an enema into the lower bowel to render it radiopaque.

barium fluoride [INORG CHEM] BaF_2 Colorless, cubic crystals, slightly soluble in water; used in enamels.

barium fuel cell [ELEC] A fuel cell in which barium is used with either oxygen or chlorine to convert chemical energy into electrical energy.

barium glass [MATER] Glass which differs from ordinary lime-soda glass in that barium oxide replaces part of the calcium oxide.

barium hydroxide [INORG CHEM] $Ba(OH)_2 \cdot 8H_2O$ Colorless, monoclinic crystals, melting at 78°C; soluble in water, insoluble in acetone; used for fat saponification and fusing of silicates.

barium hyposulfite *See* barium thiosulfate.

barium meal [MED] A suspension of barium sulfate taken orally to render the upper gastrointestinal tract radiopaque.

barium mercuric iodide [INORG CHEM] $BaHg_9I_4$ Crystals that are yellow or reddish in hue and deliquesce; used in aqueous solution as Rohrbach's solution for separation of minerals on the basis of their densities. Also known as barium tetraiodomercurate; mercuric barium iodide.

barium monosulfide [INORG CHEM] BaS A colorless, cubic crystal that is soluble in water; used in pigments.

barium monoxide *See* barium oxide.

barium nitrate [INORG CHEM] $Ba(NO_3)_2$ A toxic salt occurring as colorless, cubic crystals, melting at 592°C, and soluble in water; used as a reagent, in explosives, and in pyrotechnics. Also known as nitrobarite.

barium oxide [INORG CHEM] BaO A white to yellow powder that melts at 1923°C; it forms the hydroxide with water; may be used as a dehydrating agent. Also known as barium monoxide; barium protoxide.

barium perchlorate [INORG CHEM] $Ba(ClO_4)_2 \cdot 4H_2O$ Tetrahydrate variety which forms colorless hexagons; used in pyrotechnics.

barium peroxide [INORG CHEM] BaO_2 A compound formed as white toxic powder, insoluble in water; used as a bleach and in the glass industry. Also known as barium binoxide; barium dioxide; barium superoxide.

barium protoxide *See* barium oxide.

barium rhodanide *See* barium thiocyanate.

barium silicide [INORG CHEM] $BaSi_2$ A compound that has the appearance of metal-gray lumps; melts at white heat; used in metallurgy to deoxidize steel.

barium stearate [ORG CHEM] $Ba(C_{18}H_{35}O_2)_2$ A white, crystalline solid; melting point 160°C; used as a lubricant in manufacturing plastics and rubbers, in greases, and in plastics as a stabilizer against deterioration caused by heat and light.

barium sulfate [INORG CHEM] $BaSO_4$ A salt occurring in the form of white, rhombic crystals, insoluble in water; used as a white pigment, as an opaque contrast medium for roentgenographic processes, and as an antidiarrheal.

barium sulfocyanate *See* barium thiocyanate.

barium sulfocyanide *See* barium thiocyanate.

barium superoxide *See* barium peroxide.

barium tetraiodomercurate *See* barium mercuric iodide.

barium tetrasulfide [INORG CHEM] $BaS_4 \cdot H_2O$ Red or yellow, rhombic crystals, soluble in water.

barium thiocyanate [INORG CHEM] $Ba(SCN)_2 \cdot 2H_2O$ White crystals that deliquesce; used in dyeing and in photography. Also known as barium rhodanide; barium sulfocyanate; barium sulfocyanide.

barium thiosulfate [INORG CHEM] $BaS_2O_3 \cdot H_2O$ A white powder that decomposes upon heating; used to make explosives and in matches. Also known as barium hyposulfite.

barium titanate [INORG CHEM] $BaTiO_3$ A grayish powder that is insoluble in water but soluble in concentrated sulfuric acid; used as a ferroelectric ceramic.

bar joist [BUILD] A small steel truss with wire or rod web lacing used for roof and floor supports.

bark [BOT] The tissues external to the cambium in a stem or root. [MET] The decarburized layer formed beneath the scale on the surface of steel heated in air. [NAV ARCH] A three-masted sailing ship whose foremast and mainmast are square-rigged and whose mizzenmast is fore-and-aft-rigged.

bark cloth [TEXT] Fabric made from inner tree bark by beating it to a smooth, wearable thinness.

bar keel [NAV ARCH] A solid keel with a rectangular cross section in an iron or steel ship.

barker [DES ENG] *See* bark spud. [ENG] A machine, used mainly in pulp mills, which removes bark from logs. [FOR] 1. A worker who subjects logs and pulpwood to water pressure in a stream barker or tumbling in a drum barker, in order to free them from bark and dirt. Also known as power barker. 2. A worker who prepares or shovels bark for tanning.

Barker method [CRYSTAL] A method utilizing a number of convenient rules which allow two observers to choose the same reference system to describe the same noncubic crystal.

barkevikite [MINERAL] A brown or black member of the amphibole mineral group; looks like basaltic hornblende but differs from it in its iron concentration.

bark graft [BOT] A graft made by slipping the scion beneath a slit in the bark of the stock.

barkhan *See* barchan.

Barkhausen effect [ELECTROMAG] The succession of abrupt changes in magnetization occurring when the magnetizing force acting on a piece of iron or other magnetic material is varied.

Barkhausen interference [COMMUN] Interference caused by Barkhausen oscillations.

Barkhausen-Kurz oscillator [ELECTR] An oscillator of the retarding-field type in which the frequency of oscillation depends solely on the transit time of electrons oscillating about a highly positive grid before reaching the less positive anode. Also known as Barkhausen oscillator; positive-grid oscillator.

Barkhausen oscillation [ELECTR] Undesired oscillation in the horizontal output tube of a television receiver, causing one or more ragged dark vertical lines on the left side of the picture.

Barkhausen oscillator *See* Barkhausen-Kurz oscillator.

barkometer [CHEM ENG] A hydrometer calibrated to test the strength of tanning liquors used in tanning leather.

bark-splitting disease *See* Armillaria root rot.

bark spud [DES ENG] A tool which peels off bark. Also known as barker.

barley [BOT] A plant of the genus *Hordeum* in the order Cyperales that is cultivated as a grain crop; the seed is used to manufacture malt beverages and as a cereal.

barley coal [MIN ENG] A stream size of anthracite sized on a round punched plate; passes through quarter-inch holes. Also known as buckwheat no. 3.

barley scald [PL PATH] A fungus disease of barley caused by *Rhynchosporium secalis* and characterized by bluish-green to yellow blotches and blighting of the foliage.

barley smut [PL PATH] 1. A loose smut disease of barley caused by *Ustilago nuda.* 2. A covered smut disease of barley caused by *U. hordei.*

barley stripe [PL PATH] A fungus disease of barley characterized by light green or yellow stripes on the leaves; incited by the diffusible toxin of *Helminthosporium gramineum.*

bar linkage [MECH ENG] A set of bars joined together at pivots by means of pins or equivalent devices; used to transmit power and information.

Barlow's equation [MECH] A formula, $t = DP/2S$, used in computing the strength of cylinders subject to internal pressures, where t is the thickness of the cylinder in inches, D the outside diameter in inches, P the pressure in pounds per square inch, and S the allowable tensile strength in pounds per square inch.

Barlow's rule [PHYS CHEM] The rule that the volume occupied by the atoms in a given molecule is proportional to the valences of the atoms, using the lowest valency values.

bar magnet [ELECTROMAG] A bar of hard steel that has been strongly magnetized and holds its magnetism, thereby serving as a permanent magnet.

bar mining [MIN ENG] The mining of river bars, usually between low and high waters, although the stream is sometimes deflected and the bar worked below water level.

barn [AGR] A farm building used for storage of agricultural products and equipment or for housing farm animals. [NUC PHYS] A unit of area equal to 10^{-24} square centimeter; used in specifying nuclear cross sections. Abbreviated b.

barnacle [INV ZOO] The common name for a number of species of crustaceans which compose the subclass Cirripedia.

Barnard's star [ASTRON] A star 6.1 light-years away from earth, of visual magnitude 9.5 and proper motion of 10.31 seconds of arc annually.

Barnett effect [ELECTROMAG] The development of a slight magnetization in an initially unmagnetized iron rod when it is rotated at high speed about its axis.

Barnett method [ELETROMAG] Use of the Barnett effect to determine the gyromagnetic moment of ferromagnetic material.

barney [MIN ENG] A small car or truck, attached to a rope or cable, used to push cars up a slope or an inclined plane. Also known as bullfrog; donkey; groundhog; larry; mule; ram; truck.

baroclinic [PHYS] Of, pertaining to, or characterized by baroclinity.

baroclinic disturbance [METEOROL] Any migratory cyclone associated with strong baroclinity of the atmosphere, evidenced on synoptic charts by temperature gradients in the constant-pressure surfaces, vertical wind shear, tilt of pressure troughs with height, and concentration of solenoids in the frontal surface near the ground. Also known as baroclinic wave.

baroclinic field [METEOROL] A distribution of atmospheric pressure and mass such that the specific volume, or density, of air is a function not solely of pressure.

baroclinic instability [METEOROL] A hydrodynamic instability arising from the existence of a meridional temperature gradient (and hence of a thermal wind) in an atmosphere in quasi-geostrophic equilibrium and possessing static stability.

baroclinicity *See* baroclinity.

baroclinic model [METEOROL] A concept of stratification in the atmosphere, involving surfaces of constant pressure intersecting surfaces of constant density.

baroclinic wave *See* baroclinic disturbance.

baroclinity [PHYS] The state of stratification in a fluid in which surfaces of constant pressure (isobaric surfaces) intersect surfaces of constant density (isosteric surfaces). Also known as baroclinicity; barocliny.

barocliny *See* baroclinity.

baroduric bacteria [MICROBIO] Bacteria that can tolerate conditions of high hydrostatic pressure.

barodynamics [MECH] The mechanics of heavy structures which may collapse under their own weight.

barogram [ENG] The record of an aneroid barograph.

barometer [ENG] An absolute pressure gage specifically designed to measure atmospheric pressure.

barometer elevation [METEOROL] The vertical distance above mean sea level of the ivory point (zero point) of a weather station's mercurial barometer; frequently the same as station elevation. Also known as elevation of ivory point.

barometric [ENG] Pertaining to a barometer or to the results obtained by using a barometer. [PHYS] Loosely, pertaining to atmospheric pressure; for example, barometric gradient (meaning pressure gradient).

barometric altimeter *See* pressure altimeter.

barometric condenser [MECH ENG] A contact condenser that uses a long, vertical pipe into which the condensate and cooling liquid flow to accomplish their removal by the pressure created at the lower end of the pipe.

barometric corrections [PHYS] The corrections which must be applied to the reading of a mercury barometer in order that

BARK

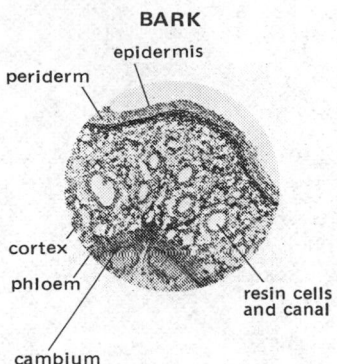

Tranverse section of young twig of balsam fir (*Abies balsamae* L.) showing tissues often considered to compose the bark. (*Forest Products Laboratory, USDA*)

BARNACLE

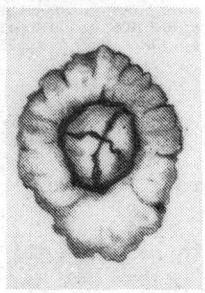

Top view of adult stage of *Balanus*, the acorn barnacle.

BAROMETRIC CONDENSER

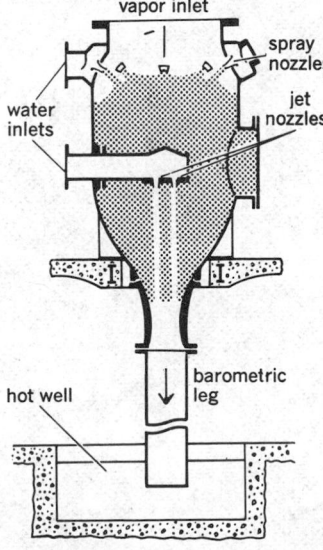

A multijet barometric condenser—one of three types of contact condensers. (*Schutte and Koerting Co.*)

the observed value may be rendered accurate. Also known as barometric errors.

barometric errors *See* barometric corrections.

barometric fuel control [AERO ENG] A device that maintains the correct flow of fuel to an engine by adjusting to atmospheric pressure at different altitudes, as well as to impact pressure.

barometric fuse [ENG] A fuse that functions as a result of change in the pressure exerted by the surrounding air.

barometric gradient *See* pressure gradient.

barometric hypsometry [ENG] The determination of elevations by means of either mercurial or aneroid barometers.

barometric leveling [ENG] The measurement of approximate elevation differences in surveying with the aid of a barometer; used especially for large areas.

barometric pressure *See* atmospheric pressure.

barometric surface [PHYS] A surface at each point of which the barometric pressure is the same.

barometric switch *See* baroswitch.

barometric tendency *See* pressure tendency.

barometric tide [GEOPHYS] A daily variation in atmospheric pressure due to the gravitational attraction of the sun and moon.

barometric wave [METEOROL] Any wave in the atmospheric pressure field; the term is usually reserved for short-period variations not associated with cyclonic-scale motions or with atmospheric tides. Also known as pressure wave.

barometrograph *See* aneroid barograph.

barometry [ENG] The study of the measurement of atmospheric pressure, with particular reference to ascertaining and correcting the errors of the different types of barometer.

baromil [MECH] The unit of length used in graduating a mercury barometer in the centimeter-gram-second system.

barophile [MICROBIO] An organism that thrives under conditions of high hydrostatic pressure.

barophobia [PSYCH] Abnormal fear of gravity.

baroscope [ENG] An apparatus which demonstrates the equality of the weight of air displaced by an object and its loss of weight in air.

barosinusitis [MED] Inflammation of the sinuses, characterized by edema and hemorrhage, due to expansion of air within the sinuses at decreased barometric pressure. Also known as aerosinusitis.

barostat [ENG] A mechanism which maintains constant pressure inside a chamber.

baroswitch [ENG] 1. A pressure-operated switching device used in a radiosonde which determines whether temperature, humidity, or reference signals will be transmitted. 2. Any switch operated by a change in barometric pressure. Also known as barometric switch.

barotaxis [BIOL] Orientation movement of an organism in response to pressure changes.

barothermogram [ENG] The record made by a barothermograph.

barothermograph [ENG] An instrument which automatically records pressure and temperature.

barothermohygrogram [ENG] The record made by a barothermohygrograph.

barothermohygrograph [ENG] An instrument that produces graphs of atmospheric pressure, temperature, and humidity on a single sheet of paper.

barotitis [MED] Inflammation of the ear, or a part of it, caused by changes in atmospheric pressure. Also known as aerootitis.

barotrauma [MED] Injury to air-containing structures, such as the middle ears, sinuses, lungs, and gastrointestinal tract, due to unequal pressure differences across their walls.

barotropic [PHYS] Of, pertaining to, or characterized by a condition of barotropy.

barotropic disturbance [METEOROL] Also known as barotropic wave. 1. A wave disturbance in a two-dimensional nondivergent flow; the driving mechanism lies in the variation of either vorticity of the basic current or the variation of the vorticity of the earth about the local vertical. 2. An atmospheric wave of cyclonic scale in which troughs and ridges are approximately vertical.

barotropic field [METEOROL] A distribution of atmospheric

pressures and mass such that the specific volume, or density, of air is a function solely of pressure.

barotropic model [METEOROL] Any of a number of model atmospheres in which some of the following conditions exist throughout the motion: coincidence of pressure and temperature surfaces, absence of vertical wind shear, absence of vertical motions, absence of horizontal velocity divergence, and conservation of the vertical component of absolute vorticity.

barotropic wave *See* barotropic disturbance.

barotropy [PHYS] The state of a fluid in which surfaces of constant density (or temperature) are coincident with surfaces of constant pressure; it is the state of zero baroclinity.

bar pattern [ELECTR] Pattern of repeating lines or bars on a television screen.

bar plain [GEOL] A plain formed by a stream without a low-water channel or an alluvial cover.

barracuda [VERT ZOO] The common name for about 20 species of fishes belonging to the genus *Sphyraena* in the order Perciformes.

barrage [CIV ENG] An artificial dam which increases the depth of water of a river or watercourse, or diverts it into a channel for navigation or irrigation. [ORD] A prearranged barrier of fire designed to protect friendly troops and installations by impeding enemy movement across defensive lines or areas.

barrage balloon [ORD] A balloon restrained from free flight by means of a cable attaching it to the earth; used to support wires or nets as protection against air attacks.

barrage jamming [COMMUN] The simultaneous jamming of a number of radio or radar bands of frequencies.

barrage reception [COMMUN] A type of radio reception in which interference from certain directions is reduced by selecting from a system of directional antennas those which yield the greatest signal-to-interference ratio.

barrage-type spillway [CIV ENG] A passage for surplus water with sluice gates across the width of the entrance.

barranca [GEOL] A hole or deep break made by heavy rain; a ravine.

Barr body [CYTOL] A condensed, inactivated X chromosome inside the nuclear membrane in interphase somatic cells of women and most female mammals.

barre [TEXT] 1. A pattern of bars or stripes parallel to the weft of a fabric. 2. A defect in the production of a fabric that results in a streak parallel to the weft.

barred basin *See* restricted basin.

barrel [DES ENG] 1. A container having a circular lateral cross section that is largest in the middle, and ends that are flat; often made of staves held together by hoops. 2. A piece of small pipe inserted in the end of a cartridge to carry the squib to the powder. [MECH] Abbreviated bbl. 1. The unit of liquid volume equal to 31.5 gallons. 2. The unit of liquid volume for petroleum equal to 42 gallons. 3. The unit of dry volume equal to 105 quarts. 4. A unit of weight that varies in sizes according to the commodity being weighed. [NAV ARCH] The central part of a windlass or capstan about which the cable is wound. [OPTICS] A tapering cylindrical housing containing the lenses of a camera and the iris diaphragm. [ORD] The cylindrical metallic part of a gun which controls the initial direction of a projectile.

barrel arch [ARCH] A plain arch with a barrellike cross section in which the length is greater than the span (diameter).

barrel assembly [ORD] The gun barrel together with the other parts necessary to attach it to the rest of the gun.

barrel chest *See* emphysematous chest.

barrel copper [MIN ENG] Copper in lumps small enough to be picked out of the mass of rock and put in the furnace without dressing.

barrel distortion [OPTICS] A defect in an optical system whereby lateral magnification decreases with object size; the image of a square then appears barrel-shaped.

barrel erosion [ORD] Wearing away of the interface of the bore due to the combined effects of gas washing, scoring, and mechanical abrasion; causes a reduction in muzzle velocity.

barrel extension [ORD] Metal projection fixed to the rear of the barrel in certain automatic guns; extends backward and

BARRACUDA

The great or predatory barracuda *(Sphyraena barracuda)*, with a long jaw and numerous sharp teeth.

holds the breech locked against the gas pressure in the chamber when the gun is fired.

barrelhead [DES ENG] The flat end of a barrel.

barreling [MET] *See* tumbling. [ORD] Expansion of the body of a cartridge case when the gun chamber recovers longitudinally following firing and the mouth of the case is not free to move in the chamber.

barrel life [ORD] As applied to small arms and automatic weapons, the number of rounds which may be fired through a barrel at a particular firing schedule before the barrel becomes unserviceable; varies with the firing schedule.

barrel plating [MET] An electroplating process by which articles are brought into contact with an electrolyte while rotating in a perforated hardwood barrel.

barrels per calendar day [CHEM ENG] A unit measuring the average rate of oil processing in a petroleum refinery, with allowances for downtime over a period of time. Abbreviated BCD.

barrels per day [CHEM ENG] A unit measuring the rate at which petroleum is produced at the refinery. Abbreviated BD; bpd.

barrels per month [CHEM ENG] A unit measuring the rate at which petroleum is produced at the refinery. Abbreviated BM; bpm.

barrels per stream day [CHEM ENG] A measurement used to denote rate of oil or oil-product flow while a fluid-processing unit is in continuous operation. Abbreviated BSD.

barrel whip [ORD] The movement of a gun barrel in a plane normal to the longitudinal axis of the gun bore, as the gun operates through a complete firing cycle.

Barremian [GEOL] Lower Cretaceous geologic age, between Hauterivian and Aptian.

barren liquor [CHEM ENG] Liquid (liquor) from filter-cake washing in which there is little or no recovery value; for example, barren cyanide liquor from washing of gold cake slimes.

barrens [GEOGR] An area that because of adverse environmental conditions is relatively devoid of vegetation compared with adjacent areas.

barretter [ELEC] Bolometer that consists of a fine wire or metal film having a positive temperature coefficient of resistivity, so that resistance increases with temperature; used for making power measurements in microwave devices.

barricade [ENG] Structure composed essentially of concrete, earth, metal, or wood, or any combination thereof, and so constructed as to reduce or confine the blast effect and fragmentation of an explosion.

barricade shield [ENG] A type of movable shield made of a material designed to absorb ionizing radiation, for protection from radiation.

barrier [ECOL] Any physical or biological factor that restricts the migration or free movement of individuals or populations. [NAV] Anything which obstructs or prevents passage of a craft. [PHYS] *See* potential barrier.

barrier bar [GEOL] Ridges whose crests are parallel to the shore and which are usually made up of water-worn gravel put down by currents in shallow water at some distance from the shore.

barrier basin [GEOL] A basin formed by natural damming, for example, by landslides or moraines.

barrier beach [GEOL] A single, long, narrow ridge of sand which rises slightly above the level of high tide and lies parallel to the shore, from which it is separated by a lagoon. Also known as offshore beach.

barrier chain [GEOL] A series of barrier spits, barrier islands, and barrier beaches extending along a coastline.

barrier curb [CIV ENG] A curb with vertical sides high enough to keep vehicles from crossing it.

barrier diode *See* Schottky barrier diode.

barrier flat [GEOL] An area which is relatively flat and frequently occupied by pools of water that separate the seaward edge of the barrier from a lagoon on the landward side.

barrier-grid storage tube *See* Radechon.

barrier ice *See* shelf ice.

barrier island [GEOL] 1. An island similar to an offshore bar but differing from it in having multiple ridges, areas of

vegetation, and swampy terraces extending toward the lagoon. 2. A detached portion of offshore bar between two inlets.

barrier lagoon [GEOGR] A shallow body of water that separates the shore and a barrier reef.

barrier lake [HYD] A small body of water that lies in a basin, retained there by a natural dam or barrier.

barrier layer *See* depletion layer.

barrier-layer cell *See* photovoltaic cell.

barrier-layer photocell *See* photovoltaic cell.

barrier-layer rectification *See* depletion-layer rectification.

barrier marsh [ECOL] A type of marsh that restricts or prevents invasion of the area beyond it by new species of animals.

barrier material [MATER] Packing material impervious to moisture, vapor, or other liquids and gases. [ORD] An inert material placed in an explosive charge to shape the detonation wave.

barrier penetration [QUANT MECH] The passage of a particle through a potential barrier, that is, through a region of finite extent in which the particle's potential energy is greater than its total energy.

barrier reef [GEOL] A coral reef that runs parallel to the coast of an island or continent, from which it is separated by a lagoon.

barrier separation [CHEM ENG] The separation of a two-component gaseous mixture by selective diffusion of one component through a separative barrier (microporous metal or nonporous polymeric).

barrier shield [ENG] A wall or enclosure made of a material designed to absorb ionizing radiation, shielding the operator from an area where radioactive material is being used or processed by remote-control equipment.

barrier spit [GEOL] A barrier of sand joined at one of its ends to the mainland.

barrier theory of cyclones [METEOROL] A theory of cyclone development, proposed by F. M. Exner, which states that a slow-moving mass of cold air in the path of rapidly eastward-moving warmer air will bring about the formation of low pressure on the lee side of the cold air; analogous to the formation of a dynamic trough on the lee side of an orographic barrier. Also known as drop theory.

Barrovian metamorphism [GEOL] A regional metamorphism that can be zoned into facies that are metamorphic.

barrow *See* handbarrow; wheelbarrow.

bar scale *See* graphic scale.

bar screen [MECH ENG] A sieve with parallel steel bars for separating small from large pieces of crushed rock.

bar sight [ORD] The rear sight of a firearm, consisting of a movable bar, usually with an open notch.

bar steel [MET] Steel formed into bars.

barstock *See* bar.

Barstovian [GEOL] Upper Miocene geologic time.

bar theory [GEOL] A theory that accounts for thick deposits of salt, gypsum, and other evaporites in terms of increased salinity of a solution in a lagoon caused by evaporation.

Barth plan [IND ENG] A wage incentive plan intended for a low task and for all efficiency points and defined as: earning = rate per hour × square root of the product (hours standard × hours actual).

Bartlett force [NUC PHYS] A force between nucleons in which spin is exchanged.

Bartonella [MICROBIO] A genus of rickettsial organisms, which contains the single species *B. bacilliformis,* in the family Bartonellaceae, and which multiplies only on erythrocytes and within fixed-tissue cells.

Bartonellaceae [MICROBIO] A family of the order Rickettsiales including parasites of red blood cells of man and animals.

bartonellosis *See* Carrion's disease.

Bartonian [GEOL] A European stage: Eocene geologic time above Auversian, below Ludian. Also known as Marinesian.

Bart reaction [ORG CHEM] Formation of an aryl arsenic acid by treating the aryl diazo compound with trivalent arsenic compounds, such as sodium arsenite.

bar turret lathe [MECH ENG] A turret lathe in which the bar stock is slid through the headstock and collet on line with the

BARRETTER

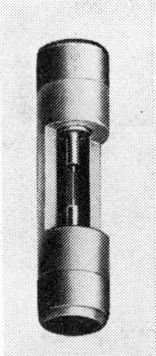

Cutaway view of commercial barretter about 1 inch (2.5 centimeters) in length. *(Sperry Gyroscope Co.)*

turning axis of the lathe and held firmly by the closed collet.

bar winding [ELEC] An armature winding made up of a series of metallic bars connected at their ends.

barycenter [MATH] The center of mass of a system of finitely many equal point masses distributed in euclidean space in such a way that their position vectors are linearly independent.

barycentric coordinates [MATH] The coefficients in the representation of a point in a simplex as a linear combination of the vertices of the simplex.

barycentric element [ASTROPHYS] An orbital element referred to the center of mass of the solar system.

barycentric energy [MECH] The energy of a system in its center-of-mass frame.

Barychilinidae [PALEON] A family of Paleozoic crustaceans in the suborder Platycopa.

barye [MECH] The pressure unit of the centimeter-gram-second system of physical units; equal to 1 dyne/cm² (0.001 millibar). Also known as microbar.

Barylambdidae [PALEON] A family of late Paleocene and early Eocene aquatic mammals in the order Pantodonta.

baryon [PARTIC PHYS] Any elementary particle which can be transformed into a nucleon and some number of mesons and lighter particles. Also known as heavy particle.

baryon number [PARTIC PHYS] A conserved quantum number, equal to the number of baryons minus the number of antibaryons in a system; neutrons and protons have baryon number one; mesons and leptons have baryon number zero.

baryon octet [PARTIC PHYS] The group of one lambda, three sigma, and two xi hyperons and two nucleons, all having spin ½ and positive parity, and forming a symmetrical pattern as suggested by SU_3 symmetry.

baryon resonance [PARTIC PHYS] A cross section anomaly indicating the existence of an unstable baryon.

baryon spectroscopy [PARTIC PHYS] The science of the energy levels and changes of state occurring among baryon particles.

barysphere *See* centrosphere.

baryta feldspar *See* hyalophane.

baryte *See* barite.

Barytheriidae [PALEON] A family of extinct proboscidean mammals in the suborder Barytherioidea.

Barytherioidea [PALEON] A suborder of extinct mammals of the order Proboscidea, in some systems of classification.

barytine *See* barite.

basal [BIOL] Pertaining to or located at the base. [PHYSIO] Being the minimal level for, or essential for maintenance of, vital activities of an organism, such as basal metabolism.

basal arkose [PETR] Partially reworked feldspathic residuum in the lower section of a sandstone that overlies granitic rock.

basal body [CYTOL] A cellular organelle that induces the formation of cilia and flagella and is similar to and sometimes derived from a centriole. Also known as kinetosome.

basal-cell carcinoma [MED] A locally invasive, rarely metastatic nevoid tumor of the epidermis. Also known as basal-cell epithelioma.

basal-cell epithelioma *See* basal-cell carcinoma.

basal cleavage [CRYSTAL] Cleavage parallel to the base of the crystal structure or to the lattice plane which is normal to one of the lattice axes.

basal conglomerate [GEOL] A coarse gravelly sandstone or conglomerate forming the lowest member of a series of related strata which lie unconformably on older rocks; records the encroachment of the seabeach on dry land.

basal coplane [GRAPHICS] The condition of exposure of a pair of photographs in which the two photographs lie in a common plane parallel to the air base.

basal disc [BIOL] The expanded basal portion of the stalk of certain sessile organisms, used for attachment to the substrate.

basal ganglia [ANAT] The corpus striatum, or the corpus striatum and the thalamus considered together as the important subcortical centers.

basalia [VERT ZOO] The cartilaginous rods that support the base of the pectoral and pelvic fins in elasmobranchs.

basalis [HISTOL] The basal portion of the endometrium; it is not shed during menstruation.

basal lamina [EMBRYO] The portion of the gray matter of the embryonic neural tube from which motor nerve roots develop.

basal membrane [ANAT] The tissue beneath the pigment layer of the retina that forms the outer layer of the choroid.

basal metabolic rate [PHYSIO] The amount of energy utilized per unit time under conditions of basal metabolism; expressed as calories per square meter of body surface or per kilogram of body weight per hour. Abbreviated BMR.

basal metabolism [PHYSIO] The sum total of anabolic and catabolic activities of an organism in the resting state providing just enough energy to maintain vital functions.

basal orientation [CRYSTAL] A crystal orientation in which the surface is parallel to the base of the lattice or to the lattice plane which is normal to one of the lattice axes.

basal plane [CRYSTAL] The plane perpendicular to the long, or *c*, axis in all crystals except those of the isometric system.

basal rot [PL PATH] Any rot that affects the basal parts of a plant, especially bulbs.

basalt [PETR] An aphanitic crystalline rock of volcanic origin, composed largely of plagioclase feldspar (labradorite or bytownite) and dark minerals such as pyroxene and olivine; the extrusive equivalent of gabbro.

basalt glass *See* tachylite.

basaltic dome *See* shield volcano.

basaltic hornblende [PETR] A black or brown variety of hornblende rich in ferric iron and occurring in basalts and other iron-rich basic igneous rocks. Also known as basaltine; lamprobolite; oxyhornblende.

basaltic lava [PETR] A volcanic fluid rock of basaltic composition.

basaltic magma [GEOL] Mobile rock material of basaltic composition.

basaltic rock [PETR] Igneous rock that is fine-grained and contains basalt, diabase, and dolerite; if andesite is included the rock is dark in color.

basaltic shell [GEOL] The lower crystal layer of basalt underlying the oceans and beneath the sialic layer of continents.

basaltiform [GEOL] Similar to basalt in form.

basaltine *See* basaltic hornblende.

basalt obsidian *See* tachylite.

basal wall [BOT] The wall dividing the oospore into an anterior and a posterior half in plants bearing archegonia.

basanite [PETR] A basaltic extrusive rock closely allied to chert, jasper, or flint. Also known as lydite; Lydian stone.

basculating fault *See* wrench fault.

bascule bridge [CIV ENG] A bridge structure with a cantilever span which can be raised for passing vessels.

bascule leaf [CIV ENG] The span of a bascule bridge.

base [ADP] *See* root. [CHEM] Any chemical species, ionic or molecular, capable of accepting or receiving a proton (hydrogen ion) from another substance; the other substance acts as an acid in giving of the proton; the hydroxyl ion is a base. [ENG] Foundation or part upon which an object or instrument rests. [ORD] Station or installation from which military forces operate and from which supplies are obtained.

base address *See* address constant.

base-altitude ratio [GRAPHICS] The ratio between the air-base length and the flight attitude of a stereoscopic pair of photographs. Also known as base-height ratio; K factor.

base apparatus [ENG] Any apparatus designed for use in measuring with accuracy and precision the length of a base line in triangulation, or the length of a line in first- or second-order traverse.

baseball [PL PHYS] A machine used in controlled fusion research to confine a plasma; consists of a linear magnetic bottle sealed by magnetic mirrors at both ends, and has current-carrying structures, which resemble the seams of a baseball in shape, to stabilize the plasma.

baseband [COMMUN] The band of frequencies occupied by all transmitted signals used to modulate the radio wave that is produced by the transmitter in the absence of the signals.

baseband frequency response [COMMUN] Frequency response characteristics of the frequency band occupied by all of the signals used to modulate a transmitted carrier.

BARYON OCTET

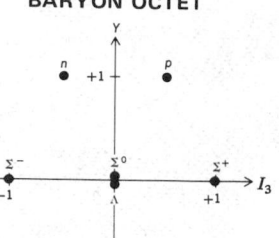

The baryon octet, arrayed with respect to I_3 as abscissa and Y as ordinate. The charge number Q is given by $Q = I_3 + Y/2$.

BASALT

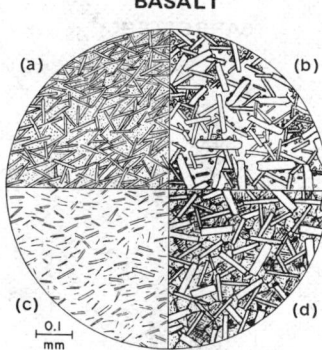

Textures of basalt.
(a) Pilotaxitic texture with feldspar microlites and some interstitial, microcrystalline material.
(b) Intersertal texture with feldspar microlites and some granular pyroxene and much interstitial glass.
(c) Hyalopilitic texture with feldspar microlites in glass.
(d) Intergranular texture with feldspar laths and some interstitial granular pyroxene.

base bias [ELECTR] The direct voltage that is applied to the majority-carrier contact (base) of a transistor.

base box [MET] A unit of area used for tin-plated steel sheet; one base box is equivalent to 112 sheets, 14 by 20 inches (35.6 by 50.8 centimeters), or 62,720 square inches of surface, coated on two sides; 1 pound (0.454 kilogram) of tin per base box is equal to a coating of tin 0.000059 inch (0.0014986 millimeter) thick.

base bullion [MET] Crude lead that has enough silver in it to make the extraction of silver worthwhile; gold may be present.

base-centered lattice [CRYSTAL] A space lattice in which each unit cell has lattice points at the centers of each of two opposite faces as well as at the vertices; in a monoclinic crystal, they are the faces normal to one of the lattice axes.

base circle [DES ENG] The circle on a gear such that each tooth-profile curve is an involute of it.

base conditions [PETRO ENG] Standard conditions of 14.65 psia pressure and 60°F (15.6°C) used to calculate the amount of gas contained in oil from a well (the gas-oil ratio).

base correction [ENG] The adjustment made to reduce measurements taken in field exploration to express them with reference to the base station values.

base course [BUILD] The lowest or first course of a wall. [CIV ENG] The first layer of material laid down in construction of a pavement.

Basedow's disease *See* exophthalmic goiter.

base drag [FL MECH] Drag owing to a base pressure lower than the ambient pressure; it is a part of the pressure drag.

base exchange [GEOCHEM] Replacement of certain ions by others in clay.

base flow [HYD] The flow of water entering stream channels from groundwater sources in the drainage of large lakes.

base fracture [MIN ENG] In quarrying, the broken condition of the base after a blast; it may be a good or bad base fracture.

base fuse [ORD] A fuse installed in the bottom of a projectile.

base-height ratio *See* base-altitude ratio.

base initiation [ORD] Detonation initiated at the base (rear) of the charge.

base insulator [ELEC] Heavy-duty insulator used to support the weight of an antenna mast and insulate the mast from the ground or some other surface.

base level [GEOL] That critical plane of erosion and deposition represented by river level on continents and by wave or current base in the sea.

base-leveled plain [GEOL] Any land surface changed almost to a plain by subaerial erosion. Also known as peneplain.

base-leveling epoch *See* gradation period.

base line Abbreviated BL. [ELECTR] The line traced on amplitude-modulated indicators which corresponds to the power level of the weakest echo detected by the radar; it is retraced with every pulse transmitted by the radar but appears as a nearly continuous display on the scope. [ENG] 1. A surveyed line, established with more than usual care, to which surveys are referred for coordination and correlation. 2. A cardinal line extending east and west along the astronomic parallel passing through the initial point, along which standard township, section, and quarter-section corners are established. [NAV] The geodesic line joining the two stations between which electrical phase or time is compared in determining navigational coordinates. [SCI TECH] A line drawn in the graphical representation of a varying physical quantity, such as a voltage or current, to indicate a reference value, such as the voltage value of a bias.

base-line break [ELECTR] Technique in radar which uses the characteristic break in the base line on an A-scope display due to a pulse signal of significant strength in noise jamming.

base-line check *See* ground check.

base-line delay [NAV] The time interval which elapses while a signal travels from a loran master station to a slave station.

base-line extension [NAV] In the loran navigation system, the extension in both directions of a line passing through a slave station and its master; on this line the maximum time interval between the reception of signals from the master and slave stations may be received.

base-line technique [ANALY CHEM] A method for measurement of absorption peaks for quantitative analysis of chemical compounds in which a base line is drawn tangent to the spectrum background; the distance from the base line to the absorption peak is the absorbence due to the sample under study.

baseload [ELEC] Minimum load of a power generator over a given period of time.

base-loaded antenna [ELECTROMAG] Vertical antenna having an impedance in series at the base for loading the antenna to secure a desired electrical length.

base map [MAP] A map having essential outlines and onto which additional geographical or topographical data may be placed for comparison or correlation. Also known as mother map.

basement [BUILD] A building story which is wholly or less than half below ground; generally used for living space. [GEOL] 1. A complex, usually of igneous and metamorphic rocks, that is overlain unconformably by sedimentary strata. 2. A crustal layer beneath a sedimentary one and above the Mohorovičić discontinuity.

basement membrane [HISTOL] A delicate connective-tissue layer underlying the epithelium of many organs.

base metal [CHEM] Any of the metals on the lower end of the electrochemical series. [MET] 1. The metal that is to be worked. 2. The principal metal of an alloy. 3. Any metal that will oxidize when heated in air. 4. The metal of parts to be welded. 5. Metal to which cladding or plating is applied. Also known as basis metal.

base modulation [ELECTR] Amplitude modulation produced by applying the modulating voltage to the base of a transistor amplifier.

base mortar [ORD] Mortar in a platoon for which initial firing data are computed and with reference to which data for other mortars in the unit are computed.

basendite [INV ZOO] In crustaceans, either of a pair of lobes at the end of each specialized paired appendage.

base net [ENG] A system, in surveying, of quadrilaterals and triangles that include and are quite close to a base line in a triangulation system.

base notation *See* radix notation.

base of a logarithm [MATH] The number of which the logarithm is the exponent.

base of a number system [MATH] The number whose powers determine place value.

base of a topological space [MATH] A collection of sets, unions of which form all open sets.

base of projectile [ORD] The rearmost section of a projectile; for projectiles having a rotating band, it is the section located to the rear thereof.

base ore [MIN ENG] Ore in which the gold is associated with sulfides, as contrasted with free-milling ores in which the sulfides have been removed by leaching.

base pairing [MOL BIO] The hydrogen bonding of complementary purine and pyrimidine bases—adenine with thymine, guanine with cytosine—in double-stranded deoxyribonucleic acids (DNA) or ribonucleic acids (RNA) or in DNA/RNA hybrid molecules.

base piece [ORD] The gun first selected after calibration fire whose center of burst or impact is taken as the reference point in determining calibration corrections for the remaining guns of the battery.

base pin *See* pin.

base plate [DES ENG] The part of a theodolite which carries the lower ends of the three foot screws and attaches the theodolite to the tripod for surveying. [ENG] A metal plate that provides support or a foundation.

base point [ORD] A well-defined point in the target area used as a point of reference from which range and direction adjustments of artillery fire are made.

base pressure [FL MECH] The pressure exerted on the base or extreme aft end of a body, as of a cylindrical or boat-tailed body or of a blunt-trailing-edge wing in fluid flow. [MECH] A pressure used as a reference base, for example, atmospheric pressure.

base rate area [COMMUN] Area within the telephone exchange in which all types of service are given without mileage charges; the area does not necessarily follow the same lines or

BASE-CENTERED LATTICE

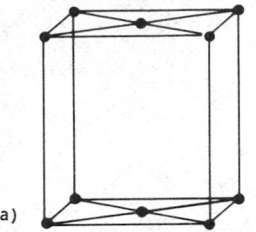

(a)

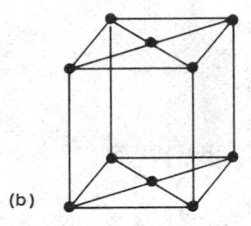

(b)

Two of fourteen Bravais lattices.
(a) Base-centered orthorhombic.
(b) Base-centered monoclinic.

the same area as city limits; an extension of city limits, for example, would have no bearing on the base rate area boundary; customers located outside the area pay for the extra cost of providing service.

base register   See index register.

base spray   [ORD] Fragments of a bursting projectile that are thrown to the rear in the line of flight, in contrast to nose spray and side spray.

base-spreading resistance   [ELECTR] Resistance which is found in the base of any transistor and acts in series with it, generally a few ohms in value.

base station   [COMMUN] **1.** A land station, in the land mobile service, carrying on a service with land mobile stations, (a base station may secondarily communicate with other base stations incident to communications with land mobile stations). **2.** A station in a land mobile system which remains in a fixed location and communicates with mobile stations. [ENG] The point from which a survey begins. [GEOD] A geographic position whose absolute gravity value is known.

base time   See normal elemental time; normal time.

base-timing sequencing   [NAV] The control of the time sharing of a single transponder between several ground transmitters through the use of suitable coded timing signals.

BASF process   [CHEM ENG] Production of acetylene by controlled burning of light hydrocarbons (for example, natural gas) with oxygen.

bashing   [MIN ENG] **1.** The building of walls and nonporous stoppings for the complete isolation of a district of a mine in which a fire has occurred. **2.** The complete stowing of old mine workings or roadways after all equipment has been removed.

basic   [CHEM] Of a chemical species that has the properties of a base. [PETR] Of igneous rocks, having low silica content (generally less than 54%) and usually rich in iron, magnesium, or calcium.

BASIC   [ADP] A procedure-level computer language well suited for conversational mode on a terminal usually connected with a remotely operated computer.

basic bismuth carbonate   See bismuth subcarbonate.

basic bismuth gallate   See bismuth subgallate.

basic bismuth nitrate   See bismuth subnitrate.

basic bismuth pyrogallate   See bismuth pyrogallate.

basic bismuth salicylate   See bismuth subsalicylate.

basic converter   See basic-lined converter.

basic dye   [MATER] Any of the dyes which are salts of colored organic bases containing amino and imino groups, combined with a colorless acid, such as hydrochloric or sulfuric.

basic feasible solution   [IND ENG] A basic solution to a linear program model in which all the variables are nonnegative.

basic frequency   [PHYS] Frequency, in any wave, which is considered the most important; in a driven system, it would generally be the driving frequency, while in most periodic waves it would correspond to the fundamental frequency.

basic front   [GEOL] An advancing zone of granitization enriched in calcium, magnesium, and iron.

basic group   [CHEM] A chemical group (for example, OH⁻) which, when freed by ionization in solution, produces a pH greater than 7.

basic hornfels   [PETR] A type of hornfels derived from a basic igneous rock.

basichromatic   [BIOL] Staining readily with basic dyes.

basic-lined   [MET] Pertaining to the walls and bottom of a melting furnace made of refractory materials, such as lime or dolomite, that have a basic reaction in the melting process.

basic-lined converter   [MET] A converter, such as the Peir-Smith copper converter, which has a basic refractory lining. Also known as basic converter.

basic motion   [IND ENG] A single, complete movement of a body member; determined by motion studies.

basic motion-time study   [IND ENG] A system of predetermined motion-time standards for basic motions. Abbreviated BMT study.

basic open-hearth process   [MET] An open-hearth process for steelmaking under basic slag; used for pig iron and scrap with a phosphorus content too low for the Bessemer process and too high for the acid open-hearth process.

basic oxide   [INORG CHEM] A metallic oxide that is a base, or

that forms a hydroxide when combined with water, such as sodium oxide to sodium hydroxide.

basic processing unit   [COMMUN] Principal controller and data processor within the communications system.

basic pulse repetition rate   [NAV] In the loran navigation system, the lowest pulse repetition rate used by a chain of stations; the chain has a number of stations, all operating on the same radio frequency; one of these stations uses the basic rate while the others use rates that differ by fractional numbers of pulses per second.

basic refractory   [MATER] Any heat-resistant material used for basic linings; examples are dolomite and magnesite.

basic research   [SCI TECH] Fundamental theoretical or experimental investigation to advance scientific knowledge, immediate practical application not being a direct objective. Also known as pure research.

basic rock   [PETR] An igneous rock with a relatively low silica content, and rich in iron, magnesium, or calcium.

basic salt   [INORG CHEM] A compound that is a base and a salt because it contains elements of both, for example, $Cu_2(OH)_2CO_3$.

basic schist   [PETR] A schistose rock that forms from the metamorphism of a basic igneous rock.

basic sediment and water   [PETRO ENG] Oil, water, and foreign matter that collects in the bottom of petroleum storage tanks. Abbreviated BS&W. Also known as bottoms; bottom settlings; sediment and water.

basic slag   [MET] A slag resulting from the steelmaking process; rich in phosphorus, it is ground and used as a nutrient in grasslands.

basic software   [ADP] Software requirements that are taken into account in the design of the data-processing hardware and usually provided by the original equipment manufacturer.

basic solution   [IND ENG] A solution to a linear program model, consisting of m equations in n variables, obtained by solving for m variables in terms of the remaining $(n - m)$ variables and setting the $(n - m)$ variables equal to zero.

basic steel   [MET] Steel made by the basic process, in a furnace with a basic lining.

basic titanium sulfate   See titanium sulfate.

basic titrant   [CHEM] A standard solution of a base used for titration.

basic variables   [ADP] The m variables in a basic feasible solution for a linear programming model.

basic zirconium chloride   See zirconium oxychloride.

basic zirconium phosphate   See zirconium phosphate.

basidiocarp   [MYCOL] The fruiting body of a fungus in the class Basidiomycetes.

Basidiolichenes   [BOT] A class of the Lichenes characterized by the production of basidia.

Basidiomycetes   [MYCOL] A class of fungi in the subdivision Eumycetes; important as food and as causal agents of plant diseases.

basidiophore   [MYCOL] A basidia-bearing sporophore.

basidiospore   [MYCOL] A spore produced by a basidium.

basidium   [MYCOL] A cell, usually terminal, of Basidiomycetes that produces spores (basidiospores) by nuclear fusion followed by meiosis.

basification   [GEOL] Development of a more basic rock, usually with more hornblende, biotite, and oligoclase, by contamination of a granitic magma in the assimilation of country rock.

basifixed   [BOT] Attached at or near the base.

basil   [BOT] The common name for any of the aromatic plants in the genus *Ocimum* of the mint family; leaves of the plant are used for food flavoring. [MATER] Sheephide tanned with bark.

basilar   [BIOL] Of, pertaining to, or situated at the base.

basilar groove   [ANAT] The cavity which is located on the upper surface of the basilar process of the brain and upon which the medulla rests.

basilar index   [ANTHRO] The ratio of the distance from the basion to the alveolar point to the total skull length multiplied by 100.

basilar membrane   [ANAT] A membrane of the mammalian inner ear supporting the organ of Corti and separating two cochlear channels, the scala media and scala tympani.

BASIDIUM

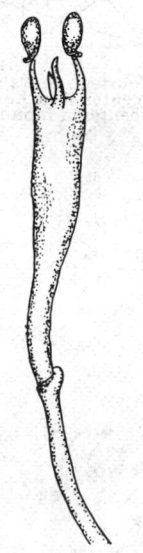

A drawing of the basidiospores and basidium in fungi.

basilar meningitis [MED] Inflammation of the meninges which affects chiefly the base of the brain, or in which exudate collects predominantly at the basal cisterns.

basilar papilla [ANAT] **1.** A sensory structure in the lagenar portion of an amphibian's membranous labyrinth between the oval and round windows. **2.** The organ of Corti in mammals.

basilar plate [EMBRYO] An embryonic cartilaginous plate in vertebrates that is formed from the parachordals and anterior notochord and gives rise to the ethmoid and other bones of the skull.

basilar process [ANAT] A strong, quadrilateral plate of bone forming the anterior portion of the occipital bone, in front of the foramen magnum.

basilic vein [ANAT] The large superficial vein of the arm on the medial side of the biceps brachii muscle.

basil oil [MATER] Any of the yellow, aromatic essential oils derived from the leaves of sweet basil; used for flavors and perfumes and in medicines. Also known as sweet basil oil.

basimesostasis [GEOL] A process of the partial or entire enclosure of plagioclase crystals in a diabase by augite.

basin [CIV ENG] **1.** A dock employing floodgates to keep water level constant during tidal variations. **2.** A harbor for small craft. [DES ENG] An open-top vessel with relatively low sloping sides for holding liquids. [GEOL] **1.** A low-lying area, wholly or largely surrounded by higher land, that varies from a small, nearly enclosed valley to an extensive, mountain-rimmed depression. **2.** An entire area drained by a given stream and its tributaries. **3.** An area in which the rock strata are inclined downward from all sides toward the center. **4.** An area in which sediments accumulate. [OCEANOGR] Deep portion of sea surrounded by shallower regions.

basin accounting *See* hydrologic accounting.

basin-and-range structure [GEOL] Regional structure dominated by fault-block mountains separated by basins filled with sediment.

basin fold [GEOL] Synclinal and anticlinal folds in structural basins.

basining [GEOL] A settlement of earth in the form of basins due to the solution and transportation of underground deposits of salt and gypsum.

basin length [GEOL] Length in a straight line from the mouth of a stream to the farthest point on the drainage divide of its basin.

basin order [GEOL] A classification of basins according to stream drainage; for example, a first-order basin contains all of the drainage area of a first-order stream.

basin peat *See* local peat.

basin perimeter [MAP] The length of a map line that encloses the catchment area of a drainage basin.

basin range [GEOL] A mountain range characteristic of the Great Basin in the western United States and formed by a faulted and tilted block of strata.

basin valley [GEOL] The filled-in depression of large intermountain areas; an example is Salt Lake Valley in Utah.

basion [ANAT] In craniometry, the point on the anterior margin of the foramen magnum where the midsagittal plane of the skull intersects the plane of the foramen magnum.

basion-bregma height [ANTHRO] Distance between basion and bregma.

basipetal [BIOL] Movement or growth from the apex toward the base.

basipodite [INV ZOO] The distal segment of the protopodite of a biramous appendage in arthropods.

basipterygium [VERT ZOO] A basal bone or cartilage supporting one of the paired fins in fishes.

basis [MATH] A set of linearly independent vectors in a vector space such that each vector in the space is a linear combination of vectors from the set.

basis metal *See* base metal.

basisternum [INV ZOO] In insects, the anterior one of the two sternal skeletal plates.

basistyle [INV ZOO] Either of a pair of flexible processes on the hypopygium of certain male Diptera.

basis weight [ENG] The weight in pounds of 500 sheets of standard-size paper; certain sizes for a given class of paper are accepted as standard.

basitarsus [INV ZOO] The basal segment of the tarsus in arthropods.

basket [DES ENG] A lightweight container with perforations. [MECH ENG] A type of single-tube core barrel made from thin-wall tubing with the lower end notched into points, which is intended to pick up a sample of granular or plastic rock material by bending in on striking the bottom of the borehole or solid layer; may be used to recover an article dropped into a borehole. Also known as basket barrel; basket tube; saw-tooth barrel.

basket barrel *See* basket.

basket cell [HISTOL] A type of cell in the cerebellum whose axis-cylinder processes terminate in a basketlike network around the cells of Purkinje.

basket coil *See* basket winding.

basket-handle arch [ARCH] An arch whose width is much greater than its height and which resembles an ellipse and is drawn from three or more centers. Also known as multicenter arch.

basket star [INV ZOO] The common name for ophiuroid echinoderms belonging to the family Gorgonocephalidae.

basket strainer [CHEM ENG] A porous-sided or screen-covered vessel used to screen solid particles out of liquid or gas streams.

basket tube *See* basket.

basket-weave [TEXT] Pertaining to fabric in which two or more yarns are worked in warp and weft.

basket winding [ELECTR] A crisscross coil winding in which successive turns are far apart except at points of crossing, giving low distributed capacitance. Also known as basket coil.

Basommatophora [INV ZOO] An order of mollusks in the subclass Pulmonata containing many aquatic snails.

basonym [SYST] The original, validly published name of a taxon.

basophil [HISTOL] A white blood cell with granules that stain with basic dyes and are water-soluble.

basophilia [BIOL] An affinity for basic dyes. [MED] An increase in the number of basophils in the circulating blood. [PATH] Stippling of the red cells with basic staining granules, representing a degenerative condition as seen in severe anemia, leukemia, malaria, lead poisoning, and other toxic states.

basophilous [BIOL] Staining readily with basic dyes. [ECOL] Of plants, growing best in alkaline soils.

basophobia [PSYCH] Abnormal fear of walking or standing erect.

bas-relief *See* low relief.

bass [ACOUS] Sounds having frequencies at the lower end of the audio range, below about 250 hertz. [VERT ZOO] The common name for a number of fishes assigned to two families, Centrarchidae and Serranidae, in the order Perciformes.

bass boost [ELECTR] A circuit that emphasizes the lower audio frequencies, generally by attenuating higher audio frequencies.

bass compensation [ELECTR] A circuit that emphasizes the low-frequency response of an audio amplifier at low volume levels to offset the lower sensitivity of the human ear to weak low frequencies.

bass control [ELECTR] A manual tone control that attenuates higher audio frequencies in an audio amplifier and thereby emphasizes bass frequencies.

basset [GEOL] The outcropping edge of a layer of rock exposed to the surface. [VERT ZOO] A French breed of short-legged hunting dogs with long ears and a typical hound coat.

basse-taille [GRAPHICS] Having a background carved in low relief.

basso [GRAPHICS] A printing plate used in electrotyping of currency, made by depositing a layer of nickel, followed by a thick layer of iron.

bassora gum [MATER] A type of high-colored gum similar to tragacanth gum; includes Indian gum.

bass reflex baffle [ENG ACOUS] A loudspeaker baffle having an opening of such size that bass frequencies from the rear of the loudspeaker emerge to reinforce those radiated directly forward.

bass response [ELECTR] A measure of the output of an

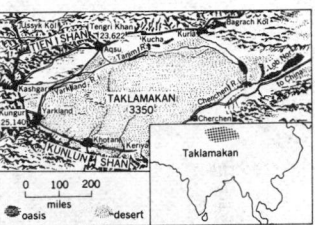

BASIN

Tarim Basin in central Asia, an example of a mountain-rimmed basin of interior drainage. The Taklamakan desert occupies the greater part of the basin. (*After E. Raisz, in P. E. James, An Outline of Geography, Ginn, 1935*)

BASS

The largemouth bass (*Micropterus salmoides*), a game fish found in parts of the United States.

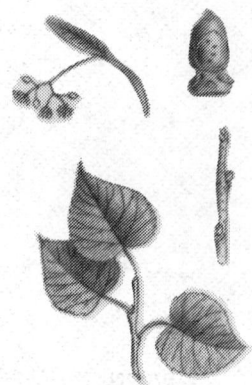

BASSWOOD

American basswood (*Tilia americana*).

BAT

Epauletted fruit bat. (*Epomophorus wahlbergi*).

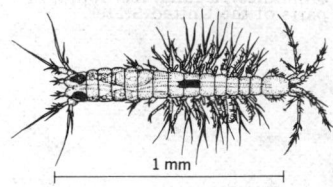

BATHYNELLACEA

1 mm

Bathynella natans, female.

electronic device or system as a function of an input of low audio frequencies.

basswood [BOT] A common name for trees of the genus *Tilia* in the linden family of the order Malvales. Also known as linden.

bassy [ENG ACOUS] Pertaining to sound reproduction that overemphasizes low-frequency notes.

bast *See* phloem.

bastard-cut file [DES ENG] A file that has coarser teeth than a rough-cut file.

bastard sawing *See* backsawing.

bastard thread [DES ENG] A screw thread that does not match any standard threads.

bast fiber [BOT] Any fiber stripped from the inner bark of plants, such as flax, hemp, jute, and ramie; used in textile and paper manufacturing.

bastite [MINERAL] A hydrated magnesium silicate, a variety of serpentine occurring from the alteration of orthorhombic pyroxenes such as enstatite.

bastnaesite [MINERAL] (Ce,La)CO(F,OH) A greasy yellow to reddish-brown fluorocarbonate rare-earth metal mineral; source of rare earths, for example, cerium and lanthanum.

bat [VERT ZOO] The common name for all members of the mammalian order Chiroptera.

Batales [BOT] A small order of dicotyledonous plants in the subclass Caryophillidae of the class Magnoliopsida containing a single family with only one genus, *Batis*.

bat bolt [DES ENG] A bolt whose butt or tang is bashed or jagged.

batch [ADP] A set of items, records, or documents to be processed as a single unit. [ENG] **1.** The quantity of material required for or produced by one operation. **2.** An amount of material subjected to some unit chemical process or physical mixing process to make the final product substantially uniform.

batch distillation [CHEM ENG] Distillation where the entire batch of liquid feed is placed into the still at the beginning of the operation, in contrast to continuous distillation, where liquid is fed continuously into the still.

batcher [MECH ENG] A machine in which the ingredients of concrete are measured and combined into batches before being discharged to the concrete mixer.

batching [ADP] Grouping records for the purpose of processing them in a computer.

Batchinsky relation [FL MECH] The relation stating that the fluidity of a liquid is proportional to the difference between the specific volume and a characteristic specific volume, approximately equal to the specific volume appearing in the van der Waals equation.

batch oil [MATER] A pale, lemon-colored, low-viscosity mineral oil used particularly in the manufacture of cordage.

batch process [ENG] A process that is not in continuous or mass production; operations are carried out with discrete quantities of material or a limited number of items.

batch processing [ADP] A technique that uses a single program loading to process many individual jobs, tasks, or requests for service.

batch production *See* series production.

batch rectification [CHEM ENG] Batch distillation in which the boiled-off vapor is re-condensed into liquid form and refluxed back into the still to make contact with the rising vapors.

batch total [ADP] The total for a specified constituent quantity in a batch; used to verify the accuracy of operations on the batch.

batea [MIN ENG] A conical-shaped wood unit (12.30 inches in diameter with about 150° apex angle) used to recover valuable metals from river channels and bars.

Batesian mimicry [ECOL] Resemblance of an innocuous species to one that is distasteful to predators.

bat-handle switch [ELEC] A toggle switch having an actuating lever shaped like a baseball bat.

bathochromatic shift [PHYS CHEM] The shift of the fluorescence of a compound toward the red part of the spectrum due to the presence of a bathochrome radical in the molecule.

batholite [GEOL] An older massive protrusion of magma that

solidifies as coarse crystalline rock in the deep horizons of the earth's crust.

batholith [GEOL] A body of igneous rock, 40 square miles or more in area, emplaced at great or intermediate depth in the earth's crust.

bathometer [ENG] A mechanism which measures depths in water.

Bathonian [GEOL] A European stage of geologic time: Middle Jurassic, below Callovian, above Bajocian. Also known as Bathian.

bathophobia [PSYCH] Abnormal fear of depths.

Bathornithidae [PALEON] A family of Oligocene birds in the order Gruiformes.

bathtub capacitor [ELEC] A capacitor enclosed in a metal housing having broadly rounded corners like those on a bathtub.

bathtub curve [IND ENG] An equipment failure-rate curve with an initial sharply declining failure rate, followed by a prolonged constant-average failure rate, after which the failure rate again increases sharply.

bathvillite [MINERAL] An oxygenated hydrocarbon mineral, found in Tortane Hill, Scotland, that is amorphous, fawn-brown, opaque, and quite friable.

bathyal zone [OCEANOGR] The biogeographic realm of the ocean depths between 100 and 1000 fathoms (180 and 1800 meters).

bathyclinograph [ENG] A mechanism which measures vertical currents in the deep sea.

bathyconductograph [ENG] A device to measure the electrical conductivity of sea water at various depths from a moving ship.

Bathyctenidae [INV ZOO] A family of bathypelagic coelenterates in the phylum Ctenophora.

Bathyergidae [VERT ZOO] A family of mammals, including the South African mole rats, in the order Rodentia.

bathygram [ENG] A graph recording the measurements of sonic sounding instruments.

Bathylaconoidei [VERT ZOO] A suborder of deep-sea fishes in the order Salmoniformes.

bathymetric biofacies [GEOL] The lateral distribution and character of underwater sedimentary strata.

bathymetric chart [MAP] A topographic map of the floor of the ocean.

bathymetry [ENG] The science of measuring ocean depths in order to determine the sea floor topography.

Bathynellacea [INV ZOO] An order of crustaceans in the superorder Syncarida found in subterranean waters in England and central Europe.

Bathynellidae [INV ZOO] The single family of the crustacean order Bathynellacea.

bathyorographical [GEOD] Concerned with depths of oceans and heights of mountains.

bathypelagic zone [OCEANOGR] The biogeographic realm of the ocean lying between depths of 900 and 3700 meters.

Bathypteroidae [VERT ZOO] A family of benthic, deep-sea fishes in the order Salmoniformes.

bathyscaph [NAV ARCH] A free, manned vehicle having a spherical cabin on the underside for exploring the deep ocean.

bathysphere [NAV ARCH] A spherical chamber in which persons are lowered for observation and study of ocean depths.

Bathysquillidae [INV ZOO] A family of mantis shrimps, with one genus (*Bathysquilla*) and two species, in the order Stomatopoda.

bathythermograph [ENG] A device for obtaining a record of temperature against depth (actually, pressure) in the ocean from a ship underway. Abbreviated BT. Also known as bathythermosphere.

bathythermosphere *See* bathythermograph.

bathyvessel [NAV ARCH] A ship, such as a bathysphere or submarine, designed to operate far below the surface of the water.

batik [TEXT] **1.** A method of dyeing fabric in which parts of the cloth not intended to be dyed are covered with removable wax. **2.** The print so produced. **3.** The dyed cloth.

bating [CHEM ENG] Cleaning of depilated leather hides by the action of tryptic enzymes.

Batoidea [VERT ZOO] The skates and rays, an order of the subclass Elasmobranchii.

batophobia [PSYCH] Abnormal fear of being near something of great height, such as a mountain or tall building.

Batrachoididae [VERT ZOO] The single family of the order Batrachoidiformes.

Batrachoidiformes [VERT ZOO] The toadfishes, an order of teleostean fishes in the subclass Actinopterygii.

batt [MATER] A blanket of insulating material usually 16 inches wide and 3 to 6 inches thick, used to insulate building walls and roofs. [MIN ENG] A thin layer of coal occurring in the lower part of shale strata that lie above and close to a coal bed.

batten [AERO ENG] Metal, wood, or plastic panels laced to the envelope of a blimp in the nose cone to add rigidity to the nose and provide a good point of attachment for mooring. [BUILD] **1.** A sawed timber strip of specific dimension (usually 7 inches broad, less than 4 inches thick, and more than 6 feet long) used for outside walls of houses, flooring, and such. **2.** A strip of wood nailed across a door or other structure made of parallel boards to strengthen it and prevent warping.

batten plate [CIV ENG] A rectangular plate used to connect two parallel structural steel members by riveting or welding.

Batten system [ADP] A system of information retrieval that uses cards in which holes have been punched. Also known as Cordonnier system; peekaboo system.

batter [CIV ENG] A uniformly steep slope in a retaining wall or pier; inclination is expressed as 1 foot horizontally per vertical unit (in feet).

battered-child syndrome [MED] A clinical condition in young children due to serious physical abuse, generally from a parent or foster parent.

battery [CHEM ENG] A series of distillation columns or other processing equipment operated as a single unit. [ELEC] A direct-current voltage source made up of one or more units that convert chemical, thermal, nuclear, or solar energy into electrical energy. [ORD] A group of guns or other weapons, such as mortars, machine guns, artillery pieces, or of searchlights, set up under one tactical commander in a certain area.

battery amalgamation [MET] Amalgamation by means of mercury placed in the mortar box of a stamp battery.

battery assay [MIN ENG] An assay of samples taken from ore as crushed in a stamp battery.

battery charger [ELEC] A rectifier unit used to change alternating to direct power for charging a storage battery. Also known as charger.

battery clip [ELEC] A terminal of a connecting wire having spring jaws that can be quickly snapped on a terminal of a device, such as a battery, to which a temporary wire connection is desired.

battery command periscope [OPTICS] An optical instrument consisting of dual telescope tubes positioned vertically on a common mounting; it provides periscopic vision for the observer, and may be used to observe artillery fire.

battery depolarizer *See* depolarizer.

battery electrolyte [PHYS CHEM] A liquid, paste, or other conducting medium in a battery, in which the flow of electric current takes place by migration of ions.

battery eliminator [ELECTR] A device which supplies electron tubes with voltage from electric power supply mains.

battery limits [CHEM ENG] An area in a refinery or chemical plant encompassing a processing unit or battery of units along with their related utilities and services.

battery manganese *See* manganese dioxide.

battery of genes [GEN] The set of producer genes which is activated when a particular sensor gene activates its set of integrator genes.

battery reefs *See* Kimberley reefs.

battery separator [ELEC] An insulating plate inserted between the positive and negative plates of a battery to prevent them from touching.

Battey disease [MED] A tuberculosislike disease of man caused by *Mycobacterium intracellulare*.

battlefield recovery [ORD] Removal of disabled or abandoned material, either enemy or friendly, from the battlefield and moving it to a recovery collecting point or to a maintenance or supply establishment.

battleship [NAV ARCH] One of a class of heaviest and most extensively armed and armored vessels, with at least 10-inch armor plating and guns of 12-inch or larger caliber.

battle short [ELEC] Switch for short-circuiting safety interlocks and lighting a red warning light.

battle sight [ORD] A predetermined sight setting, carried on a weapon, that enables the firer to engage targets effectively at battle ranges when conditions do not permit exact sight settings.

batture [GEOL] An elevation of the bed of a river under the surface of the water; sometimes used to signify the same elevation when it has risen above the surface.

baud [COMMUN] A unit of telegraph signaling speed equal to the number of code elements (pulses and spaces) per second or twice the number of pulses per second. Also known as unit pulse.

Baudot code [COMMUN] A teleprinter code that uses a combination of five or six marking and spacing intervals of equal duration for each character.

Baumé hydrometer scale [PHYS CHEM] A calibration scale for liquids that is reducible to specific gravity by the following formulas: for liquids heavier than water, specific gravity = $145 \div 145 - n$ (at 60°F); for liquids lighter than water, specific gravity = $140 \div 130 + n$ (at 60°F); n is the reading on the Baumé scale, in degrees Baumé. Baumé is abbreviated Bé.

Bauschinger effect [MET] A phenomenon by which the plastic deformation of a metal increases the tensile yield strength and decreases the compressive yield strength.

bauxite [PETR] A whitish, grayish, brown, yellow, or reddish-brown rock composed of hydrous aluminum oxides and aluminum hydroxides and containing impurities such as free silica, silt, iron hydroxides, and clay minerals; the principal commercial source of aluminum.

bauxitization [GEOL] Bauxite development from either primary aluminum silicates or secondary clay minerals.

bavenite *See* duplexite.

Baveno twin law [CRYSTAL] An uncommon twin law in feldspar, in which the twin plane and composition surface are (021); a Baveno twin usually consists of two individuals.

b axis [CRYSTAL] A crystallographic axis that is oriented horizontally, right to left. [PETR] A direction in the plane of movement that is at a right angle to the tectonic transport direction.

bay [AERO ENG] A space formed by structural partitions on an aircraft. [ARCH] Division of a building between adjacent beams or columns. [BOT] *Laurus nobilis.* An evergreen tree of the laurel family. [ELECTROMAG] One segment of an antenna array. [ENG] A housing for equipment. [GEOGR] **1.** A body of water, smaller than a gulf and larger than a cove in a recess in the shoreline. **2.** A narrow neck of water leading from the sea between two headlands.

Bayard-Alpert ionization gage [ELECTR] A type of ionization vacuum gage using a tube with an electrode structure designed to minimize x-ray-induced electron emission from the ion collector.

bay bar *See* baymouth bar.

bay barrier [GEOL] A narrow shoal or small point of land projecting from the shore across the mouth of a bay and severing the bay's connection with the main body of water.

bayberry [BOT] **1.** *Pimenta acris.* A West Indian tree related to the allspice; a source of bay oil. Also known as bay-rum tree; Jamaica bayberry; wild cinnamon. **2.** Any tree of the genus *Myrica.*

bayberry tallow *See* bayberry wax.

bayberry wax [MATER] Green, bitter-tasting wax derived from boiling of wax myrtle (*Myrica*) berries; used for candles, soaps, medicine. Also known as bayberry tallow; laurel wax; myrtle wax.

bay delta [GEOL] A usually triangular alluvial deposit formed at the point where the mouth of a stream enters the head of a drowned valley.

Bayer 205 [PHARM] Trade name under which suramin sodium was originally introduced.

Bayer letter [ASTRON] The Greek (or Roman) letter used in a Bayer name.

Bayer name [ASTRON] The Greek (or Roman) letter and the possessive form of the Latin name of a constellation, used as

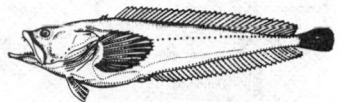

BATRACHOIDIFORMES

Atlantic midshipman (*Porichthys porosissimus*), about 8 inches (20 centimeters) long. (*After D. S. Jordan*)

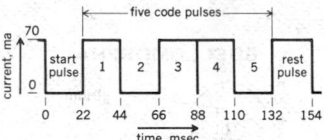

BAUDOT CODE

The five-unit stop-start telegraph code, seven-unit code pattern. This combination represents letter F.

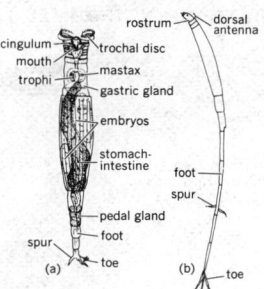

BDELLOIDEA

Bdelloid rotifers, with elongated, segmented bodies. *(a) Rotaria. (b) R. neptunia. (After Weber, 1898; from L. Hyman, The Invertebrates, vol. 3, McGraw-Hill, 1951)*

BDELLOMORPHA

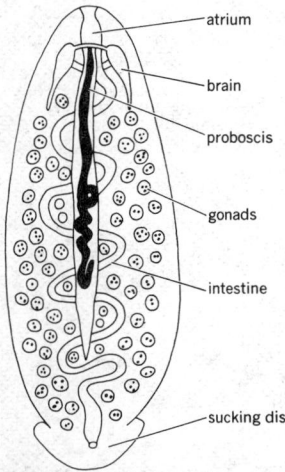

Malacobdella grossa. (From W. R. Coe, Biology of the nemerteans of the Atlantic coast of North America, Trans. Conn. Acad. Arts. Sci., 35:308, 1943)

B DISPLAY

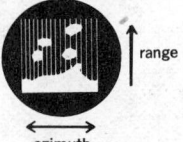

Type B radar display.

a star name; examples are α Cygni (Deneb), β Orionis (Rigel), and η Ursae Majoris (Alkaid).

Bayer process [MET] A method of producing alumina from bauxite by heating it in a sodium hydroxide solution.

Bayer's constellations [ASTRON] Thirteen constellations in the southern hemisphere named by J. Bayer.

Bayes decision rule [STAT] A decision rule under which the strategy chosen from among several available ones is the one for which the expected value of payoff is the greatest.

bay head [GEOL] A swampy region at the head of a bay.

bay head bar [GEOL] A bar formed a short distance from the shore at the head of a bay.

bay head beach [GEOL] A beach formed around a bay head by storm waves; layers of sediment cover the bay floor and bare rock benches front the headland cliffs.

bay head delta [GEOL] A delta at the head of an estuary or a bay into which a river discharges because of the margin of the land's late partial submergence.

bay ice [OCEANOGR] Sea ice that is young and flat but sufficiently thick to impede navigation.

bay leaf [FOOD ENG] An herb; the dried leaf of the bay tree (*Laurus nobilis*).

baymouth bar [GEOL] A bar extending entirely or partially across the mouth of a bay. Also known as bay bar.

bay oil [MATER] Yellow essential oil with clovelike odor and pungent taste; derived from distillation of West Indian bayberry (*Pimenta acris*) leaves and used in flavors and perfumes. Also known as myrica oil.

bayonet [ORD] An edged steel blade with a tapered point and a formed handle with an underhand grip, designed to be attached to the muzzle end of a rifle, shotgun, or the like for use in hand-to-hand combat.

bayonet base [ELEC] A tube base or lamp base having two projecting pins on opposite sides of a smooth cylindrical surface to engage in corresponding slots in a bayonet socket and hold the base firmly in the socket.

bayonet coupling [DES ENG] A coupling in which two or more pins extend out from a plug and engage in grooves in the side of a socket.

bayonet socket [DES ENG] A socket, having J-shaped slots on opposite sides, into which a bayonet base or coupling is inserted against a spring and rotated until its pins are seated firmly in the slots.

bayonet-tube exchanger [MECH ENG] A dual-tube apparatus with heating (or cooling) fluid flowing into the inner tube and out of the annular space between the inner and outer tubes; can be inserted into tanks or other process vessels to heat or cool the liquid contents.

bayou [HYD] A small, sluggish secondary stream or lake that exists often in an abandoned channel or a river delta.

bay rum [MATER] A product originally prepared from the distillation of rum and water mixed with bayberry (*Pimenta acris*) leaves but now made from a mixture of bay oil, orange oil, pimenta oil, alcohol, and water; used in shaving lotions and alcohol rubs.

bay-rum tree *See* bayberry.

bayside beach [GEOL] A beach formed at the side of a bay by materials eroded from nearby headlands and deposited by longshore currents.

bazooka [ELECTR] *See* balun. [ORD] Popular name applied to the 2.36-inch (60-millimeter) rocket launcher.

bazzite [MINERAL] $Sc_2Be_3Si_6O_{18}$ An azure-blue mineral that crystallizes in the hexagonal system; the rare scandium analog of beryl.

B battery [ELECTR] The battery that furnishes required direct-current voltages to the plate and screen-grid electrodes of the electron tubes in a battery-operated circuit.

BBC *See* bromobenzylcyanide.

B-B fraction [CHEM ENG] A mixture of butanes and butenes distilled from a solution of light liquid hydrocarbons.

BB gun *See* air rifle.

B bit [MIN ENG] A nonstandard core bit no longer in common use except in drilling deep boreholes to sample gold-bearing deposits in South Africa; the set outside and inside diameters are about $2\frac{1}{16}$ and $1\frac{3}{8}$ inches, respectively.

bbl *See* barrel.

B blasting powder [MATER] A mixture of nitrate of soda, charcoal, and sulfur; used in coal mines. Also known as soda blasting powder.

B box *See* index register.

BCD *See* barrels per calendar day.

BCD system *See* binary coded decimal system.

B cell [IMMUNOL] One of a heterogeneous population of bone-marrow-derived lymphocytes which participates in the immune responses.

b-c fracture [GEOL] A tension fracture parallel with the fabric plane and normal to the *a* axis.

BCG vaccine *See* Bacillus Calmette-Guerin vaccine.

b-c plane [GEOL] A plane that is perpendicular to the plane of movement and parallel to the *b* direction in that plane.

BCS theory *See* Bardeen-Cooper-Schrieffer theory.

BD *See* barrels per day.

B damage [ORD] Combat damage such that the aircraft is unable to return to its base.

BDC *See* bottom dead center.

Bdelloidea [INV ZOO] An order of the class Rotifera comprising animals which resemble leeches in body shape and manner of locomotion.

Bdellomorpha [INV ZOO] An order of ribbonlike worms in the class Enopla containing the single genus *Malacobdella*.

Bdellonemertini [INV ZOO] An equivalent name for the Bdellomorpha.

bdft *See* board-foot.

B display [ELECTR] The presentation of radar output data in rectangular coordinates in which range and azimuth are plotted on the coordinate axes. Also known as range-bearing display.

Bé *See* Baumé hydrometer scale.

Be *See* beryllium.

BE *See* binding energy.

beach [GEOL] The zone of unconsolidated material that extends landward from the low-water line to where there is marked change in material or physiographic form or to the line of permanent vegetation. [NAV] To intentionally run a craft ashore, as a landing ship.

beach cusp *See* cusp.

beach cycle [GEOL] Periodic retreat and outbuilding of beaches resulting from waves and tides.

beach drift [GEOL] The material transported by drifting of beach.

beach face *See* foreshore.

beach gravel [GEOL] Gravels in which most of the particles cluster about one size.

beachhead [ORD] An area on an enemy shoreline which opposing troops have landed upon and occupied.

beach marker [NAV] Usually a light, buoy, panel, or electronic device, used to identify a beach, for incoming water traffic.

beach mining [MIN ENG] The mining of the heavy minerals, such as rutile, zircon, monazite, ilmenite, and sometimes gold, which occur in sand dunes, beaches, coastal plains, and deposits located inland from the shoreline.

beach plain [GEOL] Embankments of wave-deposited material added to a prograding shoreline.

beach platform *See* wave-cut bench.

beach profile [GEOL] Intersection of a beach's ground surface with a vertical plane perpendicular to the shoreline.

beach ridge [GEOL] A continuous mound of beach material behind the beach that was heaped up by waves or other action.

beachrock [PETR] A friable to well-cemented rock made of calcareous skeletal debris that is cemented together by calcium carbonate.

beach scarp [GEOL] A nearly vertical slope along the beach caused by wave erosion.

beacon [NAV] **1.** A light, group of lights, electronic apparatus, or other device which emits identifying signals related to their positions so that the information so produced can be used by the navigator or pilots of aircraft and ships for guidance orientation or warning. **2.** A structure where such a device is mounted or located.

beaconage [NAV] A system of beacons.

beacon delay [ELECTR] The amount of transponding delay

within a beacon, that is, the time between the arrival of a signal and the response of the beacon.

beacon presentation [ELECTR] The radarscope presentation resulting from radio-frequency waves sent out by a radar beacon.

beacon skipping [ELECTR] A condition where transponder return pulses from a beacon are missing at the interrogating radar.

beacon stealing [ELECTR] Loss of beacon tracking by one radar due to stronger signals from an interfering radar.

beacon tracking [ENG] The tracking of a moving object by means of signals emitted from a transmitter or transponder within or attached to the object.

bead [DES ENG] A projecting rim or band. [ELECTROMAG] A glass, ceramic, or plastic insulator through which passes the inner conductor of a coaxial transmission line and by means of which the inner conductor is supported in a position coaxial with the outer conductor. [MET] 1. The drop of precious metal obtained by cupellation in fire assaying. 2. *See* weld bead.

beaded molding [BUILD] A molding or cornice bearing a cast plaster string of beads.

beaded transmission line [ELECTROMAG] Line using beads to support the inner conductor in coaxial transmission lines.

beading [MET] The placing of a bead on a piece of sheet metal for either decorative or strengthening purposes.

bead test [ANALY CHEM] In mineral identification, a test in which borax is fused to a transparent bead, by heating in a blowpipe flame, in a small loop formed by platinum wire; when suitable minerals are melted in this bead, characteristic glassy colors are produced in an oxidizing or reducing flame and serve to identify elements.

bead thermistor [ELEC] A thermistor made by applying the semiconducting material to two wire leads as a viscous droplet, which cements the leads upon firing.

beak [BOT] Any pointed projection, as on some fruits, that resembles a bird bill. [INV ZOO] The tip of the umbo in bivalves. [VERT ZOO] 1. The bill of a bird or some other animal, such as the turtle. 2. A projecting jawbone element of certain fishes, such as the sawfish and pike.

beaker [NAV ARCH] A shipboard vessel, usually used for storing water. [SCI TECH] A deep, open-mouthed, cylindrical vessel with thin walls, which usually has a projecting lip for pouring.

beaker sampler [PETRO ENG] A small, cylindrical vessel with a tapered top used to collect crude oil samples; it is made of low-sparking metal or glass, the bottom is weighted, and there is a small stoppered opening at the top.

beakhead [NAV ARCH] A space forward of the forecastle of a ship where latrines for crewmen are located.

beam [CIV ENG] A body, with one dimension large compared with the other dimensions, whose function is to carry lateral loads (perpendicular to the large dimension) and bending movements. [NAV ARCH] The width of a ship at its widest point. [PHYS] A concentrated, nearly unidirectional flow of particles, or a like propagation of electromagnetic or acoustic waves. [TEXT] Spool-shaped holder, 8 to 12 feet in length, on which is wrapped yarn that is to be transferred to the warp holder of a loom.

beam angle *See* beam width.

Beaman stadia arc [ENG] An attachment to an alidade consisting of a stadia arc on the outer edge of the visual vertical arc; enables the observer to determine the difference in elevation of the instrument and stadia rod without employing vertical angles.

beam antenna [ELECTROMAG] An antenna that concentrates its radiation into a narrow beam in a definite direction.

beam approach beacon system *See* blind approach beacon system.

beam attack [ORD] An attack directed against the side of an aircraft, a tank, or a ship.

beam attenuator [SPECT] An attachment to the spectrophotometer that reduces reference to beam energy to accommodate undersized chemical samples.

beam-balanced pump [PETRO ENG] An oil well pumping unit having a center-pivoted beam with the sucker rod plunger (pump) at the front end and a counterweight on the rearward extension.

beam bracketing *See* beam-rider guidance.

beam bridge [CIV ENG] A fixed structure consisting of a series of steel or concrete beams placed parallel to traffic and supporting the roadway directly on their top flanges.

beam building [MIN ENG] A process of rock bolting in flat-lying deposits where the bolts are installed in bedded rock to bind the strata together to act as a single beam capable of supporting itself, thus stabilizing the overlying rock.

beam-climber guidance *See* beam-rider guidance.

beam column [CIV ENG] A structural member subjected simultaneously to axial load and bending moments produced by lateral forces or eccentricity of the longitudinal load.

beam compass [GRAPHICS] A compass for drawing large circles, consisting of a beam with sliding sockets carrying steel or pencil points.

beam-condensing unit [SPECT] An attachment to the spectrophotometer that condenses and remagnifies the beam to provide reduced radiation at the sample.

beam-deflection amplifier [MECH ENG] A jet-interaction fluidic device in which the direction of a supply jet is varied by flow from one or more control jets which are oriented at approximately 90° to the supply jet.

beam-deflection tube [ELECTR] An electron-beam tube in which current to an output electrode is controlled by transversely moving the electron beam.

beam drop [ELECTROMAG] Distortion of the normal rectilinear fan pattern of a detection radar in which a portion of the fan is at a lower elevation than the rest of the fan.

beam-forming electrode [ELECTR] Electron-beam focusing elements in power tetrodes and cathode-ray tubes.

beam hanger [PETRO ENG] An attachment at the end of a walking beam above a well casing to lift the pump rods or sacked rods.

beam holding [ELECTR] Use of a diffused beam of electrons to regenerate the charges stored on the screen of a cathode-ray storage tube.

beam hole [NUCLEO] A hole through the shield, and usually the reflector, of a nuclear reactor which allows a beam of radiation, especially fast neutrons, to escape for experimental purposes. Also known as glory hole.

beamhouse [CHEM ENG] A place where the initial wet operations of tanning, involving soaking in water and solutions of alkali, are carried out.

beam-indexing picture tube [ELECTR] A type of color television tube in which a single electron beam scans a screen that has three color phosphors coated in vertical stripes, and index signals that indicate the beam position are fed back to the color-switching section to control the electron beam according to color signal. Also known as Apple Tube.

beam lobe switching [ELECTR] Method of determining the direction of a remote object by comparison of the signals corresponding to two or more successive beam angles, differing slightly from the direction of the object.

beam magnet *See* convergence magnet.

beam parametric amplifier [ELECTR] Parametric amplifier that uses a modulated electron beam to provide a variable reactance.

beam pattern *See* directivity pattern.

beam power tube [ELECTR] An electron-beam tube which uses directed electron beams to provide most of its power-handling capability and in which the control grid and screen grid are essentially aligned. Also known as beam tetrode.

beam rider [AERO ENG] A missile for which the guidance system consists of standard reference signals transmitted in a radar beam which enable the missile to sense its location relative to the beam, correct its course, and thereby stay on the beam.

beam-rider guidance [NAV] A system for guiding aircraft or spacecraft in which a craft follows a radar beam, light beam, or other kind of beam along the desired path. Also known as beam bracketing; beam-climber guidance.

beam riding [AERO ENG] The maneuver of a spacecraft or other vehicle as it follows a beam.

beam sea [NAV] Waves moving in a direction approximately 90° from the heading.

beam splice [CIV ENG] A connection between two lengths of a beam or girder; may be shear or moment connections.

BEAM-DEFLECTION AMPLIFIER

A typical beam-deflection amplifier.

BEAR

The polar bear *(Thalarctos maritimus)*; unlike most bears, it is somewhat slender and has good vision.

BEARING

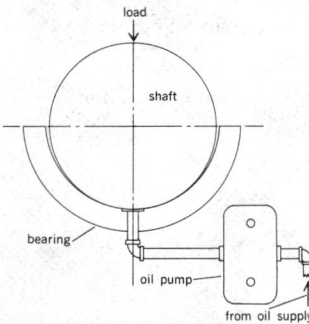

Fluid-film hydrostatic bearing. Hydrostatic oil lift can reduce starting friction drag to less than one-tenth of usual starting drag. *(From W. Staniar, ed., Plant Engineering Handbook, 2d ed., McGraw-Hill, 1959)*

BEAVER

The common beaver *(Castor canadensis)*, showing characteristic webbed hindfeet and broad, flat tail.

beam splitter [OPTICS] A mirror that reflects part of a beam of light falling on it and transmits part.

beam splitting [ELECTR] Process for increasing accuracy in locating targets by radar; by noting the azimuths at which one radar scan first discloses a target and at which radar data from it ceases, beam splitting calculates the mean azimuth for the target. [OPTICS] The division of a beam of light into two beams by placing a special type of mirror in the path of the beam that reflects part of the light falling on it and transmits part.

beam switching [ELECTR] Method of obtaining more accurately the bearing or elevation of an object by comparing the signals received when the beam is in directions differing slightly in bearing or elevation; when these signals are equal, the object lies midway between the beam axes. Also known as lobe switching.

beam-switching tube [ELECTR] An electron tube which has a series of electrodes arranged around a central cathode and in which an electron beam is switched from one electrode to another. Also known as counter tube; counting tube; cyclophon; magnetron beam-switching tube; trochotron.

beam tetrode *See* beam-power tube.

beam well [PETRO ENG] A well pumped by a walking beam.

beam width [ELECTROMAG] The angle, measured in a horizontal plane, between the directions at which the intensity of an electromagnetic beam, such as a radar or radio beam, is one-half its maximum value. Also known as beam angle.

beam wind [NAV] Nautical term for a crosswind, especially a wind blowing 90° from a ship's heading.

bean [BOT] The common name for various leguminous plants used as food for man and livestock; important commercial beans are true beans *(Phaseolus)* and California blackeye *(Vigna sinensis)*.

bean anthracnose [PL PATH] A fungus disease of the bean caused by *Colletotrichum lindemuthianum*, producing pink to brown lesions on the pod and seed and dark discolorations on the veins on the lower surface of the leaf.

bean blight [PL PATH] A bacterial disease of the bean caused by *Xanthomonas phaseoli*, producing water-soaked lesions that become yellowish-brown spots on all plant parts.

bean ore [GEOL] A lenticular, pisolitic aggregate of limonite.

bear [MIN ENG] 1. To underhole or undermine; to drive in at the top or side of a working. [VERT ZOO] The common name for a few species of mammals in the family Ursidae.

beard *See* awn.

bearding [NAV ARCH] 1. Cutting a timber so as to form an inclined surface that fits the angle of a ship's side. 2. The forward edge of a rudder or sternpost.

bearding line [NAV ARCH] A line on the side of the sternpost, keel, deadwoods, and stem of a ship that indicates the intersection of these members with the outer face of the frames.

Bear Driver *See* Boötes.

bearer [GRAPHICS] 1. In photoengraving, the metal left on a plate to protect the printing surface while molding. 2. In composition, one of the type-high slugs locked inside a chase to protect the printing surface. 3. In printing presses, one of the surface-to-surface ends of cylinders that come in contact.

bearing [MECH ENG] A machine part that supports another part which rotates, slides, or oscillates in or on it. [MIN ENG] The direction of a mine drivage, usually given in terms of the horizontal angle turned off a datum direction, such as the true north and south line. [NAV] The horizontal direction from one terrestrial point to another; basically synonymous with azimuth.

bearing angle [NAV] Horizontal direction measured from 0° at the reference direction clockwise or counterclockwise through 90° or 180°.

bearing bar *See* azimuth instrument.

bearing capacity [MECH] Load per unit area which can be safely supported by the ground.

bearing circle [ENG] A ring designed to fit snugly over a compass or compass repeater, and provided with vanes for observing compass bearings.

bearing cursor [ENG] Of a radar set, the radial line inscribed on a transparent disk which can be rotated manually about an axis coincident with the center of the plan position indicator; used for bearing determination. Also known as mechanical bearing cursor.

bearing pile [ENG] A vertical post or pile which carries the weight of a foundation, transmitting the load of a structure to the bedrock or subsoil without detrimental settlement.

bearing plate [CIV ENG] A flat steel plate used under the end of a wall-bearing beam to distribute the load over a broader area.

bearing pressure [MECH] Load on a bearing surface divided by its area. Also known as bearing stress.

bearing resolution [ELECTR] Minimum angular separation in a horizontal plane between two targets at the same range that will allow an operator to obtain data on either target.

bearing strain [MECH] The deformation of bearing parts subjected to a load.

bearing strength [MECH] The maximum load that a column, wall, footing, or joint will sustain at failure, divided by the effective bearing area.

bearing stress *See* bearing pressure.

bearing test [ENG] A test of the bearing capacities of pile foundations, such as a field loading test of an individual pile; a laboratory test of soil samples for bearing capacities.

bear trap gate [CIV ENG] A type of crest gate with an upstream leaf and a downstream leaf which rest in a horizontal position, one leaf overlapping the other, when the gate is lowered.

beat [PHYS] The periodic variation in amplitude of a wave that is the superposition of two simple harmonic waves of different frequencies.

beater [AGR] 1. A device for chopping or pulverizing unwanted parts of crops such as cornstalks or potato vines. 2. The part of a thresher that strikes the grains. [ENG] 1. A tool for packing in material to fill a blasthole containing a charge of powder. 2. A laborer who shovels or dumps asbestos fibers and sprays them with water to prepare them for the beating. [MECH ENG] A machine that cuts or beats paper stock. [TEXT] The section of a loom that drives the weft from the shed into the cloth.

beater mill *See* hammer mill.

beat frequency [ELECTR] The frequency of a signal equal to the difference in frequencies of two signals which produce the signal when they are combined in a nonlinear circuit.

beat-frequency oscillator [ELECTR] An oscillator in which a desired signal frequency, such as an audio frequency, is obtained as the beat frequency produced by combining two different signal frequencies, such as two different radio frequencies. Abbreviated BFO. Also known as heterodyne oscillator.

beating [ENG] A process that reduces asbestos fibers to pulp for making asbestos paper.

beating-in [ELECTR] Interconnecting two transmitter oscillators and adjusting one until no beat frequency is heard in a connected receiver; the oscillators are then at the same frequency.

beat note [ELECTR] The beat frequency whose signal is produced by two signals having waves that are sinusoidal.

beat reception *See* heterodyne reception.

Beattie and Bridgman equation [THERMO] An equation that relates the pressure, volume, and temperature of a real gas to the gas constant.

beat-time programming [ADP] A type of programming which requires that data be made available to the computer during some ongoing process prior to a particular point in time.

beat tone [ENG ACOUS] Musical tone due to beats, produced by the heterodyning of two high-frequency wave trains.

Beaufort force [METEOROL] A number denoting the speed (or so-called strength) of the wind according to the Beaufort wind scale. Also known as Beaufort number.

Beaufort number *See* Beaufort force.

Beaufort wind scale [METEOROL] A system of code numbers from 0 to 12 classifying wind speeds into groups from 0–1 mile per hour (Beaufort 0) to those over 75 miles per hour (Beaufort 12).

beaver [VERT ZOO] The common name for two different and unrelated species of rodents, the mountain beaver *(Aplodontia rufa)* and the true or common beaver *(Castor canadensis)*.

beavertail [ELECTROMAG] Fan-shaped radar beam, wide in the horizontal plane and narrow in the vertical plane, which is swept up and down for height finding.

bebeerine [ORG CHEM] $C_{36}H_{38}N_2O_6$ An alkaloid derived from the bark of the tropical tree *Nectandra rodiaei*; the dextro form is soluble in acetone, the levo form is soluble in benzene and is an antipyretic; the dextro form is also known as chondrodendrin; the levo, as curine.

Béchamp reduction [ORG CHEM] Reduction of nitro groups to amino groups by the use of ferrous salts or iron and dilute acid.

bêche [MECH ENG] A pneumatic forge hammer having an air-operated ram and an air-compressing cylinder integral with the frame.

Becke test [MINERAL] A microscope test in which indices of refraction are compared for minerals; the Becke line appears to move toward the material of higher refractivity as the tube of the microscope is raised.

Beckmann rearrangement [ORG CHEM] An intramolecular change of a ketoxime into its isomeric amide when treated with phosphorus pentachloride.

Beckmann thermometer [ENG] A sensitive thermometer with an adjustable range so that small differences in temperature can be measured.

Becquerel effect [ELEC] The phenomenon of a current flowing between two unequally illuminated electrodes of a certain type when they are immersed in an electrolyte.

Becquerel rays [NUC PHYS] Formerly, radiation emitted by radioactive substances; later renamed alpha, beta, and gamma rays.

bed [GEOL] **1.** The smallest division of a stratified rock series, marked by a well-defined divisional plane from its neighbors above and below. **2.** An ore deposit, parallel to the stratification, constituting a regular member of the series of formations; not an intrusion. [GRAPHICS] The surface of a flatbed printing press on which the chase of composed type is secured for printing. [HYD] The bottom of a channel for the passage of water.

Bedaux plan [IND ENG] A wage incentive plan in which work is standardized into man-minute units called bedaux (B); 60 B per hour is 100% productivity, and earnings are based on work units per length of time.

bedbug [INV ZOO] The common name for a number of species of household pests in the insect family Cimicidae that infest bedding, and by biting humans obtain blood for nutrition.

bedded [GEOL] Pertaining to rocks exhibiting depositional layering or bedding formed from consolidated sediments.

bedded vein [GEOL] A lode occupying the position of a bed that is parallel with the enclosing rock stratification.

bed detector [PETRO ENG] Apparatus to detect and measure the extent of underground formations that are potential oil and gas reservoirs; methods include induction logs, gamma-ray logs, and sonic logs.

bedding [GEOL] Condition where planes divide sedimentary rocks of the same or different lithology.

bedding cleavage [GEOL] Cleavage parallel to the rock bedding.

bedding fault [GEOL] A fault whose fault surface is parallel to the bedding plane of the constituent rocks. Also known as bedding-plane fault.

bedding fissility [GEOL] Primary foliation parallel to the bedding of sedimentary rocks.

bedding joint [GEOL] A joint parallel to the rock bedding.

bedding plane [GEOL] Any of the division planes which separate the individual strata or beds in sedimentary or stratified rock.

bedding-plane fault *See* bedding fault.

bedding-plane slip *See* flexural slip.

bedding schistosity [GEOL] Schistosity that is parallel to the rock bedding.

bedding thrust [GEOL] A thrust fault parallel to bedding.

bedding void [GEOL] A void formed between successive batches of lava that are discharged in a single short activity of a volcano, as well as between flows made a long time apart.

bede [MIN ENG] A miner's pick.

Bedford cord [TEXT] A closely woven, sturdy ribbed fabric

with a raised effect; may be made of wool, cotton, silk, acrylic, or polyesters.

Bedford limestone *See* spergenite.

bediasite [GEOL] A black to brown tektite found in Texas.

bed load [GEOL] Particles of sand, gravel, or soil carried by the natural flow of a stream on or immediately above its bed.

Bedoulian [GEOL] Lower Cretaceous (lower Aptian) geologic time in Switzerland.

bed rock [GEOL] Solid rock exposed at the surface or overlain by unconsolidated material.

Bedsonia [MICROBIO] The psittacosis-lymphogranuloma-trachoma (PLT) group of bacteria belonging to the Chlamydozoaceae; all are obligatory intracellular parasites. Also known as Chlamydia.

bedsore *See* decubitus ulcer.

bedspring array *See* billboard array.

bee [INV ZOO] Any of the membranous-winged insects which compose the superfamily Apoidea in the order Hymenoptera characterized by a hairy body and by sucking and chewing mouthparts.

beech [BOT] Any of various deciduous trees of the genus *Fagus* in the beech family (Fagaceae) characterized by smooth gray bark, triangular nuts enclosed in burs, and hard wood with a fine grain.

beech bark disease [PL PATH] A disease of beech caused by the beech scale (*Cryptococcus fagi*) and a fungus (*Nectria coccinea faginata*) acting together; bark is destroyed, foliage wilts, and the tree eventually dies.

beef [AGR] The flesh of a bovine animal, such as a cow or steer, used as food.

beef stearin *See* oleostearin.

beehive Also known as hive. [AGR] A container that is constructed to house a colony of honeybees. [INV ZOO] A colony of bees.

Beehive *See* Praesepe.

beehive oven [ENG] An arched oven that carbonizes coal into coke by using the heat of combustion of gases that are formed, and of a small part of the coke that is formed, with no recovery of by-products.

beekeeping [AGR] The management and maintenance of colonies of honeybees.

beekite [MINERAL] **1.** A concretionary form of calcite or silica that occurs in small rings on the surface of a fossil shell which has weathered out of its matrix. **2.** White, opaque accretions of silica found on silicified fossils or along joint surfaces as a replacement of organic matter.

beer [FOOD ENG] A lightly hopped, fermented malt beverage brewed by bottom fermentation.

beerbachite [PETR] A hornfels with large poikiloblastic crystals of olivine.

beer drinkers' cardiomyopathy [MED] Congestive heart failure and nonspecific cardiomyopathy presumed due to cobalt added to beer.

Beer-Lambert-Bouguer law *See* Bouguer-Lambert-Beer law.

Beer's law [PHYS CHEM] The law which states that the absorption of light by a solution changes exponentially with the concentration, all else remaining the same.

beer stone [FOOD ENG] A mixture of calcium oxalate and organic material that deposits on container surfaces during brewing operations. Also known as beer scale.

beeswax [MATER] Yellow to grayish-brown solid wax obtained from bee honeycombs by boiling and straining; used in floor waxes, waxed paper, and textile finishes and in pharmacy. Also known as yellow wax.

beet [BOT] *Beta vulgaris*. The red or garden beet, a cool-season biennial of the order Caryophyllales grown for its edible, enlarged fleshy root.

Beethoven exploder [ENG] A machine for the multishot firing of series-connected detonators in tunneling and quarrying.

beetle [INV ZOO] The common name given to members of the insect order Coleoptera. [MIN ENG] A powerful, cable-hauled propulsion unit, operated under remote control, for moving a train of wagons at the mine surface.

beetle stone *See* septarium.

beetling [TEXT] Beating cloth with hammers to force the fibers together and produce a lustrous, linenlike effect.

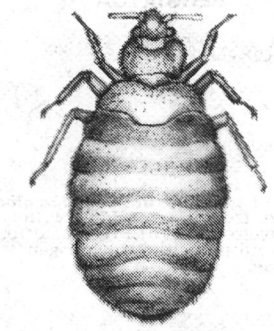

BEDBUG

Cimex lectularius, a species of bedbug parasitic on humans in many temperate and subtropical regions.

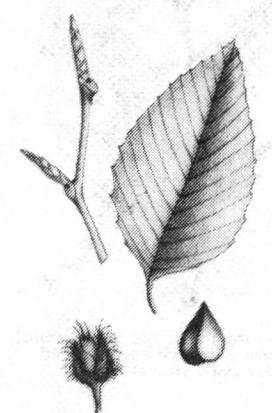

BEECH

American beech (*Fagus grandifolia*), showing slender scaly winter buds, a leaf, an open spiny involucre containing a pair of edible three-sided nuts, and a three-sided nut.

BEGGIATOA

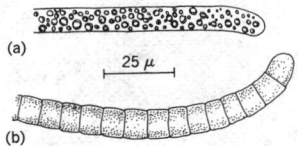

(a)

25 μ

(b)

End of trichome of *Beggiatoa alba*, drawn from life. (a) Filled with sulfur droplets. (b) Free of sulfur, as in culture without hydrogen sulfide.

BELLADONNA

Belladonna (*Atropa belladonna*), flowering branch and isolated flowers (two views).

BELLED CAISSON

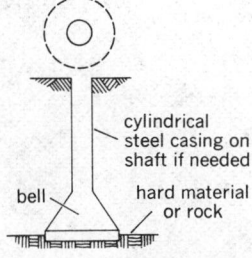

cylindrical steel casing on shaft if needed

bell

hard material or rock

A drawing of a belled caisson, showing top view and cross section from side.

beet sugar [FOOD ENG] Sugar made from sugarbeets by crystalization of the syrup extracted from the roots.

before the wind [NAV] In a direction approximating that toward which the wind is blowing; applies particularly to the situation where the wind is aft, and the craft is being aided by it.

Beggiatoa [MICROBIO] A genus of bacteria occurring as free, motile, segmented trichomes in the family Beggiatoaceae.

Beggiatoaceae [MICROBIO] A family of bacteria in the order Beggiatoales that belong to the physiological group known as the sulfur bacteria.

Beggiatoales [MICROBIO] An order of motile, filamentous, and unicellular bacteria in the class Schizomycetes.

BEGIN [ADP] An enclosing statement of ALGOL used to indicate the beginning of a block; any variable in a block enclosed by BEGIN and END is normally local to this block.

Begoniaceae [BOT] A family of dicotyledonous plants in the order Violales characterized by an inferior ovary, unisexual flowers, stipulate leaves, and two to five carpels.

behavior [PSYCH] Any overt activity of an organism.

behavioral psychophysics [PSYCH] A branch of psychology concerned primarily with the measurement of sensory capacities of normal, intact animals.

behavior disorder *See* abnormal behavior.

behaviorism [PSYCH] A school of psychology concerned with observable, tangible, and measurable data regarding behavior and human activities, but excluding ideas and emotions as purely subjective phenomena.

beheaded stream [HYD] A water course whose upper portion, through erosion, has been cut off and captured by another water course.

behenic acid *See* docosanoic acid.

Behrens-Fisher problem [STAT] The problem of calculating the probability of drawing two random samples whose means differ by some specified value (which may be zero) from normal populations, when one knows the difference of the means of these populations but not the ratio of their variances.

bei function [MATH] One of the functions defined by $ber_n(z) \pm i \ bei_n(z) = J_n(ze^{\pm 3\pi i/4})$, where J_n is the nth Bessel function.

Beilby layer [MET] An amorphous layer formed on a polished metal surface.

bejel [MED] An infectious nonvenereal treponemal disease occurring principally in children in the Middle East.

Békésy audiometry [ACOUS] A subject-controlled auditory threshold testing procedure.

bel [PHYS] A dimensionless unit expressing the ratio of two powers or intensities, or the ratio of a power to a reference power, such that the number of bels is the common logarithm of this ratio. Abbreviated b; B.

belat [METEOROL] A strong land wind from the north or northwest which sometimes blows across the southeastern coast of Arabia and is accompanied by a hazy atmosphere due to sand blown from the interior desert.

belaying pin [NAV ARCH] A pin-shaped metal rod about which ropes are secured on shipboard.

Belemnoidea [PALEON] An order of extinct dibranchiate mollusks in the class Cephalopoda.

Belgian block [MATER] **1.** A stone block used for paving, having the shape of a truncated pyramid, with a depth of 7–8 inches, a base of 5–6 inches square, and a face opposite the base that is 1 inch or less smaller than the base. **2.** Any stone block used for paving.

Belgian retort process *See* horizontal retort process.

B eliminator [ELECTR] Power pack that changes the alternating-current powerline voltage to the direct-current source required by plant circuits of vacuum tubes or semiconductor devices.

Belinuracea [PALEON] An extinct group of horseshoe crabs; arthropods belonging to the Limulida.

belite *See* larnite.

bell [ENG] **1.** A hollow metallic cylinder closed at one end and flared at the other; it is used as a fixed-pitch musical instrument or signaling device and is set vibrating by a clapper or tongue which strikes the lip. **2.** *See* bell tap. [MET] A conical device that seals the top of a blast furnace.

belladonna [BOT] *Atropa belladonna*. A perennial poisonous herb that belongs to the family Solanaceae; atropine is produced from the roots and leaves; used as an antispasmodic, as a cardiac and respiratory stimulant, and to check secretions. Also known as deadly nightshade.

Bellamy drift [NAV] The net drift angle of an aircraft calculated between any two pressure soundings; it is that drift attributable only to the cross-wind component of the wind; the actual drift may be greater or less, but the difference is small for winds normally encountered.

bell and spigot joint [ENG] A pipe joint in which a pipe ending in a bell-like shape is joined to a pipe ending in a spigotlike shape.

Bellatrix [ASTRON] A bluish-white star of stellar magnitude 1.7, spectral classification B2-III, in the constellation Orion; the star γ Orionis.

bell buoy [NAV] A channel buoy with a bell which rings as the buoy swings in the waves, usually marking rocks or shoals.

bell cap [CHEM ENG] A hemispherical or triangular metal casting used on distillation-column trays to force upflowing vapors to bubble through layers of downcoming liquid.

belled caisson [CIV ENG] A type of drilled caisson with a flared bottom.

Bellerophontacea [PALEON] A superfamily of extinct gastropod mollusks in the order Aspidobranchia.

bell glass *See* bell jar.

bell hole [MIN ENG] **1.** One of the holes or excavations made at the section joints of a pipeline for the purpose of repairs. **2.** A conical cavity in a coal mine roof caused by the falling of a large concretion.

bellite [MATER] An explosive consisting of five parts of ammonium nitrate to one of metadinitrobenzene with some potassium nitrate.

bell jar [ENG] A bell-shaped vessel, usually made of glass, which is used for enclosing a vacuum, holding gases, or covering objects. Also known as bell glass.

Bellman's principle of optimality [IND ENG] The principle that an optimal sequence of decisions in a multistage decision process problem has the property that whatever the initial state and decisions are, the remaining decisions must constitute an optimal policy with regard to the state resulting from the first decisions.

bell metal [MET] An alloy of copper and tin, containing 15–25% tin but 20–24% for best tonal quality.

bell-metal ore *See* stannite.

bell mouth [DES ENG] A flared mouth on a pipe opening or other orifice.

bellows [ENG] **1.** A mechanism that expands and contracts, or has a rising and falling top, to suck in air through a valve and blow it out through a tube. **2.** Any of several types of enclosures which have accordionlike walls, allowing one to vary the volume. [METEOROL] *See* aneroid capsule. [OPTICS] An accordionlike component of a camera which forms a passage between the lens and the film and allows one to vary the distance between them.

bellows gage [ENG] A device for measuring pressure in which the pressure on a bellows, with the end plate attached to a spring, causes a measurable movement of the plate.

bellows gas meter [ENG] A device for measuring the total volume of a continuous gas flow stream in which the motion of two bellows, alternately filled with and exhausted of the gas, actuates a register.

bell screw *See* bell tap.

Bell's law [PHYSIO] **1.** The law that in the spinal cord the ventral roots are motor and the dorsal roots sensory in function. **2.** The law that in a reflex arc the nerve impulse can be conducted in one direction only.

bell socket *See* bell tap.

bell tap [ENG] A cylindrical fishing tool having an upward-tapered inside surface provided with hardened threads; when slipped over the upper end of lost, cylindrical, downhole drilling equipment and turned, the threaded inside surface of the bell tap cuts into and grips the outside surface of the lost equipment. Also known as bell; bell screw; bell socket; box bill; die; die collar; die nipple.

bell-type manometer [ENG] A differential pressure gage in which one pressure input is fed into an inverted cuplike container floating in liquid, and the other pressure input presses down upon the top of the container so that its level in the liquid is the measure of differential pressure.

belly [ANAT] **1.** The abdominal cavity or abdomen. **2.** The most prominent, fleshy, central portion of a muscle.

belonephobia [PSYCH] Abnormal fear of sharp-pointed objects.

Beloniformes [VERT ZOO] The former ordinal name for a group of fishes now included in the order Atheriniformes.

belonite [GEOL] A rod- or club-shaped microscopic embryonic crystal in a glassy rock.

Belostomatidae [INV ZOO] The giant water bugs, a family of hemipteran insects in the subdivision Hydrocorisae.

below minimums [METEOROL] Below operational weather limits for aircraft.

belt [ECOL] **1.** Any altitudinal vegetation zone or band from the base to the summit of a mountain. **2.** Any benthic vegetation zone or band from sea level to the ocean depths. **3.** Any of the concentric vegetation zones around bodies of fresh water. [MECH ENG] A flexible band used to connect pulleys or to convey materials by transmitting motion and power.

belt conveyor [MECH ENG] A heavy-duty conveyor consisting essentially of a head or drive pulley, a take-up pulley, a level or inclined endless belt made of canvas, rubber, or metal, and carrying and return idlers.

belt drive [MECH ENG] The transmission of power between shafts by means of a belt connecting pulleys on the shafts.

belted plain [GEOL] A plain whose surface has been slowly worn down and sculptured into bands or belts of different levels.

belteroporic [GEOL] Of crystals in rocks whose growth was determined by the direction of easiest growth.

belt feeder [MECH ENG] A short belt conveyor used to transfer granulated or powdered solids from a storage or supply point to an end-use point; for example, from a bin hopper to a chemical reactor.

belt grinding [MET] Grinding with an abrasive-coated continuous belt. Also known as linishing.

belting [MATER] **1.** A sturdy fabric, usually of cotton, used in belts. **2.** A heavy leather, made from hides of cattle, used in power transmission belts. Also known as belting leather.

belting leather *See* belting.

belt of cementation *See* zone of cementation.

belt of soil moisture *See* belt of soil water.

belt of soil water [GEOL] The upper subdivision of the zone of aeration limited above by the land surface and below by the intermediate belt; this zone contains plant roots and water available for plant growth. Also known as belt of soil moisture; discrete film zone; soil-water belt; soil-water zone; zone of soil water.

belt slip [MECH ENG] The difference in speed between the driving drum and belt conveyor.

belt tightener [MECH ENG] In a belt drive, a device that takes up the slack in a belt that has become stretched and permanently lengthened.

Bemberg [MATER] Trade name of a synthetic fiber made from cellulose by Beaunit Corporation.

Bénard convection cells [PHYS] A regular array of hexagonal cells which sometimes appear in convection in a layer of liquid heated from below.

Bence-Jones protein [PATH] An abnormal group of globulins appearing in the serum and urine, usually in association with multiple myeloma and characterized by coagulation at 50–60°C.

Bence-Jones proteinuria [MED] The presence of Bence-Jones protein in the urine.

bench [GEOL] A terrace of level earth or rock that is raised and narrow and that breaks the continuity of a declivity. [MIN ENG] **1.** One of two or more divisions of a coal seam, separated by slate and so forth or simply separated by the process of cutting the coal, one bench or layer being cut before the adjacent one. **2.** A long horizontal ledge of ore in an underground working place. **3.** A ledge in an open-pit mine from which excavation takes place at a constant level.

bench blasting [MIN ENG] A mining system used either underground or in surface pits whereby a thick ore or waste zone is removed by blasting a series of successive horizontal layers called benches.

bench check [IND ENG] A workshop or servicing bay check which includes the typical check or actual functional test of an item to ascertain what is to be done to return the item to a serviceable condition or ascertain the item's temporary or permanent disposition.

bench gravel [GEOL] Gravel beds found on the sides of valleys above the present stream bottoms, representing parts of the bed of the stream when it was at a higher level.

bench height *See* bank height.

benching [MIN ENG] A method of working small quarries or opencast pits in steps or benches.

bench lathe [MECH ENG] A small engine or toolroom lathe suitable for attachment to a workbench; bed length usually does not exceed 6 feet and workpieces are generally small.

bench lava [GEOL] Semiconsolidated, crusted basaltic lava forming raised platforms and crags about the edges of lava lakes. Also known as bench magma.

bench magma *See* bench lava.

bench mark [ENG] A relatively permanent natural or artificial object bearing a marked point whose elevation above or below an adopted datum, such as sea level, is known. Abbreviated BM.

bench-mark problem [ADP] A problem to be run on computers to evaluate their performances relative to one another.

bench photometer [ENG] A device which uses an optical bench with the two light sources to be compared mounted one at each end; the comparison between the two illuminations is made by a device moved along the bench until matching brightnesses appear.

bench placer [GEOL] A placer in ancient stream deposits from 50 to 300 feet above present streams.

bench-scale testing [ENG] Testing of materials, methods, or chemical processes on a small scale, such as on a laboratory worktable.

bench slope *See* bank slope.

bend [DES ENG] **1.** The characteristic of an object, such as a machine part, that is curved. **2.** A section of pipe that is curved. **3.** A knot formed by a rope fastened to an object or another rope. [ELECTROMAG] A smooth change in the direction of the longitudinal axis of a waveguide. [NAV] The departure of a defined navigation course or bearing from a straight line.

bend allowance [DES ENG] Length of the arc of the neutral axis between the tangent points of a bend in any material.

Benday plate [GRAPHICS] A printing plate treated with the Benday process.

Ben Day process *See* Benday process.

Benday process [GRAPHICS] A process for printing shadings consisting of patterns of lines, dots, stipples, and so on, which involves inking a Benday screen (a rectangle of hardened gelatin with the pattern in relief), printing it on portions of the metal plate on which an outline drawing has been photoprinted, and then etching the metal as a line plate. Also spelled Ben Day process.

bender *See* bending machine.

bender element [ELECTR] A combination of two thin strips of different piezoelectric materials bonded together so that when a voltage is applied, one strip increases in length and the other becomes shorter, causing the combination to bend.

Bender process [CHEM ENG] Treatment of petroleum liquids with lead sulfide catalyst to sweeten (to convert mercaptans to disulfides) by oxidation.

bending [ENG] **1.** The forming of a metal part, by pressure, into a curved or angular shape, or the stretching or flanging of it along a curved path. **2.** The forming of a wooden member to a desired shape by pressure after it has been softened or plasticized by heat and moisture.

bending brake [MECH ENG] A press brake for making sharply angular linear bends in sheet metal.

bending machine [MECH ENG] A machine for bending a metal or wooden part by pressure. Also known as bender.

bending moment [MECH] The algebraic sum of all moments located between a cross section and one end of a structural member; a bending moment that bends the beam convex

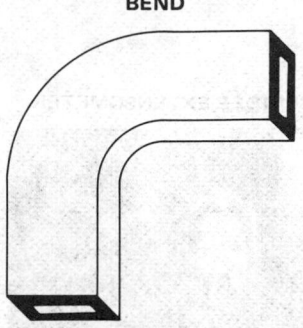

BEND

H-plane 90° bend for a rectangular waveguide.

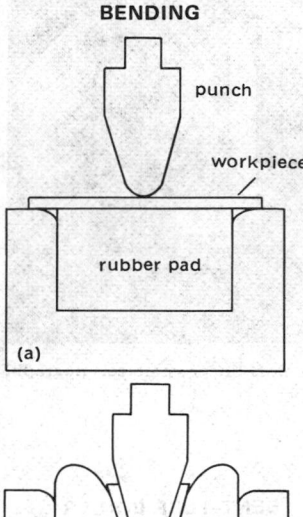

BENDING

punch

workpiece

rubber pad

(a)

(b)

The bending process involved in sheet-metal forming using a rubber pad. (a) Before forming. (b) After forming.

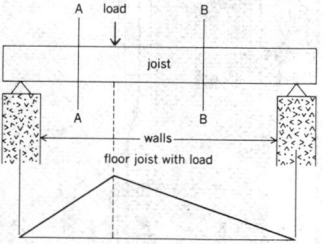

BENDING MOMENT

A load B

joist

A B

walls

floor joist with load

Schematic of bending moment on an end-supported joist with a concentrated load. Bending moment at section A-A is upward reactive force of left wall times distance from wall to A-A. Bending moment at section B-B is upward reactive force of left wall times the distance from wall to B-B plus the moment produced by the downward load acting at its distance from B-B. The resulting bending moment at each cross section is shown on the lower portion of the diagram.

downward is positive, and one that bends it convex upward is negative.

bending moment diagram [MECH] A diagram showing the bending moment at every point along the length of a beam plotted as an ordinate.

bending shackle *See* anchor shackle.

bending slab [NAV ARCH] A slab that is made up of a number of large cast-iron blocks with holes for pins about which frames and other structural members of ships are bent.

bending stress [MECH] An internal tensile or compressive longitudinal stress developed in a beam in response to curvature induced by an external load.

Bendix-Weiss universal joint [MECH ENG] A universal joint that provides for constant angular velocity of the driven shaft by transmitting the torque through a set of four balls lying in the plane that contains the bisector of, and is perpendicular to, the plane of the angle between the shafts.

bend radius [DES ENG] The radius corresponding to the curvature of a bent specimen or part, as measured at the inside surface of the bend.

bends *See* caisson disease.

bend test [MET] A ductility test in which a specimen is bent through an arc of known radius and angle.

bend wheel [MECH ENG] A wheel used to interrupt and change the normal path of travel of the conveying or driving medium; most generally used to effect a change in direction of conveyor travel from inclined to horizontal or a similar change.

Benedict equation of state [PHYS CHEM] An empirical equation relating pressures, temperatures, and volumes for gases and gas mixtures; superseded by the Benedict-Webb-Rubin equation of state.

Benedict's solution [ANALY CHEM] A solution of potassium and sodium tartrates, copper sulfate, and sodium carbonate; used to detect reducing sugars.

beneficiation [MET] Improving the chemical or physical properties of an ore so that metal can be recovered at a profit. Also known as mineral dressing.

Bengal lights [MATER] Constituent of pyrotechnics made of sulfur, sugar, and potassium nitrate, with barium or strontium salts added for green or red light.

Benger unit [BIOL] A unit for the standardization of hyaluronidase.

Benguela Current [OCEANOGR] A strong current flowing northward along the southwestern coast of Africa.

Benham top [OPTICS] A disk whose surface has black and white portions and which, when rotated at certain speeds and subjected to certain lighting, produces sensations of color.

benign [MED] Of no danger to life or health.

benign lymphoreticulosis *See* cat scratch disease.

benign myalgic encephalomyelitis *See* neuromyasthenia.

benign tertian malaria *See* vivax malaria.

benign tumor [MED] A nonmalignant neoplasm.

Benioff extensometer [ENG] A linear strainmeter for measuring the change in distance between two reference points separated by 20–30 meters or more; used to observe earth tides.

Benioff seismic zone [GEOPHYS] A plane defined by the locus of earthquake foci dipping at an angle of about 45° from ocean trenches toward and beneath adjacent continents.

benitoite [MINERAL] $BaTi(SiO_3)_3$ A blue to violet barium-titanium silicate mineral; at one time it was cut and sold as sapphire.

benjamin gum *See* benzoin gum.

benne oil *See* sesame oil.

Bennettitales [PALEOBOT] An equivalent name for the Cycadeoidales.

Bennettitatae [PALEOBOT] A class of fossil gymnosperms in the order Cycadeoidales.

ben oil [MATER] Oil from seeds of ben (*Moringa*) trees; used in foods, cosmetics, perfumes, and laxatives and as a lubricant.

bent [CIV ENG] A framework support transverse to the length of a structure.

bent-back occlusion [METEOROL] An occluded front that has reversed its direction of motion as a result of the development of a new cyclone (usually near the point of occlusion) or, less

frequently, as the result of the displacement of the old cyclone along the front. Also known as back-bent occlusion.

benthic [OCEANOGR] Of, pertaining to, or living on the bottom or at the greatest depths of a large body of water. Also known as benthonic.

benthonic *See* benthic.

benthos [ECOL] Bottom-dwelling forms of marine life. Also known as bottom fauna. [OCEANOGR] The floor or deepest part of a sea or ocean.

bentonite [GEOL] A clay formed from volcanic ash decomposition and largely composed of montmorillonite and beidellite. Also known as taylorite.

bentonite slurry [MATER] Slurry composed of water and clay powder consisting chiefly of montmorillonite minerals.

bent-tube boiler [MECH ENG] A water-tube steam boiler in which the tubes terminate in upper and lower steam-and-water drums. Also known as drum-type boiler.

bentu de soli [METEOROL] An east wind on the coast of Sardinia.

benzalacetaldehyde *See* cinnamaldehyde.

benzalacetic acid *See* cinnamic acid.

benzal chloride [ORG CHEM] $C_6H_5CHCl_2$ A colorless liquid that is refractive and fumes in air; boiling point 207°C; used to make benzaldehyde and cinnamic acid. Also known as benzylidene chloride; chlorobenzal.

benzaldehyde [ORG CHEM] C_6H_5CHO A colorless, liquid aldehyde, boiling at 170°C and possessing the odor of bitter almonds; used as a flavoring agent and an intermediate in chemical syntheses.

benzaldoxime [ORG CHEM] C_6H_5CHNOH An oxime of benzaldehyde; the antiisomeric form melts at 130°C, the syn form at 34°C; both forms are soluble in ethyl alcohol and ether; used in synthesis of other organic compounds.

benzalkonium [ORG CHEM] $C_6H_5CH_2N(CH_3)_2R^+$ An organic radical in which R may range from C_8H_{17} to $C_{18}H_{37}$; found in surfactants, as the chloride salt.

benzamide [ORG CHEM] $C_6H_5CONH_2$ A compound with melting point 132.5° to 133.5°C; slightly soluble in water, soluble in ethyl alcohol and carbon tetrachloride; used in chemical synthesis.

benzanilide [ORG CHEM] $C_6H_5CONHC_6H_5$ An antipyretic, melting point 163°C; insoluble in water, slightly soluble in alcohol. Also known as *N*-phenyl benzoic acid amide.

benzanthracene [ORG CHEM] $C_{18}H_{14}$ A weakly carcinogenic material that is isomeric with naphthacene; melting point 162°C; insoluble in water, soluble in benzene. Also known as 1,2-benzanthracene; 2,3-benzophenanthrene.

1,2-benzanthracene *See* benzanthracene.

benzanthrone [ORG CHEM] $C_{17}H_{10}O$ A compound with melting point 170°; insoluble in water; used in dye manufacture.

Benzedrine *See* amphetamine.

benzene [ORG CHEM] C_6H_6 A colorless, liquid, flammable, aromatic hydrocarbon that boils at 80.1°C and freezes at 5.4–5.5°C; used to manufacture styrene and phenol. Also known as benzol.

benzene azimide *See* 1,2,3-benzotriazole.

benzene carbonitrile *See* benzonitrile.

benzene diazonium [ORG CHEM] $C_6H_5N_2^+$ A cation found in salts such as benzenediazonium chloride, $C_6H_5N_2Cl$.

benzene-*ortho*-dicarboxylic acid *See* phthalic acid.

benzene-*para*-dicarboxylic acid *See* terephthalic acid.

benzene hexachloride [ORG CHEM] $C_6H_6Cl_6$ An alicyclic compound which forms colorless to yellow crystals or flakes; made by direct chlorination of benzene, using a mercury-vapor lamp to start the reaction.

benzene ring [ORG CHEM] The six-carbon ring structure found in benzene, C_6H_6, and in organic compounds formed from benzene by replacement of one or more hydrogen atoms by other chemical atoms or radicals.

benzene series [ORG CHEM] A series of carbon-hydrogen compounds based on the benzene ring, with the general formula C_nH_{2n-6}, where n is 6 or more; examples are benzene, C_6H_6; toluene, C_7H_8; and xylene, C_8H_{10}.

benzenesulfonate [ORG CHEM] Any salt or ester of benzenesulfonic acid.

benzenesulfonic acid [ORG CHEM] $C_6H_5SO_3H$ An organosulfur compound, strongly acidic, water soluble, nonvolatile,

BENIOFF EXTENSOMETER

Benioff linear strain seismograph.

BENT-TUBE BOILER

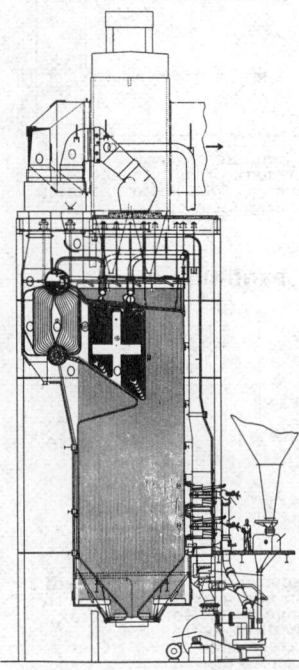

Diagram of a bent-tube boiler.

and hygroscopic; used in the manufacture of detergents and phenols.

benzene-toluene-xylene [MATER] An admixture of three aromatic chemicals formed during the catalytic reforming of petroleum naphthas. Abbreviated BTX.

benzene-1,2,3-tricarboxylic acid See hemimellitic acid.

benzenyl trichloride See benzotrichloride.

benzestrol [PHARM] $C_{20}H_{26}O_2$ A white powder; melting point 162–166°C; soluble in ethanol and acetone, insoluble in water; used as a medicine. Also known as 2,4-bis(4-hydroxyphenyl)-3-ethyl hexane.

benzhydrol [ORG CHEM] $(C_6H_5)_2CHOH$ Colorless needles; melting point 69°C; slightly soluble in water, very soluble in ethanol and ether; used in preparation of other organic compounds including antihistamines. Also known as diphenyl methanol.

benzidine [ORG CHEM] $NH_2C_6H_4C_6H_4NH_2$ An aromatic amine with a melting point of 128°C; used as an intermediate in syntheses of direct dyes for cotton.

benzil [ORG CHEM] $C_6H_5COCOC_6H_5$ A yellow powder; melting point 95°C; insoluble in water, soluble in ethanol, ether, and benzene; used in organic synthesis. Also known as dibenzoyl; diphenylglyoxal.

benzilic acid [ORG CHEM] $(C_6H_5)_2C(OH)CO_2H$ A white, crystalline acid, synthesized by heating benzil with alcohol and potassium hydroxide; used in organic synthesis.

benzimidazole [ORG CHEM] $C_7H_6N_2$ Colorless crystals; melting point 170°C; slightly soluble in water, soluble in ethanol; used in organic synthesis. Also known as 1,3-benzodiazole; benzoglioxaline.

benzin See petroleum benzin.

benzine See petroleum benzin.

benzoate [ORG CHEM] A salt or ester of benzoic acid, formed by replacing the acidic hydrogen of the carboxyl group with a metal or organic radical.

benzocaine See ethyl-*para*-amino benzoate.

1,4-benzodiazine See quinoxaline.

2,3-benzodiazine See phthalazine.

β-benzo-*ortho*-diazine See phthalazine.

benzo-*para*-diazine See quinoxaline.

1,3-benzodiazole See benzimidazole.

benzofuran See coumarone.

benzoglioxaline See benzimidazole.

benzoglycolic acid See mandelic acid.

benzoic acid [ORG CHEM] C_6H_5COOH An aromatic carboxylic acid that melts at 122.4°C, boils at 250°C, and is slightly soluble in water and relatively soluble in alcohol and ether; derivatives are valuable in industry, commerce, and medicine.

benzoic anhydride [ORG CHEM] $(C_6H_5CO)_2O$ An acid anhydride that melts at 42°C, boils at 360°C, and crystallizes in colorless prisms; used in synthesis of a variety of organic chemicals, including some dyes.

benzoic trichloride See benzotrichloride.

benzoin [MATER] A balsamic resin obtained from trees of the genus *Styrax;* used as an expectorant, as an inhalant in respiratory tract inflammations, and as an antiseptic. Also known as benjamin gum; benzoinam; gum benzoin. [ORG CHEM] $C_{14}H_{12}O_2$ An optically active compound; white or yellowish crystals, melting point 137°C; soluble in acetone, slightly soluble in water; used in organic synthesis. Also known as benzoylphenyl carbinol; 2-hydroxy-2-phenyl acetophenone; phenyl benzoyl carbinol.

benzoinam See benzoin.

benzol See benzene.

benzoline See normal benzine.

benzonitrile [ORG CHEM] C_6H_5CN A colorless liquid with an almond odor; made by heating benzoic acid with lead thiocyanate and used in the synthesis of organic chemicals. Also known as benzene carbonitrile; phenyl cyanide.

2,3-benzophenanthrene See benzanthracene.

benzophenone [ORG CHEM] $C_6H_5COC_6H_6$ A diphenyl ketone, boiling point 305.9°C, occurring in four polymorphic forms (α, β, γ, and δ) each with different melting point; used as a constituent of synthetic perfumes and as a chemical intermediate. Also known as diphenyl ketone; phenyl ketone.

1,2-benzopyrone See coumarin.

2,3-benzopyrrole See indole.

benzoquinone See quinone.

benzoresorcinol [ORG CHEM] $C_{13}H_{10}O_3$ A compound crystallizing as needles from hot-water solution; used in paints and plastics as an ultraviolet light absorber. Also known as resbenzophenone.

benzosulfimide See saccharin.

benzothiazole [ORG CHEM] C_6H_4SCHN A thiazole fused to a benzene ring; can be made by ring closure from *o*-amino thiophenols and acid chlorides; derivatives are important industrial products. Also known as benzopyrene.

1,2,3-benzotriazole [ORG CHEM] $C_6H_5N_3$ A compound with melting point 98.5°C; soluble in ethanol, insoluble in water; derivatives are ultraviolet absorbers; used as a chemical intermediate. Also known as azimidobenzene; benzene azimide.

benzotrichloride [ORG CHEM] $C_6H_5CCl_3$ A colorless to yellow liquid that fumes upon exposure to air; has penetrating odor; insoluble in water, soluble in ethanol and ether; used to make dyes. Also known as benzenyl trichloride; benzoic trichloride; phenylchloroform; toluene trichloride.

benzoyl [ORG CHEM] The radical C_6H_5CO- found, for example, in benzoyl chloride.

benzoyl chloride [ORG CHEM] C_6H_5COCl Colorless liquid whose vapor induces tears; soluble in ether, decomposes in water; used as an intermediate in chemical synthesis.

benzoyl peroxide [ORG CHEM] $(C_6H_5CO)_2O_2$ A white, crystalline solid; melting point 103–105°C; explodes when heated above 105°C; slightly soluble in water, soluble in organic solvents; used as a bleaching and drying agent and a polymerization catalyst. Also known as dibenzoyl peroxide.

3,4-benzpyrene [ORG CHEM] $C_{20}H_{12}$ A polycyclic hydrocarbon; a chemical carcinogen that will cause skin cancer in many species when applied in low dosage.

benzyl [ORG CHEM] The radical $C_6H_5CH_2-$ found, for example, in benzyl alcohol, $C_6H_5CH_2OH$.

benzyl acetate [ORG CHEM] $C_6H_5CH_2OOCCH_3$ A colorless liquid with a flowery odor; used in perfumes and flavorings and as a solvent for plastics and resins, inks, and polishes. Also known as phenylmethyl acetate.

benzyl alcohol [ORG CHEM] $C_6H_5CH_2Cl$ An alcohol that melts at 15.3°C, boils at 205.8°C, and is soluble in water and readily soluble in alcohol and ether; valued for the esters it forms with acetic, benzoic, and sebacic acids and used in the soap, perfume, and flavor industries.

benzylamine [ORG CHEM] $C_6H_5CH_2CH_2$ A liquid that is soluble in water, ethanol, and ether; boils at 185°C (770 mm Hg) and at 84°C (24 mm Hg); it is toxic; used as a chemical intermediate in dye production. Also known as aminotoluene.

benzylbenzene See diphenyl methane.

benzyl benzoate [ORG CHEM] $C_6H_5COOCH_2C_6H_5$ An oily, colorless liquid ester; used as an antispasmodic drug and as a scabicide.

benzyl carbinol See phenethyl alcohol.

benzyl chloride [ORG CHEM] C_6H_5COCl An acid halide, a colorless liquid with a pungent odor; a strong lacrimator.

benzyl cinnamate [ORG CHEM] $C_8H_7COOCH_2C_6H_5$ White crystals; melting point 39°C; insoluble in water, soluble in ethanol; used in perfumery. Also known as cinnamein.

benzyl cyanide [ORG CHEM] $C_6H_5CH_2CN$ A toxic, colorless liquid; insoluble in water, soluble in alcohol and ethanol; boils at 234°C; used in organic synthesis. Also known as phenylacetonitrile; α-tolunitrile.

benzyl dichloride See benzal chloride.

benzyleneglycol See hydrobenzoin.

benzyl ether [ORG CHEM] $(C_6H_5CH_2)_2O$ A liquid unstable at room temperature; boiling point 295–298°C; used in perfumes and as a plasticizer for nitrocellulose. Also known as dibenzyl ether.

1-benzyl-3-ethyl-6,7-dimethoxy isoquinoline See eupavarine.

benzylhydroperoxide See perbenzoic acid.

benzylidene chloride See benzal chloride.

benzyl penicillinic acid [ORG CHEM] $C_{16}H_{18}N_2O_4S$ An

BENZOTHIAZOLE

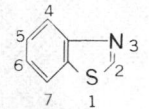

Structural formula of benzothiazole.

amorphous white powder extracted with ether or chloroform from an acidified aqueous solution of benzyl penicillin.

benzyl penicillin potassium [MICROBIO] $C_{16}H_{17}KN_2O_4S$ Moderately hygroscopic crystals; soluble in water; inactivated by acids and alkalies; obtained from fermentation of *Penicillium chrysogenum*; used as an antimicrobial drug in human and animal disease. Also known as penicillin G_1 potassium; potassium benzyl penicillinate; potassium penicillin G_1.

benzyl penicillin sodium [MICROBIO] $C_{16}H_{17}N_2NaO_4S$ Crystals obtained from a methanol-ethyl acetate acidified extract of fermentation broth of *Penicillium chrysogenum*; used as an antimicrobial in human and animal disease. Also known as penicillin; sodium benzyl penicillinate; sodium penicillin G_1.

benzyl salicylate [ORG CHEM] $C_{14}H_{12}O_3$ A thick liquid with a slight, pleasant odor; used as a fixer in perfumery and in sunburn preparations.

benzyne [ORG CHEM] C_6H_4 An unstable chemical intermediate which has not been isolated; it is inferred from indirect evidence.

Beranek scale [ACOUS] A scale which measures the subjective loudness of a noise; noises are arranged into six arbitrary categories: very quiet, quiet, moderately quiet, noisy, very noisy, and intolerably noisy.

berbamine [ORG CHEM] $C_{37}H_{40}N_2O_6$ An alkaloid; melting point 170°C; slightly soluble in water, soluble in alcohol and ether.

Berberidaceae [BOT] A family of dicotyledonous herbs and shrubs in the order Ranunculales characterized by alternate leaves, perfect, well-developed flowers, and a seemingly solitary carpel.

berberine [ORG CHEM] $C_{20}H_{19}NO_5$ A toxic compound; melting point 145°C; the anhydrous form is insoluble in water, soluble in alcohol and ether; an alkaloid whose sulfate or hydrochloride might have some use in medicine.

Berenice's Hair *See* Coma Berenices.

ber function [MATH] One of the functions defined by $ber_n(z) \pm i\ bei_n(z) = J_n(ze^{\pm 3\pi i/4})$, where J_n is the nth Bessel function.

bergamot oil [MATER] A yellow-green essential oil from the rind of bergamot (*Citrus bergamia*) fruit; it is volatile, contains linalyl acetate, limonene, and livalol, and is used in perfumes.

berg crystal *See* rock crystal.

Bergeron-Findeisen theory [METEOROL] The theoretical explanation that precipitation particles form within a mixed cloud (composed of both ice crystals and liquid water drops) because the equilibrium vapor pressure of water vapor with respect to ice is less than that with respect to liquid water at the same temperature. Also known as ice-crystal theory; Wedener-Bergeron process.

Bergius process [CHEM ENG] Treatment of carbonaceous matter, such as coal or cellulosic materials, with hydrogen at elevated pressures and temperatures in the presence of a catalyst, to form an oil similar to crude petroleum. Also known as coal hydrogenation.

Bergmann's rule [ECOL] The principle that in a polytypic wide-ranging species of warm-blooded animals the average body size of members of each geographic race varies with the mean environmental temperature.

Bergman-Turner unit [BIOL] A unit for the standardization of thyroid extract.

bergmehl *See* rock milk.

bergschrund [HYD] A type of crevice in a glacier; formed when ice and snow break away from a rock face.

Berg's diver method *See* diver method.

berg till *See* floe till.

beriberi [MED] A disorder resulting from the deficiency of vitamin B_1 and characterized by neurologic symptoms, cardiovascular abnormalities, edema, and cerebral manifestations.

Bering Sea [GEOGR] A body of water north of the Pacific Ocean, bounded by Siberia, Alaska, and the Aleutian Islands.

Berkefeld filter [MICROBIO] A diatomaceous-earth filter used for sterilization of heat-labile liquids, such as blood serum, enzyme solutions, and antibiotics.

berkelium [CHEM] A radioactive element, symbol Bk, atomic number 97, the eighth member of the actinide series; properties resemble those of the rare-earth cerium.

berkeyite *See* lazulite.

Berl saddle [CHEM ENG] A type of column packing used in distillation columns.

berm [CIV ENG] A horizontal ledge cut between the foot and top of an embankment to stabilize the slope by intercepting sliding earth. [GEOL] **1.** A narrow terrace which originates from the interruption of an erosion cycle with rejuvenation of a stream in the mature stage of its development and renewed dissection. **2.** A horizontal portion of a beach or backshore formed by deposit of material as a result of wave action. Also known as backshore terrace; coastal berm.

berm crest [GEOL] The seaward limit and usually the highest spot on a coastal berm. Also known as berm edge.

berm edge *See* berm crest.

Bermuda grass [BOT] *Cynodon dactylon*. A long-lived perennial in the order Cyperales.

Bermuda high [METEOROL] The semipermanent subtropical high of the North Atlantic Ocean, especially when it is located in the western part of that ocean area.

Bernoulli differential equation *See* Bernoulli equation.

Bernoulli distribution *See* binomial distribution.

Bernoulli effect [FL MECH] As a consequence of the Bernoulli theorem, the pressure of a stream of fluid is reduced as its speed of flow is increased.

Bernoulli equation [FL MECH] *See* Bernoulli theorem. [MATH] A nonlinear first-order differential equation of the form $(dy/dx) + yf(x) = y^n g(x)$, where n is a number different from unity and f and g are given functions. Also known as Bernoulli differential equation.

Bernoulli-Euler law [MECH] A law stating that the curvature of a beam is proportional to the bending moment.

Bernoulli law *See* Bernoulli theorem.

Bernoulli number [MATH] Numerical value of the coefficient of $x^{2n}/(2n)!$ is the expansion of $xe^x/(e^x - 1)$.

Bernoulli polynomial [MATH] The nth one is

$$\sum_{k=0}^{n} \binom{n}{k} B_k Z^{n-k}$$

where $\binom{n}{k}$ is a binomial coefficient, and B_k is a Bernoulli number.

Bernoulli's lemniscate [MATH] A curve shaped like a figure eight whose equation in rectangular coordinates is expressed as $(x^2 + y^2)^2 = a^2(x^2 - y^2)$.

Bernoulli theorem [FL MECH] An expression of the conservation of energy in the steady flow of an incompressible, inviscid fluid; it states that the quantity $(p/\rho) + gz + (v^2/2)$ is constant along any streamline, where p is the fluid pressure, v is the fluid velocity, ρ is the mass density of the fluid, g is the acceleration due to gravity, and z is the vertical height. Also known as Bernoulli equation; Bernoulli law. [STAT] *See* law of large numbers.

Beroida [INV ZOO] The single order of the class Nuda in the phylum Ctenophora.

Berriasian [GEOL] Part of or the underlying stage of the Valanginian at the base of the Cretaceous.

berry [BOT] A usually small, simple, fleshy or pulpy fruit, such as a strawberry, grape, tomato, or banana.

berrylate [INORG CHEM] A salt such as potassium beryllate, formed by action of a strong base, such as potassium hydroxide, and beryllium oxide.

berth deck [NAV ARCH] **1.** The deck on a warship where hammocks were formerly swung. **2.** A space in which lie the crew's sleeping quarters.

Berthelot equation [PHYS CHEM] A form of the equation of state which relates the temperature, pressure, and volume of a gas with the gas constant.

Berthelot relation [PHYS] A relationship between molecular attraction constants of like and unlike species.

Berthelot-Thomsen principle [PHYS CHEM] The principle that of all chemical reactions possible, the one developing the greatest amount of heat will take place, with certain obvious exceptions such as changes of state.

bertholide [CHEM] A name once proposed for a compound of variable proportions.

Bertrand curve [MATH] One of a pair of curves having the

BERKELIUM

97
Bk

Periodic table of the chemical elements showing the position of berkelium.

BERL SADDLE

The shape of the Berl saddle.

BERNOULLI'S LEMNISCATE

Curve known as Bernoulli's lemniscate.

same principal normals. Also known as associate curve; conjugate curve.

bertrandite [MINERAL] $Be_4Si_2O_7(OH)_2$ A colorless or pale-yellow mineral consisting of a beryllium silicate occurring in prismatic crystals; hardness is 6–7 on Mohs scale, and specific gravity is 2.59–2.60.

Bertrand lens [OPTICS] An auxiliary lens that can be inserted in the tube of a polarizing microscope to obtain interference figures.

Bertrand's postulate [MATH] The proposition that there exists at least one prime number between any integer greater than three and twice the integer minus two.

Bertrand's rule [MICROBIO] The rule stating that in those compounds, and only in those compounds, having cis secondary alcoholic groups containing at least one carbon atom of D configuration which is subtended by a primary alcohol group, or having a methyl-substituted primary alcohol group of D configuration, the D-carbon atom will be dehydrogenated by the vinegar bacteria *Acetobacter suboxydans,* yielding a ketone.

Beryciformes [VERT ZOO] An order of actinopterygian fishes in the infraclass Teleostei.

Berycomorphi [VERT ZOO] An equivalent name for the Beryciformes.

beryllia See beryllium oxide.

beryllide [INORG CHEM] A chemical combination of beryllium with a metal, such as zirconium or tantalum.

berylliosis [MED] Chronic lung inflammation due to inhalation of beryllium oxide dust.

beryllium [CHEM] A chemical element, symbol Be, atomic number 4, atomic weight 9.0122. [MET] A rare metal, occurring naturally in combinations, with density about one-third of aluminum; used most commonly in the manufacture of beryllium-copper alloys which find numerous industrial and scientific applications.

beryllium alloy [MET] Any dilute alloy of base metals containing a few percent of beryllium in a precipitation-hardening system.

beryllium bronze See beryllium copper.

beryllium copper [MET] **1.** An alloy of copper and beryllium containing not more than about 3% beryllium; used for springs, tools, and plastic molds. Also known as beryllium bronze. **2.** An alloy of copper and beryllium specifically for addition to metals in the foundry.

beryllium monel [MET] A nickel-copper alloy containing beryllium.

beryllium nitrate [INORG CHEM] $Be(NO_3)_2 \cdot 3H_2O$ A compound that forms colorless, deliquescent crystals that are soluble in water; used to introduce beryllium oxide into materials used in incandescent mantles.

beryllium oxide [INORG CHEM] BeO An amorphous white powder, insoluble in water; used to make beryllium salts and as a refractory. Also known as beryllia.

beryllonite [MINERAL] $NaBe(PO_4)$ A colorless or yellow mineral occurring in short, prismatic or tabular, monoclinic crystals with two good pinacoidal cleavages at right angles; hardness is 5.5–6 on Mohs scale, and specific gravity is 2.85.

Berytidae [INV ZOO] The stilt bugs, a small family of hemipteran insects in the superfamily Pentatomorpha.

berzelianite [MINERAL] Cu_2Se A silver-white mineral composed of copper selenide and found in igneous rock; specific gravity is 4.03.

Bessel ellipsoid of 1841 [GEOD] The reference ellipsoid of which the semimajor axis is 6,377,397.2 meters, the semiminor axis is 6,356,079.0 meters, and the flattening or ellipticity equals 1/299.15. Also known as Bessel spheroid of 1841.

Bessel equation [MATH] The differential equation $z^2f''(z) + zf'(z) + (z^2 - n^2)f(z) = 0$.

Bessel function [MATH] A solution of the Bessel equation. Symbolized $J_n(z)$.

Besselian elements [ASTRON] Data on a solar eclipse, giving, for selected times, the coordinates of the axis of the moon's shadow with respect to the fundamental plane, and the radii of umbra and penumbra in that plane; the data allow one to derive local circumstances of the eclipse at any point on the earth's surface.

Besselian star numbers [ASTRON] Constants used in the reduction of a mean position of a star to an apparent position;

used to account for short-term variations in precession, nutation, aberration, and parallax.

Besselian year See fictitious year.

Bessel inequality [MATH] The statement that the sum of the squares of the inner product of a vector with the members of an orthonormal set is no larger than the square of the norm of the vector.

Bessel spheroid of 1841 See Bessel ellipsoid of 1841.

Bessel transform See Hankel transform.

Bessemer converter [MET] A pear-shaped, basic-lined, cylindrical vessel for producing steel by the Bessemer process.

Bessemer iron [MET] Pig iron with about 1% silicon and a sulfur and a phosphorus content below 0.10%; used to make steel by the Bessemer or the acid open-hearth process. Also known as Bessemer pig iron.

Bessemer matte [MET] Product of the oxidation of furnace matte; contains nickel, copper, cobalt, precious metals, and about 22% sulfur.

Bessemer pig iron See Bessemer iron.

Bessemer process [MET] A steelmaking process in which carbon, silicon, phosphorus, and manganese contained in molten pig iron are oxidized by a strong blast of air.

Bessemer steel [MET] Steel manufactured by the Bessemer process.

Bessey unit [BIOL] A unit for the standardization of phosphatase.

Be star [ASTRON] A star of spectral type B in the Draper catalog that has emission lines indicating mass loss and a surrounding gaseous shell.

best commercial practice [ENG] A manufacturing standard for a process vessel which has not been designed according to standard codes, such as the ASME Boiler Code.

beta [ELECTR] The current gain of a transistor that is connected as a grounded-emitter amplifier, expressed as the ratio of change in collector current to resulting change in base current, the collector voltage being constant. [SCI TECH] The second letter of the Greek alphabet; β, B.

beta-absorption gage See beta gage.

beta brass [MET] A type of brass containing nearly equal proportions of copper and zinc.

beta carotene [BIOCHEM] $C_{40}H_{56}$ A carotenoid hydrocarbon pigment found widely in nature, always associated with chlorophylls; converted to vitamin A in the liver of many animals.

beta cell [HISTOL] **1.** Any of the basophilic chromophiles in the anterior lobe of the adenohypophysis. **2.** One of the cells of the islets of Langerhans which produce insulin.

beta coefficient [STAT] Also known as beta weight. **1.** One of the coefficients in a regression equation. **2.** A moment ratio, especially one used to describe skewness and kurtosis.

betacyanin [BIOCHEM] A group of purple plant pigments found in leaves, flowers, and roots of members of the order Caryophyllales.

beta decay [NUC PHYS] Radioactive transformation of a nuclide in which the atomic number increases or decreases by unity with no change in mass number; the nucleus emits or absorbs a beta particle (electron or positron). Also known as beta disintegration.

beta decay spectrum [NUC PHYS] The distribution in energy or momentum of the beta particles arising from a nuclear disintegration process.

beta disintegration See beta decay.

beta distribution [STAT] The probability distribution of a random variable with density function $f(x) = [x^{\alpha-1}(1-x)^{\beta-1}]/B(\alpha,\beta)$, where B represents the beta function, α and β are positive real numbers, and $0 < x < 1$. Also known as Pearson Type I distribution.

beta emitter [NUC PHYS] A radionuclide that disintegrates by emission of a negative or positive electron.

beta factor [PL PHYS] In plasma physics, the ratio of the plasma kinetic pressure to the magnetic pressure.

beta function [MATH] A function of two positive variables, defined by

$$\beta(m,n) = \int_0^1 x^{m-1}(1-x)^{n-1}dx.$$

beta gage [NUCLEO] A penetration-type thickness gage that measures the absorption of beta rays in the sample. Also known as beta-absorption gage.

beta-gamma survey meter [NUCLEO] An ionization-cham-

BERYCIFORMES

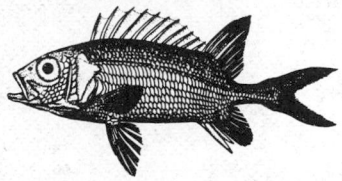

Squirrelfish (*Holocentrus ascensionis*) of the family Holocentridae that are nocturnal fishes found in shallow tropical and subtropical reefs. (*After G. B. Goode, Fishery Industries of the United States, 1884*)

BERYLLIUM

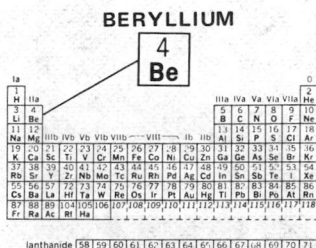

Periodic table of the chemical elements showing the position of beryllium.

BERYLLONITE

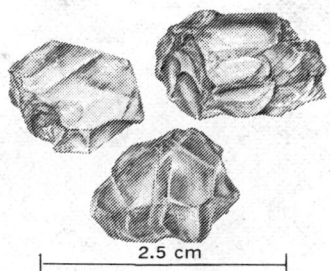

Beryllonite crystals, Stoneham, Maine. (*Specimen from Department of Geology, Bryn Mawr College*)

BESSEMER CONVERTER

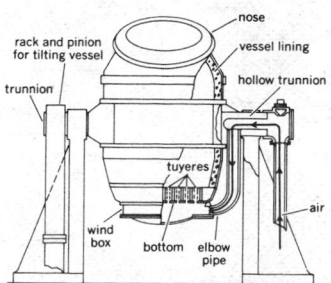

Cutaway view of Bessemer converter. (*From United States Steel Corp., The Making, Shaping, and Testing of Steel, 7th ed., 1957*)

ber type of monitor that is sensitive primarily to beta particles and gamma rays.

beta globulin [BIOCHEM] A heterogeneous fraction of serum globulins containing transferrin and various complement components.

beta hemolysis [MICROBIO] A sharply defined, clear, colorless zone of hemolysis surrounding certain streptococci colonies growing on blood agar.

betaine [ORG CHEM] $C_5H_{11}O_2N$ An alkaloid; very soluble in water, soluble in ethyl alcohol and methanol; the hydrochloride is used as a source of hydrogen chloride and in medicine. Also known as lycine; oxyneurine.

beta interaction See weak interaction.

betanin [BIOCHEM] An anthocyanin that contains nitrogen and constitutes the principal pigment of garden beets.

beta particle [NUC PHYS] An electron or positron emitted from a nucleus during beta decay.

beta plane [GEOPHYS] The model, introduced by C. G. Rossby, of the spherical earth as a plane whose rate of rotation (corresponding to the Coriolis parameter) varies linearly with the north-south direction.

beta ray [NUC PHYS] A stream of beta particles.

beta-ray spectrometer [SPECT] An instrument used to determine the energy distribution of beta particles and secondary electrons. Also known as beta spectrometer.

beta rhythm [PHYSIO] An electric current of low voltage from the brain, with a pulse frequency of 13–30/sec, encountered in a person who is aroused and anxious.

beta spectrometer See beta-ray spectrometer.

beta taxonomy [SYST] The phylogenic phase in the history of animal systematics.

betatron [NUCLEO] A device for accelerating electrons in an evacuated ring by means of a time-varying magnetic flux encircled by the ring. Also known as induction accelerator; rheotron.

beta weight See beta coefficient.

Betelgeuse [ASTRON] An orange-red giant star of stellar magnitude 0.1–1.2, 650 light-years from the sun, spectral classification M2-Iab, in the constellation Orion; the star α Orionis.

betel nut [BOT] A dried, ripe seed of the palm tree *Areca catechu* in the family Palmae; contains a narcotic.

BET equation See Brunauer-Emmett-Teller equation.

Bethell process See full-cell process.

Bethe-Salpeter equation [PARTIC PHYS] The relativistic analog of the integral form of the two-body Schrödinger equation, the two-particle interaction kernel being the analog of the potential.

Bethylidae [INV ZOO] A small family of hymenopteran insects in the superfamily Bethyloidea.

Bethyloidea [INV ZOO] A superfamily of hymenopteran insects in the suborder Apocrita.

betrunked river [GEOL] A river that is shorn of its lower course as a result of submergence of the land margin by the sea.

Betterton-Kroll process [CHEM ENG] A method for obtaining pure bismuth from softened and desilverized lead.

Betti group See homology group.

Betti number See connectivity number.

Betts' process [MET] A refining process for electrolytically purifying lead.

Betula [BOT] The birches, a genus of deciduous trees composing the family Betulaceae.

Betulaceae [BOT] A small family of dicotyledonous plants in the order Fagales characterized by stipulate leaves, seeds without endosperm, and by being monoecious with female flowers mostly in catkins.

betula oil See methyl salicylate.

betulinic acid [ORG CHEM] $C_{30}H_{48}O_3$ A dibasic acid, slightly soluble in water, ethyl alcohol, and acetone.

between decks [NAV ARCH] **1.** The space beneath the main deck of a ship. **2.** A deck beneath the main deck; in particular, a raised deck in a cargo ship's hold.

betwixt mountains See median mass.

Betz cell [HISTOL] Any of the large conical cells composing the major histological feature of the precentral motor cortex in man.

Betz momentum theory [MECH ENG] A theory of windmill

performance that considers the deceleration in the air traversing the windmill disk.

BeV [PHYS] A billion (10^9) electron volts, a unit used in the United States; international unit is GeV (giga electron volts), which has the same value.

Bevatron [NUCLEO] Name of the 6-GeV proton synchrotron at the University of California at Berkeley.

bevel [DES ENG] **1.** The angle between one line or surface and another line or surface, or the horizontal, when this angle is not a right angle. **2.** A sloping surface or line. [GRAPHICS] An instrument composed of two rules joined together, opening to any angle, which is used to draw angles and measure and lay off bevels.

bevel gear [MECH ENG] One of a pair of gears used to connect two shafts whose axes intersect.

beveling [GEOL] Planing by erosion of the outcropping edges of strata. [MECH ENG] See chamfering.

beveloid gear [DES ENG] An involute gear, of the Vinco Corporation, with tapered outside diameter, root diameter, and tooth thickness; used to reduce backlash in the drives of precision instruments. Also known as tapered-tooth gear.

Beverage antenna See wave antenna.

beyond-the-horizon communication See scatter propagation.

Beyrichacea [PALEON] A superfamily of extinct ostracods in the suborder Beyrichicopina.

Beyrichicopina [PALEON] A suborder of extinct ostracods in the order Paleocopa.

Beyrichiidae [PALEON] A family of extinct ostracods in the superfamily Beyrichacea.

bezel [DES ENG] **1.** A grooved rim used to hold a transparent glass or plastic window or lens for a meter, tuning dial, or other indicating device. **2.** A sloping face on a cutting tool. [LAP] The oblique face of a cut gem, between the table and the girdle.

BFO See beat-frequency oscillator.

B girdle [PETR] A circular pattern in petrofabric diagrams that indicates a B axis.

BHA See butylated hydroxy anisole.

Bhabba scattering [PARTIC PHYS] The scattering of positrons by electrons.

B-H curve [ELECTROMAG] A graphical curve showing the relation between magnetic induction B and magnetizing force H for a magnetic material. Also known as magnetization curve.

B-H meter [ENG] A device used to measure the intrinsic hysteresis loop of a sample of magnetic material.

B horizon [GEOL] The zone of accumulation in soil below the A horizon (zone of leaching). Also known as illuvial horizon; subsoil; zone of accumulation; zone of illuviation.

bhp See boiler horsepower; brake horsepower.

BHT See butylated hydroxytoluene.

Bi See abampere; bismuth.

biacetyl See 2,3-butanedione.

Bial's test [PATH] A test for the presence of a pentose in urine, utilizing oracin, hydrochloric acid, and ferric chloride; a green color or green precipitate indicates pentose.

biamperometry [ANALY CHEM] Amperometric titration that uses two polarizing or indicating electrodes to detect the end point of a redox reaction between the substance being titrated and the titrant.

Bianchi identity [MATH] A relation between covariant derivatives of the curvature tensor.

bias [ELEC] **1.** A direct-current voltage used on signaling or telegraph relays or electromagnets to secure desired time spacing of transitions from marking to spacing. **2.** The restraint of a relay armature by spring tension to secure a desired time spacing of transitions from marking to spacing. **3.** The effect on teleprinter signals produced by the electrical characteristics of the line and equipment. **4.** The force applied to a relay to hold it in a given position. [ELECTR] **1.** A direct-current voltage applied to a transistor control electrode to establish the desired operating point. **2.** See grid bias. [STAT] In estimating the value of a parameter of a probability distribution, the difference between the expected value of the estimator and the true value of the parameter.

bias cell [ELECTR] A small dry cell used singly or in series to provide the required negative bias for the grid circuit of an electron tube. Also known as grid-bias cell.

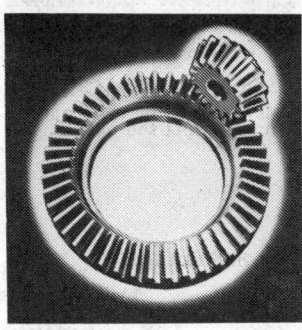

BEVEL GEAR

Types of bevel gears. (Gleason Gear Works)

bias current [ELECTR] 1. An alternating electric current above about 40,000 hertz added to the audio current being recorded on magnetic tape to reduce distortion. 2. An electric current flowing through the base-emitter junction of a transistor and adjusted to set the operating point of the transistor.

bias distortion [COMMUN] See bias telegraph distortion. [ELECTR] Distortion resulting from the operation on a nonlinear portion of the characteristic curve of a vacuum tube or other device, due to improper biasing.

biased automatic gain control See delayed automatic gain control.

biased statistic [STAT] A statistic whose expected value, as obtained from a random sampling, does not equal the parameter or quantity being estimated.

bias error [STAT] A measurement error that remains constant in magnitude for all observations; a kind of systematic error.

bias meter [COMMUN] A meter used in teletypewriter work for measuring signal bias directly in percent; a positive reading indicates a marking signal bias; a negative reading, a spacing signal bias.

bias oscillator [ELECTR] An oscillator used in a magnetic recorder to generate the alternating-current signal that is added to the audio current being recorded on magnetic tape to reduce distortion.

bias resistor [ELECTR] A resistor used in the cathode or grid circuit of an electron tube to provide a voltage drop that serves as the bias.

bias telegraph distortion [COMMUN] A distortion that causes telegraph mark-and-space pulses to be lengthened or shortened; often caused by changes in the amplitude of incoming pulses. Also known as bias distortion; spacing bias.

bias voltage [ELECTR] A voltage applied or developed between two electrodes as a bias.

bias winding [ELEC] A control winding that carries a steady direct current which serves to establish desired operating conditions in a magnetic amplifier or other magnetic device.

biaxial crystal [CRYSTAL] A crystal of low symmetry in which the index ellipsoid has three unequal axes.

biaxial indicatrix [CRYSTAL] An ellipsoid whose three axes at right angles to each other are proportional to the refractive indices of a biaxial crystal.

biaxial stress [MECH] The condition in which there are three mutually perpendicular principal stresses; two act in the same plane and one is zero.

Biazzi process [CHEM ENG] A continuous-flow process for the nitration of glycerin to nitroglycerin that is safer than the Schmid process; also used to produce glycol dinitrate and diethylene glycol nitrate.

bibb cock See bibcock.

bibcock [DES ENG] A faucet or stopcock whose nozzle is bent downward. Also spelled bibb cock.

bibenzyl [ORG CHEM] $C_{14}H_{14}$ A hydrocarbon consisting of two benzene rings attached to ethane. Also known as dibenzyl.

Bibionidae [INV ZOO] The March flies, a family of orthorrhaphous dipteran insects in the series Nematocera.

bibliophobia [PSYCH] An abnormal fear of books.

bicable tramway [MECH ENG] A tramway consisting of two stationary cables on which the wheeled carriages travel, and an endless rope, which propels the carriages.

bicameral [BIOL] Having two chambers, as the heart of a fish.

bicapsular [BIOL] 1. Having two capsules. 2. Having a capsule with two locules.

bicarbonate [INORG CHEM] A salt obtained by the neutralization of one hydrogen in carbonic acid.

bicarinate [BIOL] Having two keellike projections.

bicaudal [ZOO] Having two tails.

bicellular [BIOL] Having two cells.

bicephalous [ZOO] Having two heads.

biceps [ANAT] 1. A bicipital muscle. 2. The large muscle of the front of the upper arm that flexes the forearm; biceps brachii. 3. The thigh muscle that flexes the knee joint and extends the hip joint; biceps femoris.

bichloride of mercury See mercuric chloride.

biciliate [BIOL] Having two cilia.

bicipital [ANAT] 1. Pertaining to muscles having two origins.

2. Pertaining to ribs having double articulation with the vertebrae. [BOT] Having two heads or two supports.

bicipital groove See intertubercular sulcus.

bicipital tuberosity [ANAT] An eminence on the anterior inner aspect of the neck of the radius; the tendon of the biceps muscle is inserted here.

bicompact set See compact set.

biconcave lens See double concave lens.

biconical antenna [ELECTROMAG] An antenna consisting of two metal cones having a common axis with their vertices coinciding or adjacent and with coaxial-cable or waveguide feed to the vertices.

bicontinuous function See homeomorphism.

biconvex lens See double convex lens.

bicornuate uterus [ANAT] A uterus with two horn-shaped processes on the superior aspect.

Bicosoecida [INV ZOO] An order of colorless, free-living protozoans, each having two flagella, in the class Zoomastigophorea.

bicostate [BOT] Of a leaf, having two principal longitudinal ribs.

bicuspid [ANAT] Any of the four double-pointed premolar teeth in man. [BIOL] Having two points or prominences.

bid [ENG] An estimate of costs for specified construction, equipment, or services proposed to a customer company by one or more supplier or contractor companies.

Bidder's organ [VERT ZOO] A structure in the males of some toad species that may develop into an ovary in older individuals.

bidentate [BIOL] Having two teeth or teethlike processes.

bidirectional [ENG] Being directionally responsive to inputs in opposite directions.

bidirectional antenna [ELECTROMAG] An antenna that radiates or receives most of its energy in only two directions.

bidirectional clamping circuit [ELECTR] A clamping circuit that functions at the prescribed time irrespective of the polarity of the signal source at the time the pulses used to actuate the clamping action are applied.

bidirectional clipping circuit [ELECTR] An electronic circuit that prevents transmission of the portion of an electrical signal that exceeds a prescribed maximum or minimum voltage value.

bidirectional counter See forward-backward counter.

bidirectional microphone [ENG ACOUS] A microphone that responds equally well to sounds reaching it from the front and rear, corresponding to sound incidences of 0 and 180°.

bidirectional pulse-amplitude modulation See double-polarity pulse-amplitude modulation.

bidirectional transducer [ELECTR] A transducer capable of measuring in both positive and negative directions from a reference position. Also known as bilateral transducer.

bidirectional transistor [ELECTR] A transistor that provides switching action in either direction of signal flow through a circuit; widely used in telephone switching circuits.

bidirectional triode thyristor [ELECTR] A gate-controlled semiconductor switch designed for alternating-current power control.

Biebrich red See scarlet red.

Biedenharn identity [NUC PHYS] A relationship among the six-j symbols of Wigner.

Biela Comet [ASTRON] A comet seen in 1852 at one perihelion passage; presumed to have separated into two bodies.

Bienayme-Chebycheff inequality [STAT] The probability that the magnitude of the difference between the mean of the sample values of a random variable and the mean of the variable is less than st, where s is the standard deviation and t is any number greater than 1, is equal to or greater than $1 - (1/t^2)$.

biennial plant [BOT] A plant that requires two growing seasons to complete its life cycle.

Bierbaum scratch hardness equipment [ENG] Equipment to test the hardness of a solid sample by microscopic measurement of the width of scratch made by a diamond point under preset pressure.

biface tool [DES ENG] A tool, as an ax, made from a coil flattened on both sides to form a V-shaped cutting edge.

bifacial [BOT] Of a leaf, having dissimilar tissues on the

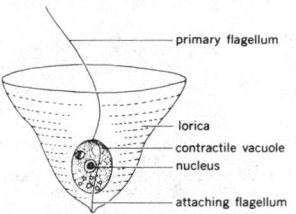

BICOSOECIDA

A bicosoecid, *Codomonas annulata.*

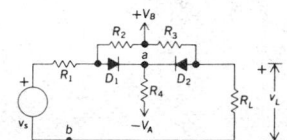

BIDIRECTIONAL CLIPPING CIRCUIT

Circuit diagram of bidirectional clipping obtained by connecting two diodes.

BIFILAR ELECTROMAGNETIC OSCILLOGRAPH

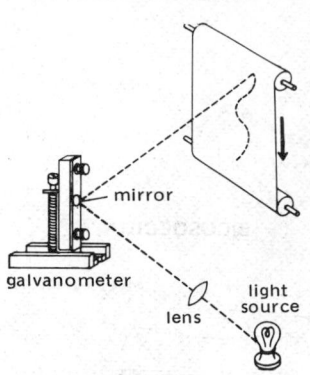

Bifilar electromagnetic oscillograph with light-beam, photographic-film recorder. *(From I. F. Kinnard, Applied Electrical Measurements, Wiley, 1956)*

upper and lower surfaces. [DES ENG] Of a tool, having both sides alike.

bifanged [ANAT] Of a tooth, having two roots.

bifid [BIOL] Divided into two equal parts by a median cleft.

bifilar electromagnetic oscillograph [ELECTROMAG] A writing low-frequency light-beam oscillograph usually using a moving coil with a single U-shaped turn (bifilar type).

bifilar resistor [ELEC] A resistor wound with a wire doubled back on itself to reduce the inductance.

bifilar transformer [ELEC] A transformer in which wires for the two windings are wound side by side to give extremely tight coupling.

bifilar winding [ELEC] A winding consisting of two insulated wires, side by side, with currents traveling through them in opposite directions.

biflabellate [INV ZOO] The shape of certain insect antennae, characterized by short joints with long, flattened processes on opposite sides.

biflagellate [BIOL] Having two flagella.

bifluoride [INORG CHEM] An acid fluoride whose formula has the form MHF_2, such as sodium bifluoride, $NaHF_2$.

bifocal lens [OPTICS] 1. A lens with two parts having different focal lengths. 2. In particular, an eyeglass lens having one part that corrects for distant vision and one part for near vision.

bifoliate [BOT] Two-leaved.

biforate [BIOL] Having two perforations.

bifurcated contact [ELEC] A contact having a forked shape such that it can slide over and interlock with an identical mating contact.

bifurcation [SCI TECH] 1. Division into two branches, parts, or aspects. 2. Point at which division occurs.

bifurcation buoy [NAV] A red and black horizontally banded buoy used to mark the point at which a channel divides into two branches when proceeding from seaward.

bifurcation ratio [HYD] The ratio of number of stream segments of one order to the number of the next higher order.

big bud [PL PATH] 1. A parasitic disease of currants caused by a gall mite (*Eriophyes ribis*) and characterized by abnormal swelling of the buds. 2. A virus disease of the tomato characterized by swelling of the buds.

Big Coal D [MATER] A nongelatinous explosive used in coal mining.

Big Dipper [ASTRON] A group of stars that is part of the constellation Ursa Major. Also known as Charles' wain.

Big Four [ASTRON] A group of large asteroids including Ceres, Pallas, Vesta, and Juno, the first four that were discovered.

bight [GEOL] 1. A long, gradual bend or recess in the coastline which forms a large, open receding bay. 2. A bend in a river or mountain range. [OCEANOGR] An indentation in shelf ice, fast ice, or a floe.

big inch pipe [PETRO ENG] A pipeline 24 inches in diameter which carries oil or gas, usually for great distances.

big M method [ADP] A technique for solving linear programming problems in which artificial variables are assigned cost coefficients which are a very large number M, say, $M = 10^{35}$.

Bignoniaceae [BOT] A family of dicotyledonous trees or shrubs in the order Scrophulariales characterized by a corolla with mostly five lobes, mature seeds with little or no endosperm and with wings, and opposite or whorled leaves.

big vein [PL PATH] A soil-borne virus disease of lettuce characterized by enlargement and yellowing of the leaf veins.

bigwoodite [PETR] A medium-grained plutonic rock consisting of microcline, microcline-microperthite, sodic plagioclase, and hornblende, aegirine-augite, or biotite.

biharmonic function [MATH] A solution to the partial differential equation $\Delta^2 u(x,y,z) = 0$, where Δ is the Laplacian operator; occurs frequently in problems in electrostatistics.

bijugate [BOT] Of a pinnate leaf, having two pairs of leaflets.

bilabial [ANTHRO] Distance between the highest point on the upper lip and lowest point on the lower lip.

bilabiate [BOT] Having two lips, such as certain corollas.

bilateral [BIOL] Of or relating to both right and left sides of an area, organ, or organism.

bilateral amplifier [ELECTR] An amplifier capable of receiving as well as transmitting signals; used primarily in transceivers.

bilateral antenna [ELECTROMAG] An antenna having maximum response in exactly opposite directions, 180° apart, such as a loop.

bilateral circuit [ELEC] Circuit wherein equipment at opposite ends is managed, operated, and maintained by different services.

bilateral cleavage [EMBRYO] The division pattern of a zygote that results in a bilaterally symmetrical embryo.

bilateral hermaphroditism [ZOO] The presence of an ovary and a testis on each side of the animal body.

bilateral Laplace transform [MATH] A generalization of the Laplace transform in which the integration is done over the negative real numbers as well as the positive ones.

bilateral observation [ORD] System for determining deviation of impacts or bursts from the target by the use of two instruments and observers located at a distance from each other.

bilateral symmetry [BIOL] Symmetry such that the body can be divided by one median, or sagittal, dorsoventral plane into equivalent right and left halves, each a mirror image of the other.

bilateral tolerance [DES ENG] The amount that the size of a machine part is allowed to vary above or below a basic dimension; for example, 3.650 ± 0.003 inches indicates a tolerance of ± 0.003 inch.

Bilateria [ZOO] A major division of the animal kingdom embracing all forms with bilateral symmetry.

bile [PHYSIO] An alkaline fluid secreted by the liver and delivered to the duodenum to aid in the emulsification, digestion, and absorption of fats. Also known as gall.

bile acid [BIOCHEM] Any of the liver-produced steroid acids, such as taurocholic acid and glycocholic acid, that appear in the bile as sodium salts.

bile duct [ANAT] Any of the major channels in the liver through which bile flows toward the hepatic duct.

bile pigment [BIOCHEM] Either of two colored organic compounds found in bile: bilirubin and biliverdin.

bile salt [BIOCHEM] The sodium salt of glycocholic and taurocholic acids found in bile.

bilge [NAV ARCH] 1. Part of the underwater body of a ship between the flat of the bottom and the straight vertical sides. 2. Internally, the lowest part of the hull, next to the keelson.

bilge block [CIV ENG] A wooden support under the turn of a ship's bilge in dry dock.

bilge board [NAV ARCH] 1. A wood or metal board that slides in a case like a centerboard, but is built into each bilge of a ship. 2. *See* limber board.

bilge keel [NAV ARCH] A keel extending outside of the hull and fastened along a ship near the turn of the bilge to reduce rolling. Also known as rolling chock.

bilge keelson [NAV ARCH] A keelson close to the turn of the bilge.

bilge pump [NAV ARCH] A pump to remove water that collects in the bottom of a ship.

bilge water [NAV ARCH] Water that builds up in a ship's bilges.

biliary atresia [MED] Failure of the bile ducts to develop in the embryo.

biliary cirrhosis [MED] A progressive inflammatory disease of the liver due to obstruction of bile ducts.

biliary colic [MED] Severe abdominal pain caused by passage of a gallstone through the bile ducts into the duodenum.

biliary diskinesia [MED] A functional spasticity of the sphincter of Oddi with disturbances in the speed of evacuation of the biliary tract.

biliary system [ANAT] The complex of canaliculi, or microscopic bile ducts, that empty into the larger intrahepatic bile ducts.

bilicyanin [BIOCHEM] A blue pigment found in gallstones; an oxidation product of biliverdin or bilirubin.

bilification [PHYSIO] Formation and excretion of bile.

bilinear concomitant [MATH] An expression $B(u,v)$, where u, v are functions of x, satisfying $vL(u) - u\bar{L}(v) = (d/dx)B(u,v)$, where L, $\bar{L}$ are given adjoint differential equations.

bilinear expression [MATH] An expression which is linear in each of two variables separately.

biliprotein [BIOCHEM] The generic name for the organic

compounds in certain algae that are composed of phycobilin and a conjugated protein.

bilirubin [BIOCHEM] $C_{33}H_{36}N_4O_6$ An orange, crystalline pigment occurring in bile; the major metabolic breakdown product of heme.

bilirubinemia [MED] The presence of bilirubin in the blood.

biliverdin [BIOCHEM] $C_{33}H_{34}N_4O_6$ A green, crystalline pigment occurring in the bile of amphibians, birds, and man; oxidation product of bilirubin in man.

bill [DES ENG] One blade of a pair of scissors. [INV ZOO] A flattened portion of the shell margin of the broad end of an oyster. [NAV ARCH] The point at the end of an anchor fluke. [VERT ZOO] The jaws, together with the horny covering, of a bird. [ZOO] Any jawlike mouthpart.

billboard array [ELECTROMAG] A broadside antenna array consisting of stacked dipoles spaced 1/4 to 3/4 wavelength apart in front of a large sheet-metal reflector. Also known as bedspring array; mattress array.

billet [MET] A semifinished, short, thick bar of iron or steel in the form of a cylinder or rectangular prism produced from an ingot; limited to 1.5 inches in width and thickness with a cross-sectional area up to 36 square inches.

Billet furnace [MET] A furnace for heating steel in sizes between a bloom and a bar.

billet mill [MET] A rolling mill for making billets from ingots.

Billet split lens [OPTICS] A lens cut into two halves, along the optic axis; used in interferometry.

billiard cloth [TEXT] A heavy-weight wool fabric with a smooth, even surface; used on billiard tables.

billibit *See* gigabit.

Billingsellacea [PALEON] A group of extinct articulate brachiopods in the order Orthida.

billow cloud [METEOROL] Broad, nearly parallel lines of cloud oriented normal to the wind direction, with cloud bases near an inversion surface. Also known as undulatus.

bilobate [BIOL] Divided into two lobes.

bilobular [BIOL] Having two lobules.

bilocular [BIOL] Having two cells or compartments.

bilophodont [ZOO] Having two transverse ridges, as the molar teeth of certain animals.

bimaceral [GEOL] A coal microlithotype that consists of a mixture of two macerals.

Bimat film [GRAPHICS] A polyester transfer film, by Kodak Company, coated with gelatin containing physical development nuclei; it is soaked in solvent developer and rolled up in contact with an exposed roll of film, which it develops to a negative, while a positive can be produced on the Bimat film.

bimaxillary [ANTHRO] Pertaining to the distance between the lower margins of the sutures of the maxilla and malar bones. [ZOO] Pertaining to the two halves of the maxilla.

bimetallic [MET] Composed of two dissimilar metals bonded together.

bimetallic corrosion [MET] Corrosion of two different metals exposed to an electrolyte while in electrical contact.

bimetallic plate [GRAPHICS] Printing plate in planographic printing composed of two metals; one rejects ink, the other does not.

bimetallic strip [ENG] A strip formed of two dissimilar metals welded together; different temperature coefficients of expansion of the metals cause the strip to bend or curl when the temperature changes.

bimetallic thermometer [ENG] A temperature-measuring instrument in which the differential thermal expansion of thin, dissimilar metals, bonded together into a narrow strip and coiled into the shape of a helix or spiral, is used to actuate a pointer. Also known as differential thermometer.

bimolecular [CHEM] Referring to two molecules.

bimolecular reaction [CHEM] A chemical transformation or change involving two molecules.

bimorph cell [ELECTR] Two piezoelectric plates cemented together in such a way that an applied voltage causes one to expand and the other to contract so that the cell bends in proportion to the applied voltage; conversely, applied pressure generates double the voltage of a single cell; used in phonograph pickups and microphones.

bin [ADP] A magnetic-tape memory in which a number of tapes are stored in a single housing. [ENG] An enclosed space, box, or frame for the storage of bulk substance.

binary [ADP] Possessing a property for which there exists two choices or conditions, one choice excluding the other. [SCI TECH] Composed of or characterized by two parts or elements.

binary alloy [MET] An alloy composed of two principal metallic components.

binary cell [ADP] An elementary unit of computer storage that can have one or the other of two stable states and can thus store one bit of information.

binary chain [ADP] A series of binary circuit elements so arranged that each can change the state of the one following it.

binary code [ADP] A code in which each allowable position has one of two possible states, commonly 0 and 1; the binary number system is one of many binary codes.

binary coded character [ADP] One element of a notation system representing alphanumeric characters such as decimal digits, alphabetic letters, and punctuation marks by a predetermined configuration of consecutive binary digits.

binary coded decimal system [ADP] A system of number representation in which each digit of a decimal number is represented by a binary number. Abbreviated BCD system.

binary compound [CHEM] A compound that has two elements; it may contain two or more atoms; examples are KCl and $AlCl_3$.

binary counter *See* binary scaler.

binary encounter approximation [ATOM PHYS] An approximation for predicting the probability that an incident proton will eject an inner shell electron from an atom; it uses a semiclassical treatment of momentum transfer from the incident proton to the ejected electron.

binary explosive [MATER] A high explosive composed of a mixture of two high explosives, to secure an explosive which is superior to its components in regard to sensitivity, fragmentation, blast, or loadability.

binary fission [BIOL] A method of asexual reproduction accomplished by the splitting of a parent cell into two equal, or nearly equal, parts, each of which grows to parental size and form.

binary granite [PETR] **1.** A granite made up of quartz and feldspar. **2.** A granite containing muscovite mica and biotite.

binary magnetic core [SOLID STATE] A ferromagnetic core that can be made to take either of two stable magnetic states.

binary notation *See* binary number system.

binary number [MATH] A number expressed in the binary number system of positional notation.

binary number system [MATH] A representation for numbers using only the digits 0 and 1 in which successive digits are interpreted as coefficients of successive powers of the base 2. Also known as binary notation; binary system.

binary operation [MATH] A rule for combining two elements of a set to obtain a third element of that set, for example, addition and multiplication.

binary point [ADP] The character, or the location of an implied symbol, that separates the integral part of a numerical expression from its fractional part in binary notation.

binary scaler [ELECTR] A scaler that produces one output pulse for every two input pulses. Also known as binary counter; scale-of-two circuit.

binary search [ADP] A dichotomizing search in which the set of items to be searched is divided at each step into two equal, or nearly equal, parts.

binary separation [CHEM ENG] Separation by distillation or solvent extraction of a fully miscible liquid mixture of two chemical compounds.

binary star *See* double star.

binary system [ENG] Any system containing two principal components. [MATH] *See* binary number system.

binary to decimal conversion [MATH] The process of converting a number written in binary notation to the equivalent number written in ordinary decimal notation.

binary weapon [ORD] A weapon that operates by storing less-than-lethal chemicals in separate compartments of a projectile; the wall between the compartments breaks after the munition is fired, allowing the chemicals to combine in a lethal mixture.

binary word [ADP] A group of bits which occupies one storage address and is treated by the computer as a unit.

BILLET SPLIT LENS

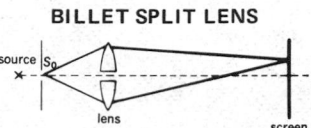

Billet split-lens interference.

BIMETALLIC THERMOMETER

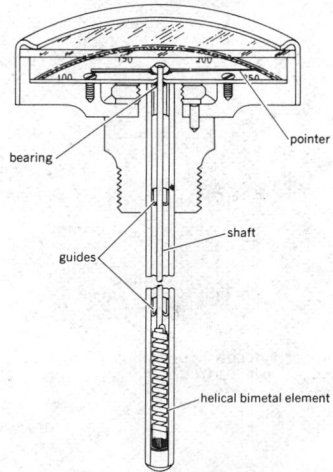

Cutaway view of bimetallic thermometer. (*Weston Instruments, Division of Daystrom, Inc.*)

binate [BOT] Growing in pairs.

binaural [ACOUS] Pertaining to sound that reaches the listener over two paths, to give the effect of auditory perspective.

binaural hearing [PHYSIO] The perception of sound by stimulation of two ears.

binaural intensity effect [ACOUS] The relationship wherein, if sound of the same frequency and phase is incident at both ears, the angle between the apparent direction of the sound and the median plane of the line joining the ears is proportional to the logarithm of the ratio of the intensities of sound received at the left and right ears.

binaural phase effect [ACOUS] A displacement in the apparent direction of a sound that results when a difference in phase is introduced between otherwise identical sound signals applied to the two ears; the angular displacement from the median plane is proportional to the phase difference.

binaural sound [ACOUS] The sound resulting from a reproduction system which has two channels, each fed into a different earphone or loudspeaker, so that a listener hears sounds coming from their original directions (with reference to the separated microphones used in recording the original sounds).

binder [MATER] A resin or other cementlike material used to hold particles together and provide mechanical strength or to ensure uniform consistency, solidification, or adhesion to a surface coating; typical binders are resin, glue, gum, and casein.

binder course [CIV ENG] Coarse aggregate with a bituminous binder between the foundation course and the wearing course of a pavement.

binderless briquetting [ENG] The briquetting of coal by the application of pressure without the addition of a binder.

bindheimite [MINERAL] $Pb_2Sb_2O_6(O,OH)$ A hydrous lead antimonate mineral produced from natural oxidation of jamesonite; found in Nevada.

B indicator See B scope.

binding energy [PHYS] Abbreviated BE. Also known as total binding energy (TBE). **1.** The net energy required to remove a particle from a system. **2.** The net energy required to decompose a system into its constituent particles.

binding post [ELEC] A manually turned screw terminal used for making electrical connections.

Binet age [PSYCH] An individual's mental age as determined by the Binet-Simon intelligence scale.

Binet's formula [PSYCH] The premise that children under 9 years of age whose mental development is retarded by 2 years are probably mentally deficient, and children of 9 years or older retarded by 3 years are definitely deficient.

Binet-Simon intelligence scale [PSYCH] Test for determining the relative mental development of children between 3 and 12 years of age; results are expressed as an intelligence quotient, the ratio of mental to chronological age.

Bingham number [FL MECH] A dimensionless number used to study the flow of Bingham plastics.

Bingham plastic [FL MECH] A non-Newtonian fluid exhibiting a yield stress which must be exceeded before flow starts; thereafter the rate-of-shear versus shear stress curve is linear.

bing ore [GEOL] The purest lead ore, with the largest crystals of galena.

binistor [ELECTR] A silicon *npn* tetrode that serves as a bistable negative-resistance device.

binnacle [NAV ARCH] A stand, case, or box for a ship's magnetic compass, which may contain nonmagnetic gear such as a lamp.

binocular [BIOL] **1.** Of, pertaining to, or used by both eyes. **2.** Of a type of visual perception providing depth-of-field focus due to angular difference between the two retinal images. [OPTICS] Any optical instrument designed for use with both eyes to give enhanced views of distant objects, whose distinguishing performance feature is the depth perception obtainable.

binocular accommodation [PHYSIO] Automatic lens adjustment by both eyes simultaneously for focusing on distant objects.

binocular microscope [OPTICS] A microscope having two oculars, allowing the use of both eyes at once.

binode [ELECTR] An electron tube with two anodes and one

BINOCULAR

Modern prism binocular. *(Bausch and Lomb Optical Co.)*

cathode used as a full-wave rectifier. Also known as double diode.

binomen [SYST] A binomial name assigned to species, as *Canis familiaris* for the dog.

binomial [MATH] A polynomial with only two terms.

binomial array See Pascal triangle.

binomial array antenna [ELECTROMAG] Directional antenna array for reducing minor lobes and providing maximum response in two opposite directions.

binomial coefficient [MATH] A coefficient in the expansion of $(x+y)^n$, where n is a positive integer; the $(k+1)$st coefficient is equal to the number of ways of choosing k objects out of n without regard for order. Symbolized $\binom{n}{k}$; $_nC_k$; $C(n,k)$; C_k^n.

binomial differential [MATH] A differential of the form $x^p(a + bx^q)^r dx$, where p, q, r are integers.

binomial distribution [STAT] The distribution of a binomial random variable; the distribution (n,p) is given by $P(B = r) = \binom{n}{r} p^r q^{n-r}$, $p+q = 1$. Also known as Bernoulli distribution.

binomial equation [MATH] An equation of the form $x^n - a = 0$.

binomial law [MATH] The law that the probability of an event occurring r times in n Bernoulli trials is equal to

$$\binom{n}{r}p^r(1-p)^{n-r}$$

where p is the probability of the event.

binomial nomenclature [SYST] The Linnean system of classification requiring the designation of a binomen, the genus and species name, for every species of plant and animal.

binomial series [MATH] The expansion of $(x + y)^n$ when n is neither a positive integer nor zero.

binomial surd [MATH] A polynomial having two terms, at least one of which is a surd.

binomial theorem [MATH] The rule for expanding $(x+y)^n$.

binomial trials [STAT] A sequence of trials, on each trial of which a certain result may or may not happen.

binomial trials model [STAT] A product model in which each factor has two simple events with probabilities p and $q = 1 - p$.

binormal [MATH] A vector on a curve at a point so that, together with the positive tangent and principal normal, it forms a system of right-handed rectangular cartesian axes.

binuclear [CYTOL] Having two nuclei.

binucleolate [CYTOL] Having two nucleoli.

bioacoustics [BIOL] The study of the relation between living organisms and sound.

bioassay [ANALY CHEM] A method for quantitatively determining the concentration of a substance by its effect on the growth of a suitable animal, plant, or microorganism under controlled conditions.

bioastronautics [BIOL] The study of biological, behavioral, and medical problems pertaining to astronautics.

bioautography [ANALY CHEM] A bioassay based upon the ability of some compounds (for example, vitamin B_{12}) to enhance the growth of some organisms or compounds and to repress the growth of others; used to assay certain antibiotics.

biobubble [ECOL] A model concept of the ecosphere in which all living things are considered as particles held together by nonliving forces.

biocatalyst [BIOCHEM] A biochemical catalyst, especially an enzyme.

biocenology [ECOL] The study of natural communities and of interactions among the members of these communities.

biocenose See biotic community.

biochemical deposit [GEOL] A precipitated deposit formed directly or indirectly from vital activities of organisms, such as bacterial iron ore and limestone.

biochemical fuel cell [ELEC] A projected fuel cell in which small amounts of electric power will be produced continuously for many years by some form of biological system.

biochemical oxygen demand [MICROBIO] The amount of dissolved oxygen required to meet the metabolic needs of anaerobic microorganisms in water rich in organic matter, such as sewage. Abbreviated BOD. Also known as biological oxygen demand.

biochemical oxygen demand test [MICROBIO] A standard laboratory procedure for measuring biochemical oxygen demand; standard measurement is made for 5 days at 20°C. Abbreviated BOD test.

biochemistry [CHEM] The study of chemical substances occurring in living organisms and the reactions and methods for identifying these substances.

biochemorphology [BIOCHEM] The science dealing with the chemical structure of foods and drugs and their reactions on living organisms.

biochore [ECOL] A group of similar biotopes.

biochrome [BIOCHEM] Any naturally occurring plant or animal pigment.

biochron [PALEON] A fossil of relatively short range of time.

biochronology [PALEON] The study of the fauna and flora of specific geologic time ranges.

biociation [ECOL] A subdivision of a biome distinguished by the predominant animal species.

biocide See pesticide.

bioclastic rock [PETR] Rock formed from material broken or arranged by animals, man, or sometimes plants; a rock composed of broken calcareous remains of organisms.

bioclimatic law [ECOL] The law which states that phenological events are altered by about 4 days for each 5° change of latitude northward or longitude eastward; events are accelerated in spring and retreat in autumn.

bioclimatograph [ECOL] A climatograph showing the relation between climatic conditions and some living organisms.

bioclimatology [ECOL] The study of the effects of the natural environment on living organisms.

biocycle [ECOL] A group of similar biotopes composing a major division of the biosphere; there are three biocycles: terrestrial, marine, and fresh-water.

biocytin [BIOCHEM] $C_{16}H_{28}N_4O_4S$ Crystals with a melting point of 241–243°C; obtained from dilute methanol or acetone solutions; characterized by its utilization by *Lactobacillus casei* and *L. delbrückii* LD5 as a biotin source, and by its unavailability as a biotin source to *L. arabinosus*. Also known as biotin complex of yeast.

biodegradability [MATER] The characteristic of a substance that can be broken down by microorganisms.

biodynamic [AGR] Of or pertaining to a system of organic farming. [BIOPHYS] Of or pertaining to the dynamic relation between an organism and its environment.

biodynamics [BIOPHYS] The study of the effects of dynamic processes (motion, acceleration, weightlessness, and so on) on living organisms.

bioelectric current [PHYSIO] A self-propagating electric current generated on the surface of nerve and muscle cells by potential differences across excitable cell membranes.

bioelectricity [PHYSIO] The generation by and flow of an electric current in living tissue.

bioelectric model [PHYSIO] A conceptual model for the study of animal electricity in terms of physical principles.

bioelectrochemistry [PHYSIO] The study of the control of biological growth and repair processes by electrical stimulation.

bioelectronics [BIOPHYS] The application of electronic theories and techniques to the problems of biology.

bioenergetics [BIOCHEM] The branch of biology dealing with energy transformations in living organisms.

bioengineering [ENG] The application of engineering knowledge to the fields of medicine and biology.

biofacies [GEOL] **1.** A rock unit differing in biologic aspect from laterally equivalent biotic groups. **2.** Lateral variation in the biologic aspect of a stratigraphic unit.

bioflavonoid [BIOCHEM] A group of compounds obtained from the rinds of citrus fruits and involved with the homeostasis of the walls of small blood vessels; in guinea pigs a marked reduction of bioflavonoids results in increased fragility and permeability of the capillaries; used to decrease permeability and fragility in capillaries in certain conditions. Also known as citrus flavonoid compound; vitamin P complex.

biofog [METEOROL] A type of steam fog caused by contact between extremely cold air and the warm, moist air surrounding human or animal bodies or generated by human activity.

biogenesis [BIOL] Development of a living organism from a similar living organism.

biogenetic law See recapitulation theory.

biogenic [BIOL] **1.** Essential to the maintenance of life. **2.** Produced by actions of living organisms.

biogenic chert [PETR] Chert derived from the tests of pelagic silica-secreting organisms, chiefly diatoms and radiolarians.

biogenic mineral [MINERAL] A mineral in sediments or sedimentary rock which represents the hard parts of dead organisms.

biogenic reef [GEOL] A mass consisting of the hard parts of organisms, or of a biogenically constructed frame enclosing detrital particles, in a body of water; most biogenic reefs are made of corals or associated organisms.

biogenic sediment [GEOL] A deposit resulting from the physiological activities of organisms.

biogeochemical cycle [GEOCHEM] The chemical interactions that exists between the atmosphere, hydrosphere, lithosphere, and biosphere.

biogeochemical prospecting [GEOCHEM] A prospecting technique for subsurface ore deposits based on interpretation of the growth of certain plants which reflect subsoil concentrations of some elements.

biogeochemistry [GEOCHEM] A branch of geochemistry that is concerned with biologic materials and their relation to earth chemicals in an area.

biogeography [ECOL] The science concerned with the geographical distribution of animal and plant life.

bioherm [GEOL] A circumscribed mass of rock exclusively or mainly constructed by marine sedimentary organisms such as corals, algae, and stromatoporoids. Also known as organic mound.

biohermal limestone [PETR] Reefs or reeflike mounds of carbonate that accumulated much in the same fashion as modern reefs and atolls of the Pacific Ocean.

biohermite [PETR] Limestone formed of debris from a bioherm.

bioinstrumentation [ENG] The use of instruments attached to animals and man to record biological parameters such as breathing rate, pulse rate, body temperature, or oxygen in the blood.

biolite [GEOL] A concretion formed of concentric layers through the action of living organisms. [PETR] See biolith.

biolith [PETR] A rock formed from or by organic material. Also known as biolite.

biolithite [PETR] An inclusive category for all organic limestone.

biological [BIOL] Of or pertaining to life or living organisms. [IMMUNOL] A biological product used to induce immunity to various infectious diseases or noxious substances of biological origin.

biological agent [ORD] Any of the viruses, microorganisms, and toxic substances derived from living organisms and used as offensive weapons to produce death or disease in man, animals, and growing plants.

biological balance [ECOL] Dynamic equilibrium that exists among members of a stable natural community.

biological clock [PHYSIO] Any physiologic factor that functions in regulating body rhythms.

biological control [ECOL] Natural or applied regulation of populations of pest organisms, especially insects, through the role or use of natural enemies.

biological corrosion [MET] Deterioration of metals as a result of the metabolic activity of microorganisms.

biological equilibrium [BIOPHYS] A state of body balance for an actively moving animal, when internal and external forces are in equilibrium.

biological half-life [PHYSIO] The time required by the body to eliminate half of the amount of an administered substance through normal channels of elimination.

biological oceanography [OCEANOGR] The study of the flora and fauna of oceans in relation to the marine environment.

biological oil-spill control [ECOL] The use of cultures of microorganisms capable of living on oil as a means of degrading an oil slick biologically.

biological oxidation [BIOCHEM] Energy-producing reactions in living cells involving the transfer of hydrogen atoms or electrons from one molecule to another.

biological oxygen demand See biochemical oxygen demand.

biological productivity [ECOL] The quantity of organic matter or its equivalent in dry matter, carbon, or energy content which is accumulated during a given period of time.

biological shield [MICROBIO] A structure designed to pre-

BIOGENIC REEF

A biogenic reef, the Ine Anchorage Reef, Arno Atoll, Marshall Islands.

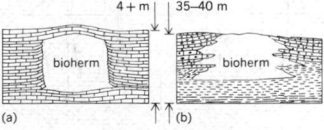

BIOHERM

Bioherms. *(a)* Small stumplike mass of cavernous dolostone in midst of thin-bedded dolomitic limestone of Middle Silurian near Lomira, Wis. *(b)* Coelenterate bioherm in midst of shale and thin-bedded cherty limestone of Middle Silurian of Wabash, Ind. *(From R. R. Shrock, Sequence in Layered Rocks, McGraw-Hill, 1948)*

vent the migration of living organisms from one part of a system to another; used on sterilized space probes.   [NUCLEO] A radiation-absorbing shield used to protect personnel from the effects of nuclear particles or radiation in the vicinity of a nuclear reactor.

biological specificity   [BIOL] The principle that defines the orderly patterns of metabolic and developmental reactions giving rise to the unique characteristics of the individual and of its species.

biological standardization   [PHARM] The standardization of drugs or biological products that cannot be chemically analyzed by studying the drugs' pharmacologic action on animals.

biological value   [BIOCHEM] A measurement of the efficiency of the protein content in a food for the maintenance and growth of the body tissues of an individual.

biological warfare   [ORD] Abbreviated BW. **1.** Employment of living microorganisms, toxic biological products, and plant growth regulators to produce death or injury in man, animals, or plants. **2.** Defense against such action.

biology   [SCI TECH] A division of the natural sciences concerned with the study of life and living organisms.

bioluminescence   [BIOL] The emission of visible light by living organisms.

biolysis   [BIOL] **1.** Death and the following tissue disintegration. **2.** Decomposition of organic materials, such as sewage, by living organisms.

biomass   [ECOL] The dry weight of living matter, including stored food, present in a species population and expressed in terms of a given area or volume of the habitat.

biome   [ECOL] A complex biotic community covering a large geographic area and characterized by the distinctive life-forms of important climax species.

biomechanics   [BIOPHYS] The study of the mechanics of living things.

biomedical engineering   [ENG] The application of engineering technology to the solution of medical problems; examples are the development of prostheses such as artificial valves for the heart, various types of sensors for the blind, and automated artificial limbs.

biomedicine   [MED] The science concerned with the study of the environment required for astronauts in space vehicles.

biomere   [ECOL] A biostratigraphic unit bounded by abrupt nonevolutionary changes in the dominant elements of a single phylum.

biometeorology   [BIOL] The study of the relationship between living organisms and atmospheric phenomena.

biometrics   [STAT] The use of statistics to analyze observations of biological phenomena.

biomicrite   [PETR] A limestone resembling biosparite except that the microcrystalline calcite matrix exceeds calcite cement.

biomicrosparite   [PETR] **1.** Biomicrite in which the micrite groundmass has recrystallized to form microspar. **2.** Microsparite containing fossil fragments or fossils.

biomicrudite   [PETR] Biomicrite with fossil fragments or fossils greater than 1 millimeter in diameter.

bion   [ECOL] An independent, individual organism.

bionavigation   [VERT ZOO] The ability of animals such as birds to find their way back to their roost, even if the landmarks on the outward-bound trip were effectively concealed from them.

bionics   [ENG] The study of systems, particularly electronic systems, which function after the manner of living systems.

biopak   [ENG] A container for housing a living organism in a habitable environment and for recording biological functions during space flight.

biopelite   *See* black shale.

biopelmicrite   [PETR] A limestone similar to biopelsparite but with a microcrystalline matrix that exceeds calcite cement.

biopelsparite   [PETR] A limestone similar to biosparite but with the ratio of fossils and fossil fragments to pellets between 3:1 and 1:3.

biophage   *See* macroconsumer.

biophagous   [ZOO] Feeding on living organisms.

biophile   [BIOCHEM] Any element concentrated or found in the bodies of living organisms and organic matter; examples are carbon, nitrogen, and oxygen.

biophysics   [SCI TECH] The hybrid science involving the application of physical principles and methods to study and explain the structures of living organisms and the mechanics of life processes.

biopotency   [BIOCHEM] Capacity of a chemical substance, as a hormone, to function in a biological system.

biopotential   [PHYSIO] Voltage difference measured between points in living cells, tissues, and organisms.

biopsy   [PATH] The removal and examination of tissues, cells, or fluids from the living body for the purposes of diagnosis.

biorheology   [BIOPHYS] The application of the principles of rheology to biological systems; involves study of biological fluids such as blood, mucus, and synovial fluid.

biosatellite   [AERO ENG] An artificial satellite designed to contain and support man, animals, or other living material in a reasonably normal manner for a period of time and to return safely to earth.

biosensor   [ENG] A sensor used to provide information about a life process.

biosonar   [PHYSIO] A guidance system in certain animals, such as bats, utilizing the reflection of sounds that they produce as they move about.

biosparite   [PETR] A limestone made up of less than 25% oolites and less than 25% intraclasts, with the ratio by volume of fossils and fragments to pellets being more than 3:1 and the calcite cement content being greater than the microcrystalline calcite content.

biosphere   [ECOL] The life zone of the earth, including the lower part of the atmosphere, the hydrosphere, soil, and the lithosphere to a depth of about 2 kilometers.

biostabilizer   [CIV ENG] A component in mechanized composting systems; consists of a drum in which moistened solid waste is comminuted and tumbled for about 5 days until the aeration and biodegradation turns the waste into a fine dark compost.

biostratigraphy   [PALEON] A part of paleontology concerned with the study of the conditions and deposition order of sedimentary rocks.

biostromal limestone   [GEOL] Biogenic carbonate accumulations that are laterally uniform in thickness, in contrast to the moundlike nature of bioherms.

biostrome   [GEOL] A bedded structure or layer (bioclastic stratum) composed of calcite and dolomitized calcarenitic fossil fragments distributed over the sea bottom as fine lentils, independent of or in association with bioherms or other areas of organic growth.

biosynthesis   [BIOCHEM] Production, by synthesis or degradation, of a chemical compound by a living organism.

biot   [ELEC] *See* abampere.   [OPTICS] A unit of rotational strength in substances exhibiting circular dichroism, equal to 10^{-40} times the corresponding centimeter-gram-second unit.

biota   [BIOL] **1.** Animal and plant life characterizing a given region. **2.** Flora and fauna, collectively.

biotar lens   [OPTICS] A modern camera lens which is a modified Gauss objective with a large aperture and a field of about 24°.

biotechnology   [ENG] The application of engineering and technological principles to the life sciences.

biotelemetry   [ENG] The use of telemetry techniques, especially radio waves, to study behavior and physiology of living things.

Biot-Fourier equation   [THERMO] An equation for heat conduction which states that the rate of change of temperature at any point divided by the thermal diffusivity equals the Laplacian of the temperature.

biotherapy   [MED] Treatment of disease with biologicals, that is, materials produced by living organisms.

biotic   [BIOL] **1.** Of or pertaining to life and living organisms. **2.** Induced by the actions of living organisms.

biotic community   [ECOL] An aggregation of organisms characterized by a distinctive combination of both animal and plant species in a particular habitat. Also known as biocenose.

biotic district   [ECOL] A subdivision of a biotic province.

biotic environment   [ECOL] That environment comprising

BIOTAR LENS

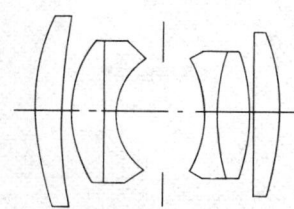

Biotar camera lens.

living organisms, which interact with each other and their abiotic environment.

biotic isolation [ECOL] The occurrence of organisms in isolation from others of their species.

biotic province [ECOL] A community, according to some systems of classification, occupying an area where similarity of climate, physiography, and soils leads to the recurrence of similar combinations of organisms.

biotin [BIOCHEM] $C_{10}H_{16}N_2O_3S$ A colorless, crystalline vitamin of the vitamin B complex occurring widely in nature, mainly in bound form.

biotin complex of yeast *See* biocytin.

biotite [MINERAL] A black, brown, or dark green, abundant and widely distributed species of rock-forming mineral of the mica group; its chemical composition is variable: $K_2[Fe(II),Mg]_{6-4}[Fe(III),Al,Ti]_{0-2}(Si_{6-5},Al_{2-3})O_{20-22}(OH,F)_{4-2}$. Also known as black mica; iron mica; magnesia mica; magnesium-iron mica.

biotite schist [PETR] A schist composed of biotite.

Biot number [FL MECH] A dimensionless group, used in the study of mass transfer between a fluid and a solid, which gives the ratio of the mass-transfer rate at the interface to the mass-transfer rate in the interior of a solid wall of specified thickness.

biotope [ECOL] An area of uniform environmental conditions and biota.

biotron [ENG] A test chamber used for biological research within which the environmental conditions can be completely controlled, thus allowing observations of the effect of variations in environment on living organisms.

Biot-Savart law [ELECTROMAG] A law that gives the intensity of the magnetic field due to a wire carrying a constant electric current.

Biot's law [OPTICS] The law that an optically active substance rotates plane-polarized light through an angle inversely proportional in its wavelength.

biotype [GEN] A group of organisms having the same genotype.

biozone [PALEON] The range of a single taxonomic entity in geologic time as reflected by its occurrence in fossiliferous rocks.

bipack [GRAPHICS] A photographic film composed of two sensitive emulsions; each is sensitive to a different color when exposed simultaneously.

biparasitic [ECOL] Parasitic upon or in a parasite.

biparietal [ANTHRO] Distance between the most distant opposite points of the parietal bone.

biparous [BOT] Having branches on dichotomous axes. [VERT ZOO] Bringing forth two young at a birth.

bipartite cubic [MATH] The points satisfying the equation $y^2 = x(x-a)(x-b)$.

bipartite graph [STAT] A linear graph (network) in which the nodes can be partitioned into two groups G_1 and G_2 such that for every arc (i,j) node i is in G_1 and node j in G_2.

bipartite uterus [ANAT] A uterus divided into two parts almost to the base.

bipectinate [INV ZOO] Of the antennae of certain moths, having two margins with comblike teeth. [ZOO] Branching like a feather on both sides of a main shaft.

biped [VERT ZOO] **1.** A two-footed animal. **2.** Any two legs of a quadruped.

bipedal [BIOL] Having two feet.

bipedal dinosaur [PALEON] A dinosaur having two long, stout hindlimbs for walking and two relatively short forelimbs.

bipeltate [BOT] Having two shield-shaped parts. [ZOO] Having a shell or other covering resembling a double shield.

bipenniform [ANAT] Of the arrangement of muscle fibers, resembling a feather barbed on both sides.

biphasic [BOT] Having both a sporophyte and a gametophyte generation in the life cycle.

biphenyl [ORG CHEM] $C_{12}H_{10}$ A white or slightly yellow crystalline hydrocarbon, melting point 70.0°C, boiling point 255.9°C, and density 1.9896, which gives plates or monoclinic prismatic crystals; used as a heat-transfer medium and as a raw material for chlorinated diphenyls. Also known as diphenyl; phenylbenzene.

biphyletic [EVOL] Descended in two branches from a common ancestry.

Biphyllidae [INV ZOO] The false skin beetles, a family of coleopteran insects in the superfamily Cucujoidea.

bipinnaria [INV ZOO] The complex, bilaterally symmetrical, free-swimming larval stage of most asteroid echinoderms.

bipinnate [BOT] Pertaining to a leaf that is pinnate for both its primary and secondary divisions.

biplane [AERO ENG] An aircraft with two wings fixed at different levels, especially one above and one below the fuselage.

bipolar [SCI TECH] Having two poles.

bipolar amplifier [ELECTR] An amplifier capable of supplying a pair of output signals corresponding to the positive or negative polarity of the input signal.

bipolar circuit [ELECTR] A logic circuit in which zeros and ones are treated in a symmetric or bipolar manner, rather than by the presence or absence of a signal; for example, a balanced arrangement in a square-loop-ferrite magnetic circuit.

bipolar electrode [ELEC] Electrode, without metallic connection with the current supply, one face of which acts as anode surface and the opposite face as a cathode surface when an electric current is passed through a cell.

bipolar flagellation [MICROBIO] The presence of flagella at both poles in certain bacteria.

Bipolarina [INV ZOO] A suborder of protozoan parasites in the order Myxosporida.

bipolar integrated circuit [ELECTR] An integrated circuit in which the principal element is the bipolar junction transistor.

bipolar magnetic driving unit [ENG ACOUS] Headphone or loudspeaker unit having two magnetic poles acting directly on a flexible iron diaphragm.

bipolar transistor [ELECTR] A transistor that uses both positive and negative charge carriers.

bipotential [BIOL] Having the potential to develop in either of two mutually exclusive directions.

bipotential electrostatic lens [ELECTR] An electron lens in which image and object space are field-free, but at different potentials; examples are the lenses formed between apertures of cylinders at different potentials. Also known as immersion electrostatic lens.

bipotentiality [BIOL] **1.** Capacity to function either as male or female. **2.** Hermaphroditism.

biprism [OPTICS] A prism with apex angle only a little less than 180°, which produces a double image of a point source, giving rise to interference fringes on a nearby screen.

biprism interference [OPTICS] Light interference fringes seen on a screen near a biprism.

bipropellant [MATER] A rocket propellant consisting of two unmixed or uncombined chemicals (fuel and oxidizer) fed to the combustion chamber separately.

biquadratic [MATH] Any fourth-degree algebraic expression. Also known as quartic.

biquinary abacus [MATH] An abacus in which the frame is divided into two parts by a bar which separates each wire into two- and five-counter segments.

biquinary notation [MATH] A mixed-base notation system in which the first of each pair of digits counts 0 or 1 unit of five, and the second counts 0, 1, 2, 3, or 4 units. Also known as biquinary number system.

biquinary number system *See* biquinary notation.

biradial symmetry [BIOL] Symmetry both radial and bilateral. Also known as disymmetry.

biradical [CHEM] A chemical species having two independent odd-electron sites.

biramous [BIOL] Having two branches, such as an arthropod appendage.

birch [BOT] The common name for all deciduous trees of the genus *Betula* that compose the family Betulaceae in the order Fagales.

birch tar oil [MATER] A toxic liquid mixture of guaiacol, phenols, cresol, xylenol, and creosol; derived from distillation of birch tar and by dry distillation of the wood of *Betula alba*; used as a disinfectant, in leather dressing, and in medicine.

bird [AERO ENG] A colloquial term for a rocket, satellite, or spacecraft. [VERT ZOO] Any of the warm-blooded vertebrates which make up the class Aves.

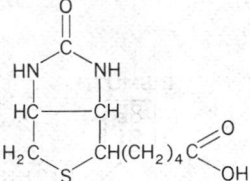

BIOTIN

Structural formula for biotin.

BIPEDAL DINOSAUR

Restored bipedal carnivorous saurischian *Allosaurus* (about 40 feet or 9 meters long), Late Jurassic, North America.

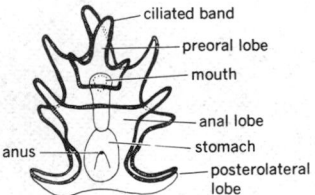

BIPINNARIA

ciliated band
preoral lobe
mouth
anal lobe
anus
stomach
posterolateral lobe

Bipinnaria larva showing component parts.

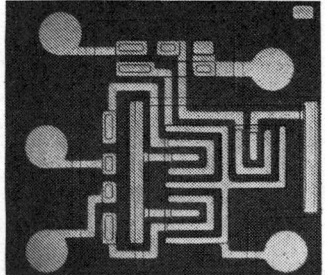

BIPOLAR INTEGRATED CIRCUIT

Photomicrograph of three-input bipolar logic gate circuit — one of the earliest digital integrated circuits. (*Fairchild Semiconductor*)

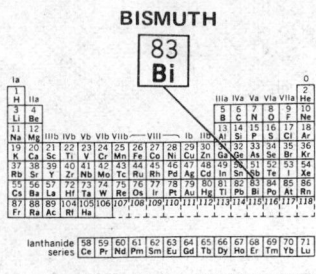

BISMUTH

83 Bi

Periodic table of the chemical elements showing the position of bismuth.

Bird centrifuge [MIN ENG] A dewatering machine for fine coal or other fine materials such as potash minerals, clays, and cement rock. Also known as Bird coal filter.

Bird coal filter *See* Bird centrifuge.

bird-foot delta [GEOL] A delta formed by the outgrowth of fingers or pairs of natural levees at the mouth of river distributaries; an example is the Mississippi delta.

bird-hipped dinosaur [PALEON] Any member of the order Ornithischia, distinguished by the birdlike arrangement of their hipbones.

bird louse [INV ZOO] The common name for any insect of the order Mallophaga. Also known as biting louse.

birdnesting [ORD] Clumping together of chaff dipoles after they have been dispensed from an aircraft.

bird of prey [VERT ZOO] Any of various carnivorous birds of the orders Falconiformes and Strigiformes which feed on meat taken by hunting.

bird's eye rot [PL PATH] A fungus disease of the grape caused by *Elsinoe ampelina* and characterized by small, dark, sunken spots with light centers on the fruit.

bird's eye spot [PL PATH] **1.** A plant disease characterized by dark, round spots surrounded by lighter tissue. **2.** A fungus disease of tea leaves caused by *Orcospora theae*. **3.** A leaf spot of the hevea rubber tree caused by the fungus *Helminthosporium heveae*.

Bird unit [BIOL] A unit for the standardization of leuteotropin.

birectangular [MATH] Property of a geometrical object that has two right angles.

birefringence [OPTICS] Splitting of a light beam into two components, which travel at different velocities, by a material. Also known as double refraction.

birefringent filter [OPTICS] A filter consisting of a multiple sandwich of alternate layers of polarizing films and plates cut from a birefringent crystal; transmits light in a series of sharp, widely spaced wavelength bands.

birefringent plate [OPTICS] A plate made of a birefringent optical material.

birimose [BOT] Opening by two slits, as an anther.

Birkeland-Eyde process [CHEM ENG] An arc process of nitrogen fixation in which air passes through an alternating-current arc flattened by a magnetic field to form about 1% nitric oxide.

Birmingham wire gage [DES ENG] A system of standard sizes of brass wire, telegraph wire, steel tubing, seamless tubing, sheet spring steel, strip steel, and steel plates, bands, and hoops. Abbreviated BWG.

birotulate [INV ZOO] A sponge spicule characterized by two wheel-shaped ends.

birth [BIOL] The emergence of a new individual from the body of its parent.

birth canal [ANAT] The channel in mammals through which the fetus is expelled during parturition; consists of the cervix, vagina, and vulva.

birth control [MED] Limitation of the number of children born by preventing or reducing the frequency of impregnation.

birth-death process [IND ENG] A simple queuing model in which units to be served arrive (birth) and depart (death) in a completely random manner.

birth defect *See* congenital anomaly.

birthmark [MED] Any abnormal cellular or vascular benign nevus that is present at birth or that appears sometime later.

birth rate [BIOL] The ratio between the number of live births and a specified number of people in a population over a given period of time.

bis- [CHEM] A prefix indicating doubled or twice.

biscuit [ENG ACOUS] *See* preform. [MATER] **1.** A clay object that has been fired once prior to glazing. **2.** Pottery that is unglazed in its final form. [MET] An upset blank for drop forging.

biscuit cutter [MIN ENG] A short (6–8 inches) core barrel that is sharpened at the bottom and forced into the rocks by the jars.

bise [METEOROL] A cold, dry wind which blows from a northerly direction in the winter over the mountainous districts of southern Europe. Also spelled bize.

bisector [MATH] The ray dividing an angle into two equal angles.

bisectrix [CRYSTAL] A line that is the bisector of the angle between the optic axes of a biaxial crystal.

biserial [BIOL] Arranged in two rows or series.

biserrate [BIOL] **1.** Having serrated serrations. **2.** Serrate on both sides.

bisexual [BIOL] Of or relating to two sexes. [PSYCH] **1.** Having mental and behavioral characteristics of both sexes. **2.** Having sexual desires for members of both sexes.

Bishop's ring [METEOROL] A faint, broad, reddish-brown corona occasionally seen in dust clouds, especially those which result from violent volcanic eruptions.

bishydroxycoumarin *See* dicoumarin.

1,3-bishydroxymethylurea *See* dimethylolurea.

bisilicate [MET] **1.** A slag whose silicate degree is 2. **2.** *See* metasilicate.

bismite [MINERAL] Bi_2O_3 A monoclinic mineral composed of bismuth trioxide; native bismuth ore, occurring as a yellow earth. Also known as bismuth ocher.

bismuth [CHEM] A metallic element, symbol Bi, of atomic number 83 and atomic weight 208.980. [MINERAL] The brittle, rhombohedral mineral form of the native element bismuth.

bismuth alloy [MET] A group of low-melting alloys (many below 100°C) of bismuth combined with lead, tin, and cadmium; used in automatic sprinklers, special solders, safety plugs in compressed-gas cylinders, automatic shutoffs for water-heating systems, castings, and type metal.

bismuthate [INORG CHEM] A compound of bismuth in which the bismuth has a valence of +5; an example is sodium bismuthate, $NaBiO_3$.

bismuth blende *See* eulytite.

bismuth carbonate *See* bismuth subcarbonate.

bismuth chloride [INORG CHEM] $BiCl_3$ A deliquescent material that melts at 230–232°C and decomposes in water to form the oxychloride; used to make bismuth salts. Also known as bismuth trichloride.

bismuth citrate [ORG CHEM] BiC_6H_5O A salt of citric acid that forms white crystals, insoluble in water; used as an astringent.

bismuth glance *See* bismuthinite.

bismuth hydroxide [INORG CHEM] $Bi(OH)_3$ A water-insoluble, white powder; precipitated by hydroxyl ion from bismuth salt solutions.

bismuthinite [MINERAL] Bi_2S_3 A mineral consisting of bismuth trisulfide, which has an orthorhombic structure and is usually found in fibrous or leafy masses that are lead gray with a yellowish tarnish and a metallic luster. Also known as bismuth glance.

bismuth iodide [INORG CHEM] BiI_3 A bismuth halide that sublimes in grayish-black hexagonal crystals melting at 408°C, insoluble in water; used in analytical chemistry.

bismuth nitrate [INORG CHEM] $Bi(NO_3)_3 \cdot 5H_2O$ White, triclinic crystals that decompose in water; used as an astringent and antiseptic.

bismuth ocher *See* bismite.

bismuth oleate [ORG CHEM] $Bi(C_{17}H_{33}COO)_3$ A salt of oleic acid obtained as yellow granules; used in medicines to treat skin diseases.

bismuth oxide *See* bismuth trioxide.

bismuth oxycarbonate *See* bismuth subcarbonate.

bismuth oxychloride [INORG CHEM] $BiOCl$ A white powder; insoluble in water, soluble in acid; a toxic material if ingested; used in pigments and cosmetics.

bismuth phenate [ORG CHEM] $C_6H_5O \cdot Bi(OH)_2$ An odorless, tasteless, gray-white powder; used in medicine. Also known as bismuth phenolate; bismuth phenylate.

bismuth phenolate *See* bismuth phenate.

bismuth phenylate *See* bismuth phenate.

bismuth pyrogallate [ORG CHEM] $Bi(OH)C_6H_3(OH)O_2$ An odorless, tasteless, yellowish-green, amorphous powder; used in medicine as intestinal antiseptic and dusting powder. Also known as basic bismuth pyrogallate; helcosol.

bismuth spar *See* bismutite.

bismuth subcarbonate [INORG CHEM] $(BiO)_2CO_3$ or $Bi_2O_3 \cdot CO_2 \cdot \frac{1}{2}H_2O$ A white powder; dissolves in hydrochloric or nitric acid, insoluble in alcohol and water; used as opacifier in x-ray diagnosis, in ceramic glass, and in enamel fluxes. Also known as basic bismuth carbonate; bismuth carbonate; bismuth oxycarbonate.

bismuth subgallate [ORG CHEM] $C_6H_2(OH)_3COOBi(OH)_2$ A yellow powder; dissolves in dilute alkali solutions, but is insoluble in water, ether, and alcohol; used in medicine. Also known as basic bismuth gallate.

bismuth subnitrate [INORG CHEM] $4BiNO_3(OH)_2 \cdot BiO(OH)$ A white, hygroscopic powder; used in bismuth salts, pefumes, cosmetics, ceramic enamels, pharmaceuticals, and analytical chemistry. Also known as basic bismuth nitrate; Spanish white.

bismuth subsalicylate [INORG CHEM] $Bi(C_7H_5O)_3Bi_2O_3$ A white powder that is insoluble in ethanol and water; used in medicine. Also known as basic bismuth salicylate.

bismuth trichloride *See* bismuth chloride.

bismuth trioxide [INORG CHEM] Bi_2O_3 A yellow powder; melting point 820°C; insoluble in water, dissolves in acid; used to make enamels and to color ceramics. Also known as bismuth oxide; bismuth yellow.

bismuth yellow *See* bismuth trioxide.

bismutite [MINERAL] $(BiO)_2CO_3$ A dull-white, yellowish, or gray, earthy, amorphous mineral consisting of basic bismuth carbonate. Also known as bismuth spar.

bison [VERT ZOO] The common name for two species of the family Bovidae in the order Artiodactyla; the wisent or European bison (*Bison bonasus*), and the American species (*B. bison*).

bisphenoid [CRYSTAL] A form apparently consisting of two sphenoids placed together symmetrically.

bisphenol [ORG CHEM] $(CH_3)_2C(C_6H_5OH)_2$ Brown crystals that are insoluble in water; used in the production of phenolic and epoxy resins. Also known as *p,p'*-dihydroxydiphenyldimethylene.

bisporangiate [BOT] Having two different types of sporangia.

bispore [BOT] In certain red algae, an asexual spore that is produced in pairs.

bistable [SCI TECH] Capable of assuming either of two stable states.

bistable circuit [ELECTR] A circuit with two stable states such that the transition between the states cannot be accomplished by self-triggering.

bistable multivibrator [ELECTR] A multivibrator in which either of the two active devices may remain conducting, with the other nonconducting, until the application of an external pulse. Also known as Eccles-Jordan circuit; Eccles-Jordan multivibrator; flip-flop circuit; trigger circuit.

bistable unit [ENG] A physical element that can be made to assume either of two stable states; a binary cell is an example.

bistatic radar [ENG] Radar system in which the receiver is some distance from the transmitter, with separate antennas for each.

bistatic reflectivity [OPTICS] The characteristic of a reflector which reflects energy along a line, or lines, different from or in addition to that of the incident ray.

bisulfate [INORG CHEM] A compound that has the HSO_4^- radical; derived from sulfuric acid.

bit [ADP] **1.** A unit of information content equal to one binary decision or the designation of one of two possible and equally likely values or states of anything used to store or convey information. **2.** A dimensionless unit of storage capacity specifying that the capacity of a storage device is expressed by the logarithm to the base 2 of the number of possible states of the device. [DES ENG] **1.** A machine part for drilling or boring. **2.** The cutting plate of a plane. **3.** The blade of a cutting tool such as an ax. **4.** A removable tooth of a saw. **5.** Any cutting device which is attached to or part of a drill rod or drill string to bore or penetrate rocks.

bitartrate [ORG CHEM] A salt that has the radical $HC_4H_4O_6^-$. Also known as acid tartrate.

bit blank [DES ENG] A steel bit in which diamonds or other cutting media may be inset by hand peening or attached by a mechanical process such as casting, sintering, or brazing. Also known as bit shank; blank; blank bit; shank.

bit buffer unit [COMMUN] A unit that terminates bit-serial communications lines coming from and going to technical control.

bit count appendage [ADP] One of the two-byte elements replacing the parity bit stripped off each byte transferred from main storage to disk volume (the other element is the cyclic check); these two elements are appended to the block during the write operation; on a subsequent read operation these elements are calculated and compared to the appended elements for accuracy.

bit density [ADP] Number of bits which can be placed, per unit length, area, or volume, on a storage medium; for example, bits per inch of magnetic tape. Also known as record density.

bit drag [DES ENG] A rotary-drilling bit with serrated teeth. Also known as drag bit.

bite [BIOL] **1.** To seize with the teeth. **2.** Closure of the lower teeth against the upper teeth. [GRAPHICS] In photoengraving, the various stages of etching accomplished through the action of acid. [MED] Skin injury produced by an animal's teeth or the mouthparts of an insect.

bitegmic [BOT] Having two integuments, especially in reference to ovules.

biternate [BOT] Of a ternate leaf, having each division ternate.

bithionol [ORG CHEM] A halogenated form of bisphenol used as an ingredient in antigermicidal soaps and as a medicine in the treatment of clonorchiases.

biting angle [ORD] Smallest angle of impact at which a projectile will penetrate or pierce armor.

biting-in [GRAPHICS] In etching, the action of dilute nitric acid or other mordant upon the bare steel or copper plate.

biting louse *See* bird louse.

bit location [ADP] Storage position on a record capable of storing one bit.

Bitot's spots [MED] The silver-gray, shiny, triangular spots on the cornea characteristic of xerosis conjunctivae.

bit pattern [ADP] A combination of binary digits arranged in a sequence.

bit rate [COMMUN] Quantity, per unit time, of binary digits (or pulses representing them) which will pass a given point on a communications line or channel in a continuous stream.

bit shank *See* bit blank.

bit synchronization [COMMUN] Element of a message header used to synchronize all of the bits and characters that follow.

bitt [NAV ARCH] **1.** A short metal or wood post on the deck of a ship used to secure mooring or other lines; usually in a pair. **2.** To secure a line about a ship's bitt.

bitter end [NAV ARCH] The end of a line or cable, especially the inboard end.

bitter lake [HYD] A lake rich in alkaline carbonates and sulfates.

bittern [CHEM ENG] Concentrated sea water or brine containing the bromides and magnesium and calcium salts left in solution after sodium chloride has been removed by crystallization. [VERT ZOO] Any of various herons of the genus *Botaurus* characterized by streaked and speckled plumage.

bitter orange oil [MATER] An essential oil from the rind or peel of the orange (*Citrus vulgaris*); insoluble in water; used as a flavoring and in some perfumes.

bitter pattern [SOLID STATE] A pattern produced when a drop of a colloidal suspension of ferromagnetic particles is placed on the surface of a ferromagnetic crystal; the particles collect along domain boundaries at the surface.

bitter pit [PL PATH] A disease of uncertain etiology affecting apple, pear, and quince; spots of dead brown tissue appear in the flesh of the fruit and discolored depressions are seen on its surface. Also known as Baldwin spot; stippen.

bitter rot [PL PATH] A fungus disease of apples, grapes, and other fruit caused by *Glomerella cingulata*.

Bittner milk factor *See* milk factor.

bitumastic [MATER] A combination of asphalt and filler used mainly to coat metals to protect them against corrosion and weathering.

bitumen [MATER] Naturally occurring or pyrolytically obtained substances of dark to black color consisting almost entirely of carbon and hydrogen, with very little oxygen, nitrogen, or sulfur.

bitumenite *See* torbanite.

bituminization *See* coalification.

BISON

North American bison (*Bison bison*), with enormous forequarter development and hump behind the head.

BISTABLE MULTIVIBRATOR

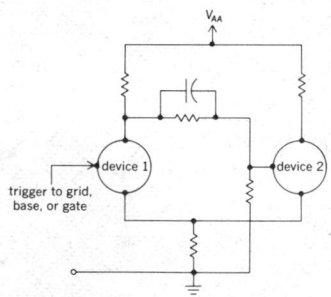

Circuit diagram of an unsymmetrical bistable multivibrator.

BITTER PATTERN

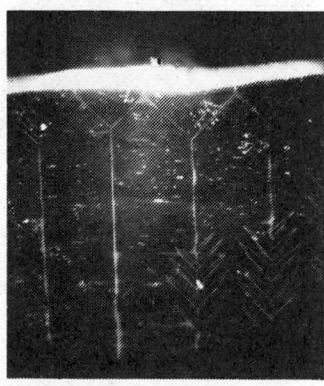

Bitter powder patterns on a (100) surface of silicon-iron. (*Photograph by H. J. Williams*)

BLACKBERRY

Thorny, biennial stem of blackberry shrub.

bituminous [MATER] **1.** Containing much organic, or at least carbonaceous, matter, mostly in the form of tarry hydrocarbons which are usually described as bitumen. **2.** Similar to bitumen. **3.** Giving off volatile bituminous substances on heating, as in bituminous coal. [MINERAL] Of a mineral, having the odor of bitumen.

bituminous cement [MATER] A bituminous material suitable for use as a binder, having cementing qualities which are dependent mainly on its bituminous character.

bituminous coal [GEOL] A dark brown to black coal that is high in carbonaceous matter and has 15–50% volatile matter. Also known as soft coal.

bituminous coating [MATER] A coating made principally of bituminous material and used as a surfacing for roads and as a water-repellent barrier in buildings.

bituminous concrete [MATER] A concrete made with bituminous material as a binder for sand and gravel.

bituminous lignite [GEOL] A brittle, lustrous bituminous coal. Also known as pitch coal.

bituminous paint [MATER] Paint with a high proportion of bitumen; usually dark in color.

bituminous rock *See* asphalt rock.

bituminous sand [GEOL] Sand containing bituminous-like material, such as the tar sands at Athabasca, Canada, from which oil is extracted commercially.

bituminous sandstone [PETR] A sandstone containing bituminous matter.

bituminous shale [PETR] A shale containing bituminous material.

bituminous wood [GEOL] A variety of brown coal having the fibrous structure of wood. Also known as board coal; wood coal; woody lignite; xyloid coal; xyloid lignite.

bit zone [ADP] **1.** One of the two left-most bits in a commonly used system in which six bits are used for each character; related to overpunch. **2.** Any bit in a group of bit positions that are used to indicate a specific class of items; for example, numbers, letters, special signs, and commands.

biuret [ORG CHEM] $NH_2CONHCONH_2$ Colorless needles that are soluble in hot water and decompose at 190°C; a condensation product of urea. Also known as allophanamide; carbamylurea; dicarbamylamine.

bivalent [CHEM] Possessing a valence of two.

bivalent chromosome [CYTOL] The structure formed following synapsis of a pair of homologous chromosomes from the zygotene stage of meiosis up to the beginning of anaphase.

bivalve [INV ZOO] The common name for a number of diverse, bilaterally symmetrical animals, including mollusks, ostracod crustaceans, and brachiopods, having a soft body enclosed in a calcareous two-part shell.

Bivalvia [INV ZOO] A large class of the phylum Mollusca containing the clams, oysters, and other bivalves.

bivane [ENG] A double-jointed vane which measures vertical as well as horizontal wind direction.

biventer [ANAT] A muscle having two bellies.

bivittate [ZOO] Having a pair of longitudinal stripes.

bivium [INV ZOO] The pair of starfish rays that extend on either side of the madreporite.

bivoltine [INV ZOO] **1.** Having two broods in a season, used especially of silkworms. **2.** Of insects, producing two generations a year.

bivouac [ORD] An encampment set up for a short period of time by using tents.

bixbyite [MINERAL] $(Mn,Fe)_2O_3$ A manganese-iron oxide mineral; black cubic crystals found in cavities in rhyolite. Also known as partridgeite; sitaparite.

bixin [ORG CHEM] $C_{25}H_{30}O_4$ A carotenoid acid occurring in the seeds of *Bixa orellana;* used as a fat and food coloring agent.

bize *See* bise.

Bk *See* berkelium.

BL *See* base line.

black [CHEM] Fine particles of impure carbon made by incomplete burning of carbon compounds, such as natural gas, naphthas, acetylene, bones, ivory, and vegetables. [COMMUN] *See* black signal. [OPTICS] Quality of an object which uniformly absorbs large percentages of light of all visible wavelengths.

black acids [MATER] Sulfonates in the sludge formed during treatment of petroleum products with sulfuric acid; soluble in water but insoluble in naphtha, benzene, and carbon tetrachloride.

black alkali [GEOL] A deposit of sodium carbonate that has formed on or near the surface in arid to semiarid areas.

black amber *See* jet coal.

black-and-white groups *See* Shubnikov groups.

black-and-white television *See* monochrome television.

black annealing [MET] A type of box annealing used to impart a black color to the metal surface; first process in tin plating.

black antimony *See* antimony trisulfide.

black area [COMMUN] Area with only encrypted signal present.

black ash [MATER] A carbon product made by furnace heating of black liquor from papermaking processes.

black balsam *See* Peru balsam.

blackband [GEOL] An earthy carbonate of iron that is present with coal beds.

blackberry [BOT] Any of the upright or trailing shrubs of the genus *Rubus* in the order Rosales; an edible berry is produced by the plant.

blackbird [VERT ZOO] Any bird species in the family Icteridae, of which the males are predominantly or totally black.

black blight [PL PATH] Any of several diseases of tropical plants caused by superficial sooty molds.

blackboard [MATER] A panel, usually black but sometimes colored, for writing on with chalk. Also known as chalkboard.

blackbody [THERMO] An ideal body which would absorb all incident radiation and reflect none. Also known as hohlraum; ideal radiator.

blackbody radiation [THERMO] The emission of radiant energy which would take place from a blackbody at a fixed temperature; it takes place at a rate expressed by the Stefan-Boltzmann law, with a spectral energy distribution described by Planck's equation.

black box [ENG] Any component, usually electronic and having known input and output, that can be readily inserted into or removed from a specific place in a larger system without knowledge of the component's detailed internal structure.

black-bulb thermometer [ENG] A thermometer whose sensitive element has been made to approximate a blackbody by covering it with lampblack.

black buran *See* karaburan.

black canker *See* ink disease.

black carbon counter [NUCLEO] The original type of radiation counter used in radiocarbon dating, in which the sample, whose carbon has first been converted to carbon black, is mounted on the inside of a steel cylinder which is inserted into a sensitive Geiger counter. Also known as Libby counter.

black chaff [PL PATH] A bacterial disease of wheat caused by *Xanthomonas translucens undulosa* and characterized by dark, longitudinal stripes on the chaff.

black cobalt *See* asbolite.

black copper [MET] The more or less impure metallic copper (70–99% copper) produced in blast furnaces when running on oxide ores or roasted sulfide material.

black coral [INV ZOO] The common name for antipatharian coelenterates having black, horny axial skeletons.

black cyanide *See* calcium cyanide.

blackdamp [MIN ENG] A nonexplosive mixture of carbon dioxide with other gases, especially with 85–90% nitrogen, which is heavier than air and cannot support flame or life. Also known as chokedamp.

black death *See* plague.

black diamond *See* carbonado.

Black Diamond Nu-Gel [MATER] An explosive used in mining.

black disease [VET MED] Necrotic hepatitis of sheep, resulting from infection with *Clostridium novyi* type B, with the necessary conditions for the growth of the clostridia provided by the damaged liver tissue produced by the fluke *Fasciola hepatica*.

black drop [ASTRON] As seen through a telescope, an apparent dark elongation of the image of Venus or Mercury when the planets' images are at the sun's limb.

black durain [GEOL] A durain that has high hydrogen content and volatile matter, many microspores, and some vitrain fragments.

black end [PL PATH] **1.** A disease of the pear marked by blackening of the epidermis and flesh in the region of the calyx; believed to be a result of a disturbed water relation. **2.** A fungus disease of the banana caused by several species, especially *Gloesporium musarum,* characterized by discoloration of the stem of the fruit.

blacker-than-black level [COMMUN] In television, a level of greater instantaneous amplitude than the black level, used for synchronization and control signals.

blackeye bean *See* cowpea.

blackfire [PL PATH] A bacterial disease of tobacco caused by *Pseudomonas angulata* and characterized by angular leaf spots which gradually darken and may fall out, leaving ragged holes.

black frost [HYD] A dry freeze with respect to its effects upon vegetation, that is, the internal freezing of vegetation unaccompanied by the protective formation of hoarfrost. Also known as hard frost.

black granite *See* diorite.

blackhead *See* comedo.

blackhead disease [PL PATH] **1.** A parasitic disease of the banana caused by eelworms of the family Tylenchidae. **2.** A rot disease of the banana rootstock caused by the fungus *Thielaviopsis paradoxa.*

blackheart malleable iron *See* whiteheart malleable iron.

black hole [ASTRON] A star with radius just outside the Schwarzchild radius; it is invisible but can capture matter and light from the outside.

black ice [HYD] A type of ice forming on lake or salt water; compact, and dark in appearance because of its transparency.

blacking [MET] Carbonaceous material, such as powdered graphite, used to coat the inner surfaces of a dry-sand mold to improve the separation and finish of a casting.

black iron oxide *See* ferrous oxide.

black kernel [PL PATH] A fungus disease of rice caused by *Curvularia lunata* and characterized by dark discoloration of the kernels.

black knot [PL PATH] A fungus disease of certain fruit and nut trees that is characterized by black excrescences on the branches; destructive to plum, cherry, gooseberry, filbert, and hazel.

black lead *See* graphite.

blackleg [VET MED] An acute, usually fatal bacterial disease of cattle, and occasionally of sheep, goats, and swine, caused by *Clostridium chauvoei.*

black level [ELECTR] The level of the television picture signal corresponding to the maximum limit of black peaks.

black light [OPTICS] Invisible light, such as ultraviolet rays which fall on fluorescent materials and cause them to emit visible light.

black lignite [GEOL] A lignite with a fixed carbon content of 35–60% and a total carbon content of 73.6–76.2% that contains between 6300 and 8300 Btu per pound; higher in rank than brown lignite. Also known as lignite A.

black line [PL PATH] A disease of walnuts, especially of English varieties grafted to black walnuts, characterized by a black line of dead tissue at the graft union, eventually leading to death of the tree.

black liquor [MATER] **1.** The liquid material remaining from pulpwood cooking in the soda or sulfate papermaking process. **2.** *See* iron acetate liquor.

blacklung *See* anthracosis.

black measles [MED] *See* hemorrhagic measles. [PL PATH] A disease of California grapevines of uncertain etiology; characterized by black spotting of the skin of the fruit, browning and dropping of the leaves, and dying back of the canes from the tip.

black mica *See* biotite.

black mold [MYCOL] Any dark fungus belonging to the order Mucorales. [PL PATH] A fungus disease of rose grafts and onion bulbs marked by black appearance due to the mold.

black mud [GEOL] A mud formed where there is poor circu-

lation or weak tides, such as in lagoons, sounds, or bays; the color is due to iron sulfides and organic matter.

blacknose [PL PATH] A physiological disease of the date, with the distal end of the fruit becoming dark, cracking, and shriveling.

black oil [MATER] Low-grade, black petroleum oil used to lubricate slow-moving or rough-surfaced machinery where high-grade lubricants are impractical or too expensive.

blackout [COMMUN] *See* radio blackout. [ELEC] Shutting off of power in an electrical power transmission system, either deliberately or through failure of the system.

black patch [PL PATH] A fungus disease of red clover characterized by simultaneous blackening of plants in groups.

black peak [COMMUN] A peak excursion of the television picture signal in the black direction.

black pepper [FOOD ENG] A spice; the dried unripe berries of *Piper nigrum,* a vine of the pepper family (Piperaceae) in the order Piperales.

black pod [PL PATH] A pod rot of cacao caused by the fungus *Phytophthora faveri.*

black point [PL PATH] A disease of wheat and other cereals characterized by blackening of the embryo ends of the grains.

black powder [MATER] A low explosive consisting of an intimate mixture of potassium nitrate or sodium nitrate, charcoal, and sulfur.

black print [GRAPHICS] The film having the black component in a four-color separation process, or the comparable plate in the four-color printing process.

black ring [PL PATH] **1.** A virus disease of cabbage and other members of the family Cruciferae characterized by dark necrotic and often sunken rings on the surface of the leaf. **2.** A virus disease of the tomato characterized in the early stage by small black rings on young leaves.

black root [PL PATH] Any plant disease characterized by black discolorations of the roots.

black root rot [PL PATH] **1.** Any of several plant diseases characterized by dark lesions of the root. **2.** A fungus disease of the apple caused by *Xylaria mali.* **3.** A fungus disease of tobacco and other plants caused by *Thielaviopsis basicola.*

black rot [PL PATH] Any fungal or bacterial disease of plants characterized by dark brown discoloration and decay of a plant part.

black sand [GEOL] Heavy, dark, sandlike minerals found on beaches and in stream beds; usually magnetite and ilmenite and sometimes gold, platinum, and monazite are present.

black scope [ELECTR] Cathode-ray tube operating at the threshold of luminescence when no video signals are being applied.

black scour [VET MED] Hemorrhagic enteritis of sheep, swine, and cattle, usually associated with a heavy worm burden but sometimes caused by bacterial infection.

Black Sea [GEOGR] A large inland sea, area 423,000 square kilometers, bounded on the north and east by the U.S.S.R., on the south and southwest by Turkey, and on the west by Bulgaria and Rumania.

black shale [PETR] Very thinly bedded shale rich in sulfides such as pyrite and organic material deposited under barred basin conditions so that there was an anaerobic accumulation. Also known as biopelite.

black shank [PL PATH] A black rot disease of tobacco caused by *Phytophthora parasitica* var. *nicotianae.*

black signal [COMMUN] Signal at any point in a facsimile system produced by the scanning of a maximum density area of the subject copy. Also known as black; picture black.

black silver *See* stephanite.

black smoke [ENG] A smoke that has many particulates in it from inefficient combustion; comes from burning fossil fuel, either coal or oil.

black snow [HYD] Snow that falls through a particulate-laden atmosphere.

black spot [PL PATH] Any bacterial or fungal disease of plants characterized by black spots on a plant part.

black stem [PL PATH] Any of several fungal diseases of plants characterized by blackening of the stem.

black storm *See* karaburan.

blackstrap molasses [FOOD ENG] Syrupy mother liquor left after all economical sucrose has been removed from cane- or

BLACK PEPPER

Berries on a branch of *Piper nigrum.*

beet-sugar juice; used as raw material for butanol, acetone, citric acid, and ethyl alcohol manufacture.

black stripe   *See* black thread.

black-surface enclosure   [THERMO] An enclosure for which the interior surfaces of the walls possess the radiation characteristics of a blackbody.

black tellurium   *See* nagyagite.

black thread   [PL PATH] A fungus disease of the para rubber tree at the point where the tree is tapped, caused by *Phytophthora meadii* and characterized by black stripes extending through the exposed bast into the cambium or wood. Also known as black stripe; stripe canker.

black tip   [PL PATH] Any of several plant diseases characterized by dark necrotic areas at the tip of the seed or fruit.

blacktongue   [VET MED] A niacin-deficiency disease of dogs characterized by black discoloration of the tongue.

blacktop   [MATER] A black bituminous material that is used to pave roadways; it is spread over a layer of crushed rocks and packed down into a level surface; it may be spread over small areas of roadways in need of repair.

blacktop paver   [MECH ENG] A construction vehicle that spreads a specified thickness of bituminous mixture over a prepared surface.

black vomit   [MED] Dark vomited matter, consisting of digested blood and gastric contents.

blackwater   [MED] Any disease characterized by dark-colored urine.

blackwater fever   [MED] A complication of falciparum malaria, characterized by intravascular hemolysis, hemoglobinuria, tachycardia, high fever, and poor prognosis.

bladder   [ANAT] Any saclike structure in man and animals, such as a swimbladder or urinary bladder, that contains a gas or functions as a receptacle for fluid. [BIOL] *See* vesicle.

bladder cell   [INV ZOO] Any of the large vacuolated cells in the outer layers of the tunic in some tunicates.

bladder press   [MECH ENG] A machine which simultaneously molds and cures (vulcanizes) a pneumatic tire.

blade   [BOT] The broad, flat portion of a leaf. [ELEC] A flat moving conductor in a switch. [ENG] **1.** A broad, flat arm of a fan, turbine, or propeller. **2.** The broad, flat surface of a bulldozer or snowplow by which the material is moved. **3.** The part of a cutting tool, such as a saw, that cuts. [VERT ZOO] A single plate of baleen.

bladed-surface aerator   [CIV ENG] A bladed, rotating component of a water treatment plant; used to infuse air into the water.

blade loading   [AERO ENG] A rotor's thrust in a rotary-wing aircraft divided by the total area of the rotor blades.

Blagden's law   [PHYS CHEM] The law that the lowering of a solution's freezing point is proportional to the amount of dissolved substance.

Blaine formation   [GEOL] A Permian red bed formation containing red shale and gypsum beds of marine origin in Oklahoma, Texas, and Kansas.

blairmorite   [PETR] A porphyritic extrusive rock consisting mainly of analcite phenocrysts in a groundmass of sanidine, analcite, and alkalic pyroxene, with accessory sphene, melanite, and nepheline.

Blake jaw crusher   [MECH ENG] A crusher with one fixed jaw plate and one pivoted at the top so as to give the greatest movement on the smallest lump.

Blake number   [FL MECH] A dimensionless number used in the study of beds of particles.

Blancan   [GEOL] Upper Pliocene or lowermost Pleistocene geologic time.

blanc fixe   [INORG CHEM] $BaSO_4$ A white sulfate compound of barium, with some use in pure form in the paint, paper, and pigment industries as a pigment extender.

blanching   [FOOD ENG] A hot-water or steam direct-scalding treatment of raw foodstuffs of particulate type to inactivate enzymes which otherwise might cause quality deterioration, particularly of flavor, during processing or storage.

Blanc rule   [ORG CHEM] The rule that glutaric and succinic acids yield cyclic anhydrides on pyrolysis, while adipic and pimolic acids yield cyclic ketones; there are certain exceptions.

blank   [DES ENG] *See* bit blank. [ELECTR] To cut off the electron beam of a television picture tube, camera tube, or

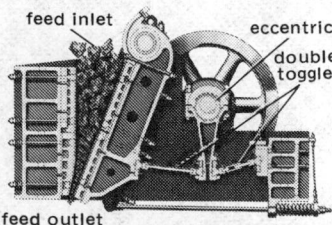

BLAKE JAW CRUSHER

feed inlet / eccentric / double toggle / feed outlet

Drawing of a Blake jaw crusher. *(Allis Chalmers Co.)*

cathode-ray oscilloscope tube during retrace by applying a rectangular pulse voltage to the grid or cathode during each retrace interval. Also known as beam blank. [ENG] **1.** The result of the final cutting operation on a natural crystal. **2.** *See* blind. [MET] **1.** A semifinished piece of metal to be stamped or forged into a tool or implement. **2.** A semifinished, pressed, compacted mass of powdered metal. **3.** Metal sheet prepared for a forming operation. [ORD] Ammunition which contains no projectile but which does contain a charge of low explosive, such as black powder, to produce a noise.

blank bit   *See* bit blank.

blank carburizing   [MET] A simulated carburizing procedure carried out without a carburizing medium. Also known as pseudocarburizing.

blank character   [ADP] A character, either printed or appearing as a blank, used to denote a blank space among printed characters. Also known as space character.

blanket   [COMMUN] To blank out or obscure weak radio signals by a stronger signal. [GRAPHICS] In offset lithography, a rubber sheet covering the cylinder of an offset press that transfers the image from the plate to the paper. [MIN ENG] A textile material used in ore treatment plants for catching coarse free gold and sometimes associated minerals, for example, pyrite. [NUCLEO] A layer of fertile uranium-238 or thorium-232 material placed around or within the core of a nuclear reactor to breed new fuel.

blanket cylinder   [GRAPHICS] The cylinder on which the blanket is mounted in offset lithography.

blanket deposit   [GEOL] A flat deposit of ore; its length and width are relatively great compared with its thickness.

blanketing   [COMMUN] Interference due to a nearby transmitter whose signals are so strong that they override other signals over a wide band of frequencies. [MIN ENG] The material caught on the blankets that are used in concentrating gold-bearing sands or slimes.

blanket sand   [GEOL] A relatively thin body of sand or sandstone covering a large area. Also known as sheet sand.

blank holder   [MET] A tool to prevent the edge of a sheet metal blank from wrinkling during deep-drawing operations.

blankholder slide   [MECH ENG] The outer slide of a double-action power press; it is usually operated by toggles or cams.

blanking   [ENG] The closing off of flow through a liquid-containing process pipe by the insertion of solid disks at joints or unions; used during maintenance and repair work as a safety precaution. Also known as blinding. [MET] Cutting of sheet metal into shapes.

blanking circuit   [ELECTR] A circuit preventing the transmission of brightness variations during the horizontal and vertical retrace intervals in television scanning.

blanking level   [ELECTR] The level that separates picture information from synchronizing information in a composite television picture signal; coincides with the level of the base of the synchronizing pulses. Also known as pedestal; pedestal level.

blanking pulse   [ELECTR] A positive or negative square-wave pulse used to switch off a part of a television or radar set electronically for a predetermined length of time.

blanking signal   [ELECTR] The signal rendering the return trace invisible on the picture tube of a television receiver.

Blasius theorem   [AERO ENG] A theorem that provides formulas for finding the force and moment on the airfoil profiler.

blast   [ENG] The setting off of a heavy explosive charge. [PHYS] **1.** The brief and rapid movement of air or other fluid away from a center of outward pressure, as in an explosion. **2.** The characteristic instantaneous rise in pressure, followed by a sudden decrease, that results from this movement, differentiated from less rapid pressure changes.

blast area   [ORD] **1.** Area affected by the blast of an explosion. **2.** Scorched area of ground around the muzzle of a gun, caused by repeated blasts.

blast burner   [ENG] A burner in which a controlled burst of air or oxygen under pressure is supplied to the illuminating gas used. Also known as blast lamp.

blast cell   [HISTOL] An undifferentiated precursor of a human blood cell in the reticuloendothelial tissue.

blast chamber   [AERO ENG] A combustion chamber, especially in a gas-turbine, jet, or rocket engine.

blast deflector [AERO ENG] A device used to divert the exhaust of a rocket fired from a vertical position.

blast ditching [CIV ENG] The use of explosives to aid in ditch excavation, such as for laying pipelines.

blast effect [PHYS] Violent air movements and pressure changes and the destruction or damage resulting therefrom, generally caused by an explosion on or above the surface of the earth.

blastema [EMBRYO] 1. A mass of undifferentiated protoplasm capable of growth and differentiation. 2. *See* anlage.

blaster [ENG] A device for detonating an explosive charge; usually consists of a machine by which an operator, by pressing downward or otherwise moving a handle of the device, may generate a powerful transient electric current which is transmitted to an electric blasting cap. Also known as blasting machine.

blast furnace [MET] A tall, cylindrical smelting furnace for reducing iron ore to pig iron; the blast of air blown through solid fuel increases the combustion rate.

blast-furnace coke [MATER] Coke which supplies carbon monoxide to reduce the ore in a blast furnace and supplies heat to melt the iron.

blast-furnace gas [MATER] The gas product from iron ore smelting when hot air passes over coke in blast ovens; contains carbon dioxide, carbon monoxide, hydrogen, and nitrogen and is used as fuel gas.

blasthole [ENG] 1. A hole that takes a heavy charge of explosive. 2. The hole through which water enters in the bottom of a pump stock.

blastic deformation [GEOL] Rock deformation involving recrystallization in which space lattices are destroyed or replaced.

blasting [COMMUN] Distortion due to overloading of any part of a radio transmitter, receiver, or audio amplifier. [ENG] 1. Cleaning materials by a blast of air that blows small abrasive particles against the surface. 2. The act of detonating an explosive. [GEOL] Abrasion caused by movement of fine particles against a stationary fragment.

blasting agent [MATER] A compound or mixture, such as ammonium nitrate or black powder, that detonates as a result of heat or shock; used in mining, blasting, pyrotechnics, and propellants.

blasting barrel [MIN ENG] A piece of iron pipe, usually about one-half inch in diameter, used to provide a smooth passageway through the stemming for the miner's squib; it is recovered after each blast and used until destroyed.

blasting cap [ENG] A copper shell closed at one end and containing a charge of detonating compound, which is ignited by electric current or the spark of a fuse; used for detonating high explosives.

blasting fuse [ENG] A core of gunpowder in the center of jute, yarn, and so on for igniting an explosive charge in a shothole.

blasting gelatin [MATER] A plastic dynamite that contains 5–10% nitrocellulose added to nitroglycerin; used principally in submarine work.

blasting machine *See* blaster.

blasting powder [MATER] A powder containing less nitrate, and in its place more charcoal than black powder; composition is 65–75% sodium or nitrate, potassium nitrate, 10–15% sulfur, and 15–20% charcoal.

blast lamp *See* blast burner; blowtorch.

blasto- [EMBRYO] A germ or bud, with reference to early embryonic stages of development. [PETR] A prefix indicating the presence in a rock of residual structures somewhat modified by metamorphism.

Blastobasidae [INV ZOO] A family of lepidopteran insects in the superfamily Tineoidea.

blastocarpous [BOT] Germinating in the pericarp.

blastochyle [EMBRYO] The fluid filling the blastocoele.

Blastocladiales [MYCOL] An order of aquatic fungi in the class Phycomycetes.

blastocoele [EMBRYO] The cavity of a blastula. Also known as segmentation cavity.

blastocone [EMBRYO] An incomplete blastomere.

blastocyst [EMBRYO] A modified blastula characteristic of placental mammals.

blastocyte [EMBRYO] An embryonic cell that is undifferenti-

ated. [INV ZOO] An undifferentiated cell capable of replacing damaged tissue in certain lower animals.

blastoderm [EMBRYO] The blastodisk of a fully formed vertebrate blastula.

blastodisk [EMBRYO] The embryo-forming, protoplasmic disk on the surface of a yolk-filled egg, such as in reptiles, birds, and some fish.

blast-off [AERO ENG] The takeoff of a rocket or missile.

blastogranitic rock [PETR] A metamorphic granitic rock which still has parts of the original granitic texture.

Blastoidea [PALEON] A class of extinct pelmatozoan echinoderms in the subphylum Crinozoa.

blastokinesis [INV ZOO] Movement of the embryo into the yolk in some insect eggs.

blastoma [MED] 1. A tumor whose parenchymal cells have certain embryonal characteristics. 2. A true tumor.

blastomere [EMBRYO] A cell of a blastula.

blastomycosis [MED] A term for two infectious, yeastlike fungus diseases of man: North American blastomycosis, caused by *Blastomyces dermatitidis,* and South American (paracoccidioidomycosis) caused by *Blastomyces brasiliensis.*

blastomylonite [PETR] Rock which has recrystallized after granulation.

blastopelitic [PETR] Descriptive of the structure of metamorphosed argillaceous rocks.

blastophitic [PETR] A metamorphosed rock which once contained lath-shaped crystals partly or wholly enclosed in augite and in which part of the original texture remains.

blastophore [CYTOL] The cytoplasm that is detached from a spermatid during its transformation to a spermatozoon. [INV ZOO] An amorphose core of cytoplasm connecting cells of the male morula of developing germ cells in oligochaetes.

blastopore [EMBRYO] The opening of the archenteron.

blastoporphyritic [PETR] Applied to the textures of metamorphic rocks that are derived from porphyritic rocks; the porphyritic character still remains as a relict feature.

blastopsammite [GEOL] A relict fragment of sandstone that is contained in a metamorphosed conglomerate.

blastopsephitic [GEOL] Descriptive of the structure of metamorphosed conglomerate or breccia.

blastospore [MYCOL] A fungal resting spore that arises by budding.

blastostyle [INV ZOO] A zooid on certain hydroids that lacks a mouth and tentacles and functions to produce medusoid buds.

blastotomy [EMBRYO] Separation of cleavage cells during early embryogenesis.

blastozooid [INV ZOO] A zooid produced by budding.

blast pressure [PHYS] The impact pressure of the air set in motion by an explosion.

blast roasting [MIN ENG] The roasting of finely divided ores by means of a blast to maintain internal combustion within the charge. Also known as roast sintering.

blastula [EMBRYO] A hollow sphere of cells characteristic of the early metazoan embryo.

blastulation [EMBRYO] Formation of a blastula from a solid ball of cleaving cells.

blast wall [ENG] A heavy wall used to isolate buildings or areas which contain highly combustible or explosive materials or to protect a building or area from blast damage when exposed to explosions.

blast wave [PHYS] The air wave set in motion by an explosion.

Blattidae [INV ZOO] The cockroaches, a family of insects in the order Orthoptera.

blaze-of-grating technique [OPTICS] A technique whereby the ruled grooves of a diffraction grating are given a controlled shape so that they reflect as much as 80% of the incoming light into one particular order for a given wavelength.

BLC *See* boundary-layer control.

bleaching [FOOD ENG] *See* decolorizing. [GRAPHICS] An afterprocess in the production of direct positive photographs, in which an oxidizing solution dissolves the negative silver. [TEXT] A process in which natural coloring matter is removed from a fiber, yarn, or fabric to make it white.

bleaching agent [CHEM] An oxidizing or reducing chemical such as sodium hypochlorite, sulfur dioxide, sodium acid

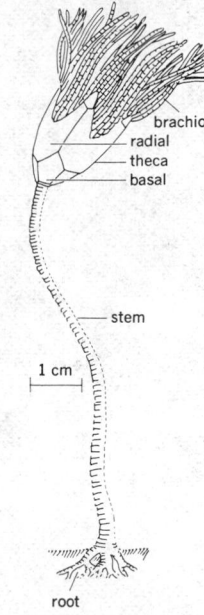

BLASTOIDEA

A drawing of a blastoid, *Pentremites.*

sulfite, or hydrogen peroxide. [FOOD ENG] A chemical, such as an aromatic acyl peroxide or monoperoxyphthalic acid, used to bleach flour, fats, oils, and other edibles.

bleaching assistant [MATER] A material added to textile bleach baths to cause better penetration by the bleach; includes pine oils, borax, and sulfonated oils.

bleaching clay [MATER] Absorbent clay contacted with petroleum and vegetable and other oils to make decolorized products.

bleaching powder [MATER] A mixture of calcium hydroxide, calcium chloride, and calcium hypochlorite that is used as a bleaching agent. Also known as chloride of lime; chlorinated lime.

bleach liquor [MATER] A water solution of calcium hypochlorite used as a laundry and textile bleach, germicide, and deodorant.

bleach spot [GEOL] A green or yellow area in red rocks formed by reduction of ferric oxide around an organic particle. Also known as deoxidation sphere.

bleb [MED] A localized collection of fluid, as serum or blood, in the epidermis.

bleed [CHEM] Diffusion of coloring matter from a substance. [ENG] To let a fluid, such as air or liquid oxygen, escape under controlled conditions from a pipe, tank, or the like through a valve or outlet. [GRAPHICS] The extension of a photograph or other artwork to the very edge of the printed page. [MED] To exude blood from a wound.

bleeder [ENG] A connection located at a low place in an air line or a gasoline container so that, by means of a small valve, the condensed water or other liquid can be drained or bled off from the line or container without discharging the air or gas. [MED] 1. A person subject to frequent hemorrhages, as a hemophiliac. 2. A blood vessel from which there is persistent uncontrolled bleeding. 3. A blood vessel which has escaped closure by cautery or ligature during a surgical procedure.

bleeder current [ELEC] Current drawn continuously from a voltage source to lessen the effect of load changes or to provide a voltage drop across a resistor.

bleeder resistor [ELEC] A resistor connected across a power pack or other voltage source to improve voltage regulation by drawing a fixed current value continuously; also used to dissipate the charge remaining in filter capacitors when equipment is turned off.

BLIMP

Later examples of nonrigid airships.
(*Goodyear Aircraft Corp.*)

bleeder turbine [MECH ENG] A multistage turbine where steam is extracted (bled) at pressures intermediate between throttle and exhaust, for process or feedwater heating purposes.

bleed illustration [GRAPHICS] An illustration printed so that it extends beyond one or more page edges.

bleeding [ENG] Natural separation of a liquid from a liquid-solid or semisolid mixture; for example, separation of oil from a stored lubricating grease, or water from freshly poured concrete. [TEXT] Referring to a fabric in which the dye is not fast and therefore comes out when the fabric is wet.

bleeding canker [PL PATH] A fungus disease of hardwoods caused by *Phytophthora cactorum* and characterized by cankers which exude a reddish ooze on the trunk and branches.

bleeding cycle [MECH ENG] A steam cycle in which steam is drawn from the turbine at one or more stages and used to heat the feedwater. Also known as regenerative cycle.

bleeding disease [PL PATH] A fungus disease of the coconut palm caused by *Ceratostomella paradoxa*, characterized by a rust-colored exudation from cracks on the stem.

bleeding time [PHYSIO] The time required for bleeding to stop after a small puncture wound.

bleed line [GRAPHICS] A line-width change usually due to overexposure or overdeveloping in photography.

bleed-through [GRAPHICS] Of records printed on both sides, the obtrusive show-through of printed matter from one side to the other.

bleed valve [ENG] A small-flow valve connected to a fluid process vessel or line for the purpose of bleeding off small quantities of contained fluid.

blend [MATER] A mixture so combined as to render the parts indistinguishable from one another.

blende See sphalerite.

blended data [ENG] Q point that is the combination of scan data and track data to form a vector.

blended fuel oil [MATER] A mixture of petroleum residual and distillate fuel oils.

blended unconformity [GEOL] An unconformity that is not sharp because the original erosion surface was covered by a thick residual soil that graded downward into the underlying rock.

blended whiskey [FOOD ENG] Whiskey containing at least 20% by volume of 100 proof straight whiskey and mixed with other whiskey or neutral spirits, the mixture being not less than 80 proof at the time of bottling.

blending [TEXT] Producing uniform yarn and fabric by bringing together two or more distinctive types of fibers of different lengths.

blending inheritance [GEN] Inheritance in which the character of the offspring is a blend of those in the parents; a common feature for quantitative characters, such as stature, determined by large numbers of genes and affected by environmental variation.

blending naphtha [MATER] A petroleum distillate used to thin or dilute heavy petroleum stocks (for example, lubricating oil) to simplify handling and processing.

blending problem [IND ENG] A linear programming problem in which it is required to find the least costly mix of ingredients which yields the desired product characteristics.

blending value [ENG] Measure of the ability of an added component (for example, tetraethyllead, isooctane, and aromatics) to affect the octane rating of a base gasoline stock.

blend stop [BUILD] A thin wood strip fastened to the exterior vertical edge of the pulley stile or jamb to hold the sash in position.

Blenniidae [VERT ZOO] The blennies, a family of carnivorous marine fishes in the suborder Blennioidei.

Blennioidei [VERT ZOO] A large suborder of small marine fishes in the order Perciformes that live principally in coral and rock reefs.

blennorrhagia [MED] Excessive discharge of mucus. Also known as blennorrhea.

blennorrhea See blennorrhagia.

Blephariceridae [INV ZOO] A family of dipteran insects in the suborder Orthorrhapha.

blepharism [MED] Spasm of the eyelids causing rapid, repetitive involuntary winking.

blepharitis [MED] Inflammation of the eyelids.

blepharoconjunctivitis [MED] Inflammation of the eyelids and the conjunctiva.

blepharospasm [MED] Spasmodic winking due to spasms of the orbicular muscle of the eyelid.

blight [PL PATH] Any plant disease or injury that results in general withering and death of the plant without rotting.

blight canker [PL PATH] A phase of fire blight characterized by cankers.

blimp [AERO ENG] A name originally applied to nonrigid, pressure-type airships, usually of small size; now applied to airships with volumes of approximately 1,500,000 cubic feet.

blind [ENG] A solid disk inserted at a pipe joint or union to prevent the flow of fluids through the pipe; used during maintenance and repair work as a safety precaution. Also known as blank.

blind approach beacon system [NAV] A pulse-type, ground-based navigation beacon used for runway approach at airports, which sends out signals that produce range and runway position information on the L-scan cathode-ray indicator of an aircraft making an instrument approach. Also known as beam approach beacon system (British usage). Abbreviated babs.

blind bombing See bombing through overcast.

blind controller system [CONT SYS] A process control arrangement that separates the in-plant measuring points (for example, pressure, temperature, and flow rate) and control points (for example, a valve actuator) from the recorder or indicator at the central control panel.

blind drainage See closed drainage.

blind drift [MIN ENG] In a mine, a horizontal passage not yet connected with the other workings.

blind hole [ENG] A type of borehole that does not have the drilling mud or other circulating medium carry the cuttings to the surface.

blinding [ENG] See blanking. [MIN ENG] Interference with the functioning of a screen mesh by a matting of fine materials during screening. Also known as blocking; plugging.

blind landing [AERO ENG] Landing an aircraft solely by the use of instruments because of poor visibility.

blindness [MED] **1.** Loss or absence of the ability to perceive visual images. **2.** The condition of a person having less than 1/10 (20/200 on the Snellen test) normal vision.

blind riser [MET] An internal riser that does not extend to the outer surface of a mold.

blind rollers [OCEANOGR] Long, high swells which have increased in height, almost to the breaking point, as they pass over shoals or run in shoaling water. Also known as blind seas.

blind seas *See* blind rollers.

blind seed [PL PATH] A fungus disease of forage grasses caused by *Phealea temulenta,* resulting in abortion of the seed.

blind spot [ENG] An area on a filter screen where no filtering occurs. Also known as dead area. [PHYSIO] A place on the retina of the eye that is insensitive to light, where the optic nerve passes through the eyeball's inner surface.

blind trial [STAT] A trial in which experimenters and subjects are kept uninformed as to whether or not treatment has been given.

blind valley [GEOL] A valley that has been made by a spring from an underground channel which emerged to form a surface stream, and that is enclosed at the head of the stream by steep walls.

blind zone [COMMUN] Area from which echoes cannot be received; generally, an area shielded from the transmitter by some natural obstruction and therefore from which there can be no return.

B line *See* index register.

blink [MECH] A unit of time equal to 10^{-5} day or to 0.864 second. [METEOROL] A brightening of the base of a cloud layer, caused by the reflection of light from a snow- or ice-covered surface.

blink comparator [OPTICS] An optical instrument used to alternately view two pictures in the same visual field in rapid succession, to detect small differences in similar images.

blinker light [COMMUN] A light that can be flashed on and off to send a coded message, especially from ship to ship.

blinking [COMMUN] Method of providing information in pulse systems by modifying the signal at its source so that signal presentation on the display scope alternately appears and disappears; in loran, this indicates that a station is malfunctioning. [ELECTR] Electronic-countermeasures technique employed by two aircraft separated by a short distance and within the same azimuth resolution so as to appear as one target to a tracking radar; the two aircraft alternately spot-jam, causing the radar system to oscillate from one place to another, making an accurate solution of a fire control problem impossible. [NAV] Regular shifting right and left or alternate appearance and disappearance of a loran signal to indicate that the signals of a pair of stations are out of synchronization.

blink microscope [OPTICS] A blink comparator which magnifies the compared pictures.

blip [ELECTR] **1.** The display of a received pulse on the screen of a cathode-ray tube. Also known as pip. **2.** An ideal infrared radiation detector that detects with unit quantum efficiency all of the radiation in the signal for which the detector was designed, and responds only to the background radiation noise that comes from the field of view of the detector.

blip-scan ratio [ELECTR] The ratio of the number of times a target appears on a radarscope to the number of times it could have been seen.

blister [ENG] A raised area on the surface of a metallic or plastic object caused by the pressure of gases developed while the surface was in a partly molten state, or by diffusion of high-pressure gases from an inner surface. [GEOL] A dome-like protuberance caused by the buckling of the cooling crust of a molten lava before the flowing mass has stopped. [MED] A local swelling of the skin resulting from the accumulation of serous fluid between the epidermis and true skin. [MIN ENG] A protrusion, more or less circular in plan, extending downward into a coal seam. [NUCLEO] A protuberance that

sometimes develops on the surface of a nuclear-reactor fuel element during use, generally because of entrapped gases.

blister blight [PL PATH] **1.** A fungus disease of the tea plant caused by *Exobasidium vexans* and characterized by blister-like lesions on the leaves. **2.** A rust disease of Scotch pine caused by *Cronartium asclepiadeum* and characterized by blisterlike lesions on the twigs.

blister canker [PL PATH] A fungus disease of the apple tree caused by *Nummularia discreta* and characterized by rough, black cankers on the trunk and large branches. Also known as apple pox.

blister copper [MET] Impure copper having a blistered appearance and formed by forcing air through molten copper matte.

blister furnace [MET] A furnace for smelting ore to produce blister copper.

blister gas [MATER] Any of several war gases, such as lewisite, which cause burning, inflammation, or tissue destruction internally or externally.

blister hypothesis [GEOL] A theory of the formation of compressional mountains by a process in which radiogenic heat expands and melts a portion of the earth's crust and subcrust, causing a domed regional uplift (blister) on a foundation of molten material that has no permanent strength.

blistering [ENG] The appearance of enclosed or broken macroscopic cavities in a body or in a glaze or other coating during firing.

blister rust [PL PATH] Any of several diseases of pines caused by rust fungi of the genus *Cronartium;* the sapwood and inner bark are affected and blisters are produced externally.

blister spot [PL PATH] A bacterial disease of the apple caused by *Pseudomonas papulans* and characterized by dark-brown blisters on the fruit and cankers on the branches.

blister steel [MET] Raw steel, made from wrought iron by cementation followed by slow cooling, in which blisters are formed by gas attempting to escape from the metal.

blizzard [METEOROL] A severe weather condition characterized by low temperatures and by strong winds bearing a great amount of snow (mostly fine, dry snow picked up from the ground).

BL method [ENG] Research test for gasoline road octanes; the material is tested at five different speeds in an engine of given compression ratio.

bloat [VET MED] Distension of the rumen in cattle and other ruminants due to excessive gas formation following heavy fermentation of legumes eaten wet.

blob [METEOROL] In radar, oscilloscope evidence of a fairly small-scale temperature and moisture inhomogeneity produced by turbulence within the atmosphere.

Bloch equations [SOLID STATE] Approximate equations for the rate of change of magnetization of a solid in a magnetic field due to spin relaxation and gyroscopic precession.

Bloch function [SOLID STATE] A wave function for an electron in a periodic lattice, of the form $u(\mathbf{r}) \exp [i\mathbf{k}\cdot\mathbf{r}]$ where $u(\mathbf{r})$ has the periodicity of the lattice.

Bloch theorem [QUANT MECH] The theorem that the lowest state of a quantum-mechanical system without a magnetic field can carry no current. [SOLID STATE] The theorem that, in a periodic structure, every electronic wave function can be represented by a Bloch function.

Bloch wall [SOLID STATE] A transition layer, with a finite thickness of a few hundred lattice constants, between adjacent ferromagnetic domains. Also known as domain wall.

block [ADP] **1.** A group of words transported as a unit or considered as a unit by virtue of their being stored in successive storage locations. **2.** The set of locations or tape positions in which a group of words is stored or recorded as a unit. [DES ENG] **1.** A metal or wood case enclosing one or more pulleys; has a hook with which it can be attached to an object. **2.** *See* cylinder block. [MIN ENG] A division of a mine, usually bounded by workings but sometimes by survey lines or other arbitrary limits.

block and block *See* chockablock.

block and fall *See* block and tackle.

block and tackle [MECH ENG] Combination of a rope or other flexible material and independently rotating frictionless pulleys. Also known as block and fall.

BLOCK BRAKE

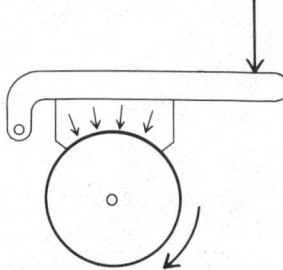

Single-block brake. Block is fixed to operating lever, where force is applied in direction of top arrow.

BLOCK CAVING

Block caving technique in underground mining showing the block on the left undercut and caved in; broken ore is drawn off through bell-shaped draw points.

BLOCKING OSCILLATOR

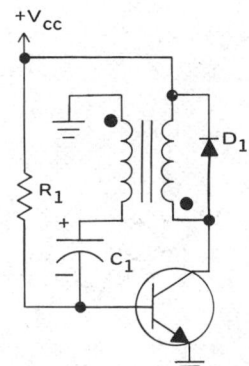

Circuit diagram of a typical free-running blocking oscillator.

block brake [MECH ENG] A brake which consists of a block or shoe of wood bearing upon an iron or steel wheel.

block brazing [MET] The process of joining metals by applying heated blocks to the joint and using a nonferrous filler metal with a melting point above 800°F (427°C).

block caving [MIN ENG] A method of caving where a block, 150–250 feet on a side and several hundred feet high, is induced to cave in after it is undercut; the broken ore is drawn off at a bell-shaped draw point.

block coal [MIN ENG] **1.** A bituminous coal that breaks into large lumps or cubical blocks. **2.** Coal that passes over 5-, 6-, and 8-inch block screens; used in smelting iron.

block coefficient [NAV ARCH] The ratio of the volume of water displaced by a ship to the product of the ship's length, breadth, and draft.

block correction [MAP] A corrected reproduction of a small area of a nautical chart that is pasted to the chart for which it is issued.

block data [ADP] A statement in FORTRAN which declares that the program following is a data specification subprogram.

block diagram [ENG] A diagram in which the essential units of any system are drawn in the form of rectangles or blocks and their relation to each other is indicated by appropriate connecting lines.

blocked F-format data set See FB data set.

blocked impedance [ELEC] The impedance at the input of a transducer when the impedance of the output system is made infinite, as by blocking or clamping the mechanical system.

blocked operation [CHEM ENG] The use of a single chemical or refinery process unit alternately in more than one operation; for example, a catalytic reactor will first produce a chemical product and then will be blocked from the main process stream during catalyst regeneration.

blocked-out ore See developed reserves.

blocked resistance [ENG ACOUS] Resistance of an audio-frequency transducer when its moving elements are blocked so they cannot move; represents the resistance due only to electrical losses.

blocker-type forging [ENG] A type of forging for designs involving the use of large radii and draft angles, smooth contours, and generous allowances.

blockette [ADP] A subdivision of a group of consecutive machine words transferred as a unit, particularly with reference to input and output.

block faulting [GEOL] A type of faulting in which fault blocks are displaced at different orientations and elevations.

block grease [MATER] High-melting-point grease that can be handled as a block or stick at normal temperatures; used for journal bearings. Also known as brick grease.

block-grid keying [COMMUN] Method of keying a continuous-wave transmitter by operating the amplifier stage as an electronic switch; during the spacing interval when the key is open, the bias on the control grid becomes highly negative and prevents the flow of plate current so that the tube has no output; during the marking interval when the key is closed, this bias is removed and full plate current flows.

block hole [ENG] A small hole drilled into a rock or boulder into which an anchor bolt or a small charge or explosive may be placed; used in quarries for breaking large blocks of stone or boulders.

blockhouse [ENG] **1.** A reinforced concrete structure, often built underground or half-underground, and sometimes dome-shaped, to provide protection against blast, heat, or explosion during rocket launchings or related activities, and usually housing electronic equipment used in launching the rocket. **2.** The activity that goes on in such a structure.

block identifier [ADP] A means of identifying an area of storage in FORTRAN so that this area may be shared by a program and its subprograms.

blocking [ADP] Combining two or more computer records into one block. [CHEM] Undesired adhesion of granular particles; often occurs with damp powders or plastic pellets in storage bins or during movement through conduits. [ELECTR] **1.** Applying a high negative bias to the grid of an electron tube to reduce its anode current to zero. **2.** Overloading a receiver by an unwanted signal so that the automatic gain control reduces the response to a desired signal. **3.** Distortion occurring in a resistance-capacitance-coupled electron tube amplifier stage when grid current flows in the following tube. [MET] **1.** A preliminary hot-forging operation which imparts an approximate shape to the rough stock. **2.** Reducing the oxygen content of the bath in an open-hearth furnace. [MIN ENG] See blinding. [STAT] The grouping of sample data into subgroups with similar characteristics.

blocking and wedging [MIN ENG] A method of holding mine timber sets in place; blocks of wood are set on the caps directly over the post supports and have a grain of block parallel with the top of the cap; wedges are driven tightly between the blocks and the roof.

blocking anticyclone See blocking high.

blocking capacitor See coupling capacitor.

blocking high [METEOROL] Any high (or anticyclone) that remains nearly stationary or moves slowly compared to the west-to-east motion upstream from its location, so that it effectively blocks the movement of migratory cyclones across its latitudes. Also known as blocking anticyclone.

blocking layer See depletion layer.

blocking oscillator [ELECTR] A relaxation oscillator that generates a short-time-duration pulse by using a single transistor or electron tube and associated circuitry. Also known as squegger; squegging oscillator.

blocking oscillator driver [ELECTR] Circuit which develops a square pulse used to drive the modulator tubes, and which usually contains a line-controlled blocking oscillator that shapes the pulse into the square wave.

block input [ADP] **1.** A block of computer words considered as a unit and intended or destined to be transferred from an internal storage medium to an external destination. **2.** See output area.

block lava [GEOL] Lava flows which occur as a tumultuous assemblage of angular blocks.

block length [ADP] The total number of records, words, or characters contained in one block.

block loading [ADP] A program loading technique in which the control sections of a program or program segment are loaded into contiguous positions in main memory.

block mountain [GEOL] A mountain formed by the combined processes of uplifting, faulting, and tilting. Also known as fault-block mountain.

block plane [DES ENG] A small type of hand plane, designed for cutting across the grain of the wood and for planing end grains.

block polymer [ORG CHEM] A copolymer whose chain is composed of alternating sequences of identical monomer units.

block printing [GRAPHICS] The earliest form of printing, involving the cutting of crude pictures and lettering on blocks of wood.

block protector [ELEC] Rectangular piece of carbon, bakelite with a metal insert, or porcelain with a carbon insert which, in combination with each other, make one element of a protector; they form a gap which will break down and provide a path to ground for excessive voltages.

block signal system [CONT SYS] An automatic railroad traffic control system in which the track is sectionalized into electrical circuits to detect the presence of trains, engines, or cars.

block standby [ADP] Locations always set aside in storage for communication with buffers in order to make more efficient use of such buffers.

block system [MIN ENG] A system of pillars in which a series of entries, rooms, and crosscuts are driven to divide the coal into blocks of about equal size which are then extracted on retreat.

blödite See bloedite.

bloedite [MINERAL] $MgSO_4 \cdot Na_2SO_4 \cdot 4H_2O$ A white or colorless monoclinic mineral consisting of magnessium sodium sulfate. Also spelled blödite. Also known as astrakanite; astrochanite.

blomstrandine See priorite.

blondel See apostilb.

Blondel-Rey law [OPTICS] A law utilized to determine the apparent point brilliance of a flashing light.

blood [HISTOL] A fluid connective tissue consisting of the plasma and cells that circulate in the blood vessels.

blood agar [MICROBIO] A nutrient microbiologic culture medium enriched with whole blood and used to detect hemolytic strains of bacteria.

blood bank [ENG] A place for storing whole blood or plasma under refrigeration.

blood blister [MED] A blister that is filled with blood.

blood cell [HISTOL] An erythrocyte or a leukocyte.

blood chimerism [GEN] Having red blood cells of two genetic types.

blood count [PATH] Determination of the number of white and red blood cells in a definite volume of blood.

blood crisis [MED] The sudden appearance of large numbers of nucleated erythrocytes in the circulating blood.

blood disease [PL PATH] A bacterial disease affecting the vascular tissues of the banana in the Celebes; believed to be caused by *Xanthomonas celebensis* and characterized by blighting of the leaves and reddish-brown rot of the fruit.

blood dyscrasia [MED] Any abnormal condition of the formed elements of blood or of the constituents required for clotting.

blood gas [ORD] War gas which, when absorbed into the body, primarily by breathing, affects body functions by interfering with transfer of oxygen from the lungs via the blood to tissues.

blood group [IMMUNOL] An immunologically distinct, genetically determined class of human erythrocyte antigens, identified as A, B, AB, and O.

blood island [EMBRYO] One of the areas in the yolk sac of vertebrate embryos allocated to the production of the first blood cells.

blood-plate hemolysis [MICROBIO] Destruction of red blood cells in a blood agar medium by a bacterial toxin.

blood platelet *See* thrombocyte.

blood poisoning *See* septicemia.

blood pressure [PHYSIO] Pressure exerted by blood on the walls of the blood vessels.

blood rain [METEOROL] Rain of a reddish color caused by dust particles containing iron oxide that were picked up by the raindrops during descent.

bloodstone [MINERAL] 1. A form of deep green chalcedony flecked with red jasper. Also known as heliotrope; oriental jasper. 2. *See* hematite.

bloodstream [PHYSIO] The flow of blood in its circulation through the body.

blood sugar [BIOCHEM] The carbohydrate, principally glucose, of the blood.

blood test [PATH] 1. A serologic test for syphilis. 2. A blood count. 3. A test for detection of blood, usually one based on the peroxidase activity of blood, such as the benzidine test or guaiac test.

blood typing [IMMUNOL] Determination of an individual's blood group.

blood vessel [ANAT] A tubular channel for blood transport.

bloom [BOT] 1. An individual flower. Also known as blossom. 2. To yield blossoms. 3. The waxy coating that appears as a powder on certain fruits, such as plums, and leaves, such as cabbage. [ECOL] Colored area on the surface of bodies of water caused by heavy planktonic growth. [ENG] 1. Fluorescence in lubricating oils or a cloudy surface on varnished or enameled surfaces. 2. To apply an antireflection coating to glass. [GEOL] *See* blossom. [GRAPHICS] A milky or foggy defect that may appear on the surface of a varnished painting; caused by moisture. [MINERAL] *See* efflorescence. [MET] 1. A semifinished bar of metal formed from an ingot and having a rectangular cross section exceeding 36 square inches (232 square centimeters). 2. To hammer or roll metal in order to make its surface bright.

bloomer [MET] A furnace used to shape wrought iron directly from ore.

blooming [ELECTR] 1. Defocusing of television picture areas where excessive brightness results in enlargement of spot size and halation of the fluorescent screen. 2. An increase in radarscope spot size due to an increase in signal intensity.

blooming mill [MET] A rolling mill for making blooms from ingots. Also known as cogging mill.

blooper [COMMUN] An oscillating radio receiver that is radiating an undesired signal.

blossom [BOT] *See* bloom. [GEOL] The oxidized or decom-posed outcrop of a vein or coal bed. Also known as bloom.

blossom-end rot [PL PATH] 1. Any rot disease of fruit that originates at the blossom end. 2. A physiological disease of tomato believed to be caused by extreme fluctuations in available moisture; characterized by shallow leathery depressions with a water-soaked appearance around the tip end of the fruit.

blotter [ENG] A disk of compressible material used between a grinding wheel and its flanges to avoid concentrated stress.

blotter model [PETRO ENG] An analysis device in which the analogous movement of copper ammonium or zinc ammonium ions in blotting paper or gelatin indicates oil-well injection-fluid movement through porous underground reservoirs.

blotter press [CHEM ENG] A plate-and-frame filter in which the filter medium is blotting paper.

blotting paper [MATER] An unsized paper used to absorb excess ink from penned letters or signatures; also used for other applications where a soft, spongy paper is required.

blow [ELEC] Opening of a circuit because of excess current, particularly when the current is heavy and a melting or breakdown point is reached.

blowback [CHEM ENG] 1. A continuous stream of liquid or gas bled through air lines from instruments and to the process line being monitored; prevents process fluid from backing up and contacting the instrument. 2. Reverse flow of fluid through a filter medium to remove caked solids. Also known as backwash. [GRAPHICS] To enlarge, or make an enlargement of, an image.

blowback gun [ORD] An automatic gun utilizing the pressure of the propellant gases to force the bolt to the rear, independently of the barrel which does not move relative to the receiver.

blowball [BOT] A fluffy seed ball, as of the dandelion.

blowby [MECH ENG] Leaking of fluid between a cylinder and its piston during operation.

blowcase [CHEM ENG] A cylindrical or spherical corrosion- and pressure-resistant container from which acid is forced by compressed air to the agitator; used in manufacture of acids but largely superseded by centrifugal pumps. Also known as acid blowcase; acid egg.

blowdown [CHEM ENG] Removal of liquids or solids from a process vessel or storage vessel or a line by the use of pressure. [METEOROL] A wind storm that causes trees or structures to be blown down.

blowdown line [CHEM ENG] A large conduit to receive and confine fluids forced by pressure from process vessels.

blowdown stack [CHEM ENG] A vertical stack or chimney into which the contents of a chemical or petroleum process unit are emptied in case of an operational emergency.

blowdown tunnel [AERO ENG] A wind tunnel in which stored compressed gas is allowed to expand through a test section to provide a stream of gas or air for model testing.

blowdown turbine [AERO ENG] A turbine attached to a reciprocating engine which receives exhaust gases separately from each cylinder, utilizing the kinetic energy of the gases.

blower [MECH ENG] A fan which operates where the resistance to gas flow is predominantly downstream of the fan.

blowhole [GEOL] A longitudinal tunnel opening in a sea cliff, on the upland side away from shore; columns of sea spray are thrown up through the opening, usually during storms. [MET] A pocket of air or gas formed in a metal during solidification. [VERT ZOO] The nostril on top of the head of cetacean mammals.

blow in [MET] To put a blast furnace into operation. [PETRO ENG] Of an oil well, to begin sending forth oil or gas.

blowing *See* blow molding.

blowing agent [MATER] A chemical added to plastics and rubbers that generates inert gases on heating, causing the resin to assume a cellular structure. Also known as foaming agent.

blowing boundary-layer control [AERO ENG] A technique that is used in addition to purely geometric means to control boundary-layer flow; it consists of reenergizing the retarded flow in the boundary layer by supplying high-velocity flow through slots or jets on the surface of the body.

blowing cave [GEOL] A cave with an alternating air movement. Also known as breathing cave.

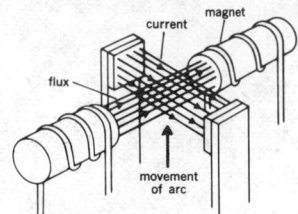

BLOWOUT COIL

The relation of direction of current, magnetic flux, and movement of arc in a blowout coil.

BLUEBERRY

A cluster of blueberries.

BLUEFISH

The bluefish, an excellent food and game fish.

blowing dust [METEOROL] Dust picked up locally from the surface of the earth and blown about in clouds or sheets.

blowing fan *See* forcing fan.

blowing pressure [ENG] Pressure of the air or other gases used to inflate the parison in blow molding.

blowing sand [METEOROL] Sand picked up from the surface of the earth by the wind and blown about in clouds or sheets.

blowing snow [METEOROL] Snow lifted from the surface of the earth by the wind to a height of 6 feet or more (higher than drifting snow) and blown about in such quantities that horizontal visibility is restricted.

blowing spray [METEOROL] Spray lifted from the sea surface by the wind and blown about in such quantities that horizontal visibility is restricted.

blowing still [CHEM ENG] A still or process column in which blown or oxidized asphalt is made.

blowing-up furnace [MET] A furnace used for sintering ore and for the volatilization of lead and zinc.

blow molding [ENG] A method of producing hollow objects, such as bottles, by injecting a parison (blob) of hot melt into a hollow, two-part mold and then inflating the parison against the cool mold surface, where it freezes into shape. Also known as blowing.

blown-fuse indicator [ELEC] A neon warning light connected across a fuse so that it lights when the fuse is blown.

blown glass [ENG] Glassware formed by blowing air into a ball of liquefied glass until it reaches the desired shape.

blown oil [MATER] A vegetable or animal oil that has been agitated and partially oxidized by a current of warm air or oxygen, including castor, whale, fish, rape, and linseed oils; used in paints, lubricants, and plasticizers.

blown tubing [ENG] A flexible thermoplastic film tube made by applying pressure inside a molten extruded plastic tube to expand it prior to cooling and winding flat onto rolls.

blowoff [AERO ENG] The action of applying an explosive force and separating a package section away from the remaining part of a rocket vehicle or reentry body, usually to retrieve an instrument or to obtain a record made during early flight.

blowout [ELEC] The melting of an electric fuse because of excessive current. [ELECTROMAG] The extinguishing of an electric arc by deflection in a magnetic field. Also known as magnetic blowout. [ENG] **1.** The bursting of a container (such as a tube pipe, pneumatic tire, or dam) by the pressure of the contained fluid. **2.** The rupture left by such bursting. **3.** The abrupt escape of air from the working chamber of a pneumatic caisson. [GEOL] Any of the various trough-, saucer-, or cuplike hollows formed by wind erosion on a dune or other sand deposit. [HYD] A bubbling spring which bursts from the ground behind a river levee when water at flood stage is forced under the levee through pervious layers of sand or silt. Also known as sand boil.

blowout coil [ELECTROMAG] A coil that produces a magnetic field in an electrical switching device for the purpose of lengthening and extinguishing an electric arc formed as the contacts of the switching device part to interrupt the current.

blowout dune [GEOL] The accumulation of sand from blowout troughs or basins, especially where the accumulation is large.

blowout magnet [ELECTROMAG] An electromagnet or permanent magnet used to deflect and extinguish the arc formed when a high-current circuit breaker or switch is opened.

blowpipe [BIOL] A small tube, tapering to a straight or slightly curved tip, used in anatomy and zoology to reveal or clean a cavity. [ENG] **1.** A long, straight tube, used in glass blowing, on which molten glass is gathered and worked. **2.** A small, tapered, and frequently curved tube that leads a jet, usually of air, into a flame to concentrate and direct it; used in flame tests in analytical chemistry and in brazing and soldering of fine work. **3.** *See* blowtorch.

blowpipe reaction analysis [ANALY CHEM] A method of analysis in which a blowpipe is used to heat and decompose a compound or mineral; a characteristic color appears in the flame or a colored crust appears on charcoal.

blowpit *See* blowtank.

blow rate [ENG] The speed of the cycle at which air or an inert gas is applied intermittently during the forming procedure of plastics blow molding.

blowtank [CHEM ENG] A tank or pit, used in papermaking, into which the contents of a digester are blown upon completion of a cook. Also known as blowpit.

blowtorch [ENG] A small, portable blast burner which operates either by having air or oxygen and gaseous fuel delivered through tubes or by having a fuel tank which is pressured by a hand pump. Also known as blast lamp; blowpipe.

blowup [GRAPHICS] An enlargement of a photographic print, or of a detail of the print.

blowup ratio [ENG] The ratio of a plastics mold cavity or die diameter to the diameter of the final product after expansion by blow molding.

blubber [INV ZOO] A large sea nettle or medusa. [VERT ZOO] A thick insulating layer of fat beneath the skin of whales and other marine mammals.

blubber oil *See* whale oil.

blue [OPTICS] The hue evoked in an average observer by monochromatic radiation having a wavelength in the approximate range from 455 to 492 nanometers; however, the same sensation can be produced in a variety of other ways.

blue annealing [MET] Softening metal sheets by heating in an open furnace and cooling in air; bluish oxide forms on the metal surface.

blue asbestos *See* crocidolite.

blue baby [MED] An infant with congenital cyanosis due to cardiac or pulmonary defect, causing shunting of unoxygenated blood into the systemic circulation.

blue band [GEOL] **1.** A layer of bubble-free, dense ice found in a glacier. **2.** A bluish clay found as a thin, persistent bed near the base of No. 6 coal everywhere in the Illinois-Indiana coal basin.

blueberry [BOT] Any of several species of plants in the genus *Vaccinium* of the order Erecales; the fruit, a berry, occurs in clusters on the plant.

blue brittleness [MET] Loss of ductility noted for some steels when heated to 400–600°F (204–316°C), the blue heat range.

blue cap [MIN ENG] The characteristic blue halo, or tip, of the flame of a safety lamp when firedamp is present in the air.

blue comb [VET MED] A disease of domestic fowl and other birds that resembles Bright's disease in man; may be a viral disease or caused by excessive salt intake.

blue copper ore *See* azurite.

bluefish [VERT ZOO] *Pomatomus saltatrix*. A predatory fish in the order Perciformes. Also known as skipjack.

blue flash *See* green flash.

blue gas [MATER] A gas consisting chiefly of carbon monoxide and hydrogen, formed by the action of steam upon hot coke; used mainly as a source of hydrogen and in synthesis of other chemical compounds. Also known as blue water gas.

blue glow [ELECTR] A glow normally seen in electron tubes containing mercury vapor, due to ionization of the mercury molecules. [MET] Luminescence emitted by certain metallic oxides when heated.

bluegrass [BOT] The common name for several species of perennial pasture and lawn grasses in the genus *Poa* of the order Cyperales.

blue-green algae [BOT] The common name for members of the Cyanophyceae.

blue-green flame *See* green flash.

blue ground [GEOL] **1.** The decomposed peridotite or kimberlite that carries the diamonds in the South African mines. **2.** Strata of the coal measures, consisting principally of beds of hard clay or shale.

blue ice [HYD] Pure ice in the form of large, single crystals that is blue owing to the scattering of light by the ice molecules; the purer the ice, the deeper the blue.

blueing *See* bluing.

blue iron earth *See* vivianite.

blue lead *See* galena.

blueline [GRAPHICS] A blue image or outline printed on paper or plastic sheeting and used as a guide for drafting, stripping, or layout; it does not reproduce when the finished work is photographed.

blue magnetism [GEOPHYS] The magnetism displayed by the south-seeking end of a freely suspended magnet; this is the magnetism of the earth's north magnetic pole.

blue malachite *See* azurite.

blue metal [GEOL] The common fine-grained blue-gray mudstone which is part of many of the coal beds of England.

blue mold [MYCOL] Any fungus of the genus *Penicillium*.

blue mud [GEOL] A combination of terrigenous and deep-sea sediments having a bluish gray color due to the presence of organic matter and finely divided iron sulfides.

blue nevus [MED] A nevus composed of spindle-shaped pigmented melanocytes in the middle and lower two-thirds of the dermis.

blueprint [GRAPHICS] A contact print, with white lines on a blue background, of a drawing; made on linen or on ferroprussiate paper and developed in water or a special solution.

blueprint machine [GRAPHICS] A machine which exposes and develops blueprint paper.

blue rot [PL PATH] A fungus disease of conifers caused by members of the genus *Ceratostomella*, characterized by blue discoloration of the wood. Also known as bluing.

blueschist facies [PETROL] High-pressure, low-temperature metamorphism associated with subduction zones which produces a broad mineral association including glaucophane, actinolite, jadeite, aegirine, lawsonite, and pumpellyite.

blue-sky scale *See* Linke scale.

blue spar *See* lazulite.

blue stain [PL PATH] A bluish stain of the sapwood of many trees caused by several fungi, especially of the genera *Fusarium*, *Ceratostomella*, and *Penicillium*.

blue star [ASTRON] A star of spectral type O, B, A, or F according to the Draper catalog.

bluestem grass [BOT] The common name for several species of tall, perennial grasses in the genus *Andropogon* of the order Cyperales.

bluestone [MINERAL] *See* chalcanthite. [PETR] 1. A sandstone that is highly argillaceous and of even texture and bedding. 2. The commercial name for a feldspathic sandstone that is dark bluish gray; it is easily split into thin slabs and used as flagstone.

bluetongue [VET MED] 1. A disease of African horses characterized by lesions that are most pronounced about the head. Also known as thickhead. 2. A virus disease of African sheep characterized by hyperemia, cyanosis, and punctate hemorrhages and by swelling and sloughing of the epithelium about the mouth and tongue.

blue vitriol *See* chalcanthite.

blue water gas *See* blue gas.

bluff [GEOGR] 1. A steep, high bank. 2. A broad-faced cliff.

bluff body [AERO ENG] A body having a broad, flattened front, as in some reentry vehicles.

bluff-bowed ship [NAV ARCH] A ship that has a full broad bow.

bluing [MET] Also spelled blueing. 1. Formation of a bluish oxide film on polished steel; improves appearance and provides some corrosion resistance. 2. Heating of formed springs to reduce internal stress. [PL PATH] *See* blue rot.

blunger [ENG] 1. A large spatula-shaped wooden implement used to mix clay with water. 2. A vat, containing a rotating shaft with fixed knives, for mixing clay and water into slip.

blunging [ENG] The mixing or suspending of ceramic material in liquid by agitation, to form slip.

blunt dissection [MED] In surgery, the exposure of structures or separation of tissues without cutting.

blunt file [DES ENG] A file whose edges are parallel.

blunting [DES ENG] Slightly rounding a cutting edge to reduce the probability of edge chipping.

BM *See* barrels per month; bench mark.

BM-AGA coal test [ENG] A laboratory test developed jointly by the United States Bureau of Mines and a committee of the American Gas Association which forecasts the quality of coke producible in commercial practice.

B meson [PARTIC PHYS] An elementary particle with strong nuclear interactions, baryon number $B = 0$, and mass 1237 MeV.

BMR *See* basal metabolic rate.

BMT *See* basic motion-time study.

boa [VERT ZOO] Any large, nonvenomous snake of the family Boidae in the order Squamata.

boar *See* wild boar.

board coal *See* bituminous wood.

board drop hammer [MECH ENG] A type of drop hammer in which the ram is attached to wooden boards which slide between two rollers; after the ram falls freely on the forging, it is raised by friction between the rotating rollers. Also known as board hammer.

board-foot [ENG] Unit of volume in measuring lumber; equals 144 cubic inches, or the volume of a board 1 foot square and 1 inch thick. Abbreviated bd-ft.

board hammer *See* board drop hammer.

boarding [ENG] 1. A batch of boards. 2. Covering with boards.

board measure [ENG] Measurement of lumber in board-feet. Abbreviated bm.

Board of Trade unit *See* kilowatt-hour.

boast [ENG] 1. To shape stone or curve furniture roughly in preparation for finer work later on. 2. To finish the face of a building stone by cutting a series of parallel grooves.

Boas' test [PATH] A test that uses an alcoholic solution of resorcinol and sucrose to determine the presence of free hydrochloric acid in gastric juices.

boasting chisel [DES ENG] A broad chisel used in boasting stone.

boat [CHEM] A platinum or ceramic vessel for holding a substance for analysis by combustion. [NAV ARCH] A small watercraft.

boat boom [NAV ARCH] A spar swung out from the side of a vessel, to which boats can be securely attached. Also known as boat spar; riding boom.

boat deck [NAV ARCH] The deck on a ship that usually is used to stow lifeboats.

boat fall [NAV ARCH] One of a pair of tackles used in lowering or hoisting a boat from or to the davits of a ship.

boat hook [NAV ARCH] A long rod with a knob on one end and a metal point and hook at the other; used to push or pull other boats, logs, or objects from or to the side of a boat or to engage lines, rings, or buoys.

boat plug [NAV ARCH] A threaded plug that stops up the drainage hole of a boat near the keel and can be removed when the boat is dry-docked to drain out bilge water.

boat spar *See* boat boom.

boat spike [DES ENG] A long, square spike used in construction with heavy timbers. Also known as barge spike.

boattail [AERO ENG] Of an elongated body such as a rocket, the rear portion having decreasing cross-sectional area.

Bobasatranidae [PALEON] A family of extinct palaeonisciform fishes in the suborder Platysomoidei.

bobbin [ELECTROMAG] An insulated spool serving as a support for a coil. [TEXT] A cylinder with projecting edges at one or both ends and a hole along the axis, used for winding twisted strands of textile fiber, thread, or yarn.

bobbin core [ELECTROMAG] A magnetic core having a form or bobbin on which the ferromagnetic tape is wrapped for support of the tape.

bobbing [ELECTR] Fluctuation of the strength of a radar echo, or its indication on a radarscope, due to alternate interference and reinforcement of returning reflected waves.

Bobillier's law [MECH] The law that, in general plane rigid motion, when *a* and *b* are the respective centers of curvature of points *A* and *B*, the angle between *Aa* and the tangent to the centrode of rotation (pole tangent) and the angle between *Bb* and a line from the centrode to the intersection of *AB* and *ab* (collineation axis) are equal and opposite.

bock beer [FOOD ENG] A dark, heavy, rich beer.

BOD *See* biochemical oxygen demand.

Bodansky unit [BIOL] A unit for the standarization of phosphatase based on the activity of phosphatase toward β-glycerophosphate.

Bode diagram [ELECTR] A diagram in which the phase shift or the gain of an amplifier, a servomechanism, or other device is plotted against frequency to show frequency response; logarithmic scales are customarily used for gain and frequency.

Bodenstein number [FL MECH] A numberless group used in the study of diffusion in reactors.

Bode's law [ASTRON] An empirical law giving mean distances of planets to the sun by the formula $a = 0.4 + 0.3 \times 2^n$, where *a* is in astronomical units and *n* equals $-\infty$ for Mercury, 0 for Venus, 1 for Earth, and so on; the asteroids are included as planets. Also known as Titius Bode law.

bodily tide *See* earth tide.

BOA

The boa constrictor (*Constrictor constrictor*), an arboreal-aquatic member of the boa family.

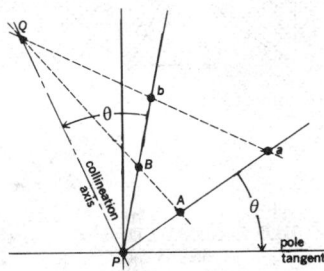

BOBILLIER'S LAW

Application of Bobillier's law of four-bar linkage consisting of fixed member *ab*, cranks *aA* and *bB* and connecting rod *AB*. Law states that angle θ between *Aa* and pole tangent equals angle θ between *Bb* and collineation axis joining centrode *P* and intersection *Q* of *AB* and *ab*.

BODONIDAE

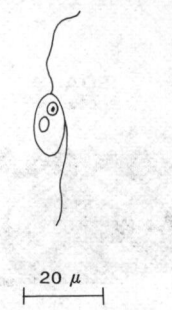

Bodo, a genus in the family Bodonidae, showing the two unequally long flagella.

BODY-CENTERED LATTICE

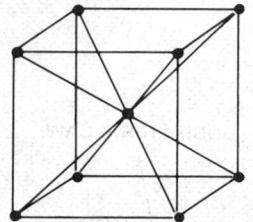

Drawing of a body-centered cubic lattice.

BOHR-SOMMERFELD THEORY

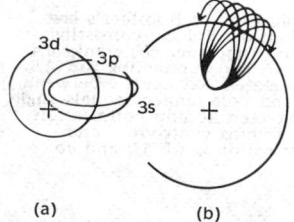

(a)   (b)

Possible elliptical orbits, according to the Bohr-Sommerfeld theory. (a) The three permitted orbits for n=3. (b) Precession of the 3s orbit caused by the relativistic variation of mass.

BOILER AIR HEATER

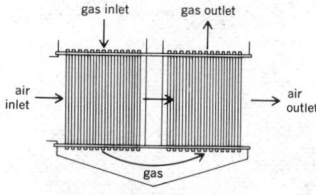

A two-gas single-air-pass tubular heater, a recuperative type of boiler air heater.

Bodonidae [INV ZOO] A family of protozoans in the order Kinetoplastida characterized by two unequally long flagella, one of them trailing.

BOD test See biochemical oxygen demand test.

body [AERO ENG] **1.** The main part or main central portion of an airplane, airship, rocket, or the like; a fuselage or hull. **2.** Any fabrication, structure, or other material form, especially one aerodynamically or ballistically designed; for example, an airfoil is a body designed to produce an aerodynamic reaction. [GEOGR] A separate entity or mass of water, such as an ocean or a lake. [GEOL] An ore body, or pocket of mineral deposit. [MATER] The consistency or viscosity of fluid materials, such as lubricating oils, paints, and cosmetics. [MECH ENG] The part of a drill which runs from the outer corners of the cutting lips to the shank or neck.

body angle [AERO ENG] The angle which the longitudinal axis of the airframe makes with some selected line.

body axis [AERO ENG] Any one of a system of mutually perpendicular reference axes fixed in an aircraft or a similar body and moving with it.

body burden [NUCLEO] The amount of radioactive material present in the body of a man or animal.

body capacitance [ELEC] Capacitance existing between the human hand or body and a circuit.

body cavity [ANAT] The peritoneal, pleural, or pericardial cavities, or the cavity of the tunica vaginalis testis.

body-centered lattice [CRYSTAL] A space lattice in which the point at the intersection of the body diagonals is identical to the points at the corners of the unit cell.

body motion [IND ENG] Motion of parts of a human body requiring a change of posture or weight distribution.

body of revolution [MATH] A symmetrical body having the form described by rotating a plane curve about an axis in its plane.

body rhythm [PHYSIO] Any bodily process having some degree of regular periodicity.

body-righting reflex [PHYSIO] A postural reflex, initiated by the asymmetric stimulation of the body surface by the weight of the body, so that the head tends to assume a horizontal position.

boehmite [MINERAL] AlO(OH) Gray, brown, or red orthorhombic mineral that is a major constituent of some bauxites.

boehm lamellae [GEOL] Lines or bands with dusty inclusions that are subparallel to the basal plane of quartz.

Boettger's test [ANALY CHEM] A test for the presence of saccharides, utilizing the reduction of bismuth subnitrate to metallic bismuth, a precipitate.

bog [ECOL] A plant community that develops and grows in areas with permanently waterlogged peat substrates. Also known as moor; quagmire.

bogen structure [GEOL] The structure of vitric tuffs composed largely of shards of glass.

bogey See bogie.

bog harrow [AGR] A type of disk harrow with extra-large notched disks.

boghead cannel shale [GEOL] A coaly shale that contains much waxy or fatty algae.

boghead coal [GEOL] Bituminous or subbituminous coal containing a large proportion of algal remains and volatile matter; similar to cannel coal in appearance and combustion.

bogie Also spelled bogey; bogy. [AERO ENG] A type of landing-gear unit consisting of two sets of wheels in tandem with a central strut. [ENG] **1.** A supporting and aligning wheel or roller on the inside of an endless track. **2.** A low truck or cart of solid build. **3.** A truck or axle to which wheels are fixed, which supports a railroad car, the leading end of a locomotive, or the end of a vehicle (such as a gun carriage) and which is allowed to swivel under it. **4.** A railroad car or locomotive supported by a bogie. [MECH ENG] The drive-wheel assembly and supporting frame comprising the four rear wheels of a six-wheel truck, mounted so that they can adjust themselves to sharp curves and irregularities in the road. [MIN ENG] A small truck or trolley upon which a bucket is carried from the shaft to the spoil bank.

bog iron ore [MINERAL] A soft, spongy, porous deposit of impure hydrous iron oxides formed in bogs, marshes, swamps, peat mosses, and shallow lakes by precipitation from iron-bearing waters and by the oxidation action of algae, iron bacteria, or the atmosphere. Also known as lake ore; limnite; marsh ore; meadow ore; morass ore; swamp ore.

bog-mine ore See bog ore.

bog moss [ECOL] Moss of the genus *Sphagnum* occurring as the characteristic vegetation of bogs.

bog ore [MINERAL] A poorly stratified accumulation of earthy metallic mineral substances, consisting mainly of oxides, that are formed in bogs, marshes, swamps, and other low-lying moist places. Also known as bog-mine ore.

bogy See bogie.

Bohemian glass See hard glass.

Bohemian ruby See rose quartz.

Bohemian topaz [MINERAL] Yellow quartz when cut as a gem.

Bohr atom [ATOM PHYS] An atomic model having the structure postulated in the Bohr theory.

Bohr-Breit-Wigner theory See Breit-Wigner theory.

Bohr magneton [ATOM PHYS] The amount $he/4\pi mc$ of magnetic moment, where h is Plank's constant, e and m are the charge and mass of the electron, and c is the speed of light.

Bohr orbit [ATOM PHYS] One of the electron paths about the nucleus in Bohr's model of the hydrogen atom.

Bohr radius [ATOM PHYS] The radius of the ground-state orbit of the hydrogen atom in the Bohr theory.

Bohr's correspondence principle See correspondence principle.

Bohr-Sommerfeld theory [ATOM PHYS] A modification of the Bohr theory in which elliptical as well as circular orbits are allowed.

Bohr theory [ATOM PHYS] A theory of atomic structure postulating an electron moving in one of certain discrete circular orbits about a nucleus with emission or absorption of electromagnetic radiation necessarily accompanied by transitions of the electron between the allowed orbits.

Bohr–van Leeuwen theorem [QUANT MECH] The theorem that magnetism is inexplicable in classical physics and is a quantum phenomenon.

Bohr-Wheeler theory of fission [NUC PHYS] A theory accounting for the stability of a nucleus against fission by treating it as a droplet of incompressible and uniformly charged liquid endowed with surface tension.

Boidae [VERT ZOO] The boas, a family of nonvenomous reptiles of the order Squamata, having teeth on both jaws and hindlimb rudiments.

boil See furuncle.

boil disease [VET MED] A protozoan disease of fish caused by *Myxobolus pfeiffer* that forms large tumorous masses in the muscles and connective tissues, finally causing death.

boiled oil [MATER] A drying oil in which metallic driers are added during the cooking; the reaction releases water, resulting in a boiling of the batch.

boiler [MECH ENG] A water heater for generating steam.

boiler air heater [MECH ENG] A component of a steam-generating unit that transfers heat from the products of combustion after they have passed through the steam-generating and superheating sections to combustion air, which recycles heat to the furnace.

boiler casing [MECH ENG] The gastight structure surrounding the component parts of a steam generator.

boiler circulation [MECH ENG] Circulation of water and steam in a boiler, which is required to prevent overheating of the heat-absorbing surfaces; may be provided naturally by gravitational forces, mechanically by pumps, or by a combination of both methods.

boiler cleaning [ENG] A mechanical or chemical process for removal of grease, scale, and other deposits from steam boiler surfaces.

boiler code [MECH ENG] A code, established by professional societies and administrative units, which contains the basic rules for the safe design, construction, and materials for steam-generating units, such as the ASME code.

boiler compound [CHEM] Any chemical used to treat boiler water to prevent corrosion, the fouling of heat-absorbing surfaces, foaming, and the contamination of steam.

boiler controls [MECH ENG] Either manual or automatic devices which maintain desired boiler operating conditions with respect to variables such as feedwater flow, firing rate, and steam temperature.

boiler draft [MECH ENG] The difference between atmospheric pressure and some lower pressure existing in the furnace or gas passages of a steam-generating unit.

boiler economizer [MECH ENG] A component of a steam-generating unit that transfers heat from the products of combustion after they have passed through the steam-generating and superheating sections to the feedwater, which it receives from the boiler feed pump and delivers to the steam-generating section of the boiler.

boiler efficiency [MECH ENG] The ratio of heat absorbed in steam to the heat supplied in fuel, usually measured in percent.

boiler feedwater [MECH ENG] Water supplied to a steam-generating unit.

boiler feedwater regulation [MECH ENG] Addition of water to the steam-generating unit at a rate commensurate with the removal of steam from the unit.

boiler fuel [MATER] Natural gas, residual oil, and various solid fuels used to heat a boiler, usually the most economical fuel available.

boiler furnace [MECH ENG] An enclosed space provided for the combustion of fuel to generate steam in a boiler. Also known as steam-generating furnace.

boiler heat balance [MECH ENG] A means of accounting for the thermal energy entering a steam-generating system in terms of its ultimate useful heat absorption or thermal loss.

boiler horsepower [MECH ENG] A measurement of water evaporation rate; 1 boiler horsepower equals the evaporation per hour of 34½ pounds (15.7 kilograms) of water at 212°F (100°C) into steam at 212°F. Abbreviated bhp.

boiler hydrostatic test [MECH ENG] A procedure that employs water under pressure, in a new boiler before use or in old equipment after major alterations and repairs, to test the boiler's ability to withstand about 1½ times the design pressure.

boiler plate [GEOL] 1. A fairly smooth surface on a cliff, consisting of flush or overlapping slabs of rock, having little or no foothold. 2. Crusty, frozen surface of snow. [MET] Flat-rolled steel, usually ¼ to ½ inch thick; used mainly for covering ships and making boilers and tanks. Also known as boiler steel.

boiler-plate model [AERO ENG] A metal copy of a flight vehicle, the structure or components of which are heavier than the flight model.

boiler scale [CHEM] Deposits from silica and other contaminants in boiler water that form on the internal surfaces of heat-absorbing components, increase metal temperatures, and result in eventual failure of the pressure parts due to overheating.

boiler setting [MECH ENG] The supporting steel and gastight enclosure for a steam generator.

boiler steel See boiler plate.

boiler storage [MECH ENG] A steam-generating unit that, when out of service, may be stored wet (filled with water) or dry (filled with protective gas).

boiler superheater [MECH ENG] A boiler component, consisting of tubular elements, in which heat is added to high-pressure steam to increase its temperature and enthalpy.

boiler tube [MECH ENG] One of the tubes in a boiler that carry water (water-tube boiler) to be heated by the high-temperature gaseous products of combustion or that carry combustion products (fire-tube boiler) to heat the boiler water that surrounds them.

boiler walls [MECH ENG] The refractory walls of the boiler furnace, usually cooled by circulating water and capable of withstanding high temperatures and pressures.

boiler water [MECH ENG] Water in the steam-generating section of a boiler unit.

boiling [PHYS CHEM] The transition of a substance from the liquid to the gaseous phase, taking place at a single temperature in pure substances and over a range of temperatures in mixtures.

boiling point [PHYS CHEM] Abbreviated bp. 1. The temperature at which the transition from the liquid to the gaseous

phase occurs in a pure substance at fixed pressure. 2. See bubble point.

boiling point elevation [CHEM] The raising of the normal boiling point of a pure liquid compound by the presence of a dissolved substance, the elevation being in direct relation to the dissolved substance's molecular weight.

boiling range [CHEM] The temperature range of a laboratory distillation of an oil from start until evaporation is complete.

boiling spring [HYD] 1. A spring which emits water at a high temperature or at boiling point. 2. A spring located at the head of an interior valley and rising from the bottom of a residual clay basin.

boiling-water reactor [NUCLEO] A nuclear reactor in which the coolant is water, maintained at such a pressure as to allow it to boil and form steam. Abbreviated BWR.

boil-off [THERMO] The vaporization of a liquid, such as liquid oxygen or liquid hydrogen, as its temperature reaches its boiling point under conditions of exposure, as in the tank of a rocket being readied for launch. [TEXT] Removal of impurities from fabric by boiling the material in a solution.

boil-off assistant [MATER] A material used to facilitate the removal of natural glue from silk during boiling in soap solution (degumming); examples are sulfonated oils, pine oils, and solvents and other wetting agents.

boil smut [PL PATH] A fungus disease of corn caused by *Ustilago maydis*, characterized by galls containing black spores.

bojite [PETR] 1. A gabbro with primary hornblende substituting for augite. 2. Hornblende diorite.

boldface [GRAPHICS] A type that has thick heavy lines so that it has a conspicuous black appearance.

bole [GEOL] Any of various red, yellow, or brown earthy clays consisting chiefly of hydrous aluminum silicates. Also known as bolus; terra miraculosa. [GRAPHICS] The foundation laid for gold leaf.

boletic acid See fumaric acid.

bolide [ASTRON] A brilliant meteor, especially one which explodes; a detonating fireball meteor.

boll [BOT] A pod or capsule (pericarp), as of cotton and flax.

bollard [CIV ENG] A heavy post on a dock or ship used in mooring ships.

boll rot [PL PATH] A fungus rot of cotton bolls caused by *Glomerella gossypii* and *Xanthomonas malvacearum*.

boll weevil [INV ZOO] A beetle, *Anthonomus grandis*, of the order Coleoptera; larvae destroy cotton plants and are the most important pests in agriculture.

boll-weevil hanger [PETRO ENG] A screw-on connector used to connect and seal the lower end of a length of oil-well casing to the next smaller casing string.

bolometer [ENG] An instrument that measures the energy of electromagnetic radiation in certain wavelength regions by utilizing the change in resistance of a thin conductor caused by the heating effect of the radiation. Also known as thermal detector.

bolson [GEOL] In the southwestern United States, a basin or valley having no outlet.

bolster [MET] A block of steel to which drop-forging dies are attached.

bolster plate [MECH ENG] A plate fixed on the bed of a power press to locate and support the die assembly.

bolt [DES ENG] A rod, usually of metal, with a square, round, or hexagonal head at one end and a screw thread on the other, used to fasten objects together. [MIN ENG] See bolthole. [ORD] The sliding part in a breechloading weapon that pushes a cartridge into position and holds it there as the gun is fired.

bolted joint [ENG] The assembly of two or more parts by a threaded bolt and nut or by a screw that passes through one member and threads into another.

bolted rail crossing [CIV ENG] A crossing whose running surfaces are made of rolled rail and whose parts are joined with bolts.

bolt face [ORD] That portion of a gun bolt that abuts the base of the cartridge case.

bolthole [MIN ENG] A short, narrow opening made to connect the main working with the airhead or ventilating drift of a coal mine. Also known as bolt.

BOLTED JOINT

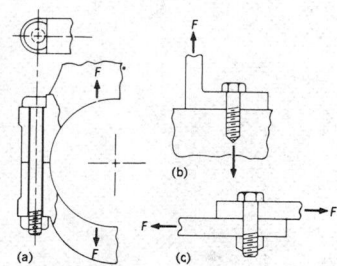

Forms of bolted joint (*F* represents force). (*a*) Direct tension. (*b*) Eccentric loading. (*c*) Single shear.

BOMB CALORIMETER

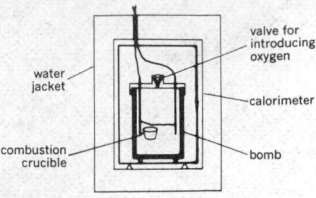

Schematic diagram of bomb calorimeter.

BOMBIDAE

A drawing of a hairy, black and yellow bumblebee. (*From T. I. Storer and R. L. Usinger, General Zoology, 3d ed., McGraw-Hill, 1957*)

bolting [ENG] A fastening system using screw-threaded devices, for example, nuts, bolts, or studs. [FOOD ENG] The process of refining or purifying, especially of sifting flour or meal through a sieve. [MIN ENG] The use of vibrating sieves to separate particles of different sizes.

bolting cloth [MATER] A sieve cloth made of wire, hair, or silk or other thread used to remove lumps from flour or to make screen prints or needlework.

boltwoodite [MINERAL] $K_2(UO_2)_2(SiO_3)_2(OH)_2 \cdot 5H_2O$ Yellow mineral consisting of hydrous potassium uranyl silicate.

Boltzmann constant [STAT MECH] The ratio of the universal gas constant to the Avogadro number.

Boltzmann distribution [STAT MECH] A function giving the probability that a molecule of a gas in thermal equilibrium will have generalized position and momentum coordinates within given infinitesimal ranges of values, assuming that the molecules obey classical mechanics.

Boltzmann H-theorem [STAT MECH] The theorem that the entropy of a system never decreases; Boltzmann proved this for a classical gas of colliding particles. Also known as H-theorem of Boltzmann.

Boltzmann statistics *See* Maxwell-Boltzmann statistics.

Boltzmann transport equation [STAT MECH] An equation used to study the nonequilibrium behavior of a collection of particles; it states that the rate of change of a function which specifies the probability of finding a particle in a unit volume of phase space is equal to the sum of terms arising from external forces, diffusion of particles, and collisions of the particles. Also known as Maxwell-Boltzmann equation.

Boltzmann-Vlasov equations [PL PHYS] The equations that govern a high-temperature plasma in which the collisional mean free path is much larger than all the characteristic lengths of the system.

bolus [GEOL] *See* bole. [PHARM] A large pill. [PHYSIO] The mass of food prepared by the mouth for swallowing.

bolus alba *See* kaolin.

Bolyai geometry *See* Lobachevski geometry.

Bolzano's theorem [MATH] The theorem that a single-valued, real-valued, continuous function of a real variable is equal to zero at some point in an interval if its values at the end points of the interval have opposite sign.

Bolzano-Weierstrass theorem [MATH] The theorem that every bounded, infinite set in finite dimensional euclidean space has a cluster point.

bomb [GEOL] A pyroclast larger than a lapillus. [ORD] An explosive or other lethal agent, together with its container or holder, which is planted or thrown by hand, dropped from an aircraft, or projected by some other slow-speed device (such as a mortar) and used to destroy, damage, injure, or kill.

Bombacaceae [BOT] A family of dicotyledonous tropical trees in the order Malvales with dry or fleshy fruit usually having woolly seeds.

bomb adapter-booster [ORD] A device consisting of an adapter and an auxiliary explosive charge, used in an explosive train to detonate a bomb.

bombard [NUCLEO] To direct a stream of particles or photons against a target. [ORD] To carry out a sustained attack upon a city, fort, or the like with bombs, projectiles, rockets, or other explosive missiles.

bombardment [ELECTR] The use of induction heating to heat electrodes of electron tubes to drive out gases during evacuation.

bombardment aircraft *See* bomber.

Bombay blood group system [IMMUNOL] A system comprising an immunologically distinct, genetically determined group of human erythrocytes characterized by the lack of A, B, or H antigens.

bomb ballistics [MECH] The special branch of ballistics concerned with bombs dropped from aircraft.

bomb bay [AERO ENG] The compartment or bay in the fuselage of a bomber where the bombs are carried for release.

bomb calorimeter [ENG] A calorimeter designed with a strong-walled container constructed of a corrosion-resistant alloy, called the bomb, immersed in about 2.5 liters of water in a metal container; the sample, usually an organic compound, is ignited by electricity, and the heat generated is measured.

bomb carpet [ORD] The fall of bombs, or the bombfall pattern produced, in carpet bombing; the area struck by carpet bombing.

bomb casing [ORD] The principal container, usually metal, for the main charge of a bomb.

bomb complete round [ORD] A complete aerial bomb, including all of the components such as arming wires and fuses, necessary to attach the bomb to a release mechanism and to make the bomb function after release.

bomb dispenser [ORD] An aerodynamically shaped unit that is externally mounted but not permanently fixed on high-speed aircraft to carry and eject bombs.

bomb ejection cartridge [ORD] An explosive item used to eject a bomb from a bomb cluster or bomb station.

bomber [AERO ENG] An airplane specifically designed to carry and drop bombs. Also known as bombardment aircraft.

bomb fin [ORD] A fin attached to a bomb in order to afford directional stability.

bombiccite *See* hartite.

Bombidae [INV ZOO] A family of relatively large, hairy, black and yellow bumblebees in the hymenopteran superfamily Apoidea.

bombing [ORD] The action of dropping bombs from an aircraft with the purpose of hitting a target.

bombing table [ORD] A table giving the bombsight settings required for dropping a particular type of bomb at various speeds and altitudes.

bombing through overcast [ORD] Blind bombing through clouds, using radar, infrared equipment, or other electronic equipment for guidance. Abbreviated BTO. Also known as blind bombing.

bomb load [ORD] 1. The weight or number of bombs carried by an aircraft. 2. The bomb or bombs carried.

bombproof [ENG] Referring to shelter, building, or other installation resistant or impervious to the effects of bomb explosions.

bomb rack [ORD] A mechanical device, fitted to an airplane in the bomb bay or suspended underneath the plane, that releases and arms bombs at the bombardier's or pilot's command.

bomb reconnaissance [ORD] Reconnoitering to determine the presence of an unexploded missile, ascertaining its nature, applying all practicable protective measures for the protection of personnel, installations, and equipment, and finally reporting essential information to the authority directing explosive ordnance disposal operations.

bomb-release line [ORD] An imaginary line around a target area at which a bomber, traveling toward it at a constant speed and altitude, releases its first bomb so that it and others will strike the target area.

bomb-release point [ORD] The point in flight on a bomb run at which a bombing airplace releases its bomb load.

bomb run [ORD] The flight course of a bombing airplane just before the release of bombs.

bomb sag [GEOL] Depressed and deranged laminae mainly found in beds of fine-grained ash or tuff around an included volcanic bomb or block which fell on and became buried in the deposit.

bomb shelter [CIV ENG] A bomb-proof structure for protection of people.

bombsight [ORD] A device which determines, or enables a bombardier to determine, the point in space at which a bomb must be released from an aircraft in order to hit a target.

Bombycidae [INV ZOO] A family of lepidopteran insects of the superorder Heteroneura that includes only the silkworms.

Bombyliidae [INV ZOO] The bee flies, a family of dipteran insects in the suborder Orthorrhapha.

Bombyx [INV ZOO] The type genus of Bombycidae.

Bombyx mori [INV ZOO] The commercial silkworm.

bond [CHEM] The strong attractive force that holds together atoms in molecules and crystalline salts. Also known as chemical bond. [ELEC] The connection made by bonding electrically. [ENG] 1. A wire rope that fixes loads to a crane hook. 2. Adhesion between cement or concrete and masonry or reinforcement. [MET] 1. Material added to molding sand to impart bond strength. 2. Junction of the base metal and filler metal, or the base metal beads, in a welded joint.

Bond and Wang theory [MECH ENG] A theory of crushing and grinding from which the energy, in horsepower-hours, required to crush a short ton of material is derived.

bond angle [PHYS CHEM] The angle between bonds sharing a common atom. Also known as valence angle.

bond clay [MATER] A type of clay with high plasticity and high dry strength used to bond nonplastic materials; may be refractory.

bond distance [PHYS CHEM] The distance separating the two nuclei of two atoms bonded to each other in a molecule. Also known as bond length.

bonded coating [MATER] A finishing or protecting layer of any compound affixed to a surface.

bonded NR diode [ELECTR] An n^+ junction semiconductor device in which the negative resistance arises from a combination of avalanche breakdown and conductivity modulation which is due to the current flow through the junction.

bonded strain gage [ENG] A strain gage in which the resistance element is a fine wire, usually in zigzag form, embedded in an insulating backing material, such as impregnated paper or plastic, which is cemented to the pressure-sensing element.

bond energy [PHYS CHEM] The heat of formation of a molecule from its constituent atoms.

bonderize [MET] To coat steel with a solution of phosphates for corrosion protection.

bond hybridization [CHEM] The linear combination of two or more simple atomic orbitals.

bonding [CHEM] The joining together of atoms to form molecules or crystalline salts. [ELEC] The use of low-resistance material to connect electrically a chassis, metal shield cans, cable shielding braid, and other supposedly equipotential points to eliminate undesirable electrical interaction resulting from high-impedance paths between them. [ENG] The fastening together of two components of a device by means of adhesives, as in anchoring the copper foil of printed wiring to an insulating baseboard.

bonding agent [MATER] Any substance that fixes one material to another.

bonding electron [PHYS CHEM] An electron whose orbit spans the entire molecule and so assists in holding it together.

bonding strength [MECH] Structural effectiveness of adhesives, welds, solders, glues, or of the chemical bond formed between the metallic and ceramic components of a cermet, when subjected to stress loading, for example, shear, tension, or compression.

bonding wire [ELEC] Wire used to connect metal objects so they have the same potential (usually ground potential).

bond length See bond distance.

Bond number [FL MECH] A dimensionless number used in the study of atomization and the study of bubbles and drops, equal to $(\rho - \rho')L^2 g / \sigma$, where ρ is the density of a bubble or drop, ρ' is the density of the surrounding medium, L is a characteristic dimension, g is the acceleration of gravity, and σ is the surface tension of the bubble or drop.

bond paper [MATER] A paper used for writing paper, business forms, and typewriter paper; the less expensive bond papers are made from wood sulfite pulps; rag-content bonds contain 25, 50, 75, or 100% of pulp made from rags, and offer greater permanence and strength.

Bond's law [MECH ENG] A statement that relates the work required for the crushing of solid materials (for example, rocks and ore) to the product size and surface area and the lengths of cracks formed. Also known as Bond's third theory.

Bond's third theory See Bond's law.

bondstone [BUILD] A stone joining the coping above a gable to the wall. [CIV ENG] A stone that passes through a masonry wall in order to hold the wall together.

bone [ANAT] One of the parts of a vertebrate skeleton. [HISTOL] A hard connective tissue that forms the major portion of the vertebrate skeleton.

Bone Age [ARCHEO] A prehistoric period of human culture characterized by the use of implements made of bone and antler.

bone ash [CHEM] A white ash consisting primarily of tribasic calcium phosphate obtained by burning bones in air; used in cleaning jewelry and in some pottery.

bone bed [GEOL] Several thin strata or layers with many fragments of fossil bones, scales, teeth, and also organic remains.

bone black [MATER] A black substance made by carbonizing crushed, defatted bones in closed vessels; used as a paint and varnish pigment, as a decolorizing absorbent in clarifying shellac, in cementation, and in gas masks. Also known as animal black; bone char.

bone char See bone black.

bone chert [PETR] A weathered residual chert that appears chalky and porous with a white color but may be stained red or other colors.

bone coal [GEOL] Argillaceous coal or carbonaceous shale that is found in coal seams.

bone conduction [BIOPHYS] Transmission of sound vibrations to the internal ear via the bones of the skull.

Bonellidae [INV ZOO] A family of wormlike animals belonging to the order Echiuroinea.

bone marrow [HISTOL] A vascular modified connective tissue occurring in the long bones and certain flat bones of vertebrates.

bone oil [MATER] Dark brown oil with a disagreeable odor, derived by destructive distillation of bones or other animal substances; used as an alcohol denaturant, an insecticide, or a source of pyrrole and for organic preparations. Also known as animal oil; Dippel's oil; hartshorn oil; Jeppel's oil.

bone seeker [NUCLEO] A radioisotope that tends to accumulate in the bones when it is introduced into the body; an example is strontium-90, which behaves chemically like calcium.

boninite [PETR] An andesitic rock that contains much glass and abundant phenocrysts of bronzite and less of olivine and augite.

Bonne projection [MAP] A type of conical map projection; meridians are plotted as curves and the parallels are spaced along them at true distances.

Bononian [GEOL] Upper Jurassic (lower Portlandian) geologic time.

bonus [PETRO ENG] Payment by a lessee of an oil- or gas-production royalty to the landowner at a rate greater than the customary one-eighth of the value of the oil or gas withdrawn.

bony fish [VERT ZOO] The name applied to all members of the class Osteichthyes.

bony labyrinth [ANAT] The system of canals within the otic bones of vertebrates that houses the membranous labyrinth of the inner ear.

Boo See Boötes.

booby trap [ORD] An explosive charge such as a mine, grenade, demolition block, shell, or bulk explosive fitted with a detonator and a firing device, and usually concealed and set to explode when an unsuspecting person touches off its firing mechanism by stepping upon, lifting, or moving a harmless-looking object.

Boodleaceae [BOT] A family of green marine algae in the order Siphonocladales.

book See mica book.

book capacitor [ELEC] A trimmer capacitor consisting of two plates which are hinged at one end; capacitance is varied by changing the angle between them.

book gill [INV ZOO] A type of gill in king crabs consisting of folds of membranous tissue arranged like the leaves of a book.

bookkeeping operation [ADP] A computer operation which does not directly contribute to the result, that is, arithmetical, logical, and transfer operations used in modifying the address section of other instructions in counting cycles and in rearranging data. Also known as red-tape operation.

book louse [INV ZOO] A common name for a number of insects belonging to the order Psocoptera; important pests in herbaria, museums, and libraries.

book lung [INV ZOO] A saccular respiratory organ in many arachnids consisting of numerous membranous folds arranged like the pages of a book.

book mold [MET] A mold made in two halves hinged together like a book.

book structure [GEOL] A rock structure of numerous parallel sheets of slate alternating with quartz.

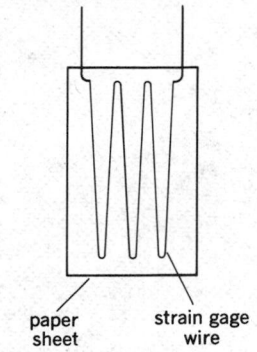

BONDED STRAIN GAGE

paper sheet · strain gage wire

Elements of a bonded strain gage.

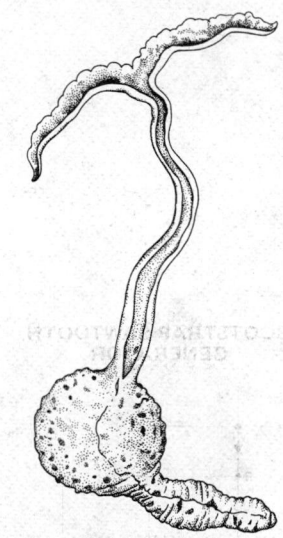

BONELLIDAE

Bonellia species, half size, showing the long and cleft (at the top) prostomium characteristic of the Bonellidae.

boolean [ADP] A scalar declaration in ALGOL defining variables similar to FORTRAN's logical variables.

Boolean algebra [MATH] An algebraic system with two binary operations and one unary operation important in representing a two-valued logic.

Boolean calculus [MATH] Boolean algebra modified to include the element of time.

Boolean function [MATH] A function $f(x, y, \ldots, z)$ assembled by the application of the operations AND, OR, NOT on the variables $x, y, \ldots, z$ and elements whose common domain is a Boolean algebra.

boom [COMMUN] A movable mechanical support, usually in a television or motion picture studio, to suspend a microphone within range of the performers but above the field of view of the camera. [NAV ARCH] A spar attached to a mast or kingpost of a ship carrying cargo-hoisting gear.

boom crutch [NAV ARCH] A movable prop for supporting the free end of the boom of a ship when it is not being used.

boomer [MIN ENG] In placer mining, an automatic gate in a dam that holds the water until the reservoir is filled, then opens automatically and allows the escape of such a volume that the soil and upper gravel of the placer are washed away.

Boopidae [INV ZOO] A family of lice in the order Mallophaga, parasitic on Australian marsupials.

Boord synthesis [CHEM ENG] A method of producing alpha olefins by the reduction of alpha bromo ethers with zinc.

boorga See burga.

boost [AERO ENG] 1. An auxiliary means of propulsion such as by a booster. 2. To supercharge. 3. To launch or push along during a portion of flight. 4. See boost pressure. [ELECTR] To augment in relative intensity, as to boost the bass response in an audio system. [ENG] To bring about a more potent explosion of the main charge of an explosive by using an additional charge to set it off.

boost charge [ELEC] Partial charge of a storage battery, usually at a high current rate for a short period.

booster [AERO ENG] See booster engine; booster rocket; launch vehicle. [ELEC] A small generator inserted in series or parallel with a larger generator to maintain normal voltage output under heavy loads. [ELECTR] 1. A separate radio-frequency amplifier connected between an antenna and a television receiver to amplify weak signals. 2. A radio-frequency amplifier that amplifies and rebroadcasts a received television or communication radio carrier frequency for reception by the general public. [IMMUNOL] The dose of an immunizing agent given to stimulate the effects of a previous dose of the same agent. [ORD] An assembly of metal parts and explosive charge provided to augment the explosive component of a fuse, to cause detonation of the main explosive charge of the munition.

booster battery [ELECTR] A battery which increases the sensitivity of a crystal detector by maintaining a certain voltage across it and thereby adjusting conditions to increase the response to a given input.

booster ejector [MECH ENG] A nozzle-shaped apparatus from which a high-velocity jet of steam is discharged to produce a continuous-flow vacuum for process equipment.

booster engine [AERO ENG] An engine, especially a booster rocket, that adds its thrust to the thrust of the sustainer engine. Also known as booster.

booster pump [MECH ENG] A machine used to increase pressure in a water or compressed-air pipe.

booster response See anamnestic response.

booster rocket [AERO ENG] Also known as booster. 1. A rocket motor, either solid- or liquid-fueled, that assists the normal propulsive system or sustainer engine of a rocket or aeronautical vehicle in some phase of its flight. 2. A rocket used to set a vehicle in motion before another engine takes over.

booster stations [ENG] Booster pumps or compressors located at intervals along a liquid-products or gas pipeline to boost the pressure of the flowing fluid to keep it moving toward its destination.

booster voltage [ELECTR] The additional voltage supplied by the damper tube to the horizontal output, horizontal oscilla-

tor, and vertical output tubes of a television receiver to give greater sawtooth sweep output.

booster well [ORD] A hollow space, in the main explosive charge of an item of ammunition, into which the booster fits.

boost-glide vehicle [AERO ENG] An air vehicle capable of aerodynamic lift which is projected to an extreme altitude by reaction propulsion and then coasts down with little or no propulsion, gliding to increase its range when it reenters the sensible atmosphere.

boost pressure [AERO ENG] Manifold pressure greater than the ambient at atmospheric pressure, obtained by supercharging. Also known as boost.

boot [ELEC] A protective covering over any portion of a cable, wire, or connector. [MIN ENG] 1. A projecting portion of a reinforced concrete beam acting as a corbel to support the facing material, such as brick or stone. 2. The lower end of a bucket elevator. [PETRO ENG] See surge column.

Boot See Boötes.

Boötes [ASTRON] A constellation which lies south and east of Ursa Major. The star Arcturus is a member of the group. Abbreviated Boo; Boot. Also known as Bear Driver.

α Boötis See Arcturus.

bootjack [ENG] A fishing tool used in drilling wells.

bootleg [MIN ENG] A hole, shaped somewhat like the leg of a boot, caused by a blast that has failed to shatter the rock properly.

bootstrap circuit [ELECTR] A single-stage amplifier in which the output load is connected between the negative end of the anode supply and the cathode, while signal voltage is applied between grid and cathode; a change in grid voltage changes the input signal voltage with respect to ground by an amount equal to the output signal voltage.

bootstrap driver [ELECTR] Electronic circuit used to produce a square pulse to drive the modulator tube; the duration of the square pulse is determined by a pulse-forming line.

bootstrap instructor technique [ADP] A technique permitting a system to bring itself into an operational state by means of its own action. Also known as bootstrap technique.

bootstrap integrator [ELECTR] A bootstrap sawtooth generator in which an integrating amplifier is used in the circuit. Also known as Miller generator.

bootstrap memory [ADP] A device that provides for the automatic input of new programs without erasing the basic instructions in the computer.

bootstrapping [ELECTR] A technique for lifting a generator circuit above ground by a voltage value derived from its own output signal.

bootstrap process [AERO ENG] A self-generating or self-sustaining process; specifically, the operation of liquid-propellant rocket engines in which, during main-stage operation, the gas generator is fed by the main propellants pumped by the turbopump, and the turbopump in turn is driven by hot gases from the gas generator system.

bootstrap program See loading program.

bootstrap sawtooth generator [ELECTR] A circuit capable of generating a highly linear positive sawtooth waveform through the use of bootstrapping.

bootstrap scheme [PARTIC PHYS] A theory of elementary particles in which the existence of each particle contributes to forces between it and other particles; these forces lead to bound systems which are the particles themselves.

bootstrap technique See bootstrap instructor technique.

Bopyridae [INV ZOO] A family of epicaridean isopods in the tribe Bopyrina known to parasitize decapod crustaceans.

Bopyrina [INV ZOO] A tribe of dioecious isopods in the suborder Epicaridea.

Bopyroidea [INV ZOO] An equivalent name for Epicaridea.

bora [METEOROL] A fall wind whose source is so cold that when the air reaches the lowlands or coast the dynamic warming is insufficient to raise the air temperature to the normal level for the region; hence it appears as a cold wind.

boracic acid See boric acid.

boracite [MINERAL] $Mg_3B_7O_{13}Cl$ A white, yellow, green, or blue orthorhombic borate mineral occurring in crystals which appear isometric in external form; it is strongly

BOOTSTRAP SAWTOOTH GENERATOR

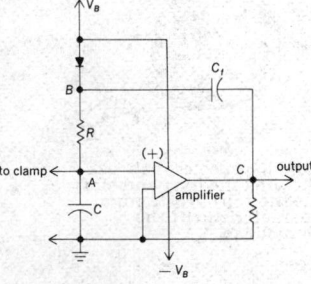

Circuit diagram of bootstrap sawtooth generator.

BORACITE

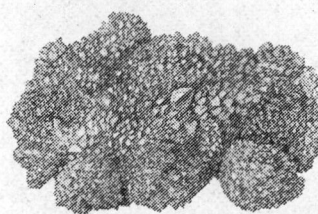

├─ 2.5 cm ─┤

Specimen of boracite from Segeberg, Germany. (Specimen from Department of Geology, Bryn Mawr College)

pyroelectric, has a hardness of 7 on Mohs scale, and a specific gravity of 2.9.

bora fog [METEOROL] A dense fog caused when the bora lifts a spray of small drops from the surface of the sea.

Boraginaceae [BOT] A family of flowering plants in the order Lamiales comprising mainly herbs and some tropical trees.

Boral [MET] An aluminum of the Aluminum Company of America containing 35% boron carbide powder and rolled into sheets which are coated on both sides with aluminum; used in thermal neutron shielding.

borane [INORG CHEM] 1. A class of binary compounds of boron and hydrogen; boranes are used as fuels. Also known as boron hydride. 2. A substance which may be considered a derivative of a boron-hydrogen compound, such as BCl_3 and $B_{10}H_{12}I_2$.

borate [CHEM] 1. A generic term referring to salts or esters of boric acid. 2. Related to boric oxide, B_2O_3, or commonly to only the salts of orthoboric acid, H_3BO_3.

borate mineral [MINERAL] Any of the large and complex group of naturally occurring crystalline solids in which boron occurs in chemical combination with oxygen.

borax [MINERAL] $Na_2B_4O_7 \cdot 10H_2O$ A white, yellow, blue, green or gray borate mineral that is an ore of boron and occurs as an efflorescence or in monoclinic crystals; when pure it is used as a cleaning agent, antiseptic, and flux. Also known as diborate; pyroborate; sodium (1:2) borate; sodium tetraborate; tincal.

borax glass [MATER] A glassy, transparent solid formed by fusing borax.

borazole [INORG CHEM] $B_3N_3H_6$ A colorless liquid boiling at 53°C; with water it hydrolyzes to form boron hydrides; the borazole molecule is the inorganic analog of the benzene molecule.

Borda mouthpiece [FL MECH] A reentrant tube in a hydraulic reservoir, whose contraction coefficient (the ratio of the cross section of the issuing jet of liquid to that of the opening) can be calculated more simply than for other discharge openings.

Bordeaux mixture [MATER] A fungicide made from a mixture of lime, copper sulfate, and water.

bordered [BOT] Having a margin with a distinctive color or texture; used especially of a leaf.

bordered pit [BOT] A wood-cell pit having the secondary cell wall arched over the cavity of the pit.

borderland [GEOL] One of the crystalline, continental landmasses postulated to have existed on the exterior (oceanward) side of geosynclines.

borderland slope [GEOL] A declivity which indicates the inner margin of the borderland of a continent.

borderline mental retardation [PSYCH] Below-normal intellectual functioning associated with an intelligence quotient of about 68–85.

borderline psychosis [PSYCH] The psychiatric diagnosis of an individual whose symptoms are severe but are not clearly psychotic or neurotic.

borderline syndrome [PSYCH] An impairment of ego function noted in individuals who have been angry most of their lives and who cannot relate meaningfully to or love other people, and who in consequence develop nonpsychotic but deviant mechanisms, such as withdrawal while expressing angry concern, or showing passive compliance at the price of real involvement, to maintain a precarious mental equilibrium.

border-punched card See edge-notched card.

Bordetella [MICROBIO] A genus of minute, gram-negative, parasitic coccobacilli in the family Brucellaceae.

Bordetella parapertussis [MICROBIO] A small, nonmotile coccobacillus that causes parapertussis.

Bordetella pertussis [MICROBIO] A small, nonmotile coccobacillus that causes whooping cough.

Bordet-Gengou bacillus See Hemophilus pertussis.

Bordini effect [MET] A phenomenon that takes place when metals having a close-packed crystal structure are subjected to an oscillating stress; there is a peak in the internal friction at a particular frequency of the stress.

bore [DES ENG] Inside diameter of a pipe or tube. [MECH ENG] 1. The diameter of a piston-cylinder mechanism as found in reciprocating engines, pumps, and compressors. 2. To penetrate or pierce with a rotary tool. 3. To machine a workpiece to increase the size of an existing hole in it. [MIN ENG] 1. A tunnel under construction. 2. To cut or drill a hole for blasting, water infusion, exploration, or water or firedamp drainage. [OCEANOGR] A high, breaking wave of water, advancing rapidly up an estuary. Also known as eager; mascaret; tidal bore. [ORD] The interior of a gun barrel or tube.

boreal [ECOL] Of or relating to northern geographic regions.

Boreal life zone [ECOL] The zone comprising the climate and biotic communities between the Arctic and Transitional zones.

bore axis [ORD] The longitudinal center line of a gunbore.

bore diameter [ORD] The interior diameter or caliber of a gun or launching tube.

bore evacuation [ORD] Clearing of the bore of propellant gases and extraneous material after firing.

bore evacuator [ORD] A device placed on the tube of artillery weapons to force propellant gases to flow outward through the muzzle end of the bore from a cylindrical chamber around the gun tube; the chamber encloses a series of jets around the cannon tube through which the gases flow.

bore expansion [ORD] Permanent enlargement of the bore of a gun due to interior pressure; it does not include enlargement due to bore erosion.

borehole [ENG] A hole made by boring into the ground to study stratification, to obtain natural resources, or to release underground pressures.

borehole mining [PETRO ENG] Extraction of minerals as liquid or gas from the earth's crust by means of boreholes and suction pumps.

borehole survey [ENG] Also known as drillhole survey. 1. Determining the course of and the target point reached by a borehole, using an azimuth-and-dip recording apparatus small enough to be lowered into a borehole. 2. The record of the information thereby obtained.

Borell unit [BIOL] A unit for the standardization of thyroid extracts.

Borel measurable function [MATH] 1. A real-valued function such that the inverse image of the set of real numbers greater than any given real number is a Borel set. 2. More generally, a function to a topological space such that the inverse image of any open set is a Borel set.

Borel set [MATH] 1. A set obtained from the collection of open sets of real numbers by taking countable unions, intersections, and complements. 2. More generally, a member of the smallest r-algebra containing compact subsets of a topological space.

bore premature [ORD] The premature explosion of a projectile occurring within the bore.

borer [INV ZOO] Any insect or other invertebrate that burrows into wood, rock, or other substances. [MECH ENG] An apparatus used to bore openings into the earth up to about 8 feet in diameter.

bore riding pin [ORD] A safety pin which is held in place in the fuse while the projectile or missile is within the gun barrel or launching tube and then ejected from the fuse by centrifugal force or spring action beyond the muzzle.

borescope [ENG] A straight-tube telescope using a mirror or prism, used to visually inspect a cylindrical cavity, such as the cannon bore of artillery weapons for defects of manufacture and erosion caused by firing.

boresight [ORD] A telescopic device utilizing a boresighting reticle to align the axis of a weapon with a target; it is attached to the exterior of a weapon barrel or tube by a mount.

boresight camera [OPTICS] A camera mounted in the optical axis of a tracking radar to photograph rockets being tracked or known, fixed targets while in camera range and thus to provide a correction for the alignment of the radar.

boresighting [ENG] Initial alignment of a directional microwave or radar antenna system by using an optical procedure or a fixed target at a known location. [ORD] The process by which the axis of a gun bore and the line of sight of a gunsight are made parallel or are made to converge on a point.

Borhyaenidae [VERT ZOO] A family of carnivorous mammals in the superfamily Borhyaenoidea.

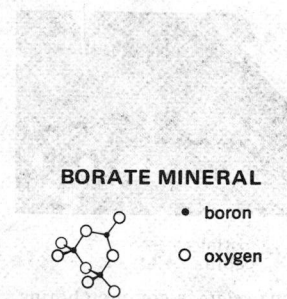

BORATE MINERAL

- boron
○ oxygen

Structural elements of meyerhofferite, a borate mineral.

BORING SPONGE

Cliona celata, a common boring sponge living on an oyster shell, showing several expanded ostial and oscular papillae. (*After W. D. Hartman, 1958*)

BORNITE

A sample of bornite from Messina, Transvaal, South Africa. (*Specimen from Department of Geology, Bryn Mawr College*)

BORON

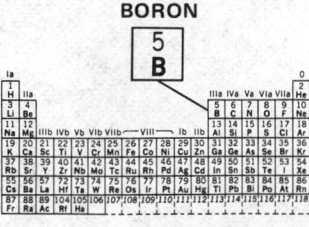

Periodic table of the chemical elements showing the position of boron.

Borhyaenoidea [VERT ZOO] A superfamily of carnivorous mammals in the order Marsupialia.

boric acid [INORG CHEM] H_3BO_3 An acid derived from boric oxide in the form of white, triclinic crystals, melting at 185°C, soluble in water. Also known as boracic acid; orthoboric acid.

boric acid ester [ORG CHEM] Any compound readily hydrolyzed to yield boric acid and the respective alcohol; for example, trimethyl borate hydrolyzes to boric acid and methyl alcohol.

boric oxide [INORG CHEM] B_2O_3 A trioxide of boron obtained as rhombic crystals melting at 460°C; used as an intermediate in the production of boron halides and metallic borides and as a thermal neutron absorber in nuclear engineering. Also known as boron oxide.

boride [INORG CHEM] A binary compound of boron and a metal formed by heating a mixture of the two elements.

boring log See drill log.

boring machine [MECH ENG] A machine tool designed to machine internal work such as cylinders, holes in castings, and dies; types are horizontal, vertical, jig, and single.

boring mill [MECH ENG] A boring machine tool used particularly for large workpieces; types are horizontal and vertical.

boring sponge [INV ZOO] Marine sponge of the family Clionidae represented by species which excavate galleries in mollusks, shells, corals, limestone, and other calcareous matter.

Born approximation [QUANT MECH] A method used for the computation of cross sections in scattering problems; the interactions are treated as perturbations of free-particle systems.

borneol [ORG CHEM] $C_{10}H_{17}OH$ White lumps with camphor odor; insoluble in water, soluble in alcohol; melting point 203°C; used in perfumes, medicine, and chemical synthesis. Also known as bornyl alcohol; 2-camphanol; 2-hydroxycamphane.

Born equation [PHYS CHEM] An equation for determining the free energy of solvation of an ion in terms of the Avogadro number, the ionic valency, the ion's electronic charge, the dielectric constant of the electrolytic, and the ionic radius.

Born-Haber cycle [SOLID STATE] A sequence of chemical and physical processes by means of which the cohesive energy of an ionic crystal can be deduced from experimental quantities; it leads from an initial state in which a crystal is at zero pressure and 0 K to a final state which is an infinitely dilute gas of its constituent ions, also at zero pressure and 0 K.

bornite [MINERAL] Cu_5FeS_4 A primary mineral in many copper ore deposits; specific gravity 5.07; the metallic and brassy color of a fresh surface rapidly tarnishes upon exposure to air to an iridescent purple.

Born-Madelung model [SOLID STATE] A classical theory of cohesive energy, lattice spacing, and compressibility of ionic crystals.

Born-Mayer equation [SOLID STATE] An equation for the cohesive energy of an ionic crystal which is deduced by assuming that this energy is the sum of terms arising from the Coulomb interaction and a repulsive interaction between nearest neighbors.

Born-Oppenheimer method [PHYS CHEM] A method for calculating the force constants between atoms by assuming that the electron motion is so fast compared with the nuclear motions that the electrons follow the motions of the nuclei adiabatically.

Born-von Kármán theory [SOLID STATE] A theory of specific heat which considers an acoustical spectrum for the vibrations of a system of point particles distributed like the atoms in a crystal lattice.

bornyl alcohol See borneol.

boroarsenate [MINERAL] One of a group of borate minerals containing arsenic; cahnite is an example.

boroethane See diborane.

borolanite [PETR] A hypabyssal rock that is essentially orthoclase and melanite with subordinate nepheline, biotite, and pyroxene.

Boroll [GEOL] A member of the soil suborder of Mollisols found in cool and cold regions, mostly in moderately high latitudes or at high altitudes, having fairly thick, nearly black

A horizons, dark grayish brown B horizons, and paler C horizons that are commonly calcareous; found extensively in Canada and the Soviet Union.

Borolon [MATER] A trade name for manufactured corundum.

boron [CHEM] A chemical element, symbol B, atomic number 5, atomic weight 10.811; it has three valence electrons and is nonmetallic.

boron-10 [NUC PHYS] A nonradioactive isotope of boron with a mass number of 10; it is a good absorber for slow neutrons, simultaneously emitting high-energy alpha particles, and is used as a radiation shield in Geiger counters.

boron alloy [MET] Alloy of boron and iron used to increase the high-temperature strength characteristics of alloy steel.

boronatrocalcite See ulexite.

boron carbide [ORG CHEM] Any compound of boron and carbon, especially B_4C (used as an abrasive, alloying agent, and neutron absorber).

boron chamber [NUCLEO] An ionization chamber that is lined with boron or boron compounds or filled with a gaseous boron compound.

boron counter tube [NUCLEO] A counter tube filled with boron fluoride or having electrodes coated with boron or boron compounds; used for detecting slow neutrons.

boron fiber [CHEM] Fiber produced by vapor-deposition methods; used in various composite materials to impart a balance of strength and stiffness. Also known as boron filament.

boron filament See boron fiber.

boron fluoride [INORG CHEM] BF_3 A colorless pungent gas in a dry atmosphere; used in industry as an acidic catalyst for polymerizations, esterifications, and alkylations. Also known as boron trifluoride.

boron fuel [MATER] The boron compounds alkyl decaborane, diborane, and pentaborane; used for ramjet engines and afterburners.

boron hydride See borane.

boron nitride [INORG CHEM] BN A binary compound of boron and nitrogen, especially a white, fluffy powder with high chemical and thermal stability and high electrical resistance.

boron nitride fiber [INORG CHEM] Inorganic, high-strength fiber, made of boron nitride, that is resistant to chemicals and electricity but susceptible to oxidation above 1600°F (870°C); used in composite structures for yarns, fibers, and woven products.

boron oxide See boric oxide.

boron polymer [ORG CHEM] Macromolecules formed by polymerization of compounds containing, for example, boron-nitrogen, boron-phosphorus, or boron-arsenic bonds.

boron steel [MET] Alloy steel with a small amount (as little as 0.0005%) boron added to increase hardenability; can be used to replace other alloys in short supply.

boron thermopile [NUCLEO] A thermopile in which alternate thermocouple junctions are coated with boron; exposure to a flux of slow neutrons generates heat in these junctions, producing an output voltage proportional to neutron flux.

boron trichloride [INORG CHEM] BCl_3 A colorless liquid used as a catalyst and in refining of aluminum, magnesium, zinc, and copper.

boron triethoxide See ethyl borate.

boron trifluoride See boron fluoride.

borosilicate glass [MATER] A type of glass containing at least 5% boric oxide; used in glassware that resists heat.

Borrelia [MICROBIO] A genus of spirochetes in the order Spirochaetales that are parasitic on many animals; many species cause relapsing fever in man.

Borrman effect [PHYS] The irregular transmission of x-rays when a single crystal of high perfection is placed in a monochromatic x-ray beam in a reflecting position.

borrow [MATH] An arithmetically negative carry; it occurs in direct subtraction by raising the low-order digit of the minuend by one unit of the next-higher-order digit; for example, when subtracting 67 from 92, a tens digit is borrowed from the 9, to raise the 2 to a factor of 12; the 7 of 67 is then subtracted from the 12 to yield 5 as the units digit of the difference; the 6 is then subtracted from 8, or 9−1, yielding 2 as the tens digit of the difference.

borrow pit [CIV ENG] An excavation dug to provide material (borrow) for fill elsewhere.

bort [MINERAL] Imperfectly crystallized diamond material unsuitable for gems because of its shape, size, or color and because of flaws or inclusions; used for abrasive and cutting purposes.

bort bit *See* diamond bit.

Bosanquet's law [ELECTROMAG] The statement that, in analogy to Ohm's law for the resistance of an electric circuit, in a magnetic circuit the ratio of the magnetomotive force to the magnetic flux is a constant known as the reluctance.

Bosch fuel injection pump [MECH ENG] A pump in the fuel injection system of an internal combustion engine, whose pump plunger and barrel are a very close lapped fit to minimize leakage.

Bose distribution *See* Bose-Einstein distribution.

Bose-Einstein condensation [CRYO] A phenomenon that occurs in the study of systems of bosons; there is a critical temperature below which the ground state is highly populated. Also known as condensation; Einstein condensation.

Bose-Einstein distribution [STAT MECH] For an assembly of independent bosons, such as photons or helium atoms of mass number 4, a function that specifies the number of particles in each of the allowed energy states. Also known as Bose distribution.

Bose-Einstein statistics [STAT MECH] The statistical mechanics of a system of indistinguishable particles for which there is no restriction on the number of particles that may exist in the same state simultaneously. Also known as Einstein-Bose statistics.

Bose gas [STAT MECH] An assemblage of noninteracting or weakly interacting bosons.

bosh [MET] **1.** Tapering lower portion of a blast furnace, from the blast holes of the hearth up to the maximum internal diameter at the bottom of the stack. **2.** Quartz deposited on the furnace lining during the smelting of copper ore.

boson [STAT MECH] A particle that obeys Bose-Einstein statistics; includes photons, pi mesons, and all nuclei having an even number of particles and all particles with integer spin.

bosporus [GEOGR] A strait connecting two seas or a lake and a sea.

bosque *See* temperate and cold scrub.

boss [DES ENG] Protuberance on a cast metal or plastic part to add strength, facilitate assembly, provide for fastenings, or so forth. [GEOL] A large, irregular mass of crystalline igneous rock that formed some distance below the surface but is now exposed by denudation. [NAV ARCH] *See* bossing.

bossing [NAV ARCH] **1.** A faired structural extension of the hull covering the propeller shaft where it emerges from the main hull and housing the main shaft bearing. **2.** The hub of a screw propeller. Also known as boss.

bostonite [PETR] A rock with coarse trachytic texture formed almost wholly of albite and microcline and with accessory pyroxene.

Bostrichidae [INV ZOO] The powder-post beetles, a family of coleopteran insects in the superfamily Bostrichoidea.

Bostrichoidea [INV ZOO] A superfamily of beetles in the coleopteran suborder Polyphaga.

botanical garden [BOT] An institution for the culture of plants collected chiefly for scientific and educational purposes.

botany [BIOL] A branch of the biological sciences which embraces the study of plants and plant life.

bothridium [INV ZOO] A muscular holdfast organ, often with hooks, on the scolex of tetraphyllidean tapeworms.

Bothriocephaloidea [INV ZOO] The equivalent name for the Pseudophyllidea.

Bothriocidaroida [PALEON] An order of extinct echinoderms in the subclass Perischoechinoidea in which the ambulacra consist of two columns of plates, the interambulacra of one column, and the madreporite is placed radially.

bothrium [INV ZOO] A suction groove on the scolex of pseudophyllidean tapeworms.

botryoid [GEOL] A formation of calcium carbonate occurring in a cave and shaped like a bunch of grapes. Also known as clusterite.

botryoidal [GEOL] Of mineral formations that are shaped like a bunch of grapes.

botryomycosis [VET MED] A chronic infectious bacterial disease of horses caused by *Staphylococcus aureus* and characterized by localized fibromatous tumors.

Botrytis disease [PL PATH] Any of various fungus diseases of plants caused by fungi of the genus *Botrytis;* characterized by soft rotting.

bottle centrifuge [ENG] A centrifuge in which the mixture to be separated is poured into small bottles or test tubes; they are then placed in a rotor assembly which is spun rapidly.

bottled gas [MATER] Butane, propane, or butane-propane mixtures liquefied and bottled under pressure for use as a domestic cooking or heating fuel. Also known as bugas; LP-gas.

bottle graft [BOT] A plant graft in which the scion is a detached branch and is protected from wilting by keeping the base of the branch in a bottle of water until union with the stock.

bottleneck assignment problem [IND ENG] A linear programming problem in which it is required to assign machines to jobs (or vice versa) so that the efficiency of the least efficient operation is maximized.

bottle thermometer [ENG] A thermoelectric thermometer used for measuring air temperature; the name is derived from the fact that the reference thermocouple is placed in an insulated bottle.

bottom [GEOL] **1.** The bed of a body of running or still water. **2.** *See* root.

bottom break [BOT] A branch that arises from the base of a plant stem. [MIN ENG] The break or crack that separates a block of stone from a quarry floor.

bottom chord [CIV ENG] Any of the bottom series of truss members parallel to the roadway of a bridge.

bottom dead center [MECH ENG] The position of the crank of a vertical reciprocating engine, compressor, or pump when the piston is at the end of its downstroke. Abbreviated BDC.

bottom dump [ENG] A construction wagon with movable gates in the bottom to allow vertical discharge of its contents.

bottomed hole [ENG] A completed borehole, or a borehole in which drilling operations have been discontinued.

bottom fauna *See* benthos.

bottom fermentation [FOOD ENG] A slow alcoholic fermentation during which yeast cells accumulate at the bottom of the fermenting liquid; occurs during fermentation of lager beer and wines of low alcoholic content.

bottom flow [ENG] A molding apparatus that forms hollow plastic articles by injecting the blowing air at the bottom of the mold. [HYD] A density current that is denser than any section of the surrounding water and that flows along the bottom of the body of water. Also known as underflow.

bottom-hole cash [PETRO ENG] Cash which is contributed by mineral-rights lessees adjacent to a drilling lessee and which is payable when the well reaches a specified depth, regardless of whether or not the completed well is a producer.

bottom-hole packer [PETRO ENG] An anchored-in-place seal used to provide liquid-proof packing in the annular space between the outside of the oil-producing tubing and the inside of the drill casing.

bottom-hole pressure [PETRO ENG] Gas-drive pressure recorded at the bottom of an oil-well shaft; used to analyze oil-reservoir performance and evaluate the performance of downhole equipment.

bottom-hole samples [PETRO ENG] Fluid samples from gas-condensate-well reservoirs; used to study the state of the hydrocarbon system under reservoir conditions and to estimate total hydrocarbons in place.

bottom ice *See* anchor ice.

bottoming drill [DES ENG] A flat-ended twist drill designed to convert a cone at the bottom of a drilled hole into a cylinder.

bottomland [GEOL] A lowland formed by alluvial deposit about a lake basin or a stream.

bottom load [GEOL] An obsolete term for the material rolled and pushed along the bottom of a stream.

bottom moraine *See* ground moraine.

bottom pillar [MIN ENG] A large block of solid coal left unworked around the shaft.

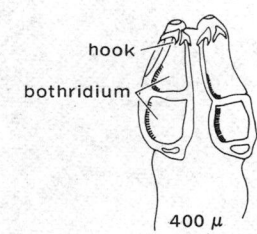

BOTHRIDIUM

hook

bothridium

400 μ

A drawing of the scolex of *Acanthobothrium* species showing the hooks on the bothridium.

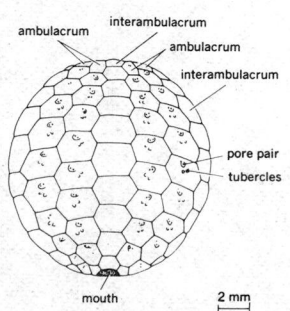

BOTHRIOCIDAROIDA

ambulacrum

interambulacrum

ambulacrum

interambulacrum

pore pair

tubercles

mouth   2 mm

A reconstruction of the test of *Bothriocidaris.*

bottom rot [PL PATH] **1.** A fungus disease of lettuce, caused by *Pellicularia filamentosa*, that spreads from the base upward. **2.** A fungus disease of tree trunks caused by pore fungi.

bottoms *See* basic sediment and water.

bottom sampler [ENG] Any instrument used to obtain a sample from the bottom of a body of water.

bottom sediment [PETRO ENG] A mixture of liquids and solids which form in the bottom of oil storage tanks.

bottomset beds [GEOL] Horizontal or gently inclined layers of finer material carried out and deposited on the bottom of a lake or sea in front of a delta.

bottom settlings *See* basic sediment and water.

bottom tap [DES ENG] A tap with a chamfer 1-1½ threads in length. [MET] A hole in the bottom of a furnace for draining out the slag.

bottom water [HYD] Water lying beneath oil or gas in productive formations. [OCEANOGR] The water mass at the deepest part of a water column in the ocean.

botulin [MICROBIO] The neurogenic toxin which is produced by *Clostridium botulinum* and *C. parabotulinum* and causes botulism. Also known as botulinus toxin.

botulinus [MICROBIO] A bacterium that causes botulism.

botulinus toxin *See* botulin.

botulism [MED] Food poisoning due to intoxication by the exotoxin of *Clostridium botulinum* and *C. parabotulinum*.

bouclé [TEXT] **1.** Yarn made with loose loops. **2.** A fabric made from bouclé yarn.

boudin [GEOL] One of a series of sausage-shaped segments found in a boudinage.

boudinage [GEOL] A structure in which beds set in a softer matrix are divided by cross fractures into segments resembling pillows.

bough [BOT] A main branch on a tree.

bougie decimale [OPTICS] Formerly, a unit of luminous intensity equal to 0.96 international standard candle.

Bouguer anomaly [GEOL] The gravity value after the Bouguer reduction is applied to a level datum.

Bouguer correction *See* Bouguer reduction.

Bouguer-Lambert-Beer law [ANALY CHEM] The intensity of a beam of monochromatic radiation in an absorbing medium decreases exponentially with penetration distance. Also known as Beer-Lambert-Bouguer law; Lambert-Beer law.

Bouguer-Lambert law [ANALY CHEM] The law that the change in intensity of light transmitted through an absorbing substance is related exponentially to the thickness of the absorbing medium and a constant which depends on the sample and the wavelength of the light. Also known as Lambert's law.

Bouguer reduction [GEOL] A correction made in gravity work to take account of the station's altitude and the rock between the station and sea level. Also known as Bouguer correction.

Bouguer's halo [METEOROL] A faint, white circular arc of light of about 39° radius around the antisolar point. Also known as Ulloa's ring.

Bouin's solution [MATER] A picric acid-acetic acid-formaldehyde fixative and preserving fluid for contractile forms.

boulder [GEOL] A worn rock with a diameter exceeding 256 millimeters. Also spelled bowlder.

boulder belt [GEOL] A long, narrow accumulation of boulders elongately transverse to the direction of glacier movement.

boulder buster [ENG] A heavy, pyramidical- or conicalpoint steel tool which may be attached to the bottom end of a string of drill rods and used to break, by impact, a boulder encountered in a borehole. Also known as boulder cracker.

boulder clay *See* till.

boulder cracker *See* boulder buster.

boulder pavement [GEOL] A surface of till with boulders; the till has been abraded to flatness by glacier movement.

boulder train [GEOL] Glacial boulders derived from one locality and arranged in a right-angled line or lines leading off in the direction in which the drift agency operated.

bounce cast [GEOL] A short ridge underneath a stratum fading out gradually in both directions.

bounce table [MECH ENG] A testing device which subjects devices and components to impacts such as might be encountered in accidental dropping.

bouncing putty [MATER] A silicone polymer in a soft elastic mass; the material's elasticity will increase as the applied force increases.

boundary [GEOL] A line between areas occupied by rocks or formations of different type and age. [SCI TECH] A line or area which determines inclusion in a system.

boundary condition [MATH] A requirement to be met by a solution to a set of differential equations on a specified set of values of the independent variables.

boundary-layer control [FL MECH] Control over the development of a boundary layer by reduction of surface roughness and choice of surface contours. Abbreviated BLC.

boundary-layer flow [FL MECH] The flow of that portion of a viscous fluid which is in the neighborhood of a body in contact with the fluid and in motion relative to the fluid.

boundary-layer photocell *See* photovoltaic cell.

boundary-layer separation [FL MECH] That point where the boundary layer no longer continues to follow the contour of the boundary because the residual momentum of the fluid (left after overcoming viscous forces) may be insufficient to allow the flow to proceed into regions of increasing pressure. Also known as flow separation.

boundary-layer theory *See* film theory.

boundary line [MAP] A line of demarcation along which two areas meet.

boundary lubrication [ENG] A lubricating condition that is a combination of solid-to-solid surface contact and liquid-film shear.

boundary map [MAP] A map constructed for the purpose of delineating a boundary line and adjacent territory.

boundary marker [NAV] A radio transmitter operating at 75 megahertz and installed near the approach end of landing runway (3.9 nautical miles ±1000 feet, or 7123 ±305 meters) and approximately on the localizer course line.

boundary monument [ENG] A material object placed on or near a boundary line to preserve and identify the location of the boundary line on the ground.

boundary of a set *See* frontier of a set in a topological space.

boundary pillar [MIN ENG] A pillar left in mines between adjoining properties.

boundary point [MATH] A boundary point of a set E is p if every neighborhood of p contains a point of E and also a point not in E.

boundary survey [ENG] A survey made to establish or to reestablish a boundary line on the ground or to obtain data for constructing a map or plat showing a boundary line.

bound barrel [ORD] A barrel that is touching parts of the stock in such a manner that expansion due to heat from firing causes the barrel to bind and bend, resulting in inaccurate fire.

bound charge [ELEC] Electric charge which is confined to atoms or molecules, in contrast to free charge, such as metallic conduction electrons, which is not. Also known as polarization charge.

bounded function [MATH] **1.** A function whose image is a bounded set. **2.** A function of a metric space to itself which moves each point no more than some constant distance.

bounded linear transformation [MATH] A linear transformation T for which there is some positive number A such that the norm of $T(x)$ is equal to or less than A times the norm of x for each x.

bounded set [MATH] **1.** A collection of numbers whose absolute values are all smaller than some constant. **2.** A set of points, the distance between any two of which is smaller than some constant.

bounded variation [MATH] A real-valued function is of bounded variation on an interval if its total variation there is bounded.

bound electron [ATOM PHYS] An electron whose wave function is negligible except in the vicinity of an atom.

bounding mine [ORD] Type of antipersonnel mine usually buried just below the surface of the ground; it has a small charge which throws the case up into the air; this case explodes at a height of 3 or 4 feet, throwing shrapnel or fragments in all directions.

bound level [NUC PHYS] An energy level in a nucleus so close to the ground state that it can only decay by gamma emission.

BOUNDARY-LAYER CONTROL

In this airfoil, retarded flow in the boundary layer is reenergized by supplying high-velocity flow through a slot in the surface.

bound particle [PHYS] A particle which is confined to some finite region.

bound symbol [COMMUN] A contextual symbol either not preceded or not followed, or neither preceded nor followed, by space.

bound symbol sequence [COMMUN] A symbol sequence either not preceded or not followed, or neither preceded nor followed, by space.

bound variable [MATH] In logic, a variable that occurs within the scope of a quantifier, and cannot be replaced by a constant.

bound water [CHEM] Water that is a portion of a system such as tissues or soil and does not form ice crystals until the material's temperature is lowered to about −20°C.

bourbon [FOOD ENG] A whiskey distilled from a corn mash containing at least 51% corn, with malt and rye composing the remaining ingredients, and aged in containers made of charred oak.

Bourdon pressure gage [ENG] A mechanical pressure-measuring instrument employing as its sensing element a curved or twisted metal tube, flattened in cross section and closed. Also known as Bourdon tube.

Bourdon tube *See* Bourdon pressure gage.

Bourges process [GRAPHICS] A process used in art and color preparation for fake color work; transparent films in a variety of colors and densities are used as overlays for the black and white art work.

Bournesville's disease *See* tuberous sclerosis.

bournonite [MINERAL] PbCuSbS$_3$ Steel-gray to black orthorhombic crystals; mined as an ore of copper, lead, and antimony. Also known as cogwheel ore.

Bourquin-Sherman unit [BIOL] A unit for the standardization of riboflavin.

bourrelet [ORD] The cylindrical surface of a projectile on which the projectile bears while in the bore of the weapon.

Boussinesq approximation [FL MECH] The assumption (frequently used in the theory of convection) that the fluid is incompressible except insofar as the thermal expansion produces a buoyancy, represented by a term $g\alpha T$, where g is the acceleration of gravity, α is the coefficient of thermal expansion, and T is the perturbation temperature.

Boussinesq equation [ENG] A relation used to calculate the influence of a concentrated load on the backfill behind a retaining wall.

Boussinesq number [FL MECH] A dimensionless number used to study wave behavior in open channels.

bouton [ANAT] A club-shaped enlargement at the end of a nerve fiber. Also known as end bulb.

boutonneuse fever *See* fièvre boutonneuse.

Bouveault-Blanc method [ORG CHEM] A laboratory method for preparing alcohols by reduction of esters utilizing sodium dissolved in alcohol.

Bovidae [VERT ZOO] A family of pecoran ruminants in the superfamily Bovoidea containing the true antelopes, sheep, and goats.

bovine [VERT ZOO] **1.** Any member of the genus *Bos*. **2.** Resembling or pertaining to a cow or ox.

bovine mastitis [VET MED] Inflammation of the udder of a cow; may result from injury or bacterial infection.

bovine staggers [VET MED] A disease of cattle in southern Africa caused by eating the poisonous herb *Matricaria nigellaefolia* and characterized by staggering, emaciation, and finally paralysis.

Bovoidea [VERT ZOO] A superfamily of pecoran ruminants in the order Artiodactyla comprising the pronghorns and bovids.

bow [AERO ENG] The forward part of an aircraft. [ARCH] A part of a building shaped as an arc or a polygon and projecting from a straight wall. [NAV ARCH] The forward part of a ship.

bow compass [GRAPHICS] A small compass whose legs are connected by a spring shaped like a bow, rather than by a joint.

Bowden cable [MECH ENG] A wire made of spring steel which is enclosed in a helical casing and used to transmit longitudinal motions over distances, particularly around corners.

bow divider [GRAPHICS] A small divider whose legs are connected by a bow-shaped spring, rather than by a joint.

bowel [ANAT] The intestine.

Bowen reaction series [MINERAL] A series of minerals wherein any early-formed phase will react with the melt later in the differentiation to yield a new mineral further in the series.

Bowen's disease [MED] **1.** Intraepithelial squamous-cell carcinoma of the skin, forming distinctive plaques. **2.** A similar carcinoma occurring in mucous membranes.

Bower-Barff process [MET] A method of coating iron or steel with magnetic oxide, such as Fe$_3$O$_4$, in order to minimize atmospheric corrosion.

bowfin [VERT ZOO] *Amia calva.* A fish recognized as the only living species of the family Amiidae. Also known as dogfish; grindle; mudfish.

bow flare slamming [NAV ARCH] An extreme pitch-and-heave motion of a ship in which the keel emerges from the water but the forefoot does not.

bow gun [ORD] Gun mounted at the front of a ship or armored vehicle, especially a semifixed, forward-firing gun in tanks.

Bowie formula [GEOPHYS] A correction used for calculation of the local gravity anomaly on earth.

bowk *See* hoppit.

bowl classifier [CHEM ENG] A shallow bowl with a concave bottom so that a liquid-solid suspension can be fed to the center; coarse particles fall to the bottom, where they are raked to a central discharge point, and liquid and fine particles overflow the edges and are collected.

bowlder *See* boulder.

bow light [NAV ARCH] A white light near the bow of a ship at anchor or under way.

bowline [NAV ARCH] **1.** A rope attached to the vertical edge of a square sail near its midpoint, and used to keep the sail's weather edge taut forward when the vessel is close-hauled. **2.** A knot forming a loop that does not slip under tension, used particularly for mooring and hauling.

bowlingite *See* saponite.

bowl mill *See* bowl-mill pulverizer.

bowl-mill pulverizer [MECH ENG] A type of pulverizer which directly feeds a coal-fired furnace, in which springs press pivoted stationary rolls against a rotating bowl grinding ring, crushing the coal between them. Also known as a bowl mill.

bowl scraper [MECH ENG] A towed steel bowl hung within a fabricated steel frame, running on four or two wheels; transports soil, in addition to spreading and leveling it.

Bowman's capsule [ANAT] A two-layered membranous sac surrounding the glomerulus and constituting the closed end of a nephron in the kidneys of all higher vertebrates.

bow-on [ORD] Facing the firer; a bow-on target presents its narrower dimension exactly toward the gun firing at it; when an enemy tank is headed exactly at the gun firing at it, the tank is a bow-on target.

bowshock [ASTROPHYS] The shock wave set up by the interaction of the supersonic solar wind with a planet's magnetic field.

Bow's notation [MECH] A graphical method of representing coplanar forces and stresses, using alphabetical letters, in the solution of stresses or in determining the resultant of a system of concurrent forces.

bowsprit [NAV ARCH] A large spar that projects forward from the forward end of a sailing ship, used to carry sails and support the masts by stays.

bowstring beam [CIV ENG] A steel, concrete, or timber beam or girder shaped in the form of a bow and string; the string resists the horizontal forces caused by loads on the arch.

bow wave [FL MECH] A shock wave occurring in front of a body, such as an airfoil, or apparently attached to the forward tip of the body.

box [DES ENG] *See* boxing. [ENG] A protective housing.

box annealing [MET] Slow heating of metal sheets in a closed metal box to prevent oxidation, followed by cooling; usually limited to iron-base alloys.

box barrage [ORD] In antiaircraft artillery fire, an antiaircraft barrage delivered in a box-shape pattern by guns surrounding a defended target.

box beam *See* box girder.

box bill *See* bell tap.

BOURNONITE

|-2.5 cm-|

Crystals of bournonite with rhodochrosite and quartz from Pribram, Bohemia, Czechoslovakia. *(Specimen from Department of Geology, Bryn Mawr College)*

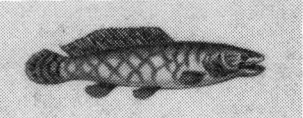

BOWFIN

Bowfin *(Amia calva)*, which may attain 2½ feet (76 centimeters) and is found only in some parts of the United States.

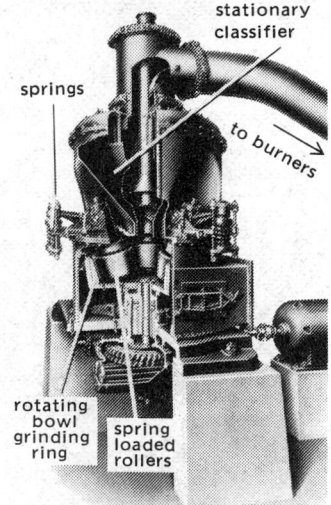

BOWL-MILL PULVERIZER

stationary classifier

springs

to burners

rotating bowl grinding ring

spring loaded rollers

Medium-speed bowl-mill coal pulverizer, which feeds burners. *(Combustion Engineering Co.)*

BOX PILES

Box-pile cross section.

BRACED-RIB ARCH

The Bayonne Bridge, a braced-rib arch, across the Kill Van Kull in New Jersey. (*Port of New York Authority*)

BRACHYGNATHA

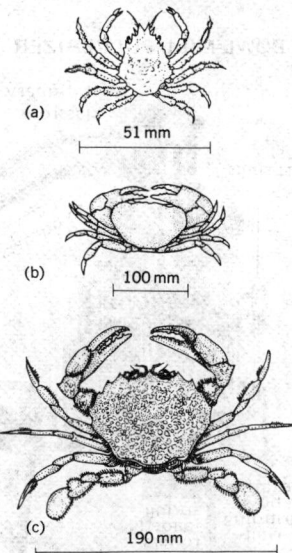

Representative crabs of the Brachygnatha: *(a)* spider crab, *Mithrax acuticornis;* *(b)* fresh-water crab, *Epilobocera sinuatifrons;* *(c)* swimming crab, *Ovalipes ocellatus.* (*Smithsonian Institution*)

boxboard [MATER] Paperboard used for making cardboard boxes.

box caisson [CIV ENG] A floating steel or concrete box with an open top which will be filled and sunk at a foundation site in a river or seaway. Also known as American caisson; stranded caisson.

box camera [OPTICS] A camera that consists of a box, an arrangement for loading and winding film, a simple lens of fixed focus, and a simple shutter with a speed of about 1/30 second.

box canyon [GEOGR] A canyon with steep rock sides and a zigzag course, that is usually closed upstream.

boxcar [COMMUN] One of a series of long signal-wave pulses that are separated by very short time intervals. [ENG] A railroad car with a flat roof and vertical sides, usually with sliding doors, which carries freight that needs to be protected from weather and theft.

boxcar circuit [ELECTR] A circuit used in radar for sampling voltage waveforms and storing the latest value sampled; the term is derived from the flat, steplike segments of the output voltage waveform.

boxcar function [MATH] A function whose value is zero except for a finite interval of its argument, for which it has a constant nonzero value.

box-coking test [ENG] A laboratory test which forecasts the quality of coke producible in commercial practice; uses a specially designed sheet-steel box containing about 60 pounds (27 kilograms) of coal in a commercial coke oven.

box fold [GEOL] A fold in which the broad, flat top of an anticline or the broad, flat bottom of a syncline is bordered by steeply dipping limbs.

box girder [CIV ENG] A hollow girder or beam with a square or rectangular cross section. Also known as box beam.

box-girder bridge [CIV ENG] A fixed structure constructed from box girders; the top flange plate of each box may be widened to serve as the deck.

Box Hole [GEOL] A meteorite crater in central Australia, 575 feet (244 meters) in diameter.

boxing [DES ENG] The threaded nut for the screw of a mounted auger drill. Also known as box. [ENG] A method of securing shafts solely by slabs and wooden pegs. [MET] Continuing a fillet weld around a corner. Also known as end turning.

boxing the compass [NAV] In nautical practice, calling out in sequence the names of the points (and sometimes the half and quarter points) indicated by the compass of a maneuvering vessel.

box loom [TEXT] A loom that uses several shuttles to weave fabric with more than one type of filling yarn or with a regular filling pattern.

box piles [CIV ENG] Pile foundations made by welding together two sections of steel sheet piling or combinations of beams, channels, and plates.

boxwork [GEOL] Limonite and other minerals which formed at one time as blades or plates along cleavage or fracture planes, after which the intervening material dissolved, leaving the intersecting blades or plates as a network.

Boyle's law [PHYS] The law that the product of the volume of a gas times its pressure is a constant at fixed temperature. Also known as Mariotte's law.

Boyle's temperature [THERMO] For a given gas, the temperature at which the virial coefficient B in the equation of state $Pv = RT [1 + (B/v) + (C/v^2) + \ldots]$ vanishes.

Boys camera [OPTICS] A type of camera used for the observation of lightning flashes.

bp *See* boiling point.

bpd *See* barrels per day.

bpm *See* barrels per month.

B power supply *See* B supply.

Br *See* bromine.

BRA *See* β-resorcylic acid.

brace [DES ENG] A cranklike device for turning a bit. [ENG] A diagonally placed structural member that withstands tension and compression, and often stiffens a structure against wind.

brace and bit [DES ENG] A small hand tool to which is attached a metal- or wood-boring bit.

braced-rib arch [CIV ENG] A type of steel arch, usually used in bridge construction, which has a system of diagonal bracing.

brace head [ENG] A cross handle attached at the top of a column of drill rods by means of which the rods and attached bit are turned after each drop in chop-and-wash operations while sinking a borehole through overburden. Also known as brace key.

brace key *See* brace head.

brace root *See* prop root.

brachial [ZOO] Of or relating to an arm or armlike process.

brachial artery [ANAT] An artery which originates at the axillary artery and branches into the radial and ulnar arteries; it distributes blood to the various muscles of the arm, the shaft of the humerus, the elbow joint, the forearm, and the hand.

brachial cavity [INV ZOO] The anterior cavity which is located inside the valves of brachiopods and into which the brachia are withdrawn.

brachial plexus [ANAT] A plexus of nerves located in the neck and axilla and composed of the anterior rami of the lower four cervical and first thoracic nerves.

Brachiata [INV ZOO] A phylum of deuterostomous, sedentary bottom-dwelling marine animals that live encased in tubes.

brachiate [BOT] Having widely divergent branches. [ZOO] Having arms.

brachiolaria [INV ZOO] A transitional larva in the development of certain starfishes that is distinguished by three anterior processes homologous with those of the adult.

Brachiopoda [INV ZOO] A phylum of solitary, marine, bivalved coelomate animals.

brachioradialis [ANAT] Pertaining to the arm and radius.

brachistochrone [MECH] The curve along which a smooth-sliding particle, under the influence of gravity alone, will fall from one point to another in the minimum time.

brachium [ANAT] The upper arm or forelimb, from the shoulder to the elbow. [INV ZOO] **1.** A ray of a crinoid. **2.** A tentacle of a cephalopod. **3.** Either of the paired appendages constituting the lophophore of a brachiopod.

brachyaxis [CRYSTAL] The shorter lateral axis, usually the a axis, of an orthorhombic or triclinic crystal. Also known as brachydiagonal.

brachyblast [BOT] A short shoot often bearing clusters of leaves.

brachycephalic [ANTHRO] Being short- or broad-headed, that is, having a cephalic index of over 80.

brachycephaly [ANTHRO] The state of being brachycephalic.

brachycerebral [ANTHRO] Having a round or short brain.

brachycranial [ANTHRO] Being short- or broad-skulled, that is, with a cephalic index of over 80.

brachydactylia [MED] Abnormal shortening of fingers or toes.

brachydiagonal *See* brachyaxis.

brachydont [ANAT] Of teeth, having short crowns, well-developed roots, and narrow root canals; characteristic of man.

brachyfacial [ANTHRO] Having a short or broad face.

Brachygnatha [INV ZOO] A subsection of brachyuran crustaceans to which most of the crabs are assigned.

brachypinacoid [GEOL] A pinacoid parallel to the vertical and the shorter lateral axis.

Brachypsectridae [INV ZOO] A family of coleopteran insects in the superfamily Cantharoidea represented by a single species.

Brachypteraciidae [VERT ZOO] The ground rollers, a family of colorful Madagascan birds in the order Coraciiformes.

brachypterous [INV ZOO] Having rudimentary or abnormally small wings, referring to certain insects.

brachysclereid [BOT] A sclereid that is more or less isodiametric and is found in certain fruits and in the pith, cortex, and bark of many stems. Also known as stone cell.

brachyskelic [ANTHRO] Having legs short in proportion to trunk length, with a skelic index of 75–80.

brachysm [BOT] Plant dwarfing in which there is shortening of the internodes only.

brachysyncline [GEOL] A broad, short syncline.

brachytherapy [MED] Radiation treatment using a solid or enclosed radioisotopic source on the surface of the body or at a short distance from the area to be treated.

Brachythoraci [PALEON] An order of the extinct joint-necked fishes.

Brachyura [INV ZOO] The section of the crustacean order Decapoda containing the true crabs.

bracing [ENG] The act or process of strengthening or making rigid.

bracket [BUILD] A vertical board to support the tread of a stair. [CIV ENG] A projecting support. [ORD] **1.** The distance between two strikes or series of strikes, one of which is over the target and the other short of it, or one of which is to the right and the other to the left of the target. **2.** A group of shots (or bombs) which fall both over and short of the target.

bracket fungus [MYCOL] A basidiomycete characterized by shelflike sporophores, sometimes seen on tree trunks.

bracketing method [ORD] A method of adjusting artillery and mortar fire in which a bracket is established by obtaining an over and a short strike, with respect to the observer, then successively splitting the bracket in half until a target hit is obtained or the smallest practicable range change has been made.

bracketing salvo [ORD] A series of shots in which the number of shots going over the target equals the number falling short of it.

brackish [HYD] **1.** Of water, having salinity values ranging from approximately 0.50 to 17.00 parts per thousand. **2.** Of water, having less salt than sea water, but undrinkable.

Braconidae [INV ZOO] The braconid wasps, a family of hymenopteran insects in the superfamily Ichneumonoidea.

bract [BOT] A modified leaf associated with plant reproductive structures.

bracteolate [BOT] Having bracteoles.

bracteole [BOT] A small bract, especially if on the floral axis. Also known as bractlet.

bractlet *See* bracteole.

Bradford breaker [MIN ENG] A machine which combines coal crushing and screening.

Bradfordian [GEOL] Uppermost Devonian geologic time.

Bradford preferential separation process [MIN ENG] A flotation process for the treatment of mixed sulfides in which certain mineral salts, such as thiosulfates, are added to the water used in the flotation cells.

Bradford spinning [TEXT] A process for spinning wool into worsted yarn, in which oil is applied to the wool before combing so that the yarn will be smooth.

Bradley aberration [ASTRON] Stellar aberration with a maximum of 20.5 seconds of arc; can be used to compute an approximate velocity for light.

bradyauxesis [BIOL] Allometric growth in which a part lags behind the body as a whole in development.

bradycardia [MED] Slow heart rate.

bradykinin [BIOCHEM] $C_{50}H_{73}N_{15}O_{11}$ A polypeptide kinin; forms an amorphous precipitate in glacial acetic acid; released from plasma precursors by plasmin. Also known as callideic I; kallidin I.

Bradyodonti [PALEON] An order of Paleozoic cartilaginous fishes (Chondrichthyes), presumably derived from primitive sharks.

Bradypodidae [VERT ZOO] A family of mammals in the order Edentata comprising the true sloths.

bradytely [EVOL] Evolutionary change that is either arrested or occurring at a very slow rate over long geologic periods.

Bragg angle [SOLID STATE] One of the characteristic angles at which x-rays reflect specularly from planes of atoms in a crystal.

Bragg curve [ATOM PHYS] **1.** A curve showing the average number of ions per unit distance along a beam of initially monoenergetic ionizing particles, usually alpha particles, passing through a gas. Also known as Bragg ionization curve. **2.** A curve showing the average specific ionization of an ionizing particle of a particular kind as a function of its kinetic energy, velocity, or residual range.

Bragg diffraction *See* Bragg scattering.

Bragg ionization curve *See* Bragg curve.

braggite [MINERAL] PtS A steel-gray platinum sulfide mineral with tetragonal crystals.

Bragg-Kleeman rule *See* Bragg rule.

Bragg-Pierce law [PHYS] A relationship for determining an element's atomic absorption coefficient for x-rays when the atomic number of the element and the wavelength of the x-rays are known.

Bragg reflection *See* Bragg scattering.

Bragg rule [ATOM PHYS] An empirical rule according to which the mass stopping power of an element for alpha particles is inversely proportional to the square root of the atomic weight. Also known as Bragg-Kleeman rule.

Bragg scattering [SOLID STATE] Scattering of x-rays or neutrons by the regularly spaced atoms in a crystal, for which constructive interference occurs only at definite angles called Bragg angles. Also known as Bragg diffraction; Bragg reflection.

Bragg's equation *See* Bragg's law.

Bragg's law [SOLID STATE] A statement of the conditions under which a crystal will reflect a beam of x-rays with maximum intensity. Also known as Bragg's equation; Bravais' law.

Bragg spectrometer [SOLID STATE] An instrument for x-ray analysis of crystal structure, in which a homogeneous beam of x-rays is directed on the known face of a crystal and the reflected beam is detected in a suitably placed ionization chamber. Also known as crystal spectrometer; crystal diffraction spectrometer; ionization spectrometer.

braided stream [HYD] A stream flowing in several channels that divide and reunite.

braided wire [ELEC] A tube of fine wires woven around a conductor or cable for shielding purposes or used alone in flattened form as a grounding strap.

Braille [COMMUN] A system of written communication for the blind in which letters are represented by raised dots over which the trained blind person moves his fingertips.

Brailsford-Morquio syndrome *See* Morquio's syndrome.

brain [ANAT] The portion of the vertebrate central nervous system enclosed in the skull. [ZOO] The enlarged anterior portion of the central nervous system in most bilaterally symmetrical animals.

braincase *See* cranium.

brain coral [INV ZOO] A reef-building coral resembling the human cerebrum in appearance.

brain hormone [INV ZOO] A neurohormone secreted by the insect brain that regulates the release of ecdysone from the prothoracic glands.

brainstem [ANAT] The portion of the brain remaining after the cerebral hemispheres and cerebellum have been removed.

brainstorming [IND ENG] A procedure used to find a solution for a problem by collecting all the ideas, without regard for feasibility, which occur from a group of people meeting together.

brain wave [PHYSIO] A rhythmic fluctuation of voltage between parts of the brain, ranging from about 1 to 60 hertz and 10 to 100 microvolts.

brake [MECH ENG] A machine element for applying friction to a moving surface to slow it (and often, the containing vehicle or device) down or bring it to rest.

brake band [MECH ENG] The contracting element of the band brake.

brake drum [MECH ENG] A rotating cylinder attached to a rotating part of machinery, which the brake band or brake shoe presses against.

brake fluid [MATER] The liquid in the cylinder of an automotive brake which, under the action of a piston, is forced through tubing into cylinders at each car wheel, moving a pair of pistons outward so that the brake shoes are thrust against the revolving brake drums.

brake horsepower [MECH ENG] The power developed by an engine as measured by the force applied to a friction brake or by an absorption dynamometer applied to the shaft or flywheel. Abbreviated bhp.

brake lining [MECH ENG] A covering, riveted or molded to the brake shoe or brake band, which presses against the rotating brake drum; made of either fabric or molded asbestos material.

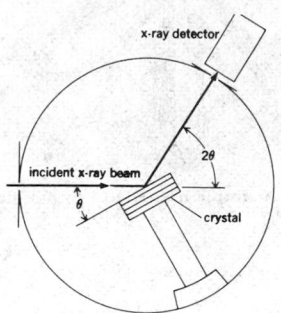

BRAGG SPECTROMETER

Schematic of Bragg spectrometer. θ is angle between incident beam and crystallographic planes; 2θ is angle between incident and diffracted beams.

BRAILLE

Sketch showing Braille alphabet. Combinations of dots (up to six per letter) represent letters. The trained blind person runs the fingers over the raised dots.

brake shoe [MECH ENG] The renewable friction element of a shoe brake. Also known as shoe.

braking ellipses [AERO ENG] A series of ellipses, decreasing in size due to aerodynamic drag, followed by a spacecraft in entering a planetary atmosphere.

braking rocket *See* retrorocket.

brale [MET] A conical diamond indenter with an angle of 120° used in the Rockwell hardness test.

bramble [BOT] **1.** A plant of the genus *Rubus.* **2.** A rough, prickly vine or shrub.

brammalite [MINERAL] A mica-type clay mineral that is different from illite because it has soda instead of potash; it is the sodium analog of illite. Also known as sodium illite.

branch [ADP] **1.** Any one of a number of instruction sequences in a program to which computer control is passed, depending upon the status of one or more variables. **2.** *See* jump. [BOT] A shoot or secondary stem on the trunk or a limb of a tree. [ELEC] A portion of a network consisting of one or more two-terminal elements in series. Also known as arm. [MATH] A complex function which is analytic in some domain and which takes on one of the values of a multiple-valued function in that domain. [NUC PHYS] A product resulting from one mode of decay of a radioactive nuclide that has two or more modes of decay. [ORG CHEM] A carbon side chain attached to a molecule's main carbon chain. [SCI TECH] An area of study representing an independent offshoot of a related basic discipline.

branch-and-bound technique [IND ENG] A technique in nonlinear programming in which all sets of feasible solutions are divided into subsets, and those having bounds inferior to others are rejected.

branch circuit [ELEC] A portion of a wiring system in the interior of a structure that extends from a final overload protective device to a plug receptable or a load such as a lighting fixture, motor, or heater.

branch-circuit distribution center [COMMUN] Distribution center at which branch circuits are supplied.

branch cut [MATH] A line or curve of singular points used in defining a branch of a multiple-valued complex function.

branched acinous gland [ANAT] A multicellular structure with saclike glandular portions connected to the surface of the containing organ or structure by a common duct.

branched-chain ketoaciduria *See* maple syrup urine disease.

branched tubular gland [ANAT] A multicellular structure with tube-shaped glandular portions connected to the surface of the containing organ or structure by a common secreting duct.

branch gain *See* branch transmittance.

branchia *See* gill.

branchial [ZOO] Of or pertaining to gills.

branchial arch [VERT ZOO] One of the series of paired arches on the sides of the pharynx which support the gills in fishes and amphibians.

branchial basket [ZOO] A cartilaginous structure that supports the gills in protochordates and certain lower vertebrates such as cyclostomes.

branchial cleft [EMBRYO] A rudimentary groove in the neck region of air-breathing vertebrate embryos. [VERT ZOO] One of the openings between the branchial arches in fishes and amphibians.

branchial heart [INV ZOO] A muscular enlarged portion of a vein of a cephalopod that contracts and forces the blood into the gills.

branchial plume [INV ZOO] An accessory respiratory organ that extends out under the mantle in certain Gastropoda.

branchial pouch [ZOO] In cyclostomes and some sharks, one of the respiratory cavities occurring in the branchial clefts.

branchial sac [INV ZOO] In tunicates, the dilated pharyngeal portion of the alimentary canal that has vascular walls pierced with clefts and serves as a gill.

branchial segment [EMBRYO] Any of the paired pharyngeal segments indicating the visceral arches and clefts posterior to and including the third pair in air-breathing vertebrate embryos.

branchiate [VERT ZOO] Having gills.

branching [ADP] The selection, under control of a computer program, of one of two or more branches. [NUC PHYS] The occurrence of two or more modes by which a radionuclide

can undergo radioactive decay. Also known as multiple decay; multiple disintegration.

branching adaptation *See* divergent adaptation.

branching bay *See* estuary.

branching fraction [NUC PHYS] That fraction of the total number of atoms involved which follows a particular branch of the disintegration scheme; usually expressed as a percentage.

branching process [STAT] A stochastic process in which the members of a population may have offspring and the lines of descent branch out as the new members are born.

branch instruction [ADP] An instruction that makes the computer choose between alternative subprograms, depending on the conditions determined by the computer during the execution of the program.

branchiocranium [VERT ZOO] The division of the fish skull constituting the mandibular and hyal regions and the branchial arches.

branchiomere [EMBRYO] An embryonic metamere that will differentiate into a visceral arch and cleft; a branchial segment.

branchiomeric musculature [VERT ZOO] Those muscles derived from branchial segments in vertebrates.

Branchiopoda [INV ZOO] A subclass of crustaceans containing small or moderate-sized animals commonly called fairy shrimps, clam shrimps, and water fleas.

branchiostegite [INV ZOO] A gill cover and chamber in certain malacostracan crustaceans, formed by lateral expansion of the carapace.

Branchiostoma [ZOO] A genus of lancelets formerly designated as amphioxus.

Branchiotremata [INV ZOO] The hemichordates, a branch of the subphylum Oligomera.

branchite *See* hartite.

Branchiura [INV ZOO] The fish lice, a subclass of fish ectoparasites in the class Crustacea.

branch joint [ELEC] Joint used for connecting a branch conductor or cable, where the latter continues beyond the branch.

branch point [ELEC] A terminal in an electrical network that is common to more than two elements or parts of elements of the network. Also known as junction point; node. [MATH] A point at which two or more sheets of a Riemann surface join together.

branch sewer [CIV ENG] A part of a sewer system that is larger in diameter than the lateral sewer system; receives sewage from both house connections and lateral sewers.

branch transmittance [CONT SYS] The amplification of current or voltage in a branch of an electrical network; used in the representation of such a network by a signal-flow graph. Also known as branch gain.

brandy [CHEM ENG] A potable alcoholic beverage distilled from wine or fermented fruit juice, usually after the aging of the wine in wooden casks; cognac is a brandy distilled from wines made from grapes from the Cognac region of France.

brannerite [MINERAL] A complex, black, opaque titanite of uranium and other elements in which the weight of uranium exceeds that of titanium; monoclinic and possibly (U,Ca, Fe,Y,Th)$_3$Ti$_5$O$_6$.

brass [GEOL] A British term for sulfides of iron (pyrites) in coal. Also known as brasses. [MET] A copper-zinc alloy of varying proportions but typically containing 67% copper and 33% zinc.

brass chills *See* metal fume fever.

brass founder's ague *See* metal fume fever.

Brassica [BOT] A large genus of herbs in the family Cruciferae of the order Capparales, including cabbage, watercress, and sweet alyssum.

Brassicaceae [BOT] An equivalent name for the Cruciferae.

brassin [BIOCHEM] Any of a class of plant hormones characterized as long-chain fatty-acid esters; brassins act to induce both cell elongation and cell division in leaves and stems.

brass ore *See* aurichalcite.

Brathinidae [INV ZOO] The grass root beetles, a small family of coleopteran insects in the superfamily Staphylinoidea.

brattice [MIN ENG] A temporary board or cloth partition in any mine passage to confine the air and force it into the working places. Also spelled brattish; brettice; brettis.

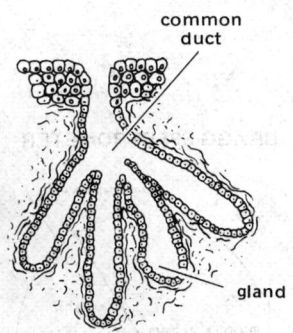
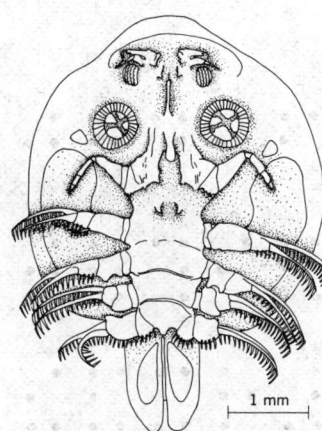

brattice cloth [MIN ENG] Fire-resistant canvas or duck used to erect a brattice.

brattish *See* brattice.

Braulidae [INV ZOO] The bee lice, a family of cyclorrhaphous dipteran insects in the section Pupipara.

braunite [MINERAL] $3Mn_2O_3 \cdot MnSiO_3$ Brittle mineral that forms tetragonal crystals; commonly found as steel-gray or brown-black masses in the United States, Europe, and South America; it is an ore of manganese.

Braun tube *See* cathode-ray tube.

Bravais indices [CRYSTAL] A modification of the Miller indices; frequently used for hexagonal and trigonal crystalline systems; they refer to four axes: the *c*-axis and three others at 120° angles in the basal plane.

Bravais lattice [CRYSTAL] One of the 14 possible arrangements of lattice points in space such that the arrangement of points about any chosen point is identical with that about any other point.

Bravais' law *See* Bragg's law.

bra vector [QUANT MECH] A vector describing the state of a dynamic system in Hilbert space; the dual of a ket vector.

brave west winds [METEOROL] A nautical term for the strong and rather persistent westerly winds over the oceans in temperate latitudes, between 40° and 50°S.

bravoite [MINERAL] $(Ni,Fe)S_2$ A yellow sulfide ore of nickel containing iron.

brayer [GRAPHICS] A soft rubber roller attached to a wooden or metal handle; used to ink blocks, stones, or printing plates.

Brayton cycle [THERMO] A thermodynamic cycle consisting of two constant-pressure processes interspersed with two constant-entropy processes. Also known as complete-expansion diesel cycle; Joule cycle.

braze [MET] To solder metals by melting a nonferrous filler metal, such as brass or brazing alloy (hard solder), with a melting point lower than that of the base metals, at the point of contact. Also known as hard-solder.

brazed joint [MET] The joining of two or more metallic components by brazing or braze welding.

braze welding [MET] A method of welding in which coalescence is produced by heating above 800°F (427°C) and by using a nonferrous filler metal having a melting point below that of the base metals; in distinction to brazing, capillary attraction does not distribute the filler metal in the joint.

Brazil Current [OCEANOGR] The warm ocean current that flows southward along the Brazilian coast below Natal; the western boundary current in the South Atlantic Ocean.

Brazil nut [BOT] *Bertholettia excelsa.* A large broad-leafed evergreen tree of the order Lecythedales; an edible seed is produced by the tree fruit.

Brazil wax *See* carnauba wax.

brazing alloy [MET] **1.** An alloy used as a filler metal for brazing; copper alloys and nickel alloys are used for brazing of steels. **2.** Solder that does not melt below red heat. Formerly known as hard solder.

brazing brass *See* spelter solder.

brazing metal [MET] A nonferrous metal to be added to a joint in braze welding; can be a pure metal such as copper, zinc, or nickel, or a brazing alloy.

brazing sheet [MET] **1.** Brazing filler metal in sheet form. **2.** Flat-rolled metal clad with brazing filler metal on one or both sides.

breached anticline [GEOL] An anticline that has been more deeply eroded in the center. Also known as scalped anticline.

breached cone [GEOL] A cinder cone in which lava has broken through the sides and broken material has been carried away.

breadboarding [ELECTR] Assembling an electronic circuit in the most convenient manner, without regard for final locations of components, to prove the feasibility of the circuit and to facilitate changes when necessary.

breadboard model [ENG] Uncased assembly of an instrument or other piece of equipment, such as a radio set, having its parts laid out on a flat surface and connected together to permit a check or demonstration of its operation.

breadcrust [GEOL] A surficial structure resembling a crust of bread, as the concretions formed by evaporation of salt water.

breadcrust bomb [GEOL] A volcanic bomb with a cracked exterior.

breadfruit [BOT] *Artocarpus altilis.* An Indo-Malaysian tree, a species of the mulberry family (Moraceae). The tree produces a multiple fruit which is edible.

bread mold [MYCOL] Any fungus belonging to the family Mucoraceae in the order Mucorales.

breadth-height index [ANTHRO] Ratio of the maximum breadth of the skull to its maximum height multiplied by 100.

break [COMMUN] An interruption of a radio, telegraph, or telephone communication, as for sending in the opposite direction. [ELEC] **1.** A fault in a circuit. **2.** The minimum distance in a circuit-opening device between the stationary and movable contacts when these contacts are in the open position. [ELECTR] A reflected radar pulse appearing on a radarscope as a line perpendicular to the base line. [GEOGR] A significant variation of topography, such as a deep valley. [GEOL] *See* knickpoint. [METEOROL] **1.** A sudden change in the weather; usually applied to the end of an extended period of unusually hot, cold, wet, or dry weather. **2.** A hole or gap in a layer of clouds. [MIN ENG] **1.** A plane of discontinuity in the coal seam such as a slip, fracture, or cleat; the surfaces are in contact or slightly separated. **2.** A fracture or crack in the roof beds as a result of mining operations.

breakage and reunion [CYTOL] The classical model of crossing over by means of physical breakage and crossways reunion of completed chromatids during the process of meiosis.

breakaway phenomenon *See* breakoff phenomenon.

break-before-make contact [ELEC] One of a pair of contacts that interrupt one circuit before establishing another.

breakbone fever *See* dengue.

break-bulk cargo [IND ENG] Miscellaneous goods packed in boxes, bales, crates, cases, bags, cartons, barrels, or drums; may also include lumber, motor vehicles, pipe, steel, and machinery.

break contact [ELEC] The contact of a switching device which opens a circuit upon the operation of the device.

breakdown [ELEC] A large, usually abrupt rise in electric current in the presence of a small increase in voltage; can occur in a confined gas between two electrodes, a gas tube, the atmosphere (as lightning), an electrical insulator, and a reverse-biased semiconductor diode. [MET] The initial process of rolling and drawing, or a series of such processes, which reduce a casting or extruded shape before its final reduction to desired size.

breakdown diode *See* Zener diode.

breakdown impedance [ELECTR] Of a semiconductor, the small-signal impedance at a specified direct current in the breakdown region.

breakdown law [STAT] The law that if the event E is broken down into the exclusive events $E_1, E_2, \ldots$ so that E is the event E_1 or E_2 or $\ldots$, then if F is any event, the probability of F is the sum of the products of the probabilities of E_i and the conditional probability of F given E_i.

breakdown potential *See* breakdown voltage.

breakdown voltage [ELEC] **1.** The voltage measured at a specified current in the electrical breakdown region of a semiconductor diode. Also known as Zener voltage. **2.** The voltage at which an electrical breakdown occurs in a dielectric. **3.** The voltage at which an electrical breakdown occurs in a gas. Also known as breakdown potential; sparking potential; sparking voltage.

breaker [MIN ENG] **1.** In anthracite mining, the structure in which the coal is broken, sized, and cleaned for market. Also known as coalbreaker. **2.** One of a row of drill holes above the mining holes in a tunnel face. [OCEANOGR] A wave breaking on a shore, over a reef, or other mass in a body of water.

breaker cam [MECH ENG] A rotating, engine-driven device in the ignition system of an internal combustion engine which causes the breaker points to open, leading to a rapid fall in the primary current.

breaker points [ELEC] Low-voltage contacts used to interrupt the current in the primary circuit of a gasoline engine's ignition system.

breaker terrace [GEOL] A type of shore found in lakes in glacial drift; the terrace is formed from stones deposited by waves.

BREADFRUIT

Breadfruit *(Artocarpus altilis)*; end of branch with multiple fruits.

break-even analysis [IND ENG] Determination of the break-even point.

break-even point [IND ENG] The point at which a company neither makes a profit nor suffers a loss from the operations of the business, and at which total costs are equal to total sales volume.

break frequency [CONT SYS] The frequency at which a graph of the logarithm of the amplitude of the frequency response versus the logarithm of the frequency has an abrupt change in slope. Also known as corner frequency; knee frequency.

break-in device [ELECTR] A device in a radiotelegraph communication system allowing an operator to receive signals in intervals between his own transmission signals.

breaking-down rolls [MET] A rolling mill unit used for breakdown operations.

breaking-drop theory [GEOPHYS] A theory of thunderstorm charge separation based upon the suggested occurrence of the Lenard effect in thunderclouds, that is, the separation of electric charge due to the breakup of water drops.

breaking load [MECH] The stress which, when steadily applied to a structural member, is just sufficient to break or rupture it.

breaking strength [MECH] The ability of a material to resist breaking or rupture from a tension force.

breaking stress [MECH] The stress required to fracture a material whether by compression, tension, or shear.

break-in operation [COMMUN] A method of radio communication in which it is possible for the receiving operator to interrupt or break into the transmission.

breakoff phenomenon [AERO ENG] The feeling which sometimes occurs during high-altitude flight of being totally separated and detached from the earth and human society. Also known as breakaway phenomenon.

breakout [ELEC] A joint at which one or more conductors are brought out from a multiconductor cable. [MIN ENG] To pull drill rods or casing from a borehole and unscrew them at points where they are joined by threaded couplings to form lengths that can be stacked in the drill tripod or derrick.

breakoutput [ADP] An ALGOL procedure which causes all bytes in a device buffer to be sent to the device rather than wait until the buffer is full.

break period [COMMUN] Of a dial telephone, the time interval during which the circuit contacts are open.

breakpoint [ADP] A point in a program where an instruction, instruction digit, or other condition enables a programmer to interrupt the run by external intervention or by a monitor routine. [CHEM ENG] See breakthrough.

breakpoint switch [ADP] A manually operated switch which controls conditional operation at breakpoints, used primarily in debugging.

breakpoint symbol [ADP] A symbol which may be optionally included in an instruction, as an indication, tag, or flag, to designate it as a breakpoint.

breaks in overcast [METEOROL] In United States weather observing practice, a condition wherein the cloud cover is more than 0.9 but less than 1.0.

breakthrough [ADP] An interruption in the intended character stroke in optical character recognition. [CHEM ENG] A localized break in a filter cake or precoat that permits fluid to pass through without being filtered. Also known as breakpoint. [MIN ENG] A passage cut through the pillar to allow the ventilating current to pass from one room to another; larger than a doghole. Also known as room crosscut.

breakthrough sweep efficiency [PETRO ENG] The completeness with which an oil-field waterflood sweeps through a reservoir area; related to the critical water saturation (point beyond which oil will not be pushed ahead of the water flow).

break thrust [GEOL] A thrust fault cutting across one limb of a fold.

breakup [HYD] The spring melting of snow, ice, and frozen ground; specifically, the destruction of the ice cover on rivers during the spring thaw.

breakwater [CIV ENG] A wall built into the sea to protect a shore area, harbor, anchorage, or basin from the action of waves.

breast [ANAT] The human mammary gland. [MIN ENG] 1.

In coal mines, a chamber driven in the seam from the gangway, for the extraction of coal. 2. See face.

breastbeam [NAV ARCH] A beam at the edge of the quarterdeck or forecastle.

breast boards [CIV ENG] Timber planks used to support the tunnel face when excavation is in loose soil.

breast drill [DES ENG] A small, portable hand drill customarily used by handsetters to drill the holes in bit blanks in which diamonds are to be set; it includes a plate that is pressed against the worker's breast.

breast hole [MET] A hole for raking cinders out of a smelting cupola.

breasting dolphin [CIV ENG] A pile or other structure against which a moored ship rests.

breast wall [CIV ENG] A low wall built to retain the face of a natural bank of earth.

breastwork [ORD] Earthwork, constructed wholly or partly above the surface of the ground, which gives protection for defenders in a standing position, firing over the crest.

breath-hold diving [ENG] A form of diving without the use of any artificial breathing mixtures.

breathing [ENG] 1. Opening and closing of a plastics mold to let gases escape during molding. Also known as degassing. 2. Movement of gas, vapors, or air in and out of a storage-tank vent line as a result of liquid expansions and contractions induced by temperature changes. [PHYSIO] Inhaling and exhaling.

breathing apparatus [ENG] An appliance that enables a man to function in irrespirable or poisonous gases or fluids; contains a supply of oxygen and a regenerator which removes the carbon dioxide exhaled.

breathing cave See blowing cave.

breathing line [CIV ENG] A level of 5 feet above the floor; suggested temperatures for various occupancies of rooms and other chambers are usually given at this level.

breccia [PETR] A rock made up of very angular coarse fragments; may be sedimentary or may be formed by grinding or crushing along faults.

breccia dike [GEOL] A dike formed of breccia injected into the country rock.

breccia marble [PETR] Any marble containing angular fragments.

breccia pipe See pipe.

breech [ORD] The rear part of the bore of a gun, especially the opening that permits the projectile to be inserted at the rear of the bore.

breechblock [ORD] A movable steel block in the mechanism of a breech-loading gun that seals the breech opening of the barrel during firing.

breech bolt [ORD] A mechanism which opens and closes the breech in a carbine, machine gun, rifle, and the like; designed to push a cartridge into the chamber by sliding action.

breeches buoy [NAV ARCH] A device for carrying men from a stranded ship to shore or between ships.

breech flash [ORD] Flames and gas flash occurring at the breech of a weapon.

breech interlock [ORD] A safety device used with weapons that are loaded or rammed automatically or in weapons in which the position of the breechblock cannot be readily seen by the loader; prevents the loading or ramming of a round when the breechblock is not fully open.

breech loading [ORD] A method of loading a weapon in which the ammunition is inserted at the rear of the bore.

breech mechanism [ORD] The assembly at the rear of a gun which receives the round of ammunition, inserts it in the chamber, fires the round by detonating the primer, and extracts the empty case.

breech preponderance [ORD] Unbalance of the tipping parts of a weapon when the weight of the breech exerts a greater moment about the trunnions than does the weight of the muzzle end.

breech pressure [ORD] In interior ballistics, the pressure from the propellant gases acting against the inner face of the breechblock.

breeder reactor [NUCLEO] A nuclear reactor that produces more fissionable material that it consumes.

breeding [AGR] The application of genetic principles to the improvement of farm animals and cultivated plants. [GEN] Controlled mating and selection, or hybridization of plants and animals in order to improve the species. [NUCLEO] The production of nuclear fuel, by absorption of neutrons in a nuclear reactor, at a rate exceeding that at which fuel is being consumed.

breeding factor *See* breeding ratio.

breeding gain [NUCLEO] The excess of fissionable atoms produced per fissionable atom consumed in a breeder reactor.

breeding ratio [NUCLEO] The ratio of the number of fissionable atoms produced in a breeder reactor to the number of fissionable atoms consumed in the reactor; breeding gain is the breeding ratio minus 1. Also known as breeding factor; conversion factor; conversion ratio.

breeze [METEROL] **1.** A light, gentle, moderate, fresh wind. **2.** In the Beaufort scale, a wind speed ranging from 4 to 31 miles per hour.

B register *See* index register.

bregma [ANAT] The point at which the coronal and sagittal sutures of the skull meet.

breithauptite [MINERAL] NiSb A nickel antimonide that is opaque violet to copper red in color.

Breit-Wigner formula [NUC PHYS] A formula which relates the cross section of a particular nuclear reaction with the energy of the incident particle, when the energy is near that required to form a discrete resonance level of the component nucleus.

Breit-Wigner theory [NUC PHYS] A theory of nuclear reactions from which the Breit-Wigner formula is derived. Also known as Bohr-Breit-Wigner theory.

bremsstrahlung [ELECTROMAG] Radiation that is emitted by an electron accelerated in its collision with the nucleus of an atom.

Brennan monorail car [MECH ENG] A type of car balanced on a single rail so that when the car starts to tip, a force automatically applied at the axle end is converted gyroscopically into a strong righting moment which forces the car back into a position of lateral equilibrium.

brennschluss [AERO ENG] **1.** The cessation of burning in a rocket, resulting from consumption of the propellants, from deliberate shutoff, or from other cause. **2.** The time at which this cessation occurs.

Brentidae [INV ZOO] The straight-snouted weevils, a family of coleopteran insects in the superfamily Curculionoidea.

Bretonian orogeny [GEOL] Post-Devonian diastrophism that is found in Nova Scotia.

Bretonian strata [GEOL] Upper Cambrian strata in Cape Breton, Nova Scotia.

brettice *See* brattice.

brettis *See* brattice.

breunnerite [MINERAL] (Mg,Fe,Mn)CO$_3$ A carbonate mineral consisting of an isomorphous system of the metallic components.

Brevibacteriaceae [MICROBIO] A family of gram-positive, rod-shaped, schizomycetous bacteria in the order Eubacteriales.

Brevibacterium [MICROBIO] A genus of short, unbranched, rod-shaped bacteria; the type genus of Brevibacteriaceae.

brevity code [COMMUN] Code which has as its sole purpose the shortening of messages rather than the concealment of their content.

Brewer anaerobic jar [MICROBIO] A glass container in which petri dish cultures are stacked and maintained under anaerobic conditions.

brewers' yeast [FOOD ENG] Dried yeast cells recovered as a by-product of the brewing of beer and used as a natural source of vitamin B and protein.

brewing [FOOD ENG] The process of making beer, ale, and other malt beverages by boiling mashed malt to produce a wort, flavoring the wort with hops, fermenting this mixture with yeast, and drawing off the fermented wort for bottling.

brewster [OPTICS] A unit of stress optical coefficient of a material; it is equal to the stress optical coefficient of a material in which a stress of 1 bar produces a relative retardation between the components of a linearly polarized light beam of 1 angstrom when the light passes through a

thickness of 1 millimeter in a direction perpendicular to the stress. Abbreviated B.

brewsterite [MINERAL] Sr(Al$_2$Si$_6$O$_{18}$)·5H$_2$O A member of the zeolite family of minerals; crystallizes in the monoclinic system and usually contains some calcium.

Brewster point [METEOROL] One of the three commonly detectable points of zero polarization of skylight (neutral points) along the vertical circle through the sun.

Brewster process [CHEM ENG] Concentration of dilute acetic acid by use of an extraction solvent (for example, isopropyl ether), followed by distillation.

Brewster's angle [OPTICS] The angle of incidence of light reflected from a dielectric surface at which the reflectivity for light whose electrical vector is in the plane of incidence becomes zero; given by Brewster's law. Also known as polarizing angle.

Brewster's law [OPTICS] The law that the index of refraction for a material is equal to the tangent of the polarizing angle for the material.

Brewster window [OPTICS] A special glass window used at opposite ends of some gas lasers to transmit one polarization of the laser output beam without loss.

Brianchon's theorem [MATH] The theorem that if a hexagon circumscribes a conic section, the three lines joining three pairs of opposite vertices are concurrent (or are parallel).

brick [MATER] A building material usually made from clay, molded as a rectangular block, and baked or burned in a kiln.

brickfielder [METEOROL] A hot, dry, dusty north wind blowing from the interior across the southern coast of Australia.

brick grease *See* block grease.

bricking curb [MIN ENG] A curb set in a circular shaft to support the brick walling.

brick walling [MIN ENG] A permanent support for circular shafts by walling or casing.

bridge [CIV ENG] A structure erected to span natural or artificial obstacles, such as rivers, highways, or railroads, and supporting a footpath or roadway for pedestrian, highway, or railroad traffic. [ELEC] **1.** An electrical instrument having four or more branches, by means of which one or more of the electrical constants of an unknown component may be measured. **2.** An electrical shunt path. [MIN ENG] A piece of timber held above the cap of a set by blocks and used to facilitate the driving of spiling in soft or running ground. [NAV ARCH] An elevated structure extending across or over the weather deck of a vessel, containing stations for control and visual communications.

bridge abutment [CIV ENG] The end foundation upon which the bridge superstructure rests.

bridge bearing [CIV ENG] The support at a bridge pier carrying the weight of the bridge; may be fixed or seated on expansion rollers.

bridge cable [CIV ENG] Cable from which a roadway or truss is suspended in a suspension bridge; may be of pencil-thick wires laid parallel or strands of wire wound spirally.

bridge circuit [ELEC] An electrical network consisting basically of four impedances connected in series to form a rectangle, with one pair of diagonally opposite corners connected to an input device and the other pair to an output device.

bridge crane [MECH ENG] A hoisting machine in which the hoisting apparatus is carried by a bridgelike structure spanning the area over which the crane operates.

bridge deck [NAV ARCH] A partial deck above the main deck on merchant ships, usually near the middle of the ship.

bridged tap [ELEC] Portion of a cable pair connected to a circuit which is not a part of the useful path.

bridged-T network [ELEC] A T network with a fourth branch connected between an input and an output terminal and across two branches of the network.

bridge foundation [CIV ENG] The piers and abutments of a bridge, on which the superstructure rests.

bridge graft [BOT] A plant graft in which each of several scions is grafted in two positions on the stock, one above and the other below an injury.

bridge house [NAV ARCH] A structure above the main deck near the middle of a ship whose top forms the bridge deck.

bridge hybrid *See* hybrid junction.

BRIDGED-T NETWORK

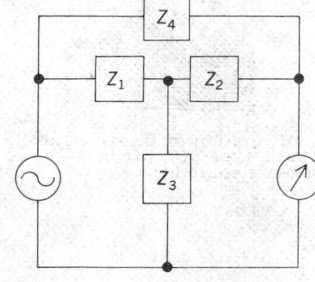

Schematic circuit of bridged-T network.

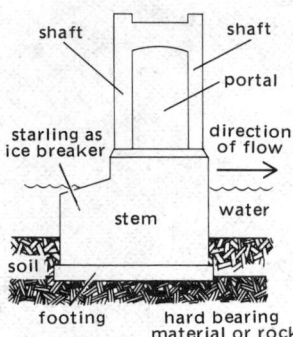

shaft   shaft

portal

starling as ice breaker   direction of flow

stem   water

soil

footing   hard bearing material or rock

Large bridge pier—vertical shafts with heavy portal and base.

BRIDGMAN ANVIL

Configuration of Bridgman anvil, a basic type of static high-pressure equipment.

bridge magnetic amplifier [ELECTR] A magnetic amplifier in which each of the gate windings is connected in series with an arm of a bridge rectifier; the rectifiers provide self-saturation and direct-current output.

bridge oscillator [ELECTR] An oscillator using a balanced bridge circuit as the feedback network.

bridge pier [CIV ENG] The main support for a bridge, upon which the bridge superstructure rests; constructed of masonry, steel, timber, or concrete founded on firm ground below river mud.

bridge rectifier [ELECTR] A full-wave rectifier with four elements connected as a bridge circuit with direct voltage obtained from one pair of opposite junctions when alternating voltage is applied to the other pair.

bridge trolley [MECH ENG] Either of the wheeled attachments at the ends of the bridge of an overhead traveling crane, permitting the bridge to move backward and forward on elevated tracks.

bridge vibration [MECH] Mechanical vibration of a bridge superstructure resulting from natural and man-made excitations.

bridging [ELEC] **1.** Connecting one electric circuit in parallel with another. **2.** The action of a selector switch whose movable contact is wide enough to touch two adjacent contacts so that the circuit is not broken during contact transfer. [MATH] The operation of carrying in addition or multiplication. [MET] **1.** Formation of arched cavities in a powder compact. **2.** Jamming of the charge in a blast or a cupola furnace due to adherence of fine ore particles to the inner walls. **3.** Formation of solidified metal over the top of the charge in a mold or crucible. [MIN ENG] The obstruction of the receiving opening in a material-crushing device by two or more pieces wedged together, each of which could easily pass through.

bridging amplifier [ELECTR] Amplifier with an input impedance sufficiently high so that its input may be bridged across a circuit without substantially affecting the signal level of the circuit across which it is bridged.

bridging connection [ELECTR] Parallel connection by means of which some of the signal energy in a circuit may be withdrawn frequently, with imperceptible effect on the normal operation of the circuit.

bridging loss [ELECTR] Loss resulting from bridging an impedance across a transmission system; quantitatively, the ratio of the signal power delivered to that part of the system following the bridging point, and measured before the bridging, to the signal power delivered to the same part after the bridging.

Bridgman anvil [PHYS] A device for producing high static pressures using two large massive opposed pistons bearing on a small thin sample confined by a gasket material.

Bridgman effect [SOLID STATE] The phenomenon that when an electric current passes through an anisotropic crystal, there is an absorption or liberation of heat due to the nonuniformity in current distribution.

Bridgman relation [SOLID STATE] $P = QT\sigma$ in a metal or semiconductor, where P is the Ettingshausen coefficient, Q the Nernst-Ettingshausen coefficient, T the temperature, and σ the thermal conductivity in a transverse magnetic field.

Bridgman sampler [MIN ENG] A mechanical device that automatically selects two samples as the ore passes through.

bridled-cup anemometer [ENG] A combination cup anemometer and pressure-plate anemometer, consisting of an array of cups about a vertical axis of rotation, the free rotation of which is restricted by a spring arrangement; by adjustment of the force constant of the spring, an angular displacement can be obtained which is proportional to wind velocity.

bridled pressure plate [METEOROL] An instrument for measuring air velocity in which the pressure on a plate exposed to the wind is balanced by the force of a spring, and the deflection of the plate is measured by an inductance-type transducer.

briefing *See* pilot briefing.

brig [PHYS] A unit to express the ratio of two quantities, as a logarithm to the base 10; that is, a ratio of 10^x is equal to x brig; it is analogous to the bel, but the latter is restricted to power ratios. Also known as dex.

Briggs equalizer [ENG] A breathing device consisting of a

head harness, mouthpiece, nose clip, corrugated breathing tube, Briggs equalizing device, 120 feet of reinforced air tubes, and a strainer and spike.

Briggs' logarithm *See* common logarithm.

Briggs pipe thread *See* American standard pipe thread.

Briggs stretcher carriage [MIN ENG] A stretcher used as an ambulance trolley in transporting casualties from underground workings.

bright [OPTICS] Attribute of an area that appears to emit a large amount of light.

bright annealing [MET] Heating and cooling a metal in an inert atmosphere to inhibit oxidation; surface remains relatively bright.

bright band [METEOROL] The enhanced echo of snow as it melts to rain, as displayed on a range-height indicator (RHI) scope.

bright-banded coal *See* bright coal.

bright coal [GEOL] A jet-black, pitchlike type of banded coal that is more compact than dull coal and breaks with a shell-shaped fracture; microscopic examination shows a consistency of more than 5% anthraxyllon and less than 20% opaque matter. Also known as bright-banded coal; brights.

bright dipping [MET] Immersing metal into an acid solution to give a bright, clean surface.

brightener [MET] Any of the agents which are employed in small concentrations in the electrolytic bath for electroplating metal to yield smoother or brighter coatings.

bright-field [OPTICS] Having a brightly lighted background.

bright-line spectrum [SPECT] An emission spectrum made up of bright lines on a dark background.

brightness [OPTICS] **1.** The characteristic of light that gives a visual sensation of more or less light. **2.** *See* luminance.

brightness control [ELECTR] A control that varies the luminance of the fluorescent screen of a cathode-ray tube, for a given input signal, by changing the grid bias of the tube and hence the beam current. Also known as brilliance control; intensity control.

brightness level *See* adaptation luminance.

bright plating [MET] An electroplating process resulting in a smooth, lustrous surface without polishing.

bright points [ASTRON] Relatively small regions on the sun, distributed uniformly over the solar disk, from which there is increased x-ray and ultraviolet emission, having lifetimes on the order of 8 hours.

brights *See* bright coal.

Bright's disease [MED] Any of several kidney diseases attended by glomerulonephritis.

bright segment [GEOPHYS] A faintly glowing band which appears above the horizon after sunset or before sunrise. Also known as crepuscular arch; twilight arch.

bright stars catalog [ASTRON] A catalog of stars brighter than 6.5 magnitude, giving positions, motions, parallaxes, and spectral classes.

bright stock [MATER] High-viscosity refined and dewaxed lubricating oils used in the compounding of motor oils.

Brikollare system [CIV ENG] A method of processing solid waste for composting that is designed to replace windrowing; after metals are separated from solid waste, it is comminuted and sent through a ballistic separator; the organic components are mixed with sewage sludge, compacted into briquets, and stacked; as biodegradation takes place, the briquets are turned; after the process is finished, the bricks are crumbled, bagged, and sold as compost.

bril [OPTICS] A unit of subjective luminance; 100 brils is the luminance level that corresponds to a luminance of 1 millilambert, and a doubling of luminance level corresponds to an increase of 1 bril.

brilliance [ELECTR] **1.** The degree of brightness and clarity of the display of a cathode-ray tube. **2.** The degree to which the higher audio frequencies of an input sound are reproduced by a radio receiver, by a public address amplifier, or by a sound-recording playback system.

brilliance control *See* brightness control.

brilliancy [LAP] A characteristic of gems that depends on the refractive index, transparency, polish, and proportions of the cut stone.

Brillouin function [SOLID STATE] A function of x with index (or parameter) n that appears in the quantum-mechanical theories of paramagnetism and ferromagnetism and is expressed as $[(2n+1)/2n]$ coth $[(2n+1)x/2n] - (1/2n) \cdot$ coth $(x/2n)$.

Brillouin scattering [SOLID STATE] Light scattering by acoustic phonons.

Brillouin zone [SOLID STATE] A fundamental region of wave vectors in the theory of the propagation of waves through a crystal lattice; any wave vector outside this region is equivalent to some vector inside it.

Brill's disease [MED] A mild recurrence of typhus some years after the initial infection.

brimstone [MINERAL] A common or commercial name for native sulfur.

brine [MATER] A liquid used in a refrigeration system, usually an aqueous solution of calcium chloride or sodium chloride, which is cooled by contact with the evaporator surface and then goes to the space which is to be refrigerated. [OCEANOGR] Sea water containing a higher concentration of dissolved salt than that of the ordinary ocean.

brine cooler [MECH ENG] The unit for cooling brine in a refrigeration system; the brine usually flows through tubes or pipes surrounded by evaporating refrigerant.

Brinell number [ENG] A hardness rating obtained from the Brinell test; expressed in kilograms per square millimeter.

Brinell test [ENG] A test to determine the hardness of a material, in which a steel ball 1 centimeter in diameter is pressed into the material with a standard force (usually 3000 kilograms); the spherical surface area of indentation is measured and divided into the load; the results are expressed as Brinell number.

brine spring [HYD] A salt-water spring.

Brinkmann number [FL MECH] A dimensionless number used to study viscous flow.

briquet [MATER] A block of some compressed substance, such as coal dust, metal powder, or sawdust, used as a fuel. Also spelled briquette.

briquette *See* briquet.

briquetting [ENG] 1. The process of binding together pulverized minerals, such as coal dust, into briquets under pressure, often with the aid of a binder, such as asphalt. 2. A process or method of mounting mineral ore, rock, or metal fragments in an embedding or casting material, such as natural or artificial resins, waxes, metals, or alloys, to facilitate handling during grinding, polishing, and microscopic examination.

brisa [METEOROL] 1. A northeast wind which blows on the coast of South America or an east wind which blows on Puerto Rico during the trade wind season. 2. The northeast monsoon in the Philippines. Also spelled briza.

brisance [ORD] The ability of an explosive to shatter the medium which confines it; the shattering effect of the explosive.

brisance index [ENG] The ratio of an explosive's power to shatter a weight of graded sand as compared to the weight of sand shattered by TNT.

Brisingidae [INV ZOO] A family of deep-water echinoderms with as many as 44 arms, belonging to the order Forcipulatida.

brisote [METEOROL] The northeast trade wind when it is blowing stronger than usual on Cuba.

bristle [BIOL] A short stiff hair or hairlike structure on an animal or plant.

Bristol board [MATER] Cardboard with a surface smooth enough for painting or writing, usually at least 0.006 inch thick.

britannia cell [MIN ENG] In mineral processing, a pneumatic flotation cell 7 to 9 feet deep.

britannia metal [MET] A silver-white tin alloy, similar to pewter, containing about 7% antimony, 2% copper, and often some zinc and bismuth; used in domestic utensils.

britholite [MINERAL] $(Na,Ce,Ca)_5(OH)[(P,Si)O_4]_3$ A rare-earth phosphate found in carbonatites in Kola Peninsula, Soviet Union.

British absolute system of units [PHYS] A measurement system based on the foot, the second, and the pound mass; force unit is the poundal.

British antilewisite *See* dimercaprol.

British engineering system of units *See* British gravitational system of units.

British gravitational system of units [PHYS] A measurement system based on the foot, the second, and the slug mass; 1 slug weighs 32.174 pounds at sea level and 45° latitude, and equals 14.594 kilograms. Also known as British engineering system of units; engineer's system of units.

British imperial pound [MECH] The British standard of mass, of which a standard is preserved by the government.

British thermal unit [THERMO] Abbreviated Btu. **1.** A unit of heat energy equal to the heat needed to raise the temperature of 1 pound of air-free water from 60° to 61°F at a constant pressure of 1 standard atmosphere; it is found experimentally to be equal to 1054.5 joules. Also known as sixty degrees Fahrenheit British thermal unit ($Btu_{60/61}$). **2.** A unit of heat energy equal to 1/180 of the heat needed to raise 1 pound of air-free water from 32°F (0°C) to 212°F (100°C) at a constant pressure of 1 of 1 standard atmosphere; it is found experimentally to be equal to 1055.79 joules. Also known as mean British thermal unit (Btu_{mean}). **3.** A unit of heat energy whose magnitude is such that 1 British thermal unit per pound equals 2326 joules per kilogram; it is equal to exactly 1055.05585262 joules. Also known as international table British thermal unit (Btu_{IT}).

brittle fracture [MET] A break in a brittle piece of metal which failed because stress exceeded cohesion.

brittle mica [MINERAL] Hydrous sodium, calcium, magnesium, and aluminum silicates; a group of more or less related minerals that resemble true micas but cleave to brittle flakes and contain calcium as the essential constituent.

brittleness [MECH] That property of a material manifested by fracture without appreciable prior plastic deformation.

brittle silver ore *See* stephanite.

brittle star [INV ZOO] The common name for all members of the echinoderm class Ophiuroidea.

brittle temperature [THERMO] The temperature point below which a material, especially metal, is brittle; that is, the critical normal stress for fracture is reached before the critical shear stress for plastic deformation.

Brix degree [CHEM ENG] A unit of the Brix scale.

Brix scale [CHEM ENG] A hydrometer scale for sugar solutions indicating the percentage by weight of sugar in the solution at a specified temperature.

briza *See* brisa.

broach [MECH ENG] A multiple-tooth, barlike cutting tool; the teeth are shaped to give a desired surface or contour, and cutting results from each tooth projecting farther than the preceding one.

broaching [ENG] **1.** The restoration of the diameter of a borehole by reaming. **2.** The breaking down of the walls between two contiguous drill holes. [MECH ENG] The machine-shaping of metal or plastic by pushing or pulling a broach across a surface or through an existing hole in a workpiece.

broaching bit *See* reaming bit.

broad band [COMMUN] A band with a wide range of frequencies.

broad-band amplifier [ELECTR] An amplifier having essentially flat response over a wide range of frequencies.

broad-band antenna [ELECTROMAG] An antenna that functions satisfactorily over a wide range of frequencies, such as for all 12 very-high-frequency television channels.

broad-band klystron [ELECTR] Klystron having three or more resonant cavities that are externally loaded and stagger-tuned to broaden the bandwidth.

broad-band path [COMMUN] A path having a bandwidth of 20 kilohertz or greater.

broadcast [COMMUN] A television or radio transmission intended for public reception.

broadcast band [COMMUN] The band of frequencies extending from 535 to 1605 kilohertz, corresponding to assigned carrier frequencies that increase in multiples of 10 kHz between 540 and 1600 kHz for the United States. Also known as standard broadcast band.

broadcaster [AGR] A machine that utilizes a rotating fanlike distributor for sowing grain, grass, and clover seed or for spreading fertilizer.

broadcast station [COMMUN] A television or radio station

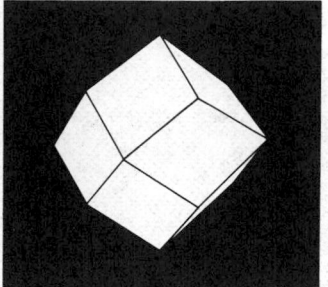

BRILLOUIN ZONE

Brillouin zone for the body-centered cubic lattice.

BROACHING

Dual ram-type vertical broaching machine with fixtures. *(Colonial Broach and Machine Co.)*

used for transmitting programs to the general public. Also known as station.

broadcast transmitter [ELECTR] A transmitter designed for use in a commercial amplitude-modulation, frequency-modulation, or television broadcast channel.

broadcloth [TEXT] A closely woven ribbed fabric having the rib running in the direction of the weft.

broadening of spectral line [SPECT] A widening of spectral lines by collision or pressure broadening, or possibly by Doppler effect.

broadleaf tree [BOT] Any deciduous or evergreen tree having broad, flat leaves.

broad on the beam [NAV] Bearing 090° (broad on the starboard beam) or 270° relative (broad on the port beam); if the bearings are approximate, the expression "on the beam" or "abeam" is used.

broad on the bow [NAV] Marine terminology indicating that a ship is bearing 045° relative (broad on the starboard bow) or 315° relative (broad on the port bow); if the bearings are approximate, the expression "on the bow" should be used.

broad on the quarter [NAV] Bearing 135° relative (broad on the starboard quarter) or 225° relative (broad on the port quarter); if the bearings are approximate, the expression "on the quarter" is used.

broadside [ELECTROMAG] Perpendicular to an axis or plane.

broadside array [ELECTROMAG] An antenna array whose direction of maximum radiation is perpendicular to the line or plane of the array.

broadside on [NAV] Beam on, as to the wind or sea.

broad-spectrum antibiotic [MICROBIO] An antibiotic that is effective against both gram-negative and gram-positive bacterial species.

broad tuning [ELECTR] Poor selectivity in a radio receiver, causing reception of two or more stations at a single setting of the tuning dial.

brocade [TEXT] Fabric made in a jacquard weave, usually with raised designs, and having a luxurious appearance; made of silk, polyester, or blends.

broccoli [BOT] *Brassica oleracea* var. *italica.* A biennial crucifer of the order Capparales which is grown for its edible stalks and buds.

brochanite *See* brochantite.

brochanthite *See* brochantite.

brochantite [MINERAL] $Cu_4(SO_4)(OH)_6$ A monoclinic copper mineral, emerald to dark green, commonly found with copper sulfide deposits; a minor copper ore. Also known as brochanite; brochanthite; warringtonite.

Brocken bow *See* anticorona.

Brocken specter [METEOROL] The illusory appearance of a gigantic figure (actually, the observer's shadow projected on cloud surfaces), observed on the Brocken peak in the Hartz Mountains of Saxony, but visible from other mountaintops under suitable conditions.

Brodmann's area 4 *See* motor area.

Brodmann's area 17 *See* visual projection area.

Brodmann's areas [PHYSIO] Numbered regions of the cerebral cortex used to identify cortical functions.

broeboe [METEOROL] A strong, dry east wind in the southwestern part of the island of Celebes.

Broenner's acid *See* Brönner's acid.

broken [METEOROL] Descriptive of a sky cover of from 0.6 to 0.9 (expressed to the nearest tenth).

broken belt [OCEANOGR] The transition zone between open water and consolidated ice.

broken bracket *See* angle bracket.

broken ground *See* loose ground.

broken line [MATH] A line which is composed of a series of line segments lying end to end, and which does not form a continuous line.

broken stone *See* crushed stone.

broken wind *See* heaves.

bromacetone [ORG CHEM] $CH_2BrCOCH_3$ A colorless liquid which is a powerful irritant and lacrimator; used as tear gas and to make other chemicals.

bromacil [ORG CHEM] 5-Bromo-3-*sec*-butyl-6-methyluracil, a soil sterilant; general at high dosage and selective at low.

bromate [CHEM] **1.** BrO_3^- A negative ion derived from bromic acid, $HBrO_3$. **2.** A salt of bromic acid.

bromatium [ECOL] A swollen hyphal tip on fungi growing in ants nests that is eaten by the ants.

bromcresol green *See* bromocresol green.

bromcresol purple *See* bromocresol purple.

bromegrass [BOT] The common name for a number of forage grasses of the genus *Bromus* in the order Cyperales.

bromelain [BIOCHEM] An enzyme that digests protein and clots milk; prepared by precipitation by acetone from pineapple juice; used to tenderize meat, to chill-proof beer, and to make protein hydrolysates. Also spelled bromelin.

Bromeliaceae [BOT] The single family of the flowering plant order Bromeliales.

Bromeliales [BOT] An order of monocotyledonous plants in the subclass Commelinidae, including terrestrial xerophytes and some epiphytes.

bromelin *See* bromelain.

bromellite [MINERAL] BeO A white hexagonal mineral consisting of beryllium oxide; it is harder than zincite.

bromeosin *See* eosin.

bromhidrosiphobia [PSYCH] Abnormal fear of offensive body odors.

bromic acid [INORG CHEM] $HBrO_3$ A liquid, colorless to slightly yellow; boils with decomposition at 100°C; used in dyes and as a chemical intermediate.

bromic ether *See* ethyl bromide.

bromide [CHEM] A compound derived from hydrobromic acid, HBr, containing the bromine atom in the minus-one oxidation state.

bromide paper [GRAPHICS] A photographic paper with high sensitivity, used for short exposures; it is coated with an emulsion of silver bromide.

brominating agent [CHEM] A compound capable of introducing bromine into a molecule; examples are phosphorus tribromide, bromine chloride, and aluminum tribromide.

bromination [CHEM] The process of introducing bromine into a molecule.

bromine [CHEM] A chemical element, symbol Br, atomic number 35, atomic weight 79.904; used to make dibromide ethylene and in organic synthesis and plastics.

bromine number [ANALY CHEM] The amount of bromine absorbed by a fatty oil; indicates the purity of the oil and degree of unsaturation.

bromine trifluoride [CHEM] BrF_3 A liquid with a boiling point of 135°C.

bromine water [CHEM] An aqueous saturated solution of bromine used as a reagent wherever a dilute solution of bromine is needed.

bromism [MED] A disease state produced by prolonged usage or overdosage of bromide compounds.

bromlite [MINERAL] $BaCa(CO_3)_2$ An orthorhombic mineral composed of a carbonate of barium and calcium. Also known as alstonite.

bromo- [CHEM] A prefix that indicates the presence of bromine in a molecule.

bromoacetone [ORG CHEM] $BrCH_2COCH_3$ A colorless liquid used as a lacrimatory agent.

bromo acid *See* eosin.

bromoalkane [ORG CHEM] An aliphatic hydrocarbon with bromine bonded to it. Also known as bromohydrocarbon.

bromobenzene [ORG CHEM] C_6H_5Br A heavy, colorless liquid with a pleasant odor; used as a solvent, in motor fuels and top-cylinder compounds, and to make other chemicals.

bromobenzylcyanide [ORG CHEM] $C_6H_5CHBrCN$ A light yellow oily compound used as a tear gas for training and for riot control. Abbreviated BBC.

bromocresol green [ORG CHEM] Tetrabromo-*m*-cresol sulfonphthalein, a gray powder soluble in water or alcohol; used as an indicator between pH 4.5 (yellow) and 5.5 (blue). Also known as bromcresol green.

bromocresol purple [ORG CHEM] Dibromo-*o*-cresol sulfonphthalein, a yellow powder soluble in water; used as an indicator between pH 5.2 (yellow) and 6.8 (purple). Also known as bromcresol purple.

bromoethane *See* ethyl bromide.

bromoform [ORG CHEM] $CHBr_3$ A colorless liquid, slightly

BROCHANTITE

Needlelike crystals from Tsumeb, Namibia. *(Specimen from Department of Geology, Bryn Mawr College)*

BROMINE

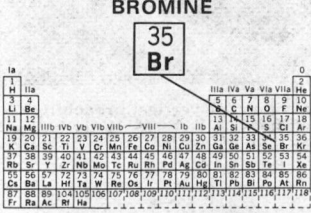

Periodic table of the chemical elements showing the position of bromine.

soluble in water; used in the separation of minerals. Also known as tribromomethane.

bromohydrocarbon　*See* bromoalkane.

bromomethane　*See* methyl bromide.

bromophosgene　*See* carbonyl bromide.

bromotrifluoromethane　[ORG CHEM] $CBrF_3$　Fluorine compound with molecular weight 148.93, melting point $-180°C$, boiling point $-59°C$; used as a fire-extinguishing agent.

bromouracil　[BIOCHEM] $C_4H_3N_2O_2Br$　5-Bromouracil, an analog of thymine that can react with deoxyribonucleic acid to produce a polymer with increased susceptibility to mutation.

bromthymol blue　[ORG CHEM] An acid-base indicator in the pH range 6.0 to 7.6; color change is yellow to blue. Also known as dibromothymolsulfonphthalein.

Bromwich contour　[MATH] A path of integration in the complex plane running from $c - i\infty$ to $c + i\infty$, where c is a real, positive number chosen so that the path lies to the right of all singularities of the analytic function under consideration.

bromyrite　[MINERAL] AgBr　A secondary ore of silver that occurs in the oxidized zone of silver deposits; exists in crusts and coatings resembling a wax.

bronchial adenoma　[MED] A low-grade malignant or potentially malignant tumor of bronchi.

bronchial asthma　[MED] Asthma usually due to hypersensitivity to an inhaled or ingested allergen.

bronchial tree　[ANAT] The arborization of the bronchi of the lung, considered as a structural and functional unit.

bronchiectasis　[MED] Dilation of the bronchi and bronchioles following a chronic inflammatory process or an infection attended by pus formation.

bronchiolar carcinoma　[MED] Adenocarcinoma of the lung characterized by mucus-producing cells which spread over the alveoli.

bronchiole　[ANAT] A small, thin-walled branch of a bronchus, usually terminating in alveoli.

bronchiolitis　[MED] Inflammation of the bronchioles.

bronchiolitis obliterans　[MED] Inflammation of the bronchioles with the formation of an exudate and fibrous tissue that obliterate the lumen.

bronchitis　[MED] An inflammation of the bronchial tubes.

bronchodilator　[MED] An instrument used to increase the caliber of the pulmonary air passages.　[PHARM] Any drug which has the property of increasing the caliber of the pulmonary air passages.

bronchogram　[MED] Radiography of the bronchial tree made after the introduction of a radiopaque substance.

bronchography　[MED] Roentgenographic visualization of the bronchial tree following injection of a radiopaque material.

bronchopneumonia　[MED] Inflammation of the lungs which has spread from infected bronchi. Also known as lobular pneumonia.

bronchorrhea　[MED] Excessive discharge of mucus from the bronchial mucous membranes.

bronchoscope　[MED] An instrument for the visual examination of the interior of the bronchi.

bronchospasm　[MED] Temporary narrowing of the bronchi due to violent, involuntary contraction of the smooth muscle of the bronchi.

bronchospirometry　[MED] The determination of various aspects of the functional capacity of a single lung or lung segment.

bronchus　[ANAT] Either of the two primary branches of the trachea or any of the bronchi's pulmonary branches having cartilage in their walls.

Brönner's acid　[ORG CHEM] $C_{10}H_6(NH_2)SO_3H$　A colorless, water-soluble naphthylamine sulfonic acid that forms needle crystals; used in dyes. Also spelled Broenner's acid. Also known as 2-napthylamine-6-sulfonic acid.

Brönsted acid　[CHEM] A chemical species which can act as a source of protons.

Brönsted-Lowry theory　[CHEM] A theory that all acid-base reactions consist simply of the transfer of a proton from one base to another. Also known as Brönsted theory.

Brönsted theory　*See* Brönsted-Lowry theory.

brontides　[GEOPHYS] Low, rumbling, thunderlike sounds of short duration, most frequently heard in active seismic regions and believed to be of seismic origin.

brontophobia　[PSYCH] An abnormal fear of thunder.

Brontotheriidae　[PALEON] The single family of the extinct mammalian superfamily Brontotherioidea.

Brontotherioidea　[PALEON] The titanotheres, a superfamily of large, extinct perissodactyl mammals in the suborder Hippomorpha.

bronze　[MET] An alloy of copper and tin in varying proportions; other elements such as zinc, nickel, and lead may be added.

bronze mica　*See* phlogopite.

bronzing liquid　[MATER] A solvent, gloss oil, or varnish containing a bronze powder; used to produce bronze-colored finishes.

bronzite　[MINERAL] $(Mg,Fe)(SiO_3)$　An orthopyroxene mineral that forms metallic green orthorhombic crystals; a form of the enstatite-hypersthene series.

bronzitfels　*See* bronzitite.

bronzitite　[PETR] A pyroxenite that is composed almost entirely of bronzite. Also known as bronzitfels.

brood　[BOT] Heavily infested by insects.　[ZOO] **1.** The young of animals. **2.** To incubate eggs or cover the young for warmth. **3.** An animal kept for breeding.

brood capsule　[INV ZOO] A secondary scolex-containing cyst constituting the infective agent of a tapeworm.

brood parasitism　[ECOL] A type of social parasitism among birds characterized by a bird of one species laying and abandoning its eggs in the nest of a bird of another species.

brood pouch　[VERT ZOO] A pouch of an animal body where eggs or embryos undergo certain stages of development.

Brookhill waffler　[MIN ENG] A coal cutter with the ordinary horizontal jib and also a shearing or mushroom jib.

brookite　[MINERAL] TiO_2　A brown, reddish, or black orthorhombic mineral; it is trimorphous with rutile and anatase, has hardness of 5.5–6 on Mohs scale, and a specific gravity of 3.87–4.08. Also known as pyromelane.

Brooks variable inductometer　[ELEC] An inductometer providing a nearly linear scale and consisting of two movable coils, side by side in a plane, sandwiched between two pairs of fixed coils.

brooming　[CIV ENG] A method of finishing uniform concrete surfaces, such as the tops of pavement slabs or floor slabs, by dragging a broom over the surface to produce a grooved texture.

broomy flow　[FL MECH] A swirling flow of a fluid in a pipe after passing through a constricted section or after a sudden change of direction.

Brotulidae　[VERT ZOO] A family of benthic teleosts in the order Perciformes.

brown acid　[CHEM ENG] Oil-soluble petroleum sulfonate found in sludge following sulfuric acid treatment of petroleum products.

brown algae　[BOT] The common name for members of the Phaeophyta.

Brown and Sharpe gage　*See* American wire gage.

brown bast　[PL PATH] A physiological disease of the para rubber tree characterized by grayish- or greenish-brown discolorations of the inner bark near the tapping cut.

brown blight　[PL PATH] A virus disease of lettuce characterized by spots and streaks on the leaves, reduction in leaf size, and gradual browning of the foliage, beginning at the base.

brown blotch　[PL PATH] **1.** A bacterial disease of mushrooms caused by *Pseudomonas tolaasi* and characterized by brown blotchy discolorations. **2.** A fungus disease of the pear characterized by brown blotches on the fruit.

brown canker　[PL PATH] A fungus disease of roses caused by *Cryptosporella embrina* and characterized by lesions that are initially purple and gradually become buff.

brown clay　*See* red clay.

brown clay ironstone　[GEOL] Limonite in the form of concrete masses, often in concretionary nodules.

brown coat　[MATER] Mortar with about 1 to 1½ bushels of hair per 200 pounds quicklime; used to make a brown coat of plaster, which is covered with a finish coat, and often covers a scratch coat.

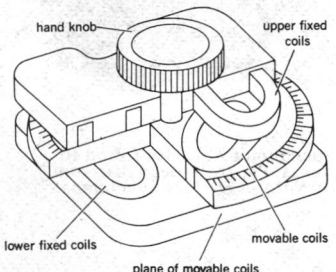

BROOKS VARIABLE INDUCTOMETER

hand knob　upper fixed coils　lower fixed coils　movable coils　plane of movable coils

Drawing of Brooks variable inductometer.

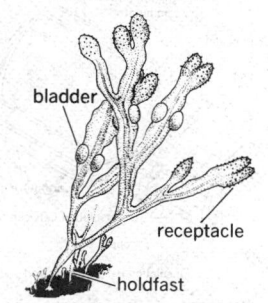

BROWN ALGAE

bladder　receptacle　holdfast

Fucus, a brown alga. (*From H. J. Fuller and O. Tippo, College Botany, rev. ed., Holt, 1954*)

brown fat cell [HISTOL] A moderately large, generally spherical cell in adipose tissue that has small fat droplets scattered in the cytoplasm.

brown felt blight [PL PATH] A fungus disease of conifers caused by several Ascomycetes, especially *Herpotrichia nigra* and *Neopeckia coulteri;* a dense felty growth of brown or black mycelia forms on the branches.

brown hematite *See* limonite.

Brownian movement [STAT MECH] Random movements of small particles suspended in a fluid, caused by the statistical pressure fluctuations over the particle.

brown induration [MED] A pathologic condition marked by acute pulmonary congestion and edema with leakage of blood into the alveoli.

browning [PL PATH] Any plant disorder or disease marked by brown discoloration of a part. Also known as stem break.

Browning automatic rifle [ORD] A gas-operated weapon whose cyclical rate of fire is adjustable but which is capable of firing 200-350 rounds per minute. Abbreviated BAR.

Browning machine gun [ORD] Any of certain caliber-.30 and caliber-.50 machine guns designed by, or modified from designs by, John M. Browning.

brown iron ore *See* limonite.

brown lead oxide *See* lead dioxide.

brown leaf rust [PL PATH] A fungus disease of rye caused by *Puccinia dispersa.*

brown lignite [GEOL] A type of lignite with a fixed carbon content ranging from 30 to 55% and total carbon from 65 to 73.6%; contains 6300 Btu per pound. Also known as lignite B.

brownline [GRAPHICS] A print with a white background and brown lines made by contact-printing a negative on sensitized paper.

brown mica *See* phlogopite.

brownout [ELEC] **1.** A restriction of electrical power usage during a power shortage, especially for advertising and display purposes. **2.** An extinguishing of some of the lights in a city as a defensive measure against enemy bombardment.

brown patch [PL PATH] A fungus disease of grasses in golf greens and lawns caused by various soil-inhabiting species, typically producing brown circular areas surrounded by a band of grayish-black mycelia.

brownprint [GRAPHICS] A print with a brown background and white lines made by contact-printing a negative on a sensitized paper.

brown-ring test [ANALY CHEM] A common qualitative test for the nitrate ion; a brown ring forms at the juncture of a dilute ferrous sulfate solution layered on top of concentrated sulfuric acid if the upper layer contains nitrate ion.

brown root [PL PATH] A fungus disease of numerous tropical plants, such as coconut and rubber, caused by *Hymenochaete noxia* and characterized by defoliation and by incrustation of the roots with earth and stones held together by brown mycelia.

brown root rot [PL PATH] **1.** A fungus disease of plants of the pea, cucumber, and potato families caused by *Thielavia basicola* and characterized by blackish discoloration and decay of the roots and stem base. **2.** A disease of tobacco and other plants comparable to the fungus disease but believed to be caused by nematodes.

brown rot [PL PATH] Any fungus or bacterial plant disease characterized by browning and tissue decay.

brown seaweed [BOT] A common name for the larger algae of the division Phaeophyta.

brown smoke [ENG] Smoke with less particulates than black smoke; comes from burning fossil fuel, usually fuel oil.

brown snow [METEOROL] Snow intermixed with dust particles.

brown soil [GEOL] Any of a zonal group of soils, with a brown surface horizon which grades into a lighter-colored soil and then into a layer of carbonate accumulation.

brown spar [GEOL] Any light-colored crystalline carbonate that contains iron, such as ankerite or dolomite, and is therefore brown.

brown spot [PL PATH] Any fungus disease of plants, especially Indian corn, characterized by brown leaf spots.

brown stem rot [PL PATH] A fungus disease of soybeans caused by *Cephalosporium gregatum* in which there is brown-

ish internal stem rot followed by discoloration and withering of leaves.

brownstone [PETR] Ferruginous sandstone with its grains coated with iron oxide.

brown stringy rot [PL PATH] A disease of conifers caused by the Indian point fungus; rusty or brown fibrous streaks appear in the heartwood.

browse [BIOL] **1.** Twigs, shoots, and leaves eaten by livestock and other grazing animals. **2.** To feed on this vegetation.

brubru [METEOROL] A squall in Indonesia.

Brucella [MICROBIO] A genus of short, rod-shaped to coccoid, encapsulated, gram-negative, parasitic and pathogenic bacteria of the family Brucellaceae.

Brucella abortus [MICROBIO] The etiological agent of contagious abortion in cattle, brucellosis in man, and a wasting disease in chickens.

Brucellaceae [MICROBIO] A family of small, coccoid to rod-shaped, gram-negative bacteria in the order Eubacteriales.

brucellergen [BIOCHEM] A nucleoprotein fraction of brucellae used in skin tests to detect the presence of *Brucella* infections.

brucellergen test [IMMUNOL] A diagnostic skin test for detection of *Brucella* infections.

brucellosis [MED] An infectious bacterial disease of man caused by *Brucella* species acquired by contact with diseased animals. Also known as Malta fever; Mediterranean fever; undulant fever. [VET MED] *See* contagious abortion.

Bruchidae [INV ZOO] The pea and bean weevils, a family of coleopteran insects in the superfamily Chrysomeloidea.

Bruch's membrane [ANAT] The membrane of the retina that separates the pigmented layer of the retina from the choroid coat of the eye.

brucine [ORG CHEM] $C_{23}H_{26}N_2O_4$ A poisonous alkaloid from the seeds of plant species such as *Nux vomica;* used in alcohol as a denaturant. Also known as dimethoxy strychnine.

brucite [MINERAL] $Mg(OH)_2$ A hexagonal mineral; native magnesium hydroxide that appears gray and occurs in serpentines and impure limestones; hardness is 2.5 on Mohs scale, and specific gravity is 2.38-2.40.

Brückner cycle [CLIMATOL] An alternation of relatively cool-damp and warm-dry periods, forming an apparent cycle of about 35 years.

bruma [METEOROL] A haze that appears in the afternoons on the coast of Chile when sea air is transported inland.

Brunauer-Emmett-Teller equation [PHYS CHEM] An extension of the Langmuir isotherm equation in the study of sorption; used for surface area determinations by computing the monolayer area. Abbreviated BET equation.

Brunner's glands [ANAT] Simple, branched, tubular mucus-secreting glands in the submucosa of the duodenum in mammals. Also known as duodenal glands; glands of Brunner.

Brunt-Douglas isallobaric wind *See* isallobaric wind.

Brunton *See* Brunton compass.

Brunton compass [ENG] A compact field compass, with sights and reflector attached, used for geological mapping and surveying. Also known as Brunton; Brunton pocket transit.

Brunton pocket transit *See* Brunton compass.

brüscha [METEOROL] A northwest wind in the Bergell Valley, Switzerland.

brush [ECOL] *See* tropical scrub. [ELEC] A conductive metal or carbon block used to make sliding electrical contact with a moving part.

brush border [CYTOL] A superficial protoplasic modification in the form of filiform processes or microvilli; present on certain absorptive cells in the intestinal epithelium and the proximal convolutions of nephrons.

brush discharge [ELEC] A luminous electric discharge that starts from a conductor when its potential exceeds a certain value but remains too low for the formation of an actual spark.

brush fire [FOR] A fire involving growth that is heavier than grass but less than full tree size.

brush hopper [IND ENG] A rotating brush that wipes quantities of eyelets, rivets, and other small special parts past shaped openings in a chute.

BRUCITE

Specimen of brucite found in Texas, Pa. *(American Museum of Natural History specimens)*

├─ 5 cm ─┤

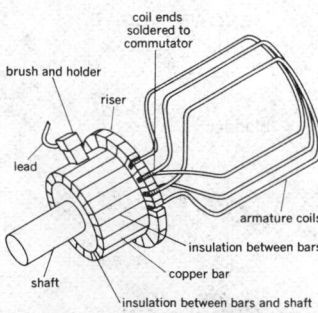

BRUSH

Commutator and brush assembly, showing stationary brush that makes sliding contact with rotating commutator.

brushing [TEXT] A process for finishing fabrics by rubbing them against a rough surface, such as a wire-covered cylinder, usually to produce a thick nap. Also known as napping.

brushing shot [MIN ENG] A charge fired in the air of a mine to blow out obnoxious gases or to start an air current.

brushite [MINERAL] $CaHPO_4 \cdot 2H_2O$ A nearly colorless mineral that is a constituent of rock phosphates that crystallizes in slender or massive crystals.

brush machine [FOOD ENG] A machine used in rice processing to rub the grains smooth; consists of a vertical cylindrical frame covered with soft leather strips revolving at high speed within a cylinder of wire screening.

brush plating [MET] Electroplating in which the anode with the solution is in the form of a brush or pad; used for plating equipment too large to be immersed.

brush-shifting motor [ENG] A category of alternating-current motor in which the brush contacts shift to modify operating speed and power factor.

brush station [ADP] A location in a device where the holes in a punched card are sensed by brushes sweeping electrical contacts.

brussels sprouts [BOT] *Brassica oleracea* var. *gemmifera.* A biennial crucifer of the order Capparales cultivated for its small, edible, headlike buds.

brute-force filter [ELEC] Type of powerpack filter depending on large values of capacitance and inductance to smooth out pulsations rather than on resonant effects of tuned filters.

Bruxellian [GEOL] Lower middle Eocene geologic time.

Bryatae *See* Bryopsida.

bryology [BOT] The study of bryophytes.

Bryophyta [BOT] A small phylum of the plant kingdom, including mosses, liverworts, and hornworts, characterized by the lack of true roots, stems, and leaves.

Bryopsida [BOT] The mosses, a class of small green plants in the phylum Bryophyta. Also known as Musci.

Bryopsidaceae [BOT] A family of green algae in the order Siphonales.

Bryozoa [INV ZOO] The moss animals, a major phylum of sessile aquatic invertebrates occurring in colonies with hardened exoskeleton.

BS&W *See* basic sediments and water.

B scan *See* B scope.

B scope [ELECTR] A cathode-ray scope on which signals appear as spots, with bearing angle as the horizontal coordinate and range as the vertical coordinate. Also known as B indicator; B scan.

BSD *See* barrels per stream day.

B size [ENG] One of a series of sizes to which trimmed paper and board are manufactured; for size BN, with N equal to any integer from 0 to 10, the length of the shorter side is $2^{-N/2}$ meters, and the length of the longer side is $2^{(1-N)/2}$ meters, with both lengths rounded off to the nearest millimeter.

BSR *See* bulk shielding reactor.

B-stage resin [ORG CHEM] A phenolic resin formed as the second stage in the alkaline condensation of an aldehyde and a phenol; it is thermoplastic. Also known as resitol.

B station [NAV] In loran, the designation applied to one transmitting station of a pair, the signal of which always occurs more than half a repetition period after the succeeding signal and less than half a repetition period before the preceding signal from the other station of the pair, designated an A station.

B store *See* index register.

B supply [ELECTR] Anode high voltage and screen-grid power source in vacuum tube circuits. Also known as B power supply.

BT *See* bathythermograph.

B tectonite [PETR] Tectonite with a fabric dominated by linear elements indicating an axial direction rather than a slip surface.

BTO *See* bombing through overcast.

B trace [ELECTR] In loran the second trace of an oscilloscope which corresponds to the signal from the B station.

Btu *See* British thermal unit.

BTX *See* benzene-toluene-xylene.

B-type star [ASTRON] A type in a classification based on stellar spectral characteristics; has strong He I absorption.

bu *See* bushel.

bubble [METEOROL] *See* bubble high. [PHYS] **1.** A small, approximately spherical body of fluid within another fluid or solid. **2.** A thin, approximately spherical film of liquid inflated with air or other gas.

bubble cap [CHEM ENG] A metal cup covering a hole in plate within a distillation tower; designed to permit vapors to rise from below the plate, pass through the cap, and make contact with liquid on the plate.

bubble-cap plate [CHEM ENG] One of the devices in large-diameter fractional distillation columns that are designed to produce a bubbling action to exchange the vapor bubbles flowing up the column.

bubble cavitation [FL MECH] **1.** Formation of vapor- or gas-filled cavities in liquids by mechanical forces. **2.** The formation of vapor-filled cavities in the interior of liquids in motion when the pressure is reduced without change in ambient temperature.

bubble chamber [NUCLEO] A chamber in which the movements and interactions of charged particles can be observed as visible tracks in a superheated liquid, the tracks being gas bubbles that form along the paths of the moving particles.

bubble high [METEOROL] A small high, complete with anticyclonic circulation, of the order of 50 to 300 miles across, often induced by precipitation and vertical currents associated with thunderstorms. Also known as bubble.

bubble horizon [NAV] An artificial horizon parallel to the celestial horizon, established by means of a bubble level on a bubble sextant.

bubble point [PHYS CHEM] In a solution of two or more components, the temperature at which the first bubbles of gas appear. Also known as boiling point.

bubble-point reservoir *See* dissolved-gas-drive reservoir.

bubble raft [SOLID STATE] A visual demonstration for the structure of dislocations in metal lattices, showing slip propagation; it consists of many identical bubbles floating on a liquid surface in something like a crystalline array.

bubble sextant [NAV] A sextant with a bubble or spirit level to indicate the horizontal.

bubble test [ENG] Measurement of the largest opening in the mesh of a filter screen; determined by the pressure needed to force air or gas through the screen while it is submerged in a liquid.

bubble tower [CHEM ENG] A plate tower used in distillation, with plates containing bubble caps.

bubble train [GEOL] A string or strings of vesicles in lava, indicating the path of rising gas escaping a flow of lava.

bubble tray [CHEM ENG] A perforated, circular plate placed within a distillation tower at specific places to collect the fractions of petroleum produced in fractional distillation.

bubble tray column [CHEM ENG] A fractionating column whose plates are formed from bubble caps.

bubble tube [ENG] The glass tube in a spirit level containing the liquid and bubble.

bubble wall fragment [GEOL] A glassy volcanic shard revealing part of a vesicle surface which may be curved or flat.

bubo [MED] An inflammatory enlargement of lymph nodes, usually of the groin or axilla; commonly associated with chancroid, lymphogranuloma venereum, and plague.

bubonic plague *See* plague.

bubulum oil *See* neatsfoot oil.

bucaramangite [MINERAL] A pale yellow variety of retinite that looks like amber but is insoluble in alcohol.

buccal cavity [ANAT] The space anterior to the teeth and gums in the mouths of all vertebrates having lips and cheeks. Also known as vestibule.

buccal gland [ANAT] Any of the mucous glands in the membrane lining the cheeks of mammals, except aquatic forms.

Buccinacea [INV ZOO] A superfamily of gastropod mollusks in the order Prosobranchia.

Buccinidae [INV ZOO] A family of marine gastropod mollusks in the order Neogastropoda containing the whelks in the genus *Buccinum.*

Bucconidae [VERT ZOO] The puffbirds, a family of neotropical birds in the order Piciformes.

Bucerotidae [VERT ZOO] The hornbills, a family of Old World tropical birds in the order Coraciiformes.

BRUSSELS SPROUTS

Brussels sprouts (*Brassica oleracea* var. *gemmifera*), Jade Cross. (*Joseph Harris Co., Rochester, N.Y.*)

BUBBLE CAP

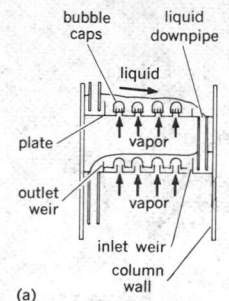

(a)

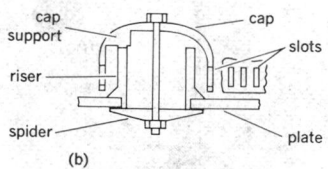

(b)

Schematic of *(a)* cross section through bubble-cap-plate column and *(b)* typical bubble cap.

BUCKEYE

Leaf of horse chestnut
(*Aesculus hippocastanum*).

BUCKWHEAT

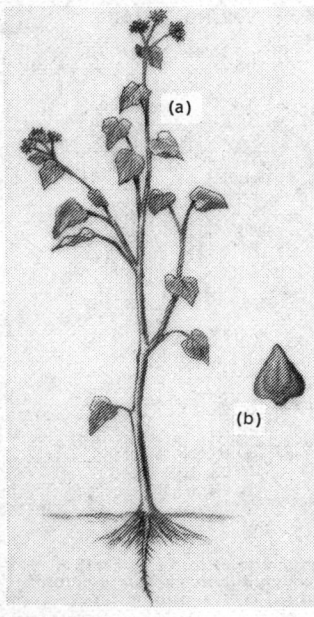

Buckwheat.
(a) Mature plant. (b) Seed.

BUFFALO

The Indian or water buffalo
(*Bubalus bubalis*), domesticated
for draft purposes as well as
milk production.

Bucherer reaction [ORG CHEM] A method of preparation of polynuclear primary aromatic amines; for example, α-naphthylamine is obtained by heating β-naphthol in an autoclave with a solution of ammonia and ammonium sulfite.

buchite [PETR] A partially vitrified inclusion of sandstone in basalt.

buchonite [PETR] An extrusive rock formed of labradorite, titanaugite, and titaniferous hornblende, with nepheline and sodic sanidine and accessory biotite, apatite, and opaque oxides.

buck [MIN ENG] **1.** To break up or pulverize ore samples. **2.** A large quartz reef in which there is little or no accessory minerals such as gold. Also known as buck quartz; bull quartz. [VERT ZOO] A male deer.

bucket [BOT] *See* calyx. [ENG] **1.** A cup on the rim of a Pelton wheel against which water impinges. **2.** A reversed curve at the toe of a spillway to deflect the water horizontally and reduce erosiveness. **3.** A container on a lift pump or chain pump. **4.** A container on some bulk-handling equipment, such as a bucket elevator, bucket dredge, or bucket conveyor. **5.** A water outlet in a turbine. **6.** *See* calyx.

bucket carrier *See* bucket conveyor.

bucket conveyor [MECH ENG] A continuous bulk conveyor constructed of a series of buckets attached to one or two strands of chain or in some instances to a belt. Also called bucket carrier.

bucket dredge [MECH ENG] A floating mechanical excavator equipped with a bucket elevator.

bucket drill [MIN ENG] An auger stem drill in which the drill bit is replaced by a bit incorporating a steel cylinder to confine the cutting.

bucket dumper *See* lander.

bucket elevator [MECH ENG] A bucket conveyor operating on a steep incline or vertical path. Also known as elevating conveyor.

bucket excavator [MECH ENG] An elevating scraper, that is, one that does the work of a conventional scraper but has a bucket elevator mounted in front of the bowl.

bucket ladder *See* bucket-ladder dredge.

bucket-ladder dredge [MECH ENG] A dredge whose digging mechanism consists of a ladderlike truss on the periphery of which is attached an endless chain riding on sprocket wheels and carrying attached buckets. Also known as bucket ladder; bucket-line dredge; ladder-bucket dredge; ladder dredge.

bucket-ladder excavator *See* trench excavator.

bucket-line dredge *See* bucket-ladder dredge.

bucket loader [MECH ENG] A form of portable, self-feeding, inclined bucket elevator for loading bulk materials into cars, trucks, or other conveyors.

bucket thermometer [ENG] A thermometer mounted in a bucket and used to measure the temperature of water drawn into the bucket from the surface of the ocean.

bucket-wheel excavator [MECH ENG] A continuous digging machine used extensively in large-scale stripping and mining. Abbreviated BWE. Also known as rotary excavator.

buckeye [BOT] The common name for deciduous trees composing the genus *Aesculus* in the order Sapindales; leaves are opposite and palmately compound, and the seed is large with a firm outer coat.

bucking [MIN ENG] A hand process for crushing ore.

bucking coil [ELECTROMAG] A coil connected and positioned in such a way that its magnetic field opposes the magnetic field of another coil; for example, the hum-bucking coil of an excited-field loudspeaker.

Buckingham's equations [MECH ENG] Equations which give the durability of gears and the dynamic loads to which they are subjected in terms of their dimensions, hardness, surface endurance, and composition.

Buckingham's π theorem [PHYS] The theorem that if there are n physical quantities, $x_1, x_2, \ldots, x_n$, which can be expressed in terms of m fundamental quantities and if there exists one and only one mathematical expression connecting them which remains formally true no matter how the units of the fundamental quantities are changed, namely $\phi(x_1, x_2, \ldots, x_n) = 0$, then the relation ϕ can be expressed by a relation of the form $F(\pi_1, \pi_2, \ldots, \pi_{n-m}) = 0$, where the πs are $n - m$ independent dimensionless products of $x_1, x_2, \ldots, x_n$. Also known as pi theorem.

bucking transformer [ELEC] A transformer whose voltage opposes that of a second transformer.

bucking voltage [ELEC] A voltage having a polarity opposite to that of another voltage against which it acts.

buckle [MET] An up-and-down wrinkle on the surface of a metal bar or sheet.

buckle fold [GEOL] A double flexure of rock beds formed by compression acting in the plane of the folded beds.

buckle plate [CIV ENG] A steel floor plate which is slightly arched to increase rigidity.

Buckley gage [ENG] A device that measures very low gas pressures by sensing the amount of ionization produced in the gas by a predetermined electric current.

buckling [MECH] Bending of a sheet, plate, or column supporting a compressive load. [NUCLEO] The size-shape factor that appears in the general nuclear reactor equation and is a measure of the curvature of the neutron density distribution in the reactor.

buckling stress [MECH] Force exerted by the crippling load.

buckwheat [BOT] The common name for several species of annual herbs in the genus *Fagopyrum* of the order Polygonales; used for the starchy seed.

buckwheat coal [GEOL] An anthracite coal that passes through 9/16-inch holes and over 5/16-inch holes in a screen.

buckwheat no. 3 *See* barley coal.

Bucky diaphragm *See* Potter-Bucky grid.

bud [BOT] An embryonic shoot containing the growing stem tip surrounded by young leaves or flowers or both and frequently enclosed by bud scales.

Budan's theorem [MATH] The theorem that the number of roots of an nth-degree polynomial lying in an open interval equals the difference in the number of sign changes induced by n differentiations at the two ends of the interval.

budbreak [BOT] Initiation of growth from a bud.

budding [BIOL] A form of asexual reproduction in which a new individual arises as an outgrowth of an older individual. Also known as gemmation. [BOT] A method of vegetative propagation in which a single bud is grafted laterally onto a stock.

budding bacteria [MICROBIO] Bacteria that reproduce by budding.

buddle [MIN ENG] A device for concentrating ore that uses a circular arrangement from which the finely divided ore is delivered in water from a central point, the heavier particles sinking and the lighter particles overflowing.

Buddy [MIN ENG] A shortwall coal cutter designed for light duty such as stabling on longwall power-loaded faces and for subsidiary developments.

budget year [METEOROL] The 1-year period beginning with the start of the accumulation season at the firn line of a glacier or ice cap and extending through the following summer's ablation season.

bud grafting [BOT] Grafting a plant by budding.

budling [BOT] The shoot that develops from the bud which was the scion of a bud graft.

bud rot [PL PATH] Any plant disease or symptom involving bud decay.

bud scale [BOT] One of the modified leaves enclosing and protecting buds in perennial plants.

Buerger's disease *See* thromboangitis obliterans.

buffalo [VERT ZOO] The common name for several species of artiodactyl mammals in the family Bovidae, including the water buffalo and bison.

buffer [CHEM] A solution selected or prepared to minimize changes in hydrogen ion concentration which would otherwise occur as a result of a chemical reaction. Also known as buffer solution. [ECOL] An animal that is introduced to serve as food for other animals to reduce the losses of more desirable animals. [ELEC] An electric circuit or component that prevents undesirable electrical interaction between two circuits or components. [ELECTR] **1.** An isolating circuit in an electronic computer used to prevent the action of a driven circuit from affecting the corresponding driving circuit. **2.** A storage device used to compensate for a difference in rate of flow of information or time of occurrence of events when information is transmitted from one computer device to another. **3.** *See* buffer amplifier.

buffer amplifier [ELECTR] An amplifier used after an oscillator or other critical stage to isolate it from the effects of load impedance variations in subsequent stages. Also known as buffer; buffer stage.

buffer cap [ORD] A light cap of ductile metal placed over, and in contact with, an armor-piercing cap in some designs of armor-piercing ammunition.

buffer capacitor [ELECTR] A capacitor connected across the secondary of a vibrator transformer or between the anode and cathode of a cold-cathode rectifier tube to suppress voltage surges that might otherwise damage other parts in the circuit.

buffered computer [ADP] A computer having a temporary storage device to compensate for differences in transmission speeds.

buffered I/O channel [ADP] A storage device located between input/output (I/O) channels and main storage control to free the channels for use by other operations.

buffering agent [FOOD ENG] A chemical, such as lactic, citric, or acetic acid or the sodium salts of various acids, added to processed food to adjust and regulate its pH.

buffer solution *See* buffer.

buffer stage *See* buffer amplifier.

buffer storage [ADP] A synchronizing element used between two different forms of storage in a computer; computation continues while transfers take place between buffer storage and the secondary or internal storage.

buffer zone [ADP] An area of main memory set aside for temporary storage.

buffeting [AERO ENG] **1.** The beating of an aerodynamic structure or surfaces by unsteady flow, gusts, and so forth. **2.** The irregular shaking or oscillation of a vehicle component owing to turbulent air or separated flow.

buffeting Mach number [AERO ENG] The free-stream Mach number over the tops of the wings as the number approaches unity.

buffing [ENG] The smoothing and brightening of a surface by an abrasive compound pressed against it by a soft wheel or belt.

buffing wheel [DES ENG] A flexible wheel with a surface of fine abrasive particles for buffing operations.

Bufonidae [VERT ZOO] A family of toothless frogs in the suborder Procoela including the true toads (*Bufo*).

bufotenine [ORG CHEM] $C_{12}H_{16}N_2O$ An active pressor agent found in the skin of the common toad; a toxic alkaloid with epinephrinelike biological activity.

bug [ADP] A defect in a program code or in designing a routine or a computer. [ELECTR] **1.** A semiautomatic code-sending telegraph key in which movement of a lever to one side produces a series of correctly spaced dots and movement to the other side produces a single dash. **2.** An electronic listening device, generally concealed, used for commercial or military espionage. [ENG] **1.** A defect or imperfection present in a piece of equipment. **2.** *See* bullet. [INV ZOO] Any insect in the order Hemiptera.

bugas *See* bottled gas.

bug dust [MIN ENG] The fine coal or other material resulting from a boring or cutting of a drill, a mining machine, or even a pick.

buggy [ENG] *See* concrete buggy. [MIN ENG] A four-wheeled steel car used for hauling coal to and from chutes.

bughole *See* vug.

buhrstone [PETR] A silicified fossiliferous limestone with abundant cavities previously occupied by fossil shells. Also known as millstone.

buhrstone mill [MECH ENG] A mill for grinding or pulverizing grain in which a flat siliceous rock (buhrstone), generally of cellular quartz, rotates against a stationary stone of the same material.

build [ELECTR] To increase in received signal strength.

builder [MATER] A material added to a soap or a synthetic surface-active agent to produce a mixture having enhanced detergency. [MET] A fire-clay brick cull used for bottom construction in kilns or for boxing brick during burning.

building [CIV ENG] A fixed structure for human occupancy and use.

building blocks [MATER] Hollow blocks, either of burned clay or concrete, used in constructing the walls of buildings, and often faced with brick or stone.

building code [CIV ENG] Local building laws to promote safe practices in the design and construction of a building.

building dock [CIV ENG] A type of graving dock or basin, usually built of concrete; in which ships are constructed and then floated out through a caisson gate after flooding the dock.

building-out circuit [ELEC] Short section of transmission line, or a network which is shunted across a transmission line, for the purpose of impedance matching.

building-out network [ELEC] Network designed to be connected to a based network so that the combination will simulate the sending-end impedance, neglecting dissipation, of a line having a termination other than that for which the basic network was designed.

building-out section [ELEC] Short section of transmission line, either open or short-circuited at the far end, shunted across another transmission line for use on an impedance-matching transformer.

buildup [MET] Deposition of excess metal, either by electro-plating or spraying, on worn or undersized machine components to restore required dimensions.

buildup curve [PETRO ENG] Graph of bottom-hole pressure buildup versus shut-in time for a gas or oil well.

buildup index *See* fire-danger meter.

buildup pressure [PETRO ENG] The increase in bottom-hole pressure up to an equilibrium value in a shut-in oil or gas well.

built-in antenna [ELETROMAG] An antenna located inside the cabinet of a radio or television receiver.

built-in check [ADP] A hardware device which controls the accuracy of data either moved or stored within the computer system.

built terrace *See* wave-built terrace.

built-up beam [ENG] A structural steel member that is fabricated by welding or riveting rather than being rolled.

built-up edge [ENG] Chip material adhering to the tool face adjacent to a cutting edge during cutting.

built-up fraction [GRAPHICS] In typesetting, a fraction set with at least two blocks of type, with each successive block to the right of the preceding one; each block contains one numeral, and the slash is between two blocks.

built-up gun [ORD] Gun tube assembled by the shrinkage method; consists of two or more concentric cylinders shrunk one over another; stronger than a simple cylinder of the same dimensions.

built-up mica [MATER] Large, laminated plates of mica made by bonding thin splittings of natural mica with shellac, glyptol, or some other suitable adhesive.

built-up roofing [MATER] A seamless piece of flexible, water-proofed roofing material consisting of plies of felt mopped with asphalt or pitch.

bulb [BOT] A short, subterranean stem with many overlapping fleshy leaf bases or scales, such as in the onion and tulip. [ELEC] *See* envelope.

bulb angle [DES ENG] A steel angle iron enlarged to a bulbous thickening at one end.

bulbar paralysis [MED] A clinical syndrome due to involvement of the nuclei of the last four or five cranial nerves, characterized principally by paralysis or weakness of the muscles which control swallowing, talking, movement of the tongue and lips, and sometimes respiratory paralysis.

bulb glacier [HYD] A glacier formed at the foot of a mountain and out into an open slope; the glacier ends spread out into an ice fan.

bulbil [BOT] A secondary bulb usually produced on the aerial part of a plant.

bulbocavernosus *See* bulbospongiosus.

bulb of percussion [ARCHEO] A cone-shaped bulge on a fractured flint surface that was made by a blow striking at an angle.

bulb of the penis [ANAT] The expanded proximal portion of the corpus spongiosum of the penis.

bulbospongiosus [ANAT] A muscle encircling the bulb and adjacent proximal parts of the penis in the male, and

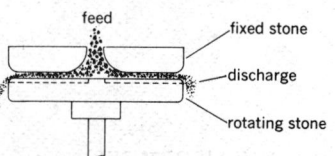

BUHRSTONE MILL

feed — fixed stone
— discharge
— rotating stone

Buhrstone mill. Material is fed at center of fixed stone and moves toward outer edge of the stones where finely ground product is discharged.

encircling the orifice of the vagina and covering the lateral parts of the vestibular bulbs in the female. Also known as bulbocavernosus.

bulbourethral gland [ANAT] Either of a pair of compound tubular glands, anterior to the prostate gland in males, which discharge into the urethra. Also known as Cowper's gland.

bulbous bow [NAV ARCH] A large, elliptical or cylindrical swelling of the lower underwater hull form protruding forward of the bow designed to cancel wave-making at high speeds.

bulimia [MED] Excessive, insatiable appetite, seen in psychotic states; a symptom of diabetes mellitus and of certain cerebral lesions. Also known as hyperphagia.

Buliminacea [INV ZOO] A superfamily of benthic, marine foraminiferans in the suborder Rotaliina.

bulk cargo [IND ENG] Cargo which is loaded into a ship's hold without being boxed, bagged, or hand stowed, or is transported in large tank spaces.

bulk carrier [NAV ARCH] A vessel designed to transport dry or liquid bulk cargo.

bulk density [IND ENG] The mass of powdered or granulated solid material per unit of volume.

bulked-down *See* solid-piled.

bulk eraser [ELECTROMAG] A device used to erase an entire reel of recorded magnetic tape at once without running it through a recorder.

bulk factor [ENG] The ratio of the volume of loose powdered or granulated solids to the volume of an equal weight of the material after consolidation into a voidless solid.

bulk flotation [MIN ENG] The rising of a mineralized froth, of more than one mineral, in a single operation.

bulk flow *See* convection.

bulk-handling machine [MECH ENG] Any of a diversified group of materials-handling machines designed for handling unpackaged, divided materials.

bulkhead [AERO ENG] A wall, partition, or such in a rocket, spacecraft, airplane fuselage, or similar structure, at right angles to the longitudinal axis of the structure and serving to strengthen, divide, or help give shape to the structure. [MIN ENG] A tight-seal partition of wood, rock, and mud or concrete in mines for protection against gas, fire, and water. [NAV ARCH] An upright partition separating compartments in a ship.

bulkhead deck [NAV ARCH] The highest continuous deck of a ship to which all transverse watertight bulkheads extend.

bulkhead line [CIV ENG] The farthest offshore line to which a structure may be constructed without interfering with navigation.

bulkhead wharf [CIV ENG] A bulkhead that may be used as a wharf by addition of mooring appurtenances, paving, and cargo-handling facilities.

bulking [MATER] The difference in volume of a given mass of sand or other fine material in moist and dry conditions.

bulking power [TEXT] The relative ability of a textile fiber to increase the finished material's volume.

bulking value [CHEM ENG] The relative ability of a pigment or other substance to increase the volume of paint.

bulk insulation [ENG] A type of insulation that retards the flow of heat by the interposition of many air spaces and, in most cases, by opacity to radiant heat.

bulk mining [MIN ENG] Mining in which large quantities of low-grade ore are taken without attempt to segregate the high-grade portions.

bulk modulus *See* bulk modulus of elasticity.

bulk modulus of elasticity [MECH] The ratio of the compressive or tensile force applied to a substance per unit surface area to the change in volume of the substance per unit volume. Also known as bulk modulus; compression modulus; hydrostatic modulus; modulus of volume elasticity.

bulk photoconductor [ELECTR] A photoconductor having high power-handling capability and other unique properties that depend on the semiconductor and doping materials used.

bulk plant [PETRO ENG] A wholesale receiving and distributing facility for petroleum products; includes storage tanks, warehouses, railroad sidings, truck loading racks, and related elements. Also known as bulk terminal.

Bulk Shielding Reactor [NUCLEO] The prototype swimming-pool reactor located at Oak Ridge, Tennessee; it uses heterogeneous enriched fuel and provides a combination of high thermal-neutron flux, ready accessibility, and versatility. Abbreviated BSR.

bulk strength [MECH] The strength per unit volume of a solid.

bulk terminal *See* bulk plant.

bulk transport [MECH ENG] Conveying, hoisting, or elevating systems for movement of solids such as grain, sand, gravel, coal, or wood chips.

Bull *See* Taurus.

bullate [BIOL] Appearing blistered or puckered, especially of certain leaves.

bull bit [MIN ENG] A flat drill bit.

bull block [MET] Power-driven machine for drawing wire through a die.

bulldozer [MECH ENG] A wheeled or crawler tractor equipped with a reinforced, curved steel plate mounted in front, perpendicular to the ground, for pushing excavated materials. [MET] A machine for bending, forging, and punching narrow plates and bars, in which a ram is pushed along a horizontal path by a pair of cranks that are linked to two bullwheels with eccentric pins.

bullet [ENG] 1. A conical-nosed, cylindrical weight, attached to a wire rope or line, either notched or seated to engage and attach itself to the upper end of a wire line core barrel or other retrievable or retractable device placed in a borehole. Also known as bug; go-devil; overshot. 2. A scraper with self-adjusting spring blades, inserted in a pipeline and carried forward by the fluid pressure, clearing away accumulations or debris from the walls of a pipe. Also known as go-devil. 3. A bullet-shaped weight or small explosive charge dropped to explode a charge of nitroglycerin placed in a borehole. Also known as go-devil. 4. An electric lamp covered by a conical metal case, usually at the end of a flexible metal shaft. [GRAPHICS] A hollow hemispherical shell, made of iron and filled with pitch, which holds small objects during the execution of artistic designs in metal. [MATER] A small, lustrous, nearly spherical industrial diamond. [ORD] The projectile fired, or intended to be fired, from a small arm.

bullet drop [MECH] The vertical drop of a bullet.

bullet group [ORD] The bullet holes in a target made by fire from a given small arm from one position and caused by ballistic variations or inaccurate aim.

bullet jacket [ORD] A metal shell surrounding a metal core, the combination composing a bullet for small arms; the jacket is either made of or coated with a relatively soft metal which engages the rifling in the bore, causing rotation of the bullet.

bulletproof armor [ORD] Light, homogeneous, hard armor plate, having a Brinell hardness of 400 to 475, resistant to small-caliber armor-piercing bullets.

bullet splash [ORD] Dispersion of finely divided or melted metal produced upon impact of a projectile with armor plate or other hard objects.

bullfrog *See* barney.

bull gear [DES ENG] A bull wheel with gear teeth.

bullhorn [COMMUN] A portable loudspeaker, generally with built-in amplifier and microphone, used for voice messages to crowds or from one ship to another at sea.

bulliform [BOT] Type of plant cell involved in tissue contraction or water storage, or of uncertain function.

bulliform cell [BOT] One of the large, highly vacuolated cells occurring in the epidermis of grass leaves. Also known as motor cell.

bulling bar [ENG] A bar for ramming clay into cracks containing blasting charges which are about to be exploded.

bullion [MET] Gold or silver in bulk in the shape of bars or ingots. [MIN ENG] A concretion found in some types of coal, composed of carbonate or silica stained by brown humic derivatives; often well-preserved plant structures form the nuclei.

bull nose [BUILD] A rounded external angle, as one used at window returns and doorframes.

bull-nose bit *See* wedge bit.

bullous emphysema [MED] Acute overinflation of the lungs due to extreme efforts to inhale to overcome a bronchial obstruction.

BULIMINACEA

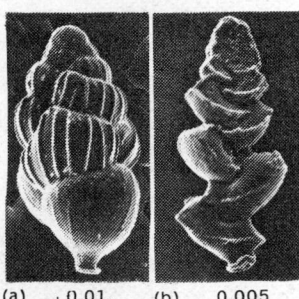

(a) $\overline{\frac{0.01}{cm}}$ (b) $\overline{\frac{0.005}{cm}}$

Scanning electron micrographs of foraminiferans, suborder Rotaliina, superfamily Buliminacea, (a) *Eouvigerina*, from Upper Cretaceous of Texas. (b) *Uvigerina*, from Miocene of California. (*R. B. MacAdam, Chevron Oil Field Research Co.*)

BULLIFORM CELL

bulliform cells

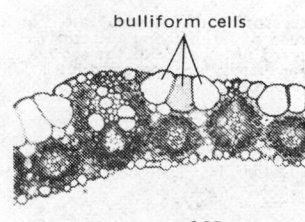

125 μ

Transection of leaf of *Andropogon* showing large bulliform cells in the upper epidermis.

bull's-eye rot [PL PATH] A fungus disease of apples caused by either *Neofabraea malicorticis* or *Gloeosporium perennans* and characterized by spots resembling eyes on the fruit.

bull's-eye squall [METEOROL] A squall forming in fair weather, characteristic of the ocean off the coast of South Africa; it is named for the peculiar appearance of the small, isolated cloud marking the top of the invisible vortex of the storm.

bull shaker [MIN ENG] A shaking chute where large coal from the dump is cleaned by hand.

bull wheel [MECH ENG] **1.** The main wheel or gear of a machine, which is usually the largest and strongest. **2.** A cylinder which has a rope wound about it for lifting or hauling. **3.** A wheel attached to the base of a derrick boom which swings the derrick in a vertical plane.

bulwark [NAV ARCH] The part of a ship's side that extends above the main deck to protect it against heavy weather.

bulwark plating [NAV ARCH] Light plating that extends above a ship's hull plating and acts as a bulwark.

Bulygen number [THERMO] A dimensionless number used in the study of heat transfer during evaporation.

bumblebee [INV ZOO] The common name for several large, hairy social bees of the genus *Bombus* in the family Apidae.

bummer [FOR] A low truck with two wheels for carrying logs, or a tracked cart for dragging them. [MIN ENG] The man who runs conveyors in a quarry or mine.

bump contact [ELECTR] A large-area contact used for alloying directly to the substrate of a transistor for mounting or interconnecting purposes.

bumper [ENG] A metal bar attached to one or both ends of a powered transportation vehicle, especially an automobile, to prevent damage to the body. [MET] A vibrating machine for ramming and consolidating sand into a mold.

bumpiness [AERO ENG] An atmospheric condition causing aircraft to experience sudden vertical jolts.

bumping [CHEM] Uneven boiling of a liquid caused by irregular rapid escape of large bubbles of highly volatile components as the liquid mixture is heated. [MECH ENG] *See* chugging. [MET] Forming a dish in metal by many repeated blows.

bumps [MIN ENG] Sudden, violent expulsion of coal from one or more pillars, accompanied by loud reports and earth tremors.

bunched pair [ELEC] Group of pairs tied together or otherwise associated for identification.

buncher *See* buncher resonator.

buncher resonator [ELECTR] The first or input cavity resonator in a velocity-modulated tube, next to the cathode; here the faster electrons catch up with the slower ones to produce bunches of electrons. Also known as buncher; input resonator.

bunching [ELECTR] The flow of electrons from cathode to anode of a velocity-modulated tube as a succession of electron groups rather than as a continuous stream.

bunching voltage [ELECTR] Radio-frequency voltage between the grids of the buncher resonator in a velocity-modulated tube such as a klystron; generally, the term implies the peak value of this oscillating voltage.

bunchy top [PL PATH] Any viral disease of plants in which there is a shortening of the internodes with crowding of leaves and shoots at the stem apex.

bund [CIV ENG] An embankment or embanked thoroughfare along a body of water; the term is used particularly for such structures in the Far East.

bundle branch [ANAT] Either of the components of the atrioventricular bundle passing to the right and left ventricles of the heart.

bundle scar [BOT] A mark within a leaf scar that shows the point of an abscised vascular bundle.

bundle sheath [BOT] A sheath around a vascular bundle that consists of a layer of parenchyma.

bundling machine [MECH ENG] A device that automatically accumulates cans, cartons, or glass containers for semiautomatic or automatic loading or for shipping cartons by assembling the packages into units of predetermined count and pattern which are then machine-wrapped in paper, film paperboard, or corrugated board.

Bundsandstein *See* Bunter.

Buniakowski's inequality *See* Cauchy-Schwarz inequality.

bunion [MED] A swelling of a bursa of the foot, especially of the metatarsophalangeal joint of the great toe; associated with thickening of the adjacent skin and a forcing of the great toe into adduction.

bunker [ORD] A fortified structure for the protection of personnel, a defended gun position, or a defensive position.

bunker C fuel oil [MATER] A special grade of bunker fuel oil. Also known as Navy Heavy; No. 6 fuel.

bunker fuel oil [MATER] A heavy residual petroleum oil used as fuel by ships, industry, and large-scale heating and power-production installations.

bunkering [ENG] Storage of solid or liquid fuel in containers from which the fuel can be continuously or intermittently withdrawn to feed a furnace, internal combustion engine, or fuel tank, for example, coal bunkering and fuel-oil bunkering.

bunodont [ANAT] Having tubercles or rounded cusps on the molar teeth, as in man.

bunolophodont [VERT ZOO] **1.** Of teeth, having the outer cusps in the form of blunt cones and the inner cusps as transverse ridges. **2.** Having such teeth, as in tapirs.

bunoselenodont [VERT ZOO] **1.** Of teeth, having the inner cusps in the form of blunt cones and the outer ones as longitudinal crescents. **2.** Having such teeth, as in the extinct titanotheres.

Bunsen burner [ENG] A type of gas burner with an adjustable air supply.

Bunsen ice calorimeter [ENG] Apparatus to gage heat released during the melting of a compound by measuring the increase in volume of the surrounding ice-water solution caused by the melting of the ice. Also known as ice calorimeter.

bunsenite [MINERAL] NiO A pistachio-green mineral consisting of nickel monoxide and occurring as octahedral crystals.

Bunsen-Kirchoff law [SPECT] The law that every element has a characteristic emission spectrum of bright lines and absorption spectrum of dark lines.

bunt [PL PATH] A fungus disease of wheat caused by two *Tilletia* species and characterized by grain replacement with fishy-smelling smut spores.

buntal [TEXT] A fine, white fiber derived from the leaves of the talipot palm. Also known as buri.

Bunter [GEOL] Lower Triassic geologic time. Also known as Bundsandstein.

bunton [MIN ENG] A steel or timber element in the lining of a rectangular shaft.

buoy [ENG] An anchored or moored floating object, other than a lightship, intended as an aid to navigation, to attach or suspend measuring instruments, or to mark the position of something beneath the water.

buoyage [NAV] A system of buoys; for example, one in which the buoys are assigned shape, color, and number distinction in accordance with location relative to the nearest obstruction is called a cardinal buoyage system.

buoyancy [FL MECH] The resultant vertical force exerted on a body by a static fluid in which it is submerged or floating.

buoyancy parameter [FL MECH] The Grashof number divided by the square of the Reynolds number.

buoyancy pontoons [PETRO ENG] Pontoons that buoy up offshore pipelines during the welding together of sections over bodies of water, after which the pontoons are removed and the pipeline is allowed to sink into position on the bottom.

buoyancy tank [NAV ARCH] An airtight tank near one of the ends of a small boat to keep it from sinking if it capsizes or fills with water.

buoyancy-type density transmitter [ENG] An instrument which records the specific gravity of a flowing stream of a liquid or gas, using the principle of hydrostatic weighing.

buoyant density [PHYS] A technique that uses the sedimentation equilibrium in a density gradient to characterize a solute.

buoyant force [FL MECH] The force exerted vertically upward by a fluid on a body wholly or partly immersed in it; its

**BUOYANCY-TYPE
DENSITY TRANSMITTER**

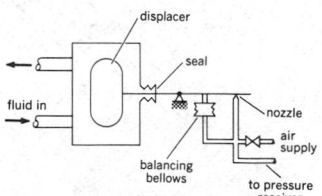

Buoyancy-type density transmitter, used to obtain a continuous record of liquid density. *(From D. M. Considine, ed., Process Instruments and Controls Handbook, McGraw-Hill, 1957)*

BUPRESTIDAE

An example of Buprestidae. *(From T. I. Storer and R. L. Usinger, General Zoology, 3d ed., McGraw-Hill, 1957)*

BURGERS VECTOR

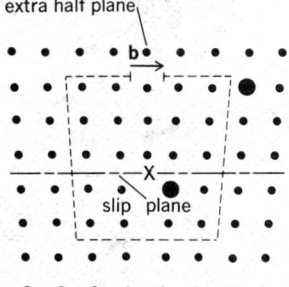

extra half plane

b

X

slip plane

An arrangement of atoms around an edge dislocation. The dashed lines represent Burgers closure circuits which define Burgers vector **b**. Dislocation is perpendicular to the paper at X. The large dots represent impurity atoms which may be present.

magnitude is equal to the weight of the fluid displaced by the body.

Buprestidae [INV ZOO] The metallic wood-boring beetles, the large, single family of the coleopteran superfamily Buprestoidea.

Buprestoidea [INV ZOO] A superfamily of coleopteran insects in the suborder Polyphaga including many serious pests of fruit trees.

Burali-Forti paradox [MATH] The order-type of the set of all ordinals is the largest ordinal, but that ordinal plus one is larger.

buran [METEOROL] A violent northeast storm of south Russia and central Siberia, similar to the blizzard.

burble [FL MECH] **1.** A separation or breakdown of the laminar flow past a body. **2.** The eddying or turbulent flow resulting from this occurrence.

burble angle *See* burble point.

burble point [FL MECH] A point reached in an increasing angle of attack at which burble begins. Also known as burble angle.

burden [GEOL] All types of rock or earthy materials overlying bedrock. [MET] **1.** The material which is melted in a direct arc furnace. **2.** In an iron blast furnace, the ratio of iron and flux to coke and other fuels in the charge.

Burdigalian [GEOL] Upper lower Miocene geologic time.

buret [CHEM] A graduated glass tube used to deliver variable volumes of liquid; usually equipped with a stopcock to control the liquid flow.

burga [METEOROL] A storm of wind and sleet in Alaska. Also spelled boorga.

Burgers vector [CRYSTAL] A translation vector of a crystal lattice representing the displacement of the material to create a dislocation.

burglar alarm [ENG] An alarm in which interruption of electric current to a relay, caused, for example, by the breaking of a metallic tape placed at an entrance to a building, deenergizes the relay and causes the relay contacts to operate the alarm indicator. Also known as intrusion alarm.

Burhinidae [VERT ZOO] The thick-knees or stone curlews, a family of the avian order Charadriiformes.

buri *See* buntal.

burial ground [NUCLEO] A place for burying unwanted radioactive objects to prevent escape of their radiations, the earth acting as a shield. Also known as graveyard.

buried hill [GEOL] A hill of resistant older rock over which later sediments are deposited.

buried placer [GEOL] Old deposit of a placer which has been buried beneath lava flows or other strata.

buried river [GEOL] A river bed which has become buried beneath streams of alluvial drifts or basalt.

buried soil *See* paleosol.

burin [GRAPHICS] An etching tool used for engraving on metal or wood.

Burkett's lymphoma [MED] A malignant lymphoma of children, typically involving the retroperitoneal area and the mandible, but sparing the peripheral lymph nodes, bone marrow, and spleen.

burlap [TEXT] A coarse cloth woven from jute fiber. Also known as hessian.

burling [TEXT] Removing loops, knots, and vegetable matter in the finishing section of a woolen or worsted mill.

Burmanniaceae [BOT] A family of monocotyledonous plants in the order Orchidales characterized by regular flowers, three or six stamens opposite the petals, and ovarian nectaries.

burn [ENG] To consume fuel. [MED] An injury to tissues caused by heat, chemicals, electricity, or irradiation effects.

burnable poison [NUCLEO] A neutron absorber that is incorporated in the fuel or fuel cladding of a nuclear reactor and gradually burns up under neutron irradiation. Also known as burnout poison.

burn cut *See* parallel cut.

burner [CHEM ENG] A furnace where sulfur or sulfide ore are burned to produce sulfur dioxide and other gases. [ENG] **1.** The part of a fluid-burning device at which the flame is produced. **2.** Any burning device used to soften old paint to aid in its removal. **3.** A worker who operates a kiln which

burns brick or tile. **4.** A worker who alters the properties of a mineral substance by burning. **5.** A worker who uses a flame-cutting torch to cut metals. [MECH ENG] A unit of a steam boiler which mixes and directs the flow of fuel and air so as to ensure rapid ignition and complete combustion.

burner fuel oil [MATER] Any of the petroleum distillate and residual oils used to heat homes and buildings.

burnettize [ENG] To saturate fabric or wood with a solution of zinc chloride under pressure to keep it from decaying.

Burnett's syndrome *See* milk-alkali syndrome.

burn-in [ELECTR] Operation of electronic components before they are applied in order to stabilize their characteristics and reveal defects.

burning [ENG] The firing of clay products in a kiln. [MET] **1.** Permanent damage to a metal caused by heating beyond temperature limits of the treatment. **2.** Deep pitting of a metal caused by excessive pickling.

burning constant [ORD] In interior ballistics, a figure expressing the relative rate of burning of a propellant; this constant has a different value for each type of propellant. Symbolized *B*.

burning glass [OPTICS] A converging lens used to produce intense heat by converging the rays of the sun on a small area.

burning index *See* fire-danger meter.

burning line [PETRO ENG] A pipeline used to convey refinery fuel gas, as distinguished from gas intended for subsequent processing.

burning oil [MATER] An oil such as kerosine or mineral seal oil suitable for burning.

burning point [ENG] The lowest temperature at which a volatile oil in an open vessel will continue to burn when ignited by a flame held close to its surface; used to test safety of kerosine and other illuminating oils.

burning quality [ENG] Rated performance for a burning oil as determined by specified ASTM tests.

burning-quality index [ENG] Prediction of burning performance of furnace and heater oils; derived from ASTM distillation, API gravity, paraffinicity, and volatility.

burning rate [MATER] The tendency and rate of materials to burn at given temperatures, in contrast to melting or disintegrating.

burning-rate constant [AERO ENG] A constant, related to initial grain temperature, used in calculating the burning rate of a rocket propellant grain.

burning train *See* igniter train.

burning velocity [CHEM] The normal velocity of the region of combustion reaction (reaction zone) relative to nonturbulent unburned gas, in the combustion of a flammable mixture.

burnish [ENG] To polish or make shiny. [MET] To develop a smooth, lustrous surface finish by tumbling with steel balls or rubbing with a hard metal pad.

burnisher [GRAPHICS] **1.** An etching tool. **2.** A tool usually made of agate for polishing gold leaf after it has been laid.

burn off [METEOROL] With reference to fog or low stratus cloud layers, to dissipate by daytime heating from the sun.

burnout [AERO ENG] **1.** An act or instance of fuel or oxidant depletion or of depletion of both at once. **2.** The time at which this depletion occurs. [ELEC] Failure of a device due to excessive heat produced by excessive current. [ENG] An instance of a device or a part overheating so as to result in destruction or damage. [NUCLEO] **1.** To receive the greatest amount of radiation permissible during a given time. **2.** The point at which the heat flux across a surface causes film-blanketing of the surface, resulting in a drop in the film heat-transfer coefficient, overheating, and possible surface failure.

burnout poison *See* burnable poison.

burnout velocity [AERO ENG] The velocity of a rocket at the time when depletion of the fuel or oxidant occurs. Also known as all-burnt velocity; burnt velocity.

Burnside boring machine [MECH ENG] A machine for boring in all types of ground with the feature of controlling water immediately if it is tapped.

burnt deposit [MET] A dark, powdery deposit obtained by excessive current density in electroplating.

burn-through *See* jammer finder.

burnt lime *See* calcium oxide.

burnt-on sand [MET] A mixture of sand and metal on the

surface of a casting due to metal penetration of the sand mold.

burnt plate oil [MATER] A material used to thin etching ink too heavy in body; made by boiling linseed oil and, at a certain temperature, igniting it.

burnt shale [MIN ENG] Carbonaceous shale which has remained for a long period in a colliery tip and undergone spontaneous combustion and converted into a coppery slag material. Also known as oxidized shale.

burnt velocity *See* burnout velocity.

burnup [NUCLEO] A measure of nuclear-reactor fuel consumption, expressed either as the percentage of fuel atoms that have undergone fission or as the amount of energy produced per unit weight of fuel.

burr [BOT] **1.** A rough or prickly envelope on a fruit. **2.** A fruit so characterized. [MET] A thin, ragged fin left on the edge of a piece of metal by a cutting or punching tool.

burr mill [FOOD ENG] A mill for grinding crops, in which two ribbed plates or disks rub or crush the material between them.

burro [VERT ZOO] A small donkey used as a pack animal.

burrow [MIN ENG] A refuse heap at a coal mine.

bursa [ANAT] A simple sac or cavity with smooth walls containing a clear, slightly sticky fluid and interposed between two moving surfaces of the body to reduce friction.

bursa of Fabricius [VERT ZOO] A thymuslike organ in the form of a diverticulum at the lower end of the alimentary canal in birds.

Burseraceae [BOT] A family of dicotyledonous plants in the order Sapindales characterized by an ovary of two to five cells, prominent resin ducts in the bark and wood, and an intrastaminal disk.

bursicle [BOT] A purse or pouchlike receptacle.

bursitis [MED] Inflammation of a bursa.

burst [COMMUN] **1.** A sudden increase in the strength of a signal being received from beyond line-of-sight range. **2.** *See* color burst. [ELECTR] **1.** An exceptionally large electric pulse in the circuit of an ionization chamber due to the simultaneous arrival of several ionizing particles. **2.** A radar term for a single pulse of radio energy. [ORD] **1.** Continuous fire from an automatic weapon, as from an aircraft machine gun, sometimes described as a long or short burst. **2.** The explosion of a projectile, bomb, or similar munition.

burst amplifier [ELECTR] An amplifier stage in a color television receiver keyed into conduction and amplification by a horizontal pulse at the instant of each arrival of the color burst. Also known as chroma band-pass amplifier.

burst center [ORD] A point about which the bursts of several projectiles from rounds fired under like conditions are evenly distributed.

burst disk [AERO ENG] A diaphragm designed to burst at a predetermined pressure differential; sometimes used as a valve, for example, in a liquid-propellant line in a rocket. Also known as rupture disk.

burster [BOT] An abnormally double flower having the calyx split or fragmented. [ORD] An explosive element used in chemical ammunition such as bombs, mines, and shells to open the container and disperse the contents.

bursting charge [ORD] The main explosive charge in a mine, bomb, projectile, or the like that breaks the casing and produces fragmentation or demolition.

bursting layer [ORD] Layer of hard material used in the roofs of dugouts or cave shelters; it sets off projectiles fused for short delay or immediate detonation before they can enter deeply enough to cause great destruction.

bursting strength [MECH] A measure of the ability of a material to withstand pressure without rupture; it is the hydraulic pressure required to burst a vessel of given thickness.

burst pedestal [COMMUN] Rectangular pulselike television signal which may be part of the color burst; the amplitude of the color burst pedestal is measured from the alternating-current axis of the sine-wave portion to the horizontal pedestal.

burst pressure [MECH] The maximum inside pressure that a process vessel can safely withstand.

burst range [ORD] Horizontal distance from the piece to the point of burst.

burst separator [ELECTR] The circuit in a color television

receiver that separates the color burst from the composite video signal.

burst slug detector [NUCLEO] A radiation detector used for detecting small leaks in a fuel element of a nuclear reactor by measuring the radiation from short-lived fission products that escape into the coolant.

burst transmission [COMMUN] Radio transmission in which messages stored for a given time are sent at from 10 to more than 100 times the normal rate, recorded when received, and then slowed down to the normal rate for the user.

burst wave [FL MECH] Wave of compressed air caused by a bursting projectile or bomb; a detonation wave; it may produce extensive local damage.

burton [MECH ENG] A small hoisting tackle with two blocks, usually a single block and a double block, with a hook block in the running part of the rope.

bus [ELEC] **1.** A set of two or more electric conductors that serve as common connections between load circuits and each of the polarities (in direct-current systems) or phases (in alternating-current systems) of the source of electric power. **2.** *See* busbar. [ELECTR] One or more conductors in a computer along which information is transmitted from any of several sources to any of several destinations. [ENG] A motor vehicle for carrying a large number of passengers.

busbar [ELEC] A heavy, rigid metallic conductor, usually uninsulated, used to carry a large current or to make a common connection between several circuits. Also known as bus.

Busch lemniscate [METEOROL] The locus in the sky, or on a diagrammatic representation thereof, of all points at which the plane of polarization of diffuse sky radiation is inclined 45° to the vertical; a polarization isocline. Also known as neutral line.

bushbaby [VERT ZOO] Any of six species of African arboreal primates in two genera (*Galago* and *Euoticus*) of the family Lorisidae. Also known as galago; night ape.

bushel Abbreviated bu. [MECH] **1.** A unit of volume (dry measure) used in the United States, equal to 2150.42 cubic inches or 35.238 liters. **2.** A unit of volume (liquid and dry measure) used in Britain, equal to 2219.36 cubic inches or 8 imperial gallons.

bushing [MECH ENG] A removable piece of soft metal or graphite-filled sintered metal, usually in the form of a bearing, that lines a support for a shaft.

buster [MET] A pair of dies used in press forging for barreling, for flattening a hot metal billet, or for loosening scale on hot, ferrous forging billets. [MIN ENG] An expanding wedge used to break down coal or rock.

bustite [GEOL] An achondritic meteorite essentially of enstatite, with small quantities of oligoclase and diopside and some nickel-iron. Also known as aubrite.

busway [ELEC] A prefabricated assembly of standard lengths of busbars rigidly supported by solid insulation and enclosed in a sheet-metal housing.

busy test [COMMUN] A test, in telephony, made to find out whether certain facilities which may be desired, such as a subscriber line or trunk, are available for use.

busy tone [COMMUN] Interrupted low tone returned to the subscriber as an indication that the party's line is busy.

1,3-butadiene [ORG CHEM] C_4H_6 A colorless gas, boiling point −4.41°C, a major product of the petrochemical industry; used in the manufacture of synthetic rubber, latex paints, and nylon.

butadiene dimer [ORG CHEM] C_8H_{12} The third ingredient in ethylene-propylene-terpolymer (EPT) synthetic rubbers; isomers include 3-methyl-1,4,6-heptatriene, vinylcyclohexene, and cyclooctadiene.

butadiene rubber *See* polybutadiene.

butadiene-styrene rubber [MATER] A synthetic rubber that is formed by copolymerization of butadiene and styrene.

butaldehyde *See* butyraldehyde.

α-butalene *See* butene-1.

Butamer process [CHEM ENG] A method of isomerizing normal butane into isobutane in the presence of hydrogen and a solid, noble-metal catalyst; used to prepare raw material in a gasoline alkylation process.

n-butanal *See* butyraldehyde.

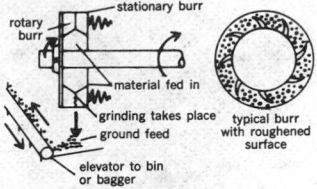

BURR MILL

Drawing of a burr mill.

BUS

Bus with rear-mounted engine. (*Mack Trucks*)

BUTTERFLY

Adult of the marsh (meadow) fritillary.

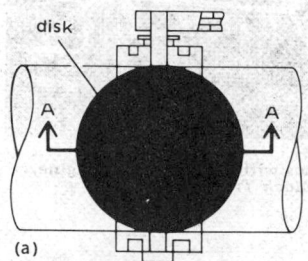

BUTTERFLY VALVE

disk

A ← → A

(a)

(b)

Butterfly-type valve for penstock is typically 25 feet (7.6 meters) in diameter. (a) Elevation. (b) Section A-A.

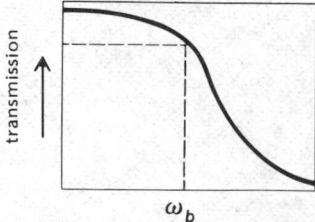

BUTTERWORTH FILTER

transmission

ω_b

Plot of the passband of the Butterworth filter; ω_b is the point where the transmission function falls below the passband tolerance.

butane [ORG CHEM] C_4H_{10} An alkane of which there are two isomers, *n* and isobutane; occurs in natural gas and is produced by cracking petroleum.

butane dehydrogenation [CHEM ENG] A process to remove hydrogen from butane to produce butene or butadiene.

butane dicarboxylic acid See adipic acid.

butanedioic acid See succinic acid.

butanedioic anhydride See succinic anhydride.

2,3-butanediol [ORG CHEM] $CH_3CHOHCHOHCH_3$ A major fermentation product of several species of bacteria. Also known as 2,3-butylene glycol; diol.

2,3-butanedione [ORG CHEM] $CH_3COCOCH_3$ A greenish-yellow liquid with boiling point 88–91°C, melting point −4°C; soluble in water, alcohol, and ether; used in foods as an aroma carrier. Also known as biacetyl; dimethylglyoxal.

butane vapor-phase isomerization [CHEM ENG] A process to isomerize normal butane into isobutane in the presence of aluminum chloride catalyst and hydrogen chloride promoter.

butanoic acid See butyric acid.

butanol [ORG CHEM] Any one of four isomeric alcohols having the formula C_4H_9OH; colorless, toxic liquids soluble in most organic liquids. Also known as butyl alcohol.

butanol fermentation [MICROBIO] Butanol production as a result of the fermentation of corn and molasses by the anaerobic bacterium *Clostridium acetobutylicum*.

2-butanone See methyl ethyl ketone.

butazolidine See phenylbutazone.

Butenandt unit [BIOL] A unit for the standardization of estrogens.

butene-1 [ORG CHEM] $CH_3CH_2CHCH_2$ A colorless, highly flammable gas; insoluble in water, soluble in organic solvents; used to produce polybutenes, butadiene aldehydes, and other organic derivatives. Also known as α-butalene; ethylethylene.

butene-2 [ORG CHEM] $CH_3CHCHCH_3$ A colorless, highly flammable gas, used to make butadiene and in the synthesis of four- and five-carbon organic molecules; the cis form, boiling point 3.7°C, is insoluble in water, soluble in organic solvents, and is also known as high-boiling butene-2; the trans form, boiling point 0.88°C, is insoluble in water, soluble in most organic solvents, and is also known as low-boiling butene-2.

α-butenic acid See crotonic acid.

trans-2-butenoic acid See ethyl crotonate.

1-butine See ethyl acetylene.

Butler finish See satin finish.

Butler oscillator [ELEC] Oscillator in which a piezoelectric crystal is connected between the cathode of two tubes, one functioning as a cathode follower, and the other as a grounded-grid amplifier.

Butomaceae [BOT] A family of monocotyledonous plants in the order Alismatales characterized by secretory canals, linear leaves, and a straight embryo.

butt [MIN ENG] Coal exposed at right angles to the face and, in contrast to the face, generally having a rough surface. [ORD] Rear end of the stock of a rifle or other small arm.

butt cable See hand cable.

butt contact [ELEC] A hemispherically shaped contact designed to mate against a similarly shaped contact.

butte [GEOGR] A detached hill or ridge which rises abruptly.

butter [FOOD ENG] A fatty food product made from milk or cream or both and consisting of a solid emulsion of fat globules, air, and water.

butterfat [FOOD ENG] A mixture of glycerides derived from fatty acids; the natural fat of milk and butter.

butter finish [MET] The semilustrous surface produced with a mildly abrasive wheel.

butterfly [INV ZOO] Any insect belonging to the lepidopteran suborder Rhopalocera, characterized by a slender body, broad colorful wings, and club-shaped antennae.

butterfly bomb [ORD] A small fragmentation or antipersonnel bomb equipped with two folding wings which rotate and arm the fuse as the bomb descends; designed to be dropped in clusters, they are frequently fitted with antidisturbance or delay fuses.

butterfly capacitor [ELEC] A variable capacitor having stator and rotor plates shaped like butterfly wings, with the stator plates having an outer ring to provide an inductance so that both capacitance and inductance may be varied, thereby giving a wide tuning range.

butterfly damper See butterfly valve.

butterfly nut See wing nut.

butterfly valve [ENG] A valve that utilizes a turnable disk element to regulate flow in a pipe or duct system, such as a hydraulic turbine or a ventilating system. Also known as butterfly damper.

buttering [MET] Coating the faces of a weld joint prior to welding to prevent cross contamination of the weld metal and base metal.

buttermilk [FOOD ENG] A fermentation food product prepared by inoculating sweet milk with cultures of *Streptococcus lactis* and *Leuconostoc citrovorum*.

butter rock See halotrichite.

Butterworth filter [ELECTR] An electric filter whose pass band (graph of transmission versus frequency) has a maximally flat shape.

Butterworth head [MECH ENG] A mechanical hose head with revolving nozzles; used to wash down shipboard storage tanks.

butt fusion [ENG] The joining of two pieces of plastic or metal pipes or sheets by heating the ends until they are molten and then pressing them together to form a homogeneous bond.

butt joint [ELEC] A connection formed by placing the ends of two conductors together and joining them by welding, brazing, or soldering. [ELECTROMAG] A connection giving physical contact between the ends of two waveguides to maintain electrical continuity. [ENG] A joint in which the parts to be joined are fastened end to end or edge to edge with one or more cover plates (or other strengthening) generally used to accomplish the joining.

buttock [MIN ENG] A corner formed by two coal faces more or less at right angles, such as the end of a working face.

buttock lines [ENG] The lines of intersection of the surface of an aircraft or its float, or of the hull of a ship, with its longitudinal vertical planes. Also known as buttocks.

buttocks [ANAT] The two fleshy parts of the body posterior to the hip joints. [ENG] See buttock lines. [NAV ARCH] The convex part of the stern end of a ship above the water line. Also known as counter.

button [ELECTR] 1. A small, round piece of metal that is alloyed to the base wafer of an alloy-junction transistor. Also known as dot. 2. The container that holds the carbon granules of a carbon microphone. Also known as carbon button. [MET] 1. Mass of metal remaining in a crucible after fusion has been completed. 2. That part of a weld which tears out in the destructive testing of spot-, seam-, or projection-welded specimens.

button balance [MIN ENG] A small, very delicate balance used for weighing assay buttons.

button die [DES ENG] A mating member, usually replaceable, for a piercing punch. Also known as die bushing.

buttonhead [DES ENG] A screw, bolt, or rivet with a hemispherical head.

buttonhook contact [ELEC] A curved, hooklike contact often used on feed-through terminals of headers to facilitate soldering or unsoldering of leads.

buttress [ARCH] An upright projection that supports or resists lateral forces in a building. [CIV ENG] A pier constructed at right angles to a restraining wall on the side opposite to the restrained material; increases the strength and thrust resistance of the wall. [PALEON] A ridge on the inner surface of a pelecypod valve which acts as a support for part of the hinge.

buttress dam [CIV ENG] A concrete dam constructed as a series of buttresses.

buttress sands [GEOL] Sandstone bodies deposited above an unconformity; the upper portion rests upon the surface of the unconformity.

buttress thread [DES ENG] A screw thread whose forward face is perpendicular to the screw axis and whose back face is at an angle to the axis, so that the thread is both efficient in transmitting power and strong.

buttress-thread casing [PETRO ENG] A drill casing in which the ends of the sections are buttressed together and held in place with a short threaded outer sleeve; used where greater than normal clearance, strength, and leak resistance are needed.

butt rot [PL PATH] A fungus decay of the base of a tree trunk caused by polypores (such as *Fomea* species).

butt weld [MET] A weld that joins the ends of two pieces of metal of similar cross section without overlapping.

butyl [ORG CHEM] Any of the four variations of the hydrocarbon radical C_4H_9: $CH_3CH_2CH_2CH_2-$, $(CH_3)_2CHCH_2-$, $CH_3CH_2CHCH_3-$, and $(CH_3)_3C-$.

butyl acetate [ORG CHEM] $CH_3COOC_4H_9$ A colorless liquid slightly soluble in water; used as a solvent.

butylacetic acid *See* caproic acid.

butyl acrylate [ORG CHEM] $CH_2CHCOOC_4H_9$ A colorless liquid that is nearly insoluble in water and polymerizes readily upon heating; used as an intermediate for organic synthesis, polymers, and copolymers.

butyl alcohol *See* butanol.

butyl aldehyde *See* butyraldehyde.

***n*-butyl aldehyde** *See* butyraldehyde.

***n*-butyl amine** [ORG CHEM] $C_4H_9NH_2$ A colorless, flammable liquid; miscible with water and ethanol; used as an intermediate in organic synthesis and to make insecticides, emulsifying agents, and pharmaceuticals. Also known as 1-aminobutane.

***sec*-butyl amine** [ORG CHEM] $CH_3CHNH_2C_2H_5$ A flammable, colorless liquid; boils in the range 63–68°C; may be used as an intermediate in organic synthesis. Also known as 2-aminobutane.

***tert*-butyl amine** [ORG CHEM] $(CH_3)_3CNH_2$ A flammable liquid; boiling range 63–68°C; may be used in organic synthesis as an intermediate.

butylated hydroxyanisole [ORG CHEM] $(CH_3)_3CC_6H_3OH-(OCH_3)$ An antioxidant consisting chiefly of a mixture of 2- and 3-*tert*-butyl-4-hydroxyanisole and used to control rancidity of lard and animal fats in foods. Abbreviated BHA.

butylated hydroxytoluene [ORG CHEM] An antioxidant used to retard rancidity in fatty foods. Abbreviated BHT.

butylbenzene [ORG CHEM] $C_6H_5C_4H_9$ A colorless liquid used as a raw material for organic synthesis, especially for insecticides; forms are normal (1-phenylbutane), secondary (2-phenylbutane), and tertiary (2-methyl-2-phenylpropane).

butyl borate *See* tributyl borate.

butyl carbinol [ORG CHEM] $(CH_3)_3CCH_2OH$ Colorless crystals that melt at 52°C; slightly soluble in water.

butyl chloride [ORG CHEM] C_4H_9Cl A colorless liquid used as an alkylating agent in organic synthesis, as a solvent, and as an anthelminthic; forms are normal (1-chlorobutane), secondary, and iso or tertiary.

butyl citrate [ORG CHEM] $C_3H_5O(COOC_4H_9)_3$ A colorless, odorless, nonvolatile liquid, almost insoluble in water; used as a plasticizer, solvent for cellulose nitrate, and antifoam agent. Also known as tributyl citrate.

4-butyl-1,2-diphenyl-3,5-pyrazolidinedione *See* phenylbutazone.

butylene [ORG CHEM] Any of three isomeric alkene hydrocarbons with the formula C_4H_8; all are flammable and easily liquefied gases.

2,3-butylene glycol *See* 2,3-butanediol.

butylene oxide [ORG CHEM] C_4H_8O An intermediate and a solvent for polyvinyl chloride. Also known as tetrahydrofuran.

butyl ether [ORG CHEM] $C_8H_{18}O$ A colorless liquid, boiling at 142°C, and soluble in water; used as an extracting agent, as a medium for Grignard and other reactions, and for purifying other solvents.

butyl ethylacetic acid *See* 2-ethyl hexoic acid.

butyl formate [ORG CHEM] $HCOOC_4H_9$ An ester of formic acid and butyric acid.

butyl lactate [ORG CHEM] $CH_3CHOHCOOC_4H_9$ A stable liquid, water-white and nontoxic, miscible with many solvents; used as a solvent for resins and gums, in lacquers and varnishes, and as a chemical intermediate.

butyl mercaptan [ORG CHEM] C_4H_9SH A colorless, odorous liquid that appears in skunk secretion; used commercially as a gas-odorizing agent.

butyl octadecanoate *See* butyl stearate.

butyl oleate [ORG CHEM] $C_{22}H_{42}O_2$ A butyl ester of oleic acid; used as a plasticizer.

butyl propionate [ORG CHEM] $C_2H_5COOC_4H_9$ A colorless aromatic liquid; used in fruit essences.

butyl rubber [MATER] A synthetic rubber made by the polymerization of isoprene and isobutylene.

butyl stearate [ORG CHEM] $C_{17}H_{35}COOC_4H_9$ A liquid that solidifies at approximately 19°C; mixes with vegetable oils and is soluble in alcohol and ethers but insoluble in water; used as a lubricant, in polishes, as a plasticizer, and as a dye solvent. Also known as butyl octadecanoate.

butyl titanate *See* tetrabutyl titanate.

1-butyne *See* ethyl acetylene.

butyraldehyde [ORG CHEM] $CH_3(CH_2)_2CHO$ A colorless liquid boiling at 75.7°C; soluble in ether and alcohol, insoluble in water; derived from the oxo process. Also known as butaldehyde; *n*-butanal; *n*-butyl aldehyde; butyric aldehyde.

butyrate [ORG CHEM] An ester or salt of butyric acid containing the $C_4H_7O_2$ radical.

Butyribacterium [MICROBIO] A genus of gram-positive, non-motile, rod-shaped bacteria of the family Propionibacteriaceae occurring in vertebrate intestines.

butyric acid [ORG CHEM] $CH_3CH_2CH_2COOH$ A colorless, combustible liquid with boiling point 163.5°C (757 mm Hg); soluble in water, alcohol, and ether; used in synthesis of flavors, in pharmaceuticals, and in emulsifying agents. Also known as butanoic acid; *n*-butyric acid; ethyl acetic acid; propylformic acid.

***n*-butyric acid** *See* butyric acid.

butyric aldehyde *See* butyraldehyde.

butyric anhydride [ORG CHEM] $C_8H_{14}O_3$ A colorless liquid that decomposes in water to form butyric acid; exists in two isomeric forms.

butyric ether *See* ethyl butyrate.

butyric fermentation [BIOCHEM] Fermentation in which butyric acid is produced by certain anaerobic bacteria acting on organic substances, such as butter; occurs in putrefaction and in digestion in herbivorous mammals.

butyrinase [BIOCHEM] An enzyme that hydrolyzes butyrin, found in the blood serum.

butyrolactone [ORG CHEM] $C_4H_6O_2$ A liquid, the anhydride of butyric acid; used as a solvent in the manufacture of plastics.

butyrone *See* 4-heptanone.

Buxbaumiales [BOT] An order of very small, atypical mosses (Bryopsida) composed of three genera and found on soil, rock, and rotten wood.

Buys-Ballot's law [METEOROL] A law describing the relationship of the horizontal wind direction in the atmosphere to the pressure distribution: if one stands with one's back to the wind, the pressure to the left is lower than to the right in the Northern Hemisphere; in the Southern Hemisphere the relation is reversed. Also known as baric wind law.

buzz [AERO ENG] Sustained oscillation of an aerodynamic control surface caused by intermittent flow separation on the surface, or by a motion of shock waves across the surface, or by a combination of flow separation and shock-wave motion on the surface. [CONT SYS] *See* dither. [ELECTR] The condition of a combinatorial circuit with feedback that has undergone a transition, caused by the inputs, from an unstable state to a new state that is also unstable. [FL MECH] In supersonic diffuser aerodynamics, a nonsteady shock motion and airflow associated with the shock system ahead of the inlet.

buzz bomb *See* V-1.

buzzer [ELECTROMAG] An electromagnetic device having an armature that vibrates rapidly, producing a buzzing sound.

B virus [VIROL] An animal virus belonging to subgroup A of the herpesvirus group.

BW *See* bandwidth; biological warfare.

BWE *See* bucket-wheel excavator.

BWG *See* Birmingham wire gage.

BWO *See* backward-wave oscillator.

BWR *See* boiling-water reactor.

BX cable [ELEC] Insulated wires in flexible metal tubing used for bringing electric power to electronic equipment.

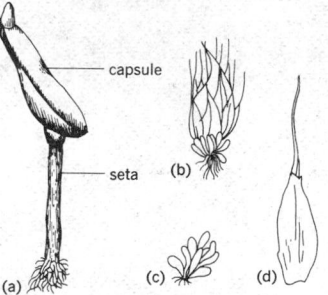

BUXBAUMIALES

Morphological features of Buxbaumiales. *(a)* Sporophyte of *Buxbaumia aphylla (from W. H. Welch, Mosses of Indiana, Indiana Department of Conservation, 1957). (b) Diphyscium foliosum,* habit sketch, *(c)* leaves, and *(d)* perichaetial leaf *(from H. S. Conard, How to Know the Mosses and Liverworts, William C. Brown, 1956).*

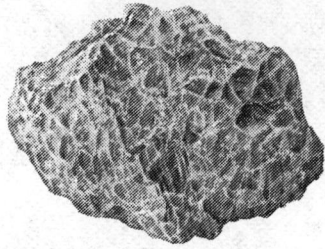

BYTOWNITE

├─2.5 cm─┤

Bytownite, from Crystal Bay, Minn. *(Specimen from Department of Geology, Bryn Mawr College)*

byerite [GEOL] Bituminous coal that does not crack in fire and melts and enlarges upon heating.

Byers process [MET] The principal process for manufacturing wrought iron; pig iron is melted in a cupola, desulfurized in a ladle, and refined in a Bessemer converter.

byon [GEOL] Gem-bearing gravel, particularly that with brownish-yellow clay in which corundum, rubies, sapphires, and so forth occur.

bypass [CIV ENG] A road which carries traffic around a congested district or a temporary obstruction. [ELEC] A shunt path around some element or elements of a circuit. [ENG] An alternating, usually smaller, diversionary flow path in a fluid dynamic system to avoid some device, fixture, or obstruction.

bypass capacitor [ELEC] A capacitor connected to provide a low-impedance path for radio-frequency or audio-frequency currents around a circuit element. Also known as bypass condenser.

bypass channel [CIV ENG] **1.** A channel built to carry excess water from a stream. Also known as flood relief channel; floodway. **2.** A channel constructed to divert water from a main channel.

bypass condenser *See* bypass capacitor.

bypass filter [ELECTR] Filter which provides a low-attenuation path around some other equipment, such as a carrier frequency filter used to bypass a physical telephone repeater station.

by-product [ENG] A product from a manufacturing process that is not considered the principal material.

Byrrhidae [INV ZOO] The pill beetles, the single family of the coleopteran insect superfamily Byrrhoidea.

Byrrhoidea [INV ZOO] A superfamily of coleopteran insects in the superorder Polyphaga.

bysmalith [GEOL] A body of igneous rock that is more or less vertical and cylindrical; it crosscuts adjacent sediments.

byssinosis [MED] A pneumoconiosis caused by the inhalation of cotton dust.

byte [ADP] A sequence of adjacent binary digits operated upon as a unit in a computer and usually shorter than a word.

by the head *See* down by the head.

by the stern *See* down by the stern.

bytownite [MINERAL] A plagioclase feldspar with a composition ranging from $Ab_{30}An_{70}$ to $Ab_{10}An_{90}$, where $Ab = NaAlSi_3O_8$ and $An = CaAl_2Si_2O_8$; occurs in basic and ultrabasic igneous rock.

Byturidae [INV ZOO] The raspberry fruitworms, a small family of coleopteran insects in the superfamily Cucujoidea.

c *See* calorie; centi-; curie.

C *See* capacitance; capacitor; carbon.

Ca *See* calcium.

Ca⁴⁵ *See* calcium-45.

caatinga [ECOL] A sparse, stunted forest in areas of little rainfall in northeastern Brazil; trees are leafless in the dry season.

cab [ENG] In a locomotive, truck, tractor, or hoisting apparatus, a compartment for the operator.

caballing [OCEANOGR] The mixing of two water masses of identical in situ densities but different in situ temperatures and salinities, such that the resulting mixture is denser than its components and therefore sinks.

cabane [AERO ENG] The arrangement of struts used on early types of airplanes to brace the wings.

Cabannes' factor [ANALY CHEM] An equational factor to correct for the depolarization effect of the horizontal components of scattered light during the determination of molecular weight by optical methods.

cabbage [BOT] *Brassica oleracea* var. *capitata*. A biennial crucifer of the order Capparales grown for its head of edible leaves.

cabbage yellows [PL PATH] A fungus disease of cabbage caused by *Fusarium conglutinans* and characterized by yellowing and dwarfing.

cabble [MET] To break up into pieces preparatory to the processes of fagoting, fusing, and rolling into bars, as is done with charcoal iron.

Cabernet [FOOD ENG] A red dry table wine made from a grape variety that corresponds to Burgundy grapes.

cabezon [VERT ZOO] *Scorpaenichthys marmoratus*. A fish that is the largest of the sculpin species, weighing as much as 25 pounds (11.3 kilograms) and reaching a length of 30 inches (76 centimeters).

Cabibbo theory [PARTIC PHYS] A theory describing baryon beta-decay processes, according to which the amplitude for such processes is given by $G \{\cos \Theta [V(\Delta s = 0) + A (\Delta s = 0)] + \sin \Theta [V(\Delta s = +1) + A (\Delta s = +1)]\}$, where Θ is the Cabbibo angle, Δs is the change in strangeness for the baryon, G is a universal beta-decay amplitude, and V and A are vector and axial vector amplitudes, respectively; it is experimentally determined that $\sin \Theta \approx 0.25$, so that $\cos \Theta = 0.97$.

cabinet file [DES ENG] A coarse-toothed file for woodworking, with flat and convex faces.

cabinet hardware [DES ENG] Parts for the final trim of a cabinet, such as fastening hinges, drawer pulls, and knobs.

cabinet saw [DES ENG] A short saw, one edge used for ripping, the other for crosscutting.

cabinet scraper [DES ENG] A steel tool with a contoured edge used to remove irregularities on a wood surface.

cable [ARCH] A convex molding within one of the vertical grooves of a column or pilaster. [DES ENG] A stranded, ropelike assembly of wire or fiber. [ELEC] Strands of insulated electrical conductors laid together, usually around a central core, and wrapped in a heavy insulation. [OCEANOGR] *See* cable length.

cable-and-trunk schematic [ELEC] A drawing which shows, in block form, the interconnection between all major electric circuits in an office.

cable armor [ELEC] One or more layers of extra-strength material, such as steel wire or tape, to reinforce the usual lead wall in cable construction.

cable bend [NAV ARCH] **1.** A small rope used to tie the end of a cable into a loop to secure an anchor. **2.** A knot used to attach a cable to an anchor.

cable buoy [ENG] A buoy used to mark one end of a submarine underwater cable during time of installation or repair.

cable chain *See* chain cable.

cable code *See* Morse cable code.

cable complement [ELEC] Group of wire pairs in a cable having some common distinguishing characteristic.

cable conveyor [MECH ENG] A powered conveyor in which a trolley runs on a flexible, torque-transmitting cable that has helical threads.

cable delay [ADP] The time required for one bit of data to go through a cable, about 1.5 nanoseconds per foot of cable.

cable duct [ENG] A pipe, either earthenware or concrete, through which prestressing wires or electric cable are pulled.

cable fill [ELEC] Ratio of the number of wire pairs in use to the total number of pairs in a cable.

cablegram [COMMUN] A message transmitted over a submarine telegraph cable.

cable lacquer [MATER] Black, colored, or clear lacquer made from synthetic resins, having a high dielectric strength and being resistant to oils and heat; used to give a tough, flexible coating.

cable-laid [DES ENG] Consisting of three ropes with a left-hand twist, each rope having three twisted strands.

cable layer [NAV ARCH] A type of ship fitted with huge reels of submarine telephone cable and devices for paying out the cable to the sea bottom.

cable length [OCEANOGR] A unit of distance, originally equal to the length of a ship's anchor cable, now variously considered to be 600 feet (183 meters), 608 feet (one-tenth of a British nautical mile), or 720 feet or 120 fathoms. Also known as cable.

cableman [ENG] A person who installs, repairs, or otherwise works with cables.

cable messenger [ELEC] Stranded group of wires supported above the ground at intervals by poles or other structures and employed to furnish, within these intervals, frequent points of support for conductors or cables.

cable paper [MATER] A paper used to insulate electrical cables.

cable railway [MECH ENG] An inclined track on which rail cars travel, with the cars fixed to an endless steel-wire rope at equal spaces; the rope is driven by a stationary engine.

cable release [ENG] A wire plunger to actuate the shutter of a camera, thus avoiding undesirable camera movement.

cable run [ELEC] Path occupied by a cable on cable racks or other support from one termination to another.

cable running list [ELEC] Drawing showing the code of cable, terminations, circuit names, and numbering of cables appearing in an office.

cable skidding [FOR] The use of a chain or cable choker to move tree lengths or tree segments for a short distance over unimproved terrain.

cable stopper [NAV ARCH] A device, such as a short chain fastened to a pad on the deck and attached to the anchor chain, to take the strain off the anchor windlass when a ship is at anchor.

cable system [MIN ENG] A drilling system involving a heavy string of tools suspended from a flexible cable.

cable-system drill *See* churn drill.

cable tier [NAV ARCH] **1.** The part of a ship used to store spare rigging and cables. **2.** *See* chain locker.

cable tools [MIN ENG] The bits and other bottom-hole tools and equipment used to drill boreholes by percussive action, using a cable, instead of rods, to connect the drilling bit with the machine on the surface.

cable vault [CIV ENG] A manhole containing electrical cables. [ELEC] Vault in which the outside plant cables are spliced to the tipping cables.

cableway [MECH ENG] A transporting system consisting of a cable extended between two or more points on which cars are

CABBAGE

Cabbage (*Brassica oleracea* var. *capitata*), cultivar Golden Acre 84. (*Joseph Harris Co., Inc., Rochester, N.Y.*)

CACTUS

Mule cactus, with closeup of flower, maximum height of 6 feet (1.8 meters).

CADDIS FLY

Adult caddis fly.

CADMIUM

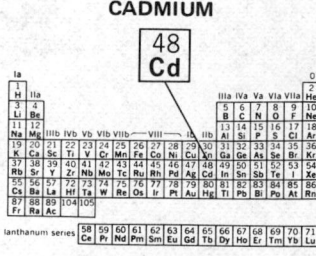

Periodic table of the chemical elements showing the position of cadmium.

propelled to transport bulk materials for construction operations. [MIN ENG] A cable system of material handling in which carriers are supported by a cable and not detached from the operating span.

cableway carriage [MECH ENG] A trolley that runs on main load cables stretched between two or more towers.

cabochon [LAP] One of two basic types of cutting styles used to fashion gemstones; the gem is cut in convex form, highly polished, but not faceted.

caboose [ENG] A car on a freight train, often the last car, usually for use by trainmen.

Cabot's ring [CYTOL] A ringlike body in immature erythrocytes that may represent the remains of the nuclear membrane.

CABRA numbers [MET] Copper and Brass Research Association number designations for various wrought copper and copper alloy grades.

cabretta [MATER] Tanned sheepskin leather.

cab signal [ENG] A signal in a locomotive that informs the engineman about conditions affecting train movement.

cacao [BOT] *Theobroma cacao.* A small tropical tree of the order Theales that bears capsular fruits which are a source of cocoa powder and chocolate. Also known as chocolate tree.

cacao butter *See* cocoa butter.

cache [ADP] A small, fast, storage buffer integrated in the central processing unit of some large computers.

cachexia [MED] Weight loss, weakness, and wasting of the body encountered in certain diseases or in terminal illnesses.

cacimbo [METEOROL] Local name in Angola (Portuguese West Africa) for the wet fogs and drizzles noted with onshore winds from the Benguela Current.

cacodyl [ORG CHEM] $(CH_3)_2As-$ A radical found in, for example, cacodylic acid, $(CH_3)_2AsOOH$.

cacodylate [ORG CHEM] Any salt of cacodylic acid.

cacodylic acid [ORG CHEM] $(CH_3)_2AsOOH$ Colorless crystals that melt at 200°C; soluble in alcohol and water; used as a herbicide. Also known as dimethylarsinic acid.

cacomistle [VERT ZOO] *Bassariscus astutus.* A raccoonlike mammal that inhabits the southern and southwestern United States; distinguished by a bushy black-and-white ringed tail. Also known as civet cat; ringtail.

caconym [SYST] A taxonomic name that is linguistically unacceptable.

cacotheline [ORG CHEM] $C_{20}H_{22}N_2O_5(NO_2)_2$ An azoic compound used as a metal indicator in chelometric titrations.

cacoxenite [MINERAL] $Fe_4(PO_4)_3(OH)_3 \cdot 12H_2O$ Yellow or brownish mineral consisting of a hydrous basic iron phosphate occurring in radiated tufts.

Cactaceae [BOT] The cactus family of the order Caryophyllales; represented by the American stem succulents, which are mostly spiny with reduced leaves.

cactus [BOT] The common name for any member of the family Cactaceae, a group characterized by a fleshy habit, spines and bristles, and large, brightly colored, solitary flowers.

cactus alkaloid *See* anhalonium alkaloid.

cadalene [ORG CHEM] $C_{15}H_{18}$ A colorless liquid which boils at 291–292°C (720 mm Hg) and which is a substituted naphthalene. Also known as 1,6-dimethyl-4-isopropyl naphthalene.

cadang-cadang [PL PATH] An infectious virus disease of the coconut palm characterized by yellow-bronzing of the leaves.

cadastral survey [CIV ENG] A survey made to establish property lines.

cadaver [MED] A dead animal or human body to be studied by dissection.

cadaverine [BIOCHEM] $C_5H_{14}N_2$ A nontoxic, organic base produced as a result of the decarboxylation of lysine by the action of putrefactive bacteria on flesh.

caddis fly [INV ZOO] The common name for all members of the insect order Trichoptera; adults are mothlike and the immature stages are aquatic.

cade oil [MATER] A brown, viscous essential oil that has a tar odor and is slightly soluble in water; it is derived by dry distillation of the wood of the European juniper (*Juniperus oxycedrus*); used in antiseptic soaps, perfumes, and pharmaceuticals. Also known as juniper tar oil.

cadinene [ORG CHEM] $C_{15}H_{24}$ A colorless liquid that boils at 274.5°C, and is a terpene derived from cubeb oil, cade oil, juniper berry oil, and other essential oils.

cadmium [CHEM] A chemical element, symbol Cd, atomic number 48, atomic weight 112.40. [MET] A tin-white, malleable, ductile metal capable of high polish; principal use is in the plating of iron and steel, and to a much less extent of copper, brass, and other alloys, to protect them from corrosion and improve solderability and surface conductivity, and as a control absorber and shield in nuclear reactors.

cadmium acetate [ORG CHEM] $Cd(OOCCH_3)_2 \cdot 3H_2O$ A compound that forms colorless monoclinic crystals, soluble in water and in alcohol; used for chemical testing for sulfides, selenides, and tellurides and for producing iridescent effects on porcelain.

cadmium blende *See* greenockite.

cadmium bromate [INORG CHEM] $Cd(BrO_3)_2$ Colorless powder, soluble in water; used as an analytical reagent.

cadmium bromide [INORG CHEM] $CdBr_2$ A compound produced as a yellow crystalline powder, soluble in water and alcohol; used in photography, process engraving, and lithography.

cadmium carbonate [INORG CHEM] $CdCO_3$ A white crystalline powder, insoluble in water, soluble in acids and potassium cyanide; used as a starting compound for other cadmium salts.

cadmium cell [ELEC] A standard cell used as a voltage reference; at 20°C its voltage is 1.0186 volt.

cadmium chlorate [INORG CHEM] $CdClO_3$ White crystals, soluble in water; a highly toxic material.

cadmium chloride [INORG CHEM] $CdCl_2$ A cadmium halide in the form of colorless crystals, soluble in water, methanol, and ethanol; used in photography, in dyeing and calico printing, and as a solution to precipitate sulfides.

cadmium hydroxide [INORG CHEM] $Cd(OH)_2$ A white powder, soluble in dilute acids; used to prepare negative electrodes for cadmium-nickel storage batteries.

cadmium iodide [INORG CHEM] CdI_2 A cadmium halide that forms lustrous, white, hexagonal scales, consisting of two water-soluble allotropes; used in photography, in process engraving, and formerly as an antiseptic.

cadmium lamp [ELEC] A lamp containing cadmium vapor; wavelength (6438.4696 international angstroms) of light emitted is a standard of length.

cadmium metallurgy [MET] The extraction of cadmium from zinc ores, or from complex ores as a by-product of zinc, lead, and copper smelting.

cadmium-nickel storage cell *See* nickel-cadmium battery.

cadmium nitrate [CHEM] $Cd(NO_3)_2 \cdot 4H_2O$ White, hygroscopic crystals, soluble in water, alcohol, and liquid ammonia; used to give a reddish-yellow luster to glass and porcelain ware.

cadmium ocher *See* greenockite.

cadmium oxide [INORG CHEM] CdO In the cubic form, a brown, amorphous powder, insoluble in water, soluble in acids and ammonia salts; used for cadmium plating baths and in the manufacture of paint pigments.

cadmium potassium iodide *See* potassium tetraiodocadinate.

cadmium red [MATER] A pigment composed of a mixture of cadmium sulfide, cadmium selenide, and barite; used as a red pigment.

cadmium silver oxide cell [ELEC] An alkaline-electrolyte cell that may be used without recharging in primary batteries or that may be recharged for secondary-battery use.

cadmium sulfate [INORG CHEM] $CdSO_4$ A compound that forms colorless, efflorescent crystals, soluble in water; used as an antiseptic and astringent, in the treatment of syphilis, gonorrhea, and rheumatism, and as a detector of hydrogen sulfide and fumaric acid.

cadmium sulfide [INORG CHEM] CdS A compound with two forms: orange, insoluble in water, used as a pigment, and also known as orange cadmium; light yellow, hexagonal crystals, insoluble in water, and also known as cadmium yellow.

cadmium sulfide cell [ELECTR] A photoconductive cell in

which a small wafer of cadmium sulfide provides an extremely high dark-light resistance ratio.

cadmium yellow *See* cadmium sulfide.

caducicorn [VERT ZOO] Having deciduous horns, as certain deer.

caducous [BOT] Lasting on a plant only a short time before falling off.

cadwaladerite [MINERAL] $Al(OH)_2Cl \cdot 4H_2O$ A mineral consisting of a hydrous basic aluminum chloride.

Cae *See* Caelum.

caecilian [VERT ZOO] The common name for members of the amphibian order Apoda.

caecum *See* cecum.

Caelum [ASTRON] A southern constellation, right ascension 5 hours, declination 40°S. Abbreviated Cae. Also known as Chisel.

Caenolestidae [PALEON] A family of extinct insectivorous mammals in the order Marsupialia.

Caenolestoidea [VERT ZOO] A superfamily of marsupial mammals represented by the single living family Caenolestidae.

caenostylic [VERT ZOO] Having the first two visceral arches attached to the cranium and functioning in food intake; a condition found in sharks, amphibians, and chimaeras.

Caesalpinoidea [BOT] A subfamily of dicotyledonous plants in the legume family, Leguminosae.

caffeic acid [ORG CHEM] $C_9H_8O_4$ A yellow crystalline acid that melts at 223–225°C with decomposition; soluble in water and alcohol. Also known as 3,4-dihydroxycinnamic acid.

caffeine [ORG CHEM] $C_8H_{10}O_2N_4 \cdot H_2O$ An alkaloid found in a large number of plants, such as tea, coffee, cowa, and mate.

caffetannin [MATER] One of the hydrolyzable tannins obtained from coffee berries and other plant products.

cage [MECH ENG] A frame for maintaining uniform separation between the balls or rollers in a bearing. Also known as separator. [MIN ENG] The car which carries men and materials in a mine hoist.

cage antenna [ELECTROMAG] Broad-band dipole antenna in which each pole consists of a cage of wires whose overall shape resembles that of a cylinder or a cone.

cage guides [MIN ENG] Directive apparatus used to guide the cages in the mine shaft and to prevent their swinging or colliding with each other.

cage mill [MECH ENG] Pulverizer used to disintegrate clay, press cake, asbestos, packing-house by-products, and various tough, gummy, high-moisture-content or low-melting-point materials.

cage shoes [MIN ENG] Fittings attached on the side of a cage by bolts so that they engage the rigid guides in a shaft.

caging [NAV] The process of orienting and mechanically locking the spin axis of a gyroscope to an internal reference position.

cahnite [MINERAL] $Ca_2B(OH)_4(AsO_4)$ A tetragonal borate mineral occurring in white, sphenoidal crystals.

Cailletet and Mathias law [PHYS CHEM] The law that describes the relationship between the mean density of a liquid and its saturated vapor at that temperature as being a linear function of the temperature.

caiman [VERT ZOO] Any of five species of reptiles of the genus *Caiman* in the family Alligatoridae, differing from alligators principally in having ventral armor and a sharper snout.

cainophobia [PSYCH] Abnormal fear of newness.

Cainotheriidae [PALEON] The single family of the extinct artiodactyl superfamily Cainotherioidea.

Cainotherioidea [PALEON] A superfamily of extinct, rabbit-sized tylopod ruminants in the mammalian order Artiodactyla.

cairn [ENG] An artificial mound of rocks, stones, or masonry, usually conical or pyramidal, whose purpose is to designate or to aid in identifying a point of surveying or of cadastral importance.

caisson [CIV ENG] 1. A watertight, cylindrical or rectangular chamber used in underwater construction to protect workers from water pressure and soil collapse. 2. A float used to raise a sunken vessel. 3. *See* dry-dock caisson. [ORD] A two-wheeled, horse-drawn vehicle used for transporting ammunition and other military equipment.

caisson disease [MED] A condition resulting from a rapid change in atmospheric pressure from high to normal, causing nitrogen bubbles to form in the blood and body tissues. Also known as bends; compressed-air illness.

caisson foundation [CIV ENG] A shaft of concrete placed under a building column or wall and extending down to hardpan or rock. Also known as pier foundation.

cajeput oil [MATER] An essential oil with a blue-green color and a camphorlike odor; chief constituents are eucalyptol, pinene, and alpha terpineol; distilled from leaves and twigs of species of East Indian trees of the genus *Melaleuca*; used in perfumes and medicines. Also spelled cajuput oil.

cajuput oil *See* cajeput oil.

caju rains [METEOROL] In northeastern Brazil, light showers that occur during the month of October.

cake [MIN ENG] 1. Solidified drill sludge. 2. That portion of a drilling mud adhering to the walls of a borehole. 3. To form into a mass, as when ore sinters together in roasting, or coal cakes in coking.

cake of gold [MET] Gold formed into a compact mass (though not melted) by distillation of the mercury from a mercury-gold amalgam. Also known as sponge gold.

caking [ENG] Changing of a powder into a solid mass by heat, pressure, or water.

caking coal [GEOL] A type of coal which agglomerates and softens upon heating; after volatile material has been expelled at high temperature, a hard, gray cellular mass of coke remains. Also known as binding coal.

cal *See* calorie.

Cal *See* kilocalorie.

CAL [ADP] A higher-level language, developed especially for time-sharing purposes, in which a user at a remote console typewriter is directly connected to the computer and can work out problems on-line with considerable help from the computer. Derived from Conversational Algebraic Language.

calabarine *See* physostigmine.

Calabar swelling [MED] Edematous, painful, subcutaneous swelling occurring in the body of natives of Calabar and of other parts of West Africa, probably due to an allergic reaction to *Loa loa* infection.

Calabrian [GEOL] Lower Pleistocene geologic time.

calaite *See* turquoise.

calamanthia oil *See* marjoram oil.

calamine [MET] An alloy of zinc, lead, and tin. [MINERAL] *See* hemimorphite; smithsonite. [PHARM] A powder mixture of zinc oxide and ferric oxide, used in skin lotions and ointments.

Calamitales [PALEOBOT] An extinct group of reedlike plants of the subphylum Sphenopsida characterized by horizontal rhizomes and tall, upright, grooved, articulated stems.

calamus oil [MATER] A yellow essential oil that is slightly soluble in water; composed mainly of eugenol and 2,4,5-trimethoxy-1-propenyl benzene; derived by steam distillation of the roots of calamus (*Acorus calamus*); used in perfumery and medicine.

calandria [CHEM ENG] One of the tubes through which the heating fluid circulates in an evaporator.

calandria evaporator *See* short-tube vertical evaporator.

Calanoida [INV ZOO] A suborder of the crustacean order Copepoda, including the larger and more abundant of the pelagic species.

Calappidae [INV ZOO] The box crabs, a family of reptantian decapods in the subsection Oxystomata of the section Brachyura.

calathiform [BIOL] Cup-shaped, being almost hemispherical.

calaverite [MINERAL] $AuTe_2$ A yellowish or tin-white, monoclinic mineral commonly containing gold telluride and minor amounts of silver.

calc-alkalic series [PETR] Series of igneous rocks in which the weight percentage of silica is 55–61.

calcaneocuboid ligament [ANAT] The ligament that joins the calcaneus and the cuboid bones.

calcaneum [ANAT] 1. A bony projection of the metatarsus in birds. 2. *See* calcaneus.

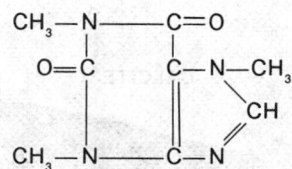

CAFFEINE

Structural formula of caffeine.

CAIMAN

The broad-nosed caiman (*Caiman latirostris*).

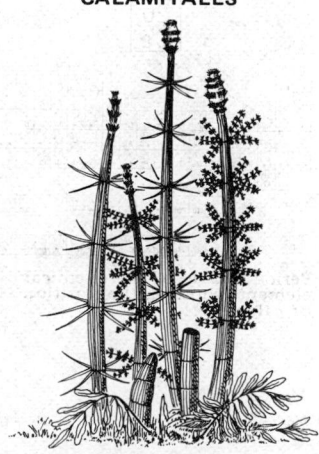

CALAMITALES

Scouring rush (*Calamites*) restored with whorls of branches at each joint of articulated stems. (*From R. C. Hussey, Historical Geology, 2d ed., McGraw-Hill, 1947*)

CALCINEA

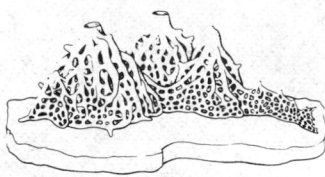

Clathrina clathrus, a calcinean sponge. *(After Minchin, 1900)*

CALCITE

├── 5 cm ──┤

Scalenohedral crystal of calcite from Joplin, Mo. *(American Museum of Natural History specimens)*

CALCIUM

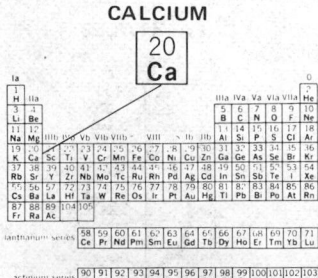

Periodic table of the chemical elements showing the position of calcium.

calcaneus [ANAT] A bone of the tarsus, forming the heel bone in man. Also known as calcaneum.

calcar [ZOO] A spur or spurlike process, especially on an appendage or digit.

calcarate [BOT] Having spurs.

Calcarea [INV ZOO] A class of the phylum Porifera, including sponges with a skeleton composed of calcium carbonate spicules.

calcarenite [PETR] A type of limestone or dolomite composed of coral or shell sand or of sand formed by erosion of older limestones, with particle size ranging from $\frac{1}{16}$ to 2 millimeters.

calcareous [SCI TECH] Resembling, containing, or composed of calcium carbonate.

calcareous algae [BOT] Algae that grow on limestone or in soil impregnated with lime.

calcareous crust *See* caliche.

calcareous duricrust *See* caliche.

calcareous ooze [GEOL] A fine-grained pelagic sediment containing undissolved sand- or silt-sized calcareous skeletal remains of small marine organisms mixed with amorphous clay-sized material.

calcareous schist [PETR] A coarse-grained metamorphic rock derived from impure calcareous sediment.

calcareous soil [GEOL] A soil containing accumulations of calcium and magnesium carbonate.

calcareous tufa *See* tufa.

calcarine fissure *See* calcarine sulcus.

calcarine sulcus [ANAT] A sulcus on the medial aspect of the occipital lobe of the cerebrum, between the lingual gyrus and the cuneus. Also known as calcarine fissure.

Calcaronea [INV ZOO] A subclass of sponges in the class Calcarea in which the larva are amphiblastulae.

calcemia *See* hypercalcemia.

calcic [SCI TECH] Derived from or containing calcium.

calcicole [BOT] Requiring soil rich in calcium carbonate for optimum growth.

calcicosis [MED] A form of pneumoconiosis caused by the inhalation of marble (calcium carbonate) dust.

calciferous [BIOL] Containing or producing calcium or calcium carbonate.

calciferous gland [INV ZOO] One of a series of glands that secrete calcium carbonate into the esophagus of certain oligochaetes.

calcification [GEOCHEM] Any process of soil formation in which the soil colloids are saturated to a high degree with exchangeable calcium, thus rendering them relatively immobile and nearly neutral in reaction. [PHYSIO] The deposit of calcareous matter within the tissues of the body.

calcifuge [ECOL] A plant that grows in an acid medium that is poor in calcareous matter.

calcilutite [PETR] 1. A dolomite or limestone formed of calcareous rock flour that is typically nonsiliceous. 2. A rock of calcium carbonate formed of grains or crystals with average diameter less than $\frac{1}{16}$ millimeter.

calcimeter [ENG] An instrument for estimating the amount of lime in soils.

calcimine [MATER] A thin paint, white or colored, made of pigment, glue, and water. Also known as kalsomine.

calcine [ENG] 1. To heat to a high temperature without fusing, as to heat unformed ceramic materials in a kiln, or to heat ores, precipitates, concentrates, or residues so that hydrates, carbonates, or other compounds are decomposed and volatile material is expelled. 2. To heat under oxidizing conditions. [MATER] Product of calcining or roasting.

Calcinea [INV ZOO] A subclass of sponges in the class Calcarea in which the larvae are parenchymulae.

calcined clay [MATER] Clay that has been heated to drive out volatile materials; a natural abrasive.

calcined coke [MATER] Coke that has been heated to expel volatile material.

calcined gypsum *See* plaster of Paris.

calcined limestone [MATER] Limestone that has been heated in a vertical-shaft kiln to drive off carbon dioxide.

calcined soda *See* soda ash.

calcining furnace [ENG] A heating device, such as a vertical-shaft kiln, that raises the temperature (but not to the melting point) of a substance such as limestone to make lime. Also known as calciner.

calcinosis [MED] Deposition of calcium salts in the skin, subcutaneous tissue, or other part of the body in certain pathologic conditions.

calciocarnotite *See* tyuyamunite.

calcioferrite [MINERAL] $Ca_2Fe_2(PO_4)OH\cdot7H_2O$ A yellow or green mineral consisting of a hydrous basic calcium iron phosphate and occurring in nodular masses.

calciovolborthite [MINERAL] $CaCu(VO_4)(OH)$ Green, yellow, or gray mineral consisting of a basic vanadate of calcium and copper. Also known as tangeite.

calcirudite [PETR] Dolomite or limestone formed of worn or broken pieces of coral or shells or of limestone fragments coarser than sand; the interstices are filled with sand, calcite, or mud, the whole bound together with a calcareous cement.

calcite [MINERAL] $CaCO_3$ One of the commonest minerals, the principal constituent of limestone; hexagonal-rhombohedral crystal structure, dimorphous with aragonite. Also known as calcspar.

calcite dolomite [PETR] A carbonate rock with a composition of 10–50% calcite and 90–50% dolomite.

calcitonin [BIOCHEM] A polypeptide, calcium-regulating hormone produced by the ultimobranchial bodies in vertebrates. Also known as thyrocalcitonin.

calcium [CHEM] A chemical element, symbol Ca, atomic number 20, atomic weight 40.08; used in metallurgy as an alloying agent for aluminum-bearing metal, as an aid in removing bismuth from lead, and as a deoxidizer in steel manufacture, and also used as a cathode coating in some types of photo tubes.

calcium-45 [NUC PHYS] A radioisotope of calcium having a mass number of 45, often used as a radioactive tracer in studying calcium metabolism in man and other organisms; half-life is 165 days. Designated Ca⁴⁵.

calcium acetate [ORG CHEM] $Ca(C_2H_3O_2)_2$ A compound that crystallizes as colorless needles that are soluble in water; formerly used as an important source of acetone and acetic acid; now used as a mordant and as a stabilizer of plastics.

calcium acrylate [ORG CHEM] $(CH_2CHCOO)_2Ca$ Free-flowing, water-soluble white powder used for soil stabilization, oil-well sealing, and ion exchange and as a binder for clay products and foundry molds.

calcium arsenate [INORG CHEM] $Ca_3(AsO_4)_2$ An arsenic compound used as an insecticide to control cotton pests.

calcium arsenite [INORG CHEM] $Ca_3(AsO_3)_2$ White granules that are soluble in water; used as an insecticide.

calcium bisulfite [INORG CHEM] $Ca(HSO_3)_2$ A white powder, used as an antiseptic and in the sulfite pulping process.

calcium-bomb process [MET] A former process to produce pellets of pure titanium metal by mixing titanium chloride with calcium in a steel bomb and heating.

calcium bromide [INORG CHEM] $CaBr_2$ A deliquescent salt in the form of colorless hexagonal crystals that are soluble in water and absolute alcohol.

calcium carbide [INORG CHEM] CaC_2 An alkaline earth carbide obtained in the pure form as transparent crystals that decompose in water; used to make acetylene gas.

calcium carbonate [INORG CHEM] $CaCO_3$ White rhombohedrons or a white powder; occurs naturally as calcite; used in paint manufacture, as a dentifrice, as an anticaking medium for table salt, and in manufacture of rubber tires.

calcium chlorate [INORG CHEM] $Ca(ClO_3)_2\cdot2H_2O$ White monoclinic crystals, decomposed by heating.

calcium chloride [INORG CHEM] $CaCl_2$ A colorless, deliquescent powder that is soluble in water and ethanol; used as an antifreeze and as an antidust agent.

calcium chromate [INORG CHEM] $CaCrO_4\cdot2H_2O$ Yellow, monoclinic crystals that are slightly soluble in water; used to make other pigments.

calcium cyanamide [INORG CHEM] $CaCN_2$ In pure form, colorless rhombohedral crystals, the commercial form being a gray material containing 55–70% $CaCN_2$; used as a fertilizer, weed killer, and defoliant.

calcium cyanide [INORG CHEM] $Ca(CN)_2$ In pure form, a white powder that gives off hydrogen cyanide in air at normal humidity; prepared commercially in impure black or gray flakes; used as an insecticide and rodenticide. Also known as black cyanide.

calcium cyclamate [ORG CHEM] $C_{12}H_{24}O_6N_2S_2Ca_2H_2O$ White crystals with a very sweet taste, soluble in water; has been used as a low-calorie sweetening agent.

calcium dihydrogen phosphate *See* calcium phosphate.

calcium fluoride [INORG CHEM] CaF_2 Colorless, cubic crystals that are slightly soluble in water and soluble in ammonium salt solutions; used in etching glass and preparing hydrofluoric acid.

calcium gluconate [ORG CHEM] $Ca(C_6H_{11}O_7)_2 \cdot H_2O$ White powder that loses water at 120°C; soluble in hot water but less soluble in cold water, insoluble in acetic acid and alcohol; used in medicine, as a foaming agent, and as a buffer in foods.

calcium hardness [CHEM] Presence of calcium ions in water, from dissolved carbonates and bicarbonates; treated in boiler water by introducing sodium phosphate.

calcium hydride [INORG CHEM] CaH_2 In pure form, white crystals that are insoluble in water; used in the production of chromium, titanium, and zirconium in the Hydromet process.

calcium hydrogen phosphate *See* calcium phosphate.

calcium hydroxide [INORG CHEM] $Ca(OH)_2$ White crystals, slightly soluble in water; used in cement, mortar, and manufacture of calcium salts. Also known as hydrated lime.

calcium hypochlorite [INORG CHEM] $Ca(OCl)_2 \cdot 4H_2O$ A white powder, used as a bleaching agent and disinfectant for swimming pools.

calcium iodide [INORG CHEM] CaI_2 A yellow, hygroscopic powder that is very soluble in water; used in photography.

calcium iodobehenate [ORG CHEM] $Ca(OOCC_{21}H_{42}I)_2$ A yellowish powder that is soluble in warm chloroform; used in feed additives.

calcium lactate [ORG CHEM] $Ca(C_3H_5O_3)_2 \cdot 5H_2O$ A salt of lactic acid in the form of white crystals that are soluble in water; used in calcium therapy and as a blood coagulant.

calcium metabolism [BIOCHEM] Biochemical and physiological processes involved in maintaining the concentration of calcium in plasma at a constant level and providing a sufficient supply of calcium for bone mineralization.

calcium naphthenate [ORG CHEM] Calcium derivative of cycloparaffin hydrocarbon (generally cyclopentane or cyclohexane base) that is a light, sticky, water-insoluble mass; used as a hardening agent in plastic compounds, in waterproofing, adhesives, wood fillers, and varnishes.

calcium nitrate [INORG CHEM] $Ca(NO_3)_2 \cdot 4H_2O$ Colorless, monoclinic crystals that are soluble in water; the anhydrous salt is very deliquescent; used as a fertilizer and in explosives. Also known as nitrocalcite.

calcium oxalate [INORG CHEM] $CaC_2O_4 \cdot H_2O$ A salt of oxalic acid in the form of white crystals that are insoluble in water.

calcium oxide [INORG CHEM] CaO A caustic white solid sparingly soluble in water; the commercial form is prepared by roasting calcium carbonate limestone in kilns until all the carbon dioxide is driven off; used as a refractory, in pulp and paper manufacture, and as a flux in manufacture of steel. Also known as burnt lime; calx; caustic lime.

calcium pantothenate [ORG CHEM] $(C_9H_{16}NO_5)_2Ca$ White slightly hygroscopic powder; soluble in water, insoluble in chloroform and ether; melts at 170–172°C; found in either the dextro or levo form or in racemic mixtures; used in nutrition and in animal feed.

calcium peroxide [INORG CHEM] CaO_2 A cream-colored powder that decomposes in water; used as an antiseptic and a detergent.

calcium phosphate [INORG CHEM] **1.** Any phosphate of calcium. **2.** Any of the following three calcium orthophosphates, all of which are white or colorless in pure form: $Ca(H_2PO_4)_2$ is used as a fertilizer, as a plastics stabilizer, and in baking powder, and is also known as acid calcium phosphate, calcium dihydrogen phosphate, monobasic calcium phosphate, monocalcium phosphate; $CaHPO_4$ is used in pharmaceuticals, animal feeds, and toothpastes, and is also known as calcium hydrogen phosphate, dibasic calcium phosphate, dicalcium orthophosphate, dicalcium phosphate; $Ca_3(PO_4)_2$ is used as a fertilizer, and is also known as tribasic calcium phosphate, tricalcium phosphate.

calcium plumbate [INORG CHEM] $Ca(PbO_3)_2$ Orange crystals that are insoluble in cold water but decompose in hot water; used as an oxidizer in the manufacture of glass and matches.

calcium plumbite [INORG CHEM] $CaPbO_2$ Colorless crystals that are slightly soluble in water.

calcium pyrophosphate [INORG CHEM] $Ca_2P_2O_7$ White, abrasive powder, used in dentifrice polishes, in metal polishes, and as a food supplement.

calcium resinate [ORG CHEM] Yellowish white, amorphous powder that is soluble in acid, insoluble in water; made by boiling rosin with calcium hydroxide and filtering, or by fusion of melted rosin with hydrated lime; used for waterproofing, leather tanning, and the manufacture of paint driers and enamels. Also known as limed rosin.

calcium silicate [INORG CHEM] Any of three silicates of calcium: tricalcium silicate, Ca_3SiO_5; dicalcium silicate, Ca_2SiO_4; calcium metasilicate, $CaSiO_3$.

calcium stearate [ORG CHEM] $Ca(C_{18}H_{35}O_2)_2$ A metallic soap produced as a white powder that is insoluble in water but slightly soluble in petroleum, benzene, and toluene.

calcium sulfate [INORG CHEM] **1.** $CaSO_4$ A white crystalline salt, insoluble in water; used in Keene's cement, in pigments, as a paper filler, and as a drying agent. **2.** Either of two hydrated forms of the salt: the dihydrate, $CaSO_4 \cdot 2H_2O$, and the hemihydrate, $CaSO_4 \cdot \frac{1}{2}H_2O$.

calcium sulfide [INORG CHEM] CaS In pure form, white cubic crystals, slightly soluble in water; used as a base for luminescent materials. Also known as hepar calcies; sulfurated lime.

calcium sulfite [INORG CHEM] $CaSO_3 \cdot 2H_2O$ A white powder that is soluble in dilute sulfurous acid; may be dehydrated at 150°C to the anhydrous salt; used in the sulfite process for the manufacture of wood pulp.

calcium tungstate [INORG CHEM] $CaWO_4$ White, tetragonal crystals, slightly soluble in water; used in manufacture of luminous paints. Also known as artificial scheelite; calcium wolframate.

calcium wolframate *See* calcium tungstate.

Calclamnidae [PALEON] A family of Paleozoic echinoderms of the order Dendrochirotida.

calclithite [PETR] Limestone with 50% or more fragments of older limestone that was redeposited after being eroded from the land.

calcrete [GEOL] A conglomerate of surficial gravel and sand cemented by calcium carbonate.

calc-silicate hornfels [PETR] A metamorphic rock with a fine grain of calc-silicate minerals.

calc-silicate marble [PETR] Marble having conspicuous calcium silicate or magnesium silicate minerals.

calcspar *See* calcite.

calcsparite *See* sparry calcite.

calculated altitude *See* computed altitude.

calculating machine *See* calculator.

calculating punch [ADP] A calculator having a card reader and a card punch.

calculator [ADP] A device that is capable of carrying out logic and arithmetic digital operations but, incapable of guiding itself, requires direction by an operator at almost every step. Also known as calculating machine.

calculus [ANAT] A small cuplike structure. [MATH] The branch of mathematics dealing with differentiation and integration and related topics. [PATH] An abnormal, solid concretion of minerals and salts formed around organic materials and found chiefly in ducts, hollow organs, and cysts.

calculus of enlargement *See* calculus of finite differences.

calculus of finite differences [MATH] A method of interpolation that makes use of formal relations between difference operators which are, in turn, defined in terms of the values of a function on a set of equally spaced points. Also known as calculus of enlargement.

calculus of tensors [MATH] The branch of mathematics dealing with the differentiation of tensors.

calculus of variations [MATH] The study of problems concerning maximizing or minimizing a given definite integral relative to the dependent variables of the integrand function.

calculus of vectors [MATH] That branch of calculus con-

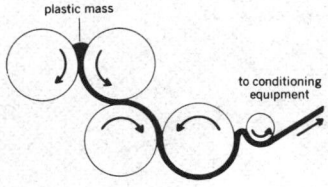

CALENDER

plastic mass

to conditioning equipment

A four-roll calender.

CALIFORNIUM

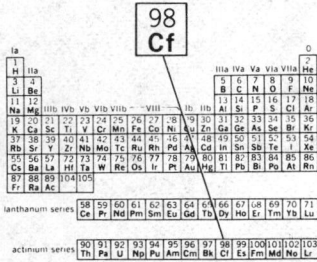

98
Cf

Periodic table of the chemical elements showing the position of californium.

CALIGOIDA

1 mm

Pandarus satyrus Dana, male, a caligoidan.

cerned with differentiation and integration of vector-valued functions.

caldera [GEOL] A more or less circular volcanic depression whose diameter is many times greater than that of the volcanic vent.

Caledonian orogeny [GEOL] Deformation of the crust of the earth by a series of diastrophic movements beginning perhaps in Early Ordovician and continuing through Silurian, extending from Great Britain through Scandinavia.

Caledonides [GEOL] A mountain system formed in Late Silurian to Early Devonian time in Scotland, Ireland, and Scandinavia.

caledonite [MINERAL] $Cu_2Pb_5(SO_4)_3CO_3(OH)_6$ A mineral occurring as green, orthorhombic crystals composed of basic copper lead sulfate; found in copper-lead deposits.

calefaction [ENG] **1.** Warming. **2.** The condition of being warmed.

calendar clock [HOROL] A clock that shows day, date, month, and phases of the moon, in addition to time of day.

calendar day [ASTRON] The period from midnight to midnight; it is 24 hours of mean solar time in length and coincides with the civil day.

calendar month [ASTRON] The month of the calendar, varying from 28 to 31 days in length.

calendar year [ASTRON] The year in the Gregorian calendar, common years having 365 days and leap years 366. Also known as civil year.

calender [ENG] **1.** To pass a material between rollers or plates to thin it into sheets or to make it smooth and glossy. **2.** The machine which performs this operation.

calendered paper [MATER] Paper that has passed through the calenders of a paper machine.

calendering [TEXT] Mechanical finishing process to produce a hard, shiny fabric.

calf [OCEANOGR] *See* calved ice. [VERT ZOO] The young of the domestic cow, elephant, rhinoceros, hippopotamus, moose, whale, and others.

calf circumference [ANTHRO] An average of three measurements taken at maximum horizontal distance around the left calf as the subject stands with weight distributed evenly on both feet.

calf diphtheria [VET MED] Disease of calves caused by species of *Sphaerophorus* which affects the mouth and pharynx and develops into pneumonia and septicemia, resulting in death.

caliber [ORD] The diameter of a projectile or the diameter of the bore of a gun or launching tube; for example, a caliber .22 cartridge has a diameter of approximately .22 inch.

calibrate [CONT SYS] To determine the settings of the control devices so that a system will operate or perform within certain limits. [SCI TECH] To determine, by measurement or comparison with a standard, the correct value of each scale reading on a meter or other device, or the correct value for each setting of a control knob.

calibrated airspeed [AERO ENG] The airspeed as read from a differential-pressure airspeed indicator which has been corrected for instrument and installation errors; equal to true airspeed for standard sea-level conditions.

calibrated altitude [NAV] Indicated altitude corrected for instrument and installation errors.

calibrating tank [ENG] A tank having known capacity used to check the volumetric accuracy of liquid delivery by positive-displacement meters. Also known as meter-proving tank.

calibration curve [ENG] A plot of calibration data, giving the correct value for each indicated reading of a meter or control dial.

calibration fire [ORD] Preparatory fire to determine the separate corrections to be applied to the individual guns or launchers of a battery in order to cause all the weapons to hit the same point, or bursts or impacts to assume a desired pattern.

calibration markers [ENG] On a radar display, electronically generated marks which provide numerical values for the navigational parameters such as bearing, distance, height, or time.

calibration point [ORD] A point at which calibration fire is directed.

calibration radio-beacon [NAV] A special radio beacon operated primarily for calibrating shipboard radio direction-finders; it transmits either continuously during scheduled hours or upon request.

calibration reference [ANALY CHEM] Any of the standards of various types that indicate whether an analytical instrument or procedure is working within prescribed limits; examples are test solutions used with pH meters, and solutions with known concentrations (standard solutions) used with spectrophotometers.

caliche [GEOL] **1.** Conglomerate of gravel, rock, soil, or alluvium cemented with sodium salts in Chilean and Peruvian nitrate deposits; contains sodium nitrate, potassium nitrate, sodium iodate, sodium chloride, sodium sulfate, and sodium borate. **2.** A thin layer of clayey soil capping auriferous veins (Peruvian usage). **3.** Whitish clay in the selvage of veins (Chilean usage). **4.** A recently discovered mineral vein. **5.** A secondary accumulation of opaque, reddish brown to buff or white calcareous material occurring in layers on or near the surface of stony soils in arid and semiarid regions of the southwestern United States; called hardpan, calcareous duricrust, and kanker in different geographic regions. Also known as calcareous crust; croute calcaire; nari; sabach; tepetate.

Caliciaceae [BOT] A family of lichens in the order Caliciales in which the disk of the apothecium is borne on a short stalk.

Caliciales [BOT] An order of lichens in the class Ascolichenes characterized by an unusual apothecium.

calico [TEXT] Any plain-weave or inexpensive figured cotton cloth.

California Current [OCEANOGR] The ocean current flowing southward along the western coast of the United States to northern Baja California.

California fog [METEOROL] Fog peculiar to the coast of California and its coastal valleys; off the coast, winds displace warm surface water, causing colder water to rise from beneath, resulting in the formation of fog; in the coastal valleys, fog is formed when moist air blown inland during the afternoon is cooled by radiation during the night.

California method [HYD] A form of frequency analysis which employs the return period, a parameter that measures the average time period between the occurrence of a quantity in hydrology and that of an equal or greater quantity, as the plotting position.

California Nebula [ASTRON] A diffuse nebula that is purely gaseous.

California polymerization [CHEM ENG] A polymerization process for converting C_3-C_4 olefins to motor fuel by utilizing a catalyst of phosphoric acid on quartz chips.

California sampler [MIN ENG] A drive sampler equipped with a piston that can be retracted mechanically to any desired point within the barrel of the sampler.

California-type dredge [MIN ENG] A single-lift dredge in which closely spaced buckets deliver to a trommel; oversize rocks are piled behind the dredge by a conveyor or stacker; the undersize are washed on gold-saving tables on the deck, and tailings discharge astern through sluices.

californite [MINERAL] $Ca_{10}Al_4(Mg,Fe)_2Si_9O_{34}(OH,F)_4$ A variety of vesuvianite that resembles jade; it is dark-, yellowish-, olive-, or grass-green and occurs in translucent to opaque compact or massive form. Also known as American jade.

californium [CHEM] A chemical element, symbol Cf, atomic number 98; all isotopes are radioactive.

Caligidae [INV ZOO] A family of fish ectoparasites belonging to the crustacean suborder Caligoida.

Caligoida [INV ZOO] A suborder of the crustacean order Copepoda, including only fish ectoparasites and characterized by a sucking mouth with styliform mandibles.

calina [METEOROL] A haze prevalent in Spain during the summer, when the air becomes filled with dust swept up from the dry ground by strong winds.

caliper [DES ENG] An instrument with two legs or jaws that can be adjusted for measuring linear dimensions, thickness, or diameter.

caliper gage [DES ENG] An instrument, such as a micrometer, of fixed size for calipering.

caliper log [MIN ENG] A graphic record showing the diameter of a drilled hole at each depth; measurements are obtained by drawing a caliper upward through the hole and recording the diameter on quadrile paper.

caliper splint [MED] A splint designed for the leg, consisting of two metal rods from a posterior-thigh band or a padded ischial ring to a metal plate attached to the sole of the shoe at the instep.

calite [MET] A heat-resistant alloy of iron, nickel, and aluminum which resists oxidation up to 2200°F (1204°C) and is practically noncorrodible under ordinary conditions of exposure.

calk See caulk.

call [ADP] To transfer control to a specified closed subroutine.

Call-A-Matic telephone set [ELECTR] An automatic dialer with Touch-Tone calling in which up to 500 numbers are stored by recording binary pulses on a magnetic tape contained in a cartridge snapped into the dialing mechanism.

call announcer [ELECTR] Device for receiving pulses from an automatic telephone office and audibly reproducing the corresponding number in words, so that it may be heard by a manual operator.

Callao painter See painter.

call circuit [ELEC] Communications circuit between switching points used by traffic forces for transmitting switching instructions.

Call Director telephone set [ELECTR] A telephone set connected to more than one telephone line and equipped with 18–30 buttons to terminate these lines.

Callendar and Barnes' continuous flow calorimeter [ENG] A calorimeter in which the heat to be measured is absorbed by water flowing through a tube at a constant rate, and the quantity of heat is determined by the rate of flow and the temperature difference between water at ends of the tube.

Callendar's equation [THERMO] **1.** An equation of state for steam whose temperature is well above the boiling point at the existing pressure, but less than the critical temperature: $(V - b) = (RT/p) - (a/T^n)$, where V is the volume, R is the gas constant, T is the temperature, p is the pressure, n equals 10/3, and a and b are constants. **2.** A very accurate equation relating temperature and resistance of platinum, according to which the temperature is the sum of a linear function of the resistance of platinum and a small correction term, which is a quadratic function of temperature.

Callendar's thermometer See platinum resistance thermometer.

call fire [ORD] Fire delivered on a specific target in response to a request from the supported unit.

Callichthyidae [VERT ZOO] A family of tropical catfishes in the suborder Siluroidei.

callideic I See bradykinin.

Callier's Q factor [GRAPHICS] A relationship between the diffuse transmission density of a photographic negative, which measures the total light transmitted, and the specular density, which measures only the light passing directly through.

calligraphy [GRAPHICS] Elegant writing or penmanship as an art and as a profession.

call in [ADP] To transfer control of a digital computer, temporarily, from a main routine to a subroutine that is inserted in the sequence of calculating operations, to fulfill an ancillary purpose.

call indicator [ELECTR] Device for receiving pulses from an automatic switching system and displaying the corresponding called number before an operator at a manual switchboard.

calling device [ELECTR] Apparatus which generates the pulses required for establishing connections in an automatic telephone switching system.

calling sequence [ADP] A specific set of instructions to set up and call a given subroutine, make available the data required by it, and tell the computer where to return after the subroutine is executed.

calling song [INV ZOO] A high-intensity insect sound which may play a role in habitat selection among certain species.

Callionymoidei [VERT ZOO] A suborder of fishes in the order Perciformes, including two families of colorful marine bottom fishes known as dragonets.

Callipallenidae [INV ZOO] A family of marine arthropods in the subphylum Pycnogonida lacking palpi and having chelifores and 10-jointed ovigers.

Calliphoridae [INV ZOO] The blow flies, a family of myodarian cyclorrhaphous dipteran insects in the subsection Calypteratae.

Callipic cycle [ASTRON] Four Metonic cycles, or 76 years.

Callisto [ASTRON] A satellite of Jupiter orbiting at a mean distance of 1,884,000 kilometers. Also known as Jupiter IV.

Callithricidae [VERT ZOO] The marmosets, a family of South American mammals in the order Primates.

call letters [COMMUN] Identifying letters, sometimes including numerals, assigned to radio and television stations by the Federal Communications Commission and other regulatory authorities throughout the world. Also known as call sign.

call number [ADP] In computer operations, a set of characters identifying a subroutine, and containing information concerning parameters to be inserted in the subroutine, or information to be used in generating the subroutine, or information related to the operands.

Callon's rule [MIN ENG] The rule stating that when a pillar is left in an inclined seam for support in a shaft or a structure on the surface, a greater width should be left on the rise side of the shaft or structure than on the dip side.

Callorhinchidae [VERT ZOO] A family of ratfishes of the chondrichthyan order Chimaeriformes.

callose [BIOCHEM] A carbohydrate component of plant cell walls; associated with sieve plates where calluses are formed. [BIOL] Having hardened protuberances, as on the skin or on leaves and stems.

Callovian [GEOL] A stage in uppermost Middle or lowermost Upper Jurassic which marks a return to clayey sedimentation.

Callow flotation cell [MIN ENG] Nonmechanical apparatus for separation of floatable solid gangue from pulverized ore, with the mixture suspended in liquid and aerated by air bubbles coming up through a porous medium, so that the lighter gangue floats away from the heavier ore.

Callow process [MIN ENG] A flotation process in which agitation is provided by the forcing of air into the pulp through the canvas-covered bottom of the cell.

Callow screen [MIN ENG] A continuous belt system formed of fine screen wire that is used to separate fine solids from coarse ones.

callus [BOT] **1.** A thickened callose deposit on sieve plates. **2.** Hard tissue that forms on a damaged plant surface. [MED] Hard, thick area on the surface of the skin.

calm [METEOROL] The absence of apparent motion of the air; in the Beaufort wind scale, smoke is observed to rise vertically, or the surface of the sea is smooth and mirrorlike; in U.S. weather observing practice, the wind has a speed under 1 mile per hour (or 1 knot).

calm belt [METEOROL] A belt of latitude in which the winds are generally light and variable; the principal calm belts are the horse latitudes (the calms of Cancer and of Capricorn) and the doldrums.

calms of Cancer [METEOROL] One of the two light, variable winds and calms which occur in the centers of the subtropical high-pressure belts over the oceans; their usual position is about latitude 30°N, the horse latitudes.

calms of Capricorn [METEOROL] One of the two light, variable winds and calms which occur in the centers of the subtropical high-pressure belts over the oceans; their usual position is about latitude 30°S, the horse latitudes.

calomel [MINERAL] Hg_2Cl_2 A colorless, white, grayish, yellowish, or brown secondary, sectile, tetragonal mineral; used as a cathartic, insecticide, and fungicide. Also known as calomelene; calomelite; horn quicksilver; mercurial horn ore.

calomel electrode [PHYS CHEM] A reference electrode of known potential consisting of mercury, mercury chloride (calomel), and potassium chloride solution; used to measure pH and electromotive force. Also known as calomel half-cell; calomel reference electrode.

calomelene See calomel.

calomel half-cell See calomel electrode.

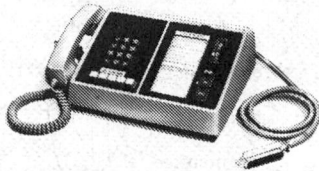

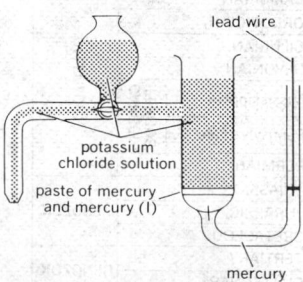

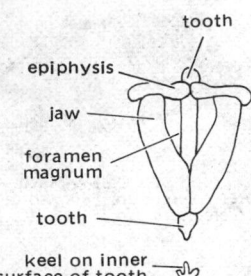

CAMARODONT DENTITION

tooth

epiphysis

jaw

foramen
magnum

tooth

keel on inner
surface of tooth

Jaw and tooth arrangement.

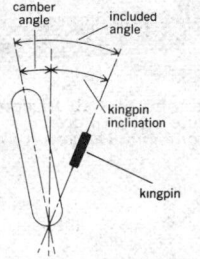

CAMBER ANGLE

camber
angle

included
angle

kingpin
inclination

kingpin

Front left wheel of an automobile
viewed from driver's position.
Angles are exaggerated.

CAMBRIAN

PRECAMBRIAN		
CAMBRIAN	PALEOZOIC	
ORDOVICIAN		
SILURIAN		
DEVONIAN		
Mississippian	CARBON-IFEROUS	
Pennsylvanian		
PERMIAN		
TRIASSIC	MESOZOIC	
JURASSIC		
CRETACEOUS		
TERTIARY	CENOZOIC	
QUATERNARY		

Position of the Cambrian
system and its relationship
to other periods.

calomelite *See* calomel.

calomel reference electrode *See* calomel electrode.

caloreceptor [PHYSIO] A cutaneous sense organ that is stimulated by heat.

calorescence [PHYS] The production of visible light by infrared radiation; the transformation is indirect, the light being produced by heat and not by any direct change of wavelength.

calorie [THERMO] Abbreviated cal; often designated c. **1.** A unit of heat energy, equal to 4.1868 joules. Also known as international table calorie (IT calorie). **2.** A unit of energy, equal to the heat required to raise the temperature of 1 gram of water from 14.5° to 15.5°C at a constant pressure of 1 standard atmosphere; equal to 4.1855 ± 0.0005 joules. Also known as fifteen-degrees calorie; gram-calorie (g-cal); small calorie. **3.** A unit of heat energy equal to 4.184 joules; used in thermochemistry. Also known as thermochemical calorie.

calorific value [ENG] Quantity of heat liberated on the complete combustion of a unit weight or unit volume of fuel.

calorifier [ENG] A device that heats fluids by circulating them over heating coils.

calorimeter [ENG] An apparatus for measuring heat quantities generated in or emitted by materials in processes such as chemical reactions, changes of state, or formation of solutions.

calorimetric test [ENG] The use of a calorimeter to determine the thermochemical characteristics of propellants and explosives; Properties normally determined are heat of combustion, heat of explosion, heat of formation, and heat of reaction.

calorimetric titration *See* thermometric titration.

calorimetry [ENG] The measurement of the quantity of heat involved in various processes, such as chemical reactions, changes of state, and formations of solutions, or in the determination of the heat capacities of substances; fundamental unit of measurement is the joule or the calorie (4.184 joules).

calorizator [FOOD ENG] A device for heating beet juice; used in beet-sugar factories.

calorize [MET] To treat by a process by which a coating of aluminum and aluminum-iron alloys is produced on iron and steel (and, less commonly, brass, copper, or nickel), which coating protects the metal from burning in temperatures up to 1800°F (982°C).

calotte [BIOL] A cap or caplike structure. [INV ZOO] **1.** The four-celled polar cap in Dicyemidae. **2.** A ciliated, retractile disc in certain bryozoan larva. **3.** A dark-colored anterior area in certain nematomorphs.

calotype [GRAPHICS] An obsolete method of photography in which paper is treated with silver iodide, silver nitrate, and acetic and gallic acids; after exposure the paper is developed in a solution of silver nitrate and gallic acid.

calthrop [INV ZOO] A sponge spicule having four axes in which the rays are equal or almost equal in length.

Calutron [NUCLEO] An electromagnetic apparatus for separating isotopes of uranium and other elements according to their masses, using the principle of the mass spectrograph.

calvarium [ANAT] A skull lacking facial parts and the lower jaw.

calved ice [OCEANOGR] A piece of ice floating in a body of water after breaking off from a mass of land ice or an iceberg. Also known as calf.

Calvé's disease *See* osteochondrosis.

calving [GEOL] The breaking off of a mass of ice from its parent glacier, iceberg, or ice shelf. Also known as ice calving. [VERT ZOO] Giving birth to a calf.

calvus *See* cumulonimbus calvus.

calx *See* calcium oxide.

calycanthemy [PL PATH] Abnormal development of calyx structures into petals or petaloid structures.

calyculate [BOT] Having bracts that imitate a second, external calyx.

calymma [INV ZOO] The outer, vacuolated protoplasmic layer of certain radiolarians.

Calymnidae [INV ZOO] A family of echinoderms in the order Holasteroida characterized by an ovoid test with a marginal fasciole.

calypter [INV ZOO] A scalelike or lobelike structure above the haltere of certain two-winged flies. Also known as alula; squama.

Calypteratae [INV ZOO] A subsection of dipteran insects in the suborder Cyclorrhapha characterized by calypters associated with the wings.

Calyptoblastea [INV ZOO] A suborder of coelenterates in the order Hydroida, including the hydroids with protective cups around the hydranths and gonozooids.

calyptra [BOT] **1.** A membranous cap or hoodlike covering, especially the remains of the archegonium over the capsule of a moss. **2.** Tissue surrounding the archegonium of a liverwort. **3.** Root cap.

calyptrate [BOT] Having a calyptra.

calyptrogen [BOT] The specialized cell layer from which a root cap originates.

Calyssozoa [INV ZOO] The single class of the bryozoan subphylum Entoprocta.

calyx [BOT] The outermost whorl of a flower; composed of sepals. [INV ZOO] A cup-shaped structure to which the arms are attached in crinoids. [ENG] A steel tube that is a guide rod and is also used to catch cuttings from a drill rod. Also known as bucket; sludge barrel; sludge bucket.

calyx drill [ENG] A rotary core drill with hardened steel shot for cutting rock. Also known as shot drill.

Calzecchi-Onesti effect [ELEC] A change in the conductivity of a loosely aggregated metallic powder caused by an applied electric field.

cam [MECH ENG] A plate or cylinder which communicates motion to a follower by means of its edge or a groove cut in its surface.

Cam *See* Camelopardalis.

cam acceleration [MECH ENG] The acceleration of the cam follower.

Camallanida [INV ZOO] An order of phasmid nematodes in the subclass Spiruria, including parasites of domestic animals.

camanchaca *See* garúa.

Camarodonta [INV ZOO] An order of Euechinoidea proposed by R. Jackson and abandoned in 1957.

camarodont dentition [INV ZOO] In echinoderms, keeled teeth meeting the epiphyses so that the foramen magnum of the jaw is closed.

Cambaridae [INV ZOO] A family of crayfishes belonging to the section Macrura in the crustacean order Decapoda.

Cambarinae [INV ZOO] A subfamily of crayfishes in the family Astacidae, including all North American species east of the Rocky Mountains.

camber [AERO ENG] The rise of the curve of an airfoil section, usually expressed as the ratio of the departure of the curve from a straight line joining the extremities of the curve to the length of this straight line. [DES ENG] Deviation from a straight line; applied to a convex, edgewise sweep or curve, or to the increase in diameter at the center of rolled materials. [GEOL] **1.** A terminal, convex shoulder of the continental shelf. **2.** A structural feature that is caused by plastic clay beneath a bed flowing toward a valley so that the bed sags downward and seems to be draped over the sides of the valley. [NAV ARCH] *See* round of beam.

camber angle [MECH ENG] The inclination from the vertical of the steerable wheels of an automobile.

camber arch [ARCH] An arch with a slightly curved interior and horizontal exterior.

camber-keeled [NAV ARCH] Property of a ship whose keel is somewhat convex.

cambium [BOT] A layer of cells between the phloem and xylem of most vascular plants that is responsible for secondary growth and for generating new cells.

Cambrian [GEOL] The lowest geologic system that contains abundant fossils of animals, and the first (earliest) geologic period of the Paleozoic era from 570 to 500 million years ago.

cambric [TEXT] **1.** Tightly woven cotton with one side calendered to resemble linen. **2.** Linen with a sheen.

cam cutter [MECH ENG] A semiautomatic or automatic machine that produces the cam contour by swinging the work as it revolves; uses a master cam in contact with a roller.

cam dwell [DES ENG] That part of a cam surface between the opening and closing acceleration sections.

camel [VERT ZOO] The common name for two species of artiodactyl mammals, the bactrian camel (*Camelus bactrianus*) and the dromedary camel (*C. dromedarius*), in the family Camelidae.

Camelidae [VERT ZOO] A family of tylopod ruminants in the superfamily Cameloidea of the order Artiodactyla, including four species of camels and llamas.

Cameloidea [VERT ZOO] A superfamily of tylopod ruminants in the order Artiodiodactyla.

Camelopardalis [ASTRON] Latin name for the Giraffe constellation of the northern hemisphere. Abbreviated Cam; Caml. Also known as Camelopardus; Giraffe.

Camelopardus *See* Camelopardalis.

camel's hair [TEXT] A fine, natural textile fiber obtained from the bactrian camel.

cam engine [MECH ENG] A piston engine in which a cam-and-roller mechanism seems to convert reciprocating motion into rotary motion.

cameo [LAP] A type of carved gemstone in which the background is cut away to leave the subject in relief.

camera [ELECTR] *See* television camera. [OPTICS] A light-tight enclosure containing an aperture (usually provided with an optical lens or system of lenses) through which the light from an object passes and forms an image, often on a light-sensitive material, inside.

camera cable [ELEC] Cable or group of wires that carries the picture from the television camera to the control room.

camera chain [COMMUN] A television camera, associated amplifiers, a monitor, and the cable needed to bring the camera output signal to the control room.

camera lucida [OPTICS] An instrument having a peculiarly shaped prism or a system of mirrors, and often a microscope, which causes a virtual image of an object to be produced on a plane surface, enabling the image's outline to be traced.

camera obscura [OPTICS] A primitive camera in which the real image of an object can be observed or traced on the wall of the enclosure opposite the aperture, rather than being recorded photographically.

camera-ready [GRAPHICS] A layout prepared for the offset printer; contains the actual material to be reproduced.

camera station [MAP] In aerial photography, the point in space occupied by the camera lens at the moment of exposure. Also known as air station.

camera study *See* memomotion study.

Camerata [PALEON] A subclass of extinct stalked echinoderms of the class Crinoidea.

camera tube [ELECTR] An electron-beam tube used in a television camera to convert an optical image into a corresponding charge-density electric image and to scan the resulting electric image in a predetermined sequence to provide an equivalent electric signal. Also known as pickup tube; television camera tube.

cam follower [MECH ENG] The output link of a cam mechanism.

Caml *See* Camelopardalis.

cam mechanism [MECH ENG] A mechanical linkage whose purpose is to produce, by means of a contoured cam surface, a prescribed motion of the output link.

Cammett table [MIN ENG] A side-moving table for concentrating ore.

camouflage [ORD] The method of concealing things or people from the enemy by trying to make them appear to be a section of the natural background.

Campanian [GEOL] European stage of Upper Cretaceous.

Campanulaceae [BOT] A family of dicotyledonous plants in the order Campanulales characterized by a style without an indusium but with well-developed collecting hairs below the stigmas, and by a well-developed latex system.

Campanulales [BOT] An order of dicotyledonous plants in the subclass Asteridae distinguished by a chiefly herbaceous habit, alternate leaves, and inferior ovary.

campanulate [BOT] Bell-shaped; applied particularly to the corolla.

cam pawl [MECH ENG] A pawl which prevents a wheel from turning in one direction by a wedging action, while permitting it to rotate in the other direction.

Campbell bridge [ELEC] **1.** A bridge designed for comparison of mutual inductances. **2.** A circuit for measuring frequencies by adjusting a mutual inductance, until the current across a detector is zero.

Campbell process [MET] An open-hearth steel manufacturing process in which ore and pig iron are used as raw materials in a tilting furnace.

Campbell's formula [ELECTROMAG] A formula which relates the propagation constant of a loaded transmission line to the propagation constant and characteristic impedance of an unloaded line and the impedance of each loading coil.

Campbell-Stokes recorder [ENG] A sunshine recorder in which the time scale is supplied by the motion of the sun and which has a spherical lens that burns an image of the sun upon a specially prepared card.

camp ceiling [BUILD] A ceiling that is flat in the center portion and sloping at the sides.

campestrian [ECOL] Of or pertaining to the northern Great Plains area.

camphane [ORG CHEM] $C_{10}H_{18}$ An alicyclic hydrocarbon; white crystals, soluble in alcohol, with melting point 158–159°C. Also known as bornylane; dihydrocamphane.

2-camphanol *See* borneol.

camphene [ORG CHEM] $C_{10}H_{16}$ A bicyclic terpene used as raw material in the synthesis of insecticides such as toxaphene and camphor. Also known as 3,3-dimethyl-2-methylenenorcamphane.

camphor [ORG CHEM] $C_{10}H_{16}O$ A bicyclic saturated terpene ketone that exists in optically active dextro and levo forms and as a racemic mixture of these forms; the dextro form is obtained from the wood and bark of the camphor tree, the levo form is found in some essential oils, and the inactive form is obtained from an Asiatic chrysanthemum or made synthetically from certain terpenes.

camphor oil [MATER] An essential oil obtained by steam distillation from the wood of the camphor tree (*Cinnamomum camphora*); used in the manufacture of camphor and safrole.

camphor tree [BOT] *Cinnamomum camphora.* A plant of the laurel family (Lauraceae) in the order Magnoliales from which camphor is extracted.

Camp-Meidell condition [STAT] For determining the distribution of a set of numbers, the guideline stating that if the distribution has only one mode, if the mode is the same as the arithmetic mean, and if the frequencies decline continuously on both sides of the mode, then more than $1 - (1/2.25t^2)$ of any distribution will fall within the closed range $\bar{X} \pm t\sigma$, where t = number of items in a set, $\bar{X}$ = average, and σ = standard deviation.

Campodeidae [INV ZOO] A family of primarily wingless insects in the order Diplura which are most numerous in the Temperate Zone of the Northern Hemisphere.

camp-on system [COMMUN] A circuit control feature whereby a user attempting to establish a telephone call and encountering a busy station will hold the connection for a preset time, to the exclusion of other callers, in case the original conversation should terminate.

campos [ECOL] The savanna of South America.

cam profile [DES ENG] The shape of the contoured cam surface by means of which motion is communicated to the follower. Also known as pitch line.

camptonite [PETR] A lamprophyre containing pyroxene, sodic hornblende, and olivine as dark constituents and labradorite as the light constituent; sodic orthoclase may be present.

campylotropous [BOT] Having the ovule symmetrical but half inverted, with the micropyle and funiculus at right angles to each other.

camshaft [MECH ENG] A rotating shaft to which a cam is attached.

can [DES ENG] A cylindrical metal vessel or container, usually with an open top or a removable cover. [NUCLEO] *See* jacket.

Canaceidae [INV ZOO] The seashore flies, a family of myodarian cyclorrhaphous dipteran insects in the subsection Acalypteratae.

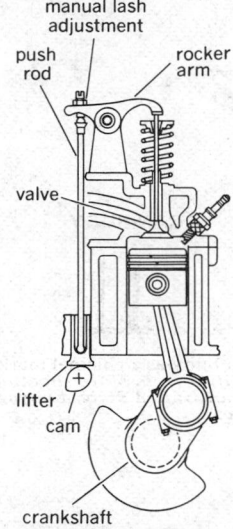

CAM MECHANISM

Cam mechanism for opening and closing valves in automotive engine. (*Texaco, Inc.*)

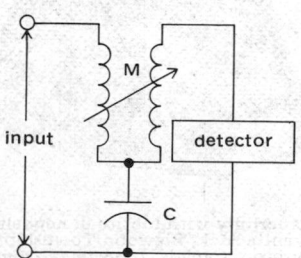

CAMPBELL BRIDGE

Circuit diagram of Campbell bridge (def. 2).

CAMPODEIDAE

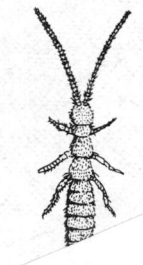

Ca
in

CAN BUOY

A can buoy as a channel marker. *(Modified from Motor Boating's chart of United States buoyage system)*

CANCRINITE

cancrinite

biotite

├─2.5 cm─┤

Cancrinite with biotite in nepheline syenite rock, Bigwood Township, Ontario, Canada. *(Specimen from Department of Geology, Bryn Mawr College)*

Canadian life zone [ECOL] The zone comprising the climate and biotic communities of the portion of the Boreal life zone exclusive of the Hudsonian and Arctic-Alpine zones.

Canadian Shield *See* Laurentian Shield.

canal [BIOL] A tubular duct or passage in bone or soft tissues. [CIV ENG] An artificial open waterway used for transportation, waterpower, or irrigation. [DES ENG] A groove on the underside of a corona. [GEOGR] A long, narrow arm of the sea extending far inland, between islands, or between islands and the mainland. [NUCLEO] A water-filled trench or conduit associated with a nuclear reactor, used for removing and sometimes storing radioactive objects taken from the reactor; the water acts as a shield against radiation.

canal boat [NAV ARCH] A long, narrow boat used on canals, with nearly vertical bow and stern giving it a large cargo capacity.

canal cell [BOT] One of the row of cells that make up the axial row within the neck of an archegonium.

canaliculate [BIOL] Having small channels, canals, or grooves.

canaliculus [HISTOL] **1.** One of the minute channels in bone radiating from a Haversian canal and connecting lacunnae with each other and with the canal. **2.** A passage between the cells of the cell cords in the liver.

canalization [ENG] Any system of distribution canals or conduits for water, gas, electricity, or steam. [MED] Surgical method of wound drainage without tubes by forming channels. [PHYSIO] The formation of new channels in tissues, such as the formation of new blood vessels through a thrombus.

canal of Schlemm [ANAT] An irregular channel at the junction of the sclera and cornea in the eye that drains aqueous humor from the anterior chamber.

canal ray [ATOM PHYS] The name given in early gaseous discharge experiments to the particles passing through a hole or canal in the cathode; the ray comprises positive ions of the gas being used in the discharge.

canal valve [ANAT] The semilunar valve in the right atrium of the heart between the orifice of the inferior vena cava and the right atrioventricular orifice. Also known as eustachian valve.

cananga oil *See* ilang-ilang oil.

canard [AERO ENG] **1.** An aerodynamic vehicle in which horizontal surfaces used for trim and control are forward of the wing or main lifting surface. **2.** The horizontal trim and control surfaces in such an arrangement.

Canary Current [OCEANOGR] The prevailing southward flow of water along the northwestern coast of Africa.

canary-pox virus [VIROL] An avian poxvirus that causes canary pox, a disease closely related to fowl pox.

Canastotan [GEOL] Lower Upper Silurian geologic time.

canavanine [BIOCHEM] $C_5H_{12}O_3N_4$ An amino acid found in the jack bean.

can buoy [NAV] Floating cylindrical unlighted buoy with a flat top, constructed of metal; depending on position, it may be painted black with or without an odd number on it, or painted with horizontal stripes.

Canc *See* Cancer.

cancellate [BIOL] Lattice-shaped. Also known as clathrate.

cancellation circuit [ELECTR] A circuit used in providing moving-target indication on a plan position indicator scope; cancels constant-amplitude fixed-target pulses by subtraction of successive pulse trains.

cancellation law [MATH] A rule which allows formal division by common factors in equal products, even in systems which have no division, as integral domains; $ab = ac$ implies $b = c$.

cancellous [BIOL] Having a reticular or spongy structure.

cancellous bone [HISTOL] A form of bone near the ends of long bones having a cancellous matrix composed of rods, plates, or tubes; spaces are filled with marrow.

cancer [MED] Any malignant neoplasm, including carcinoma and sarcoma.

Cancer [ASTRON] A constellation with right ascension 9 hours, declination 20°N. Abbreviated Canc. Also known as Crab.

cancer eye [VET MED] A malignant epithelioma of the eye of cattle, common in regions of intense sunlight.

cancerphobia [PSYCH] An abnormal fear of acquiring cancer.

cancrinite [MINERAL] $Na_3CaAl_3Si_3O_{12}CO_3(OH)_2$ A feldspathoid tectosilicate occurring in hexagonal crystals in nepheline syenites, usually in compact or disseminated masses.

cancroid [MED] A squamous-cell carcinoma.

candela [OPTICS] A unit of luminous intensity, defined as $\frac{1}{60}$ of the luminous intensity per square centimeter of a blackbody radiator operating at the temperature of freezing platinum. Formerly known as candle. Also known as new candle.

candelilla wax [MATER] A wax obtained from the wax-coated stems of candelilla shrubs, especially *Euphorbia antisyphilitica;* used for varnishes and furniture and shoe polishes.

Candida [MYCOL] A genus of yeastlike, pathogenic imperfect fungi that produce very small mycelia.

Candida albicans [MYCOL] The fungus organism that causes thrush and other types of moniliasis. Also known as *Monilia albicans.*

candidiasis [MED] A fungus infection of the skin, lungs, mucous membranes, and viscera of man caused by a species of *Candida,* usually *C. albicans.* Also known as moniliasis.

candite *See* ceylonite.

candle *See* candela.

Candlemas crack *See* Candlemas Eve winds.

Candlemas Eve winds [METEOROL] Heavy winds often occurring in Great Britain in February or March (Candlemas is February 2). Also known as Candlemas crack.

candlenut oil *See* lumbang oil.

candlepower [OPTICS] Luminous intensity expressed in candelas. Abbreviated cp.

cane [BOT] **1.** A hollow, usually slender, jointed stem, such as in sugarcane or the bamboo grasses. **2.** A stem growing directly from the base of the plant, as in most Rosaceae, such as blackberry and roses.

cane blight [PL PATH] A fungus disease affecting the canes of several bush fruits, such as currants and raspberries; caused by several species of fungi.

cane molasses [FOOD ENG] The heavy, residual syrup after the crystallization of cane syrup.

canescent [BOT] Having a grayish epidermal covering of short hairs.

cane sugar [ORG CHEM] Sucrose derived from sugarcane.

Canes Venatici [ASTRON] A northern constellation with right ascension 13 hours, declination 40°N, between Ursa Major and Boötes. Abbreviated CVn. Also known as Hunting Dogs.

canfieldite [MINERAL] Ag_8SnS_6 A black mineral of the argyrodite series consisting of silver thiostannate, with a specific gravity of 6.28; found in Germany and Bolivia.

can hoisting system [MIN ENG] A hoisting method used in shallow lead and zinc mines in which cans are loaded below and hoisted to the surface where they are capsized and the load discharged; the operation is controlled at the top of the shaft.

Canidae [VERT ZOO] A family of carnivorous mammals in the superfamily Canoidea, including dogs and their allies.

canine [ANAT] A conical tooth, such as one located between the lateral incisor and first premolar in man and many other mammals. Also known as cuspid. [VERT ZOO] Pertaining or related to dogs or to the family Canidae.

canine distemper [VET MED] A pantropic virus disease occurring among animals of the family Canidae.

Canis [VERT ZOO] The type genus of the dog family (Canidae), including dogs, wolves, and jackals.

Canis familiaris [VERT ZOO] The scientific name for all breeds of domestic dogs.

Canis Major [ASTRON] A constellation with right ascension 7 hours, declination 20°S. Abbreviated CMa. Also known as Greater Dog.

α Canis Majoris *See* Sirius.

ε Canis Majoris *See* Adhara.

Canis Minor [ASTRON] A constellation with right ascension 8 hours, declination 5°N. Abbreviated CMi. Also known as Lesser Dog.

α **Canis Minoris**   *See* Procyon.

canister   [ORD] A special short-range, antipersonnel projectile designed to be fired from rifled guns.

canker   [PL PATH] An area of necrosis on a woody stem resulting in shrinkage and cracking followed by the formation of callus tissue around the area, ultimately killing the stem. [VET MED] A localized chronic inflammation of the ear in cats, dogs, foxes, ferrets, and others caused by the mite *Otodectes cynotis*.

canker sore   [MED] Small ulceration of the mucous membrane of the mouth, sometimes caused by a food allergy.

canker stain   [PL PATH] A fungus disease of plane trees caused by *Endoconidiophora fimbriata platani* and characterized by bluish-black or reddish-brown discolorations beneath blackened cankers on the trunk and sometimes the branches.

cankerworm   [INV ZOO] Any of several lepidopteran insect larvae in the family Geometridae which cause severe plant damage by feeding on buds and foliage.

Cannabaceae   [BOT] A family of dicotyledonous herbs in the order Urticales, including Indian hemp (*Cannabis sativa*) and characterized by erect anthers, two styles or style branches, and the lack of milky juice.

cannabidiol   [ORG CHEM] $C_{21}H_{28}(OH)_2$ A constituent of cannabis which, upon isomerization to a tetrahydrocannabinol, has some of the physiologic activity of marijuana.

cannabinol   [ORG CHEM] $C_{21}H_{26}O_2$ A physiologically inactive phenol formed by spontaneous dehydrogenation of tetrahydrocannabinol from cannabis.

Cannabis   [BOT] A genus of tall annual herbs in the family Cannabaceae having erect stems, leaves with three to seven elongate leaflets, and pistillate flowers in spikes along the stem.

cannabism   [MED] Poisoning resulting from excessive or habitual use of cannabis.

Cannabis sativa   [BOT] Indian hemp, the source of a commercial fiber, hashish, and marijuana.

Cannaceae   [BOT] A family of monocotyledonous plants in the order Zingiberales characterized by one functional stamen, a single functional pollen sac in the stamen, mucilage canals in the stem, and numerous ovules in each of the one to three locules.

canned motor   [MECH ENG] A motor enclosed within a casing along with the driven element (that is, a pump) so that the motor bearings are lubricated by the same liquid that is being pumped.

canned pump   [MECH ENG] A watertight pump that can operate under water.

cannel coal   [GEOL] A fine-textured, highly volatile bituminous coal distinguished by a greasy luster and blocky, conchoidal fracture; burns with a steady luminous flame. Also known as cannelite.

cannelite   *See* cannel coal.

canneloid   [GEOL] 1. Coal that resembles cannel coal. 2. Coal intermediate between bituminous and cannel. 3. Durain laminal in banded coal. 4. Cannel coal of anthracite or semianthracite rank.

cannel shale   [GEOL] A black shale formed by the accumulation of an aquatic ooze rich in bituminous organic matter in association with inorganic materials such as silt and clay.

cannelure   [ORD] 1. A groove in a bullet which contains a lubricant, or into which the cartridge is crimped. 2. A groove in a cartridge case providing a purchase for the extractor. 3. A ringlike groove for locking the jacket of an armor-piercing bullet to the core. 4. A ringlike groove in the rotating band of a gun projectile to lessen the resistance offered to the gun rifling and to prevent fringing grooves.

canning   [FOOD ENG] Packing and preserving of food in cans or jars which are subjected to sterilizing temperatures. [NUCLEO] Placing a jacket around a slug of uranium before inserting the slug in a nuclear reactor.

Cannizzaro reaction   [ORG CHEM] The reaction in which aldehydes that do not have a hydrogen attached to the carbon adjacent to the carbonyl group, upon encountering strong alkali, readily form an alcohol and an acid salt.

cannon   [ORD] A complete assembly which consists of a tube

and a breech mechanism with a firing mechanism or base cap and which is a component of a gun, howitzer, or mortar; may include muzzle appendages; the term is generally limited to calibers greater than 1 inch (2.5 centimeters).

cannonball   [ORD] A missile that is spheroidal in shape and fired from a cannon.

cannon cradle   [ORD] A portion of a cannon assembly which is designed to support the cannon and allows the cannon to recoil and counter-recoil.

cannon pinion   [HOROL] A tube that holds the minute hand at one end and meshes with the minute wheel of a timepiece on the other end.

cannon primer   [ORD] Primer used with separate-loading ammunition.

cannula   [MED] A small tube that can be inserted into a body cavity, duct, or vessel.

cannular combustion chambers   [AERO ENG] The separate combustion chambers in an aircraft gas turbine. Also known as can-type combustors.

Canoidea   [VERT ZOO] A superfamily belonging to the mammalian order Carnivora, including all dogs and doglike species such as seals, bears, and weasels.

canonical correlation   [STAT] The maximum correlation between linear functions of two sets of random variables when specific restrictions are imposed upon the coefficients of the linear functions of the two sets.

canonical ensemble   [STAT MECH] A hypothetical collection of systems of particles used to describe an actual individual system which is in thermal contact with a heat reservoir but is not allowed to exchange particles with its environment.

canonical equations of motion   *See* Hamilton's equations of motion.

canonically conjugate variable   [MECH] A generalized coordinate and its conjugate momentum.

canonical momentum   *See* conjugate momentum.

canonical time unit   [ASTRON] For geocentric orbits, the time required by a hypothetical satellite to move one radian in a circular orbit of the earth's equatorial radius, that is, 13.447052 minutes.

canonical transformation   [MATH] Any function which has a standard form, depending on the context. [MECH] A transformation which occurs among the coordinates and momenta describing the state of a classical dynamical system and which leaves the form of Hamilton's equations of motion unchanged. Also known as contact transformation.

Canopus   [ASTRON] A star that is 180 light-years from the sun; spectral classification F0Ia. Also known as α Carinae.

canopy   [AERO ENG] 1. The umbrellalike part of a parachute which acts as its main supporting surface. 2. The overhead, transparent enclosure of an aircraft cockpit.

cant   [ORD] The leaning or tilt of an object; especially, the sidewise tilt of a gun.

cantala   [BOT] A fiber produced from agave (*Agave cantala*) leaves; used to make twine. Also known as Cebu maguey; maguey; manila maguey.

cantaloupe   [BOT] The fruit (pepo) of *Cucumis malo*, a small, distinctly netted, round to oval muskmelon of the family Cucurbitaceae in the order Violales.

cant body   [NAV ARCH] The part of a ship's body where the frames are set at an angle to the keel to form the stern and bow.

Canterbury northwester   [METEOROL] A strong northwest foehn descending the New Zealand Alps onto the Canterbury plains of South Island, New Zealand.

cant file   [DES ENG] A fine-tapered file with a triangular cross section, used for sharpening saw teeth.

canthariasis   [MED] Infection or disease caused by coleopteran insects or their larvae.

Cantharidae   [INV ZOO] The soldier beetles, a family of coleopteran insects in the superfamily Cantharoidea.

cantharides camphor   *See* cantharidin.

cantharidin   [ORG CHEM] $C_{10}H_{12}O_4$ Colorless crystals that melt at 218°C; slightly soluble in acetone, chloroform, alcohol, and water; used in veterinary medicine. Also known as cantharides camphor.

Cantharoidea   [INV ZOO] A superfamily of coleopteran insects in the suborder Polyphaga.

CANNACEAE

Flowers of a cultivated *Canna* of the order Zingiberales. These plants have been so extensively hybridized that most of them can no longer be identified with any of the wild species. *(Photograph by Luoma Photos, from National Audubon Society)*

CANTILEVER BRIDGE

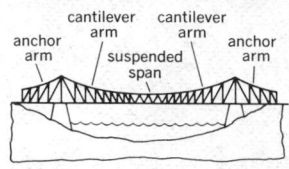

Drawing of a cantilever bridge.

CANTILEVER RETAINING WALL

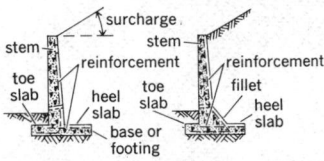

Cantilever walls, showing the parts.

cant hook [DES ENG] A lever with a hooklike attachment at one end, used in lumbering.

canthus [ANAT] Either of the two angles formed by the junction of the eyelids, designated outer or lateral, and inner or medial.

cantilever [ENG] A beam or member securely fixed at one end and hanging free at the other end.

cantilever arch [ARCH] An arch supported by flat projections on opposing walls.

cantilever bridge [CIV ENG] A three-span bridge consisting of two outer spans which are anchored to the shore and a middle span made up of two cantilever arms extending from the outer spans to a juncture at the center; sometimes a suspended span is interposed between the ends of the cantilever arms.

cantilever retaining wall [CIV ENG] A type of wall formed of three cantilever beams: stem, toe projection, and heel projection.

cantilever spring [MECH ENG] A flat spring supported at one end and holding a load at or near the other end.

cantilever vibration [MECH] Transverse oscillatory motion of a body fixed at one end.

Cantor diagonal process [MATH] A technique of proving statements about infinite sequences, each of whose terms is an infinite sequence by operation on the nth term of the nth sequence for each n; used to prove the uncountability of the real numbers.

Cantor function [MATH] A real-valued nondecreasing continuous function defined on the closed interval [0,1] which maps the Cantor ternary set onto the interval [0,1].

Cantor ternary set [MATH] A perfect, uncountable, totally disconnected subset of the real numbers having Lebesgue measure zero; it consists of all numbers between 0 and 1 (inclusive) with ternary representations containing no ones.

Cantor theorem [MATH] A theorem that there is no one-to-one correspondence between a set and the collection of its subsets.

cant strip [BUILD] **1.** A strip placed along the angle between a wall and a roof so that the roofing will not bend sharply. **2.** A strip placed under the edge of the lowest row of tiles on a roof to give them the same slope as the other tiles.

can-type combustors See cannular combustion chambers.

canvas [TEXT] A firm, closely woven fabric of plain weave made principally from hemp, but also from flax, jute, cotton, or a blend of fibers.

canvas duck [TEXT] A lightweight cotton or linen cloth; the term is occasionally used for heavier canvases as well.

canyon [GEOGR] A chasm, gorge, or ravine cut in the surface of the earth by running water; the sides are steep and form cliffs.

canyon bench [GEOL] A steplike level of hard strata in the walls of deep valleys in regions of horizontal strata.

canyon fill [GEOL] Loose, unconsolidated material which fills a canyon to a depth of 50 feet (15 meters) or more during periods between great floods.

canyon wind [METEOROL] Also known as gorge wind. **1.** The mountain wind of a canyon; that is, the nighttime down-canyon flow of air caused by cooling at the canyon walls. **2.** Any wind modified by being forced to flow through a canyon or gorge; its speed may be increased as a jet-effect wind, and its direction is rigidly controlled.

caoutchouc [MATER] Formerly, crude rubber which had been cured over a fire into a solid, dark mass for shipment.

cap [ENG] A detonating or blasting cap. [MIN ENG] **1.** A piece of timber placed on top of a prop or post in a mine. **2.** The horizontal section of a set of timber that is used as a support in a mine roadway.

Cap See Capricornus.

capacitance [ELEC] The ratio of the charge on one of the conductors of a capacitor (there being an equal and opposite charge on the other conductor) to the potential difference between the conductors. Designated by C. Formerly known as capacity.

capacitance altimeter [ENG] An absolute altimeter which determines height of an aircraft aboveground by measuring the variations in capacitance between two conductors on the aircraft when the ground is near enough to act as a third conductor.

capacitance box [ELEC] An assembly of capacitors and switches which permits adjustment of the capacitance existing at the terminals in nominally uniform steps, from a minimum value near zero to the maximum which exists when all the capacitors are connected in parallel.

capacitance bridge [ELEC] A bridge for comparing two capacitances, such as a Schering bridge.

capacitance level indicator [ENG] A level indicator in which the material being monitored serves as the dielectric of a capacitor formed by a metal tank and an insulated electrode mounted vertically in the tank.

capacitance meter [ENG] An instrument used to measure capacitance values of capacitors or of circuits containing capacitance.

capacitance-operated intrusion detector [ENG] A boundary alarm system in which the approach of an intruder to an antenna wire encircling the protected area a few feet above ground changes the antenna-ground capacitance and sets off the alarm.

capacitance relay [ELECTR] An electronic relay that responds to a small change in capacitance, such as that created by bringing a hand near a pickup wire or plate.

capacitance standard See standard capacitor.

capacitive coupling [ELEC] Use of a capacitor to transfer energy from one circuit to another.

capacitive diaphragm [ELECTROMAG] A resonant window used in a waveguide to provide the equivalent of capacitive reactance at the frequency being transmitted.

capacitive divider [ELEC] Two or more capacitors placed in series across a source, making available a portion of the source voltage across each capacitor; the voltage across each capacitor will be inversely proportional to its capacitance.

capacitive feedback [ELECTR] Process of returning part of the energy in the plate (or output) circuit of a vacuum tube (or other device) to the grid (or input) circuit by means of a capacitance common to both circuits.

capacitive load [ELECTROMAG] A load in which the capacitive reactance exceeds the inductive reactance; the load draws a leading current.

capacitive post [ELECTROMAG] Metal post or screw extending across a waveguide at right angles to the E field, to provide capacitive susceptance in parallel with the waveguide for tuning or matching purposes.

capacitive pressure transducer [ENG] A measurement device in which variations in pressure upon a capacitive element proportionally change the element's capacitive rating and thus the strength of the measured electric signal from the device.

capacitive reactance [ELECTROMAG] Reactance due to the capacitance of a capacitor or circuit, equal to the inverse of the product of the capacitance and the angular frequency.

capacitive tuning [ELECTR] Tuning involving use of a variable capacitor.

capacitive window [ELECTROMAG] Conducting diaphragm extending into a waveguide from one or both sidewalls, producing the effect of a capacitive susceptance in parallel with the waveguide.

capacitor [ELEC] A device which consists essentially of two conductors (such as parallel metal plates) insulated from each other by a dielectric and which introduces capacitance into a circuit, stores electrical energy, blocks the flow of direct current, and permits the flow of alternating current to a degree dependent on the capacitor's capacitance and the current frequency. Designated by C. Also known as condenser; electrical condenser.

capacitor antenna [ELECTROMAG] Antenna consisting of two conductors or systems of conductors, the essential characteristic of which is its capacitance. Also known as condenser antenna.

capacitor bank [ELEC] A number of capacitors connected in series or in parallel.

capacitor box [ELECTR] A box-shaped structure in which a capacitor is submerged in a heat-absorbing medium, usually water. Also known as condenser box.

capacitor color code [ELEC] A method of marking the value on a capacitor by means of dots or bands of colors as specified in the Electronic Industry Association color code.

capacitor hydrophone [ENG ACOUS] A capacitor microphone that responds to waterborne sound waves.

capacitor-input filter [ELECTR] A power-supply filter in which a shunt capacitor is the first element after the rectifier.

capacitor loudspeaker *See* electrostatic loudspeaker.

capacitor microphone [ENG ACOUS] A microphone consisting essentially of a flexible metal diaphragm and a rigid metal plate that together form a two-plate air capacitor; sound waves set the diaphragm in vibration, producing capacitance variations that are converted into audio-frequency signals by a suitable amplifier circuit. Also known as condenser microphone; electrostatic microphone.

capacitor motor [ELEC] **1.** A single-phase induction motor having a main winding connected directly to a source of alternating-current power and an auxiliary winding connected in series with a capacitor to the source of ac power. **2.** *See* capacitor-start motor.

capacitor pickup [ENG ACOUS] A phonograph pickup in which movements of the stylus in a record groove cause variations in the capacitance of the pickup.

capacitor-resistor unit *See* rescap.

capacitor-start motor [ELEC] A capacitor motor in which the capacitor is in the circuit only during the starting period; the capacitor and its auxiliary winding are disconnected automatically by a centrifugal switch or other device when the motor reaches a predetermined speed. Also known as capacitor motor.

capacity [ADP] *See* storage capacity. [ELEC] *See* capacitance. [SCI TECH] Volume, especially in reference to merchandise or containers thereof.

capacity cell [ELEC] **1.** Capacitance-type device used to measure the dielectric constants of gases, liquids, or solids. **2.** Capacitance-type device used to monitor certain composition changes in flowing streams.

capacity correction [ENG] The correction applied to a mercury barometer with a nonadjustable cistern in order to compensate for the change in the level of the cistern as the atmospheric pressure changes.

capacity factor *See* plant factor.

capacity of the wind [GEOL] The total weight of airborne particles (soil and rock) of given size, shape, and specific gravity, which can be carried in 1 cubic mile of wind blowing at a given speed.

cap cloud [METEOROL] An approximately stationary cloud, or standing cloud, on or hovering above an isolated mountain peak; formed by the cooling and condensation of humid air forced up over the peak. Also known as cloud cap.

cap crimper [ENG] A tool resembling a pliers that is used to press the open end of a blasting cap onto the safety fuse before placing the cap in the primer.

cape [GEOGR] A prominent point of land jutting into a body of water. Also known as head; headland; mull; naze; ness; point; promontory.

cape chisel [DES ENG] A chisel that tapers to a flat, narrow cutting end; used to cut flat grooves.

cape doctor [METEOROL] The strong southeast wind which blows on the South African coast.

cape foot [MECH] A unit of length equal to 1.033 feet or to 0.3148584 meter.

Cape Horn Current [OCEANOGR] That part of the west wind drift flowing eastward in the immediate vicinity of Cape Horn, and then curving northeastward to continue as the Falkland Current.

Capella [ASTRON] A star that is 45 light-years from the sun; spectral classification G0IIIp. Also known as α Aurigae.

Capell fan [MIN ENG] A centrifugal type of mine shaft fan.

caper [FOOD ENG] The buds and berries of the caper plant (*Capparis spinosa*) pickled and used as a condiment.

capillarity [FL MECH] The action by which the surface of a liquid where it contacts a solid is elevated or depressed, because of the relative attraction of the molecules of the liquid for each other and for those of the solid.

capillarity correction [ENG] As applied to a mercury barometer, that part of the instrument correction which is required by the shape of the meniscus of the mercury.

capillaroscope [MED] A microscope used for diagnostic examination of the cutaneous capillaries, as in the nail beds and conjunctiva.

capillary [ANAT] The smallest vessel of both the circulatory and lymphatic systems; the walls are composed of a single cell layer.

capillary angioma *See* hemangioma.

capillary attraction [FL MECH] The force of adhesion existing between a solid and a liquid in capillarity.

capillary bed [ANAT] The capillaries, collectively, of a given area or organ.

capillary collector [ENG] An instrument for collecting liquid water from the atmosphere; the collecting head is fabricated of a porous material having a pore size of the order of 30 micrometers; the pressure difference across the water-air interface prevents air from entering the capillary system while allowing free flow of water.

capillary condensation [PHYS CHEM] Condensation of an adsorbed vapor within the pores of the adsorbate.

capillary control [PETRO ENG] Discarded theory of reservoir oil flow to a well hole through capillary pores; it attributed flow resistance to gas bubbles within the capillaries.

capillary depression [FL MECH] The depression of the meniscus of a liquid contained in a tube where the liquid does not wet the walls of the container, as in a mercury barometer; the meniscus has a convex shape, resulting in a depression.

capillary drying [ENG] Progressive removal of moisture from a porous solid mass by surface evaporation followed by capillary movement of more moisture to the drying surface from the moist inner region, until the surface and core stabilize at the same moisture concentration.

capillary ejecta *See* Pele's hair.

capillary electrometer [ENG] An electrometer designed to measure a small potential difference between mercury and an electrolytic solution in a capillary tube by measuring the effect of this potential difference on the surface tension between the liquids.

capillary equilibrium method [PETRO ENG] Test method to predict oil and gas flow through an oil reservoir core by dethrottling the flow to hold capillary flow in equilibrium between oil and gas within the reservoir.

capillary fringe [HYD] The lower subdivision of the zone of aeration that overlies the zone of saturation and in which the pressure of water in the interstices is lower than atmospheric.

capillary migration [HYD] Movement of water produced by the force of molecular attraction between rock material and the water.

capillary pressure [PHYSIO] Pressure exerted by blood against capillary walls.

capillary pyrites *See* millerite.

capillary rise [FL MECH] The rise of a liquid within small-bore tubes or within fibers with bores when the force of intermolecular capillary attraction is greater than that of gravitational force.

capillary tube [ENG] A tube sufficiently fine so that capillary attraction of a liquid into the tube is significant.

capillary viscometer [ENG] A long, narrow tube used to measure the laminar flow of fluids.

capillary water [HYD] Soil water held by capillarity as a continuous film around soil particles and in interstices between particles above the phreactic line.

capillary wave [FL MECH] **1.** A wave occurring at the interface between two fluids, such as the interface between air and water on oceans and lakes, in which the principal restoring force is controlled by surface tension. **2.** A water wave of less than 1.7 centimeters. Also known as capillary ripple; ripple.

capillitium [MYCOL] A network of threadlike tubes or filaments in which spores are embedded within sporangia of certain fungi, such as the slime molds.

capital [ARCH] The topmost part of a column.

capital amount factor [IND ENG] Any of 20 common compound interest formulas used to calculate the equivalent uniform annual cost of all cash flows.

capital budgeting [IND ENG] Planning the most effective use of resources to obtain the highest possible level of sustained profits.

capital expenditure [IND ENG] Money spent for long-term

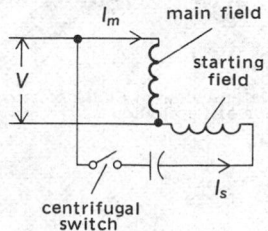

CAPACITOR MOTOR

Winding connections of the starting winding connected to the supply through a capacitor. I_m = main winding; I_s = starting winding; V = common voltage.

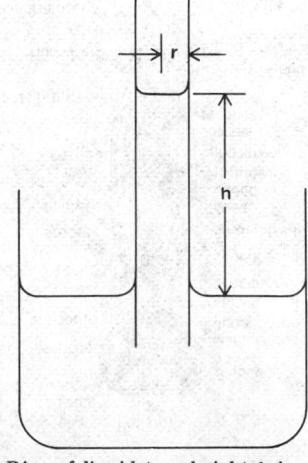

CAPILLARY TUBE

Rise of liquid to a height h in a capillary tube whose radius is r.

CAPITELLIDAE

Left lateral view of *Capitomastus*, a genus of Capitellidae.

CAPRICORNUS

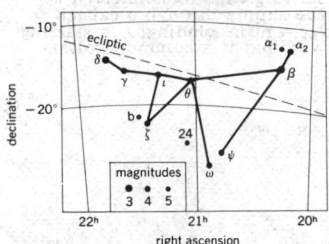

Line pattern of constellation Capricornus. Grid lines represent coordinates of the sky. Apparent brightness, or magnitude, of stars is shown by size of dots, which are graded by appropriate numbers as indicated.

CAPRELLIDEA

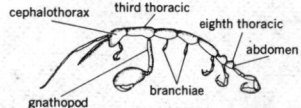

Caprella grandimana, a caprellid.

CAPSALOIDEA

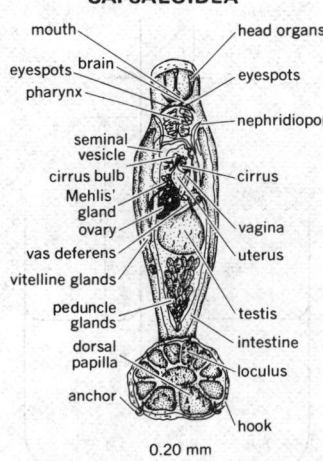

Heterocotyle acetobactis Hargis, from the spotted eagle ray, ventral view.

additions or improvements and charged to a capital assets account.

capital ship [NAV ARCH] A surface warship classified as major, for example, a battleship or aircraft carrier.

capitate [BIOL] Enlarged and swollen at the tip. [BOT] Forming a head, as certain flowers of the Compositae.

capitellate [BOT] **1.** Having a small knoblike termination. **2.** Grouped to form a capitulum.

Capitellidae [INV ZOO] A family of mud-swallowing annelid worms, sometimes called bloodworms, belonging to the Sedentaria.

capitellum [ANAT] A small head or rounded process of a bone.

Capitonidae [VERT ZOO] The barbets, a family of pantropical birds in the order Piciformes.

capitulum [BIOL] A rounded, knoblike, usually terminal protuberance on a structure. [BOT] One of the rounded cells on the manubrium in the antheridia of lichens belonging to the Caliciales.

cap lamp [MIN ENG] The lamp a miner wears on his safety hat or cap for illumination.

capon [AGR] A castrated male chicken.

Caponidae [INV ZOO] A family of arachnid arthropods in the order Araneida characterized by having tracheae instead of book lungs.

capon unit [BIOL] A unit for the standardization of androgens.

Capparaceae [BOT] A family of dicotyledonous herbs, shrubs, and trees in the order Capparales characterized by parietal placentation; hypogynous, mostly regular flowers; four to many stamens; and simple to trifoliate or palmately compound leaves.

Capparales [BOT] An order of dicotyledonous plants in the subclass Dilleniidae.

capped column [HYD] A form of ice crystal consisting of a hexagonal column with plate or stellar crystals (so-called caps) at its ends and sometimes at intermediate positions; the caps are perpendicular to the column.

capped fuse [ENG] A length of safety fuse with the cap or detonator crimped on before it is taken to the place of use.

capped steel [MET] Partially deoxidized steel cast in an open-top mold, which is capped to solidify the top metal and enforce internal pressure, resulting in a surface condition similar to that of rimming steel.

cappelenite [MINERAL] $(Ba,Ca,Na)(Y,La)_6B_6Si_{13}(O,OH)_{27}$ A greenish-brown hexagonal mineral consisting of a rare yttrium-barium borosilicate occurring in crystals.

cap piece [MIN ENG] A piece of wood fitted over a straight post or timber to provide more bearing surface.

capping [ENG] Preparation of a capped fuse. [GEOL] **1.** Consolidated barren rock overlying a mineral or ore deposit. **2.** *See* gossan. [MET] Separation of a compact into two or more portions by making diagonal cracks which originate near the edges of the punch faces. [MIN ENG] The attachment at the end of a winding rope. [PETRO ENG] **1.** The process of sealing or covering a borehole such as an oil or gas well. **2.** The material or device used to seal or cover a borehole.

capraldehyde *See* decyl aldehyde.

caprate [ORG CHEM] Any of the salts of capric acid, containing the group $C_9H_{19}COO-$.

Caprellidae [INV ZOO] The skeleton shrimps, a family of slender, cylindrical amphipod crustaceans in the suborder Caprellidea.

Caprellidea [INV ZOO] A suborder of marine and brackish-water animals of the crustacean order Amphipoda.

capric acid [ORG CHEM] $CH_3(CH_2)_8COOH$ A fatty acid found in oils and animal fats. Also known as *n*-decanoic acid; decatoic acid; *n*-decoic acid; decyclic acid; octylacetic acid.

capric aldehyde *See* decyl aldehyde.

capric anhydride [ORG CHEM] $(CH_3(CH_2)_8CO)_2O$ White crystals that are insoluble in water; used as a chemical intermediate.

Capricornus [ASTRON] A constellation with right ascension 21 hours, declination 20°S. Abbreviated Cap. Also known as Sea Goat.

Caprifoliaceae [BOT] A family of dicotyledonous, mostly

woody plants in the order Dipsacales, including elderberry and honeysuckle; characterized by distinct filaments and anthers, typically five stamens and five corolla lobes, more than one ovule per locule, and well-developed endosperm.

caprilydene *See* octyne.

Caprimulgidae [VERT ZOO] A family of birds in the order Caprimulgiformes, including the nightjars, or goatsuckers.

Caprimulgiformes [VERT ZOO] An order of nocturnal and crepuscular birds, including nightjars, potoos, and frogmouths.

capristor *See* rescap.

caproamide [ORG CHEM] $CH_3(CH_2)_4CONH_2$ An amide, melting point 100–101°C; used as a chemical intermediate.

cap rock [GEOL] **1.** An overlying, generally impervious layer or stratum of rock that overlies an oil- or gas-bearing rock. **2.** Barren vein matter, or a pinch in a vein, supposed to overlie ore. **3.** A hard layer of rock, usually sandstone, a short distance above a coal seam. **4.** An impervious body of anhydrite and gypsum in a salt dome.

caproic acid [ORG CHEM] $CH_3(CH_2)_4COOH$ A colorless liquid fatty acid found in oils and animal fats; used in synthesizing pharmaceuticals and flavors. Also known as butylacetic acid; hexanoic acid; *n*-hexoic acid; *n*-hexylic acid; pentylformic acid.

caproic aldehyde *See* *n*-hexaldehyde.

caproic anhydride [ORG CHEM] $(CH_3(CH_2)_4COO)_2$ White crystals that are insoluble in water.

caprolactam [ORG CHEM] $(CH_2)_5NH·C:O$ White flakes, melting point 68–69°C, made from cyclohexanone; used to make synthetic fiber, particularly nylon-6.

ε-caprolactone [ORG CHEM] $CH_2(CH_2)_4NHCO$ White crystals, used to make synthetic fibers, plastics, films, coatings, and plasticizers; its vapors or fine crystals are respiratory irritants.

Caprolan [TEXT] Trade name for a nylon-6 fiber manufactured by the Allied Chemical Corporation.

capryaldehyde *See* octyl aldehyde.

caprylamide [ORG CHEM] $CH_3(CH_2)_6CONH_2$ An amine, melting point 105–110°C; decomposes above 200°C; used as a chemical intermediate.

capryl compounds [ORG CHEM] A misnomer for octyl compounds; that is, the term octyl halide is preferred for caprylic halides, and octanoic acid for caprylic acid.

1-caprylene *See* 1-octene.

caprylic acid [ORG CHEM] $C_8H_{16}O_2$ A liquid fatty acid occurring in butter, coconut oil, and other fats and oils. Also known as hexylacetic acid; *n*-octanoic acid; octylic acid.

caprylic anhydride [ORG CHEM] $(CH_3(CH_2)_6CO)_2O$ A white solid that melts at −1°C; used as a chemical intermediate.

capsaicin [ORG CHEM] $C_{18}H_{27}O_3N$ A toxic material extracted from capsicum.

Capsaloidea [INV ZOO] A superfamily of ectoparasitic trematodes in the subclass Monogenea characterized by a sucker-shaped holdfast with anchors and hooks.

capsanthin [BIOCHEM] $C_{40}H_{58}O_3$ Carmine-red carotenoid pigment occurring in paprika.

cap screw [DES ENG] A screw which passes through a clear hole in the part to be joined, screws into a threaded hole in the other part, and has a head which holds the parts together.

capsicum [BOT] The fruit of a plant of the genus *Capsicum*, especially *C. frutescens*, cultivated in southern India and the tropics; a strong irritant to mucous membranes and eyes.

capsid [INV ZOO] The name applied to all members of the family Miridae. [VIROL] The protein coat of a virus particle.

capsomere [VIROL] An individual protein subunit of a capsid.

capstan [NAV ARCH] A rotating vertical spindle-mounted drum on which cable is wound for raising an anchor or other heavy weight.

capstan nut [DES ENG] A nut whose edge has several holes, in one of which a bar can be inserted for turning it.

capstan screw [DES ENG] A screw whose head has several radial holes, in one of which a bar can be inserted for turning it.

capstone [ARCH] A stone placed at the top of a stone arch.

capsular ligament [ANAT] A saclike ligament surrounding

the articular cavity of a freely movable joint and attached to the bones.

capsulate [BIOL] Enclosed in a capsule.

capsule [ANAT] A membranous structure enclosing a body part or organ. [BOT] A closed structure bearing seeds or spores; it is dehiscent at maturity. [MICROBIO] A thick, mucous envelope, composed of polypeptide or carbohydrate, surrounding certain microorganisms. [PHARM] A soluble shell in which drugs are enclosed for oral administration.

captive balloon [AERO ENG] A moored balloon, usually by steel cables.

captive fastener [DES ENG] A screw-type fastener that does not drop out after it has been unscrewed.

captive test [ENG] A hold-down test of a propulsion subsystem, rocket engine, or motor.

Captorhinimorpha [PALEON] An extinct subclass of primitive lizardlike reptiles in the order Cotylosauria.

capture [AERO ENG] The process in which a missile is taken under control by the guidance system. [ASTROPHYS] Of a central force field, as of a planet, to overcome by gravitational force the velocity of a passing body and bring the body under the control of the central force field, in some cases absorbing its mass. [HYD] The natural diversion of the headwaters of one stream into the channel of another stream having greater erosional activity and flowing at a lower level. Also known as piracy; river capture; river piracy; robbery; stream capture; stream piracy; stream robbery. [PHYS] A process in which an atomic or nuclear system acquires an additional particle; for example, the capture of electrons by positive ions, or capture of neutrons by nuclei.

capture cross section [NUC PHYS] The cross section that is effective for radiative capture.

capture effect [ELECTR] The effect wherein a strong frequency-modulation signal in an FM receiver completely suppresses a weaker signal on the same or nearly the same frequency.

capturing [ENG] The use of a torquer to restrain the spin axis of a gyro to a specified position relative to the spin reference axis.

capybara [VERT ZOO] *Hydrochoerus capybara.* An aquatic rodent (largest rodent in existence) found in South America and characterized by partly webbed feet, no tail, and coarse hair.

car *See* automobile.

Car *See* Carina.

Carabidae [INV ZOO] The ground beetles, a family of predatory coleopteran insects in the suborder Adephaga.

carabine [METEOROL] In France and Spain, a sudden and violent wind. Also known as brisa carabinera; brise carabinée.

Caracarinae [VERT ZOO] The caracaras, a subfamily of carrion-feeding birds in the order Falconiformes.

caracolite [MINERAL] A rare, colorless mineral occurring as crystalline incrustations, and consisting of a sulfate and chloride of sodium and lead.

Caradocian [GEOL] Lower Upper Ordovician geologic time.

caramel [MATER] A dark-brown mass formed by heating sugar in the presence of ammonium salts.

caranda wax [MATER] A wax similar to carnauba wax; obtained from the tropical palm caranda (*Copernicia australis*).

Carangidae [VERT ZOO] A family of perciform fishes in the suborder Percoidei, including jacks, scads, and pompanos.

carapace [GEOL] The upper normal limb of a fold having an almost horizontal axial plane. [INV ZOO] A dorsolateral, chitonous case covering the cephalothorax of many arthropods. [VERT ZOO] The bony, dorsal part of a turtle shell.

Carapidae [VERT ZOO] The pearlfishes, a family of sinuous, marine shore fishes in the order Gadiformes that live as commensals in the body cavity of holothurians.

carat [LAP] A unit of weight of gemstones, equal to 200 milligrams. Also known as metric carat.

carat count [LAP] The number of near-equal-size diamonds having a total weight of 1 carat.

carate *See* pinta.

Carathéodory outer measure [MATH] A positive, countably subadditive set function defined on the class of all subsets of a given set; used for defining measures.

Carathéodory's principle [THERMO] An expression of the second law of thermodynamics which says that in the neighborhood of any equilibrium state of a system, there are states which are not accessible by a reversible or irreversible adiabatic process.

caraway [BOT] *Carum carvi.* A white-flowered perennial herb of the family Umbelliferae; the fruit is used as a spice and flavoring agent.

caraway oil [MATER] A pale-yellow to colorless liquid that is slightly soluble in water; distilled from the dried fruit of the caraway plant (*Carum carvi*), the main ingredients being carvone and *d*-limonene; used in flavors, medicines, soaps, and perfumes. Also known as caraway seed oil; carui oil.

caraway seed oil *See* caraway oil.

carbachol [PHARM] $C_6H_{15}ClN_2O_2$ The choline ester carbamoylcholine chloride, used principally as a miotic in the local treatment of glaucoma.

carbamic acid [BIOCHEM] NH_2COOH An amino acid known for its salts, such as urea and carbamide. Also known as amidocarbonic acid; aminoformic acid.

carbamic acid β-hydroxyphenethyl ester *See* styramate.

carbamide peroxide *See* urea peroxide.

carbamidine *See* guanidine.

carbamino [BIOCHEM] A compound formed by the combination of carbon dioxide with a free amino group in an amino acid or a protein.

carbamoyl [ORG CHEM] The radical NH_2CO, formed from carbamic acid.

carbamylhydrazine hydrochloride *See* semicarbazide hydrochloride.

carbamyl phosphate [BIOCHEM] $NH_2COPO_4H_2$ The ester formed from reaction of phosphoric acid and carbamyl acid.

carbamylurea *See* biuret.

carbanilide [ORG CHEM] $(NHC_6H_5)CO(NHC_6H_5)$ Colorless crystals that are very slightly soluble in water, and dissolve in ether and alcohol; used in organic synthesis. Also known as diphenylurea.

carbanion [CHEM] One of the charged fragments which arise on heterolytic cleavage of a covalent bond involving carbon; the fragment carries an unshared pair of electrons and bears a negative charge.

carbazide *See* carbodihydrazide.

carbazole [ORG CHEM] One of a group of organic heterocyclic compounds containing a dibenzopyrrole system. Also known as 9-azafluorene.

carbazotic acid *See* picric acid.

carbene [ORG CHEM] A compound of carbon which exhibits two valences to a carbon atom; the two valence electrons are distributed in the same valence; an example is CH_2.

carbide [INORG CHEM] A binary compound of carbon with an element more electropositive than carbon; carbon-hydrogen compounds are excluded. [MATER] A cemented or compacted mixture of powdered carbides of heavy metals forming a hard material used in metal-cutting tools. Also known as cemented carbide.

carbide lamp [MIN ENG] A lamp that is charged with calcium carbide and water to form acetylene, which it burns.

carbide miner [MIN ENG] An automated coal mining machine; the unit is a continuous miner controlled from outside the coal seam.

carbide nuclear fuel [NUCLEO] A nuclear reactor fuel mixed with carbon compounds and a metal to give structural strength and oxidation resistance.

carbide tool [DES ENG] A cutting tool made of tungsten, titanium, or tantalum carbides, having high heat and wear resistance.

carbine [ORD] A rifle of short length and light weight.

carbinol [ORG CHEM] **1.** A primary alcohol with general formula RCH_2OH. **2.** The radical CH_2OH of primary alcohols. **3.** An alcohol derived from methanol.

carbocyclic compound [ORG CHEM] A compound with a homocyclic ring in which all the ring atoms are carbon, for example, benzene.

carbodihydrazide [ORG CHEM] $CO(NHNH_2)_2$ Colorless crystals that melt at 154°C; very soluble in alcohol and water; used in photographic chemicals. Also known as carbazide.

carbohumin *See* ulmin.

CAPTORHINIMORPHA

postorbital

squamosal

2.5 cm

Skull of *Captorhinus* in the subclass Captorhinimorpha. (From E. H. Colbert, *Evolution of the Vertebrates*, Wiley, 1965)

CARABIDAE

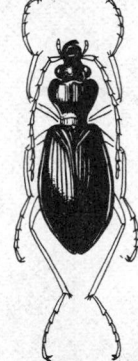

A drawing of a representative of the Carabidae. (From T. I. Storer and R. L. Usinger, *General Zoology*, 3d ed., McGraw-Hill, 1957)

carbohydrase [BIOCHEM] Any enzyme that catalyzes the hydrolysis of disaccharides and more complex carbohydrates.

carbohydrate [BIOCHEM] Any of the group of organic compounds composed of carbon, hydrogen, and oxygen, including sugars, starches, and celluloses.

carbohydrate gum [ORG CHEM] A polysaccharide which produces a gel of a viscous solution when it is dispersed in water at low concentrations; examples are agar, guar gum, xanthan gum, gum arabic, and sodium carboxymethyl cellulose.

carbohydrate metabolism [BIOCHEM] The sum of the biochemical and physiological processes involved in the breakdown and synthesis of simple sugars, oligosaccharides, and polysaccharides and in the transport of sugar across cell membranes.

carbolfuchsin [MATER] A solution of fuchsin, phenol, alcohol, and water; used as a stain in the identification of bacteria.

carbolic acid *See* phenol.

Carboloy [MET] Trade name for a bonded tungsten carbide used in manufacture of cutting tools, gages, drawing dies, and wear parts.

carbometer [ENG] An instrument for measuring the carbon content of steel by measuring magnetic properties of the steel in a known magnetic field.

carbomycin [MICROBIO] $C_{42}H_{67}O_{16}N$ A colorless, crystalline antibiotic produced by *Streptomyces halstedii;* principally active against gram-positive bacteria.

carbomycin B [MICROBIO] $C_{42}H_{67}O_{15}N$ A colorless, crystalline antibiotic differing from carbomycin only in having one less oxygen atom in its molecule.

carbon [CHEM] A nonmetallic chemical element, symbol C, atomic number 6, atomic weight 12.01115; occurs freely as diamond, graphite, and coal.

carbon-12 [NUC PHYS] A stable isotope of carbon with mass number of 12, forming about 98.9% of natural carbon; used as the basis of the newer scale of atomic masses, having an atomic mass of exactly 12 u (relative nuclidic mass unit) by definition.

carbon-13 [NUC PHYS] A heavy isotope of carbon having a mass number of 13.

carbon-14 [NUC PHYS] A naturally occurring radioisotope of carbon having a mass number of 14 and half-life of 5780 years; used in radiocarbon dating and in the elucidation of the metabolic path of carbon in photosynthesis. Also known as radiocarbon.

carbonaceous [SCI TECH] Relating to or composed of carbon.

carbonaceous meteorite [GEOL] A meteorite yielding relatively large amounts of carbon when it is analyzed.

carbonaceous rock [PETR] Rock with carbonaceous material included.

carbonaceous sandstone [PETR] Sandstone rich in carbon.

carbonaceous shale [GEOL] Shale rich in carbon.

carbonado [MINERAL] A dark-colored, fine-grained diamond aggregate; valuable for toughness and absence of cleavage planes. Also known as black diamond; carbon diamond.

carbon arc [ELEC] An electric arc between two electrodes, at least one of which is made of carbon; used in welding and high-intensity lamps, such as in searchlights and photography lamps.

carbon arc lamp [ELEC] An arc lamp in which an electric current flows between two electrodes of pure carbon, with incandescence at one or both electrodes and some light from the luminescence of the arc.

carbon arc welding [MET] Welding by maintaining an electric arc between a nonconsumable carbon electrode and the work.

carbonate [CHEM] **1.** An ester or salt of carbonic acid. **2.** A compound containing the carbonate (CO_3^{--}) radical. **3.** Containing carbonates.

carbonate cycle [GEOCHEM] The biogeochemical carbonate pathways, involving the conversion of carbonate to CO_2 and HCO_3, the solution and deposition of carbonate, and the metabolism and regeneration of it in biological systems.

carbonate mineral [MINERAL] A mineral containing considerable amounts of carbonates.

carbonate reservoir [GEOL] An underground oil or gas trap

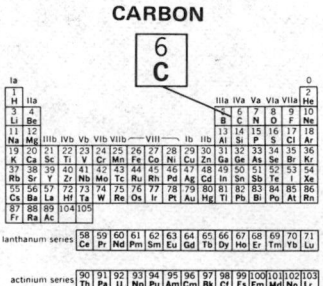

CARBON

Periodic table of the chemical elements showing the position of carbon.

formed in reefs, clastic limestones, chemical limestones, or dolomite.

carbonate rock [PETR] A rock composed principally of carbonates, especially if at least 50% by weight.

carbonate spring [HYD] A type of spring containing dissolved carbon dioxide gas.

carbonation [CHEM] Conversion to a carbonate. [CHEM ENG] The process by which a fluid, especially a beverage, is impregnated with carbon dioxide. [GEOCHEM] A process of chemical weathering whereby minerals that contain soda, lime, potash, or basic oxides are changed to carbonates by the carbonic acid in air or water.

carbonatite [PETR] **1.** Intrusive carbonate rock associated with alkaline igneous intrusive activity. **2.** A sedimentary rock composed of at least 80% calcium or magnesium.

carbon bit [DES ENG] A diamond bit in which the cutting medium is inset carbon.

carbon black [CHEM] **1.** An amorphous form of carbon produced commercially by thermal or oxidative decomposition of hydrocarbons and used principally in rubber goods, pigments, and printer's ink. **2.** *See* gas black.

carbon brush [ELEC] A rod made of carbon that bears against a commutator, collector ring, or slip ring to provide passage for the electric current from a dynamo through an outside circuit or for an external current through a motor.

carbon burning rate [CHEM ENG] The weight of carbon burned per unit time from the catalytic-cracking catalyst in the regenerator.

carbon button *See* button.

carbon cycle [GEOCHEM] The cycle of carbon in the biosphere, in which plants convert carbon dioxide to organic compounds that are consumed by plants and animals, and the carbon is returned to the biosphere in the form of inorganic compounds by processes of respiration and decay. [NUC PHYS] *See* carbon-nitrogen cycle.

carbon diamond *See* carbonado.

carbon dioxide [INORG CHEM] CO_2 A colorless, odorless, tasteless gas about 1.5 times as dense as air.

carbon dioxide absorption tube [ANALY CHEM] Absorbent-packed tube to capture the carbon dioxide formed during the microdetermination of carbon-hydrogen by the Pragl combustion procedure.

carbon dioxide fire extinguisher [CHEM ENG] A type of chemical fire extinguisher in which the extinguishing agent is liquid carbon dioxide, stored under 800–900 psi (pounds per square inch) at normal room temperature.

carbon dioxide gas laser [PHYS] A powerful, continuously operating laser in the infrared that can emit several hundred watts of power at a wavelength of 10.6 micrometers.

carbon dioxide indicator [MIN ENG] A detector of carbon dioxide in mines based on the gas's absorption by potassium hydroxide.

carbon dioxide process [MET] A casting process in which the molding material is a mixture of sand and 1.5–6% liquid silicate as a binder, and the mixture is packed around the pattern and hardened by blowing carbon dioxide gas through it.

carbon disulfide [ORG CHEM] CS_2 A sulfide, used as a solvent for oils, fats, and rubbers and in paint removers.

carbon electrode [MET] A nonfiller-metal electrode consisting of carbon or a graphite rod; sometimes contains copper powder for increased electrical conductivity; used in carbon arc welding.

carbon equivalent [MET] An empirical relationship of the total carbon (TC), silicon (Si) and phosphorus (P) content of gray iron: CE = %TC + 0.3(%Si + %P).

carbon fiber [MATER] Commercial material made by pyrolyzing any spun, felted, or woven raw material to a char at temperatures from 700 to 1800°C.

carbon film [ANALY CHEM] Carbon deposited by evaporation onto a specimen to protect and prepare it for electron microscopy.

carbon-film hygrometer element [ELEC] An electrical hygrometer element constructed of a plastic strip coated with a film of carbon black dispersed in a hygroscopic binder; variations in atmospheric moisture content vary the volume of the binder and thus change the resistance of the carbon coating.

carbon-film resistor [ELEC] A resistor made by depositing a thin carbon film on a ceramic form.

carbon-hydrogen analyzer [ANALY CHEM] A device used in the quantitative analysis of the carbon and hydrogen content of organic compounds.

carbon hydrophone [ENG ACOUS] A carbon microphone that responds to waterborne sound waves.

carbonic acid [INORG CHEM] H_2CO_3 The acid formed by combination of carbon dioxide and water.

carbonic anhydrase [BIOCHEM] An enzyme which aids carbon dioxide transport and release by catalyzing the synthesis, and the dehydration, of carbonic acid from, and to, carbon dioxide and water.

Carboniferous [GEOL] A division of late Paleozoic rocks and geologic time including the Mississippian and Pennsylvanian periods.

carbonification See coalification.

carbon isotope ratio [GEOL] Ratio of carbon-12 to either of the less common isotopes, carbon-13 or carbon-14, or the reciprocal of one of these ratios; if not specified, the ratio refers to C^{12}/C^{13}. Also known as carbon ratio.

carbonite [GEOL] A natural coke resulting from the intrusion of igneous rock into a coal seam.

carbonitrided steel [MET] Steel produced by carbonitriding.

carbonitriding [MET] Surface-hardening of low-carbon steel or other solid ferrous alloy by introducing carbon and nitrogen in a gaseous atmosphere containing carbon monoxide or hydrocarbons and ammonia at 800–875°C. Also known as gas cyaniding; nicarbing.

carbonium ion [CHEM] The positively charged fragment arising from the heterolytic cleavage of a covalent bond involving carbon.

carbonization [CHEM] Conversion of a carbon-containing substance to carbon or a carbon residue as the destructive distillation of coal by heat in the absence of air, yielding a solid residue with a higher percentage of carbon than the original coal; carried on for the production of coke and of fuel gas. [GEOCHEM] 1. In the coalification process, the accumulation of residual carbon by changes in organic material and their decomposition products. 2. Deposition of a thin film of carbon by slow decay of organic matter underwater. 3. A process of converting a carbonaceous material to carbon by removal of other components.

carbon knock [MECH ENG] Premature ignition resulting in knocking or pinging in an internal combustion engine caused when the accumulation of carbon produces overheating in the cylinder.

carbon lamp [ELEC] An arc lamp with carbon electrodes.

carbon microphone [ENG ACOUS] A microphone in which a flexible diaphragm moves in response to sound waves and applies a varying pressure to a container filled with carbon granules, causing the resistance of the microphone to vary correspondingly.

carbon monoxide [INORG CHEM] CO A colorless, odorless gas resulting from the incomplete oxidation of carbon; found, for example, in mines and automobile exhaust; poisonous to animals.

carbonmonoxyhemoglobin [BIOCHEM] A stable combination of carbon monoxide and hemoglobin formed in the blood when carbon monoxide is inhaled. Also known as carbonylhemoglobin and carboxyhemoglobin.

carbon-nitrogen cycle [NUC PHYS] A series of thermonuclear reactions, with release of energy, which presumably occurs in the sun and other stars; the net accomplishment is the synthesis of four hydrogen atoms into a helium atom, the emission of two positrons and much energy, and restoration of a carbon-12 atom with which the cycle began. Also known as carbon cycle; nitrogen cycle.

carbon-nitrogen-phosphorus ratio [OCEANOGR] The relatively constant relationship between the concentrations of carbon (C), nitrogen (N), and phosphorus (P) in plankton, and N and P in sea water, owing to removal of the elements by the organisms in the same proportions in which the elements occur and their return upon decomposition of the dead organisms.

carbon number [ANALY CHEM] The number of carbon atoms in a material under analysis; plotted against chromatographic retention volume for compound identification.

carbon paper [MATER] A paper, coated with dark waxy pigment, used to make duplicate copies while typewriting or handwriting; a sheet of carbon paper is sandwiched between two paper sheets, so that the impression made on the top sheet causes the carbon paper to transfer a pigmented impression onto the bottom sheet.

carbon pile [ELEC] A variable resistor consisting of a stack of carbon disks mounted between a fixed metal plate and a movable one that serve as the terminals of the resistor; the resistance value is reduced by applying pressure to the movable plate.

carbon pile pressure transducer [ENG] A measurement device in which variations in pressure upon a conductive carbon core proportionately change the core's electrical resistance, and thus the strength of the measured electric signal from the device.

carbon ratio [GEOL] 1. The ratio of fixed carbon to fixed carbon plus volatile hydrocarbons in a coal. 2. See carbon isotope ratio.

carbon ratio theory [GEOL] The theory that the gravity of oil in any area is inversely proportional to the carbon ratio of the coal.

carbon refractory [MATER] Carbon, generally in the form of graphite, used as a refractory in such equipment as crucibles and stopper nozzles for steel casting.

carbon replication [ANALY CHEM] A faithful carbon-film, mold of a specimen surface (for example, powders, bones, or crystals) which is thin enough to be studied by electron microscopy.

carbon residue [CHEM ENG] The quantity of carbon produced from a lubricating oil heated in a closed container under standard conditions.

carbon residue test [CHEM ENG] A destructive-distillation method for estimation of carbon residues in fuels and lubricating oils. Also known as Conradson carbon test.

carbon resistance thermometer [ENG] A highly sensitive resistance thermometer for measuring temperatures in the range 0.05–20K; capable of measuring temperature changes of the order 10^{-5} degree.

carbon resistor [ELECTR] A resistor consisting of carbon particles mixed with a binder, molded into a cylindrical shape, and baked; terminal leads are attached to opposite ends. Also known as composition resistor.

carbon star [ASTRON] Any of a class of stars with an apparently high abundance ratio of carbon to hydrogen; a majority of these are low-temperature red giants of the C class.

carbon steel [MET] Steel containing carbon, to about 2%, as the principal alloying element.

carbon suboxide [INORG CHEM] C_3O_2 A colorless lacrimatory gas having an unpleasant odor with a boiling point of −6.8°C.

carbon tetrachloride [ORG CHEM] CCl_4 A colorless, dense liquid, specific gravity 1.595, slightly soluble in water; used as a dry-cleaning agent. Also known as tetrachloromethane.

carbon tetrafluoride [ORG CHEM] CF_4 A colorless gas with a boiling point of −126°C; used as a refrigerant. Also known as fluorocarbon-14; Freon-14; tetrafluoromethane.

carbon transducer [ENG] A transducer consisting of carbon granules in contact with a fixed electrode and a movable electrode, so that motion of the movable electrode varies the resistance of the granules.

carbon transfer recording [COMMUN] Type of facsimile recording in which carbon particles are deposited on the record sheet in response to the received signal.

carbon trichloride See hexachloroethane.

carbonyl [ORG CHEM] A radical (CO) that is made up of one atom of carbon and one atom of oxygen connected by a double bond; found, for example, in aldehydes and ketones. Also known as carbonyl group.

carbonylation [CHEM] Introduction of a carbonyl radical into a molecule.

carbonyl bromide [ORG CHEM] $COBr_2$ A poisonous liquid boiling at 187.83°C; may be used by the military as a toxic suffocant. Also known as bromophosgene.

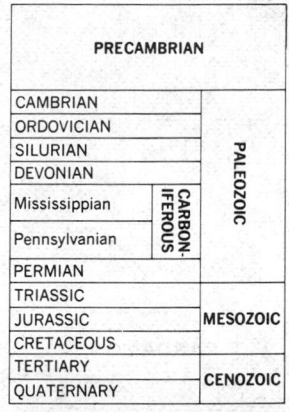

CARBONIFEROUS

PRECAMBRIAN		
CAMBRIAN		
ORDOVICIAN		PALEOZOIC
SILURIAN		
DEVONIAN		
Mississippian	CARBONIFEROUS	
Pennsylvanian		
PERMIAN		
TRIASSIC		
JURASSIC		MESOZOIC
CRETACEOUS		
TERTIARY		CENOZOIC
QUATERNARY		

Position of the Carboniferous and its relationship to the eras and periods of geologic time.

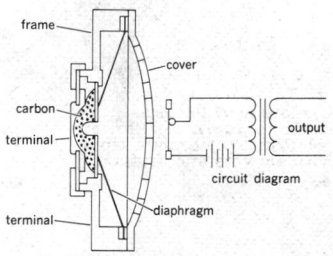

CARBON MICROPHONE

Sectional view and electrical circuit diagram of a carbon microphone, widely used in telephones.

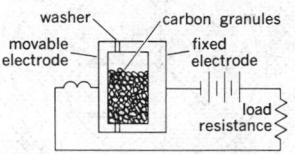

CARBON TRANSDUCER

Sectional view and electrical circuit diagram of a carbon transducer.

carbonyl chloride *See* phosgene.

carbonyl compound [ORG CHEM] A compound containing the carbonyl radical (CO).

carbonyl fluoride [ORG CHEM] COF_2 A colorless gas that is soluble in water; used in organic synthesis.

carbonyl group *See* carbonyl.

carbonylhemoglobin *See* carbonmonoxyhemoglobin.

carbonyl process [MET] **1.** A process in powder metallurgy for the production of iron, nickel, and iron-nickel alloy powders for magnetic applications. **2.** A process used in putting a metallic coating on molybdenum tungsten and other metals.

carborane [ORG CHEM] **1.** Any of a class of compounds containing boron, carbon, and hydrogen. **2.** $B_{10}C_2H_{12}$ A specific member of the class.

Carborundum [MATER] A manufactured crystalline material (silicon carbide), prepared by fusing coke and sand in an electric furnace; used as an abrasive in the grinding of low-tensile-strength materials, and as a semiconductor with a maximum operating temperature of 1300°C, to rectify and detect radio waves.

Carbowax [MATER] Trade name for a synthetic polyester wax.

carboxy *See* carboxyl.

carboxyhemoglobin *See* carbonmonoxyhemoglobin.

carboxyl [ORG CHEM] COOH The radical which determines the basicity of an organic acid. Also known as carboxy; oxatyl.

carboxylase [BIOCHEM] Any enzyme that catalyzes a carboxylation or decarboxylation reaction.

carboxylation [ORG CHEM] Addition of a carboxyl group into a molecule.

carboxylic [CHEM] Having chemical properties resembling those of carboxylic acid.

carboxylic acid [ORG CHEM] Any of a family of organic acids characterized by the presence of one or more carboxyl groups.

carboxyl terminal [BIOCHEM] The end of a polypeptide chain with a free carboxyl group.

carboxymethyl cellulose [ORG CHEM] An acid ether derivative of cellulose used as a sodium salt; a white, odorless, bulky solid used as a stabilizer and emulsifier. Also known as cellulose gum; sodium carboxymethylcellulose; sodium cellulose glycolate.

carboxypeptidase [BIOCHEM] Any enzyme that catalyzes the hydrolysis of a peptide at the end containing the free carboxyl group.

4-carboxy-resorcinol *See* β-resorcylic acid.

carbro process [GRAPHICS] A photographic method of making carbon prints in which tanning of pigmented gelatin occurs in a special bleach bath, and the gelatin yields prints made on bromide paper.

carbuncle [MED] A bacterial infection of subcutaneous tissue caused by *Staphylococcus aureus;* multiple sinuses are created by tissue destruction.

carbureted water gas [MATER] Water gas that has been enriched by hydrocarbon gases of high fuel value.

carburetion [CHEM ENG] The process of enriching a gas by adding volatile carbon compounds, such as hydrocarbons, to it, as in the manufacture of carbureted water gas. [MECH ENG] The process of mixing fuel with air in a carburetor.

carburetor [CHEM ENG] An apparatus for vaporizing, cracking, and enriching oils in the manufacture of carbureted water gas. [MECH ENG] A device that makes and controls the proportions and quantity of fuel-air mixture fed to a spark-ignition internal combustion engine.

carburetor icing [MECH ENG] The formation of ice in an engine carburetor as a consequence of expansive cooling and evaporation of gasoline.

carburize [MET] To surface-harden steel by converting the outer layer of low-carbon steel to high-carbon steel by heating the steel above the transformation range in contact with a carbonaceous material.

carcenet [METEOROL] A very cold and violent gorge wind in the eastern Pyrenees (upper Aude valley).

Carcharhinidae [VERT ZOO] A large family of sharks belonging to the charcharinid group of galeoids, including the tiger sharks and blue sharks.

Carchariidae [VERT ZOO] A family of shallow-water predatory sharks belonging to the isurid group of galeoids.

carcharodont [VERT ZOO] Having sharp, flat, triangular teeth with serrated margins, like those of the man-eating sharks.

carcinogen [MED] Any agent that incites development of a carcinoma or any other sort of malignancy.

carcinoid [MED] A potentially malignant tumor of the argentaffin cells of the stomach and intestine.

carcinoid syndrome [MED] A complex of symptoms arising from the metastasis of a carcinoid tumor to the liver.

carcinoma [MED] A malignant epithelial tumor.

carcinoma in situ [MED] A malignant tumor in the premetastatic stage, when cells are at the site of origin.

carcinomatosis [MED] Metastasis of a primary carcinoma to many sites throughout the body.

carcinotron *See* backward-wave oscillator.

card [ADP] An information-carrying medium, common to practically all computers, for the introduction of data and instructions into the computers either directly or indirectly.

cardamom *See* cardamon.

cardamon [BOT] *Elettaria cardamomum.* A perennial herbaceous plant in the family Zingiberaceae; the seed of the plant is used as a spice. Also spelled cardamom.

cardamon oil [MATER] A pale-yellow, combustible essential oil; insoluble in water, soluble in alcohol and ether; distilled from cardamon seeds, the chief known ingredients being terpinene, borneol, dipentene, limonene, and eucalyptol; used in flavoring and medicine.

Cardan joint *See* Hooke's joint.

Cardan motion [MECH ENG] The straight-line path followed by a moving centrode in a four-bar centrode linkage.

Cardan shaft [MECH ENG] A shaft with a universal joint at its end to accommodate a varying shaft angle.

Cardan's suspension [DES ENG] An arrangement of rings in which a heavy body is mounted so that the body is fixed at one point; generally used in a gyroscope.

card bed [ADP] The metal plate along which the card travels to the punching and reading stations.

cardboard [MATER] A good quality of chemical pulp or rag pasteboard made by combining two or more webs of paper, either with or without paste, while still wet; used for signs, printed material, and high-quality boxes.

card checking [ADP] The verification carried out by the computer to ensure that all the data keypunched on a card has been correctly read into the memory.

card code [ADP] The representation of characters on a punched card by means of punching one or more holes per column.

card face [ADP] The printed side of a punched card or, if printing is on both sides, the side of chief importance.

card feed [ADP] A device that inserts cards into a machine one at a time.

card field [ADP] A specified group of card columns used for a particular category of data.

card hopper [ADP] A device that holds cards and makes them available to a card-feed mechanism. Also known as hopper.

cardia [ANAT] **1.** The orifice where the esophagus enters the stomach. **2.** The large, blind diverticulum of the stomach adjoining the orifice. [INV ZOO] Anterior enlargement of the ventriculus in some insects.

cardiac [ANAT] **1.** Of, pertaining to, or situated near the heart. **2.** Of or pertaining to the cardia of the stomach.

cardiac arrest [MED] Cessation of the heartbeat.

cardiac cirrhosis [MED] Progressive fibrosis of the liver due to prolonged venous blood retention as a result of prolonged and severe heart failure.

cardiac cycle [PHYSIO] The sequence of events in the heart between the start of one contraction and the start of the next.

cardiac edema [MED] Accumulation of fluids throughout the body, as a function of cardiac failure.

cardiac electrophysiology [PHYSIO] The science that is concerned with the mechanism, spread, and interpretation of the

---

CARDAMON

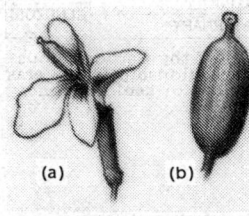

(a)   (b)

(c)

Cardamon *(Ellettaria cardamomum).* (a) Flower. (b) Fruit. (c) Branch tip with flowers and fruits.

CARDAN'S SUSPENSION

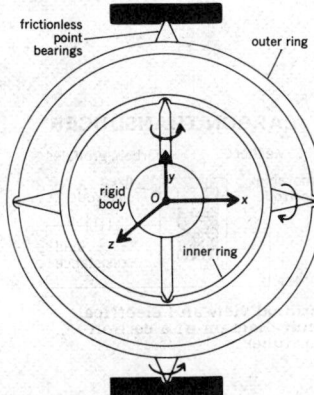

Cardan's suspension. Point *O* of the rigid body is fixed in space while the body is free to rotate about any axis in space under no external moments.

electric currents which arise within heart muscle tissue and initiate each heart muscle contraction.

cardiac failure [MED] A complex of symptoms resulting from failure of the heart to pump sufficient quantities of blood. Also known as heart failure.

cardiac gland [ANAT] Any of the mucus-secreting, compound tubular structures near the esophagus or in the cardia of the stomach of vertebrates; capable of secreting digestive enzymes.

cardiac input [PHYSIO] The amount of venous blood returned to the heart during a specified period of time.

cardiac loop [EMBRYO] The embryonic heart formed by bending and twisting of the cardiac tube.

cardiac massage [MED] Rhythmic compression of the heart by a physician or other person in the effort to maintain effective circulation following heart failure.

cardiac murmur [MED] Any adventitious sound heard in the region of the heart.

cardiac muscle [HISTOL] The principal tissue of the vertebrate heart; composed of a syncytium of striated muscle fibers.

cardiac output [PHYSIO] The total blood flow from the heart during a specified period of time.

cardiac plexus [ANAT] A network of visceral nerves situated at the base of the heart; contains both sympathetic and vagal nerve fibers.

cardiac sphincter [ANAT] The muscular ring at the orifice between the esophagus and stomach.

cardiac tamponade [MED] Cardiac compression caused by an accumulation of fluid within the pericardium.

cardiac valve [ANAT] Any of the structures located within the orifices of the heart that maintain unidirectional blood flow.

cardiectomy [MED] Excision of the cardiac end of the stomach.

card image [ADP] A one-to-one representation of the contents of a punched card, such as a matrix in which a 1 represents a hole and a 0 represents the absence of a hole.

cardinal heading [NAV] A heading in the direction of any of the cardinal points of the compass.

cardinal number [MATH] The number of members of a set; usually taken as a particular well-ordered set representative of the class of all sets which are in one-to-one correspondence with one another.

cardinal point [GEOD] Any of the four principal directions: north, east, south, or west of a compass. [OPTICS] Any one of six points in an optical system, namely, the two principal points, two nodal points, and two focal points. Also known as Gauss point.

cardinal point effect [ELECTR] The increased intensity of a line or group of returns on the radarscope occurring when the radar beam is perpendicular to the rectangular surface of a line or group of similarly aligned features in the ground pattern.

cardinal system [NAV] In marine operations, a buoyage system in which the buoys are assigned distinctive shape, color, and number in accordance with location relative to the nearest obstruction and pertinent to the cardinal points.

cardinal teeth [INV ZOO] Ridges and grooves on the inner surfaces of both valves of a bivalve mollusk near the anterior end of the hinge.

cardinal vein [EMBRYO] Any of four veins in the vertebrate embryo which run along each side of the vertebral column; the paired veins on each side discharge blood to the heart through the duct of Cuvier.

cardinal winds [METEOROL] Winds from the four cardinal points of the compass, that is, north, east, south, and west winds.

carding [TEXT] Straightening or smoothing of raw fibers in a parallel fashion, with a carding machine.

carding machine [TEXT] A machine for disentangling fibers before they are spun.

cardioblast [INV ZOO] Any of certain early embryonic cells in insects from which the heart develops.

cardiogenic plate [EMBRYO] An area of splanchnic mesoderm in the early mammalian embryo from which the heart develops.

cardiogenic shock [MED] Shock due to inadequate arterial blood flow following left-ventricular failure or pulmonary embolism.

cardiography [MED] Analysis of heart movements in the cardiac cycle by means of electronic instruments, especially by tracings.

cardioid [MATH] A heart-shaped curve generated by a point of a circle that rolls without slipping on a fixed circle of the same diameter.

cardioid condenser [OPTICS] A substage condenser that cuts off the direct light and allows only the light diffracted or dispersed from the object to enter the microscope; used in dark-field microscopes.

cardioid microphone [ENG ACOUS] A microphone having a heart-shaped, or cardioid, response pattern, so it has nearly uniform response for a range of about 180° in one direction and minimum response in the opposite direction.

cardioid pattern [ENG] Heart-shaped pattern obtained as the response or radiation characteristic of certain directional antennas, or as the response characteristic of certain types of microphones.

cardiolipin [BIOCHEM] A complex phospholipid found in the ether alcohol extract of powdered beef heart; mixed with leutin and cholesterol, it functions as the antigen in the Wassermann complement-fixation test for syphilis. Also known as diphosphatidyl glycerol.

cardiology [MED] The study of the heart.

cardiomyopathy See myocardiopathy.

cardiophobia [PSYCH] An abnormal fear of heart disease.

cardiorrhaphy [MED] Suturing of the heart muscle.

cardioscope [MED] 1. An instrument for the examination or visualization of the interior of the cardiac chambers. 2. An instrument which, by means of a cathode-ray oscillograph, projects an electrocardiographic record on a luminous screen.

cardiospasm [MED] Failure of the cardiac sphincter to relax; associated with spasm of the cardiac portion of the stomach and dilation of the esophagus.

cardiotachometer [MED] An electronic amplifier that times and records pulse rates of the heart.

cardiotomy [MED] Dissection or incision of the heart or the cardia of the stomach.

cardiotonic drug [PHARM] Any agent, such as digitalis, that increases cardiac muscle tonus.

cardiovascular system [ANAT] Those structures, including the heart and blood vessels, which provide channels for the flow of blood.

carditis [MED] Inflammation of the heart tissues.

card loader [ADP] A programming routine which permits a deck of cards to be read into a memory.

card machine [ADP] 1. Any of several small computers which perform particular operations called for by instruction cards, concurrently with the reading of data cards. 2. Loosely, any type of peripheral equipment which reads or punches cards.

card punch See key punch.

card reader [ADP] A mechanism that senses information punched on cards, using wire brushes, metal feelers, or a photoelectric system. Also known as punched card reader.

card reproducer [ADP] A machine that makes a duplicate of a punched card.

card row [ADP] A row of punching positions parallel to the long edge of a punched card.

card sorter [ADP] A machine used to arrange punched cards into an appropriate sequence for further processing. Also known as punched card sorter.

card-to-card transceiving [ADP] A system that makes possible instantaneous and accurate duplication of punched cards over telephone and telegraph networks between locations separated by either just a few miles or thousands of miles.

card-to-disk conversion [ADP] A straightforward operation which consists in loading the data in a deck of cards onto a disk by means of a utility program.

card-to-tape conversion [ADP] A straightforward operation which consists in loading the data in a deck of cards onto a magnetic tape by means of a utility program.

car dump [MECH ENG] Any one of several devices for unloading industrial or railroad cars by rotating or tilting the car.

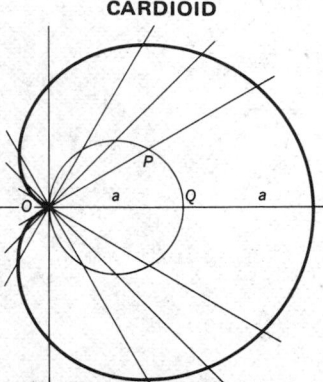

CARDIOID

Point-construction of a cardioid. The fixed circle is *OPQ* which has a diameter *a*. The cardioid is constructed by laying off, along every secant *OP* passing through the fixed point *O*, the distance *a* in both directions from *P*.

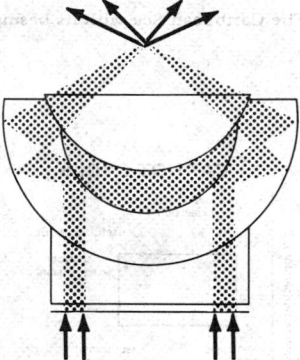

CARDIOID CONDENSER

Cardioid condenser, a dark-field device. Shaded meniscus area is air space; unshaded areas are portions of condenser through which passes the light, indicated by arrows and shaded paths. (*Photographic Service Department, Kodak Research Laboratory*)

card verifier [ADP] An electromechanical device which allows the operator to check that a card has been properly keypunched.

δ-3-carene [ORG CHEM] $C_{10}H_{16}$ A clear, colorless, combustible terpene liquid, stable to about 250°C; used as a solvent and in chemical synthesis. Also known as 3,7,7-trimethylbicyclo-[4.1.0]-hept-3-ene.

Carettochelyidae [VERT ZOO] A family of reptiles in the order Chelonia containing only one species, the New Guinea plateless turtle (*Carettochelys insculpta*).

car ferry [NAV ARCH] A ferry that is specially designed to carry railroad cars, trucks, or passenger cars.

car float [NAV ARCH] A barge with railroad tracks on its deck, used to carry railroad cars in harbors or inland waterways.

car-following theory [ENG] A mathematical model of the interactions between motor vehicles in terms of relative speed, absolute speed, and separation.

cargo boom [MECH ENG] A long spar extending from the mast of a derrick to support or guide objects lifted or suspended.

cargo liner *See* cargo ship.

cargo mill [IND ENG] A sawmill equipped with docks so the product can be loaded directly onto ships.

cargo ship [NAV ARCH] A power-driven ship employed exclusively in commercial transportation of commodities on the ocean and large inland bodies of water. Also known as cargo liner.

cargo winch [MECH ENG] A motor-driven hoisting machine for cargo having a drum around which a chain or rope winds as the load is lifted.

Cariamidae [VERT ZOO] The long-legged cariamas, a family of birds in the order Gruiformes.

Caribbean Current [OCEANOGR] A water current flowing westward through the Caribbean Sea.

Caribbean Sea [GEOGR] One of the largest and deepest enclosed basins in the world, surrounded by Central and South America and the West Indian island chains.

Caribosireninae [VERT ZOO] A subfamily of trichechiform sirenean mammals in the family Dugongidae.

Caridea [INV ZOO] A large section of decapod crustaceans in the suborder Natantia including many diverse forms of shrimps and prawns.

caries [MED] 1. Bone decay. 2. Tooth decay. Also known as dental caries.

carina [BIOL] A ridge or a keel-shaped anatomical structure. [VERT ZOO] *See* keel.

Carina [ASTRON] A constellation, right ascension 9 hours, declination 60°S. Abbreviated Car. Also known as Keel.

α Carinae *See* Canopus.

Carina Nebula [ASTRON] A gaseous nebula in the Milky Way.

carinate [BIOL] Having a ridge or keel, as the breastbone of certain birds.

Carinomidae [INV ZOO] A monogeneric family of littoral ribbonlike worms in the order Palaeonemertini.

Carinthian furnace [MET] A zinc distillation furnace with small, vertical retorts. [MIN ENG] A small reverberatory furnace with an inclined hearth, in which lead ore is treated by roasting and reaction, wood being the usual fuel.

Carius method [ANALY CHEM] A procedure used to analyze organic compounds for sulfur, halogens, and phosphorus that involves heating the sample with fuming nitric acid in a sealed tube.

carling [NAV ARCH] A short girder or timber running fore and aft, used to support or stiffen a ship's deck, or to frame an opening in the deck where beams have been cut.

Carlsbad law [CRYSTAL] A feldspar twin law in which the twinning axis is the *c* axis, the operation is rotation of 180°, and the contact surface is parallel to the side pinacoid.

Carlsbad turn [CRYSTAL] A twin crystal in the monoclinic system with the vertical axis as the turning axis.

carminic acid [ORG CHEM] $C_{22}H_{20}O_{13}$ A glucosidal hydroxyanthrapurin that is derived from cochineal; a red crystalline dye used as a stain for biological materials. Also known as cochinilin.

carminite [MINERAL] $PbFe_2(AsO_4)_2(OH)_2$ A carmine to tile-red mineral consisting of a basic arsenate of lead and iron.

carnallite [MINERAL] $KMgCl_3 \cdot 6H_2O$ A milky-white or reddish mineral that crystallizes in the orthorhombic system and occurs in deliquescent masses; it is valuable as an ore of potassium.

carnassial [ANAT] Of or pertaining to molar or premolar teeth specialized for cutting and shearing.

carnauba wax [MATER] The hardest natural wax, having a melting point of 85°C, exuded from the leaves of the carnauba palm (*Carnauba cerifera*); used for insulating purposes and in making candles, shoe polish, high-luster wax, varnishes, phonograph records, and surface coating of automobiles. Also known as Brazil wax.

carnaubic acid [ORG CHEM] $C_{24}H_{48}O_2$ An acid found in carnauba wax and beef kidney.

carnegieite [MINERAL] $NaAlSiO_4$ An artificial mineral similar to feldspar; it is triclinic at low temperatures, isometric at elevated temperatures.

Carnian [GEOL] Lower Upper Triassic geologic time. Also spelled Karnian.

carnitine [BIOCHEM] $C_7H_{15}NO_3$ α-Amino-β-hydroxybutyric acid trimethylbetaine; a constituent of striated muscle and liver, identical with vitamin B_T.

Carnivora [VERT ZOO] A large order of placental mammals, including dogs, bears, and cats, that is primarily adapted for predation as evidenced by dentition and jaw articulation.

carnivorous [BIOL] Eating flesh or, as in plants, subsisting on nutrients obtained from animal protoplasm.

carnivorous plant *See* insectivorous plant.

Carnosauria [PALEON] A group of large, predacious saurischian dinosaurs in the suborder Theropoda having short necks and large heads.

carnosine [BIOCHEM] $C_9H_{14}N_4O_3$ A colorless, crystalline dipeptide occurring in the muscle tissue of vertebrates.

Carnot cycle [THERMO] A hypothetical cycle consisting of four reversible processes in succession: an isothermal expansion and heat addition, an isentropic expansion, an isothermal compression and heat rejection process, and an isentropic compression.

Carnot efficiency [THERMO] The efficiency of a Carnot engine receiving heat at a temperature absolute T_1 and giving it up at a lower temperature absolute T_2; equal to $(T_1 - T_2)/T_1$.

Carnot engine [MECH ENG] An ideal, frictionless engine which operates in a Carnot cycle.

carnotite [MINERAL] $K(UO_2)_2(VO_4)_2 \cdot nH_2O$ A canary-yellow, fine-grained hydrous vanadate of potassium and uranium having monoclinic microcrystals; an ore of radium and uranium.

Carnot number [THERMO] A property of two heat sinks, equal to the Carnot efficiency of an engine operating between them.

Carnot's theorem [THERMO] 1. The theorem that all Carnot engines operating between two given temperatures have the same efficiency, and no cyclic heat engine operating between two given temperatures is more efficient than a Carnot engine. 2. The theorem that any system has two properties, the thermodynamic temperature T and the entropy S, such that the amount of heat exchanged in an infinitesimal reversible process is given by $dQ = TdS$; the thermodynamic temperature is a strictly increasing function of the empirical temperature measured on an arbitrary scale.

Carnoy's solution [MATER] A tissue fixative composed of mercuric salts and acetic acids; used where penetration of hard objects such as seeds is required.

caroba *See* carob wood.

carob wood [MATER] Wood from a large Brazilian tree, *Jacaranda copaia*. Also known as caroba; jacaranda.

Carolina Bays [GEOGR] Shallow, marshy, often ovate depressions on the coastal plain of the mideastern and southeastern United States of unknown origin.

Carolinian life zone [ECOL] A zone comprising the climate and biotic communities of the oak savannas of eastern North America.

Caro's acid [INORG CHEM] H_2SO_5 A white solid melting at about 45°C, formed during the acid hydrolysis of peroxydisulfates. Also known as peroxymonosulfuric acid.

carotenase [BIOCHEM] An enzyme that effects the hydrolysis of carotenoid compounds, used in bleaching of flour.

CARIBBEAN SEA

The Caribbean Sea with its basins.

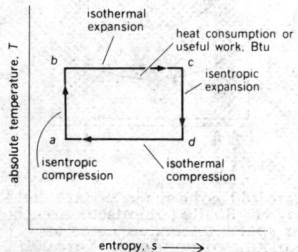

CARNOT CYCLE

Carnot cycle when air is used as a working substance, comprising the four processes *b-c*, *c-d*, *d-a*, and *a-b*.

carotene [BIOCHEM] $C_{40}H_{56}$ Any of several red, crystalline, carotenoid hydrocarbon pigments occurring widely in nature, convertible in the animal body to vitamin A, and characterized by preferential solubility in petroleum ether. Also known as carotin.

carotenemia [MED] The presence of carotene in the blood; may cause yellowing of the skin.

carotenoid [BIOCHEM] A class of labile, easily oxidizable, yellow, orange, red, or purple pigments that are widely distributed in plants and animals and are preferentially soluble in fats and fat solvents.

carotid artery [ANAT] Either of the two principal arteries on both sides of the neck that supply blood to the head and neck. Also known as common carotid artery.

carotid body [ANAT] Either of two chemoreceptors sensitive to changes in blood chemistry which lie near the bifurcations of the carotid arteries. Also known as glomus caroticum.

carotid ganglion [ANAT] A group of nerve cell bodies associated with each carotid artery.

carotid sinus [ANAT] An enlargement at the bifurcation of each carotid artery that is supplied with sensory nerve endings and plays a role in reflex control of blood pressure.

carotin *See* carotene.

carotol [BIOCHEM] $C_{15}H_{25}OH$ A sesquiterpenoid alcohol in carrots.

carp [VERT ZOO] The common name for a number of freshwater, cypriniform fishes in the family Cyprinidae, characterized by soft fins, pharyngeal teeth, and a suckerlike mouth.

carpel [BOT] The basic specialized leaf of the female reproductive structure in angiosperms; a megasporophyll.

carpenter's level [DES ENG] A bar, usually of aluminum or wood, containing a spirit level.

carpenter stopper [NAV ARCH] A device, generally made from high-carbon steel, that allows a temporary grip on a wire rope up to its breaking strength without injury or slippage.

carpet-bombing [ORD] The laying down of bombs in a creeping pattern to cover the area as with a carpet.

carpholite [MINERAL] $MnAl_2Si_2O_6(OH)_4$ A straw-yellow fibrous mineral consisting of a hydrous aluminum manganese silicate occurring in tufts; specific gravity is 2.93.

carphosiderite [MINERAL] A yellow mineral consisting of a basic hydrous iron sulfate occurring in masses and crusts.

car pincher [MIN ENG] A worker in a mine who uses a pinch bar to position cars under loading chutes.

carpincho [MATER] Processed capybara skin, noted for its elastic properties.

carpogonium [BOT] The basal, egg-bearing portion of the female reproductive organ in some thallophytes, especially red algae.

Carpoidea [PALEON] Former designation for a class of extinct homalozoan echinoderms.

carpoids [PALEON] An assemblage of three classes of enigmatic, rare Paleozoic echinoderms formerly grouped together as the class Carpoidea.

carpology [BOT] The study of the morphology of fruit and seeds.

carpophagous [ZOO] Feeding on fruits.

carpophore [BOT] The portion of a flower receptacle that extends between and attaches to the carpels. [MYCOL] The stalk of a fruiting body in fungi.

carpophyte [BOT] A thallophyte that forms a sporocarp following fertilization.

carposporangium [BOT] In red algae, a sporangium that forms the cystocarp and contains carpospores.

carpospore [BOT] In red algae, a diploid spore produced terminally by a gonimoblast, giving rise to the diploid tetrasporic plant.

carpus [ANAT] **1.** The wrist in man or the corresponding part in other vertebrates. **2.** The eight bones of the human wrist. [INV ZOO] The fifth segment from the base of a generalized crustacean appendage.

carrageen [BOT] *Chondrus crispus.* A cartilaginous red algae harvested in the Northern Atlantic as a source of carrageenin.

carrageenin [ORG CHEM] A colloid extracted from carrageen and used chiefly as an emulsifying and stabilizing agent in foods, pharmaceuticals, and cosmetics.

Carrara marble [PETR] All marble quarried near Carrara,

Italy, having a prevailing white to bluish color, or white with blue veins.

carrene *See* methylene chloride.

car retarder [ENG] A device located along the track to reduce or control the velocity of railroad or mine cars.

carriage [ENG] **1.** A device that moves in a predetermined path in a machine and carries some other part, such as a recorder head. **2.** A mechanism designed to hold a paper in the active portion of a printing or typing device, for example, a typewriter carriage. [ORD] Mobile or fixed support for a cannon; sometimes includes the elevating and traversing mechanisms.

carriage bolt [DES ENG] A round-head type of bolt with a square neck, used with a nut as a through bolt.

carriage return [ADP] The operation that causes the next character to be printed at the extreme left margin, and usually advances to the next line at the same time.

carriage stop [MECH ENG] A device added to the outer way of a lathe bed for accurately spacing grooves, turning multiple diameters and lengths, and cutting off pieces of specified thickness.

carriage tape *See* control tape.

Carribel explosive [MATER] A medium-strength explosive used in wet boreholes if immersion does not exceed 2 to 3 hours.

carrier [COMMUN] **1.** The radio wave produced by a transmitter when there is no modulating signal, or any other wave, recurring series of pulses, or direct current capable of being modulated. Also known as carrier wave; signal carrier. **2.** A wave generated locally at a receiver that, when combined with the sidebands of a suppressed-carrier transmission in a suitable detector, produces the modulating wave. **3.** *See* carrier system. [GEN] An individual who is heterozygous for a recessive gene. [MECH ENG] Any machine for transporting materials or people. [MED] A person who harbors and eliminates an infectious agent and so transmits it to others, but who may not show signs of the disease. [NAV ARCH] *See* aircraft carrier. [NUCLEO] A substance that, when associated with a radioactive trace of another substance, will carry the trace with it through a chemical or physical process; an isotope is often used for this purpose. Also known as isotopic carrier. [SOLID STATE] *See* charge carrier.

carrier amplifier [ELECTR] A direct-current amplifier in which the dc input signal is filtered by a low-pass filter, then used to modulate a carrier so it can be amplified conventionally as an alternating-current signal; the amplified dc output is obtained by rectifying and filtering the rectified carrier signal.

carrier amplitude regulation [COMMUN] Change in amplitude of the carrier wave in an amplitude-modulated transmitter when modulation is applied under conditions of symmetrical modulation.

carrier beat [COMMUN] An undesirable heterodyne of facsimile signals, each synchronous with a different stable reference oscillator, causing a pattern in received copy.

carrier channel [COMMUN] The equipment and lines that make up a complete carrier-current circuit between two or more points.

carrier chrominance signal *See* chrominance signal.

carrier-controlled approach system [NAV] An aircraft-carrier radar system providing information by which aircraft approaches may be directed by radio.

carrier current [COMMUN] A higher-frequency alternating current superimposed on ordinary telephone, telegraph, and power-line frequencies for communication and control purposes.

carrier density [SOLID STATE] The density of electrons and holes in a semiconductor.

carrier frequency [COMMUN] The frequency generated by an unmodulated radio, radar, carrier communication, or other transmitter, or the average frequency of the emitted wave when modulated by a symmetrical signal. Also known as center frequency; resting frequency.

carrier isolating choke coil [ELECTROMAG] Inductor inserted in series with a line on which carrier energy is applied to impede the flow of carrier energy beyond that point.

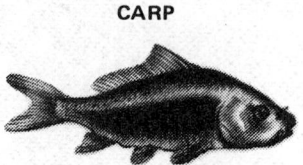

CARP

Carp *(Cyprinus carpio).*

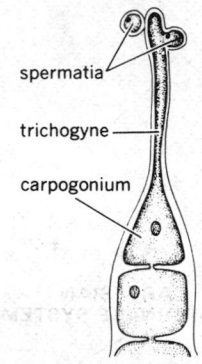

CARPOGONIUM

spermatia

trichogyne

carpogonium

Filament with terminal carpogonium (containing an egg) and bearing an elongated trichogyne. *(From H. J. Fuller and O. Tippo, College Botany, rev. ed., Holt, 1954)*

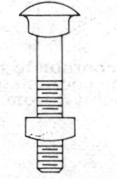

CARRIAGE BOLT

A round-head carriage bolt.

carrier leak [COMMUN] Carrier remaining after carrier suppression in a suppressed-carrier transmission system.

carrier level [COMMUN] The strength or level of an unmodulated carrier signal at a particular point in a radio system, expressed in decibels in relation to some reference level.

carrier line [ELEC] Any transmission line used for multiple-channel carrier communication.

carrier loading [ELECTROMAG] The addition of lumped inductances to the cable section of a transmission line specifically designed for carrier transmission; it serves to minimize impedance mismatch between cable and open wire and to reduce the cable attenuation.

carrier mobility [SOLID STATE] The average drift velocity of carriers per unit electric field in a homogeneous semiconductor; the mobility of electrons is usually different from that of holes.

carrier noise [COMMUN] Noise produced by undesired variation of a radio-frequency signal in the absence of any intended modulation. Also known as residual modulation.

carrier pipe [ENG] Pipe used to carry or conduct fluids, as contrasted with an exterior protective or casing pipe.

carrier power output rating [COMMUN] Power available at the output terminals of a transmitter when the output terminals are connected to the normal-load circuit or to a circuit equivalent thereto.

carrier repeater [ELECTR] Equipment designed to raise carrier signal levels to such a value that they may traverse a succeeding line section at such amplitude as to preserve an adequate signal-to-noise ratio; while the heart of a repeater is the amplifier, necessary adjuncts are filters, equalizers, level controls, and so on, depending upon the operating methods.

carrier rocket [AERO ENG] A rocket vehicle used to carry something, as the carrier rocket of the first artificial earth satellite.

carrier shift [COMMUN] **1.** Transmission of radio-teletypewriter messages by shifting the carrier frequency in one direction for a marking signal and in the opposite direction for a spacing signal. **2.** Condition resulting from imperfect modulation whereby the positive and negative excursions of the envelope pattern are unequal, thus effecting a change in the power associated with the carrier.

carrier signaling [COMMUN] Method by which busy signals, ringing, or dial signaling relays are operated by the transmission of a carrier-frequency tone.

carrier suppression [COMMUN] **1.** Suppression of the carrier frequency after conventional modulation at the transmitter, with reinsertion of the carrier at the receiving end before demodulation. **2.** Suppression of the carrier when there is no modulation signal to be transmitted; used on ships to reduce interference between transmitters.

carrier swing [COMMUN] The total deviation of a frequency-modulated or phase-modulated wave from the lowest instantaneous frequency to the highest instantaneous frequency.

carrier system [COMMUN] A system permitting a number of simultaneous, independent communications over the same circuit. Also known as carrier.

carrier telegraphy [COMMUN] Telegraphy in which a single-frequency carrier wave is modulated by the transmitting apparatus for transmission over wire lines.

carrier telephony [COMMUN] Telephony in which a single-frequency carrier wave is modulated by a voice-frequency signal for transmission over wire lines.

carrier terminal [ELECTR] Apparatus at one end of a carrier transmission system, whereby the processes of modulation, demodulation, filtering, amplification, and associated functions are effected.

carrier-to-noise ratio [COMMUN] The ratio of the magnitude of the carrier to that of the noise after selection and before any nonlinear process such as amplitude limiting and detection.

carrier transfer filters [ELECTR] Filters arranged as a carrier-frequency crossover or bridge between two transmission circuits.

carrier transmission [COMMUN] Transmission in which the transmitted electric wave is a wave resulting from the modulation of a single-frequency carrier wave by a modulating wave.

carrier wave See carrier.

Carrion's disease [MED] A bacterial infection of man endemic in the Andes caused by *Bartonella bacilliformis,* which attacks red blood cells and blood-forming organs. Also known as bartonellosis.

carrot [BOT] *Daucus carota.* A biennial umbellifer of the order Umbellales with a yellow or orange-red edible root.

carrot oil [MATER] A light-yellow oil distilled from the seeds of the carrot (*Daucus carota*), and containing carotene, pinene, palmitic acid, limonene, and butyric or isobutyric acid; used as a flavoring.

Carr-Price unit [BIOL] A unit for the standardization of vitamin A.

carry-complete signal [ADP] A signal generated by a digital parallel adder, indicating that all carries from an adding operation have been generated and propagated, and that the addition operation is completed.

carry flag [ADP] A flip-flop circuit which indicates overflow in arithmetic operations.

carry lookahead [ADP] A circuit which allows low-order carries to ripple through all the way to the highest-order bit to output a completed sum.

carry-over [CHEM ENG] Unwanted liquid or solid material carried by the overhead effluent from a fractionating column, absorber, or reaction vessel. [HYD] The portion of the stream flow during any month or year derived from precipitation in previous months or years.

carry signal [ADP] A signal produced in a computer when the sum of two digits in the same column equals or exceeds the base of the number system in use or when the difference between two digits is less than zero.

carry time [ADP] The time needed to transfer all carry digits to the next higher column.

car shaker [MECH ENG] A device consisting of a heavy yoke on an open-top car's sides that actively vibrates and rapidly discharges a load, such as coal, gravel, or sand, when an unbalanced pulley attached to the yoke is rotated fast.

carsickness [MED] Motion sickness resulting from acceleratory movements of a train or automobile.

car stop [ENG] An appliance used to arrest the movement of a mine or railroad car.

Carter chart [ELECTROMAG] An Argand diagram of the complex reflection coefficient of a waveguide junction on which are drawn lines of constant magnitude and phase of the impedance.

Carterinacea [INV ZOO] A monogeneric superfamily of marine, benthic foraminiferans in the suborder Rotaliina characterized by a test with monocrystal calcite spicules in a granular groundmass.

cartesian axis [MATH] One of a set of mutually perpendicular lines which all pass through a single point, used to define a cartesian coordinate system; the value of one of the coordinates on the axis is equal to the directed distance from the intersection of axes, while the values of the other coordinates vanish.

cartesian coordinates [MATH] The set of numbers which locate a point in space with respect to a collection of mutually perpendicular axes.

cartesian coordinate system [MATH] A coordinate system in n dimensions where n is any integer made by using n number axes which intersect each other at right angles at an origin, enabling any point within that rectangular space to be identified by the distances from the n lines. Also known as rectangular cartesian coordinate system.

cartesian diver manostat [ENG] Preset, on-off-control manometer arrangement by which a specified low pressure (high vacuum) is maintained via the rise or submergence of a marginally buoyant float within a liquid mercury reservoir.

cartesian geometry See analytic geometry.

cartesian product [MATH] In reference to the product of P and Q, the set $P \times Q$ of all pairs (p,q), where p belongs to P and q belongs to Q.

cartesian surface [MATH] A surface obtained by rotating the curve $n_0(x^2 + y^2)^{1/2} \pm n_1[(x - a)^2 + y^2]$ about the x axis.

cartesian tensor [MATH] The aggregate of the functions of position in a tensor field in an n-dimensional cartesian coordinate system.

cartilage [HISTOL] A specialized connective tissue which is bluish, translucent, and hard but yielding.

**CARTESIAN
COORDINATE SYSTEM**

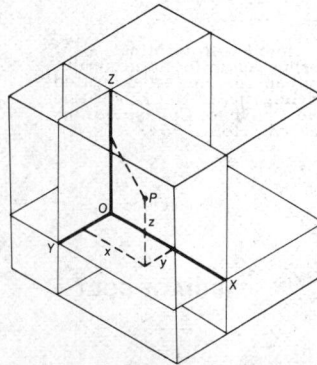

Cartesian coordinate system showing the coordinates x, y, and z of point P from the origin O.

cartilage bone [HISTOL] Bone formed by ossification of cartilage.

cartilaginous fish [VERT ZOO] The common name for all members of the class Chondrichthyes.

cartogram [MAP] A type of single-factor or topical map that is often diagrammatic to show traffic flow, movement of people or goods, or value by area, where areas of the political subdivisions are distorted so that their size is proportional to their monetary value.

cartographer [GRAPHICS] An individual who makes charts or maps.

cartographic satellite [AERO ENG] An applications satellite that is used to prepare maps of the earth's surface and of the culture on it.

cartography [GRAPHICS] The making of maps and charts for the purpose of visualizing spatial distributions over various areas of the earth.

cartoon [GRAPHICS] **1.** Animated drawings in a motion picture format. **2.** A drawing on paper that is used as a model for a final work.

cartouche [GRAPHICS] A border or scroll that is decorative and executed with a pen or brush.

cartridge [ENG] A cylindrical, waterproof, paper shell filled with high explosive and closed at both ends; used in blasting. [ENG ACOUS] *See* phonograph pickup; tape cartridge. [NUCLEO] *See* jacket. [ORD] **1.** An assemblage of the components required to function a weapon once. **2.** Ammunition for a gun which contains in a unit assembly all components required to function the gun once, and which is loaded into the gun in one operation.

cartridge actuated initiator [AERO ENG] An item designed to provide gas pressure for activating various aircraft components such as canopy removers, thrusters, and catapults.

cartridge belt [ORD] An ammunition belt with loops or pockets for carrying cartridges or clips of cartridges.

cartridge brass [MET] An alloy containing 70% copper and 30% zinc; uses include cartridge cases, automotive radiator cores and tanks, lighting fixtures, eyelets, rivets, springs, screws, and plumbing products.

cartridge case [ORD] An item which is designed to hold an ammunition primer and propellant and to which a projectile may be affixed; its profile and size conform to the chamber of the weapon in which the round is fired.

cartridge clip [ORD] A metallic device used to hold cartridges for ease of loading into a rifle or a revolver. Also known as clip.

cartridge filter [ENG] A filter for the clarification of process liquids containing small amounts of solids; turgid liquid flows between thin metal disks, assembled in a vertical stack, to openings in a central shaft supporting the disks, and solids are trapped between the disks.

cartridge fuse [ELEC] A type of electric fuse in which the fusible element is connected between metal ferrules at either end of an insulating tube.

cartridge magazine [ORD] A metallic spring-loaded container of rectangular cross section designed to hold cartridges and to be inserted into the magazine well of a pistol, rifle, or machine gun; it is an integral part of a gun, capable of being removed from the magazine well, reloaded, and reused.

cartridge receiver [ORD] An item integral to a carbine, rifle, or shotgun, designed to take the charge from a cartridge magazine or clip or a single cartridge or shell and hold it until chambered.

cartridge silk *See* powder silk.

cartridge starter [MECH ENG] An explosive device which, when placed in an engine and detonated, moves a piston, thereby starting the engine.

cartridge tape drive [ADP] A tape drive which will automatically thread the tape on the takeup reels without human assistance. Formerly known as hypertape drive.

car tunnel kiln [ENG] A long kiln with the fire located near the midpoint; ceramic ware is fired by loading it onto cars which are pushed through the kiln.

carui oil *See* caraway oil.

caruncle [ANAT] Any normal or abnormal fleshy outgrowth, such as the comb and wattles of fowl or the mass in the inner canthus of the eye. [BOT] A fleshy outgrowth developed from the seed coat near the hilum in some seeds, such as the castor bean.

carvacrol [ORG CHEM] $(CH_3)_2CHC_6H_3(CH_3)OH$ A colorless liquid, boiling at 237°C; used in perfumes, flavorings, and fungicides. Also known as 2-hydroxy-*para*-cymene; isopropyl-*ortho*-cresol; 2-methyl-5-isopropyl phenol.

Carvallo paradox [OPTICS] The absurdity that since light is composed from infinitely long wave trains of various frequencies, a spectrograph should show the spectrum of a source both before and after it is illuminated.

carved-out payment [PETRO ENG] A proportionate royalty payment based on proceeds from oil or gas production from leased property that has been assigned (carved) out of total payments for the leased property.

carvol *See* carvone.

carvone [ORG CHEM] $C_{10}H_{14}O$ A liquid ketone that boils at 231°C; soluble in water and alcohol; it is optically active and occurs naturally in both dextro and levo forms; used in flavorings and perfumery. Also known as carvol.

caryinite [MINERAL] $(Ca,Pb,Na)_5(Mn,Mg)_4(AsO_4)_5$ A mineral consisting chiefly of a calcium manganese arsenate.

Caryophanaceae [MICROBIO] A family of large, gram-negative bacteria belonging to the order Caryophanales and having disklike cells arranged in chains.

Caryophanales [MICROBIO] An order of bacteria in the class Schizomycetes occurring as trichomes; produce short structures that function as reproductive units.

Caryophyllaceae [BOT] A family of dicotyledonous plants in the order Caryophyllales differing from the other families in lacking betalains.

Caryophyllales [BOT] An order of dicotyledonous plants in the subclass Caryophyllidae characterized by free-central or basal placentation.

caryophyllene [ORG CHEM] $C_{15}H_{24}$ A liquid sesquiterpene that is found in some essential oils, particularly clove oil.

Caryophyllidae [BOT] A relatively small subclass of plants in the class Magnoliopsida characterized by trinucleate pollen, ovules with two integuments, and a multilayered nucellus.

caryophyllin [ORG CHEM] $C_{30}H_{48}O_3$ A ketone, soluble in alcohol, extracted from oil of cloves.

caryophyllus oil *See* oil of cloves.

caryopsis [BOT] A small, dry, indehiscent fruit having a single seed with such a thin, closely adherent pericarp that a single body, a grain, is formed.

Cas *See* Cassiopeia.

casaba melon [BOT] *Cucumis melo.* A winter muskmelon with a yellow rind and sweet flesh belonging to the family Cucurbitaceae of the order Violales.

Casale process [CHEM ENG] A process, of long use, that employs promoted iron oxide catalyst for synthesis of ammonia from nitrogen and hydrogen.

cascade [ELEC] An electric-power circuit arrangement in which circuit breakers of reduced interrupting ratings are used in the branches, the circuit breakers being assisted in their protection function by other circuit breakers which operate almost instantaneously. Also known as backup arrangement. [ELECTR] *See* avalanche. [ENG] An arrangement of separation devices, such as isotope separators, connected in series so that they multiply the effect of each individual device. [GEOL] A landform structure formed by gravity collapse, consisting of a bed that buckles into a series of folds as it slides down the flanks of an anticline. [HYD] A small waterfall or series of falls descending over rocks.

cascade amplifier [ELECTR] A vacuum-tube amplifier containing two or more stages arranged in the conventional series manner. Also known as multistage amplifier.

cascade-amplifier klystron A klystron having three resonant cavities to provide increased power amplification and output; the extra resonator, located between the input and output resonators, is excited by the bunched beam emerging from the first resonator gap and produces further bunching of the beam.

cascade compensation [CONT SYS] Compensation in which the compensator is placed in series with the forward transfer function. Also known as series compensation; tandem compensation.

cascade connection [ELECTR] A series connection of ampli-

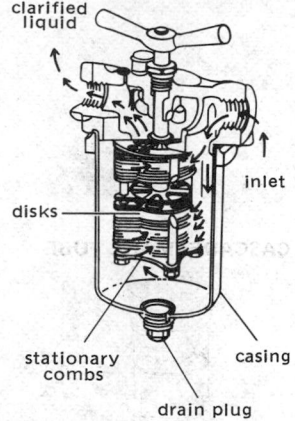

CARTRIDGE FILTER

Diagram of a cartridge filter. *(From W. L. McCabe and J. C. Smith, Unit Operations of Chemical Engineering, McGraw-Hill, 1956)*

CARTRIDGE FUSE

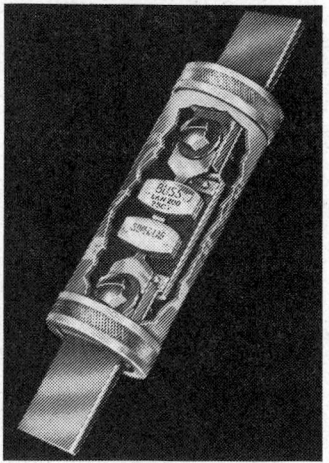

Renewable cartridge fuse, 200 amperes at 250 volts. *(Bussman Manufacturing Division, McGraw Edison Co.)*

CAR TUNNEL KILN

View of medium-sized car tunnel kiln used to fire vitrified ceramic floor tile. *(Swindell-Dressler, Inc.)*

fier stages, networks, or tuning circuits in which the output of one feeds the input of the next. Also known as tandem connection.

cascade control [CONT SYS] An automatic control system in which various control units are linked in sequence, each control unit regulating the operation of the next control unit in line.

cascade cooler [CHEM ENG] Fluid-cooling device through which the fluid flows in a series of horizontal tubes, one above the other; cooling water from a trough drips over each tube, then to a drain. Also known as serpentine cooler; trickle cooler.

cascaded [ENG] Of a series of elements or devices, arranged so that the output of one feeds directly into the input of another, as a series of dynodes or a series of airfoils.

cascaded carry [ADP] A carry process in which the addition of two numerals results in a sum numeral and a carry numeral that are in turn added together, this process being repeated until no new carries are generated.

cascaded feedback canceler [ELECTR] Sophisticated moving-target-indicator canceler which provides clutter and chaff rejection. Also known as velocity shaped canceler.

cascade gamma emission [NUC PHYS] The emission by a nucleus of two or more gamma rays in succession.

cascade hyperon See xi hyperon.

cascade image tube [ELECTR] An image tube having a number of sections stacked together, the output image of one section serving as the input for the next section; used for light detection at very low levels.

cascade impactor [ENG] A low-speed impaction device for use in sampling both solid and liquid atmospheric suspensoids; consists of four pairs of jets (each of progressively smaller size) and sampling plates working in series and designed so that each plate collects particles of one size range.

cascade junction [ELECTR] Two *pn* semiconductor junctions in tandem such that the condition of the first governs that of the second.

cascade limiter [ELECTR] A limiter circuit that uses two vacuum tubes in series to give improved limiter operation for both weak and strong signals in a frequency-modulation receiver. Also known as double limiter.

cascade mixer-settler [CHEM ENG] Series of liquid-holding vessels with stirrers, each connected to an unstirred vessel in which solids or heavy immiscible liquids settle out of suspension; light liquid moves through the mixer-settler units, counterflowing to heavy material, so that fresh liquid contacts treated heavy material, and spent (used) liquid contacts fresh (untreated) heavy material.

cascade networks [ELEC] Two networks in tandem such that the output of the first feeds the input of the second.

cascade noise [ELECTR] The noise in a communications receiver after an input signal has been subjected to two tandem stages of amplification.

cascade particle See xi hyperon; xi-minus particle.

cascade pulverizer [MECH ENG] A form of tumbling pulverizer that uses large lumps to do the pulverizing.

cascade sequence [MET] Combined longitudinal and build-up sequence in which weld beads are deposited in overlapping layers.

cascade shower [PARTIC PHYS] A cosmic-ray shower of electrons, positrons, and gamma rays which grows by pair production and bremsstrahlung events.

cascade transformer [ELEC] A source of high voltage that is made up of a collection of step-up transformers; secondary windings are in series, and primary windings, except the first, are supplied from a pair of taps on the secondary winding of the preceding transformer.

cascade tray [CHEM ENG] A fractionating apparatus that consists of a series of parallel troughs arranged in stairstep fashion.

cascade unit See radiation length.

Cascadian orogeny [GEOL] Post-Tertiary deformation of the crust of the earth in western North America.

cascading [ELEC] An effect in which a failure of an electrical power system causes this system to draw excessive amounts of power from power systems which are interconnected with it, causing them to fail, and these systems cause adjacent

systems to fail in a similar manner, and so forth. [MECH ENG] An effect in ball-mill rotating devices when the upper level of crushing bodies breaks clear and falls to the top of the crop load.

cascading glacier [HYD] A glacier broken by numerous crevasses because of passing over a steep irregular bed, giving the appearance of a cascading stream.

cascarilla oil [MATER] An essential oil from cascarilla bark (*Croton eluteria*), containing cascarillic acid, $C_{11}H_{20}O_2$.

cascode amplifier [ELECTR] An amplifier consisting of a grounded-cathode input stage that drives a grounded-grid output stage; advantages include high gain and low noise; widely used in television tuners.

case [ADP] In computers, a set of data to be used by a particular program. [ENG] An item designed to hold a specific item in a fixed position by virtue of conforming dimensions or attachments; the item which it contains is complete in itself for removal and use outside the container. [MET] Outer layer of a ferrous alloy which has been made harder than the core by case hardening. [MIN ENG] A small fissure admitting water into the mine workings. [PETRO ENG] To line a borehole with steel tubing, such as casing or pipe.

caseation necrosis [PATH] Tissue death involving loss of cellular integrity with the consequent conversion to a cheeselike substance; typical in tuberculosis.

case bay [BUILD] A division of a roof or floor, consisting of two principal rafters and the joists between them.

cased glass [MATER] Glass composed of two or more layers of different colors.

case I firing See case I pointing.

case II firing See case II pointing.

case III firing See case III pointing.

case hardening [GEOL] Formation of a mineral coating on the surface of porous rock by evaporation of a mineral-bearing solution. [MET] Process of carburizing low-carbon steel or other ferrous alloy for making the outer layer (case) harder than the core.

casein [ORG CHEM] The protein of milk; a white solid soluble in acids.

casein-formaldehyde [ORG CHEM] A modified natural polymer.

casein glue [MATER] A produce of dried curds of milk, lime, and other chemical ingredients, mixed cold; used for both plywood and assembly work.

casein paint [MATER] A paint with casein substituted for linseed oil.

casein plastic [MATER] A plastic made with casein and used for buttons, beads, knitting needles, and novelties; thin sheets and rods of casein plastic are cured (hardened) in formaldehyde baths.

casemate [ORD] A bombproof structure used as a powder magazine, gun emplacement, or the like.

caseous lymphadenitis [VET MED] A chronic bacterial disease of sheep and goats caused by *Corynebacterium pseudotuberculosis*, characterized by caseation of the lymph glands and sometimes the lungs, liver, spleen, and kidneys.

case I pointing [ORD] Direct pointing, laying, or firing; gun pointing, in which direction and elevation are set with sight or telescope pointed at the target. Also known as case I firing.

case II pointing [ORD] Combined direct and indirect pointing, laying, or firing; gun pointing, in which direction is set with a sight or telescope pointed at the target; the elevation is set with an elevation quadrant, range quadrant, or range disk. Also known as case II firing.

case III pointing [ORD] Indirect pointing, laying, or firing; gun pointing, in which direction is set with an azimuth circle or with a sight or telescope pointed at an aiming point other than the target; the elevation is set with an elevation quadrant, range quadrant, or range disk. Also known as case III firing.

case primer [ORD] Primer intended to be assembled into the cartridge case of case ammunition.

case shot [ORD] A shell loaded with shrapnel.

cashew [BOT] *Anacardium occidentale.* An evergreen tree of the order Sapindales grown for its kidney-shaped edible nuts and resinous oil.

cashmere [TEXT] A natural textile fiber obtained from the

Cashmere goat of the Himalayan region of China and India.

Casimir–du Pré theory [SOLID STATE] A theory of spin-lattice relaxation which treats the lattice and spin systems as distinct thermodynamic systems in thermal contact with one another.

casing [PETRO ENG] A special steel tubing welded or screwed together and lowered into a borehole to prevent entry of loose rock, gas, or liquid into the borehole or to prevent loss of circulation liquid into porous, cavernous, or crevassed ground.

casing hanger See hanger.

casinghead [PETRO ENG] A fitting at the head of an oil or gas well that allows the pumping operation to take place, as well as the separation of oil and gas.

casinghead gas [MATER] The natural gas that is emitted from the mouth or opening of an oil well.

casinghead gasoline [MATER] Liquid hydrocarbon product removed from casinghead gas by absorption, compression, or refrigeration.

casinghead tank [PETRO ENG] Storage tank for natural gasoline or other liquids with vapor pressures between 4 and 40 psig; intermediate between a general-purpose tank and a compressed-gas tank.

casing joint [PETRO ENG] Joint or union that connects two lengths of pipe used to form an oil-well casing.

casing log [PETRO ENG] Recorded data of a down-hole inspection of an oil or gas well made to determine some characteristic of the formations penetrated by the drill hole; types of logs include resistivity, induction, radioactivity, geologic, temperature, and acoustic.

casing nail [DES ENG] A nail about half a gage thinner than a common wire nail of the same length.

casing shoe [ENG] A ring with a cutting edge on the bottom of a well casing.

casing spear [PETRO ENG] An instrument used for recovering casing which has accidentally fallen into the well.

casing tester [PETRO ENG] A closely fitting, rubber-flanged bucket or a similar tool let down in a well to determine the location of a leak in the casing.

cask See coffin.

cask buoy [NAV] A buoy in the shape of a cask.

casket See coffin.

Casparian strip [BOT] A thin band of suberin- or lignin-like deposition in the radial and transverse walls of certain plant cells during the primary development phase of the endodermis.

Cassadagan [GEOL] Middle Upper Devonian geologic time, above Chemungian.

cassava [BOT] *Manihot esculenta*. A shrubby perennial plant grown for its starchy, edible tuberous roots. Also known as manihot; manioc.

Cassegrain antenna [ELECTROMAG] A microwave antenna in which the feed radiator is mounted at or near the surface of the main reflector and aimed at a mirror at the focus; energy from the feed first illuminates the mirror, then spreads outward to illuminate the main reflector.

Cassegrain telescope [OPTICS] A reflecting telescope in which a small hyperboloidal mirror reflects the convergent beam from the paraboloidal primary mirror through a hole in the primary mirror to an eyepiece in back of the primary mirror.

Casselian See Chattian.

cassette [ENG] A light-tight container designed to hold photographic film or plates. [ENG ACOUS] A small, compact container that holds a magnetic tape and can be readily inserted into a matching tape recorder for recording or playback; the tape passes from one hub within the container to the other hub.

cassette-cartridge system [ADP] An input system often used in minicomputers; its low cost and ease in mounting often offset its slow access time.

cassia oil [MATER] An essential oil extracted from the bark of Chinese cinnamon (*Cinnamomum cassia*), and containing cinnamaldehyde. Also known as Chinese oil of cinnamon.

Cassiar orogeny [GEOL] Orogenic episode in the Canadian Cordillera during late Paleozoic time.

Cassidulinacea [INV ZOO] A superfamily of marine, benthic

foraminiferans in the suborder Rotaliina, characterized by a test of granular calcite with monolamellar septa.

Cassiduloida [INV ZOO] An order of exocyclic Euechinoidea possessing five similar ambulacra which form petal-shaped areas (phyllodes) around the mouth.

cassidyite [MINERAL] $Ca_2(Ni,Mg)(PO_4)_2 \cdot 2H_2O$ A mineral found in meteorites.

Cassini's division [ASTRON] The dark ring, 2500 miles (4000 kilometers) wide, that separates ring A from ring B of the planet Saturn.

Cassiopeia [ASTRON] A constellation with right ascension 1 hour, declination 60°N. Abbreviated Cas.

cassiterite [MINERAL] SnO_2 A yellow, black, or brown mineral that crystallizes in the tetragonal system in prisms terminated by dipyramids; the most important ore of tin. Also known as tin stone.

cassowary [VERT ZOO] Any of three species of large, heavy, flightless birds composing the family Casuariidae in the order Casuariiformes.

cast [ENG] 1. To form a liquid or plastic substance into a fixed shape by letting it cool in the mold. 2. Any object which is formed by placing a castable substance in a mold or form and allowing it to solidify. Also known as casting. [MED] 1. A rigid dressing used to immobilize a part of the body. 2. See strabismus. [NAV] 1. To turn a ship in its own water. 2. To turn a ship to a desired direction without gaining headway or sternway. 3. To take a sounding with the lead. [OPTICS] A change in a color because of the adding of a different hue. [PALEON] A fossil reproduction of a natural object formed by infiltration of a mold of the object by waterborne minerals. [PHYSIO] A mass of fibrous material or exudate having the form of the body cavity in which it has been molded; classified from its source, such as bronchial, renal, or tracheal.

castable [MATER] A refractory aggregate mixed with a bonding agent such as aluminous hydraulic cement which, with addition of water, will develop structural strength and set in a mold.

Castaing-Slodzian mass analyzer See direct-imaging mass analyzer.

castaneous [BIOL] Chestnut-colored.

caste [INV ZOO] One of the levels of mature social insects in a colony that carry out a specific function; examples are workers and soldiers.

castellanus [METEOROL] A cloud species with at least a fraction of its upper part presenting some vertically developed cumuliform protuberances (some of which are more tall than wide) which give the cloud a crenellated or turreted appearance. Previously known as castellatus.

castellated bit [DES ENG] 1. A long-tooth, sawtooth bit. 2. A diamond-set coring bit with a few large diamonds or hard metal cutting points set in the face of each of several upstanding prongs separated from each other by deep waterways. Also known as padded bit.

castellated nut [DES ENG] A type of hexagonal nut with a cylindrical portion above through which slots are cut so that a cotter pin or safety wire can hold it in place.

castellatus See castellanus.

caster [ENG] 1. The inclination of the kingpin or its equivalent in automotive steering, which is positive if the kingpin inclines forward, negative if it inclines backward, and zero if it is vertical as viewed along the axis of the front wheels. 2. A wheel which is free to swivel about an axis at right angles to the axis of the wheel, used to support trucks, machinery, or furniture.

cast-film extrusion See chill-roll extrusion.

Castigliano's principle See Castigliano's theorem.

Castigliano's theorem [MECH] The theorem that the component in a given direction of the deflection of the point of application of an external force on an elastic body is equal to the partial derivative of the work of deformation with respect to the component of the force in that direction. Also known as Castigliano's principle.

Castile soap [MATER] A white, odorless, hard soap made from sodium hydroxide and olive oil.

casting See cast.

casting alloy [MET] An alloy which cannot be forged or rolled and can be shaped only as a casting.

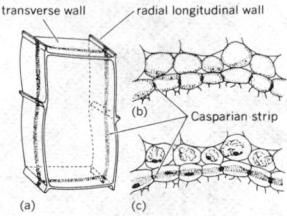
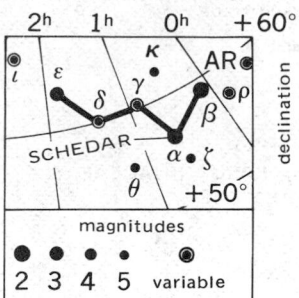

CASTOR BEAN

Ricinus communis, the castor bean plant. *(a)* Plant grows 3–40 feet (0.9–12 meters) high. *(b)* Pods of the fruit cluster liberate smooth, distinctively marked seeds.

CASUARINACEAE

5 cm

Casuarina equisetifolia, typical member of the family Casuarinaceae and order Casuarinales. *(Photograph by W. D. Brush)*

CAT

The Angora cat, noted for its fine, thick fur.

casting copper [MET] Copper used for making foundry castings; obtained from copper ores, and inferior to electrolytic copper.

casting ladle [MET] A refractory-lined steel ladle used to transport molten metal from the furnace to a mold.

casting-out nines [MATH] A method of checking the correctness of elementary arithmetical operations, based on the fact that an integer yields the same remainder as the sum of its decimal digits, when divided by 9.

casting plaster [MATER] A white plaster used for castings and carvings.

castings [ENG] Any term or symbol designating a low-quality drill diamond. [GEOL] *See* fecal pellets.

casting shrinkage [MET] **1.** Total reduction in volume of a casting due to partial reductions at each stage of solidification. **2.** Reduction in volume at each stage of solidification of a casting.

casting slip [MATER] A slurry of clay and additives mixed in water with deflocculating agents and used for casting in molds.

casting strain [MECH] Any strain that results from the cooling of a casting, causing casting stress.

casting stress [MECH] Any stress that develops in a casting due to geometry and casting shrinkage.

casting wheel [MET] A large turntable with molds mounted on the outer edge; used primarily in the base-metal industries for cast ingots, anodes, and so on.

cast iron [MET] Any carbon-iron alloy cast to shape and containing 1.8–4.5% carbon, that is, in excess of the solubility in austenite at the eutectic temperature. Abbreviated C.I.

cast-iron front [ARCH] A style of architecture characterized by large window areas and cast-iron columns and spandrels.

Castle's intrinsic factor *See* intrinsic factor.

cast loading *See* melt loading.

Castner cell [CHEM ENG] A type of mercury cell used in the commercial production of chlorine and sodium.

Castner process [CHEM ENG] A process used industrially to make high-test sodium cyanide by reacting sodium, glowed charcoal, and dry ammonia gas to form sodamide, which is converted to cyanamide immediately; the cyanamide is converted to cyanide with charcoal.

Castniidae [INV ZOO] The castniids; large diurnal, butterfly-like moths composing the single, small family of the lepidopteran superfamily Castnioidea.

Castnioidea [INV ZOO] A superfamily of neotropical and Indo-Australian lepidopteran insects in the suborder Heteroneura.

Castor [ASTRON] A multiple star of spectral classification A0 in the constellation Gemini; the star α Geminorum.

castor bean [BOT] The seed of the castor oil plant (*Ricinus communis*), a coarse, erect annual herb in the spurge family (Euphorbiaceae) of the order Geraniales.

castoreum gland [VERT ZOO] A preputial scent gland in the beaver.

castorite [MINERAL] A transparent variety of petalite occurring in crystals.

castor machine oil [MATER] Petroleum-base lubricating oil thickened with an aluminum-base soap, such as aluminum oleate.

castor oil [MATER] A colorless or greenish nondrying oil extracted from the castor bean; used as a cathartic, in soap, and after processing as a lubricant, and as a leather preservative. Also known as ricinus oil.

castor oil acid *See* ricinoleic acid.

castration [MED] Removing, or inhibiting the function or development of, the ovaries or testes.

castration anxiety [PSYCH] Anxiety due to the fear of loss of the genitals or injury to them.

cast setting *See* mechanical setting.

cast steel [MET] Steel shaped by casting.

cast stone [MATER] Building stone molded from concrete so that it resembles natural stone.

cast structure [MET] The microstructure of a casting.

cast-weld [MET] Joining parts by pouring molten metal over them in a mold.

casual carrier [MED] A person who carries an infectious microorganism but never manifests the disease.

Casuariidae [VERT ZOO] The cassowaries, a family of flightless birds in the order Casuariiformes lacking head and neck feathers and having bony casques on the head.

Casuariiformes [VERT ZOO] An order of large, flightless, ostrichlike birds of Australia and New Guinea.

Casuarinaceae [BOT] The single, monogeneric family of the plant order Casuarinales characterized by reduced flowers and green twigs bearing whorls of scalelike, reduced leaves.

Casuarinales [BOT] A monofamilial order of dicotyledonous plants in the subclass Hamamelidae.

cat [NAV ARCH] **1.** A sturdy tackle used for bringing an anchor up to a ship's cathead. **2.** To hoist an anchor up to the cathead of a ship. [VERT ZOO] The common name for all members of the carnivoran mammalian family Felidae, especially breeds of the domestic species, *Felis domestica*.

CAT *See* clear-air turbulence.

catabiosis [PHYSIO] Degenerative changes accompanying cellular senescence.

catabolism [BIOCHEM] That part of metabolism concerned with the breakdown of large protoplasmic molecules and tissues, often with the liberation of energy.

catabolite [BIOCHEM] Any product of catabolism.

catabolite repression [BIOCHEM] An intracellular regulatory mechanism in bacteria whereby glucose, or any other carbon source that is an intermediate in catabolism, prevents formation of inducible enzymes.

catachosis [GEOL] Fracturing or crushing of rock during metamorphism.

cataclasis [GEOL] Deformation of rock by fracture and rotation of aggregates or mineral grains.

cataclasite *See* cataclastic rock.

cataclastic metamorphism [PETR] Local metamorphism restricted to a region of faults and overthrusts involving purely mechanical forces resulting in cataclasis.

cataclastic rock [PETR] Rock containing angular fragments formed by cataclasis. Also known as cataclasite.

cataclastic structure *See* mortar structure.

cataclysmic variable [ASTRON] A star showing a sudden increase in the magnitude of light, followed by a slow fading of light; examples are novae and supernovae. Also known as explosive variable.

catadioptric [OPTICS] Involving both reflection and refraction of light.

catadromous [VERT ZOO] Pertaining to fishes which live in fresh water and migrate to spawn in salt water.

catagelophobia [PSYCH] An abnormal fear of ridicule.

Catalan forge [MET] A furnace or forge for making wrought iron from ore, in which the ore loaded at front and charcoal at rear are covered with fine ore and charcoal dust moistened with water.

catalase [BIOCHEM] An enzyme that catalyzes the decomposition of hydrogen peroxide into oxygen and water and the oxidation by hydrogen peroxide of alcohols to aldehydes.

catalectrotonus [PHYSIO] The negative electric potential during the passage of a current on the surface of a nerve or muscle in the region of the cathode.

catalepsy [PSYCH] Suspended animation with loss of voluntary motion associated with hysteria and the schizophrenic reactions in man, and with organic nervous system disease in animals.

catalog [ADP] **1.** All the indexes to data sets or files in a system. **2.** The index to all other indexes; the master index. **3.** To add an entry to an index or to build an entire new index.

cataloged procedure [ADP] A group of control cards (job control language statements) that has been placed in a cataloged data set.

catalog number [ASTRON] The designation of a star composed of the name of a particular star catalog and the number of the star as listed there.

catalysis [CHEM] A phenomenon in which a relatively small amount of substance augments the rate of a chemical reaction without itself being consumed.

catalyst [CHEM] Substance that alters the velocity of a chemical reaction and may be recovered essentially unaltered in form and amount at the end of the reaction.

catalyst carrier [CHEM] A neutral material used to support a catalyst, such as activated carbon, diatomaceous earth, or activated alumina.

catalyst selectivity [CHEM] **1.** The relative activity of a catalyst in reference to a particular compound in a mixture. **2.** The relative rate of a single reactant in competing reactions.

catalyst stripping [CHEM ENG] Introduction of steam to remove hydrocarbons retained on the catalyst; the steam is introduced where the spent catalyst leaves the reactor.

catalytic activity [CHEM ENG] The ratio of the space velocity of a catalyst being tested, to the space velocity required for a standard catalyst to give the same conversion as the catalyst under test.

catalytically cracked gasoline [MATER] Gasoline produced in a refinery reactor equipped with catalytic cracking equipment.

catalytic converter [CHEM ENG] A cracker utilizing catalysts.

catalytic cracking [CHEM ENG] Conversion of high-boiling hydrocarbons into lower-boiling types by a catalyst.

catalytic hydrogenation [CHEM ENG] Hydrogenating by means of catalysts such as nickel or palladium.

catalytic polymerization [CHEM ENG] Polymerization of monomers to form high-molecular-weight molecules in the presence of catalysts.

catalytic reforming [CHEM ENG] Rearranging of hydrocarbon molecules in a gasoline boiling-range feedstock to form hydrocarbons having a higher antiknock quality. Abbreviated CR.

catamaran [NAV ARCH] **1.** A sailing or powered boat having two rather slender hulls joined by a deck or other structure. **2.** A rectangular raft resting on two parallel cylindrical floats.

catamorphism *See* katamorphism.

catamount *See* puma.

cat-and-mouse engine [MECH ENG] A type of rotary engine, typified by the Tschudi engine, which is an analog of the reciprocating piston engine, except that the pistons travel in a circular motion. Also known as scissor engine.

cataphoresis *See* electrophoresis.

cataplelite [MINERAL] $(Na_2,Ca)ZrSi_3O_9 \cdot 2H_2O$ A yellow or yellowish-brown mineral crystallizing in the hexagonal system, consisting of a hydrous silicate of sodium, calcium, and zirconium, and occurring in thin tabular crystals; hardness is 6 on Mohs scale, and specific gravity is 2.8.

cataplexy [MED] **1.** Sudden loss of muscle tone provoked by exaggerated emotion, such as excessive anger or laughter, often associated with a profound desire for sleep. **2.** Prostration by the sudden onset of disease. **3.** Hypnotic sleep.

Catapochrotidae [INV ZOO] A monospecific family of coleopteran insects in the superfamily Cucujoidea.

catapult [AERO ENG] **1.** A power-actuated machine or device for hurling an object at high speed, for example, a device which launches aircraft from a ship deck. **2.** A device, usually explosive, for ejecting a person from an aircraft. [ORD] A mechanical device for hurling grenades or bombs.

cataract [HYD] A waterfall of considerable volume with the vertical fall concentrated in one sheer drop. [MED] An opacity in the crystalline lens or the lens capsule of the eye.

cataracting [MECH ENG] A motion of the crushed bodies in a ball mill in which some, leaving the top of the crop load, fall with impact to the toe of the load.

catarrh [MED] An old term for an inflammation of mucous membranes, particularly of the respiratory tract.

catarrhal conjunctivitis [MED] A usually acute inflammation of the conjunctiva with smarting of the eyes, heaviness of the lids, photophobia, and excessive mucous or mucopurulent secretion, caused by a variety of contagious organisms, but sometimes becoming chronic as a sequela of the acute form or because of irritation from polluted atmosphere or allergic factors. Also known as pinkeye.

catarrhal jaundice *See* infectious hepatitis.

catastrophism [GEOL] The theory that most features in the earth were produced by sudden, short-lived, worldwide events. [PALEON] The theory that the differences between fossils in successive stratigraphic horizons resulted from a general catastrophe followed by creation of the different organisms found in the next-younger beds.

catatonia [PSYCH] A type of schizophrenic reaction in which the individual remains speechless and motionless, assumes fixed postures, and lacks the will and resists attempts to activate speech and movement.

catazone [GEOL] The deepest zone of rock metamorphism where high temperatures and pressures prevail.

cat back [NAV ARCH] A piece of rope sometimes attached to the hook of a cat block to assist in hooking the ring of the anchor. Also known as backrope.

cat block [NAV ARCH] A heavy block with a hook, used in hoisting a ship's anchor up to the cathead.

catch basin [CIV ENG] **1.** A basin at the point where a street gutter empties into a sewer, built to catch matter that would not easily pass through the sewer. **2.** A well or reservoir into which surface water may drain off.

catcher [ELECTR] Electrode in a velocity-modulated vacuum tube on which the spaced electron groups induce a signal; the output of the tube is taken from this element.

catching diode [ELECTR] Diode connected to act as a short circuit when its anode becomes positive; the diode then prevents the voltage of a circuit terminal from rising above the diode cathode voltage.

catchlights [GRAPHICS] Reflections of light sources in a subject's eyes or eyeglasses.

catchment area *See* drainage basin.

catchment glacier *See* snowdrift glacier.

catch pit [MIN ENG] In mineral processing, a sump in a mill to which the floor slopes gently; spillage gravitates or is hosed to this area.

catchwater [CIV ENG] A ditch for catching water on sloping land.

cat cracker [CHEM ENG] A refinery unit where catalytic cracking is done.

catechol [ORG CHEM] One of a group of three isomeric dihydroxy benzenes in which the two hydroxyl groups are ortho to each other. Also known as catechin; pyrocatechol; pyrocatechuic acid.

catecholamine [BIOCHEM] Any one of a group of sympathomimetic amines containing a catechol moiety, including especially epinephrine, norepinephrine (levarterenol), and dopamine.

categorization [ADP] Process of separating multiple addressed messages to form individual messages for singular addresses.

category [MATH] A class of objects together with a set of morphisms for each pair of objects and a law of composition for morphisms; sets and functions form an important category, as do groups and homomorphisms.

catenary [MATH] The curve obtained by suspending a uniform chain by its two ends; the graph of the hyperbolic cosine function.

catenary suspension [ENG] Holding a flexible wire or chain aloft by its end points; the wire or chain takes the shape of a catenary.

catenation [CHEM] The property of an element to link to itself to form molecules as with carbon.

catenulate [BIOL] Having a chainlike form.

Catenulida [INV ZOO] An order of threadlike, colorless freshwater rhabdocoeles with a simple pharynx and a single, median protonephridium.

caterer problem [MATH] A linear programming problem in which it is required to find the optimal policy for a caterer who must choose between buying new napkins and sending them to either a fast or a slow laundry service.

caterpillar [INV ZOO] **1.** The wormlike larval stage of a butterfly or moth. **2.** The larva of certain insects, such as scorpion flies and sowflies. [MECH ENG] A vehicle, such as a tractor or army tank, which runs on two endless belts, one on each side, consisting of flat treads and kept in motion by toothed driving wheels.

caterpillar chain [DES ENG] A short, endless chain on which dogs (grippers) or teeth are arranged to mesh with a conveyor.

caterpillar gate [CIV ENG] A steel gate carried on crawler tracks that is used to control water flow through a spillway.

catfish [VERT ZOO] The common name for a number of fishes which constitute the suborder Siluroidei in the order Cypriniformes, all of which have barbels around the mouth.

catforming [CHEM ENG] A naphtha-reforming process with a catalyst of platinum-silica-alumina which results in very high hydrogen purity.

catgut [MATER] A thin cord made from the submucosa of sheep and other animal intestine; used for sutures and

CATECHOL

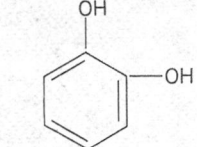

Structural formula of catechol.

CATENULIDA

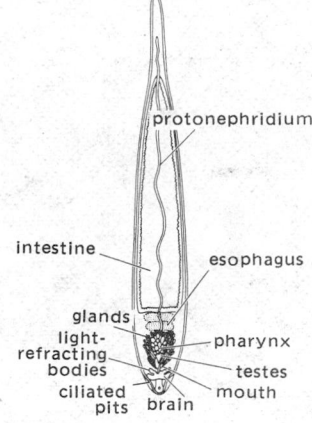

protonephridium

intestine

esophagus

glands

light-refracting bodies

pharynx

ciliated pits

testes

brain

mouth

Stenostomum grande, the commonest and most widely distributed of all freshwater rhabdocoeles, in the order Catenulida. (*After Nuttycombe and Waters, 1938*)

CATERPILLAR

Forest tent caterpillar (*Malacosoma disstria*).

CATFISH

Brown bullhead (*Ictaturus nebulosus*), a catfish found in warm mud-bottomed waters across the United States.

ligatures, for strings of musical instruments, and for tennis racket strings. Also known as gut.

catharsis [PSYCH] Release of tension by releasing deep-seated emotions or reliving a traumatic experience.

cathartic [PHARM] Any drug, such as castor oil, mineral oil, or a laxative, that causes defecation.

Cathartidae [VERT ZOO] The New World vultures, a family of large, diurnal predatory birds in the order Falconiformes that lack a voice and have slightly webbed feet.

cathead [NAV ARCH] A projection on a ship's bow which supports tackle for hoisting an anchor, and to which the anchor is secured.

cathedral glass [MATER] Unpolished translucent sheet glass.

cathepsin [BIOCHEM] Any of several proteolytic enzymes occurring in animal tissue that hydrolyze high-molecular-weight proteins to proteoses and peptones.

catheter [MED] A hollow, tubular device for insertion into a cavity, duct, or vessel to permit injection or withdrawal of fluids or to establish patency of the passageway.

catheterization [MED] Insertion or use of a catheter.

cathetometer [ENG] An instrument for measuring small differences in height, for example, between two columns of mercury.

Cathie unit [BIOL] A unit for the standardization of streptokinase.

cathode [ELEC] The positively charged pole of a primary cell or a storage battery. [ELECTR] **1.** The primary source of electrons in an electron tube; in directly heated tubes the filament is the cathode, and in indirectly heated tubes a coated metal cathode surrounds a heater. Designated K. **2.** The terminal of a semiconductor diode that is negative with respect to the other terminal when the diode is biased in the forward direction. [PHYS CHEM] The electrode at which reduction takes place in an electrochemical cell, that is, a cell through which electrons are being forced.

cathode bias [ELECTR] Bias obtained by placing a resistor in the common cathode return circuit, between cathode and ground; flow of electrode currents through this resistor produces a voltage drop that serves to make the control grid negative with respect to the cathode.

cathode cleaning [MET] Electrolytically removing soil from work connected as the cathode.

cathode copper [MET] Copper deposited at the cathode during electrolytic refining; it is melted and marketed as electrolytic copper.

cathode corrosion [MET] **1.** Corrosion of the cathode of an electrochemical circuit, usually caused by production of alkaline reaction products. **2.** Corrosion of the cathodic member of a galvanic couple.

cathode-coupled amplifier [ELECTR] A cascade amplifier in which the coupling between two stages is provided by a common cathode resistor.

cathode coupling [ELECTR] Use of an input or output element in the cathode circuit for coupling energy to another stage.

cathode dark space [ELECTR] The relatively nonluminous region between the cathode glow and the negative flow in a glow-discharge cold-cathode tube. Also known as Crookes dark space; Hittorf dark space.

cathode disintegration [ELECTR] The destruction of the active area of a cathode by positive-ion bombardment.

cathode drop [ELECTR] The voltage between the arc stream and the cathode of a glow-discharge tube. Also known as cathode fall.

cathode efficiency [CHEM ENG] The proportion of current used for completion of a given process at the cathode.

cathode emission [ELECTR] A process whereby electrons are emitted from the cathode structure.

cathode fall *See* cathode drop.

cathode follower [ELECTR] A vacuum-tube circuit in which the input signal is applied between the control grid and ground, and the load is connected between the cathode and ground. Also known as grounded-anode amplifier; grounded-plate amplifier.

cathode glow [ELECTR] The luminous glow that covers all or part of the cathode in a glow-discharge cold-cathode tube.

cathode interface capacitance [ELECTR] A capacitance

which, when connected in parallel with an appropriate resistance, forms an impedance approximately equal to the cathode interface impedance. Also known as layer capacitance.

cathode interface impedance [ELECTR] The impedance between the cathode base and coating in an electron tube, due to a high-resistivity layer or a poor mechanical bond. Also known as layer impedance.

cathode keying [ELECTR] Transmitter keying by means of a key in the cathode lead of the keyed vacuum-tube stage, opening the direct-current circuits for the grid and anode simultaneously.

cathode modulation [ELECTR] Amplitude modulation accomplished by applying the modulating voltage to the cathode circuit of an electron tube in which the carrier is present.

cathode ray [ELECTR] A stream of electrons, such as that emitted by a heated filament in a tube, or that emitted by the cathode of a gas-discharge tube when the cathode is bombarded by positive ions.

cathode-ray oscillograph [ELECTR] A cathode-ray oscilloscope in which a photographic or other permanent record is produced by the electron beam of the cathode-ray tube.

cathode-ray oscilloscope [ELECTR] A test instrument that uses a cathode-ray tube to make visible on a fluorescent screen the instantaneous values and waveforms of electrical quantities that are rapidly varying as a function of time or another quantity. Abbreviated CRO. Also known as oscilloscope; scope.

cathode-ray output [ADP] A cathode-ray tube used in a computer to display output information in graphic form or by character representation.

cathode-ray storage tube [ELECTR] A storage tube in which the information is written by means of a cathode-ray beam.

cathode-ray tube [ELECTR] An electron tube in which a beam of electrons can be focused to a small area and varied in position and intensity on a surface. Abbreviated CRT. Originally known as Braun tube; also known as electron-ray tube.

cathode-ray tuning indicator [ELECTR] A small cathode-ray tube having a fluorescent pattern whose size varies with the voltage applied to the grid; used in radio receivers to indicate accuracy of tuning and as a modulation indicator in some tape recorders. Also known as electric eye; electron-ray indicator; magic eye; tuning eye.

cathode-ray voltmeter [ELEC] An instrument consisting of a cathode-ray tube of known sensitivity, whose deflection can be used to measure voltages.

cathode resistor [ELECTR] A resistor used in the cathode circuit of a vacuum tube, having a resistance value such that the voltage drop across it due to tube current provides the correct negative grid bias for the tube.

cathode spot [ELECTR] The small cathode area from which an arc appears to originate in a discharge tube.

cathode sputtering *See* sputtering.

cathodic coating [MET] Material forming a continuous film on a base metal, deposited by mechanical coating or electroplating.

cathodic inhibitor [CHEM ENG] A compound, such as calcium bicarbonate or sodium phosphate, which is deposited on a metal surface in a thin film that operates at the cathodes to provide physical protection over the entire surface against corrosive attack in a conducting medium.

cathodic polarization [PHYS CHEM] Portion of electric cell polarization occurring at the cathode.

cathodic protection [MET] Protecting a metal from electrochemical corrosion by using it as the cathode of a cell with a sacrificial anode. Also known as electrolytic protection.

cathodoluminescence [ELECTR] Luminescence produced when high-velocity electrons bombard a metal in vacuum, thus vaporizing small amounts of the metal in an excited state, which amounts emit radiation characteristic of the metal. Also known as electroluminescence.

cathodophosphorescence [ELECTR] Phosphorescence produced when high-velocity electrons bombard a metal in a vacuum.

catholyte [CHEM] Electrolyte adjacent to the cathode in an electrolytic cell.

CATHODE FOLLOWER

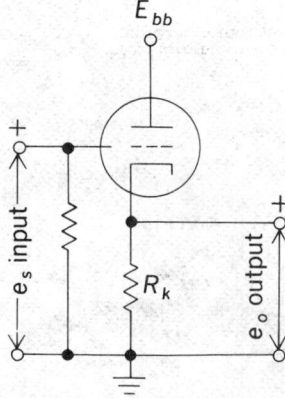

Circuit diagram of a cathode-follower amplifier.

cathomycin [MICROBIO] A trade name for novobiocin.

cation [CHEM] A positively charged atom or group of atoms, or a radical which moves to the negative pole (cathode) during electrolysis.

cation analysis [ANALY CHEM] Qualitative analysis for cations in aqueous solution.

cation exchange [CHEM] A chemical reaction in which hydrated cations of a solid are exchanged, equivalent for equivalent, for cations of like charge in solution.

cation exchange resin [ORG CHEM] A highly polymerized synthetic organic compound consisting of a large, nondiffusible anion and a simple, diffusible cation, which later can be exchanged for a cation in the medium in which the resin is placed.

cationic detergent [CHEM] A member of a group of detergents that have molecules containing a quaternary ammonium salt cation with a group of 12 to 24 carbon atoms attached to the nitrogen atom in the cation; an example is alkyltrimethyl ammonium bromide.

cationic hetero atom [CHEM] A positively charged atom, other than carbon, in an otherwise carbon atomic chain or ring.

cationic polymerization [ORG CHEM] A polymerization in which cationic catalysts are present.

cationic reagent [CHEM] A surface-active agent with active positive ions used for ore beneficiation (flotation via flocculation); an example of a cationic reagent is cetyl trimethyl ammonium bromide.

cationtrophy [CHEM] The breaking off of an ion, such as a hydrogen ion or metal ion, from a molecule so that a negative ion remains in equilibrium.

catkin [BOT] An indeterminate type of inflorescence that resembles a scaly spike and sometimes is pendant.

catoptric light [OPTICS] Light reflected from a mirror, for example, light from a filament, concentrated into a parallel beam by means of a reflector.

Catostomidae [VERT ZOO] The suckers, a family of cypriniform fishes in the suborder Cyprinoidei.

cat scratch disease [MED] A benign systematic illness in man characterized by malaise, fever, and a granulomatous lymphadenitis; the causative organism has not been identified. Also known as benign lymphoreticulosis.

cat's eye [LAP] Any of several gems cut in convex form, principally from crysoberyl, exhibiting opalescent reflections. Also known as cymophane.

cat's paw [METEOROL] A puff of wind; a light breeze affecting a small area, as one that causes patches of ripples on the surface of water.

Cattell infant intelligence scale [PSYCH] A modification of the revised Stanford-Binet test adapted for children from 2 to 30 months of age.

cattle [AGR] Domesticated bovine animals, including cows, steers, and bulls, raised and bred on a ranch or farm.

CATV *See* community antenna television system.

catwalk [ENG] A narrow, raised platform or pathway used for passage to otherwise inaccessible areas, such as a raised walkway on a ship permitting fore and aft passage when the main deck is awash, a walkway on the roof of a freight car, or a walkway along a vehicular bridge.

catwhisker [ELECTR] A sharply pointed, flexible wire used to make contact with the surface of a semiconductor crystal at a point that provides rectification. Also known as whisker.

Cauchy boundary conditions [MATH] The conditions imposed on a surface in euclidean space which are to be satisfied by a solution to a partial differential equation.

Cauchy condensation test [MATH] A monotone decreasing series of positive terms Σa_n converges or diverges, as does $\Sigma p^n a_{p^n}$, for any positive integer p.

Cauchy dispersion formula [OPTICS] A semiempirical formula for the index of refraction n of a medium as a function of wavelength λ, according to which $n = A + (B/\lambda^2)$, where A and B are constants.

Cauchy distribution [STAT] A distribution function having the form $M/[\pi M^2 + (x - a)^2]$, where x is the variable and M and a are constants. Also known as Cauchy frequency distribution.

Cauchy formula [MATH] An expression for the value of an analytic function f at a point z in terms of a line integral

$$f(z) = \frac{1}{2\pi i} \int_c \frac{f(\zeta)}{\zeta - z} d\zeta$$

where C is a simple closed curve containing z. Also known as Cauchy integral formula.

Cauchy frequency distribution *See* Cauchy distribution.

Cauchy inequality [MATH] The square of the sum of the products of two variables for a range of values is less than or equal to the product of the sums of the squares of these two variables for the same range of values.

Cauchy integral formula *See* Cauchy formula.

Cauchy integral test *See* Cauchy's test for convergence.

Cauchy integral theorem [MATH] The theorem that if γ is a closed path in a region R satisfying certain topological properties, then the integral around γ of any function analytic in R is zero.

Cauchy mean [MATH] The Cauchy mean value theorem for the ratio of two continuous functions.

Cauchy mean value theorem [MATH] If f and g are functions satisfying certain conditions on an interval $[a,b]$, then there is a point x in the interval at which the ratio of the derivatives $f'(x)/g'(x)$ equals the ratio of the net change in f, $f(b) - f(a)$, to that of g.

Cauchy number [FL MECH] A dimensionless number used in the study of compressible flow, equal to the density of a fluid times the square of its velocity divided by its bulk modulus. Also known as Hooke number.

Cauchy principal value [MATH] **1.** The Cauchy principal value of

$$\int_{-\infty}^{\infty} f(x)dx \text{ is } \lim_{s \to \infty} \int_{-s}^{s} f(x)dx$$

provided the limit exists. **2.** If a function f is bounded on an interval (a,b) except in the neighborhood of a point c, the Cauchy principal value of

$$\int_a^b f(x)\,dx$$

is

$$\lim_{\delta \to 0} \left[\int_a^{c-\delta} f(x)\,dx + \int_{c+\delta}^b f(x)\,dx \right]$$

provided the limit exists.

Cauchy problem [MATH] The problem of determining the solution of a system of partial differential equation of order m from the prescribed values of the solution and of its derivatives of order less than m on a given surface.

Cauchy product [MATH] A method of multiplying two absolutely convergent series to obtain a series which converges absolutely to the product of the limits of the original series:

$$\left(\sum_{n=0}^{\infty} a_n\right)\left(\sum_{n=0}^{\infty} b_n\right) = \sum_{n=0}^{\infty} c_n \text{ where } c_n = \sum_{k=0}^{n} a_k b_{n-k}.$$

Cauchy radical test [MATH] A test for convergence of series of positive terms: if the nth root of the nth term is less than some number less than unity, the series converges; if it remains equal to or greater than unity, the series diverges.

Cauchy ratio test [MATH] A series of nonnegative terms converges if the limit as n approaches infinity of the ratio of the $(n + 1)$st to nth term is smaller than 1, and diverges if it is greater than 1; the test fails if this limit is 1.

Cauchy residue theorem [MATH] The theorem expressing a line integral around a closed curve of a function which is analytic in a simply connected domain containing the curve, except at a finite number of poles interior to the curve, as a sum of residues of the function at these poles.

Cauchy-Riemann equations [MATH] A pair of partial differential equations satisfied by the real and imaginary parts of a complex function $f(z)$ if and only if the function is analytic: $\partial u / \partial x = \partial v / \partial y$ and $\partial u / \partial y = -\partial v / \partial x$, where $f(z) = u + iv$ and $z = x + iy$.

Cauchy-Schwarz inequality [MATH] The square of the inner-product of two vectors does not exceed the product of the squares of their norms. Also known as Buniakowski's inequality; Schwarz' inequality.

Cauchy sequence [MATH] A sequence with the property that the difference between any two terms is arbitrarily small provided they are both sufficiently far out in the sequence; more precisely stated: a sequence $\{a_n\}$ such that for every $\varepsilon > 0$ there is an integer N with the property that if n and m are

CAULERPACEAE

Caulerpa showing the stolonlike branches, rhizoidal branches, and erect featherlike frond typical of the Caulerpaceae.

CAULOBACTERACEAE

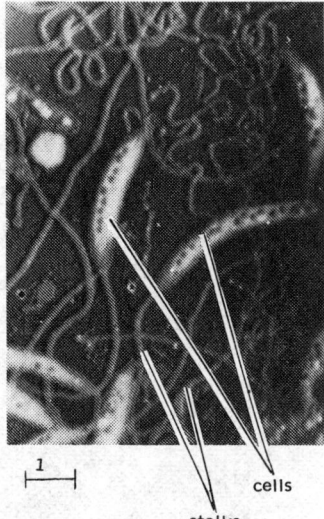

Electron micrograph of *Caulobacter* species in the Caulobacteraceae, showing cells and stalks. *(Courtesy of J. Wachsman)*

both greater than N, then $|a_n - a_m| < \varepsilon$. Also known as fundamental sequence.

Cauchy's test for convergence [MATH] **1.** A series is absolutely convergent if the limit as n approaches infinity of its nth term raised to the $1/n$ power is less than unity. Also known as Maclaurin-Cauchy test. **2.** A series a_n is convergent if there exists a monotonically decreasing function f such that $f(n) = a_n$ for n greater than some fixed number N, and if the integral of $f(x)dx$ from N to ∞ converges. Also known as Cauchy integral test.

Cauchy transcendental equation [MATH] An equation whose roots are characteristic values of a certain type of Sturm-Liouville problem: $\tan \sigma\pi = (k + K)/(\sigma^2 - kK)$, where k and K are given, and σ is to be determined.

cauda equina [ANAT] The roots of the sacral and coccygeal nerves, collectively; so called because of their resemblance to a horse's tail.

caudal [ZOO] Toward, belonging to, or pertaining to the tail or posterior end.

caudal artery [VERT ZOO] The extension of the dorsal aorta in the tail of a vertebrate.

caudal vertebra [ANAT] Any of the small bones of the vertebral column that support the tail in vertebrates; in man, three to five are fused to form the coccyx.

Caudata [VERT ZOO] An equivalent name for Urodela.

caudate [ZOO] **1.** Having a tail or taillike appendage. **2.** Any member of the Caudata.

caudate lobe [ANAT] The tailed lobe of the liver that separates the right extremity of the transverse fissure from the commencement of the fissure for the inferior vena cava.

caudate nucleus [ANAT] An elongated arched gray mass which projects into and forms part of the lateral wall of the lateral ventricle.

caudex [BOT] The main axis of a plant, including stem and roots.

caudicle [BOT] A slender appendage attaching pollen masses to the stigma in orchids.

Cauer form [ELEC] A continued fraction expansion of the impedance used in the network synthesis for a driving point function resulting in a ladder network.

cauldron subsidence [GEOL] **1.** A structure formed by the lowering along a steep ring fracture of a more or less cylindrical block, usually 1 to 10 miles (1.6 to 16 kilometers) in diameter, into a magma chamber. **2.** The process of forming such a structure.

Caulerpaceae [BOT] A family of green algae in the order Siphonales.

caulescent [BOT] Having an aboveground stem.

cauliflorous [BOT] Producing flowers on the older branches or main stem.

cauliflower [BOT] *Brassica oleracea* var. *botrytis.* A biennial crucifer of the order Capparales grown for its edible white head or curd, which is a tight mass of flower stalks.

cauliflower disease [PL PATH] **1.** A disease of the strawberry plant caused by the eelworm and manifested as clustered, puckered, and malformed leaves. **2.** A bacterial disease of the strawberry and some other plants caused by *Corynebacterium fascians.*

cauline [BOT] Belonging to or arising from the stem, particularly if on the upper portion.

caulk Also spelled calk. [ENG] To make a seam or point airtight, watertight, or steamtight by driving in caulking compound, dry pack, lead wool, or other material. [MATER] Material used to caulk seams.

caulking compound [MATER] A heavy paste, such as a synthetic, containing a polysulfide rubber and lead peroxide curing agent, or a natural product such as oakum, used for caulking.

caulking iron [DES ENG] A tool for applying caulking to a seam.

Caulobacteraceae [MICROBIO] A family of stalked, aquatic, gram-negative bacteria in the order Pseudomonadales.

caulocarpic [BOT] Having stems that bear flowers and fruit every year.

caulome [BOT] The stem structure or stem axis of a plant as a whole.

causalgia [MED] A sensation of burning pain, especially of the palms and soles, which may be of psychic or organic origin.

causality [MECH] In classical mechanics, the principle that the specification of the dynamical variables of a system at a given time, and of the external forces acting on the system, completely determines the values of dynamical variables at later times. Also known as determinism. [PHYS] **1.** The principle that an event cannot precede its cause; in a relativistic theory, an event cannot have an effect outside its future light cone. **2.** In relativistic quantum field theory, the principle that the field operators at different space-time points commute (for boson fields; anticommute in the case of fermion fields) if the separation of the points is spacelike. [QUANT MECH] The principle that the specification of the dynamical state of a system at a given time, and of the interaction of the system with its environment, determines the dynamical state of the system at later times, from which a probability distribution for the observation of any dynamical variable may be determined. Also known as determinism. [SCI TECH] The existence of regularities which control natural phenomena.

causal system [CONT SYS] A system whose response to an input does not depend on values of the input at later times. Also known as nonanticipatory system; physical system.

caustic [CHEM] **1.** Burning or corrosive. **2.** A hydroxide of a light metal. [OPTICS] A curve or surface which is tangent to the rays of an initially parallel beam after reflection or refraction in an optical system.

caustic alcohol *See* sodium ethylate.

caustic antimony *See* antimony chloride.

caustic barley *See* sabadilla.

caustic cracking *See* caustic embrittlement.

caustic dip [MET] Immersion of metal into a caustic solution such as sodium hydroxide.

caustic embrittlement [MET] Intercrystalline cracking of steel caused by exposure to caustic solutions above 70°C while under tensile stress; once common in riveted boilers. Also known as caustic cracking.

causticity [CHEM] The property of being caustic.

caustic lime *See* calcium oxide.

caustic potash *See* potassium hydroxide.

caustic soda [INORG CHEM] *See* sodium hydroxide. [MATER] Sodium hydroxide that contains 76–78% sodium oxide; the most important of the commercial caustic materials, used in chemical manufacture, petroleum refining, and pulp and paper manufacture.

caustic wash [CHEM] **1.** Treating a product with a solution of caustic soda to remove impurities. **2.** The solution itself.

caustobiolith [GEOL] Combustible organic rock formed by direct accumulation of plant materials; includes coal peat.

cauterization [MED] Use of a device or chemical agent to coagulate or destroy tissue.

caution area [NAV] An area in which there is a visible hazard to flight or navigation, such as unusual aircraft maneuvers, high density of aircraft, or parachute drops.

cautionary characteristic [NAV] In marine operations, a unique characteristic of light signifying a particular caution, for example, a quick flashing characteristic indicates a sharp turn in a channel.

cavaburd [METEOROL] Shetland Islands term for a thick fall of snow. Also spelled kavaburd.

cavaliers [METEOROL] The local term, in the vicinity of Montpelier, France, for the days near the end of March or the beginning of April when the mistral is usually strongest.

cave [ENG] A pit or tunnel under a glass furnace for collecting ashes or raking the fire. [GEOL] A natural, hollow chamber or series of chambers and galleries beneath the earth's surface, or in the side of a mountain or hill, with an opening to the surface. [MIN ENG] **1.** Fragmented rock materials, derived from the sidewalls of a borehole, that obstruct the hole or hinder drilling progress. Also known as cavings. **2.** The partial or complete failure of borehole sidewalls or mine workings. Also known as cave-in. [NUCLEO] A heavily shielded compartment in which highly radioactive material can be handled, generally by remote control. Also known as hot cell.

cave breccia [GEOL] Sharp fragments of limestone debris deposited on the floor of a cave.

cave formation *See* speleothem.

cave-in *See* cave.

Cavellinidae [PALEON] A family of Paleozoic ostracods in the suborder Platycopa.

Cavendish balance [ENG] An instrument for determining the constant of gravitation, in which one measures the displacement of two small spheres of mass m, which are connected by a light rod suspended in the middle by a thin wire, caused by bringing two large spheres of mass M near them.

cave pearl [GEOL] A small, smooth, rounded concretion of calcite or aragonite, formed by concentric precipitation about a nucleus and usually found in limestone caves.

caver [METEOROL] A gentle breeze in the Hebrides, west of Scotland. Also spelled kaver.

cavern [GEOL] An underground chamber or series of chambers of indefinite extent carved out by rock springs in limestone.

cavernicolous [BIOL] Inhabiting caverns.

cavernous [GEOL] 1. Having many caverns or cavities. 2. Producing caverns. 3. Of or pertaining to a cavern, that is, suggesting vastness.

cavernous sinus [ANAT] Either of a pair of venous sinuses of the dura mater located on the side of the body of the sphenoid bone.

Caviidae [VERT ZOO] A family of large, hystricomorph rodents distinguished by a reduced number of toes and a rudimentary tail.

caving [MIN ENG] A mining procedure, used when the surface is expendable, in which the ore body is undercut and allowed to fall, breaking into small pieces that are recovered by passages (drifts) driven for that purpose; sublevel caving, block caving, and top slicing are examples.

caving ground [MIN ENG] Rock formation that will not stand in the walls of an underground opening without support such as that offered by cementation, casing, or timber.

cavings *See* cave.

CA virus *See* croup-associated virus.

cavitation [CHEM] Emulsification produced by disruption of a liquid into a liquid-gas two-phase system, when the hydro-dynamic pressure of the liquid is reduced to the vapor pressure. [ENG] Pitting of a solid surface such as metal or concrete. [FL MECH] Formation of gas- or vapor-filled cavities within liquids by mechanical forces; broadly includes bubble formation when water is brought to a boil and effervescence of carbonated drinks; specifically, the formation of vapor-filled cavities in the interior or on the solid boundaries of vaporized liquids in motion where the pressure is reduced to a critical value without a change in ambient temperature. [PATH] The formation of one or more cavities in an organ or tissue, especially as the result of disease.

cavitation corrosion *See* cavitation erosion.

cavitation damage *See* cavitation erosion.

cavitation erosion [MET] Attack of metal surfaces caused by the collapse of cavitation bubbles on the surface of the liquid and characterized by pitting. Also known as cavitation corrosion; cavitation damage.

cavitation noise [ACOUS] Noise resulting from the formation of vapor- or gas-filled cavities in liquids by mechanical forces, as occurs near a propeller.

cavitation number [FL MECH] The excess of the local static pressure head over the vapor pressure head divided by the velocity head.

cavity [BIOL] A hole or hollow space in an organ, tissue, or other body part. [ELECTROMAG] *See* cavity resonator.

cavity charge *See* shaped charge.

cavity coupling [ELECTROMAG] The extraction of electromagnetic energy from a resonant cavity, either waveguide or coaxial, using loops, probes, or apertures.

cavity impedance [ELECTR] The impedance of the cavity of a microwave tube which appears across the gap between the cathode and the anode.

cavity magnetron [ELECTR] A magnetron having a number of resonant cavities forming the anode; used as a microwave oscillator.

cavity oscillator [ELECTR] An ultra-high-frequency oscillator whose frequency is controlled by a cavity resonator.

cavity radiator [THERMO] A heated enclosure with a small opening which allows some radiation to escape or enter; the escaping radiation approximates that of a blackbody.

cavity resonance [ELECTROMAG] The resonant vibration of a cavity. [ENG ACOUS] The natural resonant vibration of a loudspeaker baffle; if in the audio range, it is evident as unpleasant emphasis of sounds at that frequency.

cavity resonator [ELECTROMAG] A space totally enclosed by a metallic conductor and excited in such a way that it becomes a source of electromagnetic oscillations. Also known as cavity; microwave cavity; microwave resonance cavity; resonant cavity; resonant chamber; resonant element; rhumbatron; tuned cavity; waveguide resonator.

cavity tuning [ELECTROMAG] Use of an adjustable cavity resonator as a tuned circuit in an oscillator or amplifier, with tuning usually achieved by moving a metal plunger in or out of the cavity to change the volume, and hence the resonant frequency of the cavity.

cavity-type diode amplifier *See* diode amplifier.

cavity wall [BUILD] A wall constructed in two separate thicknesses with an air space between; provides thermal insulation. Also known as hollow wall.

CAVU [AERO ENG] An operational term commonly used in aviation, which designates a condition wherein the ceiling is more than 10,000 feet (3048 meters) and the visibility is more than 10 miles (16 kilometers). Derived from ceiling and visibility unlimited.

cavy [VERT ZOO] Any of the rodents composing the family Caviidae, which includes the guinea pig, rock cavies, mountain cavies, capybara, salt desert cavy, and mara.

CAW *See* channel address word.

c axis [CRYSTAL] A vertically oriented crystal axis, usually the principal axis; the unique symmetry axis in tetragonal and hexagonal crystals. [GEOL] The reference axis perpendicular to the plane of movement of rock or mineral strata.

cay [GEOL] 1. A flat coral island. 2. A flat mound of sand built up on a reef slightly above high tide. 3. A small, low coastal islet or emergent reef composed largely of sand or coral.

Cayley-Hamilton theorem [MATH] The theorem that a linear transformation or matrix is a root of its own characteristic polynomial. Also known as Hamilton-Cayley theorem.

Cayley-Klein parameters [MATH] A set of four complex numbers used to describe the orientation of a rigid body in space, or equivalently, the rotation which produces that orientation, starting from some reference orientation.

Cayley numbers [MATH] The division algebra consisting of pairs of quaternions; may be identified with an eight-dimensional vector space over the real numbers.

cay sandstone [GEOL] Firmly cemented or friable coral sand formed near the base of coral reef cays.

Caytoniales [PALEOBOT] An order of Mesozoic plants.

Cayugan [GEOL] Upper Silurian geologic time.

Cazenovian [GEOL] Lower Middle Devonian geologic time.

C-band waveguide [ELECTROMAG] A rectangular waveguide, with dimensions 3.48 by 1.58 centimeters, which is used to excite only the dominant mode (TE_{01}) for wavelengths in the range 3.7–5.1 centimeters.

C battery [ELEC] The battery that supplies the steady bias voltage required by the control-grid electrodes of electron tubes in battery-operated equipment. Also known as grid battery.

C bias *See* grid bias.

C.B. system *See* central-battery system.

cc *See* cubic centimeter.

Cc *See* cirrocumulus cloud.

C chart [IND ENG] A quality-control chart showing number of defects in subgroups of constant size; gives information concerning quality level, its variability, and evidence of assignable causes of variation.

CCIT 2 code [COMMUN] A printing-telegraph code in which each character is represented by five binary digits. Also known as international telegraph alphabet; International Telegraphic Consultative Committee code 2.

ccm *See* cubic centimeter.

C core [ELECTROMAG] A spirally wound magnetic core that is formed to a desired rectangular shape before being cut into two C-shaped pieces and placed around a transformer or magnetic amplifier coil.

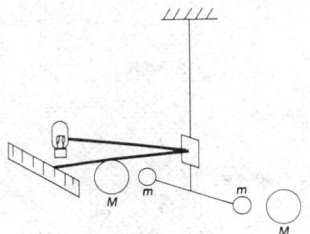
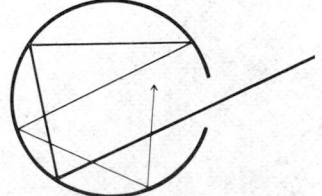

CEDAR

Leaf arrangement for the Atlas cedar (*Cedrus atlantica*).

CELERY

Celery (*Apium graveolens* var. *dulce*, cultivar Greenlight). (*Joseph Harris Co., Rochester, N.Y.*)

CCR process *See* cyclic catalytic reforming process.

CCTV *See* closed-circuit television.

CCW *See* channel command word.

Cd *See* cadmium.

CD *See* circular dichroism.

C damage [ORD] Damage of a degree or nature that prohibits an aircraft from completing its mission.

C display [ELECTR] In radar, a rectangular display in which targets appear as blips with bearing indicated by the horizontal coordinate, and angles of elevation by the vertical coordinate.

Ce *See* cerium.

Cebidae [VERT ZOO] The New World monkeys, a family of primates in the suborder Anthropoidea including the capuchins and howler monkeys.

Cebochoeridae [PALEON] A family of extinct palaeodont artiodactyls in the superfamily Entelodontoidae.

cebollite [MINERAL] $H_2Ca_4Al_2Si_3O_{16}$ A greenish to white mineral consisting of hydrous calcium aluminum silicate occurring in fibrous aggregates; hardness is 5 on Mohs scale, and specific gravity is 3.

Cebrionidae [INV ZOO] The robust click beetles, a family of cosmopolitan coleopteran insects in the superfamily Elateroidea.

cecidium [PL PATH] Plant gall produced either by insects in ovipositing or by fungi as a consequence of infection.

cecilite [PETR] A basaltic rock having few phenocrysts and consisting of at least 50% leucite with augite, melilite, nepheline, olivine, anorthite, magnetite, and apatite.

Cecropidae [INV ZOO] A family of crustaceans in the suborder Caligoida which are external parasites on fish.

cecum [ANAT] The blind end of a cavity, duct, or tube, especially the sac at the beginning of the large intestine. Also spelled caecum.

cedar [BOT] The common name for a large number of evergreen trees in the order Pinales having fragrant, durable wood.

cedarwood oil [MATER] An essential alcohol-soluble oil obtained from the heartwood of various cedars, the chief components being cedrol and cedrene.

cedricite [MINERAL] A variety of lamproite composed principally of diopside, leucite, and phlogopite and usually containing crystals of serpentine.

ceiba *See* kapok.

ceiling [BUILD] The covering made of plaster, boards, or other material constituting the overhead surface in a room. [METEOROL] In the United States, the height ascribed to the lowest layer of clouds or of obscuring phenomena when it is reported as broken, overcast, or obscuration and not classified as thin or partial.

ceiling and visibility unlimited *See* CAVU.

ceiling balloon [AERO ENG] A small balloon used to determine the height of the cloud base; the height is computed from the ascent velocity of the balloon and the time required for its disappearance into the cloud.

ceiling classification [METEOROL] In aviation weather observations, a description or explanation of the manner in which the height of the ceiling is determined.

ceiling light [ENG] A type of cloud-height indicator which uses a searchlight to project vertically a narrow beam of light onto a cloud base. Also known as ceiling projector.

ceiling projector *See* ceiling light.

ceilometer [ENG] An automatic-recording cloud-height indicator.

celadonite [MINERAL] A soft, green variety of mica having high iron content and containing silicates of magnesium and potassium.

Celanese [TEXT] Trade name for textiles manufactured by the Celanese Corporation.

Celaperm [TEXT] Trade name of the Celanese Corporation for acetate yarn with the color sealed into the fiber.

Celastraceae [BOT] A family of dicotyledonous plants in the order Celastrales characterized by erect and basal ovules, a flower disk that surrounds the ovary at the base, and opposite or sometimes alternate leaves.

Celastrales [BOT] An order of dicotyledonous plants in the subclass Rosidae marked by simple leaves and regular flowers.

Celcos [TEXT] Trade name of the Celanese Corporation for a fiber which is a combination of acetate and viscose rayon.

celerity *See* phase velocity.

celery [BOT] *Apium graveolens* var. *dulce*. A biennial umbellifer of the order Umbellales with edible petioles or leaf stalks.

celery seed oil [MATER] A colorless liquid extracted from celery seeds, containing selinene and apiol.

celestial body [ASTRON] Any aggregation of matter in space constituting a unit for astronomical study, as the sun, moon, a planet, comet, star, or nebula. Also known as heavenly body.

celestial coordinates [ASTRON] Any set of coordinates, such as zenithal distance, altitude, celestial latitude, celestial longitude, local hour angle, azimuth and declination, used to define a point on the celestial sphere.

celestial equator [ASTRON] The primary great circle of the celestial sphere in the equatorial system, everywhere 90° from the celestial poles; the intersection of the extended plane of the equator and the celestial sphere. Also known as equinoctial.

celestial equator system of coordinates *See* equatorial system.

celestial fix [NAV] A position established by means of observation on one or more celestial bodies.

celestial geodesy [GEOD] The branch of geodesy which utilizes observations of near celestial bodies and earth satellites to determine the size and shape of the earth.

celestial globe [ASTRON] A small globe representing the celestial sphere, on which the apparent positions of the stars are located. Also known as star globe.

celestial guidance [NAV] The process of directing movements of an aircraft or spacecraft, especially in the selection of a flight path, by reference to celestial bodies.

celestial horizon [ASTRON] That great circle of the celestial sphere which is formed by the intersection of the celestial sphere and a plane through the center of the earth and is perpendicular to the zenith-nadir line. Also known as rational horizon.

celestial-inertial guidance [NAV] The process of directing the movements of an aircraft or spacecraft by an inertial guidance system which receives correction inputs from observations of celestial bodies. Also known as automatic celestial navigation.

celestial latitude [ASTRON] Angular distance north or south of the ecliptic; the arc of a circle of latitude between the ecliptic and a point on the celestial sphere, measured northward or southward from the ecliptic through 90°, and labeled N or S to indicate the direction of measurement. Also known as ecliptic latitude.

celestial line of position [NAV] A line of position determined by observation of a celestial body.

celestial longitude [ASTRON] Angular distance east of the vernal equinox, along the ecliptic; the arc of the ecliptic or the angle at the ecliptic pole between the circle of latitude of the vernal equinox and the circle of latitude of a point on the celestial sphere, measured eastward from the circle of latitude of the vernal equinox, through 360°. Also known as ecliptic longitude.

celestial mechanics [ASTROPHYS] The calculation of motions of celestial bodies under the action of their mutual gravitational attractions. Also known as gravitational astronomy.

celestial meridian [ASTRON] A great circle on the celestial sphere, passing through the two celestial poles and the observer's zenith.

celestial navigation *See* astronavigation.

celestial observation [NAV] Also known as sight. **1.** The measurement of the altitude, and sometimes the azimuth, of a celestial body for a line of position. **2.** The data obtained by such observation.

celestial parallel *See* parallel of declination.

celestial pole [ASTRON] Either of the two points of intersection of the celestial sphere and the extended axis of the earth, labeled N or S to indicate the north celestial pole or the south celestial pole.

celestial sphere [ASTRON] An imaginary sphere of indefinitely large radius, which is described about an assumed center, and upon which positions of celestial bodies are projected along radii passing through the bodies.

celestine *See* celestite.

celestite [MINERAL] $SrSO_4$ A colorless or sky-blue mineral occurring in orthorhombic, tabular crystals and in compact forms; fracture is uneven and luster is vitreous; principal ore of strontium. Also known as celestine.

celiac [ANAT] Of, in, or pertaining to the abdominal cavity.

celiac syndrome [MED] A complex of symptoms produced by intestinal malabsorption of fat and marked by bulky, loose, foul-smelling stools, high in fatty-acid content. Also known as idiopathic steatorrhea.

cell [ADP] A storage or memory location capable of containing one character, one byte, or one word, the capacity being defined by the natural storage capacity of the computer. [BIOL] The microscopic functional and structural unit of all living organisms, consisting of a nucleus, cytoplasm, and a limiting membrane. [ELECTR] A single unit of a device that converts radiant energy into electrical energy, such as a nuclear cell, solar cell, or photovoltaic cell. [MATH] The homeomorphic image of the unit ball. [MIN ENG] A compartment in a flotation machine. [NUCLEO] One of a set of elementary regions in a heterogeneous reactor, all of which have the same geometrical form and the same neutron characteristics. [PHYS CHEM] A cup, jar, or vessel containing electrolyte solutions and metal electrodes to produce an electric current (conductiometric or potentiometric) or for electrolysis (electrolytic). Also known as electric cell.

cell complex [MATH] A topological space which is the last term of a finite sequence of spaces, each obtained from the previous by sewing on a cell along its boundary.

cell constancy [BIOL] The condition in which the entire body, or a part thereof, consists of a fixed number of cells that is the same for all adults of the species.

cell constant [PHYS CHEM] The ratio of distance between conductance-titration electrodes to the area of the electrodes, measured from the determined resistance of a solution of known specific conductance.

cell differentiation [CYTOL] The series of events involved in the development of a specialized cell having specific structural, functional, and biochemical properties.

cell division [CYTOL] The process by which living cells multiply; may be mitotic or amitotic.

Cellfalcicula [MICROBIO] Genus of gram-negative, spindle-shaped soil bacteria belonging to the family Spirillaceae.

cell-free extract [CYTOL] A fluid obtained by breaking open cells; contains most of the soluble molecules of a cell.

cell inclusion [CYTOL] A small, nonliving intracellular particle, usually representing a form of stored food, not immediately vital to life processes.

cell lineage [EMBRYO] The developmental history of individual blastomeres from their first cleavage division to their ultimate differentiation into cells of tissues and organs.

cell membrane [CYTOL] A thin layer of protoplasm, consisting mainly of lipids and proteins, which is present on the surface of all cells. Also known as plasmalemma; plasma membrane.

cell movement [CYTOL] **1.** Intracellular movement of cellular components. **2.** Movement of a cell relative to its environment.

cellobiose [ORG CHEM] $C_{12}H_{22}O_{11}$ A disaccharide, not known to exist in nature, obtained by partial hydrolysis of cellulose.

celloidin [MATER] A concentrated solution of pyroxylin used principally in microscopy for embedding specimens or for section-cutting.

cellophane [MATER] A thin, transparent sheeting of regenerated cellulose; it is moisture-proof, and sometimes dyed, and used chiefly as food wrapping or as bags for dialysis.

cellosolve [ORG CHEM] $C_2H_5OCH_2CH_2OH$ An important industrial chemical used in varnish removers, in cleaning solutions, and as a solvent for paints, varnishes, and plastics. Also known as 2-ethoxyethanol.

cell pathology [PATH] Abnormalities of the events taking place within cells.

cell permeability [CYTOL] The permitting or activating of the passage of substances into, out of, or through cells.

cell plate [CYTOL] A membrane-bound disk formed during cytokinesis in plant cells which eventually becomes the middle lamella of the wall formed between daughter cells.

cell sap [CYTOL] The liquid content of a plant cell vacuole.

cells of Paneth [HISTOL] Coarsely granular secretory cells found in the crypts of Lieberkühn in the small intestine. Also known as Paneth cells.

cell-surface differentiation [CYTOL] The specialization or modification of the cell surface.

cell-surface ionization [PHYSIO] The presence of a negative charge on the surface of all living cells suspended in aqueous salt solutions at neutral pH.

cell theory [BIOL] **1.** A principle that describes the cell as the fundamental unit of all living organisms. **2.** A principle that describes the properties of an organism as the sum of the properties of its component cells.

cell-type tube [ELECTR] Gas-filled radio-frequency switching tube which operates in an external resonant circuit; a tuning mechanism may be incorporated in either the external resonant circuit or the tube.

cellular [BIOL] Characterized by, consisting of, or pertaining to cells. [PETR] Pertaining to igneous rock having a porous texture, usually with the cavities larger than pore size and smaller than caverns.

cellular affinity [BIOL] The phenomenon of selective adhesiveness observed among the cells of certain sponges, slime molds, and vertebrates.

cellular cofferdam [CIV ENG] A cofferdam consisting of interlocking steel-sheet piling driven as a series of interconnecting cells; cells may be of circular type or of straight-wall diaphragm type; space between lines of pilings is filled with sand.

cellular convection [METEOROL] An organized, convective, fluid motion characterized by the presence of distinct convection cells or convective units, usually with upward motion (away from the heat source) in the central portions of the cell, and sinking or downward flow in the cell's outer regions.

cellular glass [MATER] Sheets or blocks of thermal insulating material for walls and roofs made from pulverized glass that is heated with a gas-forming chemical at the flow temperature of glass. Also known as cellulated glass; foamed glass.

cellular horn *See* multicellular horn.

cellular infiltration [MED] **1.** Passage of cells into tissues in the course of inflammation. **2.** Migration of or invasion by cells of neoplasms.

cellular plastic [MATER] A type of plastic with apparent density decreased substantially by the presence of numerous cells disposed throughout its mass.

cellular rubber *See* rubber sponge.

cellular soil *See* polygonal ground.

cellular striation [ENG] Stratum of cells inside a cellular-plastic object that differs noticeably from the cell structure of the remainder of the material.

cellulase [BIOCHEM] Any of a group of extracellular enzymes, produced by various fungi, bacteria, insects, and other lower animals, that hydrolyze cellulose.

cellulated glass *See* cellular glass.

cellulitis [MED] Diffuse inflammation of a connective tissue.

Celluloid [MATER] A trade name for a malleable substance prepared from nitrocellulose and camphor, pressed into various forms such as film or sheets or tubes and made less flammable by adding ammonium phosphate and other compounds; may be used where malleable plastics are needed and its flammability may be neglected.

cellulolytic [BIOL] Having the ability to hydrolyze cellulose; applied to certain bacteria and protozoans.

cellulose [BIOCHEM] $(C_6H_{10}O_5)_n$ The main polysaccharide in living plants, forming the skeletal structure of the plant cell wall; a polymer of β-D-glucose units linked together, with the elimination of water, to form chains of 2000–4000 units.

α-cellulose [ORG CHEM] A fraction of the cellulose polymer mixture consisting of those cellulose molecules which do not dissolve in an alkaline solution.

cellulose acetate [ORG CHEM] An acetic acid ester of cellulose; a tough, flexible, slow-burning, and longlasting thermoplastic material used as the base for magnetic tape and movie film, in acetate rayon, as a plastic film in food packaging, in lacquers, and for molded receiver cabinets.

cellulose acetate butyrate [ORG CHEM] An ester of cellulose

CELESTITE

Bladed crystals of celestite in limestone from Clay Center, Ohio. (Specimen from Department of Geology, Bryn Mawr College)

CELLOBIOSE

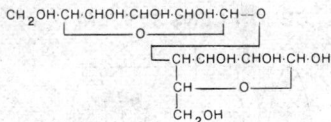

Structural formula of cellobiose.

CELLULAR COFFERDAM

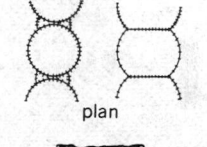

plan

section

Steel-sheet pile cellular cofferdam.

CELLULOSE

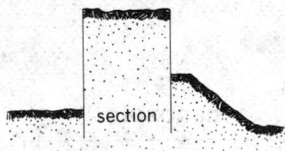

Structure of the cellulose polymer.

formed by the action of a mixture of acetic acid and butyric acid and their anhydrides on purified cellulose; has high impact resistance, clarity, and weatherability; used in making plastic film, lacquer, lenses, and outdoor signs.

cellulose acetate rayon [TEXT] The spun product of the acetic ester of cellulose.

cellulose diacetate [ORG CHEM] The ester formed by esterification of two hydroxyl groups of a cellulose molecule with acetic acid.

cellulose ester [ORG CHEM] Cellulose in which the free hydroxyl groups have been replaced wholly or in part by acidic groups.

cellulose ether [ORG CHEM] The product of the partial or complete etherification of the hydroxyl groups in a cellulose molecule.

cellulose fiber [ORG CHEM] Any fiber based on esters or ethers of cellulose.

cellulose gum See carboxymethyl cellulose.

cellulose nitrate [ORG CHEM] Any of several esters of nitric acid, produced by treating cotton or some other form of cellulose with a mixture of nitric and sulfuric acids; used as explosive and propellant. Also known as nitrocellulose.

cellulose propionate [ORG CHEM] An ester of cellulose and propionic acid.

cellulose triacetate [ORG CHEM] A cellulose resin formed by the complete esterification of the cellulose by acetic acid; used as a base in protective coatings.

cellulose xanthate [ORG CHEM] A compound formed by reaction of soda cellulose (prepared by treating cellulose with strong sodium hydroxide solution) with carbon disulfide.

cellulosic [ORG CHEM] Any of the derivatives of cellulose, such as cellulose acetate.

cellulosic resin [ORG CHEM] Any resin based on cellulose compounds such as esters and ethers.

cell wall [CYTOL] A semirigid, permeable structure that is composed of cellulose, lignin, or other substances and that envelops most plant cells.

celo [MECH] A unit of acceleration equal to the acceleration of a body whose velocity changes uniformly by 1 foot per second in 1 second.

Celor lens system [OPTICS] An anastigmatic lens system consisting of two air-spaced achromatic doublet lenses, one on each side of the stop. Also known as Gauss lens system; Gauss objective lens.

celsian [MINERAL] $BaAl_2Si_2O_8$ Colorless, monoclinic mineral consisting of barium feldspar.

Celsius degree [THERMO] Unit of temperature interval or difference equal to the kelvin.

Celsius temperature scale [THERMO] Temperature scale in which the temperature Θ_c in degrees Celsius (°C) is related to the temperature T_k in kelvins by the formula $\Theta_c = T_k - 273.15$; the freezing point of water at standard atmospheric pressure is very nearly 0°C and the corresponding boiling point is very nearly 100°C. Formerly known as centigrade temperature scale.

Celyphidae [INV ZOO] A family of myodarian cyclorrhaphous dipteran insects in the subsection Acalypteratae.

cement [GEOL] Any chemically precipitated material, such as carbonates, gypsum, and barite, occurring in the interstices of clastic rocks. [HISTOL] A calcified tissue which fastens the roots of teeth to the alveolus. Also known as cementum. [INV ZOO] Any of the various adhesive secretions, produced by certain invertebrates, that harden on exposure to air or water and are used to bind objects. [MATER] **1.** A dry powder made from silica, alumina, lime, iron oxide, and magnesia which hardens when mixed with water; used as an ingredient in concrete. **2.** An adhesive for the assembling of surfaces which are not in close contact.

cementation [CHEM] Setting of a plastic material. [ENG] **1.** Plugging a cavity or drill hole with cement. Also known as dental work. **2.** Consolidation of loose sediments or sand by injection of a chemical agent or binder. [GEOL] The precipitation of a binding material around minerals or grains in rocks. [MET] **1.** High-temperature impregnation of a metal surface with another material. **2.** Conversion of wrought iron into steel by packing layers of bars in charcoal sealed with clay and heating to 1000°C for 7–10 days.

cementation factor [PETRO ENG] Mathematical expression in oil-reservoir analysis for the degree to which precipitated minerals have bound together the grains (for example, of sand).

cementation sinking [MIN ENG] A technique of shaft sinking through strata containing water by injecting liquid cement or chemicals into the ground.

cement brick [MATER] A type of brick made from a mixture of cement and sand, molded under pressure and cured under steam at 200°F (93°C); used as backing brick and where there is no danger of attack from acid or alkaline conditions.

cement carrier [NAV ARCH] A special ship designed to carry bulk cement; usually fitted with an air-mixing system to allow pumping dry cement powder through pipes to unload or load the ship.

cement copper [MET] A precipitate of copper from copper sulfate solution by the addition of iron.

cemented carbide See carbide.

cemented lens See compound lens.

cement gland [INV ZOO] A structure in many invertebrates that produces cement.

cement gravel [GEOL] Gravel consolidated by clay, silica, calcite, or other binding material.

cement gun [MECH ENG] **1.** A machine for mixing, wetting, and applying refractory mortars to hot furnace walls. Also known as cement injector. **2.** A mechanical device for the application of cement or mortar to the walls or roofs of mine openings or building walls.

cement injector See cement gun.

cementite [MET] Fe_3C A hard, brittle, crystalline compound occurring as lamellae or plates in steel. Also known as iron carbide.

cementitious material [MATER] Any of various building materials which may be mixed with a liquid, such as water, to form a plastic paste, and to which an aggregate may be added; includes cements, limes, and mortar.

cement kiln [ENG] A kiln used to fire cement to less than complete melting.

cement-lined casing [PETRO ENG] Steel-, oil-, or gas-well casing pipe with internal lining of special cement; used to withstand severe corrosive conditions.

cement log [PETRO ENG] Gamma-ray measurement and logging of the height and condition of cement surrounding downhole oil-well casing.

cement mill [MECH ENG] A mill for grinding rock to a powder for cement.

cement mortar [MATER] A mixture of approximately four parts of sand to one part of portland cement with a small amount of lime and enough water to make it plastic.

cement paint [MATER] A mixture based on portland cement, with filler, accelerator, and water repellant added, that is combined with water and applied to masonry, concrete, or brickwork; provides a waterproof coating.

cement paste [MATER] A mixture of cement and water, hardened or unhardened.

cement plaster [MATER] A gypsum plaster used in mortar mixtures for plastering interior surfaces.

cement pump [MECH ENG] A piston device used to move concrete through pipes.

cement rock [PETR] An argillaceous limestone containing lime, silica, and alumina in variable proportions and usually some magnesia; used in the manufacture of natural hydraulic cement.

cement silo [ENG] A silo used to store dry, bulk cement.

cementum See cement.

cement valve [MECH ENG] A ball-, flapper-, or clack-type valve placed at the bottom of a string of casing, through which cement is pumped, so that when pumping ceases, the valve closes and prevents return of cement into the casing.

Cen See Centaurus.

Cenomanian [GEOL] Lower Upper Cretaceous geologic time.

cenote See pothole.

Cenozoic [GEOL] The youngest of the eras, or major subdivisions of geologic time, extending from the end of the Mesozoic Era to the present, or Recent.

cent [ACOUS] The interval between two sounds whose basic frequency ratio is the twelve-hundredth root of 2; the interval, in cents, between any two frequencies is 1200 times the

---

CELOR LENS SYSTEM

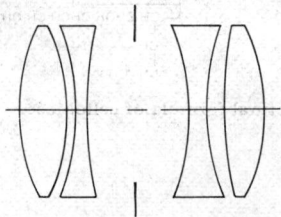

Celor lens system, an anastigmatic photographic objective.

CENOZOIC

PRECAMBRIAN		
CAMBRIAN		
ORDOVICIAN		
SILURIAN	PALEOZOIC	
DEVONIAN		
Mississippian	CARBON-IFEROUS	
Pennsylvanian		
PERMIAN		
TRIASSIC		
JURASSIC	MESOZOIC	
CRETACEOUS		
TERTIARY	CENOZOIC	
QUATERNARY		

Chart showing the relationship of the Cenozoic to other eras.

logarithm to the base 2 of the frequency ratio. [NUCLEO] A unit of nuclear reactivity equal to one-hundredth of a dollar.

cental *See* hundredweight.

α Centauri [ASTRON] A double star, the brightest in the constellation Centaurus; apart from the sun, it is the nearest bright star to earth, about 4.3 light-years away; spectral classification G2.

β Centauri [ASTRON] A first-magnitude navigational star in the constellation Centaurus; 200 light-years from the sun; spectral classification B0.

Centaurus [ASTRON] A constellation with right ascension 13 hours, declination 50°S. Abbreviated Cen.

Centaurus X-3 [ASTROPHYS] A source of x-rays that pulses with a period of 4.8 seconds and is eclipsed every 2.1 days; believed to be a binary star, one of whose members is a rotating neutron star. Abbreviated Cen X-3.

center [MATH] **1.** The point which is equidistant from all the points on a circle or sphere. **2.** For an ellipsoid or hyperboloid, the point about which the surface is symmetrical.

centerboard [NAV ARCH] A metal or wooden slab in a casing along the centerline of a sailboat which may be lowered to increase the boat's resistance to lateral motion, and raised when the boat is in shallow water or is beached.

center-coupled loop [ELECTR] Coupling loop in the center of one of the resonant cavities of a multicavity magnetron.

center-feed tape [ADP] Punched tape in which the centers of the sprocket holes are in line with the centers of the holes carrying the data or message.

center-fire [ORD] Pertaining to a cartridge with primer in the center of the head, or base.

center frequency *See* carrier frequency.

center gage [DES ENG] A gage used to check angles; for example, the angles of cutting tool points or screw threads, or the angular position of cutting tools.

centering [CIV ENG] A curved, temporary support for an arch or dome during a casting or laying operations.

centering control [ELECTR] One of the two controls used for positioning the image on the screen of a cathode-ray tube; either the horizontal centering control or the vertical centering control.

centering machine [MECH ENG] A machine for drilling and countersinking work to be turned on a lathe.

center jump [METEOROL] The formation of a second low-pressure center within an already well-developed low-pressure center; the latter diminishes in magnitude as the center of activity shifts or appears to jump to the new center.

centerless grinder [MECH ENG] A cylindrical metal-grinding machine that carries the work on a support or blade between two abrasive wheels.

center line *See* stroke center line.

center of action [METEOROL] A semipermanent high or low atmospheric pressure system at the surface of the earth; fluctuations in the intensity, position, orientation, shape, or size of such a center are associated with widespread weather changes.

center of a distribution [STAT] The expected value of any random variable which has the given distribution.

center of area [MATH] For a plane figure, the center of mass of a thin uniform plate having the same boundaries as the plane figure. Also known as center of figure; centroid.

center of attraction [MECH] A point toward which a force on a body or particle (such as gravitational or electrostatic force) is always directed; the magnitude of the force depends only on the distance of the body or particle from this point.

center of buoyancy [MECH] The point through which acts the resultant force exerted on a body by a static fluid in which it is submerged or floating; located at the centroid of displaced volume.

center of burst [ORD] A point in the air about which the bursts of several antiaircraft projectiles from rounds fired under like conditions are evenly distributed.

center of curvature [MATH] At a given point on a curve, the center of the osculating circle of the curve at that point.

center of falls *See* pressure-fall center.

center of figure *See* center of area; center of volume.

center of force [MECH] The point toward or from which a central force acts.

center of gravity [MECH] A fixed point in a material body through which the resultant force of gravitational attraction acts.

center of inversion [CRYSTAL] A point in a crystal lattice such that the lattice is left invariant by an inversion in the point.

center of lift [AERO ENG] The mean of all the centers of pressure on an airfoil.

center of mass [MECH] That point of a material body or system of bodies which moves as though the system's total mass existed at the point and all external forces were applied at the point.

center of mass coordinate system [MECH] A reference frame which moves with the velocity of the center of mass, so that the center of mass is at rest in this system, and the total momentum of the system is zero. Also known as center of momentum coordinate system.

center of momentum coordinate system *See* center of mass coordinate system.

center of oscillation [MECH] Point in a physical pendulum, on the line through the point of suspension and the center of mass, which moves as if all the mass of the pendulum were concentrated there.

center of percussion [MECH] If a rigid body, free to move in a plane, is struck a blow at a point O, and the line of force is perpendicular to the line from O to the center of mass, then the initial motion of the body is a rotation about the center of percussion relative to O; it can be shown to coincide with the center of oscillation relative to O.

center of pressure [AERO ENG] The point in the chord of an airfoil section which is at the intersection of the chord (prolonged if necessary) and the line of action of the combined air forces (resultant air force).

center of pressure coefficient [AERO ENG] The ratio of the distance of a center of pressure from the leading edge of an airfoil to its chord length.

center of rises *See* pressure-rise center.

center of similitude [MATH] A point of intersection of lines that join the ends of parallel radii of coplanar circles.

center of symmetry [SCI TECH] A point in an object through which any straight line encounters exactly similar points on opposite sides. Also known as symmetry center.

center of thrust *See* thrust axis.

center of twist [MECH] A point on a line parallel to the axis of a beam through which any transverse force must be applied to avoid twisting of the section. Also known as shear center.

center of volume [MATH] For a three-dimensional figure, the center of mass of a homogeneous solid having the same boundaries as the figures. Also known as center of figure; centroid.

center plug [DES ENG] A small diamond-set circular plug, designed to be inserted into the annular opening in a core bit, thus converting it to a noncoring bit.

center punch [DES ENG] A tool similar to a prick punch but having the point ground to an angle of about 90°; used to enlarge prick-punch marks or holes.

center square [DES ENG] A straight edge with a sliding square; used to locate the center of a circle.

center staff [HOROL] The arbor that holds the minute hand and cannon pinion of a timepiece.

center tap [ELEC] A terminal at the electrical midpoint of a resistor, coil, or other device. Abbreviated CT.

center wheel [HOROL] A wheel that drives the wheels leading to the escapement, and that has a pinion post on which the minute hand is mounted.

centesis [MED] Surgical puncture, or perforation as of a tumor or membrane.

centi- [SCI TECH] A prefix representing 10^{-2}, which is 0.01 or one-hundredth. Abbreviated c.

centiare [MECH] Unit of area equal to 1 square meter. Also spelled centare.

centibar [MECH] A unit of pressure equal to 0.01 bar or to 1000 pascals.

centigrade heat unit [THERMO] A unit of heat energy, equal to 0.01 of the quantity of heat needed to raise 1 pound of air-free water from 0 to 100°C at a constant pressure of 1 standard atmosphere; equal to 1900.44 joules. Symbolized CHU; (more correctly) CHU$_{mean}$.

centigrade temperature scale *See* Celsius temperature scale.

CENTIPEDE

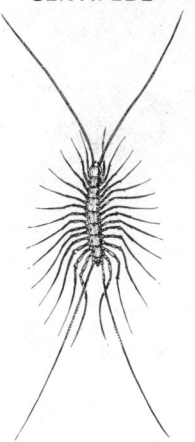

Scutigera coleoptrata (L.), the house centipede; body length about 25 millimeters. *(From R. E. Snodgrass, A Textbook of Arthropod Anatomy, Cornell University Press, 1952)*

CENTRAL PLACENTATION

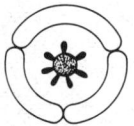 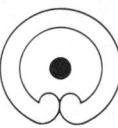

Free central placentation, ovules shown in black.

CENTRAL POCKET LOOP

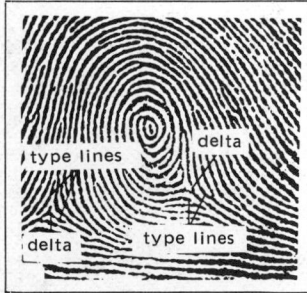

Central pocket loop, a type of whorl pattern. *(Federal Bureau of Investigation)*

centigram [MECH] Unit of mass equal to 0.01 gram or 10^{-5} kilogram. Abbreviated cg.

centihg *See* centimeter of mercury.

centiliter [MECH] A unit of volume equal to 0.01 liter or to 10^{-5} cubic meter.

centimeter [ELEC] *See* statfarad. [ELECTROMAG] *See* abhenry. [MECH] A unit of length equal to 0.01 meter. Abbreviated cm.

centimeter-candle *See* phot.

centimeter-gram-second system [PHYS] An absolute system of metric units in which the centimeter, gram mass, and the second are the basic units. Abbreviated cgs system.

centimeter of mercury [MECH] A unit of pressure equal to the pressure that would support a column of mercury 1 centimeter high, having a density of 13.5951 grams per cubic centimeter, when the acceleration of gravity is equal to its standard value (980.665 centimeters per second per second); equal to 1333. 22387415 pascals; it differs from the decatorr by less than 1 part in 7,000,000. Abbreviated cmHg. Also known as centihg.

centimetric waves [COMMUN] Microwaves having wavelengths between 1 and 10 centimeters, corresponding to frequencies between 3 and 30 gigahertz.

centimorgan [GEN] A unit of genetic map distance, equal to the distance along a chromosome that gives a recombination frequency of 1%.

centipede [INV ZOO] The common name for an arthropod of the class Chilopoda.

centipoise [FL MECH] A unit of viscosity equal to 0.01 poise. Abbreviated cp.

centistoke [FL MECH] A cgs unit of kinematic viscosity in customary use, equal to the kinematic viscosity of a fluid having a dynamic viscosity of 1 centipoise and a density of 1 gram per cubic centimeter. Abbreviated cs.

centrad [MATH] A unit of plane angle equal to 0.01 radian or to about 0.573 degree.

central apparatus [CYTOL] The centrosome or centrosomes together with the surrounding cytoplasm. Also known as cytocentrum.

central-battery system [COMMUN] A telephone or telegraph system which obtains all the energy for signaling (and for speaking, in the case of the telephone) from a single battery of secondary cells located at the main exchange. Abbreviated C.B. system.

central breaker [MIN ENG] A breaker where the coal from several mines in a district is prepared.

central canal [ANAT] The small canal running through the center of the spinal cord from the conus medullaris to the lower part of the fourth ventricle; represents the embryonic neural tube.

central control [AERO ENG] The place, facility, or activity at which the whole action incident to a test launch and flight is coordinated and controlled, from the make-ready at the launch site and on the range, to the end of the rocket flight down-range. [ORD] Fire control of weapons by a central location, not by the individual gunner; especially used in antiaircraft batteries. [SYS ENG] Control exercised over an extensive and complicated system from a single center.

central eclipse [ASTRON] An eclipse in which the eclipsing body passes centrally (midpoints in line) over the body eclipsed.

Centrales [BOT] An order of diatoms (Bacillariophyceae) in which the form is often circular and the markings on the valves are radial.

central field approximation [PHYS] The approximation that the electrons in an atom or the nucleons in a nucleus move in the potential of a central force which is the same for all the particles.

central force [MECH] A force whose line of action is always directed toward a fixed point; the force may attract or repel.

central gear [MECH ENG] The gear on the central axis of a planetary gear train, about which a pinion rotates. Also known as sun gear.

central heating [CIV ENG] The use of a single steam or hot-water heating plant to serve a group of buildings, facilities, or even a complete community through a system of distribution pipes.

centralized data processing [ADP] The processing of all the data concerned with a given activity at one place, usually with fixed equipment within one building.

centralized traffic control [CIV ENG] Control of train movements by signal indications given by a train director at a central control point. Abbreviated CTC.

centralizer [MATH] The subgroup consisting of all elements which commute with a given element of a group.

central-limit theorem [STAT] The theorem that the distribution of sample means taken from a large population approaches a normal (Gaussian) curve.

central mix concrete [MATER] A concrete prepared at a concrete mixing plant and transported to the building site.

central nervous system [ANAT] The division of the vertebrate nervous system comprising the brain and spinal cord.

central office [COMMUN] A switching unit, installed in a telephone system serving the general public, having the necessary equipment and operating arrangements for terminating and interconnecting lines and trunks. Also known as telephone central office.

central office line *See* subscriber line.

central paralysis [MED] Paralysis due to a lesion of the brain or spinal cord.

central placentation [BOT] Having the ovules located in the center of the ovary.

central pocket loop [ANAT] A whorl type of fingerprint pattern having two deltas and at least one ridge that make a complete circuit.

central pressure [METEOROL] At any given instant, the atmospheric pressure at the center of a high or low; the highest pressure in a high, the lowest pressure in a low.

central processing unit [ADP] The part of a computer containing the circuits required to interpret and execute the instructions. Abbreviated CPU. Also known as frame; main frame.

central quadric [MATH] A quadric surface that has a point about which the surface is symmetrical; namely, a sphere, ellipsoid, or hyperboloid.

central sulcus [ANAT] A groove situated about the middle of the lateral surface of the cerebral hemisphere, separating the frontal from the parietal lobe.

central terminal [ADP] A communication device which queues tellers' requests for processing and which channels answers to the consoles originating the transactions.

central valley *See* rift valley.

central water [OCEANOGR] Upper water mass associated with the central region of oceanic gyre.

Centrarchidae [VERT ZOO] A family of fishes in the order Perciformes, including the fresh-water or black basses and several sunfishes.

Centrex telephone line [COMMUN] A line in a PBX system permitting direct inward and outward dialing.

centric [ANAT] Having all teeth of both jaws meet normally with perfect distribution of forces in the dental arch.

centrifugal [MECH] Acting or moving in a direction away from the axis of rotation or the center of a circle along which a body is moving.

centrifugal atomizer [MECH ENG] Device that atomizes liquids with a spinning disk; liquid is fed onto the center of the disk, and the whirling motion (3000 to 50,000 revolutions per minute) forces the liquid outward in thin sheets to cause atomization.

centrifugal barrier [MECH] A steep rise, located around the center of force, in the effective potential governing the radial motion of a particle of nonvanishing angular momentum in a central force field, which results from the centrifugal force and prevents the particle from reaching the center of force, or causes its Schrödinger wave function to vanish there in a quantum-mechanical system.

centrifugal brake [MECH ENG] A safety device on a hoist drum that applies the brake if the drum speed is greater than a set limit.

centrifugal casting [MET] Casting method in which molten metal is allowed to solidify in a rapidly revolving mold.

centrifugal clarification [MECH ENG] The removal of solids from a liquid by centrifugal action which decreases the settling time of the particles from hours to minutes.

centrifugal classification [MECH ENG] A type of centrifugal clarification purposely designed to settle out only the large particles (rather than all particles) in a liquid by reducing the centrifuging time.

centrifugal classifier [MECH ENG] A machine that separates particles into size groups by centrifugal force.

centrifugal clutch [MECH ENG] A clutch operated by centrifugal force from the speed of rotation of a shaft, as when heavy expanding friction shoes act on the internal surface of a rim clutch, or a flyball-type mechanism is used to activate clutching surfaces on cones and disks.

centrifugal collector [MECH ENG] Device used to separate particulate matter of 0.1–1000 micrometers from an airstream; some types are simple cyclones, high-efficiency cyclones, and impellers.

centrifugal compressor [MECH ENG] A machine in which a gas or vapor is compressed by radial acceleration in an impeller with a surrounding casing, and can be arranged multistage for high ratios of compression.

centrifugal discharge elevator [MECH ENG] A high-speed bucket elevator from which free-flowing materials are discharged by centrifugal force at the top of the loop.

centrifugal distortion [PHYS] Tendency of a molecule to stretch slightly as its speed of rotation increases.

centrifugal drainage pattern *See* radial drainage pattern.

centrifugal extractor [CHEM ENG] A device for separating components of a liquid solution, consisting of a series of perforated concentric rings in a cylindrical drum that rotates at 2000–5000 revolutions per minute around a cylindrical shaft; liquids enter and leave through the shaft; they flow radially and concurrently in the rotating drum.

centrifugal fan [MECH ENG] A machine for moving a gas, such as air, by accelerating it radially outward in an impeller to a surrounding casing, generally of scroll shape.

centrifugal filter [CHEM ENG] An adaptation of the centrifugal settler; centrifugal action of a spinning container segregates heavy and light materials but heavy materials escape through nozzles as a thick slurry.

centrifugal filtration [MECH ENG] The removal of a liquid from a slurry by introducing the slurry into a rapidly rotating basket, where the solids are retained on a porous screen and the liquid is forced out of the cake by the centrifugal action.

centrifugal force [MECH] 1. An outward pseudo-force, in a reference frame that is rotating with respect to an inertial reference frame, which is equal and opposite to the centripetal force that must act on a particle stationary in the rotating frame. 2. The reaction force to a centripetal force.

centrifugal molecular still [CHEM ENG] A device used for molecular distillation; material is fed to the center of a hot, rapidly rotating cone housed in a chamber at a high vacuum; centrifugal force spreads the material rapidly over the hot surface, where the evaporable material goes off as a vapor to the condenser.

centrifugal moment [MECH] The product of the magnitude of centrifugal force acting on a body and the distance to the center of rotation.

centrifugal pump [MECH ENG] A machine for moving a liquid, such as water, by accelerating it radially outward in an impeller to a surrounding volute casing.

centrifugal sedimentation [CHEM ENG] Removing solids from liquids by causing particles to settle through the liquid radially toward or away from the center of rotation (depending on the solid-liquid relative densities) by use of a centrifuge.

centrifugal separation [MECH ENG] The separation of two immiscible liquids in a centrifuge in a much shorter period of time than could be accomplished solely by gravity. [NUCLEO] Separation of isotopes by spinning a mixture in gas or vapor form at high speed.

centrifugal settler [CHEM ENG] Spinning container that separates solid particles from liquids; centrifugal force causes suspended solids to move toward or away from the center of rotation, thus concentrating them in one area for removal.

centrifugal stretching [PHYS] Stretching of the bonds of a rotating molecule caused by centrifugal force, resulting in an increase in the molecule's moment of inertia and a modification of its energy levels.

centrifugal switch [MECH ENG] A switch opened or closed by centrifugal force; used on some induction motors to open the starting winding when the motor has almost reached synchronous speed.

centrifugal tachometer [MECH ENG] An instrument which measures the instantaneous angular speed of a shaft by measuring the centrifugal force on a mass rotating with it.

centrifuge [MECH ENG] 1. A rotating device for separating liquids of different specific gravities or for separating suspended colloidal particles, such as clay particles in an aqueous suspension, according to particle-size fractions by centrifugal force. 2. A large motor-driven apparatus with a long arm, at the end of which human and animal subjects or equipment can be revolved and rotated at various speeds to simulate the prolonged accelerations encountered in rockets and spacecraft.

centrifuge microscope [OPTICS] An instrument which permits magnification and observation of living cells being centrifuged; image of the material magnified by the objective which rotates near the periphery of the centrifuge head is brought to the axis of rotation where it is observed in a stationary occular.

centrifuge refining [CHEM ENG] The use of centrifuges for liquids processing, such as separation of solids or immiscible droplets from liquid carriers, or for liquid-liquid solvent extraction.

centrifuge tube [ANALY CHEM] Calibrated, tube-shaped glass container used with laboratory centrifuges for volumetric analysis of separable (solid-liquid or immiscible liquid) samples.

centrilobular emphysema [MED] A disorder marked by pulmonary inflation, primarily affecting the respiratory bronchioles and usually more severe in the upper lobes.

centrimetric wave [COMMUN] Radiation in the super-high-frequency band, having a wavelength between 1 and 10 centimeters.

centriole [CYTOL] A complex cellular organelle forming the center of the centrosome in most cells; usually found near the nucleus in interphase cells and at the spindle poles during mitosis.

centripetal [MECH] Acting or moving in a direction toward the axis of rotation or the center of a circle along which a body is moving.

centripetal acceleration [MECH] The radial component of the acceleration of a particle or object moving around a circle, which can be shown to be directed toward the center of the circle.

centripetal force [MECH] The radial force required to keep a particle or object moving in a circular path, which can be shown to be directed toward the center of the circle.

centrobaric [MECH] 1. Pertaining to the center of gravity, or to some method of locating it. 2. Possessing a center of gravity.

centroclinal [GEOL] Referring to geologic strata dipping toward a common center, as in a structural basin.

centrode [MECH] The path traced by the instantaneous center of a plane figure when it undergoes plane motion.

Centrohelida [INV ZOO] An order of protozoans in the subclass Heliozoia lacking a central capsule and having axopodia or filopodia, and siliceous scales and spines.

centroid *See* center of area; center of mass; center of volume.

centroid of asymptotes [CONT SYS] The intersection of asymptotes in a root-locus diagram.

centroids of areas and lines [MATH] Points positioned identically with the centers of gravity of corresponding thin homogeneous plates or thin homogeneous wires; involved in the analysis of certain problems of mechanics such as the phenomenon of bending.

centrolecithal ovum [CYTOL] An egg cell having the yolk centrally located; occurs in arthropods.

Centrolenidae [VERT ZOO] A family of arboreal frogs in the suborder Procoela characterized by green bones.

centromere [CYTOL] A specialized chromomere to which the spindle fibers are attached during mitosis. Also known as kinetochore; kinomere; primary constriction.

Centronellidina [PALEON] A suborder of extinct articulate brachiopods in the order Terebratulida.

centrosome [CYTOL] A spherical hyaline region of the cytoplasm surrounding the centriole in many cells; plays a dynamic part in mitosis as the focus of the spindle pole.

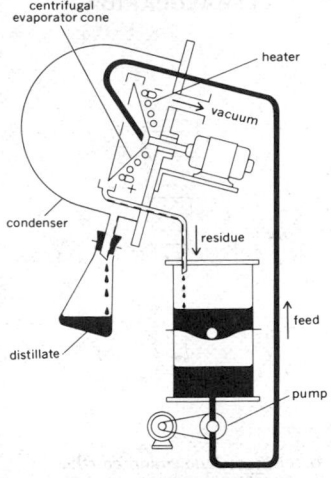

CENTRIFUGAL MOLECULAR STILL

Diagram of the centrifugal molecular still.

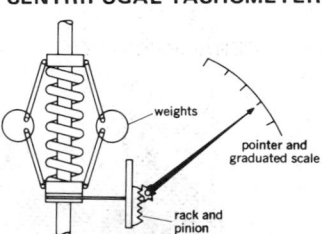

CENTRIFUGAL TACHOMETER

The centrifugal-force tachometer indicates instantaneous speed. *(From D. M. Considine, ed., Process Instruments and Controls Handbook, McGraw-Hill, 1957)*

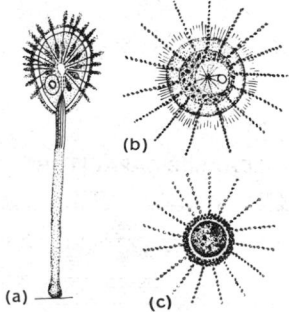

CENTROHELIDA

Some Centrohelida species. *(a) Actinolophus pedunculatus, sessile on Bryozoa (after Villeneuve), (b) Heterophrys myriopoda (after Penard), (c) Pompholyxophrys punicea (after Penard). (From R. P. Hall, Protozoology, Prentice-Hall, 1953)*

CEPHALOCARIDA

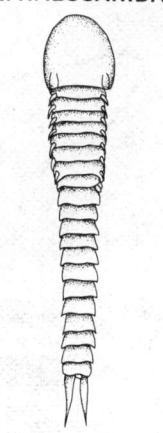

Hutchinsoniella macracantha of the Cephalocarida. (After Scourfield)

CERAMBYCIDAE

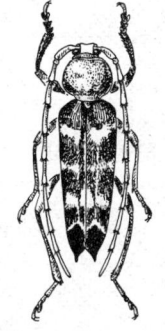

A longhorn beetle of the Cerambycidae. (From T. I. Storer and R. L. Usinger, *General Zoology*, 3d ed., McGraw-Hill, 1957)

CERAMIC CAPACITOR

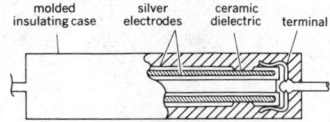

molded insulating case · silver electrodes · ceramic dielectric · terminal

Cutaway drawing of typical molded-case ceramic capacitor. (From K. Henney and C. Walsh, eds., *Electronic Components Handbook*, McGraw-Hill, 1957)

Centrospermae [BOT] An equivalent name for the Caryophyllales.

Centrospermales [BOT] An equivalent name for the Caryophyllales.

centrosphere [CYTOL] The differentiated layer of cytoplasm immediately surrounding the centriole. [GEOL] The central core of the earth. Also known as the barysphere.

centrum [ANAT] The main body of a vertebra. [BOT] The central space in hollow-stemmed plants.

Cen X-3 *See* Centaurus X-3.

Cep *See* Cepheus.

cephaeline [ORG CHEM] $C_{14}H_{19}O_2N$ An alkaloid, slightly soluble in water, extracted from the root of ipecac; used as an emetic.

Cephalaspida [PALEON] An equivalent name for the Osteostraci.

Cephalaspidomorphi [VERT ZOO] An equivalent name for Monorhina.

cephalic [ZOO] Of or pertaining to the head or anterior end.

cephalic index [ANTHRO] The ratio of maximum breadth to maximum length of the head multiplied by 100.

cephalic module [ANTHRO] A measure of absolute head size derived by averaging the length, breadth, and auricular height of the head.

cephalic vein [ANAT] A superficial vein located on the lateral side of the arm which drains blood from the radial side of the hand and forearm into the axillary vein.

cephalin [BIOCHEM] Any of several acidic phosphatides whose composition is similar to that of lecithin but having ethanolamine, serine, and inositol instead of choline; found in many living tissues, especially nervous tissue of the brain.

Cephalina [INV ZOO] A suborder of protozoans in the order Eugregarinida that are parasites of certain invertebrates.

cephalization [ZOO] Anterior specialization resulting in the concentration of sensory and neural organs in the head.

Cephalobaenida [INV ZOO] An order of the arthropod class Pentastomida composed of primitive forms with six-legged larvae.

Cephalocarida [INV ZOO] A subclass of Crustacea erected to include the primitive crustacean *Hutchinsoniella macracantha*.

Cephalochordata [VERT ZOO] A subphylum of the Chordata comprising the lancelets, including *Branchiostoma*.

Cephaloidea [INV ZOO] The false longhorn beetles, a small family of coleopteran insects in the superfamily Tenebrionoidea.

cephalomere [INV ZOO] One of the somites that make up the head of an arthropod.

cephalometry [ANTHRO] The science of measuring the head, especially for determining the characteristics of a particular race, sex, or somatotype.

cephalont [INV ZOO] A sporozoan just prior to spore formation.

Cephalopoda [INV ZOO] Exclusively marine animals constituting the most advanced class of the Mollusca, including squids, octopuses, and *Nautilus*.

cephalosporin [MICROBIO] Any of a group of antibiotics produced by strains of the imperfect fungus *Cephalosporium*.

cephalothin [MICROBIO] An antibiotic derived from the fungus *Cephalosporium*, resembling penicillin units in structure and activity, and effective against many gram-positive cocci that are resistant to penicillin.

cephalothorax [INV ZOO] The body division comprising the united head and thorax of arachnids and higher crustaceans.

Cephalothrididae [INV ZOO] A family of ribbonlike worms in the order Palaeonemertini.

cephalotrichous flagellation [CYTOL] Insertion of flagella in polar tufts.

cepheid [ASTRON] One of a subgroup of periodic variable stars whose brightness does not remain constant with time and whose period of variation is a function of intrinsic mean brightness.

Cepheus [ASTRON] A constellation with right ascension 22 hours, declination 70°N. Abbreviated Cep.

Cephidae [INV ZOO] The stem sawflies, composing the single family of the hymenopteran superfamily Cephoidea.

Cephoidea [INV ZOO] A superfamily of hymenopteran insects in the suborder Symphyta.

cepstrum [ACOUS] The Fourier transform of the logarithm of a speech power spectrum; used to separate vocal tract information from pitch excitation in voiced speech.

cepstrum vocoder [ENG ACOUS] A digital device for reproducing speech in which samples of the cepstrum of speech, together with pitch information, are transmitted to the receiver, and are then converted into an impulse response that is convolved with an impulse train generated from the pitch information.

Ceractinomorpha [INV ZOO] A subclass of sponges belonging to the class Demospongiae.

ceramagnet [ELECTROMAG] A ferrimagnet composed of the hard magnetic material $BaO \cdot 6Fe_2O_3$.

ceramal *See* cermet.

Cerambycidae [INV ZOO] The longhorn beetles, a family of coleopteran insects in the superfamily Chrysomeloidea.

ceramet *See* cermet.

ceramic [MATER] **1.** A product made by the baking or firing of a nonmetallic mineral, such as tile, cement, plaster refractories, and brick. **2.** Consisting of such a product.

ceramic aggregate [MATER] **1.** Portland cement concrete containing lumps of ceramic material. **2.** Concrete made with porous clay to reduce its weight.

ceramic amplifier [ELECTR] An amplifier that utilizes the piezoelectric properties of semiconductors such as silicon.

ceramic capacitor [ELEC] A capacitor whose dielectric is a ceramic material such as steatite or barium titanate, the composition of which can be varied to give a wide range of temperature coefficients.

ceramic cartridge [ENG ACOUS] A device containing a piezoelectric ceramic element, used in phonograph pickups and microphones.

ceramic coating [MET] A nonmetallic, inorganic coating made of sprayed aluminum oxide or of zirconium oxide, or a cemented coating of an intermetallic compound, such as aluminum disilicide, of essentially crystalline nature, applied as a protective film on metal to protect against temperatures above 1100°C.

ceramic earphones *See* crystal headphones.

ceramic fiber [MATER] A small-dimension filament or thread composed of a ceramic material, usually alumina and silica, used in lightweight units for electrical, thermal, and sound insulation, filtration at high temperatures, packing, and reinforcing other ceramic materials.

ceramic glaze [ENG] A glossy finish on a clay body obtained by spraying with metallic oxides, chemicals, and clays and firing at high temperature.

ceramicite [PETR] A porcelained pyrometamorphic rock composed of basic plagioclase and cordierite with a small amount of hypersthene and a groundmass of glass.

ceramic magnet [ELECTROMAG] A permanent magnet made from pressed and sintered mixtures of ceramic and magnetic powders. Also known as ferromagnetic ceramic.

ceramic microphone [ENG ACOUS] A microphone using a ceramic cartridge.

ceramic mold casting [MET] A precision casting process using a ceramic body fired to high temperature as the mold, and carbon, low-alloy, or stainless steel as the casting.

ceramic pickup [ENG ACOUS] A phonograph pickup using a ceramic cartridge.

ceramic reactor [NUCLEO] A nuclear reactor in which the fuel and moderator assemblies are made from high-temperature-resistant ceramic materials such as metal oxides, carbides, or nitrides.

ceramics [ENG] The art and science of making ceramic products.

ceramic tile [MATER] A burned-clay product composed of a clay body with a decorative surface glaze; used principally for decorative and sanitary effects.

ceramic tool [DES ENG] A cutting tool made from metallic oxides.

ceramic transducer *See* electrostriction transducer.

ceramic tube [ELECTR] An electron tube having a ceramic envelope capable of withstanding operating temperatures over 500°C, as required during reentry of guided missiles.

ceramoplastic [MATER] A high-temperature insulating material made by bonding synthetic mica with glass.

Ceramoporidae [PALEON] A family of extinct, marine bryozoans in the order Cystoporata.

Cerapachyinae [INV ZOO] A subfamily of predacious ants in the family Formicidae, including the army ant.

Ceraphronidae [INV ZOO] A superfamily of hymenopteran insects in the superfamily Proctotrupoidea.

cerargyrite [MINERAL] AgCl A colorless to pearl-gray mineral; crystallizes in the isometric system, but crystals, usually cubic, are rare; a secondary mineral that is an ore of silver. Also known as chlorargyrite; horn silver.

cerata [INV ZOO] Respiratory papillae of the mantle in certain nudibranchs.

cerate [ORG CHEM] A metallic salt or soap made from lard. [PHARM] An unguent made of wax, resin, or spermaceti mixed with oil, lard, and medicinal ingredients; used for external application.

ceratine [INV ZOO] A hornlike material secreted by some anthozoans.

Ceratiomyxaceae [MYCOL] The single family of the fungal order Ceratiomyxales.

Ceratiomyxales [MYCOL] An order of myxomycetous fungi in the subclass Ceratiomyxomycetidae.

Ceratiomyxomycetidae [MYCOL] A subclass of fungi belonging to the Myxomycetes.

ceratite [PALEON] A fossil ammonoid of the genus *Ceratites* distinguished by a type of suture in which the lobes are further divided into subordinate crenulations while the saddles are not divided and are smoothly rounded.

ceratitic [PALEON] Pertaining to a ceratite.

Ceratodontidae [PALEON] A family of Mesozoic lungfishes in the order Dipteriformes.

Ceratomorpha [VERT ZOO] A suborder of the mammalian order Perissodactyla including the tapiroids and the rhinoceratoids.

Ceratophyllaceae [BOT] A family of rootless, free-floating dicotyledons in the order Nymphaeales characterized by unisexual flowers and whorled, cleft leaves with slender segments.

Ceratopsia [PALEON] The horned dinosaurs, a suborder of Upper Cretaceous reptiles in the order Ornithischia.

cercaria [INV ZOO] The larval generation which terminates development of a digenetic trematode in the intermediate host.

Cercopidae [INV ZOO] A family of homopteran insects belonging to the series Auchenorrhyncha.

Cercopithecidae [VERT ZOO] The Old World monkeys, a family of primates in the suborder Anthropoidea.

cercopod [INV ZOO] 1. Either of two filamentous projections on the posterior end of notostracan crustaceans. 2. *See* cercus.

Cercospora [MYCOL] A genus of imperfect fungi having dark, elongate, multiseptate spores.

Cercospora leaf spot [PL PATH] Any of several fungus diseases of plants caused by *Cercospora* species and characterized by areas of discoloration on the leaves.

cercus [INV ZOO] Either of a pair of segmented sensory appendages on the last abdominal segment of many insects and certain other anthropods. Also known as cercopod.

cere [VERT ZOO] A soft, swollen mass of tissue at the base of the upper mandible through which the nostrils open in certain birds, such as parrots and birds of prey.

cereal [BOT] Any member of the grass family (Graminae) which produces edible, starchy grains usable as food by man and his livestock. Also known as grain.

cerebellar ataxia [MED] Incoordination of muscles due to disease of the cerebellum.

cerebellum [ANAT] The part of the vertebrate brain lying below the cerebrum and above the pons, consisting of three lobes and concerned with muscular coordination and the maintenance of equilibrium.

cerebral cortex [ANAT] The superficial layer of the cerebral hemispheres, composed of gray matter and concerned with coordination of higher nervous activity.

cerebral hemisphere [ANAT] Either of the two lateral halves of the cerebrum.

cerebral localization [ANAT] Designation of a specific region of the brain as the area controlling a specific physiologic function or as the site of a lesion.

cerebral palsy [MED] Any nonprogressive motor disorder in man caused by brain damage incurred during fetal development.

cerebral peduncle [ANAT] One of two large bands of white matter (containing descending axons of upper motor neurons) which emerge from the underside of the cerebral hemispheres and approach each other as they enter the rostral border of the pons.

cerebration [PSYCH] The act or process of using the mind; thinking.

cerebrose *See* galactose.

cerebroside [BIOCHEM] Any of a complex group of glycosides found in nerve tissue, consisting of a hexose, a nitrogenous base, and a fatty acid. Also known as galactolipid.

cerebroside lipoidosis *See* Gaucher's disease.

cerebrospinal axis [ANAT] The axis of the body composed of the brain and spinal cord.

cerebrospinal fluid [PHYSIO] A clear liquid that fills the ventricles of the brain and the spaces between the arachnoid mater and pia mater.

cerebrospinal meningitis [MED] Inflammation of the meninges of the brain and spinal cord.

cerebrotonia [PSYCH] A temperament pattern proposed by W. H. Sheldon as characteristic of an ectomorph, who is an individual with an intellectualized, solitudinous, and inhibited approach to life.

cerebrovascular accident [MED] A symptom complex resulting from cerebral hemorrhage, embolism, or thrombosis of the cerebral vessels, characterized by sudden loss of consciousness.

cerebrum [ANAT] The enlarged anterior or upper part of the vertebrate brain consisting of two lateral hemispheres.

cerelose *See* glucose.

Cerenkov counter [NUCLEO] An apparatus for detecting high-energy charged particles by observation of the Cerenkov radiation produced.

Cerenkov radiation [ELECTROMAG] Light emitted by a high-speed charged particle when the particle passes through a transparent, nonconducting material at a speed greater than the speed of light in the material.

Cerenkov rebatron radiator [ELECTR] Device in which a tightly bunched, velocity-modulated electron beam is passed through a hole in a dielectric; the reaction between the higher velocity of the electrons passing through the hole and the slower velocity of the electromagnetic energy passing through the dielectric results in radiation at some frequency higher than the frequency of modulation of the electron beam.

ceresine [MATER] An almost white wax produced by purification of ozokerite; a native mineral wax used in the manufacture of candles, shoe polishes, electrical insulation, and floor waxes.

ceresin wax [MATER] 1. A hydrocarbon wax, used as a substitute for beeswax. Also known as cerin; cerosin; purified ozokerite. 2. *See* paraffin wax.

ceria *See* ceric oxide.

Ceriantharia [INV ZOO] An order of the Zoantharia distinguished by the elongate form of the anemone-like body.

Ceriantipatharia [INV ZOO] A subclass proposed by some authorities to include the anthozoan orders Antipatharia and Ceriantharia.

ceric oxide [INORG CHEM] CeO₂ A pale-yellow to white powder; soluble in sulfuric acid, insoluble in dilute acid and water; used in ceramics and as a polish for optical glass. Also known as ceria; cerium dioxide; cerium oxide.

ceric sulfate [INORG CHEM] Ce(SO₄)₂·4H₂O Yellow needles forming a basic salt with excess water; used in waterproofing, mildew-proofing, and in dyeing and printing textiles.

cerin *See* ceresin wax.

cerinic acid *See* cerotic acid.

cerite [MINERAL] (Ca,Fe)Ce₃Si₃O₁₂·H₂O A brown rare-earth hydrous silicate of cerium and other metals found in gneiss; hardness is 5.5 on Mohs scale, and specific gravity is 4.86.

CERARGYRITE

Cerargyrite crystals from Leadville, Colorado. *(Specimen from Department of Geology, Bryn Mawr College)*

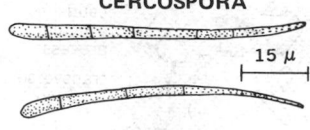

CERCOSPORA

The dark multiseptate spores of *Cercospora*.

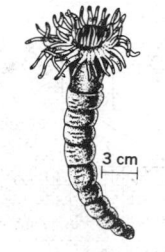

CERIANTHARIA

Cerianthus solitarius lives in sandy marine substrata. *(After Y. Delage)*

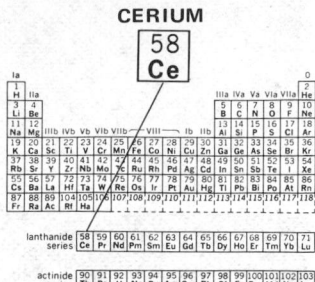

CERIUM

58
Ce

Periodic table of the chemical elements showing the position of cerium.

CERUSSITE

├── 3 cm ──┤

Cerussite crystal specimen from Ems, Nassau. *(American Museum of Natural History Specimen)*

CERVICAL VERTEBRA

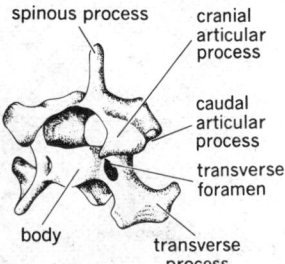

spinous process

cranial articular process

caudal articular process

transverse foramen

body

transverse process

Cervical vertebra of a dog. *(From M. E. Miller, G. C. Christensen, and H. E. Evans, Anatomy of the Dog, Saunders, 1964)*

CESIUM

55
Cs

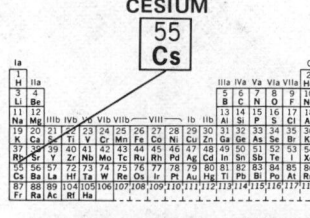

Periodic table of the chemical elements showing the position of cesium.

Cerithiacea [INV ZOO] A superfamily of gastropod mollusks in the order Prosobranchia.

cerium [CHEM] A chemical element, symbol Ce, atomic number 58, atomic weight 140.12; a rare-earth metal, used as a getter in the metal industry, as an opacifier and polisher in the glass industry, in Welsbach gas mantles, in cored carbon arcs, and as a liquid-liquid extraction agent to remove fission products from spent uranium fuel.

cerium-140 [NUC PHYS] An isotope of cerium with atomic mass number of 140, 88.48% of the known amount of the naturally occurring element.

cerium-142 [NUC PHYS] A radioactive isotope of cerium with atomic mass number of 142; emits α-particles and has a half-life of 5×10^{15} years.

cerium-144 [NUC PHYS] A radioactive isotope of the element cerium with atomic mass number of 144; a beta emitter with a half-life of 285 days.

cerium dioxide *See* ceric oxide.

cerium fluoride [INORG CHEM] CeF_3 White hexagonal crystals, melting point 1460°C; used in arc carbons to increase the brilliance of carbon-arc lamps.

cerium oxide *See* ceric oxide.

cerium stearate [ORG CHEM] $Ce(C_{18}H_{35}O_2)_2$ White, waxy, inert powder, melting point 100–110°C; used in waterproofing compounds.

Cermak-Spirek furnace [ENG] An automatic reverberatory furnace of rectangular form divided into two sections by a wall; used for roasting zinc and quicksilver ores.

cermet [MATER] Any of a group of composite materials made by mixing, pressing, and sintering metal with ceramic; examples are silicon-silicon carbide and chromium-alumina carbide. Also known as ceramal; ceramet; metal ceramic.

cermet nuclear fuel [NUCLEO] A nuclear reactor fuel mixed with a heat-resistant ceramic and a metal to give it both refractory and damage-resistant properties.

cermet resistor [ELEC] A metal-glaze resistor, consisting of a mixture of finely powdered precious metals and insulating materials fired onto a ceramic substrate.

cernuous [BOT] Drooping or inclining.

cerography [GRAPHICS] Painting in which wax is used as a binder for the pigments.

cerolite [MINERAL] A mixture of serpentine and stevensite occurring in yellow or greenish waxlike masses.

Cerophytidae [INV ZOO] A small family of coleopteran insects in the superfamily Elateroidea.

cerosin *See* ceresin wax.

cerotic acid [ORG CHEM] $CH_3(CH_2)_{24}COOH$ A fatty acid derived from carnauba wax or beeswax; melts at 87.7°C. Also known as cerinic acid; hexacosanoic acid.

cers [METEOROL] A term for the mistral in Catalonia, Narbonne, and parts of Provence (southern France and northeastern Spain).

ceruloplasmin [BIOCHEM] The copper-binding serum protein in human blood.

cerumen [PHYSIO] The waxy secretion of the ceruminous glands of the external ear. Also known as earwax.

ceruminous gland [ANAT] A modified sweat gland in the external ear that produces earwax.

cerussite [MINERAL] $PbCO_3$ A yellow or white member of the aragonite group occurring in orthorhombic crystals; produced by the action of carbon dioxide on lead ore.

cervantite [MINERAL] Sb_2O_4 A white or yellow secondary mineral crystallizing in the orthorhombic system and formed by oxidation of antimony sulfide.

cervical [ANAT] Of or relating to the neck, a necklike part, or the cervix of an organ.

cervical canal [ANAT] Canal of the cervix of the uterus.

cervical flexure [EMBRYO] A ventrally concave flexure of the embryonic brain occurring at the junction of hindbrain and spinal cord.

cervical ganglion [ANAT] Any ganglion of the sympathetic nervous system located in the neck.

cervical plexus [ANAT] A plexus in the neck formed by the anterior branches of the upper four cervical nerves.

cervical sinus [EMBRYO] A triangular depression caudal to the hyoid arch containing the posterior visceral arches and grooves.

cervical vertebra [ANAT] Any of the bones in the neck region of the vertebral column; the transverse process has a characteristic perforation by a transverse foramen.

cervicitis [MED] Inflammation of the cervix uteri.

Cervidae [VERT ZOO] A family of pecoran ruminants in the superfamily Cervoidea, characterized by solid, deciduous antlers; includes deer and elk.

cervix uteri [ANAT] The narrow outer end of the uterus.

Cervoidea [VERT ZOO] A superfamily of tylopod ruminants in infraorder Pecora, including deer, giraffes, and related species.

ceryl alcohol [ORG CHEM] $C_{26}H_{53}OH$ An alcohol derived from Chinese wax, melting at 79°C and insoluble in water.

cesarean section [MED] Delivery of the fetus through an abdominal incision.

cesarolite [MINERAL] $H_2PbMn_3O_8$ A steel-gray mineral consisting of a hydrous lead manganate occurring in spongy masses.

Cesaro summation [MATH] A method of attaching sums to certain divergent sequences and series by taking averages of the first n-terms and passing to the limit.

cesium [CHEM] A chemical element, symbol Cs, atomic number 55, atomic weight 132.905.

cesium-134 [NUC PHYS] An isotope of cesium, atomic mass number of 134; emits negative beta particles and has a half-life of 2.19 years; used in photoelectric cells and in ion propulsion systems under development.

cesium-137 [NUC PHYS] An isotope of cesium with atomic mass number of 137; emits negative beta particles and has a half-life of 30 years; offers promise as an encapsulated radiation source for therapeutic and other purposes. Also known as radiocesium.

cesium-antimonide photocathode [ELECTR] A photocathode obtained by exposing a thin layer of antimony to cesium vapor at elevated temperatures; has a maximum sensitivity in the blue and ultraviolet regions of the spectrum.

cesium-beam atomic clock [NUCLEO] An instrument, used as the primary standard of frequency and time, in which a microwave oscillator, which generates radiation in a microwave cavity, is maintained at a frequency such that a hyperfine transition is induced in cesium atoms in a beam passing through the cavity. Also known as cesium-beam atomic oscillator.

cesium-beam atomic oscillator *See* cesium beam atomic clock.

cesium beam tube *See* cesium electron tube.

cesium chloride [INORG CHEM] CsCl Colorless cuboid crystals, melting point 646°C; used in filaments of radio tubes to increase sensitivity, in photoelectric cells, and for photosensitive deposit on cathodes.

cesium electron tube [ELECTR] An electronic device used as an atomic clock, producing electromagnetic energy that is accurate and stable in frequency. Also known as cesium beam tube.

cesium hollow cathode [ELECTR] A cathode in which cesium is heated at the bottom of a cylinder serving as the cathode of an electron tube, to give current densities that can be as high as 800 amperes per square centimeter.

cesium-ion engine [AERO ENG] An ion engine that uses a stream of cesium ions to produce a thrust for space travel.

cesium phototube [ELECTR] A phototube having a cesium-coated cathode; maximum sensitivity in the infrared portion of the spectrum.

cesium thermionic converter [ELECTR] A thermionic diode in which cesium vapor is stored between the plates to neutralize space charge and to lower the work function of the emitter.

cesium-vapor lamp [ELECTR] A lamp in which light is produced by the passage of current between two electrodes in ionized cesium vapor.

cesium-vapor rectifier [ELECTR] A gas tube in which cesium vapor serves as the conducting gas and a condensed monatomic layer of cesium serves as the cathode coating.

cespitose [BOT] **1.** Tufted; growing in tufts, as grass. **2.** Having short stems forming a dense turf.

cesspit *See* cesspool.

cesspool [CIV ENG] An underground tank for raw sewage

collection; used where there is no sewage system. Also known as cesspit.

Cestida [INV ZOO] An order of ribbon-shaped ctenophores having a very short tentacular axis and an elongated pharyngeal axis.

Cestoda [INV ZOO] A subclass of tapeworms including most members of the class Cestoidea; all are endoparasites of vertebrates.

Cestodaria [INV ZOO] A small subclass of worms belonging to the class Cestoidea; all are endoparasites of primitive fishes.

Cestoidea [INV ZOO] The tapeworms, endoparasites composing a class of the phylum Platyhelminthes.

Cet *See* Cetus.

Cetacea [VERT ZOO] An order of aquatic mammals, including the whales, dolphins, and porpoises.

cetane *See* n-hexadecane.

cetane index [CHEM ENG] An empirical method for finding the cetane number of a fuel based on API gravity and the mid boiling point.

cetane number [CHEM ENG] The percentage by volume of cetane (cetane number 100) in a blend with α-methylnaphthalene (cetane number 0); indicates the ability of a fuel to ignite quickly after being injected into the cylinder of an engine.

cetane number improver [CHEM] A chemical which has the effect of increasing a diesel fuel's cetane number; examples are nitrates, nitroalkanes, nitrocarbonates, and peroxides.

cetene *See* 1-hexadecene.

cetol *See* cetyl alcohol.

cetology [VERT ZOO] The study of whales.

Cetomimiformes [VERT ZOO] An order of rare oceanic, deep-water, soft-rayed fishes that are structurally diverse.

Cetomimoidei [VERT ZOO] The whalefishes, a suborder of the Cetomimiformes, including bioluminescent, deep-sea species.

Cetorhinidae [VERT ZOO] The basking sharks, a family of large, galeoid elasmobranchs of the isurid line.

Cetus [ASTRON] A constellation with right ascension 2 hours, declination 10°S. Abbreviated Cet. Also known as Whale.

cetyl [ORG CHEM] The radical represented as $C_{16}H_{33}-$.

cetyl alcohol [ORG CHEM] $C_{15}H_{33}OH$ A colorless wax, insoluble in water although a solution in kerosine forms an insoluble film on water. Also known as cetol; ethal; 1-hexadecanol.

cetylene *See* 1-hexadecene.

cetylic acid *See* palmitic acid.

Ceva's theorem [MATH] The theorem that if three concurrent straight lines pass through the vertices *A, B,* and *C* of a triangle and intersect the opposite sides, produced if necessary at *D, E,* and *F,* then the product $\overline{AF}\cdot\overline{BD}\cdot\overline{CE}$ of the lengths of three alternate segments equals the product $\overline{FB}\cdot\overline{DC}\cdot\overline{EA}$ of the other three.

cevedilla *See* sabadilla.

cevian [MATH] A straight line that passes through a vertex of a triangle or tetrahedron and intersects the opposite side or face.

ceylonite [MINERAL] A dark-green, brown, or black iron-bearing variety of spinel. Also known as candite; pleonaste; zeylanite.

C factor [MAP] An empirical value in aerial photography which expresses the vertical measuring capability of a given stereoscopic system; the factor is the ratio of the flight height to the smallest contour interval accurately plottable; in planning for aerial photography the C factor is used to determine the flight height required for a specified contour interval, camera, and instrument system. Also known as altitude contour.

C figure *See* C index.

cfm *See* cubic feet per minute.

CFR-ASTM engine [ENG] A special engine standardized by the American Society for Testing and Materials to determine knock qualities of gasoline.

CFR engine [ENG] Cooperative Fuel Research engine, a standard test engine used to determine the octane number of motor fuels.

cfs *See* cusec.

cg *See* centigram.

cgs system *See* centimeter-gram-second system.

Cha *See* Chamaeleon.

chabazite [MINERAL] $CaAl_2Si_4O_{12}\cdot6H_2O$ A white to yellow or red member of the zeolite group occurring in glassy rhombohedral crystals; hardness is 4–5 on Mohs scale, and specific gravity is 2.08–2.16.

chad [ADP] The piece of material removed when forming a hole or notch in a punched tape or punched card. Also known as chip. [NUCLEO] **1.** A unit of neutron flux equal to 1 neutron per square centimeter per second. **2.** A unit of neutron flux equal to 10^{12} neutrons per square centimeter per second.

chadacryst *See* xenocryst.

chadless tape [ADP] Paper tape in which the perforations for code characters are made by an incomplete circular cut, with the resulting flap of material folded aside.

chad tape [ADP] Paper tape in which perforations for code characters are completely punched out.

Chadwick-Goldhaber effect *See* photodisintegration.

chaeta *See* seta.

Chaetetidae [PALEON] A family of Paleozoic corals of the order Tabulata.

Chaetodontidae [VERT ZOO] The butterflyfishes, a family of perciform fishes in the suborder Percoidei.

Chaetognatha [INV ZOO] A phylum of abundant planktonic arrowworms.

Chaetonotoidea [INV ZOO] An order of the class Gastrotricha characterized by two adhesive tubes connected with the distinctive paired, posterior tail forks.

Chaetophoraceae [BOT] A family of algae in the order Ulotrichales characterized as branched filaments which taper toward the apices, sometimes bearing terminal setae.

Chaetopteridea [INV ZOO] A family of spioniform polychaete annelids belonging to the Sedentaria.

chaff [AGR] Seed coverings and small stem pieces separated from grass and grain seeds in threshing or processing. [ORD] Thin, flat pieces of metal foil, plain or backed, designed to act as a countermeasure against enemy radar when released into the atmosphere.

chafing corrosion *See* fretting corrosion.

chafing fatigue [MET] Fatigue induced by corrosion damage between metal surfaces in close contact under pressure.

Chagas' disease [MED] An acute and chronic protozoan disease of man caused by the hemoflagellate *Trypanosoma (Schizotrypanum) cruzi.* Also known as South American trypanosomiasis.

chain [ADP] **1.** A series of data or other items linked together in some way. **2.** A sequence of binary digits used to construct a code. [DES ENG] **1.** A flexible series of metal links or rings fitted into one another; used for supporting, restraining, dragging, or lifting objects or transmitting power. **2.** A mesh of rods or plates connected together, used to convey objects or transmit power. [CIV ENG] *See* engineer's chain; Gunter's chain. [COMMUN] A network of radio, television, radar, navigation, or other similar stations connected by special telephone lines, coaxial cables, or radio relay links so all can operate as a group for broadcast purposes, communication purposes, or determination of position. [MATH] A simply linearly ordered set.

chain balance [ANALY CHEM] An analytical balance with one end of a fine gold chain suspended from the beam and the other fastened to a device which moves over a graduated vernier scale.

chain belt [DES ENG] A belt of flat links used to transmit power.

chain block [MECH ENG] A tackle which uses an endless chain rather than a rope, often operated from an overhead track to lift heavy weights especially in workshops. Also known as chain fall; chain hoist.

chain bond [CIV ENG] A masonry bond formed with a chain or bar.

chain breat machine [MIN ENG] A coal-cutting machine, so constructed that a series of cutting points attached to a circulating chain work their way for a certain distance under a seam.

chain cable [NAV ARCH] The chain to which an anchor is attached. Also known as cable chain.

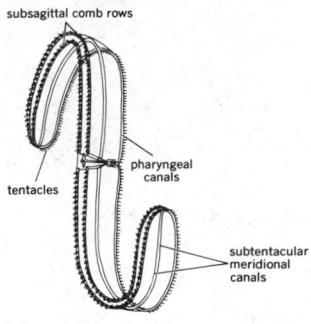

CESTIDA

Velamen species.

CHABAZITE

⊢ 2.5 cm ⊣

Crystalline specimen of the mineral chabazite found in Two Islands, Nova Scotia. *(Specimen from Department of Geology, Bryn Mawr College)*

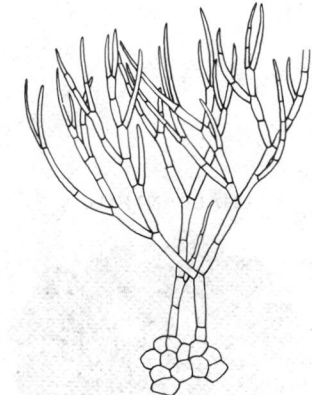

CHAETOPHORACEAE

Chaetophora showing portion of erect thallus of branched filaments which arises from a prostrate group of cells.

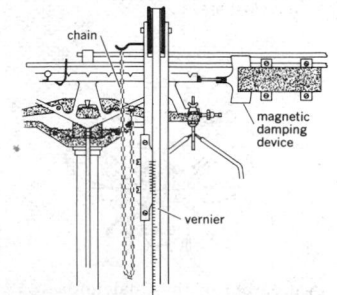

CHAIN BALANCE

Chain balance with notched beam, magnetic damper, and vernier scale.

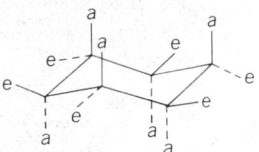

CHAIR FORM

The chair form conformation of cyclohexane; *a* represents hydrogens that are staggered and perpendicular to the mean plane of the carbon; *e* represents the hydrogens that are equatorially directed to the center of the mean plane of the carbons.

CHALCANTHITE

Crystals of chalcanthite associated with quartz. *(Specimen from Department of Geology, Bryn Mawr College)*

CHALCIDIDAE

A member of the Chalcididae. *(From T. I. Storer and R. L. Usinger, General Zoology, 3d ed., McGraw-Hill, 1957)*

chain coal cutter [MIN ENG] A cutter which makes a groove in the coal by an endless chain moving around a flat plate called a jib.

chain code [ADP] A binary code consisting of a cyclic sequence of some or all of the possible binary words at a given length such that each word is derived from the previous one by moving the binary digits one position to the left, dropping the leading bit, and inserting a new bit at the end, in such a way that no word recurs before the cycle is complete.

chain command [ADP] Any input/output command in a sequence of input/output commands such as WRITE, READ, SENSE.

chain complex [MATH] A sequence $\{C_n\}, -\infty < n < \infty$, of Abelian groups together with a sequence of boundary homomorphisms $d_n: C_n \to C_{n-1}$ such that $d_{n-1} \circ d_n = 0$ for each n.

chain conveyor [MECH ENG] A machine for moving materials that carries the product on one or two endless linked chains with crossbars; allows smaller parts to be added as the work passes.

chain course [CIV ENG] A course of stone held together by iron cramps.

chain data flag [ADP] A value of 1 given to a specific bit of a channel command word, commonly used with scatter read or scatter write operations.

chain decay *See* series disintegration.

chain disintegration *See* series disintegration.

chain drive [MECH ENG] A flexible device for power transmission, hoisting, or conveying, consisting of an endless chain whose links mesh with toothed wheels fastened to the driving and driven shafts.

chain fall *See* chain block.

chain fission yield [NUC PHYS] The sum of the independent fission yields for all isobars of a particular mass number.

chain-float liquid-level gage [ENG] Float device to measure the level of liquid in a vessel; the float, suspended from a counterweighted chain draped over a toothed sprocket, rises or falls with the liquid level, and the chain movement turns the sprocket to position a calibrated depth-indicator.

chain gear [MECH ENG] A gear that transmits motion from one wheel to another by means of a chain.

chain grate stoker [MECH ENG] A wide, endless chain used to feed, carry, and burn a noncoking coal in a furnace, control the air for combustion, and discharge the ash.

chain hoist *See* chain block.

chain homomorphism [MATH] A sequence of homomorphisms $f_n: C_n \to D_n$ between the groups of two chain complexes such that $f_{n-1} d_n = \bar{d}_n f_n$ where d_n and $\bar{d}_n$ are the boundary homomorphisms of $\{C_n\}$ and $\{D_n\}$ respectively.

chaining [ADP] A method of storing records which are not necessarily contiguous, in which the records are arranged in a sequence and each record contains means to identify its successor.

chain intermittent fillet welding [MET] **1.** The forming of two lines of intermittent fillet welds in a T joint or lap joint so that the increments of welding in one line are approximately opposite those in the other line. **2.** The forming of two lines of equal-length fillet welds concurrently on opposite sides of the perpendicular member of a T joint at intermittent intervals.

chain isomerism [ORG CHEM] A type of molecular isomerism seen in carbon compounds; as the number of carbon atoms in the molecule increases, the linkage between the atoms may be a straight chain or branched chains producing isomers that differ from each other by possessing different carbon skeletons.

chain lightning [GEOPHYS] A rare form of lightning in a long zigzag or apparently broken line.

chain locker [NAV ARCH] A compartment in the lower part of a ship toward the bow for stowing an anchor chain.

chain loom [TEXT] A knitting machine used to make flat-knit fabric.

chain of simplices [MATH] A member of the free Abelian group generated by the simplices of a given dimension of a simplicial complex.

chain pipe [NAV ARCH] A heavy steel pipe through which the anchor chain passes from the windlass to the chain locker.

chain plate [NAV ARCH] One of the metal plates bolted to the side of a ship, to which lines supporting the masts are fastened.

chain printer [ADP] A high-speed printer in which the type slugs are carried by the links of a revolving chain.

chain radar beacon [COMMUN] A beacon with a fast recovery time to permit simultaneous interrogation and tracking of the beacon by a number of radars.

chain radar system [ENG] A number of radar stations located at various sites on a missile range to enable complete radar coverage during a missile flight; the stations are linked by data and communication lines for target acquisition, target positioning, or data-recording purposes.

chain reaction [CHEM] A chemical reaction in which many molecules undergo chemical reaction after one molecule becomes activated. [NUCLEO] *See* nuclear chain reaction.

chain riveting [ENG] Riveting consisting of rivets one behind the other in rows along the seam.

chain rule [MATH] A rule for differentiating a composition of functions: $(d/dx) f(g(x)) = f'(g(x)) \cdot g'(x)$.

chain saw [MECH ENG] A gasoline-powered saw for felling and bucking timber, operated by one man; has cutting teeth inserted in a sprocket chain that moves rapidly around the edge of an oval-shaped blade.

chain scission [ORG CHEM] The cleavage of polymer chains, as in natural rubber as a result of heating.

chain stopper [NAV ARCH] A short chain with a hook attached to a ship's forecastle, used to secure the anchor chain when the anchor is raised or lowered, thus relieving the windlass from strain, or to quickly release the anchor.

chain tongs [DES ENG] A tool for turning pipe, using a chain to encircle and grasp the pipe.

chain vise [DES ENG] A vise in which the work is encircled and held tightly by a chain.

chainwall [MIN ENG] A coal mining technique in which the mine roof is supported by coal pillars, between which the coal is mined away.

chain weave [TEXT] Compact, heavyweight cloth weave made with plied yarn in both directions; its pattern is two-up and two-down broken twill, with two ends of right hand and two of left hand repeated on four threads each way.

chair form [PHYS CHEM] A particular nonplanar conformation of a cyclic molecule with more than five atoms in the ring; for example, in the chair form of cyclohexane, the hydrogens are staggered and directed perpendicularly to the mean plane of the carbons (axial conformation, *a*) or equatorially to the center of the mean plane (equatorial conformation, *e*).

chairs *See* folding boards.

chalaza [BOT] The region at the base of the nucellus of an ovule; gives rise to the integuments. [CYTOL] One of the paired, spiral, albuminous bands in a bird's egg that attach the yolk to the shell lining membrane at the ends of the egg.

chalazion [MED] A small tumor of the eyelid formed by retention of tarsal gland secretions. Also known as a Meibomian cyst.

chalazogamy [BOT] A process of fertilization in which the pollen tube passes through the chalaza to reach the embryo sac.

chalazoidite *See* mud ball.

chalcanthite [MINERAL] $CuSO_4 \cdot 5H_2O$ A blue to bluish-green mineral occurring in triclinic crystals or in massive fibrous veins or stalactites. Also known as bluestone; blue vitriol.

chalcedony [MINERAL] A cryptocrystalline variety of quartz; occurs as crusts with a rounded, mammillary, or botryoidal surface and as a major constituent of nodular and bedded cherts; varieties include carnelian and bloodstone.

chalcedonyx [MINERAL] A mineral consisting of onyx with alternating gray and white bands; valued as a semiprecious stone.

Chalcididae [INV ZOO] The chalcids, a family of hymenopteran insects in the superfamily Chalcidoidea.

Chalcidoidea [INV ZOO] A superfamily of hymenopteran insects in the suborder Apocrita, including primarily insect parasites.

chalcoalumite [MINERAL] $CuAl_4(SO_4)(OH)_{12} \cdot 3H_2O$ A turquoise-green to pale-blue mineral consisting of a hydrous basic sulfate of copper and aluminum.

chalcocite [MINERAL] Cu_2S A fine-grained, massive mineral with a metallic luster which tarnishes to dull black on exposure; crystallizes in the orthorhombic system, the crystals being rare and small usually with hexagonal outline as a result of twinning; hardness is 2.5-3 on Mohs scale, and specific gravity is 5.5-5.8. Also known as beta chalcocite; chalcosine; copper glance; redruthite; vitreous copper.

chalcocyanite [MINERAL] $CuSO_4$ A white mineral consisting of copper sulfate. Also known as hydrocyanite.

chalcography [GRAPHICS] The art of copper engraving.

chalcolite *See* torbernite.

chalcomenite [MINERAL] $CuSeO_3 \cdot 2H_2O$ A blue mineral consisting of copper selenite occurring in crystals.

chalcophanite [MINERAL] $(Zn,Mn,Fe)Mn_2O_5 \cdot nH_2O$ Black mineral with metallic luster consisting of hydrous manganese and zinc oxide.

chalcophile [GEOL] Having an affinity for sulfur and therefore massing in greatest concentration in the sulfide phase of a molten mass.

chalcophyllite [MINERAL] $Cu_{18}Al_2(AsO_4)_3(OH)_{27} \cdot 33H_2O$ A green mineral consisting of basic arsenate and sulfate of copper and aluminum occurring in tabular crystals or foliated masses. Also known as copper mica.

chalcopyrite [MINERAL] $CuFeS_2$ A major ore mineral of copper; crystallizes in the tetragonal crystal system, but crystals are generally small with diphenoidal faces resembling the tetrahedron; usually massive with a metallic luster and brass-yellow color; hardness is 3.5-4 on Mohs scale, and specific gravity is 4.1-4.3. Also known as copper pyrite; fool's gold; yellow pyrite.

chalcopyrrohite [MINERAL] $CuFe_4S_5$ A sulfide mineral occurring in meteorites.

chalcosiderite [MINERAL] $Cu(Fe,Al)_6(PO_4)_4(OH)_8 \cdot 4H_2O$ A green mineral, isomorphous with turquoise, consisting of a hydrous basic phosphate of copper, iron, and aluminum.

chalcostibite [MINERAL] $CuSbS_2$ A lead-gray mineral consisting of antimony copper sulfide.

chalcotrichite [MINERAL] A capillary variety of cuprite occurring in long needlelike crystals. Also known as hair copper; plush copper ore.

chaldron [MECH] **1.** A unit of volume in common use in the United Kingdom, equal to 36 bushels, or 288 gallons, or approximately 1.30927 cubic meters. **2.** A unit of volume, formerly used for measuring solid substances in the United States, equal to 36 bushels, or approximately 1.26861 cubic meters.

chalice cell *See* goblet cell.

chalicosis [MED] A pulmonary affection caused by inhalation of stone dust.

Chalicotheriidae [PALEON] A family of extinct perissodactyl mammals in the superfamily Chalicotherioidea.

Chalicotherioidea [PALEON] A superfamily of extinct, specialized perissodactyls having claws rather than hooves.

chalk [PETR] A variety of limestone formed from pelagic organisms; it is very fine-grained, porous, and friable; white or very light-colored, it consists almost entirely of calcite. [MATER] Artificially prepared pure calcium carbonate; used as the basis for pastels. Also known as whiting.

chalkboard *See* blackboard.

chalking [CHEM] **1.** Treating with chalk. **2.** Forming a powder which is easily rubbed off. [MET] Defect of coated metals in which a layer of powder forms between the coating and the base metal.

chalkstone [PATH] A gouty deposit, usually of sodium urate, in the hands or feet.

challenge [COMMUN] To cause an interrogator to transmit a signal which puts a transponder into operation.

challenger *See* interrogator.

challenging signal *See* interrogation.

challiho [METEOROL] Strong southerly winds which blow for about 40 days in spring in some parts of India; sometimes the winds are violent, causing blinding dust storms.

chalones [BIOCHEM] Substances thought to be molecules of the protein-polypeptide class that are produced as part of the growth-control systems of tissues; known to inhibit cell division by acting on several phases in the mitotic cycle.

chalybite *See* siderite.

Chamaeleon [ASTRON] A constellation, right ascension 11 hours, declination 80°S. Abbreviated Cha. Also spelled Chameleon.

Chamaeleontidae [VERT ZOO] The chameleons, a family of reptiles in the suborder Sauria.

Chamaemyidae [INV ZOO] The aphid flies, a family of myodarian cyclorrhaphous dipteran insects of the subsection Acalypteratae.

chamaephyte [ECOL] Any perennial plant whose winter buds are within 25 centimeters of the soil surface.

chamaerrhine [ANTHRO] Having a short broad nose with a nasal index of 51-57.9 for a skull, and 85-99.9 for a person.

Chamaesiphonales [BOT] An order of blue-green algae of the class Cyanophyceae; reproduce by cell division, colony fragmentation, and endospores.

chamber [CIV ENG] The space in a canal lock between the upper and lower gates. [GRAPHICS] A sleeve or channel of a transparent film jacket. [MIN ENG] **1.** The working place of a miner. **2.** A body of ore with definite boundaries apparently filling a preexisting cavern. [ORD] The part of the gun in which the charge is placed: in a revolver, the hole in the cylinder; in a cannon, the space between the obturator or breechblock and the forcing cone.

chamber acid [INORG CHEM] Sulfuric acid made by the obsolete chamber process.

chamber capacity *See* chamber volume.

chambering [MIN ENG] Increasing the size of a drill hole in a quarry by firing a succession of small charges, until the hole can take a proper explosive charge to bring down the face of the quarry. [ORD] That phase of small-arms operation that deals with the placing of the round into the chamber after the round has been fed into the weapon by the feeding device.

chamber kiln [ENG] A kiln consisting of a series of adjacent chambers in a ring or oval through which the fire moves, taking several days to make a circuit; waste gas from the fire preheats ware in chambers toward which the fire is moving, while combustion air is preheated by ware in chambers already fired.

chamber pressure [AERO ENG] The pressure of gases within the combustion chamber of a rocket engine.

chamber process [CHEM ENG] An obsolete method of manufacturing sulfuric acid in which sulfur dioxide, air, and steam are reacted in a lead chamber with oxides of nitrogen as the catalyst.

chamber volume [AERO ENG] The volume of the rocket combustion chamber, including the convergent portion of the nozzle up to the throat. Also known as chamber capacity.

chambray [TEXT] A cotton fabric similar to gingham; it is plain woven, lightweight, and without a pattern, and has a dyed or unbleached warp and white filling.

chamecephalic [ANTHRO] Having a flattened backward-slanting head with a length-height index of 70 or less.

chamecranial [ANTHRO] Having a flat low skull with a length-height index of less than 70.

chameleon [VERT ZOO] The common name for about 80 species of small-to-medium-sized lizards composing the family Chamaeleontidae.

Chameleon *See* Chamaeleon.

chameprosopic [ANTHRO] Having a broad low face with a facial index of 90 or less.

chamfer [DES ENG] A beveled edge.

chamfer angle [DES ENG] The angle that a beveled surface makes with one of the original surfaces.

chamfering [MECH ENG] Machining operations to produce a beveled edge. Also known as beveling.

chamfer plane [DES ENG] A plane for chamfering edges of woodwork.

chamois [VERT ZOO] *Rupicapra rupicapra.* A goatlike mammal included in the tribe Rupicaprini of the family Bovidae.

chamosite [MINERAL] A greenish-gray or black mineral consisting of silicate belonging to the chlorite group and having monoclinic crystals; found in many oolitic iron ores.

Champlainian [GEOL] Middle Ordovician geologic time.

chance-constrained programming [ADP] Type of nonlinear

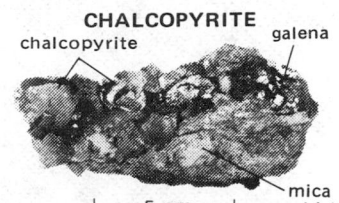

CHALCOPYRITE

Chalcopyrite with galena and mica schist from Bottino, Tuscany, Italy. *(American Museum of Natural History)*

CHAMELEON

Flap-necked chameleon *(Chamaeleo dilepis)*, with opposable toes, a prehensile tail; the neck flap is raised.

CHAMOIS

The chamois *(Rupicapra rupicapra)*, an intermediate between goats and antelopes, found only in Europe.

programming wherein the deterministic constraints are replaced by their probabilistic counterparts.

chance process [MIN ENG] A method for separating clean coal from slate and other impurities in a mixture of sand and water.

chancre [MED] **1.** A lesion or ulcer at the site of primary inoculation by an infecting organism. **2.** The initial lesion of syphilis.

chancroid [MED] A lesion of the genitalia, usually of venereal origin, caused by *Hemophilus ducreyi.* Also known as soft chancre.

Chandler motion *See* polar wandering.

Chandler period [GEOPHYS] The period of the Chandler wobble.

Chandler wobble [GEOPHYS] A movement in the earth's axis of rotation, the period of motion being about 14 months.

chandui *See* chanduy.

chanduy [METEOROL] A cool, descending wind at Guayaquil, Ecuador, which blows during the dry season (July to November). Also spelled chandui.

change chart [METEOROL] A chart indicating the amount and direction of change of some meteorological element during a specified time interval; for example, a height-change chart or pressure-change chart. Also known as tendency chart.

changed memory routine [ADP] A selective memory dump routine in which only those words that have been changed in the course of running a program are printed.

change gear [MECH ENG] A gear used to change the speed of a driven shaft while the speed of the driving remains constant.

change of tide [OCEANOGR] A reversal of the direction of motion (rising or falling) of a tide, or in the set of a tidal current. Also known as turn of the tide.

changeover switch [ELEC] A means of moving a circuit from one set of connections to another.

change tape [ADP] A paper tape or magnetic tape carrying information that is to be used to update filed information; the latter is often on a master tape. Also known as transaction tape.

changing bag [ENG] An enclosure of lightproof material used for operations such as loading of film holders in daylight.

Chanidae [VERT ZOO] A monospecific family of teleost fishes in the order Gonorynchiformes containing the milkfish (*Chanos chanos*), distinguished by the lack of teeth.

channel [ADP] **1.** A path along which digital or other information may flow in a computer. **2.** The section of a storage medium that is accessible to a given reading station in a computer, such as a path parallel to the edge of a magnetic tape or drum or a path in a delay-line memory. **3.** One of the longitudinal rows of intelligence holes punched along the length of paper tape. Also known as level. [CHEM ENG] In percolation filtration, a portion of the clay bed where there is a preponderance of flow. [CIV ENG] A natural or artificial waterway connecting two bodies of water or containing moving water. [COMMUN] **1.** A band of radio frequencies allocated for a particular purpose; a standard broadcasting channel is 10 kilohertz wide, a television channel 6 megahertz wide. **2.** A path for carrier-current signals. [ELECTR] **1.** A path for a signal, as an audio amplifier may have several input channels. **2.** The main current path between the source and drain electrodes in a field-effect transistor or other semiconductor device. [HYD] The deeper portion of a waterway carrying the main current. [NAV] Navigable portion of a body of water. [NUCLEO] A passage for fuel slugs or heat-transfer fluid in a reactor.

channel address word [ADP] A four-byte code containing the protection key and the main storage address of the first channel command word at the start of an input/output operation. Abbreviated CAW.

channel bank [ELECTR] Part of a carrier-multiplex terminal that performs the first step of modulation of the transmitting voice frequencies into a higher-frequency band, and the final step in the demodulation of the received higher-frequency band into the received voice frequencies.

channel black *See* gas black.

channel buoy [NAV] In marine operations, a buoy marking a channel.

channel capacity [COMMUN] The maximum number of bits

or other information elements that can be handled in a particular channel per unit time.

channel command [ADP] The step, equivalent to a program instruction, required to tell an input/output channel what operation is to be performed, and where the data are or should be located.

channel command word [ADP] A code specifying an operation, one or more flags, a count, and a storage location. Abbreviated CCW.

channel control [HYD] A condition whereby the stage of a stream is controlled only by discharge and the general configuration of the stream channel, that is, the contours of its bed, banks, and floodplains.

channel control command [ADP] An order to a control unit to perform a nondata input/output operation.

channel design [ADP] The type of channel, characterized by the tasks it can perform, available to a computer.

channel designator [COMMUN] Number associated with each channel, tributary, or trunk for reference purposes. Also known as channel sequence number.

channel effect [ELECTR] A leakage current flowing over a surface path between the collector and emitter in some types of transistors.

channel-end condition [ADP] A signal indicating that the use of an input/output channel is no longer required.

channeler *See* channeling machine.

channel fill [GEOL] Accumulations of sand and detritus in a stream channel where the transporting capacity of the water is insufficient to remove the material as rapidly as it is delivered.

channel frequency *See* stream frequency.

channel gradient ratio *See* stream gradient ratio.

channeling [COMMUN] A type of multiplex transmission in which the separation between communication channels is accomplished by the use of carriers or subcarriers. [NUCLEO] The transmission of extra particles through a medium in a nuclear reactor due to the presence of voids in the medium.

channeling machine [MECH ENG] An electrically powered machine that operates by a chipping action of three to five chisels while traveling back and forth on a track; used for primary separation from the rock ledge in marble, limestone, and soft sandstone quarries. Also known as channeler.

channel iron [DES ENG] A metal strip or beam with a U-shape.

channelizing [COMMUN] The process of subdividing a wideband transmission facility so as to handle a number of different circuits requiring comparatively narrow bandwidths.

channel-lag deposit [GEOL] Coarse residual material left as accumulations in the channel in the normal processes of the stream.

channel light [NAV] In marine operations, a light marking a channel.

channel marker [NAV] In marine operations, a marker indicating a channel.

channel miles [COMMUN] The summation, in miles, of the electrical path of individual channels between two points; these points may be connected by wire or radio, or a combination of both.

channel-mouth bar [GEOL] A bar formed where moving water enters a body of still water, due to decreased velocity.

channel net [HYD] Stream channel pattern within a drainage basin.

channel order *See* stream order.

channel process [CHEM ENG] A carbon-black process in which iron channel beams are used as depositing surfaces for carbon black.

channel program [ADP] The set of steps, called channel commands, by means of which an input/output channel is controlled.

channel pulse [COMMUN] Telemetering pulse representing intelligence on a channel by virtue of its time or modulation characteristics.

channel read-backward command [ADP] A command to transfer data from tape device to main storage while the tape is moving backward.

channel read command [ADP] A command to transfer data from an input/output device to main storage.

CHANIDAE

The milkfish (*Chanos chanos*), length up to 5 feet (1.5 meters), lives in marine and estuarine waters of the tropical Indo-Pacific. (*After Jordan and Evermann, The Fishes of North and Middle America, U.S. Nat. Mus. Bull., no. 47, pt. 4, 1900*)

CHANNEL BUOY

SPAR   CAN

Black buoy with odd number indicates channel lies to starboard, proceeding from seaward. (*Modified from Motor Boating's chart of United States buoyage system*)

channel reliability [COMMUN] The percent of time a channel was available for use in a specific direction during a specified period of time.

channel sample *See* groove sample.

channel sand [GEOL] A sandstone or sand deposited in a stream bed or other channel eroded into the underlying bed.

channel segment *See* stream segment.

channel selector [ELEC] A control used to tune in the desired channel in a radio or television receiver.

channel sense command [ADP] A command commonly used to denote an unusual condition existing in an input/output device and requesting more information.

channel sequence number *See* channel designator.

channel shifter [ELECTR] Radiotelephone carrier circuit that shifts one or two voice-frequency channels from normal channels to higher voice-frequency channels to reduce cross talk between channels; the channels are shifted back by a similar circuit at the receiving end.

channel spacing [COMMUN] The difference in frequency between successive radio or television channels.

channel status word [ADP] A storage register containing the status information of the input/output operation which caused an interrupt. Abbreviated CSW.

channel steamer [NAV ARCH] A special type of ship which operates across the English Channel, carrying passengers and cars.

channel synchronizer [ELECTR] An electronic device providing the proper interface between the central processing unit and the peripheral devices.

channel-to-channel adapter [ADP] A device which provides two computer systems with interchannel communications.

channel width [GEOL] The distance across a stream or channel as measured from bank to bank near bankful stage.

channel wing [AERO ENG] A wing that is trough-shaped so as to surround partially a propeller to get increased lift at low speeds from the slipstream.

channel write command [ADP] A command which transfers data from main storage to an input/output device.

Channidae [VERT ZOO] The snakeheads, a family of fresh-water perciform fishes in the suborder Anabantoidei.

Chaoboridae [INV ZOO] The phantom midges, a family of dipteran insects in the suborder Orthorrhapha.

chaparral [ECOL] A vegetation formation characterized by woody plants of low stature, impenetrable because of tough, rigid, interlacing branches, which have simple, waxy, evergreen, thick leaves.

chaplet [MET] Metal support used to space and hold the core in position within a sand mold.

Chaplygin-Kármán-Tsien relation [FL MECH] The relation that in the case of isentropic flow of an ideal gas with negligible viscosity and thermal conductivity, the sum of the pressure and a constant times the reciprocal of the density of the fluid is constant along a streamline; a useful, although physically impossible, approximation.

Chapman-Enskog approximations [STAT MECH] Approximations to a solution of the Boltzmann transport equation in the Chapman-Enskog theory.

Chapman-Enskog solution [STAT MECH] The solution of the Boltzmann transport equation according to the Chapman-Enskog theory.

Chapman-Enskog theory [STAT MECH] A method of solving the Boltzmann transport equation by successive approximations, essentially in powers of the mean free path. Also known as Enskog theory.

Chapman equation [GEOPHYS] A theoretical relation describing the distribution of electron density with height in the upper atmosphere. [STAT MECH] The relationship that the viscosity of a gas equals $(0.499)mv/[\sqrt{2}\ \pi\sigma^2\ (1\ +\ C/T)]$, where m is the mass of a molecule, v its average speed, σ its collision diameter, C the Sutherland constant, and T the absolute temperature (Kelvin scale).

chapmanite [MINERAL] $Fe_2Sb(SiO_4)_2(OH)$ A mineral consisting of a silicate of iron and antimony.

Chapman-Jouguet plane [MECH] A hypothetical, infinite plane, behind the initial shock front, in which it is variously assumed that reaction (and energy release) has effectively been completed, that reaction product gases have reached thermodynamic equilibrium, and that reaction gases, stream-ing backward out of the detonation, have reached such a condition that a forward-moving sound wave located at this precise plane would remain a fixed distance behind the initial shock.

Chapman region [GEOPHYS] A hypothetical region in the upper atmosphere in which the distribution of electron density with height can be described by Chapman's theoretical equation.

Characeae [BOT] The single family of the order Charales.

Characidae [VERT ZOO] The characins, the single family of the suborder Characoidei.

Characoidei [VERT ZOO] A suborder of the order Cypriniformes including fresh-water fishes with toothed jaws and an adipose fin.

character [ADP] **1.** An elementary mark used to represent data, usually in the form of a graphic spatial arrangement of connected or adjacent strokes, such as a letter or digit. **2.** A small collection of adjacent bits used to represent a piece of data, addressed and handled as a unit, often corresponding to a digit or letter.

character adjustment [ADP] An address modification affecting a specific number of characters of the address part of the instruction.

character boundary [ADP] In character recognition, a real or imaginary rectangle which serves as the delimiter between consecutive characters or successive lines on a source document.

character code [COMMUN] A bit pattern assigned to a particular character in a coded character set.

character density [ADP] The number of characters recorded per unit of length or area. Also known as record density.

character disorder [PSYCH] A pattern of behavior and emotional response that is socially disapproved or unacceptable, with little evidence of anxiety or other symptoms seen in neuroses.

character emitter [ADP] In character recognition, an electromechanical device which conveys a specimen character in the form of a time pulse or group of pulses.

character generator [ADP] A hard-wired subroutine which will display alphanumeric characters on a screen.

character group [MATH] The set of all continuous homomorphisms of a topological group onto the group of all complex numbers with unit norm.

characteristic [MATH] That part of the logarithm of a number which is the integral (the whole number) to the left of the decimal point in the logarithm.

characteristic chamber length [AERO ENG] The length of a straight, cylindrical tube having the same volume as that of the chamber of a rocket engine if the chamber had no converging section.

characteristic cone [MATH] A conelike region important in the study of initial value problems in partial differential equations.

characteristic curve [GRAPHICS] In photography, a graph that shows how increases in exposure increase the density of the film. [MATH] **1.** One of a pair of conjugate curves in a surface with the property that the directions of the tangents through any point of the curve are the characteristic directions of the surface. **2.** A curve plotted on graph paper to show the relation between two changing values.

characteristic distortion [COMMUN] **1.** Displacement of signal transitions resulting from the persistence of transients caused by preceding transitions. **2.** In teletypewriter transmission-systems, repetitive displacement or disruption peculiar to specific portions of a teletypewriter signal; the two types of characteristic distortions are line and equipment.

characteristic equation [MATH] **1.** Any equation which has a solution, subject to specified boundary conditions, only when a parameter occurring in it has certain values. **2.** Specifically, the equation $A\mathbf{u} = \lambda\ \mathbf{u}$, which can have a solution only when the parameter λ has certain values, where A can be a square matrix which multiplies the vector $\mathbf{u}$, or a linear differential or integral operator which operates on the function $\mathbf{u}$, or in general, any linear operator operating on the vector $\mathbf{u}$ in a finite or infinite dimensional vector space. Also known as eigenvalue equation. **3.** An equation which sets the characteristic polynomial of a given linear transformation on a finite dimensional vector space, or of its matrix representation,

CHARACTER RECOGNITION

ABCDEFGHIJKLM
NOPQRSTUVWXYZ
0123456789
• ⌐ : ⌐ = + / ⏁ * ⊓ & ⏌
' - { } % ? ∫ Ч Н

Character set approved by the American Standards Association for use in optical character recognition applications.

CHARADRIIFORMES

The common or Wilson's snipe *(Gallinago gallinago),* a typical wader. It frequents wet woodlands and marshes of North America, France, and the British Isles.

equal to zero. [PHYS] An equation relating a set of variables, such as pressure, volume, and temperature, whose values determine a substance's physical condition. [PL PHYS] An equation whose solutions give the frequencies and modes of those perturbations of a hydromagnetic system which decay or grow exponentially in time, and indicate regions of stability of such a system.

characteristic exhaust velocity [AERO ENG] Of a rocket engine, a descriptive parameter, related to effective exhaust velocity and thrust coefficient. Also known as characteristic velocity.

characteristic form [MATH] A means of classifying partial differential equations.

characteristic frequency [COMMUN] Frequency which can be easily identified and measured in a given emission.

characteristic function [MATH] The function χ_A defined for any subset A of a set by setting $\chi_A(x) = 1$ if x is in A and $\chi_A(x) = 0$ if x is not in A. Also known as indicator function. [STAT] A function that uniquely defines a probability distribution; it is equal to $\sqrt{2\pi}$ times the Fourier transform of the frequency function of the distribution.

characteristic impedance [COMMUN] The impedance that, when connected to the output terminals of a transmission line of any length, makes the line appear to be infinitely long, for there are then no standing waves on the line, and the ratio of voltage to current is the same for each point on the line. Also known as surge impedance.

characteristic length [MECH] A convenient reference length (usually constant) of a given configuration, such as overall length of an aircraft, the maximum diameter or radius of a body of revolution, or a chord or span of a lifting surface.

characteristic loss spectroscopy [SPECT] A branch of electron spectroscopy in which a solid surface is bombarded with monochromatic electrons, and backscattered particles which have lost an amount of energy equal to the core-level binding energy are detected. Abbreviated CLS.

characteristic manifold [MATH] A surface used to study the problem of existence of solutions to partial differential equations.

characteristic number *See* eigenvalue.

characteristic of a logarithm [MATH] The number preceding the decimal.

characteristic overflow [ADP] An error condition encountered when the characteristic of a floating point number exceeds the limit imposed by the hardware manufacturer.

characteristic polynomial [MATH] The polynomial whose roots are the eigenvalues of a given linear transformation on a finite dimensional vector space.

characteristic radiation [ATOM PHYS] Radiation originating in an atom following removal of an electron, whose wavelength depends only on the element concerned and the energy levels involved.

characteristic ray [MATH] For a differential equation, an integral curve which generates all the others.

characteristic root *See* eigenvalue.

characteristic temperature *See* Debye temperature.

characteristic underflow [ADP] An error condition encountered when the characteristic of a floating point number is smaller than the smallest limit imposed by the hardware manufacturer.

characteristic value *See* eigenvalue.

characteristic vector *See* eigenvector.

characteristic velocity *See* characteristic exhaust velocity.

characteristic x-rays [ATOM PHYS] Electromagnetic radiation emitted as a result of rearrangements of the electrons in the inner shells of atoms; the spectrum consists of lines whose wavelengths depend only on the element concerned and the energy levels involved.

characterization factor [CHEM ENG] A number which expresses the variations in physical properties with change in character of the paraffinic stock; ranges from 12.5 for paraffinic stocks to 10.0 for the highly aromatic stocks. Also known as Watson factor.

character of the bottom [NAV] In marine operations, the type of material of which the bottom is composed; a sounding lead is sometimes armed with tallow or similar substance so that a sample of the bottom is obtained when a sounding is made; the sample is often helpful in determining position. Also known as nature of the bottom.

character outline [ADP] The graphic pattern formed by the stroke edges of a printed or handwritten character in character recognition.

character printer [ADP] A device which prints a character at a time by using a laterally moving carriage to move the paper or the type font to successive positions on the line.

character reader [ADP] In character recognition, any device capable of locating, identifying, and translating into machine code the handwritten or printed data appearing on a source document.

character recognition [ADP] The technology of using a machine to sense and encode into a machine language the characters which are originally written or printed by human beings.

character set [COMMUN] A set of unique representations called characters, for example, the 26 letters of the English alphabet, the Boolean 0 and 1, the set of signals in Morse code, and the 128 characters of the USASCII.

character skew [ADP] In character recognition, an improper appearance of a character to be recognized, in which it appears in a tilted condition with respect to a real or imaginary horizontal base line.

character stroke *See* stroke.

character style [ADP] In character recognition, a distinctive construction that is common to all members of a particular character set.

character-writing tube [ELECTR] A cathode-ray tube that forms alphanumeric and symbolic characters on its screen for viewing or recording purposes.

Charactron gun [ELECTR] The electron gun used in a Charactron tube.

Charactron tube [ELECTR] A numerical indicator tube in which an electron beam is deflected to pass through a mask in which different numbers or letters are punched and is subsequently deflected and focused back to any desired position on a fluorescent screen.

Charadrii [VERT ZOO] The shore birds, a suborder of the order Charadriiformes.

Charadriidae [VERT ZOO] The plovers, a family of birds in the superfamily Charadrioidea.

Charadriiformes [VERT ZOO] An order of cosmopolitan birds, most of which live near water.

Charadrioidea [VERT ZOO] A superfamily of the suborder Charadrii, including plovers, sandpipers, and phalaropes.

Charales [BOT] Green algae composing the single order of the class Charophyceae.

charcoal [MATER] Also known as char. 1. A porous solid product containing 85–98% carbon and produced by heating carbonaceous materials such as cellulose, wood, or peat at 500–600°C in the absence of air. 2. The residue obtained from the carbonization of a noncoking coal, such as subbituminous coal, lignite, or anthracite. 3. *See* low-temperature coke.

charcoal rot [PL PATH] A fungus disease of potato, corn, and other plants caused by *Macrophomina phaseoli;* tissues of the root and lower stem are destroyed and blackened.

charcoal test [CHEM ENG] A determination of the natural gasoline content of natural gas by adsorbing the gasoline on activated charcoal and then recovering it by distillation.

Chareae [BOT] A tribe of green algae belonging to the family Characeae.

charge [ELEC] 1. A basic property of elementary particles of matter; the charge of an object may be a positive or negative number or zero; only integral multiples of the proton charge occur, and the charge of a body is the algebraic sum of the charges of its constituents; the value of the charge may be inferred from the Coulomb force between charged objects. Also known as electric charge. 2. To convert electrical energy to chemical energy in a secondary battery. 3. To feed electrical energy to a capacitor or other device that can store it. [ENG] 1. A unit of an explosive, either by itself or contained in a bomb, projectile, mine, or the like, or used as the propellant for a bullet or projectile. 2. To load a borehole with an explosive. 3. The material or part to be heated by induction or dielectric heating. 4. The measurement or weight of material, either liquid, preformed, or powder, used to load a mold at one time during one cycle in the manufacture of plastics or metal. [MET] Material introduced into a furnace for melting. [NUCLEO] The fissionable material or fuel placed in a reactor to produce a chain reaction.

charge carrier [SOLID STATE] A mobile conduction electron or mobile hole in a semiconductor. Also known as carrier.

charge conjugation conservation [PARTIC PHYS] The principle that the laws of motion are left unchanged by the charge conjugation operation; it is violated by the weak interactions, but no other violations have as yet been established.

charge conjugation operation [PARTIC PHYS] The operation of changing every particle into its antiparticle.

charge conjugation parity *See* charge parity.

charge conservation *See* conservation of charge.

charge-coupled devices [ELECTR] Semiconductor devices arrayed so that the electric charge at the output of one provides the input stimulus to the next.

charge coupling [ADP] Transfer of all electric charges within a semiconductor storage element to a similar, nearby element by means of voltage manipulations.

charge density [ELEC] The charge per unit area on a surface or per unit volume in space.

charged particle [PARTIC PHYS] A particle whose charge is not zero; the charge of a particle is added to its designation as a superscript, with particles of charge $+1$ and -1 (in terms of the charge of the proton) denoted by $+$ and $-$ respectively; for example, π^+, Σ^-.

charge establishment [ORD] Process of establishing the correct weight of a propelling charge to produce the prescribed muzzle velocity with the prescribed projectile in a particular weapon; performed at a proving ground.

charge independence [NUC PHYS] The principle that the nuclear (strong) force between a neutron and a proton is identical to the force between two protons or two neutrons in the same orbital and spin state. [PARTIC PHYS] As a generalization of the nuclear physics definition, the principle that the strong interactions of particles are unchanged if a particle is replaced by another particle of the same isotopic spin multiplet.

charge-mass ratio [ELEC] The ratio of the electric charge of a particle to its mass.

charge multiplet *See* isospin multiplet.

charge neutrality [PL PHYS] The near equality in the density of positive and negative charges throughout a volume, which is characteristic of a plasma. [SOLID STATE] The condition in which electrons and holes are present in equal numbers in a semiconductor.

charge parity [PARTIC PHYS] The eigenvalue of the charge conjugation operation; it exists only for a system which goes into itself under this operation. Also known as charge conjugation parity.

charge quantization [ELEC] The principle that the electric charge of an object must equal an integral multiple of a universal basic charge.

charger *See* battery charger.

charger-reader [NUCLEO] An auxiliary device used to charge and read small, portable ionization chambers.

charge-storage transistor [ELECTR] A transistor in which the collector-base junction will charge when forward bias is applied with the base at a high level and the collector at a low level.

charge-storage tube [ELECTR] A storage tube in which information is retained on a surface in the form of electric charges.

charge transfer [PHYS CHEM] The process in which an ion takes an electron from a neutral atom, with a resultant transfer of charge.

charge-weight ratio [ORD] The ratio of the weight of a charge, especially an explosive charge, to the total weight of the complete bomb or projectile that contains the charge; the term is not used in connection with propellants.

charging current [ELEC] The current that flows into a capacitor when a voltage is first applied.

charging pump [CHEM ENG] Pump that provides pressurized fluid flow for the input of another unit, such as to a triplex pump that requires positive pressure.

Charles' law [PHYS] The law that at constant pressure the volume of a fixed mass or quantity of gas varies directly with

the pressure; a close approximation. Also known as Gay-Lussac law.

Charles' Wain *See* Big Dipper.

Charlton white *See* lithopone.

charm [PARTIC PHYS] A quantum number which has been proposed to account for an apparent lack of symmetry in the behavior of hadrons relative to that of leptons, to explain why certain reactions of elementary particles do not occur, and to account for the longevity of the J-1 and J-2 particles.

Charmat process [FOOD ENG] A bulk process for making champagne in which the wine undergoes secondary fermentation in a glass-lined tank instead of in a bottle.

Charmouthian [GEOL] Middle Lower Jurassic geologic time.

charnockite [PETR] Any of various faintly foliated, nearly massive varieties of quartzofeldspathic rocks containing hypersthene.

Charophyceae [BOT] A class of green algae, division Chlorophyta.

Charophyta [BOT] A group of aquatic plants, ranging in size from a few inches to several feet in height, that live entirely submerged in water.

Charpak-Massonet current distribution system [PARTIC PHYS] An electronic data readout method used in spark chambers to locate a single spark, as determined by observing how the spark current divides between the two available paths to the ground.

Charpy test [MET] An impact test to determine the ductility of a metal; a freely swinging pendulum is allowed to strike and break a notched specimen laid loosely on a support; the work done by the pendulum is obtained by comparing the position of the pendulum before release with the position to which it swings after breaking the specimen.

charring ablator [MATER] An ablation material characterized by the formation of a carbonaceous layer at the heated surface which impedes heat flow into the material by its insulating and reradiating characteristics.

chart [MAP] **1.** A map, generally designed for navigation or other particular purposes, in which essential map information is combined with various other data critical to the intended use. **2.** To prepare a chart or to engage in a charting operation. [SCI TECH] A form, such as a graph, table, or diagram, which gives information about some variable quantity.

chartaceous [BOT] Resembling paper.

chart catalog [NAV] A list or enumeration of navigational charts, sometimes with index charts indicating the extent of coverage of the various navigational charts.

chart comparison unit [ENG] A device that permits simultaneous viewing of a radar plan position indicator display and a navigation chart so that one appears superimposed on the other. Also known as autoradar plot.

chart convergence [MAP] Convergence of the meridians as shown on a chart.

chart datum *See* datum plane.

chart desk [ENG] A flat surface on which charts are spread out, usually with storage space for charts and other navigating equipment below the plotting surface.

charted depth [OCEANOGR] The vertical distance from the tidal datum to the bottom.

charted visibility [NAV] The extreme distance, shown in numbers on a chart, at which a navigational light can be seen.

chart house [NAV ARCH] A room, usually adjacent to or on the bridge, where charts and other navigational equipment are stored, and where navigational computations, plots, and so on may be made. Also known as chart room.

chartlet [NAV] **1.** A small chart, such as one showing the coverage area of a loran radar, with the distribution of its lines of position, corrections to be applied to readings, location and identification of transmitters, and so on. **2.** A block correction.

chart projection [MAP] A map projection used for a chart.

chart reading [NAV] Interpretation of the symbols, lines, abbreviations, and terms appearing on charts.

chartreusin [MICROBIO] $C_{18}H_{18}O_{18}$ Crystalline, greenish-yellow antibiotic produced by a strain of *Streptomyces chartreusis;* active against gram-positive microorganisms, acid-fast bacilli, and phage of *Staphylococcus pyogenes.*

chart room *See* chart house.

**CHARPAK-MASSONET
CURRENT DISTRIBUTION
SYSTEM**

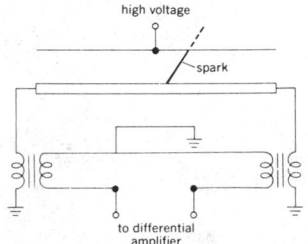

Circuit diagram of an electronic determination of spark position in a narrow gap by the Charpak-Massonet current distribution system.

chart symbol [NAV] A character, letter, or similar graphical representation used on a chart to indicate some object, characteristic, or such.

chart table [ENG] A flat surface on which charts are spread out, particularly one without storage space below the plotting surface, as in aircraft and VPR (virtual PPI reflectoscope) equipment.

Charybdis *See* Galofaro.

chase [DES ENG] A series of cuts, each having a path that follows the path of the cut before it; an example is a screw thread. [ENG] **1.** The main body of the mold which contains the molding cavity or cavities. **2.** The enclosure used to shrink-fit parts of a mold cavity in place to prevent spreading or distortion, or to enclose an assembly of two or more parts of a split-cavity block. [ORD] The exposed part of a gun (artillery) in front on the trunnion band or cradle.

chase mortise [DES ENG] A mortise with a sloping edge from bottom to surface so that a tenon can be inserted when the outside clearance is small.

chase pilot [AERO ENG] A pilot who flies an escort airplane and advises another pilot who is making a check, training, or research flight in another craft.

chaser [AERO ENG] The vehicle that maneuvers in order to effect a rendezvous with an orbiting object.

chasing tool [DES ENG] **1.** A hammer or chisel used to decorate metal surfaces. **2.** A thread-cutting tool on a metal lathe that has a number of blades instead of a single point.

chasmophyte [ECOL] A plant that grows in rock crevices.

chassignite [GEOL] An achondritic stony meteorite composed chiefly of olivine (95%); resembles dunite.

chassis [ENG] **1.** A frame on which the body of an automobile or airplane is mounted. **2.** A frame for mounting the working parts of a radio or other electronic device.

chassis lubricant [MATER] A lubricating grease of consistency to be applied with a grease gun through fittings on autos and farm and industrial equipment.

chassis punch [DES ENG] A hand tool used to make round or square holes in sheet metal.

chatoyant [MINERAL] Of a mineral or gemstone, having a changeable luster or color marked by a band of light, resembling the eye of a cat in this respect.

chatter [ELEC] Prolonged undesirable opening and closing of electric contacts, as on a relay. Also known as contact chatter. [ENG ACOUS] Vibration of a disk-recorder cutting stylus in a direction other than that in which it is driven.

chattermark [GEOL] A scar on the surface of bedrock made by the abrasive action of drift carried at the base of a glacier.

Chattian [GEOL] Upper Oligocene geologic time. Also known as Casselian.

Chauffard-Still disease *See* Still's disease.

chaulmoogra oil [MATER] Any of several fixed oils extracted from seeds of trees in the family Flacourtiaceae; widely used at one time to treat leprosy and other diseases.

Chautauquan [GEOL] Upper Devonian geologic time, below Bradfordian.

chavicol [ORG CHEM] $C_3H_5C_6H_4OH$ A colorless phenol that is liquid at room temperature; boils at 230°C; soluble in alcohol and water; found in many essential oils. Also known as 1-allyl-4-hydroxybenzene; *para*-allyl phenol.

chavicol methyl ether *See* estragole.

Chazyan [GEOL] Middle Ordovician geologic time.

chE *See* cholinesterase.

Cheadle's disease *See* infantile scurvy.

Chebyshev approximation *See* min-max technique.

Chebyshev filter [ELECTR] A filter in which the transmission frequency curve has an equal-ripple shape, with very small peaks and valleys. Also spelled Tchebycheff filter.

Chebyshev polynomials [MATH] A family of orthogonal polynomials which solve Chebyshev's differential equation. Also spelled Tchebycheff polynominals.

Chebyshev's differential equation [MATH] A special case of Gauss' hypergeometric second-order differential equation: $(1-x^2)f''(x) - xf'(x) + n^2f(x) = 0$. Also spelled Tchebycheff's differential equation.

Chebyshev's inequality [STAT] Given a nonnegative random variable $f(x)$, and $k > 0$, the probability that $f(x) \geq k$ is less than or equal to the expected value of f divided by k. Also spelled Tchebycheff's inequality.

check [ADP] A test necessary to detect a mistake in computer programming or a computer malfunction.

check bit [ADP] A binary check digit.

check character [ADP] A redundant character used to perform a check.

check dam [CIV ENG] A low, fixed structure, constructed of timber, loose rock, masonry, or concrete, to control water flow in an erodable channel or irrigation canal.

check digit [ADP] A redundant digit used to perform a check.

checkerboard regenerator [ENG] An open-checkerwork arrangement of firebrick in a high-temperature chamber that absorbs heat during a batch processing cycle, then releases it to preheat fresh combustion air during the down cycle; used, for example, in the steel industry with open-hearth and heat-treating furnaces.

checkers [ENG] Open brickwork in a checkerboard regenerator allowing for the passage of hot, spent gases.

check flight [AERO ENG] **1.** A flight made to check or test the performance of an aircraft, rocket, or spacecraft, or a piece of its equipment, or to obtain measurements or other data on performance. **2.** A familiarization flight in an aircraft, or a flight in which the pilot or the aircrew are tested for proficiency.

check indicator [ADP] A console device, usually a light, informing the operator that an error has occurred.

check indicator instruction [ADP] A computer instruction which directs that a signal device is turned on to call the operator's attention to the fact that there is some discrepancy in the instruction now in use.

checking [MATER] Fine cracks in a ceramic coating. [MET] Temporarily reducing the volume or temperature of the air blast in a blast furnace.

check ligament [ANAT] A thickening of the orbital fascia running from the insertion of the lateral rectus muscle to the medial orbital wall (medial check ligament) or from the insertion of the lateral rectus muscle to the lateral orbital wall (lateral check ligament).

checkline [NAV ARCH] A heavy line passed through a chock and fastened to a bitt of a ship that is coming alongside a wharf in order to slow the ship down or to keep it from moving away from the wharf.

check number [ADP] A number denoting a specific type of hardware malfunction.

check observation [METEOROL] An aviation weather observation taken primarily for aviation radio broadcast purposes; usually abbreviated to include just those elements of a record observation that have an important affect on aircraft operations.

checkout [ENG] A sequence of actions to test or examine a thing as to its readiness for incorporation into a new phase of use or as to the performance of its intended function.

checkpoint [ADP] That place in a routine at which the entire state of the computer (memory, registers, and so on) is written out on auxiliary storage (tape, disk, cards) from which it may be read back into the computer if the program is to be restarted later. [NAV] Geographical location on land or water above which the position of an aircraft in flight may be determined by observation or by electronic means.

check problem *See* check routine.

check rail [BUILD] A rail, thicker than the window, that spans the opening between the top and bottom sash; usually beveled and rabbeted. [CIV ENG] *See* guard rail.

check register [ADP] A register in which transferred data are temporarily stored so that they may be compared with a second transfer of the same data, to verify the accuracy of the transfer.

check routine [ADP] A routine or problem designed primarily to indicate whether a fault exists in a computer, without giving detailed information on the location of the fault. Also known as check problem; test program; test routine.

check screen *See* oversize control screen.

check sum [ADP] A sum of digits or numbers used in a summation check.

check valve [ENG] A device for automatically limiting flow in a piping system to a single direction.

Chediak-Higashi anomaly [PATH] Deeply staining, coarse, peroxidase-positive granules in the cytoplasm of neutrophils and eosinophils in certain disease states.

CHEBYSHEV FILTER

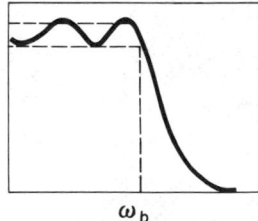

ω_b

Equal ripple shape characteristic in the transmission band of the Chebyshev filter. Frequency ω_b is lower limit of transition region from pass band to stop band.

cheek [ANAT] The wall of the mouth in man and other mammals. [MET] Portion of a three-part flask between the cope and the drag. [ZOO] The lateral side of the head in submammalian vertebrates and in invertebrates.

cheek pouch [VERT ZOO] A saclike dilation of the cheeks in certain animals, such as rodents, in which food is held.

cheese [FOOD ENG] A food produced from milk that has been clotted by acid or rennet to form a curd which is cut, shaped, pressed, and salted or brined. [TEXT] Tube of spun yarn to be put on a warp beam for weaving.

cheese antenna [ELECTROMAG] An antenna having a parabolic reflector between two metal plates, dimensioned to permit propagation of more than one mode in the desired direction of polarization.

cheesebox still [CHEM ENG] One of the first types of vertical cylindrical stills designed with a vapor dome.

cheese cement [MATER] A glue made from cheese or milk curd.

cheesecloth [TEXT] A very loosely woven, thin, lightweight cotton fabric; used in unfinished (gray) form for covering tobacco plants and for making tea bags and wiping cloths; the finished form is used for curtains, bedspreads, and such.

cheese head [DES ENG] A raised cylindrical head on a screw or bolt.

cheetah [VERT ZOO] *Acinonyx jubatus.* A doglike carnivoran mammal belonging to the cat family, having nonretractile claws and long legs.

cheiloplasty [MED] Any plastic operation upon the lip.

cheilosis [MED] Cracking at the corners of the mouth and scaling of the lips, usually associated with riboflavin deficiency.

Cheilostomata [INV ZOO] An order of ectoproct bryozoans in the class Gymnolaemata possessing delicate erect or encrusting colonies composed of loosely grouped zooecia.

cheimaphobia [PSYCH] An abnormal fear of cold.

Cheiracanthidae [PALEON] A family of extinct acanthodian fishes in the order Acanthodiformes.

cheiromegaly [MED] Enlargement of one or both hands that is not attributable to disease of the hypophysis. Also spelled chiromegaly.

cheiroplasty [MED] Any plastic operation on the hand. Also spelled chiroplasty.

chela [INV ZOO] **1.** A claw or pincer on the limbs of certain crustaceans and arachnids. **2.** A sponge spicule with talonlike terminal processes.

chelate [ORG CHEM] A molecular structure in which a heterocyclic ring can be formed by the unshared electrons of neighboring atoms.

chelate laser [OPTICS] A laser in which the coherent pulses of light are produced by a material such as europium, each atom of which is completely enclosed by a chelate molecule, which is in turn held in a transparent plastic.

chelating agent [ORG CHEM] An organic compound in which atoms form more than one coordinate bonds with metals in solution.

chelating resin [ORG CHEM] Any of the ion-exchange resins with unusually high selectivity for specific cations; for example, phenol-formaldehyde resin with 8-quinolinol replacing part of the phenol, particularly selective for copper, nickel, cobalt, and iron(III).

chelation [ORG CHEM] A chemical process involving formation of a heterocyclic ring compound which contains at least one metal cation or hydrogen ion in the ring.

chelerythrine [ORG CHEM] $C_{21}H_{17}O_4H$ A poisonous, crystalline alkaloid, slightly soluble in alcohol; it is derived from the seeds of the herb celandine (*Chelidonium majus*) and has narcotic properties.

chelicera [INV ZOO] Either appendage of the first pair in arachnids, usually modified for seizing, crushing, or piercing.

Chelicerata [INV ZOO] A subphylum of the phylum Arthropoda; chelicerae are characteristically modified as pincers.

Chelidae [VERT ZOO] The side-necked turtles, a family of reptiles in the suborder Pleurodira.

chelidonic acid [ORG CHEM] $C_7H_4O_6$ A pyran isolated from the perennial herb celandine (*Chelidonium majus*).

chelifore [INV ZOO] Either of the first pair of appendages on the cephalic segment of pycnogonids.

cheliform [INV ZOO] Having a forcepslike organ formed by a

movable joint closing against an adjacent segment; referring especially to a crab's claw.

cheliped [INV ZOO] Either of the paired appendages bearing chelae in decapod crustaceans.

chellin *See* khellin.

chelometry [ANALY CHEM] Analytical technique involving the formation of 1:1 soluble chelates when a metal ion is titrated with aminopolycarboxylate and polyamine reagents. A form of complexiometric titration.

Chelonariidae [INV ZOO] A family of coleopteran insects in the superfamily Dryopoidea.

Chelonethida [INV ZOO] An equivalent name for the Pseudoscorpionida.

Chelonia [VERT ZOO] An order of the Reptilia, subclass Anapsida, including the turtles, terrapins, and tortoises.

Cheloniidae [VERT ZOO] A family of reptiles in the order Chelonia including the hawksbill, loggerhead, and green sea turtles.

Cheluridae [INV ZOO] A family of amphipod crustaceans in the suborder Gammaridea.

Chelydridae [VERT ZOO] The snapping turtles, a small family of reptiles in the order Chelonia.

Chemag process [MET] A process for oxide-coating iron and steel by electrolytic means, the object to be coated being the anode in an alkaline solution.

chemical [CHEM] **1.** Related to the science of chemistry. **2.** A substance characterized by definite molecular composition.

chemical affinity *See* affinity.

chemical agent [MATER] A solid, liquid, or gas employed in three principal categories: war gases, smokes, and incendiaries; through its chemical properties produces lethal, injurious, or irritant effects on humans, makes a screening or colored smoke, or acts as a fire starter.

chemical ammunition [ORD] Any ammunition, such as bombs, projectiles, bullets, or flares, containing a chemical agent, such as war gases, smokes, and incendiaries.

chemical bomb [ORD] A bomb with a chemical agent for the main charge.

chemical bond *See* bond.

chemical burn [MED] Tissue destruction caused by caustic agents, irritant gases, or other chemical agents.

chemical-cartridge respirator [MIN ENG] An air purification device worn by miners that removes small quantities of toxic gases or vapors from the inspired air; the cartridge contains chemicals which operate by processes of oxidation, absorption, or chemical reaction.

chemical compound *See* compound.

chemical conversion coating [MET] A protective or decorative coating formed on the surface of a metal as the result of chemical reaction of the metal with a selected environment.

chemical crystallography [CRYSTAL] The geometric description, and study, of the internal arrangement of atoms in crystals formed from chemical compounds.

chemical dating [ANALY CHEM] The determination of the relative or absolute age of minerals and of ancient objects and materials by measurement of their chemical compositions.

chemical denudation [GEOL] Wasting of the land surface by water transport of soluble materials into the sea.

chemical deposition [CHEM] Precipitation of a metal from a solution of a salt by introducing another metal.

chemical dosimeter [NUCLEO] A dosimeter in which the accumulated radiation-exposure dose is indicated by color changes accompanying chemical reactions induced by the radiation.

chemical element *See* element.

chemical energy [PHYS CHEM] Energy of a chemical compound which, by the law of conservation of energy, must undergo a change equal and opposite to the change of heat energy in a reaction; the rearrangement of the atoms in reacting compounds to produce new compounds causes a change in chemical energy.

chemical engineering [ENG] That branch of engineering serving those industries that chemically convert basic raw materials into a variety of products, and dealing with the design and operation of plants and equipment to perform such work; all products are formed in chemical processes involving chemical reactions carried out under a wide range

CHEETAH

The cheetah (*Acinonyx jubatus*), which has been trained for hunting, especially in India.

CHEILOSTOMATA

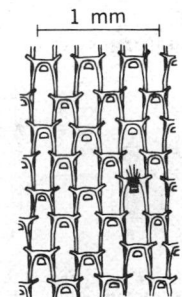

Anascan cheilostome, showing the surface of encrusting sheetlike colony of *Membranipora*, Miocene-Recent. (*T. Hincks, from L. H. Hyman, 1959*)

CHELIDONIC ACID

Structural formula of chelidonic acid.

of conditions and frequently accompanied by changes in physical state or form.

chemical equilibrium [CHEM] A condition in which a chemical reaction is occurring at equal rates in its forward and reverse directions, so that the concentrations of the reacting substances do not change with time.

chemical etching [MET] Formation of characteristic surface features when a polished metal surface is etched by suitable reagents.

chemical exchange process [CHEM] A method of separating isotopes of the lighter elements by the repetition of a process of chemical change which involves exchange of the isotopes.

chemical film dielectric [ELEC] A chemical film on one or both electrodes of an electrolytic capacitor, which constitutes the dielectric of the capacitor.

chemical fire extinguisher [CHEM ENG] Any of three types of fire extinguishers (vaporizing liquid, carbon dioxide, and dry chemical) which expel chemicals in solid, liquid, or gaseous form to blanket or smother a fire.

chemical focus See actinic focus.

chemical formula [CHEM] A notation utilizing chemical symbols and numbers to indicate the chemical composition of a pure substance; examples are CH_4 for methane and HCl for hydrogen chloride.

chemical grenade [ORD] General term for any hand grenade or rifle grenade charged with a chemical agent.

chemical hygrometer See absorption hygrometer.

chemical indicator [ANALY CHEM] A substance whose physical appearance is altered at or near the end point of a chemical titration.

chemical inhibitor [CHEM] A substance capable of stopping or retarding a chemical reaction.

chemical-ion pump [CHEM ENG] A vacuum pump whose pumping action is based on evaporation of a metal whose vapor then reacts with the chemically active molecules in the gas to be evacuated.

chemical kinetics [PHYS CHEM] That branch of physical chemistry concerned with the mechanisms and rates of chemical reactions. Also known as reaction kinetics.

chemical laser See chemically pumped laser.

chemically foamed plastic [MATER] Those cellular plastics with a structure formed by chemical foaming agents.

chemically pumped laser [OPTICS] A laser in which pumping is achieved by using a chemical action rather than electrical energy to produce the required pulses of light. Also known as chemical laser.

chemically pure [CHEM] Without impurities detectable by analysis. Abbreviated cp.

chemical machining [MET] Making of metal parts to specified dimensions by removing surface metal with chemicals (acids or alkalies). Also known as chemical milling.

chemical metallurgy [MET] The science and technology of extracting metals from ores and refining them. Also known as process metallurgy.

chemical microscopy [ANALY CHEM] Application of the microscope to the solution of chemical problems.

chemical milling See chemical machining.

chemical operations See chemical warfare.

chemical pathology [PATH] The study of disease by using chemical methods.

chemical polarity [PHYS CHEM] Tendency of a molecule, or compound, to be attracted or repelled by electrical charges because of an asymmetrical arrangement of atoms around the nucleus.

chemical polishing [MET] Smoothing and brightening the surface of a metal by treatment with a chemical agent.

chemical porcelain [MATER] High-purity, nonporous grade of porcelain used to make laboratory analysis utensils, such as crucibles, retorts, and spatulas.

chemical potential [PHYS CHEM] In a thermodynamic system of several constituents, the rate of change of the Gibbs function of the system with respect to the change in the number of moles of a particular constituent.

chemical precipitates [GEOL] A sediment formed from precipitated materials as distinguished from detrital particles that have been transported and deposited.

chemical pressurization [AERO ENG] The pressurization of propellant tanks in a rocket by means of high-pressure gases

developed by the combustion of a fuel and oxidizer or by the decomposition of a substance.

chemical pulp [MATER] Wood pulp made by separating the fibers of wood chips by the action of alkalies or acids.

chemical pulping [CHEM ENG] Separation of wood fiber for paper pulp by chemical treatment of wood chips to dissolve the lignin that cements the fibers together.

chemical pump [PETRO ENG] Skid-mounted pumping unit used to feed chemicals into the power oil (used to operate bottom-hole pumps in oil wells) to reduce corrosion in the system and to assist in water removal when the power oil and well-produced oil reach the ground-level wash tank.

chemical reaction [CHEM] A change in which a substance (or substances) is changed into one or more new substances; there is only a minute change, Δm, in the mass of the system, given by $\Delta E = \Delta mc^2$, where ΔE is the energy emitted or absorbed and c is the speed of light.

chemical reactivity [CHEM] The tendency of two or more chemicals to react to form one or more products differing from the reactants.

chemical reactor [CHEM ENG] Vessel, tube, pipe, or other container within which a chemical reaction is made to take place; may be batch or continuous, open or packed, and can use thermal, catalytic, or irradiation actuation.

chemical remanent magnetization [GEOPHYS] Permanent magnetization of rocks acquired when a magnetic material, such as hematite, is grown at low temperature through the oxidation of some other iron mineral, such as magnetite or goethite; the growing mineral becomes magnetized in the direction of any field which is present. Abbreviated CRM.

chemical reservoir [GEOL] An underground oil or gas trap formed in limestones or dolomites deposited in quiescent geologic environments.

chemical resistance [MATER] Ability of solid materials to resist damage by chemical reactivity or solvent action.

chemical sense [PHYSIO] A process of the nervous system for reception of and response to chemical stimulation by excitation of specialized receptors.

chemical shift [PHYS CHEM] Shift in a nuclear magnetic-resonance spectrum resulting from diamagnetic shielding of the nuclei by the surrounding electrons.

chemical shim [NUCLEO] A chemical, usually boric acid, that is placed in the coolant system of a nuclear reactor to serve as a neutron absorber and that compensates for fuel burnup during normal operation.

chemical shutdown [NUCLEO] Addition of a dissolved poison to the coolant of a nuclear reactor to achieve shutdown.

chemical similitude [CHEM ENG] A procedure used to ensure satisfactory operation of a full-scale chemical process by comparison with pilot plant data.

chemical spray [ORD] Aerial release, or device for aerial release, of liquid war gas for casualty effect, or of liquid smoke for aerial smoke screens.

chemical sterilization [ENG] The use of bactericidal chemicals to sterilize solutions, air, or solid surfaces.

chemical stoneware [MATER] Clay pottery material that resists acids and alkalies; used for ball mills, pipes, laboratory sinks and utensils, and so on.

chemical symbol [CHEM] A notation for one of the chemical elements, consisting of letters; for example Ne, O, C, and Na represent neon, oxygen, carbon, and sodium.

chemical synthesis [CHEM] The formation of one chemical compound from another.

chemical tanker [NAV ARCH] Ship designed with tanks of stainless steel, or of other materials, capable of containing chemicals.

chemical thermodynamics [PHYS CHEM] The application of thermodynamic principles to problems of chemical interest.

chemical thermometer [ENG] A filled-system temperature-measurement device in which gas or liquid enclosed within the device responds to heat by a volume change (rising or falling of mercury column) or by a pressure change (opening or closing of spiral coil).

chemical tracer [NUCLEO] A tracer having chemical properties similar to those of the substance with which it is mixed.

chemical warfare [ORD] Originally, the employment of poison gases as antipersonnel agents; later expanded to include flame and incendiary warfare, smoke for screening or signal-

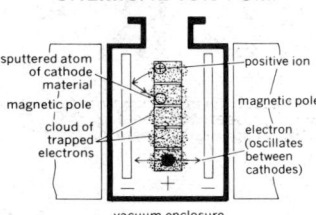

CHEMICAL-ION PUMP

sputtered atom of cathode material

magnetic pole

cloud of trapped electrons

positive ion

magnetic pole

electron (oscillates between cathodes)

vacuum enclosure

The principal features of an ion pump.

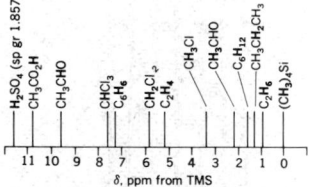

CHEMICAL SHIFT

H_2SO_4 (sp gr 1.857)

CH_3CO_2H

CH_3CHO

$CHCl_3$

C_6H_6

CH_2Cl_2

C_2H_4

CH_3Cl

CH_3CHO

C_6H_{12}

$CH_3CH_2CH_3$

C_2H_6

$(CH_3)_4Si$

11 10 9 8 7 6 5 4 3 2 1 0

δ, ppm from TMS

Chemical shifts for representative compounds. Decreasing values of δ correspond to increasing magnetic field in a constant-frequency spectrometer. The scale calibration is obtained from the resonance signal of a small amount of tetramethylsilane (TMS) placed in the sample tube to provide a zero reference point.

ing purposes, and microorganisms (bacteria and their toxins, rickettsia, viruses) for the production of casualties or destruction of crops. Also known as chemical operations.

chemical weathering [GEOCHEM] A weathering process whereby rocks and minerals are transformed into new, fairly stable chemical combinations by such chemical reactions as hydrolysis, oxidation, ion exchange, and solution. Also known as decay; decomposition.

chemiluminescence [PHYS CHEM] Emission of light as a result of a chemical reaction without an apparent change in temperature.

chemimechanical pulp [MATER] Plant material treated by the sulfite, soda, or sulfate process for papermaking.

chemisorption [CHEM] The process of chemical adsorption.

chemist [CHEM] A scientist specializing in chemistry.

chemistry [SCI TECH] The scientific study of the properties, composition, and structure of matter, the changes in structure and composition of matter, and accompanying energy changes.

chemoautotroph [MICROBIO] Any of a number of autotrophic bacteria and protozoans which do not carry out photosynthesis.

chemocline [HYD] The transition in a meromictic lake between the mixolimnion layer (at the top) and the monimolimnion layer (at the bottom).

chemodectoma [MED] A benign tumor of the carotid body.

chemodifferentiation [EMBRYO] The process of cellular differentiation at the molecular level by which embryonic cells become specialized as tissues and organs.

chemoheterotroph [BIOL] An organism that derives energy and carbon from the oxidation of preformed organic compounds.

chemonite [MATER] A wood preservative consisting of a water solution of small percentages of copper hydroxide, arsenic trioxide, ammonia, and acetic acid.

chemoprophylaxis [MED] Use of drugs to prevent the development of infectious diseases.

chemoreception [PHYSIO] Reception of a chemical stimulus by an organism.

chemoreceptor [PHYSIO] Any sense organ that responds to chemical stimuli.

chemosphere [METEOROL] The vaguely defined region of the upper atmosphere in which photochemical reactions take place; generally considered to include the stratosphere (or the top thereof) and the mesosphere, and sometimes the lower part of the thermosphere.

chemostat [MICROBIO] An apparatus, and a principle, for the continuous culture of bacterial populations in a steady state.

chemosynthesis [BIOCHEM] The synthesis of organic compounds from carbon dioxide by microorganisms using energy derived from chemical reactions.

chemotaxis [BIOL] The orientation movement of a motile organism with reference to a chemical agent.

chemotherapeutic [PHARM] Any agent used for chemotherapy.

chemotherapeutic index [PHARM] The relationship between toxicity of a compound for the body and the toxicity for parasites.

chemotherapy [MED] Administering chemical substances for treatment of disease, especially cancer and diseases caused by parasites.

chemotropism [BIOL] Orientation response of a sessile organism with reference to chemical stimuli.

Chemungian [GEOL] Middle Upper Devonian geologic time, below Cassodagan.

chemurgy [CHEM ENG] A branch of chemistry concerned with the profitable utilization of organic raw materials, especially agricultural products, for nonfood purposes such as for paints and varnishes.

chenevixite [MINERAL] $Cu_2Fe_2(AsO_4)_2(OH)_4 \cdot H_2O$ A dark-green to greenish-yellow mineral consisting of a hydrous copper iron arsenate occurring in masses.

chenier [GEOL] A continuous ridge of beach material built upon swampy deposits; often supports trees, such as pines or evergreen oaks.

Chenopodiaceae [BOT] A family of dicotyledonous plants in the order Caryophyllales having reduced, mostly greenish flowers.

chenopodium oil [MATER] An alcohol-soluble, colorless to yellow oil, derived from the herb *Chenopodium ambrosioides;* chief constituents are *para*-cymeme, *l*-limonene, and ascaridole; used in medicine. Also known as American wormseed oil; goosefoot oil.

chergui [METEOROL] An eastern or southeastern desert wind in Morocco (North Africa), especially in the north; it is persistent, very dry and dusty, hot in summer, cold in winter.

Chermidae [INV ZOO] A small family of minute homopteran insects in the superfamily Aphidoidea.

Cherminae [INV ZOO] A subfamily of homopteran insects in the family Chermidae; all forms have a beak and an open digestive tract.

Chernozem [GEOL] One of the major groups of zonal soils, developed typically in temperate to cool, subhumid climate; the Chernozem soils in modern classification include Borolls, Ustolls, Udolls, and Xerolls. Also spelled Tchernozem.

cherophobia [PSYCH] An abnormal fear of happiness.

cherry [BOT] **1.** Any trees or shrub of the genus *Prunus* in the order Rosales. **2.** The simple, fleshy, edible drupe or stone fruit of the plant.

Cherry-Burrell process [FOOD ENG] A butter manufacturing process in which high fat cream is pasteurized, cooled, and collected in vats, where color and salt are added and the fat content reduced to 80%; it is pumped continuously through a chiller-agitator and a texturator, from which it emerges as butter in a continuous ribbon.

cherry leaf spot [PL PATH] A fungus disease of the cherry caused by *Coccomyces hiemalis;* spotting and chlorosis of the leaves occurs, with consequent retardation of tree and fruit development.

cherry picker [AERO ENG] A crane used to remove the aerospace capsule containing astronauts from the top of the rocket in the event of a malfunction. [MECH ENG] Any of several small traveling cranes, especially one used to hoist a passenger on the end of a boom. [MIN ENG] A small hoist used to facilitate car changing near the loader in a mine tunnel.

chersophyte [ECOL] A plant that grows in dry waste lands.

chert [PETR] A hard, dense, micro- or cryptocrystalline rock composed of chalcedony and microcrystalline quartz. Also known as hornstone; phthanite; silexite; white chert.

chertification [GEOL] A process of replacement by silica in limestone in the form of fine-grained quartz or chalcedony.

chest breadth [ANTHRO] The measurement across the chest at nipple level.

chest circumference [ANTHRO] The horizontal circumference taken just above the nipples during a period of quiet breathing.

chest depth [ANTHRO] A measurement of the chest taken front to back from the sternum to the spinal groove.

Chesterian [GEOL] Upper Mississippian geologic time.

chestnut [BOT] The common name for several species of large, deciduous trees of the genus *Castanea* in the order Fagales, which bear sweet, edible nuts.

chestnut blight [PL PATH] A fungus disease of the chestnut caused by *Endothia parasitica*, which attacks the bark and cambium, causing cankers that girdle the stem and kill the plant. Also known as chestnut canker.

chestnut canker *See* chestnut blight.

chestnut coal [GEOL] Anthracite coal small enough to pass through a round mesh of 1⅝ inches (4.13 centimeters) but too large to pass through a round mesh of 1³⁄₁₆ inches (3.02 centimeters).

Chestnut soil [GEOL] One of the major groups of zonal soils, developed typically in temperate to cool, subhumid to semi-arid climate; the Chestnut soils in modern classification include Ustolls, Borolls, and Xerolls.

Chevalier lens [OPTICS] A type of magnifying lens composed of an achromatic negative lens combined with a distant collecting front lens; a magnifying power up to 10X with an object distance up to 3 inches (7.62 centimeters) can be obtained.

chevkinite [MINERAL] $(Fe,Ca)(Ce,La)_2(Si,Ti)_2O_8$ A mineral consisting of silicotitanate of iron, calcium, and rare-earth elements.

CHERT

Bedded chert, Monterey formation (Miocene), California. *(Photograph by M. N. Bramlette, USGS)*

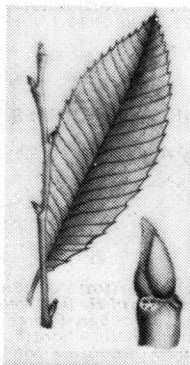

CHESTNUT

Twig, leaf, and bud of American chestnut *(Castanea dentata).*

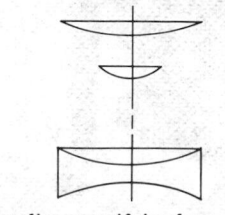

CHEVALIER LENS

Chevalier magnifying lens.

CHEVROTAIN

Indian chevrotain, distinguished by white spots on coat.

CHICLE

Fruit of *Achras zapota*. (From L. H. Bailey, *The Standard Cyclopedia of Horticulture, vol. 3, Macmillan, 1937*)

CHIMAERIFORMES

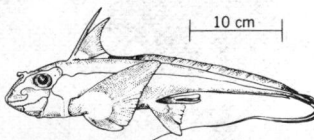

Modern chimaeriform of genus *Chimaera*. (From H. B. Bigelow and W. C. Schroeder, *Fishes of the Western North Atlantic, pt. 2, Sears Foundation for Marine Research, 1953*)

CHIMPANZEE

The chimpanzee, with a highly developed brain, and an anatomical resemblance to man.

chevron [ANAT] The bone forming the hemal arch of a caudal vertebra.

chevron fold [GEOL] An accordionlike fold with limbs of equal length.

chevrotain [VERT ZOO] The common name for four species of mammals constituting the family Tragulidae in the order Artiodactyla. Also known as mouse deer.

Cheyne-Stokes respiration [MED] Breathing characterized by periods of hyperpnea alternating with periods of apnea; rhythmic waxing and waning of respiration; occurs most commonly in older patients with heart failure and cerebrovascular disease.

Chézy formula [FL MECH] For the velocity V of open-channel flow which is steady and uniform, $V = \sqrt{8g/f} \cdot \sqrt{mS}$, where f is the Darcy-Weisbach friction coefficient, m the hydraulic radius, S the energy dissipation per unit length, and g the acceleration of gravity.

chiasma [ANAT] A cross-shaped point of intersection of two parts, especially of the optic nerves. [CYTOL] The point of junction and fusion between paired chromatids or chromosomes, first seen during diplotene of meiosis.

Chiasmodontidae [VERT ZOO] A family of deep-sea fishes in the order Perciformes.

chiastolite [MINERAL] A variety of andalusite whose crystals have a cross-shaped appearance in cross section due to the arrangement of carbonaceous impurities. Also known as macle.

chibli *See* ghibli.

Chicago boom [MECH ENG] A hoisting device that is supported on the structure being erected.

Chicago caisson [CIV ENG] A cofferdam about 4 feet (1.2 meters) in diameter lined with planks and sunk in medium-stiff clays to hard ground for pier foundations. Also known as open-well caisson.

chicken [VERT ZOO] *Galus galus.* The common domestic fowl belonging to the order Galliformes.

chickenpox [MED] A mild, highly infectious viral disease of man caused by a herpesvirus and characterized by vesicular rash. Also known as varicella.

chick unit [BIOL] A unit for the standardization of pantothenic acid.

chicle [MATER] A gummy exudate obtained from the bark of *Achras zapota*, an evergreen tree belonging to the sapodilla family (Sapotaceae); used as the principal ingredient of chewing gum.

chicory [BOT] *Cichorium intybus.* A perennial herb of the order Campanulales grown for its edible green leaves.

Chideruan [GEOL] Uppermost Permian geologic time.

chief cell [HISTOL] **1.** A parenchymal, secretory cell of the parathyroid gland. **2.** A cell in the lumen of the gastric fundic glands.

chigger [INV ZOO] The common name for bloodsucking larval mites of the Trombiculidae which parasitize vertebrates.

chilarium [INV ZOO] One of a pair of processes between the bases of the fourth pair of walking legs in the king crab.

childhood aphasia [PSYCH] Loss or impairment of the use of language in a child.

Child-Langmuir equation *See* Child's law.

Child-Langmuir-Schottky equation *See* Child's law.

childrenite [MINERAL] (Fe,Mn)AlPO$_4$(OH)$_2 \cdot$H$_2$O A pale-yellowish to dark-brown orthorhombic mineral consisting of a hydrous basic iron aluminum phosphate occurring as translucent crystals; it is isomorphous with eosphorite; hardness is 4.5–5 on Mohs scale, and specific gravity is 3.18–3.24.

Child's law [ELECTR] A law stating that the current in a thermionic diode varies directly with the three-halves power of anode voltage and inversely with the square of the distance between the electrodes, provided the operating conditions are such that the current is limited only by the space charge. Also known as Child-Langmuir equation; Child-Langmuir-Schottky equation; Langmuir-Child equation.

Chile mill [MECH ENG] A crushing mill having vertical rollers running in a circular enclosure with a stone or iron base or die. Also known as edge runner.

Chile niter *See* Chile saltpeter.

Chile saltpeter [MINERAL] Also known as Chile niter. **1.** Soda niter found in large quantities in caliche in arid regions of northern Chile. **2.** Deposits of sodium nitrate.

chili [METEOROL] A warm, dry, descending wind in Tunisia, resembling the sirocco; in southern Algeria it is called chichili.

chill [MET] **1.** A metal plate inserted in the surface of a sand mold or placed in the mold cavity to rapidly cool and solidify the casting, producing a hard surface. **2.** White or mottled iron occurring on the surface of a rapidly cooled gray iron casting.

chilled contact [PETR] The finer-grained portion of an igneous rock found near its contact with older rock.

chilled iron [MET] Cast iron made in iron- or steel-faced molds so the surface of the casting cools rapidly, retaining most of the carbon and becoming white and hard.

chilled roll [MET] A roll consisting of an outer hard layer of white (chilled) iron with a middle transitional layer of mottled iron and a core of full gray iron.

chilled shot [MET] Lead shots containing 3–6% antimony.

chiller [CHEM ENG] Oil-refining apparatus in which the temperature of paraffin distillates is lowered preparatory to filtering out the solid wax components.

chilling [MET] Rapidly removing the heat from a casting.

chill roll [ENG] A cored roll used in chill-roll extrusion of plastics.

chill-roll extrusion [ENG] Method of extruding plastic film in which the film is cooled while being drawn around two or more highly polished chill rolls, inside of which there is cooling water. Also known as cast-film extrusion.

chill wind factor [METEOROL] An arbitrary index, developed by the Canadian Army, to correlate the performance of equipment and personnel in an Arctic winter; it is equal to the sum of the wind speed in miles per hour and the negative of the Fahrenheit temperature; the term is not to be confused with wind chill.

Chilobolbinidae [PALEON] A family of extinct ostracods in the superfamily Hollinacea showing dimorphism of the velar structure.

Chilopoda [INV ZOO] The centipedes, a class of the Myriapoda that is exclusively carnivorous and predatory.

Chimaeridae [VERT ZOO] A family of the order Chimaeriformes.

Chimaeriformes [VERT ZOO] The single order of the chondrichthyan subclass Holocephali comprising the ratfishes, marine bottom-feeders of the Atlantic and Pacific oceans.

chimera [BIOL] An organism or a part made up of tissues or cells exhibiting chimerism.

chimerism [BIOL] The admixture of cell populations from more than one zygote.

chimney [BUILD] A vertical, hollow structure of masonry, steel, or concrete, built to convey gaseous products of combustion from a building. [ELECTR] A pipelike enclosure placed over a heat sink to improve natural upward convection of heat and thereby increase the dissipating ability of the sink. [GEOL] *See* pipe.

chimney cloud [METEOROL] A cumulus cloud in the tropics that has much greater vertical than horizontal extent.

chimney rock [GEOL] **1.** A chimney-shaped remnant of a rock cliff whose sides have been cut into and carried away by waves and the gravel beach. **2.** A rock column rising above its surroundings. [MATER] A porous phosphate rock used principally in chimney construction.

chimopelagic [ECOL] Pertaining to, belonging to, or being marine organisms living at great depths throughout most of the year; during the winter they move to the surface.

chimpanzee [VERT ZOO] Either of two species of Primates of the genus *Pan* indigenous to central-west Africa.

chin [ANAT] The lower part of the face, at or near the symphysis of the lower jaw.

china clay [MATER] A high-grade white kaolin composed principally of the mineral kaolinite, and often occurring as a lenticular-shaped body; used in the manufacture of ceramics, paper, rubber, catalysts, and ink.

China grass *See* ramie.

China oil *See* Peru balsam.

China wood oil *See* tung oil.

chin breadth [ANTHRO] Contact measurement of the maximum width of the chin, taken between the points of intersection of the mandible and the menton.

chinchilla [VERT ZOO] The common name for two species of rodents in the genus *Chinchilla* belonging to the family Chinchillidae.

chinchilla cloth [TEXT] A wool or other fabric that has a long nap or is tufted or nubbed.

Chinchillidae [VERT ZOO] A family of rodents comprising the chinchillas and viscachas.

chine [FOOD ENG] A part of an animal carcass backbone, with parts adjoining it, cut for cooking. [NAV ARCH] 1. A part of a ship's waterway rising above the deck and hollowed out on the inboard edge to make a path for water. 2. That point where the sides and bottom meet on a flat- or V-bottomed boat. 3. A longitudinal member that lies along the bilge where the sides and bottom of the boat meet.

Chinese binary *See* column binary.

Chinese ink *See* india ink.

Chinese oil *See* Peru balsam.

Chinese oil of cinnamon *See* cassia oil.

Chinese vermilion *See* mercuric sulfide.

Chinese wax [MATER] A white or yellowish crystalline wax formed on certain trees by the secretions of a scale insect, especially *Ceroplastes ceriferus*.

Chinese white [CHEM] A term used in the paint industry for zinc oxide and kaolin used as a white pigment. Also known as zinc white.

chinic acid *See* quinic acid.

chinidine *See* quinidine.

chin-neck projection [ANTHRO] Measurement from the tip of the thyroid cartilage to the midpoint of the menton.

chinone *See* quinone.

chinook [METEOROL] The foehn on the eastern side of the Rocky Mountains.

chinook arch [METEOROL] A foehn cloud formation appearing as a bank of clouds over the Rocky Mountains, generally a flat layer of altostratus, heralding the approach of a chinook.

chin projection [ANTHRO] Measurement of the distance from the gonion to the most forward point on the vertical midline of the menton.

chiolite [MINERAL] $Na_5Al_3F_{14}$ A snow white mineral resembling cryolite. Also known as arksutite.

Chionididae [VERT ZOO] The white sheathbills, a family of birds in the order Charadriiformes.

chionophile [ECOL] Having a preference for snow.

chionophobia [PSYCH] An abnormal fear of snow.

chip [ELECTR] 1. The shaped and processed semiconductor die that is mounted on a substrate to form a transistor, diode, or other semiconductor device. 2. An integrated microcircuit performing a significant number of functions and constituting a subsystem.

chip blower [MED] A dental instrument used to blow drilling debris from a tooth cavity that is being prepared for filling.

chipboard [MATER] A low-density paper board made from mixed waste paper and used where strength and quality are needed.

chip breaker [DES ENG] An irregularity or channel cut into the face of a lathe tool behind the cutting edge to cause removed stock to break into small chips or curls.

chip cap [DES ENG] A plate or cap on the upper part of the cutting iron of a carpenter's plane designed to give the tool rigidity and also to break up the wood shavings.

chip circuit *See* large-scale integrated circuit.

chipmunk [VERT ZOO] The common name for 18 species of rodents belonging to the tribe Marmotini in the family Sciuridae.

chipper [ENG] A tool such as a chipping hammer used for chipping.

chipping [MET] Removing seams, surface defects, or other excess fragments from semifinished metal products by using a manual or pneumatic chisel or a continuous machine.

chipping hammer [ENG] A hand or pneumatic hammer with chisel-shaped or pointed faces used to remove rust and scale from metal surfaces.

chip sampling [MIN ENG] Taking small pieces of ore or coal from the width of an ore face exposure; may be done at random or along a line.

chirality [CHEM] The handedness of an asymmetric molecule. [PARTIC PHYS] The characteristic of particles of spin $\frac{1}{2}\hbar$ that are allowed to have only one spin state with respect to an axis of quantization parallel to the particle's momentum; if the particle's spin is always parallel to its momentum, it has positive chirality; antiparallel, negative chirality. [PHYS] The characteristic of an object that cannot be superimposed upon its mirror image.

chiral twinning *See* optical twinning.

Chiridotidae [INV ZOO] A family of holothurians in the order Apodida having six-spoked, wheel-shaped spicules.

Chirodidae [PALEON] A family of extinct chondrostean fishes in the suborder Platysomoidei.

Chirognathidae [PALEON] A family of conodonts in the suborder Neurodontiformes.

chiromegaly *See* cheiromegaly.

chiroplasty *See* cheiroplasty.

chiropodist [MED] One who treats minor ailments of the feet. Also known as podiatrist.

chiropractic [MED] A system of therapeutics based upon the theory that disease is caused by abnormal function of the nervous system; attempts to restore normal function are made through manipulation and treatment of the structures of the body, especially those of the spinal column.

chiropractor [MED] One who practices the chiropractic arts.

Chiroptera [VERT ZOO] The bats, an order of mammals having the front limbs modified as wings.

chiropterophilous [BIOL] Pollinated by bats.

chiropterygium [VERT ZOO] A typical vertebrate limb, thought to have evolved from a finlike appendage.

Chirotheuthidae [INV ZOO] A family of mollusks comprising several deep-sea species of squids.

chirp radar [ENG] Radar in which a swept-frequency signal is transmitted, received from a target, then compressed in time to give a narrow pulse called the chirp signal.

chisel [AGR] A strong, heavy tool with curved points used for tilling; drawn by a tractor, it stirs the soil at an appreciable depth without turning it. [DES ENG] A tool for working the surface of various materials, consisting of a metal bar with a sharp edge at one end and often driven by a mallet.

Chisel *See* Caelum.

chisel bit *See* chopping bit.

chisel-tooth saw [DES ENG] A circular saw with chisel-shaped cutting edges.

chi-square distribution [STAT] The distribution of the sample variances indicated by

$$S^2 = \sum_{i=1}^{n} (x_i - \bar{x})^2/(n-1),$$

where $x_1, x_2, \ldots x_n$ are observations of a random sample n drawn from a normal population.

chi-square test [STAT] A generalization, and an extension, of a test for significant differences between a binomial population and a multinomial population, wherein each observation may fall into one of several classes and furnishes a comparison among several samples instead of just two.

chitin [BIOCHEM] A white or colorless amorphous polysaccharide that forms a base for the hard outer integuments of crustaceans, insects, and other invertebrates.

chitinase [BIOCHEM] An externally secreted digestive enzyme produced by certain microorganisms and invertebrates that hydrolyzes chitin.

chitinivorous bacterium [MICROBIO] Any bacterium which secretes chitinase and can digest chitin; organisms extract chitin from lobster exoskeletons, causing an infection called soft-shell disease.

Chitinozoa [PALEON] An extinct group of unicellular microfossils of the kingdom Protista.

chiton [INV ZOO] The common name for over 600 extant species of mollusks which are members of the class Polyplacophora.

Chitral fever *See* phlebotomus fever.

chiviatite [MINERAL] $Pb_2Bi_6S_{11}$ A lead-gray mineral consisting of a lead bismuth sulfide occurring in foliated masses.

Chladni's figures [MECH] Figures produced by sprinkling sand or similar material on a horizontal plate and then

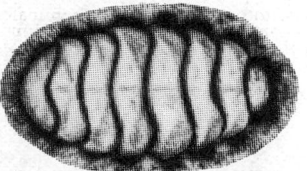

vibrating the plate while holding it rigid at its center or along its periphery; indicate the nodal lines of vibration.

chlamydeous [BOT] **1.** Pertaining to the floral envelope. **2.** Having a perianth.

Chlamydia *See* Bedsonia.

Chlamydiaceae [MICROBIO] A family of parasitic bacteria in the order Rickettsiales.

Chlamydobacteriaceae [MICROBIO] A family of gram-negative bacteria in the order Chlamydobacteriales possessing trichomes in which false branching may occur.

Chlamydobacteriales [MICROBIO] An order comprising colorless, gram-negative, algae-like bacteria of the class Schizomycetes which occur in trichomes.

Chlamydomonadidae [INV ZOO] A family of colorless, flagellated protozoans in the order Volvocida considered to be close relatives of protozoans that possess chloroplasts.

Chlamydoselachidae [VERT ZOO] The frilled sharks, a family of rare deep-water forms having a combination of hybodont and modern shark characteristics.

chlamydospore [MYCOL] A thick-walled, unicellar resting spore developed from vegetative hyphae in almost all parasitic fungi.

Chlamydozoaceae [MICROBIO] A family of small, gram-negative, coccoid bacteria in the order Rickettsiales; members are obligate intracytoplasmic parasites, or saprophytes.

chloanthite [MINERAL] NiAs$_{2-3}$ A white or gray mineral with metallic luster forming crystals in the isometric system; it is isomorphous with nickel-skutterudite.

chloasma [MED] Patchy tan, brown, and black hyperpigmentation, especially on the brow and cheek; of unknown cause, but may be due to the action of sunshine upon perfume or to endocrinopathy.

Chloracea [MICROBIO] The green sulfur bacteria, a family of photosynthetic bacteria in the suborder Rhodobacteriineae.

chloracetone *See* chloroacetone.

chloracetophenone *See* chloroacetophenone.

chloracne [MED] An acnelike eruption caused by chlorinated hydrocarbons.

chloragogen [INV ZOO] Of or pertaining to certain specialized cells forming the outer layer of the alimentary tract in earthworms and other annelids.

chloral [ORG CHEM] CCl$_3$CHO A colorless, oily liquid soluble in water; used industrially to prepare DDT; a hypnotic. Also known as trichloroacetic aldehyde; trichloroethanal.

chloralase [ORG CHEM] C$_8$H$_{11}$Cl$_3$O$_6$ Colorless, water-soluble crystals, melting at 185°C; made by heating chloral with dextrose; used as a hypnotic.

chloral hydrate [ORG CHEM] CCl$_3$CH(OH)$_2$ Colorless, deliquescent needles with slightly bitter caustic taste, soluble in water; a hypnotic. Also known as crystalline chloral; hydrated chloral.

chloralkane [ORG CHEM] Chlorinated aliphatic hydrocarbon of the methane series (C$_n$H$_{2n+2}$).

chloraluminite [MINERAL] AlCl$_3$·6H$_2$O A mineral consisting of hydrous aluminum chloride.

chloramine T [ORG CHEM] CH$_3$C$_6$H$_4$SO$_2$NClNa·3H$_2$O A white, crystalline powder that decomposes slowly in air, freeing chlorine; used as an antiseptic, a germicide, and an oxidizing agent and chlorinating agent.

chloramphenicol [MICROBIO] C$_{11}$H$_{12}$O$_2$N$_2$Cl$_2$ A colorless, crystalline, broad-spectrum antibiotic produced by *Streptomyces venezuelae;* industrial production is by chemical synthesis. Also known as chloromycetin.

Chlorangiaceae [BOT] A primitive family of colonial green algae belonging to the Tetrasporales in which the cells are directly attached to each other.

chloranil [ORG CHEM] C$_6$Cl$_4$O$_2$ Yellow leaflets melting at 290°C; soluble in organic solvents; made from phenol by treatment with potassium chloride and hydrochloric acid; used as an agricultural fungicide and as an oxidizing agent in the manufacture of dyes. Also known as tetrachloroquinone.

chloranthy [BOT] A reverting of normally colored floral leaves or bracts to green foliage leaves.

chlorapatite [MINERAL] Ca$_5$(PO$_4$)$_3$Cl An apatite mineral containing chlorine.

chlorastrolite [MINERAL] A mottled, green variety of pum-

pellyite occurring as grains or small nodules of a stellate structure in basic igneous rock in the Lake Superior region; used as a semiprecious stone.

chlorate [INORG CHEM] ClO$_3^-$ **1.** A negative ion derived from chloric acid. **2.** A salt of chloric acid.

chlorate candle [MATER] A mixture of solid chemical compounds which, when ignited, liberates free oxygen.

chlorate explosive [MATER] A type of explosive with a potassium chlorate base; a substitute for blackpowder in which potassium chlorate is used in place of potassium nitrate. Also known as chlorate powder.

chlorate powder *See* chlorate explosive.

chlordan *See* chlordane.

chlordane [ORG CHEM] C$_{10}$H$_6$Cl$_8$ A volatile liquid insecticide; a chlorinated hexahydromethanoindene. Also spelled chlordan.

chlordiazepoxide hydrochloride [PHARM] C$_{16}$H$_{14}$ON$_3$Cl A white crystalline conpound, soluble in water; the hydrochloride salt is used as a tranquilizer.

chlorenchyma [BOT] Chlorophyll-containing tissue in parts of higher plants, as in leaves.

chlorendic acid [ORG CHEM] C$_9$H$_4$Cl$_6$O$_4$ White, fine crystals used in fire-resistant polyester resins and as an intermediate for dyes, fungicides, and insecticides.

chlorendic anhydride [ORG CHEM] C$_9$H$_2$Cl$_6$O$_3$ White, fine crystals used in fire-resistant polyester resins, in hardening epoxy resins, and as a chemical intermediate.

Chlorex process [CHEM ENG] A method of extracting lubricating-oil stocks; the solvent is Chlorex, a trademark for β,β-dichlorodiethyl ether.

chlorhydrin *See* chlorohydrin.

chloric acid [INORG CHEM] HClO$_3$ A compound that exists only in solution and as chlorate salts; breaks down at 40°C.

chloride [CHEM] **1.** A compound which is derived from hydrochloric acid and contains the chlorine atom in the −1 oxidation state. **2.** In general, any binary compound containing chloride.

chloride benzilate *See* lachesne.

chloride of lime *See* bleaching powder.

chloride paper [MATER] A paper made with an emulsion of silver chloride; usually used in photography as contact paper or very-slow-speed enlarging paper.

chloride shift [PHYSIO] The reversible exchange of chloride and bicarbonate ions between erythrocytes and plasma to effect transport of carbon dioxide and maintain ionic equilibrium during respiration.

chloridization [CHEM] *See* chlorination. [MET] Treatment of mineral ores with hydrochloric acid or chlorine to form the chloride of the main metal present.

chlorimide *See* dichloramine.

chlorinated acetone *See* chloroacetone.

chlorinated lime *See* bleaching powder.

chlorinated paraffin [ORG CHEM] One of a group of chlorine derivatives of paraffin compounds.

chlorinated rubber [MATER] A nonrubbery, incombustible rubber derivative produced by the action of chlorine on rubber in solution; used in corrosion-resistant paints and varnishes, and in inks and adhesives.

chlorinated wool [TEXT] Wool treated with chlorine chemicals to minimize shrinkage and enhance dye penetration.

chlorination [CHEM] **1.** Introduction of chlorine into a compound. Also known as chloridization. **2.** Water sterilization by chlorine gas.

chlorinator [CHEM ENG] The apparatus used in chlorinating.

chlorine [CHEM] A chemical element, symbol Cl, atomic number 17, atomic weight 35.453; used in manufacture of solvents, insecticides, and many non-chlorine-containing compounds, and to bleach paper and pulp.

chlorine-36 [NUCLEO] A radioactive isotope of chlorine with atomic mass number of 36; a beta emitter with a half-life of 3 × 10^5 years.

chlorine dioxide [INORG CHEM] ClO$_2$ A green gas used to bleach cellulose and to treat water.

chlorine log *See* chlorinolog.

chlorine war gas [ORD] Chlorine gas packaged to be released against enemy troops; greenish yellow, toxic, and gaseous at normal temperatures and pressures.

CHLAMYDOSPORE

~10 μ

A chlamydospore, an asexual spore type produced by Ascomycetes and Basidiomycetes.

CHLORINE

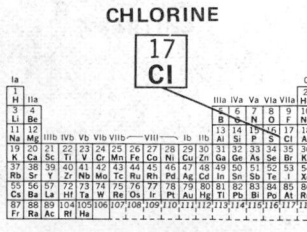

Periodic table of the chemical elements showing the position of chlorine.

chlorinity [OCEANOGR] A measure of the chloride and other halogen content, by mass, of sea water.

chlorinolog [PETRO ENG] A log (record) of the presence and concentration of chlorine in oil reservoirs, prepared as a method of locating salt-water strata. Also known as chlorine log.

chlorite [INORG CHEM] A salt of chlorous acid. [MINERAL] Any of a group of greenish, platyhydrous monoclinic silicates of aluminum, ferrous iron, and magnesium which are closely associated with and resemble the micas.

chlorite schist [PETR] A metamorphic rock whose composition is dominated by members of the chlorite group.

chlorite-sericite schist [PETR] A low-grade, fine-grained variety of mica schist without biotite.

chloritization [CHEM] The introduction of, production of, replacement by, or conversion into chlorite.

chloritoid [MINERAL] $FeAl_4Si_2O_{10}(OH)_4$ A micaceous mineral related to the brittle mica group; has both monoclinic and triclinic modifications, a gray to green color, and weakly pleochroic crystals.

chloritoid schist [PETR] A variety of mica schist whose composition is dominated by chloritoid.

chloro- [ORG CHEM] A prefix describing an organic compound which contains chlorine atoms substituted for hydrogen.

chloroacetone [ORG CHEM] CH_3COCH_2Cl Pungent, colorless liquid used as military tear gas and in organic synthesis. Also known as chloracetone; chlorinated acetone; monochloroacetone.

chloroacetophenone [ORG CHEM] $C_6H_5COCH_2Cl$ Rhombic crystals melting at 59°C; an intermediate in organic synthesis. Also known as chloracetophenone; phenacyl chloride.

Chlorobacteriaceae [MICROBIO] The equivalent name for Chlorobiaceae.

chlorobenzal See benzal chloride.

chlorobenzene [ORG CHEM] C_6H_5Cl A colorless, mobile, volatile liquid with an almondlike odor; used to produce phenol, DDT, and aniline.

Chlorobiaceae [MICROBIO] A family of green sulfur bacteria and a few chocolate-brown bacteria in the suborder Rhodobacteriineae; all members contain bacteriochlorophyll c or d as the predominant green pigment.

Chlorobium [MICROBIO] A genus of gram-negative, nonmotile, sulfur green bacteria in the family Chlorobiaceae occurring singly and in chains.

chlorobium chlorophyll [BIOCHEM] $C_{51}H_{67}O_4N_4Mg$ Either of two spectral forms of chlorophyll occurring as esters of farnesol in certain (Chlorobium) photosynthetic bacteria. Also known as bacterioviridin.

chlorobromide paper [GRAPHICS] A paper with an emulsion composed of silver bromide and silver chloride; used in photography for fast-speed contact paper, and medium-speed enlarging paper.

chlorobutadiene See chloroprene.

2-chloro-1,3-butadiene See chloroprene.

chlorocarbon [ORG CHEM] A compound of chlorine and carbon only, such as carbon tetrachloride, CCl_4.

chlorochromic anhydride See chromyl chloride.

Chlorococcales [BOT] A large, highly diverse order of unicellular or colonial, mostly fresh-water green algae in the class Chlorophyceae.

Chlorodendrineae [BOT] A suborder of colonial green algae in the order Volvocales comprising some genera with individuals capable of detachment and motility.

10-chloro-5,10-dihydrophenarsazine See phenarsazine chloride.

1-chloro-2,3-epoxypropane See epichlorohydrin.

chloroethane See ethyl chloride.

chloroethene See vinyl chloride.

chloroethyl alcohol See ethylene chlorohydrin.

chloroethylene See vinyl chloride.

chloroform [ORG CHEM] $CHCl_3$ A colorless, sweet-smelling, nonflammable liquid; used at one time as an anesthetic. Also known as trichloromethane.

chloroformyl chloride See phosgene.

chloroguanide See chloroguanide hydrochloride.

chloroguanide hydrochloride [PHARM] $C_{11}H_{16}N_5Cl$ Very effective suppressive drug in low doses, against the three kinds of malaria. Also known as chloroguanide.

chlorohydrin [ORG CHEM] Any of the compounds derived from a group of glycols or polyhydroxy alcohols by chlorine substitution for part of the hydroxyl groups. Also spelled chlorhydrin.

chlorohydrocarbon [ORG CHEM] A carbon- and hydrogen-containing compound with chlorine substituted for some hydrogen in the molecule.

6-chloro-4-isopropyl-1-methyl-3-phenol See chlorothymol.

chloroma [MED] A focal tumorous proliferation of granulocytes, with or without the blood findings of granulocytic leukemia; the sectioned surfaces of the mass are green.

chloromagnesite [MINERAL] $MgCl_2$ A mineral consisting of anhydrous magnesium chloride, found on the volcano Vesuvius.

chloromethane [ORG CHEM] CH_3Cl A colorless, noncorrosive, liquefiable gas which condenses to a colorless liquid; used as a refrigerant, and as a catalyst carrier in manufacture of butyl rubber. Also known as methyl chloride.

Chloromonadida [INV ZOO] An order of flattened, grass-green or colorless, flagellated protozoans of the class Phytamastigophorea.

Chloromonadina [INV ZOO] The equivalent name for Chloromonadida.

Chloromonadophyceae [BOT] A group of algae considered by some to be a class of the division Chrysophyta.

Chloromonadophyta [BOT] A division of algae in the plant kingdom considered by some to be a class, Chloromondophyceae.

chloromycetin See chloramphenicol.

chloronaphthalene [ORG CHEM] $C_{10}H_2Cl$ One of two isomeric, monosubstituted chlorine compounds of naphthalene; used in organic synthesis or added to gasoline as a lubricant for engine valves.

chloropal See nontronite.

chlorophenol red [ORG CHEM] $C_{19}H_{12}Cl_2O_5S$ A dye that is used as an acid-base indicator; yellow in acid solution, red in basic solution. Also spelled chlorphenol red.

chlorophoenicite [MINERAL] $(Mn,An)_5(AsO_4)(OH)_7$ Gray-green monoclinic mineral consisting of a basic arsenate of manganese and zinc occurring in crystals.

Chlorophyceae [BOT] A class of microscopic or macroscopic green algae, division Chlorophyta, composed of fresh- or salt-water, unicellular or multicellular, colonial, filamentous or sheetlike forms.

chlorophyll [BIOCHEM] The generic name for any of several oil-soluble green tetrapyrrole plant pigments which function as photoreceptors of light energy for photosynthesis.

chlorophyll a [BIOCHEM] $C_{55}H_{72}O_5N_4Mg$ A magnesium chelate of dihydroporphyrin that is esterified with phytol and has a cyclopentanone ring; occurs in all higher plants and algae.

chlorophyllase [BIOCHEM] An enzyme that splits or hydrolyzes chlorophyll.

chlorophyll b [BIOCHEM] $C_{55}H_{70}O_6N_4Mg$ An ester similar to chlorphyll a but with a −CHO substituted for a −CH₃; occurs in small amounts in all green plants and algae.

Chlorophyta [BOT] The green algae, a highly diversified plant division characterized by chloroplasts, having chlorophyll a and b as the predominating pigments.

chloropicrin [INORG CHEM] CCl_3NO_2 A colorless liquid with a sweet odor whose vapor is very irritating to the lungs and causes vomiting, coughing, and crying; used as a soil fumigant. Also known as nitrochloroform; trichloronitromethane.

Chloropidae [INV ZOO] The chloropid flies, a family of myodarian cyclorrhaphous dipteran insects in the subsection Acalypteratae.

chloroplast [BOT] A type of cell plastid occurring in the green parts of plants, containing chlorophyll pigments, and functioning in photosynthesis and protein synthesis.

chloroplatinate [INORG CHEM] 1. A double salt of platinic chloride and another chloride. 2. A salt of chloroplatinic acid. Also known as platinochloride.

chloroplatinic acid [INORG CHEM] H_2PtCl_6 An acid ob-

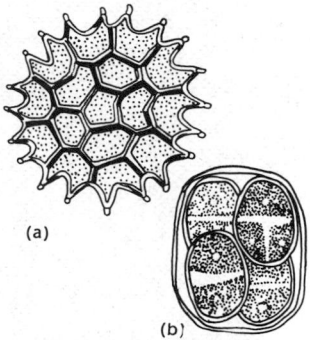

CHLOROCOCCALES

Representative colonial forms of chlorococcales. (a) Pediastrum boryanum. (b) Oöcystis borgei. (From G. M. Smith, Fresh-water Algae of the United States, 2d ed., McGraw-Hill, 1950)

CHLOROFORM

$$Cl-\underset{\displaystyle Cl}{\overset{\displaystyle Cl}{C}}-H$$

Structural formula of chloroform.

CHLOROMONADIDA

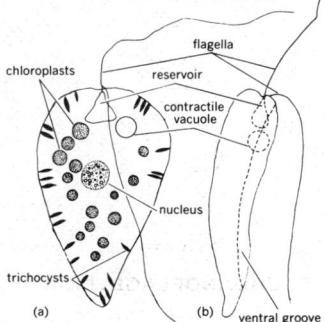

Gonyostomum semen. (a) Dorsal view. (b) Side view.

tained as red-brown deliquescent crystals; used in chemical analysis. Also known as platinic chloride.

chloroprene [ORG CHEM] C_4H_5Cl A colorless liquid which polymerizes to chloroprene resin. Also known as chlorobutadiene; 2-chloro-1,3-butadiene.

chloroprene resin [ORG CHEM] A polymer of chloroprene used to form materials resembling natural rubber.

chloropropane [ORG CHEM] Propane molecules with chlorine substituted in various amounts for the hydrogen atoms.

chloropropene [CHEM] Propene molecules with chlorine substituted for some hydrogen atoms.

3-chloropropene See allyl chloride.

chloropropylene oxide See epichlorohydrin.

chloropsia [MED] A defect of vision in which all objects appear green.

chlorosis [MED] A form of macrocytic anemia in young females characterized by marked reduction in hemoglobin and a greenish skin color. [PL PATH] A disease condition of green plants seen as yellowing of green parts of the plant.

chlorosity [OCEANOGR] The chlorine and bromide content of one liter of sea water; equals the chlorinity of the sample times its density at 20°C.

chlorosulfonic acid [INORG CHEM] $ClSO_2OH$ A fuming liquid that decomposes in water to sulfuric acid and hydrochloric acid; used in pharmaceuticals, pesticides, and dyes, and as a chemical intermediate. Also known as chlorosulfuric acid; sulfuric chlorohydrin.

chlorosulfuric acid See chlorosulfonic acid; sulfuryl chloride.

chlorothionite [MINERAL] $K_2Cu(SO_4)Cl_2$ Bright-blue secondary mineral consisting of potassium copper sulfate chloride, found on the volcano Vesuvius.

chlorothymol [ORG CHEM] $CH_3C_6H_2(OH)(C_3H_7)Cl$ White crystals melting at 59–61°C; soluble in benzene alcohol, insoluble in water; used as a bactericide. Also known as 6-chloro-4-isopropyl-1-methyl-3-phenol.

chlorotic streak [PL PATH] A systemic virus disease of sugarcane characterized by yellow or white streaks on the foliage.

chlorotrifluoroethylene polymer [ORG CHEM] A colorless, noninflammable, heat-resistant resin, soluble in most organic solvents, and with a high impact strength; can be made into transparent filling and thin sheets; used for chemical piping, fittings, and insulation for wire and cables, and in electronic components. Also known as fluorothene; polytrifluorochloroethylene resin.

chloroxiphite [MINERAL] $Pb_3CuCl_2(OH)_2O_2$ A dull-olive or pistachio-green mineral consisting of a basic chloride of lead and copper, found in the Mendip hills of England.

chlorphenol red See chlorophenol red.

chlorpromazine [PHARM] $C_{17}H_{19}ClN_2S$ A gray-white, crystalline compound used as a sedative and in preventing or relieving nausea and vomiting.

chlortetracycline [MICROBIO] $C_{22}H_{23}O_8N_2Cl$ Yellow, crystalline, broad-spectrum antibiotic produced by a strain of *Streptomyces aureofaciens*.

choana [ANAT] A funnel-shaped opening, especially the posterior nares. [INV ZOO] A protoplasmic collar surrounding the basal ends of the flagella in certain flagellates and in the choanocytes of sponges.

choanate fish [PALEON] Any of the lobefins composing the subclass Crossopterygii.

Choanichthyes [VERT ZOO] An equivalent name for the Sarcopterygii.

choanocyte [INV ZOO] Any of the choanate, flagellate cells lining the cavities of a sponge. Also known as collar cell.

Choanoflagellida [INV ZOO] An order of single-celled or colonial, colorless flagellates in the class Zoomastigophorea; distinguished by a thin protoplasmic collar at the anterior end.

choanosome [INV ZOO] The inner layer of a sponge; composed of choanocytes.

chock [MIN ENG] A square pillar for supporting the roof in a mine, constructed of prop timber laid up in alternate cross layers, in log-cabin style, the center being filled with waste. [NAV ARCH] **1.** An open or closed metal fitting through which ropes, wires, or cables are passed. **2.** A block or wedge for supporting a boat that is being repaired.

CHOANOFLAGELLIDA

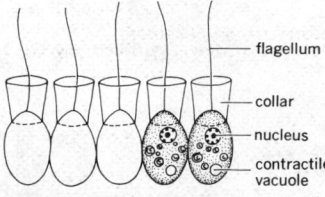

A linear colony of the choanoflagellate *Desmarella moniliformis*. (*After McCracken*)

chockablock [NAV ARCH] Also known as block and block. **1.** Brought close together, especially in reference to two blocks of tackle. **2.** Hoisted all the way up.

chocolate gale See chocolatero.

chocolatero [METEOROL] A moderate norther in the Gulf region of Mexico. Also known as chocolate gale.

chocolate spot [PL PATH] A fungus disease of legumes caused by species of *Botrytis* and characterized by brown spots on leaves and stems, with withering of shoots.

chocolate tree See cacao.

Choeropotamidae [PALEON] A family of extinct palaeodont artiodactyls in the superfamily Entelodontoidae.

choke [ELEC] An inductance used in a circuit to present a high impedance to frequencies above a specified frequency range without appreciably limiting the flow of direct current. Also known as choke coil. [ELECTROMAG] A groove or other discontinuity in a waveguide surface so shaped and dimensioned as to impede the passage of guided waves within a limited frequency range. [MECH ENG] **1.** To increase the fuel feed to an internal combustion engine through the action of a choke valve. **2.** See choke valve. [ORD] A narrowing toward the muzzle in the bore of gun, hence the choked bore; often applied to shotguns.

choke coil See choke.

choke coupling [ELECTROMAG] Coupling between two parts of a waveguide system that are not in direct mechanical contact with each other.

choke crushing [MIN ENG] A recrushing of fine ore.

chokedamp [MIN ENG] A mine atmosphere that causes choking, or suffocation, due to insufficient oxygen.

choked disk See papilledema.

choked flow [FL MECH] Flow in a duct or passage such that the flow upstream of a certain critical section cannot be increased by a reduction of downstream pressure.

choked neck [DES ENG] Container neck which has a narrowed or constricted opening.

choke filter See choke input filter.

choke flange [ELECTROMAG] A waveguide flange having in its mating surface a slot (choke) so shaped and dimensioned as to restrict leakage of microwave energy within a limited frequency range.

choke input filter [ELEC] A power-supply filter in which the first filter element is a series choke. Also known as choke filter.

choke joint [ELECTROMAG] A connection between two waveguides that uses two mating choke flanges to provide effective electrical continuity without metallic continuity at the inner walls of the waveguide.

choke piston [ELECTROMAG] A piston in which there is no metallic contact with the walls of the waveguide at the edges of the reflecting surface; the short circuit for high-frequency currents is achieved by a choke system. Also known as noncontacting piston; noncontacting plunger.

choke ring [ORD] Metal ring used in the reaction chambers of certain recoilless weapons to control gas escape; the same function is carried out by throat rings, throat blocks, and restricting plugs in other types of recoilless weapons.

choke valve [MECH ENG] A valve which supplies the higher suction necessary to give the excess fuel feed required for starting a cold internal combustion engine. Also known as choke.

choking [FL MECH] The condition which prevails in compressible fluid flow when the upper limit of mass flow is reached, or when the speed of sound is reached in a duct.

choking gas [ORD] Casualty gas which causes irritation and inflammation of the bronchial tubes and lungs; for example, phosgene.

choking Mach number [FL MECH] The Mach number at some reference point in a duct or passage (for example, at the inlet) at which the flow in the passage becomes choked.

cholagogic [PHYSIO] Inducing the flow of bile.

cholaic acid See taurocholic acid.

cholane [BIOCHEM] $C_{24}H_{42}$ A tetracyclic hydrocarbon which may be considered as the parent substance of sterols, hormones, bile acids, and digitalis aglycons.

cholangiogram [MED] The x-ray film produced by means of cholangiography.

cholangiography [MED] Roentgenography of the bile ducts.

cholangiolitis [MED] Inflammation of the bile capillaries.

cholangioma [MED] Adenocarcinoma of the bile ducts.

cholangitis [MED] Inflammation of the bile ducts.

cholate [BIOCHEM] Any salt of cholic acid.

cholecystectomy [MED] Surgical removal of the gallbladder and cystic duct.

cholecystitis [MED] Inflammation of the gallbladder.

cholecystography [MED] Radiography of the gallbladder following injection or ingestion of a radiopaque substance excreted in bile. Also known as Graham-Cole test.

cholecystokinin [BIOCHEM] A hormone produced by the mucosa of the upper intestine which stimulates contraction of the gallbladder.

cholecystostomy [MED] The establishment of an opening into the gallbladder, usually for external drainage of its contents.

choledochoduodenal junction [ANAT] The point where the common bile duct enters the duodenum.

choledocholithiasis [MED] The presence of calculi in the common bile duct.

choledochostomy [MED] The draining of the common bile duct through the abdominal wall.

choleglobin [BIOCHEM] Combined native protein (globin) and open-ring iron-porphyrin, which is bile pigment hemoglobin; a precursor of biliverdin.

cholelithiasis [MED] The production of or the condition associated with gallstones in the gallbladder or bile ducts.

cholera [MED] **1.** An acute, infectious bacterial disease of man caused by *Vibrio comma*; characterized by diarrhea, delirium, stupor, and coma. **2.** Any condition characterized by profuse vomiting and diarrhea.

cholera morbus [MED] An acute, severe gastroenteritis.

cholera vibrio [MICROBIO] *Vibrio comma,* the bacterium that causes cholera.

cholerophobia [PSYCH] Abnormal fear of cholera.

cholesteatoma [MED] An epidermal inclusion cyst of the middle ear, or mastoid bone, sometimes in the external ear canal, brain, or spinal cord. Also known as pearly tumor.

cholesterol [BIOCHEM] $C_{27}H_{46}O$ A sterol produced by all vertebrate cells, particularly in the liver, skin, and intestine, and found most abundantly in nerve tissue.

cholic acid [BIOCHEM] $C_{24}H_{40}O_5$ An unconjugated, crystalline bile acid.

choline [BIOCHEM] $C_5H_{15}O_2N$ A basic hygroscopic substance constituting a vitamin of the B complex; used by most animals as a precursor of acetylcholine and a source of methyl groups.

cholinergic [PHYSIO] Liberating, activated by, or resembling the physiologic action of acetylcholine.

cholinergic nerve [PHYSIO] Any nerve, such as autonomic preganglionic nerves and somatic motor nerves, that releases a cholinergic substance at its terminal points.

cholinesterase [BIOCHEM] An enzyme found in blood and in various other tissues that catalyzes hydrolysis of choline esters, including acetylcholine. Abbreviated chE.

choline succinate dichloride dihydrate *See* succinylcholine chloride.

choluria [MED] The presence of bile in the urine.

cholytaurine *See* taurocholic acid.

Chondrichthyes [VERT ZOO] A class of vertebrates comprising the cartilaginous, jawed fishes characterized by the absence of true bone.

chondrification [PHYSIO] Formation of or conversion into cartilage.

chondrin [BIOCHEM] A horny gelatinous protein substance obtainable from the collagen component of cartilage.

chondriome [CYTOL] Referring collectively to the chondriosomes (mitochondria) of a cell as a functional unit.

chondriosome [CYTOL] Any of a class of self-perpetuating lipoprotein complexes in the form of grains, rods, or threads in the cytoplasm of most cells; thought to function in cellular metabolism and secretion.

chondrite [GEOL] A stony meteorite containing chondrules.

chondroblast [HISTOL] A cell that produces cartilage.

Chondrobrachii [VERT ZOO] The equivalent name for Ateleopoidei.

chondroclast [HISTOL] A cell that absorbs cartilage.

chondrocranium [ANAT] The part of the adult cranium derived from the cartilaginous cranium. [EMBRYO] The cartilaginous, embryonic cranium of vertebrates.

chondrocyte [HISTOL] A cartilage cell.

chondrodendrin *See* bebeerine.

chondrodite [MINERAL] $Mg_5(SiO_4)_2(F_7OH)_2$ A monoclinic mineral of the humite group; has a resinous luster, is yellow-red in color, and occurs in contact-metamorphosed dolomites.

chondrodysplasia *See* enchondromatosis.

chondrodystrophy fetalis *See* achondroplasia.

chondrogenesis [EMBRYO] The development of cartilage.

chondroitin [BIOCHEM] A nitrogenous polysaccharide occurring in cartilage in the form of condroitinsulfuric acid.

chondrology [ANAT] The anatomical study of cartilage.

chondroma [MED] A benign tumor of bone, cartilage, or other tissue which simulates the structure of cartilage in its growth.

chondromalacia [MED] Softening of a cartilage.

chondromucoid [BIOCHEM] A mucoid found in cartilage; a glycoprotein in which chondroitinsulfuric acid is the prosthetic group.

Chondromyces [MICROBIO] A genus of myxobacteria in the family Polyangiaceae having numerous cysts grouped at the end of a colored stalk.

chondrophone [INV ZOO] In bivalve mollusks, a structure or cavity supporting the internal hinge cartilage.

Chondrophora [INV ZOO] A suborder of polymorphic, colonial, free-floating coelenterates of the class Hydrozoa.

chondroprotein [BIOCHEM] A protein (glycoprotein) occurring normally in cartilage.

chondrosarcoma [MED] A malignant tumor of cartilage.

chondroskeleton [ANAT] **1.** The parts of the bony skeleton formed from cartilage. **2.** Cartilaginous parts of a skeleton. [VERT ZOO] A cartilaginous skeleton, as in Chondrostei.

Chondrostei [PALEON] The most archaic infraclass of the subclass Actinopterygii, or rayfin fishes.

Chondrosteidae [PALEON] A family of extinct actinopterygian fishes in the order Acipenseriformes.

chondrule [GEOL] A spherically shaped body consisting chiefly of pyroxene or olivine minerals embedded in the matrix of certain stony meteorites.

Chonetidina [PALEON] A suborder of extinct articulate brachiopods in the order Strophomenida.

Chonotrichida [INV ZOO] A small order of vase-shaped ciliates in the subclass Holotrichia; commonly found as ectocommensals on marine crustaceans.

chopper [ENG] Any knife, axe, or mechanical device for chopping or cutting an object into segments. [PHYS] A device for interrupting an electric current, beam of light, or beam of infrared radiation at regular intervals, to permit amplification of the associated electrical quantity or signal by an alternating-current amplifier; also used to interrupt a continuous stream of neutrons to measure velocity.

chopper amplifier [ELECTR] A carrier amplifier in which the direct-current input is filtered by a low-pass filter, then converted into a square-wave alternating-current signal by either one or two choppers.

chopper-stabilized amplifier [ELECTR] A direct-current amplifier in which a direct-coupled amplifier is in parallel with a chopper amplifier.

chopping [ELECTR] Removal, by electronic means, of one or both extremities of a wave at a predetermined level. [PHYS] The act of interrupting an electric current, beam of light, beam of infrared radiation, or stream of neutrons at regular intervals.

chopping bit [MECH ENG] A steel bit with a chisel-shaped cutting edge, attached to a string of drill rods to break up, by impact, boulders, hardpan, and a lost core in a drill hole. Also known as chisel bit.

choppy sea [OCEANOGR] In popular usage, short, rough, irregular wave motion on a sea surface.

chop-type feeder [MECH ENG] Device for semicontinuous feed of solid materials to a process unit, with intermittent opening and closing of a hopper gate (bottom closure) by a control arm actuated by an eccentric cam.

CHONETIDINA

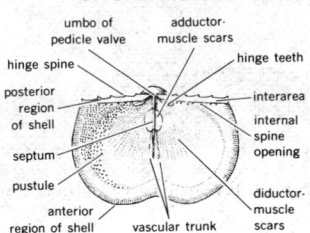

Neochonetes showing features of the pedicle valve.

CHONOTRICHIDA

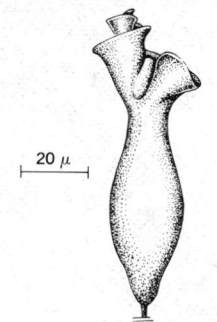

Spirochona, an example of a chonotrichid.

CHOPPER AMPLIFIER

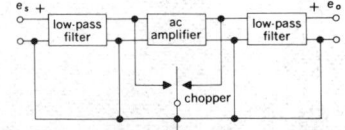

Diagram of chopper amplifier.

chord [ACOUS] A combination of two or more tones. [AERO ENG] **1.** A straight line intersecting or touching an airfoil profile at two points. **2.** Specifically, that part of such a line between two points of intersection. [CIV ENG] The top or bottom, generally horizontal member of a truss. [MATH] A line segment which intersects a curve or surface only at the endpoints of the segment.

chordal thickness [DES ENG] The tangential thickness of a tooth on a circular gear, as measured along a chord of the pitch circle.

chordamesoderm [EMBRYO] The portion of the mesoderm in the chordate embryo from which the notochord and related structures arise, and which induces formation of ectodermal neural structures.

Chordata [ZOO] The highest phylum in the animal kingdom, characterized by a notochord, nerve cord, and gill slits; includes the urochordates, lancelets, and vertebrates.

chord length [AERO ENG] The length of the chord of an airfoil section between the extremities of the section.

Chordodidae [INV ZOO] A family of worms in the order Gordioidea distinguished by a rough cuticle containing thickenings called areoles.

chordoma [MED] A rarely malignant tumor derived from persistent remnants of the notochord.

chordotomy [MED] Surgical division of a spinal nerve tract to relieve severe intractable pain.

chorea [MED] A nervous disorder seen as part of a syndrome following an organic dysfunction or an infection and characterized by irregular, involuntary movements of the body, especially of the face and extremities.

choreiform syndrome [MED] A complex of symptoms representing a form or component of minimal brain dysfunction in children, manifested by twitching movements of the face, trunk, and extremities.

chorioadenoma [MED] A tumor intermediate in malignancy between a hydatidiform mole and choriocarcinoma.

chorioallantois [EMBRYO] A vascular fetal membrane that is formed by the close association or fusion of the chorion and allantois.

choriocarcinoma [MED] A highly malignant tumor derived from chorionic tissue; found most commonly in the uterus and testis. Also known as chorioepithelioma.

chorioepithelioma *See* choriocarcinoma.

chorion [EMBRYO] The outermost of the extraembryonic membranes of amniotes, enclosing the embryo and all of its other membranes.

chorionic adenoma [MED] A benign tumor of the placenta.

chorionic gonadotropin *See* human chorionic gonadotropin.

chorionitis *See* scleroderma.

chorioretinal [ANAT] Pertaining to the choroid and retina of the eye.

chorioretinitis [MED] Inflammation of the choroid and retina of the eye.

choripetalous *See* polypetalous.

chorisepalous *See* polysepalous.

chorisis [BOT] Separation of a leaf or floral part into two or more parts during development.

chorismite [PETR] A mixed rock whose fabric is macropolyschematic and which consists of petrologically dissimilar materials of varied origins.

Choristida [INV ZOO] An order of sponges in the class Demospongiae in which at least some of the megascleres are tetraxons.

Choristodera [PALEON] A suborder of extinct reptiles of the order Eosuchia composed of a single genus, *Champsosaurus.*

C horizon [GEOL] The portion of the parent material in soils which has been penetrated with roots.

chorography [MAP] All of the methods used to map a region or district.

choroid [ANAT] The highly vascular layer of the vertebrate eye, lying between the sclera and the retina.

choroiditis [MED] Inflammation of the choroid.

choroid plexus [ANAT] Any of the highly vascular, folded processes that project into the third, fourth, and lateral ventricles of the brain.

chosen position *See* assumed position.

chrematophobia [PSYCH] An abnormal fear of money.

Christensen unit [BIOL] A unit for the standardization of streptokinase and plasmin.

Christiansen effect [ANALY CHEM] Monochromatic transparency effect when finely powdered substances, such as glass or quartz, are immersed in a liquid having the same refractive index.

Christiansen filter [OPTICS] A type of color filter, a solid-in-liquid suspension, which scatters all incident energy except that of a narrow frequency range out of the direct beam. Also known as band-pass filter.

Christmas disease [MED] A hereditary, sex-linked, hemophilia-like disease involving failure of the clotting mechanism due to a deficiency of Christmas factor.

Christmas factor [BIOCHEM] A soluble protein blood factor involved in blood coagulation. Also known as factor IX; plasma thromboplastin component (PTC).

Christmas tree [PETRO ENG] An assembly of valves, tees, crosses, and other fittings at the wellhead, used to control oil or gas production and to give access to the well tubing.

Christoffel symbols [MATH] Symbols which represent particular functions of the coefficients and their first-order derivatives of a quadratic form.

christophite *See* marmatite.

chroma [OPTICS] **1.** The dimension of the Munsell system of color that corresponds most closely to saturation, which is the degree of vividness of a hue. Also known as Munsell chroma. **2.** *See* color saturation.

chroma band-pass amplifier *See* burst amplifier.

chroma control [ELECTR] The control that adjusts the amplitude of the carrier chrominance signal fed to the chrominance demodulators in a color television receiver, so as to change the saturation or vividness of the hues in the color picture. Also known as color control; color-saturation control.

chromadizing [MET] Treating the surface of aluminum or aluminum alloys with chromic acid to improve paint adhesion.

Chromadoria [INV ZOO] A subclass of nematode worms in the class Adenophorea.

Chromadorida [INV ZOO] An order of principally aquatic nematode worms in the subclass Chromadoria.

Chromadoridae [INV ZOO] A family of soil and fresh-water, free-living nematodes in the superfamily Chromadoroidea; generally associated with algal substances.

Chromadoroidea [INV ZOO] A superfamily of small to moderate-sized, free-living nematodes with spiral, transversely ellipsoidal amphids and a striated cuticle.

chromaffin [BIOL] Staining with chromium salts.

chromaffin body *See* paraganglion.

chromaffin cell [HISTOL] Any cell of the suprarenal organs in lower vertebrates, of the adrenal medulla in mammals, of the paraganglia, or of the carotid bodies that stains with chromium salts.

chromaffin system [PHYSIO] The endocrine organs and tissues of the body that secrete epinephrine; characterized by an affinity for chromium salts.

chroma oscillator [ELECTR] A crystal oscillator used in color television receivers to generate a 3.579545-megahertz signal for comparison with the incoming 3.579545-megahertz chrominance subcarrier signal being transmitted. Also known as chrominance-subcarrier oscillator; color oscillator; color-subcarrier oscillator.

chromascope [OPTICS] An instrument used to determine the optical effects of color.

chromate [INORG CHEM] CrO_4^{--} **1.** An ion derived from the unstable acid H_2CrO_4. **2.** A salt or ester of chromic acid. [MINERAL] A mineral characterized by the cation CrO_4^{--}.

chromate treatment [MET] Treatment of metal with a solution of a hexavalent chromium compound to produce a protective coating of metal chromate.

chromatic [OPTICS] Relating to color.

chromatic aberration [ELECTR] An electron-gun defect causing enlargement and blurring of the spot on the screen of a cathode-ray tube, because electrons leave the cathode with different initial velocities and are deflected differently by the electron lenses and deflection coils. [OPTICS] An optical lens defect causing color fringes, because the lens material

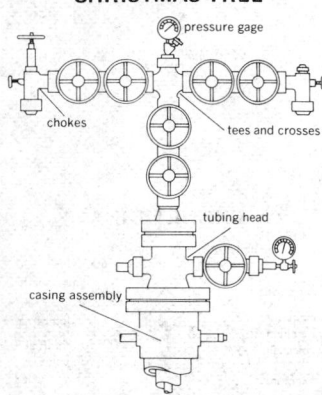

CHRISTMAS TREE

Typical layout of a Christmas tree.

(labels: pressure gage, chokes, tees and crosses, tubing head, casing assembly) brings different colors of light to focus at different points. Also known as color aberration.

chromatic diagram *See* chromaticity diagram.

chromaticity [OPTICS] The color quality of light that can be defined by its chromaticity coordinates; depends only on hue and saturation of a color, and not on its luminance (brightness).

chromaticity coordinates [OPTICS] The fractional amounts of the x, y, and z primary colors, specified by the International Committee on Illumination, in a color sample; more precisely, $x = X / (X + Y + Z)$, $y = Y / (X + Y + Z)$, $z = Z / (X + Y + Z)$, where X, Y, and Z are the integrals over wavelength λ of the product of the amount of light emerging from the sample per unit wavelength, and the tristimulus values, $\bar{x}(\lambda)$, $\bar{y}(\lambda)$, and $\bar{z}(\lambda)$ respectively.

chromaticity diagram [OPTICS] A triangular graph for specifying colors, whose ordinate is the y chromaticity coordinate and whose abscissa is the x chromaticity coordinate; the apexes of the triangle represent primary colors. Also known as chromatic diagram.

chromatic mineral [MINERAL] A mineral with color.

chromatic resolving power [OPTICS] The difference between two equally strong spectral lines that can barely be separated by a spectroscopic instrument, divided into the average wavelength of these two lines; for prisms and gratings Rayleigh's criteria are used, and the term is defined as the width of the emergent beam times the angular dispersion.

chromatics [OPTICS] **1.** The branch of optics concerned with the properties of colors. **2.** The part of colorimetry concerned with hue and saturation.

chromatic vision [PHYSIO] Vision pertaining to the color sense, that is, the perception and evaluation of the colors of the spectrum.

chromatid [CYTOL] **1.** One of the pair of strands formed by longitudinal splitting of a chromosome which are joined by a single centromere in somatic cells during mitosis. **2.** One of a tetrad of strands formed by longitudinal splitting of paired chromosomes during diplotene of meiosis.

chromatin [BIOCHEM] The deoxyribonucleoprotein complex forming the major portion of the nuclear material and of the chromosomes.

chromating [MET] Performing a chromate treatment.

chromatogram [ANALY CHEM] The pattern formed by zones of separated pigments and of colorless substance in chromatographic procedures.

chromatograph [ANALY CHEM] To employ chromatography to separate substances.

chromatographic adsorption [ANALY CHEM] Preferential adsorption of chemical compounds (gases or liquids) in an ascending molecular-weight sequence onto a solid adsorbent material, such as activated carbon, alumina, or silica gel; used for analysis and separation of chemical mixtures.

chromatography [ANALY CHEM] A method of separating and analyzing mixtures of chemical substances by chromatographic adsorption.

chromatophobia [PSYCH] An abnormal fear of colors.

chromatophore [HISTOL] A type of pigment cell found in the integument and certain deeper tissues of lower animals that contains color granules capable of being dispersed and concentrated.

chromatophorotrophin [BIOCHEM] Any crustacean neurohormone which controls the movement of pigment granules within chromatophores.

chromatoplasm [BOT] The peripheral protoplasm in bluegreen algae containing chlorophyll, accessory pigments, and stored materials.

chromatopsia [MED] A disorder of visual sensation in which color impressions are disturbed or arise subjectively, with objects appearing as unnaturally colored or colorless objects as colored; may be caused by a disturbance of the optic centers, psychic disturbance, or drugs.

chromatoscope [OPTICS] An instrument in which light beams are used to mix color stimuli.

chromatosis [MED] A pathologic process or pigmentary disease in which there is a deposit of coloring matter in a normally unpigmented site, or an excessive deposit in a normally pigmented area.

chromatron [ELECTR] A single-gun color picture tube having color phosphors deposited on the screen in strips instead of dots. Also known as Lawrence tube.

chrome alum [INORG CHEM] $KCr(SO_4)_2 \cdot 12H_2O$ An alum obtained as purple crystals and used as a mordant, in tanning, and in photography in the fixing bath. Also known as potassium chromium sulfate.

chrome brick *See* chrome refractory.

chrome dye [CHEM] One of a class of acid dyes used on wool with a chromium compound as mordant.

chrome green *See* chromic oxide.

Chromel [MET] Trade name for one of several nickel-based alloys that are resistant to oxidation and to loss of strength due to heat and that have high electrical resistivity and low temperature coefficient of resistance; grade A is 80% nickel, 20% chromium; grade B, 85% nickel, 15% chromium; grade C, 65% nickel, 25% iron, 10% chromium; grade P, 90% nickel, 10% chromium, is used in thermocouples.

Chromel-Alumel thermometer [ENG] A thermocouple type of thermometer utilizing the alloys Chromel and Alumel.

Chromel-Constantan thermocouple [ENG] A thermocouple utilizing the alloys Chromel and Constantan.

chrome leather [MATER] A leather tanned with chromium salts and used in making shoe uppers.

Chromel filament [MET] A filament composed of the alloy Chromel and used in pyrolysis of organic compounds.

chrome plating [MET] A thin plate of chromium deposited by electrolysis on a corrodible metal, giving a bright, metallic surface which is highly resistant to tarnish; used to coat automobile trimming, bathroom fixtures, and many household and other articles. Also known as chromium coating; chromium plating.

chrome red [CHEM] **1.** A pigment containing basic lead chromate. **2.** Any of several mordant acid dyes.

chrome refractory [MATER] A ceramic material made from chrome ore and used to line steel furnaces. Also known as chrome brick.

chrome spinel *See* picotite.

chrome steel *See* chromium steel.

chrome tanning [CHEM ENG] Tanning treatment of animal skin with chromium salts.

chrome-vanadium steel *See* chromium-vanadium steel.

chrome yellow [CHEM] **1.** A yellow pigment composed of normal lead chromate, $PbCrO_4$, or other lead compounds. **2.** Any of several mordant acid dyes.

chromic acid [INORG CHEM] H_2CrO_4 The hydrate of CrO_3; exists only as salts or in solution.

chromic chloride [INORG CHEM] $CrCl_3$ Crystals that are pinkish violet shimmering plates, almost insoluble in water, but easily soluble in presence of minute traces of chromous chloride; used in calico printing, as a mordant for cotton and silk.

chromic fluoride [INORG CHEM] $CrF_3 \cdot 4H_2O$ Crystals that are green, soluble in water; used in dyeing cottons.

chromic hydroxide [INORG CHEM] $Cr(OH)_3 \cdot 2H_2O$ Graygreen, gelatinous precipitate formed when a base is added to a chromic salt; the precipitate dries to a bluish, amorphous powder; prepared as an intermediate in the manufacture of other soluble chromium salts.

chromic nitrate [INORG CHEM] $Cr(NO_3)_3 \cdot 9H_2O$ Purple, rhombic crystals that are soluble in water; used as a mordant in textile dyeing.

chromic oxide [INORG CHEM] Cr_2O_3 A dark green, amorphous powder, forming hexagonal crystals on heating that are insoluble in water or acids; used as a pigment to color glass and ceramic ware and as a catalyst. Also known as chrome green.

chrominance [OPTICS] The difference between any color and a specified reference color of equal brightness; in color television, this reference color is white having coordinates $x = 0.310$ and $y = 0.316$ on the chromaticity diagram.

chrominance carrier *See* chrominance subcarrier.

chrominance-carrier reference [COMMUN] A continuous signal having the same frequency as the chrominance subcarrier in a color television system and having fixed phase with respect to the color burst; this signal is the reference with which the phase of a chrominance signal is compared for the

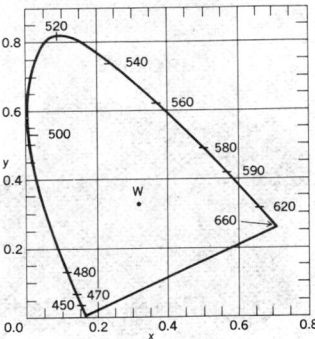

CHROMATICITY DIAGRAM

International Committee on Illumination chromaticity diagram. The colors of the spectrum are indicated along the curve. W represents a white composed of equal amounts of the three primaries. *(Adapted from A. C. Hardy, ed., Handbook of Colorimetry, MIT Press, 1936)*

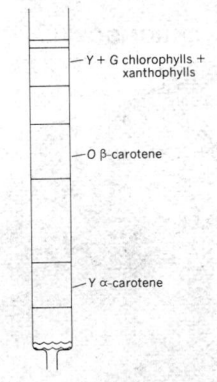

CHROMATOGRAPHIC ADSORPTION

Separation of carotenes by adsorption on activated magnesium oxide. The solvent consists of petroleum ether plus 1% acetone. *G* signifies green; *O*, orange; *Y*, yellow.

CHROMITE

⊢2.5 cm⊣

Chromite veins in a peridotite sample found in Selukwe, Rhodesia. *(Specimen from Department of Geology, Bryn Mawr College)*

CHROMIUM

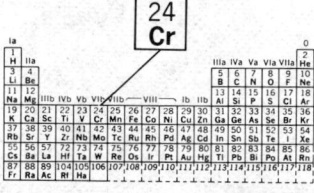

Periodic table of the chemical elements showing the position of chromium.

CHROMOCYTE

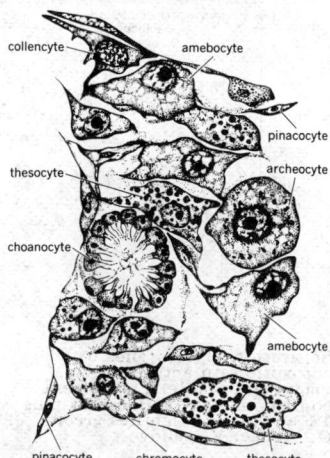

Cell types found in a fresh-water sponge as seen in a cross section through the interior of the sponge. Note chromocyte at bottom of tissue. *(After Meewis, 1936)*

purpose of modulation or demodulation. Also known as chrominance-subcarrier reference; color-carrier reference; color-subcarrier reference.

chrominance channel [COMMUN] Any path that is intended to carry the chrominance signal in a color television system.

chrominance demodulator [ELECTR] A demodulator used in a color television receiver for deriving the I and Q components of the chrominance signal from the chrominance signal and the chrominance-subcarrier frequency. Also known as chrominance-subcarrier demodulator.

chrominance frequency [COMMUN] The frequency of the chrominance subcarrier, equal to 3.579545 megahertz.

chrominance gain control [ELECTR] Variable resistors in red, green, and blue matrix channels that individually adjust primary signal levels in color television.

chrominance modulator [ELECTR] A modulator used in a color television transmitter to generate the chrominance signal from the video-frequency chrominance components and the chrominance subcarrier. Also known as chrominance-subcarrier modulator.

chrominance primary [COMMUN] The nonphysical color represented by either the I or Q chrominance signal component in a color television system.

chrominance signal [COMMUN] One of the two components, called the I signal and Q signal, that add together to produce the total chrominance signal in a color television system. Also known as carrier chrominance signal.

chrominance subcarrier [COMMUN] The 3.579545-megahertz carrier whose modulation sidebands are added to the monochrome signal to convey color information in a color television receiver. Also known as chrominance carrier; color carrier; color subcarrier; subcarrier.

chrominance-subcarrier demodulator *See* chrominance demodulator.

chrominance-subcarrier modulator *See* chrominance modulator.

chrominance-subcarrier oscillator *See* chroma oscillator.

chrominance-subcarrier reference *See* chrominance-carrier reference.

chrominance video signal [ELECTR] Voltage output from the red, green, or blue section of a color television camera or receiver matrix.

chromite [INORG CHEM] A compound of chromic oxide and a metal oxide. [MINERAL] $FeCr_2O_4$ A mineral of the spinel group; crystals and pure form are rare, and it usually is massive; the only important ore mineral of chromium. Also known as chrome iron ore.

chromium [CHEM] A metallic chemical element, symbol Cr, atomic number 24, atomic weight 51.996. [MET] A blue-white, hard, brittle metal used in chrome plating, in chromizing, and in many alloys.

chromium-51 [NUC PHYS] A radioactive isotope with atomic mass 51 made by neutron bombardment of chromium; radiates gamma rays.

chromium chloride [INORG CHEM] A group of compounds of chromium and chloride; chromium may be in +2, +3, or +6 oxidation state.

chromium coating *See* chrome plating.

chromium-iron alloy [MET] Any of several acid- and corrosion-resistant alloys containing chromium and iron.

chromium molybdenum steel [MET] Cast steel containing up to 1% carbon, 0.7–1.1% chromium, and 0.2–0.4% molybdenum; characterized by high strength and ductility.

chromium-nickel alloy [MET] Any of several alloys containing chromium and nickel in various proportions together with small amounts of other metals.

chromium oxide [INORG CHEM] A compound of chromium and oxygen; chromium may be in the +2, +3, or +6 oxidation state.

chromium oxychloride *See* chromyl chloride.

chromium plating *See* chrome plating.

chromium stearate [ORG CHEM] $Cr(C_{18}H_{35}O_2)_3$ A dark-green powder, melting at 95–100°C; used in greases, ceramics, and plastics.

chromium steel [MET] Hard, wear-resistant steel containing chromium as the predominating alloying element. Also known as chrome steel.

chromium-vanadium steel [MET] Any of several strong, hard alloy steels containing 0.15–0.25% vanadium, 0.50–1% chromium, and 0.45–0.55% carbon. Also known as chrome-vanadium steel.

chromizing [MET] Surface-alloying of metals in which an alloy is formed by diffusion of chromium into the base metal.

chromoblastomycosis [MED] A granulomatous skin disease caused by any of several fungi, usually *Hormodendrum pedrosoi,* and characterized by warty nodules which may ulcerate. Also known as chromomycosis.

chromocenter [CYTOL] An irregular, densely staining mass of heterochromatin in the chromosomes, with six armlike extensions of euchromatin, in the salivary glands of *Drosophila.*

chromocratic *See* melanocratic.

chromocyte [HISTOL] A pigmented cell.

chromogen [BIOCHEM] A pigment precursor. [MICROBIO] A microorganism capable of producing color under suitable conditions.

chromogenesis [BIOCHEM] Production of colored substances as a result of metabolic activity; characteristic of certain bacteria and fungi.

chromolipid *See* lipochrome.

chromolithography [GRAPHICS] Lithographic printing with several colors, requiring a stone for each color.

chromomere [CYTOL] Any of the linearly arranged chromatin granules in leptotene and pachytene chromosomes and in polytene nuclei.

chromometer *See* colorimeter.

chromomycin [MICROBIO] Any of five components of an antibiotic complex produced by a strain of *Streptomyces griseus;* components are designated A_1 to A_5, of which A_3 $(C_{51}H_{72}O_{32})$ is biologically active.

chromomycosis *See* chromoblastomycosis.

chromonema [CYTOL] The coiled core of a chromatid; it is thought to contain the genes.

chromophile [BIOL] Staining readily.

chromophobe [BIOL] Not readily absorbing a stain.

chromophore [CHEM] An arrangement of atoms that gives rise to color in many organic substances.

chromophyll [BIOCHEM] Any plant pigment.

chromoplasm [BOT] The pigmented, peripheral protoplasm of blue-green algae cells; contains chlorophyll, carotenoids, and phycobilins.

chromoplast [CYTOL] Any colored cell plastid, excluding chloroplasts.

chromoprotein [BIOCHEM] Any protein, such as hemoglobin, with a metal-containing pigment.

chromoscope [OPTICS] An instrument for analyzing color values and intensities.

chromosomal hybrid sterility [GEN] Sterility caused by inability of homologous chromosomes to pair during meiosis due to a chromosome aberration.

chromosome [CYTOL] Any of the complex, threadlike structures seen in animal and plant nuclei during karyokinesis which carry the linearly arranged genetic units.

chromosome aberration [GEN] Modification of the normal chromosome complement due to deletion, duplication, or rearrangement of genetic material.

chromosome complement [GEN] The species-specific, normal diploid set of chromosomes in somatic cells.

chromosome map *See* genetic map.

chromosome puff [CYTOL] Chromatic material accumulating at a restricted site on a chromosome; thought to reflect functional activity of the gene at that site during differentiation.

chromosphere [ASTRON] A transparent, tenuous layer of gas that rests on the photosphere in the atmosphere of the sun.

Chromspun [TEXT] Trade name of Eastman Chemical Products for acetate yarn and staple fiber made by adding dyes to the solution before the fiber is extruded from the spinneret.

chromyl chloride [INORG CHEM] CrO_2Cl_2 A dark-red, toxic, fuming liquid that boils at 116°C; reacts with water to form chromic acid; used to make dyes and chromium complexes. Also known as chlorochromic anhydride; chromium oxychloride.

chron [GEOL] The time unit equivalent to the stratigraphic unit, subseries, and geologic name of a division of geologic time.

chronaxie [PHYSIO] The time interval required to excite a tissue by an electric current of twice the galvanic threshold.

chronic [MED] Long-continued; of long duration.

chronic alcoholism [MED] Excessive consumption of alcohol over a prolonged period of time.

chronic appendicitis [MED] Inflammation of the vermiform appendix characterized by recurring attacks of right-sided abdominal pain over an extended period of time.

chronic carrier [MED] A person who harbors and transmits an infectious agent for an indefinite period.

chronic catarrhal enteritis [MED] Inflammation of the intestinal tract associated with vascular congestion, mucosal edema, and increased outpouring of mucus in the intestinal lumen.

chronic exposure [NUCLEO] Continued exposure to small doses of nuclear radiation.

chronic glomerulonephritis [MED] Diffuse inflammation of the glomeruli over a prolonged period of time; characterized by progressive fibrosis and associated with hypertension and uremia.

chronic hyperplastic perihepatitis *See* polyserositis.

chronic infectious arthritis *See* rheumatoid arthritis.

chronic leukemia [MED] A leukemia in which the life expectancy is prolonged; leukemias are classified as to acute or chronic, and according to the predominant cell type; the life expectancy is highly variable depending on the latter.

chronic myeloid leukemia [MED] A form of leukemia in which immature granulocytes are predominant and the life expectancy is 1–20 years or more.

chronic rhinitis [MED] Inflammation of the nasal mucous membrane due to repeated attacks of acute rhinitis; associated with membrane hypertrophy and later atrophy.

chronistor [ELECTR] A subminiature elapsed-time indicator that uses electroplating principles to totalize operating time of equipment up to several thousand hours.

chronoamperometry [ANALY CHEM] Electroanalysis by measuring at a working electrode the rate of change of current versus time during a titration; the potential is controlled.

chronocline [PALEON] A cline shown by successive morphological changes in the members of a related group, such as a species, in successive fossiliferous strata.

chronocyclegraph [IND ENG] A device used in micromotion studies to record a complete work cycle by taking still pictures with long exposures, the motion paths being traced by small electric lamps fastened to the worker's hands or fingers; time is obtained by interrupting the light circuits with a controlled frequency which produces dots on the film.

chronograph [ENG] An instrument used to register the time of an event or graphically record time intervals such as the duration of an event.

chronolith *See* time-stratigraphic unit.

chronolithologic unit *See* time-stratigraphic unit.

chronology [SCI TECH] The arrangement of data in order of time of appearance.

chronometer [HOROL] 1. Any extremely accurate watch. 2. A large, strongly built timepiece that beats half seconds and is especially designed for precise timekeeping on ships at sea.

chronometer time [HOROL] The hour of the day as indicated by a chronometer, generally set approximately to Greenwich mean time; unless the chronometer has a 24-hour dial, chronometer time is usually expressed on a 12-hour cycle and labeled A.M. or P.M.

chronometric data [ENG] Data in which the desired quantity is the time of occurrence of an event or the time interval between two or more events.

chronometric radiosonde [ENG] A radiosonde whose carrier wave is switched on and off in such a manner that the interval of time between the transmission of signals is a function of the magnitude of the meteorological elements being measured.

chronometric tachometer [ENG] A tachometer which repeatedly counts the revolutions during a fixed interval of time and presents the average speed during the last timed interval.

chronometry [HOROL] 1. The science of measuring time. 2. The measurement of time by a chronometer.

chronopher [ELECTR] Instrument for emitting standard time signal impulses from a standard clock or timing device.

chronopotentiometry [ANALY CHEM] Electroanalysis based on the measurement at a working electrode of the rate of change in potential versus time; the current is controlled.

chronoscope [HOROL] An electronic instrument used for measuring extremely short intervals of time, such as the time of passage of a rifle bullet between two points.

chronostratic unit *See* time-stratigraphic unit.

chronostratigraphic unit *See* time-stratigraphic unit.

chronothermometer [ENG] A thermometer consisting of a clock mechanism whose speed is a function of temperature; automatically calculates the mean temperature.

chronotron [ELECTR] A device that measures millimicrosecond time intervals between pulses on a transmission line to determine the time between the events which initiated the pulses.

Chroococcales [BOT] An order of blue-green algae (Cyanophyceae) that reproduce by cell division and colony fragmentation only.

Chryomyidae [INV ZOO] A family of myodarian cyclorrhaphous dipteran insects in the subsection Acalypteratae.

chrysene [ORG CHEM] $C_{18}H_{12}$ An organic, polynuclear hydrocarbon which when pure gives a bluish fluorescence; a component of short afterglow or luminescent paint.

Chrysididae [INV ZOO] The cuckoo wasps, a family of hymenopteran insects in the superfamily Bethyloidea having brilliant metallic blue and green bodies.

chrysoberyl [MINERAL] $BeAl_2O_4$ A pale green, yellow, or brown mineral that crystallizes in the orthorhombic system and is found most commonly in pegmatite dikes; used as a gem. Also known as chrysopal; cymophane; gold beryl.

chrysocarpous [BOT] Bearing yellow fruits.

Chrysochloridae [PALEON] The golden moles, a family of extinct lipotyphlan mammals in the order Insectivora.

chrysocolla [MINERAL] $CuSiO_3 \cdot 2H_2O$ A silicate mineral ordinarily occurring in impure cryptocrystalline crusts and masses with conchoidal fracture; a minor ore of copper; luster is vitreous, and color is normally emerald green to greenish-blue.

chrysoidine [ORG CHEM] $C_6H_5NNC_6H_3(NH_2)_2 \cdot HCl$ Large, black crystals or a red-brown powder that melts at 117°C; soluble in water and alcohol; used as an orange dye for silk and cotton. Also known as *meta*-diaminoazobenzene hydrochloride.

chrysolite [MINERAL] 1. A gem characterized by light-yellowish-green hues, especially the gem varieties of olivine, but also including beryl, topaz, and spinel. 2. A variety of olivine having a magnesium to magnesium-iron ratio of 0.90–0.70.

Chrysomelidae [INV ZOO] The leaf beetles, a family of coleopteran insects in the superfamily Chrysomeloidea.

Chrysomeloidea [INV ZOO] A superfamily of coleopteran insects in the suborder Polyphaga.

Chrysomonadida [INV ZOO] An order of yellow to brown, flagellated colonial protozoans of the class Phytamastigophorea.

Chrysomonadina [INV ZOO] The equivalent name for the Chrysomonadida.

Chrysopetalidae [INV ZOO] A small family of scale-bearing polychaete worms belonging to the Errantia.

chrysophanic acid [ORG CHEM] $C_{15}H_{10}O_4$ Yellow leaves that melt at 196°C; soluble in ether, chloroform, and hot alcohol; extracted from senna leaves and rhubarb root; used in medicine as a mild laxative.

Chrysophyceae [BOT] Golden-brown algae making up a class of fresh- and salt-water unicellular forms in the division Chrysophyta.

Chrysophyta [BOT] The golden-brown algae, a division of plants with a predominance of carotene and xanthophyll pigments in addition to chlorophyll.

chrysoprase [MINERAL] An apple-green variety of chalcedony that contains nickel; used as a gem. Also known as green chalcedony.

chrysotherapy [MED] The use of gold compounds in the treatment of disease.

chrysotile [MINERAL] $Mg_3Si_2O_5(OH)_4$ A fibrous form of serpentine that constitutes one type of asbestos.

CHROOCOCCALES

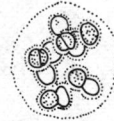

Examples of *Chroococcus*. (From E. N. Transeau, H. C. Sampson, and L. H. Tiffany, *Textbook of Botany, rev. ed.*, Harper and Row, 1953)

CHRYSOBERYL

⊢2.5 cm⊣

Chrysoberyl crystals in pegmatite, Greenwood, Maine. (*Specimen from Department of Geology, Bryn Mawr College*)

CHRYSOMELIDAE

A leaf beetle. (From T. I. Storer and R. L. Usinger, *General Zoology, 3d ed.*, McGraw-Hill, 1957)

CHRYSOMONADIDA

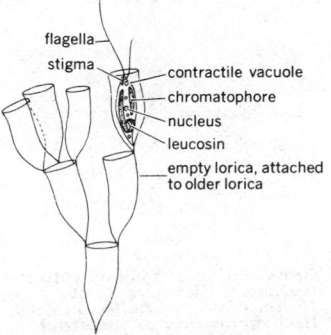

flagella
stigma
contractile vacuole
chromatophore
nucleus
leucosin
empty lorica, attached to older lorica

A chrysomonad, *Dinobryon sertularia*; cells measure 30–40 micrometers.

CICADELLIDAE

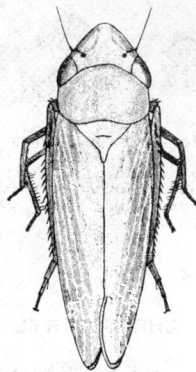

A cicadellid, dorsal view.

CICINDELIDAE

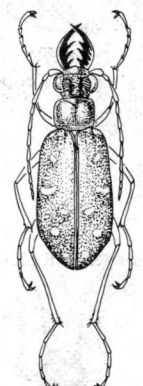

A tiger beetle. *(From T. I. Storer and R. L. Usinger, General Zoology, 3d ed., McGraw-Hill, 1957)*

CILIA

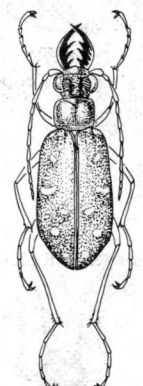

Cross section of a cilium with the peripheral filaments numbered. *(From P. Satir, Studies on Cilia, II: Examination of the distal region of the ciliary shaft and the role of the filaments in motility, J. Cell Biol, 26:805–834, 1965)*

Chthamalidae [INV ZOO] A small family of barnacles in the suborder Thoracica.

CHU *See* centigrade heat unit.

CHU_mean *See* centigrade heat unit.

chubasco [METEOROL] A severe thunderstorm with vivid lightning and violent squalls coming from the land on the west coast of Nicaragua and Costa Rica in Central America.

Chubb [GEOL] A meteorite crater in Ungava, Quebec, Canada.

chuck [DES ENG] A device for holding a component of an instrument rigid, usually by means of adjustable jaws or set screws, as the workpiece in a metalworking or woodworking machine, or the stylus or needle of a phonograph pickup. [MET] A small bar between flask bars to secure the sand in the upper box (cope) of a flask.

chucking [MECH ENG] The grasping of an outsize workpiece in a chuck or jawed device in a lathe.

chucking lug [MET] A projection forged or cast onto a piece of metal that functions as a location marker when the work is being machined.

chucking machine [MECH ENG] A lathe or grinder in which the outsize workpiece is grasped in a chuck or jawed device.

chuffing *See* chugging.

Chugaev reaction [ORG CHEM] The thermal decomposition of methyl esters of xanthates to yield olefins without rearrangement.

chugging [AERO ENG] Also known as bumping; chuffing. **1.** A form of combustion instability in a rocket engine, characterized by a pulsing operation at a fairly low frequency, sometimes defined as occurring between particular frequency limits. **2.** The noise that is made in this kind of combustion. [NUCLEO] An instability in a water-moderated reactor in which the formation of steam bubbles in the core and their subsequent collapse cause oscillations in the reactivity.

churada [METEOROL] A severe rain squall in the Mariana Islands (western Pacific Ocean) during the northeast monsoon; these squalls occur from November to April or May, but especially from January through March.

churchite *See* weinschenkite.

churn drill [MECH ENG] Portable drilling equipment, with drilling performed by a heavy string of tools tipped with a blunt-edge chisel bit suspended from a flexible cable, to which a reciprocating motion is imparted by its suspension from an oscillating beam or sheave, causing the bit to be raised and dropped. Also known as American system drill; cable-system drill.

churn hole *See* pothole.

churning [FOOD ENG] A mechanical mixing process used to separate the fat phase from a fat-water system; universally used in the manufacture of butter.

churn shot drill [MECH ENG] A boring rig with both churn and shot drillings.

chute [ENG] A conduit for conveying free-flowing materials at high velocity to lower levels. [HYD] A short channel across a narrow land area which bypasses a bend in a river; formed by the river's breaking through the land.

chute blades [ADP] Thin metal bands which form channels to the various pockets of a sorter.

chute conveyor *See* jigging conveyor.

chute spillway [CIV ENG] A spillway in which the water flow passes over a crest into a sloping, lined, open channel; used for earth and rock-fill dams.

chute system [MIN ENG] A method of mining by which ore is broken from the surface downward into chutes and is removed through passageways below. Also known as glory hole system; milling system.

chyle [PHYSIO] Lymph containing emulsified fat, present in the lacteals of the intestine during digestion of ingested fats.

chylomicron [BIOCHEM] One of the extremely small lipid droplets, consisting chiefly of triglycerides, found in blood after ingestion of fat.

chylophyllous [BOT] Having succulent or fleshy leaves.

chylothorax [MED] An accumulation of chyle in the pleural cavity.

chyluria [MED] The presence of chyle or lymph in the urine, usually caused by a fistulous communication between the urinary and lymphatic tracts or by lymphatic obstruction.

chyme [PHYSIO] The semifluid, partially digested food mass that is expelled into the duodenum by the stomach.

chymosin *See* rennin.

chymotrypsin [BIOCHEM] A proteinase in the pancreatic juice that clots milk and hydrolyzes casein and gelatin.

chymotrypsinogen [BIOCHEM] An inactive proteolytic enzyme of pancreatic juice; converted to the active form, chymotrypsin, by trypsin.

Chytridiales [MYCOL] An order of mainly aquatic fungi of the class Phycomycetes having a saclike to rhizoidal thallus and zoospores with a single posterior flagellum.

Chytridiomycetes [MYCOL] A class of true fungi.

Ci *See* cirrus cloud.

CI *See* color index; temperature humidity index.

C.I. *See* cast iron.

cibarium [INV ZOO] In insects, the space anterior to the mouth cavity in which food is chewed.

cibophobia [PSYCH] An abnormal aversion to food.

Cicadellidae [INV ZOO] A large family of homopteran insects belonging to the series Auchenorrhyncha; includes leaf hoppers.

Cicadidae [INV ZOO] A family of large homopteran insects belonging to the series Auchenorrhyncha; includes the cicadas.

cicatrix [BIOL] A scarlike mark, usually caused by previous attachment of a part or organ. [MED] The connective-tissue scar formed at the site of a healing wound.

Cichlidae [VERT ZOO] The cichlids, a family of perciform fishes in the suborder Percoidei.

Cicindelidae [INV ZOO] The tiger beetles, a family of coleopteran insects in the suborder Adephaga.

Ciconiidae [VERT ZOO] The tree storks, a family of wading birds in the order Ciconiiformes.

Ciconiiformes [VERT ZOO] An order of predominantly long-legged, long-necked birds, including herons, storks, ibises, spoonbills, and their relatives.

Cidaroida [INV ZOO] An order of echinoderms in the subclass Perischoechinoidea in which the ambulacra comprise two columns of simple plates.

CID disease *See* combined immunological deficiency disease.

Ciereszko unit [BIOL] A unit for the standardization of thyroid extracts.

Cifax [COMMUN] Enciphered facsimile communication in which the output of a keyed pulse generator is mixed with the output of the facsimile converter.

cigarette burning [CHEM] In rocket propellants, black powder, gasless delay elements, and pyrotechnic candles, the type of burning induced in a solid grain by permitting burning on one end only, so that the burning progresses in the direction of the longitudinal axis.

Ciidae [INV ZOO] The minute, tree-fungus beetles, a family of coleopteran insects in the superfamily Cucujoidea.

CIL flow test [ENG] Determination of flow properties of thermoplastic resins by measuring the amount of molten resin forced through a specific orifice per unit time by a specified variable force; developed by Canadian Industries Ltd.

cilia [ANAT] Eyelashes. [CYTOL] Relatively short, centriole-based, hairlike processes on certain anatomical cells and motile organisms.

ciliary body [ANAT] A ring of tissue lying just anterior to the retinal margin of the eye.

ciliary movement [BIOL] A type of cellular locomotion accomplished by the rhythmical beat of cilia.

ciliary muscle [ANAT] The smooth muscle of the ciliary body.

ciliary process [ANAT] Circularly arranged choroid folds continuous with the iris in front.

Ciliatea [INV ZOO] The single class of the protozoan subphylum Ciliophora.

ciliated epithelium [HISTOL] Epithelium composed of cells bearing cilia on their free surfaces.

ciliolate [BIOL] Ciliated to a very minute degree.

Ciliophora [INV ZOO] The ciliated protozoans, a homogeneous subphylum of the Protozoa distinguished principally by a mouth, ciliation, and infraciliature.

Cimbicidae [INV ZOO] The cimbicid sawflies, a family of

hymenopteran insects in the superfamily Tenthredinoidea.

Cimex [INV ZOO] The type genus of Cimicidae, including bedbugs and related forms.

Cimex lectularius [INV ZOO] The common bedbug of temperate and subtropical regions.

Cimicidae [INV ZOO] The bat, bed, and bird bugs, a family of flattened, wingless, parasitic hemipteran insects in the superfamily Cimicimorpha.

Cimicimorpha [INV ZOO] A superfamily, or group according to some authorities, of hemipteran insects in the subdivision Geocorisae.

Cimicoidea [INV ZOO] A superfamily of the Cimicimorpha in some systems of classification.

ciminite [PETR] An extrusive rock consisting essentially of olivine with sanidine and pyroxene and basic plagioclase.

cimolite [MINERAL] $2Al_2O_3 \cdot 9SiO_3 \cdot 6H_2O$ A white, grayish, or reddish mineral consisting of hydrous aluminum silicate occurring in soft, claylike masses.

cinching [GRAPHICS] The tightening of successive loops of film on a roll.

cinchocaine *See* dibucaine.

cincholepidine *See* lepidine.

cinchona [BOT] The dried, alkaloid-containing bark of trees of the genus *Cinchona*.

cinchonamine [ORG CHEM] $C_{19}H_{24}N_2O$ A yellow, crystalline, water-insoluble alkaloid that melts at 184°C; derived from the bark of *Remijia purdieana*, a member of the madder family of shrubs.

cinchonine [ORG CHEM] $C_{19}H_{22}N_2O$ A colorless, crystalline alkaloid that melts at about 245°C; extracted from cinchona bark, it is used as a substitute for quinine and as a spot reagent for bismuth.

Cincinnatian [GEOL] Upper Ordovician geologic time.

Cinclidae [VERT ZOO] The dippers, a family of insect-eating songbirds in the order Passeriformes.

cinclides [INV ZOO] Pores in the body wall of some sea anemones for the release of water and stinging cells.

cinclis [INV ZOO] Singular of cinclides.

cinder [MATER] Slag from a metal furnace. [MET] Scale cast off in forging metal.

cinder block [MATER] A hollow block made of cinder concrete. [MET] A block which closes the front of a blast furnace, containing the cinder notch.

cinder concrete [MATER] A concrete containing cinders as the aggregate.

cinder cone [GEOL] A conical elevation formed by the accumulation of volcanic debris around a vent.

cinder notch [MET] An opening in a blast furnace that allows molten slag to flow out.

cinder pig [MET] Pig iron produced from a mixture of slag in the furnace and crude metal or ore.

cinders [MATER] Incombustible residue from a burning process; in particular, small pieces of clinker from the burning of soft coal.

C index [GEOPHYS] A subjectively obtained daily index of geomagnetic activity, in which each day's record is evaluated on the basis of 0 for quiet, 1 for moderately disturbed, and 2 for very disturbed. Also known as C figure; magnetic character figure.

C indicator *See* C scope.

cinefluorography [GRAPHICS] The motion picture recording of fluoroscopic images.

cinema *See* motion picture.

cinematography [GRAPHICS] Motion picture photography.

cinemicrography [GRAPHICS] The photography of objects formed by a microscope, using a motion picture camera.

cineplasty *See* kineplasty.

cinereous [BIOL] **1.** Ashen in color. **2.** Having the inert and powdery quality of ashes.

cinetheodolite [ENG] A surveying theodolite in which 35-millimeter motion picture cameras with lenses of 60- to 240-inch focal length are substituted for the surveyor's eye and telescope; used for precise time-correlated observation of distant airplanes, missiles, and artificial satellites.

Cingulata [VERT ZOO] A group of xenarthran mammals in the order Edentata, including the armadillos.

cingulate [BIOL] Having a girdle of bands or markings.

cingulum [ANAT] **1.** The ridge around the base of the crown of a tooth. **2.** The tract of association nerve fibers in the brain, connecting the callosal and hippocampal convolutions. [BOT] The part of a plant between stem and root. [INV ZOO] **1.** Any girdlelike structure. **2.** A band of color or a raised line on certain bivalve shells. **3.** The outer zone of cilia on discs of certain rotifers. **4.** The clitellum in annelids.

cinnabar [MINERAL] HgS A vermilion-red mineral that crystallizes in the hexagonal system, although crystals are rare, and commonly occurs in fine, granular, massive form; the only important ore of mercury. Also known as cinnabarite; vermilion.

cinnamaldehyde [ORG CHEM] $C_6H_5CHCHCHO$ A yellow, volatile liquid; a constituent of cinnamon oil. Also known as benzalacetaldehyde; cinnamic aldehyde; phenylacrolein; 3-phenylpropenal.

cinnamate [ORG CHEM] A salt of cinnamic acid, containing the radical $C_9H_7O_2-$.

cinnamein *See* benzyl cinnamate.

cinnamic acid [ORG CHEM] $C_6H_5CHCHCOOH$ Colorless, monoclinic acid; forms scales, slightly soluble in water; found in natural balsams. Also known as benzalacetic acid; β-phenylacrylic acid; 3-phenyl-2-propen-1-ol; styrylformic acid.

cinnamic alcohol [ORG CHEM] $C_6H_5CH:CHCH_2OH$ White needles that congeal upon heating and are soluble in alcohol; used in perfumery. Also known as cinnamyl alcohol; phenylallylic alcohol; 3-phenyl-1-propen-1-ol; styryl carbinol.

cinnamon [BOT] *Cinnamomum zeylanicum*. An evergreen shrub of the laurel family (Lauraceae) in the order Magnoliales; a spice is made from the bark.

cinnamoyl chloride [ORG CHEM] $C_6H_5CHCHCOCl$ Yellow crystals that melt at 35°C, and decompose in water; used as a chemical intermediate. Also known as phenylacrylyl chloride.

cinnamyl alcohol *See* cinnamic alcohol.

cipher [COMMUN] A transposition or substitution code for transmitting secret messages.

cipher machine [COMMUN] Mechanical or electrical apparatus for enciphering and deciphering.

ciphony [COMMUN] A technique by which security is accomplished by converting speech into a series of on-off pulses and mixing these with the pulses supplied by a key generator; to recover the original speech, the identical key must be subtracted and the resultant on-off pulses reconverted into the original speech pattern; unauthorized listeners are unable to reconstruct the plain text unless they have an identical key generator and the daily key setting.

ciphony equipment [ELECTR] Any equipment attached to a radio transmitter, radio receiver, or telephone for scrambling or unscrambling voice messages.

Cipolletti weir [CIV ENG] Trapezoidal weir in which the sides of the notch slope are one horizontal to four vertical; used to measure water flow in open channels, especially streams and rivers.

CIPW classification [PETR] A designation for the Norm system of classifying igneous rocks; from the initial letters of the names of those who devised it: Cross, Iddings, Pirsson, and Washington.

Cir *See* Circinus.

circadian rhythm [PHYSIO] A rhythmic process within an organism occurring independently of external synchronizing signals.

circinate [BIOL] Having the form of a flat coil with the apex at the center.

circinate vernation [BOT] Uncoiling of new leaves from the base toward the apex, as in ferns.

Circinus [ASTRON] A constellation, right ascension 15 hours, declination 60°S. Abbreviated Cir. Also known as Compasses.

circle [MATH] **1.** The set of all points in the plane at a given distance from a fixed point. **2.** A unit of angular measure, equal to one complete revolution, that is, to 2π radians or 360°. Also known as turn.

circle diagram [ELEC] A diagram which gives a graphical solution of equations for a transmission line, giving the input

CINNABAR

Cinnabar on quartz crystals, Hunan Province, China. *(American Museum of Natural History)*

├── 5 cm ──┤

CINNAMON

Cinnamomum zeylanicum. *(USDA)*

impedance of the line as a function of load impedance and electrical length of the line.

circle-dot mode [ELECTR] Mode of cathode-ray storage of binary digits in which one kind of digit is represented by a small circle of excitation of the screen, and the other kind by a similar circle with a concentric dot.

circle haul [MIN ENG] A haulage system in strip mining; empty units enter the mine over one lateral and leave, loaded, over the lateral nearest the tipple.

circle of confusion [OPTICS] The blurred circular image of a point object which is formed by a camera lens, even with the best focusing. Also known as circle of least confusion.

circle of convergence [MATH] The region in which a power series possesses a limit.

circle of curvature [MATH] The circle tangent to a curve on the concave side and having the same curvature at the point of tangency as does the curve.

circle of declination *See* hour circle.

circle of equal altitude [GEOD] A circle on the surface of the earth, on every point of which the altitude of a given celestial body is the same at a given instant; the pole of this circle is the geographical position of the body, and the great-circle distance from this pole to the circle is the zenith distance of the body.

circle of equal declination *See* parallel of declination.

circle of equal probability [AERO ENG] A measure of the accuracy with which a rocket or missile can be guided; the radius of the circle at a specific distance in which 50% of the reliable shots land. Also known as circle of probable error; circular error probable.

circle of inertia *See* inertial circle.

circle of latitude Also known as parallel of latitude. [ASTRON] A great circle of the celestial sphere through the ecliptic poles, and hence perpendicular to the plane of the ecliptic. [GEOD] A meridian of the terrestrial sphere along which latitude is measured.

circle of least confusion *See* circle of confusion.

circle of longitude Also known as parallel of longitude. [ASTRON] A circle of the celestial sphere, parallel to the ecliptic. [GEOD] A circle on the surface of the earth, parallel to the plane of the Equator; a parallel, along which longitude is measured.

circle of perpetual apparition [ASTRON] That circle of the celestial sphere, centered on the polar axis and having a polar distance from the elevated pole approximately equal to the latitude of the observer, within which celestial bodies do not set.

circle of perpetual occultation [ASTRON] That circle of the celestial sphere, centered on the polar axis and having a polar distance from the depressed pole approximately equal to the latitude of the observer, within which celestial bodies do not rise.

circle of position [NAV] A circular line of position, used most frequently with reference to the circle of equal altitude surrounding the geographical position of a celestial body, but also used to refer to the lines of position produced by the DME (distance measuring equipment) system.

circle of probable error *See* circle of equal probability.

circle of right ascension *See* hour circle.

circle of uncertainty [NAV] A circle having as its center a position and as its radius the probable error of the position; the percent of the probable error must be specified error; it is a circle within which a craft is considered to be located.

circle of Willis [ANAT] A ring of arteries at the base of the cerebrum.

circle shear [MECH ENG] A shearing machine that cuts circular disks from a metal sheet rolling between the cutting wheels.

circle sheet [NAV] A chart with curves enabling a graphical solution of a three-point problem rather than using a three-arm protractor.

circling approach area [NAV] The area in which aircraft circle to land under visual conditions after completing an instrument approach.

circuit [ELEC] *See* electric circuit. [ELECTROMAG] A complete wire, radio, or carrier communications channel.

circuit analyzer *See* volt-ohm-milliammeter.

circuit breaker [ELEC] An electromagnetic device that opens a circuit automatically when the current exceeds a predetermined value.

circuit capacity [COMMUN] Number of communications channels which can be handled by a given circuit at the same time.

circuit conditioning [ELECTR] Test, analysis, engineering, and installation actions to upgrade a communications circuit to meet an operational requirement; includes the reduction of noise, the equalization of phase and level stability and frequency response, and the correction of impedance discontinuities, but does not include normal maintenance and repair activities.

circuit design [ELEC] The art of specifying the components and interconnections of an electrical network.

circuit diagram [ELEC] A drawing, using standardized symbols, of the arrangement and interconnections of the conductors and components of an electrical or electronic device or installation. Also known as schematic circuit diagram; wiring diagram.

circuit efficiency [ELECTR] Of an electron tube, the power delivered to a load at the output terminals of the output circuit at a desired frequency divided by the power delivered by the electron stream to the output circuit at that frequency.

circuit element *See* component.

circuit interrupter [ELEC] A device in a circuit breaker to remove energy from an arc in order to extinguish it.

circuit loading [ELEC] Power drawn from a circuit by an electric measuring instrument, which may alter appreciably the quantity being measured.

circuit noise [COMMUN] In telephone practice, the noise which is brought to the receiver electrically from a telephone system, excluding noise picked up acoustically by telephone transmitters.

circuit noise level [COMMUN] Ratio of the circuit noise at that point to some arbitrary amount of circuit noise chosen as a reference; usually expressed in decibels above reference noise, signifying the reading of a circuit noise meter, or in adjusted decibels, signifying circuit noise meter reading adjusted to represent interfering effect under specified conditions.

circuit protection [ELECTR] Provision for automatically preventing excess or dangerous temperatures in a conductor and limiting the amount of energy liberated when an electrical failure occurs.

circuit reliability [COMMUN] The percent of time a circuit was available to the user during a specified period of time.

circuitron [ELECTR] Combination of active and passive components mounted in a single envelope like that used for tubes, to serve as one or more complete operating stages.

circuitry [ELEC] The complete combination of circuits used in an electrical or electronic system or piece of equipment.

circuit shift *See* cyclic shift.

circuit switching [COMMUN] The method of providing communication service through a switching facility, either from local users or from other switching facilities.

circuit testing [ELEC] The testing of electric circuits to determine and locate an open circuit, or a short circuit or leakage.

circuit theory [ELEC] The mathematical analysis of conditions and relationships in an electric circuit. Also known as electric circuit theory.

circulant determinant [MATH] A determinant in which the elements of each row are the same as those of the previous row moved one place to the right, with the last element put first.

circulant matrix [MATH] A matrix in which the elements of each row are those of the previous row moved one place to the right.

circular accelerator *See* circular particle accelerator.

circular antenna [ELECTROMAG] A folded dipole that is bent into a circle, so the transmission line and the abutting folded ends are at opposite ends of a diameter.

circular arc *See* arc.

circular birefringence [OPTICS] The phenomenon in which an optically active substance transmits right circularly polarized light with a different velocity from left circularly polarized light.

circular channel [ENG] Continuous-length opening with cir-

CIRCUIT BREAKER

Bulk oil circuit breaker for 138-kilovolt application.

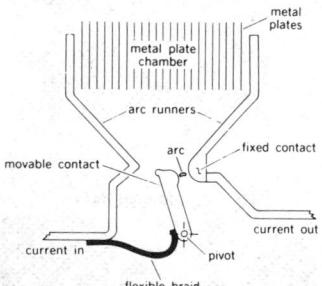

CIRCUIT INTERRUPTER

metal plates
metal plate chamber
arc runners
arc
fixed contact
movable contact
current out
current in
pivot
flexible braid

Cross section of interrupter for a typical medium-voltage circuit breaker.

cular cross section through which liquid or gas can be made to flow.

circular-chart recorder [ENG] Graphic pen-and-ink recorder where measured values are drawn onto a rotating circular chart by the backward and forward movement of a pivoted pen actuated by the input signal (such as temperature, pressure, flow, or force) from an instrument transmitter.

circular chromatography *See* radial chromatography.

circular coal *See* eye coal.

circular coil [ELECTROMAG] In eddy-current nondestructive tests, a type of test coil which surrounds an object.

circular current [ELEC] An electric current moving in a circular path.

circular cutter [MECH ENG] A rotating blade with a square or knife edge used to slit or shear metal.

circular cylinder [MATH] A solid bounded by two parallel planes and a cylindrical surface whose intersections with planes perpendicular to the straight lines forming the surface are circles.

circular deoxyribonucleic acid [BIOCHEM] A single- or double-stranded ring of deoxyribonucleic acid found in certain bacteriophages and in human wart virus. Also known as ring deoxyribonucleic acid.

circular dichroism [OPTICS] A change from planar to elliptic polarization when an initially plane-polarized light wave traverses an optically active medium. Abbreviated CD.

circular electric wave [ELECTROMAG] A transverse electric wave for which the lines of electric force form concentric circles.

circular error [ORD] **1.** A bombing error measured by the radial distance of a point of bomb impact, or mean point of impact, from the center of the target, excluding gross errors. **2.** With an airburst atomic bomb, the bombing error measured from the point on the ground immediately below the bomb burst to the desired ground zero.

circular error average [ORD] The bombing error in a given bombing attack, expressed as the average radial distance of the bomb impacts, or mean points of impact, from the center of the target.

circular error probable *See* circle of equal probability.

circular flow method [FL MECH] A method to determine viscosities of Newtonian fluids by measuring the torque from viscous drag of sample material between a closely spaced rotating plate-stationary cone assembly.

circular form tool [DES ENG] A round or disk-shaped tool with the cutting edge on the periphery.

circular functions *See* trigonometric functions.

circular horn [ELECTROMAG] A circular-waveguide section that flares outward into the shape of a horn, to serve as a feed for a microwave reflector or lens.

circular inch [MECH] The area of a circle 1 inch in diameter.

circular magnetic wave [ELECTROMAG] A transverse magnetic wave for which the lines of magnetic force form concentric circles.

circular mil [MECH] A unit equal to the area of a circle whose diameter is 1 mil (0.001 inch); used chiefly in specifying cross-sectional areas of round conductors. Abbreviated cir mil.

circular motion [MECH] **1.** Motion of a particle in a circular path. **2.** Motion of a rigid body in which all its particles move in circles about a common axis, fixed with respect to the body, with a common angular velocity.

circular nomograph [MATH] A chart with concentric circular scales for three variables, laid out so that any straight line passes through values of the variables satisfying a given equation.

circular orbit [ASTRON] An orbit comprising a complete constant-altitude revolution around the earth.

circular paper chromatography [ANALY CHEM] Paper chromatographic technique in which migration from a spot in the sheet takes place in 360° so that zones separate as a series of concentric rings.

circular particle accelerator [NUCLEO] A particle accelerator which utilizes a magnetic field to bend charged-particle orbits and confine the extent of particle motion. Also known as circular accelerator.

circular pitch [DES ENG] The linear measure in inches along the pitch circle of a gear between corresponding points of adjacent teeth.

circular plane [DES ENG] A plane that can be adjusted for convex or concave surfaces.

circular polarization [PHYS] Attribute of a transverse wave (either of electromagnetic radiation, or in an elastic medium) whose electric or displacement vector is of constant amplitude and, at a fixed point in space, rotates in a plane perpendicular to the propagation direction with constant angular velocity.

circular polarized loop vee [ELECTROMAG] Airborne communications antenna with an omnidirectional radiation pattern to provide optimum near-horizon communications coverage.

circular saw [MECH ENG] Any of several power tools for cutting wood or metal, having a thin steel disk with a toothed edge that rotates on a spindle.

circular scanning [ENG] Radar scanning in which the direction of maximum radiation describes a right circular cone.

circular segment [MATH] Portion of circle cut off from the main body of the circle by a straight line (chord) through the circle.

circular shaft [MIN ENG] A shaft excavated in a round shape.

circular shift *See* cyclic shift.

circular slide rule [MATH] A slide rule in a circular form whose advantages over a straight slide rule are its precision, because it is equivalent to a straight slide rule many times longer than the circular slide rule's diameter, and ease of multiplication, because the scale is continuous.

circular sweep generation [ELECTR] The use of electronic circuits to provide voltage or current which causes an electron beam in a device such as a cathode-ray tube to move in a circular deflection path at constant speed.

circular velocity [MECH] At any specific distance from the primary, the orbital velocity required to maintain a constant-radius orbit.

circular vortex [METEOROL] An atmospheric flow in parallel planes in which streamlines and other isopleths are concentric circles about a common axis; an atmospheric model of easterly and westerly winds is a circular vortex about the earth's polar axis.

circular waveguide [ELECTROMAG] A waveguide whose cross-sectional area is circular.

circulating fluid [ENG] A fluid pumped into a borehole through the drill stem, the flow of which cools the bit and transports the cuttings out of the borehole.

circulating memory [ELECTR] A digital computer device that uses a delay line to store information in the form of a pattern of pulses in a train; the output pulses are detected electrically, amplified, reshaped, and reinserted in the delay line at the beginning. Also known as delay-line memory; delay-line storage; circulating storage.

circulating pump [CHEM ENG] Pump used to circulate process liquid out of and back into a process system, as in the circulation of distillation column bottoms through an external heater, or the circulation of storage tank bottoms to mix tank contents.

circulating reactor [NUCLEO] A nuclear reactor in which the fissionable material circulates through the core in fluid form or as small particles suspended in a fluid.

circulating register [ADP] A shift register in which data move out of one end and reenter the other end, as in a closed loop.

circulating scrap [MET] At steelworks and founderies, scrap arising during the manufacture of finished iron and steel or of castings.

circulating storage *See* circulating memory.

circulating system [CHEM ENG] Fluid system in which the process fluid is taken from and pumped back into the system, as in the circulation of distillation column bottoms through an external heater.

circulation [FL MECH] The flow or motion of fluid in or through a given area or volume. [MATH] For the circulation of a vector field around a closed path, the line integral of the field vector around the path. [METEOROL] For an air mass, in the line integral of the tangential component of the velocity field about a closed curve. [OCEANOGR] Water current flow occurring within a large area, usually a closed circular pattern. [PHYSIO] The movement of blood through defined

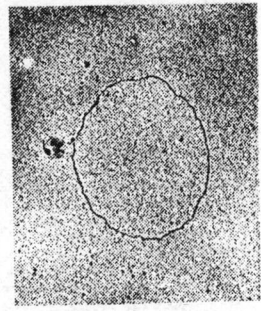

CIRCULAR DEOXYRIBONUCLEIC ACID

Electron micrograph of circular deoxyribonucleic acid extracted from the human wart virus. *(Courtesy of E. A. C. Follett)*

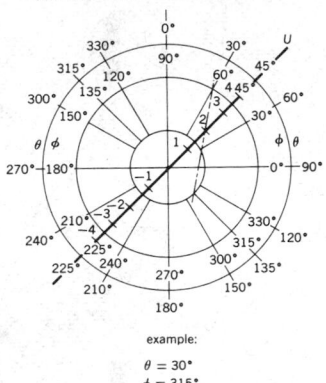

CIRCULAR NOMOGRAPH

example:

$\theta = 30°$
$\phi = 315°$
cut U at 2.1

Circular nomograph which results from a trigonometric equation expressed in the form of a determinant.

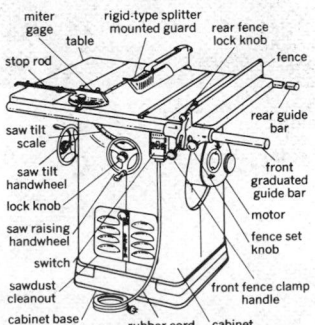

CIRCULAR SAW

Bench circular saw with tilting arbor is used for parting or slotting and can make cuts as long as working space permits. *(Delta)*

channels and tissue spaces; movement is through a closed circuit in vertebrates and certain invertebrates.

circulation area [BUILD] The area required for human traffic in a building, including permanent corridors, stairways, elevators, escalators, and lobbies.

circulation flux [METEOROL] Flux due to mean atmospheric motion as opposed to eddy flux; the dominant flux in low latitudes.

circulation index [METEOROL] A measure of the magnitude of one of several aspects of large-scale atmospheric circulation patterns; indices most frequently measured represent the strength of the zonal (east-west) or meridional (north-south) components of the wind, at the surface or at upper levels, usually averaged spatially and often averaged in time.

circulation map See traffic-circulation map.

circulation pattern [METEOROL] The general geometric configuration of atmospheric circulation usually applied, in synoptic meteorology, to the large-scale features of synoptic charts and mean charts.

circulation theorem [METEOROL] Any of several theorems concerning fluid circulation.

circulator [ELECTROMAG] A waveguide component having a number of terminals so arranged that energy entering one terminal is transmitted to the next adjacent terminal in a particular direction. Also known as microwave circulator.

circulatory system [ANAT] The vessels and organs composing the lymphatic and cardiovascular systems.

circulin [MICROBIO] Any of a group of peptide antibiotics produced by *Bacillus circulans* which are related to polymixin and are active against both gram-negative and gram-positive bacteria.

circulus [BIOL] Any of various ringlike structures, such as the vascular circle of Willis or the concentric ridges on fish scales.

circumcenter [MATH] For a triangle or a regular polygon, the center of the circle that is circumscribed about the triangle or polygon.

circumcision [MED] Surgical excision of the foreskin.

circumduction [ANAT] Movement of the distal end of a body part in the form of an arc; performed at ball-and-socket and saddle joints.

circumference [MATH] 1. The length of a circle. 2. For a sphere, the length of any great circle on the sphere.

circumferentor [ENG] A horizontal compass used in surveying that has arms diametrically placed with vertical slit sights in them.

circumflex artery [ANAT] Any artery that follows a curving or winding course.

circumhorizontal arc [OPTICS] A halo phenomenon consisting of a colored arc, red on its upper margin; it extends for about 90° parallel to the horizon and lies about 46° below the sun.

circumlunar [ASTRON] Around the moon; generally applied to trajectories.

circummeridian altitude See exmeridian altitude.

circum-Pacific province See Pacific suite.

circumpharyngeal connective [INV ZOO] One of a pair of nerve strands passing around the esophagus in annelids and anthropods, connecting the brain and subesophageal ganglia.

circumpolar [ASTRON] Revolving about the elevated pole without setting. [GEOGR] Located around one of the polar regions of earth.

circumpolar star [ASTRON] A star with its polar distance approximately equal to or less than the latitude of the observer.

circumpolar westerlies See westerlies.

circumpolar whirl See polar vortex.

circumradius [MATH] The radius of a circle that is circumscribed about a polygon.

circumscissile [BOT] Dehiscing along the line of a circumference, as exhibited by a pyxidium.

circumscribed [MATH] 1. A closed curve (or surface) is circumscribed about a polygon (or polyhedron) if every vertex of the polygon (or polyhedron) is incident upon the curve (or surface) and the polygon (or polyhedron) is contained in the curve (or surface). 2. A polygon (or polyhedron) is circumscribed about a closed curve (or surface) if every side of the polygon (or face of the polyhedron) is tangent to the curve (or surface) and the curve (or surface) is contained within the polygon (or polyhedron).

circumvallate papilla See vallate papilla.

circumzenithal arc [OPTICS] A brilliant rainbow-colored arc of about a quarter of a circle with its center at the zenith and about 46° above the sun, produced by refraction and dispersion of the sun's light striking the top of prismatic ice crystals in the atmosphere, and usually lasting only a few minutes.

ciré [TEXT] Any fabric treated with wax, heat, and pressure to produce a glossy appearance.

cir mil See circular mil.

Cirolanidae [INV ZOO] A family of isopod crustaceans in the suborder Flabellifera composed of actively swimming predators and scavengers with biting mouthparts.

cirque [GEOL] A steep elliptic to elongated enclave high on mountains in calcareous districts, usually forming the blunt end of a valley.

cirque lake [HYD] A small body of water occupying a cirque.

Cirratulidae [INV ZOO] A family of fringe worms belonging to the Sedentaria which are important detritus feeders in coastal waters.

cirrhosis [MED] A progressive, inflammatory disease of the liver characterized by a real or apparent increase in the proportion of hepatic connective tissue.

cirriform [METEOROL] Descriptive of clouds composed of small particles, mostly ice crystals, which are fairly widely dispersed, usually resulting in relative transparency and whiteness and often producing halo phenomena not observed with other cloud forms. [ZOO] Having the form of a cirrus; generally applied to a prolonged, slender process.

Cirripedia [INV ZOO] A subclass of the Crustacea, including the barnacles and goose barnacles; individuals are free-swimming in the larval stages but permanently fixed in the adult stage.

cirrocumulus cloud [METEOROL] A principal cloud type, appearing as a thin, white path of cloud without shadows, composed of very small elements in the form of grains, ripples, and so on. Abbreviated Cc.

Cirromorpha [INV ZOO] A suborder of cephalopod mollusks in the order Octopoda.

cirrostratus cloud [METEOROL] A principal cloud type, appearing as a whitish veil, usually fibrous but sometimes smooth, which may totally cover the sky and often produces halo phenomena, either partial or complete. Abbreviated Cs.

cirrus [INV ZOO] 1. The conical locomotor structure composed of fused cilia in hypotrich protozoans. 2. Any of the jointed thoracic appendages of barnacles. 3. Any hairlike tuft on insect appendages. 4. The male copulatory organ in some mollusks and trematodes. [VERT ZOO] Any of the tactile barbels of certain fishes. [ZOO] A tendrillike animal appendage.

cirrus cloud [METEOROL] A principal cloud type composed of detached cirriform elements in the form of white, delicate filaments, of white (or mostly white) patches, or narrow bands. Abbreviated Ci.

cirrus sac [INV ZOO] A pouch or channel containing the copulatory organ (cirrus) in certain invertebrates.

cis [ORG CHEM] A descriptive term indicating a form of isomerism in which atoms are located on the same side of an asymmetric molecule.

cislunar [ASTRON] Of or pertaining to phenomena, projects, or activity in the space between the earth and moon, or between the earth and the moon's orbit.

cissoid [MATH] A plane curve consisting of all points which lie on a variable line passing through a fixed point on a circle, and whose distance from the fixed point is equal to the distance from the line's intersection with the circle to its intersection with the tangent to the circle at the point diametrically opposite the fixed point; in cartesian coordinates the equation is $y^2(2a - x) = x^3$.

cistern [ANAT] A closed, fluid-filled sac or vesicle, such as the subarachnoid spaces or the vesicles comprising the dictyosomes of a Golgi apparatus. [CIV ENG] A tank for storing water or other liquid. [GEOL] A hollow that holds water.

cistern barometer [ENG] A pressure-measuring device in which pressure is read by the liquid rise in a vertical, closed-top tube as a result of system pressure on a liquid reservoir

CIRRATULIDAE

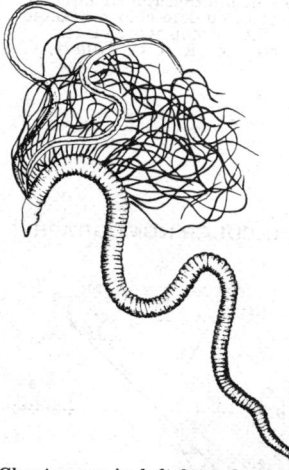

Chaetozone in left lateral view.

CIRRUS CLOUD

Cirrus, with trails of slowly falling ice crystals at a high level. (*F. Ellerman, U.S. Weather Bureau*)

(cistern) into which the bottom, open end of the tube is immersed.

cis-trans isomerism [ORG CHEM] A type of geometrical isomerism found in alkenic systems in which it is possible for each of the doubly bonded carbons to carry two different atoms or groups; two similar atoms or groups maybe on the same side (cis) or on opposite sides (trans) of a plane bisecting the alkenic carbons and perpendicular to the plane of the alkenic system.

cistron [MOL BIO] The genetic unit within which mutants do not complement each other; determined by the cis/trans complementation test. Also known as structural gene.

Citheroniinae [INV ZOO] A subfamily of lepidopteran insects in the family Saturniidae, including the regal moth and the imperial moth.

citizens' band [COMMUN] A frequency band allocated for citizens' radio service (460–470 or 27.23–27.28 megahertz). Also known as citizens' waveband.

citizens' radio service [COMMUN] A radio communication service intended for private or personal radio communication, including radio signaling and control of objects by radio.

citizens' waveband *See* citizens' band.

citraconic acid [ORG CHEM] $C_5H_6O_4$ A dicarboxylic acid; hygroscopic crystals that melt at 91°C; derived from citric acid by heating. Also known as methyl maleic acid.

citral [ORG CHEM] $C_{10}H_{16}O$ A pale-yellow liquid that in commerce is a mixture of two isomeric forms, alpha and beta; insoluble in water, soluble in glycerin or benzyl benzoate; used in perfumery and as an intermediate to form other compounds. Also known as 3,7-dimethyl-2,6-octadienal; geranial; geranialdehyde.

citramalase [BIOCHEM] An enzyme that is involved in fermentation of glutamate by *Clostridium tetanomorphum;* catalyzes the breakdown of citramalic acid to acetate and pyruvate.

citrate [BIOCHEM] A salt or ester of citric acid.

citrate test [MICROBIO] A differential cultural test to identify genera within the bacterial family Enterobacteriaceae that are able to utilize sodium citrate as a sole source of carbon.

citric acid [BIOCHEM] $C_6H_8O_7 \cdot H_2O$ A colorless crystalline or white powdery organic, tricarboxylic acid occurring in plants, especially citrus fruits, and used as a flavoring agent, as an antioxidant in foods, and as a sequestering agent; the commercially produced form melts at 153°C.

citric acid cycle *See* Krebs cycle.

citriculture [BOT] The cultivation of citrus fruits.

citrine [MINERAL] An important variety of crystalline quartz, yellow to brown in color and transparent. Also known as Bohemian topaz; false topaz; quartz topaz; topaz quartz; yellow quartz.

citron [BOT] *Citrus medica.* 1. A shrubby, evergreen citrus tree in the order Sapindales cultivated for its edible, large, lemonlike fruit.

citronella [BOT] *Cymbopogon nardus.* A tropical grass; the source of citronella oil.

citronella oil [MATER] A yellowish oil distilled from the leaves of either of two grasses, *Cymbopogon nardus* or *C. winterianus;* used as an insect repellent. Also known as Java citronella oil.

citronellol [ORG CHEM] $C_{10}H_{19}OH$ A liquid derived from citronella oil; soluble in alcohol; used in perfumery. Also known as 3,7-dimethyl-6(or 7)octen-1-ol.

citrulline [BIOCHEM] $C_6H_{13}O_3N_3$ An amino acid formed in the synthesis of arginine from ornithine.

citrus anthracnose [PL PATH] A fungus disease of citrus plants caused by *Colletotrichum gloeosporioides* and characterized by tip blight, stains on the leaves, and spots, stains, or rot on the fruit.

citrus blast [PL PATH] A bacterial disease of citrus trees caused by *Pseudomonas syringae* and marked by drying and browning of foliage and twigs and black pitting of the fruit.

citrus canker [PL PATH] A bacterial disease of citrus plants caused by *Xanthomonas citri* and producing lesions on twigs, foliage, and fruit.

citrus flavanoid compound *See* bioflavanoid.

citrus fruit [BOT] Any of the edible fruits having a pulpy

endocarp and a firm exocarp that are produced by plants of the genus *Citrus* and related genera.

citrus scab [PL PATH] A fungus disease of citrus plants caused by *Sphaceloma rosarum,* producing scablike lesions on all plant parts.

city plan [MAP] A large-scale, comprehensive map of a city delineating streets, important buildings, and other urban elements; relief is shown when important. Also known as town plan.

civet [PHYSIO] A fatty substance secreted by the civet gland; used as a fixative in perfumes. [VERT ZOO] Any of 18 species of catlike, nocturnal carnivores assigned to the family Viverridae, having a long head, pointed muzzle, and short limbs with nonretractile claws.

civet cat *See* cacomistle.

civet gland [VERT ZOO] A large anal scent gland in civet cats that secretes civet.

civetone [BIOCHEM] $C_{17}H_{30}O$ 9-Cycloheptadecen-1-one, a macrocyclic ketone component of civet used in perfumes because of its pleasant odor and lasting quality; believed to function as a sex attractant among civet cats.

civil airway [NAV] An airway designated for air commerce.

civil day [ASTRON] A mean solar day beginning at midnight instead of at noon; may be based on either apparent solar time or mean solar time.

civil engineering [ENG] The planning, design, construction, and maintenance of fixed structures and ground facilities for industry, transportation, use and control of water, or occupancy.

civil time [ASTRON] Solar time in a day (civil day) that begins at midnight; may be either apparent solar time or mean solar time.

civil twilight [ASTRON] The interval of incomplete darkness between sunrise (or sunset) and the time when the center of the sun's disk is 6° below the horizon.

civil year *See* calendar year.

Cl *See* chlorine.

cladding [ENG] Process of covering one material with another and bonding them together under high pressure and temperature. Also known as bonding. [NUCLEO] An outer jacket, usually metallic, for a nuclear fuel element; prevents corrosion of fuel and release of fission products into the coolant.

Cladistia [VERT ZOO] The equivalent name for Polypteriformes.

clad metal [MET] A metal overlaid on one or both sides with a different metal.

Cladocera [INV ZOO] An order of small, fresh-water branchiopod crustaceans, commonly known as water fleas, characterized by a transparent bivalve shell.

Cladocopa [INV ZOO] A suborder of the order Myodocopida including marine animals having a carapace that lacks a permanent aperture when the two valves are closed.

Cladocopina [INV ZOO] The equivalent name for Cladocopa.

clade *See* cladophyll.

cladodont [PALEON] Pertaining to sharks of the most primitive evolutionary level.

cladogenesis [EVOL] Evolution associated with altered habit and habitat, usually in isolated species populations.

cladogenic adaptation *See* divergent adaptation.

Cladoniaceae [BOT] A family of lichens in the order Lecanorales, including the reindeer mosses and cup lichens, in which the main thallus is hollow.

Cladophorales [BOT] An order of coarse, wiry, filamentous, branched and unbranched algae in the class Chlorophyceae.

cladophyll [BOT] A branch arising from the axil of a true leaf and resembling a foliage leaf. Also known as cladode.

cladoptosis [BOT] The annual abscission of twigs or branches instead of leaves.

Cladoselachii [PALEON] An order of extinct elasmobranch fishes including the oldest and most primitive of sharks.

cladus [BOT] A branch of a ramose spicule.

Claibornian [GEOL] Middle Eocene geologic time.

claim *See* mining claim.

clairite *See* enargite.

clairvoyance [PSYCH] A form of extrasensory perception in which there is a receiver and an extant event of which he has

CIS-TRANS ISOMERISM

cis-2-Butylene

trans-2-Butylene

Cis-trans isomerism among olefins.

CITRIC ACID

$$H_2C—COOH$$
$$HO—C—COOH$$
$$H_2C—COOH$$

Structural formula of citric acid.

CITRON

Commercial citron *(Citrus medica),* foliage and fruit. Inset shows cross section of fruit. *(From J. Horace McFarland Co.)*

CLADOSELACHII

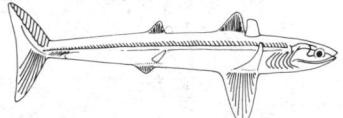

Cladoselache, a Late Devonian sharklike fish. The original specimens range from about 1½ to 4 feet (0.48–1.2 meters) in length. *(After Harris and Dean)*

knowledge that has not been conveyed to him through his sensory channels.

Claisen condensation [ORG CHEM] **1.** Condensation in the presence of sodium ethoxide of esters or of esters and ketones to form β-dicarbonyl compounds. **2.** Condensation of arylaldehydes and acylphenones with esters or ketones in the presence of sodium ethoxide to yield unsaturated esters. Also known as Claisen reaction.

Claisen flask [CHEM] A glass flask with a U-shaped neck, used for distillation.

Claisen reaction *See* Claisen condensation.

Claisen-Schmidt condensation [ORG CHEM] A reaction employed for preparation of unsaturated aldehydes and ketones by condensation of aromatic aldehydes with aliphatic aldehydes or ketones in the presence of sodium hydroxide.

clam [INV ZOO] The common name for a number of species of bivalve mollusks, many of which are important as food.

Clambidae [INV ZOO] The minute beetles, a family of coleopteran insects in the superfamily Dascilloidea.

clammy [BIOL] Moist and sticky, as the skin or a stem.

clamp [DES ENG] A tool for binding or pressing two or more parts together, holding them firmly in their relative position. [ELECTR] *See* clamping circuit.

clamper *See* direct-current restorer.

clamping [ELECTR] The introduction of a reference level that has some desired relation to a pulsed waveform, as at the negative or positive peaks. Also known as direct-current reinsertion; direct-current restoration.

clamping circuit [ELECTR] A circuit that reestablishes the direct-current level of a waveform; used in the dc restorer stage of a television receiver to restore the dc component to the video signal after its loss in capacitance-coupled alternating-current amplifiers, to reestablish the average light value of the reproduced image. Also known as clamp.

clamping coupling [MECH ENG] A coupling with a split cylindrical element which clamps the shaft ends together by direct compression, through bolts or rings, and by the wedge action of conical sections; not considered a permanent part of the shaft.

clamping diode [ELECTR] A diode used to clamp a voltage at some point in a circuit.

clamp-on [COMMUN] A method of holding a call for a line that is in use and of signaling when it becomes free.

clamp-on ammeter *See* snap-on ammeter.

clamp screw [DES ENG] A screw that holds a part by forcing it against another part.

clamp screw sextant [ENG] A marine sextant having a clamp screw for controlling the position of the tangent screw.

clamshell bucket [MECH ENG] A two-sided bucket used in a type of excavator to dig in a vertical direction; the bucket is dropped while its leaves are open and digs as they close. Also known as clamshell grab.

clamshell grab *See* clamshell bucket.

clam worm [INV ZOO] The common name for a number of species of dorsoventrally flattened annelid worms composing the large family Nereidae in the class Polychaeta; all have a distinct head, with numerous appendages.

clan [ECOL] A very small community, perhaps a few square yards, in climax formation, and dominated by one species. [PETR] A category of igneous rocks defined in terms of similarities in mineralogical or chemical composition.

clapboard [MATER] A board, thicker at one edge than the other, used to cover exterior walls.

Clapeyron-Clausius equation *See* Clausius-Clapeyron equation.

Clapeyron equation *See* Clausius-Clapeyron equation.

Clapeyron's theorem [MECH] The theorem that the strain energy of a deformed body is equal to one-half the sum over three perpendicular directions of the displacement component times the corresponding force component, including deforming loads and body forces, but not the six constraining forces required to hold the body in equilibrium.

clapper [ELEC] A hinged or pivoted relay armature.

clapper box [MECH ENG] A hinged device that permits a reciprocating cutting tool (as in a planer or shaper) to clear the work on the return stroke.

Clapp oscillator [ELECTR] A series-tuned Colpitts oscillator, having low drift.

clarain [GEOL] A coal lithotype appearing as stratifications parallel to the bedding plane and usually having a silky luster and scattered or diffuse reflection. Also known as clarite.

Clarendonian [GEOL] Lower Pliocene or upper Miocene geologic time.

clarified oil [MATER] The heavy oil which is taken from the bottom of a fractionator in a catalytic cracking process and from which residual catalyst has been removed.

clarifier [ENG] A device for filtering a liquid.

clarifying agents [FOOD ENG] Tannin, gelatin, albumin, methylcellulose, pectinases, and proteinases used to remove turbidity from fruit juices, vinegar, wine, beer, and soft drinks.

clarifying centrifuge [MECH ENG] A device that clears liquid of foreign matter by centrifugation.

clarifying filter [ENG] Any filter, such as a sand filter or a cartridge filter, used to purify liquids with a low solid-liquid ratio; in some instances color may be removed as well.

Clariidae [VERT ZOO] A family of Asian and African catfishes in the suborder Siluroidei.

clarinite [MINERAL] A heterogeneous, generally translucent material making up the major micropetrological ingredient of clarain.

clarite *See* clarain.

clarity [CHEM ENG] Measure of the amount of opaque suspended solids in a liquid, determined by visual or optical methods.

Clark cell [ELEC] An early form of standard cell, having 1.433 volts at 15°C, now largely replaced by the Weston standard cell as a voltage standard.

Clark degree *See* English degree.

clarke [GEOCHEM] A unit of the average abundance of an element in the earth's crust, expressed as a percentage. Also known as crustal abundance.

Clarkecarididae [PALEON] A family of extinct crustaceans in the order Anaspidacea.

Clarke ellipsoid of 1866 [GEOD] The reference ellipsoid adopted by the U.S. Coast and Geodetic Survey of 1880 for charting North America.

Clarke's soap solution [MATER] Reagent used in standard APHA (American Pharmaceutical Association) method to estimate hardness in water; consists of powdered castile soap in 80% ethyl alcohol solution.

Clark process [CHEM ENG] Softening of water by adding alkaline solutions of calcium hydroxide so that the acid carbonates are converted to normal carbonates.

clarodurain [GEOL] A transitional lithotype of coal composed of vitrinite and other macerals, principally micrinite and exinite.

clarofusain [GEOL] A transitional lithotype of coal composed of fusinite and vitrinite and other macerals.

clarovitrain [GEOL] A transitional lithotype of coal rock composed primarily of the maceral vitrinite, with lesser amounts of other macerals.

clasp [DES ENG] A releasable catch which holds two or more objects together.

clasper [VERT ZOO] A modified pelvic fin of male elasmobranchs and holocephalians used for the transmission of sperm.

clasp lock [DES ENG] A spring lock with a self-locking feature.

clasp nut [DES ENG] A split nut that clasps a screw when closed around it.

class [SYST] A taxonomic category ranking above the order and below the phylum or division.

class A amplifier [ELECTR] An amplifier in which the grid bias and alternating grid voltages are such that anode current in a specific tube flows at all times.

class AB amplifier [ELECTR] An amplifier in which the grid bias and alternating grid voltages are such that anode current in a specific tube flows for appreciably more than half but less than the entire electric cycle.

class A modulator [ELECTR] A class A amplifier used to supply the necessary signal power to modulate a carrier.

class A push-pull sound track [ENG ACOUS] Two single photographic sound tracks side by side, the transmission of

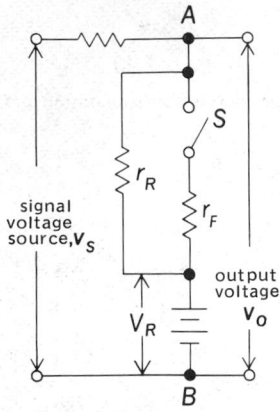
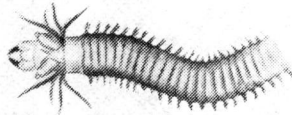
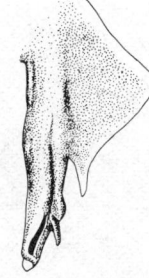

one being 180° out of phase with the transmission of the other; both positive and negative halves of the sound wave are linearly recorded on each of the two tracks.

class B amplifier [ELECTR] An amplifier in which the grid bias is approximately equal to the cutoff value, so that anode current is approximately zero when no exciting grid voltage is applied, and flows for approximately half of each cycle when an alternating grid voltage is applied.

class B auxiliary power [ELEC] Standby power plant to cover extended outages (days) of primary power.

class B modulator [ELECTR] A class B amplifier used to supply the necessary signal power to modulate a carrier; usually connected in push-pull.

class B push-pull sound track [ENG ACOUS] Two photographic sound tracks side by side, one of which carries the positive half of the signal only, and the other the negative half; during the inoperative half-cycle, each track transmits little or no light.

class C amplifier [ELECTR] An amplifier in which the bias on the control element is appreciably greater than the cutoff valve, so that the output current in each device is zero when no alternating control signal is applied, and flows for appreciably less than half of each cycle when an alternating control signal is applied.

class C auxiliary power [ELEC] Quick start (10–60 seconds) power unit to cover short-term outages (hours) of primary power.

class D auxiliary power [ELEC] Uninterruptible (no-break) power unit using stored energy to provide continuous power within specified voltage and frequency tolerances.

classical approximation [QUANT MECH] The approximation that Planck's constant may be considered infinitely small; the laws of quantum mechanics must then reduce to those of classical mechanics.

classical conductivity theory [STAT MECH] A theory which treats the system of electrons in a metal as a gas and uses the Boltzmann transport equation to calculate conductivity.

classical electron radius [ELECTROMAG] The quantity $e^2/m_e c^2$, where e is the electron's charge in electrostatic units, m_e its mass, and c the speed of light; equal to approximately 2.82×10^{-13} centimeter.

classical field theory [PHYS] The study of distributions of energy, matter, and other physical quantities under circumstances where their discrete nature is unimportant, and they may be regarded as (in general, complex) continuous functions of position. Also known as c-number theory; continuum mechanics; continuum physics.

classical mechanics [MECH] Mechanics based on Newton's laws of motion.

classical wave equation See wave equation.

classic epidemic typhus [MED] An epidemic disease caused by *Rickettsia prowazeki* var. *prowazekii*, and characterized by violent headache, a rash, neurological symptoms, and high fever. Also known as epidemic typhus.

classification [ENG] **1.** Sorting out or categorizing of particles or objects by established criteria, such as size, function, or color. **2.** Stratification of a mixture of various-sized particles (that is, sand and gravel), with the larger particles migrating to the bottom position. [IND ENG] See grading. [ORD] Placing of military documents in special groups for safeguarding defense information. [SYST] A systematic arrangement of plants and animals into categories based on a definite plan, considering evolutionary, physiologic, cytogenetic, and other relationships.

classification societies [NAV ARCH] Private organizations which issue rules for the construction, equipment, and maintenance of merchant ships.

classification track [CIV ENG] A railroad track used to separate cars from a train according to destination.

classification yard [CIV ENG] A railroad yard for separating trains according to car destination.

classifier [MECH ENG] Any apparatus for separating mixtures of materials into their constituents according to size and density.

class interval [STAT] One of several convenient intervals into which the values of the variate of a frequency distribution may be grouped.

class mark [STAT] The mid-value of a class interval, or the integral value nearest the midpoint of the interval.

classons [PARTIC PHYS] Massless bosons which are quanta of the two classical fields, gravitational and electromagnetic.

clast [GEOL] An individual grain, fragment, or constituent of detrital sediment or sedimentary rock produced by physical breakdown of a larger mass.

clastation See weathering.

clastic [GEOL] Rock or sediment composed of clasts which have been transported from their place of origin, as sandstone and shale.

clastic dike [GEOL] A tabular-shaped sedimentary dike composed of clastic material and transecting the bedding of a sedimentary formation; represents invasion by extraneous material along a crack of the containing formation.

clastic ratio [GEOL] The ratio of the percentage of clastic rocks to that of nonclastic rocks in a geologic section. Also known as detrital ratio.

clastic reservoir [GEOL] An underground oil or gas trap formed in clastic limestone.

clastic sediment [GEOL] Deposits of clastic materials transported by mechanical agents. Also known as mechanical sediment.

clastic wedge [GEOL] The sediments of the exogeosyncline, derived from the tectonic landmasses of the adjoining orthogeosyncline.

clathrate [BIOL] See cancellate. [CHEM] A well-defined addition compound formed by inclusion of molecules in cavities formed by crystal lattices or present in large molecules; examples include hydroxyquinone, urea, and cyclodextrin. [PETR] Pertaining to a condition, chiefly in leucite rock, in which clear leucite crystals are surrounded by tangential leucite crystals to give the rock an appearance of a net or a section of sponge.

Clathrinida [INV ZOO] A monofamilial order of sponges in the subclass Calcinea having an asconoid structure and lacking a true dermal membrane or cortex.

Clathrinidae [INV ZOO] The single family of the order Clathrinida.

Clauberg unit [BIOL] A unit for the standardization of progesterone.

Claude process [CHEM ENG] A process of ammonia synthesis which uses high operating pressures and a train of converters. [CRYO] A method of liquefying air or other gases in stages, in which the gas is cooled by doing work in an expansion engine and then undergoing the Joule-Thomson effect as it passes through an expansion valve.

claudetite [MINERAL] As_2O_3 A mineral containing arsenic that is dimorphous with arsenolite; crystallizes in the monoclinic system.

clause [ADP] A part of a statement in the COBOL language which may describe the structure of an elementary item, give initial values to items in independent and group work areas, or redefine data previously defined by another clause.

clausius [THERMO] A unit of entropy equal to the increase in entropy associated with the absorption of 1000 international table calories of heat at a temperature of 1K, or to 4186.8 joules per kelvin.

Clausius-Clapeyron equation [THERMO] An equation governing phase transitions of a substance, $dp/dT = \Delta H/(T\Delta V)$, in which p is the pressure, T is the temperature at which the phase transition occurs, ΔH is the change in heat content (enthalpy), and ΔV is the change in volume during the transition. Also known as Clapeyron-Clausius equation; Clapeyron equation.

Clausius-Dickel column See thermogravitational column.

Clausius equation [THERMO] An equation of state for gases which applies a correction to the van der Waals equation: $\{P + (n^2a/[T(V + c)^2])\} (V - nb) = nRT$, where P is the pressure, T the temperature, V the volume of the gas, n the number of moles in the gas, R the gas constant, a depends only on temperature, b is a constant, and c is a function of a and b.

Clausius inequality [THERMO] The principle that for any

system executing a cyclical process, the integral over the cycle of the infinitesimal amount of heat transferred to the system divided by its temperature is equal to or less than zero. Also known as Clausius theorem; inequality of Clausius.

Clausius law [THERMO] The law that an ideal gas's specific heat at constant volume does not depend on the temperature.

Clausius-Mosotti equation [ELEC] An expression for the polarizability γ of an individual molecule in a medium which has the relative dielectric constant ε and has N molecules per unit volume: $\gamma = (3/4\pi N) [(\varepsilon - 1)/(\varepsilon + 2)]$ (Gaussian units).

Clausius-Mosotti-Lorentz-Lorenz equation [ELECTROMAG] The equation that results from replacing the real relative dielectric constant in the Clausius-Mosotti equation, or the real index of refraction in the Lorentz-Lorenz equation, with its complex counterpart.

Clausius number [THERMO] A dimensionless number used in the study of heat conduction in forced fluid flow, equal to $V^3 L\rho/k\Delta T$, where V is the fluid velocity, ρ is its density, L is a characteristic dimension, k is the thermal conductivity, and ΔT is the temperature difference.

Clausius range [STAT MECH] The condition in which the mean free path of molecules in a gas is much smaller than the dimensions of the container.

Clausius' statement [THERMO] A formulation of the second law of thermodynamics, stating it is not possible that, at the end of a cycle of changes, heat has been transferred from a colder to a hotter body without producing some other effect.

Clausius theorem *See* Clausius inequality.

Clausius virial theorem [STAT MECH] The theorem that in a system of particles whose positions and velocities are bounded, the total kinetic energy of the system averaged over a long period of time equals the virial of the system. Also known as virial theorem.

Claus method [CHEM ENG] Industrial method of obtaining sulfur by a partial oxidation of gaseous hydrogen sulfide in the air to give water and sulfur.

clausthalite [MINERAL] PbSe A mineral consisting of lead selenide and resembling galena in appearance; specific gravity is 7.6–8.8.

claustrophobia [PSYCH] An abnormal fear of confined spaces.

claustrum [ANAT] A thin layer of gray matter in each cerebral hemisphere between the lenticular nucleus and the island of Reil.

clava [BIOL] A club-shaped structure, as the tip on the antennae of certain insects or the fruiting body of certain fungi.

clavate [BIOL] Club-shaped. Also known as claviform.

Clavatoraceae [PALEOBOT] A group of middle Mesozoic algae belonging to the Charophyta.

Clavaxinellida [INV ZOO] An order of sponges in the class Demospongiae; members have monaxonid megascleres arranged in radial or plumose tracts.

clavicle [ANAT] A bone in the pectoral girdle of vertebrates with articulation occurring at the sternum and scapula.

claviculate [ANAT] Having a clavicle.

claviform *See* clavate.

clavus [INV ZOO] Any of several rounded or fingerlike processes, such as the club of an insect antenna or the pointed anal portion of the hemelytron in hemipteran insects.

claw [ANAT] A sharp, slender, curved nail on the toe of an animal, such as a bird. [DES ENG] A fork for removing nails or spikes. [INV ZOO] A sharp-curved process on the tip of the limb of an insect.

claw clutch [MECH ENG] A clutch consisting of claws that interlock when pushed together.

claw hammer [DES ENG] A woodworking hammer with a flat working surface and a claw to pull nails.

clay [GEOL] **1.** A natural, earthy, fine-grained material which develops plasticity when mixed with a limited amount of water; composed primarily of silica, alumina, and water, often with iron, alkalies, and alkaline earths. **2.** The fraction of an earthy material containing the smallest particles, that is, finer than 3 micrometers. [MATER] A special grade of absorbent clay used as a filtering medium in refineries for removing solids or colorizing matter from lubricating oils.

clay atmometer [ENG] An atmometer consisting of a porous

CLAVATORACEAE

250 μ

A calcified part of the unfertilized oogonia of a Clavatoraceae fossil.

porcelain container connected to a calibrated reservoir filled with distilled water; evaporation is determined by the depletion of water.

Clay Belt [GEOL] A lowland area bordering on the western and southern portions of Hudson and James bays in Canada, composed of clays and silts recently deposited in large glacial lakes during the withdrawal of the continental glaciers.

clay bit [ENG] **1.** A bit designed for use on a clay barrel. **2.** *See* mud auger.

clay brick [MATER] Brick made from diverse types of clays and used for normal constructional purposes.

Clayden effect *See* dark lightning.

clay digger [MECH ENG] A power-driven, hand-held spade for digging hard soil or soft rock.

clay gall [GEOL] A dry, curled clay shaving derived from dried, cracked mud and embedded and flattened in a sand stratum.

clay ironstone [PETR] **1.** A clayey rock containing large quantities of iron oxide, usually limonite. **2.** A clayey-looking stone occurring among carboniferous and other rocks; contains 20–30% iron.

clay loam [GEOL] Soil containing 27–40% clay, 20–45% sand, and the remaining portion silt.

clay marl [GEOL] A chalky clay, whitish with a smooth texture.

clay mineral [MINERAL] One of a group of finely crystalline, hydrous silicates with a two- or three-layer crystal structure; the major components of clay materials; the most common minerals belong to the kaolinite, montmorillonite, attapulgite, and illite groups.

claypan [GEOL] A stratum of compact, stiff, relatively impervious noncemented clay; can be worked into a soft, plastic mass if immersed in water.

clay plug [GEOL] Sediment, with a great deal of organic muck, deposited in a cutoff river meander.

clay press [ENG] A press used to remove excess water from a pottery-clay slurry.

clay refining [CHEM ENG] A treating process for vaporized gasoline or other light petroleum product; the material is passed through a bed of granular clay, and certain olefins are polymerized to gums and absorbed by the clay.

clay regeneration [CHEM ENG] Cleaning coarse-grained absorbent clays for reuse in percolation processes by deoiling them with naphtha, steaming out excess naphtha, and roasting in a stream of air to remove carbonaceous matter.

clay shale [GEOL] **1.** Shale composed wholly or chiefly of clayey material which becomes clay again on weathering. **2.** Consolidated sediment composed of up to 10% sand and having a silt to clay ratio of less than 1:2.

clay slip [MATER] A slurry of clay and water used in glazing pottery.

clay soil [GEOL] A fine-grained inorganic soil which forms hard lumps when dry and becomes sticky when wet.

claystone [GEOL] Indurated clay, consisting predominantly of fine material of which a major proportion is clay mineral.

clay vein [GEOL] A body of clay which is similar to an ore vein in form and fills a crevice in a coal seam. Also known as dirt slip.

clay wash [MATER] A light oil such as naphtha or kerosine used to clean fuller's earth after it has been used in a filter.

clay worsted [TEXT] Worsted made in a weave of three-up and three-down right-handed twill.

clean bomb [ORD] A nuclear bomb that produces relatively little radioactive fallout.

cleanout auger *See* cleanout jet auger.

cleanout jet auger [ENG] An auger equipped with water-jet orifices designed to clean out collected material inside a driven pipe or casing before taking soil samples from strata below the bottom of the casing. Also known as cleanout auger.

clean room [ENG] A room in which elaborate precautions are employed to reduce dust particles and other contaminants in the air, as required for assembly of delicate equipment.

cleanser [MATER] Any material used to remove dirt, soil, and impurities from surfaces of all kinds.

clean ship [NAV ARCH] A tanker used to carry refined light-petroleum products instead of crude oil or heavy fuel oils.

cleanup [AERO ENG] Improving the external shape and smoothness of an aircraft to reduce its drag. [ELECTR] Gradual disappearance of gases from an electron tube during operation, due to absorption by getter material or the tube structure. [MIN ENG] **1.** The collecting of all the valuable product of a given period of operation in a stamp mill or in a hydraulic or placer mine. **2.** The valuable material resulting from a cleanup.

clear [ADP] To restore a storage device, memory device, or binary stage to a prescribed state, usually that denoting zero. Also known as reset. [METEOROL] **1.** After United States weather observing practice, the state of the sky when it is cloudless or when the sky cover is less than 0.1 (to the nearest tenth). **2.** To change from a stormy or cloudy weather condition to one of no precipitation and decreased cloudiness. [NAV] In marine navigation, to leave or pass safely, as to clear port or clear a shoal. [ORD] **1.** To give a person a security clearance. **2.** To operate a gun so as to unload it or make certain no ammunition remains; to free a gun of stoppages.

clear-air turbulence [METEOROL] A meteorological phenomenon occurring in the upper troposphere and lower stratosphere, in which high-speed aircraft are subject to violent updrafts and downdrafts. Abbreviated CAT.

clearance [ENG] Unobstructed space required for occasional removal of parts of equipment. [MECH ENG] **1.** In a piston-and-cylinder mechanism, the space at the end of the cylinder when the piston is at dead-center position toward the end of the cylinder. **2.** The ratio of the volume of this space to the piston displacement during a stroke. [MIN ENG] The space between the top or side of a car and the roof or wall. [NAV] **1.** The clear space between a vessel and an object such as a navigation light, hazard to navigation, or another vessel. **2.** A specific message from air-traffic control to a pilot of an aircraft allowing him to proceed in accordance with the flight plan which the pilot had filed, or with some modification of the original plan. **3.** In the instrument landing system, the difference in the depth of modulation which is required to produce a full-scale deflection of the course deviation indicator needle in any flight sector outside the on-course sectors. [ORD] Elevation of a gun at such an angle that a projectile will not strike an obstacle between the muzzle and the target. [PETRO ENG] The annular space between down-hole drill-string equipment, such as bits, core barrels, and casing, and the walls of the borehole with the down-hole equipment centered in the hole.

clearance angle [MECH ENG] The angle between a plane containing the end surface of a cutting tool and a plane passing through the cutting edge in the direction of cutting motion.

clearance sector [NAV] In instrument-landing-system localizers, that section which extends from the front to the back course of sectors and within which the deviation indicator provides the required off-course indications.

clearance test [PATH] The use of a substance such as urea or creatinine, or an injected foreign substance such as inulin to measure renal excretory activity; the ratio of the amount of these excreted substances in two 1-hour periods contrasted with the level of these substances in the blood is calculated in terms of the amount of blood cleared of these substances in a given unit time.

clear area [ADP] In optical character recognition, any area designated to be kept free of printing or any other extraneous markings.

clear band [ADP] In character recognition, a continuous horizontal strip of blank paper which must be obtained between consecutive code lines on a source document.

clear-cell carcinoma See renal-cell carcinoma.

clear channel [COMMUN] A standard broadcast channel in which the dominant station or stations render service over wide areas; stations are cleared of objectionable interference within their primary service areas and over all or a substantial portion of their secondary.

clear-face worsted [TEXT] Closely woven worsted fabric made of twisted yarns and with the nap removed so that the weave is visible.

clear gasoline [MATER] Gasoline free from antiknock additives.

clear ice [HYD] Generally, a layer or mass of ice which is relatively transparent because of its homogeneous structure and small number and size of air pockets.

clearing-out drop [COMMUN] Drop signal associated with a cord or trunk circuit, which is operated by ringing current to attract the attention of the operator.

clear text [COMMUN] Text or language which conveys an intelligible meaning in the language in which it is written with no hidden meaning.

cleat [CIV ENG] A strip of wood, metal, or other material fastened across something to serve as a batten or to provide strength or support. [DES ENG] A fitting having two horizontally projecting horns around which a rope may be made fast. [GEOL] Vertical breakage planes in coal. Also spelled cleet.

cleavage [CRYSTAL] Splitting, or the tendency to split, along planes determined by crystal structure and always parallel to a possible face. [EMBRYO] The subdivision of activated eggs into blastomeres. [GEOL] Splitting, or the tendency to split, along parallel, closely positioned planes in rock.

cleavage banding [GEOL] A compositional banding, usually formed from incompetent material such as argillaceous rocks, that is parallel to the cleavage rather than the bedding.

cleavage crystal [CRYSTAL] A crystal fragment bounded by cleavage faces giving it a regular form.

cleavage fracture [CRYSTAL] **1.** Manner of breaking a crystalline substance along the cleavage plane. **2.** The appearance of such a broken surface.

cleavage nucleus [EMBRYO] The nucleus of a zygote formed by fusion of male and female pronuclei.

cleavage plane [CRYSTAL] Plane along which a crystalline substance may be split.

cleavelandite [MINERAL] A white, lamellar variety of albite that is almost pure $NaAlSi_3O_8$ and has a tabular habit, with individuals often showing mosaic developments and tending to occur in fan-shaped aggregates.

Clebsch-Gordan coefficient See vector coupling coefficient.

cleet See cleat.

cleft grafting [BOT] A top-grafting method in which the scion is inserted into a cleft cut into the top of the stock.

cleft lip See harelip.

cleft palate [MED] A birth defect resulting from incomplete closure of the palate during embryogenesis.

cleft weld [MET] A weld in which a V-shaped projection on one piece is joined to a V-shaped groove in the other.

cleidocranial dysostosis [MED] A congenital defect in which there is deficient formation of bone in the skull and clavicle.

cleistocarp See cleistothecium.

cleistocarpous [BOT] Of mosses, having the capsule opening irregularly without an operculum. [MYCOL] Forming or having cleistothecia.

cleistogamy [BOT] Production of small closed flowers that are self-pollinating and contain numerous seeds.

cleistothecium [MYCOL] A closed sporebearing structure in Ascomycetes; asci and spores are freed of the fruiting body by decay or desiccation. Also known as cleistocarp.

cleithrophobia [PSYCH] An abnormal fear of being locked in.

cleithrum [VERT ZOO] A bone external and adjacent to the clavicle in certain fishes, stegocephalians, and primitive reptiles.

clerestory [ARCH] The upward extension of enclosed space achieved by bringing a windowed wall up to interrupt the slope of the roof.

Cleridae [INV ZOO] The checkered beetles, a family of coleopteran insects in the superfamily Cleroidea.

Cleroidea [INV ZOO] A superfamily of coleopteran insects in the suborder Polyphaga.

Cleveland open-cup tester [ANALY CHEM] A laboratory apparatus used to determine flash point and fire point of petroleum products.

clevis [DES ENG] A U-shaped metal fitting with holes in the open ends to receive a bolt or pin; used for attaching or suspending parts. [MIN ENG] A spring hook or snap hook which, in coal mining, is used to attach the bucket to the hoisting rope. Also known as clivvy.

clevis pin [DES ENG] A fastener with a head at one end, used to join the ends of a clevis.

CLEFT GRAFTING

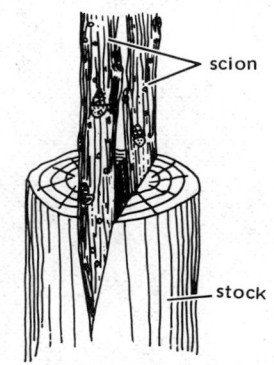

scion

stock

The components of a cleft graft. *(From H. J. Fuller and Z. B. Carothers, The Plant World, 4th ed., Holt, Rinehart, and Winston, 1963)*

CLEVIS

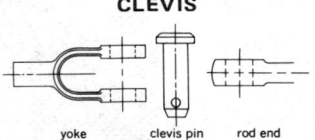

yoke   clevis pin   rod end

A clevis (yoke), with clevis pin and rod end.

cliachite [MINERAL] A group of brownish, colloidal aluminum hydroxides that constitutes most bauxite.

cliff [GEOGR] A high, steep, perpendicular or overhanging face of a rock; a precipice.

cliff of displacement *See* fault scarp.

Cliftonian [GEOL] Middle Middle Silurian geologic time.

climacophobia [PSYCH] An abnormal fear of staircases.

climacteric *See* menopause.

climagram *See* climatic diagram.

climagraph *See* climatic diagram.

climate [CLIMATOL] The long-term manifestations of weather.

climate control [CLIMATOL] Schemes for artificially altering or controlling the climate of a region. [ENG] *See* air conditioning.

climatic change [CLIMATOL] The long-term fluctuation in rainfall, temperature, and other aspects of the earth's climate.

climatic classification [CLIMATOL] The division of the earth's climates into a system of contiguous regions, each one of which is defined by relative homogeneity of the climate elements.

climatic climax [ECOL] A climax community viewed, by some authorities, as controlled by climate.

climatic controls [CLIMATOL] The relatively permanent factors which govern the general nature of the climate of a portion of the earth, including solar radiation, distribution of land and water masses, elevation and large-scale topography, and ocean currents.

climatic cycle [CLIMATOL] A long-period oscillation of climate which recurs with some regularity, but which is not strictly periodic. Also known as climatic oscillation.

climatic diagram [CLIMATOL] A graphic presentation of climatic data; generally limited to a plot of the simultaneous variations of two climatic elements, usually through an annual cycle. Also known as climagram; climagraph; climatograph; climogram; climograph.

climatic divide [CLIMATOL] A boundary between regions having different types of climate.

climatic factor [CLIMATOL] Climatic control, but regarded as including more local influences; thus city smoke and the extent of the builtup metropolitan area are climatic factors, but not climatic controls.

climatic forecast [CLIMATOL] A forecast of the future climate of a region; that is, a forecast of general weather conditions to be expected over a period of years.

climatic optimum [CLIMATOL] The period in history (about 5000–2500 B.C.) during which temperatures were warmer than at present in nearly all parts of the world.

climatic oscillation *See* climatic cycle.

climatic prediction [CLIMATOL] The estimation of the chances that specified climatic conditions will occur in a certain time interval.

climatic province [CLIMATOL] A region of the earth's surface characterized by an essentially homogeneous climate.

climatic snow line [METEOROL] The altitude above which a flat surface (fully exposed to sun, wind, and precipitation) would experience a net accumulation of snow over an extended period of time; below this altitude, ablation would predominate.

climatic zone [CLIMATOL] A belt of the earth's surface within which the climate is generally homogeneous in some respect; an elemental region of a simple climatic classification.

Climatiidae [PALEON] A family of archaic tooth-bearing fishes in the suborder Climatioidei.

Climatiiformes [PALEON] An order of extinct fishes in the class Acanthodii having two dorsal fins and large plates on the head and ventral shoulder.

Climatioidei [PALEON] A suborder of extinct fishes in the order Climatiiformes.

climatograph *See* climatic diagram.

climatography [CLIMATOL] A quantitative description of climate, particularly with reference to the tables and charts which show the characteristic values of climatic elements at a station or over an area.

climatological forecast [METEOROL] A weather forecast based upon the climate of a region instead of upon the dynamic implications of current weather, with consideration given to such synoptic weather features as cyclones and anticyclones, fronts, and the jet stream.

climatological station elevation [CLIMATOL] The elevation above mean sea level chosen as the reference datum level for all climatological records of atmospheric pressure in a given locality.

climatological station pressure [CLIMATOL] The atmospheric pressure computed for the level of the climatological station elevation, used to give all climatic records a common reference; it may or may not be the same as station pressure.

climatological substation [CLIMATOL] A weather-observing station operated (by an unpaid volunteer) for the purpose of recording climatological observations.

climatology [METEOROL] That branch of meteorology concerned with the mean physical state of the atmosphere together with its statistical variations in both space and time as reflected in the weather behavior over a period of many years.

climatopathology [MED] The study of disease in relation to the effects of the natural environment.

climatophysiology [PHYSIO] The study of the interaction of the natural environment with physiologic factors.

climatotherapy [MED] Placing a person in a suitable climate to treat a certain disease.

climax [ECOL] A mature, relatively stable community in an area, which community will undergo no further change under the prevailing climate; represents the culmination of ecological succession.

climax plant formation [ECOL] A mature, stable plant population in a climax community.

climb [AERO ENG] The gain in altitude of an aircraft.

climb cutting [MET] A milling technique in which the teeth of a cutting tool advance into the work in the same direction as the feed. Also known as climb milling; down cutting; down milling.

climbing crane [MECH ENG] A crane used on top of a high-rise construction that ascends with the building as work progresses.

climbing dune [GEOL] A dune that develops on the windward side of mountains or hills.

climbing irons [DES ENG] Spikes attached to a steel framework worn on shoes to climb wooden utility poles and trees.

climbing stem [BOT] A long, slender stem that climbs up a support or along the tops of other plants by using spines, adventitious roots, or tendrils for attachment.

climb milling *See* climb cutting.

climogram *See* climatic diagram.

climograph *See* climatic diagram.

cline [BIOL] A graded series of morphological or physiological characters exhibited by a natural group (as a species) of related organisms, generally along a line of environmental or geographic transition.

clinical genetics [GEN] The study of biological inheritance by direct observation of the living patient.

clinical pathology [PATH] A medical specialty encompassing the diagnostic study of disease by means of laboratory tests of material from the living patient.

clinical pharmacology [PHARM] The study and evaluation of the effects of drugs in man.

clinker [GEOL] Burnt or vitrified stony material, as ejected by a volcano or formed in a furnace. [MATER] An overburned brick.

clinker building [DES ENG] A method of building ships and boilers in which the edge of the wooden planks or steel plates used for the outside covering overlap the edge of the plank or plate next to it; clinched nails fasten the planks together, and rivets fasten the steel plates.

clinoamphibole [MINERAL] A group of amphiboles which crystallize in the monoclinic system.

clinoaxis [CRYSTAL] The inclined lateral axis that makes an oblique angle with the vertical axis in the monoclinic system. Also known as clinodiagonal.

clinochlore [MINERAL] $(Mg,Fe,Al)_3(Si,Al)_2O_5(OH)_4$ Green mineral of the chlorite group, occurring in monoclinic crystals, in folia or scales, or massive.

clinoclase [MINERAL] $Cu_3(AsO_4)(OH)_3$ A dark-green mineral consisting of basic copper arsenate occurring in translucent prismatic crystals or massive. Also known as clinoclasite.

clinoclasite *See* clinoclase.

clinoenstatite [MINERAL] $Mg_2(Si_2O_6)$ A monoclinic pyroxene consisting principally of magnesium silicate; occurs frequently in stony meteorites, but is rare in terrestrial environments.

clinoferrosilite [MINERAL] $Fe_2(Si_2O_6)$ A monoclinic pyroxene consisting of iron silicate.

clinoform [GEOL] A subaqueous landform, such as the continental slope of the ocean or the foreset bed of a delta.

clinograph [ENG] A type of directional surveying instrument that records photographically the direction and magnitude of deviations from the vertical of a borehole, well, or shaft; the information is obtained by the instrument in one trip into and out of the well.

clinohedral class [CRYSTAL] A rare class of crystals in the monoclinic system having a plane of symmetry but no axis of symmetry. Also known as domatic class.

clinohedrite [MINERAL] $CaZnSiO_3(OH)_2$ Colorless, white, or purplish monoclinic mineral consisting of a calcium zinc silicate occurring in crystals; hardness is 5.5 on Mohs scale, and specific gravity is 3.33.

clinohumite [MINERAL] $Mg_9(SiO_4)_4(F,OH_2)$ A monoclinic mineral of the humite group.

clinometer [ENG] A hand-held surveying device for measuring vertical angles; consists of a sighting tube surmounted by a graduated vertical arc with an attached level bubble; used in meteorology to measure cloud height at night, in conjunction with a ceiling light, and in ordnance for boresighting.

clinopinacoid [CRYSTAL] A form of monoclinic crystal whose faces are parallel to the inclined and vertical axes.

clinopyroxene [MINERAL] The general term for any of those pyroxenes that crystallize in the monoclinic system; on occasion, these pyroxenes have large amounts of calcium with or without aluminum and the alkalies. Also known as monopyroxene clinoaugite.

clinozoisite [MINERAL] $Ca_2Al_3(SiO_4)_3(OH)$ A grayish-white, pink, or green monoclinic mineral of the epidote group.

clint [GEOL] A hard or flinty rock, such as a projecting rock or ledge.

Clintonian [GEOL] Lower Middle Silurian geologic time.

clintonite [MINERAL] $Ca(Mg,Al)_3(Al,Si)O_{10}(OH)_2$ A reddish-brown, copper-red, or yellowish monoclinic mineral of the brittle mica group occurring in crystals or foliated masses. Also known as seybertite; xanthophyllite.

Clionidae [INV ZOO] The boring sponges, a family of marine sponges in the class Demospongiae.

clip [DES ENG] A device that fastens by gripping, clasping, or hooking one part to another. [ORD] *See* cartridge clip.

clip and shave [MET] Dual forging operation in which one cutter removes the flash and then another cutter shaves and sizes the piece.

clip bond [CIV ENG] A bond in which the inner edge of face brick is cut off so that bricks laid diagonal to a wall can be joined to those laid parallel to it.

clipper *See* limiter.

clipper-limiter [ELECTR] A device whose output is a function of the instantaneous input amplitude for a range of values lying between two predetermined limits but is approximately constant, at another level, for input values above the range.

clipping [COMMUN] Perceptible mutilation of signals or speech syllables during transmission. [ELECTR] *See* limiting.

clipping circuit *See* limiter.

clipping edge [MET] Area of a forging where flash is removed.

clipping level [ELECTR] The level at which a clipping circuit is adjusted; for example, the magnitude of the clipped wave shape.

clisere [ECOL] The succession of ecological communities, especially climax formations, as a consequence of intense climatic changes.

clitellum [INV ZOO] The thickened, glandular, saddlelike portion of the body wall of some annelid worms.

clitoris [ANAT] The homolog of the penis in females, located in the anterior portion of the vulva.

clivvy *See* clevis.

clo [ENG] The amount of insulation which will maintain normal skin temperature of the human body when heat production is 50 kilogram-calories per meter squared per hour, air temperature is 70°F (21°C), and the air is still.

cloaca [INV ZOO] The chamber functioning as a respiratory, excretory, and reproductive duct in certain invertebrates. [VERT ZOO] The chamber which receives the discharges of the intestine, urinary tract, and reproductive canals in monotremes, amphibians, birds, reptiles, and many fish.

cloacal bladder [VERT ZOO] A diverticulum of the cloacal wall in monotremes, amphibians, and some fish, into which urine is forced from the cloaca.

cloacal gland [VERT ZOO] Any of the sweat glands in the cloaca of lower vertebrates, as snakes or amphibians.

clock [ELECTR] A source of accurately timed pulses, used for synchronization in a digital computer or as a time base in a transmission system. [HOROL] A device for indicating the passage of time, usually containing a means for producing a regularly recurring action.

Clock *See* Horologium.

clock control system [CONT SYS] A system in which a timing device is used to generate the control function. Also known as time-controlled system.

clock frequency [ELECTR] The master frequency of the periodic pulses that schedule the operation of a digital computer.

clock method [ORD] Method of calling artillery shots by reference to the figures on an imaginary clock dial assumed to have the target at its center; thus a shot directly above the target is at 12 o'clock.

clock motor *See* timing motor.

clock oscillator [ELECTR] An oscillator that controls an electronic clock.

clock paradox [RELAT] The apparent contradiction between the principle of relativity, which asserts the equivalence of different observers, and the prediction, also part of the theory of relativity, that the clock of an observer who passes back and forth will be slower than the clock of an observer at rest. Also known as twin paradox.

clock rate [ELECTR] The rate at which bits or words are transferred from one internal element of a computer to another.

clock star [ASTRON] Any star that is used to measure time; always a bright star, whose right ascension is well known.

clock valve [ENG] A hinged valve that permits flow in one direction only.

clock watch [HOROL] A watch that strikes the hours.

clockwork [HOROL] A timing mechanism.

clog snow [HYD] A skiing term for wet, sticky, new snow.

cloister vault [ARCH] A vault resembling a pyramid with outward-curving sides.

clone [BIOL] All individuals, considered collectively, produced asexually or by parthenogenesis from a single individual.

clonorchiasis [MED] A parasitic infection of man and other fish-eating mammals caused by the trematode *Opisthorchis* (*Clonorchis*) *sinensis*, which is usually found in the bile ducts.

clonus [PHYSIO] Irregular, alternating muscular contractions and relaxations.

close [METEOROL] Colloquially, descriptive of oppressively still, warm, moist air, frequently applied to indoor conditions.

close-control radar [ENG] Ground radar used with radio to position an aircraft over a target that is normally difficult to locate or is invisible to the pilot.

close-coupled pump [MECH ENG] Pump with built-in electric motor (sometimes a steam turbine), with the motor drive and pump impeller on the same shaft.

close coupling [ELEC] **1.** The coupling obtained when the primary and secondary windings of a radio-frequency or intermediate-frequency transformer are close together. **2.** A degree of coupling that is greater than critical coupling. Also known as tight coupling.

closed aerodrome [NAV] An aerodrome at which ceiling and visibility are below prescribed minimum for ordinary landings and takeoffs.

closed-belt conveyor [MECH ENG] Solids-conveying device with zipperlike teeth that mesh to form a closed tube wrapped

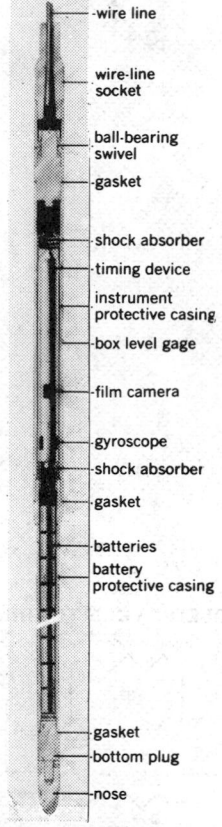

CLINOGRAPH

-wire line
wire-line socket
ball-bearing swivel
-gasket
-shock absorber
-timing device
instrument protective casing
-box level gage
-film camera
-gyroscope
-shock absorber
-gasket
-batteries
battery protective casing
-gasket
-bottom plug
-nose

Vertical section through a clinograph. *(Sperry-Sun Well Surveying Co.)*

CLINOMETER

A typical clinometer. *(Keuffel and Esser Co.)*

snugly around the conveyed material; used with fragile materials.

closed-cell foam [MATER] A cellular plastic in which there is a predominance of noninterconnecting cells.

closed circuit [COMMUN] Program source that is not broadcast for general consumption but is fed to remote monitoring units by wire. [ELEC] A complete path for current.

closed-circuit communications system [COMMUN] Certain communications systems which are entirely self-contained, and do not exchange intelligence with other facilities and systems.

closed-circuit grinder [MIN ENG] A grinder connected to a size classifier (cyclone or screen) to return oversized particles to the grinding operation in closed-circuit pulverizing.

closed-circuit pulverizing [MIN ENG] A process used in ore dressing in which the material discharged from the pulverizer is passed through an external classifier where the finished product is removed and oversized particles are returned to the pulverizer.

closed-circuit signaling [COMMUN] Signaling in which current flows in the idle condition, and a signal is initiated by increasing or decreasing the current.

closed-circuit telegraph system [COMMUN] Telegraph system in which, when no station is transmitting, the circuit is closed and current flows through the circuit.

closed-circuit television [COMMUN] Any application of television that does not involve broadcasting for public viewing; the programs can be seen only on specified receivers connected to the television camera by circuits, which include microwave relays and coaxial cables. Abbreviated CCTV.

closed cycle [THERMO] A thermodynamic cycle in which the thermodynamic fluid does not enter or leave the system, but is used over and over again.

closed-cycle fuel cell [ELEC] A fuel cell in which the reactants are regenerated by an auxiliary process, such as electrolysis.

closed-cycle reactor [NUCLEO] A nuclear reactor in which the primary coolant flows to a heat exchanger and then recirculates through the core in a completely closed circuit.

closed-cycle turbine [MECH ENG] A gas turbine in which essentially all the working medium is continuously recycled, and heat is transferred through the walls of a closed heater to the cycle.

closed die [MET] A forming or forging die in which the flow of metal is restricted to the cavity of the die set.

closed drainage [HYD] Drainage in which the surface flow of water collects in sinks or lakes having no surface outlet. Also known as blind drainage.

closed ecological system [AERO ENG] A system used in spacecraft that provides for the maintenance of life in an isolated living chamber through complete reutilization of the material available, in particular, by means of a cycle wherein exhaled carbon dioxide, urine, and other waste matter are converted chemically or by photosynthesis into oxygen, water, and food. [ECOL] A community into which a new species cannot enter due to crowding and competition.

closed fold [GEOL] A fold whose limbs have been compressed until they are parallel, and whose structure contour lines form a closed loop. Also known as tight fold.

closed frame [MIN ENG] A mine support frame that is completely closed; especially in inclined shafts, it is used to protect all sides from rock pressure.

closed graph theorem [MATH] If T is a linear transformation on a Banach space X to a Banach space Y whose domain $D(T)$ is closed and whose graph, that is, the set of pairs (x, Tx) for x in $D(T)$, is closed in $X \times Y$, then T is bounded (and hence continuous).

closed high [METEOROL] A high that may be completely encircled by an isobar or contour line.

closed linear manifold [MATH] A topologically closed vector subspace of a topological vector space.

closed loop [ADP] A loop whose execution continues indefinitely in the absence of external intervention. [CONT SYS] A family of automatic control units linked together with a process to form an endless chain; the effects of control action are constantly measured so that if the controlled quantity departs from the norm, the control units act to bring it back.

CLOSED-CYCLE TURBINE

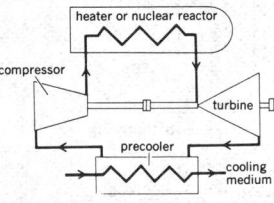

A closed-cycle gas turbine with precooler, series flow, single-shaft power plant.

closed-loop control system *See* feedback control system.

closed-loop telemetry system [ENG] **1.** A telemetry system which is also used as the display portion of a remote-control system. **2.** A system used to check out test vehicle or telemetry performance without radiation of radio-frequency energy.

closed low [METEOROL] A low that may be completely encircled by an isobar or contour line, that is, an isobar or contour line of any value, not necessarily restricted to those arbitrarily chosen for the analysis of the chart.

closed magnetic circuit [ELECTROMAG] A complete circulating path for magnetic flux around a core of ferromagnetic material.

closed operator [MATH] A linear transformation f whose domain A is contained in a normed vector space X satisfying the condition that if $\lim x_n = x$ for a sequence x_n in A, and $\lim f(x_n) = y$, then x is in A and $f(x) = y$.

closed orthonormal set *See* complete orthonormal set.

closed pair [MECH] A pair of bodies that are subject to constraints which prevent any relative motion between them.

closed pass [MET] A metal-rolling operation in which the top roll has a collar that fits a groove on the bottom roll, allowing a flash-free shape to be formed in the rolled metal.

closed reduction [MED] Reduction of fractures or dislocations by manipulation without surgical intervention.

closed respiratory gas system [ENG] A self-contained system within a sealed cabin, capsule, or spacecraft that will provide adequate oxygen for breathing, maintain adequate cabin pressure, and absorb the exhaled carbon dioxide and water vapor.

closed rotative gas lift [PETRO ENG] Oil-well control system in which high-pressure compressor gas is injected into a well to force oil fluids from the reservoir, with spent lift gas recompressed for reinjection.

closed set [MATH] A set of points which contains all its cluster points. Also known as topologically closed set.

closed shell [PHYS] An atomic or nuclear shell containing the maximum number of electrons or nucleons allowed by the Pauli exclusion principle.

closed shop [IND ENG] An establishment permitting only union members to be employed.

closed steam [ENG] Steam that flows through a heating coil or annulus so that there is no direct contact between the steam and the material being heated.

closed subroutine [ADP] A subroutine that can be stored outside the main routine and can be connected to it by linkages at one or more locations.

closed system [THERMO] A system which is isolated so that it cannot exchange matter or energy with its surroundings and can therefore attain a state of thermodynamic equilibrium.

closefile [ADP] A procedure call in time sharing which enables an ALGOL program to close a file no longer required.

close-grained [MATER] Consisting of fine, closely spaced particles, crystals, or other elements.

close-packed crystal [CRYSTAL] A crystal structure in which the lattice points are centers of spheres of equal radius arranged so that the volume of the interstices between the spheres is minimal.

closest approach [ASTRON] **1.** The event that occurs when two planets or other celestial bodies are nearest to each other as they orbit about the sun or other primary. **2.** The place or time of such an event.

close-talking microphone [ENG ACOUS] A microphone designed for use close to the mouth, so noise from more distant points is suppressed. Also known as noise-canceling microphone.

close-tolerance forging [MET] Forging in which draft angles are on the order of 1-3°, tolerances are less than half of those for commercial designs, and there is little or no allowance for finish.

close work [MIN ENG] Driving a tunnel or drifting between two coal seams.

closing line [MECH] The vector required to complete a polygon consisting of a set of vectors whose sum is zero (such as the forces acting on a body in equilibrium).

closing plug [ORD] A plug used to close openings of various components of a round of ammunition, that is, primer or nose of an unfused projectile.

closing rate [AERO ENG] The speed at which two aircraft or missiles come closer together.

Clostridium [MICROBIO] A genus of motile, spore-forming, anaerobic, rod-shaped bacteria of the family Bacillaceae commonly found in soil.

Clostridium botulinum [MICROBIO] A species of human pathogenic bacteria that produce an exotoxin, botulinin, which causes botulism.

Clostridium chauvoei [MICROBIO] A species of pathogenic bacteria that produce an exotoxin causing black leg in cattle.

Clostridium perfringens [MICROBIO] A gas-producing species of bacteria that produce several toxins and are the principal cause of gas gangrene in humans. Also known as *Clostridium welchii*.

Clostridium tetani [MICROBIO] A species of pathogenic soil bacteria which cause tetanus in humans.

Clostridium welchii *See* Clostridium perfringens.

closure [GEOL] The vertical distance between the highest and lowest point on an anticline enclosed by contour lines. [MATH] The union of a set and its cluster points; the smallest closed set containing the set.

clot [PHYSIO] A semisolid coagulum of blood or lymph.

cloth [TEXT] 1. A sheet of fibers assembled by weaving, knitting, felting, or some other similar process. 2. A nonfibrous material of similar properties.

cloth beam [TEXT] One of five essential parts of a loom, upon which the newly constructed fabric is wound.

cloth count *See* thread count.

clothing monitor [NUCLEO] An instrument designed for monitoring radioactive contamination on clothing.

clothoid *See* Cornu's spiral.

cloth print [GRAPHICS] A photo reproduction image on a woven fiber base.

cloth wheel [DES ENG] A polishing wheel made of sections of cloth glued or sewn together.

clot retraction time [PATH] The length of time required for the appearance or completion of the contraction or shrinkage of a blood clot, resulting in the extrusion of serum.

clotting time [PHYSIO] The length of time required for shed blood to coagulate under standard conditions. Also known as coagulation time.

cloud [METEOROL] Suspensions of minute water droplets or ice crystals produced by the condensation of water vapor. [NUC PHYS] The nucleons that are in the nucleus of an atom but not in closed shells. [SCI TECH] Any suspension of particulate matter, such as dust or smoke, dense enough to be seen.

cloud absorption [GEOPHYS] The absorption of electromagnetic radiation by the waterdrops and water vapor within a cloud.

cloudage *See* cloud cover.

cloud attenuation [ELECTROMAG] The attenuation of microwave radiation by clouds (for the centimeter-wavelength band, clouds produce Rayleigh scattering); due largely to scattering, rather than absorption, for both ice and water clouds.

cloud band [METEOROL] A broad band of clouds, about 10 to 100 or more miles (16 to 161 kilometers) wide, and varying in length from a few tens of miles to hundreds of miles.

cloud bank [METEOROL] A fairly well-defined mass of cloud observed at a distance; covers an appreciable portion of the horizon sky, but does not extend overhead.

cloud banner *See* banner cloud.

cloud bar [METEOROL] 1. A heavy bank of clouds that appears on the horizon with the approach of an intense tropical cyclone (hurricane or typhoon); it is the outer edge of the central cloud mass of the storm. 2. Any long, narrow, unbroken line of cloud, such as a crest cloud or an element of billow cloud.

cloud base [METEOROL] For a given cloud or cloud layer, that lowest level in the atmosphere at which the air contains a perceptible quantity of cloud particles.

cloudburst [METEOROL] In popular terminology, any sudden and heavy fall of rain, usually of the shower type, and with a fall rate equal to or greater than 100 millimeters (3.94 inches) per hour. Also known as rain gush; rain gust.

cloudburst hardness test [MET] A procedure in which a shower of steel balls, dropped from a predetermined height, dulls the surface of a hardened part in proportion to its softness and thus reveals defective areas.

cloudburst treatment [MET] Cold-working the surface of a metal by impingement of an avalanche of metal shot; a form of shot-peening.

cloud chamber [NUCLEO] A particle detector in which the path of a charged particle is made visible by the formation of liquid droplets along the trail of ions left by the particle as it passes through the gas of the chamber. Also known as expansion chamber; fog chamber.

cloud classification [METEOROL] 1. A scheme of distinguishing and grouping clouds according to their appearance and, where possible, to their process of formation. 2. A scheme of classifying clouds according to their altitudes: high, middle, or low clouds. 3. A scheme of classifying clouds according to their particulate composition: water clouds, ice-crystal clouds, or mixed clouds.

cloud column [NUCLEO] The column of smoke extending upward from the point of burst of an atomic weapon.

cloud cover [METEOROL] That portion of the sky cover which is attributed to clouds, usually measured in tenths of sky covered. Also known as cloudage; cloudiness.

cloud crest *See* crest cloud.

cloud deck [METEOROL] The upper surface of a cloud.

cloud-detection radar [ENG] A type of weather radar designed specifically for the detection of clouds (rather than precipitation).

cloud discharge [GEOPHYS] A lightning discharge occurring between a positive charge center and a negative charge center, both of which lie in the same cloud. Also known as cloud flash; intracloud discharge.

cloud droplet [METEOROL] A particle of liquid water from a few micrometers to tens of micrometers in diameter, formed by condensation of atmospheric water vapor and suspended in the atmosphere with other drops to form a cloud.

cloud-drop sampler [ENG] An instrument for collecting cloud particles, consisting of a sampling plate or cylinder and a shutter, which is so arranged that the sampling surface is exposed to the cloud for a predetermined length of time; the sampling surface is covered with a material which either captures the cloud particles or leaves an impression characteristic of the impinging elements.

cloud echo [METEOROL] The radar target signal returned from clouds alone, as detected by cloud detection radars or other very-short-wavelength equipment.

cloud flash *See* cloud discharge.

cloud forest *See* temperate rainforest.

cloud formation [METEOROL] 1. The process by which various types of clouds are formed, generally involving adiabatic cooling of ascending moist air. 2. A particular arrangement of clouds in the sky, or a striking development of a particular cloud.

cloud height [METEOROL] The absolute altitude of the base of a cloud.

cloud height indicator [ENG] General term for an instrument which measures the height of cloud bases.

cloudiness *See* cloud cover.

cloud layer [METEOROL] An array of clouds, not necessarily all of the same type, whose bases are at approximately the same level; may be either continuous or composed of detached elements.

cloud level [METEOROL] 1. A layer in the atmosphere in which are found certain cloud genera; three levels are usually defined: high, middle, and low. 2. At a particular time, the layer in the atmosphere bounded by the limits of the bases and tops of an existing cloud form.

cloud mirror *See* mirror nephoscope.

cloud modification [METEOROL] Any process by which the natural course of development of a cloud is altered by artificial means.

cloud particle [METEOROL] A particle of water, either a drop of liquid water or an ice crystal, comprising a cloud.

cloud-phase chart [METEOROL] A chart designed to indicate

CLOUD

Cumulonimbus clouds photographed over the upland adjoining the upper Colorado River Valley. Note the rain showers which appear under some of the clouds. (*Lt. B. H. Wyatt, U.S.N., U.S. Weather Bureau*)

CLOUD CHAMBER

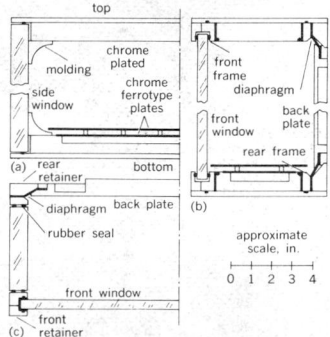

Cloud chamber designed for use in a magnetic field. (*a*) Vertical section parallel to front. (*b*) Vertical section parallel to side. (*c*) Horizontal section. The back plate moves to produce the expansion. The mechanism for compressing the chamber and producing the expansion is not shown.

and distinguish supercooled water clouds from ice-crystal clouds.

cloud physics [METEOROL] The study of the physical and dynamical processes governing the structure and development of clouds and the release from them of snow, rain, and hail.

cloud point [CHEM ENG] The temperature at which paraffin wax or other solid substance begins to separate from a solution of petroleum oil; a cloudy appearance is seen in the oil at this point.

cloud pulse [ELECTR] The output resulting from space charge effects produced by turning the electron beam on or off in a charge-storage tube.

cloud seeding [METEOROL] Any technique carried out with the intent of adding to a cloud certain particles that will alter its natural development.

cloud shield [METEOROL] The principal cloud structure of a typical wave cyclone, that is, the cloud forms found on the cold-air side of the frontal system.

cloud street [METEOROL] A line of cumuliform clouds frequently one cumulus element wide, but ranging upward in width so that it is sometimes difficult to differentiate between streets and bands.

clouds with vertical development [METEOROL] Types of clouds characterized by great vertical development, the principal types being cumulus and cumulonimbus. Also known as heap clouds (British usage).

cloud symbol [METEOROL] One of a set of specified ideograms that represent the various cloud types of greatest significance or those most commonly observed, and entered on a weather map as part of a station model.

cloud system [METEOROL] An array of clouds and precipitation associated with a cyclonic-scale feature of atmospheric circulation, and displaying typical patterns and continuity. Also known as nephsystem.

cloud test [CHEM ENG] An ASTM method for determining the cloud point of petroleum oil.

cloud-to-cloud discharge [GEOPHYS] A lightning discharge occurring between a positive charge center of one cloud and a negative charge center of a second cloud. Also known as intercloud discharge.

cloud-to-ground discharge [GEOPHYS] A lightning discharge occurring between a charge center (usually negative) in the cloud and a center of opposite charge at the ground. Also known as ground discharge.

cloud top [METEOROL] The highest level in the atmosphere at which the air contains a perceptible quantity of cloud particles for a given cloud or cloud layer.

cloud track [NUCLEO] The string of minute water droplets that forms along the path of an ionizing particle in the supersaturated vapor of a cloud chamber.

cloudy [METEOROL] The character of a day's weather when the average cloudiness, as determined from frequent observations, is more than 0.7 for the 24-hour period.

cloudy crystal-ball model [NUC PHYS] An optical analogy used in explaining scattering of nucleons by nuclei, in which the nucleus is thought of as a sphere of nuclear matter which partially refracts and partially absorbs the incident nucleon (de Broglie) wave. Also known as optical model.

cloudy swelling [PATH] A retrogressive change in the cytoplasm of parenchymatous cells, whereby the cell enlarges and its outline becomes irregular, with resultant swelling of the organ.

clough [GEOGR] A cleft in a hill; a ravine or narrow valley.

clout nail [DES ENG] A nail with a large, thin, flat head used in building.

clove [BOT] **1.** The unopened flower bud of a small, conical, symmetrical evergreen tree, *Eugenia caryophyllata,* of the myrtle family (Myrtaceae); the dried buds are used as a pungent, strongly aromatic spice. **2.** A small bulb developed within a larger bulb, as in garlic.

clove oil *See* oil of cloves.

clover [BOT] **1.** A common name designating the true clovers, sweet clovers, and other members of the Leguminosa. **2.** A herb of the genus *Trifolium.*

cloverleaf [CIV ENG] A highway intersection resembling a clover leaf and designed to allow movement and interchange of traffic without direct crossings and left turns.

cloverleaf antenna [ELECTROMAG] Antenna having radiating units shaped like a four-leaf clover.

CLS *See* characteristic loss spectroscopy.

clubfoot [MED] Congenital malpositioning of a foot such that the forefoot is inverted and rotated with a shortened Achilles tendon.

club fungi [MYCOL] The common name for members of the class Basidiomycetes.

club moss [BOT] The common name for members of the class Lycopodiatae.

clubroot [PL PATH] A disease principally of crucifers, such as cabbage, caused by the slime mold *Plasmodiophora brassicae* in which roots become enlarged and deformed, leading to plant death.

clupanodonic acid *See* docosatrienynoic acid.

Clupeidae [VERT ZOO] The herrings, a family of fishes in the suborder Clupoidea composing the most primitive group of higher bony fishes.

Clupeiformes [VERT ZOO] An order of teleost fishes in the subclass Actinopterygii, generally having a silvery, compressed body.

clupeine [BIOCHEM] A protamine found in salmon sperm, mainly composed of arginine (74.1%) and small percentages of threonine, serine, proline, alanine, valine, and isoleucine.

Clupoidea [VERT ZOO] A suborder of fishes in the order Clupeiformes comprising the herrings and anchovies.

clusec [MECH ENG] A unit of power used to measure the power of evacuation of a vacuum pump, equal to the power associated with a leak rate of 1 centiliter per second at a pressure of 1 millitorr, or to approximately 1.33322×10^{-6} watt.

Clusiidae [INV ZOO] A family of myodarian cyclorrhaphous dipteran insects in the subsection Acalypteratae.

cluster [ENG] **1.** A pyrotechnic signal consisting of a group of stars or fireballs. **2.** A grouping of rocket motors fastened together. [ORD] A collection of small bombs held together by an adapter for dropping.

cluster expansion [STAT MECH] A virial expansion in which the virial coefficients (of inverse powers of the volume of the gas in question) are obtained from integrals, over positions of a small number of molecules, of functions involving intermolecular potentials.

clusterite *See* botryoid.

cluster mill [MET] A rolling mill in which small-diameter rolls are supported by larger rolls.

cluster of stars [ASTRON] A number of stars moving in the same direction with nearly the same speed.

cluster point [MATH] A cluster point of a set in a topological space is a point *p* whose neighborhoods all contain at least one point of the set other than *p*. Also known as accumulation point; limit point.

clutch [MECH ENG] A machine element for the connection and disconnection of shafts in equipment drives, especially while running. [VERT ZOO] A nest of eggs or a brood of chicks.

clutch point [ADP] The moment in time at which the clutch is engaged in a peripheral device such as a card punch.

clutter [ELECTROMAG] Unwanted echoes on a radar screen, such as those caused by the ground, sea, rain, stationary objects, chaff, enemy jamming transmissions, and grass. Also known as background return; radar clutter.

clutter gating [ELECTR] A technique which provides switching between moving-target-indicator and normal videos; this results in normal video being displayed in regions with no clutter and moving-target-indicator video being switched in only for the clutter areas.

Clypeasteroida [INV ZOO] An order of exocyclic Euechinoidea having a monobasal apical system in which all the genital plates fuse together.

clypeus [INV ZOO] An anterior medial plate on the head of an insect, commonly bearing the labrum on its anterior margin. [MYCOL] A disk of black tissue about the mouth of the perithecia in certain ascomycetes.

clysis [MED] **1.** Administration of an enema. **2.** Subcutaneous or intravenous administration of fluids.

Clythiidae [INV ZOO] The flat-footed flies, a family of cyclorrhaphous dipteran insects in the series Aschiza characterized by a flattened distal end on the hind tarsus.

CLOVE

Closed flower buds (cloves) and open buds on a branch of the evergreen tree *Eugenia caryophyllata.*

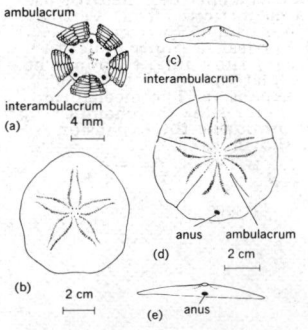

CLYPEASTEROIDA

Typical clypeasteroids. *Laganum,* (*a*) monobasal apical system, (*b*) aboral aspect, and (*c*) posterior aspect. *Arachnoides,* (*d*) aboral aspect, and (*e*) posterior aspect.

cm *See* centimeter.

Cm *See* curium.

cm³ *See* cubic centimeter.

CMa *See* Canis Major.

cmHg *See* centimeter of mercury.

CMi *See* Canis Minor.

CMOS device [ELECTR] A device formed by the combination of a PMOS (*p*-type channel metal-oxide semiconductor device) with an NMOS (*n*-type channel semiconductor device). Derived from complementary metal-oxide semiconductor device.

CM Tauri [ASTRON] A supernova observed by the Chinese and Japanese in 1054; remnants are still seen as spectacular nebulae.

C network [ELECTR] Network composed of three impedance branches in series, the free ends being connected to one pair of terminals, and the junction points being connected to another pair of terminals.

C neutron [NUCLEO] A neutron of such energy, up to about 0.3 electron-volt, that it is strongly absorbable in cadmium.

Cnidaria [INV ZOO] The equivalent name for Coelenterata.

cnidoblast [INV ZOO] A cell that produces nematocysts. Also known as nettle cell; stinging cell.

cnidocil [INV ZOO] The trigger on a cnidoblast that activates discharge of the nematocyst when touched.

cnidophore [INV ZOO] A modified structure bearing nematocysts in certain coelenterates.

Cnidospora [INV ZOO] A subphylum of spore-producing protozoans that are parasites in cells and tissues of invertebrates, fishes, a few amphibians, and turtles.

c-number theory *See* classical field theory.

Co *See* cobalt.

Co⁶⁰ *See* cobalt-60.

CoA *See* coenzyme A.

coacervate [CHEM] An aggregate of colloidal droplets bound together by the force of electrostatic attraction.

coacervation [CHEM] The separation, by addition of a third component, of an aqueous solution of a macromolecule colloid (polymer) into two liquid phases, one of which is colloid-rich (the coacervate) and the other an aqueous solution of the coacervating agent (the equilibrium liquid).

coach screw [DES ENG] A large, square-headed, wooden screw used to join heavy timbers. Also known as lag bolt; lag screw.

coagulability test [PATH] Any of several clinical tests of the ability of blood to coagulate, such as clot retraction time and quantification, prothrombin time, partial thromboplastin time, and platelet enumeration.

coagulant [CHEM] An agent that causes coagulation.

coagulase [BIOCHEM] Any enzyme that causes coagulation of blood plasma.

coagulation [CHEM] A separation or precipitation from a dispersed state of suspensoid particles resulting from their growth; may result from prolonged heating, addition of an electrolyte, or from a condensation reaction between solute and solvent; an example is the setting of a gel.

coagulation time *See* clotting time.

coal [GEOL] The natural, rocklike, brown to black derivative of forest-type plant material, usually accumulated in peat beds and progressively compressed and indurated until it is finally altered into graphite or graphite-like material.

coal auger [MIN ENG] A type of continuous miner which consists of a screw drill of large diameter and which cuts, transports, and loads the coal.

coal ball [GEOL] A subspherical mass containing mineral matter embedded with plant material, found in coal seams and overlying beds of the late Paleozoic.

coal bank [MIN ENG] A seam of coal that is exposed.

coal barrier [MIN ENG] A protective pillar composed of coal.

coal bed [GEOL] A seam or stratum of coal parallel to the rock stratification. Also known as coal rake; coal seam.

coal blasting [MIN ENG] Breaking coal with explosives.

coalbreaker *See* breaker.

coal breccia [GEOL] Angular fragments of coal within a coal bed.

coal chemicals [MATER] Chemicals obtained as by-products in the primary processing of coal to metallurgical coke; the

main source of aromatic compounds used as intermediates in the synthesis of dyes, drugs, antiseptics, and solvents.

coal clay *See* underclay.

coal cutter [MIN ENG] A power-operated machine which cuts out a thin strip of coal from the bottom of the seam; it draws itself by rope haulage along the coal face.

coal digger *See* faceman.

coal drill [MIN ENG] Usually, an electric drill of a compact, light design; however, also a light pneumatic drill.

coal dust [MIN ENG] A finely divided coal, sometimes defined as coal that will pass through 100-mesh screens (100 wires to the inch).

coalesce [SCI TECH] To come together to form a whole.

coalesced copper [MET] A mass, oxygen-free copper made by compacting and sintering cathode copper at high pressure and temperature.

coalescence [METEOROL] In cloud physics, the merging of two or more water drops into a single larger drop. [PHYS] The uniting by growth in one body, as particles, gas, or a liquid.

coalescence efficiency [METEOROL] The fraction of all collisions which occur between waterdrops of a specified size and which result in actual merging of two drops into a single larger drop.

coalescence process [METEOROL] The growth of raindrops by the collision and coalescence of cloud drops or small precipitation particles.

coalescent [CHEM] Chemical additive used in immiscible liquid-liquid mixtures to cause small droplets of the suspended liquid to unite, preparatory to removal from the carrier liquid.

coalescent pack [CHEM ENG] High-surface-area packing to consolidate liquid droplets for gravity separation from a second phase (for example, gas or immiscible liquid); packing must be wettable by the droplet phase; Berl saddles, Raschig rings, knitted wire mesh, excelsior, and similar materials are used.

coalescer [CHEM ENG] Mechanical process vessel with wettable, high-surface area packing on which liquid droplets consolidate for gravity separation from a second phase (for example, gas or immiscible liquid).

coal face [MIN ENG] The mining face from which coal is extracted.

coalfield [MIN ENG] A region containing coal deposits.

coal gas [MATER] **1.** Flammable gas derived from coal either naturally in place, or by induced methods of industrial plants and underground gasification. **2.** Specifically, fuel gas obtained from carbonization of coal.

coal gasification [CHEM ENG] The conversion of coal, char, or coke to a gaseous product by reaction with air, oxygen, steam, carbon dioxide, or mixtures of these.

coal getter *See* faceman.

coal hydrogenation *See* Bergius process.

coalification [GEOL] Formation of coal from plant material by the processes of diagenesis and metamorphism. Also known as bituminization; carbonification; incarbonization; incoalation.

coal-in-oil suspension [MATER] A fluid mixture of pulverized coal dispersed in either fuel oil or coal tar oil, used as a fuel chiefly in large installations. Also known as colloidal fuel.

coal liquefaction [CHEM ENG] The process of preparing a liquid mixture of hydrocarbons by destructive distillation of coal.

Coal Measures [GEOL] The sequence of rocks typically containing coal of the Upper Carboniferous.

coal mining [MIN ENG] The technical and mechanical job of removing coal from the earth and preparing it for market.

coal oil [MATER] **1.** Condensed liquid from coal distillation. **2.** An archaic term for kerosine made from petroleum.

coal paleobotany [PALEOBOT] A branch of the paleobotanical sciences concerned with the origin, composition, mode of occurrence, and significance of fossil plant materials that occur in or are associated with coal seams.

coal pebbles [GEOL] Rounded masses of coal occurring in sedimentary rock.

coal petrology [GEOL] The science that deals with the origin,

COAL AUGER

View from spoil bank showing operation of auger miner and coal being loaded into truck in the Pittsburgh coal bed, West Virginia. *(U.S. Bureau of Mines)*

COAL BALL

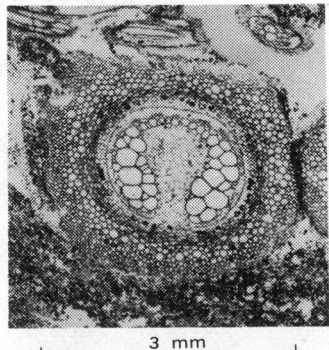

Photomicrograph of the peel made from a coal ball containing the branch of a fern.

history, occurrence, structure, chemical composition, and classification of coal.

coal planer [MIN ENG] A type of continuous coal mining machine for longwall mining; consists of a heavy steel plow with cutting knives, with power equipment to drag it back and forth across a coal face.

coal plough [MIN ENG] A device with steel blades which shears off coal and pushes it onto the face conveyor.

coal rake *See* coal bed.

Coalsack [ASTRON] An area in one of the brighter regions of the Southern Milky Way which to the naked eye appears entirely devoid of stars and hence dark with respect to the surrounding Milky Way region.

coal seam *See* coal bed.

coal-sensing probe [MIN ENG] A nucleonic instrument to measure the thickness of coal left in the seam floor by means of a gamma-ray backscattering unit.

coal sizes [MIN ENG] The sizes by which anthracite coal is marketed.

coal split *See* split.

coal tar [MATER] A tar obtained from carbonization of coal, usually in coke ovens or retorts, containing several hundred organic chemicals.

coal tar dye [ORG CHEM] Dye made from a coal tar hydrocarbon or a derivative such as benzene, toluene, xylene, naphthalene, or aniline.

coal tar enamel [MATER] Coal tar used as a paintlike coating for petroleum-product pipelines; provides protection from both water and cathodic corrosion.

coal tar pitch [MATER] Dark-brown to black amorphous residue from the redistillation of coal tar; melts at 150°F (66°C); used as a thermoplastic.

co-altitude *See* zenith distance.

coaming [NAV ARCH] A rim placed on a roof or around a hatch, deck, or bulkhead opening to stop water from entering.

Coanda effect [FL MECH] The tendency of a gas or liquid coming out of a jet to travel close to the wall contour even if the wall's direction of curvature is away from the jet's axis; a factor in the operation of a fluidic element.

coarse *See* low-grade.

coarse aggregate [MATER] Crushed stone or gravel used in concrete; will not, when dry, pass through a sieve with ¼-inch-diameter holes.

coarse-grained [MATER] Having a coarse texture. [PETR] *See* phaneritic.

coarse roll [MIN ENG] A large roll for the preliminary crushing of large pieces of ore, rock, or coal.

coarse setting [ORD] Preliminary adjustment of a sight in laying a gun, made first on the main scale; then the fine setting is made on the associated scale of smaller graduations.

coarse sighting [ORD] Adjustment of the sight of a gun so that a part of the front sight is seen through the notch in the rear sight.

coast [ENG] A memory feature on a radar which, when activated, causes the range and angle systems to continue to move in the same direction and at the same speed as that required to track an original target. [GEOGR] The general region of indefinite width that extends from the sea inland to the first major change in terrain features.

coastal berm *See* berm.

coastal dune [GEOL] A mobile mound of windblown material found along many sea and lake shores.

coastal engineering [CIV ENG] A branch of civil engineering pertaining to the study of the action of the seas on shorelines and to the design of structures to protect against this action.

coastal ice *See* fast ice.

coastal landform [GEOGR] The characteristic features and patterns of land in a coastal zone subject to marine and subaerial processes of erosion and deposition.

coastal plain [GEOGR] Broad, low-level plain between a mountain range and a seashore.

coastal refraction [ELECTROMAG] An apparent change in the direction of travel of a radio wave when it crosses a shoreline obliquely. Also known as land effect.

coastal sediment [GEOL] The mineral and organic deposits of deltas, lagoons, and bays, barrier islands and beaches, and the surf zone.

coast chart [NAV] A nautical chart for use in inshore, coastwise navigation when a course carries a vessel inside outlying reefs and shoals, for use in entering or leaving bays and harbors of considerable size, or for use in navigating larger inland waterways.

coaster [NAV ARCH] A small merchant ship, about 200 feet (61 meters) long, which operates near coasts, in rivers and estuaries, and on short ocean passages.

coast guard [ORD] A naval force which guards a coast and ensures the order, safety, and effective operation of traffic on the coastal waters.

coast guard cutter [NAV ARCH] A small, armed boat in a coast guard.

Coast Guard lines [NAV] Lines established by the U.S. Coast Guard for separating areas of the sea where the inland rules of the road apply, from those areas where the international rules apply.

Coast Guard station [NAV] In American usage, any building on the coast used to house personnel and equipment for saving life at sea. Also known as life-saving station.

coast ice *See* fast ice.

coasting [NAV] Proceeding approximately parallel to a coastline and near enough to be in pilot waters most of the time.

coasting flight [AERO ENG] The flight of a rocket between burnout or thrust cutoff of one stage and ignition of another, or between burnout and summit altitude or maximum horizontal range.

coastline [GEOGR] 1. The line that forms the boundary between the shore and the coast. 2. The line that forms the boundary between the water and the land.

coastlining [MAP] The process of obtaining data from which the coastline can be drawn on a chart.

coast pilot [NAV] A book serving as an adjunct to nautical charts, containing important information which cannot be shown conveniently on the charts, and not readily available elsewhere; prepared by the U.S. Coast and Geodetic Survey for coastal waters of continental United States, Hawaii, the Virgin Islands, and Puerto Rico; and by the U.S. Naval Oceanographic Office for foreign waters. Also known as sailing directions.

coast piloting [NAV] The directing of the movements of a vessel near a coast by means of terrestrial reference points.

coast shelf *See* submerged coastal plain.

coastwise navigation [NAV] Navigation in the vicinity of a coast, in contrast to offshore navigation at a distance from a coast.

coated abrasive [MATER] An abrasive product having the abrasive particles attached to a backing material with glue or a synthetic resin.

coated cathode [ELECTR] A cathode that has been coated with compounds to increase electron emission.

coated electrode [MET] A wire covered with metal oxides and silicates and used as a filler-metal electrode in arc welding. Also known as covered electrode.

coated filament [ELECTR] A vacuum-tube filament coated with metal oxides to provide increased electron emission.

coated lens [OPTICS] A lens whose surfaces have been coated with a thin, transparent film having an index of refraction that minimizes light loss by reflection.

coated paper [MATER] Paper with a surface coating of clay and other materials to produce a smooth, shiny surface; especially useful for fine, detailed, blur-free reproductions in color or black and white. Also known as enamel paper.

coati [VERT ZOO] The common name for three species of carnivorous mammals assigned to the raccoon family (Procyonidae) characterized by their elongated snout, body, and tail.

coating [MATER] 1. Any material that will form a continuous film over a surface. 2. The film formed by the material.

coax *See* coaxial cable.

coaxial [MECH] Sharing the same axes. [MECH ENG] Mounted on independent concentric shafts.

coaxial antenna [ELECTROMAG] An antenna consisting of a quarter-wave extension of the inner conductor of a coaxial line and a radiating sleeve that is in effect formed by folding back the outer conductor of the coaxial line for a length of approximately a quarter wavelength.

COATI

Common coati (*Nasua nasua*) with a long snout and striped bushy tail, which is held erect.

coaxial attenuator [ELECTROMAG] An attenuator that has a coaxial construction and terminations suitable for use with coaxial cable.

coaxial bolometer [ELECTR] A bolometer in which the desired square-law detection characteristic is provided by a fine Wollaston wire element that has been thoroughly cleaned before being axially located and soldered in position in its cylinder.

coaxial cable [ELECTROMAG] A transmission line in which one conductor is centered inside and insulated from an outer metal tube that serves as the second conductor. Also known as coax; coaxial line; coaxial transmission line; concentric cable; concentric line; concentric transmission line.

coaxial capacitor *See* cylindrical capacitor.

coaxial cavity [ELECTROMAG] A cylindrical resonating cavity having a central conductor in contact with its pistons or other reflecting devices.

coaxial cavity magnetron [ELECTR] A magnetron which achieves mode separation, high efficiency, stability, and ease of mechanical tuning by coupling a coaxial high Q cavity to a normal set of quarter-wavelength vane cavities.

coaxial circles [MATH] Family of circles such that any pair have the same radical axis.

coaxial connector [ELECTROMAG] An electric connector between a coaxial cable and an equipment circuit, so constructed as to maintain the conductor configuration, through the separable connection, and the characteristic impedance of the coaxial cable.

coaxial cylinder magnetron [ELECTR] A magnetron in which the cathode and anode consist of coaxial cylinders.

coaxial cylinders [MATH] Two cylinders whose cylindrical surfaces consist of the lines that pass through concentric circles in a given plane and are perpendicular to this plane.

coaxial diode [ELECTR] A diode having the same outer diameter and terminations as a coaxial cable, or otherwise designed to be inserted in a coaxial cable.

coaxial filter [ELECTROMAG] A section of coaxial line having reentrant elements that provide the inductance and capacitance of a filter section.

coaxial hybrid [ELECTROMAG] A hybrid junction of coaxial transmission lines.

coaxial isolator [ELECTROMAG] An isolator used in a coaxial cable to provide a higher loss for energy flow in one direction than in the opposite direction; all types use a permanent magnetic field in combination with ferrite and dielectric materials.

coaxial line *See* coaxial cable.

coaxial line resonator [ELECTROMAG] A resonator consisting of a length of coaxial line short-circuited at one or both ends.

coaxially fed linear array [ELECTROMAG] A beacon antenna having a uniform azimuth pattern.

coaxial machine gun [ORD] A machine gun mounted integrally with the primary gun of a tank.

coaxial relay [ELECTROMAG] A relay designed for opening or closing a coaxial cable circuit without introducing a mismatch that would cause wave reflections.

coaxial speaker [ENG ACOUS] A loudspeaker system comprising two, or less commonly three, speaker units mounted on substantially the same axis in an integrated mechanical assembly, with an acoustic-radiation-controlling structure.

coaxial stub [ELECTROMAG] A length of nondissipative cylindrical waveguide or coaxial cable branched from the side of a waveguide to produce some desired change in its characteristics.

coaxial transistor [ELECTR] A point-contact transistor in which the emitter and collector are point electrodes making pressure contact at the centers of opposite sides of a thin disk of semiconductor material serving as base.

coaxial transmission line *See* coaxial cable.

coaxing [MET] Improving fatigue strength of a metal by increasing the stress range, beginning just below the fatigue limit.

cob [MIN ENG] To chip away waste material from an ore, using hand hammers.

cobalamin *See* vitamin B_{12}.

cobalt [CHEM] A metallic element, symbol Co, atomic number 27, atomic weight 58.93; used chiefly in alloys.

cobalt-60 [NUC PHYS] A radioisotope of cobalt, symbol Co^{60}, having a mass number of 60; emits gamma rays and has many medical and industrial uses; the most commonly used isotope for encapsulated radiation sources.

cobalt-beam therapy [MED] Therapy involving the use of gamma radiation from a cobalt-60 source mounted in a cobalt bomb. Also known as cobalt therapy.

cobalt bloom *See* erythrite.

cobalt blue [CHEM] A green-blue pigment formed of alumina and cobalt oxide. Also known as cobalt ultramarine; king's blue.

cobalt bomb [NUCLEO] A quantity of cobalt-60 mounted in a housing with walls having up to 8 inches (20 centimeters) of lead for protection, and with means for removing a lead plug to release a beam of gamma rays for use in cobalt-beam therapy. [ORD] A theoretical atomic or hydrogen bomb encased in cobalt, which cobalt would be transformed into deadly radioactive dust upon detonation.

cobalt chloride *See* cobaltous chloride.

cobaltic fluoride *See* cobalt trifluoride.

cobaltite [MINERAL] CoAsS A silver-white mineral with a metallic luster that crystallizes in the isometric system, resembling crystals of pyrite; one of the chief ores of cobalt. Also known as cobalt glance; gray cobalt; white cobalt.

cobalt-molybdate desulfurization [CHEM ENG] A process for desulfurization of petroleum by using cobalt molybdate as a catalyst.

cobaltocalcite [MINERAL] A red, cobalt-bearing variety of calcite.

cobalt ocher *See* asbolite; erythrite.

cobaltomenite [MINERAL] $CoSeO_3 \cdot 2H_2O$ A mineral consisting of a hydrous cobalt selenium oxide.

cobaltous chloride [INORG CHEM] $CoCl_2$ or $CoCl_2 \cdot 6H_2O$ A compound whose anhydrous form consists of blue crystals and sublimes when heated, and whose hydrated form consists of red crystals and melts at 86.8°C; both forms are used as an absorbent for ammonia in dyes and as a catalyst. Also known as cobalt chloride.

cobaltous fluorosilicate [INORG CHEM] $CoSiF_6 \cdot H_2O$ A water-soluble, orange-red powder, used in toothpastes.

cobalt oxide [INORG CHEM] CoO A grayish brown powder that decomposes at 1935°C, insoluble in water; used as a colorant in ceramics and in manufacture of glass.

cobalt pyrites *See* linnaeite.

cobalt sulfate [INORG CHEM] Any compound of either divalent or trivalent cobalt and the sulfate group; anhydrous cobaltous sulfate, $CoSO_4$, contains divalent cobalt, has a melting point of 96.8°C, is soluble in methanol, and is utilized to prepare pigments and cobalt salts; cobaltic sulfate, $Co_2(SO_4)_3 \cdot 18H_2O$, contains trivalent cobalt, is soluble in sulfuric acid, and functions as an oxidizing agent.

cobalt therapy *See* cobalt-beam therapy.

cobalt trifluoride [INORG CHEM] CoF_3 A brownish powder that reacts with water to form a precipitate of cobaltic hydroxide; used as a fluorinating agent. Also known as cobaltic fluoride.

cobalt ultramarine *See* cobalt blue.

cobber [MIN ENG] **1.** A device used to reject waste materials from ore concentrates. **2.** A person who breaks fibers from asbestos rocks or chips low-grade material from ore.

cobble beach *See* shingle beach.

cobblestone [MATER] A rounded stone 64–256 millimeters in diameter used in construction, especially for paving.

Cobb's disease [PL PATH] A bacterial disease of sugarcane caused by *Xanthomonas vascularum* and characterized by a slime in the vascular bundles, dwarfing, streaking of leaves, and decay. Also known as sugarcane gummosis.

cobinotron [ELECTROMAG] The combination of a corbino disk and a coil arranged to produce a magnetic field perpendicular to the disk.

Cobitidae [VERT ZOO] The loaches, a family of small fishes, many eel-shaped, in the suborder Cyprinoidei, characterized by barbels around the mouth.

Coblentzian [GEOL] Upper Lower Devonian geologic time.

COBOL [ADP] A business data-processing language that can be given to a computer as a series of English statements

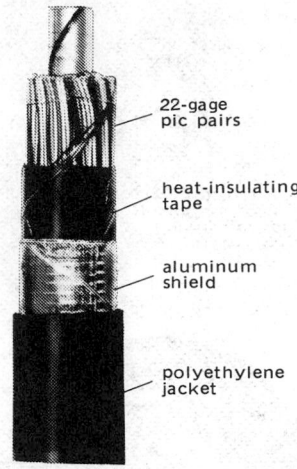

COAXIAL CABLE

22-gage pic pairs

heat-insulating tape

aluminum shield

polyethylene jacket

Cutaway view of coaxial transmission line.

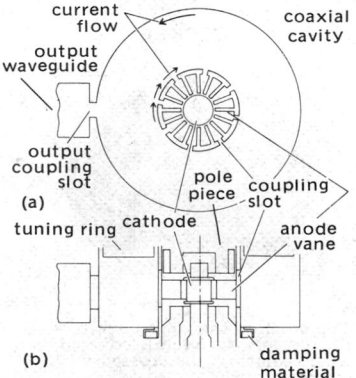

COAXIAL CAVITY MAGNETRON

current flow

coaxial cavity

output waveguide

output coupling slot

(a)

tuning ring

cathode

pole piece

coupling slot

anode vane

(b)

damping material

Coaxial cavity magnetron. (a) End view. (b) Cross-sectional view. (*After J. W. Gewartowski and H. A. Watson, Principles of Electron Tubes, Van Nostrand, 1965*)

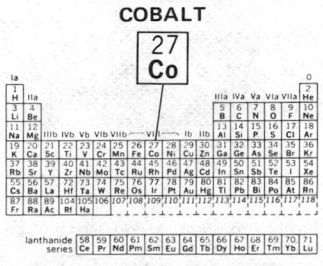

COBALT

Periodic table of the chemical elements showing the position of cobalt.

COCCINELLIDAE

A ladybird beetle. *(From T. I. Storer and R. L. Usinger, General Zoology, 3d ed., McGraw-Hill, 1957)*

COCCOSPHERE

2 μ

Scanning electron micrograph of coccosphere. *(A. McIntyre, Lamont-Doherty Geological Observatory of Columbia University)*

COCKLE

The shell of the rock cockle *(Protothaca laciniata)*, covered with radiating ribs crossed with concentric ribs.

COD

Gadus morrhua, a codfish of the northern seas.

describing a complete business operation. Derived from common business-oriented language.

coboundary [MATH] An image under the coboundary operator.

coboundary operator [MATH] If $\{C^n\}$ is a sequence of Abelian groups, coboundary operators are homomorphisms $\{\delta^n\}$ such that $\delta^n: C^n \to C^{n+1}$ and $\delta^{n+1} \circ \delta^n = 0$.

cobra [VERT ZOO] Any of several species of venomous snakes in the reptilian family Elaphidae characterized by a hoodlike expansion of skin on the anterior neck that is supported by a series of ribs.

coca [BOT] *Erythroxylon coca.* A shrub in the family Erythroxylaceae; its leaves are the source of cocaine.

cocaine [PHARM] $C_{17}H_{21}O_4N$ An alkaloid obtained from coca leaves that is used for local anesthesia and as a tonic in digestive and nervous disorders. Also known as erythroxylon; methylbenzoylecgonine.

cocarboxylase *See* thiamine pyrophosphate.

cocarcinogen [MED] A noncarcinogenic agent which augments the carcinogenic process.

Coccidia [INV ZOO] A subclass of protozoans in the class Telosporea; typically intracellular parasites of epithelial tissue in vertebrates and invertebrates.

coccidioidomycosis [MED] An infectious fungus disease of man and animals of either a pulmonary or a cutaneous nature; caused by *Coccidioides immitis.* Also known as San Joaquin Valley fever.

coccidiosis [MED] The state of or the conditions associated with being infected with coccidia.

Coccinellidae [INV ZOO] The ladybird beetles, a family of coleopteran insects in the superfamily Cucujoidea.

coccobacillus [MICROBIO] A short, thick, oval bacillus, midway between the coccus and the bacillus in appearance.

Coccoidea [INV ZOO] A superfamily of homopteran insects belonging to the Sternorrhyncha; includes scale insects and mealy bugs.

coccolith [BOT] One of the small, interlocking calcite plates covering members of the Coccolithophorida.

coccolith ooze [GEOL] A fine-grained pelagic sediment containing undissolved sand- or silt-sized particles of coccoliths mixed with amorphous clay-sized material.

Coccolithophora [INV ZOO] An order of phytoflagellates in the protozoan class Phytamastigophorea.

Coccolithophorida [BOT] A group of unicellular, biflagellate, golden-brown algae characterized by a covering of coccoliths.

Coccomyxaceae [BOT] A family of algae belonging to the Tetrasporales composed of elongate cells which reproduce only by vegetative means.

coccosphere [PALEOBOT] The fossilized remains of a member of Coccolithophorida.

Coccosteomorphi [PALEON] An aberrant lineage of the joint-necked fishes.

coccus [MICROBIO] A form of eubacteria which are more or less spherical in shape.

coccygeal body [ANAT] A small mass of vascular tissue near the tip of the coccyx.

coccygectomy [MED] Surgical excision of the coccyx.

coccyx [ANAT] The fused vestige of caudal vertebrae forming the last bone of the vertebral column in man and certain other primates.

cochain complex [MATH] A sequence of Abelian groups C^n, $-\infty < n < \infty$, together with coboundary homomorphisms $\delta^n: C^n \to C^{n+1}$ such that $\delta^{n+1} \circ \delta^n = 0$.

cochannel interference [COMMUN] Interference caused on one communication channel by a transmitter operating in the same channel.

cochineal [CHEM] A red dye made of the dried bodies of the female cochineal insect (*Coccus cacti*), found in Central America and Mexico; used as a biological stain and indicator.

cochineal solution [ANALY CHEM] An indicator in acid-base titration.

cochinilin *See* carminic acid.

cochlea [ANAT] The snail-shaped canal of the mammalian inner ear; it is divided into three channels and contains the essential organs of hearing.

cochlear duct *See* scala media.

Cochleariidae [VERT ZOO] A family of birds in the order Ciconiiformes composed of a single species, the boatbill.

cochlear nerve [ANAT] A sensory branch of the auditory nerve which receives impulses from the organ of Corti.

cochlear nucleus [ANAT] One of the two nuclear masses in which the fibers of the cochlear nerve terminate; located ventrad and dorsad to the inferior cerebellar peduncle.

cochleate [BIOL] Spiral; shaped like a snail shell.

Cochliodontidae [PALEON] A family of extinct chondrichthian fishes in the order Bradyodonti.

cock [ENG] Any mechanism which starts, stops, or regulates the flow of liquid, such as a valve, faucet, or tap. [VERT ZOO] The adult male of the domestic fowl and of gallinaceous birds.

Cockcroft-Walton accelerator [NUCLEO] An electrostatic particle accelerator utilizing as a source of high voltage a transformer and an array of rectifiers and condensers, giving voltage multiplication.

cockeyed bob [METEOROL] A thunder squall occurring during the summer, on the northwest coast of Australia.

cocking lever [ORD] A lever for drawing back or sometimes lowering the striker of an automatic firearm.

cockle [INV ZOO] The common name for a number of species of marine mollusks in the class Bivalvia characterized by a shell having convex radial ribs.

cockle finish [MATER] An irregular surface usually produced on rag bond and ledger paper, obtained by coating the paper with sizing and drying it in loop or festoon fashion in heated air.

cockpit [AERO ENG] A space in an aircraft or spacecraft where the pilot sits. [NAV ARCH] A sunken area on the deck of a small vessel, near the stern, from which the craft is steered.

cockroach *See* roach.

cocoa butter [MATER] A brown fat obtained from cacao seeds; melts at 30–35°C; used in the manufacture of chocolate, cosmetics, and pharmaceuticals. Also known as cacao butter; oleum theobromatis; theobroma oil.

cocodyl oxide [ORG CHEM] $(CH_3)_2AsOAs(CH_3)_2$ A liquid that has an obnoxious odor; slightly soluble in water, soluble in alcohol and ether; boils at 150°C. Also known as alkarsine; bisdimethyl arsenic oxide; dicacodyl oxide.

coconsciousness [PSYCH] A dissociated mental state coexisting with a person's consciousness, but without his awareness, though it is psychodynamically active and may account for various normal and abnormal mental phenomena.

coconut [BOT] *Cocos nucifera.* A large palm in the order Arecales grown for its fiber and fruit, a large, ovoid, edible drupe with a fibrous exocarp and a hard, bony endocarp containing fleshy meat (endosperm).

coconut bud rot [PL PATH] A fungus disease of the coconut palm caused by *Phytophthora palmivora* and characterized by destruction of the terminal bud and adjacent leaves.

coconut oil [MATER] A nearly colorless or yellow oil from fresh coconut (*Cocos nucifera*) or from copra (dried coconut); used in foods, in making soap, and as a raw material in fatty-acid production.

cocoon [INV ZOO] 1. A protective case formed by the larvae of many insects, in which they pass the pupa stage. 2. Any of the various protective egg cases formed by invertebrates.

cocurrent line [OCEANOGR] A line through places having the same tidal current hour.

cod [VERT ZOO] The common name for fishes of the subfamily Gadidae, especially the Atlantic cod (*Gadus morrhua*).

codan [ELECTR] A device that silences a receiver except when a modulated carrier signal is being received.

codan lamp [COMMUN] Visual indication that a usable transmitted signal has been received by a particular radio receiver.

Coddington lens [OPTICS] A magnifier consisting of a glass sphere with a deep groove cut around a great circle to serve as a stop.

Coddington shape factor *See* shape factor.

code [COMMUN] A system of symbols and rules for expressing information, such as the Morse code, EIA color code, and the binary and other machine languages used in digital computers.

code beacon [NAV] A beacon that flashes a characteristic signal by which it may be recognized.

codecarboxylase [BIOCHEM] The prosthetic component of

the enzyme carboxylase which catalyzes decarboxylation of L-amino acids. Also known as pyridoxal phosphate.

code-check [ADP] To remove mistakes from a coded routine or program.

code checking time [ADP] Time spent checking out a problem on the computer, making sure that the problem is set up correctly and that the code is correct.

codeclination [NAV] In celestial navigation, 90° minus the declination; when the declination and latitude are of the same name, codeclination is the same as polar distance measured from the elevated pole.

code converter [ADP] A converter that changes coded information to a different code system.

coded character set [ADP] A set of characters together with the code assigned to each character for computer use.

coded decimal *See* decimal-coded digit.

coded interrogator [COMMUN] An interrogator whose output signal forms the code required to trigger a specific radio or radar beacon.

coded passive reflector antenna [ELECTROMAG] An object intended to reflect Hertzian waves and having variable reflecting properties according to a predetermined code for the purpose of producing an indication on a radar receiver.

coded program [ADP] A program expressed in the required code for a computer.

coded stop [ADP] A stop instruction built into a computer routine.

code element [COMMUN] One of the separate elements or events constituting a coded message, such as the presence or absence of a pulse, dot, dash, or space.

code error [ADP] A surplus or lack of a bit or bits in a machine instruction.

code group [COMMUN] A combination of letters or numerals or both, assigned to represent one or more words of plain text in a coded message.

code holes [ADP] The informational holes in perforated tape, as opposed to the feed holes or other holes.

codehydrogenase I *See* diphosphopyridine nucleotide.

codehydrogenase II *See* triphosphopyridine nucleotide.

codeine [PHARM] $C_{18}H_{21}NO_3$ An alkaloid prepared from morphine; used as mild analgesic and cough suppressant.

code line [ADP] In character recognition, the area reserved for the inscription of the printed or handwritten characters to be recognized.

code practice oscillator [ELECTR] An oscillator used with a key and either headphones or a loudspeaker to practice sending and receiving Morse code.

coder [ADP] A person who translates a sequence of computer instructions into codes acceptable to the machine. [COMMUN] A device that generates a code by generating pulses having varying lengths or spacings, as required for radio beacons and interrogators. Also known as moder; pulse coder; pulse-duration coder.

code ringing [COMMUN] In telephone switching, party-line ringing wherein the number or duration of rings indicates which station is being called.

code-sending radiosonde [ENG] A radiosonde which transmits the indications of the meteorological sensing elements in the form of a code consisting of combinations of dots and dashes. Also known as code-type radiosonde; contracted code sonde.

code signal [COMMUN] A sequence of discrete conditions or events corresponding to a coded message.

code translation [COMMUN] Conversion of a directory code or number into a predetermined code for controlling the selection of an outgoing trunk or line.

code-type radiosonde *See* code-sending radiosonde.

Codiaceae [BOT] A family of green algae in the order Siphonales having macroscopic thalli composed of aggregates of tubes.

codimer [ORG CHEM] **1.** A copolymer formed from the polymerization of two dissimilar olefin molecules. **2.** The product of polymerization of isobutylene with one of the two normal butylenes.

coding [ADP] **1.** The process of converting a program design into an accurate, detailed representation of that program in some suitable language. **2.** A list, in computer code, of the successive operations required to carry out a given routine or solve a given problem.

coding delay [NAV] An arbitrary time delay in the transmission of pulse signals; in the loran system it is inserted between transmission of the master and slave signals to prevent zero or small readings, and thus to aid in distinguishing between master and slave station signals.

coding disk [COMMUN] Disk with small projections for operating contacts to give a certain predetermined code to a transmission.

coding line *See* instruction word.

coding sheet [ADP] A sheet of paper printed with a form on which one can conveniently write a coded program.

codistor [ELECTR] A multijunction semiconductor device which provides noise rejection and voltage regulation functions.

cod-liver oil [MATER] A yellow oil, high in vitamin D, extracted from the liver of the Atlantic cod (*Gadus morrhua*); soluble in alcohol. Also known as banks oil; oleum morrhuae.

codon [COMMUN] A basic unit in a coded message, corresponding to a single character. [GEN] The basic unit of the genetic code, comprising three-nucleotide sequences of messenger ribonucleic acid, each of which is translated into one amino acid in protein synthesis.

coefficient [MATH] A factor in a product.

coefficient A [NAV] A component of magnetic compass deviation of constant value with compass heading, resulting from mistakes in calculations, compass and pelorous misalignment, and unsymmetrical arrangements of horizontal soft iron.

coefficient B [NAV] A component of magnetic compass deviation, varying with the sine function of the compass heading, resulting from the fore-and-aft component of the craft's permanent magnetic field and induced magnetism in unsymmetrical vertical iron forward or aft of the compass.

coefficient C [NAV] A component of magnetic compass deviation, varying with the cosine function of the compass heading, resulting from the athwartship component of the craft's permanent magnetic field and induced magnetism in unsymmetrical vertical iron port or starboard of the compass.

coefficient D [NAV] A component of magnetic compass deviation, varying with the sine function of twice the compass heading, resulting from induced magnetism in all symmetrical arrangements of the craft's horizontal soft iron.

coefficient E [NAV] A component of magnetic compass deviation, varying with the cosine function of twice the compass heading, resulting from induced magnetism in all unsymmetrical arrangements of the craft's horizontal soft iron.

coefficient J [NAV] A change in magnetic compass deviation, varying with the cosine function of the compass heading for a given value of J, where J is the change of deviation for a heel of 1° on compass heading 000°.

coefficient of absorption *See* absorption coefficient.

coefficient of compressibility [MECH] The decrease in volume per unit volume of a substance resulting from a unit increase in pressure; it is the reciprocal of the bulk modulus.

coefficient of condensation [STAT MECH] The ratio of the number of molecules condensed on the surface of a solid or liquid in equilibrium with its vapor phase to the total number of vapor molecules striking the surface.

coefficient of contingency [STAT] A measure of the strength of dependence between two statistical variables, based on a contingency table.

coefficient of contraction [FL MECH] The ratio of the minimum cross-sectional area of a jet of liquid discharging from an orifice to the area of the orifice.

coefficient of coupling *See* coupling constant.

coefficient of cubical expansion [THERMO] The increment in volume of a unit volume of solid, liquid, or gas for a rise of temperature of 1° at constant pressure. Also known as coefficient of expansion; coefficient of thermal expansion; expansion coefficient; expansivity.

coefficient of discharge *See* discharge coefficient.

coefficient of eddy diffusion *See* eddy diffusivity.

coefficient of elasticity *See* modulus of elasticity.

coefficient of expansion *See* coefficient of cubical expansion.

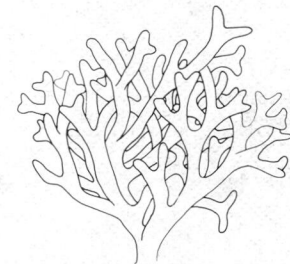

CODIACEAE

A portion of a dichotomously branched thallus of *Codium*.

coefficient of friction   [MECH] The ratio of the frictional force between two bodies in contact, parallel to the surface of contact, to the force, normal to the surface of contact, with which the bodies press against each other. Also known as friction coefficient.

coefficient of friction of rest   *See* coefficient of static friction.

coefficient of kinematic viscosity   *See* kinematic viscosity.

coefficient of kinetic friction   [MECH] The ratio of the frictional force, parallel to the surface of contact, that opposes the motion of a body which is sliding or rolling over another, to the force, normal to the surface of contact, with which the bodies press against each other.

coefficient of linear expansion   [THERMO] The increment of length of a solid in a unit of length for a rise in temperature of 1° at constant pressure.

coefficient of performance   [THERMO] In a refrigeration cycle, the ratio of the heat energy extracted by the heat engine at the low temperature to the work supplied to operate the cycle; when used as a heating device, it is the ratio of the heat delivered in the high-temperature coils to the work supplied.

coefficient of permeability   *See* permeability coefficient.

coefficient of reflection   *See* reflection coefficient.

coefficient of resistance   [FL MECH] The ratio of the loss of head of fluid, issuing from an orifice or passing over a weir, to the remaining head.

coefficient of restitution   [MECH] The constant e, which is the ratio of the relative velocity of two elastic spheres after direct impact to that before impact; e can vary from 0 to 1, with 1 equivalent to an elastic collision and 0 equivalent to a perfectly elastic collision.

coefficient of rigidity   *See* modulus of elasticity in shear.

coefficient of rolling friction   [MECH] The ratio of the frictional force, parallel to the surface of contact, opposing the motion of a body rolling over another, to the force, normal to the surface of contact, with which the bodies press against each other.

coefficient of sliding friction   [MECH] The ratio of the frictional force, parallel to the surface of contact, opposing the motion of a body sliding over another, to the force, normal to the surface of contact, with which the bodies press against each other.

coefficient of static friction   [MECH] The ratio of the maximum possible frictional force, parallel to the surface of contact, which acts to prevent two bodies in contact, and at rest with respect to each other, from sliding or rolling over each other, to the force, normal to the surface of contact, with which the bodies press against each other. Also known as coefficient of friction of rest.

coefficient of strain   [MATH] Multiplier used in transformations to elongate or compress configurations in a direction parallel to an axis.   [MECH] For a substance undergoing a one-dimensional strain, the ratio of the distance along the strain axis between two points in the body, to the distance between the same points when the body is undeformed.

coefficient of thermal expansion   *See* coefficient of cubical expansion.

coefficient of variation   [STAT] The ratio of the standard deviation of a distribution to its arithmetic mean.

coefficient of velocity   *See* velocity coefficient.

coefficient of viscosity   *See* absolute viscosity.

coefficients of form   [NAV ARCH] Parameters depending on the form of a ship, such as the block coefficient, midship section coefficient, mean length coefficient, water plane coefficient, and displacement coefficient.

coelacanth   [VERT ZOO] Any member of the Coelacanthiformes, an order of lobefin fishes represented by a single living genus, *Latimeria*.

Coelacanthidae   [PALEON] A family of extinct lobefin fishes in the order Coelacanthiformes.

Coelacanthiformes   [VERT ZOO] An order of lobefin fishes in the subclass Crossopterygii which were common fresh-water animals of the Carboniferous and Permian; one genus, *Latimeria*, exists today.

Coelacanthini   [VERT ZOO] The equivalent name for Coelacanthiformes.

Coelenterata   [INV ZOO] A phylum of the Radiata whose members typically bear tentacles and possess intrinsic nematocysts.

coelenteron   [INV ZOO] The internal cavity of coelenterates.

coeloblastula   [EMBRYO] A simple, hollow blastula with a single-layered wall.

Coelolepida   [PALEON] An order of extinct jawless vertebrates (Agnatha) distinguished by skin set with minute, close-fitting scales of dentine, similar to placoid scales of sharks.

coelom   [ZOO] The mesodermally lined body cavity of most animals higher on the evolutionary scale than flatworms and nonsegmented roundworms.

Coelomata   [ZOO] The equivalent name for Eucoelomata.

coelomocyte   [INV ZOO] A corpuscle, including amebocytes and eleocytes, in the coelom of certain animals, especially annelids.

coelomoduct   [INV ZOO] Either of a pair of ciliated excretory and reproductive channels passing from the coelom to the exterior in certain invertebrates, including annelids and mollusks.

Coelomomycetaceae   [MYCOL] A family of entomophilic fungi in the order Blastocladiales which parasitize primarily mosquito larvae.

coelomostome   [INV ZOO] The opening of a coelomoduct into the coelom.

Coelomycetes   [MYCOL] A group set up by some authorities to include the Sphaerioidaceae and the Melanconiales.

Coelopidae   [INV ZOO] The seaweed flies, a family of myodarian cyclorrhaphous dipteran insects in the subsection Acalypteratae whose larvae breed on decomposing seaweed.

coeloplanula   [INV ZOO] A hollow planula having a wall of two layers of cells.

coelostat   [ENG] A device consisting of a clockwork-driven mirror that enables a fixed telescope to continuously keep the same region of the sky in its field of view.

Coelurosauria   [PALEON] A group of small, lightly built saurischian dinosaurs in the suborder Theropoda having long necks and narrow, pointed skulls.

Coenagrionidae   [INV ZOO] A family of zygopteran insects in the order Odonata.

coenenchyme   [INV ZOO] The mesagloea surrounding and uniting the polyps in compound anthozoans. Also known as coenosarc.

Coenobitidae   [INV ZOO] A family of terrestrial decapod crustaceans belonging to the Anomura.

coenobium   [INV ZOO] A colony of protozoans having a constant size, shape, and cell number, but with undifferentiated cells.

coenocyte   [BIOL] A multinucleate mass of protoplasm formed by repeated nucleus divisions without cell fission.

Coenomyidae   [INV ZOO] A family of orthorrhaphous dipteran insects in the series Brachycera.

Coenopteridales   [PALEOBOT] A heterogeneous group of fernlike fossil plants belonging to the Polypodiophyta.

coenosarc   [INV ZOO] 1. The living axial portion of a hydroid colony. 2. *See* coenenchyme.

coenosteum   [INV ZOO] The calcareous skeleton of a compound coral or bryozoan colony.

Coenothecalia   [INV ZOO] An order of the class Alcyonaria that forms colonies; lacks spicules but has a skeleton composed of fibrocrystalline argonite.

coenotype   [BIOL] An organism having the characteristic structure of the group to which it belongs.

coenurosis   [VET MED] An infestation by a coenurus, the metacestode of *Taenia* species; most common in sheep, rabbits, and other herbivores.

coenzyme   [BIOCHEM] The nonprotein portion of an enzyme; a prosthetic group which functions as an acceptor of electrons or functional groups.

coenzyme A   [BIOCHEM] $C_{21}H_{36}O_{16}N_7P_3S$ A coenzyme in all living cells; required by certain condensing enzymes to act in acetyl or other acyl-group transfer and in fatty-acid metabolism. Abbreviated CoA.

coenzyme I   *See* diphosphopyridine nucleotide.

coenzyme II   *See* triphosphopyridine nucleotide.

coercive force   [ELECTROMAG] The magnetic field H which must be applied to a magnetic material in a symmetrical,

COELACANTH

Living coelacanth, *Latimeria chalumnae*; 5 feet (1.5 meters). (*After P. P. Grassé, ed., Traité de Zoologie, tome 13, fasc. 3, 1958*)

COELUROSAURIA

Skeleton of *Coelophysis* (about 8 feet, or 2.5 meters long), a Late Triassic theropod from New Mexico. (*After E. H. Colbert, Evolution of the Vertebrates, Wiley, 1955*)

cyclicly magnetized fashion, to make the magnetic induction B vanish. Also known as magnetic coercive force.

coercivity [ELECTROMAG] The coercive force of a magnetic material in a hysteresis loop whose maximum induction approximates the saturation induction.

coeruleolactite [MINERAL] $(Ca,Cu)Al_6(PO_4)_4(OH)_8 \cdot 4\text{-}5H_2O$ A milky-white to sky-blue mineral consisting of an aluminum phosphate.

coesite [MINERAL] A high-pressure polymorph of SiO_2 formed in nature only under unique physical conditions, requiring pressures of more than 20 kilobars; usually found in meteor impact craters.

coetaneous [SCI TECH] Contemporary.

coextrusion [ENG] Extrusion-forming of plastic or metal products in which two or more compatible feed materials are used in physical admixture through the same extrusion die.

cofactor See minor.

coffee [BOT] Any of various shrubs or small trees of the genus *Coffea* (family Rubiaceae) cultivated for the seeds (coffee beans) of its fruit; most coffee beans are obtained from the Arabian species, *C. arabica.*

cofferdam [CIV ENG] A temporary damlike structure constructed around an excavation to exclude water.

coffered ceiling [BUILD] An ornamental ceiling constructed of panels that are sunken or recessed.

coffin [NUCLEO] A box of heavy shielding material, usually lead, used for transporting radioactive objects and having walls thick enough to attenuate radiation from the contents to an allowable level. Also known as cask; casket.

coffin corner [AERO ENG] The range of Mach numbers between the buffeting Mach number and the stalling Mach number within which an aircraft must be operated.

coffinite [MINERAL] $USiO_4$ A black silicate important as a uranium ore; found in sandstone deposits and hydrothermal veins in New Mexico, Utah, and Wyoming.

cofinal [MATH] A subset C of a directed set D is cofinal if for each element of D there is a larger element in C.

cog [DES ENG] A tooth on the edge of a wheel.

cog belt [MECH ENG] A flexible device used for timing and for slip-free power transmission.

cogging mill See blooming mill.

cognac [FOOD ENG] Brandy distilled from grapes grown in the Charente and Charente-Maritime departments of France.

cognate [GEOL] Pertaining to contemporaneous fractures in a system with regard to time of origin and deformational type.

cognate ejecta [GEOL] Essential or accessory pyroclasts derived from the magmatic materials of a current volcanic eruption.

cognate inclusion See autolith.

cognition [PSYCH] The conscious faculty or process of knowing, of becoming or being aware of thoughts or perceptions, including understanding and reasoning.

COGO [ADP] A higher-level computer language oriented toward civil engineering, enabling one to write a program in a technical vocabulary familiar to engineers and feed it to the computer; several versions have been implemented. Derived from coordinated geometry.

cogon [BOT] *Imperate cylindrica.* A grass found in rainforests. Also known as alang-alang.

cog railway [CIV ENG] A steep railway that employs a cograil that meshes with a cogwheel on the locomotive to ensure traction.

cogwheel [DES ENG] A wheel with teeth around its edge.

cogwheel ore See bournonite.

cohenite [MINERAL] $(Fe,Ni,Co)_3C$ A tin-white, isometric mineral found in meteorites.

cohered video [ELECTR] The video detector output signal in a coherent moving-target indicator radar system.

coherence [PHYS] 1. The existence of a correlation between the phases of two or more waves, so that interference effects may be produced between them. 2. Property of moving in unison, such as is characteristic of the particles in a synchrotron.

coherence distance See coherence length.

coherence length [SOLID STATE] A measure of the distance through which the effect of any local disturbance is spread out

in a superconducting material. Also known as coherence distance.

coherent carrier system [NAV] Transponder system in which the interrogating carrier is retransmitted at a definite multiple frequency for comparison.

coherent detector [ELECTR] A detector used in moving-target indicator radar to give an output-signal amplitude that depends on the phase of the echo signal instead of on its strength, as required for a display that shows only moving targets.

coherent echo [ELECTR] A radar echo whose phase and amplitude at a given range remain relatively constant.

coherent interrupted waves [COMMUN] Interrupted continuous waves occurring in wave trains in which the phase of the waves is maintained through successive wave trains.

coherent light [OPTICS] Radiant electromagnetic energy of the same, or almost the same, wavelength, and with definite phase relationships between different points in the field.

coherent light communications [COMMUN] Communications using the optical band as a transmission medium by modulating a laser in amplitude or pulse frequency.

coherent noise [ENG] Noise that affects all tracks across a magnetic tape equally and simultaneously.

coherent oscillator [ELECTR] An oscillator used in moving-target indicator radar to serve as a reference by which changes in the radio-frequency phase of successively received pulses may be recognized. Abbreviated coho.

coherent precipitate [PHYS CHEM] A precipitate that is a continuation of the lattice structure of the solvent and has no phase or grain boundary.

coherent-pulse radar [ELECTR] A radar in which the radio-frequency oscillations of recurrent pulses bear a constant phase relation to those of a continuous oscillation.

coherent pulses [ELECTR] Characterizing pulses in which the phase of the radio-frequency waves is maintained through successive pulses.

coherent radiation [PHYS] Radiation in which there are definite phase relationships between different points in a cross section of the beam.

coherent reference [ELECTR] A reference signal, usually of stable frequency, to which other signals are phase-locked to establish coherence throughout a system.

coherent scattering [PHYS] Scattering in which there is a definite phase relationship between incoming and scattered particles or photons.

coherent signal [ELECTR] In a pulsed radar system, a signal having a constant phase; it is mixed with the echo signal, whose phase depends upon the range of the target, in order to detect the phase shift and measure the target's range.

coherent source [PHYS] A source in which there is a constant phase difference between waves emitted from different parts of the source.

coherent system [NAV] A navigation system in which the signal output is obtained by demodulating the received signal after mixing with a local signal having a fixed phase relation to that of the transmitted signal, to permit use of the information carrier by the phase of the received signal.

coherent transponder [ELECTR] A transponder in which a fixed relation between frequency and phase of input and output signals is maintained.

coherent video [ELECTR] The video signal produced in a moving-target indicator system by combining a radar echo signal with the output of a continuous-wave oscillator; after delay, this signal is detected, amplified, and subtracted from the next pulse train to give a signal representing only moving targets.

coherer [ELEC] A cell containing a granular conductor between two electrodes; the cell becomes highly conducting when subjected to an electric field, and conduction can then be stopped only by jarring the granules.

cohesion [BOT] The union of similar plant parts or organs, as of the petals to form a corolla. [PHYS] The tendency of parts of a body of like composition to hold together, as a result of intermolecular attractive forces. [SCI TECH] The state or process of sticking together.

cohesionless [GEOL] Referring to a soil having low shear

COFFEE

Branch of *Coffea arabica.*

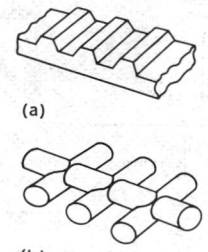

COG BELT

(a)

(b)

Cog belts for various uses. *(a)* Flat belt for timing or high-speed power transmission. *(b)* Projections integrally molded with self-lubricating plastic belt for engaging gears from either side and twisting to mesh with misaligned gears.

COINCIDENCE AMPLIFIER

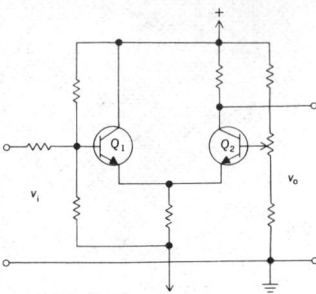

Simple coincidence amplifier.

COKE

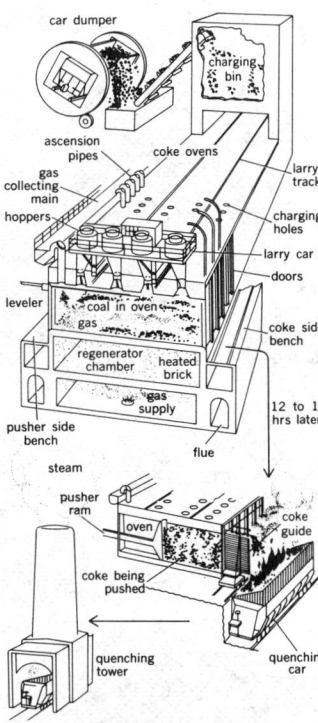

Steps in the production of coke from coal. After the coal has been in the over for 12–18 hours, the doors are removed, and a ram mounted on the same machine that operates the leveling bar shoves the coke into a quenching car for cooling. (*American Iron and Steel Institute*)

COLA

Branch of *Cola acuminata*.

strength when dry, and low cohesion when wet. Also known as frictional; noncohesive.

cohesive energy [SOLID STATE] The difference between the energy per atom of a system of free atoms at rest far apart from each other, and the energy of the solid.

cohesiveness [GEOL] Property of unconsolidated fine-grained sediments by which the particles stick together by surface forces.

cohesive soil [GEOL] A sticky soil, such as clay or silt; its shear strength equals about half its unconfined compressive strength.

cohesive strength [MECH] **1.** Strength corresponding to cohesive forces between atoms. **2.** Hypothetically, the stress causing tensile fracture without plastic deformation.

coho *See* coherent oscillator.

cohomology theory [MATH] A theory which uses algebraic groups to study the geometric properties of topological spaces; closely related to homology theory.

coil [ELECTROMAG] A number of turns of wire used to introduce inductance into an electric circuit, to produce magnetic flux, or to react mechanically to a changing magnetic flux; in high-frequency circuits a coil may be only a fraction of a turn. Also known as electric coil; inductance; inductance coil; inductor. [SCI TECH] An arrangement of flexible material into a spiral or helix.

coil antenna [ELECTROMAG] An antenna that consists of one or more complete turns of wire.

coil breaks [MET] Creases across a metal strip transverse to the direction of coiling and representing areas of reduced thickness.

coiled tubular gland [ANAT] A structure having a duct interposed between the surface opening and the coiled glandular portion; an example is a sweat gland.

coil form [ELECTROMAG] The tubing or spool of insulating material on which a coil is wound.

coil loading [COMMUN] Loading in which inductors, commonly called loading coils, are inserted in a line at intervals.

coil method [GRAPHICS] One of the methods used in terracotta sculpture; the clay is rolled into cylindrical strips about the size of an ordinary pencil and wound up to create the desired shape.

coil neutralization *See* inductive neutralization.

coil serving *See* serving.

coil spring [DES ENG] A helical or spiral spring, such as one of the helical springs used over the front wheels in an automotive suspension.

coil weld [MET] A butt weld joining the ends of two metal sheets; forms a continuous strip for coiling.

coil winder [ENG] A manual or motor-driven mechanism for winding coils individually or in groups.

coincidence [GEN] A numerical value equal to the number of double crossovers observed, divided by the number expected.

coincidence amplifier [ELECTR] An electronic circuit that amplifies only that portion of a signal present when an enabling or controlling signal is simultaneously applied.

coincidence circuit [ELECTR] A circuit that produces a specified output pulse only when a specified number or combination of two or more input terminals receives pulses within an assigned time interval. Also known as coincidence counter; coincidence gate.

coincidence correction *See* dead-time correction.

coincidence counter *See* coincidence circuit.

coincidence counting [NUCLEO] A method of distinguishing particular types of events from background events and of measuring the velocities or directions of particles, by registering the occurrence of counts in two or more particle detectors within a given time interval by means of coincidence circuits.

coincidence gate *See* coincidence circuit.

coincidence magnet [NUCLEO] An electromagnet used in one type of scram mechanism for a nuclear reactor; has three electrically independent coils and releases when any two coils are deenergized.

coincidence rangefinder [OPTICS] An optical rangefinder in which one-eyed viewing through a single eyepiece provides the basis for manipulation of the rangefinder adjustment to

cause two images of the target or parts of each, viewed over different paths, to match or coincide.

coincident-current selection [ELECTR] The selection of a particular magnetic cell, for reading or writing in computer storage, by simultaneously applying two or more currents.

coin gold [MET] Gold of the legal fineness for coins.

coining [MET] **1.** A process of forming metals by squeezing between two dies so as to impress well-defined imprints on both surfaces of the work; usually performed cold. **2.** Final pressing of a sintered compact in powder metallurgy.

coin silver [MET] An alloy of 90% silver, 10% copper; has been used for coining American currency.

coir [MATER] A coarse, brown fiber obtained from the husk of the coconut.

coitophobia [PSYCH] An abnormal fear of coitus.

coitus [ZOO] The act of copulation. Also known as intercourse.

coke [MATER] A coherent, cellular, solid residue remaining from the dry (destructive) distillation of a coking coal or of pitch, petroleum, petroleum residues, or other carbonaceous materials; contains carbon as its principal constituent, together with mineral matter and volatile matter.

coke drum [CHEM ENG] A vessel in which coke is produced.

cokeite [GEOL] Naturally occurring coke formed by the action of magma on coal or by natural combustion of coal.

coke knocker [MECH ENG] A mechanical device used to break loose coke within a drum or tower.

coke number [CHEM ENG] A number used to report the results of the Ramsbottom carbon residue test.

coke oven [CHEM ENG] A retort in which coal is converted to coke by carbonization.

coke-oven gas [MATER] A gas produced during carbonization of coal to form coke.

coke-oven regenerator [CHEM ENG] Arrangement of refractory blocks in the flue system of a coke oven to recover waste heat from hot, exiting combustion gases; the blocks, in turn, release heat to warm, incoming fuel gas.

coker [CHEM ENG] The processing unit in which coking occurs.

coking [CHEM ENG] **1.** Destructive distillation of coal to make coke **2.** A process for thermally converting the heavy residual bottoms of crude oil entirely to lower-boiling petroleum products and by-product petroleum coke.

coking coal [GEOL] A very soft bituminous coal suitable for coking.

coking still [CHEM ENG] A still in which coking is done; usually, it is a batch still.

col [GEOL] A high, sharp-edged pass in a mountain ridge, usually produced by the headward erosion of opposing cirques. [METEOROL] The point of intersection of a trough and a ridge in the pressure pattern of a weather map; it is the point of relatively lowest pressure between two highs and the point of relatively highest pressure between two lows. Also known as neutral point; saddle point.

Col *See* Columba.

cola [BOT] *Cola acuminata.* A tree of the sterculia family (Sterculiaceae) cultivated for cola nuts, the seeds of the fruit; extract of cola nuts is used in the manufacture of soft drinks.

colatitude [GEOD] Ninety degrees minus the latitude.

Colburn analogy [FL MECH] Dimensionless Reynolds equation for fluid-flow resistance modified to be analogous to the Colburn j factor heat-transfer equation.

Colburn j factor equation [THERMO] Dimensionless heat-transfer equation to calculate the natural convection movement of heat from vertical surfaces or horizontal cylinders to fluids (gases or liquids) flowing past these surfaces.

Colburn method [CHEM ENG] Graphical method, and equations to calculate the theoretical number of plates (trays) needed to separate light and heavy liquids in a distillation column.

colchicine [ORG CHEM] $C_{22}H_{25}O_6N$ An alkaloid extracted from the stem of the autumn crocus; used experimentally to inhibit spindle formation and delay centromere division, and medicinally in the treatment of gout.

colcothar [INORG CHEM] Red ferric oxide made by heating ferrous sulfate in the air; used as a pigment and as an abrasive in polishing glass.

cold [ELEC] Pertaining to electrical circuits that are disconnected from voltage supplies and at ground potential; opposed to hot, pertaining to carrying an electrical charge.

cold agglutination phenomenon [IMMUNOL] Clumping of human blood group O erythrocytes at 0–4°C, but not at body temperature; occurs in primary atypical pneumonia, trypanasomiasis, and other unidentifiable states.

cold agglutinin [IMMUNOL] A nonspecific panagglutinin found in many normal human serums which produce maximum clumping of erythrocytes at 4° and none at 37°C.

cold-air drop *See* cold pool.

cold-air machine [MECH ENG] A refrigeration system in which air serves as the refrigerant in a cycle of adiabatic compression, cooling to ambient temperature, and adiabatic expansion to refrigeration temperature; air is customarily reused in a closed superatmospheric pressure system. Also known as dense-air system.

cold-air outbreak *See* polar outbreak.

cold anticyclone *See* cold high.

cold area [SCI TECH] Laboratory area that is continuously held at a specified degree of coldness to facilitate the conduct of physical and chemical tests.

cold bending [MET] The bending of metal rods, especially concrete-reinforcing rods, without heat.

cold-blooded [PHYSIO] Having body temperature approximating that of the environment and not internally regulated.

cold cathode [ELECTR] A cathode whose operation does not depend on its temperature being above the ambient temperature.

cold-cathode counter tube [ELECTR] A counter tube having one anode and three sets of 10 cathodes; two sets of cathodes serve as guides that direct the flow discharge to each of the 10 output cathodes in correct sequence in response to driving pulses. Also known as decade glow counting tube.

cold cathode discharge *See* glow discharge.

cold-cathode ionization gage *See* Philips ionization gage.

cold-cathode rectifier [ELECTR] A cold-cathode gas tube in which the electrodes differ greatly in size so electron flow is much greater in one direction than in the other. Also known as gas-filled rectifier.

cold-cathode tube [ELECTR] An electron tube containing a cold cathode, such as a cold-cathode rectifier, mercury-pool rectifier, neon tube, phototube, or voltage regulator.

cold-chamber machine [MET] Machine for die casting in which no heat is applied to the metal chamber or the plunger.

cold chisel [DES ENG] A chisel specifically designed to cut or chip cold metal; made of specially tempered tool steel machined into various cutting edges. Also known as cold cutter.

cold-core cyclone *See* cold low.

cold-core high *See* cold high.

cold-core low *See* cold low.

cold cure [CHEM ENG] Vulcanization of rubber at nonelevated temperatures with a solution of a sulfur compound.

cold cutter *See* cold chisel.

cold dome [METEOROL] A cold air mass, considered as a three-dimensional entity.

cold drawing [MET] Drawing a tube or wire through a series of successively smaller dies, without the application of heat, to reduce its diameter. [TEXT] Drawing a textile, as nylon, when cold.

cold drop *See* cold pool.

cold emission *See* field emission.

cold extrusion [MET] Shaping cold metal by striking a slug in a closed cavity with a punch so that the metal is forced up around the punch. Also known as cold forging; cold pressing; extrusion pressing; impact extrusion.

cold flow [MECH] Creep in polymer plastics.

cold-flow test [AERO ENG] A test of a liquid rocket without firing it to check or verify the integrity of a propulsion subsystem, and to provide for the conditioning and flow of propellants (including tank pressurization, propellant loading, and propellant feeding).

cold forging *See* cold extrusion.

cold forming [MET] Any forging operation performed cold, such as cold extrusion, cold drawing, or coining, which enables close dimensional accuracy to be achieved.

cold front [METEOROL] Any nonoccluded front, or portion thereof, that moves so that the colder air replaces the warmer air; the leading edge of a relatively cold air mass.

cold-front-like sea breeze [METEOROL] A sea breeze which forms over the ocean, moves slowly toward the land, and then moves inland quite suddenly. Also known as sea breeze of the second kind.

cold-front thunderstorm [METEOROL] A thunderstorm attending a cold front.

cold galvanizing [MET] Painting iron with a suspension of zinc particles in an organic solvent, so that a zinc coating remains following evaporation of the solvent.

cold hemagglutination [IMMUNOL] A phenomenon caused by the presence of cold agglutinin.

cold high [METEOROL] At a given level in the atmosphere, any high that is generally characterized by colder air near its center than around its periphery. Also known as cold anticyclone; cold-core high.

cold injury [MED] Physical trauma following exposure to very low temperatures.

cold junction [ELECTR] The reference junction of thermocouple wires leading to the measuring instrument; normally at room temperature.

cold lap *See* cold shut.

cold light [PHYS] 1. Light emitted in luminescence. 2. Visible light which is accompanied by little or no infrared radiation, and therefore has little heating effect.

cold lime-soda process [CHEM ENG] A water-softening process in which water is treated with hydrated lime (sometimes in combination with soda ash), which reacts with dissolved calcium and magnesium compounds to form precipitates that can be removed as sludge.

cold low [METEOROL] At a given level in the atmosphere, any low that is generally characterized by colder air near its center than around its periphery. Also known as cold-core cyclone; cold-core low.

cold molding [ENG] Shaping of an unheated compound in a mold under pressure, followed by heating the article to cure it.

cold neutron [SOLID STATE] A very-low-energy neutron in a reactor, used for research into solid-state physics because it has a wavelength of the order of crystal lattice spacings and can therefore be diffracted by crystals.

cold nosing *See* wildcat drilling.

cold plate [MECH ENG] An aluminum or other plate containing internal tubing through which a liquid coolant is forced, to absorb heat transferred to the plate by transistors and other components mounted on it. Also known as liquid-cooled dissipator.

cold pole [CLIMATOL] The location which has the lowest mean annual temperature in its hemisphere.

cold pool [METEOROL] A region of relatively cold air surrounded by warmer air; the term is usually applied to cold air of appreciable vertical extent that has been isolated in lower latitudes as part of the formation of a cutoff low. Also known as cold-air drop; cold drop.

cold pressing *See* cold extrusion.

cold rolling [MET] Rolling metal at room temperature to reduce thickness or harden the surface; results in a smooth finish and improved resistance to fatigue.

cold rubber [MATER] Butadiene-styrene type of synthetic rubber produced by polymerization at about 40°F (4°C), instead of the conventional 120°F (49°C); has improved strength and abrasion resistance.

cold saw [MECH ENG] 1. Any saw for cutting cold metal, as opposed to a hot saw. 2. A disk made of soft steel or iron which rotates at a speed such that a point on its edge has a tangential velocity of about 15,000 feet per minute (75 meters per second), and which grinds metal by friction.

cold settling [CHEM ENG] A process that removes wax from high-viscosity stocks.

cold-short [MET] Pertaining to lack of ductility in some metals at temperatures below the recrystallization temperature.

cold shot [MET] Intensely hard, globular portions of the surface of an ingot or casting formed by premature solidification upon first contact with the cold sand during pouring.

cold shut [MET] 1. A surface defect of a metal casting in the form of a discontinuity where two streams failed to unite. Also known as cold lap. **2.** Freezing of the top surface of an ingot before the mold is full.

cold slug [ENG] The first material to enter an injection mold in plastics manufacturing.

cold soldering [MET] Soldering of parts without heat.

cold-spot hygrometer *See* dewpoint hygrometer.

cold storage [ENG] The storage of perishables at low temperatures produced by refrigeration, usually above freezing, to increase storage life.

cold-storage locker plant [ENG] A plant with many rental steel lockers, each with a capacity of about 6 cubic feet (0.17 cubic meter) and generally for food storage by an individual family, placed in refrigerated rooms, at about 0°F (-18°C).

cold stress [MECH] Forces tending to deform steel, cement, and other materials, resulting from low temperatures.

cold stretch [ENG] A pulling operation, usually on extruded filaments, to improve tensile operations.

cold test [CHEM ENG] A test for one of an oil's characteristics: the temperature at which it becomes a solid.

cold tongue [METEOROL] In synoptic meteorology, a pronounced equatorward extension or protrusion of cold air.

cold torpor [PHYSIO] Condition of reduced body temperature in poikilotherms.

cold treatment [MET] Subzero cooling, to a temperature of -100°F (-73°C).

cold-type composition [GRAPHICS] Any typesetting method which produces copy suitable for offset lithography; copy may be obtained from a typewriter or photocomposition equipment.

cold-water sphere [OCEANOGR] Those portions of the ocean water having a temperature below 8°C. Also known as oceanic stratosphere.

cold wave [METEOROL] A rapid fall in temperature within 24 hours to a level requiring substantially increased protection to agriculture, industry, commerce, and social activities.

cold welding [MET] Welding in which a molecular bond is obtained by a cold flow of metal under extremely high pressures, without heat; widely used for sealing transistors and quartz crystal holders.

cold working [MET] Plastic deformation of a metal below the annealing temperature to cause permanent strain hardening.

Colebrook equation [FL MECH] An empirical equation for the flow of liquids in ducts, relating the friction factor to the Reynolds number and the relative roughness of the duct.

colectomy [MED] Excision of all or a portion of the colon.

colemanite [MINERAL] $Ca_2B_6O_{11} \cdot 5H_2O$ A colorless or white hydrated borate mineral that crystallizes in the monoclinic system and occurs in massive crystals or as nodules in clay.

Coleochaetaceae [BOT] A family of green algae in the suborder Ulotrichineae; all occur as attached, disklike, or parenchymatous thalli.

Coleodontidae [PALEON] A family of conodonts in the suborder Neurodontiformes.

Coleoidea [INV ZOO] A subclass of cephalopod mollusks including all cephalopods except *Nautilus,* according to certain systems of classification.

Coleophoridae [INV ZOO] The case bearers, narrow-winged moths composing a family of lepidopteran insects in the suborder Heteroneura; named for the silk-and-leaf shell carried by larvae.

coleopter [AERO ENG] An aircraft having an annular (barrel-shaped) wing, the engine and body being mounted within the circle of the wing.

Coleoptera [INV ZOO] The beetles, holometabolous insects making up the largest order of the animal kingdom; general features of the Insecta are found in this group.

coleoptile [BOT] The first leaf of a monocotyledon seedling.

coleorhiza [BOT] The sheath surrounding the radicle in monocotyledons.

Coleorrhyncha [INV ZOO] A monofamilial group of homopteran insects in which the beak is formed at the anteroventral extremity of the face and the propleura form a shield for the base of the beak.

Coleosporaceae [MYCOL] A family of parasitic fungi in the order Uredinales.

colic [MED] **1.** Acute paroxysmal abdominal pain usually caused by smooth muscle spasm, obstruction, or twisting. **2.** In early infancy, paroxysms of pain, crying, and irritability caused by swallowing air, overfeeding, intestinal allergy, and emotional factors.

colic artery [ANAT] Any of the three arteries that supply the colon.

colicin [MICROBIO] A highly specific protein, released by some enteric bacilli, which kills other enteric bacilli.

colidar *See* ladar.

coliform bacteria [MICROBIO] Colon bacilli, or forms which resemble or are related to them.

Coliidae [VERT ZOO] The colies or mousebirds, composing the single family of the avian order Coliiformes.

Coliiformes [VERT ZOO] A monofamilial order of birds distinguished by long tails, short legs, and long toes, all four of which are directed forward.

coliphage [VIROL] Any bacteriophage able to infect *Escherichia coli.*

colistin [MICROBIO] $C_{45}H_{85}O_{10}N_{13}$ A basic polypeptide antibiotic produced by *Bacillus colistinus;* consists of an A and B component, active against a broad spectrum of gram-positive microorganisms and some gram-negative microorganisms.

colitis [MED] Inflammation of the large bowel, or colon.

colk *See* pothole.

colla [METEOROL] In the Philippines, a fresh or strong (less than 39–46 miles per hour, or 63–74 kilometers per hour) south to southwest wind, accompanied by heavy rain and severe squall. Also known as colla tempestade.

collada [METEOROL] A strong wind (35–50 miles per hour, or 56–80 kilometers per hour) in the Gulf of California, blowing from the north or northwest in the upper part, and from the northeast in the lower part of the gulf.

collage [GRAPHICS] A composition consisting of paper, cloth, wood, photographs, and so on, pasted together to form a texture or pattern.

collagen [BIOCHEM] A fibrous protein found in all multicellular animals, especially in connective tissue.

collagenase [BIOCHEM] Any proteinase that decomposes collagen and gelatin.

collagen disease [MED] Any of various clinical syndromes characterized by widespread alterations of connective tissue, including inflammation and fibrinoid degeneration.

collagraph [GRAPHICS] A print made from a plate produced by the collage method.

collapse breccia [GEOL] Angular rock fragments derived from the collapse of rock overlying a hollow space.

collapse caldera [GEOL] A caldera formed primarily as a result of collapse due to withdrawal of magmatic support.

collapse properties [MECH] Strength and dimensional attributes of piping, tubing, or process vessels, related to the ability to resist collapse from exterior pressure or internal vacuum.

collapse sink [GEOL] A sinkhole resulting from local collapse of a cavern that has been enlarged by solution and erosion.

collapse structure [GEOL] A structure resulting from rock slides under the influence of gravity. Also known as gravity-collapse structure.

collapsing pressure [MECH] The external pressure which causes a thin-walled body or structure to collapse.

collar [DES ENG] A ring placed around an object to restrict its motion, hold it in place, or cover an opening. [MIN ENG] The mouth of a mine shaft. [NAV ARCH] **1.** An opening in the end or bight of a rope or cable supporting a mast that goes over the masthead. **2.** A ring or loop of metal, rope, or other material, used to secure a heart or deadeye. **3.** A fitting over a structural part passing through a bulkhead or deck.

collar beam [BUILD] A tie beam in a roof truss connecting the rafters well above the wall plate.

collar bearing [MECH ENG] A bearing that resists the axial force of a collar on a rotating shaft.

collar cell *See* choanocyte.

collard [BOT] *Brassica oleracea* var. *acephala.* A biennial crucifer of the order Capparales grown for its rosette of edible leaves.

collared hole [ENG] A started hole drilled sufficiently deep to

---

COLEODONTIDAE

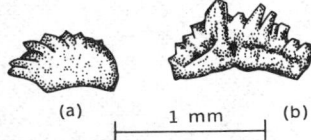

(a)   1 mm   (b)

Representatives of Coleodontidae. *(a) Coleodus,* a typical denticulate blade. *(b) Erismodus. (After illustration in R. R. Shrock and W. H. Twenhofel, Principles of Invertebrate Paleontology, McGraw-Hill, 2d ed., 1953)*

COLLARD

Collard *(Brassica oleracea var. acephala).*

confine the drill bit and prevent slippage of the bit from normal position.

collar locator log [PETRO ENG] Down-hole nuclear-log measurement to locate drill-hole casing collars, usually for precise location of perforating points.

collar vortex *See* vortex ring.

collate [ADP] To combine two or more similarly ordered sets of values into one set that may or may not have the same order as the original sets. [GRAPHICS] To assemble in proper sequence all the sheets, signatures, or insertions for a printed piece.

colla tempestade *See* colla.

collateral bud [BOT] An accessory bud produced beside an axillary bud.

collateral series [NUC PHYS] A radioactive decay series, initiated by transmutation, that eventually joins into one of the four radioactive decay series encountered in natural radioactivity.

collating sequence [ADP] The ordering of a set of items such that sets in that assigned order can be collated.

collating unit [ADP] An electromechanical device capable of performing singly or simultaneously the merging, sequence-checking, selection, and matching of punched cards. Also known as collator.

collator *See* collating unit.

collecting tubule [ANAT] One of the ducts conveying urine from the renal tubules (nephrons) to the minor calyces of the renal pelvis.

collection trap [ANALY CHEM] Cooled device to collect gas-chromatographic eluent, holding it for subsequent compound-identification analysis.

collective [METEOROL] In aviation weather observations, a group of observations transmitted in prescribed order by stations on the same long-line teletypewriter circuit. Also known as sequence.

collective bargaining [IND ENG] The negotiation for mutual agreement in the settlement of a labor contract between an employer or his representatives and a labor union or its representatives.

collective fire [ORD] Combined fire of various small arms concentrated on a given target or area.

collective motion [NUC PHYS] Motion of nucleons in a nucleus correlated so that their overall space pattern is essentially constant or undergoes changes which are slow compared to the motions of individual nucleons.

collective transition [NUC PHYS] A nuclear transition from one state of collective motion to another.

collector [ELECTR] **1.** A semiconductive region through which a primary flow of charge carriers leaves the base of a transistor; the electrode or terminal connected to this region is also called the collector. **2.** An electrode that collects electrons or ions which have completed their functions within an electron tube; a collector receives electrons after they have done useful work, whereas an anode receives electrons whose useful work is to be done outside the tube. Also known as electron collector. [ENG] A class of instruments employed to determine the electric potential at a point in the atmosphere, and ultimately the atmospheric electric field; all collectors consist of some device for rapidly bringing a conductor to the same potential as the air immediately surrounding it, plus some form of electrometer for measuring the difference in potential between the equilibrated collector and the earth itself; collectors differ widely in their speed of response to atmospheric potential changes.

collector capacitance [ELECTR] The depletion-layer capacitance associated with the collector junction of a transistor.

collector cutoff [ELECTR] The reverse saturation current of the collector-base junction.

collector junction [ELECTR] A semiconductor junction located between the base and collector electrodes of a transistor.

collector modulation [ELECTR] Amplitude modulation in which the modulator varies the collector voltage of a transistor.

collector plate [ELEC] One of several metal inserts that are sometimes embedded in the lining of an electrolyte cell to make the resistance between the cell lining and the current leads as small as possible.

collector resistance [ELECTR] The back resistance of the collector-base diode of a transistor.

collector ring *See* slip ring.

collector voltage [ELECTR] The direct-current voltage, obtained from a power supply, that is applied between the base and collector of a transistor.

Collembola [INV ZOO] The springtails, an order of primitive insects in the subclass Apterygota having six abdominal segments.

collenchyma [BOT] A primary, or early-differentiated, subepidermal supporting tissue in leaf petioles and vein ribs formed before vascular differentiation.

collenchyme [INV ZOO] A loose mesenchyme that fills the space between ectoderm and endoderm in the body wall of many lower invertebrates, such as sponges.

collenia [PALEOBOT] A convex, slightly arched, or turbinate stromatolite produced by late Precambrian blue-green algae of the genus *Collenia*.

Colles' fracture [MED] A fracture of the radius about 1 inch (2.5 centimeters) above the wrist with dorsal displacement of the distal fragment.

collet [DES ENG] A split, coned sleeve to hold small, circular tools or work in the nose of a lathe or other machine. [ENG] **1.** The glass neck remaining on a bottle after it is taken off the glass-blowing iron. **2.** Pieces of glass, ordinarily discarded, that are added to a batch of glass. Also spelled cullet. [HOROL] A small, friction-tight collar on a balance staff which holds the inner end of a balance spring. [LAP] The small, horizontal face at the bottom of a brilliant-cut gemstone.

Colletidae [INV ZOO] The colletid bees, a family of hymenopteran insects in the superfamily Apoidea.

colliculus [ANAT] **1.** Any of the four prominences of the corpora quadrigemina. **2.** The elevation of the optic nerve where it enters the retina. **3.** The anterolateral, apical elevation of the arytenoid cartilages.

2,4,6-collidine [ORG CHEM] $(CH_3)_3C_5H_2N$ A liquid boiling at 170.4°C; slightly soluble in water, soluble in alcohol; used as a chemical intermediate. Also known as 2,4,6-trimethylpyridine.

colliery [MIN ENG] A whole coal mining plant; generally the term is used in connection with anthracite mining but sometimes to designate the mine, shops, and preparation plant of a bituminous operation.

colligative properties [PHYS CHEM] Those properties that are related by a mathematical function, that is, dependent on the number of particles.

collimate [PHYS] To render parallel to a certain line or direction; paths of electrons in a flooding beam, or paths of various rays of a scanning beam are collimated to cause them to become more nearly parallel as they approach the storage assembly of a storage tube.

collimating lens [OPTICS] A lens on a collimator used to focus light from a source near one of its focal points into a parallel beam.

collimating sight [ORD] Sight equipped with a collimator, set parallel with the axis of the bore of the gun in horizontal direction, but adjustable in elevation, so that it can be kept focused on an aiming point while the gun is raised or lowered.

collimation error [ENG] **1.** Angular error in magnitude and direction between two nominally parallel lines of sight. **2.** Specifically, the angle by which the line of sight of a radar differs from what it should be.

collimation tower [ENG] Tower on which a visual and a radio target are mounted to check the electrical axis of an antenna.

collimator [OPTICS] An instrument which produces parallel rays of light. [PHYS] A device for confining the elements of a beam within an assigned solid angle.

collinear array *See* linear array.

collinear vectors [MATH] Two vectors one of which is a nonzero scaler multiple of the other.

collineation [MATH] A mapping which transforms points into points, lines into lines, and planes into planes. Also known as collineatory transformation.

collineatory transformation *See* collineation.

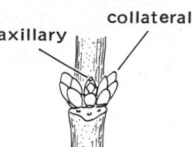

COLLATERAL BUD

Collateral bud in a red maple.

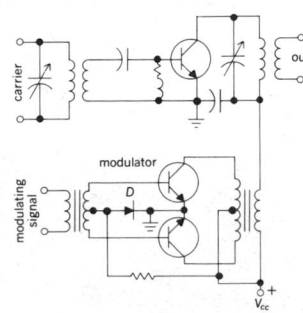

COLLECTOR MODULATION

Circuit diagram of a collector-modulated transistor.

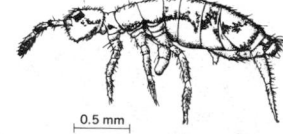

COLLEMBOLA

A collembolan, *Entomobrya cubensis.* (From J. W. Folsom, Proc. U.S. Nat. Mus., 72(6), plate 6, 1927)

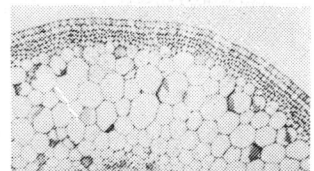

COLLENCHYMA

Collenchyma in a transverse section of Jimsonweed stem. (W. S. Walker)

collinite [GEOL] The maceral, of collain consistency, of jellified plant material precipitated from solution and hardened; a variety of euvitrinite.

Collins helium liquefier [CRYO] A machine which uses the Joule-Thomson effect and work done by helium gas in expansion against a movable piston to liquefy helium.

collinsite [MINERAL] $Ca_2(Mg,Fe)(PO_4)_2$ A phosphate mineral occurring in concentric layers in phosphoric nodules; found in meteorites.

Collins miner [MIN ENG] A type of remote-controlled continuous miner for thin-seam extraction.

Collip unit [BIOL] A unit for the standardization of parathyroid extract.

collision [PHYS] An interaction resulting from the close approach of two or more bodies, particles, or systems of particles, and confined to a relatively short time interval during which the motion of at least one of the particles or systems changes abruptly.

collision-avoidance radar [ENG] Radar equipment utilized in a collision-avoidance system.

collision-avoidance system [ENG] Electronic devices and equipment used by a pilot to perform the functions of conflict detection and avoidance.

collision bearing [NAV] A constant bearing maintained while the distance between two craft is decreasing.

collision blasting [ENG] The blasting out of different sections of rocks against each other.

collision broadening *See* collision line-broadening.

collision bulkhead [NAV ARCH] A watertight partition in a ship, perpendicular to the fore and aft centerline of the ship, usually near the bow, for keeping out water in the event of a collision.

collision course [NAV] A course which if followed will bring two craft together.

collision-course homing [ORD] Homing in which an offset antenna is used on the missile in conjunction with built-in computers that anticipate the motion of the target and direct the missile ahead of the present position of the target on a converging course that gives a collision in minimum missile travel time.

collision cross section *See* cross section.

collision efficiency [METEOROL] The fraction of all waterdrops which, initially moving on a collision course with respect to other drops, actually collide (make surface contact) with the other drops.

collision excitation [ATOM PHYS] The excitation of a gas by collisions of moving charged particles.

collision frequency [PHYS] The average number of collisions undergone by a particle traveling through a material, such as an electron traveling through a gas, in a unit time.

collision ionization [ATOM PHYS] The ionization of atoms or molecules of a gas or vapor by collision with other particles.

collisionless Boltzmann equation *See* Vlasov equation.

collision line-broadening [SPECT] Spreading of a spectral line due to interruption of the radiation process when the radiator collides with another particle. Also known as collision broadening.

collision matrix *See* scattering matrix.

collision of the first kind [PHYS] An inelastic collision in which some of the kinetic energy of translational motion is converted to internal energy of the colliding systems. Also known as endoergic collision.

collision of the second kind [PHYS] An inelastic collision in which some of the internal energy of the colliding systems is converted to kinetic energy of translation. Also known as exoergic collision.

collision parameter [AERO ENG] In orbit computation, the distance between a center of attraction of a central force field and the extension of the velocity vector of a moving object at a great distance from the center.

collision-radiative recombination [ATOM PHYS] The capture of an electron by an ion in a gas, accompanied by the emission of one or more photons.

collision theory [PHYS CHEM] Theory of chemical reaction proposing that the rate of product formation is equal to the number of reactant-molecule collisions multiplied by a factor that corrects for low-energy-level collisions. [QUANT MECH]

Theory to describe collisions of simple or complex particles, the derivation of collision cross sections from postulated interactions and the study of properties of collision amplitudes which follow from invariance principles such as conservation of probability and time-reversal invariance.

colloblast [INV ZOO] An adhesive cell on the tentacles of ctenophores.

collodion [ORG CHEM] Cellulose nitrate deposited from a solution of 60% ether and 40% alcohol, used for making fibers and film and in membranes for dialysis.

collodion cotton *See* pyroxylin.

collodion replication [ANALY CHEM] Production of a faithful collodion-film mold of a specimen surface (for example, powders, bones, microorganisms, crystals) which is thin enough to be studied by electron microscopy. [GRAPHICS] A process of image formation utilizing a negative that is a glass plate coated with a collodion containing iodides.

colloform [GEOL] Pertaining to the rounded, globular texture of mineral formed by colloidal precipitation.

colloid [CHEM] A system of which one phase is made up of particles having dimensions of 10–10,000 angstroms (1–1000 nanometers) and which is dispersed in a different phase.

colloidal dispersion *See* colloidal system.

colloidal fuel *See* coal-in-oil suspension.

colloidal graphite [MATER] Extremely fine flakes of graphite suspended in water, petroleum oil, castor oil, glycerin, or other liquids; used to provide conductive shields on the inside or outside surfaces of electron tubes.

colloidal instability [METEOROL] A property attributed to clouds, by which the particles of the cloud tend to aggregate into masses large enough to precipitate.

colloidal osmotic pressure *See* oncotic pressure.

colloidal suspension *See* colloidal system.

colloidal system [CHEM] An intimate mixture of two substances, one of which, called the dispersed phase (or colloid), is uniformly distributed in a finely divided state through the second substance, called the dispersion medium (or dispersing medium); the dispersion medium or dispersed phase may be a gas, liquid, or solid. Also known as colloidal dispersion; colloidal suspension.

colloid chemistry [PHYS CHEM] The scientific study of matter whose size is approximately 10 to 10,000 angstroms (1 to 1000 nanometers), and which exists as a suspension in a continuous medium, especially a liquid, solid, or gaseous substance.

colloider [CIV ENG] A device that removes colloids from sewage.

colloid goiter [MED] A soft, diffuse enlargement of the thyroid gland in which colloid fills the acinar spaces.

colloid mill [MECH ENG] A grinding mill for the making of very fine dispersions of liquids or solids by breaking down particles in an emulsion or paste.

collophane [MINERAL] A massive, cryptocrystalline, carbonate-containing variety of apatite and a principal source of phosphates for fertilizers. Also known as collophanite.

collophanite *See* collophane.

Collothecacea [INV ZOO] A monofamilial suborder of mostly sessile rotifers in the order Monogonata; many species are encased in gelatinous tubes.

Collothecidae [INV ZOO] The single family of the Collothecacea.

collotype *See* photogelatin process.

colluvium [GEOL] Loose, incoherent deposits at the foot of a slope or cliff, brought there principally by gravity.

Collyritidae [PALEON] A family of extinct, small, ovoid, exocyclic Euechinoidea with fascioles or a plastron.

Colmol miner [MIN ENG] A continuous miner, in which the coal is completely augered by two banks of cutting arms fitted with picks; the arms rotate in opposite directions to assist in gathering up the cuttings for the central conveyor.

coloboma [MED] A congenital, pathologic, or operative fissure, especially of the eye or eyelid.

colog *See* cologarithm.

cologarithm [MATH] The cologarithm of a number is the logarithm of the reciprocal of that number. Abbreviated colog.

colon [ANAT] The portion of the human intestine extending from the cecum to the rectum; it is divided into four sections:

COLLOTHECIDAE

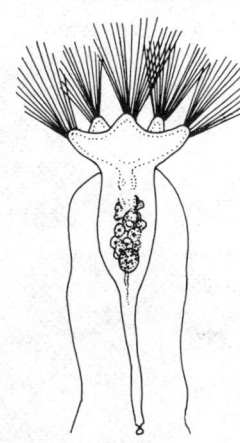

Collotheca species.

ascending, transverse, descending, and sigmoid. Also known as large intestine.

colon bacillus *See* Escherichia coli.

colonnade [ARCH] A series of columns placed at regular intervals.

colony [BIOL] A localized population of individuals of the same species which are living either attached or separately. [MICROBIO] A cluster of microorganisms growing on the surface of or within a solid medium; usually cultured from a single cell.

colony count [MICROBIO] The number of colonies of bacteria growing on the surface of a solid medium.

color [OPTICS] A general term that refers to the wavelength composition of light, with particular reference to its visual appearance.

color aberration *See* chromatic aberration.

Coloradoan [GEOL] Middle Upper Cretaceous geologic time.

coloradoite [MINERAL] HgTe A grayish-black, isometric telluride mineral with a metallic luster; specific gravity is 8.6.

Colorado low [METEOROL] A low which makes its first appearance as a definite center in the vicinity of Colorado on the eastern slopes of the Rocky Mountains; analogous to the Alberta low.

Colorado tick fever [MED] A nonexanthematous acute viral disease of man occurring in the western United States and transmitted by a bite of the tick *Dermacentor andersoni*; characterized by a short course, intermittent fever, leukopenia, and occasionally meningoencephalitis. Also known as mountain tick fever.

color balance [ELECTR] Adjustment of the circuits feeding the three electron guns of a television color picture tube to compensate for differences in light-emitting efficiencies of the three color phosphors on the screen of the tube.

color-bar generator [ELECTR] A signal generator that delivers to the input of a color television receiver the signal needed to produce a color-bar test pattern on one or more channels.

color-blind [GRAPHICS] Of a photographic emulsion, sensitive only to blue, violet, and ultraviolet light.

color blindness [MED] Inability to perceive one or more colors.

color burst [ELECTR] The portion of the composite color television signal consisting of a few cycles of a sine wave of chrominance subcarrier frequency. Also known as burst; reference burst.

color carrier *See* chrominance subcarrier.

color-carrier reference *See* chrominance-carrier reference.

color center [SOLID STATE] A point lattice defect which produces optical absorption bands in an otherwise transparent crystal.

color circle [OPTICS] An arrangement of hues about the circumference of a circle in the order in which they appear in the electromagnetic spectrum, with pairs of complementary colors at opposite ends of diameters.

color code [ELEC] A system of colors used to indicate the electrical value of a component or to identify terminals and leads. [ENG] **1.** Any system of colors used for purposes of identification, such as to identify dangerous areas of a factory. **2.** A system of colors used to identify the type of material carried by a pipe; for example, dangerous materials, protective materials, extra valuable materials.

color coder *See* matrix.

color comparator [ANALY CHEM] A photoelectric instrument that compares an unknown color with that of a standard color sample for matching purposes. Also known as photoelectric color comparator.

color control *See* chroma control.

color correction [OPTICS] The construction of an optical system so that the image positions of an object are the same for two or more wavelengths, and chromatic aberration is thus minimized.

color decoder *See* matrix.

color-difference signal [ELECTR] A signal that is added to the monochrome signal in a color television receiver to obtain a signal representative of one of the three tristimulus values needed by the color picture tube.

color disk [OPTICS] A rotating circular disk having three

filter sections to produce the individual red, green, and blue pictures in a field-sequential color television system.

colored smoke [MATER] Gaseous products of a distinctive color which forms a colored cloud and may be used for signals or target markers.

color encoder *See* matrix.

color equation [ASTRON] In astronomy, a measure of the color sensitivity and response of a method of observation; photographic, visual, or photoelectric techniques may be employed.

color facsimile [COMMUN] A facsimile system for transmission of color photographs, in which three separate facsimile transmissions are made from the original color print, using color-separation filters in the optical system of the facsimile transmitter.

color film [GRAPHICS] A sensitized film used for three-color photography; consists of three emulsions coated one above another and sensitive respectively to blue, green, and red light.

color filter [OPTICS] An optical element that partially absorbs incident light, consisting of a pane of glass or other partially transparent material, or of films separated by narrow layers; the absorption may be either selective or nonselective with respect to wavelength. Also known as light filter.

color fringing [ELECTR] Spurious chromaticity at boundaries of objects in a television picture.

colorimeter [ANALY CHEM] A device for measuring concentration of a known constituent in solution by comparison with colors of a few solutions of known concentration of that constituent. Also known as chromometer. [OPTICS] An instrument that measures color by determining the intensities of the three primary colors that will give that color.

colorimetric photometer [OPTICS] A photometer that can measure light intensities in several spectral regions, using color filters placed in the path of the light.

colorimetry [OPTICS] Any technique by which an unknown color is evaluated in terms of standard colors; the technique may be visual, photoelectric, or indirect by means of spectrophotometry; used in chemistry and physics.

color index Abbreviated CI. [ASTRON] Of a star, the numerical difference between the apparent photographic magnitude and the apparent photovisual magnitude. [PATH] The amount of hemoglobin per erythrocyte relative to normal, equal to the percent normal hemoglobin concentration divided by percent normal erythrocyte count.

coloring agent [FOOD ENG] Any substance of natural origin, such as tumeric, annatto, caramel, carmine, and carotine, or a synthetic certified food color added to food to compensate for color changes during processing or to give an appetizing color.

color killer circuit [ELECTR] The circuit in a color television receiver that biases chrominance amplifier tubes to cutoff during reception of monochrome programs. Also known as killer stage.

color kinescope *See* color picture tube.

color lake *See* lake.

color media [OPTICS] Any colored, transparent material that is placed in front of a lighting unit to color the light transmitted.

color oscillator *See* chroma oscillator.

color phase [COMMUN] The difference in phase between components (I or Q) of a chrominance signal and the chrominance-carrier reference in a color television receiver.

color-phase alternation [COMMUN] The periodic changing of the color phase of one or more components of the chrominance subcarrier between two sets of assigned values after every field in a color television system. Abbreviated cpa.

color-phase detector [ELECTR] The color television receiver circuit that compares the frequency and phase of the incoming burst signal with those of the locally generated 3.579545-megahertz chroma oscillator and delivers a correction voltage to a reactance tube to ensure that the color portions of the picture will be in exact register with the black-and-white portions on the screen.

color photography [GRAPHICS] A film process that uses color film to produce three-color positive pictures.

color picture signal [COMMUN] The electric signal that rep-

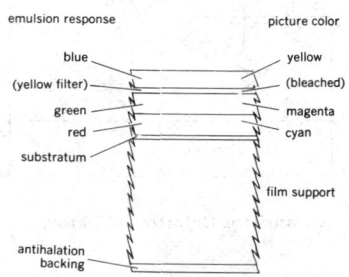

COLOR FILM

Cross section of Kodachrome film showing the various emulsion layers with (left) the emulsion response and (right) the developed color of each layer.

resents complete color picture information, excluding all synchronizing signals.

color picture tube [ELECTR] A cathode-ray tube having three different colors of phosphors, so that when these are appropriately scanned and excited in a color television receiver, a color picture is obtained. Also known as color kinescope; color television picture tube; tricolor picture tube.

color printing [GRAPHICS] The art and craft of embellishing designs, pictures, and typographic pages with color for a more pleasing effect than obtained in black and white; in addition, pictures more closely represent the original object or painting.

color saturation [OPTICS] The degree to which a color is mixed with white; high saturation means little white, low saturation means much white. Also known as chroma; saturation.

color-saturation control *See* chroma control.

color-sensitive [GRAPHICS] Referring to a photographic emulsion which is not color-blind.

color separation [GRAPHICS] The process of preparing a separate drawing, engraving, or negative for each color required in the reproduction of a colored picture.

color signal [COMMUN] Any signal that controls the chromaticity values of a color television picture, such as the color picture signal and the chrominance signal.

color solid [OPTICS] A three-dimensional diagram which represents the relationship of three attributes of surface color: hue, saturation, and brightness.

color stability [CHEM] Resistance of materials to change in color that can be caused by light or aging, as of petroleum or whiskey.

color standard [ANALY CHEM] Liquid solution of known chemical composition and concentration, hence of known and standardized color, used for optical analysis of samples of unknown strength.

color subcarrier *See* chrominance subcarrier.

color-subcarrier oscillator *See* chroma oscillator.

color-subcarrier reference *See* chrominance-carrier reference.

color television [COMMUN] A television system that reproduces an image approximately in its original colors.

color television picture tube *See* color picture tube.

color temperature [STAT MECH] Of a solid surface, that temperature of a blackbody from which the radiant energy has essentially the same spectral distribution as that from the surface.

color test [ANALY CHEM] The quantitative analysis of a substance by comparing the intensity of the color produced in a sample by a reagent with a standard color produced similarly in a solution of known strength.

color-translating microscope [OPTICS] A type of compound microscope that employs three different wavelengths of light to reveal details produced by ultraviolet or other nonvisible radiation.

color transmission [COMMUN] In television, the transmission of a signal wave which represents both the brightness values and the chromaticity values in the picture.

color vision [PHYSIO] The ability to discriminate light on the basis of wavelength composition.

Colossendeidae [INV ZOO] A family of deep-water marine arthropods in the subphylum Pycnogonida, having long palpi and lacking chelifores, except in polymerous forms.

colostomy [MED] Surgical formation of an artificial anus by joining the colon to an opening in the anterior abdominal wall.

colostrum [PHYSIO] The first milk secreted by the mammary gland during the first days following parturition.

colotomy [MED] Incision of the colon; may be abdominal, lateral, lumbar, or iliac, according to the region of entrance.

Colpitts oscillator [ELECTR] An electron-tube oscillator in which a parallel-tuned tank circuit is connected between grid and anode; the tank capacitance consists of two voltage-dividing capacitors in series, with their common connection at cathode potential.

colposcope [MED] An instrument for the visual examination of the vagina and cervix; a vaginal speculum.

colpotomy [MED] Incision of the vagina.

Colubridae [VERT ZOO] A family of cosmopolitan snakes in the order Squamata.

Columba [ASTRON] A constellation, right ascension 6 hours, declination 35°S. Abbreviated Col. Also known as Dove.

Columbia-Gel [MATER] Trade name for a gelatinous explosive whose use is permitted by the U.S. Bureau of Mines regulations.

Columbia unit [BIOL] A unit for the standardization of bacitracin.

Columbidae [VERT ZOO] A family of birds in the order Columbiformes composed of the pigeons and doves.

Columbiformes [VERT ZOO] An order of birds distinguished by a short, pointed bill, imperforate nostrils, and short legs.

columbite [MINERAL] $(Fe,Mn)(Cb,Ta)_2O_6$ An iron-black mineral with a submetallic luster that crystallizes in the orthorhombic system; the chief ore mineral of niobium (columbium); hardness is 6 on Mohs scale, and specific gravity is 5.4–6.5. Also known as dianite; greenlandite; niobite.

columbium *See* niobium.

columella [ANAT] *See* stapes. [BIOL] Any part shaped like a column. [BOT] A sterile axial body within the capsules of certain mosses, liverworts, and many fungi.

column [ADP] A vertical arrangement of characters or other expressions, usually referring to a specific print position on a printer or a vertical area on a card. [CHEM ENG] *See* tower. [ENG] A vertical shaft designed to bear axial loads in compression. [MATH] *See* place. [NUCLEO] A hollow cylinder of water and spray thrown up from an underwater burst of an atomic weapon, through which hot, high-pressure gases are vented to the atmosphere; a somewhat similar column of dirt is formed in an underground explosion. Also known as plume.

columnar epithelium [HISTOL] Epithelium distinguished by elongated, columnar, or prismatic cells.

columnar jointing [GEOL] Parallel, prismatic columns that are formed as a result of contraction during cooling in basaltic flow and other extrusive and intrusive rocks. Also known as columnar structure; prismatic jointing; prismatic structure.

columnar resistance [GEOPHYS] The electrical resistance of a column of air 1 centimeter square, extending from the earth's surface to some specified altitude.

columnar section [GEOL] A vertical strip or scale drawing of the strip taken from a given area or locality showing the sequence of the rock units and their stratigraphic relationship, and indicating the thickness, lithology, age, classification, and fossil content of the rock units. Also known as section.

columnar stem [BOT] An unbranched, cylindrical stem bearing a set of large leaves at its summit, as in palms, or no leaves, as in cacti.

columnar structure [GEOL] *See* columnar jointing. [MINERAL] Mineral structure consisting of parallel columns of slender prismatic crystals. [PETR] A primary sedimentary structure consisting of columns arranged perpendicular to the bedding.

column binary [ADP] The binary representation of data on punched cards in which adjacent positions in a column correspond to adjacent bits of data. Also known as Chinese binary.

column binary card [ADP] A punched card containing data in column binary representation.

column bleed [ANALY CHEM] The loss of carrier liquid during gas chromatography due to evaporation into the gas under analysis.

column chromatography [ANALY CHEM] Chromatographic technique of two general types: packed columns usually contain either a granular adsorbent or a granular support material coated with a thin layer of high-boiling solvent (partitioning liquid); open-tubular columns contain a thin film of partitioning liquid on the column walls and have an opening so that gas can pass through the center of the column.

column crane [MECH ENG] A jib crane whose boom pivots about a post attached to a building column.

column development chromatography [ANALY CHEM] Columnar apparatus for separating or concentrating one or more components from a physical mixture by use of adsorbent packing; as the specimen percolates along the length of the

---

COLPITTS OSCILLATOR

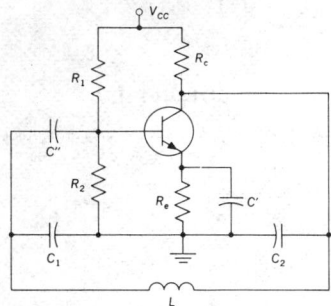

A transistor Colpitts oscillator.

COLUMBITE

Columbite associated with muscovite from Minas Gerais, Brazil. *(American Museum of Natural History)*

adsorbent, its various components are preferentially held at different rates, effecting a separation.

column drill [MECH ENG] A tunnel rock drill supported by a vertical steel column.

column formation *See* trail formation.

column indicator [ADP] The visible column on the program card which shows to the punch card operator the next column to be encountered.

column operations [MATH] A set of rules for manipulating the columns of a matrix so that the image of the corresponding linear transformation remains unchanged.

column order [ADP] The storage of a matrix $a(m,n)$ as $a(1,1)$, $a(2,1), \ldots, a(m,1), a(1,2), \ldots$.

column pipe [MIN ENG] The large cast-iron (or wooden) pipe through which the water is conveyed from the mine pumps to the surface.

column rank [MATH] The number of linearly independent columns of a matrix; the dimension of the image of the corresponding linear transformation.

column space [MATH] The vector space spanned by the columns of a matrix.

column splice [CIV ENG] A connection between two lengths of a compression member (column); an erection device rather than a stress-carrying element.

column vector [MATH] A matrix consisting of only one column.

colure [ASTRON] A great circle of the celestial sphere through the celestial poles and either the equinoxes or solstices, called respectively the equinoctial colure or the solstitial colure.

Colydiidae [INV ZOO] The cylindrical bark beetles, a large family of coleopteran insects in the superfamily Cucujoidea.

colza oil *See* rape oil.

Com *See* Coma Berenices.

coma [ASTRON] The gaseous envelope that surrounds the nucleus of a comet. [ELECTR] A cathode-ray tube image defect that makes the spot on the screen appear comet-shaped when away from the center of the screen. [MED] Unconsciousness from which the patient cannot be aroused. [OPTICS] A manifestation of errors in an optical system, so that a point has an asymmetrical image (that is, appears as a pear-shaped spot).

Coma Berenices [ASTRON] A constellation, right ascension 13 hours, declination 20°N. Abbreviated Com. Also known as Berenice's Hair.

Coma cluster [ASTRON] A specific group of several thousand galaxies.

comagmatic province *See* petrographic province.

coma lobe [ELECTROMAG] Side lobe that occurs in the radiation pattern of a microwave antenna when the reflector alone is tilted back and forth to sweep the beam through space because the feed is no longer always at the center of the reflector; used to eliminate the need for a rotary joint in the feed waveguide.

Comanchian [GEOL] Lower Cretaceous geologic time.

Comasteridae [INV ZOO] A family of radially symmetrical Crinozoa in the order Comatulida.

comatose [MED] In a condition of coma; resembling coma.

Comatulida [INV ZOO] The feather stars, an order of free-living echinoderms in the subclass Articulata.

comb [INV ZOO] 1. A system of hexagonal cells constructed of beeswax by a colony of bees. 2. A comblike swimming plate in ctenophores. [VERT ZOO] A crest of naked tissue on the head of many male fowl.

comb antenna [ELECTROMAG] A broad-band antenna for vertically polarized signals, in which half of a fishbone antenna is erected vertically and fed against ground by a coaxial line.

combat analysis [ORD] A theoretical analysis conducted to determine the probable effectiveness of an existing or projected weapons system under combat conditions.

combatant ship [NAV ARCH] A ship whose main function is combat with an enemy.

combat chart [MAP] A chart overprinted with a system of grid lines as an aid in fire control.

combat development [ORD] The research, development, and testing of new doctrine and organization, and their integra-

tion with new materiel into the army services to obtain the greatest combat effectiveness.

combat development field experiment [ORD] A field trial usually under controlled conditions designed to collect data on operations, organizations, or materiel for use in preparing new or modified operational or organizational concepts or qualitative materiel requirements.

combat development study project [ORD] A study directed toward a determination of operational concepts and techniques, new organizations, or qualitative materiel requirements for the army in the field, or a study which contributes to such determination.

combat development test project [ORD] A troop test directed at the determination of the feasibility and desirability of operational or organizational concepts, the suitability of materiel, or a combination of these tasks.

combat information center [COMMUN] A shipboard location at which tactical information from radar, sonar, and other equipment is received, displayed for rapid analysis, and evaluated.

combat loading [ORD] The division of military cargo into categories for loading aboard ships.

combat serviceable item [ORD] An item of equipment which is complete, ready to perform immediately at its rated capacity, and will remain serviceable under severe operating conditions for a reasonable length of time.

combat tire [ORD] Pneumatic tire of heavy construction which is designed to operate without air pressure for a limited distance in an emergency.

combat vehicle [ORD] A land or amphibious vehicle, with or without armor or armament, designed for specific functions in combat or battle.

combat zone [ORD] A region in a theater of operations where fighting takes place, or where space is designated for the operations of friendly combat forces, extending from the front line to a line or boundary designated by the theater commander.

combescure transformation [MATH] A one-to-one continuous mapping of one space curve onto another space curve so that tangents to corresponding points are parallel.

comb filter [ELECTR] A wave filter whose frequency spectrum consists of a number of equispaced elements resembling the teeth of a comb.

comb growth unit [BIOL] A unit for the standardization of male sex hormones.

combinational circuit [ELECTR] A switching circuit whose outputs are determined only by the concurrent inputs.

combination buoy [NAV] A buoy in which a light and a sound signal are combined such as a lighted bell buoy, lighted gong buoy, lighted whistle buoy, or lighted horn buoy.

combination cable [ELEC] Cable having conductors grouped in both quads and pairs.

combination chuck [DES ENG] A chuck which is used in a lathe and whose jaws either move independently or simultaneously.

combination coefficient [GEOPHYS] A measure of the specific rate of disappearance of small ions in the atmosphere due to either union with neutral Aitken nuclei to form new large ions, or union with large ions of opposite sign to form neutral Aitken nuclei.

combination cycle *See* mixed cycle.

combination die [MET] A die having more than one cavity for different castings.

combination distributing frame [ELEC] Frame which combines the functions of a main distributing frame and an intermediate distributing frame.

combination-drive reservoir [PETRO ENG] A type of reservoir in which hydrocarbons are swept (displaced) toward the drill hole by injection of water followed by liquefied-petroleum-gas or gas injection.

combination fuse [ORD] A fuse combining two different types of fuse mechanisms, especially one combining impact and time mechanisms.

combination lock [ENG] A lock that can be opened only when its dial has been set to the proper combination of symbols, in the proper sequence.

combination mill [MET] A rolling mill arranged with con-

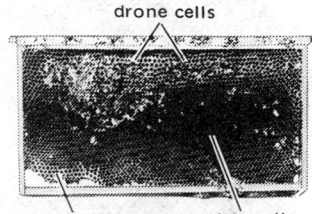

COMB

drone cells

drone cells   worker cells

Drone and worker bee cells in comb. *(Walter T. Kelley Co., Clarkson, Ky.)*

tinuous rolls for roughing and a guide or looping mill for shaping.

combination pliers [DES ENG] Pliers that can be used either for holding objects or for cutting and bending wire.

combination principle *See* Ritz's combination principle.

combination saw [MECH ENG] A saw made in various tooth arrangement combinations suitable for ripping and crosscut mitering.

combination square [DES ENG] A square head and steel rule that when used together have both a 45 and 90° face to allow the testing of the accuracy of two surfaces intended to have these angles.

combination tone [ACOUS] A subjective tone produced by simultaneously sounding two pure tones whose frequencies differ by a large amount.

combination trap [GEOL] Underground reservoir structure closure, deformation, or fault where reservoir rock covers only part of the structure.

combination unit [CHEM ENG] A processing unit that combines more than one process, such as straight-run distillation together with selective cracking.

combination vibration [SPECT] A vibration of a polyatomic molecule involving the simultaneous excitation of two or more normal vibrations.

combination wrench [DES ENG] A wrench that is an open-end wrench at one end and a socket wrench at the other.

combinatorial analysis [MATH] **1.** The determination of the number of possible outcomes in ideal games of chance by using formulas for computing numbers of combinations and permutations. **2.** The study of large finite problems.

combinatorial theory [MATH] The branch of mathematics which studies the arrangements of elements into sets.

combinatorial topology [MATH] The study of polyhedrons, simplicial complexes, and generalizations of these. Also known as piecewise-linear topology.

combinatorics [MATH] Combinatorial topology, which studies geometric forms by breaking them into simple geometric figures.

combined carbon [CHEM] Carbon that is chemically combined within a compound, as contrasted with free or uncombined elemental carbon.

combined cyanide [ORG CHEM] The cyanide portion of a complex ion composed of cyanide and a metal.

combined flexure [MECH] The flexure of a beam under a combination of transverse and longitudinal loads.

combined footing [CIV ENG] A footing consisting of a large concrete slab under two or more columns; also used under pairs of columns subjected to uplift or to opposing or different horizontal stresses.

combined head *See* read/write head.

combined immunological deficiency disease [MED] A severe and usually fatal disease in which the individual lacks not only the T (thymus-derived) cells, which are responsible for graft rejection and for defense against viruses, but also the B (bone-marrow-derived) cells, which are responsible for production of globulins and antibodies. Also known as CID disease.

combined moisture [MIN ENG] Moisture in coal that cannot be removed by ordinary drying.

combined sewers [CIV ENG] A drainage system that receives both surface runoff and sewage.

combined stresses [MECH] Bending or twisting stresses in a structural member combined with direct tension or compression.

combine-harvester [AGR] A machine for harvesting legumes and grasses; cuts off the plant tops, then beats and cleans the grain while the machine continues to move across the field.

combiner circuit [ELECTR] The circuit that combines the luminance and chrominance signals with the synchronizing signals in a color television camera chain.

combing [TEXT] Elimination of short, curly fibers by means of a machine process.

combing machine [TEXT] A machine that combs fibers, separating the longer ones from the shorter.

combining-volumes principle [CHEM] The principle that when gases take part in chemical reactions the volumes of the reacting gases and those of the products (if gaseous) are in the ratio of small whole numbers, provided that all measurements are made at the same temperature and pressure. Also known as Gay-Lussac law.

combining weight [CHEM] The weight of an element that chemically combines with 8 grams of oxygen or its equivalent.

comb nephoscope [ENG] A direct-vision nephoscope constructed with a comb (a crosspiece containing equispaced vertical rods) attached to the end of a column 8-10 feet (2.4-3 meters) long and supported on a mounting that is free to rotate about its vertical axis; in use, the comb is turned so that the cloud appears to move parallel to the tips of the vertical rods.

combustible gas [MATER] A gas that burns, including the fuel gases, hydrogen, hydrocarbon, carbon monoxide, or a mixture of these.

combustible shale *See* tasmanite.

combustion [CHEM] The burning of gas, liquid, or solid, in which the fuel is oxidized, evolving heat and often light.

combustion chamber [AERO ENG] That part of the rocket engine in which the combustion of propellants takes place at high pressure. Also known as firing chamber. [ENG] Any chamber in which a fuel such as oil, coal, or kerosine is burned to provide heat. [MECH ENG] The space at the head end of an internal combustion engine cylinder where most of the combustion takes place.

combustion deposit [ENG] A layer of ash on the heat-exchange surfaces of a combustion chamber, resulting from the burning of a fuel.

combustion efficiency [CHEM] The ratio of heat actually developed in a combustion process to the heat that would be released if the combustion were perfect.

combustion engine [MECH ENG] An engine that operates by the energy of combustion of a fuel.

combustion engineering [MECH ENG] The design of combustion furnaces for a given performance and thermal efficiency, involving study of the heat liberated in the combustion process, the amount of heat absorbed by heat elements, and heat-transfer rates.

combustion furnace [ANALY CHEM] A heating device used in the analysis of organic compounds for elements. [ENG] A furnace whose source of heat is the energy released in the oxidation of fossil fuel.

combustion instability [AERO ENG] Unsteadiness or abnormality in the combustion of fuel, as may occur in a rocket engine.

combustion knock *See* engine knock.

combustion nucleus [METEOROL] A condensation nucleus formed as a result of industrial or natural combustion processes.

combustion rate [CHEM] The rate of burning of any substance.

combustion shock [ENG] Shock resulting from abnormal burning of fuel in an internal combustion engine, caused by preignition or fuel-air detonation; or in a diesel engine, the uncontrolled burning of fuel accumulated in the combustion chamber.

combustion train [ANALY CHEM] The arrangement of apparatus for elementary organic analysis.

combustion tube [ANALY CHEM] A glass, silica, or porcelain tube, resistant to high temperatures, that is a component of a combustion train.

combustion turbine *See* gas turbine.

combustion wave [CHEM] **1.** A zone of burning propagated through a combustible medium. **2.** The zoned, reacting, gaseous material formed when an explosive mixture is ignited.

combustor [MECH ENG] The combustion chamber together with burners, igniters, and injection devices in a gas turbine or jet engine.

come-along [DES ENG] **1.** A device for gripping and effectively shortening a length of cable, wire rope, or chain by means of two jaws which close when one pulls on a ring. **2.** *See* puller.

comedo [MED] A collection of sebaceous material and keratin retained in the hair follicle and excretory duct of the sebaceous gland, whose surface is covered with a black dot

---

COMBINED FOOTING

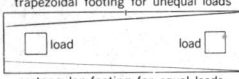

trapezoidal footing for unequal loads

load · load

rectangular footing for equal loads

Rectangular and trapezoidal shapes of combined footing.

COMBUSTION CHAMBER

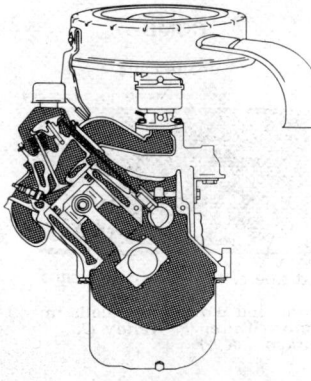

Combustion chamber as found in a spark-ignition engine. (*From L. C. Lichty, Internal-Combustion Engines, 6th ed., McGraw-Hill, 1951*)

caused by oxidation of sebum at the follicular orifice. Also known as blackhead.

comedocarcinoma [MED] A type of adenocarcinoma of the breast in which the ducts are filled with cells which, when expressed from the cut surface, resemble comedos.

comendite [GEOL] A white, sodic rhyolite containing alkalic amphibole or pyroxene.

comes [ASTRON] The smaller star in a binary system. Also known as companion.

Comesomatidae [INV ZOO] A family of free-living nematodes in the superfamily Chromadoroidea found as deposit feeders on soft bottom sediments.

comet [ASTRON] A nebulous celestial body having a fuzzy head surrounding a bright nucleus, one of two major types of bodies moving in closed orbits about the sun; in comparison with the planets, the comets are characterized by their more eccentric orbits and greater range of inclination to the ecliptic.

comet family [ASTRON] Those short-period comets whose aphelia correspond closely to Jupiter's orbit.

comet group [ASTRON] A division of comets on the basis of their period; the short-period group (periods of less than 200 years) contains orbits that show a strong preference for the plan of the solar system; the planes of long-period comets are randomly distributed.

comfort chart [ENG] A diagram showing curves of relative humidity and effective temperature superimposed upon rectangular coordinates of wet-bulb temperature and dry-bulb temperature.

comfort control [ENG] Control of temperature, humidity, flow, and composition of air by using heating and air-conditioning systems, ventilators, or other systems to increase the comfort of people in an enclosure.

comfort curve [ENG] A line drawn on a graph of air temperature versus some function of humidity (usually wet-bulb temperature or relative humidity) to show the varying conditions under which the average sedentary person feels the same degree of comfort; a curve of constant comfort.

comfort index See temperature humidity index.

comfort standard See comfort zone.

comfort zone [ENG] The ranges of indoor temperature, humidity, and air movement, under which most persons enjoy mental and physical well-being. Also known as comfort standard.

COMIT [ADP] A user-oriented, general-purpose, symbol-manipulation programming language for computers.

Comleyan [GEOL] Lower Cambrian geologic time.

comlognet [COMMUN] A long-line data-transmission network for rapid processing and distribution of logistic information at a computer-equipped automatic switching center.

command [ADP] A signal that initiates a predetermined type of computer operation that is defined by an instruction. [CONT SYS] An independent signal in a feedback control system, from which the dependent signals are controlled in a predetermined manner.

command code See operation code.

command control See command guidance.

command control program [ADP] The interface between a time-sharing computer and its users by means of which they can create, edit, save, delete, and execute their programs.

command destruct [CONT SYS] A command control system that destroys a flightborne test rocket or a guided missile, actuated by the safety officer whenever the vehicle's performance indicates a safety hazard.

command guidance [ENG] A type of electronic guidance of guided missiles or other guided aircraft wherein signals or pulses sent out by an operator cause the guided object to fly a directed path. Also known as command control.

command mode [ADP] The status of a terminal in a time-sharing environment enabling the programmer to use the command control program.

command net [ORD] The several points that may exercise control over airborne guided missiles.

command pulses [ELECTR] The electrical representations of bit values of 1 or 0 which control input/output devices.

command set [COMMUN] A radio set used to receive or give commands, as between one aircraft and another or between an aircraft and the ground.

Commelinaceae [BOT] A family of monocotyledonous plants in the order Commelinales characterized by differentiation of the leaves into a closed sheath and a well-defined, commonly somewhat succulent blade.

Commelinales [BOT] An order of monocotyledonous plants in the subclass Commelinidae marked by having differentiated sepals and petals but lacking nectaries and nectar.

Commelinidae [BOT] A subclass of flowering plants in the class Liliopsida.

commensal [ECOL] An organism living in a state of commensalism.

commensalism [ECOL] An interspecific, symbiotic relationship in which two different species are associated, wherein one is benefited and the other neither benefited nor harmed.

comment [ADP] An expression identifying or explaining one or more steps in a routine, which has no effect on execution of the routine.

comment code [ADP] One or more characters identifying a comment.

commercial diesel cycle See mixed cycle.

commercial harbor [CIV ENG] A harbor in which docks are provided with cargo-handling facilities.

commercial lecithin See lecithin.

commercial mine [MIN ENG] A coal mine operated to supply purchasers in general, as contrasted with a captive mine.

commercial ore [MIN ENG] Mineralized material profitable at prevailing metal prices.

commercial-type vehicle [ORD] A motor vehicle designed to meet civilian requirements and used by the armed services, without major modification, for routine purposes in connection with the transportation of supplies, personnel, and equipment.

comminution [MECH ENG] Breaking up or grinding into small fragments. Also known as pulverization.

comminutor [MECH ENG] A machine that breaks up solids.

commissioned [NAV] Pertaining to an aid to navigation which, after it has been constructed, tested, and placed in operation, is declared suitable for use by vessels or crafts.

commission ore [MIN ENG] Uranium-bearing ore of 0.10% U_3O_8 or higher, for which the U.S. Atomic Energy Commission has an established price.

commissure [BIOL] A joint, seam, or closure line where two structures unite.

commissurotomy [MED] The surgical destruction of a commissure, usually the anterior commissure, particularly in the treatment of certain psychiatric disorders.

committed [ORD] Pertaining to the condition of a fuse when the arming process has reached the point from which arming will continue to completion in spite of the absence of any arming forces.

common area [ADP] An area of storage which two or more routines share.

common-base connection See grounded-base connection.

common battery [COMMUN] System of current supply where all direct current energy for a unit of a telephone system is supplied by one source in a central office or exchange.

common bile duct [ANAT] The duct formed by the union of the hepatic and cystic ducts.

common bond See American bond.

common brick [MATER] Brick made from natural clay.

common business-oriented language See COBOL.

common carotid artery See carotid artery.

common carrier [IND ENG] A company recognized by an appropriate regulatory agency as having a vested interest in furnishing communications services or in transporting commodities or people.

common cold [MED] A viral disease of humans most frequently caused by the rhinovirus and accompanied by inflammation of the mucous membranes of the nose, throat, and eyes.

common-collector connection See grounded-collector connection.

common control unit [ADP] Control unit that is shared by more than one machine.

common declaration statement [ADP] A nonexecutable statement in FORTRAN which allows specified arrays or variables to be stored in an area available to other programs.

common denominator [MATH] Any common multiple of the denominators of a collection of fractions.

common-drain amplifier [ELECTR] An amplifier using a field-effect transistor so that the input signal is injected between gate and drain, while the output is taken between the source and drain. Also known as source-follower amplifier.

common-emitter connection *See* grounded-emitter connection.

common establishment *See* high-water full and change.

common feldspar *See* orthoclase.

common fraction [MATH] A fraction whose numerator and denominator are both integers.

common-gate amplifier [ELECTR] An amplifier using a field-effect transistor in which the gate is common to both the input circuit and the output circuit.

common hepatic duct *See* hepatic duct.

common iliac artery *See* iliac artery.

common-ion effect [CHEM] The lowering of the degree of ionization of a compound when another ionizable compound is added to a solution; the compound added has a common ion with the other compound.

common item [ORD] Piece of equipment or materiel of supply, used or procured by more than one technical service.

common joist [BUILD] An ordinary floor beam to which floor boards are attached.

common labor [IND ENG] Unskilled workers.

common language [ADP] A machine-readable language that is common to a group of computers and associated equipment.

common language family [ADP] Various types of business machines, capable of being operated automatically and interchangeably by the same piece of punched paper tape, are said to be of a common language family.

common logarithm [MATH] The exponent in the representation of a number as a power of 10. Also known as Briggs' logarithm.

common mica *See* muscovite.

common-mode rejection [ELECTR] The ability of an amplifier to cancel a common-mode signal while responding to an out-of-phase signal. Also known as in-phase rejection.

common-mode signal [ELECTR] A signal applied equally to both ungrounded inputs of a balanced amplifier stage or other differential device. Also known as in-phase signal.

common multiple [MATH] A quantity (polynomial number) divisible by all quantities in a given set.

common projectile [ORD] A penetrating type of projectile containing a bursting charge of high explosive, intended to explode after passing through the lighter protective armor of a tank or naval vessel.

common pyrite *See* pyrite.

common rafter [BUILD] A rafter to which roofing is attached.

common return [ELECTR] A return conductor that serves two or more circuits.

common salt *See* halite; sodium chloride.

common-source amplifier [ELECTR] An amplifier stage using a field-effect transistor in which the input signal is applied between gate and source and the output signal is taken between drain and source.

common storage [ADP] A section of memory in certain computers reserved for temporary storage of program outputs to be used as input for other programs.

common-user channel [COMMUN] Any of the communications channels which are available to all authorized agencies for transmission of command, administrative, and logistic traffic.

common user circuit [ELEC] A circuit designated to furnish a communications service to a number of users.

common year [ASTRON] A calendar year of 365 days.

communicable disease [MED] An infectious disease that can be transmitted from one individual to another either directly by contact or indirectly by fomites and vectors.

communicating hydrocephaly [MED] A form of hydrocephaly in which there is normal communication between the ventricles and the subarachnoid space of the brain.

communication [COMMUN] The transmission of intelligence between two or more points over wires or by radio; the terms telecommunication and communication are often used inter-

changeably, but telecommunication is usually the preferred term when long distances are involved.

communication band [COMMUN] The band of frequencies effectively occupied by a radio transmitter for the type of transmission and the speed of signaling used.

communication cable [COMMUN] A uniform conductive circuit used in the telephone industry to connect customers to their local switching centers and to interconnect local and long-distance switching centers.

communication channel [COMMUN] The wire or radio channel that serves to convey intelligence between two or more terminals.

communication countermeasure [COMMUN] Any electronic countermeasure against communications, such as jamming.

communication engineering [COMMUN] The design, construction, and operation of all types of equipment used for radio, wire, or other types of communication.

communication link *See* data link.

communication receiver [ELECTR] A receiver designed especially for reception of voice or code messages transmitted by radio communication systems.

communications [ENG] The science and technology by which information is collected from an originating source, transformed into electric currents or fields, transmitted over electrical networks or space to another point, and reconverted into a form suitable for interpretation by a receiver.

communications intelligence [COMMUN] Technical and intelligence information derived from foreign communications by other than the intended recipients.

communications language [COMMUN] A language structure complete with conventions, syntax, and character set, used primarily for conveying knowledge of processes between two participants.

communications network [COMMUN] Organization of stations capable of intercommunications but not necessarily on the same channel.

communication speed [COMMUN] The rate at which information is transmitted over a communications channel, adjusted for redundancies.

communications relay station [COMMUN] Facility for rapidly passing message traffic from one tributary to another by automatic, semiautomatic, or manual means, or by electrically connecting circuits (circuit switching) between two tributaries for direct transmission.

communications satellite [AERO ENG] An orbiting, artificial earth satellite that relays radio, television, and other signals between ground terminal stations thousands of miles apart. Also known as radio relay satellite; relay satellite.

Communications Satellite Corporation [COMMUN] A common carrier created under the provisions of the Communications Satellite Act of 1962, to provide communications satellite service. Abbreviated COMSAT.

communications traffic [COMMUN] All transmitted and received messages.

communication system [COMMUN] A telephone, telegraph, teletypewriter, picture, data transmission, or other system in which electrical impulses originated at one place are faithfully reproduced at a distant point.

communications zone indicator [ELECTR] Device to indicate whether or not long-distance high-frequency broadcasts are successfully reaching their destinations.

communication theory [COMMUN] The mathematical theory of the communication of information from one point to another.

community [ECOL] Aggregation of organisms characterized by a distinctive combination of two or more ecologically related species; an example is a deciduous forest. Also known as ecological community.

community antenna television system [COMMUN] A television program distribution system for poor-reception areas, in which signals are picked up by a high and sensitive antenna array, amplified for all receivable channels, and fed through coaxial cables to receivers in individual homes. Abbreviated CATV.

community classification [ECOL] Arrangement of communities into classes with respect to their complexity and extent, their stage of ecological succession, or their primary production.

COMMUNICATIONS SATELLITE

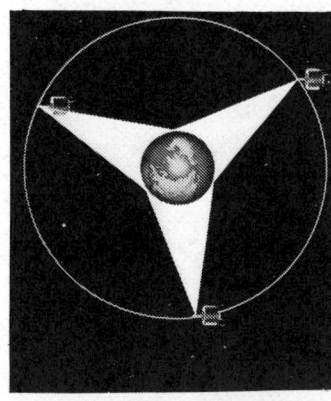

Three satellites in synchronous equatorial orbits to provide nearly worldwide coverage. Each satellite is about 22,300 miles (35,900 kilometers) above the surface of the earth.

community dial office [COMMUN] Small dial office with no employees located in the building serving an exchange area.

commutating capacitor [ELECTR] A capacitor used in gas-tube rectifier circuits to prevent the anode from going highly negative immediately after extinction.

commutating pole [ELECTROMAG] One of several small poles between the main poles of a direct-current generator or motor, which serves to neutralize the flux distortion in the neutral plane caused by armature reaction. Also known as compole; interpole.

commutating reactance [ELECTR] An inductive reactance placed in the cathode lead of a three-phase mercury-arc rectifier to ensure that tube current holds over during transfer of conduction from one anode to the next.

commutation [COMMUN] The sampling of various quantities in a repetitive manner for transmission over a single channel in telemetering. [ELECTR] The transfer of current from one channel to another in a gas tube. [ELECTROMAG] The process of current reversal in the armature windings of a direct-current rotating machine to provide direct current at the brushes.

commutation rules [QUANT MECH] The specification of the commutators of operators corresponding to the dynamical variables of a system, which are equal to $i\hbar$ times the Poisson brackets of the classical variables to which the operators correspond.

commutative group *See* Abelian group.

commutative law [MATH] A rule which requires that the result of a binary operation be independent of order; that is, $ab = ba$.

commutative ring [MATH] A ring in which the multiplication obeys the commutative law.

commutator [ELECTROMAG] That part of a direct-current motor or generator which serves the dual function, in combination with brushes, of providing an electrical connection between the rotating armature winding and the stationary terminals, and of permitting reversal of the current in the armature windings. [MATH] The commutator of a and b is the element c of a group such that $bac = ab$. [QUANT MECH] The commutator of a and b is $[a,b] = ab - ba$.

commutator-controlled welding [MET] Spot or projection welding in which several electrodes in contact with the work simultaneously are operated under the control of an electrical commutating device. Also known as ultraspeed welding.

commutator head [ELEC] The butt end of a commutator.

commutator motor [ELEC] An electric motor having a commutator.

commutator switch [ELEC] A switch, usually rotary and mechanically driven, that performs a set of switching operations in repeated sequential order, such as is required for telemetering many quantities. Also known as sampling switch; scanning switch.

comose [BOT] Having a tuft of soft hairs.

compact [MET] A briquette made by the compression of metal powder, with or without the addition of nonmetallic constituents.

compaction [ENG] Increasing the dry density of a granular material, particularly soil, by means such as impact or by rolling the surface layers. [GEOL] Process by which soil and sediment mass loses pore space in response to the increasing weight of overlying material.

compact-open topology [MATH] A topology on the space of all continuous functions from one topological space into another; a subbase for this topology is given by the sets $W(K,U) = \{f : f(K) \subset U\}$, where K is compact and U is open.

compact operator [MATH] A linear transformation from one normed vector space to another, with the property that the image of every bounded set has a compact closure.

compactor [MECH ENG] Machine designed to consolidate earth and paving materials by kneading, weight, vibration, or impact, to sustain loads greater than those sustained in an uncompacted state.

compact set [MATH] A set in a topological space with the property that every open cover has a finite subset which is also a cover. Also known as bicompact set.

compact space [MATH] A topological space which is a compact set.

companding [ELECTR] A process in which compression is followed by expansion; often used for noise reduction in equipment, in which case compression is applied before noise exposure and expansion after exposure.

compandor [ELECTR] A system for improving the signal-to-noise ratio by compressing the volume range of the signal at a transmitter or recorder by means of a compressor and restoring the normal range at the receiving or reproducing apparatus with an expander.

companion *See* comes.

companion body [AERO ENG] A nose cone, last-stage rocket, or other body that orbits along with an earth satellite or follows a space probe.

companion cell [BOT] A specialized parenchyma cell occurring in close developmental and physiologic association with a sieve-tube member.

companion flange [DES ENG] A pipe flange that can be bolted to a similar flange on another pipe.

companionway [NAV ARCH] A stairway that runs from one deck of a ship to another.

compar [MATER] Generic term for a family of compounded, modified, and plasticized polyvinyl alcohol resins; used for making full- and solvent-resistant tubing and hose, and printing rolls.

comparative embryology [EMBRYO] A branch of embryology that deals with the similarities and differences in the development of animals or plants of different orders.

comparative experiments [STAT] Experiments conducted to determine statistically whether one procedure is better than another.

comparative pathology [MED] Investigation and comparison of disease in various animals, including man, to arrive at resemblances and differences which may clarify disease as a phenomenon of nature.

comparative rabal [ENG] A rabal observation (that is, a radiosonde balloon tracked by theodolite) taken simultaneously with the usual rawin observation (tracking by radar or radio direction-finder), to provide a rough check on the alignment and operating accuracy of the electronic tracking equipment.

comparator [ADP] A device that compares two transcriptions of the same information to verify the accuracy of transcription, storage, arithmetical operation, or some other process in a computer, and delivers an output signal of some form to indicate whether or not the two sources are equal or in agreement. [CONT SYS] A device which detects the value of the quantity to be controlled by a feedback control system and compares it continuously with the desired value of that quantity. [ENG] A device used to inspect a gaged part for deviation from a specified dimension, by mechanical, electrical, pneumatic, or optical means.

comparator circuit [ELECTR] An electronic circuit that produces an output voltage or current whenever two input levels simultaneously satisfy predetermined amplitude requirements; may be linear (continuous) or digital (discrete).

comparator-densitometer [ANALY CHEM] Device that projects a labeled spectrum onto a screen adjacent to an enlarged image of the spectrum to be analyzed, allowing visual comparison.

comparing brushes [ADP] Sets of metallic brushes which verify that all the cards in a gang-punching operation have been properly punched.

comparing unit [ELECTR] An electromechanical device which compares two groups of timed pulses and signals to establish either identity or nonidentity.

comparing watch [HOROL] A hack watch, particularly one having an error determined by comparison with a chronometer.

comparison [ADP] A computer operation in which two numbers are compared as to identity, relative magnitude, or sign.

comparison bridge [ELECTR] A bridge circuit in which any change in the output voltage with respect to a reference voltage creates a corresponding error signal, which, by means of negative feedback, is used to correct the output voltage and thereby restore bridge balance.

comparison indicators [ADP] Registers, one of which is activated during the comparison of two quantities to indicate

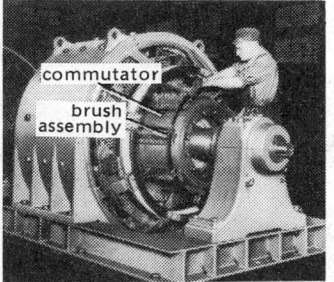
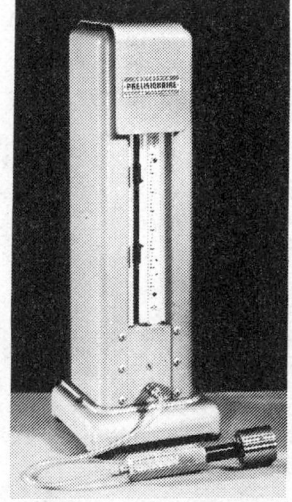

whether the first quantity is lower than, equal to, or greater than the second quantity.

comparison lamp [OPTICS] An incandescent lamp whose luminous intensity is constant (although not necessarily known), and which is compared against other lamps in a photometer.

comparison microscope [OPTICS] 1. An arrangement of two microscopes connected by a special receiving ocular so that the field of one microscope is seen at one side of a dividing line and the field of the other microscope at the opposite side. 2. A projection type of microscope in which the image is compared with a template or known pattern.

comparison spectrum [SPECT] A line spectrum whose wavelengths are accurately known, and which is matched with another spectrum to determine the wavelengths of the latter.

compartment [MIN ENG] A section of a mine shaft separated by framed timbers and planking.

compartment mill [MECH ENG] A multisection pulverizing device divided by perforated partitions, with preliminary grinding at one end in a short ball-mill operation, and finish grinding at the discharge end in a longer tube-mill operation.

compass [ENG] An instrument for indicating a horizontal reference direction relative to the earth. [GRAPHICS] An instrument used for describing arcs or circles with pencil or pen; has two legs hinged together at the top.

compass adjustment [NAV] The process of neutralizing the magnetic effect a craft exerts on a magnetic compass: permanent magnets and soft iron correctors are arranged about the binnacle so that their effects are nearly equal and opposite to the magnetic material in the craft, thus reducing the deviations and eliminating the sectors of sluggishness and unsteadiness.

compass amplitude [NAV] In marine navigation, amplitude relative to compass east or west.

compass azimuth [NAV] Azimuth relative to compass north.

compass bearing [NAV] Direction relative to north as indicated by a compass.

compass bowl [ENG] That part of a compass in which the compass card is mounted.

compass calibration See swinging ship.

compass card [DES ENG] The part of a compass on which the direction graduations are placed, it is usually in the form of a thin disk or annulus graduated in degrees, clockwise from 0° at the reference direction to 360°, and sometimes also in compass points.

compass card axis [DES ENG] The line joining 0° and 180° on a compass card.

compass compensation [NAV] 1. The process of neutralizing the effects degaussing currents exert on a marine magnetic compass. 2. The process of neutralizing the magnetic effect an aircraft exerts on a magnetic compass.

compass correction card [NAV] A card posted in a holder near the magnetic compass, on which there is recorded the difference between the readings of the compass and the correct geomagnetic directions; these errors (deviations) are given for at least the four cardinal points; sometimes the card lists the compass bearings to be flown when it is desired to fly corresponding magnetic headings.

compass corrector [NAV] A magnet or piece of iron placed close to a ship's or aircraft's compass to neutralize the effect of the ship's or aircraft's magnetic field. Also known as Flinder's bar.

compass course [NAV] Course relative to compass north.

compass declinometer [ENG] An instrument used for magnetic distribution surveys; employs a thin compass needle 6 inches (15 centimeters) long, supported on a sapphire bearing and steel pivot of high quality; peep sights serve for aligning the compass box on an azimuth mark.

compass deviation [NAV] The difference between the readings of a compass which is without mechanical defects and is held motionless in space, and the same instrument when it is installed in the same geographic position but is mounted on a ship or aircraft; deviation is a systematic error which is compensated by placing iron bars in places about the compass; residual deviation errors are calibrated and noted on a card so it can be used by the pilot or navigator.

compass direction [NAV] Horizontal direction expressed as angular distance from compass north.

compass error [NAV] The angle by which a compass direction differs from the true geographical direction; assuming that there is no lubber line error, it is the algebraic sum of the declination, variation, and motion errors.

Compasses See Circinus.

compass heading [NAV] Heading relative to compass north.

compass locator [NAV] A low-power, low-frequency, nondirectional radio beacon installed near an airfield to facilitate instrument approaches.

compass meridian [NAV] A line through the north-south points of a magnetic compass, in which the compass card axis lies.

compass motion error [NAV] Errors in compass reading when the craft or vessel is under way; these errors are caused by northerly turning, speed acceleration and deceleration, the liquid in the compass swirling on turns, and vibration caused by engines; these are all random errors, and compensation is not attained in the simple compass.

compass north [NAV] The direction of north as indicated by a magnetic compass; the reference direction for measurement of compass directions.

compass points [GEOD] The 32 divisions of a compass at intervals of $11\frac{1}{4}°$, with each division further divided into quarter points. Also known as points of the compass.

compass prime vertical [NAV] The vertical circle through the compass east and west points of the horizon.

compass repeater [NAV] That part of a remote-indicating compass system which repeats at a distance the indications of the master compass. Also known as repeater compass.

compass roof [BUILD] A roof in which each truss is in the form of an arch.

compass rose [NAV] A graduated circle, usually marked in degrees, indicating directions north, south, east, and west, and inscribed on a chart; used for the calibration of compasses on crafts.

compass saw [DES ENG] A handsaw which has a handle with several attachable thin, tapering blades of varying widths, making it suitable for a variety of work, such as cutting circles and curves.

compass track [NAV] The direction of the track relative to compass north.

compass transmitter [NAV] That part of a remote-indicating compass system which sends the direction indications to the repeaters.

compass variation error [NAV] The angular difference between the geographic and magnetic meridians at any place as read on the magnetic compass; variation errors vary in time and place, but they are systematic and can readily be canceled out merely by referring to government-published variation charts; the importance of this error may be judged by this example: the variation in northern California is 20° east while in northern New York it is 10° west, thus a pilot attempting to fly between the two places who failed to correct would be off by 30°. Also known as declination error.

compatibility [ADP] The ability of one device to accept data handled by another device without conversion of data or modification of the code. [IMMUNOL] Ability of two bloods or other tissues to unite and function together. [ORD] In ammunition, the ability of a given material to exist unchanged under certain conditions of temperature and moisture when in the presence of some other material. [PHARM] The capacity of two or more ingredients in a medicine to mix without chemical change or loss of therapeutic effectiveness. [SYS ENG] The ability of a new system to serve users of an old system.

compatible color television system [COMMUN] A color television system that permits the substantially normal monochrome reception of the transmitted color picture signal on a typical unaltered monochrome receiver.

compatible monolithic integrated circuit [ELECTR] Device in which passive components are deposited by thin-film techniques on top of a basic silicon-substrate circuit containing the active components and some passive parts.

compatible single-sideband system [COMMUN] A single-sideband system that can be received by an ordinary amplitude-modulation radio receiver without distortion.

compensated amplifier [ELECTR] A broad-band amplifier in which the frequency range is extended by choice of circuit constants.

compensated ionization chamber [NUCLEO] An arrangement of two ionization chambers in parallel, with potentials reversed, used as a radiation null indicator.

compensated-loop direction finder [ELECTR] A direction finder employing a loop antenna and a second antenna system to compensate for polarization error.

compensated pendulum [DES ENG] A pendulum made of two materials with different coefficients of expansion so that the distance between the point of suspension and center of oscillation remains nearly constant when the temperature changes.

compensated semiconductor [ELECTR] Semiconductor in which one type of impurity or imperfection (for example, donor) partially cancels the electrical effects on the other type of impurity or imperfection (for example, acceptor).

compensated volume control *See* loudness control.

compensated winding *See* pole-face winding.

compensating capacitor *See* balancing capacitor.

compensating eyepiece [OPTICS] A type of Huygens eyepiece in which the eye lens is achromatized to compensate for the color errors of the objective.

compensating network [CONT SYS] A network used in a low-energy-level method for suppression of excessive oscillations in a control system.

compensating plate [OPTICS] The first of two plates in a Brace compensator, which covers the entire field of view.

compensation [CONT SYS] Introduction of additional equipment into a control system in order to reshape its root locus so as to improve system performance. Also known as stabilization. [ELECTR] Modification of the amplitude-frequency response of an amplifier to broaden the bandwidth or to make the response more nearly uniform over the existing bandwidth. Also known as frequency compensation. [PSYCH] Counterbalancing a weakness or failure in one area by stressing or substituting a strength or success in another area.

compensation balance [HOROL] A balance that compensates for changes caused by temperature variations.

compensation pendulum [HOROL] A clock pendulum which has compensating mechanisms that are designed to maintain constant pendulum length in spite of temperature changes.

compensation point [BOT] The light intensity at which the amount of carbon dioxide released in respiration equals the amount used in photosynthesis, and the amount of oxygen released in photosynthesis equals the amount used in respiration.

compensation signals [ENG] In telemetry, signals recorded on a tape, along with the data and in the same track as the data, used during the playback of data to correct electrically the effects of tape-speed errors.

compensator [CONT SYS] A device introduced into a feedback control system to improve performance and achieve stability. Also known as filter. [ELECTR] A component that offsets an error or other undesired effect. [OPTICS] A device, usually consisting of two quartz wedges, for determining the phase difference between the two components of elliptically polarized light.

compensatory emphysema [MED] Simple, nonobstructive overdistension of lung segments or an entire lung in intrathoracic adaptation to collapse, destruction, or removal of portions of the lung or the opposite lung.

compensatory hypertrophy [MED] An increase in the size of an organ following injury or removal of the opposite paired organ or of part of the same organ.

competence [EMBRYO] The ability of a reacting system to respond to the inductive stimulus during early developmental stages. [GEOL] The ability of the wind to transport solid particles either by rolling, suspension, or saltation (intermittent rolling and suspension); usually expressed in terms of the weight of a single particle. [HYD] The ability of a stream, flowing at a given velocity, to move the largest particles.

competent beds [GEOL] Beds or strata capable of withstanding the pressures of folding without flowing or changing in original thickness.

competing equilibria condition [CHEM] The competition for a reactant in a complex chemical system in which several reactions are taking place at the same time.

competition [ECOL] The inter- or intraspecific interaction resulting when several individuals share an environmental necessity.

competitive enzyme inhibition [BIOCHEM] Prevention of an enzymatic process resulting from the reversible interaction of an inhibitor with a free enzyme.

compile [ADP] To prepare a machine-language program automatically from a program written in a higher programming language, usually generating more than one machine instruction for each symbolic statement.

compiler [ADP] A program to translate a higher programming language into machine language. Also known as compiling routine.

compiler-level language [ADP] A higher-level language normally supplied by the computer manufacturer.

compiler system [ADP] The set consisting of a higher-level language, such as FORTRAN, and its compiler which translates the program written in that language into machine-readable instructions.

compiling routine *See* compiler.

compital [BOT] **1.** Of the vein of a leaf, intersecting at a wide angle. **2.** Of a fern, bearing sori at the intersection of two veins.

complement [IMMUNOL] A heat-sensitive, complex system in fresh human and other sera which, in combination with antibodies, is important in the host defense mechanism against invading microorganisms. [MATH] **1.** The complement of a number A is another number B such that the sum $A + B$ will produce a specified result. **2.** *See* radix complement.

complemental air [PHYSIO] The amount of air that can still be inhaled after a normal inspiration.

complementarity [QUANT MECH] The principle that nature has complementary aspects, particle and wave; the two aspects are related by $p = h/\lambda$ and $E = h\nu$, where p and E are the momentum and energy of the particle, λ and ν are the wavelength and frequency of the wave, and h is Planck's constant.

complementary angle [MATH] One of a pair of angles whose sum is 90°.

complementary colors [OPTICS] Two colors which lie on opposite sides of the white point in the chromaticity diagram so that an additive mixture of the two, in appropriate proportions, can be made to yield an achromatic mixture.

complementary function [MATH] Any solution of the equation obtained from a given linear differential equation by replacing the inhomogeneous term with zero.

complementary genes [GEN] Nonallelic genes that complement one another's expression in a trait.

complementary metal-oxide semiconductor device *See* CMOS device.

complementary minor *See* minor.

complementary rocks [GEOL] Rocks which are differentiated from the same magma, and whose average composition is the same as the parent magma.

complementary symmetry [ELECTR] A circuit using both *pnp* and *npn* transistors in a symmetrical arrangement that permits push-pull operation without an input transformer or other form of phase inverter.

complementary transistors [ELECTR] Two transistors of opposite conductivity (*pnp* and *npn*) in the same functional unit.

complementary wave [ELECTROMAG] Wave brought into existence at the ends of a coaxial cable, or two-conductor transmission lines, or any discontinuity along the line.

complementary wavelength [OPTICS] The wavelength of light that, when combined with a sample color in suitable proportions, matches a reference standard light.

complementation [GEN] The action of complementary genes. [MATH] The act of replacing a set by its complement.

complementation law [STAT] The law that the probability of an event E is 1 minus the probability of the event not E.

complemented lattice [MATH] A lattice with distinguished elements *a* and *b*, and with the property that corresponding to each point *x* of the lattice, there is a *y* such that the greatest lower bound of *x* and *y* is *a*, and the least upper bound of *x* and *y* is *b*.

complement fixation [IMMUNOL] The binding of complement to an antigen-antibody complex so that the complement is unavailable for subsequent reaction.

complement-fixation test [IMMUNOL] A diagnostic test to determine the presence of antigen or antibody in the blood by adding complement to the test system; used especially in diagnosing syphilis.

complement number system [ADP] System of number handling in which the complement of the actual number is operated upon; used in some computers to facilitate arithmetic operations.

complement of a subset of a set [MATH] The collection of all members of the set which are not in the given subset.

complete blood count [PATH] Differential and absolute determinations of the numbers of each type of blood cell in a sample and, by extrapolation, in the general circulation.

complete degeneracy [QUANT MECH] The condition in which all the states of interest have the same energy.

complete-expansion diesel cycle *See* Brayton cycle.

complete flower [BOT] A flower having all four floral parts, that is, having sepals, petals, stamens, and carpels.

complete fusion [MET] Fusion which has occurred over all surfaces of the base metal exposed for welding.

complete induction *See* mathematical induction.

complete integral [MATH] **1.** A solution of an *n*th order ordinary differential equation which depends on *n* arbitrary constants as well as the independent variable. Also known as complete primitive. **2.** A solution of a first-order partial differential equation with *n* independent variables which depends upon *n* arbitrary parameters as well as the independent variables.

complete lattice [MATH] A partially ordered set in which every subset has both a supremum and an infimum.

complete leaf [BOT] A dicotyledon leaf consisting of three parts: blade, petiole, and a pair of stipules.

complete lubrication [ENG] Lubrication taking place when rubbing surfaces are separated by a fluid film, and frictional losses are due solely to the internal fluid friction in the film. Also known as viscous lubrication.

completely inelastic collision *See* perfectly inelastic collision.

completely normal space [MATH] A topological space with the property that any pair of sets with disjoint closures can be separated by open sets.

completely regular space [MATH] A topological space *X* where for every point *x* and neighborhood *U* of *x* there is a continuous function from *X* to [0,1] with $f(x) = 1$ and $f(y) = 0$, if *y* is not in *U*. Also known as Tychonoff space.

complete metric space [MATH] A metric space in which every Cauchy sequence converges to a point of the space.

complete normed linear space *See* Banach space.

complete operation [ADP] An operation which includes obtaining all operands from storage, performing the operation, returning resulting operands to storage, and obtaining the next instruction.

complete orthonormal set [MATH] A set of mutually orthogonal unit vectors in a (possibly infinite dimensional) vector space which is contained in no larger such set, that is no nonzero vector is perpendicular to all the vectors in the set. Also known as closed orthonormal set.

complete primitive *See* complete integral.

complete round [ORD] All components of ammunition that are needed to fire a given gun or firearm once.

completion of a metric space [MATH] A complete metric space obtained from another metric space by formally adding limits to Cauchy sequences.

complex [GEOL] An assemblage of rocks that has been folded together, intricately mixed, involved, or otherwise complicated. [MATH] A space which is represented as a union of simplices which intersect only on their faces. [MED] *See* syndrome. [MINERAL] Composed of many ingredients. [PSYCH] A group of associated ideas with strong emotional tones, which have been transferred from the conscious mind into the unconscious and which influence the personality.

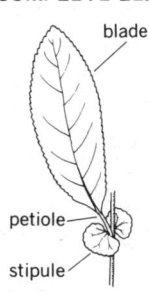
complexation *See* complexing.

complex chemical reaction [CHEM] A chemical system in which a number of chemical reactions take place simultaneously, including reversible reactions, consecutive reactions, and concurrent or side reactions.

complex climatology [CLIMATOL] Analysis of the climate of a single space, or comparison of the climates of two or more places, by the relative frequencies of various weather types or groups of such types; a type is defined by the simultaneous occurrence within specified narrow limits of each of several weather elements.

complex compound [CHEM] Any of a group of chemical compounds in which a part of the molecular bonding is of the coordinate type.

complex conjugate [MATH] One of a pair of complex numbers with identical real parts and with imaginary parts differing only in sign. Also known as conjugate.

complex declaration statement [ADP] A nonexecutable statement in FORTRAN used to specify that the type of identifier appearing in the program is of the form $a + bi$, where *i* is the square root of -1.

complex fold [GEOL] A fold whose axial line is also folded.

complex fraction [MATH] A fraction whose numerator or denominator is a fraction.

complex impedance [ELEC] *See* electrical impedance. [PHYS] *See* impedance.

complexing [CHEM] Formation of a complex compound. Also known as complexation.

complexing agent [CHEM] A substance capable of forming a complex compound with another material in solution.

complex ion [CHEM] A complex, electrically charged group of atoms or radical, for example, $Cu(NH_3)_2^{++}$.

complex low [METEOROL] An area of low atmospheric pressure within which more than one low-pressure center is found.

complex notation [PHYS] The representation of a physical quantity by a complex number whose real component equals the instantaneous value of the physical quantity, a sinusoidally varying quantity thus being represented by a point rotating in a circle centered at the origin of the complex plane with uniform speed.

complex number [MATH] Any number of the form $a + bi$, where *a* and *b* are real numbers, and $i^2 = -1$.

complexometric titration [ANALY CHEM] A technique of volumetric analysis in which the formation of a colored complex is used to indicate the end point of a titration. Also known as chelatometry.

complex permeability [ELECTROMAG] A property, designated by μ^*, of a magnetic material, equal to $\mu_0 (L/L_0)$, where *L* is the complex inductance of an inductance coil in which the magnetic material forms the core when the coil is connected to a sinusoidal voltage source, and L_0 is the vacuum inductance of the coil.

complex permittivity [ELEC] A property of a dielectric, equal to $\varepsilon_0(C/C_0)$, where *C* is the complex capacitance of a capacitor in which the dielectric is the insulating material when the capacitor is connected to a sinusoidal voltage source, and C_0 is the vacuum capacitance of the capacitor.

complex potential [FL MECH] An analytic function in ideal aerodynamics whose real part is the velocity potential and whose imaginary part is the stream function. [NUC PHYS] Generalization of the potential in the Schrödinger equation describing the scattering of a nucleon by a nucleus in the cloudy crystal-ball model.

complex reflector [ENG] A structure or group of structures having many radar-reflecting surfaces facing in different directions.

complex relative attenuation [ELECTR] The ratio of the peak output voltage, in complex notation, of an electric filter to the output voltage at the frequency being considered.

complex salt [INORG CHEM] A class of salts in which there are no detectable quantities of each of the metal ions existing in solution; an example is $K_3Fe(CN)_6$, which in solution has K^+ but no Fe^{3+} because Fe is strongly bound in the complex ion, $Fe(CN)_6^{3-}$.

complex sphere [MATH] Unit sphere on which the complex plane is projected stereographically.

complex target [ENG] A radar target composed of a number of reflecting surfaces that, in the aggregate, are smaller in all dimensions than the resolution capabilities of the radar.

complex tombolo [GEOL] A system resulting when several islands and the mainland are interconnected by a complex series of tombolos. Also known as tombolo cluster; tombolo series.

complex tone [ACOUS] A sound wave produced by the combination of simple sinusoidal components of different frequencies.

complex unit [MATH] Any complex number, $x + iy$, whose absolute value, $\sqrt{x^2 + y^2}$, equals 1.

complex variable [MATH] A variable which assumes complex numbers for values.

complex velocity [FL MECH] In ideal aerodynamic flow, the derivative of the complex potential with respect to $z = x + iy$, where x and y are the chosen coordinates.

compliance [MECH] The displacement of a linear mechanical system under a unit force.

compliance constant [MECH] Any one of the coefficients of the relations in the generalized Hooke's law used to express strain components as linear functions of the stress components. Also known as elastic constants.

complicate [INV ZOO] Folded lengthwise several times, as applied to insect wings.

compo board *See* composition board.

compole *See* commutating pole.

component [CHEM] **1.** A part of a mixture. **2.** The smallest number of chemical substances able to form all the constituents of a system in whatever proportion they may be present. [ELEC] Any electric device, such as a coil, resistor, capacitor, generator, line, or electron tube, having distinct electrical characteristics and having terminals at which it may be connected to other components to form a circuit. Also known as circuit element; element. [SCI TECH] A constituent part of a system; examples are a vector term which when added to others gives a vector sum, an ingredient of a chemical system, or the mineral portion of a rock.

component substances law [CHEM] The law that each substance, singly or in mixture, composing a material exhibits specific properties that are independent of the other substances in that material.

component symbol [ELEC] A graphical design used to represent a component in a circuit diagram.

composing rule [GRAPHICS] A piece of brass or steel against which type is placed in setting.

Compositae [BOT] A family of dicotyledonous plants in the order Campanulales distinguished by a large calyx.

composite [MATER] A structural material composed of combinations of metal alloys or plastics, usually with the addition of strengthening agents.

composite balance [ELEC] An electric balance made by modifying the Kelvin balance to measure amperage, voltage, or wattage.

composite beam [CIV ENG] Beam action of two materials joined to act as a unit, especially that developed by a concrete slab resting on a steel beam and joined by shear connectors.

composite cable [ELEC] Cable in which conductors of different gages or types are combined under one sheath.

composite circuit [ELECTR] A circuit used simultaneously for voice communication and telegraphy, with frequency-discriminating networks serving to separate the two types of signals.

composite color signal [COMMUN] The color television picture signal plus all blanking and synchronizing signals. Also known as composite picture signal.

composite color sync [COMMUN] The signal comprising all the synchronization signals necessary for proper operation of a color television receiver.

composite column [CIV ENG] A concrete column having a structural-steel or cast-iron core with a maximum core area of 20%.

composite compact [MET] A powder compact composed of more than one layer of different components with each layer retaining its identity.

composite cone [GEOL] A large volcanic cone constructed of lava and pyroclastic material in alternating layers.

composite defense [ORD] In antiaircraft artillery, a defense that employs two or more types of fire units which are integrated into a single defense.

composite dialing [COMMUN] Method of dialing between distant offices over one leg of a composite set.

composite dike [GEOL] A dike consisting of several intrusions differing in chemical and mineralogical composition.

composite electrode [MET] A filler-metal electrode composed of more than one metal.

composite explosive [MATER] A mixture of substances which consume and give off oxygen, together with one or several simple explosives; dynamite is an example.

composite filter [ELECTR] A filter constructed by linking filters of different kinds in series.

composite flash [GEOPHYS] A lightning discharge which is made up of a series of distinct lightning strokes with all strokes following the same or nearly the same channel, and with successive strokes occurring at intervals of about 0.05 second. Also known as multiple discharge.

composite fold [GEOL] A fold having smaller folds on its limbs.

composite fuel [MATER] A broad class of solid chemical fuels composed of a fuel and oxidizer and used as propellants in rockets; an example of a fuel is phenol formaldehyde, and an oxidizer is ammonium perchlorate. Also known as composite propellant.

composite gneiss [PETR] A banded rock formed by intimate penetration of magma into country rocks.

composite grain [GEOL] A sedimentary clast formed of two or more original particles.

composite I-beam bridge [CIV ENG] A beam bridge in which the concrete roadway is mechanically bonded to the I beams by means of shear connectors.

composite joint [MET] A joint connected by welding in conjunction with one or more mechanical means.

composite map [MIN ENG] A map in which several levels of a mine are shown on a single sheet.

composite nerve [PHYSIO] A nerve containing both sensory and motor fibers.

composite number [MATH] Any positive integer which is not prime. Also known as composite quantity.

composite picture signal *See* composite color signal.

composite pile [CIV ENG] A pile in which the upper and lower portions consist of different types of piles.

composite plate [MET] A layer of electrodeposited material consisting of at least two different constituents.

composite propellant *See* composite fuel.

composite pulse [ELECTR] A pulse composed of a series of overlapping pulses received from the same source over several paths in a pulse navigation system.

composite quantity *See* composite number.

composite sailing [NAV] In marine operations, a modification of great-circle sailing, used when limiting the highest latitude.

composite sampler [ENG] A hydrometer cylinder equipped with sample cocks at regular intervals along its vertical height; used to take representative (vertical composite) samples of oil from storage tanks.

composite sequence [GEOL] An ideal sequence of cyclic sediments containing all the lithological types in their proper order.

composite set [ELECTR] Assembly of apparatus designed to provide one end of a composite circuit.

composite sill [GEOL] A sill consisting of several intrusions differing in chemical and mineralogical compositions.

composite steel [MET] Bar steel machined along the entire length which is cast around an insert of tool steel welded to the backing of mild steel; used for shear blades and die parts.

composite track [NAV] A modified great-circle track consisting of an initial great-circle track from the point of departure with its vertex on a limiting parallel of latitude; a parallel-sailing track from this vertex along the limiting parallel to the vertex of a final great-circle track passing through the destination.

composite vein [GEOL] A large fracture zone composed of parallel ore-filled fissures and converging diagonals, whose walls and intervening country rock have been replaced to a certain degree.

composite volcano *See* stratovolcano.

COMPOSITE COLUMN

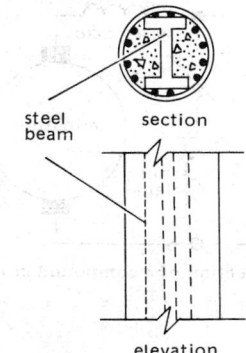

steel beam

section

elevation

Cross section and elevation of a composite column.

COMPOUND LEAF

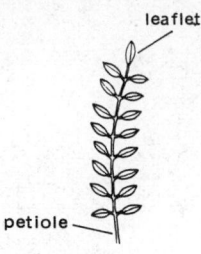

leaflet

petiole

Odd-pinnately compound leaf.

COMPOUND MICROSCOPE

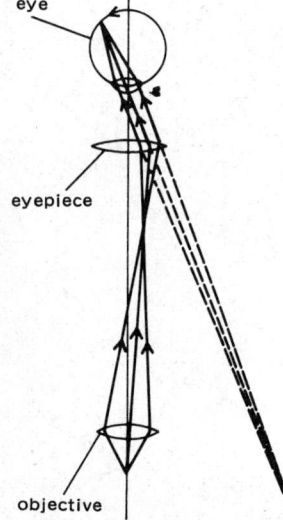

eye

eyepiece

objective

Compound microscope diagram.
(From F. A. Jenkins and H. E.
White, Fundamentals of Optics,
3d ed., McGraw-Hill, 1957)

COMPOUND MOTOR

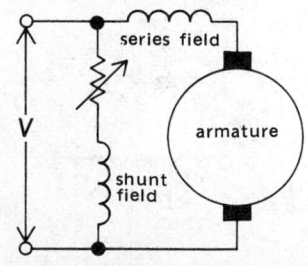

series field

armature

V

shunt
field

Connection of a compound motor.

COMPOUND TUBULAR-ACINOUS GLAND

One of the distal secreting units
of a compound tubular-acinous
gland.

composite wave filter [ELECTR] A combination of two or more low-pass, high-pass, band-pass, or band-elimination filters.

composition [CHEM] The elements or compounds making up a material or produced from it by analysis. [GRAPHICS] The act of composing or combining type for printing, either by hand or by machine. [SCI TECH] The elements or compounds making up a material or produced from it by analysis.

compositional maturity [GEOL] Concept of a type of maturity in sedimentary rocks in which a sediment approaches the compositional end product to which formative processes drive it.

composition board [MATER] A sheet product composed of vegetable fibers mechanically or chemically formed into a pulp which is rolled and pressed. Also known as compo board.

composition diagram [CHEM ENG] Graphical plots to show the solvent-solute concentration relationships during various stages of extraction operations (leaching, or solid-liquid extraction; and liquid-liquid extraction).

composition face See composition surface.

composition metal [MET] A cast copper alloy having a composition of more than 80% copper, with tin, zinc, and lead.

composition of forces [MECH] The determination of a force whose effect is the same as that of two or more given forces acting simultaneously; all forces are considered acting at the same point.

composition of vectors See addition of vectors.

composition of velocities law [MECH] A law relating the velocities of an object in two references frames which are moving relative to each other with a specified velocity.

composition plane [CRYSTAL] A planar composition surface in a crystal uniting two individuals of a contact twin.

composition resistor See carbon resistor.

composition series [MATH] A normal series $G_1, G_2, \ldots$, of a group, where each G_i is a proper normal subgroup of G_{i-1} and no further normal subgroups both contain G_i and are contained in G_{i-1}.

composition surface [CRYSTAL] The surface uniting individuals of a crystal twin; may or may not be planar. Also known as composition face.

compost [MATER] A mixture of decaying organic matter used to fertilize and condition the soil.

compound [CHEM] A substance whose molecules consist of unlike atoms and whose constituents cannot be separated by physical means. Also known as chemical compound.

compound acinous gland [ANAT] A structure with spherical secreting units connected to many ducts that empty into a common duct.

compound alluvial fan [GEOL] Structure formed by the lateral growth and merger of fans made by neighboring streams.

compound compact [MET] A powder compact made from a mixture of metals, with each particle retaining its original composition.

compound cryosar [ELECTR] A cryosar consisting of two normal cryosars with different electrical characteristics in series.

compound curve [MATH] A curve made up of two arcs of differing radii whose centers are on the same side, connected by a common tangent; used to lay out railroad curves because curvature goes from nothing to a maximum gradually, and vice versa.

compound die [MET] A die designed to perform more than one operation on the work with each stroke of the press.

compound elastic scattering [NUC PHYS] Scattering in which the final state is the same as the initial state, but there is an intermediate state with the colliding systems amalgamating to form a compound system.

compound engine [MECH ENG] A multicylinder-type displacement engine, using steam, air, or hot gas, where expansion proceeds successively (sequentially).

compound eye [INV ZOO] An eye typical of crustaceans, insects, centipedes, and horseshoe crabs, constructed of many functionally independent photoreceptor units (ommatidia) separated by pigment cells.

compound generator [ELEC] A direct-current generator which has both a series field winding and a shunt field

winding, both on the main poles with the shunt field winding on the outside.

compound gland [ANAT] A secretory structure with many ducts.

compounding [MECH ENG] The series placing of cylinders in an engine (such as steam) for greater ratios of expansion and consequent improved engine economy.

compound leaf [BOT] A type of leaf with the blade divided into two or more separate leaflets such as the rose.

compound lens [OPTICS] 1. A combination of two or more lenses in which the second surface of one lens has the same radius as the first surface of the following lens, and the two lenses are cemented together. Also known as cemented lens. 2. Any optical system consisting of more than one element, even when they are not in contact.

compound lever [MECH ENG] A train of levers in which motion or force is transmitted from the arm of one lever to that of the next.

compound microscope [OPTICS] A microscope which utilizes two lenses or lens systems; one lens forms an enlarged image of the object, and the second magnifies the image formed by the first.

compound modulation See multiple modulation.

compound motor [ELEC] A direct-current motor with two separate field windings, one connected in parallel with the armature circuit, the other connected in series with the armature circuit.

compound nucleus [NUC PHYS] An intermediate state in a nuclear reaction in which the incident particle combines with the target nucleus and its energy is shared among all the nucleons of the system.

compound number [MATH] A quantity which is expressed as the sum of two or more quantities in terms of different units, for example, 3 feet 10 inches, or 2 pounds 5 ounces.

compound ripple marks [GEOL] Complex ripple marks of great diversity which originate by simultaneous interference of wave oscillation with current action.

compound screw [DES ENG] A screw having different or opposite pitches on opposite ends of the shank.

compound shaft [MIN ENG] A shaft in which the upper stage is often a vertical shaft, while the lower stage, or stages, may be inclined and driven into the deposit.

compound sugar See oligosaccharide.

compound tubular-acinous gland [ANAT] A structure in which the secreting units are simple tubes with acinous side chambers and all are connected to a common duct.

compound tubular gland [ANAT] A structure with branched ducts between the surface opening and the secreting portion.

compound twins [CRYSTAL] Individuals of one mineral group united in accordance with two or more different twin laws.

compound valley glacier [HYD] A glacier composed of several ice streams emanating from different tributary valleys.

compound volcano [GEOL] 1. A volcano consisting of a complex of two or more cones. 2. A volcano with an associated volcanic dome.

compound wave [FL MECH] A plane wave of finite amplitude in which neither the sum of the velocity potential and the component of velocity in the direction of wave motion, nor the difference of these two quantities, is constant.

compound winding [ELEC] A winding that is a combination of series and shunt winding.

compregnate [ENG] Compression of materials into a dense, hard substance with the aid of heat.

compressed air [MECH] Air whose density is increased by subjecting it to a pressure greater than atmospheric pressure.

compressed-air blasting [MIN ENG] A method for breaking down coal by compressed-air power.

compressed-air diving [ENG] Any form of diving in which air is supplied under high pressure to prevent lung collapse.

compressed-air illness See caisson disease.

compressed-air loudspeaker [ENG ACOUS] A loudspeaker having an electrically actuated valve that modulates a stream of compressed air.

compressed-air power [MECH ENG] The power delivered by the pressure of compressed air as it expands, utilized in tools such as drills, in hoists, grinders, riveters, diggers, pile drivers, motors, locomotives, and mine ventilating systems.

compressibility [MECH] The property of a substance capable of being reduced in volume by application of pressure; quantitively, the reciprocal of the bulk modulus.

compressibility burble [FL MECH] A region of disturbed flow, produced by and rearward of a shock wave.

compressibility correction [FL MECH] The correction of the calibrated airspeed caused by compressibility error.

compressibility error [FL MECH] The error in the readings of a differential-pressure-type airspeed indicator due to compression of the air on the forward part of the pitot tube component moving at high speeds.

compressibility factor [THERMO] The product of the pressure and the volume of a gas, divided by the product of the temperature of the gas and the gas constant; this factor may be inserted in the ideal gas law to take into account the departure of true gases from ideal gas behavior. Also known as deviation factor; gas-deviation factor; supercompressibility factor.

compressible flow [FL MECH] Flow in which the fluid density varies.

compressible-flow principle [FL MECH] The principle that when flow velocity is large, it is necessary to consider that the fluid is compressible rather than to assume that it has a constant density.

compressible fluid flow [CHEM ENG] Gas flow when the pressure drop due to the flow of a gas through a system is large enough, compared with the inlet pressure, to cause a 10% or greater decrease in gas density.

compression [ADP] *See* data compression. [ELECTR] 1. Reduction of the effective gain of a device at one level of signal with respect to the gain at a lower level of signal, so that weak signal components will not be lost in background and strong signals will not overload the system. 2. *See* compression ratio. [GEOD] *See* flattening. [GEOL] A system of forces which tend to decrease the volume or shorten rocks. [MECH] Reduction in the volume of a substance due to pressure; for example in building, the type of stress which causes shortening of the fibers of a wooden member. [MECH ENG] *See* compression ratio.

compressional wave [PHYS] A disturbance traveling in an elastic medium; characterized by changes in volume and by particle motion parallel with the direction of wave movement. Also known as dilatational wave; irrotational wave; P wave; pressure wave.

compression cable *See* pressure cable.

compression coupling [MECH ENG] 1. A means of connecting two perfectly aligned shafts in which a slotted tapered sleeve is placed over the junction and two flanges are drawn over the sleeve so that they automatically center the shafts and provide sufficient contact pressure to transmit medium loads. 2. A type of tubing fitting.

compression cup [ENG] A cup from which lubricant is forced to a bearing by compression.

compression failure [ENG] Buckling or collapse caused by compression, as of a steel or concrete column or of wood fibers.

compression ignition [MECH ENG] Ignition produced by compression of the air in a cylinder of an internal combustion engine before fuel is admitted.

compression-ignition engine *See* diesel engine.

compression machine *See* compressor.

compression member [ENG] A beam or other structural member which is subject to compressive stress.

compression mold [ENG] A mold for plastics which is open when the material is introduced and which shapes the material by heat and by the pressure of closing.

compression plant [PETRO ENG] Gas-compression facility used to produce a high-pressure gas stream for injection into reservoir formations to increase oil yield; when the injected gas is that recovered from the well during oil production, the facility is called a gas-cycling plant.

compression process [CHEM ENG] The recovery of natural gasoline from gas containing a high proportion of hydrocarbons.

compression ratio [ELECTR] The ratio of the gain of a device at a low power level to the gain at some higher level, usually expressed in decibels. Also known as compression. [MECH ENG] The ratio in internal combustion engines be-

tween the volume displaced by the piston plus the clearance space, to the volume of the clearance space. Also known as compression. [MET] Ratio of the volume of loose metal powder to the volume of the compact made from it.

compression refrigeration [MECH ENG] The cooling of a gaseous refrigerant by first compressing it to liquid form (with resultant heat buildup), cooling the liquid by heat exchange, then releasing pressure to allow the liquid to vaporize (with resultant absorption of latent heat of vaporization and a refrigerative effect).

compression spring [ENG] A spring, usually a coil spring, which resists a force tending to compress it.

compression strength [MECH] Property of a material to resist rupture under compression.

compression stroke [MECH ENG] The phase of a positive displacement engine or compressor in which the motion of the piston compresses the fluid trapped in the cylinder.

compression syndrome *See* crush syndrome.

compression test [ENG] A test to determine compression strength, usually applied to materials of high compression but low tensile strength, in which the specimen is subjected to increasing compressive forces until failure occurs.

compression wave [FL MECH] A wave in a fluid in which a compression is propagated.

compression wood [BOT] Dense wood found at the base of some tree trunks and on the undersides of branches.

compressive strength [MECH] The maximum compressive stress a material can withstand without failure.

compressive stress [MECH] A stress which causes an elastic body to shorten in the direction of the applied force.

compressor [ELECTR] The part of a compandor that is used to compress the intensity range of signals at the transmitting or recording end of a circuit. [MECH ENG] A machine used for increasing the pressure of a gas or vapor. Also known as compression machine.

compressor blade [MECH ENG] The vane components of a centrifugal or axial-flow, air or gas compressor.

compressor valve [MECH ENG] A valve in a compressor, usually automatic, which operates by pressure difference (less than 5 pounds per square inch) on the two sides of a movable, single-loaded member and which has no mechanical linkage with the moving parts of the compressor mechanism.

compromise joint [CIV ENG] 1. A joint bar used for joining rails of different height or section. 2. A rail that has different joint drillings from that of the same section.

compromise network [ELEC] 1. Network employed in conjunction with a hybrid coil to balance a subscriber's loop; adjusted for an average loop length or an average subscriber's set, or both, to secure compromise (not precision) isolation between the two directional paths of the hybrid. 2. Hybrid balancing network which is designed to balance the average of the impedances that may be connected to the switchboard side of a hybrid arrangement of a repeater.

compromise rail [CIV ENG] A short rail having different sections at the ends to correspond with the rail ends to be joined, thus providing a transition between rails of different sections.

compromising emanations [COMMUN] Unintentional data-related or intelligence-bearing signals which, if intercepted and analyzed by any technique, could disclose the classified information transmitted, received, handled, or otherwise processed by equipments.

Compton absorption [QUANT MECH] The absorption of an x-ray or gamma-ray photon in Compton scattering, accompanied by the emission of another photon of lower energy.

Compton cross section [QUANT MECH] The differential cross section for the elastic scattering of photons by electrons.

Compton-Debye effect *See* Compton effect.

Compton effect [QUANT MECH] The increase in wavelength of electromagnetic radiation in the x-ray and gamma-ray region on being scattered by material objects; the scattering is due to the interaction of the protons with electrons that are effectively free. Also known as Compton-Debye effect.

Compton electron *See* Compton recoil electron.

Compton-Getting effect [ASTROPHYS] The sidereal diurnal variation of the intensity of cosmic rays which would be expected from the rotation of the galaxy if cosmic radiation

COMPRESSOR

Mobile air compressor unit which is driven by engine in the forward compartment. *(Ingersoll-Rand)*

originated in extragalactic regions and was isotropic in intergalactic space, and if this radiation was unaffected at entry to and passage through the galaxy.

Compton incoherent scattering [NUC PHYS] Scattering of gamma rays by individual nucleons in a nucleus or electrons in an atom when the energy of the gamma rays is large enough so that binding effects may be neglected.

Compton meter [NUCLEO] An ionization chamber having a balance chamber with a uranium source that is adjusted until it balances out normal cosmic radiation; variations in cosmic radiation are then shown on an electrometer.

Compton process *See* Compton scattering.

Compton recoil electron [QUANT MECH] An electron set in motion by its interaction with a photon in Compton scattering. Also known as Compton electron.

Compton recoil particle [QUANT MECH] Any particle that has acquired its momentum in a scattering process similar to Compton scattering.

Compton rule [PHYS CHEM] An empirical law stating that the heat of fusion of an element times its atomic weight divided by its melting point in degrees Kelvin equals approximately 2.

Compton scattering [QUANT MECH] The elastic scattering of photons by electrons. Also known as Compton process; gamma-ray scattering.

Compton shift [QUANT MECH] The change in wavelength of scattered radiation due to the Compton effect.

Compton wavelength [QUANT MECH] A property of an elementary particle, equal to Planck's constant divided by the product of 2π, the particle's mass, and the speed of light.

compulsion [PSYCH] An irresistible, impulsive act performed by an individual against his conscious will and usually arising from an obsession.

compulsive reaction [PSYCH] Behavior disorder in which a person is uncomfortable with conditions of ambiguity and uncertainty.

compulsory reporting points [NAV] In air operations, geographical points for which an aircraft must report; these points are designated by regulations and can be approved or deleted only by rule-making action.

computation [MATH] 1. The act or process of calculating. 2. The result so obtained.

computed altitude [NAV] 1. In celestial navigation, tabulated altitude interpolated for increments of latitude, declination, or hour angle; if no interpolation is required, the tabulated altitude and computed altitude are identical. 2. Altitude determined by computation, table, mechanical computer, or graphics, particularly such an altitude of the center of a celestial body measured as an arc on a vertical circle of the celestial sphere from the celestial horizon. Also known as calculated altitude.

computed azimuth [NAV] In celestial navigation, an azimuth determined by computation, table, mechanical device, or graphics for a given place and time.

computed azimuth angle [NAV] Azimuth angle determined by computation, table, mechanical device, or graphics for a given place and time.

computed go to [ADP] A control procedure in FORTRAN which allows the transfer of control to the *i*th label of a set of *n* labels used as statement numbers in the program.

computer [ADP] A device that receives, processes, and presents data; the two types are analog and digital. Also known as computing machine.

computer-aided design [ADP] The generation of computer automated designs for display on cathode-ray tubes.

computer analyst [ADP] A person who defines a problem, determines exactly what is required in the solution, and defines the outlines of the machine solution; generally, an expert in automatic data processing applications.

computer animation [ADP] The use of a computer to present, either continuously or in rapid succession, pictures on a cathode-ray tube or other device, graphically representing a time developing system at successive times.

computer center *See* electronic data-processing center.

computer code [ADP] The code representing the operations built into the hardware of a particular computer.

computer control [ADP] Parts of a digital computer which effect the carrying out of instructions in proper sequence, the interpretation of each instruction, and the application of the

proper signals to the arithmetic unit and other parts according to this interpretation.

computer control counter [ADP] Counter which stores the next required address; any counter which furnishes information to the control unit.

computer-controlled system [CONT SYS] A feedback control system in which a computer operates on both the input signal and the feedback signal to effect control.

computer control register *See* program register.

computer efficiency [ADP] 1. The ratio of actual operating time to scheduled operating time of a computer. 2. In time-sharing, the ratio of user time to the sum of user time plus system time.

computer entry punch [ADP] Combination card-reader and key punch that enters data directly onto a computer's memory drum.

computer graphics [ADP] The process of pictorial communication between men and computers, in which the computer input and output have the form of charts, drawings, or appropriate pictorial representation; such devices as cathode-ray tubes, mechanical plotting boards, curve tracers, coordinate digitizers, and light pens are employed.

computer-limited [ADP] Pertaining to a situation in which the time required for computation exceeds the time required to read inputs and write outputs.

computer memory *See* memory.

computer operation [ADP] The electronic action required in a computer to give a desired computation.

computer performance evaluation [ADP] The measurement and evaluation of the performance of a computer system, aimed at ensuring that a minimum amount of effort, expense, and waste is incurred in the production of data-processing services, and encompassing such tools as canned programs, source program optimizers, software monitors, hardware monitors, simulation, and bench-mark problems. Abbreviated CPE.

computer programming *See* programming.

computer storage device *See* storage device.

computer system [ADP] 1. A set of related but unconnected components (hardware) of a computer or data-processing system. 2. A set of hardware parts that are related and connected, and thus form a computer.

computer theory [ADP] A discipline covering the study of circuitry, logic, microprogramming, compilers, programming languages, file structures, and system architectures.

computer word *See* word.

computing gunsight [ORD] A sight which includes an electrical or mechanical means for computing the proper angle between the line of sight to the target and the line of departure for the projectile.

computing machine *See* computer.

computing unit [ADP] The section of a computer that carries out arithmetic, logical, and decision-making operations.

COMSAT *See* Communications Satellite Corporation.

Comstock refraction formula [ASTROPHYS] A formula for the apparent angular displacement of an object outside the earth's atmosphere due to refraction, in terms of the barometric pressure, the temperature of the atmosphere, and the observed zenith distance.

conarium *See* pineal body.

concatenate [ADP] To unite in a sequence, link together, or link to a chain.

concatenation [ELEC] A method of speed control of induction motors in which the rotors of two wound-rotor motors are mechanically coupled together and the stator of the second motor is supplied with power from the rotor slip rings of the first motor.

Concato's disease *See* polyserositis.

concave [SCI TECH] Having a curved form which bulges inward resembling the interior of a sphere or cylinder or a section of these bodies.

concave bit [DES ENG] A type of tungsten carbide drill bit having a concave cutting edge; used for percussive boring.

concave fillet weld [MET] A fillet weld having a concave surface.

concave function [MATH] A function $f(x)$ is said to be concave over the interval a,b if for any three points x_1, x_2, x_3 such that $a < x_1 < x_2 < x_3 < b$, $f(x_2) \geq L(x_2)$, where $L(x)$ is the equa-

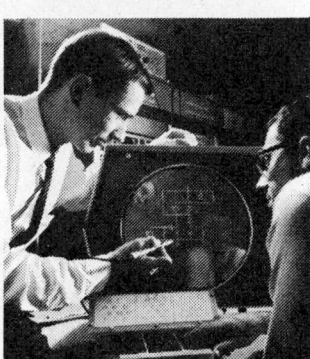

tion of the straight line passing through the points x_1,x_3.

concave grating [SPECT] A reflection grating which both collimates and focuses the light falling upon it, made by spacing straight grooves equally along the chord of a concave spherical or paraboloid mirror surface.

concave polygon [MATH] A polygon at least one of whose angles is greater than 180°.

concave spherical mirror [OPTICS] A round mirror having a concavely curved surface, in the form of a portion of a sphere.

concentrate [CHEM] To increase the amount of a dissolved substance by evaporation. [MIN ENG] **1.** To separate ore or metal from its containing rock or earth. **2.** The clean product recovered in froth flotation or other methods of mineral separation.

concentrated load [MECH] A force that is negligible because of a small contact area; a beam supported on a girder represents a concentrated load on the girder.

concentrating table [MIN ENG] A device consisting of a riffled deck to which a reciprocating motion in a horizontal direction is imparted; the material to be separated is fed in a stream of water, the heavy particles collect between the riffles and are conveyed in the direction of the reciprocating motion, while the lighter particles are borne by the water over the riffles to be discharged laterally from the table.

concentration [CHEM] In solutions, the mass, volume, or number of moles of solute present in proportion to the amount of solvent or total solution. [MIN ENG] Separation and accumulation of economic minerals from gangue.

concentration cell [PHYS CHEM] **1.** Electrochemical cell for potentiometric measurement of ionic concentrations where the electrode potential electromotive force produced is determined as the difference in emf between a known cell (concentration) and the unknown cell. **2.** An electrolytic cell in which the electromotive force is due to a difference in electrolyte concentrations at the anode and the cathode.

concentration-dilution test [PATH] A renal function test to measure the ability of the kidney to concentrate and dilute urine under stress; specific gravity for urine from a normal kidney fluctuates from 1.030 on a restricted fluid intake to 1.003 on a high water intake.

concentration gradient [CHEM] The graded difference in the concentration of a solute throughout the solvent phase.

concentration polarization [PHYS CHEM] That part of the polarization of an electrolytic cell resulting from changes in the electrolyte concentration due to the passage of current through the solution.

concentration potential [CHEM] Tendency for a univalent electrolyte to concentrate in a specific region of a solution.

concentration scale [CHEM] Any of several numerical systems defining the quantitative relation of the components of a mixture; for solutions, concentration is expressed as the mass, volume, or number of moles of solute present in proportion to the amount of solvent or total solution.

concentration time [HYD] The time required for water to travel from the most remote portion of a river basin to the basin outlet; it varies with the quantity of flow and channel conditions.

concentrator [ELECTR] Buffer switch (analog or digital) that serves to reduce the number of trunks required. [ENG] **1.** An apparatus used to concentrate materials. **2.** A plant where materials are concentrated.

concentric [SCI TECH] Pertaining to the relationship between two different-sized circular, cylindrical, or spherical shapes when the smaller one is exactly centered within the larger one.

concentric bundle [BOT] A vascular bundle in which xylem surrounds phloem, or phloem surrounds xylem.

concentric cable See coaxial cable.

concentric circles [MATH] A family of coplanar circles with the same center.

concentric faults [GEOL] Faults that are arranged concentrically.

concentric fold [GEOL] A fold in which the original thickness of the strata is unchanged during deformation. Also known as parallel fold.

concentric fractures [GEOL] A system of fractures concentrically arranged about a center.

concentric groove See locked groove.

concentric lens [OPTICS] A lens whose two spherical surfaces have the same center.

concentric line See coaxial cable.

concentric locating [DES ENG] The process of making the axis of a tooling device coincide with the axis of the workpiece.

concentric orifice plate [DES ENG] A fluid-meter orifice plate whose edges have a circular shape and whose center coincides with the center of the pipe.

concentric slip ring [ELEC] A large slip-ring assembly consisting of concentrically arranged insulators and conducting materials.

concentric transmission line See coaxial cable.

concentric tube column [CHEM ENG] A carefully insulated distillation apparatus which is capable of very high separating power, and in which the outer vapor-rising annulus of the column is concentric around an inner, bottom-discharging reflux return.

concentric weathering See spheroidal weathering.

concentric windings [ELEC] Transformer windings in which the low-voltage winding is in the form of a cylinder next to the core, and the high-voltage winding, also cylindrical, surrounds the low-voltage winding.

conceptacle [BOT] A cavity which is shaped like a flask with a pore opening to the outside, contains reproductive structures, and is bound in a thallus such as in the brown algae.

concept coordination [ADP] The basic principles of various punched-card, aspect, and mechanized information retrieval systems in which independently assigned concepts are used to characterize the subject content of documents, and the latter are identified during searching by means of either such assigned concepts or their combination.

conception [BIOL] Fertilization of an ovum by the sperm resulting in the formation of a viable zygote. [PSYCH] The mental process of forming ideas, especially abstract ideas.

conceptual modeling [ADP] Writing a program by means of which a given result will be obtained, although the result is incapable of proof. Also known as heuristic programming.

concertina wire [ORD] A barbed wire coiled for easier handling and emplacement; when uncoiled, it forms an entanglement for the enemy infantry.

concession lease [MIN ENG] A lease form that conveys specified national or state permission to a lessee to explore for or produce minerals (such as oil, gas, or uranium) from specified properties.

conch [INV ZOO] The common name for several species of large, colorful gastropod mollusks of the family Strombidae; the shell is used to make cameos and porcelain.

conchiolin [BIOCHEM] A nitrogenous substance that is the organic basis of many molluscan shells.

conchoid [MATH] A plane curve consisting of the locus of both ends of a line segment of constant length on a line which rotates about a fixed point, while the midpoint of the segment remains on a fixed line which does not contain the fixed point.

conchoidal [GEOL] Having a smoothly curved surface; used especially to describe the fracture surface of a mineral or rock.

Conchorhagae [INV ZOO] A suborder of benthonic wormlike animals in the class Kinorhyncha.

Conchostraca [INV ZOO] An order of mussellike crustaceans of moderate size belonging to the subclass Branchiopoda.

concordance [GEN] Similarity in appearance of members of a twin pair with respect to one or more specific traits.

concordant body [GEOL] An intrusive igneous body whose contacts are parallel to the bedding of the country rock. Also known as concordant injection; concordant pluton.

concordant coastline [GEOL] A coastline parallel to the land structures which form the margin of an ocean basin.

concordant injection See concordant body.

concordant pluton See concordant body.

concrescence [BIOL] Convergence and fusion of parts originally separate, as the lips of the blastopore in embryogenesis.

concrete [MATER] A mixture of aggregate, water, and a binder, usually portland cement; hardens to stonelike condition when dry.

concrete beam [CIV ENG] A structural member of reinforced concrete, placed horizontally over openings to carry loads.

concrete block [MATER] A solid or hollow block of precast concrete.

CONCENTRIC SLIP RING

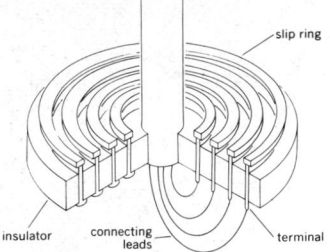

Concentric slip-ring assembly configuration.

CONCENTRIC WINDINGS

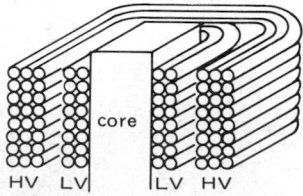

Section through the core and concentric winding of a transformer. LV = low-voltage winding; HV = high-voltage winding.

CONCH

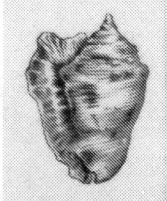

Shell of hawk-wing conch.

CONCHOSTRACA

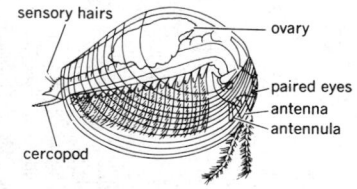

Limnadia lenticularis female, lateral aspect.

concrete bridge [CIV ENG] A bridge constructed of prestressed or reinforced concrete.

concrete bucket [ENG] A container with movable gates at the bottom that is attached to power cranes or cables to transport concrete.

concrete buggy [ENG] A cart which carries up to 6 cubic feet (0.17 cubic meter) of concrete from the mixer or hopper to the forms. Also known as buggy; concrete cart.

concrete caisson sinking [CIV ENG] A shaft-sinking method similar to caisson sinking except that reinforced concrete rings are used and an airtight working chamber is not adopted.

concrete cart *See* concrete buggy.

concrete chute [ENG] A long metal trough with rounded bottom and open ends used for conveying concrete to a lower elevation.

concrete column [CIV ENG] A vertical structural member made of reinforced or unreinforced concrete.

concrete dam [CIV ENG] A dam built of concrete.

concrete finish [MATER] The texture or smoothness on the surface of hardened concrete.

concrete form oil [MATER] Nonviscous, neutral mineral oil used on wooden or metal forms to allow easy removal from set concrete.

concrete hardener [MATER] An admixture such as calcium chloride, sodium chloride, or sodium hydroxide that hastens or decreases the hydration rate of cementing material; the concrete takes less time to set and has earlier higher strength.

concrete masonry [MATER] Building units composed of block, brick, or tile laid by masons.

concrete mixer [MECH ENG] A machine with a rotating drum in which the components of concrete are mixed.

concrete pile [CIV ENG] A reinforced pile made of concrete, either precast and driven into the ground, or cast in place in a hole bored into the ground.

concrete pipe [CIV ENG] A porous pipe made of concrete and used principally for subsoil drainage; diameters over 15 inches (38 centimeters) are usually reinforced.

concrete pump [MECH ENG] A device which drives concrete to the placing position through a pipeline of 6-inch (15-centimeter) diameter or more, using a special type of reciprocating pump.

concrete retarder [MATER] A material added to concrete that decreases the hydration rate of cement, thereby increasing the setting time and decreasing the strengthening rate during the early age.

concrete slab [CIV ENG] A flat, reinforced-concrete structural member, relatively sizable in length and width, but shallow in depth; used for floors, roofs, and bridge decks.

concrete steel [MET] Steel used in reinforced concrete, which should comply with standard specifications for prestressed concrete.

concrete thinking [PSYCH] Mental processes characterized by literalness and the tendency to be bound to the most immediate and obvious sense impressions, as well as by a lack of generalization and abstraction.

concrete vibrator [MECH ENG] Vibrating device used to achieve proper consolidation of concrete; the three types are internal, surface, and form vibrators.

concretion [GEOL] A hard, compact mass of mineral matter in the pores of sedimentary or fragmental volcanic rock; represents a concentration of a minor constituent of the enclosing rock or of cementing material.

concretionary [GEOL] Tending to grow together, forming concretions. [PATH] A compact mass of inorganic material formed in a body cavity or in tissue.

concretioning [GEOL] The process of forming concretions.

concurrent conversion [ADP] The transfer of data from one medium to another, such as card to tape, under computer control while programs are being run on the same computer.

concurrent heating [MET] Application of supplemental heat in metal cutting or welding.

concurrent infection [MED] Two or more forms of an infection existing simultaneously.

concurrent input/output [ADP] The simultaneous reading from and writing on different media by a computer.

concurrent operations control [ADP] The supervisory capability required by a computer to handle more than one program at a time.

concurrent processing [ADP] The capability of a computer to process more than one program at the time.

concurrent real-time processing [ADP] The capability of a computer to process simultaneously several programs, each of which requires responses within a time span related to its particular time frame.

concussion [MED] A state of shock following traumatic injury, especially cerebral trauma, in which there is temporary functional impairment without physical evidence of damage to impaired tissues.

concussion fracture [GEOL] Radiating system of fractures in a shock-metamorphosed rock.

concussion fuse [ORD] A bomb fuse designed to function in the air in response to the concussion produced by the explosion of a preceding bomb.

concussion table [MIN ENG] An inclined table, agitated by a series of shocks and operating like a buddle. Also known as percussion table.

condensable vapors [CHEM] Gases or vapors which when subjected to appropriately altered conditions of temperature or pressure become liquids.

condensate [MATER] **1.** The liquid product from a condenser. Also known as condensate liquid. **2.** A light hydrocarbon mixture formed as a liquid product in a gas-recycling plant through expansion and cooling of the gas.

condensate flash [CHEM ENG] Partial evaporation (flash) of hot condensed liquid by a stepwise reduction in system pressure, the hot vapor supplying heat to a cooler evaporator step (stage).

condensate liquid *See* condensate.

condensate well [MECH ENG] A chamber into which condensed vapor falls for convenient accumulation prior to removal. [PETRO ENG] Well that produces a natural gas highly saturated with condensable hydrocarbons heavier than methane and ethane.

condensation [ACOUS] A measure of the increase in the instantaneous density at a given point resulting from a sound wave, namely $(\rho - \rho_0)/\rho_0$, where ρ is the density and ρ_0 is the constant mean density at the point. [CHEM] Transformation from a gas to a liquid. [ELEC] An increase of electric charge on a capacitor conductor. [MECH] An increase in density. [METEOROL] The process by which water vapor becomes a liquid such as dew, fog, or cloud or a solid like snow; condensation in the atmosphere is brought about by either of two processes: cooling of air to its dew point, or addition of enough water vapor to bring the mixture to the point of saturation (that is, the relative humidity is raised to 100%). [OPTICS] Focusing or collimation of light. [STAT MECH] *See* Bose-Einstein condensation.

condensation cloud [METEOROL] A mist or fog of minute water droplets that temporarily surrounds the fireball following an atomic detonation in a comparatively humid atmosphere.

condensation nucleus [METEOROL] A particle, either liquid or solid, upon which condensation of water vapor begins in the atmosphere.

condensation polymer [ORG CHEM] A high-molecular-weight compound formed by condensation polymerization.

condensation polymerization [ORG CHEM] The formation of high-molecular-weight polymers from monomers by chemical reactions of the condensation type.

condensation pressure [METEOROL] The pressure at which a parcel of moist air expanded dry adiabatically reaches saturation. Also called adiabatic condensation pressure; adiabatic saturation pressure.

condensation reaction [CHEM] One of a class of chemical reactions involving a combination between molecules or between parts of the same molecule.

condensation resin [ORG CHEM] A resin formed by polycondensation.

condensation shock wave [FL MECH] A sheet of discontinuity associated with a sudden condensation and fog formation in a field of flow; it occurs, for example, on a wing where a rapid drop in pressure causes the temperature to drop considerably below the dew point.

condensation temperature [ANALY CHEM] In boiling-point

determination, the temperature established on the bulb of a thermometer on which a thin moving film of liquid coexists with vapor from which the liquid has condensed, the vapor phase being replenished at the moment of measurement from a boiling-liquid phase. [METEOROL] The temperature at which a parcel of moist air expanded dry adiabatically reaches saturation. Also known as adiabatic condensation temperature; adiabatic saturation temperature.

condensation trail [METEOROL] A visible trail of condensed water vapor or ice particles left behind an aircraft, an airfoil, or such, in motion through the air. Also known as contrail; vapor trail.

condensed instruction deck [ADP] The card output from an assembly program in which several instructions per card are punched in machine language; input to the assembly program may consist of one instruction per card, so the output is condensed.

condenser [ELEC] *See* capacitor. [MECH ENG] A heat-transfer device that reduces a thermodynamic fluid from its vapor phase to its liquid phase, as in a vapor-compression refrigeration plant or a condensing steam power plant. [OPTICS] A system of lenses or mirrors in an optical projection system, which gathers as much of the light from the source as possible and directs it through the projection lens.

condenser bushing [ELEC] An insulation made up of alternate layers of insulating material and metal foil placed between the conductor and outer casing in terminals of transformers and other high-voltage equipment such as switchgears.

condenser-discharge anemometer [ENG] A contact anemometer connected to an electrical circuit which is so arranged that the average wind speed is indicated.

condenser microphone *See* capacitor microphone.

condenser transducer *See* electrostatic transducer.

condenser tubes [MECH ENG] Metal tubes used in a heat-transfer device, with condenser vapor as the heat source and flowing liquid such as water as the receiver.

condensing engine [MECH ENG] A steam engine in which the steam exhausts from the cylinder to a vacuum space, where the steam is liquefied.

condensing flow [FL MECH] The flow and simultaneous condensation (partial or complete) of vapor through a cooled pipe or other closed conduit or container.

condensing gas drive [PETRO ENG] Reservoir-oil displacement by gas where hydrocarbon components of the injected gas condense in the oil that it is displacing.

condensing routine [ADP] A routine that converts a program format having one instruction per card to a program format having several instructions per card.

conditional [ADP] Subject to the result of a comparison made during computation in a computer, or subject to human intervention.

conditional assembly [ADP] A feature of some assemblers which suppresses certain sections of code if stated program conditions are not met at assembly time.

conditional branch *See* conditional jump.

conditional breakpoint [ADP] A conditional jump that, if a specified switch is set, will cause a computer to stop; the routine may then be continued as coded or a jump may be forced.

conditional convergence [MATH] The property of a series that is convergent but not absolutely convergent.

conditional distribution [STAT] If W and Z are random variables with discrete values $w_1, w_2, \ldots$, and $z_1, z_2, \ldots$, the conditional distribution of W given $Z = z$ is the distribution which assigns to w_i, $i = 1, 2, \ldots$, the conditional probability of $W = w_i$ given $Z = z$.

conditional expectation [MATH] If X is a random variable on a probability space (Ω, F, P), the conditional expectation of X with respect to a given sub σ-field F' of F is an F'-measurable random variable whose expected value over any set in F' is equal to the expected value of X over this set. [STAT] The expected value of a conditional distribution.

conditional expression [ADP] A COBOL language expression which is either true or false, depending upon the status of the variables within the expression.

conditional frequency [STAT] If r and s are possible outcomes of an experiment which is performed n times, the conditional frequency of s given that r has occurred is the ratio of the number of times both r and s have occurred to the number of times r has occurred.

conditional instability [METEOROL] The state of a column of air in the atmosphere when its lapse rate of temperature is less than the dry adiabatic lapse rate but greater than the saturation adiabatic lapse rate.

conditional jump [ADP] A computer instruction that will cause the proper one of two or more addresses to be used in obtaining the next instruction, depending on some property of a numerical expression that may be the result of some previous instruction. Also known as conditional branch; conditional transfer; decision instruction; discrimination.

conditional lethal mutant [GEN] A mutant gene that has lethal expression only under particular growth conditions, such as a specific temperature range.

conditionally compact set [MATH] A set whose closure is compact. Also known as relatively compact set.

conditionally periodic motion [MECH] Motion of a system in which each of the coordinates undergoes simple periodic motion, but the associated frequencies are not all rational fractions of each other so that the complete motion is not simply periodic.

conditionally stable circuit [ELECTR] A circuit which is stable for certain values of input signal and gain, and unstable for other values.

conditional probability [STAT] The probability that a second event will be B if the first event is A, expressed as $P(B/A)$.

conditional transfer *See* conditional jump.

conditioned reflex [PSYCH] Response of an organism to a stimulus which was inadequate to elicit the response until paired for one or more times with an adequate stimulus.

conditioning [ELECTR] Equipment modifications or adjustments necessary to match transmission levels and impedances or provide equalization between facilities. [SCI TECH] Subjecting a material or organism to a stipulated treatment or stimulus so that it will respond in a uniform and desired manner to subsequent testing or processing.

Condon-Shortley-Wigner phase convention [QUANT MECH] Convention relating the phases of states having the same eigenvalue of $J^2 = J_x^2 + J_y^2 + J_z^2$, and different eigenvalues of J_z, where $\mathbf{J}$ is the total angular momentum, according to which the matrix elements of $\mathbf{J}_+ = J_x + iJ_y$ and $\mathbf{J}_- = J_x - iJ_y$ between such states are real.

condor [NAV] A continuous-wave navigation system, similar to benito, that automatically measures bearing and distance from a single ground station; the distance is determined by phase comparison and the bearing by automatic direction finding. [VERT ZOO] *Vultur gryphus.* A large American vulture having a bare head and neck, dull black plumage, and a white neck ruff.

conductance [ELEC] The real part of the admittance of a circuit; when the impedance contains no reactance, as in a direct-current circuit, it is the reciprocal of resistance, and is thus a measure of the ability of the circuit to conduct electricity. Also known as electrical conductance. Designated G. [THERMO] *See* thermal conductance.

conductance-variation method [ELEC] A technique for measuring low admittances; measurements in a parallel-resonance circuit with the terminals open-circuited, with the unknown admittance connected, and then with the unknown admittance replaced by a known conductance standard are made; from them the unknown can be calculated.

conducted interference [COMMUN] Interfering signals arriving on power lines.

conductimetry [CHEM] The scientific study of conductance measurements of solutions; to avoid electrolytic complications, conductance measurements are usually taken with alternating current.

conducting polymer [MATER] A plastic having high conductivity, approaching that of metals.

conduction [ELEC] The passage of electric charge, which can occur by a variety of processes, such as passage of electrons or ionized atoms. Also known as electrical conduction. [PHYS] Transmission of energy by a medium which does not involve movement of the medium itself.

conduction band [SOLID STATE] An energy band in which

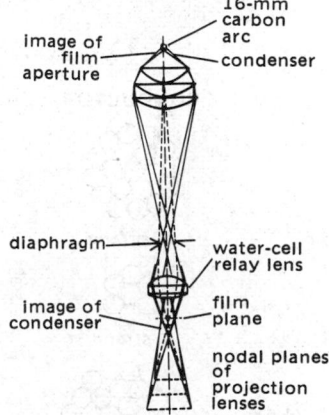

CONDENSER

image of film aperture — 16-mm carbon arc — condenser

diaphragm — water-cell relay lens

image of condenser — film plane

nodal planes of projection lenses

A relay condenser system having a water cell incorporated in second stage.

electrons can move freely in a solid, producing net transport of charge.

conduction current [SOLID STATE] A current due to a flow of conduction electrons through a body.

conduction deafness [MED] Deafness due to abnormal conduction of energy to the cochlea.

conduction electron [SOLID STATE] An electron in the conduction band of a solid, where it is free to move under the influence of an electric field. Also known as outer-shell electron; valence electron.

conduction field [ELECTROMAG] Energy surrounding a conductor when an electric current is passed through the conductor, which, because of the difference in phase between the electrical field and magnetic field set up in the conductor, cannot be detached from the conductor.

conduction pump [ENG] A pump in which liquid metal or some other conductive liquid is moved through a pipe by sending a current across the liquid and applying a magnetic field at right angles to current flow.

conductive coating [MATER] A coating used to reduce surface resistance and thus prevent the accumulation of static electric charges.

conductive coupling [ELEC] Electric connection of two electric circuits by their sharing the same resistor.

conductive equilibrium See isothermal equilibrium.

conductive gasket [ELEC] A flexible metallic gasket used to reduce radio-frequency leakage at joints in shielding.

conductivity [ELEC] The ratio of the electric current density to the electric field in a material. Also known as electrical conductivity; specific conductance.

conductivity bridge [ELEC] A modified Kelvin bridge for measuring very low resistances.

conductivity cell [ELEC] A glass vessel with two electrodes at a definite distance apart and filled with a solution whose conductivity is to be measured.

conductivity current See air-earth conduction current.

conductivity modulation [ELECTR] Of a semiconductor, the variation of the conductivity of a semiconductor through variation of the charge carrier density.

conductivity modulation transistor [ELECTR] Transistor in which the active properties are derived from minority carrier modulation of the bulk resistivity of the semiconductor.

conductivity tensor [ELEC] A tensor which, when multiplied by the electric field vector according to the rules of matrix multiplication, gives the current density vector.

conductivity theory [STAT MECH] Theory which treats the system of electrons in a metal as a gas and uses the Boltzmann transport equation to calculate conductivity.

conductometer [ENG] An instrument designed to measure thermal conductivity; in particular, one that compares the rates at which different rods transmit heat.

conductometric titration [ANALY CHEM] A titration in which electrical conductance of a solution is measured during the course of the titration.

conductor [ELEC] A wire, cable, or other body or medium that is suitable for carrying electric current.

conductor skin effect See skin effect.

conduit [ELEC] Solid or flexible metal or other tubing through which insulated electric wires are run. [ENG] Any channel or pipe for conducting the flow of water or other fluid.

conduplicate [BOT] Folded lengthwise and in half with the upper faces together, applied to leaves and petals in the bud.

Condylarthra [PALEON] A mammalian order of extinct, primitive, hoofed herbivores with five-toed plantigrade to semidigitigrade feet.

condyle [ANAT] A rounded bone prominence that functions in articulation. [BOT] The antheridium of certain stoneworts. [INV ZOO] A rounded, articular process on arthropod appendages.

condyloid articulation [ANAT] A joint, such as the wrist, formed by an ovoid surface that fits into an elliptical cavity, permitting all movement except rotation.

condyloma acuminata [MED] A venereal disease characterized by wartlike growths on the genital organs; thought to be of viral origin.

cone [BOT] Either the ovulate or staminate strobilus of a gymnosperm. [ENG ACOUS] The cone-shaped paper or fiber diaphragm of a loudspeaker. [HISTOL] A photoceptor of the vertebrate retina that responds differentially to light across the visible spectrum, providing color vision and visual acuity in bright light. [MATH] A solid bounded by a region enclosed in a closed curve on a plane and a surface formed by the segments joining each point of the closed curve to a point not in the plane. [TEXT] A bobbin on which yarn is wound for weaving.

cone antenna See conical antenna.

cone bearing [MECH ENG] A cone-shaped journal bearing running in a correspondingly tapered sleeve.

cone-bottom tank [ENG] Liquids-storage tank with downward-pointing conical bottom to facilitate drainage of bottom, as of water or sludge.

cone brake [MECH ENG] A type of friction brake whose rubbing parts are cone-shaped.

cone classifier [MECH ENG] Inverted-cone device for the separation of heavy particulates (such as sand, ore, or other mineral matter) from a liquid stream; feed enters the top of the cone, heavy particles settle to the bottom where they can be withdrawn, and liquid overflows the top edge, carrying the smaller particles or those of lower gravity over the rim; used in the mining and chemical industries.

cone clutch [MECH ENG] A clutch which uses the wedging action of mating conical surfaces to transmit friction torque.

cone crusher [MECH ENG] A machine that reduces the size of materials such as rock by crushing in the tapered space between a truncated revolving cone and an outer chamber.

conehead rivet [DES ENG] A rivet with a head shaped like a truncated cone.

cone-in-cone structure [GEOL] The structure of a concretion characterized by the development of a succession of cones one within another.

cone key [DES ENG] A taper saddle key placed on a shaft to adapt it to a pulley with a too-large hole.

cone loudspeaker [ENG ACOUS] A loudspeaker employing a magnetic driving unit that is mechanically coupled to a paper or fiber cone. Also known as cone speaker.

Conelrad [COMMUN] A system for providing official civil defense information and instructions by radio in an emergency without providing radio homing guidance for the enemy. Derived from control of electromagnetic radiation.

cone mandrel [DES ENG] A mandrel in which the diameter can be changed by moving conical sleeves.

Conemaughian [GEOL] Upper Middle Pennsylvanian geologic time.

cone nozzle [DES ENG] A cone-shaped nozzle that disperses fluid in an atomized mist.

cone of ambiguity [NAV] The conical volume of airspace above the beacon in which bearing information is unreliable in VOR and Tacan.

cone of depression [HYD] The depression in the water table around a well defining the area of influence of the well. Also known as cone of influence.

cone of dispersion [ORD] The pattern in space formed by phenomena from point sources, such as shots on a target from the same gun, that spread out in conical form.

cone of escape [GEOPHYS] A hypothetical cone in the exosphere, directed vertically upward, through which an atom or molecule would theoretically be able to pass to outer space without a collision.

cone of influence See cone of depression.

cone of nulls [ELECTROMAG] In antenna practice, a conical surface formed by directions of negligible radiation.

cone of revolution [MATH] The surface obtained by rotating a line around another line which it intersects, using the intersection point as a pivot.

cone of silence [NAV] A cone-shaped region, directly over the antenna of a radio-beacon transmitter, in which no signal is detected.

cone-of-silence marker [NAV] A marker beacon, usually a Z (for zone) marker beacon, located at a radio-range station; operates at 75 megahertz and produces a light signal accompanied by a 3000-hertz tone.

cone of visibility [AERO ENG] Generally, the right conical space which has its apex at some ground target and within which an aircraft must be located if the pilot is to be able to discern the target while flying at a specified altitude.

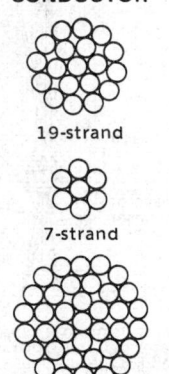
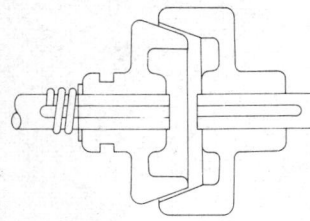

cone pulley *See* step pulley.

cone rock bit [MECH ENG] A rotary drill with two hardened knurled cones which cut the rock as they roll. Also known as roller bit.

cone-roof tank [ENG] Liquids-storage tank with flattened conical roof to allow a vapor reservoir at the top for filling operations.

cone settler [MIN ENG] A conical vessel fed centrally with fine ore pulp, in which the apex discharge carries the larger-sized particles, and the peripheral top overflow carries the finer fraction of the solids.

cone sheet [GEOL] An accurate dike forming part of a concentric set that dips inward toward the center of the arc. Also known as cone dike.

cone speaker *See* cone loudspeaker.

cone valve [CIV ENG] A divergent valve whose cone-shaped head in a fixed cylinder spreads water around the wide, downstream end of the cone in spillways of dams or hydroelectric facilities. Also known as Howell-Bunger valve.

Conewangoan [GEOL] Upper Upper Devonian geologic time.

conference communications [COMMUN] Communications facilities whereby direct speech conversation may be conducted between three or more locations simultaneously.

confidence [STAT] The degree of assurance that a specified failure rate is not exceeded.

confidence coefficient [STAT] The probability associated with a confidence interval; that is, the probability that the interval contains a given parameter or characteristic. Also known as confidence level.

confidence interval [STAT] An interval which has a specified probability of containing a given parameter or characteristic.

confidence level [IND ENG] The probability in acceptance sampling that the quality of accepted lots manufactured will be better than the rejectable quality level (RQL); 90% level indicates that accepted lots will be better than the RQL 90 times in 100. [STAT] *See* confidence coefficient.

confidence limit [STAT] One of the end points of a confidence interval.

configuration [AERO ENG] A particular type of specific aircraft, rocket, or such, which differs from others of the same model by the arrangement of its components or by the addition or omission of auxiliary equipment; for example, long-range configuration or cargo configuration. [ELEC] A group of components interconnected to perform a desired circuit function. [MATH] An arrangement of geometric objects. [SYS ENG] A group of machines interconnected and programmed to operate as a system.

configurational free energy [STAT MECH] The free energy of a solid lattice associated with the interaction between neighboring atoms, and with external electric and magnetic fields.

configuration interaction [PHYS CHEM] Interaction between two different possible arrangements of the electrons in an atom (or molecule); the resulting electron distribution, energy levels, and transitions differ from what would occur in the absence of the interaction.

confined charge [ORD] An explosive charge loaded in a resistant container, as opposed to a bare charge.

confined explosion [ORD] Explosion occurring in a closed chamber, where the volume is constant.

confined flow [ENG] The flow of any fluid (liquid or gas) through a continuous container (process vessel) or conduit (piping or tubing).

confinement [ENG] Physical restriction, or degree of such restriction, to passage of detonation wave or reaction zone, for example, that of a resistant container which holds an explosive charge.

confinement of plasma [PL PHYS] Restriction of a hot plasma to a given volume as long as possible, by such means as magnetic mirrors and pinch effect.

confining bed [GEOL] An impermeable bed adjacent to an aquifer.

confining liquid [CHEM ENG] A liquid seal (most often mercury or sodium sulfate brine) that is displaced during the no-loss transfer of a gas sample from one container to another.

confining pressure [GEOL] An equal, all-sided pressure, such as lithostatic pressure produced by overlying rocks in the crust of the earth.

confluence [HYD] **1.** A stream formed from the flowing together of two or more streams. **2.** The place where such streams join.

confluent hypergeometric function [MATH] A solution to the differential equation $z(d^2w/dz^2) + (\rho - z)(dw/dz) - \alpha w = 0$.

confocal coordinates [MATH] Coordinates of a point in the plane with norm greater than 1 in terms of the system of ellipses and hyperbolas whose foci are at $(1,0)$ and $(-1,0)$.

confocal resonator [ELECTROMAG] A wavemeter for millimeter wavelengths, consisting of two spherical mirrors facing each other; changing the spacing between the mirrors affects propagation of electromagnetic energy between them, permitting direct measurement of free-space wavelength.

conformable [GEOL] **1.** Pertaining to the contact of an intrusive body when it is aligned with the internal structures of the intrusion. **2.** Referring to strata in which layers are formed above one another in an unbroken, parallel order.

conformable optical mask [ELECTR] An optical mask made on a flexible glass substrate so that it can be pulled down under vacuum into intimate contact with the substrate for accurate circuit fabrication.

conformal chart [MAP] A chart on a conformal map projection.

conformality [MAP] The retention of angular relationships at each point on a map projection.

conformal mapping [MATH] An angle-preserving analytic function of a complex variable.

conformal map projection [MAP] A map projection on which the shape of any small area of the surface mapped is preserved unchanged. Also known as orthomorphic map projection.

conformal reflection chart [ELECTROMAG] An Argand diagram for plotting the complex reflection coefficient of a waveguide junction and its image, the two being related by a conformal transformation.

conformation [PHYS CHEM] The spatial arrangement of the atoms in a molecule; mainly employed when a given molecule has two or more stable arrangements with the same set of chemical bonds.

conformational analysis [PHYS CHEM] The determination of the arrangement in space of the constituent atoms of a molecule that may rotate about a single bond.

confusion reflector [ORD] An electromagnetic-wave reflector dropped from aircraft to create false signals on enemy radarscopes, consisting of strips of aluminum foil or metallized paper, such as chaff or window, cut to lengths that are multiples or submultiples of enemy radar frequencies. Also known as radar confusion reflector.

conge [FOOD ENG] A machine in which the ingredients for sweet chocolate are worked to a smooth and viscosity-stable paste.

congelifraction [GEOL] The splitting or disintegration of rocks as the result of the freezing of the water contained. Also known as frost bursting; frost riving; frost shattering; frost splitting; frost weathering; frost wedging; gelifraction; gelivation.

congeliturbate [GEOL] Soil or unconsolidated earth which has been moved or disturbed by frost action.

congeliturbation [GEOL] The churning and stirring of soil as a result of repeated cycles of freezing and thawing; includes frost heaving and surface subsidence during thaws. Also known as cryoturbation; frost churning; frost stirring; geliturbation.

congenital [MED] Dating from or existing before birth.

congenital agammaglobulinemia [MED] A congenital deficiency (in serum) of immunoglobulins, characterized clinically by increased susceptibility to bacterial infections; may be a sex-linked recessive trait, affecting male infants, or sporadic, affecting both sexes.

congenital anomaly [MED] A structural or functional abnormality of the human body that develops before birth. Also known as birth defect.

congenital disease [MED] Any disorder or disease state that is present at birth.

congenital pathology [MED] The study of diseases and defects existing at birth.

congestin [BIOCHEM] A toxin produced by certain sea anemones.

congestion [MED] An abnormal accumulation of fluid, usu-

CONICAL BALL MILL

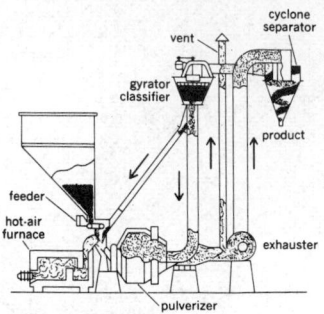

Conical ball mill pulverizer in closed circuit with classifier. (*Hardinge Co., Inc.*)

CONICONCHIA

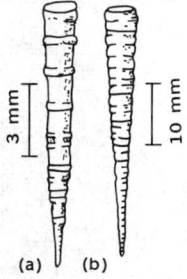

Early Paleozoic *Tentaculites*. (a) *T. gyracanthus* (Eaton), Upper Silurian. (b) *T. scalariforms* Hall, Middle Devonian. (*Adapted from R. C. Moore, C. G. Lalicker, and A. G. Fischer, Invertebrate Fossils, McGraw-Hill, 1952*)

CONIC PROJECTION

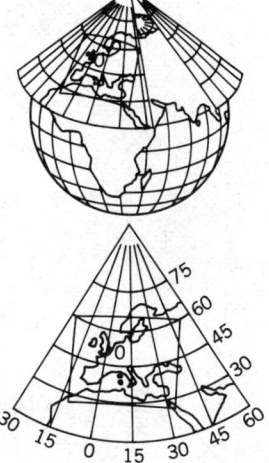

Conic projection based on origin 45°N 10°E. (*From American Oxford Atlas, Oxford University Press, 1951*)

CONIDIUM

Two conidia.

ally blood, but occasionally bile or mucus, within the vessels of an organ or part.

congestive heart failure [MED] A state in which circulatory congestion exists as a result of heart failure.

conging process [FOOD ENG] In making sweet chocolate, working a ground mixture of chocolate liquor, sugar, and cocoa butter with additional cocoa butter in conges.

conglomerate [GEOL] Cemented, rounded fragments of water-worn rock or pebbles, bound by a siliceous or argillaceous substance.

conglutination [IMMUNOL] The completion of an agglutinating system, or the enhancement of an incomplete one, by the addition of certain substances. [MED] Abnormal union of two contiguous surfaces or bodies.

conglutination phenomenon [IMMUNOL] Clumping of cells or particles, such as red cells or bacteria, when treated with conglutinin in the presence of antibody and nonhemolytic complement.

conglutinin [IMMUNOL] A heat-stable substance in bovine and other serums that aids or causes agglomeration or lysis of certain sensitized cells or particles.

congo red [ORG CHEM] $C_{32}H_{22}N_6Na_2O_6S_2$ An azo dye, sodium diphenyldiazo-bis-α-naphthylamine sulfonate, used as a biological stain and as an acid-base indicator; it is red in alkaline solution and blue in acid solution.

congo red test [PATH] Diagnostic test for amyloidosis, in which congo red is injected intravenously; 30% disappears within 1 hour in normal individuals, but in amyloidosis 40-100% disappears.

congruence [MATH] 1. The property of geometric figures that can be made to coincide by a rigid transformation. 2. The property of two integers having the same remainder on division by another integer.

congruent matrices [MATH] Two matrices A and B related by the transformation $B = SAT$, where S and T are nonsingular matrices and T is the transpose of S.

congruent melting point [THERMO] A point on a temperature composition plot of a nonstoichiometric compound at which the one solid phase and one liquid phase are adjacent.

congruent numbers [MATH] Two numbers having the same remainder when divided by a given quantity called the modulus.

congruent transformation [MET] An isothermal or isobaric phase change in an alloy where the integrity of both phases is maintained throughout the process.

Coniacian [GEOL] Lower Senonian geologic time.

conic [MATH] A curve which may be represented as the intersection of a cone with a plane; the four types of conics are circle, ellipse, parabola, and hyperbola. Also known as conic section.

conical [SCI TECH] Having the shape of or pertaining to a cone.

conical antenna [ELECTROMAG] A wide-band antenna in which the driven element is conical in shape. Also known as cone antenna.

conical ball mill [MECH ENG] A cone-shaped tumbling pulverizer in which the steel balls are classified, with the larger balls at the feed end where larger lumps are crushed, and the smaller balls at the discharge end where the material is finer.

conical beam [ELECTR] The radar beam produced by conical scanning methods.

conical bearing [MECH ENG] An antifriction bearing employing tapered rollers.

conical buoy [NAV] A buoy whose part above water is in the shape of a cone.

conical helimagnet [SOLID STATE] A helimagnet in which the directions of atomic magnetic moments all make the same angle with a specified axis of the crystal, this angle is greater than 0° and less than 90°, moments of atoms in successive basal planes are separated by equal azimuthal angles, and all moments have the same magnitude.

conical horn [ACOUS] A horn having a circular cross section and straight sides.

conical-horn antenna [ELECTROMAG] A horn antenna having a circular cross section and straight sides.

conical pendulum [MECH] A weight suspended from a cord or light rod and made to rotate in a horizontal circle about a vertical axis with a constant angular velocity.

conical refiner [MECH ENG] In paper manufacture, a cone-shaped continuous refiner having two sets of bars mounted on the rotating plug and fixed shell for beating unmodified cellulose fibers.

conical refraction [OPTICS] Phenomenon in which a ray incident on the surface of a biaxial crystal at a certain direction splits into a family of rays which lie along a cone.

conical scanning [ELECTR] Scanning in radar in which the direction of maximum radiation generates a cone, the vertex angle of which is of the order of the beam width; may be either rotating or nutating, according to whether the direction of polarization rotates or remains unchanged.

conical surface [MATH] A surface formed by the lines which pass through each of the points of a closed plane curve and a fixed point which is not in the plane of the curve.

conical vault [ARCH] A vault whose inner surface is conical.

conic chart [MAP] A chart on a conic projection.

conic chart with two standard parallels [MAP] A chart on the conic projection with two standard parallels. Also known as secant conic chart.

Coniconchia [PALEON] A class name proposed for certain extinct organisms thought to have been mollusks; distinguished by a calcareous univalve shell that is open at one end and by lack of a siphon.

conic projection [MAP] A map deformation pattern resulting from the transfer of the map to a tangent or intersecting cone.

conic projection with two standard parallels [MAP] A conic map projection in which the surface of a sphere or spheroid, such as the earth, is conceived as developed on a cone which intersects the sphere or spheroid along two standard parallels, the cone being spread out to form a plane; for example, the Lambert conformal projection. Also known as secant conic projection.

conic section *See* conic.

Conidae [INV ZOO] A family of marine gastropod mollusks in the order Neogastropoda containing the poisonous cone shells.

conidiophore [MYCOL] A specialized aerial hypha that produces conidia in certain ascomycetes and imperfect fungi.

conidiospore *See* conidium.

conidium [MYCOL] A unicellular, asexual reproductive spore produced externally upon a conidiophore. Also known as conidiospore.

conifer [BOT] The common name for plants of the order Pinales.

Coniferales [BOT] The equivalent name for Pinales.

Coniferophyta [BOT] The equivalent name for Pinicae.

coniferous forest [ECOL] An area of wooded land predominated by conifers.

coniscope *See* koniscope.

conjoint tendon [ANAT] The common tendon of the transverse and internal oblique muscles of the abdomen.

Conjugales [BOT] An order of fresh-water green algae in the class Chlorophyceae distinguished by the lack of flagellated cells, and conjugation being the method of sexual reproduction.

conjugase [BIOCHEM] Any of a group of enzymes which catalyze the breakdown of pteroylglutamic acid.

conjugate [GEOL] 1. Pertaining to fractures in which both sets of veins or joints show the same strike but opposite dip. 2. Pertaining to any two sets of veins or joints lying perpendicular. [MATH] *See* complex conjugate.

conjugate acid-base pair [CHEM] An acid and a base related by the ability of the acid to generate the base by loss of a proton.

conjugate binomial surds *See* conjugate radicals.

conjugate branches [ELEC] Any two branches of an electrical network such that a change in the electromotive force in either branch does not result in a change in current in the other. Also known as conjugate conductors.

conjugate bridge [ELECTR] A bridge in which the detector circuit and the supply circuits are interchanged, as compared with a normal bridge of the given type.

conjugate conductors *See* conjugate branches.

conjugate convex functions [MATH] Two functions $f(x)$ and $g(y)$ are conjugate convex functions if the derivative of $f(x)$ is 0 for $x = 0$ and constantly increasing for $x > 0$, and the derivative of $g(y)$ is the inverse of the derivative of $f(x)$.

conjugate curve *See* Bertrand curve.

conjugate diameters [MATH] **1.** For a conic section, any pair of straight lines either of which bisects all the chords that are parallel to the other. **2.** For an ellipsoid or hyperboloid, any three lines passing through the point of symmetry of the surface such that the plane containing the conjugate diameters (first definition) of one of the lines also contains the other two lines.

conjugate division [MYCOL] Division of dikaryotic cells in certain fungi in which the two haploid nuclei divide independently, each daughter cell receiving one product of each nuclear division.

conjugated protein [BIOCHEM] A protein combined with a nonprotein group, other than a salt or a simple protein.

conjugate foci *See* conjugate points.

conjugate hyperbolas [MATH] Two hyperbolas having the same asymptotes with semiaxes interchanged.

conjugate impedances [ELEC] Impedances having resistance components that are equal, and reactance components that are equal in magnitude but opposite in sign.

conjugate joint system [GEOL] Two joint sets with a symmetrical pattern arranged about another structural feature or an inferred stress axis.

conjugate lines [MATH] **1.** For a conic section, two lines each of which passes through the intersection of the tangents to the conic at its points of intersection with the other line. **2.** For a quadric surface, two lines each of which intersects the polar line of the other.

conjugate momentum [MECH] If q_j ($j = 1,2, \ldots$) are generalized coordinates of a classical dynamical system, and L is its Lagrangian, the momentum conjugate to q_j is $p_j = \partial L/\partial \dot{q}_j$. Also known as canonical momentum; generalized momentum.

conjugate particles [PARTIC PHYS] A particle and its antiparticle.

conjugate planes [MATH] For a quadric surface, two planes each of which contains the pole of the other.

conjugate points [MATH] For a conic section, two points either of which lies on the line that passes through the points of contact of the two tangents drawn to the conic from the other. [OPTICS] Any pair of points such that all rays from one are imaged on the other within the limits of validity of Gaussian optics. Also known as conjugate foci.

conjugate radicals [MATH] Binomial surds that are of the type $a\sqrt{b} + c\sqrt{d}$ and $a\sqrt{b} - c\sqrt{d}$, where a, b, c, d are rational but $\sqrt{b}$ and $\sqrt{d}$ are not both rational. Also known as conjugate binomial surds.

conjugate roots [MATH] Conjugate complex numbers which are roots of a given equation.

conjugate space [MATH] The set of all continuous linear functions defined on a normed linear space.

conjugate triangles [MATH] Two triangles in which the poles of the sides of each with respect to a given curve are the vertices of the other.

conjugation [BOT] Sexual reproduction by fusion of two protoplasts in certain thallophytes to form a zygote. [INV ZOO] Sexual reproduction by temporary union of cells with exchange of nuclear material between two individuals, principally ciliate protozoans. [MICROBIO] Reproduction among colon bacilli by temporary union of cells and transfer of chromosome material.

conjunction [ASTRON] **1.** The situation in which two celestial bodies have either the same celestial longitude or the same sidereal hour angle. **2.** The time at which this conjunction takes place. [MATH] The connection of two statements by the word "and."

conjunctiva [ANAT] The mucous membrane covering the eyeball and lining the eyelids.

conjunctive search [ADP] A search to identify items having all of a certain set of characteristics.

conjunctive transformation [MATH] The transformation $B = SAT$, where S is the Hermitian conjugate of T, and matrices A and B are equivalent.

conjunctivitis [MED] Inflammation of the conjunctiva.

Conklin process [MIN ENG] A dense-media coal cleaning process.

conn [NAV] To direct or conduct the steering of a vessel; to give orders to the helmsman on steering the ship.

connarite [MINERAL] A green mineral consisting of hydrous nickel silicate occurring as small crystals or grains.

connate [SCI TECH] Born, originated, or produced in a united or fused condition.

connate leaf [BOT] A leaf shaped as though the bases of two opposite leaves had fused around the stem.

connate water [HYD] Water entrapped in the interstices of igneous rocks when the rocks were formed; usually highly mineralized.

connected load [ELEC] The sum of the continuous power ratings of all load-consuming apparatus connected to an electric power distribution system or any part thereof.

connected set [MATH] A set which cannot be contained completely in each of two disjoint open sets; intuitively, a set with only one piece.

connected surface [MATH] A surface between any two points of which there is a continuous path that does not cross the surface's boundary.

connecting bar *See* tombolo.

connecting circuit [ELECTR] A functional switching circuit which directly couples other functional circuit units to each other to exchange information as dictated by the momentary needs of the switching system.

connecting rod [MECH ENG] Any straight link that transmits motion or power from one linkage to another within a mechanism, especially linear to rotary motion, as in a reciprocating engine or compressor.

connective tissue [HISTOL] A primary tissue, distinguished by an abundance of fibrillar and nonfibrillar extracellular components.

connectivity number [MATH] **1.** The number of points plus 1 which can be removed from a curve without separating the curve into more than one piece. **2.** The number of closed cuts or cuts joining points of previous cuts (or joining points on the boundary) plus 1 which can be made on a surface without separating the surface. Also known as Betti number.

connector [ELECTR] A switch, or relay group system, which finds the telephone line being called as a result of digits being dialed; it also has the function of causing interrupted ringing voltage to be placed on the called line or of returning a busy tone to the calling party if the line is busy. [ENG] **1.** A detachable device for connecting electrical conductors. **2.** A metal part for joining timbers.

connellite [MINERAL] $Cu_{19}(SO_4)Cl_4(OH)_{32} \cdot 3H_2O$ A deep-blue striated copper mineral; crystals are in the hexagonal system. Also known as footeite.

conning tower [NAV ARCH] **1.** The raised observation post of a submarine, which is in addition usually used as an entrance or exit. **2.** The armored pilothouse of a warship.

connivent [BIOL] Converging so as to meet, but not fused into a single part.

Conoclypidae [PALEON] A family of Cretaceous and Eocene exocyclic Euechinoidea in the order Holectypoida having developed aboral petals, internal partitions, and a high test.

Conocyeminae [PALEON] A subfamily of Mesozoan parasites in the family Dicyemidae.

conode *See* tie line.

conodont [PALEON] A minute, toothlike microfossil, composed of translucent amber-brown, fibrous or lamellar calcium phosphate; taxonomic identity is controversial.

Conodontiformes [PALEON] A suborder of conodonts from the Ordovician to the Triassic having a lamellar internal structure.

Conodontophoridia [PALEON] The ordinal name for the conodonts.

conoid [SCI TECH] Shaped somewhat like a cone, but not quite conical.

Conopidae [INV ZOO] The wasp flies, a family of dipteran insects in the suborder Cyclorrhapha.

conoplain *See* pediment.

conoscope [OPTICS] An instrument, essentially a wide-angle microscope, used for study and observation of interference figures and related phenomena of specially cut crystal plates, especially for measuring the axial angle. Also known as hodoscope.

Conrad discontinuity [GEOPHYS] A relatively abrupt discontinuity in the velocity of elastic waves in the earth, increasing from 6.1 to 6.4–6.7 kilometers per second; occurs at various

CONNATE LEAF

Shape of a connate leaf.

CONNECTING CIRCUIT

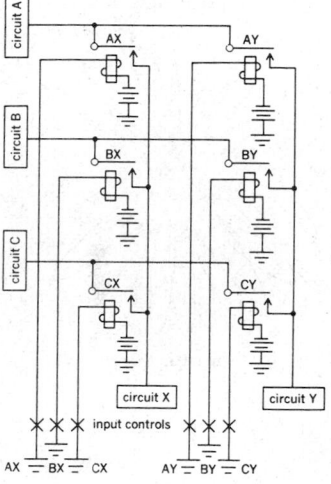

Diagram of a connecting circuit.

depths and marks contact of granitic and basaltic layers.

Conrad-Limpach method [CHEM ENG] Synthesis of 2-methyl-4-quinolone from aniline and acetoacetic ester.

Conrad machine [MIN ENG] Mechanized pit digger used in checking of alluvial boring; sections of tubing, 5 feet (152 centimeters) long and 2 feet (61 centimeters) in inside diameter, are worked into the ground while spoil is removed by means of a bucket or grab.

Conradson carbon test *See* carbon residue test.

consanguineous [GEOL] Of a natural group of sediments or sedimentary rocks, having common or related origin.

consanguinity [GEN] Blood relationship arising from common parentage. [PETR] The genetic relationship between igneous rocks in a single petrographic province which are presumably derived from a common parent magma.

conscience [PSYCH] The moral, self-critical part of oneself wherein have developed, and reside, standards of behavior and performance and value judgments.

consciousness [PSYCH] State of being aware of one's own existence, of one's mental states, and of the impressions made upon one's senses.

consensus [SCI TECH] A method of checking or confirming the correctness of an observation or report, based on agreement between different observers.

consequent [GEOL] Of, pertaining to, or characterizing movements of the earth resulting from the external transfer of material in the process of gradation.

consequent stream [GEOL] A stream whose course is determined by the slope of the land. Also known as superposed stream.

consequent valley [GEOL] 1. A valley whose direction depends on corrugation. 2. A valley formed by the widening of a trench cut by a consequent stream.

conservation [ECOL] Those measures concerned with the preservation, restoration, beneficiation, maximization, reutilization, substitution, allocation, and integration of natural resources.

conservation law [PHYS] A law which states that some physical quantity associated with an isolated system is constant.

conservation of charge [ELEC] A law which states that the total charge of an isolated system is constant; no violation of this law has been discovered. Also known as charge conservation.

conservation of energy [PHYS] The principle that energy cannot be created or destroyed, although it can be changed from one form to another; no violation of this principle has been found. Also known as energy conservation.

conservation of mass [PHYS] The notion that mass can neither be created nor destroyed; it is violated by many microscopic phenomena.

conservation of matter [PHYS] The notion that matter can be neither created nor destroyed; it is violated by microscopic phenomena.

conservation of momentum [MECH] The principle that, when a system of masses is subject only to internal forces that masses of the system exert on one another, the total vector momentum of the system is constant; no violation of this principle has been found. Also known as momentum conservation.

conservation of orbital symmetry *See* Woodward-Hoffmann rule.

conservation of parity [QUANT MECH] The law that, if the wave function describing the initial state of a system has even (odd) parity, the wave function describing the final state has even (odd) parity; it is violated by the weak interactions. Also known as parity conservation.

conservation of probability [QUANT MECH] The requirement that the sum of the probabilities of finding a system in each of its possible states is constant.

conservation of vorticity [FL MECH] 1. The principle that the vertical component of the absolute vorticity of each particle in an inviscid, autobarotropic fluid flowing horizontally remains constant. 2. The hypothesis that the vorticity of fluid particles remains constant during the turbulent mixing of the fluid.

conservative force field [MECH] A field of force in which the work done on a particle in moving it from one point to

another depends only on the particle's initial and final positions.

conservative property [THERMO] A property of a system whose value remains constant during a series of events.

conserved vector current [PARTIC PHYS] The hypothesis that the weak hadronic vector current is identical to the conserved isotopic-spin current. Abbreviated CVC.

consistency [MATER] The degree of solidity or fluidity of a material such as grease, pulp, or slurry.

consistency condition [MATH] The requirement that a mathematical theory be free from contradiction.

consistency routine [ADP] A debugging routine which is used to determine whether the program being checked gives consistent results at specified check points; for example, consistent between runs or with values calculated by other means.

consistent estimate [STAT] A method of estimation which has the property that the estimate is practically certain to fall very close to a parameter being estimated, provided there are sufficient observations.

Consol [NAV] A radio navigation aid that provides a number of characteristic signal zones which rotate in a time sequence; a bearing may be determined by observation of the instant at which transition occurs from one zone to the following zone. Also known as Sonne.

Consolan [NAV] A long-range directional navigation system that transmits a slowly rotating keyed radio field pattern; an American version of the German Sonne and British Consol, using two radiators instead of three to minimize night-effect errors.

Consol chart [NAV] A chart showing the lines of position of the Consol navigation system.

console [ADP] The section of a computer that is used to control the machine manually, correct errors, manually revise the contents of storage, and provide communication in other ways between the operator or service engineer and the central processing unit. [ENG] 1. A main control desk for electronic equipment, as at a radar station, radio or television station, or airport control tower. Also known as control desk. 2. A large cabinet for a radio or television receiver, standing on the floor rather than on a table. 3. A grouping of controls, indicators, and similar items contained in a specially designed model cabinet for floor mounting; constitutes an operator's permanent working position.

console display [ADP] The visible representation of information, whether in words, numbers, or drawings, on a console screen connected to a computer.

console file adapter [ADP] A special input/output device which allows the operator to load reloadable control storage from the system console.

console receiver [ELECTR] A television or radio receiver in a console.

console typewriter [ADP] A typewriter by means of which the computer operator can monitor system and program operations.

consolidated ice [OCEANOGR] Ice which has been compacted into a solid mass by wind and ocean currents and covers an area of the ocean.

consolidation [GEOL] 1. Processes by which loose, soft, or liquid earth become coherent and firm. 2. Adjustment of a saturated soil in response to increased load; involves squeezing of water from the pores and a decrease in void ratio.

consolidation test [MIN ENG] A test in which the specimen is confined laterally in a ring and is compressed between porous plates which are saturated with water.

Consol station [NAV] A short-base-line directional-antenna system consisting of three low-frequency/medium-frequency vertical antennas in a line on the ground, evenly spaced at a distance of about three times the length of the transmitted continuous wave; coverage is about 240° over long distances, in the form of 22 lobes; the lobes are areas where the signals corresponding to the Morse E and T are heard; these patterns rotate in space and the line of position is determined by counting the numbers of E's and T's that are heard.

consolute [CHEM] Of or pertaining to liquids that are perfectly miscible in all proportions under certain conditions.

consolute temperature [THERMO] The upper temperature of immiscibility for a two-component liquid system. Also

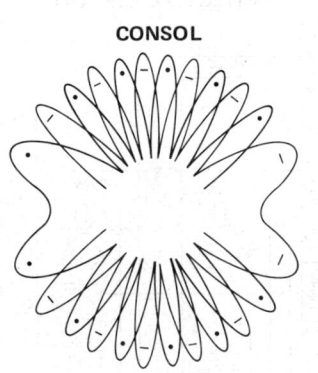

CONSOL

Field pattern of a Consol system.

known as upper consolute temperature; upper critical solution temperature.

consonance [ACOUS] The interval between two tones whose frequencies are in a ratio approximately equal to the quotient of two whole numbers, each equal to or less than 6, or to such a quotient multiplied or divided by some power of 2.

consortism *See* symbiosis.

constant [SCI TECH] A value that does not change during a particular process.

constant-amplitude recording [ENG ACOUS] A sound-recording method in which all frequencies having the same intensity are recorded at the same amplitude.

Constantan [MET] An alloy containing 55% copper and 45% nickel, used in making precision wire-wound resistors because of its constant temperature coefficient of resistance, and used with iron or copper in thermocouples.

constant-angle fringes *See* Haidinger fringes.

Constantan thermocouple [ENG] A dual-wire-junction type of heat-detecting device in which one wire is Constantan alloy, the other being iron, copper, or Chromel-P alloy.

constant area [ADP] A part of storage used for constants.

constant-bandwidth analyzer [ACOUS] A tunable sound analyzer which has a fixed pass band that is swept through the frequency range of interest. Also known as constant-bandwidth filter.

constant-bandwidth filter *See* constant-bandwidth analyzer.

constant-conductance network *See* constant-resistance network.

constant-current characteristic [ELECTR] The relation between the voltages of two electrodes in an electron tube when the current to one of them is maintained constant and all other electrode voltages are constant.

constant-current dc potentiometer [ELEC] A potentiometer in which the unknown electromotive force is balanced by a constant current times the resistance of a calibrated resistor or slide-wire. Also known as Poggendorff's first method.

constant-current electrolysis [CHEM] Electrolysis in which a constant current flows through the cell; used in electrodeposition analysis.

constant-current filter [ELECTR] A filter network intended to be connected to a source whose internal impedance is so high it can be assumed as infinite.

constant-current generator [ELECTR] A vacuum-tube circuit, generally containing a pentode, in which the alternating-current anode resistance is so high that anode current remains essentially constant despite variations in load resistance.

constant-current modulation [COMMUN] A system of amplitude modulation in which the output circuits of the signal amplifier and the carrier-wave generator or amplifier are connected through a common coil to a constant-current source. Also known as Heising modulation.

constant-current supply [ELEC] The power supply for repeatered submarine telephone cables; the voltage is varied automatically to maintain a constant current through the use of variable-voltage rectifiers and constant-current regulators at each shore station.

constant-current titration *See* potentiometric titration.

constant-current transformer [ELEC] A transformer that automatically maintains a constant current in its secondary circuit under varying loads, when supplied from a constant-voltage source.

constant-delay discriminator *See* pulse demoder.

constant-deviation fringes *See* Haidinger fringes.

constant-effect model [STAT] A model of a test in which the effect of a treatment is the same for all subjects.

constant-false-alarm rate [ELECTR] Radar system devices used to prevent receiver saturation and overload so as to present clean video information to the display, and to present a constant noise level to an automatic detector.

constant field *See* stationary field.

constant-force spring [MECH ENG] A spring which has a constant restoring force, regardless of displacement.

constant-gradient synchrotron [NUCLEO] A synchrotron in which the radial gradient of the magnetic field is constant as a function of angle around the orbit.

constant-head meter [ENG] A flow meter which maintains a constant pressure differential but varies the orifice area with flow, such as a rotameter or piston meter.

constant-height chart [METEOROL] A synoptic chart for any surface of constant geometric altitude above mean sea level (a constant-height surface), usually containing plotted data and analyses of the distribution of such variables as pressure, wind, temperature, and humidity at that altitude. Also known as constant-level chart; fixed-level chart; isohypsic chart.

constant-height surface [METEOROL] A surface of constant geometric or geopotential altitude measured with respect to mean sea level. Also known as constant-level surface; isohypsic surface.

constant instruction [ADP] A nonexecutable instruction.

constant-k filter [ELECTR] A filter in which the product of the series and shunt impedances is a constant that is independent of frequency.

constant-k lens [ELECTROMAG] A microwave lens that is constructed as a solid dielectric sphere; a plane electromagnetic wave brought to a focus at one point on the sphere emerges from the opposite side of the sphere as a parallel beam.

constant-k network [ELECTR] A ladder network in which the product of the series and shunt impedances is independent of frequency within the operating frequency range.

constant-level balloon [AERO ENG] A balloon designed to float at a constant pressure level. Also known as constant-pressure balloon.

constant-level chart *See* constant-height chart.

constant-level surface *See* constant-height surface.

constant-load balance [ENG] An instrument for measuring weight or mass which consists of a single pan, together with a set of weights that can be suspended from a counterpoised beam, that has a constant load (200 grams for the microbalance).

constant-luminance transmission [COMMUN] Type of transmission in which the transmission primaries are a luminance primary and two chrominance primaries.

constant of aberration [ASTRON] The maximum aberration of a star observed from the surface of the earth, equal to 20.49 seconds of arc.

constant of gravitation *See* gravitational constant.

constant of motion [MECH] A dynamical variable of a system which remains constant in time.

constant of the cone [MAP] The chart convergence factor for a conic projection.

constant-potential accelerator [NUCLEO] An accelerator in which constant direct-current voltage is applied to an accelerating tube to produce high-energy ions or electrons.

constant-potential electrolysis [CHEM] Electrolysis in which a constant voltage is applied to the cell; used in electrodeposition analysis.

constant-pressure balloon *See* constant-level balloon.

constant-pressure chart [METEOROL] The synoptic chart for any constant-pressure surface, usually containing plotted data and analyses of the distribution of height of the surface, wind temperature, humidity, and so on. Also known as isobaric chart; isobaric contour chart.

constant-pressure-pattern flight [NAV] A technique of pressure-pattern flight whereby an aircraft is navigated, in the direction of wind flow, along a height contour line on a constant-pressure surface, thereby assuring a continuous, nearly direct tail wind.

constant-pressure surface *See* isobaric surface.

constant radio code [COMMUN] Code in which all characters are represented by combinations having a fixed ratio of ones to zeros.

constant-resistance dc potentiometer [ELEC] A potentiometer in which the ratio of an unknown and a known potential are set equal to the ratio of two known constant resistances. Also known as Poggendorff's second method.

constant-resistance network [ELECTR] A network having at least one driving-point impedance that is a positive constant. Also known as constant-conductance network.

constant series *See* displacement series.

constant-speed drive [MECH ENG] A mechanism transmitting motion from one shaft to another that does not allow the velocity ratio of the shafts to be varied, or allows it to be varied only in steps.

constant-speed propeller [AERO ENG] A variable-pitch pro-

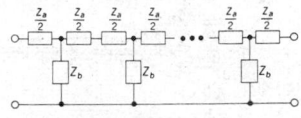

CONSTANT-K NETWORK

T sections, one form of constant-*k* recurrent ladder network.

peller having a governor which automatically changes the pitch to maintain constant engine speed.

constant-velocity recording [ENG ACOUS] A sound-recording method in which, for input signals of a given amplitude, the resulting recorded amplitude is inversely proportional to the frequency; the velocity of the cutting stylus is then constant for all input frequencies having that given amplitude.

constant-velocity universal joint [MECH ENG] A universal joint that transmits constant angular velocity from the driving to the driven shaft, such as the Bendix-Weiss universal joint.

constant-voltage generator [ELEC] An axle generator that is equipped with a regulator which keeps voltage constant.

Constellariidae [PALEON] A family of extinct, marine bryozoans in the order Cystoporata.

constellation [ASTRON] **1.** Any one of the star groups interpreted as forming configurations in the sky; examples are Orion and Leo. **2.** Any one of the definite areas of the sky.

constipation [MED] The passage of hard, dry stools.

constituent [SCI TECH] An essential part or component of a system or group: examples are an ingredient of a chemical system, or a component of an alloy.

constituent day [ASTRON] The duration of one rotation of the earth on its axis with respect to an astre fictif, that is, a fictitious star representing one of the periodic elements in the tidal forces; approximates the length of a lunar or solar day.

constituent number [OCEANOGR] One of the harmonic elements in a mathematical expression for the tide-producing force, and in corresponding formulas for the tide or tidal current.

constitutional anthropology [ANTHRO] A branch of physical anthropology concerned with body composition and constitution of an individual.

constitution diagram [MET] Graphical representation of phase-stability relationships in an alloy system as a function of temperature. Also known as phase diagram.

constitutive enzyme [BIOCHEM] A type of enzyme always produced by the cell regardless of environmental conditions.

constitutive equations [ELECTROMAG] The equations $D = \varepsilon E$ and $B = \mu H$, which relate the electric displacement D with the electric field intensity E, and the magnetic induction B with the magnetic field intensity H.

constitutive property [CHEM] Any physical or chemical property that depends on the constitution or structure of the molecule.

constrained mechanism [MECH ENG] A mechanism in which all members move only in prescribed paths.

constraint [ENG] Anything that restricts the transverse contraction which normally occurs in a solid under longitudinal tension. [MECH] A restriction on the natural degrees of freedom of a system; the number of constraints is the difference between the number of natural degrees of freedom and the number of actual degrees of freedom. [SCI TECH] A condition imposed on a system which limits the freedom of the system; may be physical or mathematical, necessary or incidental.

constraint matrix [ADP] The set of equations and inequalities defining the set of admissible solutions in linear programming.

constriction [SCI TECH] Narrowing of a channel or cylindrical member.

constriction disease [PL PATH] A fungus disease of peach trees caused by a species of *Phomopsis* and marked by death of peripheral structures.

constrictive pericarditis [MED] Inflammation and fibrosis of the pericardium resulting in constriction of the heart and restriction of contraction and blood flow. Also known as Pick's disease.

constrictor [AERO ENG] The exit portion of the combustion chamber in some designs of ramjets, where there is a narrowing of the tube at the exhaust.

constringence *See* nu value.

construction [DES ENG] The number of strands in a wire rope and the number of wires in a strand; expressed as two numbers separated by a multiplication sign. [ENG] **1.** Putting parts together to form an integrated object. **2.** The manner in which something is put together. [TEXT] A fabric

formula, being the number of warp and filling threads per square inch and the weight of the yarns.

constructional apraxia *See* optic apraxia.

construction area [BUILD] The area of exterior walls and permanent interior walls and partitions.

construction engineering [CIV ENG] A specialized branch of civil engineering concerned with the planning, execution, and control of construction operations for projects such as highways, dams, utility lines, and buildings.

construction equipment [MECH ENG] Heavy power machines which perform specific construction or demolition functions.

construction joint [CIV ENG] A vertical or horizontal surface in reinforced concrete where concreting was stopped and continued later.

construction paper [MATER] A heavy paper made from mechanical pulp and available in a large range of colors, most of which are not lightfast.

construction survey [CIV ENG] A survey that gives locations for construction work.

construction weight [AERO ENG] The weight of a rocket exclusive of propellant, load, and crew if any. Also known as structural weight.

construction wrench [DES ENG] An open-end wrench with a long handle; the handle is used to align matching rivet or bolt holes.

constructive interference [PHYS] Phenomenon in which the phases of waves arriving at a specified point over two or more paths of different lengths are such that the square of the resultant amplitude is greater than the sum of the squares of the component amplitudes.

consumable electrode [MET] A metal electrode that supplies the filler for welding.

consumer [ECOL] A nutritional grouping in the food chain of an ecosystem, composed of heterotrophic organisms, chiefly animals, which ingest other organisms or particulate organic matter.

consumer's risk [IND ENG] The probability that a lot whose quality equals the poorest quality that a consumer is willing to tolerate in an individual lot will be accepted by a sampling plan.

consummatory behavior [PSYCH] Actions that fulfill a motive and cause appetitive behavior to end.

consumption *See* tuberculosis.

consumptive coagulopathy [MED] Reduction in one or more of the blood elements involved in coagulation as the result of marked blood clotting.

consumptive use [HYD] The total annual land water loss in an area, due to evaporation and plant use.

contact [ELEC] *See* electric contact. [ENG] Initial detection of an aircraft, ship, submarine, or other object on a radarscope or other detecting equipment. [FL MECH] The surface between two immiscible fluids in a reservoir. [GEOL] The surface between two different kinds of rocks.

contact acid [INORG CHEM] Sulfuric acid produced by the contact process.

contact adsorption [CHEM ENG] Process for removal of minor constituents from fluids by stirring in direct contact with powdered or granulated adsorbents, or by passing the fluid through fixed-position adsorbent beds (activated carbon or ion-exchange resin); used to decolorize petroleum lubricating oils and to remove solvent vapors from air.

contact aerator [CIV ENG] A tank in which sewage that is settled on a bed of stone, cement-asbestos, or other surfaces is treated by aeration with compressed air.

contact anemometer [ENG] An anemometer which actuates an electrical contact at a rate dependent upon the wind speed. Also known as contact-cup anemometer.

contact aureole *See* aureole.

contact bed [CIV ENG] A bed of coarse material such as coke, used to purify sewage.

contact block [ELEC] A block of conducting material such as carbon, used in a relay.

contact bounce [ELEC] The uncontrolled making and breaking of contact one or more times, but not continuously, when relay contacts are moved to the closed position.

contact breccia [PETR] Angular rock fragments resulting

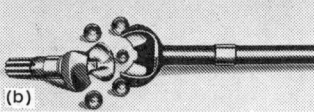

Constant-velocity universal joint. (a) Partially separated cutaway. (b) Disassembled joint showing the arrangement; the right-hand member has been cut away. (*Bendix Aviation Corp.*)

from shattering of wall rocks around laccolithic and other igneous masses.

contact catalysis [CHEM ENG] Process of change in the structure of gas molecules adsorbed onto solid surfaces; the basis of many industrial processes.

contact ceiling [BUILD] A ceiling in which the lath and construction are in direct contact, without use of furring or runner channels.

contact chart *See* aeronautical pilotage chart.

contact chatter *See* chatter.

contact clip [ELEC] The clip which the blade of a knife switch is clamped to in the closed condition.

contact condenser [MECH ENG] A device in which a vapor, such as steam, is brought into direct contact with a cooling liquid, such as water, and is condensed by giving up its latent heat to the liquid. Also known as direct-contact condenser.

contact corrosion *See* crevice corrosion.

contact-cup anemometer *See* contact anemometer.

contact dermatitis [MED] An acute or chronic inflammation of the skin resulting from irritation by or sensitization to some substance coming in contact with the skin.

contact electricity [ELEC] An electric charge at the surface of contact of two different materials.

contact electromotive force *See* contact potential difference.

contact filtration [CHEM ENG] A process in which finely divided adsorbent clay is mixed with oil to remove color bodies and to improve the oil's stability.

contact flight [NAV] Flight in which visual contact is maintained with the surface of the earth.

contact follow [ELEC] The distance two contacts travel together after just touching. Also known as contact overtravel.

contact force [ELEC] The force exerted by the moving contact of a switch or relay on a stationary contact.

contact fuse *See* impact fuse.

contact gear ratio *See* contact ratio.

contact inhibition [CYTOL] Cessation of cell division when cultured cells are in physical contact with each other.

contact-initiated discharge machining [MECH ENG] An electromachining process in which the discharge is initiated by allowing the tool and workpiece to come into contact, after which the tool is withdrawn and an arc forms.

contact lens [OPTICS] **1.** A thin lens fitted over the cornea to correct defects of vision. **2.** A similar lens or prism used with a gonioscope in eye examinations.

contact log [PETRO ENG] Record of electrical-resistivity data pertaining to strata structures along the depth of a drill hole.

contact-making meter *See* instrument-type relay.

contact material [MET] A metal having high electrical and thermal conductivity, low contact resistance, minimum sticking or welding tendencies, and high corrosion resistance.

contact metamorphic rock [PETR] A rock formed by the processes of contact metamorphism.

contact metamorphism [PETR] Metamorphism that is genetically related to the intrusion or extrusion of magmas and takes place in rocks at or near their contact.

contact metasomatism [GEOL] One of the main local processes of thermal metamorphism that is related to intrusion of magmas; takes place in rocks or near their contact with a body of igneous rock.

contact microphone [ENG ACOUS] A microphone designed to pick up mechanical vibrations directly and convert them into corresponding electric currents or voltages.

contact mine [ORD] A mine that is fired when struck by the hull of a passing ship.

contact mineral [MINERAL] A mineral formed by the processes of contact metamorphism.

contactor [ELEC] A heavy-duty relay used to control electric power circuits. Also known as electric contactor.

contactor control system [CONT SYS] A feedback control system in which the control signal is a discontinuous function of the sensed error and may therefore assume one of a limited number of discrete values.

contact overtravel *See* contact follow.

contact paper [GRAPHICS] Photographic paper designed to be exposed when directly in contact with a reproducible document.

contact piston [ELECTROMAG] A waveguide piston that makes contact with the walls of the waveguide. Also known as contact plunger.

contact plunger *See* contact piston.

contact potential *See* contact potential difference.

contact potential difference [ELEC] The potential difference that exists across the space between two electrically connected materials. Also known as contact electromotive force; contact potential; Volta effect.

contact pressure [ELEC] The amount of pressure holding a set of contacts together.

contact-pressure resin *See* contact resin.

contact print [GRAPHICS] A photographic image produced by the exposure of a sensitized emulsion in direct contact with a negative or positive transparency.

contact printer [GRAPHICS] **1.** A device which provides a light source and a means for holding the negative and the sensitive material in contact during exposure. **2.** A specialized device for exposing diapositive plates at the same scale as that of the negative.

contact process [CHEM ENG] Catalytic manufacture of sulfuric acid from sulfur dioxide and oxygen.

contact protection [ELEC] Any method for suppressing the surge which results when an inductive circuit is suddenly interrupted; the break would otherwise produce arcing at the contacts, leading to their deterioration.

contact ratio [DES ENG] The ratio of the length of the path of contact of two gears to the base pitch, equal to approximately the average number of pairs of teeth in contact. Also known as contact gear ratio.

contact rectifier *See* metallic rectifier.

contact resin [MATER] A liquid resin which thickens or polymerizes on heating and, when used for bonding laminates, requires little or no pressure for adherence. Also known as contact-pressure resin.

contact resistance [ELEC] The resistance in ohms between the contacts of a relay, switch, or other device when the contacts are touching each other.

contact thermography [ENG] A method of measuring surface temperature in which a thin layer of luminescent material is spread on the surface of an object and is excited by ultraviolet radiation in a darkened room; the brightness of the coating indicates the surface temperature.

contact time [ENG] The length of time a substance is held in direct contact with a treating agent.

contact transformation *See* canonical transformation.

contact twin [CRYSTAL] Twinned crystals whose members are symmetrically arranged about a twin plane.

contact vein [GEOL] **1.** A variety of fissure vein formed by deposition of minerals in a fault fissure at a rock contact. **2.** A replacement vein formed by mineralized solutions percolating along the more permeable surface areas of the contact.

contact zone *See* aureole.

contagion [MED] **1.** The process whereby disease spreads from one person to another, by direct or indirect contact. **2.** The bacterium or virus which transmits disease.

contagious abortion [VET MED] Brucellosis in cattle caused by *Brucella abortus* and inducing abortion. Also known as Bang's disease; infectious abortion.

contagious disease [MED] An infectious disease communicable by contact with one suffering from it, with his bodily discharge, or with an object touched by him.

contagious distribution [STAT] A probability distribution which is dependent on a parameter that itself has a probability distribution.

contagious polyarthritis [VET MED] An infectious bacterial disease of mice caused by members of the genus *Bacteroides*; characterized by inflammation and abcess formation in the joints.

contagious pustular dermatitis [VET MED] An infectious disease of sheep and goats characterized by vesicles on the skin which are transformed into pustules.

container [IND ENG] A portable compartment of standard, uniform size, used to hold cargo for air, sea, or ground transport.

container car [ENG] A railroad car designed specifically to hold containers.

containerization [IND ENG] The practice of placing cargo in

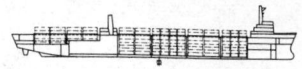

CONTAINER SHIP

Inboard profile of a container ship.

large containers such as truck trailers to facilitate loading on and off ships and railroad flat cars.

container ship [NAV ARCH] A cargo ship which carries its cargo in weatherproof boxes (usually metal) of standard size, called containers, which need not be opened and are rapidly loaded or unloaded from the ship.

containment [NUCLEO] **1.** Provision of a gastight enclosure around the highly radioactive components of a nuclear power plant, to contain the radioactivity released by a possible major accident. **2.** The use of remote-control devices (slave apparatus) to remove spent cores from nuclear power plants or, in shielded laboratory hoods, to perform chemical studies of dangerous radioactive materials.

containment vessel [NUCLEO] A gas-tight shell or other enclosure around a reactor.

contaminate [SCI TECH] To render unfit or to soil by the introduction of foreign or unwanted material.

contamination [MICROBIO] Process or act of soiling with bacteria. [NUCLEO] The deposit of radioactive materials, such as fission fragments or radiological warfare agents, on any objective or surface or in the atmosphere. [PSYCH] The fusion of words, resulting in a new word. [SCI TECH] Something that contaminates.

contamination monitor [NUCLEO] A radiation counter used to detect radioactive contamination of surface areas or of the atmosphere.

contemporaneous [GEOL] **1.** Formed, existing, or originating at the same time. **2.** Of a rock, developing during formation of the enclosing rock.

contemporaneous deformation [GEOL] Deformation that takes place during or immediately following deposition of a sediment. Also known as penecontemporaneous deformation.

content-addressed storage See associative memory.

content indicator [ADP] Display unit that indicates the content in a computer, and the program or mode being used.

contents [ADP] The information stored at any address or in any register of a computer.

continent [GEOGR] A protuberance of the earth's crustal shell, with an area of several million square miles and sufficient elevation so that much of it is above sea level.

continental accretion [GEOL] The theory that continents have grown by the addition of new continental material around an original nucleus, mainly through the processes of geosynclinal sedimentation and orogeny.

continental air [METEOROL] A type of air whose characteristics are developed over a large land area and which therefore has relatively low moisture content.

continental anticyclone See continental high.

continental borderland [GEOL] The area of the continental margin between the shoreline and the continental slope.

continental climate [CLIMATOL] Climate characteristic of the interior of a landmass of continental size, marked by large annual, daily, and day-to-day temperature ranges, low relative humidity, and a moderate or small irregular rainfall; annual extremes of temperature occur soon after the solstices.

continental code [COMMUN] The code commonly used for manual telegraph communication, consisting of short (dot) and long (dash) symbols, but not the various-length spaces used in the original Morse code. Also known as international Morse code.

continental crust [GEOL] The basement complex of rock, that is, metamorphosed sedimentary and volcanic rock with associated igneous rocks mainly granitic, that underlies the continents and the continental shelves.

continental deposits [GEOL] Sedimentary deposits laid down within a general land area.

continental drift [GEOL] The concept of continent formation by the fragmentation and movement of land masses on the surface of the earth.

continental geosyncline [GEOL] A geosyncline filled with nonmarine sediments.

continental glacier [HYD] A sheet of ice covering a large tract of land, such as the ice caps of Greenland and the Antarctic.

continental growth [GEOL] The processes contributing to growth of continents at the expense of ocean basins.

continental heat flow [GEOPHYS] The amount of thermal energy escaping from the earth through the continental crust per unit area and unit time.

continental high [METEOROL] A general area of high atmospheric pressure which on mean charts of sea-level pressure is seen to overlie a continent during the winter. Also known as continental anticyclone.

continentality [CLIMATOL] The degree to which a point on the earth's surface is in all respects subject to the influence of a land mass.

continental margin [GEOL] Those provinces between the shoreline and the deep-sea bottom; generally consists of the continental borderland, shelf, slope, and rise.

continental mass [GEOGR] The continental land rising more or less abruptly from the ocean floor and also the shallow submerged areas surrounding this land.

continental nucleus [GEOL] A large area of basement rock consisting of basaltic and more mafic oceanic crust and periodotitic mantle from which it is postulated that continents have grown. Also known as continental shield; cratogene; shield.

continental plate [GEOL] Thick continental crust.

continental plateau See tableland.

continental platform See continental shelf.

continental polar air [METEOROL] Polar air having low surface temperature, low moisture content, and (especially in its source regions) great stability in the lower layers.

continental rise [GEOL] A transitional part of the continental margin; a gentle slope with a generally smooth surface, built up by the shedding of sediments from the continental block, and located between the continental slope and the abyssal plain.

continental shelf [GEOL] The zone around a continent, that part of the continental margin extending from the shoreline and the continental slope; composes with the continental slope the continental terrace. Also known as continental platform; shelf.

continental slope [GEOL] The part of the continental margin consisting of the declivity from the edge of the continental shelf extending down to the continental rise.

continental terrace [GEOL] The continental shelf and slope together.

continental tropical air [METEOROL] A type of tropical air produced over subtropical arid regions; it is hot and very dry.

continent formation [GEOL] A series of six or seven major episodes, resulting from the buildup of radioactive heat and then the melting or partial melting of the earth's interior; the molten rock melt rises to the surface, differentiating into less primitive lavas; the continent then nucleates, differentiates, and grows from oceanic crust and mantle.

contingency interrupt [ADP] A processing interruption due to an operator's action or due to an abnormal result from the system or from a program.

contingency table [STAT] A table for classifying elements of a population according to two variables, the rows corresponding to one variable and the columns to the other.

continued fraction [MATH] The sum of a number and a fraction whose denominator is the sum of a number and a fraction, and so forth; it may have either a finite or an infinite number of terms.

continued-fraction expansion [MATH] An expansion of a driving-point function about infinity (or zero) in a continued fraction, in which the terms are alternately constants and multiples of the complex frequency (or multiples of the reciprocal of the complex frequency).

continued product [MATH] A product of three or more factors, or of an infinite number of factors.

continue statement [ADP] A nonexecutable statement in FORTRAN used principally as a target for transfers, particularly as the last statement in the range of a do statement.

continuity [CIV ENG] Joining of structural members to each other, such as floors to beams, and beams to beams and to columns, so they bend together and strengthen each other when loaded. Also known as fixity. [ELEC] Continuous effective contact of all components of an electric circuit to give it high conductance by providing low resistance.

continuity chart [METEOROL] A chart maintained for weather analysis and forecasting upon which are entered the positions of significant features (pressure centers, fronts, instability lines, through lines, ridge lines) of the regular synoptic charts at regular intervals in the past.

continuity equation [PHYS] An equation obeyed by any conserved, indestructible quantity such as mass, electric charge, thermal energy, electrical energy, or quantum-mechanical probability, which is essentially a statement that the rate of increase of the quantity in any region equals the total current flowing into the region.

continuity test [ELEC] An electrical test used to determine the presence and location of a broken connection.

continuous at a point [MATH] A function f is continuous at a point x if for every sequence $\{x_n\}$ whose limit is x, the sequence $f(x_n)$ converges to $f(x)$; in a general topological space, for every neighborhood W of $f(x)$, there is a neighborhood N of x such that $f^{-1}(W)$ is contained in N.

continuous beam [CIV ENG] **1.** A beam resting upon several supports, which may be in the same horizontal plane. **2.** A beam having several spans in one straight line; generally has at least three supports.

continuous brake [MECH ENG] A train brake that operates on all cars but is controlled from a single point.

continuous bridge [CIV ENG] A bridge structure supported at three or more points that extends past its supports and is capable of resisting bending moments and shear at all sections.

continuous bucket elevator [MECH ENG] A bucket elevator on an endless chain or belt.

continuous bucket excavator [MECH ENG] A bucket excavator with a continuous bucket elevator mounted in front of the bowl.

continuous casting [MET] A technique in which an ingot, billet, tube, or other shape is continuously solidified and withdrawn while it is being poured, so that its length is not determined by mold dimensions.

continuous clamp *See* voltage-amplitude-controlled clamp.

continuous coal cutter [MIN ENG] A coal mining machine that cuts the coal face without being withdrawn from the cut.

continuous comparator *See* linear comparator.

continuous contact coking [CHEM ENG] A thermal conversion process using the mass-flow lift principle to give continuous coke circulation; oil-wetted particles of coke move downward into the reactor in which cracking, coking, and drying take place; pelleted coke, gas, gasoline, and gas oil are products of the process.

continuous control [CONT SYS] Automatic control in which the controlled quantity is measured continuously and corrections are a continuous function of the deviation.

continuous countercurrent leaching [CHEM ENG] Leaching by the use of continuous equipment in which the solid and liquid are both moved mechanically, and by the use of a series of leach tanks and the countercurrent flow of solvent through the tanks in reverse order to the flow of solid.

continuous distillation [CHEM ENG] Separation by boiling of a liquid mixture with different component boiling points; feed is introduced continuously, with continuous removal of overhead vapors and high-boiling bottoms liquids.

continuous distribution [STAT] Distribution of a continuous population, which is a class of pairs such that the second member of each pair is a value, and the first member of the pair is a proportion density for that value.

continuous dryer [ENG] An apparatus in which drying is accomplished by passing wet material through without interruption.

continuous-duty rating [ELEC] The rating that defines the load which can be carried for an indefinite time without exceeding a specified temperature rise.

continuous dyeing [TEXT] The application of color-producing agents to textiles by impregnating the cloth with dye and then passing it through a series of developing, washing, and drying zones to a final take-up roll.

continuous equilibrium vaporization *See* equilibrium flash vaporization.

continuous extension [MATH] A continuous function which is equal to another continuous function defined on a smaller domain.

continuous film scanner [ELECTR] A television film scanner in which the motion picture film moves continuously while being scanned by a flying-spot kinescope.

continuous fire [ORD] **1.** Fire conducted at a normal rate without interruption, for application of adjustment corrections or for other causes. **2.** In field artillery, a succession of salvos, the pieces being fired consecutively at the interval designated in the command.

continuous-flow conveyor [MECH ENG] A totally enclosed, continuous-belt conveyor pulled transversely through a mass of granular, powdered or small-lump material fed from an overhead hopper.

continuous footing [CIV ENG] A concrete footing that may extend along the entire length of a wall or single row of columns.

continuous forms [ADP] In character recognition, any batch of source information that exists in reel form, such as tally rolls or cash-register receipts.

continuous function [MATH] A function which is continuous at each point of its domain. Also known as continuous transformation.

continuous furnace [MET] A type of reheating furnace in which the charge introduced at one end moves continuously through the furnace and is discharged at the other end.

continuous gas lift [PETRO ENG] Oil production in which reservoir gas pressure (natural or injected) is sufficient to provide a continuous upward flow of oil through the well tubing.

continuous geometry [MATH] A generalization of projective geometry.

continuous image [MATH] The image of a set under a continuous function.

continuous industry [IND ENG] An industry in which raw material is subjected to successive operations, turning it into a finished product.

continuous kiln [ENG] **1.** A long kiln through which ware travels on a moving device, such as a conveyor. **2.** A kiln through which the fire travels progressively.

continuous leader *See* dart leader.

continuous loading [ELEC] Loading in which the added inductance is distributed uniformly along a line by wrapping magnetic material around each conductor.

continuously adjustable transformer *See* variable transformer.

continuous mill [MET] A rolling mill in which metal is successively rolled thinner as it passes through a series of synchronized rolls in tandem.

continuous miner [MIN ENG] Machine designed to remove coal or other soft minerals from the face and to load it into cars or conveyors continuously, without the use of cutting machines, drills, or explosives.

continuous mining [MIN ENG] A type of mining in which the continuous miner cuts or rips coal or other soft minerals from the face and loads it in a continuous operation.

continuous mixer [MECH ENG] A mixer in which materials are introduced, mixed, and discharged in a continuous flow.

continuous operation [ENG] A process that operates on a continuous flow (materials or time) basis, in contrast to batch, intermittent, or sequenced operations.

continuous operator [MATH] A linear transformation of Banach spaces which is continuous with respect to their topologies.

continuous permafrost zone [GEOL] Regional zone predominantly underlain by permanently frozen subsoil that is not interrupted by pockets of unfrozen ground.

continuous phase [MET] The matrix or background phase of a multiphasic alloy.

continuous precipitation [MET] Precipitation that is characteristic of certain alloys, from a supersaturated solid solution, involving a gradual change of the lattice parameter of the matrix with aging time.

continuous production [IND ENG] Manufacture of products, such as chemicals or paper, involving a sequence of processes performed by a series of machines receiving the materials through a closed channel of flow.

continuous profiling [GEOL] A method of shooting in seismic exploration in which uniformly placed seismometer stations

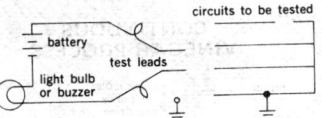
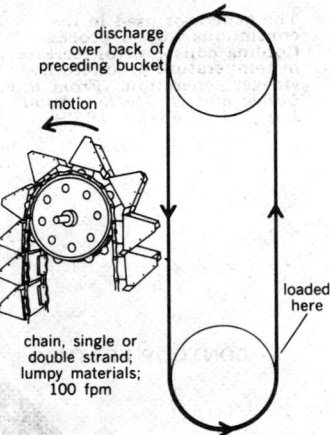
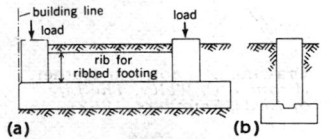

along a line are shot from holes spaced along the same line so that each hole records seismic ray paths geometrically identical with those from adjacent holes.

continuous radio beacon [NAV] A single marine radio beacon operating on a frequency without interruption; used specifically with automatic direction finders.

continuous-rail frog [ENG] A metal fitting that holds continuous welded rail sections to railroad ties.

continuous reaction series [MINERAL] A branch of Bowen's reaction series comprising the plagioclase mineral group in which reaction of early-formed crystals with water takes place continuously, without abrupt changes in crystal structure.

continuous recorder [ENG] A recorder whose record sheet is a continuous strip or web rather than individual sheets.

continuous sintering [MET] Sintering process in which materials are moved through the furnace at a fixed rate without interruption.

continuous spectrum [MATH] The portion of the spectrum of a linear operator which is a continuum. [SPECT] A radiation spectrum which is continuously distributed over a frequency region without being broken up into lines or bands.

continuous spinning [TEXT] Spinning in which fiber is extruded, coagulated, washed, and wound in a continuous operation on a single machine.

continuous stationery reader [ADP] A type of character reader which processes only continuous forms of predefined dimensions.

continuous still [FOOD ENG] A type of still in which rectification is accomplished, allowing for the collection of several relatively pure fractions of distilled spirits.

continuous system [CONT SYS] A system whose inputs and outputs are capable of changing at any instant of time. Also known as continuous-time signal system.

continuous-time signal system See continuous system.

continuous titrator [ANALY CHEM] A titrator so equipped that a reservoir refills the buret.

continuous tone [GRAPHICS] An image which has not been screened and contains unbroken gradient tones from black to white, and may be either in negative or positive form.

continuous transformation See continuous function.

continuous tube process [ENG] Plastics blow-molding process that uses a continuous extrusion of plastic tubing as feed to a series of blow molds as they clamp in sequence.

continuous vinegar process [FOOD ENG] Continuous, rather than batch, bacterial fermentation of apple cider, wine, other fruit juice, malt, or barley.

continuous wave [ELECTROMAG] A radio or radar wave whose successive sinusoidal oscillations are identical under steady-state conditions. Abbreviated CW. Also known as type A wave.

continuous-wave Doppler radar See continuous-wave radar.

continuous-wave gas laser [OPTICS] A laser having a quartz envelope filled with a mixture of helium and neon at low pressure, with Brewster-angle mirrors at opposite ends and an external optical system.

continuous-wave jammer [ELECTR] An electronic jammer that emits a single frequency which gives the appearance of a picket or rail fence on an enemy's radarscope. Also known as rail-fence jammer.

continuous-wave modulation [COMMUN] Modulation of a continuous wave by modification of its amplitude or phase, in contrast to pulse modulation.

continuous-wave radar [ENG] A radar system in which a transmitter sends out a continuous flow of radio energy; the target reradiates a small fraction of this energy to a separate receiving antenna. Also known as continuous-wave Doppler radar.

continuous-wave tracking system [ELECTR] Tracking system which operates by keeping a continuous radio beam on a target and determining its behavior from changes in the antenna necessary to keep the beam on the target.

continuous weld [MET] A weld that is continuous along the entire length of the joint.

continuous x-rays [ELECTROMAG] The electromagnetic radiation, having a continuous spectral distribution, that is produced when high-velocity electrons strike a target.

continuum [MATH] A compact, connected set.

continuum mechanics See classical field theory.

continuum physics See classical field theory.

contorted [BOT] Twisted; applied to proximate leaves whose margins overlap.

contour [MAP] See contour line. [PHYS] A curve drawn on a two-dimensional diagram through points satisfying $f(x,y) = c$, where c is a constant and f is some function, such as the field strength for a transmitter. [SCI TECH] The periphery of a figure or body.

contour analysis [ADP] In optical character recognition, a reading technique that employs a roving spot of light which searches out the character's outline by bouncing around its outer edges.

contour-change line See height-change line.

contour code [METEOROL] A code in which data on the topography of constant-pressure surfaces are transmitted; a modification of the international analysis code.

contour feather [VERT ZOO] Any of the large flight feathers or long tail feathers of a bird. Also known as penna; vane feather.

contour finder [GRAPHICS] An optical instrument of simple design for use with photographic prints; used to produce contour lines from differences in relief evident in the print.

contour forming [MET] Shaping a sheet of metal onto a shaped die.

contouring control [ADP] The guidance by a computer of a machine tool along a programmed path by interpolating many intermediate points between selected points.

contour integral [MATH] A line integral of a complex function, usually over a simple closed curve.

contour interval [MAP] The difference in elevation between adjacent contours.

contour line [MAP] A map line representing a contour, that is, connecting points of equal elevation above or below a datum plane, usually mean sea level. Also known as contour; isoheight; isohypse. [METEOROL] A line on a weather map connecting points of equal atmospheric pressure, temperature, or such.

contour machining [MECH ENG] Machining of an irregular surface.

contour map [MAP] A map displaying topographic or structural contour lines.

contour microclimate [CLIMATOL] That portion of the microclimate which is directly attributable to the small-scale variations of ground level.

contour milling [MET] Milling of an irregular surface.

contourograph [ELECTR] Device using a cathode-ray oscilloscope to produce imagery that has a three-dimensional appearance.

contour plan [MIN ENG] A plan showing surface contours or calculated contours of coal seams to be developed.

contour plowing [AGR] Cultivation of land along lines connecting points of equal elevation, to prevent water erosion. Also known as terracing.

contour turning [MECH ENG] Making a three-dimensional reproduction of the shape of a template by controlling the cutting tool with a follower that moves over the surface of a template.

contour value [MAP] A numerical value placed upon a contour line to denote its elevation relative to a given datum, usually mean sea level.

contraception [MED] Prevention of impregnation.

contraceptive [MED] Any mechanical device or chemical agent used to prevent conception.

contracted code sonde See code-sending radiosonde.

contracted curvature tensor [MATH] A symmetric tensor of second order, obtained by summation on two indices of the Riemann curvature tensor which are not antisymmetric. Also known as contracted Riemann-Christoffel tensor; Ricci tensor.

contracted pelvis [MED] A pelvis having one or more major diameters reduced in size, interfering with parturition.

contracted Riemann-Christoffel tensor See contracted curvature tensor.

contractile [BIOL] Displaying contraction; having the property of contracting.

CONTINUOUS VINEGAR PROCESS

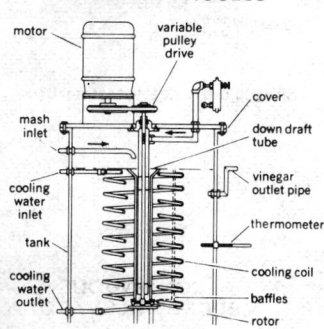

motor · variable pulley drive · cover · mash inlet · down draft tube · vinegar outlet pipe · cooling water inlet · thermometer · tank · cooling coil · cooling water outlet · baffles · rotor

The generator used in the continuous vinegar process. Cooling coils control increase in temperature produced by vinegar generation. *(From R. F. Cohee and G. Steffen, Food Eng., 31(3):58–59, 1959)*

CONTOUR FEATHER

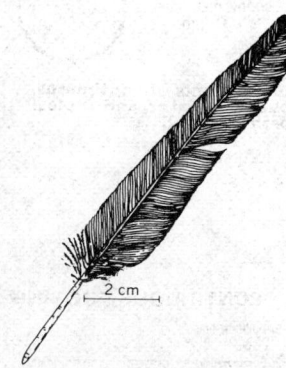

2 cm

Drawing of a contour feather. *(From J. C. Welty, The Life of Birds, Saunders, 1962)*

contractile vacuole [CYTOL] A tiny, intracellular, membranous bladder that functions in maintaining intra- and extracellular osmotic pressures in equilibrium, as well as excretion of water, such as occurs in protozoans.

contracting stitching [TEXT] Arranging fibers so that they are parallel, and removing those fibers that are shorter than a specified length.

contraction [MATH] A continuous function of a metric space to itself which moves each pair of points closer together. [MECH] The action or process of becoming smaller or pressed together, as a gas on cooling. [PHYSIO] Shortening of the fibers of muscle tissue.

contraction hypothesis [GEOL] Theory that shrinking of the earth is the cause of compression folding and thrusting.

contraction joint [CIV ENG] A break designed in a structure to allow for drying and temperature shrinkage of concrete, brickwork, or masonry, thereby preventing the formation of cracks.

contraction loss [FL MECH] In fluid flow, the loss in mechanical energy in a stream flowing through a closed duct or pipe when there is a sudden contraction of the cross-sectional area of the passage.

contraction rule [MET] A measuring rule having larger divisions than standard measures to allow for shrinkage of a metal casting. Also known as shrinkage rule; shrink rule.

contracture [ARCH] Narrowing of a section of a column. [MED] **1.** Shortening, as of muscle or scar tissue, producing distortion or deformity or abnormal limitation of movement of a joint. **2.** Retarded relaxation of muscle, as when it is injected with veratrine.

contracurrent system *See* katoptric system.

contraflexure point [CIV ENG] The point in a structure where bending occurs in opposite directions.

contrail *See* condensation trail.

contrail-formation graph [METEOROL] A graph containing the parameters pressure, temperature, and relative humidity for critical values at which condensation trails (contrails) form; used as an aid in forecasting the formation of condensation trails.

contraindication [MED] A symptom, indication, or condition in which a remedy or a method of treatment is inadvisable or improper.

contralateral [PHYSIO] Opposite; acting in unison with a similar part on the opposite side of the body.

CONTRAN [ADP] Computer programming language in which instructions are written at the compiler level, thereby eliminating the need for translation by a compiling routine.

contrapositive [MATH] The contrapositive of the statement "*if p*, then *q*" is the equivalent statement "*if not q*, then not *p*."

contrarotating propellers [MECH ENG] A pair of propellers on concentric shafts, turning in opposite directions.

contrarotation [ENG] Rotation in the direction opposite to another rotation.

contra solem [METEOROL] Characterizing air motion that is counterclockwise in the Northern Hemisphere and clockwise in the Southern Hemisphere; literally, against the sun.

contrast [COMMUN] The degree of difference in tone between the lightest and darkest areas in a television or facsimile picture.

contrast control [ELECTR] A manual control that adjusts the range of brightness between highlights and shadows on the reproduced image in a television receiver.

contrastes [METEOROL] Winds a short distance apart blowing from opposite quadrants, frequent in the spring and fall in the western Mediterranean.

contrast ratio [ELECTR] The ratio of the maximum to the minimum luminance values in a television picture.

contrast threshold *See* threshold contrast.

contravariant functor [MATH] A functor which reverses the sense of morphisms.

contravariant index [MATH] A tensor index such that, under a transformation of coordinates, the procedure for obtaining a component of the transformed tensor for which this index has the value p involves taking a sum over q of the product of a component of the original tensor for which the index has the value q times the partial derivative of the pth transformed coordinate with respect to the qth original coordinate; it is written as a superscript.

contravariant tensor [MATH] A tensor with only contravariant indices.

contravariant vector [MATH] A contravariant tensor of degree 1, such as the tensor whose components are differentials of the coordinates.

contributory *See* tributary.

control [ADP] **1.** The section of a digital computer that carries out instructions in proper sequence, interprets each coded instruction, and applies the proper signals to the arithmetic unit and other parts in accordance with this interpretation. **2.** A mathematical check used in some computer operations. [CONT SYS] A means or device to direct and regulate a process or sequence of events. [ELECTR] An input element of a cryotron. [STAT] **1.** A test made to determine the extent of error in experimental observations or measurements. **2.** A procedure carried out to give a standard of comparison in an experiment. **3.** Observation made on subjects which have not undergone treatment, to use in comparison with observations made on subjects which have undergone treatment.

control accuracy [CONT SYS] The degree of correspondence between the ultimately controlled variable and the ideal value in a feedback control system.

control agent [CHEM ENG] In process automatic-control work, material or energy within a process system of which the manipulated (controlled) variable is a condition or characteristic.

control area [NAV] An area over which air-traffic control is exercised by a ground-located traffic-control center.

control block [ADP] A storage area containing (in condensed, formalized form) the information required for the control of a task, function, operation, or quantity of information.

control board [ELEC] A panel at which one can make circuit changes, as in lighting a theater. [ENG] A panel in which meters and other indicating instruments display the condition of a system, and dials, switches, and other devices are used to modify circuits to control the system. Also known as control panel; panel board.

control card [ADP] A punched card containing input data or parameters which are necessary to begin or modify a program, or containing instructions needed for the specific application of a general routine.

control character [ADP] A character whose occurrence in a particular context initiates, modifies, or stops a control operation in a computer or associated equipment.

control characteristic [ELECTR] **1.** The relation, usually shown by a graph, between critical grid voltage and anode voltage of a gas tube. **2.** The relation between control ampere-turns and output current of a magnetic amplifier.

control chart [IND ENG] A chart in which quantities of data concerning some property of a product or process are plotted and used specifically to determine the variation in the process.

control circuit [ADP] One of the circuits that responds to the instructions in the program for a digital computer. [ELEC] A circuit that controls some function of a machine, device, or piece of equipment. [ELECTR] The circuit that feeds the control winding of a magnetic amplifier.

control column [AERO ENG] A cockpit control lever pivoted or sliding in front of the pilot; controls operation of the elevator and aileron.

control computer [ADP] A computer which uses inputs from sensor devices and outputs connected to control mechanisms to control physical processes.

control counter [ADP] A counter providing data used to control the execution of a computer program.

control data [ADP] Data used for identifying, selecting, executing, or modifying another set of data, a routine, a record, or the like.

control day [METEOROL] One of several days on which the weather is supposed (according to folklore) to provide the key for the weather of a subsequent period. Also known as key day.

control desk *See* console.

control diagram *See* flow chart.

control drive [NUCLEO] The system of control rods which regulate the reaction rate of a nuclear reactor.

control electrode [ELECTR] An electrode used to initiate or

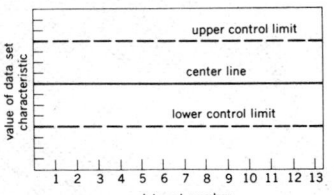

CONTROL CHART

Control chart, consisting of a center line and two sets of control lines.

vary the current between two or more electrodes in an electron tube.

control element [CONT SYS] The portion of a feedback control system that acts on the process or machine being controlled.

control feel [AERO ENG] The impression of the stability and control of an aircraft that a pilot receives through the cockpit controls, either from the aerodynamic forces acting on the control surfaces or from forces simulating these aerodynamic forces.

control field [ADP] A constant location where information for control purposes is placed, such as specified columns on punched cards.

control grid [ELECTR] A grid, ordinarily placed between the cathode and an anode, that serves to control the anode current of an electron tube.

control-grid bias [ELECTR] Average direct-current voltage between the control grid and cathode of a vacuum tube.

control-grid plate transconductance [ELECTR] Ratio of the amplification factor of a vacuum tube to its plate resistance, combining the effects of both into one term.

control head gap [ADP] The distance maintained between the read/write head of a disk drive and the disk surface.

control hole *See* designation punch.

control inductor *See* control winding.

controllability [AERO ENG] The quality of an aircraft or guided weapon which determines the ease of producing changes in flight direction or altitude by operation of its controls. [CONT SYS] Property of a system for which, given any initial state and any desired state, there exists a time interval and an input signal which brings the system from the initial state to the desired state during the time interval.

controllable-pitch propeller [MECH ENG] An aircraft or ship's propeller in which the pitch of the blades can be changed while the propeller is in motion; five types used for aircraft are two-position, variable-pitch, constant-speed, feathering, and reversible-pitch. Abbreviated CP propeller.

controlled aerodrome [NAV] An aerodrome for which air-traffic control facilities are provided.

controlled airspace [NAV] An airspace of defined dimensions within which air-traffic control is provided.

controlled atmosphere [SCI TECH] A specified gas or mixture of gases at a predetermined temperature, and sometimes humidity, in which selected processes take place.

controlled avalanche device [ELECTR] A semiconductor device that has rigidly specified maximum and minimum avalanche voltage characteristics and is able to operate and absorb momentary power surges in this avalanche region indefinitely without damage.

controlled avalanche rectifier [ELECTR] A silicon rectifier in which carefully controlled, nondestructive internal avalanche breakdown across the entire junction area protects the junction surface, thereby eliminating local heating that would impair or destroy the reverse blocking ability of the rectifier.

controlled carrier modulation [COMMUN] System of modulation wherein the carrier is amplitude-modulated by the signal frequencies and, in addition, the carrier is amplitude-modulated according to the envelope of the signal so that the modulation factor remains constant regardless of the amplitude of the signal. Also known as floating carrier modulation; variable carrier modulation.

controlled cooling [MET] Process by which an object is cooled from an elevated temperature in a predetermined manner to avoid cracking, internal damage, or hardening, or to produce a desired microstructure.

controlled fragment [ORD] One of the pieces produced from warhead casings that have been designed to break up in specific patterns; the fragment takes on its final shape during the detonation of the explosive charge. Also known as fire-formed fragment.

controlled fusion [NUCLEO] The use of thermonuclear fusion reactions in a controlled manner to generate power.

controlled-leakage system [AERO ENG] A system that provides for the maintenance of life in an aircraft or spacecraft cabin by a controlled escape of carbon dioxide and other waste from the cabin, with replenishment provided by stored oxygen and food.

controlled medium [CHEM ENG] In process automatic-control work, material within a process system in which a variable (for example, concentration) is controlled.

controlled mercury-arc rectifier [ELECTR] A mercury-arc rectifier in which one or more electrodes control the start of the discharge in each cycle and thereby control output current.

controlled mine [ORD] A mine fitted with firing devices capable of being activated by an electrical system leading to a central control station; may be an underwater or land mine.

controlled parameter [ENG] In the formulation of an optimization problem, one of the parameters whose values determine the value of the criterion parameter.

controlled rectifier [ELECTR] A three-terminal semiconductor junction device with four regions of alternating conductivity type (*pnpn*).

controlled thermonuclear reaction [NUCLEO] A fusion reaction generated in a controlled manner for research purposes or for production of useful power.

controlled thermonuclear reactor [NUCLEO] The heart of a fusion spacecraft propulsion system, based on the thermonuclear reaction of deuterium with a helium-3 isotope to produce helium-4 and protons. Abbreviated CTR.

controlled time of arrival [NAV] A time of arrival determined in advance, requiring careful planning, accurate navigation, and frequent checking of position and speed.

controlled variable [CONT SYS] In process automatic-control work, that quantity or condition of a controlled system that is directly measured or controlled. [SCI TECH] The quantity or condition that is measured and controlled.

control limits [ELECTR] In radar evaluation, upper and lower control limits are established at those performance figures within which it is expected that 95% of quality-control samples will fall when the radar is performing normally. [IND ENG] In statistical quality control, the limits of acceptability placed on control charts; parts outside the limits are defective.

controlling depth [NAV] The least depth in the approach channel to an area, such as a port or anchorage, governing the maximum draft of vessels that can enter.

controlling obstacle [NAV] In terminal instrument procedures, the highest obstacle relative to a prescribed plane within a specified area; in precision approach procedures where obstacles penetrate the approach surface, the controlling obstacle is the one which results in the requirements for the highest decision height.

control logic [ADP] The sequence of steps required to perform a specific function.

control module [ADP] The set of registers and circuitry required to carry out a specific function.

control-moment gyro [AERO ENG] An internal momentum storage device that applies torques to the attitude-control system through large rotating gyros.

control of electromagnetic radiation *See* Conelrad.

control panel [ADP] **1.** An array of jacks or sockets in which wires (or other elements) may be plugged to control the action of an electromechanical device in a data-processing system such as a printer. Also known as plugboard; wiring board. **2.** *See* panel. [ENG] *See* control board.

control plane [AERO ENG] An aircraft from which the movements of another craft are controlled remotely.

control point [ADP] The numerical value of the controlled variable (speed, temperature, and so on) which, under any fixed set of operating conditions, an automatic controller operates to maintain. [MAP] Any station in a horizontal and vertical grid that is identified on a photograph and used for correlating the data shown on that photograph. [NAV] A position marked by a buoy, boat, aircraft, electronic device, conspicuous terrain feature, or other identifiable object which is given a name or number and used as an aid for navigation or control of ships, boats, or aircraft.

control program [ADP] A program which carries on input/output operations, loading of programs, detection of errors, communication with the operator, and so forth.

control punch *See* designation punch.

control register [ADP] Any one of the registers in a computer used to control the execution of a computer program.

control rocket [AERO ENG] A vernier engine, retrorocket, or

CONTROLLED RECTIFIER

collector (anode) | p | n | p | n | emitter (cathode)

base (gate)

Controlled rectifier diagram.

other such rocket used to change the attitude of, guide, or make small changes in the speed of a rocket, spacecraft, or the like.

control rod [NUCLEO] Any rod used to control the reactivity of a nuclear reactor; may be a fuel rod or part of the moderator; in a thermal reactor, commonly a neutron absorber. Also known as absorbing rod.

control room [COMMUN] A room from which engineers and production men control and direct a television or radio program or a sound-recording session; the room is adjacent to the main studios and separated from them by large sound-proof glass windows. [ENG] A room from which space flights are directed.

control section [ADP] The smallest integral subsection of a program, that is, the smallest unit of code that can be separately relocated during loading.

control sequence [ADP] The order in which a set of executions are carried to perform a specific function.

control signal [ADP] A set of pulses used to identify the channels to be followed by transferred data. [CONT SYS] The signal applied to the device that makes corrective changes in a controlled process or machine.

control state [ADP] The operating mode of a system which permits it to override its normal sequence of operations.

control statement *See* job control statement.

control supervisor [ADP] The computer software which controls the processing of the system.

control surface [AERO ENG] **1.** Any movable airfoil used to guide or control an aircraft, guided missile, or the like in the air, including the rudder, elevators, ailerons, spoiler flaps, and trim tabs. **2.** In restricted usage, one of the main control surfaces, such as the rudder, an elevator, or an aileron.

control switching point [COMMUN] A telephone office which is an important switching center in the routing of long-distance calls in the direct distance dialing system. Abbreviated CSP.

control symbol [ADP] A symbol which, coded into the machine memory, controls certain steps in the mechanical translation process; since control symbols are not contextual symbols, they appear neither in the input nor in the output.

control synchro *See* control transformer.

control system [ENG] A system in which one or more outputs are forced to change in a desired manner as time progresses.

control system feedback [CONT SYS] A signal obtained by comparing the output of a control system with the input, which is used to diminish the difference between them.

control systems equipment [ADP] Computers which are an integral part of a total facility or larger complex of equipment and have the primary purpose of controlling, monitoring, analyzing, or measuring a process or other equipment.

control tape [ADP] Loop of paper tape to control the carriage operation of character printers. Also known as carriage tape.

control total [ADP] The sum of the numbers in a specified record field of a batch of records, determined repetitiously during computer processing so that any discrepancy from the control indicates an error.

control tower *See* airport traffic control tower.

control-tower visibility [METEOROL] The visibility observed from an airport control tower.

control track [ENG ACOUS] A supplementary sound track, usually containing tone signals that control the reproduction of the sound track, such as by changing feed levels to loudspeakers in a theater to achieve stereophonic effects.

control transformer [ELEC] A synchro in which the electrical output of the rotor is dependent on both the shaft position and the electric input to the stator. Also known as control synchro.

control unit [ADP] An electronic device containing data buffers and logical circuitry, situated between the computer channel and the input/output device, and controlling data transfers and such operations as tape rewind.

control valve [ENG] A valve which controls pressure, volume, or flow direction in a fluid transmission system.

control vane [AERO ENG] A movable vane used for control, especially a movable air vane or jet vane on a rocket used to control flight altitude.

control winding [ELECTR] A winding used on a magnetic

amplifier or saturable reactor to apply control magnetomotive forces to the core. Also known as control inductor.

control word [ADP] A computer word specifying a certain action to be taken.

control zone [NAV] An airspace of defined dimensions extended upward from the surface to include one or more airports, and within which rules additional to those governing flight in control areas apply for the protection of traffic. Also known as air-traffic control zone.

contusion [MED] A subcutaneous bruise caused by an injury in which the skin is not broken.

Conularida [PALEON] A small group of extinct invertebrates showing a narrow, four-sided, pyramidal-shaped test.

Conulata [INV ZOO] A subclass of free-living coelenterates in the class Scyphozoa; individuals are described as tetraramous cones to elongate pyramids having tentacles on the oral margin.

Conulidae [PALEON] A family of Cretaceous exocyclic Euechinoidea characterized by a flattened oral surface.

conus arteriosus [EMBRYO] The cone-shaped projection from which the pulmonary artery arises on the right ventricle of the heart in man and mammals.

convalescence [MED] The period and process of recovery after an illness or injury.

convalescent carrier [MED] A person who harbors an infectious agent after recovery from a clinical attack of a disease.

convalescent serum [IMMUNOL] The serum of the blood of one or more patients recovering from an infectious disease; used for prophylaxis of the particular infection.

convection [FL MECH] Diffusion in which the fluid as a whole is moving in the direction of diffusion. Also known as bulk flow. [METEOROL] Atmospheric motions that are predominantly vertical, resulting in vertical transport and mixing of atmospheric properties. [OCEANOGR] Movement and mixing of ocean water masses. [PHYS] Transmission of energy or mass by a medium involving movement of the medium itself.

convectional stability *See* static stability.

convection cell [METEOROL] An atmospheric unit in which organized convective fluid motion occurs. [GEOPHYS] A concept in plate tectonics that accounts for the lateral or the upward and downward movement of subcrustalar mantle material as due to heat variation in the earth.

convection coefficient *See* film coefficient.

convection cooling [ENG] Heat transfer by natural, upward flow of hot air from the device being cooled.

convection current [ELECTR] The time rate at which the electric charges of an electron stream are transported through a given surface. [GEOPHYS] Mass movement of subcrustal or mantle material as a result of temperature variations. [METEOROL] Any current of air involved in convection; usually, the upward-moving portion of a convection circulation, such as a thermal or the updraft in cumulus clouds. Also known as convective current.

convection modulus [FL MECH] An intrinsic property of a fluid which is important in determining the Nusselt number, equal to the acceleration of gravity times the volume coefficient of thermal expansion divided by the product of the kinematic viscosity and the thermal diffusivity.

convection section [ENG] That portion of the furnace in which tubes receive heat from the flue gases by convection.

convection theory of cyclones [METEOROL] A theory of cyclone development proposing that the upward convection of air (particularly of moist air) due to surface heating can be of sufficient magnitude and duration that the surface inflow of air will attain appreciable cyclonic rotation.

convective activity [METEOROL] Generally, manifestations of convection in the atmosphere, alluding particularly to the development of convective clouds and resulting weather phenomena, such as showers, thunderstorms, squalls, hail, and tornadoes.

convective cloud [METEOROL] A cloud which owes its vertical development, and possibly its origin, to convection.

convective-cloud-height diagram [METEOROL] A graph used as an aid in estimating the altitude of the base of convective clouds; since its basis is the same as that for the dew-point

CONULARIDA

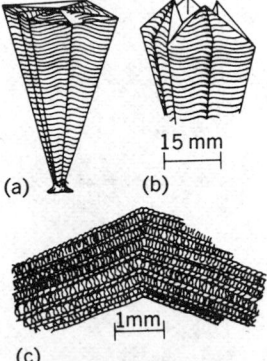

15 mm

(a)   (b)

1mm

(c)

Conularida restorations. *(a)* Attachment disk. *(b)* Distal part of same individual with triangular flaps raised. *(c)* Part of exterior showing the ornamentation. *(Adapted from R. C. Moore, C. G. Lalicker, and A. G. Fischer, Invertebrate Fossils, McGraw-Hill, 1952)*

formula, only the surface temperature and dew point need be known to use the diagram.

convective condensation level [METEOROL] On a thermodynamic diagram, the point of intersection of a sounding curve (representing the vertical distribution of temperature in an atmospheric column) with the saturation mixing-ratio line corresponding to the average mixing ratio in the surface layer (that is, approximately the lowest 1500 feet, or 450 meters).

convective current See convection current.

convective discharge [ELECTR] The movement of a visible or invisible stream of charged particles away from a body that has been charged to a sufficiently high voltage. Also known as electric wind; static breeze.

convective equilibrium See adiabatic equilibrium.

convective instability [METEOROL] The state of an unsaturated layer or column of air in the atmosphere whose wet-bulb potential temperature (or equivalent potential temperature) decreases with elevation. Also known as potential instability.

convective overturn See overturn.

convective precipitation [METEOROL] Precipitation from convective clouds, generally considered to be synonymous with showers.

convective region [METEOROL] An area particularly favorable for the formation of convection in the lower atmosphere, or one characterized by convective activity at a given time.

convector [ENG] A heat-emitting unit for the heating of room air; it has a heating element surrounded by a cabinet-type enclosure with openings below and above for entrance and egress of air.

convenience receptacle See outlet.

conventional bomb [ORD] A nonatomic bomb designed primarily for explosive effect.

conventional grouping [ADP] When a unit record containing a coding field is used for a single item code and the set of codes of the terms which describe the item, the grouping is conventional; for example, a personnel file in which each individual is represented by a card on which are punched codes for his or her age, sex, education, and salary.

conventional milling [MET] Milling in which the cutter and feed move in opposite directions from the point of contact.

conventional mining [MIN ENG] The cycle which includes cutting the coal, drilling the shot holes, charging and shooting the holes, loading the broken coal, and installing roof support. Also known as cyclic mining.

conventional spinning See manual spinning.

convergence [ANAT] The coming together of a group of afferent nerves upon a motoneuron of the ventral horn of the spinal cord. [ANTHRO] Independent development of similarities between unrelated cultures. [EVOL] Development of similarities between animals or plants of different groups resulting from adaptation to similar habitats. [ELECTR] A condition in which the electron beams of a multibeam cathode-ray tube intersect at a specified point, such as at an opening in the shadow mask of a three-gun color television picture tube; both static and dynamic convergence are required. [GEOL] Diminution of the interval between geologic horizons. [HYD] The line of demarcation between turbid river water and clear lake water. [MATH] The property of having a limit for infinite series, sequences, products, and so on. [METEOROL] The increase in wind setup observed beyond that which would take place in an equivalent rectangular basin of uniform depth, due to changes in platform or depth.

convergence circuit [ELECTROMAG] An auxiliary deflection system in a color television receiver which maintains convergence, having separate convergence coils for electromagnetic controls of the positions of the three beams in a convergence yoke around the neck of the kinescope.

convergence coil [ELECTR] One of the coils used to obtain convergence of electron beams in a three-gun color television picture tube.

convergence constant See convergence factor.

convergence control [ELECTR] A control used in a color television receiver to adjust the potential on the convergence electrode of the three-gun color picture tube to achieve convergence.

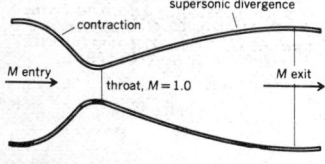

CONVERGENT-DIVERGENT NOZZLE

Diagram of convergent-divergent nozzle. M = local Mach number.

convergence electrode [ELECTR] An electrode whose electric field converges two or more electron beams.

convergence factor [MAP] For a specific chart or map projection, the factor which, when multiplied by the difference of longitude between points on two meridians, will give the convergence of the meridians. Also known as convergence constant.

convergence in measure [MATH] A sequence of functions $f_n(x)$ converges in measure to $f(x)$ if given any $\varepsilon > 0$, the measure of the set of points at which $|f_n(x) - f(x)| > \varepsilon$ is less than ε, provided n is sufficiently large.

convergence line [METEOROL] Any horizontal line along which horizontal convergence of the airflow is occurring. Also known as asymptote of convergence.

convergence magnet [ELECTR] A magnet assembly whose magnetic field converges two or more electron beams; used in three-gun color picture tubes. Also known as beam magnet.

convergence of meridians [MAP] **1.** The angular drawing together of the geographic meridians in passing from the equator to the poles. **2.** The relative difference of directions of meridians at specific points on the meridians; it is equal to the product of the difference of longitude and the convergence factor.

convergence pressure [PHYS CHEM] The pressure at which the different constant-temperature K (liquid-vapor equilibrium) factors for each member of a two-component system converge to unity.

convergence zone [ACOUS] A sound transmission channel produced in sea water by a combination of pressure and temperature changes in the depth range between 2500 and 15,000 feet (750 and 4500 meters); utilized by sonar systems.

convergent-divergent nozzle [DES ENG] A nozzle in which supersonic velocities are attained; has a divergent portion downstream of the contracting section. Also known as supersonic nozzle.

convergent integral [MATH] An improper integral which has a finite value.

convergent precipitation [METEOROL] A synoptic type of precipitation caused by local updrafts of moist air.

convergent sequence [MATH] A sequence which has a limit.

convergent series [MATH] A series whose sequence of partial sums has a limit.

converging lens [OPTICS] A lens that has a positive focal length, and therefore causes rays of light parallel to its axis to converge.

Conversational Algebraic Language See CAL.

conversational processing [ADP] The operating mode of a computer system which enables a user to have each statement he keys into the system processed immediately.

conversational time-sharing [ADP] The simultaneous utilization of a computer system by multiple users, each user being equipped with a remote terminal with which he communicates with the computer in conversational mode.

converse [MATH] The converse of the statement "if p, then q" is the statement "if q, then p."

conversion [ADP] See data conversion. [CHEM] Change of a compound from one isomeric form to another. [NAV] Determination of the rhumb-line direction of one point from another when the initial great-circle direction is known, or vice versa, the difference between the two directions being the conversion angle; used in connection with radio bearings, Consol, Consolan, and in great-circle sailing. [NUC PHYS] Nuclear transformation of a fertile substance into a fissile substance. [PHYS] Change in a quantity's numerical value as a result of using a different unit of measurement. [PSYCH] A defense mechanism whereby unconscious emotional conflict is transformed into physical disability, the affected part always having symbolic meaning pertinent to the nature of the conflict.

conversion angle [NAV] The angle between the rhumb line and the great circle between two points. Also known as arc to chord correction; half-convergency.

conversion coating [MET] A metal-surface coating consisting of a compound of the base metal.

conversion coefficient Also known as conversion fraction; internal conversion coefficient. [NUC PHYS] **1.** The ratio of the number of conversion electrons emitted per unit time to the number of photons emitted per unit time in the de-excitation of a nucleus between two given states. **2.** In older literature, the ratio of the number of conversion electrons emitted per unit time to the number of conversion electrons plus the number of photons emitted per unit time in the de-excitation of a nucleus between two given states.

conversion electron [NUC PHYS] An electron which receives energy directly from a nucleus in an internal conversion process and is thereby expelled from the atom.

conversion equipment [ADP] Equipment used for conversion of data from one recording medium to another, as from card to tape.

conversion factor [MATH] The numerical factor by which one must multiply (or divide) a quantity expressed in terms of one unit to express the quantity in terms of another unit. Also known as conversion ratio. [NUCLEO] *See* breeding ratio.

conversion fraction *See* conversion coefficient.

conversion gain [ELECTR] **1.** Ratio of the intermediate-frequency output voltage to the input signal voltage of the first detector of a superheterodyne receiver. **2.** Ratio of the available intermediate-frequency power output of a converter or mixer to the available radio-frequency power input. [NUCLEO] The conversion ratio minus one in a nuclear reactor.

conversion program [ADP] A set of instructions which allows a program written for one system to be run on a different system.

conversion ratio [MATH] *See* conversion factor. [NUCLEO] The number of fissionable atoms produced per fissionable atom fissioned in a converter type of nuclear reactor.

conversion reaction [PSYCH] The form of hysterical neurosis in which the impulse causing anxiety is converted into functional symptoms of the special senses or voluntary nervous system.

conversion routine [ADP] A flexible, self-contained, and generalized program used for data conversion, which only requires specifications about very few facts in order to be used by a programmer.

conversion table [SCI TECH] A list of equivalent values for converting from one set of units to another.

conversion time [ADP] The time required to read in data from one code into another code.

conversive heating [MED] The conversion of some form of energy, especially radio waves, into heat for use in thermotherapy.

convert [ADP] To transform the representation of data.

converter [ADP] A computer unit that changes numerical information from one form to another, as from decimal to binary or vice versa, from fixed-point to floating-point representation, or from punched cards to magnetic tape. [ELECTR] **1.** The section of a superheterodyne radio receiver that converts the desired incoming radio-frequency signal to the intermediate-frequency value; the converter section includes the oscillator and the mixer-first detector. Also known as heterodyne conversion transducer; oscillator-mixer-first detector. **2.** An auxiliary unit used with a television or radio receiver to permit reception of channels or frequencies for which the receiver was not originally designed. **3.** In facsimile, a device that changes the type of modulation delivered by the scanner. **4.** Unit of a radar system in which the mixer of a superheterodyne receiver and usually two stages of intermediate-frequency amplification are located; performs a preamplifying operation. **5.** *See* remodulator. [MET] A type of furnace in which impurities are oxidized out by blowing air through or across a path of molten metal or matte. [NUCLEO] Also known as nuclear converter. **1.** A nuclear reactor that converts fertile atoms into fuel by neutron capture, using one kind of fuel and producing another. **2.** A nuclear reactor that produces some fissionable fuel, but less than it consumes; the fuel produced may be the same as that consumed or different.

converter tube [ELECTR] An electron tube that combines the mixer and local-oscillator functions of a heterodyne conversion transducer.

convertiplane [AERO ENG] A hybrid form of heavier-than-air craft capable, because of one or more horizontal rotors or units acting as rotors, of taking off, hovering, and landing in a fashion similar to a helicopter; and once aloft and moving forward, capable, by means of a mechanical conversion, of flying purely as a fixed-wing aircraft, especially in higher speed ranges.

convex [SCI TECH] Having a curved form which bulges outward, resembling the exterior of a sphere or cylinder or a section of these bodies.

convex combination [MATH] A linear combination of vectors in which the sum of the coefficients is 1.

convex curve [MATH] A plane curve for which any straight line that crosses the curve crosses it at just two points.

convex fillet weld [MET] A fillet weld having a convex surface.

convex function [MATH] A function $f(x)$ is said to be convex over the interval a,b if for any three points x_1, x_2, x_3 such that $a < x_1 < x_2 < x_3 < b$, $f(x_2) \leq L(x_2)$, where $L(x)$ is the equation of the straight line passing through the points x_1, x_3.

convex hull [MATH] The smallest convex set containing a given collection of points in a real linear space. Also known as convex linear hull.

convex linear hull *See* convex hull.

convexo-concave [SCI TECH] Having one side convex and the other concave, usually with greater curvature on the convex side.

convex polyhedron [MATH] A polyhedron in the plane which is a convex set, for example, any regular polyhedron.

convex set [MATH] A set which contains the entire line segment joining any pair of its points.

conveyor [MECH ENG] Any materials-handling machine designed to move individual articles such as solids or free-flowing bulk materials over a horizontal, inclined, declined, or vertical path of travel with continuous motion.

convolute [BIOL] Twisted or rolled together, specifically referring to leaves, mollusk shells, and renal tubules.

convolute bedding [GEOL] The extremely contorted laminae usually confined to a single layer of sediment, resulting from subaqueous slumping.

convolution [ANAT] A fold, twist, or coil of any organ, especially any one of the prominent convex parts of the brain, separated from each other by depressions or sulci.

convolution family *See* faltung.

convolution of two functions [MATH] The convolution of the functions f and g is the function F, defined by

$$F(x) = \int_0^x f(t)g(x-t)\,dt.$$

Convolvulaceae [BOT] A large family of dicotyledonous plants in the order Polemoniales characterized by internal phloem, the presence of chlorophyll, two ovules per carpel, and plicate cotyledons.

convoy [ORD] An accompanying and protective force on sea or land, such as a fleet or troops.

convulsion [MED] An episode of involuntary, generally violent muscular contractions, rhythmically alternated with periods of relaxation; associated with many systematic and neurological diseases.

Conwell-Weisskopf equation [SOLID STATE] An equation for the mobility of electrons in a semiconductor in the presence of donor or acceptor impurities, in terms of the dielectric constant of the medium, the temperature, the concentration of ionized donors (or acceptors), and the average distance between them.

Cooke objective [OPTICS] A three-lens objective consisting of one biconcave lens, the dispersive component, between two biconvex lens, the collective components; used in astronomical cameras.

Cooke unit [BIOL] A unit for the standardization of pollen antigenicity.

cooking snow *See* water snow.

cook-off [ORD] A cartridge fired by environmental heat, because it remained in the chamber of an overheated weapon.

coolant [MATER] **1.** A cutting fluid for machine operations, which keeps the tool cool to prevent reduction in hardness and resistance to abrasion, and prevents distortion of the work. **2.** A substance, ordinarily fluid, used for cooling any part of a reactor in which heat is generated. **3.** In general, any cooling agent, usually a fluid.

cooled infrared detector [ELECTR] An infrared detector that

COOKE OBJECTIVE

Cooke objective, a type of wide-field lens objective for astronomical cameras.

must be operated at cryogenic temperatures, such as at the temperature of liquid nitrogen, to obtain the desired infrared sensitivity.

cooled-tube pyrometer [ENG] A thermometer for high-temperature flowing gases that uses a liquid-cooled tube inserted in the flowing gas; gas temperature is deduced from the law of convective heat transfer to the outside of the tube and from measurement of the mass flow rate and temperature rise of the cooling liquid.

cooler nail [DES ENG] A thin, cement-coated wire nail.

cool flame [CHEM] A faint, luminous phenomenon observed when, for example, a mixture of ether vapor and oxygen is slowly heated; it proceeds by diffusion of reactive molecules which initiate chemical processes as they go.

Coolidge tube [ELECTROMAG] An x-ray tube in which the needed electrons are produced by a hot cathode.

cooling [NUCLEO] Setting aside a highly radioactive material until the radioactivity has diminished to a desired level.

cooling coil [MECH ENG] A coiled arrangement of pipe or tubing for the transfer of heat between two fluids.

cooling curve [THERMO] A curve obtained by plotting time against temperature for a solid-liquid mixture cooling under constant conditions.

cooling degree day [MECH ENG] A unit for estimating the energy needed for cooling a building; one unit is given for each degree Fahrenheit that the daily mean temperature exceeds 75°F (24°C).

cooling fin [MECH ENG] The extended element of a heat-transfer device that effectively increases the surface area.

cooling load [MECH ENG] The total amount of heat energy that must be removed from a system by a cooling mechanism in a unit time, equal to the rate at which heat is generated by people, machinery, and processes, plus the net flow of heat into the system not associated with the cooling machinery.

cooling pond [CHEM ENG] Outdoor depression into which hot process water is pumped for purposes of cooling by evaporation, convection, and radiation.

cooling power [MECH ENG] A parameter devised to measure the air's cooling effect upon a human body; it is determined by the amount of heat required by a device to maintain the device at a constant temperature (usually 34°C); the entire system should be made to correspond, as closely as possible, to the external heat exchange mechanism of the human body.

cooling-power anemometer [ENG] Any anemometer operating on the principle that the heat transfer to air from an object at an elevated temperature is a function of airspeed.

cooling process [ENG] Physical operation in which heat is removed from process fluids or solids; may be by evaporation of liquids, expansion of gases, radiation or heat exchange to a cooler fluid stream, and so on.

cooling range [MECH ENG] The difference in temperature between the hot water entering and the cold water leaving a cooling tower.

cooling stress [MECH] Stress resulting from uneven contraction during cooling of metals and ceramics due to uneven temperature distribution.

cooling table *See* hotbed.

cooling tower [ENG] A towerlike device in which atmospheric air circulates and cools warm water, generally by direct contact (evaporation).

coolometer [ENG] An instrument which measures the cooling power of the air, consisting of a metal cylinder electrically heated to maintain a constant temperature; the electrical heating power required is taken as a measure of the air's cooling power.

cool star [ASTROPHYS] A low-temperature star, generally visible in the infrared range of the electromagnetic spectrum.

cool time [MET] The period of time between successive heat times in pulsation and seam welding.

Coombs serum [IMMUNOL] An immune serum containing antiglobulin that is used in testing for Rh and other sensitizations.

cooperative observer [METEOROL] An unpaid observer who maintains a meteorological station for the U.S. National Weather Service.

cooperative phenomenon [SOLID STATE] A process involving a simultaneous collective interaction among many atoms or electrons in a crystal, such as ferromagnetism, superconductivity, and order-disorder transformations.

cooperative system [ENG] A missile guidance system that requires transmission of information from a remote ground station to a missile in flight, processing of the information by the missile-borne equipment, and retransmission of the processed data to the originating or other remote ground stations, as in azusa and dovap.

cooperite [MINERAL] (Pt,Pd)S A steel-gray tetragonal mineral of metallic luster consisting of a sulfide of platinum, occurring in irregular grains in igneous rock.

Cooper pairs [SOLID STATE] Pairs of bound electrons which occur in a superconducting medium according to the Bardeen-Cooper-Schrieffer theory.

coordinate axes [MATH] One of a set of lines or curves used to define a coordinate system; the value of one of the coordinates uniquely determines the location of a point on the axis, while the values of the other coordinates vanish on the axis.

coordinate bond *See* coordinate valence; dative bond.

coordinate conversion [MAP] Changing the map coordinate values from one system to those of another system.

coordinate data receiver [ELECTR] A receiver specifically designed to accept the signal of a coordinate data transmitter and reconvert this signal into a form suitable for input to associated equipment such as a plotting board, computer, or radar set.

coordinate data transmitter [ELECTR] A transmitter that accepts two or more coordinates, such as those representing a target position, and converts them into a form suitable for transmission.

coordinated complex *See* coordination compound.

coordinated geometry *See* COGO.

coordinated transpositions [ELEC] Transpositions which are installed in either electric supply or communications circuits or in both, for the purpose of reducing inductive coupling, and which are located effectively with respect to the discontinuities in both the electric supply and communications circuits.

coordinate indexing [ADP] An indexing scheme in which equal-rank descriptors are used to describe a document, for information retrieval by a computer or other means.

coordinate plotter [GRAPHICS] An automated drafting device in which a transverse beam and drafting head are driven over a drawing surface by a computer and tape readers to produce highly precise drawings at high speed. Also known as mechanical plotting board; XY coordinate plotter; XY plotter.

coordinates [MAP] **1.** Linear or angular quantities which designate the position that a point occupies in a given reference frame or system. **2.** A general term to designate the particular kind of reference frame or system, such as plane rectangular coordinates or spherical coordinates. [MATH] A set of numbers which locate a point in space.

coordinate systems [MATH] A rule for designating each point in space by a set of numbers.

coordinate transformation [MATH] A mathematical or graphic process of obtaining a modified set of coordinates by performing some nonsingular operation on the coordinate axes, such as rotating or translating them.

coordinate valence [CHEM] A chemical bond between two atoms in which a shared pair of electrons forms the bond and the pair has been supplied by one of the two atoms. Also known as coordinate bond.

coordinating holes [DES ENG] Holes in two parts of an assembly which form a single continuous hole when the parts are joined.

coordination chemistry [CHEM] The chemistry of metal ions in their interactions with other molecules or ions.

coordination compound [CHEM] A compound with a central atom or ion and a group of ions or molecules surrounding it. Also known as coordinated complex; Werner complex.

coordination lattice [CRYSTAL] The crystal structure of a coordination compound.

coordination number [PHYS] The number of nearest neighbors of a point in a space lattice, of an atom or an ion in a solid, or of an anion or cation in a solution.

coordination polygon [CHEM] The symmetrical polygonal

COOLING TOWER

Natural-draft cooling towers. *(Haman, Inc.)*

chemical structure of simple polyatomic aggregates having coordination numbers of 4 or less.

coordination polyhedron [CHEM] The symmetrical polyhedral chemical structure of relatively simple polyatomic aggregates having coordination numbers of 4 to 8.

coordination polymer [ORG CHEM] Organic addition polymer that is neither free-radical nor simply ionic; prepared by catalysts that combine an organometallic (for example, triethyl aluminum) and a transition metal compound (for example, $TiCl_4$).

coorongite [GEOL] A boghead coal in the peat stage.

copaiba balsam [MATER] An oleoresin extracted from trees of the genus *Copaifera* of South America; used as a plasticizer and in medicine.

copal [MATER] Hard, resinous substance exuded from certain trees in the East Indies, South America, and Africa and used in varnish and printing ink.

cope [MET] The upper portion of a flask, mold, or pattern.

cope chisel [DES ENG] A chisel used to cut grooves in metal.

Copenhagen water *See* normal water.

Copeognatha [INV ZOO] An equivalent name for Psocoptera.

Copepoda [INV ZOO] An order of Crustacea commonly included in the Entomostraca; contains free-living, parasitic, and symbiotic forms.

Copernican system [ASTRON] The system of planetary motions according to Copernicus, who maintained that the earth revolves about an axis once every day and revolves around the sun once every year while the other planets also move in orbits centered near the sun.

copiapite [MINERAL] 1. $Fe_5(SO_4)_6(OH)_2 \cdot 20H_2O$ A yellow mineral occurring in granular or scalar aggregates. Also known as ihleite; knoxvillite; yellow copperas. 2. A group of minerals containing hydrous iron sulfates.

coping [BUILD] A covering course on a wall. [MECH ENG] Shaping stone or other nonmetallic substance with a grinding wheel. [MIN ENG] 1. Process of cutting and trimming the edges of stone slabs. 2. Process of cutting a stone slab into two pieces.

coping saw [DES ENG] A type of handsaw that has a narrow blade, usually about ⅛ inch (3 millimeters) wide, held taut by a U-shaped frame equipped with a handle; used for shaping and cutout work.

coplanar electrodes [ELECTR] Electrodes mounted in the same plane.

coplanar forces [MECH] Forces that act in a single plane; thus the forces are parallel to the plane and their points of application are in the plane.

Copodontidae [PALEON] An obscure family of Paleozoic fishes in the order Bradyodonti.

copolymer [ORG CHEM] A mixed polymer, the product of polymerization of two or more substances at the same time.

copolymerization [CHEM] A polymerization reaction that forms a copolymer.

copper [CHEM] A chemical element, symbol Cu, atomic number 29, atomic weight 63.546. [MET] One of the most important nonferrous metals; a ductile and malleable metal found in various ores and used in industry, engineering, and the arts in both pure and alloyed form.

copper-64 [NUC PHYS] Radioactive isotope of copper with mass number of 64; derived from pile-irradiation of metallic copper; used as a research aid to study diffusion, corrosion, and friction wear in metals and alloys.

copper acetate *See* cupric acetate.

copper alloy [MET] A solid solution of one or more metals in copper.

copper amalgam [MET] An alloy of copper and mercury.

copper arsenate [INORG CHEM] $Cu_3(AsO_4)_2 \cdot 4H_2O$ or $Cu_5H_2(AsO_4)_4 \cdot 2H_2O$ Bluish powder, soluble in ammonium hydroxide and dilute acids, insoluble in water and alcohol; used as a fungicide and insecticide.

copper blight [PL PATH] A leaf spot disease of tea caused by the fungus *Guignardia camelliae*.

copper blue *See* mountain blue.

copper brazing [MET] Brazing by using copper as the filler metal.

copper bromide *See* cupric bromide; cuprous bromide.

copper cable [ELEC] A mechanically assembled group of copper wires, used in place of a single, large wire for increased flexibility.

copper chloride *See* cupric chloride; cuprous chloride.

copper chromate *See* cupric chromate.

copper converter [MET] A converter for purifying copper.

copper cyanide *See* cupric cyanide.

copper dish gum [CHEM ENG] The milligrams of gum found in 100 milliliters of gasoline when evaporated under controlled conditions in a polished copper dish.

copper fluoride *See* cupric fluoride; cuprous fluoride.

copperhead [VERT ZOO] *Agkistrodon contortrix*. A pit viper of the eastern United States; grows to about 3 feet (90 centimeters) in length and is distinguished by its coppery-brown skin with dark transverse blotches.

copper hydroxide *See* cupric hydroxide.

coppering [ORD] Metal fouling accumulated in the bore of a weapon because of repeated firing; deposited from the rotating bands or jackets of the projectiles.

copperite [MINERAL] An important platinum mineral, composed of platinum sulfide.

copper loss [ELEC] Power loss in a winding due to current flow through the resistance of the copper conductors. Also known as I^2R loss.

copper mica *See* chalcophyllite.

copper nickel *See* niccolite.

copper nitrite *See* cupric nitrate.

copper number [ANALY CHEM] The number of milligrams of copper obtained by the reduction of Benedict's or Fehling's solution by 1 gram of carbohydrate.

copperon *See* cupferron.

copper ore [GEOL] Rock containing copper minerals.

copper oxide *See* cupric oxide; cuprous oxide.

copper oxide photovoltaic cell [ELECTR] A photovoltaic cell in which light acting on the surface of contact between layers of copper and cuprous oxide causes a voltage to be produced.

copper oxide rectifier [ELECTR] A metallic rectifier in which the rectifying barrier is the junction between metallic copper and cuprous oxide.

copperplate engraving [GRAPHICS] A thin, rigid plate of copper, with the lines of a picture cut, or engraved, into it, used for printing purposes; it is inked over, the ink is removed so that it is retained only in the engraved lines, and the plate is placed in a handpress where ink is transferred from engraved lines to overlying paper.

copper plating [MET] Coating of a substance with copper by an electrolytic process, to minimize corrosion.

copper powder [MET] A bronzing powder made by saturating nitrous acid with copper and precipitating the latter by the addition of iron.

copper pressure gage *See* Nobel's copper cylinder.

copper resinate [ORG CHEM] Poisonous green powder, soluble in oils and ether, insoluble in water; made by heating rosin oil with copper sulfate, followed by filtering and drying of the resultant solids; used as a metal-paint preservative and insecticide.

copper spot [PL PATH] A fungus disease of lawn grasses caused by *Gloeocercospora sorghi* and marked by coppery-red areas.

copper steel [MET] Low-carbon steel containing up to 0.25% copper.

copper strip corrosion [ENG] A qualitative method of determining the corrosivity of a petroleum product by observing its effect on a strip of polished copper suspended or placed in the product.

copper sulfate *See* cupric sulfate.

copper sulfide [INORG CHEM] CuS Black, monoclinic or hexagonal crystals that break down at 220°C; used in paints on ship bottoms to prevent fouling.

copper sulfide rectifier [ELECTR] A semiconductor rectifier in which the rectifying barrier is the junction between magnesium and copper sulfide.

copper sweetening [CHEM ENG] Those refining processes using cupric chloride to oxidize mercaptans in petroleum.

copper uranite *See* torbernite.

copperweld [MET] Copper-covered steel, used as a conductor for high-voltage transmission spans where tensile strength is more important than high conductance.

**COORDINATION
POLYHEDRON**

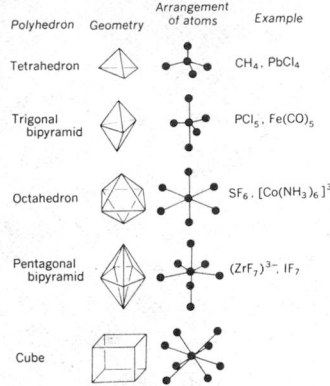

Types of coordination polyhedrons.

COPEPODA

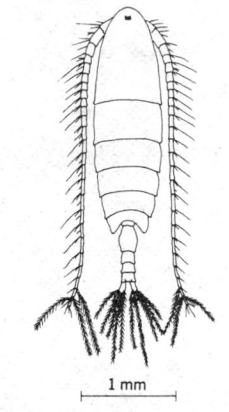

Calanus finmarchicus, a calcanoid free-living copepod.

COPPER

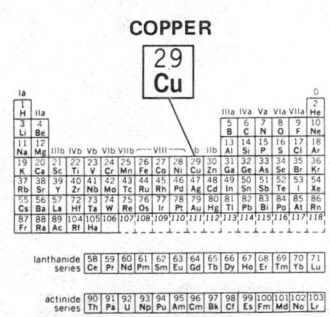

Periodic table of the chemical elements showing the position of copper.

COPPER OXIDE RECTIFIER

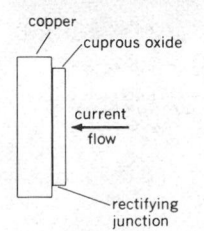

Cross section of a copper oxide rectifier.

copper wire [MET] Wire commonly made from copper by drawing from a hot-rolled rod without annealing; however, the smaller sizes may involve intermediate anneals.

coppice [ECOL] A growth of small trees that are repeatedly cut down at short intervals; the new shoots are produced by the old stumps.

coprecipitation [CHEM] Simultaneous precipitation of more than one substance.

coprolite [GEOL] Petrified excrement.

coprophagy [ZOO] Feeding on dung or excrement.

coprophilous [ECOL] Living in dung.

coprophobia [PSYCH] An abnormal fear of fecal matter.

copulation [ZOO] The sexual union of two individuals, resulting in insemination or deposition of the male gametes in close proximity to the female gametes.

copulatory bursa [INV ZOO] **1.** A sac that receives the sperm during copulation in certain insects. **2.** The caudal expansion of certain male nematodes that functions as a clasper during copulation.

copulatory organ [ANAT] An organ employed by certain male animals for insemination.

copulatory spicule *See* spiculum.

copy [ADP] A string procedure in ALGOL by means of which a new byte string can be generated from an existing byte string. [COMMUN] **1.** To transcribe Morse code signals into written form. **2.** To reproduce graphical material usually by an electrostatic device. **3.** To reproduce information in a new location and possibly in a different form, leaving the source of the information unchanged. [GRAPHICS] *See* subject copy.

copying program [ADP] A system program which copies a data or program file from one peripheral device onto another.

coquimbite [MINERAL] $Fe_2(SO_4)_3 \cdot 9H_2O$ A white mineral that crystallizes in the hexagonal system; it is dimorphous with paracoquimbite. Also known as white copperas.

coquina [INV ZOO] A small marine clam of the genus *Donax*. [PETR] A coarse-grained, porous, easily crumbled variety of limestone composed principally of mollusk shell and coral fragments cemented together as rock.

coquinoid [PETR] **1.** Of or pertaining to coquina. **2.** Lithified coquina. **3.** An autochthonous deposit of limestone made up of more or less whole mollusk shells.

Coraciidae [VERT ZOO] The rollers, a family of Old World birds in the order Coraciiformes.

Coraciiformes [VERT ZOO] An order of predominantly tropical and frequently brightly colored birds.

coracite *See* uraninite.

coracoid [ANAT] One of the paired bones on the posterior-ventral aspect of the pectoral girdle in vertebrates.

coracoid ligament [ANAT] The transverse ligament of the scapula which crosses over the suprascapular notch.

coracoid process [ANAT] The beak-shaped process of the scapula.

coral [INV ZOO] The skeleton of certain solitary and colonial anthozoan coelenterates; composed chiefly of calcium carbonate.

coral head [GEOL] A small reef patch of coralline material. Also known as coral knoll.

coral knoll *See* coral head.

Corallanidae [INV ZOO] A family of sometimes parasitic, but often free-living, isopod crustaceans in the suborder Flabellifera.

Corallidae [INV ZOO] A family of dimorphic coelenterates in the order Gorgonacea.

Corallimorpharia [INV ZOO] An order of solitary sea anemones in the subclass Zoantharia resembling coral in many aspects.

Corallinaceae [BOT] A family of red algae, division Rhodophyta, having compact tissue with lime deposits within and between the cell walls.

coralline [INV ZOO] Any animal that resembles coral, such as a bryozoan or hydroid.

coralline algae [BOT] Red algae belonging to the family Corallinaceae.

corallite [INV ZOO] Skeleton of an individual polyp in a compound coral.

coralloid [BIOL] Resembling coral, or branching like certain coral.

corallum [INV ZOO] Skeleton of a compound coral.

coral mud [GEOL] Fine-grade deposits of coral fragments formed around coral islands and coasts bordered by coral reefs.

coral pinnacle [GEOL] A sharply upward-projecting growth of coral rising from the floor of an atoll lagoon.

coral reef [GEOL] A ridge or mass of limestone built up of detrital material deposited around a framework of skeletal remains of mollusks, colonial coral, and massive calcareous algae.

coral-reef lagoon [GEOGR] The central, shallow body of water of an atoll or the water separating a barrier reef from the shore.

coral-reef shoreline [GEOL] A shoreline formed by reefs composed of coral polyps. Also known as coral shoreline.

coral rock *See* reef limestone.

coral sand [GEOL] Coarse-grade deposits of coral fragments formed around coral islands and coasts bordered by coral reefs.

coral shoreline *See* coral-reef shoreline.

corbel [ARCH] An architectural bracket projecting from within a wall, formed by extensions of the masonry and wood beyond the wall surface, and supporting a burdensome weight above.

corbel arch [ARCH] A structure resembling an arch, in which successive courses project farther into a gap until they close.

Corbiculidae [INV ZOO] A family of fresh-water bivalve mollusks in the subclass Eulamellibranchia; an important food in the Orient.

corbino disk [ELECTROMAG] A variable-resistance device utilizing the effect of a magnetic field on the flow of carriers from the center to the circumference of a disk made of semiconducting or conducting material.

cord [ELEC] A small, very flexible insulated cable. [MATER] **1.** A unit of measure for wood stacked for fuel or pulp; equals $4 \times 4 \times 8$, or 128 cubic feet (approximately 3.6246 cubic meters). **2.** A long, flexible, cylindrical construction of natural or synthetic fibers twisted or woven together.

cordage [ENG] Number of cords of lumber per given area. [MATER] Ropes or cords, especially those in the rigging of a ship.

Cordaitaceae [PALEOBOT] A family of fossil plants belonging to the Cordaitales.

Cordaitales [PALEOBOT] An extensive natural grouping of forest trees of the late Paleozoic.

cordate [BOT] Heart-shaped; generally refers to a leaf base.

cord circuit [ELEC] Connecting circuit terminating in a plug at one or both ends and used at switchboard positions in establishing telephone connections.

cord factor [MICROBIO] A toxic glycolipid found as a surface component of tubercle bacilli that is responsible for virulence and serpentine growth.

cord foot [ENG] A stack of wood measuring 16 cubic feet (approximately 0.45307 cubic meter).

cordierite [MINERAL] $Mg_2(Al_4Si_5O_{18})$ A blue, orthorhombic magnesium aluminosilicate mineral frequently occurring associated with thermally metamorphosed rocks derived from argillaceous sediments.

cordillera [GEOGR] A mountain range or group of ranges, including valleys, plains, rivers, lakes, and so on, forming the main mountain axis of a continent.

cordilleran geosyncline [GEOL] The Devonian geosynclinal region of western North America.

Cordirie process [MET] The refining of lead by conducting steam through it, while molten, to oxidize certain metallic impurities.

cordite [MATER] A trinitrate cellulose derivative prepared by treating cotton fiber or purified wood pulp with a mixture of nitric and sulfuric acids; an explosive powder.

cordless switchboard [COMMUN] Manual telephone switchboard which uses manually operated keys to make connections.

cord of ore [MIN ENG] A unit of about 7 tons (6.35 metric tons), but measured by wagonloads and not by weight.

cordon [BOT] A plant trained to grow flat against a vertical

COPULATORY ORGAN

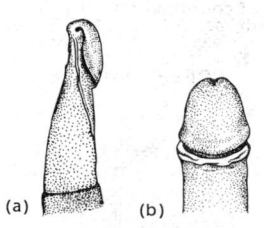

Glans penis of *(a)* bull, *(b)* man.

CORDIERITE

├─2.5 cm─┤

Cordierite from Tsilaizina, Madagascar. *(Specimen from Department of Geology, Bryn Mawr College)*

structure, in a single horizontal shoot or two opposed horizontal shoots.

cordonazo [METEOROL] A southerly wind of hurricane force generated along the western coast of Mexico when a tropical cyclone passes offshore in a northerly direction.

Cordonnier system *See* Batten system.

cordovan [MATER] Nonporous leather made from split horsehide and tanned with vegetable materials.

Cordtex [MIN ENG] A trademark for a detonating fuse suitable for opencut and quarry mining; consists of an explosive core of pentaerythritol tetranitrate (PETN) contained within plastic covering.

cord tire [DES ENG] A pneumatic tire made with cords running parallel to the tread.

Cordulegasteridae [INV ZOO] A family of anisopteran insects in the order Odonata.

corduroy [TEXT] Cotton, rayon, or other fabric with a cut-pile surface of wales; may be either plain or twill weave.

cordwood [MATER] Wood stacked and sold in cords.

cordwood module [ELECTR] High-density circuit module in which discrete components are mounted between and perpendicular to two small, parallel printed circuit boards to which their terminals are attached.

cordylite [MINERAL] $(Ce,La)_2Ba(CO_3)_3F_2$ A colorless to wax-yellow mineral consisting of a carbonate and fluoride of cerium, lanthanum, and barium.

core [ELECTR] *See* magnetic core. [ELECTROMAG] *See* magnetic core. [ENG] The inner material of a wall, column, veneered door, or similar structure. [GEOL] **1.** Center of the earth, beginning at a depth of 2900 kilometers. Also known as earth core. **2.** A vertical, cylindrical boring of the earth from which composition and stratification may be determined; in oil or gas well exploration the presence of hydrocarbons or water are items of interest. [MET] A specially formed part of a mold used to form internal holes in a casting. [NUCLEO] The active portion of a nuclear reactor, containing the fissionable material. [SCI TECH] The central part of a body or structure.

core analysis [GEOL] The use of core samples taken from the borehole during drilling to give information on strata age, composition, and porosity, and the presence of hydrocarbons or water along the length of the borehole.

core array [ELECTR] A rectangular grid arrangement of magnetic cores.

core bank [ELECTR] A stack of core arrays and associated electronics, the stack containing a specific number of core arrays.

core barrel [DES ENG] A hollow cylinder attached to a specially designed bit; used to obtain a continuous section of the rocks penetrated in drilling.

core barrel rod *See* guide rod.

core binder [MATER] A substance for binding core sand together; an example is core oil.

core bit [DES ENG] The hollow, cylindrical cutting part of a core drill.

core blower [MET] A machine using compressed air to blow and pack sand into a core box.

core box [MET] A container for shaping a sand core for a casting.

core catcher *See* split-ring core lifter.

core-catcher case *See* lifter case.

cored ammonium nitrate dynamite [MATER] A class of dynamite that has good water resistance and exhibits velocities higher than straight ammonia dynamite; the gelatin core provides for detonation in the complete explosive column.

cored bar [MET] A bar-shaped powder compact which has been heated by electricity to melt the interior.

cored electrode [MET] An electrode made of metal with a core of flux or other material.

core drier [MET] A light, skeleton cast-iron or aluminum box, whose internal shape conforms closely to the cope portion of a core for molding, used to support, during baking, a core which cannot be placed on a flat plate.

core drill [MECH ENG] A mechanism designed to rotate and to cause an annular-shaped rock-cutting bit to penetrate rock formations, produce cylindrical cores of the formations penetrated, and lift such cores to the surface, where they may be collected and examined.

core-dump [ADP] To copy the contents of all or part of core storage, usually into an external storage device.

core gripper *See* split-ring core lifter.

core-gripper case *See* lifter case.

core hitch [ELEC] Attachment to a cable core to permit pulling it into a duct without damaging the sheath.

Coreidae [INV ZOO] The squash bugs and leaf-footed bugs, a family of hemipteran insects belonging to the superfamily Coreoidea.

core-image library [ADP] A collection of computer programs residing on mass-storage device in ready-to-run form.

core intersection [GEOL] **1.** The point in a borehole where an ore vein or body is encountered as shown by the core. **2.** The width or thickness of the ore body, as shown by the core. Also known as core interval.

core interval *See* core intersection.

core iron [MET] A grade of soft iron suitable for cores of chokes, transformers, and relays.

coreless-type induction heater [ENG] A device in which a charge is heated directly by induction, with no magnetic core material linking the charge. Also known as coreless-type induction furnace.

core lifter *See* split-ring core lifter.

core-lifter case *See* lifter case.

core logging [GEOL] The analysis of the strata through which a borehole passes by the taking of core samples at predetermined depth intervals as the well is drilled.

core loss [ELECTROMAG] The rate of energy conversion into heat in a magnetic material due to the presence of an alternating or pulsating magnetic field. Also known as excitation loss; iron loss.

core memory *See* magnetic core storage.

core memory resident [ADP] A control program which is in the main memory of a computer at all times to supervise the processing of the computer.

coremium [MYCOL] A small bundle of conidiophores in certain imperfect fungi.

core molding [MET] A molding process which makes use of assembled cores to construct the mold.

core oil [MATER] An oil compound used with sand to make foundry core.

core oven [MET] An oven used for baking cores for molding; the walls are constructed of inner and outer layers of sheet metal separated by rock wool or Fiberglas insulation, with interlocked joints.

core print [MET] A projection on a cylindrical casting pattern which supports a core.

corer [ENG] An instrument used to obtain cylindrical samples of geological materials or ocean sediments.

core rod [MET] The part of a die used to make a hole in a compact.

core rope storage [ADP] Direct-access storage consisting of a large number of doughnut-shaped ferrite cores arranged on a common axis, with sense, inhibit, and set wires threaded through or around individual cores in a predetermined manner to provide fixed storage of digital data; each core rope stores one or more complete words, rather than just a single bit.

core sample [GEOL] A sample of rock, soil, snow, or ice obtained by driving a hollow tube into the undisturbed medium and withdrawing it with its contained sample or core.

core sand [MATER] Sand used in a core for molding, made from standard molding-sand mixtures or from silica sand, usually with a binder. Also known as foundry core sand.

core-spring case *See* lifter case.

core-spun yarn [TEXT] Yarn made by wrapping fibers around an inner fiber.

core stack [ELECTR] A number of core arrays, next to one another and treated as a unit.

core storage *See* magnetic core storage.

core test [TEXT] A test for grading wool in which a core sample is withdrawn mechanically from a bag or bale and tested for moisture, ash, vegetable matter, and grease content.

core-type induction heater [ELECTROMAG] A device in

CORE TEST

Wool coring to test wool content.

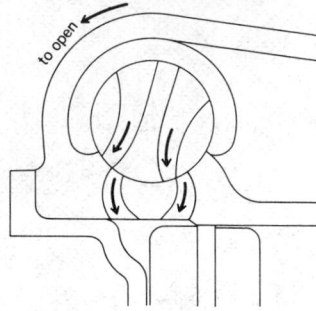

CORLISS VALVE

Corliss steam valve in closed position; arrows indicate path steam will take when the valves are opened.

CORM

A corm, a specialized stem.

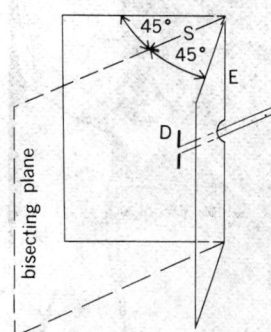

CORNER REFLECTOR

Configuration of a 90° corner reflector. Distance S from the driven radiator D to edge E need not be critically chosen with respect to wavelength; reflector D may lie between 0.25 and 0.7 wavelength.

which a charge is heated by induction, with a magnetic core being used to link the induction coil to the charge.

core wall *See* cutoff wall.

core wire [MET] Copper wire having a steel core, often used for antennas.

coriaceous [BIOL] Leathery, applied to leaves and certain insects.

coriander [BOT] *Coriandrum sativum.* A strong-scented perennial herb in the order Umbellales; the dried fruit is used as a flavoring.

coriandrol *See* linalool.

Cori ester *See* glucose-1-phosphate.

coring [MET] A variable composition of individual crystals across a casting, due to nonequilibrium growth over a range of temperature; the purest material is near the center. [PETRO ENG] The use of a core barrel (hollow length of tubing) to take samples from the underground formation during the drilling operation; used for core analysis.

Coriolis acceleration [MECH] 1. An acceleration which, when added to the acceleration of an object relative to a rotating coordinate system and to its centripetal acceleration, gives the acceleration of the object relative to a fixed coordinate system. 2. A vector which is equal in magnitude and opposite in direction to that of the first definition.

Coriolis correction [NAV] A correction applied to an assumed position, celestial line of position, or celestial fix or to a computed or observed altitude to allow for apparent acceleration due to the Coriolis force.

Coriolis deflection *See* Coriolis effect.

Coriolis effect [MECH] Also known as Coriolis deflection. 1. The deflection relative to the earth's surface of any object moving above the earth, caused by the Coriolis force; an object moving horizontally is deflected to the right in the Northern Hemisphere, to the left in the Southern. 2. The effect of the Coriolis force in any rotating system. [PHYSIO] The physiological effects (nausea, vertigo, dizziness, and so on) felt by a person moving radially in a rotating system, as a rotating space station.

Coriolis force [MECH] A velocity-dependent pseudoforce in a reference frame which is rotating with respect to an inertial reference frame; it is equal and opposite to the product of the mass of the particle on which the force acts and its Coriolis acceleration.

Coriolis operator [SPECT] An operator which gives a large contribution to the energy of an axially symmetric molecule arising from the interaction between vibration and rotation when two vibrations have equal or nearly equal frequencies.

Coriolis parameter [GEOPHYS] Twice the component of the earth's angular velocity about the local vertical $2\Omega \sin \phi$, where Ω is the angular speed of the earth and ϕ is the latitude; the magnitude of the Coriolis force per unit mass on a horizontally moving fluid parcel is equal to the product of the Coriolis parameter and the speed of the parcel.

Coriolis resonance interactions [SPECT] Perturbation of two vibrations of a polyatomic molecule, having nearly equal frequencies, on each other, due to the energy contribution of the Coriolis operator.

corium [ANAT] *See* dermis. [INV ZOO] Middle portion of the forewing of hemipteran insects.

Corixidae [INV ZOO] The water boatmen, the single family of the hemipteran superfamily Corixoidea.

Corixoidea [INV ZOO] A superfamily of hemipteran insects belonging to the subdivision Hydrocorisae that lack ocelli.

cork [BOT] A protective layer of cells that replaces the epidermis in older plant stems.

corkboard [MATER] Board made of compressed cork.

cork paint [MATER] A paint containing fine cork particles; used on steel parts on ships to prevent sweating.

corkscrew rule [ELECTROMAG] The rule that the direction of the current and that of the resulting magnetic field are related to each other as the forward travel of a corkscrew and the direction in which it is rotated.

cork tile [MATER] Floor tile made of compressed cork bound with phenolic or other resin binders; used on moisture-free rigid subfloors or on plywood or hardboard.

Corliss valve [MECH ENG] An oscillating type of valve gear with a trip mechanism for the admission and exhaust of steam to and from an engine cylinder.

corm [BOT] A short, erect, fleshy underground stem, usually broader than high and covered with membranous scales.

cormatose [BOT] Having or producing a corm.

cormidium [INV ZOO] The assemblage of individuals dangling in clusters from the main stem of pelagic siphonophores.

corn [BOT] *Zea mays.* A grain crop of the grass order Cyperales grown for its edible seeds (technically fruits).

Cornaceae [BOT] A family of dicotyledonous plants in the order Cornales characterized by perfect or unisexual flowers, a single ovule in each locule, as many stamens as petals, and opposite leaves.

Cornales [BOT] An order of dicotyledonous plants in the subclass Rosidae marked by a woody habit, simple leaves, well-developed endosperm, and fleshy fruits.

cornea [ANAT] The transparent anterior portion of the outer coat of the vertebrate eye covering the iris and the pupil. [INV ZOO] The outer transparent portion of each ommatidium of a compound eye.

corneite [GEOL] A biotite-hornfels formed during deformation of shale by folding.

Corner-Allen unit [BIOL] A unit for the standardization of progesterone.

corner chisel [DES ENG] A chisel with two cutting edges at right angles.

corner cut [ADP] A corner cut off a punched card at an oblique angle to aid in orientation.

corner frequency *See* break frequency.

cornering tool [DES ENG] A cutting tool with a curved edge, used to round off sharp corners.

cornerite [BUILD] A corner reinforcement for interior plastering.

corner joint [ENG] An L-shaped joint formed by two members positioned perpendicular to each other.

corner reflector [ELECTROMAG] An antenna consisting of two conducting surfaces intersecting at an angle that is usually 90°, with a dipole or other antenna located on the bisector of the angle. [OPTICS] A reflector which returns a laser beam in the direction of its source, consisting of perpendicular reflecting surfaces; used to make precise determinations of distances in surveying.

cornerstone [BUILD] An inscribed stone laid at the corner of a building, usually at a ceremony.

cornetite [MINERAL] $Cu_3(PO_4)(OH)_3$ A peacock-blue mineral consisting of basic copper phosphate.

cornice [ARCH] The crowning, overhanging part of an architectural structure.

cornice brake [MECH ENG] A machine used to bend sheet metal into different forms.

cornicle [INV ZOO] Either of two protruding horn-shaped dorsal tubes in aphids which secrete a waxy fluid.

corniculate [BIOL] Having small horns or hornlike processes.

corniculate cartilage [ANAT] The cartilaginous nodule on the tip of the arytenoid cartilage.

cornification [PHYSIO] Conversion of stratified squamous epithelial cells into a horny layer and into derivatives such as nails, hair, and feathers.

Cornish rolls [MIN ENG] A geared pair of horizontal cylinders, one fixed in a frame and the other held by strong springs; used for grinding.

corn oil [MATER] A semidrying, fatty oil of yellowish color, extracted from germs of corn kernels; used mainly as a salad oil, in soft soaps, and in compounded petroleum lubricants.

corn smut [PL PATH] A fungus disease of corn caused by *Ustilago maydis.*

corn snow *See* spring snow.

corn sugar *See* dextrose.

cornu [ANAT] A horn or hornlike structure.

Cornu-Hartmann formula [OPTICS] An empirical formula, $\lambda = \lambda_0 + C(D - D_0)^{-1}$, expressing the relation between the deviation D produced by a prism and the wavelength λ of the light; λ_0, C, and D_0 are constants for a specific case. Also known as Hartmann dispersion formula.

Cornu quartz prism [OPTICS] A prism constructed of two 30° quartz prisms, left- and right-handed, used in conjunction with left- and right-handed lenses, so that the rotation of polarization occurring in one half of the optical path is exactly compensated by the reverse rotation in the other; used in a quartz spectrograph.

Cornu's spiral [MATH] The graph of the function

$$f(x,y) = \int_{-\infty}^{+\infty} \exp\left[-y^2 z^2 - z - xe^{-z}\right] dz.$$

Also known as clothoid.

corn whiskey [FOOD ENG] Whiskey distilled from corn mash containing at least 80% corn, and aged in uncharred oak containers.

corolla [BOT] A collective term for the petals of a flower.

corollate [BOT] Having a corolla.

corolline [BOT] Relating to, resembling, or being borne on a corolla.

coromant cut [MIN ENG] A drill hole pattern; two overlapping holes about 2¼ inches (about 5.5 centimeters) in diameter are drilled in the tunnel center and left uncharged; they form a slot roughly 4 by 2 inches (10 by 5 centimeters) to which the easers can break.

coromell [METEOROL] A land breeze from the south at La Paz, Mexico, near the mouth of the Gulf of California, prevailing from November to May; it sets in at night and usually persists until 8 or 10 A.M.

corona [BOT] 1. An appendage or series of fused appendages between the corolla and stamens of some flowers. 2. The region where stem and root of a seed plant merge. Also known as crown. [ELEC] *See* corona discharge. [GEOL] A mineral zone that is usually radial about another mineral or at the area between two minerals. Also known as kelyphite. [INV ZOO] 1. The anterior ring of cilia in rotifers. 2. A sea urchin test. 3. The calyx and arms of a crinoid. [MET] An area sometimes surrounding the nugget at the faying surfaces of a spot weld which provides a degree of bond strength. [METEOROL] A set of one or more prismatically colored rings of small radii, concentrically surrounding the disk of the sun, moon, or other luminary when veiled by a thin cloud; due to diffraction by numerous waterdrops. [MINERAL] An annular zone of minerals disposed either around another mineral or at the contact between two minerals.

Corona Australis [ASTRON] A constellation, right ascension 19 hours, declination 40°S. Abbreviated CrA. Also known as Southern Crown.

Corona Borealis [ASTRON] A constellation, right ascension 16 hours, declination 30°N. Abbreviated CrB. Also known as Northern Crown.

corona current [ELEC] The current of electricity equivalent to the rate of charge transferred to the air from an object experiencing corona discharge.

corona discharge [ELEC] A discharge of electricity appearing as a bluish-purple glow on the surface of and adjacent to a conductor when the voltage gradient exceeds a certain critical value; due to ionization of the surrounding air by the high voltage. Also known as corona; electric corona.

coronadite [MINERAL] $Pb(Mn^{2+},Mn^{4+})_8O_{16}$ A black mineral consisting of a lead and manganese oxide, occurring in massive form with fibrous structure; an important constituent of manganese ore.

coronagraph [ASTRON] An instrument for photographing the corona and prominences of the sun at times other than at solar eclipse.

coronal hole [ASTRON] A large-scale, apparently open structure in the solar corona, devoid of any soft x-ray emission and surrounded by diverging boundary structures.

coronal suture [ANAT] The union of the frontal with the parietal bones transversely across the vertex of the skull.

corona method [GEOPHYS] A method of estimating drop sizes in clouds by utilizing measurements of the angular radii of the rings of a corona.

corona radiata [HISTOL] The layer of cells immediately surrounding a mammalian ovum.

corona resistance [ELEC] Ability of a conductor to resist destruction when a high-voltage electrostatic field ionizes within insulation voids.

coronary artery [ANAT] Either of two arteries arising in the aortic sinuses that supply the heart tissue with blood.

coronary disease [MED] Any condition that reduces the flow of blood to the heart muscles.

coronary failure [MED] Prolonged precordial pain or discomfort without conventional evidence of myocardial infarction; subendocardial ischemia caused by a disparity between coronary blood flow and myocardial needs; this condition is more commonly referred to as coronary artery insufficiency.

coronary occlusion [MED] Complete blockage of a coronary artery.

coronary sinus [ANAT] A venous sinus opening into the heart's right atrium which drains the cardiac veins.

coronary stenosis [MED] Narrowing of the lumen of a coronary artery.

coronary sulcus [ANAT] A groove in the external surface of the heart separating the atria from the ventricles, containing the trunks of the nutrient vessels of the heart.

coronary thrombosis [MED] Formation of a thrombus in a coronary artery.

coronary valve [ANAT] A semicircular fold of the endocardium of the right atrium at the orifice of the coronary sinus.

coronary vein [ANAT] 1. Any of the blood vessels that bring blood from the heart and empty into the coronary sinus. 2. A vein along the lesser curvature of the stomach.

corona shield [ELEC] A shield placed about a point of high potential to redistribute electrostatic lines of force.

Coronatae [INV ZOO] An order of the class Scyphozoa which includes mainly abyssal species having the exumbrella divided into two parts by a coronal furrow.

corona tube [ELEC] A gas-discharge voltage-reference tube employing a corona discharge.

coronavirus [VIROL] A major group of animal viruses including avian infectious bronchitis virus and mouse hepatitis virus.

corona voltmeter [ELEC] A voltmeter in which the crest value of a voltage is indicated by the inception of corona at a known electrode spacing.

coronizing [MET] Process to electroplate zinc on nickel, thermally treated at 375°C; coating is used on ferrous and copper-based alloys to give resistance to sulfur dioxide, SO_2, and sulfur trioxide, SO_3.

coronoid [BIOL] Shaped like a beak.

coronoid fossa [ANAT] A depression in the humerus into which the apex of the coronoid process of the ulna fits in extreme flexion of the forearm.

coronoid process [ANAT] 1. A thin, flattened process projecting from the anterior portion of the upper border of the ramus of the mandible, and serving for the insertion of the temporal muscle. 2. A triangular projection from the upper end of the ulna, forming the lower part of the radial notch.

coronule [INV ZOO] A peripheral ring of spines on some diatom shells.

Corophiidae [INV ZOO] A family of amphipod crustaceans in the suborder Gammaridea.

corpora quadrigemina [ANAT] The inferior and superior colliculi collectively. Also known as quadrigeminal body.

cor pulmonale [MED] Hypertrophy and dilation of the right ventricle secondary to obstruction to the pulmonary blood flow and consequent pulmonary hypertension.

corpus albicans [HISTOL] The white fibrous scar in an ovary; produced by the involution of the corpus luteum.

corpus allatum [INV ZOO] An endocrine structure near the brain of immature arthropods that secretes a juvenile hormone, neotenin.

corpus callosum [ANAT] A band of nerve tissue connecting the cerebral hemispheres in man and higher mammals.

corpus cavernosum [ANAT] The cylinder of erectile tissue forming the clitoris in the female and the penis in the male.

corpus cerebelli [ANAT] The central lobe or zone of the cerebellum; regulates reflex tonus of postural muscles in mammals.

corpuscle [ANAT] 1. A small, rounded body. 2. An encapsulated sensory-nerve end organ. [OPTICS] A particle of light in the corpuscular theory, corresponding to the photon in the quantum theory.

corpuscular radiation [PHYS] Radiation consisting of subatomic particles, such as electrons, protons, deuterons, and neutrons, as distinguished from electromagnetic radiation.

corpuscular theory of light [OPTICS] Theory that light consists of a stream of particles; now considered a limiting case of the quantum theory. Also known as Newton's theory of light.

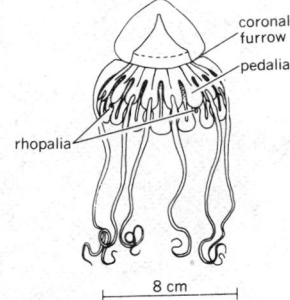

CORONATAE

Periphylla. (From L. H. Hyman, The Invertebrates, vol. 1, McGraw-Hill, 1940)

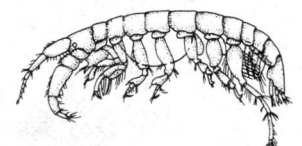

COROPHIIDAE

Lateral view of female *Corophium crassicorne* Bruzelius, a tube-builder. (From G. O. Sars, An Account of the Crustacea of Norway, vol. 1, 1895)

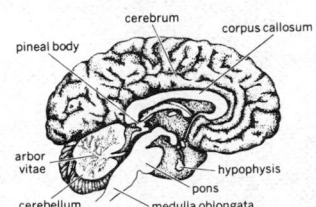

CORPUS CALLOSUM

A sagittal section through the human brain showing corpus callosum. (From H. E. Walter and L. P. Sayles, Biology of the Vertebrates, 3d ed., Macmillan, 1949)

corpus luteum [HISTOL] The yellow endocrine body formed in the ovary at the site of a ruptured Graafian follicle.

corpus striatum [ANAT] The caudate and lenticular nuclei, together with the internal capsule which separates them.

corrasion [GEOL] Mechanical wearing away of rock and soil by the action of solid materials moved along by wind, waves, running water, glaciers, or gravity. Also known as mechanical erosion.

corrected altitude [METEOROL] The indicated altitude corrected for temperature deviation from the standard atmosphere. Also known as true altitude.

corrected azimuth [ORD] Azimuth of the axis of the bore of a gun firing on a moving target, after allowances have been made for atmospheric, material, and other variable conditions.

corrected compass course [NAV] Compass course with deviation applied.

corrected compass heading [NAV] Compass heading with deviation applied.

corrected establishment *See* mean high water lunitidal interval.

corrected sextant altitude [NAV] Sextant altitude for which index error, height of eye, parallax, refraction, and so on have been applied. Also known as observed altitude.

correcting [NAV] The process of applying corrections, particularly that of converting compass to magnetic direction, or compass, magnetic, or gyro to true direction.

correcting wedge [MIN ENG] A deflection wedge used to deflect a crooked borehole back into its intended course.

correction [SCI TECH] A quantity added to a calculated or observed value to obtain the true value.

correction of soundings [NAV] In marine operations, the adjustment of soundings for any departure from true depth because of the method of sounding or any fault in the measuring apparatus.

correction time [CONT SYS] The time required for the controlled variable to reach and stay within a predetermined band about the control point following any change of the independent variable or operating condition in a control system. Also known as settling time.

corrective action [CONT SYS] The act of varying the manipulated process variable by the controlling means in order to modify overall process operating conditions.

corrective network [ELEC] An electric network inserted in a circuit to improve its transmission properties, impedance properties, or both. Also known as shaping circuit; shaping network.

corrective operation *See* remedial operation.

corrective therapy [MED] A program, and the techniques, designed to improve or maintain the health of a patient by improving neuromuscular activities and personal health habits and promoting relaxation by adjustment to stresses.

corrector [ENG] A magnet, piece of soft iron, or device used in the adjustment or compensation of a magnetic compass.

correed relay [ELEC] Hermetically sealed reed capsule surrounded by a coil winding, used as a switching device with telephone equipment.

correlated orientation tracking and range *See* cotar.

correlation [GEOL] **1.** The determination of the equivalence or contemporaneity of geologic events in separated areas. **2.** As a step in seismic study, the selecting of corresponding phases, taken from two or more separated seismometer spreads, of seismic events seemingly developing at the same geologic formation boundary.

correlation coefficient [STAT] A measurement, which is unchanged by both addition and multiplication of the random variable by positive constants, of the tendency of two random variables X and Y to vary together; it is given by the ratio of the covariance of X and Y to the square root of the product of the variance of X and the variance of Y.

correlation curve *See* correlogram.

correlation detection [ENG] A method of detection of aircraft or space vehicles in which a signal is compared, point to point, with an internally generated reference. Also known as cross-correlation detection.

correlation direction finder [ENG] Satellite station separated from a radar to receive jamming signals; by correlating the signals received from several such stations, range and azimuth of many jammers may be obtained.

correlation distance [COMMUN] In tropospheric scatter propagation, the minimum spatial separation between antennas which will give rise to independent fading of the received signals.

correlation tracking and triangulation *See* cotat.

correlation tracking system [ENG] A trajectory-measuring system utilizing correlation techniques where signals derived from the same source are correlated to derive the phase difference between the signals.

correlation-type receiver *See* correlator.

correlative rights [PETRO ENG] Legal rights protecting property over a portion of a gas or oil reservoir from excessive or wasteful withdrawal of hydrocarbons by adjoining properties overlying the same reservoir.

correlator [ELECTR] A device that detects weak signals in noise by performing an electronic operation approximating the computation of a correlation function. Also known as correlation-type receiver.

correlogram [MATH] A curve showing the assumed correlation between two mathematical variables. Also known as correlation curve.

correspondence principle [QUANT MECH] The principle that quantum mechanics has a classical limit in which it is equivalent to classical mechanics. Also known as Bohr's correspondence principle.

corresponding points [PHYSIO] Any two retinal areas in the respective eyes so that the area in one eye has an identical direction in the opposite retina.

corresponding states [PHYS CHEM] The condition when two or more substances are at the same reduced pressures, the same reduced temperatures, and the same reduced volumes.

corridor [ECOL] A land bridge that allows free migration of fauna in both directions.

Corrigan's pulse [MED] A pulse characterized by a rapid, forceful ascent (water-hammer quality) and rapid downstroke or descent (collapsing quality); seen with aortic regurgitation and hyperkinetic circulatory states.

Corrodentia [INV ZOO] The equivalent name for Psocoptera.

corroding lead [MET] Lead that can be corroded to make white lead.

Corrodkote test [MET] An accelerated corrosion test in which the article is coated with a slurry of clay and a salt solution and then exposed to a high humidity for a specified period.

corrosion [GEOCHEM] Chemical erosion by motionless or moving agents. [MET] Gradual destruction of a metal or alloy due to chemical processes such as oxidation or the action of a chemical agent.

corrosion border *See* corrosion rim.

corrosion fatigue [MET] Damage to or failure of a metal due to corrosion combined with fluctuating fatigue stresses.

corrosion fatigue limit [MET] The maximum stress that a corroded material can withstand for a given number of stress reversals.

corrosion number *See* acid number.

corrosion protection [MET] The minimization of corrosion by coating with a protective metal, with an oxide or phosphide or similar substance, or with a protective paint, or by rendering the metal passive.

corrosion rim [MINERAL] A modification of the outlines of a porphyritic crystal due to the corrosive action of a magma on previously stable minerals. Also known as corrosion border.

corrosion test [MET] Any of various tests to determine the resistance of a metal to chemical attack.

Corrosiron [MET] Trade name for a silicon cast iron.

corrosive [MATER] A substance that causes corrosion.

corrosive flux [MET] A soldering flux, usually composed of inorganic salts and acids, which provides oxide removal of the base metal upon application of solder; flux remaining on the base metal is corrosive and should be removed.

corrosiveness [MET] The tendency of a metal to wear away another by chemical attack.

corrosive sublimate *See* mercuric chloride.

corrugated bar [DES ENG] Steel bar with transverse ridges; used in reinforced concrete.

corrugated fastener [DES ENG] A thin corrugated strip of steel that can be hammered into a wood joint to fasten it.

corrugated lens [OPTICS] A lens having circular sections cut out from the surface to reduce its weight without lowering its focal power.

corrugating [DES ENG] Forming straight, parallel, alternate ridges and grooves in sheet metal, cardboard, or other material.

corsite [PETR] A spheroidal variety of gabbro. Also known as miagite; napoleonite.

cortex [ANAT] The outer portion of an organ or structure, such as of the brain and adrenal glands. [BOT] A primary tissue in roots and stems of vascular plants that extends inward from the epidermis to the phloem. [INV ZOO] The peripheral layer of certain protozoans.

cortical stimulator [MED] An electronic instrument used in nerve and mental therapy to deliver an electric shock of prescribed strength by means of a pulsating current.

corticoid *See* adrenal cortex hormone.

corticosteroid [BIOCHEM] **1.** Any steroid hormone secreted by the adrenal cortex of vertebrates. **2.** Any steroid with properties of an adrenal cortex steroid.

corticosterone [BIOCHEM] $C_{21}H_{30}O_4$ A steroid hormone produced by the adrenal cortex of vertebrates that stimulates carbohydrate synthesis and protein breakdown and is antagonistic to the action of insulin.

corticotrophic [PHYSIO] Having an effect on the adrenal cortex.

corticotropin [BIOCHEM] A hormonal preparation having adrenocorticotropic activity, derived from the adenohypophysis of certain domesticated animals.

cortin unit [BIOL] A unit for the standardization of adrenal cortical hormones.

cortisol *See* hydrocortisone.

cortisone [BIOCHEM] $C_{21}H_{28}O_5$ A steroid hormone produced by the adrenal cortex of vertebrates that acts principally in carbohydrate metabolism.

cortlandite [PETR] A peridotite consisting of large crystals of hornblende with poikilitically included crystals of olivine. Also known as hudsonite.

corundum [MINERAL] Al_2O_3 A hard mineral occurring in various colors and crystallizing in the hexagonal system; crystals are usually prismatic or in rounded barrel shapes; gem varieties are ruby and sapphire.

corvette [NAV ARCH] **1.** A warship with a continuous deck from fore to stern, usually with no structure above, and usually with only one row of guns. **2.** A very maneuverable escort ship having antisubmarine and antiaircraft guns, depth charges, and detection equipment.

Corvidae [VERT ZOO] A family of large birds in the order Passeriformes having stout, long beaks; includes the crows, jays, and magpies.

Corvus [ASTRON] A constellation, right ascension 12 hours, declination 20°S. Abbreviated Crv. Also known as Crow.

Corylophidae [INV ZOO] The equivalent name for Orthoperidae.

corymb [BOT] An inflorescence in which the flower stalks arise at different levels but reach the same height, resulting in a flat-topped cluster.

corymbose [BOT] Resembling or pertaining to a corymb.

Corynebacteriaceae [MICROBIO] A family of nonsporeforming, usually nonmotile rod-shaped bacteria in the order Eubacteriales including animal and plant parasites and pathogens.

corynebacteriophage [VIROL] Any bacteriophage able to infect *Corynebacterium diphtheriae*.

Corynebacterium [MICROBIO] A genus of gram-positive, slightly curved rod-shaped bacteria, sometimes containing granules, in the family Corynebacteriaceae.

Corynebacterium diphtheriae [MICROBIO] A facultatively aerobic, nonmotile species of bacteria that causes diphtheria in man. Also known as *Bacillus diphtheriae*; Klebs-Loeffler bacillus.

Coryphaenidae [VERT ZOO] A family of pelagic fishes in the order Perciformes characterized by a blunt nose and deeply forked tail.

Coryphodontidae [PALEON] The single family of the Coryphodontoidea, an extinct superfamily of mammals.

Coryphodontoidea [PALEON] A superfamily of extinct mammals in the order Pantodonta.

coryza [MED] Inflammation of the mucous membranes of the nose, usually marked by sneezing and discharge of watery mucus.

cos *See* cosine function.

cosalite [MINERAL] $Pb_2Bi_2S_5$ A lead-gray or steel-gray mineral consisting of lead, bismuth, and sulfur; specific gravity is 6.39–6.75

cosecant [MATH] The reciprocal of the sine. Denoted csc.

cosecant antenna [ELECTROMAG] An antenna that gives a beam whose amplitude varies as the cosecant of the angle of depression below the horizontal; used in navigation radar.

cosecant-squared antenna [ELECTROMAG] An antenna having a cosecant-squared pattern.

cosecant-squared pattern [ELECTROMAG] A ground radar-antenna radiation pattern that sends less power to nearby objects than to those farther away in the same sector; the field intensity varies as the square of the cosecant of the elevation angle.

cosh *See* hyperbolic cosine.

cosine emission law [OPTICS] The law that the energy emitted by a radiating surface in any direction is proportional to the cosine of the angle which that direction makes with the normal.

cosine function [MATH] In a right triangle having an angle θ, the cosine function gives the ratio of adjacent side to hypotenuse; more generally, it is the function which assigns to any real number θ the abscissa of the point on the unit circle obtained by moving from (1,0) counterclockwise θ units along the circle, or clockwise $|\theta|$ units if θ is less than 0. Denoted cos.

cosine pulse [PHYS] A pulse whose amplitude varies during some time interval in proportion to the cosine function over the range from $-\pi/2$ to $\pi/2$, and vanishes outside this time interval.

cosine-squared pulse [PHYS] A pulse whose amplitude varies during some time interval in proportion to the square of the cosine function over the range from $-\pi/2$ to $\pi/2$, and vanishes outside this time interval.

cosine winding [ELECTR] A winding used in the deflection yoke of a cathode-ray tube to prevent changes in focus as the beam is deflected over the entire area of the screen.

cosmic [ASTRON] Pertaining to the cosmos, the vast extraterrestrial regions of the universe.

cosmic dust [ASTRON] Fine particles of solid matter forming clouds in interstellar space.

cosmic electrodynamics [ASTROPHYS] The science concerned with electromagnetic phenomena in ionized media encountered in interstellar space, in stars, and above the atmosphere.

cosmic expansion [ASTRON] The recession of all distant galaxies from each other, as manifested in the red shift of their spectral lines.

cosmic noise [COMMUN] Radio static caused by a phenomenon outside the earth's atmosphere, such as sunspots.

cosmic radiation *See* cosmic rays.

cosmic radio waves [ASTRON] Radio waves reaching the earth from interstellar or intergalactic sources.

cosmic rays [NUC PHYS] Electrons and the nuclei of atoms, largely hydrogen, that impinge upon the earth from all directions of space with nearly the speed of light. Also known as cosmic radiation; primary cosmic rays.

cosmic-ray shower [NUC PHYS] The simultaneous appearance of a number of downward-directed ionizing particles, with or without accompanying photons, caused by a single cosmic ray. Also known as shower.

cosmic-ray telescope [NUCLEO] An array of counters, sensitive to the direction of the rays detected.

cosmic year [ASTRON] The period of rotation of the Milky Way Galaxy, about 220 million years.

Cosmocercidae [INV ZOO] A group of nematodes assigned to the suborder Oxyurina by some authorities and to the suborder Ascaridina by others.

cosmochemistry [ASTROPHYS] The branch of science which treats of the chemical composition of the universe and its origin.

CORUNDUM

A specimen of corundum from Steinkopf, South Africa. (*American Museum of Natural History*)

├── 4.4 cm ──┤

COSMOID SCALE

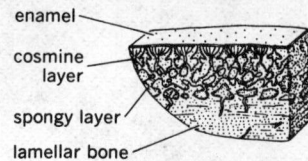

enamel
cosmine layer
spongy layer
lamellar bone

Cross section through a cosmoid scale. (*From A. S. Romer, The Vertebrate Body, Saunders, 1962*)

COSSURIDAE

anterior

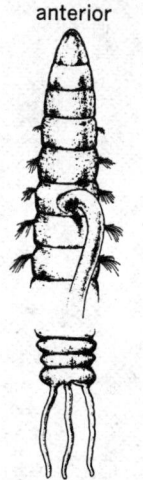

Cossura, anterior and posterior ends in dorsal view.

COTTER PIN

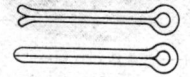

Two common forms of cotter pin.

COTTON EFFECT

rotation or ellipticity

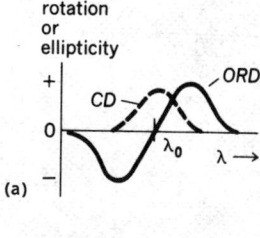

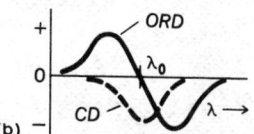

Behavior of optical rotatory dispersion (ORD) and circular dichroism (CD) curves in the vicinity of an absorption band at wavelength λ_0 (idealized). (*a*) Positive Cotton effect. (*b*) Negative Cotton effect.

cosmogony [ASTROPHYS] Study of the origin and evolution of specific astronomical systems and of the universe as a whole.

cosmoid scale [VERT ZOO] A structure in the skin of primitive rhipidistians and dipnoans that is composed of enamel, a dentine layer (cosmine), and laminated bone.

cosmological constant [RELAT] The multiplicative constant for a term proportional to the metric in Einstein's equation relating the curvature of space to the energy-momentum tensor.

cosmology [ASTRON] The study of the overall structure of the physical universe.

cosmonaut [AERO ENG] A Soviet astronaut.

cosmopolitan [ECOL] Having a worldwide distribution wherever the habitat is suitable, with reference to the geographical distribution of a taxon.

Cosmos satellites [AERO ENG] A series of earth satellites launched by the U.S.S.R. starting in 1962, ostensibly for geophysical experiments.

cospectrum [PHYS] **1.** The spectral decomposition of the in-phase components of the covariance of two functions of time. **2.** The real part of the cross spectrum of two functions.

Cossidae [INV ZOO] The goat or carpenter moths, a family of heavy-bodied lepidopteran insects in the superfamily Cossoidea having the abdomen extending well beyond the hindwings.

Cosslett process [MET] A process in which iron or steel articles immersed for 3 or 4 hours in a boiling solution, made by mixing iron filings with concentrated H_3PO_4 (sufficient to form a paste) and then adding to weak phosphoric acid, become coated with a rust-resisting deposit of basic ferrous phosphate.

Cossoidea [INV ZOO] A monofamilial superfamily of lepidopteran insects belonging to suborder Heteroneura.

Cossuridae [INV ZOO] A family of fringe worms belonging to the Sedentaria.

Cossyphodidae [INV ZOO] The lively ant guest beetles, a small family of coleopteran insects in the superfamily Tenebrionoidea.

costa [BIOL] A rib or riblike structure. [BOT] The midrib of a leaf. [INV ZOO] The anterior vein of an insect's wing.

costa bulb [NAV ARCH] A streamlined body of revolution integral with a rudder and directly in line with the propeller, designed to improve propulsion efficiency of ships.

cost accounting [IND ENG] The branch of accounting in which one records, analyzes, and summarizes costs of material, labor, and burden, and compares these actual costs with predetermined budgets and standards.

Costaceae [BOT] A family of monocotyledonous plants in the order Zingiberales distinguished by having one functional stamen with two pollen sacs and spirally arranged leaves and bracts.

costal cartilage [ANAT] The cartilage occupying the interval between the ribs and the sternum or adjacent cartilages.

costal process [ANAT] An anterior or ventral projection on the lateral part of a cervical vertebra. [EMBRYO] An embryonic rib primordium, the ventrolateral outgrowth of the caudal, denser half of a sclerotome.

cost analysis [IND ENG] Analysis of the factors contributing to the costs of operating a business and of the costs which will result from alternative procedures, and of their effects on profits.

costate [BIOL] Having ribs or ridges.

cost control *See* industrial cost control.

cost function [SYS ENG] In decision theory, a loss function which does not depend upon the decision rule.

cost-plus contract [ENG] A contract under which a contractor furnishes all material, construction equipment, and labor at actual cost, plus an agreed-upon fee for his services.

cotangent [MATH] The reciprocal of the tangent. Denoted cot; ctn.

cotar [ENG] A passive system used for tracking a vehicle in space by determining the line of direction between a remote ground-based receiving antenna and a telemetering transmitter in the missile, using phase-comparison techniques. Derived from correlated orientation tracking and range.

cotat [ENG] A trajectory-measuring system using several antenna base lines, each separated by large distances, to measure direction cosines to an object; then the object's space position is computed by triangulation. Derived from correlation tracking and triangulation.

coth *See* hyperbolic cotangent.

cotidal chart [MAP] A chart of cotidal lines that show approximate locations of high water at hourly intervals as measured from a reference meridian, usually Greenwich.

cotidal hour [ASTRON] The average interval expressed in solar or lunar hours between the moon's passage over the meridian of Greenwich and the following high water at a specified place.

cotidal line [MAP] A line on a chart passing through all points where high water occurs at the same time.

Cotingidae [VERT ZOO] The cotingas, a family of neotropical suboscine birds in the order Passeriformes.

cotter [DES ENG] A tapered piece that can be driven in a tapered hole to hold together an assembly of machine or structural parts.

cottered joint [MECH ENG] A joint in which a cotter, usually a flat bar tapered on one side to ensure a tight fit, transmits power by shear on an area at right angles to its length.

cotter pin [DES ENG] A split pin, inserted into a hole, to hold a nut or cotter securely to a bolt or shaft, or to hold a pair of hinge plates together.

Cottidae [VERT ZOO] The sculpins, a family of perciform fishes in the suborder Cottoidei.

Cottiformes [VERT ZOO] An order set up in some classification schemes to include the Cottoidei.

Cottoidei [VERT ZOO] The mail-cheeked fishes, a suborder of the order Perciformes characterized by the expanded third infraorbital bone.

cotton [BOT] Any plant of the genus *Gossypium* in the order Malvales; cultivated for the fibers obtained from its encapsulated fruits or bolls. [TEXT] The most economical natural fiber, obtained from plants of the genus *Gossypium*, used in making fabrics, cordage, and padding and for producing man-made fibers and cellulose.

cotton anthracnose [PL PATH] A fungus disease of cotton caused by *Glomerella gossypii* and characterized by reddish-brown to light-colored or necrotic spots.

cotton ball *See* ulexite.

cotton-belt climate [CLIMATOL] A type of warm climate characterized by dry winters and rainy summers; that is, a monsoon climate, in contrast to a Mediterranean climate.

Cotton effect [ANALY CHEM] The characteristic wavelength dependence of the optical rotatory dispersion curve or the circular dichroism curve or both in the vicinity of an absorption band.

cotton gin [TEXT] A machine that separates cottonseed from the fibers.

cottonmouth *See* water moccasin.

Cotton-Mouton birefringence *See* Cotton-Mouton effect.

Cotton-Mouton constant [OPTICS] A constant giving the strength of the Cotton-Mouton effect in a liquid; when multiplied by the path length and the square of the magnetic field, it gives the phase difference between the components of light parallel and perpendicular to the field.

Cotton-Mouton effect [OPTICS] Double refraction (birefringence) of light in a liquid in a magnetic field at right angles to the direction of light propagation. Also known as Cotton-Mouton birefringence.

cotton oil [MATER] The yellow, viscous fixed oil, containing principally linoleic acid, pressed from the seeds of various *Gossypium* species; the refined oil is colorless and used in foods and some pharmaceutical preparations. Also known as cottonseed oil; oleum gossypii seminis.

cotton root rot [PL PATH] A fungus disease of cotton caused by *Phymatotrichum omnivorum* and marked by bronzing of the foliage followed by sudden wilting and death of the plant.

cotton rust [PL PATH] A fungus disease of cotton caused by *Puccinia stakmanii* producing low, greenish-yellow or orange elevations on the undersurface of leaves.

cottonseed oil *See* cotton oil.

cotton staple [TEXT] Natural cotton fiber.

cotton wax [MATER] The wax composing cottonseed coating.

cotton wilt [PL PATH] **1.** A fungus disease of cotton caused by

Fusarium vasinfectum growing in the water-conducting vessels and characterized by wilt, yellowing, blighting, and death. **2.** A fungus blight of cotton caused by *Verticillium albo-atrum* and characterized by yellow mottling of the foliage.

cottonwood [BOT] Any of several poplar trees *(Populus)* having hairy, encapsulated fruit.

cottony rot [PL PATH] A fungus disease of many plants, especially citrus trees, marked by fluffy white growth caused by *Sclerotinia sclerotiorum*, in which there is stem wilt and rot.

Cottrell hardening [SOLID STATE] Hardening of a material caused by locking of its dislocations when impurity atoms whose size differs from that of the solvent cluster around them.

Cottrell precipitator [ENG] A machine for removing dusts and mists from gases, in which the gas passes through a grounded pipe with a fine axial wire at a high negative voltage, and particles are ionized by the corona discharge of the wire and migrate to the pipe.

cotunnite [MINERAL] $PbCl_2$ An alteration product of galena; a soft, white to yellowish mineral that crystallizes in the orthorhombic crystal system.

cotyledon [BOT] The first leaf of the embryo of seed plants.

cotylocercous cercaria [INV ZOO] A digenetic trematode larva characterized by a sucker or adhesive gland on the tail.

Cotylosauria [PALEON] An order of primitive reptiles in the subclass Anapsida, including the stem reptiles, ancestors of all of the more advanced Reptilia.

cotype *See* syntype.

coudé focus [OPTICS] Focus achieved with a coudé telescope.

coudé spectrograph [SPECT] A stationary spectrograph that is attached to the tube of a coudé telescope.

coudé spectroscopy [SPECT] The production and investigation of astronomical spectra using a coudé spectrograph.

coudé telescope [OPTICS] An instrument in which light is reflected along the polar axis to come to focus at a fixed place where it is viewed through a fixed eyepiece or where a spectrograph can be mounted.

Couette flow [FL MECH] Low-speed, steady motion of a viscous fluid between two infinite plates moving parallel to each other.

Couette viscometer [ENG] A viscometer in which the liquid whose viscosity is to be measured fills the space between two vertical coaxial cylinders, the inner one suspended by a torsion wire; the outer cylinder is rotated at a constant rate, and the resulting torque on the inner cylinder is measured by the twist of the wire. Also known as rotational viscometer.

cougar *See* puma.

cough [MED] A sudden, violent expulsion of air after deep inspiration and closure of the glottis.

Couinae [VERT ZOO] The couas, a subfamily of Madagascan birds in the family Cuculidae.

coul *See* coulomb.

coulee [GEOL] **1.** A thick, solidified sheet or stream of lava. **2.** A steep-sided valley or ravine, sometimes with a stream at the bottom.

coulisse [ENG] A piece of wood that has a groove cut in it to enable another piece of wood to slide in it. Also known as cullis.

coulomb [ELEC] A unit of electric charge, defined as the amount of electric charge that crosses a surface in 1 second when a steady current of 1 absolute ampere is flowing across the surface; this is the absolute coulomb and has been the legal standard of quantity of electricity since 1950; the previous standard was the international coulomb, equal to 0.999835 absolute coulomb. Abbreviated coul.

Coulomb attraction [ELEC] The electrostatic force of attraction exerted by one charged particle on another charged particle of opposite sign. Also known as electrostatic attraction.

Coulomb barrier [NUC PHYS] **1.** The Coulomb repulsion which tends to keep positively charged bombarding particles out of the nucleus. **2.** Specifically, the Coulomb potential associated with this force.

Coulomb excitation [NUC PHYS] Inelastic scattering of a positively charged particle by a nucleus and excitation of the nucleus, caused by the interaction of the nucleus with the rapidly changing electric field of the bombarding particle.

Coulomb field [ELEC] The electric field created by a stationary charged particle.

Coulomb force [ELEC] The electrostatic force of attraction or repulsion exerted by one charged particle on another, in accordance with Coulomb's law.

Coulomb friction [MECH] Friction occurring between dry surfaces.

Coulomb interactions [ELEC] Interactions of charged particles associated with the Coulomb forces they exert on one another. Also known as electrostatic interactions.

Coulomb potential [ELEC] A scalar point function equal to the work per unit charge done against the Coulomb force in transferring a particle bearing an infinitesimal positive charge from infinity to a point in the field of a specific charge distribution.

Coulomb repulsion [ELEC] The electrostatic force of repulsion exerted by one charged particle on another charged particle of the same sign. Also known as electrostatic repulsion.

Coulomb scattering [PHYS] A collision of two charged particles in which the Coulomb force is the dominant interaction.

Coulomb's law [ELEC] The law that the attraction or repulsion between two electric charges acts along the line between them, is proportional to the product of their magnitudes, and is inversely proportional to the square of the distance between them. Also known as law of electrostatic attraction.

coulometer [PHYS CHEM] An electrolytic cell for the precise measurement of electrical quantities or current intensity by quantitative determination of chemical substances produced or consumed.

coulometric analysis [ANALY CHEM] A technique in which the amount of a substance is determined quantitatively by measuring the total amount of electricity required to deplete a solution of the substance.

coulometric titration [ANALY CHEM] The slow electrolytic generation of a soluble species which is capable of reacting quantitatively with the substance sought; some independent property must be observed to establish the equivalence point in the reaction.

coulometry [ANALY CHEM] A determination of the amount of an electrolyte released during electrolysis by measuring the number of coulombs used.

coulostatic analysis [PHYS CHEM] An electrochemical technique involving the application of a very short, large pulse of current to the electrode; the pulse charges the capacitive electrode-solution interface to a new potential, then the circuit is opened, and the return of the working electrode potential to its initial value is monitored; the current necessary to discharge the electrode interface comes from the electrolysis of electroactive species in solution; the change in electrode potential versus time results in a plot, the shape of which is proportional to concentration.

Coulter counter [MICROBIO] An electronic device for counting the number of cells in a liquid culture.

coumarin [ORG CHEM] $C_9H_6O_2$ The anhydride of *o*-coumaric acid; a toxic, white, crystalline lactone found in many plants and made synthetically; used in making perfume and soap. Also known as 1,2-benzopyrone.

coumarin glycoside [BIOCHEM] Any of several glycosidic aromatic principles in many plants; contains coumaric acid as the aglycon group.

coumarone [ORG CHEM] C_8H_6O A colorless liquid, boiling point 169°C. Also known as benzofuran.

coumarone-indene resin [ORG CHEM] A synthetic resin prepared by polymerization of coumarone and indene.

count [AERO ENG] **1.** To proceed from one point to another in a countdown or plus count, normally by calling a number to signify the point reached. **2.** To proceed in a countdown, as in T minus 90 and counting. [CHEM] An ionizing event. [NUCLEO] **1.** A single response of the counting system in a radiation counter. **2.** The total number of events indicated by a counter. [TEXT] The number of warp and filling threads per square inch of fabric.

countable [MATH] Either finite or denumerable. Also known as enumerable.

COTTONWOOD

Cottonwood poplar *(Populus deltoides)* showing leaf, leaf scar, terminal bud, and twig.

countably additive [MATH] Given a measure m, and a sequence of pairwise disjoint measurable sets, the property that the measure of the union is equal to the sum of the measures of the sets.

countably compact set [MATH] A set with the property that every cover with countably many open sets contains a finite number of sets which is also a cover.

countably infinite set *See* denumerable set.

countably subadditive [MATH] A set function m is countably subadditive if, given any sequence of sets, the measure of the union is less than or equal to the sum of the measures of the sets.

count cyle [ADP] An increase or decrease of the cycle index by unity or by an arbitrary integer.

countdown [AERO ENG] **1.** The process in the engineering definition, used in leading up to the launch of a large or complicated rocket vehicle, or in leading up to a captive test, a readiness firing, a mock firing, or other firing test. **2.** The act of counting inversely during this process. [COMMUN] The ratio of the number of interrogation pulses not answered by a transponder to the total number received. [ENG] A step-by-step process that culminates in a climatic event, each step being performed in accordance with a schedule marked by a count in inverse numerical order.

counter [ADP] **1.** A register or storage location used to represent the number of occurrences of an event. **2.** *See* accumulator. [ELECTR] *See* scaler. [ENG] A complete instrument for detecting, totalizing, and indicating a sequence of events. [NAV ARCH] *See* buttocks. [NUCLEO] *See* radiation counter.

counterbalance *See* counterweight.

counterblow hammer [MECH ENG] A forging hammer in which the ram and anvil are driven toward each other by compressed air or steam.

counterbore [DES ENG] A flat-bottom enlargement of the mouth of a cylindrical bore to enlarge a borehole and give it a flat bottom. [ENG] To enlarge a borehole by means of a counterbore.

counter circuit *See* counting circuit.

counter-countermeasures *See* electronic counter-countermeasures.

counter coupling [ADP] The technique of combining two or more counters into one counter of larger capacity in electro-mechanical devices by means of control panel wiring.

countercurrent [SCI TECH] A current flowing adjacent to the main current but in the opposite direction.

countercurrent distribution [CHEM ENG] A profile of a compound's concentration in different ratios of two immiscible liquids.

countercurrent extraction [CHEM ENG] A liquid-liquid extraction process in which the solvent and the process stream in contact with each other flow in opposite directions. Also known as countercurrent separation.

countercurrent flow [MECH ENG] A sensible heat-transfer system in which the two fluids flow in opposite directions.

countercurrent leaching [CHEM ENG] A process utilizing a series of leach tanks and countercurrent flow of solvent through them in reverse order to the flow of solid.

countercurrent separation *See* countercurrent extraction.

countercurrent spray dryer [ENG] A dryer in which drying gases flow in a direction opposite to that of the spray.

counter dead time [NUCLEO] The time interval between the start of a counted event and the earliest instant at which a new event can be counted by a radiation counter.

counter decade *See* decade scaler.

counterelectromotive cell [ELEC] Cell of practically no ampere-hour capacity, used to oppose the line voltage.

counterelectromotive force [ELECTROMAG] The voltage developed in an inductive circuit by a changing current; the polarity of the induced voltage is at each instant opposite that of the applied voltage. Also known as back electromotive force.

counterfire [ORD] **1.** Fire delivered in answer to the fire of an attacker. **2.** Fire intended to destroy or neutralize enemy weapons.

counterflow [ENG] Fluid flow in opposite directions in adjacent parts of an apparatus, as in a heat exchanger.

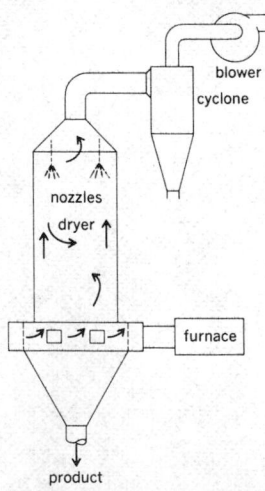

**COUNTERCURRENT
SPRAY DRYER**

Liquid feed is atomized from nozzles at the top; arrows show the path that the drying gases take.

counterfort [CIV ENG] A strengthening pier perpendicular and bonded to a retaining wall.

counterfort wall [CIV ENG] A type of retaining wall that resembles a cantilever wall but has braces at the back; the toe slab is a cantilever and the main steel is placed horizontally.

counterglow *See* gegenschein.

counterlath [BUILD] **1.** A strip placed between two rafters to support crosswise laths. **2.** A lath placed between a timber and a sheet lath. **3.** A lath nailed at a more or less random spacing between two precisely spaced laths. **4.** A lath put on one side of a partition after the other side has been finished.

countermeasures [ORD] Devices and techniques intended to impair the operational effectiveness of enemy activity.

countermeasures set [ELECTR] A complete electronic set specifically designed to provide facilities for intercepting and analyzing electromagnetic energy propagated by transmitter and to provide a source of radio-frequency signals which deprive the enemy of effective use of his electronic equipment.

counterpoise [ELEC] A system of wires or other conductors that is elevated above and insulated from the ground to form a lower system of conductors for an antenna. Also known as antenna counterpoise. [MECH ENG] *See* counterweight.

counterpoise gun carriage [ORD] The mechanism that counterbalances the weight of the breechblock of a large gun.

counterradiation [GEOPHYS] The downward flux of atmospheric radiation passing through a given level surface, usually taken as the earth's surface. Also known as back radiation.

counterrecoil [ORD] Forward movement of a gun returning to firing position after recoil.

countershaft [MECH ENG] A secondary shaft that is driven by a main shaft and from which power is supplied to a machine part.

countersink [DES ENG] The tapered and relieved cutting portion in a twist drill, situated between the pilot drill and the body.

countersinking [MECH ENG] Drilling operation to form a flaring depression around the rim of a hole.

counterstain [BIOL] A second stain applied to a biological specimen to color elements other than those demonstrated by the principal stain.

counter sun *See* anthelion.

countertransference [PSYCH] The conscious or unconscious emotional reaction of the therapist to the patient, which may interfere with psychotherapy.

counter tube [ELECTR] An electron tube having one signal-input electrode and 10 or more output electrodes, with each input pulse serving to transfer conduction sequentially to the next output electrode; beam-switching tubes and cold-cathode counter tubes are examples. Also known as counting tube; decade counter tube. [NUCLEO] An electron tube that converts an incident particle or burst of incident radiation into a discrete electric pulse, generally by utilizing the current flow through a gas that is ionized by the radiation; used in radiation counters. Also known as radiation counter tube.

counterweight [MECH ENG] **1.** A device which counterbalances the original load in elevators and skip and mine hoists, going up when the load goes down, so that the engine must only drive against the unbalanced load and overcome friction. **2.** Any weight placed on a mechanism which is out of balance so as to maintain static equilibrium. Also known as counterbalance; counterpoise.

counting chamber [MICROBIO] An accurately dimensioned chamber in a microslide which can hold a specific volume of fluid and which is usually ruled into units to facilitate the counting under the microscope of cells, bacteria, or other structures in the fluid.

counting circuit [ELECTR] A circuit that counts pulses by frequency-dividing techniques, by charging a capacitor in such a way as to produce a voltage proportional to the pulse count, or by other means. Also known as counter circuit.

counting-down circuit *See* frequency divider.

counting glass [TEXT] A magnifying glass for counting the number of threads per inch in textile fabrics.

counting rate [PHYS] The average rate of occurrence of events as observed by means of a counting system.

counting rate meter [NUCLEO] An instrument that indicates the time rate of occurrence of input pulses to a radiation counter, averaged over a time interval. Also known as rate meter.

counting rate-voltage characteristic *See* plateau characteristic.

counting tube *See* counter tube.

country rock [GEOL] **1.** Rock that surrounds and is penetrated by mineral veins. **2.** Rock that surrounds and is invaded by an igneous intrusion.

count system *See* numerical system.

couple [CHEM] Joining of two molecules. [ELEC] To connect two circuits so signals are transferred from one to the other. [ELECTR] Two metals placed in contact, as in a thermocouple. [ENG] To connect with a coupling, such as of two pipes or two pipes. [MECH] A system of two parallel forces of equal magnitude and opposite sense.

coupled antenna [ELECTROMAG] An antenna electromagnetically coupled to another.

coupled circuits [ELEC] Two or more electric circuits so arranged that energy can transfer electrically or magnetically from one to another.

coupled column [ARCH] One of two columns used as a grouped pair.

coupled engine [MECH ENG] A locomotive engine having the driving wheels connected by a rod.

coupled field vectors [ELECTROMAG] The electric- and magnetic-field vectors, which depend upon each other according to Maxwell's field equations.

coupled harmonic oscillators [PHYS] Linear oscillators with an interaction, often also linear or weak.

coupled modes [ACOUS] Modes of acoustic transmission along a duct having a discontinuity, so that the reflected and transmitted waves contain modes other than the incident ones.

coupled oscillators [ELECTROMAG] A set of alternating-current circuits which interact with each other, for example, through mutual inductances or capacitances. [MECH] A set of particles subject to elastic restoring forces and also to elastic interactions with each other.

coupled transistors [ELECTR] Transistors connected in series by transformers or resistance-capacitance networks, in much the same manner as electron tubes.

coupler [ELEC] A component used to transfer energy from one circuit to another. [ELECTROMAG] **1.** A passage joining two cavities or waveguides, allowing them to exchange energy. **2.** A passage joining the ends of two waveguides, whose cross section changes continuously from that of one to that of the other. [ENG] A device that connects two railroad cars. [NAV] The portion of a navigation system that receives signals of one type from a sensor and transmits signals of a different type to an actuator.

coupling [ELEC] **1.** A mutual relation between two circuits that permits energy transfer from one to another, through a wire, resistor, transformer, capacitor, or other device. **2.** A hardware device used to make a temporary connection between two wires. [ENG] Any device that serves to connect the ends of adjacent parts, as railroad cars. [MECH ENG] The mechanical fastening that connects shafts together for power transmission. Also known as shaft coupling.

coupling aperture [ELECTROMAG] An aperture in the wall of a waveguide or cavity resonator, designed to transfer energy to or from an external circuit. Also known as coupling hole; coupling slot.

coupling capacitor [ELECTR] A capacitor used to block the flow of direct current while allowing alternating or signal current to pass; widely used for joining two circuits or stages. Also known as blocking capacitor; stopping capacitor.

coupling coefficient [ELECTR] The ratio of the maximum change in energy of an electron traversing an interaction space to the product of the peak alternating gap voltage and the electronic charge. [PHYS] *See* coupling constant.

coupling constant [PARTIC PHYS] A measure of the strength of a type of interaction between particles, such as the strong interaction between mesons and nucleons, and the weak interaction between four fermions; analogous to the electric charge, which is the coupling constant between charged

particles and electromagnetic radiation. [PHYS] **1.** A measure of the strength of the coupling between two systems, especially electric circuits; maximum coupling is 1 and no coupling is 0. Also known as coefficient of coupling; coupling coefficient. **2.** A measure of the dependence of one physical quantity on another.

coupling hole *See* coupling aperture.

coupling loop [ELECTROMAG] A conducting loop projecting into a waveguide or cavity resonator, designed to transfer energy to or from an external circuit.

coupling probe [ELECTROMAG] A probe projecting into a waveguide or cavity resonator, designed to transfer energy to or from an external circuit.

coupling slot *See* coupling aperture.

coupons [CHEM ENG] Polished metal strips of specified size and weight used to detect the corrosive action of liquid or gas products or to test the efficiency of corrosion-inhibitor additives.

course [NAV] The intended direction of travel expressed as an angle in the horizontal plane between a reference line (true magnetic north) and the course line (the line connecting the point of origin and the point of destination), usually measured clockwise from the reference line. Also known as desired track.

course angle [NAV] Course measured from 0° at the reference direction clockwise or counterclockwise through 90° or 180°; it is labeled with the reference direction as a prefix and the direction of measurement from the reference direction as a suffix; for example, course angle S 21° E is 21° east of south, or course 159°.

coursed rubble [CIV ENG] Masonry in which rough stones are fitted into approximately level courses.

course error [NAV] Angular difference between the course and the course made good.

course line [NAV] **1.** A line of position plotted on a chart, parallel or substantially parallel to the intended course of a craft, showing whether the craft is to the right or the left of its course. **2.** Any line representing a course.

course linearity [NAV] In equisignal radio-range-type beacons, the change in the difference in the depth of modulation (DDM) of the two signals which produce the course with respect to displacement of the measuring position from the course line but within the course sector.

course-line computer [NAV] An airborne computer that accepts bearing and distance information derived from ground facilities and uses this data to compute a course from the aircraft's present position to any other point which the pilot selects, providing only that it be within the coverage of the ground facilities; the steering information is displayed on a right-left indicator, while additional displays such as distance to go are also provided. Also known as arbitrary course computer; bearing distance computer; off-line computer; parallel course computer; rho-theta computer.

course-line deviation [NAV] The angular or linear difference between the actual track of a vehicle and the intended course line.

course-line deviation indicator [NAV] An instrument that indicates deviation from a desired course line.

course-line selector [NAV] A device providing means for selecting a course to be followed automatically, usually by means of an electronic system of navigation such as omnirange and distance-measuring equipment.

course made good [NAV] The resultant direction of actual travel of a vehicle, equivalent to its bearing from the point of departure.

course over the ground [NAV] The direction of the track that a vessel has actually made, measured clockwise from north through 360°.

course programmer [CONT SYS] An item which initiates and processes signals in a manner to establish a vehicle in which it is installed along one or more projected courses.

course recorder [NAV] A device which makes a graphic record of the headings and distances traveled.

course scalloping [NAV] Irregularities in the field pattern produced by a ground-based beacon caused by terrain features; appears while in flight as cyclical variation in the course error.

COUPLED CIRCUITS

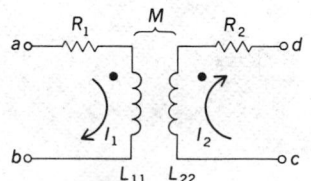

A pair of coupled circuits.

COURSE-LINE COMPUTER

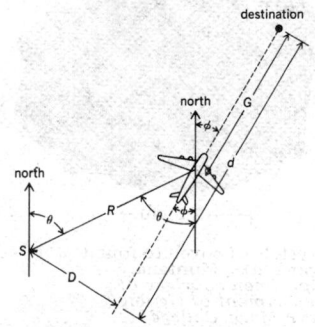

D = shortest distance from transponder to desired course
R = distance from transponder to aircraft
d = distance from perpendicular to destination
G = distance to go
ϕ = bearing of desired course to destination
θ = aircraft bearing from transponder

Arbitrary-course problem. With facility at S, it is desired to generate courses in any arbitrary direction.

course softening [NAV] An intentional decrease in course sensitivity as the navigation aid is approached.

coursing joint [CIV ENG] A mortar joint connecting two courses of brick or pebble.

Couvinian [GEOL] Lower Middle Devonian geologic time.

covalence [CHEM] The number of covalent bonds which an atom can form.

covalent bond [CHEM] A bond in which each atom of a bound pair contributes one electron to form a pair of electrons. Also known as electron pair bond.

covalent crystal [CRYSTAL] A crystal held together by covalent bonds. Also known as valence crystal.

covalent hydride [INORG CHEM] A compound formed from a nonmetal and hydrogen, for example, H_2S and NH_3.

covalent radius [PHYS CHEM] The effective radius of an atom in a covalent bond.

covariance [STAT] A measurement of the tendency of two random variables, X and Y, to vary together, given by the expected value of the variable $(X-\overline{X})(Y-\overline{Y})$, where $\overline{X}$ and $\overline{Y}$ are the expected values of the variables X and Y respectively.

covariant [RELAT] A scalar, vector, or higher-order tensor.

covariant components [MATH] Vector or tensor components which, in a transformation from one set of basis vectors to another, transform in the same manner as the basis vectors.

covariant functor [MATH] A functor which does not change the sense of morphisms.

covariant index [MATH] A tensor index such that, under a transformation of coordinates, the procedure for obtaining a component of the transformed tensor for which this index has value p involves taking a sum over q of the product of a component of the original tensor for which the index has the value q times the partial derivative of the qth original coordinate with respect to the pth transformed coordinate; it is written as a subscript.

covariant tensor [MATH] A tensor with only covariant indices.

covariant theory [PHYS] A theory in which the equations have the same form in any inertial reference frame, the frames being related to each other by Lorentz transformations.

covariant vector [MATH] A covariant tensor of degree 1, such as the gradient of a function.

covellite [MINERAL] CuS An indigo-blue mineral of metallic luster that crystallizes in the hexagonal system; it is usually massive or occurs in disseminations through other copper minerals and represents an ore of copper. Also known as indigo copper.

cover [MIN ENG] The thickness of rock between the mine workings and the surface.

coverage [COMMUN] *See* service area. [GRAPHICS] In microfilming, the portion of a document plane included in the lens field.

cover crops [AGR] Crops, especially grasses, grown for the express purpose of preventing and protecting a bare soil surface.

covered electrode *See* coated electrode.

covered smut [PL PATH] A seed-borne smut of certain grain crops caused by *Ustilago hordei* in barley and *U. avenae* in oats.

cover half [MET] The stationary portion of a die.

cover hole [MIN ENG] One of a group of boreholes drilled in advance of mine workings to probe for and detect water-bearing fissures or structures.

covering [MATH] A set of subsets such that any point of the set belongs to at least one of the subsets.

covering of a set [MATH] A collection of sets whose union contains the given set. Also known as cover of a set.

covering power [ENG] The degree to which a coating obscures the underlying material. [MET] The ability of an electroplating bath to produce a coating at low current density. [OPTICS] The field of view over which a camera lens can produce a sharp image, frequently expressed as an angle.

cover of a set *See* covering of a set.

covers *See* coversed sine.

coversed sine [MATH] The coversed sine of A is $1 -$ sine A. Denoted covers. Also known as coversine; versed cosine.

coversine *See* coversed sine.

covert [ECOL] A refuge or shelter, such as a coppice, for game animals. [TEXT] A tightly woven woolen twill fabric

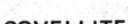

COVELLITE

$\vdash$ 2.5 cm $\dashv$

Crystals of covellite found near Butte, Montana. *(Specimen courtesy of Department of Geology, Bryn Mawr College)*

made by using a single-color yarn for filling threads and yarns in two different shades in the warp.

covey [VERT ZOO] **1.** A brood of birds. **2.** A small flock of birds of one kind, used typically of partridge and quail.

covite [PETR] A rock of igneous origin composed of sodic orthoclase, hornblende, sodic pyroxene, nepheline, and accessory sphene, apatite, and opaque oxides.

cow [AGR] A domestic bovine of any sex or age. [VERT ZOO] A mature female cattle of the genus *Bos*.

Cowell method [AERO ENG] A method of orbit computation using direct step-by-step integration in rectangular coordinates of the total acceleration of the orbiting body.

cowling [AERO ENG] The streamlined metal cover of an aircraft engine. [ENG] A metal cover that houses an engine.

cowpea [BOT] *Vigna sinensis.* An annual legume in the order Rosales cultivated for its edible seeds. Also known as black-eye bean.

Cowper's gland *See* bulbourethral gland.

cowpox *See* vaccinia.

cowpox virus [VIROL] The causative agent of cowpox in cattle.

cowshee *See* kaus.

coxa [INV ZOO] The proximal or basal segment of the leg of insects and certain other arthropods which articulates with the body.

coxal cavity [INV ZOO] A cavity in which the coxa of an arthropod limb articulates.

coxal gland [INV ZOO] One of certain paired glands with ducts opening in the coxal region of arthropods.

Cox chart [CHEM] A straight-line graph of the logarithm of vapor pressure against a special nonuniform temperature scale; vapor pressure–temperature lines for many substances intersect at a common point on the Cox chart.

Coxiella [MICROBIO] A genus of small, gram-negative, rod-shaped or coccoid intracellular parasitic bacteria in the family Rickettsiaceae; includes the etiological agent of Q fever.

coxitis [MED] Inflammation of the hip joint.

coxopodite [INV ZOO] The basal joint of a crustacean limb.

coxsackie disease [MED] A variety of syndromes resulting from a coxsackievirus infection.

coxsackievirus [VIROL] A large subgroup of the enteroviruses in the picornavirus group including various human pathogens.

coyote [VERT ZOO] *Canis latrans.* A small wolf native to western North America but found as far eastward as New York State. Also known as prairie wolf.

coyote blasting [MIN ENG] A method of blasting in which large charges are fired in small adits or tunnels driven at the level of the floor, in the face of a quarry or the slope of an open-pit mine. Also known as coyote-hole blasting; gopher-hole blasting; heading blasting.

coyote hole *See* gopher hole.

coyote-hole blasting *See* coyote blasting.

COZI [COMMUN] An ionospheric sounding system for determining propagation characteristics of the ionosphere at various angles at any instant; used to determine how well long-distance, high-frequency broadcasts are reaching their intended destinations. Derived from communications zone indicator.

cp *See* candlepower; centipoise; chemically pure.

cpa *See* color-phase alternation.

CPE *See* computer performance evaluation.

CP invariance [PARTIC PHYS] The principle that the laws of physics are left unchanged by a combination of the operations of charge conjugation C and space inversion P; a small violation of this principle has been observed in the decay of neutral K-mesons.

cpm *See* cycle per minute.

CPM *See* critical path method.

C power supply [ELECTR] A device connected in the circuit between the cathode and grid of a vacuum tube to apply grid bias.

CP propeller *See* controllable-pitch propeller.

cps *See* hertz.

CPT theorem [PARTIC PHYS] A theorem which states that a Lorentz invariant field theory is invariant to the product of charge conjugation C, space inversion P, and time reversal T.

CPU *See* central processing unit.

Cr *See* chromium.

CR *See* catalytic reforming.

CrA *See* Corona Australis.

crab [INV ZOO] 1. The common name for a number of crustaceans in the order Decapoda having five pairs of walking legs, with the first pair modified as chelipeds. 2. The common name for members of the Merostoma. [NAV] To drift sideways or to leeward, as a ship.

crabapple [BOT] Any of several trees of the genus *Malus*, order Rosales, cultivated for their small, edible pomes.

crabbing [NAV] The horizontal attitude of an aircraft in flight when a crosswind causes its heading to differ from the course. [TEXT] A finishing process that sets warp and weft threads by winding the fabric on a roller under tension and then boiling or steaming.

crab locomotive [MIN ENG] A type of trolley locomotive equipped with an electric motor, a drum, and haulage cable mounted on a small truck; used to haul mine cars from workings.

Crab Nebula [ASTRON] A gaseous nebula in the constellation Taurus; an amorphous mass which radiates a continuous spectrum involved in a mesh of filaments that radiate a bright-line spectrum.

crachin [METEOROL] A period of light rain accompanied by low stratus clouds and poor visibility which frequently occurs in the China Sea between January and April.

Cracidae [VERT ZOO] A family of New World tropical upland game birds in the order Galliformes; includes the chachalacas, guans, and curassows.

crack [CHEM] To break a compound into simpler molecules. [ENG] To open slightly, for instance, a valve. [SCI TECH] A fissure.

crack arrester [NAV ARCH] 1. On a ship, a plate riveted over another plate where the latter has a crack or a stressed area where a crack might begin. 2. A hole or slot formed at the end of a crack to keep it from spreading, or in a stressed area to prevent a crack from beginning.

cracked [MATER] Applied to those oils produced by the cracking process rather than straight distillation.

cracked gasoline [MATER] Gasoline manufactured by heating crude petroleum distillation fractions or residues under pressure, or by heating with or without pressure in the presence of a catalyst, so that heavier hydrocarbons are broken into others, some of which distill in the gasoline range.

cracked stem [PL PATH] A boron-deficiency disease of celery characterized by brown mottling of leaves and brittleness and cracking of leaf stalks.

cracking [CHEM ENG] A process used to reduce the molecular weight of hydrocarbons by breaking molecular bonds by thermal, catalytic, or hydrocracking methods. [ENG] Presence of relatively large cracks extending into the interior of a structure, usually produced by overstressing the structural material.

cracking coil [CHEM ENG] A coil used for cracking heavy petroleum products.

cracking still [CHEM ENG] The furnace, reaction chamber, and fractionator for thermal conversion of heavier charging stock to gasoline.

cradle [ENG] A framework or other resting place for supporting or restraining objects. [ORD] The nonrecoiling structure of a weapon that houses the recoiling parts and rotates to elevate the gun. [TEXT] A device that catches the cards as they fall from a jacquard head.

cradle cap [MED] Heavy, greasy crusts on the scalp of an infant; seborrheic dermatitis of infants.

cradle dump [MIN ENG] A tipple which dumps cars with a rocking motion.

cradle printing [GRAPHICS] Early printing of a crude sort done from movable types. Also known as incunabula printing.

crag [GEOL] A steep, rugged point or eminence of rock, as one projecting from the side of a mountain.

Cramblinae [INV ZOO] The snout moths, a subfamily of lepidopteran insects in the family Pyralididae containing small marshland and grassland forms.

Cramer's rule [MATH] The method of solving a system of linear equations by means of determinants.

cramp [DES ENG] A metal plate with bent ends used to hold blocks together. [MED] 1. Painful, involuntary contraction of a muscle, such as a leg or foot cramp that may occur in normal individuals at night or in swimming. 2. Any cramplike pain, as of the intestine, or that accompanying dysmenorrhea. 3. Spasm of certain muscles, which may be intermittent or constant, from excessive use.

crampon [DES ENG] A device for holding heavy objects such as rock or lumber to be lifted by a crane or hoist; shaped like scissors, with points bent inward for grasping the load. Also spelled crampoon.

crampoon *See* crampon.

cranberry [BOT] Any of several plants of the genus *Vaccinium*, especially *V. macrocarpon*, in the order Ericales, cultivated for its small, edible berries.

Cranchiidae [INV ZOO] A family of cephalopod mollusks in the subclass Dibranchia.

crandallite [MINERAL] $CaAl_3(PO_4)_2(OH)_5 \cdot H_2O$ A white to light-grayish mineral consisting of a hydrous phosphate of calcium and aluminum occurring in fine, fibrous masses.

crane [MECH ENG] A hoisting machine with a power-operated inclined or horizontal boom and lifting tackle for moving loads vertically and horizontally. [VERT ZOO] The common name for the long-legged wading birds composing the family Gruidae of the order Gruiformes.

Crane *See* Grus.

crane hoist [MECH ENG] A mobile construction machine built principally for lifting loads by means of cables and consisting of an undercarriage on which the unit moves, a cab or house which envelops the main frame and contains the power units and controls, and a movable boom over which the cables run.

crane hook [DES ENG] A hoisting fixture designed to engage a ring or link of a lifting chain, or the pin of a shackle or cable socket.

crane ship [NAV ARCH] A ship or barge that carries cranes for handling heavy loads.

crane truck [MECH ENG] A crane with a jiblike boom mounted on a truck. Also known as yard crane.

Craniacea [INV ZOO] A family of inarticulate branchiopods in the suborder Craniidina.

cranial capacity [ANAT] The volume of the cranial cavity.

cranial flexure [EMBRYO] A flexure of the embryonic brain.

cranial fossa [ANAT] Any of the three depressions in the floor of the interior of the skull.

cranial index [ANTHRO] The ratio of maximum skull height to maximum skull breadth multiplied by 100.

cranial nerve [ANAT] Any of the paired nerves which arise in the brainstem of vertebrates and pass to peripheral structures through openings in the skull.

Craniata [VERT ZOO] A major subdivision of the phylum Chordata comprising the vertebrates, from cyclostomes to mammals, distinguished by a cranium.

craniectomy [MED] Surgical removal of strips or pieces of the cranial bones.

Craniidina [INV ZOO] A subdivision of inarticulate branchiopods in the order Acrotretida known to possess a pedicle; all forms are attached by cementation.

craniobuccal pouch [EMBRYO] A diverticulum from the buccal cavity in the embryo from which the anterior lobe of the hypophysis is developed. Also known as Rathke's pouch.

cranioclasis [MED] The operation of breaking the fetal head by means of a cranioclast.

cranioclast [MED] A heavy forceps for crushing the fetal head.

craniofacial index [ANTHRO] The ratio of the width of the cranium to the width of the face.

craniometer [ANTHRO] An instrument used to measure a skull.

craniometry [ANTHRO] The science of measuring the skull, especially for determining characteristics of a particular race, sex, or somatotype.

craniopharyngioma [MED] An epithelial tumor of the craniopharyngeal canal, usually in children.

cranioplasty [MED] Surgical correction of defects in the cranial bones, usually by implants of metal, plastic material, or bone.

cranioscopy [MED] Examination of the human skull.

CRAB

Male fiddler crab (*Uca pugnax*), with an enlarged claw.

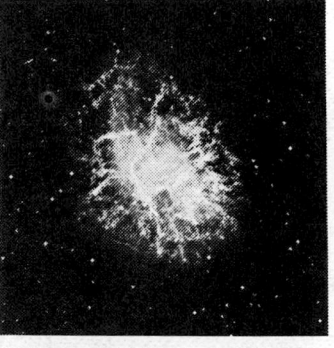

CRAB NEBULA

Crab Nebula, in the constellation Taurus, emitter of strong radio waves and of x-rays.

CRANE HOOK

A typical crane hook.

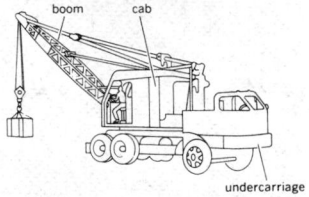

CRANE TRUCK

A truck crane lifting a load. (Link-Belt Co.)

CRANIACEA

Pedicle valve of *Crania*, a genus in Craniacea. (From R. C. Moore, ed., Treatise on Invertebrate Paleontology, pt. H, Geological Society of America, Inc., and University of Kansas Press, 1965)

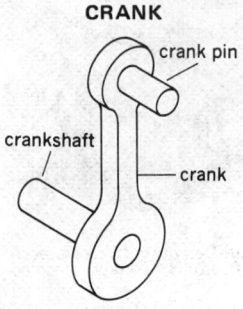

CRANK

crank pin

crankshaft

crank

A crank for changing radius of rotation.

cranium [ANAT] That portion of the skull enclosing the brain. Also known as braincase.

crank [MECH ENG] A link in a mechanical linkage or mechanism that can turn about a center of rotation.

crank angle [MECH ENG] 1. The angle between a crank and some reference direction. 2. Specifically, the angle between the crank of a slider crank mechanism and a line from crankshaft to the piston.

crank arm [MECH ENG] The arm of a crankshaft attached to a connecting rod and piston.

crank axle [MECH ENG] 1. An axle containing a crank. 2. An axle bent at both ends so that it can accommodate a large body with large wheels.

crankcase [MECH ENG] The housing for the crankshaft of an engine, where, in the case of an automobile, oil from hot engine parts is collected and cooled before returning to the engine by a pump.

crankpin [DES ENG] A cylindrical projection on a crank which holds the connecting rod.

crank press [MECH ENG] A punch press that applies power to the slide by means of a crank.

crankshaft [MECH ENG] The shaft about which a crank rotates.

crank throw [MECH ENG] 1. The web or arm of a crank. 2. The displacement of a crankpin from the crankshaft.

crank web [MECH ENG] The arm of a crank connecting the crankshaft to crankpin, or connecting two adjacent crankpins.

crash [TEXT] Rough-textured wool, cotton, polyester, linen, or other fabric made by weaving together thick, uneven yarns.

crash locator beacon [COMMUN] An automatic radio beacon carried in aircraft to guide searching forces in the event of a crash.

craspedon [INV ZOO] A coelenterate medusa stage possessing a velum.

craspedote [INV ZOO] Having a velum, used specifically for velate hydroid medusae.

Crassulaceae [BOT] A family of dicotyledonous plants in the order Rosales notable for their succulent leaves and resistance to desiccation.

crater [GEOL] 1. A large, bowl-shaped topographic depression with steep sides. 2. A rimmed structure at the summit of a volcanic cone; the floor is equal to the vent diameter. [MECH ENG] A depression in the face of a cutting tool worn down by chip contact. [MET] A depression at the end of the weld head or under the electrode during welding.

Crater [ASTRON] A constellation, right ascension 11 hours, declination 15°S. Abbreviated Crt. Also known as Cup.

crater cone [GEOL] A cone built around a volcanic vent by lava extruded from the vent.

crater cuts [MIN ENG] Cuts with one or more fully charged holes in which blasting is conducted toward the face of the tunnel.

crater lake [HYD] A fresh-water lake formed by the accumulation of rain and groundwater in a caldera or crater.

crater lamp [ELECTR] A glow-discharge tube used as a point source of light whose brightness is proportional to the signal current sent through the tube; used for photographic recording of facsimile signals.

craton [GEOL] The large, relatively immobile portion of continents consisting of both shield and platform areas.

craw [ZOO] 1. The crop of a bird or insect. 2. The stomach of a lower animal.

crawler [MECH ENG] 1. One of a pair of an endless chain of plates driven by sprockets and used instead of wheels by certain power shovels, tractors, bulldozers, drilling machines, and such, as a means of propulsion. 2. Any machine mounted on such tracks.

crawler crane [MECH ENG] A self-propelled crane mounted on two endless tracks that revolve around wheels.

crawler tractor [MECH ENG] A tractor that propels itself on two endless tracks revolving around wheels.

crawler wheel [MECH ENG] A wheel that drives a continuous metal belt, as on a crawler tractor.

crawl space [BUILD] A shallow space in a building which workers can enter to gain access to pipes, wires, and equipment.

crayfish [INV ZOO] The common name for a number of lobsterlike fresh-water decapod crustaceans in the section Astacura.

crayon [GRAPHICS] A small stick for drawing, usually made of a combination of pigments or dyes in a wax or oil medium.

crazing [ENG] Cracking of an enamel, glaze, or varnish surface. [MET] Development of a network of cracks on a metal surface.

CrB See Corona Borealis.

CRC method [CHEM ENG] A research technique formulated, approved, and published by the Coordinating Research Council, Inc.

cream ice See sludge.

creatine [BIOCHEM] $C_4H_9O_2N_3$ α-Methylguanidine-acetic acid; a compound present in vertebrate muscle tissue, principally as phosphocreatine.

creatinine [BIOCHEM] $C_4H_7ON_3$ A compound present in urine, blood, and muscle that is formed from the dehydration of creatine.

creatinuria [MED] The occurrence of creatine in the urine.

creation operator [QUANT MECH] An operator which increases the occupation number of a single state by unity and leaves all the other occupation numbers unchanged.

credence [ELECTROMAG] In radar, a measure of confidence in a target detection, generally proportional to target return amplitude.

Credé procedure [MED] Installation of silver nitrate drops into the eyes of a newborn infant to prevent ophthalmia neonatorum.

crednerite [MINERAL] $CuMn_2O_4$ A steel-gray to iron-black foliated mineral consisting of copper, manganese, and oxygen.

creedite [MINERAL] $Ca_3Al_2(SO_4)(F,OH)_{10} \cdot 2H_2O$ A white or colorless monoclinic mineral consisting of hydrous calcium aluminum fluoride with calcium sulfate, occurring in grains and radiating crystalline masses; hardness is 2 on Mohs scale, and specific gravity is 2.7.

creep [ELECTR] A slow change in a characteristic with time or usage. [GEOL] A slow, imperceptible downward movement of slope-forming rock or soil under sheer stress. [MECH] A time-dependent strain of solids caused by stress. [MIN ENG] See squeeze.

creepage [ELEC] The conduction of electricity across the surface of a dielectric.

creeper [MIN ENG] An endless chain that catches mine car axles on projecting bars.

creeping barrage [ORD] An artillery barrage that precedes infantry soldiers at a predetermined rate in their attack to protect them and facilitate their advance.

creeping disk [INV ZOO] The smooth and adhesive undersurface of the foot or body of a mollusk or of certain other invertebrates, on which the animal creeps.

creeping eruption [MED] A red line of eruption on the skin produced by larva burrowing in the dermis; characterizes the condition of larva migrans.

creep limit [MECH] The maximum stress a given material can withstand in a given time without exceeding a specified quantity of creep.

creep recovery [MECH] Strain developed in a period of time after release of load in a creep test.

creep rupture strength [MECH] The stress which, at a given temperature, will cause a material to rupture in a given time.

creep strength [MECH] The stress which, at a given temperature, will result in a creep rate of 1% deformation in 100, 000 hours.

creep test [ENG] Any one of a number of methods of measuring creep, for example, by subjecting a material to a constant stress or deforming it at a constant rate.

cremnophobia [PSYCH] An abnormal fear of steep places or precipices.

cremocarp [BOT] A dry dehiscent fruit consisting of two indehiscent one-seeded mericarps which separate at maturity and remain pendant from the carpophore.

cremone bolt [DES ENG] A fastening for double doors or casement windows; employs vertical rods that move up and down to engage the top and bottom of the frame.

crenate [BIOL] Having a scalloped margin; used specifically for foliar structures, shrunken erythrocytes, and shells of certain mollusks.

crenation [PHYSIO] A notched appearance of shrunken erythrocytes; seen when they are exposed to the air or to strong saline solutions.

crenitic [GEOL] Relating to or resulting from the raising of subterranean minerals by the action of spring water.

Crenotrichaceae [MICROBIO] A family of bacteria in the order Chlamydobacteriales having trichomes that are differentiated at the base and tip and attached to a firm substrate.

crenulate [BIOL] Having a minutely crenate margin.

crenulation cleavage See slip cleavage.

Creodonta [PALEON] A group formerly recognized as a suborder of the order Carnivora.

creosol [ORG CHEM] $CH_3C_6H_4OH$ A combination of isomers, derived from coal tar or petroleum; a yellowish liquid with a phenolic odor; used as a disinfectant, in the manufacture of resins, and flotation of ore. Also known as hydroxymethylbenzene; methyl phenol.

creosote [MATER] A colorless or yellowish oily liquid containing a mixture of phenolic compounds obtained by distillation of tar; commercial creosote is distilled from coal tar, and pharmaceutical creosote is distilled from wood tar.

creosote bush [BOT] Larrea divaricata. A bronze-green, xerophytic shrub characteristic of all the American warm deserts.

creosote oil [MATER] A coal tar fraction, boiling between 240 and 270°F (116–132°C); used for producing materials such as creosote and tar acids and used directly as a germicide, insecticide, or pesticide.

crepe [TEXT] A silk, polyester, wool, rayon, or other fabric with a crinkled surface obtained by using yarns twisted alternately right and left in the filling.

crepe paper [MATER] A lightweight, crinkled paper available in many colors and used for displays, floats, and decorations; it has no strength when wet, and the colors run.

crepitation [MED] A noise produced by the rubbing of fractured ends of bones, by cracking joints, and by pressure upon tissues containing abnormal amounts of air, as in cellular emphysema.

crepuscular [ZOO] Active during the hours of twilight or preceding dawn.

crepuscular arch See bright segment.

crepuscular rays [ASTRON] Streaks of light radiating from the sun shortly before and after sunset which shine through breaks in the clouds or through irregular spaces along the horizon.

crescent beach [GEOL] A crescent-shaped beach at the head of a bay or the mouth of a stream entering the bay, with the concave side facing the sea.

crescent beam [ENG] A beam bounded by arcs having different centers of curvature, with the central section the largest.

crescentic dune See barchan.

crescentic lake See oxbow lake.

Creslan [TEXT] Trade name of the American Cyanamid Company for an acrylic fiber which comes in staple and continuous filaments and to which basic materials are introduced to form sites for anionic dyes.

cresol [ORG CHEM] $CH_3O(CH_3)C_6H_3OH$ One of three poisonous, colorless isomeric methyl phenols: o-cresol, m-cresol, p-cresol; used in the production of phenolic resins, tricresyl phosphate, disinfectants, and solvents.

cresol red [ORG CHEM] $C_{21}H_{18}O_5S$ A compound derived from o-cresol and used as an acid-base indicator; color change is yellow to red at pH 0.4 to 1.8, or 7.0 to 8.8, depending on preparation. Also known as ortho-cresolsulfonphthalein.

ortho-cresolsulfonphthalein See cresol red.

cress [BOT] Any of several prostrate crucifers belonging to the order Capparales and grown for their flavorful leaves; includes watercress (Nasturtium officinale), garden cress (Lepidium sativum), and upland or spring cress (Barbarea verna).

crest [DES ENG] The top of a screw thread. [SCI TECH] The highest point of a structure or natural formation, such as the top edge of a dam, the ridge of a roof, the highest point of a gravity wave, or the highest natural projection of a hill or mountain.

crestal injection See external gas injection.

crestal plane [GEOL] The plane formed by joining the crests of all beds of an anticline.

crest clearance [DES ENG] The clearance, in a radial direction, between the crest of the thread of a screw and the root of the thread with which the screw mates.

crest cloud [METEOROL] A type of standing cloud which forms along a mountain ridge, either on the ridge, or slightly above and leeward of it, and remains in the same position relative to the ridge. Also known as cloud crest.

crest factor [PHYS] The ratio of the peak value to the effective value of any periodic quantity such as a sinusoidal alternating current. Also known as amplitude factor; peak factor.

crest gate [CIV ENG] A gate in the spillway of a dam which functions to maintain or change the water level.

crest line [GEOL] The line connecting the highest points on the same bed of an anticline in an infinite number of cross sections.

crest stage [HYD] The highest stage reached at a point along a stream culminating a rise by waters of that stream.

crest value See peak value.

crest voltmeter [ELEC] A voltmeter reading the peak value of the voltage applied to its terminals.

cresylic acid [MATER] 1. A mixture of phenols containing varying amounts of xylenols, cresols, and other high-boiling fractions. 2. A crude mixture of the three cresol isomers.

para-cresyl methyl ether See methyl para-cresol.

Cretaceous [GEOL] The latest system of rocks or period of the Mesozoic Era, between 136 and 65 million years ago.

cretin [MED] An individual afflicted with cretinism.

cretinism [MED] A type of dwarfism caused by hypothyroidism and associated with generalized body changes, including mental deficiency.

crevasse [GEOL] An open, nearly vertical fissure in a glacier or other mass of land ice or the earth, especially after earthquakes.

crevasse deposit [GEOL] Kame deposited in a crevasse.

crevasse hoar [HYD] Ice crystals which form and grow in glacial crevasses and in other cavities where a large cooled space is formed and in which water vapor can accumulate under calm, still conditions.

crevice [SCI TECH] A deep, narrow opening.

crevice corrosion [MET] Corrosive degradation of metal parts at the crevices left at rolled joints or from other forming procedures; common in stainless steel heat exchangers in contact with chloride-containing fluids or other dissolved corrosives. Also known as contact corrosion.

crew-served [ORD] Of or pertaining to anything served or operated by a crew as distinguished from an individual; for example, a weapon operated by a crew of two or more persons.

criador [METEOROL] The rain-bringing west wind of northern Spain.

crib [CIV ENG] The space between two successive ties along a railway track. [ENG] 1. Any structure composed of a layer of timber or steel joists laid on the ground, or two layers across each other, to spread a load. 2. Any structure composed of frames of timber placed horizontally on top of each other to form a wall. [GEOL] See arête.

crib death [MED] Sudden death of a sleeping infant without apparent cause. Also known as sudden death syndrome.

cribellum [INV ZOO] 1. A small accessory spinning organ located in front of the ordinary spinning organ in certain spiders. 2. A chitinous plate perforated with the openings of certain gland ducts in insects.

Cribrariaceae [BOT] A family of true slime molds in the order Liceales.

cribriform [BIOL] Perforated, like a sieve.

cribriform fascia [ANAT] The sievelike covering of the fossa ovalis of the thigh.

cribriform plate [ANAT] 1. The horizontal plate of the ethmoid bone, part of the floor of the anterior cranial fossa. 2. The bone lining a dental alveolus.

Cricetidae [VERT ZOO] A family of the order Rodentia including hamsters, voles, and some mice.

Cricetinae [VERT ZOO] A subfamily of mice in the family Cricetidae.

CRESOL

(a)

(b)

(c)

Structural formulas for the three isomeric methyl phenols, (a) o-cresol, (b) m-cresol, (c) p-cresol.

CRETACEOUS

PRECAMBRIAN		
CAMBRIAN		PALEOZOIC
ORDOVICIAN		
SILURIAN		
DEVONIAN		
Mississippian	CARBON-IFEROUS	
Pennsylvanian		
PERMIAN		
TRIASSIC		MESOZOIC
JURASSIC		
CRETACEOUS		
TERTIARY		CENOZOIC
QUATERNARY		

Chart showing relationship of Cretaceous to other geologic periods.

CRICKET

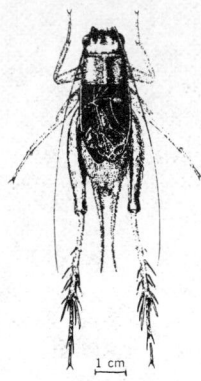

Grass cricket (*Nemobius fasciatus*), a tree cricket of the family Gryllidae. (*From Illinois Natural History Survey*)

CRICOID

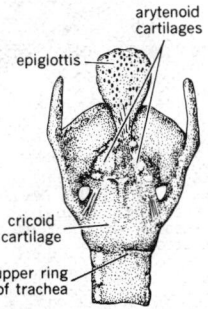

arytenoid cartilages

epiglottis

cricoid cartilage

upper ring of trachea

Back view of human laryngeal cartilages and ligaments. (*After Sappey, from J. Symington, ed., Quains' Elements of Anatomy, vol. 2, pt. 2, Longmans, Green, 1914*)

CRINOIDEA

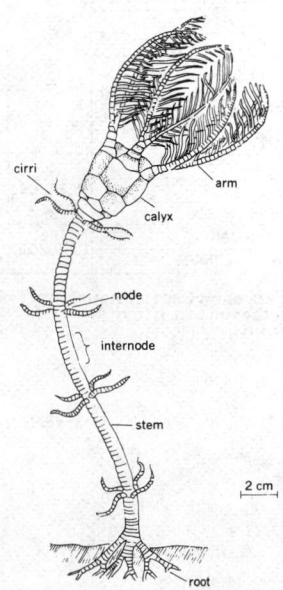

cirri

arm

calyx

node

internode

stem

root

Schematic diagram of crinoid.

cricket [INV ZOO] **1.** The common name for members of the insect family Gryllidae. **2.** The common name for any of several related species of orthopteran insects in the families Tettigoniidae, Gryllotalpidae, and Tridactylidae.

cricoid [ANAT] The signet-ring-shaped cartilage forming the base of the larynx in man and most other mammals.

cricondenbar [PHYS CHEM] Maximum pressure at which two phases (for example, liquid and vapor) can coexist.

cricondentherm [PHYS CHEM] Maximum temperature at which two phases (for example, liquid and vapor) can coexist.

criminal abortion [MED] Illegal interruption of pregnancy.

crimp [ENG] **1.** To cause to become wavy, crinkled, or warped, as lumber. **2.** To pinch or press together, especially a tubular or cylindrical shape, in order to seal or unite. [TEXT] To give curl to synthetic fibers.

crimp contact [ELEC] A contact whose back portion is a hollow cylinder that will accept a wire; after a bared wire is inserted, a swaging tool is applied to crimp the contact metal firmly against the wire. Also known as solderless contact.

Crimplene [TEXT] Trade name of a stable, modified, high-bulk yarn of the polyester Terylene, characterized by low stretchability; manufactured by Imperial Chemical Industries, Ltd.

crinal [MECH] A unit of force equal to 0.1 newton.

crinion-menton [ANTHRO] The measurement from the hairline in the center part of the forehead to the midpoint of the lower edge of the chin.

crinoidal limestone [PETR] A rock composed predominantly of crystalline joints of crinoids, with foraminiferans, corals, and mollusks.

Crinoidea [INV ZOO] A class of radially symmetrical crinozoans in which the adult body is flower-shaped and is either carried on an anchored stem or is free-living.

Crinozoa [INV ZOO] A subphylum of the Echinodermata comprising radially symmetrical forms that show a partly meridional pattern of growth.

crippled leap-frog test [ADP] A variation of the leap-frog test, modified so the computer tests are repeated from a single set of storage locations rather than a changing set of locations.

crisis [MED] A turning point in the course of a disease. [PSYCH] The psychological events associated with a specific stage of life, as an identity crisis or developmental crisis.

crispation number [PHYS] A dimensionless number used in the study of convection currents, equal to the product of a fluid's dynamic viscosity and its thermal diffusivity, divided by the product of its undisturbed surface tension and a layer thickness.

crissum [VERT ZOO] **1.** The region surrounding the cloacal opening in birds. **2.** The vent feathers covering the circumcloacal region.

crista [BIOL] A ridge or crest.

cristate [BIOL] Having a crista.

cristobalite [MINERAL] SiO_2 A silicate mineral that is a high-temperature form of quartz; stable above 1470°C; crystallizes in the tetragonal system at low temperatures and the isometric system at high temperatures.

crit [NUCLEO] The mass of fissionable material that is critical under a given set of conditions; sometimes applied to the mass of an untamped critical sphere of fissionable material.

crith [MECH] A unit of mass, used for gases, equal to the mass of 1 liter of hydrogen at standard pressure and temperature; it is found experimentally to equal 8.9885×10^{-5} kilogram.

critical [NUCLEO] Capable of sustaining a chain reaction at a constant level.

critical absorption wavelength [SPECT] The wavelength, characteristic of a given electron energy level in an atom of a specified element, at which an absorption discontinuity occurs.

critical altitude [AERO ENG] The maximum altitude at which a supercharger can maintain a pressure in the intake manifold of an engine equal to that existing during normal operation at rated power and speed at sea level without the supercharger. [ORD] The maximum altitude at which the propulsion system of a missile performs satisfactorily.

critical angle [OPTICS] An angle associated with the total reflection of electromagnetic radiation back into a medium from the boundary with another medium with a lower index of refraction; it is the smallest angle with the normal to the boundary at which total reflection occurs.

critical angle of attack [AERO ENG] The angle of attack of an airfoil at which the flow of air about the airfoil changes abruptly so that lift is sharply reduced and drag is sharply increased. Also known as stalling angle of attack.

critical anode voltage [ELECTR] The anode voltage at which breakdown occurs in a gas tube.

critical area *See* picture element.

critical assembly [NUCLEO] An assembly of sufficient fissionable and moderator material to sustain a fission chain reaction at a low power level.

critical bottom slope [GEOL] The depth distribution in which depth d of an ocean increases with latitude ϕ according to an equation of the form $d = d_0 \sin \phi + $ constant.

critical condensation temperature [PHYS CHEM] The temperature at which the sublimand of a sublimed solid recondenses; used to analyze solid mixtures, analogous to liquid distillation. Also known as true condensing point.

critical constant [PHYS CHEM] A characteristic temperature, pressure, and specific volume of a gas above which it cannot be liquefied.

critical cooling rate [MET] The minimum cooling rate that will suppress undesired transformations in a metal.

critical coupling [ELEC] The degree of coupling that provides maximum transfer of signal energy from one radio-frequency resonant circuit to another when both are tuned to the same frequency. Also known as optimum coupling.

critical current [SOLID STATE] The current in a superconductive material above which the material is normal and below which the material is superconducting, at a specified temperature and in the absence of external magnetic fields.

critical current density [PHYS CHEM] The amount of current per unit area of electrode at which an abrupt change occurs in a variable of an electrolytic process.

critical damping [PHYS] Damping in a linear system on the threshold between oscillatory and exponential behavior.

critical density [CIV ENG] For a highway, the density of traffic when the volume equals the capacity.

critical elevation [MAP] That elevation which is the high point within the area of a chart.

critical equation [NUCLEO] Any equation relating parameters of a reactor that must be satisfied for the reactor to be critical.

critical experiment [NUCLEO] An experiment in which fissionable material is assembled gradually until the arrangement will support a self-sustaining chain reaction.

critical facility [NUCLEO] A facility where critical experiments are conducted.

critical field [ELECTR] The smallest theoretical value of steady magnetic flux density that would prevent an electron emitted from the cathode of a magnetron at zero velocity from reaching the anode. Also known as cutoff field.

critical flicker frequency [OPTICS] That frequency of an intermittent light source at which the light appears half the time as flickering and half the time as continuous.

critical flow [FL MECH] The rate of flow of a fluid equivalent to the speed of sound in that fluid.

critical flow prover [PETRO ENG] Device used to measure the velocity of gas flow during open-flow testing of gas wells.

critical frequency [ELECTR] *See* cutoff frequency. [ELECTROMAG] The limiting frequency below which a radio wave will be reflected by an ionospheric layer at vertical incidence at a given time. [GEOPHYS] The minimum frequency of a vertically directed radio wave which will penetrate a particular layer in the ionosphere; for example, all vertical radio waves with frequencies greater than the E-layer critical frequency will pass through the E layer. Also known as penetration frequency.

critical function [MATH] A function satisfying the Euler equations in the calculus of variations.

critical gas saturation *See* equilibrium gas saturation.

critical grid current [ELECTR] Instantaneous value of grid current when the anode current starts to flow in a gas-filled vacuum tube.

critical grid voltage [ELECTR] The grid voltage at which anode current starts to flow in a gas tube. Also known as firing point.

critical gun pull [ORD] The maximum pulling force in pounds developed by the gun feeder mechanism during operation of automatic guns.

critical humidity [CHEM ENG] The humidity of a system's atmosphere above which a crystal of a water-soluble salt will always become damp (absorb moisture from the atmosphere) and below which it will always stay dry (release moisture to the atmosphere). [MET] The atmospheric humidity above which the corrosion rate increases rapidly for a particular metal.

criticality [NUCLEO] The condition in which a nuclear reactor is just self-sustaining.

critical level of escape [GEOPHYS] **1.** That level, in the atmosphere, at which a particle moving rapidly upward will have a probability of $1/e$ (e is base of natural logarithm) of colliding with another particle on its way out of the atmosphere. **2.** The level at which the horizontal mean free path of an atmospheric particle equals the scale height of the atmosphere.

critical locus [PHYS CHEM] The line connecting the critical points of a series of liquid-gas phase-boundary loops for multicomponent mixtures plotted on a pressure versus temperature graph.

critical Mach number [AERO ENG] The free-stream Mach number at which a local Mach number of 1.0 is attained at any point on the body under consideration.

critical magnetic field [SOLID STATE] The field below which a superconductive material is superconducting and above which the material is normal, at a specified temperature and in the absence of current.

critical mass [NUCLEO] The mass of fissionable material of a particular shape that is just sufficient to sustain a nuclear chain reaction.

critical micelle concentration [PHYS CHEM] The concentration of a micelle (oriented molecular arrangement of an electrically charged colloidal particle or ion) at which the rate of increase of electrical conductance with increase in concentration levels off or proceeds at a much slower rate.

critical moisture content [CHEM ENG] The average moisture throughout a solid material being dried, its value being related to drying rate, thickness of material, and the factors that influence the movement of moisture within the solid.

critical opalescence [OPTICS] Extreme opalescence resulting from strong density fluctuations in a medium near a critical point.

critical path method [SYS ENG] A systematic procedure for detailed project planning and control. Abbreviated CPM.

critical phenomena [PHYS CHEM] Physical properties of liquids and gases at the critical point (conditions at which two phases are just about to become one); for example, critical pressure is that needed to condense a gas at the critical temperature, and above the critical temperature the gas cannot be liquefied at any pressure.

critical point [MATH] A point at which the first derivative of a function is either 0 or does not exist. [PHYS CHEM] **1.** The temperature and pressure at which two phases of a substance in equilibrium with each other become identical, forming one phase. **2.** The temperature and pressure at which two ordinarily partially miscible liquids are consolute.

critical potential [ATOM PHYS] The energy needed to raise an electron to a higher energy level in an atom (resonance potential) or to remove it from the atom (ionization potential). [ELEC] A potential which results in sudden change in magnitude of the current.

critical pressure [FL MECH] For a nozzle whose cross section at each point is such that a fluid in isentropic flow just fills it, the pressure at the section of minimum area of the nozzle; if the nozzle is cut off at this point with no diverging section, decrease in discharge pressure below the critical pressure (at constant admission pressure) does not result in increased flow. [THERMO] The pressure of the liquid-vapor critical point.

critical pressure ratio [FL MECH] The ratio of the critical pressure of a nozzle to the admission pressure of the nozzle (equals 0.53 for gases).

critical properties [PHYS CHEM] Physical and thermodynamic properties of materials at conditions of critical temperature, pressure, and volume, that is, at the critical point.

critical range [MET] The temperature range for the reversible change of austenite to ferrite, pearlite, and cementite.

critical ratio [STAT] The ratio of a particular deviation from the mean value to the standard deviation.

critical reactor [NUCLEO] A nuclear reactor in which the ratio of moderator to fuel is either subcritical or just critical; used to study the properties of the system and determine critical size.

critical Reynolds number [FL MECH] The Reynolds number at which there is a transition from laminar to turbulent flow.

critical size [NUCLEO] A set of physical dimensions for the core and reflector of a nuclear reactor at which a critical chain reaction is maintained.

critical solution temperature [PHYS CHEM] The temperature at which a mixture of two liquids, immiscible at ordinary temperatures, ceases to separate into two phases.

critical speed [CRYO] *See* critical velocity. [FL MECH] *See* critical velocity. [MECH ENG] The angular speed at which a rotating shaft becomes dynamically unstable with large lateral amplitudes, due to resonance with the natural frequencies of lateral vibration of the shaft.

critical state [PHYS CHEM] Unique condition of pressure, temperature, and composition wherein all properties of coexisting vapor and liquid become identical.

critical strain [MET] The strain at which heating causes rapid growth of large grains in many metals and alloys; phase transformations do not occur.

critical temperature [PHYS CHEM] The temperature of the liquid-vapor critical point, that is, the temperature above which the substance has no liquid-vapor transition.

critical value [MATH] The value of the independent variable at a critical point of a function. [STAT] A number which causes rejection of the null hypothesis if a given test statistic is this number or more, and acceptance of the null hypothesis if the test statistic is smaller than this number.

critical velocity [AERO ENG] In rocketry, the speed of sound at the conditions prevailing at the nozzle throat. Also known as throat velocity. [CRYO] The velocity of a superfluid in very narrow channels (on the order of 10^{-5} centimeter), which is nearly constant. Also known as critical speed. [FL MECH] **1.** The speed of flow equal to the local speed of sound. Also known as critical speed. **2.** The speed of fluid flow through a given conduit above which it becomes turbulent.

critical voltage [ELECTR] The highest theoretical value of steady anode voltage, at a given steady magnetic flux density, at which electrons emitted from the cathode of a magnetron at zero velocity would fail to reach the anode. Also known as cutoff voltage.

critical volume [PHYS] The volume occupied by one mole of a substance at the liquid-vapor critical point, that is, at the critical temperature and pressure.

critical wavelength [COMMUN] The free-space wavelength corresponding to the critical frequency.

critical zone [FL MECH] In fluid flow, the area on a graph of the Reynolds number versus friction factor indicating unstable flow (Reynolds number 2000 to 4000) between laminar flow and the transition to turbulent flow. [ORD] Area over which a bombing plane in horizontal-flight or glide bombing must maintain straight flight so that the bombsight can be operated properly and bombs dropped accurately.

crivetz [METEOROL] A wind blowing from the northeast quadrant in Rumania and southern Russia, especially a cold boralike wind from the north-northeast, characteristic of the climate of Rumania.

CRM *See* chemical remanent magnetization.

CRO *See* cathode-ray oscilloscope.

Crocco's equation [FL MECH] A relationship, expressed as $\mathbf{v} \times \omega = -T \operatorname{grad} S$, between vorticity and entropy gradient for the steady flow of an inviscid compressible fluid; $\mathbf{v}$ is the fluid velocity vector, ω ($=$ curl $\mathbf{v}$) is the vorticity vector, T is the fluid temperature, and S is the entropy per unit mass of the fluid.

CRITICAL PATH METHOD

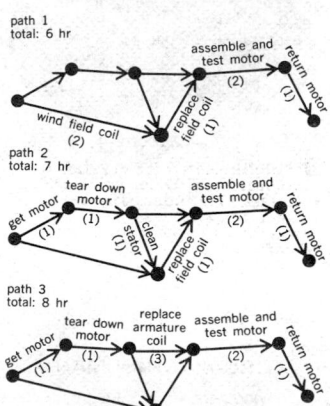

The critical path method for motor maintenance job, showing activities involved, duration of the activity in hours (numbers in parentheses), and how the activities dovetail together.

CROCODILE

A typical representative of the Crocodylidae, with short, powerful legs and a long, narrow head.

CROCOITE

⊢— 2.5 cm —⊣

Needlelike crocoite crystals in association with cerussite and dundasite, Dundas, Tasmania. (*Specimen courtesy of Department of Geology, Bryn Mawr College*)

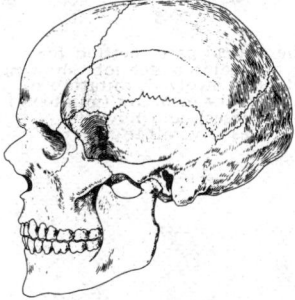

CRO-MAGNON MAN

Skull of a Cro-Magnon male. (*From M. F. Ashley Montagu, An Introduction to Physical Anthropology, 2d ed., Charles C. Thomas, 1951*)

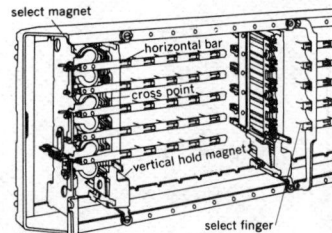

CROSSBAR SWITCH

select magnet
horizontal bar
cross point
vertical hold magnet
select finger

Drawing of a crossbar switch.

crochet file [DES ENG] A thin, flat, round-edged file that tapers to a point.

crocidolite [MINERAL] A lavender-blue, indigo-blue, or leek-green asbestiform variety of riebeckite; occurs in fibrous, massive, and earthy forms. Also known as blue asbestos; krokidolite.

Crockett magnetic separator [MIN ENG] An assembly consisting of a continuous belt submerged in a tank through which ore pulp flows; magnetic solids adhere to the belt, which has a series of flat magnets attached to it, and the solids are dragged clear.

crocodile [ELEC] A unit of potential difference or electromotive force, equal to 10^6 volts; used informally at some nuclear physics laboratories. [VERT ZOO] The common name for about 12 species of aquatic reptiles included in the family Crocodylidae.

crocodile shears *See* lever shears.

Crocodilia [VERT ZOO] An order of the class Reptilia which is composed of large, voracious, aquatic species, including the alligators, caimans, crocodiles, and gavials.

crocodiling *See* alligatoring.

Crocodylidae [VERT ZOO] A family of reptiles in the order Crocodilia including the true crocodiles, false gavial, alligators, and caimans.

Crocodylinae [VERT ZOO] A subfamily of reptiles in the family Crocodylidae containing the crocodiles, *Osteolaemus*, and the false gavial.

crocoite [MINERAL] $PbCrO_4$ A yellow to orange or hyacinth-red secondary mineral occurring as monoclinic, prismatic crystals; it is also massive granular. Also known as crocoisite; red lead ore.

crocus [BOT] A plant of the genus *Crocus*, comprising perennial herbs cultivated for their flowers. [MATER] Finely powdered oxide of iron, of dark red color, used for buffing and polishing.

crofting [TEXT] The whitening of linen by soaking it in an alkaline solution, then drying it in the sun.

Croixian [GEOL] Upper Cambrian geologic time.

Cro-Magnon man [PALEON] 1. A race of tall, erect Caucasoid men having large skulls; identified from skeletons found in southern France. 2. A general term to describe all fossils resembling this race that belong to the upper Paleolithic (35,000–8000 B.C.) in Europe.

cromfordite *See* phosgenite.

Cromwell Current [OCEANOGR] An eastward-setting subsurface current that extends about 1½° north and south of the equator, and from about 150°E to 92°W.

croning process [MET] A shell-molding process.

cronstedtite [MINERAL] $Fe_4^{2+}Fe_2^{3+}(Fe_2^{3+}Si_2)O_{10}(OH)_8$ A black to brownish-black mineral consisting of a hydrous iron silicate crystallizing in hexagonal prisms; specific gravity is 3.34–3.35.

crooked hole [PETRO ENG] A borehole drilled at an angle, often because of steeply dipping formations; not to be confused with holes deliberately deviated from the vertical to avoid obstacles or to tap otherwise unavailable reservoirs.

Crookes dark space *See* cathode dark space.

crookesite [MINERAL] $(Cu,Tl,Ag)_2Se$ An important selenium mineral occurring in lead-gray masses and having a metallic appearance.

Crookes radiometer [PHYS] A radiometer used to demonstrate that radiant energy from the sun can produce motion; a miniature four-vane windmill is mounted in a glass-envelope vacuum tube, with each vane polished on one side and black on the other.

Crookes tube [ELECTR] An early form of low-pressure discharge tube whose cathode was a flat aluminum disk at one end of the tube, and whose anode was a wire at one side of the tube, outside the electron stream; used to study cathode rays.

crop [AGR] A plant or animal grown for its commercial value. [MET] Defective end portion of an ingot which is removed for scrap before rolling the ingot. [VERT ZOO] A distensible saccular diverticulum near the lower end of the esophagus of birds which serves to hold and soften food before passage into the stomach.

crop coal [MIN ENG] Coal of inferior quality found near the surface.

crop dusting [AGR] Applying fungicides or insecticides in powder form to a crop; usually done from a low-flying aircraft.

crop out *See* outcrop.

cropping [GRAPHICS] The elimination of unwanted details in a picture by removing portions on the edges of the picture.

crop rotation [AGR] A method of protecting the soil and replenishing its nutrition by planting a succession of different crops on the same land.

Cross *See* Crux.

cross axle [MECH ENG] 1. A shaft operated by levers at its ends. 2. An axle with cranks set at 90°.

crossband [COMMUN] Two-way communication in which one radio frequency is used in one direction and a frequency having different propagation characteristics is used in the opposite direction.

crossbar switch [ELEC] A switch having a three-dimensional arrangement of contacts and a magnet system that selects individual contacts according to their coordinates in the matrix.

crossbar system [COMMUN] Automatic telephone switching system which is generally characterized by the following features: selecting mechanisms are crossbar switches, common circuits select and test the switching paths and control the operation of the selecting mechanisms, and method of operations is one in which the switching information is received and stored by controlling mechanisms that determine the operations necessary in establishing a telephone connection.

cross-bedding [GEOL] The condition of having laminae lying transverse to the main stratification planes of the strata; occurs only in granular sediments. Also known as cross-lamination; cross-stratification.

crossbolt [DES ENG] A lock bolt with two parts which can be moved in opposite directions.

cross bond [CIV ENG] A masonry bond in which a course of alternating lengthwise and endwise bricks (Flemish bond) alternates with a course of bricks laid lengthwise.

crossbreed [BIOL] To propagate new individuals by breeding two distinctive varieties of a species.

cross-color [ELECTR] In color television, the interference in the receiver chrominance channel caused by cross talk from monochrome signals.

cross-correlation [STAT] 1. Correlation between corresponding members of two or more series: if $q_1, \ldots q_n$ and $r_1, \ldots r_n$ are two series, correlation between q_i and r_i, or between q_i and r_{i+j} (for fixed j), is a cross correlation. 2. Correlation between or expectation of the inner product of two series of random variables, where the difference in indices between the corresponding values of the two series is fixed.

cross-correlation detection *See* correlation detection.

cross-correlation function [COMMUN] A function, $\phi_{12}(\tau)$, where τ is a time-delay parameter, equal to the limit, as T approaches infinity, of the reciprocal of $2T$ times the integral over t from $-T$ to T of $f_1(t)f_2(t-\tau)$, where f_1 and f_2 are functions of time, such as the input and output of a communication system.

cross-correlator [ELECTR] A correlator in which a locally generated reference signal is multiplied by the incoming signal and the result is smoothed in a low-pass filter to give an approximate computation of the cross-correlation function. Also known as synchronous detector.

cross-country mill [MET] A rolling mill in which the tables of the mill stands are parallel with a crossover table that connects them; used to produce special forms of bar stock.

cross-coupling [COMMUN] A measure of the undesired power transferred from one channel to another in a transmission medium.

crosscurrent [FL MECH] A current that flows across or opposite to another current.

crosscurrent extraction [ANALY CHEM] Procedure of batch-wise liquid-liquid extraction in a separatory funnel; solvent is added to the sample in the funnel, which is then shaken, and the extract phase is allowed to coalesce, then is drawn off.

crosscut [ENG] A cut made through wood across the grain. [MIN ENG] 1. A small passageway driven at right angles to the main entry of a mine to connect it with a parallel entry of air course. **2.** A passageway in a mine that cuts across the geological structure.

crosscut file [DES ENG] A file with a rounded edge on one side and a thin edge on the other; used to sharpen straight-sided saw teeth with round gullets.

crosscut saw [DES ENG] A type of saw for cutting across the grain of the wood; designed with about eight teeth per inch.

crossed belt [MECH ENG] A pulley belt arranged so that the sides cross, thereby making the pulleys rotate in opposite directions.

crossed-field accelerator [AERO ENG] A plasma engine for space travel in which plasma serves as a conductor to carry current across a magnetic field, so that a resultant force is exerted on the plasma.

crossed-field amplifier [ELECTR] A forward-wave, beam-type microwave amplifier that uses crossed-field interaction to achieve good phase stability, high efficiency, high gain, and wide bandwidth for most of the microwave spectrum.

crossed-field backward-wave oscillator [ELECTR] One of several types of backward-wave oscillators that utilize a crossed field, such as the amplitron and carcinotron.

crossed-field device [ELECTR] Any instrument which uses the motion of electrons in perpendicular electric and magnetic fields to generate microwave radiation, either as an amplifier or oscillator.

crossed-field multiplier phototube [ELECTR] A multiplier phototube in which repeated secondary emission is obtained from a single active electrode by the combined effects of a strong radio-frequency electric field and a perpendicular direct-current magnetic field.

crossed lens [OPTICS] A lens designed with radii of curvature which give minimum spherical aberration for parallel incident rays.

crossed paralysis [MED] Paralysis of the arm and leg on one side, associated with contralateral cranial nerve palsies caused by a brainstem lesion involving cranial nerve nuclei and the ipsilateral pyramidal tract.

crossed-pointer indicator [NAV] A two-pointer indicator used with an instrument-landing system to indicate the position of an airplane with respect to the glide path.

crossed prisms [OPTICS] A pair of Nicol prisms whose principal planes are perpendicular to each other, so that light passing through one is extinguished by the other.

cross-eye *See* esotropia.

cross fault [GEOL] **1.** A fault whose strike is perpendicular to the general trend of the regional structure. **2.** A minor fault that intersects a major fault.

cross-fertilization [BOT] Fertilization between two separate plants. [ZOO] Fertilization between different kinds of individuals.

cross fire [COMMUN] Interfering current in one telegraph or signaling channel resulting from telegraph or signaling currents in another channel.

cross-flow [AERO ENG] A flow going across another flow, as a spanwise flow over a wing.

cross-flow plane [AERO ENG] A plane at right angles to the free-stream velocity.

cross flux [ELECTROMAG] A component of magnetic flux perpendicular to that produced by the field magnets in an electrical rotating machine.

cross-fold [GEOL] A secondary fold whose axis is perpendicular or oblique to the axis of another fold. Also known as subsequent fold; superimposed fold; transverse fold.

crossfoot [ADP] To add numbers in several different ways in a computer, for checking purposes.

cross furring ceiling [BUILD] A ceiling in which furring members are attached perpendicular to the main runners or other structural members.

cross gateway *See* cross heading.

cross hair [ENG] An inscribed line or a strand of hair, wire, silk, or the like used in an optical sight, transit, or similar instrument for accurate sighting.

crosshatch generator [ELECTR] A signal generator that generates a crosshatch pattern for adjusting color television receiver circuits.

crosshaul [MECH ENG] A device for loading objects onto vehicles, consisting of a chain that is hooked on opposite sides

of a vehicle, looped under the object, and connected to a power source and that rolls the object onto the vehicle.

crosshead [MECH ENG] A block sliding between guides and containing a wrist pin for the conversion of reciprocating to rotary motion, as in an engine or compressor. [MET] A device generally employed in wire coating which is attached to the discharge end of the extruder cylinder; designed to facilitate extruding material at an angle. [MIN ENG] A runner or guide positioned just above a sinking bucket to restrict excessive swinging.

cross heading [MIN ENG] Mine passage driven for ventilation from the airway to the gangway, or from one breast through the pillar to the adjoining working. Also known as cross gateway; cross hole; headway.

cross hole *See* cross heading.

crossing angle [NAV] The angle at which two lines of position intersect. Also known as angle of cut.

crossing over [GEN] The exchange of genetic material between paired homologous chromosomes during meiosis. Also known as crossover.

crossing plates [CIV ENG] Plates placed between a crossing and the ties to support the crossing and protect the ties.

crossing symmetry [PARTIC PHYS] The amplitude for a process that involves creation of a particle with four-momentum P_μ is equal to the amplitude for a process which is the same except it involves destruction of the antiparticle with four-momentum $-P_\mu$.

cross joint [GEOL] A fracture in igneous rock perpendicular to the lineation caused by flow magma. Also known as transverse joint.

cross-lamination *See* cross-bedding.

cross-level [ENG] To level at an angle perpendicular to the principal line of sight.

crosslights [OPTICS] Lights arranged to balance the effects of key lighting on the subject's face. Also known as modeling lights.

cross-linking [ORG CHEM] The setting up of chemical links between the molecular chains of polymers.

cross-magnetizing effect [ELECTROMAG] The distortion in the flux-density distribution in the air gap of an electric rotating machine caused by armature reaction.

cross matching [IMMUNOL] Determination of blood compatibility for transfusion by mixing donor cells with recipient serum, and recipient cells with donor serum, and examining for an agglutination reaction.

cross modulation [COMMUN] A type of interference in which the carrier of a desired signal becomes modulated by the program of an undesired signal on a different carrier frequency; the program of the undesired station is then heard in the background of the desired program.

cross multiplication [MATH] Multiplication of the numerator of each of two fractions by the denominator of the other, as when eliminating fractions from an equation.

cross-neutralization [ELECTR] Method of neutralization used in push-pull amplifiers, whereby a portion of the plate-cathode alternating-current voltage of each vacuum tube is applied to the grid-cathode circuit of the other vacuum tube through a neutralizing capacitor.

cross office switching time [COMMUN] Time required to connect any input through the switching center to any selected output.

Crossopterygii [PALEON] A subclass of the class Osteichthyes comprising the extinct lobefins or choanate fishes and represented by one extant species; distinguished by two separate dorsal fins.

Crossosomataceae [BOT] A monogeneric family of xerophytic shrubs in the order Dilleniales characterized by perigynous flowers, seeds with thin endosperm, and small, entire leaves.

crossover [CIV ENG] An S-shaped section of railroad track joining two parallel tracks. [GEN] *See* crossing over.

crossover distortion [ELECTR] Amplitude distortion in a class B transistor power amplifier which occurs at low values of current, when input impedance becomes appreciable compared with driver impedance.

crossover experiment [MED] An experiment or clinical investigation in which subjects are divided randomly into at

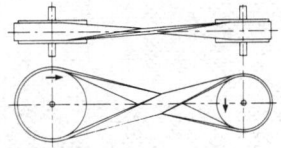

CROSSED BELT

Arrangement of crossed pulley belt showing the directions in which the pulleys rotate.

least as many groups as there are kinds of treatment to be given, and then the groups are interchanged until every subject has received each treatment.

crossover flange [ENG] Intermediate pipe flange used to connect flanges of different working pressures.

crossover frequency [ENG ACOUS] **1.** The frequency at which a dividing network delivers equal power to the upper and lower frequency channels when both are terminated in specified loads. **2.** *See* transition frequency.

crossover network [ENG ACOUS] A selective network used to divide the audio-frequency output of an amplifier into two or more bands of frequencies. Also known as dividing network; loudspeaker dividing network.

cross-over spiral *See* lead-over groove.

crossover voltage [ELECTR] In a cathode-ray storage tube, the voltage of a secondary writing surface, with respect to cathode voltage, on which the secondary emission is unity.

cross-pointer indicator [NAV] A flight instrument having two needles which cross in the center when the aircraft is on course; the vertical needle indicates the position of the aircraft with respect to the localizer course, and the horizontal needle serves a similar purpose for the glide slope of an instrument landing system.

cross-polarization [ELECTROMAG] The component of the electric field vector normal to the desired polarization component.

cross-pollination [BOT] Transfer of pollen from the anthers of one plant to the stigmata of another plant.

cross product [MATH] An anticommutative multiplication on the vectors of euclidean three-dimensional space. Also known as vector product.

cross-reaction [IMMUNOL] Reaction between an antibody and a closely related, but not complementary, antigen.

cross-rolling [MET] **1.** Straightening metal sheets by passing them through rolls at right angles to the principal direction of rolling. **2.** Straightening round bars or tubes by passing the work through parallel to the axes of rolls.

cross section [GRAPHICS] A diagram or drawing representing a cut at right angles to an axis. [MAP] A horizontal grid system laid out on the ground for determining contours, quantities of earthwork, and so on, by means of elevations of the grid points. [MATH] **1.** The intersection of an *n*-dimensional geometric figure in some euclidean space with a lower dimensional hyperplane. **2.** A right inverse for the projection of a fiber bundle. [PHYS] An area characteristic of a collision reaction between atomic or nuclear particles or systems, such that the number of reactions which occur equals the product of the number of target particles or systems and the number of incident particles or systems which would pass through this area if their velocities were perpendicular to it. Also known as collision cross section.

cross section per atom [NUC PHYS] The microscopic cross section for a given nuclear reaction referred to the natural element, even though the reaction involves only one of the natural isotopes.

cross-section symbols [GRAPHICS] Standardized shadings used to represent various materials in a cross section.

cross slide [MECH ENG] A part of a machine tool that allows the tool carriage to move at right angles to the main direction of travel.

cross spectrum [PHYS] The complex vector sum of the cospectrum and quadrature spectrum.

cross-staff [NAV] A forerunner of the modern sextant, used for measuring altitudes of celestial bodies, consisting essentially of a wooden rod with one or more perpendicular crosspieces free to slide along the main rod. Also known as forestaff; Jacob's staff.

cross-stone *See* harmotome; staurolite.

cross-stratification *See* cross-bedding.

crosstalk [COMMUN] **1.** The sound heard in a receiver along with a desired program because of cross modulation or other undesired coupling to another communication channel; it is also observed between adjacent pairs in a telephone cable. **2.** Interaction of audio and video signals in a television system, causing video modulation of the audio carrier or audio modulation of the video signal at some point. **3.** Interaction of chrominance and luminance signals in a color television receiver. [ENG ACOUS] *See* magnetic printing.

crosstalk coupling [COMMUN] Cross coupling between speech communications channels or their component parts. Also known as crosstalk loss.

crosstalk level [COMMUN] Volume of crosstalk energy, measured in decibels, referred to a base.

crosstalk loss *See* crosstalk coupling.

crosstalk unit [COMMUN] A measure of the coupling between two circuits; the number of crosstalk units is 1 million times the ratio of the current or voltage at the observing point to the current or voltage at the origin of the disturbing signal, the impedances at these points being equal. Abbreviated cu.

crosstie [ENG] A timber or metal sill placed transversely under the rails of a railroad, tramway, or mine-car track.

cross-tolerance [MED] Tolerance or resistance to the action of a drug brought about by continued use of another drug of similar pharmacologic action.

cross trail error [NAV] The error introduced in a drift observation made by means of an object dropped from an aircraft when the object lands to one side of the track.

cross turret [MECH ENG] A turret that moves horizontally and at right angles to the lathe guides.

cross valley *See* transverse valley.

crossvein [MIN ENG] A vein that intersects an older or larger vein.

crosswind [METEOROL] A wind which has a component directed perpendicularly to the course (or heading) of an exposed, moving object.

cross-wire weld [MET] A projection weld localized by crossed wires or bars.

Crotalidae [VERT ZOO] A family of proglyphodont venomous snakes in the reptilian suborder Serpentes.

crotch [SCI TECH] The angular form made by the parting of two branches, legs, or members.

crotch height [ANTHRO] The measure of the vertical distance from the crotch of a standing subject to the floor.

crotonaldehyde [ORG CHEM] C_3H_5CHO A colorless liquid boiling at 104°C, soluble in water; vapors are lacrimatory; used as an intermediate in manufacture of *n*-butyl alcohol and quinaldine. Also known as propylene aldehyde.

crotonic acid [ORG CHEM] C_3H_5COOH An unsaturated acid, with colorless, monoclinic crystals, soluble in water; used in the preparation of synthetic resins, plasticizers, and pharmaceuticals. Also known as α-butenic acid.

croton oil [MATER] A yellow-brown oil obtained from the seeds of the *Croton tiglium;* used as a purgative and as a substitute for castor oil.

croup [MED] Any condition of upper-respiratory pathway obstruction in children, especially acute inflammation of the pharynx, larynx, and trachea, characterized by a hoarse, brassy, and stridulent cough and difficulties in breathing.

croup-associated virus [VIROL] A virus belonging to subgroup 2 of the parainfluenza viruses and found in children with croup. Also known as CA virus; laryngotracheobronchitis virus.

croute calcaire *See* caliche.

Crout reduction [MATH] Modification of the Gauss procedure for numerical solution of simultaneous linear equations; adapted for use on desk calculators and digital computers.

Crova wavelength [STAT MECH] The wavelength in the spectrum of a radiator whose intensity divided by the intensity of the total radiation equals the derivative of the intensity of the wavelength with respect to temperature divided by the derivative of the total intensity with respect to temperature.

crow [VERT ZOO] The common name for a number of predominantly black birds in the genus *Corvus* comprising the most advanced members of the family Corvidae.

Crow *See* Corvus.

crowbar [DES ENG] An iron or steel bar that is usually bent and has a wedge-shaped working end; used as a lever and for prying. [ELEC] A device or action that in effect places a high overload on the actuating element of a circuit breaker or other protective device, thus triggering it.

Crowe process [MET] In cyanidation, after extraction of the gold, separation of the solution from the ore tailings by

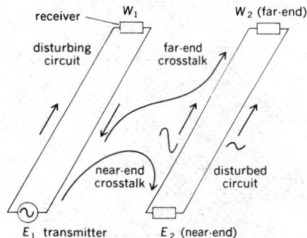

CROSSTALK

Crosstalk arising in short section of two parallel circuits (def. 1). Conductor closer to disturbing circuit has larger crosstalk voltage than farther conductor; voltage difference is effective at E_2 and W_2 terminals. Energy in disturbing circuit is assumed to flow from east to west.

filtration or countercurrent decantation and by passage through a vacuum chamber where deaeration occurs.

crown [ANAT] **1.** The top of the skull. **2.** The portion of a tooth above the gum. [BOT] **1.** The topmost part of a plant or plant part. **2.** See corona. [CIV ENG] Center of a roadway elevated above the sides. [ENG] **1.** The part of a drill bit inset with diamonds. **2.** The vertex of an arch or arched surface. **3.** The top or dome of a furnace or kiln. [LAP] The portion of a faceted gem above the girdle. [MET] That part of the sheet or roll where the thickness or diameter increases from edge to center. [MIN ENG] A horizontal roof member of a timber up to 16 feet (4.9 meters) long and supported at each end by an upright.

crown block [PETRO ENG] A wooden or steel beam joined to the tops of derrick posts of an oil well to support pulleys.

crown cell [ELEC] Generic name for alkaline zinc-manganese dioxide dry-cell battery; manganese dioxide-graphite cathode mix is pressed into a steel can onto which a steel cap is spot-welded to contain the amalgamated powdered-zinc anode.

crown-cut See backsawing.

crown fire [FOR] A forest fire burning primarily in the tops of trees and shrubs.

crown gall [PL PATH] A bacterial disease of many plants induced by Bacterium tumefaciens and marked by abnormal enlargement of the stem near the root crown.

crown glass [MATER] A soda-lime glass, typically having 72% SiO_2, 13% CaO, and 15% Na_2O, which is hard and will take a simple polish; highly transparent for visible light.

crown grafting [BOT] A method of vegetative propagation whereby a scion 3–6 inches (8–15 centimeters) long is grafted at the root crown, just below ground level.

crown rot [PL PATH] Any plant disease or disorder marked by deterioration of the stem at or near ground level.

crown rust [PL PATH] A rust disease of oats and certain other grasses caused by varieties of Puccinia coronata and marked by light-orange masses of fungi on the leaves.

crown saw [DES ENG] A saw consisting of a hollow cylinder with teeth around its edge; used for cutting round holes. Also known as hole saw.

crown wheel [DES ENG] A light, crown-shaped gear. [HOROL] **1.** The horizontal escape wheel of a verge escapement timepiece. **2.** The wheel in the winding mechanism of a watch that drives the ratchet wheel and is itself driven by the winding-stem pinion.

crow quill pen [GRAPHICS] An artist's pen made from a crow quill and used to form an extremely fine line.

Crt See Crater.

CRT See cathode-ray tube.

Cru See Crux.

cruciate [ANAT] Resembling a cross.

crucible [SCI TECH] A refractory vessel or pot, varying in size from a small laboratory utensil to large industrial equipment for melting or calcining.

crucible steel See drill steel.

Cruciferae [BOT] A large family of dicotyledonous herbs in the order Capparales characterized by parietal placentation; hypogynous, mostly regular flowers; and a two-celled ovary with the ovules attached to the margins of the partition.

cruciform [SCI TECH] Resembling or arranged like a cross.

cruciform core [ELEC] A transformer core in which all windings are on one center leg, and four additional legs arranged in the form of a cross serve as return paths for magnetic flux.

cruciform wing [AERO ENG] An aircraft wing in the shape of a cross.

α Crucis [ASTRON] A double star in the constellation Crux that is 220 light-years from the sun; spectral classification BO.5V.

crude assay [CHEM ENG] A procedure for determining the general distillation characteristics and other quality information of crude oil.

crude drug [PHARM] **1.** A plant or animal drug containing all principles characteristic of the drug. **2.** The dried leaves, bark, or rhizome of a plant containing therapeutically active principles.

crude lecithin See lecithin.

crude naphtha [MATER] A light distillate made in the fractionation of crude oil.

crude oil [GEOL] A comparatively volatile liquid bitumen composed principally of hydrocarbon, with traces of sulfur, nitrogen, or oxygen compounds; can be removed from the earth in a liquid state.

crude ore [MIN ENG] The ore as it leaves the mine in an unconcentrated form.

crude scale See scale wax.

crude still [CHEM ENG] The distillation equipment in which crude oil is separated into various products.

crude yellow scale [MATER] The trade name for a low-grade paraffin wax.

cruise control [NAV] The act or practice of operating an aircraft so as to achieve the most efficient performance on a given flight under available conditions; it may be instituted to obtain maximum economy, endurance, speed, or range, or maximum efficiency at a predetermined airspeed or power setting.

cruiser [NAV ARCH] A type of large warship, but smaller than a battleship, having a displacement of 6000 to 15,000 tons (5442–13,605 metric tons), moderately armed and armored, and capable of any naval duty except combat with battleships.

cruising altitude [NAV] An indicated pressure altitude maintained in cruising.

cruising radius [NAV] **1.** The maximum distance that an aircraft, starting with full fuel tanks, can cruise under given or specified conditions from its takeoff point before returning with a specified fuel reserve remaining. **2.** The distance a craft can travel at cruising speed without refueling. Also known as cruising range.

cruising range See cruising radius.

crunode [MATH] A point on a curve through which pass two branches of the curve with different tangents. Also known as node.

crura [ANAT] Plural of crus.

crus [ANAT] **1.** The shank of the hindleg, that portion between the femur and the ankle. **2.** Any of various parts of the body resembling a leg or root.

crush [MET] Casting defect caused by damage to the mold before pouring the metal. [MIN ENG] **1.** A general settlement of the strata above a coal mine due to failure of pillars; generally accompanied by numerous local falls of roof in mine workings. **2.** To reduce ore or quartz by stamps, crushers, or rolls.

crush breccia [GEOL] A breccia formed in place by mechanical fragmentation of rock during movements of the earth's crust.

crush conglomerate [GEOL] Beds similar to a fault breccia, except that the fragments are rounded by attrition. Also known as tectonic conglomerate.

crushed steel [MATER] An abrasive used in the stone, brick, glass, and metal trades, made by heating high-grade crucible steel to white heat, quenching in a bath of cold water, and crushing the fragments.

crushed stone [MIN ENG] Irregular fragments of rock crushed or ground to smaller sizes after quarrying. Also known as broken stone.

crusher [MECH ENG] A machine for crushing rock and other bulk materials.

crush fold [GEOL] A fold of large dimensions that may involve considerable minor folding and faulting such as would produce a mountain chain or an oceanic deep.

crush-forming [ENG] Shaping the face of a grinding wheel by forcing a rotating metal roll into it.

crushing [MIN ENG] The quantity of ore pulverized or crushed at a single operation in processing.

crushing mill See stamping mill.

crushing strength [MECH] The compressive stress required to cause a solid to fail by fracture; in essence, it is the resistance of the solid to vertical pressure placed upon it.

crushing test [ENG] A test of the suitability of stone that might be mined for roads or building use. [MET] A test to determine quality of tubing, especially welded tubing, by applying compression parallel to the axis.

crush kidney See lower nephron nephrosis.

crush syndrome [MED] A severe, often fatal condition that

CROWN FIRE

Characteristic crown fire.
(U.S. Forest Service)

follows a severe crushing injury, particularly involving large muscle masses, characterized by fluid and blood loss, shock, hematuria, and renal failure. Also known as compression syndrome.

crush zone [GEOL] A zone of fault breccia on fault gouge.

crust [GEOL] The outermost solid layer of the earth, mostly consisting of crystalline rock and extending no more than a few miles from the surface to the Mohorovičić discontinuity. Also known as earth crust. [HYD] A hard layer of snow lying on top of a soft layer.

Crustacea [INV ZOO] A class of arthropod animals in the subphylum Mandibulata having jointed feet and mandibles, two pairs of antennae, and segmented, chitin-encased bodies.

crustal motion [GEOL] Movement of the earth's crust.

crustal plate [GEOL] One of the few major 100-kilometer-thick blocks into which the lithosphere can be divided according to plate tectonics.

crustecdysone [BIOCHEM] $C_{27}H_{44}O_7$ 20-Hydroxyecdysone, the molting hormone produced by Y organs in crustaceans.

crustose [BOT] Of a lichen, forming a thin crustlike thallus which adheres closely to the substratum of rock, bark, or soil.

crust vegetation [ECOL] Zonal growths of algae, mosses, lichens, or liverworts having variable coverage and a thickness of only a few centimeters.

crutter [MIN ENG] **1.** A worker who drills blasting holes and prepares the blasting charge. **2.** A worker who removes blasted rock.

Crux [ASTRON] A constellation having four principal bright stars which form the figure of a cross; right ascension 12 hours, declination 60°S. Abbreviated Cru. Also known as Cross; Southern Cross.

Crv *See* Corvus.

cry-, cryo- [SCI TECH] Combining form meaning cold, freezing.

cryobiology [BIOL] The use of low-temperature environments in the study of living plants and animals.

cryochem process [CHEM ENG] A freeze-drying technique involving conduction heat transfer to the frozen solid held on a metallic surface.

cryoconite [GEOL] A dark, powdery dust transported by wind and deposited on the surface of snow or ice; found, however, mainly in cryoconite holes. [MINERAL] A mixture of garnet, sillimanite, zircon, pyroxene, quartz, and various other minerals.

cryoconite hole [GEOL] A cylindrical dust well filled with cryoconite; absorbs solar radiation, causing melting of glacier ice around and below it.

cryogen *See* cryogenic fluid.

cryogenic coil [CRYO] A high-purity coil refrigerated to very low temperatures to reduce effective coil resistivity.

cryogenic conductor *See* superconductor.

cryogenic device [CRYO] A device whose operation depends on superconductivity as produced by temperatures near absolute zero. Also known as superconducting device.

cryogenic engineering [ENG] A branch of engineering specializing in technical operations at very low temperatures (about 200 to 400°R, or −160 to −50°C).

cryogenic film [ADP] A storage element using superconducting thin films of lead at liquid-helium temperature.

cryogenic fluid [CRYO] A liquid which boils at temperatures of less than about 110K at atmospheric pressure, such as hydrogen, helium, nitrogen, oxygen, air, or methane. Also known as cryogen.

cryogenic freezing [FOOD ENG] A freezing technique for preserving shrimp and other foods that are high-priced or have low moisture content by spraying them with liquid nitrogen as they pass through a tunnel on a conveyor belt.

cryogenic gyroscope [ENG] A gyroscope in which a spherical rotor of superconducting niobium spins while in levitation at cryogenic temperatures. Also known as superconducting gyroscope.

cryogenic period [GEOL] A time period in geologic history during which large bodies of ice appeared at or near the poles and climate favored the formation of continental glaciers.

cryogenic propellant [MATER] A rocket fuel, oxidizer, or propulsion fluid which is liquid only at very low temperatures.

cryogenic pump [CRYO] A high-speed vacuum pump that can produce an extremely low vacuum and has a low power consumption; to reduce the pressure, gases are condensed on surfaces within an enclosure at extremely low temperatures, usually attained by using liquid helium or liquid or gaseous hydrogen. Also known as cryopump.

cryogenics [PHYS] The production and maintenance of very low temperatures, and the study of phenomena at these temperatures.

cryogenic temperature [CRYO] A temperature within a few degrees of absolute zero.

cryogenic transformer [ELECTR] A transformer designed to operate in digital cryogenic circuits, such as a controlled-coupling transformer.

cryoglobulin [PATH] An abnormal protein, usually an immunoglobulin, which precipitates from plasma between 40 and 70°F (4.4 and 21°C).

cryohydrate [CHEM] A salt that contains water of crystallization at low temperatures. Also known as cryosel.

cryolaccolith *See* hydrolaccolith.

cryoelectronics [ELECTR] Technology concerning the characteristics of electronic components at cryogenic temperatures.

cryolite [MINERAL] Na_3AlF_6 A white or colorless mineral that crystallizes in the monoclinic system but has a pseudocubic aspect; found in masses of waxy luster; hardness is 2.5 on Mohs scale, and specific gravity is 2.95–3.0; used chiefly as a flux in producing aluminum from bauxite and for making salts of sodium and aluminum and porcelaneous glass. Also known as Greenland spar; ice stone.

cryolithionite [MINERAL] $Na_3Li_3Al_2F_{12}$ A colorless mineral that crystallizes in the isometric system; found in the Ural Mountains.

cryology [HYD] The study of ice and snow. [MECH ENG] The study of low-temperature (about 200°R, or −160°C) refrigeration.

cryomagnetic [CRYO] Pertaining to production of very low temperatures by adiabatic demagnetization of paramagnetic salts.

cryometer [ENG] A thermometer for measuring low temperatures.

cryopedology [GEOL] A branch of geology that deals with the study of intensive frost action and permanently frozen ground.

cryophilic *See* cryophilous.

cryophilous [ECOL] Having a preference for low temperatures. Also known as cryophilic.

cryophysics [CRYO] Physics as restricted to phenomena occurring at very low temperatures, approaching absolute zero.

cryophyte [ECOL] A plant that forms winter buds below the soil surface.

cryoplanation [GEOL] Land erosion at high latitudes or elevations due to processes of intensive frost action.

cryoprecipitate [BIOCHEM] The precipitate of a cryoglobulin.

cryopreservation [ENG] Preservation of food, biologicals, and other materials at extremely low temperatures.

cryopump *See* cryogenic pump.

cryoresistive transmission line [ELEC] An electric power transmission line whose conducting cables are cooled to the temperature of liquid nitrogen, 77K (−196°C), resulting in a reduction of the resistance of the conductor by a factor of approximately 10, leading to increased transmission capacity.

cryosar [ELECTR] A cryogenic, two-terminal, negative-resistance semiconductor device, consisting essentially of two contacts on a germanium wafer operating in liquid helium.

cryoscope [ENG] A device to determine the freezing point of a liquid.

cryoscopic constant [ANALY CHEM] Equation constant expressed in degrees per mole of pure solvent; used to calculate the freezing-point-depression effects of a solute.

cryoscopy [ANALY CHEM] A phase-equilibrium technique to determine molecular weight and other properties of a solute by dissolving it in a liquid solvent and then ascertaining the solvent's freezing point.

cryosel *See* cryohydrate.

cryosistor [ELECTR] A cryogenic semiconductor device in which a reverse-biased *pn* junction is used to control the ionization between two ohmic contacts.

cryosorption pump [MECH ENG] A high-vacuum pump that employs a sorbent such as activated charcoal or synthetic

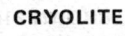

CRYOLITE

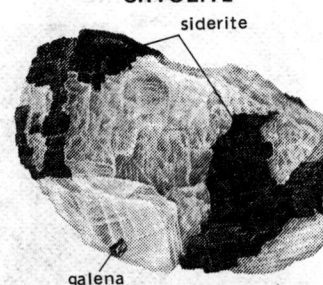

siderite

galena

|—2.5 cm—|

White, translucent cryolite in association with siderite and galena, Ivigtut, Greenland. *(Specimen from Department of Geology, Bryn Mawr College)*

zeolite cooled by nitrogen or some other refrigerant; used to reduce pressure from atmospheric pressure to a few milliton.

cryostat [ENG] An apparatus used to provide low-temperature environments in which operations may be carried out under controlled conditions.

cryosurgery [MED] Selective destruction of tissue by freezing, as the use of a liquid nitrogen probe to the brain in parkinsonism.

cryotherapy [MED] A form of therapy which consists of local or general use of cold.

cryotron [ELECTR] A switch that operates at very low temperatures at which its components are superconducting; when current is sent through a control element to produce a magnetic field, a gate element changes from a superconductive zero-resistance state to its normal resistive state.

cryotronics [ELECTR] The branch of electronics that deals with the design, construction, and use of cryogenic devices.

cryoturbation *See* congeliturbation.

Cryphaeaceae [BOT] A family of mosses in the order Isobryales distinguished by a rough calyptra.

crypt [ANAT] 1. A follicle or pitlike depression. 2. A simple glandular cavity.

cryptanalysis [COMMUN] Steps and operations performed in converting encrypted messages into plain text without previous knowledge of the key employed.

cryptic coloration [ZOO] A phenomenon of protective coloration by which an animal blends with the background through color matching or countershading.

Cryptobiidae [INV ZOO] A family of flagellate protozoans in the order Kinetoplastida including organisms with two flagella, one free and one with an undulating membrane.

cryptobiotic [ECOL] Living in concealed or secluded situations.

Cryptobranchidae [VERT ZOO] The giant salamanders and hellbenders, a family of tailed amphibians in the suborder Cryptobranchoidea.

Cryptobranchoidea [VERT ZOO] A primitive suborder of amphibians in the order Urodela distinguished by external fertilization and aquatic larvae.

Cryptocerata [INV ZOO] A division of hemipteran insects in some systems of classification that includes the water bugs (Hydrocorisae).

Cryptochaetidae [INV ZOO] A family of myodarian cyclorrhaphous dipteran insects in the subsection Acalypteratae.

cryptoclastic [GEOL] Composed of extremely fine, almost submicroscopic, broken or fragmental particles.

cryptoclimate [ENG] The climate of a confined space, such as inside a house, barn, or greenhouse, or in an artificial or natural cave; a form of microclimate. Also spelled kryptoclimate.

cryptoclimatology [CLIMATOL] The science of climates of confined spaces (cryptoclimates); basically, a form of microclimatology. Also spelled kryptoclimatology.

Cryptococcaceae [MYCOL] A family of imperfect fungi in the order Moniliales in some systems of classification; equivalent to the Cryptococcales in other systems.

Cryptococcales [MYCOL] An order of imperfect fungi, in some systems of classification, set up to include the yeasts or yeastlike organisms whose perfect or sexual stage is not known.

cryptococcal meningitis [MED] Inflammation of the meninges due to yeasts of the genus *Cryptococcus*.

cryptococcosis [MED] A yeast infection of man, primarily of the central nervous system, caused by *Cryptococcus neoformans*. Also known as torulosis.

Cryptococcus [MYCOL] A genus of encapsulated pathogenic yeasts in the order Moniliales.

cryptocrystalline [GEOL] Having a crystalline structure but of such a fine grain that individual components are not visible with a magnifying lens.

Cryptodira [VERT ZOO] A suborder of the reptilian order Chelonia including all turtles in which the cervical spines are uniformly reduced and the head folds directly back into the shell.

cryptogam [BOT] An old term for nonflowering plants.

cryptogram [COMMUN] Information written in code or cipher.

cryptography [COMMUN] The science of preparing messages

in a form which cannot be read by those not privy to the secrets of the form.

cryptolite *See* monazite.

cryptology [COMMUN] The science of preparing messages in forms which are intended to be unintelligible to those not privy to the secrets of the form, and of deciphering such messages.

cryptomedusa [INV ZOO] The final stage in the reduction of a hydroid medusa to a rudiment having sex cells within the gonophore.

cryptomelane [MINERAL] $KMn_8O_{16} \cdot H_2O$ A usually massive mineral, common in manganese ores; contains an oxide of manganese and potassium and crystallizes in the monoclinic system.

cryptomitosis [INV ZOO] Cell division in certain protozoans in which a modified spindle forms, and chromatin assembles with no apparent chromosome differentiation.

Cryptomonadida [BIOL] An order of the class Phytamastigophorea considered to be protozoans by biologists and algae by botanists.

Cryptomonadina [BIOL] The equivalent name for Cryptomonadida.

cryptopart [COMMUN] One of several portions of a cryptotext; each cryptopart bears a different message indicator.

cryptoperthite [MINERAL] A fine-grained, submicroscopic variety of perthite consisting of an intergrowth of potassic and sodic feldspar, detectable only by means of x-rays or with the aid of an electron microscope.

Cryptophagidae [INV ZOO] The silken fungus beetles, a family of coleopteran insects in the superfamily Cucujoidea.

Cryptophyceae [BOT] A class of algae of the Pyrrhophyta in some systems of classification; equivalent to the division Cryptophyta.

Cryptophyta [BOT] A division of the algae in some classification schemes; equivalent to the Cryptophyceae.

cryptophyte [BOT] A plant that produces buds either underwater or underground on corms, bulbs, or rhizomes.

Cryptopidae [INV ZOO] A family of epimorphic centipedes in the order Scolopendromorpha.

cryptorchidism *See* cryptorchism.

cryptorchism [MED] Failure of the testes to descend into the scrotum from the abdomen or inguinal canals. Also known as cryptorchidism.

Cryptostomata [PALEON] An order of extinct bryozoans in the class Gymnolaemata.

cryptotext [COMMUN] In cryptology, a text of visible writing which conveys no intelligible meaning in any language, or which apparently conveys an intelligible meaning that is not the real meaning.

cryptotope [IMMUNOL] A determinant (or epitope) of an immunological antigen or immunogen which is initially hidden and becomes functional only when the molecule is broken or degraded.

cryptovolcanic [GEOL] A small, nearly circular area of highly disturbed strata in which there is no evidence of volcanic materials to confirm the origin as being volcanic.

cryptoxanthin [BIOCHEM] $C_{40}H_{57}O$ A xanthophyll carotenoid pigment found in plants; convertible to vitamin A by many animal livers.

cryptozoon [PALEOBOT] A hemispherical or cabbagelike reef-forming fossil algae, probably from the Cambrian and Ordovician.

crypts of Lieberkühn [ANAT] Simple, tubular glands which arise as evaginations into the mucosa of the small intestine.

crystal [CRYSTAL] A homogeneous solid made up of an element, chemical compound or isomorphous mixture throughout which the atoms or molecules are arranged in a regularly repeating pattern. [ELECTR] A natural or synthetic piezoelectric or semiconductor material whose atoms are arranged with some degree of geometric regularity. [MINERAL] *See* rock crystal.

crystal activity [ELECTR] A measure of the amplitude of vibration of a piezoelectric crystal plate under specified conditions.

crystal aerugo *See* cupric acetate.

crystal-audio receiver [ELECTR] Similar to the crystal-video receiver, except for the path detection bandwidth which is audio rather than video.

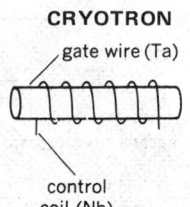

CRYOTRON

gate wire (Ta)

control coil (Nb)

The wire-wound cryotron operated at temperatures close to absolute zero.

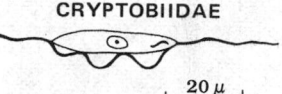

CRYPTOBIIDAE

20 μ

Cryptobia, a genus in Cryptobiidae showing the characteristic two flagella, one free and one with undulating membrane.

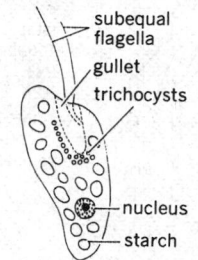

CRYPTOMONADIDA

subequal flagella
gullet
trichocysts
nucleus
starch

Chilomonas paramecium, an example of a cryptomonad.

CRYSTAL COUNTER

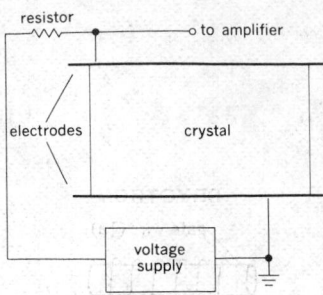

Circuit diagram of a crystal counter.

CRYSTAL GROWTH

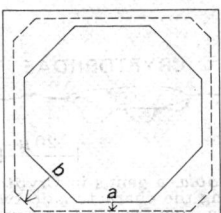

Schematic representation of cross section of crystal at three stages of growth; *a* represents slower growing faces, *b* faster growing faces.

CRYSTAL HEADPHONES

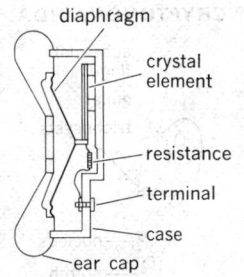

Section through a crystal headphone.

CRYSTALLITE

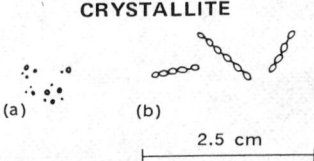

2.5 cm

Examples of crystallites. *(a)* Globulites and *(b)* margarites.

crystal axis [CRYSTAL] A reference axis used for the vectoral properties of a crystal.

crystal blank [ELECTR] The result of the final cutting operation on a piezoelectric or semiconductor crystal.

crystal calibrator [ELECTR] A crystal-controlled oscillator used as a reference standard to check frequencies.

crystal cartridge [ENG ACOUS] A piezoelectric unit used with a stylus in a phonograph pickup to convert disk recordings into audio-frequency signals, or used with a diaphragm in a crystal microphone to convert sound waves into af signals.

crystal chemistry [CRYSTAL] The study of the crystalline structure and properties of a mineral or other solid.

crystal class [CRYSTAL] One of 32 categories of crystals according to the inversions, rotations about an axis, reflections, and combinations of these which leaves the crystal invariant. Also known as symmetry class.

crystal clock [HOROL] A clock which uses the mechanical resonance of a crystal plate coupled piezoelectrically into an electronic circuit.

crystal control [ELECTR] Control of the frequency of an oscillator by means of a quartz crystal unit.

crystal-controlled oscillator [ELECTR] An oscillator whose frequency of operation is controlled by a crystal unit.

crystal-controlled transmitter [ELECTR] A transmitter whose carrier frequency is directly controlled by the electromechanical characteristics of a quartz crystal unit.

crystal counter [NUCLEO] A particle detector in which the sensitive material is a dielectric (nonconducting) crystal mounted between two metallic electrodes.

crystal current [ELECTR] The actual alternating current flowing through a crystal unit.

crystal cutter [ENG ACOUS] A cutter in which the mechanical displacements of the recording stylus are derived from the deformations of a crystal having piezoelectric properties.

crystal defect [CRYSTAL] Any departure from crystal symmetry caused by free surfaces, disorder, impurities, vacancies and interstitials, dislocations, lattice vibrations, and grain boundaries. Also known as lattice defect.

crystal detector [ELECTR] 1. A crystal diode, or an equivalent earlier crystal-catwhisker combination, used to rectify a modulated radio-frequency signal to obtain the audio or video signal directly. 2. A crystal diode used in a microwave receiver to combine an incoming radio-frequency signal with a local oscillator signal to produce an intermediate-frequency signal.

crystal diffraction [SOLID STATE] Diffraction by a crystal of beams of x-rays, neutrons, or electrons whose wavelengths (or de Broglie wavelengths) are comparable with the interatomic spacing of the crystal.

crystal diffraction spectrometer [PHYS] An instrument utilizing diffraction of waves by a crystal lattice to measure wavelengths or crystal properties. Also known as crystal spectrometer. [SOLID STATE] See Bragg spectrometer.

crystal diode See semiconductor diode.

crystal dynamics See lattice dynamics.

crystal face [CRYSTAL] One of the outward planar surfaces which define a crystal and reflect its internal structure. Also known as face.

crystal field theory [PHYS CHEM] The theory which assumes that the ligands of a coordination compound are the sources of negative charge which perturb the energy levels of the central metal ion and thus subject the metal ion to an electric field analogous to that within an ionic crystalline lattice.

crystal filter [ELECTR] A highly selective tuned circuit employing one or more quartz crystals; sometimes used in intermediate-frequency amplifiers of communication receivers to improve the selectivity.

crystal glass [MATER] A water-clear lead glass which polishes readily and has a high index of refraction.

crystal gliding [CRYSTAL] Slip along a crystal plane due to plastic deformation; often produces crystal twins. Also known as translation gliding.

crystal grating [SPECT] A diffraction grating for gamma rays or x-rays which uses the equally spaced lattice planes of a crystal.

crystal growth [CRYSTAL] The growth of a crystal, which involves diffusion of the molecules of the crystallizing sub-

stance to the surface of the crystal, diffusion of these molecules over the crystal surface to special sites on the surface, incorporation of molecules into the surface at these sites, and diffusion of heat away from the surface.

crystal habit [CRYSTAL] The size and shape of the crystals in a crystalline solid. Also known as habit.

crystal harmonic generator [ELECTR] A crystal-controlled oscillator which produces an output rich in harmonics (overtones or multiples) of its fundamental frequency.

crystal headphones [ENG ACOUS] Headphones using Rochelle salt or other crystal elements to convert audio-frequency signals into sound waves. Also known as ceramic earphones.

crystal holder [DES ENG] A housing designed to provide proper support, mechanical protection, and connections for a quartz crystal plate.

crystal hydrophone [ENG ACOUS] A crystal microphone that responds to waterborne sound waves.

crystal indices See Miller indices.

crystal laser [OPTICS] A laser that uses a pure crystal of ruby or other material for generating a coherent beam of output light.

crystal lattice [CRYSTAL] A lattice from which the structure of a crystal may be obtained by associating with every lattice point an assembly of atoms identical in composition, arrangement, and orientation. Also known as lattice; space lattice.

crystalliferous bacteria [MICROBIO] *Bacillus thuringiensis* and related species characterized by the formation of a protein crystal in the sporangium at the time of spore formation.

crystalline [CRYSTAL] Of, pertaining to, resembling, or composed of crystals.

crystalline alumina [MATER] An abrasive which consists of essentially the same mineral as corundum, but whose physical properties such as crystal structure, size, and shape of grain are so controlled as to produce the most desirable abrasives for specific types of grinding.

crystalline anisotropy [SOLID STATE] The tendency of crystals to have different properties in different directions; for example, a ferromagnet will spontaneously magnetize along certain crystallographic axes.

crystalline chloral See chloral hydrate.

crystalline double refraction [OPTICS] The splitting which a wavefront experiences when a wave disturbance propagates through an anisotropic crystal.

crystalline field [SOLID STATE] The internal electric field in a solid due to localized charges, especially ions, inside.

crystalline fracture [MET] A break in a polycrystalline metal, with the fractured surface having a grainy appearance.

crystalline frost [HYD] Hoarfrost that exhibits a relatively simple macroscopic crystalline structure.

crystalline-granular texture [PETR] A primary texture of an igneous rock due to crystallization from a fluid medium.

crystalline lens See lens.

crystalline polymer [CHEM] A polymer whose sections of adjacent chains are packed in a regular array.

crystalline porosity [GEOL] Porosity in crystalline limestone and dolomite, making possible underground oil reservoirs.

crystalline rock [PETR] 1. Rock made up of minerals in a clearly crystalline state. 2. Igneous and metamorphic rock, as opposed to sedimentary rock.

crystallinity [CRYSTAL] The quality or state of being crystalline. [PETR] Degree of crystallization exhibited by igneous rock.

crystallite [GEOL] A small, rudimentary form of crystal which is of unknown mineralogic composition and which does not polarize light.

crystallization [CRYSTAL] The formation of crystalline substances from solutions or melts.

crystallization differentiation See fractional crystallization.

crystallizer [CHEM ENG] Process vessel within which dissolved solids in a supersaturated solution are forced out of solution by cooling or evaporation, and then recovered as solid crystals.

crystalloblast [MINERAL] A mineral crystal produced by metamorphic processes.

crystalloblastic series [GEOL] A series of metamorphic minerals ordered according to decreasing formation energy, so crystals of a listed mineral have a tendency to form idioblastic outlines at surfaces of contact with simultaneously developed crystals of all minerals in lower positions.

crystalloblastic texture [GEOL] A crystalline texture resulting from metamorphic recrystallization under conditions of high viscosity and directed pressure.

crystallogram [CRYSTAL] A photograph of the x-ray diffraction pattern of a crystal.

crystallographic axis [CRYSTAL] One of three lines (sometimes four, in the case of a hexagonal crystal), passing through a common point, that are chosen to have definite relation to the symmetry properties of a crystal, and are used as a reference in describing crystal symmetry and structure.

crystallography [PHYS] The branch of science that deals with the geometric description of crystals and their internal arrangement.

crystallomagnetic [SOLID STATE] Pertaining to magnetic properties of crystals.

crystallophobia [PSYCH] An abnormal fear of glass.

crystal loudspeaker [ENG ACOUS] A loudspeaker in which movements of the diaphragm are produced by a piezoelectric crystal unit that twists or bends under the influence of the applied audio-frequency signal voltage. Also known as piezoelectric loudspeaker.

crystal microphone [ENG ACOUS] A microphone in which deformation of a piezoelectric bar by the action of sound waves or mechanical vibrations generates the output voltage between the faces of the bar. Also known as piezoelectric microphone.

crystal mixer [ELECTR] A mixer that uses the nonlinear characteristic of a crystal diode to mix two frequencies; widely used in radar receivers to convert the received radar signal to a lower intermediate-frequency value by mixing it with a local oscillator signal.

crystal monochromator [SPECT] A spectrometer in which a collimated beam of slow neutrons from a reactor is incident on a single crystal of copper, lead, or other element mounted on a divided circle.

crystal operation [ELECTR] Operation using crystal-controlled oscillators.

crystal optics [OPTICS] The study of the propagation of light, and associated phenomena, in crystalline solids.

crystal oscillator [ELECTR] An oscillator in which the frequency of the alternating-current output is determined by the mechanical properties of a piezoelectric crystal. Also known as piezoelectric oscillator.

crystal oven [ENG] A temperature-controlled oven in which a crystal unit is operated to stabilize its temperature and thereby minimize frequency drift.

crystal phase [MET] A crystal structure formed by an alloy over a certain range of values of the relative proportions of its constituents.

crystal pickup [ENG ACOUS] A phonograph pickup in which movements of the needle in the record groove cause deformation of a piezoelectric crystal, thereby generating an audio-frequency output voltage between opposite faces of the crystal. Also known as piezoelectric pickup.

crystal plate [ELECTR] A precisely cut slab of quartz crystal that has been lapped to final dimensions, etched to improve stability and efficiency, and coated with metal on its major surfaces for connecting purposes. Also known as quartz plate.

crystal pulling [CRYSTAL] A method of crystal growing in which the developing crystal is gradually withdrawn from a melt.

crystal rectifier *See* semiconductor diode.

crystal resonator [ELECTR] A precisely cut piezoelectric crystal whose natural frequency of vibration is used to control or stabilize the frequency of an oscillator. Also known as piezoelectric resonator.

crystal sandstone [GEOL] Siliceous sandstone in which deposited silica is precipitated upon the quartz grains in crystalline position.

crystal set [ELECTR] A radio receiver having a crystal detector stage for demodulation of the received signals, but no amplifier stages.

crystal settling [GEOL] Sinking of crystals in magma from the liquid in which they formed, by the action of gravity.

crystal shutter [ELECTROMAG] Mechanical waveguide or coaxial-cable shorting switch that, when closed, prevents undesired radio-frequency energy from reaching and damaging a crystal detector.

crystals of Venus *See* cupric acetate.

crystal spectrometer *See* crystal diffraction spectrometer.

crystal-stabilized transmitter [ELECTR] A transmitter employing automatic frequency control, in which the reference frequency is that of a crystal oscillator.

crystal structure [CRYSTAL] The arrangement of atoms or ions in a crystalline solid.

crystal symmetry [CRYSTAL] The existence of nontrivial operations, consisting of inversions, rotations around an axis, reflections, and combinations of these, which bring a crystal into a position indistinguishable from its original position.

crystal system [CRYSTAL] One of seven categories (cubic, hexagonal, tetragonal, trigonal, orthorhombic, monoclinic, and triclinic) into which a crystal may be classified according to the shape of the unit cell of its Bravais lattice, or according to the dominant symmetry elements of its crystal class.

crystal transducer [ELECTR] A transducer in which a piezoelectric crystal serves as the sensing element.

crystal tuff [GEOL] Consolidated volcanic ash in which crystals and crystal fragments predominate.

crystal twin *See* twin.

crystal unit [ELECTR] A complete assembly of one or more quartz plates in a crystal holder.

crystal video receiver [ELECTR] A broad-tuning radar or other microwave receiver consisting only of a crystal detector and a video or audio amplifier.

crystal video rectifier [ELECTR] A crystal rectifier transforming a high-frequency signal directly into a video-frequency signal.

crystal-vitric tuff [GEOL] Consolidated volcanic ash composed of 50–75% crystal fragments and 25–50% glass fragments.

crystal whiskers [CRYSTAL] Single crystals that have grown in a filamentary form.

crystogen *See* cystamine.

Crystolon [MATER] Trade name for a brand of silicon carbide used as an abrasive.

crystosphene [HYD] A buried sheet or mass of ice, as in the tundra of northern America, formed by the freezing of rising and spreading springwater beneath alluvial deposits.

cs *See* centistoke.

Cs *See* cesium; cirrostratus cloud.

csc *See* cosecant.

C scan *See* C scope.

csch *See* hyperbolic cosecant.

C scope [ELECTR] A cathode-ray scope on which signals appear as spots, with bearing angle as the horizontal coordinate and elevation angle as the vertical coordinate. Also known as C indicator; C scan.

C size [ENG] One of a series of sizes to which trimmed paper and board are manufactured; for size CN, with N equal to any integer, the length of the longer side is $2^{3/8-N/2}$ meters, while the length of the shorter side is $2^{1/8-N/2}$ meters, with both lengths rounded off to the nearest millimeter.

CSP *See* control switching point.

C-stage resin [ORG CHEM] A material used in the final stage of preparing certain thermosetting resins, at which time the material is relatively insoluble and infusible; the resin in a fully cured thermoset molding is in this stage. Also known as resite.

CSW *See* channel status word.

CT *See* center tap.

CTC *See* centralized traffic control.

CTD recorder *See* salinity-temperature-depth recorder.

ctenidium [INV ZOO] 1. The comb- or featherlike respiratory apparatus of certain mollusks. 2. A row of spines on the head or thorax of some fleas.

Ctenodrilidae [INV ZOO] A family of fringe worms belonging to the Sedentaria.

ctenoid scale [VERT ZOO] A thin, acellular structure composed of bonelike material and characterized by a serrated margin; found in the skin of advanced teleosts.

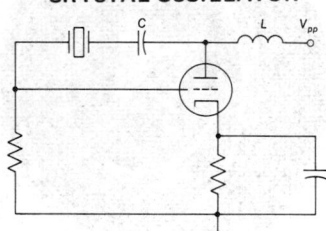

CRYSTAL OSCILLATOR

Circuit diagram of Pierce crystal oscillator; C is capacitor, L is inductor, and V_{pp} is plate voltage.

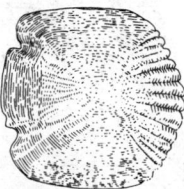

CTENOID SCALE

Ctenoid scale from carp. *(From K. F. Lagler et al., Ichthyology, Wiley, 1962)*

CTENOPHORA

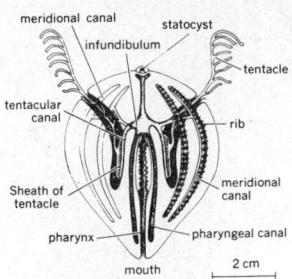

Structure of a cydippid ctenophore.

CTENOTHRISSIFORMES

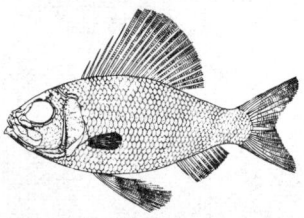

Cretaceous ctenothrissid, *Ctenothrissus radians*; length 10–12 inches (25–30 centimeters). (After C. Patterson, Phil. Trans. Roy. Soc. London Ser. B. Biol. Sci. no. 739, vol. 247, 1964)

C-TUBE BOURDON ELEMENT

C-tube bourdon element with mechanism composing an industrial pressure gage. (Foxboro Co.)

CUCKOO

Black-billed cuckoo, brown above and white underneath with thin red rings around eyes, maximum length 12 inches (30 centimeters).

Ctenophora [INV ZOO] The comb jellies, a phylum of marine organisms having eight rows of comblike plates as the main locomotory structure.

Ctenostomata [INV ZOO] An order of bryozoans in the class Gymnolaemata recognized as inconspicuous, delicate colonies made up of relatively isolated, short, tubular zooecia with chitinous walls.

Ctenostomatida [INV ZOO] The equivalent name for Odontostomatida.

Ctenothrissidae [PALEON] A family of extinct teleostean fishes in the order Ctenothrissiformes.

Ctenothrissiformes [PALEON] A small order of extinct teleostean fishes; important as a group on the evolutionary line leading from the soft-rayed to the spiny-rayed fishes.

CTR *See* controlled thermonuclear reactor.

C-tube bourdon element [ENG] Hollow tube of flexible (elastic) metal shaped like the arc of a circle; changes in internal gas or liquid pressure flexes the tube to a degree related to the pressure change; used to measure process-stream pressures.

cu *See* crosstalk unit; cubic.

Cu *See* copper.

cubanite [MINERAL] $CuFe_2S_3$ Bronze-yellow mineral that crystallizes in the orthorhombic system. Also known as chalmersite.

cube [MATH] Regular polyhedron whose faces are all square.

cubeb [BOT] The dried, nearly ripe fruit (berries) of a climbing vine, *Piper cubeba*, of the pepper family (Piperaceae).

cubeb oil *See* oil of cubeb.

cube of a number [MATH] The new number obtained by taking the three-fold product of the given number a with itself: $a \times a \times a$.

cube ore *See* pharmacosiderite.

cube root of a number [MATH] Another number whose cube is the original number.

cubic [MECH] Denoting a unit of volume, so that if x is a unit of length, a cubic x is the volume of a cube whose sides have length $1x$; for example, a cubic meter, or a meter cubed, is the volume of a cube whose sides have a length of 1 meter. Abbreviated cu.

cubical antenna [ELECTROMAG] An antenna array, the elements of which are positioned to form a cube.

cubical dilation [MECH] The isotropic part of the strain tensor describing the deformation of an elastic solid, equal to the fractional increase in volume.

cubical expansion [PHYS] The increase in volume of a substance with a change in temperature or pressure.

cubic centimeter [MECH] A unit of volume equal to the volume of a cube whose edges are 1 centimeter long. Abbreviated cc; ccm; cm³.

cubic crystal [CRYSTAL] A crystal whose lattice has a unit cell with perpendicular axes of equal length.

cubic curve [MATH] A plane curve which has an equation of the form $f(x, y) = 0$, where $f(x,y)$ is a polynomial of degree three in x and y.

cubic determinant [MATH] A mathematical form analogous to an ordinary determinant, with the elements forming a cube instead of a square.

cubic equation [MATH] A polynomial equation with no exponent larger than 3.

cubic foot [MECH] A unit of volume equal to the volume of a cube whose edges are 1 foot long; equal to $2.8316846592 \times 10^{-2}$ cubic meter.

cubic foot per minute [MECH] A unit of volume flow rate, equal to a uniform flow of 1 cubic foot in 1 minute; equal to 1/60 cusec. Abbreviated cfm.

cubic foot per second *See* cusec.

cubic inch [MECH] A unit of volume equal to the volume of a cube whose edges are 1 inch long; equal to 1.6387064×10^{-5} cubic meter.

cubicle [BUILD] Any small, approximately square room or compartment. [ENG] An enclosure for high-voltage equipment.

cubic measure [MECH] A unit or set of units to measure volume.

cubic meter [MECH] A unit of volume equal to the volume of a cube whose edges are 1 meter long.

cubic packing [CRYSTAL] The spacing pattern of uniform solid spheres in a clastic sediment or crystal lattice in which the unit cell is a cube.

cubic plane [CRYSTAL] A plane that is at right angles to any one of the three crystallographic axes of the cubic system.

cubic polynomial [MATH] A polynomial in which all exponents are no greater than 3.

cubic spline [MATH] One of a collection of cubic polynomials used in interpolating a function whose value is specified at each of a collection of distinct ordered values, X_i ($i = 1, \ldots, n$), and whose slope is specified at X_1 and X_n; one cubic polynomial is found for each interval, such that the interpolating system has the prescribed values at each of the X_i, the prescribed slope at X_1 and X_n, and a continuous slope at each of the X_i.

cubic system *See* isometric system.

cuboctahedron [MATH] A polyhedron whose faces consist of six equal squares and eight equal equilateral triangles, and which can be formed by cutting the corners off a cube; it is one of the 13 Archimedean solids. Also spelled cubooctahedron.

cuboid [ANAT] The outermost distal tarsal bone in vertebrates. [INV ZOO] Main vein of the wing in many insects, particularly the flies (Diptera). [SCI TECH] Nearly cubic in shape.

cuboidal epithelium [HISTOL] A single-layered epithelium made up of cubelike cells.

Cubomedusae [INV ZOO] An order of coelenterates in the class Scyphozoa distinguished by a cubic umbrella.

cubooctahedron *See* cuboctahedron.

Cuccia coupler *See* electron coupler.

cuckoo [VERT ZOO] The common name for about 130 species of primarily arboreal birds in the family Cuculidae; some are social parasites.

cucoloris [GRAPHICS] Material having a cutout that allows light to pass through it and project a form.

Cucujidae [INV ZOO] The flat-back beetles, a family of predatory coleopteran insects in the superfamily Cucujoidea.

Cucujoidea [INV ZOO] A large superfamily of coleopteran insects in the suborder Polyphaga.

Cuculidae [VERT ZOO] A family of perching birds in the order Cuculiformes, including the cuckoos and the roadrunner, characterized by long tails, heavy beaks and conspicuous lashes.

Cuculiformes [VERT ZOO] An order of birds containing the cuckoos and allies, characterized by the zygodactyl arrangement of the toes.

Cucumariidae [INV ZOO] A family of dendrochirotacean holothurian echinoderms in the order Dendrochirotida.

cucumber [BOT] *Cucumis sativus*. An annual cucurbit, in the family Cucurbitaceae grown for its edible, immature fleshy fruit.

cucumber mildew [PL PATH] **1.** A downy mildew of cucumbers and melons caused by *Peronoplasmopara cubensis*. **2.** A powdery mildew of cucumbers and melons caused by *Erysiphe cichoracearum*.

cucumber mosaic [PL PATH] A virus disease of cucumbers and related fruits, producing mottling of terminal leaves and fruits and dwarfing of vines.

cucurbit wilt [PL PATH] A bacterial disease of cucumbers and related plants caused by *Erwinia tracheiphila*, characterized by sudden wilting of the plant.

cue circuit [ELECTR] A one-way communication circuit used to convey program control information.

cuesta [GEOGR] A gently sloping plain which terminates in a steep slope on one side.

cul-de-sac [ANAT] Blind pouch or diverticulum. [CIV ENG] A dead-end street with a circular area for turning around. [MIN ENG] *See* dead end.

culdoscope [MED] An instrument used to visualize female pelvic organs, introduced through the vagina or a perforation into the retrouterine pouch.

Culex [INV ZOO] A genus of mosquitoes important as vectors for malaria and several filarial parasites.

Culicidae [INV ZOO] The mosquitoes, a family of slender, orthorrhaphous dipteran insects in the series Nematocera having long legs and piercing mouthparts.

Culicinae [INV ZOO] A subfamily of the dipteran family Culicidae.

Cullender isochronal method [PETRO ENG] Procedure for back-pressure testing to analyze gas wells that produce from low-permeability reservoirs with a consequent slow approach to stabilized producing conditions.

cullet *See* collet.

cullis *See* coulisse.

culm [BOT] **1.** A jointed, usually hollow grass stem. **2.** The solid stem of certain monocotyledons, such as the sedges. [MIN ENG] Fine, refuse coal, screened and separated from larger pieces.

culmen [VERT ZOO] The edge of the upper bill in birds.

culmination [ASTRON] **1.** The position of a heavenly body when at highest apparent altitude. **2.** For a heavenly body which is continually above the horizon, the position of lowest apparent altitude. [GEOL] A high point on the axis of a fold.

cultellation [ENG] Transferring a surveyed point from a high level (such as on overhang) to a lower level by dropping a marking pin.

cultigen [BIOL] A cultivated variety or species of organism for which there is no known wild ancestor. Also known as cultivar.

cultivar *See* cultigen.

cultivate [AGR] To prepare soil for the raising of crops.

cultivator [AGR] A farm implement pulled behind a powered machine that is used to break up soil, kill weeds, and create a surface mulch for moisture.

cultural anthropology [ANTHRO] The division of anthropology dealing with the study of all aspects of culture.

cultural-familial mental retardation [PSYCH] Subnormal general intellectual functioning, usually borderline or mild, presumably on the basis of some degree of environmental deprivation resulting from familial retardation as evidenced by its presence in one parent and one or more siblings.

culture [ANTHRO] The complex pattern of behavior that distinguishes a social, ethnic, or religious group. [BIOL] A growth of living cells or microorganisms in a controlled artificial environment.

cultured pearl [INV ZOO] A natural pearl grown by controlled stimulation of the oyster.

culture medium [MICROBIO] The nutrients and other organic and inorganic materials used for the growth of microorganisms and plant and animal tissue in culture.

culvert [ENG] A covered channel or a large-diameter pipe that takes a watercourse below ground level.

Cumacea [INV ZOO] An order of the class Crustacea characterized by a well-developed carapace which is fused dorsally with at least the first three thoracic somites and overhangs the sides.

cumatophyte [ECOL] A plant that grows under surf conditions.

cumberlandite [PETR] A coarse-grained, ultramafic, ultrabasic rock composed principally of olivine crystals in a ground mass of magnetite and ilmenite with minor plagioclase.

cumbraite [PETR] A variety of dacite or rhyodacite containing very calcic plagioclase and pyroxene in a glassy groundmass.

cumec [MECH] A unit of volume flow rate equal to 1 cubic meter per second.

cumene [ORG CHEM] $C_6H_5CH(CH_3)_2$ A colorless, oily benzenoid hydrocarbon cooling at 152.4°C; used as an additive for high-octane motor fuel. Also known as isopropylbenzene.

cumene hydroperoxide [ORG CHEM] $C_6H_5C(CH_3)_2OOH$ An isopropyl hydroperoxide of cumene; an oily liquid, used to make phenol and acetone.

cumengite [MINERAL] $Pb_4Cu_4Cl_8(OH)_8 \cdot H_2O$ A deep-blue or light-indigo-blue tetragonal mineral consisting of a basic lead-copper chloride occurring in crystals.

cumidine [ORG CHEM] $C_9H_{13}N$ A colorless, water-insoluble liquid, boiling at 225°C. Also known as *para*-isopropylaniline.

cumin [BOT] *Cuminum cyminum.* An annual herb in the family Umbelliferae; the fruit is valuable for its edible, aromatic seeds.

cummingtonite [MINERAL] $(Fe,Mg)_7Si_8O_{22}(OH)_2$ A brownish mineral that crystallizes in the monoclinic system; usually occurs as lamellae or fibers in metamorphic rocks.

cum sole [GEOPHYS] With the sun; hence anticyclonic or clockwise in the Northern Hemisphere.

cumulants [STAT] A set of parameters k_h $(h = 1, \ldots, r)$ of a one-dimensional probability distribution defined by

$$ln\, \chi_x(q) = \sum_{h=1}^{r} k_h[(iq)^h/h!] + o(q^r)$$

where $\chi_x(q)$ is the characteristic function of the probability distribution of x. Also known as semi-invariants.

cumulate [PETR] Any igneous rock formed by the accumulation of crystals settling out of a magma.

cumulated double bond [CHEM] Two double bonds on the same carbon atom, as in $>C=C=C<$.

cumulative compound generator [ELEC] A compound generator in which the series field is connected to aid the shunt field magnetomotive force.

cumulative compound motor [MECH ENG] A motor with operating characteristics between those of the constant-speed (shunt-wound) and the variable-speed (series-wound) types.

cumulative dose [NUCLEO] The total dose resulting from repeated exposures to radiation.

cumulative error [STAT] An error whose magnitude does not approach zero as the number of observations increases. Also known as accumulative error.

cumulative excitation [ATOM PHYS] Process by which the atom is raised from one excited state to a higher state by collision, for example, with an electron.

cumulative gas [PETRO ENG] Measurement of total gas produced from a reservoir, usually expressed in graphical relationship to total (cumulative) oil produced from the same reservoir.

cumulative ionization [ATOM PHYS] Ionization of an excited atom in the metastable state by cumulative excitation. [ELECTR] *See* avalanche.

cumulative sum chart [IND ENG] A statistical control chart on which the cumulative sum of deviations is plotted over a period of time and which often has a sliding V-shaped mask for comparing the plot with allowable limits. Also known as cusum chart.

cumuliform cloud [METEOROL] A fundamental cloud type, showing vertical development in the form of rising mounds, domes, or towers.

cumulonimbus calvus [METEOROL] A species of cumulonimbus cloud which is characterized by the thunderstorm or shower it causes or by the wisps or falling trails of precipitation, but in which at first no cirriform parts can be made out. Also known as calvus.

cumulonimbus calvus cloud [METEOROL] A species of cumulonimbus cloud evolving from cumulus congestus: the protuberances of the upper portion have begun to lose the cumuliform outline; they loom and usually flatten, then transform into a whitish mass with a more or less diffuse outline and vertical striation; cirriform cloud is not present, but the transformation into ice crystals often proceeds with great rapidity.

cumulonimbus capillatus cloud [METEOROL] A species of cumulonimbus cloud characterized by the presence of distinct cirriform parts, frequently in the form of an anvil, a plume, or a vast and more or less disorderly mass of hair, and usually accompanied by a thunderstorm.

cumulonimbus cloud [METEOROL] A principal cloud type, exceptionally dense and vertically developed, occurring either as isolated clouds or as a line or wall of clouds with separated upper portions.

cumulus cloud [METEOROL] A principal cloud type in the form of individual, detached elements which are generally dense and possess sharp nonfibrous outlines; these elements develop vertically, appearing as rising mounds, domes, or towers, the upper parts of which often resemble a cauliflower.

cumulus congestus cloud [METEOROL] A strongly sprouting cumulus species with generally sharp outline and sometimes a great vertical development, and with cauliflower or tower aspect.

cumulus humilis cloud [METEOROL] A species of cumulus

CULTIVATOR

Coil-shank field cultivator.
(Allis-Chalmers)

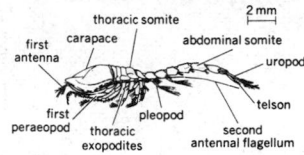

CUMACEA

Typical adult male cumacean.

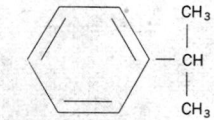

CUMENE

Cumene structural formula.

CUMULUS CLOUD

Small cumulus cloud formation.
(U.S. Weather Bureau)

cloud characterized by small vertical development and a generally flattened appearance, vertical growth is usually restricted by the existence of a temperature inversion in the atmosphere, which in turn explains the unusually uniform height of the cloud. Also known as fair-weather cumulus.

cumulus mediocris cloud [METEOROL] A cloud species unique to the species cumulus, of moderate vertical development, the upper protuberances or sproutings being not very marked; there may be a small cauliflower aspect; while this species does not give any precipitation, it frequently develops into cumulus congestus and cumulonimbus.

cumulus oophorus [HISTOL] The layer of gelatinous, follicle cells surrounding the ovum in a Graafian follicle.

cuneate [BIOL] Wedge-shaped with the acute angle near the base, as in certain insect wings and the leaves of various plants.

cuneiform [ANAT] **1.** Any of three wedge-shaped tarsal bones. **2.** Either of a pair of cartilages lying dorsal to the thyroid cartilage of the larynx. **3.** Wedge-shaped, chiefly referring to skeletal elements.

cunnus [ANAT] The vulva.

cup [DES ENG] A cylindrical part with only one end open. [MET] Sheet metal part formed during the first deep-drawing operation.

Cup *See* Crater.

cup-and-ball joint [GEOL] A dish-shaped transverse fracture which divides a basalt column into segments. Also known as ball-and-socket jointing.

cup-and-cone fracture *See* cup fracture.

cup anemometer [ENG] A rotation anemometer, usually consisting of three or four hemispherical or conical cups mounted with their diametral planes vertical and distributed symmetrically about the axis of rotation; the rate of rotation of the cups, which is a measure of the wind speed, is determined by a counter.

cup barometer [ENG] A barometer in which one end of a graduated glass tube is immersed in a cup, both cup and tube containing mercury.

cup-case thermometer [ENG] Total-immersion type of thermometer with a cup container at the bulb end to hold a specified amount and depth of the material whose temperature is to be measured.

cup crystal [HYD] A crystal of ice in the form of a hollow hexagonal cup; a common form of depth hoar.

Cupedidae [INV ZOO] The reticulated beetles, the single family of the coleopteran suborder Archostemata.

cupel [MET] Cup made of bone ash used in assaying precious metals.

cupellation [MET] **1.** Method using a cupel for assaying precious metals. **2.** Process for refining gold and silver by alloying them with lead and then oxidizing the molten lead to separate the base metal from the precious metal.

cupferron [ORG CHEM] $NH_4ONONC_6H_5$ A colorless salt that forms crystals melting at 164°C; its acid solution is a precipitating reagent. Also known as copperon; nitrosophenylhydroxylamine.

cup fracture [MET] A break in a ductile material under tensile stress in which the surface of failure on one piece has a central flat area with an exterior extended rim. Also known as cup-and-cone fracture.

cup grease [MATER] A lubricating grease, usually lime base, for many applications.

cupola [GEOL] An isolated, upward-projecting body of plutonic rock that lies near a larger body; both bodies are presumed to unite at depth. [MET] A vertical cylindrical furnace for melting gray iron for foundry use; the metal, coke, and flux are put into the top of the furnace onto a bed of coke through which air is blown. Also known as furnace cupola.

cupping [MET] **1.** First operation of a deep-drawing process. **2.** Fracture of a wire or rod in which one fracture surface is conical and the other concave.

cup product [MATH] A multiplication defined on cohomology classes; it gives cohomology a ring structure.

cuprammonium cellulose [TEXT] The cellulose of cotton linters treated with copper sulfate and ammonium.

cuprammonium process [TEXT] A process by which rayon is made from regenerated cellulose in a solution of ammoniacal copper oxide.

cuprammonium rayon [TEXT] Rayon made from cuprammonium cellulose.

cupreous [SCI TECH] Containing or resembling copper.

cupric [CHEM] The divalent ion of copper.

cupric acetate [ORG CHEM] $Cu(C_2H_3O_2)_2 \cdot H_2O$ Blue-green crystals, soluble in water; used as a raw material to make paris green. Also known as copper acetate; crystal aerugo; crystals of Venus; verdigris.

cupric bromide [INORG CHEM] $CuBr_2$ Black prismatic crystals; used in photography as an intensifier and in organic synthesis as a brominating agent. Also known as copper bromide.

cupric chloride [INORG CHEM] Also known as copper chloride. **1.** $CuCl_2$ Yellowish-brown, deliquescent powder soluble in water, alcohol, and ammonium chloride. **2.** $CuCl_2 \cdot H_2O$ A dihydrate of cupric chloride forming green crystals soluble in water; used as a mordant in dyeing and printing textile fabrics and in the refining of copper, gold, and silver.

cupric chromate [INORG CHEM] $CuCrO_4$ A yellow liquid, used as a mordant. Also known as copper chromate.

cupric cyanide [INORG CHEM] $Cu(CN)_2$ A green powder, insoluble in water; used in electroplating copper on iron. Also known as copper cyanide.

cupric fluoride [INORG CHEM] CuF_2 White crystalline powder used in ceramics and in the preparation of brazing and soldering fluxes. Also known as copper fluoride.

cupric hydroxide [INORG CHEM] $Cu(OH)_2$ Blue macro- or microscopic crystals; used as a mordant and pigment, in manufacture of many copper salts, and for staining paper. Also known as copper hydroxide.

cupric nitrate [INORG CHEM] $Cu(NO_3)_2 \cdot 3H_2O$ Green powder or blue crystals soluble in water; used in electroplating copper on iron. Also known as copper nitrate.

cupric oxide [INORG CHEM] CuO Black, monoclinic crystals, insoluble in water; used in making fibers and ceramics, and in organic and gas analyses. Also known as copper oxide.

cupric sulfate [INORG CHEM] $CuSO_4$ A water-soluble salt used in copper-plating baths; crystallizes as hydrous copper sulfate, which is blue. Also known as copper sulfate.

cuprite [MINERAL] Cu_2O A red mineral that crystallizes in the isometric system and is found in crystals and fine-grained aggregates or is massive; a widespread supergene copper ore. Also known as octahedral copper ore; red copper ore; ruby copper ore.

cuprodescloizite *See* mottramite.

cupronickel [MET] A copper-base alloy with 10–30% nickel and small amounts of manganese and iron; used in industrial and marine installations as condenser and heat-exchanger tubing.

cuprotungstite [MINERAL] $Cu_2(WO_4)(OH)_2$ A green mineral that forms compact masses; soluble in acids; the crystal system is not known.

cuprouranite *See* torbernite.

cuprous bromide [INORG CHEM] Cu_2Br_2 White or gray crystals slightly soluble in cold water. Also known as copper bromide.

cuprous chloride [INORG CHEM] $CuCl$ or Cu_2Cl_2 Green, tetrahedral crystals, insoluble in water. Also known as copper chloride; resin of copper.

cuprous fluoride [INORG CHEM] Cu_2F_2 Red, crystalline powder, melting point 908°C. Also known as copper fluoride.

cuprous oxide [INORG CHEM] Cu_2O An oxide of copper found in nature as cuprite and formed on copper by heat; used chiefly as a pigment and as a fungicide. Also known as copper oxide.

cupule [BOT] **1.** The cup-shaped involucre characteristic of oaks. **2.** A cup-shaped corolla. **3.** The gemmae cup of the Marchantiales. [INV ZOO] A small sucker on the feet of certain male flies.

curare [MATER] A poisonous extract from the plant *Strychnos toxifera* containing alkaloids that produce paralysis of the voluntary muscles by acting on synaptic junctions; used as an adjunct to anesthesia in surgery.

CUPRITE

Crystals and veins of cuprite in limonite-bearing rocks. *(Specimen from Department of Geology, Bryn Mawr College)*

curb [CIV ENG] A border of concrete or row of joined stones forming part of a gutter along a street edge. [MIN ENG] A timber frame, circular or square, wedged in a shaft to make a foundation for walling or tubbing, or to support, with or without other timbering, the walls of the shaft.

curb roof [ARCH] A roof with a ridge at its center and a parallel lower ridge on both sloping sides.

Curculionidae [INV ZOO] The true weevils or snout beetles, a family of coleopteran insects in the superfamily Curculionoidea.

Curculionoidea [INV ZOO] A superfamily of coleopteran insects in the suborder Polyphaga.

Curcurbitaceae [BOT] A family of dicotyledonous herbs or herbaceous vines in the order Violales characterized by an inferior ovary, unisexual flowers, one to five stamens but typically three, and a sympetalous corolla.

Curcurbitales [BOT] The ordinal name assigned to the Cucurbitaceae in some systems of classification.

curd [BOT] The edible flower heads of members of the mustard family such as broccoli. [FOOD ENG] **1.** The clotted portion of soured milk or milk treated with an acid or enzyme; used in making cheese. **2.** Any food resembling milk curd.

curet [MED] An instrument, shaped like a spoon or scoop, for scraping away tissue.

curettage [MED] Scraping of the inside of a body cavity or the hollow of an organ with a curet.

curie [NUCLEO] A unit of radioactivity, defined as that quantity of any radioactive nuclide which has 3.700×10^{10} disintegrations per second. Abbreviated c.

Curie constant [ELECTROMAG] The electric or magnetic susceptibility at some temperature times the difference of the temperature and the Curie temperature, which is a constant at temperatures above the Curie temperature according to the Curie-Weiss law.

Curie point See Curie temperature.

Curie principle [THERMO] The principle that a macroscopic cause never has more elements of symmetry than the effect it produces; for example, a scalar cause cannot produce a vectorial effect.

Curie's law [ELECTROMAG] The law that the magnetic susceptibilities of most paramagnetic substances are inversely proportional to their absolute temperatures.

Curie temperature [ELECTROMAG] The temperature marking the transition between ferromagnetism and paramagnetism, or between the ferroelectric phase and the paraelectric phase. Also known as Curie point.

Curie-Weiss law [ELECTROMAG] A relation between magnetic or electric susceptibilities and the absolute temperatures which is followed by ferromagnets, antiferromagnets, nonpolar ferroelectrics, antiferroelectrics, and some paramagnets.

curine See bebeerine.

curing [CHEM] Changing the properties of a resin by chemical reaction, which may be condensation or addition, usually accomplished by heat or catalyst or both, and with or without pressure. [ENG] Process of keeping concrete moist for its first week or month to provide enough water for the cement to harden. Also known as maturing.

curing agent See hardener.

curing temperature [CHEM] That temperature at which a resin or adhesive is subjected to curing.

curing time [CHEM] The period of time in which a part is subjected to heat or pressure to cure the resin.

curite [MINERAL] $Pb_2U_5O_{17} \cdot 4H_2O$ An orange-red radioactive mineral, occurring in acicular crystals, an alteration product of uraninite.

curium [CHEM] An element, symbol Cm, atomic number 96; the isotope of mass 244 is the principal source of this artificially produced element.

curium-242 [NUC PHYS] An isotope of curium, mass number 242; half-life is 165.5 days for α-particle emission; 7.2×10^6 years for spontaneous fission.

curium-244 [NUC PHYS] An isotope of curium, mass number 244; half-life is 16.6 years for α-particle emission; 1.4×10^7 years for spontaneous fission; potential use as compact thermoelectric power source.

curl [FOR] A block of timber cut from a crotch for cutting into veneers. [MATER] A defect of paper caused by unequal alteration in the dimensions of the top and underside of the sheet. [MATH] The curl of a vector function is a vector which is formally the cross product of the del operator and the vector. Also known as rotation.

curling dies [MECH ENG] A set of tools that shape the ends of a piece of work into a form with a circular cross section.

curling factor See griseofulvin.

curling machine [MECH ENG] A machine with curling dies; used to curl the ends of cans.

Curling's ulcer [MED] An acute gastric ulcer associated with severe skin burns.

curly top [PL PATH] A virus disease of sugarbeets and certain other plants that is transmitted by a leafhopper; affected plants are dwarfed and have curled, upturned leaves.

currant [BOT] A shrubby, deciduous plant of the genus *Ribes* in the order Rosales; the edible fruit, a berry, is borne in clusters on the plant.

currant leaf spot [PL PATH] **1.** An angular leaf spot of currants caused by the fungus *Cercospora angulata*. **2.** An anthracnose of currants caused by *Pseudopeziza ribis* and characterized by brown or black spots.

current [ELEC] The net transfer of electric charge per unit time; a specialization of the physics definition. Also known as electric current. [PHYS] **1.** The rate of flow of any conserved, indestructible quantity across a surface per unit time. **2.** See current density.

current algebra [PARTIC PHYS] The application of algebraic relationships among currents derived from approximate symmetries, such as broken SU_3 symmetry, to the study of hadrons.

current amplification [ELECTR] The ratio of output-signal current to imput-signal current for an electron tube, transistor, or magnetic section of a multiplier phototube, the multiplier section of a multiplier phototube, or any other amplifying device; often expressed in decibels by multiplying the common logarithm of the ratio by 20.

current amplifier [ELECTR] An amplifier capable of delivering considerably more signal current than is fed in.

current antinode [ELEC] A point at which current is a maximum along a transmission line, antenna, or other circuit element having standing waves. Also known as current loop.

current attenuation [ELECTR] The ratio of input-signal current for a transducer to the current in a specified load impedance connected to the transducer; often expressed in decibels.

current balance [ELEC] An apparatus with which force is measured between current-carrying conductors, with the purpose of assigning the value of the ampere.

current-bedding [GEOL] Cross-bedding resulting from water or air currents.

current-carrying capacity [ELEC] The maximum current that can be continuously carried without causing permanent deterioration of electrical or mechanical properties of a device or conductor.

current chart [MAP] A map of a water area depicting current speeds and directions by current roses, vectors, or other means.

current constants [OCEANOGR] Tidal current relations that remain practically constant for any particular locality.

current curve [OCEANOGR] In marine operations, a graphic representation of the flow of a current, consisting of a rectangular-coordinate graph on which speed is represented by the ordinates and time by the abscissas.

current cycle [OCEANOGR] A complete set of tidal current conditions, as those occurring during a tidal day, lunar month, or Metonic cycle.

current decay [MET] In certain welding operations, controlled reduction of the welding impulse over a predetermined time interval to prevent rapid cooling of the weld nugget.

current density [ELEC] The current per unit cross-sectional area of a conductor; a specialization of the physics definition. Also known as electric current density. [PHYS] A vector quantity whose component perpendicular to any surface equals the rate of flow of some conserved, indestructible quantity across that surface per unit area per unit time. Also known as current.

CURCULIONIDAE

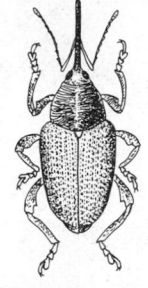

Snout beetle. *(From T. I. Storer and R. L. Usinger, General Zoology, 3d ed., McGraw-Hill, 1957)*

CURIUM

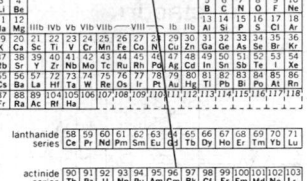

Periodic table of the chemical elements showing the position of curium.

CURRANT

The black currant *(Ribes nigrum)*.

CURRENT BALANCE

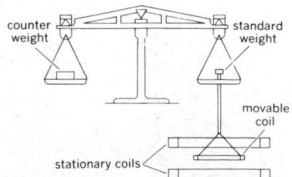

Rayleigh current balance.

current diagram [OCEANOGR] A graph showing the average speeds of flood and ebb currents throughout the current cycle for a considerable part of a tidal waterway.

current difference [OCEANOGR] In marine operations, the difference between the time of slack water or strength of current at a subordinate station and its reference station.

current divider [ELEC] A device used to deliver a desired fraction of a total current to a circuit.

current drain [ELEC] The current taken from a voltage source by a load. Also known as drain.

current drogue [ENG] A current-measuring assembly consisting of a weighted current cross, sail, or parachute, and an attached surface buoy.

current efficiency [PHYS CHEM] The ratio of the amount of electricity, in coulombs, theoretically required to yield a given quantity of material in an electrochemical process, to the amount actually consumed.

current ellipse [OCEANOGR] In marine operations, a graphic representation of a rotary current, in which the speed and direction of the current at various hours of the current cycle are represented by radius vectors; a line connecting the ends of the radius vectors approximates an ellipse.

current feed [ELECTR] Feed to a point where current is a maximum, as at the center of a half-wave antenna.

current feedback [ELECTR] Feedback introduced in series with the input circuit of an amplifier.

current feedback circuit [ELECTR] A circuit used to eliminate effects of amplifier gain instability in an indirect-acting recording instrument, in which the voltage input (error signal) to an amplifier is the difference between the measured quantity and the voltage drop across a resistor.

current function See Lagrange stream function.

current gain [ELECTR] The fraction of the current flowing into the emitter of a transistor which flows through the base region and out the collector.

current generator [ELECTR] A two-terminal circuit element whose terminal current is independent of the voltage between its terminals.

current hogging [ELECTR] A condition in which the largest fraction of a current passes through one of several parallel logic circuits because it has a lower resistance than the others.

current hour [OCEANOGR] The average time interval between the moon's transit over the meridian of Greenwich and the time of the following strength of flood current modified by the times of slack water and strength of ebb.

current-instruction register See instruction register.

current intensity [ELEC] The magnitude of an electric current. Also known as current strength.

current interrupter [ELEC] Mechanism connected into a current-carrying line to periodically interrupt current flow to allow no-current tests of system components.

current limiter [ELECTR] A device that restricts the flow of current to a certain amount, regardless of applied voltage. Also known as demand limiter.

current-limiting reactor See series reactor.

current-limiting resistor [ELEC] A resistor inserted in an electric circuit to limit the flow of current to some predetermined value; used chiefly to protect tubes and other components during warm-up.

current line [ENG] In marine operations, a graduated line attached to a current pole, used to measure the speed of a current; as the pole moves away with the current, the speed of the current is determined by the amount of line paid out in a specified time. Also known as log line.

current lineation See parting lineation.

current location reference [ADP] A symbolic expression, such as a star, which indicates the current location reached by the program; a transfer to * + 2 would bring control to the second statement after the current statement.

current loop See current antinode.

current margin [COMMUN] Difference between the steady-state currents flowing through a telegraph receiving instrument corresponding respectively to the two positions of the telegraph transmitter.

current measurement [ELEC] The measurement of the flow of electric current.

current meter See ammeter; velocity-type flowmeter.

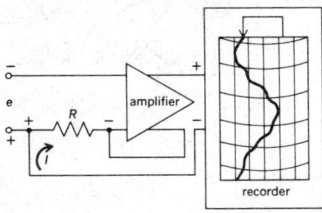
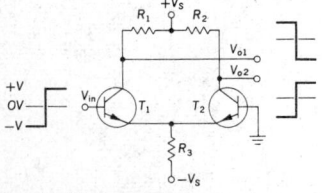
current-mode logic [ELECTR] Integrated-circuit logic in which transistors are paralleled so as to eliminate current hogging.

current node [ELEC] A point at which current is zero along a transmission line, antenna, or other circuit element having standing waves.

current phasor [ELEC] A line referenced to a point, whose length and angle represent the magnitude and phase of a current.

current pole [ENG] A pole used to determine the direction and speed of a current; the direction is determined by the direction of motion of the pole, and the speed by the amount of an attached current line paid out in a specified time.

current ratio [ELECTROMAG] In a waveguide, the ratio of maximum to minimum current.

current regulator [ELECTR] A device that maintains the output current of a voltage source at a predetermined, essentially constant value despite changes in load impedance.

current relay [ELEC] A relay that operates at a specified current value rather than at a specified voltage value.

current ripple [GEOL] A type of ripple mark having a long, gentle slope toward the direction from which the current flows, and a shorter, steeper slope on the lee side.

current rips [OCEANOGR] Small waves formed on the surface of water by the meeting of opposing ocean currents; vertical oscillation, rather than progressive waves, is characteristic of current rips.

current rose [MAP] A graphic presentation of ocean currents for specified areas, utilizing arrows at the cardinal and intercardinal compass points to show the direction toward which the prevailing current flows and the present frequency of set for a given period of time.

current saturation See anode saturation.

current strength See current intensity.

current tables [OCEANOGR] Tables listing predictions of the time and speeds of tidal currents at various places.

current tap See multiple lamp holder; plug adapter lamp holder.

current transformer [ELEC] An instrument transformer intended to have its primary winding connected in series with a circuit carrying the current to be measured or controlled; the current is measured across the secondary winding.

current transformer phase angle [ELEC] Angle between the primary current vector and the secondary current vector reversed; it is conveniently considered as positive when the reversed secondary current vector leads the primary current vector.

current-type flowmeter [ENG] A mechanical device to measure liquid velocity in open and closed channels; similar to the vane anemometer (where moving liquid turns a small windmill-type vane), but more rugged.

current-type telemeter [COMMUN] A telemeter in which the magnitude of a single current is the translating means.

current-voltage dual [ELEC] A circuit which is equivalent to a specified circuit when one replaces quantities with dual quantities; current and voltage impedance and admittance, and meshes and nodes are examples of dual quantities.

curry [FOOD ENG] A mixture of plant spices including turmeric, coriander, cinnamon, cumin, ginger, cardamom, cayenne pepper, cloves, and nutmeg.

cursor [DES ENG] A clear or amber-colored filter that can be placed over a radar screen and rotated until an etched diameter line on the filter passes through a target echo; the bearing from radar to target can then be read accurately on a stationary 360° scale surrounding the filter.

cursorial [VERT ZOO] Adapted for running.

curtain [NUCLEO] A thin shield, usually cadmium, used in a nuclear reactor to shut off a flow of slow neutrons.

curtain array [ELECTROMAG] An antenna array consisting of vertical wire elements stretched between two suspension cables.

curtain board [BUILD] A fire-retardant partition applied to a ceiling.

curtain coating [CHEM ENG] A method in which the substrate to be coated with low-viscosity resins or solutions is passed through, and is perpendicular to, a freely falling liquid curtain.

curtain rhombic antenna [ELECTROMAG] A multiple-wire rhombic antenna having a constant input impedance over a wide frequency range; two or more conductors join at the feed and terminating ends but are spaced apart vertically from 1 to 5 feet (30 to 150 centimeters) at the side poles.

curtain wall [CIV ENG] An external wall that is not load-bearing.

Curtis-Doisy unit [BIOL] A unit for the standardization of estrogens.

Curtis stage [MECH ENG] A velocity-staged impulse turbine using reversing buckets between stages.

Curtis turbine [MECH ENG] A velocity-staged, impulse-type steam engine.

Curtius reaction [ORG CHEM] A laboratory method for degrading a carboxylic acid to a primary amine by converting the acid to an acyl azide to give products which can be hydrolyzed to amines.

curvature [MATH] The reciprocal of the radius of the circle which most nearly approximates a curve at a given point; the rate of change of the unit tangent vector to a curve with respect to arc length of the curve.

curvature correction [ASTRON] A correction applied to the mean of a series of observations on a star or planet to take account of the divergence of the apparent path of the star or planet from a straight line. [GEOD] The correction applied in some geodetic work to take account of the divergence of the surface of the earth (spheroid) from a plane.

curvature of field [OPTICS] Error in the image of a plane object formed on a flat screen by an optical system when the best image lies on a curved surface.

curvature of space [RELAT] 1. The deviation of a spacelike three-dimensional subspace of curved space-time from euclidean geometry. 2. The Gaussian curvature of a spacelike three-dimensional subspace of curved space-time.

curvature tensor See Riemann-Christoffel tensor.

curve [MATH] The continuous image of the unit interval.

curved beam [ENG] A beam bounded by circular arcs.

curved space-time [RELAT] A four-dimensional Riemannian space, in which there are no straight lines but only curves, which is a generalization of the Minkowski universe in the general theory of relativity.

curve fitting [STAT] The calculation of a curve of some particular character (as a logarithmic curve) that most closely approaches a number of points in a plane.

curve follower [ADP] A device in which a photoelectric, capacitive or inductive pick-off guided by a servomechanism reads data in the form of a graph, such as a curve drawn on paper with suitable ink. Also known as graph follower.

curve of constant bearing See curve of equal bearing.

curve of equal bearing [NAV] On a map plot, a curve connecting all points at which the great-circle bearing of a given point is the same. Also known as curve of constant bearing.

curve resistance [MECH] The force opposing the motion of a railway train along a track due to track curvature.

curves of form [NAV ARCH] Graphs of properties of a vessel's form, such as the displacement and the area of wetted surface, versus the vessel's draft.

curve tracing [MATH] The method of graphing a function by plotting points and analyzing symmetries, derivatives, and so on.

curvilinear [SCI TECH] Pertaining to curved lines, as in curvilinear coordinates or curvilinear motion.

curvilinear coordinates [MAP] Any linear coordinates which are not cartesian coordinates; frequently used curvilinear coordinates are polar coordinates and cylindrical coordinates.

curvilinear motion [MECH] Motion along a curved path.

curvilinear regression [STAT] Regression study of jointly distributed random variables where the function measuring their statistical dependence is analyzed in terms of curvilinear coordinates.

curvilinear transformation [MATH] A transformation from one coordinate system to another in which the coordinates in the new system are arbitrary twice-differentiable functions of the coordinates in the old system.

cuscus oil See vetiver oil.

Cuscutaceae [BOT] A family of parasitic dicotyledonous plants in the order Polemoniales which lack internal phloem and chlorophyll, have capsular fruit, and are not rooted to the ground at maturity.

cusec [MECH] A unit of volume flow rate, used primarily to describe pumps, equal to a uniform flow of 1 cubic foot in 1 second. Also known as cubic foot per second (cfs).

Cushing's syndrome [MED] A complex of symptoms including facial and truncal obesity, hypertension, edema, and osteoporosis, resulting from oversecretion of adrenocortical hormones.

cusp [ANAT] 1. A pointed or rounded projection on the masticating surface of a tooth. 2. One of the flaps of a heart valve. [ARCH] A pointed projection or peak created by the intersection of two arcs. [GEOL] One of a series of low, crescent-shaped mounds of beach material separated by smoothly curved, shallow troughs spaced at more or less regular intervals along and generally perpendicular to the beach face. Also known as beach cusp. [MATH] A singular point of a curve at which the limits of the tangents of the portions of the curve on either side of the point coincide. Also known as spinode.

cuspate bar [GEOL] A crescentic bar joining with the shore at each end.

cuspate ripple mark See linguoid ripple mark.

cusped magnetic field [ELECTROMAG] A magnetic field created by adjacent parallel coils that carry current in opposite directions; used in fusion research, to contain a plasma of high-energy deuterium ions.

cuspid See canine.

cuspidal cubic [MATH] A cubic curve that has one cusp, one point of inflection, and no node.

cuspidal locus [MATH] A curve consisting of the cusps of some family of curves.

cuspidate [BIOL] Having a cusp; terminating in a point.

custard winds [METEOROL] Cold easterly winds on the northeastern coast of England.

custodial area [BUILD] Area of a building designated for service and custodial personnel; includes rooms, closets, storage, toilets, and lockers.

cusum chart See cumulative sum chart.

cut [CHEM ENG] A fraction obtained by a separation process. [CRYSTAL] A section of a crystal having two parallel major surfaces; cuts are specified by their orientation with respect to the axes of the natural crystal, such as X cut, Y cut, BT cut, and AT cut. [LAP] The style in which a gem is cut, such as brilliant cut, single cut, or rose cut. [MET] See fraction. [MIN ENG] 1. To intersect a vein or working. 2. To excavate coal. 3. To shear one side of an entry or crosscut by digging out the coal from floor to roof with a pick. [NUCLEO] The fraction that is removed as product or advanced to the next separative element in an isotope separation process.

cut-and-carry method [MET] A die-fabricating method in which the part remains attached to the strip or is forced back into the strip to be fed through the succeeding stations of a progressive die.

cut and fill [CIV ENG] Construction of a road, railway, or canal which is partly embanked and partly below ground. [GEOL] 1. Lateral corrosion of one side of a meander accompanied by deposition on the other. 2. A sedimentary structure consisting of a small filled-in channel.

cutaneous anaphylaxis [IMMUNOL] Hypersensitivity that is marked by an intense skin reaction following parenteral contact with a sensitizing agent.

cutaneous anthrax See malignant pustule.

cutaneous appendage [ANAT] Any of the epidermal derivatives, including the nails, hair, sebaceous glands, mammary glands, and sweat glands.

cutaneous blastomycosis [MED] A form of North American blastomycosis considered by some to be a clinical manifestation of the systemic form.

cutaneous coccidioidomycosis [MED] A primary skin infection by the fungus *Coccidioides immitis;* a skin infection secondary to a pulmonary lesion.

cutaneous leishmaniasis [MED] A parasitic skin infection by *Leishmania tropica* characterized by deep ulcers of the skin and subcutaneous tissue.

cutaneous pain [PHYSIO] A sensation of pain arising from the skin.

cutaneous reaction [MED] 1. Any change in the outer layers of the skin, as in sunburn or the rash in measles. **2.** Any immediate or delayed immune reaction in the skin resulting from antigen-antibody interaction.

cutaneous sensation [PHYSIO] Any feeling originating in sensory nerve endings of the skin, including pressure, warmth, cold, and pain.

cutaway [GRAPHICS] An illustration of an object with a part of its covering or surface removed to show its interior construction or movement.

cutback [CHEM ENG] Blending of heavier oils with lighter ones to bring the heavier to desired specifications.

cutback asphalt [MATER] Asphalt which has been softened or liquefied by blending with petroleum distillates.

cutbank [GEOL] The concave bank of a winding stream that is maintained as a steep or even overhanging cliff by the action of water at its base.

cut capacity [MATH] For a network whose points have been partitioned into two specified classes, C_1 and C_2, the sum of the capacities of all the segments directed from a point in C_1 to a point in C_2. Also known as cut value.

cutch [MATER] Tannin extracted from mangrove bark.

Cuterebridae [INV ZOO] The robust botflies, a family of myodarian cyclorrhaphous dipteran insects in the subsection Calypteratae.

cut form [ADP] In optical character recognition, any document form, receipt, or such, of standard dimensions which must be issued a separate read command in order to be recognized.

cut glass [MATER] Flint glass ornamented with patterns cut into its surface.

cuticle [ANAT] The horny layer of the nail fold attached to the nail plate at its margin. [BIOL] A noncellular, hardened or membranous secretion of an epithelial sheet, such as the integument of nematodes and annelids, the exoskeleton of arthropods, and the continuous film of cutin on certain plant parts.

cutie pie [NUCLEO] A radiation dose-rate meter having a pistol grip, a plastic cylinder or barrel containing an ionization chamber, and an indicating meter mounted above the grip.

cut in [NAV] To observe and plot lines of position locating an object or craft, particularly by bearings.

cutin [BIOCHEM] A mixture of fatty substances characteristically found in epidermal cell walls and in the cuticle of plant leaves and stems.

cutinite [GEOL] A variety of exinite consisting of plant cuticles.

cutis _See_ dermis.

Cutler feed [ELECTROMAG] A resonant cavity that transfers radio-frequency energy from the end of a waveguide to the reflector of a radar spinner assembly.

cut nail [DES ENG] A flat, tapered nail sheared from steel plate; it has greater holding power than a wire nail and is generally used for fastening flooring.

cutoff [ELECTR] **1.** The minimum value of negative grid bias that will prevent the flow of anode current in an electron tube. **2.** _See_ cutoff frequency. [ENG] A misfire in a round of shots because of severance of fuse owing to rock shear as adjacent charges explode. [GEOL] A new, relatively short channel formed when a stream cuts through the neck of an oxbow or horseshoe bend. [MECH ENG] **1.** The shutting off of the working fluid to an engine cylinder. **2.** The time required for this process. [MIN ENG] **1.** A quarryman's term for the direction along which the granite must be channeled, because it will not split. **2.** The number of feet a bit may be used in a particular type of rock (as specified by the drill foreman). **3.** Minimum percentage of mineral in an ore that can be mined profitably. [PHYS] Technique used when the contribution to the value of a physical quantity given by integration over a certain variable is absurd (in particular, when the contribution is infinite); involves cutting off the integral at some limit.

cutoff attenuator [ELECTROMAG] Variable length of waveguide used below its cutoff frequency to introduce variable nondissipative attenuation.

cutoff bias [ELECTR] The direct-current bias voltage that must be applied to the grid of an electron tube to stop the flow of anode current.

cutoff field _See_ critical field.

cutoff frequency [ELECTR] A frequency at which the attenuation of a device begins to increase sharply, such as the limiting frequency below which a traveling wave in a given mode cannot be maintained in a waveguide, or the frequency above which an electron tube loses efficiency rapidly. Also known as critical frequency; cutoff.

cutoff high [METEOROL] A warm high which has become displaced out of the basic westerly current, and lies to the north of this current.

cutoff lake _See_ oxbow lake.

cutoff limiting [ELECTR] Limiting the maximum output voltage of a vacuum tube circuit by driving the grid beyond cutoff.

cutoff low [METEOROL] A cold low which has become displaced out of the basic westerly current, and lies to the south of this current.

cutoff point [MECH ENG] **1.** The point at which there is a transition from spiral flow in the housing of a centrifugal fan to straight-line flow in the connected duct. **2.** The point on the stroke of a steam engine where admission of steam is stopped.

cutoff tool [MECH ENG] A tool used on bar-type lathes to separate the finished piece from the bar stock.

cutoff trench [CIV ENG] A trench which is below the foundation base line of a dam or other structure and is filled with an impervious material, such as clay or concrete, to form a watertight barrier.

cutoff valve [MECH ENG] A valve used to stop the flow of steam to the cylinder of a steam engine.

cutoff voltage [ELECTR] **1.** The electrode voltage value that reduces the dependent variable of an electron-tube characteristic to a specified low value. **2.** _See_ critical voltage.

cutoff wall [CIV ENG] A thin, watertight wall of clay or concrete built up from a cutoff trench to reduce seepage. Also known as core wall.

cutoff wavelength [ELECTROMAG] **1.** The ratio of the velocity of electromagnetic waves in free space to the cutoff frequency in a uniconductor waveguide. **2.** The wavelength corresponding to the cutoff frequency.

cutoff wheel [MECH ENG] A thin wheel impregnated with an abrasive used for severing or cutting slots in a material or part.

cut oil [MATER] An oil which has been partially emulsified with water in the presence of air.

cutout [ELEC] **1.** Pairs brought out of a cable and terminated at some place other than at the end of the cable. **2.** Electrical device used to interrupt the flow of current through any particular apparatus or instrument, either automatically or manually. Also known as electric cutout. [GEOL] _See_ horseback.

cutout box [ELEC] A fireproof cabinet or box with one or more hinged doors that contains fuses and switches for various leads in an electrical wiring system. Also known as fuse box.

cut over [FOREST] To cut marketable timber.

cut platform _See_ wave-cut platform.

cut point [CHEM ENG] The boiling-temperature division between cuts of a crude oil or base stock.

cut-set [ELEC] A set of branches of a network such that the cutting of all the branches of the set increases the number of separate parts of the network, but the cutting of all the branches except one does not.

cut shot [MIN ENG] A shot designed to bring down coal which has been sheared or opened on one side.

cut-signal-branch operation [ELECTR] In systems where radio reception continues without cutting off the carrier, the cut-signal-branch operation technique disables a signal branch in one direction when it is enabled in the other to preclude unwanted signal reflections.

cutter [ENG ACOUS] An electromagnetic or piezoelectric device that converts an electric input to a mechanical output, used to drive a stylus that cuts a wavy groove in the highly polished wax surface of a recording disk. Also known as cutting head; head; phonograph cutter; recording head. [MECH ENG] _See_ cutting tool. [MIN ENG] **1.** An operator of a coal-cutting or rock-cutting machine, or a worker engaged in underholing by pick or drill. **2.** A joint, usually a dip joint, running in the direction of working; usually in the plural.

cutter bar [MECH ENG] The bar that supports the cutting tool in a lathe or other machine.

cutterhead [MECH ENG] A device on a machine tool for holding a cutting tool.

cutter sweep [MECH ENG] The section that is cut off or eradicated by the milling cutter or grinding wheel in entering or leaving the flute.

cutting [BOT] A piece of plant stem with one or more nodes, which, when placed under suitable conditions, will produce roots and shoots resulting in a complete plant.

cutting angle [MECH ENG] The angle that the cutting face of a tool makes with the work surface back of the tool.

cutting down [MET] Removing surface roughness or irregularities from metal by the use of an abrasive.

cutting drilling [MECH ENG] A rotary drilling method in which drilling occurs through the action of the drill steel rotating while pressed against the rock.

cutting edge [DES ENG] **1.** The point or edge of a diamond or other material set in a drill bit. Also known as cutting point. **2.** The edge of a lathe tool in contact with the work during a machining operation.

cutting fluid [MATER] A fluid flowed over the tool and work in metal cutting to reduce heat generated by friction, lubricate, prevent rust, and flush away chips.

cutting head *See* cutter.

cutting machine [MIN ENG] A power-driven apparatus used to undercut or shear the coal to help in its removal from the face.

cutting-off machine [MECH ENG] A machine for cutting off metal bars and shapes; includes the lathe type using single-point cutoff tools, and several types of saws.

cutting-off process [METEOROL] A sequence of events by which a warm high or cold low, originally within the westerlies, becomes displaced either poleward (cutoff high) or equatorward (cutoff low) out of the westerly current; this process is evident at very high levels in the atmosphere, and it frequently produces, or is part of the production of, a blocking situation.

cutting oil [MATER] Oil (rather than an oil emulsion) used as a cutting fluid; used when lubricating, rather than cooling, is most desired.

cutting pliers [DES ENG] Pliers with cutting blades on the jaws.

cutting point *See* cutting edge.

cutting rule [ENG] A sharp steel rule used in a machine for cutting paper or cardboard.

cuttings [MIN ENG] Rock fragments broken from the penetrated rock during drilling operations.

cutting speed [MECH ENG] The speed of relative motion between the tool and workpiece in the main direction of cutting. Also known as feed rate; peripheral speed.

cutting stylus [ENG ACOUS] A recording stylus with a sharpened tip that removes material to produce a groove in the recording medium.

cutting tip [ENG] The end of the snout of a cutting torch from which gas flows.

cutting tool [MECH ENG] The part of a machine tool which comes into contact with and removes material from the workpiece by the use of a cutting medium. Also known as cutter.

cutting torch [ENG] A torch that preheats metal while the surface is rapidly oxidized by a jet of oxygen issuing through the flame from an additional feed line.

cuttlefish [INV ZOO] An Old World decapod mollusk of the genus *Sepia*; shells are used to manufacture dentifrices and cosmetics.

cut value *See* cut capacity.

cutwater [CIV ENG] A sharp-edged structure built around a bridge pier to protect it from the flow of water and material carried by the water.

Cuvieroninae [PALEON] A subfamily of extinct proboscidean mammals in the family Gomphotheriidae.

CVC *See* conserved vector current.

CVn *See* Canes Venatici.

CW *See* continuous wave.

cwt *See* hundredweight.

Cyamidae [INV ZOO] The whale lice, a family of amphipod crustaceans in the suborder Caprellidea that bear a resemblance to insect lice.

cyanalcohol *See* cyanohydrin.

cyanamide [INORG CHEM] NHCNH An acidic compound that forms colorless needles, melting at 46°C, soluble in water. Also known as urea anhydride.

cyanate [INORG CHEM] A salt or ester of cyanic acid containing the radical CNO.

cyanic acid [ORG CHEM] HCNO A colorless, poisonous liquid, which polymerizes to cyamelide and fulminic acid.

4-cyanic acid *See* fulminic acid.

cyanidation [CHEM] Joining of cyanide to an atom or molecule. [MET] *See* cyanide process.

cyanide [INORG CHEM] Any of a group of compounds containing the CN group and derived from hydrogen cyanide, HCN.

cyanide copper [MET] **1.** An electrolytic solution containing a complex of copper and the cyanide radical. **2.** Copper electrodeposited from the solution.

cyanide process [MET] Process of dissolving powdered gold and silver ores in a weak solution of sodium cyanide or potassium cyanide; the precious metals are precipitated from solution by zinc. Also known as cyanidation.

cyanide pulp [MET] The mixture resulting from grinding of gold and silver ore, then dissolving out the precious-metal content in a solution of sodium cyanide.

cyanide slime [MET] Minute particles of precious metals precipitated from cyanide solutions used in extracting the metals from ore.

cyanine dye [ORG CHEM] $C_{29}H_{35}N_2I$ Green metallic crystals, soluble in water; unstable to light, the dye is used in the photography industry as a chemical sensitizer for film. Also known as iodocyanin; quinoline blue.

cyanite *See* kyanite.

cyano- [CHEM] Combining form indicating the radical CN.

cyanocarbon [ORG CHEM] A derivative of hydrocarbon in which all of the hydrogen atoms are replaced by the CN group.

cyanochroite [MINERAL] $K_2Cu(SO_4)_2 \cdot 6H_2O$ A blue mineral consisting of a hydrous sulfate of potassium and copper.

cyanocobalamin *See* vitamin B_{12}.

cyano complex [CHEM] A coordination compound containing the CN group.

2-cyanoethanol *See* ethylene cyanohydrin.

cyanoethylation [ORG CHEM] A chemical reaction involving the addition of acrylonitrile to compounds with a reactive hydrogen.

Cyanogas [INORG CHEM] Trade name for a powdered calcium cyanide, used as an insect and rodent killer.

cyanogen [CHEM] A univalent radical, CN. [INORG CHEM] C_2N_2 A colorless, highly toxic gas with a pungent odor; a starting material for the production of complex thiocyanates used as insecticides. Also known as dicyanogen.

cyanogen bromide [INORG CHEM] CNBr White crystals melting at 52°C, vaporizing at 61.3°C, and having toxic fumes that affect nerve centers; used in the synthesis of organic compounds and as a fumigant.

cyanogen chloride [INORG CHEM] ClCN A poisonous, colorless gas or liquid, soluble in water; used in organic synthesis.

cyanogen iodide *See* iodine cyanide.

cyanohydrin [ORG CHEM] A compound containing the radicals CN and OH. Also known as cyanalcohol.

cyanometer [OPTICS] An instrument designed to measure or estimate the degree of blueness of light, as of the sky.

cyanometry [OPTICS] The study and measurement of the blueness of light.

cyanonitroacetamide *See* fulminuric acid.

cyanophilous [BIOL] Having an affinity for blue or green dyes.

Cyanophyceae [BOT] The blue-green algae, a class of plants in the division Schizophyta.

cyanophycin [BIOCHEM] A granular protein food reserve in the cells of blue-green algae, especially in the peripheral cytoplasm.

CUTTLEFISH

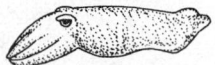

Drawing of cuttlefish. (*From T. I. Storer and R. L. Usinger, General Zoology, 3d ed., McGraw-Hill, 1957*)

Cyanophyta [BOT] An equivalent name for the Cyanophyceae.

cyanoplatinate *See* platinocyanide.

cyanosis [MED] A bluish coloration in the skin and mucous membranes due to deficient levels of oxygen in the blood.

cyanotrichite [MINERAL] $Cu_4Al_2(SO_4)(OH)_{12} \cdot 2H_2O$ A bright-blue or sky-blue mineral consisting of a hydrous basic copper aluminum sulfate.

cyanuric acid [ORG CHEM] $HOC(NCOH)_2N \cdot 2H_2O$ Colorless, monoclinic crystals, slightly soluble in water; formed by polymerization of cyanic acid. Also known as pyrolithic acid; pyrouric acid; pyruric acid; *s*-triazinetriol; trihydroxycyanidine; trioxycyanidine.

cyanurtriamide *See* melamine.

Cyatheaceae [BOT] A family of tropical and pantropical tree ferns distinguished by the location of sori along the veins.

cyathium [BOT] An inflorescence in which the flowers arise from the base of a cuplike involucre.

Cyathoceridae [INV ZOO] The equivalent name for the Lepiceridae.

cybernation [IND ENG] The use of computers in connection with automation.

cybernetics [SCI TECH] 1. The science of control and communication in all of their manifestations within and between machines, animals, and organizations. 2. Specifically, the interaction between automatic control and living organisms, especially humans and animals.

cybotaxis [PHYS] A transient molecular orientation in a liquid evidenced by x-ray diffraction effects.

Cycadales [BOT] An ancient order of plants in the class Cycadopsida characterized by tuberous or columnar stems that bear a crown of large, usually pinnate leaves.

Cycadatae *See* Cycadopsida.

Cycadeoidaceae [PALEOBOT] A family of extinct plants in the order Cycadeoidales characterized by sparsely branched trunks and a terminal crown of leaves.

Cycadeoidales [PALEOBOT] An order of extinct plants that were abundant during the Triassic, Jurassic, and Cretaceous periods.

Cycadicae [BOT] A subdivision of large-leaved gymnosperms with stout stems in the plant division Pinophyta; only a few species are extant.

Cycadofilicales [PALEOBOT] The equivalent name for the extinct Pteridospermae.

Cycadophyta [BOT] An equivalent name for Cycadecae elevated to the level of a division.

Cycadophytae [BOT] An equivalent name for Cycadicae.

Cycadopsida [BOT] A class of gymnosperms in the plant subdivision Cycadicae.

cyclamate [ORG CHEM] The calcium or sodium salt of cyclohexylsulfamate, an artificial sweetener.

Cyclanthaceae [BOT] The single family of the order Cyclanthales.

Cyclanthales [BOT] An order of monocotyledonous plants composed of herbs; or, seldom, composed of more or less woody plants with leaves that usually have a bifid, expanded blade.

cycle [ENG] To run a machine through an operating cycle. [FL MECH] A system of phases through which the working substance passes in an engine, compressor, pump, turbine, power plant, or refrigeration system. [MATH] A member of the kernel of a boundary homomorphism. [SCI TECH] 1. One complete sequence of values of an alternating quantity, or of a sequence of process operations. 2. A set of operations that is repeated as a unit.

cycle count [ADP] The operation of keeping track of the number of cycles a computer system goes through during processing time.

cycle criterion [ADP] Total number of times a cycle in a computer program is to be repeated.

cycle delay selector [ADP] An electromechanical device in a sorter which causes a cycle to be skipped so that the card out of sequence may be directed to a different pocket.

cyclegraph technique [IND ENG] Recording a brief work cycle by attaching small lights to various parts of a worker and then exposing the work motions on a still-film time plate; motion will appear on the plate as superimposed streaks of light; constituting a cyclegraph.

CYCADALES

ovulate cones

Sago palm (*Cycas revoluta*) showing megasporophylls grouped into cones. (*New York Botanical Garden*)

cycle index [ADP] 1. The number of times a cycle has been carried out by a computer. 2. The difference, or its negative, between the number of executions of a cycle which are desired and the number which have actually been carried out.

cycle-matching loran *See* low-frequency loran.

cycle of erosion *See* geomorphic cycle.

cycle of sedimentation [GEOL] Also known as sedimentary cycle. 1. A series of related processes and conditions appearing repeatedly in the same sequence in a sedimentary deposit. 2. The sediments deposited from the beginning of one cycle to the beginning of a second cycle of the spread of the sea over a land area, consisting of the original land sediments, followed by those deposited by shallow water, then deep water, and then the reverse process of the receding water. 3. *See* cyclothem.

cycle per minute [PHYS] A unit of frequency of action, equal to 1/60 hertz. Abbreviated cpm.

cycle per second *See* hertz.

cycle plant [CHEM ENG] A plant in which the liquid hydrocarbons are removed from natural gas and then the gas is put back into the earth to maintain pressure in the oil reservoir.

cycle reset [ADP] The resetting of a cycle index to its initial or other specified value.

cycle skip *See* skip logging.

cycle stock [CHEM ENG] The unfinished product taken from a stage of a refinery process and recharged to the process at an earlier stage in the operation.

cycle time [ADP] The shortest time elapsed between one store (or fetch) and the next store (or fetch) in the same memory unit. Also known as memory cycle. [SCI TECH] The time required to carry out a cycle; used principally for time-and-motion studies.

cycle timer [ELECTR] A timer that opens or closes circuits according to a predetermined schedule.

Cycleweld [MATER] Trade name for a heat-applied cement consisting of rubber and other plastics.

cyclic [SCI TECH] 1. Pertaining to some cycle. 2. Repeating itself in some manner in space or time.

cyclic amide [ORG CHEM] An amide arranged in a ring of carbon atoms.

cyclic anhydride [ORG CHEM] A ring compound formed by the removal of water from a compound; an example is phthalic anhydride.

cyclic catalytic reforming process [CHEM ENG] A method for the production of low-Btu reformed gas consisting of the conversion of carbureted water-gas sets by installing a bed of nickel catalyst in the superheater and using the carburetor as a combustion chamber and process steam superheater. Abbreviated CCR process.

cyclic check [ADP] One of the two-byte elements replacing the parity bit stripped off each byte transferred from main storage to disk volume (the other element is the bit count appendage); these two elements are appended to the block during the write operation; on a subsequent read operation these elements are calculated and compared to the appended elements for accuracy.

cyclic chronopotentiometry [ANALY CHEM] An analytic electrochemical method in which instantaneous current reversal is imposed at the working electrode, and its potential is monitored with time.

cyclic code [ADP] A code, such as a binary code, that changes only in one digit when going from one number to the number immediately following, and in that digit by only one unit.

cyclic compound [ORG CHEM] A compound that contains a ring of atoms.

cyclic coordinate [MECH] A generalized coordinate on which the Lagrangian of a system does not depend explicitly.

cyclic curve [MATH] 1. A curve (such as a cycloid, cardioid, or epicycloid) generated by a point of a circle that rolls (without slipping) on a given curve. 2. The intersection of a quadric surface with a sphere. Also known as spherical cyclic curve. 3. The stereographic projection of a spherical cyclic curve. Also known as plane cyclic curve.

cyclic feeding [ADP] In character recognition, a system employed by character readers in which each input document is issued to the document transport in a predetermined and constant period of time.

cyclic identity [MATH] The principle that the sum of any component of the Riemann-Christoffel tensor and two other components obtained from it by cyclic permutation of any three indices, while the fourth is held fixed, is zero.

cyclic magnetization [ELECTROMAG] A magnetizing force varying between two specific limits long enough so that the magnetic induction has the same value for corresponding points in successive cycles.

cyclic mining *See* conventional mining.

cyclic permutation [MATH] A permutation of an ordered set of symbols which sends the first to the second, the second to the third, . . . , the last to the first.

cyclic salt [OCEANOGR] Salt removed from the sea as spray, blown inland, and returned to its source by land drainage.

cyclic shift [ADP] A computer shift in which the digits dropped off at one end of a word are returned at the other end of the word. Also known as circuit shift; circular shift; end-around shift; nonarithmetic shift; ring shift.

cyclic train [MECH ENG] A set of gears, such as an epicyclic gear system, in which one or more of the gear axes rotates around a fixed axis.

cyclic transfer [ADP] The automatic transfer of data from some medium to memory or from memory to some medium until all the data are read.

cyclic twinning [CRYSTAL] Repeated twinning of three or more individuals in accordance with the same twinning law but without parallel twinning axes.

cyclic voltammetry [PHYS CHEM] An electrochemical technique for studying variable potential at an electrode involving application of a triangular potential sweep, allowing one to sweep back through the potential region just covered.

cycling [CHEM ENG] A series·of operations in petroleum refining or natural-gas processing in which the steps are repeated periodically in the same sequence. [CONT SYST] A periodic change of the controlled variable from one value to another in an automatic control system. Also known as oscillation.

cyclitis [MED] Inflammation of the ciliary body of the eye.

cyclization [ORG CHEM] Changing an open-chain hydrocarbon to a closed ring.

cyclized rubber [MATER] A thermoplastic, nonrubbery, tough or hard rubber derivative formed by the action of certain agents, such as sulfonic acid and chlorostannic acid, on rubber; used in paints and adhesives and for insulation.

cycloaddition [ORG CHEM] A reaction in which unsaturated molecules combine to form a cyclic compound.

cycloaliphatic *See* alicyclic.

cycloalkane *See* alicyclic.

cycloalkene [ORG CHEM] An unsaturated, monocyclic hydrocarbon having the formula C_nH_{2n-2}. Also known as cycloolefin.

cycloalkyne [ORG CHEM] A cyclic compound containing one or more triple bonds between carbon atoms.

cyclobarbital [PHARM] $C_{12}H_{16}N_2O_3$ A hypnotic and sedative of short duration.

cyclobutadiene [ORG CHEM] C_4H_4 A cyclic compound containing two alternate double bonds; used in organic synthesis. Also known as butene; tetramethyldiene; $\Delta^{1,3}$-tetramethylene.

cyclobutane [ORG CHEM] C_4H_8 An alicyclic hydrocarbon, boiling point 11°C; synthesized as a condensable gas; used in organic synthesis. Also known as tetramethylene.

cyclobutene [ORG CHEM] C_4H_6 An asymmetrical cyclic hydrocarbon occurring in several isomeric forms. Also known as cyclobutylene.

cycloconverter [ELEC] A device that produces an alternating current of constant or precisely controllable frequency from a variable-frequency alternating-current input, with the output frequency usually one-third or less of the input frequency.

Cyclocystoidea [PALEON] A class of small, disk-shaped, extinct echinozoans in which the lower surface of the body probably consisted of a suction cup.

cyclodextrin [ORG CHEM] Cyclic degradation products of starch which contain six, seven, or eight glucose residues and have the shape of large-ring molecules.

cyclodialysis [MED] Detaching the ciliary body from the sclera in order to effect reduction of intraocular tension in certain cases of glaucoma, especially in aphakia.

cyclodiathermy [MED] Destruction, by diathermy, of the ciliary body.

cyclodiolefin [ORG CHEM] A cycloalkene with two double bonds; sometimes included with alkenes, cycloalkenes, and hydrocarbons containing more than one ethylene bond as olefins in a generic sense.

cyclododecatriene [ORG CHEM] $C_{12}H_{18}$ One of two cyclic hydrocarbons with three double bonds; the two forms are stereoisomeric; used to make nylon-6 and nylon-12.

cyclogenesis [METEOROL] Any development or strengthening of cyclonic circulation in the atmosphere.

1,3-cyclohexadiene [ORG CHEM] C_6H_8 A partly saturated benzene compound with two double bonds; used in organic synthesis. Also known as dihydrobenzene.

cyclohexane [ORG CHEM] C_6H_{12} A colorless liquid that is a cyclic hydrocarbon synthesized by hydrogenation of benzene; used in organic synthesis. Also known as hexamethylene.

cyclohexanone [ORG CHEM] $C_6H_{10}O$ A ketone, a colorless liquid boiling at 157°C; used as an industrial solvent. Also known as pimelinketone.

cycloid [MATH] The curve traced by a point on the circumference of a circle as the circle rolls along a straight line.

cycloidal gear teeth [DES ENG] Gear teeth whose profile is formed by the trace of a point on a circle rolling without slippage on the outside or inside of the pitch circle of a gear; now used only for clockwork and timer gears.

cycloidal mass spectrometer [SPECT] Small mass spectrometer of limited mass range fitted with a special-type analyzer that generates a cycloidal-path beam of the sample mass.

cycloidal pendulum [MECH] A modification of a simple pendulum in which a weight is suspended from a cord which is slung between two pieces of metal shaped in the form of cycloids; as the bob swings, the cord wraps and unwraps on the cycloids; the pendulum has a period that is independent of the amplitude of the swing.

cycloidal propeller [NAV ARCH] A type of vertical-axis propeller; used especially on shallow-draft craft.

cycloid scale [VERT ZOO] A thin, acellular structure which is composed of a bonelike substance and shows annual growth rings; found in the skin of soft-rayed fishes.

cyclolysis [METEOROL] The weakening or decay of cyclonic circulation in the atmosphere.

cyclomorphosis [ECOL] Cyclic recurrent polymorphism in certain planktonic fauna in response to seasonal temperature or salinity changes.

cyclone [CHEM ENG] A static reaction vessel in which fluids under pressure form a vortex. [MECH ENG] Any cone-shaped air-cleaning apparatus operated by centrifugal separation that is used in particle collecting and fine grinding operations. [METEOROL] A low-pressure region of the earth's atmosphere with roundish to elongated-oval ground plan, in-moving air currents, centrally upward air movement, and generally outward movement at various higher elevations in the troposphere.

cyclone cellar [CIV] An underground shelter, often built in areas frequented by tornadoes. Also known as storm cellar; tornado cellar.

cyclone classifier *See* cyclone separator.

cyclone family [METEOROL] A series of wave cyclones occurring in the interval between two successive major outbreaks of polar air, and traveling along the polar front, usually eastward and poleward.

cyclone furnace [ENG] A water-cooled, horizontal cylinder in which fuel is fired cyclonically and heat is released at extremely high rates.

cyclone separator [MECH ENG] A funnel-shaped device for removing particles from air or other fluids by centrifugal means; used to remove dust from air or other fluids, steam from water, and water from steam, and in certain applications to separate particles into two or more size classes. Also known as cyclone classifier.

cyclone wave [METEOROL] 1. A disturbance in the lower troposphere, of wavelength 1000-2500 kilometers; cyclone waves are recognized on synoptic charts as migratory high-

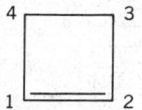

Structural formula of cyclobutane.

Structural formula of cyclobutene showing numbered carbons.

Cycloid scale, a primitive vertebrate scale. (*From K. F. Lagler et al., Ichthyology, Wiley, 1962*)

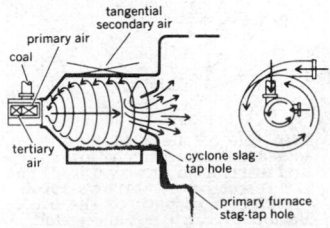

Twin cycloidal propellers, rated at 1100 horsepower each, on U.S. Army towboat. Blade length is 4½ feet (1.4 meters). (*Pacific Car and Foundry Co.*)

Schematic diagram of cyclone furnace. (*From T. Baumeister, ed., Standard Handbook for Mechanical Engineers, 7th ed., McGraw-Hill, 1967*)

CYCLORHAGAE

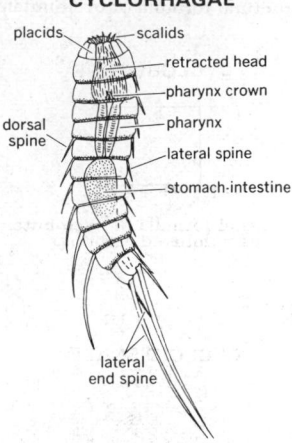

placids — scalids — retracted head — pharynx crown

dorsal spine — pharynx — lateral spine — stomach-intestine

lateral end spine

Side view of *Echinoderella* species, a cyclorhagous kinorhynch.

CYCLOTRON

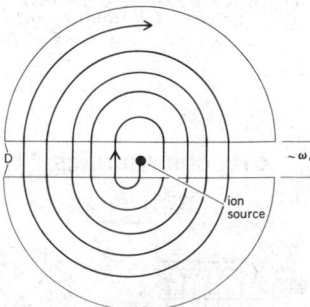

Principle of the cyclotron. The ions are formed in the ion source and are drawn out by one of the D's (electrodes). Arrows show circular orbit path of the ions. Period of orbit equals period $(2\pi/\omega_c)$ of applied D voltage.

and low-pressure systems. **2.** A frontal wave at the crest of which there is a center of cyclonic circulation, that is, the frontal wave of a wave cyclone.

cyclonic [GEOPHYS] Having a sense of rotation about the local vertical that is the same as that of the earth's rotation: as viewed from above, counterclockwise in the Northern Hemisphere, clockwise in the Southern Hemisphere, undefined at the Equator.

cyclonic scale [METEOROL] The scale of the migratory high- and low-pressure systems (or cyclone waves) of the lower troposphere, with wavelengths of 1000–2500 kilometers. Also known as synoptic scale.

cyclonic shear [METEOROL] Horizontal wind shear of such a nature that it contributes to the cyclonic vorticity of the flow; that is, it tends to produce cyclonic rotation of the individual air particles along the line of flow.

cyclonite [ORG CHEM] $(CH_2)_3N_3(NO_2)_3$ A white, crystalline explosive, consisting of hexahydro-trinitro-triazine, and having high sensitivity and brisance; mixed with other explosives or substances. Abbreviated RDX. Also known as cyclotrimethylenetrinitramine.

1,5-cyclooctadiene [ORG CHEM] C_8H_{12} A cyclic hydrocarbon with two double bonds; prepared from butadiene and used to make cyclooctene and cyclooctane, which are intermediates for the production of plastics, fibers, and so on.

cyclooctane [ORG CHEM] $(CH_2)_8$ A cyclic alkane melting at 9.5°C; used as an intermediate in production of plastics, fibers, adhesives, and coatings. Also known as octomethylene.

cyclooctatetraene [ORG CHEM] C_8H_8 A cyclic olefin with alternate double bonds; highly reactive; rearranges to styrene.

cycloolefin See cycloalkene.

cycloparaffin See alicyclic.

cyclopean [MATER] Mass concrete with aggregate larger than 6 inches; used for thick structures such as dams. [PETR] See mosaic.

1,3-cyclopentadiene [ORG CHEM] C_5H_6 A colorless liquid boiling at 41.5°C; used to make resins.

cyclopentadienyl anion [ORG CHEM] $C_5H_5^-$ A radical formed from cyclopentadiene.

cyclopentane [ORG CHEM] C_5H_{10} A cyclic hydrocarbon that is a colorless liquid; present in crude petroleum, it is converted during refining to aromatics which improve antiknock and combustion properties of gasoline.

cyclopentanol [ORG CHEM] C_5H_9OH A colorless liquid boiling at 139°C; used as a solvent for perfumes and pharmaceuticals. Also known as cyclopentyl alcohol.

cyclopentanone [ORG CHEM] C_5H_8O A saturated monoketone; a colorless liquid boiling at 130°C; used as an intermediate in pharmaceutical preparation. Also known as adipinketone.

cyclopentene [ORG CHEM] $(CH_2)_3CHCH$ A colorless liquid boiling at 45°C; used as a chemical intermediate in petroleum chemistry. Also known as Δ^1-pentamethylene.

cyclopentenylundecylic acid See hydnocarpic acid.

cyclopentyl alcohol See cyclopentanol.

cyclophon See beam-switching tube.

Cyclophoracea [INV ZOO] A superfamily of gastropod mollusks in the order Prosobranchia.

Cyclophoridae [INV ZOO] A family of land snails in the order Pectinibranchia.

Cyclophyllidea [INV ZOO] An order of platyhelminthic worms comprising most tapeworms of warm-blooded vertebrates.

cyclopia [MED] A congenital anomaly characterized by fusion of the eye sockets with various degrees of fusion of the eyes, to the occurrence of a single median eye.

Cyclopinidae [INV ZOO] A family of copepod crustaceans in the suborder Cyclopoida, section Gnathostoma.

Cyclopoida [INV ZOO] A suborder of small copepod crustaceans.

cyclopropane [ORG CHEM] C_3H_6 A colorless gas, insoluble in water; used as an anesthetic.

Cyclopteridae [VERT ZOO] The lumpfishes and snailfishes, a family of deep-sea forms in the suborder Cottoidei of the order Perciformes.

cyclorama [GRAPHICS] A vertical surface, often curved, used to form the background for theatrical settings; an illusion of depth is achieved by even lighting.

Cyclorhagae [INV ZOO] A suborder of benthonic, microscopic marine animals in the class Kinorhyncha of the phylum Aschelminthes.

Cyclorrhapha [INV ZOO] A suborder of true flies, order Diptera, in which developing adults are always formed in a puparium from which they emerge through a circular opening.

cycloserine [MICROBIO] $C_3H_6O_2N_2$ Broad-spectrum, crystalline antibiotic produced by several species of *Streptomyces*; useful in the treatment of tuberculosis and urinary-tract infections caused by resistant gram-negative bacteria.

cyclosilicate [MINERAL] A silicate having the SiO_4 tetrahedra linked to form rings, with a silicon-oxygen ratio of 1:3, such as $Si_3O_9^{6-}$ or $Si_6O_{18}^{12-}$. Also known as ring silicate.

cyclosis [CYTOL] Massive rotational streaming of cytoplasm in certain vacuolated cells, such as the stonewort *Nitella* and *Paramecium*.

Cyclosporeae [BOT] A class of brown algae, division Phaeophyta, in which there is only a free-living diploid generation.

Cyclosteroidea [PALEON] A class of Middle Ordovician to Middle Devonian echinoderms in the subphylum Echinozoa.

Cyclostomata [INV ZOO] An order of bryozoans in the class Stenolaemata. [VERT ZOO] A subclass comprising the simplest and most primitive of living vertebrates characterized by the absence of jaws and the presence of a single median nostril and an uncalcified cartilaginous skeleton.

cyclostrophic flow [FL MECH] A form of gradient flow in which the centripetal acceleration exactly balances the horizontal pressure force.

cyclostrophic wind [METEOROL] The horizontal wind velocity for which the centripetal acceleration exactly balances the horizontal pressure force.

cyclothem [GEOL] A rock stratigraphic unit associated with unstable shelf of interior basin conditions, in which the sea has repeatedly covered the land.

cyclothymia [PSYCH] A disposition marked by alterations of mood between elation and depression out of proportion to apparent external events and stimulated, rather, by internal factors.

cyclotol [MATER] High explosive composed of RDX (cyclonite) and TNT.

cyclotomic equation [MATH] An equation which has the form $x^{n-1} + x^{n-2} + \ldots + x + 1 = 0$, where n is a prime number.

cyclotomy [MATH] Theory of dividing the circle into equal parts or constructing regular polygons or, analytically, of finding the nth roots of unity.

cyclotrimethylenetrinitramine See cyclonite.

cyclotron [NUCLEO] An accelerator in which charged particles are successively accelerated by a constant-frequency alternating electric field that is synchronized with movement of the particles on spiral paths in a constant magnetic field normal to their path. Also known as phasotron.

cyclotron cataract See irradiation cataract.

cyclotron D See dee.

cyclotron emission See cyclotron radiation.

cyclotron frequency [ELECTROMAG] The frequency at which an electron traverses an orbit when moving subject to a uniform magnetic field, at right angles to the field. Also known as gyrofrequency.

cyclotron-frequency magnetron [ELECTR] A magnetron whose frequency of operation depends on synchronism between the alternating-current electric field and the electrons oscillating in a direction parallel to this field.

cyclotron magnets [NUCLEO] The magnets which bend charged-particle orbits and confine the extent of particle motion in a cyclotron.

cyclotron radiation [ELECTROMAG] The electromagnetic radiation emitted by charged particles as they orbit in a magnetic field, at a speed which is not close to the speed of light. Also known as cyclotron emission.

cyclotron resonance [PHYS] Resonance absorption of energy from an alternating-current electric field by electrons in a uniform magnetic field when the frequency of the electric field equals the cyclotron frequency, or the cyclotron frequency corresponding to the electron's effective mass if the electrons are in a solid. Also known as diamagnetic resonance.

cyclotron resonance heating [PL PHYS] A modification of magnetic pumping that involves compressing and expanding plasma at a frequency approximating the cyclotron frequency of the ions in the plasma; the goal is temperatures above several million degrees.

cyclotron wave [ELECTROMAG] A wave associated with the electron beam of a traveling-wave tube.

Cycloversion desulfurization [CHEM ENG] A continuous, fixed-bed catalytic process to desulfurize gasoline, kerosine, and other light fractions.

Cydippidea [INV ZOO] An order of the Ctenophora having well-developed tentacles.

Cydnidae [INV ZOO] The ground or burrower bugs, a family of hemipteran insects in the superfamily Pentatomorpha.

Cyg See Cygnus.

Cygnus [ASTRON] A conspicuous northern summer constellation; the five major stars are arranged in the form of a cross, but the constellation is represented by a swan with widespread wings flying southward; right ascension 21 hours, declination 40°N. Abbreviated Cyg. Also known as Northern Cross; Swan.

Cygnus X-1 [ASTROPHYS] A source of x-rays whose intensity varies in an irregular manner, associated with a weak variable radio source and a ninth-magnitude spectroscopic binary star, designated HDE226868, that consists of a blue supergiant and an invisible companion, which may be a black hole. Abbreviated Cyg X-1.

Cygnus X-3 [ASTROPHYS] A variable source of x-rays, with a period of 4.8 hours, associated with a variable radio source that flared up to enormous levels in September 1972 with no observed increase in x-ray emission. Abbreviated Cyg X-3.

Cyg X-1 See Cygnus X-1.

Cyg X-3 See Cygnus X-3.

cylinder [ADP] The virtual cylinder represented by the tracks of equal radius of a set of disks on a disk drive. [CIV ENG] **1.** A steel tube 10–60 inches (25–152 centimeters) in diameter with a wall at least 1/8 inch (3 millimeters) thick that is driven into bedrock, excavated inside, filled with concrete, and used as a pile foundation. **2.** A domed, closed tank for storing hot water to be drawn off at taps. Also known as storage calorifier. [MATH] **1.** A solid bounded by a cylindrical surface and two parallel planes, or the surface of such a solid. **2.** See cylindrical surface. [MECH ENG] See engine cylinder.

cylinder actuator [MECH ENG] A device that converts hydraulic power into useful mechanical work by means of a tight-fitting piston moving in a closed cylinder.

cylinder block [DES ENG] The metal casting comprising the piston chambers of a multicylinder internal combustion engine. Also known as block; engine block.

cylinder bore [DES ENG] The internal diameter of the tube in which the piston of an engine or pump moves.

cylinder head [MECH ENG] The cap that serves to close the end of the piston chamber of a reciprocating engine, pump, or compressor.

cylinder machine [ENG] A paper-making machine consisting of one or a series of rotary cylindrical filters on which wet paper sheets are formed.

cylinder oil [MATER] A viscous lubricating oil for the cylinders and valves of steam engines.

cylinder press [GRAPHICS] A large printing press in which paper is rolled against a flat, reciprocating printing surface by a rotating cylinder.

cylinder stock [MATER] **1.** Residual material in a still after vaporization of lighter petroleum stock. **2.** Compounded or straight oil used for lubricating steam cylinders.

cylinder stop [ORD] Device on a pistol that checks the action of the cylinder and acts as a lock to prevent it from moving.

cylindrarthrosis [ANAT] A joint characterized by rounded articular surfaces.

cylindrical antenna [ELECTROMAG] An antenna in which hollow cylinders serve as radiating elements.

cylindrical bow [NAV ARCH] A type of bow used principally on full, slow-speed, bulk-cargo-carrying ships; in general, the bow ends in a large-radius, vertical cylinder.

cylindrical cam [MECH ENG] A cam mechanism in which the cam follower undergoes translational motion parallel to the camshaft as a roller attached to it rolls in a groove in a circular cylinder concentric with the camshaft.

cylindrical capacitor [ELEC] A capacitor made of two concentric metal cylinders of the same length, with dielectric filling the space between the cylinders. Also known as coaxial capacitor.

cylindrical cavity [ELECTROMAG] A cavity resonator in the shape of a right circular cylinder.

cylindrical coordinates [MATH] A system of curvilinear coordinates in which the position of a point in space is determined by its perpendicular distance from a given line, its distance from a selected reference plane perpendicular to this line, and its angular distance from a selected reference line when projected onto this plane.

cylindrical cutter [DES ENG] Any cutting tool with a cylindrical shape, such as a milling cutter.

cylindrical-film storage [ELECTR] A computer storage in which each storage element consists of a short length of glass tubing having a thin film of nickel-iron alloy on its outer surface.

cylindrical grinder [MECH ENG] A machine for doing work on the peripheries or shoulders of workpieces composed of concentric cylindrical or conical shapes, in which a rotating grinding wheel cuts a workpiece rotated from a power headstock and carried past the face of the wheel.

cylindrical helix [MATH] A curve lying on a cylinder which intersects the elements of the cylinder at a constant angle.

cylindrical map projection [MAP] A map projection produced by projecting the geographic meridians and parallels onto a cylinder which is tangent to (or intersects) the surface of a sphere, and then developing the cylinder into a plane.

cylindrical pinch See pinch effect.

cylindrical reflector [ELECTROMAG] A reflector that is a portion of a cylinder; this cylinder is usually parabolic.

cylindrical surface [MATH] A surface consisting of each of the straight lines which are parallel to a given straight line and pass through a given curve. Also known as cylinder.

cylindrical wave [ELECTROMAG] A wave whose equiphase surfaces form a family of coaxial cylinders.

cylindrical winding [ELEC] The current-carrying element of a core-type transformer, consisting of a single coil of one or more layers wound concentrically with the iron core.

cylindrite [MINERAL] $Pb_3Sn_4Sb_2S_{14}$ A blackish-gray mineral consisting of sulfur, lead, antimony, and tin, occurring in cylindrical forms that separate under pressure into distinct sheets or folia.

Cylindrocapsaceae [BOT] A family of green algae in the suborder Ulotrichineae comprising thick-walled, sheathed cells having massive chloroplasts.

cymbocephalic [ANTHRO] Of a head or skull, having an unusually prolonged receding forehead and a protrudent occiput.

cyme [BOT] An inflorescence in which each main axis terminates in a single flower; secondary and tertiary axes may also have flowers, but with shorter flower stalks.

cymene [ORG CHEM] Any of the isomeric hydrocarbons metacymene, paracymene, and orthocymene; paracymene is a liquid that is colorless, has a pleasant odor, and is made from oil of cumin or oil of wild thyme.

cymophane See cat's eye.

cymose [BOT] Of, pertaining to, or resembling a cyme.

Cymothoidae [INV ZOO] A family of isopod crustaceans in the suborder Flabellifera; members are fish parasites with reduced maxillipeds ending in hooks.

cymrite [MINERAL] $Ba_2Al_5Si_5O_{19}(OH)\cdot 3H_2O$ Zeolite mineral consisting of a basic aluminosilicate of barium.

Cynipidae [INV ZOO] A family of hymenopteran insects in the superfamily Cynipoidea.

Cynipoidea [INV ZOO] A superfamily of hymenopteran insects in the suborder Apocrita.

Cynoglossidae [VERT ZOO] The tonguefishes, a family of Asiatic flatfishes in the order Pleuronectiformes.

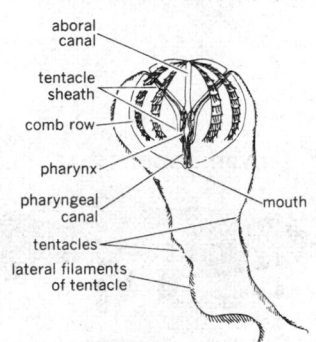

CYDIPPIDEA

aboral canal
tentacle sheath
comb row
pharynx
pharyngeal canal
mouth
tentacles
lateral filaments of tentacle

Pleurobranchia, a cydippid ctenophore. (*From L. H. Hyman, The Invertebrates, vol. 1, McGraw-Hill, 1940*)

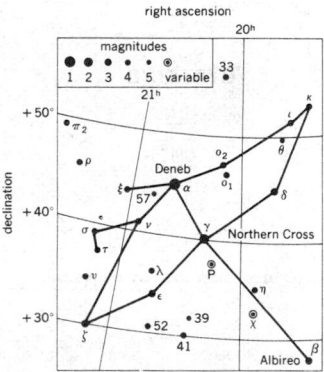

CYGNUS

right ascension

Line pattern of constellation Cygnus. Grid lines represent coordinates of sky. Apparent brightness, or magnitude, of stars is shown by sizes of dots, which are graded by appropriate numbers as indicated.

CYPRIDACEA

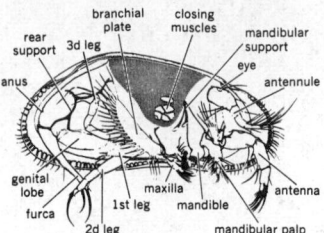

Female *Candona suburbana* Hoff, an ostracod of the Cypridacea. (*After D. J. McGregor and R. V. Kesling, Contributions from the Museum of Paleontology of the University of Michigan, 1969*)

CYPRINODONTIDAE

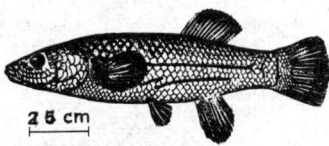

Striped killifish (*Fundulus majalis*). (*After G. B. Goode, Fishery Industries of the U.S., 1884*)

CYSTEINE

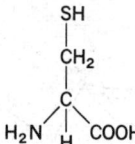

Structural formula of cysteine.

CYSTICERCUS

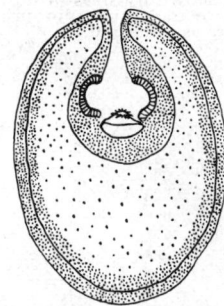

Cysticercus, the metacestode larva of cyclophyllidean tapeworms.

cynophobia [PSYCH] An abnormal fear of dogs.

Cyperaceae [BOT] The sedges, a family of monocotyledonous plants in the order Cyperales characterized by spirally arranged flowers on a spike or spikelet; a usually solid, often triangular stem; and three carpels.

Cyperales [BOT] An order of monocotyledonous plants in the subclass Commelinidae with reduced, mostly wind-pollinated or self-pollinated flowers that have a unilocular, two- or three-carpellate ovary bearing a single ovule.

Cypheliaceae [BOT] A family of typically crustose lichens with sessile apothecia in the order Caliciales.

cyphonautes [INV ZOO] The free-swimming bivalve larva of certain bryozoans.

Cyphophthalmi [INV ZOO] A family of small, mitelike arachnids in the order Phalangida.

Cypraeacea [INV ZOO] A superfamily of gastropod mollusks in the order Prosobranchia.

Cypraeidae [INV ZOO] A family of colorful marine snails in the order Pectinibranchia.

cypress [BOT] The common name for members of the genus *Cupressus* and several related species in the order Pinales.

Cypridacea [INV ZOO] A superfamily of mostly fresh-water ostracods in the suborder Podocopa.

Cypridinacea [INV ZOO] A superfamily of ostracods in the suborder Myodocopa characterized by a calcified carapace and having a round back with a downward-curving rostrum.

cypridophobia [PSYCH] An abnormal fear of acquiring venereal disease.

Cyprinidae [VERT ZOO] The largest family of fishes, including minnows and carps in the order Cypriniformes.

Cypriniformes [VERT ZOO] An order of actinopterygian fishes in the suborder Ostariophysi.

Cyprinodontidae [VERT ZOO] The killifishes, a family of actinopterygian fishes in the order Atheriniformes that inhabit ephemeral tropical ponds.

Cyprinoidei [VERT ZOO] A suborder of primarily fresh-water actinopterygian fishes in the order Cypriniformes having toothless jaws, no adipose fin, and faliciform lower pharyngeal bones.

cypris [INV ZOO] An ostracod-like, free-swimming larval stage in the development of Cirripedia.

Cyrtophorina [INV ZOO] The equivalent name for Gymnostomatida.

cyrtopia [INV ZOO] A type of crustacean larva (Ostracoda) characterized by an elongation of the first pair of antennae and loss of swimming action in the second pair.

cyrtosis [PL PATH] A virus disease of cotton characterized by stunting, distortion, and abnormal branching and coloration.

cyst [MED] A normal or pathologic sac with a distinct wall, containing fluid or other material.

cystacanth [INV ZOO] The infective larva of the Acanthocephala; lies in the hemocele of the intermediate host.

cystamine [ORG CHEM] $(CH_2)_6N_4$ A white, crystalline powder, melting at 280°C; used to make synthetic resins. Also known as aminiform; crystogen; cystamine methenamine; hexamethylene tetramine; urotropin.

cystamine methenamine *See* cystamine.

L-cystathionine [BIOCHEM] $C_7H_{14}N_2O_4S$ An amino acid formed by condensation of homocysteine with serine, catalyzed by an enzyme transsulfurase; found in high concentration in the brain of primates.

cystectomy [MED] **1.** Excision of the gallbladder, or part of the urinary bladder. **2.** Removal of a cyst. **3.** Removal of a piece of the anterior capsule of the lens for the extraction of a cataract.

cysteine [BIOCHEM] $C_3H_7O_2NS$ A crystalline amino acid occurring as a constituent of glutathione and cystine.

cystic disease [MED] A disorder of women, usually at or near menopause, characterized by the development of large cysts in the breast.

cystic duct [ANAT] The duct of the gallbladder.

cysticercosis [MED] The infestation in man by cysticerci of the genus *Taenia*.

cysticercus [INV ZOO] A larva of tapeworms in the order Cyclophyllidea that has a bladder with a single invaginated scolex.

cystic fibrosis [MED] A hereditary disease of the pancreas transmitted as an autosomal recessive; involves obstructive lesions, atrophy, and fibrosis of the pancreas and lungs, and the production of mucus of high viscosity. Also known as mucoviscidosis.

cystic kidney [MED] A kidney with one or more cysts.

cystine [BIOCHEM] $C_6H_{12}N_2S_2$ A white, crystalline amino acid formed biosynthetically from cysteine.

cystinosis [MED] A congenital metabolic disorder involving sulfur-containing amino acids, usually cystine; characterized by deposits of cystine crystals in the body organs.

cystinuria [MED] The presence in the urine of crystals of cystine together with some lysine, arginine, and ornithine.

cystitis [MED] Inflammation of a fluid-filled organ, especially the urinary bladder.

cystocarp [BOT] A fruiting structure with a special protective envelope, produced after fertilization in red algae.

cystocele [MED] Herniation of the urinary bladder into the vagina.

cystocercous cercariae [INV ZOO] A digenetic trematode larva that can withdraw the body into the tail.

cystography [MED] Radiography of the urinary bladder after the injection of a contrast medium.

Cystoidea [PALEON] A class of extinct crinozoans characterized by an ovoid body that was either sessile or attached by a short aboral stem.

cystolith [BOT] A concretion of calcium carbonate arising from the cell walls of modified epidermal cells in some flowering plants.

cystoma [MED] A cystic mass, especially in or near the ovary.

cystometer [MED] An instrument used to determine pressure in the urinary bladder under standard conditions.

Cystoporata [PALEON] An order of extinct, marine bryozoans characterized by cystopores and minutopores.

cystopyelitis [MED] Inflammation of the urinary bladder and the renal pelvis.

cystopyelography [MED] Radiography of the urinary bladder, the ureter, and the renal pelvis after injection of a radiopaque material.

cystopyelonephritis [MED] Inflammation of the urinary bladder, renal pelvis, and renal parenchyma.

cystoscope [MED] An optical instrument for visual examination of the urinary bladder, ureters, and kidneys.

cystoureteritis [MED] Inflammation of the urinary bladder and the ureters.

cytac [NAV] A system in which hyperbolic lines of position are determined by measuring the time relationship between two synchronized radio signals; similar to loran but capable of higher accuracy and greater range.

cytase [BIOCHEM] Any of several enzymes in the seeds of cereals and other plants, which hydrolyze the cell-wall material.

Cytheracea [INV ZOO] A superfamily of ostracods in the suborder Podocopa comprising principally crawling and digging marine forms.

Cytherellidae [INV ZOO] The family comprising all living members of the ostracod suborder Platycopa.

cytidine [BIOCHEM] $C_9H_{13}N_3O_5$ Cytosine riboside, a nucleoside composed of one molecule each of cytosine and D-ribose.

cytidylic acid [BIOCHEM] $C_9H_{14}O_8N_3P$ A nucleotide synthesized from the base cytosine and obtained by hydrolysis of nucleic acid. Also known as cytidine monophosphate.

cytocentrum *See* central apparatus.

cytochalasin [BIOCHEM] One of a series of structurally related fungal metabolic products which, among other effects on biological systems, selectively and reversibly block cytokinesis while not affecting karyokinesis; the molecule with minor variations consists of a benzyl-substituted hydroaromatic isoindolone system, which in turn is fused to a small macrolide-like cyclic ring.

cytochemistry [CYTOL] The science concerned with the chemistry of cells and cell components, primarily with the location of chemical constituents and enzymes.

cytochrome [BIOCHEM] Any of the complex protein respiratory pigments occurring within plant and animal cells, usually in mitochondia, that function as electron carriers in biological oxidation.

cytochrome a₃   *See* cytochrome oxidase.

cytochrome oxidase   [BIOCHEM] Any of a family of respiratory pigments that react directly with oxygen in the reduced state. Also known as cytochrome a_3.

cytocidal unit   [BIOL] A unit for the standardization of adrenal cortical hormones.

cytocrine gland   [CYTOL] A cell, especially a melanocyte, that passes its secretion directly to another cell.

cytodiagnosis   [PATH] The determination of the nature of an abnormal liquid by the study of cells it contains.

cytogenetics   [CYTOL] A branch of molecular pathology combining the methods of cytology and genetics.

cytogenous gland   [PHYSIO] A structure that secretes living cells; an example is the testis.

cytokinesis   [CYTOL] Division of the cytoplasm following nuclear division.

cytokinin   [BIOCHEM] Any of a group of plant hormones which elicit certain plant growth and development responses, especially by promoting cell division.

cytology   [BIOL] A branch of the biological sciences which deals with the structure, behavior, growth, and reproduction of cells and the function and chemistry of cell components.

cytolysis   [PATH] Disintegration or dissolution of cells, usually associated with a pathologic process.

cytolysosome   [CYTOL] An enlarged lysosome that contains organelles such as mitochondria.

cytomegalic   [MED] Of, pertaining to, or characterizing the greatly enlarged cells with enlarged nuclei and inclusion bodies found in tissues in cytomegalic inclusion disease.

cytomegalic inclusion disease   [MED] A virus infection primarily of infants characterized by jaundice, liver enlargement, and circulatory disturbances.

cytomegalovirus   [VIROL] An animal virus belonging to subgroup B of the herpesvirus group; causes cytomegalic inclusion disease and pneumonia.

cytomorphosis   [CYTOL] All the structural alterations which cells or successive generations of cells undergo from the earliest undifferentiated stage to their final destruction.

cytopathology   [PATH] A branch of pathology concerned with abnormalities within cells.

cytopenia   [PATH] A blood cell count below normal.

Cytophagaceae   [MICROBIO] A family of slime bacteria, order Myxobacterales, having spherical microcysts surrounded by a thick wall.

cytopharynx   [INV ZOO] A channel connecting the surface with the protoplasm in certain protozoans; functions as a gullet in ciliates.

cytophilic antibody   [IMMUNOL] A substance capable of combining directly with the receptors of a corresponding antigenic cell. Also known as cytotropic antibody.

cytoplasm   [CYTOL] The protoplasm of an animal or plant cell external to the nucleus.

cytoplasmic inheritance   [GEN] The control of genetic difference by hereditary units carried in cytoplasmic organelles. Also known as extrachromosomal inheritance.

cytoplasmic streaming   [CYTOL] Intracellular movement involving irreversible deformation of the cytoplasm produced by endogenous forces.

cytopyge   [INV ZOO] A fixed point for waste discharge in the body of a protozoan, especially a ciliate.

cytosine   [BIOCHEM] $C_4H_5ON_3$ A pyrimidine occurring as a fundamental unit or base of nucleic acids.

cytoskeleton   [CYTOL] Protein fibers composing the structural framework of a cell.

cytosome   [CYTOL] The cytoplasm of the cell, as distinct from the nucleus.

cytostome   [INV ZOO] The mouth-like opening in many unicellular organisms, particularly Ciliophora.

cytotaxis   [PHYSIO] Attraction of motile cells by specific diffusible stimuli emitted by other cells.

cytotechnologist   [PATH] A person trained to prepare smears of and examine exfoliated cells, referring abnormalities to a physician.

cytotrophoblast   [EMBRYO] The inner, cellular layer of a trophoblast, covering the chorion and the chorionic villi during the first half of pregnancy.

cytotropic antibody   *See* cytophilic antibody.

cytotropism   [BIOL] The tendency of individual cells and groups of cells to move toward or away from each other.

Czapek's agar   [MICROBIO] A nutrient culture medium consisting of salt, sugar, water, and agar; used for certain mold cultures.

Czochralski method   [ECOL] Alteration of the original state of a community by human agencies.

CYTOSINE

Structural formula of cytosine.

D   *See* dee; diopter.

2,4-D   *See* dichlorophenoxyacetic acid.

DABS   *See* Digital Address Beacon System.

dac   *See* digital-to-analog converter.

dachiardite   [MINERAL] $(Na_2Ca)_2(Al_4Si_{20}O_{48})\cdot 12H_2O$ A white to colorless mineral in the mordenite group of the zeolite family that crystallizes in the monoclinic system.

Dacian   [GEOL] Lower upper Pliocene geologic time.

dacite   [GEOL] Very fine crystalline or glassy rock of volcanic origin, composed chiefly of sodic plagioclase and free silica with subordinate dark-colored minerals.

dacite glass   [GEOL] A natural glass formed by rapid cooling of dacite lava.

Dacromycetales   [MYCOL] An order of jelly fungi in the subclass Heterobasidiomycetidae having branched basidia with the appearance of a tuning fork.

Dacron   [TEXT] Trade name of a DuPont polyester fiber.

dacryoblennorrhea   [MED] Chronic inflammation of the lacrimal sac of the eye accompanied by discharge of mucus.

dacryocyst   *See* lacrimal sac.

dacryocystitis   [MED] Inflammation of the lacrimal sac.

dacryon   [ANAT] The point of the face where the frontomaxillary, the maxillolacrimal, and frontolacrimal sutures meet.

Dactinomycin   [MICROBIO] A trade name for actinomycin D.

dactylitic   [GEOL] Of a rock texture, characterized by fingerlike projections of a mineral that penetrate another mineral.

Dactylochirotida   [INV ZOO] An order of dendrochirotacean holothurians in which there are 8–30 digitate or digitiform tentacles, which sometimes bifurcate.

dactylognathite   [INV ZOO] The distal segment of a maxilliped in crustaceans.

dactylography   [SCI TECH] The scientific study of fingerprints as a device for identifying people.

Dactylogyroidea   [INV ZOO] A superfamily of trematodes in the subclass Monogenea; all are fish ectoparasites.

dactylopodite   [INV ZOO] The distal segment of ambulatory limbs in decapods and of certain limbs in other arthropods.

dactylopore   [INV ZOO] Any of the small openings on the surface of Milleporina through which the bodies of the polyps are extended.

Dactylopteridae   [VERT ZOO] The flying gurnards, the single family of the perciform suborder Dactylopteroidei.

Dactylopteroidei   [VERT ZOO] A suborder of marine shore fishes in the order Perciformes, characterized by tremendously expansive pectoral fins.

Dactyloscopidae   [VERT ZOO] The sand stargazers, a family of small tropical and subtropical perciform fishes in the suborder Blennioidei.

dactylosternal   [VERT ZOO] Of turtles, having marginal fingerlike processes joining the plastron to the carapace.

dactylozooid   [INV ZOO] One of the long, defensive polyps of the Milleporina, armed with stinging cells.

dactylus   [INV ZOO] The structure of the tarsus of certain insects which follows the first joint; usually consists of one or more joints.

dado   [ARCH] **1.** The lower portion of an interior wall set off by molding or other decoration. **2.** The portion of a pedestal between surbase and base. **3.** The portion of a wall basement between surbase and base course.

dado head   [MECH ENG] A machine consisting of two circular saws with one or more chippers in between; used for cutting flat-bottomed grooves in wood.

dado plane   [DES ENG] A narrow plane for cutting flat grooves in woodwork.

dadur   [METEOROL] In India, a wind blowing down the Ganges Valley from the Siwalik hills at Hardwar.

dagger board   [NAV ARCH] A device used in a sailing vessel to increase the area of lateral resistance; it is a narrow metal flat or wooden board that slides up and down in the trunk, which extends from inside the hull to the bottom.

Dagor lens   [OPTICS] An anastigmatic lens consisting of two lens systems that are nearly symmetrical with respect to the stop, each system containing three or more lenses.

daguerreotype   [GRAPHICS] A photograph produced on a silver plate or a copper plate coated with silver sensitized by the action of iodine; after exposure of the plate in a camera, a latent image is developed by use of mercury vapor.

daily aberration   *See* diurnal aberration.

daily forecast   [METEOROL] A forecast for periods of from twelve to forty-eight hours in advance.

daily keying element   [COMMUN] Part of a specific cipher key that changes at predetermined intervals, usually daily.

daily mean   [METEOROL] The average value of a meteorological element over a period of twenty-four hours.

daily rate   *See* watch rate.

dairy   [AGR] A farm concerned with the production of milk. [FOOD ENG] An establishment where milk products are made.

dairy cattle   [AGR] A cattle breed selected, raised, and bred for ability to produce large quantities of milk.

dairy machinery   [FOOD ENG] Equipment, such as pasteurizers, used in the production and processing of milk and milk products.

dairy product   [FOOD ENG] Any food made from milk or milk products.

daisy chain   [ADP] A means of connecting devices (readers, printers, and so on) to a central processor by party-line input/output buses which join these devices by male and female connectors, the last female connector being shorted by a suitable line termination.

Dakotan   [GEOL] Lower Upper Cretaceous geologic time.

Dalatiidae   [VERT ZOO] The spineless dogfishes, a family of modern sharks belonging to the squaloid group.

d'Alembertian   [MATH] A differential operator in four-dimensional space,

$$\frac{\partial^2}{\partial x^2} + \frac{\partial^2}{\partial y^2} + \frac{\partial^2}{\partial z^2} - \frac{1}{c^2}\frac{\partial^2}{\partial t^2}$$

which is used in the study of relativistic mechanics.

d'Alembert's paradox   [FL MECH] The paradox that no forces act on a body moving at constant velocity in a straight line through a large mass of incompressible, inviscid fluid which was initially at rest, or in uniform motion.

d'Alembert's principle   [MECH] The principle that the resultant of the external forces and the kinetic reaction acting on a body equals zero.

d'Alembert's test for convergence   [MATH] A series Σa_n converges if there is an N such that the absolute value of the ratio a_n/a_{n-1} is always less than some fixed number smaller than 1, provided n is at least N, and diverges if the ratio is always greater than 1.

d'Alembert's wave equation   *See* wave equation.

Dalitz pair   [PARTIC PHYS] The electron and positron resulting from the decay of a neutral pion to these particles and a photon.

Dalitz plot   [PARTIC PHYS] Pictorial representation for data on the distribution of certain three-particle configurations.

Dallis grass   [BOT] The common name for the tall perennial forage grasses composing the genus *Paspalum* in the order Cyperales.

Dall tube   [MECH ENG] Fluid-flow measurement device, similar to a venturi tube, inserted as a section of a fluid-carrying pipe; flow rate is measured by pressure drop across a restricted throat.

Dalmatian wettability   [PETRO ENG] Theory of wettability that some in situ reservoir rocks are partly preferentially oil-wet and partly preferentially water-wet.

dalmatian sage oil   *See* oil of sage.

DACTYLOGYROIDEA

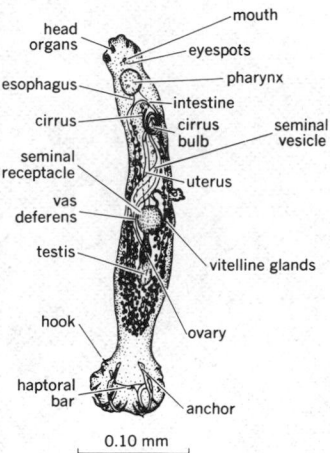

Pseudohaliotrema carbunculus Hargis from the pinfish.

DAGOR LENS

Diagram of Dagor lens.

DALITZ PLOT

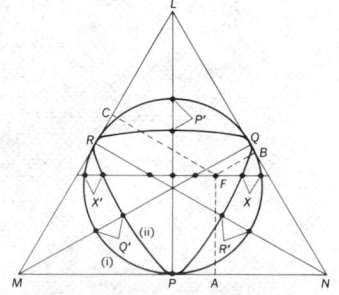

Configuration of a three-particle system *(abc)* in its barycentric frame is specified by a point F such that the three perpendiculars FA, FB, and FC to the sides of an equilateral triangle LMN (of height Q) are equal in magnitude to the kinetic energies T_a, T_b, T_c, where Q denotes the sum.

dalton *See* atomic mass unit.

Dalton's atomic theory [CHEM] Theory forming the basis of accepted modern atomic theory, according to which matter is made of particles called atoms, reactions must take place between atoms or groups of atoms, and atoms of the same element are all alike but differ from atoms of another element.

Dalton's law [PHYS] The law that the pressure of a gas mixture is equal to the sum of the partial pressures of the gases composing it. Also known as law of partial pressures.

Dalton's temperature scale [THERMO] A scale for measuring temperature such that the absolute temperature T is given in terms of the temperature on the Dalton scale τ by $T = 273.15(373.15/273.15)^{\tau/100}$.

dam [CIV ENG] 1. A barrier constructed to obstruct the flow of a watercourse. 2. A pair of cast-steel plates with interlocking fingers built over an expansion joint in the road surface of a bridge.

damage [ORD] 1. An injury short of complete destruction inflicted upon persons, equipment, or installations. 2. To cause damage.

damage assessment [ORD] The result of examination of combat material, particularly aircraft, ships, and armored vehicles, after a simulated attack, to determine the category in which the damage resulting from the attack would be placed.

damage control [ORD] Procedures for maintaining or restoring integrity, stability, or weapon power of a ship or aircraft.

damage potential [ORD] The damage which a projectile or explosive can be expected to do to a specific target.

damage radius [ORD] 1. The distance at which, in terms of experience or theoretical calculations, certain types of damage can be expected from a specified type of explosive item. 2. In atomic explosion, the distance from ground zero at which there is a 50% probability that a target element susceptible to the weapon effect considered will be damaged.

damaging stress [MECH] The minimum unit stress for a given material and use that will cause damage to the member and make it unfit for its expected length of service.

damask [TEXT] A fabric with a satin-weave pattern against a plainwoven background, made on a jacquard loom.

damkjernite [PETR] A melanocratic dike rock composed of biotite and pyroxene phenocrysts in a groundmass of pyroxene, biotite, and magnetite.

Damköhler number I [PHYS] A dimensionless number, equal to the ratio of the time it takes a fluid to flow some characteristic distance, to the time it takes some chemical reaction or other physical process to be completed. Symbolized Da I. Also known as Damköhler's ratio.

Damköhler number II [PHYS] A measure of the ratio of the rate of a chemical reaction to the rate of molecular diffusion, equal to the square of a characteristic length divided by the product of the diffusivity and the time it takes for a chemical reaction or other physical process to be completed. Symbolized Da II.

Damköhler number III [PHYS] A measure of the ratio of the heat liberated by a chemical reaction to the bulk transport of heat in a fluid, equal to the time it takes the fluid to travel a characteristic length, divided by the product of the fluid temperature and the time it would take for the chemical reaction to raise this temperature one unit if all the heat liberated by it were immediately absorbed by the fluid. Symbolized Da III.

Damköhler number IV [PHYS] A measure of the ratio of the heat liberated by a chemical reaction to the conductive heat transfer, equal to a characteristic length times the heat liberated per unit volume per unit time divided by the product of the thermal conductivity and the temperature. Symbolized Da IV.

Damköhler number V *See* Reynold's number.

Damköhler's ratio *See* Damköhler number I.

dammar [MATER] 1. A type of hard resin obtained from evergreen trees of the genus *Agathis*. 2. A type of soft, clear to yellow East Indian resin derived from several trees of the family Dipterocarpaceae. Also known as gum dammar.

dammar varnish [MATER] Varnish made from East Indian dammar.

damp [ENG] To reduce the fire in a boiler or a furnace by putting a layer of damp coals or ashes on the fire bed. [MIN ENG] A poisonous gas in a coal mine. [PHYS] To gradually diminish the amplitude of a vibration or oscillation.

damp air [METEOROL] Air that has a high relative humidity.

damp course [CIV ENG] A layer of impervious material placed horizontally in a wall to keep out water.

damp down [MET] To stop the blast in a blast furnace by closing the openings in the furnace.

damped harmonic motion [PHYS] Also known as damped oscillation; damped vibration. 1. The linear motion of a particle subject both to an elastic restoring force proportional to its displacement and to a frictional force in the direction opposite to its motion and proportional to its speed. 2. A similar variation in a quantity analogous to the displacement of a particle, such as the charge on a capacitor in a simple series circuit containing a resistance.

damped oscillation [PHYS] 1. Any oscillation in which the amplitude of the oscillating quantity decreases with time. Also known as damped vibration. 2. *See* damped harmonic motion.

damped vibration *See* damped harmonic motion; damped oscillation.

damped wave [PHYS] 1. A wave whose amplitude drops exponentially with distance because of energy losses which are proportional to the square of the amplitude. 2. A wave in which the amplitudes of successive cycles progressively diminish at the source.

dampener [ENG] A device for damping spring oscillations after abrupt removal or application of a load. [GRAPHICS] In offset printing, one of the rollers that transfers water to the nonprinting areas of the total surface that is printed.

damper [ELECTR] A diode used in the horizontal deflection circuit of a television receiver to make the sawtooth deflection current decrease smoothly to zero instead of oscillating at zero; the diode conducts each time the polarity is reversed by a current swing below zero. [MECH ENG] A valve or movable plate for regulating the flow of air or the draft in a stove, furnace, or fireplace.

damper pedal [ENG] A pedal that controls the damping of piano strings.

damper winding [ELEC] A winding consisting of several conducting bars on the field poles of a synchronous machine, short-circuited by conducting rings or plates at their ends, and used to prevent pulsating variations of the position or magnitude of the magnetic field linking the poles. Also known as amortisseur winding.

damp haze [METEOROL] Small water droplets or very hygroscopic particles in the air, reducing the horizontal visibility somewhat, but to not less than 1¼ miles (2 kilometers); similar to a very thin fog, but the droplets or particles are more scattered than in light fog and presumably smaller.

damping [ENG] Reducing or eliminating reverberation in a room by placing sound-absorbing materials on the walls and ceiling. Also known as soundproofing. [PHYS] 1. The dissipation of energy in motion of any type, especially oscillatory motion and the consequent reduction or decay of the motion. 2. The extent of such dissipation and decay.

damping capacity [MECH] A material's capability in absorbing vibrations.

damping coefficient *See* damping factor; resistance.

damping constant *See* resistance.

damping factor [PHYS] 1. The ratio of the logarithmic decrement of any underdamped harmonic motion to its period. Also known as damping coefficient. 2. *See* decrement.

damping magnet [ELECTROMAG] A permanent magnet used in conjunction with a disk or other moving conductor to produce a force that opposes motion of the conductor and thereby provides damping.

damping-off [PL PATH] A fungus disease of seedlings and cuttings in which the parasites invade the plant tissues near the ground level, causing wilting and rotting.

damping ratio [PHYS] The ratio of the actual resistance in damped harmonic motion to that necessary to produce critical damping. Also known as relative damping ratio.

damp sheet [MIN ENG] A curtain used in a mine to direct airflow, thus preventing gas accumulation.

Danaidae [INV ZOO] A family of large tropical butterflies, order Lepidoptera, having the first pair of legs degenerate.

danalite [MINERAL] $(Fe,Mn,Zn)_4Be_3(SiO_4)_3S$ A mineral consisting of a silicate and sulfide of iron and beryllium; it is isomorphous with helvite and genthelvite.

dan buoy [NAV] A floating marker buoy used temporarily, as on fishing grounds and in minesweeping operations.

danburite [MINERAL] $CaB_2(SiO_4)_2$ An orange-yellow, yellowish-brown, grayish, or colorless transparent to translucent borosilicate mineral with a feldspar structure crystallizing in the orthorhombic system; it resembles topaz and is used as an ornamental stone.

dancing dervish See dust whirl.

dancing devil See dust whirl.

Danckwerts model [CHEM ENG] Theory applied to liquid flow across packing in a liquid-gas absorption tower; allows for liquid eddies that bring fresh liquid from the interior of the liquid body to the surface, thus contacting the gas in the column.

Dancool [TEXT] Trade name of a permanent-press fabric of Dan River Mills, made of 50% cotton and 50% polyester.

dandruff [MED] Scales of dry sebum formed on the scalp in seborrhea.

dandy fever See dengue.

dandy roll [MECH ENG] A roll in a fourdrinier paper-making machine; used to compact the sheet and sometimes to imprint a watermark.

Danel [TEXT] Trade name of a yarn-dyed, durable-press fabric of Dan River Mills, made of 50% cotton and 50% polyester.

danger angle [NAV] The angle between two known points as measured by an observer at a third point (the vessel), which is the safety limit for a vessel approaching a dangerous obstacle such as a reef.

danger area [NAV] A specified area within, below, or over which there may exist activities constituting potential danger to aircraft flying over it, or to persons, property, and traffic on land or sea.

danger bearing [NAV] The bearing of any object or obstruction as measured on board a vessel which will put a ship in jeopardy.

danger buoy [NAV] A buoy marking an isolated danger to navigation.

danger coefficient [NUCLEO] The change in reactivity per unit mass of a substance resulting from inserting the substance in a particular nuclear reactor.

danger line [NAV] A line on a chart representing a boundary, beyond which some hazard will be encountered.

dangerous semicircle [METEOROL] The half of the circular area of a tropical cyclone having the strongest winds and heaviest seas, where a ship tends to be drawn into the path of the storm.

danger sounding [NAV] A minimum sounding chosen for a vessel of specific draft in a given area to indicate the limit of safe navigation.

danger space [ORD] **1.** That portion of the range within which a target of given dimensions could be hit by a projectile with a given angle of fall. **2.** Space around the bursting point of an antiaircraft projectile.

dangler [MET] The flexible electrode used in barrel plating.

dangling bond [SOLID STATE] A chemical bond associated with an atom in the surface layer of a solid that does not join the atom with a second atom but extends in the direction of the solid's exterior.

Danian [GEOL] Lowermost Paleocene or uppermost Cretaceous geologic time.

Daniell cell [PHYS CHEM] A primary cell with a constant electromotive force of 1.1 volts, having a copper electrode in a copper sulfate solution and a zinc electrode in dilute sulfuric acid or zinc sulfate, the solutions separated by a porous partition or by gravity.

Daniell hygrometer [ENG] An instrument for measuring dew point; dew forms on the surface of a bulb containing ether which is cooled by evaporation into another bulb, the second bulb being cooled by the evaporation of ether on its outer surface.

dannemorite [MINERAL] $(Fe,Mn,Mg)_7Si_8O_{22}(OH)_2$ A yellowish-brown to greenish-gray monoclinic mineral consisting of a columnar or fibrous amphibole.

Dantwill [TEXT] Trade name for a piece-dyed, durable-press fabric of Dan River Mills, made of 50% cotton and 50% polyester.

Danysz reaction [IMMUNOL] A toxin-antitoxin reaction that occurs when an exact equivalence of toxin is added to antitoxin, not in one portion but in successive increments.

DAP See diaminopimelate.

daphnite [MINERAL] $(MgFe)_3(Fe,Al)_3(Si,Al)_4O_{10}(OH)_8$ A mineral of the chlorite group consisting of a basic aluminosilicate of magnesium, iron, and aluminum.

Daphoenidae [PALEON] A family of extinct carnivoran mammals in the superfamily Miacoidea.

dapsone See 4,4'-sulfonyldianiline.

daraf [ELEC] The unit of elastance, equal to the reciprocal of 1 farad.

darapskite [MINERAL] $Na_3(NO_3)(SO_4) \cdot H_2O$ A naturally occurring hydrate mineral consisting of a hydrous nitrate and sulfate of sodium.

darby [ENG] A flat-surfaced tool for smoothing plaster.

darcy [PHYS] A unit of permeability, equivalent to the passage of 1 cubic centimeter of fluid of 1 centipoise viscosity flowing in 1 second under a pressure of 1 atmosphere through a porous medium having a cross-sectional area of 1 square centimeter and a length of 1 centimeter.

Darcy number 1 [FL MECH] A dimensionless group, equal to four times the Fanning friction factor. Symbolized Da_1. Also known as Darcy-Weisbach coefficient; resistance coefficient 2.

Darcy number 2 [FL MECH] A dimensionless group used in the study of the flow of fluids in porous media, equal to the fluid velocity times the flow path divided by the permeability of the medium. Symbolized Da_2.

Darcy's law [FL MECH] The law that the rate at which a fluid flows through a permeable substance per unit area is equal to the permeability, which is a property only of the substance through which the fluid is flowing, times the pressure drop per unit length of flow, divided by the viscosity of the fluid.

Darcy-Weisbach coefficient See Darcy number 1.

Darcy-Weisbach equation [FL MECH] An equation for the loss of head due to friction h_f during turbulent flow of a fluid through a duct of any shape; for a circular pipe, it is $h_f = f(L/d)(V^2/2g)$, where L and d are the length and diameter of the pipe, V is the fluid velocity, g the acceleration of gravity, and f a dimensionless number called Darcy number 1.

dark box [GRAPHICS] A light-proof box used to store certain light-sensitive photographic papers.

dark conduction [ELECTR] Residual conduction in a photosensitive substance that is not illuminated.

dark current See electrode dark current.

dark-current pulse [ELECTR] A phototube dark-current excursion that can be resolved by the system employing the phototube.

dark discharge [ELECTR] An invisible electrical discharge in a gas.

dark-eclipsing variables [ASTRON] A binary star system, comprising a bright star and an almost dark companion that revolve about each other.

dark-field illumination [OPTICS] A method of microscope illumination in which the illuminating beam is a hollow cone of light formed by an opaque stop at the center of the condenser large enough to prevent direct light from entering the objective; the specimen is placed at the concentration of the light cone, and is seen with light scattered or diffracted by it.

dark lightning [GRAPHICS] A photographic effect in which lightning gives black photographic streaks instead of white due to multiple exposures caused by successive members of a composite flash. Also known as Clayden effect.

dark-line spectrum [SPECT] The absorption spectrum that results when white light passes through a substance, consisting of dark lines against a bright background.

dark nebula [ASTRON] A cloud of solid particles which absorbs or scatters away radiation directed toward an observer and becomes apparent when silhouetted against a bright nebula or rich star field.

dark of the moon [ASTRON] **1.** The time period of approximately a week at the time of a new moon, when the light of the moon is absent at night. **2.** Any period in which the light of the moon is obscured.

dark plaster [MATER] A plaster made from calcined, unground gypsum.

DARK-FIELD ILLUMINATION

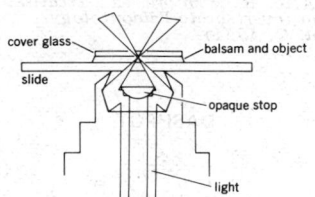

cover glass
slide
balsam and object
opaque stop
light

Diagram of dark-field illumination. *(American Optical Co., Instrument Division)*

D'ARSONVAL GALVANOMETER

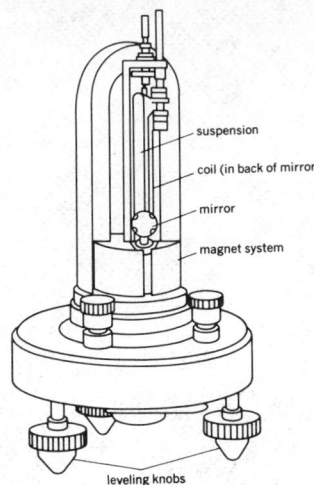

suspension

coil (in back of mirror)

mirror

magnet system

leveling knobs

D'Arsonval galvanometer showing the detached sacale and the light source as used in laboratory work. *(From D. M. Considine, ed., Process Instruments and Control Handbook, McGraw-Hill, 1957)*

DARWINULACEA

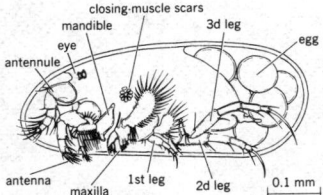

closing-muscle scars
mandible   3d leg
eye
antennule   egg

antenna
maxilla   1st leg   2d leg   0.1 mm

Darwinula stevensoni (Brady and Robertson), a podocopan ostracod of the Darwinulacea. In this parthenogenetic species, eggs and young are protected in a brood space behind the body. *(After R. C. Moore, ed., Treatise on Invertebrate Paleontology, pt. Q, 1961)*

DASHPOT

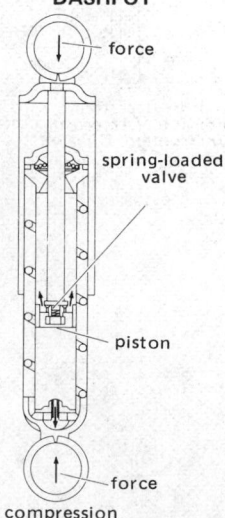

force

spring-loaded valve

piston

force

compression

A dashpot-type shock absorber. *(Plymouth Division, Chrysler Corp.)*

dark-red silver ore   *See* pyrargyrite.

dark resistance   [ELECTR] The resistance of a selenium cell or other photoelectric device in total darkness.

darkroom   [GRAPHICS] A light-free room illuminated by a safelight for processing light-sensitive materials.

dark-ruby silver   *See* pyrargyrite.

dark satellite   [AERO ENG] Satellite that gives no information to a friendly ground environment, either because it is controlled or because the radiating equipment is inoperative.

dark segment   [METEOROL] A bluish-gray band appearing along the horizon opposite the rising or setting sun and lying just below the antitwilight arch. Also known as earth's shadow.

dark space   [ELECTR] A region in a glow discharge that produces little or no light.

dark spot   [ELECTR] A spot on a television receiver tube that results from a spurious signal generated in the television camera tube during rescan, generally from the redistribution of secondary electrons over the mosaic in the tube.

dark star   [ASTRON] A star that is not visible but is a part of a binary star system; in particular, a star which causes, in an eclipsing variable, a primary eclipse.

dark-trace tube   [ELECTR] A cathode-ray tube having a bright face that does not necessarily luminesce, on which signals are displayed as dark traces or dark blips where the potassium chloride screen is hit by the electron beam. Also known as skiatron.

Darling shower   [METEOROL] A dust storm caused by cyclonic winds in the vicinity of the River Darling in Australia.

Darlington amplifier   [ELECTR] A current amplifier consisting essentially of two separate transistors and often mounted in a single transistor housing.

d'Arsonval current   [ELEC] A current consisting of isolated trains of heavily damped high-frequency oscillations of high voltage and relatively low current, used in diathermy.

d'Arsonval galvanometer   [ENG] A galvanometer in which a light coil of wire, suspended from thin copper or gold ribbons, rotates in the field of a permanent magnet when current is carried to it through the ribbons; the position of the coil is indicated by a mirror carried on it, which reflects a light beam onto a fixed scale. Also known as light-beam galvanometer.

dart   [INV ZOO] A small sclerotized structure ejected from the dart sac of certain snails into the body of another individual as a stimulant before copulation.

dart configuration   [AERO ENG] An aerodynamic configuration in which the control surfaces are at the tail of the vehicle.

dart leader   [GEOPHYS] The leader which, after the first stroke, initiates each succeeding stroke of a composite flash of lightning. Also known as continuous leader.

dart sac   [INV ZOO] A dart-forming pouch associated with the reproductive system of certain snails.

darwin   [EVOL] A unit of evolutionary rate of change; if some dimension of a part of an animal or plant, or of the whole animal or plant, changes from l_0 to l_t over a time of t years according to the formula $l_t = l_0 \exp(Et/10^6)$, its evolutionary rate of change is equal to E darwins.

Darwin-Doodson system   [GEOPHYS] A method for predicting tides by expressing them as sums of harmonic functions of time.

Darwin glass   [GEOL] A highly siliceous, vesicular glass shaped in smooth blobs or twisted shreds, found in the Mt. Darwin range in western Tasmania. Also known as queenstownite.

Darwinism   [BIOL] The theory of the origin and perpetuation of new species based on natural selection of those offspring best adapted to their environment because of genetic variation and consequent vigor. Also known as Darwin's theory.

Darwin's finch   [VERT ZOO] A bird of the subfamily Fringillidae; Darwin studied the variation of these birds and used his data as evidence for his theory of evolution by natural selection.

Darwin's theory   *See* Darwinism.

Darwinulacea   [INV ZOO] A small superfamily of nonmarine, parthenogenetic ostracods in the suborder Podocopa.

Darzen's procedure   [ORG CHEM] Preparation of alkyl halides by refluxing a molecule of an alcohol with a molecule of thionyl chloride in the presence of a molecule of pyridine.

Darzen's reaction   [ORG CHEM] Condensation of aldehydes and ketones with α-haloesters to produce glycidic esters.

Dasayatidae   [VERT ZOO] The stingrays, a family of modern sharks in the batoid group having a narrow tail with a single poisonous spine.

Dascillidae   [INV ZOO] The soft-bodied plant beetles, a family of coleopteran insects in the superfamily Dascilloidea.

Dascilloidea   [INV ZOO] A superfamily of coleopteran insects in the suborder Polyphaga.

dasheen   [BOT] *Colocasia esculenta.* A plant in the order Arales, grown for its edible corm.

dashkesanite   [MINERAL] $(Na,K)Ca_2(Fe,Mg)_5(Si,Al)_8O_{22}Cl_2$ A monoclinic mineral of the amphibole group consisting of a chloroaluminosilicate of sodium, potassium, iron, and magnesium.

dashpot   [MECH ENG] A device used to dampen and control a motion, in which an attached piston is loosely fitted to move slowly in a cylinder containing oil.

Dasycladaceae   [BOT] A family of green algae in the order Dasycladales comprising plants formed of a central stem from which whorls of branches develop.

Dasycladales   [BOT] An order of lime-encrusted marine algae in the division Chlorophyta, characterized by a thallus composed of nonseptate, highly branched tubes.

dasymeter   [PHYS] A thin glass globe used to measure the density of gas by weighing the globe in the gas.

Dasyonygidae   [INV ZOO] A family of biting lice, order Mallophaga, that are confined to rodents of the family Procaviidae.

Dasypodidae   [VERT ZOO] The armadillos, a family of edentate mammals in the infraorder Cingulata.

Dasytidae   [INV ZOO] An equivalent name for Melyridae.

Dasyuridae   [VERT ZOO] A family of mammals in the order Marsupialia characterized by five toes on each hindfoot.

Dasyuroidea   [VERT ZOO] A superfamily of marsupial mammals.

data   [ADP] **1.** General term for numbers, letters, symbols, and analog quantities that serve as input for computer processing. **2.** Any representations of characters or analog quantities to which meaning, if not information, may be assigned.   [SCI TECH] Numerical or qualitative values derived from scientific experiments.

data acquisition   [COMMUN] The phase of data handling that begins with the sensing of variables and ends with a magnetic recording or other record of raw data; may include a complete radio telemetering link.

data analysis   [ADP] The evaluation of digital data.

data automation   [ADP] The use of electronic, electromechanical, or mechanical equipment and associated techniques to automatically record, communicate, and process data and to present the resultant information.

data bank   [ADP] A complete collection of information such as contained in automated files, a library, or a set of computer disks. Also known as data base.

data base   *See* data bank.

data cell drive   [ADP] A large-capacity storage device consisting of strips of magnetic tape which can be individually transferred to the read-write head.

data chain   [ADP] Any combination of two or more data elements, data items, data codes, and data abbreviations in a prescribed sequence to yield meaningful information; for example, "date" consists of data elements year, month, and day.

data chaining   [ADP] A technique used in scatter reading or scatter writing in which new storage areas are defined for use as soon as the current data transfer is completed.

data channel   [ADP] A bidirectional data path between input/output devices and the main memory of a digital computer permitting one or more input/output operations to proceed concurrently with computation.

data circuit   [ELECTR] A telephone facility allowing transmission of digital data pulses with minimum distortion.

data code   [ADP] A number, letter, character, symbol, or any combination thereof, used to represent a data item.

data collection   [ADP] The process of sending data to a central point from one or more locations.

data communications   [COMMUN] The conveying from one location to another by electrical means of information that originates or is recorded in alphabetic, numeric, or pictorial form, or as a signal that represents a measurement; includes telemetering, telegraphy, and facsimile but not voice or television. Also known as data transmission.

data compression [ADP] Any means of increasing the quantity of data that can be stored in a given space, or decreasing the space needed to store a given quantity of data. Also known as compression.

data conversion [ADP] The changing of the representation of data from one form to another, as from binary to decimal, or from one physical recording medium to another, as from card to disk. Also known as conversion.

data conversion line [ADP] The channel, electronic or manual, through which data elements are transferred between data banks.

data descriptor [ADP] A pointer indicating the memory location of a data item.

data display [ADP] Visual presentation of processed data by specially designed electronic or electromechanical devices through interconnection (either on- or off-line) with digital computers or component equipments; although line printers and punch cards may display data, they are not usually categorized as displays but as output equipments.

data division [ADP] The section of a program (written in the COBOL language) which describes each data item used for input, output, and storage.

data element [ADP] A set of data items pertaining to information of one kind, such as months of a year.

data error [ADP] A deviation from correctness in data, usually an error, which occurred prior to processing the data.

data field [ADP] An area in the main memory of the computer in which a data record is contained.

data flow [ADP] The transfer of data from an external storage device, through the processing unit and memory, and out to an external storage device.

data formatting [ADP] Structuring the presentation of data as numerical or alphabetic and specifying the size and type of each datum.

data-handling system [ADP] Automatically operated equipment used to interpret data gathered by instrument installations. Also known as data-reduction system.

data-initiated control [ADP] The automatic handling of a program dependent only upon the value of input data fed into the computer.

data interchange [ADP] Switching of data in and out of storage units.

data item [ADP] A single member of a data element. Also known as datum.

data link [COMMUN] The physical equipment for automatic transmission and reception of information. Also known as communication link; information link; tie line; tie-link.

data logging [ADP] Conversion of electrical impulses from process instruments into digital data to be recorded, stored, and periodically tabulated.

data manipulation [ADP] The standard operations of sorting, merging, input/output, and report generation.

datamation [ADP] A shortened term for automatic data processing; taken from data and automation.

data organization [ADP] Any one of the data-management conventions for physical and spatial arrangement of the physical records of a data set. Also known as data-set organization.

data origination [ADP] The process of putting data in a form that can be read by a machine.

Data-Phone [ELECTR] A Bell System device that permits transmission of data over telephone channels at speeds up to 2000 bits per second.

data plotter [ADP] A device which plots digital information in a continuous fashion.

data processing [ADP] Any operation or combination of operations on data, including everything that happens to data from the time they are observed or collected to the time they are destroyed. Also known as information processing.

data-processing center [ADP] A computer installation providing data-processing service for others, sometimes called customers, on a reimbursable or nonreimbursable basis.

data-processing inventory [ADP] An identification of all major data-processing areas in an agency for the purpose of selecting and focusing upon those in which the use of automatic data-processing (ADP) techniques appears to be potentially advantageous, establishing relative priorities and schedules for embarking on ADP studies, and identifying significant relationships among areas to pinpoint possibilities for the integration of systems.

data-processing machine [ADP] A computer or a component of a data-processing system, such as a card reader or tape unit.

data processor [ADP] **1.** Any device capable of performing operations on data, for instance, a desk calculator, an analog computer, or a digital computer. **2.** Person engaged in processing data.

data purification [ADP] The process of removing as many inaccurate or incorrect items as possible from a mass of data before automatic data processing is begun.

data record [ADP] A collection of data items related in some fashion and usually contiguous in location.

data reduction [ADP] The transformation of raw data into a more useful form.

data-reduction system *See* data-handling system.

data retrieval [ADP] The searching, selecting, and retrieving of actual data from a personnel file, data bank, or other file.

data rules [ADP] Conditions which must be met by data to be processed by a computer program.

data set [ADP] **1.** A named collection of similar and related data records recorded upon some computer-readable medium. **2.** A data file in IBM 360 terminology.

data set coupler [ADP] The interface between a parallel computer input-output bus and the serial input-output of a modem.

data set migration [ADP] The process of moving inactive data sets from on-line storage to back up storage in a time-sharing environment.

data-set organization *See* data organization.

data stabilization [ELECTR] Stabilization of the display of radar signals with respect to a selected reference, regardless of changes in radar-carrying vehicle attitude, as in azimuth-stabilized plan-position indicator.

data station [ADP] A remote input/output device which handles a variety of transmissions to and from certain centralized computers.

data station control [ADP] The supervision of a data station by means of a program resident in the central computer.

data system [ADP] The means, either manual or automatic, of converting data into action or decision information, including the forms, procedures, and processes which together provide an organized and interrelated means of recording, communicating, processing, and presenting information relative to a definable function or activity.

data system interface [ADP] **1.** A common aspect of two or more data systems involving the capability of intersystem communications. **2.** A common boundary between automatic data-processing systems or parts of a single system.

data systems integration [ADP] Achievement through systems design of an improved or broader capability by functionally or technically relating two or more data systems, or by incorporating a portion of the functional or technical elements of one data system into another.

data system specifications [ADP] **1.** The delineation of the objectives which a data system is intended to accomplish. **2.** The data-processing requirements underlying that accomplishment; includes a description of the data output, the data files and record content, the volume of data, the processing frequencies, training, and such other facts as may be necessary to provide a full description of the system.

data tracks [ADP] Information storage positions on drum storage devices; information is stored on the drum surface in the form of magnetized or nonmagnetized areas.

data transcription equipment [ADP] Those devices or equipment designed to convey data from its original state to a data-processing media.

data transfer [ADP] The technique used by the hardware manufacturer to transmit data from computer to storage device or from storage device to computer; usually under specialized program control.

data transmission   *See* data communications.

data transmission equipment   [ADP] The communications equipment used in direct support of data-processing equipment.

data transmission line   [ELEC] A system of electrical conductors, such as a coaxial cable or pair of wires, used to send information from one place to another or one part of a system to another.

data transmission-utilization measure   [COMM] The ratio of useful data output to the sum total of data input.

data unit   [ADP] A set of digits or characters treated as a whole.

data validation   [ADP] The checking of data for correctness, or the determination of compliance with applicable standards, rules, and conventions.

data word   [ADP] A computer word that is part of the data which the computer is manipulating, in contrast with an instruction word. Also known as information word.

date time group   [COMMUN] The date and time, expressed in digits and zone suffix, at which the message was prepared for transmission (expressed as six digits followed by the zone suffix; first pair of digits denoting the date, second pair the hours, third pair the minutes).

dating   [SCI TECH] The use of methods and techniques to fix dates, assign periods of time, and determine age in archeology, biology, and geology.

dative bond   *See* coordinate valence.

datolite   [MINERAL] $CaBSiO_4(OH)$ A mineral nesosilicate crystallizing in the monoclinic system; luster is vitreous, and crystals are colorless or white with a greenish tinge. Also known as dystome spar; humboldtite.

datum   [ADP] *See* data item.   [ENG]   **1.** A direction, level, or position from which angles, heights, speeds or distances are conveniently measured.   **2.** Any numerical or geometric quantity or value that serves as a base reference for other quantities or values (such as a point, line, or surface in relation to which others are determined).   [GEOD] The latitude and longitude of an initial point; the azimuth of a line from this point.   [GEOL] The top or bottom of a bed of rock on which structure contours are drawn.

datum line   *See* reference line.

datum plane   [ENG] A permanently established horizontal plane, surface, or level to which soundings, ground elevations, water surface elevations, and tidal data are referred. Also known as chart datum; datum level; reference level; reference plane.

datum point   [MAP] Any reference point of known or assumed coordinates from which calculation or measurements may be taken.

Daubentoniidae   [VERT ZOO] A family of Madagascan prosimian primates containing a single species, the aye-aye.

Daubenton's plane   [ANTHRO] A plane passing through the opisthion and the orbital points on the skull.

daubreelite   [MINERAL] $FeCr_2S_4$ A mineral composed of a black chromium iron sulfide; occurs in some meteors.

daughter   [NUC PHYS] The immediate product of radioactive decay of an element, such as uranium. Also known as decay product; radioactive decay product.

Dauphine law   [CRYSTAL] A twin law in which the twinned parts are related by a rotation of 180° around the *c* axis.

Davian   [GEOL] A subdivision of the Upper Cretaceous in Europe; a limestone formation with abundant hydrocorals, bryozoans, and mollusks in Denmark; marine limestone and nonmarine rocks in southeastern France; and continental formations in the Davian of Spain and Portugal.

davidite   [MINERAL] A black primary pegmatite uranium mineral of the general formula $A_6B_{15}(O,OH)_{36}$, where $A = Fe^{2+}$, rare earths, uranium, calcium, zirconium, and thorium, and $B =$ titanium, Fe^{3+}, vanadium, and chromium.

Davidson Current   [OCEANOGR] A coastal countercurrent of the Pacific Ocean running north, inshore of the California Current, along the western coast of the United States (from northern California to Washington to at least latitude 48°N) during the winter months.

daviesite   [MINERAL] An orthorhombic mineral consisting of a lead oxychloride, occurring in minute crystals.

Davis correction   [FL MECH] Empirical relation of flow-line diameters used to correct data calculated from the Atherton equation (friction loss in annular passages).

Davis magnetic tester   [MIN ENG] An instrument used to determine the magnetic contents of ores.

davisonite   [MINERAL] $Ca_3Al(PO_4)_2(OH)_3H_2O$ White mineral consisting of a hydrous basic phosphate of calcium and aluminum.

Davis wing   [AERO ENG] A narrow-chord wing that has comparatively low drag and a stable center of pressure and develops lift at relatively small angles of attack.

davit   [NAV ARCH] A fixed or movable shipboard crane projecting over the side or over a hatchway, which hoists and lowers boats, anchors, or cargo.

Davy lamp   [MIN ENG] An early safety lamp with a mantle of wire gauze around the flame to dissipate the heat from the flame to below the ignition temperature of methane.

Dawes' limit   [OPTICS] The resolving power of a telescope, limited by diffraction effects, is $4.5/a$ seconds of arc, where a is the aperture in inches.

dawn   [ASTRON] The first appearance of light in the eastern sky before sunrise, or the time of that appearance. Also known as daybreak.

dawn side   [ASTRON] That side of a celestial object, such as a planet, which points in the direction of its orbital movement.

Dawsoniales   [BOT] An order of mosses comprising rigid plants with erect stems rising from a rhizomelike base.

dawsonite   [MINERAL] $NaAl(OH)_2CO_3$ A white, bladed mineral found in certain oil shales that contains large quantities of alumina; specific gravity is 2.40.

day   [ASTRON] One of various units of time equal to the period of rotation of the earth with respect to one or another direction in space; specific examples are the mean solar day and the sidereal day.

day beacon   [NAV] An unlighted structure serving as an aid to navigation in the daytime.

daybreak   *See* dawn.

day clock   [ADP] An internal binary counter, with a resolution usually of a microsecond and a cycle measured in years, providing an accurate measure of elapsed time independent of system activity.

day drift   [MIN ENG] A mine passageway that has one end at the surface.

dayglow   [ASTRON] Airglow of the day sky.

daylight   [ASTRON] Light of the day, from sun and sky.

daylight controls   [ENG] Special devices which automatically control the electric power to the lamp, causing the light to operate during hours of darkness and to be extinguished during daylight hours.

daylight glass   [MATER] A glass that absorbs red light; used in incandescent lamps to remove excess red emission so that the light spectrum resembles natural daylight.

daylighting   [CIV ENG] To light an area with daylight.

daylight lamp   [ELEC] An incandescent or fluorescent lamp that emits light whose spectral distribution is approximately that of daylight.

daylight saving meridian   [ASTRON] The meridian used for reckoning daylight saving time; generally 15° east of the zone of standard meridian.

daylight saving noon   [ASTRON] Twelve o'clock daylight saving time, or the instant the mean sun is over the upper branch of the daylight saving meridian; during a war, when daylight saving time may be used throughout the year and called war time, the expression war noon applies. Also known as summer noon.

daylight saving time   [ASTRON] A variation of zone time, usually 1 hour more advanced than standard time, frequently kept during the summer to make better use of daylight. Also known as summer time.

daymark   [NAV] The daytime-identifying characteristics of an aid to navigation, unique and distinctive to facilitate its recognition; a conspicuous target added to a day beacon or light.

day neutral   [BOT] Reaching maturity regardless of relative length of light and dark periods.

Day's multiplexing system   [COMMUN] A system of phase-discrimination multiplexing applicable to pairs of channels;

used to multiplex the two so-called color components in color television broadcasting.

day's run [NAV] The distance traveled by a vessel in 1 day, usually reckoned from noon to noon.

day's work [NAV] In marine operations, the daily routine of the navigation of a vessel at sea, usually consisting principally of the dead reckoning from noon to noon, evening and morning twilight observations, a morning sun observation for a line of position and for checking the compass, and a sun observation at or near noon to obtain a running fix.

day wage [IND ENG] A fixed rate of pay per shift or per daily hours of work, irrespective of the amount of work completed.

dB See decibel.

dBa See adjusted decibel.

dBk See decibels above 1 kilowatt.

dBm See decibels above 1 milliwatt.

dBp See decibels above 1 picowatt.

dBrn See decibels above reference noise.

dBV See decibels above 1 volt.

dBW See decibels above 1 watt.

dBx See decibels above reference coupling.

dc See direct current.

D cable [ELEC] Two-conductor cable, each conductor having the shape of the letter D, with insulation between the conductors and between the conductors and the sheath.

dc casting See direct-chill casting.

4-D chart [METEOROL] A chart showing the field of D values (deviations of the actual altitudes along a constant-pressure surface from the standard atmosphere altitude of that surface) in terms of the three dimensions of space and one of time; it is a form of a four-dimensional display of pressure altitude; the space dimensions are represented by D-value contours, and the time dimension is provided by tau-value lines.

DCTL See direct-coupled transistor logic.

dc-to-ac converter See inverter.

dc-to-ac inverter See inverter.

dc-to-dc converter [ELEC] An electronic circuit which converts one direct-current voltage into another, consisting of an inverter followed by a step-up or step-down transformer and rectifier.

dcwv See direct-current working volts.

DDA See digital differential analyzer.

DDA value See depth-duration-area value.

DDBS See Digital Data Broadcast System.

D display [ELECTR] In radar, a C display in which the blips extend vertically to give a rough estimate of distance.

DDT [ORG CHEM] Common name for an insecticide; melting point 108.5°C, insoluble in water, very soluble in ethanol and acetone, colorless, and odorless; especially useful against agricultural pests, flies, lice, and mosquitoes. Also known as dichlorodiphenyltrichloroethane.

DDTA See derivative differential thermal analysis.

DEA See diethanolamine.

deaccentuator [ELECTR] A circuit used in a frequency-modulation receiver to offset the preemphasis of higher audio frequencies introduced at the transmitter.

deacetylate [ORG CHEM] The removal of an acetyl grouping from a molecule.

deacidify [CHEM] 1. Removal of acid. 2. A process of reducing acidity.

Deacon process [CHEM ENG] A method of chlorine production by passing a hot mixture of gaseous hydrochloric acid with oxygen over a cuprous chloride catalyst.

deactivation [CHEM] 1. Rendering inactive, as of a catalyst. 2. Loss of radioactivity. [MET] Chemical removal of the active constituents of a corrosive liquid. [MIN ENG] Treatment of one or more species of mineral particles to reduce floating during froth flotation.

dead [ELEC] Free from any electric connection to a source of potential difference from electric charge; not having a potential different from that of earth; the term is used only with reference to current-carrying parts which are sometimes alive or charged.

dead ahead [NAV] In marine operations, a position directly ahead of the ship; a bearing relative 000°.

dead air [MIN ENG] Air in a mine when it is stagnant or contains carbonic acid.

dead-air space [BUILD] A sealed air space, such as in a hollow wall.

dead area See blind spot.

dead arm [PL PATH] A fungus disease caused by *Cryptosporella viticola* in which the main lateral shoots are destroyed; common in grapes.

dead astern [NAV] In marine navigation, a bearing 180° relative; if the bearing is approximate, the term astern should be used. Also known as right astern.

dead axle [MECH ENG] An axle that carries a wheel but does not drive it.

dead band [ELEC] The portion of a potentiometer element that is shortened by a tap; when the wiper traverses this area, there is no change in output. [ENG] The range of values of the measured variable to which an instrument will not effectively respond. Also known as dead zone; neutral zone.

deadbeat [MECH] Coming to rest without vibration or oscillation, as when the pointer of a meter moves to a new position without overshooting. Also known as deadbeat response.

deadbeat compass See aperiodic compass.

deadbeat escapement [HOROL] A watch escapement without recoil, having arresting faces of the pallets described by a circular arc whose center is at the pivot point of the anchor; escape-wheel teeth are contoured to give impulses to these pallet faces, over which they slide without recoil.

deadbeat response See deadbeat.

dead block [ENG] A device placed on the ends of railroad passenger cars to absorb the shock of impacts.

dead bolt [DES ENG] A lock bolt that is moved directly by the turning of a knob or key, not by spring action.

dead-bright [MET] Of a metal, polished to remove tool marks.

dead-burn [MIN ENG] A calcination to produce a dense refractory substance; done at a higher temperature and for a longer time than for normal calcination.

dead center [MECH ENG] 1. A position of a crank in which the turning force applied to it by the connecting rod is zero; occurs when the crank and rod are in a straight line. 2. A support for the work on a lathe which does not turn with the work.

dead-center position [ELEC] Position in which a brush would be placed on the commutator of a direct-current motor or generator if the field flux were not distorted by armature reaction.

dead earth [ELEC] A connection between a line conductor and earth by means of a path of low resistance.

dead end [ACOUS] The end of a sound studio that has the greater sound-absorbing characteristics. [ELEC] The portion of a tapped coil through which no current is flowing at a particular switch position. [SCI TECH] The end of a conduit, passage, power line, or similar system having no exit or continuation. Also known as cul-de-sac.

dead-end effect [ELEC] Absorption of energy by unused portions of a tapped coil.

dead-end switch [ELEC] A switch used to short-circuit unused portions of a tapped coil to prevent dead-end effects.

dead-end tower [CIV ENG] Antenna or transmission line tower designed to withstand unbalanced mechanical pull from all the conductors in one direction together with the wind strain and vertical loads.

deadeye [NAV ARCH] A flat, rounded wooden block usually pierced with three holes to receive the lanyard; used to fasten shrouds and stays.

dead flat [NAV ARCH] A portion of length in the middle of a ship in which all sections have the same shape. Also known as parallel middle body.

dead ground [ELEC] A low-resistance connection between the ground and an electric circuit.

dead halt See drop-dead halt.

deadhead [MET] The portion of a casting that fills up the ingate. [MIN ENG] To begin a new cut without excavating the material from the preceding cut.

deadlight [NAV ARCH] 1. A strong plate or a cover with light-obstructing baffles fitted over ventilation ports or windows in stormy weather. 2. A strong glass or plastic window set in the deck or side of a ship to admit light.

dead load See static load.

DEADMAN

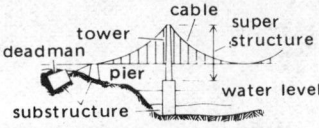

A drawing of a bridge showing buried block used as a deadman.

deadly nightshade   *See* belladonna.

deadman   [CIV ENG]   **1.** A buried plate, wall, or block attached at some distance from and forming an anchorage for a retaining wall. Also known as anchorage; anchor block; anchor wall.   **2.** *See* anchor log.

deadman's brake   [MECH ENG]   An emergency device that automatically is activated to stop a vehicle when the driver removes his foot from the pedal.

deadman's handle   [MECH ENG]   A handle on a machine designed so that the operator must continuously press on it in order to keep the machine running.

dead oil   [MATER]   An oil, of density greater than water, distilled from tar.

dead rail   [CIV ENG]   One of two rails on a railroad weighing platform that permit an excessive load to leave the platform.

dead reckoning   [NAV]   Determination of position of a craft by advancing a previous position to a new one on the basis of assumed distance and direction traveled.

dead-reckoning analyzer   *See* dead-reckoning equipment.

dead-reckoning equipment   [NAV]   A computing device which continuously indicates the dead-reckoning position of a craft, vessel, or vehicle and which may provide a graphical record of the dead-reckoning track. Also known as dead-reckoning analyzer.

dead-reckoning plot   [NAV]   **1.** In marine navigation, the graphic plot of consecutive course lines representing courses steered or planned to be steered, originating from a fix or running fix position, and suitably labeled as to course, speeds, and times of periodic dead-reckoning positions.   **2.** In air navigation, the graphic plot of track and ground speed, suitably labeled. Also known as ground plot.

dead-reckoning position   [NAV]   A position determined by advancing a known position of the past to the position of the present time by adding one or more vectors representing known or assumed speed, bearing, and wind or current.

dead-reckoning tracer   [NAV]   A mechanical device used to produce a continuous plot of all ships within range of a ship's radar.

dead-reckoning track   [NAV]   A line representing successive dead-reckoning positions of a craft. Also known as DR track line.

dead-rise   [NAV ARCH]   The vertical distance between the bottom of a vessel's keel and the intersection of a line drawn along the bottom of a vessel's midship section with a vertical line tangent to the widest part of the underwater body. Also known as rise of bottom; rise of floor.

dead rising   [NAV ARCH]   In a wooden ship's profile or sheer plan, a curved fore-and-aft line that passes through the upper end of the floor timbers.

dead-roast   [MIN ENG]   **1.** A roasting process for driving off sulfur. Also known as sweet roast.   **2.** Removing volatiles by roasting within a specified temperature range.

dead room   *See* anechoic chamber.

dead sea   [HYD]   A body of water that has undergone precipitation of its rock salt, gypsum, or other evaporites.

Dead Sea   [GEOGR]   A salt lake between Jordan and Israel.

dead short   [ELEC]   A short-circuit path that has extremely low resistance.

dead soft steel   [MET]   **1.** Steel very low in carbon.   **2.** Steel annealed until it is very soft.

dead space   [ANAT]   The space in the trachea, bronchi, and other air passages which contains air that does not reach the alveoli during respiration, the amount of air being about 140 milliliters. Also known as anatomical dead space.   [MED]   A cavity left after closure of a wound.   [PHYSIO]   A calculated expression of the anatomical dead space plus whatever degree of overventilation or underperfusion is present; it is alleged to reflect the relationship of ventilation to pulmonary capillary perfusion. Also known as physiological dead space.   [THERMO]   A space filled with gas whose temperature differs from that of the main body of gas, such as the gas in the capillary tube of a constant-volume gas thermometer.

dead spot   [COMMUN]   A geographic location in which signals from a radio, television, or radar transmitter are received poorly or not at all.   [ELECTR]   A portion of the tuning range of a receiver in which stations are heard poorly or not at all, due to improper design of tuning circuits.

dead stick   [AERO ENG]   The propeller of an airplane that is not rotating because the engine has stopped.

dead-stroke   [MECH ENG]   Having a recoilless or nearly recoilless stroke.

dead-stroke hammer   [MECH ENG]   A power hammer provided with a spring on the hammer head to reduce recoil.

dead time   [CONT SYS]   The time interval between a change in the input signal to a process control system and the response to the signal.   [ENG]   The time interval, after a response to one signal or event, during which a system is unable to respond to another. Also known as insensitive time.

dead-time correction   [ENG]   A correction applied to an observed counting rate to allow for the probability of the occurrence of events within the dead time. Also known as coincidence correction.

dead track   [CIV ENG]   **1.** Railway track that is no longer used.   **2.** A section of railway track that is electrically isolated from the track signal circuits.

dead water   [OCEANOGR]   The mass of eddying water associated with formation of internal waves near the keel of a ship; forms under a ship of low propulsive power when it negotiates water which has a thin layer of fresher water over a deeper layer of more saline water.

deadweight capacity   *See* deadweight tonnage.

deadweight gage   [ENG]   An instrument used as a standard for calibrating pressure gages in which known hydraulic pressures are generated by means of freely balanced (dead) weights loaded on a calibrated piston.

deadweight tonnage   [NAV ARCH]   The total carrying capacity of a ship expressed in long tons (2240 pounds); displacement of a fully loaded ship, less the weight of the ship itself. Abbreviated dwt. Also known as deadweight capacity.

dead work   [MIN ENG]   Preparatory work which is for future operations and not directly productive.

dead zone   *See* dead band.

deaeration   [ENG]   Removal of gas or air from a substance, as from feedwater or food.

deaerator   [MECH ENG]   A device in which oxygen and carbon dioxide are removed from boiler feedwater.

deafness   [MED]   Temporary or permanent impairment or loss of hearing.

deagglomeration   [CHEM ENG]   Size-reduction process in which loosely adhered clumps (agglomerates) of powders or crystals are broken apart without further disintegration of the powder or crystal particles themselves.

deaister   *See* doister.

deal   [DES ENG]   **1.** A face on which numbers are registered by means of a pointer.   **2.** A disk usually with a series of markings around its border, which can be turned to regulate the operation of a machine or electrical device.

dealkalize   [CHEM]   **1.** The removal of alkali.   **2.** The reduction of alkalinity as in the neutralization process.

dealkylate   [CHEM]   To remove alkyl groups from a compound.

dealuminization   [CHEM]   Removal of aluminum.

deamidase   [BIOCHEM]   An enzyme that catalyzes the removal of an amido group from a compound.

deamidate   [ORG CHEM]   Removal from a molecule of the amido group.

deaminase   [BIOCHEM]   An enzyme that catalyzes the hydrolysis of amino compounds, removing the amino group.

deaminate   [ORG CHEM]   Removal from a molecule of the amino group.

Dean number   [FL MECH]   A dimensionless number giving the ratio of the viscous force acting on a fluid flowing in a curved pipe to the centrifugal force; equal to the Reynolds number times the square root of the ratio of the radius of the pipe to its radius of curvature. Symbolized N_D.

deasphalting   [CHEM ENG]   The process of removing asphalt from petroleum fractions.

death   [MED]   Cessation of all life functions; can involve the whole organism, an organ, individual cells, or cell parts.

death instinct   [PSYCH]   In psychoanalytic theory, the unconscious drive which leads the individual toward dissolution and death, and which coexists with the life instinct.

death point   [PHYSIO]   The limit (as of extremes of temperature) beyond which an organism cannot survive.

death rate   *See* mortality rate.

debenzylation [ORG CHEM] Removal from a molecule of the benzyl group.

de Beurmann–Gougerot disease *See* sporotrichosis.

deblocking [ADP] Breaking up a block of records into individual records.

deblooming [CHEM ENG] The process by which the fluorescence, or bloom, is removed from petroleum oils by exposing them in shallow tanks to the sun and atmospheric conditions or by using chemicals.

Deborah number [MECH] A dimensionless number used in rheology, equal to the relaxation time for some process divided by the time it is observed. Symbolized D.

debossed [GRAPHICS] Having a depressed pattern on the surface of a material.

Debot effect [GRAPHICS] A variation of the Herschel effect, in which red or infrared radiation converts an internal latent image into a surface latent image.

debriefing [AERO ENG] The relating of factual information by a flight crew at the termination of a flight, consisting of flight weather encountered, the condition of the aircraft, or facilities along the airways or at the airports.

debris [GEOL] Large fragments arising from disintegration of rocks and strata.

debris avalanche [GEOL] The sudden and rapid downward movement of incoherent mixtures of rock and soil on deep slopes.

debris cone [GEOL] **1.** A mound of fine-grained debris piled atop certain boulders moved by a landslide. **2.** A mound of ice or snow on a glacier covered with a thin layer of debris.

debris dam [CIV ENG] A fixed dam across a stream channel for the retention of sand, gravel, driftwood, or other debris.

debris fall [GEOL] A relatively free downward or forward falling of unconsolidated or poorly consolidated earth or rocky debris from a cliff, cave, or arch.

debris-flow [GEOL] Rapid flowage of debris resulting from heavy saturation by rains; specifically, a high-density mudflow.

debris glacier [HYD] A glacier formed from ice fragments that have fallen from a larger and taller glacier.

debris line *See* swash mark.

debris slide [GEOL] A type of landslide involving a rapid downward sliding and forward rolling of comparatively dry, unconsolidated earth and rocky debris.

debris slope *See* talus slope.

de Broglie equation *See* de Broglie relation.

de Broglie relation [QUANT MECH] The relation in which the de Broglie wave associated with a free particle of matter, and the electromagnetic wave in a vacuum associated with a photon, has a wavelength equal to Planck's constant divided by the particle's momentum and a frequency equal to the particle's energy divided by Planck's constant. Also known as de Broglie equation.

de Broglie's theory [QUANT MECH] The theory that particles of matter have wavelike properties which can give rise to interference effects, and electrons in an atom are associated with standing waves on a Bohr orbit.

de Broglie wave [QUANT MECH] The quantum-mechanical wave associated with a particle of matter.

de Broglie wavelength [QUANT MECH] The wavelength of the wave associated with a particle as given by the de Broglie relation.

de Brun–van Eckstein rearrangement [ORG CHEM] The isomerization of an aldose or ketose when mixed with aqueous calcium hydroxide to form a mixture of various monosaccharides and unfermented ketoses; used to prepare certain ketoses.

debubblizer [ENG] A worker who removes bubbles from plastic rods and tubing.

debug [ADP] To test for, locate, and remove mistakes from a program or malfunctions from a computer. [ELECTR] To detect and remove secretly installed listening devices popularly known as bugs. [ENG] To eliminate from a newly designed system the components and circuits that cause early failures.

debugging routine [ADP] A routine to aid programmers in the debugging of their routiness; some typical routines are storage printout, tape printout, and drum printout routines.

debugging statement [ADP] Temporary instructions inserted into a program being tested so as to pinpoint problem areas.

debunching [ELECTR] A tendency for electrons in a beam to spread out both longitudinally and transversely due to mutual repulsion; the effect is a drawback in velocity modulation tubes.

deburr [MET] To remove burrs, sharp edges, or fins from metal parts by placing them in a revolving barrel containing abrasives suspended in a liquid.

debutanization [CHEM ENG] Removal of butane and lighter components in a natural-gasoline plant.

debutanizer [CHEM ENG] The fractionating column in a natural-gasoline plant in which butane and lighter components are removed.

debye [ELEC] A unit of electric dipole moment, equal to 10^{-18} Franklin centimeter.

Debye effect [ELECTROMAG] Selective absorption of electromagnetic waves by a dielectric, due to molecular dipoles.

Debye equation [SOLID STATE] The equation for the Debye specific heat, which satisfies the Dulong and Petit law at high temperatures and the Debye T^3 law at low temperatures.

Debye equation for polarization [STAT MECH] The Langevin-Debye formula for the polarization of a dielectric material, relating the total polarization for n molecules to the permanent moment of the specific molecule and its polarizability.

Debye-Falkenhagen effect [PHYS CHEM] The increase in the conductance of an electrolytic solution when the applied voltage has a very high frequency.

Debye frequency [SOLID STATE] The maximum allowable frequency in the computation of the Debye specific heat.

Debye-Hückel screening radius *See* Debye shielding length.

Debye-Hückel theory [PHYS CHEM] A theory of the behavior of strong electrolytes, according to which each ion is surrounded by an ionic atmosphere of charges of the opposite sign whose behavior retards the movement of ions when a current is passed through the medium.

Debye-Jauncey scattering [SOLID STATE] Incoherent background scattering of x-rays from a crystal in directions between those of the Bragg reflections.

Debye length *See* Debye shielding length.

Debye potentials [ELECTROMAG] Two scalar potentials, designated Π_e and Π_m, in terms of which one can express the electric and magnetic fields resulting from radiation or scattering of electromagnetic waves by a distribution of localized sources in a homogeneous isotropic medium.

Debye relaxation time [PHYS CHEM] According to the Debye-Hückel theory, the time required for the ionic atmosphere of a charge to reach equilibrium in a current-carrying electrolyte, during which time the motion of the charge is retarded.

Debye-Scherrer method [SOLID STATE] An x-ray diffraction method in which the sample, consisting of a powder stuck to a thin fiber or contained in a thin-walled silica tube, is rotated in a monochromatic beam of x-rays, and the diffraction pattern is recorded on a cylindrical film whose axis is parallel to the axis of rotation of the sample.

Debye-Sears ultrasonic cell [ACOUS] A process in ultrasonic imaging for which the acoustic wavefronts act as optical gratings to diffract the light on either side of the central spot.

Debye shielding length [PL PHYS] A characteristic distance in a plasma beyond which the electric field of a charged particle is shielded by particles having charges of the opposite sign. Also known as Debye-Hückel screening radius; Debye length; shielding distance.

Debye specific heat [SOLID STATE] The specific heat of a solid under the assumption that the energy of the lattice arises entirely from acoustic lattice vibration modes which all have the same sound velocity, and that frequencies are cut off at a maximum such that the total number of modes equals the number of degrees of freedom of the solid.

Debye T^3 law [SOLID STATE] The law that the specific heat of a solid at constant volume varies as the cube of the absolute temperature T at temperatures which are small with respect to the Debye temperature.

Debye temperature [SOLID STATE] The temperature Θ arising in the computation of the Debye specific heat, defined by $k\Theta = h\nu$, where k is the Boltzmann constant, h is Planck's constant, and ν is the Debye frequency. Also known as characteristic temperature.

Debye unit [ELEC] A unit of electric moment that is equal to 10^{-18} statcoulomb-centimeter.

Debye-Waller factor [SOLID STATE] A reduction factor for the intensity of coherent (Bragg) scattering of x-rays, neutrons, or electrons by a crystal, arising from thermal motion of the atoms in the lattice.

deca- [SCI TECH] A prefix denoting 10.

decaborane (14) [INORG CHEM] $B_{10}H_{14}$ A binary compound of boron and hydrogen that is relatively stable at room temperature; melting point 99.5°C, boiling point 213°C.

decade [ELEC] A group or assembly of 10 units; for example, a decade counter counts 10 in one column, and a decade box inserts resistance quantities in multiples of powers of 10. [SCI TECH] The interval between any two quantities having the ratio of 10 to 1.

decade box [ELEC] An assembly of precision resistors, coils, or capacitors whose individual values vary in submultiples and multiples of 10; by appropriately setting a 10-position selector switch for each section, the decade box can be set to any desired value within its range.

decade bridge [ELECTR] Electronic apparatus for measurement of unknown values of resistances or capacitances by comparison with known values (bridge); one secondary section of the oscillator-driven transformer is tapped in decade steps, the other in 10 uniform steps.

decade counter See decade scaler.

decade counter tube See counter tube.

decade glow counting tube See cold-cathode counter tube.

decade scaler [ELECTR] A scaler that produces one output pulse for every 10 input pulses. Also known as counter decade; decade counter; scale-of-ten circuit.

decagon [MATH] A ten-sided polygon.

decahedron [MATH] A polyhedron that has 10 faces.

decahydrate [CHEM] A compound that has 10 water molecules.

decahydronaphthalene [ORG CHEM] $C_{10}H_{18}$ A liquid hydrocarbon, used in some paints and lacquers as a solvent.

decalcification [CHEM] Loss or removal of calcium or calcium compounds from a calcified material such as bone or soil.

decalescence [MET] Darkening of a metal surface due to isothermal absorption of the latent heat of phase transformation.

decaliter [MECH] A unit of volume, equal to 10 liters, or to 0.01 cubic meters.

decameter [MECH] A unit of length in the metric system equal to 10 meters.

decametric wave [ELECTROMAG] British term for a radio wave ranging from 10 to 100 meters long.

decanal See decyl aldehyde.

decane [ORG CHEM] $C_{10}H_{22}$ Any of several saturated aliphatic hydrocarbons, especially $CH_3(CH_2)_8CH_3$.

decanedioic acid See sebacic acid.

decanning [NUCLEO] Removing the outer container of an enriched uranium fuel rod, in preparation for reprocessing of the fuel.

n-decanoic acid See capric acid.

decanol See decyl alcohol.

decantation [ENG] A method for mechanical dewatering of a wet solid by pouring off the liquid without disturbing underlying sediment or precipitate.

decanter [ENG] Tank or vessel in which solids or immiscible dispersions in a carrier liquid settle or coalesce, with clear upper liquid withdrawn (decanted) as overflow from the top.

decanth larva See lycophore larva.

Decapoda [INV ZOO] 1. A diverse order of the class Crustacea including the shrimps, lobsters, hermit crabs, and true crabs; all members have a carapace, well-developed gills, and the first three pairs of thoracic appendages specialized as maxillipeds. 2. An order of dibranchiate cephalopod mollusks containing the squids and cuttle fishes, characterized by eight arms and two long tentacles.

decarbonize [CHEM] To remove carbon by chemical means.

decarboxylase [BIOCHEM] An enzyme that hydrolyzes the carboxyl radical, COOH.

decarboxylate [ORG CHEM] To remove the carboxyl radical, especially from amino acids and protein.

decarburize [MET] To remove carbon from the surface of a ferrous alloy, particularly steel, by heating in a medium that reacts with carbon.

decastere [MECH] A unit of volume, equal to 10 cubic meters.

decating [TEXT] A finishing process in which fabric is sponged or steamed in rollers to set the width and length as well as to achieve smoothness. Also known as decatizing.

decatizing See decating.

decatoic acid See capric acid.

decatyl alcohol See decyl alcohol.

decay [GEOCHEM] See chemical weathering. [MATER] To undergo decomposition. [NUC PHYS] See radioactive decay. [OCEANOGR] In ocean-wave studies, the loss of energy from wind-generated ocean waves after they have ceased to be acted on by the wind; this process is accompanied by an increase in length and a decrease in height of the wave. [PHYS] Gradual reduction in the magnitude of a quantity, as of current, magnetic flux, a stored charge, or phosphorescence.

decay area [OCEANOGR] The area into which ocean waves travel (as swell) after leaving the generating area.

decay chain See radioactive series.

decay coefficient See decay constant.

decay constant [PHYS] The constant c in the equation $I = I_0 e^{-ct}$, for the time dependence of rate of decay of a radioactive species; here, I is the number of disintegrations per unit time. Also known as decay coefficient; disintegration constant; radioactive decay constant; transformation constant.

decay curve [NUC PHYS] A graph showing how the activity of a radioactive sample varies with time; alternatively, it may show the amount of radioactive material remaining at any time.

decay distance [OCEANOGR] The distance through which ocean waves pass after leaving the generating area.

decay family See radioactive series.

decay gammas [NUC PHYS] The characteristic gamma rays emitted during the decay of most radioisotopes.

decay heat [NUCLEO] Heat produced by the decay of radioactive nuclides.

decay mode [NUC PHYS] A possible type of decay of a radionuclide or elementary particle.

decay product See daughter.

decay rate [NUC PHYS] The time rate of disintegration of radioactive material, generally accompanied by emission of particles or gamma radiation.

decay series See radioactive series.

decay time [PHYS] The time taken by a quantity to decay to a stated fraction of its initial value; the fraction is commonly $1/e$. Also known as storage time (deprecated).

Decca [NAV] A hyperbolic navigation system which establishes a line of position from measurement of the phase difference between two continuous-wave signals; the intersection of the two lines of position from two pairs of transmitting stations establishes a navigational fix, or location.

Decca chain [NAV] A system or combination of three slave radio transmitting stations disposed about a master Decca station. Also known as star chain.

Decca chart [NAV] A chart showing Decca lines of position.

Deccan basalt [GEOL] Fine-grained, nonporphyritic, tholeiitic basaltic lava consisting essentially of labradorite, clinopyroxene, and iron ore; found in the Deccan region of southeastern India. Also known as Deccan trap.

Deccan trap See Deccan basalt.

decelerating electrode [ELECTR] Of an electron-beam tube, an electrode to which a potential is applied to decrease the velocity of the electrons in the beam.

deceleration [MECH] The rate of decrease of speed of a motion.

deceleration parachute See drogue.

decelerometer [ENG] An instrument that measures the rate at which the speed of a vehicle decreases.

deceleron [AERO ENG] A lateral control surface of an airplane that is divided so as to combine the functions of an airbrake and an aileron.

December solstice [ASTRON] Winter solstice in the Northern Hemisphere.

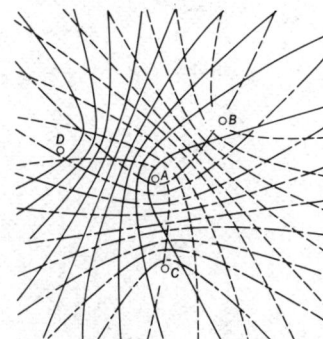

decentralized data processing [ADP] An arrangement comprising a data-processing center for each division or location of a single organization.

deception [ELECTR] The deliberate radiation, reradiation, alteration, absorption, or reflection of electromagnetic energy in a manner intended to mislead an enemy in the interpretation of information received by his electronic systems.

decerebellate [MED] Lacking the cerebellum, generally by experimental removal.

decerebrate [MED] Lacking the cerebrum either by experimental removal or by disconnection.

decerebrate rigidity [MED] Exaggerated postural tone in the antigravity muscles due to release of vestibular nuclei from cerebral control.

dechlorination [CHEM] Removal of chlorine from a substance.

deci- [SCI TECH] A prefix indicating 10^{-1}, 0.1, or a tenth.

deciare [MECH] A unit of area, equal to 0.1 are or 10 square meters.

decibar [MECH] A metric unit of pressure equal to one-tenth bar.

decibel [PHYS] A unit for describing the ratio of two powers or intensities, or the ratio of a power to a reference power; in the measurement of sound intensity, the pressure of the reference sound is usually taken as 2×10^{-4} dynes per square centimeter; equal to one-tenth bel; if P_1 and P_2 are two amounts of power, the first is said to be n decibels greater, where $n = 10 \log_{10}(P_1/P_2)$. Abbreviated dB.

decibel loss [COMMUN] Loss of a signal over a conductor expressed in decibels.

decibel meter [ENG] An instrument calibrated in logarithmic steps and labeled with decibel units and used for measuring power levels in communication circuits.

decibels above 1 kilowatt [ELEC] A measure of power equal to 10 times the common logarithm of the ratio of a given power to 1000 watts. Abbreviated dBk.

decibels above 1 milliwatt [ELEC] A measure of power equal to 10 times the common logarithm of the ratio of a given power to 0.001 watt; a negative value, such as −2.7 dBm, means decibels below 1 milliwatt. Abbreviated dBm.

decibels above 1 picowatt [ELEC] A measure of power equal to 10 times the common logarithm of the ratio of a given power to 1 picowatt. Abbreviated dBp.

decibels above 1 volt [ELEC] A measure of voltage equal to 20 times the common logarithm of the ratio of a given voltage to 1 volt. Abbreviated dBV.

decibels above 1 watt [ELEC] A measure of power equal to 10 times the common logarithm of the ratio of a given power to 1 watt. Abbreviated dBW.

decibels above reference coupling [ELEC] A measure of the coupling between two circuits, expressed in relation to a reference value of coupling that gives a specified reading on a specified noise-measuring set when a test tone of 90 dBa is impressed on one circuit. Abbreviated dBx.

decibels above reference noise [ELEC] Units used to show the relationship between the interfering effect of a noise frequency, or band of noise frequencies, and a fixed amount of noise power commonly called reference noise; a 1000-hertz tone having a power level of −90 dBm was selected as the reference noise power; superseded by the adjusted decibel unit. Abbreviated dBrn.

decicaine *See* tetracaine hydrochloride.

decidua [MED] The endometrium of pregnancy and associated fetal membranes which are cast off at parturition.

deciduous [BIOL] Falling off or being shed at the end of the growing period or season.

deciduous teeth [ANAT] Teeth of a young mammal which are shed and replaced by permanent teeth. Also known as milk teeth.

decigram [MECH] A unit of mass, equal to 0.1 gram.

decile [STAT] Any of the points which divide the total number of items in a frequency distribution into 10 equal parts.

deciliter [MECH] A unit of volume, equal to 0.1 liter, or 10^{-4} cubic meter.

decimal [MATH] A number expressed in the scale of tens.

decimal attenuator [ELECTR] System of attenuators arranged so that a voltage or current can be reduced decimally.

decimal-binary switch [ELEC] A switch that connects a single input lead to appropriate combinations of four output leads (representing 1, 2, 4, and 8) for each of the decimal-numbered settings of its control knob; thus, for position 7, output leads 1, 2, and 4 would be connected to the input.

decimal code [ADP] A code in which each allowable position has one of 10 possible states; the conventional decimal number system is a decimal code.

decimal-coded digit [ADP] One of 10 arbitrarily selected patterns of 1 and 0 used to represent the decimal digits. Also known as coded decimal.

decimal fraction [MATH] Any number written in the form: an integer followed by a decimal point followed by a (possibly infinite) string of digits.

decimal number [MATH] A number signifying a decimal fraction by a decimal point to the left of the numerator with the number of figures to the right of the point equal to the power of 10 of the denominator.

decimal number system [MATH] A representational system for the real numbers in which place values are read in powers of 10.

decimal place [MATH] Reference to one of the digits following the decimal point in a decimal fraction; the kth decimal place registers units of 10^{-k}.

decimal point [MATH] A dot written either on or slightly above the line; used to mark the point at which place values change from positive to negative powers of 10 in the decimal number system.

decimal processor [ADP] A digital computer organized to calculate by decimal arithmetic.

decimal system [MATH] A number system based on the number 10; in theory, each unit is 10 times the next smaller one.

decimal-to-binary conversion [ADP] The mathematical process of converting a number written in the scale of 10 into the same number written in the scale of 2.

decimeter [MECH] A metric unit of length equal to one-tenth meter.

decimetric wave [ELECTROMAG] An electromagnetic wave having a wavelength between 0.1 and 1 meter, corresponding to a frequency between 300 and 3000 megahertz.

decineper [PHYS] One-tenth of a neper.

decinormal [CHEM] Pertaining to a chemical solution that is one-tenth normality in reference to a 1 normal solution.

decision [ADP] The computer operation of determining if a certain relationship exists between words in storage or registers, and taking alternative courses of action; this is effected by conditional jumps or equivalent techniques.

decision box [ADP] A flow-chart symbol indicating a decision instruction; usually diamond-shaped.

decision calculus [SYS ENG] A guide to the process of decision-making, often outlined in the following steps: analysis of the decision area to discover applicable elements; location or creation of criteria for evaluation; appraisal of the known information pertinent to the applicable elements and correction for bias; isolation of the unknown factors; weighting of the pertinent elements, known and unknown, as to relative importance; and projection of the relative impacts on the objective, and synthesis into a course of action.

decision element [ELECTR] A circuit that performs a logical operation such as "and," "or," "not," or "except" on one or more binary digits of input information representing "yes" or "no" and that expresses the result in its output. Also known as decision gate.

decision gate *See* decision element.

decision height [NAV] A height specified in MSL (mean sea level) above the highest runway elevation in the touchdown zone at which a missed approach shall be initiated if the required visual reference has not been established; this term is used only in procedure where an electronic glide slope provides a reference for descent or in ILS (instrument landing systems) or PAR (precision approach radar). Abbreviated d.h.

decision instruction *See* conditional jump.

decision mechanism [ADP] In character recognition, that component part of a character reader which accepts the finalized version of the input character and makes an assessment as to its most probable identity.

decision rule [SYS ENG] In decision theory, the mathematical representation of a physical system which operates upon the observed data to produce a decision.

decision table [ADP] A table of contingencies to be considered in the definition of a problem, together with the actions to be taken; sometimes used in place of a flow chart for program documentation.

decision theory [SYS ENG] A broad spectrum of concepts and techniques which have been developed to both describe and rationalize the process of decision making, that is, making a choice among several possible alternatives.

decision tree [IND ENG] Graphic display of the underlying decision process involved in the introduction of a new product by a manufacturer.

deck [ADP] A set of punched cards. [CIV ENG] 1. A floor, usually of wood, without a roof. 2. The floor or roadway of a bridge. [ENG] A magnetic-tape transport mechanism. [NAV ARCH] Horizontal or cambered and sloping surfaces on a ship, corresponding to the floors of a building.

deck beam [NAV ARCH] A strengthening member of the deck that may run the length or width of the deck and is connected at both ends to the ship's frames or bulkheads by girders.

deck bridge [CIV ENG] A type of bridge in which the supporting members are all beneath the roadway.

deck charge [MIN ENG] A charge that is separated into several smaller components and placed along a quarry borehole.

deck erection [NAV ARCH] A forecastle, bridge, poop, or deckhouse erected on the upper deck of a ship.

deck fitting [NAV ARCH] 1. A fitting where a pipeline penetrates the deck to maintain watertightness. 2. Any fitting attached to the deck.

deck height [NAV ARCH] Vertical distance between the molded lines of two adjacent decks.

deckhouse [NAV ARCH] A low building or superstructure, such as a cabin, constructed on the top deck of a ship which may or may not extend to the edges of the deck.

decking [CIV ENG] Surface material on a deck. [MIN ENG] Changing tubs on a cage at both ends of a shaft.

deckle edge [MATER] The unfinished edge of paper having a characteristic appearance as a result of leakage under the frame (deckle) in which the paper is made; handmade paper has a deckle edge on all four sides, machine-made paper only on two sides.

deck light [NAV ARCH] Any opening in the deck covered with glass; allows daylight to come in below deck.

deck line [NAV ARCH] A line that passes through the intersection of the molded line of the deck beams and the molded line of the frames.

deck loading [MIN ENG] The method of loading deck charges in a quarry borehole.

deck log [NAV] A record of significant navigational data and events of a ship's voyage written (usually by the quartermaster) in chronological order.

deck machinery [NAV ARCH] Any machinery on the decks of a ship; includes capstans, windlasses, and winches.

deck roof [BUILD] A roof that is nearly flat and without parapet walls.

deck stringer plate [NAV ARCH] The outboard fore-and-aft plate on any deck.

deck switch *See* gang switch.

deck truss [CIV ENG] The frame of a deck.

deckzelle [INV ZOO] In certain hydroids, one of the supporting or epithelial cells which are usually columnar or cuboidal.

declarative statement [ADP] Any program statement describing the data which will be used or identifying the memory locations which will be required.

declinate [BIOL] Curved toward one side or downward.

declination [GEOPHYS] The angle between a magnetic needle and the geographical meridian. Also known as magnetic deviation; variation. [NAV] 1. In a system of polar or spherical coordinates, the angle at the origin between a line to a point and the equatorial plane, measured in a plane perpendicular to the equatorial plane. 2. The arc between the equator and the point measured on a great circle perpendicular to the equator.

declination axis [ENG] For an equatorial mounting of a telescope, an axis of rotation that is perpendicular to the polar axis and allows the telescope to be pointed at objects of different declinations.

declination circle [ENG] For a telescope with an equatorial mounting, a setting circle attached to the declination axis that shows the declination to which the telescope is pointing.

declination compass *See* declinometer.

declination difference [NAV] The difference between two declinations, particularly between the declination of a celestial body and the value used as an argument for entering a table.

declination of grid north *See* grid declination.

declination variometer [ENG] An instrument that measures changes in the declination of the earth's magnetic field, consisting of a permanent bar magnet, usually about 1 centimeter long, suspended with a plane mirror from a fine quartz fiber 5–15 centimeters in length; a lens focuses to a point a beam of light reflected from the mirror to recording paper mounted on a rotating drum. Also known as D variometer.

declinometer [ENG] A magnetic instrument similar to a surveyor's compass, but arranged so that the line of sight can be rotated to conform with the needle or to any desired setting on the horizontal circle; used in determining magnetic declination. Also known as declination compass.

declivity [GEOL] 1. A slope descending downward from a point of reference. 2. A downward deviation from the horizontal.

Declomycin [MICROBIO] Trade name for demethylchlortetracycline.

decode [COMMUN] 1. To translate coded characters into a more understandable form. 2. *See* demodulate.

decoder [COMMUN] A device that decodes. [ELECTR] 1. A matrix of logic elements that selects one or more output channels, depending on the combination of input signals present. 2. *See* decoder circuit; matrix; tree.

decoder circuit [ELECTR] A circuit that responds to a particular coded signal while rejecting others. Also known as decoder.

decoding gate [ADP] The use of combinatorial logic in circuitry to select a device identified by a binary address code. Also known as recognition gate.

decoking [CHEM ENG] Removal of petroleum coke from equipment.

décollement [GEOL] Folding or faulting of sedimentary beds by sliding over the underlying rock.

decolorize [CHEM ENG] To remove the color from, as from a liquid.

decolorizer [CHEM ENG] An agent used to decolorize; the removal of color may occur by a chemical reaction or a physical reaction.

decolorizing carbon [CHEM] Porous or finely divided carbon (activated or bone) with large surface area; used to adsorb colored impurities from liquids, such as lube oils.

decometer [ELECTR] An adding-type phasemeter which rotates continuously and adds up the total number of degrees of phase shift between two signals, such as those received from two transmitters in the Decca navigation system.

decommutation [ELECTR] The process of recovering a signal from the composite signal previously created by a commutation process.

decommutator [ELECTR] The section of a telemetering system that extracts analog data from a time-serial train of samples representing a multiplicity of data sources transmitted over a single radio-frequency link.

decomposable process [MATH] A process which can be reduced to several basic events.

decomposer [ECOL] A heterotrophic organism (including bacteria and fungi) which breaks down the complex compounds of dead protoplasm, absorbs some decomposition products, and releases substances usable by consumers. Also known as microcomposer; microconsumer; reducer.

decomposition [CHEM] The more or less permanent breakdown of a molecule into simpler molecules or atoms. [GEOCHEM] *See* chemical weathering.

decomposition potential [PHYS CHEM] The electrode potential at which the electrolysis current begins to increase appreciably. Also known as decomposition voltage.

DECLINATION VARIOMETER

Declination variometer equipped with Helmholtz coil for calibration. *(U.S. Coast and Geodetic Survey)*

decomposition voltage See decomposition potential.

decompound [BOT] Divided or compounded several times, with each division being compound.

decompression [ENG] Any procedure for the relief of pressure or compression.

decompression chamber [ENG] **1.** A steel chamber fitted with auxiliary equipment to raise its air pressure to a value two to six times atmospheric pressure; used to relieve a diver who has decompressed too quickly in ascending. **2.** Such a chamber in which conditions of high atmospheric pressure can be simulated for experimental purposes.

decompression illness See aeroembolism.

deconcentrator [ENG] An apparatus for removing dissolved or suspended material from feedwater.

decontamination [ENG] The removing of chemical, biological, or radiological contamination from, or the neutralizing of it on, a person, object, or area.

decontamination factor [NUCLEO] The ratio of initial specific radioactivity to final specific radioactivity resulting from a separation process.

decontamination index [NUCLEO] The logarithm of the ratio of initial specific radioactivity to final specific radioactivity resulting from a separation process.

decoppering agent [ORD] Material, such as finely divided tin, included in a propelling charge, or inserted in the chamber with the propelling charge, for the purpose of removing the coppering from the surface of the bore.

decorticate [BIOL] Lacking a cortical layer.

decouple [ENG] **1.** To minimize or eliminate airborne shock waves of a nuclear or other explosion by placing the explosives deep under the ground. **2.** To minimize the seismic effect of an underground explosion by setting it off in the center of an underground cavity.

decoupling [ELEC] Preventing transfer or feedback of energy from one circuit to another.

decoupling filter [ELECTR] One of a number of low-pass filters placed between each of several amplifier stages and a common power supply.

decoupling network [ELEC] Any combination of resistors, coils, and capacitors placed in power supply leads or other leads that are common to two or more circuits, to prevent unwanted interstage coupling.

decoy [ORD] **1.** An object with reflective characteristics of a target, used in radar deception. **2.** Underwater device which reflects acoustic radiation, used by submarines in the deception of acoustic listening devices and acoustic homing torpedoes.

decoy target [ORD] A device assembled from prefabricated materials, designed to simulate miscellaneous types of field equipment.

decoy transponder [ELECTR] A transponder that returns a strong signal when triggered directly by a radar pulse, to produce large and misleading target signals on enemy radar screens.

decreasing function [MATH] **1.** A function of x whose value gets smaller as x gets larger, that is, if $x < y$, then $f(x) > f(y)$. **2.** See monotone nonincreasing function.

decrement [ADP] A specific part of an instruction word in some binary computers, thus a set of digits. [HYD] See groundwater discharge. [MATH] The quantity by which a variable is decreased. [PHYS] The ratio of the amplitudes of an underdamped harmonic motion during two successive oscillations. Also known as damping factor; numerical decrement.

decrement field [ADP] That part of an instruction word which is used to modify the contents of a storage location or register.

decremeter [ENG] An instrument for measuring the logarithmic decrement (damping) of a train of waves.

decrepitation [GEOPHYS] Breaking up of mineral substances when exposed to heat; usually accompanied by a crackling noise.

decrypt [ELECTR] To convert a cryptogram or series of electronic pulses into plain text by electronic means.

Dectra [NAV] A radio navigation aid that provides coverage over a specific section of a long ocean route, using equipment and techniques similar to Decca; a master station and a slave station are located at each end of the route; by comparing the phase difference of a master and its slave, guidance is provided to the right or left of the course; comparison of transmission with those of a high-precision quartz clock provides distance information along the course.

decubitus ulcer [MED] An ulcer of the skin and subcutaneous tissues following prolonged lying down, due to pressure on bony protuberances. Also known as bedsore; pressure ulcer.

decumbent [BOT] Lying down on the ground but with an ascending tip, specifically referring to a stem.

decurrent [BOT] Running downward, especially of a leaf base extended past its insertion in the form of a winged expansion.

decussate [BOT] Of the arrangement of leaves, occurring in alternating pairs at right angles. [SCI TECH] To intersect in the form of an X.

decussate structure [GEOL] A crisscross microstructure of certain minerals; most noticeable in rocks composed predominantly of minerals with a columnar habit.

decyclic acid See capric acid.

decyl [ORG CHEM] An isomeric grouping of univalent radicals, all with formulas $C_{10}H_{21}$, and derived from the decanes by removing one hydrogen.

decyl acetate [ORG CHEM] $CH_3(CH_2)_9OOCCH_3$ Perfumery liquid with a floral orange-rose aroma.

decyl alcohol [ORG CHEM] $C_{10}H_{21}OH$ A colorless oil, boiling at 231°C; used in plasticizers, synthetic lubricants, and detergents. Also known as decanol; decatyl alcohol; nonyl carbinol.

decyl aldehyde [ORG CHEM] $CH_3(CH_2)_8CHO$ A liquid aldehyde, found in essential oils; used in flavorings and perfumes. Also known as capraldehyde; capric aldehyde; decanal.

decylene [ORG CHEM] Any of a group of isomeric hydrocarbons with formula $C_{10}H_{20}$; the group is part of the ethylene series.

decylethylene See 1-dodecene.

decyltrichlorosilane [ORG CHEM] $n\text{-}C_{10}H_{21}SiCl_3$ An organochlorosilane that boils at 183°C at 84 mm Hg; used in coupling agents or primers to obtain improved bonding between organic polymers and mineral surfaces.

Dedekind cut [MATH] A set of rational numbers satisfying certain properties, with which a unique real number may be associated; used to define the real numbers as an extension of the rationals.

Dedekind test [MATH] If the series $\sum_i (b_i - b_{i+1})$ converges absolutely, the b_i converge to zero, and the series $\sum_i a_i$ has bounded partial sums, then the series $\sum_i a_i b_i$ converges.

dedendum [DES ENG] The difference between the radius of the pitch circle of a gear and the radius of its root circle.

dedendum circle [DES ENG] A circle tangent to the bottom of the spaces between teeth on a gear wheel.

dedifferentiation [BIOL] Disintegration of a specialized habit or adaptation. [PHYSIO] Return of a specialized cell or structure to a more general or primitive condition.

dedolomitization [GEOL] Destruction of dolomite to form calcite and periclase, usually by contact metamorphism at low pressures.

deduction [MATH] The process of deriving a statement from certain assumed statements by applying the rules of logic.

dedusting [MIN ENG] Cleaning ore, using pneumatic means and screening, to remove dust and other fine impurities. Also known as aspirating.

dee [NUCLEO] A hollow accelerating-cyclotron electrode in the shape of the letter D. Also known as cyclotron D; D.

dee line [NUCLEO] A structural member that supports the dee of a cyclotron and acts with the dee to form the resonant circuit.

deemphasis [ENG ACOUS] A process for reducing the relative strength of higher audio frequencies before reproduction, to complement and thereby offset the preemphasis that was introduced to help override noise or reduce distortion. Also known as postemphasis; postequalization.

deemphasis network [ENG ACOUS] An *RC* filter inserted in a system to restore preemphasized signals to their original form.

deenergize [ELEC] To disconnect from the source of power.

deep [OCEANOGR] An area of great depth in the ocean, representing a depression in the ocean floor.

deep-casting [OCEANOGR] Sampling ocean water at great depths by lowering a number of self-sealing bottles, usually made of brass or bronze, on a cable.

deep-draw [MET] To form shapes with large depth-diameter ratios in sheet or strip metal by considerable plastic distortion in dies.

deep easterlies *See* equatorial easterlies.

deepening [METEOROL] A decrease in the central pressure of a pressure system on a constant-height chart, or an analogous decrease in height on a constant-pressure chart.

deep-etch [GRAPHICS] The etching of an offset printing plate so that the printing area becomes slightly recessed, thereby leading to sharper definition and longer life on press. [MET] Severe etching of a metal surface to reveal gross features, such as abnormal grain size, segregation, or cracks, at magnifications of 10 diameters or less. Also known as macroetching.

deep-etch printing [GRAPHICS] Printing with plates produced by deep-etching.

deep fascia [ANAT] The fibrous tissue between muscles and forming the sheaths of muscles, or investing other deep, definitive structures, as nerves and blood vessels.

deep fording [ORD] The ability of a gun or vehicle, equipped with built-in waterproofing with its suspension in contact with the ground, to negotiate a water obstacle by application of a special waterproofing kit.

deep hibernation [PHYSIO] Profound decrease in metabolic rate and physiological function during winter, with a body temperature near 0°C, in certain warm-blooded vertebrates. Also known as hibernation.

deep inelastic transfer *See* quasifission.

deep inland sea [GEOGR] A sea adjacent to but in restricted communication with the sea; depth exceeds 200 meters.

deep pain [PHYSIO] Pattern of somesthetic sensation of pain, usually indefinitely localized, originating in the viscera, muscles, and other deep tissues.

deep palmar arch [ANAT] The anastomosis between the terminal part of the radial artery and the deep palmar branch of the ulnar artery. Also known as deep volar arch; palmar arch.

deep-penetration bomb [ORD] A bomb designed to enter deeply into a target before it explodes.

deep scattering layer [OCEANOGR] The stratified populations of organisms which scatter sound in most oceanic waters.

deep-sea basin [GEOL] A depression of the sea floor more or less equidimensional in form and of variable extent.

deep-sea channel [GEOL] A trough-shaped valley of low relief beyond the continental rise on the deep-sea floor. Also known as mid-ocean canyon.

deep-sea plain [GEOL] A broad, almost level area forming the predominant portion of the ocean floor.

deep-seated *See* plutonic.

deep-sea trench [GEOL] A long, narrow depression of the deep-sea floor having steep sides and containing the greatest ocean depths; formed by depression, to several kilometers' depth, of the high-velocity crustal layer and the mantle.

deep sleep [PSYCH] The third and fourth stage of the sleep cycle, determined by electroencephalographic recording and characterized by slow brain waves. Also known as slow wave sleep (SWS).

deep space [ASTRON] Space beyond the gravitational influence of the earth.

Deep Space Network [AERO ENG] A spacecraft network operated by NASA which tracks, commands, and receives telemetry for all types of spacecraft sent to explore deep space, the moon, and solar system planets. Abbreviated DSN.

deep-space probe [AERO ENG] A spacecraft designed for exploring space beyond the gravitational and magnetic fields of the earth.

deep-submergence rescue vehicle [NAV ARCH] A small rescue and research submarine; intended primarily to rescue the crew of another submarine to depths of 5000 feet (1500 meters). Abbreviated DSRV.

deep tank [NAV ARCH] A tank that extends from the bottom of a ship up to or higher than the lower deck.

deep trades *See* equatorial easterlies.

deep volar arch *See* deep palmar arch.

deep water [OCEANOGR] An ocean area where depth of the water layer is greater than one-half the wave length.

deep waterline [NAV ARCH] The height of the line on a ship's hull that is reached by the water when the ship is loaded to its maximum safe capacity.

deep-water wave [OCEANOGR] A surface wave whose length is less than twice the depth of the water.

deep well [CIV ENG] A well that draws its water from beneath shallow impermeable strata, at depths exceeding 22 feet (6.7 meters).

deep-well pump [MECH ENG] A multistage centrifugal pump for lifting water from deep, small-diameter wells; a surface electric motor operates the shaft. Also known as vertical turbine pump.

deer [VERT ZOO] The common name for 41 species of even-toed ungulates that compose the family Cervidae in the order Artiodactyla; males have antlers.

deerhorn antenna [ELECTROMAG] A dipole antenna whose ends are swept back to reduce wind resistance when mounted on an airplane.

deethanize [CHEM ENG] To separate and remove ethane and sometimes lighter fractions from heavy substances, such as propane, by distillation.

deethanizer [CHEM ENG] The equipment used to deethanize.

defecation [CHEM ENG] Industrial purification, or clarification, of sugar solutions. [PHYSIO] The process by which fecal wastes that reach the lower colon and rectum are evacuated from the body.

defect [SCI TECH] An irregularity that spoils the appearance or impairs the usefulness or effectiveness of an object or a material by causing weakness or failure.

defect cluster [CRYSTAL] A macroscopic cluster of crystal defects which can arise from attraction among defects.

defect conduction [SOLID STATE] Electric conduction in a semiconductor by holes in the valence band.

defective [SCI TECH] An item or product which has at least one defect.

defective number *See* deficient number.

defective track [ADP] Any circular path on the surface of a magnetic disk which is detected by the system as unable to accept one or more bits of data.

defective virus [VIROL] A virus, such as adeno-associated satellite virus, that can grow and reproduce only in the presence of another virus.

defect motion [CRYSTAL] Movement of a point defect from one lattice point to another.

defeminization [PHYSIO] Loss or reduction of feminine attributes, usually due to ovarian dysfunction or removal. [PSYCH] Psychic process involving a deep and permanent change in the character of a woman, resulting in a giving up of feminine feelings, and the assumption of masculine qualities.

defender [IND ENG] A machine or facility which is being considered for replacement.

defense information [ORD] Official information which requires protection in the interests of national defense, which is not common knowledge, and which would be of intelligence value to an enemy or potential enemy in the planning or waging of war against a nation.

defense mechanism [PSYCH] Any psychic device, such as rationalization, denial, or repression, for concealing unacceptable feelings or for protecting oneself against unpleasant feelings, memories, or experiences.

defensive minefield [ORD] A minefield so situated that an adversary's attack will be delayed or repulsed.

deferent [ASTRON] An imaginary circle around the earth, postulated by Ptolemy, in whose circumference a celestial body or its epicycle is supposed to move.

deferment factor *See* discount factor.

deferred entry [ADP] The passing of control of the central

DEER

North American elk *(Cervus canadensis)*, a deer common in Canada, stands about 5½ feet (1.7 meters) at shoulder.

processing unit to a subroutine or to an entry point as the result of an asynchronous event.

deferred processing [ADP] The making of computer runs which are postponed until nonpeak periods.

defervescence *See* lysis.

defibrillation [MED] Stopping a local quivering of muscle fibers, especially of the heart.

defibrillator [MED] An electronic instrument used for stopping fibrillation during a heart attack by applying controlled electric pulses to the heart muscles.

deficiency disease [MED] Any disease resulting from a dietary deficiency of minerals, vitamins, or essential nutrients.

deficiency index [MATH] For a curve or equation involving two complex variables this is the genus of the Riemann surface associated to the equation.

deficient number [MATH] A positive integer the sum of whose divisors, including 1 but excluding itself, is less than itself. Also known as defective number.

defilade [ORD] An arrangement of fortifications that minimizes the effect of frontal or enfilading fire and plunging or reverse fire.

definite composition law [CHEM] The law that a given chemical compound always contains the same elements in the same fixed proportions by weight. Also known as definite proportions law.

definite proportions law *See* definite composition law.

definite Riemann integral [MATH] A number associated with a function defined on an interval [a,b] which is

$$\lim_{N\to\infty} \sum_{k=0}^{N-1} f\left(a +\frac{k}{N}\right) \cdot \frac{b-a}{N},$$

if f is bounded and continuous; denoted by

$$\int_a^b f(x)\,dx;$$

if f is a positive function, the definite integral measures the area between the graph of f and the x axis.

definition [COMMUN] The fidelity with which a television or facsimile receiver forms an image. [ELECTR] The extent to which the fine-line details of a printed circuit correspond to the master drawing. [OPTICS] Lens image clarity or discernible detail.

definitive host [BIOL] The host in which a parasite reproduces sexually.

deflagrating spoon [CHEM] A long-handled spoon used in chemistry to demonstrate deflagration.

deflagration [CHEM] A chemical reaction accompanied by vigorous evolution of heat, flame, sparks, or spattering of burning particles.

deflashing [ENG] Finishing technique to remove excess material (flash) from a plastic or metal molding.

deflation [GEOL] The sweeping erosive action of the wind over the ground.

deflation basin [GEOL] A topographic depression formed by deflation.

deflecting torque [MECH] An instrument's moment, resulting from the quantity measured, that acts to cause the pointer's deflection.

deflection [ELECTR] The displacement of an electron beam from its straight-line path by an electrostatic or electromagnetic field. [ENG] 1. Shape change or reduction in diameter of a conduit, produced without fracturing the material. 2. Elastic movement or sinking of a loaded structural member, particularly of the mid-span of a beam. [ORD] 1. Horizontal clockwise angle between the axis of the bore and the line of sighting. 2. The setting on the scale to compensate for deflection.

deflection angle [GEOD] The angle at a point on the earth between the direction of a plumb line (the vertical) and the perpendicular (the normal) to the reference spheroid; this difference seldom exceeds 30 seconds of arc. [ORD] 1. An angle that measures the departure of a moving object from its directed course. 2. The angle of a deflection shot in gunnery, measured between the line of sight to the target and the line of sight to the aiming point.

deflection bit [DES ENG] A long, cone-shaped, noncoring bit used to drill past a deflection wedge in a borehole.

deflection board [ORD] Instrument used in artillery for figuring azimuth or deflection corrected for wind, drift, and other factors. Also known as gun deflection board.

deflection change [ORD] Change in azimuth setting applying to all guns in a battery when the target moves, or when a shift is made from one target to another; it does not include the deflection difference, which allows for the difference in positions of the various guns firing at the same target.

deflection circuit [ELECTR] A circuit which controls the deflection of an electron beam in a cathode-ray tube.

deflection coil [ELECTR] One of the coils in a deflection yoke.

deflection correction [ORD] A correction that must be applied to the azimuth or shift measured on a firing chart so that the line of fire will pass through the target.

deflection curve [MECH] The curve, generally downward, described by a shot deviating from its true course.

deflection defocusing [ELECTR] Defocusing that becomes greater as deflection is increased in a cathode-ray tube, because the beam hits the screen at a greater slant and the beam spot becomes more elliptical as it approaches the edges of the screen.

deflection difference [ORD] The amount that an artillery piece in a battery is traversed toward or away from a given piece.

deflection electrode [ELECTR] An electrode whose potential provides an electric field that deflects an electron beam. Also known as deflection plate.

deflection error [ORD] 1. In artillery, the distance to the right or left of a target between the point aimed at and the burst of the projectile. 2. In bombing, the distance between the point of impact and the mean point of impact and the center of the target measured at right angles from the line of the aircraft's approach.

deflection factor [ELECTR] The reciprocal of the deflection sensitivity in a cathode-ray tube.

deflection-modulated indicator *See* amplitude-modulated indicator.

deflection plate *See* deflection electrode.

deflection polarity of an oscilloscope [ELECTR] Relationship between the direction of a displacement of the cathode beam and the polarity of the applied signal wave.

deflection probable error [ORD] The directional error, caused by dispersion, which will be exceeded, as often as not, in an infinite number of rounds fired at a single deflection; it is one-eighth the width of the dispersion pattern at its greatest width; this value is given in the firing tables.

deflection scale [ORD] Scale on a sight, marked in mils or degrees, for applying corrections in deflection or for moving the piece in direction.

deflection sensitivity [ELECTR] The displacement of the electron beam at the target or screen of a cathode-ray tube per unit of change in the deflection field; usually expressed in inches per volt applied between deflection electrodes or inches per ampere in a deflection coil.

deflection voltage [ELECTR] The voltage applied between a pair of deflection electrodes to produce an electric field.

deflection wedge [DES ENG] A wedge-shaped tool inserted into a borehole to direct the drill bit.

deflection yoke [ELECTR] An assembly of one or more electromagnets that is placed around the neck of an electron-beam tube to produce a magnetic field for deflection of one or more electron beams. Also known as scanning yoke; yoke.

deflectometer [ENG] An instrument for measuring minute deformations in a structure under transverse stress.

deflector [ENG] A plate, baffle, or the like that diverts the flow of a forward-moving stream.

deflexed [BIOL] Turned sharply downward.

deflocculant [CHEM] An agent that causes deflocculation; examples are sodium carbonate and other basic materials used to deflocculate clay slips.

deflocculate [CHEM] To break up and disperse agglomerates and form a stable colloid.

defluorination [CHEM] Removal of fluorine.

defluvium [PATH] The pathological loss of a part of an animal or plant, as nails or bark.

defoaming [CHEM ENG] Reduction or elimination of foam.

defocus-dash mode [ELECTR] A mode of cathode-ray tube storage of binary digits in which the writing beam is initially defocused so as to excite a small circular area on the screen; for one kind of binary digit it remains defocused, and for the other kind it is suddenly focused to a concentric dot and drawn out into a dash.

defocus-focus mode [ELECTR] A variation of the defocus-dash mode in which the focused dot is drawn out into a dash.

defoliant [MATER] A chemical sprayed on plants that causes leaves to fall off prematurely.

defoliate [BOT] To remove leaves or cause leaves to fall, especially prematurely.

deforestation [FOR] The act or process of removing trees from or clearing a forest.

deformation [MATH] A homotopy of the identity map to some other map. [MECH] Any alteration of shape or dimensions of a body caused by stresses, thermal expansion or contraction, chemical or metallurgical transformations, or shrinkage and expansions due to moisture change.

deformation bands *See* Lüder's lines.

deformation curve [MECH] A curve showing the relationship between the stress or load on a structure, structural member, or a specimen and the strain or deformation that results. Also known as stress-strain curve.

deformation ellipsoid *See* strain ellipsoid.

deformation energy [NUC PHYS] The energy which must be supplied to an initially spherical nucleus to give it a certain deformation in the Bohr-Wheeler theory.

deformation fabric [GEOL] The space orientation of rock elements produced by external stress on the rock.

deformation lamella [GEOL] A type of slipband in the crystalline grains of a material (particularly quartz) produced by intracrystalline slip during tectonic deformation.

deformation potential [SOLID STATE] The effective electric potential experienced by free electrons in a semiconductor or metal resulting from a local deformation in the crystal lattice.

deformation thermometer [ENG] A thermometer with transducing elements which deform with temperature; examples are the bimetallic thermometer and the Bourdon-tube type of thermometer.

deformed bar [CIV ENG] A steel bar with projections or indentations to increase mechanical bonding; used to reinforce concrete.

deformeter [ENG] An instrument used to measure minute deformations in materials in structural models.

defrost [ENG] To keep free of or to remove ice. [THERMO] To thaw out from a frozen state.

defruit [ELECTR] To remove random asynchronous replies from the video input of a display unit in a radar beacon system by comparing the video signals on successive sweeps.

defuse [ORD] To remove the fuse from a munition.

degas [ELECTR] To drive out and exhaust the gases occluded in the internal parts of an electron tube or other gastight apparatus, generally by heating during evacuation. [ENG] To remove gas from a liquid or solid.

degasifier [MET] An alloy added to molten metal to facilitate the removal of dissolved gases.

degasser *See* getter.

degassing *See* breathing.

degaussing [ELECTROMAG] A method of neutralizing (demagnetizing) a magnetic field of, for example, a ship hull or television tube; a direct current of the correct value is sent through a cable around the ship hull; a current-carrying coil is brought up to and then removed from the television tube. Also known as deperming.

degaussing cable [ELECTROMAG] A single-conductor or multiple-conductor cable used on ships for degaussing.

degaussing coil [ELECTROMAG] A plastic-encased coil, about 1 foot in diameter, that can be plugged into a 120-volt alternating-current wall outlet and moved slowly toward and away from a color television picture tube to demagnetize adjacent parts.

degaussing control [ELECTROMAG] A control that automatically varies the current in the degaussing cable as a ship changes heading or rolls and pitches.

degeneracy [MATH] The condition in which two characteristic functions of an operator have the same characteristic value. [PHYS] The condition in which two or more modes of a vibrating system have the same frequency; a special case of

the mathematics definition. [QUANT MECH] The condition in which two or more stationary states of the same system have the same energy even though their wave functions are not the same; a special case of the mathematics definition.

degenerate amplifier [ELECTR] Parametric amplifier with a pump frequency exactly twice the signal frequency, producing an idler frequency equal to that of the signal input; it is considered as a single-frequency device.

degenerate codons [MOL BIO] Two or more codons that act as initiators for the same amino acid.

degenerate conduction band [SOLID STATE] A band in which two or more orthogonal quantum states exist that have the same energy, the same spin, and zero mean velocity.

degenerate electron gas [STAT MECH] An electron gas that is far below its Fermi temperature and is therefore described in first approximation by the Fermi distribution; most of the electrons completely fill the lower energy levels and are unable to take part in physical processes until excited out of these levels.

degenerate matter [PHYS] Matter that has been stripped of its orbital electrons, so the nuclei are packed close together.

degenerate semiconductor [SOLID STATE] A semiconductor in which the number of electrons in the conduction band approaches that of a metal.

degeneration [ELECTR] The loss or gain in an amplifier through unintentional negative feedback. [MED] **1.** Deterioration of cellular integrity with no sign of response to injury or disease. **2.** General deterioration of a physical, mental, or moral state. [STAT MECH] A phenomenon which occurs in gases at very low temperatures when the molecular heat drops to less than $\frac{3}{2}$ the gas constant.

degenerative arthritis *See* degenerative joint disease.

degenerative disease [MED] General debility and diseases associated with advancing age.

degenerative joint disease [MED] A chronic joint disease characterized pathologically by degeneration of articular cartilage and hypertrophy of bone, clinically by pain on activity which subsides with rest. Also known as degenerative arthritis; hypertrophic arthritis; osteoarthritis; senescent arthritis.

Degeneriaceae [BOT] A family of dicotyledonous plants in the order Magnoliales characterized by laminar stamens; a solitary, pluriovulate, unsealed carpel; and ruminate endosperm.

deglaciation [HYD] Exposure of an area from beneath a glacier or ice sheet as a result of shrinkage of the ice by melting.

deglutition [PHYSIO] Act of swallowing.

degradation [ADP] Condition under which a computer operates when some area of memory or some units of peripheral equipment are not available to the user. [GEOL] The wearing down of the land surface by processes of erosion and weathering. [HYD] **1.** Lowering of a steam bed. **2.** Shrinkage or disappearance of permafrost. [ORG CHEM] Conversion of an organic compound to one containing a smaller number of carbon atoms. [PHYS] Loss of energy of a particle, such as a neutron or photon, through a collision. [THERMO] The conversion of energy into forms that are increasingly difficult to convert into work, resulting from the general tendency of entropy to increase.

degraded illite [MINERAL] Illite with a depleted potassium content because of prolonged leaching. Also known as stripped illite.

degras [MATER] **1.** A semioxidized fat obtained from sheep skins by subjecting them to the action of oxidized fish oil and pressing them; used to dress leather. Also known as moellen. **2.** A mixture of this material with other fatty oils or fats or with wool grease. **3.** *See* wool grease.

degrease [CHEM ENG] **1.** To remove grease from wool with chemicals. **2.** To remove grease from hides or skins in tanning by tumbling them in solvents. [MET] To remove grease, oil, or fatty material from a metal surface with fumes from a hot solvent.

degreaser [ENG] A machine designed to clean grease and foreign matter from mechanical parts and like items, usually metallic, by exposing them to vaporized or liquid solvent solutions confined in a tank or vessel. [MATER] A solvent, such as a polyhalogenated hydrocarbon, that removes fat or oil in many industrial processes.

degree [CHEM] Any one of several units for measuring hardness of water, such as the English or Clark degree, the French degree, and the German degree. [FL MECH] One of the units in any of various scales of specific gravity, such as the Baumé scale. [MATH] A unit for measurement of plane angles, equal to 1/360 of a complete revolution, or 1/90 of a right angle. Symbolized °. [THERMO] One of the units of temperature or temperature difference in any of various temperature scales, such as the Celsius, Fahrenheit, and Kelvin temperature scales (the Kelvin degree is now known as the kelvin).

degree-day [MECH ENG] A measure of the departure of the mean daily temperature from a given standard; one degree-day is recorded for each degree of departure above (or below) the standard during a single day; used to estimate energy requirements for building heating and, to a lesser extent, for cooling.

degree Engler [FL MECH] A measure of viscosity; the ratio of the time of flow of 200 milliliters of the liquid through a viscometer devised by Engler, to the time for the flow of the same volume of water.

degree of current rectification [ELECTR] Ratio between the average unidirectional current output and the root mean square value of the alternating current input from which it was derived.

degree of curve [CIV ENG] A measure of the curvature of a railway or highway, equal to the angle subtended by a 100-foot chord (railway) or by a 100-foot arc (highway).

degree of degeneracy [MATH] The number of characteristic functions of an operator having the same characteristic value. Also known as order of degeneracy.

degree of freedom [MECH] Of a gyro, the number of orthogonal axes about which the spin axis is free to rotate, the spin axis freedom not being counted; this is not a universal convention; for example, the free gyro is frequently referred to as a three-degree-of-freedom gyro, the spin axis being counted. [MECH ENG] Any one of the number of ways in which the space configuration of a mechanical system may change. [PHYS CHEM] Any one of the variables, including pressure, temperature, composition, and specific volume, which must be specified to define the state of a system. [STAT] A number one less than the number of frequencies being tested with a chi-square test.

degree of polymerization [ORG CHEM] The number of structural units in the average polymer molecule in a particular sample. Abbreviated D.P.

degree of voltage rectification [ELECTR] Ratio between the average unidirectional voltage and the root mean square value of the alternating voltage from which it was derived.

degrees of frost [METEOROL] In England, the number of degrees Fahrenheit that the temperature falls below the freezing point; thus a day with a minimum temperature of 27°F may be designated as a day of five degrees of frost.

degritting [CHEM ENG] Removal of fine solid particles (grit) from a liquid carrier by gravity separation (settling) or centrifugation.

de Gua's rule [MATH] The rule that if, in a polynomial equation $f(x) = 0$, a group of r consecutive terms is missing, then the equation has at least r imaginary roots if r is even, or the equation has at least $r + 1$ or $r - 1$ imaginary roots if r is odd (depending on whether the terms immediately preceding and following the group have like or unlike signs).

degumming [TEXT] Removing sericin, a gluelike substance, from silk by using a solution of soap, to improve the quality of the silk.

de Haas–Van Alphen effect [SOLID STATE] An effect occurring in many complex metals at low temperatures, consisting of a periodic variation in the diamagnetic susceptibility of conduction electrons with changes in the component of the applied magnetic field at right angles to the principal axis of the crystal.

dehiscence [BOT] Spontaneous bursting open of a mature plant structure, such as fruit, anther, or sporangium, to discharge its contents. [MED] A defect in the boundary of a bony canal or cavity.

dehumidification [MECH ENG] The process of reducing the moisture in the air; serves to increase the cooling power of air.

dehumidifier [MECH ENG] Equipment designed to reduce the amount of water vapor in the ambient atmosphere.

dehydrase [BIOCHEM] An enzyme which catalyzes the removal of water from a substrate.

dehydration [CHEM] Removal of water from any substance.

dehydrator [CHEM] A substance that removes water from a material; an example is sulfuric acid. [CHEM ENG] Vessel or process system for the removal of liquids from gases or solids by the use of heat, absorbents, or adsorbents.

dehydroacetic acid [ORG CHEM] $C_8H_8O_4$ Crystals that melt at 108.5°C and are insoluble in water, soluble in acetone; used as a fungicide and bactericide. Also known as DHA; methyl-acetopyranone.

dehydroascorbic acid [ORG CHEM] $C_6H_6O_6$ A relatively inactive acid resulting from elimination of two hydrogen atoms from ascorbic acid when the latter is oxidized by air or other agents; has potential ascorbic acid activity.

dehydrocholesterol [BIOCHEM] $C_{27}H_{43}OH$ A provitamin of animal origin found in the skin of man, in milk, and elsewhere, which upon irradiation with ultraviolet rays becomes vitamin D.

dehydrocholic acid [ORG CHEM] $C_{24}H_{34}O_5$ A white powder melting at 231–240°C, very slightly soluble in water; used as a pharmaceutical intermediate and in medicine.

dehydrocyclization [CHEM ENG] Any process involving both dehydrogenation and cyclization, as in petroleum refining.

dehydrofreezing [FOOD ENG] A dual process for preserving food which involves partial dehydration followed by quick freezing.

dehydrogenase [BIOCHEM] An enzyme which removes hydrogen atoms from a substrate and transfers it to an acceptor other than oxygen.

dehydrogenation [CHEM] Removal of hydrogen from a compound.

dehydrohalogenation [CHEM] Removal of hydrogen and a halogen from a compound.

deicer [AERO ENG] Any device to keep the wings and propeller of an airplane free of ice. [MATER] Any substance used to keep a surface free of ice or to rid it of ice; ethylene glycol is used to deice windshields of automobiles and airplanes.

deicing [ENG] The removal of ice deposited on any object, especially as applied to aircraft icing, by heating, chemical treatment, and mechanical rupture of the ice deposit.

Deimos [ASTRON] A satellite of Mars orbiting at a mean distance of 23,500 kilometers.

Deinotheriidae [PALEON] A family of extinct proboscidean mammals in the suborder Deinotherioidea; known only by the genus *Deinotherium*.

Deinotherioidea [PALEON] A monofamilial suborder of extinct mammals in the order Proboscidea.

deion circuit breaker [ELEC] Circuit breaker built so that the arc that forms when the circuit is broken is magnetically blown into a stack of insulated copper plates, giving the effect of a large number of short arcs in series; each arc becomes almost instantly deionized when the current drops to zero in the alternating current cycle, and the arc cannot reform.

deionization [ELECTR] The return of an ionized gas to its neutral state after all sources of ionization have been removed, involving diffusion of ions to the container walls and volume recombination of negative and positive ions.

deionization potential [ELECTR] The potential at which ionization of the gas in a gas-filled tube ceases and conduction stops.

deionization time [ELECTR] The time required for a gas tube to regain its preconduction characteristics after interruption of anode current, so that the grid regains control. Also called recontrol time.

Deister phase [GEOL] A subdivision of the late Ammerian phase of the Jurassic period between the Kimmeridgian and lower Portlandian.

dekapoise [FL MECH] A unit of absolute viscosity, equal to 10 poises.

Delaborne prism [OPTICS] A special compound prism which, when rotated about an axis parallel to the reflecting face and lying in a plane perpendicular to the refracting faces, rotates

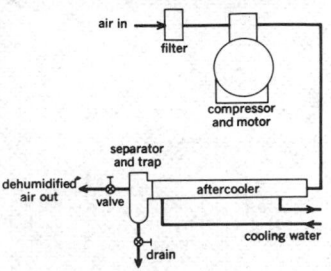

DEHUMIDIFIER

Dehumidifying by compression and aftercooling.

DELABORNE PRISM

Diagram of Delaborne prism.

the image through twice the angle. Also known as Dove prism.

delafossite [MINERAL] $CuFeO_2$ A mineral consisting of an oxide of copper and iron.

delamination [BIOL] The separation of cells into layers. [EMBRYO] Gastrulation in which the endodermal layer splits off from the inner surface of the blastoderm and the space between this layer and the yolk represents the archenteron. [ENG] Separation of a laminate into its constituent layers.

de la Rue and Miller's law [ELECTR] The law that in a field between two parallel plates, the sparking potential of a gas is a function of the product of gas pressure and sparking distance only.

de la Tour method [ANALY CHEM] Measurement of critical temperature, involving sealing the sample in a tube and heating it; the temperature at which the meniscus disappears is the critical temperature.

Delaunay orbit element [MECH] In the n-body problem, certain functions of variable elements of an ellipse with a fixed focus along which one of the bodies travels; these functions have rates of change satisfying simple equations.

de Laval nozzle [AERO ENG] A converging-diverging nozzle used in certain rockets. Also known as Laval nozzle.

de Lavaud process [MET] A centrifugal casting process employing water-cooled metal molds, used to produce cast iron pipe, gun barrels, and other cylindrical objects.

delay [COMMUN] 1. Time required for a signal to pass through a device or a conducting medium. 2. Time which elapses between the instant at which any designated point of a transmitted wave passes any two designated points of a transmission circuit; such delay is primarily determined by the constants of the circuit. [IND ENG] Interruption of the normal tempo of an operation; may be avoidable or unavoidable.

delay-action detonator See delay blasting cap.

delay allowance [IND ENG] A percentage of the normal operating time added to the normal time to allow for delays.

delay blasting cap [ENG] A blasting cap which explodes at a definite time interval after the firing current has been passed by the exploder. Also known as delay-action detonator.

delay circuit See time-delay circuit.

delay counter [ADP] A counter which inserts a time delay in a sequence of events.

delay distortion [ELECTR] Phase distortion in which the rate of change of phase shift with frequency of a circuit or system is not constant over the frequency range required for transmission. Also called envelope delay distortion.

delayed action bomb [ORD] A bomb with a delay fuse; the delay action may vary from a fraction of a second to several days after impact.

delayed action mine [ORD] A mine that explodes some time after being activated.

delayed alpha particle [NUC PHYS] An alpha particle emitted by an excited nucleus that was formed an appreciable time after a beta disintegration process.

delayed automatic gain control [ELECTR] An automatic gain control system that does not operate until the signal exceeds a predetermined magnitude; weaker signals thus receive maximum amplification. Also known as biased automatic gain control; delayed automatic volume control; quiet automatic volume control.

delayed automatic volume control See delayed automatic gain control.

delayed coincidence [NUCLEO] Occurrence of a count in one detector at a short but measurable time later than a count in another detector, the two counts being due to successive events in the same nucleus.

delayed coking [CHEM ENG] A semicontinuous thermal process for converting heavy petroleum stock to lighter material.

delayed critical [NUCLEO] The condition in which a nuclear reactor is critical because of delayed neutrons alone, without requiring the contribution of prompt neutrons.

delayed development well See step-out well.

delayed hypersensitivity [IMMUNOL] Abnormal reactivity in a sensitized individual beginning several hours after contact with the allergen.

delayed neutron [NUC PHYS] A neutron emitted spontane-

DELAY LINE

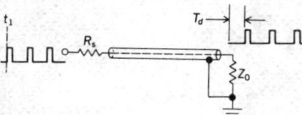

Circuit diagram of a transmission line as a delay line. R_s is series resistance of line; Z_0 is its characteristic impedance. Graphs at left and right represent input and output pulses respectively. Series of pulses starting at time t_1 require time T_d to propagate down line.

ously from a nucleus as a consequence of excitation left from a preceding radioactive decay event; in particular, a delayed fission neutron.

delayed neutron fraction [NUC PHYS] The ratio of the mean number of delayed fission neutrons per fission to the mean total number of neutrons (prompt plus delayed) per fission.

delayed-opening chaff [ORD] Chaff packages, with parachutes and time fuses, deployed from aircraft or ballistic missiles; intended to make the enemy believe there are flying targets far removed from the dispensing aircraft or missiles. Abbreviated DOC.

delayed PPI [ELECTR] A plan position indicator in which initiation of the time base is delayed a fixed time after each transmitted pulse, to give expansion of the range scale for distant targets so that they show more clearly on the screen.

delayed proton [NUC PHYS] A proton emitted spontaneously from a nucleus as a consequence of excitation left from a previous radioactive decay event.

delayed repeater satellite [AERO ENG] Satellite which stores information obtained from a ground terminal at one location, and upon interrogation by a terminal at a different location, transmits the stored message.

delayed speech [MED] A speech disorder characterized by a complete absence of vocalization or vocalization with no communicative value; speech is considered delayed when it fails to develop by the second year, caused by impaired hearing, severe childhood illness, or emotional disturbance.

delayed sweep [ELECTR] A sweep whose beginning is delayed for a definite time after the pulse that initiates the sweep.

delayed yield [MET] Time delay between the sudden application of a yield stress and the appearance of yielding.

delay element [ORD] A component which provides a specified delay between actuation of the propellant-actuated devices and ignition of the propellant.

delay equalizer [ELECTR] A corrective network used to make the phase delay or envelope delay of a circuit or system substantially constant over a desired frequency range.

delayer [MATER] A substance mixed with solid rocket propellants to decrease the rate of combustion.

delay line [ELECTR] A transmission line (as dissipationless as possible), or an electric network approximation of it, which, if terminated in its characteristic impedance, will reproduce at its output a waveform applied to its input terminals with little distortion, but at a time delayed by an amount dependent upon the electrical length of the line. Also known as artificial delay line.

delay-line memory See circulating memory.

delay-line storage See circulating memory.

delay multivibrator [ELECTR] A monostable multivibrator that generates an output pulse a predetermined time after it is triggered by an input pulse.

delay time [CONT SYS] The amount of time by which the arrival of a signal is retarded after transmission through physical equipment or systems. [ELECTR] The time taken for collector current to start flowing in a transistor that is being turned on from the cutoff condition.

delay unit [ELECTR] Unit of a radar system in which pulses may be delayed a controllable amount.

Delbrück scattering [NUC PHYS] Elastic scattering of gamma rays by a nucleus caused by virtual electron-positron pair production.

Delco Remy distributor [ELEC] A distributor in which two breaker arms are connected in parallel, one coil and one capacitor are used, and one set of contacts closes a few degrees after the other is broken.

d electron [ATOM PHYS] An atomic electron that has an orbital angular momentum of 2 in the central field approximation.

Delepine reaction [ORG CHEM] Slow ammonolysis of alkyl halides in acid to primary amines in the presence of hexamethylenetetramine.

deletion [GEN] Loss of a chromosome segment of any size, down to a part of a single gene.

deletion record [ADP] A record which removes and replaces an existing record when it is added to a file.

deliberate fire [ORD] Fire which is conducted at a rate

salvo, for tactical reasons, or for conserving ammunition.

delimiter [ADP] A character that separates items of data.

delinquency [PSYCH] **1.** The tendency to commit legal or moral offenses, especially by a minor. **2.** The offense committed.

delinquency proneness [PSYCH] The likelihood of becoming delinquent, measured by estimates of the probability that an individual will become delinquent if exposed to fairly commonplace temptations and opportunities.

deliquescence [BOT] The condition of repeated divisions ending in fine divisions; seen especially in venation and stem branching. [PHYS CHEM] The absorption of atmospheric water vapor by a crystalline solid until the crystal eventually dissolves into a saturated solution.

delirium [MED] Severely disordered mental state associated with fever, intoxication, head trauma, and other encephalopathies.

delirium tremens [MED] Delirium associated with tremors, insomnia, and other physical and neurological symptoms frequently following chronic alcoholism.

delivery system [ORD] The means of delivering atomic weapons to the target.

dell [GEOGR] A small, secluded valley or vale.

dellenite *See* rhyodacite.

Dellinger fadeout [COMMUN] Type of fadeout that occurs during shortwave reception, believed to be caused by rapid shifting of ionosphere layers during solar eruptions.

Delmontian [GEOL] Upper Miocene or lower Pliocene geologic time.

delomorphous cell *See* parietal cell.

del operator [MATH] The rule which replaces the function f of three variables, x, y, z, by the vector valued function whose components in the x, y, z directions are the respective partial derivatives of f. Written ∇f. Also known as nabla.

delorenzite *See* tanteuxenite.

Delphinidae [VERT ZOO] A family of aquatic mammals in the order Cetacea; includes the dolphins.

delphinidin [BIOCHEM] $C_{15}H_{11}O_7Cl$ A purple or brownish-red anthocyanin compound occurring widely in plants.

delphinium blue [OPTICS] A variable color averaging a brilliant blue.

Delphinus [ASTRON] A northern constellation, right ascension, 21 hours, declination, 10° north. Also known as Dolphin. [VERT ZOO] A genus of cetacean mammals, including the dolphin.

Delrac [NAV] An electronic long-range navigation system in which the position of a vessel is ascertained by comparing the phase of signal bursts from different transmission stations.

delta [ELECTR] The difference between a partial-select output of a magnetic cell in a one state and a partial-select output of the same cell in a zero state. [GEOL] An alluvial deposit, usually triangular in shape, at the mouth of a river, stream, or tidal inlet.

delta baryon [PARTIC PHYS] **1.** Any excited baryon state belonging to a multiplet having a total isospin of 3/2, a hypercharge of +1, a spin of 3/2, positive parity, and an approximate mass of 1236 Mev. Designated Δ(1236). **2.** Any excited baryon state belonging to any multiplet having a total isospin of 3/2 and a hypercharge of +1.

delta connection [ELEC] A combination of three components connected in series to form a triangle like the Greek letter delta. Also known as mesh connection.

delta current [ELEC] Electricity going through a delta connection.

delta ferrite *See* delta iron.

delta function [MATH] A distribution δ such that

$$\int_{-\infty}^{\infty} f(t)\,\delta(x-t)\,dt \text{ is } f(x).$$

Also known as Dirac delta function.

delta geosyncline *See* exogeosyncline.

delta-gun tube [ELECTR] A color television picture tube in which three electron guns, arranged in a triangle, provide electron beams that fall on phosphor dots on the screen, causing them to emit light in three primary colors; a shadow mask located just behind the screen ensures that each beam excites only dots of one color.

deltaic deposits [GEOL] Sedimentary deposits in a delta.

delta iron [MET] The nonmagnetic polymorphic form of iron stable between about 1403°C and the melting point, about 1535°C. Also known as delta ferrite.

deltaite [MINERAL] A mixture of crandallite and hydroxylapatite.

delta matching transformer [ELEC] Impedance device used to match the impedance of an open-wire transmission line to an antenna; the two ends of the transmission line are fanned out so that the impedance of the line gradually increases; the ends of the transmission line are attached to the antenna at points of equal impedance, symmetrically located with respect to the center of the antenna.

delta meson [PARTIC PHYS] Any scalar meson resonance, with positive charge conjugation parity, belonging to a multiplet with a total isospin of 1, a hypercharge of zero, a mass of 962 ± 5 MeV, and a width <5 MeV. Designated δ(962).

delta modulation [ELECTR] A pulse-modulation technique in which a continuous signal is converted into a binary pulse pattern, for transmission through low-quality channels.

delta moraine *See* ice-contact delta.

delta network [ELEC] A set of three branches connected in series to form a mesh.

delta plain [GEOL] A plain formed by deposition of silt at the mouth of a stream or by overflow along the lower stream courses.

delta pulse code modulation [ELECTR] A modulation system that converts audio signals into corresponding trains of digital pulses to give greater freedom from interference during transmission over wire or radio channels.

delta ray [ATOM PHYS] An electron or proton ejected by recoil when a rapidly moving alpha particle or other primary ionizing particle passes through matter.

delta region [METEOROL] A region in the atmosphere characterized by difluence.

delta rhythm [PHYSIO] An electric current generated in slow waves with frequencies of 0.5–3 per second from the forward portion of the brain of normal subjects when asleep.

Deltatheridia [PALEON] An order of mammals that includes the dominant carnivores of the early Cenozoic.

delta wing [AERO ENG] A triangularly shaped wing of an aircraft.

delta-Y transformation *See* Y-delta transformation.

deltic method [ELECTR] A method of sampling incoming radar, sonar, seismic, speech, or other waveforms along with reference signals, compressing the samples in time, and comparing them by autocorrelation.

deltohedron [CRYSTAL] A polyhedron which has 12 quadrilateral faces, and is the form of a crystal belonging to the cubic system and having hemihedral symmetry. Also known as deltoid dodecahedron; tetragonal tristetrahedron.

deltoid [ANAT] The large triangular shoulder muscle; originates on the pectoral girdle and inserts on the humerus. [BIOL] Triangular in shape.

deltoid dodecahedron *See* deltohedron.

deltoid ligament [ANAT] The ligament on the medial side of the ankle joint; the fibers radiate from the medial malleolus to the talus, calcaneus, and navicular bones.

delusion [PSYCH] A conviction based on faulty perceptions, feelings, and thinking.

deluster [TEXT] To reduce brightness, for example, by adding pigment to the spinning solution of a yarn.

delvauxite [MINERAL] A mineral, with the approximate formula $Fe_4(PO_4)_2(OH)_6\cdot nH_2O$, consisting of a hydrous phosphate of iron.

demagnetization [ELECTROMAG] **1.** The process of reducing or removing the magnetism of a ferromagnetic material. **2.** The reduction of magnetic induction by the internal field of a magnet. [MIN ENG] Deflocculation in dense-media process using ferrosilicon by passing the fluid through an alternating-current field.

demagnetizer [ELECTR] A device for removing undesired magnetism, as from the playback head of a tape recorder or from a recorded reel of magnetic tape that is to be erased.

demal [CHEM] A unit of concentration, equal to the concentration of a solution in which 1 gram-equivalent of solute is dissolved in 1 cubic decimeter of solvent.

demand *See* demand factor.

demand factor [ELEC] The ratio of the maximum demand of

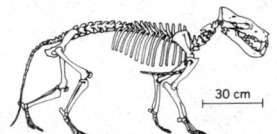

DELTATHERIDIA

The deltatheridian *Hyaenodon*, original about 4 feet (1.2 meters) long. (*After Scott, from A. S. Romer, Vertebrate Paleontology, 3d ed., University of Chicago Press, 1966*)

a building for electric power to the total connected load. Also known as demand.

demand limiter *See* current limiter.

demand meter [ENG] Any of several types of instruments used to determine a customer's maximum demand for electric power over an appreciable time interval; generally used for billing industrial users.

demand paging [ADP] The characteristic of a virtual memory system which retrieves only that part of a user's program which is required during execution.

demand rate [ELEC] The maximum amount of electric power that must be kept available to a customer.

demand system [ENG] A system in an airplane that automatically dispenses oxygen according to the demand of the flyer's body.

demantoid [MINERAL] A lustrous, green variety of andradite; used as a gem.

demarcation potential *See* injury potential.

De Marre formula [ORD] A formula expressing the relationship between projectile characteristics and armor-plate penetration capabilities.

Dematiaceae [MYCOL] A family of fungi in the order Moniliales; sporophores are not grouped, hyphae are always dark, and the spores are hyaline or dark.

Dember effect [ELECTR] Creation of a voltage in a conductor or semiconductor by illumination of one surface. Also known as photodiffusion effect.

deme [ECOL] A local population in which the individuals freely interbreed among themselves but not with those of other demes.

dementia [PSYCH] Deterioration of intellectual and other mental processes due to organic brain disease.

dementia praecox *See* schizophrenia.

dementia simplex [PSYCH] A subtype of schizophrenia broadly characterized by a slow, progressive deterioration, often combined with mental deficiency.

demersal [BIOL] Living at or near the bottom of the sea.

demethanation *See* demethanization.

demethanator [CHEM ENG] The apparatus in which demethanization is conducted.

demethanization [CHEM ENG] The process of distillation in which methane is separated from the heavier components. Also known as demethanation.

demethylation [ORG CHEM] Removal of the methyl group from a compound.

demethylchlortetracycline [MICROBIO] $C_{21}H_{21}O_8N_2Cl$ A broad-spectrum tetracycline antibiotic produced by a mutant strain of *Streptomyces aureofaciens*.

demilune [BIOL] Crescent-shaped.

demineralization [CHEM ENG] Removal of mineral constituents from water. [MED] Removal or loss of minerals and salts from the body, especially by disease.

demister [MECH ENG] A series of ducts in automobiles arranged so that hot, dry air directed from the heat source is forced against the interior of the windscreen or windshield to prevent condensation.

Demjanov rearrangement [ORG CHEM] A structural rearrangement that accompanies treatment of certain primary aliphatic amines with nitrous acid; the amine will undergo a ring contraction or expansion.

Demodicidae [INV ZOO] The pore mites, a family of arachnids in the suborder Trombidiformes.

demodulate [COMMUN] To recover the modulating wave from a modulated carrier. Also known as decode; detect.

demodulator *See* detector.

demography [ECOL] The statistical study of populations with reference to natality, mortality, migratory movements, age, and sex, among other social, ethnic, and economic factors.

De Moivre's theorem [MATH] The nth power of the quantity $\cos \theta + i \sin \theta$ is $\cos n\theta + i \sin n\theta$ for any integer n.

demolition [CIV ENG] The act or process of tearing down a building or other structure. [ORD] Destroying a structure or an area by the use of explosives.

demolition block [ORD] An explosive charge, usually in a nonmetallic container, used for demolition purposes.

demolition bomb [ORD] Former classification for a bomb that explodes after a short penetration, accomplishing dam-

age and destruction by both blast and underground explosion.

demolition kit [ORD] A group of items of an explosive nature, with the necessary nonexplosive accessories and tools, specially designed containers, and carrying attachments, to enable efficient performance of designated demolition tasks.

demon of Maxwell [THERMO] Hypothetical creature who controls a trapdoor over a microscopic hole in an adiabatic wall between two vessels filled with gas at the same temperature, so as to supposedly decrease the entropy of the gas as a whole and thus violate the second law of thermodynamics. Also known as Maxwell's demon.

demonophobia [PSYCH] An abnormal fear of devils and demons.

Demon Star *See* Algol.

De Morgan's rules [MATH] The complement of the union of two sets equals the intersection of their respective complements; the complement of the intersection of two sets equals the union of their complements.

De Morgan's test [MATH] A series with term u_n, for which $|u_{n+1}/u_n|$ converges to 1, will converge absolutely if there is $c > 0$ such that the limit superior of $n(|u_{n+1}/u_n| - 1)$ equals $-1 - c$.

demorphism *See* weathering.

Demospongiae [INV ZOO] A class of the phylum Porifera, including sponges with a skeleton of one- to four-rayed siliceous spicules, or of spongin fibers, or both.

demountable tube [ELECTR] High-power radio tube having a metal envelope with porcelain insulation; can be taken apart for inspection and for renewal of electrodes.

demulsification [CHEM ENG] Prevention or breaking of liquid-liquid emulsions by chemical, mechanical or electrical demulsifiers.

demulsifier [CHEM ENG] A chemical, mechanical, or electrical system that either breaks liquid-liquid emulsions or prevents them from forming.

demultiplexer [ELECTR] A device used to separate two or more signals that were previously combined by a compatible multiplexer and transmitted over a single channel.

demultiplexing circuit [ELECTR] A circuit used to separate the signals that were combined for transmission by multiplex.

demyelinating disease [MED] Any disease associated with the destruction or removal of myelin from nerves.

demyelination [PATH] Destruction of myelin; loss of myelin from nerve sheaths or nerve tracts.

denaturant [CHEM] An inert, bad-tasting, or poisonous chemical substance added to a product such as ethyl alcohol to make it unfit for human consumption. [NUCLEO] A nonfissionable isotope that can be added to fissionable material to make it unsuitable for use in atomic weapons without extensive processing.

denature [CHEM] **1.** To change a protein by heating it or treating it with alkali or acid so that the original properties such as solubility are changed as a result of the protein's molecular structure being changed in some way. **2.** To add a denaturant, such as methyl alcohol, to grain alcohol to make the grain alcohol poisonous and unfit for human consumption.

denatured alcohol [CHEM] Ethyl alcohol containing a poisonous substance, such as methyl alcohol or benzene, which makes it unfit for human consumption.

dendrite [ANAT] The part of a neuron that carries the unidirectional nerve impulse toward the cell body. Also known as dendron. [CRYSTAL] A crystal having a treelike structure.

dendritic [SCI TECH] Having a branching, treelike structure or pattern.

dendritic drainage [HYD] Irregular stream branching, with tributaries joining the main stream at all angles.

dendritic powder [MET] Fine metal particles having a dendritic structure; usually of electrolytic origin. Also known as arborescent powder.

dendritic valleys [GEOL] Treelike extensions of the valleys in a region lying upon horizontally bedded rock.

Dendrobatinae [VERT ZOO] A subfamily of anuran amphibians in the family Ranidae, including the colorful poisonous frogs of Central and South America.

dendrobranchiate gill [INV ZOO] A respiratory structure of

DENDROBRANCHIATE GILL

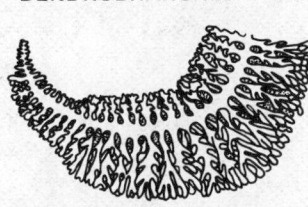

Dendrobranchiate gill of penaeid shrimp, *Benthesicymus*. (*Smithsonian Institution*)

certain decapod crustaceans, characterized by extensive branching of the two primary series.

Dendroceratida [INV ZOO] A small order of sponges of the class Demospongiae; members have a skeleton of spongin fibers or lack a skeleton.

Dendrochirotacea [INV ZOO] A subclass of echinoderms in the class Holothuroidea.

Dendrochirotida [INV ZOO] An order of dendrochirotacean holothurian echinoderms with 10–30 richly branched tentacles.

dendrochronology [GEOL] The science of measuring time intervals and dating events and environmental changes by reading and dating growth layers of trees as demarcated by the annual rings.

Dendrocolaptidae [VERT ZOO] The woodcreepers, a family of passeriform birds belonging to the suboscine group.

dendrogram [BIOL] A genealogical tree; the trunk represents the oldest common ancestor, and the branches indicate successively more recent divisions of a lineage for a group.

dendrohydrology [HYD] The science of determining hydrologic occurrences by the comparison of tree ring thickness with streamflow or precipitation.

dendroid [BIOL] Branched or treelike in form.

Dendroidea [PALEON] An order of extinct sessile, branched colonial animals in the class Graptolithina occurring among typical benthonic fauna.

dendrology [FOR] The division of forestry concerned with the classification, identification, and distribution of trees and other woody plants.

dendrometer [FOR] A device used to measure a tree's height and diameter using principles based on the relation of the sides of similar triangles.

Dendromurinae [VERT ZOO] The African tree mice and related species, a subfamily of rodents in the family Muridae.

dendron *See* dendrite.

dendrophagous [ZOO] Feeding on trees, referring to insects.

dendrophysis [MYCOL] A hyphal thread with arboreal branching in certain fungi.

Deneb [ASTRON] A white star of spectral classification A2-Ia in the constellation Cygnus; the star α Cygni.

Denebola [ASTRON] A white star of stellar magnitude 2.2, spectral classification A2, in the constellation Leo; the star β Leonis.

denervate [MED] To interfere with or cut off the nerve supply to a part of the body, or to remove a nerve; may occur by excision, drugs, or a disease process.

denervation hypersensitivity [PHYSIO] Extreme sensitivity of an organ that has recovered from the removal or interruption of its nerve supply.

dengue [MED] An acute viral disease of man characterized by fever, rash, prostration, and lymphadenopathy; transmitted by the mosquito *Aedes aegypti*. Also known as breakbone fever; dandy fever.

denial [PSYCH] An unconscious defense mechanism in which an individual denies himself recognition of an observation in order to avoid pain or anxiety.

denier [TEXT] A unit expressing the mass of a fiber divided by its length, equal to 1 gram for 9000 meters of fiber.

denim [TEXT] A sturdy twill-weave cotton fabric having a solid-colored warp and a white filling.

Denison sampler [ENG] A soil sampler consisting of a central nonrotating barrel which is forced into the soil as friction is removed by a rotating external barrel; the bottom can be closed to retain the sample during withdrawal.

denitration [CHEM] Removal of nitrates or nitrogen. Also known as denitrification.

denitrification [CHEM] *See* denitration. [MICROBIO] The reduction of nitrate or nitrite to gaseous products such as nitrogen, nitrous oxide, and nitric oxide; brought about by denitrifying bacteria.

denitrifying bacteria [MICROBIO] Bacteria that reduce nitrates to nitrites or nitrogen gas; most are found in soil.

denitrogenate [PHYSIO] To remove nitrogen from the body by breathing nitrogen-free gas.

denominator [MATH] In a fraction, the term that divides the other term (called the numerator), and is written below the line.

De Nora cell [CHEM ENG] Mercury-cathode cell used for production of chlorine and caustic soda by electrolysis of sodium chloride brine.

dense [GRAPHICS] Very opaque because of a concentration of material, as pertaining to a negative or transparency that has been overdeveloped or overexposed.

dense-air refrigeration cycle *See* reverse Brayton cycle.

dense-air system *See* cold-air machine.

dense binary code [ADP] A code in which all possible states of the binary pattern are used.

dense connective tissue [HISTOL] A fibrous connective tissue with an abundance of enlarged collagenous fibers which tend to crowd out the cells and ground substance.

dense-in-itself set [MATH] A set every point of which is an accumulation point; a set without any isolated points.

dense-media separator [MIN ENG] A device in which a heavy mineral is dispersed in water, causing heavier ores to sink and lighter ores to float.

dense subset [MATH] A subset of a topological space whose closure is the entire space.

densify [ENG] To increase the density of a material such as wood by subjecting it to pressure or impregnating it with another material.

densimeter [ENG] An instrument which measures the density or specific gravity of a liquid, gas, or solid. Also known as densitometer; density gage; density indicator; gravitometer.

densitometer [ENG] 1. An instrument which measures optical density by measuring the intensity of transmitted or reflected light; used to measure photographic density. 2. *See* densimeter.

density [MATER] Closeness of texture or consistency. [MECH] The mass of a given substance per unit volume. [OPTICS] 1. The degree of opacity of a translucent material. 2. The common logarithm of opacity. [PHYS] The total amount of a quantity, such as energy, per unit of space.

density airspeed [AERO ENG] Calibrated airspeed corrected for pressure altitude and true air temperature.

density altitude [METEOROL] The altitude, in the standard atmosphere, at which a given density occurs.

density bombing [ORD] Dropping a given tonnage of bombs onto an area in order to make certain of striking particular targets.

density channel [METEOROL] A channel used to investigate a density current; for example, in experiments relating to the behavior of cold masses of air in the atmosphere and related frontal structures.

density correction [AERO ENG] A correction made necessary because the airspeed indicator is calibrated only for standard air pressure; it is applied to equivalent airspeed to obtain true airspeed, or to calibrated airspeed to obtain density airspeed. [ENG] 1. The part of the temperature correction of a mercury barometer which is necessitated by the variation of the density of mercury with temperature. 2. The correction, applied to the indications of a pressure-tube anemometer or pressure-plate anemometer, which is necessitated by the variation of air density with temperature.

density current [METEOROL] Intrusion of a dense air mass beneath a lighter air mass; the usage applies to cold fronts. [OCEANOGR] *See* turbidity current.

density error [AERO ENG] The error in the indications of a differential-pressure-type airspeed indicator due to nonstandard atmospheric density.

density function [MATH] A function which when integrated with respect to a measure gives another measure as the density function for the new measure. [STAT] *See* probability density function.

density gage *See* densimeter.

density indicator *See* densimeter.

density log [PETRO ENG] Radioactivity logging of reservoir structure densities down an oil-well bore by emission and detection of gamma rays.

density matrix [QUANT MECH] A matrix ρ_{mn} describing an ensemble of quantum-mechanical systems in a representation based on an orthonormal set of functions ϕ_n; for any operator G with representation G_{mn}, the ensemble average of the expectation value of G is the trace of ρG.

density modulation [ELECTR] Modulation of an electron

DENDROCERATIDA

Dendritic skeleton of *Dendrilla cactus*. (*After Lendenfeld, 1889*)

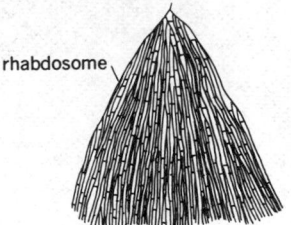

DENDROIDEA

rhabdosome

Whole colony of a dendroid graptolite.

beam by making the density of the electrons in the beam vary with time.

density of states [SOLID STATE] A function of energy E equal to the number of quantum states in the energy range between E and $E + dE$ divided by the product of dE and the volume of the substance.

density packing [ADP] In computers, the number of binary digit magnetic pulses stored on tape or drum per linear inch on a single track by a single head.

density ratio [METEOROL] The ratio of the density of the air at a given altitude to the air density at the same altitude in a standard atmosphere.

density rule [ENG] A grading system for lumber based on the width of annual rings.

density specific impulse [AERO ENG] The product of the specific impulse of a propellant combination and the average specific gravity of the propellants.

density step tablet [COMMUN] Facsimile test chart consisting of a series of areas; density of the areas increases from a low value to a maximum value in steps. Also known as step tablet.

density transmitter [ENG] An instrument used to record the density of a flowing stream of liquid by measuring the buoyant force on an air-filled chamber immersed in the stream.

density-wave theory [ASTROPHYS] A theory explaining the spiral structure of galaxies by a periodic variation in space in the density of matter which rotates with a fixed angular velocity while the angular velocity of the matter itself varies with distance from the galaxy's center.

densofacies *See* metamorphic facies.

dental [ANAT] Pertaining to the teeth.

dental arch [ANAT] The parabolic curve formed by the cutting edges and masticating surfaces of the teeth.

dental calculi [MED] Calcareous deposits of organic and mineral matter on the teeth. Also known as tartar.

dental caries *See* caries.

dental coupling [MECH ENG] A type of flexible coupling used to join a steam turbine to a reduction-gear pinion shaft; consists of a short piece of shaft with gear teeth at each end, and mates with internal gears in a flange at the ends of the two shafts to be joined.

dental epithelium [HISTOL] The cells forming the boundary of the enamel organ.

dental follicle *See* dental sac.

dental formula [VERT ZOO] An expression of the number and kind of teeth in each half jaw, both upper and lower, of mammals.

dental gold [MET] An alloy composed of 5 to 12% silver, 4 to 10% copper, with the balance gold.

Dentaliidae [INV ZOO] A family of mollusks in the class Scaphopoda; members have pointed feet.

dental index [ANTHRO] A ratio of the length of the teeth to the distance from the nasion to the basion multiplied by 100, used to determine the relative size of teeth.

dental pad [VERT ZOO] A firm ridge that replaces incisors in the maxilla of cud-chewing herbivores.

dental papilla [EMBRYO] The mass of connective tissue located inside the enamel organ of a developing tooth, and forming the dentin and dental pulp of the tooth.

dental plate [INV ZOO] A flat plate that replaces teeth in certain invertebrates, such as some worms. [VERT ZOO] A flattened plate that represents fused teeth in parrot fishes and related forms.

dental pulp [HISTOL] The vascular connective tissue of the roots and pulp cavity of a tooth.

dental ridge [EMBRYO] An elevation of the embryonic jaw that forms a cusp or margin of a tooth.

dental sac [EMBRYO] The connective tissue that encloses the developing tooth. Also known as dental follicle.

dental work *See* cementation.

dentate [BIOL] **1.** Having teeth. **2.** Having toothlike or conical marginal projections.

dentate fissure *See* hippocampal sulcus.

dentate nucleus [ANAT] An ovoid mass of nerve cells located in the center of each cerebellar hemisphere, which give rise to fibers found in the superior cerebellar peduncle.

denticle [ZOO] A small tooth or toothlike projection, as the type of scale of certain elasmobranchs.

denticulate [ZOO] Having denticles; serrate.

dentigerous [BIOL] Having teeth or toothlike structures.

dentil [ARCH] One of a series of small rectangular blocks under a cornice.

dentil band [ARCH] A molding or band on a cornice resembling a row of dentils, but with the spaces between dentils filled.

dentin [HISTOL] A bonelike tissue composing the bulk of a vertebrate tooth; consists of 70% inorganic materials and 30% water and organic matter.

dentinoblast [HISTOL] A mesenchymal cell that forms dentin.

dentinogenesis [PHYSIO] The formation of dentin.

dentinoma [MED] A benign odontogenic tumor made up of dentin.

dentist [MED] One who practices dentistry.

dentistry [MED] A branch of medical science concerned with the prevention, diagnosis, and treatment of diseases of the teeth and adjacent tissues and the restoration of missing dental structures.

dentition [VERT ZOO] The arrangement, type, and number of teeth which are variously located in the oral or in the pharyngeal cavities, or in both, in vertebrates.

denture [MED] A partial or complete prosthetic appliance to replace one or more missing teeth.

denudation [GEOL] General wearing away of the land; laying bare of subjacent lands.

denumerable set [MATH] A set which may be put in one-to-one correspondence with the positive integers. Also known as countably infinite set.

Denver cell [MIN ENG] A subaeration type of flotation cell, mechanized and self-aerating.

Denver jig [MIN ENG] A pulsion-suction diaphragm for separating sulfur from coal before flotation; hydraulic water is admitted through a rotary valve.

deodorant [MATER] A substance used to remove, correct, or repress undesirable odors.

deodorized kerosine [MATER] A highly refined petroleum kerosine that has very little odor; used as an illuminant for wick lamps. Also known as refined kerosine.

deodorizing [CHEM ENG] A process for removing odor-creating substances from oil or fat, in which the oil or fat is held at high temperatures and low pressure while steam is blown through.

deoil [CHEM ENG] To reduce the amount of liquid oil entrained in solid wax.

deoperculate [BOT] Of mosses and liverworts, to shed the operculum.

deorbit [AERO ENG] To recover a spacecraft from earth orbit by providing a new orbit which intersects the earth's atmosphere.

deoxidant *See* deoxidizer.

deoxidation [CHEM] **1.** The condition of a molecule's being deoxidized. **2.** The process of deoxidizing.

deoxidation sphere *See* bleach spot.

deoxidize [CHEM] **1.** To remove oxygen by any of several processes. **2.** To reduce from the state of an oxide. [MET] To remove an oxide film from a metal surface.

deoxidized copper [MET] Pure copper deoxidized with phosphorus to reduce cuprous oxide and eliminate porosity.

deoxidizer [CHEM] Any substance which reduces the amount of oxygen in a substance, especially a metal, or reduces oxide compounds. Also known as deoxidant.

deoxycholate [BIOCHEM] A salt or ester of deoxycholic acid.

deoxycholic acid [BIOCHEM] $C_{24}H_{40}O_4$ One of the unconjugated bile acids; in bile it is largely conjugated with glycine or taurine.

deoxycorticosterone [BIOCHEM] $C_{21}H_{30}O_3$ A steroid hormone secreted in small amounts by the adrenal cortex.

6-deoxy-L-galactose *See* L-fucose.

deoxygenation [CHEM] Removal of oxygen from a substance, such as blood or polluted water.

deoxyribonuclease [BIOCHEM] An enzyme that catalyzes the hydrolysis of deoxyribonucleic acid to nucleotides.

deoxyribonucleic acid [BIOCHEM] A linear polymer made up

DENSITY TRANSMITTER

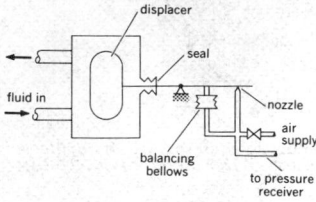

Buoyancy-type density transmitter which is used to obtain a continuous record of liquid density. *(From D. M Considine, ed., Process Instruments and Controls Handbook, McGraw-Hill, 1957)*

DEOXYCORTICOSTERONE

Structural formula of deoxycorticosterone.

of deoxyribonucleotide repeating units (composed of the sugar 2-deoxyribose, phosphate, and a purine or pyrimidine base) linked by the phosphate group joining the 3′ position of one sugar to the 5′ position of the next; most molecules are double-stranded and antiparallel, resulting in a right-handed helix structure kept together by hydrogen bonds between a purine on one chain and a pyrimidine on another; carrier of genetic information, which is encoded in the sequence of bases; present in chromosomes and chromosomal material of cell organelles such as mitochondria and chloroplasts, and also present in some viruses. Abbreviated DNA.

deoxyribonucleoprotein [BIOCHEM] A protein containing molecules of deoxyribonucleic acid in close association with protein molecules.

deoxyribose [BIOCHEM] $C_5H_{10}O_4$ A pentose sugar in which the hydrogen replaces the hydroxyl groups of ribose; a major constituent of deoxyribonucleic acid.

deoxy sugar [BIOCHEM] A substance which has the characteristics of a sugar, but which shows a deviation from the required hydrogen-to-oxygen ratio.

departure [METEOROL] The amount by which the value of a meteorological element differs from the normal value. [NAV] **1.** The distance between two meridians at any given parallel of latitude, expressed in linear units, usually nautical miles; the distance to the east or west made good by a craft in proceeding from one point to another. **2.** The point at which reckoning of a voyage begins; usually established by bearings of prominent landmarks as the vessel clears a harbor and proceeds to sea; when a person establishes this point, he is said to take departure. Also known as point of departure. **3.** Act of departing or leaving.

departure point [NAV] A navigational checkpoint used by aircraft as a marker for setting course.

departure track [CIV ENG] A railroad yard track for combining freight cars into outgoing trains.

depauperate [BIOL] Inferiority of natural development or size.

depegram [METEOROL] On a diagram having entropy and temperature as coordinates, a curve representing the distribution of the dew point as a function of pressure for a given sounding of the atmosphere.

dependence [STAT] The existence of a relationship between frequencies obtained from two parts of an experiment which does not arise from the direct influence of the result of the first part on the chances of the second part but indirectly from the fact that both parts are subject to influences from a common outside factor.

dependency needs [PSYCH] The vital, originally infantile needs for mothering, love, affection, shelter, protection, security, food, and warmth, which are also present in some degree even in adult life.

dependent variable [MATH] If y is a function of x, that is, if the function assigns a single value of y to each value of x, then y is the dependent variable.

depentanizer [CHEM ENG] A fractionating column for removal of pentane and lighter fractions from a hydrocarbon mixture.

depéq [METEOROL] Strong winds over Loet Tawar (Sumatra, East Indies) during the southwest monsoon.

depergelation [HYD] The act or process of thawing permafrost.

deperming See degaussing.

depersonalization [PSYCH] Loss of the sense of one's identity or of reality concerning the self.

Depertellidae [PALEON] A family of extinct perissodactyl mammals in the superfamily Tapiroidea.

dephlegmation [CHEM ENG] In a distillation operation, the partial condensation of vapor to form a liquid richer in higher boiling constituents than the original vapor.

dephlegmator [CHEM ENG] An apparatus used in fractional distillation to cool the vapor mixture, thereby condensing higher-boiling fractions.

dephosphorize [MET] Removal of phosphorus from a molten metal such as steel.

depilation [ENG] Removal of hair from animal skins in processing leather.

depilatory [MATER] A chemical that removes hairs from skin.

depleted material [NUCLEO] Material in which the amount of one or more isotopes of a constituent has been reduced by an isotope separation process or by a nuclear reaction.

depleted uranium [NUCLEO] Uranium having a smaller percentage of uranium-235 than the 0.7% found in natural uranium.

depletion [ECOL] Using a resource, such as water or timber, faster than it is replenished. [NUCLEO] The percentage reduction in the quantity of fissionable atoms in the fuel assemblies or fuel mixture that occurs during operation of a nuclear reactor.

depletion drive [PETRO ENG] Displacement mechanism (type of drive) to expel hydrocarbons from porous reservoir formations, that is, to remove more hydrocarbon from the reservoir; types of drives are gas or water (natural or injected) and injected LPG.

depletion layer [ELECTR] An electric double layer formed at the surface of contact between a metal and a semiconductor having different work functions, because the mobile carrier charge density is insufficient to neutralize the fixed charge density of donors and acceptors. Also known as barrier layer (deprecated); blocking layer (deprecated); space-charge layer.

depletion-layer rectification [ELECTR] Rectification at the junction between dissimilar materials, such as a *pn* junction or a junction between a metal and a semiconductor. Also known as barrier-layer rectification.

depletion-layer transistor [ELECTR] A transistor that relies directly on motion of carriers through depletion layers, such as spacistor.

depletion-mode field-effect transistor See junction-gate field-effect transistor.

depletion region [ELECTR] The portion of the channel in a metal oxide field-effect transistor in which there are no charge carriers.

depletion-type reservoir [PETRO ENG] Oil reservoir which is initially in (and during depletion remains in) a state of equilibrium between the gas and liquid phases; includes single-phase gas, two-phase bubble-point, and retrograde-gas-condensate (or dew-point) reservoirs.

depocenter [GEOL] A site of maximum deposition.

depolarization [ELEC] The removal or prevention of polarization in a substance (for example, through the use of a depolarizer in an electric cell) or of polarization arising from the field due to the charges induced on the surface of a dielectric when an external field is applied. [OPTICS] The resolution of polarized light in an optical depolarizer.

depolarization factor [ELEC] The ratio of the internal electric field induced by the charges on the surface of a dielectric when an external field is applied to the polarization of the dielectric.

depolarizer [PHYS CHEM] A substance added to the electrolyte of a primary cell to prevent excessive buildup of hydrogen bubbles by combining chemically with the hydrogen gas as it forms. Also known as battery depolarizer.

depolymerization [ORG CHEM] Decomposition of macromolecular compounds into relatively simple compounds.

deposit [GEOL] Consolidated or unconsolidated material that has accumulated by a natural process or agent. [MATER] Any material applied to a base by means of vacuum, electrical, chemical, screening, or vapor methods. [SCI TECH] Any solid matter which is gradually laid down on a surface by a natural process.

deposit attack [MET] Corrosion under or around the edge of a noncontinuous local deposit on a metal surface.

deposit dose [NUCLEO] The residual radioactivity deposited on the surface after a nuclear explosion, as by water falling as rain from the base surge of an underwater atomic explosion.

deposited carbon resistor [ELECTR] A resistor in which the resistive element is a carbon film pyrolytically deposited on a ceramic substrate.

deposit feeder [INV ZOO] Any animal that feeds on the detritus that collects on the substratum at the bottom of water. Also known as detritus feeder.

deposit gage [ENG] The general name for instruments used in air pollution studies for determining the amount of material deposited on a given area during a given time.

depositional dip See primary dip.

depositional fabric [PETR] Arrangement of detrital particles settled from suspension or of crystals from a differentiating magma determined by the plane of the surface on which they come to rest.

depositional remanent magnetization [GEOPHYS] Remanent magnetization occurring in sedimentary rock following the depositional alignment of previously magnetized grains. Abbreviated DRM.

depositional strike [GEOL] Sedimentary deposits that are continuous laterally on a gently sloping surface.

deposition efficiency [MET] The ratio of the weight of deposited metal to the net weight of the consumed electrodes, exclusive of stubs, in welding.

deposition potential [PHYS CHEM] The smallest potential which can produce electrolytic deposition when applied to an electrolytic cell.

deposition sequence [MET] The order of deposition of weld-metal increments.

depot [ORD] **1.** An establishment for storing supplies or for maintaining equipment. **2.** The installation for this establishment.

deprecated usage [SCI TECH] Word usage which is disapproved by experts in the pertinent field because the term in question has misleading connotations; for example, a capacitor has frequently been called a "condenser," but it does not condense anything.

depreciation [IND ENG] Loss of value due to physical deterioration.

depressed center car [ENG] A flat railroad car having a low center section; used to provide adequate tunnel clearance for oversized loads.

depressed fracture [MED] A fracture of the skull in which the fractured part is depressed below the normal level.

depression [GEOL] **1.** A hollow of any size on a plain surface having no natural outlet for surface drainage. **2.** A structurally low area in the crust of the earth. [METEOROL] An area of low pressure; usually applied to a certain stage in the development of a tropical cyclone, to migratory lows and troughs, and to upper-level lows and troughs that are only weakly developed. Also known as low. [PSYCH] A mood provoked by conscious awareness of an idea or feeling that was previously pushed into the unconscious.

depression angle *See* angle of depression.

depression spring [HYD] A type of gravity spring that flows onto the land surface because the surface slopes down to the water table.

depression storage [HYD] Water retained in puddles, ditches, and other depressions in the surface of the ground.

depressor [ANAT] A muscle that draws a part down. [CHEM ENG] An agent that prevents or retards a chemical reaction or process.

depressor nerve [PHYSIO] A nerve which, upon stimulation, lowers the blood pressure either in a local part or throughout the body.

depropanize [CHEM ENG] In petroleum processing, the removal of propane and sometimes lighter fractions from the petroleum.

depropanized material [MATER] Material that has undergone distillation to remove lighter components from butanes and heavier material.

depropanizer [CHEM ENG] A fractionating column in a gasoline plant for removal of propane and lighter components.

deproteinize [ORG CHEM] To remove protein from a substance.

depside [ORG CHEM] One of a class of esters that form from the joining of two or more molecules of phenolic carboxylic acid.

depsidone [ORG CHEM] One of a class of compounds that consists of esters such as depsides, but are also cyclic ethers.

depth [NAV ARCH] The vertical distance amidships from the upper surface of the flat plate keel to the underside of the plating of a specified deck at ship's side. [OCEANOGR] The vertical distance from a specified sea level to the sea floor.

depth bomb [ORD] An explosive item designed to be dropped from an aircraft for use against underwater targets.

depth charge [ORD] A cylindrical or teardrop-shaped container holding a charge of TNT or other explosive, dropped from the deck of a ship, and detonated at a preset depth as an antisubmarine weapon.

depth contour *See* isobath.

depth curve *See* isobath.

depth dose [NUCLEO] The radiation dose delivered at a particular depth beneath the surface of a body; usually expressed as percent of surface dose or of air dose.

depth-duration-area value [METEOROL] The average depth of precipitation that has occurred within a specified time interval over an area of given size. Abbreviated DDA value.

depth finder [ENG] A radar or ultrasonic instrument for measuring the depth of the sea.

depth gage [DES ENG] An instrument or tool for measuring the depth of depression to a thousandth inch.

depth hoar [HYD] Ice crystals formed beneath the snow surface by sublimation. Also known as sugar snow.

depth marker [ENG] A thin board or other lightweight substance used as a means of identifying the surface of snow or ice which has been covered by a more recent snowfall.

depth micrometer [DES ENG] A micrometer used to measure the depths of holes, slots, and distances of shoulders and projections.

depth of compensation [GEOPHYS] That depth at which density differences occurring in the earth's crust are compensated isostatically; calculated to be between 100 and 113–117 kilometers. [HYD] The depth in a body of water at which illuminance has diminished to the extent that oxygen production through photosynthesis and oxygen consumption through respiration by plants are equal; it is the lower boundary of the euphotic zone.

depth of engagement [DES ENG] The depth of contact, in a radial direction, between mating threads.

depth of field [OPTICS] The range of distances over which a camera gives satisfactory definition, its lens in the best focus for a certain specific distance.

depth of fusion [MET] The distance that fusion extends from the original surface into the base metal in a welding operation.

depth of thread [DES ENG] The distance, in a radial direction, from the crest of a screw thread to the base.

depth perception [PHYSIO] Ability to judge spatial relationships.

depth sounder [ENG] An instrument for mechanically measuring the depth of the sea beneath a ship.

depth-type filtration [CHEM ENG] Removal of solids by passing the carrier fluid through a mass-filter medium that provides a tortuous path with many entrapments to catch the solids.

depth zone of earth [GEOL] A zone within the earth giving rise to different metamorphic assemblages.

depurination [BIOCHEM] Detachment of guanine from sugar in a deoxyribonucleic acid molecule.

derail [ENG] **1.** To cause a railroad car or engine to run off the rails. **2.** A device to guide railway cars or engines off the tracks to avoid collision or other accident.

derating [ELECTR] The reduction of the rating of a device to improve reliability or to permit operation at high ambient temperatures.

derby [MET] A large, usually cylindrical piece of primary metal, whose weight may exceed 100 pounds (45 kilograms), formed by bomb reduction.

derbylite [MINERAL] $Fe_6Ti_6Sb_2O_{23}$ A black or brown orthorhombic mineral occurring in cinnabar-bearing gravels.

Derbyshire spar *See* fluorite.

derealization [PSYCH] Loss of the sense of the reality of people or objects in one's environment.

derelict [NAV] Any property abandoned at sea, often of sufficient size as to constitute a menace to navigation; especially an abandoned vessel.

derepression [MICROBIO] Transfer of microbial cells from an enzyme-repressing medium to a nonrepressing medium.

derichment [ANALY CHEM] In gravimetric analysis by coprecipitation of salts, a system with λ less than unity, when λ is the logarithmic distribution coefficient expressed by the ratio of the logarithms of the ratios of the initial and final solution concentrations of the two salts.

derivation [MATH] **1.** The process of deducing a formula. **2.**

A function D on an algebra which satisfies the equation $D(uv) = uD(v) + vD(u)$.

derivative [CHEM] A substance that is made from another substance. [MATH] The slope of a graph $y=f(x)$ at a given point c; more precisely, it is the limit as h approaches zero of $f(c+h) - f(c)$ divided by h. Also known as differential coefficient.

derivative action [CONT SYS] Control action in which the speed at which a correction is made depends on how fast the system error is increasing. Also known as derivative compensation; rate action.

derivative compensation See derivative action.

derivative differential thermal analysis [ANALY CHEM] A method for precise determination in thermograms of slight temperature changes by taking the first derivative of the differential thermal analysis curve (thermogram) which plots time versus differential temperature as measured by a differential thermocouple. Also known as DDTA.

derivative network [CONT SYS] A compensating network whose output is proportional to the sum of the input signal and its derivative. Also known as lead network.

derivative polarography [ANALY CHEM] Polarography technique in which the rate of change of current with respect to applied potential is measured as a function of the applied potential (di/dE vs. E, where i is current and E is applied potential).

derivative rock See sedimentary rock.

derivative thermometric titration [ANALY CHEM] The use of a special resistance-capacitance network to record first and second derivatives of a thermometric titration curve (temperature versus weight change upon heating) to produce a sharp end-point peak.

derived curve [MATH] A curve whose ordinate, for each value of the abscissa, is equal to the slope of some given curve. Also known as first derived curve.

derived gust velocity [METEOROL] The maximum velocity of a sharp-edged gust that would produce a given acceleration on a particular airplane flown in level flight at the design cruising speed of the aircraft at a given air density.

derived set [MATH] The set of cluster points of a given set.

Dermacentor [INV ZOO] A genus of ticks, important as vectors of disease.

Dermacentor andersoni [INV ZOO] The wood tick, which is the vector of Rocky Mountain spotted fever and tularemia.

Dermacentor variabilis [INV ZOO] A North American tick, parasitic primarily on dogs but may attack man and other mammals.

dermal [ANAT] Pertaining to the dermis.

dermal bone [ANAT] A type of bone that ossifies directly from membrane without a cartilaginous predecessor; occurs only in the skull and shoulder region. Also known as investing bone; membrane bone.

dermal denticle [VERT ZOO] A toothlike scale composed mostly of dentine with a large central pulp cavity, found in the skin of sharks.

dermalia [INV ZOO] Dermal microscleres in sponges.

dermal pore [INV ZOO] One of the minute openings on the surface of poriferans leading to the incurrent canals.

Dermaptera [INV ZOO] An order of small or medium-sized, slender insects having incomplete metamorphosis, chewing mouthparts, short forewings, and cerci.

Dermatemydinae [VERT ZOO] A family of reptiles in the order Chelonia; includes the river turtles.

dermatitis [MED] Inflammation of the skin.

Dermatocarpaceae [BOT] A family of lichens in the order Pyrenulales having an umbilicate or squamulose growth form; most members grow on limestone or calcareous soils.

dermatocranium [ANAT] Bony parts of the skull derived from ossifications in the dermis of the skin.

dermatogen [BOT] The outer layer of primary meristem or the primordial epidermis in embryonic plants. Also known as protoderm.

dermatoglyphics [ANAT] **1.** The integumentary patterns on the surface of the fingertips, palms, and soles. **2.** The study of these patterns.

dermatograph [MED] A crayonlike or similar instrument, used to mark the skin before surgery to outline positions of organs.

dermatologist [MED] A physician who specializes in diseases of the skin.

dermatology [MED] The science of the structure, function, and diseases of the skin.

dermatome [ANAT] An area of skin delimited by the supply of sensory fibers from a single spinal nerve. [EMBRYO] Lateral portion of an embryonic somite from which the dermis will develop. [MED] Instrument for cutting skin for grafting.

dermatomyositis [MED] An inflammatory reaction of unknown cause involving degenerative changes of skin and muscle. Also known as pseudotrichinosis.

dermatopathic lymphadenitis See lipomelanotic reticulosis.

dermatopathology [MED] A branch of pathology concerned with diseases of the skin.

dermatopathophobia [PSYCH] An abnormal fear of contracting a skin disease.

dermatophyte [MYCOL] A fungus parasitic on skin or its derivatives.

dermatophytosis [MED] A fungus infection, such as ringworm, of the skin of humans caused by the organism living in the keratinized tissues, characterized by vesicles, cracking, and scaling.

dermatosclerosis See scleroderma.

dermatosis [MED] Any skin disease.

Dermestidae [INV ZOO] The skin beetles, a family of coleopteran insects in the superfamily Dermestoidea, including serious pests of stored agricultural grain products.

Dermestoidea [INV ZOO] A superfamily of coleopteran insects in the suborder Polyphaga.

dermis [ANAT] The deep layer of the skin, a dense connective tissue richly supplied with blood vessels, nerves, and sensory organs. Also known as corium; cutis.

Dermitron [MET] An electrical instrument which applies a high-frequency current and the resulting eddy current series as a measure of thickness of nonmagnetic coatings on nonmagnetic basis metals, provided the conductivity of the coating and that of the basis metal are different.

Dermochelidae [VERT ZOO] A family of reptiles in the order Chelonia composed of a single species, the leatherback turtle.

dermoepidermal junction [HISTOL] The area of separation between the stratum basale of the epidermis and the papillary layer of the dermis.

dermographia [MED] A condition in which the skin is peculiarly susceptible to irritation, characterized by elevations or wheals with surrounding erythematous axon reflex flare, caused by tracing a fingernail or a blunt instrument over the skin.

dermoid cyst [MED] A benign cystic teratoma with skin, skin appendages, and their products as the most prominent components, usually involving the ovary or the skin.

Dermoptera [VERT ZOO] The flying lemurs, an ancient order of primatelike herbivorous and frugivorous gliding mammals confined to southeastern Asia and eastern India.

de Robertis unit [BIOL] A unit for the standardization of thyroid extracts.

Derodontidae [INV ZOO] The tooth-necked fungus beetles, a small family of coleopteran insects in the superfamily Dermestoidea.

derosination [CHEM ENG] Removing excess resins from wood by saponification with alkaline aqueous solutions or organic solvents.

derrick [MECH ENG] A hoisting machine consisting usually of a vertical mast, a slanted boom, and associated tackle; may be operated mechanically or by hand.

derrick crane See stiffleg derrick.

derrick post See king post.

derris [BOT] Any of certain tropical shrubs in the genus *Derris* in the legume family (Leguminosae), having long climbing branches.

Deryagin number [PHYS] A dimensionless group equal to the ratio of the thickness of a film coating a liquid to the capillary length of the liquid. Symbolized *De*.

desalination [CHEM ENG] Removal of salt, as from water or soil. Also known as desalting.

DERMESTIDAE

A drawing of a skin beetle. (*From T. I. Storer and R. L. Usinger, General Zoology, 3d ed., McGraw-Hill, 1957*)

desalting *See* desalination.

Desargues' theorem [MATH] If the three lines passing through corresponding vertices of two triangles are concurrent, then the intersections of the three pairs of corresponding sides lie on a straight line, and conversely.

DeSauty's bridge [ELEC] A four-arm bridge used to compare two capacitances; two adjacent arms contain capacitors in series with resistors, while the other two arms contain resistors only. Also known as Wien–De Sauty bridge.

descaling [ENG] Removing scale, usually oxides, from the surface of a metal or the inner surface of a pipe, boiler, or other object.

Descartes laws of refraction *See* Snell laws of refraction.

Descartes ray [OPTICS] A ray of light incident on a sphere of transparent material, such as a water droplet, which after one internal reflection leaves the drop at the smallest possible angle of deviation from the direction of the incident ray; these rays make the primary rainbow.

Descartes' rule of signs [MATH] A polynomial with real coefficients has at most k real positive roots, where k is the number of sign changes in the polynomial.

Descemet's membrane [HISTOL] A layer of the cornea between the posterior surface of the stroma and the anterior surface of the endothelium which contains collagen arranged on a crystalline lattice.

descending [ANAT] Extending or directed downward or caudally, as the descending aorta. [PHYSIO] In the nervous system, efferent; conducting impulses or progressing down the spinal cord or from central to peripheral.

descending branch [MECH] That portion of a trajectory which is between the summit and the point where the trajectory terminates, either by impact or air burst, and along which the projectile falls, with altitude constantly decreasing. Also known as descent trajectory.

descending chromatography [ANALY CHEM] A type of paper chromatography in which the sample-carrying solvent mixture is fed to the top of the developing chamber, being separated as it works downward.

descending node [AERO ENG] That point at which an earth satellite crosses to the south side of the equatorial plane of its primary. Also known as southbound node. [ASTRON] The point at which a planet, planetoid, or comet crosses the ecliptic from north to south.

descending vertical angle *See* angle of depression.

descent [AERO ENG] Motion of a craft in which the path is inclined with respect to the horizontal.

descent trajectory *See* descending branch.

describing function [CONT SYS] A function used to represent a nonlinear transfer function by an approximately equivalent linear transfer function; it is the ratio of the phasor representing the fundamental component of the output of the nonlinearity, determined by Fourier analysis, to the phasor representing a sinusoidal input signal.

descriptive anatomy [ANAT] Study of the separate and individual portions of the body, with regard to form, size, character, and position.

descriptive astronomy [ASTRON] Astronomy as presented by graphic and verbal description.

descriptive botany [BOT] The branch of botany that deals with diagnostic characters or systematic description of plants.

descriptive climatology [CLIMATOL] Climatology as presented by graphic and verbal description, without going into causes and theory.

descriptive geometry [MATH] The application of graphical methods to the solution of three-dimensional space problems.

descriptive meteorology [METEOROL] A branch of meteorology which deals with the description of the atmosphere as a whole and its various phenomena, without going into theory. Also known as aerography.

deseaming [MET] Removing defects from the surfaces of ingots, blooms, or semifinished products, usually by means of a chipping hammer or an oxy-gas flame.

desensitization [COMMUN] Reduction in receiver sensitivity due to the presence of a high-level off-channel signal overloading the radio-frequency amplifier or mixer stages, or causing automatic gain control action. [IMMUNOL] Loss or reduction of sensitivity to infection or an allergen accomplished by means of frequent, small doses of the antigen. Also known as hyposensitization. [PSYCH] Relief from or removal of a mental complex.

desert [GEOGR] **1.** A wide, open, comparatively barren tract of land with few forms of life and little rainfall. **2.** Any waste, uninhabited tract, such as the vast expanse of ice in Greenland.

desert climate [CLIMATOL] A climate type which is characterized by insufficient moisture to support appreciable plant life; that is, a climate of extreme aridity.

desert crust *See* desert pavement.

desert devil *See* dust whirl.

deserticolous [ECOL] Living in a desert.

desert pavement [GEOL] A mosaic of pebbles and large stones which accumulate as the finer dust and sand particles are blown away by the wind. Also known as desert crust.

desert peneplain *See* pediplain.

desert plain *See* pediplain.

desert polish [GEOL] A smooth, shining surface imparted to rocks and other hard substances by the action of windblown sand and dust of desert regions.

desert soil [GEOL] In early United States classification systems, a group of zonal soils that have a light-colored surface soil underlain by calcareous material and a hardpan.

desert varnish [GEOL] A brown or black stain or crust of manganese or iron oxide characterizing many exposed rock surfaces in the desert.

desert wind [METEOROL] A wind blowing off the desert, which is very dry and usually dusty, hot in summer but cold in winter, and with a large diurnal range of temperature.

desexualization [PHYSIO] Depriving an organism of sexual characters or power, as by spaying or castration. [PSYCH] Repression of the sexual drive or rechanneling of sexual energy into areas considered by the individual to be more socially acceptable.

desiccant *See* drying agent.

desiccation [HYD] The permanent decrease or disappearance of water from a region, caused by a decrease of rainfall, a failure to maintain irrigation, or deforestation or overcropping. [SCI TECH] Thorough removal of water from a substance, often with the use of a desiccant.

desiccation breccia [GEOL] Fragments of a mud-cracked layer of sediment deposited with other sediments.

desiccation crack [GEOL] A crack formed by shrinkage of clay or clayey beds under the influence of heat from the sun.

desiccator [CHEM ENG] A closed vessel, usually made of glass and having an airtight lid, used for drying solid chemicals by means of a desiccant.

design [SCI TECH] The act of conceiving and planning the structure and parameter values of a system, device, process, or work of art.

designation hole *See* designation punch.

designation punch [ADP] A hole in a punched card indicating what is the nature of the data on the card or which functions are to be performed by the computer. Also known as control hole; control punch; designation hole; function hole.

design engineering [ENG] A branch of engineering concerned with the creation of systems, devices, and processes useful to and sought by society.

design flood [CIV ENG] The flood, either observed or synthetic, which is chosen as the basis for the design of a hydraulic structure.

design gross weight [AERO ENG] The gross weight at takeoff that an aircraft, rocket, or such is expected to have, used in design calculations.

design head [CIV ENG] The planned elevation between the free level of a water supply and the point of free discharge or the level of free discharge surface.

design load [DES ENG] The most stressful combination of weight or other forces a building, structure, or mechanical system or device is designed to sustain.

design of experiments [STAT] A pattern for setting up experiments and making observations about the relationship between several variables in which one attempts to obtain as much information as possible for a fixed expenditure level.

design pressure [CIV ENG] **1.** The force exerted by a body of still water on a dam. **2.** The pressure which the dam can withstand.

design speed [CIV ENG] The highest continuous safe vehicular speed as governed by the design features of a highway.

design standards [DES ENG] Generally accepted uniform procedures, dimensions, materials, or parts that directly affect the design of a product or facility.

design storm [CIV ENG] A storm whose magnitude, rate, and intensity do not exceed the design load for a storm drainage system or flood protection project.

design stress [DES ENG] A permissible maximum stress to which a machine part or structural member may be subjected, which is large enough to prevent failure in case the loads exceed expected values, or other uncertainties turn out unfavorably.

design waterline [NAV ARCH] The waterline on a ship when it is floating freely at rest in still water in its normally loaded condition. Abbreviated DWL.

desilication [GEOCHEM] Removal of silica, as from rock or a magma.

desilker [FOOD ENG] A machine consisting of a series of revolving rolls and brushes for removing silk from ears of corn.

desilter [MECH ENG] Wet, mechanical solids classifier (separator) in which silt particles settle as the carrier liquid is slowly stirred by horizontally revolving rakes; solids are plowed outward and removed at the periphery of the container bowl.

desilting basin [CIV ENG] A space or structure constructed just below a diversion structure of a canal to remove bed, sand, and silt loads. Also known as desilting works.

desilting works *See* desilting basin.

desilverization [MET] The act or process of removing silver; specifically, the process used to remove silver and gold from ore after softening.

desired ground zero [ORD] For a surface burst, the point on the earth's surface where atomic detonation is desired; for an air burst or underground burst, the point is on the earth's surface directly below or directly above the desired point of detonation.

desired track *See* course.

de Sitter space [RELAT] The four-dimensional surface of a sphere in five-dimensional space, used as a model of the universe.

desize [TEXT] To remove size or sizing agents from warp yarns prior to weaving to protect them against the abrasive action of loom parts.

desk calculator [ADP] A device that is used to perform arithmetic operations and is small enough to be conveniently placed on a desk.

deslimer [MECH ENG] Apparatus, such as a bowl-type centrifuge, used to remove fine, wet particles (slime) from cement rocks and slurry to use as size pigments and abrasives.

desma [INV ZOO] A branched, knobby spicule in some Demospongiae.

desmacyte [INV ZOO] A bipolar collencyte found in the cortex of certain sponges.

Desmidiaceae [BOT] A family of desmids, mostly unicellular algae in the order Conjugales.

desmine *See* stilbite.

desmochore [ECOL] A plant having sticky or barbed disseminules.

Desmodonta [PALEON] An order of extinct bivalve, burrowing mollusks.

Desmodontidae [VERT ZOO] A small family of chiropteran mammals comprising the true vampire bats.

Desmodoroidea [INV ZOO] A superfamily of marine- and brackish-water-inhabiting nematodes with an annulated, usually smooth cuticle.

Des Moinesian [GEOL] Lower Middle Pennsylvanian geologic time.

Desmokontae [BOT] The equivalent name for Desmophyceae.

desmolase [BIOCHEM] Any of a group of enzymes which catalyze rupture of atomic linkages that are not cleaved through hydrolysis, such as the bonds in the carbon chain of D-glucose.

desmoneme [INV ZOO] A nematocyst having a long coiled tube which is extruded and wrapped around the prey.

desmopelmous [VERT ZOO] A type of bird foot in which the hindtoe cannot be bent independently because planter tendons are united.

Desmophyceae [BOT] A class of rare, mostly marine algae in the division Pyrrhophyta.

desmoplasia [MED] **1.** The formation and proliferation of connective tissue, frequently in the growth of tumors. **2.** The formation of adhesions.

Desmoscolecida [INV ZOO] An order of the class Nematoda.

Desmoscolecidae [INV ZOO] A family of nematodes in the superfamily Desmoscolecoidea; individuals resemble annelids in having coarse annulation.

Desmoscolecoidea [INV ZOO] A small superfamily of free-living nematodes characterized by a ringed body, an armored head set, and hemispherical amphids.

desmose [INV ZOO] A fibril connecting the centrioles during mitosis in certain protozoans.

desmosome [CYTOL] A surface modification of stratified squamous epithelium formed by the apposition of short cytoplasmic processes of neighboring cells.

Desmostylia [PALEON] An extinct order of large hippopotamuslike, amphibious, gravigrade, shellfish-eating mammals.

Desmostylidae [PALEON] A family of extinct mammals in the order Desmostylia.

Desmothoracida [INV ZOO] An order of sessile and free-living protozoans in the subclass Heliozoia having a spherical body with a perforate, chitinous test.

desorption [PHYS CHEM] The process of removing a sorbed substance by the reverse of adsorption or absorption.

Desor's larva [INV ZOO] An oval, ciliated larva of certain nemertineans in which the gastrula remains inside the egg membrane.

despun antenna [ELECTROMAG] Satellite directional antenna pointed continuously at earth by electrically or mechanically despinning the antenna at the same rate that the satellite is spinning for stabilization.

desquamation [PHYSIO] Shedding; a peeling and casting off, as of the superficial epithelium, mucous membranes, renal tubules, and the skin.

destearinate [CHEM ENG] A process of removing from a fatty oil the lower melting point compounds.

destination address [ADP] The location to which a jump instruction passes control in a program.

destination time [ADP] The time involved in a memory access plus the time required for indirect addressing.

destressing [MIN ENG] Relieving stress on the abutments of an excavation by drilling and blasting to loosen peak stress zones.

Destriau effect [SOLID STATE] Sustained emission of light by suitable phosphor powders that are embedded in an insulator and subjected only to the action of an alternating electric field.

destroyed [ORD] Of aircraft, material, installations, or the like, ruined beyond repair and unfit for further military use.

destroyer [NAV ARCH] A small, fast, lightly armored warship capable of a variety of functions, usually armed with 5-inch guns, torpedoes, depth charges, and mines, and sometimes with guided missiles.

destruct [AERO ENG] The deliberate action of destroying a rocket vehicle after it has been launched, but before it has completed its course.

destruction operator *See* annihilation operator.

destructive distillation [ORG CHEM] Decomposition of organic compounds by heat without the presence of air.

destructive interference [OPTICS] The interaction of superimposed light from two different sources when the phase relationship is such as to reduce or cancel the resultant intensity to less than the sum of the individual lights.

destructive read [ADP] Reading that partially or completely erases the stored information as it is being read.

destructive testing [ENG] Intentional operation of equipment until it fails, to reveal design weaknesses.

destruct line [AERO ENG] On a rocket test range, a boundary line on each side of the down-range course beyond which a rocket cannot fly without being destroyed under destruct

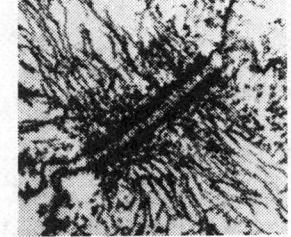

DESMOSOME

Electron micrograph of desmosome in skin of newt. *(Courtesy of D. Kelly)*

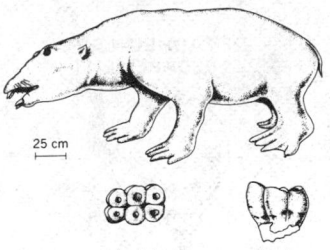

DESMOSTYLIA

25 cm

occlusal view   side view

Restoration of *Desmostylus* from the Miocene of California. *(Redrawn from R. A. Stirton, 1959)*

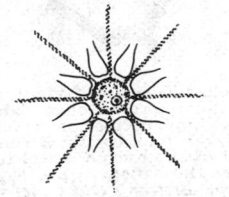

DESMOTHORACIDA

Choanocystis lepidula, a floating species. *(From R. P. Hall, Protozoology, Prentice-Hall, 1953)*

procedures; or a line beyond which the impact point cannot pass.

destructor [ORD] An explosive device used intentionally by the control center to destroy a missile after launching. Also known as destruct system.

destruct system [ORD] **1.** An explosive device enabling destruction of materials, weapons, or other objects to prevent acquisition by the enemy. **2.** *See* destructor.

desulfonate [CHEM] Removal from a sulfonated molecule of the sulfonate group.

Desulfovibrio [MICROBIO] A genus of strictly anaerobic, motile, rod-shaped bacteria in the family Spirillaceae which reduce sulfates to hydrogen sulfide.

desulfurization [CHEM ENG] The removal of sulfur, as from molten metals or petroleum oil.

desulfurization unit [CHEM ENG] A unit in petroleum refining for removal of sulfur compounds or sulfur.

desyl [ORG CHEM] The radical $C_6H_5COCH(C_6H_5) - $; may be formed from desoxybenzoin. Also known as α-phenyl phenacyl.

detachable bit [ENG] An all-steel drill bit that can be removed from the drill steel, and can be resharpened. Also known as knock-off bit; rip bit.

detached core [GEOL] The inner bed or beds of a fold that may become separated or pinched off from the main body of the strata due to extreme folding and compression.

detached-lever escapement [HOROL] A watch escapement whose regulating device is given an impulse during only a small part of its operating cycle.

detached meristem [BOT] A meristematic region originating from apical meristem but becoming discontinuous with it because of differentiation of intervening tissue.

detached shock wave [FL MECH] A shock wave not in contact with the body which originates it.

detail [GRAPHICS] The extent to which image elements that are close together can be individually distinguished.

detail chart [ADP] A flow chart representing every single step of a program.

detail drawing [GRAPHICS] A large-scale drawing of a small part of a structure or machine.

detailed balance [STAT MECH] The hypothesis that when a system is in equilibrium any process occurs with the same frequency as the reverse process.

detail file [ADP] A file containing current or transient data used to update a master file or processed with the master file to obtain a specific result. Also known as transaction file.

detailing *See* screening.

detail printing [ADP] The printing of information for each card as the card passes through the machine; the function is used to prepare reports that show complete detail about each card; during this listing operation, the machine adds, subtracts, cross-adds, or cross-subtracts, and prints many combinations of totals.

det drill *See* fusion-piercing drill.

detect *See* demodulate.

Detectophone [ENG ACOUS] An audio system used to listen to conversations secretly, in which a sensitive microphone is concealed in a room and connected to an audio amplifier feeding headphones or a magnetic tape recorder, or connected to a radio transmitter.

detector [ELECTR] The stage in a receiver at which demodulation takes place; in a superheterodyne receiver this is called the second detector. Also known as demodulator; envelope detector. [SCI TECH] Apparatus or system used to detect the presence of an object, radiation, chemical compound, or such.

detector balanced bias [ELECTR] Controlling circuit used in radar systems for anticlutter purposes.

detector bar [CIV ENG] A device that keeps a railroad switch locked while a train is passing over it.

detector car [ENG] A railroad car used to detect flaws in rails.

detent [MECH ENG] A catch or lever in a mechanism which initiates or locks movement of a part, especially in escapement mechanisms.

detention basin [CIV ENG] A reservoir without control gates for storing water over brief periods of time until the stream

has the capacity for ordinary flow plus released water; used for flood regulation.

detergent [MATER] A synthetic cleansing agent resembling soap in the ability to emulsify oil and hold dirt, and containing surfactants which do not precipitate in hard water; may also contain protease enzymes and whitening agents.

detergent additive [MATER] A substance incorporated in lubricating oils which gives them the property of keeping insoluble material in suspension.

detergent alkylate *See* dodecylbenzene.

detergent oil [MATER] A lubricating oil with special sludge-dispersing properties for use in internal combustion engines.

deteriorating supplies [ORD] Those items that may reasonably be expected to become unusable within 1 or 2 years, whether used or not.

deterioration [ENG] Decline in the quality of equipment or structures over a period of time due to the chemical or physical action of the environment.

determinant [CONT SYS] The product of the partial return differences associated with the nodes of a signal-flow graph. [MATH] A certain real-valued function of the column vectors of a square matrix which is zero if and only if the matrix is singular; used to solve systems of linear equations and to study linear transformations.

determinant tensor [MATH] A tensor whose components are each equal to the corresponding component of the Levi-Civita tensor density times the square root of the determinant of the metric tensor, and whose contravariant components are each equal to the corresponding component of the Levi-Civita density divided by the square root of the metric tensor. Also known as permutation tensor.

determinate [SCI TECH] Bounded by definite limits.

determinate cleavage [EMBRYO] A type of cleavage which separates portions of the zygote with specific and distinct potencies for development as specific parts of the body.

determinate growth [BOT] Growth in which the axis, or central stem, being limited by the development of the floral reproductive structure, does not grow or lengthen indefinitely.

determinate structure [MECH] A structure in which the equations of statics alone are sufficient to determine the stresses and reactions.

determination [ANALY CHEM] The finding of the value of a chemical or physical property of a compound, such as reaction-rate determination or specific-gravity determination.

determinism *See* causality.

detonating agent [MATER] An explosive, such as PETN, contained in the blasting cap or detonator.

detonating fuse [ENG] A device consisting of a core of high explosive within a waterproof textile covering and set off by an electrical blasting cap fired from a distance by means of a fuse line; used in large, deep boreholes.

detonating net [ORD] Network of detonating cord that is interlaced in a mesh design; used for clearing paths through mine fields by exploding the mines over which the nets are placed and detonated.

detonating primer [MATER] A primer used to fire high explosives that is exploded by a fuse.

detonating rate [MECH] The velocity at which the explosion wave passes through a cylindrical charge.

detonating relay [ENG] A device used in conjunction with the detonating fuse to avoid short-delay blasting.

detonation [CHEM] An exothermic chemical reaction that propagates with such rapidity that the rate of advance of the reaction zone into the unreacted material exceeds the velocity of sound in the unreacted material; that is, the advancing reaction zone is preceded by a shock wave. [MECH ENG] Spontaneous combustion of the compressed charge after passage of the spark in an internal combustion engine; it is accompanied by knock.

detonation front [ENG] The reaction zone of a detonation.

detonation wave [FL MECH] A shock wave that accompanies detonation and has a shock front followed by a region of decreasing pressure in which the reaction occurs.

detonator [ENG] A device, such as a blasting cap, employing

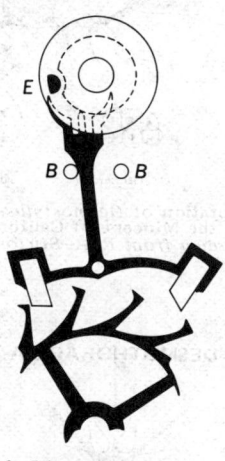

DETACHED-LEVER ESCAPEMENT

Detached-lever escapement, often used in modern watches. Banking pins *B* limit oscillation of anchor and lever; impulse pin *E* causes recoil of the escape wheel to release the pallet. (*From F. J. Britten, Britten's Old Clocks and Watches and Their Makers, 7th ed., Dutton, 1956*)

a sensitive primary explosive to detonate a high-explosive charge.

detonator safety [ENG] A fuse has detonator safety or is detonator safe when the functioning of the detonator cannot initiate subsequent explosive train components.

detonics [ENG] The study of detonating and explosives performance.

detorsion [INV ZOO] Untwisting of the 180° visceral twist imposed by embryonic torsion on many gastropod mollusks. [MED] Untwisting of an abnormal torsion, as of a ureter or intestine.

detoxification [BIOCHEM] The act or process of removing a poison or the toxic properties of a substance in the body.

detrainment [METEOROL] The transfer of air from an organized air current to the surrounding atmosphere.

detrital fan *See* alluvial fan.

detrital minerals [MINERAL] Grains of heavy minerals found in sediment, resulting from mechanical disintegration of the parent rock.

detrital reservoir [GEOL] A clastic or detrital-granular reservoir, classified by rock type and other factors such as sediments (quartzose-type, graywacke, or arkose sediments).

detrital sediment [GEOL] Accumulations of the organic and inorganic fragmental products of the weathering and erosion of land transported to the place of deposition.

detritus [GEOL] Any loose material removed directly from rocks and minerals by mechanical means, such as disintegration or abrasion.

detritus feeder *See* deposit feeder.

detritus tank [CIV ENG] A tank in which heavy suspended matter is removed in sewage treatment.

Detroit rocking furnace [ENG] An indirect arc type of rocking furnace having graphite electrodes entering horizontally from opposite ends.

detune [ELECTR] To change the inductance or capacitance of a tuned circuit so its resonant frequency is different from the incoming signal frequency.

detuning stub [ELECTROMAG] Quarter-wave stub used to match a coaxial line to a sleeve-stub antenna; the stub detunes the outside of the coaxial feed line while tuning the antenna itself.

deutencephalon *See* epichordal brain.

deuteranomaly [MED] A partial deuteranopia.

deuteranopia [MED] Defective vision consisting of red-green color confusion, with no marked reduction in the brightness of any color.

deuteration [CHEM] The addition of deuterium to a chemical compound.

deuteric [GEOL] Of or pertaining to alterations in igneous rock during the later stages and as a direct result of consolidation of magma or lava. Also known as epimagmatic; paulopost.

deuteride [CHEM] A hydride in which the hydrogen is deuterium.

deuterium [CHEM] The isotope of the element hydrogen with one neutron and one proton in the nucleus; atomic weight 2.0144. Designated D, d, H², or ²H.

deuterium cycle *See* proton-proton chain.

deuterium discharge tube [ELECTR] A tube similar to a hydrogen discharge lamp, but with deuterium replacing the hydrogen; source of high-intensity ultraviolet radiation for spectroscopic microanalysis.

deuterium oxide *See* heavy water.

deuterogamy [BOT] Secondary pairing of sexual cells or nuclei replacing direct copulation in many fungi, algae, and higher plants.

Deuteromycetes [MYCOL] The equivalent name for Fungi Imperfecti.

deuteron [NUC PHYS] The nucleus of a deuterium atom, consisting of a neutron and a proton. Designated d. Also known as deuton.

deuteron accelerator [NUCLEO] An accelerator that produces a flux of slow neutrons by bombarding a metal-tritium target with deuterons.

deuteron capture [NUC PHYS] The absorption of a deuteron by a nucleus, giving rise to a compound nucleus which subsequently decays.

Deuterophlebiidae [INV ZOO] A family of dipteran insects in the suborder Cyclorrhapha.

Deuterostomia [ZOO] A division of the animal kingdom which includes the phyla Echinodermata, Chaetognatha, Hemichordata, and Chordata.

deutocerebrum [INV ZOO] The median lobes of the insect brain.

deuton *See* deuteron.

deutoplasm [EMBRYO] The nutritive yolk granules in egg cells.

DeVecchis process [MIN ENG] A smelting process for pyrites in which the raw material is roasted, concentrated magnetically, and then reduced in a rotary kiln or electric furnace.

developed blank [MET] A blank requiring little or no trimming after being formed.

developed dye [CHEM] A direct azo dye that can be further diazotized by a developer after application to the fiber; it couples with the fiber to form colorfast shades. Also known as diazo dye.

developed muzzle velocity [ORD] The actual muzzle velocity produced by any gun.

developed ore *See* developed reserves.

developed planform [AERO ENG] The plan of an airfoil as drawn with the chord lines at each section rotated about the airfoil axis into a plane parallel to the plane of projection and with the airfoil axis rotated or developed and projected into the plane of projection.

developed reserves [MIN ENG] Ore that is exposed on three sides and for which tonnage yield and quality estimates have been made. Also known as assured mineral; blocked-out ore; developed ore; measured ore; ore in sight.

developer [CHEM] An organic compound which interacts on a textile fiber to develop a dye. [GRAPHICS] A chemical solution used to develop exposed photographic materials by reducing silver salts to metallic silver.

developing [GRAPHICS] The complete process of producing an image on sensitized material by means of chemical agents.

development [GEOL] The progression of changes in fossil groups which have succeeded one another during deposition of the strata of the earth. [METEOROL] The process of intensification of an atmospheric disturbance, most commonly applied to cyclones and anticyclones. [MIN ENG] Opening of a coal seam or ore body by sinking shafts or driving levels, as well as installing equipment, for proving ore reserves and exploiting them. [SCI TECH] The work required to determine the best production techniques to bring a new process or piece of equipment to the production stage.

development age [PSYCH] An index of physical, mental, social, or emotional development stated in age equivalent as determined by standardized methods.

developmental alexia [PSYCH] A deficiency in learning to read in the absence of physical defects or mental deficiency.

developmental aphasia [PSYCH] A deficiency in learning to speak, in contrast with the child's mental or development age.

developmental crisis [PSYCH] A period of childhood stress related to unsuccessful attempts to establish trust, identity, autonomy, or initiative.

developmental psychology [PSYCH] The branch of psychology that deals with changes in behavior occurring with changes in age.

development drift [MIN ENG] A tunnel dug in a mine either from the surface or a point underground to get to coal or ore for exploitation or mining purposes.

development drilling [MIN ENG] Drilling boreholes to locate, identify, and prove an ore body or coal seam.

development index [METEOROL] An index used as an aid in forecasting cyclogenesis; the development index I is defined most frequently as the difference in divergence between two well-separated, tropospheric, constant-pressure surfaces. Also known as relative divergence.

development rock [MIN ENG] Rock containing both barren and valuable rock, broken during development work.

Devereaux agitator [MIN ENG] An agitator that utilizes an upthrust propeller to stir pulp; used in leach agitation of minerals.

devernalization [BOT] Annulment of the vernalization effect.

deviation [ENG] The difference between the actual value of a

DEVONIAN

PRECAMBRIAN		
CAMBRIAN		
ORDOVICIAN		
SILURIAN		PALEOZOIC
DEVONIAN		
Mississippian	CARBON-IFEROUS	
Pennsylvanian		
PERMIAN		
TRIASSIC		
JURASSIC		MESOZOIC
CRETACEOUS		
TERTIARY		CENOZOIC
QUATERNARY		

Chart showing relationship of Devonian to other periods.

DEWAR FLASK

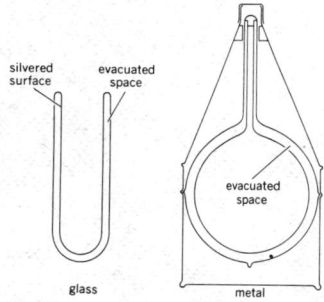

Typical Dewar containers.

DEW-POINT HYGROMETER

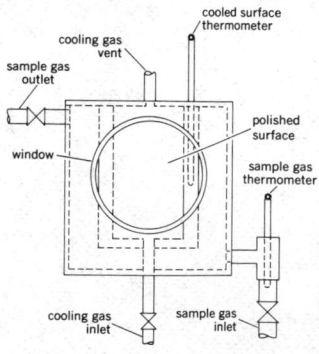

Dew-point type of hygrometer.

controlled variable and the desired value corresponding to the set point. [EVOL] Evolutionary differentiation involving interpolation of new stages in the ancestral pattern of morphogenesis. [STAT] The difference between any given number in a set and the mean average of those numbers.

deviation absorption [COMMUN] Distortion in a frequency-modulated receiver due to inadequate bandwidth, inadequate amplitude-modulation rejection, or inadequate discriminator linearity.

deviation card [NAV] A card having a table of the deviation of an aircraft magnetic compass on various headings; this card is usually mounted on the compass.

deviation factor *See* compressibility factor.

deviation hole [PETRO ENG] Drilled hole with deviation from true vertical, usually limited by contract to 3–5°; not to be confused with a crooked hole, resulting from carelessness or a steeply dipping formation.

deviation ratio [COMMUN] Ratio of the maximum frequency deviation to the maximum modulating frequency of a frequency-modulated system under specified conditions.

deviation sensitivity [NAV] A value expressed as the ratio of the rate of change in course indication to the deviation from the course line.

deviation table [NAV] A table of the deviation of a magnetic compass on various headings, magnetic or compass; for an aircraft compass, this information is usually placed on a card called a deviation card. Also known as magnetic compass table.

device [ADP] A general-purpose term used, often indiscriminately, to refer to a computer component or the computer itself. [ELECTR] An electronic element that cannot be divided without destroying its stated function; commonly applied to active elements such as transistors and transducers. [ENG] A mechanism, tool, or other piece of equipment designed for specific uses.

device address [ADP] The binary code which corresponds to a unique device, referred to when selecting this specific device.

device assignment [ADP] The use of a logical device number used in conjunction with an input/output instruction, and made to refer to a specific device.

device driver [ADP] A subroutine which handles a complete input/output operation.

device-end condition [ADP] The completion of an input/output operation, such as the transfer of a complete data block, recognized by the hardware in the absence of a byte count.

device flag [ADP] A flip-flop output which indicates the ready status of an input/output device.

device number [ADP] The physical or logical number which refers to a specific input/output device.

device selector [ADP] A circuit which gates data-transfer or command pulses to a specific input/output device.

devillite [MINERAL] $Cu_4Ca(SO_4)_2(OH)_6 3H_2O$ A dark-green mineral consisting of a hydrous basic sulfate of copper and calcium, occurring in six-sided platy crystals.

devitrification [CHEM] The process by which the glassy texture of a material is converted into a crystalline texture.

devitrified glass [MATER] A glassy material which has been changed from a vitreous to a brittle crystalline state during manufacture.

Devonian [GEOL] The fourth period of the Paleozoic Era.

dew [HYD] Water condensed onto grass and other objects near the ground, the temperatures of which have fallen below the dew point of the surface air because of radiational cooling during the night but are still above freezing.

Dewar flask [PHYS] A vessel having double walls, the space between being evacuated to prevent the transfer of heat and the surfaces facing the vacuum being heat-reflective; used to hold liquid gases and to study low-temperature phenomena.

dewaterer [MECH ENG] Wet-type mechanical classifier (solids separator) in which solids settle out of the carrier liquid and are concentrated for recovery.

dewatering [ENG] 1. Removal of water from solid material by wet classification, centrifugation, filtration, or similar solid-liquid separation techniques. 2. Removing or draining water from an enclosure or a structure, such as a riverbed, caisson, or mine shaft, by pumping or evaporation.

dewaxed oil [MATER] Lubricating oil that has had a portion of the wax removed.

dewaxing [CHEM ENG] Removing wax from a material or object; a process used to separate solid hydrocarbons from petroleum.

dew cell [ENG] An instrument used to determine the dew point, consisting of a pair of spaced, bare electrical wires wound spirally around an insulator and covered with a wicking wetted with a water solution containing an excess of lithium chloride; an electrical potential applied to the wires causes a flow of current through the lithium chloride solution, which raises the temperature of the solution until its vapor pressure is in equilibrium with that of the ambient air.

dewclaw [VERT ZOO] 1. A vestigial digit on the foot of a mammal which does not reach the ground. 2. A claw or hoof terminating such a digit.

deweylite [MINERAL] A mixture of clinochrysotile and stevensite. Also known as gymnite.

dewindtite [MINERAL] $Pb(UO_2)_2(PO_4)_2 3H_2O$ A canary-yellow secondary mineral consisting of a hydrous phosphate of lead and uranium.

de Witte relation [GEOPHYS] Graphical plot of the relation between electrical conductivity and distance over which the conductivity is measured through reservoir rock with clay minerals, (the effect is similar to two parallel electrical circuits), the current passing through the conducting clay minerals and the water-filled pores.

dewlap [ANAT] A fleshy or fatty fold of skin on the throat of some humans. [BOT] One of a pair of hinges at the joint of a sugarcane leaf blade. [VERT ZOO] A fold of skin hanging from the neck of some reptiles and bovines.

dew point [CHEM] The temperature at which water vapor begins to condense. [METEOROL] The temperature at which air becomes saturated when cooled without addition of moisture or change of pressure; any further cooling causes condensation. Also known as dew-point temperature.

dew-point boundary [CHEM ENG] On a phase diagram for a gas-condensate reservoir (pressure versus temperature with constant gas-oil ratios), the area along which the gas-oil ratio approaches zero.

dew-point composition [CHEM ENG] The water vapor–air composition at saturation, that is, at the temperature at which water exerts a vapor pressure equal to the partial pressure of water vapor in the air-water mixture.

dew-point curve [CHEM ENG] On a PVT phase diagram, the line that separates the two-phase (gas-liquid) region from the one-phase (gas) region, and indicates the point at a given gas temperature or pressure at which the first dew or liquid phase occurs.

dew-point depression [CHEM ENG] Reduction of the liquid-vapor dew point of a gas by removal of a portion of the liquid (such as water) from the gas (such as air). [METEOROL] The number of degrees the dew point is found to be lower than the temperature.

dew-point formula [METEOROL] A formula for the calculation of the approximate height of the lifting condensation level; employed to estimate the height of the base of convective clouds, under suitable atmospheric and topographic conditions.

dew-point hygrometer [CHEM ENG] An instrument for determining the dew point by measuring the temperature at which vapor being cooled in a silver vessel begins to condense. Also known as cold-spot hygrometer.

dew-point pressure [CHEM ENG] The gas pressure at which a system is at its dew point, that is, the conditions of gas temperature and pressure at which the first dew or liquid phase occurs.

dew-point recorder [ENG] An instrument which gives a continuous recording of the dew point; it alternately cools and heats the target and uses a photocell to observe and record the temperature at which the condensate appears and disappears. Also known as mechanized dew-point meter.

dew-point reservoir [PETRO ENG] A hydrocarbon reservoir in which the temperature lies between the critical temperature and the cricondentherm (maximum temperature and pressure at which two phases can coexist) and in the one-phase region. Also known as retrograde gas-condensate reservoir.

dew-point spread [METEOROL] The difference in degrees between the air temperature and the dew point.

dew-point temperature *See* dew point.

dex *See* brig.

Dexaminidae [INV ZOO] A family of amphipod crustaceans in the suborder Gammeridea.

dexterotropic [BIOL] Turning toward the right; applied to cleavage, shell formation, and whorl patterns.

dextral drag fold [GEOL] A drag fold in which the trace of a given surface bed is displaced to the right.

dextral fault [GEOL] A strike-slip fault in which an observer approaching the fault sees the opposite block as having moved to the right. Also known as right-lateral fault; right-lateral slip fault; right-slip fault.

dextral fold [GEOL] An asymmetric fold in which the long limb appears to be offset to the right to an observer looking along the long limb.

dextran [BIOCHEM] Any of the several polysaccharides, $(C_5H_{10}O_5)_n$, that yield glucose units on hydrolysis.

dextrin [BIOCHEM] A polymer of D-glucose which is intermediate in complexity between starch and maltose.

dextrinization [ORG CHEM] Any process that involves dextrinizing.

dextrinize [ORG CHEM] To convert a starch into dextrins.

dextro *See* dextrorotatory.

dextrocardia [MED] The presence of the heart in the right hemithorax, with the cardiac apex directed to the right.

dextrophobia [PSYCH] An abnormal fear of objects to the right of the body.

dextropimeric acid [ORG CHEM] $C_{19}H_{29}COOH$ A compound found in particular in oleoresins of pine trees.

dextrorotatory [OPTICS] Rotating clockwise the plane of polarization of a wave traveling through a medium in a clockwise direction, as seen by an eye observing the light. Abbreviated dextro.

dextrorse [BOT] Twining toward the right.

dextrose [BIOCHEM] $C_6H_{12}O_6 \cdot H_2O$ A dextrorotatory monosaccharide obtained as a white, crystalline, odorless, sweet powder, which is soluble in about one part of water; an important intermediate in carbohydrate metabolism; used for nutritional purposes, for temporary increase of blood volume, and as a diuretic. Also known as corn sugar; grape sugar.

dextrotopic cleavage [EMBRYO] A clockwise spiral cleavage pattern.

dezincification [CHEM] Removal of zinc. [MET] Corrosion of brass in which both components of the alloy are dissolved and the copper is redeposited as a porous surface residue.

DG synchro amplifier [ELECTR] Synchro differential generator driven by servosystem.

d.h. *See* decision height.

DHA *See* dehydroacetic acid.

D horizon [GEOL] A soil horizon sometimes occurring below a B or C horizon, consisting of unweathered rock.

di- [SCI TECH] Prefix meaning two.

Di *See* didymium.

diabantite [MINERAL] $(Mg,Fe^{2+},Al)_6(Si,Al)_4O_{10}(OH)_8$ Mineral of the chlorite group consisting of a basic silicate of magnesium, iron, and aluminum, occurring in cavities in basic igneous rock.

diabase [PETR] An intrusive rock consisting principally of labradorite and pyroxene.

diabase amphibolite [PETR] Amphibolite formed by dynamic metamorphism of diabase.

diabasic [PETR] Denoting igneous rock in which the interstices between the feldspar crystals are filled with discrete crystals or grains of pyroxene.

diabatic [THERMO] Involving a thermodynamic change of state of a system in which there is a transfer of heat across the boundaries of the system.

diabetes [MED] Any of various abnormal conditions characterized by excessive urinary output, thirst, and hunger; usually refers to diabetes mellitus.

diabetes insipidus [MED] A form of diabetes due to a disfunction of the hypothalamus.

diabetes mellitus [MED] A metabolic disorder arising from a defect in carbohydrate utilization by the body, related to inadequate or abnormal insulin production by the pancreas.

diabetic acidosis [MED] Metabolic acidosis seen in diabetes mellitus, due to an excess of ketone bodies.

diabetic gangrene [MED] A moist form of gangrene occurring in persons with diabetes mellitus, often following a minor injury.

diabetic glomerulosclerosis *See* intercapillary glomerulosclerosis.

diabetophobia [PSYCH] An abnormal fear of becoming diabetic.

diablastic [PETR] Pertaining to a texture in metamorphic rock that consists of intergrown and interpenetrating rod-shaped components.

diaboleite [MINERAL] $Pb_2CuCl_2(OH)_4$ A sky-blue mineral consisting of a basic chloride of lead and copper.

Diac [ELECTR] Two-lead alternating-current switch semiconductor.

diacetate [ORG CHEM] An ester or salt that contains two acetate groups.

diacetic acid *See* acetoacetic acid.

diacetic ether *See* ethyl acetoacetate.

diacetin [ORG CHEM] $C_3H_5(OH)(CH_3COO)_2$ A colorless, hygroscopic liquid that is soluble in water, alcohol, ether, and benzene; boiling point 259°C; used as a plasticizer and softening agent and as a solvent. Also known as glyceryl diacetate.

diacetone alcohol [ORG CHEM] $CH_3COCH_2C(CH_3)_2OH$ A colorless liquid used as a solvent for nitrocellulose and resins.

diacetyl- [CHEM] **1.** Prefix indicating two acetyl radicals. **2.** *See* biacetyl-.

diacetylurea [ORG CHEM] $C_5H_8O_3N_2$ An acyl derivative of urea containing two acetyl groups.

diachronous [GEOL] Of a rock unit, varying in age in different areas or cutting across time planes or biostratigraphic zones. Also known as time-transgressive.

diacid [CHEM] An acid that has two acidic hydrogen atoms; an example is oxalic acid.

diaclinal [GEOL] Pertaining to a stream crossing a fold, perpendicular to the strike of the underlying strata it traverses.

Diacodectidae [PALEON] A family of extinct artiodactyl mammals in the suborder Palaeodonta.

diactine [INV ZOO] A type of sponge spicule which develops in two directions from a central point.

diactor [ELEC] Direct-acting automatic regulator for control of shunt generator voltage output.

diadelphous stamen [BOT] A stamen that has its filaments united into two sets.

Diadematacea [INV ZOO] A superorder of Euchinoidea having a rigid or flexible test, perforate tubercles, and branchial slits.

Diadematidae [INV ZOO] A family of large euechinoid echinoderms in the order Diadematoida having crenulate tubercles and long spines.

Diadematoida [INV ZOO] An order of echinoderms in the superorder Diadematacea with hollow primary radioles and crenulate tubercles.

diadochite [MINERAL] $Fe_2(PO_4)(SO_4)(OH)5H_2O$ A brown or yellowish mineral consisting of a basic hydrous ferric phosphate and sulfate.

diadochy [CRYSTAL] Replacement or ability to be replaced of one atom or ion by another in a crystal lattice.

diadromous [BOT] Having venation in the form of fanlike radiations. [VERT ZOO] Of fish, migrating between salt and fresh waters.

Diadumenidae [INV ZOO] A family of anthozoans in the order Actiniaria.

diafocal point [OPTICS] For a ray of light refracted by a lens, a point on the ray which lies on a plane passing through the axis of the lens which is parallel to the ray on the opposite side of the lens.

diagenesis [GEOL] Chemical and physical changes occurring in sediments during and after their deposition but before consolidation.

DIADELPHOUS STAMEN

staminode

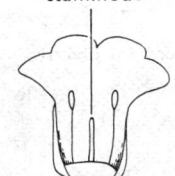

Cutaway drawing of a flower showing diadelphous stamen.

diageotropism [BIOL] Growth orientation of a sessile organism or structure perpendicular to the line of gravity.

diagnosis [ADP] The process of locating and explaining detectable errors in a computer routine or hardware component. [MED] Identification of a disease from its signs and symptoms.

diagnostic check *See* diagnostic routine.

diagnostic equation [METEOROL] Any equation governing a system which contains no time derivative and therefore specifies a balance of quantities in space at a moment of time; examples are a hydrostatic equation or a balance equation.

diagnostic routine [ADP] A routine designed to locate a computer malfunction or a mistake in coding. Also known as diagnostic check; diagnostic subroutine; diagnostic test; error detection routine.

diagnostic subroutine *See* diagnostic routine.

diagnostic test *See* diagnostic routine.

diagnotor [ADP] A combination diagnostic and edit routine which questions unusual situations and notes the implied results.

diagonal [CIV ENG] A sloping structural member, under compression or tension or both, of a truss or bracing system. [MATH] 1. The set of points all of whose coordinates are equal to one another in an *n*-dimensional coordinate system. 2. A line joining opposite vertices of a polygon with an even number of sides. [TEXT] A heavy twilled fabric.

diagonal bond [CIV ENG] A masonry bond with diagonal headers.

diagonal fault [GEOL] A fault whose strike is diagonal or oblique to the strike of the adjacent strata. Also known as oblique fault.

diagonal horn antenna [ELECTROMAG] Horn antenna in which all cross sections are square and the electric vector is parallel to one of the diagonals; the radiation pattern in the far field has almost perfect circular symmetry.

diagonal joint [GEOL] A joint having its strike oblique to the strike of the strata of the sedimentary rock, or to the cleavage plane of the metamorphic rock in which it occurs. Also known as oblique joint.

diagonal pitch [ENG] In rows of staggered rivets, the distance between the center of a rivet in one row to the center of the adjacent rivet in the next row.

diagram [ADP] A schematic representation of a sequence of subroutines designed to solve a problem; it is a coarser and less symbolic representation than a flow chart, frequently including descriptions in English words. [GRAPHICS] 1. A line drawing that represents an object or area according to a scale. 2. A graph which shows the relation between two variables or which plots the occurrence of events or objects as a function of two variables.

diagram factor [MECH ENG] The ratio of the actual mean effective pressure, as determined by an indicator card, to the map of the ideal cycle for a steam engine.

diagram on the plane of the celestial equator *See* time diagram.

diagram on the plane of the equinoctial *See* time diagram.

diakinesis [CYTOL] The last stage of meiotic prophase, when the chromatids attain maximum contraction and the bivalents move apart and position themselves against the nuclear membrane.

dial [DES ENG] A separate scale or other device for indicating the value to which a control is set. [ELECTR] In automatic telephone switching, a type of calling device which, when wound up and released, generates pulses required for establishing connections.

dial cable [DES ENG] Braided cord or flexible wire cable used to make a pointer move over a dial when a separate control knob is rotated, or used to couple two shafts together mechanically.

dial central office [COMMUN] Telephone or teletypewriter office where necessary automatic equipment is located for connecting two or more users together by wires for communications purposes.

dial cord [DES ENG] A braided cotton, silk, or glass fiber cord used as a dial cable.

dialdehyde [ORG CHEM] A molecule that has two aldehyde groups, such as dialdehyde starch.

dial exchange [COMMUN] A telephone exchange area in which all subscribers originate their calls by dialing.

dial feed [MECH ENG] A device that rotates workpieces into position successively so they can be acted on by a machine.

dial indicator [DES ENG] Meter or gage with a calibrated circular face and a pivoted pointer to give readings.

dialing key [COMMUN] Method of dialing in which a set of numerical keys is used to originate dial pulses instead of a dial; generally used in connection with voice-frequency dialing.

dial jacks [ELEC] Strip of jacks associated with and bridged to a regular out-going trunk jack circuit to provide a connection between the dial cords and the outgoing trunks.

dial key [ELEC] Key unit of the subscriber's cord circuit used to connect the dial into the line.

dialkyl [ORG CHEM] A molecule that has two alkyl groups.

dialkyl amine [ORG CHEM] An amine that has two alkyl groups bonded to the amino nitrogen.

diallage [MINERAL] A green, brown, gray, or bronze-colored clinopyroxene characterized by prominent parting parallel to the front pinacoid *a* (100).

dial lamp [ELEC] A small lamp used to illuminate a dial.

dial leg [ELEC] Conductor in a circuit brought out for direct-current dial signaling.

dial office [COMMUN] Central office operating on dial signals.

dial press [MECH ENG] A punch press with dial feed.

dial pulsing *See* loop pulsing.

dial telegraph [COMMUN] A telegraph system in which there is a rotating circular dial plate at each station with letters, numbers, and other symbols along the edge, and the movements of the dial at the receiving station copy those at the transmitting station.

dial telephone system [COMMUN] A telephone system in which telephone connections between customers are ordinarily established by electronic and mechanical apparatus, controlled by manipulations of dials operated by calling parties.

dial tone [COMMUN] A tone employed in a dial telephone system to indicate that the equipment is ready for dialing operation.

dial-up [COMMUN] 1. The service whereby a dial telephone can be used to initiate and effect station-to-station telephone calls. 2. In computer networks, pertaining to terminals which must dial up to receive service, as contrasted with those hand-wired or permanently connected into the network.

dialuric acid [ORG CHEM] $C_4H_4N_2O$ An acid that is derived by oxidation of uric acid or by the reduction of alloxan; may be used in organic synthesis. Also known as 5-hydroxy-barbituric acid.

dialysis [PHYS CHEM] A process of selective diffusion through a membrane; usually used to separate low-molecular-weight solutes which diffuse through the membrane from the colloidal and high-molecular-weight solutes which do not.

dialyzate [CHEM] The material that does not diffuse through the membrane during dialysis; alternatively, it may be considered the material that has diffused.

dialyzer [CHEM ENG] 1. The semipermeable membrane used for dialyzing liquid. 2. The container used in dialysis; it is separated into compartments by membranes.

diamagnet [ELECTROMAG] A substance which is diamagnetic, such as the alkali and alkaline earth metals, the halogens, and the noble gases.

diamagnetic [ELECTROMAG] Having a magnetic permeability less than 1; materials with this property are repelled by a magnet and tend to position themselves at right angles to magnetic lines of force.

diamagnetic Faraday effect [OPTICS] Faraday effect at frequencies near an absorption line which is split due to the splitting of the upper level only.

diamagnetic resonance *See* cyclotron resonance.

diamagnetic susceptibility [ELECTROMAG] The susceptibility of a diamagnetic material, which is always negative and usually on the order of -10^{-5} cm³/mole.

diamagnetism [ELECTROMAG] The property of a material which is repelled by magnets.

diamantine [MINERAL] Consisting of or resembling diamond.

diameter [MATH] 1. A line segment which passes through the

DIALYSIS

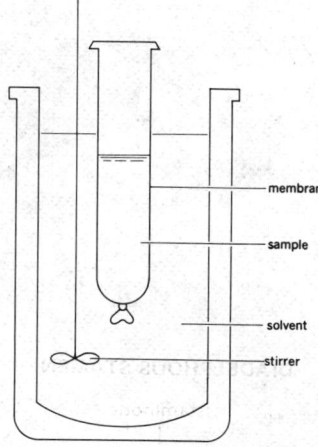

membrane

sample

solvent

stirrer

Equipment for dialysis.

**DIAMAGNETIC
FARADAY EFFECT**

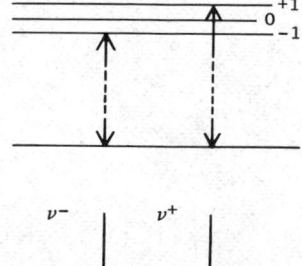

ν^- ν^+

Diamagnetic Faraday effect which is temperature-independent; ν = frequency of light. Splitting of the absorption line is due to the splitting of the upper level only, and the lower level of the line is not split.

center of a circle, and whose end points lie on the circle. **2.** The length of such a line.

diameter group [MECH ENG] A dimensionless group, used in the study of flow machines such as turbines and pumps, equal to the fourth root of pressure number 2 divided by the square root of the delivery number.

diameter of a conic [MATH] Any straight line that passes through the midpoints of all the chords of the conic that are parallel to a given chord.

diameter of a set [MATH] The smallest number which is greater than or equal to the distance between every pair of points of the set.

diameter tape [ENG] A tape for measuring the diameter of trees; when warpped around the circumference of a tree, it reads the diameter directly.

diametral curve [MATH] A curve that passes through the midpoints of a family of parallel chords of a given curve.

diametral pitch [DES ENG] A gear tooth design factor expressed as the ratio of the number of teeth to the diameter of the pitch circle measured in inches.

diametral plane [MATH] **1.** A plane that passes through the center of a sphere. **2.** A plane that passes through the midpoints of a family of parallel chords of a quadric surface that are parallel to a given chord.

diametral surface [MATH] A surface that passes through the midpoints of a family of parallel chords of a given surface that are parallel to a given chord.

diamictite [PETR] A calcareous, terrigenous sedimentary rock that is not sorted or poorly sorted and contains particles of many sizes. Also known as mixtite.

diamicton [PETR] A nonlithified diamictite. Also known as symmicton.

diamide [ORG CHEM] A molecule that has two amide ($-CONH_2$) groups.

diamidine [ORG CHEM] A molecule that has two amidine ($-C=NHNH_2$) groups.

diamine [ORG CHEM] Any compound containing two amino groups.

diamine oxidase [BIOCHEM] A flavoprotein which catalyzes the aerobic oxidation of amines to the corresponding aldehyde and ammonia.

diamino [ORG CHEM] A term used in chemical nomenclature to indicate the presence in a molecule of two amino ($-NH_2$) groups.

meta-**diaminoazobenzene hydrochloride** *See* chrysoidine.

diaminobenzene *See* phenylenediamine.

ortho-**diaminobenzene** *See* phenylenediamine.

diaminoditolyl *See ortho*-tolidine.

diaminoethane *See* ethylenediamine.

2,4-diaminophenolhydrochloride *See* amidol.

diaminopimelate [BIOCHEM] $C_7H_{14}O_4N_2$ A compound that serves as a component of cell wall mucopeptide in some bacteria and as a source of lysine in all bacteria. Abbreviated DAP. Also known as diaminopimelic acid.

diaminopimelic acid *See* diaminopimelate.

diamond [MINERAL] A colorless mineral composed entirely of carbon crystallized in the isometric system as octahedrons, dodecahedrons, and cubes; the hardest substance known; used as a gem and in cutting tools.

diamond antenna *See* rhombic antenna.

diamond bit [DES ENG] A rotary drilling bit crowned with bort-type diamonds, used for rock boring. Also known as bort bit.

diamond boring [ENG] Boring with a diamond tool.

diamond canker [PL PATH] A virus disease that affects the bark of certain stone-fruit trees, resulting in weakening of the trunk and limbs.

diamond chisel [DES ENG] A chisel having a V-shaped or diamond-shaped cutting edge.

diamond circuit [ELECTR] A gate circuit that provides isolation between input and output terminals in its off state, by operating transistors in their cutoff region; in the on state the output voltage follows the input voltage as required for gating both analog and digital signals, while the transistors provide current gain to supply output current on demand.

diamond coring [ENG] Obtaining core samples of rock by using a diamond drill.

diamond count [DES ENG] The number of diamonds set in a diamond crown bit.

diamond crossing [CIV ENG] An oblique railroad crossing that forms a diamond shape between the tracks.

diamond crown [DES ENG] The cutting bit used in diamond drilling; it consists of a steel shell set with black diamonds on the face and cutting edges.

diamond drill [DES ENG] A drilling machine with a hollow, diamond-set bit for boring rock and yielding continuous and columnar rock samples.

Diamond-Hinman radiosonde [ENG] A variable audio-modulated radiosonde used by United States weather services; the carrier signal from the radiosonde is modulated by audio signals determined by the electrical resistance of the humidity- and temperature-transducing elements and by fixed reference resistors; the modulating signals are transmitted in a fixed sequence at predetermined pressure levels by means of a baroswitch.

diamond indenter [ENG] An instrument that measures hardness by indenting a material with a diamond point.

diamond matrix [DES ENG] The metal or alloy in which diamonds are set in a drill crown. [GEOL] The rock material in which diamonds are formed.

diamond orientation [DES ENG] The set of a diamond in a cutting tool so that the crystal face will be in contact with the material being cut.

diamond-particle bit [DES ENG] A diamond bit set with small fragments of diamonds.

diamond paste [MATER] An abrasive consisting of diamond dust in a viscous material.

diamond pattern [DES ENG] The arrangement of diamonds set in a diamond crown.

diamond plate [LAP] A plate with a layer of diamond dust and oil; used for rubbing down gems.

diamond point [DES ENG] A cutting tool with a diamond tip.

diamond-point bit *See* mud auger.

diamond-pyramid hardness number [MET] The quotient of the load applied in the diamond-pyramid hardness test divided by the pyramidal area of the impression.

diamond-pyramid hardness test [MET] An indentation hardness test in which a diamond-pyramid indenter, with a 136° angle between opposite faces, is forced under variable loads into the surface of a test specimen. Also known as Vickers hardness test.

diamond reamer [DES ENG] A diamond-inset pipe behind, and larger than, the drill bit and core barrel that is used for enlarging boreholes.

diamonds [FL MECH] The pattern of shock waves often visible in a rocket exhaust which resembles a series of diamond shapes placed end to end.

diamond saw [DES ENG] A circular, band, or frame saw inset with diamonds or diamond dust for cutting sections of rock and other brittle substances.

diamond setter [ENG] A person skilled at setting diamonds by hand in a diamond bit or a bit mold.

diamond size [ENG] In the bit-setting and diamond-drilling industries, the number of equal-size diamonds having a total weight of 1 carat; a 10-diamond size means 10 stones weighing 1 carat.

diamond structure [CRYSTAL] A crystal structure in which each atom is the center of a tetrahedron formed by its nearest neighbors.

diamond stylus [ENG ACOUS] A stylus having a ground diamond as its point.

diamond tool [DES ENG] **1.** Any tool using a diamond-set bit to drill a borehole. **2.** A diamond shaped to the contour of a single-pointed cutting tool, used for precision machining.

diamond washer [MIN ENG] An apparatus for shaking and separating rock gravel containing diamonds, utilizing a vertical series of screens with 8-, 4-, 2-, and 1-millimeter mesh.

diamond wheel [DES ENG] A grinding wheel in which synthetic diamond dust is bonded as the abrasive to cut very hard materials such as sintered carbide or quartz.

diandrous [BOT] Having two stamens.

Dianemaceae [MICROBIO] A family of slime molds in the order Trichales.

DIAMOND

Diamond crystals from Kimberley, South Africa. *(American Museum of Natural History Specimens)*

DIAMOND STRUCTURE

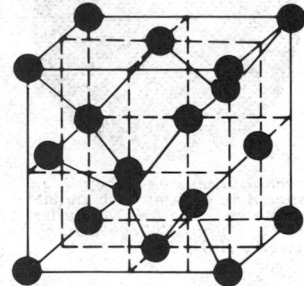

Diamond-type crystal structure.

Dianulitidae [PALEON] A family of extinct, marine bryozoans in the order Cystoporata.

diapause [PHYSIO] A period of spontaneously suspended growth or development in certain insects, mites, crustaceans, and snails.

diapedesis [MED] Hemorrhage of blood cells, especially erythrocytes, through an intact vessel wall into the tissues.

Diapensiaceae [BOT] The single family of the Diapensiales, an order of flowering plants.

Diapensiales [BOT] A monofamilial order of dicotyledonous plants in the subclass Dilleniidae comprising certain herbs and dwarf shrubs in temperate and arctic regions of the Northern Hemisphere.

Diaphanocephalidae [INV ZOO] A family of parasitic roundworms belonging to the Strongyloidea; snakes are the principal host.

diaphone [NAV] A fog signal device that produces sounds similar to a siren and uses a reciprocating piston actuated by compressed air.

diaphorase [BIOCHEM] Mitochondrial flavoprotein enzymes which catalyze the reduction of dyes, such as methylene blue, by reduced pyridine nucleotides such as reduced diphosphopyridine nucleotide.

diaphorite [MINERAL] $PB_2Ag_3Sb_3S_8$ A gray-black orthorhombic mineral consisting of sulfide of lead, silver, and antimony, occurring in crystals. Also known as ultrabasite.

diaphragm [ANAT] The dome-shaped partition composed of muscle and connective tissue that separates the abdominal and thoracic cavities in mammals. [ELECTROMAG] *See* iris. [ENG] A thin sheet placed between parallel parts of a member of structural steel to increase its rigidity. [ENG ACOUS] A thin, flexible sheet that can be moved by sound waves, as in a microphone, or can produce sound waves when moved, as in a loudspeaker. [OPTICS] Any opening in an optical system which controls the cross section of a beam of light passing through it, to control light intensity, reduce aberration, or increase depth of focus. [PHYS] **1.** A separating wall or membrane, especially one which transmits some substances and forces but not others. **2.** In general, any opening, sometimes adjustable in size, which is used to control the flow of a substance or radiation.

diaphragmatic hernia [MED] Protrusion of an abdominal organ through the diaphragm into the thoracic cavity.

diaphragmatic respiration [PHYSIO] Respiration effected primarily by movement of the diaphragm, changing the intrathoracic pressure.

diaphragm cell [CHEM ENG] An electrolytic cell used to produce sodium hydroxide and chlorine from sodium chloride brine; porous diaphragm separates the anode and cathode compartments.

diaphragm compressor [MECH ENG] Device for compression of small volumes of a gas by means of a reciprocally moving diaphragm, in place of pistons or rotors.

diaphragm gage [ENG] Pressure- or vacuum-sensing instrument in which pressures act against opposite sides of an enclosed diaphragm that consequently moves in relation to the difference between the two pressures, actuating a mechanical indicator or electric-electronic signal.

diaphragm horn [ENG ACOUS] A horn that produces sound by means of a diaphragm vibrated by compressed air, steam, or electricity.

diaphragm jig [MIN ENG] A jig having a flexible diaphragm to pulse water; used in gravity concentration of minerals.

diaphragm meter [ENG] A flow meter which uses the movement of a diaphragm in the measurement of a difference in pressure created by the flow, such as a force-balance-type or a deflection-type meter.

diaphragm pump [MECH ENG] A metering pump which uses a diaphragm to isolate the operating parts from pumped liquid in a mechanically actuated diaphragm pump, or from hydraulic fluid in a hydraulically actuated diaphragm pump.

diaphragm setting [OPTICS] The position of a camera's diaphragm after opening or closing it.

diaphragm valve [ENG] A fluid valve in which the open-close element is a flexible diaphragm; used for fluids containing suspended solids, but limited to low-pressure systems.

diaphthoresis *See* retrograde metamorphism.

diaphthorite [PETR] Schistose rocks in which minerals have formed by retrograde metamorphism.

diaphyseal aclasis *See* multiple hereditary exostoses.

diaphysis [ANAT] The shaft of a long bone.

diapir [GEOL] A dome or anticlinal fold in which a mobile plastic core has ruptured the more brittle overlying rock. Also known as diapiric fold; piercement dome; piercing fold.

diapiric fold *See* diapir.

diapophysis [ANAT] The articular portion of a transverse process of a vertebra.

diapositive [GRAPHICS] A positive formed on a transparent support such as glass.

Diapriidae [INV ZOO] A family of hymenopteran insects in the superfamily Proctotrupoidea.

diarrhea [MED] The passage of loose or watery stools, usually at more frequent than normal intervals.

diarthrosis [ANAT] A freely moving articulation, characterized by a synovial cavity between the bones.

diarylamine [ORG CHEM] A molecule that contains an amine group and two aryl groups joined to the amino nitrogen.

diaspore [MINERAL] $AlO(OH)$ A mineral composed of some bauxites occurring in white, lamellar masses; crystallizes in the orthorhombic system.

diasporometer [OPTICS] Oppositely rotating wedges used in optical rangefinders to obtain deviation of the axis of the image.

diastase [BIOCHEM] An enzyme that catalyzes the hydrolysis of starch to maltose. Also known as vegetable diastase.

diastasis [MED] Any simple separation of parts normally joined together, as the separation of an epiphysis from the body of a bone without true fracture, or the dislocation of an amphiarthrosis. [PHYSIO] The final phase of diastole, the phase of slow ventricular filling.

diastem [GEOL] A temporal break between adjacent geologic strata that represents nondeposition or local erosion but not a change in the general regimen of deposition.

diastema [ANAT] A space between two types of teeth, as between an incisor and premolar. [CYTOL] Modified cytoplasm of the equatorial plane prior to cell division.

diastereoisomer [ORG CHEM] One of a pair of optical isomers which are not mirror images of each other. Also known as diastereomer.

diastereomer *See* diastereoisomer.

diastole [PHYSIO] The rhythmic relaxation and dilation of a heart chamber, especially a ventricle.

diastolic pressure [PHYSIO] The lowest arterial blood pressure during the cardiac cycle; reflects relaxation and dilation of a heart chamber.

diastrophism [GEOL] **1.** The general process or combination of processes by which the earth's crust is deformed. **2.** The results of this deforming action.

diathermy [MED] The therapeutic use of high-frequency electric currents to produce localized heat in body tissues.

diathermy interference [COMMUN] Television interference caused by diathermy equipment; produces a herringbone pattern in a dark horizontal band across the picture.

diathermy machine [ELECTR] A radio-frequency oscillator, sometimes followed by rf amplifier stages, used to generate high-frequency currents that produce heat within some part of the body for therapeutic purposes.

diathesis [MED] A constitutional or hereditary predisposition to some disease or a structural or mental abnormality.

diatom [INV ZOO] The common name for algae composing the class Bacillariophyceae; noted for the symmetry and sculpturing of the siliceous cell walls.

diatomaceous earth [GEOL] A yellow, white, or light-gray, siliceous, porous deposit made of the opaline shells of diatoms; used as a filter aid, paint filler, adsorbent, abrasive, and thermal insulator. Also known as kieselguhr; tripolite.

diatomaceous ooze [GEOL] A pelagic, siliceous sediment composed of more than 30% diatom tests, up to 40% calcium carbonate, and up to 25% mineral grains.

diatomic [CHEM] Consisting of two atoms.

diatomite [GEOL] Dense, chert-like, consolidated diatomaceous earth.

diatonic scale [ACOUS] A musical scale in which the octave

DIATOM

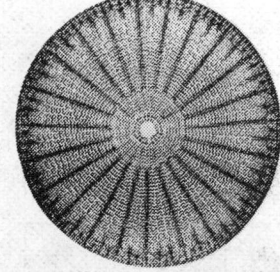

Arachnoidiscus ehrenbergii, a concentric diatom with radial symmetry. (*From H. J. Fuller and O. Tippo, College Botany, rev. ed., Holt, 1954*)

is divided into intervals of two different sizes, five of one and two of the other, with adjustments in tuning systems other than equal temperament.

diatreme [GEOL] A circular volcanic vent produced by the explosive energy of gas-charged magmas.

diatropism [BOT] Growth orientation of certain plant organs that is transverse to the line of action of a stimulus.

Diatrymiformes [PALEON] An order of extinct large, flightless birds having massive legs, tiny wings, and large heads and beaks.

Diatype [GRAPHICS] A small, desk-model film-lettering machine that produces display composition from master negatives either in sheet form or on rolls.

diauxic growth [MICROBIO] The diphasic response of a culture of microorganisms based on a phenotypic adaptation to the addition of a second substrate; characterized by a growth phase followed by a lag after which growth is resumed.

diazine [ORG CHEM] 1. A hydrocarbon consisting of an unsaturated hexatomic ring of two nitrogen atoms and four carbons. 2. Suffix indicating a ring compound with two nitrogen atoms.

diazinon [ORG CHEM] $C_{12}H_{21}O_3N_2PS$ Generic name for O,O-diethyl-O-(isopropyl-4-methyl-6-pyrimidinyl)-phosphorothioate; an organophosphorus insecticide; if ingested, it is highly toxic.

diazoamine [ORG CHEM] The grouping $-N=NNH-$. Also known as azimino.

diazoate [ORG CHEM] A salt with molecular formula of the type $C_6H_5N=NOOM$, where M is a nonvalent metal.

diazo compound [ORG CHEM] An organic compound containing the radical $-N=N-$.

diazo dye See developed dye.

diazoic acid [ORG CHEM] $C_6H_5N=NOOH$ An isomeric form of phenylnitramine.

diazoimide See hydrazoic acid.

diazole [ORG CHEM] A cyclic hydrocarbon with five atoms in the ring, two of which are nitrogen atoms and three are carbon.

1,3-diazole See imidazole.

diazomethane [ORG CHEM] CH_2N_2 A poisonous gas used in organic synthesis to methylate compounds. Also known as azimethane; azimethylene.

diazonium [ORG CHEM] The grouping $=N\equiv N$.

diazonium salts [ORG CHEM] Compounds of the type $R\cdot X\cdot N:N$, with X being an acid radical such as chlorine.

diazo oxide [ORG CHEM] An organic molecule or a grouping of organic molecules that have a diazo group and an oxygen atom joined to ortho positions of an aromatic nucleus. Also known as diazophenol.

diazophenol See diazo oxide.

diazo print [GRAPHICS] A reduction made by the whiteprint process.

diazo process See diazotization.

diazosulfonate [ORG CHEM] A salt formed from diazosulfonic acid.

diazosulfonic acid [ORG CHEM] $C_6H_5N=NSO_3H$ Any of a group of aromatic acids containing the diazo group bonded to the sulfonic acid group.

diazotization [ORG CHEM] Reaction between a primary aromatic amine and nitrous acid to give a diazo compound. Also known as diazo process.

diazotype [GRAPHICS] A photograph or photocopy produced on a surface by coating with a photosensitive solution containing a diazo compound.

Dibamidae [VERT ZOO] The flap-legged skinks, a small family of lizards in the suborder Sauria comprising three species confined to southeastern Asia.

dibasic [CHEM] 1. Compounds containing two hydrogens that may be replaced by a monovalent metal or radical. 2. An alcohol that has two hydroxyl groups, for example, ethylene glycol.

dibasic acid [CHEM] An acid having two hydrogen atoms capable of replacement by two basic atoms or radicals.

dibasic calcium phosphate See calcium phosphate.

dibasic magnesium citrate [ORG CHEM] $MgHC_6H_5O_7\cdot 5H_2O$ A white or yellowish powder soluble in water; used as a

dietary supplement or in medicine. Also known as acid magnesium citrate.

dibenzopyrone See xanthone.

dibenzothiazine See phenothiazine.

dibenzoyl See benzil.

dibenzoyl peroxide See benzoyl peroxide.

dibenzyl See bibenzyl.

dibenzyl ether See benzyl ether.

diborane [INORG CHEM] B_2H_6 A colorless, volatile compound that is soluble in ether; boiling point $-92.5°C$, melting point $-165.5°C$; can be used to produce pentaborane and decaborane, proposed for use as rocket fuels; also used to synthesize organic boron compounds. Also known as boroethane; diboron hexahydride.

diborate See borax.

diboron hexahydride See diborane.

Dibothriocephalus See Diphyllobothrium.

Dibranchia [INV ZOO] A subclass of the Cephalopoda containing all living cephalopods except *Nautilus*; members possess two gills and, when present, an internal shell.

dibromide [CHEM] Indicating the presence of two bromine atoms in a molecule.

dibromo- [CHEM] A prefix indicating two bromine atoms.

1,2-dibromoethane See ethylene dibromide.

dibromomethane See methylene bromide.

dibromothymolsulfonphthalein See bromthymol blue.

dibucaine [ORG CHEM] $C_{20}H_{29}O_2N_3$ A local anesthetic used both as the base and the hydrochloride salt. Also known as cinchocaine.

dibutyl [ORG CHEM] Indicating the presence of two butyl groupings bonded through a third atom or group in a molecule.

dibutyl amine [ORG CHEM] $C_8H_{19}N$ A colorless, clear liquid with amine aroma; either di-n-butylamine, $(C_4H_9)_2NH$, boiling at 160°C, insoluble in water, soluble in hydrocarbon solvents, or di-sec-butylamine, $(CH_3CHCH_2CH_3)_2NH$, boiling at 133°C, flammable; used in the manufacture of dyes.

dibutyl maleate [ORG CHEM] $C_4H_9OOCCHCHCOOC_4H_9$ Oily liquid used for copolymers and plasticizers and as a chemical intermediate.

dibutyl oxalate [ORG CHEM] $(COOC_4H_9)_2$ High-boiling, water-white liquid with mild odor, used as a solvent and in organic synthesis.

dibutyl phthalate [ORG CHEM] $C_{16}H_{22}O_4$ A colorless liquid, used as a plasticizer and insect repellent.

dibutyl tartrate [ORG CHEM] $(COOC_4H_9)_2(CHOH)_2$ Liquid used as a solvent and plasticizer for cellulosics and as a lubricant.

dicalcium [CHEM] A molecule containing two atoms of calcium.

dicalcium orthophosphate See calcium phosphate.

dicalcium phosphate See calcium phosphate.

di-cap storage [ELECTR] Device capable of holding data in the form of an array of charged capacitors and using diodes for controlling information flow.

dicarbamylamine See biuret.

dicarbocyanine [ORG CHEM] 1. A member of a group of dyes termed the cyanine dyes; structure consists of two heterocyclic rings joined to the five-carbon chain: $=CH-CH=CH-CH=CH-$. 2. A particular dicarbocyanine dye containing two quinoline heterocyclic rings. Also known as pentamethine.

dicarboxylic acid [ORG CHEM] A compound with two carboxyl groups.

dicarpellate [BOT] Having two carpels.

dicaryon See dikaryon.

dice See die.

DICE See digital intercontinental conversion equipment.

dicentric [CYTOL] Having two centromeres.

dicephaly [MED] A severe congenital anomaly in which the infant is born with two distinct heads.

dicerous [INV ZOO] Having two tentacles or two antennae.

Dice's life zones [ECOL] Biomes proposed by L. R. Dice based on the concept of the biotic province.

dichasium [BOT] A cyme producing two main axes from the primary axis or shoot.

Dichelesthiidae [INV ZOO] A family of parasitic copepods in

Diatryma steini (reconstruction). *(American Museum of Natural History)*

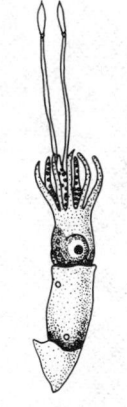

The luminous deep-sea squid *Lycoteuthis diadema*.

the suborder Caligoida; individuals attach to the gills of various fishes.

dichlamydeous [BOT] Having both calyx and corolla.

dichlone [ORG CHEM] $C_{10}H_4O_2Cl_2$ A yellow, crystalline compound, used as a fungicide for foliage and as an algicide. Also known as 2,3-dichloro-1,4-naphthoquinone.

dichloramine [INORG CHEM] **1.** NH_2Cl_2 An unstable molecule considered to be formed from ammonia by action of chlorine. Also known as chlorimide. **2.** Any chloramine with two chlorine atoms joined to the nitrogen atom.

dichloride [CHEM] Any inorganic salt or organic compound that has two chloride atoms in its molecule.

dichloroacetic acid [ORG CHEM] $CHCl_2COOH$ A strong liquid acid, formed by chlorinating acetic acid; used in organic synthesis.

dichlorobenzene [ORG CHEM] $C_6H_4Cl_2$ Any of a group of substitution products of benzene and two atoms of chlorine; the three forms are *meta*-dichlorobenzene, colorless liquid boiling at 172°C, soluble in alcohol and ether, insoluble in water, or *ortho*-, colorless liquid boiling at 179°C, used as a solvent and chemical intermediate, or *para*-, volatile white crystals, insoluble in water, soluble in organic solvents, used as a germicide, insecticide, and chemical intermediate.

1,4-dichlorobutane [ORG CHEM] $Cl(CH_2)_4Cl$ A liquid; used in the manufacture of adiponitrile.

2,2′-dichlorodiethyl ether *See* dichloroethyl ether.

dichlorodiethylsulfide *See* mustard gas.

dichlorodifluoromethane [ORG CHEM] CCl_2F_2 A nontoxic, nonflammable, colorless gas made from carbon tetrachloride; boiling point $-30°C$; used as a refrigerant and as a propellant in aerosols.

dichlorodiphenyltrichloroethane *See* DDT.

dichloroether *See* dichloroethyl ether.

sym-dichloroethylene [ORG CHEM] $CHClCHCl$ Colorless, toxic liquid with pleasant aroma, boiling at 59°C; decomposes in light, air, and moisture; soluble in organic solvents, insoluble in water; exists in cis and trans forms; used as solvent, in medicine, and for chemical synthesis. Also known as acetylene dichloride.

dichloroethyl ether [ORG CHEM] $ClCH_2CH_2OCH_2CH_2Cl$ A colorless liquid insoluble in water, soluble in organic solvents; used as a solvent in paints, varnishes, lacquers, and as a soil fumigant. Also known as 2,2′-dichlorodiethyl ether; dichloroether; dichloroethyl oxide.

dichloroethyl oxide *See* dichloroethyl ether.

α-dichlorohydrin [ORG CHEM] $CH_2ClCHOHCH_2Cl$ Unstable liquid, the commercial product consisting of a mixture of two isomers; used as a solvent and a chemical intermediate. Also known as GDCH; glycerol dichlorohydrin; α-propenyldichlorohydrin.

dichloromethane *See* methylene chloride.

2,3-dichloro-1,4-naphthoquinone *See* dichlone.

dichloropentane [ORG CHEM] $C_5H_{10}Cl_2$ Mixed dichloro derivatives of normal pentane and isopentane; clear, light-yellow liquid used as solvent, paint and varnish remover, insecticide, and soil fumigant.

dichlorophenoxyacetic acid [ORG CHEM] Yellow crystals, melting at 142°C; used as a herbicide and pesticide. Also known as 2,4-D.

1,2-dichloropropane *See* propylene dichloride.

dichlorotoluene [ORG CHEM] $C_7H_6Cl_2$ A colorless liquid, soluble in organic solvents, insoluble in water; isomers are 2,4-$CH_3C_6H_3Cl_2$, boiling at 200–202°C, and 3,4-(CH_3-$C_6H_3Cl_2$), boiling at 209°C; used as solvent and chemical intermediate.

Dichobunidae [PALEON] A family of extinct artiodactyl mammals in the superfamily Dichobunoidea.

Dichobunoidea [PALEON] A superfamily of extinct artiodactyl mammals in the suborder Paleodonta composed of small- to medium-size forms with tri- to quadrituberular bunodont upper teeth.

dichogamy [BIOL] Producing mature male and female reproductive structures at different times.

dichoptous [INV ZOO] Having the margins of the compound eyes separate.

dichotomic variable [QUANT MECH] A variable with a range consisting of two values; used, for example, to describe a particle with spin ½.

dichotomizing search [ADP] A procedure for searching an item in a set, in which, at each step, the set is divided into two parts, one part being then discarded if it can be logically shown that the item could not be in that part.

dichotomy [ADP] A division into two subordinate classes; for example, all white and all nonwhite, or all zero and all nonzero. [ASTRON] A configuration of three bodies so that they form a right triangle; specifically, such a configuration in the solar system with the sun at the apex of the 90° angle. [BIOL] **1.** Divided in two parts. **2.** Repeated branching or forking.

dichotriaene [INV ZOO] A type of sponge spicule with three rays.

dichroic mirror [OPTICS] A glass surface coated with a special metal film that reflects certain colors of light while allowing others to pass through.

dichroism [OPTICS] In certain anisotropic materials, the property of having different absorption coefficients for light polarized in different directions.

dichromate [INORG CHEM] A salt of dichromic acid, usually orange or red.

dichromate treatment [MET] Processing technique involving the formation of a corrosion-resistant film on the surface of a magnesium alloy by boiling the alloy in a sodium dichromate solution.

dichromatic [BIOL] Having or exhibiting two color phases independently of age or sex.

dichromatic dye [CHEM] Dye or indicator in which different colors are seen, depending upon the thickness of the solution.

dichromatism [MED] Partial color blindness in which vision is apparently based on two primary colors rather than the normal three.

dichromic [CHEM] Pertaining to a molecule with two atoms of chromium.

dichromic acid [INORG CHEM] $H_2Cr_2O_7$ An acid known only in solution, especially in the form of dichromates.

dicing [ELECTR] Sawing or otherwise machining a semiconductor wafer into small squares, or dice, from which transistors and diodes can be fabricated.

dicing cutter [MECH ENG] A cutting mill for sheet material; sheet is first slit into horizontal strands by blades, then fed against a rotating knife for dicing.

Dicke fix [ELECTR] Technique designed to protect a receiver from fast sweep jamming.

Dicke radiometer [ELECTR] A radiometer-type receiver that detects weak signals in noise by modulating or switching the incoming signal before it is processed by conventional receiver circuits.

Dickinsoniidae [PALEON] A family that comprises extinct flat-bodied, mutisegmented coelomates; identified as ediacaran fauna.

dickinsonite [MINERAL] $H_2Na_6(Mn,Fe,Ca,Mg)_{14}(PO_4)_{12} \cdot H_2O$ A green mineral consisting of foliated hydrous acid phosphate, chiefly of manganese, iron, and sodium, and is isostructural with arrojadite; specific gravity is 3.34.

dickite [MINERAL] $Al_2Si_2O_5(OH)_4$ A mineral of the kaolin group found crystallized in clay in hydrothermal veins; it is polymorphous with kaolinite and nacrite.

Dicksoniaceae [BOT] A family of tree ferns characterized by marginal sori which are terminal on the veins and protected by a bivalved indusium.

Dick test [IMMUNOL] A skin test to determine susceptibility or immunity to scarlet fever.

diclinous [BOT] Having stamens and pistils on different flowers.

dicoccous [BOT] Composed of two adherent one-seeded carpels.

dicotyledon [BOT] Any plant of the class Magnoliopsida, all having two cotyledons.

Dicotyledoneae [BOT] The equivalent name for Magnoliopsida.

Dicranales [BOT] An order of mosses having erect stems, dichotomous branching, and dense foliation.

dicrotic [MED] Pertaining to a secondary pressure wave in an artery on the descending limb of a main wave during diastole of the heart.

Dictaphone [ENG ACOUS] Trademark for an instrument that can both record and reproduce sound, used in dictation.

Dictaphone reception [COMMUN] Recording of high-speed radio-telegraph signals with a Dictaphone or phonograph, which afterward can be run slower for reading or copying the message.

dictionary [ADP] A table establishing the correspondence between specific words and their code representations.

dictionary code [ADP] An alphabetical arrangement of English words and terms, associated with their code representations.

dictyoblastospore [MYCOL] A blastospore with both cross and longitudinal septa.

Dictyoceratida [INV ZOO] An order of sponges of the class Demospongiae; includes the bath sponges of commerce.

Dictyonellidina [PALEON] A suborder of extinct articulate brachiopods.

dictyonema bed [GEOL] A thin shale bed rich in remains of graptolites of the genus *Dictyonema*.

dictyosome [CYTOL] A stack of two or more cisternae; a component of the Golgi apparatus.

Dictyospongiidae [PALEON] A family of extinct sponges in the subclass Amphidiscophora having spicules resembling a one-ended amphidisc (paraclavule).

Dictyosporae [MYCOL] A spore group of the imperfect fungi characterized by multicelled spores with cross and longitudinal septae.

dictyospore [MYCOL] A multicellular spore in certain fungi characterized by longitudinal walls and cross septa.

dictyostele [BOT] A modified siphonostele in which the vascular tissue is dissected into a network of distinct strands; found in certain fern stems.

Dictyosteliaceae [MICROBIO] A family of microorganisms belonging to the Acrasiales and characterized by strongly differentiated fructifications.

dicyanide [CHEM] A salt that has two cyanide groups.

dicyanoargentates I *See* argentocyanides.

dicyclopentadienyl iron *See* ferrocene.

Dicyemida [INV ZOO] An order of mesozoans comprising minute, wormlike parasites of the renal organs of cephalopod mollusks.

didelphic [ANAT] Having a double uterus or genital tract.

Didelphidae [VERT ZOO] The opossums, a family of arboreal mammals in the order Marsupialia.

Didolodontidae [PALEON] A family of extinct medium-sized herbivores in the order Condylarthra.

Didymiaceae [MICROBIO] A family of slime molds in the order Physarales.

didymium [CHEM] A mixture of the rare-earth elements praeseodymium and neodymium. Abbreviated Di.

didymolite [MINERAL] $Ca_2Al_6Si_9O_{29}$ A dark-gray monoclinic mineral consisting of a calcium aluminum silicate, occurring in twinned crystals.

Didymosporae [MYCOL] A spore group of the imperfect fungi characterized by two-celled spores.

didymous [BIOL] Occurring in pairs.

didynamous [BOT] Having four stamens occurring in two pairs, one pair long and the other short.

die [DES ENG] A tool or mold used to impart shapes to or to form impressions on materials such as metals and ceramics. [ELECTR] The tiny, sawed or otherwise machined piece of semiconductor material used in the construction of a transistor, diode, or other semiconductor device; plural is dice. [MED] To pass from physical life. [MIN ENG] *See* bell tap.

die adapter [ENG] That part of an extrusion die which holds the die block.

dieback [PL PATH] Of a plant, to die from the top or peripheral parts.

die blade [ENG] A deformable member attached to a die body which determines the slot opening and is adjusted to produce uniform thickness across plastic film or sheet.

die block [ENG] A tool-steel block which is bolted to the bed of a punch press and into which the desired impressions are machined.

die body [ENG] The stationary part of an extrusion die, used to separate and form material.

die bushing *See* button die.

die-cast [ENG] To force molten material, such as metal, into a die for casting.

die casting [ENG] A casting formed in a die.

die chaser [ENG] One of the cutting parts of a composite die or a die used to cut threads.

Dieckman condensation [CHEM ENG] Any condensation of esters of dicarboxylic acids which produce cyclic β-ketoesters.

die clearance [ENG] The distance between die members that meet during an operation.

die collar *See* bell tap.

die cushion [ENG] A device located in or under a die block or bolster to provide additional pressure or motion for stamping.

die-cut [ENG] To cut shapes from sheet stock by striking it sharply with a shaped knife edge.

die cutting [ENG] An object made by cutting with a die.

die down [BOT] Normal seasonal death of aboveground parts of herbaceous perennials.

die drawing [MET] Reducing the diameter of wire or tubing by pulling it through a die.

die forging [MET] Shaping metal by plastic deformation in a die.

die forming [MET] Shaping metal by means of a die under pressure.

die gap [ENG] In plastics and metals forming, the distance between the two opposing metal faces forming the opening of a die.

Diego blood group [IMMUNOL] A genetically determined, immunologically distinct group of human erythrocyte antigens recognized by reaction with a specific antibody.

die holder [ENG] A plate or block on which the die block is mounted; it is fastened to the bolster or press bed.

dieing machine [MECH ENG] A vertical press with the slide activated by pull rods attached to the drive mechanism below the bed of the press.

die insert [ENG] A removable part or the liner of a die body or punch.

dieldrin [ORG CHEM] $C_{12}H_8Cl_6O$ A white, crystalline contact insecticide obtained by oxidation of aldrin; used in mothproofing carpets and other furnishings.

dielectric [MATER] A material which is an electrical insulator or in which an electric field can be sustained with a minimum dissipation in power.

dielectric absorption [ELEC] The persistence of electric polarization in certain dielectrics after removal of the electric field. [ELECTROMAG] *See* dielectric loss.

dielectric amplifier [ELECTR] An amplifier using a ferroelectric capacitor whose capacitance varies with applied voltage so as to give signal amplification.

dielectric antenna [ELECTROMAG] An antenna in which a dielectric is the major component used to produce a desired radiation pattern.

dielectric breakdown [ELECTR] Breakdown which occurs in an alkali halide crystal at field strengths on the order of 10^6 volts/cm.

dielectric circuit [ELEC] Any electric circuit which has capacitors.

dielectric constant [ELEC] **1.** For an isotropic medium, the ratio of the capacitance of a capacitor filled with a given dielectric to that of the same capacitor having only a vacuum as dielectric. **2.** More generally, $1 + \gamma\chi$, where γ is 4π in Gaussian and cgs electrostatic units or 1 in rationalized mks units, and χ is the electric susceptibility tensor. Also known as relative dielectric constant; relative permittivity; specific inductive capacity (SIC).

dielectric crystal [ELEC] A crystal which is electrically nonconducting.

dielectric current [ELEC] The current flowing at any instant through a surface of a dielectric that is located in a changing electric field.

dielectric displacement *See* electric displacement.

dielectric ellipsoid [ELEC] For an anisotropic medium in which the dielectric constant is a tensor quantity **K**, the locus of points **r** satisfying **r** · **K** · **r** = 1.

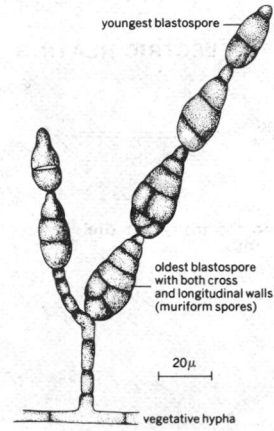

DICTYOBLASTOSPORE

youngest blastospore

oldest blastospore with both cross and longitudinal walls (muriform spores)

20μ

vegetative hypha

Alternaria tenuis. Sporophore with branched chain of dictyoblastospores. *(After G. Goidanich, 1938)*

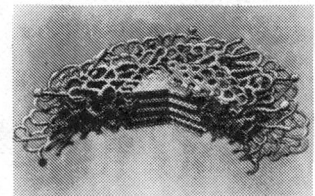

DICTYOSOME

Cutaway view of a model of a portion of a dictyosome composed of six cisternae. Cisternal maturation is depicted from top to bottom.

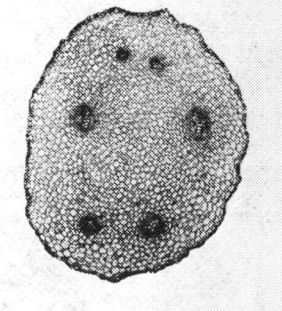

DICTYOSTELE

Dictyostele of *Polypodium*.

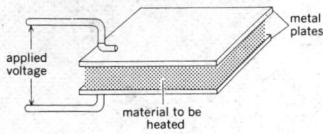

DIELECTRIC HEATING

Basic assembly for dielectric heating.

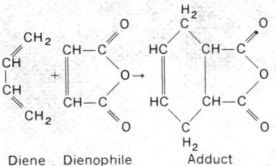

DIELS-ALDER REACTION

Diene   Dienophile   Adduct

Equation demonstrating Diels-Alder reaction.

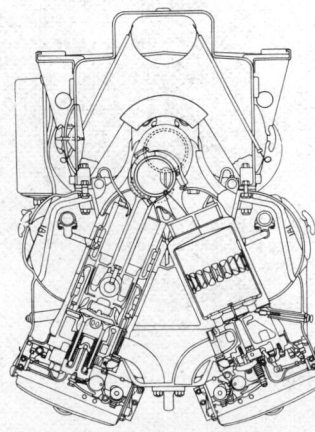

DIESEL ENGINE

Section through a locomotive diesel engine. *(General Motors, Electromotive Division)*

dielectric fatigue [ELECTR] The property of some dielectrics in which resistance to breakdown decreases after a voltage has been applied for a considerable time.

dielectric film [ELEC] A film possessing dielectric properties; used as the central layer of a capacitor.

dielectric flux density *See* electric displacement.

dielectric gas [ELEC] A gas having a high dielectric constant, such as sulfur hexafluoride.

dielectric heating [ELEC] Heating of a nominally electrical insulating material due to its own electrical (dielectric) losses, when the material is placed in a varying electrostatic field.

dielectric hysteresis *See* ferroelectric hysteresis.

dielectric lens [ELECTROMAG] A lens made of dielectric material so that it refracts radio waves in the same manner that an optical lens refracts light waves; used with microwave antennas.

dielectric-lens antenna [ELECTROMAG] An aperture antenna in which the beam width is determined by the dimensions of a dielectric lens through which the beam passes.

dielectric loss [ELECTROMAG] The electric energy that is converted into heat in a dielectric subjected to a varying electric field. Also known as dielectric absorption.

dielectric loss angle [ELEC] Difference between 90° and the dielectric phase angle.

dielectric loss factor [ELEC] Product of the dielectric constant of a material and the tangent of its dielectric loss angle.

dielectric matching plate [ELECTROMAG] In waveguide technique, a dielectric plate used as an impedance transformer for matching purposes.

dielectric phase angle [ELEC] Angular difference in phase between the sinusoidal alternating potential difference applied to a dielectric and the component of the resulting alternating current having the same period as the potential difference.

dielectric polarization *See* polarization.

dielectric power factor [ELEC] Cosine of the dielectric phase angle (or sine of the dielectric loss angle).

dielectric-rod antenna [ELECTROMAG] A surface-wave antenna in which an end-fire radiation pattern is produced by propagation of a surface wave on a tapered dielectric rod.

dielectric shielding [ELEC] The reduction of an electric field in some region by interposing a dielectric substance, such as polystyrene, glass, or mica.

dielectric soak *See* absorption.

dielectric strength [ELEC] The maximum electrical potential gradient that a material can withstand without rupture; usually specified in volts per millimeter of thickness. Also known as electric strength.

dielectric susceptibility *See* electric susceptibility.

dielectric test [ELEC] A test involving application of a voltage higher than the rated value for a specified time, to determine the margin of safety against later failure of insulating materials.

dielectric vapor detector [ANALY CHEM] Apparatus to measure the change in the dielectric constant of gases or gas mixtures; used as a detector in gas chromatographs to sense changes in carrier gas.

dielectric waveguide [ELEC] A waveguide consisting of a dielectric cylinder surrounded by air.

dielectric wedge [ELECTROMAG] A wedge-shaped piece of dielectric used in a waveguide to match its impedance to that of another waveguide.

dielectric wire [ELECTROMAG] A dielectric waveguide used to transmit ultra-high-frequency radio waves short distances between parts of a circuit.

dielectronic recombination [ATOM PHYS] The combination of an electron with a positive-ion in a gas, so that the energy released is taken up by two electrons of the resulting atom.

Dieline [ORG CHEM] $C_2H_2Cl_2$ Trade name for dichloroethylene, used as a refrigerant in some vapor compression systems.

die lines [MET] Lines or markings on the surface of a drawn, formed, or extruded metal product due to imperfections in the surface of the die.

Diels-Alder reaction [ORG CHEM] The 1,4 addition of a conjugated diolefin to a compound, known as a dienophile, containing a double or triple bond; the dienophile may be activated by conjugation with a second double bond or with an electron acceptor.

die lubricant [MATER] Any material applied to a die to facilitate movement of the work in the die in certain die-forming operations.

diencephalon [EMBRYO] The posterior division of the embryonic forebrain in vertebrates.

diene [ORG CHEM] One of a class of organic compounds containing two ethylenic linkages (carbon-to-carbon double bonds) in the molecules. Also known as diolefin.

diene resin [ORG CHEM] Material containing the diene group of double bonds that may polymerize.

diene value [ORG CHEM] A number that represents the amount of conjugated bonds in a fatty acid or fat.

die nipple *See* bell tap.

die opening [MET] The distance between electrodes in flash or upset welding; it is measured with parts in contact but before the beginning or immediately after completion of the weld cycle.

die radius [MET] The radius on the exposed edge of a deep-drawing die.

die scalping [MET] Drawing wire, tubing, bars, or rods through a sharp-edged die to remove surface layers containing defects.

diesel cycle [THERMO] An internal combustion engine cycle in which the heat of compression ignites the fuel.

diesel electric locomotive [MECH ENG] A locomotive with a diesel engine driving an electric generator which supplies electric power to traction motors for propelling the vehicle. Also known as diesel locomotive.

diesel electric power generation [MECH ENG] Electric power generation in which the generator is driven by a diesel engine.

diesel engine [MECH ENG] An internal combustion engine operating on a thermodynamic cycle in which the ratio of compression of the air charge is sufficiently high to ignite the fuel subsequently injected into the combustion chamber. Also known as compression-ignition engine.

diesel fuel [MATER] Fuel used for internal combustion in diesel engines; usually that fraction of crude oil that distills after kerosine.

diesel fuel additives [MATER] Compounds added to diesel fuels to improve performance, such as cetane number improvers, metal deactivators, corrosion inhibitors, antioxidants, rust inhibitors, and dispersants.

diesel fuel grades [MATER] Fuels suitable for various classes of engines and service, which must meet specifications for flash point temperature, distillation temperature, cetane number, and concentrations of impurities.

diesel fuel water and sediment [MATER] Undesirable constituents of diesel fuel which should not exceed certain limits to ensure clean fuel.

diesel index [CHEM ENG] An empirical expression for the correlation between the aniline number of a diesel fuel and its ignitability.

diesel knock [MECH ENG] A combustion knock caused when the delayed period of ignition is long so that a large quantity of atomized fuel accumulates in the combustion chamber; when combustion occurs, the sudden high pressure resulting from the accumulated fuel causes diesel knock.

diesel locomotive *See* diesel electric locomotive.

diesel oil [MATER] Heavy oil residue used as fuel for certain types of diesel engines.

diesel rig [MECH ENG] Any diesel engine apparatus or machinery.

die set [ENG] A tool or tool holder consisting of a die base for the attachment of a die and a punch plate for the attachment of a punch.

die shoe [MECH ENG] A block placed beneath the lower part of a die upon which the die holder is mounted; spreads the impact over the die bed, thereby reducing wear.

diesinking [ENG] Making a depressed pattern in a die by forming or machining.

die slide [MECH ENG] A device in which the lower die of a power press is mounted; it slides in and out of the press for easy access and safety in feeding the parts.

die steel [MET] Plain carbon steel or alloy steel used in making tools for cutting, machining, shearing, stamping, punching, and chipping.

diester [ORG CHEM] A compound containing two ester groupings.

diesterase [BIOCHEM] An enzyme such as a nuclease which splits the linkages binding individual nucleotides of a nucleic acid.

diestrus [PHYSIO] The long, quiescent period following ovulation in the estrous cycle in mammals; the stage in which the uterus prepares for the reception of a fertilized ovum.

Diesulforming [CHEM ENG] A fixed-bed catalytic process, employing hydrogen, for desulfurizing naphtha, middle distillate, and gas oil; the catalyst is pelleted molybdenum oxide.

die swell ratio [ENG] The ratio of the outer parison diameter (or parison thickness) to the outer diameter of the die (or die gap).

diet [BIOL] Food or drink regularly consumed. [MED] Food prescribed, regulated, or restricted as to kind and amount, for therapeutic or other purpose.

Dieterici equation of state [THERMO] An empirical equation of state for gases, $pe^{a/vRT}(v - b) = RT$, where p is the pressure, T is the absolute temperature, v is the molar volume, R is the gas constant, and a and b are constants characteristic of the substance under consideration.

Dietert tester [MET] An apparatus for reading Brinell hardness directly from the impression made in the part being tested by means of a depth pin pressed into the depression.

dietetics [MED] The science concerned with applying the principle of nutrition to the feeding of people under various economic conditions or for therapeutic purposes.

diethanolamine [ORG CHEM] $(HOCH_2CH_2)_2NH$ Colorless, water-soluble, deliquescent crystals, or liquid boiling at 217°C; soluble in alcohol and acetone, insoluble in ether and benzene; used in detergents, as an absorbent of acid gases, and as a chemical intermediate. Also known as DEA.

diether [ORG CHEM] A molecule that has two oxygen atoms with ether bonds.

1,1-diethoxyethane See acetal.

diethyl [ORG CHEM] Pertaining to a molecule with two ethyl groups.

diethyl adipate [ORG CHEM] $C_2H_5OCO(CH_2)_4OCOC_2H_5$ Water-insoluble, colorless liquid, boiling at 245°C; used as a plasticizer.

diethylamine [ORG CHEM] $(C_2H_5)_2NH$ Water-soluble, colorless liquid with ammonia aroma, boiling at 56°C; used in rubber chemicals and pharmaceuticals and as a solvent and flotation agent.

5,5-diethylbarbituric acid See barbital.

diethylbarbituric acid See barbital.

diethylbenzene [ORG CHEM] $C_6H_4(C_2H_5)_2$ Colorless liquid, boiling at 180–185°C; soluble in organic solvents, insoluble in water; usually a mixture of three isomers, which are 1,2- (or *ortho*-diethylbenzene), boiling at 183°C, and 1,3- (or *meta*-), boiling at 181°C, and 1,4- (or *para*-), boiling at 184°C; used as a solvent.

diethylcarbamazine [ORG CHEM] $C_{16}H_{29}O_8N_3$ White, water-soluble, hygroscopic crystals, melting at 136°C; used as an anthelminthic.

diethyl carbinol [ORG CHEM] $(CH_3CH_2)_2CHOH$ Colorless, alcohol-soluble liquid, boiling at 116°C; slightly soluble in water; used in pharmaceuticals and as a solvent and flotation agent. Also known as sec-*n*-amyl alcohol.

diethyl carbonate [ORG CHEM] $(C_2H_5)_2CO_3$ Stable, colorless liquid with mild aroma, boiling at 126°C; soluble with most organic solvents; used as a solvent and for chemical synthesis. Also known as ethyl carbonate.

diethylenediamine See piperazine.

diethylene dioxide See 1,4-dioxane.

diethylene glycol [ORG CHEM] $CH_2OHCH_2OCH_2CH_2OH$ Clear, hygroscopic, water-soluble liquid, boiling at 245°C; soluble in many organic solvents; used as a softener, conditioner, lubricant, and solvent, and in antifreezes and cosmetics.

diethyl ether [ORG CHEM] $C_4H_{10}O$ A colorless liquid, slightly soluble in water; used as a reagent and solvent. Also known as ethyl ether; ethyl oxide; ethylic ether.

diethyl ketone See pentanone.

diethyl maleate [ORG CHEM] $(HCCOOC_2H_5)_2$ Clear, colorless liquid, boiling at 225°C; slightly soluble in water, soluble in most organic solvents; used as a chemical intermediate.

diethylmalonylurea See barbital.

diethylmethylmethane See 3-methylpentane.

diethyl para-nitrophenyl phosphate See para-oxon.

diethyl phthalate [ORG CHEM] $C_6H_4(CO_2C_2H_5)_2$ Clear, colorless, odorless liquid with bitter taste, boiling at 298°C; soluble in alcohols, ketones, esters, and aromatic hydrocarbons, partly soluble in aliphatic solvents; used as a cellulosic solvent, wetting agent, alcohol denaturant, mosquito repellent, and in perfumes.

diethylstilbesterol [BIOCHEM] $C_{18}H_{20}O_2$ A white, crystalline, nonsteroid estrogen that is used therapeutically as a substitute for natural estrogenic hormones. Also known as stilbestrol.

diethyl succinate [ORG CHEM] $(CH_2COOC_2H_5)_2$ Water-white liquid with pleasant aroma, boiling at 216°C; soluble in alcohol and ether, slightly soluble in water; used as a chemical intermediate and plasticizer.

diethyl sulfate [ORG CHEM] $(C_2H_5)_2SO_4$ A colorless oil with a peppermint odor, and boiling at 208°C; used as an intermediate in organic synthesis. Also known as ethyl sulfate.

diethyl sulfide See ethyl sulfide.

dietician [MED] A person trained in dietetics, or the scientific management of meals for individuals or groups.

Dietl's crisis [MED] Recurrent attacks of radiating pain in the costovertebral angle, accompanied by nausea, vomiting, tachycardia, and hypotension, caused by kinking or twisting of the ureter with intermittent obstructive dilation.

dietrichite [MINERAL] $(Zn,Fe,Mn)Al_2(SO_4)_4 \cdot 22H_2O$ Mineral consisting of a hydrous sulfate of aluminum and one or more of the metals zinc, iron, and manganese.

dietzeite [MINERAL] $Ca_2(IO_3)_2(CrO_4)$ A dark-golden-yellow iodate mineral commonly in fibrous or columnar form as a component of caliche.

die welding [MET] Forge welding in which the weld is completed under pressure between dies.

difference [MATH] The result of subtracting one number from another.

difference amplifier See differential amplifier.

difference channel [ENG ACOUS] An audio channel that handles the difference between the signals in the left and right channels of a stereophonic sound system.

difference detector [ELECTR] A detector circuit in which the output is a function of the difference between the amplitudes of the two input waveforms.

difference equation [MATH] An equation expressing a functional relationship of one or more independent variables, one or more functions dependent on these variables, and successive differences of these functions.

difference in depth modulation [COMMUN] In directive systems employing overlapping lobes with modulated signals, a ratio obtained by subtracting from the percentage of modulation of the larger signal the percentage of modulation of the smaller signal and dividing by 100.

difference limen See just-noticeable difference.

difference number See neutron excess.

difference of latitude [GEOD] The shorter arc of any meridian between the parallels of two places, expressed in angular measure.

difference of longitude [GEOD] The smaller angle at the pole or the shorter arc of a parallel between the meridians of two places, expressed in angular measure.

difference of meridional parts See meridional difference.

difference spectrophotometer See absorption spectrophotometer.

difference threshold See just-noticeable difference.

difference tone [ACOUS] A combination tone whose frequency equals the difference of the frequencies of the pure tones producing it.

differentiable atlas [MATH] A family of embeddings $h_i:E^n \rightarrow M$ of euclidean space into a topological space M with the property that $h_i^{-1}h_j:E^n \rightarrow E^n$ is a differentiable map for each pair of indices, i, j.

differentiable function [MATH] A function which has a derivative at each point of its domain.

differentiable manifold [MATH] A topological space with a maximal differentiable atlas; roughly speaking, a smooth surface.

differential [CONT SYS] The difference between levels for turn-on and turn-off operation in a control system. [MATH] **1.** The differential of a real-valued function $f(x)$, where x is a vector, evaluated at a given vector c, is the linear, real-valued function whose graph is the tangent hyperplane to the graph of $f(x)$ at $x=c$; if x is a real number, the usual notation is $df=f'(c)dx$. **2.** *See* total differential. [MECH ENG] Any arrangement of gears forming an epicyclic train in which the angular speed of one shaft is proportional to the sum or difference of the angular speeds of two other gears which lie on the same axis; allows one shaft to revolve faster than the other, the speed of the main driving member being equal to the algebraic mean of the speeds of the two shafts. Also known as differential gear.

differential amplifier [ELECTR] An amplifier whose output is proportional to the difference between the voltages applied to its two inputs. Also called difference amplifier.

differential analysis [METEOROL] Synoptic analysis of change charts or of vertical differential charts (such as thickness charts) obtained by the graphical or numerical subtraction of the patterns of some meteorological variable at two times or two levels.

differential analyzer [ADP] A mechanical or electromechanical device designed primarily to solve differential equations.

differential ballistic wind [ORD] In bombing, a hypothetical wind equal to the difference in velocity between the ballistic wind and the actual wind at release altitude.

differential blood count *See* differential leukocyte count.

differential brake [MECH ENG] A brake in which operation depends on a difference between two motions.

differential calculus [MATH] The study of the manner in which the value of a function changes as one changes the value of the independent variable; includes maximum-minimum problems and expansion of functions into Taylor series.

differential calorimetry [THERMO] Technique for measurement of and comparison (differential) of process heats (reaction, absorption, hydrolysis, and so on) for a specimen and a reference material.

differential capacitance [ELECTR] The derivative with respect to voltage of a charge characteristic, such as an alternating charge characteristic or a mean charge characteristic, at a given point on the characteristic.

differential capacitor [ELEC] A two-section variable capacitor having one rotor and two stators so arranged that as capacitance is reduced in one section it is increased in the other.

differential centrifugation [CYTOL] The separation of mixtures such as cellular particles in a medium at various centrifugal forces to separate particles of different density, size, and shape from each other.

differential chart [METEOROL] A chart showing the amount and direction of change of a meteorological quantity in time or space.

differential chemical reactor [CHEM ENG] A flow reactor operated at constant temperature and very low concentrations (resulting from very short residence times), with product and reactant concentrations essentially constant at the levels in the feed.

differential coefficient *See* derivative.

differential compaction [GEOL] Compression in sediments, such as sand or limestone, as the weight of overburden causes reduction in pore space and forcing out of water.

differential compound motor [ELEC] A direct-current motor whose speed may be made nearly constant or may be adjusted to increase with increasing load.

differential correction [ASTRON] A method for finding from the observed residuals minus the computed residuals $(O - C)$ small corrections which, when applied to the orbital elements or constants, will reduce the deviations from the observed motion to a minimum.

differential cross section [PHYS] The cross section for a collision process resulting in the emission of particles or photons at a specified angle relative to the direction of the incident particles, per unit angle or per unit solid angle.

differential delay [COMMUN] The difference between the maximum and minimum frequency delays occurring across a band.

differential diagnosis [MED] Distinguishing between diseases of similar character by comparing their signs and symptoms.

differential discriminator [ELECTR] A discriminator that passes only pulses whose amplitudes are between two predetermined values, neither of which is zero.

differential duplex system [ELECTR] System in which the sent currents divide through two mutually inductive sections of a receiving apparatus, connected respectively to the line and to a balancing artificial line in opposite directions, so that there is substantially no net effect on the receiving apparatus; the received currents pass mainly through one section, or through the two sections in the same direction, and operate the apparatus.

differential ebuliometer [ANALY CHEM] Apparatus for precise and simultaneous measurement of both the boiling temperature of a liquid and the condensation temperature of the vapors of the boiling liquid.

differential effects [MECH] The effects upon the elements of the trajectory due to variations from standard conditions.

differential equation [MATH] An equation expressing a relationship between functions and their derivatives.

differential erosion [GEOL] Rapid erosion of one area of the earth's surface relative to another.

differential extraction [CHEM ENG] Theoretical limiting case of crosscurrent extraction in a single vessel where feed is continuously extracted with infinitesimal amounts of fresh solvent; true differential extraction cannot be achieved.

differential fault *See* scissor fault.

differential flotation [MIN ENG] Separation of a complex ore into two or more mineral components and gangue by flotation.

differential form [MATH] A homogeneous polynomial in differentials.

differential gain control [ELECTR] Device for altering the gain of a radio receiver according to expected change of signal level, to reduce the amplitude differential between the signals at the output of the receiver. Also known as gain sensitivity control.

differential galvanometer [ELEC] A galvanometer having a magnetic needle which is free to rotate in the magnetic field produced by currents flowing in opposite directions through two separate identical coils, so that there is no deflection when the currents are equal.

differential game [ADP] A two-sided optimal control problem. [MATH] A game in which the describing equations are differential equations.

differential gap controller [CONT SYS] A two-position (on-off) controller that actuates when the manipulated variable reaches the high or low value of its range (differential gap).

differential gear *See* differential.

differential geometry [MATH] The study of curves and surfaces using the methods of differential calculus.

differential heat of dilution *See* heat of dilution.

differential heat of solution [THERMO] The partial derivative of the total heat of solution with respect to the molal concentration of one component of the solution, when the concentration of the other component or components, the pressure, and the temperature are held constant.

differential indexing [MECH ENG] A method of subdividing a circle based on the difference between movements of the index plate and index crank of a dividing engine.

differential input [ELECTR] Amplifier input circuit that rejects voltages that are the same at both input terminals and amplifies the voltage difference between the two input terminals.

differential instrument [ENG] Galvanometer or other measuring instrument having two circuits or coils, usually identical, through which currents flow in opposite directions; the difference or differential effect of these currents actuates the indicating pointer.

differential ionization chamber [NUCLEO] A two-section ionization chamber in which electrode potentials are such that output current is equal to the difference between the separate ionization currents of the two sections.

differential keying [ELECTR] Method for obtaining chirp-free break-in keying of continuous wave transmitters by using circuitry that arranges to have the oscillator turn on fast before the keyed amplifier stage can pass any signal, and turn off fast after the keyed amplifier stage has cut off.

differential leukocyte count [PATH] The percentage of each variety of leukocytes in the blood, usually based on counting 100 leukocytes. Also known as differential blood count.

differential leveling [ENG] A surveying process in which a horizontal line of sight of known elevation is intercepted by a graduated standard, or rod, held vertically on the point being checked.

differential manometer [ENG] An instrument in which the difference in pressure between two sources is determined from the vertical distance between the surfaces of a liquid in two legs of an erect or inverted U-shaped tube when each of the legs is connected to one of the sources.

differential microphone See double-button microphone.

differential-mode signal [ELECTR] A signal that is applied between the two ungrounded terminals of a balanced three-terminal system.

differential modulation [COMMUN] Modulation in which the choice of the significant condition for any signal element is dependent on the choice for the previous signal element.

differential motion [MECH ENG] A mechanism in which the follower has two driving elements; the net motion of the follower is the difference between the motions that would result from either driver acting alone.

differential operator [MATH] An operator on a space of functions which maps a function f into a linear combination of higher-order derivatives of f.

differential permeability [ELECTROMAG] The slope of the magnetization curve for a magnetic material.

differential phase [ELECTR] Difference in output phase of a small high-frequency sine-wave signal at two stated levels of a low-frequency signal on which it is superimposed in a video transmission system.

differential phase-shift keying [COMMUN] Form of phase-shift keying in which the reference phase for a given keying interval is the phase of the signal during the preceding keying interval.

differential piece-rate system [IND ENG] A wage plan based on a standard task time whereby the worker receives increased or decreased piece rates as his production varies from that expected for the standard time. Also known as accelerating incentive.

differential polarography [ANALY CHEM] Technique of polarographic analysis which measures the difference in current flowing between two identical dropping-mercury electrodes at the same potential but in different solutions.

differential pressure [PHYS] The difference in pressure between two points of a system, such as between the well bottom and wellhead or between the two sides of an orifice.

differential pressure gage [ENG] Apparatus to measure pressure differences between two points in a system; it can be a pressured liquid column balanced by a pressured liquid reservoir, a formed metallic pressure element with opposing force, or an electrical-electronic gage (such as strain, thermal-conductivity, or ionization).

differential pressure pickup [ELEC] An instrument that measures the difference in pressure between two pressure sources and translates this difference into a change in inductance, resistance, voltage, or some other electrical quality.

differential process [CHEM ENG] A process in which a system is caused to move through a bubble point and as a result to form two phases, the minor phase being removed from further contact with the major phase; thus the system continuously changes in quantity and composition.

differential psychology [PSYCH] The comparative study of the psychological differences between groups of people, for example, men and women.

differential pulley [MECH ENG] A tackle in which an endless cable passes through a movable lower pulley, which carries the load, and two fixed coaxial upper pulleys having different diameters; yields a high mechanical advantage.

differential pulse-code modulation [COMMUN] A type of pulse-code modulation proposed for television transmission, in which only the differences between the continuous picture elements on the scanning lines are transmitted, enabling the bandwidth of the signal to be reduced. Abbreviated DPCM.

differential reaction rate [PHYS CHEM] The order of a chemi-

cal reaction expressed as a differential equation with respect to time; for example, $dx/dt = k(a - x)$ for first order, $dx/dt = k(a - x)(b - x)$ for second order, and so on, where k is the specific rate constant, a is the concentration of reactant A, b is the concentration of reactant B, and dx/dt is the rate of change in concentration for time t.

differential relay [ELEC] A two-winding relay that operates when the difference between the currents in the two windings reaches a predetermined value.

differential screw [MECH ENG] A type of compound screw which produces a motion equal to the difference in motion between the two component screws.

differential selection [STAT] A biased selection of a conditioned sample.

differential selsyn [ELEC] Selsyn in which both rotor and stator have similar windings that are spread 120° apart; position of the rotor corresponds to the algebraic sum of the fields produced by the stator and rotor.

differential separation [CHEM ENG] Release of gas (vapor) from liquids by a reduction in pressure that allows the vapor to come out of the solution, so that the vapor can be removed from the system; differs from flash separation, in which the vapor and liquid are kept in contact following pressure reduction.

differential spectrophotometry [SPECT] Spectrophotometric analysis of a sample when a solution of the major component of the sample is placed in the reference cell; the recorded spectrum represents the difference between the sample and the reference cell.

differential synchro See synchro differential receiver; synchro differential transmitter.

differential temperature survey [PETRO ENG] Well-temperature logging method that detects very small temperature anomalies; two thermometers, 6 feet (1.8 meters) apart, record the temperature gradient down the well bore, with small difference changes showing up anomalies.

differential thermal analysis [THERMO] A method of determining the temperature at which thermal reactions occur in a material undergoing continuous heating to elevated temperatures; also involves a determination of the nature and intensity of such reactions.

differential thermogravimetric analysis [THERMO] Thermal analysis in which the rate of material weight change upon heating versus temperature is plotted; used to simplify reading of weight-versus-temperature thermogram peaks that occur close together.

differential thermometer See bimetallic thermometer.

differential thermometric titration [ANALY CHEM] Thermometric titration in which titrant is added simultaneously to the reaction mixture and to a blank in identically equipped cells.

differential topology [MATH] The branch of mathematics dealing with differentiable manifolds.

differential transducer [ELEC] A transducer that simultaneously senses two separate sources and provides an output proportional to the difference between them.

differential transformer [ELEC] A transformer used to join two or more sources of signals to a common transmission line.

differential-transformer transducer [ELEC] A transducer in which movement of the iron core of a transformer varies the output voltage across two series-opposing secondary windings.

differential voltmeter [ELEC] A voltmeter that measures only the difference between a known voltage and an unknown voltage.

differential wind [ORD] In bomb ballistics, the vector difference between the wind at the bomb-release altitude and the wind at some other specific lower altitude; differential winds are required for the computation of the differential ballistic wind.

differential winding [ELEC] A winding whose magnetic field opposes that of a nearby winding.

differential windlass [MECH ENG] A windlass in which the barrel has two sections, each having a different diameter; the rope winds around one section, passes through a pulley (which carries the load), then winds around the other section of the barrel.

differential wound field [ELEC] Type of motor or generator field having both series and shunt coils that are connected to oppose each other.

differentiating circuit [ELEC] A circuit whose output voltage is proportional to the rate of change of the input voltage. Also known as differentiating network.

differentiating network *See* differentiating circuit.

differentiation [MATH] The act of taking a derivative.

differentiator [ELECTR] A device whose output function is proportional to the derivative, or rate of change, of the input function with respect to one or more variables.

diffluence [FL MECH] A region of fluid flow in which the fluid is diverging from the direction of flow.

diffracted wave [PHYS] A wave whose front has been changed in direction by an obstacle or other nonhomogeneity in a medium, other than by reflection or refraction.

diffraction [PHYS] Any redistribution in space of the intensity of waves that results from the presence of an object causing variations of either the amplitude or phase of the waves; found in all types of wave phenomena.

diffractional pulse-height discriminator *See* pulse-height selector.

diffraction grating [SPECT] An optical device which consists of an assembly of narrow slits or grooves which produce a large number of beams that can interfere to produce spectra. Also known as grating.

diffraction instrument *See* diffractometer.

diffraction pattern [PHYS] Pattern produced on a screen or plate by waves which have undergone diffraction.

diffraction propagation [ELECTROMAG] Propagation of electromagnetic waves around objects or over the horizon by diffraction.

diffraction ring [OPTICS] Circular light pattern which appears to surround particles in a microscope field.

diffraction scattering [PHYS] Elastic scattering that occurs when inelastic processes remove particles from a beam.

diffraction spectrum [SPECT] Parallel light and dark or colored bands of light produced by diffraction.

diffraction symmetry [CRYSTAL] Any symmetry in a crystal lattice which causes the systematic annihilation of certain beams in x-ray diffraction.

diffraction velocimeter [OPTICS] A velocity-measuring instrument that uses a continuous-wave laser to send a beam of coherent light at objects moving at right angles to the beam; the needlelike diffraction lobes reflected by the moving objects sweep past the optical grating in the receiver, thereby generating in a photomultiplier a series of impulses from which velocity can be determined and read out. Also known as laser velocimeter; optical diffraction velocimeter.

diffraction zone [ELECTROMAG] The portion of a radio propagation path which lies outside a line-of-sight path.

diffractometer [PHYS] An instrument used to study the structure of matter by means of the diffraction of x-rays, electrons, neutrons, or other waves. Also known as diffraction instrument.

diffractometry [CRYSTAL] The science of determining crystal structures by studying the diffraction of beams of x-rays or other waves.

diffuse aurora [GEOPHYS] A widespread and relatively uniform type of aurora which is easily overlooked from the ground but is prominent in satellite pictures.

diffuse-cutting filter [OPTICS] A color filter that gradually changes in absorption with wavelength.

diffused-alloy transistor [ELECTR] A transistor in which the semiconductor wafer is subjected to gaseous diffusion to produce a nonuniform base region, after which alloy junctions are formed in the same manner as for an alloy-junction transistor; it may also have an intrinsic region, to give a *pnip* unit. Also known as drift transistor.

diffused-base transistor [ELECTR] A transistor in which a nonuniform base region is produced by gaseous diffusion; the collector-base junction is also formed by gaseous diffusion, while the emitter-base junction is a conventional alloy junction.

diffused emitter-collector transistor [ELECTR] A transistor in which both the emitter and collector are produced by diffusion.

diffused junction [ELECTR] A semiconductor junction that

has been formed by the diffusion of an impurity within a semiconductor crystal.

diffused-junction rectifier [ELECTR] A semiconductor diode in which the *pn* junction is produced by diffusion.

diffused-junction transistor [ELECTR] A transistor in which the emitter and collector electrodes have been formed by diffusion by an impurity metal into the semiconductor wafer without heating.

diffused-mesa transistor [ELECTR] A diffused-junction transistor in which an *n*-type impurity is diffused into one side of a *p*-type wafer; a second *pn* junction, required for the emitter, is produced by alloying or diffusing a *p*-type impurity into the newly formed *n*-type surface; after contacts have been applied, undesired diffused areas are etched away to create a flat-topped peak called a mesa.

diffuse front [METEOROL] A front across which the characteristics of wind shift and temperature change are weakly defined.

diffuse hypergammaglobulinemia [MED] General increase in serum immunoglobulins due to infection, hepatic disease, collagen diseases, and advanced sarcoidosis.

diffuse illumination [OPTICS] Lighting so arranged that the object is illuminated from many directions or sources.

diffuse nebula [ASTRON] A type of nebula ranging from huge masses presenting relatively high surface brightness down to faint, milky structures that are detectable only with long exposures and special filters; may contain both dust and gas or may be purely gaseous.

diffuse placenta [EMBRYO] A placenta having villi diffusely scattered over most of the surface of the chorion; found in whales, horses, and other mammals.

diffuser [ENG] A duct, chamber, or section in which a high-velocity, low-pressure stream of fluid (usually air) is converted into a low-velocity, high-pressure flow.

diffuse radiation [PHYS] Radiant energy propagating in many different directions through a given small volume of space.

diffuse reflection [PHYS] Reflection of light, sound, or radio waves from a surface in all directions according to the cosine law.

diffuse reflector [OPTICS] Any surface whose irregularities are so large compared to the wavelength of the incident radiation that the reflected rays are sent back in a multiplicity of directions.

diffuse series [SPECT] A series occurring in the spectra of many atoms having one, two, or three electrons in the outer shell, in which the total orbital angular momentum quantum number changes from 2 to 1.

diffuse skylight *See* diffuse sky radiation.

diffuse sky radiation [ASTROPHYS] Solar radiation reaching the earth's surface after having been scattered from the direct solar beam by molecules or suspensoids in the atmosphere. Also known as diffuse skylight; skylight; sky radiation.

diffuse sound [ACOUS] Sound that has uniform energy density in a given region so that all directions of energy flux at all parts of the region are equally probable.

diffuse spectrum [SPECT] Any spectrum having lines which are very broad even when there is no possibility of line broadening by collisions.

diffuse transmission [PHYS] Transmission of electromagnetic or acoustic radiation in all directions by a transmitting body.

diffuse transmission density [OPTICS] The value of the photographic transmission density obtained when light flux impinges normally on the sample and all the transmitted flux is collected and measured.

diffusing disk *See* diffusion disk.

diffusing screen [GRAPHICS] A translucent screen used in contact printing to ensure even diffusion of light; the screen is sometimes denser at those points immediately facing the electric lamps, to avoid increased light in those areas.

diffusiometer [PHYS] An instrument which measures diffusion in liquids.

diffusion [ELECTR] A method of producing a junction by diffusing an impurity metal into a semiconductor at a high temperature. [MECH ENG] The conversion of air velocity into static pressure in the diffuser casing of a centrifugal fan, resulting from increases in radius of the air spin and in area.

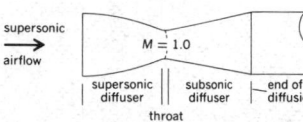
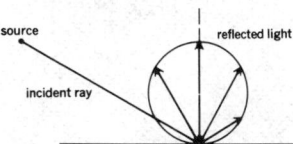

[METEOROL] The exchange of fluid parcels (and hence the transport of conservative properties) between regions in space, in the apparently random motions of the parcels on a scale too small to be treated by the equations of motion; the diffusion of momentum (viscosity), vorticity, water vapor, heat (conduction), and gaseous components of the atmospheric mixture have been studied extensively. [OPTICS] **1.** The distribution of incident light by reflection. **2.** Transmission of light through a translucent material. [PHYS] The spontaneous movement and scattering of particles (atoms and molecules), of liquids, gases, and solids. [SOLID STATE] **1.** The actual transport of mass, in the form of discrete atoms, through the lattice of a crystalline solid. **2.** The movement of carriers in a semiconductor.

diffusion annealing [MET] Heat treatment of metal to promote homogeneity by diffusion of components.

diffusion barrier [CHEM ENG] Porous barrier through which gaseous mixtures are passed for enrichment of the lighter-molecular-weight constituent of the diffusate; used as a many-stage cascade system for the recovery of $U^{235}F_6$ isotopes from a $U^{238}F_6$ stream.

diffusion bonding [MET] A solid-state process for joining metals by using only heat and pressure to achieve atomic bonding.

diffusion capacitance [ELECTR] The rate of change of stored minority-carrier charge with the voltage across a semiconductor junction.

diffusion cloud chamber [NUCLEO] A cloud chamber in which vapor diffuses from a source near a hot plate and condenses on a cold plate; the resulting layer of supersaturated vapor between the plates is sensitive to the passage of ionizing particles.

diffusion coating [MET] An alloy coating produced by allowing the coating material to diffuse into the base at high temperature.

diffusion coefficient [PHYS] The weight of a material, in grams, diffusing across an area of 1 square centimeter in 1 second in a unit concentration gradient. Also known as diffusivity.

diffusion constant [SOLID STATE] The diffusion current density in a homogeneous semiconductor divided by the charge carrier concentration gradient.

diffusion current [ANALY CHEM] In polarography with a dropping-mercury electrode, the flow that is controlled by the rate of diffusion of the active solution species across the concentration gradient produced by the removal of ions or molecules at the electrode surface.

diffusion diagram [METEOROL] A diagram for displaying the comparative properties of various diffusion processes, with coordinates of the mean free path or mixing length and mean molecular speed or diffusion velocity, for molecular or eddy diffusion, respectively; each point of the diagram determines diffusivity.

diffusion disk [OPTICS] A piece of transparent material that is marked or embossed, and is used with a camera lens to give the image a hazy softened quality. Also known as diffusing disk.

diffusion equation [PHYS] **1.** An equation for diffusion which states that the rate of change of the density of the diffusing substance, at a fixed point in space, equals the sum of the diffusion coefficient times the Laplacian of the density, the amount of the quantity generated per unit volume per unit time, and the negative of the quantity absorbed per unit volume per unit time. **2.** More generally, any equation which states that the rate of change of some quantity, at a fixed point in space, equals a positive constant times the Laplacian of that quantity.

diffusion flame [CHEM] A long gas flame that radiates uniformly over its length and precipitates free carbon uniformly.

diffusion gradient [PHYS] The graphed distance of penetration (diffusion) versus concentration of the material (or effect) diffusing through a second material; applies to heat, liquids, solids, or gases.

diffusion hygrometer [ENG] A hygrometer based upon the diffusion of water vapor through a porous membrane; essentially, it consists of a closed chamber having porous walls and containing a hygroscopic compound, whose absorption of

water vapor causes a pressure drop within the chamber that is measured by a manometer.

diffusion kernel [NUCLEO] The neutron flux resulting from a point source emitting one neutron per second; it is a function of the distance between the source and the point where the flux is measured.

diffusion length [PHYS] The average distance traveled by a particle, such as a minority carrier in a semiconductor or a thermal neutron in a nuclear reactor, from the point at which it is formed to the point at which it is absorbed.

diffusion number [FL MECH] A dimensionless number used in the study of mass transfer, equal to the diffusivity of a solute through a stationary solution contained in the solid, times a characteristic time, divided by the square of the distance from the midpoint of the solid to the surface. Symbolized β.

diffusion plant [NUCLEO] A plant which separates isotopes by isotopic diffusion or thermal diffusion.

diffusion pump [ENG] A vacuum pump in which a stream of heavy molecules, such as mercury vapor, carries gas molecules out of the volume being evacuated; also used for separating isotopes according to weight, the lighter molecules being pumped preferentially by the vapor stream.

diffusion respiration [PHYSIO] Exchange of gases through the cell membrane, between the cells of unicellular or other simple organisms and the environment.

diffusion theory [ELEC] The theory that in semiconductors, where there is a variation of carrier concentration, a motion of the carriers is produced by diffusion in addition to the drift determined by the mobility and the electric field.

diffusion-transfer process [GRAPHICS] Any of several photographic processes for copying documents in which the copy is produced by developing a photographic image, transferring by diffusion the silver salts in undeveloped areas of the receiving paper, and developing the transferred image.

diffusion transistor [ELECTR] A transistor in which current flow is a result of diffusion of carriers, donors, or acceptors, as in a junction transistor.

diffusion velocity [FL MECH] **1.** The relative mean molecular velocity of a selected gas undergoing diffusion in a gaseous atmosphere, commonly taken as a nitrogen (N_2) atmosphere; a molecular phenomenon that depends upon the gaseous concentration as well as upon the pressure and temperature gradients present. **2.** The velocity or speed with which a turbulent diffusion process proceeds as evidenced by the motion of individual eddies.

diffusive equilibrium [METEOROL] The steady state resulting from the diffusion process, primarily of interest when external forces and sources and sinks exist within the field; in such a state the constituent gases of the atmosphere would be distributed independently of each other, the heavier decreasing more rapidly with height than the lighter; but the presence of turbulent mixing precludes establishment of complete diffusive equilibrium.

diffusivity [PHYS] *See* diffusion coefficient. [THERMO] The quantity of heat passing normally through a unit area per unit time divided by the product of specific heat, density, and temperature gradient. Also known as thermometric conductivity.

diffusivity analysis [ANALY CHEM] Analysis of difficult-to-separate materials in solution by diffusion effects, using, for example, dialysis, electrodialysis, interferometry, amperometric titration, polarography, or voltammetry.

digallic acid *See* tannic acid.

digastric [ANAT] Of a muscle, having a fleshy part at each end and a tendinous part in the middle.

Digenea [INV ZOO] A group of parasitic flatworms or flukes constituting a subclass or order of the class Trematoda and having two types of generations in the life cycle.

digenite [MINERAL] Cu_9S_5 A blue to black mineral consisting of an isometric copper sulfide having a variable deficiency in copper. Also known as alpha chalcocite; blue chalcocite.

Di George's syndrome *See* thymic aplasia.

digested sludge [CIV ENG] Sludge or thickened mixture of sewage solids with water that has been decomposed by anaerobic bacteria.

digester [CHEM ENG] A vessel used to produce cellulose pulp from wood chips by cooking under pressure. [CIV ENG] A

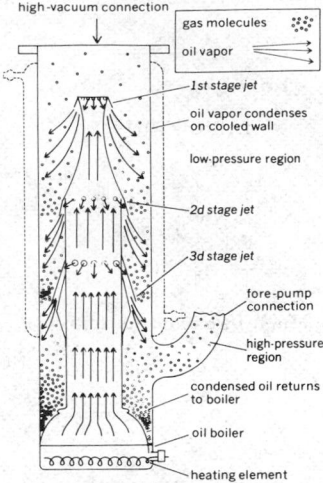

DIFFUSION PUMP

high-vacuum connection

gas molecules
oil vapor

1st stage jet
oil vapor condenses on cooled wall
low-pressure region
2d stage jet
3d stage jet
fore-pump connection
high-pressure region
condensed oil returns to boiler
oil boiler
heating element

Main operating features of diffusion pump.

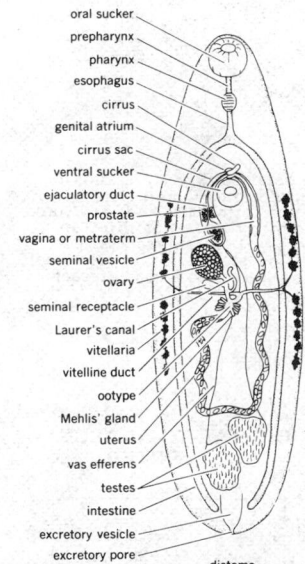

DIGENEA

oral sucker
prepharynx
pharynx
esophagus
cirrus
genital atrium
cirrus sac
ventral sucker
ejaculatory duct
prostate
vagina or metraterm
seminal vesicle
ovary
seminal receptacle
Laurer's canal
vitellaria
vitelline duct
ootype
Mehlis' gland
uterus
vas efferens
testes
intestine
excretory vesicle
excretory pore
distome

Diagram of an adult digenetic trematode. (*From R. M. Cable, An Illustrated Laboratory Manual of Parasitology, Burgess, 1940*)

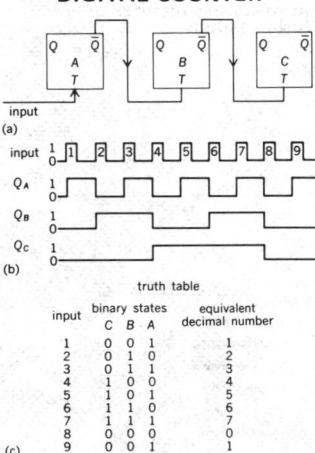

DIGICITRIN

Structural formula of digicitrin.

DIGITAL COUNTER

An octal counter. *T* stands for trigger input, *Q* and *Q̄* represent output terminals, and *A*, *B*, and *C* identify different flip-flop stages. (*a*) Three successive flip-flop stages. (*b*) Input signal and *Q*-terminal states of each flip-flop. (*c*) Truth table.

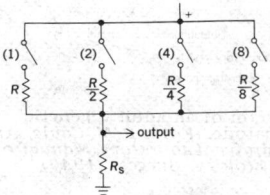

DIGITAL-TO-ANALOG CONVERTER

Circuit diagram for simple digital-to-analog converter. Each of the four switches represents binary 1 when closed and 0 when open. Current flowing through summing resistor R_s and voltage at output are proportional to value of number represented.

sludge-digestion tank containing a system of hot water or steam pipes for heating the sludge.

digestion [CHEM ENG] **1.** Preferential dissolving of mineral constituents in concentrations of ore. **2.** Liquefaction of organic waste materials by action of microbes. **3.** Separation of fabric from tires by the use of hot sodium hydroxide. **4.** Removing lignin from wood in manufacture of chemical cellulose paper pulp. [CIV ENG] The process of sewage treatment by the anaerobic decomposition of organic matter. [PHYSIO] The process of converting food to an absorbable form by breaking it down to simpler chemical compounds.

digestive enzyme [BIOCHEM] Any enzyme that causes or aids in digestion.

digestive gland [PHYSIO] Any structure that secretes digestive enzymes.

digestive system [ANAT] A system of structures in which food substances are digested.

digestive tract [ANAT] The alimentary canal.

digger [ENG] A tool or apparatus for digging in the ground. [MIN ENG] A person who digs in the ground; usually refers to a coal miner.

digging [ENG] A sudden increase in cutting depth of a cutting tool due to an erratic change in load.

digging height *See* bank height.

digging line *See* inhaul cable.

diggings [SCI TECH] **1.** Excavated materials. **2.** A place of excavating.

digicitrin [BIOCHEM] $C_{21}H_{21}O_{10}$ A flavone compound that is found in foxglove leaves.

digicom [COMMUN] A wire communication system that transmits speech signals in the form of corresponding trains of pulses and transmits digital information directly from computers, radar, tape readers, teleprinters, and telemetering equipment.

digit [ADP] In a decimal digital computer, the space reserved for the storage of one digit of information. [MATH] A character used to represent one of the nonnegative integers smaller than the base of a system of positional notation. Also known as numeric character.

digit absorbing selector [ELECTR] Dial switch arranged to set up and then fall back on the first one of two digits dialed; it then operates on the next digit dialed.

digital [ADP] Pertaining to data in the form of digits.

Digital Address Beacon System [NAV] The future discrete address beacon system, adopted by the Federal Aviation Agency, intended to augment the presently used air-traffic control radar beacon system for gathering information for the control of air traffic; the system provides for surveillance by the discrete addressing of identification-interrogation to specific aircraft; it will add a new ground-to-air digital link for passing routine air-traffic control and flight information. Abbreviated DABS.

digital communications [COMMUN] System of telecommunications employing a nominally discontinuous signal that changes in frequency, amplitude, or polarity.

digital comparator [ELECTR] A comparator circuit operating on input signals at discrete levels. Also known as discrete comparator.

digital computer [ADP] A computer operating on discrete data by performing arithmetic and logic processes on these data.

digital converter [ELECTR] A device that converts voltages to digital form; examples include analog-to-digital converters, pulse-code modulators, encoders, and quantizing encoders.

digital counter [ELECTR] A discrete-state device (one with only a finite number of output conditions) that responds by advancing to its next output condition.

digital data [ADP] Data represented by a sequence of code characters.

Digital Data Broadcast System [NAV] A system that will provide information aiding air-traffic control; digital data to aircraft over vortac channels will carry information on the geographic location, elevation, magnetic variation, and related data of the vortac station being received. Abbreviated DDBS.

digital data recorder [ADP] Electronic device that converts continuous electrical analog signals into number (digital)

values and records these values onto a data log via a high-speed typewriter.

digital differential analyzer [ADP] A differential analyzer which uses numbers to represent analog quantities. Abbreviated DDA.

digital display [ADP] A display in which the result is indicated in directly readable numerals.

digital filter [ELECTR] An electrical filter that responds to an input which has been quantified, usually as pulses.

digital frequency meter [ELECTR] A frequency meter in which the value of the frequency being measured is indicated on a digital display.

digital intercontinental conversion equipment [ELECTR] Equipment which uses pulse-code modulation to convert a 525-line, 60-frame-per-second television signal used in the United States into a 625-line, 50-frame-per-second phase-alternation line signal used in Europe; the 525-line signal is sampled and quantized into a pulse-code modulation signal which is stored in shift registers from which the phase-alternation line signal is read out. Abbreviated DICE.

digitalis [PHARM] The dried leaf of the purple foxglove plant (*Digitalis purpurea*), containing digitoxin and gitoxin; constitutes a powerful cardiac stimulant and diuretic.

Digitalis [BOT] A genus of herbs in the figwort family, Scrophulariaceae.

digital message entry system [ELECTR] A system that encodes formatted messages in digital form; it enters the encoded digital information into a voice communications transceiver by frequency shift techniques.

digital phase shifter [ELECTR] Device which provides a signal phase shift by the application of a control pulse; a reversal or phase shift requires a control pulse of opposite polarity.

digital recording [ELECTR] Magnetic recording in which the information is first coded in a digital form, generally with a binary code that uses two discrete values of residual flux.

digital simulation [ADP] The representation of a system in a form acceptable to a digital computer as opposed to an analog computer.

digital speech communications [COMMUN] Transmission of voice in digitized or binary form via landline or radio.

digital system [ADP] Any of the levels of operation for a digital computer, including the wires and mechanical parts, the logical elements, and the functional units for reading, writing, storing, and manipulating information.

digital telemetering [ADP] Conversion of a continuous electrical analog signal into a digital (number system) code prior to transmitting the signal to a receiver (such as digital readout, card punch, or tape).

digital television [COMMUN] Television in which picture redundancy is reduced or eliminated by transmitting only the data needed to define motion in the picture, as represented by changes in the areas of continuous white or continuous black.

digital-to-analog converter [ELECTR] A converter in which digital input signals are changed to essentially proportional analog signals. Abbreviated dac.

digital transducer [ELECTR] A transducer that measures physical quantities and transmits the information as coded digital signals rather than as continuously varying currents or voltages.

digital voltmeter [ELECTR] A voltmeter in which the unknown voltage is compared with an internally generated analog voltage, the result being indicated in digital form rather than by a pointer moving over a meter scale.

digitate [ANAT] Having digits or digitlike processes.

digitation [GEOL] A secondary recumbent anticline emanating from a larger recumbent anticline.

digit-coded voice [ADP] A limited, spoken vocabulary, each word of which corresponds to a code and which, upon keyed inquiry, can be strung in meaningful sequence and can be outputted as audio response to the inquiry.

digit compression [ADP] Any process which increases the number of digits stored at a given location.

digitellum [INV ZOO] A tentaclelike gastric filament in scyphozoans.

digit emitter [ADP] A character emitter limited to the twelfth-row pulses in a punched card.

digitigrade [VERT ZOO] Pertaining to animals, such as dogs and cats, which walk on the digits with the posterior part of the foot raised from the ground.

digitinervate [BOT] Having straight veins extending from the petiole like fingers.

digitipinnate [BOT] Having digitate leaves with pinnate leaflets.

digitize [ADP] To convert an analog measurement of a quantity into a numerical value.

digitonin [PHARM] $C_{41}H_{64}O_{13}$ A glycoside derived from the purple foxglove plant (*Digitalis purpurea*); a white powder melting at 255–256°C, used as a medicine for cardiac conditions.

digitoxigenin [ORG CHEM] $C_{23}H_{34}O_4$ The steroid aglycone formed by removal of three molecules of the sugar digitoxose from digitoxin.

digitoxin [ORG CHEM] $C_{41}H_{64}O_{13}$ A poisonous steroid glycoside found as the most active principle of digitalis, from the foxglove leaf.

digit selector [ADP] A device which separates a card column into individual pulses corresponding to punched row positions.

digitus [INV ZOO] In insects, the claw-bearing terminal segment of the tarsus.

diglucoside [BIOCHEM] A compound containing two glucose molecules.

diglycerol [ORG CHEM] A compound that is a diester of glycerol.

diglycolic acid [ORG CHEM] $O(CH_2COOH)_2$ A white powder that forms a monohydrate; used in the manufacture of plasticizers and in organic synthesis, and to break emulsions.

digoxin [ORG CHEM] $C_{41}H_{64}O_{14}$ A crystalline steroid obtained from a foxglove leaf (*Digitalis lanata*); similar to digitalis in pharmacological effects.

digraph *See* directed graph.

dihalide [CHEM] A molecule containing two atoms of halogen combined with a radical or element.

dihedral [AERO ENG] The upward or downward inclination of an airplane's wing or other supporting surface in respect to the horizontal; in some contexts, the upward inclination only. [MATH] *See* dihedron.

dihedral angle [MATH] The angle between two planes; it is said to be zero if the planes are parallel; if the planes intersect, it is the plane angle between two lines, one in each of the planes, which pass through a point on the line of intersection of the two planes and are perpendicular to it.

dihedral reflector [OPTICS] A corner reflector having two sides meeting at a line.

dihedron [MATH] A geometric figure formed by two half planes that are bounded by the same straight line. Also known as dihedral.

diheptal base [ELECTR] A tube base having 14 pins or 14 possible pin positions; used chiefly on television cathode-ray tubes.

dihexagonal [CRYSTAL] Of crystals, having a symmetrical form with 12 sides.

dihexagonal-dipyramidal [CRYSTAL] Characterized by the class of crystals in the hexagonal system in which any section perpendicular to the sixfold axis is dihexagonal.

dihexahedron [CRYSTAL] A type of crystal that has 12 faces, such as a double six-sided pyramid.

dihexy *See* dodecane.

dihydrate [CHEM] A compound with two molecules of water of hydration.

dihydrazone [ORG CHEM] A molecule containing two hydrazone radicals.

dihydro- [CHEM] A prefix indicating combination with two atoms of hydrogen.

dihydrobenzene *See* 1,3-cyclohexadiene.

dihydrocamphane *See* camphane.

dihydrochloride [CHEM] A compound containing two molecules of hydrochloric acid.

dihydrodiketoanthracene *See* anthraquinone.

dihydrostreptomycin [MICROBIO] $C_{21}H_{41}O_{12}N_7$ A hydrogenated derivative of streptomycin having the same action as streptomycin.

dihydroxy [CHEM] A molecule containing two hydroxyl groups.

dihydroxyacetone phosphate *See* dihydroxyacetonephosphoric acid.

dihydroxyacetonephosphoric acid [BIOCHEM] $C_3H_7O_6P$ A phosphoric acid ester of dehydroxyacetone, produced as an intermediate substance in the conversion of glycogen to lactic acid during muscular contraction. Also known as dihydroxyacetone phosphate.

dihydroxy alcohol *See* glycol.

1,2-dihydroxyanthraquinone *See* alizarin.

2,4-dihydroxybenzene carboxylic acid *See* β-resorcylic acid.

2,4-dihydroxybenzoic acid *See* β-resorcylic acid.

3,4-dihydroxycinnamic acid *See* caffeic acid.

p,p′-dihydroxydiphenyldimethyane *See* bisphenol.

dihydroxyethylsulfide *See* thiodiglycol.

3,5-dihydroxy-3-methyl valeric acid *See* mevalonic acid.

dihydroxyphenylalanine [BIOCHEM] $C_9H_{11}NO_4$ An amino acid that can be formed by oxidation of tyrosine; it is converted by a series of biochemical transformations, utilizing the enzyme dopa oxidase, to melanins. Also known as dopa.

2,3-dihydroxypropanal *See* glyceraldehyde.

1,2-dihydroxypropane *See* propylene glycol.

dihydroxysuccinic acid *See* tartaric acid.

diiodomethane *See* methylene iodide.

di-iron enneacarbonyl *See* iron nonacarbonyl.

diisobutylene [ORG CHEM] C_8H_{16} Any one of a number of isomers, but most often 2,4,4-trimethylpentene-1 and 2,4,4-trimethylpentene-2; used in alkylation and as a chemical intermediate.

diisobutyl ketone [ORG CHEM] $(CH_3)_2CHCH_2COCH_2-CH(CH_3)_2$ Stable liquid, boiling at 168°C; soluble in most organic liquids; toxic and flammable; used as a solvent, in lacquers and coatings, and as a chemical intermediate.

diisobutyl phenol *See* octyl phenol.

diisocyanate [ORG CHEM] A compound that contains two NCO (isocyanate) groups; used to produce polyurethane foams, resins, and rubber.

diisopropyl [ORG CHEM] A molecule containing two isopropyl groups.

diisopropyl ether *See* isopropyl ether.

diisopropyl phosphorofluoridate [PHARM] $[(CH_3)_2CHO]_2-FPO$ A colorless, oily liquid that inhibits cholinesterase, prolongs meiosis, and is effective in treating glaucoma.

dikaryon [MYCOL] Also spelled dicaryon. **1.** A pair of distinct, unfused nuclei in the same cell brought together by union of plus and minus hyphae in certain mycelia. **2.** Mycelium containing dikaryotic cells.

dike [CIV ENG] An embankment constructed on dry ground along a riverbank to prevent overflow of lowlands and to retain floodwater. [GEOL] A tabular body of igneous rock that cuts across adjacent rocks or cuts massive rocks.

dike ridge [GEOL] Any small wall-like ridge created by differential erosion.

dike set [GEOL] A small group of dikes arranged linearly or parallel to each other.

dike swarm [GEOL] A large group of parallel, linear, or radially oriented dikes.

diketene [ORG CHEM] $CH_3COCHCO$ A colorless, readily polymerized liquid with pungent aroma; insoluble in water, soluble in organic solvents; used as a chemical intermediate. Also known as acetyl ketene.

diketone [ORG CHEM] A molecule containing two ketone carbonyl groups.

diketopiperazine [ORG CHEM] **1.** $C_4H_6N_2O_2$ A compound formed by dehydration of two molecules of glycine. Also known as 2,5-piperazine-dione. **2.** Any of the cyclic molecules formed from α-amino acids other than glycine or by partial hydrolysis of protein.

2,5-diketopyrrolidine *See* succinimide.

2,5-diketotetrahydrofurane *See* succinic anhydride.

diktoma *See* neuroepithelioma.

dilactone [ORG CHEM] A molecule that contains two lactone groups.

Dilantin [PHARM] A trademark for diphenlhydantoin, an anticonvulsant in the treatment of epilepsy; the compound is also used as the sodium derivative.

dilatancy [CHEM] The property of a viscous suspension

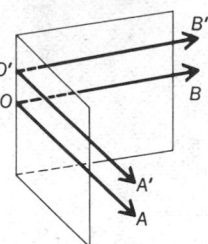

DIHEDRON

Dihedron is formed by the half-planes bounded by the line passing through O and O'. AOB and $A'O'B'$ are dihedral angles.

under the influence of pressure. [GEOL] ... formed masses of granular material, such as ... rrangement of the component grains.

... A material with the ability to increase in ... shape is changed.

...] The increase in volume per unit volume of ... substance, caused by deformation.

... *See* compressional wave.

... A transformation which changes the size, ... , of a geometric figure. [SCI TECH] The act ... etching or expanding.

... G] An instrument for measuring thermal ex- ... ation of liquids or solids.

... YS] The measurement of changes in the vol- ... l or dimensions of a solid which occur in ... h as allotropic transformations, thermal ex- ... ession, creep, or magnetostriction.

dilator [PHYSIO] Any muscle, instrument, or drug causing dilation of an organ or part.

dill [BOT] *Anethum graveolens.* A small annual or biennial herb in the family Umbelliferae; the aromatic leaves and seeds are used for food flavoring.

Dilleniaceae [BOT] A family of dicotyledonous trees, woody vines, and shrubs in the order Dilleniales having hypogynous flowers and mostly entire leaves.

Dilleniales [BOT] An order of dicotyledonous plants in the subclass Dilleniidae characterized by separate carpels and numerous stamens.

Dilleniidae [BOT] A subclass of plants in the class Magnoliopsida distinguished by being syncarpous, having centrifugal stamens, and usually having bitegmic ovules and binucleate pollen.

dill oil [MATER] A yellowish essential oil, soluble in propylene glycol, slightly soluble in glycerine, obtained by steam distillation of the dill plant *Anethum graveolens;* chief ingredient is carvone. Also known as American dillweed oil; anethum oil.

diluent [CHEM] An inert substance added to some other substance or solution so that the volume of the latter substance is increased and its concentration per unit volume is decreased.

dilute [CHEM] To make less concentrated.

dilute phase [CHEM ENG] In liquid-liquid extraction, the liquid phase that is dilute with respect to the material being extracted.

dilution [CHEM] Increasing the proportion of solvent to solute in any solution and thereby decreasing the concentration of the solute per unit volume. [OPTICS] Reducing the intensity of a color by adding white.

dilution gene [GEN] Any modifier gene that acts to reduce the effect of another gene.

dilution method [MICROBIO] A technique in which a series of cultures is tested with various concentrations of an antibiotic to determine the minimum inhibiting concentration of antibiotic.

DIM *See* nonthermal decimetric emission.

dimension [GRAPHICS] In a mechanical drawing, a labeled measure in a straight line of the breadth, height, or thickness of a part, the angular position of a line, or the location of a detail such as a hole or boss.

dimensional analysis [PHYS] A technique that involves the study of dimensions of physical quantities, used primarily as a tool for obtaining information about physical systems too complicated for full mathematical solutions to be feasible.

dimensional constant [PHYS] A physical quantity whose numerical value depends on the units chosen for fundamental quantities but not on the system being considered.

dimensional formula [PHYS] The expression of a derived quantity as a product of powers of the fundamental quantities.

dimensional stability [GRAPHICS] The percentage of change in size of paper under two different conditions of temperature and humidity. [TEXT] Shape-retaining quality of material.

dimension declaration statement [ADP] A FORTRAN statement identifying arrays and specifying the number and bounds of the subscripts.

DIMENSIONING

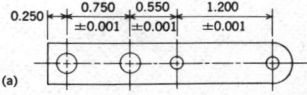

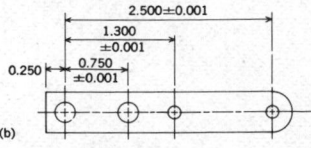

Two standard methods for marking dimensions on mechanical drawings. (a) Point-to-point dimensioning. (b) Datum dimensioning.

dimensioning [GRAPHICS] Assigning of dimensions to a mechanical drawing.

dimensionless group [PHYS] Any combination of dimensional or dimensionless quantities possessing zero overall dimensions; an example is the Reynold's number.

dimensionless number [MATH] A ratio of various physical properties (such as density or heat capacity) and conditions (such as flow rate or weight) of such nature that the resulting number has no defining units of weight, rate, and so on. Also known as nondimensional parameter.

dimension line [GRAPHICS] A line on a drawing pointing to another line or part to which the dimensions relate.

dimension of a simplex [MATH] One less than the number of vertices of the simplex.

dimension of a vector space [MATH] The number of vectors in any basis of the vector space.

dimensions [PHYS] The product of powers of fundamental quantities (or of convenient derived quantities) which are used to define a physical quantity; the fundamental quantities are often mass, length, and time.

dimension stone [MATER] Large, sound, relatively flawless blocks of stone used as building stone, monumental stone, paving stone, curbing, and flagging.

dimension theory [MATH] The study of abstract notions of dimension, which are topological invariants of a space.

dimer [CHEM] A condensation product consisting of two molecules.

dimercaprol [PHARM] $C_3H_8OS_2$ 2,3-Dimercapto-1-propanol, a colorless, water-soluble oily liquid with a mercaptanlike odor; used as an antidote for arsenic, gold, and mercury poisoning. Also known as British antilewisite (BAL); dithiol.

dimeric water [INORG CHEM] Water in which pairs of molecules are joined by hydrogen bonds.

dimerization [CHEM] Formation of dimers.

dimerous [BIOL] Composed of two parts.

dimetan [ORG CHEM] The generic name for 5,5-dimethyldehydroresorcinol dimethylcarbamate, a synthetic carbamate insecticide.

dimethoate [ORG CHEM] $C_5H_{12}NO_3PS_2$ A crystalline compound, soluble in most organic solvents; used as an insecticide.

dimethoxymethane *See* methylal.

dimethoxy strychnine *See* brucine.

6,7-dimethoxy-1-veratrylisoquinoline *See* papaverine.

dimethyl [ORG CHEM] A compound that has two methyl groups.

dimethylamine [ORG CHEM] $(CH_3)_2NH$ Flammable gas with ammonia aroma, boiling at 7°C; soluble in water, ether, and alcohol; used as an acid-gas absorbent, solvent, and flotation agent, in pharmaceuticals and electroplating, and in dehairing hides.

N,N-dimethylaniline [ORG CHEM] $C_6H_5N(CH_3)_2$ A yellowish liquid slightly soluble in water; used in dyes and solvent and in the manufacture of vanillin. Also known as aniline N,N-dimethyl.

dimethylarsinic acid *See* cacodylic acid.

dimethylbenzene *See* xylene.

1,2-dimethylbenzene *See* ortho-xylene.

1,3-dimethylbenzene *See* meta-xylene.

1,4-dimethylbenzene *See* para-xylene.

dimethyl diaminophenazine chloride *See* neutral red.

dimethyl ether [ORG CHEM] CH_3OCH_3 A flammable, colorless liquid, boiling at −25°C; soluble in water and alcohol; used as a solvent, extractant, reaction medium, and refrigerant. Also known as methyl ether; wood ether.

dimethylethylene *See* butylene.

N,N-dimethylformamide [ORG CHEM] $HCON(CH_3)_2$ A liquid that boils at 152.8°C; extensively used as a solvent for organic compounds. Abbreviated DMF.

dimethylglyoxime [ORG CHEM] $(CH_3)_2C_2(NOH)_2$ White, crystalline or powdered solid, used in analytical chemistry as a reagent for nickel.

1,1-dimethylhydrazine *See* uns-dimethylhydrazine.

uns-dimethylhydrazine [ORG CHEM] $(CH_3)_2NNH_2$ A flammable, highly toxic, colorless liquid; used as a component of rocket and jet fuels and as a stabilizer for organic peroxide fuel additives. Also known as 1,1-dimethylhydrazine; UDMH.

dimethylhydroxybenzene *See* xylenol.

1,6-dimethyl-4-isopropyl naphthalene *See* cadalene.

dimethylmethane *See* propane.

3,3-dimethyl-2-methylenenorcamphane *See* camphene.

3,7-dimethyl-2,6-octadienal *See* citral.

dimethyloctadiene-2,7-ol-6 *See* linalool.

3,7-dimethyl-6(or 7)octen-1-ol *See* citronellol.

dimethylolurea [ORG CHEM] $CO(NHCH_2OH)_2$ Colorless crystals melting at 126°C, soluble in water; used to increase fire resistance and hardness of wood, and in textiles to prevent wrinkles. Also known as 1,3-bis-hydroxymethylurea; DMU.

dimethylphenol *See* xylenol.

dimethyl phthalate [ORG CHEM] $C_6H_4(COOCH_3)_2$ Odorless, colorless liquid, boiling at 282°C; soluble in organic solvents, slightly soluble in water; used as a plasticizer, in resins, lacquers, and perfumes, and as an insect repellent.

2,2-dimethylpropane *See* neopentane.

dimethyl sebacate [ORG CHEM] $[(CH_2)_4COOCH_3]_2$ Clear, colorless liquid, boiling at 294°C; used as a vinyl resin, nitrocellulose solvent, or plasticizer.

dimethyl sulfate [ORG CHEM] $(CH_3)_2SO_4$ Poisonous, corrosive, colorless liquid, boiling at 188°C; slightly soluble in water, soluble in ether and alcohol; used to methylate amines and phenols. Also known as methyl sulfate.

dimethyl sulfide *See* methyl sulfide.

dimethyl sulfoxide [ORG CHEM] $(CH_3)_2SO$ A colorless liquid used as a local analgesic and anti-inflammatory agent, as a solvent in industry, and in laboratories as a medium for carrying out chemical reactions. Abbreviated DMSO.

dimethyl terephthalate [ORG CHEM] $C_6H_4(COOCH_3)_2$ Colorless crystals, melting at 140°C and subliming above 300°; slightly soluble in water, soluble in hot alcohol and ether; used to make polyester fibers and film. Abbreviated DMT.

3,7-dimethylxanthine *See* theobromine.

dimetilan [ORG CHEM] A generic name for 1-(dimethylcarbamoyl)-5-methyl-3-pyrazolyl dimethylcarbamate, a synthetic carbamate insecticide.

dimictic lake [HYD] A lake which circulates twice a year.

diminution [BOT] Increasing simplification of inflorescences on successive branches.

dimmer [ELEC] An electrical or electronic control for varying the intensity of a lamp or other light source.

dimmerfoehn [METEOROL] A rare form of foehn where, during a very strong upper wind from the south, a pressure difference of 12 millibars or more exists between the south and north sides of the Alps; a stormy foehn wind then overleaps the upper valleys in the northern slopes, reaches the ground in the lower parts of the valleys, and enters the foreground as a very strong wind; the foehn wall and the precipitation area extend beyond the crest across the almost calm surface area in the upper valleys.

dimorphism [CHEM] Having crystallization in two forms with the same chemical composition. [SCI TECH] Existing in two distinct forms, with reference to two members expected to be identical.

dimorphite [MINERAL] As_4S_3 An orange-yellow mineral consisting of arsenic sulfide.

dimpling [ENG] Forming a conical depression in a metal surface in order to countersink a rivet head.

Dimylidae [PALEON] A family of extinct lipotyphlan mammals in the order Insectivora; a side branch in the ancestry of the hedgehogs.

dina [ELECTR] An airborne radar-jamming transmitter operating in the band from 92 to 210 megahertz with an output of 30 watts, radiating noise in one side band for spot or barrage jamming; the carrier and the other side band are suppressed.

Dinantian [GEOL] Lower Carboniferous geologic time. Also known as Avonian.

Dinarides [GEOGR] A mountain system, east of the Adriatic Sea, in Yugoslavia.

D indicator *See* D scope.

Dines anemometer [ENG] A pressure-tube anemometer in which the pressure head on a weather vane is kept facing into the wind, and the suction head, near the bearing which supports the vane, develops a suction independent of wind direction; the pressure difference between the heads is proportional to the square of the wind speed and is measured by a float manometer with a linear wind scale.

dineutron [NUC PHYS] **1.** A hypothetical bound state of two neutrons, which probably does not exist. **2.** A combination of two neutrons which has a transitory existence in certain nuclear reactions.

dinghy [NAV ARCH] A boat, less than 20 feet (6 meters) long, propelled by oars or sails, that may be used as a tender to a ship or yacht.

dingot [MET] A massive derby, usually a ton or more, produced in a bomb reaction.

Dings magnetic separator [MECH ENG] A device which is suspended above a belt conveyor to pull out and separate magnetic material from burden as thick as 40 inches (1 meter) and at belt speeds up to 750 feet (229 meters) per minute.

Dinidoridae [INV ZOO] A family of hemipteran insects in the superfamily Pentatomoidea.

dinitrate [CHEM] A molecule that contains two nitrate groups.

dinitrite [CHEM] A molecule that has two nitrite groups.

4,6-dinitro-2-aminophenol *See* picramic acid.

dinitrobenzene [ORG CHEM] Any one of three isomeric substitution products of benzene having the empirical formula $C_6H_4(NO_2)_2$.

dinitrogen fixation *See* nitrogen fixation.

dinitrogen tetroxide *See* nitrogen dioxide.

dinitrophenol [ORG CHEM] Any one of six isomeric substituent products of benzene having the empirical formula $(NO_2)_2C_6H_3OH$.

dinitrotoluene [ORG CHEM] Any one of six isomeric substitution products of benzene having the empirical formula $CH_3C_6H_3(NO_2)_2$; they are high explosives formed by nitration of toluene. Abbreviated DNT.

dinking [MECH ENG] Using a sharp, hollow punch for cutting light-gage soft metals or nonmetallic materials.

Dinocerata [PALEON] An extinct order of large, herbivorous mammals having semigraviportal limbs and hoofed, five-toed feet; often called uintatheres.

Dinoflagellata [INV ZOO] The equivalent name for Dinoflagellida.

Dinoflagellida [INV ZOO] An order of flagellate protozoans in the class Phytamastigophorea; most members have fixed shapes determined by thick covering plates.

Dinophilidae [INV ZOO] A family of annelid worms belonging to the Archiannelida.

Dinophyceae [BOT] The dinoflagellates, a class of thallophytes in the division Pyrrhophyta.

Dinornithiformes [PALEON] The moas, an order of extinct birds of New Zealand; all had strong legs with four-toed feet.

dinosaur [PALEON] The name, meaning terrible lizard, applied to the fossil bones of certain large, ancient bipedal and quadripedal reptiles placed in the orders Saurischia and Ornithischia.

DIN system [GRAPHICS] A system in photography used to find the speed of photographic emulsions; it is stated in terms of the logarithm of the reciprocal of the exposure needed to obtain a density of 0.1 above fog density.

diocoel [EMBRYO] The cavity of the diencephalon, which becomes the third brain ventricle.

dioctahedral [CRYSTAL] Pertaining to a crystal structure in which only two of the three available octahedrally coordinated positions are occupied. [MATH] Having 16 faces.

Dioctophymatida [INV ZOO] An order of parasitic nematode worms in the subclass Enoplia.

Dioctophymoidea [INV ZOO] An order or superfamily of parasitic nematodes characterized by the peculiar structure of the copulatory bursa of the male.

dioctyl [ORG CHEM] A compound that has two octyl groups.

dioctylmethane *See* heptadecane.

dioctyl phthalate [ORG CHEM] $(C_8H_{17}OOC)_2C_6H_4$ Pale, viscous liquid, boiling at 384°C; insoluble in water; used as a plasticizer for acrylate, vinyl, and cellulosic resins. Abbreviated DOP.

dioctyl phthalate test [ENG] A method used to evaluate air filters to be used in critical air-cleaning applications; a light-scattering technique counts the number of particles of con-

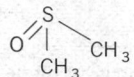

DIMETHYL SULFOXIDE

Structural formula for dimethyl sulfoxide.

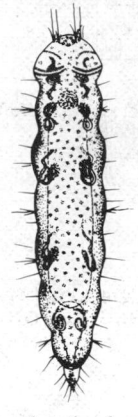

DINOPHILIDAE

Dinophilus female, dorsal view. *(From Ruebush, Trans. Amer. Micr. Sci., vol. 59, Fig. 1, 1940)*

DIODE DEMODULATOR

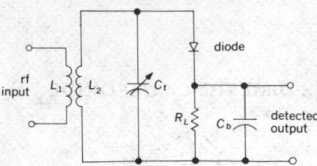

Circuit diagram of diode demodulator.

DIODE TRANSISTOR LOGIC

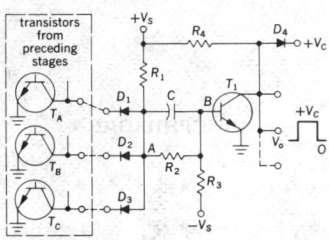

Circuit diagram of a diode transistor logic.

DIODE VOLTAGE REGULATOR

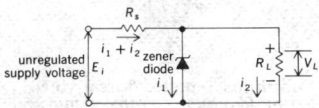

A Zener diode voltage regulator.

DIOPSIDE

Diopside crystals from St. Lawrence County, New York. (Specimen from Department of Geology, Bryn Mawr College)

1,4-DIOXANE

Structural formula of 1,4-dioxane.

---

trolled size (0.3 micrometer) entering and emerging from the test filter. Abbreviated DOP test.

dioctyl sebacate [ORG CHEM] $(CH_2)_8(COOC_8H_{17})_2$ Water-insoluble, straw-colored liquid, boiling at 248°C; used as a plasticizer for vinyl, cellulosic, and styrene resins.

diode [ELECTR] **1.** A two-electrode electron tube containing an anode and a cathode. **2.** *See* semiconductor diode.

diode amplifier [ELECTR] A parametric amplifier that uses a special diode in a cavity to amplify signals at frequencies as high as 6000 megahertz. Also called cavity-type diode amplifier.

diode-capacitor transistor logic [ELECTR] A circuit that uses diodes, capacitors, and transistors to provide logic functions.

diode characteristic [ELECTR] The composite electrode characteristic of an electron tube when all electrodes except the cathode are connected together.

diode clamp *See* diode clamping circuit.

diode clamping circuit [ELECTR] A clamping circuit in which a diode provides a very low resistance whenever the potential at a certain point rises above a certain value in some circuits or falls below a certain value in others. Also known as diode clamp.

diode clipping circuit [ELECTR] A clipping circuit in which a diode is used as a switch to perform the clipping action.

diode demodulator [ELECTR] A demodulator using one or more crystal or electron tube diodes to provide a rectified output whose average value is proportional to the original modulation. Also known as diode detector.

diode detector *See* diode demodulator.

diode function generator [ELECTR] A function generator that uses the transfer characteristics of resistive networks containing biased diodes; the desired function is approximated by linear segments.

diode gate [ELECTR] An AND gate that uses diodes as switching elements.

diode laser [OPTICS] A laser in which an active *pn* semiconductor junction converts direct-current, forward-bias input pump power into coherent, optical output power by a process of stimulated emission in a region near the junction. Also known as laser diode.

diode limiter [ELECTR] A peak-limiting circuit employing a diode that becomes conductive when signal peaks exceed a predetermined value.

diode mixer [ELECTR] A mixer that uses a crystal or electron tube diode; it is generally small enough to fit directly into a radio-frequency transmission line.

diode modulator [ELECTR] A modulator using one or more diodes to combine a modulating signal with a carrier signal; used chiefly for low-level signaling because of inherently poor efficiency.

diode pack [ELECTR] Combination of two or more diodes integrated into one solid block.

diode peak detector [ELECTR] Diode used in a circuit to indicate when peaks exceed a predetermined value.

diode-pentode [ELECTR] Vacuum tube having a diode and a pentode in the same envelope.

diode rectifier-amplifier meter [ELECTR] The most widely used vacuum tube voltmeter for measurement of alternating-current voltage; has separate tubes for rectification and direct-current amplification, permitting an optimum design for each.

diode-switch [ELECTR] Diode which is made to act as a switch by the successive application of positive and negative biasing voltages to the anode (relative to the cathode), thereby allowing or preventing, respectively, the passage of other applied waveforms within certain limits of voltage.

diode theory [ELEC] The theory that in a semiconductor, when the barrier thickness is comparable to or smaller than the mean free path of the carriers, then the carriers cross the barrier without being scattered, much as in a vacuum tube diode.

diode transistor logic [ELECTR] A circuit that uses diodes, transistors, and resistors to provide logic functions. Abbreviated DTL.

diode-triode [ELECTR] Vacuum tube having a diode and a triode in the same envelope.

diode voltage regulator [ELECTR] A voltage regulator with a Zener diode, making use of its almost constant voltage over a range of currents. Also known as Zener diode voltage regulator.

diodide [CHEM] A molecule that contains two iodine atoms bonded to an element or radical.

dioecious [BIOL] Having the male and female reproductive organs on different individuals. Also known as dioic.

d-i offset system *See* direct-image offset system.

diogenite [MINERAL] An achondritic stony meteorite composed essentially of iron-rich pyroxene minerals. Also known as rodite.

dioic *See* dioecious.

diolefin *See* diene.

diolefin hydrogenation [CHEM ENG] A fixed-bed catalytic process used to hydrogenate diolefins in C_4 and C_5 fractions to mono-olefin in alkylation feedstocks.

Diomedeidae [VERT ZOO] The albatrosses, a family of birds in the order Procellariiformes.

-dione [ORG CHEM] Suffix indicating the presence of two keto groups.

diophantine analysis [MATH] A means of determining integer solutions for certain algebraic equations.

diophantine equations [MATH] Equations with more than one independent variable and with integer coefficients for which integer solutions are desired.

Diopsidae [INV ZOO] The stalk-eyed flies, a family of myodarian cyclorrhaphous dipteran insects in the subsection Acalypteratae.

diopside [MINERAL] $CaMg(SiO_3)_2$ A white to green monoclinic pyroxene mineral which forms gray to white, short, stubby, prismatic, often equidimensional crystals. Also known as malacolite.

dioptase [MINERAL] $CuSiO_2(OH)_2$ A rare emerald-green mineral that forms hexagonal, hydrous crystals.

diopter [OPTICS] A measure of the power of a lens or a prism, equal to the reciprocal of its focal length in meters. Abbreviated D.

dioptometer [OPTICS] An instrument for determining ocular refraction.

dioptric [OPTICS] **1.** Serving in or effecting refraction. **2.** Produced by means of refraction.

dioptrics [OPTICS] The branch of optics that treats of the refraction of light, especially by the transparent medium of the eye, and by lenses.

diorchism [ANAT] Having two testes.

diorite [PETR] A phaneritic plutonic rock with granular texture composed largely of plagioclase feldspar with smaller amounts of dark-colored minerals; used occasionally as ornamental and building stone. Also known as black granite.

Dioscoreaceae [BOT] A family of monocotyledonous, leafy-stemmed, mostly twining plants in the order Liliales, having an inferior ovary and septal nectaries and lacking tendrils.

diose [BIOCHEM] A member of a class of monosaccharides that have two carbon atoms; the only compound that belongs to this class is glycolaldehyde.

1,4-dioxane [ORG CHEM] $C_4H_8O_2$ The cyclic ether of ethylene glycol; it is soluble in water in all proportions and is used as a solvent. Also known as diethylene dioxide; *para*-dioxane; glycol ethylene ether.

***para*-dioxane** *See* 1,4-dioxane.

dioxide [CHEM] A compound containing two atoms of oxygen.

dioxolane [ORG CHEM] $C_3H_6O_2$ A cyclic acetal that is a liquid; used as a solvent and extractant.

dioxopurine *See* xanthine.

dip [ENG] The vertical angle between the sensible horizon and a line to the visible horizon at sea, due to the elevation of the observer and to the convexity of the earth's surface. Also known as dip of horizon. [GEOL] **1.** The angle that a stratum or fault plane makes with the horizontal. Also known as angle of dip; true dip. **2.** A pronounced depression in the land surface.

dip brazing [MET] Soldering by dipping the work into a hot, molten salt or metal bath and by using a nonferrous metal with a melting point above 800°F (427°C).

dip coating [ENG] A coating applied to ceramic ware or metal by immersion into a tank of melted nonmetallic material, such as resin or plastic, then chilling the adhering melt.

dip correction *See* height-of-eye correction.

dip-dye [TEXT] To dye after knitting.

dipentene [ORG CHEM] The racemic mixture of dextro and levo isomers of limonene.

dipentene glycol *See* terpin hydrate.

dipentene hydrochloride *See* terpene hydrochloride.

dipeptidase [BIOCHEM] An enzyme that hydrolyzes a dipeptide.

dip fault [GEOL] A type of fault that strikes parallel with the dip of the strata involved.

diphase cleaning [MET] Removing soilage from metal surfaces in a cleaning tank incorporating a solvent phase and an aqueous phase.

diphead [MIN ENG] A passage that follows the inclination of a coal seam.

diphenol [ORG CHEM] A compound that has two phenol groups, for example, resorcinol.

diphenyl *See* biphenyl.

diphenylamine [ORG CHEM] $(C_6H_5)_2NH$ Colorless leaflets, sparingly soluble in water; melting point 54°C; used as an additive in propellants to increase the storage life by neutralizing the acid products formed upon decomposition of the nitrocellulose. Also known as phenylaniline.

diphenylaminechloroarsine *See* adamsite.

1,4-diphenylbenzene *See para*-terphenyl.

1,3-diphenyl-3-butene-1-one *See* dipnone.

diphenylcarbazide [ORG CHEM] $CO(NHNHC_6H_5)_2$ White powder, melting point 170°C; used as an indicator, pink for alkalies, colorless for acids.

diphenyl carbonate [ORG CHEM] $(C_6H_5O)_2CO$ Easily hydrolyzed, white crystals, melting at 78°C; soluble in organic solvents, insoluble in water; used as a solvent, plasticizer, and chemical intermediate.

diphenylchloroarsine [ORG CHEM] $(C_6H_5)_2AsCl$ Colorless crystals used during World War I as an antipersonnel device to generate a smoke causing sneezing and vomiting.

diphenyldihydroxyethane *See* hydrobenzoin.

diphenyldiimide *See* azobenzene.

diphenylenedianilinohydrotriazole *See* nitron.

diphenylethylene *See* stilbene.

diphenylglyoxal *See* benzil.

diphenylguanidine [ORG CHEM] $HNC(NHC_6H_5)_2$ A white powder, melting at 147°C; used as a rubber accelerator. Also known as DPG; melaniline.

***N,N'*-diphenylhydrazine** *See* hydrazobenzene.

diphenyl ketone *See* benzophenone.

diphenylmethane [ORG CHEM] $(C_6H_5)_2CH_2$ Combustible, colorless crystals melting at 26.5°C; used in perfumery, dyes, and organic synthesis. Also known as benzylbenzene.

diphenyl methanol *See* benzhydrol.

4,7-diphenyl-1,10-phenanthroline *See* phenanthroline indicator.

diphenyl phthalate [ORG CHEM] $C_6H_4(COOC_6H_5)_2$ White powder, melting at 80°C; soluble in chlorinated hydrocarbons, esters, and ketones, insoluble in water; used as a plasticizer for cellulosic and other resins.

diphenylurea *See* carbanilide.

diphosgene [INORG CHEM] $C_2O_2Cl_4$ A volatile liquid used as a poison gas during World War I.

diphosphate [CHEM] A salt that has two phosphate groups.

diphosphatidyl glycerol *See* cardiolipin.

diphosphoglyceric acid [ORG CHEM] $C_3H_8O_9P_2$ An ester of glyceric acid, with two molecules of phosphoric acid, characterized by a high-energy phosphate band.

diphosphopyridine nucleotide [BIOCHEM] $C_{21}H_{27}O_{14}N_7P_2$ An organic coenzyme that functions in enzymatic systems concerned with oxidation-reduction reactions. Abbreviated DPN. Also known as codehydrogenase 1; coenzyme 1; nicotinamide adenine dinucleotide (NAD).

diphtheria [MED] A communicable bacterial disease of man caused by the growth of *Corynebacterium diphtheriae* on any mucous membrane, especially of the throat.

diphtheritic myocarditis [MED] Inflammation of the cardiac muscle arising from local or generalized diphtheria.

diphycercal [VERT ZOO] Pertaining to a tail fin, having symmetrical upper and lower parts, and with the vertebral column extending to the tip without upturning.

diphyletic [EVOL] Originating from two lines of descent.

Diphyllidea [INV ZOO] A monogeneric order of tapeworms in the subclass Cestoda; all species live in the intestine of elasmobranch fishes.

Diphyllobothrium [INV ZOO] A genus of tapeworms; including parasites of man, dogs, and cats. Formerly known as *Dibothriocephalus*.

Diphyllobothrium latum [INV ZOO] A large tapeworm that infects man, dogs, and cats; causes anemia and disorders of the nervous and digestive systems in man.

diphyllous [BOT] Having two leaves.

diphyodont [ANAT] Having two successive sets of teeth, deciduous followed by permanent, as in man.

dipicolinic acid [BIOCHEM] $C_7H_5O_4N \cdot 1\frac{1}{2}H_2O$ A chelating agent composing 5–15% of the dry weight of bacterial spores.

dip inductor *See* earth inductor.

dip joint [GEOL] A joint that strikes approximately at right angles to the cleavage or bedding of the constituent rock.

Diplacanthidae [PALEON] A family of extinct acanthodian fishes in the suborder Diplacanthoidei.

Diplacanthoidei [PALEON] A suborder of extinct acanthodian fishes in the order Climatiiformes.

diplacusis [MED] A difference in the pitch perceptions of the two ears when stimulated by the same sound frequency.

Diplasiocoela [VERT ZOO] A suborder of amphibians in the order Anura typically having the eighth vertebra biconcave.

diplegia [MED] Paralysis of similar parts on the two sides of the body.

dipleurula [INV ZOO] **1.** A hypothetical bilaterally symmetrical larva postulated to be an ancestral form of echinoderms and chordates. **2.** Any bilaterally symmetrical, ciliated echinoderm larva.

diplexer [ELECTR] A coupling system that allows two different transmitters to operate simultaneously or separately from the same antenna.

diplex operation [COMMUN] Simultaneous transmission or reception of two signals using a specified common feature, such as a single antenna or a single carrier.

diplex radio transmission [COMMUN] The simultaneous transmission of two signals by using a common carrier wave.

diplex reception [ELEC] Simultaneous reception of two signals which have some features in common, such as a single receiving antenna or a single carrier frequency.

Diplobathrida [PALEON] An order of extinct, camerate crinoids having two circles of plates beneath the radials.

diplobiont [BIOL] An organism characterized by alternating, morphologically dissimilar haploid and diploid generations.

diploblastic [ZOO] Having two germ layers, referring to embryos and certain lower invertebrates.

diploblastula [INV ZOO] A two-layered, flagellated larva of certain ceractinomorph sponges. Also known as parenchymella.

diplococci [MICROBIO] A pair of micrococci.

Diplococcus [MICROBIO] A genus of gram-positive, parasitic bacteria in the family Lactobacillaceae; cells usually occur in pairs, sometimes in chains.

Diplococcus gonorrhoeae *See* Neisseria gonorrhoeae.

Diplococcus intracellularis meningitidis *See* Neisseria meningitidis.

Diplococcus pneumoniae [MICROBIO] A species of encapulated, nonmotile bacteria that causes pneumonia and other infectious diseases of humans; the type species of the genus. Also known as pneumococci.

dip log [GEOL] A log of the dips of formations traversed by boreholes.

diploglossate [VERT ZOO] Pertaining to certain lizards, having the ability to retract the end of the tongue into the basal portion.

diplohaplont [BIOL] An organism characterized by alternating, morphologically similar haploid and diploid generations.

diploid [CRYSTAL] A crystal form in the isometric system having 24 similar quadrilateral faces arranged in pairs.

DIPHENYLMETHANE

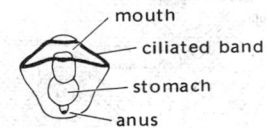

Structural formula of diphenylmethane showing numbered carbons.

DIPLEURULA

mouth
ciliated band
stomach
anus

Dipleurula larva.

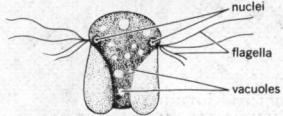

DIPLOMONADIDA

A diplomonad, *Trepomonas rotans. (After McCracken)*

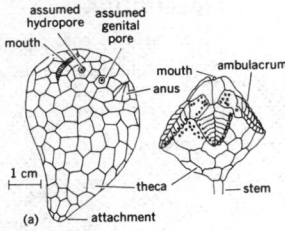

DIPLOPORITA

Morphological features of representative Diploporita. *(a) Aristocystitis*, from the Ordovician *(after J. Barrande). (b) Asteroblastus*, from the Silurian *(after O. Jaekel).*

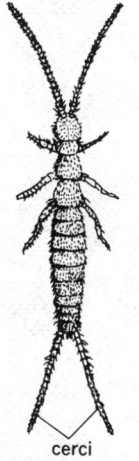

DIPLURA

cerci

Campodea folsomi Silvestri (Campodeidae). *(From E. O. Essig, College Entomology, Macmillan, 1942)*

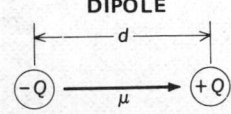

DIPOLE

Electric dipole with moment $\mu = Qd$.

DIPTERIFORMES

African lungfish *(Protopterus annectens)*, length to 32 inches (81 centimeters). *(After G. A. Boulenger, Catalogue of the Fresh Water Fishes of Africa in the British Museum, vol. 1, 1909)*

diploidization [GEN] The process of attaining the diploid state.

diploid merogony [EMBRYO] Development of a part of an egg in which the nucleus is the normal diploid fusion product of egg and sperm nuclei.

diploid state [GEN] A condition in which a chromosome set is present in duplicate in a nucleus (2N).

diplolepidious [BOT] Double-scaled, specifically referring to the peristome of mosses with two rows of scales on the outside and one row on the inner.

Diplomonadida [INV ZOO] An order of small, colorless protozoans in the class Zoomastigophorea, having a bilaterally symmetrical body with four flagella on each side.

Diplomystidae [VERT ZOO] A family of catfishes in the suborder Siluroidei confined to the waters of Chile and Argentina.

diplont [BIOL] An organism with diploid somatic cells and haploid gametes.

diplopia [MED] A disorder characterized by double vision.

Diplopoda [INV ZOO] The millipeds, a class of terrestrial tracheate, oviparous arthropods; each body segment except the first few bears two pairs of walking legs.

Diploporita [PALEON] An extinct order of echinoderms in the class Cystoidea in which the thecal canals were associated in pairs.

Diplorhina [VERT ZOO] The subclass of the class Agnatha that includes the jawless vertebrates with paired nostrils.

diplosome [CYTOL] A double centriole.

diplospondyly [ANAT] Having two centra in one vertebra.

diplotene [CYTOL] The stage of meiotic prophase during which pairs of nonsister chromatids of each bivalent repel each other and are kept from falling apart by the chiasmata.

Diplura [INV ZOO] An order of small, primarily wingless insects of worldwide distribution.

dipmeter [ENG] An instrument used to measure the direction and angle of dip of geologic formations.

dipmeter log [GEOL] A dip log produced by reading of the direction and angle of formation dip as analyzed from impulses from a dipmeter consisting of three electrodes 120° apart in a plane perpendicular to the borehole.

dip mold [ENG] A one-piece glassmaking mold with an open top; used to mold patterns.

dip needle [ENG] An obsolete type of magnetometer consisting of a magnetized needle that rotates freely in the vertical plane, with an adjustable weight on one side of the pivot.

Dipneumonomorphae [INV ZOO] A suborder of the order Araneida comprising the spiders common in the United States, including grass spiders, hunting spiders, and black widows.

Dipneusti [VERT ZOO] The equivalent name for Dipnoi.

Dipnoi [VERT ZOO] The lungfishes, a subclass of the Osteichthyes having lungs that arise from a ventral connection in the gut.

dipnone [ORG CHEM] $C_{16}H_{14}O$ A liquid ketone, formed by condensation of two acetophenone molecules; used as a plasticizer. Also known as 1,3-diphenyl-3-butene-1-one; β-methylchalcone.

Dipodidae [VERT ZOO] The Old World jerboas, a family of mammals in the order Rodentia.

dip of horizon *See* dip.

dip oil [MATER] Oil containing about 25% tar acids; used as dip for animals to kill insect parasites.

dipolar ion [CHEM] An ion carrying both a positive and a negative charge. Also known as zwitterion.

dipole [ELECTROMAG] Any object or system that is oppositely charged at two points, or poles, such as a magnet or a polar molecule; more precisely, the limit as either charge goes to infinity, the separation distance to zero, while the product remains constant. Also known as doublet; electric doublet.

dipole antenna [ELECTROMAG] An antenna approximately one-half wavelength long, split at its electrical center for connection to a transmission line whose radiation pattern has a maximum at right angles to the antenna. Also known as doublet antenna; half-wave dipole.

dipole disk feed [ELECTROMAG] Antenna, consisting of a dipole near a disk, used to reflect energy to the disk.

dipole moment *See* electric dipole moment; magnetic dipole moment.

dipole polarization *See* orientation polarization.

dipole radiation [ELECTROMAG] The electromagnetic radiation generated by an oscillating electric or magnetic dipole.

dipole relaxation [ELEC] The process, occupying a certain period of time after a change in the applied electric field, in which the orientation polarization of a substance reaches equilibrium.

dipole sound field [ACOUS] A sound field generated by an oscillating dipole source.

dipole transition [ATOM PHYS] A transition of an atom or nucleus from one energy state to another in which dipole radiation is emitted or absorbed.

diporpa larva [INV ZOO] A developmental stage of a monogenean trematode.

Dippel's oil *See* bone oil.

dipper dredge [MECH ENG] A power shovel resembling a grab crane mounted on a flat-bottom boat for dredging under water. Also known as dipper shovel.

dipper stick [MECH ENG] A straight shaft connecting the digging bucket of an excavating machine or power shovel with the boom.

dipper trip [MECH ENG] A device which releases the door of a shovel bucket.

dipping acid *See* sulfuric acid.

dipping refractometer *See* immersion refractometer.

dipping sonar [ENG] A sonar transducer that is lowered into the water from a hovering antisubmarine-warfare helicopter and recovered after the search is complete. Also known as dunking sonar.

dip plating *See* immersion plating.

dip pole *See* magnetic pole.

dip reversal *See* reversal of dip.

Diprionidae [INV ZOO] The conifer sawflies, a family of hymenopteran insects in the superfamily Tenthredinoidea.

dipropyl [ORG CHEM] A compound containing two propyl groups.

dipropyl ketone *See* 4-heptanone.

Diprotodonta [VERT ZOO] A proposed order of marsupial mammals to include the phalangers, wombats, koalas, and kangaroos.

Diprotodontidae [PALEON] A family of extinct marsupial mammals.

diproton [NUC PHYS] A hypothetical bound state of two protons, which probably does not exist.

Dipsacales [BOT] An order of dicotyledonous herbs and shrubs in the subclass Asteridae characterized by an inferior ovary and usually opposite leaves.

dip slip [GEOL] The component of a fault parallel to the dip of the fault. Also known as normal displacement.

dip slope [GEOL] A slope of the surface of the land determined by and conforming approximately to the dip of the underlying rocks. Also known as back slope; outface.

Dipsocoridae [INV ZOO] A family of hemipteran insects in the superfamily Dipsocoroidea; members are predators on small insects under bark or in rotten wood.

Dipsocoroidea [INV ZOO] A superfamily of minute, ground-inhabiting hemipteran insects belonging to the subdivision Geocorisae.

dip soldering [MET] A method similar to dip brazing but using a filler metal having a melting point below 800°F (427°C).

dipsophobia [PSYCH] An abnormal fear of drinking.

dipstick [ENG] A graduated rod to measure depth when dipped in a liquid; used, for example, to measure the oil in an automobile engine crankcase.

dip stream [HYD] A consequent stream that flows in the direction of the dip of the strata it traverses.

dip-strike symbol [GEOL] A geologic symbol used on maps to show the strike and dip of a planar feature.

Diptera [INV ZOO] The true flies, an order of the class Insecta characterized by possessing only two wings and a pair of balancers.

Dipteriformes [VERT ZOO] The single order of the subclass Dipnoi, the lungfishes.

Dipterocarpaceae [BOT] A family of dicotyledonous plants in the order Theales having mostly stipulate, alternate leaves, a prominently exserted connective, and a calyx that is mostly winged in fruit.

dipterous [BIOL] **1.** Of, related to, or characteristic of Diptera. **2.** Having two wings or winglike structures.

dipulse [COMMUN] Transmission of a binary code in which the presence of one cycle of a sine-wave tone represents a binary "1" and the absence of one cycle represents a binary "0."

dipyramid [CRYSTAL] A crystal having the form of two pyramids that melt at a plane of symmetry.

dipyre *See* mizzonite.

Dirac covariants [QUANT MECH] Quantities which behave as a scalar, a pseudoscalar, a vector, an axial vector, or a second-rank tensor under Lorentz transformations, and whose elements consist of basis elements of the Dirac gamma algebra multiplied by the Dirac wave function on the right and its adjoint on the left.

Dirac delta function *See* delta function.

Dirac electron theory *See* Dirac theory.

Dirac equation [QUANT MECH] A relativistic wave equation for an electron in an electromagnetic field, in which the wave function has four components corresponding to four internal states specified by a two-valued spin coordinate and an energy coordinate which can have a positive or negative value.

Dirac fields [QUANT MECH] Operators, arising in the second quantization of the Dirac theory, which correspond to the Dirac wave functions in the original theory.

Dirac gamma algebra [QUANT MECH] An algebra whose basis consists of 16 linearly independent 4×4 matrices constructed from products of the four basic Dirac matrices.

Dirac h *See* h-bar.

Dirac matrix [QUANT MECH] Any one of four matrices, designated γ_u ($u = 1, 2, 3, 4$), each having four rows and four columns and satisfying $\gamma_u \gamma_v + \gamma_v \gamma_u = \delta uv$, where δuv is the Kronecker delta function, which matrices operate on the four-component wave function in the Dirac equation. Also known as gamma matrix.

Dirac moment [QUANT MECH] Magnetic moment of the electron according to the Dirac theory, equal to $eh/2mc$, where e and m are the charge and mass of the positron respectively, h is Planck's constant divided by 2π, and c is the speed of light.

Dirac monopole [QUANT MECH] A magnetic monopole whose magnetic charge is an integral multiple of $hc/2e$, where h is Planck's constant divided by 2π, c is the speed of light, and e is the charge of the electron.

Dirac particle [PARTIC PHYS] A particle behaving according to the Dirac theory, which describes the behavior of electrons and muons except for radiative corrections, and is envisaged as describing a central core of a hadron of spin ½ℏ which remains when the effects of nuclear forces are removed.

Dirac quantization [QUANT MECH] The condition, arising from conservation of angular momentum, that for any electric charge q and magnetic monopole with magnetic charge m, one has $2qm = n\hbar c$, where n is an integer, $\hbar$ is Planck's constant divided by 2π, and c is the speed of light (gaussian units).

Dirac spinor *See* spinor.

Dirac theory [QUANT MECH] Theory of the electron based on the Dirac equation, which accounts for its spin angular momentum and gives its magnetic moment and its behavior in an electromagnetic field (except for higher-order corrections). Also known as Dirac electron theory.

Dirac wave function [QUANT MECH] A function appropriate for describing a spin ½ particle and antiparticle; it is a column matrix with four entries, each of which is a function of the space and time coordinates; the four-components form two first-rank Lorentz spinors.

direct access [ADP] The ability to read or write information anywhere within a storage device in an amount of time that is constant regardless of the location of the information accessed and of the location of the information previously accessed.

direct-access library [ADP] A disk-stored set of programs, each of which is directly accessible without sequential search.

direct-access storage [ADP] A method of storing data such that any datum can be accessed without the need for a sequential search.

direct-acting pump [MECH ENG] A displacement reciprocating pump in which the steam or power piston is connected to the pump piston by means of a rod, without crank motion or flywheel.

direct-acting recorder [ENG] A recorder in which the marking device is mechanically connected to or directly operated by the primary detector.

direct address [ADP] Any address specifying the location of an operand.

direct-address processing [ADP] Any computer operation during which data are accessed by means of addresses rather than contents.

direct air cycle [AERO ENG] A thermodynamic propulsion cycle involving a nuclear reactor and gas turbine or ramjet engine, in which air is the working fluid. Also known as direct cycle.

direct-aperture antenna [ELECTROMAG] An antenna whose conductor or dielectric is a surface or solid, such as a horn, mirror, or lens.

direct-arc furnace [ENG] A furnace in which a material in a refractory-lined shell is rapidly heated to pour temperature by an electric arc which goes directly from electrodes to the material.

direct bearing [CIV ENG] A direct vertical support in a structure.

direct cell [METEOROL] A closed thermal circulation in a vertical plane in which the rising motion occurs at higher potential temperature than the sinking motion.

direct-chill casting [MET] A continuous ingot- or billet-casting process in which metal is poured into short molds on a platform and then cooled when the platform is lowered into a water bath. Abbreviated dc casting. Also known as semicontinuous casting.

direct code [ADP] A code in which instructions are written in the basic machine language.

direct command guidance [ENG] Control of a missile or drone entirely from the launching site by radio or by signals sent over a wire.

direct-connected [MECH ENG] The connection between a driver and a driven part, as a turbine and an electric generator, without intervening speed-changing devices, such as gears.

direct-contact condenser *See* contact condenser.

direct cost [IND ENG] The cost in goods and labor to produce a product which would not be spent if the product were not made.

direct-coupled [MECH ENG] Joined without intermediate connections.

direct-coupled amplifier [ELECTR] A direct-current amplifier in which a resistor or a direct connection provides the coupling between stages, so small changes in direct currents can be amplified.

direct-coupled transistor logic [ELECTR] Integrated-circuit logic using only resistors and transistors, with direct conductive coupling between the transistors; speed can be up to 1 megahertz. Abbreviated DCTL.

direct coupling [ELEC] Coupling of two circuits by means of a non-frequency-sensitive device, such as a wire, resistor, or battery, so both direct and alternating current can flow through the coupling path. [MECH ENG] The direct connection of the shaft of a prime mover (such as a motor) to the shaft of a rotating mechanism (such as a pump or compressor).

direct current [ELEC] Electric current which flows in one direction only, as opposed to alternating current. Abbreviated dc.

direct-current amplifier [ELECTR] An amplifier that is capable of amplifying dc voltages and slowly varying voltages.

direct-current circuit [ELEC] Any combination of dc voltage or current sources, such as generators and batteries, in conjunction with transmission lines, resistors, and power converters such as motors.

direct-current circuit theory [ELEC] An analysis of relationships within a dc circuit.

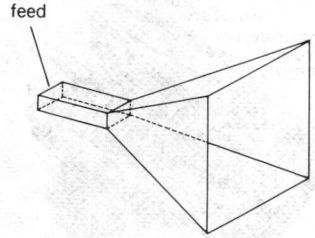

**DIRECT-APERTURE
ANTENNA**

feed

Pyramidal horn, a type of direct-aperture antenna.

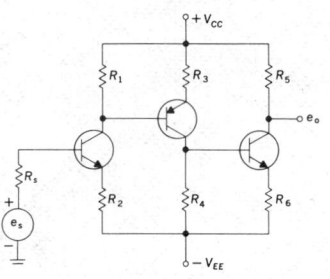

**DIRECT-COUPLED
AMPLIFIER**

Direct-coupled direct-current amplifier of the type which uses *npn* and *pnp* transistors.

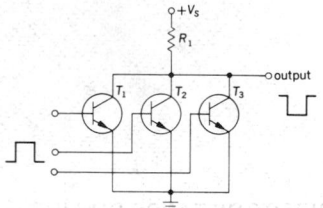

**DIRECT-COUPLED
TRANSISTOR LOGIC**

Direct-coupled transistor logic circuit.

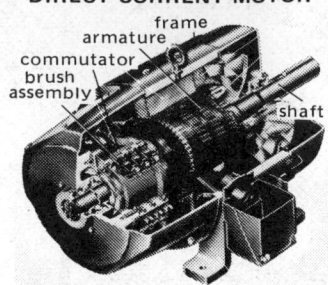

DIRECT-CURRENT MOTOR

Cutaway view of typical direct-current motor. *(General Electric)*

direct-current component [COMMUN] The average value of a signal; in television, it represents the average luminance of the picture being transmitted; in radar, the level from which the transmitted and received pulses rise.

direct-current coupling [ELECTR] That type of coupling in which the zero-frequency term of the Fourier series representing the input signal is transmitted.

direct-current discharge [ELECTR] The passage of a direct current through a gas.

direct-current dump [ELECTR] Removal of all direct-current power from a computer system or component intentionally, accidentally, or conditionally; in some types of storage, this results in loss of stored information.

direct-current erase [ELECTR] Use of direct current to energize an erasing head of a tape recorder.

direct-current generator [ELEC] A rotating electric machine that converts mechanical power into dc power.

direct-current inserter [ELECTR] A television transmitter stage that adds to the video signal a dc component known as the pedestal level.

direct-current motor [ELEC] An electric rotating machine energized by direct current and used to convert electric energy to mechanical energy.

direct-current motor control *See* electronic motor control.

direct-current picture transmission [COMMUN] Television transmission in which the signal contains a dc component that represents the average illumination of the entire scene. Also known as direct-current transmission.

direct-current plate resistance [ELECTR] Value or characteristic used in vacuum-tube computations; it is equal to the direct-current plate voltage divided by the direct-current plate current.

direct-current power [ELEC] The power delivered by a dc power system, equal to the line voltage times the load current.

direct-current power supply [ELEC] A power supply that provides one or more dc output voltages, such as a dc generator, rectifier-type power supply, converter, or dynamotor.

direct-current quadruplex system [COMMUN] Direct-current telegraph system which affords simultaneous transmission of two messages in each direction over the same line, achieved by superimposing neutral telegraph upon polar telegraph.

direct-current receiver [ELECTR] A radio receiver designed to operate directly from a 115-volt dc power line.

direct-current reinsertion *See* clamping.

direct-current restoration *See* clamping.

direct-current restorer [ELECTR] A clamp circuit used to establish a dc reference level in a signal without modifying to any important degree the waveform of the signal itself. Also known as clamper; reinserter.

direct-current signaling [ELEC] A transmission method that uses direct current.

direct-current tachometer [ELEC] A dc generator operating with negligible load current and with constant field flux provided by a permanent magnet, so its dc output voltage is proportional to speed.

direct-current telegraphy [COMMUN] Telegraphy in which direct current controlled by the transmitting apparatus is supplied to the line to form the transmitted signal.

direct-current transducer [ELECTR] A transducer that requires dc excitation and provides a dc output that varies with the parameter being sensed.

direct-current transmission *See* direct-current picture transmission.

direct-current vacuum-tube voltmeter [ELECTR] The amplifying and indicating portions of the diode rectifier-amplifier meter, which are usually designed so that the diode rectifier can be disconnected for dc measurements.

direct-current voltage *See* direct voltage.

direct-current working volts [ELEC] The maximum continuously applied dc voltage for which a capacitor is rated. Abbreviated dcwV.

direct cycle *See* direct air cycle.

direct-cycle reactor [NUCLEO] A nuclear power plant in which the heat-transfer fluid circulates through the reactor and then passes directly to the turbine in a continuous cycle.

direct digital control [CONT SYS] The use of a digital computer generally on a time-sharing or multiplexing basis, for process control in petroleum, chemical, and other industries.

direct distance dialing [COMMUN] A telephone exchange service that allows a telephone user to dial subscribers outside the local area using a standard routing pattern from the local or end office.

direct drive [MECH ENG] A drive in which the driving part is directly connected to the driven part.

direct-drive vibration machine [MECH ENG] A vibration machine in which the vibration table is forced to undergo a displacement by a positive linkage driven by a direct attachment to eccentrics or camshafts.

direct dye [MATER] A group of coal tar dyes that act without mordants, for example, benzidine dyes. Also known as substantive dye.

directed angle [MATH] An angle for which one side is designated as initial, the other as terminal.

directed graph [MATH] A graph in which a direction is shown for every arc. Also known as digraph.

directed line [MATH] A line on which a positive direction has been specified.

directed number [MATH] A number together with a sign.

directed set [MATH] A partially ordered set with the property that for every pair of elements a,b in the set, there is a third element which is larger than both a and b.

direct electromotive force [ELEC] Unidirectional electromotive force in which the changes in values are either zero or so small that they may be neglected.

direct energy conversion [ENG] Conversion of thermal or chemical energy into electric power by means of direct-power generators.

direct-expansion coil [MECH ENG] A finned coil, used in air cooling, inside of which circulates a cold fluid or evaporating refrigerant. Abbreviated DX coil.

direct feedback system [CONT SYS] A system in which electrical feedback is used directly, as in a tachometer.

direct fire [ORD] Fire delivered on a target in which the weapon sights are brought directly on the target.

direct-fire [ENG] To fire a furnace without preheating the air or gas.

direct-fired evaporator [CHEM ENG] An evaporator in which the flame and combustion gases are separated from the boiling liquid by a metal wall, or other heating surface.

direct-gap semiconductor [SOLID STATE] A semiconductor in which the minimum of the conduction band occurs at the same wave vector as the maximum of the valence band, and recombination radiation consequently occurs with relatively large intensity.

direct-geared [MECH ENG] Joined by a gear on the shaft of one machine meshing with a gear on the shaft of another machine.

direct grid bias *See* grid bias.

direct-image offset system [GRAPHICS] A system for converting letterpress color plates to offset that involves pulling a proof of the letterpress form directly onto a special paper-backed aluminum foil plate, which then receives a slight treatment. Also known as d-i offset system.

direct-imaging mass analyzer [ENG] A type of secondary ion mass spectrometer in which secondary ions pass through an electrostatic immersion lens which forms an image that bears a point-to-point relation to the ion's place of origin on the sample surface, and then traverse magnetic sectors which effect mass separation. Also known as Castaing-Slodzian mass analyzer.

direct indicating compass [NAV] A compass in which the dial, scale, or index is carried on the sensing element.

direct-insert subroutine [ADP] A body of coding or a group of instructions inserted directly into the logic of a program, often in multiple copies, whenever required.

direct interelectrode capacitance *See* interelectrode capacitance.

direct inward dialing [COMMUN] The capability for dialing individual telephone extensions in a large organization directly from outside, without going through a central switchboard.

DIRECT DISTANCE DIALING

symbols	names	abbreviation	class number
to other regional centers	regional center	RC	1
	sectional center	SC	2
	primary center	PC	3
	toll center	TC	4
	end office	EO	5

Standard routing pattern for direct distance dialing. *(American Telephone and Telegraph Co.)*

direction [ENG] The position of one point in space relative to another without reference to the distance between them; may be either three-dimensional or two-dimensional, the horizontal being the usual plane of the latter; usually indicated in terms of its angular distance from a reference direction. [GEOL] *See* trend.

directional antenna [ELECTROMAG] An antenna that radiates or receives radio waves more effectively in some directions than others.

directional beacon *See* direction-giving beacon.

directional beam [ELECTROMAG] A radio or radar wave that is concentrated in a given direction.

directional control [ENG] Control of motion about the vertical axis; in an aircraft, usually by the rudder.

directional control valve [ENG] A control valve serving primarily to direct hydraulic fluid to the point of application.

directional counter [NUCLEO] A counter that is more sensitive to nuclear radiation from some directions than from others.

directional coupler [ELECTR] A device that couples a secondary system only to a wave traveling in a particular direction in a primary transmission system, while completely ignoring a wave traveling in the opposite direction. Also known as directive feed.

directional derivative [MATH] The rate of change of a function in a given direction; more precisely, if f maps an n-dimensional euclidean space into the real numbers, and $\mathbf{x} = (x_1, \ldots, x_n)$ is a vector in this space, and $\mathbf{u} = (u_1, \ldots, u_n)$ is a unit vector in the space (that is, $u_1^2 + \ldots + u_n^2 = 1$), then the directional derivative of f at $\mathbf{x}$ in the direction of $\mathbf{u}$ is the limit as h approaches zero of $[f(\mathbf{x} + h\mathbf{u}) - f(\mathbf{x})]/h$.

directional drilling [ENG] A method of drilling in which the direction of the hole is planned before.

directional filter [ELECTR] A low-pass, band-pass, or high-pass filter that separates the bands of frequencies used for transmission in opposite directions in a carrier system. Also known as directional separation filter.

directional gain *See* directivity index.

directional gyro [AERO ENG] A flight instrument incorporating a gyro that holds its position in azimuth and thus can be used as a directional reference. Also known as direction indicator. [MECH] A two-degrees-of-freedom gyro with a provision for maintaining its spin axis approximately horizontal.

directional homing [NAV] Homing in which the bearing is the parameter which is maintained constant.

directional hydrophone [ENG ACOUS] A hydrophone whose response varies significantly with the direction of sound incidence.

directional log [PETRO ENG] A record of the wellhole drift, from the vertical, and the direction of that drift.

directional microphone [ENG ACOUS] A microphone whose response varies significantly with the direction of sound incidence.

directional pattern *See* radiation pattern.

directional phase shifter [ELEC] Passive phase shifter in which the phase change for transmission in one direction differs from that for transmission in the opposite direction.

directional property [MET] Any property of a metal whose magnitude varies with the orientation of the test axis to a specific direction within the metal.

directional relay [ELEC] Relay which functions in conformance with the direction of power, voltage, current, pulse, rotation, and so on.

directional response pattern *See* directivity pattern.

directional separation filter *See* directional filter.

directional solidification [MET] Controlled solidification of molten metal in a casting so as to provide feed metal to the solidifying front of the casting.

directional stability [AERO ENG] The property of an aircraft, rocket, or such, enabling it to restore itself from a yawing or side-slipping condition. Also known as weathercock stability.

directional structure [GEOL] Any sedimentary structure having directional significance; examples are cross-bedding and ripple marks. Also known as vectorial structure.

directional well [PETRO ENG] A well drilled at an angle up to 70° from the vertical to avoid obstacles over the reservoir, such as towns, beaches, or bodies of water.

direction angles [MATH] The three angles which a line in space makes with the positive x, y, and z axes.

direction cosine [ENG] In tracking, the cosine of the angle between a baseline and the line connecting the center of the baseline with the target. [MATH] The cosine of one of the direction angles of a line in space.

direction finder *See* radio direction finder.

direction-finder-bearing indicator [NAV] That portion of a direction-finder on which the bearing (relative, magnetic, true bearing or reciprocal) of the received transmission is indicated.

direction finder deviation [NAV] The angular difference between a bearing observed by a radio direction finder and the correct bearing, caused by disturbances due to the characteristics of the receiving craft or the station site.

direction-finding net [NAV] Two or more direction-finding facilities furnishing coordinated information to a designated control-evaluation agency facility for the purpose of providing position or course information to a pilot.

direction-giving beacon [NAV] A beacon (light or radio) which, when observed or received by suitable equipment on the craft or vehicle, gives the observer the bearing (usually magnetic) from the beacon; although it operates in a different mode from that of a direction-finding station, the information produced by both types of stations is identical; however, a direction-giving station (which is shore- or ground-based) produces navigational parameters that differ from those of a craft or vehicle-borne direction finder. Also known as directional beacon.

direction-independent radar [ENG] Doppler radar used in sentry applications.

direction indicator *See* directional gyro.

direct ionization *See* extrinsic photoemission.

direction of propagation [PHYS] **1.** The normal to a surface of constant phase, in a propagating wave. **2.** The direction of the group velocity. **3.** The direction of time-average energy flow. (In a homogeneous isotropic medium, these three directions coincide.)

direction of relative movement [NAV] The direction of motion relative to a reference point, itself usually in motion.

direction rectifier [ELECTR] A rectifier that supplies a direct-current voltage whose magnitude and polarity vary with the magnitude and relative polarity of an alternating-current synchro error voltage.

directive [INV ZOO] Any of the dorsal and ventral paired mesenteries of certain anthozoan coelenterates.

directive feed *See* directional coupler.

directive gain [ELECTROMAG] Of an antenna in a given direction, 4π times the ratio of the radiation intensity in that direction to the total power radiated by the antenna.

directive therapy [PSYCH] A method of psychiatric treatment by which the therapist, assuming complete understanding of the patient's needs, endeavors to change the patient's attitudes, behavior, or mode of living.

directivity [ELECTROMAG] **1.** The value of the directive gain of an antenna in the direction of its maximum value. **2.** The ratio of the power measured at the forward-wave sampling terminals of a directional coupler, with only a forward wave present in the transmission line, to the power measured at the same terminals when the direction of the forward wave in the line is reversed; the ratio is usually expressed in decibels.

directivity factor [ENG ACOUS] **1.** The ratio of radiated sound intensity at a remote point on the principal axis of a loudspeaker or other transducer, to the average intensity of the sound transmitted through a sphere passing through the remote point and concentric with the transducer; the frequency must be stated. **2.** The ratio of the square of the voltage produced by sound waves arriving parallel to the principal axis of a microphone or other receiving transducer, to the mean square of the voltage that would be produced if sound waves having the same frequency and mean-square pressure were arriving simultaneously from all directions with random phase; the frequency must be stated.

directivity index [ENG ACOUS] The directivity factor expressed in decibels; it is 10 times the logarithm to the base 10 of the directivity factor. Also known as directional gain.

directivity pattern [ENG ACOUS] A graphical or other description of the response of a transducer used for sound emission or reception as a function of the direction of the transmitted or incident sound waves in a specified plane and at a specified frequency. Also known as beam pattern; directional response pattern.

direct labor [IND ENG] The labor or effort actually producing goods or services.

direct labor standard *See* standard time.

direct-line drive [PETRO ENG] Waterflood operation involving a network of wells in a direct (straight) line.

directly heated cathode *See* filament.

direct material [IND ENG] Any raw or semifinished material which will be incorporated into the product.

direct memory access [ADP] The use of special hardware for direct transfer of data to or from memory to minimize the interruptions caused by program-controlled data transfers. Abbreviated dma.

direct motion [ASTRON] Eastward, or counterclockwise, motion of a planet or other object as seen from the North Pole (motion in the direction of increasing right ascension).

direct nuclear reaction [NUC PHYS] A nuclear reaction which is completed in the time required for the incident particle to transverse the target nucleus, so that it does not combine with the nucleus as a whole but interacts only with the surface or with some individual constituent.

direct numerical control [ADP] The use of a computer to program, service, and log a process such as a machine-tool cutting operation.

director [ELECTR] Telephone switch which translates the digits dialed into the directing digits actually used to switch the call. [ELECTROMAG] A parasitic element placed a fraction of a wavelength ahead of a dipole receiving antenna to increase the gain of the array in the direction of the major lobe. [ORD] Electromechanical equipment which is used to track a moving target in azimuth and angular height and which, with the addition of other necessary information from an outside source, such as a radar set or a range finder, continuously computes firing data and transmits them to the guns.

direct organization [ADP] A type of processing in which records within data sets stored on direct-access devices may be fetched directly if their physical locations are known.

director-sight system [ORD] A sighting system used on airborne flexible guns in which the gunner has direct control of the tracking line (line of aim) and indirect control of the gun or guns.

director-type computer [ADP] A gunsight computer used in the director-sight system which, in response to the gunner's action of tracking, computes the angle at which a gun must be fired in order to hit a target.

director-type fire control [ORD] The control of gun fire by use of a director-type computer.

directory [ADP] The listing and description of all the fields of the records making up a file.

direct pickup [COMMUN] The transmission of television images without intermediate photographic or magnetic recording.

direct piezoelectricity [SOLID STATE] Name sometimes given to the piezoelectric effect in which an electric charge is developed on a crystal by the application of mechanical stress.

direct pointing [ORD] Pointing a piece either in a direction or in a range and direction by means of a sight directed at the target.

direct point repeater [ELECTR] Telegraph repeater in which the receiving relay controlled by the signals received over a line repeats corresponding signals directly into another line or lines without the interposition of any other repeating or transmitting apparatus.

direct positive [GRAPHICS] A photographic positive made by exposure to light and developed without a negative.

direct-power generator [ENG] Any device which converts thermal or chemical energy into electric power by methods more direct than the conventional thermal cycle.

direct printing [TEXT] Printing by passing the textile through a series of rollers, each printing a different color or different part of the pattern.

direct process [MET] A process in which the metal is produced from the ore in a single step (for example, steel without intermediate pig iron).

direct product of a finite family of sets [MATH] Given sets $A_1, \ldots, A_n$, the direct product is the set of all n-tuples $(a_1, \ldots, a_n)$, where a_i belongs to A_i for $i = 1, \ldots, n$.

direct proportion [MATH] A statement that the ratio of two variable quantities is equal to a constant.

direct quenching [MET] Rapid cooling of carburized parts directly from the carburizing process.

direct-radiator speaker [ENG ACOUS] A loudspeaker in which the radiating element acts directly on the air, without a horn.

direct reading [GRAPHICS] In microfilm technology, that image which is legible in a normal reading position.

direct-reading gage [ENG] Gage that records directly (instead of inferentially) measured values, for example, a liquid-level gage pointer actuated by direct linkage with a float.

direct recording [ENG ACOUS] Recording in which a record is produced immediately, without subsequent processing, in response to received signals.

direct-reduction process [MET] Any of several methods for extracting iron ore below the melting point of iron, to produce solid reduced iron that may be converted to steel with little further refining.

direct reflection *See* specular reflection.

direct resistance-coupled amplifier [ELECTR] Amplifier in which the plate of one stage is connected either directly or through a resistor to the control grid of the next stage, with the plate-load resistor being common to both stages; used to amplify small changes in direct current.

directrix [MATH] A fixed line used in one method of defining a conic; the distance from this line divided by the distance from a fixed point (called the focus) is the same for all points on the conic.

direct route [ELEC] In wire communications, the trunks that connect a pair of switching centers, regardless of the geographical direction the actual trunk facilities may follow. [NAV] The shortest navigational distance between two points on the earth's surface, for example, the great circle.

direct scanning [COMMUN] A scanning method in which the subject is illuminated at all times and only one elemental area of the subject is viewed at a time by the television camera.

direct solar radiation [ASTROPHYS] That portion of the radiant energy received at the actinometer direct from the sun, as distinguished from diffuse sky radiation, effective terrestrial radiation, or radiation from any other source.

direct sum [MATH] If each of the sets in a finite direct product of sets has a group structure, this structure may be imposed on the direct product by defining the composition "componentwise"; the resulting group is called the direct sum.

direct support artillery [ORD] Artillery assigned the mission of providing the fire requested by the supported unit.

direct symbol recognition [ADP] Recognition by sensing the unique geometrical properties of symbols.

direct tide [GEOPHYS] A gravitational solar or lunar tide in the ocean or atmosphere which is in phase with the apparent motions of the attracting body, and consequently has its local maxima directly under the tide-producing body, and on the opposite side of the earth.

direct viewfinder [OPTICS] A viewfinder in which the user views the subject directly through a lens or sight.

direct-view storage tube [ELECTR] A cathode-ray tube in which secondary emission of electrons from a storage grid is used to provide an intensely bright display for long and controllable periods of time. Also known as display storage tube; viewing storage tube.

direct-vision nephoscope [OPTICS] A type of nephoscope in which the cloud motion is observed by looking directly into the instrument.

direct-vision spectroscope [SPECT] A spectroscope that allows the observer to look in the direction of the light source by means of an Amici prism.

direct voltage [ELEC] A voltage that forces electrons to move through a circuit in the same direction continuously, thereby producing a direct current. Also known as direct-current voltage.

direct wave [COMMUN] A radio wave that is propagated directly through space from transmitter to receiver without being refracted by the ionosphere.

direct-wire circuit [ELEC] Supervised protective signaling circuit usually consisting of one metallic conductor and a ground return and having signal-receiving equipment responsive to either an increase or a decrease in current.

direct-writing galvanometer [ENG] A direct-writing recorder in which the stylus or pen is attached to a moving coil positioned in the field of the permanent magnet of a galvanometer.

direct-writing recorder [ENG] A recorder in which the permanent record of varying electrical quantities or signals is made on paper, directly by a pen attached to the moving coil of a galvanometer or indirectly by a pen moved by some form of motor under control of the galvanometer. Also known as mechanical oscillograph.

Dirichlet conditions [MATH] The requirement that a function be bounded, and have finitely many maxima, minima, and discontinuities on the closed interval $[-\pi, \pi]$.

Dirichlet problem [MATH] To determine a solution to Laplace's equation which satisfies certain conditions in a region and on its boundary.

Dirichlet series [MATH] A series whose nth term is a complex number divided by n to the zth power.

Dirichlet test for convergence [MATH] If Σb_n is a series whose sequence of partial sums is bounded, and if $\{a_n\}$ is a monotone decreasing null sequence, then the series

$$\sum_{n=1}^{\infty} a_n b_n \text{ converges.}$$

Dirichlet transform [MATH] For a function $f(x)$, this is the integral of $f(x) \cdot \sin (kx)/x$; its convergence determines the convergence of the Fourier series of $f(x)$.

dirigible [AERO ENG] A lighter-than-air craft equipped with means of propelling and steering for controlled flight.

dirt band [GEOL] A dark layer in a glacier representing a former surface, usually a summer surface, where silt and debris accumulated.

dirt bed [GEOL] A buried soil containing partially decayed organic material; sometimes occurs in glacial drift.

dirty bomb [ORD] An explosive based on nuclear fission that emits many long-lived radioactive isotopes.

dirty ship [NAV ARCH] Tanker carrying oil or heavy petroleum products.

dirty snowball model [ASTRON] A model of comet structure in which the nucleus of the comet resembles a large dirty snowball.

disability glare See glare.

disaccharide [BIOCHEM] Any of the class of compound sugars which yield two monosaccharide units upon hydrolysis.

disappearing carriage [ORD] A movable part for raising a heavy gun above a parapet and lowering it automatically after firing.

disappearing stair [BUILD] A stair that can be swung up into a ceiling space.

disappearing target [ORD] Target that is exposed to the firer's view for a short time; for example, bobbing targets or targets raised from target pits for short periods of time.

disarm [ORD] To remove the detonating device or fuse of a bomb, mine, or other piece of explosive ordnance, or otherwise render it incapable of exploding in its usual manner.

disassemble [ENG] To take apart into constituent parts.

Disasteridae [PALEON] A family of extinct burrowing, exocyclic Euechinoidea in the order Holasteroida comprising mainly small, ovoid forms without fascioles or a plastron.

disc See disk.

discarding petal [ORD] A part of a discarding sabot that is composed of a base and attached pieces, or petals, which surround the core, and are peeled back under centrifugal and aerodynamic forces and discarded just in front of the gun muzzle.

Discellaceae [MYCOL] A family of fungi of the order Sphaeropsidales, including saprophytes and some plant pathogens.

discharge [ELEC] To remove a charge from a battery, capacitor, or other electric-energy storage device. [ELECTR] The passage of electricity through a gas, usually accompanied by a glow, arc, spark, or corona. Also known as electric discharge. [FL MECH] The flow rate of a fluid at a given instant expressed as volume per unit of time.

discharge coefficient [FL MECH] In a nozzle or other constriction, the ratio of the mass flow rate at the discharge end of the nozzle to that of an ideal nozzle which expands an identical working fluid from the same initial conditions to the same exit pressure. Also known as coefficient of discharge.

discharge head [MECH ENG] Vertical distance between the intake level of a water pump and the level at which it discharges water freely to the atmosphere.

discharge hydrograph [CIV ENG] A graph showing the discharge or flow of a stream or conduit with respect to time.

discharge key [ELEC] Device for switching a capacitor suddenly from a charging circuit to a load through which it can discharge.

discharge lamp [ELECTR] A lamp in which light is produced by an electric discharge between electrodes in a gas (or vapor) at low or high pressure. Also known as electric-discharge lamp; gas-discharge lamp; vapor lamp.

discharge liquor [CHEM ENG] Liquid that has passed through a processing operation. Also known as effluent; product.

discharge printing [GRAPHICS] A method of printing in which an electric discharge is shaped to produce characters. [TEXT] Using bleaching chemicals on a previously dyed fabric to remove the dye and thus imprint a pattern.

discharger [ELEC] A silver-impregnated cotton wick encased in a flexible plastic tube with an aluminum mounting lug, used on aircraft to reduce precipitation static.

discharge tube [ELECTR] An evacuated enclosure containing a gas at low pressure, through which current can flow when sufficient voltage is applied between metal electrodes in the tube. Also known as electric-discharge tube.

discharging agent [TEXT] A stripping agent such as sodium hyposulfite which is used to remove dyes from fabric that has been vat-dyed or printed.

discharging arch [CIV ENG] A support built over, and not touching, a weak structural member, such as a wooden lintel, to carry the main load. Also known as relieving arch.

discifloral [BOT] Having flowers with enlarged, disklike receptacles.

disciform [BIOL] Disk-shaped.

Discinacea [INV ZOO] A family of inarticulate brachiopods in the suborder Acrotretidina.

disclimax [ECOL] A climax community that includes foreign species following a disturbance of the natural climax by man or domestic animals. Also known as disturbance climax.

Discoaster [BOT] A star-shaped coccolith.

discoblastula [EMBRYO] A blastula formed by cleavage of a meroblastic egg; the blastoderm is disk-shaped.

discocephalous [INV ZOO] Having a sucker on the head.

discoctaster [INV ZOO] A type of spicule with eight rays terminating in discs in hexactinellid sponges.

discodactylous [VERT ZOO] Having sucking disks on the toes.

discogastrula [EMBRYO] A gastrula formed from a blastoderm.

Discoglossidae [VERT ZOO] A family of anuran amphibians in and typical of the suborder Opisthocoela.

discoid [BIOL] **1.** Being flat and circular in form. **2.** Any structure shaped like a disc.

discoidal cleavage [EMBRYO] A type of cleavage producing a disc of cells at the animal pole.

Discoidiidae [PALEON] A family of extinct conical or globular, exocyclic Euechinoidea in the order Holectypoida distinguished by the rudiments of internal skeletal partitions.

Discolichenes [BOT] The equivalent name for Lecanorales.

Discolomidae [INV ZOO] The tropical log beetles, a family of coleopteran insects in the superfamily Cucujoidea.

discomfort glare See glare.

discomfort index See temperature-humidity index.

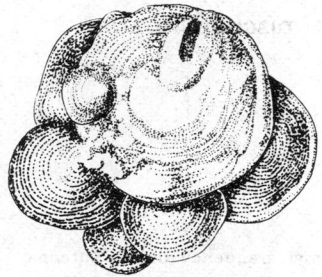

DISCINACEA

A cluster of *Discinisca*. (*From R. C. Moore, ed.,* Treatise on Invertebrate Paleontology, *pt. H, Geological Society of America, Inc., and University of Kansas Press, 1965*)

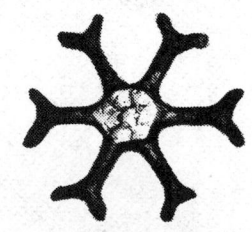

DISCOASTER

3 µ

Discoaster, an example of Coccolithophorida. (*A. McIntyre, Lamont-Doherty Geological Observatory of Columbia University*)

discomposition [NUCLEO] The process in which an atom is knocked out of its position in a crystal lattice by direct nuclear impact, as by fast neutrons or by fast ions that have been previously knocked out of their lattice positions.

discomposition effect [NUCLEO] Changes in physical or chemical properties of a substance caused by discomposition. Also known as Wigner effect.

Discomycetes [MYCOL] A group of fungi in the class Ascomycetes in which the surface of the fruiting body is exposed during maturation of the spores.

discone antenna [ELECTROMAG] A biconical antenna in which one of the cones is spread out to 180° to form a disk; the center conductor of the coaxial line terminates at the center of the disk, and the cable shield terminates at the vertex of the cone.

disconformity [GEOL] Unconformity between parallel beds or strata.

disconnect [ELEC] To open a circuit by removing wires or connections, as distinguished from opening a switch to stop current flow. [ENG] To sever a connection.

disconnect fitting [ELEC] An electrical connection that can be disconnected without tools.

disconnecting switch [ELEC] A switch that isolates a circuit or piece of electrical apparatus after interruption of the current. Also known as disconnector.

disconnector *See* disconnecting switch.

disconnector release [ELEC] Device which disengages the apparatus used in a telephone connection to restore it to its original condition when not in use.

discontinuity [ELECTROMAG] An abrupt change in the shape of a waveguide. [GEOL] **1.** An interruption in sedimentation. **2.** A surface that separates unrelated groups of rocks. [GEOPHYS] A boundary at which the velocity of seismic waves changes abruptly. [MATH] A point at which a function is not continuous. [PHYS] A break in the continuity of a medium or material at which a reflection of wave energy can occur.

discontinuous amplifier [ELECTR] Amplifier in which the input waveform is reproduced on some type of averaging basis.

discontinuous phase *See* disperse phase.

discontinuous precipitation [MET] Precipitation principally at and away from the grain boundaries in a supersaturated solid solution; diffraction patterns show two lattice parameters, the solute in solution and the precipitate.

discontinuous reaction series [GEOL] The branch of Bowen's reaction series that include olivine, pyroxene, amphibole, and biotite; each change in the series represents an abrupt change in phase.

discontinuous yielding [MET] The nonuniform plastic deformation of a metal along the length strained in tension.

discopodous [INV ZOO] Having a disk-shaped foot.

Discorbacea [INV ZOO] A superfamily of foraminiferan protozoans in the suborder Rotaliina characterized by a radial, perforate, calcite test and a monolamellar septa.

discord *See* dissonance.

discordance [GEOL] An unconformity characterized by lack of parallelism between strata which touch without fusion.

discordant pluton [GEOL] An intrusive igneous body that cuts across the bedding or foliation of the intruded formations.

discount [IND ENG] A reduction from the gross amount, price, or value.

discount factor [PETRO ENG] The ratio of the present worth of one or a series of future payments to the total undiscounted amount of such future payments. Also known as average discount factor; deferment factor; present-worth factor.

discovery [MIN ENG] Finding of a valuable mineral deposit.

discovery claim [MIN ENG] The first claim for the finding of a mineral deposit.

discovery vein [MIN ENG] The vein on which a mining claim is based.

discrete [SCI TECH] **1.** Composed of separate and distinct parts. **2.** Having an individually distinct identity.

discrete comparator *See* digital comparator.

discrete-film zone *See* belt of soil water.

discrete radio source [ASTROPHYS] A source of radio waves coming from a small area of the sky.

discrete sampling [ELECTR] Sampling in which the individual samples are of such long duration that the frequency response of the channel is not deteriorated by the sampling process.

discrete set [MATH] A set with no cluster points.

discrete spectrum [SPECT] A spectrum in which the component wavelengths constitute a discrete sequence of values rather than a continuum of values.

discrete system [CONT SYS] A control system in which signals at one or more points may change only at discrete values of time. Also known as discrete-time system.

discrete-time system *See* discrete system.

discrete transfer function *See* pulsed transfer function.

discrete variable [MATH] A variable for which the possible values form a discrete set.

discrete-word intelligibility [COMMUN] The percent of intelligibility obtained when the speech units under consideration are words, usually presented so as to minimize the contextual relation between them.

discriminant [MATH] The quantity $b^2 - 4ac$ where a, b, c are the coefficients of a given quadratic polynomial: $ax^2 + bx + c$.

discrimination [ADP] *See* conditional jump. [COMMUN] **1.** In frequency-modulated systems, the detection or demodulation of the imposed variations in the frequency of the carriers. **2.** In a tuned circuit, the degree of rejection of unwanted signals. **3.** Of any system or transducer, the difference between the losses at specified frequencies with the system or transducer terminated in specified impedances.

discrimination learning [PSYCH] A learning procedure in which a response is reinforced in the presence of one stimulus but not in the presence of others, enabling the experimental animal to discriminate between the one stimulus and the others.

discriminator [ELECTR] A circuit in which magnitude and polarity of the output voltage depend on how an input signal differs from a standard or from another signal.

discriminator transformer [ELECTR] A transformer designed to be used in a stage where frequency-modulated signals are converted directly to audio-frequency signals or in a stage where frequency changes are converted to corresponding voltage changes.

disdrometer [ENG] Equipment designed to measure and record the size distribution of raindrops as they occur in the atmosphere.

disease [MED] An alteration of the dynamic interaction between an individual and his environment which is sufficient to be deleterious to the well-being of the individual and produces signs and symptoms.

disengage [ENG] To break the contact between two objects.

dish *See* parabolic reflector.

disharmonic fold [GEOL] A fold in which changes in form or magnitude occur with depth.

dishing [MET] Producing a shallow, concave surface in metal-forming operations.

dishpan experiment [METEOROL] A model experiment carried out by differential heating of fluid in a flat, rotating pan; it establishes similarity with the atmosphere and is used to reproduce many important features of the general circulation and, on a smaller scale, atmospheric motion.

disilane [INORG CHEM] Si_2H_6 A spontaneously flammable compound of silicon and hydrogen; it exists as a liquid at room temperature.

disilicate [CHEM] A silicate compound that has two silicon atoms in the molecule.

disilicide [CHEM] A compound that has two silicon atoms joined to a radical or another element.

disinfectant [MATER] A chemical agent that destroys microorganisms but not bacterial spores.

disintegration [NUC PHYS] Any transformation of a nucleus, whether spontaneous or induced by irradiation, in which particles or photons are emitted.

disintegration chain *See* radioactive series.

disintegration constant *See* decay constant.

disintegration energy [NUC PHYS] The energy released, or the negative of the energy absorbed, during a nuclear or

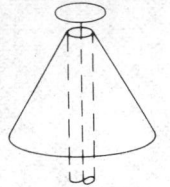

DISCONE ANTENNA

A high-frequency discone antenna.

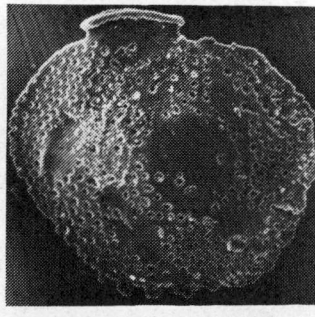

DISCORBACEA

Scanning electron micrograph of a spiral view of coarsely perforate, trochospiral test with terminal aperture and everted lip of *Siphonina* from upper Eocene of Mississippi. *(R. B. MacAdam, Chevron Oil Field Research Co.)*

particle reaction. Designated Q. Also known as Q value; reaction energy.

disintegration family *See* radioactive series.

disintegration of measure [MATH] The representation of a measure as an integral of a family of positive measures.

disintegration rate [NUCLEO] **1.** The absolute rate of decay of a radioactive substance, usually expressed in terms of disintegrations per unit of time. **2.** The absolute rate of transformation of a nuclide under bombardment.

disintegration series *See* radioactive series.

disintegration voltage [ELECTR] The lowest anode voltage at which destructive positive-ion bombardment of the cathode occurs in a hot-cathode gas tube.

disintegrator [MECH ENG] An apparatus used for pulverizing or grinding substances, consisting of two steel cages which rotate in opposite directions.

disjoint sets [MATH] Sets with no elements in common.

disjunct endemism [PALEON] A type of regionally restricted distribution of a fossil taxon in which two or more component parts are separated by a major physical barrier and hence not readily explicable in terms of present-day geography.

disjunction [CYTOL] Separation of chromatids or homologous chromosomes during anaphase.

disjunctive search [ADP] A search to find items that have at least one of a given set of characteristics.

disjunctor [MYCOL] A small cellulose body between the conidia of certain fungi, which eventually breaks down and thus frees the conidia.

disk Also spelled disc. [ADP] A rotating circular plate having a magnetizable surface on which information may be stored as a pattern of polarized spots on concentric recording tracks. [BIOL] Any of various rounded and flattened animal and plant structures. [ENG ACOUS] *See* phonograph record. [MATH] The region in the plane consisting of all points with norm less than 1 (sometimes less than or equal to 1).

disk-and-doughnut [CHEM ENG] A type of fractionating tower construction of alternating disks and plates that are doughnut-shaped, to provide mixing.

disk armature [ELEC] The armature in a motor that has a disk winding or is made up of a metal disk.

disk attrition mill *See* disk mill.

disk brake [MECH ENG] A type of brake in which disks attached to a fixed frame are pressed against disks attached to a rotating axle or against the inner surfaces of a rotating housing.

disk cam [MECH ENG] A disk with a contoured edge which rotates about an axis perpendicular to the disk, communicating motion to the cam follower which remains in contact with the edge of the disk.

disk canvas wheel [DES ENG] A polishing wheel made of disks of canvas sewn together with heavy twine or copper wire, and reinforced by steel side plates and side rings with bolts or screws.

disk centrifuge [MECH ENG] A centrifuge with a large bowl having a set of disks that separate the liquid into thin layers to create shallow settling chambers.

disk clutch [MECH ENG] A clutch in which torque is transmitted by friction between friction disks with specially prepared friction material riveted to both sides and contact plates keyed to the inner surface of an external hub.

disk colorimeter [ANALY CHEM] A device for comparing standard and sample colors by means of rotating color disks.

disk coupling [MECH ENG] A flexible coupling in which the connecting member is a flexible disk.

disk cultivator [AGR] A cultivator consisting of pairs of oppositely inclined disks.

disk engine [MECH ENG] A rotating engine in which the piston is a disk.

disk filter [ENG] A filter in which the substance to be filtered is drawn through membranes stretched on segments of revolving disks by a vacuum inside each disk; the solids left on the membrane are lifted from the tank and discharged. Also known as American filter.

disk flower [BOT] One of the flowers on the disk of a composite plant.

disk furrower [AGR] A furrower in which concave disks, at an angle to the direction of motion, are used to cut the soil.

disk grinder [MECH ENG] A grinding machine that employs abrasive disks.

disk grinding [MECH ENG] Grinding with the flat side of a rigid, bonded abrasive disk or segmental wheel.

disk harrow [AGR] A harrow which has two or more opposed gangs of 3–12 disks for cutting clods and trash, destroying weeds, cutting in cover crops, and smoothing and preparing the surface for various farming operations.

disk leather wheel [DES ENG] A polishing wheel made of leather disks glued together.

disk loading [AERO ENG] A measure which expresses the design gross weight of a helicopter as a function of the swept areas of the lifting rotor.

disk memory *See* disk storage.

disk meter [ENG] A positive displacement meter to measure flow rate of a fluid; consists of a disk that wobbles or nutates within a chamber so that each time the disk nutates a known volume of fluid passes through the meter.

disk mill [MECH ENG] Size-reduction apparatus in which grinding of feed solids takes place between two disks, either or both of which rotate. Also known as disk attrition mill.

disk pack [ADP] A set of magnetic disks that can be removed from a disk drive as a unit.

disk plow [AGR] A plow consisting of a number of disk blades attached to one axle or gang bolt; used for rapid, shallow plowing.

disk recording [ENG ACOUS] **1.** The process of inscribing suitably transformed acoustical or electrical signals on a phonograph record. **2.** *See* phonograph record.

disk sander [MECH ENG] A machine that uses a circular disk coated with abrasive to smooth or shape surfaces.

disk-seal tube [ELECTR] An electron tube having disk-shaped electrodes arranged in closely spaced parallel layers, to give low interelectrode capacitance along with high power output, up to 2500 megahertz. Also known as lighthouse tube; megatron.

disk signal [CIV ENG] An automatic block signal with colored disks that indicate train movements.

disk spring [MECH ENG] A mechanical spring which consists of a disk or washer supported by one force (distributed by a suitable chuck or holder) at the periphery and by an opposing force on the center or hub of the disk.

disk storage [ELECTR] An external computer storage device consisting of one or more disks spaced on a common shaft, and magnetic heads mounted on arms that reach between the disks to read and record information on them. Also known as disk memory; magnetic disk storage.

disk telescope [OPTICS] A telescope designed for observations of the brilliant solar disk; examples are the tower telescope and the horizontal fixed telescope.

disk thermistor [ELECTR] A thermistor which is produced by pressing and sintering an oxide binder mixture into a disk, 0.2–0.6 inch (5–15 millimeters) in diameter and 0.04–0.5 inch (1.0–13 millimeters) thick, coating the major surfaces with conducting material, and attaching leads.

disk-wall packer [PETRO ENG] A disklike seal between the outside of the well tubing and the inside of the well casing; used to prevent fluid movement from the pressure differential above and below the sealing point.

disk wheel [DES ENG] A wheel in which a solid metal disk, rather than separate spokes, joins the hub to the rim.

dislocation [CRYSTAL] A defect occurring along certain lines in the crystal structure and present as a closed ring or a line anchored at its ends to other dislocations, grain boundaries, the surface, or other structural feature. Also known as line defect. [GEOL] Relative movement of rock on opposite sides of a fault. Also known as displacement. [MED] Displacement of one or more bones of a joint.

dislocation breccia *See* fault breccia.

dismicrite [GEOL] Fine-grained limestone of obscure origin, resembling micrite but containing sparry calcite bodies.

dismount [ORD] To remove a weapon or piece of equipment from its setting, mount, or carriage.

disodium citrate *See* sodium citrate.

disodium ethylene-bis-dithiocarbamate *See* nabam.

disodium hydrogen citrate *See* sodium citrate.

disodium hydrogen phosphate *See* disodium phosphate.

DISK BRAKE

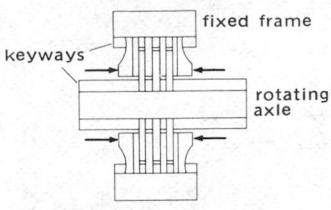

Forces in direction of arrows press disks keyed to axle against disks keyed to frame.

DISK CAM

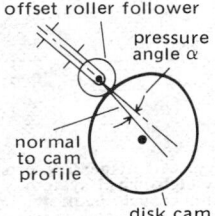

Diagram of disk cam.

DISK CENTRIFUGE

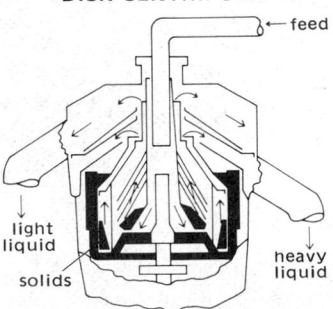

Cutaway view of disk centrifuge bowl.

DISK CLUTCH

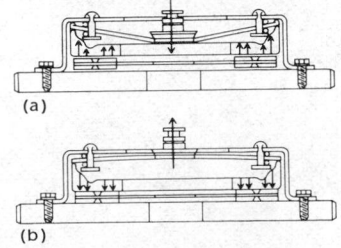

Disk clutch (a) in disengaged position, (b) in engaged position.

disodium phosphate [INORG CHEM] Na_2HPO_4 Transparent crystals, soluble in water; used in the textile processing and other industries to control pH in the range 4–9, as an additive in processed cheese to maintain spreadability, and as a laxative and antacid. Also known as disodium hydrogen phosphate.

disomaty [CYTOL] Duplication of chromosomes unaccompanied by nuclear division.

Disomidae [INV ZOO] A family of spioniform annelid worms belonging to the Sedentaria.

disorder [CRYSTAL] Departures from regularity in the occupation of lattice sites in a crystal containing more than one element.

disorientation [MED] Mental confusion as to one's normal relationship to his environment, especially time, place, and people; associated with organic brain disorders.

dispatching [ADP] The control of priorities in a queue of requests in a multiprogramming or multitasking environment.

dispatching priority [ADP] In a multiprogramming or multitasking environment, the priority assigned to an active (non-real time, nonforeground) task.

dispenser [ENG] Device that automatically dispenses radar chaff from an aircraft.

dispenser cathode [ELECTR] An electron tube cathode having provisions for continuously replacing evaporated electron-emitting material.

dispermy [PHYSIO] Entrance of two spermatozoa into an ovum.

dispersal [CIV ENG] The practice of building or establishing industrial plants, government offices, or the like, in separated areas, to reduce vulnerability to enemy attack. [ORD] The spreading out of equipment, supplies, or personnel, especially for protection against enemy action.

dispersal pattern [GEOCHEM] Distribution pattern of metals in soil, rock, water, or vegetation.

disperse [ADP] A data-processing operation in which grouped input items are distributed among a larger number of groups in the output.

dispersed elements [GEOCHEM] Elements which form few or no independent minerals but are present as minor ingredients in minerals of abundant elements.

dispersed gas injection [PETRO ENG] Gas-injection pressure maintenance of an oil reservoir in which the injection wells are arranged geometrically to distribute the gas uniformly throughout the oil-productive portions of the reservoir.

disperse dye [MATER] A very slightly water-soluble, colored material for use on cellulose acetate and other synthetic fibers; color is transferred to the fiber as extremely finely divided particles, resulting in a solution of the dye in the solid fiber.

disperse phase [CHEM] The phase of a disperse system consisting of particles or droplets of one substance distributed through another system. Also known as discontinuous phase; internal phase.

disperser [MATER] Material added to solid-in-liquid or liquid-in-liquid suspensions to separate the individual suspended particles; used in pigment grinding and dye dispersion. Also known as dispersing agent; emulsifier; emulsifying agent.

disperse system [CHEM] A two-phase system consisting of a dispersion medium and a disperse phase.

dispersing agent See disperser.

dispersing prism [OPTICS] An optical prism which deviates light of different wavelengths by different amounts and can therefore be used to separate white light into its monochromatic parts.

dispersion [AERO ENG] Deviation from a prescribed flight path; specifically, circular dispersion especially as applied to missiles. [CHEM] A distribution of finely divided particles in a medium. [COMMUN] The entropy of the output of a communications channel when the input is known. [ELECTROMAG] Scattering of microwave radiation by an obstruction. [MINERAL] In optical mineralogy, the constant optical values at different positions on the spectrum. [PHYS] **1.** The separation of a complex of electromagnetic or sound waves into its various frequency components. **2.** Quantita-

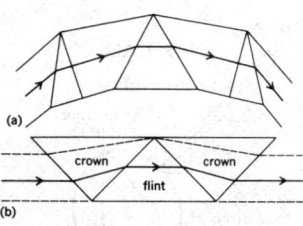

DISPERSING PRISM

Two types of dispersing prisms.
(a) Rayleigh prism system.
(b) Amici direct-vision system consisting of flint-glass prism and two crown-glass prims.

tively, the rate of change of refractive index with wavelength or frequency at a given wavelength or frequency. **3.** The rate of change of deviation with wavelength or frequency. **4.** In general, any process separating radiation into components having different frequencies, energies, velocities, or other characteristics, such as the sorting of electrons according to velocity in a magnetic field.

dispersion equation See dispersion formula.

dispersion error [ORD] Chance variation in a series of shots even though firing conditions are kept as constant as possible.

dispersion force [PHYS CHEM] The force of attraction that exists between molecules that have no permanent dipole.

dispersion formula [PHYS] Any formula which gives the refractive index as a function of wavelength of electromagnetic radiation. Also known as dispersion equation.

dispersion ladder [ORD] Table showing the probable distribution of a succession of shots made with the same firing data; specifically, a diagram made up of eight zones, showing the percentage of shots which may be expected to fall within each zone, based on direction (deflection) or range.

dispersion mill [MECH ENG] Size-reduction apparatus that disrupts clusters or agglomerates of solids, rather than breaking down individual particles; used for paint pigments, food products, and cosmetics.

dispersion of a random variable [STAT] The spread of a random variable's distribution about its mean.

dispersion pattern [ORD] The distribution of a series of shots by using coordinate settings as nearly identical as possible.

dispersion relation [NUC PHYS] A relation between the cross section for a given effect and the de Broglie wavelength of the incident particle, which is similar to a classical dispersion formula. [PHYS] An integral formula relating the real and imaginary parts of some function of frequency or energy, such as a refractive index or scattering amplitude, based on the causality principle and the Cauchy integral formula. [PL PHYS] A relation between the radian frequency and the wave vector of a wave motion or instability in a plasma.

dispersion zone [ORD] The area over which shots scatter when fired with the same sight setting.

dispersive line [ELECTROMAG] A delay line that delays each frequency a different length of time.

dispersive medium [ELECTROMAG] A medium in which the phase velocity of an electromagnetic wave is a function of frequency.

dispersive power [OPTICS] A measure of the power of a medium to separate different colors of light, equal to $(n_2 - n_1)/(n - 1)$, where n_1 and n_2 are the indices of refraction at two specified widely differing wavelengths, and n is the index of refraction for the average of these wavelengths, or for the D line of sodium.

dispersoid [CHEM] Matter in a form produced by a disperse system.

disphenoid [CRYSTAL] **1.** A crystal form with four similar triangular faces combined in a wedge shape; can be tetragonal or orthorhombic. **2.** A crystal form with eight scalene triangles combined in pairs.

displaced ore body [GEOL] An ore body which has been subjected to displacement or disruption after its initial deposition.

displacement [CHEM] A chemical reaction in which an atom, radical, or molecule displaces and sets free an element of a compound. [ELEC] See electric displacement. [FL MECH] **1.** The weight of fluid which is displaced by a floating body, equal to the weight of the body and its contents; the displacement of a ship is generally measured in long tons (1 long ton = 2240 pounds). **2.** The volume of fluid which is displaced by a floating body. [GEOL] See dislocation. [MECH] **1.** The linear distance from the initial to the final position of an object moved from one place to another, regardless of the length of path followed. **2.** The distance of an oscillating particle from its equilibrium position. [MECH ENG] The volume swept out in one stroke by a piston moving in a cylinder as for an engine, pump, or compressor.

displacement angle [ELEC] The change in the phase of an alternator's terminal voltage when a load is applied.

displacement boat [NAV ARCH] Any ship or boat which travels immersed and operates at relatively lower speeds than craft, such as hydroplanes, which plane on the surface of the water at high speed.

displacement chromatography [ANALY CHEM] Variation of column-development or elution chromatography in which the solvent is sorbed more strongly than the sample components; the freed sample migrates down the column, pushed by the solvent.

displacement compressor [MECH ENG] A type of compressor that depends on displacement of a volume of air by a piston moving in a cylinder.

displacement current [ELECTROMAG] The rate of change of the electric displacement vector, which must be added to the current density to extend Ampère's law to the case of time-varying fields (meter-kilogram-second units). Also known as Maxwell's displacement current.

displacement curve [NAV ARCH] A graph of the displacement of a vessel versus its draft; it is a curve of form.

displacement efficiency [PETRO ENG] In a gas condensate reservoir, the proportion (by volume) of wet hydrocarbons swept out of pores during dry-gas cycling.

displacement engine See piston engine.

displacement gyroscope [ENG] A gyroscope that senses, measures, and transmits angular displacement data.

displacement law See radioactive displacement law; Wien's displacement law.

displacement length coefficient [NAV ARCH] The displacement, in tons of sea water, of a ship divided by the length over 100 cubed.

displacement manometer [ENG] A differential manometer which indicates the pressure difference across a solid or liquid partition which can be displaced against a restoring force.

displacement meter [ENG] A water meter that measures water flow quantitatively by recording the number of times a vessel of known capacity is filled and emptied.

displacement pump [MECH ENG] A pump that develops its action through the alternate filling and emptying of an enclosed volume as in a piston-cylinder construction.

displacement series [CHEM] The elements in decreasing order of their negative potentials. Also known as constant series; electromotive series; Volta series.

displacer-type meter [ENG] Apparatus to detect liquid level or gas density by measuring the effect of the fluid (gas or liquid) on the buoyancy of a displacer unit immersed within the fluid.

display [ELECTR] **1.** A visible representation of information, in words, numbers, or drawings, as on the cathode-ray tube screen of a radar set, navigation system, or computer console. **2.** The device on which the information is projected. Also known as display device. **3.** The image of the information.

display device See display.

display information processor [ADP] Computer used to generate situation displays in a combat operations center.

display loss See visibility factor.

display packing [ADP] An efficient means of transmitting the x and y coordinates of a point packed in a single word to halve the time required to freshen the spot on a cathode-ray tube display.

display primary [COMMUN] One of the primary colors produced in a television receiver that, when mixed in proper proportions, serve to produce the other desired colors. Also known as receiver primary.

display processor [ADP] A section of a computer, or a minicomputer which handles the routines required to display an output on a cathode-ray tube.

display storage tube See direct-view storage tube.

display system [ADP] The total system, combining hardware and software, needed to achieve a visible representation of information in a data-processing system.

display tube [ELECTR] A cathode-ray tube used to provide a visual display.

display window [COMMUN] Width of the portion of the frequency spectrum presented on panoramic presentation; expressed in frequency units, usually megahertz.

disposable [ENG] Within a manufacturing system, designed to be discarded after use and replaced by an identical item, such as a filter element.

disposal area [NAV] On U.S. Coast and Geodetic Survey charts, an area established or approved by the Corps of Engineers for depositing dredged material where existing depths indicate that the intent is not to cause sufficient shoaling to create a danger to surface navigation; soundings and curves are retained.

disruptive discharge [ELEC] A sudden and large increase in current through an insulating medium due to complete failure of the medium under electrostatic stress.

disruptive strength [MET] Failure stress caused by hydrostatic tension.

dissect [BIOL] To divide, cut, and separate into different parts.

dissected topography [GEOGR] Physical features marked by erosive cutting.

dissecting microscope [OPTICS] Either of two types of optical microscope used to magnify materials undergoing dissection.

dissection [GEOL] Destruction of the continuity of the land surface by erosive cutting of valleys or ravines into a relatively even surface.

dissector tube [ELECTR] Camera tube having a continuous photo cathode on which is formed a photoelectric emission pattern which is scanned by moving its electron-optical image over an aperture.

disseminated necrotizing periarteritis See polyarteritis nodosa.

disseminule [BIOL] An individual organism or part of an individual adapted for the dispersal of a population of the same organisms.

dissepiment [BOT] A partition which divides a fruit or an ovary into chambers. [PALEON] One of the vertically positioned thin plates situated between the septa in extinct corals of the order Rugosa.

dissipation [PHYS] Any loss of energy, generally by conversion into heat; quantitatively, the rate at which this loss occurs. Also known as energy dissipation.

dissipation coefficient See scattering coefficient.

dissipation constant [GEOPHYS] In atmospheric electricity, a measure of the rate at which a given electrically charged object loses its charge to the surrounding air.

dissipation factor [ELEC] The inverse of Q, the storage factor.

dissipation function See Rayleigh's dissipation function; viscous dissipation function.

dissipation line [ELECTROMAG] A length of stainless steel or Nichrome wire used as a noninductive terminating impedance for a rhombic transmitting antenna when several kilowatts of power must be dissipated.

dissipation trail [FL MECH] A clear rift left behind an aircraft as it flies in a thin cloud layer; the opposite of a condensation trail. Also known as distrail.

dissipative muffler [ENG] A device which absorbs sound energy as the gas passes through it; a duct lined with sound-absorbing material is the most common type.

dissipative tunneling [SOLID STATE] Quantum-mechanical tunneling of individual electrons, rather than pairs, across a thin insulating layer separating two superconducting metals when there is a voltage across this layer, resulting in partial disruption of cooperative motion.

dissipator See heat sink.

dissociation [MED] Independent, uncoordinated functioning of the atria and ventricles. [MICROBIO] The appearance of a novel colony type on solid media after one or more subcultures of the microorganism in liquid media. [PHYS CHEM] Separation of a molecule into two or more fragments (atoms, ions, radicals) by collision with a second body or by the absorption of electromagnetic radiation. [PSYCH] The segregation of ideas from their affects or feelings, resulting in independent functioning of these components of a person's mental processes.

dissociation constant [PHYS CHEM] A constant whose numerical value depends on the equilibrium between the undissociated and dissociated forms of a molecule; a higher value indicates greater dissociation.

dissociation energy [PHYS CHEM] The energy required for complete separation of the atoms of a molecule.

dissociation pressure [PHYS CHEM] The pressure, for a given temperature, at which a chemical compound dissociates.

dissociative reaction [PSYCH] A neurotic disorder leading to gross disorganization of the personality.

dissociative recombination [ATOM PHYS] The combination of an electron with a positive molecular ion in a gas followed by dissociation of the molecule in which the resulting atoms carry off the excess energy.

dissogeny [ZOO] Having two sexually mature stages, larva and adult, in the life of an individual.

dissolution [CHEM] Dissolving of a material.

dissolve [CHEM] **1.** To cause to disperse. **2.** To cause to pass into solution. [GRAPHICS] A superimposing of one television or motion picture shot upon another, the emergent shot gradually brightening and the overlapped shot gradually darkening, so that as one scene disappears, another gradually appears. Also known as lap dissolve.

dissolved air floatation [CHEM ENG] A liquid-solid separation process wherein the main mechanism of suspended-solids removal is the change of apparent specific gravity of those suspended solids in relation to that of the suspending liquid by the attachment of small gas bubbles formed by the release of dissolved gas to the solids.

dissolved gas *See* solution gas.

dissolved-gas drive *See* internal gas drive.

dissolved-gas-drive reservoir [PETRO ENG] Oil reservoir in which the temperature of the liquid phase is below critical, and the liquid is driven from the reservoir by the expansion of dissolved gas. Also known as a bubble-point reservoir.

dissolved-gas reservoir *See* solution-gas reservoir.

dissonance [ACOUS] An unpleasant combination of harmonics heard when certain musical tones are played simultaneously. Also known as discord.

dissymmetrical network *See* dissymmetrical transducer.

dissymmetrical transducer [ELECTR] A transducer whose input and output image impedances are not equal. Also known as dissymmetrical network.

dissymmetry [SCI TECH] Lack of symmetry.

dissymmetry coefficient [ANALY CHEM] Ratio of the intensities of scattered light at 45 and 135°, used to correct for destructive interference encountered in light-scattering-photometric analyses of liquid samples.

dissymmetry factor [OPTICS] A quantity which expresses the strength of circular dichroism, equal to the difference in the absorption indices for left and right circularly polarized light divided by the absorption index for ordinary light of the same wavelength. Also known as anisotropy factor.

Distacodidae [PALEON] A family of conodonts in the suborder Conodontiformes characterized as simple curved cones with deeply excavated attachment scars.

distal [BIOL] Located away from the point of origin or attachment.

distal convoluted tubule [ANAT] The portion of the nephron in the vertebrate kidney lying between the loop of Henle and the collecting tubules.

distance [MATH] A nonnegative number associated with pairs of geometric objects. [MECH] The spatial separation of two points, measured by the length of a hypothetical line joining them.

distance finding station [NAV] A radio beacon equipped with a synchronized sound signal to provide the pilot or marine with a means of determining distance from the source of the sound, by measuring the difference in the time of reception of the two signals; the sound may be transmitted through air or water and from the same location as the radio signal or a location remote from it.

distance-luminosity relation [ASTRON] The relation in which the light intensity from a star is inversely proportional to the square of its distance.

distance mark [ELECTR] A movable point produced on a radar display by a special signal generator, so that when the mark is moved to a target position on the screen the range to the target can be read on the calibrated dial of the signal generator; usually used for gun laying where highly accurate distance is important.

distance marker [ENG] One of a series of concentric circles, painted or otherwise fixed on the screen of a plan position indicator, from which the distance of a target from the radar antenna can be read directly; used for surveillance and navigation where the relative distances between a number of targets are required simultaneously. Also known as radar range marker; range marker.

distance marking light [NAV] An approach light indicating distance from the end of a runway, landing strip, or channel.

distance-measuring equipment [NAV] A radio aid to navigation that provides distance information by measuring total round-trip time of transmission from an airborne interrogator to a ground-based transponder and return. Abbreviated DME.

distance modulus *See* modulus of distance.

distance protection [ELEC] Effect of a device operative within a predetermined electrical distance on the protected circuit to cause and maintain an interruption of power in a faulty circuit.

distance reception [COMMUN] Reception of messages from, or communication with, distant radio stations. Abbreviated DX.

distance relay [ELEC] Protective relay, the operation of which is a function of the distance between the relay and the point of fault.

distance resolution [ENG] The minimum radial distance by which targets must be separated to be separately distinguishable by a particular radar. Also known as range discrimination; range resolution.

distance/velocity lag [CONT SYS] The delay caused by the amount of time required to transport material or propagate a signal or condition from one point to another. Also known as transport lag; transportation lag.

distant early warning line [ORD] Defense line of radar stations at about the 70th parallel on the North American continent.

distant field [ELECTROMAG] The electromagnetic field at a distance of five wavelengths or more from a transmitter, where the radial electric field becomes negligible.

distant signal [CIV ENG] A signal placed at a distance from a block of track to give advance warning when the block is closed.

distemper [VET MED] Any of several contagious virus diseases of mammals, especially the form occurring in dogs, marked by fever, respiratory inflammation, and destruction of myelinated nerve tissue.

disthene *See* kyanite.

distichous [BIOL] Occurring in two vertical rows.

distillate [CHEM] The products of distillation formed by condensing vapors.

distillate fuel [MATER] Any one of the wide variety of fuels obtained from fractions boiling above the temperature at which gasoline comes off in the distillation of petroleum.

distillate fuel oil [MATER] A classification for one of the overhead fractions produced from crude oil in conventional distillation operations.

distillation [CHEM] The process of producing a gas or vapor from a liquid by heating the liquid in a vessel and collecting and condensing the vapors into liquids.

distillation column [CHEM] A still for fractional distillation.

distillation curve [CHEM] The graphical plot of temperature versus overhead product (distillate) volume or weight for a distillation operation.

distillation loss [CHEM] In a laboratory distillation, the difference between the volume of liquid introduced into the distilling flask and the sum of the residue and condensate received.

distillation range [CHEM] The difference between the temperature at the initial boiling point and at the end point of a distillation test.

distillation test [CHEM ENG] A standardized procedure for finding the initial, intermediate, and final boiling points in the boiling range of petroleum products.

distilled liquor [FOOD ENG] Alcoholic beverages obtained by distilling an alcohol-containing liquid such as wine or fermented fruit juice and then further treating the distillate to obtain a beverage of specific character. Also known as hard liquor.

DISTACODIDAE

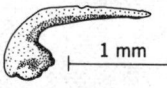

|← 1 mm →|

Multioistodus, a typical conodont of the family Distacodidae. *(After illustration in R. R. Shrock and W. H. Twenhofel, Principles of Invertebrate Paleontology, McGraw-Hill, 2d ed., 1953)*

distilled mustard gas [ORG CHEM] A delayed-action casualty gas (mustard gas) that has been distilled, or purified, to greatly reduce the odor and thereby increase its difficulty of detection.

distilled water [CHEM] Water that has been freed of dissolved or suspended solids and organisms by distillation.

distillery [FOOD ENG] The building where distillation of alcoholic beverages occurs.

distilling flask [CHEM] A round-bottomed glass flask that is capable of holding a liquid to be distilled.

distocclusion [MED] Malocclusion of the teeth in which those of the lower jaw are in distal relation to the upper teeth.

distome [INV ZOO] A digenetic trematode characterized by possession of an oral and a ventral sucker.

distorted water [METEOROL] A multimolecular layer of water, at the boundary between a mass of liquid water and the surrounding vapor, whose structure is not identical with that of bulk water.

distortion [ELECTR] Any undesired change in the waveform of an electric signal passing through a circuit or other transmission medium.

distortional wave *See* S wave.

distortion factor [COMMUN] Ratio of the effective value of the residue of a wave after elimination of the fundamental to the effective value of the original wave.

distortion meter [ENG] An instrument that provides a visual indication of the harmonic content of an audio-frequency wave.

distrail *See* dissipation trail.

distress frequency [COMMUN] A frequency allotted to distress calls, generally by international agreement; for ships at sea and aircraft over the sea, it is 500 kilohertz.

distress signal [COMMUN] An international signal used when a ship, aircraft, or other vehicle is threatened by grave and imminent danger and requests immediate assistance; examples are special radiotelegraph and radiotelephone signals or special signal flags or flares.

distributary [HYD] An irregular branch flowing out from a main stream and not returning to it, as in a delta. Also known as distributary channel.

distributary channel *See* distributary.

distributed amplifier [ELECTR] A wide-band amplifier in which tubes are distributed along artificial delay lines made up of coils acting with the input and output capacitances of the tubes.

distributed capacitance [ELEC] The capacitance that exists along the entire length of a transmission line.

distributed communications [COMMUN] Information transfer beyond the local level that may involve the originating source to transmit information to all communications centers on any one network, and may also cause an interchange of communications among several whole networks.

distributed constant [ELECTROMAG] A circuit parameter that exists along the entire length of a transmission line.

distributed-emission photodiode [ELECTR] A broad-band photodiode proposed for detection of modulated laser beams at millimeter wavelengths; incident light falls on a photocathode strip that generates a traveling wave of photocurrent having the same wave velocity as the transmission line which the photodiode feeds.

distributed fault *See* fault zone.

distributed inductance [ELECTROMAG] The inductance that exists along the entire length of a conductor, as distinguished from inductance concentrated in a coil.

distributed-parameter system *See* distributed system.

distributed paramp [ELECTR] Paramagnetic amplifier that consists essentially of a transmission line shunted by uniformly spaced, identical varactors; the applied pumping wave excites the varactors in sequence to give the desired traveling-wave effect.

distributed system [SYS ENG] A system whose behavior is governed by partial differential equations, and not merely ordinary differential equations. Also known as distributed-parameter system.

distributing frame [ELECTR] Structure for terminating permanent wires of a central office, private branch exchange, or private exchange, and for permitting the easy change of connections between them by means of cross-connecting wires.

distributing point [ORD] Point at which supplies, obtained from the supply point by a division or other unit, are broken down for immediate distribution to subordinate units.

distributing terminal assembly [ELECTR] Frame situated between each pair of selector bays to provide terminal facilities for the selector bank wiring and facilities for cross-connection to trunks running to succeeding switches.

distribution [MATH] An abstract object which generalizes the idea of function; used in applied mathematics, quantum theory, and probability theory; the delta function is an example. Also known as generalized function.

distribution amplifier [ELECTR] A radio-frequency power amplifier used to feed television or radio signals to a number of receivers, as in an apartment house or a hotel. [ENG ACOUS] An audio-frequency power amplifier used to feed a speech or music distribution system and having sufficiently low output impedance so changes in load do not appreciably affect the output voltage.

distribution box [CIV ENG] A device that equalizes the disposal of septic tank effluent through various lines.

distribution cable [ELEC] Cable extending from a feeder cable into a specific area for the purpose of providing service to that area.

distribution center [ELEC] In an alternating-current power system, the point at which control and routing equipment is installed.

distribution coefficient [OPTICS] One of the tristimulus values of monochromatic radiations having equal power, usually denoted by $\bar{x}, \bar{y}, \bar{z}$. [PHYS CHEM] The ratio of the amounts of solute dissolved in two immiscible liquids at equilibrium.

distribution control *See* linearity control.

distribution curve [STAT] The graph of the distribution function of a random variable.

distribution deck [ADP] A card file which duplicates all or part of a master card file and used for disseminating or decentralizing.

distribution factor [NUCLEO] A term used to express the modification of the effect of radiation in a biological system attributable to the nonuniform distribution of an internally deposited isotope, such as radium's being concentrated in bones.

distribution function *See* distribution of random variable.

distribution graph [HYD] A statistically derived hydrograph for a storm of specified duration, graphically representing the percent of total direct runoff passing a point on a stream, as a function of time; usually presented as a histogram or table of percent runoff within each of successive short time intervals.

distribution law [ANALY CHEM] The law stating that if a substance is dissolved in two immiscible liquids, the ratio of its concentration in each is constant. [STAT MECH] A law which gives a density function specifying the probability of finding a particle in a unit volume of phase space, or the number of particles in each of the states which a particle may occupy, or the number of particles per unit volume of phase space.

distribution of a random variable [STAT] For a discrete random variable, a function (or table) which assigns to each possible value of the random variable the probability that this value will occur; for a continuous random variable x, the monotone nondecreasing function which assigns to each real t the probability that x is less than or equal to t. Also known as distribution function; probability distribution; statistical distribution.

distribution photometer *See* light-distribution photometer.

distribution reservoir [CIV ENG] A service reservoir connected with the conduits of a primary water supply; used to supply water to consumers according to fluctuations in demand over short time periods and serves for local storage in case of emergency.

distribution substation [ELEC] An electric power substation associated with the distribution system and the primary feeders for supply to residential, commercial, and industrial loads.

distribution switchboard [ELEC] Power switchboard used for the distribution of electrical energy at the voltage common for each distribution within a building.

distribution system [ELEC] Circuitry involving high-voltage switchgear, step-down transformers, voltage dividers, and related equipment used to receive high-voltage electricity from a primary source and redistribute it at lower voltages. Also known as electric distribution system.

distribution transformer [ELEC] An element of an electric distribution system located near consumers which changes primary distribution voltage to secondary distribution voltage.

distributive fault *See* step fault.

distributive lattice [MATH] A lattice in which "greatest lower bound" obeys a distributive law with respect to "least upper bound," and vice versa.

distributive law [MATH] A rule which stipulates how two binary operations on a set shall behave with respect to one another; in particular, if $+$, $\circ$ are two such operations then $\circ$ distributes over $+$ means $a \circ (b + c) = (a \circ b) + (a \circ c)$ for all a,b,c in the set.

distributor [ELEC] **1.** Any device which allocates a telegraph line to each of a number of channels, or to each row of holes on a punched tape, in succession. **2.** A rotary switch that directs the high-voltage ignition current in the proper firing sequence to the various cylinders of an internal combustion engine. [ELECTR] The electronic circuitry which acts as an intermediate link between the accumulator and drum storage.

distributor gear [MECH ENG] A gear which meshes with the camshaft gear to rotate the distributor shaft.

distributor points [ELEC] Cam-operated contacts, the opening of which triggers the ignition pulse in an internal combustion engine.

district forecast [METEOROL] In U.S. Weather Bureau usage, a general weather forecast for conditions over an established geographical "forecast district."

district heating [MECH ENG] The supply of heat, either in the form of steam or hot water, from a central source to a group of buildings.

disturbance [COMMUN] An undesired interference or noise signal affecting radio, television, or facsimile reception. [CONT SYS] An undesired command signal in a control system. [GEOL] Folding or faulting of rock or a stratum from its original position. [METEOROL] **1.** Any low or cyclone, but usually one that is relatively small in size and effect. **2.** An area where weather, wind, pressure, and so on show signs of the development of cyclonic circulation. **3.** Any deviation in flow or pressure that is associated with a disturbed state of the weather, such as cloudiness and precipitation. **4.** Any individual circulatory system within the primary circulation of the atmosphere.

disturbance climax *See* disclimax.

disturbed-one output [ELECTR] One output of a magnetic cell to which partial-read pulses have been applied since that cell was last selected for writing.

disturbed sun noise [ASTROPHYS] Noise at times of sunspot or solar flare activity.

disulfate [CHEM] A compound that has two sulfate radicals.

disulfide [CHEM] **1.** A compound that has two sulfur atoms bonded to a radical or element. **2.** One of a group of organosulfur compounds RSSR′ that may be symmetrical ($R = R′$) or unsymmetrical (R and R′, different).

disulfide oil [MATER] One of the oils obtained by oxidizing to disulfides the mercaptans extracted from light petroleum distillates; the disulfides separate from the extract as an oily layer.

disulfonate [CHEM] A molecule that has two sulfonate groups.

disulfonic acid [CHEM] A molecule that has two sulfonic acid groups.

ditch [CIV ENG] A small artificial channel cut through earth or rock to carry water for irrigation or drainage.

ditch check [CIV ENG] A small dam positioned at intervals in a road ditch to prevent erosion.

ditcher *See* trench excavator.

ditching [AERO ENG] A forced landing on water, or the process of making such a landing. [ENG] The digging of ditches, as around storage tanks or process areas to hold liquids in the event of a spill or along the sides of a roadway for drainage.

diterpene [ORG CHEM] $C_{20}H_{32}$ **1.** A group of terpenes that have twice as many atoms in the molecule as monoterpenes. **2.** Any derivative of diterpene.

dither [CONT SYS] A force having a controlled amplitude and frequency, applied continuously to a device driven by a servomotor so that the device is constantly in small-amplitude motion and cannot stick at its null position. Also known as buzz.

dithiocarbamate [ORG CHEM] **1.** A salt of dithiocarbamic acid. **2.** Any other derivative of dithiocarbamic acid.

dithiocarbamic acid [ORG CHEM] NH_2CS_2H A colorless, unstable powder; various metal salts are readily obtained, and used as strong accelerators for rubber. Also known as amino-dithioformic acid.

dithiol *See* dimercaprol.

dithionate [CHEM] Any salt formed from dithionic acid.

dithionic acid [INORG CHEM] $H_2S_2O_6$ A strong acid formed by the oxidation of sulfurous acid, and known only by its salts and in solution.

ditokous [VERT ZOO] Producing two eggs or giving birth to two young at one time.

Dittus-Boelter equation [FL MECH] An equation used to calculate the surface coefficient of heat transfer for fluids in turbulent flow inside clean, round pipes.

ditty box [NAV ARCH] A small box with a hinged lid and lock used by the crew for personal belongings.

ditungsten carbide [INORG CHEM] W_2C A gray powder having hardness approaching that of diamond; forms hexagonal crystals with specific gravity 17.2; melting point 2850°C.

diuresis [MED] Increased excretion of urine.

diuretic [PHARM] Any agent that increases the volume and flow of urine.

diuretic hormone [BIOCHEM] A neurohormone that promotes water loss in insects by increasing the volume of fluid secreted into the Malpighian tubules.

diurnal [BIOL] Active during daylight hours. [METEOROL] Pertaining to meteorological actions which are completed within 24 hours and which recur every 24 hours.

diurnal aberration [ASTRON] Aberration caused by the rotation of the earth; its value varies with the latitude of the observer and ranges from zero at the poles to 0.31 second of arc. Also known as daily aberration.

diurnal age *See* age of diurnal inequality.

diurnal arc [ASTRON] That part of a celestial body's diurnal circle which lies above the horizon of the observer.

diurnal circle [ASTRON] The apparent daily path of a celestial body, approximating a parallel of declination.

diurnal inequality [OCEANOGR] The difference between the heights of the two high waters or the two low waters of a lunar day.

diurnal migration [BIOL] The daily rhythmic movements of organisms in the sea from deeper water to the surface at the approach of darkness and their return to deeper water before dawn.

diurnal motion [ASTRON] The apparent daily motion of a celestial body as observed from a rotating body.

diurnal parallax *See* geocentric parallax.

diurnal range *See* great diurnal range.

diurnal tide [OCEANOGR] A tide in which there is only one high water and one low water each lunar day.

diurnal variation [GEOPHYS] Daily variations of the earth's magnetic field at a given point on the surface, with both solar and lunar periods having their source in the horizontal movements of air in the ionosphere.

diuron [ORG CHEM] $C_9H_{10}Cl_2N_2O$ A water-soluble, crystalline compound; used as a preemergence herbicide.

divalent metal [CHEM] A metal whose atoms are each capable of chemically combining with two atoms of hydrogen.

divanadyl tetrachloride *See* vanadyl chloride.

divaricate [BIOL] Broadly divergent and spread apart.

divaricator [ZOO] A muscle that causes separation of parts, as of brachiopod shells.

dive [AERO ENG] A rapid descent by an aircraft or missile, nose downward, with or without power or thrust. [ENG] To submerge into an underwater environment so that it may be studied or utilized; includes the use of specialized equipment such as scuba, diving helmets, diving suits, diving bells, and underwater research vessels.   [NAV] To submerge a submarine under power.

dive bomber   [AERO ENG] An aircraft designed to release bombs during a steep dive.

dive brake   [AERO ENG] An air brake designed for operation in a dive; flaps at the following edge of one wing that can be extended into the airstream to increase drag and hold the aircraft to its "never exceed" dive speed in a vertical dive; used on dive bombers and sailplanes.

divergence   [ELECTR] The spreading of a cathode-ray stream due to repulsion of like charges (electrons).   [FL MECH] The ratio of the area of any section of fluid emerging from a nozzle to the area of the throat of the nozzle.   [METEOROL] The two-dimensional horizontal divergence of the velocity field.   [NUCLEO] In a nuclear reactor, the condition wherein the number of neutrons produced increases in each succeeding generation.   [OCEANOGR] A horizontal flow of water, in different directions, from a common center or zone.

divergence of a vector-valued function   [MATH] The sum of the diagonal entries of the Jacobian matrix; it is the scalar product of the del operator and the vector.

divergence theorem   *See* Gauss' theorem.

divergent adaptation   [EVOL] Adaptation to different kinds of environment that results in divergence from a common ancestral form. Also known as branching adaptation; cladogenic adaptation.

divergent die   [ENG] A die with the internal channels that lead to the orifice diverging, such as the dies used for manufacture of hollow-body plastic items.

divergent integral   [MATH] An improper integral which does not have a finite value.

divergent nozzle   [DES ENG] A nozzle whose cross section becomes larger in the direction of flow.

divergent sequence   [MATH] A sequence which does not converge.

divergent series   [MATH] An infinite series whose sequence of partial sums does not converge.

diverging duct   [DES ENG] Fluid-flow conduit whose internal cross-sectional area increases in the direction of flow.

diverging lens   [OPTICS] A lens whose focal length is negative, so that light incident parallel to its axis diverges after passing through it. Also known as negative lens.

diverging meniscus lens   *See* negative meniscus lens.

diverging yaw   [AERO ENG] In the flight of a projectile, an angle of yaw increasing from the initial yaw, so that the projectile is unstable.

diver method   [PHYS CHEM] Measure of the size of suspended solid particles; small glass divers of known density sink to the level where the liquid-suspension density is equal to that of the diver, allowing calculation of particle size. Also known as Berg's diver method.

diverse vector   [NAV] In air operations, an instruction issued by a radar controller to fly a specific course which is not a part of the predetermined radar pattern. Also known as random vector.

diversion canal   [CIV ENG] An artificial channel for diverting water from one place to another.

diversion chamber   [ENG] A chamber designed to direct a stream into a channel or channels.

diversion dam   [CIV ENG] A fixed dam for diverting stream water away from its course.

diversion gate   [CIV ENG] A gate which may be closed to divert water from the main conduit or canal to a lateral or some other channel.

diversion tunnel   [CIV ENG] An underground passageway used to divert flowing water around a construction site.

diversity   [COMMUN] Method of signal extraction by which an optimum resultant signal is derived from a combination of, or selection from, a plurality of transmission paths, channels, techniques, or physical arrangements; the system may employ space diversity, polarization diversity, frequency diversity, or any other arrangement by which a choice can be made between signals.

diversity factor   [ELEC] Ratio of the sum of the individual maximum demands to total maximum demand, as applied to an electrical distribution system.

diversity gain   [COMMUN] Gain in reception as a result of the use of two or more receiving antennas.

diversity radar   [ENG] A radar that uses two or more transmitters and receivers, each pair operating at a slightly different frequency but sharing a common antenna and video display, to obtain greater effective range and reduce susceptibility to jamming.

diversity receiver   [ELECTR] A radio receiver designed for space or frequency diversity reception.

diversity reception   [COMMUN] Radio reception in which the effects of fading are minimized by combining two or more sources of signal energy carrying the same modulation.

diverter   [ELEC] A low resistance which is connected in parallel with the series or compole winding of a direct-current machine and diverts current from it. causing the magnetomotive force produced by the winding to vary.

diverter-pole generator   [ELEC] Compound wound direct-current generator with the series winding of the diverter pole opposing the flux generated by the shunt wound main pole; provides a close voltage regulation.

diverticulitis   [MED] Inflammation of a diverticulum.

diverticulosis   [MED] Presence of many diverticula in the intestine.

diverticulum   [MED] An abnormal outpocketing or sac on the wall of a hollow organ.

diverting agent   [PETRO ENG] A viscous gel or suspension of graded solids used during acidizing of an oil reservoir to temporarily block off the most permeable sections of the pay zone to force the acid into less permeable sections.

Divesian   *See* Oxfordian.

divide   [GEOGR] A ridge or section of high ground between drainage systems.   [SCI TECH] A point or line of division.

divide check   [ADP] An error signal indicating that an illegal division (such as dividing by zero) was attempted.

divided lane   [CIV ENG] A highway divided into lanes by a median strip.

divided pitch   [DES ENG] In a screw with multiple threads, the distance between corresponding points on two adjacent threads measured parallel to the axis.

divided slit scan   [ADP] In optical character recognition, a device consisting of a narrow column of photoelectric cells which scans an input character at given intervals for the purpose of obtaining its horizontal and vertical components.

divider   [DES ENG] A tool like a compass, used in metalworking to lay out circles or arcs and to space holes or other dimensions.

divides   [MATH] One object (integer, polynomial) divides another if their quotient is an object of the same type.

dividing head   *See* indexing head.

dividing network   *See* crossover network.

diving bell   [ENG] An early diving apparatus constructed in the shape of a box or cylinder without a bottom and connected to a compressed-air hose.

diving bird   [VERT ZOO] Any bird adapted for diving and swimming, including loons, grebes, and divers.

diving plane   *See* diving rudder.

diving rudder   [NAV ARCH] One of the movable plane surfaces attached to the hull of a submarine that control the vertical motion of a submarine. Also known as diving plane; hydroplane.

diving suit   [ENG] A waterproof outfit designed for diving, especially one with a helmet connected to a compressed-air hose.

divining   [MIN ENG] An unscientific method for searching for subsurface water or minerals by means of a divining rod. Also known as dowsing.

divining rod   [MIN ENG] An unscientific device in the form of a forked rod or tree branch that is supposed to dip when held over water or minerals, depending on the specialty of the operator, or dowser. Also known as dowsing rod; wiggle stick.

divinyl   [ORG CHEM] **1.** A molecule that has two vinyl groups. **2.** *See* 1,3-butadiene.

divinyl acetylene [ORG CHEM] C_6H_6 A linear trimer of acetylene, made by passing acetylene into a hydrochloric acid solution that has metallic catalysts; used as an intermediate in neoprene manufacture.

divinylbenzene [ORG CHEM] $C_6H_4(CHCH_2)_2$ Polymerizable, water-white liquid used to make rubbers, drying oils, and ion-exchange resins and other polymers; forms include ortho, meta, and para isomers. Also known as vinylstyrene.

divinyl ether See vinyl ether.

divinyl oxide See vinyl ether.

division [MATH] The inverse operation of multiplication; the number a divided by the number b is the number c such that b multiplied by c is equal to a.

division algebra [MATH] A hypercomplex system that is also a skew field.

division modulo p [MATH] Division in the finite field with p elements, where p is a prime number.

division subroutine [ADP] A built-in program which achieves division by methods such as repetitive subtraction.

division wall [BUILD] A wall used to create major subdivisions in a building.

dixenite [MINERAL] $Mn_5(SiO_3)(AsO_3)(OH)_2$ A black hexagonal mineral consisting of a manganese arsenite and silicate, occurring in scales.

Dixidae [INV ZOO] A family of orthorrhaphous dipteran insects in the series Nematocera.

dizygotic twins [BIOL] Twins derived from two eggs. Also known as fraternal twins.

djerfisherite [MINERAL] $K_3CuFe_{12}S_{14}$ A sulfide mineral found only in meteorites.

Djulfian [GEOL] Upper upper Permian geologic time.

D layer [GEOL] The lower mantle of the earth, between a depth of 1000 and 2900 km. [GEOPHYS] The lowest layer of ionized air above the earth, occurring in the D region only in the daytime hemisphere; reflects frequencies below about 50 kilohertz and partially absorbs higher-frequency waves.

D line [SPECT] The yellow line that is the first line of the major series of the sodium spectrum; the doublet in the Fraunhofer lines whose almost equal components have wavelengths of 5895.93 and 5889.96 angstroms respectively.

DM See adamsite.

dma See direct memory access.

DMDT See methoxychlor.

DME See distance-measuring equipment.

D meson [PARTIC PHYS] A neutral pseudovector meson resonance having a mass of 1285 ± 4 MeV, width of about 30 MeV, and positive charge conjugation parity; the only singlet state consistent with the $(1^+)^+$ nonet.

DMF See N,N-dimethylformamide.

DMSO See dimethyl sulfoxide.

DMT See dimethyl terephthalate.

DMU See dimethylolurea.

DNA See deoxyribonucleic acid.

dneprovskite See wood tin.

D nickel [MET] A nickel-manganese (4.75%) alloy of medium strength that resists spark erosion and hence is used for sparkplug electrodes and for some electronic applications.

Dobbin's reagent [ANALY CHEM] A mercuric chloride-potassium iodide reagent used to test for caustic alkalies in soap.

dobby loom [TEXT] A weaving loom on which small all-over designs, dots, and stripes can be made in raised woven or self-color effects by means of special attachments.

Dobrowolsky generator [ELEC] Three-wire, direct-current generator with a balance coil connected across the armature; the coil's midpoint produces the midpoint voltage for the system.

Dobson prop [MIN ENG] A hydraulic supporting post used in mine tunnel construction.

Dobson spectrophotometer [SPECT] A photoelectric spectrophotometer used in the determination of the ozone content of the atmosphere; compares the solar energy at two wavelengths in the absorption band of ozone by permitting the radiation of each to fall alternately upon a photocell.

Dobson support system [MIN ENG] A self-advancing unit consisting of three Dobson props used to support longwall faces.

DOC See delayed-opening chaff.

dock [CIV ENG] **1.** The slip or waterway that is between two piers or cut into the land for the berthing of ships. **2.** A basin or enclosure for reception of vessels, provided with means for controlling the water level.

docking [AERO ENG] The mechanical coupling of two or more man-made orbiting objects.

docking block [CIV ENG] A timber used to support a ship in dry dock.

docking bridge [NAV ARCH] A platform that rises above the deck of a large ship at the stern.

docking keel [NAV ARCH] A keel placed on some ships parallel to and between the bilge keels and the main keel to support the vessel while in dry dock.

dock landing ship [NAV ARCH] An amphibious warfare ship which transports and supports landing craft and can quickly unload troops and equipment onto a beach. Designated LSD.

dockyard [CIV ENG] A yard utilized for ship construction and repair.

Docodonta [PALEON] A primitive order of Jurassic mammals of North America and England.

docosane [ORG CHEM] $C_{22}H_{46}$ A paraffin hydrocarbon, especially the normal isomer $CH_3(CH_2)_{20}CH_3$.

docosanoic acid [ORG CHEM] $CH_3(CH_2)_{20}CO_2H$ A crystalline fatty acid, melting at 80°C, slightly soluble in water and alcohol, and found in the fats and oils of some seeds such as peanuts. Also known as behenic acid.

docosapentanoic acid [ORG CHEM] $C_{21}H_{33}CO_2H$ A paleyellow liquid, boils at 236°C (5 mm Hg), insoluble in water, soluble in ether, and found in fish blubber. Also known as clupanodonic acid.

cis-13-docosenoic acid See erucic acid.

doctor [METEOROL] A cooling sea breeze in the tropics.

doctor blade [CHEM ENG] A device for regulating the amount of liquid material on the rollers of a spreader. Also known as doctor knife.

doctor knife See doctor blade.

doctor roll [CHEM ENG] Roller device used to remove accumulated filter cake from rotary filter drums.

doctor solution [CHEM ENG] Sodium plumbite solution used to remove mercaptan sulfur from gasoline and other light petroleum distillates; used in doctor treatment.

doctor test [CHEM ENG] A procedure using doctor solution (sodium plumbite) to detect sulfur compounds in light petroleum distillates which react with the sodium plumbite.

doctor treatment [CHEM ENG] Refining process to sweeten (reduce the odor) of gasoline, solvents, and kerosine; sodium plumbite and sulfur convert the odoriferous mercaptans into disulfides.

document [ADP] **1.** Any record, printed or otherwise, that can be read by man or machine. **2.** To prepare a written text and charts describing the purpose, nature, usage, and operation of a program or a system of programs.

document alignment [ADP] The phase of the reading process in which a transverse force is applied to a document to line up its reference edge with that of the reading station.

documentation [ADP] The collection, organized and stored, of records that describe the purpose, use, structure, details, and operational requirements of a program, for the purpose of making this information easily accessible to the user.

document flow [ADP] The path taken by documents as they are processed through a record handling system.

document handling [ADP] In character recognition, the process of loading, feeding, transporting, and unloading a cutform document that has been submitted for character recognition.

document leading edge [ADP] In character recognition, that edge which is the foremost one encountered during the reading process and whose relative position defines the document's direction of travel.

document misregistration [ADP] In character recognition, the improper state of appearance of a document, on site in a character reader, with respect to real or imaginary horizontal baselines.

document number [ADP] The number given to a document by its originators to be used as a means for retrieval; this number will follow any one of various systems, such as chronological, subject area, or accession.

document plane [GRAPHICS] In microfilm technology, that surface or area in space at which the document is positioned during exposure.

document platen [GRAPHICS] A device for positioning a document in the document plane.

document reference edge [ADP] In character recognition, that edge of a source document which provides the basis of all subsequent reading processes, insofar as it indicates the relative position of registration marks, and the impending text.

docuterm [ADP] A word or phrase descriptive of the subject matter or concept of an item of information and considered important for later retrieval of information.

dodecagon [MATH] A 12-sided polygon.

dodecahedron [MATH] A polyhedron with 12 faces.

dodecahydrate [CHEM] A hydrated compound that has a total of 12 water molecules associated with it.

dodecamerous [BOT] Having the whorls of floral parts in multiples of 12.

dodecane [ORG CHEM] $CH_3(CH_2)_{10}CH_3C_{12}H_{26}$ An oily paraffin compound, a colorless liquid, boiling at 214.5°C, insoluble in water; used as a solvent and in jet fuel research. Also known as dihexyl; propylene tetramer; tetrapropylene.

dodecanoic acid *See* lauric acid.

n-**dodecanol** *See* lauryl alcohol.

1-dodecene [ORG CHEM] $CH_2CH(CH_2)_9CH_3$ A colorless liquid, boiling at 213°C, insoluble in water; used in flavors, dyes, perfumes, and medicines. Also known as decylethylene; dodecylene.

dodecyl [ORG CHEM] $C_{12}H_{25}$ A radical derived from dodecane by removing one hydrogen atom; in particular, the normal radical, $CH_3(CH_2)_{10}CH_2-$.

dodecyl alcohol *See* lauryl alcohol.

dodecylbenzene [ORG CHEM] Blend of isomeric (mostly monoalkyl) benzenes with saturated side chains averaging 12 carbon atoms; used in the alkyl amyl sulfonate type of detergents. Also known as detergent alkylate.

dodecylene *See* 1-dodecene.

dodecyl sodium sulfate *See* sodium lauryl sulfate.

dodge chain [DES ENG] A chain with detachable bearing blocks between the links.

Dodge crusher [MIN ENG] A type of jaw crusher with the movable jaw hinged at the bottom, allowing a highly uniform product to be discharged.

Dodge pulverizer [MIN ENG] A hexagonal drum-shaped pulverizer that rotates on a horizontal axis and contains steel balls for reducing rock and ore.

Dodge-Romig tables [IND ENG] Tabular data for acceptance sampling, including lot tolerance and AOQL tables.

dodging [GRAPHICS] A technique of shading a portion of the photographic paper during exposure so that the tone of certain portions of the print is modified.

dodo [VERT ZOO] *Raphus calculltus.* A large, flightless, extinct bird of the family Raphidae.

doe [VERT ZOO] The adult female deer, antelope, goat, rabbit, or any other mammal of which the male is referred to as buck.

Doebner-Miller synthesis [CHEM ENG] Synthesis of methylquinoline by heating aniline with paraldehyde in the presence of hydrochloric acid.

doffer [TEXT] **1.** A device that strips material from another part on a textile machine. **2.** A worker who replaces full bobbins or cones with empty ones.

doffing [TEXT] Removing fibers from a cotton carding machine by means of a toothed bar or cylinder.

dog [DES ENG] **1.** Any of various simple devices for holding, gripping, or fastening, such as a hook, rod, or spike with a ring, claw, or lug at the end. **2.** An iron for supporting logs in a fireplace. **3.** A drag for the wheel of a vehicle. [VERT ZOO] Any of various wild and domestic animals identified as *Canis familiaris* in the family Canidae; all are carnivorous and digitigrade, are adapted to running, and have four toes with nonretractable claws on each foot.

dog clutch [DES ENG] A clutch in which projections on one part fit into recesses on the other part.

dog days [CLIMATOL] The period of greatest heat in summer.

dogfish *See* bowfin.

dogfish oil *See* shark liver oil.

dogger [GEOL] Concretionary masses of calcareous sandstone or ironstone.

doghole [MIN ENG] A small opening in a mine.

doghole mine [MIN ENG] A small coal mine employing 15 or less miners.

doghouse [ELECTR] Small enclosure placed at the base of a transmitting antenna tower to house antenna tuning equipment.

dog iron [DES ENG] **1.** A short iron bar with ends bent at right angles. **2.** An iron pin that can be inserted in stone or timber in order to lift it.

dogleg [NAV] That portion of a flight which does not lead directly to the destination or way point, followed to comply with established flight procedures, avoid possible dangers or bad weather areas, or delay time of arrival. [PETRO ENG] Bend or sudden direction change in a wellhole that can cause tubing wear and failure.

dogs *See* folding boards.

dog screw [DES ENG] A screw with an eccentric head; used to mount a watch in its case.

Dog Star *See* Sirius.

dog's tooth [CIV ENG] A masonry string course in which the brick corner projects.

Doherty amplifier [ELECTR] A linear radio-frequency power amplifier that is divided into two sections whose inputs and outputs are connected by quarter-wave networks; for all values of input signal voltage up to one-half maximum amplitude, section no. 1 delivers all the power to the load; above this level, section no. 2 comes into operation.

doister [METEOROL] In Scotland, a severe storm from the sea. Also known as deaister; dyster.

Doisy unit [BIOL] A unit for standardization of vitamin K.

dolabriform [BIOL] Shaped like an ax head.

Dolan equation [PETRO ENG] Empirical equation for reservoir-permeability damage factor by the invasion of drilling mud or other foreign materials.

doldrums [METEOROL] A nautical term for the equatorial trough, with special reference to the light and variable nature of the winds. Also known as equatorial calms.

dolerophanite [MINERAL] $Cu_2(SO_4)O$ A brown, monoclinic mineral consisting of a basic copper sulfate, occurring in crystals.

Dolichopodidae [INV ZOO] The long-legged flies, a family of orthorrhaphous dipteran insects in the series Brachycera.

Dolichothoraci [PALEON] A group of joint-necked fishes assigned to the Arctolepiformes in which the pectoral appendages are represented solely by large fixed spines.

dolioform [BIOL] Barrel-shaped.

doliolaria larva [INV ZOO] A free-swimming larval stage of crinoids and holothurians having an apical tuft and four or five bands of cilia.

Doliolida [INV ZOO] An order of pelagic tunicates in the class Thaliacea; transparent forms, partly or wholly ringed by muscular bands.

dollar [NUCLEO] A unit of reactivity, equal to the difference between the reactivities for delayed critical and prompt critical conditions in a given nuclear reactor.

dollar spot [PL PATH] A fungus disease of lawn grasses caused by *Sclerotinia homeocarpa* and characterized by small, round, brownish areas which gradually coalesce.

dolly [ENG] Any of several types of industrial hand trucks consisting of a low platform or specially shaped carrier mounted on rollers or combinations of fixed and swivel casters; used to carry such things as furniture, milk cans, paper rolls, machinery weighing up to 80 tons, and television cameras short distances.

dolocast [GEOL] The cast or impression of a dolomite crystal.

dolomite [MINERAL] $CaMg(CO_3)_2$ The carbonate mineral; white or colorless with hexagonal symmetry and a structure similar to that of calcite, but with alternate layers of calcium ions being completely replaced by magnesium.

dolomite rock *See* dolomitic limestone.

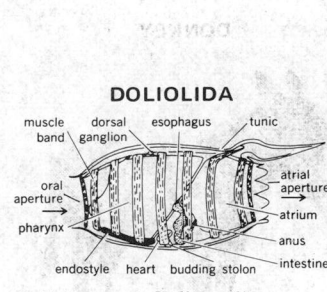

DOLIOLIDA

muscle band, dorsal ganglion, esophagus, tunic, oral aperture, pharynx, atrial aperture, atrium, anus, intestine, endostyle, heart, budding stolon

Doliolum, a doliolid, solitary asexual form. *(After Uljanin)*

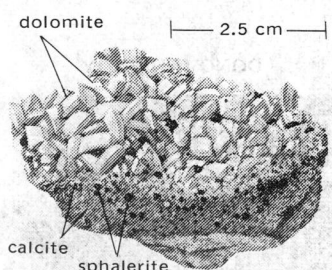

DOLOMITE

dolomite, 2.5 cm, calcite, sphalerite

Crystals of dolomite with calcite and sphalerite on limestone, found in Joplin, Missouri. *(Specimen from Department of Geology, Bryn Mawr College)*

DOLPHIN

The common dolphin (*Delphinus delphis*), found in many seas throughout the world.

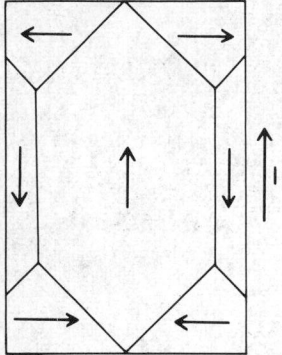

DOMAIN GROWTH

Domain growth, a stage of magnetization in an applied magnetic field. Arrow at right with symbol "||" represents direction of applied magnetic field. Other arrows represent direction of magnetization of domains.

DONKEY

The donkey, sometimes referred to as the burro.

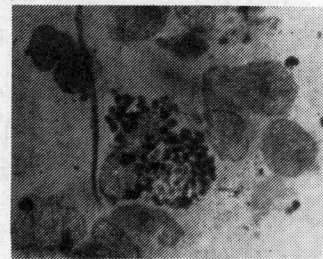

DONOVAN BODY

Mononuclear cell, with intramonocytic Donovan body.

dolomitic limestone [PETR] A limestone whose carbonate fraction contains more than 50% dolomite. Also known as dolomite rock; dolostone.

dolomitization [GEOL] Conversion of limestone to dolomite rock by replacing a portion of the calcium carbonate with magnesium carbonate.

dolomol *See* magnesium stearate.

do loop [ADP] A FORTRAN iterative technique which enables any number of instructions to be executed repeatedly.

dolostone *See* dolomitic limestone.

dolphin [CIV ENG] **1.** A group of piles driven close and tied together to provide a fixed mooring in the open sea or a guide for ships coming into a narrow harbor entrance. **2.** A mooring post on a wharf. [VERT ZOO] The common name for about 33 species of cetacean mammals included in the family Delphinidae and characterized by the pronounced beak-shaped mouth.

Dolphin *See* Delphinus.

domain [MATH] **1.** A nonempty open connected set in euclidean space. Also known as region. **2.** *See* Abelian field. [SOLID STATE] A region in a solid within which elementary atomic or molecular magnetic or electric moments are uniformly arrayed.

domain growth [SOLID STATE] A stage in the process of magnetization in which there is a growth of those magnetic domains in a ferromagnet oriented most nearly in the direction of an applied magnetic field.

domain of a function [MATH] The set of values of the independent variable.

domain rotation [SOLID STATE] The stage in the magnetization process in which there is rotation of the direction of magnetization of magnetic domains in a ferromagnet toward the direction of a magnetic applied field and against anisotropy forces.

domain theory [SOLID STATE] A theory of the behavior of ferromagnetic and ferroelectric crystals according to which changes in the bulk magnetization and polarization arise from changes in size and orientation of domains that are each polarized to saturation but which point in different directions.

domain wall *See* Bloch wall.

domatic class *See* clinohedral class.

domatophobia [PSYCH] An abnormal fear of being in a house.

dome [ARCH] A hemispherical roof. [CRYSTAL] An open crystal form consisting of two faces astride a symmetry plane. [ENG ACOUS] An enclosure for a sonar transducer, projector, or hydrophone and associated equipment; designed to have minimum effect on sound waves traveling underwater. [GEOL] **1.** A circular or elliptical, almost symmetrical upfold or anticlinal type of structural deformation. **2.** A large igneous intrusion whose surface is convex upward. [ORD] The mound of water spray created in air when the shock wave from an underwater detonation of an atomic weapon reaches the surface.

Domerian [GEOL] Upper Charmouthian geologic time.

domestication [BIOL] The adaptation of an animal or plant through breeding in captivity to a life intimately associated with and advantageous to man.

domestic coke [MATER] Coke for residential heating, which must have as low an ash content and as high a softening temperature (preferably above 2300°F, or 1260°C) for the ash as possible.

domestic induction heater [ENG] A cooking utensil heated by current (usually of commercial power line frequency) induced in it by a primary inductor.

domestic public-frequency bands [COMMUN] Radio-frequency bands reserved for public service within the United States.

domestic refrigerator [MECH ENG] A refrigeration system for household use which typically has a compression machine designed for continuous automatic operation and for conservation of the charges of refrigerant and oil, and is usually motor-driven and air-cooled. Also known as refrigerator.

dome theory [MIN ENG] The theory that the movements of strata resulting from underground excavations are limited by a dome whose base is the area of excavation, and that the movements decrease in intensity as they extend upward from the center of the base.

domeykite [MINERAL] Cu_3As A tin-white or steel-gray mineral consisting of copper arsenide; specific gravity is 7.2–7.75.

dominance [GEN] The expression of a heritable trait in the heterozygote such as to make it phenotypically indistinguishable from the homozygote.

dominant allele [GEN] The member of a pair of alleles which is phenotypically indistinguishable in both the homozygous and heterozygous condition.

dominant hemisphere [PHYSIO] The cerebral hemisphere which controls certain motor activities; usually the left hemisphere in right-handed individuals.

dominant mode *See* fundamental mode.

dominant wave [ELECTROMAG] The electromagnetic wave that has the lowest cutoff frequency in a given uniconductor waveguide.

dominant wavelength [OPTICS] The single wavelength of light that, when combined in suitable proportions with a reference standard light, matches the color of a given sample.

dominated convergence theorem [MATH] If a sequence $\{f_n\}$ of Lebesgue measurable functions converges to f and if each f_n is dominated by the same measurable function, then $\lim \int f_n dm = \int f dm$.

dominating integral [MATH] An improper integral whose nonnegative, nonincreasing integrand function has the property that its value for all sufficiently large positive integers n is no smaller than the nth term of a given series of positive terms; used in the integral test for convergence.

dominating series [MATH] A series, each term of which is larger than the respective term in some other given series; used in the comparison test for convergence of series.

doming [MIN ENG] Setting up a domelike region above the open space created by stope excavation.

Donau glaciation [GEOL] A Pleistocene glacial time unit in the Alps region in Europe.

donkey [MIN ENG] *See* barney. [VERT ZOO] A domestic ass (*Equus asinus*); a perissodactyl mammal in the family Equidae.

donkey boiler [NAV ARCH] A steam boiler on a ship deck used to supply steam to deck machinery when main boilers are shut down.

donkey engine [MECH ENG] A small auxiliary engine which is usually portable or semiportable and powered by steam, compressed air, or other means, particularly one used to power a windlass to lift cargo on shipboard or to haul logs.

donkey power [PHYS] A unit of power equal to 250 watts; it is approximately 1/3 horsepower.

Donnan equilibrium [PHYS CHEM] The particular equilibrium set up when two coexisting phases are subject to the restriction that one or more of the ionic components cannot pass from one phase into the other; commonly, this restriction is caused by a membrane which is permeable to the solvent and small ions but impermeable to colloidal ions or charged particles of colloidal size. Also known as Gibbs-Donnan equilibrium.

Donohue equation [THERMO] Equation used to determine the heat-transfer film coefficient for a fluid on the outside of a baffled shell-and-tube heat exchanger.

donor [ELECTR] An impurity that is added to a pure semiconductor material to increase the number of free electrons. Also known as donor impurity; electron donor.

donor impurity *See* donor.

donor level [SOLID STATE] An intermediate energy level close to the conduction band in the energy diagram of an extrinsic semiconductor.

Donovan body [MED] The causative microorganism of granuloma inguinale, demonstrated in stained mononuclear cells and characterized by one or two opposite polar chromatin masses.

donut *See* doughnut.

doodlebug [GEOL] Also known as douser. **1.** Any unscientific device or apparatus, such as a divining rod, used to locate subsurface water, minerals, gas, or oil. **2.** A scientific instrument used for locating minerals. [INV ZOO] The larva of an ant lion. [MECH ENG] **1.** A small tractor. **2.** A motor-driven railcar used for maintenance and repair work. [MIN ENG] The treatment plant or washing unit of a dredge which is mounted on a pontoon and can be floated in an excavation dug by a dragline. [ORD] **1.** A small military tank or utility truck. **2.** An airborne, magnetic submarine-detecting device.

door [ENG] A piece of wood, metal, or other firm material pivoted or hinged on one side, sliding along grooves, rolling up and down, revolving, or folding, by means of which an opening into or out of a building, room, or other enclosure is open or closed to passage.

doorknob capacitor [ELEC] A high-voltage, plastic-encased capacitor resembling a doorknob in size and shape.

DOP *See* dioctyl phthalate.

dopa *See* dihydroxyphenylalanine.

dopamine [BIOCHEM] $C_8H_{11}O_2N$ An intermediate in epinephrine and norepinephrine biosynthesis; the decarboxylation product of dopa.

dopa oxidase [BIOCHEM] An enzyme that catalyzes the oxidation of dihydroxyphenylalanine to melanin; occurs in the skin.

dope [ELECTR] *See* doping agent. [MATER] A cellulose ester lacquer used as an adhesive or a coating.

doped junction [ELECTR] A junction produced by adding an impurity to the melt during growing of a semiconductor crystal.

dope dyeing *See* solution dyeing.

doping [ELECTR] The addition of impurities to a semiconductor to achieve a desired characteristic, as in producing an *n*-type or *p*-type material. Also known as semiconductor doping. [ENG] Coating the mold or mandrel with a substance which will prevent the molded plywood part from sticking to it and will facilitate removal.

doping agent [ELECTR] An impurity element added to semiconductor materials used in crystal diodes and transistors. Also known as dope.

doping compensation [ELECTR] The addition of donor impurities to a *p*-type semiconductor or of acceptor impurities to an *n*-type semiconductor.

Doppler broadening [SPECT] Frequency spreading that occurs in single-frequency radiation when the radiating atoms, molecules, or nuclei do not all have the same velocity and may each give rise to a different Doppler shift.

Doppler effect [PHYS] The change in the observed frequency of a wave due to relative motion of source and observer.

Doppler frequency *See* Doppler shift.

Doppler-inertial navigation equipment [NAV] Navigation equipment utilizing an inertial navigator and a Doppler navigator in combination so that inherent weaknesses of both systems will be bolstered to increase the reliability of the resultant navigational information.

dopplerite [GEOL] A naturally occurring gel of humic acids found in peat bags or where an aqueous extract from a low-rank coal can collect.

Doppler navigation [NAV] Dead reckoning performed automatically by a device which gives a continuous indication of position by integrating the speed and the crab angle of the aircraft as derived from measurement of the Doppler effect of echoes from directed beams of radiant energy transmitted from the craft.

Doppler radar [ENG] A radar that makes use of the Doppler shift of an echo due to relative motion of target and radar to differentiate between fixed and moving targets and measure target velocities.

Doppler range *See* doran.

Doppler shift [PHYS] The amount of the change in the observed frequency of a wave due to Doppler effect, usually expressed in hertz. Also known as Doppler frequency.

Doppler sonar [ENG] Sonar based on Doppler shift measurement technique. Abbreviated DS.

Doppler tracking [ENG] Tracking of a target by using Doppler radar.

Doppler velocity and position [NAV] A continuous-wave trajectory-measuring system using the Doppler effect caused by a target moving relative to a ground transmitter and receiving station. Abbreviated Dovap.

Doppler VOR [NAV] A ground-based navigational aid operating at very high frequency and using a wide-aperture

radiation system to reduce azimuth errors caused by reflection from terrain and other obstacles; makes use of the Doppler principle to solve the problem of ambiguity that arises from the use of a radiation system with apertures that exceed one-half wavelength; the system is so designed that its signals may be received on the equipment used for the narrow-aperture VOR (very-high-frequency omnidirectional radio range).

DOP test *See* dioctyl phthalate test.

Doradidae [VERT ZOO] A family of South American catfishes in the suborder Siluroidei.

Dorado [ASTRON] A constellation of the southern hemisphere, right ascension 5 hours, declination 65° south. Also known as Swordfish.

30 Doradus *See* Loop Nebula.

doran [ENG] A Doppler ranging system that uses phase comparison of three different modulation frequencies on the carrier wave, such as 0.01, 0.1, and 1 megahertz, to obtain missile range data with high accuracy. Derived from Doppler range.

doraphobia [PSYCH] An abnormal fear of touching animal skin or fur.

dore [MET] Gold and silver bullion remaining in a cupeling furnace after removal of the oxidized lead.

Dorilaidae [INV ZOO] The big-headed flies, a family of cyclorrhaphous dipteran insects in the series Aschiza.

Dorippidae [INV ZOO] The mask crabs, a family of brachyuran decapods in the subsection Oxystomata.

dormancy [BOT] A state of quiescence during the development of many plants characterized by their inability to grow, though continuing their morphological and physiological activities.

dormant bud *See* latent bud.

dormer window [BUILD] An extension of an attic room through a sloping roof to accommodate a vertical window.

dormouse [VERT ZOO] The common name applied to members of the family Gliridae; they are Old World arboreal rodents intermediate between squirrels and rats.

Dorn effect [PHYS CHEM] A difference in a potential resulting from the motions of particles through water; the potential exists between the particles and the water.

Dorngeholz *See* thornbush.

Dorngestrauch *See* thornbush.

dornveld *See* thornbush.

doroid [ELECTROMAG] A coil resembling half a toroid, using a removable core segment to simplify the winding process.

Dorr agitator [MECH ENG] A tank used for batch washing of precipitates which cannot be leached satisfactorily in a tank; equipped with a slowly rotating rake at the bottom, which moves settled solids to the center, and an air lift that lifts slurry to the launders. Also known as Dorr thickener.

Dorr classifier [MECH ENG] A horizontal flow classifier consisting of a rectangular tank with a sloping bottom, a rake mechanism for moving sands uphill along the bottom, an inlet for feed, and outlets for sand and slime.

Dorr thickener *See* Dorr agitator.

dorsal [ANAT] Located near or on the back of an animal or one of its parts.

dorsal aorta [ANAT] The portion of the aorta extending from the left ventricle to the first branch. [INV ZOO] The large, dorsal blood vessel in many invertebrates.

dorsal column [ANAT] A column situated dorsally in each lateral half of the spinal cord which receives the terminals of some afferent fibers from the dorsal roots of the spinal nerves.

dorsal fin [VERT ZOO] A median longitudinal vertical fin on the dorsal aspect of a fish or other aquatic vertebrate.

dorsal lip [EMBRYO] In an amphibian embryo, the margin or lip of the fold of blastula wall marking the dorsal limit of the blastopore during gastrulation and constituting the primary organizer, is necessary to the development of neural tissue, and forms the originating point of chordamesoderm.

dorsiferous [BOT] Of ferns, bearing sori on the back of the frond. [ZOO] Bearing the eggs or young on the back.

dorsiflex [ZOO] To flex or cause to flex in a dorsal direction.

dorsiflexion sign *See* Homan's sign.

dorsigrade [VERT ZOO] Walking on the back of the toes.

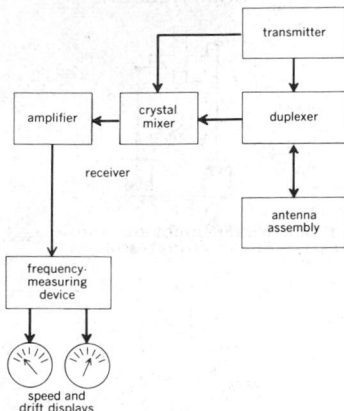

DOPPLER RADAR

Block diagram of a simple Doppler system.

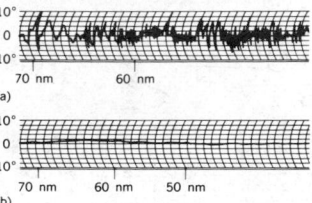

DOPPLER VOR

Comparison of radial flight recordings *(a)* from VOR installation, *(b)* from Doppler VOR installation. Radial flight flight recording revealed for conventional VOR course some fluctuations which exceed International Civil Aviation Organization limits; maximum deviation of Doppler VOR was about 0.5°.

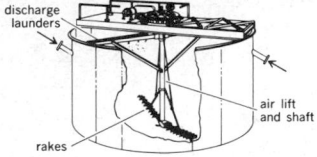

DORR AGITATOR

Dorr agitator for batch washing of precipitates. *(From W. L. Badger and J. T. Banchero, Introduction to Chemical Engineering, McGraw-Hill, 1955)*

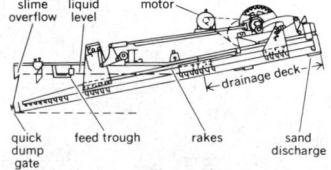

DORR CLASSIFIER

Diagram of Dorr classifier. *(From J. H. Perry, ed., Chemical Engineers' Handbook, 4th ed., McGraw-Hill, 1963)*

DOUBLE-ACTING COMPRESSOR

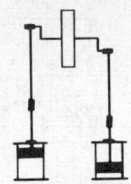

Frame arrangement of double-acting piston compressor.

DOUBLE-ACTING PAWL

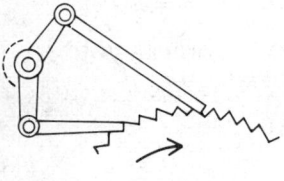

Elements of a double-acting pawl.

dorsocaudad [ANAT] To or toward the dorsal surface and caudal end of the body.

dorsomedial [ANAT] Located on the back, toward the midline.

dorsoposteriad [ANAT] To or toward the dorsal surface and posterior end of the body.

dorsum [ANAT] **1.** The entire dorsal surface of the animal body. **2.** The upper part of the tongue, opposite the velum.

Dorvilleidae [INV ZOO] A family of minute errantian annelids in the superfamily Eunicea.

Dorylaimoidea [INV ZOO] An order or superfamily of nematodes inhabiting soil and fresh water.

Dorylinae [INV ZOO] A subfamily of predacious ants in the family Formicidae, including the army ant (*Eciton hamatum*).

Dorypteridae [PALEON] A family of Permian palaeonisciform fishes sometimes included in the suborder Platysomoidei.

dosage [GEN] The number of genes with a similar action that control a given character. [MED] The prescribed or correct amount of medicine or other therapeutic agent administered to treat a given illness. Also known as dose. [NUCLEO] See absorbed dose.

dosage compensation [GEN] A mechanism that equalizes the phenotypic effects of genes located on the X chromosome, which genes therefore exist in two doses in the homogametic sex and one dose in the heterogametic sex.

Dosco miner [MIN ENG] A large crawler-tracked cutter-loader that is designed for longwall faces in seams over 4½ feet (1.4 meters) thick and takes a buttock 0.5 foot (0.15 meter) wide; rated at 200 horsepower; has seven cutter chains mounted on the cutterhead, with a capacity of over 400 tons per machine.

dose [MED] **1.** The measure, expressed in number of roentgens, of a property of x-rays at a particular place; used in radiology. **2.** See dosage. [NUCLEO] See absorbed dose.

dose equivalent [NUCLEO] The product of absorbed dose, in rads, and a number of modifying factors due to nonuniform distribution of internally deposited isotopes in radiobiology; the unit is the rem.

dosemeter See dosimeter.

dose rate [NUCLEO] The rate at which nuclear radiation is delivered.

dose-rate meter [NUCLEO] An instrument that measures radiation dose rate.

dosimeter [NUCLEO] An instrument that measures the total dose of nuclear radiation received in a given period. Also spelled dosemeter.

dosimetry [NUCLEO] The measurement of radiation doses. Also known as radiation dosimetry.

dosing tank [CIV ENG] A holding tank that discharges sewage at a rate required by treatment processes.

dot See button.

dot angel [ELECTROMAG] An angel that appears on the screens of vertically pointing radars, often on clear cloudless days, as a bright dot; believed to be produced by a vertical column of rising air passing through air layers having different indices of refraction.

dot cycle [COMMUN] In teletypewriter systems, refers to an on-off or mark-space cycle in which both mark and space have the same length as the unit pulse.

dot etching [GRAPHICS] A technique in correcting the color of a positive or halftone negative employing the chemical reduction of halftone dots.

dot generator [ELECTR] A signal generator that produces a dot pattern on the screen of a three-gun color television picture tube, for use in convergence adjustments.

dot product See inner product.

dotriacontane [ORG CHEM] $C_{32}H_{66}$ A paraffin hydrocarbon, in particular, the normal isomer $CH_3(CH_2)_{30}CH_3$, which is crystalline.

dot-sequential color television [ELECTR] A color television system in which the red, blue, and green primary-color dots are formed in rapid succession along each scanning line.

dot system [ELECTR] Manufacturing technique for producing microelectronic circuitry.

dotter [OPTICS] A worker who uses a centering machine to locate the optical center and optical axis of a lens, for the guidance of workers who will cut, edge, trim, and mount the lens. Also known as spotter.

double-acting [MECH ENG] Acting in two directions, as with a reciprocating piston in a cylinder with a working chamber at each end.

double-acting compressor [MECH ENG] A reciprocating compressor in which both ends of the piston act in working chambers to compress the fluid.

double-acting hammer [MET] A forging hammer in which the ram is raised and forced down by a charge of air or steam.

double-acting pawl [MECH ENG] A double pawl which can drive in either direction.

double action [ORD] Method of fire in a revolver and in old-style rifles and shotguns in which a single pull of the trigger both cocks and fires the weapon.

double-action die [MET] A die designed to perform more than one operation with each stroke of the press.

double-action forming [MET] Forming in which more than one shape is imparted by each stroke of the press.

double-action mechanical press [MECH ENG] A press having two slides which move one within the other in parallel movements.

double altitudes See equal altitudes.

double-amplitude-modulation multiplier [ELECTR] Multiplier in which one variable is amplitude-modulated by a carrier, and the modulated signal is again amplitude-modulated by the other variable; the resulting double-modulated signal is applied to a balanced demodulator to obtain the product of the two variables.

double-base diode See unijunction transistor.

double-base junction diode See unijunction transistor.

double-base junction transistor [ELECTR] A tetrode transistor that is essentially a junction triode transistor having two base connections on opposite sides of the central region of the transistor. Also known as tetrode junction transistor.

double-base rocket propellant [MATER] A solid rocket propellant using two unstable compounds, such as nitrocellulose and nitroglycerine.

double-beam cathode-ray tube [ELECTR] A cathode-ray tube having two beams and capable of producing two independent traces that may overlap; the beams may be produced by splitting the beam of one gun or by using two guns.

double-beam spectrophotometer [SPECT] An instrument that uses a photoelectric circuit to measure the difference in absorption when two closely related wavelengths of light are passed through the same medium.

double beta decay [NUC PHYS] A nuclear transformation in which the atomic number changes by 2 and the mass number does not change; either two electrons are emitted or two orbital electrons are captured.

double-bevel groove weld [MET] A type of groove weld in which one member has a joint edge beveled on both sides. Also known as double-V groove weld.

double-block brake [MECH ENG] Two single-block brakes in symmetrical opposition, where the operating force on one lever is the reaction on the other.

double blossom [PL PATH] A fungus disease of dewberry and blackberry caused by *Fusarium rubi* and characterized by witches'-brooms and enlargement and malformation of the flowers.

double bond [PHYS CHEM] A type of linkage between atoms in which two pair of electrons are shared equally.

double-bond isomerism [PHYS CHEM] Isomerism in which two or more substances possess the same elementary composition but differ in having double bonds in different positions.

double-bond shift [ORG CHEM] In an organic molecular structure, the occurrence when a pair of valence bonds that join a pair of carbons (or other atoms) shifts, via chemical reaction, to a new position, for example, $H_2C=C-C-CH_2$ (butene-1) to $H_2C-C=C-CH_2$ (butene-2).

double bottom [NAV ARCH] A ship bottom with space between the inner and outer plating.

double-bottom cellular [NAV ARCH] Pertaining to division of the double bottom of a ship into numerous rectangular compartments by transverse and longitudinal framing.

double-bounce calibration [ELECTR] Method of radar calibration which is used to determine the zero set error by using round-trip echoes; the correct range is the difference between the first and second echoes.

double-break switch [ELEC] Switch which opens the connected circuit at two points.

double bridge *See* Kelvin bridge.

double brilliant [LAP] A brilliant cut with 40 facets above the girdle and 32 below.

double-buffered data transfer [ADP] The transmission of data into the buffer register and from there into the device register proper.

double-button microphone [ENG ACOUS] A carbon microphone having two carbon-filled buttonlike containers, one on each side of the diaphragm, to give twice the resistance change obtainable with a single button. Also known as differential microphone.

double-channel duplex [COMMUN] A method that provides for simultaneous communication between two stations through use of two radio-frequency channels, one in each direction.

double-channel simplex [COMMUN] A method that provides for nonsimultaneous communication between two stations through use of two radio-frequency channels, one in each direction.

double circulation [PHYSIO] A circulatory system in which blood flows through two separate circuits, as pulmonary and systemic.

double-concave lens [OPTICS] A lens having surfaces that are adjacent portions of nonintersecting spheres whose centers lie on opposite sides of the plane of the lens. Also known as biconcave lens.

double-cone bit [DES ENG] A type of roller bit having only two cone-shaped cutting members.

double-convex lens [OPTICS] A lens having surfaces that are adjacent portions of intersecting spheres whose centers lie on opposite sides of the plane of the lens. Also known as biconvex lens.

double-core barrel drill [DES ENG] A core drill consisting of an inner and an outer tube; the inner member can remain stationary while the outer one revolves.

double-coursed [BUILD] Covered with a material such as shingles in such a way that no area is covered with less than two thicknesses.

double-crank press [MECH ENG] A mechanical press with a single wide slide operated by a crankshaft having two crank pins.

double crossover *See* scissors crossover.

double-current cable code [COMMUN] A cable code in which characters are determined by bipolar characters of equal length.

double-current generator [ELEC] Machine which supplies both direct and alternating current from the same armature winding.

double-current signaling [COMMUN] A system of telegraph signaling that uses both positive and negative currents.

double-cut file [DES ENG] A file covered with two series of parallel ridges crossing at angles to each other.

double-cut planer [MECH ENG] A planer designed to cut in both the forward and reverse strokes of the table.

double-cut saw [DES ENG] A saw with teeth that cut during the forward and return strokes.

double decomposition [CHEM] The simple exchange of elements of two substances to form two new substances; for example, $CaSO_4 + 2NaCl \rightarrow CaCl_2 + Na_2SO_4$.

double-diffused transistor [ELECTR] A transistor in which two pn junctions are formed in the semiconductor wafer by gaseous diffusion of both p-type and n-type impurities; an intrinsic region can also be formed.

double diode *See* binode; duodiode.

double-diode limiter [ELECTR] Type of limiter which is used to remove all positive signals from a combination of positive and negative pulses, or to remove all the negative signals from such a combination of positive and negative pulses.

double distribution [CHEM ENG] The product distribution resulting from counter double-current extraction, a scheme in which each of the two liquid phases is transferred simultaneously and continuously in opposite directions through an interconnected train of contact vessels.

double-doped transistor [ELECTR] The original grown-junction transistor, formed by successively adding p-type and n-type impurities to the melt during growing of the crystal.

double-doublet antenna [ELECTROMAG] Two half-wave doublet antennas criss-crossed at their center, one being shorter than the other to give broader frequency coverage.

double drift [NAV] A method of determining the speed and direction of the wind by observing the drift angle on two aircraft headings and deriving the wind information therefrom.

double drill column [MIN ENG] Two drill columns connected by a horizontal bar on which a drill machine can be mounted. Also known as double jack.

double-drum hoist [MECH ENG] A hoisting device consisting of two cable drums which rotate in opposite directions and can be operated separately or together.

double-entry method [MIN ENG] A mining arrangement involving twin entries in flat or gently dipping coal, so that rooms can be extended from both entryways.

double exposure [GRAPHICS] The act of recording two images on top of each other completely or in part.

double fertilization [BOT] In most seed plants, fertilization involving fusion between the egg nucleus and one sperm nucleus, and fusion between the other sperm nucleus and the polar nuclei.

double floor [BUILD] A floor in which binding joists support the ceiling joists below as well as the floor joists above.

double frequency shift keying [COMMUN] Multiplex system in which two telegraph signals are combined and transmitted simultaneously by a method of frequency shifting between four radio frequencies.

double group [QUANT MECH] A type of group useful in studying systems of half-integral spin; it is formed by modifying a finite point group by introducing an element which is a rotation through an angle of 2π about an arbitrary axis and which is not the unit element but gives the unit element when applied twice.

doublehand drilling [ENG] A rock-drilling method performed by two men, one striking the rock with a long-handled sledge hammer while a second holds the drill and twists it between strokes. Also known as double jacking.

double headings [MIN ENG] A pair of coal headings driven parallel to each other and positioned side by side about 10–20 yards (9–18 meters) apart.

double Hooke's joint [MECH ENG] A universal joint which eliminates the variation in angular displacement and angular velocity between driving and driven shafts, consisting of two Hooke's joints with an intermediate shaft.

double-housing planer [MECH ENG] A planer having two housings to support the cross rail, with two heads on the cross rail and one sidehead on each housing.

double-hung [BUILD] Counterweighted or equipped with a spring on each side of a window for easier raising and lowering.

double image [ELECTR] A television picture consisting of two overlapping images due to reception of the signal over two paths of different length so that signals arrive at slightly different times.

double impeller breaker *See* impact breaker.

double integral [MATH] The Riemann integral of functions of two variables.

double-integrating gyro [MECH] A single-degree-of-freedom gyro having essentially no restraint of its spin axis about the output axis.

double jack [DES ENG] A heavy hammer, weighing about 10 pounds, requiring the use of both hands. [MIN ENG] *See* double drill column.

double jacking *See* doublehand drilling.

double-J groove weld [MET] A groove weld in which one member has a joint edge in the form of a double J or two half U's, one from each side. Also known as double-U groove weld.

double layer *See* electric double-layer.

double-length number [ADP] A number having twice as many digits as are ordinarily used in a given computer. Also known as double-precision number.

double limiter *See* cascade limiter.

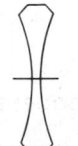

DOUBLE-CONCAVE LENS

Double-concave lens, a type of diverging or negative lens. *(From F. A. Jenkins and H. E. White, Fundamentals of Optics, 3d ed., McGraw-Hill, 1957)*

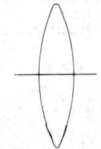

DOUBLE-CONVEX LENS

A double-convex lens, a type of collecting or positive lens. *(From F. A. Jenkins and H. E. White, Fundamentals of Optics, 3d ed., McGraw-Hill, 1957)*

DOUBLE HOOKE'S JOINT

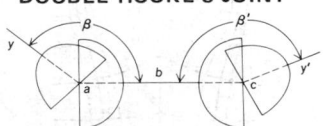

The arrangement in a double Hooke's joint; axes y and y' of the two yokes attached to intermediate shaft b lie respectively in planes containing axes of adjoining shafts (a, y, b in the same plane; b, y', c in the same plane) and the angle β between the driving and intermediate shafts equal angle β' between the intermediate and driven shafts. *(From C. W. Ham, E. J. Crane, and W. L. Rogers, Mechanics of Machinery, McGraw-Hill, 1958)*

DOUBLE-LOOP PATTERN

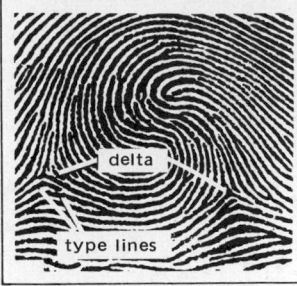

Double-loop pattern in a fingerprint. *(Federal Bureau of Investigation)*

DOUBLET

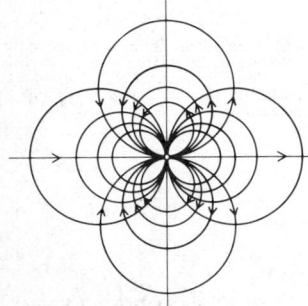

Equipotential lines and streamlines for the two-dimensional doublet.

DOUBLET FLOW

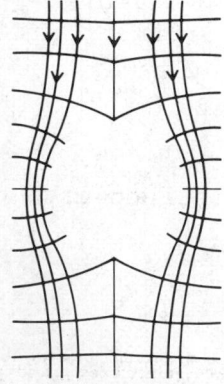

Streamlines and equipotential lines for uniform flow about a sphere at rest.

double load [ENG] A charge separated by inert material in a borehole.

double-loop pattern [ANAT] A whorl type of fingerprint pattern consisting of two separate loop formations and two deltas.

double mast *See* A frame.

double mirror [OPTICS] Two plane mirrors inclined at an angle to each other.

double moding [ELECTR] Undesirable shifting of a magnetron from one frequency to another at irregular intervals.

double modulation [COMMUN] A method of modulation in which a subcarrier is first modulated with the desired intelligence, and the modulated subcarrier is then used to modulate a second carrier having a higher frequency.

double pendulum [MECH] Two masses, one suspended from a fixed point by a weightless string or rod of fixed length, and the other similarly suspended from the first; often the system is constrained to remain in a vertical plane.

double-pipe exchanger [CHEM ENG] Fluid-fluid heat exchanger made of two concentric pipe sections; one fluid (such as a coolant) flows in the annular space between pipes, and the other fluid (such as hot process stream) flows through the inner pipe.

double point [MATH] A point on a curve at which a curve crosses or touches itself, or has a cusp; that is, a point at which the curve has two tangents (which may be coincident).

double-polarity pulse-amplitude modulation [COMMUN] Pulse-amplitude modulation employing pulses of positive and negative polarity, the average value being equal to zero. Also known as bidirectional pulse-amplitude modulation.

double-pole double-throw switch [ELEC] A six-terminal switch or relay contact arrangement that simultaneously connects one pair of terminals to either of two other pairs of terminals. Abbreviated dpdt switch.

double-pole single-throw switch [ELEC] A four-terminal switch or relay contact arrangement that simultaneously opens or closes two separate circuits or both sides of the same circuit. Abbreviated dpst switch.

double-pole switch [ELEC] A switch that operates simultaneously in two separate electric circuits or in both lines of a single circuit.

double precision [ADP] The use of two computer words to represent a number.

double-precision number *See* double-length number.

double pulsing [NAV] The transmitting of loran signals of two rates by a single station.

double-pulsing station [NAV] A master loran station that controls two slaves and therefore emits pulses at two different rates.

double-quantum stimulated-emission device [OPTICS] A laser in which the crystal contains two species of fluorescent ions whose fluorescence frequencies are so related that, when the flash lamp coils produce pumping action, the ions of one species contribute photons to the fluorescence of the other species, which is the active ion.

doubler *See* frequency doubler; voltage doubler.

double-recoil system [ORD] A system in which the gun recoils on the top carriage and the top carriage recoils on the bottom carriage or trail.

double refraction *See* birefringence.

double-rivet [ENG] To rivet a lap joint with two rows of rivets or a butt joint with four rows.

double-roll crusher [MECH ENG] A machine which crushes materials between teeth on two roll surfaces; used mainly for coal.

double root [MATH] A number a such that $(x-a)^2p(x) = 0$ where $p(x)$ is a polynomial of which a is not a root.

double rose [LAP] A gem cut to resemble two roses base to base, with 48 facets.

double salt [INORG CHEM] **1.** A salt that upon hydrolysis forms two different anions and cations. **2.** A salt that is a molecular combination of two other salts.

double sampling [IND ENG] Inspecting one sample and then deciding whether to accept or reject the lot or to defer action until a second sample is inspected.

double screen [ELECTR] Three-layer cathode-ray tube screen consisting of a two-layer screen with the addition of a second long-persistence coating having a different color and different persistence from the first.

double sextant [NAV] A sextant designed to enable the observer to measure simultaneously the left and right horizontal sextant angles of a three-point problem; this sextant differs from a standard sextant in that it comprises a circle with two indices inscribed and two coaxial index mirrors; the horizon glass is silvered top and bottom but is clear in the middle to permit direct observation of the center object ashore.

double-shield enclosure [ELEC] Type of shielded enclosure or room in which the inner wall is partially isolated electrically from the outer wall.

double-shot molding [ENG] A means of turning out two-color parts in thermoplastic materials by successive molding operations.

double-sideband modulation [COMMUN] Amplitude modulation in which the modulated wave is composed of a carrier, an upper sideband whose frequency is the sum of the carrier and modulation frequencies, and a lower sideband whose frequency is the difference between the carrier and modulation frequencies. Abbreviated DSB.

double-sideband transmission [COMMUN] The transmission of a modulated carrier wave accompanied by both of the sidebands resulting from modulation; the upper sideband corresponds to the sum of the carrier and modulation frequencies, whereas the lower sideband corresponds to the difference between the carrier and modulation frequencies.

double-slider coupling *See* slider coupling.

double-solvent refining [CHEM ENG] Petroleum-refining process using two solvents to simultaneously deasphalt and solvent-treat lubricating-oil stocks.

double-spot tuning [COMMUN] Superheterodyne reception of a given station at two different local oscillator frequency values with the local oscillator adjusted above the incoming signal frequency by the intermediate-frequency value or, with the local oscillator adjusted below the incoming signal frequency by the intermediate-frequency value. Also known as repeat point.

double star [ASTRON] A star which appears as a single point of light to the eye but which can be resolved into two points by a telescope. Also known as binary star.

double-stream amplifier [ELECTR] Microwave traveling-wave amplifier in which amplification occurs through interaction of two electron beams having different average velocities.

double-stub tuner [ELECTROMAG] Impedance-matching device, consisting of two stubs, usually fixed three-eighths of a wavelength apart, in parallel with the main transmission lines.

double-superheterodyne reception [COMMUN] Method of reception in which two frequency converters are employed before final detection. Also known as triple detection.

doublet [ATOM PHYS] Two stationary states which have the same orbital and spin angular momentum but which have different total angular momenta, and therefore have slightly different energies due to spin-orbit coupling. [ELECTROMAG] *See* dipole. [FL MECH] A source and a sink separated by an infinitesimal distance, each having an infinitely large strength so that the product of this strength and the separation is finite. [OPTICS] A lens made up of two components, especially an achromat. [PARTIC PHYS] Two elementary particles which have slightly differing masses and the same baryon number, spin, parity, and charge conjugation parity (if self-conjugate), but have different charges. [PHYS CHEM] Two electrons which are shared between two atoms and give rise to a nonpolar valence bond. [SPECT] Two closely separated spectral lines arising from a transition between a single state and a pair of states forming a doublet as described in the atomic physics definition.

doublet antenna *See* dipole antenna.

doublet flow [FL MECH] The motion of a fluid in the vicinity of a doublet; can be superposed with uniform flow to yield flow around a cylinder or a sphere.

double-theodolite observation [ENG] A technique for making winds-aloft observations in which two theodolites located at either end of a base line follow the ascent of a pilot balloon; synchronous measurements of the elevation and azimuth angles of the balloon, taken at periodic intervals, permit computation of the wind vector as a function of height.

double-throw circuit breaker [ELEC] Circuit breaker by means of which a change in the circuit connections can be obtained by closing either of two sets of contacts.

double-throw switch [ELEC] A switch that connects one set of two or more terminals to either of two other similar sets of terminals.

double-track tape recorder [ENG ACOUS] A tape recorder with a recording head that covers half the tape width, so two parallel tracks can be recorded on one tape. Also known as dual-track tape recorder; half-track tape recorder.

double triode [ELECTR] An electron tube having two triodes in the same envelope. Also known as duotriode.

doublet trigger [ELECTR] A trigger signal consisting of two pulses spaced a predetermined amount for coding purposes.

double-tuned amplifier [ELECTR] Amplifier of one or more stages in which each stage uses coupled circuits having two frequencies of resonance, to obtain wider bands than those obtainable with single tuning.

double-tuned circuit [ELECTR] A circuit that is resonant to two adjacent frequencies, so that there are two approximately equal values of peak response, with a dip between.

double-tuned detector [ELECTR] A type of frequency-modulation discriminator in which the limiter output transformer has two secondaries, one tuned above the resting frequency and the other tuned an equal amount below.

double-U groove weld See double-J groove weld.

double valves [PETRO ENG] Two valves in series used as subsurface traveling or standing valves in wells, the dual arrangement being more reliable than a single valve.

double-V groove weld See double-bevel groove weld.

double-wall cofferdam [CIV ENG] A cofferdam consisting of two lines of steel piles tied to each other, and having the space between filled with sand.

double-wedge cut [MIN ENG] A drill-hole pattern composed of a shallow wedge within a larger, outer wedge.

double-welded joint [MET] Any joint that has been welded from both sides.

double-winding synchronous generator [ELEC] Synchronous generator which has two similar windings, in phase with one another, mounted on the same magnetic structure but not connected electrically, designed to supply power to two independent external circuits.

double word [ADP] A unit containing twice as many bits as a word.

double-word addressing [ADP] An addressing mode in computers with short words (less than 16 bits) in which the second of two consecutive instruction words contains the address of a location.

double-work [BOT] In plant propagation, to graft or bud a scion to an intermediate variety that is itself grafted on a stock of still another variety.

doubling the angle on the bow [NAV] A method of obtaining a running fix by measuring the distance a craft travels while the relative bearing (right or left) of a fixed object doubles; the distance from the object at the time of the second bearing is equal to the run between bearings, neglecting drift.

doubling time [NUCLEO] The time required for a breeding reactor to double its fuel inventory.

doubly plunging fold [GEOL] A fold that plunges in opposite directions, either away from or toward a central point.

doughnut [NUCLEO] Also spelled donut. **1.** The toroidal vacuum chamber in which electrons are accelerated in a betatron or synchrotron. Also known as toroid. **2.** An assembly of enriched fissionable material, often doughnut-shaped, used in a thermal reactor to provide a local increase in fast neutron flux for experimental purposes.

Douglas fir [BOT] *Pseudotsuga menziesii.* A large coniferous tree in the order Pinales characterized by bracts extending beyond the cone scales. Also known as red fir.

Douglas fir oil See pine needle oil.

douglasite [MINERAL] $K_2FeCl_4 \cdot 2H_2O$ Ore from Stassfurt, Germany; a member of the erythrosiderite group; orthorhombic, in the isomorphous series.

douse [MIN ENG] To locate and delineate subsurface resources such as water, oil, gas, or minerals.

douser See doodlebug.

Dovap See Doppler velocity and position.

dove [VERT ZOO] The common name for a number of small birds of the family Columbidae.

Dove See Columba.

Dove prism See Delaborne prism.

dovetail joint [DES ENG] A joint consisting of a flaring tenon in a fitting mortise.

dovetail molding [ARCH] A molding in a zig-zag pattern resembling a series of dovetails.

dovetail saw [DES ENG] A short stiff saw with a thin blade and fine teeth; used for accurate woodwork.

dowel [DES ENG] **1.** A headless, cylindrical pin which is sunk into corresponding holes in adjoining parts, to locate the parts relative to each other or to join them together. **2.** A round wooden stick from which dowel pins are cut.

dowel screw [DES ENG] A dowel with threads at both ends.

down [GEOL] **1.** Hillock of sand thrown up along the coast by the sea or the wind. **2.** A flat eminence on the top of a hill or mountain.

down by the head [NAV ARCH] Having greater draft at the bow than at the stern. Also known as by the head.

down by the stern [NAV ARCH] Having a greater draft at the stern than at the bow. Also known as by the stern.

downcast [MIN ENG] Intake shaft for air in a mine.

downcomer [CHEM ENG] A method of conveying liquid from one tray to the one below in a bubble-tray column.

down cutting See climb cutting.

downdip [GEOL] Pertaining to a position parallel to or in the direction of the dip of a stratum or bed.

down-Doppler [ACOUS] The sonar situation wherein the target is moving away from the transducer, so that the frequency of the echo is less than the frequency of the reverberations received immediately after the end of the outgoing ping; opposite of up-Doppler.

downdraft [PHYS] A current of air or other gas that travels downward, as during a thunderstorm or in a mine shaft.

downdraft carburetor [MECH ENG] A carburetor in which the fuel is fed into a downward current of air.

downhand welding See flat-position welding.

downhole drill [MIN ENG] A hammer or percussive drill in which a reciprocating pneumatic piston is located immediately behind the drill bit and can follow and enter the bit down the hole, for minimizing energy losses.

downhole equipment See drill fittings.

down-lead See lead-in.

down lock [AERO ENG] An airplane mechanism that locks the landing gear in a down position after the gear is lowered.

down milling See climb cutting.

downrange [AERO ENG] Any area along the flight course of a rocket or missile test range.

downrush [METEOROL] A term sometimes applied to the strong downward-flowing air current that marks the dissipating stages of a thunderstorm.

Downs cell [CHEM ENG] A brick-lined steel vessel with four graphite anodes projecting upward from the bottom, with cathodes in the form of steel cylinders concentric with the anodes, containing an electrolyte which is 40% NaCl and 60% $CaCl_2$ at 590°C; used to make sodium.

downslope time [MET] Time necessary for current decrease when using slope control in resistance welding.

Down's process [CHEM ENG] A method for producing sodium and chlorine from sodium chloride; potassium chloride and fluoride are added to the sodium chloride to reduce the melting point; the fused mixture is electrolyzed, with sodium forming at the cathode and chlorine at the anode.

Down's syndrome [MED] A syndrome of congenital defects, especially mental retardation, typical facies responsible for the term mongoloid idiocy, or mongolism, and cytogenetic abnormality consisting of trisomy 21 or its equivalent in the form of an unbalanced translocation. Also known as mongolism; trisomy 21 syndrome.

downstream [CHEM ENG] Portion of a product stream that has already passed through the system; that portion located after a specific process unit. [HYD] In the direction of flow, as a current or waterway.

DOUGLAS FIR

Cone on a branch of Douglas fir (*Pseudotsuga menziesii*).

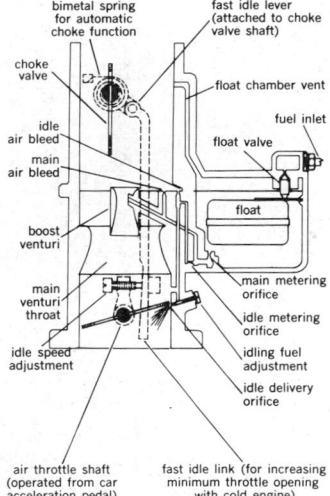

DOWNDRAFT CARBURETOR

bimetal spring for automatic choke function

fast idle lever (attached to choke valve shaft)

choke valve

float chamber vent

fuel inlet

idle air bleed

float valve

main air bleed

float

boost venturi

main venturi throat

main metering orifice

idle metering orifice

idling fuel adjustment

idle speed adjustment

idle delivery orifice

air throttle shaft (operated from car acceleration pedal)

fast idle link (for increasing minimum throttle opening with cold engine)

Idling fuel circuit and choke device for starting and warm-up in typical downdraft carburetor.

DRAG-CUP GENERATOR

Schematic circuit and components of a drag-cup generator.

downthrow [GEOL] The side of a fault whose relative movement appears to have been downward.

downtime [IND ENG] The lost production time during which a piece of equipment is not operating correctly due to a breakdown, maintenance, necessities, or power failure.

downward modulation [COMMUN] Modulation in which the instantaneous amplitude of the modulated wave is never greater than the amplitude of the unmodulated carrier.

downwash [FL MECH] The downward deflection of air, relative to the direction of motion of an airfoil.

downwelling See sinking.

downwind [NAV] In the direction toward which the wind is blowing; applies particularly to moving downwind, whether desired or not.

downy mildew [PL PATH] A fungus disease of higher plants caused by members of the family Peronosporaceae and characterized by a white, downy growth on the diseased plant parts.

Dow oscillator See electron-coupled oscillator.

Dow process [CHEM ENG] A process to produce magnesium by electrolysis of its fused chloride.

dowsing See divining.

dowsing rod See divining rod.

Dowtherm [MATER] A trade name for any of several eutectic mixtures of diphenyl oxide and diphenyl; used in liquid form as a heat-transfer fluid at elevated temperatures (300–500°F; 150–260°C), and in vapor form to heat the deck of a vibrating conveyor dryer by passing through a jacket fastened to the deck.

Dowtonian [GEOL] Uppermost Silurian or lowermost Devonian geologic time.

Dowty prop [MIN ENG] A self-contained hydraulic supporting post consisting of two telescoping tubes; the upper (inner) tube contains the oil, pump, yield valve, and other accessories.

doz See dozen.

dozen [SCI TECH] A group of 12 items. Abbreviated doz.

dozer See bulldozer.

D.P. See degree of polymerization.

DPCM See differential pulse-code modulation.

dpdt switch See double-pole double-throw switch.

DPG See diphenylguanidine.

DPN See diphosphopyridine nucleotide.

dpst switch See double-pole single-throw switch.

dr See dram.

Dra See Draco.

Drac See Draco.

drachm See dram.

Draco [ASTRON] A long, serpentine constellation that surrounds half of the Little Dipper in the north. Abbreviated Dra; Drac. Also known as Dragon.

Draconids [ASTRON] Several meteor showers whose radiants lie in the constellation Draco.

draconitic month See nodical month.

Dracunculoidea [INV ZOO] An order or superfamily of parasitic nematodes characterized by their habitat in host tissues and by the way larvae leave the host through a skin lesion.

Draeger breathing apparatus [MIN ENG] A long-service, self-contained oxygen-breathing apparatus with the oxygen feed governed by the lungs; allows the user to do hard work for up to 5 hours and normal work for 7 hours, and can sustain a resting individual for 18 hours.

Draeger escape apparatus [MIN ENG] A portable, self-contained oxygen-breathing apparatus that is carried on the back of the user; protects against poisonous gases or oxygen shortages for 1 hour.

draft Also spelled draught. [CIV ENG] A line of a traverse survey. [ENG] 1. In molds, the degree of taper on a side wall or the angle of clearance present to facilitate removal of cured or hardened parts from a mold. 2. The area of a water discharge opening. [FL MECH] 1. An air current in a confined space, such as that in a cooling tower or chimney. 2. The difference between atmospheric pressure and some lower pressure in a confined space that causes air to flow, such as exists in the furnace or gas passages of a steam-generating unit or in a chimney. [MET] 1. The act or process of drawing, with dies. 2. The work or quantity of work drawn. [NAV ARCH] The vertical distance from the top of the keel plate or bar keel to the load waterline.

draft gage [ENG] A modified U-tube manometer used to measure draft of low gas heads, such as draft pressure in a furnace, or small differential pressures, for example, less than 2 inches (5 centimeters) of water.

drafting [GRAPHICS] The making of drawings of objects, structures, or systems that have been visualized by engineers, scientists, and others. [TEXT] 1. The process of lengthening raw fibers, in the form of slubbing, sliver, or roving, to make the stock look more like yarn. 2. Plotting directions for weaving on cross-section paper, showing the movement of the threads.

drafting machine [GRAPHICS] A mechanical device to aid in drafting, having two straight edges fixed perpendicularly to each other.

drafting machine See parallel motion protractor.

drafting paper [MATER] A fine white or cream-colored paper that is hard-surfaced and has good erasing characteristics.

draft marks [NAV ARCH] Numbers at the bow and stern of a ship to indicate the height from the bottom of keel or lowest projection.

draftsman [ENG] An individual skilled in drafting, especially of machinery and structures.

draft tube [MECH ENG] The piping system for a reaction-type hydraulic turbine that allows the turbine to be set safely above tail water and yet utilize the full head of the site from head race to tail race.

drag [FL MECH] Resistance caused by friction in the direction opposite to that of the motion of the center of gravity of a moving body in a fluid. [MET] The bottom part of a flask used in casting. [MIN ENG] Movement of the hanging wall with respect to the foot wall due to the weight of the arch block in an inclined slope.

drag bit See bit drag.

drag-body flowmeter [ENG] Device to meter liquid flow; measures the net force parallel to the direction of flow; the resulting pressure difference is used to solve flow equations.

drag chain [ENG] 1. A chain dragged along the ground from a motor vehicle chassis to prevent the accumulation of static electricity. 2. A chain for coupling rail cars.

drag-chain conveyor [MECH ENG] A conveyor in which the open links of a chain drag material along the bottom of a hard-faced concrete or cast iron trough. Also known as dragline conveyor.

drag chute See drag parachute.

drag classifier [MECH ENG] A continuous belt containing transverse rakes, used to separate coarse sand from fine; the belt moves up through an inclined trough, and fast-settling sands are dragged along by the rakes.

drag coefficient [FL MECH] A characteristic of a body in a flowing inviscous fluid, equal to the ratio of twice the force on the body in the direction of flow to the product of the density of the fluid, the square of the flow velocity, and the effective cross-sectional area of the body.

drag conveyor See flight conveyor.

drag-cup generator [ENG] A type of tachometer which uses eddy currents and functions in control systems; it consists of two stationary windings, positioned so as to have zero coupling, and a nonmagnetic metal cup, which is revolved by the source whose speed is to be measured; one of the windings is used for excitation, inducing eddy currents in the rotating cup. Also known as drag-cup tachometer.

drag-cup motor [ELEC] An induction motor having a cup-shaped rotor or conducting material, inside of which is a stationary magnetic core.

drag-cup tachometer See drag-cup generator.

drag cut [ENG] A drill hole pattern for breaking out rock, in which angled holes are drilled along a floor toward a parting, or on a free face and then broken by other holes drilled into them.

drag direction [AERO ENG] In stress analysis of a given airfoil, the direction of the relative wind.

drag factor [CHEM ENG] Ratio of hindered diffusion rate to unhindered rate through a swollen dialysis membrane. Also known as Faxen drag factor; hindrance factor.

drag fold [GEOL] A minor fold formed in an incompetent bed by movement of a competent bed so as to subject it to couple; the axis is at right angles to the direction in which the beds slip.

drag force [PL PHYS] A force on an electrically conducting fluid arising from inelastic collisions of electrons and ions and proportional to the fluid velocity.

drag-in [MET] Solution carried by the work and handling equipment to another solution.

dragline [MECH ENG] An excavator operated by pulling a bucket on ropes towards the jib from which it is suspended. Also known as dragline excavator.

dragline conveyor *See* drag-chain conveyor.

dragline excavator *See* dragline.

dragline scraper [MECH ENG] A machine with a flat, plow-like blade or partially open bucket pulled on rope for withdrawing piled material, such as stone or coal, from a stockyard to the loading platform; the empty bucket is subsequently returned to the pile of material by means of a return rope.

drag link [MECH ENG] A four-bar linkage in which both cranks traverse full circles; the fixed member must be the shortest link.

drag mark [GEOL] Long, even mark usually having longitudinal striations produced by current drag of an object across a sedimentary surface.

Dragon *See* Draco.

dragonfly [INV ZOO] Any of the insects composing six families of the suborder Anisoptera and having four large, membranous wings and compound eyes that provide keen vision.

drag-out [MET] Solution taken from a bath by the work or handling equipment.

drag parachute [AERO ENG] Any of various types of parachutes that can be deployed from the rear of an aircraft, especially during landings, to decrease speed and also, under certain flight conditions, to control and stabilize the aircraft. Also known as drag chute.

drag rope [AERO ENG] A long, heavy rope carried in the basket of a balloon and permitted to hang over the side and drag on the ground in order to lighten the basket.

dragsaw [DES ENG] A saw that cuts on the pulling stroke; used in power saws for cutting felled trees.

drag-stone mill [MIN ENG] A mill in which a heavy stone is dragged over ore to grind it.

drag technique [MET] An arc-welding method in which the electrode is in contact with the joint being welded without being in short circuit.

drag truss [AERO ENG] A truss that is positioned horizontally between the wing spars; used to stiffen the wing structure and as a resistance for drag forces acting on the airplane wing.

drag-type tachometer *See* eddy-current tachometer.

drag-weight ratio [AERO ENG] The ratio of the drag of a missile to its total weight.

drag wire [AERO ENG] A part of the truss in an airplane wing and also in the wing support; used to sustain the backward reaction due to the wing's drag.

drain [CIV ENG] **1.** A channel which carries off surface water. **2.** A pipe which carries off liquid sewage. [ELEC] *See* current drain. [ELECTR] One of the electrodes in a thin-film transistor.

drainage [CIV ENG] Removal of groundwater or surface water, or of water from structures, by gravity or pumping.

drainage basin [HYD] An area in which surface runoff collects and from which it is carried by a drainage system, as a river and its tributaries. Also known as catchment area; drainage area; feeding ground; gathering ground; hydrographic basin.

drainage canal [CIV ENG] An artificial canal built to drain water from an area having no natural outlet for precipitation accumulation.

drainage density [HYD] Ratio of the total length of all channels in a drainage basin and the basin area.

drainage divide [GEOL] **1.** The border of a drainage basin. **2.** The boundary separating adjacent drainage basins.

drainage gallery [CIV ENG] A gallery in a masonry dam parallel to the top of the dam, to intercept seepage from the upstream face and conduct it away from the downstream face.

drainage pattern [HYD] The configuration of a natural or artificial drainage system; stream patterns reflect the topography and rock patterns of the area.

drainage well [CIV ENG] A vertical shaft in a masonry dam to intercept seepage before it reaches the downstream side.

drainage wind *See* gravity wind.

drain valve [CHEM ENG] A valve used to drain off material that has separated from a fluid or gas stream, or one used to empty a process line, vessel, or storage tank.

drain wire [ELEC] Metallic conductor frequently used in contact with foil-type signal-cable shielding to provide a low-resistance ground return at any point along the shield.

dram [MECH] **1.** A unit of mass, used in the apothecaries' system of mass units, equal to $\frac{1}{8}$ apothecaries' ounce or 60 grains or 3.8879346 grams. Also known as apothecaries' dram (dram ap); drachm (British). **2.** A unit of mass, formerly used in the United Kingdom, equal to $\frac{1}{16}$ ounce (avoirdupois) or approximately 1.77185 grams. Abbreviated dr.

dram ap *See* dram.

drape forming [ENG] A method of forming thermoplastic sheet in which the sheet is clamped into a movable frame, heated, and draped over high points of a male mold; vacuum is then applied to complete the forming operation.

Draper catalog [ASTRON] A nine-volume catalog of stars completed in 1924; it gives positions, magnitudes, and spectral classes of 225,300 stars.

Draper effect [CHEM ENG] The increase in volume at constant pressure at the start of the reaction of hydrogen and chlorine to form hydrogen chloride; the volume increase is caused by an increase in temperature of the reactants, due to heat released in the reaction.

draping [GEOL] Structural concordance of the strata overlying a limestone reef or other hard core to the surface of the reef or core.

draught *See* draft.

draught stop *See* fire stop.

draw [ENG] To haul a load. [MET] **1.** A fissure or pocket in a casting formed when the supply of molten metal is inadequate during solidification. **2.** To remove a pattern from a foundry flask. [MIN ENG] **1.** To remove timber supports, allowing overhanging coal to fall down for collection. **2.** To allow ore to run down chutes from stopes, chambers, or ore bins. **3.** To collect broken coal in trucks. **4.** To hoist coal, rock, ore, or other materials to the surface. **5.** The horizontal distance to which creep extends on the surface beyond the stopes.

drawability [MET] The ability of a metal to be deep-drawn.

drawbar [ENG] **1.** A bar used to connect a tender to a steam locomotive. **2.** A beam across the rear of a tractor for coupling machines or other loads. **3.** A clay block submerged in a glassmaking furnace to define the point at which sheet glass is drawn.

drawbar horsepower [MECH ENG] The horsepower available at the drawbar in the rear of a locomotive or tractor to pull the vehicles behind it.

drawbar pull [MECH ENG] The force with which a locomotive or tractor pulls vehicles on a drawbar behind it.

draw bead [MET] A projection on the surface of a metal sheet to control its flow during drawing.

drawbench [MET] A stand on which metal is drawn through dies; used in wire-making, or for drawing of rods and tubing.

drawbridge [CIV ENG] Any bridge that can be raised, lowered, or drawn aside to provide clear passage for ships.

drawdown [HYD] The magnitude of the change in water surface level in a well, reservoir, or natural body of water resulting from the withdrawal of water.

drawer [ENG] A box or receptacle that slides or rolls on tracks within a cabinet.

draw-filing [ENG] Filing by pushing and pulling a file sideways across the work.

drawhead [MET] A group of rollers through which strip tubing or solid stock is drawn to form angled sections.

DRAGONFLY

Winged adult dragonfly.

drawhole [MIN ENG] The aperture in a battery through which coal or ore is drawn.

drawing [GRAPHICS] A surface portrayal of a form or figure in line. [MET] **1.** Pulling a wire or tube through a die to reduce the cross section. **2.** Forcing plastic deformation of metal in a die to form recessed parts.

drawing bristol [MATER] A cardboard made of 100% cotton in the higher grades; has good characteristics of permanence, strength, and erasability.

drawing cloth [MATER] A linen cloth that is specially treated to be smooth and translucent so that it may be used for ink tracings.

drawing compound [MET] A material applied to the work during drawing or pressing operations to eliminate draw marks by preventing direct contact between the work and die.

drawing die [MET] A die that forms sheet metal into cuplike, wrinkle-free shapes.

drawing out [TEXT] The action of pulling staple textile fibers lengthwise over each other, producing longer and thinner slivers.

drawing paper [MATER] One of a wide variety of papers used for pen-and-pencil drawing by artists and architects.

drawing timber [MIN ENG] The act of withdrawing timber and other supports from abandoned or worked-out mines.

draw mark [MET] An impairment of the die or metal surface caused during drawing due to friction or a defect in the die; examples are scoring and die lines.

drawn glass [MATER] Glass made automatically by drawing the molten material through rollers.

draw piece [MET] Any part made by drawing.

drawplate [MET] A circular plate having a central hole through which wire is drawn by a punch. Also known as draw ring.

drawpoint [ENG] A steel point used to scratch lines or to pierce holes.

draw radius [MET] A measure of cutting edge of a die or punch over which the metal is drawn.

draw ring *See* drawplate.

draw works [PETRO ENG] An oil-well drilling mechanism used to supply driving power and to lift heavy objects; consists of a countershaft and drum.

dream [PSYCH] An involuntary series of visual, auditory, or kinesthetic imagery, emotions, and thoughts occurring in the mind during sleep or a sleeplike state, which take the form of a sequence of events or of a story, having a feeling of reality but totally lacking a feeling of free will.

dredge [ENG] A cylindrical or rectangular device for collecting samples of bottom sediment and benthic fauna. [MECH ENG] A floating excavator used for widening or deepening channels, building canals, constructing levees, raising material from stream or harbor bottoms to be used elsewhere as fill, or mining.

dredge ship [NAV ARCH] A watercraft used as a dredge.

dredging [ENG] Removing solid matter from the bottom of a water area.

dredging buoy [NAV] A buoy marking the limit of an area where dredging is being performed.

D region [GEOPHYS] The region of ionosphere up to about 60 miles above the earth, below the E and F regions, in which the D layer forms.

dreikanter [GEOL] A pebble shaped with three facets by sandblasting.

Drepanellacea [PALEON] A monomorphic superfamily of extinct paleocopan ostracods in the suborder Beyrichicopina having a subquadrate carapace, many with a marginal rim.

Drepanellidae [PALEON] A monomorphic family of extinct ostracods in the superfamily Drepanellacea.

Drepanidae [INV ZOO] The hooktips, a small family of lepidopteran insects in the suborder Heteroneura.

Dresbachian [GEOL] Lower Croixan geologic time.

dress [ELECTR] The arrangement of connecting wires in a circuit to prevent undesirable coupling and feedback. [MECH ENG] **1.** To shape a tool. **2.** To restore a tool to its original shape and sharpness. [MIN ENG] To sort, grind, clean, and concentrate ore.

dresser [ENG] Any tool or apparatus used for dressing something.

Dresser coupling [ENG] A type of coupling for unthreaded pipe.

dressing [AGR] Manure or compost used as a fertilizer. [MED] **1.** Application of various materials for protecting a wound and encouraging healing. **2.** Material so applied.

Dressler kiln [MECH ENG] The first successful muffle-type tunnel kiln.

Drew number [PHYS CHEM] A dimensionless group used in the study of diffusion of a solid material A into a stream of vapor initially composed of substance B, equal to

$$\frac{Z_A(M_A - M_B) + M_B}{(Z_A - Y_{AW})(M_B - M_A)} \cdot \ln\frac{M_V}{M_W},$$

where M_A and M_B are the molecular weights of components A and B, M_V and M_W are the molecular weights of the mixture in the vapor and at the wall, and Y_{AW} and Z_A are the mole fractions of A at the wall and in the diffusing stream, respectively. Symbolized N_D.

drewite [GEOL] Calcareous ooze composed of impalpable calcareous material.

drex [TEXT] A unit of yarn density (mass per unit length), equal to 1 gram per 10 kilometers of yarn fiber.

dribbling [MIN ENG] Fall of debris from the roof of an excavation, usually preceding a heavy fall or cave-in.

drier [ENG] A device to remove water. [MATER] **1.** A substance that absorbs water. **2.** A substance that is used to hasten solidification. **3.** Material, such as salts of lead, manganese, and cobalt, which facilitates the oxidation of oils; used in paints and varnishes to speed drying.

Drierite [CHEM] A trade name for an anhydrous calcium sulfate used as a desiccant.

drift [ENG] **1.** A gradual deviation from a set adjustment, such as frequency or balance current, or from a direction. **2.** The deviation, or the angle of deviation, of a borehole from the vertical or from its intended course. [GEOL] **1.** Rock material picked up and transported by a glacier and deposited elsewhere. **2.** Detrital material moved and deposited on a beach by waves and currents. [MECH ENG] The water lost in a cooling tower as mist or droplets entrained by the circulating air, not including the evaporative loss. [MIN ENG] A horizontal mine opening which follows a vein or lies within the trend of an ore body. [NAV] **1.** The movement of a craft caused by the action of wind or current. **2.** To move gradually from a set position without control. [OCEANOGR] *See* drift current. [SOLID STATE] The movement of current carriers in a semiconductor under the influence of an applied voltage.

drift angle [NAV] **1.** The horizontal angle between the axis of a ship and the tangent to its path. Also known as drift correction angle. **2.** The angle between the longitudinal axis of an aircraft and its path relative to the ground.

drift axis [NAV] Of a gyroscope, the axis about which drift occurs; for example, in a directional gyro with the spin axis mounted horizontally, the drift axis is the vertical axis.

drift bolt [ENG] **1.** A bolt used to force out other bolts or pins. **2.** A metal rod used to secure timbers.

drift bottle [OCEANOGR] A bottle which is released into the sea for studying currents; contains a card, identifying the date and place of release, to be returned by the finder with date and place of recovery. Also known as floater.

drift card [OCEANOGR] A card, such as is used in a drift bottle, encased in a buoyant, waterproof envelope and released in the same manner as a drift bottle.

drift correction angle *See* drift angle.

drift current [OCEANOGR] **1.** A wide, slow-moving ocean current principally caused by winds. Also known as drift; wind drift; wind-driven current. **2.** Current determined from the differences between dead reckoning and a navigational fix.

drift dam [GEOL] A dam formed by glacial drift in a stream valley.

drifter [MECH ENG] A rock drill, similar to but usually larger than a jack hammer, mounted for drilling holes up to $4\frac{1}{2}$ inches (11.4 centimeters) in diameter. [MIN ENG] **1.** A person who excavates mine drifts. **2.** An air-driven rock drill used for excavating mine drifts and crosscuts.

drift glacier *See* snowdrift glacier.

drift ice [OCEANOGR] Sea ice that has drifted from its place of formation.

drift indicator [ENG] Device used to record directional logs; records only the amount of drift (deviation from the vertical), and not the direction. [NAV] *See* drift meter.

drifting [MIN ENG] Tunneling along the strike of a lode.

drifting mine [ORD] An underwater mine adjusted to float unanchored on or just below the surface of the water.

drifting snow [METEOROL] Wind-driven snow raised from the surface of the earth to a height of less than 6 feet (1.8 meters).

drift lead [NAV] A lead placed on the bottom to indicate movement of a vessel; at anchor the lead line is usually secured to the rail with a little slack and if the ship drags anchor, the line tends forward; also used to indicate when a vessel coming to anchor is dead in the water or when it is moving astern, or to indicate current if a ship is dead in the water.

drift meter [NAV] An instrument for measuring drift angle. Also known as drift indicator; drift sight.

drift mining [MIN ENG] Working of shallow veins or beds through drifts or shafts from the surface.

drift mobility [SOLID STATE] The average drift velocity of carriers per unit electric field in a homogeneous semiconductor. Also known as mobility.

drift observation [NAV] Also known as drift sight. **1.** The process of observing drift or leeway. **2.** The value obtained by such an observation.

driftpin [DES ENG] A round, tapered metal rod that is driven into matching rivet holes of two metal parts for stretching the parts and bringing them into alignment.

drift plug [ENG] A plug that can be driven into a pipe to straighten it or to flare its opening.

drift sight *See* drift meter; drift sight.

drift space [ELECTR] A space in an electron tube which is substantially free of externally applied alternating fields and in which repositioning of electrons takes place.

drift speed [ELEC] Average speed at which electrons or ions progress through a medium.

drift station [OCEANOGR] **1.** A scientific station established on the ice of the Arctic Ocean, generally based on an ice flow. **2.** A set of observations made over a period of time from a drifting vessel.

drift transistor [ELECTR] **1.** A transistor having two plane parallel junctions, with a resistivity gradient in the base region between the junctions to improve the high-frequency response. **2.** *See* diffused-alloy transistor.

drift tube [NUCLEO] A tubular electrode placed in the vacuum chamber of a circular accelerator, to which radio-frequency voltage is applied to accelerate the particles.

drift velocity [SOLID STATE] The average velocity of a carrier that is moving under the influence of an electric field in a semiconductor, conductor, or electron tube.

drift wave [PL PHYS] An oscillation in a magnetically confined plasma which arises in the presence of density gradients, for example, at the plasma's surface, and which resembles the waves that propagate at the interface of two fluids of different density in a gravity field.

Drilidae [INV ZOO] The false firefly beetles, a family of coleopteran insects in the superfamily Cantharoidea.

drill [ENG] A rotating-end cutting tool for creating or enlarging holes in a solid material. Also known as drill bit.

drillability [ENG] Fitness for being drilled, denoting ease of penetration.

drill angle gage *See* drill grinding gage.

drill bit *See* drill.

drill cable [ENG] A cable used to pull up drill rods, casing, and other drilling equipment used in making a borehole.

drill capacity [MECH ENG] The length of drill rod of specified size that the hoist on a diamond or rotary drill can lift or that the brake can hold on a single line.

drill carriage [MECH ENG] A platform or frame on which several rock drills are mounted and which moves along a track, for heavy drilling in large tunnels. Also known as jumbo.

drill chuck [DES ENG] A chuck for holding a drill or other cutting tool on a spindle.

drill circuit [COMMUN] A telegraph circuit used only to practice sending and receiving.

drill collar [DES ENG] A ring which holds a drill bit and gives it radial location with respect to a bearing.

drill column [MIN ENG] A steel pipe that can be wedged across an underground opening in a vertical or horizontal position to serve as a base on which to mount a diamond or rock drill.

drill cuttings [ENG] Cuttings of rock and other subterranean materials brought to the surface during the drilling of well-holes.

drill doctor [MIN ENG] **1.** A person who services drill bits, tools, and steels. **2.** A shop where the mechanic works.

drilled caisson [CIV ENG] A drilled hole filled with concrete and lined with a cylindrical steel casing if needed.

drilled extrusion ingot [MET] A hollow extrusion ingot made from a solid cast extrusion ingot by drilling.

driller [ENG] A person who operates a drilling machine. [MECH ENG] *See* drilling machine.

drill extractor [ENG] A tool for recovering broken drill pieces or a detached drill from a borehole.

drill feed [MECH ENG] The mechanism by which the drill bit is fed into the borehole during drilling.

drill fittings [ENG] All equipment used in a borehole during drilling. Also known as downhole equipment.

drill floor [ENG] A work area covered with planks around the collar of a borehole at the base of a drill tripod or derrick.

drill footage [ENG] The lineal feet of borehole drilled.

drill gage [DES ENG] A thin, flat steel plate that has accurate holes for many sizes of drills; each hole, identified as to drill size, enables the diameter of a drill to be checked. [ENG] Diameter of a borehole.

drill grinding gage [DES ENG] A tool that checks the angle and length of a twist drill while grinding it. Also known as drill angle gage; drill point gage.

drill hole [ENG] A hole created or enlarged by a drill or auger.

drillhole pattern [ENG] The number, position, angle, and depth of the shot holes forming the round in the face of a tunnel or sinking pit.

drillhole survey *See* borehole survey.

drill-in [MIN ENG] The act or process of setting casting through overburden by using a drill machine.

drilling [ENG] The creation or enlarging of a hole in a solid material with a drill.

drilling column [ENG] The column of drill rods, with the drill bit attached to the end.

drilling fluid *See* drilling mud.

drilling machine [MECH ENG] A device, usually motor-driven, fitted with an end cutting tool that is rotated with sufficient power either to create a hole or to enlarge an existing hole in a solid material. Also known as driller.

drilling mud [MATER] A suspension of finely divided heavy material, such as bentonite and barite, pumped through the drill pipe during rotary drilling to seal off porous zones and flush out chippings, and to lubricate and cool the bit. Also known as drilling fluid.

drilling platform [ENG] The structural base upon which the drill rig and associated equipment is mounted during the drilling operation.

drilling rate [MECH ENG] The number of lineal feet drilled per unit of time.

drilling time [ENG] **1.** The time required in rotary drilling for the bit to penetrate a specified thickness (usually 1 foot) of rock. **2.** The actual time the drill is operating.

drilling time log [ENG] Foot-by-foot record of how fast a formation is drilled.

drill jig [MECH ENG] A device fastened to the work in repetition drilling to position and guide the drill.

drill log [ENG] **1.** A record of the events and features of the formations penetrated during boring. Also known as boring log. **2.** A record of all occurrences during drilling that might help in a complete logging of the hole or in determining the cost of the drilling.

drill out [ENG] **1.** To complete one or more boreholes. **2.** To penetrate or remove a borehole obstruction. **3.** To locate and delineate the area of a subsurface ore body or of petroleum by a series of boreholes.

drill-over [ENG] The act or process of drilling around a casing lodged in a borehole.

drill pipe [MIN ENG] A pipe used for driving a revolving drill bit, used especially in drilling wells; consists of a casing within which tubing is run to conduct oil or gas to ground level; drilling mud flows in the annular space between casing and tubing during the drilling operation.

drill point gage *See* drill grinding gage.

drill press [MECH ENG] A drilling machine in which a vertical drill moves into the work, which is stationary.

drill rod [ENG] The long rod that drives the drill bit in drilling boreholes.

drill runner [MIN ENG] A tunnel miner who operates rock drills.

drill steel [MET] Steel with at least 0.85% carbon content made by the electric furnace process. Formerly known as crucible steel, when made by the crucible process.

drill stem [MECH ENG] 1. In standard drilling, a cylindrical steel or iron bar attached to the cable tool bit to give it weight. 2. In rotary drilling, a length of steel pipe joined together and reaching from the rig floor to the bit at the hole bottom; it transmits power to the bit.

drill-stem test [PETRO ENG] Bottom-hole pressure information obtained and used to determine formation productivity.

drill string [MECH ENG] The assemblage of drill rods, core barrel, and bit, or of drill rods, drill collars, and bit in a borehole, which is connected to and rotated by the drill collar of the borehole. Also known as drill stem.

drill weave [TEXT] Special fabric sometimes used in filtration; a three-harness, warp-face twill weave, having the two-up and one-down twill effect.

drip [HYD] Condensed or otherwise collected moisture falling from leaves, twigs, and so forth. [MATER] 1. Oil which comes through the cloth of a paraffin wax press. 2. Filter drainings too dark to be included in filter stock. [PETRO ENG] A discharge mechanism installed at a low point in a gas transmission line to collect and remove liquid accumulations.

drip cap [BUILD] A molding on top of the head casing of a window frame to direct water away from the window.

drip-dry [TEXT] Of a fabric, shedding water or moisture rapidly without squeezing, spinning, or wringing.

dripping drop atomization [HYD] A type of natural gravitational atomization process in which there is periodic emission of drops from the bottom side of a surface to which a liquid is fed continuously, as in dripping of water from leaves.

dripstone cave [GEOL] Any cave of calcium carbonate or other mineral formed by the action of dripping water.

drive [ELECTR] *See* excitation. [MECH ENG] The means by which a machine is given motion or power (as in steam drive, diesel-electric drive), or by which power is transferred from one part of a machine to another (as in gear drive, belt drive). [MIN ENG] 1. To excavate in a horizontal or inclined plane. 2. A horizontal underground tunnel along or parallel to a lode, vein, or ore body. [PSYCH] A strong impetus to behavior or active striving.

drive chuck [MECH ENG] A mechanism at the lower end of a diamond-drill drive rod on the swivel head by means of which the motion of the drive rod can be transmitted to the drill string.

drive control *See* horizontal drive control.

drivehead [ENG] A cap fitted over the end of a mechanical part to protect it while it is being driven.

driven array [ELECTROMAG] An antenna array consisting of a number of driven elements, usually half-wave dipoles, fed in phase or out of phase from a common source.

driven blocking oscillator *See* monostable blocking oscillator.

driven caisson [CIV ENG] A caisson formed by driving a cylindrical steel shell into the ground with a pile-driving hammer and then placing concrete inside; the shell may be removed when concrete sets.

driven element [ELECTROMAG] An antenna element that is directly connected to the transmission line.

driven gear [MECH ENG] The member of a pair of gears to which motion and power are transmitted by the other.

driven snow [METEOROL] Snow which has been moved by wind and collected into snowdrifts.

drive pattern [COMMUN] In a facsimile system, undesired pattern of density variations caused by periodic errors in the position of the recording spot.

drivepipe [ENG] A thick-walled casing pipe that is driven through overburden or into a deep drill hole to prevent caving.

drive pulley [MECH ENG] The pulley that drives a conveyor belt.

driver [ELECTR] The amplifier stage preceding the output stage in a receiver or transmitter. [ENG ACOUS] The portion of a horn loudspeaker that converts electrical energy into acoustical energy and feeds the acoustical energy to the small end of the horn.

driver element [ELECTROMAG] Antenna array element that receives power directly from the transmitter.

drive rod [ENG] Hollow shaft in the swivel head of a diamond-drill machine through which energy is transmitted from the drill motor to the drill string. Also known as drive spindle.

driver sweep [ELECTR] Sweep triggered only by an incoming signal or trigger.

driver transformer [ELECTR] A transformer in the input circuit of an amplifier, especially in the transmitter.

drive sampling [ENG] The act or process of driving a tubular device into soft rock material for obtaining dry samples.

drivescrew [DES ENG] A screw that is driven all the way in, or nearly all the way in, with a hammer.

drive shaft [MECH ENG] A shaft which transmits power from a motor or engine to the rest of a machine.

drive shoe [DES ENG] A sharp-edged steel sleeve attached to the bottom of a drivepipe or casing to act as a cutting edge and protector.

drive spindle *See* drive rod.

driving clock [ENG] A mechanism for driving an instrument at a required rate.

driving pinion [MECH ENG] The input gear in the differential of an automobile.

driving point impedance [ELECTR] The complex ratio of applied alternating voltage to the resulting alternating current in an electron tube, network, or other transducer.

driving resistance [MECH] The force exerted by soil on a pile being driven into it.

driving signal [ELECTR] Television signal that times the scanning at the pickup point.

driving wheel [MECH ENG] A wheel that supplies driving power.

drizzle [METEOROL] Very small, numerous, and uniformly dispersed water drops that may appear to float while following air currents; unlike fog droplets, drizzle falls to the ground; it usually falls from low stratus clouds and is frequently accompanied by low visibility and fog.

drizzle drop [METEOROL] A drop of water of diameter 0.2 to 0.5 millimeter falling through the atmosphere; however, all water drops of diameter greater than 0.2 millimeter are frequently termed raindrops, as opposed to cloud drops.

DRM *See* depositional remanent magnetization.

drogue [AERO ENG] 1. A small parachute attached to a body for stabilization and deceleration. Also known as deceleration parachute. 2. A funnel-shaped device at the end of the hose of a tanker aircraft in flight, to receive the probe of another aircraft that will take on fuel. [ENG] 1. A device, such as a sea anchor, usually shaped like a funnel or cone and dragged or towed behind a boat or seaplane for deceleration, stabilization, or speed control. 2. A current-measuring assembly consisting of a weighted current cross, sail, or parachute and an attached surface buoy. Also known as drag anchor; sea anchor.

Dromadidae [VERT ZOO] A family of the avian order Charadriiformes containing a single species, the crab plover (*Dromas ardeola*).

dromedary [VERT ZOO] *Camelus dromedarius* The Arabian camel, distinguished by a single hump.

Dromiacea [INV ZOO] The dromiid crabs, a subsection of the Brachyura in the crustacean order Decapoda.

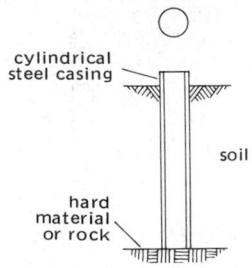

DRIVEN CAISSON

cylindrical
steel casing

soil

hard
material
or rock

Driven caisson, top view and cross-section from the side.

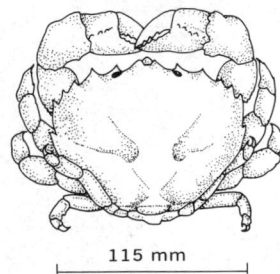

DROMIACEA

115 mm

Dromia erythropus, a dromiid crab. *(Smithsonian Institution)*

Dromiceidae [VERT ZOO] The emus, a monospecific family of flightless birds in the order Casuariiformes.

dromophobia [PSYCH] An abnormal fear of walking about.

drone [AERO ENG] A pilotless aircraft usually subordinated to the controlling influences of a remotely located command station, but occasionally preprogrammed. [INV ZOO] A haploid male bee or ant; one of the three castes in a colony.

drooped ailerons [AERO ENG] Ailerons that are of the hinged trailing-edge type and are so arranged that both the right and left one have a 10 to 15° positive downward deflection with the control column in a neutral position.

droop governor [MECH ENG] A governor whose equilibrium speed decreases as the load on the machinery controlled by the governor increases.

drop [FL MECH] The quantity of liquid that coalesces into a single globule; sizes vary according to physical conditions and the properties of the fluid itself. [MET] A casting defect due to the falling of a portion of sand from an overhanging section of the mold. [PL PATH] A fungus disease of various vegetables caused by *Sclerotinia sclerotiorum* and characterized by wilt and stem rot.

drop ball [ENG] A ball, weighing 3000-4000 pounds (1400-1800 kilograms), dropped from a crane through about 20-33 feet (6-10 meters) onto oversize quarry stones left after blasting; this method is used to avoid secondary blasting.

drop bar [ELEC] Protective device used to ground a high-voltage capacitor when opening a door. [MECH ENG] A bar that guides sheets of paper into a printing or folding machine.

drop black [MATER] Black pigment shaped into droplets.

drop-bottom car [MIN ENG] A mine car designed so that flaps drop open in the bottom to allow the coal to fall out as the car passes over the dump; flaps close as the car leaves.

drop bracket transposition [ELEC] Reversal of the relative positions of two parallel wire conductors while depressing one, so that the crossover is in a vertical plane.

drop-dead halt [ADP] A machine halt from which there is no recovery; such a halt may occur through a logical error in programming; examples in which a drop-dead halt could occur are division by zero and transfer to a nonexistent instruction word. Also known as dead halt.

drop forging [MET] Plastic deformation of hot metal under a falling weight, such as a drop hammer.

drop hammer [MECH ENG] *See* pile hammer. [MET] A hammer used in forging that is raised and then dropped on the metal resting on an anvil or on a die.

drop indicator [COMMUN] Indicator for signaling, consisting of a hinged flap normally held up by a catch; the catch is released by an electromagnet, allowing the flap to drop when a signal is received.

droplet [MED] A tiny drop of matter consisting of water, mucus, and bacterial products, released through the nasal passages or expectorated. [METEOROL] A water droplet in the atmosphere; there is no defined size limit separating droplets from drops of water, but sometimes a maximum diameter of 0.2 millimeter is the limit for droplets.

droplet infection [MED] Infection by contact with airborne droplets of sputum carrying infectious agents.

drop log [MIN ENG] A timber which can be dropped across a mine track by remote control to derail cars.

drop model of nucleus *See* liquid-drop model of nucleus.

dropout [ELEC] Of a relay, the maximum current, voltage, power, or such, at which it will release from its energized position. [ELECTR] A reduction in output signal level during reproduction of recorded data, sufficient to cause a processing error.

dropout current [ELEC] The maximum current at which a relay or other magnetically operated device will release to its deenergized position.

dropout error [ELECTR] Loss of a recorded bit or any other error occurring in recorded magnetic tape due to foreign particles on or in the magnetic coating or to defects in the backing.

dropout fuse [ELEC] A fuse used on utility line poles which springs open when the fuse metal melts to provide rapid arc extinction, and which drops to an open-circuit position readily distinguishable from the ground. Also known as flip-open cutout fuse.

dropout voltage [ELEC] The maximum voltage at which a relay or other magnetically operated device will release to its deenergized position.

dropping angle *See* range angle.

dropping fraction [ADP] In punched cards, the chance that a given sorting operation will cause a card taken at random to be selected.

dropping-mercury electrode [PHYS CHEM] An electrode consisting of a fine-bore capillary tube above which a constant head of mercury is maintained; the mercury emerges from the tip of the capillary at the rate of a few milligrams per second and forms a spherical drop which falls into the solution at the rate of one every 2-10 seconds.

dropping point [CHEM] The temperature at which grease changes from a semisolid to a liquid state under standardized conditions.

dropping resistor [ELEC] A resistor used in series with a load to decrease the voltage applied to the load.

dropping test [MET] A chemical method for determining thickness of zinc and cadmium plated coatings on metal in which a reagent is dropped on the surface until the basis metal is exposed.

drop press *See* punch press.

drop relay [ELEC] Relay activated by incoming ringing current to call an operator's attention to a subscriber's line.

drop repeater [ELECTR] Microwave repeater that is provided with the necessary equipment for local termination of one or more circuits.

drop siding [BUILD] Building siding with a shiplap joint.

drop-size distribution [METEOROL] The frequency distribution of drop sizes (diameters, volumes) that is characteristic of a given cloud or rainfall.

dropsonde [ENG] A radiosonde dropped by parachute from a high-flying aircraft to measure weather conditions and report them back to the aircraft.

dropsonde dispenser [ENG] A chamber from which dropsonde instruments are released from weather reconnaissance aircraft; used only for some models of equipment, ejection chambers being used for others.

dropsonde observation [METEOROL] An evaluation of the significant radio signals received from a descending dropsonde, and usually presented in terms of height, temperature, and dew point at the mandatory and significant pressure levels; it is comparable to a radiosonde observation.

drop spillway [CIV ENG] A spillway usually less than 20 feet (6 meters) high having a vertical downstream face, and water drops over the face without touching the face.

dropsy *See* edema.

drop tank [AERO ENG] A fuel tank on an airplane that may be jettisoned.

drop theory *See* barrier theory of cyclones.

drop wire [ELEC] Wire suitable for extending an open wire or cable pair from a pole or cable terminal to a building.

Droseraceae [BOT] A family of dicotyledonous plants in the order Sarraceniales, distinguished by leaves that do not form pitchers, parietal placentation, and several styles.

drosometer [ENG] An instrument used to measure the amount of dew deposited on a given surface.

Drosophilidae [INV ZOO] The vinegar flies, a family of myodarian cyclorrhaphous dipteran insects in the subsection Acalypteratae, including the fruit fly (*Drosophila melanogaster*).

dross [MET] An impurity, usually an oxide, formed on the surface of a molten metal.

drossing [MET] A process used in nonferrous pyrometallurgy for removing solid oxide deposits on the surface of a molten metal.

drought [CLIMATOL] A period of abnormally dry weather sufficiently prolonged so that the lack of water causes a serious hydrologic imbalance (such as crop damage, water supply shortage, and so on) in the affected area; in general, the term should be reserved for relatively extensive time periods and areas.

drowned atoll [GEOL] An atoll which has not reached the water surface.

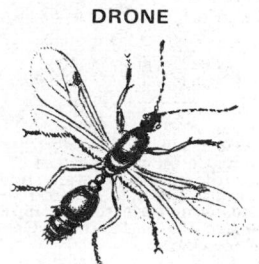

DRONE

Drone, about ³/₁₆ inch (0.5 centimeter).

DROPOUT FUSE

A dropout fuse rated 50 amperes at 7.8 kilovolts used on utility-line poles. *(General Electric Co.)*

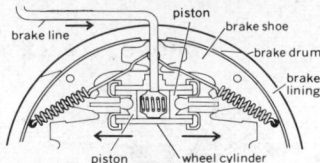

DRUM BRAKE

Schematic view of part of hydraulic braking system showing the brake line leading into the brakes which are shown applied. (*Pontiac Motor Division, General Motors Corp.*)

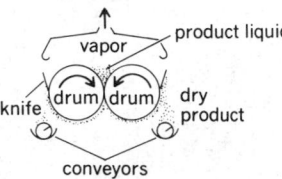

DRUM DRYER

A top-fed double drum dryer.

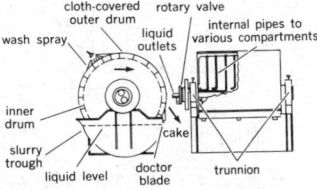

DRUM FILTER

Continuous vacuum drum filter. (*From W. L. McCabe and J. C. Smith, Unit Operations of Chemical Engineering, McGraw-Hill, 2d ed., 1967*)

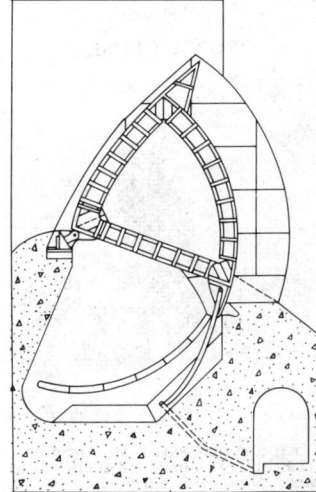

DRUM GATE

Drum gate, a type of spillway gate. (*U.S. Army Corps of Engineers and U.S. Bureau of Reclamation*)

drowned coast [GEOL] A shoreline transformed from a hilly land surface to an archipelago of small islands by inundation by the sea.

drowned river mouth *See* estuary.

drowned valley [GEOL] A valley whose lower part has been inundated by the sea due to submergence of the land margin.

droxtal [HYD] An ice particle measuring 10–20 micrometers in diameter, formed by direct freezing of supercooled water droplets at temperatures below −30°C.

DR track line *See* dead-reckoning track.

Drude equation [OPTICS] An equation which states that the rotation of the plane of polarization of plane-polarized light passing through an optically active substance is inversely proportional to the difference between the square of the wavelength of the light and the square of a constant wavelength.

Drude's theory of conduction [SOLID STATE] A theory which treats the electrons in a metal as a gas of classical particles.

drug [PHARM] **1.** Any substance used internally or externally as a medicine for the treatment, cure, or prevention of a disease. **2.** A narcotic preparation.

drug idiosyncrasy [MED] A peculiarity of constitution that makes an individual respond differently to a drug or treatment than do most people.

drug resistance [MICROBIO] A decreased reactivity of living organisms to the injurious actions of certain drugs and chemicals.

drug tolerance [MED] Condition that may follow repeated ingestion of a drug in so that the effect produced by the original dose no longer occurs.

drum [CHEM ENG] Tower or vessel in a refinery into which heated products are conducted so that volatile portions can separate. [DES ENG] **1.** A hollow, cylindrical container. **2.** A metal cylindrical shipping container for liquids having a capacity of 12–110 gallons (45–416 liters). [ELECTR] A computer storage device consisting of a rapidly rotating cylinder with a magnetizable external surface on which data can be read or written by many read/write heads floating a few millionths of an inch off the surface. Also known as drum memory; drum storage; magnetic drum; magnetic drum storage. [MECH ENG] A horizontal cylinder about which rope or wire rope is wound in a hoisting mechanism. Also known as hoisting drum.

drum armature [ELEC] An armature that has a drum winding.

drum brake [MECH ENG] A brake in which two curved shoes fitted with heat- and wear-resistant linings are forced against the surface of a rotating drum.

drum cam [MECH ENG] A device consisting of a drum with a contoured surface which communicates motion to a cam follower as the drum rotates around an axis.

drum controller [ELEC] An electric device that has a drum switch for its main switching element; used to govern the way electric power is delivered to a motor.

drum disk rectifier [ELEC] A mechanical rectifier using synchronous contacts and a copper oxide dry disk.

drum dryer [MECH ENG] A machine for removing water from substances such as milk, in which a thin film of the product is moved over a turning steam-heated drum and a knife scrapes it from the drum after moisture has been removed.

drum feeder [MECH ENG] A rotating drum with vanes or buckets to lift and carry parts and drop them into various orienting or chute arrangements. Also known as tumbler feeder.

drum filter [MECH ENG] A cylindrical drum that rotates through thickened ore pulp, extracts liquid by a vacuum, and leaves solids, in the form of a cake, on a permeable membrane on the drum end. Also known as rotary filter; rotary vacuum filter.

drum gate [CIV ENG] A movable crest gate in the form of an arc hinged at the apex and operated by reservoir pressure to open and close a spillway.

drumlin [GEOL] A hill of glacial drift or bedrock having a half-ellipsoidal streamline form like the inverted bowl of a spoon, with its long axis paralleling the direction of movement of the glacier that fashioned it.

drum mark [ADP] A character indicating the termination of a record on a magnetic drum.

drum memory *See* drum.

drummy [MIN ENG] Loose rock or coal, especially in a mine roof, that produces a hollow, weak sound when tapped with a bar.

drum parity error [ADP] Parity error occurring during transfer of information onto or from drums.

drum printing [TEXT] Dyeing yarn by winding it on a drum and applying color in bands of various widths.

drum recorder [ELECTR] A facsimile recorder in which the record sheet is mounted on a rotating drum or cylinder.

drum separator [MIN ENG] A cylindrical vessel which rotates slowly and separates run-of-mine coal into clean coal, middlings, and refuse, can be adjusted for different specific gravities.

drum storage *See* drum.

drum switch [ELEC] A switch in which the electrical contacts are made on pins, segments, or surfaces on the periphery of a rotating cylinder or sector, or by the operation of a rotating cam.

drum transmitter [ELECTR] A facsimile transmitter in which the subject copy is mounted on a rotating drum or cylinder.

drum-type boiler *See* bent-tube boiler.

drum winding [ELEC] A type of winding in electric machines in which coils are housed in long, narrow gaps either in the outer surface of a cylindrical core or in the inner surface of a core with a cylindrical bore.

drupaceous [BOT] Of, pertaining to, or characteristic of a drupe.

drupe [BOT] A fruit, such as a cherry, having a thin or leathery exocarp, a fleshy mesocarp, and a single seed with a stony endocarp. Also known as stone fruit.

drupelet [BOT] An individual drupe of an aggregate fruit. Also known as grain.

druse [GEOL] A small cavity in a rock or vein encrusted with aggregates of crystals of the same minerals which commonly constitute the enclosing rock.

drusy [GEOL] Of or pertaining to rocks containing numerous druses.

dry [SCI TECH] Free from or deficient in moisture.

dry abrasive cutting [MECH ENG] Frictional cutting using a rotary abrasive wheel without the use of a liquid coolant.

dry acid [CHEM] Nonaqueous acetic acid used for oil-well reservoir acidizing treatment.

dry adiabat [METEOROL] A line of constant potential temperature on a thermodynamic diagram.

dry-adiabatic lapse rate [METEOROL] A special process lapse rate of temperature, defined as the rate of decrease of temperature with height of a parcel of dry air lifted adiabatically through an atmosphere in hydrostatic equilibrium. Also known as adiabatic lapse rate.

dry-adiabatic process [METEOROL] An adiabatic process in a system of dry air.

dry air [METEOROL] Air that contains no water vapor.

dry ashing [ORG CHEM] The conversion of an organic compound into ash (decomposition) by a burner or in a muffle furnace.

dry assay [MET] Determination of the amount of a desired constituent in ores, metallurgical residues, and alloys by methods other than those involving liquid means of separation.

dry battery [ELEC] A battery made up of a series, parallel, or series-parallel arrangement of dry cells in a single housing to provide desired voltage and current values.

dry blast cleaning [ENG] Cleaning of metallic surfaces by blasting with abrasive material traveling at a high velocity; abrasive may be accelerated by an air nozzle or a centrifugal wheel.

dry-bone ore *See* smithsonite.

dry box [CHEM] A container or chamber filled with argon, or sometimes dry air or air with no CO_2, to provide an inert atmosphere in which manipulation of very reactive chemicals is carried out in the laboratory.

dry-bulb temperature [PHYS] The actual air temperature as measured by a dry-bulb thermometer.

dry-bulb thermometer [ENG] An ordinary thermometer, especially one with an unmoistened bulb; not dependent upon atmospheric humidity.

dry cargo [IND ENG] Nonliquid cargo, including minerals, grain, boxes, and drums.

dry-cargo ship [NAV ARCH] A ship which carries miscellaneous dry cargo, such as boxes, bales, bags, or lumber, which is normally hand-stowed.

dry cell [ELEC] A voltage-generating cell having an immobilized electrolyte.

dry-cell cap light [MIN ENG] A headlamp with a focusing lens lamp and a dry-cell battery unit clipped to the belt; to prevent explosion in a mine, the bulb is ejected automatically in case of its breakage.

dry-charged battery [ELEC] A storage battery in which the electrolyte is drained from the battery for storage, and which is filled with electrolyte and charged for a few minutes to prepare for use.

dry-chemical fire extinguisher [CHEM ENG] A dry powder, consisting principally of sodium bicarbonate, which is used for extinguishing small fires, especially electrical fires.

dry circuit [ELEC] A relay circuit in which open-circuit voltages are very low and closed-circuit currents extremely small, so there is no arcing to roughen the contacts.

dry cleaning [ENG] To utilize dry-cleaning fluid to remove stains from textile.

dry-cleaning fluid [MATER] An organic solvent such as chlorinated hydrocarbons or petroleum naphtha with narrow, carefully selected boiling points; used in dry cleaning.

dry climate [CLIMATOL] **1.** In W. Köppen's climatic classification, the major category which includes steppe climate and desert climate, defined strictly by the amount of annual precipitation as a function of seasonal distribution and of annual temperature. **2.** In C. W. Thornwaite's climatic classification, any climate type in which the seasonal water surplus does not counteract seasonal water deficiency, and having a moisture index of less than zero; included are the dry subhumid, semiarid, and arid climates.

dry coloring [MATER] A powdered form of pigment.

dry contact [ELEC] A contact that does not break or make current.

dry cooling tower [MECH ENG] A structure in which water is cooled by circulation through finned tubes, transferring heat to air passing over the fins; there is no loss of water by evaporation because the air does not directly contact the water.

dry corrosion [MET] Destruction of a metal or alloy by chemical processes resulting from attack by gases in the atmosphere above the dew point.

dry course [BUILD] An initial roofing course of felt or paper not bedded in tar or asphalt.

dry criticality [NUCLEO] Reactor criticality achieved without a coolant.

dry delta *See* alluvial fan.

dry-desiccant dehydration [CHEM ENG] Use of silica gel or other solid absorbent to remove liquids from gases, such as water from air, or liquid hydrocarbons from natural gas.

dry-disk rectifier *See* metallic rectifier.

dry distillation [CHEM] Distillation of materials that are dry.

dry dock [CIV ENG] A dock providing support for a vessel and a means for removing the water so that the bottom of the vessel can be exposed.

dry-dock caisson [CIV ENG] The floating gate to a dry dock. Also known as caisson.

drydock iceberg *See* valley iceberg.

dry drilling [MIN ENG] Drilling in which chippings and cuttings are lifted out of a borehole by a current of air or gas.

dry electrolytic capacitor [ELEC] An electrolytic capacitor in which the electrolyte is a paste rather than a liquid; the dielectric is a thin film of gas formed on one of the plates by chemical action.

dry flashover voltage [ELECTR] Voltage at which the air surrounding a clean dry insulator or shell completely breaks down between electrodes.

dry fog [METEOROL] A fog that does not moisten exposed surfaces.

dry freeze [HYD] The freezing of the soil and terrestrial objects caused by a reduction of temperature when the adjacent air does not contain sufficient moisture for the formation of hoarfrost on exposed surfaces.

dry friction [MECH] Resistance between two dry solid surfaces, that is, surfaces free from contaminating films or fluids.

dry gangrene [MED] Local death of a part caused by arterial obstruction without associated venous obstruction or infection.

dry gas [MATER] A gas that does not contain fractions which may easily condense under normal atmospheric conditions, for example, natural gas with methane and ethane.

dry grinding [ENG] Reducing particle sizes without a liquid medium.

dry haze [METEOROL] Fine dust or salt particles in the air, too small to be individually apparent but in sufficient number to reduce horizontal visibility, and to give the atmosphere a characteristic hazy appearance.

dry hole [ENG] A hole driven without the use of water. [PETRO ENG] A well in which no oil or gas is found.

dry ice [INORG CHEM] A trade name for carbon dioxide in the solid form, usually made in blocks to be used as a coolant; changes directly to a gas at $-78.5°C$ as heat is absorbed.

drying [CHEM] **1.** An operation in which a liquid, usually water, is removed from a wet solid in equipment termed a dryer. **2.** A process of oxidation whereby a liquid such as linseed oil changes into a solid film.

drying agent [CHEM] Soluble or insoluble chemical substance that has such a great affinity for water that it will abstract water from a great many fluid materials; soluble chemicals are calcium chloride and glycerol, and insoluble chemicals are bauxite and silica gel. Also known as desiccant.

drying oil [MATER] Relatively highly unsaturated oil, such as cottonseed, soybean, and linseed oil, that is easily oxidized and polymerized to form a hard, dry film on exposure to air; used in paints and varnish.

drying oven [ENG] A closed chamber for drying an object by heating at relatively low temperatures.

Dryinidae [INV ZOO] A family of hymenopteran insects in the superfamily Bethyloidea.

dry kiln [ENG] A heated room or chamber used to dry and season cut lumber.

dry measure [MECH] A measure of volume for commodities that are dry.

dry mill [FOOD ENG] A machine for processing corn consisting of a horizontal, revolving conical drum covered with metal projections, in a housing also studded with metal projections and having small perforations through which pass fine particles of hull and germ. [MECH ENG] Grinding device used to powder or pulverize solid materials without an associated liquid.

dry mining [MIN ENG] Mining operation in which there is no moisture in the ventilating air.

dry mounting [GRAPHICS] A method for mounting photographs and other paper materials on cardboard without paste or rubber cement; a light, thin tissue (mounting tissue) is placed over the photograph and heat is applied with slight pressure so that the photo will adhere to the cardboard.

dry offset *See* letterset.

Dryomyzidae [INV ZOO] A family of myodarian cyclorrhaphous dipteran insects in the subsection Acalypteratae.

Dryopidae [INV ZOO] The long-toed water beetles, a family of coleopteran insects in the superfamily Dryopoidea.

Dryopoidea [INV ZOO] A superfamily of coleopteran insects in the suborder Polyphaga, including the nonpredatory aquatic beetles.

dry ore [MIN ENG] An ore of gold or silver which requires added lead and fluxes for treatment.

dry permafrost [GEOL] A loose and crumbly permafrost which contains little or no ice.

dry pint *See* pint.

dry-pipe system [ENG] A sprinkler system that admits water only when the air it normally contains has been vented; used for systems subjected to freezing temperatures.

dry-pit pump [MECH ENG] A pump operated with the liquid conducted to and from the unit by piping.

dry placer [MIN ENG] A gold-bearing alluvial deposit found in arid regions; it cannot be mined due to lack of water.

dry plate [GRAPHICS] A photographic plate that has a sensitized coating of an emulsion of silver halide in gelatin which is dried before exposure to light in the photographic process.

dry-plate rectifier *See* metallic rectifier.

dry point [ANALY CHEM] The temperature at which the last drop of liquid evaporates from the bottom of the flask.

drypoint etching [GRAPHICS] Etching in which a sharp tool (an etching needle) scratches through only the etching ground that is placed on the surface of the copper plate; the plate is then placed in an acid bath, and the chemical action produces a line deep enough to hold ink.

dry pressing [ENG] Molding clayware by compressing moist clay powder in metal dies.

dry pt *See* pint.

dry-reed relay [ELEC] Reed-type relay which does not use mercury at the relay contacts.

dry reed switch [ELEC] A switch having contacts mounted on magnetic reeds in a vacuum enclosure, designed for reliable operation in dry circuits.

dry-relief offset [GRAPHICS] Referring to plates made for use on offset-lithographic presses, but printed without the use of dampeners and water; the plates are made photomechanically, and nonprinting areas are etched from 0.008 to 0.015 inch (0.2 to 0.4 millimeter) below the surface. Also known as high etch.

dry rot [MICROBIO] A rapid decay of seasoned timber caused by certain fungi which cause the wood to be reduced to a dry, friable texture. [PL PATH] Any of various rot diseases of plants characterized by drying of affected tissues.

dry run [ENG] Any practice test or session. [ORD] Any simulated firing practice, particularly a dive-bombing approach made without the release of a bomb.

dry sample [MIN ENG] A sample of ore obtained by dry drilling.

dry sand mold [MET] A mold made of greensand and then dried in an oven to increase its strength.

Drysdale ac polar potentiometer [ENG] A potentiometer for measuring alternating-current voltages in which the voltage is applied across a slide-wire supplied with current by a phase-shifting transformer; this current is measured by an ammeter and brought into phase with the unknown voltage by adjustment of the transformer rotor, and the unknown voltage is measured by observation of the slide-wire setting for a null indication of a vibration galvanometer.

dry season [CLIMATOL] In certain types of climate, an annually recurring period of one or more months during which precipitation is at a minimum for the region.

dry sieving [ENG] Particle-size distribution analysis of powdered solids; the sample is placed on the top sieve screen of a nest (stack), with mesh openings decreasing in size from the top to the bottom of the nest.

dry socket [MED] Inflammation of the dental alveolus, especially the inflamed condition following the removal of a tooth. Also known as alveolitis.

dry spell [CLIMATOL] A period of abnormally dry weather, generally reserved for a less extensive, and therefore less severe, condition than a drought; in the United States, describes a period lasting not less than 2 weeks, during which no measurable precipitation was recorded.

dry spot [ENG] An open area of an incomplete surface film on laminated plastic.

drystone [GEOL] A stalagmite or stalactite formed by dropping water.

dry storage [MECH ENG] Cold storage in which refrigeration is provided by chilled air.

dry strength [ENG] The strength of an adhesive joint determined immediately after drying under specified conditions or after a period of conditioning in the standard laboratory atmosphere.

dry tabling [MIN ENG] A process similar to wet tabling, but without the water; used to separate two or more minerals based on specific gravity differences.

dry-tape fuel cell [ELEC] A fuel cell in which the fuel is in the form of a dry tape, coated with fuel, oxidant, and electrolyte, which is fed into the cell at a rate corresponding to the demand for electric energy.

dry test meter [ENG] Gas-flow rate meter with two compartments separated by a movable diaphragm which is connected to a series of gears that actuate a dial; when one chamber is full, a valve switches to the other, empty chamber; used to measure household gas-flow rates and to calibrate flow-measurement instruments.

dry ticket [IND ENG] Tank inspection form signed by shore and ship inspectors before loading and after discharging the ship.

dry tongue [METEOROL] In synoptic meteorology, a pronounced protrusion of relatively dry air into a region of higher moisture content.

dry wall [BUILD] A wall covered with wallboard, in contrast to plaster. [ENG] A wall constructed of rock without cementing material.

dry wash [GEOL] A wash, arroyo, or coulee whose bed lacks water.

dry washer [MIN ENG] A machine for extracting gold mined from dry placers.

dry well [CIV ENG] **1.** A well that has been completely drained. **2.** An excavated well filled with broken stone and used to receive drainage when the water percolates into the soil. **3.** Compartment of a pumping station in which the pumps are housed. [NUCLEO] The first containment tank surrounding a water-cooled nuclear reactor that uses the pressure-suppressing containment system.

dry wire drawing [MET] The drawing of dry steel process wire, pretreated by acid cleaning, lime coating, and baking, through a lubricant and wire drawing frame.

Drzwiecki theory [MECH ENG] In theoretical investigations of windmill performance, a theory concerning the air forces produced on an element of the blade.

DS *See* Doppler sonar.

DSB *See* double-sideband modulation.

D scan *See* D scope.

D scope [ELECTR] A cathode-ray scope which combines the features of B and C scopes, the signal appearing as a spot with bearing angle as the horizontal coordinate and elevation angle as the vertical coordinate, but with each spot expanded slightly in a vertical direction to give a rough range indication. Also known as D indicator; D scan.

DSM screens [MIN ENG] Dutch State Mines screens for dewatering very fine materials; the slurry is sluiced over a concave screen surface with slots that flare out radially to promote egress of particles.

DSN *See* Deep Space Network.

D sounding *See* D value.

DSRV *See* deep-submergence rescue vehicle.

8D technique [METEOROL] A technique for using the radiosonde observation to determine the presence of liquid water-droplets in supercooled clouds in saturated or nearly saturated layers of air; for each reported level in the sounding, the negative value of eight times the dew-point spread ($-8D$) is plotted on the pseudoadiabatic chart (or equivalent chart); where the temperature sounding lies to the left of the $-8D$ curve, liquid droplet clouds are considered to be present, and icing is possible on aircraft flying in the cloud layer. Also known as frost-point technique.

DTL *See* diode transistor logic.

D₁ trisomy [MED] A syndrome resulting from the presence in triplicate of chromosomes 13–15; manifested in severe congenital anomalies and usually resulting in death in infancy. Also known as trisomy 13–15.

Dualayer distillate process [CHEM ENG] A process for the removal of mercaptan and oxygenated compounds from distillate fuel oils; treatment is with concentrated caustic Dualayer solution and electrical precipitation of the impurities.

Dualayer solution [CHEM ENG] A concentrated potassium or sodium hydroxide solution containing a solubilizer; used in the Dualayer distillate process.

dual-bed dehumidifier [MECH ENG] A sorbent dehumidifier with two beds, one bed dehumidifying while the other bed is reactivating, thus providing a continuous flow of air.

DRYSDALE AC POLAR POTENTIOMETER

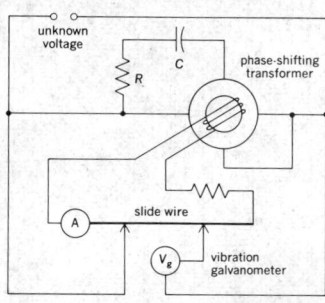

Drysdale ac potentiometer circuit. *(From I. F. Kinard, Applied Electrical Measurements, Wiley, 1956)*

DUAL-BED DEHUMIDIFIER

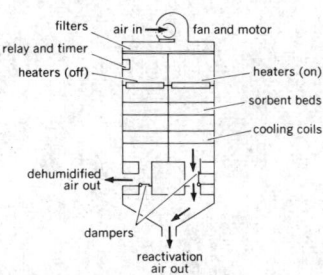

Dual-bed solid-sorbent dehumidifier. Air is being dehumidified through the left bed at the same time that the right bed is being reactivated.

dual-channel amplifier [ENG ACOUS] An audio-frequency amplifier having two separate amplifiers for the two channels of a stereophonic sound system, usually operating from a common power supply mounted on the same chassis.

dual completion well [PETRO ENG] Single well casing containing two production tubing strings, each in a different zone of the reservoir (one higher, one lower) and each separately controlled.

dual coordinates [MATH] Point coordinates and plane coordinates are dual in geometry since an equation about one determines an equation about the other.

dual-cycle boiling-water reactor [NUCLEO] A boiling-water reactor in which part of the steam used to run the steam turbine is generated in the reactor core and part is generated in an external heat exchanger. Also known as dual-cycle reactor system.

dual-cycle reactor system *See* dual-cycle boiling-water reactor.

dual diversity receiver [ELECTR] A diversity radio receiver in which the two antennas feed separate radio-frequency systems, with mixing occurring after the converter.

dual-emitter transistor [ELECTR] A passivated *pnp* silicon planar epitaxial transistor having two emitters, for use in low-level choppers.

dual-fuel engine [MECH ENG] Internal combustion engine that can operate on either of two fuels, such as natural gas or gasoline.

dual graph [MATH] A planar graph corresponding to a planar map obtained by replacing each country with its capital and each common boundary by an arc joining the two countries.

dual-gravity valve [CHEM ENG] A float-operated valve that operates on the interface between two immiscible liquids of different specific gravities.

dual group [MATH] The group of all homomorphisms of an Abelian group G into the cyclic group of order n, where n is the smallest integer such that g^n is the identity element of G.

dual-gun cathode-ray tube [ELECTR] A dual-trace oscilloscope in which beams from two electron guns are controlled by separate balanced vertical-deflection plates and also have separate brightness and focus controls.

dual in-line package [ELECTR] Microcircuit package with two rows of seven vertical leads that are easily inserted into an etched circuit board.

duality principle Also known as principle of duality. [ELEC] The principle that for any theorem in electrical circuit analysis there is a dual theorem in which one replaces quantities with dual quantities; current and voltage, impedance and admittance, and meshes and nodes are examples of dual quantities. [ELECTR] The principle that analogies may be drawn between a transistor circuit and the corresponding vacuum tube circuit. [ELECTROMAG] The principle that one can obtain new solutions of Maxwell's equations from known solutions by replacing **E** with **H**, **H** with −**E**, ε with μ, and μ with ε. [MATH] A principle that if a theorem is true, it remains true if each object and operation is replaced by its dual; important in projective geometry and Boolean algebra. [QUANT MECH] *See* wave-particle duality.

duality theorem [MATH] **1.** A theorem which asserts that for a given space, the $(n − p)$ dimensional homology group is isomorphic to a p-dimensional cohomology group for each $p = 0, \ldots, n$, provided certain conditions are met. **2.** Let G be either a compact group or a discrete group, let X be its character group, and let G' be the character group of X; then there is an isomorphism of G onto G' so that the groups G and G' may be identified.

dual laser [OPTICS] A gas laser having Brewster windows and concave mirrors at opposite ends, the mirrors having different reflectivities so as to produce two different visible or infrared wavelengths from a helium-neon laser beam.

dual linear programming [MATH] Linear programming in which the maximum and minimum number are the same number.

dual meter [ENG] Meter constructed so that two aspects of an electric circuit may be read simultaneously.

dual modulation [COMMUN] The process of modulating a common carrier wave or subcarrier with two different types of modulation, each conveying separate information.

dual network [ELEC] A network which has the same number of terminal pairs as a given network, and whose open-circuit impedance network is the same as the short-circuit admittance matrix of the given network, and vice versa.

dual-purpose gun [ORD] Gun so designed and constructed that effective fire may be delivered against either aerial or surface targets.

dual-purpose reactor [NUCLEO] Any nuclear reactor which both acts as a source of heat energy for a power plant and produces fissionable material.

dual radioactive decay [NUC PHYS] Property exhibited by a nucleus which has two or more independent and alternative modes of decay.

dual-seal tubing joint [PETRO ENG] Tubing connection joint with two sealing surfaces to assure a leak-free connection between sections.

dual space [MATH] The vector space consisting of all linear transformations from a given vector space into its scalar field.

dual tensor [MATH] The product of a given tensor, covariant in all its indices, with the contravariant form of the determinant tensor, contracting over the indices of the given tensor.

dual thrust [AERO ENG] A rocket thrust derived from two propellant grains and using the same propulsion section of a missile.

dual-thrust motor [AERO ENG] A solid-propellant rocket engine built to obtain dual thrust.

dual-tone multifrequency [COMMUN] Signaling method employing set combinations of two specific frequencies used by subscribers and telephone private branch exchange attendants, if their switchboard positions are so equipped, to indicate telephhone address digits, precedence ranks, and end of signaling.

dual-trace amplifier [ELECTR] An oscilloscope amplifier that switches electronically between two signals under observation in the interval between sweeps, so that waveforms of both signals are displayed on the screen.

dual-trace oscilloscope [ELECTR] An oscilloscope which can compare two waveforms on the face of a single cathode-ray tube, using any one of several methods.

dual-track tape recorder *See* double-track tape recorder.

dual-use line [COMMUN] Communications link normally used for more than one mode of transmission, such as voice and data.

dual variables [MATH] Mutually dependent variables.

Duane-Hunt law [QUANT MECH] The law that the frequency of x-rays resulting from electrons striking a target cannot exceed eV/h, where e is the charge of the electron, V is the exciting voltage, and h is Planck's constant.

Duane-Hunt limit [QUANT MECH] The upper limit on the frequency of radiation from an x-ray tube given by the Duane-Hunt law.

dub [ENG ACOUS] **1.** To transfer recorded material from one recording to another, with or without the addition of new sounds, background music, or sound effects. **2.** To combine two or more sources of sound into one record. **3.** To add a new sound track or new sounds to a motion picture film, or to a recorded radio or television production.

Dubbs cracking [CHEM ENG] A continuous, liquid-phase, thermal cracking process.

Duchemin's formula [PHYS] An expression for normal wind pressure per square foot on an inclined surface, $N = F[(2 \sin a)/(1 + \sin^2 a)]$, where F = normal wind force in pounds per square foot on a vertical surface, and a = angle of inclination of inclined surface.

Duchenne's dystrophy [MED] A sex-linked or autosomal recessive form of muscular dystrophy, which is progressive with pseudohypertrophy.

duck [ORD] *See* DUKW. [TEXT] Close-woven, heavy fabric made of cotton and used for liquid filtration in the process industries, as well as for making sails, tents, and clothing. [VERT ZOO] The common name for a number of small waterfowl in the family Anatidae, having short legs, a broad, flat bill, and a dorsoventrally flattened body.

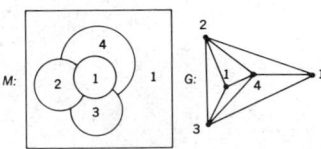

DUAL GRAPH

A map requiring four colors, indicated by numbers, and its planar graph.

DUCK-BILLED DINOSAUR

Restoration of the duck-billed dinosaur *Anatosaurus* (*Trachodon*) from the Late Cretaceous of North America; this dinosaur was 30–40 feet (9–12 meters) long.

duckbill [MECH ENG] A shaking type of combination loader and conveyor whose loading end is generally shaped like a duck's bill.

duck-billed dinosaur [PALEON] Any of several herbivorous, bipedal ornithopods having the front of the mouth widened to form a ducklike beak.

duckbill platypus *See* platypus.

duck wheat *See* tartary buckwheat.

Ducrey bacillus *See* Hemophilus ducreyi.

Ducrey test [IMMUNOL] A skin test to determine past or present infection with *Hemophilus ducreyi*.

duct [ANAT] An enclosed tubular channel for conducting a glandular secretion or other body fluid. [COMMUN] An enclosed runway for cables. [GEOPHYS] The space between two air layers, or between an air layer and the earth's surface, in which microwave beams are trapped in ducting. Also known as radio duct; tropospheric duct. [MECH ENG] A fluid flow passage which may range from a few inches in diameter to many feet in rectangular cross section, usually constructed of galvanized steel, aluminum, or copper, through which air flows in a ventilation system or to a compressor, supercharger, or other equipment at speeds ranging to thousands of feet per minute.

ducted fan [MECH ENG] A propeller or multibladed fan inside a coaxial duct or cowling. Also known as ducted propeller; shrouded propeller.

ducted-fan engine [AERO ENG] An aircraft engine incorporating a fan or propeller enclosed in a duct; especially, a jet engine in which an enclosed fan or propeller is used to ingest ambient air to augment the gases of combustion in the jetstream.

ducted propeller *See* ducted fan.

ducted rocket *See* rocket ramjet.

ductile fracture *See* fibrous fracture.

ductile iron *See* nodular cast iron.

ductility [MATER] The ability of a material to be plastically deformed by elongation, without fracture.

ducting [GEOPHYS] An atmospheric condition in the troposphere in which temperature inversions cause microwave beams to refract up and down between two air layers, so that microwave signals travel 10 or more times farther than the normal line-of-sight limit. Also known as superrefraction; tropospheric ducting.

ductless gland *See* endocrine gland.

duct of Cuvier [EMBRYO] Either of the paired common cardinal veins in a vertebrate embryo.

duct of Santorini [ANAT] The dorsal pancreatic duct in a vertebrate embryo; persists in adult life in some species and serves as the pancreatic duct in the adult elasmobranch, pig, and ox.

duct propulsion [AERO ENG] A means of propelling a vehicle by ducting a surrounding fluid through an engine, adding momentum by mechanical or thermal means, and ejecting the fluid to obtain a reactive force.

ductus arteriosus [EMBRYO] Blood shunt between the pulmonary artery and the aorta of the mammalian embryo.

ductus deferens *See* vas deferens.

ductus venosus [EMBRYO] Blood shunt between the left umbilical vein and the right sinus venosus of the heart in the mammalian embryo.

dud [ORD] An explosive munition that has failed to explode, although such was intended.

Duddell oscillograph [ELECTROMAG] A moving-coil oscillograph; the current to be observed passes through a coil in a magnetic field and a mirror attached to the coil reveals its movement.

Duffy blood group [IMMUNOL] A genetically determined, immunologically distinct group of human erythrocyte antigens defined by their reaction with anti-Fy^a serum.

Dufour effect [THERMO] Energy flux due to a mass gradient occurring as a coupled effect of irreversible processes.

Dufour number [THERMO] A dimensionless number used in studying thermodiffusion, equal to the increase in enthalpy of a unit mass during isothermal mass transfer divided by the enthalpy of a unit mass of mixture. Symbol Du_2.

dufrenite [MINERAL] A blackish-green, fibrous ferric phosphate mineral; commonly massive or in nodules.

duftite [MINERAL] $PbCu(AsO_4)(OH)$ Orthorhombic mineral that is composed of a basic arsenate of lead and copper.

Dugongidae [VERT ZOO] A family of aquatic mammals in the order Sirenia comprising two species, the dugong and the sea cow.

Dugonginae [VERT ZOO] The dugongs, a subfamily of sirenian mammals in the family Dugongidae characterized by enlarged, sharply deflected premaxillae and the absence of nasal bones.

dugout [ORD] Underground shelter built to protect troops, ammunition, and material from gunfire.

Duhamel's theorem [MATH] If f and g are continuous functions, then

$$\lim_{|\Delta x| \to 0} \sum_{i=1}^{n} f(x_i') g(x_i'') \Delta x_i = \int_a^b f(x) g(x) \, dx,$$

where x_i' and x_i'' are between x_{i-1} and x_i, $i = 1, \ldots, n$, and $|\Delta x| = \max \{x_i - x_{i-1}\}$ for a partition $a = x_0 < x_1 < \cdots < x_n = b$.

Duhem-Margules equation [THERMO] An equation showing the relationship between the two constituents of a liquid-vapor system and their partial vapor pressures:

$$\frac{d \ln p_A}{d \ln x_A} = \frac{d \ln p_B}{d \ln x_B}$$

where x_A and x_B are the mole fractions of the two constituents, and p_A and p_B are the partial vapor pressures.

Duhem's equation *See* Gibbs-Duhem equation.

Dühring's rule [PHYS CHEM] The rule that a plot of the temperature at which a liquid exerts a particular vapor pressure against the temperature at which a similar reference liquid exerts the same vapor pressure produces a straight or nearly straight line.

Dukler theory [CHEM ENG] Relationship of velocity and temperature distribution in thin films on vertical walls; used to calculate eddy viscosity and thermal conductivity near the solid boundary.

DUKW [ORD] A 2½-ton amphibious truck used by the U.S. Army to transport cargo on land or water; 36 feet (11 meters) long, with a cargo capacity of 5175 pounds (2347 kilograms). Also known as duck.

dull coal [GEOL] A component of banded coal with a grayish color and dull appearance, consisting of small anthraxylon constituents in addition to cuticles and barklike constituents embedded in the attritus.

dull emitter [ELECTR] An electron tube whose cathode is a filament that does not glow brightly.

Dulong number *See* Eckert number.

Dulong-Petit law [THERMO] The law that the product of the specific heat per gram and the atomic weight of many solid elements at room temperature has almost the same value, about 6.3 cal/°C.

Dulong's formula [ENG] A formula giving the gross heating value of coal in terms of the weight fractions of carbon, hydrogen, oxygen, and sulfur from the ultimate analysis.

dulse [BOT] Any of several species of red algae of the genus *Rhodymenia* found below the intertidal zone in northern latitudes; an important food plant.

Dultgen process [GRAPHICS] A halftone intaglio process for color work which makes it possible to vary the size as well as the depth of the dots.

Dumas method [ANALY CHEM] A procedure for the determination of nitrogen in organic substances by combustion of the substance.

dumb iron [ENG] **1.** A rod for opening seams prior to caulking. **2.** A rigid connector between the frame of a motor vehicle and the spring shackle.

dumbwaiter [MECH ENG] An industrial elevator which carries small objects but is not permitted to carry people.

dumdum [ORD] A bullet that flattens excessively on contact, or one especially designed to flatten excessively.

Dumet wire [MET] An iron-nickel alloy wire containing 42% nickel, and covered with copper; used to replace platinum as the seal-in wire in incandescent lamps and vacuum tubes; the copper prevents gassing at the seal.

dummy [ADP] An artificial address, instruction, or other unit of information inserted in a digital computer solely to fulfill prescribed conditions (such as word length or block length) without affecting operations. [COMMUN] Telegraphy network simulating a customer's loop for adjusting a telegraph repeater; the dummy side of the repeater is that toward the customer. [ENG] Simulating device with no operating features, as a dummy heat coil. [MET] A cathode that undergoes electroplating at low current densities. [ORD] **1.** A nonexplosive bomb, projectile, or the like, or an object made to appear as one of these. **2.** An object made to appear as an airplane, gun emplacement, or the like from the air.

dummy antenna [ELECTR] A device that has the impedance characteristic and power-handling capacity of an antenna but does not radiate or receive radio waves; used chiefly for testing a transmitter. Also known as artificial antenna.

dummy argument [ADP] The variable appearing in the definition of a macro or function which will be replaced by an address at call time.

dummy block [MET] A thick disk positioned between the ram and billet in extrusion working to prevent the ram from overheating.

dummy deck [ADP] Complete set of tabulating cards containing only punched, coded information (nonaperture); used as machine-handling set for sorting, reproducing, and interpreting in conjunction with an aperture card system.

dummy instruction [ADP] An artificial instruction or address inserted in a list to serve a purpose other than the execution as an instruction.

dummy joint [ENG] A groove cut into the top half of a concrete slab, sometimes packed with filler, to form a line where the slab can crack with only minimum damage.

dummy load [ELECTR] A dissipative device used at the end of a transmission line or waveguide to convert transmitted energy into heat, so that essentially no energy is radiated outward or reflected back to its source.

dummy message [COMMUN] A message sent for some purpose other than its content, which may consist of dummy groups or may have a meaningless text.

dumontite [MINERAL] $Pb_2(UO_2)_3(PO_4)_2(OH)_4 \cdot 3H_2O$ Yellow orthorhombic mineral consisting of a hydrated phosphate of uranium and lead, occurring in crystals.

dumortierite [MINERAL] $Al_8BSi_3O_{19}(OH)$ A pink, green, blue, or violet mineral that crystallizes in the orthorhombic system but commonly occurs in parallel or radiating fibrous aggregates; mined for the manufacture of high-grade porcelain.

dump [ADP] To copy the contents of all or part of a storage, usually from an internal storage device into an external storage device. [ELECTR] To withdraw all power from a system or component accidentally or intentionally. [ORD] A temporary storage area, usually in the open, for bombs, ammunition, equipment, or supplies.

dump bucket [MECH ENG] A large bucket with movable discharge gates at the bottom; used to move soil or other construction materials by a crane or cable.

dump car [MECH ENG] Any of several types of narrow-gage rail cars with bodies which can easily be tipped to dump material.

dump check [ADP] A computer check that usually consists of adding all the digits during dumping, and verifying the sum when retransferring.

dumping syndrome [MED] An imperfectly understood symptom complex of disagreeable or painful epigastric fullness, nausea, weakness, giddiness, sweating, palpitations, and diarrhea, occurring after meals in patients who have gastric surgery which interferes with the function of the pylorus.

dump power [ELEC] Electric power, generated by any source, which is in excess of the needs of the electric system and which cannot be stored or conserved.

dump scow [NAV ARCH] A craft used for transporting rubbish, normally equipped with doors in the bottom for dumping.

dump truck [ENG] A motor or hand-propelled truck for hauling and dumping loose materials, equipped with a body that discharges its contents by gravity.

dump valve [ENG] A large valve located at the bottom of a tank or container used in emergency situations to empty the tank quickly; for example, to jettison fuel from an airplane fuel tank.

dumpy level [ENG] A surveyor's level which has the telescope with its level tube rigidly attached to a vertical spindle and is capable only of horizontal rotary movement.

dundasite [MINERAL] $PbAl_2(CO_3)_2(OH)_4 \cdot 2H_2O$ A white mineral consisting of a basic lead aluminum carbonate, occurring in spherical aggregates.

dune [GEOL] A mobile mound, hill, or ridge of windblown material, usually sand.

dune complex [GEOGR] The totality of topographic forms, especially dunes, which comprise the moving landscape.

dunite [PETR] An ultrabasic rock consisting almost solely of a magnesium-rich olivine with some chromite and picotite; an important source of chromium.

dunking sonar *See* dipping sonar.

dunnage [IND ENG] **1.** Padding material placed in a container to protect shipped goods from damage. **2.** Loose wood or waste material placed in the ship's hold to protect the cargo from shifting and damage.

dunnite *See* ammonium picrate.

duo [GRAPHICS] Recording of images on one-half of the film width during one passage of the film, then turning the film end for end and rerunning it to utilize the unused half of the film width.

duodecimal number system [MATH] A representation system for real numbers using 12 as the base.

duodenal glands *See* Brunner's glands.

duodenal ulcer [MED] A peptic ulcer occurring in the wall of the duodenum, the first portion of the small intestine.

duodenum [ANAT] The first section of the small intestine of mammals, extending from the pylorus to the jejunum.

duodiode [ELECTR] An electron tube having two diodes in the same envelope, with either a common cathode or separate cathodes. Also known as double diode.

duodiode-pentode [ELECTR] An electron tube having two diodes and a pentode in the same envelope, generally with a common cathode.

duodiode-triode [ELECTR] An electron tube having two diodes and a triode in the same envelope, generally with a common cathode.

duolateral coil *See* honeycomb coil.

duoplasmatron [ELECTR] An ion-beam source in which electrons from a hot filament are accelerated sufficiently to ionize a gas by impact; the resulting positive ions are drawn out by high-voltage electrons and focused into a beam by electrostatic lens action.

duoprimed word [ADP] A computer word containing a representation of the sixth, seventh, eighth, and ninth rows of information from an 80-column card.

duotone [GRAPHICS] A process in which two halftone cuts, one with a screen angle 30° different from the other, are made from the same black-and-white photograph, and the picture is printed in two tones, usually black and a color such as blue or green.

duotriode *See* double triode.

duotype [GRAPHICS] A process in which two halftone plates, for letterpress, are produced from a black-and-white original, each plate being etched differently; one plate is etched for detail and printed in a dark color, and the other is etched for a flat effect and printed in a light color.

Duovac method [MET] Technique for testing for defects in magnetic parts; a moving magnetic field magnetizes the part in many directions, and the part is then sprayed with fluorescent magnetic particles and examined under ultraviolet light so that defects become apparent.

dupion [TEXT] A thick, irregular, double silk fiber obtained from two cocoons nested together.

duplet lens system [OPTICS] A system of lenses in which there are two groups of lenses separated by a space, and successive lens in each group are in contact.

duplex [ENG] Consisting of two parts working together or in a similar fashion.

duplex artificial line [ELEC] A balancing network, simulating the impedance of the real line and distant terminal apparatus,

DUMORTIERITE

dumortierite

quartz schist

├─2.5 cm─┤

Dumortierite in quartz schist, Dehesa, Calif. *(Specimen courtesy of Department of Geology, Bryn Mawr College)*

DUMPY LEVEL

Engineer's dumpy level. *(W. and L. E. Gurley Co.)*

DUNE

Seif dune or linear ridge, located in the Algodones dunes, Imperial County, California.

which is employed in a duplex circuit for the purpose of making the receiving device unresponsive to outgoing signal currents.

duplex cable [ELEC] Two insulated stranded conductors twisted together; they may have a common insulating covering.

duplex channel [COMMUN] A communication channel providing simultaneous transmission in both directions.

duplexer [ELECTR] A switching device used in radar to permit alternate use of the same antenna for both transmitting and receiving; other forms of duplexers serve for two-way radio communication using a single antenna at lower frequencies. Also known as duplexing assembly.

duplexing See duplex operation; duplex process.

duplexing assembly See duplexer.

duplex iron [MET] Cast iron heated in an electric furnace after it has been melted in a cupola.

duplexite [MINERAL] Ca₄BeAl₂Si₉O₂₄(OH)₂ $Ca_4BeAl_2Si_9O_{24}(OH)_2$ A white fibrous mineral consisting of hydrous beryllium calcium aluminosilicate. Also known as bavenite.

duplex lock [DES ENG] A lock with two independent pin-tumbler cylinders on the same bolt.

duplex operation [COMMUN] The operation of associated transmitting and receiving apparatus concurrently, as in ordinary telephones, without manual switching between talking and listening periods. Also known as duplexing; duplex transmission. [ENG] In radar, a condition of operation when two identical and interchangeable equipments are provided, one in an active state and the other immediately available for operation.

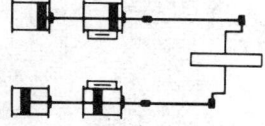

DUPLEX TANDEM COMPRESSOR

Frame arrangement of duplex tandem piston steam compressor.

duplex paper [GRAPHICS] Photosensitized paper with emulsion on both sides, one side having a smooth finish and the other a rough finish, or having sides of different color.

duplex practice See duplex process.

duplex process [MET] A two-step procedure in which steel is refined by one process (usually the Bessemer process) and finished by another process (usually open-hearth or electric-furnace). Also known as duplexing; duplex practice.

duplex pump [MECH ENG] A reciprocating pump with two parallel pumping cylinders.

duplex tandem compressor [MECH ENG] A compressor having cylinders on two parallel frames connected through a common crakshaft.

duplex transmission See duplex operation.

duplex tube [ELECTR] Combination of two vacuum tubes in one evelope.

duplex uterus [ANAT] A condition in certain primitive mammals, such as rodents and bats, that have two distinct uteri opening separately into the vagina.

duplicate field [ADP] A series of 12 punches in a program card.

duplicate film [GRAPHICS] Film generated in a camera containing a dual supply of film.

duplicate key [ADP] A key on the card punch which, when depressed, will copy a card in the reading station onto a card in the write station.

duplication check [ADP] A check based on the identity in results of two independent performances of the same task.

duplicatus [METEOROL] A cloud variety composed of superposed layers, sheets, or patches, at slightly different levels and sometimes partly merged.

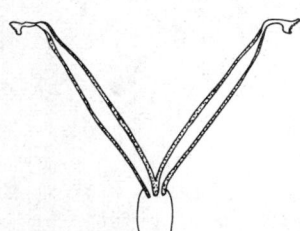

DUPLEX UTERUS

Duplex uterus of the rat. (From C. K. Weichert, Elements of Chordate Anatomy, 3d ed., McGraw-Hill, 1967)

Dupont process [MIN ENG] A method for separation of minerals in which organic liquids of high specific gravity and low viscosity are used.

Dupré equation [THERMO] The work W_{LS} done by adhesion at a gas-solid-liquid interface, expressed in terms of the surface tensions γ of the three phases, is $W_{LS} = \gamma_{GS} + \gamma_{GL} - \gamma_{LS}$.

durability [ENG] The quality of equipment, structures, or goods of continuing to be useful after an extended period of time and usage.

durain [GEOL] A hard, granular ingredient of banded coal which occurs in lenticels and shows a close, firm texture. Also known as durite.

Duralumin [MET] A trade name for an aluminum alloy containing about 4% copper, 0.5% magnesium, and 0.5% manganese.

dura mater [ANAT] The fibrous membrane forming the outermost covering of the brain and spinal cord. Also known as endocranium.

Durargid [GEOL] A great soil group constituting a subdivision of the Argids, indicating those soils with a hardpan cemented by silica and called a duripan.

duration [MECH] A basic concept of kinetics which is expressed quantitatively by time measured by a clock or comparable mechanism. [OCEANOGR] The interval of time of the rising or falling tide, or the length of time of flood or ebb tidal currents.

duration control [ELECTR] Control for adjusting the time duration of reduced gain in a sensitivity-time control circuit.

durene [ORG CHEM] $C_6H_2(CH_3)_4$ Colorless crystals with camphor aroma; boiling point 190°C; soluble in organic solvents, insoluble in water; used as a chemical intermediate. Also known as durol.

Durene [TEXT] Trade name of Durene Association of America for a plied, mercerized cotton.

Durham fermentation tube [MICROBIO] A test tube containing lactose or lauryl tryptose and an inverted vial for gas collection; used to test for the presence of coliform bacteria.

duricrust [GEOL] The case-hardened soil crust formed in semiarid climates by precipitation of salts; contains aluminous, ferruginous, siliceous, and calcareous material.

durinite [GEOL] The principal maceral of durain; a heterogeneous material, semiopaque in section (including all parts of plants); micrinite, exinite, cutinite, resinite, collinite, xylinite, suberinite, and fusinite may be present.

duripan [GEOL] A horizon in mineral soil characterized by cementation by silica.

Duriron [MET] Trade name for a type of silicon cast iron.

durite See durain.

durol See durene.

durometer [ENG] An instrument consisting of a small drill or blunt indenter point under pressure; used to measure hardness of metals and other materials.

durometer hardness [ENG] The hardness of a material as measured by a durometer.

Durville process [MET] A casting process involving the attachment of an inverted mold to the top of a crucible; the metal is melted in the bottom of the crucible, and then the molten metal is decanted into the mold by inverting the entire apparatus.

düsenwind [METEOROL] The mountain-gap wind of the Dardanelles; a strong east-northeast wind which blows out of the Dardanelles into the Aegean Sea, penetrating as far as the island of Lemnos, and caused by a ridge of high pressure over the Black Sea.

dusk [ASTRON] That part of either morning or evening twilight between complete darkness and civil twilight.

dusk side [ASTRON] The side of a planet or other celestial body pointing away from its orbital movement direction.

dussertite [MINERAL] BaFe₃(AsO₄)₂(OH)₅ $BaFe_3(AsO_4)_2(OH)_5$ A mineral consisting of a hydrous basic arsenate of barium and iron.

dust [GEOL] Dry solid matter of silt and clay size (less than 1/16 millimeter). [PHYS] A loose term applied to solid particles predominantly larger than colloidal size and capable of temporary gas suspension.

dust and fume monitor [MIN ENG] An instrument designed to measure and record concentrations of dust, fume, and gas in mine environments over an extended period of time.

dust avalanche [GEOL] An avalanche of dry, loose snow.

dust bowl [CLIMATOL] A name given, early in 1935, to the region in the south-central United States afflicted by drought and dust storms, including parts of Colorado, Kansas, New Mexico, Texas, and Oklahoma, and resulting from a long period of deficient rainfall combined with loosening of the soil by destruction of the natural vegetation; dust bowl describes similar regions in other parts of the world.

dust chamber [ENG] A chamber through which gases pass to permit deposition of solid particles for collection. Also known as ash collector; dust collector.

dust collector See dust chamber.

dust control system [ENG] System to capture, settle, or inert dusts produced during handling, drying, or other process operations; considered important for safety and health.

dust core   *See* ferrite core.

dust counter   [ENG] A photoelectric apparatus which measures the size and number of dust particles per unit volume of air. Also known as Kern counter.

dust-counting microscope   [ENG] A microscope equipped for quantitative dust sample analysis; magnification is usually 100X.

dust devil   [METEOROL] A small but vigorous whirlwind, usually of short duration, rendered visible by dust, sand, and debris picked up from the ground; diameters range from about 10–100 feet (3–30 meters), and average height is about 600 feet (180 meters).

dust-devil effect   [GEOPHYS] In atmospheric electricity, rather sudden and short-lived change (positive or negative) of the vertical component of the atmospheric electric field that accompanies passage of a dust devil near an instrument sensitive to the vertical gradient.

dust explosion   [ENG] An explosion following the ignition of flammable dust suspended in the air.

dust extinction   [OPTICS] The contribution to total extinction of light made by scattering and absorption by dust particles in the path of a light beam.

dust filter   [ENG] A gas-cleaning device using a dry or viscous-coated fiber or fabric for separation of particulate matter.

dust horizon   [METEOROL] The top of a dust layer which is confined by a low-level temperature inversion and has the appearance of the horizon when viewed from above, against the sky; the true horizon is usually obscured by the dust layer.

dusting   [MET] Spontaneous disintegration of a material on cooling due to expansion or inversion.

dusting clay   [MATER] Finely pulverized clay used as an extender or carrier in insecticide dust formulations.

dust separator   [ENG] Device or system to remove dust from a flowing stream of gas; includes electrostatic precipitators, wet scrubbers, bag filters, screens, and cyclones.

dust storm   [METEOROL] A strong, turbulent wind carrying large clouds of dust.

dust well   [HYD] A pit in an ice surface produced when small, dark particles on the ice are heated by sunshine and sink down into the ice.

dust whirl   [METEOROL] A rapidly rotating column of air over a dry and dusty or sandy area, carrying dust, leaves, and other light material picked up from the ground; when well developed, it is known as a dust devil. Also known as dancing dervish; dancing devil; desert devil; sand auger; sand devil.

Dutch door   [BUILD] A door with upper and lower parts that can be opened and closed independently.

Dutch elm disease   [PL PATH] A lethal fungus disease of elm trees caused by *Graphium ulmi*, which releases a toxic substance that destroys vascular tissue; transmitted by a bark beetle.

Dutch liquid   *See* ethylene chloride.

Dutchman's log   [ENG] A buoyant object thrown overboard to determine the speed of a vessel; the time required for a known length of the vessel to pass the object is measured, and the speed can then be computed.

Dutch metal   [MET] An alloy of 80% copper and 20% zinc that is ductile, is easily drawn, and takes a high polish; used for low-priced jewelry.

Dutch process   [CHEM ENG] A process for making white lead; metallic lead is placed in vessels containing a dilute acetic acid, and the vessels are stacked in bark or manure.

duty classification of a relay   [ELEC] Expression of the frequency with which the relay may be required to operate without exceeding prescribed limitations.

duty cycle   [COMMUN] The product of the pulse duration and pulse frequency of a pulse carrier, equal to the time per second that pulse power is applied. Also known as duty factor. [ELECTR] *See* duty ratio. [ENG] **1.** The time intervals devoted to starting, running, stopping, and idling when a device is used for intermittent duty. **2.** The ratio of working time to total time for an intermittently operating device, usually expressed as a percent. Also known as duty factor. [MET] The percentage of time that current flows in equipment over a specific period during electric resistance welding. [NUCLEO] The fraction of time during which a pulsed accel-erator beam is on target, usually expressed as a percent. Also known as duty factor.

duty cyclometer   [ENG] Test meter which gives direct reading of duty cycle.

duty factor   *See* duty cycle.

duty of water   [HYD] The total volume of irrigation water required to mature a particular type of crop, including consumptive use, evaporation and seepage from ditches and canals, and the water eventually returned to streams by percolation and surface runoff.

duty ratio   [ELECTR] In a pulse radar or similar system, the ratio of average to peak pulse power. Also known as duty cycle.

D value   [NAV] The difference between pressure altitude and absolute altitude, as determined at a given time in flight, expressed algebraically; the absolute altitude is always minuend. Also known as D sounding.

D variometer   *See* declination variometer.

dwarf   [BIOL] Being an atypically small form or variety of something. [MED] An abnormally small individual; especially one whose bodily proportions are altered.

dwarf disease   [PL PATH] A virus disease marked by the inhibition of fruit production; common in plum trees.

dwarfism   [MED] Underdevelopment of the body due to surgical removal of the pituitary gland or hyposecretion of growth hormone.

dwarf mouse unit   [BIOL] A unit for the standardization of somatotropin.

dwarf star   [ASTRON] A star that typically has surface temperature of 5730K, radius of 690,000 kilometers, mass of 2×10^{33}g, and luminosity of 4×10^{33}ergs/sec. Also known as main sequence star.

dwell   [DES ENG] That part of a cam that allows the cam follower to remain at maximum lift for a period of time. [ENG] A pause in the application of pressure to a mold.

dwey   *See* dwigh.

dwigh   [METEOROL] In Newfoundland, a sudden shower or snow storm. Also known as dwey; dwoy.

Dwight-Lloyd machine   [MIN ENG] A continuous sintering machine in which the feed is moved on articulated plates pulled by chains in conveyor-belt fashion.

Dwight-Lloyd process   [MIN ENG] Blast roasting, with air currents being drawn downward through the ore.

DWL   *See* design waterline.

dwoy   *See* dwigh.

dwt   *See* deadweight tonnage; pennyweight.

Dwyka tillite   [GEOL] A glacial Permian deposit that is widespread in South Africa.

DX   *See* distance reception.

DX coil   *See* direct-expansion coil.

Dy   *See* dysprosium.

dyad   [CYTOL] Either of the two pair of chromatids produced by separation of a tetrad during the first meiotic division. [MATH] An abstract object which is a pair of vectors **AB** in a given order on which certain operations are defined.

dyadic expansion   [MATH] The representation of a number in the binary number system.

dyadic rational   [MATH] A fraction whose denominator is a power of 2.

Dycril plate   [GRAPHICS] A flexible, photopolymer plastic duplicate plate made by DuPont.

dye   [CHEM] A colored substance which imparts more or less permanent color to other materials. Also known as dyestuff.

dyecrete process   [ENG] A process of adding permanent color to concrete with organic dyes.

dyeing   [CHEM ENG] The application of color-producing agents to material, usually fibrous or film, in order to impart a degree of color permanence demanded by the projected end use.

dyeing assistant   [CHEM] Material such as sodium sulfate added to a dye bath to control or promote the action of a textile dye.

dye laser   [OPTICS] A type of tunable laser in which the active material is a dye such as acridine red or esculin, with very large molecules, and laser action takes place between the first excited and ground electronic states, each of which comprises a broad vibrational-rotational continuum.

dye penetrant [MET] A dye-containing liquid used for detecting cracks or other surface defects in nonmagnetic materials.

dye-retarding agent [MATER] Materials that decrease the rate of dye absorption, preventing rapid exhaustion of dye baths.

dyestuff *See* dye.

dye toning [GRAPHICS] The process whereby the color of a developing image is altered by changing the color into a mordant and then placing the film in a suitable dye solution.

dynamical friction [PHYS] **1.** The drag force between electrons and ions drifitng with respect to each other. **2.** Sliding friction, in contrast to static friction.

dynamical parallax [ASTRON] A parallax of binary stars that is computed from the sum of the masses of the binary system.

dynamical similarity [MECH] Two flow fields are dynamically similar if one can be transformed into the other by a change of length and velocity scales. All dimensionless numbers of the flows must be the same.

dynamical system [MATH] An abstraction of the concept of a family of solutions to an ordinary differential equation; namely, an action of the real numbers on a topological space satisfying certain "flow" properties.

dynamical variable [MECH] One of the quantities used to describe a system in classical mechanics, such as the coordinates of a particle, the components of its velocity, the momentum, or functions of these quantities.

dynamic analogies [PHYS] Analogies that make it possible to convert the differential equations for mechanical and acoustical systems to equivalent electrical equations that can be represented by electric networks and solved by circuit theory.

dynamic augment [MECH ENG] Force produced by unbalanced reciprocating parts in a steam locomotive.

dynamic balance [MECH] The condition which exists in a rotating body when the axis about which it is forced to rotate, or to which reference is made, is parallel with a principal axis of inertia; no products of inertia about the center of gravity of the body exist in relation to the selected rotational axis.

dynamic boundary condition [FL MECH] The condition that the pressure must be continuous across an internal boundary or free surface in a fluid.

dynamic braking [MECH] A technique of electric braking in which the retarding force is supplied by the same machine that originally was the driving motor.

dynamic breccia *See* tectonic breccia.

dynamic capillary pressure [PETRO ENG] Capillary-pressure saturation curves of a core sample determined by the simultaneous steady-state flow of two fluids through the sample; capillarity pressures are determined by the difference in the pressures of the two fluids.

dynamic characteristic *See* load characteristic.

dynamic check [ENG] Check used to ascertain the correct performance of some or all components of equipment or a system under dynamic or operating conditions.

dynamic circuit [ELECTR] An MOS circuit designed to make use of its high input impedance to store charge temporarily at certain nodes of the circuit and thereby increase the speed of the circuit.

dynamic climatology [CLIMATOL] The climatology of atmospheric dynamics and thermodynamics, that is, a climatological approach to the study and explanation of atmospheric circulation.

dynamic compressor [MECH ENG] A compressor which uses rotating vanes or impellers to impart velocity and pressure to the fluid.

dynamic condenser electrometer [ELEC] A sensitive voltage-measuring instrument in which an object carrying charge resulting from the voltage is moved back and forth in an electrostatic field and the resulting alternating-current signal is observed.

dynamic convergence [ELECTR] The process whereby the locus of the point of convergence of electron beams in a color-television or other multibeam cathode-ray tube is made to fall on a specified surface during scanning.

dynamic creep [MECH] Creep resulting from fluctuations in a load or temperature.

dynamic debugging routine [ADP] A debugging routine which operates in conjunction with the program being checked and interacts with it while the program is running.

dynamic dump [ADP] A dump performed during the execution of a program.

dynamic error [ELECTR] Error in a time-varying signal resulting from inadequate dynamic response of a transducer.

dynamic factor [AERO ENG] A ratio formed from the load carried by any airplane part when the airplane is accelerating or subjected to abnormal conditions to the load carried in the conditions of normal flight.

dynamic fluidity [FL MECH] The reciprocal of the dynamic viscosity.

dynamic focusing [ELECTR] The process of varying the focusing electrode voltage for a color picture tube automatically so the electron-beam spots remain in focus as they sweep over the flat surface of the screen.

dynamic forecasting *See* numerical forecasting.

dynamic height [PHYS] The amount of work done when a water particle of unit mass is moved vertically from one level to another. Also known as geodynamic height.

dynamic-height anomaly [OCEANOGR] The excess of the actual geopotential difference, between two given isobaric surfaces, over the geopotential difference in a homogeneous water column of salinity 35 per mille and temperature 0°C. Also known as anomaly of geopotential difference.

dynamic holdup [CHEM ENG] Liquid held by a tank or process vessel, with constant introduction of fresh material and counteracting withdrawal of held material to maintain a constant liquid level.

dynamic ileus *See* spastic ileus.

dynamic impedance [ELEC] The impedance of a circuit having an inductance and a capacitance in parallel at the frequency at which this impedance has a maximum value. Also known as rejector impedance.

dynamic instability *See* inertial instability.

dynamic load [AERO ENG] With respect to aircraft, rockets, or spacecraft, a load due to an acceleration of craft, as imposed by gusts, by maneuvering, by landing, by firing rockets, and so on. [CIV ENG] A force exerted by a moving body on a resisting member, usually in a relatively short time interval. Also known as energy load.

dynamic loudspeaker [ENG ACOUS] A loudspeaker in which the moving diaphragm is attached to a current-carrying voice coil that interacts with a constant magnetic field to give the in-and-out motion required for the production of sound waves. Also known as dynamic speaker; moving-coil loudspeaker.

dynamic memory *See* dynamic storage.

dynamic memory allocation [ADP] The allocation of a specific memory area for any subroutine called by a program.

dynamic metamorphism [GEOL] Metamorphism resulting exclusively or largely from rock deformation, principally faulting and folding. Also known as dynamometamorphism.

dynamic meteorology [METEOROL] The study of atmospheric motions as solutions of the fundamental equations of hydrodynamics or other systems of equations appropriate to special situations, as in the statistical theory of turbulence.

dynamic meter [PHYS] The standard unit of dynamic height expressed as 10 square meters per second per second.

dynamic microphone [ENG ACOUS] A moving-conductor microphone in which the flexible diaphragm is attached to a coil positioned in the fixed magnetic field of a permanent magnet. Also known as moving-coil microphone.

dynamic model [ENG] A model of an aircraft or other object which has its linear dimensions and its weight and moments of inertia reproduced in scale in proportion to the original.

dynamic noise suppressor [ENG ACOUS] An audio-frequency filter circuit that automatically adjusts its band-pass limits according to signal level, generally by means of reactance tubes; at low signal levels, when noise becomes more noticeable, the circuit reduces the low-frequency response and sometimes also reduces the high-frequency response.

dynamic packing [ENG] Any packing that operates on moving surfaces; in functioning, to retain fluid under pressure, they carry the hydraulic load and therefore operate like bearings.

dynamic parallax [ASTRON] A value for the parallax of a binary star computed from the observations of the period and angular dimensions of the orbit by assuming a value for the mass of the binary system. Also known as hypothetical parallax.

dynamic pickup [ELECTR] A pickup in which the electric output is due to motion of a coil or conductor in a constant magnetic field. Also known as dynamic reproducer; moving-coil pickup.

dynamic plate impedance [ELECTR] Internal resistance to the flow of alternating current between the cathode and plate of a tube.

dynamic plate resistance [ELECTR] Opposition that the plate circuit of a vacuum tube offers to a small increment of plate voltage; it is the ratio of a small change in plate voltage to the resulting change in the plate current, other tube voltages remaining constant.

dynamic pressure [FL MECH] **1.** The pressure that a moving fluid would have if it were brought to rest by isentropic flow against a pressure gradient. **2.** The difference between the quantity in the first definition and the static pressure.

dynamic printout [ADP] A printout of data which occurs during the machine run as one of the sequential operations.

dynamic problem check [ADP] Any dynamic check used to ascertain that the computer solution satisfies the given system of equations in an analog computer operation.

dynamic programming [MATH] A mathematical technique, more sophisticated than linear programming, for solving a multidimensional optimization problem, which transforms the problem into a sequence of single-stage problems having only one variable each.

dynamic program relocation [ADP] The act of moving a partially executed program to another location in main memory, without hindering its ability to finish processing normally.

dynamic range [ELECTR] The ratio of the specified maximum signal level capability of a system or component to its noise level; usually expressed in decibels.

dynamic regulator [ELECTR] Transmission regulator in which the adjusting mechanism is in self-equilibrium at only one or a few settings and requires control power to maintain it at any other setting.

dynamic reproducer *See* dynamic pickup.

dynamic resistance [ELEC] A device's electrical resistance when it is in operation.

dynamic roughness [OCEANOGR] A quantity, designated z_o, dependent on the shape and distribution of the roughness elements of the sea surface, and used in calculations of wind at the surface. Also known as roughness length.

dynamics [MECH] That branch of mechanics which deals with the motion of a system of material particles under the influence of forces, especially those which originate outside the system under consideration.

dynamic scattering device [OPTICS] A type of numerical display device in which a voltage is applied to a cell containing a nematic liquid crystal of negative dielectric anisotropy; opposing influences from the electric field and electrical conduction produce turbulence, causing the cell to become milk white.

dynamic sequential control [ADP] Method of operation of a digital computer through which it can alter instructions as the computation proceeds, or the sequence in which instructions are executed, or both.

dynamic similarity [MECH ENG] A relation between two mechanical systems (often referred to as model and prototype) such that by proportional alterations of the units of length, mass, and time, measured quantities in the one system go identically (or with a constant multiple for each) into those in the other; in particular, this implies constant ratios of forces in the two systems.

dynamic speaker *See* dynamic loudspeaker.

dynamic stability [MECH] The characteristic of a body, such as an aircraft, rocket, or ship, that causes it, when disturbed from an original state of steady motion in an upright position, to damp the oscillations set up by restoring moments and gradually return to its original state. Also known as stability.

dynamic storage [ADP] Computer storage in which information at a certain position is not always available instantly because it is moving, as in an acoustic delay line or magnetic drum. Also known as dynamic memory.

dynamic subroutine [ADP] Subroutine that involves parameters, such as decimal point position or item size, from which a relatively coded subroutine is derived by the computer itself.

dynamic test [ENG] A test conducted under active or simulated load.

dynamic thickness [OCEANOGR] The vertical separation between two isobaric surfaces in the ocean.

dynamic topography [MAP] A topographic map indicating the dynamic depth of an isobaric surface.

dynamic trough [METEOROL] A pressure trough formed on the lee side of a mountain range across which the wind is blowing almost at right angles. Also known as lee trough.

dynamic unbalance [MECH ENG] Failure of the rotation axis of a piece of rotating equipment to coincide with one of the principal axes of inertia due to forces in a single axial plane and on opposite sides of the rotation axis, or in different axial planes.

dynamic vertical *See* apparent vertical.

dynamic viscosity *See* absolute viscosity.

dynamite [MATER] A generic term covering a class of nitroglycerin-sensitized mixtures of carbonaceous materials (wood, flour, starch) and oxygen-supplying salts, used as explosives for blasting and mining.

dynamo *See* generator.

dynamo effect [GEOPHYS] A process in the ionosphere in which winds and the resultant movement of ionization in the geomagnetic field give rise to induced current.

dynamoelectric [PHYS] Pertaining to the conversion of mechanical energy to electric energy, or vice versa.

dynamoelectric amplifier generator [ELEC] A generator that serves as a power amplifier at low frequencies or direct current; the input signal is applied to the stationary field to change the excitation, and the amplified output is taken from the rotating armature.

dynamometamorphism *See* dynamic metamorphism.

dynamometer [ENG] **1.** An instrument in which current, voltage, or power is measured by the force between a fixed coil and a moving coil. **2.** A special type of electric rotating machine used to measure the output torque or driving torque of rotating machinery by the elastic deformation produced.

dynamometer multiplier [ELEC] A multiplier in which a fixed and a moving coil are arranged so that the deflection of the moving coil is proportional to the product of the currents flowing in the coils.

dynamostatic [ELEC] Pertaining to a machine that uses direct or alternating current to produce static electricity.

dynamo theory [GEOPHYS] The hypothesis which explains the regular daily variations in the earth's magnetic field in terms of electrical currents in the lower ionosphere, generated by tidal motions of the ionized air across the earth's magnetic field.

dynamotor [ELEC] A rotating electric machine having two or more windings on a single armature containing a commutator for direct-current operation and slip rings for alternating-current operation; when one type of power is fed in for motor operation, the other type is delivered by generator action. Also known as rotary converter; synchronous inverter.

dynatron [ELECTR] A screen-grid tube in which secondary emission of electrons from the anode causes the anode current to decrease as anode voltage increases, resulting in a negative resistance characteristic. Also known as negatron.

dynatron oscillator [ELECTR] An oscillator in which secondary emission of electrons from the anode of a screen-grid tube causes the anode current to decrease as anode voltage is increased, giving the negative resistance characteristic required for oscillation.

dyne [MECH] The unit of force in the centimeter-gram-second system of units, equal to the force which imparts an acceleration of 1 cm/sec² to a 1 gram mass.

dyne-centimeter *See* erg.

DYSODONTA

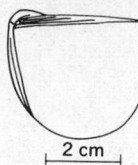

Dysodont hinge, *Ambonychinia*, Upper Ordovician, Sweden. *(After O. Isberg, 1934)*

DYSPROSIUM

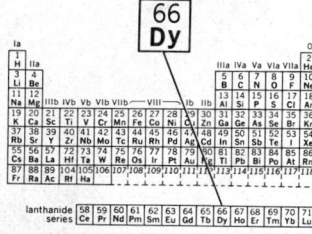

Periodic table of the chemical elements showing the position of dysprosium.

DYTISCIDAE

A diving beetle. *(From T. I. Storer and R. L. Usinger, General Zoology, 3d ed., McGraw-Hill, 1957)*

dyne-cm   *See* erg.

dynode   [ELECTR] An electrode whose primary function is secondary emission of electrons; used in multiplier phototubes and some types of television camera tubes. Also known as electron mirror.

dysanalyte   [MINERAL] A variety of the mineral perovskite in which Nb^{5+} substitutes for Ti^{5+}, and Na^+ for Ca^{++} in the formula $Ca[TiO_3]$.

dysarthria   [MED] Impairment of articulation caused by any disorder or lesion affecting the tongue or speech muscles.

dysarthrosis   [MED] **1.** Deformity, dislocation, or disease of a joint. **2.** A false joint.

dyschondroplasia   *See* enchondromatosis.

dyscrasia   [MED] An abnormal state of the body.

dyscrasite   [MINERAL] Ag_2Sb A gray mineral that forms rhombic crystals.

dysentery   [MED] Inflammation of the intestine characterized by pain, intense diarrhea, and the passage of mucus and blood.

dysgammaglobulinemia   [MED] A quantitative or qualitative abnormality of serum globulins.

dysgerminoma   [MED] An ovarian tumor composed of large polygonal cells of germ-cell origin, resembling seminoma of the testis, but less malignant. Also known as embryoma of the ovary.

dyshidrosis   [MED] Any disturbance in sweat production or excretion.

Dysideidae   [INV ZOO] A family of sponges in the order Dictyoceratida.

dyskaryosis   [PATH] Any abnormality of the nuclei of exfoliated cells, without significant change in cell integrity.

dyskeratosis   [MED] **1.** Imperfect keratinization of individual epidermal cells. **2.** Keratinization of corneal epithelium.

dyskinesia   [MED] **1.** Disordered movements of voluntary or involuntary muscles, particularly those seen in disorders of the extrapyramidal system. **2.** Impaired voluntary movements.

dyslexia   [MED] Impairment of the ability to read.

dyslogia   [MED] **1.** Difficulty in the expression of ideas by speech. **2.** Impairment of reasoning or the faculty to think logically.

dysmenorrhea   [MED] Difficult or painful menstruation.

dysmorphophobia   [PSYCH] An abnormal fear of deformity.

Dysodonta   [PALEON] An order of extinct bivalve mollusks with a nearly toothless hinge and a ligament in grooves or pits.

Dyson microscope   [OPTICS] A type of interference microscope, now obsolete, in which a light ray is split into two parallel beams and then recombined by reflections from surfaces of parallel plates, and one of the beams passes through the object under observation.

Dyson notation   [ORG CHEM] A notation system for representing organic chemicals developed by G. Malcolm Dyson; the compound is described on a single line, symbols are used for the chemical elements involved as well as for the functional groups and various ring systems; for example, methyl alcohol is C.Q and phenol is B6.Q.

dysostosis   [MED] Defective formation of bone.

dyspepsia   [MED] Disturbed digestion.

dysphagia   [MED] Difficulty in swallowing, or inability to swallow, of organic or psychic causation.

dysphasia   [MED] Partial aphasia due to a brain lesion.

dysphonia   [MED] An impairment of the voice.

dysphoria   [MED] **1.** The condition of not feeling well or of being ill at ease. **2.** Morbid impatience and restlessness, anxiety, or fidgetiness.

dysplasia   [PATH] Abnormal development or growth, especially of cells.

dyspnea   [MED] Difficult or labored breathing.

dysprosium   [CHEM] A metallic rare-earth element, symbol Dy, atomic number 66, atomic weight 162.50.

dysrhythmia   [MED] Disordered rhythm of the brain waves.

dyssebacia   [MED] Plugging of the sebaceous glands, especially around the nose, mouth, and forehead, with a dry, yellowish material.

dyssocial behavior   [PSYCH] A behavior pattern characteristic of persons who manifest disregard for social codes by following illegal pursuits such as gambling or pushing dope.

dyster   *See* doister.

dysthymia   [MED] Any childhood condition caused by malfunction of the thymus.   [PSYCH] Any despondent mood or depressive tendency, often associated with hypochondriasis.

dystonia   [PHYSIO] Disorder or lack of muscle tonicity.

dystophic   [BIOL] Pertaining to an environment that does not supply adequate nutrition.

dystrophy   [MED] **1.** Defective nutrition. **2.** Defective or abnormal development or degeneration.

Dytiscidae   [INV ZOO] The predacious diving beetles, a family of coleopteran insects in the suborder Adephaga.

e [MATH] The base of the natural logarithms; the number defined by the equation

$$\int_1^e \frac{1}{x}\,dx = 1.$$

E *See* electric field vector.

eager *See* bore.

eagle [VERT ZOO] Any of several large, strong diurnal birds of prey in the family Accipitridae.

Eagle's medium [MICROBIO] A tissue-culture medium, developed by H. Eagle, containing vitamins, amino acids, inorganic salts and serous enrichments, and dextrose.

EAM *See* electric accounting machine.

E and M lead signaling [COMMUN] Communications between a trunk circuit and a separate signaling unit over two leads: an M lead that transmits battery or ground signals to the signaling equipment, and an E lead which receives open or ground signals from the signaling unit.

ear [ANAT] The receptor organ that sends both auditory information and space orientation information to the brain in vertebrates.

eardrum *See* tympanic membrane.

ear implantation length [ANTHRO] A measure of the distance from the otobasion superior to the otobasion inferior.

earing [MET] Formation of scallops around the top edge of a deep-drawn product due to differences in the directional properties of the metal sheet.

earlandite [MINERAL] $Ca_3(C_6H_5O_7)_2 \cdot 4H_2O$ A mineral consisting of a hydrous citrate of calcium; found in sediments in the Weddell Sea.

ear length [ANTHRO] A measure of the maximum distance along the anterior-posterior axis of the ear.

earliest finish time [IND ENG] The earliest time for completion of an activity of a project; for the entire project, it equals the earliest start time of the final event included in the schedule.

earliest start time [IND ENG] The earliest time at which an activity may begin in the schedule of a project; it equals the earliest time that all predecessor activities can be completed.

earlobe [ANAT] The pendulous, fleshy lower portion of the auricle or external ear.

early-warning/control and reporting post [ORD] Long-range supplementary radar surveillance data collection and reporting facility for a sector system and a long-range supplementary control facility for offensive, defensive, and return-to-base-type missions.

early-warning radar [ORD] A line of air defense radar units along the perimeter of a defended area to provide the earliest possible warning of approaching aircraft.

early-warning station *See* aircraft early-warning station.

Earnshaw's theorem [ELEC] The theorem that a charge cannot be held in stable equilibrium by an electrostatic field.

earphone [ENG ACOUS] 1. An electroacoustical transducer, such as a telephone receiver or a headphone, actuated by an electrical system and supplying energy to an acoustical system of the ear, the waveform in the acoustical system being substantially the same as in the electrical system. 2. A small, lightweight electroacoustic transducer that fits inside the ear, used chiefly with hearing aids.

earplug [ENG] A device made of a pliable substance which fits into the ear opening; used to protect the ear from excessive noise or from water.

ear protector [ENG] A device, such as a plug or ear muff, used to protect the human ear from loud noise that may be injurious to hearing, such as that of jet engines.

ear rot [PL PATH] Any of several fungus diseases of corn, occurring both in the field and in storage and marked by decay and molding of the ears.

ear shell *See* abalone.

earth [ASTRON] The third planet in the solar system, lying between Venus and Mars; sometimes capitalized. [ELEC] *See* ground. [GEOL] 1. Solid component of the globe, distinct from air and water. 2. Soil; loose material composed of disintegrated solid matter.

earth connection *See* ground.

earth core *See* core.

earth crust *See* crust.

earth current [ELEC] Return, fault, leakage, or stray current passing through the earth from electrical equipment. Also known as ground current. [GEOPHYS] A current flowing through the ground and due to natural causes, such as the earth's magnetic field or auroral activity. Also known as telluric current.

earth-current storm [GEOPHYS] Irregular fluctuations in an earth current in the earth's crust, often associated with electric field strengths as large as several volts per kilometer, and superimposed on the normal diurnal variation of the earth currents.

earth dam [CIV ENG] A dam having the main section built of earth, sand, or rock, and a core of impervious material such as clay or concrete.

earth detector *See* leakage indicator.

earthed system *See* grounded system.

earthenware [ENG] Ceramic products of natural clay, fired at 1742-2129°F (950-1165°C), that is slightly porous, opaque, and usually covered with a nonporous glaze.

earth figure [GEOD] The shape of the earth.

earthflow [GEOL] A mass-movement landform, or the process producing it, involving downslope translation of soil and weathered rock over a discrete basal shear surface within well-defined lateral limits.

earth hummock [GEOL] A small, dome-shaped uplift of soil caused by the pressure of groundwater. Also known as earth mound.

earth inductor [ENG] A type of inclinometer that has a coil which rotates in the earth's field and in which a voltage is induced when the rotation axis does not coincide with the field direction; used to measure the dip angle of the earth's magnetic field. Also known as dip inductor; earth inductor compass; induction inclinometer.

earth inductor compass *See* earth inductor.

earthing reactor *See* grounding reactor.

earth interior [GEOL] The portion of the earth beneath the crust.

earth-layer propagation [GEOPHYS] 1. Propagation of electromagnetic waves through layers of the earth's atmosphere. 2. Electromagnetic wave propagation through layers below the earth's surface.

earthlight [ASTRON] The illumination of the dark part of the moon's disk, produced by sunlight reflected onto the moon from the earth's surface and atmosphere. Also known as earthshine.

earth mound *See* earth hummock.

earth movements [GEOPHYS] Movements of the earth, comprising revolution about the sun, rotation on the axis, precession of equinoxes, and motion of the surface of the earth relative to the core and mantle.

earthmover [MECH ENG] A machine used to excavate, transport, or push earth.

earth-nut oil *See* peanut oil.

earth orbit [ASTRON] The elliptical motion of the earth about the sun (eccentricity 0.01675, average radius 1.496×10^8 km) in a little over a year.

earth oscillations [GEOPHYS] Any rhythmic deformations of the earth as an elastic body; for example, the gravitational attraction of the moon and sun excite the oscillations known as earth tides.

EAR

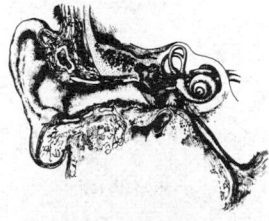

Schematic drawing of the human ear. (*Drawing by M. Brödel, Three Unpublished Drawings of the Anatomy of the Human Ear, Saunders*)

EAR PROTECTOR

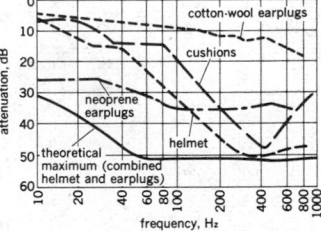

Comparison of various ear protectors in the attenuation of sound. (*After C. M. Harris*)

EARTH INDUCTOR

Photograph of earth inductor. (*U.S. Coast and Geodetic Survey*)

EARTHWORM

External features of earthworm (*Lumbricus terrestris*). (From T. I. Storer, General Zoology, 3d ed., McGraw-Hill, 1957)

EARWIG

Adult earwig with pincers.

EBONY

Twig, leaf, and bud of persimmon (*Diospyros virginiana*).

earth pig   See aardvark.

earthquake   [GEOPHYS] A series of suddenly generated elastic waves in the earth occurring in shallow depths to about 700 kilometers.

earthquake-resistant   [CIV ENG] Of a structure or building, able to withstand lateral seismic stresses at the base.

earthquake tremor   See tremor.

earthquake zone   [GEOL] An area of the earth's crust in which movements, sometimes with associated volcanism, occur. Also known as seismic area.

earth radiation   See terrestrial radiation.

earth rate   [ASTRON] The angular velocity or rate of the earth's rotation.

earth-rate correction   [NAV] A rate applied to a gyroscope to compensate for the apparent precession of the spin axis caused by the rotation of the earth.

earth resources satellite   [AERO ENG] An earth satellite with sensors that collect information concerning agriculture, forest resources, mineral and land resources, marine resources, and land use.

earth rotation   [ASTRON] Motion about the earth's axis that occurs 365.2422 times over a year's period.

earth satellite   [AERO ENG] An artificial satellite placed into orbit about the earth. [ASTRON] A natural body that revolves about the earth, such as the moon.

earth science   [SCI TECH] The science that deals with the earth or any part thereof; includes the disciplines of geology, geography, oceanography, and meteorology, among others.

earth shadow   [METEOROL] Any shadow projecting into a hazy atmosphere from mountain peaks at times of sunrise or sunset.

earthshine   See earthlight.

earth's shadow   See dark segment.

earthstar   [MYCOL] A fungus of the genus *Geastrum* that resembles a puffball with a double peridium, the outer layer of which splits into the shape of a star.

earth thermometer   See soil thermometer.

earth tide   [GEOPHYS] The periodic movement of the earth's crust caused by forces of the moon and sun. Also known as bodily tide.

earth tremor   See tremor.

earth wax   See ozocerite.

earthwork   [CIV ENG] **1.** Any operation involving the excavation or construction of earth embankments. **2.** Any construction made of earth. [ORD] A temporary or permanent fortification for attack or defense, made chiefly of earth.

earthworm   [INV ZOO] The common name for certain terrestrial members of the class Oligochaeta, especially forms belonging to the family Lumbricidae.

earthy cobalt   See asbolite.

earthy manganese   See wad.

earwax   See cerumen.

earwig   [INV ZOO] The common name for members of the insect order Dermaptera.

easel   [GRAPHICS] A standing frame, often adjustable, used to hold a painting in process, or to display a chart in meetings.

easement   [CIV ENG] The right held by one person over another person's land for a specific use; rights of tenants are excluded.

easement curve   [CIV ENG] A curve, as on a highway, whose degree of curvature is varied to provide a gradual transition between a tangent and a simple curve, or between two simple curves which it connects. Also known as transition curve.

easin   [ORG CHEM] $C_{20}H_6O_5I_4Na_2$ The sodium salt of tetraiodofluorescein; a brown powder, insoluble in water; used as a dye and a pH indicator (hydrogen ion) at pH 2.0. Also known as iodoeasin; sodium tetrafluorescein.

east   [GEOD] The direction 90° to the right of north.

East Africa Coast Current   [OCEANOGR] A current that is influenced by the monsoon drifts of the Indian Ocean, flowing southwestward along the Somalia coast in the Northern Hemisphere winter and northeastward in the Northern Hemisphere summer. Also known as Somali Current.

East African sleeping sickness   See Rhodesian trypanosomiasis.

East Australia Current   [OCEANOGR] The current which is formed by part of the South Equatorial Current and flows southward along the eastern coast of Australia.

Easter-egging   [ELECTR] An undirected procedure for checking electronic equipment, which derives its name from the children's activity of searching for hidden eggs at Eastertime.

eastern equine encephalitis   [MED] A mosquito-borne virus infection of horses and mules in the eastern and southern United States caused by a member of arbovirus group A.

Eastern Hemisphere   [GEOGR] The half of the earth lying mostly to the east of the Atlantic Ocean, including Europe, Africa, and Asia.

East Greenland Current   [OCEANOGR] A current setting south along the eastern coast of Greenland and carrying water of low salinity and low temperature.

East Indian geranium oil   See palmarosa oil.

East Indian sandalwood oil   See sandalwood oil.

easting   [NAV] The distance a craft makes good to the east.

eastonite   [MINERAL] $K_2Mg_5AlSi_5Al_3O_{20}(OH_4)$ A mineral consisting of basic silicate of potassium, magnesium, and aluminum; it is an end member of the biotite system.

east point   [GEOD] That intersection of the prime vertical with the horizon which lies to the right of the observer when facing north.

east-west effect   [ASTRON] The phenomenon due to the fact that a greater number of cosmic-ray particles approach the earth from a westerly direction than from an easterly.

Eaton agent   [MICROBIO] The name applied to *Mycoplasma pneumoniae* when it was regarded as a virus.

Eaton agent pneumonia   [MED] Pneumonitis in man, caused by *Mycoplasma pneumoniae*. Also known as primary atypical pneumonia.

eave   [BUILD] The border of a roof overhanging a wall.

eaves board   [BUILD] A strip nailed along the eaves of a building to raise the end of the bottom course of tile or slate on the roof.

eaves molding   [BUILD] A cornicelike molding below the eaves of a building.

ebb-and-flow structure   [GEOL] Rock strata with alternating horizontal and cross-bedded layers, believed to have been produced by ebb and flow of tides.

ebb current   [OCEANOGR] The tidal current associated with the decrease in the height of a tide.

ebb tide   [OCEANOGR] The portion of the tide cycle between high water and the following low water. Also known as falling tide.

Ebenaceae   [BOT] A family of dicotyledonous plants in the order Ebenales, in which a latex system is absent and flowers are mostly unisexual with the styles separate, at least distally.

Ebenales   [BOT] An order of woody, sympetalous dicotyledonous plants in the subclass Dilleniidae, having axile placentation and usually twice as many stamens as corolla lobes.

E bend   [ELECTROMAG] A smooth change in the direction of the axis of a waveguide, throughout which the axis remains in a plane parallel to the direction of polarization. Also known as E-plane bend.

Eberhard effect   [GRAPHICS] The phenomenon due to the fact that the density of a photographic plate, given a uniform exposure through a metal plate with an opening, varies with the size of the opening, when an organic developer is used.

Eberthella   See Salmonella.

Eberthella typhosa   See Salmonella typhosa.

Ebert ion counter   [ENG] An ion counter of the aspiration condenser type, used for the measurement of the concentration and mobility of small ions in the atmosphere.

EBM   See electron-beam machining.

ebonite   See hard rubber.

ebony   [BOT] Any of several African and Asian trees of the genus *Diospyros*, providing a hard, durable wood.

ebracteate   [BOT] Without bracts, or much reduced leaves.

ebracteolate   [BOT] Without bracteoles.

Ebriida   [INV ZOO] An order of flagellate protozoans in the class Phytamastigophorea characterized by a solid siliceous skeleton.

ebulliometer   [PHYS CHEM] An instrument used to measure precisely the absolute or differential boiling points of solutions. Also known as ebullioscope.

ebulliometry [PHYS CHEM] The precise measurement of the absolute or differential boiling points of solutions.

ebullioscope *See* ebulliometer.

ebullition [PHYS] The process or state of a liquid bubbling up or boiling.

EC blank fire *See* EC smokeless powder.

EC blank powder *See* EC smokeless powder.

eccentric [SCI TECH] Situated to one side with reference to a center.

eccentric angle [MATH] For an ellipse having semimajor and semiminor angles of lengths *a* and *b* respectively, lying along the *x* and *y* axes of a coordinate system respectively, and for a point (*x*,*y*) on the ellipse, the angle arc cos (*x/a*) = arc sin (*y/b*).

eccentric anomaly [ASTRON] For a planet in an elliptical orbit, the eccentric angle corresponding to the planet's location.

eccentric cam [DES ENG] A cylindrical cam with the shaft displaced from the geometric center.

eccentric gear [DES ENG] A gear whose axis deviates from the geometric center.

eccentricity [MATH] The ratio of the distance of a point on a conic from the focus to the distance from the directrix. [MECH] The distance of the geometric center of a revolving body from the axis of rotation.

eccentric load [ENG] A load imposed on a structural member at some point other than the centroid of the section.

eccentric orbit [ASTRON] An orbit of a celestial body that deviates markedly from a circle.

eccentric rotor engine [MECH ENG] A rotary engine, such as the Wankel engine, wherein motion is imparted to a shaft by a rotor eccentric to the shaft.

eccentric valve [ENG] A rubber-lined slurry or fluid valve with an eccentric rotary cut-off body to reduce corrosion and wear on mechanical moving valve parts.

ecchymosis [MED] A subcutaneous hemorrhage marked by purple discoloration of the skin.

Eccles-Jordan circuit *See* bistable multivibrator.

Eccles-Jordan multivibrator *See* bistable multivibrator.

eccrine gland [PHYSIO] One of the small sweat glands distributed all over the human body surface; they are tubular coiled merocrine glands that secrete clear aqueous sweat.

ecdysis [INV ZOO] Molting of the outer cuticular layer of the body, as in insects and crustaceans.

ecdysone [BIOCHEM] The molting hormone of insects.

ecesis [ECOL] Successful naturalization of a plant or animal population in a new environment.

ECG *See* electrocardiogram.

ecgonine [ORG CHEM] $C_9H_{15}NO_3$ An alkaloid obtained in crystalline form by the hydrolysis of cocaine.

echelette grating [SPECT] A diffraction grating with coarse groove spacing, designed for the infrared region; has grooves with comparatively flat sides and concentrates most of the radiation by reflection into a small angular coverage.

echelle grating [SPECT] A diffraction grating designed for use in high orders and at angles of illumination greater than 45° to obtain high dispersion and resolving power by the use of high orders of interference.

echelon [ORD] **1.** A formation of troops; the units are parallel but unaligned, in a steplike manner. **2.** A similar arrangement of planes or ships, as planes flying in a V formation.

echelon faults [GEOL] Separate, parallel faults having steplike trends.

echelon grating [SPECT] A diffraction grating which consists of about 20 plane-parallel plates about 1 centimeter thick, cut from one sheet, each plate extending beyond the next by about 1 millimeter, and which has a resolving power on the order of 10^6.

Echeneidae [VERT ZOO] The remoras, a family of perciform fishes in the suborder Percoidei.

echidna [VERT ZOO] A spiny anteater; any member of the family Tachyglossidae.

Echinacea [INV ZOO] A suborder of echinoderms in the order Euechinoidea; individuals have a rigid test, keeled teeth, and branchial slits.

echinate [ZOO] Having a dense covering of spines or bristles.

Echinidae [INV ZOO] A family of echinacean echinoderms in the order Echinoida possessing trigeminate or polyporous plates with the pores in a narrow vertical zone.

Echiniscoidea [INV ZOO] A suborder of tardigrades in the order Heterotardigrada characterized by terminal claws on the legs.

echinococcosis [MED] Infestation by the larva (hydatid) of *Echinococcus granulosis* in man, and in some canines and herbivores. Also known as hydatid disease; hydatidosis.

Echinococcus [INV ZOO] A genus of tapeworms.

echinococcus cyst [INV ZOO] A cyst formed in host tissues by the larva of *Echinococcus granulosus*.

Echinocystitoida [PALEON] An order of extinct echinoderms in the subclass Perischoechinoidea.

Echinodera [INV ZOO] The equivalent name for Kinorhyncha.

Echinodermata [INV ZOO] A phylum of exclusively marine coelomate animals distinguished from all others by an internal skeleton composed of calcite plates, and a water-vascular system to serve the needs of locomotion, respiration, nutrition, or perception.

Echinoida [INV ZOO] An order of Echinacea with a camarodont lantern, smooth test, and imperforate noncrenulate tubercles.

Echinoidea [INV ZOO] The sea urchins, a class of Echinozoa having a compact body enclosed in a hard shell, or test, formed by regularly arranged plates which bear movable spines.

Echinometridae [INV ZOO] A family of echinoderms in the order Echinoida, including polyporous types with either an oblong or a spherical test.

echinomycin [MICROBIO] $C_{50}H_{60}O_{12}N_{12}S_2$ A toxic polypeptide antibiotic produced by species of *Streptomyces*.

echinopluteus [INV ZOO] The bilaterally symmetrical larva of sea urchins.

echinopsine [ORG CHEM] $C_{10}H_9O$ An alkaloid obtained from *Echinops* species; crystallizes as needles from benzene solution, melts at 152°C; physiological action is similar to that of brucine and strychnine.

Echinosteliaceae [MYCOL] A family of slime molds in the order Echinosteliales.

Echinosteliales [MYCOL] An order of slime molds in the subclass Myxogastromycetidae.

echinostome cercaria [INV ZOO] A digenetic trematode larva characterized by the large anterior acetabulum and a collar with spines.

Echinothuriidae [INV ZOO] A family of deep-water echinoderms in the order Echinothurioida in which the large, flexible test collapses into a disk at atmospheric pressure.

Echinothurioida [INV ZOO] An order of echinoderms in the superorder Diadematacea with solid or hollow primary radioles, diademoid ambulacral plates, noncrenulate tubercles, and the anus within the apical system.

Echinozoa [INV ZOO] A subphylum of free-living echinoderms having the body essentially globoid with meridional symmetry and lacking appendages.

Echiurida [INV ZOO] A small group of wormlike organisms regarded as a separate phylum of the animal kingdom; members have a saclike or sausage-shaped body with an anterior, detachable prostomium.

Echiuridae [INV ZOO] A small family of the order Echiuroinea characterized by a flaplike prostomium.

Echiuroidea [INV ZOO] A phylum of schizocoelous animals.

Echiuroinea [INV ZOO] An order of the Echiurida.

echo [ELECTR] **1.** The signal reflected by a radar target, or the trace produced by this signal on the screen of the cathode-ray tube in a radar receiver. Also known as radar echo; return. **2.** *See* ghost signal. [PHYS] A wave packet that has been reflected or otherwise returned with sufficient delay and magnitude to be perceived as a signal distinct from that directly transmitted.

echo amplitude [ELECTR] In radar, an empirical measure of the strength of a target signal as determined from the appearance of the echo; the amplitude of the echo waveform usually is measured by the deflection of the electron beam from the base line of an amplitude-modulated indicator.

echo area [ELECTROMAG] In radar, the area of a fictitious perfect reflector of electromagnetic waves that would reflect

ECCENTRIC CAM

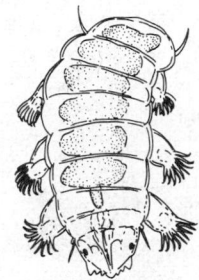

An eccentric cam with a flat-face follower.

ECHINISCOIDEA

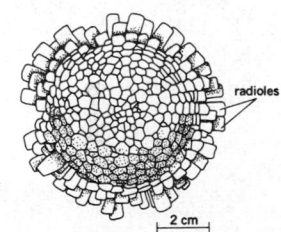

Echiniscoides sigismundi.

ECHINOIDA

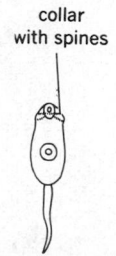

Colobocentrotus atratus, aboral aspect, a Pacific species adapted for life on wave-exposed coral reefs.

ECHINOSTOME CERCARIA

A drawing of an echinostome cercaria. *(From R. M. Cable, An Illustrated Laboratory Manual of Parasitology, Burgess, 1958)*

the same amount of energy back to the radar as the actual target. Also known as radar cross section; target cross section.

echo attenuation [ELECTR] The power transmitted at an output terminal of a transmission line, divided by the power reflected back to the same output terminal.

echo box [ELECTR] A calibrated high-Q resonant cavity that stores part of the transmitted radar pulse power and gradually feeds this energy into the receiving system after completion of the pulse transmission; used to provide an artificial target signal for test and tuning purposes. Also known as phantom target.

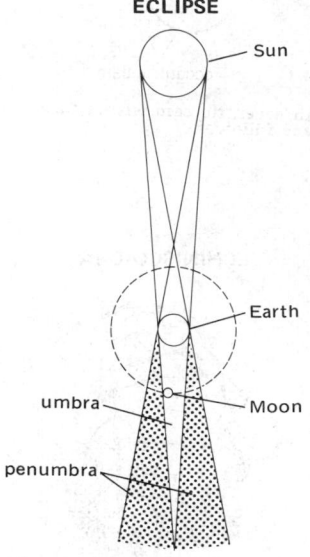

ECLIPSE

Diagram of a lunar eclipse.

echocardiography [MED] A diagnostic technique for the heart that uses a transducer held against the chest to send high-frequency sound waves which pass harmlessly into the heart; as they strike structures within the heart, they are reflected back to the transducer and recorded on an oscilloscope.

echo chamber [ACOUS] A reverberant room or enclosure used in a studio to add echo effects to sounds for radio or television programs.

echo check [ADP] A method of ascertaining the accuracy of transmission of data in which the transmitted data are returned to the sending end for comparison with original data. Also known as read-back check.

echo contour [ELECTR] A trace of equal signal intensity of the radar echo displayed on a range height indicator or plan position indicator scope.

echo frequency [ELECTR] The number of fluctuations, per unit time, in the power or amplitude of a radar target signal.

echogram [ENG] The graphic presentation of echo soundings recorded as a continuous profile of the sea bottom.

echograph [ENG] An instrument used to record an echogram.

echo intensity [ELECTR] The brightness or brilliance of a radar echo as displayed on an intensity-modulated indicator; echo intensity is, within certain limits, proportional to the voltage of the target signal or to the square root of its power.

echolalia [MED] The purposeless, often seemingly involuntary repetition of words spoken by another person; a disorder seen in certain psychotic states and in certain organic brain syndromes. Also known as echophrasia.

echolic [MED] Producing abortion or accelerating labor.

echo location *See* echo ranging.

echo matching [ENG] Rotating an antenna to a position in which the pulse indications of an echo-splitting radar are equal.

echophrasia *See* echolalia.

echo power [ELECTR] The electrical strength, or power, of a radar target signal, normally measured in watts or dBm (decibels referred to 1 milliwatt).

echo pulse [ELECTR] A pulse of radio energy received at the radar after reflection from a target; that is, the target signal of a pulse radar.

echo ranging [ENG] Active sonar, in which underwater sound equipment generates bursts of ultrasonic sound and picks up echoes reflected from submarines, fish, and other objects within range, to determine both direction and distance to each target. Also known as echo location. [VERT ZOO] An auditory feedback mechanism in bats, porpoises, seals, and certain other animals whereby reflected ultrasonic sounds are utilized in orientation.

echo recognition [ENG] Identification of a sonar reflection from a target, as distinct from energy returned by other reflectors.

Echo satellite [AERO ENG] An aluminized-surface, Mylar balloon about 100 feet (30 meters) in diameter, placed in orbit as a passive communications satellite for reflecting microwave signals from a transmitter to receivers beyond the horizon.

echo signal *See* target signal.

echo sounder *See* sonic depth finder.

echo sounding [ENG] Determination of the depth of water by measuring the time interval between emission of a sonic or ultrasonic signal and the return of its echo from the sea bottom.

echo-splitting radar [ENG] Radar in which the echo is split by special circuits associated with the antenna lobe-switching mechanism, to give two echo indications on the radarscope screen; when the two echo indications are equal in height, the target bearing is read from a calibrated scale.

echo suppressor [ELECTR] **1.** A circuit that desensitizes radar navigation equipment for a fixed period after the reception of one pulse, for the purpose of rejecting delayed pulses arriving from longer, indirect reflection paths. **2.** A relay or other device used on a transmission line to prevent a reflected wave from returning to the sending end of the line.

echo talker [ADP] The interference created by the retransmission of a message back to its source while the source is still transmitting.

echovirus [VIROL] A division of enteroviruses in the picornavirus group; the name is derived from the group designation enteric cytopathogenic human orphan virus.

eckermannite [MINERAL] $Na_3(Mg,Li)_4(Al,Fe)Si_8O_{22}(OH,F)_2$ Mineral of the amphibole group containing magnesium, lithium, iron, and fluorine.

Eckert number [PHYS] A dimensionless group used in the study of compressible flow around a body, equal to the square of the fluid velocity far from the body divided by the product of the specific heat of the fluid at constant temperature and the difference between the temperatures of the fluid and the body. Symbolized N_E. Also known as Dulong number.

eclampsia [MED] A disorder occurring during the latter half of pregnancy, characterized by elevated blood pressure, edema, proteinuria, and convulsions or coma.

eclipse [ASTRON] **1.** The reduction in visibility or disappearance of a body by passing into the shadow cast by another body. **2.** The apparent cutting off, wholly or partially, of the light from a luminous body by a dark body coming between it and the observer. Also known as astronomical eclipse.

eclipsed conformation [PHYS CHEM] A particular arrangement of constituent atoms that may rotate about a single bond in a molecule; for ethane it is such that when viewed along the axis of the carbon-carbon bond the hydrogen atoms of one methyl group are exactly in line with those of the other methyl group.

eclipse period [VIROL] A phase in the proliferation of viral particles during which the virus cannot be detected in the host cell.

eclipse seasons [ASTRON] The two times when the sun is near enough to one of the nodes of the moon's orbit for eclipses to occur; this positioning occurs at nearly opposite times of the year, and the eclipse seasons vary yearly because of westward regression of the nodes.

eclipse year [ASTRON] The interval between two successive conjunctions of the sun with the same node of the moon's orbit, equal to 346.62 sidereal days.

eclipsing variable star [ASTRON] A binary star whose orbit is such that every time one star passes between the observer and its companion an eclipse results.

ecliptic [ASTRON] The apparent annual path of the sun among the stars; the intersection of the plane of the earth's orbit with the celestial sphere.

ecliptic coordinate system [ASTRON] A celestial coordinate system in which the ecliptic is taken as the primary and the great circles perpendicular to it are then taken as secondaries.

ecliptic diagram [ASTRON] A diagram of the zodiac indicating positions of certain celestial bodies in the ecliptic region.

ecliptic latitude *See* celestial latitude.

ecliptic limits [ASTRON] The distance of the sun from a node of the moon's orbit such that a solar eclipse cannot occur, or the greatest distance of the moon from a node such that an eclipse of the moon cannot occur.

ecliptic longitude *See* celestial longitude.

ecliptic pole [ASTRON] On the celestial sphere, either of two points 90° from the ecliptic.

eclogite [PETR] A class of metamorphic rocks distinguished by their composition, consisting essentially of omphacite and pyrope with small amounts of diopside, enstatite, olivine, kyanite, rutile, and rarely, diamond.

eclogite facies [PETR] A type of facies composed of eclogite and formed by regional metamorphism at extremely high temperature and pressure.

ECM *See* electrochemical machining.

ecnephias [METEOROL] A squall or thunderstorm in the Mediterranean.

eco *See* electron-coupled oscillator.

ecocline [ECOL] A genetic gradient of adaptability to an environmental gradient; formed by the merger of ecotypes.

ecological association [ECOL] A complex of communities, such as an elm-hackberry association, which develops in accord with variations in physiography, soil, and successional history within the major subdivision of a biotic realm.

ecological climatology [BIOL] A branch of bioclimatology, including the physiological adaptation of plants and animals to their climate, and the geographical distribution of plants and animals in relation to climate.

ecological community *See* community.

ecological interaction [ECOL] The relation between species that live together in a community; specifically, the effect an individual of one species may exert on an individual of another species.

ecological pyramid [ECOL] A pyramid-shaped diagram representing quantitatively the numbers of organisms, energy relationships, and biomass of an ecosystem; numbers are high for the lowest trophic levels (plants) and low for the highest trophic level (carnivores).

ecological succession [ECOL] A gradual process incurred by the change in the number of individuals of each species of a community and by establishment of new species populations that may gradually replace the original inhabitants.

ecological system *See* ecosystem.

ecology [BIOL] A study of the interrelationships which exist between organisms and their environment. Also known as environmental biology.

ecomania [PSYCH] A symptom complex characterized by a domineering, haughty, and irritable attitude toward members of one's own family, but an attitude of humility toward those in authority.

econometrics [IND ENG] The application of mathematical and statistical techniques to the estimation of mathematical relationships for testing of economic theories and the solution of economic problems.

economic entomology [BIOL] A branch of entomology concerned with the study of economic losses of commercially important animals and plants due to insect predation.

economic geography [GEOGR] A branch of geography concerned with the relations of physical environment and economic conditions to the manufacture and distribution of commodities.

economic geology [GEOL] **1.** Application of geologic knowledge to materials usage and principles of engineering. **2.** The study of metallic ore deposits.

economic life [IND ENG] The number of years after which a capital good should be replaced in order to minimize the long-run annual cost of operation, repair, depreciation, and capital.

economic lot size [IND ENG] The number of units of a product or item to be manufactured at each setup or purchased on each order so as to minimize the cost of purchasing or setup, and the cost of holding the average inventory over a given period, usually annual.

economic mineral [MINERAL] Mineral of commercial value.

economic mobilization [ORD] The process of preparing for and carrying out such changes in the organization and functioning of the national economy as are necessary to provide the most effective use of resources in a national emergency.

economic order quantity [IND ENG] The number of orders required to fulfill the economic lot size.

economic purchase quantity [IND ENG] The economic lot size for a purchased quantity.

economics [IND ENG] A social science that deals with production, distribution, and consumption of commodities, or wealth.

economic warfare [ORD] The defensive use in peacetime, as well as during a war, of any means by military and civilian agencies to maintain or expand the economic potential for war of a nation and its (probable) allies; and, conversely, the offensive use of any measure in peace or war to diminish or neutralize the economic potential for war of the (likely) enemy nation and its accomplices.

economic war potential [ORD] The segment of the economic capacity of a nation which can be used for purposes of conducting war.

economizer [ENG] A reservoir in a continuous-flow oxygen system in which oxygen exhaled by the user is collected for recirculation in the system. [MECH ENG] A forced-flow, once-through, convection-heat-transfer tube bank in which feedwater is raised in temperature on its way to the evaporating section of a steam boiler, thus lowering flue gas temperature, improving boiler efficiency, and saving fuel.

economy [ADP] The ratio of the number of characters to be coded to the maximum number available with the code; for example, binary-coded decimal using 4 bits provides 16 possible characters but uses only 10 of them.

E core [ELECTROMAG] A transformer core made from E-shaped laminations and used in conjunction with I-shaped laminations.

ecosystem [ECOL] A functional system which includes the organisms of a natural community together with their environment. Derived from ecological system.

ecosystem mapping [ECOL] The drawing of maps that locate different ecosystems in a geographic area.

ecotone [ECOL] A zone of intergradation between ecological communities.

ecotype [ECOL] A subunit, race, or variety of a plant ecospecies that is restricted to one habitat; equivalent to a taxonomic subspecies.

EC smokeless powder [MATER] An explosive powder used chiefly in blank cartridges, but also in some .22-caliber and shotgun ammunition, and formerly in fragmentation grenades. Also known as EC blank fire; EC blank powder.

ecstasy [MED] A trancelike state with loss of sensory perception and voluntary control.

ECSW *See* extended channel status word.

ectasia [MED] Dilation, especially of a hollow organ.

Ecterocoelia [INV ZOO] The equivalent name for Protostomia.

ectethmoid [ANAT] Either one of the lateral cellular masses of the ethmoid bone.

ecthyma [MED] An inflammatory skin disease characterized by large flat pustules that ulcerate and become crusted, and are surrounded by a distinct inflammatory areola.

ectinites [PETR] One of two major groups of metamorphic rocks comprising those formed with no accession or introduction of feldspathic material.

ECTL *See* emitter-coupled transistor logic.

ectocardia [MED] An abnormal position of the heart; it may be outside the thoracic cavity or misplaced within the thorax.

ectocommensal [ECOL] An organism living on the outer surface of the body of another organism, without affecting its host.

ectocornea [ANAT] The outer layer of the cornea.

ectocyst [INV ZOO] **1.** The outer layer of the wall of a zooecium. **2.** *See* epicyst.

ectoderm [EMBRYO] The outer germ layer of an animal embryo. Also known as epiblast. [INV ZOO] The outer layer of a diploblastic animal.

ectogenesis [EMBRYO] Development of an embryo or of embryonic tissue outside the body in an artificial environment.

ectogony [BOT] The influence of pollination and fertilization on structures outside the embryo and endosperm; effect may be on color, chemical composition, ripening, or abscission.

ectohumus [GEOL] An accumulation of organic matter on the soil surface with little or no mixing with mineral material. Also known as mor; raw humus.

ectomere [EMBRYO] A blastomere that will differentiate into ectoderm.

ectomesoblast [EMBRYO] An undifferentiated layer of embryonic cells from which arises the epiblast and mesoblast.

ectomesoderm [EMBRYO] Mesoderm which is derived from ectoderm and is always mesenchymal; a type of primitive connective tissue.

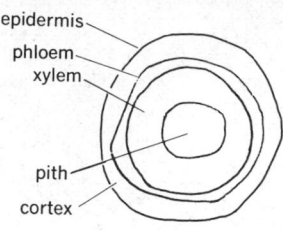

ECTOPHLOIC SIPHONOSTELE

epidermis
phloem
xylem
pith
cortex

Cross section of ectophloic siphonostele.

EDDY CURRENT

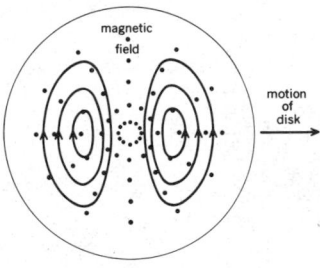

Eddy currents which are induced in a disk moving through a nonuniform magnetic field.

EDENTATA

The giant anteater (*Myrmecophaga tridactyla*), the largest of all the edentates and a native of the tropical grasslands of Central South America.

ectomorph [PSYCH] A somatotype suggested by W. H. Sheldon to describe a person with a thin physique.

ectoparasite [ECOL] A parasite that lives on the exterior of its host.

ectophagous [INV ZOO] The larval stage of a parasitic insect which is in the process of development externally on a host.

ectophloic siphonostele [BOT] A type of stele with pith that has the phloem only on the outside of the xylem.

ectophyte [ECOL] A plant which lives externally on another organism.

ectopia [MED] A congenital or acquired positional abnormality of an organ or other part of the body.

ectopic pairing [CYTOL] Pairing between nonhomologous segments of the salivary gland chromosomes in *Drosophila*, presumably involving mainly heterochromatic regions.

ectoplasm [CYTOL] The outer, gelled zone of the cytoplasmic ground substance in many cells. Also known as ectosarc.

Ectoprocta [INV ZOO] A subphylum of colonial bryozoans having eucoelomate visceral cavities and the anus opening outside the circlet of tentacles.

ectopterygoid [VERT ZOO] A membrane bone located ventrally on the skull, situated behind the palate and extending to the quadrate; found in some fishes and reptiles.

ectosarc *See* ectoplasm.

ectosome [INV ZOO] The outer, cortical layer of a sponge.

ectostosis [PHYSIO] Formation of bone immediately beneath the perichondrium and surrounding and replacing underlying cartilage.

ectosymbiont [ECOL] A symbiont that lives on the surface of or is physically separated from its host.

ectotherm [PHYSIO] A cold-blooded animal.

ectotrophic [BIOL] Obtaining nourishment from outside; applied to certain parasitic fungi that live on and surround the roots of the host plant.

ectozoa [ECOL] Animals which live externally on other organisms.

Ectrephidae [INV ZOO] An equivalent name for Ptinidae.

ectrodactylia [MED] Congenital absence of any of the fingers or toes or parts of them.

ectromelia [MED] A congenital absence or an anomaly of one or more limbs.

ectromelia virus [VIROL] A member of subgroup I of the poxvirus group; causes mousepox.

eczema [MED] Any skin disorder characterized by redness, thickening, oozing from blisters or papules, and occasional formation of fissures and crusts.

eczematoid reaction [MED] A dermal and epidermal inflammatory response characterized by erythema, edema, vesiculation, and exudation in the acute stage, and by erythema, edema, thickening of the epidermis, and scaling in the chronic stage.

ED *See* electronic dummy.

ED$_{50}$ *See* effective dose 50.

edaphic community [ECOL] A plant community that results from or is influenced by soil factors such as salinity and drainage.

edaphon [BIOL] Flora and fauna in soils.

Edaphosuria [PALEON] A suborder of extinct, lowland, terrestrial, herbivorous reptiles in the order Pelycosauria.

eddy [FL MECH] A vortexlike motion of a fluid running contrary to the main current.

eddy coefficient *See* exchange coefficient.

eddy conduction *See* eddy heat conduction.

eddy conductivity *See* austausch coefficient.

eddy correlation [METEOROL] A method of studying the effects of sea surface on the air above it by measuring simultaneous fluctuations of the horizontal and vertical components of the airflow from the mean.

eddy current [ELECTROMAG] An electric current induced within the body of a conductor when that conductor either moves through a nonuniform magnetic field or is in a region where there is a change in magnetic flux. Also known as Foucault current.

eddy-current brake [MECH ENG] A control device or dynamometer for regulating rotational speed, as of flywheels, in which energy is converted by eddy currents into heat.

eddy-current clutch [MECH ENG] A type of electromagnetic clutch in which torque is transmitted by means of eddy currents induced by a magnetic field set up by a coil carrying direct current in one rotating member.

eddy-current loss [ELECTROMAG] Energy loss due to undesired eddy currents circulating in a magnetic core.

eddy-current tachometer [ENG] A type of tachometer in which a rotating permanent magnet induces currents in a spring-mounted metal cylinder; the resulting torque rotates the cylinder and moves its attached pointer in proportion to the speed of the rotating shaft. Also known as drag-type tachometer.

eddy-current test [ELECTROMAG] A nondestructive test in which the change of impedance of a test coil brought close to a conducting specimen indicates the eddy currents induced by the coil, and thereby indicates certain properties or defects of the specimen.

eddy diffusion [FL MECH] Diffusion which occurs in turbulent flow, by the rapid process of mixing of the swirling eddies of fluid. Also known as turbulent diffusion.

eddy diffusion coefficient *See* eddy diffusivity.

eddy diffusivity [FL MECH] The exchange coefficient for the diffusion of a conservative property by eddies in a turbulent flow. Also known as coefficient of eddy diffusion; eddy diffusion coefficient.

eddy flux [FL MECH] The rate of transport (or flux) of fluid properties such as momentum, mass heat, or suspended matter by means of eddies in a turbulent motion; the rate of turbulent exchange. Also known as moisture flux; turbulent flux.

eddy heat conduction [THERMO] The transfer of heat by means of eddies in turbulent flow, treated analogously to molecular conduction. Also known as eddy heat flux; eddy conduction.

eddy heat flux *See* eddy heat conduction.

eddy kinetic energy [FL MECH] The kinetic energy of that component of fluid flow which represents a departure from the average kinetic energy of the fluid, the mode of averaging depending on the particular problem. Also known as turbulence energy.

eddy mill *See* pothole.

eddy resistance [FL MECH] Resistance or drag of a ship resulting from eddies that are shed from the hull or appendages of the ship and carry away energy.

eddy spectrum [FL MECH] **1.** The distribution of frequencies of rotation of eddies in a turbulent flow, or of those eddies having some range of sizes. **2.** The distribution of kinetic energy among eddies with various frequencies or sizes.

eddy stress *See* Reynolds stress.

eddy velocity [FL MECH] The difference between the mean velocity of fluid flow and the instantaneous velocity at a point. Also known as fluctuation velocity.

eddy viscosity [FL MECH] The turbulent transfer of momentum by eddies giving rise to an internal fluid friction, in a manner analogous to the action of molecular viscosity in laminar flow, but taking place on a much larger scale.

Edeleanu process [CHEM ENG] A process for removal of compounds of sulfur from petroleum fractions by an extraction procedure utilizing liquid sulfur dioxide, or liquid sulfur dioxide and benzene.

edema [MED] An excessive accumulation of fluid in the cells, tissue spaces, or body cavities due to a disturbance in the fluid exchange mechanism. Also known as dropsy.

Edenian [GEOL] Lower Cincinnatian geologic stage in North America, above the Mohawkian and below Maysvillian.

Edentata [VERT ZOO] An order of mammals characterized by the absence of teeth or the presence of simple prismatic, unspecialized teeth with no enamel.

edentate [VERT ZOO] **1.** Lacking teeth. **2.** Any member of the Edentata.

edentulous [VERT ZOO] Having no teeth; especially, having lost teeth that were present.

edge [MATH] **1.** A line along which two plane faces of a solid intersect. **2.** A line segment connecting nodes or vertices in a graph (a geometric representation of the relation among situations). Also known as arc.

edge dislocation [CRYSTAL] A dislocation which may be regarded as the result of inserting an extra plane of atoms, terminating along the line of the dislocation. Also known as Taylor-Orowan dislocation.

edge effect [ELEC] An outward-curving distortion of lines of force near the edges of two parallel metal plates that form a capacitor.

edge focusing [ELECTROMAG] Axial focusing of a stream of ions which occurs when it crosses a fringe magnetic field obliquely; used in mass spectrometers and cyclotrons.

edge fog [GRAPHICS] The light fog which appears along the edge of roll film, generally from exposure during loading or unloading.

edge-notched card [ADP] A card with a series of holes along one or more edges, and notches which open one or more holes, so that long needles inserted in specific holes in a deck of such cards let fall only cards with a desired type of data.

edge printing [GRAPHICS] A preprinted identification along the edge of film.

edge-punched card [ADP] A card with one or more rows of holes, representing binary data, punched along the edges.

edger [MET] The part of a forging die which portions out the quantity of metal needed for shaping.

edge runner *See* Chile mill.

edge wave [OCEANOGR] An ocean wave moving parallel to the coast, with crests normal to the coastline; maximum amplitude is at shore, with amplitude falling off exponentially farther from shore.

edging [MET] Controlling the plate width or the edge shape during rolling operations.

ediacaran fauna [PALEON] The oldest known assemblage of fossil remains of soft-bodied marine animals; first discovered in the Ediacara Hills, Australia.

edible oil [FOOD ENG] A liquid fat that is capable of being eaten as a food or food accessory.

edingtonite [MINERAL] $BaAl_2Si_3O_{10}\cdot4H_2O$ Gray zeolite mineral that forms rhombic crystals; sometimes contains large amounts of calcium.

Edison battery [ELEC] A storage battery composed of cells having nickel and iron in an alkaline solution. Also known as nickel-iron battery.

Edison distribution system [ELEC] Three-wire direct-current distribution system, usually 120 to 240 volts, for combined light and power service from a single set of mains.

Edison effect *See* thermionic emission.

E display [ELECTR] A rectangular radar display in which targets appear as blips with distance indicated by the horizontal coordinate, and elevation by the vertical coordinate.

edit [ADP] **1.** To modify the form or format of an output or input by inserting or deleting characters such as page numbers or decimal points. **2.** A computer instruction directing that this step be performed.

edit capability [ADP] The degree of sophistication available to the programmer to modify his statements while in the time-sharing mode.

edit mask [ADP] The receiving word through which a source word is filtered, allowing for the suppression of leading zeroes, the insertion of floating dollar signs and decimal points, and other such formatting.

editor program [ADP] A special program by means of which a user can easily perform corrections, insertions, modifications, or deletions in an existing program or data file.

EDM *See* electron discharge machining.

Edman degradation technique [BIOCHEM] In protein analysis, an approach to amino-end-group determination involving the use of a reagent, phenylisothiocyanate, that can be applied to the liberation of a derivative of the amino-terminal residue without hydrolysis of the remainder of the peptide chain.

EDP *See* electronic data processing.

EDP center *See* electronic data-processing center.

Edrioasteroidea [PALEON] A class of extinct Echinozoa having ambulacral radial areas bordered by tube feet, and the mouth and anus located on the upper side of the theca.

Edser and Butler's bands [OPTICS] Dark bands at intervals of equal frequency in a spectrum of light which has passed through a thin plate of transparent material with parallel sides.

EDTA *See* ethylenediaminetetraacetic acid.

educational age [PSYCH] The average achievement of a pupil or student in school subjects based on average performance for a given chronological age as measured by standard educational tests.

educational psychology [PSYCH] A field of psychology that deals with the psychological aspects of teaching and formal learning processes.

eductor [ENG] **1.** An ejectorlike device for mixing two fluids. **2.** *See* ejector.

eductor pump [MIN ENG] A pump which removes slurried material from a hydraulically disseminated subsurface ore matrix.

edulcorate [ADP] To eliminate irrelevant data from a data file.

EDVAC [ADP] The first stored program computer, built in 1952. Derived from electron discrete variable automatic compiler.

Edwards' syndrome *See* trisomy 18 syndrome.

EEG *See* electroencephalogram.

eel [VERT ZOO] The common name for a number of unrelated fishes included in the orders Anguilliformes and Cypriniformes; all have an elongate, serpentine body.

eel grass *See* tape grass.

eff *See* efficiency.

effective address [ADP] The address that is obtained by applying any specified indexing or indirect addressing rules to the specified address; the effective address is then used to identify the current operand.

effective air path [NAV] A straight line on a navigation chart connecting two air positions, commonly used between the air positions of two pressure soundings in order to determine effective true airspeed between the two soundings.

effective ampere [ELEC] The amount of alternating current flowing through a resistance that produces heat at the same average rate as 1 ampere of direct current flowing in the same resistance.

effective angle of attack [AERO ENG] That part of a given angle of attack that lies between the chord of an airfoil and a line representing the resultant velocity of the disturbed airflow.

effective antenna length [ELECTROMAG] Electrical length of an antenna, as distinguished from its physical length.

effective aperture [OPTICS] The diameter of the image of the aperture stop of an optical system, as viewed from the object.

effective area [CHEM ENG] Absolute or cross-sectional area of process media involved in the process, such as the actual area of filter media through which a fluid passes, or the available surface area of absorbent contacted by a gas or liquid. [ELECTROMAG] Of an antenna in any specified direction, the square of the wavelength multiplied by the power gain (or directive gain) in that direction and divided by 4π (12.57).

effective atmosphere *See* optically effective atmosphere.

effective bandwidth [ELECTR] The bandwidth of an assumed rectangular band-pass having the same transfer ratio at a reference frequency as a given actual band-pass filter, and passing the same mean-square value of a hypothetical current having even distribution of energy throughout that bandwidth.

effective capacitance [ELEC] Total capacitance existing between any two given points of an electric circuit.

effective confusion area [ENG] Amount of chaff whose radar cross-sectional area equals the radar cross-sectional area of the particular aircraft at a particular frequency.

effective current [ELEC] The value of alternating current that will give the same heating effect as the corresponding value of direct current. Also known as root-mean-square current.

effective decline rate [PETRO ENG] The drop in oil or gas production rate over a period of time; equal to unity (1 month or 1 year) divided by the production rate at the beginning of the period.

effective dose 50 [PHARM] The amount of a drug required to produce a response in 50% of the subjects to whom the drug is given. Abbreviated ED_{50}. Also known as median effective dose.

effective earth radius [COMMUN] A radius value used in place of the geometric radius to correct for atmospheric refraction in estimating ranges of antennas when the index of

EEL

American eel, showing typical fin structure.

refraction in the atmosphere changes linearly with height; under conditions of standard refraction it is 4/3 the geometric radius. Also known as effective radius of the earth.

effective exhaust velocity [AERO ENG] A fictitious exhaust velocity that yields the observed value of jet thrust in calculations.

effective facsimile band [COMMUN] Frequency band of a facsimile signal wave equal in width to that between zero frequency and maximum keying frequency.

effective field intensity [ELECTROMAG] Root-mean-square value of the inverse distance fields at a distance of 1 mile (1.6 kilometers) from the transmitting antenna in all directions in the horizontal plane.

effective firing time [ORD] The period of time during which an aircraft can deliver effective fire at a moving or stationary target.

effective force *See* inertial force.

effective fragment [ORD] In terminal ballistics, a fragment whose mass, velocity, and form, upon impact with the target are such as to enable the fragment to accomplish the desired effect.

effective gun bore line [MECH] The line which a projectile should follow when the muzzle velocity of the antiaircraft gun is vectorially added to the aircraft velocity.

effective gust velocity [METEOROL] The vertical component of the velocity of a sharp-edged gust that would produce a given acceleration on a particular airplane flown in level flight at the design cruising speed of the aircraft and at a given air density.

effective half-life [NUCLEO] The half-life of a radioisotope in a biological organism, resulting from a combination of radioactive decay and biological elimination.

effective height [ELECTROMAG] The height of the center of radiation of a transmitting antenna above the effective ground level.

effective horizon [COMMUN] A horizon whose distance at a given height above sea level is the distance to the horizon of a fictitious earth, having a radius 4/3 times the earth's true radius; used to estimate ranges of antennas, taking atmospheric refraction into account.

effective horsepower [NAV ARCH] The power necessary to overcome the resistance of water to motion of a ship towed at a given speed, measured in horsepower. Abbreviated ehp. Also known as towrope horsepower.

effective launcher line [MECH] The line along which the aircraft rocket would go if it were not affected by gravity.

effective length [NAV ARCH] Mean length of that portion of the ship's hull below the waterline.

effective molecular diameter [PHYS CHEM] The general extent of the electron cloud surrounding a gas molecule as calculated in any of several ways.

effective molecular weight [PETRO ENG] Empirical relationship of oil graphed against API gravity to give the effective (pseudoaverage) molecular weight of the oil for reservoir calculations.

effective multiplication factor [NUCLEO] The multiplication factor of an actual reactor, in which there is leakage of neutrons.

effectiveness level [ADP] A measure of the effectiveness of data-processing equipment, equal to the ratio of the operational use time to the total performace period, expressed as a percentage. Also known as average effectiveness level.

effective percentage modulation [COMMUN] For a single sinusoidal input component, the ratio of the peak value of the fundamental component of the envelope to the average amplitude of the modulated wave expressed in percent.

effective permeability [PHYS CHEM] The observed permeability exhibited by a porous medium to one fluid phase when there is physical interaction between this phase and other fluid phases present.

effective pitch [AERO ENG] The distance traveled by an airplane along its flight path for one complete turn of the propeller.

effective porosity [GEOL] A property of earth containing interconnecting interstices, expressed as a percent of bulk volume occupied by the interstices.

effective precipitable water [METEOROL] That part of the

precipitable water which, in theory, can actually fall as precipitation.

effective precipitation [HYD] **1.** The part of precipitation that reaches stream channels as runoff. Also known as effective rainfall. **2.** In irrigation, the portion of the precipitation which remains in the soil and is available for consumptive use.

effective radiated power [ELECTROMAG] The product of antenna input power and antenna power gain, expressed in kilowatts. Abbreviated ERP.

effective radiation *See* effective terrestrial radiation.

effective radius [ORD] The radius of a circle having its center at the center of impact of a nuclear weapon, and within which a specified dosage of radiation is equaled or exceeded within a given period of time.

effective radius of the earth *See* effective earth radius.

effective rainfall *See* effective precipitation.

effective rake [MECH ENG] The angular relationship between the plane of the tooth face of the cutter and the line through the tooth point measured in the direction of chip flow.

effective range [ORD] Distance at which a weapon may be expected to fire accurately to inflict damage or casualties.

effective resistance *See* high-frequency resistance.

effective snowmelt [HYD] The part of snowmelt that reaches stream channels as runoff.

effective sound pressure [ACOUS] The root-mean-square value of the instantaneous sound pressure at a point during a complete cycle, expressed in dynes per square centimeter. Also known as root-mean-square sound pressure; sound pressure.

effective temperature [ASTROPHYS] A measure of the temperature of a star, deduced by means of the Stefan-Boltzmann law, from the total energy emitted per unit area. [METEOROL] The temperature at which motionless, saturated air would induce, in a sedentary worker wearing ordinary indoor clothing, the same sensation of comfort as that induced by the actual conditions of temperature, humidity, and air movement.

effective terrestrial radiation [GEOPHYS] The amount by which outgoing infrared terrestrial radiation of the earth's surface exceeds downcoming infrared counterradiation from the sky. Also known as effective radiation; nocturnal radiation.

effective thermal resistance [ELECTR] Of a semiconductor device, the effective temperature rise per unit power dissipation of a designated junction above the temperature of a stated external reference point under conditions of thermal equilibrium. Also known as thermal resistance.

effective thrust [AERO ENG] In a rocket motor or engine, the theoretical thrust less the effects of incomplete combustion and friction flow in the nozzle.

effective transformation group [MATH] A transformation group in which the identity element is the only element to leave all points fixed.

effective true airspeed [NAV] The effective air distance divided by the elapsed time between two pressure soundings.

effective value *See* root-mean-square value.

effector [BIOCHEM] Activator of an allosteric enzyme. [CONT SYS] A motor, solenoid, or hydraulic piston that turns commands to a teleoperator into specific manipulatory actions. [PHYSIO] A structure that is sensitive to a stimulus and causes an organism or part of an organism to react to the stimulus, either positively or negatively.

effector organ [PHYSIO] Any muscle or gland that mediates overt behavior, that is, movement or secretion.

effector system [PHYSIO] A system of effector organs in the animal body.

efferent [PHYSIO] Carrying or conducting away, as the duct of an exocrine gland or a nerve.

effervescence [CHEM] The bubbling of a solution of an element or chemical compound as the result of the emission of gas without the application of heat; for example, the escape of carbon dioxide from carbonated water.

efficiency Abbreviated eff. [ENG] **1.** Measure of the degree of heat output per unit of fuel when all available oxidizable materials in the fuel have been burned. **2.** Ratio of useful energy provided by a dynamic system to the energy supplied to it during a specific period of operation. [NUCLEO] The probability that a count will be produced in a counter tube by a specified particle or quantum incident. [PHYS] The ratio, usually expressed as a percentage, of the useful power output to the power input of a device. [THERMO] The ratio of the work done by a heat engine to the heat energy absorbed by it. Also known as thermal efficiency.

efficiency expert [IND ENG] An individual who analyzes procedures, productivity, and jobs in order to recommend methods for achieving maximum utilization of resources and equipment.

efficient estimator [STAT] A statistical estimator that has minimum variance.

efflorescence [BOT] The period or process of flowering. [CHEM] The property of hydrated crystals to lose water of hydration and crumble when exposed to air. [MINERAL] A whitish powder, consisting of one or several minerals produced as an incrustation on the surface of a rock in an arid region. Also known as bloom.

effluent [CHEM ENG] See discharge liquor. [CIV ENG] The liquid waste of sewage and industrial processing. [HYD] 1. Flowing outward or away from. 2. Liquid which flows away from a containing space or a main waterway.

effluent refrigeration alkylation See Stratco alkylation.

effluent weir [CIV ENG] A dam at the outflow end of a watercourse.

effluvium [IND ENG] By-products of food and chemical processes, in the form of wastes.

effort-controlled cycle [IND ENG] A work cycle which is performed entirely by hand or in which the hand time controls the place. Also known as manually controlled work.

effort rating [IND ENG] Assessing the level of manual effort expended by the operator, based on the observer's concept of normal effort, in order to adjust time-study data. Also known as pace rating; performance rating.

effuse [BOT] Expanded; spread out in a definite form.

effusion [MED] A pouring out of any fluid into a body cavity or tissue. [PHYS CHEM] The movement of a gas through an opening which is small as compared with the average distance which the gas molecules travel between collisions. [SCI TECH] 1. The act or process of leaking or pouring out. 2. Any material that is effused.

effusive stage [GEOL] The second cooling stage for volcanic rocks.

EFL See error frequency limit.

e format [ADP] A decimal, normalized form of a floating point number in FORTRAN in which a number such as 18.756 appears as .18756E + 02, which stands for .18756 × 10².

Egerov's therorem [MATH] If a sequence of measurable functions converges almost everywhere on a set of finite measure to a real-valued function, then given any $\varepsilon > 0$ there is a set of measure smaller than ε on whose complement the sequence converges uniformly.

egest [PHYSIO] 1. To discharge indigestible matter from the digestive tract. 2. To rid the body of waste.

egg [CYTOL] 1. A large, female sex cell enclosed in a porous, calcareous or leathery shell, produced by birds and reptiles. 2. See ovum.

egg apparatus [BOT] A group of three cells, consisting of the egg and two synergid cells, in the micropylar end of the embryo sac in seed plants.

egg capsule See egg case.

egg case [INV ZOO] 1. A protective capsule containing the eggs of certain insects and mollusks. Also known as egg capsule. 2. A silk pouch in which certain spiders carry their eggs. Also known as egg sac. [VERT ZOO] A soft, gelatinous (amphibians) or strong, horny (skates) envelope containing the egg of certain vertebrates.

Eggertz's method [MET] A colorimetric estimation of the carbon content of steel by dissolving the metal in nitric acid and comparing the color with that produced by a similar metal of known carbon content.

eggplant [BOT] Solanum melongena. A plant of the order Polemoniales grown for its edible egg-shaped, fleshy fruit.

egg raft [ZOO] A floating mass of eggs; produced by a variety of aquatic organisms.

egg sac [INV ZOO] 1. The structure containing the eggs of certain microcrustaceans. 2. See egg case.

eggstone See oolite.

egg tempera [GRAPHICS] A painting process in which the color is bound with egg instead of oil.

egg tooth [VERT ZOO] A toothlike prominence on the tip of the beak of a bird embryo and the tip of the nose of an oviparous reptile, which is used to break the eggshell.

eglandular [BIOL] Without glands.

eglestonite [MINERAL] Hg_4Cl_2O Rare mercuric oxide mineral; forms yellow-brown isometric crystals upon exposure to air.

Egnell's law [METEOROL] The rule stating that above any fixed place the velocity of straight or nearly straight winds in the upper half of the troposphere increases with height at roughly the same rate that the density of the air decreases.

ego [PSYCH] 1. The self. 2. The conscious part of the personality that is in contact with reality.

Egyptian asphalt [GEOL] A glance pitch (bituminous mixture similar to asphalt) found in the Arabian desert.

Egyptian cotton [BOT] Long-staple, high-quality cotton grown in Egypt.

Egyptian henna See henna.

Egyptian wind [METEOROL] A westerly wind, frequent during winter and accompanied by fog and dust, blowing in Egypt and the Gulf of Suez.

EHF See extremely high frequency.

ehp See effective horsepower.

Ehrenfest's adiabatic law [QUANT MECH] The law that, if the Hamiltonian of a system undergoes an infinitely slow change, and if the system is initially in an eigenstate of the Hamiltonian, then at the end of the change it will be in the eigenstate of the new Hamiltonian that derives from the original state by continuity, provided certain conditions are met. Also known as Ehrenfest's theorem.

Ehrenfest's equations [THERMO] Equations which state that for the phase curve $P(T)$ of a second-order phase transition the derivative of pressure P with respect to temperature T is equal to $(C_p{}^f - C_p{}^i)/TV(\gamma^f - \gamma^i) = (\gamma^f - \gamma^i)/(K^f - K^i)$, where i and f refer to the two phases, γ is the coefficient of volume expansion, K is the compressibility, C_p is the specific heat at constant pressure, and V is the volume.

Ehrenfest's theorem [QUANT MECH] 1. The theorem that a quantum-mechanical wave packet obeys the equations of motion of the corresponding classical particle when the position, momentum, and force acting on the particle are replaced by the expectation values of these quantities. 2. See Ehrenfest's adiabatic law.

Ehrenhaft effect [ELECTROMAG] A helical motion of fine particles along the lines of force of a magnetic field during exposure to light, resulting from radiometer effects.

Ehrlich's 606 See arsphenamine.

eht See extra-high tension.

E-H T junction [ELECTROMAG] In microwave waveguides, a combination of E- and H-plane T junctions forming a junction at a common point of intersection with the main waveguide.

E-H tuner [ELECTROMAG] Tunable E-H T junction having two arms terminated in adjustable plungers used for impedance transformation.

ehv See extra-high voltage.

eicosane [MATER] A mixture of saturated hydrocarbons mostly straight-chained and averaging 20 carbons in the chain; for this reason, the formula $C_{20}H_{42}$ is given to the technical mixture; used in lubricants and plasticizers.

eicosanoic acid [ORG CHEM] $CH_3(CH_2)_{18}COOH$ A white, crystalline, saturated fatty acid, melting at 75.4°C; a constituent of butter. Also known as arachic acid; arachidic acid.

eidetic imagery [PSYCH] Imagery in extreme detail; a sort of projection of an image on a mental screen.

eigenfrequency [PHYS] One of the frequencies at which an oscillatory system can vibrate.

eigenfunction [MATH] 1. An eigenvector for a linear operator on a vector space whose vectors are functions. 2. A solution to the Sturm-Liouville partial differential equation.

eigenfunction expansion [MATH] By using spectral theory for linear operators defined on spaces comprised of functions,

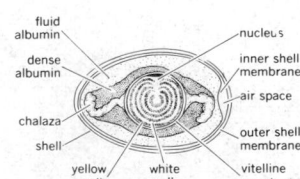

EGG

The egg of a bird. (After Schimkewitsch, from L.P. Sayles, ed., Biology of the Vertebrates, 3d ed., Macmillan, 1949)

EGGPLANT

Eggplant (Solanum melongena), cultivar Black Magic. (Joseph Harris Co., Rochester, N.Y.)

in certain cases the operator equals an integral or series involving its eigenvectors; this is known as its eigenfunction expansion and is particularly useful in studying linear partial differential equations.

eigenmatrix [MATH] Corresponding to a diagonalizable matrix or linear transformation, this is the matrix all of whose entries are 0 save those on the principal diagonal where appear the eigenvalues.

eigenstate [QUANT MECH] **1.** A dynamical state whose state vector (or wave function) is an eigenvector (or eigenfunction) of an operator corresponding to a specified physical quantity. **2.** *See* energy state.

eigenvalue [MATH] The one of the scalars λ such that $T(v) = \lambda v$, where T is a linear operator on a vector space, and v is an eigenvector. Also known as characteristic number; characteristic root; characteristic value.

eigenvalue equation *See* characteristic equation.

eigenvalue problem *See* Sturm-Lionville problem.

eigenvector [MATH] A nonzero vector v whose direction is not changed by a given linear transformation T; that is, $T(v) = \lambda v$ for some scalar λ. Also known as characteristic vector.

eight-D technique *See* 8D technique (in the D's).

eight-level code [COMMUN] A teletypewriter code that uses eight impulses, in addition to the start and stop impulses, to define a character.

eighty-column card [ADP] A card with 80 columns of punch positions, and 12 punch positions in each column.

eikonal equation [PHYS] An equation for propagation of electromagnetic or acoustic waves in a nonhomogeneous medium; it is valid only when the variation of the properties of the medium is small over the distance of a wavelength.

eikonometer [OPTICS] A scale used to measure sizes of objects viewed through a microscope, usually attached to the eyepiece so that it is seen superimposed on the image.

Eimeriina [INV ZOO] A suborder of coccidian protozoans in the order Eucoccida in which there is no syzygy and the microgametocytes produce a large number of microgametes.

Einchluss thermometer [ANALY CHEM] All-glass, liquid-filled thermometer, temperature range -201 to $+360°C$, used for laboratory test work.

E indicator *See* E scope.

einkanter [GEOL] A stone shaped by windblown sand only upon one facet.

einstein [PHYS] A unit of light energy used in photochemistry, equal to Avogadro's number times the energy of one photon of light of the frequency in question.

Einstein-Bohr equation [QUANT MECH] In a system undergoing a transition between two states so that it emits or absorbs radiation, that equation indicating that the radiation frequency equals the difference in energy between the two states divided by Planck's constant.

Einstein-Bose statistics *See* Bose-Einstein statistics.

Einstein condensation *See* Bose-Einstein condensation.

Einstein-de Haas effect [ELECTROMAG] A freely suspended body consisting of a ferromagnetic material acquires a rotation when its magnetization changes.

Einstein-de Haas method [ELECTROMAG] Method of measuring the gyromagnetic ratio of a ferromagnetic substance; one measures the angular displacement induced in a ferromagnetic cylinder suspended from a torsion fiber when magnetization of the object is reversed, and the magnetization change is measured with a magnetometer.

Einstein-de Sitter model [RELAT] A model of the universe in which ordinary euclidean geometry holds good, the distribution of matter extends infinitely at all times, and the universe expands from an infinitely condensed state at such a rate that the density is inversely proportional to the square of the time elapsed since the beginning of the expansion.

Einstein diffusion equation [STAT MECH] An equation which gives the mean square displacement caused by Brownian movement of spherical, colloidal particles in a gas or liquid.

Einstein elevator [RELAT] A windowless elevator freely falling in its shaft, inside of which conditions resemble interstellar space; used to elucidate the principle of equivalence.

Einstein equations [STAT MECH] Equations for the density and pressure of a Bose-Einstein gas in terms of power series

in a parameter which appears in the Bose-Einstein distribution law.

Einstein frequency [SOLID STATE] Single frequency with which each atom vibrates independently of other atoms, in a model of lattice vibrations; equal to the frequency observed in infrared absorption studies.

Einstein frequency condition [SOLID STATE] The assumption that all vibrations of a crystal lattice are harmonic with the same characteristic frequency.

einsteinium [CHEM] Synthetic radioactive element, symbol Es, atomic number 99; discovered in debris of 1952 hydrogen bomb explosion; now made in cyclotrons.

Einstein mass-energy relation [RELAT] The relation in which the energy of a system is equivalent to its mass times the square of the speed of light.

Einstein number [PL PHYS] A dimensionless number used in magnetofluid dynamics, equal to the ratio of the velocity of a fluid to the speed of light.

Einstein partition function [STAT MECH] The partition function for a solid, based on the Einstein frequency condition.

Einstein photochemical equivalence law [PHYS CHEM] The law that each molecule taking part in a chemical reaction caused by electromagnetic radiation absorbs one photon of the radiation. Also known as Stark-Einstein law.

Einstein photoelectric law [QUANT MECH] The law that the energy of an electron emitted from a system in the photoelectric effect is $h\nu - W$, where h is Planck's constant, ν is the frequency of the incident radiation, and W is the energy needed to remove the electron from the system; if $h\nu$ is less than W, no electrons are emitted.

Einstein-Planck law [QUANT MECH] The law that the energy of a photon is given by Planck's constant times the frequency. [RELAT] The equation of motion of a charged particle in an electromagnetic field, according to which its rate of change of momentum is equal to the Lorentz force, where the magnitude of the momentum is $mv/(1 - v^2/c^2)^{1/2}$, where m and v are the particle's mass and velocity, and c is the speed of light.

Einstein relation [PHYS] The relation in which the mobility of charges in an ionic solution or semiconductor is equal to the magnitude of the charge times the diffusion coefficient divided by the product of the Boltzmann constant and the absolute temperature.

Einstein-Rosen waves [RELAT] Gravitational waves produced by oscillating ponderable matter, along an infinitely long cylindrical axis, in an exact solution of Einstein's field equations.

Einstein's absorption coefficient [ATOM PHYS] Proportionality constant governing the absorption of electromagnetic radiation by atoms, equal to the number of quanta absorbed per second divided by the product of the energy of radiation per unit volume per unit wave number and the number of atoms in the ground state.

Einstein's equation for specific heat [SOLID STATE] The earliest equation based on quantum mechanics for the specific heat of a solid; uses the assumption that each atom oscillates with the same frequency.

Einstein's equivalency principle *See* equivalence principle.

Einstein's field equations [RELAT] Those equations relevant to the relationship in which the Einstein tensor equals -8π times the energy momentum tensor times the gravitational constant divided by the square of the speed of light. Also known as Einstein's law of gravitation.

Einstein shift [RELAT] A shift toward longer wavelengths of spectral lines emitted by atoms in strong gravitational fields.

Einstein's law of gravitation *See* Einstein's field equations.

Einstein's principle of relativity [RELAT] The principle that all the laws of physics must assume the same mathematical form in any inertial frame of reference; thus, it is impossible to determine the absolute motion of a system by any means.

Einstein's summation convention [MATH] A rotational convenience used in tensor analysis whereupon it is agreed that any term in which an index appears twice will stand for the sum of all such terms as the index assumes all of a preassigned range of values.

Einstein's unified field theories [RELAT] A series of theories attempting to express a general unifying principle underlying electromagnetism and gravity.

EINSTEINIUM

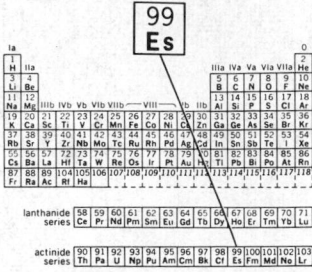

Periodic table of the chemical elements showing the position of einsteinium.

Einstein tensor [RELAT] The tensor expressed as $E_{\mu\nu} = R_{\mu\nu} - \frac{1}{2}(g_{\mu\nu}R - 2\Lambda)$, where $R_{\mu\nu}$ is the contracted curvature tensor, R is the curvature of space-time, $g_{\mu\nu}$ is the metric tensor, and Λ is the cosmological constant.

Einstein universe [RELAT] A model of the universe which is a four-dimensional cylindrical surface in a five-dimensional space.

Einstein viscosity equation [PHYS CHEM] An equation which gives the viscosity of a sol in terms of the volume of dissolved particles divided by the total volume.

ejaculation [PHYSIO] The act or process of suddenly discharging a fluid from the body; specifically, the ejection of semen during orgasm.

ejaculatory duct [ANAT] The terminal part of the ductus deferens after junction with the duct of a seminal vesicle, embedded in the prostate gland and opening into the urethra.

ejecta [GEOL] Material discharged by a volcano. [PHYSIO] Excrement. [SCI TECH] Material which is cast out.

ejection [ENG] The process of removing a molding from a mold impression by mechanical means, by hand, or by compressed air. [ORD] The expelling, by the ejector, of the empty cartridge case from small arms and rapid-fire guns.

ejection capsule [AERO ENG] In an aircraft or spacecraft, a detachable compartment (serving as a cockpit or cabin) or a payload capsule which may be ejected as a unit and parachuted to the ground.

ejection port [ORD] The opening in the receiver portion of a firearm through which the empty shell cases are thrown from the firearm after firing.

ejection seat [AERO ENG] Emergency device which expels the pilot safely from a high-speed airplane.

ejector [ENG] **1.** Any of various types of jet pumps used to withdraw fluid materials from a space. Also known as eductor. **2.** A device that ejects the finished casting from a mold. [ORD] A device in the breech mechanism of a gun, rifle, or other firearm which automatically throws out an empty cartridge case, or unfired cartridge, from the breech or receiver.

ejector condenser [MECH ENG] A type of direct-contact condenser in which vacuum is maintained by high-velocity injection water; condenses steam and discharges water, condensate, and noncondensables to the atmosphere.

ejector half [MET] The movable half of a casting die.

ejector rod [MET] A rod used to eject a piece of work made by forming.

EKG *See* electrocardiogram.

Ekman convergence [OCEANOGR] A zone of convergence of warm surface water caused by Ekman transport, creating a marked depression of the ocean's thermocline in the affected area.

Ekman current meter [ENG] A mechanical device for measuring ocean current velocity which incorporates a propeller and a magnetic compass and can be suspended from a moored ship.

Ekman layer [METEOROL] The layer of transition between the surface boundary layer of the atmosphere, where the shearing stress is constant, and the free atmosphere, which is treated as an ideal fluid in approximate geostrophic equilibrium. Also known as spiral layer.

Ekman spiral [METEOROL] A theoretical representation that a wind blowing steadily over an ocean of unlimited depth and extent and uniform viscosity would cause, in the Northern Hemisphere, the immediate surface water to drift at an angle of 45° to the right of the wind direction, and the water beneath to drift further to the right, and with slower and slower speeds, as one goes to greater depths.

Ekman transport [OCEANOGR] The movement of ocean water caused by wind blowing steadily over the surface; occurs at right angles to the wind direction.

Ekman water bottle [ENG] A cylindrical tube fitted with plates at both ends and used for deep-water samplings; when hit by a messenger it turns 180°, closing the plates and capturing the water sample.

Elaeagnaceae [BOT] A family of dicotyledonous plants in the order Proteales, noted for peltate leaf scales which often give the leaves a silvery-gray appearance.

elaidic acid [ORG CHEM] $CH_3(CH_2)_7CH:CH(CH_2)_7COOH$ A transisomer of an unsaturated fatty acid, oleic acid; crystallizes as colorless leaflets, melts at 44°C, boils at 288°C (100 mm Hg), insoluble in water, soluble in alcohol and ether; used in chromatography as a reference standard. Also known as *trans*-9-octadecenoic acid.

elaidinization [ORG CHEM] The process of changing the geometric cis form of an unsaturated fatty acid or a compound related to it into the trans form, resulting in an acid that is more resistant to oxidation.

elaidin reaction [ANALY CHEM] A test that differentiates nondrying oils such as olein from semidrying oils and drying oils; nitrous acid converts olein into its solid isomer, while semidrying oils in contact with nitrous acid thicken slowly, and drying oils such as tung oil become hard and resinous.

elaioplast [HISTOL] An oil-secreting leucoplast.

Elaphomycetaceae [MYCOL] A family of underground, saprophytic or mycorrhiza-forming fungi in the order Eurotiales characterized by ascocarps with thick, usually woody walls.

Elapidae [VERT ZOO] A family of poisonous reptiles, including cobras, kraits, mambas, and coral snakes; all have a pteroglyph fang arrangement.

Elasipodida [INV ZOO] An order of deep-sea aspidochirotacean holothurians in which there are no respiratory trees and bilateral symmetry is often quite conspicuous.

Elasmidae [INV ZOO] A family of hymenopteran insects in the superfamily Chalcidoidea.

Elasmobranchii [VERT ZOO] The sharks and rays, a subclass of the class Chondrichthyes distinguished by separate gill openings, amphistylic or hyostylic jaw suspension, and ampullae of Lorenzini in the head region.

Elassomatidae [VERT ZOO] The pygmy sunfishes, a family of the order Perciformes.

elastance [ELEC] The reciprocal of capacitance.

elastase [BIOCHEM] An enzyme which acts on elastin to change it chemically and render it soluble.

elastic [MECH] Capable of sustaining deformation without permanent loss of size or shape.

elastica [MECH] The elastic curve formed by a uniform rod that is originally straight, then is bent in a principal plane by applying forces, and couples only at its ends.

elastic aftereffect [MECH] The delay of certain substances in regaining their original shape after being deformed within their elastic limits. Also known as elastic lag.

elastic axis [MECH] The lengthwise line of a beam along which transverse loads must be applied in order to produce bending only, with no torsion of the beam at any section.

elastic body [MECH] A solid body for which the additional deformation produced by an increment of stress completely disappears when the increment is removed. Also known as elastic solid.

elastic buckling [MECH] An abrupt increase in the lateral deflection of a column at a critical load while the stresses acting on the column are wholly elastic.

elastic cartilage [HISTOL] A type of cartilage containing elastic fibers in the matrix.

elastic center [MECH] That point of a beam in the plane of the section lying midway between the flexural center and the center of twist in that section.

elastic collision [MECH] A collision in which the sum of the kinetic energies of translation of the participating systems is the same after the collision as before.

elastic constant *See* compliance constant; stiffness constant.

elastic cross section [PHYS] The cross section for an elastic collision between two particles or systems.

elastic curve [MECH] The curved shape of the longitudinal centroidal surface of a beam when the transverse loads acting on it produced wholly elastic stresses.

elastic deformation [MECH] Reversible alteration of the form or dimensions of a solid body under stress or strain.

elastic failure [MECH] Failure of a body to recover its original size and shape after a stress is removed.

elastic fiber [HISTOL] A homogeneous, fibrillar connective tissue component that is highly refractile and appears yellowish when arranged in bundles.

elastic flow [MECH] Return of a material to its original shape following deformation.

elastic force [MECH] A force arising from the deformation of

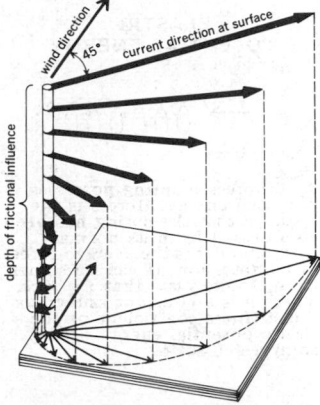

EKMAN SPIRAL

Schematic of pure wind current in deep water, showing decrease in velocity and change of direction at regular intervals of depth (the Ekman spiral).

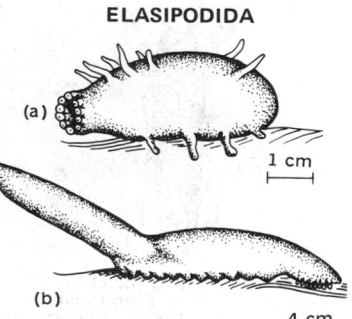

ELASIPODIDA

Two examples of bottom-dwelling Elasipodida. *(a) Elpidia. (b) Psychropotes.*

a solid body which depends only on the body's instantaneous deformation and not on its previous history, and which is conservative.

elastic hysteresis [MECH] Phenomenon exhibited by some solids in which the deformation of the solid depends not only on the stress applied to the solid but also on the previous history of this stress; analogous to magnetic hysteresis, with magnetic field strength and magnetic induction replaced by stress and strain respectively.

elasticity [MECH] **1.** The property whereby a solid material changes its shape and size under action of opposing forces, but recovers its original configuration when the forces are removed. **2.** The existence of forces which tend to restore to its original position any part of a medium (solid or fluid) which has been displaced.

elasticity modulus *See* modulus of elasticity.

elasticity number 1 [FL MECH] A dimensionless number which is a measure of the ratio of elastic forces to inertial forces on a viscoelastic fluid flowing in a pipe, and is equal to the product of the fluid's relaxation time and its dynamic viscosity, divided by the product of the fluid's density and the square of the radius of the pipe. Symbolized N_{El_1}.

elasticity number 2 [FL MECH] A dimensionless number used in studying the effect of elasticity on a flow process, equal to the fluid's density times its specific heat at constant pressure, divided by the product of its coefficient of bulk expansion and its bulk modulus. Symbolized $\bar{K}_E$.

elastic lag *See* elastic aftereffect.

elastic limit [MECH] The maximum stress a solid can sustain without undergoing permanent deformation.

elastic modulus *See* modulus of elasticity.

elasticoviscosity [FL MECH] That property of a fluid whose rate of deformation under stress is the sum of a part corresponding to a viscous Newtonian fluid and a part obeying Hooke's law.

elastic potential energy [MECH] Capacity that a body has to do work by virtue of its deformation.

elastic ratio [MECH] The ratio of the elastic limit to the ultimate strength of a solid.

elastic rebound theory [GEOL] A theory which attributes faulting to stresses (in the form of potential energy) which are being built up in the earth and which, at discrete intervals, are suddenly released as elastic energy; at the time of rupture the rocks on either side of the fault spring back to a position of little or no strain.

elastic recovery [MECH] That fraction of a given deformation of a solid which behaves elastically.

elastic scattering [MECH] Scattering due to an elastic collision.

elastic solid *See* elastic body.

elastic strain energy [MECH] The work done in deforming a solid within its elastic limit.

elastic theory [MECH] Theory of the relations between the forces acting on a body and the resulting changes in dimensions.

elastic tissue [HISTOL] A type of connective tissue having a predominance of elastic fibers, bands, or lamellae.

elastic wave [PHYS] A wave propagated by a medium having inertia and elasticity (the existence of forces which tend to restore any part of a medium to its original position), in which displaced particles transfer momentum to adjoining particles, and are themselves restored to their original position.

elastin [BIOCHEM] An elastic protein composing the principal component of elastic fibers.

elastomer [MATER] A material, such as a synthetic rubber or plastic, which at room temperature can be stretched under low stress to at least twice its original length and, upon immediate release of the stress, will return with force to its approximate original length.

elastoplasticity [MECH] State of a substance subjected to a stress greater than its elastic limit but not so great as to cause it to rupture, in which it exhibits both elastic and plastic properties.

elastoresistance [ELEC] The change in a material's electrical resistance as it undergoes a stress within its elastic limit.

elastosis [MED] **1.** Retrogressive change in elastic tissue. **2.** Retrogressive change in cutaneous connective tissue result-ing in excessive amounts of material which give the staining reactions for elastin.

elater [BOT] A spiral, filamentous structure that functions in the dispersion of spores in certain plants, such as liverworts and slime molds.

Elateridae [INV ZOO] The click beetles, a large family of coleopteran insects in the superfamily Elateroidea; many have light-producing organs.

elaterite [GEOL] A light-brown to black asphaltic pyrobitumen that is moderately soft and elastic. Also known as elastic bitumen; mineral caoutchouc.

Elateroidea [INV ZOO] A superfamily of coleopteran insects in the suborder Polyphaga.

elaterophore [BOT] A tissue bearing elaters, found in some liverworts.

E layer [GEOPHYS] A layer of ionized air occurring at altitudes between 100 and 120 kilometers in the E region of the ionosphere, capable of bending radio waves back to earth. Also known as Heaviside layer; Kennelly-Heaviside layer.

elbow [ANAT] The arm joint formed at the junction of the humerus, radius, and ulna. [DES ENG] **1.** A fitting that connects two pipes at an angle, often of 90°. **2.** A sharp corner in a pipe. [ELECTROMAG] In a waveguide, a bend of comparatively short radius, normally 90°, and sometimes for acute angles down to 15°. [GEOGR] A sharp change in direction of a coast line, channel, bank, or so on.

elbow meter [ENG] Pipe elbow used as a liquids flowmeter; flow rate is measured by determining the differential pressure developed between the inner and outer radii of the bend by means of two pressure taps located midway on the bend.

Elbs reaction [ORG CHEM] The formation of anthracene derivatives by dehydration and cyclization of diaryl ketone compounds which have a methyl group or methylene group; heating to an elevated temperature is usually required.

elective culture [MICROBIO] A type of microorganism grown selectively from a mixed culture by culturing in a medium and under conditions selective for only one type of organism.

electra [NAV] A continuous-wave radio navigation aid that uses special radio beacons to provide a number (usually 24) of equisignal zones.

Electra complex [PSYCH] The female analog of the Oedipus complex, that is, the attraction and attachment of a female child to the father.

electret [ELEC] A solid dielectric possessing persistent electric polarization, by virtue of a long time constant for decay of a charge instability.

electret microphone [ENG ACOUS] A microphone consisting of an electret transducer in which the foil electret diaphragm is placed next to a perforated, ridged, metal or metal-coated backplate, and output voltage, taken between diaphragm and backplate, is proportional to the displacement of the diaphragm.

electret transducer [ELECTR] An electroacoustic or electromechanical transducer in which a foil electret, stretched out to form a diaphragm, is placed next to a metal or metal-coated plate, and motion of the diaphragm is converted to voltage between diaphragm and plate, or vice versa.

electric [ELEC] Containing, producing, arising from, or actuated by electricity; often used interchangeably with electrical.

electric accounting machine [ADP] Data-processing equipment that is predominantly electromechanical in nature, such as sorters, collectors, and tabulators. Abbreviated EAM.

electrical [ELEC] Related to or associated with electricity, but not containing it or having its properties or characteristics; often used interchangeably with electric.

electrical analog [PHYS] An electric circuit whose behavior may be described by the same mathematical equations as some physical system under study.

electrical angle [ELEC] An angle that specifies a particular instant in an alternating-current cycle or expresses the phase difference between two alternating quantities; usually expressed in electrical degrees.

electrical axis [SOLID STATE] The x axis in a quartz crystal; there are three such axes in a crystal, each parallel to one pair of opposite sides of the hexagon; all pass through and are perpendicular to the optical, or z, axis.

ELASTIC POTENTIAL ENERGY

A compressed spring possesses potential energy. Here s is the distance that the spring has been compressed from its normal length, and $\bar{f}$ is the average force of compression. It can be shown from Hooke's law that $f = \frac{1}{2}ks$, where k is a constant known as the stiffness coefficient of the elastic potential energy of the compressed spring is $\frac{1}{2}ks^2$.

ELATERIDAE

External features of the click beetle. *(From T. I. Storer and R. L. Usinger, General Zoology, 3d ed., McGraw-Hill, 1957)*

electrical blasting cap [ENG] A blasting cap ignited by electric current and not by a spark.

electrical breakdown *See* breakdown.

electrical calorimeter [ANALY CHEM] Device to measure heat evolved (from fusion or vaporization, for example); measured quantities of heat are added electrically to the sample, and the temperature rise is noted.

electrical center [ELEC] Point approximately midway between the ends of an inductor or resistor that divides the inductor or resistor into two equal electrical values.

electrical circuit theory *See* circuit theory.

electrical code [ELEC] A systematic body of rules governing the practical application and installation of electrically operated equipment and devices and electric wiring systems.

electrical conductance *See* conductance.

electrical conduction *See* conduction.

electrical conductivity *See* conductivity.

electrical conductivity analyzer [ELEC] Alternating-current, resistance-bridge device used to measure the electrical conductivity of solutions, slurries, or wet solids.

electrical degree [ELEC] A unit equal to $\frac{1}{360}$ cycle of an alternating quantity.

electrical discharge machining *See* electron discharge machining.

electrical disintegration [MET] Removing excess metal by using an electric spark in air.

electrical distance [ELECTROMAG] The distance between two points, expressed in terms of the duration of travel of an electromagnetic wave in free space between the two points.

electrical drainage [ELEC] Diversion of electric currents from subterranean pipes to prevent electrolytic corrosion.

electrical engineer [ENG] An engineer whose training includes a degree in electrical engineering from an accredited college or university (or who has comparable knowledge and experience), to prepare him for dealing with the generation, transmission, and utilization of electric energy.

electrical engineering [ENG] Engineering that deals with practical applications of electricity; generally restricted to applications involving current flow through conductors, as in motors and generators.

electrical equipment [ELEC] Apparatus, appliances, devices, wiring, fixtures, fittings, and material used as a part of or in connection with an electrical installation.

electrical equivalent [ANALY CHEM] In conductometric analyses of electrolyte solutions, an outside, calibrated current source as compared to (equivalent to) the current passing through the sample under analysis; for example, a Wheatstone-bridge balanced reading.

electrical fault *See* fault.

electrical impedance Also known as impedance. [ELEC] **1.** The total opposition that a circuit presents to an alternating current, equal to the complex ratio of the voltage to the current in complex notation. Also known as complex impedance. **2.** The ratio of the maximum voltage in an alternating-current circuit to the maximum current; equal to the magnitude of the quantity in the first definition.

electrical impedance meter [ELEC] An instrument which measures the complex ratio of voltage to current in a given circuit at a given frequency. Also known as impedance meter.

electrical instability [ELEC] A persistent condition of unwanted self-oscillation in an amplifier or other electric circuit.

electrical insulating paper *See* insulating paper.

electrical insulation *See* insulation.

electrical insulator *See* insulator.

electrical interference *See* interference.

electrical length [ELECTROMAG] The length of a conductor expressed in wavelengths, radians, or degrees.

electrical loading *See* loading.

electrical log [ENG] Recorded measurement of the conductivities and resistivities down the length of an uncased borehole; gives a complete record of the formations penetrated.

electrical logging [ENG] The recording in uncased sections of a borehole of the conductivities and resistivities of the penetrated formations; used for geological correlation of the strata and evaluation of possibly productive horizons. Also known as electrical well logging.

electrically connected [ELEC] Connected by means of a conducting path, or through a capacitor, as distinguished from connection merely through electromagnetic induction.

electrically suspended gyro [ENG] A gyroscope in which the main rotating element is suspended by an electromagnetic or an electrostatic field.

electrical measurement [ELEC] The measurement of any one of the many quantities by which electricity is characterized.

electrical noise [ELEC] Noise generated by electrical devices, for example, motors, engine ignition, power lines, and so on, and propagated to the receiving antenna direct from the noise source.

electrical oil *See* insulating oil.

electrical porcelain *See* insulation porcelain.

electrical potential energy [ELEC] Energy possessed by electric charges by virtue of their position in an electrostatic field.

electrical properties [ELEC] Properties of a substance which determine its response to an electric field, such as its dielectric constant or conductivity.

electrical prospecting [ENG] The use of downhole electrical logs to obtain subsurface information for geological analysis.

electrical resistance meter *See* resistance meter.

electrical-resistance strain gage [ENG] A vibration-measuring device consisting of a grid of fine wire cemented to the vibrating object to measure fluctuating strains.

electrical resistance thermometer *See* resistance thermometer.

electrical resistivity [ELEC] The electrical resistance offered by a material to the flow of current, times the cross-sectional area of current flow and per unit length of current path; the reciprocal of the conductivity. Also known as resistivity; specific resistance.

electrical resistor *See* resistor.

electrical resonator *See* tank circuit.

electrical steel [MET] Low carbon–iron alloy containing 0.5–5% silicon, produced in an electric-arc furnace and used for the cores of transformers, alternators, and other iron-core electric machines.

electrical storm [METEOROL] A popular term for a thunderstorm.

electrical symbol [ELEC] A simple geometrical symbol used to represent a component of a circuit in a schematic circuit diagram.

electrical system [ELEC] System of wiring, switches, relays, and other equipment associated with receiving and distributing electricity.

electrical tape *See* insulating tape.

electrical thickness [OCEANOGR] The vertical measure between the surface of an ocean current and an isokinetic point having a value of about one-tenth the surface speed.

electrical transcription *See* transcription.

electrical unit [ELEC] A standard in terms of which some electrical quantity is evaluated.

electrical weighing system [ENG] An instrument which weighs an object by measuring the change in resistance caused by the elastic deformation of a mechanical element loaded with the object.

electrical well logging *See* electrical logging.

electrical zero [ELEC] A standard reference position from which rotor angles are measured in synchros and other rotating devices.

electric anesthesia [MED] Anesthesia produced by electrical means, as with interrupted direct current.

electric arc [ELEC] A discharge of electricity through a gas, normally characterized by a voltage drop approximately equal to the ionization potential of the gas. Also known as arc.

electric-arc furnace *See* arc furnace.

electric-arc heating *See* arc heating.

electric-arc lamp *See* arc lamp.

electric-arc welding [MET] Welding in which the joint is heated to fusion by an electric arc or by a large electric current. Also known as arc welding.

electric battery *See* battery.

electric boiler [MECH ENG] A steam generator using electric

ELECTRICAL
IMPEDANCE METER

The Madsen impedance meter uses an acoustic bridge as the basic component for testing deafness.

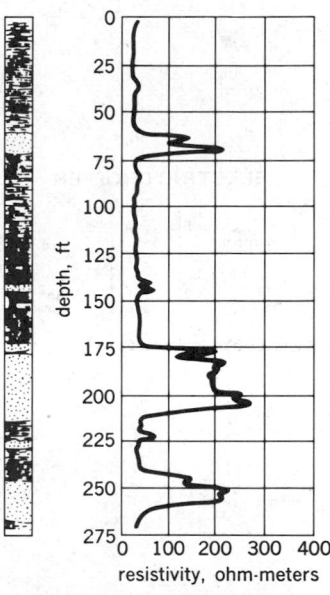

ELECTRICAL LOG

Earth resistivity measurements in a well penetrating a sequence of sandy and shaley beds lying above the permanent water table. The shaley beds are fully saturated with water, while the sandstone is partially desaturated. *(After George V. Keller and Frank C. Frischknecht, Electrical Methods in Geophysical Prospecting, Pergamon Press, 1966)*

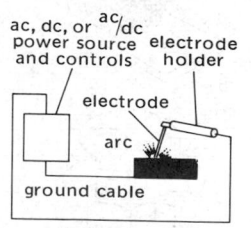

ELECTRIC-ARC WELDING

Fundamental connections of electric equipment and workpiece for arc welding.

ELECTRIC BRAKE

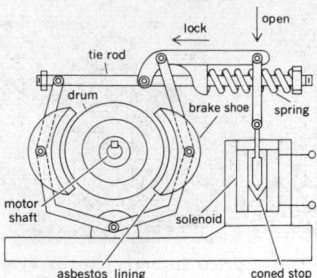

Actuating force applied to a double-block electric brake.

ELECTRIC CHOPPER

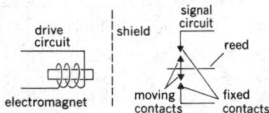

Basic chopper structure.

ELECTRIC CONNECTOR

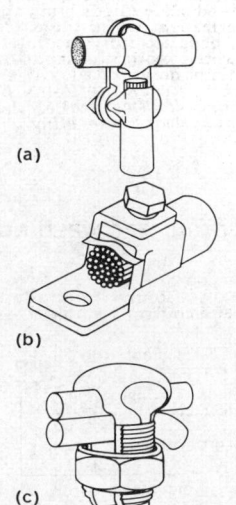

Types of electric connectors. (a) T connector. (b) Terminal connector. (c) Split-bolt connector.

energy, in immersion, resistor, or electrode elements, as the source of heat.

electric brake [MECH ENG] An actuator in which the actuating force is supplied by current flowing through a solenoid, or through an electromagnet which is thereby attracted to disks on the rotating member, actuating the brake shoes; this force is counteracted by the force of a compression spring. Also known as electromagnetic brake.

electric bridge *See* bridge.

electric burn [MED] A burn caused by electric current.

electric calamine *See* hemimorphite.

electric cell *See* cell.

electric charge *See* charge.

electric chopper [ELECTROMAG] A chopper in which an electromagnet driven by a source of alternating current sets into vibration a reed carrying a moving contact that alternately touches two fixed contacts in a signal circuit, thus periodically interrupting the signal.

electric circuit [ELEC] Also known as circuit. **1.** A path or group of interconnected paths capable of carrying electric currents. **2.** An arrangement of one or more complete, closed paths for electron flow.

electric circuit theory *See* circuit theory.

electric clock [HOROL] **1.** Any clock that is operated by electric power. **2.** Specifically, a clock driven by an alternating-current motor whose current has a definite frequency, controlled at the generator.

electric coil *See* coil.

electric comparator [ELEC] A comparator in which movement results in a change in some electrical quantity, which is then amplified by electrical means.

electric condenser *See* capacitor.

electric conductor *See* conductor.

electric connection [ELEC] A direct wire path for current between two points in a circuit.

electric connector [ELEC] A device that joins electric conductors mechanically and electrically to other conductors and to the terminals of apparatus and equipment.

electric contact [ELEC] A physical contact that permits current flow between conducting parts. Also known as contact.

electric contactor *See* contactor.

electric control [ELEC] The control of a machine or device by switches, relays, or rheostats, as contrasted with electronic control by electron tubes or by devices that do the work of electron tubes.

electric controller [ELEC] A device that governs in some predetermined manner the electric power delivered to apparatus.

electric converter *See* synchronous converter.

electric corona *See* corona discharge.

electric coupling [MECH ENG] Magnetic-field coupling between the shafts of a driver and a driven machine.

electric current *See* current.

electric current density *See* current density.

electric current meter *See* ammeter.

electric cutout *See* cutout.

electric delay line [ELECTR] A delay line using properties of lumped or distributed capacitive and inductive elements; can be used for signal storage by recirculating information-carrying wave patterns.

electric desalting [CHEM ENG] A process to remove impurities such as inorganic salts from crude oil by settling out in an electrostatic field.

electric detonator [ENG] A detonator ignited by a fuse wire which serves to touch off the primer.

electric dipole [ELEC] A localized distribution of positive and negative electricity, without net charge, whose mean positions of positive and negative charges do not coincide.

electric dipole moment [ELEC] A quantity characteristic of a charge distribution, equal to the vector sum over the electric charges of the product of the charge and the position vector of the charge.

electric dipole transition [ATOM PHYS] A transition of an atom or nucleus from one energy state to another, in which electric dipole radiation is emitted or absorbed.

electric discharge *See* discharge.

electric-discharge lamp *See* discharge lamp.

electric-discharge tube *See* discharge tube.

electric displacement [ELEC] The electric field intensity multiplied by the permittivity. Symbolized D. Also known as dielectric displacement; dielectric flux density; displacement; electric displacement density; electric flux density; electric induction.

electric displacement density *See* electric displacement.

electric distribution system *See* distribution system.

electric double layer [PHYS CHEM] An electric double layer at a solid-liquid interface; this electric double layer is made up of ions of one charge type which are fixed to the surface of the solid and an equal number of mobile ions of the opposite charge which are distributed through the neighboring region of the liquid; in such a system the movement of liquid causes a displacement of the mobile counterions with respect to the fixed charges on the solid surface. Also known as double layer.

electric doublet *See* dipole.

electric drive [MECH ENG] A mechanism which transmits motion from one shaft to another and controls the velocity ratio of the shafts by electrical means.

electric eel [VERT ZOO] *Electrophorus electricus.* An eellike cypriniform electric fish of the family Gymnotidae.

electric energy [ELECTROMAG] **1.** Energy of electric charges by virtue of their position in an electric field. **2.** Energy of electric currents by virtue of their position in a magnetic field.

electric energy measurement [ELEC] The measurement of the integral, with respect to time, of the power in an electric circuit.

electric energy meter [ELEC] A device which measures the integral, with respect to time, of the power in an electric circuit.

electric engine [AERO ENG] A rocket engine in which the propellant is accelerated by some electric device. Also known as electric propulsion system; electric rocket.

electric eye *See* cathode-ray tuning indicator; photocell; phototube.

electric fence [ENG] A fence consisting of one or more lengths of wire energized with high-voltage, low-current pulses, and giving a warning shock when touched.

electric field [ELEC] **1.** One of the fundamental fields in nature, causing a charged body to be attracted to or repelled by other charged bodies; associated with an electromagnetic wave or a changing magnetic field. **2.** Specifically, the electric force per unit test charge.

electric field effect *See* Stark effect.

electric field intensity *See* electric field vector.

electric field strength *See* electric field vector.

electric field vector [ELEC] The force on a stationary positive charge per unit charge at a point in an electric field. Designated **E**. Also known as electric field intensity; electric field strength; electric vector.

electric filter [ELECTR] **1.** A network that transmits alternating currents of desired frequencies while substantially attenuating all other frequencies. Also known as frequency-selective device. **2.** *See* filter.

electric firing mechanism [ORD] Firing mechanism using a firing magneto, battery, or alternating-current power in circuit with an electric primer; one side of the line is connected by an insulated wire to the primer, and the other side is grounded to the frame of the weapon.

electric fish [VERT ZOO] Any of several fishes capable of producing electric discharges from an electric organ.

electric flowmeter [ELEC] Fluid-flow measurement device relying on an inductance or impedance bridge or on electrical-resistance rod elements to sense flow-rate variations.

electric flux [ELEC] **1.** The integral over a surface of the component of the electric displacement perpendicular to the surface; equal to the number of electric lines of force crossing the surface. **2.** The electric lines of force in a region.

electric flux density *See* electric displacement.

electric flux line *See* electric line of force.

electric forming [ELECTR] The process of applying electric energy to a semiconductor or other device to modify permanently its electrical characteristics.

electric furnace [ENG] A furnace which uses electricity as a source of heat.

electric-furnace steel [MET] Steel produced in an electric furnace.

electric fuse *See* fuse.

electric gathering locomotive *See* gathering motor.

electric generator *See* generator.

electric guitar [ENG ACOUS] A guitar in which a contact microphone placed under the strings picks up the acoustic vibrations for amplification and for reproduction by a loudspeaker.

electric hammer [MECH ENG] An electric-powered hammer; often used for riveting or caulking.

electric heating [ENG] Any method of converting electric energy to heat energy by resisting the free flow of electric current.

electric hygrometer [ENG] An instrument for indicating by electrical means the humidity of the ambient atmosphere; usually based on the relation between the electric conductance of a film of hygroscopic material and its moisture content.

electric hysteresis *See* ferroelectric hysteresis.

electrician [ENG] A skilled worker who installs, repairs, maintains, or operates electric equipment.

electric ignition [MECH ENG] Ignition of a charge of fuel vapor and air in an internal combustion engine by passing a high-voltage electric current between two electrodes in the combustion chamber.

electric image [ELEC] A fictitious charge used in finding the electric field set up by fixed electric charges in the neighborhood of a conductor; the conductor, with its distribution of induced surface charges, is replaced by one or more of these fictitious charges. Also known as image.

electric induction *See* electric displacement.

electric instrument [ENG] An electricity-measuring device that indicates, such as an ammeter or voltmeter, in contrast to an electric meter that totalizes or records.

electricity [PHYS] Physical phenomenon involving electric charges and their effects when at rest and when in motion.

electric lamp [ELEC] A lamp in which light is produced by electricity, as the incandescent lamp, arc lamp, glow lamp, mercury-vapor lamp, and fluorescent lamp.

electric line of force [ELEC] An imaginary line drawn so that each segment of the line is parallel to the direction of the electric field or of the electric displacement at that point, and the density of the set of lines is proportional to the electric field or electrical displacement. Also known as electric flux line.

electric locomotive [MECH ENG] A locomotive operated by electric power picked up from a system of continuous overhead wires, or, sometimes, from a third rail mounted alongside the track.

electric main *See* power transmission line.

electric meter [ENG] An electricity-measuring device that totalizes with time, such as a watthour meter or ampere-hour meter, in contrast to an electric instrument.

electric moment [ELEC] One of a series of quantities characterizing an electric charge distribution; an l-th moment is given by integrating the product of the charge density, the l-th power of the distance from the origin, and a spherical harmonic Y^*_{lm} over the charge distribution.

electric monopole [ELEC] A distribution of electric charge which is concentrated at a point or is spherically symmetric.

electric motor *See* motor.

electric multipole [ELECTROMAG] One of a series of types of static or oscillating charge distributions; the multipole of order 1 is a point charge or a spherically symmetric distribution, and the electric and magnetic fields produced by an electric multipole of order 2^n are equivalent to those of two electric multipoles of order 2^{n-1} of equal strengths, but opposite sign, separated from each other by a short distance.

electric multipole field [ELECTROMAG] The electric and magnetic fields generated by a static or oscillating electric multipole.

electric network *See* network.

electric octupole moment [ELEC] A quantity characterizing an electric charge distribution; obtained by integrating the product of the charge density, the third power of the distance from the origin, and a spherical harmonic Y^*_{3m} over the charge distribution.

electric organ [VERT ZOO] An organ consisting of rows of electroplaques which produce an electric discharge.

electric outlet *See* outlet.

electric polarizability [ELEC] Induced dipole moment of an atom or molecule in a unit electric field.

electric polarization *See* polarization.

electric potential [ELEC] The work which must be done against electric forces to bring a unit charge from a reference point to the point in question; the reference point is located at an infinite distance, or, for practical purposes, at the surface of the earth or some other large conductor. Also known as electrostatic potential; potential.

electric power [ELEC] The rate at which electric energy is converted to other forms of energy, equal to the product of the current and the voltage drop.

electric power generation [MECH ENG] The large-scale production of electric power for industrial, residential, and rural use, generally in stationary plants designed for that purpose.

electric power line *See* power line.

electric power meter [ENG] A device that measures electric power consumed, either at an instant, as in a wattmeter, or averaged over a time interval, as in a demand meter. Also known as power meter.

electric power plant [MECH ENG] A power plant that converts a form of raw energy into electricity, for example, a hydro, steam, diesel, or nuclear generating station for stationary or transportation service.

electric power station [ELEC] A generating station or an electric power substation.

electric power substation [ELEC] An assembly of equipment in an electric power system through which electric energy is passed for transmission, transformation, distribution, or switching. Also known as substation.

electric power system [MECH ENG] A complex assemblage of equipment and circuits for generating, transmitting, transforming, and distributing electric energy.

electric power transmission [ELEC] Process of transferring electric energy from one point to another in an electric power system.

electric precipitation [CHEM ENG] A process that utilizes an electric field to improve the separation of hydrocarbon reagent dispersions.

electric pressure transducer *See* pressure transducer.

electric probe [PL PHYS] A device used to measure electron temperatures, electron and ion densities, space and wall potentials, and random electron currents in a plasma; consists substantially of one or two small collecting electrodes to which various potentials are applied, with the corresponding collection currents being measured. Also known as electrostatic probe.

electric propulsion [AERO ENG] A general term encompassing all the various types of propulsion in which the propellant consists of electrically charged particles which are accelerated by electric or magnetic fields, or both.

electric propulsion system *See* electric engine.

electric protective device [ELEC] A particular type of equipment used in electric power systems to detect abnormal conditions and to initiate appropriate corrective action. Also known as protective device.

electric quadrupole [ELEC] A charge distribution that produces an electric field equivalent to that produced by two electric dipoles whose dipole moments have the same magnitude but point in opposite directions and which are separated from each other by a small distance.

electric quadrupole lens [ELECTR] A device for focusing beams of charged particles which has four electrodes with alternately positive and negative polarity; used in electron microscopes and particle accelerators.

electric quadrupole moment [ELEC] A quantity characterizing an electric charge distribution, obtained by integrating the product of the charge density, the second power of the distance from the origin, and a spherical harmonic Y^*_{2m} over the charge distribution.

electric quadrupole transition [ATOM PHYS] A transition of an atom or molecule from one energy state to another, in

ELECTRIC IMAGE

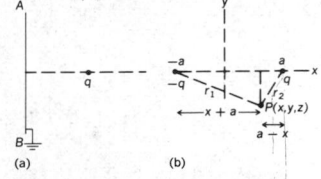

Illustration of the method of electric images. (a) Point charge q at distance a in front of an infinite grounded conducting plane AB. The problem is to find the potential function and equation for electric field intensity in the region to the right of AB, the field being caused by point charge q, and its equal distributed induced charge on the conducting plane. (b) Physically simpler but mathematically equivalent situation for the region of interest. Infinite conducting plane is replaced by point charge $-q$. Potential at P with coordinates (x,y,z) is that resulting from charges $-q$ and q at distances r_1 and r_2 respectively.

ELECTRIC POWER GENERATION

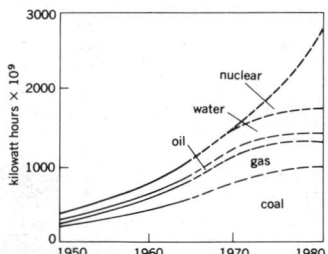

Energy sources for electric power generation in the United States. (USAEC)

ELECTRIC PROBE

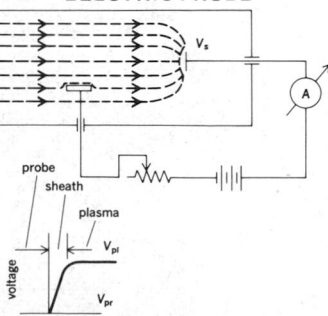

Schematic diagram of a single electric probe.

which electric quadrupole radiation is emitted or absorbed.

electric raceway *See* raceway.

electric railroad [MECH ENG] A railroad which has a system of continuous overhead wires or a third rail mounted alongside the track to supply electric power to the locomotive and cars.

electric reactor *See* reactor.

electric relay *See* relay.

electric resistance *See* resistance.

electric resistance furnace *See* resistance furnace.

electric rocket *See* electric engine.

electric rotating machinery [ELEC] Any form of apparatus which has a rotating member and generates, converts, transforms, or modifies electric power, such as a motor, generator, or synchronous converter.

electric scanning [ELECTR] Scanning in which the required changes in radar beam direction are produced by variations in phase or amplitude of the currents fed to the various elements of the antenna array.

electric shielding [ELECTROMAG] Any means of avoiding pickup of undesired signals or noise, suppressing radiation of undesired signals, or confining wanted signals to desired paths or regions, such as electrostatic shielding or electromagnetic shielding. Also known as screening; shielding.

electric shock [PHYSIO] The sudden pain, convulsion, unconsciousness, or death produced by the passage of electric current through the body.

electric shock tube [PL PHYS] A gas-filled tube used in plasma physics to ionize a gas suddenly; a capacitor bank charged to a high voltage is discharged into the gas at one tube end to ionize and heat the gas, producing a shock wave that may be studied as it travels down the tube.

electric shunt *See* shunt.

electric solenoid *See* solenoid.

electric spark *See* spark.

electric spark machining *See* electron discharge machining.

electric stacker [MECH ENG] A stacker whose carriage is raised and lowered by a winch powered by electric storage batteries.

electric strength *See* dielectric strength.

electric surface-recording thermometer [PETRO ENG] Device to measure temperatures during oil-well temperature surveying; has a thermocouple, resistance wire, or thermistor as the temperature-sensitive element.

electric susceptibility [ELEC] A dimensionless parameter measuring the ease of polarization of a dielectric, equal (in meter-kilogram-second units) to the ratio of the polarization to the product of the electric field strength and the vacuum permittivity. Also known as dielectric susceptibility.

electric switch *See* switch.

electric switchboard *See* switchboard.

electric tachometer [ENG] An instrument for measuring rotational speed by measuring the output voltage of a generator driven by the rotating unit.

electric tank *See* electrolytic tank.

electric telemetering [COMMUN] System to transmit electric impulses from the primary detector to a remote receiving station, with or without wire interconnections.

electric terminal *See* terminal.

electric thermometer [ENG] An instrument that utilizes electrical means to measure temperature, such as a thermocouple or resistance thermometer.

electric transducer [ELECTR] A transducer in which all of the waves are electric.

electric transient [ELEC] A temporary component of current and voltage in an electric circuit which has been disturbed.

electric tuning [ELECTR] Tuning a receiver to a desired station by switching a set of preadjusted trimmer capacitors or coils into the tuning circuits.

electric twinning [SOLID STATE] A defect occurring in natural quartz crystals, in which adjacent regions of quartz have their electric axes oppositely poled.

electric typewriter [MECH ENG] A typewriter having an electric motor that provides power for all operations initiated by the touching of the keys.

electric vector *See* electric field vector.

electric vehicle [MECH ENG] Any ground vehicle whose original source of energy is electric power, such as an electric car or electric locomotive.

electric wave [ELECTROMAG] An electromagnetic wave, especially one whose wavelength is at least a few centimeters. Also known as Hertzian wave.

electric-wave filter *See* filter.

electric wind *See* convective discharge.

electric wire *See* wire.

electric wiring *See* wiring.

electrification [ELEC] **1.** The process of establishing a charge in an object. **2.** The generation, distribution, and utilization of electricity.

electrification ice nucleus [METEOROL] An ice nucleus that is formed by the fragmentation of dendritic crystals exposed to an electric field strength of several hundred volts per centimeter; it is a type of fragmentation nucleus.

electrization [ELEC] The electric polarization divided by the permittivity of empty space.

electroacoustics [ENG ACOUS] The conversion of acoustic energy and waves into electric energy and waves, or vice versa.

electroacoustic transducer [ENG ACOUS] A transducer that receives waves from an electric system and delivers waves to an acoustic system, or vice versa. Also known as sound transducer.

electrobalance [ANALY CHEM] Analytical microbalance utilizing electromagnetic weighing; the sample weight is balanced by the torque produced by current in a coil in a magnetic field, with torque proportional to the current.

electrocapillarity [PHYS] A change in the surface tension of a liquid caused by an electric field at the surface.

electrocardiogram [MED] A graphic recording of the electrical manifestations of the heart action as obtained from the body surfaces. Abbreviated ECG; EKG.

electrocardiograph [MED] The instrument used to obtain an electrocardiogram.

electrocardiography [MED] The medical specialty concerned with the production and interpretation of electrocardiograms.

electrocauterization [MED] The application of a direct galvanic current to tissues to cause destruction or coagulation.

electroceramics [MATER] Ceramic materials having electrical and other properties which make them suitable for use as insulators for power lines and in electrical components.

electrochemical cell [PHYS CHEM] A combination of two electrodes arranged so that an overall oxidation-reduction reaction produces an electromotive force; includes dry cells, wet cells, standard cells, fuel cells, solid-electrolyte cells, and reserve cells.

electrochemical cleaning [MET] Removing soil by the chemical action caused or sustained by a current of electricity in an electrolyte. Also known as electrolytic cleaning.

electrochemical coating [MET] A coating formed by chemical action on the metal surface and effected by a current of electricity through an electrolyte.

electrochemical corrosion [MET] Corrosion of a metal associated with the flow of electric current in an electrolyte. Also known as electrolytic corrosion.

electrochemical effect [PHYS CHEM] Conversion of chemical to electric energy, as in electrochemical cells; or the reverse process, used to produce elemental aluminum, magnesium, and bromine from compounds of these elements.

electrochemical emf [PHYS CHEM] Electrical force generated by means of chemical action, in man-made cells (such as dry batteries) or by natural means (galvanic reaction).

electrochemical equivalent [PHYS CHEM] The weight in grams of a substance produced or consumed by electrolysis with 100% current efficiency during the flow of a quantity of electricity equal to 1 faraday ($96,487.0 \pm 1.6$ coulombs).

electrochemical machining [MET] Removing excess metal by electrolytic dissolution, effected by the tool acting as the cathode against the workpiece acting as the anode. Abbreviated ECM. Also known as electrolytic machining.

electrochemical potential [PHYS CHEM] The difference in potential that exists when two dissimilar electrodes are connected through an external conducting circuit and the two electrodes are placed in a conducting solution so that electrochemical reactions occur.

electrochemical power generation [ENG] The direct conversion of chemical energy to electric energy, as in a battery or fuel cell.

electrochemical process [PHYS CHEM] **1.** A chemical change accompanying the passage of an electric current, especially as used in the preparation of commercially important quantities of certain chemical substances. **2.** The reverse change, in which a chemical reaction is used as the source of energy to produce an electric current, as in a battery.

electrochemical recording [ELECTR] Recording by means of a chemical reaction brought about by the passage of signal-controlled current through the sensitized portion of the record sheet.

electrochemical reduction cell [PHYS CHEM] The cathode component of an electrochemical cell, at which chemical reduction occurs (while at the anode, chemical oxidation occurs).

electrochemical series [PHYS CHEM] A series in which the metals and other substances are listed in the order of their chemical reactivity or electrode potentials, the most reactive at the top and the less reactive at the bottom. Also known as electromotive series.

electrochemical techniques [PHYS CHEM] The experimental methods developed to study the physical and chemical phenomena associated with electron transfer at the interface of an electrode and solution.

electrochemical thermodynamics [THERMO] The application of the laws of thermodynamics to electrochemical systems.

electrochemical valve [ELEC] Electric valve consisting of a metal in contact with a solution or compound, across the boundary of which current flows more readily in one direction than in the other direction, and in which the valve action is accompanied by chemical changes.

electrochemistry [PHYS CHEM] A branch of chemistry dealing with chemical changes accompanying the passage of an electric current; or with the reverse process, in which a chemical reaction is used to produce an electric current.

electrochromatography [ANALY CHEM] Type of chromatography that utilizes application of an electric potential to produce an electric differential. Also known as electropherography.

electrocoagulation [MED] The coagulation of tissue by means of a high-frequency electric current.

electroconvulsive shock [MED] The technique of eliciting convulsions by applying an electric current through the brain for a brief period.

electrocyclic reaction [PHYS CHEM] The interconversion of a linear π-system containing n π-electrons and a cyclic molecule containing $n - 2$ π-electrons which is formed by joining the ends of the linear molecule.

electrode [ELEC] **1.** An electric conductor through which an electric current enters or leaves a medium, whether it be an electrolytic solution, solid, molten mass, gas, or vacuum. **2.** One of the terminals used in dielectric heating or diathermy for applying the high-frequency electric field to the material being heated.

electrode admittance [ELECTR] Quotient of dividing the alternating component of the electrode current by the alternating component of the electrode voltage, all other electrode voltages being maintained constant.

electrode capacitance [ELECTR] Capacitance between one electrode and all the other electrodes connected together.

electrode characteristic [ELECTR] Relation between the electrode voltage and the current to an electrode, all other electrode voltages being maintained constant.

electrode conductance [ELECTR] Quotient of the inphase component of the electrode alternating current by the electrode alternating voltage, all other electrode voltage being maintained constant; this is a variational and not a total conductance. Also known as grid conductance.

electrode couple [ELEC] The pair of electrodes in an electric cell, between which there is a potential difference.

electrode current [ELECTR] Current passing to or from an electrode, through the interelectrode space within a vacuum tube.

electrode dark current [ELECTR] The electrode current that

flows when there is no radiant flux incident on the photocathode in a phototube or camera tube. Also known as dark current.

electrode dissipation [ELECTR] Power dissipated in the form of heat by an electrode as a result of electron or ion bombardment.

electrode drop [ELECTR] Voltage drop in the electrode due to its resistance.

electrode force [MET] The force that occurs between electrodes during seam, spot, and projection welding. Also known as welding force.

electrode impedance [ELECTR] Reciprocal of the electrode admittance.

electrode inverse current [ELECTR] Current flowing through an electrode in the direction opposite to that for which the tube is designed.

electrodeposition [MET] Electrolytic process in which a metal is deposited at the cathode from a solution of its ions; includes electroplating and electroforming. Also known as electrolytic deposition.

electrodeposition analysis [ANALY CHEM] An electroanalytical technique in which an element is quantitatively deposited on an electrode.

electrode potential Also known as electrode voltage. [ELECTR] The instantaneous voltage of an electrode with respect to the cathode of an electron tube. [PHYS CHEM] The voltage existing between an electrode and the solution or electrolyte in which it is immersed; usually, electrode potentials are referred to a standard electrode, such as the hydrogen electrode.

electrode resistance [ELECTR] Reciprocal of the electrode conductance; this is the effective parallel resistance and is not the real component of the electrode impedance.

electrode skid [MET] Sliding of an electrode over the work surface in spot, seam, or projection welding.

electrode-type liquid-level meter [ENG] Device that senses liquid level by the effect of the liquid-gas interface on the conductance of an electrode or probe.

electrode voltage *See* electrode potential.

electrodiagnosis [MED] Diagnosis of disease states by recording the spontaneous electrical activity of tissue or organs, or by the response to stimulation of electrically excitable tissue.

electrodialysis [PHYS CHEM] Dialysis that is conducted with the aid of an electromotive force applied to electrodes adjacent to both sides of the membrane.

electrodialyzer [PHYS CHEM] An instrument used to conduct electrodialysis.

electrodrill [MECH ENG] A drilling machine driven by electric power.

electrodynamic ammeter [ENG] Instrument which measures the current passing through a fixed coil and a movable coil connected in series by balancing the torque on the movable coil (resulting from the magnetic field of the fixed coil) against that of a spiral spring.

electrodynamic drift [GEOPHYS] Motion of ions in the upper atmosphere due to the electric or geomagnetic field.

electrodynamic instrument [ENG] An instrument that depends for its operation on the reaction between the current in one or more movable coils and the current in one or more fixed coils. Also known as electrodynamometer.

electrodynamic loudspeaker [ENG ACOUS] Dynamic loudspeaker in which the magnetic field is produced by an electromagnet, called the field coil, to which a direct current must be furnished.

electrodynamic machine [ELEC] An electric generator or motor in which the output load current is produced by magnetomotive currents generated in a rotating armature.

electrodynamics [ELECTROMAG] The study of the relations between electrical, magnetic, and mechanical phenomena.

electrodynamic shaker *See* shaker.

electrodynamic wattmeter [ENG] An electrodynamic instrument connected as a wattmeter, with the main current flowing through the fixed coil, and a small current proportional to the voltage flowing through the movable coil. Also known as moving-coil wattmeter.

electrodynamometer *See* electrodynamic instrument.

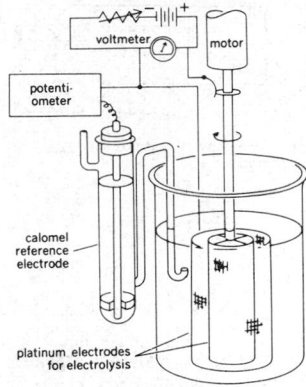

ELECTRODEPOSITION ANALYSIS

Diagram of the electrodeposition apparatus. *(From G. W. Ewing, Instrumental Methods of Chemical Analysis, 2d ed., McGraw-Hill, 1960)*

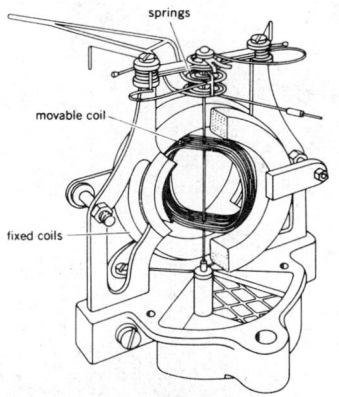

ELECTRODYNAMIC INSTRUMENT

Electrodynamic instrument mechanism. *(Weston Instruments, Inc.)*

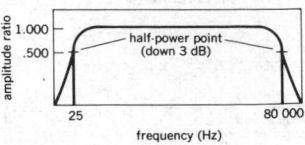

ELECTROKINETIC TRANSDUCER

Typical response curve of unit-cell transducer.

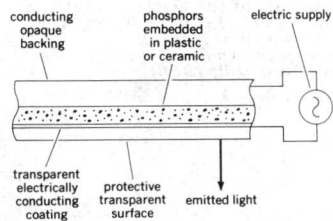

ELECTROLUMINESCENT PANEL

Simplified diagram of electroluminescent panel, not drawn to scale.

ELECTROLYSIS

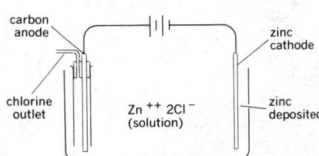

Electrolysis of zinc chloride solution.

electroencephalogram [MED] A graphic recording of the electric discharges of the cerebral cortex as detected by electrodes on the surface of the scalp. Abbreviated EEG.

electroencephalograph [MED] An instrument used to make electroencephalograms.

electroencephalography [MED] The medical specialty concerned with the production and interpretation of electroencephalograms.

electroendosmosis [PHYS] The production of an endosmosis effect by an electrical potential; that is, the use of electricity to cause diffusion of a liquid through an organic membrane.

electroerosive machining See electron discharge machining.

electroexplosive [ENG] An initiator or a system in which an electric impulse initiates detonation or deflagration of an explosive. [MATER] The explosive substance so detonated or deflagrated.

Electrofax [GRAPHICS] Trademark for an electrostatic reproduction process in which an image is formed on a zinc oxide-coated paper after it has been given a negative electrostatic charge.

electrofiltration [GEOL] Counterprocess during electrical logging of well boreholes, in which mud filtrate forced through the mud cake produces an emf in the mud cake opposite a permeable bed, positive in the direction of filtrate flow.

electrofluid [FL MECH] Newtonian (or shear-thinning) fluid whose rheological or flow properties are changed into those of a viscoplastic type by the addition of electric-field modulation.

electroformed mold [MET] A mold made by electroplating metal on the reverse pattern on the cavity; molten steel may be then sprayed on the back of the mold to increase its strength.

electroforming [MET] Shaping components by electrodeposition of the metal on a pattern.

electrogalvanizing [MET] Coating of a metal, especially iron or steel, with zinc by electroplating.

electrogram [ELECTR] A record of an image of an object made by sparking, usually on paper. [METEOROL] A record, usually automatically produced, showing the time variations of the atmospheric electric field at a given point. [PHYSIO] The graphic representation of electric events in living tissues; commonly, an electrocardiogram or electroencephalogram.

electrograph [COMMUN] Facsimile transmission equipment. [ENG] Any plot, graph, or tracing produced by the action of an electric current on prepared sensitized paper (or other chart material) or by means of an electrically controlled stylus or pen.

electrographic recording [GRAPHICS] Type of electrography in which the electrostatic image is formed by one or more rows of closely spaced parallel wires to which voltages are applied at appropriate instants to form the desired image charge pattern.

electrography [GRAPHICS] The branch of electrostatography in which electrostatic images are formed on an insulating medium without the aid of electromagnetic radiation.

electrogravimetry [ANALY CHEM] Electrodeposition analysis in which the quantities of metals deposited may be determined by weighing a suitable electrode before and after deposition.

electrojet [GEOPHYS] A stream of electricity moving in the upper atmosphere around the equator and in polar regions, where it produces auroras.

electrokinetic phenomena [PHYS CHEM] The phenomena associated with movement of charged particles through a continuous medium or with the movement of a continuous medium over a charged surface.

electrokinetic potential See zeta potential.

electrokinetics [ELECTROMAG] The study of the motion of electric charges, especially of steady currents in electric circuits, and of the motion of electrified particles in electric or magnetic fields.

electrokinetic transducer [ELEC] An instrument which converts dynamic physical forces, such as vibration and sound, into corresponding electric signals by measuring the streaming potential generated by passage of a polar fluid through a

permeable refractory-ceramic or fritted-glass member between two chambers.

electrokinetograph [ENG] An instrument used to measure ocean current velocities based on their electrical effects in the magnetic field of the earth.

electroless plating [MET] Deposition of a metal coating by immersion of a metal or nonmetal in a suitable bath containing a chemical reducing agent.

electroluminescence [ELECTR] The emission of light, not due to heating effects alone, resulting from application of an electric field to a material, usually solid.

electroluminescent cell See electroluminescent panel.

electroluminescent display [ELECTR] A display in which various combinations of electroluminescent segments may be activated by applying voltages to produce any desired numeral or other character.

electroluminescent lamp See electroluminescent panel.

electroluminescent panel [ELECTR] A surface-area light source employing the principle of electroluminescence; consists of a suitable phosphor placed between sheet-metal electrodes, one of which is essentially transparent, with an alternating current applied between the electrodes. Also known as electroluminescent cell; electroluminescent lamp; light panel; luminescent cell.

electroluminescent phosphor [MATER] Zinc sulfide powder, with small additions of copper or manganese, which emits light when suspended in an insulator in an intense alternating electric field. Also known as electroluminor.

electroluminor See electroluminescent phosphor.

electrolysis [PHYS CHEM] A method by which chemical reactions are carried out by passage of electric current through a solution of an electrolyte or through a molten salt.

electrolyte [PHYS CHEM] A chemical compound which when molten or dissolved in certain solvents, usually water, will conduct an electric current.

electrolyte acid See sulfuric acid.

electrolyte-activated battery [ELEC] A reserve battery in which an aqueous electrolyte is stored in a separate chamber, and a mechanism, which may be operated from a remote location, drives the electrolyte out of the reservoir and into the cells of the battery for activation.

electrolytic analysis [ANALY CHEM] Basic electrochemical technique for quantitative analysis of conducting solutions containing oxidizable or reducible material; measurement is based on the weight of material plated out onto the electrode.

electrolytic arrester See aluminum-cell arrester.

electrolytic brightening See electropolishing.

electrolytic capacitor [ELEC] A capacitor consisting of two electrodes separated by an electrolyte; a dielectric film, usually a thin layer of gas, is formed on the surface of one electrode. Also known as electrolytic condenser.

electrolytic cell [PHYS CHEM] A cell consisting of electrodes immersed in an electrolyte solution, for carrying out electrolysis.

electrolytic cleaning See electrochemical cleaning.

electrolytic condenser See electrolytic capacitor.

electrolytic conductance [PHYS CHEM] The transport of electric charges, under electric potential differences, by charged particles (called ions) of atomic or larger size.

electrolytic conductivity [PHYS CHEM] The conductivity of a medium in which the transport of electric charges, under electric potential differences, is by particles of atomic or larger size.

electrolytic copper [MET] Metallic copper produced by electrochemical deposition from a copper ion-containing electrolyte.

electrolytic corrosion See electrochemical corrosion.

electrolytic deposition See electrodeposition.

electrolytic dissociation [CHEM] The ionization of a compound in a solution.

electrolytic etching [MET] Engraving the surface of a metal by electrolysis.

electrolytic grinding [MECH ENG] A combined grinding and machining operation in which the abrasive, cathodic grinding wheel is in contact with the anodic workpiece beneath the surface of an electrolyte.

electrolytic interrupter [ELEC] An interrupter that consists of two electrodes in an electrolytic solution; bubbles formed in the solution continually interrupt the passage of current between the electrodes.

electrolytic mercaptan process [CHEM ENG] A process in which an aqueous caustic solution is used to extract mercaptans from refinery streams.

electrolytic model [PETRO ENG] Laboratory simulation of steady-state fluid flow through porous reservoir mediums; depends on the mobility of ions in absorbent mediums (gelatin or blotter), or through a liquid (potentiometric technique). Also known as gelatin model; oil-field model; potentiometric model.

electrolytic photocopying [GRAPHICS] A photocopying process in which an image is projected on a sheet consisting of a paper support, a thin aluminum laminate, and a coating of a white photoconductive substance in contact with an electrolyte, and electrolysis takes place in the exposed areas when a direct current is applied across the electrolyte and the aluminum underlayer.

electrolytic pickling [MET] Removal of metal by electrolysis using the metal as an electrode in a suitable electrolyte.

electrolytic polishing See electropolishing.

electrolytic potential [PHYS CHEM] Difference in potential between an electrode and the immediately adjacent electrolyte, expressed in terms of some standard electrode difference.

electrolytic powder [MET] Metal powder produced directly or indirectly by electrodeposition.

electrolytic process [PHYS CHEM] An electrochemical process involving the principles of electrolysis, especially as relating to the separation and deposition of metals.

electrolytic protection See cathodic protection.

electrolytic recording [ELECTR] Electrochemical recording in which the chemical change is made possible by the presence of an electrolyte.

electrolytic rectifier [ELEC] A rectifier consisting of metal electrodes in an electrolyte, in which rectification of alternating current is accompanied by electrolytic action; polarizing film formed on one electrode permits current flow in one direction but not the other.

electrolytic refining See electrorefining.

electrolytic rheostat [ELEC] A rheostat that consists of a tank of conducting liquid in which electrodes are placed, and resistance is varied by changing the distance between the electrodes, the depth of immersion of the electrodes, or the resistivity of the solution. Also known as water rheostat.

electrolytic solution [PHYS CHEM] A solution made up of a solvent and an ionically dissociated solute; it will conduct electricity, and ions can be separated from the solution by deposition on an electrically charged electrode.

electrolytic strip See humidity strip.

electrolytic switch [ELEC] A switch having two electrodes projecting into a chamber partly filled with electrolyte, leaving an air bubble of predetermined width; the bubble shifts position and changes the amount of electrolyte in contact with the electrodes when the switch is tilted from true horizontal.

electrolytic tank [ENG] A tank in which voltages are applied to an enlarged scale model of an electron-tube system or a reduced scale model of an aerodynamic system immersed in a poorly conducting liquid, and equipotential lines between electrodes are traced; used as an aid to electron-tube design or in computing ideal fluid flow; the latter application is based on the fact that the velocity potential in ideal flow and the stream function in planar flow satisfy the same equation, Laplace's equation, as an electrostatic potential. Also known as electric tank; potential flow analyzer.

electrolytic tough pitch [MET] Copper which has been refined electrolytically, containing mostly oxygen as an impurity.

electromachining [MECH ENG] The application of electric or ultrasonic energy to a workpiece to effect removal of material.

electromagnet [ELECTROMAG] A magnet consisting of a coil wound around a soft iron or steel core; the core is strongly magnetized when current flows through the coil, and is almost completely demagnetized when the current is interrupted.

electromagnetic [PHYS] Pertaining to phenomena in which electricity and magnetism are related.

electromagnetic amplifying lens [ELECTROMAG] Large numbers of waveguides symmetrically arranged with respect to an excitation medium in order to become excited with equal amplitude and phase to provide a net gain in energy.

electromagnetic brake See electric brake.

electromagnetic cathode-ray tube [ELECTR] A cathode-ray tube in which electromagnetic deflection is used on the electron beam.

electromagnetic clutch [MECH ENG] A clutch based on magnetic coupling between conductors, such as a magnetic fluid and powder clutch, an eddy-current clutch, or a hysteresis clutch.

electromagnetic compatibility [ELECTR] The capability of electronic equipment or systems to be operated in the intended electromagnetic environment at design levels of efficiency.

electromagnetic complex [ELECTROMAG] Electromagnetic configuration of an installation, including all significant radiators of energy.

electromagnetic constant See speed of light.

electromagnetic coupling [ELECTROMAG] Coupling that exists between circuits when they are mutually affected by the same electromagnetic field.

electromagnetic crack detector [MET] An instrument that detects cracks in iron or steel objects by applying a strong magnetizing force and measuring the resulting magnetic flux through the object.

electromagnetic current [ELECTR] Motion of charged particles (for example, in the ionosphere) giving rise to electric and magnetic fields.

electromagnetic deflection [ELECTR] Deflection of an electron stream by means of a magnetic field.

electromagnetic energy [ELECTROMAG] The energy associated with electric or magnetic fields.

electromagnetic environment [COMMUN] Radio-frequency fields existing in a given area.

electromagnetic field [ELECTROMAG] An electric or magnetic field, or a combination of the two, as in an electromagnetic wave.

electromagnetic field equations See Maxwell field equations.

electromagnetic field tensor [ELECTROMAG] An antisymmetric, second-rank Lorentz tensor, whose elements are proportional to the electric and magnetic fields; the Maxwell field equations can be expressed in a simple form in terms of this tensor.

electromagnetic flowmeter [ENG] A flowmeter that offers no obstruction to liquid flow; two coils produce an electromagnetic field in the conductive moving fluid; the current induced in the liquid, detected by two electrodes, is directly proportional to the rate of flow. Also known as electromagnetic meter.

electromagnetic focusing [ELECTR] Focusing the electron beam in a television picture tube by means of a magnetic field parallel to the beam; the field is produced by sending an adjustable value of direct current through a focusing coil mounted on the neck of the tube.

electromagnetic horn See horn antenna.

electromagnetic induction [ELECTROMAG] The production of an electromotive force either by motion of a conductor through a magnetic field so as to cut across the magnetic flux or by a change in the magnetic flux that threads a conductor. Also known as induction.

electromagnetic inertia [ELECTROMAG] **1.** Characteristic delay of a current in an electric circuit in reaching its maximum value, or in returning to zero, after the source voltage has been removed or applied. **2.** Property of self-induction.

electromagnetic interference [ELEC] Interference, generally at radio frequencies, that is generated inside systems, as contrasted to radio-frequency interference coming from sources outside a system. Abbreviated emi.

electromagnetic lens [ELECTR] An electron lens in which electron beams are focused by an electromagnetic field.

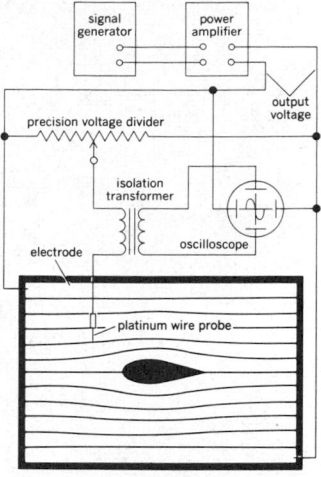

ELECTROLYTIC TANK

Setup to trace streamlines of nonlift airfoil with an electrolytic tank.

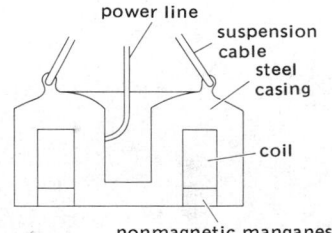

ELECTROMAGNET

Cross section of circular lifting electromagnet.

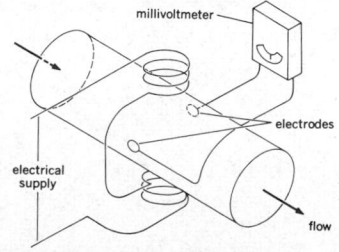

ELECTROMAGNETIC FLOWMETER

Drawing of electromagnetic flowmeter showing sensing electrodes and other components.

electromagnetic log [ENG] A log containing an electromagnetic sensing element extended below the hull of the vessel; this device produces a voltage directly proportional to speed through the water.

electromagnetic logging [ENG] A method of well logging in which a transmitting coil sets up an alternating electromagnetic field, and a receiver coil, placed in the drill hole above the transmitter coil, measures the secondary electromagnetic field induced by the resulting eddy currents within the formation. Also known as electromagnetic well logging.

electromagnetic mass [ELECTROMAG] The contribution to the mass of an object from its electric and magnetic field energy.

electromagnetic meter *See* electromagnetic flowmeter.

electromagnetic mirror [ELECTROMAG] Surface or region capable of reflecting radio waves, such as one of the ionized layers in the upper atmosphere.

electromagnetic mixing [MET] Mixing of molten alloys by exposing the melt to a strong magnetic field while passing direct current between electrodes at opposite ends of the crucible; stirring action results from interaction of the magnetic field of the current-carrying molten alloy with the external transverse magnetic field.

electromagnetic momentum [ELECTROMAG] The momentum transported by electromagnetic radiation; its volume density equals the Poynting vector divided by the square of the speed of light.

electromagnetic noise [ELEC] Noise in a communications system resulting from undesired electromagnetic radiation. Also known as radiation noise.

electromagnetic oscillograph [ELECTROMAG] An oscillograph in which the recording mechanism is controlled by a moving-coil galvanometer, such as a direct-writing recorder or a light-beam oscillograph.

electromagnetic potential [ELECTROMAG] Collective name for a scalar potential, which reduces to the electrostatic potential in a time-independent system, and the vector potential for the magnetic field; the electric and magnetic fields can be written in terms of these potentials.

electromagnetic properties [ELECTROMAG] The response of materials or equipment to electromagnetic fields, and their ability to produce such fields.

electromagnetic propulsion [AERO ENG] Motive power for flight vehicles produced by electromagnetic acceleration of a plasma fluid.

electromagnetic prospecting *See* electromagnetic surveying.

electromagnetic pulse [ELECTROMAG] The pulse of electromagnetic radiation generated by a large thermonuclear explosion.

electromagnetic pump [ELEC] A pump in which a conductive liquid is made to move through a pipe by sending a large current transversely through the liquid; this current reacts with a magnetic field that is at right angles to the pipe and to current flow, to move the current-carrying liquid conductor.

electromagnetic radiation [ELECTROMAG] Electromagnetic waves and, especially, the associated electromagnetic energy.

electromagnetic reconnaissance [ELECTR] Reconnaissance for the purpose of locating and identifying potentially hostile transmitters of electromagnetic radiation, including radar, communication, missile-guidance, and navigation-aid equipment.

electromagnetic relay [ELECTROMAG] A relay in which current flow through a coil produces a magnetic field that results in contact actuation.

electromagnetic scattering [PHYS] The process in which energy is removed from a beam of electromagnetic radiation and reemitted without appreciable changes in wavelength.

electromagnetic separator [ELECTROMAG] Device in which ions of varying mass are separated by a combination of electric and magnetic fields.

electromagnetic shielding [ELECTROMAG] Means, similar to electrostatic or magnetostatic shielding, for suppressing changing magnetic fields or electromagnetic radiation at a device.

electromagnetic shock wave [ELECTROMAG] Electromagnetic wave of great intensity which results when waves with different intensities propagate with different velocities in a nonlinear optical medium, and faster-traveling waves from a pulse of light catch up with preceding, slower traveling waves.

electromagnetic spectrum [ELECTROMAG] The total range of wavelengths or frequencies of electromagnetic radiation, extending from the longest radio waves to the shortest known cosmic rays.

electromagnetic surveying [ENG] Underground surveying carried out by generating electromagnetic waves at the surface of the earth; the waves penetrate the earth and induce currents in conducting ore bodies, thereby generating new waves that are detected by instruments at the surface or by a receiving coil lowered into a borehole. Also known as electromagnetic prospecting.

electromagnetic susceptibility [ELECTR] The tolerance of circuits and components to all sources of interfering electromagnetic energy.

electromagnetic system of units [ELECTROMAG] A centimeter-gram-second system of electric and magnetic units in which the unit of current is defined as the current which, if maintained in two straight parallel wires having infinite length and being 1 centimeter apart in vacuum, would produce between these conductors a force of 2 dynes per centimeter of length; other units are derived from this definition by assigning unit coefficients in equations relating electric and magnetic quantities. Also known as electromagnetic units (emu).

electromagnetic theory of light [ELECTROMAG] Theory according to which light is an electromagnetic wave whose electric and magnetic fields obey Maxwell's equations.

electromagnetic transducer *See* electromechanical transducer.

electromagnetic units *See* electromagnetic system of units.

electromagnetic wave [ELECTROMAG] A disturbance which propagates outward from any electric charge which oscillates or is accelerated; far from the charge it consists of vibrating electric and magnetic fields which move at the speed of light and are at right angles to each other and to the direction of motion.

electromagnetic-wave filter [ELECTROMAG] Any device to transmit electromagnetic waves of desired frequencies while substantially attenuating all other frequencies.

electromagnetic well logging *See* electromagnetic logging.

electromagnetism [PHYS] **1.** Branch of physics relating electricity to magnetism. **2.** Magnetism produced by an electric current rather than by a permanent magnet.

electromechanical [MECH ENG] Pertaining to a mechanical device, system, or process which is electrostatically or electromagnetically actuated or controlled.

electromechanical circuit [ELEC] A circuit containing both electrical and mechanical parameters of consequence in its analysis.

electromechanical dialer [ELECTR] Telephone dialer which activates one of a set of desired numbers, precoded into it, when the user selects and presses a start button.

electromechanical recording [ELECTR] Recording by means of a signal-actuated mechanical device, such as a pen arm or mirror attached to the moving coil of a galvanometer.

electromechanical relay [ELECTROMAG] A protective relay operating on the principle of electromagnetic attraction, as a plunger relay, or of electromagnetic induction.

electromechanical transducer [ELECTR] A transducer for receiving waves from an electric system and delivering waves to a mechanical system, or vice versa. Also known as electromagnetic transducer.

electromechanics [MECH ENG] The technology of mechanical devices, systems, or processes which are electrostatically or electromagnetically actuated or controlled.

electrometallurgy [MET] Industrial recovery and processing of metals by electrical and electrolytic procedures.

electrometer [ENG] An instrument for measuring voltage without drawing appreciable current.

electrometer tube [ELECTR] A high-vacuum electron tube having a high input impedance (low control-electrode conductance) to facilitate measurement of extremely small direct currents or voltages.

electromigration [ANALY CHEM] A process used to separate isotopes or ionic species by the differences in their ionic mobilities in an electric field.

electromodulation [SPECT] Modulation spectroscopy in which changes in transmission or reflection spectra induced by a perturbing electric field are measured.

electromotance *See* electromotive force.

electromotive force [PHYS CHEM] **1.** The difference in electric potential that exists between two dissimilar electrodes immersed in the same electrolyte or otherwise connected by ionic conductors. **2.** The resultant of the relative electrode potential of the two dissimilar electrodes at which electrochemical reactions occur. Abbreviated emf. Also known as electromotance.

electromotive series *See* electrochemical series.

electromyogram [MED] **1.** A graphic recording of the electrical response of a muscle to electrical stimulation. **2.** A graphic recording of eye movements during reading.

electromyograph [MED] An instrument used for making electromyograms.

electromyography [MED] A medical specialty concerned with the production and study of electromyograms.

electron [PHYS] **1.** A stable elementary particle which is the negatively charged constituent of ordinary matter, having a mass of about 9.11×10^{-28}g (equivalent to 0.511 MeV), a charge of about -1.602×10^{-19} coulomb, and a spin of ½. Also known as negative electron; negatron. **2.** Collective name for the electron, as in the first definition, and the positron.

electron accelerator [NUCLEO] A device which accelerates electrons to high energies.

electron acceptor [PHYS CHEM] An atom or part of a molecule joined by a covalent bond to an electron donor. [SOLID STATE] *See* acceptor.

electron affinity [ATOM PHYS] The work needed in removing an electron from a negative ion, thus restoring the neutrality of an atom or molecule.

electronarcosis [MED] Profound stupor or unconsciousness produced by passing an electric current through the brain.

electron avalanche *See* avalanche.

electron beam [ELECTR] A narrow stream of electrons moving in the same direction, all having about the same velocity.

electron-beam drilling [ELECTR] Drilling of tiny holes in a ferrite, semiconductor, or other material by using a sharply focused electron beam to melt and evaporate or sublimate the material in a vacuum.

electron-beam generator [ELECTR] Velocity-modulated generator, such as a klystron tube, used to generate extremely high frequencies.

electron-beam machining [MET] A machining process in which heat is produced by a focused electron beam at a sufficiently high temperature to volatilize and thereby remove metal in a desired manner; takes place in a vacuum. Abbreviated EBM.

electron-beam magnetometer [ENG] A magnetometer that depends on the change in intensity or direction of an electron beam that passes through the magnetic field to be measured.

electron-beam melting [MET] A melting process in which an electron beam provides the necessary heat.

electron-beam parametric amplifier [ELECTR] A parametric amplifier in which energy is pumped from an electrostatic field into a beam of electrons traveling down the length of the tube, and electron couplers impress the input signal at one end of the tube and translate spiraling electron motion into electric output at the other.

electron-beam tube [ELECTR] An electron tube whose performance depends on the formation and control of one or more electron beams.

electron-beam welding [MET] A technique for joining materials in which highly collimated electron beams are used at a pressure below 10^{-3} mm Hg to produce a highly concentrated heat source; used in outer space.

electron-bombardment-induced-conductivity [ELECTR] In a multimode display-storage tube, a process using an electron gun to erase the image on the cathode-ray tube interface.

electron bunching *See* bunching.

electron capture [ATOM PHYS] The process in which an atom or ion passing through a material medium either loses or gains one or more orbital electrons. [NUC PHYS] A radioactive transformation of nuclide in which a bound electron merges with its nucleus.

electron capture detector [ANALY CHEM] Extremely sensitive gas chromatography detector that is a modification of the argon ionization detector, with conditions adjusted to favor the formation of negative ions.

electron charge [PHYS] The charge carried by an electron, equal to about -1.602×10^{-19} coulomb, or -4.803×10^{-10} statcoulomb.

electron cloud [ATOM PHYS] Picture of an electron state in which the charge is thought of as being smeared out, with the resulting charge density distribution corresponding to the probability distribution function associated with the Schrödinger wave function.

electron collector *See* collector.

electron compound [MET] Alloy of two metals in which a progressive change in composition is accompanied by a progression of phases, differing in crystal structure. Also known as Hume-Rothery compound; intermetallic compound.

electron conduction [ELEC] Conduction of electricity resulting from motion of electrons, rather than from ions in a gas or solution, or holes in a solid.

electron configuration [ATOM PHYS] The orbital and spin arrangement of an atom's electrons, specifying the quantum numbers of the atom's electrons in a given state.

electron-coupled oscillator [ELECTR] An oscillator employing a multigrid tube in which the cathode and two grids operate as an oscillator; the anode-circuit load is coupled to the oscillator through the electron stream. Abbreviated eco. Also known as Dow oscillator.

electron coupler [ELECTR] A microwave amplifier tube in which electron bunching is produced by an electron beam projected parallel to a magnetic field and, at the same time, subjected to a transverse electric field produced by a signal generator. Also known as Cuccia coupler.

electron coupling [ELECTR] A method of coupling two circuits inside an electron tube, used principally with multigrid tubes; the electron stream passing between electrodes in one circuit transfers energy to electrodes in the other circuit. Also known as electronic coupling.

electron cyclotron wave [PL PHYS] A wave in a plasma which propagates parallel to the magnetic field produced by currents outside the plasma at frequencies less than that of the electron cyclotron resonance, and which is circularly polarized, rotating in the same sense as electrons in the plasma; responsible for whistlers. Also known as whistler wave.

electron density [PHYS] **1.** The number of electrons in a unit volume. **2.** When quantum-mechanical effects are significant, the total probability of finding an electron in a unit volume.

electron device [ELECTR] A device in which conduction is principally by electrons moving through a vacuum, gas, or semiconductor, as in a crystal diode, electron tube, transistor, or selenium rectifier.

electron diffraction [PHYS] The phenomenon associated with the interference processes which occur when electrons are scattered by atoms in crystals to form diffraction patterns.

electron diffraction analysis [PHYS] Examination of solid surfaces by observing the diffraction of a stream of electrons by the surface.

electron diffraction camera [OPTICS] A camera used to obtain a photographic record of the position and intensity of the diffracted beams produced when a specimen is irradiated by a beam of electrons.

electron diffractograph [PHYS] A device, allied to the electron microscope, in which a beam of electrons strikes the sample, showing crystal pattern and other physical attributes on the resulting diffraction pattern; used for chemical analysis, atomic structure determination, and so on.

electron dipole moment *See* electron magnetic moment.

electron discharge machining [MET] A process by which materials that conduct electricity can be removed from a metal by an electric spark; used to form holes of varied shapes in materials of poor machinability. Abbreviated EDM. Also

ELECTRON ACCELERATOR

A type of electron accelerator, the Stanford Linear Accelerator, installed in underground housing.

known as electrical discharge machining; electric spark machining; electroerosive machining; electrospark machining.

electron distribution [PHYS] A function which gives the number of electrons per unit volume of phase space.

electron distribution curve [PHYS CHEM] A curve indicating the electron distribution among the different available energy levels of a solid substance.

electron donor [PHYS CHEM] An atom or part of a molecule which supplies both electrons of a duplet forming a covalent bond. [SOLID STATE] *See* donor.

electron efficiency [ELECTR] The power which an electron stream delivers to the circuit of an oscillator or amplifier at a given frequency, divided by the direct power supplied to the stream. Also known as electronic efficiency.

electronegative potential [PHYS CHEM] Potential of an electrode expressed as negative with respect to the hydrogen electrode.

electronegativity [ELEC] 1. The carrying of a negative electric charge. 2. The capability of acting as the negative electrode in an electric cell. [PHYS CHEM] The relative ability of an atom or group of atoms to attract electrons to itself.

electron emission [ELECTR] The liberation of electrons from an electrode into the surrounding space, usually under the influence of heat, light, or a high electric field.

electron emitter [ELECTR] The electrode from which electrons are emitted.

electron energy level [ATOM PHYS] A quantum-mechanical concept for energy levels of electrons about the nucleus; electron energies are functions of each particular atomic species.

electron energy loss spectroscopy [SPECT] A technique for studying atoms, molecules, or solids in which a substance is bombarded with monochromatic electrons, and the energies of scattered electrons are measured to determine the distribution of energy loss.

electroneutrality principle [PHYS CHEM] The principle that in an electrolytic solution the concentrations of all the ionic species are such that the solution as a whole is neutral.

electron flow [ELEC] A current produced by the movement of free electrons toward a positive terminal; the direction of electron flow is opposite to that of current.

electron gas [PHYS] A concentration of electrons whose behavior is, in first approximation, not governed by forces.

electron gun [ELECTR] An electrode structure that produces and may control, focus, deflect, and converge one or more electron beams in an electron tube.

electron-gun density multiplication [ELECTR] Ratio of the average current density at any specified aperture through which the electron stream passes to the average current density at the cathode surface.

electron hole *See* hole.

electron hole droplets [SOLID STATE] A form of electronic excitation observed in germanium and silicon at sufficiently low cryogenic temperatures; it is associated with a liquid-gas phase transition of the charge carriers, and consists of regions of conducting electron-hole Fermi liquid coexisting with regions of insulating exciton gas.

electronic [ELECTR] Pertaining to electron devices or to circuits or systems utilizing electron devices, including electron tubes, magnetic amplifiers, transistors, and other devices that do the work of electron tubes.

electronic absorption spectrum [SPECT] Spectrum resulting from absorption of electromagnetic radiation by atoms, ions, and molecules due to excitations of their electrons.

electronic ac voltmeter [ELECTR] A voltmeter consisting of a direct-current milliammeter calibrated in volts and connected to an amplifier-rectifier circuit.

electronic altimeter *See* radio altimeter.

electronic angular momentum [ATOM PHYS] The total angular momentum associated with the orbital motion of the spins of all the electrons of an atom.

electronic azimuth marker [ELECTR] On an airborne radar plan position indicator (PPI) a bright rotatable radial line used for bearing determination. Also known as azimuth marker.

electronic band spectrum [SPECT] Bands of spectral lines

associated with a change of electronic state of a molecule; each band corresponds to certain vibrational energies in the initial and final states and consists of many rotational lines.

electronic bearing cursor [ELECTR] Of a marine radar set, the bright rotatable radial line on the plan position indicator used for bearing determination. Also known as electronic bearing marker.

electronic bearing marker *See* electronic bearing cursor.

electronic calculating punch [ADP] A card-handling machine that reads a punched card, performs a number of sequential operations, and punches the result on the card.

electronic calculator [ADP] A calculator that uses electron devices.

electronic camouflage [ELECTR] Use of electronic means, or exploitation of electronic characteristics to reduce, submerge, or eliminate the radar echoing properties of a target.

electronic carillon [ENG ACOUS] A carillon that uses electric and electronic circuits to generate, amplify, and reproduce musical tones approximating those of bells.

electronic circuit [ELECTR] An electric circuit in which the equilibrium of electrons in some of the components (such as electron tubes, transistors, or magnetic amplifiers) is upset by means other than an applied voltage.

electronic clock [HOROL] A clock that uses a ferrite rod and coil to pick up the electromagnetic field of 60-hertz power line wiring in a house; this 60-hertz voltage is amplified by a transistor amplifier and used to control a transistor oscillator that drives a tiny permanent-magnet synchronous clock motor; two mercury cells provide power for the transistors.

electronic commutator [ELECTR] An electron-tube or transistor circuit that switches one circuit connection rapidly and successively to many other circuits, without the wear and noise of mechanical switches.

electronic component [ELECTR] A component which is able to amplify or control voltages or currents without mechanical or other nonelectrical command, or to switch currents or voltages without mechanical switches; examples include electron tubes, transistors, and other solid-state devices.

electronic computing units [ELECTR] The sensing sections of tabulating equipment which enable the machine to handle the contents of punched cards in a prescribed manner.

electronic confusion area [ELECTROMAG] Amount of space that a target appears to occupy in a radar resolution cell, as it appears to that radar beam.

electronic control [ELECTR] The control of a machine or process by circuits using electron tubes, transistors, magnetic amplifiers, or other devices having comparable functions.

electronic controller [ELECTR] Electronic device incorporating vacuum tubes or solid-state devices and used to control the action or position of equipment; for example, a valve operator.

electronic counter [ELECTR] A circuit using electron tubes or equivalent devices for counting electric pulses. Also known as electronic tachometer.

electronic countermeasure [ELECTR] An offensive or defensive tactic or device using electronic and reflecting apparatus to reduce the military effectiveness of enemy equipment involving electromagnetic radiation, such as radar, communication, guidance, or other radio-wave devices. Abbreviated ECM. Also known as electromagnetic countermeasure.

electronic coupling *See* electron coupling.

electronic data processing [ADP] Processing data by using equipment that is predominantly electronic in nature, such as an electronic digital computer. Abbreviated EDP.

electronic data-processing center [ADP] The complex formed by the computer, its peripheral equipment, the personnel related to the operation of the center and control functions, and, usually, the office space housing hardware and personnel. Abbreviated EDP center. Also known as computer center.

electronic data-processing management science [ADP] The field consisting of a class of management problems capable of being handled by computer programs.

electronic data-processing system [ADP] A system for data processing by means of machines using electronic circuitry at electronic speed, as opposed to electromechanical equipment.

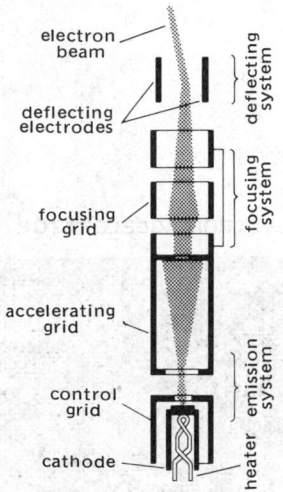

ELECTRON GUN

electron beam
deflecting electrodes
focusing grid
accelerating grid
control grid
cathode
deflecting system
focusing system
accelerating system
emission system
heater

Simplified electron gun employing electrostatic focus and deflection.

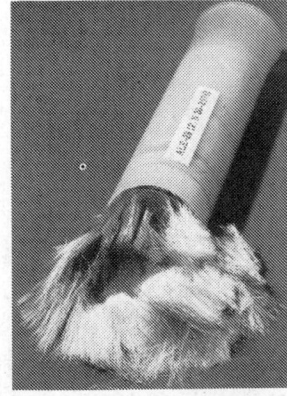

ELECTRONIC COUNTERMEASURE

Metallized glass chaff which returns spurious radar echoes.
(Lundy Electronics and Systems)

electronic deception [ORD] Radiation or reradiation of electromagnetic waves in a manner intended to mislead the enemy in the interpretation of data received by the enemy's electronic equipment.

electronic defense evaluation [ELECTR] A mutual evaluation of radar and aircraft, with the aircraft trying to penetrate the radar's area of coverage in an electronic countermeasure environment.

electronic differential analyzer [ADP] A form of analog computer using interconnected electronic integrators to solve differential equations.

electronic distance-measuring equipment [NAV] A navigation system consisting of airborne devices that transmit microsecond pulses to special ground beacons, which retransmit the signals to the aircraft; the length of expired time between transmission and reception is measured, converted to kilometers or miles, and presented to the pilot.

electronic driftmeter [NAV] An electronic instrument for measuring drift angle; may be an attachment to an airborne radar; is an integral part of a Doppler navigator.

electronic dummy [ENG ACOUS] A vocal simulator which is a replica of the head and torso of a person, covered with plastisol flesh that simulates the acoustical and mechanical properties of real flesh, and possessing an artificial voice and two artificial ears. Abbreviated ED.

electronic efficiency [ELECTR] Ratio of the power at the desired frequency, delivered by the electron stream to the circuit in an oscillator or amplifier, to the average power supplied to the stream.

electronic emission spectrum [SPECT] Spectrum resulting from emission of electromagnetic radiation by atoms, ions, and molecules following excitations of their electrons.

electronic energy curve [PHYS CHEM] A graph of the energy of a diatomic molecule in a given electronic state as a function of the distance between the nuclei of the atoms.

electronic engineering [ENG] Engineering that deals with practical applications of electronics.

electronic fix [NAV] A position fix established by means of electronic equipment.

electronic fuse [ENG] A fuse, such as the radio proximity fuse, set off by an electronic device incorporated in it.

electronic heating [ENG] Heating by means of radio-frequency current produced by an electron-tube oscillator or an equivalent radio-frequency power source. Also known as high-frequency heating; radio-frequency heating.

electronic humidistat [ENG] A humidistat in which a change in the relative humidity causes a change in the electrical resistance between two sets of alternate metal conductors mounted on a small flat plate with plastic coating, and this change in resistance is measured by a relay amplifier.

electronic interference [ELECTR] Any electrical or electromagnetic disturbance that causes undesirable response in electronic equipment.

electronic jammer *See* jammer.

electronic jamming *See* jamming.

electronic line scanning [ELECTR] Method which provides motion of the scanning spot along the scanning line by electronic means.

electronic listening device [ELECTR] A device used to capture the sound waves of conversation originating in an ostensibly private setting in a form, usually as a magnetic tape recording, which can be used against the target by adverse interests.

electronic locator *See* metal detector.

electronic locking [ELECTR] A technique for preventing the operation of a switch until a specific electrical signal (the unlocking signal) is introduced into circuitry associated with the switch; usually, but not necessarily, the unlocking signal is a binary sequence.

electronic logger *See* Geiger-Müller probe.

electronic magnetic moment [ATOM PHYS] The total magnetic dipole moment associated with the orbital motion of all the electrons of an atom and the electron spins; opposed to nuclear magnetic moment.

electronic microradiography [ELECTR] Microradiography of very thin specimens in which the emission of electrons from an irradiated object, either the specimen or a lead screen behind it, is used to produce a photographic image of the specimen, which is then enlarged.

electronic motor control [ELECTR] A control circuit used to vary the speed of a direct-current motor operated from an alternating-current power line. Also known as direct-current motor control; motor control.

electronic navigation [NAV] Navigation by means of any electronic device or instrument.

electronic noise jammer [ELECTR] An electronic jammer which emits a radio-frequency carrier modulated with a white noise signal usually derived from a gas tube; used against enemy radar.

Electronic Numerical Integrator and Calculator *See* ENIAC.

electronic packaging [ENG] The technology of packaging electronic equipment; in current usage it refers to inserting discrete components, integrated circuits, and MSI and LSI chips (usually attached to a lead frame by beam leads) into plates through holes on multilayer circuit boards (also called cards), where they are soldered in place.

electronic phase-angle meter [ELECTR] A phasemeter that makes use of electronic devices, such as amplifiers and limiters, that convert the alternating-current voltages being measured to square waves whose spacings are proportional to phase.

electronic photometer *See* photoelectric photometer.

electronic polarization [ELEC] Polarization arising from the displacement of electrons with respect to the nuclei with which they are associated, upon application of an external electric field.

electronic position indicator [NAV] A radio navigation system used in hydrographic surveying which provides circular lines of position. Abbreviated EPI.

electronic power supply *See* power supply.

electronic pumping *See* pumping.

electronic radiography [ELECTR] Radiography in which the image is detached by direct image converter tubes or by the use of television pickup or electronic scanning, and the resulant signals are amplified and presented for viewing on a kinescope.

electronic-raster scanning *See* electronic scanning.

electronic reconnaissance [ELECTR] The detection, identification, evaluation, and location of foreign, electromagnetic radiations emanating from other than nuclear detonations or radioactive sources.

electronic recording [ELECTR] The process of making a graphical record of a varying quantity or signal (or the result of such a process) by electronic means, involving control of an electron beam by electric or magnetic fields, as in a cathode-ray oscillograph, in contrast to light-beam recording.

electronics [PHYS] Study, control, and application of the conduction of electricity through gases or vacuum or through semiconducting or conducting materials.

electronic scanning [ELECTR] Scanning in which an electron beam, controlled by electric or magnetic fields, is swept over the area under examination, in contrast to mechanical or electromechanical scanning. Also known as electronic-raster scanning.

electronic sculpturing [ADP] Procedure for constructing a model of a system by using an analog computer, in which the model is devised at the console by interconnecting components on the basis of analogous configuration with real system elements; then, by adjusting circuit gains and reference voltages, dynamic behavior can be generated that corresponds to the desired response, or is recognizable in the real system.

electronic security [ELECTR] Protection resulting from all measures designed to deny to unauthorized persons information of value which might be derived from the possession and study of electromagnetic radiations.

electronic sky screen equipment [ELECTR] Electronic device that indicates the departure of a missile from a predetermined trajectory.

electronic specific heat [SOLID STATE] Contribution to the specific heat of a metal from the motion of conduction electrons.

electronic spectrum [SPECT] Spectrum resulting from emis-

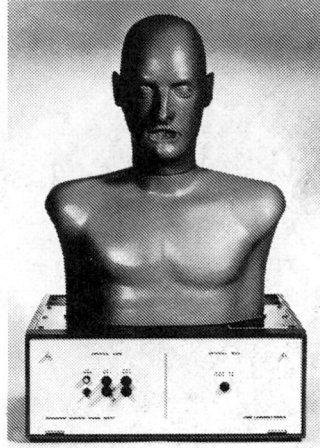

ELECTRONIC DUMMY

Electronic dummy which is used for acoustical testing. *(CBS Laboratories)*

ELECTRONIC HUMIDISTAT

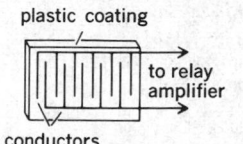

Simple diagram of electronic humidistat. *(Honeywell Inc.)*

sion or absorption of electromagnetic radiation during changes in the electron configuration of atoms, ions, or molecules, as opposed to vibrational, rotational, fine-structure, or hyperfine spectra.

electronic state [QUANT MECH] The physical state of electrons of a system, as specified, for example, by a Schrödinger-Pauli wave function of the positions and spin orientations of all the electrons.

electronic structure [PHYS] The arrangement of electrons in an atom, molecule, or solid, specified by their wave functions, energy levels, or quantum numbers.

electronic surge arrester [ELECTR] Device used to switch to ground high-energy surges, thereby reducing transient energy to a level safe for secondary protectors, for example, Zener diodes, silicon rectifiers and so on.

electronic switch [ELECTR] **1.** Vacuum tube, crystal diodes, or transistors used as an on and off switching device. **2.** Test instrument used to present two wave shapes on a single gun cathode-ray tube.

electronic switching [COMMUN] Telephone switching using a computer with a storage containing program switching logic, whose output actuates reed switches that set up telephone connections automatically. [ELECTR] The use of electronic circuits to perform the functions of a high-speed switch.

electronic tachometer *See* electronic counter.

electronic tonometer *See* tonometer.

electronic voltage regulator [ELECTR] A device which maintains the direct-current power supply voltage for electronic equipment nearly constant in spite of input alternating-current line voltage variations and output load variations.

electronic voltmeter [ENG] Voltmeter which uses the rectifying and amplifying properties of electron devices and their associated circuits to secure desired characteristics, such as high-input impedance, wide-frequency range, crest indications, and so on.

electronic warfare [ELECTR] Military action involving the use of electromagnetic energy to determine, exploit, reduce, or prevent hostile use of the electromagnetic spectrum, and action which retains friendly use of electromagnetic spectrum.

electronic warfare support measures [ELECTR] That division of electronic warfare involving actions taken to search for, intercept, locate, record, and analyze radiated electromagnetic energy for the purpose of exploiting such radiations in support of military operations.

electronic work function [SOLID STATE] The energy required to raise an electron with the Fermi energy in a solid to the energy level of an electron at rest in vacuum outside the solid.

electronic writing [ELECTR] The use of electronic circuits and electron devices to reproduce symbols, such as an alphabet, in a prescribed order on an electronic display device for the purpose of transferring information from a source to a viewer of the display device.

electron image tube *See* image tube.

electron injection [ELECTR] **1.** The emission of electrons from one solid into another. **2.** The process of injecting a beam of electrons with an electron gun into the vacuum chamber of a mass spectrometer, betatron, or other large electron accelerator.

electron lens [ELECTR] An electric or magnetic field, or a combination thereof, which acts upon an electron beam in a manner analogous to that in which an optical lens acts upon a light beam. Also known as lens.

electron lepton number [PARTIC PHYS] The number of electrons and electron-associated neutrinos minus the number of positrons and electron-associated antineutrinos; believed to be an exactly conserved quantity.

electron linear accelerator [NUCLEO] A linear accelerator used to accelerate electrons in a straight line, usually by means of radio-frequency fields which are produced in a loaded waveguide and travel with the electrons.

electron magnetic moment [ATOM PHYS] The magnetic dipole moment which an electron possesses by virtue of its spin. Also known as electron dipole moment.

electron mass [PHYS] The mass of an electron, equal to about 9.11×10^{-28}g, equivalent to 0.511 MeV. Also known as electron rest mass.

electron metallography [MET] The study of the microscopic structure of metals employing an electron microscope.

electron micrograph [GRAPHICS] Photograph of a sample taken with an electron microscope.

electron microprobe [PHYS] An x-ray machine in which electrons emitted from a hot-filament source are accelerated electrostatically, then focused to an extremely small point on the surface of a specimen by an electromagnetic lens; nondestructive analysis of the specimen can then be made by measuring the backscattered electrons, the specimen current, the resulting x-radiation, or any other resulting phenomenon. Also known as electron probe.

electron microscope [ELECTR] A device for forming greatly magnified images of objects by means of electrons, usually focused by electron lenses.

electron mirror *See* dynode.

electron mobility [SOLID STATE] The drift mobility of electrons in a semiconductor, being the electron velocity divided by the applied electric field.

electron multiplicity [ATOM PHYS] In an atom with Russell-Saunders coupling, the quantity $2S + 1$, where S is the total spin quantum number.

electron multiplier [ELECTR] An electron-tube structure which produces current amplification; an electron beam containing the desired signal is reflected in turn from the surfaces of each of a series of dynodes, and at each reflection an impinging electron releases two or more secondary electrons, so that the beam builds up in strength. Also known as multiplier.

electron-multiplier phototube *See* multiplier phototube.

electron number [ATOM PHYS] The number of electrons in an ion or atom.

electronographic tube [ELECTR] An image tube used in astronomy in which the electron image formed by the tube is recorded directly upon film or plates.

electronography [ELECTR] The use of image tubes to form intensified electron images of astronomical objects and record them directly on film or plates.

electronoluminescence *See* cathodoluminescence.

electron optics [ELECTR] The study of the motion of free electrons under the influence of electric and magnetic fields.

electron orbit [PHYS] The path described by an electron.

electron pair [PHYS CHEM] A pair of valence electrons which form a nonpolar bond between two neighboring atoms.

electron pair bond *See* covalent bond.

electron paramagnetic resonance [PHYS] Magnetic resonance arising from the magnetic moment of unpaired electrons in a paramagnetic substance or in a paramagnetic center in a diamagnetic substance. Abbreviated EPR. Also known as electron spin resonance (ESR); paramagnetic resonance.

electron paramagnetism [PHYS] Paramagnetism in a substance whose atoms or molecules possess a net electronic magnetic moment; arises because of the tendency of a magnetic field to orient the electronic magnetic moments parallel to itself.

electron probe *See* electron microprobe.

electron probe microanalysis [ANALY CHEM] A technique in analytical chemistry in which a finely focused beam of electrons is used to excite an x-ray spectrum characteristic of the elements in the sample; can be used with samples as small as 10^{-11} cubic centimeter.

electron radius [PHYS] The classical value r of 2.81777×10^{-13} centimeter for the radius of an electron; obtained by equating mc^2 for the electron to e^2/r, where e and m are the charge and mass of the electron respectively; any classical model for an electron will have approximately this radius.

electron-ray indicator *See* cathode-ray tuning indicator.

electron-ray tube *See* cathode-ray tube.

electron rest mass *See* electron mass.

electron ring accelerator [NUCLEO] Proposed particle accelerator in which protons to be accelerated are trapped by the space charge of a ring of relativisitic electrons which is then accelerated. Abbreviated ERA.

electron shell [ATOM PHYS] **1.** The collection of all the electron states in an atom which have a given principal quantum number. **2.** The collection of all the electron states in an atom which have a given principal quantum number and a given orbital angular momentum quantum number.

electron spectroscopy [SPECT] The study of the energy spectra of photoelectrons or Auger electrons emitted from a substance upon bombardment by electromagnetic radiation, electrons, or ions; used to investigate atomic, molecular, or solid-state structure, and in chemical analysis.

electron spectrum [SPECT] Visual display, photograph, or graphical plot of the intensity of electrons emitted from a substance bombarded by x-rays or other radiation as a function of the kinetic energy of the electrons.

electron spin [QUANT MECH] That property of an electron which gives rise to its angular momentum about an axis within the electron.

electron spin density [PHYS] The vector sum of the spin angular momenta of electrons at each point in a substance per unit volume.

electron spin resonance *See* electron paramagnetic resonance.

electron-stream potential [ELECTR] At any point in an electron stream, the time average of the potential difference between that point and the electron-emitting surface.

electron-stream transmission efficiency [ELECTR] At an electrode through which the electron stream (beam) passes, the ratio of the average stream current through the electrode to the stream current approaching the electrode.

electron synchrotron [NUCLEO] A circular electron accelerator in which the frequency of the accelerating system is constant, the strength of the magnetic guide field increases, and the electrons move in orbits of nearly constant radius.

electron telescope [ELECTR] A telescope in which an infrared image of a distant object is focused on the photosensitive cathode of an image converter tube; the resulting electron image is enlarged by electron lenses and made visible by a fluorescent screen.

electron temperature [PL PHYS] The temperature at which ideal gas molecules would have an average kinetic energy equal to that of electrons in a plasma under consideration.

electron transfer [PHYS] The passage of an electron from one constituent of a system to another.

electron transition [QUANT MECH] Change of an electron from one state to another, accompanied by emission or absorption of electromagnetic radiation.

electron transport system [BIOCHEM] The components of the final sequence of reactions in biological oxidations; composed of a series of oxidizing agents arranged in order of increasing strength and terminating in oxygen.

electron trap [SOLID STATE] A defect or chemical impurity in a semiconductor or insulator which captures mobile electrons in a special way.

electron tube [ELECTR] An electron device in which conduction of electricity is provided by electrons moving through a vacuum or gaseous medium within a gastight envelope. Also known as radio tube; tube; valve (British usage).

electron-tube amplifier [ELECTR] An amplifier in which electron tubes provide the required increase in signal strength.

electron-tube generator [ELECTR] A generator in which direct-current energy is converted to radio-frequency energy by an electron tube in an oscillator circuit.

electron-tube heater *See* heater.

electron tube static characteristic [ELECTR] Relation between a pair of variables such as electrode voltage and electrode current with all other voltages maintained constant.

electron tunneling [QUANT MECH] The passage of electrons through a potential barrier which they would not be able to cross according to classical mechanics, such as a thin insulating barrier between two superconductors.

electron vacuum gage [ENG] An instrument used to measure vacuum by the ionization effect that an electron flow (from an incandescent filament to a charged grid) has on gas molecules.

electron volt [PHYS] A unit of energy equal to the energy acquired by an electron when it passes through a potential difference of 1 volt in a vacuum; it is equal to $(1.602192 \pm 0.000007) \times 10^{-19}$ volt. Abbreviated eV.

electron wave [QUANT MECH] The de Broglie wave or probability amplitude wave of an electron.

electron wave function [QUANT MECH] Function of the spin orientation and position of one or more electrons, specifying the dynamical state of the electrons; the square of the function's modulus gives the probability per unit volume of finding electrons at a given position.

electron wavelength [QUANT MECH] The de Broglie wavelength of an electron, given by Planck's constant divided by the momentum.

electrooptical birefringence *See* electrooptical Kerr effect.

electrooptical character recognition *See* optical character recognition.

electrooptical Kerr effect [OPTICS] Birefringence induced by an electric field. Also known as electrooptical birefringence; Kerr effect.

electrooptic material [OPTICS] A material in which the indices of refraction are changed by an applied electric field.

electrooptic radar [ENG] Radar system using electrooptic techniques and equipment instead of microwave to perform the acquisition and tracking operation.

electrooptics [OPTICS] The study of the influence of an electric field on optical phenomena, as in the electrooptical Kerr effect and the Stark effect. Also known as optoelectronics.

electroosmosis [PHYS CHEM] The movement in an electric field of liquid with respect to colloidal particles immobilized in a porous diaphragm or a single capillary tube.

electroosmotic driver [ELECTR] A type of solion for converting voltage into fluid pressure, which uses depolarizing electrodes sealed in an electrolyte and operates through the streaming potential effect. Also known as micropump.

electropainting [ENG] Electrolytic deposition of a thin layer of paint on a metal surface which is made an anode.

electropherography *See* electrochromatography.

electrophilic [PHYS CHEM] Any chemical process in which electrons are acquired from or shared with other molecules or ions.

electrophilic reagent [PHYS CHEM] A reactant which accepts an electron pair from a molecule, with which it forms a covalent bond.

electrophonic effect [BIOPHYS] The sensation of hearing produced when an alternating current of suitable frequency and magnitude is passed through a person.

electrophoresis [PHYS CHEM] An electrochemical process in which colloidal particles or macromolecules with a net electric charge migrate in a solution under the influence of an electric current. Also known as cataphoresis.

electrophoretic coating [MET] A surface coating on a metal deposited by electric discharge of particles from a colloidal solution.

electrophoretic effect [PHYS CHEM] Retarding effect on the characteristic motion of an ion in an electrolytic solution subjected to a potential gradient, which results from motion in the opposite direction by the ion atmosphere.

electrophoretic mobility [BIOCHEM] A characteristic of living cells in suspension and biological compounds (proteins) in solution to travel in an electric field to the positive or negative electrode, because of the charge on these substances.

electrophoretic variants [BIOCHEM] Phenotypically different proteins that are separable into distinct electrophoretic components due to differences in mobilities; an example is erythrocyte acid phosphatase.

electrophorus [ELEC] A device used to produce electric charges; it consists of a hard-rubber disk, which is negatively charged by rubbing with fur, and a metal plate, held by an insulating handle, which is placed on the disk; the plate is then touched with a grounded conductor, so that negative charge is removed and the plate has net positive charge.

electrophotograph [GRAPHICS] An image formed by means of an electrostatic copying system.

electrophotography [GRAPHICS] An electrostatic image-forming process in which light, x-rays, or gamma rays form an electrostatic image on a photoconductive, insulating medium; the charged image areas attract and hold a fine powder called a toner, and the powder image is then transferred to paper or fused there by heat.

electrophotoluminescence [ELECTR] Emission of light re-

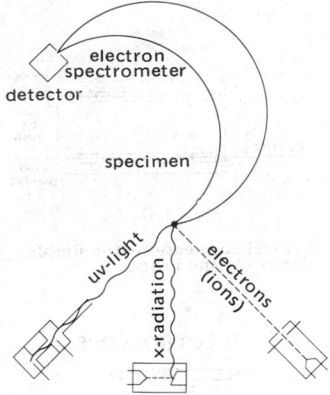

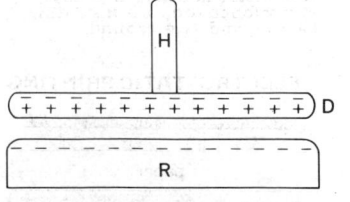

ELECTROPLATING

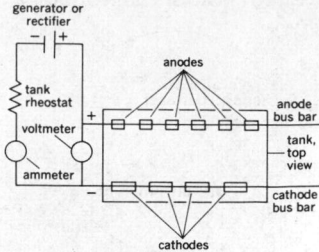

Typical connections for simple electroplating process.

ELECTROSCOPE

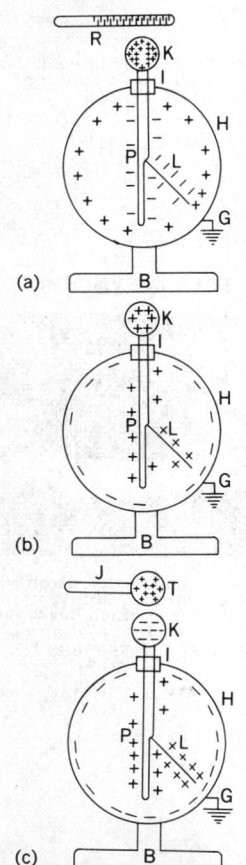

Simple gold-leaf electroscope. *(a)* An electroscope being charged by induction by negative charge on hard-rubber rod R. *(b)* Positive charge left on its leaf after induction process is complete. *(c)* Testing the sign of an unknown charge on test ball T. L = gold leaf, P = metal post, I = insulator, K = metal knob, H = metal housing, B = base, R = rubber rod, J = insulating handle, and G = ground.

ELECTROSTATIC PRINTING

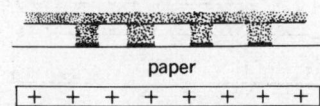

The openings in the screen (image area) are electrostatically filled with powdered ink.

sulting from application of an electric field to a phosphor which is concurrently, or has been previously, excited by other means.

electrophrenic respiration [MED] Artificial respiration in which the nerves that control breathing are stimulated electrically through appropriately placed electrodes.

electrophysiology [PHYSIO] The branch of physiology concerned with determining the basic mechanisms by which electric currents are generated within living organisms.

electroplating [MET] Electrodeposition of a metal or alloy from a suitable electrolyte solution; the article to be plated is connected as the cathode in the electrolyte solution; direct current is introduced through the anode which consists of the metal to be deposited.

electroplax [VERT ZOO] One of the structural units of an electric organ of some fishes, composed of thin, flattened plates of modified muscle that appear as two large, waferlike, roughly circular or rectangular surfaces.

electropolishing [MET] Smoothing and enhancing the appearance of a metal surface by making it an anode in a suitable electrolyte. Also known as electrolytic brightening; electrolytic polishing.

electropositive [ELEC] **1.** Carrying a positive electric charge. **2.** Capable of acting as the positive electrode in an electric cell. [PHYS CHEM] Pertaining to elements, ions, or radicals that tend to give up or lose electrons.

electropositive potential [PHYS CHEM] Potential of an electrode expressed as positive with respect to the hydrogen electrode.

electropulse engine [AERO ENG] An engine, for propelling a flight vehicle, that is based on the use of spark discharges through which intense electric and magnetic fields are established for periods ranging from microseconds to a few milliseconds; a resulting electromagnetic force drives the plasma along the leads and away from the spark gap.

electrorefining [CHEM ENG] Petroleum refinery process for light hydrocarbon streams in which an electrostatic field is used to assist in separation of chemical treating agents (acid, caustic, doctor) from the hydrocarbon phase. [MET] Purifying metals by electrolysis using an impure metal as anode from which the pure metal is dissolved and subsequently deposited at the cathode. Also known as electrolytic refining.

electroreflectance [SPECT] Electromodulation in which reflection spectra are studied. Abbreviated ER.

electroresistive effect [ELECTR] The change in the resistivity of certain materials with changes in applied voltage.

electroretinogram [MED] A graphic recording of the electric discharges of the retina. Abbreviated ERG.

electroscope [ENG] An instrument for detecting an electric charge by means of the mechanical forces exerted between electrically charged bodies.

electrosensitive recording [ELECTR] Recording in which the image is produced by passing electric current through the record sheet.

electroshock therapy [MED] Treatment of mental patients by passing an electric current of 85–110 volts through the brain.

electroslag welding [MET] A welding process in which consumable electrodes are fed into a joint containing flux; the current melts the flux, and the flux in turn melts the faces of the joint and the electrodes, allowing the weld metal to form a continuously cast ingot between the joint faces.

electrospark machining *See* electron discharge machining.

electrostatic [ELEC] Pertaining to electricity at rest, such as an electric charge on an object.

electrostatic accelerator [ELECTR] Any instrument which uses an electrostatic field to accelerate charged particles to high velocities in a vacuum.

electrostatic actuator *See* actuator.

electrostatic analyzer [ELECTR] A device which filters an electron beam, permitting only electrons within a very narrow velocity range to pass through.

electrostatic atomization [MECH ENG] Atomization in which a liquid jet or film is exposed to an electric field, and forces leading to atomization arise from either free charges on the surface or liquid polarization.

electrostatic attraction *See* Coulomb attraction.

electrostatic bond [PHYS CHEM] A valence bond in which two atoms are kept together by electrostatic forces caused by transferring one or more electrons from one atom to the other.

electrostatic cathode-ray tube [ELECTR] A cathode-ray tube in which electrostatic deflection is used on the electron beam.

electrostatic coalescence [METEOROL] **1.** The coalescence of cloud drops induced by electrostatic attractions between drops of opposite charges. **2.** The coalescence of two cloud or rain drops induced by polarization effects resulting from an external electric field.

electrostatic copier [GRAPHICS] A copier that employs principles of electrostatography.

electrostatic copying *See* electrostatography.

electrostatic deflection [ELECTR] The deflection of an electron beam by means of an electrostatic field produced by electrodes on opposite sides of the beam; used chiefly in cathode-ray tubes for oscilloscopes.

electrostatic detection [ELECTR] The detection and location of any type of solid body, such as a mineral deposit or a mine, by measuring the associated electrostatic field which arises spontaneously or is induced by the detection equipment.

electrostatic energy [ELEC] The potential energy which a collection of electric charges possesses by virtue of their positions relative to each other.

electrostatic error *See* antenna effect.

electrostatic field [ELEC] A time-independent electric field, such as that produced by stationary charges.

electrostatic focus [ELECTR] Production of a focused electron beam in a cathode-ray tube by the application of an electric field.

electrostatic force [ELEC] Force on a charged particle due to an electrostatic field, equal to the electric field vector times the charge of the particle.

electrostatic generator [ELEC] Any machine which produces electric charges by friction or (more commonly) electrostatic induction.

electrostatic gyroscope [ENG] A gyroscope in which a small beryllium ball is electrostatically suspended within an array of six electrodes in a vacuum inside a ceramic envelope.

electrostatic induction [ELEC] The process of charging an object electrically by bringing it near another charged object, then touching it to ground. Also known as induction.

electrostatic instrument [ELEC] A meter that depends for its operation on the forces of attraction and repulsion between electrically charged bodies.

electrostatic interactions *See* Coulomb interactions.

electrostatic lens [ELECTR] An arrangement of electrostatic fields which acts upon beams of charged particles similar to the way a glass lens acts on light beams.

electrostatic loudspeaker [ENG ACOUS] A loudspeaker in which the mechanical forces are produced by the action of electrostatic fields; in one type the fields are produced between a thin metal diaphragm and a rigid metal plate. Also known as capacitor loudspeaker.

electrostatic memory *See* electrostatic storage.

electrostatic microphone *See* capacitor microphone.

electrostatic octupole lens [ELECTR] A device for controlling beams of electrons or other charged particles, consisting of eight electrodes arranged in a circular pattern with alternating polarities; commonly used to correct aberrations of quadrupole lens systems.

electrostatic painting [ENG] A painting process that uses the particle-attracting property of electrostatic charges; direct current of about 100,000 volts is applied to a grid of wires through which the paint is sprayed to charge each particle; the metal objects to be sprayed are connected to the opposite terminal of the high-voltage circuit, so that they attract the particles of paint.

electrostatic potential *See* electric potential.

electrostatic precipitator [ENG] A device which removes dust or other finely divided particles from a gas by charging the particles inductively with an electric field, then attracting them to highly charged collector plates. Also known as precipitator.

electrostatic printing [GRAPHICS] An ink transfer process used in copying and printing; powdered ink is electrostatically attracted through openings (image area) in a screen to the paper; there is no contact between screen and paper.

electrostatic probe *See* electric probe.

electrostatic quadrupole lens [ELECTR] A device for focusing beams of electrons or other charged particles, consisting of four electrodes arranged in a circular pattern with alternating polarities.

electrostatic reprography [GRAPHICS] The replication of a document image as a pattern of electric charge on a metal plate; carbon particles attracted to the charge pattern are transferred to and bonded on an accepting surface; for example, multilith mats.

electrostatic repulsion *See* Coulomb repulsion.

electrostatics [ELEC] The study of electric charges at rest, their electric fields, and potentials.

electrostatic scanning [ELECTR] Scanning that involves electrostatic deflection of an electron beam.

electrostatic separation [ENG] Separation of finely pulverized materials by placing them in electrostatic separators. Also known as high-tension separation.

electrostatic separator [ENG] A separator in which a finely pulverized mixture falls through a powerful electric field between two electrodes; materials having different specific inductive capacitances are deflected by varying amounts and fall into different sorting chutes.

electrostatic shielding [ELEC] The placing of a grounded metal screen, sheet, or enclosure around a device or between two devices to prevent electric fields from interacting.

electrostatic storage [ELECTR] A storage in which information is retained as the presence or absence of electrostatic charges at specific spot locations, generally on the screen of a special type of cathode-ray tube known as a storage tube. Also known as electrostatic memory.

electrostatic storage tube *See* storage tube.

electrostatic stress [ELEC] An electrostatic field acting on an insulator, which produces polarization in the insulator and causes electrical breakdown if raised beyond a certain intensity.

electrostatic tape camera [OPTICS] A camera in which images are stored electrostatically on a plastic tape; designed for use in satellites, where the stored image is not damaged by Van Allen or other radiation.

electrostatic transducer [ENG ACOUS] A transducer consisting of a fixed electrode and a movable electrode, charged electrostatically in opposite polarity; motion of the movable electrode changes the capacitance between the electrodes and thereby makes the applied voltage change in proportion to the amplitude of the electrode's motion. Also known as condenser transducer.

electrostatic tweeter [ENG ACOUS] A tweeter loudspeaker in which a flat metal diaphragm is driven directly by a varying high voltage applied between the diaphragm and a fixed metal electrode.

electrostatic units [ELEC] A centimeter-gram-second system of electric and magnetic units in which the unit of charge is that charge which exerts a force of 1 dyne on another unit charge when separated from it by a distance of 1 centimeter in vacuum; other units are derived from this definition by assigning unit coefficients in equations relating electric and magnetic quantities. Abbreviated esu.

electrostatic valence rule [PHYS CHEM] The postulate that in a stable ionic structure the valence of each anion, with changed sign, equals the sum of the strengths of its electrostatic bonds to the adjacent cations.

electrostatic voltmeter [ENG] A voltmeter in which the voltage to be measured is applied between fixed and movable metal vanes; the resulting electrostatic force deflects the movable vane against the tension of a spring.

electrostatic wattmeter [ENG] An adaptation of a quadrant electrometer for power measurements in which two quadrants are charged by the voltage drop across a noninductive shunt resistance through which the load current passes, and the line voltage is applied between one of the quadrants and a moving vane.

electrostatic wave [PL PHYS] Wave motion of a plasma whose restoring forces are primarily electrostatic.

electrostatography [GRAPHICS] A generic term covering all processes involving the forming and use of electrostatic charged patterns for recording and reproducing images; the field is divided into electrophotography and electrography. Also known as electrostatic copying.

electrostriction [MECH] A form of elastic deformation of a dielectric induced by an electric field, associated with those components of strain which are independent of reversal of field direction, in contrast to the piezoelectric effect. Also known as electrostrictive strain.

electrostriction transducer [ENG ACOUS] A transducer which depends on the production of an elastic strain in certain symmetric crystals when an electric field is applied, or, conversely, which produces a voltage when the crystal is deformed. Also known as ceramic transducer.

electrostrictive strain *See* electrostriction.

electrosurgery [MED] The use of electricity to perform surgical procedures, as the use of electricity to simultaneously cut tissue and arrest bleeding.

electrosynthesis [CHEM] A reaction in which synthesis occurs as the result of an electric current.

electrotaxis [BIOL] Movement of an organism in response to stimulation by electric charges.

electrotherapy [MED] The therapeutic use of electricity.

electrothermal [PHYS] **1.** Pertaining to both heat and electricity. **2.** In particular, pertaining to conversion of electrical energy into heat energy.

electrothermal ammeter *See* thermoammeter.

electrothermal energy conversion [ENG] The direct conversion of electric energy into heat energy, as in an electric heater.

electrothermal process [ENG] Any process which uses an electric current to generate heat, utilizing resistance, arcs, or induction; used to achieve temperatures higher than can be obtained by combustion methods.

electrothermal propulsion [AERO ENG] Propulsion of spacecraft by using an electric arc or other electric heater to bring hydrogen gas or other propellant to the high temperature required for maximum thrust; an arc-jet engine is an example.

electrothermal recording [ELECTR] Type of electrochemical recording, used in facsimile equipment, wherein the chemical change is produced principally by signal-controlled thermal action.

electrothermal voltmeter [ENG] An electrothermal ammeter employing a series resistor as a multiplier, thus measuring voltage instead of current.

electrotinning [MET] Electroplating an object with tin.

electrotonus [PHYSIO] The change of condition in a nerve or a muscle during the passage of a current of electricity.

electrotropism [BIOL] Orientation response of a sessile organism to stimulation by electric charges.

electrotype [GRAPHICS] A duplicate printing surface prepared by making a mold of the type page or halftone plate, then suspending this mold in a bath of copper sulfate and sulfuric acid where, by electrolytic action, a thin shell of copper is deposited on it, and finally pouring molten type metal into this shell to strengthen it for use on the press.

electrotyping [GRAPHICS] The process of making an electrotype.

electrovalence [PHYS CHEM] The valence of an atom that has formed an ionic bond.

electrovalent bond *See* ionic bond.

electroviscous effect [FL MECH] Change in a liquid's viscosity induced by a strong electrostatic field.

electrowinning [MET] Extracting metal from solutions by electrochemical processes.

Elektrion process [CHEM ENG] A process of condensation and polymerization in which a mixture of a relatively light mineral oil and a fatty oil is subjected to an electric discharge in an atmosphere of hydrogen; the product is a very viscous oil used for blending with lighter lubricating oils.

element [ADP] A circuit or device performing some specific elementary data-processing function. [CHEM] A substance made up of atoms with the same atomic number; common examples are hydrogen, gold, and iron. Also known as chemical element. [ELEC] **1.** A part of an electron tube, semiconductor device, or antenna array that contributes directly to the electrical performance. **2.** *See* component. [ELECTROMAG] Radiator, active or parasitic, that is a part of an antenna. [IND ENG] A brief, relatively homogeneous part of a work cycle that can be described and identified. [MATH] **1.** In an array such as a matrix or determinant, a quantity identified by the intersection of a given row or column. **2.** In network topology, an edge.

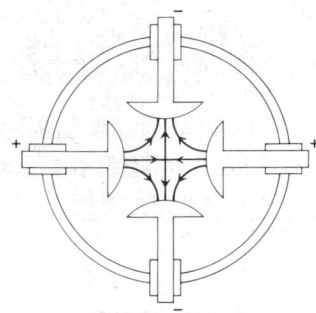

**ELECTROSTATIC
QUADRUPOLE LENS**

Electrostatic quadrupole lens showing the four electrodes.

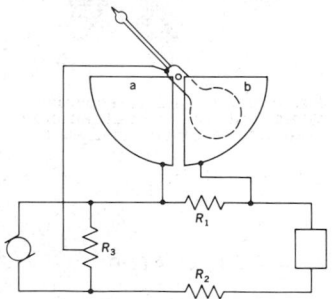

**ELECTROSTATIC
WATTMETER**

Electrostatic wattmeter circuit diagram showing two quadrants *a* and *b*.

ELEMENT 104

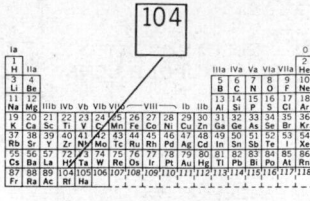

Periodic table of the chemical elements showing the position of element 104.

ELEMENT 105

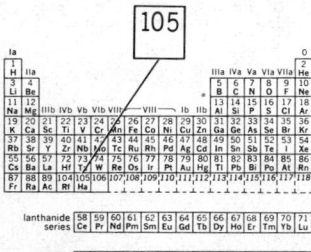

Periodic table of the chemical elements showing the position of element 105.

ELEPHANT

The Indian elephant, recognized by its concave forehead, knobby head, and relatively small ears.

ELEVATOR

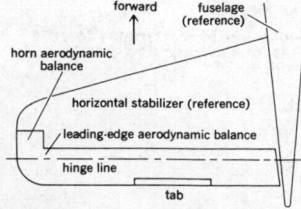

A typical elevator control surface on the left-hand tail plane of an airplane.

element 104 [CHEM] The first element beyond the actinide series, and the twelfth transuranium element; the atoms of element 104, of mass number 260, were first produced by irradiating plutonium-242 with neon-22 ions in a heavy-ion cyclotron.

element 105 [CHEM] An artificial element whose isotope of mass number 260 was discovered by bombarding californium-249 with nitrogen-15 ions in a heavy-ion linear accelerator.

elemental area *See* picture element.

elemental motion [IND ENG] In time-and-motion study, a fundamental subdivision of the hand movements in manipulating an object. Also known as basic element; fundamental motion; therblig.

elementary charge [PHYS] An electric charge such that the electric charge of any body is an integral multiple of it, equal to the electron charge.

elementary excitation [QUANT MECH] The quantum of energy of some vibration or wave, such as a photon, phonon, plasmon, magnon, polaron, or exciton.

elementary item [ADP] An item considered to have no subordinate item in the COBOL language.

elementary particle [PHYS] A particle which, in the present state of knowledge, cannot be described as compound, and is thus one of the fundamental constituents of all matter. Also known as fundamental particle; particle.

elementary process [PHYS CHEM] In chemical kinetics, the particular events at the atomic or molecular level which make up an overall reaction.

element breakdown [IND ENG] Separation of a work cycle into elemental motions.

elements of a fix [NAV] The specific values of the coordinates used to define a position.

elements of an orbit [ASTRON] A set of quantities specifying the orbit of a member of the solar system or of a binary star system, used to calculate the body's position at any time.

elements of the trajectory [MECH] The various features of the trajectory such as the angle of departure, maximum ordinate, angle of fall, and so on.

element time [IND ENG] The time to complete a specific motion element.

eleolite *See* nepheline.

eleostearic acid [ORG CHEM] $CH_3(CH_2)_7(CH:CH)_3(CH_2)_3$-COOH A colorless, water-insoluble, crystalline, unsaturated fatty acid; the glycerol ester is a chief component of tung oil. Also known as octadeca-9,11,13-trienoic acid.

elephant [VERT ZOO] The common name for two living species of proboscidean mammals in the family Elephantidae; distinguished by the elongation of the nostrils and upper lip into a sensitive, prehensile proboscis.

elephanta [METEOROL] A strong southeasterly wind on the Malabar coast of southwest India in September and October, at the end of the southwest monsoon, bringing thundersqualls and heavy rain. Also known as elephant; elephanter.

elephanter *See* elephanta.

elephant-hide pahoehoe [GEOL] A type of pahoehoe on whose surface are innumerable tummuli, broad swells, and pressure ridges which impart the appearance of elephant hide.

elephantiasis [MED] A parasitic disease of man caused by the filarial nematode *Wuchereria bancrofti;* characterized by cutaneous and subcutaneous tissue enlargement due to lymphatic obstruction.

Elephantidae [VERT ZOO] A family of mammals in the order Proboscidea containing the modern elephants and extinct mammoths.

elerwind [METEOROL] A wind of Sun Valley north of Kufstein, in the Tyrol.

elevate [ENG] To increase the angle of elevation of a gun, launcher, optical instrument, or the like.

elevated pole [ASTRON] The celestial pole that appears above the horizon.

elevating machine *See* elevator.

elevating mechanism [ORD] Mechanism on a gun carriage or launcher by which the weapon is elevated or depressed.

elevation [ENG] Vertical distance to a point or object from sea level or some other datum. [GRAPHICS] A graphic projection of a machine or structure on a vertical plane without perspective. [ORD] In antiaircraft artillery, a term sometimes applied to the angular height.

elevation angle [ELECTROMAG] The angle that a radio, radar, or other such beam makes with the horizontal. [PHYS] *See* angle of elevation.

elevation-angle error [ELECTROMAG] In radar, the error in the measurement of the elevation angle of a target resulting from the vertical bending or refraction of radio energy in traveling through the atmosphere. Also known as elevation error.

elevation difference [ORD] The change in elevation which must be applied to a particular gun when firing data are being received from a base piece or other directing point.

elevation error *See* elevation-angle error.

elevation indicator [ORD] A component which presents visually the angle between a fixed reference point and a target in a vertical plane.

elevation meter [ENG] An instrument that measures the change of elevation of a vehicle.

elevation of ivory point *See* barometer elevation.

elevation prediction correction [ORD] In antiaircraft artillery terminology, the future elevation minus the present elevation.

elevation rate [ORD] Rate of change of present elevation of artillery; it is equal to minus the zenith distance rate of the present line, and may be expressed in mils per second or degrees per second.

elevation scale [ORD] Scale on a gun carriage that shows the quadrant elevation of the gun.

elevation stop [ENG] Structural unit in a gun or other equipment that prevents it from being elevated or depressed beyond certain fixed limits.

elevation table [ORD] Firing table giving a list of ranges, with the corresponding quadrant elevation settings to be applied to a gun.

elevator [AERO ENG] The hinged rear portion of the longitudinal stabilizing surface or tail plane of an aircraft, used to obtain longitudinal or pitch-control moments. [MECH ENG] Also known as elevating machine. **1.** Vertical, continuous-belt, or chain device with closely spaced buckets, scoops, arms, or trays to lift or elevate powders, granules, or solid objects to a higher level. **2.** Pneumatic device in which air or gas is used to elevate finely powdered materials through a closed conduit. **3.** An enclosed platform or car that moves up and down in a shaft for transporting people or materials. Also known as lift.

elevator angle [AERO ENG] The angular displacement of the elevator from its neutral position; it is positive when the trailing edge of the elevator is below the neutral position, and negative when it is above.

elevator dredge [MECH ENG] A dredge which has a chain of buckets, usually flattened across the front and mounted on a nearly vertical ladder; used principally for excavation of sand and gravel beds under bodies of water.

eleven punch *See* X punch.

elevon [AERO ENG] The hinged rear portion of an aircraft wing, moved in the same direction on each side of the aircraft to obtain longitudinal control and differentially to obtain lateral control; elevon is a combination of the words elevator and aileron to denote that an elevon combines the functions of aircraft elevators and ailerons.

ELF *See* extremely low frequency.

elfinwood *See* krummholz.

Elgin extractor [CHEM ENG] Spray-tower, multistage, counterflow extractor in which the diameter of the base section is expanded to eliminate flow restriction at the light-liquid distribution location.

elimination [MATH] A process of deriving from a system of equations a new system with fewer variables, but with precisely the same solutions.

elimination factor [ADP] In information retrieval, the ratio obtained in dividing the number of documents that have not been retrieved by the total number of documents in the file.

elimination reaction [ORG CHEM] A chemical reaction involving elimination of some portion of a reactant compound, with the production of a second compound.

eliminator [ELECTR] Device that takes the place of batteries, generally consisting of a rectifier operating from alternating current.

E lines [ELEC] Contour lines of constant electrostatic field strength referred to some reference base.

E link [NAV ARCH] A bracket attached to one of the arms of the stand that the compass is on (binnacle); it permits the mounting of a quadrantal error corrector.

elinvar [MET] A nickel-chromium steel alloy containing manganese and tungsten in varying amounts and having a low thermal expansion and almost invariable modulus of elasticity; used for chronometer balances and springs for gages and other instruments.

elixir [PHARM] A sweetened, aromatic solution, usually hydroalcoholic, sometimes containing soluble medicants; intended for use only as a flavor or vehicle.

elk [VERT ZOO] *Alces alces.* A mammal (family Cervidae) in Europe and Asia that resembles the North American moose but is smaller; it is the largest living deer.

ellagic acid [ORG CHEM] $C_{14}H_6O_8$ A compound isolated from tannins as yellow crystals that are minimally soluble in hot water. Also known as gallogen.

ellipse [MATH] The locus of all points in the plane at which the sum of the distances from a fixed pair of points, the foci, is a given constant.

ellipsograph [GRAPHICS] A type of compass that draws ellipses.

ellipsoid [MATH] A surface whose intersection with every plane is an ellipse (or circle).

ellipsoidal coordinates [MATH] Coordinates in space determined by confocal quadrics.

ellipsoidal floodlight [ELEC] A lighting unit used in theatrical lighting consisting of an ellipsoidal reflector with fixed spacing and a lamp; power requirements are 250–5000 watts and the reflector diameter is 10–24 inches (25–61 centimeters). Also known as scoop.

ellipsoidal lava *See* pillow lava.

ellipsoidal of wave normals *See* index ellipsoid.

ellipsoidal reflector [OPTICS] A concave ellipsoidal surface from which light is specularly reflected; used in a light projector to focus rays from a light source at the near focal point onto the opposite focal point of the ellipse.

ellipsoidal spotlight [ELEC] A lighting unit consisting of a reflector, lamp, single or multiple lens system, and framing device; power requirements are 250–2000 watts.

ellipsoid of revolution [MATH] An ellipsoid generated by rotation of an ellipse about one of its axes. Also known as spheroid.

ellipsometer [OPTICS] An instrument for determining the degree of ellipticity of polarized light; used to measure the thickness of very thin transparent films by observing light reflected from the film.

ellipsometry [OPTICS] Techniques for measuring the degree of ellipticity of polarized light.

elliptic [SCI TECH] Oval-shaped.

elliptical coordinates [MATH] The coordinates of a point in the plane determined by confocal ellipses and hyperbolas.

elliptical galaxy [ASTRON] A galaxy whose overall shape ranges from a spheroid to an ellipsoid, without any noticeable structural features. Also known as spheroidal galaxy.

elliptical orbit [MECH] The path of a body moving along an ellipse, such as that described by either of two bodies revolving under their mutual gravitational attraction but otherwise undisturbed.

elliptical point [MATH] A point on a surface at which the total curvature is positive.

elliptical polarization [ELECTROMAG] Polarization of an electromagnetic wave in which the electric field vector at any point in space describes an ellipse in a plane perpendicular to the propagation direction.

elliptical projection [MAP] A map of the surface of the earth formed on an ellipse's interior.

elliptical system [ENG] A tracking or navigation system where ellipsoids of position are determined from time or phase summation relative to two or more fixed stations which are the focuses for the ellipsoids.

elliptic arch [ARCH] An arch whose interior curve resembles an ellipse.

elliptic differential equation [MATH] A general type of second-order partial differential equation which includes Laplace's equation, and having the form

$$\sum_{i,j=1}^{N} A_{ij}\,(\partial^2 u/\partial x_i\,\partial x_j) + B = D$$

where the A_{ij} and B are suitably differentiable real functions of the x_i, u and the $\partial u/\partial x_i$, and there exists at each point $(x_1, \ldots, x_n)$ a real linear transformation on the x_i which reduces the quadratic form

$$\sum_{i,j=1}^{N} A_{ij}\,x_i\,x_j$$

to a sum of n squares, all of the same sign. Also known as elliptic partial differential equation.

elliptic function [MATH] An inverse function of an elliptic integral; alternatively, a doubly periodic, meromorphic function of a complex variable.

elliptic gear [MECH ENG] A change gear composed of two elliptically shaped gears, each rotating about one of its focal points.

elliptic geometry [MATH] The geometry obtained from euclidean geometry by replacing the parallel line postulate with the postulate that infinitely many lines may be drawn through a given point, parallel to a given line. Also known as Riemannian geometry.

elliptic integral [MATH] An integral over x whose integrand is a rational function of x and the square root of $p(x)$, where $p(x)$ is a third- or fourth-degree polynomial without multiple roots.

ellipticity *See* axial ratio.

elliptic partial differential equation *See* elliptic differential equation.

elliptic spring [DES ENG] A spring made of laminated steel plates, arched to resemble an ellipse.

elliptocytosis [MED] A rare hereditary disease of man characterized by the presence of large numbers of oval or elliptic erythrocytes in the circulating blood.

ellsworthite [MINERAL] $(Ca,Na,U)_2(Nb,Ta)_2O_6(O,OH)$ A yellow, brown, greenish or black mineral of the pyrochlore group occurring in isometric crystals and consisting of an oxide of niobium, titanium, and uranium. Also known as betafite; hatchettolite.

elm [BOT] The common name for hardwood trees composing the genus *Ulmus*, characterized by simple, serrate, deciduous leaves.

Elmidae [INV ZOO] The drive beetles, a small family of coleopteran insects in the superfamily Dryopoidea.

El Niño [OCEANOGR] A warm current setting south along the coast of Peru; generally develops during February and March concurrently with a southerly shift in the tropical rain belt.

elongation [COMMUN] The extension of the envelope of a signal due to delayed arrival of multipath components. [MECH] The fractional increase in a material's length due to stress in tension or to thermal expansion.

Elopidae [VERT ZOO] A family of fishes in the order Elopiformes, including the tarpon, ladyfish, and machete.

Elopiformes [VERT ZOO] A primitive order of actinopterygian fishes characterized by a single dorsal fin composed of soft rays only, cycloid scales, and toothed maxillae.

elpasolite [MINERAL] K_2NaAlF_6 Mineral composed of sodium potassium aluminum fluoride.

elpidite [MINERAL] $Na_2ZrSi_6O_{15}\cdot3H_2O$ A white to brick-red mineral composed of hydrated sodium zirconium silicate.

ELR scale *See* equal listener response scale.

Elsasser's radiation chart [METEOROL] A radiation chart developed by W. M. Elsasser for the graphical solution of the radiative transfer problems of importance in meteorology: given a radiosonde record of the vertical variation of tempera-

ELM

Twig, leaf, fruit, and terminal and lateral buds of American elm (*Ulmus americana*).

ture and water vapor content, one can find with this chart such quantities as the effective terrestrial radiation, net flux of infrared radiation at a cloud base or a cloud top, and radiative cooling rates.

else rule [ADP] A convention in decision tables which spells out which action to take in the case specified conditions are not met.

Elster-Geitel effect [PHYS] The phenomenon in which a heated conductor acquires a positive or negative electric charge in the presence of a gas, while in a vacuum it always acquires a negative charge.

El Tor vibrio [MICROBIO] Any of the rod-shaped paracholera vibrios; many strains can be agglutinated with anticholera serum.

eluant [CHEM] A liquid used to extract one material from another, as in chromatography.

eluate [CHEM] The solution that results from the elution process.

elusive ulcer *See* Hunner's ulcer.

elution [CHEM] The process of extracting one material from another. Also known as elutriation; levigation.

elutriate [ENG] To separate or purify; for example, to separate ore by washing, decanting, and settling.

elutriation *See* elution.

elutriator [ENG] An apparatus used to separate suspended solid particles according to size by the process of elution.

eluvial [GEOL] Of, composed of, or relating to eluvium.

eluviation [HYD] The process of transporting dissolved or suspended materials in the soil by lateral or downward water flow when rainfall exceeds evaporation.

eluvium [GEOL] Disintegrated rock material formed and accumulated in situ or moved by the wind alone.

elvegust [METEOROL] A cold descending squall in the upper parts of Norwegian fjords. Also known as sno.

elytron [INV ZOO] **1.** One of the two sclerotized or leathery anterior wings of beetles which serve to cover and protect the membranous hindwings. **2.** A dorsal scale of certain Polychaeta.

em [GRAPHICS] The space occupied by the letter M; column width is measured in em picas, the space (about 4 millimeters) occupied by an M in pica-size type.

emaciation [MED] A wasted condition of the body; the process of losing flesh so as to become extremely lean.

emagram [THERMO] A graph of the logarithm of the pressure of a substance versus its temperature, when it is held at constant volume; in meteorological investigations, the potential temperature is often the parameter.

emanating power [NUCLEO] The fraction of radon atoms, formed in a solid or solution, which escape.

emanation *See* radioactive emanation.

emanation security [ELECTR] The protection resulting from all measures designed to deny unauthorized persons information of value which might be derived from unintentional emissions from other than telecommunications systems.

emanometer [ENG] An instrument for the measurement of the radon content of the atmosphere: radon is removed from a sample of air by condensation or adsorption on a surface, and is then placed in an ionization chamber and its activity determined.

emanometry [NUCLEO] The collective techniques for ionization-chamber determination of the amounts of radioactive gases escaping into the lower atmosphere from the earth's surface; in emanometric measurements, the objective is to count, by typical ionization-chamber methods, all of the ions produced by the alpha particles emitted by the one or more radioactive gases contained in the chamber.

emarginate [BIOL] Having a margin that is notched or slightly forked.

embacle [HYD] The piling up of ice in a stream after a refreeze, and the pile so formed.

EMB agar [MICROBIO] A culture medium containing sugar, eosin, and methylene blue, used in the confirming test for coliform bacteria.

Emballonuridae [VERT ZOO] The sheath-tailed bats, a family of mammals in the order Chiroptera.

embalm [MED] To treat a cadaver with antiseptics and preservatives to prevent decay, before burial or dissection.

embankment [CIV ENG] **1.** A ridge constructed of earth, stone, or other material to carry a roadway or railroad at a level above that of the surrounding terrain. **2.** A ridge of earth or stone to prevent water from passing beyond desirable limits. Also known as bank.

embarkation [ORD] The loading of troops with their supplies and equipment into ships or aircraft.

embata [METEOROL] A local onshore southwest wind caused by the reversal of the northeast trade winds in the lee of the Canary Islands.

embatholithic [GEOL] Pertaining to ore deposits associated with a batholith where exposure of the batholith and country rock is about equal.

embayed [GEOGR] Formed into a bay. [NAV] Pertaining to a vessel in a bay unable to put to sea or to put to sea safely because of wind, current, or sea.

embayed coastal plain [GEOL] A coastal plain that has been partly sunk beneath the sea, thereby forming a bay.

embayed mountain [GEOL] A mountain that has been depressed enough for sea water to enter the bordering valleys.

embayment [GEOGR] Indentation in a shoreline forming a bay. [GEOL] **1.** Act or process of forming a bay. **2.** A reentrant of sedimentary rock into a crystalline massif.

Embden-Meyerhof pathway *See* glycolytic pathway.

embed Also spelled imbed. [BIOL] To prepare a specimen for sectioning for microscopic examination by infiltrating with or enclosing in paraffin or other supporting material. [SCI TECH] **1.** To enclose in a matrix. **2.** To closely surround.

embedment anchor [NAV ARCH] An anchor that uses a charge of powder or hydrostatic or pneumatic pressure to drive the fluke into the sea floor.

Embiidina [INV ZOO] An equivalent name for Embioptera.

Embioptera [INV ZOO] An order of silk-spinning, orthopteroid insects resembling the grasshoppers; commonly called the embiids or web spinners.

Embiotocidae [VERT ZOO] The surfperches, a family of perciform fishes in the suborder Percoidei.

embolectomy [MED] Surgical removal of an embolus.

embolism [MED] The blocking of a blood vessel by an embolus.

embolite [MINERAL] Ag(Cl,Br) A yellow-green mineral resembling cerargyrite; composed of native silver chloride and silver bromide.

Embolomeri [PALEON] An extinct side branch of slender-bodied, fish-eating aquatic anthracosaurs in which intercentra as well as centra form complete rings.

embolus [MED] A clot or other mass of particulate matter foreign to the bloodstream which lodges in a blood vessel and causes obstruction.

emboly [EMBRYO] Formation of a gastrula by the process of invagination.

embossed plate printer [ADP] In character recognition, a data preparation device which accomplishes printing by allowing a raised character behind the paper to push the paper against the printing ribbon in front of the paper.

embossing [GRAPHICS] Producing a raised pattern on the surface of paper or wood by means of a die.

embossing die [GRAPHICS] A die used for embossing.

embossing stylus [ENG ACOUS] A recording stylus with a rounded tip that forms a groove by displacing material in the recording medium.

embouchure [GEOL] **1.** The mouth of a river. **2.** A river valley widened into a plain.

embrasure [ARCH] **1.** Opening in a wall or parapet, especially one through which a gun is fired. **2.** An opening such as for a door or window with sloping or beveled sides.

embrechites [PETR] A type of migmatite in which structural features of crystalline shifts are preserved but often partially obliterated by metablastesis.

Embrithopoda [PALEON] An order established for the unique Oligocene mammal *Arsinoitherium*, a herbivorous animal that resembled the modern rhinoceros.

embrittlement [MECH] Reduction or loss of ductility in a metal or plastic.

embryo [BOT] Young sporophyte of a seed plant. [EMBRYO] **1.** An early stage of development in multicellular organisms.

EMBIOPTERA

Body form of a typical embiid (*Pararhagadochir trachelia* Navas, female). (*From E. S. Ross, Insects Close Up; Univeristy of California Press, 1953*)

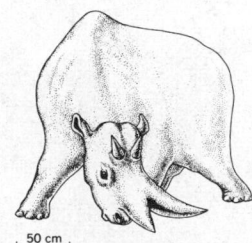

EMBRITHOPODA

Arsinoitherium, the early Oligocene embrithopod from Egypt. (*After C. R. Knight*)

50 cm

2. The product of conception up to the third month of human pregnancy.

Embryobionta [BOT] The land plants, a subkingdom of the Plantae characterized by having specialized conducting tissue in the sporophyte (except bryophytes), having multicellular sex organs, and producing an embryo.

embryogenesis [EMBRYO] The formation and development of an embryo from an egg.

embryology [BIOL] The study of the development of the organism from the zygote, or fertilized egg.

embryoma of the ovary *See* dysgerminoma.

embryonal-cell lipoma *See* liposarcoma.

embryonate [EMBRYO] **1.** To differentiate into a zygote. **2.** Containing an embryo.

embryonated egg culture [VIROL] Embryonated hen's eggs inoculated with animal viruses for the purpose of identification, isolation, titration, or for quantity cultivation in the production of viral vaccines.

embryonic differentiation [EMBRYO] The process by which specialized and diversified structures arise during embryogenesis.

embryonic inducer [EMBRYO] The acting system in embryos, which contributes to the formation of specialized tissues by controlling the mode of development of the reacting system.

embryonic induction [EMBRYO] The influence of one cell group (inducer) over a neighboring cell group (induced) during embryogenesis. Also known as induction.

embryopathy [MED] Any abnormal development of an embryo, either morphological or biochemical.

Embryophyta [BOT] The equivalent name for Embryobionta.

embryo sac [BOT] The female gametophyte of a seed plant, containing the egg, synergids, and polar and antipodal nuclei; fusion of the antipodals and a pollen generative nucleus forms the endosperm.

embryotomy [MED] Any mutilation of the fetus in the uterus to aid in its removal when natural delivery is impossible.

emerald [MINERAL] $Al_2(Be_3Si_6O_{18})$ A brilliant-green to grass-green gem variety of beryl that crystallizes in the hexagonal system; green color is caused by varying amounts of chromium. Also known as smaragd.

emerald cut [LAP] A step cut in which the gem girdle outline is square or rectangular and the steps of elongated facets are parallel to the girdle, with sets on each of the four sides and at the corners.

emerged shoreline *See* shoreline of emergence.

emergence [GEOL] **1.** Dry land which was part of the ocean floor. **2.** The act or process of becoming an emergent land mass.

emergency anchorage [NAV] An anchorage, which may have a limited defense organization, for naval vessels, mobile support units, auxiliaries, or merchant ships.

emergency brake [MECH ENG] A brake that can be set by hand and, once set, continues to hold until released; used as a parking brake in an automobile.

emergency broadcast system [COMMUN] A system of broadcast stations and interconnecting facilities authorized by the U.S. Federal Communications Commission to operate in a controlled manner during a war, threat of war, state of public peril or disaster, or other national emergency.

emergency power supply [ELEC] A source of power that becomes available, usually automatically, when normal power line service fails.

emergency radio channel [COMMUN] Any radio frequency reserved for emergency use, particularly for distress signals.

emergency receiver [COMMUN] Receiver immediately available in a station for emergency communications and capable of being energized solely by self-contained or emergency power supply.

emergency recorder plot [NAV] An emergency substitute for a tactical range recorder, consisting of a plastic sheet representing space and time coordinates on which speed and distance data are marked with a grease pencil as they are called off verbally. Abbreviated ERP.

emersion [ASTRON] The reappearance of a celestial body after an eclipse or occultation.

Emerson wage incentive plan [IND ENG] A plan comprising time wages to 66⅔% of standard performance, empiric bo-

nuses from there to standard performance, ending at 120% time wages, and thereafter a straight-line earning which is 20% above and parallel to basic piece rate.

emery [MATER] An abrasive composed of pulverized, impure corundum; used in polishing and grinding. [MINERAL] A fine, granular, gray-black, impure variety of corundum containing iron oxides, either hematite or magnetite; occurs as masses in limestone and as segregations in igneous rock.

emery cake [MATER] Caked, powdered emery in a binding material.

Emery-Dietz gravity corer [ENG] A tube, with weights attached, which forces sediment samples into its interior as it is dropped on the ocean bottom.

emery paper [MATER] An abrasive paper or cloth with an adherent surface layer of emery powder; used for polishing and cleaning metal.

emery rock [PETR] A rock that contains corundum and iron ores.

emery stone [MATER] **1.** A sharpening stone. **2.** A mixture of powdered emery and a binder which can be molded into grinding devices.

emery wheel [DES ENG] A grinding wheel made of or having a surface of emery powder; used for grinding and polishing.

emesis [MED] The act of vomiting.

emetic [PHARM] Any agent that induces emesis.

emetine [PHARM] $C_{29}H_{40}N_2O_4$ Cephaeline methyl ether, the principal alkaloid of ipecac; a white powder, sparingly soluble in water; it is emetic, diaphoretic, and expectorant, but its chief utility is as an amebicide.

emf *See* electromotive force.

emi *See* electromagnetic interference.

emiocytosis [CYTOL] Fusion of intracellular granules with the cell membrane, followed by discharge of the granules outside of the cell; applied chiefly to the mechanism of insulin secretion. Also known as reverse pinocytosis.

emissary sky [METEOROL] A sky of cirrus clouds which are either isolated or in small, separated groups; so called because this formation often is one of the first indications of the approach of a cyclonic storm.

emission [ELECTROMAG] Any radiation of energy by means of electromagnetic waves, as from a radio transmitter.

emission characteristics [ELECTR] Relation, usually shown by a graph, between the emission and a factor controlling the emission, such as temperature, voltage, or current of the filament or heater.

emission electron microscope [ELECTR] An electron microscope in which thermionic, photo, secondary, or field electrons emitted from a metal surface are projected on a fluorescent screen, with or without focusing.

emission flame photometry [ANALY CHEM] A form of flame photometry in which a sample solution to be analyzed is aspirated into a hydrogen-oxygen or acetylene-oxygen flame; the line emission spectrum is formed, and the line or band of interest is isolated with a monochromator and its intensity measured photoelectrically.

emission lines [SPECT] Spectral lines resulting from emission of electromagnetic radiation by atoms, ions, or molecules during changes from excited states to states of lower energy.

emission security [ELECTR] That component of communications security which results from all measures taken to protect any unintentional emissions of a telecommunications system from any form of exploitation other than cryptanalysis.

emission spectrum [SPECT] Electromagnetic spectrum produced when radiations from any emitting source, excited by any of various forms of energy, are dispersed.

emissive power *See* emittance.

emissivity [THERMO] The ratio of the radiation emitted by a surface to the radiation emitted by a perfect blackbody radiator at the same temperature. Also known as thermal emissivity.

emittance [THERMO] The power radiated per unit area of a radiating surface. Also known as emissive power; radiating power.

emitter [ADP] A time pulse generator found in some equipment, such as a card punch. [ELECTR] A transistor region from which charge carriers that are minority carriers in the base are injected into the base, thus controlling the current

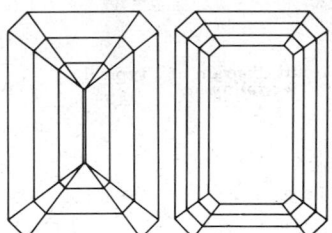

EMERALD CUT

Emerald-cut gemstones. *(Gemological Institute of America)*

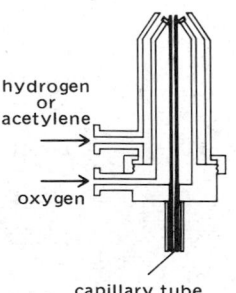

EMISSION FLAME PHOTOMETRY

hydrogen or acetylene

oxygen

capillary tube for introduction of sample solution

Beckman aspirator-burner used in emission flame photometry.

flowing through the collector; corresponds to the cathode of an electron tube. Symbolized E. Also known as emitter region.

emitter barrier [ELECTR] One of the regions in which rectification takes place in a transistor, lying between the emitter region and the base region.

emitter bias [ELECTR] A bias voltage applied to the emitter electrode of a transistor.

emitter-coupled transistor logic [ELECTR] A logic operator circuit consisting of transistors and resistors formed simultaneously on a silicon wafer; it is a negative logic system based on emitter-follower gate inputs, with a two-phase clock when timing is required. Abbreviated ECTL.

emitter follower [ELECTR] A grounded-collector transistor amplifier which provides less than unity voltage gain but high input resistance and low output resistance, and which is similar to a cathode follower in its operations.

emitter junction [ELECTR] A transistor junction normally biased in the low-resistance direction to inject minority carriers into a base.

emitter region *See* emitter.

emitter resistance [ELECTR] The resistance in series with the emitter lead in an equivalent circuit representing a transistor.

emmetropia [MED] Normal vision.

emmonsite [MINERAL] $FE_2Te_3O_9 \cdot 2H_2O$ Yellow-green mineral composed of a hydrous oxide of iron and tellurium.

E mode *See* transverse magnetic mode.

emodin [ORG CHEM] $C_{14}H_4O_2(OH)_3CH_3$ Orange needles crystallizing from alcohol solution, melting point 256–257°C, practically insoluble in water, soluble in alcohol and aqueous alkali hydroxide solutions, occurs as the rhamnoside in plants such as rhubarb root and alder buckthorn; used as a laxative. Also known as archen; frangula emodin; frangulic acid; rheum emodin.

emollient [PHARM] A softening agent, especially for use on skin and mucous membranes; lanolin is widely used as a base.

emotion [PSYCH] A strong mental feeling or affect of the consciousness involving visceral and other physiologic changes.

empennage [AERO ENG] The assembly at the rear of an aircraft; it comprises the horizontal and vertical stabilizers. Also known as tail assembly.

emphasizer *See* preemphasis network.

emphysema [MED] A pulmonary disorder characterized by overdistention and destruction of the air spaces in the lungs.

emphysematous chest [MED] The altered contour of the chest seen in pulmonary emphysema, with increased antero-posterior diameter, flaring at the lower rib margins, low position of the diaphragm, and minimal respiratory motion. Also known as barrel chest.

Empididae [INV ZOO] The dance flies, a family of orthorrhaphous dipteran insects in the series Nematocera.

empire cloth [MATER] Cotton cloth coated with oxidized oil; used as an electrical insulator.

empirical [SCI TECH] Based on actual measurement, observation, or experience, rather than on theory.

empirical curve [MATH] A smooth curve drawn through or close to points representing measured values of two variables on a graph.

empirical formula [CHEM] A chemical formula indicating the variety and relative proportions of the atoms in a molecule but not showing the manner in which they are linked together.

empirical probability [STAT] The ratio of the number of times an event has occurred to the total number of trials performed. Also known as a posteriori probability.

empirical rule [SCI TECH] A rule which is derived from measurements or observations, and is not based on any theory.

emplacement [GEOL] Intrusion of igneous rock or development of an ore body in older rocks. [ORD] **1.** Prepared position for one or more weapons or pieces of equipment, protecting against hostile fire or bombardment but permitting execution of the mission of the weapons. **2.** Act of fixing a gun in a prepared position from which it may be fired.

emplectite [MINERAL] $CuBiS_3$ A grayish or white mineral

that crystallizes in the orthorhombic system; occurs in masses.

employment test [IND ENG] Any of a wide variety of tests to measure intelligence, personality traits, skills, interests, aptitudes, or other characteristics; used to supplement interviews, physical examinations, and background investigations before employment.

empodium [INV ZOO] A small peripheral part located between the claws of the tarsi of many insects and arachnids.

empressite [MINERAL] AgTe An opaque, pale-bronze mineral whose crystal system is unknown.

empty [ORD] In ammunition nomenclature, indicates that the munition does not contain a payload, but is designed to contain one at the time of final use.

empty-cell process [ENG] A wood treatment in which the preservative coats the cells without filling them.

empty set [MATH] The set with no elements.

empyema [MED] The presence of pus in the body cavity, hollow organ, or tissue space; when the term is used without qualification, it generally refers to pus in the pleural space.

emu [ELECTROMAG] *See* electromagnetic system of units. [VERT ZOO] *Dromiceius novae-hollandiae.* An Australian rat-ite bird, the second largest living bird, characterized by rudimentary wings and a feathered head and neck without wattles.

emulation mode [ADP] A method of operation in which a computer actually executes the instructions of a different (simpler) computer, in contrast to normal mode.

emulator [ADP] The microprogram-assisted macroprogram which allows a computer to run programs written for another computer.

emulsification [CHEM] The process of dispersing one liquid in a second immiscible liquid; the largest group of emulsifying agents are soaps, detergents, and other compounds, whose basic structure is a paraffin chain terminating in a polar group.

emulsification test [CHEM ENG] Standard laboratory procedure for evaluating the resistance of insulating oils, turbine oils, and other lubricating oils to emulsification.

emulsified asphalt [MATER] An emulsion of asphalt cement and water with a small quantity of an emulsifying agent.

emulsifier *See* disperser.

emulsifying agent *See* disperser.

emulsifying oil *See* soluble oil.

emulsion [CHEM] A stable dispersion of one liquid in a second immiscible liquid, such as milk (oil dispersed in water). [GRAPHICS] In photography, the photosensitized material on film, plates, and various photographic papers.

emulsion breaking [CHEM] In an emulsion, the combined sedimentation and coalescence of emulsified drops of the dispersed phase so that they will settle out of the carrier liquid; can be accomplished mechanically (in settlers, cyclones, or centrifuges) with or without the aid of chemical additives to increase the surface tension of the droplets.

emulsion cleaner [CHEM ENG] A cleaner composed of organic solvents dispersed in an aqueous solution with the aid of an emulsifying agent.

emulsion paint [MATER] Paint whose vehicle is an emulsion of a binder (oil, resin, latex, and so on) in water.

emulsion polymerization [ORG CHEM] A polymerization reaction that occurs in one phase of an emulsion.

emulsion speed [GRAPHICS] Sensitivity of a photographic emulsion to light, under standard conditions of exposure and development.

Emydidae [VERT ZOO] A family of aquatic and semiaquatic turtles in the suborder Cryptodira.

enable [ADP] To authorize an activity which would otherwise be suppressed, such as to write on a tape.

enabling pulse [ELECTR] A pulse that prepares a circuit for some subsequent action.

Enaliornithidae [PALEON] A family of extinct birds assigned to the order Hesperornithiformes, having well-developed teeth found in grooves in the dentary and maxillary bones of the jaws.

enamel [MATER] **1.** A finely ground, resin-containing oil paint that dries relatively harder, smoother, and glossier than ordinary paint. **2.** *See* glaze.

EMITTER FOLLOWER

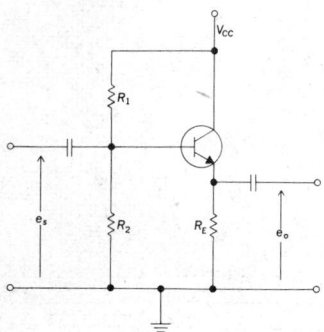

Circuit diagram of a typical emitter follower.

enamel clay [MATER] A ball clay able to float nonplastic enamel slips to make them spray or dip more evenly.

enameled brick [MATER] Brick with a smooth hard surface acquired from the application of a special wash before burning.

enameling [ENG] The application of a vitreous glaze to pottery or metal surfaces, followed by fusing in a kiln or furnace.

enamel kiln [ENG] A kiln in which enamel colors are fired.

enamel organ [EMBRYO] The epithelial ingrowth from the dental lamina which covers the dental papilla, furnishes a mold for the shape of a developing tooth, and forms the dental enamel.

enamel oxide [MATER] Any of the mixtures of calcined oxides used to color vitreous enamels used on sheet steel or cast iron.

enamel paper See coated paper.

enantiomer See enantiomorph.

enantiomorph [CHEM] One of an isomeric pair of crystalline forms or compounds whose molecules are nonsuperimposable mirror images. Also known as enantiomer; optical antipode; optical isomer.

enantiomorphism [CHEM] A phenomenon of mirror-image relationship exhibited by right-handed and left-handed crystals or by the molecular structures of two stereoisomers.

enantiotropy [CHEM] The relation of crystal forms of the same substance in which one form is stable above the transition-point temperature and the other stable below it, so that the forms can change reversibly one into the other.

Enantiozoa [INV ZOO] The equivalent name for Parazoa.

enargite [MINERAL] A lustrous, grayish-black mineral which is found in orthorhombic crystals but is more commonly columnar, bladed, or massive; hardness is 3 on Mohs scale, specific gravity is 4.44; in some places enargite is a valuable copper ore. Also known as clairite; luzonite.

enarthrosis [ANAT] A freely movable joint that allows a wide range of motion on all planes. Also known as ball-and-socket joint.

encapsulate [SCI TECH] To surround, encase, or enclose as if in a capsule; for example, the formation of a protective coating around a bacterium, or the enclosure of an item such as an electronic component in plastic.

encaustic [GRAPHICS] A method of painting in which the pigment is carried in hot wax.

Enceladus [ASTRON] A satellite of Saturn orbiting at a mean distance of 238,000 kilometers.

encephalitis [MED] Inflammation of the brain.

encephalitis lethargica [MED] Epidemic encephalitis, probably of viral etiology, characterized by lethargy, ophthalmoplegia, hyperkinesia, and at times residual neurologic disability, particularly parkinsonism with oculogyric crisis. Also known as epidemic encephalitis; sleeping sickness; von Economo's disease.

encephalocele [MED] Hernia of the brain through a congenital or traumatic opening in the cranium.

encephalogram [MED] A roentgenogram of the brain made in encephalography.

encephalography [MED] Roentgenography of the brain following removal of cerebrospinal fluid, by lumbar or cisternal puncture, and its replacement by air or other gas.

encephaloid carcinoma See medullary carcinoma.

encephalomalacia [MED] 1. Infarction of the brain. 2. Any softening or fragmentation of the brain.

encephalomyelitis [MED] Inflammation of the brain and spinal cord.

encephalomyocarditis [MED] An acute febrile RNA virus disease accompanied by pharyngitis, stiff neck, and hyperactive deep reflexes; certain species of wild rats are the reservoir; human infections range from a mild febrile illness to a severe encephalomyelitis.

encephalopathy [MED] Any disease of the brain.

enchondroma [MED] A benign tumor composed of dysplastic cartilage cells, occurring in the metaphysis of cylindric bones, especially of the hands and feet.

enchondromatosis [MED] A rare disorder principally involving tubular bones, especially those of the feet and hands, characterized by hamartomatous proliferation of cartilage in the metaphysis, indistinguishable in single lesions from enchondromas. Also known as chondrodysplasia; dyschondroplasia; Ollier's disease.

enchymatous [PHYSIO] Of gland cells, distended with secreted material.

encipher [COMMUN] To convert a plain-text message into unintelligible language by means of a cryptosystem. Also known as encrypt.

enciphered facsimile communications [COMMUN] Communications in which security is accomplished by mixing pulses produced by a key generator with the output of the facsimile converter; plain text is recovered by subtracting the identical key at the receiving terminal; unauthorized listeners are unable to reconstruct the plain text unless they have an identical key generator and the daily key setting.

Encke division [ASTRON] A faint line that splits the outer ring of Saturn into two.

Encke roots [MATH] For any two numbers a_1 and a_2, the numbers $-x_1$ and $-x_2$, where x_1 and x_2 are the roots of the equation $x^2 + a_1 x + a_2 = 0$, with $|x_1| < |x_2|$.

Encke's Comet [ASTRON] A very faint comet with the shortest period of any known comet, 3.3 years.

enclosed arc lamp [ELEC] An arc lamp in which the arc produced by carbon electrodes is protected from the atmosphere by a translucent enclosure.

encode [ADP] To prepare a routine in machine language for a specific computer. [COMMUN] To express given information by means of a code.

encoded abstract [ADP] An abstract prepared to be scanned by automatic electronic machines.

encoded question [ADP] A question set up and encoded in the form appropriate for operating, programming, or conditioning a searching device.

encoder [ADP] In character recognition, that class of printer which is usually designed for the specific purpose of printing a particular type font in predetermined positions on certain size forms. [ELECTR] 1. In an electronic computer: a network or system in which only one input is excited at a time and each input produces a combination of outputs. 2. See matrix.

encoding strip [ADP] In character recognition, the area reserved for the inscription of magnetic-ink characters, as in bank checks.

encrinal limestone [GEOL] A limestone consisting of more than 10% but less than 50% of fossil crinoidal fragments.

encrinite [PALEON] One of certain fossil crinoids, especially of the genus *Encrinus*.

encrustation [ENG] The buildup of slag or other material inside furnaces and kilns.

encrypt See encipher.

encrypted message part [COMMUN] Portion of a long message which is sent as a separate complete message with its own date-time group, station serial number, and group count.

Encyrtidae [INV ZOO] A family of hymenopteran insects in the superfamily Chalcidoidea.

encystment [BIOL] The process of forming or becoming enclosed in a cyst or capsule.

end See warp.

Endamoeba [INV ZOO] The type genus of the Endamoebidae comprising insect parasites and, in some systems of classification, certain vertebrate parasites.

Endamoeba coli [INV ZOO] A nonpathogenic ameba that inhabits the human intestinal tract.

end-and-end [TEXT] Pertaining to a weave with a minuscule check effect made by alternating white and colored warp yarns. Also known as end-to-end.

endarch [BOT] Formed outward from the center, referring to xylem or its development.

end-around carry [ADP] A carry from the most significant digit place to the least significant digit place.

end-around shift See cyclic shift.

endarteritis [MED] Inflammation of the lining (tunica intima) of an artery.

endarteritis obliterans [MED] Endarteritis, particularly of small arteries, accompanied by degeneration of the intima, leading to occlusion of the blood vessel. Also known as obliterating endarteritis.

ENARGITE

|–2.5 cm–|

Enargite crystals, Silver Bow County, Montana. (*Specimen from Department of Geology, Bryn Mawr College*)

end-bearing pile [CIV ENG] A bearing pile that is driven down to hard ground so that it carries the full load at its point. Also known as a point-bearing pile.

end bulb See bouton.

end bulb of Krause See Krause's corpuscle.

end cell [ELEC] One of a group of cells in series with a storage battery, which can be switched in to maintain the output voltage of the battery when it is not being charged.

end-cell rectifier [ELECTR] Small trickle charge rectifier used to maintain voltage of the storage battery end cells.

end construction [CIV ENG] Structural blocks or tiles laid so that the hollow cells run vertically.

end-construction tile [MATER] A type of structural clay tile designed to receive its principal stress parallel to the axis of the cells.

end distortion [COMMUN] The displacement of trailing edges of marking pulses transmitted over a teletypewriter circuit relative to the leading edge of the start pulse.

end effect [ELECTROMAG] The effect of capacitance at the ends of an antenna; it requires that the actual length of a half-wave antenna be about 5% less than a half wavelength.

Endeidae [INV ZOO] A family of marine arthropods in the subphylum Pycnogonida.

endellite [MINERAL] $Al_2Si_2O_5(OH)_4 \cdot 4H_2O$ Term used in the United States for a clay mineral, the more hydrous form of halloysite. Also known as hydrated halloysite; hydrohalloysite; hydrokaolin.

endemic [MED] Peculiar to a certain region, specifically referring to a disease which occurs more or less constantly in any locality.

endemic goiter [MED] Goiter peculiar to areas that are iodine-poor in food, water, or soil.

endemic rural plague See sylvatic plague.

endemic typhus See murine typhus.

endemism [MED] The state or quality of being endemic.

endergonic [BIOCHEM] Of or pertaining to a biochemical reaction in which the final products possess more free energy than the starting materials; usually associated with anabolism.

endermic [MED] Acting through the skin by absorption, such as medication applied to the skin.

end-fire antenna See end-fire array.

end-fire array [ELECTROMAG] A linear array whose direction of maximum radiation is along the axis of the array; it may be either unidirectional or bidirectional; the elements of the array are parallel and in the same plane, as in a fishbone antenna. Also known as end-fire antenna.

end instrument [ELECTR] A pickup used in telemetering to convert a physical quantity to an inductance, resistance, voltage, or other electrical quantity that can be transmitted over wires or by radio.

endite [INV ZOO] **1.** One of the appendages on the inner aspect of an arthropod limb. **2.** A ridgelike chewing surface on the inner part of the pedipalpus or maxilla of many arachnids.

end item [ENG] A final combination of end products, component parts, or materials which is ready for its intended use; for example, ship, tank, mobile machine shop, or aircraft.

end lap [DES ENG] A joint in which two joining members are made to overlap by removal of half the thickness of each.

endless tangent screw [NAV] That portion of the mechanism of a marine sextant which moves the index arm.

endless tangent screw sextant [NAV] A marine sextant having an endless tangent screw for controlling the position of the index arm and the vernier or micrometer drum; the index arm may be moved over the entire arc without resetting, by means of the endless tangent screw.

endlichite [MINERAL] A mineral similar to vanadinite, but with the vanadium replaced by arsenic.

end loader [MECH ENG] A platform elevator at the rear of a truck.

end loss [ELECTROMAG] The difference between the actual and effective lengths of a radiating antenna element.

end mark [ADP] A mark which signals the end of a unit of information.

end member [MINERAL] One of the two or more pure chemical compounds that enters into solid solution with other pure chemical compounds to make up a series of minerals of similar crystal structure (that is, an isomorphous, solid-solution series).

end mill [MET] A machine which has a rotating shank with cutting teeth at the end and spiral blades on the peripheral surface; used for shaping and cutting metal.

end-milled keyway See profiled keyway.

end moraine [GEOL] An accumulation of drift in the form of a ridge along the border of a valley glacier or ice sheet.

endo- [ORG CHEM] Prefix that denotes inward-directed valence bonds of a six-membered ring in its boat form. [SCI TECH] Prefix denoting within or inside.

endobasion [ANAT] The anteriormost point of the margin of the foramen magnum at the level of its smallest diameter.

endobatholithic [GEOL] Pertaining to ore deposits along projecting portions of a batholith.

endobiotic [ECOL] Referring to an organism living in the cells or tissues of a host.

endobranchiate [ZOO] Animal form with endodermal gills.

endocardial fibroelastosis [MED] Fibrous or fibroelastic thickening of the endocardium, of unknown cause.

endocarditis [MED] Inflammation of the endocardium.

endocardium [ANAT] The membrane lining the heart.

endocarp [BOT] The inner layer of the wall of a fruit or pericarp.

endocast See steinkern.

endocervicitis [MED] Inflammation of the mucous membrane of the uterine cervix.

endocervix [ANAT] The glandular mucous membrane of the cervix uteri.

endochondral ossification [PHYSIO] The conversion of cartilage into bone. Also known as intracartilaginous ossification.

endocommensal [ECOL] A commensal that lives within the body of its host.

endocorpuscular [CYTOL] Located within an erythrocyte.

endocranium [ANAT] **1.** The inner surface of the cranium. **2.** See dura mater. [INV ZOO] The processes on the inner surface of the head capsule of certain insects.

endocrine gland [PHYSIO] A ductless structure whose secretion (hormone) is passed into adjacent tissue and then to the bloodstream either directly or by way of the lymphatics. Also known as ductless gland.

endocrine system [PHYSIO] The chemical coordinating system in animals, that is, the endocrine glands that produce hormones.

endocrinology [PHYSIO] The study of the endocrine glands and the hormones that they synthesize and secrete.

endocuticle [INV ZOO] The inner, elastic layer of an insect cuticle.

endocyst [INV ZOO] The soft layer consisting of ectoderm and mesoderm, lining the ectocyst of bryozoans.

endocytic vacuole [CYTOL] A membrane-bound cellular organelle containing extracellular particles engulfed by the mechanisms of endocytosis.

endocytosis [CYTOL] An active process in which extracellular materials are introduced into the cytoplasm of cells by either phagocytosis or pinocytosis.

endoderm [EMBRYO] The inner, primary germ layer of an animal embryo; sometimes referred to as the hypoblast. Also known as entoderm; hypoblast.

endodermis [BOT] The innermost tissue of the cortex of most plant roots and certain stems consisting of a single layer of at least partly suberized or cutinized cells; functions to control the movement of water and other substances into and out of the stele.

endoenzyme [BIOCHEM] An intracellular enzyme, retained and utilized by the secreting cell.

endoergic See endothermic.

endoergic collision See collision of the first kind.

end of file [ADP] **1.** Termination or point of completion of a quantity of data; end of file marks are used to indicate this point. **2.** Automatic procedures to handle tapes when the end of an input or output tape is reached; a reflective spot, called a record mark, is placed on the physical end of the tape to signal the end.

end-of-file gap [ADP] A gap of precise dimension to indicate the end of a file on tape. Abbreviated EOF gap.

end-of-file indicator [ADP] **1.** A device that indicates the end of a file on tape. **2.** *See* end-of-file mark.

end-of-file mark [ADP] A control character which signifies that the last record of a file has been read. Also known as end-of-file indicator.

end-of-file routine [ADP] A program which checks that the contents of a file read into the computer were correctly read; may also start the rewind procedure.

end-of-file spot [ADP] A reflective piece of tape indicating the end of the tape.

end-of-job control card [ADP] The last card in a deck, normally punched with a distinctive code indicating that no additional cards are required for the job.

end-of-message [COMMUN] A character or series of characters signifying the end of a message or record, such as a message sent by teletypewriter.

end-of-record gap [ADP] A gap of precise dimension (shorter than the end-of-file gap) which indicates the physical end of a record on a magnetic tape. Abbreviated EOR gap.

end-of-tape routine [ADP] A program which is brought into play when the end of a tape is reached; may involve a series of validity checks and initiate the tape rewind.

end-of-transmission card [COMMUN] Last card of each message; used to signal the end of a transmission and contains the same information as the header card, plus additional data for traffic analysis.

end-of-transmission recognition [ADP] The capability of a computer to recognize the end of transmission of a data string even if the buffer area is not filled.

endogamy [BIOL] Sexual reproduction between organisms which are closely related. [BOT] Pollination of a flower by another flower of the same plant.

endogenetic *See* endogenic.

endogenic [GEOL] Of or pertaining to a geologic process, or its resulting feature such as a rock, that originated within the earth. Also known as endogenetic; endogenous.

endogenote [MICROBIO] The genetic complement of the partial zygote formed as a result of gene transfer during the process of recombination in bacteria.

endogenous [BIOCHEM] Relating to the metabolism of nitrogenous tissue elements. [GEOL] *See* endogenic. [MED] Pertaining to diseases resulting from internal causes. [PSYCH] Pertaining to mental disorders caused by hereditary or constitutional factors.

endogenous pyrogen [BIOCHEM] A fever-inducing substance (protein) produced by cells of the host body, such as leukocytes and macrophages.

endogenous variables [MATH] In a mathematical model, the dependent variables; their values are to be determined by the solution of the model equations.

endognath [INV ZOO] The inner and main branch of a crustacean's oral appendage.

endolecithal [INV ZOO] A type of egg found in turbellarians with yolk granules in the cytoplasm of the egg. Also spelled entolecithal.

endolithic [ECOL] Living within rocks, as certain algae and coral.

endolymph [PHYSIO] The lymph fluid found in the membranous labyrinth of the ear.

endolymphatic strommyosis *See* interstitial endometriosis.

endomeninx [EMBRYO] The internal part of the meninx primitiva that differentiates into the pia mater and arachnoid membrane.

endomere [EMBRYO] A blastomere that forms endoderm.

endometamorphism [GEOL] A phase of contact metamorphism involving changes in an igneous rock due to assimilation of portions of the rocks invaded by its magma.

endometrioma [MED] Endometriosis in which there is a discrete tumor mass.

endometriosis [MED] The presence of endometrial tissue in abnormal locations, including the uterine wall, ovaries, or extragenital sites.

endometritis [MED] Inflammation of the endometrium.

endometrium [ANAT] The mucous membrane lining the uterus.

endomitosis [CYTOL] Division of the chromosomes without dissolution of the nuclear membrane; results in polyploidy or polyteny.

endomixis [INV ZOO] Periodic division and reorganization of the nucleus in certain ciliated protozoans.

endomorph [PSYCH] A somatotype suggested by W. H. Sheldon to describe a person with a rounded physique; associated with viscerotonia.

endomorphism [MATH] **1.** A function from a set to itself. **2.** Often, a homomorphism of a group to itself.

Endomycetales [MICROBIO] Former designation for Saccharomycetales.

Endomycetoideae [MICROBIO] A subfamily of ascosporogenous yeasts in the family Saccharomycetaceae.

Endomychidae [INV ZOO] The handsome fungus beetles, a family of coleopteran insects in the superfamily Cucujoidea.

endomysium [HISTOL] The connective tissue layer surrounding an individual skeletal muscle fiber.

endoneural fibroma *See* neurofibroma.

endoneurium [HISTOL] Connective tissue fibers surrounding and joining the individual fibers of a nerve trunk.

endonuclease [BIOCHEM] An enzyme that severs the backbone chain of a strand of deoxyribonucleic acid.

endoparasite [ECOL] A parasite that lives inside its host.

endopeptidase [BIOCHEM] An enzyme that acts upon the centrally located peptide bonds of a protein molecule.

endophagous [INV ZOO] Of an insect larva, living within and feeding upon the host tissues.

endophallus [INV ZOO] Inner wall of the phallus of insects.

endophyte [ECOL] A plant that lives within, but is not necessarily parasitic on, another plant.

endoplasm [CYTOL] The inner, semifluid portion of the cytoplasm.

endoplasmic reticulum [CYTOL] A vacuolar system of the cytoplasm in differentiated cells that functions in protein synthesis and sequestration. Abbreviated ER.

endopleurite [INV ZOO] **1.** The portion of a crustacean apodeme which develops from the interepimeral membrane. **2.** One of the laterally located parts on the thorax of an insect which fold inward, extending into the body cavity.

endopodite [INV ZOO] The inner branch of a biramous crustacean appendage.

Endoprocta [INV ZOO] The equivalent name for Entoprocta.

endoprosthesis [MED] A prosthesis that is used internally.

endopterygoid [VERT ZOO] A paired dermal bone of the roof of the mouth in fishes.

Endopterygota [INV ZOO] A division of the insects in the subclass Pterygota, including those orders which undergo a holometabolous metamorphosis.

endoreduplication [CYTOL] Appearance in mitotic cells of certain chromosomes or chromosome sets in the form of multiples.

end organ [ANAT] The expanded termination of a nerve fiber in muscle, skin, mucous membrane, or other structure.

endorser [ADP] A special feature available on most magnetic-ink character-recognition readers that imprints a bank's endorsement on successful document reading.

endosalpingioma *See* serous cystadenoma.

endosalpinx [ANAT] The mucous membrane that lines the fallopian tube.

endosarc [INV ZOO] The inner, relatively fluid part of the protoplasm of certain unicellular organisms.

endoscope [MED] An instrument used to visualize the interior of a body cavity or hollow organ.

endosepsis [PL PATH] A fungus disease of figs caused by *Fusarium moniliforme fici*; fruits rot internally.

endoskeleton [ZOO] An internal skeleton or supporting framework in an animal.

endosmosis [PHYSIO] The passage of a liquid inward through a cell membrane.

endosome [CYTOL] A mass of chromatin near the center of a vesicular nucleus. [INV ZOO] The inner layer of certain sponges.

endosperm [BOT] **1.** The nutritive protein material within the embryo sac of seed plants. **2.** Storage tissue in the seeds of gymnosperms.

endosperm nucleus [BOT] The triploid nucleus formed

ENDOTHELIOCHORIAL
PLACENTA

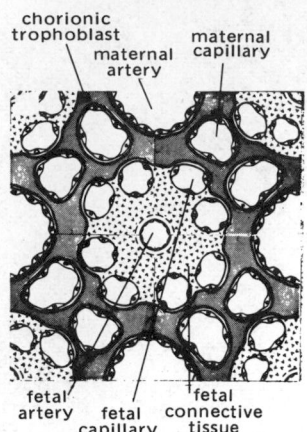

chorionic
trophoblast
maternal
artery
maternal
capillary

fetal
artery
fetal
capillary
fetal
connective
tissue

Section through an
endotheliochorial placenta.

ENDOTHYRACEA

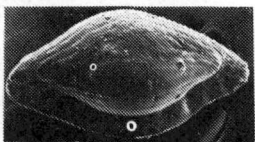

Scanning electron micrograph of
Triticites from the Pennsylvanian
formation of Texas showing the
exterior of the fusuline test.
*(R. B. MacAdam, Chevron Oil
Field Research Co.)*

ENERGY-LEVEL DIAGRAM

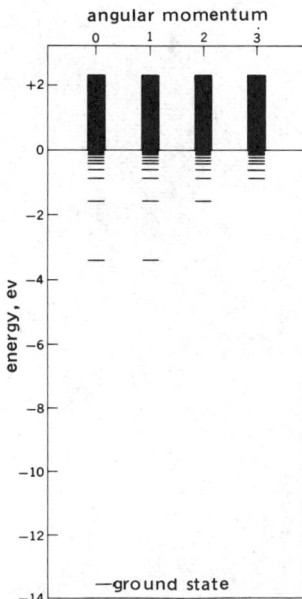

Energy levels of the hydrogen
atom, classified by the orbital
angular momentum of the
electron, expressed in units of ħ.

within the embryo sac of most seed plants by fusion of the polar nuclei with one sperm nucleus.

endospore [BIOL] An asexual spore formed within a cell.

endostome [BOT] The opening in the inner integument of a bitegmic ovule.

endostyle [INV ZOO] A ciliated groove or pair of grooves in the pharynx of lower chordates.

endosymbiont [ECOL] A symbiont that lives within the body of the host without deleterious effect on the host.

endotergite [INV ZOO] A dorsal plate to which muscles are attached in the insect skeleton.

endothecium [BOT] The middle of three layers that make up an immature anther; becomes the inner layer of a mature anther.

endothelial cell [HISTOL] A type of squamous epithelial cell composing the endothelium.

endotheliochorial placenta [EMBRYO] A type of placenta in which the maternal blood is separated from the chorion by the maternal capillary endothelium; occurs in dogs.

endothelioma [MED] Any tumor arising from, or resembling, endothelium; usually a benign growth, but occasionally a malignant tumor.

endothelium [HISTOL] The epithelial layer of cells lining the heart and vessels of the circulatory system.

Endotheriidae [PALEON] A family of Cretaceous insectivores from China belonging to the Proteutheria.

endotherm [PHYS CHEM] In differential thermal analysis, a graph of the temperature difference between a sample compound and a thermally inert reference compound (commonly aluminum oxide) as the substances are simultaneously heated to elevated temperatures at a predetermined rate, and the sample compound undergoes endothermal or exothermal processes.

endothermic [PHYS CHEM] Pertaining to a reaction which absorbs heat. Also known as endoergic.

Endothyracea [PALEON] A superfamily of extinct benthic marine foraminiferans in the suborder Fusulinina, having a granular or fibrous wall.

endotoxin [MICROBIO] A toxin that is produced within a microorganism and can be isolated only after the cell is disintegrated.

endotracheal [ANAT] Within the trachea.

endotrophic [BIOL] Obtaining nourishment from within; applied to certain parasitic fungi that live in the root cortex of the host plant.

end point [ANALY CHEM] That stage in the titration at which an effect, such as a color change, occurs, indicating that a desired point in the titration has been reached. [CHEM ENG] In the distillation analysis of crude petroleum and its products, the highest reading of a thermometer when a specified proportion of the liquid has boiled off. Also known as final boiling point.

end product [PHYS] The final product of a chemical or nuclear reaction or process.

endrin [ORG CHEM] $C_{12}H_8OCl_6$ Poisonous, white crystals that are insoluble in water; it is used as a pesticide and is a stereoisomer of dieldrin, another pesticide.

end section [COMMUN] Additional portion of switchboard added to each end of a large multiple switchboard and used to extend some of the trunks or locals to these end positions to place all jacks within easy reach of the first and last operator. Also known as head section.

end stone [HOROL] A flat jewel in a timepiece which acts as a bearing for a pivot.

end turning *See* boxing.

endurance [ENG] The time an aircraft, vehicle, or ship can continue operating under given conditions without refueling.

endurance limit *See* fatigue limit.

endurance ratio *See* fatigue ratio.

endurance strength *See* fatigue strength.

en echelon [GEOL] Referring to an overlapped or staggered arrangement of geologic features.

en echelon fault blocks [GEOL] A belt in which the individual fault blocks trend approximately 45° to the trend of the entire fault belt.

enema [MED] A rectal injection of liquid for therapeutic, diagnostic, or nutritive purposes.

energetics [PHYS] The study of energy and of its transformation from one form to another.

energetic solar particles [ASTROPHYS] Electrons and atomic nuclei produced in association with solar flares, with energies mostly in the range 1–100 million electron volts, but occasionally as high as 15 billion electron volts. Also known as solar cosmic rays.

energized [ELEC] Electrically connected to a voltage source. Also known as alive; hot; live.

energy [PHYS] The capacity for doing work.

energy absorption [PHYS] Conversion of mechanical or radiant energy into the internal potential energy or heat energy of a system.

energy balance [PHYS] The arithmetic balancing of energy inputs versus outputs for an object, reactor, or other processing system; it is positive if energy is released, and negative if it is absorbed. [PHYSIO] The relation of the amount of utilizable energy taken into the body to that which is employed for internal work, external work, and the growth and repair of tissues.

energy band *See* band.

energy band theory of solids *See* band theory of solids.

energy beam [ENG] An intense beam of light, electrons, or other nuclear particles; used to cut, drill, form, weld, or otherwise process metals, ceramics, and other materials.

energy conservation *See* conservation of energy.

energy conversion [PHYS] The process of changing energy from one form to another.

energy conversion efficiency [MECH ENG] The efficiency with which the energy of the working substance is converted into kinetic energy.

energy density [PHYS] The energy per unit volume of a medium; in the case of an electric or magnetic field, the energy needed to set up the field is thought of as residing in the field.

energy diagram *See* energy level diagram.

energy dissipation *See* dissipation.

energy efficiency ratio [ELEC] A value that represents the relative electrical efficiency of air conditioners; it is the quotient obtained by dividing Btu-per-hour output by electrical-watts input during cooling.

energy eigenstate *See* energy state.

energy flux [PHYS] A vector quantity whose component perpendicular to any surface equals the energy transported across that surface by some medium per unit area per unit time.

energy gap [SOLID STATE] A range of forbidden energies in the band theory of solids.

energy gradient [PHYS] Any change in energy over time or space.

energy head [FL MECH] The elevation of the hydraulic grade line at any section of a waterway plus the velocity head of the mean velocity of the water in that section.

energy integral [MECH] A constant of integration resulting from Newton's second law of motion in the case of a conservative force; equal to the sum of the kinetic energy of the particle and the potential energy of the force acting on it.

energy level [QUANT MECH] An allowed energy of a physical system; there may be several allowed states at one level. Also known as level.

energy-level diagram [QUANT MECH] A diagram in which the energy levels of a quantized system are indicated by distances of horizontal lines from a zero energy level. Also known as energy diagram; level scheme.

energy load *See* dynamic load.

energy management [AERO ENG] In rocketry, the monitoring of the expenditure of fuel for flight control and navigation.

energy metabolism [BIOCHEM] The chemical reactions involved in energy transformations within cells.

energy momentum tensor [PHYS] A tensor whose 16 elements give the energy density, momentum density, and stresses in a distribution of matter or radiation.

energy of a charge [ELEC] Charge energy measured in ergs according to the equation $E = \frac{1}{2}QV$, where Q is the charge and V is the potential in electrostatic units.

energy of activation *See* activation energy.

energy of rotation [PHYS] Kinetic energy of a mass with moment of inertia I rotating with angular velocity ω about the axis, expressed as $E = \frac{1}{2}I\omega^2$.

energy operator [QUANT MECH] The operator corresponding to the energy or Hamiltonian of a classical system. Also known as Hamiltonian operator.

energy product curve [ELECTROMAG] Curve obtained by plotting the product of the values of magnetic induction B and demagnetizing force H for each point on the demagnetization curve of a permanent magnet material; usually shown with the demagnetization curve.

energy pyramid [ECOL] An ecological pyramid illustrating the energy flow within an ecosystem.

energy spectrum [PHYS] Any plot, display, or photographic record of the intensity of some type of radiation as a function of its energy.

energy spread [QUANT MECH] The width in energy of a wave packet or metastable state.

energy state [QUANT MECH] An eigenstate of the energy (Hamiltonian) operator, so that the energy has a definite stationary value. Also known as eigenstate; energy eigenstate; quantum state; stationary state.

energy transfer [METEOROL] The transfer of energy of a given form among different scales of motion; for example, kinetic energy may be transferred between the zonal and meridional components of the wind, or between the mean and eddy components of the wind.

energy-variant sequential detection [COMMUN] Technique of sequential detection consisting of the transmission of a fixed number of pulses of varying energy, received with a single (upper) threshold device.

energy winds [PHYS] A group of winds which contain the bulk of recoverable kinetic energy for each month.

Enfield rifle [ORD] The popular name for the United States rifle, caliber .30, model 1917, a copy of the British Enfield rifle, a bolt-type, breech-loading magazine rifle; no longer standard for service use, it is used as a sporting weapon.

enfilade [ORD] To rake with gunfire; to fire down the length of a trench or line of troops.

enfleurage [CHEM ENG] Removal of the odoriferous components from flowers by placing them near an odorless mixture of lard and tallow; this mixture absorbs the perfume, which is subsequently extracted.

Engel-Recklinghausen disease See osteitis fibrosa cystica.

engine [MECH ENG] A machine in which power is applied to do work by the conversion of various forms of energy into mechanical force and motion.

engine balance [MECH ENG] Arrangement and construction of moving parts in reciprocating or rotating machines to reduce dynamic forces which may result in undesirable vibrations.

engine block See cylinder block.

engine cooling [MECH ENG] Controlling the temperature of internal combustion engine parts to prevent overheating and to maintain all operating dimensions, clearances, and alignment by a circulating coolant, oil, and a fan.

engine cycle [THERMO] Any series of thermodynamic phases constituting a cycle for the conversion of heat into work; examples are the Otto cycle, Stirling cycle, and Diesel cycle.

engine cylinder [MECH ENG] A cylindrical chamber in an engine in which the energy of the working fluid, in the form of pressure and heat, is converted to mechanical force by performing work on the piston. Also known as cylinder.

engine distillate [MATER] A heavy naphtha-kerosine distillate fuel of low octane number. Also known as tractor fuel.

engine efficiency [MECH ENG] Ratio between the energy supplied to an engine to the energy output of the engine.

engineer [ENG] An individual who specializes in one of the branches of engineering.

engineering [SCI TECH] The science by which the properties of matter and the sources of power in nature are made useful to man in structures, machines, and products.

engineering channel circuit [COMMUN] Auxiliary circuit or channel (radio or wire) for use by operating or maintenance personnel for communications incident to the establishment, operation, maintenance, and control of communications facilities.

engineering economy [IND ENG] **1.** Application of engineering or mathematical analysis and synthesis to decision making in economics. **2.** The knowledge and techniques concerned evaluating the worth of commodities and services relative to their cost. **3.** Analysis of the economics of engineering alternatives.

engineering geology [CIV ENG] The application of education and experience in geology and other geosciences to solve geological problems posed by civil engineering structures.

engineering plastics [MATER] Plastics that lend themselves to use for engineering design, such as gears and structural members.

engineering time [ADP] The nonproductive time of a computer, reserved for maintenance and servicing.

engineer's chain [CIV ENG] A surveyor's measuring instrument consisting of 1-foot (30.48-centimeter) steel links joined together by rings, 100 (30.5 meters) or 50 feet long. Also known as chain.

engineer's scale [GRAPHICS] A rule having a triangular cross section and different measurement scales on the two edges of each face.

engineer's system of units See British gravitational system of units.

engine fuel [MATER] Any of various substances, usually fluid, which provide heat, chemical, or pressure energy for engine operation.

engine inlet [MECH ENG] A place of entrance for engine fuel.

engine knock [MECH ENG] In spark ignition engines, the sound and other effects associated with ignition and rapid combustion of the last part of the charge to burn, before the flame front reaches it. Also known as combustion knock.

engine lathe [MECH ENG] A manually operated lathe equipped with a headstock of the back-geared, cone-driven type or of the geared-head type.

engine oil [MATER] Oil used for the bearing lubrication of all types of engines, machines, and shafting and for cylinder lubrication in other than steam engines.

engine performance [MECH ENG] Relationship between power output, revolutions per minute, fuel or fluid consumption, and ambient conditions in which an engine operates.

engine revolution counter [NAV ARCH] An instrument for registering the number of revolutions of a propeller shaft of a vessel; this information is useful in estimating a vessel's speed through the water.

engine sludge [ENG] The insoluble products of degradation of lubricating oils and fuels formed during the operation of an internal combustion engine.

engine starter [ELEC] The electric motor in the electric system of an automobile that cranks the engine for starting. Also known as starter; starting motor.

englacial [HYD] Of or pertaining to the inside of a glacier.

Engler distillation test [CHEM ENG] A standard test for determination of the volatility characteristics of a gasoline by the measurement of the percent of gasoline distilled at various specific temperatures.

Engler flask [CHEM ENG] A standardized flask of 100-milliliter volume used in the Engler distillation test.

Engler viscometer [ENG] An instrument used in the measurement of the degree Engler, a measure of viscosity; the kinematic viscosity ν in stokes for this instrument is obtained from the equation $\nu = 0.00147t - 3.74/t$, where t is the efflux time in seconds.

English degree [CHEM] A unit of water hardness, equal to 1 part calcium carbonate to 70,000 parts water; equivalent to 1 grain of calcium carbonate per gallon of water. Also known as Clark degree.

English garden-wall bond [CIV ENG] A masonry bond in which there are three courses of stretchers to one of headers.

englishite [MINERAL] $K_2Ca_4Al_8(PO_4)_8(OH)_{10}\cdot 9H_2O$ A white mineral composed of hydrous basic phosphate of potassium, calcium, and aluminum.

English red [MATER] Pigment consisting mostly of red iron oxide.

English vermilion [INORG CHEM] Bright vermilion pigment of precipitated mercury sulfide; in paints, it tends to darken when exposed to light.

engram [PHYSIO] A memory imprint; the alteration that has

ENGINE KNOCK

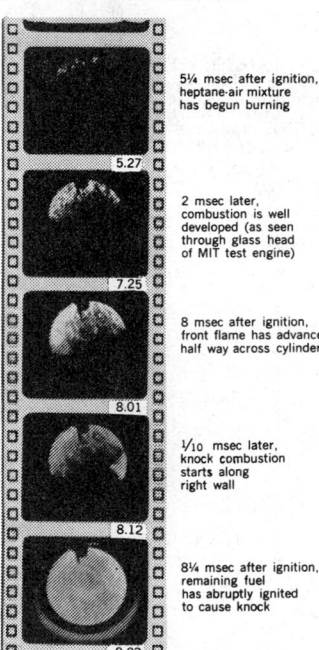

5¼ msec after ignition, heptane-air mixture has begun burning

2 msec later, combustion is well developed (as seen through glass head of MIT test engine)

8 msec after ignition, front flame has advanced half way across cylinder

¹/₁₀ msec later, knock combustion starts along right wall

8¼ msec after ignition, remaining fuel has abruptly ignited to cause knock

Photographic sequence showing (top to bottom) slow normal fuel burning ending in rapid knock combustion.

occurred in nervous tissue as a result of an excitation from a stimulus, which hypothetically accounts for retention of that experience. Also known as memory trace.

Engraulidae [VERT ZOO] The anchovies, a family of herring-like fishes in the suborder Clupoidea.

engraved-roll coating [GRAPHICS] Pattern coating engraved on a surface by a webbed roll onto which the coating substance is carefully metered. Also known as gravure coating.

engraver [GRAPHICS] One who makes engravings, either manually or by machine.

engraving [GRAPHICS] A photomechanical process in which lines are scribed on line negatives by means of a needle that produces an even transparent printing line by removal of a thin layer of photographic emulsion.

engrossing [GRAPHICS] The production of decorative designs of letters and illuminations by an artist on citations and diplomas.

engysseismology [GEOPHYS] Seismology dealing with earthquake records made close to the disturbance.

enhanced carrier demodulation [COMMUN] Amplitude demodulation system in which a synchronized local carrier of proper phase is fed into the demodulator to reduce demodulation distortion.

enhanced line *See* enhanced spectral line.

enhanced spectral line [SPECT] A spectral line of a very hot source, such as a spark, whose intensity is much greater than that of a line in a flame or arc spectrum. Also known as enhanced line.

enhancer gene [GEN] Any modifier gene that acts to enhance the action of a nonallelic gene.

ENIAC [ADP] The first digital computer in the modern sense of the word, built 1942-1945. Derived from Electronic Numerical Integrator and Calculator.

Enicocephalidae [INV ZOO] The gnat bugs, a family of hemipteran insects in the superfamily Enicocephaloidea.

Enicocephaloidea [INV ZOO] A superfamily of the Hemiptera in the subdivision Geocorisae containing a single family.

enigmatite [MINERAL] $Na_2Fe_5TiSi_6O_{20}$ A black amphibole mineral occurring in triclinic crystals; specific gravity is 3.14–3.80. Also spelled aenigmatite.

Enka [TEXT] Trade name for a viscose rayon of the American Enka Corporation.

enlargement [GRAPHICS] Photographic print made in an enlarger so that the print is bigger than the negative.

enlargement loss [FL MECH] Energy loss by friction in a flowing fluid when it moves into a cross-sectional area of sudden enlargement.

enlarger [OPTICS] An optical projector used to project an enlarged image of a photograph's negative onto photosensitized film or paper. Also known as photoenlarger.

enlarging [GRAPHICS] The process of reproducing an image from a smaller image through a projection process.

enol [ORG CHEM] An organic compound with a hydroxide group adjacent to a double bond; varies with a ketone form in the effect known as enol-keto tautomerism; an example is the compound $CH_3COH = CHCO_2C_2H_5$.

enolase [BIOCHEM] An enzyme that catalyzes the reversible dehydration of phosphoglyceric acid to phosphopyruvic acid.

enol-keto tautomerism [ORG CHEM] The tautomeric migration of a hydrogen atom from an adjacent carbon atom to a carbonyl group of a keto compound to produce the enol form of the compound; the reverse process of hydrogen atom migration also occurs.

enophthalmos [MED] Recession of the eyeball into the orbital cavity.

Enopla [INV ZOO] A class or subclass of ribbonlike worms of the phylum Rhynchocoela.

Enoplia [INV ZOO] A subclass of nematodes in the class Adenophorea.

Enoplida [INV ZOO] An order of nematodes in the subclass Enoplia.

Enoplidae [INV ZOO] A family of free-living marine nematodes in the superfamily Enoploidea, characterized by a complex arrangement of teeth and mandibles.

Enoploidea [INV ZOO] A superfamily of small to very large free-living marine nematodes having pocketlike amphids opening to the exterior via slitlike apertures.

Enoploteuthidae [INV ZOO] A molluscan family of deep-sea squids in the class Cephalopoda.

enphytotic [PL PATH] **1.** A disease that occurs regularly among plants of a specific region. **2.** An outbreak of such a disease.

enriched gas [MATER] A motor gasoline containing additives to improve combustion characteristics, as by limiting knock.

enriched material [NUC ENG] Material in which the amount of one or more isotopes has been increased above that occurring in nature, such as uranium in which the abundance of U^{235} is increased.

enriched reactor [NUCLEO] A nuclear reactor in which the fuel is an enriched material.

enriching column [CHEM ENG] The portion of a countercurrent contractor (liquid-liquid extraction or vapor-liquid distillation) above the feed point in which an upward-moving, product-rich stream from the stripping column is further purified by countercurrent contact with a downward-flowing reflux stream from the overhead product-recovery vessel.

enrichment [NUCLEO] A process that changes the isotopic ratio in a material; for uranium, for example, the ratio of U^{235} to U^{238} may be increased by gaseous diffusion of uranium hexafluoride.

enrichment culture [MICROBIO] A medium of known composition and specific conditions of incubation which favors the growth of a particular type or species of bacterium.

enroute chart [NAV] A chart of air routes in specific areas that shows the exact location of electronic aids to navigation, such as radio direction-finder stations, radio and radar beacons, and radio-range stations. Formerly known as radio facility chart.

enroute turning area [NAV] An area of specified dimensions used by an aircraft when a change of course is necessary; the area extends beyond the primary and secondary obstruction clearance areas and is provided with adequate protection; turning area criteria supplement the basic airways and route segment criteria to protect the aircraft. Also known as turning area.

ensemble [STAT MECH] A collection of systems of particles used to describe an individual system; time averages of quantities describing the individual system are found by averaging over the systems in the ensemble at a fixed time.

ensialic geosyncline [GEOL] A geosyncline whose geosynclinal prism accumulates on a sialic crust and contains clastics.

ensiform [BIOL] Sword-shaped.

ensiling [AGR] The anaerobic fermentation process used to preserve immature green corn, legumes, grasses, and grain plants; the crop is chopped and packed while at about 70–80% moisture and put into silos or other containers to exclude air.

ensimatic geosyncline [GEOL] A geosyncline whose geosynclinal prism accumulates on a simatic crust and is composed largely of volcanic rock or sediments of volcanic debris.

Enskog theory *See* Chapman-Enskog theory.

enstatite [MINERAL] $MgOSiO_2$ A member of the pyroxene mineral group that crystallizes in the orthorhombic system; usually yellowish gray but becomes green when a little iron is present.

entablature [ARCH] A unit consisting of the architrave, frieze, and cornice of a wall.

Entamoeba [INV ZOO] A genus of parasite amebas in the family Endamoebidae, including some species of the genus *Endamoeba* which are parasites of humans and other vertebrates.

Entamoeba histolytica [INV ZOO] A pathogenic ameba, causing amebic dysentery.

entanglement [ORD] An obstacle, such as barbed wire, that is utilized to stop or hamper the forward movement of troops.

Enteletacea [PALEON] A group of extinct articulate brachiopods in the order Orthida.

Entelodontidae [PALEON] A family of extinct palaeodont artiodactyls in the superfamily Entelodontoidea.

Entelodontoidea [PALEON] A superfamily of extinct piglike mammals in the suborder Palaeodonta having huge skulls and enlarged incisors.

enteralgia [MED] Pain in the intestine.

enterectomy [MED] Excision of a part of the intestine.

enteric bacilli [MICROBIO] Microorganisms, especially the

ENTELETACEA

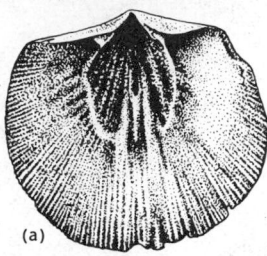

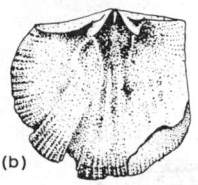

Pionodema, (a) pedicle and (b) brachial valve interiors. (From R. C. Moore, ed., Treatise on Invertebrate Paleontology, pt. H, Geological Society of America and University of Kansas Press, 1965)

ENTELODONTIDAE

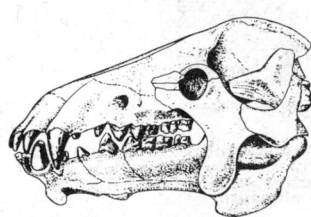

Archaeothesium, Oligocene entelodont, skull length about 24 inches (61 centimeters). (After Marsh, 1893)

gram-negative rods, found in the intestinal tract of man and animals.

enteric cytopathogenic human orphan virus *See* echovirus.

entering angle [MECH ENG] The angle between the side-cutting edge of a tool and the machined surface of the work; angle is 90° for a tool with 0° side-cutting edge angle effective.

enteritis [MED] Inflammation of the intestinal tract.

Enterobacteriaceae [MICROBIO] A family of bacteria in the order Eubacteriales consisting of straight, rod-shaped, gram-negative, nonsporeforming cells, nonmotile or motile with peritrichous flagella; includes important human and plant pathogens.

enterobiasis [MED] Infestation of the intestinal tract of man with the nematode *Enterobius vermacularis* (pinworm); characterized by mild enteritis.

enterocoel [ZOO] A coelom that arises by mesodermal outpocketing of the archenteron.

Enterocoela [SYST] A section of the animal kingdom that includes the Echinodermata, Chaetognatha, Hemichordata, and Chordata.

enterocolitis [MED] Inflammation of the small intestine and colon.

enterohydrocoel [INV ZOO] In crinoids, an anterior cavity derived from the archenteron.

enterokinase [BIOCHEM] An enzyme which catalyzes the conversion of trypsinogen to trypsin.

enterolith [PATH] A concretion formed in the intestine.

enterolithic [GEOL] Of or pertaining to structures, such as small folds, formed in evaporites due to flowage or hydration.

enteron [ANAT] The alimentary canal.

Enteropneusta [INV ZOO] The acorn worms or tongue worms, a class of the Hemichordata; free-living solitary animals with no exoskeleton and with numerous gill slits and a straight gut.

enteroptosis *See* visceroptosis.

enterorrhagia [MED] Intestinal hemorrhage.

enterotoxin [MICROBIO] A toxin produced by *Micrococcus pyogenes* var. *aureus* (*Staphylococcus aureus*) which gives rise to symptoms of food poisoning in man and monkeys.

enterovirus [VIROL] One of the two subgroups of human picornaviruses; includes the polioviruses, the coxsackieviruses, and the echoviruses.

Enterozoa [ZOO] Animals with a digestive tract or cavity; includes all animals except Protozoa, Mesozoa, and Parazoa.

enthalpimetric analysis [ANALY CHEM] Generic designation for a group of modern thermochemical methodologies such as thermometric enthalpy titrations which rely on monitoring the temperature changes produced in adiabatic calorimeters by heats of reaction occurring in solution; in contradistinction, classical methods of thermoanalysis such as thermogravimetry focus primarily on changes occurring in solid samples in response to externally imposed programmed alterations in temperature.

enthalpy [THERMO] The sum of the internal energy of a system plus the product of the system's volume multiplied by the pressure exerted on the system by its surroundings. Also known as heat content; sensible heat; total heat.

enthalpy-entropy chart [THERMO] A graph of the enthalpy of a substance versus its entropy at various values of temperature, pressure, or specific volume; useful in making calculations about a machine or process in which this substance is the working medium.

enthalpy of reaction [PHYS CHEM] The change in enthalpy accompanying a chemical reaction.

enthalpy of transition [PHYS CHEM] The change of enthalpy accompanying a phase transition.

enthalpy of vaporization *See* heat of vaporization.

enthalpy-pressure chart *See* pressure-enthalpy chart.

enthalpy titration *See* thermometric titration.

entire [BIOL] Having a continuous, unimpaired margin.

entire function [MATH] A function of a complex variable which is analytic throughout the entire complex plane.

entire series [MATH] A power series which converges for all values of its variable; a power series with an infinite radius of convergence.

Entisol [GEOL] An order of soil having few or faint horizons.

Entner-Doudoroff pathway [BIOCHEM] A sequence of reactions for glucose degradation, with the liberation of energy;

the distinguishing feature is the formation of 2-keto-3-deoxy-6-phosphogluconate from 6-phosphogluconate and the cleaving of this compound to yield pyruvate and glyceraldehyde-3-phosphate.

entoblast [EMBRYO] A blastomere that differentiates into endoderm.

entoderm *See* endoderm.

Entodiniomorphida [INV ZOO] An order of highly evolved ciliated protozoans in the subclass Spirotrichia, characterized by a smooth, firm pellicle and the lack of external ciliature.

entolecithal *See* endolecithal.

Entomoconchacea [PALEON] A superfamily of extinct marine ostracods in the suborder Myodocopa that are without a rostrum above the permanent aperture.

entomogenous [BIOL] Growing on or in an insect body, as certain fungi.

entomology [INV ZOO] A branch of the biological sciences that deals with the study of insects.

entomophagous [ZOO] Feeding on insects.

entomophilic fungi [MYCOL] Species of fungi that are insect pathogens.

entomophilous [ECOL] Pollinated by insects.

entomophobia [PSYCH] An abnormal fear of insects.

Entomophthoraceae [MYCOL] The single family of the order Entomophthorales.

Entomophthorales [MYCOL] An order of mainly terrestrial fungi in the class Phycomycetes having a hyphal thallus and nonmotile sporangiospores, or conidia.

Entomostraca [INV ZOO] A group of Crustacea comprising the orders Cephalocarida, Branchiopoda, Ostracoda, Copepoda, Branchiura, and Cirripedia.

Entoniscidae [INV ZOO] A family of isopod crustaceans in the tribe Bopyrina that are parasitic in the visceral cavity of crabs and porcellanids.

entoplastron [VERT ZOO] The anterior median bony plate of the plastron of chelonians.

Entoprocta [INV ZOO] A group of bryozoans, sometimes considered to be a subphylum, having a pseudocoelomate visceral cavity and the anus opening inside the circlet of tentacles.

entrail pahoehoe [GEOL] A type of pahoehoe having a surface that resembles an intertwined mass of entrails.

entrained fluid [FL MECH] Fluid in the form of mist, fog, or droplets that is carried out of a column or vessel by a rising gas or vapor stream.

entrainer [CHEM ENG] An additive that forms an azeotrope with one component of a liquid mixture to aid in otherwise difficult separations by distillation, as in azeotropic distillation.

entrainment [CHEM ENG] A process in which the liquid boils so violently that suspended droplets of liquid are carried in the escaping vapor. [HYD] The pickup and movement of sediment as bed load or in suspension by current flow. [METEOROL] The mixing of environmental air into a preexisting organized air current so that the environmental air becomes part of the current. [OCEANOGR] The transfer of fluid by friction from one water mass to another, usually occurring between currents moving in respect to each other.

entrance [ADP] The location of a program or subroutine at which execution is to start. Also known as entry point. [CIV ENG] The seaward end of a channel, harbor, and so on. [ENG] A place of physical entering, such as a door or passage. [NAV ARCH] The part of a ship's underwater hull which is forward of the amidships.

entrance cable [ELEC] Cable that brings power from an outside power line into a building.

entrance lock [CIV ENG] A lock between the tideway and an enclosed basin made necessary because the levels of the two bodies of water vary; by means of this lock, vessels can pass either way at all states of the tide. Also known as guard lock; tidal lock; tide lock.

entrance loss [FL MECH] Energy loss by friction in a flowing fluid when it moves into a cross-sectional area of sudden contraction, as at the entrance of a pipe or a suddenly reduced area of a duct.

entrance pupil [OPTICS] The image of the aperture stop of an optical system, as seen from the object.

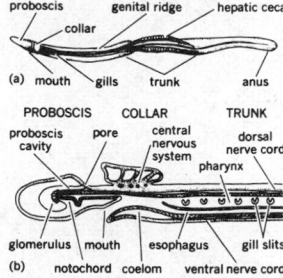

ENTEROPNEUSTA

Saccoglossus, the tongue worm or acorn worm. (a) Dorsal view. (b) Median section of anterior portion. (*From T. I. Storer and R. L. Usinger, General Zoology, 4th ed., McGraw-Hill, 1965*)

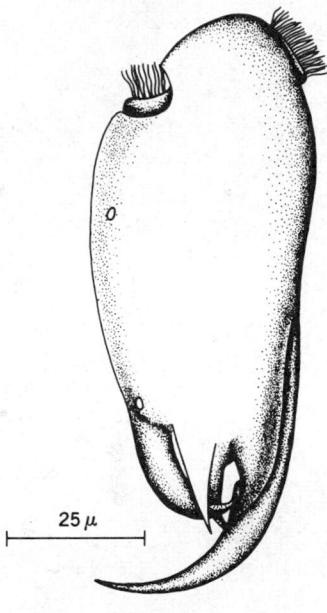

ENTODINIOMORPHIDA

25 μ

Ophryoscolex, an entodiniomorphid.

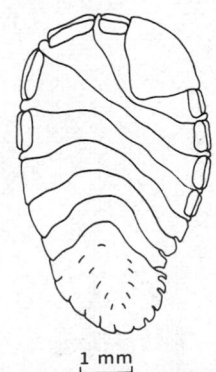

ENTONISCIDAE

1 mm

Female *Bopyrus squillarum* Latreille.

entrance region [METEOROL] The region of confluence at the upwind extremity of a jet stream.

entrance slit [SPECT] Narrow slit through which passes the light entering a spectrometer.

entrapment [GEOL] The underground trapping of oil or gas reserves by folds, faults, domes, asphaltic seals, unconformities, and such.

entrenched meander [HYD] A deepened meander of a river which is carried downward further below the valley surface in which the meander originally formed. Also known as inherited meander.

entrenched stream [HYD] A stream that flows in a valley or narrow trench cut into a plain or relatively level upland. Also spelled intrenched stream.

entropy [COMMUN] A measure of the absence of information about a situation, or, equivalently, the uncertainty associated with the nature of a situation. [MATH] In a mathematical context, this concept is attached to dynamical systems, transformations between measure spaces, or systems of events with probabilities; it expresses the amount of disorder inherent or produced. [STAT MECH] Measure of the disorder of a system, equal to the Boltzmann constant times the natural logarithm of the number of microscopic states corresponding to the thermodynamic state of the system; this statistical-mechanical definition can be shown to be equivalent to the thermodynamic definition. [THERMO] Function of the state of a thermodynamic system whose change in any differential reversible process is equal to the heat absorbed by the system from its surroundings divided by the absolute temperature of the system. Also known as thermal charge.

entropy of activation [PHYS CHEM] The difference in entropy between the activated complex in a chemical reaction and the reactants.

entropy of a partition [MATH] If ξ is a finite partition of a probability space, the entropy of ξ is the negative of the sum of the products of the probabilities of elements in ξ with the logarithm of the probability of the element.

entropy of a transformation *See* Kolmogorov-Sinai invariant.

entropy of a transformation given a partition [MATH] If T is a measure preserving transformation on a probability space and ξ is a finite partition of the space, the entropy of T given ξ is the limit as $n \to \infty$ of $1/n$ times the entropy of the partition which is the common refinement of ξ, $T^{-1}\xi$, ..., $T^{-n+1}\xi$.

entropy of mixing [PHYS CHEM] After mixing substances, the difference between the entropy of the mixture and the sum of the entropies of the components of the mixture.

entropy of transition [PHYS CHEM] The heat absorbed or liberated in a phase change divided by the absolute temperature at which the change occurs.

entry [ADP] Input data fed during the execution of a program by means of a terminal.

entry ballistics [MECH] That branch of ballistics which pertains to the entry of a missile, spacecraft, or other object from outer space into and through an atmosphere.

entry block [ADP] The area of main memory reserved for the data which will be introduced at execution time.

entry corridor [AERO ENG] Depth of the region between two trajectories which define the design limits of a vehicle about to enter a planetary atmosphere, or define the desired landing area (footprint).

entry point *See* entrance.

entypy [EMBRYO] The formation of the amnion in certain mammals by the invagination of the embryonic knob into the yolk sac, without the formation of any amniotic folds.

enucleate [MED] To remove an organ or a tumor in its entirety, as an eye from its socket.

enumerable *See* countable.

enuresis [MED] Urinary incontinence, especially in the absence of organic cause.

envelope [COMMUN] A curve drawn to pass through the peaks of a graph, such as that of a modulated radio-frequency carrier signal. [ENG] The glass or metal housing of an electron tube or the glass housing of an incandescent lamp. Also known as bulb.

envelope delay [COMMUN] The time required for the envelope of a wave to travel between two points in a system.

envelope delay distortion *See* delay distortion.

envelope detector *See* detector.

envelope of a family of curves [MATH] A curve which is tangent to every member of the given family.

envenomation [MATER] The process by which the surface of a plastic close to or in contact with another surface is deteriorated.

environment [ECOL] The sum of all external conditions and influences affecting the development and life of organisms. [ENG] The aggregate of all natural, operational, or other conditions that affect the operation of equipment or components. [PHYS] The aggregate of all the conditions and the influences that determine the behavior of a physical system.

environmental biology *See* ecology.

environmental cab [ENG] Operator's compartment in earthmovers equipped with tinted safety glass, soundproofing, air conditioning, and cleaning units.

environmental control [ENG] Modification and control of soil, water, and air environments of man and other living organisms.

environmental control system [ENG] A system used in a closed area, especially a spacecraft or submarine, to permit life to be sustained; the system provides the occupants with a suitably controlled atmosphere to permit them to live and work in the area.

environmental engineering [ENG] The technology concerned with the reduction of pollution, contamination, and deterioration of the surroundings in which humans live.

environmental impact analysis [IND ENG] Predetermination of the extent of pollution or environmental degradation which will be involved in a mining or processing project.

environmental impact statement [ENG] A report of the potential effect of plans for land use in terms of the environmental, engineering, esthetic, and economic aspects of the proposed objective.

environmental lapse rate [METEOROL] The rate of decrease of temperature with elevation in the atmosphere. Also known as atmospheric lapse rate.

environmental protection [ENG] The protection of man and equipment against stresses of climate and other elements of the environment.

environmental range [ENG] The range of environment throughout which a system or portion thereof is capable of operation at not less than the specified level of reliability.

environmental stress cracking [MECH] The susceptibility of a material to crack or craze in the presence of surface-active agents or other factors.

environmental survey satellite [AERO ENG] One of a series of meteorological satellites which completely photographs the earth each day. Abbreviated ESSA.

environmental test [ENG] A laboratory test conducted to determine the functional performance of a component or system under conditions that simulate the real environment in which the component or system is expected to operate.

environment division [ADP] The section of a program written in COBOL which defines the hardware and files to be used by the program.

environment of sedimentation [GEOL] A more or less destructive geomorphologic setting in which sediments are deposited as beach environment.

environment simulator [ENG] Any machine or artificial device that simulates all or some of the attributes of an environment, such as the solar simulators with artificial suns used in testing spacecraft.

enzootic [VET MED] **1.** A disease affecting animals in a limited geographic region. **2.** Pertaining to such a disease.

enzyme [BIOCHEM] Any of a group of catalytic proteins that are produced by living cells and that mediate and promote the chemical processes of life without themselves being altered or destroyed.

enzyme induction [MICROBIO] The process by which a microbial cell synthesizes an enzyme in response to the presence of a substrate or of a substance closely related to a substrate in the medium.

enzyme inhibition [BIOCHEM] Prevention of an enzymic process as a result of the interaction of some substance with the enzyme so as to decrease the rate of reaction.

enzyme repression [BIOCHEM] The process by which the rate of synthesis of an enzyme is reduced in the presence of a metabolite, often the end product of a chain of reactions in which the enzyme in question operates near the beginning.

enzyme unit [BIOCHEM] The amount of an enzyme that will catalyze the transformation of 10^{-6} mole of substrate per minute or, when more than one bond of each substrate is attacked, 10^{-6} of 1 gram equivalent of the group concerned, under specified conditions of temperature, substrate concentration, and pH number.

enzymology [BIOCHEM] A branch of science dealing with the chemical nature, biological activity, and biological significance of enzymes.

Eocambrian [GEOL] Pertaining to the thick sequences of strata conformably underlying Lower Cambrian fossils. Also known as Infracambrian.

Eocanthocephala [INV ZOO] An order of the Acanthocephala characterized by the presence of a small number of giant subcuticular nuclei.

Eocene [GEOL] The next to the oldest of the five major epochs of the Tertiary period (in the Cenozoic era).

Eocrinoidea [PALEON] A class of extinct echinoderms in the subphylum Crinozoa that had biserial brachioles like those of cystoids combined with a theca like that of crinoids.

EOD See explosive ordnance disposal.

EOF gap See end-of-file gap.

Eogene See Paleogene.

Eohippus [PALEON] The earliest, primitive horse, included in the genus *Hyracotherium;* described as a small, four-toed species.

eolation [GEOL] Any action of wind on the land.

eolian [METEOROL] Pertaining to the action or the effect of the wind, as in eolian sounds or eolian deposits (of dust). Also spelled aeolian.

eolian anemometer [ENG] An anemometer which works on the principle that the pitch of the eolian tones made by air moving past an obstacle is a function of the speed of the air.

eolian erosion [GEOL] Erosion due to the action of wind.

eolianite [GEOL] A sedimentary rock consisting of clastic material which has been deposited by wind.

eolian ripple mark [GEOL] A mark made in sand by the wind.

eolian sand [GEOL] Deposits of sand arranged by the wind.

eolian soil [GEOL] A type of soil ranging from sand dunes to loess deposits whose particles are predominantly of silt size.

eolian sounds [ACOUS] Sounds produced by eddying motions of air in the lee of obstacles, such as wires, twigs, and even the ear itself, when wind blows over those obstacles.

eolotropy See anisotropy.

Eomoropidae [PALEON] A family of extinct perissodactyl mammals in the superfamily Chalicotherioidea.

eon [MECH] A unit of time, equal to 10^9 years.

EOR See explosive ordnance reconnaissance.

EORA See explosive ordnance reconnaissance agent.

EOR gap See end-of-record gap.

Eosentomidae [INV ZOO] A family of primitive wingless insects in the order Protura that possess spiracles and tracheae.

eosin [ORG CHEM] $C_{20}H_8O_5Br_4$ **1.** A red fluorescent dye in the form of triclinic crystals that are insoluble in water; used chiefly in cosmetics and as a toner. Also known as bromeosin; bromo acid; eosine; tetrabromofluorescein. **2.** The red to brown crystalline sodium or potassium salt of this dye; used in organic pigments, as a biological stain, and in pharmaceuticals. Also known as eosine; eosine G; eosine Y; eosine yellowish.

eosine See eosin.

eosine G See eosin.

eosine Y See eosin.

eosine yellowish See eosin.

eosinophil [HISTOL] A granular leukocyte having cytoplasmic granules that stain with acid dyes and a nucleus with two lobes connected by a thin thread of chromatin.

eosinophilia [MED] A greater than average number of circulating eosinophils. Also known as acidophilia; oxyphilia.

eosinophilic erythroblast See normoblast.

eosinophilic granuloma [MED] A disease, principally of childhood, characterized by foci of bone inflammation and granulation containing lipids, mononuclear cells, and eosinophils.

eosinophilic pneumonitis See Loeffler's syndrome.

eosphorite [MINERAL] $(Mn,Fe)Al(PO_4)(OH)_2 \cdot H_2O$ A usually rose-pink mineral composed of hydrous aluminum manganese phosphate, found massive or in prismatic crystals.

Eosuchia [PALEON] The oldest, most primitive, and only extinct order of lepidosaurian reptiles.

eötvös [GEOPHYS] A unit of horizontal gradient of gravitational acceleration, equal to a change in gravitational acceleration of 10^{-9} galileo over a horizontal distance of 1 centimeter.

Eötvös experiment [RELAT] An experiment which tests the equality of inertial mass and gravitational mass by balancing on a given body the earth's gravitational attraction against the kinetic reaction arising from the rotation of the earth.

Eötvös number See Bond number.

Eötvös rule [THERMO] The rule that the rate of change of molar surface energy with temperature is a constant for all liquids; deviations are encountered in practice.

Eötvös torsion balance [ENG] An instrument which records the change in the acceleration of gravity over the horizontal distance between the ends of a beam; used to measure density variations of subsurface rocks.

Epacridaceae [BOT] A family of dicotyledonous plants in the order Ericales, distinguished by palmately veined leaves, and stamens equal in number with the corolla lobes.

epaulette [INV ZOO] **1.** Any of the branched or knobbed processes on the oral arms of many Scyphozoa. **2.** The first haired scale at the base of the costal vein in Diptera.

epaxial [BIOL] Above or dorsal to an axis.

epaxial muscle [ANAT] Any of the dorsal trunk muscles of vertebrates.

epeiric sea See epicontinental sea.

epeirogeny [GEOL] Movements which affect large tracts of the earth's crust.

ependyma [HISTOL] The layer of epithelial cells lining the cavities of the brain and spinal cord. Also known as ependymal layer.

ependymal layer See ependyma.

ependymoma [MED] A tumor of the central nervous system whose essential portion consists of cells derived from and resembling ependymal cells. Also known as medulloepithelioma.

ephapse [ANAT] A contact point between neurons.

ephaptic transmission [PHYSIO] Electrical transfer of activity to a postephaptic unit by the action current of a preephaptic cell.

Ephedra [BOT] A genus of low, leafless, green-stemmed shrubs belonging to the order Ephedrales; source of the drug ephedrine.

Ephedrales [BOT] A monogeneric order of gymnosperms in the subdivision Gneticae.

ephedrine [ORG CHEM] $C_{10}H_{15}NO$ A white, crystalline, water-soluble alkaloid present in several *Ephedra* species and also produced synthetically; a sympathomimetic amine, it is used for its action on the bronchi, blood pressure, blood vessels, and central nervous system.

ephemeral plant [BOT] An annual plant that completes its life cycle in one short moist season; desert plants are examples.

ephemeral stream [HYD] A stream channel which carries water only during and immediately after periods of rainfall or snowmelt.

Ephemerida [INV ZOO] An equivalent name for Ephemeroptera.

ephemeris [ASTRON] A periodical publication tabulating the predicted positions of celestial bodies at regular intervals, such as daily, and containing other data of interest to astronomers. Also known as astronomical ephemeris.

ephemeris day [ASTRON] A unit of time equal to 86,400 ephemeris seconds (International System of Units).

ephemeris second [ASTRON] The fundamental unit of time of the International System of Units of 1960, equal to 1/31556925.9747 of the tropical year defined by the mean motion of the sun in longitude at the epoch 1900 January 0 day 12 hours.

ephemeris time [ASTRON] The uniform measure of time defined by the laws of dynamics and determined in principle from the orbital motions of the planets, specifically the orbital

EOCENE

	PALEOZOIC								MESOZOIC			CENOZOIC
PRECAMBRIAN	CAMBRIAN	ORDOVICIAN	SILURIAN	DEVONIAN	CARBON-IFEROUS Mississippian	Pennsylvanian	PERMIAN	TRIASSIC	JURASSIC	CRETACEOUS	TERTIARY	QUATERNARY

TERTIARY					QUATERNARY	
Paleocene	Eocene	Oligocene	Miocene	Pliocene	Pleistocene	Recent

A chart showing the position of the Eocene epoch in geologic time.

EOSINOPHIL

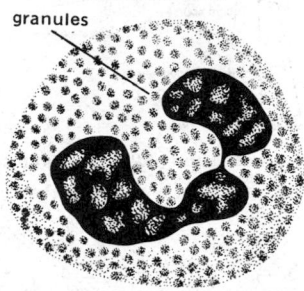

granules

Diagram of an eosinophil showing the typical two-lobed nucleus.

EPHEDRA

A species of *Ephedra* growing in Utah. *(Courtesy of Tony Gauba, from National Audubon Society)*

motion of the earth as represented by Newcomb's Tables of the Sun. Abbreviated E.T.

Ephemeroptera [INV ZOO] The mayflies, an order of exopterygote insects in the subclass Pterygota.

ephidrosis *See* hyperhidrosis.

Ephydridae [INV ZOO] The shore flies, a family of myodarian cyclorrhaphous dipteran insects in the subsection Acalypteratae.

ephyra [INV ZOO] A larval, free-swimming medusoid stage of scyphozoans; arises from the scyphistoma by transverse fission. Also known as ephyrula.

ephyrula *See* ephyra.

epi- [ORG CHEM] A prefix used in naming compounds to indicate the presence of a bridge or intramolecular connection. [SCI TECH] Prefix denoting upon, beside, near to, over, outer, anterior, prior to, or after.

EPI *See* electronic position indicator.

epiandrum [INV ZOO] The genital orifice of a male arachnid.

epibasidium [MYCOL] A lengthening of the upper part of each cell of the basidium of various heterobasidiomycetes.

epibiotic [ECOL] Living, usually parasitically, on the surface of plants or animals; used especially of fungi.

epiblast *See* ectoderm.

epiblem [BOT] A tissue that replaces the epidermis in most roots and in stems of submerged aquatic plants.

epiboly [EMBRYO] The growing or extending of one part, such as the upper hemisphere of a blastula, over and around another part, such as the lower hemisphere, in embryogenesis.

epibranchial [ANAT] Of or pertaining to the segment below the pharyngobranchial region in a branchial arch.

epicadmium [NUCLEO] Energy above the greatest level at which cadmium shows a large neutron cross section, about 0.3 electron volt.

epicalyx [BOT] A ring of fused bracts below the calyx forming a structure that resembles the calyx.

epicardium [ANAT] The inner, serous portion of the pericardium that is in contact with the heart. [INV ZOO] A tubular prolongation of the branchial sac in certain ascidians which takes part in the process of budding.

Epicaridea [INV ZOO] A suborder of the Isopoda whose members are parasitic on various marine crustaceans.

epicarp [BOT] The outer layer of the pericarp. Also known as exocarp.

epicenter [GEOL] A point on the surface of the earth which is directly above the seismic focus of an earthquake and where the earthquake vibrations reach first.

epichlorohydrin [ORG CHEM] C_3H_5OCl A colorless, unstable liquid, insoluble in water; used as a solvent for resins. Also known as 1-chloro-2,3-epoxypropane; chloropropylene oxide.

epichordal [VERT ZOO] Located upon or above the notochord.

epichordal brain [EMBRYO] The area of origin of the hindbrain or rhombencephalon, located on the dorsal side of the notochord. Also known as deutencephalon.

epiclastic [GEOL] Pertaining to the texture of mechanically deposited sediments consisting of detrital material from pre-existent rocks.

epicnemial [ANAT] Of or pertaining to the anterior portion of the tibia.

epicondyle [ANAT] An eminence on the condyle of a bone.

epicone [INV ZOO] The part anterior to the equatorial groove in a dinoflagellate.

epicontinental [GEOL] Located upon a continental plateau or platform.

epicontinental sea [OCEANOGR] That portion of the sea lying upon the continental shelf, and the portions which extend into the interior of the continent with similar shallow depths. Also known as epeiric sea; inland sea.

epicotyl [BOT] The embryonic plant stem above the cotyledons.

epicranium [INV ZOO] The dorsal wall of an insect head. [VERT ZOO] The structures covering the cranium in vertebrates.

epicuticle [INV ZOO] The outer, waxy layer of an insect cuticle or exoskeleton.

epicycle [MATH] The circle which generates an epicycloid or hypocycloid.

epicyclic gear [MECH ENG] A system of gears in which one or more gears travel around the inside or the outside of another gear whose axis is fixed.

epicyclic train [MECH ENG] A combination of epicyclic gears, usually connected by an arm, in which some or all of the gears have a motion compounded of rotation about an axis and a translation or revolution of that axis.

epicycloid [MATH] The curve traced by a point on a circle as it rolls along the outside of a fixed circle.

epicyst [INV ZOO] The outer layer of a cyst wall in encysted protozoans. Also known as ectocyst.

epidemic [MED] A sudden increase in the incidence rate of a disease to a value above normal, affecting large numbers of people and spread over a wide area.

epidemic diarrhea of the newborn [MED] Contagious, fulminating diarrhea with high mortality, seen in newborns; caused by enteropathogenic strains of *Escherichia coli*, strains of *Staphylococcus*, other bacteria, and possibly viruses.

epidemic encephalitis *See* encephalitis lethargica.

epidemic gastroenteritis [MED] Inflammation of the stomach and intestine, of viral origin; considered to be epidemic when symptoms are manifested by a member of the patient's family within 10 days of the patient's recovery.

epidemic hepatitis *See* infectious hepatitis.

epidemic jaundice *See* infectious hepatitis.

epidemic keratoconjunctivitis [MED] Inflammation of the cornea and conjunctiva, caused by a virus; epidemic by nature.

epidemic neuromyasthenia [MED] A prolonged, debilitating disease of the nervous system of adults; characterized by fatigue, headache, muscle pain, paresis, and emotional and mental disturbances; no etiologic agent has been isolated. Also known as acute infective encephalomyelitis; Akureyri disease; benign myalgic encephalomyelitis; epidemic vegetative neuritis; Iceland disease.

epidemic pleurodynia [MED] An acute epidemic disease of man, caused by coxsackie B virus; characterized by severe pain in the lower thorax and upper abdomen, and associated with fever and malaise.

epidemic roseola *See* rubella.

epidemic typhus *See* classic epidemic typhus.

epidemic vegetative neuritis *See* epidemic neuromyasthenia.

epidemiology [MED] The study of the mass aspects of disease.

epidermal ridge [ANAT] Any of the minute corrugations of the skin on the palmar and plantar surfaces of man and other primates.

epidermis [BOT] The outermost layer (sometimes several layers) of cells on the primary plant body. [HISTOL] The outer nonsensitive, nonvascular portion of the skin comprising two strata of cells, the stratum corneum and the stratum germinativum.

epidermoid carcinoma *See* squamous-cell carcinoma.

epidermoid cyst [MED] A cyst lined by stratified squamous epithelium without associated cutaneous glands.

epidermolysis [MED] The easy separation of various layers of skin, primarily of the epidermis from the corium, observed in certain pathological conditions.

epidermolysis bullosa [MED] A congenital skin disease characterized by the development of vesicles and bullae upon slight, or even without, trauma.

epidiascope [OPTICS] 1. An optical projection system for forming an enlarged real image of a flat opaque object, in which light is reflected from the object and then from a mirror before being focused by a projection lens. Also known as episcope. 2. An optical projection system which can easily be altered to project either transparent or opaque objects.

epididymis [ANAT] The convoluted efferent duct lying posterior to the testis and connected to it by the efferent ductules of the testis.

epididymitis [MED] Inflammation of the epididymis.

epidiorite [PETR] A dioritic rock formed by alteration of pyroxenic igneous rocks.

epidosite [PETR] A rare metamorphic rock composed of epidote and quartz.

epidote [MINERAL] A pistachio-green to blackish-green cal-

EPIDIASCOPE

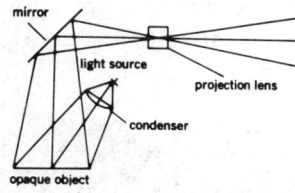

Diagram of an epidiascope (def. 1).

EPIDOTE

├─ 2.5 cm ─┤

Crystals of epidote with quartz on the right side, from Ketchikan, Alaska. *(American Museum of Natural History Specimen)*

cium aluminum sorosilicate mineral that crystallizes in the monoclinic system; the luster is vitreous, hardness is 6½ on Mohs scale, and specific gravity is 3.35-3.45.

epidote-amphibolite facies [PETR] Metamorphic rocks formed under pressures of 3000-7000 bars and temperatures of 250-450°C with conditions intermediate between those that formed greenschist and amphibolite, or with characteristics intermediate.

epidotization [GEOL] The introduction of epidote into, or the formation of epidote from, rocks.

epidural [ANAT] Located on or over the dura mater.

epieugeosyncline [GEOL] Deep troughs formed by subsidence which have limited volcanic power and overlie a eugeosyncline.

epigaster [EMBRYO] The portion of the intestine in vertebrate embryos which gives rise to the colon.

epigastric region [ANAT] The upper and middle part of the abdominal surface between the two hypochondriac regions. Also known as epigastrium.

epigastrium [ANAT] *See* epigastric region. [INV ZOO] The ventral side of mesothorax and metathorax in insects.

epigean [BOT] Pertaining to a plant or plant part that grows above the ground surface. [ZOO] Living near or on the ground surface, applied especially to insects.

epigene [GEOL] **1.** A geologic process originating at or near the earth's surface. **2.** A structure formed at or near the earth's surface.

epigenesis [EMBRYO] Development in gradual stages of differentiation. [GEOL] Alteration of the mineral content of rock due to outside influences.

epigenetic [GEOL] Produced or formed at or near the surface of the earth.

epigenous [BOT] Developing or growing on a surface, especially of a plant or plant part.

epiglottis [ANAT] A flap of elastic cartilage covered by mucous membrane that protects the glottis during swallowing.

epigynous [BOT] Having the perianth and stamens attached near the top of the ovary; that is, the ovary is inferior.

epigynum [INV ZOO] **1.** The genital pore of female arachnids. **2.** The plate covering this opening.

epihydrin alcohol *See* glycidol.

epilation [MED] Removal of the hair by the roots by the use of forceps, chemical means, or roentgenotherapy.

epilemma [HISTOL] The perineurium of very small nerves.

epilepsy [MED] A condition characterized by the paroxysmal recurrence of transient, uncontrollable episodes of abnormal neurological or mental function, or both.

epilimnion [HYD] A fresh-water zone of relatively warm water in which mixing occurs as a result of wind action and convection currents.

epimagma [GEOL] A gas-free, vesicular to semisolid magmatic residue of pasty consistency formed by cooling and loss of gas from liquid lava in a lava lake.

epimagmatic *See* deuteric.

epimer [ORG CHEM] A type of isomer in which the difference between the two compounds is the relative position of the H (hydrogen) group and OH (hydroxyl) group on the last asymmetric C (carbon) atom of the chain, as in the sugars D-glucose and D-mannose.

epimerase [BIOCHEM] A type of enzyme that catalyzes the rearrangement of hydroxyl groups on a substrate.

epimere [ANAT] The dorsal muscle plate of the lining of a coelomic cavity. [EMBRYO] The dorsal part of a mesodermal segment in the embryo of chordates.

epimerization [ORG CHEM] A process in which in an optically active compound that contains two or more asymmetric centers, only one of these is altered by some reaction to form an epimer.

epimeron [INV ZOO] **1.** The posterior plate of the pleuron in insects. **2.** The portion of a somite between the tergum and the insertion of a limb in arthropods.

epimorphosis [PHYSIO] Regeneration in which cell proliferation precedes differentiation.

epimyocardium [EMBRYO] The layer of the embryonic heart from which both the myocardium and epicardium develop.

epimysium [ANAT] The connective-tissue sheath surrounding a skeletal muscle.

epinasty [BOT] Growth changes in which the upper surface of a leaf grows, thus bending the leaf downward.

epinephrine [BIOCHEM] $C_9H_{13}O_3N$ A hormone secreted by the adrenal medulla that acts to increase blood pressure due to stimulation of heart action and constriction of peripheral blood vessels. Also known as adrenaline.

epineural [ANAT] **1.** Arising from the neural arch. **2.** Any process arising from the neural arch. [INV ZOO] The nervous tissue dorsal to the ventral nerve cord in arthropods.

epineural canal [INV ZOO] A canal that runs between the radial nerve and the epithelium in echinoids and ophiuroids.

epineurium [ANAT] The connective-tissue sheath of a nerve trunk.

epipelagic [OCEANOGR] Of or pertaining to the portion of oceanic zone into which enough light penetrates to allow photosynthesis.

epipelagic zone [OCEANOGR] The region of an ocean extending from the surface to a depth of about 200 meters; light penetrates this zone, allowing photosynthesis.

epipetalous [BOT] Having stamens located on the corolla.

epiphallus [INV ZOO] A sclerite in some orthopterans in the floor of the genital chamber.

epipharynx [INV ZOO] An organ attached beneath the labrium of many insects.

epiphragm [BOT] A membrane covering the aperture of the capsule in certain mosses. [INV ZOO] A membranous or calcareous partition that covers the aperture of certain hibernating land snails.

epiphyll [ECOL] A plant that grows on the surface of leaves.

epiphyseal arch [EMBRYO] The arched structure in the third ventricle of the embryonic brain, which marks the site of development of the pineal body.

epiphyseal plate [ANAT] **1.** The broad, articular surface on each end of a vertebral centrum. **2.** The thin layer of cartilage between the epiphysis and the shaft of a long bone. Also known as metaphysis.

epiphysiolysis [MED] The separation of an epiphysis from the shaft of a bone.

epiphysis [ANAT] **1.** The end portion of a long bone in vertebrates. **2.** *See* pineal body.

epiphyte [ECOL] A plant which grows nonparasitically on another plant or on some nonliving structure, such as a building or telephone pole, deriving moisture and nutrients from the air. Also known as aerophyte.

epiphytotic [PL PATH] **1.** Any infectious plant disease that occurs sporadically in epidemic proportions. **2.** Of or pertaining to an epidemic plant disease.

epiplankton [BIOL] Plankton occurring in the sea from the surface to a depth of about 100 fathoms.

epipleural [ANAT] Arising from a rib. [VERT ZOO] An intramuscular bone arising from and extending between some of the ribs in certain fishes.

epiploic foramen [ANAT] An aperture of the peritoneal cavity, formed by folds of the peritoneum and located between the liver and the stomach. Also known as foramen of Winslow.

epipodite [INV ZOO] A branch of the basal joint of the protopodite of thoracic limbs of many arthropods.

epipodium [BOT] The apical portion of an embryonic phyllopodium. [INV ZOO] **1.** A ridge or fold on the lateral edges of each side of the foot of certain gastropod mollusks. **2.** The elevated ring on an ambulacral plate in Echinoidea.

Epipolasina [INV ZOO] A suborder of sponges in the order Clavaxinellida having radially arranged monactinal or diactinal megascleres.

epiproct [INV ZOO] A plate above the anus forming the dorsal part of the tenth or eleventh somite of certain insects.

epipubis [VERT ZOO] A single cartilage or bone located in front of the pubis in some vertebrates, particularly in some amphibians.

episclera [ANAT] The loose connective tissue lying between the conjunctiva and the sclera.

episcope *See* epidiascope.

episcotister [OPTICS] A device for reducing the intensity of light by a known fraction, consisting of a rapidly rotating disk with transparent and opaque sectors.

episepalous [BOT] Having stamens growing on or adnate to the sepals.

EPIGYNOUS

The arrangement of the sepals, petals, and stamens in an epigynous flower.

episiotomy [MED] Medial or lateral incision of the vulva during childbirth, to avoid undue laceration.

episode [GEOL] A distinctive event or series of events in the geologic history of a region or feature.

episome [GEN] A genetic element in bacteria, presumably a deoxyribonucleic acid fragment, which is not necessary for survival of the organism and which can be integrated in the bacterial chromosome or remain free.

epispadias [MED] A congenital defect of the anterior urethra in which the canal terminates on the dorsum of the penis and posterior to its normal opening.

episperm *See* testa.

epistasis [GEN] The suppression of the effect of one gene by another. [MED] A checking or stoppage of a hemorrhage or other discharge. [PATH] A scum or film of substance floating on the surface of urine.

episternum [VERT ZOO] A dermal bone or pair of bones ventral to the sternum of certain fishes and reptiles.

epistilbite [MINERAL] $CaAl_2Si_6O_{16} \cdot 5H_2O$ A mineral of the zeolite family that contains calcium and aluminosilicate and crystallizes in the monoclinic system; occurs in white prismatic crystals or granular forms.

epistome [INV ZOO] **1.** The area between the mouth and the second antennae in crustaceans. **2.** The plate covering this region. **3.** The area between the labrum and the epicranium in many insects. **4.** A flap covering the mouth of certain bryozoans. **5.** The area just above the labrum in certain dipterans.

epitaxial diffused-junction transistor [ELECTR] A junction transistor produced by growing a thin, high-purity layer of semiconductor material on a heavily doped region of the same type.

epitaxial diffused-mesa transistor [ELECTR] A diffused-mesa transistor in which a thin, high-resistivity epitaxial layer is deposited on the substrate to serve as the collector.

epitaxial layer [SOLID STATE] A semiconductor layer having the same crystalline orientation as the substrate on which it is grown.

epitaxial transistor [ELECTR] Transistor with one or more epitaxial layers.

epitaxy [CRYSTAL] Growth of one crystal on the surface of another crystal in which the growth of the deposited crystal is oriented by the lattice structure of the substrate.

epithalamus [ANAT] A division of the vertebrate diencephalon including the habenula, the pineal body, and the posterior commissure.

epitheca [INV ZOO] **1.** An external, calcareous layer around the basal portion of the theca of many corals. **2.** A protective covering of the epicone. **3.** The outer portion of a diatom frustule.

epitheliochorial placenta [EMBRYO] A type of placenta in which the maternal epithelium and fetal epithelium are in contact. Also known as villous placenta.

epithelioid cell [HISTOL] A macrophage that resembles an epithelial cell. Also known as alveolated cell.

epithelioma [MED] A tumor derived from epithelium; usually a skin cancer, occasionally cancer of a mucous membrane.

epitheliomuscular cell [INV ZOO] An epithelial cell with an elongate base that contains contractile fibrils; common among coelenterates.

epithelium [HISTOL] A primary animal tissue, distinguished by cells being close together with little intercellular substance; covers free surfaces and lines body cavities and ducts.

epithema [VERT ZOO] A horny outgrowth on the beak of certain birds.

epithermal [GEOL] Pertaining to mineral veins and ore deposits formed from warm waters at shallow depth, at temperatures ranging from 50-200°C, and generally at some distance from the magnetic source.

epithermal deposit [GEOL] Ore deposit formed in and along openings in rocks by deposition at shallow depths from ascending hot solutions.

epithermal neutron [NUCLEO] A neutron having an energy in the range immediately above the thermal range, roughly between 0.02 and 100 electron volts.

epithermal reactor [NUCLEO] A nuclear reactor in which a substantial fraction of fissions is induced by neutrons having more than thermal energy.

epithermal thorium reactor [NUCLEO] A sodium-cooled reactor based on operation with neutrons in the high epithermal energy range; a uranium-thorium fuel mixture is used, with graphite or beryllium as moderator.

epitoke [INV ZOO] The posterior portion of marine polychaetes; contains the gonads.

epitope [IMMUNOL] A single determinant of an antigen or immunogen which influences its specificity (immunology).

epitrichium [EMBRYO] The outer layer of the fetal epidermis of many mammals.

epitrochlear [ANAT] Of or pertaining to a lymph node that lies above the trochlea of the elbow joint.

epitrochoid [MATH] A curve traced by a point rigidly attached to a circle at a point other than the center when the circle rolls without slipping on the outside of a fixed circle.

epituberculosis [MED] A massive pulmonary shadow seen in x-ray films in active juvenile tuberculosis, probably caused by bronchial obstruction.

epitympanum [ANAT] The attic of the middle ear, or tympanic cavity.

epivalve [INV ZOO] **1.** The upper or apical shell of certain dinoflagellates. **2.** The upper shell of a diatom.

epixylous [ECOL] Growing on wood; used especially of fungi.

epizoic [BIOL] Living on the body of an animal.

epizone [GEOL] **1.** The zone of metamorphism characterized by moderate temperature, low hydrostatic pressure, and powerful stress. **2.** The outer depth zone of metamorphic rocks.

epizootic [VET MED] **1.** Affecting many animals of one kind in one region simultaneously; widely diffuse and rapidly spreading. **2.** An extensive outbreak of an epizootic disease.

epizootiology [VET MED] The study of epizootics.

E-plane antenna [ELECTROMAG] An antenna which lies in a plane parallel to the electric field vector of the radiation that it emits.

E-plane bend *See* E bend.

E-plane T junction [ELECTROMAG] Waveguide T junction in which the change in structure occurs in the plane of the electric field. Also known as series T junction.

EP lubricant [MATER] A lubricating oil or grease that contains additives to improve ability to adhere to the surfaces of metals under high bearing pressures. Derived from extreme-pressure lubricant.

epoch [ASTRON] A particular instant for which certain data are valid; for example, star positions in an astronomical catalog, epoch 1950.0. [GEOL] A major subdivision of a period of geologic time. [PHYS] *See* time.

eponychium [ANAT] The horny layer of the nail fold attached to the nail plate at its margin; represents the remnant of the embryonic condition. [EMBRYO] A horny condition of the epidermis from the second to the eighth month of fetal life, indicating the position of the future nail.

epoophoron [ANAT] A blind longitudinal duct and 10-15 transverse ductules in the mesosalpinx near the ovary which represent remnants of the reproductive part of the mesonephros in the female; homolog of the head of the epididymis in the male. Also known as parovarium; Rosenmueller's organ.

epoxidation [ORG CHEM] Reaction yielding an epoxy compound, such as the conversion of ethylene to ethylene oxide.

epoxide [ORG CHEM] A reactive group in which an oxygen atom is joined to each of two carbon atoms which are already bonded.

epoxy- [ORG CHEM] A prefix indicating presence of an epoxide group in a molecule.

epoxy adhesive [MATER] An adhesive material made of epoxy resin.

1,2-epoxyethane *See* ethylene oxide.

2,3-epoxy-1-propanol *See* glycidol.

epoxy resin [ORG CHEM] A polyether resin formed originally by the polymerization of bisphenol A and epichlorohydrin, having high strength, and low shrinkage during curing; used as a coating, adhesive, casting, or foam.

Eppley pyrheliometer [ENG] A pyrheliometer of the thermoelectric type; radiation is allowed to fall on two concentric silver rings, the outer covered with magnesium oxide and the inner covered with lampblack; a system of thermocouples

EPITHELIOMUSCULAR CELL

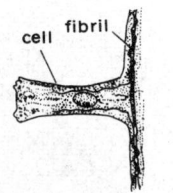

Microscopic view of an epitheliomuscular cell of *Hydra*. (After L. H. Hyman)

(thermopile) is used to measure the temperature difference between the rings; attachments are provided so that measurements of direct and diffuse solar radiation may be obtained.

EPR See electron paramagnetic resonance.

epsilon meson [PARTIC PHYS] Neutral, scalar, meson resonance having positive charge conjugation parity and G-parity, a mass of about 730 MeV, and a width of about 600 MeV; decays to two pions.

epsilon neighborhood [MATH] The set of all points in a metric space whose distance from a given point is less than some number; this number is designated ε.

epsilon structure [SOLID STATE] The hexagonal close-packed structure of the ε-phase of an electron compound.

epsomite [MINERAL] $MgSO_4 \cdot 7H_2O$ A mineral that occurs in clear, needlelike, orthorhombic crystals; commonly, it is massive or fibrous; luster varies from vitreous to milky, hardness is 2-2.5 on Mohs scale, and specific gravity is 1.68; it has a salty bitter taste and is soluble in water. Also known as epsom salt.

epsom salt See epsomite.

Epstein-Barr virus [VIROL] Herpeslike virus particles first identified in cultures of cells from Burkett's malignant lymphoma.

epulis [MED] A benign tumorlike lesion of the gingiva.

equal [MATH] Being the same in some sense determined by context.

equal altitudes [NAV] In celestial navigation, two altitudes numerically the same; the expression applies particularly to the now obsolescent practice of determining the instant of local apparent noon by observing the altitude of the sun a short time before it reaches the meridian and again at the same altitude after transit; the time of local apparent noon is midway between the times of the two observations; however the second observation must be corrected for the run of the ship which took place between the times of the two readings. Also known as double altitudes.

equal-area latitude See authalic latitude.

equal-area map projection [MAP] A map projection having a constant area scale; it is not conformal and is not used for navigation. Also known as authalic map projection; equivalent map projection.

equal-arm balance [MECH] A simple balance in which the distances from the point of support of the balance-arm beam to the two pans at the end of the beam are equal.

equal-energy source [PHYS] Electromagnetic or sound source of energy which emits the same amount of energy for each frequency of the spectrum.

equaling file [DES ENG] A slightly bulging double-cut file used in fine toolmaking.

equality [MATH] The state of being equal.

equalization [ELECTR] The effect of all frequency-discriminating means employed in transmitting, recording, amplifying, or other signal-handling systems to obtain a desired overall frequency response. Also known as frequency-response equalization.

equalizer [ELECTR] A network designed to compensate for an undesired amplitude-frequency or phase-frequency response of a system or component; usually a combination of coils, capacitors, and resistors. Also known as equalizing circuit. [MECH ENG] **1.** A bar to which one attaches a vehicle's whiffletrees to make the pull of draft animals equal. Also known as equalizing bar. **2.** A bar which joins a pair of axle springs on a railway locomotive or car for equalization of weight. Also known as equalizing bar. **3.** A device which distributes braking force among independent brakes of an automotive vehicle. Also known as equalizer brake. **4.** A machine which saws wooden stock to equal lengths. [ORD] A device attached to the carriage of those artillery weapons that, when emplaced, rest on two wheels and two trail ends; it is a compensating mechanism to transmit equally the weapon weight and firing shock.

equalizer brake See equalizer.

equalizing bar See equalizer.

equalizing circuit See equalizer.

equalizing current [ELEC] Current that circulates between two parallel-connected compound generators to equalize their output.

equalizing line [CHEM ENG] A pipe or tubing interconnection

between two closed vessels, containers, or process systems to allow pressure equalization.

equalizing pulses [ELECTR] In television, pulses at twice the line frequency, occurring just before and after the vertical synchronizing pulses, which minimize the effect of line frequency pulses on the interlace.

equalizing reservoir [CIV ENG] A reservoir located between a primary water supply and the consumer for the purpose of maintaining equilibrium between different portions of the distribution system.

equalizing support [ORD] The crossbeam support of an axle-support-type equalizer, on which is mounted the trails and upper carriage; it is connected to the axle by means of a horizontal pintle, which permits a vertical angular difference between the front and rear axis.

equal listener response scale [ACOUS] An arbitrary scale of noisiness which measures the average response of a listener to a noise when allowance is made for the apparent increase of intensity of a noise as its frequency increases. Abbreviated ELR scale.

equal loudness contour [ACOUS] A curve on a graph of sound intensity in decibels versus frequency at each point along which sound appears to be equally loud to a listener. Also known as Fletcher-Munson contour.

equally likely cases [STAT] All simple events in a trial have the same probability.

equal ripple property [MATH] For any continuous function $f(x)$ on the interval $[-1,1]$, and for any positive integer n, a property of the polynomial $p_n(x)$, which is the best possible approximation to $f(x)$ in the sense that the maximum absolute value of $e_n(x) = f(x) - p_n(x)$ is as small as possible; namely, that $e_n(x)$ assumes its extreme values at least $n + 2$ times, with the consecutive extrema having opposite signs.

equal sets [MATH] Sets with precisely the same elements.

equal tails test [STAT] A technique for choosing two critical values for use in a two-sided test; it consists of selecting critical values c and d so that the probability of acceptance of the null hypothesis if the test statistic does not exceed c is the same as the probability of acceptance of the null hypothesis if the test statistic is not smaller than d.

equal-zero indicator [ADP] A circuit component which is on when the result of an operation is zero.

equate [MATH] To state algebraically that two expressions are equal to one another.

equation [CHEM] A symbolic expression that represents in an abbreviated form the laboratory observations of a chemical change; an equation (such as $2H_2 + O_2 \rightarrow 2H_2O$) indicates what reactants are consumed (H_2 and O_2) and what products are formed (H_2O), the correct formula of each reactant and product, and satisfies the law of conservation of atoms in that the symbols for the number of atoms reacting equals the number of atoms in the products. [MATH] A statement that each of two expressions is equal to the other.

equation clock [HOROL] A timepiece that indicates the difference between mean solar time and apparent time.

equation of motion [FL MECH] One of a set of hydrodynamical equations representing the application of Newton's second law of motion to a fluid system; the total acceleration on an individual fluid particle is equated to the sum of the forces acting on the particle within the fluid. [MECH] **1.** Equation which specifies the coordinates of particles as functions of time. **2.** A differential equation, or one of several such equations, from which the coordinates of particles as functions of time can be obtained if the initial positions and velocities of the particles are known. [QUANT MECH] A differential equation which enables one to predict the statistical distribution of the results of any measurement upon a system at any time if the initial dynamical state of the system is known.

equation of piezotropy [THERMO] An equation obeyed by certain fluids which states that the time rate of change of the fluid's density equals the product of a function of the thermodynamic variables and the time rate of change of the pressure.

equation of state [PHYS CHEM] A mathematical expression which defines the physical state of a homogeneous substance (gas, liquid, or solid) by relating volume to pressure and absolute temperature for a given mass of the material.

equation of time [ASTRON] The addition of a quantity to

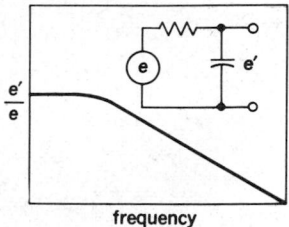

EQUALIZER

Circuit diagram and frequency response characteristics of a type of equalizer that employs a combination of resistance and capacitance; e' = output voltage, e = input voltage.

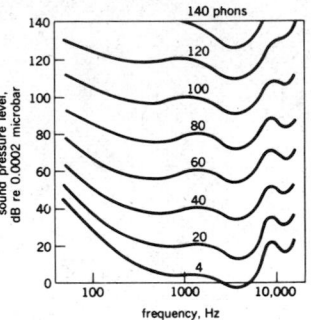

EQUAL LOUDNESS CONTOUR

Equal loudness level contours.

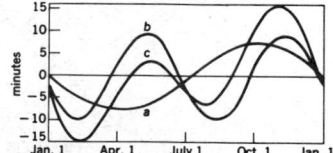

EQUATION OF TIME

The equation of time; c is the sum of the elliptical motion effect a and the inclination effect b.

mean solar time to obtain apparent solar time; formerly, when apparent solar time was in common use, the opposite convention was used; apparent solar time has annual variation as a result of the sun's inclination in the ecliptic and the eccentricity of the earth's elliptical orbit.

equation solver [ADP] A machine, usually analog, for solving systems of simultaneous equations, which may be linear, nonlinear, or differential, and for finding roots of polynomials.

equator [GEOD] The great circle around the earth, equally distant from the North and South poles, which divides the earth into the Northern and Southern hemispheres; the line from which latitudes are reckoned.

equatorial acceleration [ASTROPHYS] A state in which the equatorial atmosphere of a celestial body has a larger absolute angular velocity than the more poleward portions of the atmosphere; exhibited by the sun, Jupiter, and Saturn.

equatorial air [METEOROL] The air of the doldrums or the equatorial trough; distinguished somewhat vaguely from the tropical air of the trade-wind zones.

equatorial axis [GEOD] The diameter of the earth described between two points on the equator.

equatorial bulge [GEOD] The excess of the earth's equatorial diameter over the polar diameter.

equatorial calms *See* doldrums.

equatorial chart [MAP] A chart on an equatorial projection.

equatorial convergence zone *See* intertropical convergence zone.

Equatorial Countercurrent [OCEANOGR] An ocean current flowing eastward (counter to and between the westward-flowing North Equatorial Current and South Equatorial Current) through all the oceans.

Equatorial Current *See* North Equatorial Current; South Equatorial Current.

equatorial cylindrical orthomorphic chart *See* Mercator chart.

equatorial dry zone [CLIMATOL] An arid region existing in the equatorial trough; the most famous dry zone is situated a little south of the equator in the central Pacific. Also known as arid zone.

equatorial easterlies [METEOROL] The trade winds in the summer hemisphere when they are very deep, extending at least 8 to 10 kilometers in altitude, and generally not topped by upper westerlies; if upper westerlies are present, they are too weak and shallow to influence the weather. Also known as deep easterlies; deep trades.

equatorial electrojet [GEOPHYS] A concentration of electric current in the atmosphere found in the magnetic equator.

equatorial front *See* intertropical front.

equatorial horizontal parallax [ASTRON] The parallax of a member of the solar system measured from positional observations made at the same time at two stations on earth, whose distance apart is the earth's equatorial radius.

equatorial mounting [ENG] The mounting of an equatorial telescope; it has two perpendicular axes, the polar axis (parallel to the earth's axis) that turns on fixed bearings, and the declination axis, supported by the polar axis.

equatorial orbit [ASTRON] An orbit in the plane of the earth's equator.

equatorial plane [ASTRON] The plane passing through the equator of the earth, or of other celestial body, perpendicular to its axis of rotation and equidistant from its poles. [CYTOL] The plane in a cell undergoing mitosis that is midway between the centrosomes and perpendicular to the spindle fibers. [MECH] A plane perpendicular to the axis of rotation of a rotating body and equidistant from the intersections of this axis with the body's surface, provided that the body is symmetric about the axis of rotation and is symmetric under reflection through this plane.

equatorial projection [MAP] A map projection centered on the equator.

equatorial radius [GEOD] The radius assigned to the great circle making up the terrestrial equator; approximately $6,378,099 \pm 116$ meters.

equatorial system [ASTRON] A set of celestial coordinates based on the celestial equator as the primary great circle; usually declination and hour angle or sidereal hour angle.

Also known as celestial equator system of coordinates; equinoctial system of coordinates.

equatorial telescope [ENG] An astronomical telescope that revolves about an axis parallel to the earth's axis and automatically keeps a star on which it has been fixed in its field of view.

equatorial tide [OCEANOGR] **1.** A lunar fortnightly tide. **2.** A tidal component with a period of 328 hours.

equatorial trough [METEOROL] The quasicontinuous belt of low pressure lying between the subtropical high-pressure belts of the Northern and Southern hemispheres. Also known as meteorological equator.

Equatorial Undercurrent [OCEANOGR] **1.** A subsurface current flowing from west to east in the Indian Ocean near the 150-meter depth at the equator during the time of the Northeast Monsoon. **2.** A permanent subsurface current in the equatorial region of the Atlantic and Pacific oceans.

equatorial vortex [METEOROL] A closed cyclonic circulation with the equatorial trough.

equatorial wave [METEOROL] A wavelike disturbance of the equatorial easterlies that extends across the equatorial trough.

equatorial westerlies [METEOROL] The westerly winds occasionally found in the equatorial trough and separated from the mid-latitude westerlies by the broad belt of easterly trade winds.

equiangular polygon [MATH] A polygon all of whose interior angles are equal.

equiangular spiral *See* logarithmic spiral.

equiangular spiral antenna [ELECTROMAG] A frequency-independent broad-band antenna, cut from sheet metal, that radiates a very broad, circularly polarized beam on both sides of its surface; this bidirectional radiation pattern is its chief limitation.

equicontinuous family of functions [MATH] A family of functions with the property that given any $\varepsilon > 0$ there is a $\delta > 0$ such that whenever $|x - y| < \delta$, $|f(x) - f(y)| < \varepsilon$ for every function $f(x)$ in the family.

Equidae [VERT ZOO] A family of perissodactyl mammals in the superfamily Equoidea, including the horses, zebras, and donkeys.

equidensity technique [ANALY CHEM] Interference microscopy technique utilizing the Sabattier-effect in photographic emulsions; the equidensities (lines of equal density in a photographic emulsion) are produced by exactly superimposing a positive and a negative of the same interferogram, and making a copy; used to measure photographic film emulsion density.

equidistant [MATH] Being the same distance from some given object.

equigeopotential surface *See* geopotential surface.

equigranular [PETR] Pertaining to the texture of rocks whose essential minerals are all of the same order of size.

equilateral arch [ARCH] An arch described by two circular curves intersecting at the peak of the arch, each curve having a chord equal to the span.

equilateral polygon [MATH] A polygon all of whose sides are the same length.

equilateral polyhedron [MATH] A polyhedron all of whose faces are identical.

equilibrator [ORD] Force-producing mechanism designed to provide a moment about the trunnions of a gun cradle or launcher which is equal and opposite to that caused by the unbalanced weight of the tipping parts, thus making it easier to elevate the weapon.

equilibristat [ENG] A device for measuring the deviation from equilibrium of a railroad car as it goes around a curve.

equilibrium [MECH] Condition in which a particle, or all the constituent particles of a body, are at rest or in unaccelerated motion in an inertial reference frame. Also known as static equilibrium. [PHYS] Condition in which no change occurs in the state of a system as long as its surroundings are unaltered. [STAT MECH] Condition in which the distribution function of a system is time-independent.

equilibrium brightness [ELECTR] Viewing screen brightness occurring when a display storage tube is in a fully written condition.

equilibrium constant [CHEM] A constant at a given temperature such that when a reversible chemical reaction $cC + bB = gG + hH$ has reached equilibrium, the value of this constant K^0 is equal to

$$\frac{a_G^g a_H^h}{a_C^c a_B^b}$$

where a_G, a_H, a_C, and a_B represent chemical activities of the species G, H, C, and B at equilibrium.

equilibrium diagram [PHYS CHEM] A phase diagram of the equilibrium relationship between temperature, pressure, and composition in any system.

equilibrium dialysis [ANALY CHEM] A technique used to determine the degree of ion bonding by protein; the protein solution, placed in a bag impermeable to protein but permeable to small ions, is immersed in a solution containing the diffusible ion whose binding is being studied; after equilibration of the ion across the membrane, the concentration of ion in the protein-free solution is determined; the concentration of ion in the protein solution is determined by subtraction; if binding has occurred, the concentration of ion in the protein solution must be greater.

equilibrium flash vaporization [CHEM ENG] Process in which a continuous liquid-mixture feed stream is partly vaporized in a column or vessel, with continuous withdrawal of vapor and liquid portions, the vapor and liquid in equilibrium. Also known as continuous equilibrium vaporization; simple continuous distillation.

equilibrium gas saturation [PETRO ENG] Condition of zero relative permeability of a nonwetting/wetting phase system in a reservoir; relation to the nonwetting phase (for example, oil) to the wetting phase (for example, water) when the nonwetting-phase saturation is so small that relatively few pores contain it. Also known as critical gas saturation.

equilibrium moisture content [PHYS CHEM] The moisture content in a hydroscopic material that is being dried by contact with air at constant temperature and humidity when a definite, fixed (equilibrium) moisture content in the solid is reached.

equilibrium orbit [NUCLEO] A path, such as a circular path in a synchrotron, or a point moving through space, such as in a linear accelerator, about which the particles in a particle accelerator oscillate, experiencing an effective restoring force toward the path or point. Also known as stable orbit.

equilibrium prism [PHYS CHEM] Three-dimensional (solid) diagram for multicomponent mixtures to show the effects of composition changes on some key property, such as freezing point.

equilibrium profile *See* profile of equilibrium.

equilibrium ratio [PHYS CHEM] **1.** In any system, relation of the proportions of the various components (gas, liquid) at equilibrium conditions. **2.** *See* equilibrium vaporization ratio.

equilibrium solar tide [GEOPHYS] The form of the atmosphere which is determined solely by gravitational forces in the absence of any rotation of the earth relative to the sun.

equilibrium solubility [PHYS CHEM] The maximum solubility of one material in another (for example, water in hydrocarbons) for specified conditions of temperature and pressure.

equilibrium spheroid [GEOPHYS] The shape that the earth would attain if it were entirely covered by a tideless ocean of constant depth.

equilibrium state [IND ENG] A state in which the numbers of customers or items waiting in a queue varies in such a way that the mean and distribution remain constant over a long period.

equilibrium still [ANALY CHEM] Recirculating distillation apparatus (no product withdrawal) used to determine vapor-liquid equilibria data.

equilibrium tide [OCEANOGR] The hypothetical tide due to the tide-producing forces of celestial bodies, particularly the sun and moon.

equilibrium vaporization ratio [PHYS CHEM] In a liquid-vapor equilibrium mixture, the ratio of the mole fraction of a component in the vapor phase (y) to the mole fraction of the

same component in the liquid phase (x), or $y/x = K$ (the K factor). Also known as equilibrium ratio.

equilibrium vapor pressure [PHYS] The vapor pressure of a system in which two or more phases of water coexist in equilibrium.

equine [VERT ZOO] **1.** Resembling a horse. **2.** Of or related to the Equidae.

equine encephalitis [MED] A disease of equines and man caused by one of three viral strains: eastern, western, and Venezuelan equine viruses. Also known as equine encephalomyelitis.

equine encephalomyelitis *See* equine encephalitis.

equinoctial *See* celestial equator.

equinoctial colure [ASTRON] The great circle of the celestial sphere through the celestial poles and the equinoxes; the hour circle of the vernal equinox.

equinoctial point *See* equinox.

equinoctial rains [METEOROL] Rainy seasons which occur regularly at or shortly after the equinoxes in many places within a few degrees of the equator.

equinoctial storm [METEOROL] In semipopular belief, a violent storm of wind and rain which is supposed, both in the United States and in Britain, to occur at or near the time of the equinox. Also known as line gale; line storm.

equinoctial tide [OCEANOGR] A tide occurring near an equinox.

equinox [ASTRON] **1.** Either of the two points of intersection of the ecliptic and the celestial equator, occupied by the sun when its declination is 0°. Also known as equinoctial point. **2.** That instant when the sun occupies one of the equinoctial points.

equiparte [METEOROL] In Mexico, heavy cold rains during October to January, which last for several days. Also known as equipatos.

equipartition [CHEM] **1.** The condition in a gas where under equal pressure the molecules of the gas maintain the same average distance between each other. **2.** The equal distribution of a compound between two solvents. **3.** The distribution of the atoms in an orderly fashion, such as in a crystal.

equipartition law [STAT MECH] In a classical ideal gas, the average kinetic energy per molecule associated with any degree of freedom which occurs as a quadratic term in the expression for the mechanical energy, is equal to half of Boltzmann's constant times the absolute temperature.

equipatos *See* equiparte.

equiphase wave surface [PHYS] Any surface in a wave over which the field vectors at the same instant are in the same phase or 180° out of phase.

equiphase zone [GEOPHYS] That region in space where the difference in phase of two radio signals is indistinguishable.

equipment [ENG] One or more assemblies capable of performing a complete function.

equipment augmentation [ADP] **1.** Procuring additional automatic data-processing equipment capability to accommodate increased work load within an established data system. **2.** Obtaining additional sites or locations.

equipment chain [ENG] Group of equipments that are functionally in series; the failure of one or more of the equipments results in loss of the function.

equipment characteristic distortion [COMMUN] Teletypewriter transmission repetitive display or disruption peculiar to specific portions of a signal, normally caused by maladjusted or dirty contacts of the sending or receiving equipment.

equipment compatibility [ADP] The ability of a device to handle data prepared or handled by other equipment, without alteration of the code or of the form of the data.

equipment failure [ADP] A fault in equipment that results in its improper behavior or prevents the execution of a job as scheduled.

equipment-misuse error [ADP] An erroneous programming instruction, such as a read command addressed to a card punch.

equipment replacement study [IND ENG] A cost analysis based on estimates of operating costs over a stated time for the old facility compared with the new facility.

equipotent [SCI TECH] Equal in capacity or effect.

equipotential cathode *See* indirectly heated cathode.

EQUISETALES

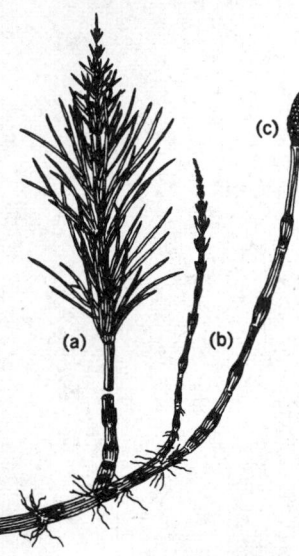

Equisetum arvense, (a) sterile
shoot, *(b)* fertile shoot growing
from underground rootstock,
(c) cone. (From E. W. Sinnot and
K. S. Wilson, Botany: Principles
and Problems, 6th ed., McGraw-
Hill, 1963)

equipotential surface [ELEC] A surface on which the electric potential is the same at every point. [GEOPHYS] A surface characterized by the potential being constant everywhere on it for the attractive forces concerned. [MECH] A surface which is always normal to the lines of force of a field and on which the potential is everywhere the same.

equipressure contour [PETRO ENG] Within a reservoir, a plot or map of the equal isopressure flow network; used to locate sites for water-injection wells for flood coverage of an areal reservoir pattern.

Equisetales [BOT] The horsetails, a monogeneric order of the class Equisetopsida; the only living genus is *Equisetum.*

Equisetatae *See* Equisetopsida.

Equisetineae [BOT] The equivalent name for Equisetophyta.

Equisetophyta [BOT] A division of the subkingdom Embryobionta represented by a single living genus, *Equisetum.*

Equisetopsida [BOT] A class of the division Equisetophyta whose members made up a major part of the flora, especially in moist or swampy places, during the Carboniferous Period.

equisignal [COMMUN] **1.** Pertaining to two signals of equal intensity, used particularly with reference to the signals of a radio range station. **2.** Referring to a radio system in which two identifiable separate radio signals are received with the same intensity.

equisignal localizer [NAV] An aircraft guidance localizer in which the localizer on-course line is centered in a zone of equal amplitude of two transmitted signals; deviations from this zone are detectable as unbalance in the levels of the two signals. Also known as equisignal radio-range beacon; tone localizer.

equisignal radio-range beacon *See* equisignal localizer.

equisignal surface [ELECTROMAG] Surface around an antenna formed by all points at which, for transmission, the field strength (usually measured in volts per meter) is constant.

equisignal zone [NAV] That region in space where the difference in amplitude of two radio signals (usually emitted by a signal station) is indistinguishable.

equitant [BOT] Of leaves, overlapping transversely at the base.

equivalence [MATH] A logic operator having the property that if P, Q, R, etc., are statements, then the equivalence of P, Q, R, etc., is true if and only if all statements are true or all statements are false.

equivalence classes [MATH] The collection of pairwise disjoint subsets determined by an equivalence relation on a set; two elements are in the same equivalence class if and only if they are equivalent under the given relation.

equivalence law of ordered sampling [STAT] If a random ordered sample of size s is drawn from a population of size N, then on any particular one of the s draws each of the N items has the same probability, $1/N$, of appearing.

equivalence point [CHEM] The point in a titration where the amounts of titrant and material being titrated are equivalent chemically.

equivalence principle [RELAT] In general relativity, the principle that the observable local effects of a gravitational field are indistinguishable from those arising from acceleration of the frame of reference. Also known as Einstein's equivalency principle; principle of equivalence.

equivalence relation [MATH] A relation which is reflexive, symmetric, and transitive.

equivalence zone *See* zone of optimal proportion.

equivalent absorption area [ACOUS] Area of perfectly absorbing surface that will absorb sound energy at the same rate as the given object under the same conditions; the acoustic unit of equivalent absorption is the sabin.

equivalent airspeed [AERO ENG] The product of the true airspeed and the square root of the density ratio; used in structural design work to designate various design conditions.

equivalent annual rate [IND ENG] A measure used in setting up a monthly rate on a comparable basis for each of the months regardless of their variation in working days, or for making the rate comparable with an annual rate regardless of the variation in working days during each month.

equivalent-barotropic model [METEOROL] A model atmosphere characterized by frictionless and adiabatic flow and by hydrostatic quasigeostrophic equilibrium, and in which the vertical shear of the horizontal wind is assumed to be proportional to the horizontal wind itself.

equivalent bending moment [MECH] A bending moment which, acting alone, would produce in a circular shaft a normal stress of the same magnitude as the maximum normal stress produced by a given bending moment and a given twisting moment acting simultaneously.

equivalent binary digits [ADP] The number of binary positions required to enumerate the elements of a given set.

equivalent circuit [ELEC] A circuit whose behavior is identical to that of a more complex circuit or device over a stated range of operating conditions.

equivalent conductance [PHYS CHEM] Property of an electrolyte, equal to the specific conductance divided by the number of gram equivalents of solute per cubic centimeter of solvent.

equivalent diameter *See* nominal diameter.

equivalent electrons [ATOM PHYS] Electrons in an atom which have the same principal and orbital quantum numbers, but not necessarily the same magnetic orbital and magnetic spin quantum numbers.

equivalent focal length [OPTICS] The focal length of a thin lens which forms images that most nearly duplicate those of a given compound lens, thick lens, or system of lenses.

equivalent footcandle *See* foot-lambert.

equivalent four-wire system [COMMUN] A transmission system in which multiplex techniques are used to carry on duplex operation over a single pair of wires.

equivalent head wind [NAV] A fictitious wind blowing along the track of an aircraft in the opposite direction to that of motion of the aircraft and of such speed that it would result in the same ground speed as that actually attained.

equivalent height *See* virtual height.

equivalent map projection *See* equal-area map projection.

equivalent noise conductance [ELECTR] Spectral density of a noise current generator measured in conductance units at a specified frequency.

equivalent noise resistance [ELECTR] Spectral density of a noise voltage generator measured in ohms at a specified frequency.

equivalent noise temperature [ELECTR] Absolute temperature at which a perfect resistor, of equal resistance to the component, would generate the same noise as does the component at room temperature.

equivalent nuclei [PHYS CHEM] A set of nuclei in a molecule which are transformed into each other by rotations, reflections, or combinations of these operations, leaving the molecule invariant.

equivalent orifice [MECH ENG] An expression of fan performance as the theoretical sharp-edge orifice area which would offer the same resistance to flow as the system resistance itself.

equivalent periodic line [ELEC] Of a uniform line, a periodic line having the same electrical behavior, at a given frequency, as the uniform line when measured at its terminals or at corresponding section junctions.

equivalent potential temperature [METEOROL] The potential temperature corresponding to the adiabatic equivalent temperature.

equivalent resistance [ELEC] Concentrated or lumped resistance that would cause the same power loss as the actual small resistance values distributed throughout a circuit.

equivalent tail wind [NAV] A fictitious wind blowing along the track of an aircraft in the same direction as that of motion of the aircraft and of such speed that it would result in the same ground speed as that actually attained.

equivalent temperature [METEOROL] **1.** The temperature that an air parcel would have if all water vapor were condensed out at constant pressure, the latent heat released being used to heat the air. Also known as isobaric equivalent temperature. **2.** The temperature that an air parcel would have after undergoing the following theoretical process: dry-adiabatic expansion until saturated, pseudoadiabatic expansion until all moisture is precipitated out, and dry adiabatic compression to the initial pressure; this is the equivalent temperature as read from a thermodynamic chart and is always greater than the isobaric equivalent temperature. Also known as adiabatic equivalent temperature; pseudoequivalent temperature. [THERMO] A term used in British engineering for that temperature of a uniform enclosure in which, in still air, a sizable blackbody at 75°F (23.9°C) would lose heat at the same rate as in the environment.

equivalent twisting moment [MECH] A twisting moment which, if acting alone, would produce in a circular shaft a shear stress of the same magnitude as the shear stress produced by a given twisting moment and a given bending moment acting simultaneously.

equivalent vapor volume [PETRO ENG] The volume occupied by a barrel of oil if all the oil were to become a vapor; expressed as cubic foot per barrel at 60°F (15.6°C).

equivalent weight [CHEM] The number of parts by weight of an element or compound which will combine with or replace, directly or indirectly, 1.008 parts by weight of hydrogen, 8.00 parts of oxygen, or the equivalent weight of any other element or compound.

equivalent width [PHYS] A measure of the total absorption of radiant energy as indicated by an absorption line or absorption band.

equiviscous temperature [CHEM ENG] A measure of viscosity used in the tar industry, equal to the temperature in degrees Celsius at which the viscosity of tar is 50 seconds as measured in a standard tar efflux viscometer. Abbreviated EVT.

equivocation [COMMUN] The entropy of the input of a communications channel when the output is known.

equivoluminal wave See S wave.

Equi See Equuleus.

Equoidea [VERT ZOO] A superfamily of perissodactyl mammals in the suborder Hippomorpha comprising the living and extinct horses and their relatives.

Equuleus [ASTRON] A northern constellation near Aquarius, right ascension 21 hours, declination 10° north. Abbreviated Equl. Also known as Little Horse.

Equus [VERT ZOO] The genus comprising the large, one-toed modern horses, including donkeys and zebras.

Er See erbium.

ER See electroreflectance; endoplasmic reticulum.

era [GEOL] A division of geologic time of the highest order, comprising one or more periods.

ERA See electron ring accelerator.

eradiation See terrestrial radiation.

erasability of storage [ADP] Ability to erase data that are recorded in a particular location and replace them with new data; storage media are said to be erasable (for example, magnetic tape) or nonerasable (for example, punched cards).

erasable storage [ADP] Any storage medium which permits new data to be written in place of the old, such as magnetic disk or tape, but not punched card or punched tape.

erase [ADP] To change all the binary digits in a digital computer storage device to binary zeros. [ELECTR] **1.** To remove recorded material from magnetic tape by passing the tape through a strong, constant magnetic field (dc erase) or through a high-frequency alternating magnetic field (ac erase). **2.** To eliminate previously stored information in a charge-storage tube by charging or discharging all storage elements.

erase oscillator [ELECTR] The oscillator used in a magnetic recorder to provide the high-frequency signal needed to erase a recording on magnetic tape; the bias oscillator usually serves also as the erase oscillator.

erasing head [ELECTR] A magnetic head used to obliterate material previously recorded on magnetic tape.

erasing speed [ELECTR] In charge-storage tubes, the rate of erasing successive storage elements.

erbia See erbium oxide.

erbium [CHEM] A trivalent metallic rare-earth element, symbol Er, of the yttrium subgroup, found in euxenite, gadolinite, fergusonite, and xenotine; atomic number 68, atomic weight 167.26, specific gravity 9.051; insoluble in water, soluble in acids; melts at 1400–1500°C.

erbium halide [INORG CHEM] A compound of erbium and one of the halide elements.

erbium nitrate [INORG CHEM] $Er(NO_3)_3 \cdot 5H_2O$ Pink crystals that are soluble in water, alcohol, and acetone; may explode if it is heated or shocked.

erbium oxalate [ORG CHEM] $Er_2(C_2O_4)_3 \cdot 10H_2O$ A red powder that decomposes at 575°C; used to separate erbium from common metals.

erbium oxide [INORG CHEM] Er_2O_3 Pink powder that is insoluble in water; used as an actuator for phosphors and in manufacture of glass that absorbs in the infrared. Also known as erbia.

erbium sulfate [INORG CHEM] $Er_2(SO_4)_3 \cdot 8H_2O$ Red crystals that are soluble in water.

erection [CIV ENG] Positioning and fixing the frame of a structure. [PHYSIO] The enlarged state of erectile tissue when engorged with blood, as of the penis or clitoris.

erection bolt [CIV ENG] A threaded rod with a head at one end, used to temporarily join parts of a structure during construction.

erection stress [MECH] The internal forces exerted on a structural member during construction.

erection tower [CIV ENG] A temporary framework built at a construction site for hoisting equipment.

erector [PHYSIO] Any muscle that produces erection of a part.

erect stem [BOT] A stem that stands, having a vertical or upright habit.

Eremascoideae [BOT] A monogeneric subfamily of ascosporogenous yeasts characterized by mostly septate mycelia, and spherical asci with eight oval to round ascospores.

eremophobia [PSYCH] An abnormal fear of being lonely.

Erethizontidae [VERT ZOO] The New World porcupines, a family of rodents characterized by sharply pointed, erectile hairs and four functional digits.

ereuthophobia See erythrophobia.

erg [PHYS] A unit of energy or work in the centimeter-gram-second system of units, equal to the work done by a force of magnitude of 1 dyne when the point at which the force is applied is displaced 1 centimeter in the direction of the force. Also known as dyne centimeter (dyne-cm).

ERG See electroretinogram.

Ergasilidae [INV ZOO] A family of copepod crustaceans in the suborder Cyclopoida in which the females are parasitic on aquatic animals, while the males are free-swimming.

ergastic [CYTOL] Pertaining to the nonliving components of protoplasm.

ergastoplasm [CYTOL] A cytoplasm component which shows an affinity for basic dyes; a form of the endoplasmic reticulum.

ergodic [STAT] **1.** Property of a system or process in which averages computed from a data sample over time converge, in a probabilistic sense, to ensemble or special averages. **2.** Pertaining to such a system or process.

ergodic theory [MATH] The study of measure-preserving transformations. [STAT MECH] Mathematical theory which attempts to show that the various possible microscopic states of a system are equally probable, and that the system is therefore ergodic.

ergodic transformation [MATH] A measure-preserving transformation on X with the property that whenever X is written as a union of two disjoint invariant subsets, one of these must have measure zero.

ergograph [ENG] An instrument with a recording device used to measure work capacity of muscles.

ergometer [ENG] An instrument with a recording device used to measure work performed by muscles under control conditions.

ergon [QUANT MECH] A quantum of energy; for any oscillator it is equal to the product of the oscillator's frequency and Planck's constant.

ergonometrics [IND ENG] The application of various procedures for determining the time for an operator to perform a task satisfactorily, using the standard method in the usual environmental conditions.

ergonomics [IND ENG] The study of human capability and psychology in relation to the working environment and the equipment operated by the worker.

ergosterin See ergosterol.

ergosterol [BIOCHEM] $C_{28}H_{44}O$ A crystalline, water-insolu-

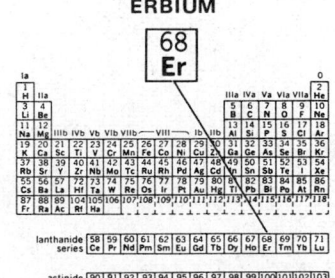

ERBIUM

Periodic table of the chemical elements showing the position of erbium.

ble, unsaturated sterol found in ergot, yeast, and other fungi, and which may be converted to vitamin D_2 on irradiation with ultraviolet light or activation with electrons. Also known as ergosterin.

ergot [MYCOL] The dark purple or black sclerotium of the fungus *Claviceps purpurea*. [ORG CHEM] Any of the five optically isomeric pairs of alkaloids obtained from this fungus; only the levorotatory isomers are physiologically active.

ergotinine [ORG CHEM] An alkaloid and an isomer of ergotoxine that is a 1:1:1 mixture of ergocornine, ergocristine, and ergocryptine; crystallizes in long needles from acetone solutions, melting point 229°C, and soluble in chloroform, alcohol, and absolute ether.

ergotism [MED] Acute or chronic intoxication resulting from ingestion of grain infected with ergot fungus, or from chronic use of drugs containing ergot.

ergotoxine [ORG CHEM] An alkaloid and an isomer of ergotinine that is a 1:1:1 mixture of ergocornine, ergocristine, and ergocryptine; crystallizes in orthorhombic crystals, melts at 190°C, and is soluble in methyl alcohol, ethyl alcohol, acetone, and chloroform.

Eri *See* Eridanus.

Erian [GEOL] Middle Devonian geologic time; a North American provincial series.

Erian orogeny [GEOL] One of the orogenies during Phanerozoic geologic time, at the end of the Silurian; the last part of the Caledonian orogenic era. Also known as Hibernian orogeny.

Ericaceae [BOT] A large family of dicotyledonous plants in the order Ericales distinguished by having twice as many stamens as corolla lobes.

Ericales [BOT] An order of dicotyledonous plants in the subclass Dilleniidae; plants are generally sympetalous with unitegmic ovules and they have twice as many stamens as petals.

Erichsen test [MET] A cupping test to measure the ductility of a piece of sheet metal and to determine its suitability for deep drawing.

Erichsen value [MET] The depth of impression in millimeters required to fracture a cupped sheet metal supported on a ring and deformed at the center by a spherically shaped tool.

ericophyte [ECOL] A plant that grows on a heath or moor.

Ericsson cycle [THERMO] An ideal thermodynamic cycle consisting of two isobaric processes interspersed with processes which are, in effect, isothermal, but each of which consists of an infinite number of alternating isentropic and isobaric processes.

Erid *See* Eridanus.

Eridanus [ASTRON] A southern constellation made up of a long, crooked line of stars beginning near Rigel in the foot of Orion, and winding west and south to the first-magnitude star Achernar. Abbreviated Eri; Erid. Also known as River Po.

erigeron oil [MATER] A volatile oil, whose components are gallic and tannic acid, that is distilled from fleabane and horseweed.

erikite [MINERAL] A brown mineral consisting of a silicate and phosphate of cerium metals; occurs in orthorhombic crystals.

Erinaceidae [VERT ZOO] The hedgehogs, a family of mammals in the order Insectivora characterized by dorsal and lateral body spines.

erineum [PL PATH] An abnormal growth of hairs induced on the epidermis of a leaf by certain mites.

erinite [MINERAL] $Cu_5(OH)_4(AsO_4)_2$ Emerald-green mineral composed of basic copper arsenate.

Erinnidae [INV ZOO] A family of orthorrhaphous dipteran insects in the series Brachycera.

Eriocaulaceae [BOT] The single family of the order Eriocaulales.

Eriocaulales [BOT] An order of monocotyledonous plants in the subclass Commelinidae, having a perianth reduced or lacking and having unisexual flowers aggregated on a long peduncle.

Eriococcinae [INV ZOO] A family of homopteran insects in the superfamily Coccoidea; adult females and late instar nymphs have an anal ring.

Eriocraniidae [INV ZOO] A small family of lepidopteran insects in the superfamily Eriocranioidea.

Eriocranioidea [INV ZOO] A superfamily of lepidopteran insects in the suborder Homoneura comprising tiny moths with reduced, untoothed mandibles.

eriodictyol [ORG CHEM] $C_{15}H_{22}O_6$ A compound isolated from *Eriodictyon californicum* as needlelike crystals from a dilute alcohol solution, sparingly soluble in boiling water, hot alcohol, and glacial acetic acid; used in medicine as an expectorant. Also known as 3',4',5,7-tetrahydroxyflavanone.

eriometer [OPTICS] A device used to measure diameters of small particles or fibers by observing the diameter of the diffraction pattern produced by them in light coming from a small hole in a metal plate.

erionite [MINERAL] A chabazite mineral of the zeolite family that contains calcium ions and crystallizes in the hexagonal system.

Eriophyidae [INV ZOO] The bud mites or gall mites, a family of economically important plant-feeding mites in the suborder Trombidiformes.

eriophyllous [BOT] Having leaves covered by a cottony pubescence.

erlang [COMMUN] A unit of communication traffic load, equal to the traffic load whose calls, if placed end to end, will keep one path continuously occupied.

Erlenmeyer flask [CHEM] A conical glass laboratory flask, with a broad bottom and a narrow neck.

Erlenmeyer synthesis [ORG CHEM] Preparation of cyclic ethers by the condensation of an aldehyde with an α-acylamino acid in the presence of acetic anhydride and sodium acetate.

eroding velocity [GEOL] The minimum average velocity required for eroding homogeneous material of a given particle size.

Eros [ASTRON] An asteroid that is about 32 kilometers in diameter; its closest approach to the earth is every 44 years.

erose [BIOL] Having an irregular margin.

erosion [GEOL] **1.** The loosening and transportation of rock debris at the earth's surface. **2.** The wearing away of the land, chiefly by rain and running water. [MED] **1.** Surgical removal of tissues by scraping. **2.** Excision of a joint.

erosional unconformity [GEOL] The surface that separates older, eroded rocks from younger, overlying sediments.

erosion-corrosion [MET] Attack on a metal surface resulting from the combined effects of erosion and corrosion.

erosion cycle [GEOL] A postulated sequence of conditions through which a new landmass proceeds as it wears down, classically the concept of youth, maturity, and old age, as stated by W. M. Davis; an original landmass is uplifted above base level, cut by canyons, gradually converted into steep hills and wide valleys, and is finally reduced to a flat lowland at or near base level.

erosion pavement [GEOL] A layer of pebbles and small rocks that prevents the soil underneath from eroding.

erosion platform *See* wave-cut platform.

erosion ridge [HYD] One of a group of ridges on the surface of snow; formed by the corrosive action of wind-blown snow.

erosion surface [GEOL] A land surface shaped by agents of erosion.

erotic [PSYCH] **1.** Pertaining to the libido or sexual passion. **2.** Moved by or arousing sexual desire.

erotophobia [PSYCH] An abnormal fear of love.

Erotylidae [INV ZOO] The pleasing fungus beetles, a family of coleopteran insects in the superfamily Cucujoidea.

ERP *See* effective radiated power; emergency recorder plot.

Errantia [INV ZOO] A group of 34 families of polychaete annelids in which the anterior region is exposed and the linear body is often long and is dorsoventrally flattened.

erratic [GEOL] A rock fragment that has been transported a great distance, generally by glacier ice or floating ice, and differs from the bedrock on which it rests.

error [ADP] An incorrect result arising from approximations used in numerical methods, rather than from a human mistake or computer malfunction. [SCI TECH] Any discrepancy between a computed, observed, or measured quantity and the true, specified, or theoretically correct value of that quantity.

error burst [ADP] The condition when more than one bit is in error in a given number of bits.

error checking and recovery [ADP] An automatic procedure which checks for parity and will proceed with the execution after error correction.

error-checking code *See* self-checking code.

error code [ADP] A specific character punched into a card or tape to indicate that a conscious error was made in the associated block of data; machines reading the error code may be programmed to throw out the entire block automatically.

error coefficient [CONT SYS] The steady-state value of the output of a control system, or of some derivative of the output, divided by the steady-state actuating signal. Also known as error constant.

error constant *See* error coefficient.

error-correcting code [ADP] Data representation that allows for error detection and error correction if the error is of a specific kind.

error-correcting telegraph system [COMMUN] System employing an error-detecting code, and so conceived that any false signal initiates a repetition of the transmission of the character incorrectly received.

error correction [ADP] Computer device for automatically locating and correcting a machine error of dropping a bit or picking up an extraneous bit, without stopping the machine or having it go to a programmed recovery routine. [COMMUN] Any system for reducing errors in an incoming message, such as sending redundant signals as a check. [ELEC] Correction of time errors in interconnected alternating-current power systems resulting from deviations from normal frequency, in order to make all areas synchronous.

error correction routine [ADP] A program which corrects specific error conditions in another program, routine, or subroutine.

error-detecting code *See* self-checking code.

error-detecting system [ADP] An automatic system which detects an error due to a lack of data, or erroneous data during transmission.

error detection and feedback system [ADP] An automatic system which retransmits a piece of data detected by the computer as being in error.

error detection routine *See* diagnostic routine.

error diagnostic [ADP] A computer printout of an instruction or data statement, pinpointing an error in the instruction or statement and spelling out the type of error involved.

error frequency limit [ADP] The maximum number of single bit errors per unit of time that a computer will accept before a machine check interrupt is initiated. Abbreviated EFL.

error function [MATH] The real function defined as the integral from 0 to x of $e^{-t^2} dt$ or $e^{t^2} dt$, or the integral from x to ∞ of $e^{-t^2} dt$.

error-indicating system [ADP] Built-in circuits designed to indicate automatically that certain computational errors have occurred.

error interrupt [ADP] The halt in execution of a program because of errors which the computer is not capable of correcting.

error message [ADP] A message indicating detection of an error.

error of closure [ENG] Also known as angular error of closure. **1.** The amount by which the measurement of the azimuth of the first line of a traverse, made after completing the circuit, fails to equal the initial measurement. **2.** The amount by which the sum of the angles measured around the horizon differs from 360°.

error of perpendicularity [NAV] That error in the reading of a marine sextant due to nonperpendicularity of the index mirror to the frame.

error range [STAT] The difference between the highest and lowest error values; a measure of the uncertainty associated with a number.

error rate [COMMUN] The number of erroneous bits or characters received for some fixed number of bits transmitted (such as 100,000).

error routine [ADP] A routine which takes control of a program and initiates corrective actions when an error is detected.

error signal [CONT SYS] In an automatic control device, a signal whose magnitude and sign are used to correct the alignment between the controlling and the controlled elements. [ELEC] *See* error voltage. [ELECTR] A voltage that depends on the signal received from the target in a tracking system, having a polarity and magnitude dependent on the angle between the target and the center of the scanning beam.

error tape [ADP] The magnetic tape on which erroneous records are stored during processing.

error voltage [ELEC] A voltage, usually obtained from a selsyn, that is proportional to the difference between the angular positions of the input and output shafts of a servosystem; this voltage acts on the system to produce a motion that tends to reduce the error in position. Also known as error signal.

ertor [METEOROL] The effective (radiational) temperature of the ozone layer (region).

erucic acid [ORG CHEM] $C_{22}H_{42}O_2$ A monoethenoid acid that is the cis isomer of brassidic acid and makes up 40 to 50% of the total fatty acid in rapeseed, wallflower seed, and mustard seed; crystallizes as needles from alcohol solution, insoluble in water, soluble in ethanol and methanol. Also known as *cis*-13-docosenoic acid.

eruption [GEOL] The ejection of solid, liquid, or gaseous material from a volcano.

eruptive prominence [ASTRON] A prominence on the sun that is formed from active material above the chromosphere and reaches high altitudes on the sun at great speed.

eruptive rock [PETR] **1.** Rock formed from a volcanic eruption. **2.** Igneous rock that reaches the earth's surface in a molten condition.

eruptive star [ASTRON] A star that has a rapid change in its intensity because of the physical change it undergoes; examples are flare stars, recurrent novae, novae, supernovae, and nebular variables.

Erwinia [MICROBIO] A genus of motile, rod-shaped bacteria in the tribe Erwinieae; these organisms invade living plant tissues and cause dry necroses, galls, wilts, and soft rots.

Erwinieae [MICROBIO] A tribe of phytopathogenic bacteria in the family Enterobacteriaceae, including the single genus *Erwinia*.

erysipelas [MED] An acute, infectious bacterial disease caused by *Streptococcus pyogenes* and characterized by inflammation of the skin and subcutaneous tissues.

erysipeloid [MED] A bacterial infection caused by *Erysipelothrix rhuscopathiae* and occurring on the hands of people who handle infected meat or fish.

Erysipelothrix [MICROBIO] A genus of gram-positive, rod-shaped bacteria of the family Corynebacteriaceae that have a tendency to form long filaments; parasites of mammals, birds, and fish.

Erysiphaceae [MYCOL] The powdery mildews, a family of ascomycetous fungi in the order Erysiphales having light-colored mycelia and conidia.

Erysiphales [MYCOL] An order of ascomycetous fungi which are obligate parasites of seed plants, causing powdery mildew and sooty mold.

erythema [MED] Localized redness of the skin in areas of variable size.

erythema multiforme [MED] An acute inflammatory skin disease characterized by red macules, papules, or tubercles on the extremities, neck, and face.

erythema nodosum [MED] The occurrence of pink to blue, tender nodules on the anterior surfaces of the lower legs; more frequent in women than men.

erythremia *See* erythrocytosis; polycythemia vera.

erythrine *See* erythrite.

erythrite [MINERAL] $Co_3(AsO_4)_2 \cdot 8H_2O$ A crimson, peach, or pink-red secondary oxidized cobalt mineral that occurs in monoclinic crystals, in globular and reniform masses, or in earthy forms. Also known as cobalt bloom; cobalt ocher; erythrine; peachblossom ore; red cobalt. [ORG CHEM] *See* erythritol.

erythritol [ORG CHEM] $H(CHOH)_4H$ A tetrahydric alcohol; occurs as tetragonal prisms, melting at 121°C, soluble in water; used in medicine as a vasodilator. Also known as antierythrite; erythrite; ethroglucin; erythrol; phycite; tetrahydroxy butane.

erythroblast [HISTOL] A nucleated cell occurring in bone marrow as the earliest recognizable cell of the erythrocytic series.

erythroblastosis [MED] The abnormal presence of erythroblasts in the blood. [VET MED] A virus disease of birds; considered to be part of the avian leukosis complex in which there is an abnormal number of erythroblasts in the blood.

erythroblastosis fetalis [MED] A form of hemolytic anemia affecting the fetus and newborn infant when a mother is Rh-negative and has developed antibodies against an Rh-positive fetus. Also known as hemolytic disease of newborn.

erythrocruorin [BIOCHEM] Any of the iron-porphyrin protein respiratory pigments found in the blood and tissue fluids of certain invertebrates; corresponds to hemoglobin in vertebrates.

erythrocyte [HISTOL] A type of blood cell that contains a nucleus in all vertebrates but man and that has hemoglobin in the cytoplasm. Also known as red blood cell.

erythrocytopoiesis *See* erythropoiesis.

erythrocytosis [MED] An increase in the number of circulating erythrocytes of more than two standard deviations above the mean normal, usually occurring secondary to hypoxia. Also known as erythremia; polycythemia.

erythroidine [ORG CHEM] $C_{16}H_{19}NO_3$ An alkaloid existing in two forms: α-erythroidine and β-erythroidine, isolated from *Erythrina* species; β-erythroidine has an action similar to that of curare as a skeletal muscle relaxant.

erythrol *See* erythritol.

erythromelalgia [MED] A cutaneous vasodilation of the feet or, more rarely, of the hands; characterized by redness, mottling, changes in skin temperature, and neuralgic pains. Also known as acromelalgia; Mitchell's disease.

erythromycin [MICROBIO] A crystalline antibiotic produced by *Streptomyces erythreus* and used in the treatment of gram-positive bacterial infections.

D-erythropentose *See* ribulose.

erythrophilous [BIOL] Having an affinity for red dyes and other coloring matter.

erythrophleine [ORG CHEM] $C_{24}H_{39}NO_5$ An alkaloid isolated from the bark of *Erythrophleum guineense*; used in medicine experimentally for its digitalislike action.

erythrophobia [PSYCH] **1.** An abnormal fear of red colors; may be associated with a fear of blood. **2.** Fear of blushing. Also known as ereuthophobia.

erythrophore [ZOO] A chromatophore containing a red pigment, especially a carotenoid.

erythropia *See* erythropsia.

erythropoiesis [PHYSIO] The process by which erythrocytes are formed. Also known as erythrocytopoiesis.

erythropoietin [BIOCHEM] A hormone, thought to be produced by the kidneys, that regulates erythropoiesis, at least in higher vertebrates.

erythropsia [MED] An abnormality of vision in which all objects appear red; red vision. Also known as erythropia.

erythrose [ORG CHEM] $HOCH_2(CHOH)_2CHO$ A tetrose sugar obtained from erythrol; a syrupy liquid at room temperature.

erythrosiderite [MINERAL] $K_2FeCl_5 \cdot H_2O$ Mineral composed of hydrous potassium iron chloride; occurs in lavas.

erythrosin [ORG CHEM] $C_{13}H_{18}O_6N_2$ A red compound obtained by reacting tyrosine with nitric acid.

erythrosis [MED] **1.** Overproliferation of erythropoietic tissue, as found in polycythemia. **2.** The unusual red skin color of individuals with polycythemia.

Erythroxylaceae [BOT] A homogeneous family of dicotyledonous woody plants in the order Linales characterized by petals that are internally appendiculate, three carpels, and flowers without a disk.

erythroxylon *See* cocaine.

erythrulose [BIOCHEM] $C_4H_8O_4$ A ketose sugar occurring as an oxidation product of erythritol due to the action of certain bacteria.

Erzgebirgian orogeny [GEOL] Diastrophism of the early Late Carboniferous.

Es *See* einsteinium.

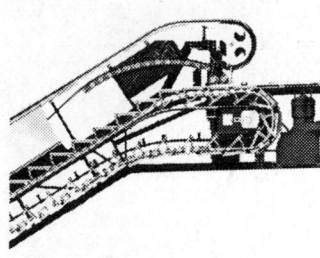

ESCALATOR

Cutaway view of escalator. (*Otis Elevator Co.*)

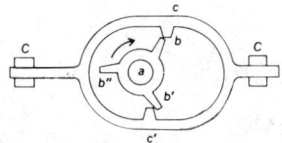

ESCAPEMENT

Simple escapement. Oscillating member cc' is an open bar arranged to slide longitudinally in bearings CC; wheel a turns continuously in direction of arrow; wheel has teeth b, b', b''; tooth b is just ceasing to drive pallet c to the right.

Esaki tunnel diode *See* tunnel diode.

Esbach's reagent [PATH] A solution of 1 gram of trinitrophenol and 2 grains of citric acid in 100 milliliters of water; used in determining albumin in urine.

escalation [IND ENG] Provision in actual or estimated costs for inflational increases in the costs of equipment, materials, labor, and so on, over those specified in an original contract.

escalator [MECH ENG] A continuously moving stairway and handrail.

E scan *See* E scope.

escape character [ADP] A character used to indicate that the succeeding character or characters are expressed in a code different from the code currently in use.

escape hatch [ENG] A hatch which permits men to escape from a compartment, such as the interior of a submarine or aircraft, when normal means of exiting are blocked.

escapement [HOROL] A device in a timepiece consisting of a toothed wheel which engages alternate pallets attached to an oscillating member. [MECH ENG] A ratchet device that permits motion in one direction slowly.

escape rocket [AERO ENG] A small rocket engine attached to the leading end of an escape tower, to provide additional thrust to the capsule in an emergency; it helps separate the capsule from the booster vehicle and carries it to an altitude where parachutes can be deployed.

escape tower [AERO ENG] A trestle tower placed on top of a space capsule, connecting the capsule to the escape rocket on top of the tower; used for emergencies.

escape trunk [NAV ARCH] An escape compartment in a submarine, designed to receive a rescue chamber or deep-submergence rescue vehicle.

escape velocity [AERO ENG] In space flight, the speed at which an object is able to overcome the gravitational pull of the earth.

escape wheel [HOROL] The final wheel in a timepiece train, designed to transmit impulses to a lever.

escaping tendency [PHYS CHEM] The tendency of a solute species to escape from solution; related to the chemical potential of the solute.

escar *See* esker.

escarpment [GEOL] A cliff or steep slope of some extent, generally separating two level or gently sloping areas, and produced by erosion or faulting. Also known as scarp. [ORD] The ground surrounding a fortified place which has been cut away nearly vertically to prevent an enemy's approach.

eschar [GEOL] *See* esker. [MED] A dry crust or slough, especially one formed after a thermal or chemical burn.

escharotic [MED] **1.** Caustic. **2.** Producing an eschar.

Escherichia [MICROBIO] A genus of gram-negative, rod-shaped bacteria in the family Enterobacteriaceae that are usually motile by peritrichous flagella.

Escherichia coli [MICRO] The type species of the genus, occurring as part of the normal intestinal flora in vertebrates. Also known as colon bacillus.

Escherichieae [MICROBIO] A tribe of bacteria in the family Enterobacteriaceae defined by the ability to ferment lactose, with the rapid production of acid and visible gas.

Eschka mixture [ANALY CHEM] A mixture of two parts magnesium oxide and one part anhydrous sodium carbonate; used as a fusion mixture for determining sulfur in coal.

Eschweiler-Clarke modification [ORG CHEM] A modification of the Leuckart reaction, involving reductive alkylation of ammonia or amines (except tertiary amines) by formaldehyde and formic acid.

eschynite [MINERAL] $(Ce,Ca,Fe,Th)(Ti,Cb)_2O_6$ A black mineral, occurring in prismatic crystals; a rare oxide of cesium, titanium, and other metals, which is isomorphous with priorite.

Esclangon effect [OPTICS] Bending of a reflected light ray caused by movement of the mirror in a direction making an acute angle with its surface.

E scope [ELECTR] A cathode-ray scope on which signals appear as spots, with range as the horizontal coordinate and elevation angle or height as the vertical coordinate. Also known as E indicator; E scan.

escort fighter [AERO ENG] A fighter designed or equipped for long-range missions, usually to accompany heavy bombers on raids.

escutcheon [DES ENG] An ornamental shield, flange, or border used around a dial, window, control knob, or other panel-mounted part. Also known as escutcheon plate.

escutcheon plate *See* escutcheon.

ESD *See* external symbol dictionary.

eserine *See* physostigmine.

eskar *See* esker.

eskebornite [MINERAL] $CuFeSe_2$ The selenium analog of the mineral pyrrhotite ($Fe_{1-x}S$).

esker [GEOL] A sinuous ridge of constructional form, consisting of stratified accumulations, glacial sand, and gravel. Also known as asar; back furrow; escar; eschar; eskar; osar; serpent kame.

Esocidae [VERT ZOO] The pikes, a family of fishes in the order Clupeiformes characterized by an elongated beaklike snout and sharp teeth.

Esocoidei [VERT ZOO] A small suborder of fresh-water fishes in the order Salmoniformes; includes the pikes, mudminnows, and pickerels.

esophageal diverticulum [MED] An outpocketing of the wall of the esophagus, or of the pharynx just above the opening to the esophagus.

esophageal fistula [MED] Congenitally, an abnormal tube communicating between the esophagus and an internal organ, usually the trachea; an acquired esophageal fistula usually communicates between the esophagus and the skin through an external opening, or may communicate with internal organs.

esophageal gland [ANAT] Any of the digestive glands within the submucosa of the esophagus; secretions are chiefly mucus and serve to lubricate the esophagus.

esophageal hiatus [ANAT] The opening in the diaphragm for passage of the esophagus.

esophageal teeth [VERT ZOO] Enamel-tipped hypapophyses of the posterior cervical vertebrae of certain snakes, which penetrate the esophagus and function to break eggshells.

esophagitis [MED] Inflammation of the esophagus.

esophagogastrostomy [MED] Establishment, by surgery, of an anastomosis between the esophagus and the stomach; may be performed by the abdominal route or by transpleural operation.

esophagus [ANAT] The tubular portion of the alimentary canal interposed between the pharynx and the stomach. Also known as gullet.

esotropia [MED] Convergent strabismus, occurring when one eye fixes upon an object and the other deviates inward. Also known as cross-eye.

ESP *See* extrasensory perception.

espalier drainage *See* trellis drainage.

esparto wax [MATER] Vegetable wax extracted from esparto grass; it is hard, blends well, and is easy to emulsify; used to give smoothness to polishes.

ESR *See* electron paramagnetic resonance.

ESSA *See* environmental survey satellite.

Essen coefficient [ELEC] The torque exerted on the moving part of an electric rotating machine divided by the volume enclosed by the air gap.

essential amino acid [BIOCHEM] Any of eight of the 20 naturally occurring amino acids that are indispensable for optimum animal growth but cannot be formed in the body and must be supplied in the diet.

essential hypertension [MED] Elevation of the systemic blood pressure, of unknown origin. Also known as primary hypertension.

essential oil [MATER] Any of the odoriferous oily products of plant origin which are distillable; the principal constituents are terpenes, but benzenoid and aliphatic compounds may also be present. Also known as ethereal oil.

essential singularity [MATH] An isolated singularity of a complex function which is neither removable nor a pole.

essexite [PETR] A rock of igneous origin composed principally of plagioclase hornblende, biotite, and titanaugite.

established flow [FL MECH] The flow when the boundary layer of a fluid flowing in a duct completely fills the duct; that is, when the effect of the wall shearing stress extends completely across the duct.

establishment [OCEANOGR] The interval of time between the transit (upper or lower) of the moon and the next high water at a place.

ester [ORG CHEM] The compound formed by the elimination of water and the bonding of an alcohol and an organic acid.

esterase [BIOCHEM] Any of a group of enzymes that catalyze the synthesis and hydrolysis of esters.

ester gum [ORG CHEM] A compound obtained by forming an ester of a natural resin with a polyhydric alcohol; used in varnishes, paints, and cellulosic lacquers. Also known as rosin ester.

esterification [ORG CHEM] A chemical reaction whereby esters are formed.

estersil [ORG CHEM] Hydrophobic silica powder, an ester of $-SiOH$ with a monohydric alcohol; used as a filler in silicone rubbers, plastics, and printing inks.

esthacyte [INV ZOO] A simple sensory cell occurring in certain lower animals, such as sponges. Also spelled aesthacyte.

esthesia [PHYSIO] The capacity for sensation, perception, or feeling. Also spelled aesthesia.

esthesiometer [ENG] An instrument used to measure tactile sensibility by determining the distance by which two points pressed against the skin must be separated in order that they be felt as separate. Also spelled aesthesiometer.

esthesioneuroblastoma *See* neuroepithelioma.

esthesioneuroepithelioma *See* neuroepithelioma.

esthiomene [MED] The chronic ulcerative lesion of the vulva in lymphogranuloma venereum.

estimate [SCI TECH] A statement of the value of a quantity or function based on incomplete data or evidence, or on a rough or approximate calculation.

estimated position [NAV] The most probable position of a craft determined from incomplete data or data of questionable accuracy.

estimated time [IND ENG] A predicted element or operation time.

estimated time of arrival [NAV] The predicted time of reaching a destination or way point. Abbreviated ETA.

estimated time of departure [NAV] The predicted time of leaving a place. Abbreviated ETD.

estimated time of interception [NAV] The predicted time of intercepting a craft by another craft.

estival [ASTRON] Of or pertaining to the summer. Also spelled aestival.

estivation [PHYSIO] **1.** The adaptation of certain animals to the conditions of summer, or the taking on of certain modifications, which enables them to survive a hot, dry summer. **2.** The dormant condition of an organism during the summer.

estivoautumnal malaria *See* falciparum malaria.

estradiol [BIOCHEM] $C_{18}H_{24}O_2$ An estrogenic hormone produced by follicle cells of the vertebrate ovary; provokes estrus and proliferation of the human endometrium, and stimulates ICSH (interstitial-cell-stimulating hormone) secretion.

estragole [ORG CHEM] $C_6H_4(C_3H_5)(OCH_3)$ A colorless liquid with the odor of anise, found in basil oil, estragon oil, and anise bark oil; used in perfumes and flavorings. Also known as chavicol methyl ether; methyl chavicol.

estragon oil [MATER] Essential oil, colorless to yellowish green, with aniselike odor and aromatic flavor; distilled from flowering herb tarragon, *Artemisia dracunculus;* used chiefly as flavoring. Also known as tarragon oil.

estriol [BIOCHEM] $C_{18}H_{24}O_3$ A crystalline estrogenic hormone obtained from human pregnancy urine.

estrogen [BIOCHEM] Any of various natural or synthetic substances possessing the biologic activity of estrus-producing hormones.

estrogenic hormone [BIOCHEM] A hormone, found principally in ovaries and also in the placenta, which stimulates the accessory sex structures and the secondary sex characteristics in the female.

Estron [TEXT] Trade name of Eastman Chemical Products for an acetate yarn and staple fiber.

estrone [BIOCHEM] $C_{18}H_{22}O_2$ An estrogenic hormone pro-

ESKER

View along the crest of an esker in southeastern Wisconsin. (*W. C. Alden, USDG*)

duced by follicle cells of the vertebrate ovary; functions the same as estradiol.

estrous cycle [PHYSIO] The physiological changes that take place between periods of estrus in the female mammal.

estrus [PHYSIO] The period in female mammals during which ovulation occurs and the animal is receptive to mating.

estuarine deposit [GEOL] A sediment deposited at the heads and floors of estuaries.

estuarine environment [OCEANOGR] The physical conditions and influences of an estuary.

estuarine oceanography [OCEANOGR] The study of the chemical, physical, biological, and geological properties of estuaries.

estuary [GEOGR] A semienclosed coastal body of water which has a free connection with the open sea and within which sea water is measurably diluted with fresh water. Also known as branching bay; drowned river mouth; firth.

esu *See* electrostatic units.

E.T. *See* ephemeris time.

ETA *See* estimated time of arrival.

etalon [OPTICS] **1.** Two adjustable parallel mirrors mounted so that either one may serve as one of the mirrors in a Michelson interferometer; used to measure distances in terms of wavelengths of spectral lines. **2.** An instrument similar to the Fabry-Perot interferometer, except that the distance between the plates is fixed. Also known as Fabry-Perot etalon.

eta meson [PARTIC PHYS] Neutral pseudoscalar meson having zero isotopic spin and hypercharge, positive charge parity and G parity, and a mass of about 549 MeV; decays via electromagnetic interactions.

etamine [TEXT] Cotton material that is loosely woven and lightweight but has a coarse texture.

Etard reaction [ORG CHEM] Direct oxidation of an aromatic or heterocyclic bound methyl group to an aldehyde by utilizing chromyl chloride or certain metallic oxides.

etch [GRAPHICS] To incise lines on a plate of metal, glass, or other material by covering it with an acid-resistant coating, scratching through the coating, and then permitting an acid bath to erode exposed parts of the plate. [MET] To corrode the surface of a metal in order to reveal its composition and structure.

etch cleaning [MET] Removing soil by electrolytic or chemical action; removes some of the surface metal along with the dirt.

etch cracks [MET] Shallow cracks in the surface of hardened steel that result from reaction with an acid, causing hydrogen cracking.

etched circuit [ENG] A printed circuit formed by chemical or electrolytic removal of unwanted portions of a layer of conductive material bonded to an insulating base.

etch figures [MET] Minute, faceted pits or surfaces produced on a metal surface by chemical reaction with etchant. [MINERAL] A minute pit produced by a solvent on the crystal face of a mineral which reveals its molecular structure.

etch printing [GRAPHICS] Printing with a printing plate that has an image formed by chemical or electrolytic action.

ETD *See* estimated time of departure.

etesian climate *See* Mediterranean climate.

etesians [METEOROL] The prevailing northerly winds in summer in the eastern Mediterranean, and especially the Aegean Sea; basically similar to the monsoon and equivalent to the maestro of the Adriatic Sea.

ethal *See* cetyl alcohol.

ethamine *See* ethyl amine.

ethanal *See* acetaldehyde.

ethanamide *See* acetamide.

ethane [ORG CHEM] CH_3CH_3 A colorless, odorless gas belonging to the alkane series of hydrocarbons, with freezing point of $-183.3°C$ and boiling point of $-88.6°C$; used as a fuel and refrigerant and for organic synthesis.

1,2-ethanediamine *See* ethylenediamine.

ethanedioyl chloride *See* oxalyl chloride.

ethane nitrile *See* acetonitrile.

ethanethiol *See* ethyl mercaptan.

ethanethiolic acid *See* thioacetic acid.

ethanoic acid *See* acetic acid.

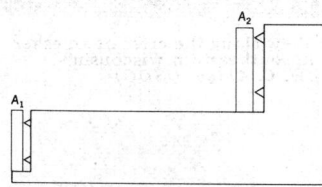

ETALON

An etalon with two adjustable mirrors A_1 and A_2 which are to be used with the Michelson interferometer in distance measurement.

ethanol [ORG CHEM] C_2H_5OH A colorless liquid, miscible with water, boiling point 78.32°C; used as a reagent and solvent. Also known as alcohol; ethyl alcohol; grain alcohol.

ethanol acid *See* glyoxalic acid.

ethanolamine [ORG CHEM] $NH_2(CH_2)_2OH$ A colorless liquid, miscible in water; used in scrubbing H_2S and CO_2 from petroleum gas streams, for dry cleaning, in paints, and in pharmaceuticals. Also known as 2-aminoethanol; 2-hydroxyethylamine; monoethanolamine (MEA).

ethanolurea [ORG CHEM] $NH_2CONHCH_2CH_2OH$ A white solid; its formaldehyde condensation products are thermoplastic and water-soluble.

ethene *See* ethylene.

ethenol *See* vinyl alcohol.

ether [ELECTROMAG] The medium postulated to carry electromagnetic waves, similar to the way a gas carries sound waves. [ORG CHEM] **1.** One of a class of organic compounds characterized by the structural feature of an oxygen linking two hydrocarbon groups (such as $R-O-R$). **2.** $(C_2H_5)_2O$ A colorless liquid, slightly soluble in water; used as a reagent, intermediate, anesthetic, and solvent. Also known as ethyl ether.

ether drag [ELECTROMAG] The hypothesis, advanced unsuccessfully to account for results of the Michelson-Morley experiment, that ether is dragged along with matter.

ether drift [ELECTROMAG] Hypothetical motion of the ether relative to the earth.

ethereal oil *See* essential oil.

etherification [ORG CHEM] The process of making an ether from an alcohol.

etherize [MED] To produce anesthesia by administration of ether.

ether thermoscope [PHYS] A device for detecting radiant heat; consists of an evacuated U-shaped tube, with ether at the bottom of the tube and a bulb at each end, one bulb being blackened.

ethidine *See* ethylidene.

ethinyl [ORG CHEM] The $CH_3:C-$ radical from acetylene. Also known as acetenyl; acetylenyl; ethynyl.

ethionic acid [ORG CHEM] $HO\cdot SO_2\cdot CH_2\cdot CH_2\cdot SO_2OH$ An unstable diacid, known only in solution. Also known as ethylene sulfonic acid.

ethionine [BIOCHEM] $C_5H_{13}O_2N$ An amino acid that is the ethyl analog of and the biological antagonist of methionine.

Ethiopian zoogeographic region [ECOL] A geographic unit of faunal homogeneity including all of Africa south of the Sahara.

ethmoid bone [ANAT] An irregularly shaped cartilage bone of the skull, forming the medial wall of each orbit and part of the roof and lateral walls of the nasal cavities.

ethmolith [GEOL] A downward tapering, funnel-shaped, discordant intrusion of igneous rocks.

ethmoturbinate [ANAT] Of or pertaining to the masses of ethmoid bone which form the lateral and superior portions of the turbinate bones in mammals.

ethnogeography [ANTHRO] The scientific study of the geographic distribution of races, peoples, or cultural groups and their adaptation and relation to the environments in which they live.

ethnology [ANTHRO] The science that deals with the study of the origin, distribution, and relations of races or ethnic groups of mankind.

ethnopsychology [PSYCH] The study of the psychology of races or ethnic groups.

ethology [VERT ZOO] The study of animal behavior in a natural context.

ethoxide [ORG CHEM] A compound formed from ethanol by replacing the hydrogen of the hydroxy group by a monovalent metal. Also known as ethylate.

ethoxy [ORG CHEM] The C_2H_5O- radical from ethyl alcohol. Also known as ethyoxyl.

2-ethoxyethanol *See* cellosolve.

ethroglucin *See* erythritol.

ethyl [ORG CHEM] **1.** The hydrocarbon radical C_2H_5. **2.** Trade name for the tetraethyllead antiknock compound in gasoline.

ethyl acetate [ORG CHEM] $CH_3COOC_2H_5$ A colorless liquid, slightly soluble in water; boils at 77°C; a medicine, reagent, and solvent. Also known as acetic ester; acetic ether; acetidin.

ethyl acetic acid *See* butyric acid.

ethyl acetoacetate [ORG CHEM] $CH_3COCH_2COOC_2H_5$ A colorless liquid, boiling at 181°C; used as a reagent, intermediate, and solvent. Also known as acetoacetic ester; diacetic ether.

ethyl acetone *See* pentanone.

ethyl acetylene [ORG CHEM] Compound with boiling point 8.1°C; insoluble in water, soluble in alcohol; used in organic synthesis. Also known as 1-butine; 1-butyne.

ethyl acrylate [ORG CHEM] $C_5H_8O_2$ A colorless liquid, boiling at 99°C; used to manufacture chemicals and resins.

ethyl alcohol *See* ethanol.

ethyl amine [ORG CHEM] A colorless liquid, boiling at 15°C, water-soluble; used as a solvent, as a dye intermediate, and in organic synthesis. Also known as aminoethane; ethamine.

ethyl-*para*-aminobenzoate [ORG CHEM] $C_6H_4NH_2CO_2C_2H_5$ A white powder, melting point 88–92°C, slightly soluble in ethanol and ether, very slightly soluble in water; used as a local anesthetic. Also known as benzocaine.

ethyl amyl ketone [ORG CHEM] $C_8H_{16}O$ A colorless liquid, almost insoluble in water; used in perfumery. Also known as octanone-3.

ethylate *See* ethoxide.

ethylation [ORG CHEM] Formation of a new compound by introducing the ethyl radical ($C_2H_5^-$).

ethyl benzene [ORG CHEM] $C_6H_5C_2H_5$ A colorless liquid that boils at 136°C, insoluble in water; used in organic synthesis, as a solvent, and in making styrene.

ethyl benzoate [ORG CHEM] $C_6H_5COOCH_2CH_3$ Colorless, aromatic liquid, boiling at 213°C, insoluble in water; used as a solvent, in flavoring extracts, and in perfumery.

ethyl borate [ORG CHEM] $B(OC_2H_5)_3$ A salt of ethanol and boric acid; colorless, flammable liquid; used in antiseptics, disinfectants, and fireproofing. Also known as boron triethoxide; triethylic borate.

ethyl bromide [ORG CHEM] C_2H_5Br A colorless liquid, boiling at 39°C; used as a refrigerant and in organic synthesis. Also known as bromic ether; bromoethane.

2-ethylbutene [ORG CHEM] $CH_3CH_2(C_2H_5)CCH_2$ Colorless liquid, soluble in alcohol and organic solvents, insoluble in water; used in organic synthesis.

2-ethylbutyl acetate [ORG CHEM] $C_2H_5CH(C_2H_5)CH_2$-O_2CCH_3 Colorless liquid with mild odor; used as a solvent for resins, lacquers, and nitrocellulose.

2-ethylbutyl alcohol [ORG CHEM] $(C_2H_5)_2CHCH_2OH$ Stable, colorless liquid, miscible in most organic solvents, slightly water-soluble; used as a solvent for resins, waxes, and dyes, and in the synthesis of perfumes, drugs, and flavorings.

ethyl butyl ketone [ORG CHEM] $C_2H_5COC_4H_9$ A colorless liquid, boiling at 147°C; used in solvent mixtures. Also known as 3-heptanone.

ethyl butyrate [ORG CHEM] $C_3H_7COOC_2H_5$ A colorless liquid, boiling at 121°C; used in flavoring extracts and perfumery. Also known as butyric ether.

ethyl caproate [ORG CHEM] $C_5H_{11}COOC_2H_5$ A colorless to yellow liquid, boiling at 167°C, soluble in ether and alcohol, and having a pleasant odor; used as a chemical intermediate and in the food industry as an artificial fruit essence. Also known as ethyl hexanoate; ethyl hexoate.

ethyl carbamate *See* urethane.

ethyl carbinol *See* propyl alcohol.

ethyl carbonate *See* diethyl carbonate.

ethyl cellulose [ORG CHEM] The ethyl ester of cellulose; it has film-forming properties and is inert to alkalies and dilute acids; used in adhesives, lacquers, and coatings.

ethyl chloride [ORG CHEM] C_2H_5Cl A colorless gas, liquefying at 12.2°C, slightly soluble in water; used as a solvent, in medicine, and as an intermediate. Also known as chloroethane; hydrochloric ether; monochloroethane; muriatic ether.

ethyl chloroacetate [ORG CHEM] $CH_2ClCOOC_2H_5$ ·A colorless liquid, boiling at 145°C; used as a poison gas, solvent, and chemical intermediate.

ethyl crotonate [ORG CHEM] $CH_3CHCHCO_2C_2H_5$ A com-
pound with a pungent aroma; boiling point of 143–147°C, soluble in water, soluble in ether; one of two isomeric forms used as an organic intermediate, a solvent for cellulose esters, and as a plasticizer for acrylic resins. Also known as *trans*-2-butenoic acid.

ethyl crotonic acid [ORG CHEM] $CH_3CHCC_2H_5COOH$ Colorless monoclinic crystals, subliming at 40°C; used as a peppermint flavoring.

ethyl cyanide [ORG CHEM] C_2H_5CN A colorless liquid that boils at 97.1°C; poisonous. Also known as hydrocyanic ether; propionitrile.

ethylene [ORG CHEM] C_2H_4 A colorless, flammable gas, boiling at −102.7°C; used as an agricultural chemical, in medicine, and for the manufacture of organic chemicals and polyethylene. Also known as ethene; olefiant gas.

ethyleneamine *See* piperazine.

ethylene bromide *See* ethylene dibromide.

ethylene carbonate [ORG CHEM] $(CH_2)_2CO$ Odorless, colorless solid with low melting point; soluble in water and organic solvents; used as a polymer and resin solvent, in solvent extraction, and in organic syntheses.

ethylene chloride [ORG CHEM] $ClCH_2CH_2Cl$ A colorless, oily liquid, boiling at 83.7°C; used as a solvent and fumigant, for organic synthesis, and for ore flotation. Also known as Dutch liquid; ethylene dichloride.

ethylene chlorobromide [ORG CHEM] CH_2BrCH_2Cl Volatile, colorless liquid with chloroformlike odor; soluble in ether and alcohol but not in water; general-purpose solvent for cellulosics; used in organic synthesis.

ethylene chlorohydrin [ORG CHEM] $ClCH_2CH_2OH$ A colorless, poisonous liquid, boiling at 129°C; used as a solvent and in organic synthesis. Also known as chloroethyl alcohol.

ethylene cyanide [ORG CHEM] $C_2H_4(CN)_2$ Colorless crystals, melting at 57°C; used in organic synthesis. Also known as succinonitrite.

ethylene cyanohydrin [ORG CHEM] C_3H_5ON A colorless liquid that is miscible with water and boils at 221°C. Also known as 2-cyanoethanol; 1-hydroxy-2-cyanoethane.

ethylenediamine [ORG CHEM] $NH_2CH_2CH_2NH_2$ Colorless liquid, melting at 8.5°C, soluble in water; used as a solvent, corrosion inhibitor, and resin and in adhesive manufacture. Also known as diaminoethane; 1,2-ethanediamine.

ethylenediaminetetraacetic acid [ORG CHEM] $(HOOC-CH_2)_2NCH_2CH_2N(CH_2COOH)$ White crystals, slightly soluble in water and decomposing above 160°C; the sodium salt is a strong chelating agent, reacting with many metallic ions to form soluble nonionic chelate. Abbreviated EDTA.

ethylene dibromide [ORG CHEM] $BrCH_2CH_2Br$ A colorless, poisonous liquid, boiling at 131°C; insoluble in water; used in medicine, as a solvent in organic synthesis, and in antiknock gasoline. Also known as 1,2-dibromoethane; ethylene bromide.

ethylene dichloride *See* ethylene chloride.

ethylene glycol *See* glycol.

ethylenehydrinsulfonic acid *See* isethionic acid.

ethyleneimine [ORG CHEM] C_2H_4NH Highly corrosive liquid, colorless and clear; miscible with organic solvents and water; used as an intermediate in fuel oil production, refining lubricants, textiles, and pharmaceuticals. Also known as aziridine.

ethylene nitrate [ORG CHEM] $(CH_2NO_3)_2$ An explosive yellow liquid, insoluble in water. Also known as glycoldinitrate.

ethylene oxide [ORG CHEM] $(CH_2)_2O$ A colorless gas, soluble in organic solvents and miscible in water, boiling point 11°C; used in organic synthesis, for sterilizing, and for fumigating. Also known as 1,2-epoxyethane.

ethylene resin [ORG CHEM] A thermoplastic material composed of polymers of ethylene; the resin is synthesized by polymerization of ethylene at elevated temperatures and pressures in the presence of catalysts. Also known as polyethylene; polyethylene resin.

ethylene sulfonic acid *See* ethionic acid.

ethylethanolamine [ORG CHEM] $C_2H_5NHCH_2CH_2OH$ Water-white liquid with amine odor; soluble in alcohol, ether, and water; used in dyes, insecticides, fungicides, and surface-active agents.

ethyl ether *See* ether.

ethylethylene   *See* butene-1.

ethyl formate   [ORG CHEM] $HCOOC_2H_5$ A colorless liquid, boiling at 54.4°C; used as a solvent, fumigant, and larvacide and in flavors, resins, and medicines.

ethyl hexanoate   *See* ethyl caproate.

ethyl hexoate   *See* ethyl caproate.

2-ethyl hexoic acid   [ORG CHEM] $C_4H_9CH(C_2H_5)COOH$ A liquid that is slightly soluble in water, boils at 226.9°C, and has a mild odor; used as an intermediate to make metallic salts for paint and varnish driers, esters for plasticizers, and light metal salts for conversion of some oils to grease. Also known as butyl ethylacetic acid.

2-ethylhexyl acetate   [ORG CHEM] CH_3COOCH_2CH-$C_2H_5C_4H_9$ Water-white, stable liquid; used as a solvent for nitrocellulose, resins, and lacquers. Also known as octyl acetate.

2-ethylhexyl acrylate   [ORG CHEM] $CH_2CHCOOCH_2CH$-$(C_2H_5)C_4H_9$ Pleasant-smelling liquid; used as monomer for plastics, protective coatings, and paper finishes.

2-ethylhexyl alcohol   [ORG CHEM] $C_4H_9CH(C_2H_5)CH_2OH$ Colorless, slightly viscous liquid; used as a defoaming or wetting agent, as a solvent for protective coatings, waxes, and oils, and as a raw material for plasticizers. Also known as octyl alcohol.

2-ethylhexylamine   [ORG CHEM] $C_4H_9CH(C_2H_5)CH_2NH_2$ Water-white liquid with slight ammonia odor; slightly water-soluble; used to synthesize detergents, rubber chemicals, and oil additives.

2-ethylhexyl bromide   [ORG CHEM] $C_4H_9CH(C_2H_5)CH_2Br$ Water-white, water-insoluble liquid; used to prepare pharmaceuticals and disinfectants.

2-ethylhexyl chloride   [ORG CHEM] $C_4H_9CH(C_2H_5)CH_2Cl$ Colorless liquid; used to synthesize cellulose derivatives, pharmaceuticals, resins, insecticides, and dyestuffs.

ethylic compound   [ORG CHEM] Generic term for ethyl compounds.

ethylic ether   *See* diethyl ether.

ethylidine   [ORG CHEM] The $CH_3 \cdot CH=$ radical from ethane, C_2H_5. Also known as ethidine.

ethyl iodide   [ORG CHEM] C_2H_5I A colorless liquid, boiling at 72.3°C; used in medicine and in organic synthesis. Also known as hydroiodic ether; iodoethane.

ethyl isobutylmethane   *See* 2-methylhexane.

ethyl lactate   [ORG CHEM] $CH_3CHOHCOOC_2H_5$ A colorless liquid that boils at 154°C, has a mild odor, and is miscible with water and organic solvents such as alcohols, ketones, esters, and hydrocarbons; used as a flavoring and as a solvent for cellulose compounds such as nitrocellulose, cellulose acetate, and cellulose ethers. Also known as ethyl-2-hydroxypropionate.

ethyl malonate   [ORG CHEM] $CH_2(COOC_2H_5)_2$ A colorless liquid, boiling at 198°C; used as an intermediate and a plasticizer. Also known as malonic ester.

ethyl mercaptan   [ORG CHEM] C_2H_5SH A colorless liquid, boiling at 36°C. Also known as ethanethiol; ethyl sulfhydrate; thioethyl alcohol.

ethyl methacrylate   [ORG CHEM] $CH_2CCH_3COOC_2H_5$ Colorless, easily polymerized liquid, water-insoluble; used to produce polymers and chemical intermediates.

ethylmethylacetylene   *See* pentyne.

ethyl methyl ketone   *See* methyl ethyl ketone.

ethyl nitrate   [ORG CHEM] $C_2H_5NO_3$ A colorless, flammable liquid, boiling at 87.6°C; used in perfumes, drugs, and dyes and in organic synthesis.

ethyl nitrite   [ORG CHEM] $C_2H_5NO_2$ A colorless liquid, boiling at 16.4°C; used in medicine and in organic synthesis. Also known as sweet spirits of niter.

ethyl oleate   [ORG CHEM] $C_{20}H_{38}O_2$ A yellow oil, insoluble in water; used as a solvent, plasticizer, and lubricant.

ethyl orthosilicate   *See* ethyl silicate.

ethyl oxalate   [ORG CHEM] $(COOC_2H_5)_2$ Oily, unstable, colorless liquid that is combustible; miscible with organic solvents, very slightly soluble in water; used as a solvent for cellulosics and resins, and as an intermediate for dyes and pharmaceuticals.

ethyl oxide   *See* diethyl ether.

ethyl propionate   [ORG CHEM] $C_2H_5COOC_2H_5$ A colorless liquid, slightly soluble in water, boiling at 99°C; used as solvent and pyroxylin cutting agent. Also known as propionic ether.

ethyl silicate   [ORG CHEM] $(C_2H_5)_4SiO_4$ A colorless, flammable liquid, hydrolyzed by water; used as a preservative for stone, brick, and masonry, in lacquers, and as a bonding agent. Also known as ethyl orthosilicate.

ethyl sulfate   *See* diethyl sulfate.

ethyl sulfhydrate   *See* ethyl mercaptan.

ethyl sulfide   [ORG CHEM] $(C_2H_5)_2S$ A colorless, oily liquid, boiling at 92°C; used as a solvent and in organic synthesis. Also known as diethyl sulfide; ethylthioethane.

ethylthioethane   *See* ethyl sulfide.

ethyl urethane   *See* urethane.

ethyl vanillin   [ORG CHEM] $C_2H_5O(OH)C_6H_3CHO$ A compound, crystallizing in fine white crystals that melt at 76.5°C, has a strong vanilla odor and four times the flavor of vanilla, soluble in organic solvents such as alcohol, chloroform, and ether; used in the food industry as a flavoring agent to replace or fortify vanilla.

ethyne   *See* acetylene.

ethynyl   *See* ethinyl.

ethynylation   [ORG CHEM] Production of an acetylenic derivative by the condensation of acetylene with a compound such as an aldehyde; for example, production of butynediol from the union of formaldehyde with acetylene.

ethyoxyl   *See* ethoxy.

etioallocholane   *See* androstane.

etioblast   [BOT] An immature chloroplast, containing prolamellar bodies.

etiolation   [BOT] The yellowing or whitening of green plant parts grown in darkness.

etiology   [MED] Any factors which cause disease.   [SCI TECH] A branch of science dealing with the causes of phenomena.

etioporphyrin   [ORG CHEM] $C_{31}H_{34}N_4$ A synthetic porphyrin that has four ethyl and four methyl groups in a red-pigmented compound whose crystals melt at 280°C.

E transformer   [ELECTROMAG] A transformer consisting of two coils wound around a laminated iron core in the shape of an E, with the primary and secondaries occupying the center and outside legs respectively.

E trisomy   *See* trisomy 18 syndrome.

Ettingshausen coefficient   [PHYS] A measure of the strength of the Ettingshausen effect, equal to the ratio of the temperature gradient to the product of the current density and magnetic field strength which produce this gradient.

Ettingshausen effect   [PHYS] The phenomenon that, when a metal strip is placed with its plane perpendicular to a magnetic field and an electric current is sent longitudinally through the strip, corresponding points on opposite edges of the strip have different temperatures.

Ettingshausen-Nernst coefficient   [PHYS] A measure of the strength of the Ettingshausen-Nernst effect, equal to the ratio of the electric field to the product of the temperature gradient and magnetic field strength which produce this field.

Ettingshausen-Nernst effect   [PHYS] The phenomenon that, when a conductor or semiconductor is subjected to a temperature gradient and to a magnetic field perpendicular to the temperature gradient, an electric field arises perpendicular to both the temperature gradient and the magnetic field. Also known as Nernst effect.

ettringite   [MINERAL] $Ca_6Al_2(SO_4)_3(OH)_{12} \cdot 26H_2O$ A mineral composed of hydrous basic calcium and aluminum sulfate.

Eu   *See* europium.

EU   *See* expected value.

Eubacteriales   [MICROBIO] The true bacteria, an order of the class Schizomycetes; characterized by simple, undifferentiated, rigid cells which are either spherical forms or straight rods.

Eubasidiomycetes   [MYCOL] An equivalent name for Homobasidiomycetidae.

Eubrya   [BOT] A subclass of the mosses (Bryopsida); the leafy gametophytes arise from buds on the protonemata, which are nearly always filamentous or branched green threads attached to the substratum by rhizoids.

Eubryales [BOT] An order of mosses (Bryatae); plants have the sporophyte at the end of a stem, vary in size from small to robust, and generally grow in tufts.

eucairite [MINERAL] CuAgSe A white, native selenide that crystallizes in the isometric crystal system.

Eucalyptus [BOT] A large genus of evergreen trees belonging to the myrtle family (Myrtaceae) and occurring in Australia and New Guinea.

eucalyptus gum [PHARM] The dried gummy exudate from *Eucalyptus longirostris* of Australia, composed of kinotannic acid, kino red, glucoside, catechol, and pyrocatechol; used in medicine as an astringent and antidiarrheal agent. Also known as eucalyptus kino; red gum.

eucalyptus kino *See* eucalyptus gum.

eucalyptus oil [MATER] Any of various essential oils of various eucalyptus leaves, a yellow liquid miscible with alcohol; varies from peppermint to turpentine in odor; used in perfumery, medicine, and ore flotation.

eucalyptus resin oil *See* eucalyptus tar.

eucalyptus tar [MATER] Residue from caustic treating of oil from distilled eucalyptus leaves; used to perfume soaps and as a disinfectant. Also known as eucalyptus resin oil.

Eucarida [INV ZOO] A large superorder of the decapod crustaceans, subclass Malacostraca, including shrimps, lobsters, hermit crabs, and crabs; characterized by having the shell and thoracic segments fused dorsally and the eyes on movable stalks.

Eucaryota [BIOL] Primitive, unicellular organisms having a well-defined nuclear membrane, chromosomes, and mitotic cell division.

eucaryote *See* eukaryote.

eucatropine hydrochloride [PHARM] $C_{17}H_{25}O_3N \cdot HCl$ A white, odorless powder that melts at 183–186°C, is soluble in alcohol and chloroform, and insoluble in ether; used in medicine as a mydriatic.

Eucestoda [INV ZOO] The true tapeworms, a subclass of the class Cestoda.

Eucharitidae [INV ZOO] A family of hymenopteran insects in the superfamily Chalcidoidea.

euchlorin [MINERAL] $(K,Na)_8Cu_9(SO_4)_{10}(OH)_6$ An emerald-green mineral consisting of a basic sulfate of potassium, sodium, and copper; found in lava at Vesuvius.

euchromatin [CYTOL] The portion of the chromosomes that stains with low intensity, uncoils during interphase, and condenses during cell division.

Eucinetidae [INV ZOO] The plate thigh beetles, a family of coleopteran insects in the superfamily Dascilloidea.

euclase [MINERAL] $BeAlSiO_4(OH)$ A brittle, pale green, blue, yellow, or violet monoclinic mineral, occurring as prismatic crystals.

Euclasterida [INV ZOO] An order of asteroid echinoderms in which the arms are sharply distinguished from a small, central disk-shaped body.

Eucleidae [INV ZOO] The slug moths, a family of lepidopteran insects in the suborder Heteroneura.

euclidean algorithm [MATH] A method of finding the greatest common divisor of a pair of integers.

euclidean geometry [MATH] The study of the properties preserved by isometries of two- and three-dimensional euclidean space.

euclidean space [MATH] A space consisting of all ordered sets $(x_1, \ldots, x_n)$ of n numbers with the distance between $(x_1, \ldots, x_n)$ and $(y_1, \ldots, y_n)$ being given by

$$\left[\sum_{i=1}^{n}(x_i - y_i)^2\right]^{\frac{1}{2}};$$

the number n is called the dimension of the space.

Euclymeninae [INV ZOO] A subfamily of annelids in the family Maldonidae of the Sedentaria, having well-developed plaques and an anal pore within the plaque.

Eucnemidae [INV ZOO] The false click beetles, a family of coleopteran insects in the superfamily Elateroidea.

Eucoccida [INV ZOO] An order of parasitic protozoans in the subclass Coccidia characterized by alternating sexual and asexual phases; stages of the life cycle occur intracellularly in vertebrates and invertebrates.

Eucoelomata [ZOO] A large sector of the animal kingdom including all forms in which there is a true coelom or body cavity; includes all phyla above Aschelminthes.

Eucommiales [BOT] A monotypic order of dicotyledonous plants in the subclass Hamamelidae; plants have two, unitegmic ovules and lack stipules.

eucrite [MINERAL] An olivine-bearing gabbro containing unusually calcic plagiocase; a meteorite component.

eucryptite [MINERAL] $LiAlSiO_4$ A colorless or white lithium aluminum silicate mineral, crystallizing in the hexagonal system; specific gravity is 2.67.

Eudactylinidae [INV ZOO] A family of parasitic copepod crustaceans in the suborder Caligoida; found as ectoparasites on the gills of sharks.

eudialite [MINERAL] $(Na,Ca,Fe)_6ZrSi_6O_{18}(OH,Cl)$ Hexagonal-crystalline silicate chloride mineral; color is red to brown.

eudidymite [MINERAL] $NaBeSi_3O_7(OH)$ A glassy white mineral composed of sodium beryllium silicate.

eudiometer [ENG] An instrument for measuring changes in volume during the combustion of gases, consisting of a graduated tube that is closed at one end and has two wires sealed into it, between which a spark may be passed.

eudoxid *See* eudoxome.

eudoxome [INV ZOO] Cormidium of most calycophoran siphonophores which lead a free existence. Also known as eudoxid.

Euechinoidea [INV ZOO] A subclass of echinoderms in the class Echinoidea; distinguished by the relative stability of ambulacra and interambulacra.

eugenics [GEN] The use of practices that influence the hereditary qualities of future generations, with the aim of improving man's genetic future.

eugenol [ORG CHEM] $CH_2CHCH_2C_6H_3(OCH_3)OH$ A colorless or yellowish aromatic liquid with spicy odor and taste, soluble in organic solvents, and extracted from clove oil; used in flavors, perfumes, medicines, and the manufacture of vanilla. Also known as 4-allyl-2-methoxyphenol.

eugeosyncline [GEOL] The internal volcanic belt of an orthogeosyncline.

Euglena [BIOL] A genus of organisms with one or two flagella, chromatophores in most species, and a generally elongate, spindle-shaped body; classified as algae by botanists (Euglenophyta) and as protozoans by zoologists (Euglenida).

Euglenida [INV ZOO] An order of protozoans in the class Phytamastigophorea, including the largest green, noncolonial flagellates.

Euglenidae [INV ZOO] The antlike leaf beetles, a family of coleopteran insects in the superfamily Tenebrionoidea.

Euglenoidina [INV ZOO] The equivalent name for Euglenida.

Euglenophyceae [BOT] The single class of the plant division Euglenophyta.

Euglenophyta [BOT] A division of the plant kingdom including one-celled, chiefly aquatic flagellate organisms having a spindle-shaped or flattened body, naked or with a pellicle.

euglobulin [BIOCHEM] True globulin; a simple protein that is soluble in distilled water and dilute salt solutions.

Eugregarinida [INV ZOO] An order of protozoans in the subclass Gregarinia; parasites of certain invertebrates.

euhedral *See* automorphic.

eukaryote [BIOL] A cell with a definitive nucleus. Also spelled eucaryote.

Eulamellibranchia [INV ZOO] The largest subclass of the molluscan class Bivalvia, having a heterodont shell hinge, leaflike gills, and well-developed siphons.

Eulenburg's disease *See* paramyotonia congenita.

Euler characteristic of a topological space X [MATH] The number $\chi(x) = \sum(-1)^q\beta_q$, where β_q is the qth-Betti number of X.

Euler diagram [MATH] A diagram consisting of closed curves, used to represent relations between logical propositions or sets; similar to a Venn diagram.

Euler equation [MECH] Expression for the energy removed from a gas stream by a rotating blade system (as a gas turbine), independent of the blade system (as a radial- or axial-flow system).

Euler equations of motion [MECH] A set of three differential

EUCALYPTUS

Eucalyptus globulus showing spray of mature foliage.

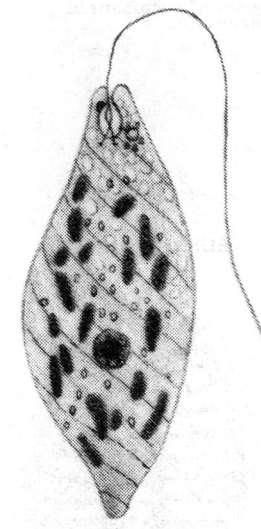

EUGLENA

Euglena showing many chloroplasts and spiral torsion.

equations expressing relations between the force moments, angular velocities, and angular accelerations of a rotating rigid body.

Euler force [MECH] The greatest load that a long, slender column can carry without buckling, according to the Euler formula for long columns.

Euler formula for long columns [MECH] A formula which gives the greatest axial load that a long, slender column can carry without buckling, in terms of its length, Young's modulus, and the moment of inertia about an axis along the center of the column.

Euler transformation [MATH] A method of obtaining from a given convergent series a new series which converges faster to the same limit, and for defining sums of certain divergent series; the transformation carries the series $a_0 - a_1 + a_2 - a_3 + \ldots$ into a series whose nth term is

$$\sum_{r=0}^{n-1} (-1)^r \binom{n-1}{r} a_r/2^n.$$

Eulerian coordinates [FL MECH] Any system of coordinates in which properties of a fluid are assigned to points in space at each given time, without attempt to identify individual fluid parcels from one time to the next; a sequence of synoptic charts is a Eulerian representation of the data.

Eulerian correlation [FL MECH] The correlation between the properties of a flow at various points in space at a single instant of time. Also known as synoptic correlation.

Eulerian equation [FL MECH] A mathematical representation of the motions of a fluid in which the behavior and the properties of the fluid are described at fixed points in a coordinate system.

Eulerian wind [METEOROL] A wind motion only in response to the pressure force; the cyclostrophic wind is a special case of the Eulerian wind, which is limited in its meteorological applicability to those situations in which the Coriolis effect is negligible.

Euler-Lagrange equation [MATH] A partial differential equation arising in the calculus of variations, which provides a necessary condition that $y(x)$ minimize the integral over some finite interval of $f(x,y,y')dx$, where $y' = dy/dx$; the equation is $(\partial f(x,y,y')/\partial y) - (d/dx)(\partial f(x,y,y')/\partial y') = 0$. Also known as Euler's equation.

Euler method [MECH] A method of studying fluid motion and the mechanics of deformable bodies in which one considers volume elements at fixed locations in space, across which material flows; the Euler method is in contrast to the Lagrangian method.

Euler number 1 [FL MECH] A dimensionless number used in the study of fluid friction in conduits, equal to the pressure drop due to friction divided by the product of the fluid density and the square of the fluid velocity.

Euler number 2 [FL MECH] A dimensionless number equal to two times the Fanning friction factor.

Euler-Rodrigues parameter [MECH] One of four numbers which may be used to specify the orientation of a rigid body; they are components of a quaternion.

Euler's expansion [FL MECH] The transformation of a derivative (d/dt) describing the behavior of a moving particle with respect to time, into a local derivative $(\delta/\delta t)$ and three additional terms that describe the changing motion of a fluid as it passes through a fixed point.

Euler's formula [MATH] The formula $e^{ix} = \cos x + i \sin x$, where $i = \sqrt{-1}$.

Euler's numbers [MATH] The numbers E_{2n} defined by the equation

$$\frac{1}{\cos z} = \sum_{n=0}^{\infty} (-1)^n \frac{E_{2n}}{(2n)!} z^{2n}.$$

Euler's theorem [MATH] For any polyhedron, $V - E + F = 2$, where V, E, F represent the number of vertices, edges, and faces respectively.

eulittoral [OCEANOGR] A subdivision of the benthic division of the littoral zone of the marine environment, extending from high-tide level to about 60 meters, the lower limit for abundant growth of attached plants.

Eulophidae [INV ZOO] A family of hymenopteran insects in the superfamily Chalcidoidea including species that are parasitic on the larvae of other insects.

eulytine *See* eulytite.

eulytite [MINERAL] $Bi_4Si_3O_{12}$ A bismuth silicate mineral usually found as minute dark-brown or gray tetrahedral crystals; specific gravity is 6.11. Also known as agricolite; bismuth blende; eulytine.

Eumalacostraca [INV ZOO] A series of the class Crustacea comprising shrimplike crustaceans having eight thoracic segments, six abdominal segments, and a telson.

Eumetazoa [ZOO] A section of the animal kingdom that includes the phyla above the Porifera; contains those animals which have tissues or show some tissue formation and organ systems.

eumitosis [CYTOL] Typical mitosis.

Eumycetes [MYCOL] The true fungi, a large group of microorganisms characterized by cell walls, lack of chlorophyll, and mycelia in most species; includes the unicellular yeasts.

Eumycetozoida [INV ZOO] An order of protozoans in the subclass Mycetozoia; includes slime molds which form a plasmodium.

Eumycophyta [MYCOL] An equivalent name for Eumycetes.

Eumycota [MYCOL] An equivalent name for Eumycetes.

Eunicea [INV ZOO] A superfamily of polychaete annelids belonging to the Errantia.

Eunicidae [INV ZOO] A family of polychaete annelids in the superfamily Eunicea having characteristic pharyngeal armature consisting of maxillae and mandibles.

eunuch [MED] An individual who has undergone complete loss of testicular function.

Euomphalacea [PALEON] A superfamily of extinct gastropod mollusks in the order Aspidobranchia characterized by shells with low spires, some approaching bivalve symmetry.

eupavarine [PHARM] $C_{20}H_{21}NO_2$ A crystalline alkaloid that melts at 214°C and is soluble in many organic solvents and hot water; used in medicine as an antispasmodic. Also known as 1-benzyl-3-ethyl-6,7,-dimethoxy isoquinoline.

eupelagic *See* pelagic.

Eupelmidae [INV ZOO] A family of hymenopteran insects in the superfamily Chalcidoidea.

Euphausiacea [INV ZOO] An order of planktonic malacostracans in the class Crustacea possessing photophores which emit a brilliant blue-green light.

euphenics [GEN] The production of a satisfactory phenotype by means other than eugenics.

Eupheterochlorina [INV ZOO] A suborder of flagellate protozoans in the order Heterochlorida.

Euphorbiaceae [BOT] A family of dicotyledonous plants in the order Euphorbiales characterized by dehiscent fruit having more than one seed and by epitropous ovules.

Euphorbiales [BOT] An order of dicotyledonous plants in the subclass Rosidae having simple leaves and unisexual flowers that are aggregated and reduced.

euphotic [OCEANOGR] Of or constituting the upper levels of the marine environment down to the limits of effective light penetration for photosynthesis.

Euphrosinidae [INV ZOO] A family of amphinomorphan polychaete annelids with short, dorsolaterally flattened bodies.

Euplexoptera [INV ZOO] The equivalent name for Dermaptera.

euploid [GEN] Having a chromosome complement that is an exact multiple of the haploid complement.

eupnea [PHYSIO] Normal or easy respiration rhythm.

Eupodidae [INV ZOO] A family of mites in the suborder Trombidiformes.

Euproopacea [PALEON] A group of Paleozoic horseshoe crabs belonging to the Limulida.

Europa [ASTRON] A satellite of Jupiter; its diameter is about 3200 miles.

Europe [GEOGR] A great western peninsula of the Eurasian landmass, usually called a continent; its eastern limits are arbitrary and are conventionally drawn along the water divide of the Ural Mountains, the Ural River, the Caspian Sea, and the Caucasus watershed to the Black Sea.

European boreal faunal region [ECOL] A zoogeographic region describing marine littoral faunal regions of the northern

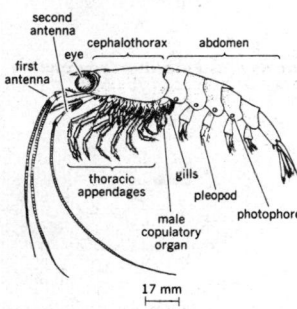
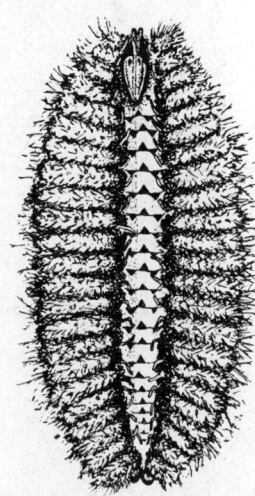

Atlantic Ocean between Greenland and the northwestern coast of Europe.

European canker [PL PATH] **1.** A fungus disease of apple, pear, and other fruit and shade trees caused by *Nectria galligena* and characterized by cankers with concentric rings of callus on the trunk and branches. **2.** A fungus disease of poplars caused by *Dothichiza populea*.

European porcelain *See* porcelain.

europium [CHEM] A member of the rare-earth elements in the cerium subgroup, symbol Eu, atomic number 63, atomic weight 151.96, steel gray and malleable, melting at 1100–1200°C.

europium halide [INORG CHEM] Any of the compounds of the element europium and the halogen elements; for example, europium chloride, $EuCl_3 \cdot xH_2O$.

europium oxide [INORG CHEM] Eu_2O_3 A white powder, insoluble in water; used in red- and infrared-sensitive phosphors.

Eurotiaceae [MYCOL] A family of ascomycetous fungi of the order Eurotiales in which the asci are borne in cleistothecia or closed fruiting bodies.

Eurotiales [MYCOL] An order of fungi in the class Ascomycetes bearing ascospores in globose or broadly oval, delicate asci which lack a pore.

Euryalae [INV ZOO] The basket fishes, a family of echinoderms in the subclass Ophiuroidea.

Euryalina [INV ZOO] A suborder of ophiuroid echinoderms in the order Phrynophiurida characterized by a leathery integument.

Euryapsida [PALEON] A subclass of fossil reptiles distinguished by an upper temporal opening on each side of the skull.

eurybathic [ECOL] Living at the bottom of a body of water.

eurycephalic [ANTHRO] Having a cephalic index of 80–84.

Eurychilinidae [PALEON] A family of extinct dimorphic ostracods in the superfamily Hollinacea.

euryene [ANTHRO] Having a short or broad forehead or both with an upper facial index of 45–50.

eurygamous [INV ZOO] Mating in flight, as in many insect species.

Eurylaimi [VERT ZOO] A monofamilial suborder of suboscine birds in the order Passeriformes.

Eurylaimidae [VERT ZOO] The broadbills, the single family of the avian suborder Eurylaimi.

eurymeric [ANTHRO] Having a broad femur with a platymeric index of 85–100.

Eurymylidae [PALEON] A family of extinct mammals presumed to be the ancestral stock of the order Lagomorpha.

euryon [ANAT] One of the two lateral points functioning as end points to measure the greatest transverse diameter of the skull.

Euryphoridae [INV ZOO] A family of copepod crustaceans in the order Caligoida; members are fish ectoparasites.

euryprosopic [ANTHRO] Having a short or broad face or both with a facial index of 80–85.

Eurypterida [PALEON] A group of extinct aquatic arthropods in the subphylum Chelicerata having elongate-lanceolate bodies encased in a chitinous exoskeleton.

Eurypygidae [VERT ZOO] The sun bitterns, a family of tropical and subtropical New World birds belonging to the order Gruiformes.

eurypylous [INV ZOO] Having a wide opening; applied to sponges with wide apopyles opening directly into excurrent canals, and wide prosopyles opening directly from incurrent canals.

eurysome [ANTHRO] Having a broad, thickset body build.

eurytherm [BIOL] An organism that is tolerant of a wide range of temperatures.

Eurytomidae [INV ZOO] The seed and stem chalcids, a family of hymenopteran insects in the superfamily Chalcidoidea.

Eusiridae [INV ZOO] A family of pelagic amphipod crustaceans in the suborder Gammaridea.

eusporangiate [BOT] Having sporogenous tissue derived from a group of epidermal cells.

eustachian tube [ANAT] A tube composed of bone and cartilage that connects the nasopharynx with the middle ear cavity.

eustachian valve *See* canal valve.

eustacy [OCEANOGR] Worldwide fluctuations of sea level due to changing capacity of the ocean basins or the volume of ocean water.

eustele [BOT] A modified siphonostele containing collateral or bicollateral vascular bundles; found in most gymnosperm and angiosperm stems.

eusternum [INV ZOO] The anterior sternal plate in insects.

Eusuchia [VERT ZOO] The modern crocodiles, a suborder of the order Crocodilia characterized by a fully developed secondary palate and procoelous vertebrae.

Eusyllinae [INV ZOO] A subfamily of polychaete annelids in the family Syllidae having a thick body and unsegmented cirri.

Eutardigrada [INV ZOO] An order of tardigrades which lack both a sensory cephalic appendage and a club-shaped appendage.

eutectic [MET] The microstructure that results when a metal of eutectic composition solidifies. [PHYS CHEM] An alloy or solution that has the lowest possible constant melting point.

eutectic crystallization [MET] Simultaneous crystallization of the constituents of a eutectic alloy during cooling of the melt.

eutectic melting [MET] Melting of isolated, microscopic areas of an alloy that correspond to the location of the eutectic of the system.

eutectic point [PHYS CHEM] The point in the constitutional diagram indicating the composition and temperature of the lowest melting point of a eutectic.

eutectic system [PHYS CHEM] The particular composition and temperature of materials at the eutectic point.

eutectic temperature [PHYS CHEM] The temperature at the lowest melting point of a eutectic.

eutectogenic system [PHYS CHEM] A multicomponent liquid-solid mixture in which pure solid phases of each component are in equilibrium with the remaining liquid mixture at a specific (usually minimum) temperature for a given composition, that is, the eutectic point.

eutectoid [MET] A mixture of phases whose composition is determined by the eutectoid point in the solid region of an equilibrium diagram and whose constituents are formed by the eutectoid reaction. [PHYS CHEM] The point in an equilibrium diagram for a solid solution at which the solution on cooling is converted to a mixture of solids.

eutely [BIOL] Having the body composed of a constant number of cells, as in certain rotifers.

Euthacanthidae [PALEON] A family of extinct acanthodian fishes in the order Climatiiformes.

euthenics [BIOL] The science that deals with the improvement of the future of man by changing his environment.

Eutheria [VERT ZOO] An infraclass of therian mammals including all living forms except the monotremes and marsupials.

Eutrichosomatidae [INV ZOO] Small family of hymenopteran insects in the superfamily Chalcidoidea.

eutrophic [HYD] Pertaining to a lake containing a high concentration of dissolved nutrients; often shallow, with periods of oxygen deficiency.

eutrophication [ECOL] Of bodies of water, the process of becoming better nourished either naturally by processes of maturation or artificially by fertilization.

euxenite [MINERAL] A brownish-black rare-earth mineral that crystallizes in the orthorhombic system, contains oxide of calcium, cerium, columbium, tantalum, titanium, and uranium, and has a metallic luster; hardness is 6.5 on Mohs scale, and specific gravity is 4.7–5.0.

euxinic [HYD] Of or pertaining to an environment of restricted circulation and stagnant or anaerobic conditions.

eV *See* electron volt.

EV *See* expected value.

evacuate [ENG] To remove something, especially gases and vapors, from an enclosure, such as from the envelope of an electron tube, or from a well. Also known as exhaust.

Evaniidae [INV ZOO] The ensign flies, a family of hymenopteran insects in the superfamily Proctotrupoidea.

Evans-Burr unit [BIOL] A unit for the standardization of vitamin E.

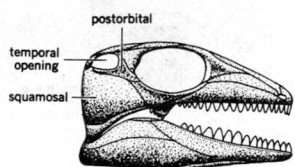

EUROPIUM

Periodic table of the chemical elements showing the position of europium.

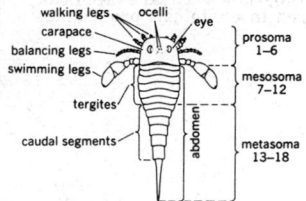

EURYAPSIDA

Lateral view of *Araeoscelis* skull showing temporal opening. *(From A. S. Romer, Vertebrate Paleontology, 3d ed., University of Chicago Press, 1966)*

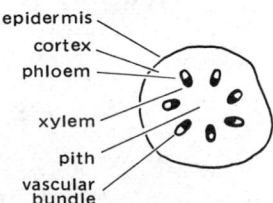

EURYPTERIDA

Dorsal view of features of a typical eurypterid. *(Adapted from Clarke and Ruedemann, 1912, from R. R. Shrock and W. H. Twenhofel, Principles of Invertebrate Paleontology, McGraw-Hill, 2d ed., 1953)*

EUSTELE

Diagram of a stem in cross section showing tissue arrangements of a eustele.

evaporation [PHYS] Conversion of a liquid to the vapor state by the addition of latent heat.

evaporation capacity *See* evaporative power.

evaporation current [OCEANOGR] An ocean current resulting from the accumulation of water through precipitation and river runoff at one point, and loss by evaporation at another point.

evaporation gage *See* atmometer.

evaporation loss [CHEM ENG] The loss of a stored volatile liquid component or mixture by evaporation; controlled by temperature, pressure, and the presence or absence of vapor-recovery systems.

evaporation pan [ENG] A type of atmometer consisting of a pan, used in the measurement of the evaporation of water into the atmosphere.

evaporation power *See* evaporative power.

evaporation tank [ENG] A tank used to measure the evaporation of water under controlled conditions.

evaporative capacity *See* evaporative power.

evaporative condenser [MECH ENG] An apparatus in which vapor is condensed within tubes that are cooled by the evaporation of water flowing over the outside of the tubes.

evaporative cooling [ENG] 1. Lowering the temperature of a large mass of liquid by utilizing the latent heat of vaporization of a portion of the liquid. 2. Cooling air by evaporating water into it. 3. *See* vaporization cooling.

evaporative cooling tower *See* wet cooling tower.

evaporative heat regulation [PHYSIO] The composite process by which an animal body is cooled by evaporation of sensible perspiration; this avenue of heat loss serves as a physical means of regulating the body temperature.

evaporative power [METEOROL] A measure of the degree to which the weather or climate of a region is favorable to the process of evaporation; it is usually considered to be the rate of evaporation, under existing atmospheric conditions, from a surface of water which is chemically pure and has the temperature of the lowest layer of the atmosphere. Also known as evaporation capacity; evaporation power; evaporative capacity; evaporativity; potential evaporation.

evaporativity *See* evaporative power.

evaporator [CHEM ENG] A device used to vaporize part or all of the solvent from a solution; the valuable product is usually either a solid or concentrated solution of the solute. [MECH ENG] Any of many devices in which liquid is changed to the vapor state by the addition of heat, for example, distiller, still, dryer, water purifier, or refrigeration system element where evaporation proceeds at low pressure and consequent low temperature.

evaporimeter *See* atmometer.

evaporite [GEOL] Deposits of mineral salts from sea water or salt lakes due to evaporation of the water.

evaporite pond [IND ENG] Any containment area for brines or solution-mined effluents constructed to permit solar evaporation and harvesting of dewatered evaporite concentrates.

evapotranspiration [HYD] Discharge of water from the earth's surface to the atmosphere by evaporation from lakes, streams, and soil surfaces and by transpiration from plants. Also known as fly-off; total evaporation; water loss.

evapotranspirometer [ENG] An instrument which measures the rate of evapotranspiration; consists of a vegetation soil tank so designed that all water added to the tank and all water left after evapotranspiration can be measured.

Evasé stack [CIV ENG] In tunnel engineering, an exhaust stack for air having a cross section that increases in the direction of airflow at a rate to regain pressure.

evection [ASTROPHYS] A perturbation of the moon in its orbit due to the attraction of the sun.

E vector [ELECTROMAG] Vector representing the electric field of an electromagnetic wave.

even-even nucleus [NUC PHYS] A nucleus which has an even number of neutrons and an even number of protons.

even function [MATH] A function with the property that $f(x) = f(-x)$ for each number x.

even harmonic [PHYS] A harmonic that is an even multiple of the fundamental frequency.

evening gun [ORD] Firing of a gun as a signal for the lowering of the flag at retreat; the gun is fired after the sounding of the last note of the bugle call at retreat. Also known as retreat gun.

evening star [ASTRON] A misnomer for a planet that can be seen without a telescope when it sets after the sun.

evening twilight [ASTRON] The period of time between sunset and darkness.

even keel [NAV ARCH] Floating level, without inclination.

even number [MATH] An integer which is a multiple of 2.

even-odd nucleus [NUC PHYS] A nucleus which has an even number of neutrons and an odd number of protons, or vice versa.

even permutation [MATH] A permutation which may be represented as a product of an even number of transpositions.

even pitch [DES ENG] The pitch of a screw in which the number of threads per inch is a multiple (or submultiple) of the threads per inch of the lead screw of the lathe on which the screw is cut.

event [ADP] The moment of time at which a specified change of state occurs; usually marks the completion of an asynchronous input/output operation. [GEOL] An incident of probable tectonic significance, but whose full implications are unknown. [PHYS] A point in space-time. [STAT] A mathematical model of the result of a conceptual experiment; this model is a measurable subset of a probability space.

Eventognathi [VERT ZOO] The equivalent name for Cypriniformes.

event recorder [ENG] A recorder that plots on-off information against time, to indicate when events start, how long they last, and how often they recur.

Everdur [MET] Trade name for a group of copper-silicon-manganese alloys characterized by high tensile strength and conductivity; Everdur 1000 contains 94.9% copper, 4% silicon, and 1.1% manganese.

evergreen [BOT] Pertaining to a perennially green plant. Also known as aiophyllous.

Eve's constant [NUCLEO] A measure of a substance's intensity of radioactivity, equal to the number of ions produced per cubic centimeter per second in air by 1 gram of the substance at a distance of 1 centimeter.

Evjen method [SOLID STATE] Method of calculating lattice sums in which groups of charges whose total charge is zero are taken together, so that the contribution of each group is small and the series rapidly converges.

evjite [PETR] A gabbro of hornblende in which the only light-colored mineral is labradorite or bytownite; hornblende must be primary, not uralitic.

evoked potential [PHYSIO] Electrical response of any neuron to stimuli.

evolute [MATH] The locus of the centers of curvature of a curve.

evolution [BIOL] The processes of biological and organic change in organisms by which descendants come to differ from their ancestors.

evolutionary operation [IND ENG] An iterative technique for optimizing a production process by systematically introducing small changes in the process and then observing and evaluating the results.

evorsion [GEOL] The process of pothole formation in riverbeds; plays an important role in denudation.

evorsion hollow *See* pothole.

EVT *See* equiviscous temperature.

Ewald-Kornfeld method [SOLID STATE] An extension of the Ewald method to calculate Coulomb energies of dipole arrays.

Ewald method [SOLID STATE] Method of calculating lattice sums in which certain mathematical techniques are employed to make series converge rapidly.

Ewald sphere [SOLID STATE] A sphere superimposed on the reciprocal lattice of a crystal, used to determine the directions in which an x-ray or other beam will be reflected by a crystal lattice.

E wave *See* transverse magnetic wave.

ewe [VERT ZOO] A mature female sheep, goat, or related animal, as the smaller antelopes.

Ewing's sarcoma [MED] A primary malignant tumor of bone, usually arising as a central tumor in long bone.

Ewing theory of ferromagnetism [SOLID STATE] Theory of

EVAPORATOR

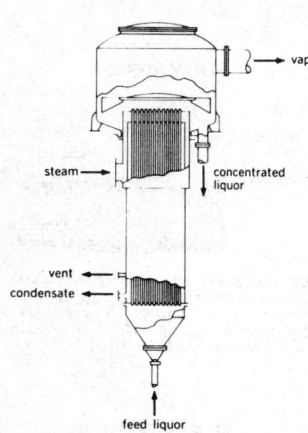

Long-tube vertical evaporator used in chemicals processing.

ferromagnetic phenomena which assumes each atom is a permanent magnet which can turn freely about its center under the influence of applied fields and other magnets.

exact differential equation [MATH] A differential equation obtained by setting the exact differential of some function equal to zero.

exact differential form [MATH] A differential form which is the differential of some other form.

exact sequence [MATH] A sequence of homomorphisms with the property that the kernel of each homomorphism is precisely the image of the previous homomorphism.

exalate [BOT] Being without winglike appendages.

exalbuminous *See* exendospermous.

exalted-carrier receiver [ELECTR] Receiver that counteracts selective fading by maintaining the carrier at a high level at all times; this minimizes the second harmonic distortion that would otherwise occur when the carrier drops out while leaving most of the sidebands at their normal amplitudes.

exanthema [MED] **1.** An eruption on the skin. **2.** Any disease or fever accompanied by a skin eruption.

exanthem subitum [MED] A mild, sometimes epidemic viral disease of young children, with abrupt onset, high fever, and rash. Also known as roseola infantum.

exarch [BOT] A vascular bundle in which the primary wood is centripetal.

exasperate [BIOL] Having a surface roughened by stiff elevations or bristles.

exaspidean [VERT ZOO] Of the tarsal envelope of birds, being continuous around the outer edge of the tarsus.

excavation [ARCHEO] Process of removing earth, stone, or other materials covering the remains of ancient civilizations. [CIV ENG] **1.** The process of digging a hollow in the earth. **2.** An uncovered cavity in the ground.

excavator [MECH ENG] A machine for digging and removing earth.

excelsior [MATER] Fine, curled wood shavings, used as packing material.

except [MATH] A logical operator which has the property that if P and Q are two statements, then the statement "P except Q" is true only when P alone is true; it is false for the other three combinations (P false Q false, P false Q true, and P true Q true).

except gate [ELECTR] A gate that produces an output pulse only for a pulse on one or more input lines and the absence of a pulse on one or more other lines.

exception-item encoding [ADP] A technique which allows the uninterrupted flow of a process by the automatic shunting of erroneous records to an error tape for later corrections.

exception-principle system [ADP] A technique which assumes no printouts except when an error is encountered.

Excepulaceae [MYCOL] The equivalent name for Discellaceae.

excess coefficient [MECH ENG] The ratio $(A - R)/R$, where A is the amount of air admitted in the combustion of fuel and R is the amount required.

excess electron [SOLID STATE] Electron introduced into a semiconductor by a donor impurity and available for conduction.

excess-fifty code [ADP] A number code in which the number n is represented by the binary equivalent of $n + 50$.

excessive precipitation [METEOROL] Precipitation (generally in the form of rain) of an unusually high rate of fall; although often used qualitatively, several meteorological services have adopted quantitative limits.

excess of arc [NAV] The part of a sextant arc beinning at zero and extending in the direction opposite to that part usually considered positive.

excess reactivity [NUCLEO] The amount of surplus reactivity over that needed to achieve criticality; it is built into a reactor (by using extra fuel) in order to compensate for fuel burnup and the accumulation of fission-product poisons during operation.

excess-three code [ADP] A number code in which the decimal digit n is represented by the four-bit binary equivalent of $n + 3$.

exchange [ADP] The interchange of contents between two locations. [COMMUN] **1.** A unit established by a telephone company for the administration of telephone service in a specified area, usually a town, a city, or a village and its environs, and consisting of one or more central offices together with the associated plant used in furnishing telephone service in that area. Also known as local exchange. **2.** Room or building equipped so telephone lines terminating there may be interconnected as required; equipment may include a switchboard or automatic switching apparatus. [QUANT MECH] **1.** Operation of exchanging the space and spin coordinates in a Schrödinger-Pauli wave function representing two identical particles; this operation must leave the wave function unchanged, except possibly for sign. **2.** Process of exchanging a real or virtual particle between two other particles.

exchange adsorption [CHEM ENG] Ion exchange process in which the fluid phase contains (or consists of) two adsorbable components which together entirely saturate the surfaces of the adsorbent.

exchange anisotropy [ELECTROMAG] Phenomenon observed in certain mixtures of magnetic materials under certain conditions, in which magnetization is favored in some direction (rather than merely along some axis); thought to be caused by exchange coupling across the interface between compounds when one is ferromagnetic and one is antiferromagnetic.

exchange broadening [SPECT] The broadening of a spectral line by some type of chemical or spin exchange process which limits the lifetime of the absorbing or emitting species and produces the broadening via the Heisenberg uncertainty principle.

exchange buffering [ADP] An input/output buffering technique that avoids the internal moving of data.

exchange cable [ELEC] Lead covered, nonquadded, paper-insulated cable used within a given area to provide cable pairs between local subscribers and a central office.

exchange coefficient [FL MECH] A coefficient of eddy flux in turbulent flow, defined in analogy to those coefficients of the kinetic theory of gases. Also known as austausch coefficient; eddy coefficient; interchange coefficient.

exchange current [ELEC] The magnitude of the current which flows through a galvanic cell when it is operating in a reversible manner.

exchange degeneracy [PARTIC PHYS] Coincidence of two Regge trajectories for particles having the same quantum numbers (except for parity, charge parity, and G parity) where one would have expected separate trajectories for alternate Regge recurrences. [QUANT MECH] An exchange process that leads back to the original configuration. Also known as exchange symmetry.

exchange force [QUANT MECH] The force arising in an exchange interaction.

exchange integral [QUANT MECH] Integral over the coordinates of two identical particles which can be thought of as the interaction between a given state and a second state in which the coordinates of the particles are exchanged.

exchange interaction [QUANT MECH] **1.** An interaction represented by a potential involving exchange of space or spin coordinates, or both, of the particles involved; can be visualized physically in terms of exchange of particles. **2.** Any interaction which can be looked upon as due to exchange of particles.

exchange line [ELEC] Line joining a subscriber or switchboard to a commercial exchange.

exchange message [ADP] A device, placed between a communication line and a computer, in order to take care of certain communication functions and thereby free the computer for other work.

exchange narrowing [SPECT] The phenomenon in which, when a spectral line is split and thereby broadened by some variable perturbation, the broadening may be narrowed by a dynamic process that exchanges different values of the perturbation.

exchange operator [QUANT MECH] An operator which exchanges the spatial coordinates of the particles in a wave function, or their spins, or both positions and spins.

exchange plant [COMMUN] Plant used to serve subscriber's needs as distinguished from that used for long-distance communication.

exchanger *See* heat exchanger.

exchange reaction [CHEM] Reaction in which two atoms or ions exchange places either in two different molecules or in the same molecule.

exchange symmetry *See* exchange degeneracy.

exchange transfusion [MED] The replacement of most or all of the recipient's blood in small amounts at a time by blood from a donor, a technique used particularly in cases of erythroblastosis fetalis, in certain types of poisoning such as salicylism, and occasionally in liver failure. Also known as replacement transfusion.

excipient [PHARM] Any inert substance combined with an active drug for preparing an agreeable or convenient dosage form.

Excipulaceae [MYCOL] The equivalent name for Discellaceae.

excision [MED] The cutting out of a part; removal of a foreign body or growth from a part, organ, or tissue.

excision enzyme [BIOCHEM] A bacterial enzyme that removes damaged dimers from the deoxyribonucleic acid molecule of a bacterial cell following light or ultraviolet radiation or nitrogen mustard damage.

excitable [BIOL] Describing a tissue or organism that exhibits its irritability.

EXCITATION

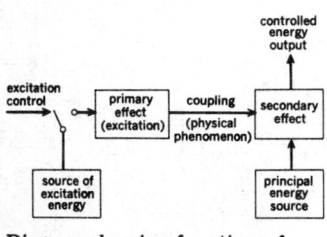

Diagram showing function of excitation in a control system.

excitation [CONT SYS] The application of energy to one portion of a system or apparatus in a manner that enables another portion to carry out a specialized function; a generalization of the electricity and electronics definitions. [ELEC] The application of voltage to field coils to produce a magnetic field, as required for the operation of an excited-field loudspeaker or a generator. [ELECTR] 1. The signal voltage that is applied to the control electrode of an electron tube. Also known as drive. 2. Application of signal power to a transmitting antenna. [QUANT MECH] The addition of energy to a particle or system of particles at ground state to produce an excited state.

excitation anode [ELECTR] An anode used to maintain a cathode spot on a pool cathode of a gas tube when output current is zero.

excitation curve [NUC PHYS] A curve showing the relative yield of a specified nuclear reaction as a function of the energy of the incident particles or photons. Also known as excitation function.

excitation energy [QUANT MECH] The minimum energy required to change a system from its ground state to a particular excited state.

excitation function [ATOM PHYS] 1. The cross section for an incident electron to excite an atom to a particular excited state expressed as a function of the electron energy. 2. *See* excitation curve.

excitation index [SPECT] In emission spectroscopy, the ratio of intensities of a pair of extremely nonhomologous spectra lines; used to provide a sensitive indication of variation in excitation conditions.

excitation loss *See* core loss.

excitation potential [QUANT MECH] Electric potential which gives the excitation energy when multiplied by the magnitude of the electron charge.

excitation purity [ANALY CHEM] The ratio of the departure of the chromaticity of a specified color to that of the reference source, measured on a chromaticity diagram; used as a guide of the wavelength of spectrum color needed to be mixed with a reference color to give the specified color. Also known as purity.

excitation spectrum [SPECT] The graph of luminous efficiency per unit energy of the exciting light absorbed by a photoluminescent body versus the frequency of the exciting light.

excitation voltage [ELEC] Nominal voltage required for excitation of a circuit.

excitation volume [PHYS] In electron-probe microanalysis, the volume of the x-ray source used to penetrate and diffuse into the target sample.

excited state [QUANT MECH] A stationary state of higher energy than the lowest stationary state or ground state of a particle or system of particles.

exciter [ELEC] 1. A small auxiliary generator that provides field current for an alternating-current generator. 2. *See* exciter lamp. [ELECTR] A crystal oscillator or self-excited oscillator used to generate the carrier frequency of a transmitter. [ELECTROMAG] 1. The portion of a directional transmitting antenna system that is directly connected to the transmitter. 2. A loop or probe extending into a resonant cavity or waveguide.

exciter lamp [ELEC] A bright incandescent lamp having a concentrated filament, used to excite a phototube or photocell in sound movie and facsimile systems. Also known as exciter.

exciter response [ELEC] In electrical rotating machinery, the rate of increase or decrease of the main exciter voltage when resistance is suddenly removed from or inserted in the main exciter field circuit.

exciting current *See* magnetizing current.

exciting line [SPECT] The frequency of electromagnetic radiation, that is, the spectral line from a noncontinuous source, which is absorbed by a system in connection with some particular process.

exciton [SOLID STATE] An excited state of an insulator or semiconductor which allows energy to be transported without transport of electric charge; may be thought of as an electron and a hole in a bound state.

excitron [ELECTR] A single-anode mercury-pool tube provided with means for maintaining a continuous cathode spot.

exclusion area [NUCLEO] The area around a nuclear operation (reactor, bomb test, and so on) where human habitation is restricted.

exclusion principle [QUANT MECH] The principle that no two fermions of the same kind may simultaneously occupy the same quantum state. Also known as Pauli exclusion principle.

exclusive or [ADP] An instruction which performs the "exclusive or" operation on a bit-by-bit basis for its two operand words, usually storing the result in one of the operand locations. [MATH] A logic operator which has the property that if P is a statement and Q is a statement, then P exclusive or Q is true if either but not both statements are true, false if both are true or both are false.

exclusive segments [ADP] Parts of an overlay program structure that cannot be resident in main memory simultaneously.

Excorallanidae [INV ZOO] A family of free-living and parasitic isopod crustaceans in the suborder Flabellifera which have mandibles and first maxillae modified as hooklike piercing organs.

excoriation [MED] Abrasion of a portion of the skin.

excrement [PHYSIO] An excreted substance; the feces.

excrescence [BIOL] 1. Abnormal or excessive increase in growth. 2. An abnormal outgrowth.

excretion [PHYSIO] The removal of unusable or excess material from a cell or a living organism.

excretory system [ANAT] Those organs concerned with solid, fluid, or gaseous excretion.

excurrent [BIOL] Flowing out. [BOT] 1. Having an undivided main stem or trunk. 2. Having the midrib extending beyond the apex.

excursion [NUCLEO] A sudden, very rapid rise in the power level of a nuclear reactor caused by supercriticality.

excursion steamer [NAV ARCH] A steam-powered ship which carries passengers on recreational trips.

execute [ADP] Usually, to run a compiled or assembled program on the computer; by extension, to compile or assemble and to run a source program.

execution control program [ADP] The program delivered by the manufacturer which permits the computer to handle the programs fed to it.

execution cycle [ADP] The time during which an elementary operation takes place.

execution error detection [ADP] The detection of errors which become apparent only during execution time.

execution time [ADP] The time during which actual work, such as addition or multiplication, is carried out in the execution of a computer instruction.

executive communications [ADP] The routine informations transmitted to the operator on the status of programs being executed and of the requirements made by these programs of the various components of the system.

executive-control language [ADP] The generic term for a finite set of instructions which enables the programmer to run a program more efficiently.

executive file-control system [ADP] The assignment of intermediate storage devices performed by the computer, and over which the programmer has no control.

executive guard mode [ADP] A protective technique which prevents the programmer from accessing, or using, the executive instructions.

executive instruction [ADP] Instruction to determine how a specially written computer program is to operate.

executive logging [ADP] The automatic bookkeeping of time utilization by programs of the various components of a computer system.

executive routine [ADP] A digital computer routine designed to process and control other routines. Also known as master routine; monitor routine.

executive schedule maintenance [ADP] The scheduling of jobs to be run according to priorities as established and maintained by the control executive or executive supervisor.

executive supervisor [ADP] The component of the computer system which controls the sequencing, setup, and execution of the jobs presented to it.

executive system [ADP] A set of programs and routines which guides a computer in the performance of its tasks, assists the programs (and programmers) with certain supporting functions, and increases the usefulness of the computer's hardware. Also known as monitor system; operating system.

executive system concurrency [ADP] The capability of the executive of a computer system to handle more than one job at the same time if these jobs do not require the same components at the same time.

executive-system control [ADP] The control exerted over the executive system by means of job control cards or commands issued at a terminal.

executive-system utilities [ADP] The set of programs, such as diagnostic programs or file utility programs, which enables the executive to handle the jobs efficiently and completely.

exendospermous [BOT] Lacking endosperm. Also known as exalbuminous.

exergonic [BIOCHEM] Of or pertaining to a biochemical reaction in which the end products possess less free energy than the starting materials; usually associated with catabolism.

exfoliation [MED] **1.** The separation of bone or other tissue in thin layers. **2.** A peeling and shedding of the horny layer of the skin. [MET] Peeling off or separation of metal at its surface in the form of thin, parallel scales or lamellae. [SCI TECH] Flaking away or peeling off in scales.

exfoliation corrosion [MET] A type of corrosion that progresses parallel to the metal surface in such a manner that underlying layers are gradually separated.

exfoliation joint *See* sheeting structure.

exfoliative cytology [PATH] The study of cells shed spontaneously from the body surfaces; used principally in the diagnosis of cancer.

exhalation [GEOPHYS] The process by which radioactive gases escape from the surface layers of soil or loose rock, where they are formed by decay of radioactive salts. [PHYSIO] The giving off or sending forth in the form of vapor; expiration.

exhaust [MECH ENG] **1.** The working substance discharged from an engine cylinder or turbine after performing work on the moving parts of the machine. **2.** The phase of the engine cycle concerned with this discharge. **3.** A duct for the escape of gases, fumes, and odors from an enclosure, sometimes equipped with an arrangement of fans. [SCI TECH] *See* evacuate.

exhaust deflecting ring [MECH ENG] A type of jetavator consisting of a ring so mounted at the end of a nozzle as to permit it to be rotated into the exhaust stream.

exhaust gas [MECH ENG] Spent gas leaving an internal combustion engine or gas turbine.

exhaust-gas analyzer [ENG] An instrument that analyzes the gaseous products to determine the effectiveness of the combustion process.

exhaust head [ENG] A device placed on the end of an exhaust pipe to remove oil and water and to reduce noise.

exhaustion delirium [MED] Acute, confusional, delirious reactions brought about by extreme fatigue, long wasting illness, or prolonged insomnia.

exhaustion region [ELECTR] A layer in a semiconductor, adjacent to its contact with a metal, in which there is almost complete ionization of atoms in the lattice and few charge carriers, resulting in a space-charge density.

exhaust manifold [MECH ENG] A branched system of pipes to carry waste emissions away from the piston chambers of an internal combustion engine.

exhaust nozzle [AERO ENG] The terminal portion of a jet engine tail pipe.

exhaust pipe [MECH ENG] The duct through which engine exhaust is discharged.

exhaust scrubber [ENG] A purifying device on internal combustion engines which removes noxious gases from engine exhaust.

exhaust stream [AERO ENG] The stream of matter or radiation emitted from the nozzle of a rocket or other reaction engine.

exhaust stroke [MECH ENG] The stroke of an engine, pump, or compressor that expels the fluid from the cylinder.

exhaust suction stroke [MECH ENG] A stroke of an engine that simultaneously removes used fuel and introduces fresh fuel to the cylinder.

exhaust trail [METEOROL] A visible condensation trail (contrail) that forms when the water vapor of an aircraft exhaust is mixed with and saturates (or slightly supersaturates) the air in the wake of the aircraft.

exhaust valve [MECH ENG] The valve on a cylinder in an internal combustion engine which controls the discharge of spent gas.

exhaust velocity [FL MECH] The velocity of gaseous or other particles in the exhaust stream of the nozzle of a reaction engine, relative to the nozzle.

exhibitionism [PSYCH] **1.** A sexual perversion in which pleasure is obtained by exposing the genitalia. **2.** In psychoanalysis, gratification of early sexual impulses in young children by physical activity, such as dancing. **3.** Any attracting of attention to oneself.

Exide ironclad battery [ELEC] A portable storage battery designed for propelling electric vehicles; a lead-antimony frame (positive plate) supports perforated hard-rubber tubes containing irregular lead-antimony cores packed with lead peroxide paste.

exine *See* exosporium.

exinite [GEOL] A hydrogen-rich maceral group consisting of spore exines, cuticular matter, resins, and waxes; includes sporinite, cutinite, alginite, and resinite. Also known as liptinite.

existence doubtful [NAV] In marine operations, expression used principally on charts to indicate the possible existence of a rock, shoal, and so on, the actual existence of which has not been established.

exit [ADP] **1.** A way of terminating a repeated cycle of operations in a computer program. **2.** A place at which such a cycle can be stopped. [ENG] A door, passage, or place of egress.

exite [INV. ZOO] A movable appendage or lobe located on the external side of the limb of a generalized arthropod.

exit region [METEOROL] The region of difluence at the downwind extremity of a jet stream.

ex lighterage [IND ENG] Price quoted exclusive of lighterage fees.

exline correction [FL MECH] Calculation of fluid-flow friction loss through annular sections with a correction for the flow eccentricity in the laminar-flow range.

exmeridian altitude [ASTRON] An altitude of a celestial body near the celestial meridian of the observer to which a correction is to be applied to determine the meridian altitude. Also known as circummeridian altitude.

exmeridian observation [ASTRON] **1.** Measurement of the altitude of a celestial body near the celestial meridian of the observer, for conversion to a meridian altitude. **2.** The altitude so measured.

exo- [ORG CHEM] A conformation of carbon bonds in a six-membered ring such that the molecule is boat-shaped with one or more substituents directed outward from the ring. [SCI TECH] A prefix denoting outside or outer.

exobiology [BIOL] The search for and study of extraterrestrial life.

exocarp *See* epicarp.

exoccipital [ANAT] Lying to the side of the foramen magnum, as the exoccipital bone.

exochorion [INV ZOO] The outer of two layers forming the covering of an insect egg.

exocline [GEOL] An inverted anticline or syncline.

exocoel [INV ZOO] The space between pairs of adjacent mesenteries in anthozoan polyps.

Exocoetidae [VERT ZOO] The halfbeaks, a family of actinopterygian fishes in the order Atheriniformes.

exocrine gland [PHYSIO] A structure whose secretion is passed directly or by ducts to its exterior surface, or to another surface which is continuous with the external surface of the gland.

exocuticle [INV ZOO] The middle layer of the cuticle of insects.

exocytosis [CYTOL] Discharge of the contents of old lysosomes or digestive vacuoles into the surrounding environment; characteristic of cells in lower organisms.

exodermis *See* hypodermis.

exoenzyme [BIOCHEM] An enzyme that functions outside the cell in which it was synthesized.

exoergic *See* exothermic.

exoergic collision *See* collision of the second kind.

exogamy [GEN] Union of gametes from organisms that are not closely related. Also known as outbreeding.

exogastrula [EMBRYO] An abnormal gastrula that is unable to undergo invagination or further development because of a quantitative increase of presumptive endoderm.

exogenote [GEN] The genetic fragment transferred from the donor to the recipient cell during the process of recombination in bacteria.

exogenous [BIOL] **1.** Due to an external cause; not arising within the organism. **2.** Growing by addition to the outer surfaces. [PHYSIO] Pertaining to those factors in the metabolism of nitrogenous substances obtained from food.

exogenous electrification [ELEC] The separation of electric charge in a conductor placed in a preexisting electric field, especially applied to the charge separation observed on metal-covered aircraft, resulting from induction effects, and by itself does not create any net total charge on the conductor.

exogenous inclusion *See* xenolith.

exogenous variables [MATH] In a mathematical model, the independent variables, which are predetermined and given outside the model.

exogeosyncline [GEOL] A parageosyncline that lies along the cratonal border and obtains its clastic sediments from erosion of the adjacent orthogeosynclinal belt outside the craton. Also known as deltageosyncline; foredeep; transverse basin.

exognathite [INV ZOO] The external branch of an oral appendage of a crustacean.

Exogoninae [INV ZOO] A subfamily of polychaete annelids in the family Syllidae having a short, small body of few segments.

exogynous [BOT] Having the style longer than and exserted beyond the corolla.

exomorphic zone *See* aureole.

exomorphism [PETR] A change in a rock mass caused by intrusion of external igneous material; in the usual sense, contact metamorphism.

exonephric [INV ZOO] Having the excretory organs discharge through the body wall.

exopeptidase [BIOCHEM] An enzyme that acts on the terminal peptide bonds of a protein chain.

exophoria [MED] A type of heterophoria in which the visual lines tend outward.

exophthalmic goiter *See* hyperthyroidism.

exophthalmos [MED] Abnormal protrusion of the eyeball from the orbit.

exopodite [INV ZOO] The outer branch of a biramous crustacean appendage.

exoprosthesis [MED] An externally applied prosthesis.

Exopterygota [INV ZOO] A division of the insect subclass Pterygota including those insects which undergo a hemimetabolous metamorphosis.

exoskeleton [INV ZOO] The external supportive covering of certain invertebrates, such as arthropods. [VERT ZOO] Bony or horny epidermal derivatives, such as nails, hoofs, and scales.

exosmosis [PHYSIO] Passage of a liquid outward through a cell membrane.

exosphere [METEOROL] An outermost region of the atmosphere, estimated at 500–1000 kilometers, where the density is so low that the mean free path of particles depends upon their direction with respect to the local vertical, being greatest for upward-traveling particles. Also known as region of escape.

exospore [MYCOL] An asexual spore formed by abstriction, as in certain Phycomycetes.

exosporium [BOT] The outer of two layers forming the wall of spores such as pollen and bacterial spores. Also known as exine.

exostome [BOT] The opening through the outer integument of a bitegmic ovule.

exostosis [MED] A benign cartilage-capped protuberance from the surface of long bones but also seen on flat bones, caused by chronic irritation as from infection, trauma, or osteoarthritis.

exotheca [INV ZOO] The tissue external to the theca of corals.

exotherm [CHEM ENG] The graphical plotting of heat rise and fall versus time for an exothermic reaction or process system.

exothermic [PHYS] Indicating liberation of heat. Also known as exoergic.

exothermic reaction [PHYS] A reaction in which heat is evolved.

exotic [ECOL] Not endemic to an area.

exotic fuels [MATER] The hydroborons which have higher calorific values than do the carbon-hydrogen fuels, once proposed as high-energy fuels for aircraft and missiles; include borane (BH_3), borobutane (B_4H_{10}), and borodecane ($B_{10}H_{14}$).

exotic stream [HYD] A stream that crosses a desert as it flows to the sea, or any stream which derives most of its water from the drainage system of another region.

exotoxin [MICROBIO] A toxin that is excreted by a microorganism.

expandable space structure [AERO ENG] A structure which can be packaged in a small volume for launch and then erected to its full size and shape outside the earth's atmosphere.

expanded clay [MATER] A material made from common brick clays by grinding, screening, and then feeding through a gas burner at about 2700°F (1482°C), thus changing the ferric oxide to ferrous oxide and causing the formation of bubbles.

expanded metal [MET] An alloy which has expanded following cooling and solidification.

expanded plastic [MATER] A light, spongy plastic made by introducing pockets of air or gas. Also known as foamed plastic; plastic foam.

expanded position indicator display [ELECTR] Display of an expanded sector from a plan position indicator presentation.

expanded scope [ELECTR] Magnified portion of a given type of cathode-ray tube presentation.

expanded slag [MATER] Slag formed by running slag from phosphate rock onto a forehearth at about 2000°F (1093°C) and then treating it with water, high-pressure steam, and air; used to make lightweight concrete blocks.

expanded sweep [ELECTR] A cathode-ray sweep in which the movement of the electron beam across the screen is speeded up during a selected portion of the sweep time.

expander [ELECTR] A transducer that, for a given input amplitude range, produces a larger output range.

expanding [MET] A process used to increase the inside diameter of a hollow piece, such as a tube, cup, or shell.

EXPANDER

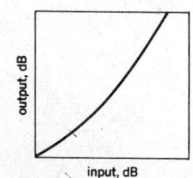

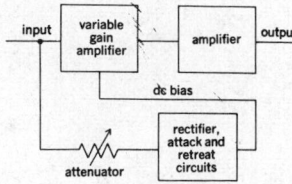

Input-output characteristics and schematic block diagram of an expander.

expanding brake [MECH ENG] A brake that operates by moving outward against the inside rim of a drum or wheel.

expanding square search *See* square search.

expanding universe [ASTROPHYS] Explanation of the red shift observed in spectral lines from distant galaxies as due to a mutual recession of galaxies away from each other. [RELAT] A model of the universe describing the process defined in the astronomy definition, in which the universe is nonstatic, homogeneous, and isotropic; based on Einstein's field equations with a nonvanishing cosmical constant.

expandor [ELECTR] The part of a compandor that is used at the receiving end of a circuit to return the compressed signal to its original form; attenuates weak signals and amplifies strong signals.

expansion [ELECTR] A process in which the effective gain of an amplifier is varied as a function of signal magnitude, the effective gain being greater for large signals than for small signals; the result is greater volume range in an audio amplifier and greater contrast range in facsimile. [MECH ENG] Increase in volume of working material with accompanying drop in pressure of a gaseous or vapor fluid, as in an internal combustion engine or steam engine cylinder. [PHYS] Process in which the volume of a constant mass of a substance increases.

expansion bolt [DES ENG] A bolt having an end which, when embedded into masonry or concrete, expands under a pull on the bolt, thereby providing anchorage.

expansion chamber *See* cloud chamber.

expansion chucking reamer [DES ENG] A machine reamer with an expansion screw at the end which increases the diameter.

expansion coefficient *See* coefficient of cubical expansion.

expansion cooling [MECH ENG] Cooling of a substance by having it undergo adiabatic expansion.

expansion ellipsoid [SOLID STATE] An ellipsoid whose axes have lengths which are proportional to the coefficient of linear expansion in the corresponding direction in a crystal.

expansion engine [MECH ENG] Piston-cylinder device that cools compressed air via sudden expansion; used in production of pure gaseous oxygen via the Claude cycle.

expansion factor [NAV] In radio hyperbolic navigation, a factor representing the degree to which the hyperbolas of a given group of hyperbolic lines of position diverge as the distance from the base line increases.

expansion fissures [GEOL] A system of fissures which radiate randomly and pass through feldspars and other minerals adjacent to olivine crystals that have been replaced by serpentine.

expansion fit [DES ENG] A condition of optimum clearance between certain mating parts in which the cold inner member is placed inside the warmer outer member and the temperature is allowed to equalize.

expansion joint [GEOL] *See* sheeting structure. [MECH ENG] **1.** A joint between parts of a structure or machine to avoid distortion when subjected to temperature change. **2.** A pipe coupling which, under temperature change, allows movement of a piping system without hazard to associated equipment.

expansion opening [ENG] A chamber in line with a pipe or tunnel and of larger diameter than the conduit containing liquid or gas, to allow lowering of pressure within the conduit by expansion of the fluid.

expansion ratio [FL MECH] For the calculation of the mass flow of a gas out of a nozzle or other expanding duct, the ratio of the nozzle exit section area to the nozzle throat area, or the ratio of final to initial volume.

expansion rollers [CIV ENG] Rollers fitted to one support of a bridge or truss to allow for thermal expansion and contraction.

expansion shield [DES ENG] An anchoring device that expands as it is driven into masonry or concrete, pressing against the sides of the hole.

expansion system [PETRO ENG] Gas-liquid recovery system in which the refrigeration effect of rapidly depressurized wellstream effluent through a wellhead choke is used to obtain maximum removal of liquefiable hydrocarbons from the gas stream.

expansion valve [MECH ENG] A valve in which fluid flows under falling pressure and increasing volume.

expansion wave [FL MECH] A pressure wave or shock wave that decreases the density of air as the air passes through it.

expansive bit [DES ENG] A bit in which the cutting blade can be set at various sizes.

expansive cement [MATER] A type of hydraulic cement, usually of high sulfate and alumina content, that expands after hardening to compensate for drying shrinkage.

expansivity *See* coefficient of cubical expansion.

expectation *See* expected value.

expectation value [QUANT MECH] The average of the results of a large number of measurements of a quantity made on a system in a given state; in case the measurement disturbs the state, the state is reprepared before each measurement.

expected approach clearance time [NAV] The anticipated time that an arriving aircraft will be cleared to commence approach for a landing.

expected utility *See* expected value.

expected value [MATH] **1.** For a random variable x with probability density function $f(x)$, this is the integral from $-\infty$ to ∞ of $xf(x)dx$. Also known as expectation. **2.** For a random variable x on a probability space (Ω, P), the integral of x with respect to the probability measure P. [SYS ENG] In decision theory, a measure of the value or utility expected to result from a given strategy, equal to the sum over states of nature of the product of the probability of the state times the consequence or outcome of the strategy in terms of some value or utility parameter. Abbreviated EV. Also known as expected utility (EU).

expectorant [PHARM] **1.** Tending to promote expectoration. **2.** An agent that promotes expectoration.

expectorate [PHYSIO] To eject phlegm or other material from the throat or lungs.

experiment [SCI TECH] The test of a hypothesis under controlled conditions.

experimental breeder reactor [NUCLEO] A fast, heterogeneous nuclear reactor used for research and breeding; its core consists of enriched U^{235} surrounded by a blanket of natural uranium.

experimental petrology [PETR] A branch of petrology in which phenomena that occur during petrological processes are reproduced and studied in the laboratory.

experimental psychology [PSYCH] The study of psychological phenomena by experimental methods.

experimental reactor [NUCLEO] A reactor to test the design of a new reactor concept.

expiration date [MATER] The anticipated date when a material may go from usable to unusable.

exploded view [GRAPHICS] A drawing· or picture of any article or piece of equipment in which the component parts are separated but so arranged to show their relationship to the whole.

exploding bridge wire [ENG] An initiator or system in which a very high energy electrical impulse is passed through a bridge wire, literally exploding the bridge wire and releasing thermal and shock energy capable of initiating a relatively insensitive explosive in contact with the bridge wire.

exploitation [MIN ENG] The extraction from the earth and utilization of ore, gas, oil, and minerals found by exploration.

exploration [MIN ENG] The search for economic deposits of minerals, ore, gas, oil, or coal by geological surveys, geophysical prospecting, boreholes and trial pits, or surface or underground headings, drifts, or tunnels.

exploratory well [PETRO ENG] An oil well drilled for purposes of exploration for underlying petroleum.

Explorer program [AERO ENG] A series of earth satellites begun by the U.S. Army and continued by NASA; *Explorer 1*, the first U.S. satellite, went into orbit on January 31, 1958.

exploring coil [ELECTROMAG] A small coil used to measure a magnetic field or to detect changes produced in a magnetic field by a hidden object; the coil is connected to an indicating instrument either directly or through an amplifier. Also known as magnetic test coil; search coil.

explosion [CHEM] A chemical reaction or change of state which is effected in an exceedingly short space of time with the generation of a high temperature and generally a large quantity of gas.

explosion breccia [PETR] Breccia resulting from volcanic eruption or a phreatic explosion.

explosion crater [GEOL] A volcanic crater formed by explosion and commonly developed along rift zones on the flanks of large volcanoes.

explosion method [THERMO] Method of measuring the specific heat of a gas at constant volume by enclosing the gas with an explosive mixture, whose heat of reaction is known, in a chamber closed with a corrugated steel membrane which acts as a manometer, and by deducing the maximum temperature reached on ignition of the mixture from the pressure change.

explosion tuff [GEOL] A tuff whose constituent ash particles are in the place they fell after being ejected from a volcanic vent.

explosive [MATER] A substance, such as trinitrotoluene, or a mixture, such as gunpowder, that is characterized by chemical stability but may be made to undergo rapid chemical change without an outside source of oxygen, whereupon it produces a large quantity of energy generally accompanied by the evolution of hot gases.

explosive-actuated device [ENG] Any of various devices actuated by means of explosive; includes devices actuated either by high explosives or low explosives, whereas propellant-actuated devices include only the latter.

explosive bolt [AERO ENG] A bolt designed to contain a remote-initiated explosive charge which, upon detonation, will shear the bolt or cause it to fail otherwise; applicable to such uses as stage separation of rockets, jettison of expended fuel tanks, and ejection of parachutes.

explosive cladding [MET] Bonding of a metal coating or metal cladding to a base metal by using the force of an explosive charge.

explosive D *See* ammonium picrate.

explosive decompression [AERO ENG] A sudden loss of pressure in a pressurized cabin, cockpit, or the like, so rapid as to be explosive, as when punctured by gunfire.

explosive disintegration [ENG] Explosive shattering when pressure is suddenly released on a pressured, permeable material (wood, mineral, and such) containing gas or liquid; the rupture of wood by this process is used to manufacture Masonite.

explosive echo ranging [ENG] Sonar in which a charge is exploded underwater to produce a shock wave that serves the same purpose as an ultrasonic pulse; the elapsed time for return of the reflected wave gives target range.

explosive evolution [EVOL] Rapid diversification of a group of fossil organisms in a short geological time.

explosive filler [MATER] Main explosive charge contained in a projectile, missile, bomb, or the like.

explosive fog signal [NAV] A fog signal produced by detonating an explosive charge.

explosive forming [MET] Shaping metal parts in dies by using an explosive charge to generate forming pressure.

explosive fuel [MATER] Any substance which combines with oxygen and other explosive ingredients to produce explosion energy, including aluminum, silicon, carbon, sulfur, glycerol, glucol, paraffin wax, diesel oil, and guar gum.

explosive limits [CHEM ENG] The upper and lower limits of percentage composition of a combustible gas mixed with other gases or air within which the mixture explodes when ignited.

explosive ordnance [ORD] Ordnance materiel containing or consisting of explosives, such as bombs, mines, torpedoes, missiles, or projectiles.

explosive ordnance disposal [ORD] The handling, disarming, or destroying of unexploded missiles and other explosive ordnance. Abbreviated EOD.

explosive ordnance disposal unit [ORD] Organization or personnel with special training or equipment who render safe explosive ordnance, make intelligence reports on such ordnance, and supervise the safe removal and disposal therof.

explosive ordnance reconnaissance [ORD] Act of reconnoitering to determine the presence of an unexploded missile, ascertaining its nature, applying all practicable protective measures for the protection of personnel, installations, and equipment, and, finally, reporting essential information to the authority directing explosive ordnance disposal operations. Abbreviated EOR.

explosive ordnance reconnaissance agent [ORD] A person trained in explosive ordnance reconnaissance techniques who takes required actions and renders the needed reports so that the unexploded ordnance in question can be effectively neutralized by the military. Abbreviated EORA.

explosive oxidizer [MATER] Any substance which yields oxygen to combine with fuels or other explosive ingredients to produce explosive energy, such as nitrates, chlorates, and perchlorates.

explosive rivet [ENG] A rivet holding a charge of explosive material; when the charge is set off, the rivet expands to fit tightly in the hole.

explosive train [ORD] A train of combustible and explosive elements arranged in order of decreasing sensitivity, inside a fuse, projectile, bomb, gun chamber, or the like; the function is to accomplish the controlled augmentation of a small impulse into one of suitable energy to cause the main charge of the munition to function.

explosive variable *See* cataclysmic variable.

explosive welding [MET] A welding process accomplished by use of an explosive charge.

exponent [MATH] A number or symbol placed to the right and above some given mathematical expression.

exponential amplifier [ELECTR] An amplifier capable of supplying an output signal proportional to the exponential of the input signal.

exponential atmosphere *See* isothermal atmosphere.

exponential curve [MATH] A graph of the function $y = a^x$, where a is a positive constant.

exponential density function [MATH] A probability density function obtained by integrating a function of the form $\exp(-|x - m|/\sigma)$, where m is the mean and σ the standard deviation.

exponential distribution [STAT] A continuous probability distribution whose density function is given by $f(x) = ae^{-ax}$, where $a > 0$ for $x > 0$, and $f(x) = 0$ for $x \leq 0$; the mean and standard deviation are both $1/a$.

exponential equation [MATH] An equation containing e^x (the Naperian base raised to a power) as a term.

exponential experiment [NUCLEO] A nuclear experiment involving a subcritical assembly of fissionable and moderator material.

exponential function [MATH] The function $f(x) = e^x$, written $f(x) = \exp(x)$.

exponential growth [MICROBIO] The period of bacterial growth during which cells divide at a constant rate. Also known as logarithmic growth.

exponential horn [ENG ACOUS] A horn whose cross-sectional area increases exponentially with axial distance.

exponential integral [MATH] The function defined to be the integral from x to ∞ of $(e^{-t}/t)\,dt$ for x positive.

exponential law [MATH] *See* law of exponents. [PHYS] The principle that growth or decay of some physical quantity is at a rate such that its value at a certain time or place is the initial value times e raised to a power equal to a constant times some convenient coordinate, such as the elapsed time or the distance traveled by a wave; there is growth if the constant is positive, decay if it is negative.

exponential pile *See* exponential reactor.

exponential pulse [PHYS] Variation of some quantity with time similar to the displacement of a critically damped harmonic oscillator which is initially given an impulse in its equilibrium position.

exponential reactor [NUCLEO] A nuclear reactor which is designed specifically for exponential experiments, as well as for determining critical size. Also known as exponential pile.

exponential series [MATH] The Maclaurin series expansion of e^x, namely,

$$e^x = 1 + \sum_{n=1}^{\infty} \frac{x^n}{n!}.$$

exponential smoothing [IND ENG] A mathematical-statistical method of forecasting used in industrial engineering which assumes that demand for the following period is some weighted average of the demands for the past periods.

exponential transmission line [ELEC] A two-conductor transmission line whose characteristic impedance varies exponentially with electrical length along the line.

export kerosine [MATER] A grade of kerosine once used for export; has the darkest shade of the standard kerosine colors, namely standard white (also called export white). Also known as standard white kerosine.

exposure [GRAPHICS] The act of permitting light to fall upon a photosensitive material. [MED] The state of being open to some action or influence that may affect detrimentally, as cold, disease, or wetness. [METEOROL] The general surroundings of a site, with special reference to its openness to winds and sunshine. [NUCLEO] 1. The total quantity of radiation at a given point, measured in air. 2. The cumulative amount of radiation exposure to which nuclear fuel has been subjected in a nuclear reactor; usually expressed in terms of the thermal energy produced by the reactor per ton of fuel initially present, as megawatt days per ton. [OPTICS] In photography, the amount of time that the image of a subject is permitted to be on the photographic sensitized film or paper.

exposure factor [NUCLEO] A quantity f used to specify radiographic exposure equal to st/d^2, where s is the intensity of radioactive source, t is the time, and d is the source to film distance.

exposure meter [OPTICS] An instrument used to measure the intensity of light reflected from an object, for the purpose of determining proper camera exposure.

exposure voltage [ELEC] The voltage at which the document-illuminating lamps are operated during exposure.

expression [CHEM ENG] Separation of liquid from a two-phase solid-liquid system by compression under conditions that permit liquid to escape while the solid is retained between the compressing surfaces. Also known as mechanical expression.

expressway [CIV ENG] A limited-access, high-speed, divided highway having grade separations at points of intersection with other roads. Also known as limited-access highway.

expulsion fuse See expulsion-fuse unit.

expulsion-fuse unit [ELEC] A vented fuse unit in which the arc is extinguished by the expulsion of gases generated by the arc and lining of the fuse holder, sometimes with the aid of a spring. Also known as expulsion fuse.

exsecant [MATH] The trigonometric function defined by subtracting unity from the secant, that is exsec $\theta = \sec \theta - 1$.

exserted [BIOL] Protruding beyond the enclosing structure, such as stamens extending beyond the margin of the corolla.

exsheath [INV ZOO] To escape from the residual membrane of a previous developmental stage, used of the larva of certain nematodes, microfilaria, and so on.

exsiccate [SCI TECH] To dry by driving off, or draining of, moisture.

exsolution [GEOL] A phenomenon during which molten rock solutions separate when cooled.

exsolution lamellae [GEOL] Layers of sedimentary rock that solidify from solution by either precipitation or secretion.

exstipulate [BOT] Lacking stipules.

exstrophy [MED] Eversion; the turning inside out of a part.

extended area [DES ENG] An engineering surface that has been extended areawise without increasing diameter, as by using pleats (as in filter cartridges) or fins (as in heat exchangers).

extended-area service [COMMUN] Telephone exchange service, without toll charges, that extends over an area where there is a community of interest, in return for a somewhat higher exchange service rate.

extended channel status word [ADP] Stored information which follows an input/output interrupt. Abbreviated ECSW.

extended dislocation [CRYSTAL] A dislocation in a close-packed structure consisting of a strip of stacking fault edged by two lines across which slip through a fraction of a lattice constant, into one of the alternative stacking positions, has occurred.

extended forecast [METEOROL] In general, a forecast of

weather conditions for a period extending beyond 2 days from the day of issue. Also known as extended-range forecast; long-range forecast.

extended-interaction tube [ELECTR] Microwave tube in which a moving electron stream interacts with a traveling electric field in a long resonator; bandwidth is between that of klystrons and traveling-wave tubes.

extended-precision word [ADP] A piece of data of 16 bytes in floating-point arithmetic when additional precision is required.

extended-range Dovap [NAV] A baseline extension of the Dovap system to provide a coherent reference to the ground transmitter and all Dovap receivers located beyond line-of-sight to the ground transmitter. Abbreviated Extradop.

extended-range forecast See long-range forecast.

extended state [QUANT MECH] A state of motion in which an electron may be found anywhere within a region of a material of linear extent equal to that of the material itself.

extended stream [HYD] A stream lengthened by the extension of its downstream course; the course is through a newly emerged land such as a coastal plain.

extender [CHEM] A material used to dilute or extend or change the properties of resins, ceramics, paints, rubber, and so on. [ELEC] A male or female receptacle connected by a short cable to make a test point more conveniently accessible to a test probe.

extender plasticizer See secondary plasticizer.

extend flip-flop [ADP] A special flag set when there is a carry-out of the most significant bit in the register after an addition or a subtraction.

extensibility [MECH] The amount to which a material can be stretched or distorted without breaking.

extensional fault See tension fault.

extension bolt [DES ENG] A vertical bolt that can be slid into place by a long extension rod; used at the top of doors.

extension cord [ELEC] A line cord having a plug at one end and an outlet at the other end.

extension jamb [BUILD] A jamb that extends past the head of a door or window.

extension joints [GEOL] Fractures that form parallel to a compressive force.

extension ladder [DES ENG] A ladder of two or more nesting sections which can be extended to almost the combined length of the sections.

extension of a field [MATH] Any field containing the original field.

extension ore See possible ore.

extension spring [DES ENG] A tightly coiled spring designed to resist a tensile force.

extensive air shower See Auger shower.

extensive property [PHYS CHEM] A noninherent property of a system, such as volume or internal energy, that changes with the quantity of material in the system; the quantitative value equals the sum of the values of the property for the individual constituents.

extensive shower See Auger shower.

extensometer [ENG] 1. A strainometer that measures the change in distance between two reference points separated 20–30 meters or more; used in studies of displacements due to seismic activities. 2. An instrument designed to measure minute deformations of small objects subjected to stress.

extent [ADP] The physical locations in a mass-storage device or volume allocated for use by a particular data set.

exterior algebra [MATH] An algebra whose structure is analogous to that of the collection of differential forms on a Riemannian manifold.

exterior angle [MATH] 1. An angle between one side of a polygon and the prolongation of an adjacent side. 2. An angle made by a line (the transversal) that intersects two other lines, and either of the latter on the outside.

exterior ballistics [MECH] The science concerned with behavior of a projectile after leaving the muzzle of the firing weapon.

exterior of a set [MATH] The largest open set contained in the complement of a given set.

extern [ADP] A pseudoinstruction found in several assembly

EXPOSURE METER

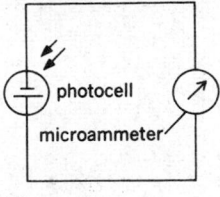

(a)

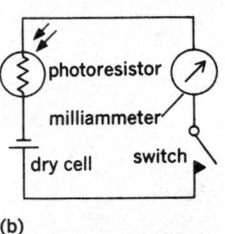

(b)

Basic exposure meter circuits.
(a) Photovoltaic-type circuit.
(b) Photoconductive-type circuit.

languages which explicitly tells an assembler that a symbol is external, that is, not defined in the program module.

external aileron [AERO ENG] An aileron offset from the wing; that is, not forming a part of the wing.

external angle [MATH] The angle defined by an arc around the boundaries of an internal angle or included angle.

external ankle height [ANTHRO] A measure of the vertical distance taken from the lower end of the fibula to the floor.

external armature [ELEC] Armature for a machine of special design in which the armature is a ring which rotates around the magnetic poles.

external auditory meatus [ANAT] The external passage of the ear, leading to the tympanic membrane in reptiles, birds, and mammals.

external beam [NUCLEO] A beam of particles which originate in a particle accelerator and are directed outside the accelerator so that they can be used for experiments with external apparatus.

external brake [MECH ENG] A brake that operates by contacting the outside of a brake drum.

external carotid artery [ANAT] An artery which originates at the common carotid and distributes blood to the anterior part of the neck, face, scalp, side of the head, ear, and dura mater.

external combustion engine [MECH ENG] An engine in which the generation of heat is effected in a furnace or reactor outside the engine cylinder.

external-device address [ADP] The address of a component such as a tape drive.

external-device control [ADP] The capability of an external device to create an interrupt during the execution of a job.

external-device operands [ADP] The part of an instruction referring to an external device such as a tape drive.

external-device response [ADP] The signal from an external device, such as a tape drive, that it is not busy.

external ear [ANAT] The portion of the ear that receives sound waves, including the pinna and external auditory meatus.

external error [ADP] An error sensed by the computer when this error occurs in a device such as a disk drive.

external fertilization [PHYSIO] Those processes involved in the union of male and female sex cells outside the body of the female.

external galaxy [ASTRON] Any galaxy known to exist, besides the Milky Way.

external gas injection [PETRO ENG] Pressure-maintenance gas injection with wells located in the structurally higher positions of the reservoir, usually in the primary or secondary gas cap. Also known as crestal injection; gas-cap injection.

external gill [ZOO] A gill that is external to the body wall, as in certain larval fishes and amphibians, and in many aquatic insects.

external grinding [MECH ENG] Grinding the outer surface of a rotating piece of work.

external interrupt [ADP] Any interrupt caused by the operator or by some external device such as a tape drive.

external-interrupt status word [ADP] The content of a special register which indicates, among other things, the source of the interrupt.

external label [ADP] A reference to a variable not defined in a program segment.

externally stored program [ADP] Program achieved by wiring plugboards, as in some tabulating equipment.

external memory [ADP] Any storage device not an integral part of a computer system, such as a magnetic tape or a deck of cards.

external photoelectric effect *See* photoemission.

external Q [ELECTR] The inverse of the difference between the loaded and unloaded Q values of a microwave tube.

external respiration [PHYSIO] The processes by which oxygen is carried into living cells from the outside environment and by which carbon dioxide is carried in the reverse direction.

external shoe brake [MECH ENG] A friction brake operated by the application of externally contracting elements.

external signal [ADP] Any message to an operator for which no printout is required but which is self-explanatory, such as

a light condition indicating whether the equipment is on or off.

external storage [ADP] Large-capacity, slow-access data storage attached to a digital computer and used to store information that exceeds the capacity of main storage.

external symbol dictionary [ADP] A list of external symbols and their relocatable addresses which allows the linkage editor to resolve interprogram references. Abbreviated ESD.

external thread [DES ENG] A screw thread cut on an outside surface.

external time [IND ENG] The time used to perform work by the operator outside the machine cycle, resulting in a loss of potential machine operating time.

external upset casing [PETRO ENG] Special oil- or gas-well casing designed for extreme conditions requiring greater than usual strength and leak resistance. Also known as extreme line casing.

external wave [FL MECH] 1. A wave in fluid motion having its maximum amplitude at an external boundary such as a free surface. 2. Any surface wave on the free surface of a homogeneous incompressible fluid is an external wave.

exteroceptor [PHYSIO] Any sense receptor at the surface of the body that transmits information about the external environment.

extinction [EVOL] The worldwide death and disappearance of a specific organism or group of organisms. [OPTICS] Phenomenon in which plane polarized light is almost completely absorbed by a polarizer whose axis is perpendicular to the plane of polarization. [PHYS CHEM] *See* absorbance. [PSYCH] Decrease in frequency and elimination of a conditioned response if reinforcement of the response is withheld.

extinction coefficient *See* absorptivity.

extinction meter [OPTICS] An exposure meter in which light intensity is measured by gradually attenuating the light by a known fraction until a selected design is just visible or disappears.

extinction voltage [ELECTR] The lowest anode voltage at which a discharge is sustained in a gas tube.

extirpate [BIOL] To uproot, destroy, make extinct, or exterminate.

extracellular [BIOL] Outside the cell.

extrachromosomal inheritance *See* cytoplasmic inheritance.

extract [ADP] 1. To form a new computer word by extracting and putting together selected segments of given words. 2. To remove from a computer register or memory all items that meet a specified condition. [CHEM] Material separated from liquid or solid mixture by a solvent. [PHARM] 1. A pharmaceutical preparation obtained by dissolving the active constituents of a drug with a suitable menstruum, evaporating the solvent, and adjusting to prescribed standards. 2. A preparation, usually in a concentrated form, obtained by treating plant or animal tissue with a solvent to remove desired odiferous, flavorful, or nutritive components of the tissue.

extractant [CHEM] The liquid used to remove a solute from another liquid.

extract a root of a number [MATH] To determine a root of the number, usually a positive real root, or a negative real odd root of a negative number.

extract instruction [ADP] An instruction that requests the formation of a new expression from selected parts of given expressions.

extraction [CHEM] A method of separation in which a solid or solution is contacted with a liquid solvent (the two being essential mutually insoluble) to transfer one or more components into the solvent. [MED] The act or process of pulling out a tooth.

extraction column [CHEM ENG] Vertical-process vessel in which a desired product is separated from a liquid by countercurrent contact with a solvent in which the desired product is preferentially soluble.

extraction parachute [AERO ENG] An auxiliary parachute designed to release and extract cargo from aircraft in flight and to deploy cargo parachutes.

extraction turbine [MECH ENG] A steam turbine equipped with openings through which partly expanded steam is bled at one or more stages.

extractive distillation [CHEM ENG] A distillation process to

EXTERNAL SHOE BRAKE

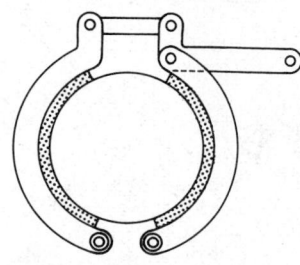

External shoe brake; shoes are lined with frictional material (shaded area).

separate components from eutectic mixtures; a solution of the mixture is cooled, causing one component to crystallize out and the other to remain in solution; used to separate p-xylene and m-xylene, using n-pentane as the solvent.

extractive metallurgy [MET] Extraction of metals from ore by various chemical and mechanical methods.

extract oil [MATER] The less desirable portion of the oil being solvent-refined; it is dissolved in and selectively removed by the solvent.

extractor [ADP] See mask. [CHEM ENG] An apparatus for solvent-contact with liquids or solids for removal of specified components. [ENG] **1.** A machine for extracting a substance by a solvent or by centrifugal force, squeezing, or other action. **2.** An instrument for removing an object. [ORD] A device in the breech mechanism of a gun, rifle, or the like, for pulling an empty cartridge case or an unfired cartridge out of the chamber of a gun, rifle, or the like.

extractor groove [ORD] Groove machined in the base of a cartridge case, a short distance above the head, which receives the extractor of the breech mechanism and permits the case to be withdrawn by the extractor; used in automatic weapons, in preference to extractor rims (flanges) formed on the cartridge case base.

Extradop See extended-range Dovap.

extrados [ARCH] The upper or outer curve of an arch.

extraembryonic coelom [EMBRYO] The cavity in the extraembryonic mesoderm; it is continuous with the embryonic coelom in the region of the umbilicus, and is obliterated by growth of the amnion.

extraembryonic membrane See fetal membrane.

extragalactic [ASTRON] Beyond the Milky Way.

extragalactic radio source [ASTROPHYS] A source of radio emission outside the Milky Way.

extra-high tension [ELECTR] British term for the high direct-current voltage applied to the second anode in a cathode-ray tube, ranging from about 4000 to 50,000 volts in various sizes of tubes. Abbreviated eht.

extra-high voltage [ELEC] A voltage above 345 kilovolts used for power transmission. Abbreviated ehv.

extraneous emission [ELECTR] Any emission of a transmitter or transponder, other than the output carrier fundamental, plus only those sidebands intentionally employed for the transmission of intelligence.

extraneous light [GRAPHICS] Undesired illumination striking photographic area or copy field.

extraneous response [ELECTR] Any undersired response of a receiver, recorder, or other susceptible device, due to the desired signals, undersired signals, or any combination or interaction among them.

extraordinary component See extraordinary wave.

extraordinary index [OPTICS] The index of refraction of the extraordinary wave propagating in a direction perpendicular to the optical axis of a uniaxial crystal.

extraordinary ray [OPTICS] One of two rays into which a ray incident on an anisotropic uniaxial crystal is split; its deviation at the crystal's surface depends on the orientation of the crystal, and it is deviated even in the case of normal incidence.

extraordinary wave [GEOPHYS] Magnetoionic wave component which, when viewed below the ionosphere in the direction of propagation, has clockwise or counterclockwise elliptical polarization respectively, accordingly as the earth's magnetic field has a positive or negative component in the same direction. Also known as X wave. [OPTICS] Component of electromagnetic radiation propagating in an anisotropic uniaxial crystal whose electric displacement vector lies in the plane containing the optical axis and the direction normal to the wavefront; it gives rise to the extraordinary ray. Also known as extraordinary component.

extrapolation [MATH] Estimating a function at a point which is larger than (or smaller than) all the points at which the value of the function is known.

extrapolation ionization chamber [NUCLEO] An ionization chamber so designed that volume, electrode separation, or some other factor can be varied in suitable steps for measurement purposes; the resulting measured values are plotted in appropriate form and the desired result is obtained by extrapolation of the curve.

extrapyramidal system [ANAT] Descending tracts of nerve fibers arising in the cortex and subcortical motor areas of the brain.

extrasensory perception [PSYCH] The alleged phenomenon of perception or awareness of external events in the absence of any sensory stimulation arising from the events. Abbreviated ESP.

extrasystole [MED] Premature beat of the heart.

extraterrestrial noise [ELECTROMAG] Cosmic and solar noise; radio disturbances from other sources other than those related to the earth.

extraterrestrial radiation [ASTROPHYS] Electromagnetic radiation which originates outside the earth or its atmosphere, as in the sun or stars.

extratropical cyclone [METEOROL] Any cyclone-scale storm that is not a tropical cyclone. Also known as extratropical low; extratropical storm.

extratropical low See extratropical cyclone.

extratropical storm See extratropical cyclone.

extrauterine pregnancy [MED] Gestation outside the uterus.

extravasation [GEOL] The eruption of lava from a vent in the earth. [MED] The pouring out or eruption of a body fluid from its proper channel or vessel into the surrounding tissue.

extravehicular activity [AERO ENG] Activity conducted outside a spacecraft during space flight.

extremals [MATH] For a variational problem in the calculus of variations entailing use of the Euler-Lagrange equation, the extremals are the solutions of this equation.

extreme [CLIMATOL] The highest, and in some cases the lowest, value of a climatic element observed during a given period or during a given month or season of that period; if this is the whole period for which observations are available, it is the absolute extreme. [MATH] See extremum.

extreme and mean ratio See golden section.

extreme close-up [GRAPHICS] A motion picture shot taken at extremely close range; for example, an extreme close-up of a face might show only an eye.

extreme line casing See external upset casing.

extreme line tubing [PETRO ENG] Special oil- or gas-well tubing designed for extreme conditions requiring greater than usual strength and leak resistance.

extremely high frequency [COMMUN] The frequency band from 30,000 to 300,000 megahertz in the radio spectrum. Abbreviated EHF.

extremely low frequency [COMMUN] A frequency below 300 hertz in the radio spectrum. Abbreviated ELF.

extreme narrowing approximation [SPECT] A mathematical approximation in the theory of spectral-line shapes to the effect that the exchange narrowing of a perturbation is complete.

extreme point [MATH] **1.** A maximum or minimum value of a function. **2.** A point in a convex subset K of a vector space is called extreme if it does not lie on the interior of any line segment contained in K.

extreme-pressure lubricant See EP lubricant.

extreme range [ORD] Greatest range of a weapon; for example, the greatest distance a gun will shoot.

extreme relativistic limit [PHYS] Limit which a formula describing a particle's behavior approaches when the speed of the particle approaches the speed of light.

extreme spread [ORD] In a firing accuracy test, the distance between the two shots farthest from each other.

extreme terms [MATH] The first and last terms in a proportion.

extreme value problem [MATH] A set of mathematical conditions which may be met by values that are less than or greater than an upper or a lower bound, that is, an extreme value.

extremum [MATH] A maximum or minimum value of a function. Also known as extreme.

extrinsic factor See vitamin B_{12}.

extrinsic photoemission [ELECTR] Photoemission by an alkali halide crystal in which electrons are ejected directly from negative ion vacancies, forming color centers. Also known as direct ionization.

extrinsic properties [ELECTR] The properties of a semicon-

EXTRAORDINARY RAY

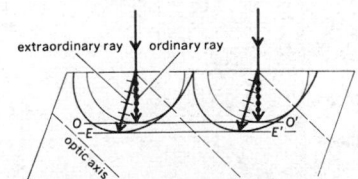

Huygens construction for a plane wave incident normally on transparent calcite showing the ordinary and extraordinary ray. (From F. A. Jenkins and H. E. White, Fundamentals of Optics, 3d ed., McGraw-Hill, 1957)

ductor as modified by impurities or imperfections within the crystal.

extrinsic semiconductor [ELECTR] A semiconductor whose electrical properties are dependent on impurities added to the semiconductor crystal, in contrast to an intrinsic semiconductor, whose properties are characteristic of an ideal pure crystal.

extrinsic sol [PHYS CHEM] A colloid whose stability is attributed to electric charge on the surface of the colloidal particles.

extrophy [MED] Malformation of an organ.

extrorse [BIOL] Directed outward or away from the axis of growth.

extroversion [BIOL] A turning outward. [PSYCH] Turning to things and persons outside oneself rather than to one's own thoughts and feelings.

extrudate [ENG] Ductile metal, plastic, or other semisoft solid material that has been shaped into a continuous form (such as fiber, film, pipe, or wire coating) by forcing the semisolid material through a die opening of appropriate shape.

extruder [ENG] A device that forces ductile or semisoft solids through die openings of appropirate shape to produce a continuous film, strip, or tubing.

extrusion [ENG] A process in which a hot or cold semisoft solid material, such as metal or plastic, is forced through the orifice of a die to produce a continuously formed piece in the shape of the desired product. [GEOL] Emission of magma or magmatic materials at the surface of the earth. [TEXT] A process for making continuous-filament synthetic fibers by forcing a syruplike liquid through minute holes of a spinneret.

extrusion billet [MET] A slug of heated metal that is forced through a die by a hydraulic ram in direct extrusion operations.

extrusion coating [ENG] A process of placing resin on a substrate by extruding a thin film of molten resin and pressing it onto or into the substrates, or both, without the use of adhesives.

extrusion defect [MET] Impaired flow of an extrusion product due to surface oxidation of the ingot or billet.

extrusion ingot [MET] A cylindrical casting used to form extruded products.

extrusion metal [MET] Any of numerous nonferrous metals, alloys, and other materials used in extrusion operations.

extrusion pressing See cold extrusion.

extrusive rock See volcanic rock.

exudate [MED] **1.** A proteinaceous, cellular material that passes through blood vessel walls into the surrounding tissue in inflammation or a superficial lesion. **2.** Any substance that is exuded.

exudation See sweating.

exudation vein See segregated vein.

exumbrella [INV ZOO] The outer, convex surface of the umbrella of jellyfishes.

eye [ZOO] A photoreceptive sense organ that is capable of forming an image in vertebrates and in some invertebrates such as the squids and crayfishes.

eye assay [MIN ENG] An estimate of the valuable mineral content of a core or ore sample as based on visual inspection. Also known as eyeball assay.

eyeball [ANAT] The globe of the eye.

eyeball assay See eye assay.

eyebar [DES ENG] A metal bar having a hole or eye through each enlarged end.

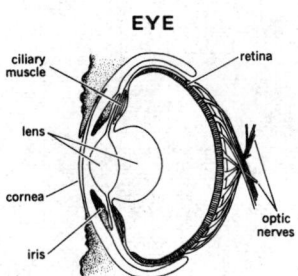

EYE

ciliary muscle
retina
lens
cornea
optic nerves
iris

Octopus eye (after J. Wells). *(From R. D. Barnes, Invertebrate Zoology, 2d ed., W. B. Saunders Co., 1968)*

eyebolt [DES ENG] A bolt with a loop at one end.

eye coal [GEOL] Coal characterized by small, circular or elliptic structural disks that reflect light and are arranged in parallel planes either in or normal to the bedding. Also known as circular coal.

eye-ear plane [ANTHRO] In craniometric study, a position for placing a human skull so that the lower margins of the orbits and the upper margin of the auditory meatus are on the same horizontal plane. Also known as Frankfurt horizontal.

eyeglasses [OPTICS] Optical devices containing corrective lenses for defects of vision or for special purposes.

eyelet [DES ENG] A small ring or barrel-shaped piece of metal inserted into a hole for reinforcement.

eyeleting [ENG] Forming a lip around the rim of a hole.

eyelid [ANAT] A movable, protective section of skin that covers and uncovers the eyeball of many terrestrial animals.

eyelights [GRAPHICS] Low-intensity light sources used to add sparkle to the eyes or teeth and reduce shadows on the face; usually placed at eye level.

eye of the storm [METEOROL] The center of a tropical cyclone, marked by relatively light winds, confused seas, rising temperature, lowered relative humidity, and often by clear skies.

eye of the wind [METEOROL] The point or direction from which the wind is blowing.

eyepiece [OPTICS] A lens or optical system which offers to the eye the image originating from another system (the objective) at a suitable viewing distance. Also known as ocular.

eye screw [DES ENG] A screw with an open loop head.

eye socket See orbit.

eyespot [BOT] **1.** A small photosensitive pigment body in certain unicellular algae. **2.** A dark area around the hilum of certain seeds, as some beans. [INV ZOO] A simple organ of vision in many invertebrates consisting of pigmented cells overlying a sensory termination. [PL PATH] A fungus disease of sugarcane and certain other grasses caused by *Helminthosporium sacchari* and characterized by yellowish oval lesions on the stems and leaves.

eyestalk [INV ZOO] A movable peduncle bearing a terminal eye in decapod crustaceans.

Eykman formula [OPTICS] An empirical formula which relates the molal refraction of a liquid at a given optical frequency to its index of refraction, density, and molecular weight.

Eyring equation [PHYS CHEM] An equation, based on statistical mechanics, which gives the specific reaction rate for a chemical reaction in terms of the heat of activation, entropy of activation, the temperature, and various constants.

Eyring formula [FL MECH] A formula, based on the Eyring theory of rate processes, which relates shear stress acting on a liquid and the resulting rate of shear.

Eyring molecular system [FL MECH] Theory to account for liquid properties; assumes that each liquid molecule can move freely within a certain free volume. Also known as Eyring theory.

Eyring theory See Eyring molecular system.

E zone [COMMUN] One of the three zones into which the earth is divided to show the variations of the F_2 layer in respect of longitude when making frequency predictions; it roughly covers Asia, Australia, the Philippines, and Japan.

E-Z Sort card [ADP] Trade name for a marginal punched card.

f [PHYS] Notation representing the Coriolis parameter.

F *See* farad; fluorine.

F_1 [GEN] Notation for the first filial generation resulting from a cross.

F_2 [GEN] Notation for the progeny produced by intercrossing members of the F_1. Also known as second generation.

F-6 *See* Skyray.

F-11 *See* Tiger.

F-84 *See* Thunderstreak.

F-86 *See* Sabrejet.

F-100 *See* Super Sabre.

F-105 *See* Thunderchief.

fa *See* femtoampere.

Faber flaw [SOLID STATE] A deformation in a superconducting material that acts as a nucleation center for the growth of a superconducting region.

Fabian system [MIN ENG] The free-fall drilling system from which all other free-fall systems have originated.

fabric [ARCH] The framework of a building. [GEOL] The spatial orientation of the elements of a sedimentary rock. [PETR] The sum of all the structural and textural features of a rock. Also known as petrofabric; rock fabric; structural fabric. [SCI TECH] **1.** Arrangement or pattern of constituent parts. **2.** Materials used in fabrication. [TEXT] A textile material.

fabric analysis *See* structural petrology.

fabrication [ENG] **1.** The manufacture of parts, usually structural or electromechanical parts. **2.** The assembly of parts into a structure.

fabric diagram [PETR] In structural petrology, a graphic representation of the data of fabric elements. Also known as petrofabric diagram.

fabric element [PETR] A surface or line of structural discontinuity in a rock fabric.

Fabriciinae [INV ZOO] A subfamily of small to minute, colonial, sedentary polychaete annelids in the family Sabellidae.

fabric laminate [MATER] Layers of fabric alternating with plastic, used as insulation in electrical equipment.

fabric-type dust collector [MIN ENG] A collector which removes dust particles from ore by means of a filter made of fabric.

fabric weight [TEXT] The number of ounces per square yard of a fabric.

Fabry-Barot method [OPTICS] Method of determining the index of refraction of a prism in which the prism is set up so that the incident beam is perpendicular to the emergent face, and the index of refraction is calculated from the angle of the prism and the angle of deviation.

Fabry-Perot etalon *See* etalon.

Fabry-Perot filter [OPTICS] An optical interference filter, similar to the Fabry-Perot interferometer except that the space between the partially reflecting surfaces is only a few thousand angstroms.

Fabry-Perot fringes [OPTICS] Series of rings observed when a monochromatic light source is viewed through a Fabry-Perot interferometer.

Fabry-Perot interferometer [OPTICS] An interferometer having two parallel glass plates (whose separation of a few centimeters may be varied), silvered on their inner surfaces so that the incoming wave is multiply reflected between them and ultimately transmitted.

facade [ARCH] The front of a building or a face of a building, given special architectural treatment.

face [ANAT] The anterior portion of the head, including the forehead and jaws. [CRYSTAL] *See* crystal face. [ELECTR] *See* faceplate. [GEOL] **1.** The main surface of a landform. **2.** The original surface of a layer of rock. [GRAPHICS] **1.** A particular style or size of letter as distinguished from another style or size. Also known as typeface. **2.** The printing surface of a printing plate or the front surface of a piece of paper. [MIN ENG] A surface on which mining operations are being performed. Also known as breast; highwall.

face angle [MATH] An angle between two successive edges of a polyhedral angle.

face area [MIN ENG] The working area toward the interior of the last open crosscut in an entry or room.

face belt conveyor [MIN ENG] A lightweight belt conveyor used at the working face in a mine.

face-bonding [ELECTR] Method of assembling hybrid microcircuits wherein semiconductor chips are provided with small mounting pads, turned facedown, and bonded directly to the ends of the thin-film conductors on the passive substrate.

face boss [MIN ENG] A foreman in charge of operations at the working face in a bituminous coal mine.

face brick [MATER] A brick of some esthetic quality to be used on the exposed surface of a building wall or other structure.

face-centered cubic lattice [CRYSTAL] A lattice whose unit cells are cubes, with lattice points at the center of each face of the cube, as well as at the vertices. Abbreviated fcc lattice.

face-centered orthorhombic lattice [CRYSTAL] An orthorhombic lattice which has lattice points at the center of each face of a unit cell, as well as at the vertices.

face conveyor [MIN ENG] Any type of mine conveyor used at and parallel to a working face.

face-discharge bit [MECH ENG] A liquid-coolant bit designed for drilling in soft formations and for use on a double-tube core barrel, the inner tube of which fits snugly into a recess cut into the inside wall of the bit directly above the inside reaming stones; the coolant flows through the bit and is ejected at the cutting face. Also known as bottom-discharge bit; face-ejection bit.

faced wall [BUILD] A wall whose masonry facing and backing are of different materials.

face-ejection bit *See* face-discharge bit.

face gear [DES ENG] A gear having teeth cut on the face.

face haulage *See* primary haulage.

face height [MIN ENG] The vertical distance between the top and toe of a quarry or opencast face.

facellite *See* kaliophillite.

faceman [MIN ENG] A coal miner who performs the duties involved in drilling underground openings into which explosives are charged and set off, to extract coal, slate, and rock. Also known as coal digger; coal getter.

face mechanization [MIN ENG] The use of a cutter-loader on a longwall face.

face milling [MECH ENG] Milling flat surfaces perpendicular to the rotational axis of the cutting tool.

face mold [ENG] A pattern for cutting forms out of sheets of wood, metal, or other material.

face nailing [ENG] Nailing of facing wood to a base, leaving the nailheads exposed.

face of a geometric simplex [MATH] Any simplex whose vertices are also vertices of the original simplex.

face of weld [MET] The exposed surface on the welded side of an arc weld or gas weld.

faceplate [ELECTR] The transparent or semitransparent glass front of a cathode-ray tube, through which the image is viewed or projected; the inner surface of the face is coated with fluorescent chemicals that emit light when hit by an electron beam. Also known as face. [ENG] **1.** A disk fixed perpendicularly to the spindle of a lathe and used for attachment of the workpiece. **2.** A protective plate used to cover holes in machines or other devices.

face sampling [MIN ENG] Taking random samples of ore and rock from exposed faces of ore and waste.

face shield [ENG] A detachable wraparound guard fitted to a

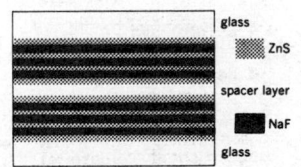

FABRY-PEROT FILTER

Schematic diagram of seven-layer solid Fabry-Perot filter. (*From D. E. Gray, ed., American Institute of Physics Handbook, McGraw-Hill, 1957*)

worker's helmet to protect the face from flying particles.

face signal [MIN ENG] A wire stretched along the face and connected to a panel near the main gate to control the running of a face conveyor.

facet [ANAT] A small plane surface, especially on a bone or a hard body; may be produced by wear, as a worn spot on the surface of a tooth. [INV ZOO] The surface of a simple eye in the compound eye of arthropods and certain other invertebrates. [MATER] The plane surface of a crystal, a cut precious stone, or other fractured surface.

faceted pebble [GEOL] A pebble with three or more faces naturally worn flat and meeting at sharp angles.

face tile [MATER] Tile with one finished surface, intended for use on a face.

face timbering [MIN ENG] Positioning of safety posts at the working portion of a coal face to support the roof of the mine.

facework [CIV ENG] Ornamental or otherwise special material on the front side or outside of a wall.

face worker [MIN ENG] A miner who works regularly at the face.

facial angle [ANTHRO] The angle formed by the union of a line connecting nasion and gnathion with the Frankfort horizontal plane of the head.

facial artery [ANAT] The external branch of the external carotid artery.

facial bone [ANAT] The bone comprising the nose and jaws, formed by the maxilla, zygoma, nasal, lacrimal, palatine, inferior nasal concha, vomer, mandible, and parts of the ethmoid and sphenoid.

facial index [ANTHRO] The ratio of the breadth of the face to its length multiplied by 100.

facial nerve [ANAT] The seventh cranial nerve in vertebrates; a paired composite nerve, with motor elements supplying muscles of facial expression and with sensory fibers from the taste buds of the anterior two-thirds of the tongue and from other sensory endings in the anterior part of the throat.

facies [ANAT] Characteristic appearance of the face in association with a disease or abnormality. [ECOL] The makeup or appearance of a community or species population. [GEOL] Any observable attribute of a rock or stratigraphic unit, such as overall appearance or composition.

facies map [GEOL] A stratigraphic map indicating distribution of sedimentary facies within a specific geologic unit.

facility assignment [ADP] The allocation of core memory and external devices by the executive as required by the program being executed.

facility dispersion [COMMUN] The distribution of circuits between two points over more than one physical or geographic route to reduce the likelihood of a trunk group being put completely out of service by facility damage or other circuit failure.

facility security clearance [ORD] An administrative determination that, from a security viewpoint, a facility is eligible for access to classified information of a certain category (and all lower categories).

facing [CIV ENG] A covering or casting of some material applied to the outer face of embankments, buildings, and other structures. [MECH ENG] Machining the end of a flat rotating surface by applying a tool perpendicular to the axis of rotation in a spiral planar path. [MET] A fine molding sand applied to the face of a mold.

facing-point lock [CIV ENG] A lock used on a railroad track, such as a switch track, which contains a plunger that engages a rod on the switch point to lock the device.

facing wall [CIV ENG] Concrete lining against the earth face of an excavation; used instead of timber sheeting.

facsimile [COMMUN] **1.** A system of communication in which a transmitter scans a photograph, map, or other fixed graphic material and converts the information into signal waves for transmission by wire or radio to a facsimile receiver at a remote point. Also known as fax; phototelegraphy; radiophoto; telephoto; telephotography; wirephoto. **2.** A photograph transmitted by radio to a facsimile receiver. Also known as radiophoto. [GRAPHICS] An exact copy of a book, document, painting, or other material.

facsimile chart [METEOROL] Any graphic form of weather information, usually a type of synoptic chart, which has been

reproduced by facsimile equipment. Also known as fax chart; fax map.

facsimile modulation [COMMUN] Process in which the amplitude, frequency, or phase of a transmitted wave is varied with time in accordance with a facsimile transmission signal.

facsimile receiver [ELECTR] The receiver used to translate the facsimile signal from a wire or radio communication channel into a facsimile record of the subject copy.

facsimile recorder [ELECTR] The section of a facsimile receiver that performs the final conversion of electric signals to an image of the subject copy on the record medium.

facsimile signal [COMMUN] The picture signal produced by scanning the subject copy in a facsimile transmitter.

facsimile signal level [ELECTR] Maximum facsimile signal power or voltage (root mean square or direct current) measured at any point in a facsimile system.

facsimile synchronizing [ELECTR] Maintenance of predetermined speed relations between the scanning spot and the recording spot within each scanning line.

facsimile telegraph [COMMUN] A telegraph system designed to transmit pictures.

facsimile transmitter [ELECTR] The apparatus used to translate the subject copy into facsimile signals suitable for delivery over a communication system.

factor I *See* fibrinogen.

factor II *See* prothrombin.

factor III *See* thromboplastin.

factor IV [BIOCHEM] Calcium ions involved in the mechanism of blood coagulation.

factor V *See* proaccelerin.

factor VI [BIOCHEM] An unidentified substance believed to be derived from factor V during blood coagulation.

factor VII [BIOCHEM] A procoagulant, related to prothrombin, that is involved in the formation of a prothrombin-converting principle which transforms prothrombin to thrombin. Also known as stable factor.

factor VIII *See* antihemophilic factor.

factor IX *See* Christmas factor.

factor X *See* Stuart factor.

factor XI [BIOCHEM] A procoagulant present in normal blood but deficient in hemophiliacs. Also known as plasma thromboplastin antecedent (PTA).

factor XII [BIOCHEM] A blood clotting factor effective experimentally only in vitro; deficient in hemophiliacs. Also known as Hageman factor.

factorable polynomial [MATH] A polynomial which has polynomial factors other than itself.

factor analysis [MATH] Given sets of variables which are related linearly, factor analysis studies techniques of approximating each set relative to the others; usually the variables denote numbers.

factor comparison [IND ENG] A quantitative system of job evaluation in which jobs are given relative positions on a rating scale based on a comparison of factors composing the job with certain previously selected key jobs.

factor group *See* quotient group.

factorial design [STAT] A design for an experiment that allows the experimenter to find out the effect levels of each factor on levels of all the other factors.

factorial of a positive integer n [MATH] The product of all positive integers less than or equal to n; written $n!$; by convention $0! = 1$.

factoring [MATH] Finding the factors of an integer or polynomial.

factoring of the secular equation [MATH] Factoring the polynomial that results from expanding the secular determinant of a matrix, in order to find the roots of this polynomial, which are the eigenvalues of the matrix.

factor model [STAT] Any one of the probability models which goes into the construction of a product model.

factor of an integer [MATH] Any integer which when multiplied by another integer gives the original integer.

factor of a polynomial [MATH] Any polynomial which when multiplied by another polynomial gives the original polynomial.

factor of proportionality [MATH] Two quantities A and B are related by a factor of proportionality μ if either $A = \mu B$ or $B = \mu A$.

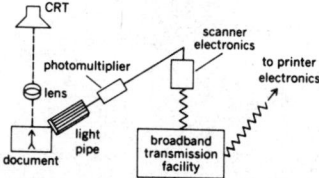

FACSIMILE

The scanning and transmission components of a facsimile system that uses a cathode-ray-tube (CRT) flying-spot scanner.

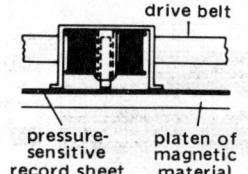

FACSIMILE RECORDER

Magnetically driven stylus for recording on a pressure-sensitive record sheet.

factor of safety [MECH] **1.** The ratio between the breaking load on a member, appliance, or hoisting rope and the safe permissible load on it. Also known as safety factor. **2.** *See* factor of stress intensity.

factor of stress concentration [MECH] Any irregularity producing localized stress in a structural member subject to load. Also known as fatigue-strength reduction factor.

factor of stress intensity [MECH] The ratio of the maximum stress to which a structural member can be subjected, to the maximum stress to which it is likely to be subjected. Also known as factor of safety.

factor of subdivision [NAV ARCH] An arbitrary factor used in computing allowable floodable length of ships after damage, set up by regulations and international convention.

factor space *See* quotient space.

factor theorem of algebra [MATH] A polynomial $f(x)$ has $(x - a)$ as a factor if and only if $f(a) = 0$.

factory [IND ENG] A building or group of buildings where goods are manufactured.

factory-data collection [ADP] The continuous input of data achieved in a working area by having the worker insert a precoded card into a device connected to a computer.

factory lumber [MATER] Softwood lumber graded and used in the factory for the manufacture of such items as doors, sashes, moldings, and so on.

factory ship [NAV ARCH] A ship equipped both to catch and to process fish into products such as frozen filet, frozen whole fish, and fish meal.

facula [ASTRON] Any of the large patches of bright material forming a veined network in the vicinity of sunspots; faculae appear to be more permanent than sunspots and are probably due to elevated clouds of luminous gas.

facultative aerobe [MICROBIO] An anaerobic microorganism which can grow under aerobic conditions.

facultative anaerobe [MICROBIO] A microorganism that grows equally well under aerobic and anaerobic conditions.

facultative parasite [ECOL] An organism that can exist independently but may be parasitic on certain occasions, such as the flea.

facultative photoheterotroph [MICROBIO] Any bacterium that usually grows anaerobically in light but can also grow aerobically in the dark.

FAD *See* flavin adenine dinucleotide.

fade chart [ELECTROMAG] Graph on which the null areas of an air-search radar antenna are plotted as an aid to estimating target altitude.

fade-in [COMMUN] A gradual increase in signal strength, as at the start of a radio or television program or when changing to a new scene, to make sound volume and picture brightness increase gradually. [GRAPHICS] In motion pictures, the gradual emergence of a screen image from black.

fade-out [COMMUN] A gradual and temporary loss of a received radio or television signal due to magnetic storms, atmospheric disturbances, or other conditions along the transmission path.

fader [ELECTR] A multiple-unit level control used for gradual changeover from one microphone, audio channel, or television camera to another.

fading [COMMUN] Variations in the field strength of a radio signal, usually gradual, that are caused by changes in the transmission medium.

fading margin [COMMUN] **1.** Number of decibels of attenuation which may be added to a specified radio-frequency propagation path before the signal-to-noise ratio of a specified channel falls below a specified minimum. **2.** Allowance made in radio system planning to accommodate estimated fading.

Fagaceae [BOT] A family of dicotyledonous plants in the order Fagales characterized by stipulate leaves, seeds without endosperm, female flowers generally not in catkins, and mostly three styles and locules.

Fagales [BOT] An order of dicotyledonous woody plants in the subclass Hamamelidae having simple leaves and much reduced, mostly unisexual flowers.

Fagergren cell [MIN ENG] A froth-flotation cell in which a squirrel-cage rotor is driven concentrically in a vertical stator, so that air is drawn down the rotor shaft and dispersed into the pulp.

Fagersta cut [MIN ENG] A cut drilled with handheld equipment in two steps, first as a pilot hole and then as an enlargement of this hole.

fagopyrism [VET MED] Photosensitization of the skin and mucous membranes, accompanied by convulsions; produced especially in sheep and swine by feeding on the buckwheat plant, *Fagopyrum sagittatum*, or clovers and grasses containing flavin or carotene and xanthophyll.

fahlband [GEOL] A stratum containing metal sulfides; occurs in crystalline rock.

fahlore *See* tetrahedrite.

Fahnestock clip [ELEC] A spring-type terminal to which a temporary connection can readily be made.

Fahrenheit scale [THERMO] A temperature scale; the temperature in degrees Fahrenheit (°F) is the sum of 32 plus 9/5 the temperature in degrees Celsius; water at 1 atmosphere pressure freezes very near 32°F and boils very near 212°F.

failed hole [ENG] A drill hole loaded with dynamite which did not explode. Also known as missed hole.

fail-safe system [ENG] A system designed so that failure of power, control circuits, structural members, or other components will not endanger people operating the system or other people in the vicinity.

failsafe tape *See* incremental dump tape.

failure [MECH] Condition caused by collapse, break, or bending, so that a structure or structural element can no longer fulfill its purpose.

fair [METEOROL] Generally descriptive of pleasant weather conditions, with regard for location and time of year; it is subject to popular misinterpretation, for it is a purely subjective description; when this term is used in forecasts of the U.S. Weather Bureau, it is meant to imply no precipitation, less than 0.4 sky cover of low clouds, and no other extreme conditions of cloudiness or windiness.

fairchildite [MINERAL] $K_2Ca(CO_3)_2$ A mineral composed of potassium calcium carbonate; occurs in partly burned trees.

fair curves [NAV ARCH] Curves which are smooth without sharp changes in direction over any portion of their length.

faired cable [DES ENG] A trawling cable covered by streamlined surfaces to reduce hydrodynamic drag.

fairfieldite [MINERAL] $Ca_2Mn(PO_4)_2 \cdot 2H_2O$ A white or pale-yellow mineral composed of hydrous calcium manganese phosphate and occurring in foliated or fibrous form.

fair game [MATH] A game in which all of the participants have equal expectation of gain.

fairing [AERO ENG] A structure or surface on an aircraft or rocket that functions to reduce drag, such as the streamlined nose of a satellite-launching rocket.

fairlead [AERO ENG] A guide through which an airplane antenna or control cable passes. [MECH ENG] A group of pulleys or rollers used in conjunction with a winch or similar apparatus to permit the cable to be reeled from any direction. [NAV ARCH] A block, ring, or other fitting through which passes a line or the running rigging on a ship to prevent chafing.

fair line [NAV ARCH] A line formed by the intersection of a plane with a ship's surface, which surface is smooth and is such as to minimize resistance to the ship's motion.

fair tide [NAV] A tidal current setting in such a direction as to increase the speed of a vessel.

fairwater [NAV ARCH] A device for making fair the lines of an underwater fitting.

fairway [NAV] Open water of sufficient depth for navigation; a marine thoroughfare.

fair-weather cumulus *See* cumulus humilis cloud.

fair wind *See* favorable wind.

fairy ring spot [PL PATH] A fungus disease of carnations caused by *Heterosporium echinulatum*, producing bleached spots with concentric dark zones on the leaves.

fairy stone *See* staurolite.

fake [NAV ARCH] To lay a rope or chain down in long bights side by side or in coils in regular order.

fake set *See* false set.

falcate [ASTRON] Crescent-shaped; applied usually to the appearance of the moon, Venus, and Mercury during their crescent phases. [BIOL] Shaped like a sickle.

falciform [BIOL] Sickle-shaped.

falciform ligament [ANAT] The ventral mesentery of the liver; its peripheral attachment extends from the diaphragm to the umbilicus and contains the round ligament of the liver.

falciparum malaria [MED] A severe form of malaria caused by *Plasmodium falciparum* and characterized by sudden attacks of chills, fever, and sweating at irregular intervals; the infecting organism usually localizes in a specific organ, causing capillary blockage. Also known as alged malaria; estivoautumnal malaria; malignant malaria; pernicious malaria.

falcon [VERT ZOO] Any of the highly specialized diurnal birds of prey composing the family Falconidae; these birds have been captured and trained for hunting.

Falcon [ORD] A U.S. Air Force air-to-air guided missile having either radar or infrared homing guidance, a speed of about Mach 2, and a range of about 5 miles (8 kilometers); can be carried in quantity by interceptor aircraft.

Falconbridge process [MET] Recovery of nickel from a nickel-copper matte; the matte is first crushed and roasted to remove sulfur, and the copper is acid-leached, filtered off, and electrolyzed; the residual solids are melted, cast as anodes, and refined electrolytically to produce nickel.

Falconidae [VERT ZOO] The falcons, a family of long-winged predacious birds in the order Falconiformes.

Falconiformes [VERT ZOO] An order of birds containing the diurnal birds of prey, including falcons, hawks, vultures, and eagles.

falculate [ZOO] Curved and with a sharp point.

Fales-Stuart windmill [MECH ENG] A windmill developed for farm use from the two-blade airfoil propeller. Also known as Stuart windmill.

Falk flexible coupling [MECH ENG] A spring coupling in which a continuous steel spring is threaded back and forth through axial slots in the periphery of two hubs on the shaft ends.

Falkland Current [OCEANOGR] An ocean current flowing northward along the Argentine coast.

fall [ASTRON] 1. Of a spacecraft or spatial body, to drop toward a spatial body under the influence of its gravity. 2. *See* autumn. [MECH ENG] The rope or chain of a hoisting tackle. [MIN ENG] A mass of rock, coal, or ore which has fallen from the roof or side in any subterranean working or gallery.

fallaway section [AERO ENG] A section of a rocket vehicle that is cast off from the vehicle during flight, especially such a section that falls back to the earth.

fallback [ADP] The system, electronic or manual, which is substituted for the computer system in case of breakdown. [NUCLEO] That part of the material carried into the air by an atomic explosion which ultimately drops back to the earth or water at the site of the explosion.

fall block [MECH ENG] A pulley block that rises and falls with the load on a lifting tackle.

faller [MECH ENG] A machine part whose operation depends on a falling action.

falling-ball viscometer *See* falling-sphere viscometer.

falling body [MECH] A body whose motion is accelerated toward the center of the earth by the force of gravity, other forces acting on it being negligible by comparison.

falling disease [VET MED] A terminal manifestation of copper deficiency in which the animal collapses and dies because of heart failure.

falling-drop method [PHYS] Technique for measurement of liquid densities in which the time of fall of a drop of the sample liquid through a reference liquid is measured.

falling film [FL MECH] A theoretical liquid film that moves downward in even flow on a vertical surface in laminar flow; the concept is used for heat- and mass-transfer calculations.

falling-film cooler [ENG] Liquid cooling system in which the cooling liquid flows down vertical tube exterior surfaces in a thin film, and hot process fluid flows upward through the tubes.

falling-film evaporator [ENG] Liquid evaporator system with heated vertical tubes; liquid to be evaporated flows down the inside tube surfaces as a film, evaporating as it flows.

falling-film molecular still *See* falling-film still.

falling-film still [CHEM ENG] Special molecular distillation apparatus designed for high evaporative and separation efficiency. Also known as falling-film molecular still.

falling-sphere viscometer [ENG] A viscometer which measures the speed of a spherical body falling with constant velocity in the fluid whose viscosity is to be determined. Also known as falling-ball viscometer.

falling tide *See* ebb tide.

fall line [GEOL] 1. The zone or boundary between resistant rocks of older land and weaker strata of plains. 2. The line indicated by the edge over which a waterway suddenly descends, as in waterfalls.

falloff curve [PETRO ENG] Graphical representatation of bottom-hole pressure falloff for a shut-in well as the reservoir drainage area expands.

fall of ground [MIN ENG] The fall of rock from the roof into a mine opening.

Fallopian tube [ANAT] Either of the paired oviducts that extend from the ovary to the uterus for conduction of the ovum in mammals.

fallout [NUCLEO] The material that descends to the earth or water well beyond the site of a surface or subsurface nuclear explosion. Also known as atomic fallout; radioactive fallout.

fallout area [NUCLEO] The area on which radioactive materials have settled out, or the area on which it is predicted from weather conditions that radioactive materials may settle out.

fallout shelter [CIV ENG] A structure that affords some protection against fallout radiation and other effects of nuclear explosion; maximum protection is in reinforced concrete shelters below the ground. Also known as radiation shelter.

fallout winds [METEOROL] Tropospheric winds that carry the radioactive fallout materials, observed by standard winds-aloft observation techniques.

fallow [AGR] Pertaining to land normally used for crop production but left unsown for one or more growing seasons.

fall streaks *See* virga.

Fallstreifen *See* virga.

fall time [ELEC] Measure of time required for a circuit to change its output from a high level to a low level.

fall wind [METEOROL] A strong, cold, downslope wind, differing from a foehn in that the initially cold air remains relatively cold despite adiabatic warming upon descent, and from the gravity wind in that it is a larger-scale phenomenon prerequiring an accumulation of cold air at high elevations.

false acceptance [STAT] Accepting on the basis of a statistical test a hypothesis which is wrong.

false alarm [ELECTR] In radar, an indication of a detected target even though one does not exist, due to noise or interference levels exceeding the set threshold of detection.

false attic [BUILD] A section under a roof normally occupied by an attic, but which has no windows and does not enclose rooms.

false bedding [GEOL] An inclined bedding produced by currents.

false blossom [PL PATH] 1. A fungus disease of the cranberry caused by *Exobasidium oxycocci*; erect flower buds are formed which produce malformed flowers that set no fruit. Also known as rosebloom. 2. A similar virus disease of the cranberry transmitted by the leafhopper, *Scleroracus vaccinii*. Also known as Wisconsin false blossom.

false bottom [CIV ENG] A temporary bottom installed in a caisson to add to its buoyancy. [MET] An insert put in either member of a die set to increase the strength and improve the life of the die. [MIN ENG] A flat, hexagonal or cylindrical iron die upon which ore is crushed in a stamp mill.

false cleavage [GEOL] 1. A weak cleavage at an angle to the slaty cleavage. 2. Spaced surfaces about a millimeter apart along which a rock splits.

false drop *See* false retrieval.

false form *See* pseudomorph.

false galena *See* sphalerite.

false header [CIV ENG] A half brick used to complete a visible bond; it is not a header.

false horizon [NAV] A line resembling the visible horizon but above or below it.

false lapis *See* lazulite.

false ligament [ANAT] Any peritoneal fold which is not a true supporting ligament.

false light [NAV] A light which is unavoidably exhibited by an aid to navigation and which is not intended to be a part of

FALES-STUART WINDMILL

Fales-Stuart windmill developed for farm use from airplane propeller techniques.

FALL LINE

more resistant rocks of old land

zone of fall line

less resistant strata of coastal plain

Diagram showing zone of fall line.

the proper characteristic of the light, such as reflections from storm panes.

false ogive [ORD] A rounded or pointed hollow cup added to the nose of a projectile to improve streamlining. Also known as ballistic cap.

false pyroelectricity *See* tertiary pyroelectricity.

false rejection [STAT] Rejecting on the basis of a statistical test a hypothesis which is correct.

false retrieval [ADP] An item retrieved in an automatic library search which is unrelated or vaguely related to the subject of the search. Also known as false drop.

false rib [ANAT] A rib that is not attached to the sternum directly; any of the five lower ribs on each side in humans.

false ring [BOT] A layer of wood that is less than a full season's growth and often does not form a complete ring.

false set [MATER] Rapid hardening of freshly mixed cement paste, mortar, or concrete with minimum evolution of heat; plasticity can be restored by mixing without addition of water. [MIN ENG] A light, temporary lagging set of timber supporting the side and roof lagging until the drive is advanced sufficiently to allow the heavy permanent set to be put, at which time the false set is taken out and used again in advance of the next permanent set. Also known as fake set.

false smut [PL PATH] **1.** A fungus disease of palm caused by *Graphiola phoenicis* and characterized by small cylindrical protruding pustules, often surrounded by yellowish leaf tissue. **2.** *See* green smut.

false sorts [ADP] Entries irrelevant to the subject sought which are retrieved in a search.

false stull [MIN ENG] A stull so placed as to offer support or reinforcement for a stull, prop, or other timber.

false target [ELECTR] A nonexistent target which shows up on a radar scope as the result of time delay.

false target generator [ELECTR] An electronic countermeasure device that generates a delayed return signal on an enemy radar frequency to give erroneous position information.

false twist [TEXT] A method by which certain synthetic yarns are given stretch characteristics; yarns are wound under heat to eliminate the twist, but because of the yarns' tendency to retain the twist, fabrics made from them have elasticity. Also known as memory twist.

false warm sector [METEOROL] The sector, in a horizontal plane, between the occluded front and a secondary cold front of an occluded cyclone.

false white rainbow *See* fogbow.

falsework [CIV ENG] A temporary support used until the main structure is strong enough to support itself.

faltung [MATH] A family of functions where the convolution of any two members of the family is also a member of the family. Also known as convolution family.

falx [ANAT] A sickle-shaped structure.

famatinite [MINERAL] Cu_3SbS_4 A reddish-gray mineral composed of copper antimony sulfide.

familial [BIOL] Of, pertaining to, or occurring among the members of a family.

familial aldosterone deficiency [MED] A hereditary metabolic disorder, probably due to a defect in the enzyme involved in dehydrogenation of 18-hydroxycorticosterone to aldosterone, characterized by growth retardation and hypoaldosteronism.

familial dysautonomia [MED] A hereditary disease transmitted as an autosomal recessive and characterized from infancy by evidence of autonomic nervous system dysfunction, including feeding difficulties, absence of overflow tears, indifference to pain, absent corneal reflexes and deep tendon reflexes, and absence of fungiform papillae on the tongue; most common in Jewish children.

familial Mediterranean fever [MED] A hereditary disease of unknown cause characterized by recurrent fever, abdominal and chest pain, arthralgia, and rash, sometimes terminating in renal failure. Abbreviated FMF. Also known as familial recurring polyserositis; periodic disease; periodic peritonitis.

familial osteochondrodystrophy *See* Morquio's syndrome.

familial polyposis [MED] A hereditary condition transmitted as an autosomal dominant and characterized by the appearance of polyps in the small intestine and colon; malignant degeneration is common.

familial recurring polyserositis *See* familial Mediterranean fever.

familial splenic anemia *See* Gaucher's disease.

family [CHEM] A group of elements whose chemical properties, such as valence, solubility of salts, and behavior toward reagents, are similar. [SYST] A taxonomic category based on the grouping of related genera.

fan [AGR] A mechanical device used for winnowing grain. [BIOL] Any structure, such as a leaf or the tail of a bird, resembling an open fan. [ELECTROMAG] Volume of space periodically energized by a radar beam (or beams) repeatedly traversing an established pattern. [GEOL] A gently sloping, fan-shaped feature usually found near the lower termination of a canyon. Also known as submarine bulge; submarine delta. [MECH ENG] **1.** A device, usually consisting of a rotating paddle wheel or an airscrew, with or without a casing, for producing currents in order to circulate, exhaust, or deliver large volumes of air or gas. **2.** A vane to keep the sails of a windmill facing the direction of the wind.

fan antenna [ELECTROMAG] An array of folded dipoles of different length forming a wide-band ultra-high-frequency or very-high-frequency antenna.

fan beam [ELECTROMAG] **1.** A radio beam having an elliptically shaped cross section in which the ratio of the major to the minor axis usually exceeds 3 to 1; the beam is broad in the vertical plane and narrow in the horizontal plane. **2.** A radar beam having the shape of a fan.

fan brake [MECH ENG] A fan used to provide a load for a driving mechanism.

Fanconi's anemia [MED] An infantile anemia that resembles pernicious anemia; related to excessive chromosomal breakage and associated with the risk of developing leukemia.

Fanconi's syndrome *See* amino diabetes.

fan cut [ENG] A cut in which holes of equal or increasing length are drilled in a pattern on a horizontal plane or in a selected stratum to break out a considerable part of the plane or stratum before the rest of the round is fired.

fan drift [MIN ENG] The short tunnel connecting the upcast shaft with the exhaust fan.

fan-drift doors [MIN ENG] Isolation doors for each drift leading to each fan, when there are two fans at a mine.

fan drilling [ENG] **1.** Drilling boreholes in different vertical and horizontal directions from a single-drill setup. **2.** A radial pattern of drill holes from a setup.

fan efficiency [MECH ENG] The ratio obtained by dividing a fan's useful power output by the power input (the power supplied to the fan shaft); it is expressed as a percentage.

fan fold [GEOL] A fold of strata in which both limbs are overturned, forming a syncline or anticline.

fang [ANAT] The root of a tooth. [VERT ZOO] A long, pointed tooth, especially one of a venomous serpent.

fang bolt [DES ENG] A bolt having a triangular nut with sharp projections at its corners; used to attach metal pieces to wood.

fanglomerate [GEOL] Coarse material in an alluvial fan, with the rock fragments being only slightly worn.

fan-in [ELECTR] The number of inputs that can be connected to a logic circuit.

fanjet [AERO ENG] A turbojet engine whose performance has been improved by the addition of a fan which operates in an annular duct surrounding the engine.

fanlight [ARCH] A segmented semicircular window over a door or window.

fan marker *See* fan-marker beacon.

fan-marker beacon [NAV] A very-high frequency radio facility having a vertically directed fan beam interesecting an airway to provide a fix. Also known as fan marker; radio fan-marker beacon.

fanned-beam antenna [ELECTROMAG] Unidirectional antenna so designed that transverse cross sections of the major lobe are approximately elliptical.

fanning beam [ELECTROMAG] Narrow antenna beam which is repeatedly scanned over a limited arc.

Fanning friction factor [FL MECH] A dimensionless number used in studying fluid friction in pipes, equal to the pipe diameter times the drop in pressure in the fluid due to friction as it passes through the pipe, divided by the product of the

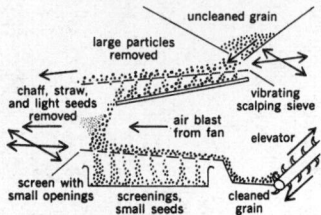

FANNING MILL

uncleaned grain

large particles
removed

chaff, straw,
and light seeds
removed

air blast
from fan

vibrating
scalping sieve

elevator

screen with
small openings

screenings,
small seeds

cleaned
grain

Fanning mill operation.

FARNESOL

Structural formula of farnesol.

pipe length and the kinetic energy of the fluid per unit volume. Symbolized f.

fanning mill [FOOD ENG] A device consisting of two vibrating screens and utilizing an air blast to clean and separate grain.

Fanning's equation [FL MECH] The equation expressing that frictional pressure drop of fluid flowing in a pipe is a function of the Reynolds number, rate of flow, acceleration due to gravity, and length and diameter of the pipe.

fanning strip [ELEC] Insulated board, often of wood, which serves to spread out the wires of a cable for distribution to a terminal board.

Fanno flow [FL MECH] An ideal flow used to study the flow of fluids in long pipes; the flow obeys the same simplifying assumptions as Rayleigh flow except that the assumption there is no friction is replaced by the requirement the flow be adiabatic.

fan-out [ELECTR] The number of parallel loads that can be driven from one output mode of a logic circuit.

fan rating [MECH ENG] The head, quantity, power, and efficiency expected from a fan operating at peak efficiency.

fan ring [DES ENG] Circular metallic collar encircling (but spaced away from) the tips of the fan blade in process equipment, such as air-cooled heat exchangers; ring design is critical to the efficiency of fan performance.

fan shaft [DES ENG] The spindle on which a fan impeller is mounted. [MIN ENG] The ventilating shaft to which a mine fan is connected.

fan static pressure [MECH ENG] The total pressure rise diminished by the velocity pressure in the fan outlet.

fantail [NAV ARCH] The area of the upper deck of a ship which is nearest the stern.

fan test [MECH ENG] Observations of the quantity, total pressure, and power of air circulated by a fan running at a known constant speed.

Fanti unit [BIOL] A unit for the standardization of thrombin.

fan total head [MECH ENG] The sum of the fan static head and the velocity head at the fan discharge corresponding to a given quantity of airflow.

fan total pressure [MECH ENG] The algebraic difference between the mean total pressure at the fan outlet and the mean total pressure at the fan inlet.

fan truss [CIV ENG] A truss with struts arranged as radiating lines.

fan vaulting [ARCH] Vaulting in which the ribs diverge like the rays of a fan.

fan velocity pressure [MECH ENG] The velocity pressure corresponding to the average velocity at the fan outlet.

farad [ELEC] The unit of capacitance in the meter-kilogram-second system, equal to the capacitance of a capacitor which has a potential difference of 1 volt between its plates when the charge on one of its plates is 1 coulomb, there being an equal and opposite charge on the other plate. Symbolized F.

faraday [PHYS] The electric charge required to liberate 1 gram-equivalent of a substance by electrolysis; experimentally equal to 96,487.0 ± 1.6 coulombs. Also known as Faraday constant.

Faraday birefringence [OPTICS] Difference in the indices of refraction of left and right circularly polarized light passing through matter parallel to an applied magnetic field; it is responsible for the Faraday effect.

Faraday cage See Faraday shield.

Faraday constant See faraday.

Faraday dark space [ELECTR] The relatively nonluminous region that separates the negative glow from the positive column in a cold-cathode glow-discharge tube.

Faraday disk machine [ELECTROMAG] A device for demonstrating electromagnetic induction, consisting of a copper disk in which a radial electromotive force is induced when the disk is rotated between the poles of a magnet. Also known as Faraday generator.

Faraday effect [OPTICS] Rotation of polarization of a beam of linearly polarized light when it passes through matter in the direction of an applied magnetic field; it is the result of Faraday birefringence. Also known as Faraday rotation; Kundt effect; magnetic rotation.

Faraday generator See Faraday disk machine.

Faraday ice bucket experiment [ELEC] Experiment in which

one lowers a charged metal body into a pail and observes the effect on an electroscope attached to the pail, with and without contact between body and pail; the experiment shows that charge resides on a conductor's outside surface.

Faraday rotation See Faraday effect.

Faraday rotation isolator See ferrite isolator.

Faraday screen See Faraday shield.

Faraday shield [ELEC] Electrostatic shield composed of wire mesh or a series of parallel wires, usually connected at one end to another conductor which is grounded. Also known as Faraday cage; Faraday screen.

Faraday's law of electromagnetic induction [ELECTROMAG] The law that the electromotive force induced in a circuit by a changing magnetic field is equal to the negative of the rate of change of the magnetic flux linking the circuit. Also known as law of electromagnetic induction.

Faraday's laws of electrolysis [PHYS CHEM] 1. The amount of any substance dissolved or deposited in electrolysis is proportional to the total electric charge passed. 2. The amounts of different substances dissolved or desposited by the passage of the same electric charge are proportional to their equivalent weights.

faradic current [ELEC] An intermittent and nonsymmetrical alternating current like that obtained from the secondary winding of an induction coil; used in electrobiology.

faradization [PHYSIO] Use of a faradic current to stimulate muscles and nerves.

farcy See glanders.

far-end crosstalk [COMMUN] Crosstalk that travels along the disturbed circuit in the same direction as desired signals in that circuit.

farewell buoy See sea buoy.

far field See Fraunhofer region.

farinaceous [BIOL] Having a mealy surface covering. [FOOD ENG] 1. Containing starch or flour. 2. Having the texture of meal.

Farinales [BOT] An order that includes several groups regarded as orders of the Commelinidae in other systems of classification.

far-infrared maser [ENG] A gas maser that generates a beam having a wavelength well above 100 micrometers, and ranging up to the present lower wavelength limit of about 500 micrometers for microwave oscillators.

far-infrared radiation [ELECTROMAG] Infrared radiation the wavelengths of which are the longest of those in the infrared region, about 50–1000 micrometers; requires diffraction gratings for spectroscopic analysis.

Farinosae [BOT] The equivalent name for Farinales.

farinose [AGR] Yielding farina, a fine meal of vegetable matter. [BIOL] Covered with a white powdery substance.

farm [AGR] A tract of land used for cultivating crops or raising animals.

farmer's lung [MED] An acute pulmonary disorder caused by the inhalation of spores from moldy hay or straw.

farmer's year [CLIMATOL] In Great Britain, the 12-month period starting with the Sunday nearest March 1.

farming [AGR] The skills and practices of agriculture.

farnesol [BIOCHEM] $C_{15}H_{25}OH$ A colorless liquid extracted from oils of plants such as citronella, neroli, cyclamen, and tuberose; it has a delicate floral odor, and is an intermediate step in the biological synthesis of cholesterol from mevalonic acid in vertebrates; used in perfumery.

Farnsworth image dissector tube See image dissector tube.

far point [OPTICS] The farthest point from an eye at which an object is distinctly seen; for a normal eye it is theoretically at infinity.

far region See Fraunhofer region.

farringtonite [MINERAL] $Mg_3(PO_4)_2$ Colorless, wax-white, or yellow phosphate mineral known only in meteorites.

farsightedness See hypermetropia.

far-ultraviolet radiation [ELECTROMAG] Ultraviolet radiation in the wavelength range of 200–300 nanometers; germicidal effects are greatest in this range.

far vane [NAV] In marine operations, the instrument sighting vane on the opposite side of the instrument from the observer's eye.

far zone See Fraunhofer region.

fascia [BUILD] A wide board fixed vertically on edge to the rafter ends or wall which carries the gutter around the eaves of a roof. [HISTOL] Layers of areolar connective tissue under the skin and between muscles, nerves, and blood vessels.

fasciate [BOT] Having bands or stripes.

fasciation [PL PATH] Malformation of plant parts resulting from disorganized tissue growth.

fascicle [BOT] A small bundle, as of fibers or leaves.

fasciculate [BOT] Arranged in tufts or fascicles.

fasciculation potential [PHYSIO] An action potential which is quantitatively comparable to that of a motor unit and which represents spontaneous contraction of a bundle of muscle fibers.

fasciculus [ANAT] A bundle or tract of nerve, muscle, or tendon fibers isolated by a sheath of connective tissues and having common origins, innervation, and functions.

fascine [CIV ENG] A cylindrical bundle of brushwood 1-3 feet (30-90 centimeters) in diameter and 10-20 feet (3-6 meters) long, used as a facing for seawalls on riverbanks, as a foundation mat, as a dam in an estuary, or to protect bridge, dike, and pier foundations from erosion.

Fasciola hepatica [INV ZOO] A digenetic trematode which parasitizes sheep, cattle, and occasionally humans.

fasciole [INV ZOO] A band of cilia on the test of certain sea urchins.

fascioliasis [MED] The infection of humans with *Fasciola hepatica*.

fasciolopsiasis [MED] The presence of the parasite *Fasciolopsis buski* in a person's small intestine.

Fasciolopsis buski [INV ZOO] A large, fleshy trematode, native to eastern Asia and the southwestern Pacific, which parasitizes humans.

fascioscapulohumeral dystrophy [MED] A progressive hereditary form of muscular dystrophy involving atrophy of the muscles of the face, pectoral girdle, and upper arm.

fast [GRAPHICS] A relative term given to the speed of emulsion.

fast-access storage [ADP] The section of a computer storage from which data can be obtained most rapidly.

fast automatic gain control [ELECTR] Radar automatic gain control method characterized by a response time that is long with respect to a pulse width, and short with respect to the time on target.

fast axis [OPTICS] The direction of the electrical displacement vector of light propagating in an anisotropic crystal with the greatest possible phase velocity corresponding to a specified direction of propagation.

fast break [MET] Interruption of the current in the magnetizing coil during nondestructive testing of magnetic particles; induces eddy currents and strong magnetization.

fast breeder reactor [NUCLEO] A type of fast reactor using highly enriched fuel in the core, fertile material in the blanket, and a liquid-metal coolant, such as sodium; high-speed neutrons fission the fuel in the compact core, and the excess neutrons convert fertile material to fissionable isotopes; the breeding ratio is 1.0 or larger. Abbreviated FBR.

fast-burst reactor [NUCLEO] A nuclear reactor that supplies microsecond pulses of fast neutrons for use in biomedical research.

fast chemical reaction [PHYS CHEM] A reaction with a half-life of milliseconds or less; such reactions occur so rapidly that special experimental techniques are required to observe their rate.

fast coupling [MECH ENG] A flexible geared coupling that uses two interior hubs on the shafts with circumferential gear teeth surrounded by a casing having internal gear teeth to mesh and connect the two hubs.

fast-delay detonation [ENG] The firing of blasts by means of a blasting timer or millisecond delay caps.

fast effect [NUCLEO] The reactivity change (increase in neutrons) due to fissions caused by fast neutrons in a thermal reactor.

fastener [DES ENG] **1.** A device for joining two separate parts of an article or structure. **2.** A device for holding closed a door, gate, or similar structure.

fastening [DES ENG] A spike, bolt, nut, or other device to connect rails to ties.

fastest mile [METEOROL] Over a specified period (usually the 24-hour observational day), the fastest speed, in miles per hour, of any mile of wind, with its accompanying direction.

fast Fourier transform [MATH] A Fourier transform employing the Cooley-Tukey algorithm to reduce the number of operations.

fast ice [HYD] Any type of sea, river, or lake ice attached to the shore (ice foot, ice shelf), beached (shore ice), stranded in shallow water, or frozen to the bottom of shallow waters (anchor ice). Also known as landfast ice. [OCEANOGR] Sea ice generally remaining in the position where originally formed and sometimes attaining a considerable thickness; it is attached to the shore or over shoals where it may be held in position by islands, grounded icebergs, or polar ice. Also known as coastal ice; coast ice.

fastigiate [BOT] **1.** Having erect branches that are close to the stem. **2.** Becoming narrower at the top. [ZOO] Arranged in a conical bundle.

fast ion *See* small ion.

fast-joint [ENG] Pertaining to a joint with a permanently secured pin.

fast neutron [NUCLEO] A neutron having energy much greater than some arbitrary lower limit (that may be only a few thousand electron volts).

fast-neutron spectrometry [NUC PHYS] Neutron spectrometry in which nuclear reactions are produced by or yield fast neutrons; such reactions are more varied than in the slow-neutron case.

fast pin [ENG] A pin that fastens immovably, particularly the pin in a fast joint.

fast powder [MATER] Any explosive having a high-speed detonation.

fast reactor [NUCLEO] A nuclear reactor in which most of the fissions are produced by fast neutrons, with little or no moderator to slow down the neutrons.

fast-spiral drill *See* high-helix drill.

fast time constant [ELEC] An electric circuit which combines resistance and capacitance to give a short time constant for capacitor discharge through the resistor. [ELECTR] Circuit with short time constant used to emphasize signals of short duration to produce discrimination against low-frequency components of clutter in radar.

fat [ANAT] Pertaining to an obese person. [BIOCHEM] Any of the glyceryl esters of fatty acids which form a class of neutral organic compounds. [PHYSIO] The chief component of fat cells and other animal and plant tissues.

fatal accident [MIN ENG] A coal mine accident in which less than five persons are killed and property damage is slight; excludes ignitions and mine fires.

Fata Morgana [OPTICS] A complex mirage characterized by multiple distortions of images, generally in the vertical, so that such objects as cliffs or cottages are distorted and magnified into fantastic castles.

fat body [INV ZOO] A nutritional reservoir of fatty tissue surrounding the viscera or forming a layer beneath the integument in the immature larval stages of many insects. [VERT ZOO] A mass of adipose tissue attached to each genital gland in amphibians.

fat cell [HISTOL] The principal component of adipose connective tissue; two types are yellow fat cells and brown fat cells.

fat dye [MATER] A type of oil-soluble dye used in the coloring of candles and other wax products.

fate map [EMBRYO] A graphic scheme indicating the definite spatial arrangement of undifferentiated embryonic cells in accordance with their destination to become specific tissues.

fat embolus [MED] An embolus composed principally of fatty substances.

fathom [OCEANOGR] The common unit of depth in the ocean, equal to 6 feet (1.8288 meters).

fathom curve *See* isobath.

Fathometer [ENG] Trade name for a sonic depth finder.

fatigue [ELECTR] The decrease of efficiency of a luminescent or light-sensitive material as a result of excitation. [MECH] Failure of a material by cracking resulting from repeated or cyclic stress. [PHYSIO] Exhaustion of strength or reduced

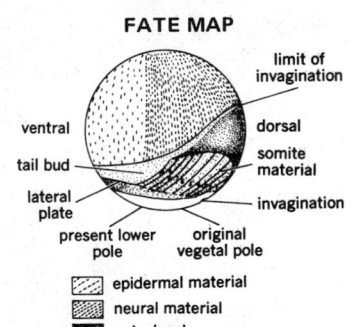

FATE MAP

A fate map for the beginning gastrula of a urodele shown in lateral view. *(After W. Vogt, 1926)*

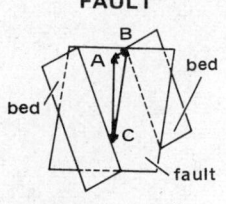

FAULT

BC = dip separation
AB = heave
AC = throw

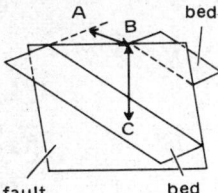

AB = normal horizontal
separation
BC = vertical separation

Disruption of a rock bed by a fault.

capacity to respond to stimulation following a period of activity.

fatigue allowance [IND ENG] An adjustment to normal time to compensate for production time lost due to exhaustion of the worker.

fatigue factor [IND ENG] The element of physical and mental exhaustion in a time-motion study; the multiplier used to add the fatigue allowance to the normal time.

fatigue life [MECH] The number of applied repeated stress cycles a material can endure before failure.

fatigue limit [MECH] The maximum stress that a material can endure for an infinite number of stress cycles without breaking. Also known as endurance limit.

fatigue notch factor [MET] A notch, scratch, or other impairment on the surface of a metal resulting in premature failure of the metal.

fatigue notch sensitivity [MET] A measure of the reduction of fatigue strength of a metal resulting from a notch.

fatigue ratio [MECH] The ratio of the fatigue limit or fatigue strength to the static tensile strength. Also known as endurance ratio.

fatigue strength [MECH] The maximum stress a material can endure for a given number of stress cycles without breaking. Also known as endurance strength.

fatigue-strength reduction factor *See* factor of stress concentration.

fatigue test [ENG] Test to determine the range of alternating stress which a material can withstand without breaking.

fat liquoring agents [MATER] Oil-in-water emulsions used to replace oils in tanned (deoiled) leather hides.

fat-metabolizing hormone *See* ketogenic hormone.

fat mortar [MATER] Mortar that adheres to the trowel.

fat necrosis [MED] Pathologic death of adipose tissue often accompanied by soap production from the hydrolyzed fat; associated with pancreatitis.

fat oil [MATER] Enriched absorber oil that is drawn off from the absorber column after being saturated by hydrocarbon values stripped from a wet natural-gas stream.

Fatou-Lebesgue lemma [MATH] Given a sequence f_n of positive measurable functions on a measure space (X, μ), then

$$\int_X (\lim_n \inf f_n)\, d\mu \leq \lim_n \inf \int_X f_n\, d\mu.$$

fatty acid [ORG CHEM] An organic monobasic acid of the general formula $C_nH_{2n+1}COOH$ derived from the saturated series of aliphatic hydrocarbons; examples are palmitic acid, stearic acid, and oleic acid; used as a lubricant in cosmetics and nutrition, and for soaps and detergents.

fatty-acid pitch *See* packing house pitch.

fatty alcohol [ORG CHEM] A high-molecular-weight, straight-chain primary alcohol derived from natural fats and oils; includes lauryl, stearyl, oleyl, and linoleyl alcohols; used in pharmaceuticals, cosmetics, detergents, plastics, and lube oils and in textile manufacture.

fatty amine [ORG CHEM] RCH_2NH_2 A normal aliphatic amine from oils and fats; used as a plasticizer, in medicine, as a chemical intermediate, and in rubber manufacture.

fatty ester [ORG CHEM] $RCOOR'$ A fatty acid in which the alkyl group (R') of a monohydric alcohol replaces the active hydrogen; for example, $RCOOCH_3$ from reaction of $RCOOH$ with methane.

fatty infiltration [PHYSIO] Infiltration of an organ or tissue with excessive amounts of fats.

fatty metamorphosis [MED] Fatty degeneration, fatty infiltration, or both.

fatty nitrile [ORG CHEM] RCN An ester of hydrogen cyanide derived from fatty acid; used in lube oil additives and plasticizers, and as a chemical intermediate.

faucal [BIOL] Of or pertaining to the fauces. [INV ZOO] The opening of a spiral shell.

fauces [ANAT] The passage in the throat between the soft palate and the base of the tongue. [BOT] The throat of a calyx, corolla, or similar part.

faucet [ENG] A fixture through which water is drawn from a pipe or vessel.

faucial tonsil *See* palatine tonsil.

Faugeron kiln [ENG] A coal-fired tunnel kiln for firing feldspathic porcelain; the distinctive feature is the separation of

the tunnel into a series of chambers by division walls on the cars and drop arches in the roof.

faujasite [MINERAL] $(Na_2,Ca)Al_2Si_4O_{12}\cdot 6H_2O$ Zeolite mineral of the sodalite group, crystallizing in the cubic system.

fault [CRYSTAL] *See* stacking fault. [ELEC] A defect, such as an open circuit, short circuit, or ground, in a circuit, component, or line. Also known as electrical fault; faulting. [ELECTR] Any physical condition that causes a component of a data-processing system to fail in performance. [GEOL] A fracture in rock along which the adjacent rock surfaces are differentially displaced.

fault basin [GEOL] A region depressed in relation to surrounding regions and separated from them by faults.

fault block [GEOL] A rock mass that is bounded by faults; the faults may be elevated or depressed and not necessarily the same on all sides.

fault-block mountain *See* block mountain.

fault breccia [GEOL] The assembly of angular fragments found frequently along faults. Also known as dislocation breccia.

fault cliff *See* fault scarp.

fault current *See* fault electrode current.

fault electrode current [ELEC] The current to an electrode under fault conditions, such as during arc-backs and load short circuits. Also known as fault current; surge electrode current.

fault escarpment *See* fault scarp.

fault finder [ENG] Test set for locating trouble conditions in communications circuits or systems.

faulting [ELEC] *See* fault. [GEOL] The fracturing and displacement processes which produce a fault.

fault ledge *See* fault scarp.

fault line [GEOL] Intersection of the fault surface with the surface of the earth or any other horizontal surface of reference. Also known as fault trace.

fault-line scarp [GEOL] A cliff produced when a soft rock erodes against hard rock at a fault.

fault plane [GEOL] A planar fault surface.

fault rock [GEOL] A rock often found along a fault plane and made up of fragments formed by the crushing and grinding which accompany a dislocation.

fault scarp [GEOL] A steep cliff formed by movement along one side of a fault. Also known as cliff of displacement; fault cliff; fault escarpment; fault ledge.

fault strike [GEOL] The angular direction, with respect to north, of the intersection of the fault surface with a horizontal plane.

fault system [GEOL] Two or more fault sets which interconnect.

fault terrace [GEOL] A step on a slope, produced by displacement of two parallel faults.

fault throw [GEOL] The amount of vertical displacement of rocks due to faulting.

fault trace *See* fault line.

fault trap [GEOL] Oil or gas reservoir formed by a structural trap limited in one or more directions by subterranean geological faulting.

fault vein [GEOL] A mineral vein deposited in a fault fissure.

fault wall [GEOL] The mass of rock on a particular side of a fault.

fault zone [GEOL] A fault expressed as an area of numerous small fractures. Also known as distributed fault.

fauna [ZOO] **1.** Animals. **2.** The animal life characteristic of a particular region or environment.

faunal extinction [EVOL] The worldwide death and disappearance of diverse animal groups under circumstances that suggest common and related causes. Also known as mass extinction.

faunal region [ECOL] A division of the zoosphere, defined by geographic and environmental barriers, to which certain animal communities are bound.

faunizone [GEOL] A bed characterized by fossils of a particular assemblage of fauna.

Faure storage battery [ELEC] A storage battery in which the plates consist of lead-antimony supporting grids covered with a lead oxide paste, immersed in weak sulfuric acid. Also known as pasted-plate storage battery.

Faust jig [MIN ENG] A plunger-type jig, usually built with multiple compartments; distinguished by synchronized plungers on both sides of the screen plate, withdrawal of refuse through kettle valves in each compartment, and discharge of the hutch periodically by means of hand valves.

favism [MED] An acute hemolytic anemia, usually in persons of Mediterranean area descent, occurring when an individual with glucose-6-phosphate dehydrogenase deficiency of erythrocytes eats the beans or inhales the pollen of *Vicia faba*.

favorable current [NAV] A current flowing in such a direction as to increase the speed of a vessel over the ground.

favorable wind [NAV] A wind which aids a craft in making progress in a desired direction; usually used in plural form and chiefly in connection with sailing vessels. Also known as fair wind.

Favorskii rearrangement [ORG CHEM] A reaction in which α-halogenated ketones undergo rearrangement in the presence of bases, with loss of the halogen and formation of carboxylic acids or their derivatives with the same number of carbon atoms.

Favositidae [PALEON] A family of extinct Paleozoic corals in the order Tabulata.

favus [MED] A fungal infection of the scalp, usually caused by *Trichophyton schoenleini*, characterized by round, yellow, cup-shaped crusts having a peculiar mousy odor. Also known as tinea favosa.

fax *See* facsimile.

fax chart *See* facsimile chart.

Faxen drag factor *See* drag factor.

fax map *See* facsimile chart.

fayalite [MINERAL] Fe_2SiO_4 A brown to black mineral of the olivine group, consisting of iron silicate and found either massive or in crystals; specific gravity is 4.1.

faying surface [ENG] The surfaces of materials in contact with each other and joined or about to be joined together.

F band [SOLID STATE] The optical absorption band arising from F centers.

FB data set [ADP] A data set which has F-format logical records and whose physical records are all some multiple of the size of the logical record, except possibly for a few truncated blocks. Also known as blocked F-format data set.

FBM data set [ADP] An FB data set which has a machine-control (M) character in its first byte of information.

FBR *See* fast breeder reactor.

FBSA data set [ADP] An FBS data set which has an ASCII (American Standard Code for Information Interchange) control (A) character in its first byte of information.

FBS data set [ADP] An FB data set which has at most one truncated block, which must be the last one in the data set. Also known as standard blocked F-format data set.

fcc lattice *See* face-centered cubic lattice.

F center [SOLID STATE] A color center consisting of an electron trapped by a negative ion vacancy in an ionic crystal, such as an alkali halide or an alkaline-earth fluoride or oxide.

F corona [ASTRON] The outer layer of the sun's corona. Also known as Fraunhofer corona.

F display [ELECTR] A rectangular display in which a target appears as a centralized blip when the radar antenna is aimed at it; horizontal and vertical aiming errors are respectively indicated by the horizontal and vertical displacement of the blip.

fdm *See* frequency-division multiplexing.

Fe *See* iron.

feasibility study [SYS ENG] **1.** A study of applicability or desirability of any management or procedural system from the standpoint of advantages versus disadvantages in any given case. **2.** A study to determine the time at which it would be practicable or desirable to install such a system when determined to be advantageous. **3.** A study to determine whether a plan is capable of being accomplished successfully.

feasibility test [SYS ENG] A test conducted to obtain data in support of a feasibility study or to demonstrate feasibility.

feasible ground [MIN ENG] Ground that can be easily worked and yet will stand without the support of timber or boards.

feasible solution [ADP] In linear programming, any set of

values for the variables x_j, $j = 1, 2, \ldots, n$, that (1) satisfy the set of restrictions

$$\sum_{j=1}^{n} a_{ij}x_j \leq b_i, \ i = 1, 2, \ldots, m$$

(alternatively, $\sum_{j=1}^{n} a_{ij}x_j \geq b_i$, or $\sum_{j=1}^{n} a_{ij}x_j = b_i$)

where the b_i are numerical constants known collectively as the right-hand side and the a_{ij} are coefficients of the variables x_j, and (2) satisfy the restrictions $x_j \geq 0$.

feather [MECH ENG] To change the pitch on a propeller. [METEOROL] *See* barb. [VERT ZOO] An ectodermal derivative which is a specialized keratinous outgrowth of the epidermis of birds; functions in flight and in providing insulation and protection.

feather alum *See* halotrichite.

Feather analysis [NUCLEO] A technique for determining the range in aluminum of the beta rays of a species by comparing the absorption curve of that species with the absorption curve of a reference species.

featheredge [CIV ENG] The thin edge of a gravel-surfaced road. [DES ENG] A wood tool with a level edge used to straighten angles in the finish coat of plaster.

feathering [FOOD ENG] Flocculation of the cream (fat) in homogenized milk when added to hot coffee or tea due to a defect in the chemistry of the cream. [MECH ENG] A pitch position in a controllable-pitch propeller; it is used in the event of engine failure to stop the windmilling action, and occurs when the blade angle is about 90° to the plane of rotation. Also known as full feathering. [VERT ZOO] Plumage.

feathering propeller [MECH ENG] A variable-pitch marine or airscrew propeller capable of increasing pitch beyond the normal high pitch value to the feathered position.

feather joint [ENG] A joint made by cutting a mating groove in each of the pieces to be joined and inserting a feather in the opening formed when the pieces are butted together. Also known as ploughed-and-tongued joint. [GEOL] One of a series of joints in a fault zone formed by shear and tension. Also known as pinnate joint.

feather ore *See* jamesonite.

feather rot [PL PATH] A fungus rot of both dead and living tree trunks caused by *Poria subacida* and characterized by the white stringy or spongy nature of the rotted tissue.

febrile disease [MED] Any disease associated with or characterized by fever.

febriphobia [PSYCH] An abnormal fear of fever.

fecal pellets [GEOL] Mainly the excreta of invertebrates occurring in marine deposits and as fossils in sedimentary rocks. Also known as castings.

feces [PHYSIO] The waste material eliminated by the gastrointestinal tract.

Fechner color [OPTICS] A sensation of color caused by achromatic stimuli at intervals in time.

Fechner law [PHYSIO] The intensity of a sensation produced by a stimulus varies directly as the logarithm of the numerical value of that stimulus.

fecundity [BIOL] The innate potential reproductive capacity of the individual organism, as denoted by its ability to form and separate from the body the mature germ cells.

Federal Telecommunications System [COMMUN] System of commercial telephone lines, leased by the government, for use between major government installations for official telecommunications.

Fedorov stage *See* universal stage.

fedsim star [ADP] The starlike shape that is characteristic of the Kiviat graph of a well-balanced computer system.

feed [ADP] **1.** To supply the material to be operated upon to a machine. **2.** A device capable of so feeding. [AGR] Any crops or other food substances for livestock. [ELECTR] To supply a signal to the input of a circuit, transmission line, or antenna. [ELECTROMAG] The part of a radar antenna that is connected to or mounted on the end of the transmission line and serves to radiate radio-frequency electromagnetic energy to the reflector or receive energy therefrom. [ENG] **1.** Process or act of supplying material to a processing unit for treatment. **2.** The material supplied to a processing unit for treatment. [FOOD ENG] The fermenting wort that is removed

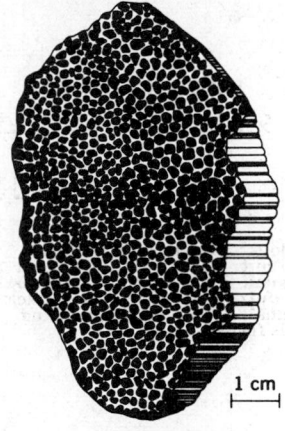
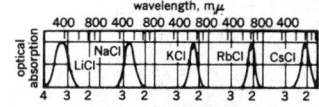
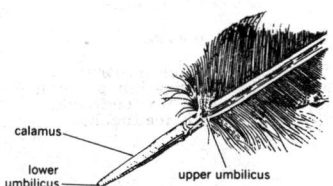

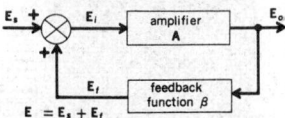

FEEDBACK CIRCUIT

Block diagram of a feedback circuit: E_s is sinusoidal input signal; E_i is actuating signal; E_o is output signal; E_f is feedback signal; A is amplifier gain; and β is feedback function.

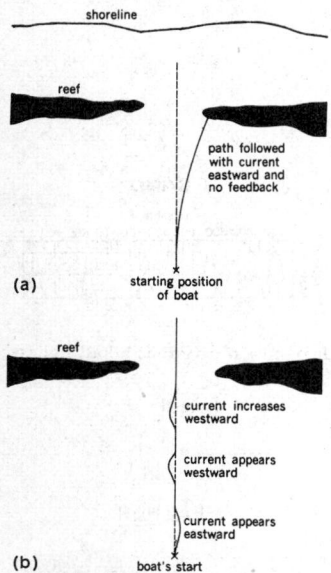

FEEDBACK CONTROL SYSTEM

Use of a feedback control system in a navigation problem. (a) Course with no feedback. (b) Course with feedback.

from the yeast troughs during brewing processes. [MECH ENG] Forward motion imparted to the cutters or drills of cutting or drilling machinery.

feedback [ELECTR] The return of a portion of the output of a circuit or device to its input. [SCI TECH] The control of input as a function of output by returning a portion of the output to the input.

feedback admittance [ELECTR] Short-circuit transadmittance from the output electrode to the input electrode of an electron tube.

feedback amplifier [ELECTR] An amplifier in which a passive network is used to return a portion of the output signal to its input so as to change the performance characteristics of the amplifier.

feedback branch [CONT SYS] A branch in a signal-flow graph that belongs to a feedback loop.

feedback circuit [ELECTR] A circuit that returns a portion of the output signal of an electronic circuit or control system to the input of the circuit or system.

feedback compensation [CONT SYS] Improvement of the response of a feedback control system by placing a compensator in the feedback path, in contrast to cascade compensation. Also known as parallel compensation.

feedback control loop See feedback loop.

feedback control signal [CONT SYS] The portion of an output signal which is retransmitted as an input signal.

feedback control system [CONT SYS] A system in which the value of some output quantity is controlled by feeding back the value of the controlled quantity and using it to manipulate an input quantity so as to bring the value of the controlled quantity closer to a desired value. Also known as closed-loop control system.

feedback factor [ELECTR] The fraction of the output voltage of an oscillator which is applied to the feedback network.

feedback inhibition [BIOCHEM] A cellular control mechanism by which the end product of a series of metabolic reactions inhibits the activity of the first enzyme in the sequence.

feedback loop [CONT SYS] A closed transmission path or loop that includes an active transducer and consists of a forward path, a feedback path, and one or more mixing points arranged to maintain a prescribed relationship between the loop input signal and the loop output signal. Also known as feedback control loop.

feedback oscillator [ELECTR] An oscillating circuit, including an amplifier, in which the output is fed back in phase with the input; oscillation is maintained at a frequency determined by the values of the components in the amplifier and the feedback circuits.

feedback regulator [CONT SYS] A feedback control system that tends to maintain a prescribed relationship between certain system signals and other predetermined quantities.

feedback transfer function [CONT SYS] In a feedback control loop, the transfer function of the feedback path.

feedback winding [ELECTR] A winding to which feedback connections are made in a magnetic amplifier.

feed chute [ORD] A chute or passage through which ammunition is guided into the breech mechanism of a machine gun.

feed-control valve [MECH ENG] A small valve, usually a needle valve, on the outlet of the hydraulic-feed cylinder on the swivel head of a diamond drill, used to control minutely the speed of the hydraulic piston travel and hence the rate at which the bit is made to penetrate the rock.

feeder [ELEC] 1. A transmission line used between a transmitter and an antenna. 2. A conductor, or several conductors, connecting generating stations, substations, or feeding points in an electric power distribution system. 3. A group of conductors in an interior wiring system which link a main distribution center with secondary or branch-circuit distribution centers. [GEOL] A small ore-bearing vein which merges with a larger one. [HYD] See tributary. [MECH ENG] 1. A conveyor adapted to control the rate of delivery of bulk materials, packages, or objects, or a control device which separates or assembles objects. 2. A device for delivering materials to a processing unit. [MET] A runner or riser so placed that it can feed molten metal to the contracting mass of the casting as it cools in its flask, therefore preventing

formation of cavities or porous structure. [ORD] A device that supplies ammunition to a weapon, usually actuated by an automatic or semiautomatic mechanism.

feeder-breaker [MECH ENG] A unit that breaks and feeds ore or crushed rock to a materials-handling system at a required rate.

feeder cable [COMMUN] In communications practice, a cable extending from the central office along a primary route (main feeder cable) or from a main feeder cable along a secondary route (branch feeder cable) and providing connections to one or more distribution cables.

feeder canal [CIV ENG] A canal serving to conduct water to a larger canal.

feeder conveyor [MECH ENG] A short auxiliary conveyor designed to transport materials to another conveyor. Also known as stage loader.

feeder distribution center [COMMUN] Distribution center at which feeders or subfeeders are connected.

feeder panel [ELEC] The part of a switchboard in an electric power distribution system where feeder connections are made.

feeder road [CIV ENG] A road that feeds traffic to a more important road.

feeder trough [MIN ENG] The trough connected to the conveyor pan line in a duckbill.

feedforward control [CONT SYS] Process control in which changes are detected at the process input and an anticipating correction signal is applied before process output is affected.

feedhead [MET] A reservoir of molten metal that is left above a casting in order to supply additional metal as the casting solidifies and shrinks. Also known as riser; sinkhead.

feed hole [DES ENG] One of a series of holes punched in paper tape at fixed intervals to be engaged by the sprocket teeth of a drive wheel.

feeding mechanism [ZOO] A mechanism by which an animal obtains and utilizes food materials.

feeding rod [MET] A rod used by working up and down to keep the passage clear between riser and casting.

feed lines [MET] The pattern produced on the surface of a piece of metal by machine grinding.

feed materials [NUCLEO] Refined uranium or thorium metal or their pure compounds in a form suitable for use in nuclear reactor fuel elements or as feed for uranium-enrichment processes.

feed nut [MECH ENG] The threaded sleeve fitting around the feed screw on a gear-feed drill swivel head, which is rotated by means of paired gears driven from the spindle or feed shaft.

feed pitch [DES ENG] The distance between the centers of adjacent feed holes in punched paper tape.

feed preparation unit [CHEM ENG] A processing unit (such as distillation or desulfurization units) providing feedstock for subsequent processing.

feed pressure [MECH ENG] Total weight or pressure, expressed in pounds or tons, applied to the drilling stem to make the drill bit cut and penetrate the geologic, rock, or ore formation.

feed pump [MECH ENG] A pump used to supply water to a steam boiler.

feed rate See cutting speed.

feed ratio [MECH ENG] The number of revolutions a drill stem and bit must turn to advance the drill bit 1 inch when the stem is attached to and rotated by a screw- or gear-feed type of drill swivel head with a particular pair of the set of gears engaged. Also known as feed speed.

feed reel [ENG] The reel from which paper tape or magnetic tape is being fed.

feed screw [MECH ENG] The externally threaded drill-rod drive rod in a screw- or gear-feed swivel head on a diamond drill; also used on percussion drills, lathes, and other machinery.

feed shaft [MECH ENG] A short shaft or countershaft in a diamond-drill gear-feed swivel head which is rotated by the drill motor through gears or a fractional drive and by means of which the engaged pair of feed gears is driven.

feed shelf [ADP] 1. A device for supporting documents for manual sensing. 2. The first few feet of a tape reel, used to prime the tape drive.

feed speed *See* feed ratio.

feedstock [ENG] The raw material furnished to a machine or process.

feed-tape [ADP] A mechanism which will feed tape to be read or sensed.

feedthrough [ELEC] A conductor that connects patterns on opposite sides of a printed circuit board. Also known as interface connection.

feedthrough capacitor [ELEC] A feedthrough terminal that provides a desired value of capacitance between the feedthrough conductor and the metal chassis or panel through which the conductor is passing; used chiefly for bypass purposes in ultra-high-frequency circuits.

feedthrough insulator *See* feedthrough terminal.

feedthrough terminal [ELEC] An insulator designed for mounting in a hole in a panel, wall, or bulkhead, with a conductor in the center on the insulator to permit feeding electricity through the partition. Also known as feedthrough insulator.

feed travel [MECH ENG] The distance a drilling machine moves the steel shank in traveling from top to bottom of its feeding range.

feed tray [CHEM ENG] For a tray-type distillation column, that tray on which fresh feedstock is introduced into the system.

feedwater [MECH ENG] The water supplied to a boiler or still.

feedwater heater [MECH ENG] An apparatus that utilizes steam extracted from an engine or turbine to heat boiler feedwater.

feeler pin [MECH ENG] A pin that allows a duplicating machine to operate only when there is a supply of paper.

feel the bottom [NAV] The effect on a ship underway in shallow water which tends to reduce its speed, make it slow in answering the helm, and often make it sheer off course; the speed reduction is largely due to increased wave making resistance resulting from higher pressure differences due to restriction of flow around the hull; the increased velocity of the water flowing past the hull results in an increase in squat. Also known as smell the bottom.

Fehling's reagent [ANALY CHEM] A solution of cupric sulfate, sodium potassium tartrate, and sodium hydroxide, used to test for the presence of reducing compounds such as sugars.

Feinc filter [MIN ENG] A vacuum-type drum filter in which a system of parallel strings is used to carry the filter cake away from the drum, instead of the usual filter cloth.

feldspar [MINERAL] A group of silicate minerals that make up about 60% of the outer 15 kilometers of the earth's crust; they are silicates of aluminum with the metals potassium, sodium, and calcium, and rarely, barium.

feldspathic graywacke [GEOL] Sandstone containing less than 75% quartz and chert and 15-75% detrital clay matrix, and having feldspar grains in greater abundance than rock fragments. Also known as high-rank graywacke.

feldspathic sandstone [GEOL] Sandstone rich in feldspar; intermediate in composition between arkosic sandstone and quartz sandstone, made up of 10-25% feldspar and less than 20% matrix material.

feldspathization [GEOL] Formation of feldspar in a rock usually as a result of metamorphism leading toward granitization.

feldspathoid [GEOL] Aluminosilicates of sodium, potassium, or calcium that are similar in composition to feldspars but contain less silica than the corresponding feldspar.

Felidae [VERT ZOO] The cats and saber-toothed cats, a family of mammals in the superfamily Feloidea.

feline [VERT ZOO] **1.** Of or relating to the genus *Felis*. **2.** Catlike.

Felis [VERT ZOO] The type genus of the Felidae, comprising the true or typical cats, both wild and domestic.

fell [FOR] The timber cut in a given season.

fell-field [ECOL] A culture community of dwarfed, scattered plants or grasses above the timberline.

Fell system [CIV ENG] A method of traction intended for steep railroad slopes; a central rail is gripped between horizontal wheels on the locomotive.

Feloidea [VERT ZOO] A superfamily of catlike mammals in the order Carnivora.

Felon's unit [BIOL] A unit for the standardization of antipneumococcic serum.

felsenmeer [GEOL] A flat or gently sloping veneer of angular rock fragments occurring on moderate mountain slopes above the timber line.

felsic [MINERAL] A light-colored mineral. [PETR] Of an igneous rock, having a mode containing light-colored minerals.

felsite [PETR] **1.** A light-colored, fine-grained igneous rock composed chiefly of quartz or feldspar. **2.** A rock characterized by felsitic texture.

felsophyric *See* aphaniphyric.

felt [MATER] A fibrous, watertight heavy paper of organic or asbestos fibers impregnated with asphalt and used as an overlining or an underlining for roofs. Also known as felt paper. [TEXT] A compressed, densely matted unwoven fabric of wool, sometimes with rayon or hair.

felt side [MATER] The upper side of a sheet of paper which was not in contact with the wire in a papermaking machine.

Felty's syndrome [MED] A complex of symptoms involving rheumatoid arthritis, splenomegaly, lymphadenopathy, and anemia.

female [BOT] A flower lacking stamens. [ZOO] An individual that bears young or produces eggs.

female connector [ELEC] A connector having one or more contacts set into recessed openings; jacks, sockets, and wall outlets are examples.

female fitting [DES ENG] In a paired pipe or an electrical or mechanical connection, the portion (fitting) that receives, contrasted to the male portion (fitting) that inserts.

female heterogamety [GEN] The occurrence, in females of a species, of an unequal pair of sex chromosomes.

female homogamety [GEN] The occurrence, in females of a species, of an equal pair of sex chromosomes.

female pseudohermaphroditism *See* gynandry.

feminizing syndrome [MED] Any of a number of symptom complexes in which males tend to take on feminine characteristics due to alterations of adrenocorticotropin output.

femitrons [ELECTR] Class of field-emission microwave devices.

femoral artery [ANAT] The principal artery of the thigh; originates as a continuation of the external iliac artery.

femoral hernia [MED] A hernia that occurs at the passage of the arteries and veins from the abdomen into the legs below the inguinal ligament.

femoral nerve [ANAT] A mixed nerve of the leg; the motor portion innervates muscles of the thigh, and the sensory portion innervates portions of the skin of the thigh, leg, hip, and knee.

femoral ring [ANAT] The abdominal opening of the femoral canal.

femoral vein [ANAT] A vein accompanying the femoral artery.

femorotibial index [ANTHRO] The ratio of the length of the femur to the length of the tibia multiplied by 100.

femto- [SCI TECH] A prefix representing 10^{-15}, which is 0.000 000 000 000 001, or one-thousandth of a millionth of a millionth.

femtoampere [ELEC] A unit of current equal to 10^{-15} ampere. Abbreviated fa.

femtometer [MECH] A unit of length, equal to 10^{-15} meter; used particularly in measuring nuclear distances. Abbreviated fm. Also known as fermi.

femtovolt [ELEC] A unit of voltage equal to 10^{-15} volt. Abbreviated fv.

femur [ANAT] **1.** The proximal bone of the hind or lower limb in vertebrates. **2.** The thigh bone in humans, articulating with the acetabulum and tibia.

fen [GEOGR] Peat land covered by water, especially in the upper regions of old estuaries and around lakes, that can be drained only artificially.

fence [AERO ENG] A stationary plate or vane projecting from the upper surface of an airfoil, substantially parallel to the airflow, used to prevent spanwise flow. [ENG] **1.** A line of data-acquisition or tracking stations used to monitor orbiting satellites. **2.** A line of radar or radio stations for detection of satellites or other objects in orbit. **3.** A line or network of early-warning radar stations. **4.** A concentric steel fence erected around a ground radar transmitting antenna to serve as an artificial horizon and suppress ground clutter that would otherwise drown out weak signals returning at a low angle from a target.

fenchol *See* fenchyl alcohol.

fenchone [ORG CHEM] $C_{10}H_{16}O$ An isomer of camphor; a colorless oil that boils at 193°C and is soluble in ether; a constituent of fennel oil; used as a flavoring.

fenchyl alcohol [ORG CHEM] $C_{10}H_{18}O$ A colorless solid or oily liquid, boiling at 198-204°C, isolated from pine oil and turpentine and also made synthetically; used as a solvent, an intermediate in organic synthesis, and as a flavoring. Also known as fenchol; 1-hydroxy fenchane.

fender [CIV ENG] A timber, cluster of piles, or bag of rope placed along dock or bridge pier to prevent damage by docking ships or floating objects. [ENG] A cover over the upper part of a wheel of an automobile or other vehicle. [MIN ENG] A thin pillar of coal adjacent to the gob, left for protection while driving a lift through the mine pillar. [NAV ARCH] A padded device acting as a buffer to prevent damage between two ships or between a ship and dock.

Fenestellidae [PALEON] A family of extinct fenestrated, cryptostomatous bryozoans which abounded during the Silurian.

fenestra [ANAT] An opening in the medial wall of the middle ear. [MED] An opening in a bandage or plaster splint for examination or drainage.

fenestrated membrane [HISTOL] One of the layers of elastic tissue in the tunica media and tunica intima of large arteries.

fenestration [ARCH] The arrangement of openings, especially windows, in the wall of a building. [BIOL] **1.** A transparent or windowlike break or opening in the surface. **2.** The presence of windowlike openings.

fennel [BOT] *Foeniculum vulgare.* A tall perennial herb of the family Umbelliferae; a spice is derived from the fruit.

fennel oil [MATER] The essential oil obtained from fennel; a colorless liquid with aromatic scent and bitter taste, insoluble in water and boiling at 160-220°C; used in medicine, perfumes, and liqueurs. Also known as oil of fennel.

fenolactine *See para*-lactophenetide.

fen peat *See* lowmoor peat.

Fenske equation *See* Fenske-Underwood equation.

Fenske-Underwood equation [CHEM ENG] Equation in plate-to-plate distillation-column calculations relating the number of theoretical plates needed at total reflux to overall relative volatility and the liquid-vapor composition ratios on upper and lower plates. Also known as Fenske equation.

fenster *See* window.

FEP resin *See* fluorinated ethylene propylene resin.

ferberite [MINERAL] $FeNO_4$ A black mineral of the wolframite solid-solution series occurring as monoclinic, prismatic crystals and having a submetallic luster; hardness is 4.5 on Mohs scale, and specific gravity is 7.5.

ferghanite [MINERAL] $U_3(VO_4)_2 \cdot 6H_2O$ Sulfur-yellow mineral composed of hydrated uranium vanadate, occurring in scales.

fergusonite [MINERAL] $Y_2O_3 \cdot (Nb,Ta)_2O_5$ Brownish-black rare-earth mineral with a tetragonal crystal form; it is isomorphous with formanite.

Fermat's last theorem [MATH] The conjecture that there are no positive integer solutions of the equation $x^n + y^n = z^n$ for $n \geq 3$.

Fermat's principle [OPTICS] The principle that an electromagnetic wave will take a path that involves the least travel time when propagating between two points. Also known as least-time principle.

ferment [BIOCHEM] An agent that can initiate fermentation and other metabolic processes.

fermentation [MICROBIO] An enzymatic transformation of organic substrates, especially carbohydrates, generally accompanied by the evolution of gas; a physiological counterpart of oxidation, permitting certain organisms to live and grow in the absence of air; used in various industrial processes for the manufacture of products such as alcohols, acids, and cheese by the action of yeasts, molds, and bacteria; alcoholic fermentation is the best-known example. Also known as zymosis.

fermentation accelerator [MATER] Substance that speeds chemical fermentation (as for wines) without participating in the resulting chemical changes; can be an enzyme or other catalytic agent.

fermentation tube [MICROBIO] A culture tube with a vertical closed arm to collect gas formed in a broth culture by microorganisms.

fermenter [FOOD ENG] A vessel used for fermenting, such as a vat for fermenting mash in brewing.

ferment oil [MATER] A volatile oil formed by the fermentation of plant material in which the oil was not present originally.

fermi *See* femtometer.

Fermi age [NUCLEO] The value calculated for the slowing-down area in the Fermi age model; it has the dimensions of area, not time. Also known as neutron age; symbolic age of neutrons.

Fermi age model [NUCLEO] A model used in studying the slowing down of neutrons by elastic collisions; it is assumed that the slowing down takes place by a very large number of very small energy changes.

Fermi beta-decay theory [NUC PHYS] Theory in which a nucleon source current interacts with an electron-neutrino field to produce beta decay, in a manner analogous to the interaction of an electric current with an electromagnetic field during the emission of a photon of electromagnetic radiation.

Fermi constant [NUC PHYS] A universal constant, introduced in beta-disintegration theory, that expresses the strength of the interaction between the transforming nucleon and the electron-neutrino field.

Fermi-Dirac distribution function [STAT MECH] A function specifying the probability that a member of an assembly of independent fermions, such as electrons in a semiconductor or metal, will occupy a certain energy state when thermal equilibrium exists.

Fermi-Dirac statistics [STAT MECH] The statistics of an assembly of identical half-integer spin particles; such particles have wave functions antisymmetrical with respect to particle interchange and satisfy the Pauli exclusion principle.

Fermi distribution [SOLID STATE] Distribution of energies of electrons in a semiconductor or metal as given by the Fermi-Dirac distribution function; nearly all energy levels below the Fermi level are filled, and nearly all above this level are empty.

Fermi energy [STAT MECH] **1.** The average energy of electrons in a metal, equal to $\frac{3}{5}$ of the Fermi level. **2.** *See* Fermi level.

Fermi hole [SOLID STATE] A region surrounding an electron in a solid in which the energy band theory predicts that the probability of finding other electrons is less than the average over the volume of the solid.

Fermi level [STAT MECH] The energy level at which the Fermi-Dirac distribution function of an assembly of fermions is equal to one-half. Also known as Fermi energy.

Fermi liquid [CRYO] A liquid of particles which have Fermi-Dirac statistics; an example is the liquid phase of helium-3, in which the atoms belong to the isotope with mass number 3.

fermion [QUANT MECH] A particle, such as the electron, proton, or neutron, which obeys the rule that the wave function of several identical particles changes sign when the coordinates of any pair are interchanged; it therefore obeys the Pauli exclusion principle.

fermion field [QUANT MECH] An operator defined at each point in space-time that creates or annihilates a particular type of fermion and its antiparticle.

Fermi plot *See* Kurie plot.

Fermi resonance [PHYS CHEM] In a polyatomic molecule, the relationship of two vibrational levels that have in zero approximation nearly the same energy; they repel each other, and the eigenfunctions of the two states mix.

Fermi selection rules [NUC PHYS] Selection rules for beta decay in a Fermi transition; that is, there is no change in total angular momentum or parity of the nucleus in an allowed transition.

Fermi's golden rules [QUANT MECH] The equations giving the first-order (rule number 2) and second-order (rule number 1) contributions to the transition probability per unit time

Fennel (*Foeniculum vulgare*). (USDA)

├─2.5 cm─┤

Black, prismatic ferberite crystals on granite, found in Boulder County, Colorado. (*Specimen from Department of Geology, Bryn Mawr College*)

induced by a perturbation Hamiltonian, in terms of matrix elements of the perturbation Hamiltonian.

Fermi sphere [STAT MECH] The Fermi surface of an assembly of fermions in the approximation that the fermions are free particles.

Fermi surface [SOLID STATE] A constant-energy surface in the space containing the wave vectors of states of members of an assembly of independent fermions, such as electrons in a semiconductor or metal, whose energy is that of the Fermi level.

Fermi temperature [STAT MECH] The energy of the Fermi level of an assembly of fermions divided by Boltzmann's constant, which appears as a parameter in the Fermi-Dirac distribution function.

Fermi transition [NUC PHYS] Beta decay subject to Fermi selection rules.

fermium [CHEM] A synthetic radioactive element, symbol Fm, with atomic number 100; discovered in debris of the 1952 hydrogen bomb explosion, and now made in nuclear reactors.

fermorite [MINERAL] $(Ca,Sr)_5[(As,P)O_4]_3$ A white mineral composed of arsenate, phosphate, and fluoride of calcium and strontium, occurring in crystalline masses.

fern [BOT] Any of a large number of vascular plants composing the division Polypodiophyta.

fernandinite [MINERAL] A dull green mineral composed of hydrous calcium vanadyl vanadate.

fernico [MET] An iron-nickel-cobalt alloy used for metal-to-glass seals.

Ferranti effect [ELEC] A rise in voltage occurring at the end of a long transmission line when its load is disconnected.

ferrate [INORG CHEM] A multiple iron oxide with another oxide, for example, Na_2FeO_4.

ferredoxins [BIOCHEM] Iron-containing proteins that transfer electrons, usually at a low potential, to flavoproteins; the iron is not present as a heme.

ferreed switch [ELEC] A switch whose contacts are mounted on magnetic blades or reeds sealed into an evacuated tubular glass housing, the contacts being operated by external electromagnets or permanent magnets.

ferret [ORD] An aircraft, ship, or vehicle especially equipped for the detection, location, recording, and analyzing of electromagnetic radiation. [VERT ZOO] *Mustela nigripes*. The largest member of the weasel family, Mustelidae, and a relative of the European polecat; has yellowish fur with black feet, tail, and mask.

ferriamphibole [MINERAL] The ferric ion equivalent of the amphibole group of minerals.

ferric [INORG CHEM] The term for a compound of trivalent iron, for example, ferric bromide, $FeBr_3$.

ferric acetate [ORG CHEM] $Fe_2(C_2H_3O_2)_3$ A brown compound, soluble in water; used as a tonic and dye mordant.

ferric ammonium alum *See* ferric ammonium sulfate.

ferric ammonium citrate [ORG CHEM] $Fe(NH_4)_3(C_6H_5O_7)_2$ Red, deliquescent scales or granules; odorless, water soluble, and affected by light; used in medicine and blueprint photography.

ferric ammonium oxalate [ORG CHEM] $(NH_4)_3Fe(C_2O_4)_3 \cdot 3H_2O$ Green, crystalline material, soluble in water and alcohol, sensitive to light; used in blueprint photography.

ferric ammonium sulfate [INORG CHEM] $FeNH_4(SO_4)_2 \cdot 12H_2O$ Efflorescent, water-soluble crystals; used in medicine, in analytical chemistry, and as a mordant in textile dyeing. Also known as ferric ammonium alum; iron ammonium sulfate.

ferric arsenate [INORG CHEM] $FeAsO_4 \cdot 2H_2O$ A green or brown powder, insoluble in water, soluble in dilute mineral acids; used as an insecticide.

ferric bromide [INORG CHEM] $FeBr_3$ Red, deliquescent crystals that decompose upon heating; soluble in water, ether, and alcohol; used in medicine and analytical chemistry. Also known as ferric sesquibromide; ferric tribromide; iron bromide.

ferric chloride [INORG CHEM] $FeCl_3$ Brown crystals, melting at 300°C, that are soluble in water, alcohol, and glycerol; used as a coagulant for sewage and industrial wastes, as an oxidizing and chlorinating agent, as a disinfectant, in copper etching, and as a mordant. Also known as anhydrous ferric chloride; ferric trichloride; flores martis; iron chloride.

ferric citrate [ORG CHEM] $FeC_6H_5O_7 \cdot 3H_2O$ Red scales that react to light; soluble in water, insoluble in alcohol; used as a medicine for certain blood disorders, and for blueprint paper. Also known as iron citrate.

ferric dichromate [INORG CHEM] $Fe_2(CrO_4)_3$ A red-brown, granular powder, miscible in water; used as a mordant.

ferric ferrocyanide [INORG CHEM] $Fe_4[Fe(CN)_6]_3$ Dark-blue crystals, used as a pigment, and with oxalic acid in blue ink. Also known as iron ferrocyanide.

ferric fluoride [INORG CHEM] FeF_3 Green, rhombohedral crystals, soluble in water and acids; used in porcelain and pottery manufacture. Also known as iron fluoride.

ferrichrome [MICROBIO] A cyclic hexapeptide that is a microbial hydroxamic acid and is involved in iron transport and metabolism in microorganisms.

ferric hydrate *See* ferric hydroxide.

ferric hydroxide [INORG CHEM] $Fe(OH)_3$ A brown powder, insoluble in water; used as arsenic poisoning antidote, in pigments, and in pharmaceutical preparations. Also known as ferric hydrate; iron hydroxide.

ferric nitrate [INORG CHEM] $Fe(NO_3)_3 \cdot 9H_2O$ Colorless crystals, soluble in water and decomposed by heat; used as a dyeing mordant, in tanning, and in analytical chemistry. Also known as iron nitrate.

ferric oxalate [ORG CHEM] $Fe_2(COO)_3$ Yellow scales, soluble in water, decomposing when heated at about 100°C; used as a catalyst and in photographic printing papers.

ferric oxide [INORG CHEM] Fe_2O_3 Red, hexagonal crystals or powder, insoluble in water and soluble in acids, melting at 1565°C; used as a catalyst and pigment for metal polishing, in metallurgy, and in medicine. Also known as ferric oxide red; jeweler's rouge; red ocher.

ferric oxide red *See* ferric oxide.

ferric phosphate [INORG CHEM] $FePO_4 \cdot 2H_2O$ Yellow, rhombohedral crystals, insoluble in water, soluble in acids; used in medicines and fertilizers. Also known as iron phosphate.

ferric resinate [ORG CHEM] Reddish-brown, water-insoluble powder; used as a drier for paints and varnishes. Also known as iron resinate.

ferric sesquibromide *See* ferric bromide.

ferric stearate [ORG CHEM] $Fe(C_{18}H_{35}O_2)_3$ A light-brown, water-insoluble powder; used as a varnish drier. Also known as iron stearate.

ferric sulfate [INORG CHEM] $Fe_2(SO_4)_3 \cdot 9H_2O$ Yellow, water-soluble, rhombohedral crystals, decomposing when heated; used as a chemical intermediate, disinfectant, soil conditioner, pigment, and analytical reagent, and in medicine. Also known as iron sulfate.

ferric tribromide *See* ferric bromide.

ferric trichloride *See* ferric chloride.

ferric vanadate [INORG CHEM] $Fe(VO_3)_3$ Grayish-brown powder, insoluble in water and alcohol; used in metallurgy. Also known as iron metavanadate.

ferricyanic acid [INORG CHEM] $H_3Fe(CN)_6$ A red-brown unstable solid.

ferricyanide [INORG CHEM] A salt containing the radical $Fe(CN)_6^{3-}$.

ferrierite [MINERAL] $(Na,K)_2MgAl_3Si_{15}O_{36}(OH) \cdot 9H_2O$ A zeolite mineral crystallizing in the orthorhombic system.

ferriferous [GEOL] Of a sedimentary rock, iron-rich. [MINERAL] Of a mineral, iron-bearing.

ferrihemoglobin [BIOCHEM] Hemoglobin in the oxidized state. Also known as methemoglobin.

ferrimagnet *See* ferrimagnetic material.

ferrimagnetic amplifier [ELECTR] A microwave amplifier using ferrites.

ferrimagnetic limiter [ELECTROMAG] Power limiter used in microwave systems to replace transmit-receive tubes; uses ferrimagnetic material (such as a piece of ferrite or garnet) that exhibits nonlinear properties.

ferrimagnetic material [SOLID STATE] A material displaying ferrimagnetism; the ferrites are the principal example. Also known as ferrimagnet.

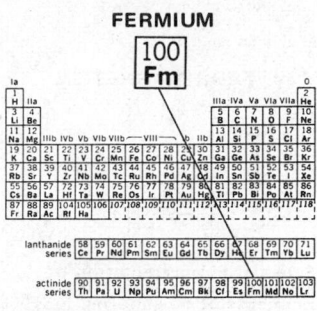

Periodic table of the chemical elements showing the position of fermium.

FERN

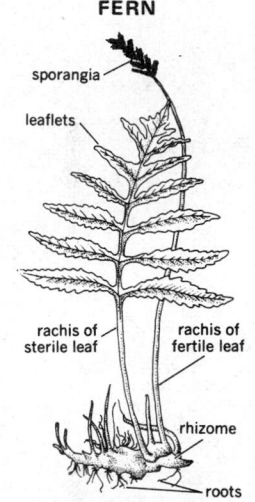

The sensitive fern (*Onoclea sensibilis*), a representative of the Polypodiatae. (*From W. W. Robbins, T. E. Weier, and C. R. Stocking, Botany; An Introduction to Plant Science, 3d ed., Wiley, 1964*)

FERRET

Black-footed ferret (*Mustela nigripes*) of North America.

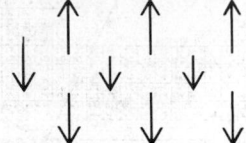

FERRIMAGNETISM

Schematic representation of arrangement of magnetic moments of neighboring ions in collinear ferrimagnetism.

FERROCENE

Fe

Structural diagram of ferrocene.

ferrimagnetic resonance [PHYS] Magnetic resonance of a ferrimagnetic material.

ferrimagnetism [SOLID STATE] A type of magnetism in which the magnetic moments of neighboring ions tend to align nonparallel, usually antiparallel, to each other, but the moments are of different magnitudes, so there is an appreciable resultant magnetization.

ferrimycin [MICROBIO] The representative antibiotic of the sideromycin group; a hydroxamic acid compound.

ferrinatrite [MINERAL] $Na_3Fe(SO_4)_3 \cdot 3H_2O$ A greenish or white mineral composed of sodium ferric iron double sulfate; usually occurs in spherical forms.

ferriporphyrin [BIOCHEM] A red-brown to black complex of iron and porphyrin in which the iron is in the 3+ oxidation state.

ferrisicklerite [MINERAL] $(Li,Fe,Mn)(PO_4)$ Mineral composed of phosphate of lithium, ferric iron, and manganese, more iron being present than manganese; it is isomorphous with sicklerite.

ferristor [ELECTR] A miniature, two-winding, saturable reactor that operates at a high carrier frequency and may be connected as a coincidence gate, current discriminator, free-running multivibrator, oscillator, or ring counter.

ferrisulphas *See* ferrous sulfate.

ferrite [INORG CHEM] An unstable compound of a strong base and ferric oxide which exists in alkaline solution, such as $NaFeO_2$. [MET] Iron that has not combined with carbon in pig iron or steel. [PETR] Grains or scales of unidentifiable, generally transparent amorphous iron oxide in the matrix of a porphyritic rock. [SOLID STATE] Any ferrimagnetic material having high electrical resistivity which has a spinel crystal structure and the chemical formula XFe_2O_4, where X represents any divalent metal ion whose size is such that it will fit into the crystal structure.

ferrite attenuator *See* ferrite limiter.

ferrite banding [MET] The formation of faint bands (flow lines) of free ferrite in rolled steel, running in the direction of working. Also known as ferrite ghosts; ferrite streaks; ghost lines; ghost structure.

ferrite bead [ELECTR] Magnetic information storage device consisting of ferrite powder mixtures in the form of a bead fired on the current-carrying wires of a memory matrix.

ferrite circulator [ELECTROMAG] A combination of two dual-mode transducers and a 45° ferrite rotator, used with rectangular waveguides to control and switch microwave energy. Also known as ferrite phase-differential circulator.

ferrite core [ELECTR] A magnetic core made of ferrite material. Also known as dust core; powdered-iron core.

ferrite-core memory [ELECTR] A magnetic memory consisting of a matrix of tiny toroidal cores molded from a square-loop ferrite, through which are threaded the pulse-carrying wires and the sense wire.

ferrite device [ELEC] An electrical device whose principle of operation is based upon the use of ferrities in powdered, compressed, sintered form, making use of their ferrimagnetism and their high electrical resistivity, which makes eddy-current losses extremely low at high frequencies.

ferrite ghosts *See* ferrite banding.

ferrite isolator [ELECTROMAG] A device consisting of a ferrite rod, centered on the axis of a short length of circular waveguide, located between rectangular-waveguide sections displaced 45° with respect to each other, which passes energy traveling through the waveguide in one direction while absorbing energy from the opposite direction. Also known as Faraday rotation isolator.

ferrite limiter [ELECTROMAG] A passive, low-power microwave limiter having an insertion loss of less than 1 decibel when operating in its linear range, with minimum phase distortion; the input signal is coupled to a single-crystal sample of either yttrium iron garnet or lithium ferrite, which is biased to resonance by a magnetic field. Also known as ferrite attenuator.

ferrite phase-differential circulator *See* ferrite circulator.

ferrite-rod antenna [ELECTROMAG] An antenna consisting of a coil wound on a rod of ferrite; used in place of a loop antenna in radio receivers. Also known as ferrod; loopstick antenna.

ferrite rotator [ELECTROMAG] A gyrator consisting of a ferrite cylinder surrounded by a ring-type permanent magnet, inserted in a waveguide to rotate the plane of polarization of the electromagnetic wave passing through the waveguide.

ferrite streaks *See* ferrite banding.

ferrite switch [ELECTROMAG] A ferrite device that blocks the flow of energy through a waveguide by rotating the electric field vector 90°; the switch is energized by sending direct current through its magnetizing coil; the rotated electromagnetic wave is then reflected from a reactive mismatch or absorbed in a resistive card.

ferrite-tuned oscillator [ELECTR] An oscillator in which the resonant characteristic of a ferrite-loaded cavity is changed by varying the ambient magnetic field, to give electronic tuning.

ferritic stainless steel [MET] Any magnetic iron alloy containing more than 12% chromium having a body-centered cubic structure. Also known as stainless iron.

ferritin [BIOCHEM] An iron-protein complex occurring in tissues, probably as a storage form of iron.

ferritremolite [MINERAL] The ferric ion equivalent of the monoclinic amphibole, tremolite.

ferritungstite [MINERAL] $Fe_2(WO_4)(OH)_4 \cdot 4H_2O$ A yellow ocher mineral composed of hydrous ferric tungstate, occurring as a powder.

ferroacoustic storage [ELECTR] A delay-line type of storage consisting of a thin tube of magnetostrictive material, a central conductor passing through the tube, and an ultrasonic driving transducer at one end of the tube.

ferroalloy [MET] Any alloy containing iron, usually in major amount. Also known as ferrous alloy.

ferroaluminum [MET] An alloy of iron and aluminum; added to molten steel as a deoxidizer or as an alloying component.

ferroamphibole [MINERAL] The ferrous iron equivalent of the amphibole group of minerals.

ferroan dolomite [MINERAL] A species of ankerite having less than 20% of the manganese positions occupied by iron.

ferroaugite [MINERAL] A form of monoclinic pyroxene.

Ferrobacillus [MICROBIO] A genus of obligate or facultative chemoautotrophic bacteria which obtain metabolic energy from the oxidation of ferrous iron to the ferric state by oxygen.

ferroboron [MET] An alloy of iron and boron that is added to steel to form hardened special steels; two grades are used, 10% boron and 17% boron.

ferrocarbon titanium [MET] An alloy of iron with 15–20% titanium and 3–8% carbon; may be added to molten steel as a component of low-alloy steel.

ferrocene [ORG CHEM] $(CH_2)_5Fe(CH_2)_5$ Orange crystals that are soluble in ether, melting point 174°C; used as a combustion control additive in fuels, and for heat stabilization in greases and plastics. Also known as dicyclopentadienyl iron.

ferrocerium [MET] An alloy of iron with a high percentage of cerium; used to make cigarette lighter flints.

ferrochromium [MET] A crude ferroalloy containing chromium.

ferrocolumbium [MET] An alloy of iron and columbium (niobium); used to add columbium to certain alloy steels.

ferrocyanic acid [INORG CHEM] $H_4Fe(CN)_6$ A white solid obtained by treating ferrocyanides with acid.

ferrocyanide [INORG CHEM] A salt containing the radical $Fe(CN)_6^{4-}$.

ferrocyanide process [CHEM ENG] A regenerative chemical treatment for removal of mercaptans from petroleum fuels; uses caustic–sodium ferrocyanide reagent.

ferrod *See* ferrite-rod antenna.

ferrodolomite [MINERAL] $CaFe(CO_3)_2$ A mineral composed of calcium iron carbonate, isomorphous with dolomite, and occurring in ankerite.

ferroelectric [SOLID STATE] A crystalline substance displaying ferroelectricity, such as barium titanate, potassium dihydrogen phosphate, and Rochelle salt; used in ceramic capacitors, acoustic transducers, and dielectric amplifiers. Also known as seignette-electric.

ferroelectric converter [ELEC] A converter that transforms thermal energy into electric energy by utilizing the change in the dielectric constant of a ferroelectric material when heated beyond its Curie temperature.

ferroelectric crystal [SOLID STATE] A crystal of a ferroelectric material.

ferroelectric domain [SOLID STATE] A region of a ferroelectric material within which the spontaneous polarization is constant.

ferroelectric hysteresis [ELEC] The dependence of the polarization of ferroelectric materials not only on the applied electric field but also on their previous history; analogous to magnetic hysteresis in ferromagnetic materials. Also known as dielectric hysteresis; electric hysteresis.

ferroelectric hysteresis loop [ELEC] Graph of polarization or electric displacement versus applied electric field of a material displaying ferroelectric hysteresis.

ferroelectricity [SOLID STATE] Spontaneous electric polarization in a crystal; analogous to ferromagnetism.

ferroelectric shutter [OPTICS] A shutter consisting of a slab of ferroelectric crystal located between polarizers whose planes are at right angles; opens to pass light when activated by a pulse of up to 100 volts.

ferrogabbro [PETR] A gabbro rock in which the pyroxene and olivine constituents have an unusually high iron content.

ferromagnetic amplifier [ELECTR] A parametric amplifier based on the nonlinear behavior of ferromagnetic resonance at high radio-frequency power levels; incorrectly known as garnet maser.

ferromagnetic crystal [SOLID STATE] A crystal of a ferromagnetic material. Also known as polar crystal.

ferromagnetic domain [SOLID STATE] A region of a ferromagnetic material within which atomic or molecular magnetic moments are aligned parallel. Also known as magnetic domain.

ferromagnetic film See magnetic thin film.

ferromagnetic material [SOLID STATE] A material displaying ferromagnetism, such as the various forms of iron, steel, cobalt, nickel, and their alloys.

ferromagnetic resonance [SOLID STATE] Magnetic resonance of a ferromagnetic material.

ferromagnetics [ELECTR] The science that deals with the storage of binary information and the logical control of pulse sequences through the utilization of the magnetic polarization properties of materials.

ferromagnetic tape [ELECTROMAG] A tape made of magnetic material for use in winding closed magnetic cores of toroids and transformers.

ferromagnetism [SOLID STATE] A property, exhibited by certain metals, alloys, and compounds of the transition (iron group) rare-earth and actinide elements, in which the internal magnetic moments spontaneously organize in a common direction; gives rise to a permeability considerably greater than that of vacuum, and to magnetic hysteresis.

ferromagnetography [GRAPHICS] A printing technique in which magnetic iron particles are applied to a latent image magnetized onto a metal sheet or drum, then transferred to ordinary paper and fixed in position by heat, pressure, or other means.

ferromanganese [MET] A ferroalloy containing about 80% manganese and used in steelmaking.

ferrometer [ENG] An instrument used to make permeability and hysteresis tests of iron and steel.

ferromolybdenum [MET] A molybdenum-iron alloy produced in the electric furnace or by a thermite process; used to introduce molybdenum into iron or steel alloys, and as a coating material on welding rods.

ferronickel [MET] A crude ferroalloy containing nickel.

ferrophosphorus [MET] A by-product formed in the heating of iron, phosphate rock, silica, and coke; this alloy is used to increase fluidity in steel casting.

ferroporphyrin [BIOCHEM] A red complex of porphyrin and iron in which the iron is in the 2+ oxidation state.

ferroprussiate paper [MATER] A paper used in a blueprint process to reproduce plans and drawings.

ferroresonant circuit [ELECTR] A resonant circuit in which a saturable reactor provides nonlinear characteristics, with tuning being accomplished by varying circuit voltage or current.

ferrosilicon [MET] A crude ferroalloy containing 15–95% silicon and used in steelmaking.

ferrosilicon process See Pidgeon process.

ferrosilite [MINERAL] A mineral in the orthopyroxene group; the iron analog of enstatite; occurs in hypersthene, but is not found separately in nature.

ferrospinel See hercynite.

ferrotitanium [MET] A ferroalloy containing 15–45% titanium and used in steelmaking.

ferrotremolite [MINERAL] The ferrous iron equivalent of the monoclinic amphibole, tremolite.

ferrotungsten [MET] A crude ferroalloy containing tungsten and used in steelmaking.

ferrotype [GRAPHICS] A photograph formed on a metal plate coated with collodion and sensitive salts.

ferrouranium [MET] An alloy of iron and uranium.

ferrous [CHEM] The term or prefix used to denote compounds of iron in which iron is in the divalent (2+) state.

ferrous acetate [ORG CHEM] $Fe(CH_3COO)_2 \cdot 4H_2O$ Soluble green crystals, soluble in water and alcohol, that are combustible and that oxidize to basic ferric acetate in air; used as textile dyeing mordant, as wood preservative, and in medicine. Also known as iron acetate.

ferrous alloy See ferroalloy.

ferrous ammonium sulfate [INORG CHEM] $Fe(SO_4) \cdot (NH_4)SO_4 \cdot 6H_2O$ Light-green, water-soluble crystals; used in medicine, analytical chemistry, and metallurgy. Also known as iron ammonium sulfate; Mohr's salt.

ferrous arsenate [INORG CHEM] $Fe_3(AsO_4)_2 \cdot 6H_2O$ Water-insoluble, toxic green amorphous powder, soluble in acids; used in medicine and as an insecticide. Also known as iron arsenate.

ferrous carbonate [INORG CHEM] $FeCO_3$ Green rhombohedral crystals that are soluble in carbonated water and decompose when heated; used in medicine.

ferrous chloride [INORG CHEM] $FeCl_2 \cdot 4H_2O$ Green, monoclinic crystals, soluble in water; used as a mordant in dyeing, for sewage treatment, in metallurgy, and in pharmaceutical preparations. Also known as iron chloride; iron dichloride.

ferrous hydroxide [INORG CHEM] $Fe(OH)_2$ A white, water-insoluble, gelatinous solid that turns reddish-brown as it oxidizes to ferric hydroxide.

ferrous oxalate [ORG CHEM] $Fe(COO)_2$ A water-soluble, yellow powder; used in photography and medicine. Also known as iron oxalate.

ferrous oxide [INORG CHEM] FeO A black powder, soluble in water, melting at 1419°C. Also known as black iron oxide; iron monoxide.

ferrous sulfate [INORG CHEM] $FeSO_4 \cdot 7H_2O$ Blue-green, water-soluble, monoclinic crystals; used as a mordant in dyeing wool, in the manufacture of ink, and as a disinfectant. Also known as ferrisulphas; green copperas; green vitriol; iron sulfate.

ferrous sulfide [INORG CHEM] FeS Black crystals, insoluble in water, soluble in acids, melting point 1195°C; used to generate hydrogen sulfide in ceramics manufacture. Also known as iron sulfide.

ferrovanadium [MET] An iron alloy high in vanadium (35–55%); used to add 0.1–2.5% vanadium during the manufacture of engineering steels and high-strength steels.

ferruccite [MINERAL] $NaBF_4$ An orthorhombic boron mineral consisting of sodium fluoborate.

ferruginous [SCI TECH] **1.** Pertaining to or containing iron. **2.** Having the appearance or color of iron rust (ferric oxide).

ferrule [DES ENG] **1.** A metal ring or cap attached to the end of a tool handle, post, or other device to strengthen and protect it. **2.** A bushing inserted in the end of a boiler flue to spread and tighten it. [ENG] See stabilizer.

ferrum [CHEM] Latin term for iron; derivation of the symbol Fe.

ferry [NAV ARCH] A boat which carries people, automotive vehicles, or goods across a river or other body of water, usually traveling back and forth on a regular schedule. [ORD] **1.** To deliver aircraft or ships by operating them under their own power. **2.** To transport personnel and materiel by air.

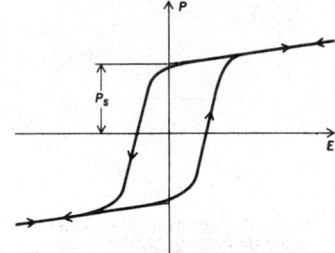

FERROELECTRIC HYSTERESIS LOOP

Hysteresis loop showing the relation between the resulting polarization P of the ferroelectric crystal and the externally applied electric field E; P_s is permanent or spontaneous magnetization.

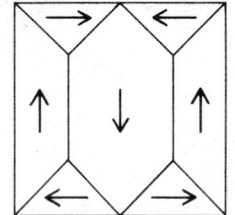

FERROMAGNETIC DOMAIN

Drawing of a uniaxial crystal showing the ferromagnetic domains. The arrows show the direction of the magnetic field in each domain.

FEYNMAN DIAGRAM

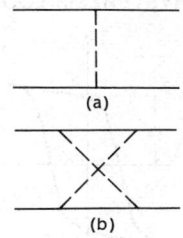

(a)

(b)

Typical Feynman diagrams for two-particle scattering. *(a)* Second-order diagram. *(b)* fourth-order diagram. Solid lines represent scattered particles; broken lines, particles which transmit force between them.

FIBER FLAX

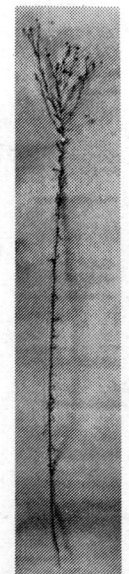

The type of flax plant grown for fiber, usually taller than the seed flax plant.

fersmanite [MINERAL] $(Na,Ca)_2(TI,Cb)Si(O,F)_6$ A brown mineral composed of a silicate fluoride of sodium, calcium, titanium, and columbium.

fersmite [MINERAL] $(Ca,Ce)(Cb,Ti)_2(O,F)_6$ A black mineral composed of an oxide and fluoride of calcium and columbium with cerium and titanium.

fertile material [NUCLEO] A material, such as thorium-232 or uranium-238, which is capable of being transformed into a fissionable material by capture of a neutron.

fertility [BIOL] The state of or capacity for abundant productivity.

fertility factor [GEN] An episomal bacterial sex factor which determines the role of a bacterium as either a male donor or as a female recipient of genetic material. Also known as F factor; sex factor.

fertilization [PHYSIO] The physicochemical processes involved in the union of the male and female gametes to form the zygote.

fertilization membrane [CYTOL] A membrane that separates from the surface of and surrounds many eggs following activation by the sperm; prevents multiple fertilization.

fertilizer [MATER] Material that is added to the soil to supply chemical elements needed for plant nutrition.

fertilizin [BIOCHEM] A mucopolysaccharide, derived from the jelly coat of an egg, that plays a role in sperm recognition and the stimulation of sperm motility and metabolic activity.

ferulic acid [ORG CHEM] $C_{10}H_{10}O_4$ A compound widely distributed in small amounts in plants, having two isomers: the cis form is a yellow oil, and the trans form is obtained from water solutions as orthorhombic crystals.

fervanite [MINERAL] $Fe_4V_4O_{16} \cdot 5H_2O$ Golden-brown mineral composed of a hydrated iron vanadate; although itself not radioactive, it occurs with radioactive minerals.

fescue [BOT] A group of grasses of the genus *Festuca*, used for both hay and pasture.

Fessler compound [FOOD ENG] Flocculating-salt mixture used to remove copper and iron from wines.

FET *See* field-effect transistor.

fetal asphyxia [MED] Deprivation of oxygen to the fetus due to interference with its blood supply.

fetal fat-cell lipoma *See* liposarcoma.

fetal hemoglobin [BIOCHEM] A normal embryonic hemoglobin having alpha chains identical to those of normal adult human hemoglobin, and gamma chains similar to adult beta chains.

fetal membrane [EMBRYO] Any one of the membranous structures which surround the embryo during its development period. Also known as extraembryonic membrane.

fetch [ADP] To locate and load into main memory a requested load module, relocating it as necessary and leaving it in a ready-to-execute condition. [OCEANOGR] **1.** The distance traversed by waves without obstruction. **2.** An area of the sea surface over which seas are generated by a wind having a constant speed and direction. **3.** The length of the fetch area, measured in the direction of the wind in which the seas are generated. Also known as generating area.

fetch bit [ADP] The fifth bit in a storage key; the value of the fetch bit can protect a stored block from destruction or from being accessed by unauthorized programs.

fetometamorphism [INV ZOO] A life cycle variation in the Cantharidae (Coleoptera); the larvae hatch prematurely as legless, immature prelarvae.

fettling knife [GRAPHICS] A sharp instrument with a flexible blade tapering to a point; used in ceramics for carving of clay models, sgraffito, removing mold marks, and miscellaneous other purposes.

fetus [EMBRYO] **1.** The unborn offspring of viviparous mammals in the later stages of development. **2.** In human beings, the developing body in utero from the beginning of the ninth week after fertilization through the fortieth week of intrauterine gestation, or until birth.

Feulgen reaction [ANALY CHEM] An aldehyde specific reaction based on the formation of a purple-colored compound when aldehydes react with fuchsin-sulfuric acid; deoxyribonucleic acid gives this reaction after removal of its purine bases by acid hydrolysis; used as a nuclear stain.

fever [MED] An elevation in the central body temperature of warm-blooded animals caused by abnormal functioning of the thermoregulatory mechanisms.

Feyliniidae [VERT ZOO] The limbless skinks, a family of reptiles in the suborder Sauria represented by four species in tropical Africa.

Feynman diagram [QUANT MECH] A diagram which gives an intuitive picture of a term in a perturbation expansion of a scattering matrix element or other physical quantity associated with interactions of particles; each line represents a particle, each vertex an interaction.

Feynman integral [QUANT MECH] A term in a perturbation expansion of a scattering matrix element; it is an integral over the Minkowski space of various particles (or over the corresponding momentum space) of the product of propagators of these particles and quantities representing interactions between the particles.

Feynman's superfluidity theory [CRYO] Microscopic theory of superfluid helium which accounts for the spectrum of elementary excitations assumed by Landau's superfluidity theory.

F factor *See* fertility factor.

F format [ADP] **1.** In data management, a fixed-length logical record format. **2.** In FORTRAN, a real variable formatted as $F\mu.d$, where μ is the width of the field and d represents the number of digits to appear after the decimal point.

fiber [BOT] **1.** An elongate, thick-walled, tapering plant cell that lacks protoplasm and has a small lumen. **2.** A very slender root. [MET] **1.** The characteristic of wrought metal that indicates directional properties as revealed by etching or by fracture appearance. **2.** The pattern of preferred orientation of metal crystals after a deformation process, usually wiredrawing. [OPTICS] A transparent threadlike object made of glass or clear plastic, used to conduct light along selected paths. [TEXT] An extremely long, pliable, cohesive natural or man-made threadlike object from which yarns are spun to be woven into textiles.

fiberboard [MATER] A hard isotropic board made by compressing wood chips or other vegetable fibers.

fiber bundle [MATH] Consists of topological spaces X, B, and F, with a continuous onto map $p: X \to B$, where each point x_0 in B has a neighborhood U such that UxF is homeomorphic to $p^{-1}(U)$. [OPTICS] A flexible bundle of glass or other transparent fibers, parallel to each other, used in fiber optics to transmit a complete image from one end of the bundle to the other.

fiber crops [AGR] Plants, such as flax, hemp, jute, and sisal, cultivated for their content or yield of fibrous material.

fiber flax [BOT] The flax plant grown in fertile, well-drained, well-prepared soil and cool, humid climate; planted in the early spring and harvested when half the seed pods turn yellow; used in the manufacture of linen.

Fiberglas [MATER] Trade name of Owens-Corning Fiberglas Company for glass fibers made in diameters from 0.0002 to 0.008 inch (0.005 to 0.2 millimeter).

fiber grease [MATER] Solid-base lubricating grease; contains soap fibers 1–1000 micrometers long and 0.1–1.0 micrometer wide, which stabilize lubricating action by immobilizing fluid lubricating components.

fiberizer [MIN ENG] A hammer mill which cracks open asbestos-bearing rock to yield a fibrous product.

fiber metal [MET] Any material composed of metal fibers that are pressed or sintered together or infiltrated with resin or other material.

fiber metallurgy [MET] A branch of metallurgy concerned with the study of metal fibers.

fiber optics [OPTICS] The technique of transmitting light through long, thin, flexible fibers of glass, plastic, or other transparent materials; bundles of parallel fibers can be used to transmit complete images.

fiber plaster [MATER] Gypsum plaster containing hair or wood fiber as a binder.

fiberscope [OPTICS] An arrangement of parallel glass fibers with an objective lens on one end and an eyepiece at the other; the assembly can be bent as required to view objects that are inaccessible for direct viewing.

fiber stress [MECH] **1.** The tensile or compressive stress on the fibers of a fiber metal or other fibrous material, especially when fiber orientation is parallel with the neutral axis. **2.**

Local stress through a small area (a point or line) on a section where the stress is not uniform, as in a beam under bending load.

Fibonacci sequence [MATH] The sequence 1,1,2,3,5,8,13,21, ..., or any sequence where each entry is the sum of the two previous entries.

fibratus [METEOROL] A cloud species characterized by a fine hairlike or striated composition, the filaments of which are usually distinctly separated from each other; the extremities of these filaments are always thin and never terminated by tufts or hooks. Also known as filosus.

fibril [BIOL] A small thread or fiber, as a root hair or one of the structural units of a striated muscle. [MATER] One of the minute threadlike elements of a natural or synthetic fiber.

fibrillation [PHYSIO] An independent, spontaneous, local twitching of muscle fibers.

fibrillose [BIOL] Having fibrils.

fibrin [BIOCHEM] The fibrous, insoluble protein that forms the structure of a blood clot; formed by the action of thrombin.

fibrinase [BIOCHEM] An enzyme that catalyzes the formation of covalent bonds between fibrin molecules. Also known as fibrin-stabilizing factor.

fibrinogen [BIOCHEM] A plasma protein synthesized by the parenchymal cells of the liver; the precursor of fibrin. Also known as factor I.

fibrinogenopenia [MED] A congenital hemorrhagic diathesis in which there is a decrease in fibrinogen in the plasma.

fibrinoid [BIOCHEM] A homogeneous, refractile, oxyphilic substance occurring in degenerating connective tissue, as in term placentas, rheumatoid nodules, and Aschoff bodies, and in pulmonary alveoli in some prolonged pneumonitides.

fibrinolysin See plasmin.

fibrinolysis [PHYSIO] Liquefaction of coagulated blood by the action of plasmin on fibrin.

fibrinous pericarditis [MED] Inflammation of the pericardium involving the deposition of fibrin and leukocytes between the layers of the pericardium; seen in rheumatic carditis and acute infectious diseases.

fibrin-stabilizing factor See fibrinase.

Fibro [TEXT] Trade name for a viscose rayon fiber made by Courtaulds North America.

fibroadenoma [MED] A benign tumor containing both fibrous and glandular elements.

fibroblast [HISTOL] A stellate connective tissue cell found in fibrous tissue. Also known as a fibrocyte.

fibroblastic [PETR] Of a metamorphic rock, having a texture that is homeoblastic as a result of the development of minerals with a fibrous habit during recrystallization.

fibrocartilage [HISTOL] A form of cartilage rich in dense, closely opposed bundles of collagen fibers; occurs in intervertebral disks, in the symphysis pubis, and in certain tendons.

fibrocyte See fibroblast.

fibroferrite [MINERAL] Fe(SO₄)(OH)·5H₂O A yellowish mineral composed of a hydrous basic ferric sulfate, occurring in fibrous form.

fibroid [HISTOL] Composed of fibrous tissue.

fibroid tumor See fibroma.

fibrolite See sillimanite.

fibroma [MED] A benign tumor composed primarily of fibrous connective tissue. Also known as fibroid tumor.

fibroma molluscum See neurofibromatosis.

fibromatosis [MED] **1.** The occurrence of multiple fibromas. **2.** Localized proliferation of fibroblasts without apparent cause.

fibromyoma [MED] A benign tumor, usually of smooth muscle, with a prominent fibrous stroma; commonly a uterine leiomyoma.

fibromyosis See interstitial endometrosis.

fibromyositis See myositis.

fibroplasia [MED] The growth of fibrous tissue, as in the second phase of wound healing.

fibrosarcoma [MED] A sarcoma composed of spindle cells that produce collagenous fibrils.

fibrosing adenomatosis See sclerosing adenomatosis.

fibrosis [MED] Growth of fibrous connective tissue in an organ or part in excess of that naturally present.

fibrositis [MED] Inflammation of white fibrous connective tissue, usually in a joint region.

fibrous dysplasia [MED] **1.** Extensive formation of fibrous tissue and transformation of bony tissue in one or more bones. **2.** Development of abnormal amounts of fibrous tissue in the mammary glands.

fibrous fracture [MECH] Failure of a material resulting from a ductile crack; broken surfaces are dull and silky. Also known as ductile fracture.

fibrous ice See acicular ice.

fibrous osteoma See ossifying fibroma.

fibrous plaster [MATER] Gypsum plaster reinforced or backed with sisal or canvas.

fibrous protein [BIOCHEM] Any of a class of highly insoluble proteins representing the principal structural elements of many animal tissues.

fibrous structure [MATER] A ropy surface on a fractured material. [MET] **1.** A lamination on an etched section of a forging. **2.** In wrought iron, a structure consisting of slag fibers embedded in ferrite.

fibula [ANAT] The outer and usually slender bone of the hind or lower limb below the knee in vertebrates; it articulates with the tibia and astragalus in humans, and is ankylosed with the tibia in birds and some mammals.

Fick's law [PHYS] The law that the rate of diffusion of matter across a plane is proportional to the negative of the rate of change of the concentration of the diffusing substance in the direction perpendicular to the plane.

fictitious [NAV] Pertaining to or measured from an arbitrary reference line as in fictitious equator, fictitious latitude, or fictitious longitude.

fictitious craft [NAV] In marine usage, an imaginary craft used in the solution of certain maneuvering problems, as when a ship to be intercepted is expected to change course or speed during the interception run.

fictitious equator [NAV] A reference line serving as the origin for measurement of a fictitious latitude.

fictitious graticule [NAV] A graticule formed from a set of fictitious parallels and meridians to resemble the geographic graticule but offset from it; coordinates taken from the fictitious graticule are used with a transverse or oblique map projection or with a navigation grid.

fictitious latitude [NAV] Angular distance from the equator as shown on a fictitious graticule.

fictitious longitude [NAV] The arc of the equator between the prime meridian and any given meridian derived from the fictitious graticule.

fictitious loxodrome See fictitious rhumb line.

fictitious loxodromic curve See fictitious rhumb line.

fictitious meridian [NAV] One of the coordinates of a fictitious graticule.

fictitious parallel [NAV] One of the coordinates derived from a fictitious graticule.

fictitious pole [NAV] One of the two points 90° from the equator on a fictitious graticule; called transverse or oblique pole, depending upon the type of fictitious equator.

fictitious rhumb line [NAV] A line on a fictitious graticule making the same oblique angle with all meridians. Also known as fictitious loxodrome; fictitious loxodromic curve.

fictitious ship [NAV] An imaginary ship serving as a fictitious craft.

fictitious vehicle [NAV] An imaginary vehicle serving as a fictitious craft.

fictitious year [ASTRON] The period between successive returns of the fictitious mean sun to a sidereal hour angle of 80° (right ascension 18 hours 40 minutes; about January 1); the length of the fictitious year is the same as that of the tropical year, since both are based upon the position of the sun with respect to the vernal equinox. Also known as Besselian year.

Ficus [BOT] A genus of tropical trees in the family Moraceae including the rubber tree and the fig tree. [INV ZOO] A genus of gastropod mollusks having pear-shaped, spirally ribbed sculptured shells.

fiddley See fidley.

fidelity [COMMUN] The degree to which a system accurately reproduces at its output the essential characteristics of the signal impressed on its input.

fidley [NAV ARCH] Also spelled fiddley. **1.** A wide opening

FIELD COIL

A rotor of a synchronous motor showing the field coils. (Allis-Chalmers)

FIELD-EFFECT TRANSISTOR

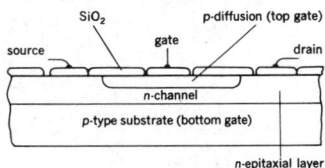

One type of field-effect transistor: the junction-gate field-effect transistor.

FIELD EMISSION

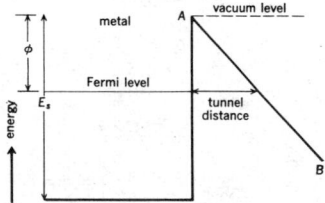

Diagram of energy level scheme for field emission from a metal at absolute zero. Dashed vacuum level represents energy of an electron at rest in free space. Energy of conduction electron at rest in metal lies below vacuum level by amount E_s. Energy levels below Fermi level are occupied by conduction electrons whereas those above are empty. AB is the potential of an electron outside the metal in presence of a strong constant field produced by a positive electrode. ϕ is the work function.

FIELD-EMISSION MICROSCOPE

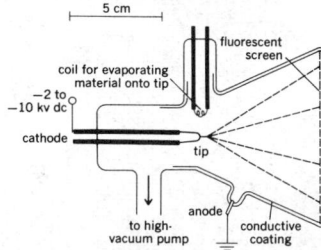

Electron-operated field-emission microscope.

above a fire room through which pass ventilators. **2.** A framework of iron about the ladder of a hatch in a ship's deck leading below the deck.

FIDO [METEOROL] A system for artificially dissipating fog, in which gasoline or other fuel is burned at intervals along an airstrip to be cleared. Derived from fog investigation dispersal operations.

fiducial point [OPTICS] A mark, or one of several marks, visible in the field of view of an optical instrument, used as a reference or for measurement. Also known as fiduciary point.

fiducial temperature [METEOROL] That temperature at which, in a specified latitude, the reading of a particular barometer requires no temperature or latitude correction. [THERMO] Any of the temperatures assigned to a number of reproducible equilibrium states on the International Practical Temperature Scale; standard instruments are calibrated at these temperatures.

fiduciary point *See* fiducial point.

fiedlerite [MINERAL] $Pb_3(OH)_2Cl_4$ A colorless mineral composed of a hydroxychloride of lead, occurring as monoclinic crystals.

Fiedler's myocarditis *See* interstitial myocarditis.

field [ADP] A specified area, such as a group of card columns or a set of bit locations in a computer word, used for a particular category of data. [ELEC] That part of an electric motor or generator which produces the magnetic flux which reacts with the armature, producing the desired machine action. [ELECTR] One of the equal parts into which a frame is divided in interlaced scanning for television; includes one complete scanning operation from top to bottom of the picture and back again. [MATH] An algebraic system possessing two operations which have all the properties that addition and multiplication of real numbers have. [OPTICS] *See* field of view. [PHYS] **1.** An entity which acts as an intermediary in interactions between particles, which is distributed over part or all of space, and whose properties are functions of space coordinates and, except for static fields, of time; examples include gravitational field, sound field, and the strain tensor of an elastic medium. **2.** The quantum-mechanical analog of this entity, in which the function of space and time is replaced by an operator at each point in space-time.

field artillery [ORD] Artillery mounted on carriages, and mobile enough to accompany infantry or armored units in the field.

field artillery observer [ORD] A person who watches the effects of artillery fire, adjusts the center of impact of that fire onto a target, and reports results to the firing agency.

field artillery trainer [ORD] A small practice gun and carriage unit with a telescope and a mechanism for adjusting elevation and deflection; used in training field artillery personnel.

fieldata code [COMMUN] A standardized military data transmission code, seven data bits plus one parity bit.

field brightness *See* adaptation luminance.

field changes [METEOROL] With regard to thunderstorm electricity, the rapid variations in the vertical component of the electric field strength at the earth's surface.

field coil [ELECTROMAG] A coil used to produce a constant-strength magnetic field in an electric motor, generator, or excited-field loudspeaker; depending on the type of motor or generator, the field core may be on the stator or the rotor. Also known as field winding.

field delimiter [ADP] Any symbol, such as a slash, colon, tab, or space, which enables an assembler to recognize the end of a field.

field desorption [SOLID STATE] A technique which tears atoms from a surface by an electric field applied at a sharp dip to produce very well-ordered, clean, plane surfaces of many crystallographic orientations.

field-desorption microscope [ELECTR] A type of field-ion microscope in which the tip specimen is imaged by ions that are field-desorbed or field-evaporated directly from the surface rather than by ions obtained from an externally supplied gas.

field discharge [ELECTR] A spark discharge due to high potential across a gap.

field distortion [ELECTROMAG] Any alteration in the direction of an electric or magnetic field; in particular, distortion

of the magnetic fields between the north and south poles of a generator due to the counter electromotive force in the armature winding.

field-effect diode [ELECTR] A semiconductor diode in which the charge carriers are of only one polarity.

field-effect display [OPTICS] A type of numerical display device in which a liquid-crystal cell is sandwiched between polarizers; the cell is treated so that it normally rotates light 90°, but ceases to rotate light when an electric field is applied to it, altering the transmission of the device.

field-effect tetrode [ELECTR] Four-terminal device consisting of two independently terminated semiconducting channels so displaced that the conductance of each is modulated along its length by the voltage conditions in the other.

field-effect transistor [ELECTR] A transistor in which the resistance of the current path from source to drain is modulated by applying a transverse electric field between grid or gate electrodes; the electric field varies the thickness of the depletion layer between the gates, thereby reducing the conductance. Abbreviated FET.

field-effect varistor [ELECTR] A passive, two-terminal, nonlinear semiconductor device that maintains constant current over a wide voltage range.

field emission [ELECTR] The emission of electrons from the surface of a metallic conductor into a vacuum (or into an insulator) under influence of a strong electric field; electrons penetrate through the surface potential barrier by virtue of the quantum-mechanical tunnel effect. Also known as cold emission.

field-emission microscope [ELECTR] A device that uses field emission of electrons or of positive ions (field-ion microscope) to produce a magnified image of the emitter surface on a fluorescent screen.

field-emission tube [ELECTR] A vacuum tube within which field emission is obtained from a sharp metal point; must be more highly evacuated than an ordinary vacuum tube to prevent contamination of the point.

field emplacement [ORD] Platform, support, or other position for artillery, machine guns, and so forth, in the field.

field engineer [ENG] **1.** An engineer who is in charge of directing civil, mechanical, and electrical engineering activities in the production and transmission of petroleum and natural gas. **2.** An engineer who operates at a construction site.

field-enhanced emission [ELECTR] An increase in electron emission resulting from an electric field near the surface of the emitter.

field excitation [MECH ENG] Control of the speed of a series motor in an electric or diesel-electric locomotive by changing the relation between the armature current and the field strength, either through a reduction in field current by shunting the field coils with resistance, or through the use of field taps.

field fortification [ORD] Fortification constructed in the field to strengthen the natural defenses of the ground features, including foxholes, obstacles, trenches, gun emplacements, and so forth.

field-free emission current [ELECTR] Electron current emitted by a cathode when the electric field at the surface of the cathode is zero. Also known as zero-field emission.

field frequency [ELECTR] The number of fields transmitted per second in television; equal to the frame frequency multiplied by the number of fields that make up one frame. Also known as field repetition rate.

field geology [GEOL] The study of rocks and rock materials in their environment and in their natural relations to one another.

field gradient [PHYS] **1.** A vector obtained by applying the del operator to a scalar field. **2.** A tensor obtained by dyadic multiplication of the del operator with a vector field.

field gun [ORD] Any artillery piece mounted on a carriage for use in the field, for example, a cannon.

field index [NUCLEO] The constant n for a betatron in which the magnetic field strength at radius r is equal to $B_0 (r/R)^{-n}$, where R is the radius of equilibrium orbit of an electron, and B_0 is the corresponding magnetic field. Also known as n value.

field intensity [COMMUN] In Federal Communications Commission regulations, the electric field intensity in the horizontal direction. .[PHYS] *See* field strength.

field investigation [SCI TECH] An investigation carried out in the field; usually applied to an investigation made by someone not domiciled at the site.

field-ion microscope [ELECTR] A microscope in which atoms are ionized by an electric field near a sharp tip; the field then forces the ions to a fluorescent screen, which shows an enlarged image of the tip, and individual atoms are made visible; this is the most powerful microscope yet produced. Also known as ion microscope.

field laboratory [SCI TECH] Usually a temporary or portable laboratory facility set up at the site of an operation to conduct chemical or physical evaluations.

field length [ADP] The number of columns, characters, or bits in a specified field.

field lens [OPTICS] The lens in a two-lens eyepiece which is farther from the eye.

field luminance *See* adaptation luminance.

field magnet [ELECTROMAG] The magnet which creates a magnetic field in an electric machine or device.

field map [MAP] A map made in the field and bearing observations of various kinds upon which the final map is based.

field of fire [ORD] The area which a weapon or group of weapons may cover effectively with fire from a given position.

field of planes on a manifold [MATH] A continuous assignment of a vector subspace of tangent vectors to each point in the manifold. Also known as plane field.

field of search [ELECTR] The space that a radar set or installation can cover effectively.

field of vectors on a manifold [MATH] A continuous assignment of a tangent vector to each point in the manifold. Also known as vector field.

field of view [OPTICS] The area or solid angle which can be viewed through an optical instrument. Also known as field.

field operator [QUANT MECH] An operator function of space and time for the annihilation or creation of a particle.

field pattern *See* radiation pattern.

field piece [ORD] An artillery gun or howitzer mounted on a carriage, for use in the field.

field pole [ELECTROMAG] A structure of magnetic material on which a field coil of a loudspeaker, motor, generator, or other electromagnetic device may be mounted.

field quenching [MET] The quench cooling and tempering of a heated metal object at the site of construction or operation by using portable equipment rather than fixed manufacturing facilities. [SOLID STATE] Decrease in the emission of light of a phosphor excited by ultraviolet radiation, x-rays, alpha particles, or cathode rays when an electric field is simultaneously applied.

field repetition rate *See* field frequency.

field rheostat [ELEC] A rheostat used to adjust the current in the field winding of an electric machine.

field scan [ELECTR] Television term denoting the vertical excursion of an electron beam downward across a cathode-ray tube face, the excursion being made in order to scan alternate lines.

field section [ADP] A portion of a field, such as the section formed by the second and third character of a 10-character field.

field stars [ASTRON] Background stars when a specific object is being observed.

field strength [PHYS] A vector characterizing a field. Also known as field intensity.

field-strength meter [ENG] A calibrated radio receiver used to measure the field strength of radiated electromagnetic energy from a radio transmitter.

field-strip [ORD] To disassemble the major components of a machine gun, cannon, or other firearm for cleaning, inspection, or the like.

field telephone [COMMUN] A portable telephone designed for field or combat use.

field test [SCI TECH] A nonformal experiment, that is, one with fewer controls than a laboratory experiment, conducted under field conditions.

field theory [MATH] The study of fields and their extensions.

[PHYS] A theory in which the basic quantities are fields; classically the equations governing the fields may be given; in quantum field theory the commutation rules satisfied by the field operators also are specified. [PSYCH] A psychological theory that emphasizes the importance of interactions between events in an individual's environment.

field waveguide [ELECTROMAG] A single wire, threaded or coated with dielectric, which guides an electromagnetic field. Also known as G string.

field weapons [ORD] Weapons designed or intended for actual use in the field.

field weld [MET] A weld made at the construction site.

field winding *See* field coil.

field wire [ELEC] An insulated flexible wire or cable used in field telephone and telegraph systems.

fieldwork [SCI TECH] Work done, such as surveying or making geological observations, in the field.

Fierz interference [NUC PHYS] Interference between the axial vector and tensor parts of the weak interaction of nucleon and lepton (electron-neutrino) fields in beta decay; measurements of the beta-particle energy spectrum indicate that it vanishes.

fièvre boutonneuse [MED] A mild febrile rickettsial disease of man caused by *Rickettsia conori;* characterized by a rash, tache noire (primary ulcer), and swollen lymph glands. Also known as boutonneuse fever; Marseilles fever.

fife rail [NAV ARCH] A metal or wooden rail with holes for belaying pins.

FIFO *See* first-in, first-out.

fifteen-degrees calorie *See* calorie.

fifty-percent zone [ORD] Area enclosing the center of dispersion or impact within which one half of all shots fired with the same setting will fall.

fig [BOT] *Ficus carica.* A deciduous tree of the family Moraceae cultivated for its edible fruit, which is a syconium, consisting of a fleshy hollow receptacle lined with pistillate flowers.

fighter aircraft [AERO ENG] A military aircraft designed primarily to destroy other aircraft in the air; may also be used to bomb military targets; it is maneuverable and has a high rate of climb.

fighter bomber [AERO ENG] A fighter aircraft that is designed to have bombs, or rockets, added to it so that it may be used as a bomber.

fighter interceptor [AERO ENG] A fighter aircraft designed to intercept and shoot down enemy aircraft.

fighting compartment [ORD] A portion of a fighting vehicle in which the occupants service and fire the principal armament, occupying a part of the hull and all of the turret, if any.

Figitidae [INV ZOO] A family of hymenopteran insects in the superfamily Cynipoidea.

figurative constant [ADP] A predefined constant in COBOL which does not require a description in data division, such as ZERO which stands for 0.

figure [MATER] The natural grain of wood, especially when it is cut as a veneer.

figurehead [NAV ARCH] An ornament placed on the foremost edge of the stem just below the bowsprit.

figure of merit [ELECTR] A performance rating that governs the choice of a device for a particular application; for example, the figure of merit of a magnetic amplifier is the ratio of usable power gain to the control time constant.

figure of the earth [GEOD] A precise geometric shape of the earth.

figures shift [COMMUN] A physical movement that permits a teletypewriter to print uppercase characters, numbers, symbols, and the like.

figure stone *See* agalmatolite.

Fiji disease [PL PATH] A virus disease of sugarcane; elongated swellings on the underside of leaves precede death of the plant.

filament [BOT] **1.** The stalk of a stamen which supports the anther. **2.** A chain of cells joined end to end, as in certain algae. [ELEC] Metallic wire or ribbon which is heated in an incandescent lamp to produce light, by passing an electric current through the filament. [ELECTR] A cathode made of resistance wire or ribbon, through which an electric current

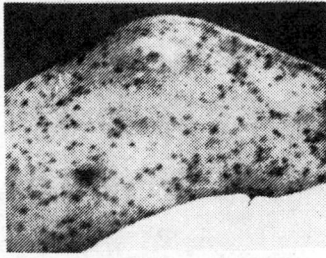

Fièvre boutonneuse, showing rash and tache noire in popliteal area. *(From C. B. Philip, in R. L. Pullen, ed., Communicable Diseases, Lea and Febiger, 1950)*

FIG

Fig *(Ficus carica)*, details of branch and fruit.

FILAMENT

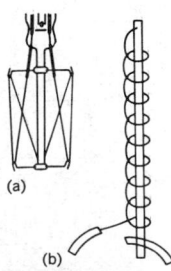

(a)

(b)

Two types of filaments: *(a)* in incandescent lamp; *(b)* in thermionic tube. *(General Electric Co.)*

is sent to produce the high temperature required for emission of electrons in a thermionic tube. Also known as directly heated cathode; filamentary cathode; filament-type cathode. [INV ZOO] A single silk fiber in the cocoon of a silkworm. [MET] A long, flexible metal wire drawn very fine. [SCI TECH] A long, flexible object with a small cross section. [TEXT] A single continuous man-made fiber which is extruded from a spinneret and joined with others to make a thread.

filamentary cathode *See* filament.

filament current [ELECTR] The current supplied to the filament of an electron tube for heating purposes.

filament drawing [MET] Reducing the cross section of wire by pulling it through a die to form a filament.

filament emission [ELECTR] Liberation of electrons from a heated filament wire in an electron tube.

filament lamp *See* incandescent lamp.

filamentous bacteria [MICROBIO] Bacteria, especially in the order Actinomycetales, whose cells resemble filaments and are often branched.

filament saturation *See* temperature saturation.

filament transformer [ELECTR] A small transformer used exclusively to supply filament or heater current for one or more electron tubes.

filament-type cathode *See* filament.

filament winding [ELECTR] The secondary winding of a power transformer that furnishes alternating-current heater or filament voltage for one or more electron tubes. [ENG] A process for fabricating a composite structure in which continuous fiber reinforcement (glass, boron, silicon carbide), either previously impregnated with a matrix material or impregnated during winding, are wound under tension over a rotating core.

filaria [INV ZOO] A parasitic filamentous nematode belonging to the order Filaroidea.

filariasis [MED] A disease due to the presence of hairlike nematodes (filariae) in humans, including *Wuchereria bancrofti*, *W. pacifica*, and *Onchocerca volvulus*.

Filarioidea [INV ZOO] An order of the class Nematoda comprising highly specialized parasites of humans and domestic animals.

filar micrometer [DES ENG] An instrument used to measure small distances in the field of an eyepiece by using two parallel wires, one of which is fixed while the other is moved at right angles to its length by means of an accurately cut screw.

filature [TEXT] A factory where silk is unwound from cocoons and the strands are collected into skeins.

filbert [BOT] Either of two European plants belonging to the genus *Corylus* and producing a thick-shelled, edible nut. Also known as hazelnut.

file [ADP] A collection of related records treated as a unit. [DES ENG] A steel bar or rod with cutting teeth on its surface; used as a smoothing or forming tool.

file control system [ADP] Software package which handles the transfer of data from any device into any device.

file event [ADP] A single access to any storage device for either input or output.

file gap [ADP] An area in a data storage medium which is used mainly to indicate the end of a file and sometimes the beginning of another.

file hardness [ENG] Hardness of a material as determined by testing with a file of standardized hardness; a material which cannot be cut with the file is considered as hard as or harder than the file.

file header [ADP] A set of words comprising the file name and various characteristics of the file, found at the beginning of a file stored on magnetic tape or disk.

file identification [ADP] A device, such as a label or tag, used to identify, describe, or name a physical medium, such as a reel of digital magnetic tape or a box of punched cards, which contains data.

file maintenance [ADP] Data-processing operation in which a master file is updated on the basis of one or more transaction files.

file name [ADP] The name given by the programmer to a specific set of data.

file organization [ADP] The structure of a file meeting two requirements: to minimize the running time of the program, and to simplify the work involved in modifying the contents of the file.

file organization routine [ADP] A program which allocates data files into random-access storage devices.

file-oriented system [ADP] A computer configuration which considers a heavy, or exclusive, usage of data files.

file processing [ADP] The job of updating, sorting, or validating a data file.

file protection [ADP] A mechanical device or a computer command which prevents erasing of or writing upon a magnetic tape but allows a program to read the data from the tape.

file reference [ADP] An operation involving looking up and retrieving the information on file for a specified item or items.

file search [ADP] An operation involving looking through the file for information on all items falling in a specified category, extracting the information for any item where the information recorded meets certain criteria, and determining whether or not there exists a specified pattern of information anywhere in the file.

Filicales [BOT] The equivalent name for Polypodiales.

Filicineae [BOT] The equivalent name for Polypodiatae.

Filicornia [INV ZOO] A group of hyperiid amphipod crustaceans in the suborder Genuina having the first antennae inserted anteriorly.

filiform [BIOL] Threadlike or filamentous.

filiform corrosion [MET] A random threadlike deterioration of a painted or lacquered metal caused by superficial corrosion of the base metal.

filiform lapilli *See* Pele's hair.

filiform papilla [ANAT] Any one of the papillae occurring on the dorsum and margins of the oral part of the tongue, consisting of an elevation of connective tissue covered by a layer of epithelium.

fill [CIV ENG] Earth used for embankments or as backfill. [MIN ENG] *See* pack.

filled band [SOLID STATE] An energy band, each of whose energy levels is occupied by an electron.

filled composite [MATER] Mixture (composite) of thermoplastic or thermosetting resin and granular or short-strand fiber fill.

filled stopes [MIN ENG] Stopes filled with barren stone, low-grade ore, sand, or tailings (mill waste) after the ore has been extracted.

filled-system thermometer [ENG] A thermometer which has a bourdon tube connected by a capillary tube to a hollow bulb; the deformation of the bourdon tube depends on the pressure of a gas (usually nitrogen or helium) or on the volume of a liquid filling the system. Also known as filled thermometer.

filled thermometer *See* filled-system thermometer.

filled thermoplastic [MATER] A thermoplastic resin material that has been extended (filled) with an inert filler powder or fibers before curing.

filled thermoset [MATER] Thermosetting resin material that has been extended (filled) with an inert filler powder or fibers before curing.

filler [MATER] 1. An inert material added to paper, resin, bituminous material, and other substances to modify their properties and improve quality. 2. A material used to fill holes in wood, plaster, or other surfaces before applying a coating such as paint or varnish. [MET] The rod used to deposit metal in a joint in brazing, soldering, or welding.

filler specks [MATER] In a cast plastic object, visible specks of a filler such as wood flour or asbestos that stand out in color contrast against the surface of the object.

fillet [BUILD] 1. A flat molding that separates rounded or angular moldings. 2. A narrow, concavely curved strip of material in the angle between two surfaces. [ENG] Any narrow, flat metal or wood member. [FOOD ENG] A boneless slice of meat or fish.

fillet gage [DES ENG] A gage for measuring convex or concave surfaces.

fillet lightning *See* ribbon lightning.

fillet weld [MET] A weld joining two edges at right angles; cross-sectional configuration is approximately triangular.

fill factor [MECH ENG] The approximate load that the dipper of a shovel is carrying, expressed as a percentage of the rated capacity.

filling [ENG] The loading of trucks with any material. [METEOROL] An increase in the central pressure of a pressure system on a constant-height chart, or an analogous increase in height on a constant-pressure chart; the term is commonly applied to a low rather than to a high. [MIN ENG] Allowing a mine to fill with water. [TEXT] The yarn running perpendicular to the lengthwise, or warp, yarn in weaving. Also known as pick; weft; woof.

fillowite [MINERAL] $H_2Na_6(Mn,Fe,Ca)_{14}(PO_4)_{12} \cdot H_2O$ A brown, yellow, or colorless mineral composed of a hydrous phosphate of manganese, iron, sodium, and other metals.

fill-up work *See* internal work.

film [BIOL] A thin, membranous skin, such as a pellicle. [ELEC] The layer adjacent to the valve metal in an electrochemical valve, in which is located the high voltage drop when current flows in the direction of high impedance. [GRAPHICS] Plastic material, such as cellulose acetate or cellulose nitrate, coated with a light-sensitive emulsion, used to make negatives or transparencies in radiography or photography. [MATER] A thin, flexible, transparent sheet of plastic, adhesive, rubber, or other material. [MED] A pathological opacity, as of the cornea. [MET] Oxide coating on a metal.

film badge [NUCLEO] A device worn for the purpose of indicating the absorbed dose of radiation received by the wearer; usually made of metal, plastic, or paper and loaded with one or more pieces of x-ray film. Also known as badge meter.

film base [GRAPHICS] The celluloid component which supports the emulsion of photographic film.

film coefficient [THERMO] For a fluid confined in a vessel, the rate of flow of heat out of the fluid, per unit area of vessel wall divided by the difference between the temperature in the interior of the fluid and the temperature at the surface of the wall. Also known as convection coefficient.

film cooling [THERMO] The cooling of a body or surface, such as the inner surface of a rocket combustion chamber, by maintaining a thin fluid layer over the affected area.

film density [GRAPHICS] Degree of film image opacity.

film-development chromatography [ANALY CHEM] Liquid-analysis chromatographic technique in which the stationary phase (adsorbent) is a strip or layer, as in paper or thin-layer chromatography.

film jacket [GRAPHICS] A transparent acetate single or multiple sleeve or pocket made to hold microfilm in flat strips.

film mottle [GRAPHICS] Cloudy or blotchy appearance of photographic film; the uneven density is generally caused by insufficient agitation.

filmogen [MATER] The film-forming material or binder in paint which imparts continuity.

film opaque [GRAPHICS] Any of the liquids applied with small brushes to processed film to cover imperfections known as pinholes, very small transparent dots in an otherwise black emulsion area.

film optical-sensing device [ADP] A device capable of digitizing the information stored on a film.

film plane [GRAPHICS] The position in a camera occupied by the film or plate.

film platen [ENG] A device which holds film in the focal plane during exposure.

film pressure [PHYS] The difference between the surface tension of a pure liquid and the surface tension of the liquid with a unimolecular layer of a given substance adsorbed on it. Also known as surface pressure.

film reader [ELECTR] A device for converting a pattern of transparent or opaque spots on a photographic film into a series of electric pulses. [OPTICS] A device for projecting or displaying microfilm so that an operator can read the data on the film; usually provided with equipment for moving or holding the film.

film recorder [ELECTR] A device which places data, usually in the form of transparent and opaque spots or light and dark spots, on photographic film.

film resistor [ELEC] A fixed resistor in which the resistance element is a thin layer of conductive material on an insulated form; the conductive material does not contain binders or insulating material.

film scanning [ELECTR] The process of converting motion picture film into corresponding electric signals that can be transmitted by a television system.

film size [GRAPHICS] Generally, an indication of film width, for example, 16 millimeter, or 5 inch, 8 inch, and so forth.

film sizing table [MIN ENG] A table used in ore dressing for sorting fine material by means of a film of flowing water.

film strength [MATER] **1.** The measurement of a lubricant's ability to keep an unbroken film over surfaces. **2.** The resistance to disruption by films of all types, such as plastic films or surface-coating films.

filmstrip [GRAPHICS] A continuous length of 35-millimeter film containing a number of still photographs, drawings, or charts, which are projected on a screen one at a time.

film theory [PHYS] A theory of the transfer of material or heat across a phase boundary, where one or both of the phases are flowing fluids, the main controlling factor being resistance to heat conduction or mass diffusion through a relatively stagnant film of the fluid next to the surface. Also known as boundary-layer theory.

film transport [MECH ENG] **1.** The mechanism for moving photographic film through the region where light strikes it in recording film tracks or sound tracks of motion pictures. **2.** The mechanism which moves the film print past the area where light passes through it in reproduction of picture and sound.

film vault [ENG] A place for safekeeping of film.

film weld [GRAPHICS] A butt splice of photographic film made with a heat splicer.

filoplume [VERT ZOO] A specialized feather that may be decorative, sensory, or both; it is always associated with papillae of contour feathers.

filopodia [INV ZOO] Filamentous pseudopodia.

filoreticulopodia [INV ZOO] Branched, filamentous pseudopodia.

Filosia [INV ZOO] A subclass of the class Rhizopodea characterized by slender filopodia which rarely anastomose.

filosus *See* fibratus.

filter [ADP] A device or program that separates data or signals in accordance with specified criteria. [CONT SYS] *See* compensator. [ELECTR] Any transmission network used in electrical systems for the selective enhancement of a given class of input signals. Also known as electric filter; electric-wave filter. [ENG] A porous article or material for separating suspended particulate matter from liquids by passing the liquid through the pores in the filter and sieving out the solids. [ENG ACOUS] A device employed to reject sound in a particular range of frequencies while passing sound in another range of frequencies. Also known as acoustic filter. [MATH] A family of subsets of a set S: it does not include the empty set, the intersection of any two members of the family is also a member, and any subset of S containing a member is also a member. [OPTICS] An optical element that partially absorbs incident electromagnetic radiation in the visible, ultraviolet, or infrared spectra, consisting of a pane of glass or other partially transparent material, or of films separated by narrow layers; the absorption may be either selective or nonselective with respect to wavelength. Also known as optical filter. [SCI TECH] In general, a selective device that transmits a desired range of matter or energy while substantially attenuating all other ranges.

filterability [ENG] The adaptability of a liquid-solid system to filtration; system is not filterable if it is too viscous to be forced through a filter medium, or if the solids are too small to be stopped by the filter medium.

filterable virus [VIROL] Virus particles that remain in a fluid after passing through a diatomite or glazed porcelain filter with pores too minute to allow the passage of bacterial cells.

filter aid [MATER] An inert powder or granules such as diatomaceous earth, fly ash, or sand added to a solution that is to be filtered in order to form a porous bed on the filter and increase the rate and improve the quality of filtration.

filter bed [CIV ENG] A fill of pervious soil that provides a site

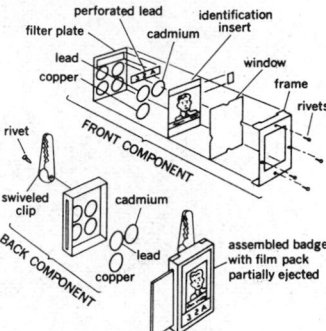

Film badge from Oak Ridge National Laboratory.

FILM TRANSPORT

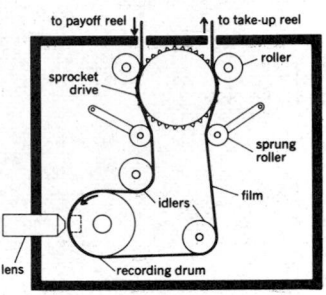

Photographic film-transport mechanism in a motion picture film sound recorder. (*After H. F. Olson, Acoustical Engineering, 3d ed., Van Nostrand, 1957*)

FILOPLUME

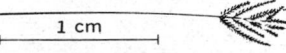

Drawing of a filoplume showing reduced hairlike vane. (*From J. C. Welty, The Life of Birds, Saunders, 1962*)

FILTER

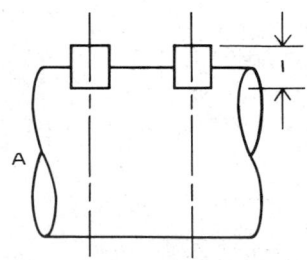

A simple high-pass acoustic filter which rejects low frequencies; sound entering the open-ended tank at A will be transmitted through the filter only if the frequency is higher than the cutoff frequency; l represents the length of the tubes.

for a septic field. [ENG] A contact bed used for filtering purposes.

filter cake [MATER] A concentrated solid or semisolid material that is separated from a liquid and remains on the filter after pressure filtration.

filter cake washing [CHEM ENG] An operation performed at the end of a filtration, in which residual liquid impurities are washed out of the cake by the flow of another liquid through the cake.

filter capacitor [ELEC] A capacitor used in a power-supply filter system to provide a low-reactance path for alternating currents and thereby suppress ripple currents, without affecting direct currents.

filter center [ORD] An information center at which all radar and other observed information concerning movements of friendly and enemy planes within a certain sector is screened and disseminated.

filter choke [ELEC] An iron-core coil used in a power-supply filter system to pass direct current while offering high impedance to pulsating or alternating current.

filter cloth [MATER] A fabric used as a medium for filtration.

filter crystal [ELECTR] Quartz crystal which is used in an electrical circuit designed to pass energy of certain frequencies.

filter design [ELECTR] The design of electrical networks in which the principle of electrical resonance is used to make the network accept wanted frequencies while rejecting unwanted ones.

filter discrimination [ELECTR] Difference between the minimum insertion loss at any frequency in a filter attenuation band and the maximum insertion loss at any frequency in the operating range of a filter transmission band.

filtered particle testing [ENG] A penetrant method of nondestructive testing by which cracks in porous objects (100 mesh or smaller) are indicated: a fluid containing suspended particles is sprayed on a test object; if a crack exists, particles are filtered out and concentrate at the surface as liquid flows into the crack.

filtered radar data [ELECTR] Radar data from which unwanted returns have been removed by mapping.

filter factor [OPTICS] The number of times the exposure must be increased when a filter is used on a camera, because the filter absorbs some of the light.

filter feeder [INV ZOO] A microphagous organism that uses complex filtering mechanisms to trap particles suspended in water.

filter flask [CHEM] A flask with a side arm to which a vacuum can be applied; usually filter flasks have heavy side walls to withstand high vacuum.

filter impedance compensator [ELECTR] Impedance compensator which is connected across the common terminals of electric wave filters when the latter are used in parallel to compensate for the effects of the filters on each other.

filtering [ENG] The process of interpreting reported information on movements of aircraft, ships, and submarines in order to determine their probable true tracks and, where applicable, heights or depths.

filter leaf [CHEM ENG] The frame or structure in a filter press that holds the filter cloth or other filter medium; a number of leaves in series usually comprises a filter press.

filter loss [SCI TECH] The amount of fluid passed through a permeable membrane in a given time.

filter medium [MATER] That portion of a filtration system that provides the liquid-solid separation, such as close-woven textiles or metal screens, papers, nonwoven fabrics, granular beds, or porous media.

filter paper [MATER] Porous cellulose paper used for filtering, especially for quantitative purposes.

filter pass band *See* filter transmission band.

filter photometry [ANALY CHEM] **1.** Colorimetric analysis of solution colors with a filter applied to the eyepiece of a conventional colorimeter. **2.** Inspection of a pair of Nessler tubes through a filter.

filter press [ENG] A metal frame on which iron plates are suspended and pressed together by a screw device; liquid to be filtered is pumped into canvas bags between the plates, and the screw is tightened so that pressure is furnished for filtration.

filter-press cell [PHYS CHEM] An electrolytic cell consisting of several units in series, as in a filter press, in which each electrode, except the two end ones, acts as an anode on one side and a cathode on the other, and the space between electrodes is divided by porous asbestos diaphragms.

filter pump [MECH ENG] An aspirator or vacuum pump which creates a negative pressure on the filtrate side of the filter to hasten the process of filtering.

filter reactor [ELEC] A reactor used for reducing the harmonic components of voltage in an alternating-current or direct-current circuit.

filter sand [MATER] Graded sand used for filtering suspended matter from a flowing liquid stream.

filter screen [ENG] A fine-pored medium through which a liquid will pass and on which solids deposit; the medium may be a metal sieve screen or a woven fabric of metal or of natural or synthetic fibers.

filter section [ELEC] A simple RC, RL, or LC network used as a broad-band filter in a power supply, grid-bias feed, or similar device.

filter slot [ELECTROMAG] Choke in the form of a slot designed to suppress unwanted modes in a waveguide.

filter spectrophotometer [SPECT] Spectrophotographic analyzer of spectral radiations in which a filter is used to isolate narrow portions of the spectrum.

filter thickener [ENG] Device that thickens a liquid-solid mixture by removing a portion of the liquid by filtration, rather than by settling.

filter transmission band [ELECTR] Frequency band of free transmission; that is, frequency band in which, if dissipation is neglected, the attenuation constant is zero. Also known as filter pass band.

filter-type respirator [ENG] A protective device which removes dispersoids from the air by physically trapping the particles on the fibrous material of the filter.

filtrate [SCI TECH] The discharge liquor in filtration. Also known as mother liquor; strong liquor.

filtration [SCI TECH] A process of separating particulate matter from a fluid, such as air or a liquid, by passing the fluid carrier through a medium that will not pass the particulates.

fimbria *See* pilus.

fimbriate [BIOL] Having a fringe along the edge.

fin [AERO ENG] A fixed or adjustable vane or airfoil affixed longitudinally to an aerodynamically or ballistically designed body for stabilizing purposes. [DES ENG] A projecting flat plate or structure, as a cooling fin. [VERT ZOO] A paddle-shaped appendage on fish and other aquatic animals that is used for propulsion, balance, and guidance.

final amplifier [ELECTR] The transmitter stage that feeds the antenna.

final boiling point *See* end point.

final common pathway *See* lower motor neuron.

final diameter [NAV] The diameter of the circle traversed by a vessel after turning through 360° and maintaining the same speed and rudder angle; it is always less than the tactical diameter, and is measured perpendicular to the original course and between the tangents at the points where 180° and 360° of the turn have been completed.

final drawing [GRAPHICS] A completed mechanical drawing from which fabrication of an item can take place.

final great-circle course [NAV] The direction, at the destination, of the great circle through that point and the point of departure, expressed as the angular distance from a reference direction, usually north, to that part of the great circle extending beyond the destination.

final heading [NAV] The aircraft heading at the end of a rating period while using gyro steering.

final lock mechanism [ORD] A device for locking the stroking member of a cartridge-actuated device in final position.

final mass [AERO ENG] The mass of a rocket after its propellants are consumed.

final-value theorem [MATH] The theorem that if $f(t)$ is a function which has a Laplace transform $F(s)$, and if the derivative of $f(t)$ with respect to t is also Laplace transformable, and if the limit of $f(t)$ as t approaches infinity exists, then this limit is equal to the limit of $sF(s)$ as s approaches zero.

fin assembly [ORD] An assembly of metal blades, usually mounted lengthwise on a sleeve, used on a missile, such as bomb or rifle grenade, to give directional stability; now seldom used.

finch [VERT ZOO] The common name for birds composing the family Fringillidae.

find [IND ENG] The therblig representing the mental reaction which occurs on recognizing an object at the end of the elemental motion search; now seldom used.

finder [COMMUN] **1.** An optical or electronic device that shows the field of action covered by a television camera. **2.** Switch or relay group in telephone switching systems that selects the path which the call is to take through the system; operates under the control of the calling station's dial. [OPTICS] A small telescope having a wide-angle lens and low power, which is attached to a larger telescope and points in the same direction; used to locate objects that are to be viewed in the larger telescope.

finder light [GRAPHICS] Light beam projected to show outline of photographic field.

F indicator *See* F scope.

finding circuit *See* lockout circuit.

findspot [ARCHEO] The place where an archeological object has been found.

fine admixture [GEOL] The smaller size grades of a sediment of mixed size grades.

fine delay [NAV] A dial on a loran navigation system receiver, for controlling relatively small changes in the position of the B trace pedestal, and serving as a vernier for the coarse delay.

fin efficiency [ENG] In extended-surface heat-exchange equations, the ratio of the mean temperature difference from surface-to-fluid divided by the temperature difference from fin-to-fluid at the base or root of the fin.

fine gold [MET] Almost pure gold; the value of bullion gold depends on its percentage of fineness. [MIN ENG] In placer mining, gold in exceedingly small particles.

fine gravel [GEOL] Gravel consisting of particles with a diameter range of 1 to 2 millimeters.

fine grinding [MECH ENG] Grinding performed in a mill rotating on a horizontal axis in which the material undergoes final size reduction, to -100 mesh.

fine index [ADP] The more specific of two indices consulted to gain access to a record.

fine-line printer [GRAPHICS] An accessory device mounted in a microfilmer to mark each filmed document.

fineness [MET] Degree of purity of gold or silver in parts per thousand.

fineness modulus [ENG] A number denoting the fineness of a fine aggregate or other fine material such as sand or paint.

fineness ratio [AERO ENG] The ratio of the length of a streamlined body, as that of a fuselage or airship hull, to its maximum diameter.

fines [MATER] **1.** Particles smaller than average in a mixture of particles varying in size. **2.** Fine material which passes through a standard screen on which coarser fragments are retained. [MET] That portion of a metal powder consisting of particles smaller than a specified size.

fine sand [GEOL] Sand grains between 0.25 and 0.125 millimeter in diameter.

fine-screen halftone [GRAPHICS] An illustration (photograph or artwork) reproduced for printing in a continuous tone by photographing the illustration with a 120-line to the inch screen between the illustration and the camera.

fine silver [MET] Silver having a minimum fineness of 999; considered to be pure silver.

fine structure [ATOM PHYS] The splitting of spectral lines in atomic and molecular spectra caused by the spin angular momentum of the electrons and the coupling of the spin to the orbital angular momentum.

fine-structure constant [PHYS] A fundamental dimensionless constant, equal to 2π times the square of the electron charge in electrostatic units, divided by the product of the speed of light and Planck's constant; mathematically, equal to $(7.297351 \pm 0.000011) \times 10^{-3}$, approximately $1/137$. Also known as Sommerfeld fine-structure constant.

fin fold [EMBRYO] A median integumentary fold extending along the body of a fish embryo which gives rise to the dorsal, caudal, and anal fins.

finger [ANAT] Any of the four digits on the hand other than the thumb. [GEOL] The tendency for gas which is displacing liquid hydrocarbons in a heterogeneous reservoir rock system to move forward irregularly (in fingers), rather than on a uniform front. [PETRO ENG] A pair or set of bracketlike projections placed at a strategic point in a drill tripod or derrick to keep a number of lengths of drill rods or casing in place when they are standing in the tripod or derrick.

finger bit [DES ENG] A steel rock-cutting bit having fingerlike, fixed or replaceable steel-cutting points.

finger board [PETRO ENG] A board with projecting dowels or pipe fingers located in the upper part of the drill derrick or tripod to support stands of drill rod, drill pipe, or casing.

finger chute [MIN ENG] Steel rails hinged independently over an ore chute, to control rate of flow of rock.

fingerprint [ANAT] A pattern of distinctive epidermal ridges on the bulbs of the inside of the end joints of fingers and thumbs. [GRAPHICS] An impression of a human fingerprint.

finger raise [MIN ENG] Steeply sloping openings permitting caved ore to flow down raises through grizzlies to chutes on the haulage level.

finish [MATER] **1.** A chemical or other material applied to surfaces to protect them, to alter their appearances, or to modify their physical properties; finishes can be physically, chemically, or electrolytically applied and have value for fabrics and fibers, metals, paper products, plastics, woods, and so on. **2.** The ultimate quality, condition, or appearance of the surface of a material.

finished goods [IND ENG] Manufactured products in inventory ready for packaging, shipment, or sale.

finished steel [MET] Steel that has undergone final processing and is ready for market.

finisher [CIV ENG] A construction machine used to smooth the freshly placed surface of a roadway, or to prepare the foundation for a pavement.

finish grinding [MECH ENG] The last action of a grinding operation to achieve a good finish and accurate dimensions.

finishing compound [MATER] A substance used to impart surface properties to textiles or leather, such as softness, flexibility, or fire resistance.

finishing hardware [BUILD] Items, such as hinges, door pulls, and strike plates, made in attractive shapes and finishes, and usually visible on the completed structure.

finishing hydrated lime [MATER] Any hydrated lime suitable for use in the finishing coat of plaster; characterized by a high degree of whiteness and plasticity.

finishing mill [MET] A rolling mill in which sheet, plate, and other mill products are subjected to final rolling operations.

finishing nail [DES ENG] A wire nail with a small head that can easily be concealed.

finishing roll [MIN ENG] The last roll, or the one that does the finest crushing in ore dressing.

finishing temperature [MET] The temperature at which hot-working is completed.

finish turning [MECH ENG] The operation of machining a surface to accurate size and producing a smooth finish.

finite clipping [ELECTR] Clipping in which the threshold level is large but is below the peak input signal amplitude.

finite closed aquifer [HYD] The part of a subterranean reservoir containing water (aquifer) in which the aquifer is limited (finite), with no water flow across the exterior reservoir boundary.

finite difference [MATH] The difference between the values of a function at two discrete points, used to approximate the derivative of the function.

finite difference equations [MATH] Equations arising from differential equations by substituting difference quotients for derivatives, and then using these equations to approximate a solution.

finite intersection property of a family of sets [MATH] If the intersection of any finite number of them is nonempty, then the intersection of all the members of the family is nonempty.

finite matrix [MATH] A matrix with a finite number of rows and columns.

finite moment theorem [MATH] The theorem that if $f(x)$ is a continuous function, and if the integral of $f(x)\, x^n$ over a finite interval is zero for all positive integers n, then $f(x)$ is identically zero in that interval.

finite quantity [MATH] Any bounded quantity.

finite set [MATH] A set whose elements can be indexed by integers $1, 2, 3, \ldots, n$ inclusive.

fin keel [NAV ARCH] A metal plate or thin fairing attached to the keel of sailing craft to give resistance to lateral motion, and frequently having a cigar-shaped lead bulb at the bottom to give transverse stability. Also known as ballast fin.

Fink truss [CIV ENG] A symmetrical steel roof truss suitable for spans up to 50 feet (15 meters).

finned surface [MECH ENG] A tubular heat-exchange surface with extended projections on one side.

finnemanite [MINERAL] $Pb_5Cl(AsO_3)_3$ A gray, olive-green, or black hexagonal mineral composed of arsenite and chloride of lead.

fin reinforcing assembly [ORD] An assemblage of components required to reinforce a bomb fin assembly.

fin rot [VET MED] A bacterial disease of hatchery fishes characterized by necrosis and erosion of the fin tissue.

Finsen lamp [ELEC] A high-temperature carbon arc or mercury arc lamp that produces a mixture of blue, violet, and near-ultraviolet light; used to treat certain skin disorders and to test paints and other protective coatings.

finsen unit [ELECTROMAG] A unit of intensity of ultraviolet radiation, equal to the intensity of ultraviolet radiation at a specified wavelength whose energy flux is 100,000 watts per square meter; the wavelength usually specified is 296.7 nanometers. Abbreviated FU.

Finsler geometry [MATH] The study of the geometry of a manifold in terms of the various possible metrics on it by means of Finsler structures.

Finsler structure on a manifold [MATH] A family of metrics varying continuously from point to point.

fin spine [VERT ZOO] A bony process that supports the fins of certain fishes.

fin stabilization [ORD] Method of stabilizing a projectile (as a rocket, bomb, or missile) during flight by the aerodynamic use of protruding fins.

fin waveguide [ELECTROMAG] Waveguide containing a thin longitudinal metal fin that serves to increase the wavelength range over which the waveguide will transmit signals efficiently; usually used with circular waveguides.

fiord *See* fjord.

fiorite *See* siliceous sinter.

Fior process [MET] The prereduction of high-grade iron particles or concentrates in a hot gaseous reactor to produce low-oxygen fines for partially metallized briquettes suitable for electric-arc steelmaking furnaces.

fir [BOT] The common name for any tree of the genus *Abies* in the pine family; needles are characteristically flat.

fire [CHEM] The manifestation of rapid combustion, or combination of materials with oxygen. [ENG] To blast with gunpowder or other explosives. [MIN ENG] A warning that a shot is being fired. [ORD] **1.** The discharge of a gun, launching of a missile, or the like. **2.** The projectiles or missiles fired. **3.** To discharge a weapon.

fire adjustment [ORD] Correcting the elevation and direction of a weapon, or regulating the explosion time of its projectile, so that the projectile will strike or burst at the desired point; for automatic weapons, it is a continuous operation from the instant the first rounds reach the vicinity of the target until the command "cease firing" is given.

firearm [ORD] **1.** In a general sense, a gun. **2.** A small arm, as a pistol or rifle, designed to be carried and used by an individual.

fire assay [MET] Analysis of a metal-bearing material, especially gold and silver, by assaying a sample with a suitable flux and measuring the content of resulting metal leads by weighing or atomic absorption techniques.

fireball [ASTRON] A bright meteor with luminosity equal to or exceeding that of the brightest planets. [NUCLEO] The luminous sphere of hot gases that forms a few millionths of a second after a nuclear explosion.

fire blight [PL PATH] A bacterial disease of apple, pear, and related pomaceous fruit trees caused by *Erwinia amylovora*;

FIR

Cone and branches with flat needles of balsam fir *(Abies balsamea)*.

leaves are blackened, cankers form on the trunk, and flowers and fruits become discolored.

fireboat [NAV ARCH] A vessel similar to a tug but fitted with fire-fighting apparatus.

fire bomb [ORD] An item designed to be dropped from an aircraft to destroy or reduce the utility of a target by the effects of combustion.

fire boss [MIN ENG] An individual who examines a mine for gas and other dangers. Also known as mine examiner.

firebreak [FOR] A cleared area of land intended to check the spread of forest or prairie fire. [MIN ENG] A strip across an area in which either no combustible material is employed or in which, if timber supports are used, sand is filled and packed tightly around them.

firebrick [MATER] A refractory brick, often made of fireclay, that is able to withstand high temperature (up to 1500–1600°C) without fusion; used to line furnaces, fireplaces, and chimneys.

fire bridge [ENG] A low wall separating the hearth and the grate in a reverberatory furnace.

fireclay [GEOL] **1.** A clay that can resist high temperatures without becoming glassy. **2.** Soft, embedded, white or gray clay rich in hydrated aluminum silicates or silica and deficient in alkalies and iron.

fire control [ORD] Control over the direction, volume, and time of fire of guns or launchers by the use of certain electrical or electronic devices or aids, or by optical or mechanical systems.

fire-control circuit [ELECTR] An electric circuit in a fire-control system.

fire-control computer [ORD] A computer used to guide a fire-control system.

fire-control grid [ORD] A system of lines that divide a military map into squares, the distance between any two parallel lines representing 1000 yards or 1000 meters, depending on the type of map; maps using the fire-control grid are of sufficiently large scale to be useful in fire control.

fire-control instrument [ORD] An aiming circle, range finder, compass, telescope, or other instrument used in fire control.

fire-control quadrant [ORD] A chemical device with a scale to measure angles and a leveling adjustment, used to measure the elevation angles of a weapon for obtaining the horizontal range of a weapon; it is attached to the gun, gun mount, or gun carriage.

fire-control radar [ORD] Radar equipment used in a fire-control system.

fire-control sonar [ORD] Sonar used in determining the aiming of antisubmarine weapons.

fire-control system [ORD] A setup for control of the aiming and firing of guns, rockets, or guided missiles.

fire coordination [ORD] The planning and execution of fire so that targets are adequately covered by a suitable weapon or groups of weapons.

firecracker [ENG] A cylindrically shaped item containing an explosive and a fuse; used to simulate the noise of an explosive charge.

firedamp [MIN ENG] **1.** A gas formed in mines by decomposition of coal or other carbonaceous matter; consists chiefly of methane and is combustible. **2.** An airtight stopping to isolate an underground fire and to prevent the inflow of fresh air and the outflow of foul air. Also known as fire wall.

firedamp alarm [MIN ENG] An instrument which gives a warning signal when the methane content in the mine atmosphere exceeds a known value.

firedamp detector [MIN ENG] A portable device to detect the presence and determine the percentage of firedamp in mine air.

firedamp drainage [MIN ENG] The collection of firedamp from coal strata, generally into pipes, with or without the use of suction. Also known as methane drainage.

firedamp drainage drill [MIN ENG] A heavy, compressed-air-operated, percussive, rotary or rotary-percussive drilling machine for putting up the boreholes in firedamp drainage.

firedamp explosion [MIN ENG] An explosion of a mixture of firedamp and air.

firedamp fringe [MIN ENG] The zone of contact between the coal gases and the ventilation air current at the face of the mine.

firedamp layer [MIN ENG] An accumulation of firedamp under the roof of a mine roadway where the ventilation is insufficient to dilute and remove the gas.

firedamp migration [MIN ENG] The movement of firedamp through the strata or coal of a mine.

firedamp pressure-chamber method [MIN ENG] A method of firedamp drainage in coal mines; pressure chambers built at the intake and the return of a worked-out area are used to trap firedamp, which is drawn off in pipes.

firedamp probe [MIN ENG] A flexible rubber tube connected to a rod, which can be thrust into roof cavities and breaks so that a sample of the air may be transferred to a methanometer and its firedamp content determined.

firedamp reforming process [CHEM ENG] A process in which methane (firedamp) is mixed with steam and passed over a nickel catalyst for conversion to a mixture of hydrogen and carbon monoxide; this mixture is blended with pure methane, and the result is a fuel of high calorific value.

fire-danger meter [ENG] A graphical aid used in fire-weather forecasting to calculate the degree of forest-fire danger (or burning index): commonly in the form of a circular slide rule, it relates numerical indices of the seasonal stage of foliage, the cumulative effect of past precipitation or lack thereof (build-up index), the measured fuel moisture, and the speed of the wind in the woods; the fuel moisture is determined by weighing a special type of wooden stick that has been exposed in the woods, its weight being proportional to its contained water; the calculated burning index falls on a scale of 1 to 100: 1 to 11 is no fire danger; 12 to 35 medium danger; 40 to 100 high danger.

fire detector [ENG] A temperature-sensing device designed to sound an alarm, to turn on a sprinkler system, or to activate some other fire preventive measure at the first signs of fire.

fire direction [ORD] Tactical employment of fire power, exercise of tactical command of one or more units in the selection of targets, concentration or distribution of fire, and allocation of ammunition for each mission.

fire direction net [COMMUN] A communication system linking observers, liaison officers, air observers, and firing batteries with the fire direction center for the purposes of fire control.

fire disclimax [ECOL] A community that is perpetually maintained at an early stage of succession through recurrent destruction by fire followed by regeneration.

fire distribution [ORD] **1.** The use of weapons of a unit to cover a target most effectively. **2.** The application of fire on targets, or on subdivisions of a target, in order of their importance. **3.** The systematic assignment of targets to the batteries of a number of vessels by message.

fire door [ENG] **1.** The door or opening through which fuel is supplied to a furnace or stove. **2.** A door that can be closed to prevent the spreading of fire, as through a building or mine.

fired process equipment [ENG] Heaters, furnaces, reactors, incinerators, vaporizers, steam generators, boilers, and other process equipment for which the heat input is derived from fuel combustion (flames); can be direct-fired (flame in contact with the process stream) or indirect-fired (flame separated from the process fluid by a metallic wall).

fire effect [ORD] Result of firing on enemy personnel and materiel.

fire escape [BUILD] An outside stairway usually made of steel and used to escape from a building in case of fire.

fire extinguisher [ENG] Any of various portable devices used to extinguish a fire by the ejection of a fire-inhibiting substance, such as water, carbon dioxide, gas, or chemical foam.

firefinder [ENG] An instrument consisting of a map and a sighting device; used in fire towers to locate forest fires.

fire flooding [PETRO ENG] A method to improve secondary recovery in an oil reservoir; a combustion process is started in the reservoir at an injection well by continued introduction of gas containing oxygen or other material to support combustion, and the combustion wave is driven through the reservoir toward the production well.

firefly [INV ZOO] Any of various flying insects which produce light by bioluminescence.

fire foam [MATER] A colloidal solution of small gas bubbles produced by chemical reaction or mechanical agitation and used to extinguish hydrocarbon fire.

fire for adjustment [ORD] Fire delivered for the purpose of determining firing data that will place the center of impact or burst on the desired portion of the target.

fire for effect [ORD] Fire which is delivered after the center of impact or burst is within the desired distance of the target or adjusting point.

fire-formed fragment *See* controlled fragment.

fire fountain *See* lava fountain.

fire hose [ENG] A collapsible, flameproof hose that can be attached to a hydrant, standpipe, or similar outlet to supply water to extinguish a fire.

fire hydrant [CIV ENG] An outlet from a water main provided inside buildings or outdoors to which fire hoses can be connected. Also known as fire plug; hydrant.

fire interrupter [ORD] For aircraft guns, a device (usually an electrical switch mechanically actuated) which interrupts firing of a weapon.

fire load [CIV ENG] The load of combustible material per square foot of floor space.

fire opal [MINERAL] A translucent or transparent, orangy-yellow, brownish-orange, or red variety of opal that gives out fiery reflections in bright light and that may have a play of colors. Also known as pyrophane; sun opal.

fire partition [BUILD] A wall inside a building intended to retard fire.

fire plug *See* fire hydrant.

fire point [CHEM] The lowest temperature at which a volatile combustible substance vaporizes rapidly enough to form above its surface an air-vapor mixture which burns continuously when ignited by a small flame.

firepower [ORD] **1.** The capability to deliver fire. **2.** The fire itself, or the quantity or effectiveness of fire delivered. **3.** Any explosive or missile that wreaks damage upon the target against which it is directed; for example, guns or rockets, or aircraft units and so forth armed with guns or rockets.

fireproof [BUILD] Having noncombustible walls, stairways, and stress-bearing members, and having all steel and iron structural members which could be damaged by heat protected by refractory materials. [MATER] The property of being relatively resistant to combustion.

fireproofing compound *See* fire retardant.

fire protection [CIV ENG] Measures for reducing injury and property loss by fire.

fire pump [MECH ENG] A pump for fire protection purposes usually driven by an independent, reliable prime mover and approved by the National Board of Fire Underwriters.

fire refining [MET] The refining of blister copper by treatment in a furnace under oxidizing conditions to remove the impurities and under reducing conditions to remove the excess oxygen.

fire-resistant [CIV ENG] Of a structural element, able to resist combustion for a specified time under conditions of standard heat intensity without burning or failing structurally.

fire retardant [MATER] A chemical used as a coating for or a component of a combustible material to reduce or eliminate a tendency to burn; used with textiles, plastics, rubbers, paints, and other materials. Also known as fireproofing compound.

fire retardant paint [MATER] A paint applied as a thin coating to reduce the rate of flame spread of a combustible material; based on silicone, casein, polyvinylchloride, or other substance.

fire scale [MET] Copper oxide remaining below the surface of silver-copper alloys after annealing and pickling.

fire sprinkling system *See* sprinkler system.

fire standpipe [CIV ENG] A high, vertical pipe or tank that holds water to assure a positive, relatively uniform pressure, particularly to provide fire protection to upper floors of tall buildings.

firestone *See* flint.

fire stop [BUILD] An incombustible, horizontal or vertical barrier, as of brick across a hollow wall or across an open room, to stop the spread of fire. Also known as draught stop.

FIRE EXTINGUISHER

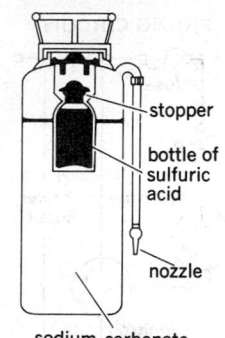

— stopper

bottle of sulfuric acid

nozzle

sodium carbonate solution

Soda-acid fire extinguisher. When it is inverted, the stopper falls out of place allowing sulfuric acid to react with sodium carbonate to produce carbon dioxide, which propels the solution through the nozzle.

FIRE-TUBE BOILER

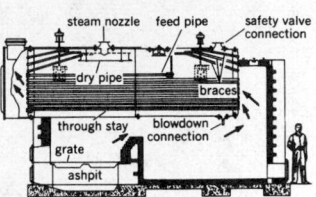

A horizontal-return-tube fire-tube boiler. Part of the heat from the combustion gases is transferred directly to the lower portion of the shell of the boiler before making a return pass through the horizontal tubes, and then going up the stack.

FIRING CIRCUIT

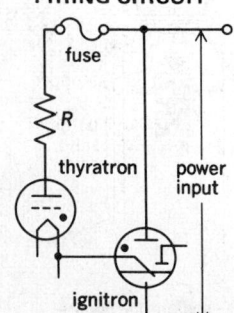

Elementary anode firing circuit.

fire storm [ORD] A fire engulfing a city, produced by an air raid with incendiary or fire bombs.

fire superiority [ORD] Fire with greater effect than that of the enemy because of its greater accuracy and volume, making possible advances against the enemy without heavy losses.

fire support [ORD] The support or protection given forces in direct contact with the enemy by ground or naval guns or by aircraft engaging in close air support.

fire tower [BUILD] A fireproof and smokeproof stairway compartment running the height of a building. [FOR] A tower used to watch for fires, especially forest fires.

fire-tube boiler [MECH ENG] A steam boiler in which hot gaseous products of combustion pass through tubes surrounded by boiler water.

fire unit analyzer [ORD] An instrument for analyzing the effectiveness of an antiaircraft fire unit against hostile aircraft or missiles under the conditions stated on the face of the analyzer.

fire wall [CIV ENG] 1. A fire-resisting wall separating two parts of a building from the lowest floor to several feet above the roof to prevent the spread of fire. 2. A fire-resisting wall surrounding an oil storage tank to retain oil that may escape and to confine fire. [MIN ENG] See firedamp.

fire weather [METEOROL] The state of the weather with respect to its effect upon the kindling and spreading of forest fires.

fire welding See forge welding.

firing [ELECTR] 1. The gas ionization that initiates current flow in a gas-discharge tube. 2. Excitation of a magnetron or transmit-receive tube by a pulse. [ENG] 1. The act or process of adding fuel and air to a furnace. 2. Igniting an explosive mixture. 3. Treating a ceramic product with heat. [NUCLEO] The transition from the unsaturated to the saturated state of a saturable reactor.

firing azimuth [ORD] Horizontal direction in which a gun or launcher is pointed for firing, expressed as an azimuth.

firing base [ORD] Part of the mechanism in some cannon that supports the gun carriage when it is in position for firing.

firing box [ELEC] A boxlike item in which are mounted switches, cables, fuses, plugs, indicator lights, batteries, and the like, specifically designed for firing a rocket or guided missile from a remote position.

firing button [ELEC] A button or switch for firing guns or rockets.

firing cable See shot-firing cable.

firing chamber See combustion chamber.

firing circuit [ELECTR] 1. Circuit used with an ignitron to deliver a pulse of current of 5-50 amperes in the forward direction, from the igniter to the mercury, to start a cathode spot and to control the time of firing. 2. By analogy, a similar control circuit of silicon-controlled rectifiers and like devices.

firing data [ORD] All data necessary for firing a weapon at a given objective, which may be determined by computation and then transmitted as verbal commands, or may be applied electromechanically by one of the several types of directing devices.

firing hammer [ORD] A metallic pivoted item, part of the firing mechanism of a firearm, designed to strike a firing pin or percussion cap and fire a gun.

firing interval [ORD] Period of time between firing one shot and the next.

firing jack [ORD] Adjustable device which stabilizes and levels certain mobile artillery weapons while the weapons are in firing position.

firing lanyard [ORD] A cord or cable of specific length, usually with a hook on one end and a handle on the opposite end, designed to be attached to a component of the firing mechanism of a gun, rocket launcher, or the like, and used to fire the weapon.

firing lock [ORD] A removable part of the firing mechanism in some weapons, incorporating the firing pin and the mechanism which drives it against the primer.

firing machine [ENG] Term used for an electric blasting machine. [MECH ENG] A mechanical stoker used to feed coal to a boiler furnace.

firing mechanism [ENG] A mechanism for firing a primer; the primer may be for initiating the propelling charge, in

which case the firing mechanism forms a part of the weapon; if the primer is for the purpose of initiating detonation of the main charge, the firing mechanism is a part of the ammunition item and performs the function of a fuse.

firing pin [ORD] A device used in the firing mechanism of a gun, mine, bomb, fuse, projectile, or the like, which strikes and detonates a sensitive explosive to initiate an explosive train or a propelling charge.

firing point [ELECTR] See critical grid voltage. [ORD] Location from which fire is delivered in target practice.

firing position [ORD] Position of a weapon ready for firing.

firing potential [ELECTR] Controlled potential at which conduction through a gas-filled tube begins.

firing table [ORD] A table or chart giving data needed for firing a gun accurately on a target under standard conditions, and also the corrections that must be made for special conditions such as winds or variations of temperature.

firing table elevation [ORD] The angle between the axis of the bore and the horizontal plane when the piece is laid to fire at a given range under conditions that are accepted as standard.

firing time [ORD] The period of time during which a weapon is fired.

firmer chisel [DES ENG] A small hand chisel with a flat blade; used in woodworking.

firm-joint caliper [DES ENG] An outside or inside caliper whose legs are jointed together at the top with a nut and which must be opened and closed by hand pressure.

firmoviscosity [MECH] Property of a substance in which the stress is equal to the sum of a term proportional to the substance's deformation, and a term proportional to its rate of deformation.

firn [HYD] Material transitional between snow and glacier ice; it is formed from snow after existing through one summer melt season and becomes glacier ice when its permeability to liquid water drops to zero. Also known as firn snow.

firn basin See firn field.

firn field [HYD] The accumulation area or upper region of a glacier where snow accumulates and firn is secreted. Also known as firn basin.

firn ice See iced firn.

firnification [HYD] The process of firn formation from snow and of transformation of firn into glacier ice.

firn limit See firn line.

firn line [GEOL] 1. The regional snow line on a glacier. 2. The line that divides the ablation area of a glacier from the accumulation area. Also known as firn limit.

firn snow See firn; old snow.

first answer print [GRAPHICS] The motion picture composite print, with the mixed sound track and the optical effects.

first bottom [GEOL] The floodplain of a river, below the first terrace.

first countable topological space [MATH] A topological space in which every point has a countable number of open neighborhoods so that any neighborhood of this point contains one of these.

first-degree burn [MED] A mild burn characterized by pain and reddening of the skin.

first derived curve See derived curve.

first detector See mixer.

first estimate–second estimate method [NAV] The method of determining time of meridian transit (especially local apparent noon) for a moving craft; the time of transit is computed for an estimated longitude of the craft, and the longitude estimate is then revised to agree with the time determined by the first estimate, and a second computation is made; the process is repeated as many times as necessary to obtain an answer of the desired precision.

first-filial generation [GEN] The first generation resulting from a cross with all members being heterozygous for characters which differ from those of the parents. Symbolized F_1.

first fire [ENG] The igniter used with pyrotechnic devices, consisting of first fire composition, loaded in direct contact with the main pyrotechnic charge; the ignition of the igniter or first fire is generally accomplished by fuse action.

first fire composition [MATER] A pyrotechnic composition (readily ignitable and easily pressed into a strong, solid mass), compounded to produce a high temperature, preferably with creation of slag to give heat capacity.

first Fresnel zone [ELECTROMAG] Circular portion of a wavefront transverse to the line between an emitter and a more distant point, where the resultant disturbance is being observed, whose center is the intersection of the front with the direct ray, and whose radius is such that the shortest path from the emitter through the periphery to the receiving point is one-half wavelength longer than the direct ray.

first generation [ADP] Term denoting electronic hardware, logical organization, and software characteristic of a first-generation computer.

first-generation computer [ADP] A computer from the earliest stage of computer development, ending in the early 1960s, characterized by the use of vacuum tubes, the performance of one operation at a time in strictly sequential fashion, and elementary software, usually including a program loader, simple utility routines, and an assembler to assist in program writing.

first gust [METEOROL] The sharp increase in wind speed often associated with the early mature stage of a thunderstorm cell; it occurs with the passage of the discontinuity zone which is the boundary of the cold-air downdraft.

first harmonic *See* fundamental.

first-in, first-out [IND ENG] An inventory cost evaluation method which transfers costs of material to the product in chronological order. Abbreviated FIFO.

first law of motion *See* Newton's first law.

first law of the mean [MATH] If a function $f(x)$ is continuous on the closed interval $[a,b]$ and differentiable on the open interval (a,b) and $f(a) = f(b)$, then there exists x_0, $a < x_0 < b$, such that $f'(x_0) = 0$. Also known as mean value theorem; Rolle's theorem.

first law of thermodynamics [THERMO] The law that heat is a form of energy, and the total amount of energy of all kinds in an isolated system is constant; it is an application of the principle of conservation of energy.

first level address [ADP] The location of a referenced operand.

first light [NAV] The beginning of morning nautical twilight, that is, when the center of the morning sun is 12° below the horizon.

first-order climatological station [METEOROL] A meteorological station at which autographic records or hourly readings of atmospheric pressure, temperature, humidity, wind, sunshine, and precipitation are made, together with observations at fixed hours of the amount and form of clouds and notes on the weather.

first-order reaction [PHYS CHEM] A chemical reaction in which the rate of decrease of concentration of component A with time is proportional to the concentration of A.

first-order spectrum [SPECT] A spectrum, produced by a diffraction grating, in which the difference in path length of light from adjacent slits is one wavelength.

first-order station [METEOROL] After U.S. National Weather Service practice, any meteorological station that is staffed in whole or in part by National Weather Service (Civil Service) personnel, regardless of the type or extent of work required of that station.

first-order theory [OPTICS] *See* Gaussian optics. [PHYS] A theory which takes into account only the most important terms, such as the term proportional to the independent variable in the series expansion of a function appearing in the theory.

first-order transition [THERMO] A change in state of aggregation of a system accompanied by a discontinuous change in enthalpy, entropy, and volume at a single temperature and pressure.

first point of Aries *See* vernal equinox.

first point of Cancer *See* summer solstice.

first point of Capricorn *See* winter solstice.

first point of Libra *See* autumnal equinox.

first quarter [ASTRON] The phase of the moon when it is near east quadrature, when the western half of it is visible to an observer on the earth.

first radiation constant [STAT MECH] A constant appearing in the Planck radiation formula; its value depends on the form of the formula used; in the formula for power emitted by a blackbody per unit area per unit wavelength interval, it is 2π times Planck's constant, times the square of the speed of light, or approximately 3.7415×10^{-16} watt (meter)^2. Symbolized c_1; C_1.

first selector [ELECTR] Selector which immediately follows a line finder in a switch train and which responds to dial pulses of the first digit of the called telephone number.

firth *See* estuary.

fir wood oil *See* pine needle oil.

Fischer ellipsoid of 1960 [GEOD] The reference ellipsoid of which the semimajor axis is 6,378,166.000 meters, the semiminor axis is 6,356,784.298 meters, and the flattening or ellipticity is 1/298.3. Also known as Fischer spheroid of 1960.

Fischer-Hepp rearrangement [ORG CHEM] The rearrangement of a nitroso derivative of a secondary aromatic amine to a *p*-nitrosoarylamine; the reaction is brought about by an alcoholic solution of hydrogen chloride.

Fischer-Hinnen method [ELEC] Method of analysis of a complex waveform which has like loops above and below the time axis, in which the amplitude and phase of the *n*-th harmonic is determined from the ordinates of the resultant wave at a series of times which divide the half wave into $2n$ equal time intervals.

Fischer indole synthesis [ORG CHEM] A reaction to form indole derivatives by means of a ring closure of aromatic hydrazones.

fischerite [MINERAL] A green mineral composed of a basic aluminum phosphate; may be identical to wavellite.

Fischer polypeptide synthesis [ORG CHEM] A synthesis of peptides in which α-amino acids or those peptides with a free amino group react with acid halides of α-haloacids, followed by amination with ammonia.

Fischer's distribution [STAT] Given data from a normal population with S_1^2 and S_2^2 two independent estimates of variance, the distribution $\frac{1}{2} \log (S_1^2/S_2^2)$.

Fischer spheroid of 1960 *See* Fischer ellipsoid of 1960.

Fischer-Tropsch process [CHEM ENG] A catalytic process to synthesize hydrocarbons and their oxygen derivatives by the controlled reaction of hydrogen and carbon monoxide.

Fischer-Yates test [STAT] A test of independence of data arranged in a 2×2 contingency table.

fish [VERT ZOO] The common name for the cold-blooded aquatic vertebrates belonging to the groups Cyclostomata, Chondrichthyes, and Osteichthyes.

fish-bone antenna [ELECTROMAG] **1.** Antenna consisting of a series of coplanar elements arranged in collinear pairs, loosely coupled to a balanced transmission line. **2.** Directional antenna in the form of a plane array of doublets arranged transversely along both sides of a transmission line.

fished joint [CIV ENG] A structural joint made with fish plates.

fisher [VERT ZOO] *Martes pennanti.* An arboreal, carnivorous mammal of the family Mustelidae; a relatively large weasel-like animal with dark fur, found in northern North America.

fisheries conservation [ECOL] Those measures concerned with the protection and preservation of fish and other aquatic life, particularly in sea waters.

Fisher-Irwin test [STAT] A method for testing the null hypothesis in an experiment with quantal response.

fishery [ECOL] A place for harvesting fish or other aquatic life, particularly in sea waters.

Fishes *See* Pisces.

fisheye [MATER] A small globular mass which has not blended completely into the surrounding material and is particularly evident in a transparent or translucent material, such as a plastic coating or surface coating. [MET] *See* flake.

fish gelatin *See* isinglass.

fish glue *See* isinglass.

fishing [ENG] In drilling, the operation by which lost or damaged tools are secured and brought to the surface from the bottom of a well or drill hole.

fishing boat [NAV ARCH] A ship having the necessary equipment to catch fish, such as a fathometer and fishing nets, but not the equipment to process fish.

fishing grounds [NAV] Water areas in which fishing is frequently carried on.

FISHER

The fisher *(Martes pennanti)*, an arboreal, nocturnal carnivorous mammal of North America.

FISSIDENTALES

sporophyte

rhizoids

Entire plant of *Fissidens adiantoides*.

FISSION

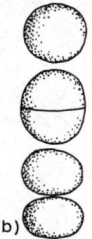

(a)

(b)

Fission in bacteria:
(a) in a bacillus; *(b)* in a coccus.
(From H. J. Fuller and O. Tippo, College Botany, rev. ed., Holt, 1954)

FISSION-TRACK DATING

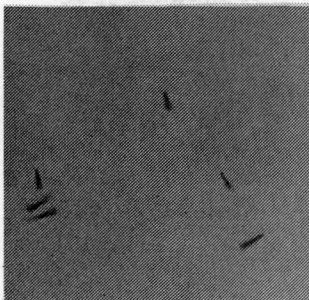

Etched natural fission tracks in muscovite mica crystal. The sample was etched in concentrated hydrofluoric acid.

fishing space [CIV ENG] The space between base and head of a rail in which a joint bar is placed.

fishing tool [ENG] A device for retrieving objects from inaccessible locations.

fish ladder [CIV ENG] A type of fishway that carries water around a dam through a series of stepped baffles or boxes and thus facilitates the migration of fish.

fish lead [ENG] A type of sounding lead used without removal from the water between soundings.

fish lice [INV ZOO] The common name for all members of the crustacean group Arguloida.

fish meal [FOOD ENG] A protein-rich, dried food product produced from the inedible portions of fishes by dry or wet rendering. Also known as fish protein concentrate.

fishmouthing *See* alligatoring.

fish net buoy [NAV] A buoy marking the limit of a fish net area.

fish oil [MATER] Oil obtained from fish such as menhaden, pilchard, herring, and sardine; used as a drying oil in paint and as a raw material for detergents, resins, margarine, and so on.

fish paper [MATER] A type of fiber used in sheet form for insulating purposes where high mechanical strength is required, as in insulating transformer windings from the transformer core.

fish plate [CIV ENG] One of a pair of steel plates bolted to the sides of a rail or beam joint, to secure the joint.

fishpole antenna *See* whip antenna.

fish protein concentrate *See* fish meal.

fish screen [CIV ENG] 1. A screen set across a water intake canal or pipe to prevent fish from entering. 2. Any similar barrier to prevent fish from entering or leaving a pond.

fish stakes [NAV] Poles or stakes placed in shallow water to outline fishing areas.

fishtail [MET] Excess metal trailing on the end of a roll forging.

fishtail bit [DES ENG] A drilling bit shaped like the tail of a fish.

fishtail burner [ENG] A burner in which two jets of gas impinge on each other to form a flame shaped like a fish's tail.

Fissidentales [BOT] An order of the Bryopsida having erect to procumbent, simple or branching stems and two rows of leaves arranged in one plane.

fissile [GEOL] Capable of being split along the line of the grain or cleavage plane. [NUCLEO] *See* fissionable.

fission [BIOL] A method of asexual reproduction among bacteria, algae, and protozoans by which the organism splits into two or more parts, each part becoming a complete organism. [NUC PHYS] The division of an atomic nucleus into parts of comparable mass; usually restricted to heavier nuclei such as isotopes of uranium, plutonium, and thorium. Also known as atomic fission; nuclear fission.

fissionable [NUCLEO] 1. A property of material whose nuclei are capable of undergoing fission. Also known as fissile. 2. A material capable of fission.

fission bomb *See* atomic bomb.

fission chamber [NUCLEO] An ionization chamber used to detect slow neutrons; the inside wall has a thin coating of uranium, in which a slow neutron produces a fission; the resulting highly ionizing fission fragments produce a count in the chamber. Also known as fission counter.

fission counter *See* fission chamber.

fission cross section [NUC PHYS] The cross section for a bombarding neutron, gamma ray, or other particle to induce fission of a nucleus.

fission detector [NUCLEO] Device for detecting spontaneous fission, consisting of a mica or special glass which is placed near the sample and which is subsequently chemically etched, making fission tracks visible.

fission fraction [NUCLEO] The fraction of the total yield of a nuclear weapon that is due to fission; for thermonuclear weapons the average value is about 50%.

fission fragments [NUCLEO] The nuclear species first produced when an atom such as uranium-238 or plutonium-239 undergoes fission. Also known as primary fission products.

fission fuel *See* nuclear fuel.

fission fungi [MICROBIO] A misnomer once used to describe the Schizomycetes.

fission-fusion bomb [NUCLEO] An explosive device which derives its energy in comparable amounts from nuclear fission and nuclear fusion.

fission neutron [NUC PHYS] A neutron emitted as a result of nuclear fission.

fission product [NUC PHYS] Any radioactive or stable nuclide resulting from fission, including both primary fission fragments and their radioactive decay products.

fission-product poisoning [NUCLEO] Inhibition of a nuclear chain reaction by fission products which have large cross sections for slow neutrons, and thus capture these neutrons before they can cause fission.

fission reactor *See* nuclear reactor.

fission spectrum [NUC PHYS] The energy distribution of neutrons arising from fission.

fission threshold [NUC PHYS] The minimum kinetic energy of a bombarding neutron required to induce fission of a nucleus.

fission-track dating [GEOL] A method of dating geological specimens by counting the radiation-damage tracks produced by spontaneous fission of uranium impurities in minerals and glasses.

fission yield [NUCLEO] The amount of energy released by fission in a nuclear explosion, as distinct from that released by fusion. [NUC PHYS] The percent of fissions that gives a particular nuclide or group of isobars.

fissiped [VERT ZOO] 1. Having the toes separated to the base. 2. Of or relating to the Fissipeda.

Fissipeda [VERT ZOO] Former designation for a suborder of the Carnivora.

fissium [NUCLEO] An equilibrium mixture of fission products in reactor fuel that can improve the stability of uranium and uranium-plutonium fuel alloys under fast-neutron irradiation.

fissure [GEOL] 1. A high, narrow cave passageway. 2. An extensive crack in a rock.

Fissurellidae [INV ZOO] The keyhole limpets, a family of gastropod mollusks in the order Archeogastropoda.

fissure vein [GEOL] A mineral deposit in a cleft or crack in the rock material of the earth's crust.

fistula [MED] An abnormal congenital or acquired communication between two surfaces or between a viscus or other hollow structure and the exterior.

Fistuliporidae [PALEON] A diverse family of extinct marine bryozoans in the order Cystoporata.

fistulous withers [VET MED] A chronic inflammation of the withers of a horse accompanied by fluid discharge, which may be initiated by mechanical injury but depends on bacterial (*Brucella abortus*) infection for development.

fit [DES ENG] The dimensional relationship between mating parts, such as press, shrink, or sliding fit.

fitness figure [METEOROL] In Great Britain, a measure of the "fitness" of the weather at an airport for the safe landing of aircraft; the figure F is computed on the basis of corrected values of visibility and cloud height; observed visibility is adjusted according to intensity of precipitation, and cloud height is corrected for height of nearby obstructions and cloud amount; further corrections are applied for the cross-runway component of the wind. Also known as fitness number.

fitness number *See* fitness figure.

fitter [ENG] One who maintains, repairs, and assembles machines in an engineering shop.

Fittig's synthesis [ORG CHEM] The synthesis of aromatic hydrocarbons by the condensation of aryl halides with alkyl halides, using sodium as a catalyst.

fitting [ENG] A small auxiliary part of standard dimensions used in the assembly of an engine, piping system, machine, or other apparatus.

fitting out [NAV ARCH] The preparation of a ship for active service by placing on board equipment and consumables required by the allowance list or for operation.

FitzGerald-Lorentz contraction [RELAT] The contraction of a moving body in the direction of its motion when its speed is comparable to the speed of light. Also known as Lorentz contraction; Lorentz-FitzGerald contraction.

five-and-ten system [METEOROL] The most common system for representing wind speed, to the nearest 5 knots, in symbolic form on synoptic charts, consisting of drawing the appropriate number of half-barbs, barbs, and pennants from the end of the wind-direction shaft; in this system, a half-barb represents 5 knots, a barb 10 knots, a pennant 50 knots.

five-day forecast [METEOROL] A forecast of the average weather conditions and large-scale synoptic features in a 5-day period; a type of extended forecast.

five-dimensional space [MATH] A vector space whose basis has five vectors.

five-level code [ADP] A code which uses five bits to specify each character.

five-level start-stop operation [COMMUN] Simplex mode of operation used in teletypewriter circuits; each code character is divided into five electrical units; the machine distributor unit makes a positive start and stop for the transmission of each character.

five-spot well pattern [PETRO ENG] A symmetrical network pattern of five wells (one in center, four equally spaced in a square pattern) as used in water-injection pressure maintenance of reservoirs.

five-wire line [ELEC] A transmission line which has four conductors, all in phase, at the corners of a square and a fifth conductor at the center of the square which is out of phase with the others.

fix [BIOL] To kill, harden, or preserve a tissue, organ, or organism by immersion in dilute acids, alcohol, or solutions of coagulants. [NAV] A position of a vessel or craft determined by its master, pilot, or navigator through the use of some or all of the equipments and techniques available.

fixation [PSYCH] A rigid habit developed as a consequence of repeated reinforcement, or of frustration.

fixative [MATER] 1. A chemical or a mixture of chemicals used to treat biological specimens before preservation so as to retain a reasonable facsimile of their appearance when alive. 2. A substance used to increase the durability of another substance; used to fix dye mordants, hold textile dyes and pigments, and slow the rate of perfume evaporation. Also known as fixing agent.

fixed ammunition [ORD] Ammunition with primer and propellant contained in a cartridge case, permanently crimped or attached to a projectile, and loaded into the weapon as a unit.

fixed and flashing light [NAV] A fixed light varied at regular intervals by one or more flashes of greater brilliance.

fixed and group flashing light [NAV] A fixed light varied at regular intervals by a group of two or more flashes of greater brilliance.

fixed arch [CIV ENG] A stiff arch having rotation prevented at its supports.

fixed area [ADP] That portion of the main storage occupied by the resident portion of the control program.

fixed artillery [ORD] Artillery weapons permanently installed on land and sea frontiers for the protection of important areas.

fixed attenuator See pad.

fixed-bed hydroforming [CHEM ENG] A cyclic petroleum process that utilizes a fixed bed of molybdenum oxide catalyst deposited on activated alumina.

fixed-bed operation [CHEM ENG] An operation in which the additive material (catalyst, absorbent, filter media, ion-exchange resin) remains stationary in the chemical reactor.

fixed bias [ELECTR] A constant value of bias voltage, independent of signal strength.

fixed-bias transistor circuit [ELECTR] A transistor circuit in which a current flowing through a resistor is independent of the quiescent collector current.

fixed bridge [CIV ENG] A rigid bridge structure having no movable parts, as distinct from a bascule or lift bridge.

fixed capacitor [ELEC] A capacitor having a definite capacitance value that cannot be adjusted.

fixed carbon [CHEM] Solid, combustible residue remaining after removal of moisture, ash, and volatile materials from coal, coke, and bituminous materials; expressed as a percentage.

fixed-charge problem [IND ENG] A linear programming problem in which each variable has a fixed-charge coefficient in addition to the usual cost coefficient; the fixed charge (for example, a setup time charge) is a nonlinear function and is incurred only when the variable appears in the solution with a positive level.

fixed contact [ELEC] A relatively immovable contact that is engaged and disengaged by a moving contact to make and break a circuit, as in a switch or relay.

fixed cost [IND ENG] A cost that remains unchanged during short-term changes in production level. Also known as overhead cost.

fixed-cycle operation [ADP] An operation completed in a specified number of regularly timed execution cycles.

fixed echo [ELECTR] An echo indication that remains stationary on a radar plan-position indicator display, indicating the presence of a fixed target.

fixed-electrode method [ENG] A geophysical surveying method used in a self-potential system of prospecting in which one electrode remains stationary while the other is grounded at progressively greater distances from it.

fixed emplacement [ORD] A fixed setting for a gun, usually made of reinforced concrete, with the base plate and base ring set in the concrete and bolted down. Also known as permanent emplacement.

fixed end beam [CIV ENG] A beam that is supported at both free ends and is restrained against rotation and vertical movement.

fixed-end column [CIV ENG] A column with the end fixed so that it cannot rotate.

fixed end moment See fixing moment.

fixed-feed grinding [MECH ENG] Feeding processed material to a grinding wheel, or vice versa, in predetermined increments or at a given rate.

fixed field [ADP] A field in computers, film selection devices, or punched cards, or a given number of holes along the edge of a marginal punched card, set aside, or "fixed," for the recording of a given type of characteristic.

fixed-field accelerator [NUCLEO] A circular particle accelerator whose magnetic fields do not vary with time, such as an ordinary cyclotron or a fixed-field, alternating-gradient synchrotron.

fixed-field method [ADP] A method of data storage in which the same type of data is always placed in the same relative position.

fixed fin [AERO ENG] A nonadjustable vane or airfoil affixed longitudinally to an aerodynamically or ballistically designed body for stabilizing purposes.

fixed-focus lens [OPTICS] A lens whose focus is invariable, as on inexpensive cameras with no mechanism for adjusting focus but so designed that all objects from a few feet away to infinity are tolerably in focus.

fixed gun [ORD] An aircraft machine gun mounted rigidly to the aircraft, and aimed by moving the aircraft.

fixed inductor [ELEC] An inductor whose coils are wound in such a manner that the turns remain fixed in position with respect to each other, and which either has no magnetic core or has a core whose air gap and position within the coil are fixed.

fixed-length record [ADP] One of a file of records, each of which must have the same specified number of data units, such as blocks, words, characters, or digits.

fixed-level chart See constant-height chart.

fixed light [NAV] A light used in navigation having constant luminous intensity. [NAV ARCH] A thick, usually circular glass fitted in a frame and fixed in an opening in the side of a ship or deckhouse or in a bulkhead, to provide light.

fixed logic [ADP] Circuit logic of computers or peripheral devices that cannot be changed by external controls; connections must be physically broken to arrange the logic.

fixed memory [ADP] Of a computer, a nondestructive readout memory that is only mechanically alterable.

fixed mooring berth [CIV ENG] A marine structure consisting of dolphins for securing a ship and a platform to support cargo-handling equipment.

fixed-needle traverse [ENG] In surveying, a traverse with a compass fitted with a sight line which can be moved above a

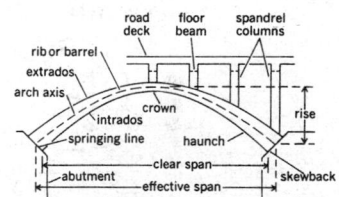

FIXED ARCH

An open-spandrel, concrete, fixed arch bridge.

FIZEAU FRINGES

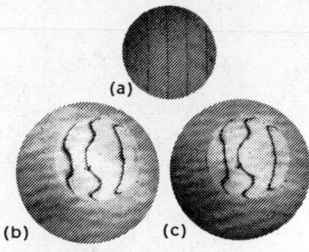

Fizeau multiple-beam fringe patterns (def. 2). *(a)* Narrow-gap plane mirrors. *(b)* Wide-gap test lens without end mirrors imaged on each other; *(c)* with mirrors imaged on each other.

FJORD

Fjord in Alaska. *(W. W. Atwood, USGS)*

FLABELLIFERA

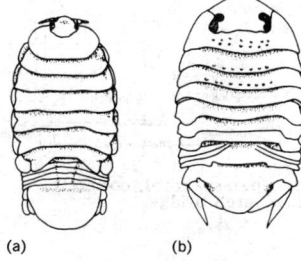

(a)   (b)

Two examples of Flabellifera showing the caudal fan. *(a) Lironeca epimerias* Richardson. *(b) Tecticeps renoculis* Richardson.

FLABELLIGERIDAE

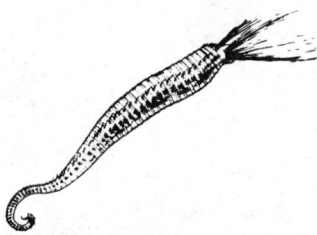

Pheurusa, a cage worm, shown in right lateral view.

---

graduated horizontal circle, so that the azimuth angle can be read, as with a theodolite.

fixed oil [MATER] A nonvolatile fatty oil of vegetable origin.

fixed point [ENG] A reproducible value, as for temperature, used to standardize measurements; derived from intrinsic properties of pure substances. [MATH] For a function f mapping a set S to itself, any element of S which f sends to itself.

fixed-point arithmetic [ADP] **1.** A method of calculation in which the computer does not consider the location of the decimal or radix point because the point is given a fixed position. **2.** A type of arithmetic in which the operands and results of all arithmetic operations must be properly scaled so as to have a magnitude between certain fixed values.

fixed point calculation [ADP] A calculation made with fixed point arithmetic.

fixed-point computer [ADP] A computer in which numbers in all registers and storage locations must have an arithmetic point which remains in the same fixed location.

fixed-point part *See* mantissa.

fixed-point representation [ADP] Any method of representing a number in which a fixed-point convention is used.

fixed-point system [ADP] A number system in which the location of the point is fixed with respect to one end of the numerals, according to some convention.

fixed-position addressing [ADP] Direct access to an item in a data file on disk or drum, as opposed to a sequential search for this item starting with the first item in the file.

fixed-position welding [MET] A welding operation in which the work is stationary.

fixed-product area [ADP] The area in core memory where multiplication takes place for certain types of computers.

fixed-program computer [ADP] A special-purpose computer having a program permanently wired in.

fixed rent *See* minimum rent.

fixed resistor [ELEC] A resistor that has no provision for varying its resistance value.

fixed screen [MIN ENG] A stationary panel, commonly of wedge wire, used to remove a large proportion of water and fines from a suspension of coal in water.

fixed service [COMMUN] Service providing radio communications between fixed points.

fixed sonar [ENG] Sonar in which the receiving transducer is not constantly rotated, in contrast to scanning sonar.

fixed square search *See* geographic square search.

fixed star [ASTRON] A misnomer to indicate those stars which kept apparently the same position with respect to other stars, in contrast to the planets which were termed wandering stars.

fixed storage [ADP] A storage for data not alterable by computer instructions, such as magnetic-core storage with a lockout feature.

fixed target [ORD] A nonmovable or immobilized target, such as a city, factory, or surrounded troops.

fixed transmitter [ELECTR] Transmitter that is operated in a fixed or permanent location.

fixed word length [ADP] The length of a computer machine word that always contains the same number of characters or digits.

fixer network [NAV] A combination of radio or radar direction-finding installations which, operating in conjunction, are capable of plotting the position relative to the ground of an aircraft in flight. Also known as fixer system.

fixer system *See* fixer network.

fixing agent *See* fixative.

fixing moment [MECH] The bending moment at the end support of a beam necessary to fix it and prevent rotation. Also known as fixed end moment.

fixity *See* continuity.

fixture [CIV ENG] An object permanently attached to a structure, such as a light or sink. [MECH ENG] A device used to hold and position a piece of work without guiding the cutting tool.

Fizeau fringes [OPTICS] **1.** Interference fringes of monochromatic light from interference in a geometrical situation other than plane parallel plates. Also known as fringes of equal

---

thickness. **2.** Interference fringes in light from a Fizeau interferometer.

Fizeau interferometer [OPTICS] Interferometer in which light from a point source is collimated and multiply reflected between a plane mirror and the partially silvered inner surface of a parallel plane plate, and is viewed in reflection.

Fizeau toothed wheel [OPTICS] Rapidly rotating toothed wheel which was used to measure the speed of light by adjusting the rotation speed until light passing through one tooth opening and reflected from a distant mirror would pass through the next tooth opening on return.

fizelyite [MINERAL] A metallic, lead-gray mineral composed of a lead silver antimony sulfide, occurring as prisms.

fjord [GEOGR] A narrow, deep inlet of the sea between high cliffs or steep slopes. Also spelled fiord.

fjord valley [GEOGR] A deep, narrow channel occupied by the sea and extending inland about 50–100 miles (80–160 kilometers).

flabellate [BIOL] Fan-shaped.

Flabellifera [INV ZOO] The largest and morphologically most generalized suborder of isopod crustaceans; the biramous uropods are attached to the sides of the abdomen and may form, with the last abdominal fragment, a caudal fan.

Flabelligeridae [INV ZOO] The cage worms, a family of spioniform worms belonging to the Sedentaria; the anterior part of the body is often concealed by a cage of setae arising from the first few segments.

flabellum [INV ZOO] Any structure resembling a fan, as the epipodite of certain crustacean limbs.

flaccid [BOT] Deficient in turgor. [PHYSIO] Soft, flabby, or relaxed.

flacherie [INV ZOO] A fatal bacterial disease of caterpillars, especially silkworms, marked by loss of appetite, dysentery, and flaccidity of the body; after death the body darkens and liquefies.

Flacourtiaceae [BOT] A family of dicotyledonous plants in the order Violales having the characteristics of the more primitive members of the order.

Flade potential [MET] The potential of a passive metal immediately preceding a final steep fall from the passive to the active region.

flag [ADP] Any of various types of indicators used for identification, such as a work mark, or a character that signals the occurrence of some condition, such as the end of a word. [ELECTR] A small metal tab that holds the getter during assembly of an electron tube. [ENG] **1.** A piece of fabric used as a symbol or as a signaling or marking device. **2.** A large sheet of metal or fabric used to shield television camera lenses from light when not in use.

flag alarm [ENG] A semaphore-type flag in the indicator of an instrument to serve as a signal, usually to warn that the indications are unreliable.

flag bridge [NAV ARCH] A separate structure built above the deck of a ship, designed for the use of a naval officer higher than the rank of captain; the visual control and communications center from which the officer exercises command of the tactical units and is separate from and at a different level from the navigation bridge.

flagella [BIOL] Relatively long, whiplike, centriole-based locomotor organelles on some motile cells.

Flagellata [INV ZOO] The equivalent name for Mastigophora.

flagellate [BIOL] **1.** Having flagella. **2.** An organism that propels itself by means of flagella. **3.** Resembling a flagellum. [INV ZOO] Any member of the protozoan superclass Mastigophora.

flagellated chamber [INV ZOO] An outpouching of the wall of the central cavity in Porifera that is lined with choanocytes; connects with incurrent canals through prosophyles.

flagellation [BIOL] The arrangement of flagella on an organism. [PSYCH] Beating or whipping as a means of producing sexual gratification.

flagellin [MICROBIO] The protein component of bacterial flagella.

flag flip-flop [ADP] A one-bit register which indicates overflow, carry, or sign bit from past or current operations.

flag float [ENG] A pyrotechnic device that floats and burns upon the water, used for marking or signaling.

flaggy [GEOL] **1.** Of bedding, consisting of strata 10–100 centimeters in thickness. **2.** Of rock, tending to split into layers of suitable thickness (1–5 centimeters) for use as flagstones.

flagman [CIV ENG] A range-pole carrier in a surveying party.

flag operand [ADP] A part of the instruction of some assembly languages denoting which elements of the object instruction will be flagged.

flagpole [ENG] A single staff or pole rising from the ground and on which flags or other signals are displayed; on charts the term is used only when the pole is not attached to a building.

flag smut [PL PATH] A smut affecting the leaves and stems of cereals and other grasses, characterized by formation of sori within the tissues, which rupture releasing black spore masses and causing fraying of the infected area.

flagstaff [ENG] A pole or staff on which flags or other signals are displayed; on charts this term is used only when the pole is attached to a building.

flagstone [GEOL] **1.** A hard, thin-bedded sandstone, firm shale, or other rock that splits easily along bedding planes or joints into flat slabs. **2.** A piece of flagstone used for making pavement or covering the side of a house.

flag tower [NAV] A scaffoldlike tower from which flags are displayed.

flail tank [ORD] A specially constructed tank equipped with a flailing device, consisting of chain flails attached to a roller powered by the tank engine, employed to detonate antitank mines.

flair [CIV ENG] A gradual widening of the flangeway near the end of a guard line of a track or rail structure.

flajolotite [MINERAL] $4FeSbO_4 \cdot 3H_2O$ A claylike, lemon-yellow mineral composed of a hydrous iron antimonate, occurring in nodular masses.

flak [ORD] **1.** Explosive or exploding missile fired from anti-aircraft cannon. **2.** Antiaircraft cannon, as in flak battery or flak installation.

flak analysis [ORD] The examination and study of flak intelligence to determine the nature, effectiveness, or probable effectiveness of enemy antiaircraft defenses.

flake [MATER] **1.** Dry, unplasticized, cellulosic plastics base. **2.** Plastic chip used as feed in molding operations. [MET] **1.** Discontinuous, internal cracks formed in steel during cooling due usually to the release of hydrogen. Also known as fisheye; snowflake; shattercrack. **2.** Fish-scale, flat particles in powder metallurgy. Also known as flake powder.

flake powder See flake.

flaking [CHEM ENG] Continuous process operation to remove heat from material in the liquid state to cause its solidification. [ENG] **1.** Reducing or separating into flakes. **2.** See frosting. [MIN ENG] Breaking small chips from the face of a refractory, particularly chrome ore containing refractories.

flaking mill [MECH ENG] A machine for converting material to flakes.

flak jacket [ENG] A jacket or vest of heavy fabric containing metal, nylon, or ceramic plates, designed especially for protection against flak; usually covers the chest, abdomen, back, and genitals, leaving the arms and legs free. Also known as flak vest.

flak vest See flak jacket.

flamboyant structure [GEOL] The optical continuity of crystals or grains as disturbed by a structure that is divergent.

flame [CHEM] A hot, luminous reaction front (or wave) in a gaseous medium into which the reactants flow and out of which the products flow.

flame annealing [MET] The careful heating of a metal part by flames, before or after working.

flame arc lamp [ELEC] An arc lamp in which carbon electrodes are impregnated with chemicals, such as calcium, barium, or titanium, which are more volatile than the carbon and radiate light when driven into the arc.

flame arrester [ENG] An assembly of screens, perforated plates, or metal-gauze packing attached to the breather vent on a flammable-product storage tank.

flame bucket [AERO ENG] A deep, cavelike construction built beneath a rocket launchpad, open at the top to receive the hot gases of the rocket, and open on one or three sides below, with a thick metal fourth side bent toward the open side or sides so as to deflect the exhaust gases.

flame bulb [INV ZOO] The enlarged terminal part of the flame cell of a protonephridium, consisting of a tuft of cilia.

flame cell [INV ZOO] A hollow cell that contains the terminal branches of excretory vessels in certain flatworms and rotifers and some other invertebrates.

flame cleaning [MET] Removing scale, rust, and dirt from metal surfaces by using a broad flame.

flame coating See flame plating.

flame collector [ENG] A device used in atmospheric electrical measurements for the removal of induction charge on apparatus; based upon the principle that products of combustion are ionized and will consequently conduct electricity from charged bodies.

flame cultivator [AGR] A flamethrower for destroying weeds between rows of crops.

flame cutting [MET] Use of an oxyacetylene, oxyhydrogen, or oxycoal gas flame to cut thick metal sections.

flame deflector [AERO ENG] **1.** In a vertical launch, any of variously designed obstructions that intercept the hot gases of the rocket engine so as to deflect them away from the ground or from a structure. **2.** In a captive test, an elbow in the exhaust conduit or flame bucket that deflects the flame into the open.

flame emission spectroscopy [SPECT] A flame photometry technique in which the solution containing the sample to be analyzed is optically excited in an oxyhydrogen or oxyacetylene flame.

flame excitation [SPECT] Use of a high-temperature flame (such as oxyacetylene) to excite spectra emission lines from alkali and alkaline-earth elements and metals.

flame gouging [MET] A form of oxygen cutting by means of a cutting torch with a slightly curved tip, enabling the flame to strike the metal surface at a low angle and making shallow cuts possible. Also known as oxygen gouging.

flame hardening [MET] A method for local surface hardening of steel by passing an oxyacetylene or similar flame over the work at a predetermined rate.

flameholder [AERO ENG] A device that sustains combustion in a flowing mixture within the combustion chamber of some types of jet engines.

flame ionization detector [ANALY CHEM] A device in which the measured change in conductivity of a standard flame (usually hydrogen) due to the insertion of another gas or vapor is used to detect the gas or vapor.

flameout [AERO ENG] The extinguishing of the flame in a reaction engine, especially in a jet engine.

flame photometer [SPECT] One of several types of instruments used in flame photometry, such as the emission flame photometer and the atomic absorption spectrophotometer, in each of which a solution of the chemical being analyzed is vaporized; the spectral lines resulting from the light source going through the vapors enters a monochromator that selects the band or bands of interest.

flame photometry [SPECT] A branch of spectrochemical analysis in which samples in solution are excited to produce line emission spectra by introduction into a flame.

flame plate [ENG] One of the plates on a boiler firebox which are subjected to the maximum furnace temperature.

flame plating [MET] Coating a thin layer of refractory material on a surface by exploding a mixture of plating powder, oxygen, and acetylene. Also known as flame coating.

flame propagation [CHEM] The spread of a flame in a combustible environment outward from the point at which the combustion started.

flame-retarded resin [MATER] A resin which is compounded with certain chemicals to reduce or eliminate its tendency to burn.

flame spectrometry [SPECT] A procedure used to measure the spectra or to determine wavelengths emitted by flame-excited substances.

flame spectrophotometry [SPECT] A method used to determine the intensity of radiations of various wavelengths in a spectrum emitted by a chemical inserted into a flame.

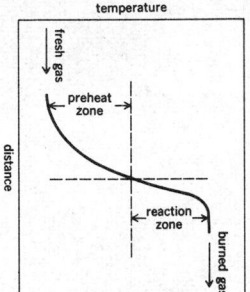

FLAME

Temperature distribution through a flame front.

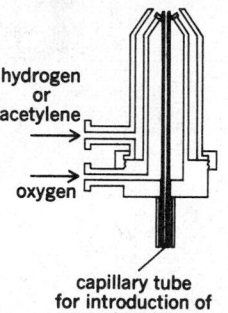

FLAME EMISSION SPECTROSCOPY

Beckman aspirator burner used to aspirate the solution directly into oxyhydrogen or oxyacetylene flame.

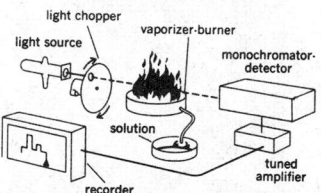

FLAME PHOTOMETER

Schematic diagram of atomic absorption spectrophotometer for determining metal concentrations.

flame spectrum [SPECT] An emission spectrum obtained by evaporating substances in a nonluminous flame.

flame speed [CHEM] The rate at which combustion moves through an explosive mixture.

flame spraying [ENG] **1.** A method of applying a plastic coating onto a surface in which finely powdered fragments of the plastic, together with suitable fluxes, are projected through a cone of flame. **2.** Deposition of a conductor on a board in molten form, generally through a metal mask or stencil, by means of a spray gun that feeds wire into a gas flame and drives the molten particles against the work.

flame straightening [MET] Correcting distorted structural metal to a straight form by local application of a gas-flame heat.

flamethrower [ENG] A device used to project ignited fuel from a nozzle so as to cause casualties to personnel or to destroy material such as weeds or insects.

flame trap [ENG] A device that prevents a gas flame from entering the supply pipe.

flame treating [ENG] A method of rendering inert thermoplastic objects receptive to inks, lacquers, paints, or adhesives, in which the object is bathed in an open flame to promote oxidation of the surface.

flamingo [VERT ZOO] Any of various long-legged and long-necked aquatic birds of the family Phoenicopteridae characterized by a broad bill resembling that of a duck but abruptly bent downward and rosy-white plumage with scarlet coverts.

flammability [CHEM] A measure of the extent to which a material will support combustion. Also known as inflammability.

flammability limits [CHEM] The stoichiometric composition limits (maximum and minimum) of an ignited oxidizer-fuel mixture what will burn indefinitely at given conditions of temperature and pressure without further ignition.

flammable [MATER] Of a material, capable of supporting combustion.

flammable liquid [MATER] A liquid which gives off combustible vapors.

Flamsteed's number [ASTRON] A number sometimes used with the possessive form of the Latin name of the constellation to identify a star, for example, 72 Ophiuchi.

flan [METEOROL] In Scotland, a sudden gust or squall of wind from land.

Flanders storm [METEOROL] In England, a heavy fall of snow coming with the south wind.

Flandrian transgression [OCEANOGR] The rapid rise of the North Sea between 8000 and 3000 B.C. from about 180 feet (55 meters) below to about 20 feet (6 meters) below its present level.

flange [SCI TECH] A projecting rim of an organism or mechanical part.

flanged pipe [DES ENG] A pipe with flanges at the ends; can be bolted end to end to another pipe.

flangeway [CIV ENG] Open way through a rail or track structure that provides a passageway for the flange of a wheel.

flank [CIV ENG] The outer edge of a carriageway. [DES ENG] **1.** The end surface of a cutting tool, adjacent to the cutting edge. **2.** The side of a screw thread. [GEOL] See limb. [VERT ZOO] The part of a quadruped mammal between the ribs and the pelvic girdle.

flank angle [DES ENG] The angle made by the flank of a screw thread with a line perpendicular to the axis of the screw.

flank hole [MIN ENG] **1.** A hole bored in advance of a working place when approaching old workings. **2.** A borehole driven from the side of an underground excavation, not parallel with the center line of the excavation, to detect water, gas, or other danger.

flank observation [ORD] Observation of fire from a place on, or near, the flank of the target; the angle at the target between the gun and the observer is between 75 and 105°.

flank wear [ENG] Loss of relief on the flank of a tool behind the cutting edge.

flannel [TEXT] A loosely woven, generally wool fabric with the weave concealed by a napped surface.

flap [AERO ENG] **1.** Any control surface, such as a speed brake, dive brake, or dive-recovery brake, used primarily to increase the lift or drag on an airplane, or to aid in recovery from a dive. **2.** Any rudder attached to a rocket and acting either in the air or within the jet stream.

flap attenuator [ELECTROMAG] A waveguide attenuator in which a contoured sheet of dissipative material is moved into the guide through a nonradiating slot to provide a desired amount of power absorption. Also known as vane attenuator.

flaperon [AERO ENG] A control surface used both as a flap and as an aileron.

flap gate [CIV ENG] A gate that opens or closes by rotation around hinges at the top of the gate. Also known as pivot leaf gate.

flapping [MET] Striking through the surface of molten copper to hasten oxidation by increasing the exposure to air.

flap valve [MECH ENG] A valve fitted with a hinged flap or disk that swings in one direction only.

flare [AERO ENG] To descend in a smooth curve, making a transition from a relatively steep descent to a direction substantially parallel to the surface, when landing an aircraft. [ASTRON] A bright eruption from the sun's chromosphere; flares may appear within minutes and fade within an hour, cover a wide range of intensity and size, and tend to occur between sunspots or over their penumbrae. [CHEM ENG] A device for disposing of combustible gases from refining or chemical processes by burning in the open, in contrast to combustion in a furnace or closed vessel or chamber. [DES ENG] An expansion at the end of a cylindrical body, as at the base of a rocket. [ELECTR] A radar screen target indication having an enlarged and distorted shape due to excessive brightness. [ELECTROMAG] See horn antenna. [ENG] A pyrotechnic item designed to produce a single source of intense light for such purposes as target or airfield illumination. [NAV ARCH] A concave curve of a boat's or ship's sides away from the center line, above the waterline, normally at the bow.

flareback [ORD] A rearward escapement of flame or gas from a gun.

flare chute [ENG] A flare attached to a parachute.

flare factor [ENG ACOUS] Number expressing the degree of outward curvature of the horn of a loudspeaker.

flareout [AERO ENG] That portion of the approach path of an aircraft in which the vertical component is modified to lessen the impact of landing.

flare stars See UV Ceti stars.

flare-type bucket [MIN ENG] A dragline bucket that has flared sides to allow heaped loading.

flaring [MET] Increasing the diameter at the end of a pipe or tube.

flaser [GEOL] Streaky layer of parallel, scaly aggregates that surrounds the lenticular bodies of granular material in flaser structure; caused by pressure and shearing during metamorphism.

flaser gabbro [GEOL] A cataclastic gabbro that contains augen of feldspar or quartz surrounded by flakes of mica or chlorite.

flaser structure [GEOL] **1.** A metamorphic structure in which small lenses and layers of granular material are surrounded by a matrix of sheared, crushed material, resembling a crude flow structure. Also known as pachoidal structure. **2.** A primary sedimentary structure consisting of fine-sand or silt lenticles that are aligned and cross-bedded.

flash [ENG] In plastics or rubber molding or in metal casting, that portion of the charge which overflows from the mold cavity at the joint line. [MET] A fin of excess metal along the mold joint line of a casting, occurring between mating die faces of a forging or expelled from a joint in resistance welding.

flash arc [ELECTR] A sudden increase in the emission of large thermionic vacuum tubes, probably due to irregularities in the cathode surface.

flashback See backfire.

flashback voltage [ELECTR] Inverse peak voltage at which ionization takes place in a gas tube.

flash-bang [ORD] The time interval between visual observation of the flash of a weapon being fired and the auditory perception of the sound of the discharge.

flashboard [CIV ENG] A relatively low, temporary barrier constructed of a series of boards along the top of a dam spillway to increase storage capacity.

flash boiler [MECH ENG] A boiler with hot tubes of small capacity; designed to immediately convert small amounts of water to superheated steam.

flash bomb [ENG] A bomb that illuminates the ground for night aerial photography.

flash burn [MED] Tissue injury resulting from exposure to high-intensity radiant heat.

flash butt welding [MET] Resistance welding to produce a butt joint by passing an electric current through two pieces of metal in light contact to create an arc which causes flashing and consequent heating; the weld is completed by applying pressure at the joint.

flash chamber [CHEM ENG] A conventional oil-and-gas separator operated at low pressure, with the liquid from a higher-pressure vessel being flashed into it. Also known as flash trap; flash vessel.

flash depressor [MATER] A substance used to reduce the flash from a rocket motor.

flash dry [CHEM ENG] The rapid evaporation of moisture from a porous or granular solid by a sudden reduction in pressure or by placing the material in an updraft of warm air.

flasher [ELEC] A switch, generally either motor-driven or using a combination heater element and bimetallic strip, that turns lamps on and off rapidly.

flash factor [OPTICS] In photography using a photoflash lamp, a number dependent on the lamp and the film speed, equal to the product of the distance of the lamp from the subject and the correct *f*-number for that distance.

flash flood [HYD] A sudden local flood of short duration and great volume; usually caused by heavy rainfall in the immediate vicinity.

flash groove [ENG] A groove in a casting die so that excess material can escape during casting.

flash hider [ORD] A metallic cone or flat disks attached to the muzzle of the gun to conceal the flash when the gun is fired and to prevent temporary blindness of the gun crew while firing.

flashing [BUILD] A strip of sheet metal placed at the junction of exterior building surfaces to render the joint watertight. [CHEM ENG] Vaporization of volatile liquids by either heat or vacuum. [ENG] Burning brick in an intermittent air supply in order to impart irregular color to the bricks. [MET] The violent expulsion of small metal particles due to arcing during flash butt welding.

flashing flow [CHEM ENG] The condition when a liquid at its boiling point flows through a heated conduit and is further heated to cause partial vaporization (flashing), with a resultant two-phase (vapor-liquid) flow.

flashing light [NAV] A light showing one or more flashes at regular intervals, the duration of the light period is less than that of the dark period; in particular, a light showing a single flash at regular intervals; distinctive names are used to indicate different combinations of flashes.

flashing over [ELEC] Accidental formation of an arc over the surface of a rotating commutator from brush-to-brush; usually caused by faulty insulation between commutator segments.

flashing ring [ENG] A ring around a pipe that holds it in place as it passes through a partition such as a floor or wall.

flash lamp [ELECTR] A gaseous-discharge lamp used in a photoflash unit to produce flashes of light of short duration and high intensity for stroboscopic photography. Also known as stroboscopic lamp.

flashless [ORD] Of a propellant or a propelling charge that does not produce a muzzle flash in the weapon for which intended.

flash magnetization [ELECTROMAG] Magnetization of a ferromagnetic object by a current impulse of short duration.

flash message [COMMUN] A category of precedence reserved for initial enemy contact messages or operational combat messages of extreme urgency; brevity is mandatory.

flashover [ELEC] An electric discharge around or over the surface of an insulator.

flash pasteurization [MICROBIO] A pasteurization method in which a heat-labile liquid, such as milk, is briefly subjected to temperatures of 230°F (110°C).

flash photography *See* stroboscopic photography.

flash photolysis [PHYS CHEM] A method of studying fast photochemical reactions in gas molecules; a powerful lamp is discharged in microsecond flashes near a reaction vessel holding the gas, and the products formed by the flash are observed spectroscopically.

flash plating [MET] Electrodeposition of a thin film of metal.

flash point [CHEM] The temperature at which the vapors of a volatile liquid, mixed with air, spontaneously ignite; this is determined for petroleum products by the open cup and closed cup techniques.

flash process [CHEM ENG] Liquid-vapor system in which the composition remains constant, but the proportion of gas and liquid phases changes as pressure or temperature change.

flash radiography [GRAPHICS] A radiography technique employing very short exposure times, such as 1 microsecond, to give unblurred pictures of moving objects.

flash ranging [ORD] Finding the position of the burst of a projectile or of an enemy gun by observing its flash.

flash ranging adjustment [ORD] Correcting friendly artillery fire on the basis of observation and location of the flash of shell bursts.

flash reconnaissance [ORD] Observation from ground posts or from aircraft to locate enemy gun positions by the flashes of enemy guns.

flash roast [MIN ENG] Rapid removal of sulfur from ore by having finely divided sulfide mineral fall through a heated oxidizing atmosphere. Also known as suspension roast.

flash separation [CHEM ENG] Process for separation of gas (vapor) from liquid components under reduced pressure; the liquid and gas remain in contact as the gas evolves from the liquid.

flash set [MATER] Rapid hardening of freshly mixed cement paste, mortar, or concrete with considerable evolution of heat; plasticity cannot be restored.

flash smelting [MET] Production of molten metal or matte in a vertical furnace in which concentrates are reacted with hot gases; the molten product is collected in a horizontal refractory-lined accumulator at the base of the furnace.

flash spectroscopy [SPECT] The study of the electronic states of molecules after they absorb energy from an intense, brief light flash.

flash test [ELEC] A method of testing insulation by applying momentarily a voltage much higher than the rated working voltage.

flash trap *See* flash chamber.

flash vessel *See* flash chamber.

flash welding [MET] A form of resistance butt welding used to weld wide, thin members or members with irregular faces, and tubing to tubing.

flask [MET] A frame used to hold molding sand in foundry work.

flat [ACOUS] A musical note that is a half step lower than a specified note. [ENG] A nonglossy painted surface. [GEOGR] A level tract of land. [GEOL] *See* mud flat. [GRAPHICS] **1.** The sheet of glass on which negative films are placed close together for printing on sensitized metal in the photoengraving process. **2.** An assemblage of negative or positive films used in preparing a photo-offset plate. [MINERAL] An inferior grade of rough diamonds. [NAV] **1.** A place covered with water too shallow for ordinary navigation. **2.** The area between high- and low-water marks along the edge of an arm of the sea, a bay, or tidal river; the term is usually used in the plural. [NAV ARCH] A partial deck below the main deck, constructed without any camber. [SCI TECH] **1.** A smooth, even surface. **2.** An object with a broad, shallow or thin form.

flat arch [ARCH] **1.** A straight horizontal arch consisting of mutually supportive wedge-shaped blocks. **2.** Any arch with a small rise-to-span ratio.

flat-back stope [MIN ENG] An overhand stoping method in which the ore is broken in slices parallel with the levels.

flatbed press [GRAPHICS] A press whose printing plates for type and cuts are flat; it and the rotary press are the two major kinds of presses.

FLASH LAMP

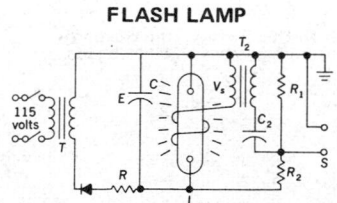

Elementary circuit of electronic flash lamp L.

FLAT-BELT PULLEY

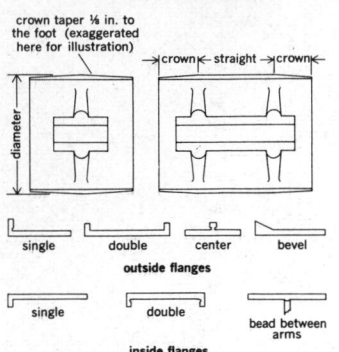

Diagrams of crowning and flanging on flat-belt pulleys. (*From D. C. Greenwood, Mechanical Power Transmission, McGraw-Hill, 1962*)

FLATFISH

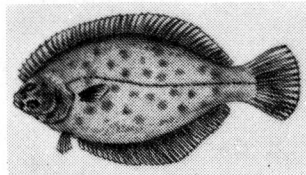

A typical left-eye flounder, with both eyes on left side.

FLATHEAD RIVET

A flathead rivet, one of many standard types of rivet heads.

FLAVONE

Structural formula of flavone (def. 2).

flatbed truck [ENG] A truck whose body is in the form of a platform.

flat-belt conveyor [MECH ENG] A conveyor belt in which the carrying run is supported by flat-belt idlers or pulleys.

flat-belt pulley [DES ENG] A smooth, flat-faced pulley made of cast iron, fabricated steel, wood, and paper and used with a flat-belt drive.

flat-blade turbine [MECH ENG] An impeller with flat blades attached to the margin.

flat-bottom crown *See* flat-face bit.

flatcar [ENG] A railroad car without fixed walls or a cover.

flat chisel [DES ENG] A steel chisel used to obtain a flat and finished surface.

flat coaxial transmission line *See* strip transmission line.

flat-crested weir [CIV ENG] A type of measuring weir whose crest is in the horizontal plane and whose length is great compared with the height of water passing over it.

flat cut [MIN ENG] A manner of placing the boreholes, for the first shot in a tunnel, in which they are started about 2 or 3 feet (60–90 centimeters) above the floor and pointed downward so that the bottom of the hole will be about level with the floor.

flat drill [DES ENG] A type of rotary drill constructed from a flat piece of material.

flat edge trimmer [MECH ENG] A machine designed to trim the notched edges of metal shells.

flat-face bit [DES ENG] A diamond core bit whose face in cross section is square. Also known as flat-bottom crown; flat-nose bit; square-nose bit.

flat fading [COMMUN] Type of fading in which all components of the received radio signal fluctuate in the same proportion simultaneously.

flat fillet weld [MET] A fillet weld having a face that is relatively flat.

flatfish [VERT ZOO] Any of a number of asymmetrical fishes which compose the order Pleuronectiformes; the body is laterally compressed, and both eyes are on the same side of the head.

flat form tool [DES ENG] A tool having a square or rectangular cross section with the form along the end.

flathead rivet [DES ENG] A small rivet with a flat manufactured head used for general-purpose riveting.

flat jack [CIV ENG] A hollow steel cushion which is made of two nearly flat disks welded around the edge and which can be inflated with oil or cement under controlled pressure; used at the arch abutments and crowns to relieve the load on the formwork at the moment of striking the formwork.

flat line [ELECTROMAG] A radio-frequency transmission line, or part thereof, having essentially 1-to-1 standing wave ratio.

flat-lying [GEOL] Of mineral deposits and coal seams, having a relatively flat dip, up to 5°.

flat nose [ORD] A missile used against submarines, designed to prevent ricocheting on water impact.

flat-nose bit *See* flat-face bit.

flat of bottom [NAV ARCH] That portion of a ship's bottom without rise or having a rise with little or no curvature.

flat pack [ELECTR] Semiconductor network encapsulated in a thin, rectangular package, with the necessary connecting leads projecting from the edges of the unit.

flat-position welding [MET] Welding above the joint with the face of the weld in the horizontal reference plane. Also known as downhand welding.

flat rope [DES ENG] A steel or fiber rope having a flat cross section and composed of a number of loosely twisted ropes placed side by side, the lay of the adjacent strands being in opposite directions to secure uniformity in wear and to prevent twisting during winding.

flats [GRAPHICS] Stage constructions, used in series, to produce a painted background, usually architectural details; they are canvas-covered frames and are used for all flat surfaces on the stage, such as room interiors or exterior walls of buildings.

flat slab [CIV ENG] A flat plate of reinforced concrete designed to span in two directions.

flat space-time [RELAT] Space-time in which the Riemann-Christoffel tensor vanishes; geometry is then equivalent to that of the Minkowski universe used in special relativity.

flat spin [MECH] Motion of a projectile with a slow spin and a very large angle of yaw, happening most frequently in fin-

stabilized projectiles with some spin-producing moment, when the period of revolution of the projectile coincides with the period of its oscillation; sometimes observed in bombs and in unstable spinning projectiles.

flat spring *See* leaf spring.

flattening [GEOD] The ratio of the difference between the equatorial and polar radii of the earth; the flattening of the earth is the ellipticity of the spheroid; the magnitude of the flattening is sometimes expressed as the numerical value of the reciprocal of the flattening. Also known as compression. [MET] Straightening of metal sheet by passing it through special rollers which flatten it without changing its thickness. Also known as roll flattening.

flattening test [MET] Quality test performed by flattening metal tubing between parallel plates that are a specified distance from each other.

flatting agent [MATER] Additive substance for paints or varnishes to disperse incident light rays to give the dried surface a nonglossy matte finish.

flat-top antenna [ELECTROMAG] An antenna having two or more lengths of wire parallel to each other and in a plane parallel to the ground, each fed at or near its midpoint.

flat top response *See* band-pass response.

flat trajectory [MECH] A trajectory which is relatively flat, that is, described by a projectile of relatively high velocity; used to describe the trajectory of a rifle or gun as opposed to that of howitzers and mortars.

flat trajectory fire [ORD] Gunfire delivered at such a range, or with such velocity and elevation setting, that the path of the projectile in flight is almost flat, rather than curved.

flat trajectory weapon [ORD] A gun that fires projectiles at such velocity that they travel in an almost straight path with little curve above the ground, for example, a machine gun or a rifle at common battle ranges.

flat tuning [ELECTR] Tuning of a radio receiver in which a change in frequency of the received waves produces only a small change in the current in the tuning apparatus.

flat-turret lathe [MECH ENG] A lathe with a low, flat turret on a power-fed cross-sliding headstock.

flatworm [INV ZOO] The common name for members of the phylum Platyhelminthes; individuals are dorsoventrally flattened.

flat yard [CIV ENG] A switchyard in which railroad cars are moved by locomotives, not by gravity.

flavan [BIOCHEM] $C_{15}H_{14}O$ 2-Phenylbenzopyran, an aromatic heterocyclic compound from which all flavonoids are derived.

flavanone [BIOCHEM] $C_{15}H_{12}O_2$ A colorless crystalline derivative of flavone.

flavescence [PL PATH] Yellowing or blanching of green plant parts due to diminution of chlorophyll accompanying certain virus disease.

flavin [BIOCHEM] 1. A yellow dye obtained from the bark of quercitron trees. 2. Any of several water-soluble yellow pigments occurring as coenzymes of flavoproteins.

flavin adenine dinucleotide [BIOCHEM] $C_{27}H_{33}N_9O_{15}P_2$ A coenzyme that functions as a hydrogen acceptor in aerobic dehydrogenases (flavoproteins). Abbreviated FAD.

flavin mononucleotide *See* riboflavin 5'-phosphate.

flavin phosphate *See* riboflavin 5'-phosphate.

Flavobacterium [MICROBIO] A genus of gram-negative, rod-shaped, motile bacteria in the family Achromobacteraceae that characteristically produce yellow, orange, or red pigmentation.

flavone [BIOCHEM] 1. Any of a number of ketones composing a class of flavonoid compounds. 2. $C_{15}H_{10}O_2$ A colorless crystalline compound occurring as dust on the surface of many primrose plants.

flavonoid [BIOCHEM] Any of a series of widely distributed plant constituents related to the aromatic heterocyclic skeleton of flavan.

flavonol [BIOCHEM] 1. Any of a class of flavonoid compounds that are hydroxy derivatives of flavone. 2. $C_{16}H_{10}O_2$ A colorless, crystalline compound from which many yellow plant pigments are derived.

flavoprotein [BIOCHEM] Any of a number of conjugated protein dehydrogenases containing flavin that play a role in biological oxidations in both plants and animals; a yellow enzyme.

flaw [METEOROL] An English nautical term for a sudden gust or squall of wind. [OCEANOGR] **1.** The seaward edge of fast ice. **2.** A shore lead just outside fast ice.

flax [BOT] *Linum usitatissimum.* An erect annual plant with linear leaves and blue flowers; cultivated as a source of flaxseed and fiber.

flax rust [PL PATH] A disease of flax caused by the rust fungus *Melampsora lini.*

flaxseed [BOT] The seed obtained from the seed flax plant; a source of linseed oil.

flaxseed ore [GEOL] Iron ore composed of disk-shaped oölites that have been partially flattened parallel to the bedding plane.

flax wilt [PL PATH] A fungus disease of flax caused by *Fusarium oxysporum lini;* diseased plants wilt, yellow, and die.

F layer [GEOPHYS] An ionized layer in the F region of the ionosphere which consists of the F_1 and F_2 layers in the day hemisphere, and the F_2 layer alone in the night hemisphere; it is capable of reflecting radio waves to earth at frequencies up to about 50 megahertz.

F_1 layer [GEOPHYS] The ionosphere layer beneath the F_2 layer during the day, at a virtual height of 200–300 kilometers, being closest to earth around noon; characterized by a distinct maximum of free-electron density, except at high latitudes during winter, when the layer is not detectable.

F_2 layer [GEOPHYS] The highest constantly observable ionosphere layer, characterized by a distinct maximum of free-electron density at a virtual height from about 225 kilometers in the polar winter to more than 400 kilometers in daytime near the magnetic equator. Also known as Appleton layer.

fl dr *See* fluid dram.

flea [INV ZOO] Any of the wingless insects composing the order Siphonaptera; most are ectoparasites of mammals and birds.

flea-borne typhus *See* murine typhus.

flechette [ORD] **1.** A small fin-stabilized missile, a large number of which can be loaded in an artillery canister or in a warhead. **2.** *See* aerial dart.

fleet [MECH ENG] Sidewise movement of a rope or cable when winding on a drum. [ORD] **1.** An organization of ships, aircraft, marine forces, and shore-based fleet activities, all under a commander who may exercise operational as well as administrative control. **2.** All naval operating forces.

fleet angle [MECH ENG] In hoisting gear, the included angle between the rope, in its position of greatest travel across the drum, and a line drawn perpendicular to the drum shaft, passing through the center of the head sheave or lead sheave groove.

fleet broadcast [COMMUN] The radio broadcast (in addition to the general broadcast) to all U.S. Navy ships and merchant ships, in which storm warnings are given.

fleeting target [ORD] Moving or transient target that remains within observing or firing distance for such a short period that it affords little time for deliberate adjustment and fire against it; for example, aircraft, vehicles, marching troops, and so forth.

Fleming cracking process [CHEM ENG] An obsolete liquid-phase thermal cracking process for heavy petroleum fractions; the charge was heated under pressure in a vertical shell still.

Fleming's rule *See* left-hand rule; right-hand rule.

Fleming's solution [MATER] A tissue fixative made up of a mixture of osmic, chromic and acetic acids.

Fleming tube [ELECTR] The original diode, consisting of a heated filament and a cold metallic electrode in an evacuated glass envelope; negative current flows from the filament to the cold electrode, but not in the reverse direction.

Flemish bond [CIV ENG] A masonry bond consisting of alternating stretchers and headers in each course, laid with broken joints.

Flemish garden wall bond [CIV ENG] A masonry bond consisting of headers and stretchers in the ratio of one to three or four in each course, with joints broken to give a variety of patterns.

flesh [ANAT] The soft parts of the body of a vertebrate, especially the skeletal muscle and associated connective tissue and fat.

Flesh-Demag process [CHEM ENG] A gas-making process in which a cyclic water-gas apparatus is used for feeding and charring the coal charge and for gas generation, with periodic automatic removal of the resultant ash.

fleshing machine [ENG] A machine that removes flesh from hides in a tannery.

fleshy fruit [BOT] A fruit having a fleshy pericarp that is usually soft and juicy, but sometimes hard and tough.

Fletcher-Munson contour *See* equal loudness contour.

Fletcher radial burner [ENG] A burner with gas jets arranged radially.

Flettner windmill [MECH ENG] An inefficient windmill with four arms, each consisting of a rotating cylinder actuated by a Savonius rotor.

flex [SCI TECH] To bend.

flexibacteria [MICROBIO] A general term to describe filamentous, gliding, nonphotosynthetic bacteria.

Flexibilia [PALEON] A subclass of extinct stalked or creeping Crinoidea; characteristics include a flexible tegmen with open ambulacral grooves, uniserial arms, a cylindrical stem, and five conspicuous basals and radials.

flexibility [MECH] The quality or state of being able to be flexed or bent repeatedly.

flexibilizer [MATER] An additive that gives an otherwise rigid plastic flexibility. Also known as plasticizer.

flexible collodion [MATER] A collodion which has two additives (2% camphor and 3% castor oil) to make a pliable film.

flexible coupling [ELECTROMAG] A coupling designed to allow a limited angular movement between the axes of two waveguides. [MECH ENG] A coupling used to connect two shafts and to accommodate their misalignment.

flexible glue [MATER] A type of glue used for pliable molds and printers' rollers, for example, a mixture of glue, glycerol, and water.

flexible gun [ORD] A gun, especially a machine gun, mounted in an aircraft turret or on a post, tripod, or other mount in such a manner that the gun may be swung in both a vertical and horizontal plane.

flexible-joint pipe [ENG] Cast-iron pipe adapted to laying under water and capable of motion through several degrees without leakage.

flexible mold [ENG] A coating mold made of flexible rubber or other elastomeric materials; used mainly for casting plastics.

flexible pavement [CIV ENG] A road or runway made of bituminous material which has little tensile strength and is therefore flexible.

flexible resistor [ELEC] A wire-wound resistor having the appearance of a flexible lead; made by winding the Nichrome resistance wire around a length of asbestos or other heat-resistant cord, then covering the winding with asbestos and braided insulating covering.

flexible sandstone [GEOL] A variety of itacolumite that consists of fine grains and occurs in thin layers.

flexible shaft [MECH ENG] **1.** A shaft that transmits rotary motion at any angle up to about 90°. **2.** A shaft made of flexible material or of segments. **3.** A shaft whose bearings are designed to accommodate a small amount of misalignment.

flexible ventilation ducting [MIN ENG] Flexible fabric tubes covered with rubber or polyvinyl chloride, used for auxiliary ventilation.

flexible waveguide [ELECTROMAG] A waveguide that can be bent or twisted without appreciably changing its electrical properties.

flexion [BIOL] Act of bending, especially of a joint.

flexional symbols [ADP] Symbols in which the meaning of each component digit is dependent on those which precede it.

flexion reflex [PHYSIO] An unconditioned, segmental reflex elicited by noxious stimulation and consisting of contraction of the flexor muscles of all joints on the same side. Also known as the nocioceptive reflex.

flexographic printing *See* flexography.

flexography [GRAPHICS] Relief printing with plates fastened to a cylinder and with a single inking roller supplied with

FLAX

Flowering top of flax plant showing buds and seed capsules. (USDA)

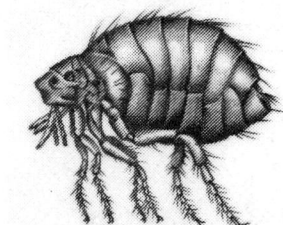

FLEA

The sticktight flea, a bird parasite.

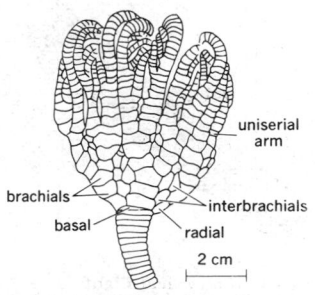

FLEXIBILIA

uniserial arm

brachials

interbrachials

basal

radial

2 cm

Talanterocrinus species. (After F. Bather, 1900)

aniline ink from two rollers in the ink fountain. Also known as aniline printing; aniline process; flexographic printing.

flexometer [ENG] An instrument for measuring the flexibility of materials.

flexor [PHYSIO] A muscle that bends or flexes a limb or a part.

flexowriter [ADP] A typewriterlike device to read in manually or to read out information of a computer to which it is connected; it can also be used to punch paper tape.

flexuous [BIOL] **1.** Flexible. **2.** Bending in a zigzag manner. **3.** Wavy.

flexural slip [GEOL] The slipping of sedimentary strata along bedding planes during folding, producing disharmonic folding and, when extreme, découllement. Also known as bedding-plane slip.

flexural strength [MECH] A measure of the resistance to fracture when the load is applied across a simple beam.

flexure [EMBRYO] A sharp bend of the anterior part of the primary axis of the vertebrate embryo. [GEOL] **1.** A broad, domed structure. **2.** A fold. [MECH] **1.** The deformation of any beam subjected to a load. **2.** Any deformation of an elastic body in which the points originally lying on any straight line are displaced to form a plane curve. [VERT ZOO] The last joint of a bird's wing.

flexure theory [MECH] Theory of the deformation of a prismatic beam having a length at least 10 times its depth and consisting of a material obeying Hooke's law, in response to stresses within the elastic limit.

flicker [OPTICS] A visual sensation produced by periodic fluctuations in light at rates ranging from a few cycles per second to a few tens of cycles per second.

flicker control [AERO ENG] Control of an aircraft, rocket, or such in which the control surfaces are deflected to their maximum degree with only a slight motion of the controller.

flicker effect [ELECTR] Random variations in the output current of an electron tube having an oxide-coated cathode, due to random changes in cathode emission.

flicker photometer [OPTICS] A photometer in which a single field of view is alternately illuminated by the light sources to be compared, and the rate of alternation is such that color flicker is absent but brightness flicker is not; disappearance of flicker signifies equality of luminance.

flight [AERO ENG] The movement of an object through the atmosphere or through space, sustained by aerodynamic reaction or other forces. [CIV ENG] A series of stairs between landings or floors.

flight briefing *See* pilot briefing.

flight characteristic [AERO ENG] A characteristic exhibited by an aircraft, rocket, or the like in flight, such as a tendency to stall or to yaw, or an ability to remain stable at certain speeds.

flight control system *See* vehicle control system.

flight conveyor [MECH ENG] A conveyor in which paddles, attached to single or double strands of chain, drag or push pulverized or granulated solid materials along a trough. Also known as drag conveyor.

flight deck [AERO ENG] In certain airplanes, an elevated compartment occupied by the crew for operating the airplane in flight. [NAV ARCH] The topmost complete deck of an aircraft carrier, used mainly for takeoff and landing of planes.

flight dynamics [AERO ENG] The study of the motion of an aircraft or missile; concerned with transient or short-term effects relating to the stability and control of the vehicle, rather than to calculating such performance as altitude or velocity.

flight envelope [AERO ENG] The boundary depicting, for a specific aircraft, the limits of speed, altitude, and acceleration which that aircraft cannot safely exceed.

flighter [FOOD ENG] A rotating vane used to cool brewers' wort.

flight feather [VERT ZOO] Any of the long contour feathers on the wing of a bird. Also known as remex.

flight feeder [MECH ENG] Short-length flight conveyor used to feed solids materials to a process vessel or other receptacle at a preset rate.

flight forecast [METEOROL] An aviation weather forecast for a specific flight.

flight inspection [NAV] Flight investigation certification of certain operational performance characteristics of electronic and visual navigation facilities by an authorized inspector in conformance with the U.S. Standard Flight Inspection Manual.

flight instrument [AERO ENG] An aircraft instrument used in the control of the direction of flight, attitude, altitude, or speed of an aircraft, for example, the artificial horizon, airspeed indicator, altimeter, compass, rate-of-climb indicator, accelerometer, turn-and-bank indicator, and so on.

flight level [AERO ENG] A surface of constant atmospheric pressure which is related to the standard pressure datum.

flight log [NAV] **1.** A complete written record of a flight, normally showing flight planning information together with actual data recorded during the flight. **2.** A device that automatically records on a screen or map the flight path flown by an aircraft.

flight path [AERO ENG] The path made or followed in the air or in space by an aircraft, rocket, or such.

flight-path angle [AERO ENG] The angle between the horizontal (or some other reference angle) and a tangent to the flight path at a point. Also known as flight-path slope.

flight-path computer [ADP] A computer that includes all of the functions of a course-line computer and also provides means for controlling the altitude of an aircraft in accordance with a desired plan of flight.

flight-path deviation [NAV] The difference between the flight track of an aircraft and the flight path, expressed in terms of either angular or linear measurement.

flight-path deviation indicator [NAV] An instrument providing visual indication of deviation from a flight path.

flight-path selector [NAV] An instrument used with a flight path computer to preset the values defining the flight path to a way point or terminal.

flight-path slope *See* flight-path angle.

flight plan [NAV] Information provided to air-traffic service units, giving in detail the proposed plan of flight, including times, altitudes, way points, and so on, and submitted for approval.

flight profile [AERO ENG] A graphic portrayal or plot of the flight path of an aeronautical vehicle in the vertical plane.

flight recorder [ENG] Any instrument or device that records information about the performance of an aircraft in flight or about conditions encountered in flight, for future study and evaluation.

flight rules [NAV] Rules established by competent authority to govern flights; the type of flight involved determines whether instrument flight rules or visual flight rules apply.

flight science [AERO ENG] The sum total of all knowledge that enables man to accomplish flight; it is compounded of both science and engineering, and is concerned with airplanes, missiles, and manned and unmanned space vehicles.

flight simulator [AERO ENG] A training device or apparatus that simulates certain conditions of actual flight or of flight operations.

flight stability [AERO ENG] The property of an aircraft or missile to maintain its attitude and to resist displacement, and, if displaced, to tend to restore itself to the original attitude.

flight test [AERO ENG] **1.** A test by means of actual or attempted flight to see how an aircraft, spacecraft, space-air vehicle, or missile flies. **2.** A test of a component part of a flying vehicle, or of an object carried in such a vehicle, to determine its suitability or reliability in terms of its intended function by making it endure actual flight.

flight time [NAV] The elapsed time from the moment an aircraft first moves under its own power for the purpose of taking off until the moment it comes to rest at the end of the flight.

flight track *See* track.

flight visibility [METEOROL] Average visibility in a forward direction from an aircraft in flight.

flight-weather briefing *See* pilot briefing.

flinching [IND ENG] In inspection, failure to call a borderline defect a defect.

Flinders bar [NAV] A bar of soft unmagnetized iron placed vertically near a magnetic compass to counteract deviation caused by magnetic induction in vertical soft iron of the craft.

FLIGHT FEATHER

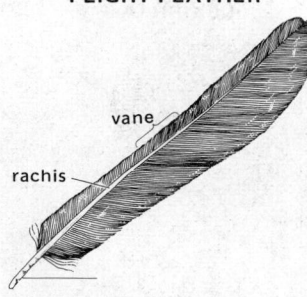

vane

rachis

Full view of a typical flight feather. *(From J. Van Tyne and A. J. Berger, Fundamentals of Ornithology, Wiley, 1959)*

flinkite [MINERAL] $Mn_3(AsO_4)(OH)_4$ Greenish-brown mineral composed of basic manganese arsenate, occurring in feathery forms.

flint [MINERAL] A black or gray, massive, hard, somewhat impure variety of chalcedony, breaking with a conchoidal fracture. Also known as firestone.

flint clay [GEOL] A hard, smooth, flintlike fireclay; when it is ground, it develops no platicity, and it breaks with conchoidal fracture.

flint glass [MATER] **1.** Heavy, colorless, brilliant glass that contains lead oxide. **2.** Any high-quality glass.

flint mill [MECH ENG] A mill employing pebbles to pulverize materials (for example, in cement manufacture).

FLIP See floating instrument platform.

flip chip [ELECTR] A tiny semiconductor die having terminations all on one side in the form of solder pads or bump contacts; after the surface of the chip has been passivated or otherwise treated, it is flipped over for attaching to a matching substrate.

flip coil [ELECTROMAG] A small coil used to measure the strength of a magnetic field; it is placed in the field, connected to a ballistic galvanometer or other instrument, and suddenly flipped over 180°; alternatively, the coil may be held stationary and the magnetic field reversed.

flip-flop amplifier See wall-attachment amplifier.

flip-flop circuit See bistable multivibrator.

flip-open cutout fuse See dropout fuse.

flip-over process See Umklapp process.

flipper [VERT ZOO] A broad, flat appendage used for locomotion by aquatic mammals and sea turtles.

flist [METEOROL] In Scotland, a keen blast or shower accompanied by a squall.

flitch beam [CIV ENG] A built-up beam composed of a metal plate sandwiched between flitches, that is, other strips of metal or other material.

flitch girder [CIV ENG] A girder composed of a metal plate sandwiched between flitches.

flitch plate [CIV ENG] The metal plate in a flitch beam or girder.

float [AGR] A device consisting of one or more blades used to level a seedbed. [BIOL] An air-filled sac in many pelagic flora and fauna that serves to buoy up the body of the organism. [DES ENG] A file having a single set of parallel teeth. [ENG] **1.** A flat, rectangular piece of wood with a handle, used to apply and smooth coats of plaster. **2.** A mechanical device to finish the surface of freshly placed concrete paving. **3.** A marble-polishing block. **4.** Any structure that provides positive buoyancy such as a hollow, watertight unit that floats or rests on the surface of a fluid. **5.** See plummet. [GEOL] An isolated, displaced rock or ore fragment. [TEXT] **1.** A thread used to create patterns in fabric by passing over other threads. **2.** A fabric defect caused by passing a thread over other threads where it should be interwoven.

floatability [MIN ENG] Response of a specific mineral to the flotation process.

float-and-sink analysis [MIN ENG] Use of a series of heavy liquids diminishing (or increasing) in density by accurately controlled stages in order to divide a sample of crushed coal or other minerals or metals into fractions that are either equal-settling or equal-floating at each stage.

float chamber [ENG] A vessel in which a float regulates the level of a liquid.

float coal [GEOL] Small, irregularly shaped, isolated deposits of coal embedded in sandstone or in siltstone. Also known as raft.

float control [ENG] Floating device used to transmit a liquid-level reading to a control apparatus, such as an on-off switch controlling liquid flow into and out of a storage tank.

float-cut file [DES ENG] A coarse file used on soft materials.

float finish [CIV ENG] A rough concrete finish, obtained by using a wooden float for finishing.

float gage [ENG] Any one of several types of instruments in which the level of a liquid is determined from the height of a body floating on its surface, by using pullies, levers, or other mechanical devices.

floating action [ENG] Controller action in which there is a predetermined relation between the deviation and the speed of a final control element; a neutral zone, in which no motion of the final control element occurs, is often used.

floating address [ADP] The symbolic address used prior to its conversion to a machine address.

floating aid [NAV] A buoy serving as an aid to navigation, secured in its charted position by a mooring.

floating arithmetic See floating-point arithmetic.

floating axle [MECH ENG] A live axle used to turn the wheels of an automotive vehicle; the weight of the vehicle is borne by housings at the ends of a fixed axle.

floating battery [ELEC] A storage battery connected permanently in parallel with another power source; the battery normally handles only small charging or discharging currents, but takes over the entire load upon failure of the main supply.

floating block See traveling block.

floating carrier modulation See controlled carrier modulation.

floating charge [ELEC] Application of a constant voltage to a storage battery, sufficient to maintain an approximately constant state of charge while the battery is idle or on light duty.

floating control [ENG] Control device in which the speed of correction of the control element (such as a piston in a hydraulic relay) is proportional to the error signal. Also known as proportional-speed control.

floating crane [CIV ENG] A crane having a barge or scow for an undercarriage and moved by cables attached to anchors set some distance off the corners of the barge; used for water work and for work on waterfronts.

floating-decimal arithmetic See floating-point arithmetic.

floating dock [CIV ENG] **1.** A form of dry dock for repairing ships; it can be partly submerged by controlled flooding to receive a vessel, then raised by pumping out the water so that the vessel's bottom can be exposed. Also known as floating dry dock. **2.** A barge or flatboat which is used as a wharf.

floating dollar sign [ADP] A dollar sign used with an edit mask, allowing the sign to be inserted before the nonzero leading digit of a dollar amount.

floating dry dock See floating dock.

floating floor [BUILD] A floor constructed so that the wearing surface is separated from the supporting structure by an insulating layer of mineral wool, resilient quilt, or other material to provide insulation against impact sound.

floating foundation [CIV ENG] A reinforced concrete slab that distributes the concentrated load from columns; used on soft soil.

floating grid [ELECTR] Vacuum-tube grid that is not connected to any circuit; it assumes a negative potential with respect to the cathode. Also known as free grid.

floating ice [OCEANOGR] Any form of ice floating in water, including grounded ice and drifting land ice.

floating input [ELEC] Isolated input circuit not connected to ground at any point.

floating instrument platform [NAV ARCH] A manned, spar-buoy-type oceanographic platform with marine instruments mounted on the submerged portions. Abbreviated FLIP.

floating lever [MECH ENG] A horizontal brake lever with a movable fulcrum; used under railroad cars.

floating neutral [ELEC] Neutral conductor whose voltage to ground is free to vary when circuit conditions change.

floating pan [ENG] An evaporation pan in which the evaporation is measured from water in a pan floating in a larger body of water.

floating plug [MET] A plug or mandrel attached to a rod and used in plug drawing. Also known as a plug die.

floating-point arithmetic [MATH] A method of performing arithmetical operations, used especially by automatic computers, in which numbers are expressed as integers multiplied by the radix raised to an integral power, as 87×10^{-4} instead of 0.0087. Also known as floating arithmetic; floating-decimal arithmetic.

floating-point calculation [ADP] A calculation made with floating point arithmetic.

floating-point coefficient See mantissa.

floating-point package [ADP] A program which enables a computer to perform arithmetic operations when such capa-

FLOAT GAGE

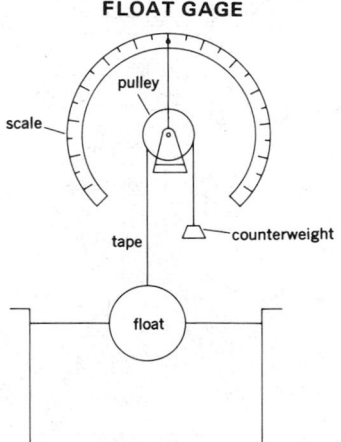

Float, tape, and pulley gage for measuring large changes in liquid level. *(From D. M. Considine ed., Process Instruments and Controls Handbook, McGraw-Hill, 1957)*

bilities are not wired into the computer. Also known as floating-point routine.

floating-point routine *See* floating-point package.

floating-point system [ADP] A number system in which the location of the point does not remain fixed with respect to one end of the numerals.

floating reticle [OPTICS] A reticle the image of which is movable within the field of view.

floating rib [ANAT] One of the last two ribs in humans which have the anterior end free.

floating roof [ENG] A type of tank roof (steel, plastic, sheet, or microballoons) which floats upon the surface of the stored liquid; used to decrease the vapor space and reduce the potential for evaporation.

floating scraper [MECH ENG] A balanced scraper blade that rests lightly on a drum filter; removes solids collected on the rotating drum surface by riding on the drum's surface contour.

floating zone refining [MET] A variation of the zone-refining technique in which the molten zone is held in place by its own surface tension between two collinear rods; since no container is needed, contamination of the pure metal is avoided.

floatless level control [ENG] Any nonfloat device for measurement and control of liquid levels in storage tanks or process vessels; includes use of manometers, capacitances, electroprobes, nuclear radiation, and sonics.

float mineral [GEOL] Small ore fragments carried from the ore bed by the action of water or by gravity; a float mineral often leads to discovery of mines.

floatoblast [INV ZOO] A free-floating statoblast having a float of air cells.

float switch [ENG] A switch actuated by a float at the surface of a liquid.

float-type rain gage [ENG] A class of rain gage in which the level of the collected rainwater is measured by the position of a float resting on the surface of the water; frequently used as a recording rain gage by connecting the float through a linkage to a pen which records on a clock-driven chart.

float valve [ENG] A valve whose on-off action is controlled directly by the fall or rise of a float concurrent with the fall or rise of liquid level in a liquid-containing vessel.

floc [CHEM] Small masses formed in a fluid through coagulation, agglomeration, or biochemical reaction of fine suspended particles.

floccose [BOT] Covered with tufts of woollike hairs.

flocculant *See* flocculating agent.

flocculate [BIOL] Having small tufts of hairs. [CHEM] To cause to aggregate or coalesce into a flocculent mass.

flocculating agent [CHEM] A reagent added to a dispersion of solids in a liquid to bring together the fine particles to form flocs. Also known as flocculant.

flocculent [CHEM] Pertaining to a material that is cloudlike and noncrystalline.

flocculonodular lobes [ANAT] The pair of lateral cerebellar lobes in vertebrates which function to regulate vestibular reflexes underlying posture; referred to functionally as the vestibulocerebellum.

flocculus [ANAT] A prominent lobe of the cerebellum situated behind and below the middle cerebellar peduncle on each side of the median fissure. [ASTRON] A patch in the sun's surface seen in the light of calcium or hydrogen; the patch may be bright or dark and is usually in the vicinity of sunspots.

floccus [BOT] A tuft of woolly hairs. [METEOROL] A cloud species in which each element is a small tuft with a rounded top and a ragged bottom.

flock [TEXT] **1.** Pulverized wool, cotton, silk, or rayon fiber used to form velvety patterns on cloth. **2.** Woolen or cotton refuse reduced by machinery and used to stuff furniture.

flocking [TEXT] A design or a surface finish made by spraying short fibers so that they adhere electrostatically to a material.

floc point [ANALY CHEM] The temperature at which wax or solids separate from kerosine and other illuminating oils as a definite floc.

floc test [ANALY CHEM] A quantitative test applied to kerosine and other illuminating oils to detect substances rendered insoluble by heat.

floe [OCEANOGR] A piece of floating sea ice other than fast

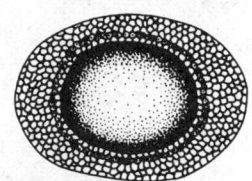

FLOATOBLAST

Floatoblast of *Plumatella repens*, in Phylactolaemata, a class of fresh-water Bryozoans.

ice or glacier ice; may consist of a single fragment or of many consolidated fragments, but is larger than an ice cake and smaller than an ice field. Also known as ice floe.

floeberg [OCEANOGR] A mass of hummocked ice formed by the piling up of many ice floes by lateral pressure; an extreme form of pressure ice; may be more than 50 feet (15 meters) high and resemble an iceberg.

floe till [GEOL] **1.** A glacial till resulting from the intact deposition of a grounded iceberg in a lake bordering an ice sheet. **2.** A lacustrine clay with boulders, stones, and other glacial matter dropped into it by melting icebergs. Also known as berg till.

flokite *See* mordenite.

flood [ELECTR] To direct a large-area flow of electrons toward a storage assembly in a charge storage tube. [ENG] To cover or fill with fluid. [HYD] The condition that occurs when water overflows the natural or artificial confines of a stream or other body of water, or accumulates by drainage over low-lying areas. [MECH ENG] To supply an excess of fuel to a carburetor so that the level rises above the nozzle. [OCEANOGR] The highest point of a tide.

floodable length [NAV ARCH] At any point in the length of a ship, the greatest part of the length centered at that point which can be flooded at the prescribed permeability without submerging the margin line.

flood basalt *See* plateau basalt.

flood control [CIV ENG] Use of levees, walls, reservoirs, floodways, and other means to protect land from water overflow.

flood coverage [PETRO ENG] The extent of subterranean coverage within an oil reservoir by the injection of pressure-maintenance (or water-drive) water.

flood current [OCEANOGR] The tidal current associated with the increase in the height of a tide.

flood dam [CIV ENG] A dam for storing floodwater, or for supplying a flood of water.

flood flow [HYD] Stream discharge during a flood.

floodgate [CIV ENG] **1.** A gate used to restrain a flow or, when opened, to allow a flood flow to pass. **2.** The lower gate of a lock.

flood icing *See* icing.

flooding [AGR] Filling of ditches or covering of land with water during the raising of crops; rice, for example, must have occasional flooding to grow properly. [CHEM ENG] Condition in a liquid-vapor counterflow device (such as a distillation column) in which the rate of vapor rise is such as to prevent liquid downflow, causing a buildup of the liquid (flooding) within the device. [PETRO ENG] Technique of increasing recovery of oil (secondary recovery) from a reservoir by injection of water into the formation to drive the oil toward producing wellholes. Also known as waterflooding.

flooding ice *See* icing.

floodlight [ELEC] A light projector used for outdoor lighting of buildings, parking lots, sports fields, and the like, usually having a filament lamp or mercury-vapor lamp and a parabolic reflector.

flood-out pattern [PETRO ENG] Pattern of subterranean water penetration and spread in an oil reservoir as a result of water injection.

floodplain [GEOL] The relatively smooth valley floors adjacent to and formed by alluviating rivers which are subject to overflow.

flood pot test [PETRO ENG] Laboratory simulation of an oil reservoir to appraise the residual reservoir saturation after waterflooding.

flood projection [COMMUN] In facsimile transmission, optical method in which the subject facsimile copy is illuminated and the scanning spot is delineated by an aperture between the subject copy and the light-sensitive device.

flood relief channel *See* bypass channel.

flood routing [HYD] The process of computing the progressive time and shape of a flood wave at successive points along a river. Also known as storage routing; streamflow routing.

Flood's equation [PHYS CHEM] A relation used to determine the liquidus temperature in a binary fused salt system.

flood stage [HYD] The stage, on a fixed river gage, at which overflow of the natural banks of the stream begins to cause damage in any portion of the reach for which the gage is used as an index.

flood tide [OCEANOGR] **1.** That period of tide between low water and the next high water. **2.** A tide at its highest point.

flood tuff *See* ignimbrite.

flood wall [CIV ENG] A levee or similar wall for the purpose of protecting the land from inundation by flood waters.

floodway *See* bypass channel.

floor [ENG] The bottom, horizontal surface of an enclosed space. [GEOL] **1.** The rock underlying a stratified or nearly horizontal deposit, corresponding to the footwall of more steeply dipping deposits. **2.** A horizontal, flat ore body. [MIN ENG] Boards laid at the heading to receive blasted rocks and to facilitate ore loading. [NAV ARCH] One of a series of vertical plates extending across the bottom of a ship at right angles to the center line and forming part of the bottom framing of the hull.

floor beam [BUILD] A beam used in the framing of floors in buildings. [CIV ENG] A large beam used in a bridge floor at right angles to the direction of the roadway, to transfer loads to bridge supports.

floorboard [MIN ENG] A thick wooden plank constituting part of a drill platform or other work platform.

floor burst [MIN ENG] A type of outburst in longwall faces which is preceded by heavy weighting due to floor lift; gas evolved below the seam collects beneath an impervious layer of rock, and a gas blister forms beneath the face, giving the observed floor lift; later, the floor fractures and the firedamp escapes into the mine atmosphere.

floor cut [MIN ENG] A machine-made cut in the floor dirt just below the coal seam.

floor drain [CIV ENG] A pipe or channel to remove water from under a floor in contact with soil.

floor framing [BUILD] Floor joists together with their strutting and supports.

flooring [MATER] Material suitable for use as a floor.

flooring saw [DES ENG] A pointed saw with teeth on both edges; cuts its own entrance into a material.

floor light [BUILD] A window set in a floor that is adapted for walking on and admitting light to areas below.

floor outlet [ELEC] An electrical outlet whose face is level with or recessed into a floor. Also known as floor plug.

floor plan [ARCH] A diagram of a floor showing partitions, doors, windows, and other features.

floor plate [BUILD] A flat board on a floor used to support wall studs. [ENG] A plate in a floor to which heavy work or machine tools can be bolted.

floor plug *See* floor outlet.

floor sill [MIN ENG] A large timber laid flat on the ground or in a level, shallow ditch, to which are fastened the drill-platform boards or planking, or which is used as the base for a full timber set.

floor system [CIV ENG] The structural floor assembly between supporting beams or girders in buildings and bridges.

flop gate [MIN ENG] An automatic gate used in placer mining when there is a shortage of water; the gate closes a reservoir until it is filled with water, then automatically opens and allows the water to flow into the sluices; when the reservoir is empty, the gate closes, and the operation is repeated.

flopover [ELECTR] A defect in television reception in which a series of frames move vertically up or down the screen, caused by lack of synchronization between the vertical and horizontal sweep frequencies.

Floquet theorem [MATH] A second-order linear differential equation whose coefficients are periodic single-valued functions of an independent variable x, has a solution of the form $e^{\mu x}P(x)$ where μ is a constant and $P(x)$ a periodic function.

flora [BOT] **1.** Plants. **2.** The plant life characterizing a specific geographic region or environment.

floral axis [BOT] A flower stalk.

floral diagram [BOT] A diagram of a flower in cross section showing the number and arrangement of floral parts.

Florence oil *See* olive oil.

florencite [MINERAL] $CeAl_3(PO_4)_2(OH)_6$ Pale-yellow mineral composed of basic phosphate of cerium and aluminum.

florentium *See* promethium-147.

flores [CHEM] A form of a chemical compound made by the process of sublimation.

flores martis *See* ferric chloride.

floret [BOT] A small individual flower that is part of a compact group of flowers, such as the head of a composite plant or inflorescence.

Florey unit [BIOL] A unit for the standardization of penicillin.

floricome [INV ZOO] A type of branched hexaster spicule.

floriculture [AGR] A segment of horticulture concerned with commercial production, marketing, and retail sale of cut flowers and potted plants, as well as home gardening and flower arrangement.

Florida Current [OCEANOGR] A fast current that sets through the Straits of Florida to a point north of Grand Bahama Island, where it joins the Antilles Current to form the Gulf Stream.

Florideophyceae [BOT] A class of red algae, division Rhodophyta, having prominent pit connections between cells.

floriferous [BOT] Blooming freely, used principally of ornamental plants.

florigen [BIOCHEM] A plant hormone that stimulates buds to flower.

florivorous [ZOO] Feeding on flowers.

flor process [FOOD ENG] A technique used to make Spanish sherries involving the participation of a film-forming yeast growing on the surface of the wine in partially filled oak barrels; the yeast imparts the characteristic sherry flavor.

floscelle [INV ZOO] A flowerlike structure around the mouth of some echinoids.

Flosculariacea [INV ZOO] A suborder of rotifers in the order Monogononta having a malleoramate mastax.

Flosculariidae [INV ZOO] A family of sessile rotifers in the suborder Flosculariacea.

flosculous [BOT] **1.** Composed of florets. **2.** Of a floret, tubular in form.

flosculus [BOT] A floret.

flospinning [MET] Power-spinning or flowing metal over a rotating bar for shaping into cylindrical, conical, and curvilinear parts.

floss [MET] Molten or solid slag floating on the surface of a metal melt.

floss hole [MET] A small door or opening of the bottom of a smokestack or flue for removal of ash.

flotation [ENG] A process used to separate particulate solids by causing one group of particles to float; utilizes differences in surface chemical properties of the particles, some of which are entirely wetted by water, others are not; the process is primarily applied to treatment of minerals but can be applied to chemical and biological materials; in mining engineering it is referred to as froth flotation.

flotation agent [CHEM] A chemical which alters the surface tension of water or which makes it froth easily.

flotation analysis [PHYS] Technique to measure liquid density in which a float of known density is adjusted with weights to match that of the liquid.

flotation cell [MIN ENG] The device in which froth flotation of ores is performed.

flotation collar [ENG] A buoyant bag carried by a spacecraft and designed so that it inflates and surrounds part of the outer surface if the spacecraft lands in the sea.

flotsam [ENG] Floating articles, particularly those that are thrown overboard to lighten a vessel in distress.

flounder [VERT ZOO] Any of a number of flatfishes in the families Pleuronectidae and Bothidae of the order Pleuronectiformes.

flour [FOOD ENG] A powdery meal obtained by milling wheat and other cereal grains or dry food products such as potato or banana.

flour gold [MET] The finest-size gold dust, much of which will float on water.

flour mill [FOOD ENG] A machine or factory that processes cereal grains such as wheat and rye into flour.

flow [ADP] The sequence in which events take place or operations are carried out. [ENG] A forward movement in a continuous stream or sequence of fluids or discrete objects or materials, as in a continuous chemical process or solids-

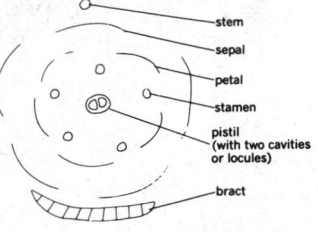

FLORAL DIAGRAM

Graphic diagram of a cross section of a flower.

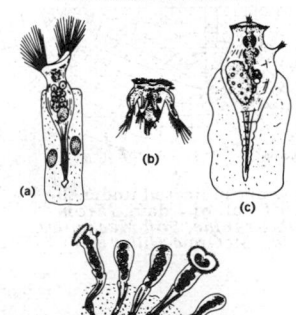

FLOSCULARIACEA

Sessile rotifers of the Flosculariacea. (a) *Floscularia mutabilis.* (b) *Pedalia mira.* (c) *Conchiloides* species. (d) *Lacinularia socialis.*

conveying or production-line operations. [FL MECH] The forward continuous movement of a fluid, such as gases, vapors, or liquids, through closed or open channels or conduits. [PHYS] The movement of electric charges, gases, liquids, or other materials or quantities.

flowability [FL MECH] Capability of a liquid or loose particulate solid to move by flow.

flowage [HYD] Flooding of water onto adjacent land.

flowage line [GEOL] A contour line at the edge of a body of water, such as a reservoir, representing a given water level.

flow banding [GEOL] An igneous rock structure resulting from flowing of magmas or lavas and characterized by alternation of mineralogically unlike layers.

flow brazing [MET] A brazing process in which coalescence is produced by the heat of molten filler metal that is poured over a joint.

flow breccia [GEOL] A breccia formed with the movement of lava flow while the flow is still in motion.

flow cast [PETR] One of a group of bedding plane structures formed in graywacke.

flow chart [ENG] A graphical representation of the progress of a system for the definition, analysis, or solution of a data-processing or manufacturing problem in which symbols are used to represent operations, data or material flow, and equipment, and lines and arrows represent interrelationships among the components. Also known as control diagram; flow diagram; flow sheet.

flow-chart symbol [ENG] Any of the existing symbols normally used to represent operations, data or materials flow, or equipment in a data-processing problem or manufacturing-process description.

flow cleavage [GEOL] Rock cleavage in which solid flow of rock accompanies recrystallization. Also known as slaty cleavage.

flow coat [ENG] A coating formed by pouring a liquid material over the object and allowing it to flow over the surface and drain off.

flow coefficient [FL MECH] An experimentally determined proportionality constant, relating the actual velocity of fluid flow in a pipe, duct, or open channel to the theoretical velocity expected under certain assumptions. [MECH ENG] A dimensionless number used in studying the power required by fans, equal to the volumetric flow rate through the fan divided by the product of the rate of rotation of the fan and the cube of the impeller diameter.

flow control [ENG] Any system used to control the flow of gases, vapors, liquids, slurries, pastes, or solid particles through or along conduits or channels.

flow control valve [ENG] A valve whose flow opening is controlled by the rate of flow of the fluid through it; usually controlled by differential pressure across an orifice at the valve. Also known as rate-of-flow control valve.

flow counter *See* gas-flow counter tube.

flow diagram *See* flow chart.

flow direction [ENG] The antecedent-to-successor relation, indicated by arrows or other conventions, between operations on a flow chart.

flow distribution *See* flow field.

flow earth *See* solifluction mantle.

flow equation [FL MECH] Equation for the calculation of fluid (gas, vapor, liquid) flow through conduits or channels; consists of an interrelation of fluid properties (such as density or viscosity), environmental conditions (such as temperature or pressure), and conduit or channel geometry and conditions (such as diameter, cross-sectional shape, or surface roughness).

flower [BOT] The characteristic reproductive structure of a seed plant, particularly if some or all of the parts are brightly colored.

flowers of sulfur [PHARM] One of three forms of pharmaceutical sulfur, made by sublimation; the other two forms are precipitated sulfur and washed sulfur. Also known as sublimed sulfur.

flowers of tin *See* stannic oxide.

flow field [FL MECH] The velocity and the density of a fluid as functions of position and time. Also known as flow distribution.

flow fold [GEOL] Folding in beds, composed of relatively plastic rock, that assume any shape impressed upon them by the more rigid surrounding rocks or by the general stress pattern of the deformed zone; there are no apparent surfaces of slip.

flow graph *See* signal-flow graph.

flowing-film concentration [MIN ENG] A concentration based on the fact that liquid films in laminar flow possess a velocity which is not the same in all depths of the film; by this principle lighter particles of ore may be washed off while the heavier particles accumulate and are intermittently removed.

flowing furnace [MET] A furnace from which molten metal can be tapped or drawn.

flowing pressure [PETRO ENG] Pressure at the bottom of an oil-well bore (bottom-hole pressure) during normal oil production.

flowing-pressure gradient [PETRO ENG] The slope of decreasing pressure plotted against distance measured for upward liquid flow in a continuous-flow gas-lift oil well.

flowing-temperature factor [THERMO] Calculation correction factor for gases flowing at temperatures other than that for which a flow equation is valid, that is, other than 60°F (15.5°C).

flowing well [PETRO ENG] Oil reservoir in which gas-drive pressure is sufficient to force oil flow up through and out of a wellhole.

flow line [ENG] **1.** The connecting line or arrow between symbols on a flow chart or block diagram. **2.** Mark on a molded plastic or metal article made by the meeting of two input-flow fronts during molding. Also known as weld mark. [HYD] A contour of the water level around a body of water. [PETR] In an igneous rock, any internal structure produced by parallel orientation of crystals, mineral streaks, or inclusions.

flow measurement [ENG] The determination of the quantity of a fluid, either a liquid, a vapor, or a gas, that passes through a pipe, duct, or open channel.

flowmeter [ENG] An instrument used to measure pressure, flow rate, and discharge rate of a liquid, vapor, or gas flowing in a pipe. Also known as fluid meter.

flow mixer [MECH ENG] Liquid-liquid mixing device in which the mixing action occurs as the liquids pass through it; includes jet nozzles and agitator vanes. Also known as line mixer.

flow net [FL MECH] A diagram used in studying the flow of a fluid through a permeable substance (such as water through a soil structure) having two nests of curves, one representing the flow lines, which follow the path of the fluid, and the other the equipotential lines, which connect points of equal head.

flow nozzle [ENG] A flowmeter in a closed conduit, consisting of a short flared nozzle of reduced diameter inset into the inner diameter of a pipe; used to cause a temporary pressure drop in flowing fluid to determine flow rate via measurement of static pressures before and after the nozzle.

flow pattern [FL MECH] Pattern of two-phase flow in a conduit or channel pipe, taking into consideration the ratio of gas to liquid and conditions of flow resistance and liquid holdup.

flow process [ENG] System in which fluids or solids are handled in continuous movement during chemical or physical processing or manufacturing.

flow rate [FL MECH] Also known as rate of flow. **1.** Time required for a given quantity of flowable material to flow a measured distance. **2.** Weight or volume of flowable material flowing per unit time.

flow reactor [CHEM ENG] A dynamic reactor system in which reactants flow continuously into the vessel and products are continuously removed, in contrast to a batch reactor.

flow resistance [FL MECH] **1.** Any factor within a conduit or channel that impedes the flow of fluid, such as surface roughness or sudden bends, contractions, or expansions. **2.** *See* viscosity.

flow rock [PETR] An igneous rock that had been liquid.

flow separation *See* boundary-layer separation.

flow sheet *See* flow chart.

flow soldering [ENG] Soldering of printed circuit boards by moving them over a flowing wave of molten solder in a solder bath; the process permits precise control of the depth of

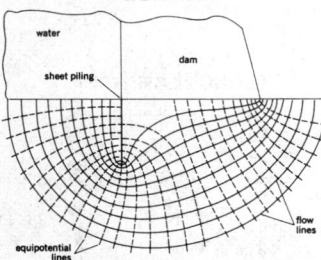

FLOW NET

Flow net indicated under the cutoff wall of a dam. (*From D. P. Krynine, Soil Mechanics, 2d ed., McGraw-Hill, 1947*)

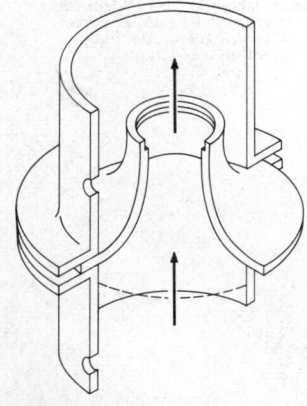

FLOW NOZZLE

Diagram of a flow nozzle. Arrows indicate direction of fluid flow.

immersion in the molten solder and minimizes heating of the board. Also known as wave soldering.

flowstone [GEOL] Deposits of calcium carbonate that accumulated against the walls of a cave where water flowed on the rock.

flow stress [MECH] The stress along one axis at a given value of strain that is required to produce plastic deformation.

flow string [PETRO ENG] Total length of oil- or gas-well tubing made up of a string of interconnected tubing sections.

flow texture [PETR] A pattern of an igneous rock that is formed when the stream or flow lines of a once-molten material have a subparallel arrangement of prismatic or tabular cyrstals or microlites. Also known as fluidal texture.

flow transmitter [ENG] A device used to measure the flow of liquids in pipelines and convert the results into proportional electric signals that can be transmitted to distant receivers or controllers.

flow valve [ENG] A valve that closes itself when the flow of a fluid exceeds a particular value.

flow visualization [ENG] Method of making visible the disturbances that occur in fluid flow, using the fact that light passing through a flow field of varying density exhibits refraction and a relative phase shift among different rays.

flow welding [MET] A welding process in which coalescence is produced by heating with molten filler metal, which is poured over the joint until the welding temperature is attained and the required amount of filler metal is added.

fl oz *See* fluid ounce.

fluctuating current [ELEC] Direct current that changes in value but not at a steady rate.

fluctuation [OCEANOGR] **1.** Wavelike motion of water. **2.** The variations of water-level height from mean sea level that are not due to tide-producing forces. [SCI TECH] **1.** Variation, especially back and forth between successive values in a series of observations. **2.** Variation of data points about a smooth curve passing among them.

fluctuation noise *See* random noise.

fluctuation test [MICROBIO] A method of demonstrating that bacterial mutations preexist in a population before they are selected; a large parent population is divided into small parts which are grown independently and the number of mutants in each subculture determined; the number of mutants in the subculture will fluctuate because in some a mutant arises early (giving a large number of progeny), while in others the mutant arises late and gives few progeny.

fluctuation theory [OPTICS] The theory proposed by M. von Smoluchowski and A. Einstein which states that the scattering of light occurs in pure water because random molecular motion causes density variations which effect changes in the refraction of light.

fluctuation velocity *See* eddy velocity.

flue [ENG] A channel or passage for conveying combustion products from a furnace, boiler, or fireplace to or through a chimney.

flue dust [MET] Fine particles of metal or alloy emitted with the gases of a smelter or metallurgical furnace.

flue gas [ENG] Gaseous combustion products from a furnace.

flue gas analyzer [ENG] A device that monitors the composition of the flue gas of a boiler heating unit to determine if the mixture of air and fuel is at the proper ratio for maximum heat output.

fluellite [MINERAL] $AlF_3 \cdot H_2O$ A colorless or white mineral composed of aluminum fluoride, occurring in crystals.

fluid [PHYS] An aggregate of matter in which the molecules are able to flow past each other without limit and without fracture planes forming.

fluidal texture *See* flow texture.

fluid amplifier [ENG] An amplifier in which all amplification is achieved by interaction between jets of fluid, with no electronic circuit and usually no moving parts.

fluid-bed process [CHEM ENG] A type of process based on the tendency of finely divided powders to behave in a fluidlike manner when supported and moved by a rising gas or vapor stream; used mainly for catalytic cracking of petroleum distillates.

fluid catalyst [CHEM ENG] Finely divided solid particles utilized as a catalyst in a fluid-bed process.

fluid catalytic cracking [CHEM ENG] An oil refining process in which the gas-oil is cracked by a catalyst bed fluidized by using oil vapors.

fluid clutch *See* fluid drive.

fluid coal [MATER] Pulverized coal that, when mixed with air, can be transported through pipes.

fluid coefficient [PETRO ENG] A measure of the flow resistance to the leaking off of reservoir fracturing fluids into the formation during the fracturing operation.

fluid coking [CHEM ENG] A thermal process utilizing the fluidized solids technique for continuous conversion of heavy, low-grade petroleum oils into petroleum coke and lighter hydrocarbon products.

fluid-compressed [MET] Pertaining to steel that has been compressed while still fluid to remove gases and make the material more homogeneous.

fluid computer [ADP] A digital computer constructed entirely from air-powered fluid logic elements; it contains no moving parts and no electronic circuits; all logic functions are carried out by interaction between jets of air.

fluid-controlled valve [MECH ENG] A valve for which the valve operator is activated by a fluid energy, in contrast to electrical, pneumatic, or manual energy.

fluid coupling [MECH ENG] A device for transmitting rotation between shafts by means of the acceleration and deceleration of a fluid such as oil. Also known as hydraulic coupling.

fluid density [FL MECH] The mass of a fluid per unit volume.

fluid die [MECH ENG] A die for shaping parts by liquid pressure; a plunger forces the liquid against the part to be shaped, making the part conform to the shape of a die.

fluid distributor [ENG] Device for the controlled distribution of fluid feed to a process unit, such as a liquid-gas or liquid-solids contactor, reactor, mixer, burner, or heat exchanger; can be a simple perforated-pipe sparger, spray head, or such.

fluid dram [MECH] Abbreviated fl dr. **1.** A unit of volume used in the United States for measurement of liquid substances, equal to 1/8 fluid ounce, or $3.6966911953125 \times 10^{-6}$ cubic meter. **2.** A unit of volume used in the United Kingdom for measurement of liquid substances and occasionally of solid substances, equal to 1/8 fluid ounce or approximately 3.55163×10^{-6} cubic meter.

fluid drive [MECH ENG] A power coupling operated on a hydraulic turbine principle in which the engine flywheel has a set of turbine blades which are connected directly to it and which are driven in oil, thereby turning another set of blades attached to the transmission gears of the automobile. Also known as fluid clutch; hydraulic clutch.

fluid duplicating [GRAPHICS] A system in which a master is produced by placing a special aniline-dye carbon paper under a sheet of coated paper; the carbon is placed face up, so that when the coated paper is typed or written on, a reversed copy is produced on the back of the sheet; a certain amount of dye from the carbon paper is transferred to the master, from which the copies are printed directly.

fluid dynamics [FL MECH] The science of fluids in motion.

fluid-energy mill [ENG] A size-reduction unit in which grinding is achieved by collision between the particles being ground and the energy supplied by a compressed fluid entering the grinding chamber at high speed. Also known as jet mill.

fluid-film bearing [MECH ENG] An antifriction bearing in which rubbing surfaces are kept apart by a film of lubricant such as oil.

fluid friction [FL MECH] Conversion of mechanical energy in fluid flow into heat energy.

fluid fuel reactor [NUCLEO] A type of reactor (for example, a fused-salt reactor) whose fuel is in fluid form.

fluid geometry [GEOL] Fluid distribution in reservoir strata controlled by rock effective pore-size distribution, rock wettability characteristics in relation to the fluids present, method of producing saturation, and rock heterogeneity.

fluid hydroforming [CHEM ENG] A type of fluid catalytic cracking process used by petroleum refineries to upgrade low-octane-number stocks.

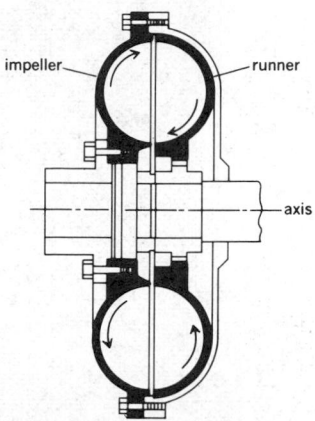

FLUID COUPLING

impeller — runner
axis

Basic fluid coupling. Fluid is contained in the impeller and the runner, both bladed rotors; the impeller acts as a pump and the runner reacts as a turbine.

fluidic device [ENG] A device that operates by the interaction of streams of fluid.

fluidics [ENG] A control technology that employs fluid dynamic phenomena to perform sensing, control, information, processing, and actuation functions without the use of moving mechanical parts.

fluid inclusion [PETR] A tiny fluid-filled cavity in an igneous rock that forms by the entrapment of the liquid from which the rock crystallized.

fluidity [FL MECH] The reciprocal of viscosity; expresses the ability of a substance to flow.

fluidization [CHEM ENG] A roasting process in which finely divided solids are suspended in a rising current of air (or other fluid), producing a fluidized bed; used in the calcination of various minerals, in Fischer-Tropsch synthesis, and in the coal industry.

fluidized adsorption [CHEM ENG] Method of vapor- or gas-fractionation (separation via adsorption-desorption cycles) in a fluidized bed of adsorbent material.

fluidized bed [ENG] A cushion of air or hot gas blown through the porous bottom slab of a container which can be used to float a powdered material as a means of drying, heating, quenching, or calcining the immersed components.

fluidized-bed coating [ENG] Method for plastic-coating of objects; the heated object is immersed into the fluidized bed of a thermoplastic resin that then fuses into a continuous uniform coating over the immersed object.

fluidized-bed reactor *See* fluidized reactor.

fluidized reactor [NUCLEO] A nuclear reactor in which the fuel has been given the properties of a quasi-fluid, such as by suspension of fine fuel particles in a carrying gas or liquid. Also known as fluidized-bed reactor.

fluid-loss agent [MATER] Material used to thicken or gel crude oil, light oil, and water or acid fracturing fluids to seal off pores and flow channels in the reservoir matrix.

fluid-loss test [PETRO ENG] Measure of fracturing fluid loss versus time (spurt loss) before the fluid-loss agent forms a nonpermeable layer in the reservoir pore matrix.

fluid mechanics [MECH] The science concerned with fluids, either at rest or in motion, and dealing with pressures, velocities, and accelerations in the fluid, including fluid deformation and compression or expansion.

fluid meter *See* flow meter.

fluid ounce [MECH] Abbreviated fl oz. **1.** A unit of volume used in the United States for measurement of liquid substances, equal to 1/16 liquid pint, or 231/128 cubic inches, or $2.95735295625 \times 10^{-5}$ cubic meter. **2.** A unit of volume used in the United Kingdom for measurement of liquid substances, and occasionally of solid substances, equal to 1/20 pint or approximately 2.84130×10^{-5} cubic meter.

fluid resistance [FL MECH] The force exerted by a gas or liquid opposing the motion of a body through it. Also known as resistance.

fluid saturation [GEOL] Measure of the gross void space in a reservoir rock that is occupied by a fluid.

fluid statics [FL MECH] The determination of pressure intensities and forces exerted by fluids at rest.

fluid stress [MECH] Stress associated with plastic deformation in a solid material.

fluid ton [MECH] A unit of volume equal to 32 cubic feet or approximately 9.0614×10^{-2} cubic meter; used for many hydrometallurgical, hydraulic, and other industrial purposes.

fluid transmission [MECH ENG] Automotive transmission with fluid drive.

fluid viscosity ratio [PETRO ENG] Ratio of viscosity of a displacing gas to that of oil in a gas-drive reservoir; used in unit displacement efficiency calculations.

fluke [INV ZOO] The common name for more than 40,000 species of parasitic flatworms that form the class Trematoda. [NAV ARCH] The broad end of each arm of an anchor. [VERT ZOO] A flatfish, especially summer flounder.

flume [ENG] **1.** An open channel constructed of steel, reinforced concrete, or wood and used to convey water to be utilized for power, to transport logs, and so on. **2.** To divert by a flume, as the waters of a stream, in order to lay bare the auriferous sand and gravel forming the bed. [GEOL] A ravine with a stream flowing through it.

flumed [MIN ENG] In hydraulic mining, pertaining to the

FLUKE

Liver fluke *Opisthorchis sinensis*, a trematode characterized by well-developed suckers.

transportation of solids by suspension or flotation in flowing water.

fluoborate [INORG CHEM] **1.** Any of a group of compounds related to the borates in which one or more oxygens have been replaced by fluorine atoms. **2.** The BF_4^- ion, which is derived from fluoboric acid, HBF_4.

fluoboric acid [INORG CHEM] HBF_4 Colorless, clear, water-miscible acid; used for electrolytic brightening of aluminum and for forming stabilized diazo salts.

fluoborite [MINERAL] $Mg_3(BO_3)(F,OH)_3$ A colorless mineral composed of magnesium fluoborate; occurs in hexagonal prisms.

fluocerite [MINERAL] $(Ce,La,Nd)F_3$ A reddish-yellow mineral composed of fluoride of cerium and related elements.

fluolite *See* pitchstone.

fluor *See* fluorite.

fluoranthene [ORG CHEM] $C_{16}H_{10}$ A tetracyclic hydrocarbon found in coal tar fractions and petroleum, forming needlelike crystals, boiling point 250°C, and soluble in organic solvents such as ether and benzene.

fluorapatite [MINERAL] **1.** $Ca_5(PO_4)_3F$ A mineral of the solid-solution series of the apatite group; common accessory mineral in igneous rocks. **2.** An apatite mineral in which the fluoride member dominates.

fluorene [ORG CHEM] $C_{13}H_{10}$ A hydrocarbon chemical present in the middle oil fraction of coal tar; insoluble in water, soluble in ether and acetone, melting point 116–117°C; used as the basis for a group of dyes. Also known as 2,3-benzindene; diphenylenemethane.

fluorescein [ORG CHEM] $C_{20}H_{12}O_5$ A yellowish to red powder, melts and decomposes at 290°C, insoluble in water, benzene, and chloroform, soluble in glacial acetic acid, boiling alcohol, ether, dilute acids, and dilute alkali; used in medicine, in oceanography as a marker in sea water, and in textiles to dye silk and wool.

fluorescence [ATOM PHYS] **1.** Emission of electromagnetic radiation that is caused by the flow of some form of energy into the emitting body and which ceases abruptly when the excitation ceases. **2.** Emission of electromagnetic radiation that is caused by the flow of some form of energy into the emitting body and whose decay, when the excitation ceases, is temperature-independent. [NUC PHYS] Gamma radiation scattered by nuclei which are excited to and radiate from an excited state.

fluorescence analysis *See* fluorometric analysis.

fluorescence method [GRAPHICS] A method of ultraviolet photography in which the subject is illuminated by ultraviolet light, and a filter is used on the camera to absorb the reflected ultraviolet light and permit only the visible fluorescence to reach the film.

fluorescence microscope [OPTICS] A variation of the compound laboratory light microscope which is arranged to admit ultraviolet, violet, and sometimes blue radiations to a specimen, which then fluoresces.

fluorescence spectra [SPECT] Emission spectra of fluorescence in which an atom or molecule is excited by absorbing light and then emits light of characteristic frequencies.

fluorescent antibody [IMMUNOL] An antibody labeled by a fluorescent dye, such as fluorescein.

fluorescent antibody test [IMMUNOL] A clinical laboratory test based on the antigen used in the diagnosis of syphilis and lupus erythematosus and for identification of certain bacteria and fungi, including the tubercle bacillus.

fluorescent dye [CHEM] A highly reflective dye that serves to intensify color and add to the brilliance of a fabric.

fluorescent lamp [ELECTR] A tubular discharge lamp in which ionization of mercury vapor produces radiation that activates the fluorescent coating on the inner surface of the glass.

fluorescent pigment [CHEM] A pigment capable of absorbing both visible and nonvisible electromagnetic radiations and releasing them quickly as energy of desired wavelength; examples are zinc sulfide or cadmium sulfide.

fluorescent screen [ENG] A sheet of material coated with a fluorescent substance so as to emit visible light when struck by ionizing radiation such as x-rays or electron beams.

fluorescent staining [CYTOL] The use of fluorescent dyes to mark specific cell structures, such as chromosomes.

fluoridation [ENG] The addition of the fluorine ion (F⁻) to municipal water supplies in a final concentration of 0.8–1.6 ppm (parts per million) to help prevent dental caries in children. [GEOCHEM] Formation in rocks of fluorine-containing minerals such as fluorite or topaz.

fluoride [INORG CHEM] A salt of hydrofluoric acid, HF, in which the fluorine atom is in the minus-one oxidation state.

fluorimeter *See* fluorometer.

fluorinated ethylene propylene resin [ORG CHEM] Copolymers of tetrafluoroethylene and hexafluoropropylene. Abbreviated FEP resin.

fluorination [CHEM] A chemical reaction in which fluorine is introduced into a chemical compound.

fluorine [CHEM] A gaseous or liquid chemical element, symbol F, atomic number 9, atomic weight 18.998; a member of the halide family, it is the most electronegative element and the most chemically energetic of the nonmetallic elements; highly toxic, corrosive, and flammable; used in rocket fuels and as a chemical intermediate.

fluorite [MINERAL] CaF₂ A transparent to translucent, often blue or purple mineral, commonly found in crystalline cubes in veins and associated with lead, tin, and zinc ores; hardness is 4 on Mohs scale; the principal ore of fluorine. Also known as Derbyshire spar; fluor; fluorspar.

fluoroacetate [ORG CHEM] Acetate in which carbon-connected hydrogen atoms are replaced by fluorine atoms.

fluoroacetic acid [ORG CHEM] CH₂FCOOH A poisonous, crystalline compound obtained from plants, such as those of the Dichapetalaceae family, South Africa, soluble in water and alcohol, and burns with a green flame; the sodium salt is used as a water-soluble rodent poison. Also known as fluoroethanoic acid; gifblaar poison.

fluoroalkane [ORG CHEM] Straight-chain, saturated hydrocarbon compound (or analog thereof) in which some of the hydrogen atoms are replaced by fluorine atoms.

fluorocarbon [ORG CHEM] A hydrocarbon such as Freon in which part or all hydrogen atoms have been replaced by fluorine atoms; can be liquid or gas and is nonflammable and heat-stable; used as refrigerant, aerosol propellant, and solvent. Also known as fluorohydrocarbon.

fluorocarbon-11 *See* trichlorofluoromethane.

fluorocarbon-14 *See* carbon tetrafluoride.

fluorocarbon fiber [ORG CHEM] Fiber made from a fluorocarbon resin, such as polytetrafluoroethylene resin.

fluorocarbon resin [ORG CHEM] Polymeric material made up of carbon and fluorine with or without other halogens (such as chlorine) or hydrogen; the resin is extremely inert and more dense than corresponding fluorocarbons such as Teflon.

fluorochemical [CHEM] Any chemical compound containing fluorine; usually refers to the fluorocarbons.

fluorochromasia [CYTOL] The immediate appearance of fluorescence inside viable cells on exposure to a fluorogenic substrate.

fluorocummingtonite [MINERAL] Cummingtonite with a high content of fluorine.

fluorod [NUCLEO] A rod made from silver-activated phosphate glass and used in solid-state dosimeters; under irradiation the rod absorbs ultraviolet light and emits orange fluorescent light; measurement of the intensity of the emitted light with a photomultiplier gives a measure of the absorbed dose of radiation.

fluoroethanoic acid *See* fluoroacetic acid.

fluoroform [ORG CHEM] CHF₃ A colorless, nonflammable gas, boiling point 84°C at 1 atmosphere, freezing point 160°C at 1 atmosphere; used in refrigeration and as an intermediate in organic synthesis. Also known as propellant 23; refrigerant 23; trifluoromethane.

fluorogenic substrate [CHEM] A nonfluorescent material that is acted upon by an enzyme to produce a fluorescent compound.

fluorography [GRAPHICS] Photography of an image produced on a fluorescent screen. Also known as photofluorography.

fluorohydrocarbon *See* fluorocarbon.

fluorometer [ENG] An instrument that measures the fluorescent radiation emitted by a sample which is exposed to monochromatic radiation, usually radiation from a mercury-arc lamp or a tungsten or molybdenum x-ray source that has

passed through a filter; used in chemical analysis, or to determine the intensity of the radiation producing fluorescence. Also spelled fluorimeter.

fluorometric analysis [ANALY CHEM] A method of chemical analysis in which a sample, exposed to radiation of one wavelength, absorbs this radiation and reemits radiation of the same or longer wavelength in about 10⁻⁹ second; the intensity of reemited radiation is almost directly proportional to the concentration of the fluorescing material. Also known as fluorescence analysis; fluorometry.

fluorometry *See* fluorometric analysis.

fluorophosphoric acid [INORG CHEM] H₂PO₃F A colorless, viscous liquid that is miscible with water; used in metal cleaners and as a catalyst.

fluoroplastics [MATER] A family of plastics based on fluorine replacement of hydrogen atoms in hydrocarbon molecules; includes polytetrafluoroethylene (PTFE), polychlorotrifluoroethylene (PCTFE), polyvinylidene fluoride, and fluorinated ethylene propylene (FEP).

fluoroscope [ENG] A fluorescent screen designed for use with an x-ray tube to permit direct visual observation of x-ray shadow images of objects interposed between the x-ray tube and the screen.

fluoroscopic image intensifier [ELECTR] An electron-beam tube that converts a relatively feeble fluoroscopic image on the fluorescent input phosphor into a much brighter image on the output phosphor.

fluoroscopy [ENG] Use of a fluoroscope for x-ray examination.

fluorothene *See* chlorotrifluoroethylene polymer.

fluorotrichloromethane *See* trichlorofluoromethane.

fluorspar *See* fluorite.

fluosilicate [INORG CHEM] A salt derived from fluosilicic acid, H₂SiF₆, and containing the SiF₆⁻⁻ ion.

fluosilicic acid [INORG CHEM] H₂SiF₆ A colorless acid, soluble in water, which attacks glass and stoneware; highly corrosive and toxic; used in water fluoridation and electroplating. Also known as hydrofluorosilicic acid; hydrofluosilicic acid.

fluosolids system [MET] In pyrometallurgy, a roasting method for finely divided solids, in which air under pressure is blown through a heated bed of mineral to keep it fluid.

fluosulfonic acid [INORG CHEM] HSO₃F Colorless, corrosive, fuming liquid; soluble in water with partial decomposition; used as organic synthesis catalyst and in electroplating.

flurry [METEOROL] A brief shower of snow accompanied by a gust of wind, or a sudden, brief wind squall.

flush [GRAPHICS] A printing term that means no indention; headings are often run flush left, that is, they align at the left margin; flush-right lines align at the right.

flush coat [CIV ENG] A coating of bituminous material, used to waterproof a surface.

flushed-zone resistivity [PETRO ENG] Electrical resistivity of the reservoir area which surrounds a borehole to a distance of at least 3 inches (7.6 centimeters) and for which the original interstitial fluids have been flushed out by drilling-mud filtrate.

flush gate [CIV ENG] A gate for flushing a channel that lies below the gate of a dam.

flushing [CIV ENG] The removal or reduction to a permissible level of dissolved or suspended contaminants in an estuary or harbor. [ENG] Removing lodged deposits of rock fragments and other debris by water flow at high velocity; used to clean water conduits and drilled boreholes.

flushing oil [MATER] A solvent oil designed to remove used lubricating oil, decomposition products, and accumulated dirt from lubrication passages, crankcase surfaces, and lubricated moving parts of automotive engines.

flush-joint casing [PETRO ENG] Lengths of casing that when connected end to end form a smooth joint flush with the outer diameter of the remainder of the section length.

flushometer [ENG] A valve that discharges a fixed quantity of water when a handle is operated; used to flush toilets and urinals.

flush tank [CIV ENG] **1.** A tank in which water or sewage is retained for periodic release through a sewer. **2.** A small water-filled tank for flushing a water closet.

flush valve [ENG] A valve used for flushing toilets.

FLUORINE

Periodic table of the chemical elements showing the position of fluorine.

FLUORITE

Fluorite crystals from Elizabethtown, Illinois. (*American Museum of Natural History specimen*)

flute [DES ENG] A groove having a curved section, especially when parallel to the main axis, as on columns, drills, and other cylindrical or conical shaped pieces. [GEOL] **1.** A natural groove running vertically down the face of a rock. **2.** A groove in a sedimentary structure formed by the scouring action of a turbulent, sediment-laden water current, and having a steep upcurrent end.

flute cast [GEOL] A raised, oblong, or subconical welt on the bottom surface of a siltstone or sandstone bed formed by the filling of a flute.

fluted chucking reamer [DES ENG] A machine reamer with a straight or tapered shank and with straight or spiral flutes; the ends of the teeth are ground on a slight chamfer for end cutting.

fluted coupling *See* stabilizer.

flute length [DES ENG] On a twist drill, the length measured from the outside corners of the cutting lips to the farthest point at the back end of the flutes.

flute storage [ELECTR] Ferrite storage consisting of a number of parallel lengths of fine prism-shaped tubing, each surrounding an insulated axial conductor that acts as a word line; the lengths of tubing are intersected at right angles by parallel sets of insulated wire bit lines that are displaced slightly from the word lines; each intersection stores one bit.

fluting [MECH ENG] A machining operation whereby flutes are formed parallel to the main axis of cylindrical or conical parts.

flutter [ACOUS] Distortion that occurs in sound reproduction as a result of undesired speed variations during the recording, duplicating, or reproducing process. [ELECTROMAG] A fast-changing variation in received signal strength, such as may be caused by antenna movements in a high wind or interaction with a signal or another frequency. [FL MECH] An aeroelastic self-excited vibration in which the external source of energy is the airstream and which depends on the elastic, inertial, and dissipative forces of the system in addition to the aerodynamic forces.

flutter echo [ACOUS] A multiple echo in which the reflections rapidly follow each other. [ELECTROMAG] A radar echo consisting of a rapid succession of reflected pulses resulting from a single transmitted pulse.

flutter valve [ENG] A valve that is operated by fluctuations in pressure of the material flowing over it; used in carburetors.

fluvarium [ENG] A large aquarium in which the tanks contain flowing stream water maintained by gravity, not pumps.

fluvial [HYD] **1.** Pertaining to or produced by the action of a stream or river. **2.** Existing, growing, or living in or near a river or stream.

fluvial cycle of erosion *See* normal cycle.

fluvial sand [GEOL] Sand laid down by a river or stream.

fluvial soil [GEOL] Soil laid down by a river or stream.

fluviatile [GEOL] Resulting from river action.

flux [ELECTROMAG] The electric or magnetic lines of force in a region. [MATER] **1.** In soldering, welding, and brazing, a material applied to the pieces to be united to reduce the melting point of solders and filler metals and to prevent the formation of oxides. **2.** A substance used to promote the fusing of minerals or metals. **3.** Additive for plastics composition to improve flow during physical processing. **4.** In enamel work, a substance composed of silicates and other materials that forms a colorless, transparent glass when fired. Also know as fondant. [NUCLEO] The product of the number of particles per unit volume and their average velocity; a special case of the physics definition. Also known as flux density. [PHYS] **1.** The integral over a given surface of the component of a vector field (for example, the magnetic flux density, electric displacement, or gravitational field) perpendicular to the surface; by definition, it is proportional to the number of lines of force crossing the surface. **2.** The amount of some quantity flowing across a given area (often a unit area perpendicular to the flow) per unit time; the quantity may be, for example, mass or volume of fluid, electromagnetic energy, or number of particles.

flux-cored welding [MET] Welding with a metal electrode that has a flux core.

flux density [NUCLEO] *See* flux. [PHYS] Any vector field whose flux is a significant physical quantity; examples are magnetic flux density, electric displacement, gravitational field, and the Poynting vector.

flux density threshold *See* threshold illuminance.

flux factor [MET] A factor for assessing the quality of steel-works-grade silica refractories.

flux gate [ENG] A detector that gives an electric signal whose magnitude and phase are proportional to the magnitude and direction of the external magnetic field acting along its axis; used to indicate the direction of the terrestrial magnetic field.

flux-gate magnetometer [ELECTROMAG] A magnetometer in which the degree of saturation of the core by an external magnetic field is used as a measure of the strength of the earth's magnetic field; the essential element is the flux gate.

flux guide [MET] A shaped piece of magnetic material used to guide magnetic flux in induction heating; may be used either to direct the flux to preferred locations or to prevent the flux from spreading beyond definite regions.

fluxing [MET] The development of the liquid phase in a ceramic body under heat treatment by the melting of low-fusion components; used in steel manufacture, metal smelting, and assaying.

fluxing ore [MET] An ore containing usually an appreciable amount of valuable metal, but smelted mainly because it contains fluxing agents which are required in the reduction of other ores.

flux jumping *See* Meissner effect.

flux leakage [ELECTROMAG] Magnetic flux that does not pass through an air gap or other part of a magnetic circuit where it is required.

flux line *See* line of force.

flux linkage [ELECTROMAG] The product of the number of turns in a coil and the magnetic flux passing through the coil. Also known as linkage.

flux mapping [NUCLEO] The process of measuring the radiation flux at representative points within a nuclear reactor or around some other radiation source.

fluxmeter [ENG] An instrument for measuring magnetic flux.

flux of energy [PHYS] The energy which passes through a surface per unit area per unit time.

flux oil [MATER] An oil suitable for blending with bitumen or asphalt to form a product of greater fluidity or softer consistency.

flux oxygen cutting [MET] Oxygen cutting of metal with the aid of a flux to reduce the temperature.

flux path [ELECTROMAG] A path which is followed by magnetic lines of force and in which the magnetic flux density is significant.

flux pump [CRYO] A cryogenic direct-current generator that converts a small alternating-current input to a large direct-current output when cooled to about 4K; the output current builds up in a series of steps, much like the action of a pump.

flux refraction [ELECTROMAG] The abrupt change in direction of magnetic flux lines at the boundary between two media having different permeabilities, or of the electric flux lines at the boundary between two media having different dielectric constants, when these lines are oblique to the boundary.

flux unit [ASTROPHYS] A unit of energy flux density of radio-astronomical sources, equal to 10^{-26} watt per square meter per hertz. Abbreviated fu.

fly [INV ZOO] The common name for a number of species of the insect order Diptera characterized by a single pair of wings, antennae, compound eyes, and hindwings modified to form knoblike balancing organs, the halters. [MECH ENG] A fan with two or more blades used in timepieces or light machinery to govern speed by air resistance.

Fly *See* Musca.

fly ash [ENG] Fine particulate, essentially noncombustible refuse, carried in a gas stream from a furnace.

flyback [ELECTR] The time interval in which the electron beam of a cathode-ray tube returns to its starting point after scanning one line or one field of a television picture or after completing one trace in an oscilloscope. Also known as retrace; return trace. [HOROL] The return to zero of the timing hand of a stopwatch or chronograph.

flyback power supply [ELECTR] A high-voltage power supply used to produce the direct-current voltage of about 10,000–25,000 volts required for the second anode of a

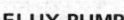

FLUX PUMP

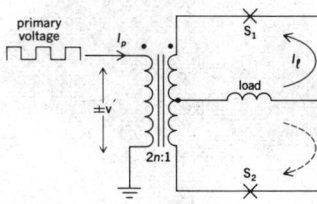

A flux-pump circuit (simplified).

FLY

A common black fly *(Simulium).*

cathode-ray tube in a television receiver or oscilloscope.

flyback transformer *See* horizontal output transformer.

fly cutter [MECH ENG] A cutting tool that revolves with the arbor of a lathe.

fly cutting [MECH ENG] Cutting with a milling cutter provided with only one tooth.

Flygt pump [MECH ENG] A submersible pump developed in Sweden; the discharge hose may be 1 or 3 inches (2.5 or 7.6 centimeters) in diameter, and operation is on a 110-volt, single-phase alternating current.

flying angle [AERO ENG] The acute angle between the longitudinal axis of an aircraft and the horizontal axis in normal level flight, or the angle of attack of a wing in normal level flight.

flying boat [AERO ENG] A seaplane with a fuselage that acts as a hull and is the means of the plane's support on water.

flying bomb [ORD] Popularly, any explosive robot plane, guided missile, rocket bomb, or the like; specifically, the German V-1 explosive robot plane of World War II.

flying bridge [NAV ARCH] A narrow walkway or platform built at the level of the top of the pilothouse and containing a duplicate set of controls for steering gear and engine-room signals and for navigating instruments; the platform extends from one side of the ship to the other. Also known as navigating bridge.

flying buttress [ARCH] A buttress connected to the building it supports by an arch.

flying-crane helicopter [AERO ENG] A heavy-lift helicopter used in rapid loading and unloading of, for example, cargo ships.

flying fish [VERT ZOO] Any of about 65 species of marine fishes which form the family Exocoetidae in the order Atheriniformes; characteristic enlarged pectoral fins are used for gliding.

Flying Fish *See* Volan.

flying head [ELECTR] A read/write head used on magnetic disks and drums, so designed that it flies a microscopic distance off the moving magnetic surface and is supported by a film of air.

flying hour rate *See* utilization rate.

flying saucer *See* unidentified flying object.

flying shear [MET] A machine which cuts lengths of rolled products and allows for continuous production by reciprocating with the product while cutting.

flying spot [ELECTR] A small point of light, controlled mechanically or electrically, which moves rapidly in a rectangular scanning pattern in a flying-spot scanner.

flying-spot scanner [ELECTR] A scanner used for television film and slide transmission, electronic writing, and character recognition, in which a moving spot of light, controlled mechanically or electrically, scans the image field, and the light reflected from or transmitted by the image field is picked up by a phototube to generate electric signals. Also known as optical scanner.

flying switch [ENG] Disconnection of railroad cars from a locomotive while they are moving and switching them to another track under their own momentum.

fly-off *See* evapotranspiration.

fly rock [ENG] The fragments of rock thrown and scattered during quarry or tunnel blasting.

flysch [GEOL] Deposits of dark, fine-grained, thinly bedded sandstone shales and of clay, thought to be deposited by turbidity currents and originally defined as rock formations on the northern and southern borders of the Alps.

flyway [VERT ZOO] A geographic migration route for birds, including the breeding and wintering areas that it connects.

flywheel [MECH ENG] A rotating element attached to the shaft of a machine for the maintenance of uniform angular velocity and revolutions per minute. Also known as balance wheel.

flywheel synchronization [ELECTR] Automatic frequency control of a scanning system by using the average timing of the incoming sync signals, rather than by making each pulse trigger the scanning circuit; used in high-sensitivity television receivers designed for fringe-area reception, when noise pulses might otherwise trigger the sweep circuit prematurely.

fm *See* femtometer.

Fm *See* fermium.

FM *See* frequency modulation.

FM/AM multiplier [ELECTR] Multiplier in which the frequency deviation from the central frequency of a carrier is proportional to one variable, and its amplitude is proportional to the other variable; the frequency-amplitude-modulated carrier is then consecutively demodulated for frequency modulation (FM) and for amplitude modulation (AM); the final output is proportional to the product of the two variables.

F martingale [MATH] A stochastic process $\{X_t, t>0\}$ such that the conditional expectation of X_t given F_s equals X_s whenever $s<t$, where $F=\{F_t, t\geq0\}$ is an increasing family of sigma algebras that represents the amount of information increasing with time.

FMF *See* familial Mediterranean fever.

FMN *See* riboflavin 5′-phosphate.

fnp *See* fusion point.

f number [OPTICS] A lens rating obtained by dividing the lens's focal length by its effective maximum diameter; the larger the *f* number, the less exposure is given. Also known as focal ratio.

foal [VERT ZOO] A young horse, especially one under 1 year of age.

foam [FL MECH] A froth of bubbles on the surface of a liquid, often stabilized by organic contaminants, as found at sea or along shore. [GEOL] *See* pumice.

foam-back fabric [TEXT] A fabric to the back of which is banded a thin layer of polyether or polyester foam. Also known as foam-laminated fabric.

foam crust [HYD] A snow surface feature that looks like small overlapping waves, like sea foam on a beach, occurring during the ablation of the snow surface and may further develop into a more pronounced wedge-shaped form, known as plowshares.

foam drilling [MIN ENG] A method of dust suppression in which thick foam is forced through the drill by means of compressed air, and the foam-and-dust mixture emerges from the mouth of the hole in the form of a thick sludge.

foamed plastic *See* expanded plastic.

foam glass [MATER] A light, black, opaque, cellular glass made by adding powdered carbon to crushed glass and firing the mixture.

foaminess [PHYS] The volume of foam produced in a liquid, in cubic centimeters, produced by passing air through it divided by the rate of flow of air, in cubic centimeters per second.

foaming [ENG] Any of various processes by which air or gas is introduced into a liquid or solid to produce a foam material. [GRAPHICS] A froth in a photographic emulsion containing minute bubbles caused by excessive turbulence.

foaming agent *See* blowing agent.

foam-in-place [ENG] The deposition of reactive foam ingredients onto the surface to be covered, allowing the foaming reaction to take place upon that surface, as with polyurethane foam; used in applying thermal insulation for homes and industrial equipment.

foam-laminated fabric *See* foam-back fabric.

foam metal [MET] Cast metal with finely divided gas bubbles evenly distributed throughout the body of the metal; an example is foam aluminum.

foam rubber *See* rubber sponge.

focal distance *See* focal length.

focal infection [MED] Infection in a limited area, such as the tonsils, teeth, sinuses, or prostate.

focal length [OPTICS] The distance from the focal point of a lens or curved mirror to the principal point; for a thin lens it is approximately the distance from the focal point to the lens. Also known as focal distance.

focal plane [OPTICS] A plane perpendicular to the axis of an optical system and passing through the focal point of the system.

focal-plane shutter [OPTICS] A camera shutter consisting of a blind containing a slot; the blind is pulled rapidly across the film, exposing it through the slot.

focal point [OPTICS] The point to which rays that are initially parallel to the axis of a lens, mirror, or other optical system are converged or from which they appear to diverge. Also known as principal focus.

focal ratio *See* f number.

FLYING FISH

The four-winged flying fish (*Parexocoetus mesogaster*), with pectoral fins extended for gliding.

FLYING-SPOT SCANNER

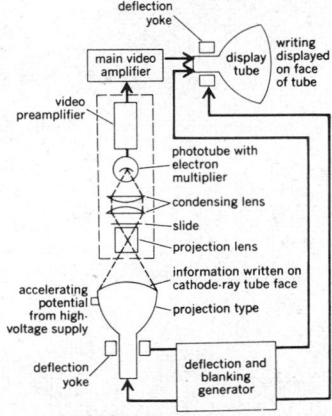

The flying-spot scanner for electronic writing.

focal seizure [MED] An epileptic manifestation of a restricted nature, usually without loss of consciousness, due to irritation of a localized area of the brain.

focus [ELECTR] To control convergence or divergence of the electron paths within one or more beams, usually by adjusting a voltage or current in a circuit that controls the electric or magnetic fields through which the beams pass, in order to obtain a desired image or a desired current density within the beam. [GEOPHYS] The center of an earthquake and the origin of its elastic waves within the earth. [MATH] A point in the plane which together with a line (directrix) defines a conic section. [NUCLEO] To guide particles along a desired path in a particle accelerator by means of electric or magnetic fields. [OPTICS] **1.** The point or small region at which rays converge or from which they appear to diverge. **2.** To move an optical lens toward or away from a screen or film to obtain the sharpest possible image of a desired object.

focus control [ELECTR] A control that adjusts spot size at the screen of a cathode-ray tube to give the sharpest possible image; it may vary the current through a focusing coil or change the position of a permanent magnet. [OPTICS] A device to adjust a lens system to produce a sharp image.

focusing anode [ELECTR] An anode used in a cathode-ray tube to change the size of the electron beam at the screen; varying the voltage on this anode alters the paths of electrons in the beam and thus changes the position at which they cross or focus.

focusing coil [ELECTR] A coil that produces a magnetic field parallel to an electron beam for the purpose of focusing the beam.

focusing electrode [ELECTR] An electrode to which a potential is applied to control the cross-sectional area of the electron beam in a cathode-ray tube.

focusing glass [OPTICS] A magnifying glass designed to enlarge the image thrown on the ground glass of the viewfinder of a camera, to help achieve exact focusing.

focusing magnet [ELECTR] A permanent magnet used to produce a magnetic field for focusing an electron beam.

focusing scale [OPTICS] A graduated scale to indicate appropriate lens-to-image plane positions for given lens-to-object plane distances.

focus lamp [ELEC] **1.** A lamp whose filament has a spiral or zigzag form in order to reduce its size, so that it can be brought into the focus of a lens or mirror. **2.** An arc lamp whose feeding mechanism is designed to hold the arc in a constant position with respect to an optical system that is used to focus its rays.

focus projection and scanning [ELECTR] Method of magnetic focusing and electrostatic deflection of the electron beam of a hybrid vidicon; a transverse electrostatic field is used for beam deflection; this field is immersed with an axial magnetic field that focuses the electron beam.

foehn [METEOROL] A warm, dry wind on the lee side of a mountain range, the warmth and dryness being due to adiabatic compression as the air descends the mountain slopes. Also spelled föhn.

foehn air [METEOROL] The air associated with foehn winds, very warm and dry.

foehn cloud [METEOROL] Any cloud form associated with a foehn, but usually signifying only those clouds of the lenticularis species formed in the lee wave parallel to the mountain ridge.

foehn cyclone [METEOROL] A cyclone formed (or at least enhanced) as a result of the foehn process on the lee side of a mountain range.

foehn island [METEOROL] An isolated area where the foehn has reached the ground, in contrast to the surrounding area where foehn air has not replaced colder surface air.

foehn nose [METEOROL] As seen on a synoptic surface chart, a typical deformation of the isobars in connection with a well-developed foehn situation; a ridge of high pressure is produced on the windward slopes of the mountain range, while a foehn trough forms on the lee side; the isobars "bulge" correspondingly, giving a noselike configuration.

foehn pause [METEOROL] **1.** A temporary cessation of the foehn at the ground, due to the formation or intrusion of a cold air layer which lifts the foehn above the valley floor. **2.** The boundary between foehn air and its surroundings.

foehn period [METEOROL] The duration of continuous foehn conditions at a given location.

foehn phase [METEOROL] One of three stages to describe the development of the foehn in the Alps: the preliminary phase, when cold air at the surface is separated from warm dry air aloft by a subsidence inversion; the anticyclonic phase, when the warm air reaches a station as the result of the cold air flowing out from the plain; and the stationary phase or cyclonic phase, when the foehn wall forms and the downslope wind becomes appreciable.

foehn sickness [MED] A phenomenon in humans in alpine regions, marked by adverse psychological and physiological effects during prolonged periods of foehn wind.

foehn storm [METEOROL] A type of destructive storm which frequently occurs in October in the Bavarian Alps.

foehn trough [METEOROL] The dynamic trough formed in connection with the foehn.

foehn wall [METEOROL] The steep leeward boundary of flat, cumuliform clouds formed on the peaks and upper windward sides of mountains during foehn conditions.

fog [GRAPHICS] A dark, hazy deposit or veil of uniform density over all or parts of a piece of film or paper; can be caused by light other than that forming the image, lens flare, aged materials, or chemical impurities. [METEOROL] Water droplets or, rarely, ice crystals suspended in the air in sufficient concentration to reduce visibility appreciably.

fogbank [METEOROL] A fairly well-defined mass of fog observed in the distance, most commonly at sea.

fogbound [NAV] Surrounded by fog; the term refers to vessels which are unable to proceed because of the fog.

fogbow [OPTICS] A faintly colored circular arc similar to a rainbow but formed on fog layers containing drops whose diameters are of the order of 100 micrometers or less. Also known as false white rainbow; mistbow; white rainbow.

fog chamber *See* cloud chamber.

fog climax [ECOL] A community that deviates from a climatic climax because of the persistent occurrence of a controlling fog blanket.

fog dectector light [NAV] A light fitted in certain lighthouses for the automatic detection of fog, in addition to the main light.

fog deposit [HYD] The deposit of an ice coating on exposed surfaces by a freezing fog.

fog dispersal [METEOROL] Artificial dissipation of a fog by means such as seeding or heating.

fog drip [HYD] Water dripping to the ground from trees or other objects which have collected the moisture from drifting fog; the dripping can be as heavy as light rain, as sometimes occurs among the redwood trees along the coast of northern California.

fog drop [METEOROL] An elementary particle of fog, physically the same as a cloud drop. Also known as fog droplet.

fog droplet *See* fog drop.

fog forest [ECOL] The dense, rich forest growth which is found at high or medium-high altitudes on tropical mountains; occurs when the tropical rain forest penetrates altitudes of cloud formation, and the climate is excessively moist and not too cold to prevent plant growth.

fogged metal [MET] A metal whose luster has been highly reduced by corrosion products.

fog gong [NAV] A gong used as a fog signal.

fog gun [NAV] A gun used as a fog signal.

fog horizon [METEOROL] The top of a fog layer which is confined by a low-level temperature inversion so as to give the appearance of the horizon when viewed from above against the sky; the true horizon is usually obscured by the fog in such instances.

foghorn [NAV] A horn used as a fog signal.

fog quenching [MET] Rapid cooling of a metal piece in a fine vapor or mist.

fog scale [METEOROL] A classification of fog intensity based on its effectiveness in decreasing horizontal visibility; such practice is not current in United States weather observing procedures.

fog signal [NAV] **1.** A warning signal transmitted by a vessel or aid to navigation during periods of low visibility. **2.** The device for producing such a signal.

fog siren [NAV] A siren used as a fog signal.

fog track [NUCLEO] A line of condensation, produced in supersaturated water vapor by the passage of charged particles; used in studying the courses and collisions of particles in cloud chambers.

fog trumpet [NAV] A trumpet used as a fog signal.

fog whistle [NAV] A whistle used as a fog signal.

fog wind [METEOROL] Humid east wind which crosses the divide of the Andes east of Lake Titicaca and descends on the west in violent squalls; probably the same as puelche.

föhn *See* foehn.

foil [MET] A thin sheet of metal, usually less than 0.006 inch (0.15 millimeter) thick.

foil decorating [ENG] The molding of paper, textile, or plastic foil, printed with compatible inks, into a plastic part so that the foil is visible below the surface of the part as a decoration.

foil dosimeter [NUCLEO] A device for measuring the amount of radiation exposure by means of the degree of activation created in a metal foil inserted in the radiation field.

foil electret [ELEC] A thin film of strongly insulating material capable of trapping charge carriers, such as polyfluoroethyleneproplylene, that is electrically charged to produce an external electric field; in the conventional design, charge carriers of one sign are injected into one surface, and a compensation charge of opposite sign forms on the opposite surface or an adjacent electrode.

Fokker-Planck equation [STAT MECH] An equation for the distribution function of a gas, analogous to the Boltzmann equation but applying where the forces are long-range and the collisions are not binary.

fold [ANAT] A plication or doubling, as of various parts of the body such as membranes and other flat surfaces. [GEOL] A bend in rock strata or other planar structure, usually produced by deformation; folds are recognized where layered rocks have been distorted into wavelike form. [MET] *See* lap.

fold belt *See* orogenic belt.

folded cavity [ELECTR] Arrangement used in a klystron repeater to make the incoming wave act on the electron stream from the cathode at several places and produce a cumulative effect.

folded dipole *See* folded-dipole antenna.

folded-dipole antenna [ELECTROMAG] A dipole antenna whose outer ends are folded back and joined together at the center; the impedance is about 300 ohms, as compared to 70 ohms for a single-wire dipole; widely used with television and frequency-modulation receivers. Also known as folded dipole.

folded horn [ENG ACOUS] An acoustic horn in which the path from throat to mouth is folded or curled to give the longest possible path in a given volume.

folded-plate roof [BUILD] A roof constructed of flat plates, usually of reinforced concrete, joined at various angles.

folding [GEOL] Compression of planar structure in the formation of fold structures.

folding boards [MIN ENG] A shifting frame on which the cage rests in a mine. Also known as chairs; dogs; keeps; keps.

folding door [ENG] A door in sections that can be folded back or can be moved apart by sliding.

folding fin [AERO ENG] A fin hinged at its base to lie flat, especially a fin on a rocket that lies flat until the rocket is in flight.

foldover [ELECTR] Picture distortion seen as a white line on the side, top, or bottom of a television picture; generally caused by nonlinear operation in either the horizontal or vertical deflection circuits of a receiver.

fold system [GEOL] A group of folds with common trends and characteristics.

foliaceous [BOT] Consisting of or having the form or texture of a foliage leaf. [GEOL] Having a leaflike or platelike structure composed of thin layers of minerals. [ZOO] Resembling a leaf in growth form or mode.

foliage [BOT] The leaves of a plant.

foliar [BOT] Of, pertaining to, or consisting of leaves.

foliate papilla [VERT ZOO] One of the papillae found on the posterolateral margin of the tongue of many mammals, but vestigial or absent in humans.

foliated ice [HYD] Large masses of ice which grow in thermal contraction cracks in permafrost. Also known as ice wedge.

foliation [BOT] **1.** The process of developing into a leaf. **2.** The state of being in leaf. [GEOL] A laminated structure formed by segregation of different minerals into layers that are parallel to the schistosity. [MET] Beating metal into thin sheets.

folic acid [BIOCHEM] $C_{19}H_{19}N_7O_6$ A yellow, crystalline vitamin of the B complex; it is slightly soluble in water, usually occurs in conjugates containing glutamic acid residues, and is found especially in plant leaves and vertebrate livers. Also known as pteroylglutamic acid (PGA).

folic acid sodium salt *See* sodium folate.

foliferous [BOT] Producing leaves.

foliicolous [BIOL] Growing or parasitic upon leaves, as certain fungi.

Folin solution [ANALY CHEM] An aqueous solution of 500 grams of ammonium sulfate, 5 grams of uranium acetate, and 6 grams of acetic acid in a volume of 1 liter; used to test for uric acid.

foliobranchiate [VERT ZOO] Having leaflike gills.

foliolate [BOT] Having leaflets.

follicle [BIOL] A deep, narrow sheath or a small cavity. [BOT] A type of dehiscent fruit composed of one carpel opening along a single suture.

follicle-stimulating hormone [BIOCHEM] A protein hormone released by the anterior pituitary of vertebrates which stimulates growth and secretion of the Graafian follicle and also promotes spermatogenesis. Abbreviated FSH.

follicular lymphoma [MED] A premalignant lymphoma in which the lymph nodes show enlarged follicles composed predominantly of closely packed, large reticuloendothelial cells.

folliculate [BIOL] Having or composed of follicles.

folliculitis [MED] Inflammation of a follicle or group of follicles.

folliculitis keloidalis *See* keloid acne.

follow current [ELEC] The current at power frequency that passes through a surge diverter or other discharge path after a high-voltage surge has started the discharge.

follower [ENG] A drill used for making all but the first part of a hole, the first part being made with a drill of larger gage.

follower rail [MIN ENG] The rail of a mine switch on the other side of the turnout corresponding to the lead rail.

following sea [NAV] A sea in which the waves move in the general direction of the heading.

following wind [METEOROL] **1.** A wind blowing in the direction of ocean-wave advance. **2.** *See* tailwind. [NAV] Wind blowing in the general direction of a vessel's course.

follow spot [ELEC] A high-intensity spotlight used to follow action in arenas and stadiums and on large stages; it is equipped with adjustable iris and shutter controls, and its light source is either a carbon arc or an incandescent bulb.

follow-the-pointer indicator [ORD] Scale, on the mount of some types of artillery, that receives and registers firing data transmitted over a remote control system; the gun is kept properly aimed when its adjustment dials are matched with the readings on the indicator.

follow through [ORD] Material which follows the jet of a shaped charge through the hole formed in the target.

fomite [MED] An inanimate object which may be contaminated with infectious organisms and thus serve to transmit disease.

fondant *See* flux.

font [GRAPHICS] A particular typeface and size, including all the uppercase and lowercase letters, punctuation marks, numerals, and so forth.

fontanelle [ANAT] A membrane-covered space between the bones of a fetal or young skull. [INV ZOO] A depression on the head of termites.

Fontéchevade man [PALEON] A fossil man representing the third interglacial *Homo sapiens* and having browridges and a cranial vault similar to those of modern *Homo sapiens*.

food [BIOL] A material that can be ingested and utilized by the organism as a source of nutrition and energy.

food additive [FOOD ENG] A substance added to foods during processing to improve color, texture, flavor, or keeping qualities; examples are antioxidants, emulsifiers, thickeners, preservatives, and colorants.

food allergy [IMMUNOL] A hypersensitivity to certain foods.

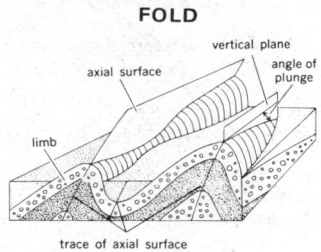

FOLD

Elements of folds.

FOLLICLE

Follicles of the milkweed plant.

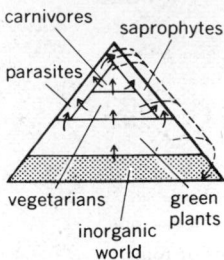

FOOD PYRAMID

carnivores

saprophytes

parasites

vegetarians

green plants

inorganic world

Diagram of the food pyramid showing the general trends in food circulation.

food-borne disease [MED] Any disease transmitted by contaminated foods.

food chain [ECOL] The scheme of feeding relationships by trophic levels which unites the member species of a biological community.

food chemistry [FOOD ENG] A specialized phase of food technology concerned with an understanding of the fundamental changes of composition and the physical condition of foodstuffs which may occur during and subsequent to industrial processing.

food engineering [ENG] The technical discipline involved in food manufacturing and processing.

food infection [MED] A type of bacterial food poisoning in which the host is infected by organisms carried by food.

food manufacturing [FOOD ENG] The commercial production and packaging of foods that are fabricated by processing, by combining various ingredients, or both.

food microbiology [FOOD ENG] The science that deals with the microorganisms involved in the spoilage, contamination, and preservation of food.

food poisoning [MED] Poisoning due to intake of food contaminated by bacteria or poisonous substances produced by bacteria.

food preservation [FOOD ENG] Processing designed to protect food from spoilage caused by microbes, enzymes, and autooxidation.

food pyramid [ECOL] An ecological pyramid representing the food relationship among the animals in a community.

food science [FOOD ENG] The applied science which deals with the chemical, biochemical, physical, physiochemical, and biological properties of foods.

food spoilage [FOOD ENG] Deterioration in the color, flavor, odor, or consistency of a food product.

food technology [FOOD ENG] The application of science and engineering to the refining, manufacturing, and handling of foods; many food technologists are food scientists rather than engineers.

food vacuole [CYTOL] A membrane-bound organelle in which digestion occurs in cells capable of phagocytosis. Also known as heterophagic vacuole; phagocytic vacuole.

food web [ECOL] A modified food chain that expresses feeding relationships at various, changing trophic levels.

fool's gold *See* pyrite.

foot [ANAT] Terminal portion of a vertebrate leg. [INV ZOO] An organ for locomotion or attachment. [SCI TECH] The unit of length in the British systems of units, equal to exactly 0.3048 meter. Abbreviated ft.

footage [ENG] The extent or length of a material expressed in feet. [MIN ENG] **1.** The number of feet of borehole drilled per unit of time, or that required to complete a specific project or contract. **2.** The payment of miners by the running foot of work.

foot-and-mouth disease [VET MED] A highly contagious virus disease of cattle, pigs, sheep, and goats that is transmissible to man; characterized by fever, salivation, and formation of vesicles in the mouth and pharynx and on the feet. Also known as hoof-and-mouth disease.

foot block [ENG] Flat pieces of wood placed under props in tunneling to give a broad base and thus prevent the superincumbent weight from pressing the props down.

foot breadth [ANTHRO] The measure of the maximum distance across the left foot, when the subject stands with his weight evenly distributed on both feet.

foot bridge [CIV ENG] A bridge structure used only for pedestrian traffic.

footcandle [OPTICS] A unit of illumination, equal to the illumination of a surface, 1 square foot in area, on which there is a luminous flux of 1 lumen uniformly distributed, or equal to the illumination of a surface all points of which are at a distance of 1 foot from a uniform point source of 1 candela; equal to approximately 10.7639 lux. Abbreviated ftc.

foot gland [INV ZOO] A glandular structure which secretes an adhesive substance in many animals. Also known as pedal gland.

foot guard [CIV ENG] A filler placed on the space between converging rails to prevent a foot from being wedged between the rails.

foot holes [MIN ENG] Holes cut in the sides of shafts or winzes to enable miners to climb up or down.

footing [CIV ENG] The widened base or substructure forming the foundation for a wall or a column.

footlambert [OPTICS] A unit of luminance (photometric brightness), equal to $1/\pi$ candela per square foot, or to the uniform luminance of a perfectly diffusing surface emitting or reflecting light at the rate of 1 lumen per square foot; equal to approximately 3.42625 nit. Abbreviated ft-L. Also known as equivalent footcandle.

foot length [ANTHRO] A measure of the distance from the heel to the longest toe of the left foot, when the subject stands with the weight evenly distributed on both feet.

foot per minute [MECH] A unit of speed equal to the speed of a body which travels uniformly 1 foot in 1 minute; equal to 0.00508 meter per second. Abbreviated fpm.

foot per second [MECH] A unit of speed equal to the speed of a body which travels uniformly 1 foot in 1 second; equal to 0.3048 meter per second. Abbreviated fps.

foot-pound [MECH] **1.** Unit of energy or work in the English gravitational system, equal to the work done by 1 pound of force when the point at which the force is applied is displaced 1 foot in the direction of the force; equal to approximately 1.355818 joules. Abbreviated ft-lb; ft-lbf. **2.** Unit of torque in the English gravitational system, equal to the torque produced by 1 pound of force acting at a perpendicular distance of 1 foot from an axis of rotation. Also known as pound-foot. Abbreviated lbf-ft.

foot-poundal [MECH] **1.** A unit of energy or work in the English absolute system, equal to the work done by a force of magnitude 1 poundal when the point at which the force is applied is displaced 1 foot in the direction of the force; equal to approximately 0.04214011 joules. Abbreviated ft-pdl. **2.** A unit of torque in the English absolute system, equal to the torque produced by a force of magnitude 1 poundal acting at a perpendicular distance of 1 foot from the axis of rotation. Also known as poundal-foot. Abbreviated pdl-ft.

foot rot [PL PATH] Any disease that involves rotting of the stem or trunk of a plant. [VET MED] *See* foul foot.

foot screw [ENG] **1.** One of the three screws connecting the tribach of a theodolite or other level with the plate screwed to the tripod head. **2.** An adjusting screw that serves also as a foot.

foot section [MECH ENG] In both belt and chain conveyors that portion of the conveyor at the extreme opposite end from the delivery point.

foot's oil [CHEM] The oil sweated out of slack wax; it takes its name from the fact that it goes to the bottom, or foot, of the pan when sweated.

foot valve [MECH ENG] A valve in the bottom of the suction pipe of a pump which prevents backward flow of water.

footwall [GEOL] The mass of rock that lies beneath a fault, an ore body, or a mine working. Also known as heading side; heading wall; lower plate.

footwall shaft *See* underlay shaft.

forage [AGR] A vegetable food for domestic animals.

Foraky boring method [MIN ENG] A percussive boring system; a closed-in derrick contains the crown pulley, over which a steel rope with the boring tools is passed from a drum; the drum moves the tools, which are vibrated by a walking beam.

Foraky freezing process [MIN ENG] A method of shaft sinking through heavily watered sands by freezing the sands.

foramen [BIOL] A small opening, orifice, pore, or perforation.

foramen magnum [ANAT] A large oval opening in the occipital bone at the base of the cranium that allows passage of the spinal cord, accessory nerves, and vertebral arteries.

foramen of Magendie [ANAT] The median aperture of the fourth ventricle of the brain.

foramen of Monro *See* interventricular foramen.

foramen of Winslow *See* epiploic foramen.

foramen ovale [ANAT] An opening in the sphenoid for the passage of nerves and blood vessels. [EMBRYO] An opening in the fetal heart partition between the two atria.

foramen primum [EMBRYO] A temporary embryonic interatrial opening.

Foraminiferida [INV ZOO] An order of dominantly marine protozoans in the subclass Granuloreticulosia having a secreted or agglutinated shell enclosing the ameboid body.

forb [BOT] A weed or broadleaf herb.

forbesite [MINERAL] $H(Ni,Co)AsO_4 \cdot 3\frac{1}{2}H_2O$ A grayish-white mineral composed of hydrous nickel cobalt arsenate; occurs in fibrocrystalline form.

Forbes log [NAV] A log consisting essentially of a small rotator in a tube projecting below the bottom of a vessel, with suitable registering devices.

forbidden band [SOLID STATE] A range of unallowed energy levels for an electron in a solid.

forbidden-character code [ADP] A bit code which exists only when an error occurs in the binary coding of characters.

forbidden-combination check [ADP] A test for the occurrence of a nonpermissible code expression in a computer; used to detect computer errors.

forbidden line [ATOM PHYS] A spectral line associated with a transition forbidden by selection rules; optically this might be a magnetic dipole or electric quadrupole transition.

forbidden transition [QUANT MECH] A transition between two states of a quantum-mechanical system which is considerably less probable than a competing allowed transition.

Forbush decrease [ASTROPHYS] A sudden decrease in cosmic-ray intensity which occurs a day or two after a solar flare, and at the same time as the commencement of magnetic storms and auroral activity.

force [ADP] To intervene manually in a computer routine and cause the computer to execute a jump instruction. [MECH] That influence on a body which causes it to accelerate; quantitatively it is a vector, equal to the body's mass times its time rate of change of momentum.

force constant [MECH] The ratio of the force to the deformation of a system whose deformation is proportional to the applied force. [PHYS CHEM] An expression for the force acting to restrain the relative displacement of the nuclei in a molecule.

forced-air heating [MECH ENG] A warm-air heating system in which positive air circulation is provided by means of a fan or a blower.

forced auxiliary ventilation [MIN ENG] A system in which the duct delivers the intake air to the face.

forced-caving system [MIN ENG] A stoping system in which the ore is broken down by large blasts into the stopes that are kept partly full of broken ore.

forced circulation [MECH ENG] The use of a pump or other fluid-movement device in conjunction with liquid-processing equipment to move the liquid through pipes and process vessels; contrasted to gravity or thermal circulation.

forced-circulation boiler [MECH ENG] A once-through steam generator in which water is pumped through successive parts.

forced draft [MECH ENG] Air under positive pressure produced by fans at the point where air or gases enter a unit, such as a combustion furnace.

forced oscillation [MECH] An oscillation produced in a simple oscillator or equivalent mechanical system by an external periodic driving force. Also known as forced vibration.

forced programming *See* minimum-access programming.

forced ventilation [MECH ENG] A system of ventilation in which air is forced through ventilation ducts under pressure.

forced vibration *See* forced oscillation.

forced wave [FL MECH] Any wave which is required to fit irregularities at the boundary of a system or satisfy some impressed force within the system; the forced wave will not in general be a characteristic mode of oscillation of the system.

force feedback [CONT SYS] A method of error detection in which the force exerted on the effector is sensed and fed back to the control, usually by mechanical, hydraulic, or electric transducers.

force fit *See* press fit.

force gage [ENG] An instrument which measures the force exerted on an object.

force main [CIV ENG] The discharge pipeline of a pumping station.

force piece *See* foreset.

force plate [ENG] A plate that carries the plunger or force plug of a mold and the guide pins on bushings.

force polygon [MECH] A closed polygon whose sides are

vectors representing the forces acting on a body in equilibrium.

forceps [DES ENG] A pincerlike instrument for grasping objects.

force pump [MECH ENG] A pump fitted with a solid plunger and a suction valve which draws and forces a liquid to a considerable height above the valve or puts the liquid under a considerable pressure.

forcing [GRAPHICS] The attempt to bring out detail in an underexposed negative by extending the development time or by using an accelerator.

forcing cone [ORD] Tapered beginning of the lands at the origin of the rifling of a gun tube; the forcing cone allows the rotating band of the projectile to be gradually engaged by the rifling, thereby centering the projectile in the bore.

forcing fan [MIN ENG] A fan which forces the intake air into mine workings. Also known as blowing fan.

forcipate [BIOL] Shaped like forceps; deeply forked.

forcipate trophus [INV ZOO] A type of masticatory apparatus in certain predatory rotifers which resembles forceps and is used for grasping.

Forcipulatida [INV ZOO] An order of echinoderms in the subclass Asteroidea characterized by crossed pedicellariae.

fording depth [ENG] Maximum depth at which a particular vehicle can operate in water.

fore [NAV ARCH] **1.** The front part of a ship. **2.** In the direction of or toward the bow.

forearm [ANAT] The part of the upper extremity between the wrist and the elbow. Also known as antebrachium.

forearm circumference [ANTHRO] The measure of the circumference taken halfway between the elbow and the wrist.

forearm length [ANTHRO] A measure of the distance from the tip of the elbow to the tip of the middle finger, with the arm flexed at the elbow.

forebay [CIV ENG] **1.** A small reservoir at the head of the pipeline that carries water to the consumer; it is the last free water surface of a distribution system. **2.** A reservoir feeding the penstocks of a hydro-power plant.

forebody [NAV ARCH] The part of a ship's hull in front of the amidships.

forebrain [EMBRYO] The most anterior expansion of the neural tube of a vertebrate embryo. [VERT ZOO] The part of the adult brain derived from the embryonic forebrain; includes the cerebrum, thalamus, and hypothalamus.

forecast [METEOROL] A statement of expected future meteorological occurrences.

forecasting [COMMUN] The prediction of conditions of radio propagation for a period extending anywhere from a few hours to a few months. [METEOROL] Procedures for extrapolation of the future characteristics of weather on the basis of present and past conditions.

forecastle deck [NAV ARCH] A deck extending from the stem aft over a forecastle erection.

forecast period [METEOROL] The time interval for which a forecast is made.

forecast-reversal test [METEOROL] A test used to evaluate the adequacy of a given method of forecast verification; the same verification method is applied, simultaneously, to a given forecast and to a fabricated forecast of opposite conditions; comparison of the verification scores gives an indication of the value of the verification method.

forecast verification [METEOROL] Any process for determining the accuracy of a weather forecast by comparing the predicted weather with the observed weather of the forecast period; used to test forecasting skills and methods.

foredeep [GEOL] **1.** A long, narrow depression that borders an orogenic belt, such as an island arc, on the convex side. **2.** *See* exogeosyncline.

fore drift [MIN ENG] That one of a pair of parallel headings which is kept a short distance in advance of the other.

foredune [GEOL] A coastal dune or ridge that is parallel to the shoreline of a large lake or ocean and is stabilized by vegetation.

forefinger [ANAT] The index finger; the first finger next to the thumb.

forefoot [NAV ARCH] The extreme forward end of the bottom of a ship. [VERT ZOO] An anterior foot of a quadruped.

foreground [ADP] A program or process of high priority that

FORCIPATE TROPHUS

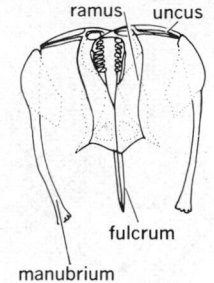

ramus   uncus

fulcrum

manubrium

Ventral view of *Dicranophorus* forcipate trophus. *(After Hauer)*

utilizes machine facilities as needed, with less critical, background work performed in otherwise unused time.

foregut [EMBRYO] The anterior alimentary canal in a vertebrate embryo, including those parts which will develop into the pharynx, esophagus, stomach, and anterior intestine.

forehand welding [MET] Welding in which the flame is directed against the base metal ahead of the weld and is moved in the direction of welding. Also known as forward welding.

forehead [ANAT] The part of the face above the eyes.

forehearth [MET] **1.** A bay in front of the hearth of a furnace. **2.** A receptacle in front of a hearth to receive the molten products.

foreign-body locator [ENG] A device for locating foreign metallic bodies in tissue by means of suitable probes that generate a magnetic field; the presence of a magnetic body within this field is indicated by a meter or a sound signal.

foreign element [IND ENG] A work element which is not a part of the normal work cycle, either because it is accidental or because it occurs only occasionally.

foreign matter [SCI TECH] Any substance not belonging naturally in the place where found.

foreland [GEOGR] An extensive area of land jutting out into the sea. [GEOL] **1.** A lowland area onto which piedmont glaciers have moved from adjacent mountains. **2.** A stable part of a continent bordering an orogenic or mobile belt.

foreland facies *See* shelf facies.

forelimb [ANAT] An appendage (as a wing, fin, or arm) of a vertebrate that is, or is homologous to, the foreleg of a quadruped.

forellenstein *See* troctolite.

Forel scale [OCEANOGR] A scale of yellows, greens, and blues for recording the color of sea water as seen against the white background of a Secchi disk.

forensic chemistry [CHEM] The application of chemistry to the study of materials or problems in cases where the findings may be presented as technical evidence in a court of law.

forensic engineering [ENG] The application of accepted engineering practices and principles for discussion, debate, argumentative, or legal purposes.

forensic medicine [MED] Application of medical evidence or medical opinion for purposes of civil or criminal law.

forensic physics [PHYS] The application of physics for discussion, debate, argumentative, or legal purposes.

forensic science [SCI TECH] The application of science for discussion, debate, argumentative or legal purposes.

forepeak [NAV ARCH] A tank in the extreme forward end of a ship that is usually used for fresh water, but can carry ballast or bunkers.

forepeak bulkhead [NAV ARCH] The partition wall nearest the stem; it forms the after-boundary of the forepeak tank.

forepoling [MIN ENG] A timbering method for a very weak roof in which a bench of timbers is set and boards or long wedges are placed above the header; as the next bench of timbers is placed at the inbye end of the wedges, other like wedges are driven in under the first wedges and over the second header. Also known as spiling.

forerunner [OCEANOGR] Low, long-period ocean swell which commonly precedes the main swell from a distant storm, especially a tropical cyclone.

foreset [MIN ENG] **1.** To place a prop under the coal-face end of a bar. **2.** Timber set used for roof support at the working face. Also known as force piece.

foreset bed [GEOL] One of a series of inclined symmetrically arranged layers of a cross-bedding unit formed by deposition of sediments that rolled down a steep frontal slope of a delta or dune.

foreshaft sinking [MIN ENG] The first 150 feet (46 meters) of shaft sinking from the surface; the plant and services for the main shaft are installed during this step.

foreshock [GEOPHYS] A tremor which precedes a larger earthquake or main shock.

foreshore [GEOL] The zone that lies between the ordinary high- and low-watermarks and is daily traversed by the rise and fall of the tide. Also known as beach face.

forest [ECOL] An ecosystem consisting of plants and animals and their environment, with trees as the dominant form of vegetation.

forestaff *See* cross-staff.

forestay [NAV ARCH] In sailing vessels, a stay extending from the foremast to the bow deck, bowsprit, or jibboom.

forest climate *See* humid climate.

forest conservation [ECOL] Those measures concerned with the protection and preservation of forest lands and resources.

forest ecology [ECOL] The science that deals with the relationship of forest trees to their environment, to one another, and to other plants and to animals in the forest.

forest fire [FOR] Uncontrolled combustion of forest fuels.

forest management [FOR] Measures concerned with the effective organization of a forest to ensure continued production of its goods and services.

forest mapping [FOR] The branch of forestry dealing with the preparation of maps showing the distribution and conformation of individual forest stands.

forest measurement [FOR] The branch of forestry concerned with the measurement of standing trees, cut roundwood, and lumber products.

forest product [FOR] Any material afforded by a forest for commercial use, such as tree products and forage.

forest resources [FOR] Forest land and the trees on it.

forestry [ECOL] The management of forest lands for wood, forages, water, wildlife, and recreation.

forest soil [FOR] The natural medium for growth of tree roots and associated forest vegetation.

forest stand [FOR] The basic unit of forest mapping; a group of trees that are more or less homogeneous with regard to species composition, density, size, and sometimes habitat.

forest-tundra [ECOL] A temperate and cold savanna which occurs at high altitudes and consists of scattered or clumped trees and a shrub layer of varying coverage.

forest wind [METEOROL] A light breeze which blows from forests toward open country on calm clear nights.

forfeiture [MIN ENG] Loss of a mining claim by operation of the law, without regard to the intention of the locator, whenever he fails to preserve his right by complying with the conditions imposed by law.

forge [MET] **1.** To form a metal, usually hot, into desirable shapes by employing compressive forces. **2.** A machine or place in which metal is formed hot, or where iron is produced from its ore.

forgeability [MET] Suitability of a material for forging.

forge delay time [MET] The time between the start of weld time and the time when forging pressure is reached by the electrode force.

forge welding [MET] A group of welding processes in which the parts to be joined, usually iron, are heated to about 1000°C and then hammered or pressed together. Also known as fire welding.

forging [MET] **1.** Using compressive force to shape metal by plastic deformation; dies may be used. **2.** A piece of work made by forging.

forging brass [MET] Brass composed of 60% copper, 38% zinc, and 2% lead, used for hot forgings, hardware, and plumbing supplies; it is extremely plastic when hot, is corrosion-resistant, and has excellent mechanical properties.

forging hammer [MET] A hammer used to pound metal into forgings.

forging press [MET] A press designed to operate dies in die forging.

forging range [MET] Optimum temperature range in which a metal can be forged.

forging rolls [MET] A machine used in making forgings by rolling the metal.

forging stock [MET] A section or a piece of metal used to make a forging.

forked lightning [GEOPHYS] A common form of lightning, in a cloud-to-ground discharge, which exhibits downward-directed branches from the main lightning channel.

forklift [MECH ENG] A machine, usually powered by hydraulic means, consisting of two or more prongs which can be raised and lowered and are inserted under heavy materials or objects for hoisting and moving them.

forklift truck *See* fork truck.

fork oscillator [ELECTR] An oscillator that uses a tuning fork as the frequency-determining element.

fork truck [MECH ENG] A vehicle equipped with a forklift. Also known as forklift truck.

FORKLIFT

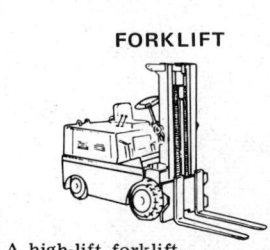

A high-lift forklift.

form [GRAPHICS] Type and material that is secured in a chase and is ready for printing or for producing an electrotype plate.

formability [MATER] Capability of a material to be shaped by plastic deformation.

formal charge [PHYS CHEM] The apparent charge of an element in a compound; for example, magnesium has a formal charge of $+2$ in MgO and oxygen has a charge of -2.

formaldehyde [ORG CHEM] HCHO The simplest aldehyde; a gas at room temperature, and a poisonous, clear, colorless liquid solution with pungent odor; used to make synthetic resins by reaction with phenols, urea, and melamine, as a chemical intermediate, as an embalming fluid, and as a disinfectant. Also known as formol; methylene oxide.

formalin [MATER] An aqueous solution of formaldehyde, usually 37% formaldehyde by weight.

formal logic [MATH] The study of the permissible relationships between propositions, a study that concerns the form rather than the content.

formamide [ORG CHEM] **1.** A compound containing the radical HCONH. **2.** $HCONH_2$ A clear, colorless hygroscopic liquid, boiling at 200–212°C; soluble in water and alcohol; used as a solvent, softener, and chemical intermediate. Also known as formylamine; methanamide.

formanite [MINERAL] A mineral composed of an oxide of uranium, zirconium, thorium, calcium, tantalum, and niobium with some rare-earth metals.

format [ADP] The specific arrangement of data on a printed page, punched card, or such to meet established presentation requirements.

formate [ORG CHEM] A compound containing the HCOO— radical.

formation [GEOL] Any assemblage of rocks which have some common character and are mappable as a unit.

formation bombing [ORD] Bombing by aircraft in formation.

formation factor [GEOL] A function of the porosity and internal geometry of a reservoir rock system, expressed as $F = \phi^{-m}$, where ϕ is the fractional porosity of the rock, and m is the cementation factor (pore-opening reduction).

formation fracturing [PETRO ENG] Method of applying hydraulic pressure to a reservoir formation to cause the rock to split open, that is to fracture; used to increase oil production.

formation resistivity [GEOPHYS] Electrical resistivity of reservoir formations measured by electrical log sondes; used for clues to formation lithography and fluid content.

formation solubility [PETRO ENG] Measure of formation rock solubility in oil-well acidizing solution (hydrochloric acid or hydrochloric-hydrofluoric acids).

formation tester [PETRO ENG] Device for retrieval of samples of fluid from an oil-reservoir formation.

formation water [HYD] Water present with petroleum or gas in reservoirs. Also known as interstitial water; oil-reservoir water.

formatted tape [ADP] A magnetic tape which employs a prerecorded timing track by means of which blocks of data can be found after reference to a directory table.

form clamp [CIV ENG] An adjustable metal clamp used to secure planks of wooden forms for concrete columns or beams.

form drag [FL MECH] **1.** The drag from all causes resulting from the particular shape of a body relative to its direction of motion, as of fuselage, wing, or nacelle. **2.** At supersonic speed, the drag caused by losses due to shock waves, exclusive of losses due to skin friction.

form factor [ELEC] **1.** The ratio of the effective value of a periodic function, such as an alternating current, to its average absolute value. **2.** A factor that takes the shape of a coil into account when computing its inductance. Also known as shape factor. [MECH] The theoretical stress concentration factor for a given shape, for a perfectly elastic material. [PHYS] A function which describes the internal structure of a particle, allowing calculations to be made even though the structure is unknown. [QUANT MECH] An expression used in studying the scattering of electrons or radiation from atoms, nuclei, or elementary particles, which gives the deviation from point particle scattering due to the distribution of charge and current in the target.

form feeding [ADP] The positioning of documents in order to move them past printing or sensing devices, either singly or in continuous rolls.

form function [ORD] The mathematical expression for the relationship between the fraction of the propellant burned and the distance that each burning surface has regressed.

form grinding [MECH ENG] Grinding by use of a wheel whose cutting face is contoured to the reverse shape of the desired form.

formic acid [ORG CHEM] HCOOH A colorless, pungent, toxic, corrosive liquid melting at 8.4°C; soluble in water, ether, and alcohol; used as a chemical intermediate and solvent, in dyeing and electroplating processes, and in fumigants. Also known as methanoic acid.

Formicariidae [VERT ZOO] The antbirds, a family of suboscine birds in the order Passeriformes.

formication [MED] An abnormal sensation as of insects crawling in or upon the skin; a common symptom in diseases of the spinal cord and the peripheral nerves; may be a hallucination.

formic ether *See* ethyl formate.

Formicidae [INV ZOO] The ants, social insects composing the single family of the hymenopteran superfamily Formicoidea.

formicivorous [ZOO] Feeding on ants.

Formicoidea [INV ZOO] A monofamilial superfamily of hymenopteran insects in the suborder Apocrita, containing the ants.

forming [ELEC] Application of voltage to an electrolytic capacitor, electrolytic rectifier, or semiconductor device to produce a desired permanent change in electrical characteristics as a part of the manufacturing process. [ENG] A process for shaping or molding sheets, rods, or other pieces of hot glass, ceramic ware, plastic, or metal by the application of pressure.

forming die [ENG] A die like a drawing die, but without a blank holder.

forming press [MECH ENG] A punch press for forming metal parts.

forming rolls [MECH ENG] Rolls contoured to give a desired shape to parts passing through them.

forming tool [DES ENG] A nonrotating tool that produces its inverse form on the workpiece.

form line [MAP] An approximation of a contour line without a definite elevation value, as one derived by visual observation, sometimes supplemented by measured elevations but not in sufficient quantity to produce accurate results; used principally to indicate the appearance of terrain which has not been accurately surveyed.

form of the physical store [ADP] The code store considered as a physical structure, which can exhibit many different forms: discrete (cards), continuous (tapes), linear (tapes), cylindrical (drums), three-dimensional (array of cores), disks, strips, sheets, reels, and so on.

form oil [MATER] An oil utilized on the contact surface of wooden or metal concrete forms to prevent concrete from sticking.

formol *See* formaldehyde.

formonitrile *See* hydrocyanic acid.

form process chart [IND ENG] A graphic representation of the process flow of paperwork forms. Also known as forms analysis chart; functional forms analysis chart; information process analysis chart.

forms analysis chart *See* form process chart.

forms control buffer [ADP] A reserved storage containing coordinates for a page position on the printer; earlier printers utilized a carriage control tape, allowing the page to be set at a specific position.

form stop [ADP] A device which stops a machine when its supply of paper has run out.

formula [CHEM] **1.** A combination of chemical symbols that expresses a molecule's composition. **2.** A reaction formula showing the interrelationship between reactants and products. [MATH] An equation or rule relating mathematical objects or quantities.

formulation [CHEM] The particular mixture of base chemicals and additives required for a product.

formula transition *See* FORTRAN.

formula weight [CHEM] **1.** The gram-molecular weight of a substance. **2.** In the case of a substance of uncertain molecular weight such as certain proteins, the molecular weight calculated from the composition, assuming that the element present in the smallest proportion is represented by only one atom.

Formvar [ORG CHEM] Trade name for certain thermoplastic resins derived from polyvinyl alcohol which are used in coatings, adhesives, and molding materials.

formwork [CIV ENG] A temporary wooden casing used to contain concrete during its placing and hardening. Also known as shuttering.

form-wound coil [ELEC] Armature coil that is formed or shaped over a fixture before being placed on the armature of a motor or generator.

formyl [ORG CHEM] The formic acid radical, $HCO-$; it is characteristic of aldehydes.

formylamine *See* formamide.

2-formyl-3,4-dihydro-2H-pyran *See* acrolein dimer.

fornix [ANAT] A structure that is folded or arched. [BOT] A small scale, especially in the corolla tube of some plants.

Forrel cell [METEOROL] A type of atmospheric circulation in which air moves away from the thermal equator at low latitude levels and in the opposite direction in higher latitudes.

Forrester machine [MIN ENG] A pneumatic flotation cell in which pulp is aerated by low-pressure air, delivering a mineralized froth along the overflow, and tailings to the end weir.

forril farina *See* rock milk.

fors *See* G; gram-force.

Forssman antibody [IMMUNOL] A heterophile antibody that reacts with Forssman antigen.

Forssman antigen [IMMUNOL] Any of a large group of heterophile antigens occurring in a wide variety of unrelated animals, including horses, dogs, and fowl, and in certain bacteria.

forsterite [MINERAL] Mg_2SiO_4 A whitish or yellowish, magnesium-rich variety of olivine. Also known as white olivine.

fort [ORD] **1.** Permanent post as opposed to a camp, which is a temporary installation. **2.** Land area within which harbor defense units are located.

fortification [ORD] **1.** A structure or earthworks, usually heavily armed, constructed as a defense; a fortified place or position. **2.** The act or art of fortifying.

fortification agate *See* landscape agate.

Fortin barometer [ENG] A type of cistern barometer; provision is made to increase or decrease the volume of the cistern so that when a pressure change occurs, the level of the cistern can be maintained at the zero of the barometer scale (the ivory point).

Fortisan [TEXT] Trade name for cellulose acetate manufactured by the Celanese Corporation.

fortnightly tide [OCEANOGR] A tide occurring at intervals of one-half the period of oscillation of the moon, approximately 2 weeks.

FORTRAN [ADP] A family of procedure-oriented languages used mostly for scientific or algebraic applications; derived from formula translation.

Fortrat parabola [SPECT] Graph of wave numbers of lines in a molecular spectral band versus the serial number of the successive lines.

Fortrel [TEXT] Trade name for a polyester fiber made by the Celanese Corporation.

fortuitous distortion [COMMUN] Distortion in a telegraph system which includes effects that cannot be classified as bias or characteristic distortion; it is a departure (for one occurrence of a particular signal pulse) from the average combined effects of bias and characteristic distortion; the direct opposite of systematic distortion.

forty-four-type repeater [ELECTR] Type of telephone repeater employing two amplifiers and no hybrid arrangements; used in a four-wire system.

Forty Saints' storm [METEOROL] A southerly gale in Greece, occurring a little before the equinox in March.

forward [NAV] In a direction nearer dead ahead than dead astern.

forward-acting regulator [ELECTR] Transmission regulator in which the adjustment made by the regulator does not affect the quantity which caused the adjustment.

forward-backward counter [ADP] A counter that has both an add and a subtract input so as to count in either an increasing or a decreasing direction. Also known as bidirectional counter.

forward bias [ELECTR] A bias voltage that is applied to a *pn*-junction in the direction that causes a large current flow; used in some semiconductor diode circuits.

forward coupler [ELECTR] Directional coupler used to sample incident power.

forward current [ELECTR] Current which flows upon application of forward voltage.

forward direction [ELECTR] Of a semiconductor diode, the direction of lower resistance to the flow of steady direct current.

forward drop [ELECTR] The voltage drop in the forward direction across a rectifier.

forward extrusion [MET] A cold extrusion process in which a formed blank is placed in a die cavity and struck by a punch; the metal is extruded through an annular space between the die and the end of the punch, moving in the same direction as the punch.

forward of the beam [NAV] Any direction between broad on the beam and ahead.

forward path [CONT SYS] The transmission path from the loop actuating signal to the loop output signal in a feedback control loop.

forward perpendicular [NAV ARCH] A vertical line through the intersection of the design waterline and the forward end of a ship. Abbreviated FP.

forward propagation by ionospheric scatter [COMMUN] Radio communications technique using the scattering phenomenon exhibited by electromagnetic waves in the 30–100-megahertz region when passing through the ionosphere at an elevation of about 85 kilometers.

forward propagation by tropospheric scatter [COMMUN] Radio communications technique using high transmitting power levels, large antenna arrays, and the scattering phenomenon of the troposphere to permit communications far beyond line-of-sight distances.

forward quarter [NAV ARCH] The portions of the sides of a ship immediately abaft the stem.

forward recovery time [ELECTR] Of a semiconductor diode, the time required for the forward current or voltage to reach a specified value after instantaneous application of a forward bias in a given circuit.

forward scatter [COMMUN] **1.** Propagation of electromagnetic waves at frequencies above the maximum usable high frequency through use of the scattering of a small portion of the transmitted energy when the signal passes from an unionized medium into a layer of the ionosphere. **2.** Collectively, the very-high-frequency forward propagation by ionospheric scatter and ultra-high-frequency forward propagation by tropospheric scatter communications techniques. [METEOROL] The scattering of radiant energy into the hemisphere of space bounded by a plane normal to the direction of the incident radiation and lying on the side toward which the incident radiation was advancing.

forward scattering [PHYS] **1.** Scattering in which there is no change in the direction of motion of the scattered particles. **2.** Scattering in which the angle between the initial and final directions of motion of the scattered particles is less than 90°.

forward-scatter propagation *See* scatter propagation.

forward transfer function [CONT SYS] In a feedback control loop, the transfer function of the forward path.

forward wave [ELECTR] Wave whose group velocity is the same direction as the electron stream motion.

forward welding *See* forehand welding.

FOSDIC II [ADP] An electronic scanner which reads filmed images of punched cards, searches for cards containing specified information, and copies the selected information onto new cards for computer input. Derived from film optical scanning device for input to computers.

foshagite [MINERAL] $Ca_5Si_3O_{10}(OH)_2 \cdot 2H_2O$ A white mineral composed of a basic hydrous calcium silicate.

fossa [ANAT] A pit or depression. [VERT ZOO] *Cryptoprocta ferox.* A Madagascan carnivore related to the civets.

fossil [PALEON] The organic remains, traces, or imprint of an organism preserved in the earth's crust since some time in the geologic past.

fossil dune [GEOL] An ancient desert dune.

fossil fuel [GEOL] Any hydrocarbon deposit that may be used for fuel; examples are petroleum, coal, and natural gas.

fossil ice [HYD] **1.** Relatively old ground ice found in regions of permafrost. **2.** Underground ice in regions where present-day temperatures are not low enough to have formed it.

fossil man [PALEON] Ancient man identified from prehistoric skeletal remains which are archeologically earlier than the Neolithic.

fossil permafrost *See* passive permafrost.

fossil reef [GEOL] An ancient reef.

fossil soil *See* paleosol.

fossil wax *See* ozocerite.

fossorial [VERT ZOO] Adapted for digging.

Foster-Seely discriminator *See* phase-shift discriminator.

Foster's formula [MIN ENG] The empirical formula $R = 3\sqrt{DT}$ for determining the radius R of a shaft pillar, where D = depth in feet and T = thickness of lode in feet.

Foster's reactance theorem [CONT SYS] The theorem that the most general driving point impedance or admittance of a network, in which every mesh contains independent inductance and capacitance, is a meromorphic function whose poles and zeros are all simple and occur in conjugate pairs on the imaginary axis, and in which these poles and zeros alternate.

Fotoceram [MATER] A high-strength crystalline material forming a fluidic element or circuit, manufactured in the Fotoform process.

Fotoform process [ENG] A method of constructing fluidic elements and modules in which fluidic circuits and interconnections are etched on separate glass sheets, which are then assembled in alternate layers, thermally bonded to form a monolithic structure, and further processed to form Fotoceram.

Foucault current *See* eddy current.

Foucault knife-edge test [OPTICS] Test of a lens or a concave mirror in which a pinhole source is placed at twice the focal length behind the lens or at the mirror's center of curvature, the eye is placed at the image of the pinhole, and defects in the lens or mirror result in irregular darkening of the image when a knife edge is moved across the image immediately in front of the eye.

Foucault mirror [OPTICS] Experiment for measuring the speed of light in which light is reflected from a rapidly rotating mirror to a distant mirror and back, and the speed of light is deduced from the displacement of the beam after its second reflection from the rotating mirror, the angular speed of the rotating mirror, and the distance the light travels.

Foucault pendulum [MECH] A swinging weight supported by a long wire, so that the wire's upper support restrains the wire only in the vertical direction, and the weight is set swinging with no lateral or circular motion; the plane of the pendulum gradually changes, demonstrating the rotation of the earth on its axis.

fougasse [ORD] A mine constructed so that upon explosion of the charge, pieces of metal, rock, gasoline, or other substances are blown in a predetermined direction.

foul berth [NAV] A berth in which a vessel at anchor is in danger of striking or fouling another vessel, the ground, or an obstruction.

foul bottom [CIV ENG] A hard, uneven, rocky or obstructed bottom having poor holding qualities for anchors, or one having rocks or wreckage that would endanger an anchored vessel. [NAV ARCH] Referring to the underwater portion of the hull when covered with foreign matter such as barnacles or grass.

foulbrood [INV ZOO] The common name for three destructive bacterial diseases of honeybee larvae.

foul foot [VET MED] A feedlot disease of cattle and sheep marked by inflammation and ulceration of the feet; common in wet feedlots. Also called foot rot.

Foulger's test [ANALY CHEM] A test for fructose in which urea, sulfuric acid, and stannous chloride are added to the solution to be tested, the solution is boiled, and in the presence of fructose a blue coloration forms.

fouling [NAV ARCH] The adhesion of different marine organ-

isms to the underwater parts of ships, causing the ships to lose speed. [ORD] The deposit that remains on the bore of a gun after firing.

fouling factor [CHEM ENG] In heat transfer, the lowering of clear-film transfer rates resulting from corrosion, dirt, or roughness of the surface of tube walls of heat exchangers.

fouling organism [ECOL] Any aquatic organism with a sessile adult stage that attaches to and fouls underwater structures of ships.

fouling point [CIV ENG] **1.** The point at a switch or turnout beyond which railroad cars must be placed so as not to interfere with cars on the main track. **2.** The location of insulated joints in a turnout on signaled tracks.

foundation [CIV ENG] **1.** The ground that supports a building or other structure. **2.** The portion of a structure which transmits the building load to the ground.

foundation mat *See* raft foundation.

foundry [ENG] A building where metal or glass castings are produced.

foundry alloy *See* master alloy.

foundry core sand *See* core sand.

foundry engineering [ENG] The science and practice of melting and casting glass or metal.

foundry facing [MET] A material applied to a sand mold to improve the surface quality of a casting.

foundry proofs [GRAPHICS] Proofs of type pages that have been locked up within frames prior to casting; identified by black bands around each page of type, caused by bearers that lend support to the type page when stereotypes or electrotypes are made.

foundry sand [MET] Sand used in foundries to make molds for the casting of metal shapes.

foundry type [GRAPHICS] Type cast as single characters. Also known as hand type.

fountain [GRAPHICS] **1.** In printing, a container or reservoir on a press that contains an ink supply. **2.** In offset lithography, a fountain solution (usually a water-alcohol mixture) that wets the nonprinting areas of the plate.

fountain effect [FL MECH] The effect occurring when two containers of superfluid helium are connected by a capillary tube and one of them is heated, so that helium flows through the tube in the direction of higher temperature.

four-address [ADP] Pertaining to an instruction address which contains four address parts.

four-ball tester [ENG] A machine designed to measure the efficiency of lubricants by driving one ball against three stationary balls clamped together in a cup filled with the lubricant; performance is evaluated by measuring wear-scar diameters on the stationary balls.

four-bar linkage [MECH ENG] A plane linkage consisting of four links pinned tail to head in a closed loop with lower, or closed, joints.

Fourcault process [ENG] A process for forming sheet glass in which the molten glass is drawn vertically upward.

four-channel stereophonic sound *See* quadraphonic sound.

fourchite [PETR] A monchiquite that lacks feldspar and olivine.

four-color printing [GRAPHICS] A method of reproducing full-color originals, such as paintings and color photographs, by overprinting a series of four plates in yellow, magenta, cyan, and black ink.

four-color separation process [GRAPHICS] Conversion of a color illustration into four negative films from which the four printing plates [yellow, magenta, cyan (blue) and black] that will be used in the printing process are made; the negative film for yellow is made by photographing the illustration through a blue filter, the magenta through a green filter, the cyan through a red filter, and the black through a yellow filter.

four-course radio range station [NAV] Radio navigation land station in the aeronautical radio navigation service providing radio equisignal zones.

four-current density [RELAT] A four-vector whose three space components are those of the ordinary current density and whose time component is the charge density.

four-degree calorie [CHEM] The heat needed to change the temperature of 1 gram of water from 3.5 to 4.5°C.

Fourdrinier machine [MECH ENG] A papermaking machine; a paper web is formed on an endless wire screen; the screen

FOTOCERAM

Monolithic Fotoceram circuit module, with the bottom view showing the cross section. *(Courtesy of the Corning Glass Works)*

FOUR-BAR LINKAGE

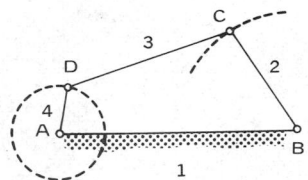

Four-bar linkage, a common form of bar linkage. Bars join and pivot at points A, B, C, and D.

passes through presses and over dryers to the calenders and reels.

four-factor formula [NUCLEO] The principle that the multiplication factor of a thermal reactor with no leakage is the product of the average number of fast neutrons emitted when a nucleus in the fuel material captures a thermal neutron, the fast fission factor, the fraction of neutrons which are not captured while being slowed down, and the number of thermal neutrons absorbed in the fuel divided by the total number of neutrons absorbed in the fuel and the moderator.

four-force [RELAT] A four-vector equal to the product of the rest mass of a particle and the rate of change of its four-momentum with respect to its proper time.

four-frequency diplex telegraphy [COMMUN] Frequency-shift telegraphy in which each of the four possible signal combinations corresponding to two telegraph channels is represented by a separate frequency.

four-hundred-day clock *See* anniversary clock.

fourier *See* thermal ohm.

Fourier analysis [MATH] The study of convergence of Fourier series and when and how a function is approximated by its Fourier series or transform.

Fourier-Bessel integrals [MATH] Given a function $F(v,\theta)$ independent of θ where v,θ are the polar coordinates in the plane, these integrals have the form

$$\int_0^\infty u\,du \int_0^\infty F(r) J_m(ur)r\,dr,$$

where J_m is a Bessel function order m.

Fourier-Bessel series [MATH] For a function $f(x)$, the series whose mth term is $a_m J_0(j_m x)$, where $j_1, j_2, \ldots$ are positive zeros of the Bessel function J_0 arranged in ascending order, and a_m is the product of $2/J_1{}^2(j_m)$ and the integral over t from 0 to 1 of $tf(t)J_0(j_m t)$; J_1 is a Bessel function.

Fourier-Bessel transform *See* Hankel transform.

Fourier expansion *See* Fourier series.

Fourier heat equation *See* Fourier law of heat conduction.

Fourier integrals [MATH] For a function $f(x)$ the Fourier integrals are

$$\frac{1}{\pi}\int_0^\infty du \int_{-\infty}^\infty f(t) \cos u(x-t)\,dt,$$

$$\frac{1}{\pi}\int_0^\infty du \int_{-\infty}^\infty f(t) \sin u(x-t)\,dt.$$

Fourier law of heat conduction [THERMO] The law that the rate of heat flow through a substance is proportional to the area normal to the direction of flow and to the negative of the rate of change of temperature with distance along the direction of flow. Also known as Fourier heat equation.

Fourier-Legendre series [MATH] Given a function $f(x)$, the series from $n = 0$ to infinity of $a_n P_n(x)$, where $P_n(x)$, $n = 0,1,2, \ldots$ are the Legendre polynomials, and a_n is the product of $(2n+1)/2$ and the integral over x from -1 to 1 of $f(x)P_n(x)$.

Fourier number [FL MECH] A dimensionless number used in unsteady-state flow problems, equal to the product of the dynamic viscosity and a characteristic time divided by the product of the fluid density and the square of a characteristic length. Symbolized Fo_f. [PHYS] A dimensionless number used in the study of unsteady-state mass transfer, equal to the product of the diffusion coefficient and a characteristic time divided by the square of a characteristic length. Symbolized N_{Fo_m}. [THERMO] A dimensionless number used in the study of unsteady-state heat transfer, equal to the product of the thermal conductivity and a characteristic time, divided by the product of the density, the specific heat at constant pressure, and the distance from the midpoint of the body through which heat is passing to the surface. Symbolized N_{Fo_h}.

Fourier series [MATH] The Fourier series of a function $f(x)$ is

$$\frac{1}{2}a_0 + \sum_{n=1}^{\infty}(a_n \cos nx + b_n \sin nx)\text{ with}$$

$$a_n = \frac{1}{\pi}\int_{-\pi}^{\pi} f(x) \cos nx\,dx,$$

$$b_n = \frac{1}{\pi}\int_{-\pi}^{\pi} f(x) \sin nx\,dx.$$

Also known as Fourier expansion.

Fourier space [MATH] The space in which the Fourier transform of a function is defined.

Fourier spectrum [PHYS] A plot of the magnitude and phase of the Fourier transform of a function.

Fourier's theorem [MATH] If $f(x)$ satisfies the Dirichlet conditions on the interval $-\pi < x < \pi$, then its Fourier series converges to $f(x)$ for all values of x in this interval at which $f(x)$ is continuous, and approaches $\frac{1}{2}[f(x+0) + f(x-0)]$ at points at which $f(x)$ is discontinuous, where $f(x-0)$ is the limit on the left of f at x and $f(x+0)$ is the limit on the right of f at x.

Fourier-Stieltjes series [MATH] For a function $f(x)$ of bounded variation on the interval $[0,2\pi]$, the series from $n = 0$ to infinity of $c_n \exp(inx)$, where c_n is $1/2\pi$ times the integral from $x = 0$ to $x = 2\pi$ of $\exp(-inx)df(x)$.

Fourier-Stieltjes transform [MATH] For a function $f(y)$ of bounded variation on the interval $(-\infty, \infty)$, the function $F(x)$ equal to $1/\sqrt{2\pi}$ times the integral from $y = -\infty$ to $y = \infty$ of $\exp(-ixy)df(y)$.

Fourier synthesis [MATH] The determination of a periodic function from its Fourier components.

Fourier transform [MATH] For a function $f(t)$, the function $F(x)$ equal to $1/\sqrt{2\pi}$ times the integral over t from $-\infty$ to ∞ of $f(t) \exp(itx)$.

four-layer diode [ELECTR] A semiconductor diode having three junctions, terminal connections being made to the two outer layers that form the junctions; a Shockley diode is an example.

four-layer transistor [ELECTR] A junction transistor having four conductivity regions but only three terminals; a thyristor is an example.

four-level laser [PHYS] A laser in which the lowest level for a laser transition is an excited state rather than the ground level.

fourmarierite [MINERAL] An orange-red to brown mineral composed of a hydrous oxide of lead and uranium.

four-piece set [MIN ENG] Squared timber frame used in underground driving to give all-around support to weak ground.

four-plus-one address [ADP] An instruction that contains four operand addresses and a control address.

four-point bearing [NAV] A relative bearing of 045° or 315°.

four-quadrant multiplier [ADP] The multiplier of an analog computer in which operation is unrestricted as to the sign of the input variables.

fourré *See* temperate and cold scrub; tropical scrub.

four-stroke cycle [MECH ENG] An internal combustion engine cycle completed in four piston strokes; includes a suction stroke, compression stroke, expansion stroke, and exhaust stroke.

four-tape [ADP] To sort input data, supplied on two tapes, into incomplete sequences alternately on two output tapes; the output tapes are used for input on the succeeding pass, resulting in longer and longer sequences after each pass, until the data are all in one sequence on one output tape.

fourth dimension [RELAT] Time in the theory of relativity, in which space and time are conceived as particular aspects of a four-dimensional world.

four-track tape [ENG ACOUS] Magnetic tape on which two tracks are recorded for each direction of travel, to provide stereo sound reproduction or to double the amount of source material that can be recorded on a given length of 1/4-inch tape.

four-vector [RELAT] A set of four quantities which transform under a Lorentz transformation in the same way as the three space coordinates and the time coordinate of an event. Also known as Lorentz four-vector.

four-vector potential [ELECTROMAG] A four-vector whose space components are the magnetic vector potential and whose time component is the electric scalar potential.

four-velocity [RELAT] A four-vector whose components are the rates of change of the space and time coordinates of a particle with respect to the particle's proper time.

four-way reinforcing [CIV ENG] A system of reinforcing rods in concrete slab construction in which the rods are placed parallel to two adjacent edges and to both diagonals of a rectangular slab.

four-way switch [ELEC] An electric switch employed in

house wiring, that makes it possible to turn a light on or off at three or more places.

four-way valve [MECH ENG] A valve at the junction of four waterways which allows passage between any two adjacent waterways by means of a movable element operated by a quarter turn.

four-wheel drive [MECH ENG] An arrangement in which the drive shaft acts on all four wheels of the automobile.

four-wire circuit [COMMUN] A two-way circuit using two paths so arranged that communication currents are transmitted in one direction only on one path, and in the opposite direction on the other path; the transmission path may or may not employ four wires.

four-wire line [ELECTROMAG] A transmission line in which four conductors lie at the corners of a rectangle, and each conductor is in phase with the conductor at the opposite corner and out of phase with the conductors at adjacent corners.

four-wire repeater [ELECTR] Telephone repeater for use in a four-wire circuit and in which there are two amplifiers, one serving to amplify the telephone currents in one side of the four-wire circuit, and the other serving to amplify the telephone currents in the other side of the four-wire circuit.

four-wire subscriber line [COMMUN] Four-wire circuit connecting a subscriber directly to a switching center.

four-wire terminating set [ELECTR] Hybrid arrangement by which four-wire circuits are terminated on a two-wire basis for interconnection with two-wire circuits.

fovea [BIOL] A small depression or pit.

fovea centralis [ANAT] A small, rodless depression of the retina in line with the visual axis, which affords acute vision.

foveal vision [PHYSIO] Vision achieved by looking directly at objects in the daylight so that the image falls on or near the fovea centralis. Also known as photopic vision.

foveola [BIOL] A small pit, especially one in the embryonic gastric mucosa from which gastric glands develop.

foveolate [BIOL] Having small depressions; pitted.

fowan [METEOROL] A dry, scorching wind of Great Britain and the Isle of Man.

fowl [AGR] A domestic cock or hen, especially an adult hen, as chicken or several gallinaceous birds.

Fowler-DuBridge theory [SOLID STATE] Theory of photoelectric emission from a metal based on the Sommerfeld model, which takes into account the thermal agitation of electrons in the metal and predicts the photoelectric yield and the energy spectrum of photoelectrons as functions of temperature and the frequency of incident radiation.

Fowler function [SOLID STATE] A mathematical function used in the Fowler-DuBridge theory to calculate the photoelectric yield.

fowlerite [MINERAL] A zinc-bearing variety of rhodonite.

fowl pox [VET MED] A disease of birds caused by a virus and characterized by wartlike nodules on the skin, particularly on the head.

fox [VERT ZOO] The common name for certain members of the dog family (Canidae) having relatively short legs, long bodies, large erect ears, pointed snouts, and long bushy tails.

Fox broadcast [COMMUN] Radio broadcast of messages for which receiving stations make no acknowledgment.

foxhole [ORD] A small pit used for cover, usually for one or two soldiers, and so constructed that an occupant can fire effectively from it.

fox lathe [MECH ENG] A lathe with chasing bar and leaders for cutting threads; used for turning brass.

foyaite [PETR] A nepheline syenite composed chiefly of potassium feldspar.

fp *See* freezing point.

FP *See* forward perpendicular.

fpm *See* foot per minute.

F process [MATH] A stochastic process $\{X_t, t > 0\}$ whose value at time t is determined by the information up to time t; more precisely, the events $\{X_t \leq a\}$ belong to F_t for every t and a, where $F = \{F_t, t \geq 0\}$ is an increasing family of sigma algebras that represents the amount of information increasing with time.

fps *See* foot per second.

Fr *See* francium; statcoulomb.

fraction [CHEM] One of the portions of a volatile liquid within certain boiling point ranges, such as petroleum naphtha fractions or gas-oil fractions. [MATH] An expression which is the product of a real number or complex number with the multiplicative inverse of a real or complex number. [MET] In powder metallurgy, that portion of sample that lies between two stated particle sizes. Also known as cut. [SCI TECH] A portion of a mixture which represents a discrete unit and can be isolated from the whole system.

fractional condensation [CHEM] Separation of components of vaporized liquid mixtures by condensing the vapors in stages (partial condensation); highest-boiling-point components condense in the first condenser stage, allowing the remainder of the vapor to pass on to subsequent condenser stages.

fractional crystallization [PETR] Separation of a cooling magma into multiple minerals as the different minerals cool and congeal at progressively lower temperatures. Also known as crystallization differentiation; fractionation.

fractional distillation [CHEM] A method to separate a mixture of several volatile components of different boiling points; the mixture is distilled at the lowest boiling point, and the distillate is collected as one fraction until the temperature of the vapor rises, showing that the next higher boiling component of the mixture is beginning to distill; this component is then collected as a separate fraction.

fractional equation [MATH] 1. Any equation that contains fractions. 2. An equation in which the unknown variable appears in the denominator of one or more terms.

fractional gas-flow curve [PETRO ENG] Graph of the fraction of free injected gas flowing through a reservoir formation versus the liquid saturation of the gas for various parameter values of oil viscosity; used to calculate displacement efficiency during gas injection.

fractional horsepower motor [ELEC] Any motor built into a frame smaller than that for a motor having an open construction and a continuous rating of 1 horsepower (745.7 watts) at 1800 revolutions per minute.

fractional precipitation [ANALY CHEM] Method for separating elements or compounds with similar solubilities by a series of analytical precipitations, each one improving the purity of the desired element.

fractional sampling [MIN ENG] Mechanical selection of samples of uniformly graded material without segregation.

fractional sine wave [PHYS] A pulse train whose waveform is a truncated sine wave.

fractionating column [CHEM] An apparatus used widely for separation of fluid (gaseous or liquid) components by vapor-liquid fractionation or liquid-liquid extraction or liquid-solid adsorption.

fractionation [CHEM] Separation of a mixture in successive stages, each stage removing from the mixture some proportion of one of the substances, as by differential solubility in water-solvent mixtures. [NUCLEO] Alterations in the isotopic composition of substances found in nature or in radioactive weapon debris, which result from small differences in the physical and chemical properties of isotopes of an element. [PETR] *See* fractional crystallization.

fraction defective [IND ENG] The number of units per 100 pieces which are defective in a lot; expressed as a decimal.

fractography [MET] The microscopic examination of fractured metal surfaces.

fracture [GEOL] A crack, joint, or fault in a rock due to mechanical failure by stress. Also known as rupture. [MED] The breaking of bone, cartilage, or teeth. [MINERAL] A break in a mineral other than along a cleavage plane. [SCI TECH] 1. The act, process, or state of being broken. 2. The surface appearance of a freshly broken material. 3. The break produced by fracturing.

fracture cleavage [GEOL] Cleavage that occurs in deformed but only slightly metamorphosed rocks along closely spaced, parallel joints and fractures.

fractured formation [PETRO ENG] Reservoir formation in which rock has been split by hydraulic pressure produced by injected fluids.

fracture dome [MIN ENG] The zone of loose or semiloose rock which exists in the immediate hanging or footwall of a stope.

FOX

The gray fox (*Urocynon cineroargenteus*), similar to the red fox but flecked with gray, found from the Great Lakes to South America.

FRANCIS TURBINE

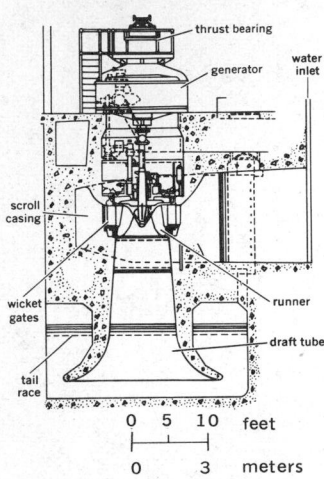

Cross section of a Francis turbine installation.

FRANCIUM

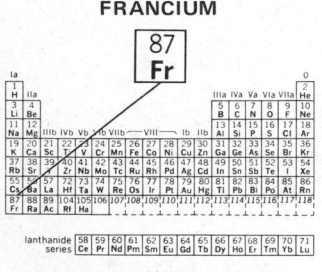

Periodic table of the chemical elements showing the position of francium.

fracture plane inclination [GEOL] Gradient or inclination of the plane of fracture formed in a reservoir formation.

fracture strength *See* fracture stress.

fracture stress [MECH] The minimum tensile stress that will cause fracture. Also known as fracture strength.

fracture system [GEOL] A stress-related group of contemporaneous fractures.

fracture test [ENG] 1. Macro- or microscopic examination of a fractured surface to determine characteristics such as grain pattern, composition, or the presence of defects. 2. A test designed to evaluate fracture stress.

fracture wear [MECH] The wear on individual abrasive grains on the surface of a grinding wheel caused by fracture.

fracture zone [GEOL] An elongate zone on the deep-sea floor that is of irregular topography and often separates regions of different depths; frequently crosses and displaces the midoceanic ridge by faulting.

fractus [METEOROL] A cloud species in which the cloud elements are irregular but generally small in size, and which presents a ragged, shredded appearance, as if torn; these characteristics change ceaselessly and often rapidly.

fragility [SCI TECH] The state or quality of being fragile, that is, brittle or easily broken.

fragility test [PATH] A measure of the resistance of red blood cells to osmotic hemolysis in hypotonic salt solutions of graded dilutions.

fragment [ORD] 1. A piece of an exploding or exploded bomb, projectile, or the like. 2. To break into fragments.

fragmentation [CYTOL] Amitotic division; a type of asexual reproduction. [MIN ENG] The blasting of coal, ore, or rock into pieces small enough to load, handle, and transport without the need for hand-breaking or secondary blasting. [PSYCH] Disordered behavior and mental processes.

fragmentation ammunition [ORD] Ammunition that is primarily intended to produce a fragmentation effect.

fragmentation bomb [ORD] An item designed to be dropped from aircraft to produce many small, high-velocity fragments when detonated.

fragmentation bomb cluster [ORD] Multiple fragmentation bombs suspended and dropped from a single station of a bomb rack on an airplane.

fragmentation grenade [ORD] A hand grenade designed to fragment, an effective weapon against personnel; since the thrower needs protective cover, it is used primarily for defensive operations, and is often called a defensive grenade.

fragmentation nucleus [METEOROL] A tiny ice particle broken from a large ice crystal, serving as an ice nucleus; that is, a growth center for a new ice crystal.

fragmentation protective body armor [ORD] Armor designed to provide fragmentation protection to vital areas of the body; usually provided in the form of garments which contain steel, nylon, or other resistant materials.

fragmentation test [ORD] A test conducted to determine the number and weight distribution, and (where the method used permits) the velocity and spatial distribution of fragments produced by a projectile or other munition upon detonation.

fragment emission [ORD] In terminal ballistics, the pattern of the fragments upon leaving the exploded projectile or other munition, including the number of fragments and the direction, weight, and velocity of each fragment.

fragmenting [ADP] The breaking up of a document into its various components.

fragment simulator projectile [ORD] Projectile which simulates the action of a fragment; used in ballistic tests at the proving ground.

Frahm frequency meter *See* vibrating-reed frequency meter.

frambesia *See* yaws.

framboid [GEOL] A microscopic aggregate of pyrite grains, often occurring in spheroidal clusters.

frame [ADP] 1. A row of recording or punch positions extending across a magnetic or paper tape in a direction at right angles to its motion. 2. *See* central processing unit. [BUILD] The skeleton structure of a building. Also known as framing. [ELECTR] 1. One complete coverage of a television picture. 2. A rectangular area representing the size of copy handled by a facsimile system. [GRAPHICS] A single complete picture on motion picture film.

frame frequency [ELECTR] The number of times per second that the frame is completely scanned in television. Also known as picture frequency.

frame of reference [PHYS] A coordinate system for the purpose of assigning positions and times to events. Also known as reference frame.

frame period [ELECTR] A time interval equal to the reciprocal of the frame frequency.

framer [ELECTR] Device for adjusting facsimile equipment so the start and end of a recorded line are the same as on the corresponding line of the subject copy.

frame set [MIN ENG] The arrangement of the legs and cap or crossbar so as to provide support for the roof of an underground passage. Also known as framing; set.

framework [ENG] The load-carrying frame of a structure; may be of timber, steel, or concrete.

framework silicate *See* tectosilicate.

framing [BUILD] *See* frame. [ELECTR] 1. Adjusting a television picture to a desired position on the screen of the picture tube. 2. Adjusting a facsimile picture to a desired position in the direction of line progression. Also known as phasing. [MIN ENG] *See* frame set.

framing camera [OPTICS] A motion picture camera that automatically controls the position of successive still photographs on the film so that, when the film is subsequently projected, the image will appear steady on the screen.

framing control [ELECTR] 1. A control that adjusts the centering, width, or height of the image on a television receiver screen. 2. A control that shifts a received facsimile picture horizontally.

framing plan [NAV ARCH] A diagram showing positions and type of construction of the framing members of a ship.

framing square [DES ENG] A graduated carpenter's square used for cutting off and making notches.

framing table [MIN ENG] An inclined table on which ore slimes are separated by running water.

Francis formula [FL MECH] An equation for the calculation of water flow rate over a rectangular weir in terms of length and head.

Francis turbine [MECH ENG] A reaction hydraulic turbine of relatively medium speed with radial flow of water in the runner.

francium [CHEM] A radioactive alkali-metal element, symbol Fr, atomic number 87, atomic weight distinguished by nuclear instability; exists in short-lived radioactive forms, the chief isotope being francium-223.

Franck-Condon principle [PHYS CHEM] The principle that in any molecular system the transition from one energy state to another is so rapid that the nuclei of the atoms involved can be considered to be stationary during the transition.

franckeite [MINERAL] A dark-gray or black massive mineral composed of lead antimony tin sulfide.

Franck-Hertz experiment [ELECTR] Experiment for measuring the kinetic energy lost by electrons in inelastic collisions with atoms; it established the existence of discrete energy levels in atoms, and can be used to determine excitation and ionization potentials.

francolite [MINERAL] $Ca_5(PO_4,CO_3)_3(F,OH)$ Colorless fluoride-bearing carbonate-apatite.

Franconian [GEOL] A North American stage of geologic time; the middle Upper Cambrian.

frangible [MECH] Breakable, fragile, or brittle.

frangible bullet [ORD] A brittle plastic or other nonmetallic bullet for firing practice which, upon striking a target, breaks into powder or small fragments without penetrating.

frangible grenade [ORD] Improvised incendiary hand grenade consisting of a glass container filled with a flammable liquid, with an igniter attached, and which breaks and ignites upon striking a resistant target, such as a tank.

frangula emodin *See* emodin.

frangulic acid *See* emodin.

Frankfurt horizontal *See* eye-ear plane.

frankincense *See* olibanum.

frankincense oil *See* olibanum oil.

Frankland's method [ORG CHEM] Reaction of dialkyl zinc compounds with alkyl halides to form hydrocarbons; may be used to form paraffins containing a quaternary carbon atom.

franklin   *See* statcoulomb.

Franklin centimeter   [ELEC] A unit of electric dipole moment, equal to the dipole moment of a charge distribution consisting of positive and negative charges of 1 statcoulomb separated by a distance of 1 centimeter.

Franklin equation   [ENG ACOUS] An equation for intensity of sound in a room as a function of time after shutting off the source, involving the volume and exposed surface area of the room, the speed of sound, and the mean sound-absorption coefficient.

franklinite   [MINERAL] $ZnFe_2O_4$ Black, slightly magnetic mineral member of the spinel group; usually possesses extensive substitution of divalent manganese and iron for the divalent zinc, and limited trivalent manganese for the trivalent iron.

Frank partial dislocation   [CRYSTAL] A partial dislocation whose Burger's vector is not parallel to the fault plane, so that it can only diffuse and not glide, in contrast to a Schockley partial dislocation.

Frank-Read source   [MET] **1.** The creation of dislocations by application of shear stress to an edge dislocation anchored terminally, causing formation of an unstable loop form followed by formation of a closed dislocation line and the establishment of the original condition. **2.** One of the sources of dislocations in a plastically deforming metal.

Franz-Keldysh effect   [OPTICS] A shift to longer wavelength in the spectrum transmitted by a semiconductor when a strong electric field is applied.

Frary metal   [MET] Metal containing 97-98% lead alloyed with 1-2% barium and calcium; used for bearings.

Frasch process   [MIN ENG] A process to remove sulfur from sulfur beds; superheated water is forced under pressure into the sulfur bed, and the molten sulfur is thus forced to the surface.

Fraser's air-sand process   [MIN ENG] A process in which dry, specific-gravity separation of coal from refuse is achieved by utilizing a flowing dense medium intermediate in density between coal and refuse.

fraternal twins   *See* dizygotic twins.

Fraude's reagent   *See* perchloric acid.

fraunhofer   [SPECT] A unit for measurement of the reduced width of a spectrum line such that a spectrum line's reduced width in fraunhofers equals 10^6 times its equivalent width divided by its wavelength.

Fraunhofer corona   *See* F corona.

Fraunhofer diffraction   [OPTICS] Diffraction of a beam of parallel light observed at an effectively infinite distance from the diffracting object, usually with the aid of lenses which collimate the light before diffraction and focus it at the point of observation.

Fraunhofer lines   [SPECT] The dark lines constituting the Fraunhofer spectrum.

Fraunhofer region   [ELECTROMAG] The region far from an antenna compared to the dimensions of the antenna and the wavelength of the radiation. Also known as far field; far region; far zone; radiation zone.

Fraunhofer spectrum   [SPECT] The absorption lines in sunlight, due to the cooler outer layers of the sun's atmosphere.

Frazer-Brace extraction method   [CHEM ENG] A method used to extract oil from citrus fruit; utilizes a machine which has abrasive carborundum rolls to rasp the peel from the fruit under a water spray; the water-and-peel mixture is screened and settled to allow oil separation.

frazil   [HYD] Ice crystals which form in supercooled water that is too turbulent to permit coagulation of the crystals into sheet ice.

frazil ice   [HYD] A spongy or slushy accumulation of frazil in a body of water. Also known as needle ice.

freckle   [MED] A pigmented macule resulting from focal increase in melanin, usually associated with exposure to sunlight, commonly on the face.

Fredholm determinant   [MATH] A power series obtained from the function $K(x,y)$ of the Fredholm equation which provides solutions to the equation under certain conditions.

Fredhom integral equations   [MATH] Given functions $f(x)$

and $K(x,y)$, the Fredholm integral equations with unknown function y are

$$\text{type 1: } f(x) = \int_a^b K(x, t)\, y\,(t)\, dt, \text{ and}$$

$$\text{type 2: } y(x) = f(x) + \lambda \int_a^b K(x, t)\, y\,(t)\, dt.$$

Fredholm operator   [MATH] A linear operator between Banach spaces which has closed range, and both the Fredholm operator and its adjoint have finite dimensional null space.

Fredholm theorem   [MATH] A Fredholm equation of type 2 with continuous $f(x)$ has a unique continuous solution, or else the corresponding equation of type 1 has a positive number of linearly independent solutions.

Fredholm theory   [MATH] The study of the solutions of the Fredholm equations.

free air   *See* free atmosphere.

free-air anomaly   [GEOPHYS] A gravity anomaly calculated as the difference between the measured gravity and the theoretical gravity at sea level and a free-air coefficient determined by the elevation of the measuring station.

free-air temperature   [METEOROL] Temperature of the atmosphere, obtained by a thermometer located so as to avoid as completely as practicable the effects of extraneous heating.

free association   [PSYCH] **1.** Spontaneous, consciously unrestricted association of ideas or mental images. **2.** A method used in psychoanalysis to gain an understanding of the organization of the content of the mind.

free atmosphere   [GEOPHYS] That portion of the earth's atmosphere, above the planetary boundary layer, in which the effect of the earth's surface friction on the air motion is negligible and in which the air is usually treated (dynamically) as an ideal fluid. Also known as free air.

free atom   [ATOM PHYS] An atom, as in a gas, whose properties, such as spectrum and magnetic moment, are not significantly affected by other atoms, ions, or molecules nearby.

free balloon   [AERO ENG] A balloon that ascends without a tether, propulsion or guidance; it is made to descend by the release of gas.

freeboard   [CIV ENG] The height between normal water level and the crest of a dam or the top of a flume.   [NAV ARCH] The vertical distance from the intersection of the top of the freeboard deck amidships with the outer surface of the side plating to the upper edge of the summer load line, or in general terms, the distance from the waterline to the deck.

freeboard deck   [NAV ARCH] The lowest exposed deck of a ship, below which all bulkheads must be made watertight.

free-burning coal   *See* noncaking coal.

free carbon   [MET] Elemental carbon present in a metal in an uncombined state.

free charge   [ELEC] Electric charge which is not bound to a definite site in a solid, in contrast to the polarization charge.

free convection   *See* natural convection.

free convection number   *See* Grashof number.

free crushing   [MIN ENG] Crushing under conditions of speed and feed so that there is ample room for the fine ore to fall away from the coarser material and thereby escape further crushing.

free cyanide   [CHEM] Cyanide not combined as part of an ionic complex.

free diving   [ENG] Diving with the use of scuba equipment to allow freedom and maneuverability.

freedom to mine   [MIN ENG] The law by which anybody has the right to mine certain minerals when he has prospected for them and has filed a proper application for the right to mine them.

free-drop   [ENG] To air-drop supplies or equipment without parachute.

free electromagnetic field   [ELECTROMAG] An electromagnetic field in empty space that does not interact with matter.

free electron   [PHYS] An electron that is not constrained to remain in a particular atom, and is therefore able to move in matter or in a vacuum when acted on by external electric or magnetic fields.

free-electron theory of metals   [SOLID STATE] A model of a metal in which the free electrons, that is, those giving rise to

FRANKLINITE

Franklinite crystals in calcite, Franklin, New Jersey. *(Specimen from Department of Geology, Bryn Mawr College)*

⊢ 2.5 cm ⊣

FRAUNHOFER DIFFRACTION

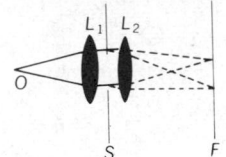

Diagram of Fraunhofer diffraction, with circular aperture. Light source O lies at the principal focus of lens L_1, which renders light parallel as it falls on aperture S. The second lens L_2 focuses parallel diffracted beams on observing screen F.

FRAUNHOFER SPECTRUM

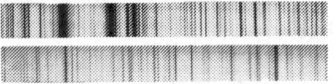

Two sections of the Fraunhofer spectrum, showing bright continuum and dark absorption lines. The wavelength range covered by each strip is approximately 85 angstroms. *(Sacramento Peak Observatory)*

the conductivity, are regarded as moving in a potential (due to the metal ions in the lattice and to all the remaining free electrons) which is approximated as constant everywhere inside the metal. Also known as Sommerfeld model; Sommerfeld theory.

free end   *See* free face.

free energy   [THERMO] **1.** The internal energy of a system minus the product of its temperature and its entropy. Also known as Helmholtz free energy; Helmholtz function; Helmholtz potential; thermodynamic potential at constant volume; work function. **2.** *See* Gibbs free energy.

free enthalpy   *See* Gibbs free energy.

free face   [MIN ENG] The exposed surface of a mass of rock or of coal. Also known as free end.

free fall   [MECH] The ideal falling motion of a body acted upon only by the pull of the earth's gravitational field.

free falling   [MECH ENG] In ball milling, the peripheral speed at which part of the crop load breaks clear on the ascending side and falls clear to the toe of the charge.

free-fed   [MIN ENG] In comminution, pertaining to rolls fed only enough ore to maintain a ribbon of material between them.

free ferrite   [MET] Relatively pure metallic iron phase present in steel or cast iron.

free field   [ACOUS] An isotropic, homogeneous sound field that is free from all bounding surfaces.   [ADP] A property of information retrieval devices which permits recording of information in the search medium without regard to preassigned fixed fields.   [PHYS] A field in empty space not interacting with other fields or sources.

free-field room   *See* anechoic chamber.

free-field storage   [ADP] Data storage that allows recording of the data without regard for fixed or preassigned fields.

free flight   [MECH] Unconstrained or unassisted flight.

free-flight angle   [MECH] The angle between the horizontal and a line in the direction of motion of a flying body, especially a rocket, at the beginning of free flight.

free-flight trajectory   [MECH] The path of a body in free fall.

free foehn   *See* high foehn.

free gas   [PETRO ENG] A hydrocarbon that exists in the gaseous phase at reservoir pressure and temperature and remains a gas when produced under normal conditions. [PHYS] Any gas at any pressure not in solution, or mechanically held in the liquid hydrocarbon phase.

free-gas saturation   [PETRO ENG] Proportion of oil-reservoir pore structure saturated by free (undissolved) gas.

free gold   [MET] Gold that is in the free state, that is, not combined with other substances.

free grid   *See* floating grid.

free group   [MATH] A group whose generators satisfy the equation $x \circ y = e$ (e the identity element in the group) only when $x = y^{-1}$ or $y = x^{-1}$.

free gyroscope   [ENG] A gyroscope that uses the property of gyroscopic rigidity to sense changes in altitude of a machine, such as an airplane; the spinning wheel or rotor is isolated from the airplane by gimbals; when the plane changes from level flight, the gyro remains vertical and gives the pilot an artificial horizon reference.

free hole   [SOLID STATE] Any hole which is not bound to an impurity or to an exciton.

free impedance   [ELECTR] Impedance at the input of the transducer when the impedance of its load is made zero. Also known as normal impedance.

freeing port   [NAV ARCH] An opening in a ship's side, at the level of the deck or in the bulwark, to allow water to escape.

free ion   [PHYS CHEM] An ion, such as found in an ionized gas, whose properties, such as spectrum and magnetic moment, are not significantly affected by other atoms, ions, or molecules nearby.

free-machining steel   [MET] Steel to which impurities have been added to improve machinability.

Freeman-Nichols roaster   [MIN ENG] A unit in which pyrite flotation concentrates are flash-roasted.

freemartin   [VERT ZOO] An intersexual, usually sterile female calf twinborn with a male.

free meander   [HYD] A stream meander that displaces itself very easily by lateral corrasion.

free milling   [MIN ENG] A process applied to ores which

contain free gold or silver and can be reduced by crushing and amalgamation (by gravity or on blankets), without roasting or other chemical treatment.

free-milling gold   [MET] Gold that has a clean surface so that it readily amalgamates with mercury (by gravity or on blankets) after liberation by comminution.

free-milling ore   [MIN ENG] Ore containing gold which can be caught with mercury by a variety of gravity processes or on blankets.

free module   [MATH] A module which is a free group with respect to its additive group.

free molecule   [PHYS CHEM] A molecule, as in a gas, whose properties, such as spectrum and magnetic moment, are not affected by other atoms, ions, and molecules nearby.

free molecule flow   [PHYS] Flow of a gas in which the mean free path of the molecules is long compared to a characteristic dimension of the flow field, such as the diameter of a tube through which gas is flowing. Also known as Knudsen flow.

free motional impedance   [ELECTR] Of a transducer, the complex remainder after the blocked impedance has been subtracted from the free impedance.

free oscillation   [PHYS] The oscillation of a physical system with no externally applied stimuli. Also known as free vibration.

free-pendulum clock   *See* Shortt clock.

free-piston engine   [MECH ENG] A prime mover utilizing free-piston motion controlled by gas pressure in the cylinders.

free port   [CIV ENG] An isolated, enclosed, and policed port in or adjacent to a port of entry, without a resident population.

free progressive wave   [PHYS] A wave in a medium or in vacuum, free from boundary effects. Also known as free wave.

free radical   [CHEM] An atom or a diatomic or polyatomic molecule which possesses at least one unpaired electron.

free recoil   [ORD] The movement in recoil of the recoiling parts of a gun, if unimpeded by resistances such as springs or pneumatic pressure; it is a theoretical term used in recoil mechanism design.

free-recoil mount   [ORD] A proving ground gun mount designed to closely approximate free-recoil conditions; used for determining design information.

free recombination   [GEN] Genetic recombination having a frequency of 50%, that is, occurring by independent reassortment.

free rocket   [ORD] A rocket having fixed fins but no control surface, that is, no provision for guidance.

free run   [ORD] As applied to guns, the travel of a projectile from its original position in the gun chamber until it engages with the rifling in the gun bore.

free-running frequency   [ELECTR] Frequency at which a normally driven oscillator operates in the absence of a driving signal.

free-running multivibrator   *See* astable multivibrator.

free-running sweep   [ELECTR] Sweep triggered continuously by an internal trigger generator.

free settling   [MIN ENG] In classification, the free fall of particles through fluid media.

free space   [COMMUN] A region high enough so that the radiation pattern of an antenna is not affected by surrounding objects such as buildings, trees, hills, and the earth.

free-space field intensity   [ELECTROMAG] Radio field intensity that would exist at a point in a uniform medium in the absence of waves reflected from the earth or other objects.

free space loss   [ELECTROMAG] The theoretical radiation loss, depending only on frequency and distance, that would occur if all variable factors were disregarded when transmitting energy between two antennas.

free-space propagation   [ELECTROMAG] Propagation of electromagnetic radiation over a straight-line path in a vacuum or ideal atmosphere, sufficiently removed from all objects that affect the wave in any way.

free-space radar equation   [ELECTROMAG] Equation that governs a radar signal characteristic when it is propagated between a radar set and a reflecting object or target in otherwise empty space.

free-space radiation pattern   [ELECTROMAG] Radiation pattern that an antenna would have if it were in free space where there is nothing to reflect, refract, or absorb the radiated waves.

free space wave [ELECTROMAG] An electromagnetic wave propagating in a vacuum, free from boundary effects.

freestone [BOT] A fruit stone to which the fruit does not cling, as in certain varieties of peach.

free surface [FL MECH] A boundary between two homogeneous fluids.

free-swelling index [ENG] A test for measuring the free-swelling properties of coal; consists of heating 1 gram of pulverized coal in a silica crucible over a gas flame under prescribed conditions to form a coke button, the size and shape of which are then compared with a series of standard profiles numbered 1 to 9 in increasing order of swelling.

free symbol [ADP] A contextual symbol preceded and followed by a space; it is always meaningful and always used to symbolize both grammatical and nongrammatical meaning; an example is the English "I."

free symbol sequence [ADP] A symbol sequence either not preceded, or not followed, or neither preceded nor followed by space.

Free test [PATH] Demonstration of a delayed hypersensitivity reaction to Bedsonia antigen in the diagnosis of lymphogranuloma venereum.

free-traveling wave *See* progressive wave.

free turbine [MECH ENG] In a turbine engine, a turbine wheel that drives the output shaft and is not connected to the shaft driving the compressor.

free vibration *See* free oscillation.

free volume [STAT MECH] In a lattice theory of a dense gas or liquid, the volume of the cage in which a given molecule is free to wander when its nearest neighbors are fixed at their lattice positions.

free vortex [FL MECH] Two-dimensional fluid flow in which the fluid moves in concentric circles at speeds inversely proportional to the radii of the circles.

free wall [MIN ENG] The wall of an ore vein filling which scales off cleanly from the gouge.

free-water content *See* water content.

free-water elevation *See* water table.

free-water surface *See* water table.

free wave *See* free progressive wave.

freeze [ENG] **1.** To permit drilling tools, casing, drivepipe, or drill rods to become lodged in a borehole by reason of caving walls or impaction of sand, mud, or drill cuttings, to the extent that they cannot be pulled out. Also known as bind-seize. **2.** To burn in a bit. Also known as burn-in. **3.** The premature setting of cement, especially when cement slurry hardens before it can be ejected fully from pumps or drill rods during a borehole cementation operation. **4.** The act or process of drilling a borehole by utilizing a drill fluid chilled to minus 30–40°F, (minus 34–40°C) as a means of consolidating, by freezing, the borehole wall materials or core as the drill penetrates a water-saturated formation, such as sand or gravel. [PHYS CHEM] To solidify a liquid by removal of heat.

freeze drying [ENG] A method of drying materials, such as certain foods, that would be destroyed by the loss of volatile ingredients or by drying temperatures above the freezing point; the material is frozen under high vacuum so that ice or other frozen solvent will quickly sublime and a porous solid remain.

freeze etching [CRYO] A method using cryogenics to prepare specimens for study with a microscope.

freezer [MECH ENG] An insulated unit, compartment, or room in which perishable foods are quick-frozen and stored.

freeze sinking [MIN ENG] A method of shaft sinking in waterlogged strata by the use of cold brine circulating through a system of pipes until an ice wall is formed. Also known as freezing method.

freeze-up [HYD] The formation of a continuous ice cover on a body of water.

freezing drizzle [METEOROL] Drizzle that falls in liquid form but freezes upon impact with the ground to form a coating of glaze.

freezing level [METEOROL] The lowest altitude in the atmosphere over a given location, at which the air temperature is 0°C; the height of the 0°C constant-temperature surface.

freezing-level chart [METEOROL] A synoptic chart showing the height of the 0°C constant-temperature surface by means of contour lines.

freezing method *See* freeze sinking.

freezing microtome [ENG] A microtome used to cut frozen tissue.

freezing nucleus [METEOROL] Any particle which, when present within a mass of supercooled water, will initiate growth of an ice crystal about itself.

freezing point [PHYS CHEM] The temperature at which a liquid and a solid may be in equilibrium. Abbreviated fp.

freezing-point depression [PHYS CHEM] The lowering of the freezing point of a solution compared to the pure solvent; the depression is proportional to the active mass of the solute in a given amount of solvent.

freezing precipitation [METEOROL] Any form of liquid precipitation that freezes upon impact with the ground or exposed objects; that is, freezing rain or freezing drizzle.

freezing rain [METEOROL] Rain that falls in liquid form but freezes upon impact to form a coating of glaze upon the ground and on exposed objects.

Fregatidae [VERT ZOO] Frigate birds or man-o'-war birds, a family of fish-eating birds in the order Pelecaniformes.

F region [GEOPHYS] The general region of the ionosphere in which the F_1 and F_2 layers tend to form.

freibergite [MINERAL] A steel-gray, silver-bearing variety of tetrahedrite.

freieslebenite [MINERAL] $Pb_3Ag_5Sb_5S_{12}$ A steel-gray to dark mineral composed of a sulfide of antimony, lead, and silver.

freight car [ENG] A railroad car in or on which freight is transported.

freighter [ENG] A ship or aircraft used mainly for carrying freight.

freight ton *See* ton.

freirinite [MINERAL] $Na_3Cu_3(AsO_4)_2(OH)_3 \cdot H_2O$ A lavender to turquoise-blue mineral composed of a basic hydrous arsenate of sodium and copper.

fremontite *See* natromontebrasite.

Frenatae [INV ZOO] The equivalent name for Heteroneura.

french [MECH] A unit of length used to measure small diameters, especially those of fiber optic bundles, equal to $\frac{1}{3}$ millimeter.

French chalk [MATER] Finely ground talc.

french curve [GRAPHICS] A guide, usually made of clear plastic, used for making regular, irregular, and reverse curves in mechanical drawings and illustrations.

French drain [CIV ENG] An underground passage for water, consisting of loose stones covered with earth.

French measles *See* rubella.

French polish [MATER] Shellac dissolved in methylated spirits.

Frenet-Serret formulas [MATH] Formulas in the theory of space curves, which give the directional derivatives of the unit vectors along the tangent, principal normal and binormal of a space curve in the direction tangent to the curve. Also known as Serret-Frenet formulas.

Frenkel defect [SOLID STATE] A crystal defect consisting of a vacancy and an interstitial which arise when an atom is plucked out of a normal lattice site and forced into an interstitial position. Also known as Frenkel pair.

Frenkel exciton [SOLID STATE] A tightly bound exciton in which the electron and the hole are usually on the same atom, although the pair can travel anywhere in the crystal.

Frenkel-Halsey-Hill isotherm equation [PHYS] An equation for the volume v of a gas adsorbed on a surface at a given temperature, $\ln (p/p_o) = k/v^s$, where p is the pressure of the gas, p_o is the vapor pressure, and k and s are constants.

Frenkel pair *See* Frenkel defect.

frenulum [ANAT] **1.** A small fold of integument or mucous membrane. **2.** A small ridge on the upper part of the anterior medullary velum. [INV ZOO] A spine on most moths that projects from the hindwings and is held to the forewings by a clasp, thus coupling the wings together.

frenum [ANAT] A fold of tissue that restricts the movements of an organ.

Freon [ORG CHEM] Trade name for a group of polyhalogenated hydrocarbons containing fluorine and chlorine; an example is trichlorofluoromethane (Freon-11).

FREQUENCY COUNTER

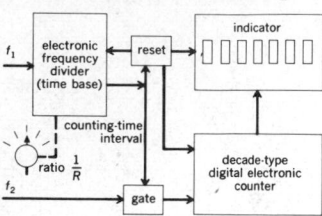

Block diagram of electronic digital counter.

FREQUENCY DIVIDER

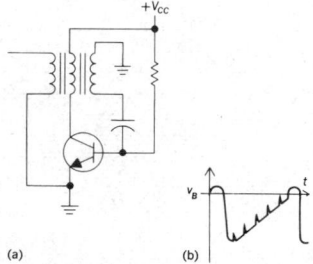

The blocking oscillator as a frequency divider. (a) Circuit. (b) Base voltage.

FREQUENCY METER

Cutaway view of an indirect-drive switchboard-type frequency meter (def. 1). (James G. Biddle Co.)

frequency [PHYS] The number of cycles completed by a periodic quantity in a unit time.

frequency agility [ORD] The ability to shift the frequency of a radar transmitter rapidly and continually in order to avoid jamming by the enemy, to reduce mutual interference with friendly sources, to enhance echoes from targets, or to provide the required patterns of electronic countermeasures or electronic counter-countermeasures radiation.

frequency allocation [COMMUN] Assignment of available frequencies in the radio spectrum to specific stations and for specific purposes, to give maximum utilization of frequencies with minimum interference between stations.

frequency analyzer [ELECTR] A device which measures the intensity of many different frequency components in some oscillation, as in a radio band; used to identify transmitting sources.

frequency-azimuth intensity [ELECTR] Type of radar display in which frequency, azimuth, and strobe intensity are correlated.

frequency band [PHYS] A continuous range of frequencies extending between two limiting frequencies.

frequency bridge [ELECTR] A bridge in which the balance varies with frequency in a known manner, such as the Wien bridge; used to measure frequency.

frequency carrier system [COMMUN] A form of frequency division multiplex in which intelligence is carried on subcarriers.

frequency changer See frequency converter.

frequency characteristic See frequency-response curve.

frequency coding [PHYSIO] A means by which the central nervous system analyzes the content of a receptor message; the frequency of discharged impulses is a function of the rate of rise of the generator current and indirectly of the stimulus strength; the greater the stimulus intensity, the higher the impulse frequency of the message.

frequency compensation See compensation.

frequency conversion [ELECTR] Converting the carrier frequency of a received signal from its original value to the intermediate frequency value in a superheterodyne receiver.

frequency converter [ELEC] A circuit, device, or machine that changes an alternating current from one frequency to another, with or without a change in voltage or number of phases. Also known as frequency changer; frequency translator.

frequency counter [ELECTR] An electronic counter used to measure frequency by counting the number of cycles in an electric signal during a preselected time interval.

frequency curve [STAT] A graphical representation of a continuous frequency distribution; the value of the variable is the abscissa and the frequency is the ordinate.

frequency cutoff [ELECTR] The frequency at which the current gain of a transistor drops 3 decibels below the low-frequency gain value.

frequency deviation [COMMUN] The peak difference between the instantaneous frequency of a frequency-modulated wave and the carrier frequency.

frequency discriminator [ELECTR] A discriminator circuit that delivers an output voltage which is proportional to the deviations of a signal from a predetermined frequency value.

frequency distortion [ELECTR] Distortion in which the relative magnitudes of the different frequency components of a wave are changed during transmission or amplification. Also known as amplitude distortion; amplitude-frequency distortion; waveform-amplitude distortion.

frequency distribution [MATH] A function which measures the relative frequency or probability that a variable can take on a set of values.

frequency diversity [COMMUN] Diversity reception involving the use of carrier frequencies separated 500 hertz or more and having the same modulation, to take advantage of the fact that fading does not occur simultaneously on different frequencies.

frequency divider [ELECTR] A harmonic conversion transducer in which the frequency of the output signal is an integral submultiple of the input frequency. Also known as counting-down circuit.

frequency division data link [COMMUN] Data link using frequency division techniques for channel spacing.

frequency-division multiplexing [COMMUN] A multiplex system for transmitting two or more signals over a common path by using a different frequency band for each signal. Abbreviated fdm. Also known as frequency multiplexing.

frequency domain [COMMUN] A plane on which signal strength can be represented graphically as a function of frequency, instead of a function of time. [CONT SYS] Pertaining to a method of analysis, particularly useful for fixed linear systems in which one does not deal with functions of time explicitly, but with their Laplace or Fourier transforms, which are functions of frequency.

frequency doubler [ELECTR] An amplifier stage whose resonant anode circuit is tuned to the second harmonic of the input frequency; the output frequency is then twice the input frequency. Also known as doubler.

frequency drift [ELECTR] A gradual change in the frequency of an oscillator or transmitter due to temperature or other changes in the circuit components that determine frequency.

frequency-exchange signaling [COMMUN] Signaling in which the change from one signaling condition to another is accompanied by decay in amplitude of one or more other frequencies.

frequency factor [PHYS CHEM] The constant A (or ν) in the Arrhenius equation, which is the relation between reaction rate and absolute temperature T; the equation is $k = Ae - \Delta H_{act}/RT$, where k is the specific rate constant, ΔH_{act} is the heat of activation, and R is the gas constant.

frequency frogging [COMMUN] Interchanging of frequency allocations for carrier channels to prevent singing, reduce crosstalk, and reduce the need for equalization; modulators in each repeater translate a low-frequency group to a high-frequency group, and vice versa.

frequency function See probability density function.

frequency interlace [COMMUN] Carrier chrominance signal frequency chosen so I and J sidebands are interwoven with luminance sidebands in the same bandwidth and in a manner that causes no mutual interference.

frequency locus [CONT SYS] The path followed by the frequency transfer function or its inverse, either in the complex plane or on a graph of amplitude against phase angle; used in determining zeros of the describing function.

frequency meter [ENG] 1. An instrument for measuring the frequency of an alternating current; the scale is usually graduated in hertz, kilohertz, and megahertz. 2. A device calibrated to indicate frequency of a radio wave.

frequency-modulated carrier current telephony [COMMUN] Telephony involving the use of a frequency-modulated carrier signal transmitted over power-line wires or other wires.

frequency-modulated cyclotron See synchrocyclotron.

frequency-modulated jamming [ELECTR] Jamming technique consisting of a constant amplitude radio-frequency signal that is varied in frequency about a center frequency to produce a signal over a band of frequencies.

frequency-modulated radar [ENG] Form of radar in which the radiated wave is frequency modulated, and the returning echo beats with the wave being radiated, thus enabling range to be measured.

frequency modulation [COMMUN] Modulation in which the instantaneous frequency of the modulated wave differs from the carrier frequency by an amount proportional to the instantaneous value of the modulating wave. Abbreviated FM.

frequency-modulation broadcast band [COMMUN] The band of frequencies extending from 88 to 108 megahertz; used for frequency-modulation radio broadcasting in the United States.

frequency-modulation detector [ELECTR] A device, such as a Foster-Seely discriminator, for the detection or demodulation of a frequency-modulated wave.

frequency-modulation Doppler [ENG] Type of radar involving frequency modulation of both carrier and modulation on radial sweep.

frequency modulation–frequency modulation [COMMUN] System in which frequency modulated subcarriers are used to frequency modulate a second carrier.

frequency-modulation laser [OPTICS] Conventional laser containing a phase modulator inside its Fabry-Perot cavity;

characterized by the lack of noise resulting from the random fluctuation in the phase in the various modes.

frequency-modulation noise level on carrier [COMMUN] Residual frequency modulation resulting from disturbance produced in an aural transmitter operating within the band of 50 to 15,000 hertz.

frequency modulation–phase modulation [COMMUN] System in which the several frequency-modulated subcarriers are used to phase modulate a second carrier.

frequency-modulation receiver [ELECTR] A radio receiver that receives frequency-modulated waves and delivers corresponding sound waves.

frequency-modulation receiver deviation sensitivity [ELECTR] Least frequency deviation that produces a specified output power.

frequency-modulation transmitter [ELECTR] A radio transmitter that transmits a frequency-modulated wave.

frequency-modulation tuner [ELECTR] A tuner containing a radio-frequency amplifier, converter, intermediate-frequency amplifier, and demodulator for frequency-modulated signals, used to feed a low-level audio-frequency signal to a separate af amplifer and loudspeaker.

frequency modulator [ELECTR] A circuit or device for producing frequency modulation.

frequency monitor [ELECTR] An instrument for indicating the amount of deviation of the carrier frequency of a transmitter from its assigned value.

frequency multiplexing See frequency-division multiplexing.

frequency multiplier [ELECTR] A harmonic conversion transducer in which the frequency of the output signal is an exact integral multiple of the input frequency. Also known as multiplier.

frequency offset [COMMUN] **1.** Change in the frequency received on a transmission line from that which was transmitted, which can occur if suppressed-carrier multiplexing is used, and may result in errors. **2.** A small difference in the carrier frequencies of television stations in adjacent cities operating on the same channel.

frequency-offset transponder [ELECTR] Transponder that changes the signal frequency by a fixed amount before retransmission.

frequency optimum traffic See optimum working frequency.

frequency polygon [STAT] A graph obtained from a frequency distribution by joining with straight lines points whose abscissae are the midpoints of successive class intervals and whose ordinates are the corresponding class frequencies.

frequency prediction chart [COMMUN] Graph showing curve for the maximum usable frequency, frequency optimum traffic, and lowest usable frequency between two specific points for various times throughout a 24-hour period.

frequency pulling [ELECTR] A change in the frequency of an oscillator due to a change in load impedance.

frequency recorder [ELEC] An instrument which uses a frequency bridge to sense the frequency of an alternating current, and which makes a graphical record of this frequency as a function of time.

frequency regulator [ELEC] A device that maintains the frequency of an alternating-current generator at a predetermined value.

frequency relay [ELECTR] Relay which functions at a predetermined value of frequency; may be an over-frequency relay, an under-frequency relay, or a combination of both.

frequency response [ENG] A measure of the effectiveness with which a circuit, device, or system transmits the different frequencies applied to it; it is a phasor whose magnitude is the ratio of the magnitude of the output signal to that of a sine-wave input, and whose phase is that of the output with respect to the input. Also known as amplitude-frequency response; response; sine-wave response.

frequency-response curve [ENG] A graph showing the magnitude or the phase of the frequency response of a device or system as a function of frequency. Also known as frequency characteristic.

frequency-response equalization See equalization.

frequency-response trajectory [CONT SYS] The path followed by the frequency-response phasor in the complex plane as the frequency is varied.

frequency run [ELECTR] A series of tests made to determine the amplitude-frequency response characteristic of a transmission line, circuit, or device.

frequency scan antenna [ELECTROMAG] A radar antenna similar to a phased array antenna in which one dimensional scanning is accomplished through frequency variation.

frequency scanning [ELECTR] Type of system in which output frequency is made to vary at a mechanical rate over a desired frequency band.

frequency-selective device See electric filter.

frequency separation multiplier [ELECTR] Multiplier in which each of the variables is split into a low-frequency part and a high-frequency part that are multiplied separately, and the results added to give the required product; this system makes it possible to get high accuracy and broad bandwidth.

frequency separator [ELECTR] The circuit that separates the horizontal and vertical synchronizing pulses in a monochrome or color television receiver.

frequency shift [ELECTR] A change in the frequency of a radio transmitter or oscillator. Also known as radio-frequency shift.

frequency-shift converter [ELECTR] A device that converts a received frequency-shift signal to an amplitude-modulated signal or a direct-current signal.

frequency-shift keyer [ELECTR] A lever to effect a frequency shift, that is, a change in the frequency of a radio transmitter, oscillator, or receiver.

frequency-shift keying [COMMUN] A form of frequency modulation used especially in telegraph and facsimile transmission, in which the modulating wave shifts the output frequency between predetermined values corresponding to the frequencies of correlated sources. Abbreviated FSK. Also known as frequency-shift modulation; frequency-shift transmission.

frequency-shift modulation See frequency-shift keying.

frequency-shift transmission See frequency-shift keying.

frequency-slope modulation [COMMUN] Type of modulation in which the carrier signal is swept periodically over the entire width of the band, much as in chirp radar; modulation of the carrier with a voice or other communication signal changes the bandwidth of the system without affecting the uniform distribution of energy over the band.

frequency spectrum [PHYS] A plot of the distribution of the intensity of some type of electromagnetic or acoustic radiation as a function of frequency. [SYS ENG] In the analysis of a random function of time, such as the amplitude of noise in a system, the limit as T approaches infinity of $1/2\pi T$ times the ensemble average of the squared magnitude of the amplitude of the Fourier transform of the function from $-T$ to T. Also known as power-density spectrum; power spectrum; spectral density.

frequency splitting [ELECTR] One condition of operation of a magnetron which causes rapid alternating from one mode of operation to another; this results in a similar rapid change in oscillatory frequency and consequent loss in power at the desired frequency.

frequency stability [ELECTR] The ability of an oscillator to maintain a desired frequency; usually expressed as percent deviation from the assigned frequency value.

frequency stabilization [COMMUN] Process of controlling the center or carrier frequency so that it differs from that of a reference source by not more than a prescribed amount.

frequency standard [ELECTR] A stable oscillator, usually controlled by a crystal or tuning fork, that is used primarily for frequency calibration.

frequency study See work sampling.

frequency swing [COMMUN] **1.** Peak difference between the maximum and the minimum values of the instantaneous frequency. **2.** In frequency modulation, a term loosely used to describe the change in frequency resulting from the modulation.

frequency synthesizer [ELECTR] A device that provides a choice of a large number of different frequencies by combining frequencies selected from groups of independent crystals, frequency dividers, and frequency multipliers.

frequency telemetering [COMMUN] The transmittal of an alternating-current signal from a primary element by variations in the signal frequency, instead of intensity.

FREQUENCY MODULATOR

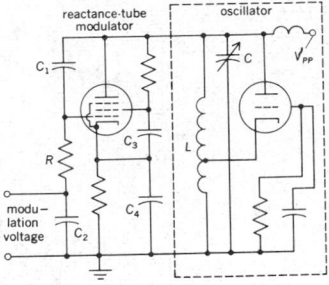

Basic circuit of reactance-tube type of frequency modulator.

FREQUENCY POLYGON

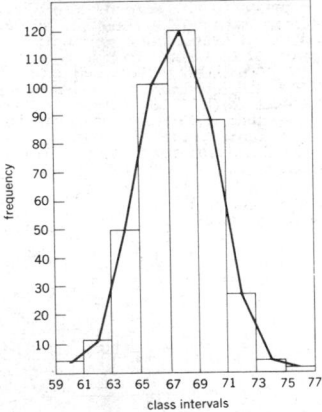

The conversion of a histogram into a frequency polygon by connecting the midpoint value (at the top of each rectangle) with the adjacent midpoint value by straight lines.

FREQUENCY-SHIFT KEYING

The radio waveform used in frequency-shift keying.

frequency theory [PHYSIO] A theory of human hearing according to which every specific frequency of sound energy is represented by nerve impulses of the same frequency, and pitch differentiation and analysis are carried out by the brain centers. Also known as telephone theory.

frequency-time-intensity [ELECTR] Type of radar display in which the frequency, time, and strobe intensity are correlated.

frequency tolerance [ELECTR] Of a radio transmitter, extent to which the carrier frequency of the transmitter may be permitted to depart from the frequency assigned.

frequency transformation [CONT SYS] A transformation used in synthesizing a band-pass network from a low-pass prototype, in which the frequency variable of the transfer function is replaced by a function of the frequency. Also known as low-pass band-pass transformation.

frequency translator *See* frequency converter.

frequency-type telemeter [ELECTR] Telemeter that employs frequency of an alternating current or voltage as the translating means.

frequency variation [ELECTR] The change over time of the deviation from assigned frequency of a radio-frequency carrier (or power supply system); usually tightly controlled because of national or industry standards.

fresco [GRAPHICS] Painting of two types, buon fresco and fresco secco; in buon fresco, dry pigments are ground with water and applied on a wet lime plaster wall, and as the plaster dries, the pigment is permanently bound to it; fresco secco utilizes pigments ground in glue, casein, or polymer emulsions and applied on a dry plaster wall that has been wetted with limewater.

fresh [GEOL] Unweathered in reference to a rock or rock surface. [METEOROL] Pertaining to air which is stimulating and refreshing.

fresh breeze [METEOROL] In the Beaufort wind scale, a wind whose speed is 17 to 21 knots (19 to 24 miles per hour, or 31 to 39 kilometers per hour).

fresh-core technique [PETRO ENG] A method in which a core sample, fresh from the field, is subjected to waterflooding in the laboratory, and the resulting residual oil is determined; used to calculate total waterflood recovery of oil from a reservoir formation.

freshet [HYD] **1.** The annual spring rise of streams in cold climates as a result of melting snow. **2.** A flood resulting from either rain or melting snow; usually applied only to small streams and to floods of minor severity. **3.** A small freshwater stream.

fresh gale [METEOROL] In the Beaufort wind scale, a wind whose speed is from 34 to 40 knots (39 to 46 miles per hour, or 63 to 74 kilometers per hour).

fresh ice *See* newly formed ice.

fresh water [HYD] Water containing no significant amounts of salts, such as in rivers and lakes.

fresh-water ecosystem [ECOL] The living organisms and nonliving materials of an inland aquatic environment.

fresnel [PHYS] A unit of frequency, equal to 10^{12} hertz.

Fresnel-Arago laws [OPTICS] The three laws stating that two rays of polarized light interfere in the same way as ordinary light if they are polarized in the same plane, but do not interfere if they are polarized at right angles; two rays polarized from ordinary light at right angles do not interfere in the ordinary sense when they are brought into the same plane of polarization; and two rays polarized at right angles from plane polarized light, and then brought into the same polarization plane, interfere.

Fresnel biprism [OPTICS] A very flat triangular prism which has two very acute angles and one very obtuse angle; used to observe the interference of light from a slit passing through the two halves of the prism.

Fresnel diffraction [OPTICS] Diffraction in which the source of light or the observing screen are at a finite distance from the aperture or obstacle.

Fresnel drag coefficient [OPTICS] The quantity $1 - (1/n^2)$, where n is the index of diffraction of a transparent medium, believed by Fresnel to be the ratio of the velocity with which ether was dragged along in the medium to the velocity of the medium itself.

Fresnel ellipsoid [OPTICS] An ellipsoid whose three perpendicular axes are proportional to the principal values of the wave velocity of light in an anisotropic medium. Also known as ray ellipsoid.

Fresnel equations [OPTICS] Equations which give the intensity of each of the two polarization components of light which is reflected or transmitted at the boundary between two media with different indices of refraction.

Fresnel fringe [OPTICS] One of a series of light and dark bands that appear near the edge of a shadow in Fresnel diffraction.

Fresnel integrals [MATH] Given a parameter x, the integrals over t from 0 to x of $\sin t^2$ and of $\cos t^2$ or from x to ∞ of $(\cos t)/t^{1/2}$ and of $(\sin t)/t^{1/2}$.

Fresnel lens [OPTICS] A thin lens constructed with stepped setbacks so as to have the optical properties of a much thicker lens.

Fresnel mirrors [OPTICS] Two plane mirrors which are inclined to each other on the order of a degree and used to observe the interference of light which originates from a slit and is reflected from both mirrors.

Fresnel ovaloid [OPTICS] For an anisotropic crystal, an ovaloid whose central section normal to the propagation direction of an electromagnetic wave gives the axes of polarization of the displacement vector and the associated wave velocities.

Fresnel reflection formula [OPTICS] The Fresnel equations for light reflected from a boundary.

Fresnel region [ELECTROMAG] The region between the near field of an antenna (close to the antenna compared to a wavelength) and the Fraunhofer region.

Fresnel rhomb [OPTICS] A glass rhomb which has an acute angle of about 52°; light which is incident normal to the end of the rhomb undergoes two internal reflections, and if it is initially linearly polarized at an angle of 45° to the plane of incidence, it emerges circularly polarized.

Fresnel spotlight [ELEC] A lighting instrument that is composed of a lamp and a Fresnel (stepped planoconvex) lens; the unit can be made with or without reflectors and has a system to adjust the spacing between the lamp and the lens so as to control the light beam; models range from 100 to 5000 watts.

Fresnel theory of double refraction [OPTICS] The theory which explains double refraction of a crystal in terms of nonspherical wave surfaces.

Fresnel zones [ELECTROMAG] Circular portions of a wavefront transverse to a line between an emitter and a point where the disturbance is being observed; the nth zone includes all paths whose lengths are between $n-1$ and n half-wavelengths longer than the line-of-sight path.

fretting corrosion [MET] Surface damage usually in an air environment between two surfaces, one or both of which are metals, in close contact under pressure and subject to a slight relative motion. Also known as chafing corrosion.

Freudianism [PSYCH] The psychoanalytic school of psychiatry founded by Sigmund Freud.

Freundlich isotherm equation [PHYS] Equation which states that the volume of gas adsorbed on a surface at a given temperature is proportional to the pressure of the gas raised to a constant power.

Freund method [ORG CHEM] A method for preparation of cycloparaffins in which dihalo derivatives of the paraffins are treated with zinc to produce the cycloparaffin.

Freund's adjuvant [IMMUNOL] A water-oil emulsion containing a killed microorganism (usually *Mycobacterium tuberculosis*) which enhances antigenicity.

friability [MATER] The ease with which a material is crumbled, pulverized, or reduced to powder.

friable [MATER] Referring to the property of a substance capable of being easily rubbed, crumbled, or pulverized into powder.

friagem [METEOROL] A period of cold weather in the middle and upper parts of the Amazon Valley and in eastern Bolivia, occurring during the dry season in the Southern Hemisphere winter. Also known as vriajem.

friar's cloth *See* monk's cloth.

friar's cowl *See* aconite.

Fricke dosimeter [NUCLEO] A radiation dosimeter in which the energy of ionizing radiation is determined from the amount of ferrous ions converted to ferric ions in an aerated acidic ferrous sulfate solution.

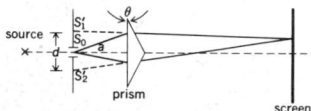
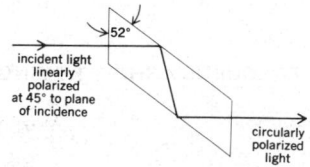

friction [MECH] A force which opposes the relative motion of two bodies whenever such motion exists or whenever there exist other forces which tend to produce such motion.

frictional See cohesionless.

frictional electricity [ELEC] The electric charges produced on two different objects, such as silk and glass or catskin and ebonite, by rubbing them together.

frictional grip [MECH] The frictional grip of a locomotive or adhesion between the wheels and the rails of the railroad track.

frictional secondary flow See secondary flow.

friction bearing [MECH ENG] A solid bearing that directly contacts and supports an axle end.

friction blocks [PETRO ENG] Thin blocks with cylindrical surfaces that drag on the inside of the well casing to prevent rotation of the packer (seal between the outside of tubing and inside of casing).

friction brake [MECH ENG] A brake in which the resistance is provided by friction.

friction calendering [ENG] Process wherein an elastomeric compound is forced into the interstices of woven or cord fabrics while passing between calender rolls.

friction clutch [MECH ENG] A clutch in which torque is transmitted by pressure of the clutch faces on each other.

friction coefficient See coefficient of friction.

friction crack [GEOL] A short, crescent-shaped crack in glaciated rock produced by a localized increase in friction between rock and ice, oriented transverse to the direction of ice flow.

friction damping [MECH] The conversion of the mechanical vibrational energy of solids into heat energy by causing one dry member to slide on another.

friction depth [OCEANOGR] The depth at which the velocity of wind-driven current becomes neglibible compared to the surface velocity; sometimes referred to as the depth of the Ekman layer.

friction drive [MECH ENG] A drive that operates by the friction forces set up when one rotating wheel is pressed against a second wheel.

friction factor [FL MECH] Any of several dimensionless numbers used in studying fluid friction in pipes, equal to the Fanning friction factor times some dimensionless constant.

friction feed [MIN ENG] Longitudinal movement or advance of a drill stem and bit accomplished by friction devices in a diamond-drill swivel head, as opposed to a system consisting entirely of meshing gears.

friction flow [FL MECH] Fluid flow in which a significant amount of mechanical energy is dissipated into heat by action of viscosity.

friction gear [MECH ENG] Gearing in which motion is transmitted through friction between two surfaces in rolling contact.

friction head [FL MECH] The head lost by the flow in a stream or conduit due to frictional disturbances set up by the moving fluid and its containing conduit and by intermolecular friction.

friction horsepower [MECH ENG] Power dissipated in a machine through friction.

friction layer See surface boundary layer.

frictionless flow See inviscid flow.

friction loss [MECH] Mechanical energy lost because of mechanical friction between moving parts of a machine.

friction pile [CIV ENG] A bearing pile surrounded by earth and supported entirely by friction; carries no load at its end.

friction ridge [ANAT] One of the integumentary ridges on the plantar and palmar surfaces of primates.

friction saw [MECH ENG] A toothless circular saw used to cut materials by fusion due to frictional heat.

friction sawing [MECH ENG] A burning process to cut stock to length by using a blade saw operating at high speed; used especially for the structural parts of mild steel and stainless steel.

friction tape [MATER] Cotton tape impregnated with a sticky moisture-repelling compound; used chiefly to hold rubber-tape insulation in position over a joint or splice.

friction torque [MECH] The torque which is produced by frictional forces and opposes rotational motion, such as that associated with journal or sleeve bearings in machines.

friction-tube viscometer [ENG] Device to determine liquid viscosity by measurement of pressure drop through a friction tube with the liquid in viscous flow; gives direct solution to Poiseuille's equation.

friction velocity [METEOROL] A reference wind velocity defined by the relation $u = \sqrt{\tau/\rho}$, where τ is the Reynolds stress, ρ the density, and u the friction velocity.

friction welding [MET] Welding process in which two members are joined by rubbing the mating faces together under high pressure.

friction yielding prop See mechanical yielding prop.

Friedel-Crafts reaction [ORG CHEM] A substitution reaction, catalyzed by aluminum chloride in which an alkyl $(R-)$ or an acyl $(RCO-)$ group replaces a hydrogen atom of an aromatic nucleus to produce hydrocarbon or a ketone.

friedelite [MINERAL] $Mn_8Si_6O_{18}(OH,Cl)_4 \cdot 3H_2O$ A rose-red mineral composed of manganese silicate with chlorine.

Friedel's law [CRYSTAL] The law that x-ray or electron diffraction measurements cannot determine whether or not a crystal has a center of symmetry.

Friedlander's bacillus See Klebsiella pneumoniae.

Friedlander's pneumonia [MED] Inflammation of the lungs caused by *Klebisiella pneumoniae*.

Friedlander synthesis [ORG CHEM] A synthesis of quinolines; the method is usually catalyzed by bases and consists of condensation of an aromatic o-amino-carbonyl derivative with a compound containing a methylene group in the alpha position to the carbonyl.

Friedman test [PATH] A pregnancy test in which a female rabbit is given an intravenous injection of urine from the patient; formation of corpora lutea in the ovaries indicates a positive test.

Friedrich's ataxia [MED] A hereditary sclerosis of the spine with speech impairment, lateral curvature of the spine, and palsy of the lower limbs. Also known as hereditary spinal ataxia.

Fries rearrangement [ORG CHEM] The conversion of a phenolic ester into the corresponding o- and p-hydroxyketone by treatment with catalysts of the type of aluminum chloride.

Fries' rule [ORG CHEM] The rule that the most stable form of the bonds of a polynuclear compound is that arrangement which has the maximum number of rings in the benzenoid form, that is, three double bonds in each ring.

frieze [ARCH] A decorated band immediately below the cornice on an interior wall.

frigate [NAV ARCH] **1.** In the U.S. Navy, a ship larger than a destroyer and smaller than a cruiser, having a displacement of 4000–9000 tons, designed mainly as an escort ship for an attack aircraft carrier. **2.** In the British and Canadian navies, an escort ship larger than a corvette and smaller than a destroyer, having a displacement of 1200–2500 tons, corresponding in size to a destroyer escort in the U.S. Navy.

frigidoreceptor [PHYSIO] A cutaneous sense organ which is sensitive to cold.

frigorie [THERMO] A unit of rate of extraction of heat used in refrigeration, equal to 1000 fifteen-degree calories per hour, or 1.16264 ± 0.00014 watt.

frigorimeter [ENG] A thermometer which measures low temperatures.

frilling [GRAPHICS] Emulsion at the edges of a photographic film loosened from base of film.

fringe [OPTICS] One of the light or dark bands produced by interference or diffraction of light.

fringe area [COMMUN] An area just beyond the limits of the reliable service area of a television transmitter, in which signals are weak and erratic.

fringe howl [ENG ACOUS] Squeal or howl heard when some circuit in a receiver is on the verge of oscillation.

fringe magnetic field [ELECTROMAG] The part of the magnetic field of a horseshoe magnet that extends outside the space between its poles.

fringe region [METEOROL] The upper portion of the exosphere, where the cone of escape equals or exceeds 180°; in this region the individual atoms have so little chance of collision that they essentially travel in free orbits, subject to

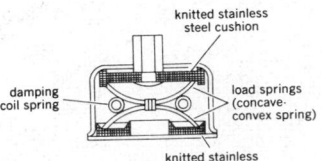

FRICTION DAMPING

knitted stainless
steel cushion

damping
coil spring

load springs
(concave-
convex spring)

knitted stainless
steel cushion

A frictional damper, effective in vertical and horizontal directions.

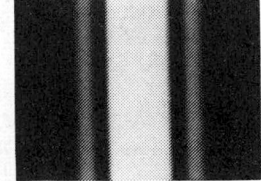

FRINGE

A Fraunhofer diffraction pattern, for a slit, or fringe, photographed with visible light. (*F. S. Harris*)

the earth's gravitation, at speeds imparted by the last collision. Also known as spray region.

fringes of equal thickness   *See* Fizeau fringes.

fringe value   [OPTICS] A quantity used in photoelastic work, equal to the stress which must be applied to a material, in pounds per square inch, to produce a relative retardation between the components of a linearly polarized light beam of 1 wavelength when the light passes through a thickness of 1 inch in a direction perpendicular to the stress.

Fringillidae   [VERT ZOO] The finches, a family of oscine birds in the order Passeriformes.

fringing groove   [ORD] A groove cut into a rotating band to collect metal from the band while it travels through the bore, to prevent its forming a fringe in rear of the rotating band; causes excess dispersion and short range.

fringing reef   [GEOL] A coral reef attached directly to or bordering the shore of an island or continental landmass.

Frise aileron   [AERO ENG] A type of aileron having its leading edge projecting well ahead of the hinge axis.

frisket   [GRAPHICS] A mask used to protect the portions of photographs and artwork that are not be be airbrushed.

frit   [MATER] Fusible ceramic mixture used to make glazes and enamels for dinnerware and metallic surfaces, as on stoves and metal-base basins and tubs.

fritillary   [BOT] The common name for plants of the genus *Fritillaria*.   [INV ZOO] The common name for butterflies in several genera of the subfamily Nymphalinae.

fritting   [ENG] Fusing materials for glass by application of heat.   [MET] The pasty condition, usually occurring a little below the melting point, of the powdered ore, flux, and other reagents in fire assaying.

Fritz process   [FOOD ENG] Treatment of cream containing up to 40–50% fat by continuous feeding to a cooled horizontal cylindrical tank, where it is distributed in a film and beaten by blades at 3000 revolutions per minute; the formed mass falls into a trough, where buttermilk drains off, and butter is moved out for kneading and packing.

Frobenius method   [MATH] A method of finding a series solution near a point for a linear homogeneous ordinary differential equation.

Froehlich's syndrome   *See* adiposogenital dystrophy.

frog   [DES ENG] A hollow on one or both of the larger faces of a brick or block; reduces weight of the brick or block; may be filled with mortar.   [ENG] A device which permits the train or tram wheels on one rail of a track to cross the rail of an intersecting track.   [VERT ZOO] The common name for a number of tailless amphibians in the order Anura; most have hindlegs adapted for jumping, scaleless skin, and large eyes.

frogeye   [PL PATH] Any of various leaf diseases characterized by formation of concentric rings around the diseased spots.

frogging repeater   [ELECTR] Carrier repeater having provisions for frequency frogging to permit use of a single multipair voice cable without having excessive crosstalk.

frog storm   [METEOROL] The first bad weather in spring after a warm period. Also known as whippoorwill storm.

frohbergite   [MINERAL] FeTe$_2$ A mineral composed of iron telluride; it is isomorphous with marcasite.

from-to tester   [ENG] Test equipment which checks continuity or impedance between points.

frond   [BOT] **1.** The leaf of a palm or fern. **2.** A foliaceous thallus or thalloid shoot.

frondelite   [MINERAL] MnFe$_4$(PO$_4$)$_5$(OH)$_5$ A mineral composed of basic phosphate of manganese and iron; it is isomorphous with rockbridgeite.

front   [METEOROL] A sloping surface of discontinuity in the troposphere, separating air masses of different density or temperature.

front abutment pressure   [GEOPHYS] The release of energy in the superincumbent strata above the seam induced by the extraction of the seam.

frontal-advance performance   [PETRO ENG] The theory that, during the waterflood of a formation reservoir, displacement of oil causes a desaturation of the displaced fluids in accordance with relative-permeability relationships, and the displacement is linear.

frontal angle   [ANTHRO] The angle formed by the intersection

of lines from the bregma and glabella to the auricular point.

frontal apron   *See* outwash plain.

frontal bone   [ANAT] Either of a pair of flat membrane bones in vertebrates, and a single bone in humans, forming the upper frontal portion of the cranium; the forehead bone.

frontal contour   [METEOROL] The line of intersection of a front (frontal surface) with a specified surface in the atmosphere, usually a constant-pressure surface; with respect to only one surface, this line is usually called the front.

frontal crest   [ANAT] A median ridge on the internal surface of the frontal bone in humans.

frontal cyclone   [METEOROL] Any cyclone associated with a front; often used synonymously with wave cyclone or with extratropical cyclone (as opposed to tropical cyclones, which are nonfrontal).

frontal drive   [PETRO ENG] In an oil reservoir with constant pressure (by gas injection or from a large gas-to-oil ratio), the driving of oil fluids into the wellbore by the free gas.

frontal eminence   [ANAT] The prominence of the frontal bone above each superciliary ridge in humans.

frontal fire   [ORD] Fire delivered at right angles to the front of the target.

frontal fog   [METEOROL] Fog associated with frontal zones and frontal passages.

frontal index   [ANTHRO] The ratio of the least to the greatest breadth of the forehead multiplied by 100.

frontal inversion   [METEOROL] A temperature inversion in the atmosphere, encountered upon vertical ascent through a sloping front (or frontal zone).

frontal lifting   [METEOROL] The forced ascent of the warmer, less-dense air at and near a front, occurring whenever the relative velocities of the two air masses are such that they converge at the front.

frontal lobe   [ANAT] The anterior portion of a cerebral hemisphere, bounded behind by the central sulcus and below by the lateral cerebral sulcus.

frontal nerve   [ANAT] A somatic sensory nerve, attached to the ophthalmic nerve, which innervates the skin of the upper eyelid, the forehead, and the scalp.

frontal occlusion   *See* occluded front.

frontal passage   [AERO ENG] The transit of an aircraft through a frontal zone.   [METEOROL] The passage of a front over a point on the earth's surface.

frontal plain   *See* outwash plain.

frontal plane   [ANAT] Any plane parallel with the long axis of the body and perpendicular to the sagittal plane.   [MED] In electrocardiography and vectorcardiography, the projection of the vertical axis.

frontal precipitation   [METEOROL] Any precipitation attributable to the action of a front; used mainly to distinguish this type from air-mass precipitation and orographic precipitation.

frontal profile   [METEOROL] The outline of a front as seen on a vertical cross section oriented normal to the frontal surface.

frontal sinus   [ANAT] Either of a pair of air spaces within the frontal bone above the nasal bridge.

frontal strip   [METEOROL] The presentation of a front, on a synoptic chart, as a frontal zone; that is, two lines, rather than a single line, are drawn to represent the boundaries of the zone; a rare usage.

frontal system   [METEOROL] A system of fronts as they appear on a synoptic chart.

frontal thunderstorm   [METEOROL] A thunderstorm associated with a front; limited to thunderstorms resulting from the convection induced by frontal lifting.

frontal wave   [METEOROL] A horizontal, wavelike deformation of a front in the lower levels, commonly associated with a maximum of cyclonic circulations in the adjacent flow; it may develop into a wave cyclone.

frontal zone   [METEOROL] The three-dimensional zone or layer of large horizontal density gradient, bounded by frontal surfaces and surface front.

front-end   [ADP] Of a minicomputer, under programmed instructions, performing data transfers and control operations to relieve a larger computer of these routines.

front-end loader   [MECH ENG] An excavator consisting of an

FROG

Common frog (*Rana temporaria*).

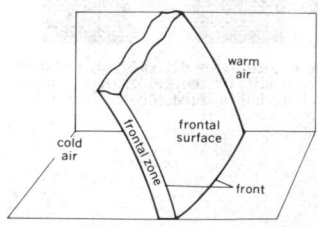

FRONTAL ZONE

Schematic diagram of the frontal zone (the angle with earth's surface is much exaggerated).

articulated bucket mounted on a series of movable arms at the front of a crawler or rubber-tired tractor.

front-end volatility [CHEM ENG] The volatility of the lower-boiling fractions of gasoline, such as butanes.

frontier of a set in a topological space [MATH] All points in the closure of the set but not in its interior. Also known as boundary of a set.

frontogenesis [METEOROL] **1.** The initial formation of a frontal zone or front. **2.** The increase in the horizontal gradient of an air mass property, mainly density, and the formation of the accompanying features of the wind field that typify a front.

frontogenetic function [METEOROL] A kinematic measure of the tendency of the flow in an air mass to increase the horizontal gradient of a conservative property.

frontolysis [METEOROL] **1.** The dissipation of a front or frontal zone. **2.** In general, a decrease in the horizontal gradient of an air mass property, principally density, and the dissipation of the accompanying features of the wind field.

front pinacoid [CRYSTAL] The {100} pinacoid in an orthorhombic, monoclinic, or triclinic crystal. Also known as macropinacoid; orthopinacoid.

front porch [COMMUN] Portion of a composite picture signal which lies between the leading edge of the horizontal blanking pulse and the leading edge of the corresponding synchronizing pulse.

front slope _See_ scarp slope.

front-to-back ratio [ELECTROMAG] Ratio of the effectiveness of a directional antenna, loudspeaker, or microphone toward the front and toward the rear. [SOLID STATE] Ratio of resistance of a crystal to current flowing in the normal direction to current flowing in the opposite direction.

frost [HYD] A covering of ice in one of its several forms, produced by the sublimation of water vapor on objects colder than 32°F (0°C).

frost action [GEOL] **1.** The weathering process caused by cycles of freezing and thawing of water in surface pores, cracks, and other openings. **2.** Alternate or repeated cycles of freezing and thawing of water contained in materials; the term is especially applied to disruptive effects of this action.

frostbite [MED] Injury to skin and subcutaneous tissues, and in severe cases to deeper tissues also, from exposure to extreme cold.

frost boil [GEOL] **1.** An accumulation of water and mud released from ground ice by accelerated spring thawing. **2.** A low mound formed by local differential frost heaving at a location most favorable for the formation of segregated ice and accompanied by the absence of an insulating cover of vegetation.

frost bursting _See_ congelifraction.

frost churning _See_ congeliturbation.

frost climate [CLIMATOL] The coldest temperature province in C. W. Thornthwaite's climatic classification: the climate of the ice cap regions of the earth, that is, those regions perennially covered with snow and ice.

frost day [METEOROL] An observational day on which frost occurs.

frosted glass [MATER] Glass that has been etched with sand, or appears to have been so treated.

frost feathers _See_ ice feathers.

frost flakes _See_ ice fog.

frost flowers _See_ ice flowers.

frost fog _See_ ice fog.

frost hazard [METEOROL] The risk of damage by frost, expressed as the probability or frequency of killing frost on different dates during the growing season, or as the distribution of dates of the last killing frost of spring or the first of autumn.

frost heaving [GEOL] The lifting and distortion of a surface due to internal action of frost resulting from subsurface ice formation; affects soil, rock, pavement, and other structures.

frosting [ENG] Decorating a scraped metal surface with a handscraper. Also known as flaking.

frostless zone [METEOROL] The warmest part of a slope above a valley floor, lying between the layer of cold air which forms over the valley floor on calm clear nights and the cold hill tops or plateaus; the air flowing down the slopes is warmed by mixing with the air above ground level, and to some extent also by adiabatic compression. Also known as green belt; verdant zone.

frost line [GEOL] **1.** The maximum depth of frozen ground during the winter. **2.** The lower limit of permafrost.

frost mound [GEOL] A hill and knoll associated with frozen ground in a permafrost region, containing a core of ice. Also known as soffosian knob; soil blister.

frost point [METEOROL] The temperature to which atmospheric moisture must be cooled to reach the point of saturation with respect to ice.

frost-point hygrometer [ENG] An instrument for measuring the frost point of the atmosphere; air under test is passed continuously across a polished surface whose temperature is adjusted so that a thin deposit of frost is formed which is in equilibrium with the air.

frost-point technique _See_ 8D technique (in the D's).

frost ring [BOT] A false annual growth ring in the trunk of a tree due to out-of-season defoliation by frost and subsequent regrowth of foliage.

frost riving _See_ congelifraction.

frost smoke [METEOROL] **1.** A rare type of fog formed in the same manner as a steam fog, but at colder temperatures so that it is composed of ice particles instead of water droplets. **2.** _See_ steam fog.

frost splitting _See_ congelifraction.

frost table [GEOL] An irregular surface in the ground which, at any given time, represents the penetration of thawing into seasonally frozen ground.

frosty mildew [PL PATH] A leaf spot caused by fungi of the genus _Cercosporella_ and characterized by pale to white lesions.

frost zone _See_ seasonally frozen ground.

frother [CHEM] Substance used in flotation processes to make air bubbles sufficiently permanent, principally by reducing surface tension.

froth flotation [ENG] A process for recovery of particles of ore or other material, in which the particles adhere to bubbles and can be removed as part of the froth.

frothing [ENG] The producing of relatively stable bubbles at an air-liquid interface as the result of agitation, aeration, ebulliation, or chemical reaction; it can be an undesired side effect, but in minerals beneficiation it is the basis of froth flotation.

frothing collector [MIN ENG] An ore collector which in addition produces a stable foam.

froth promoter [CHEM] A chemical compound used with a frothing agent.

frottage [GRAPHICS] A technique in which a material, such as paper, is placed on a rough or irregular surface and is rubbed with a pencil or paint; the approximate image of the peaks and valleys results; the method is used to copy bas-reliefs, tombstones, and bronzes.

Froude number 1 [FL MECH] A dimensionless number used in studying the motion of a body floating on a fluid with production of surface waves and eddies; equal to the ratio of the square of the relative speed to the product of the acceleration of gravity and a characteristic length of the body. Symbolized N_{Fr_1}.

Froude number 2 [FL MECH] A dimensionless number, equal to the ratio of the speed of flow of a fluid in an open channel to the speed of very small gravity waves, the latter being equal to the square root of the product of the acceleration of gravity and a characteristic length. Symbolized N_{Fr_2}.

frozen flux [PL PHYS] The lines of force of a frozen-in field.

frozen fog _See_ ice fog.

frozen ground [GEOL] Soil having a temperature below freezing, generally containing water in the form of ice. Also known as gelisol; merzlota; taele; tjaele.

frozen-in field [PL PHYS] A magnetic field in a plasma which has negligible electrical resistance; it can be shown that the lines of force of this field are constrained to move with the material.

frozen precipitation [METEOROL] Any form of precipitation that reaches the ground in frozen form; that is, snow, snow pellets, snow grains, ice crystals, ice pellets, and hail.

frozen section [BIOL] A thin slice of material cut from a frozen sample of tissue or organ.

fructescence [BOT] The period of fruit maturation.

fructification [BOT] **1.** The process of producing fruit. **2.** A fruit and its appendages. [MYCOL] A sporanginous structure.

fructivorous *See* frugivorous.

D-fructopyranose *See* fructose.

fructose [BIOCHEM] $C_6H_{12}O_5$ The commonest of ketoses and the sweetest of sugars, found in the free state in fruit juices, honey, and nectar of plant glands. Also known as D-fructopyranose.

Frue vanner [MIN ENG] A side-shake type of ore-dressing apparatus consisting of an inclined rubber belt on which the material is washed by a constant flow of water.

frugivorous [ZOO] Fruit-eating. Also known as fructivorous.

fruit [BOT] A fully matured plant ovary with or without other floral or shoot parts united with it at maturity. [NAV] Radar-beacon-system video display of a synchronous beacon return which results when several interrogator stations are located within the same general area; each interrogator receives its own interrogated reply as well as many synchronous replies resulting from interrogation of the airborne transponders by other ground stations.

fruit bud [BOT] A fertilized flower bud that matures into a fruit.

fruit fly [INV ZOO] **1.** The common name for those acalypterate insects composing the family Tephritidae. **2.** Any insect whose larvae feed on fruit or decaying vegetable matter.

fruiting body [BOT] A specialized, spore-producing organ.

fruiting myxobacteria *See* Myxobacterales.

frustrated internal reflectance *See* attenuated total reflectance.

frustration [PSYCH] The experience of nonfulfillment of some wish or need.

frustration threshold [PSYCH] The point at which an individual feels or shows frustration over inability to achieve an objective.

frustule [INV ZOO] **1.** The shell and protoplast of a diatom. **2.** A nonciliated planulalike bud in some hydrozoans.

frustum [MATH] The part of a solid between two cutting parallel planes.

frutescent [BIOL] *See* fruticose. [BOT] Shrublike in habit.

fruticose [BIOL] Resembling a shrub; applied especially to lichens. Also known as frutescent.

frying noise [ELEC] Noise in telephone transmission even when no conversation is taking place; caused by signal current flowing across a resistance element having multiple intermittent paths. Also known as transmitter noise.

F scan *See* F scope.

F scope [ELECTR] A cathode-ray scope on which a single signal appears as a spot with bearing error as the horizontal coordinate and elevation angle error as the vertical coordinate, with cross hairs on the scope face to assist in bringing the system to bear on the target. Also known as F indicator; F scan.

FSH *See* follicle-stimulating hormone.

FSK *See* frequency-shift keying.

F star [ASTRON] A star whose spectral type is F; surface temperature is 7000K, and color is yellowish.

f stop [OPTICS] An aperture setting for a camera lens; indicated by the *f* number.

f-sum rule [ATOM PHYS] The rule that the sum of the *f* values (or oscillator strengths) of absorption transitions of an atom in a given state, minus the sum of the *f* values of the emission transitions in that state, equals the number of electrons which take part in these transitions. Also known as Thomas-Reiche-Kuhn sum rule.

ft *See* foot.

ftc *See* footcandle.

F test *See* variance ratio test.

ft-L *See* foot lambert.

ft-lb *See* foot-pound.

ft-lbf *See* foot-pound.

ft-pdl *See* foot-poundal.

fu *See* flux unit.

FU *See* finsen unit.

Fubini's theorem [MATH] The theorem stating conditions under which $\iint f(u,v)dudv = \int du \int f(u,v)dv = \int dv \int f(u,v) \cdot du.$

Fucales [BOT] An order of brown algae composing the class Cyclosporeae.

fuchsin [ORG CHEM] $C_{20}H_{19}N_3HCl$ A dark-green powder, soluble in water and alcohol; used as a red dye in pharmaceuticals and leather and textile processing. Also known as magenta.

fuchsinophile [BIOL] Having an affinity for the dye fuchsin.

fuchsite [MINERAL] A bright-green variety of muscovite rich in chromium.

fucoid [GEOL] A tunnellike marking on a sedimentary structure identified as a trace fossil but not referred to a described genus.

fucoidin [BIOCHEM] A gum composed of L-fucose and sulfate acid ester groups obtained from *Fucus* species and other brown algae.

Fucophyceae [BOT] A class of brown algae.

L-fucopyranose *See* L-fucose.

L-fucose [BIOCHEM] $C_6H_{12}O_5$ A methyl pentose present in some algae and a number of gums and identified in the polysaccharides of blood groups and certain bacteria. Also known as 6-deoxy-L-galactose; L-fucopyranose; L-galactomethylose; L-rhodeose.

fucoxanthin [BIOCHEM] $C_{40}H_{60}O_6$ A carotenoid pigment; a partial xanthophyll ester found in diatoms and brown algae.

Fucus [BOT] A genus of dichotomously branched brown algae; it is harvested in the kelp industry as a source of algin.

fuel [MATER] A material that is burnt to release heat energy, for example, coal, oil, or uranium.

fuel assembly [NUCLEO] A combination of fuel and structural materials, used in some nuclear reactors to facilitate assembly of the core.

fuel cell [ELEC] A cell that converts chemical energy directly into electric energy, with electric power being produced as a part of a chemical reaction between the electrolyte and a fuel such as kerosine or industrial fuel gas.

fuel-cell catalyst [CHEM] A substance, such as platinum, silver, or nickel, from which the electrodes of a fuel cell are made, and which speeds the reaction of the cell; it is especially important in a fuel cell which does not operate at high temperatures.

fuel-cell electrolyte [CHEM] The substance which conducts electricity between the electrodes of a fuel cell.

fuel-cell fuel [CHEM] A substance, such as hydrogen, carbon monoxide, sodium, alcohol, or a hydrocarbon, which reacts with oxygen to generate energy in a fuel cell.

fuel-cooled [MECH ENG] Cooled by fuel, as in a rocket engine or an oil cooler.

fuel cycle *See* reactor fuel cycle.

fuel decanner [NUCLEO] A machine used for removing the stainless steel or other metal cans that enclose the enriched uranium fuel rods of a nuclear reactor; the cans are removed by a chipless machining operation in which the tubing is sheared into a spiral strip.

fuel element [NUCLEO] A rod, tube, plate, or other geometrical form into which nuclear fuel is fabricated for use in a reactor.

fuel filter [ENG] A device, as in an internal combustion engine, that removes particles from the fuel.

fuel gas [MATER] A gaseous fuel used to provide heat energy when burned with oxygen.

fuel injection [MECH ENG] The delivery of fuel to an internal combustion engine cylinder by pressure from a mechanical pump.

fuel injector [MECH ENG] A pump mechanism that sprays fuel into the cylinder of an internal combustion engine at the appropriate part of the cycle.

fuel oil [MATER] A liquid product burned to generate heat, exclusive of oils with a flash point below 100°F (38°C); includes heating oils, stove oils, furnace oils, bunker fuel oils.

fuel pellet [NUCLEO] A small pellet of frozen deuterium and tritium that would be used as fuel in a laser-induced fusion power plant.

fuel plate [NUCLEO] A form of nuclear fuel element consisting of a flat or slightly curved sheet of fuel, which is usually

FUEL ELEMENT

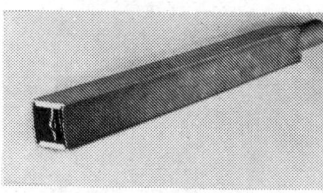

A plate-type fuel element.
(From Robert Laws, Salon of Photography)

a sandwich of uranium fuel protected by metallic cladding.

fuel pump [MECH ENG] A pump for drawing fuel from a storage tank and delivering it to an engine or furnace.

fuel reprocessing [NUCLEO] The processing of nuclear reactor fuel to recover the unused fissionable material.

fuel rod [NUCLEO] A long, rod-shaped fuel assembly.

fuel shutoff [AERO ENG] **1.** The action of shutting off the flow of liquid fuel into a combustion chamber or of stopping the combustion of a solid fuel. **2.** The event or time marking this action.

fuel structure ratio *See* fuel-weight ratio.

fuel system [MECH ENG] A system which stores fuel for present use and delivers it as needed.

fuel tank [MECH ENG] The operating, fuel-storage component of a fuel system.

fuel-weight ratio [AERO ENG] The ratio of the weight of a rocket's fuel to the weight of the unfueled rocket. Also known as fuel structure ratio.

fugacious [BOT] Lasting a short time; used principally to describe plant parts that fall soon after being formed.

fugacity [THERMO] A function used as an analog of the partial pressure in applying thermodynamics to real systems; at a constant temperature it is proportional to the exponential of the ratio of the chemical potential of a constituent of a system divided by the product of the gas constant and the temperature, and it approaches the partial pressure as the total pressure of the gas approaches zero.

fugitive air [MIN ENG] Air which moves through the ventilation fan but never reaches the mine workings.

fugitive dye [CHEM] A dye that is unstable, that is, not fast; used in the textile processing for purposes of identity.

fulchronograph [ENG] An instrument for recording lightning strokes, consisting of a rotating aluminum disk with several hundred steel fins on its rim; the fins are magnetized if they pass between two coils when these are carrying the surge current of a lightning stroke.

fulcrate [BIOL] Having a fulcrum.

fulcrate trophus [INV ZOO] A type of masticatory apparatus in certain rotifers characterized by an elongate fulcrum.

fulcrum [MECH] The rigid point of support about which a lever pivots.

Fuld-Gross unit [BIOL] A unit for the standardization of trypsin.

Fulgoroidea [INV ZOO] The lantern flies, a superfamily of homopteran insects in the series Auchenorrhyncha distinguished by the anterior and middle coxae being of equal length and joined to the body at some distance from the median line.

fulgurator [ENG] An atomizer used to spray salt solutions into a flame for analysis.

fulgurite [GEOL] A glassy, rootlike tube formed when a lightning stroke terminates in dry sandy soil; the intense heating of the current passing down into the soil along an irregular path fuses the sand.

full adder [ELECTR] A logic element which operates on two binary digits and a carry digit from a preceding stage, producing as output a sum digit and a new carry digit.

full annealing [MET] Heating steel to a high temperature and then cooling to ambient or near-ambient temperatures.

full automatic [ORD] A weapon that provides continuous fire as long as the trigger is depressed; used to distinguish from semiautomatic.

full automatic plating [MET] Electroplating a piece of work that is carried through the full cycle automatically.

full-cell process [ENG] A process of preservative treatment of wood that uses a pressure vessel and first draws a vacuum on the charge of wood and then introduces the preservative without breaking the vacuum. Also known as Bethell process.

full duplex [ADP] The complete duplication of any data-processing facility. [COMMUN] Of a telegraph or other data channel, able to operate in both directions simultaneously.

full-duplex operation [COMMUN] Simultaneous communications in both directions between two points.

full-ended [NAV ARCH] The condition when the extremities of the waterlines in the vicinity of the load line are strongly convex to the surrounding water and the ends of the sectional area curve are full indicating that the displacement is carried well forward and aft toward the ends of the vessel.

fuller [MET] A die or portion of a die used in preliminary forging operations to reduce the cross section somewhere between the ends of a piece of stock.

fuller's earth [GEOL] A natural, fine-grained earthy material, such as a clay, with high adsorptive power; consists principally of hydrated aluminum silicates; used as an adsorbent in refining and decolorizing oils, as a catalyst, and as a bleaching agent.

full-face firing [MIN ENG] Drilling of small-diameter holes from top to bottom of the face.

full-face tunneling [CIV ENG] A system of tunneling in which the tunnel opening is enlarged to desired diameter before extension of the tunnel face.

full feathering *See* feathering.

full-gear [MECH ENG] The condition of a steam engine when the valve is operated to the maximum extent by the link motion.

fulling [TEXT] A process for finishing woolen fabric in which the material is dampened and then beaten under heat to make the weave less visible.

fulling agent [TEXT] A soap solution used during the fulling of wool.

full linear group [MATH] The group of all nonsingular linear transformations of a vector space whose group operation is composition.

full load [ELEC] The greatest load that a circuit or piece of equipment is designed to carry under specified conditions.

full-load current [ELEC] The greatest current that a circuit or piece of equipment is designed to carry under specified conditions.

full-mill [BUILD] A type of construction in which all vertical apertures open onto shafts of brick or other fireproof material; used for fire retardance.

full moon [ASTRON] The moon at opposition, with a phase angle of 0°, when it appears as a round disk to an observer on the earth because the illuminated side is toward the observer.

full period allocated circuit [COMMUN] Communications link (allocated circuit) assigned for the exclusive use of previously defined users at two or more terminal points.

full-pitch winding [ELEC] An armature winding in which the distance between two active conductors of a coil equals the pole pitch.

full plate [HOROL] A watch with train wheels and escapement under a single plate; only the balance is exposed.

full-scale [ORD] Of an attack or other operation, all-out or maximum.

full-seam mining [MIN ENG] A mining system in which the entire section is dislodged together and the coal is separated from the rock outside the mine by the cleaning plant.

full section filter [ELECTR] A filter network whose graphical representation has the shape of the Greek letter pi, connoting capacitance in the upright legs and inductance or reactance in the horizontal member.

full subsidence [MIN ENG] The greatest amount of subsidence occurring as a result of mine workings.

full-track vehicle [MECH ENG] A vehicle entirely supported, driven, and steered by an endless belt, or track, on each side; for example, a tank.

full-wave amplifier [ELECTR] An amplifier without any clipping.

full-wave bridge [ELECTR] A circuit having a bridge with four diodes, which provides full-wave rectification and gives twice as much direct-current output voltage for a given alternating-current input voltage as a conventional full-wave rectifier.

full-wave rectification [ELECTR] Rectification in which output current flows in the same direction during both half cycles of the alternating input voltage.

full-wave rectifier [ELECTR] A double-element rectifier that provides full-wave rectification; one element functions during positive half cycles and the other during negative half cycles.

full-wave vibrator [ELEC] A vibrator having an armature that moves back and forth between two fixed contacts so as to change the direction of direct-current flow through a transformer at regular intervals and thereby permit voltage stepup

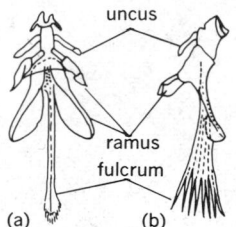

FULCRATE TROPHUS

Fulcrate trophus of *Seison*.
(a) Dorsal view. *(b)* Lateral view.
(After de Beauchamp)

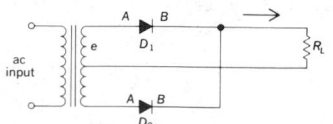

FULL-WAVE RECTIFIER

Circuit diagram of a full-wave
diode rectifier.

by the transformer; used in battery-operated power supplies for mobile and marine radio equipment.

full wires [MATER] In wire-mesh cloth, wires running the short way of the cloth as woven. Also known as shute wires.

full-word boundary [ADP] In the IBM 360 system, any address which ends in 00, and is therefore a natural boundary for a four-byte machine word.

fully arisen sea *See* fully developed sea.

fully developed mine [MIN ENG] In coal mining, a mine where all development work has reached the boundaries and further extraction will be done on the retreat.

fully developed sea [OCEANOGR] The maximum ocean waves or sea state that can be produced by a given wind force blowing over sufficient fetch, regardless of duration. Also known as fully arisen sea.

fulmar [VERT ZOO] Any of the oceanic birds composing the family Procellariidae; sometimes referred to as foul gulls because of the foul-smelling substance spat at intruders upon their nests.

fulminate [MED] Of a disease, to come suddenly and follow a severe, intense, and rapid course. [ORG CHEM] **1.** A salt of fulminic acid. **2.** $HgC_2N_2O_2$ An explosive mercury compound derived from the fulminic acid; used for the caps or exploders by means of which charges of gunpowder, dynamite, and other explosives are fired. Also known as mercury fulminate.

fulminic acid [ORG CHEM] CNOH An unstable isomer of cyanic acid, whose salts are known for their explosive characteristics. Also known as 4-cyanic acid; paracyanic acid.

fulminuric acid [ORG CHEM] $CN \cdot CH(NO_2) \cdot CONH_2$ A trimer of cyanuric acid; a water-soluble compound, crystallizing in colorless needles, melting at 138°C, and exploding at 145°C. Also known as cyanonitroacetamide; 2-cyano-2-nitroethanamide; isocyanuric acid.

fulvene [ORG CHEM] C_6H_6 A yellow oil, an isomer of benzene. Also known as methylenecyclopentadiene.

Fulvicin [MICROBIO] A trade name for the antibiotic griseofulvin.

fumagillin [MICROBIO] $C_{26}H_{34}O_7$ An insoluble, crystalline antibiotic produced by a strain of the fungus *Aspergillus fumigatus*.

fumarase [BIOCHEM] An enzyme that catalyzes the hydration of fumaric acid to malic acid, and the reverse dehydration.

Fumariaceae [BOT] A family of dicotyledonous plants in the order Papaverales having four or six stamens, irregular flowers, and no latex system.

fumaric acid [ORG CHEM] $C_4H_4O_4$ A dicarboxylic organic acid produced commercially by synthesis and fermentation; the trans isomer of maleic acid; colorless crystals, melting point 287°C; used to make resins, paints, varnishes, and inks, in foods, as a mordant, and as a chemical intermediate. Also known as boletic acid.

fumarole [GEOL] A hole, usually found in volcanic areas, from which vapors or gases escape.

fumble [IND ENG] An unintentional sensory-motor error that may be unavoidable.

fume hood [CHEM] A fume-collection device over an enclosed shelf or table, so that experiments involving poisonous or unpleasant fumes or gases may be conducted away from the experimental area.

fumes [CHEM] Particulate matter consisting of the solid particles generated by condensation from the gaseous state, generally after volatilization from melted substances, and often accompanied by a chemical reaction, such as oxidation.

fumigant [CHEM] A chemical compound which acts in the gaseous state to destroy insects and their larvae and other pests; examples are dichlorethyl ether, *p*-dichlorobenzene, and ethylene oxide.

fumigating [ENG] The use of a chemical compound in a gaseous state to kill insects, nematodes, arachnids, rodents, weeds, and fungi in confined or inaccessible locations; also used to control weeds, nematodes, and insects in the field.

fuming nitric acid [INORG CHEM] Concentrated nitric acid containing dissolved nitrogen dioxide; may be prepared by adding formaldehyde to concentrated nitric acid.

FULVENE

Structural formula of fulvene.

fumulus [METEOROL] A very thin cloud veil at any level, so delicate that it may be almost invisible.

funal *See* sthène.

Funariales [BOT] An order of mosses; plants are usually annual, are terrestrial, and have stems that are erect, short, simple, or sparingly branched.

function [ADP] In FORTRAN, a subroutine of a particular kind which returns a computational value whenever it is called. [MATH] A mathematical rule between two sets which assigns to each member of the first, exactly one member of the second.

functional [ADP] In a linear programming problem involving a set of variables x_j, $j = 1, 2, \ldots, n$, a function of the form $c_1x_1 + c_2x_2 + \ldots + c_nx_n$ (where the c_j are constants) which one wishes to optimize (maximize or minimize, depending on the problem) subject to a set of restrictions. [MATH] Any function from a vector space into its scalar field.

functional analysis [MATH] A branch of analysis which studies the properties of mappings of classes of functions from one topological vector space to another.

functional bombing [ORD] The bombing of a specially selected class of key targets, such as dams or marshaling yards, that function within an industrial complex or transportation system, as opposed to the bombing of an entire system.

functional constraint [MATH] A mathematical equation which must be satisfied by the independent parameters in an optimization problem, representing some physical principle which governs the relationship among these parameters.

functional design [ADP] A level of the design process in which subtasks are specified and the relationships among them defined, so that the total collection of subsystems performs the entire task of the system.

functional diagram [ADP] A diagram that indicates the functions of the principal parts of a total system and also shows the important relationships and interactions among these parts.

functional forms analysis chart *See* form process chart.

functional generator *See* function generator.

functional group [ORG CHEM] A chemical radical or structure that has characteristic properties; examples are a hydroxyl group and a carboxyl group.

functional interleaving [ADP] Alternating the parts of a number of sequences in a cyclic fashion, such as a number of accesses to memory followed by an access to a data channel.

functionality [CHEM] Ability of a compound to form covalent bonds; compounds may be mono-, di-, tri-, or polyfunctional, that is, one, two, three, or many functional groups may participate in a reaction.

functional multiplier *See* function multiplier.

functional paralysis *See* hysterical paralysis.

functional requirement [ADP] The documentation which accompanies a program and states in detail what is to be performed by the system.

functional switching circuit [ELECTR] One of a relatively small number of types of circuits which implements a Boolean function and constitutes a basic building block of a switching system; examples are the AND, OR, NOT, NAND, and NOR circuits.

functional unit [ADP] The part of the computer required to perform an elementary process such as an addition or a pulse generation.

function code [ADP] Special code which appears on a medium such as a paper tape and which controls machine functions such as a carriage return.

function-evaluation routine [ADP] A canned routine such as a log function or a sine function.

function generator [ELECTR] Also known as functional generator. **1.** An analog computer device that indicates the value of a given function as the independent variable is increased. **2.** A signal generator that delivers a choice of a number of different waveforms, with provisions for varying the frequency over a wide range.

function hole *See* designation punch.

function multiplier [ELECTR] An analog computer device that takes in the changing values of two functions and puts out the changing value of their product as the independent variable is changed. Also known as functional multiplier.

function switch [ELECTR] A network having a number of inputs and outputs so connected that input signals expressed in a certain code will produce output signals that are a function of the input information but in a different code.

function table [ADP] **1.** Sets of computer information arranged so an entry in one set selects one or more entries in the other sets. **2.** A computer device that converts multiple inputs into a single output or encodes a single input into multiple outputs. [MATH] A table that lists the values of a function for various values of the variable.

function unit [ADP] In computer systems, a device which can store a functional relationship and release it continuously or in increments.

functor [ADP] *See* logic element. [MATH] A function between categories which associates objects with objects and morphisms with morphisms.

fundamental [PHYS] The lowest frequency component of a complex wave. Also known as first harmonic; fundamental component.

fundamental affine connection [MATH] An affine connection whose coefficients arise from the covariant and contravariant metric tensors of a space.

fundamental circle *See* primary great circle.

fundamental complex [GEOL] An agglomeration of metamorphic rocks underlying sedimentary or unmetamorphosed rocks; specifically, an agglomeration of Archean rocks supporting a geological column.

fundamental component *See* fundamental.

fundamental forms of a surface [MATH] Differential forms which express the area and curvature of the surface.

fundamental frequency [PHYS] **1.** The lowest frequency at which a system vibrates freely. **2.** The lowest frequency in a complex wave.

fundamental group [COMMUN] In wire communications, a group of trunks that connect each local or trunk switching center to a trunk switching center of higher rank on which it homes; the term also applies to groups that interconnect zone centers.

fundamental group of a topological space [MATH] The group of homotopy classes of all closed paths about a point; this group yields information about the number and type of "holes" in a surface.

fundamental jelly *See* ulmin.

fundamental mode [ELECTROMAG] The waveguide mode having the lowest critical frequency. Also known as dominant mode; principal mode. [PHYS] The normal mode of vibration having the lowest frequency.

fundamental motion *See* elemental motion.

fundamental particle *See* elementary particle.

fundamental region [MATH] Any region in the complex plane that can be mapped conformally onto all of the complex plane.

fundamental sequence *See* Cauchy sequence.

fundamental series [SPECT] A series occurring in the line spectra of many atoms and ions having one, two, or three electrons in the outer shell, in which the total orbital angular momentum quantum number changes from 3 to 2.

fundamental star places [ASTRON] The apparent right ascensions and declinations of 1535 standard comparison stars obtained by leading observatories and published annually under the sponsorship of the International Astronomical Union.

fundamental substance *See* ulmin.

fundamental theorem of algebra [MATH] Every polynomial of degree n with complex coefficients has exactly n roots counted according to multiplicity.

fundamental theorem of arithmetic [MATH] Every positive integer greater than 1 can be factored uniquely into the form $P_1^{n_1} \ldots P_k^{n_1} \ldots P_k^{n_k}$, where the P_i are primes, the n_i positive integers.

fundamental theorem of calculus [MATH] Given a continuous function $f(x)$ on the closed interval $[a,b]$ the functional

$$F(x) = \int_a^x f(t)\,dt$$

is differentiable on $[a,b]$ and $F'(x) = f(x)$ for every x in $[a,b]$,

and if G is any function on $[a,b]$ such that $G'(x) = f(x)$ for all x in $[a,b]$, then

$$\int_a^b f(t)\,dt = G(b) - G(a).$$

fundamental tone [ACOUS] The component tone of lowest pitch in a complex tone.

fundamental wavelength [PHYS] Of an oscillatory device, that wavelength corresponding to its fundamental frequency.

fundic gland [ANAT] Any of the glands of the corpus and fundus of the stomach.

fundus [ANAT] The bottom of a hollow organ.

fungi [MYCOL] Nucleated, usually filamentous, sporebearing organisms devoid of chlorophyll.

fungicide [MATER] An agent that kills or destroys fungi.

fungiform [BIOL] Mushroom-shaped.

fungiform papilla [ANAT] One of the low, broad papillae scattered over the dorsum and margins of the tongue.

Fungi Imperfecti [MYCOL] A class of the subdivision Eumycetes; the name is derived from the lack of a sexual stage.

fungi-proofing [ENG] Application of a protective chemical coating that inhibits growth of fungi.

fungistat [MATER] A compound that inhibits or prevents growth of fungi.

Fungivoridae [INV ZOO] The fungus gnats, a family of orthorrhaphous dipteran insects in the series Nematocera; the larvae feed on fungi.

fungivorous [ZOO] Feeding on or in fungi.

Fungizone [MICROBIO] The trade name for the antibiotic amphotericin.

fungus [MYCOL] Singular of fungi.

fungus gall [PL PATH] A plant gall resulting from an attack of a parasitic fungus.

funicle *See* funiculus.

funicular [ENG] Cable railway up a mountain, with cars simultaneously ascending and descending to counterbalance.

funicular distribution [CHEM] The distribution of a two-phase, immiscible liquid mixture (such as oil and water, one a wetting phase, the other nonwetting) in a porous system when the wetting phase is continuous over the surface of the solids.

funicular polygon [MECH] **1.** The figure formed by a light string hung between two points from which weights are suspended at various points. **2.** A force diagram for such a string, in which the forces (weights and tensions) acting on points of the string from which weights are suspended are represented by a series of adjacent triangles.

funicular railroad [ENG] A railroad system used primarily to ascend and descend mountains; the weight of the descending train helps to move the ascending train up the mountain.

funiculus [ANAT] Also known as funicle. **1.** Any structure in the form of a chord. **2.** A column of white matter in the spinal cord. [BOT] The stalk of an ovule. [INV ZOO] A band of tissue extending from the adoral end of the coelom to the adoral body wall in bryozoans.

funnel [DES ENG] A tube with one conical end that sometimes holds a filter; the function is to direct flow of a liquid or, if a filter is present, to direct a flow that was filtered. [NAV ARCH] A smokestack of a ship.

funnel cloud [METEOROL] The popular term for the tornado cloud, often shaped like a funnel with the small end nearest the ground.

fur [MATER] The dressed pelt of a mammal. [VERT ZOO] The coat of a mammal.

Furadantin [MICROBIO] Trade name for nitrofurantoin.

2-furaldehyde *See* furfural.

furan [ORG CHEM] **1.** One of a group of organic heterocyclic compounds containing a diunsaturated ring of four carbon atoms and one oxygen atom. **2.** $C_4H_4O_4$ The simplest furan type of molecule; a colorless, mildly toxic liquid, boiling at 32°C, insoluble in water, soluble in alcohol and ether; used as a chemical intermediate. Also known as furfuran; tetrol.

2-furancarbinol *See* furfuryl alcohol.

furancarboxylic acid *See* furoic acid.

2,5-furandione *See* maleic anhydride.

furanose [BIOCHEM] A sugar whose cyclic or ring structure resembles that of furan.

furanoside [ORG CHEM] A glycoside whose cyclic sugar component resembles that of furan.

furan resin [ORG CHEM] A liquid, thermosetting resin in which the furan ring is an integral part of the polymer chain, made by the condensation of furfuryl alcohol; used as a cement and adhesive, casting resin, coating, and impregnant.

furca [INV ZOO] A forked process as the last abdominal segment of certain crustaceans, and as part of the spring in collembolans.

furcate [BIOL] Forked.

furcocercous cercaria [INV ZOO] A free-swimming, digenetic trematode larva with a forked tail.

furcula [ZOO] A forked structure, especially the wishbone of fowl.

furfural [ORG CHEM] C_4H_3OCHO When pure, a colorless liquid, soluble in organic solvents, slightly soluble in water; used as a lube oil–refining solvent, in cellulosic formulations, in making resins, as a weed killer, as a fungicide, and as a chemical intermediate. Also known as 2-furaldehyde; furfuraldehyde; furfurol; furol.

furfur alcohol *See* furfuryl alcohol.

furfuraldehyde *See* furfural.

furfural extraction [CHEM ENG] Process for the refining of lubricating oils and other organic materials by contact with furfural.

furfuran *See* furan.

furfurol *See* furfural.

furfuryl [ORG CHEM] The univalent radical C_5H_5O- from furfural.

furfuryl alcohol [ORG CHEM] $C_5H_6O_2$ A liquid with a faint burning odor and bitter taste, soluble in alcohol and ether, usually prepared from furfural; used as a solvent in the manufacturing of wetting agents and resins. Also known as 2-furancarbinol; furfur alcohol; furylcarbinol; 2-furylcarbinol; 2-hydroxy methyl furan.

furiani [METEOROL] A southwest wind that blows in the vicinity of the Po River, Italy, and is vehement and short-lived, followed by a gale from the south or southeast.

Furipteridae [VERT ZOO] The smoky bats, a family of mammals in the order Chiroptera having a vestigial thumb and small ears.

furlong [MECH] A unit of length, equal to 1/8 mile, 660 feet, or 201.168 meters.

furnace [ENG] An apparatus in which heat is liberated and transferred directly or indirectly to a solid or fluid mass for the purpose of effecting a physical or chemical change.

furnace black [CHEM] A carbon black formed by partial combustion of liquid and gaseous hydrocarbons in a closed furnace with a deficiency of oxygen; used as a reinforcing filler for synthetic rubber.

furnace brazing [MET] Joining two metals by mechanical union of the filler metal and joint, then heating the composite in a furnace.

furnace cupola *See* cupola.

furnace lining [ENG] The interior part of a furnace in contact with a molten charge and hot gases; constructed of heat-resistant material.

furnace oil [MATER] Distillate fuel oil intended primarily for domestic central heating systems; usually No. 1 fuel oil.

furnace refining [MET] Purification of molten metal by treatment in a reverberatory furnace.

Furnariidae [VERT ZOO] The oven birds, a family of perching birds in the superfamily Furnarioidea.

Furnarioidea [VERT ZOO] A superfamily of birds in the order Passeriformes characterized by a predominance of gray, brown, and black plumage.

furoic acid [ORG CHEM] $C_5H_4O_3$ Long monoclinic prisms crystallized from the water solution, soluble in ether and alcohol; used as a preservative and bactericide. Also known as furancarboxylic acid; pyromucic acid.

furol *See* furfural.

furred ceiling [BUILD] A ceiling in which the furring units are attached directly to the structural units of the building.

furring [BUILD] Thin strips of wood or metal applied to the joists, studs, or wall of a building to level the surface, create an air space, or add thickness.

furring brick [MATER] Hollow brick grooved for plastering.

furring tile [MATER] Non-load-bearing clay tile used for lining interior walls.

furrow [ENG] A trench plowed in the ground.

furrow irrigation [AGR] Irrigation via furrows between rows of crops.

furrow press [AGR] A device that firms the earth in a furrow after plowing.

Furry theorem [QUANT MECH] In quantum electrodynamics, the theorem that the contribution of a Feynman diagram, consisting of a closed polygon of fermion lines connected to an odd number of photon lines, vanishes.

furuncle [MED] A small cutaneous abscess, usually resulting from infection of a hair follicle by *Staphylococcus aureus*. Also known as boil.

furunculosis [MED] A condition marked by numerous furuncles, or the recurrence of furuncles following healing of a preceding crop.

furylcarbinol *See* furfuryl alcohol.

2-furylcarbinol *See* furfuryl alcohol.

fusain [GEOL] The local lithotype strands or patches, characterized by silky luster, fibrous structure, friability, and black color. Also known as mineral charcoal.

Fusarium [MYCOL] A genus of fungi in the family Tuberculariaceae having sickle-shaped, multicelled conidia; includes many important plant pathogens.

Fusarium oxysporum [MYCOL] A pathogenic fungus causing a variety of plant diseases, including cabbage yellows and wilt of tomato, flax, cotton, peas, and muskmelon.

Fusarium solani [MYCOL] A pathogenic fungus implicated in root rot and wilt diseases of several plants, including sisal and squash.

fuse [ELEC] An expandable device for opening an electric circuit when the current therein becomes excessive, containing a section of conductor which melts when the current through it exceeds a rated value for a definite period of time. Also known as electric fuse. [ENG] Also spelled fuze. **1.** A device with explosive components designed to initiate a train of fire or detonation in an item of ammunition by an action such as hydrostatic pressure, electrical energy, chemical energy, impact, or a combination of these. **2.** A nonexplosive device designed to initiate an explosion in an item of ammunition by an action such as continuous or pulsating electromagnetic waves or acceleration.

fuse alarm [ELEC] Circuit that produces a visual or audible signal to indicate a blown fuse.

fuse blasting cap [ENG] A small copper cylinder closed at one end and charged with a fulminate.

fuse block [ELEC] An insulating base on which are mounted fuse clips or other contacts for fuses. Also known as fuseboard.

fuseboard *See* fuse block.

fuse body [ENG] The part of a fuse contributing the major portion of the total weight, and which houses the majority of the functioning parts, and to which smaller parts are attached.

fuse box *See* cutout box.

fuse clip [ELEC] A spring contact used to hold and make connection to a cartridge-type fuse.

fuse cutout [ELEC] Assembly of a fuse support and a fuse holder which may or may not include the fuse link.

fused-electrolyte battery *See* thermal battery.

fuse disconnecting switch [ELEC] Disconnecting switch in which a fuse unit forms a part of the blade.

fused junction *See* alloy junction.

fused-junction diode *See* alloy junction diode.

fused-junction transistor *See* alloy junction transistor.

fused potassium sulfide *See* potassium sulfide.

fused quartz [MATER] A glasslike insulating material made by melting crushed crystals of natural quartz or a certain type of quartz sand.

fused-salt electrolysis [PHYS CHEM] Electrolysis with use of purified fused salts as raw material and as an electrolyte.

fused-salt reactor *See* molten-salt reactor.

fused semiconductor [ELECTR] Junction formed by recrystallization on a base crystal from a liquid phase of one or more components and the semiconductor.

fused silica *See* silica glass.

FURCOCERCOUS CERCARIA

Furcocercous cercaria, obtained either by crushing and examining freshly collected snails or by placing snails in water to permit larvae to escape.

labels: tail, stem, furca, caudal-fin fold

FURFURAL

Structural formula of furfural.

fuse gage [ENG] An instrument for slicing time fuses to length.

fusehead [ENG] That part of an electric detonator consisting of twin metal conductors, bridged by fine resistance wire, and surrounded by a bead of igniting compound which burns when the firing current is passed through the bridge wire.

fuselage [AERO ENG] In an airplane, the central structure to which wings and tail are attached; it accommodates flight crew, passengers, and cargo.

fuse lighter [ENG] A device for facilitating the ignition of the powder core of a fuse.

fuse link [ELEC] Part of a fuse that carries the current of the circuit and all or part of which melts when the current exceeds a predetermined value.

fusel oil [MATER] A volatile, poisonous mixture of isoamyl, butyl, propyl, and heptyl alcohols produced as by-products in alcoholic fermentation of starches, grains, or fruits to produce ethyl alcohol.

fuse wire [ELEC] Wire made from an alloy that melts at a relatively low temperature and overheats to this temperature when carrying a particular value of overload current.

fusibility [THERMO] The quality or degree of being capable of being liquefied by heat.

fusible alloy [MET] A low melting alloy, usually of bismuth, tin, cadmium, and lead, which melts at temperatures as low as 70°C (160°F).

fusible plug *See* safety plug.

fusible resistor [ELEC] A resistor designed to protect a circuit against overload; its resistance limits current flow and thereby protects against surges when power is first applied to a circuit; its fuse characteristic opens the circuit when current drain exceeds design limits.

fusiform [BIOL] Spindle-shaped; tapering toward the ends.

fusiform bacillus [MICROBIO] A bacillus having one blunt and one pointed end, as *Fusobacterium fusiforme.*

fusiform initial cell [BOT] A cell type of the vascular cambium that gives rise to all cells in the vertical system of secondary xylem and phloem.

fusimotoneuron [PHYSIO] One of the small motor fibers, composing about 30% of the fibers in the ventral root of the spinal cord, which innervate intrafusal fibers.

fusing disk [MECH ENG] A rapidly spinning disk that cuts metal by melting it.

fusinite [GEOL] The micropetrological constituent of fusain which consists of carbonized woody tissue.

fusinization [GEOL] The process of formation of fusain in coal.

fusion [NUC PHYS] Combination of two light nuclei to form a heavier nucleus (and perhaps other reaction products) with release of some binding energy. Also known as atomic fusion; nuclear fusion. [PHYS CHEM] A change of the state of a substance from the solid phase to the liquid phase. Also known as melting.

fusion bomb [ORD] A bomb that depends upon nuclear fusion for release of energy. Also known as fusion weapon.

fusion fuel [NUCLEO] A substance which may generate energy in a fusion reaction, such as deuterium, deuterium and tritium, or deuterium and helium-3.

fusion nucleus [BOT] The triploid, or $3n$, nucleus which results from double fertilization and which produces the endosperm in some seed plants.

fusion piercing [ENG] A method of producing vertical blastholes by virtually burning holes in rock. Also known as piercing.

fusion-piercing drill [ENG] A machine designed to use the fusion-piercing mode of producing holes in rock. Also known as det drill; jet-piercing drill; Linde drill.

fusion point [NUCLEO] The temperature of a plasma above which the rate of energy generation by nuclear fusion reactions exceeds the rate of energy loss from the plasma, so that the fusion reaction can be self-sustaining. Abbreviated fnp.

fusion reactor [NUCLEO] Proposed device in which controlled, self-sustaining nuclear fusion reactions would be carried out in order to produce useful power.

fusion tube [ANALY CHEM] Device used for the analysis of the elements in a compound by fusing them with another compound; for example, analysis of nitrogen in organic compounds by fusing the compound with sodium and analyzing for sodium cyanide.

fusion weapon *See* fusion bomb.

fusion welding [MET] Any welding operation involving melting of the base or parent metal.

fusion zone [MET] The volume of base or parent metal melted during a welding operation.

Fusobacterium [MICROBIO] A genus of gram-negative, straight or curved fusiform rods, occurring singly, in pairs, or in short chains, in the family Bacteroidaceae; pathogenicity is limited.

fusula [INV ZOO] A spindle-shaped, terminal projection of the spinneret of a spider.

Fusulinacea [PALEON] A superfamily of large, marine extinct protozoans in the order Foraminiferida characterized by a chambered calcareous shell.

Fusulinidae [PALEON] A family of extinct protozoans in the superfamily Fusulinacea.

Fusulinina [PALEON] A suborder of extinct rhizopod protozoans in the order Foraminiferida having a monolamellar, micrograular calcite wall.

future address patch [ADP] A computer output containing the address of a symbol and the address of the last reference to that symbol.

future label [ADP] An address referenced in the operand field of an instruction, but which has not been previously defined.

future of an event [RELAT] Those events which can be reached by a signal emitted at the event and which move at a speed less than or equal to the speed of light in a vacuum.

fuze *See* fuse.

fv *See* femtovolt.

f value *See* oscillator strength.

Fyrite [ENG] Trade name for a portable instrument for measuring the carbon dioxide content of air; air is pumped through a solution of caustic soda which combines with CO_2, causing a decrease in gas volume in the unit so that the fluid is drawn up into a graduated tube in a quantity proportionate with the CO_2 absorbed.

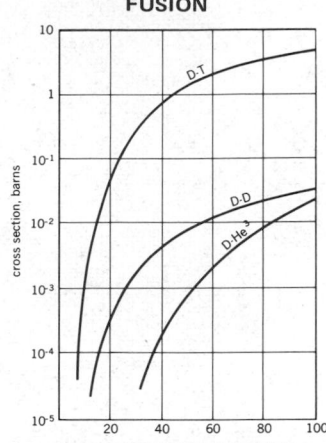

FUSION

Cross sections versus bombarding energy for three simple fusion reactions. D-T = deuterium-tritium reaction; D-D = deuterium-deuterium reaction; D-He3 = deuterium-helium 3 reaction. *(From R. F. Post, Fusion power, Sci. Amer., 197(6):73–84, 1957)*

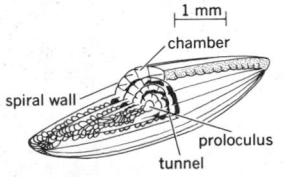

FUSULINIDAE

A representative of the Fusulinidae shown in a cutaway diagram.

FUSULININA

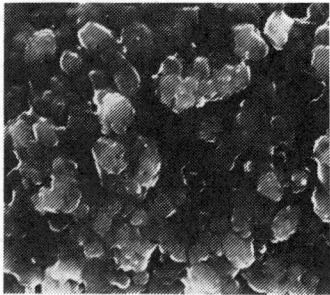

Scanning electron micrograph of the surface of the wall of *Earlandia* showing the randomly packed tiny angular calcite grains. *(R. B. MacAdam, Chevron Oil Field Research Co.)*

g   *See* gram.

G   [ELEC] *See* conductance.   [MECH] A unit of acceleration equal to the standard acceleration of gravity, 9.80665 meters per second per second, or approximately 32.1740 feet per second per second. Also known as fors; grav.   [SCI TECH] *See* giga-.

Ga   *See* gallium.

GA agent   [CHEM] $(CH_3)_2NP(CN)O(OC_2H_5)$ A phosphorus ester; a nerve gas, ethylphosphorodimethylamidocyanidate; absorbed into the body through the lungs or skin. Also known as tabun.

gabardine   [TEXT] Steep twill fabric made of closely woven single or two-ply fibers of wool, rayon, polyester, or other yarn.

gabbro   [PETR] A group of dark-colored, intrusive igneous rocks with granular texture, composed largely of basic plagioclase and clinopyroxene.

gabion   [ENG] A bottomless basket of wickerwork or strap iron filled with earth or stones; used in building fieldworks or as revetments in mining. Also known as pannier.

gable   [ARCH] The upper, triangular portion of the terminal wall of a building under the ridge of a sloped roof.

gable roof   [ARCH] A sloping peaked roof that forms a gable at each end.

Gabriel's synthesis   [ORG CHEM] A synthesis of primary amines by the hydrolysis of N-alkylphthalimides; the latter are obtained from potassium phthalimide and alkyl halides.

gad   [MIN ENG] **1.** A heavy steel wedge, 6 or 8 inches (15 or 20 centimeters) long, with a narrow chisel point used in mining to cut samples, break out pieces of loose rock, and so on. **2.** A small iron punch with a wooden handle used to break up ore.

gadder   [MIN ENG] A small car or platform with a drilling machine attached, to make a straight line of holes along its course in getting out dimension stone. Also known as gadding car; gadding machine.

gadding car   *See* gadder.

gadding machine   *See* gadder.

Gadidae   [VERT ZOO] A family of fishes in the order Gadiformes, including cod, haddock, pollock, and hake.

Gadiformes   [VERT ZOO] An order of actinopterygian fishes that lack fin spines and a swim bladder duct and have cycloid scales and many-rayed pelvic fins.

gadoleic acid   [ORG CHEM] $C_{20}H_{38}O_2$ A fatty acid derived from cod liver oil, and melting at 20°C.

gadolinite   [MINERAL] $Be_2FeY_2Si_2O_{10}$ A black, greenish-black, or brown rare-earth mineral; hardness is 6.5–7 on Mohs scale, and specific gravity is 4–4.5.

gadolinium   [CHEM] A rare-earth element, symbol Gd, atomic number 64, atomic weight 157.25; highly magnetic, especially at low temperatures.

gaffkemia   [INV ZOO] A septicemic bacterial disease of lobsters caused by *Gaffkya homari*.

Gaffkya   [MICROBIO] A genus of gram-positive, parasitic, coccoid bacteria in the family Micrococcaceae occurring singly and in tetrads; included by many in the genus *Micrococcus*.

gage   Also spelled gauge.   [CIV ENG] The distance between the inner faces of the rails of railway track; standard gage in the United States is 4 feet 8 1/2 inches (1.44 meters).   [DES ENG] **1.** A device for determining the relative shape or size of an object. **2.** The thickness of a metal sheet, a rod, or a wire.   [ELECTROMAG] One of the family of possible choices for the electric scalar potential and magnetic vector potential, given the electric and magnetic fields.   [ENG] The minimum sieve size through which most (95% or more) of an aggregate will pass.   [ORD] The interior diameter of the barrel of a shotgun expressed by the number of spherical lead bullets fitting it that are required to make a pound.

gage block   [DES ENG] A chrome steel block having two flat, parallel surfaces with the parallel distance between them being the size marked on the block to a guaranteed accuracy of a few millionths of an inch; used as the standard of precise lineal measurement for most manufacturing processes. Also known as precision block; size block.

gage complete penetration   [ORD] Penetration in which a hole of sufficient size is made through the plate so as to fully admit a plug gage of a designated percentage of diameter; if the projectile remains in the plate and prevents the complete insertion of the plug gage through the hole, the round is disregarded.

gage glass   [ENG] A glass, plastic, or metal tube, usually equipped with shutoff valves, that is connected by a suitable fitting to a tank or vessel, for the measurement of liquid level.

gage invariance   [ELECTROMAG] The invariance of electric and magnetic fields and electrodynamic interactions under gage transformations.

gageite   [MINERAL] $(Mn,Mg,Zn)_8Si_3O_{14} \cdot 2H_2O(or\ 3H_2O)$ A mineral composed of a hydrous silicate of manganese, magnesium, and zinc.

gage length   [ENG] Original length of the portion of a specimen measured for strain, length changes, and other characteristics.

gage loss   [MIN ENG] The diametrical reduction in the size of a bit or reaming shell caused by wear through use.

gage plate   [CIV ENG] A plate inserted between the parallel rails of a railroad track to maintain the gage.

gage point   [DES ENG] A point used to position a part in a jig, fixture, or qualifying gage.

gage pressure   [MECH ENG] The amount by which the total absolute pressure exceeds the ambient atmospheric pressure.

gage theory   [PHYS] Any field theory in which, as the result of the conservation of some quantity, it is possible to perform a transformation in which the phase of the fields is altered by a function of space and time without altering any measurable physical quantity, so that the fields obtained by any such transformation give a valid description of a given physical situation.

gage transformation   [ELECTROMAG] The addition of the gradient of some function of space and time to the magnetic vector potential, and the addition of the negative of the partial derivative of the same function with respect to time, divided by the speed of light, to the electric scalar potential; this procedure gives different potentials but leaves the electric and magnetic fields unchanged.

gagger   [MET] An irregular-shaped piece of metal used in a sand mold to reinforce and support a metal casting.

gaging   [NUCLEO] The measurement of the thickness, density, or quantity of material by the amount of radiation it absorbs; this is the most common use of radioactive isotopes in industry. Also spelled gauging.

gahnite   [MINERAL] $ZnAl_2O_4$ A usually dark-green, but sometimes yellow, gray, or black spinel mineral consisting of an oxide of zinc and aluminum. Also known as zinc spinel.

gaign   [METEOROL] A cross-mountain wind that causes clouds to form on the crests of mountains in Italy.

gain   [ELECTR] **1.** The increase in signal power that is produced by an amplifier; usually given as the ratio of output to input voltage, current, or power, expressed in decibels. Also known as transmission gain. **2.** *See* antenna gain.

gain asymptotes   [CONT SYS] Asymptotes to a logarithmic graph of gain as a function of frequency.

gain-bandwidth product   [ELECTR] The midband gain of an amplifier stage multiplied by the bandwidth in megacycles.

gain control   [ELECTR] A device for adjusting the gain of a system or component.

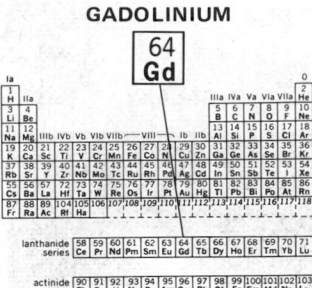
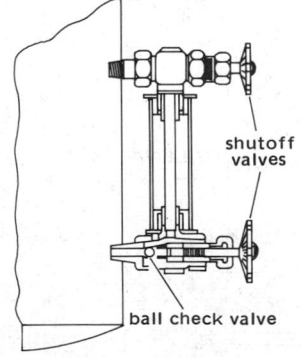

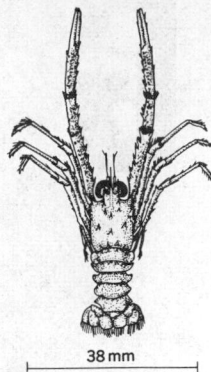

GALATHEIDEA

Munida evermanni.
(Smithsonian Institution)

GALAXY

Great Spiral NGC 224 in Andromeda, which resembles the Milky Way Galaxy. Above it is elliptical galaxy NGC 205. Photograph is by the 48-inch (1.2-meter) Schmidt telescope. *(Hale Observatories)*

GALENA

(a)

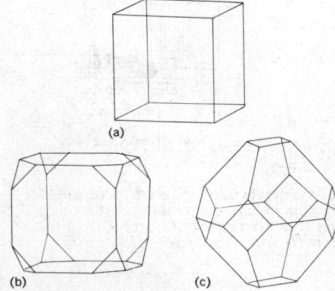

(b)   (c)

Forms of galena crystals: usually *(a)* cubic but sometimes modified to *(b,c)* octahedral truncations. *(From C. S. Hurlbut, Jr., Dana's Manual of Mineralogy, 17th ed., Wiley, 1959)*

gain-crossover frequency [CONT SYS] The frequency at which the magnitude of the loop ratio is unity.

gain margin [CONT SYS] The reciprocal of the magnitude of the loop ratio at the phase crossover frequency, frequently expressed in decibels.

gain reduction [ELECTR] Diminution of the output of an amplifier, usually achieved by reducing the drive from feed lines by use of equalizer pads or reducing amplification by a volume control.

gain sensitivity control *See* differential gain control.

gain turndown [ELEC] A receiver gain control incorporated in a transponder to protect the transmitter from overload.

gain twist [ORD] A type of rifling in which the twist is more rapid at the muzzle than at the breech, thus gradually increasing the rotation of the projectile.

gal *See* galileo; gallon.

galactan [BIOCHEM] Any of a number of polysaccharides composed of galactose units. Also known as galactosan.

galactic center [ASTRON] The gravitational center of the Milky Way galaxy; the sun and other stars of the galaxy revolve about this center.

galactic circle *See* galactic equator.

galactic cluster [ASTRON] Group of stars, not very concentrated in their central region, traveling in space as a group and having had a common origin; examples are the Pleiades and the Hyades. Also known as open cluster.

galactic coordinates *See* galactic system.

galactic disk [ASTRON] The flat distribution of stars and interstellar matter in the spiral arms and plane of the Milky Way galaxy.

galactic equator [ASTRON] A great circle of the celestial sphere, inclined 62° to the celestial equator, coinciding approximately with the center line of the Milky Way, and constituting the primary great circle for the galactic system of coordinates; it is everywhere 90° from the galactic poles. Also known as galactic circle.

galactic halo [ASTRON] The spherical distribution of oldest stars that are centered about the galactic center of the Milky Way galaxy.

galactic latitude [ASTRON] Angular distance north or south of the galactic equator; the arc of a great circle through the galactic poles, between the galactic equator and a point on the celestial sphere, measured northward or southward from the galactic equator through 90° and labeled N or S to indicate the direction of measurement.

galactic longitude [ASTRON] Angular distance east of sidereal hour angle 94°.4 along the galactic equator; the arc of the galactic equator or the angle at the galactic pole between the great circle through the intersection of the galactic equator and the celestial equator in Sagittarius (SHA 94°.4) and a great circle through the galactic poles, measured eastward from the great circle through SHA 94°.4 through 360°.

galactic nebula [ASTRON] A nebula that is in or near the galactic system known as the Milky Way.

galactic noise [ASTRON] Radio-frequency noise that originates outside the solar system; it is similar to thermal noise and is strongest in the direction of the Milky Way.

galactic nova [ASTRON] One of the novae that are concentrated largely in a band 10° on each side of the plane of the galaxy and are most frequent toward the center of the galaxy.

galactic nucleus [ASTRON] The center area in the galaxy about which there is a large spherical distribution of stars and from which the spiral arms emanate.

galactic plane [ASTRON] The plane that may be drawn through the galactic equator; the plane of the Milky Way galaxy.

galactic pole [ASTRON] On the celestial sphere, either of the two points 90° from the galactic equator.

galactic radiation [ASTROPHYS] Radiation emanating from the Milky Way galaxy.

galactic radio waves [ELECTROMAG] Radio waves emanating from the Milky Way galaxy.

galactic rotation [ASTRON] The rotation of the Milky Way about an axis through the center and perpendicular to the plane of the galaxy; the rotation is apparent from the highly flattened shape and from relative stellar motion.

galactic system [ASTRON] An astronomical coordinate system using latitude measured north and south from the galactic equator, and longitude measured in the sense of increasing right ascension from 0 to 360°. Also known as galactic coordinates.

galactic windows [ASTROPHYS] The regions near the equator of the Milky Way where there is low absorption of light by interstellar clouds so that some distant external galaxies may be seen through them.

galactocele [MED] **1.** A retention cyst caused by obstruction of one or more of the mammary ducts. **2.** A hydrocele with milky contents.

galactogen [BIOCHEM] A polysaccharide, in snails, that yields galactose on hydrolysis.

galactoglucomannan [BIOCHEM] Any of a group of polysaccharides which are prominent components of coniferous woods; they are soluble in alkali and consist of D-glucopyranose and D-mannopyranose units.

galactolipid *See* cerebroside.

galactomannan [BIOCHEM] Any of a group of polysaccharides which are composed of D-galactose and D-mannose units, are soluble in water, and form highly viscous solutions; they are plant mucilages existing as reserve carbohydrates in the endosperm of leguminous seeds.

L-galactomethylose *See* L-fucose.

galactonic acid [BIOCHEM] $C_6H_{12}O_7$ A monobasic acid derived from galactose, occurring in three optically different forms, and melting at 97°C. Also known as pentahydroxyhexoic acid.

galactophore [ANAT] A duct that carries milk.

galactorrhea [MED] Excessive flow of milk.

galactosamine [BIOCHEM] $C_6H_{14}O_5N$ A crystalline amino acid derivative of galactose; found in bacterial cell walls.

galactosan *See* galactan.

galactose [BIOCHEM] $C_6H_{12}O_6$ A monosaccharide occurring in both levo and dextro forms as a constituent of plant and animal oligosaccharides (lactose and raffinose) and polysaccharides (agar and pectin). Also known as cerebrose.

galactosemia [MED] A congenital metabolic disorder caused by an enzyme deficiency and marked by high blood levels of galactose.

galactosidase [BIOCHEM] An enzyme that hydrolyzes galactosides.

galactoside [BIOCHEM] A glycoside formed by the reaction of galactose with an alcohol; yields galactose on hydrolysis.

galactosuria [MED] Passage of urine containing galactose.

galacturonic acid [BIOCHEM] The monobasic acid resulting from oxidation of the primary alcohol group of D-galactose to carboxyl; it is widely distributed as a constituent of pectins and many plant gums and mucilages.

galago *See* bushbaby.

Galatheidea [INV ZOO] A group of decapod crustaceans belonging to the Anomura and having a symmetrical abdomen bent upon itself and a well-developed tail fan.

Galaxioidei [VERT ZOO] A suborder of mostly small, freshwater fishes in the order Salmoniformes.

galaxite [MINERAL] $MnAl_2O_4$ A black mineral of the spinel series composed of an oxide of manganese and aluminum.

galaxy [ASTRON] A large-scale aggregate of stars, gas, and dust; the aggregate is a separate system of stars covering a mass range from 10^7 to 10^{12} solar masses and ranging in diameter from 1500 to 300,000 light-years.

galbanum [MATER] A yellowish to brownish gum resin derived from *Ferula galbaniflua*, a perennial herb of western Asia; used in medicine.

Galbulidae [VERT ZOO] The jacamars, a family of highly iridescent birds of the order Piciformes that resemble giant hummingbirds.

gale [METEOROL] **1.** An unusually strong wind. **2.** In storm-warning terminology, a wind of 28–47 knots (52–87 kilometers per hour). **3.** In the Beaufort wind scale, a wind whose speed is 28–55 knots (52–102 kilometers per hour).

galea [ANAT] The epicranial aponeurosis linking the occipital and frontal muscles. [BIOL] A helmet-shaped structure. [BOT] A helmet-shaped petal near the axis. [INV ZOO] **1.** The endopodite of the maxilla of certain insects. **2.** A spinning organ on the movable digit of chelicerae of pseudoscorpions.

galeate [BIOL] **1.** Shaped like a helmet. **2.** Having a galea.

galena [MINERAL] PbS A bluish-gray to lead-gray mineral

with brilliant metallic luster, specific gravity 7.5, and hardness 2.5 on Mohs scale; occurs in cubic or octahedral crystals, in masses, or in grains. Also known as blue lead; lead glance.

galenobismutite [MINERAL] $PbBi_2S_4$ A lead-gray or tin-white mineral consisting of bismuth sulfide; specific gravity is 6.9.

Galen's vein [ANAT] One of the two veins running along the roof the third ventricle that drain the interior of the brain.

Galeritidae [PALEON] A family of extinct exocyclic Euechinoidea in the order Holectypoida, characterized by large ambulacral plates with small, widely separated pore pairs.

galerna *See* galerne.

galerne [METEOROL] A squally northwesterly wind that is cold, humid, and showery, occurring in the rear of a low-pressure area over the English Channel and off the Atlantic coast of France and northern Spain. Also known as galerna; galerno; giboulee.

galerno *See* galerne.

gale warning [METEOROL] A storm warning for marine interests of impending winds from 28 to 47 knots (52–87 kilometers per hour), signaled by two triangular red pennants by day, and a white lantern over a red lantern by night.

Galilean glass *See* Galilean telescope.

Galilean satellites [ASTRON] The four largest and brightest satellites of Jupiter (Io, Europa, Ganymede, and Callisto).

Galilean telescope [OPTICS] A refracting telescope whose objective is a converging (convex) lens and whose eyepiece is a diverging (concave) lens; it forms erect images. Also known as Galilean glass.

Galilean transformation [MECH] A mathematical transformation used to relate the space and time variables of two uniformly moving (inertial) reference systems in nonrelativistic kinematics.

galileo [MECH] The unit of acceleration in the centimeter-gram-second system, equal to 1 centimeter per second squared; commonly used in geodetic measurement. Abbreviated gal.

Galileo number [FL MECH] A dimensionless number used in studying the circulation of viscous liquids, equal to the cube of a characteristic dimension, times the acceleration of gravity, times the square of the liquid's density, divided by the square of its viscosity. Symbol N_{Ga}.

Galileo's law of inertia *See* Newton's first law.

galipol [ORG CHEM] $C_{15}H_{26}O$ A terpene alcohol derived from the oil of the angostura bark; colorless crystals that melt at 89°C.

gall [MED] A sore on the skin caused by chafing. [PHYSIO] *See* bile. [PL PATH] A large swelling on plant tissues caused by the invasion of parasites, such as fungi or bacteria, following puncture by an insect; insect oviposit and larvae of insects are found in galls.

gallacetophenone [ORG CHEM] $C_8H_8O_4$ A white to brownish-gray, crystalline powder, melting at 173°C, soluble in water, alcohol, and ether; used as an antiseptic. Also known as alizarin yellow C.

gallbladder [ANAT] A hollow, muscular organ in humans and most vertebrates which receives dilute bile from the liver, concentrates it, and discharges it into the duodenum.

gallego [METEOROL] A cold, piercing, northerly wind in Spain and Portugal.

gallein [ORG CHEM] $C_{20}H_{10}O_7$ A brown powder or green scales, broken down by heat; used as a pH indicator in the analysis of phosphates in urine and as an intermediate in the manufacture of dyes. Also known as anthracene violet; gallin; pyrogallolphthalein.

galleria forest [ECOL] A modified tropical deciduous forest occurring along stream banks.

Galleriinae [INV ZOO] A monotypic subfamily of lepidopteran insects in the family Pyralididae; contains the bee moth or wax worm (*Galleria mellonella*), which lives in beehives and whose larvae feed on beeswax.

gallery [GEOL] **1.** A horizontal, or nearly horizontal, underground passage, either natural or artificial. **2.** A subsidiary passage in a cave at a higher level than the main passage. [MIN ENG] A level or drift.

gallery deck [NAV ARCH] A partial deck below a flight deck on an aircraft carrier.

gallery practice ammunition [ORD] Small arms ammunition

with a reduced charge, used for practice shooting in a gallery and also for guard purposes.

gallery testing [MIN ENG] A method of testing explosives; a test condition is achieved by firing light charges without any stemming, and heavier charges with only 1 inch (2.5 centimeters) of stemming.

galley [ENG] The kitchen of a ship, airplane, or trailer. [GRAPHICS] A flat, oblong, open-ended tray into which the letters assembled by hand in a composing stick are transferred after the composing stick is full.

galley proof [GRAPHICS] A reproduction taken from type while the type is still in the galley; the reproduction is reviewed for errors.

gallic acid [ORG CHEM] $C_7H_6O_5$ A crystalline compound that forms needles from solutions of absolute methanol or chloroform, dissolves in water, alcohol, ether, and glycerol; obtained from nutgall tannins or from *Penicillium notatum* fermentation; used to make antioxidants and ink dyes and in photography.

gallicolous [BIOL] Producing or inhabiting galls.

Galliformes [VERT ZOO] An order of birds that includes important domestic and game birds, such as turkeys, pheasants, and quails.

gallin *See* gallein.

gallinaceous [VERT ZOO] Of, pertaining to, or resembling birds of the order Galliformes.

galling [MET] Surface damage on mating, moving metal parts due to friction caused by local welding of high spots.

gallium [CHEM] A chemical element, symbol Ga, atomic number 31, atomic weight 69.72. [MET] A silvery-white metal, melting at 29.7°C, boiling at 1983°C.

gallium arsenide [INORG CHEM] GaAs A crystalline material, melting point 1238°C; frequently alloys of this material are formed with gallium phosphide or indium arsenide. [SOLID STATE] A semiconductor having a forbidden-band gap of 1.4 electron volts and a maximum operating temperature of 400°C when used in a transistor.

gallium arsenide laser [OPTICS] A laser that emits light at right angles to a junction region in gallium arsenide, at a wavelength of 9000 angstroms; can be modulated directly at microwave frequencies; cryogenic cooling is required.

gallium halide [INORG CHEM] A compound formed by bonding of gallium to either chlorine, bromine, iodine, fluorine, or astatine.

gallium phosphide [INORG CHEM] GaP Transparent crystals made by reacting phosphorus and gallium suboxide at low temperature. [SOLID STATE] A semiconductor having a forbidden-band gap of 2.4 electron volts and a maximum operating temperature of 870°C when used in a transistor.

gallivorous [ZOO] Feeding on the tissues of galls, especially certain insect larvae.

gallnut [PL PATH] A gall resembling a nut.

gallogen *See* ellagic acid.

gallon [MECH] Abbreviated gal. **1.** A unit of volume used in the United States for measurement of liquid substances, equal to 231 cubic inches, or to 3.785 411 784 × 10⁻³ cubic meter. **2.** A unit of volume used in the United Kingdom for measurement of liquid and solid substances, usually the former; equal to the volume occupied by 10 pounds of weight of water of density 0.998859 gram per milliliter in air of density 0.001217 gram per milliliter against weights of density 8.136 grams per milliliter (the milliliter here has its old definition of 1.000028 × 10⁻⁶ cubic meter); approximately equal to 277.420 cubic inches, or to 4.54609 × 10⁻³ cubic meter.

gallop rhythm [MED] A three-sound sequence resulting from the intensification of the normal third or fourth heart sounds, occurring usually with a rapid ventricular rate.

gallotannic acid *See* tannic acid.

gallotannin *See* tannic acid.

Galloway sinking and walling stage *See* sinking and walling scaffold.

Galloway stage [MIN ENG] A platform of several decks suspended near the shaft during the sinking operation.

gallows frame [MIN ENG] *See* headframe. [MIN ARCH] A deck-mounted frame consisting of one U-shaped structural member or two legs and a header to provide large spacing at the head of the frame; accommodates securing points for a

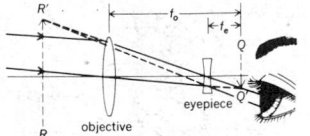

GALILEAN TELESCOPE

Diagram of Galilean telescope. QQ' is real inverted image which would be formed by objective alone. Eyepiece forms enlarged, erect virtual image, RR', which is viewed by observer. f_o and f_e are focal lengths of objective and eyepiece respectively. For distant object magnification is f_o/f_e.

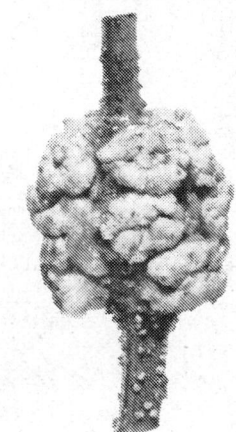

GALL

Gall formation on tomato caused by *Agrobacterium tumefaciens*. (Courtesy of Oscar N. Allen)

GALLIUM

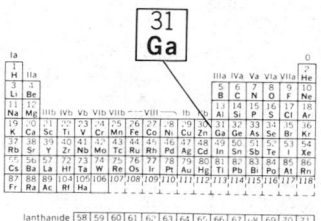

Periodic table of the chemical elements showing the position of gallium.

number of head blocks that can swing transversely without interference from each other.

gallstone [PATH] A nodule formed in the gallbladder or biliary tubes and composed of calcium, cholesterol, or bilirubin, or a combination of these.

galmei *See* hemimorphite.

Galofaro [OCEANOGR] A whirlpool in the Strait of Messina, between Sicily and Italy; formerly called Charybdis.

Galois field [MATH] A type of field extension obtained from considering the coefficients and roots of a given polynomial. Also known as root field.

Galois group [MATH] A group of isomorphisms of a particular field extension associated with a polynomial's roots.

Galois theory [MATH] The study of the Galois field and Galois group corresponding to a polynomial.

Galtonian curve [STAT] A graph showing the variation of any quantity from its normal value.

Galumnidae [INV ZOO] A family of oribatid mites in the suborder Sarcoptiformes.

galvanic [ELEC] Pertaining to electricity flowing as a result of chemical action.

galvanic battery [ELEC] A galvanic cell, or two or more such cells electrically connected to produce energy.

galvanic cell [ELEC] An electrolytic cell that is capable of producing electric energy by electrochemical action.

galvanic corrosion [MET] Electrochemical corrosion associated with the current in a galvanic cell, caused by dissimilar metals in an electrolyte because of the difference in potential (emf) of the two metals.

galvanic couple [ELEC] A pair of unlike substances, such as metals, which generate a voltage when brought in contact with an electrolyte.

galvanic current [ELEC] A steady direct current.

galvanic series [CHEM] The relative hierarchy of metals arranged in order from magnesium (least noble) at the anodic, corroded end through platinum (most noble) at the cathodic, protected end.

galvanic skin response [PHYSIO] The electrical reactions of the skin to any stimulus as detected by a sensitive galvanometer; most often used experimentally to measure the resistance of the skin to the passage of a weak electric current.

galvanism [BIOL] The use of a galvanic current for medical or biological purposes.

galvanize [MET] To deposit zinc on the surface of metal by the processes of hot dipping, sherardizing, or sometimes electroplating.

galvanoluminescence [PHYS] Light emission which may occur when electrodes of certain metals, such as aluminum or tantalum, are immersed in suitable electrolytes and current is passed between them.

galvanomagnetic effect [ELECTROMAG] One of the electrical or thermal phenomena occurring when a current-carrying conductor or semiconductor is placed in a magnetic field; examples are the Hall effect, Ettingshausen effect, transverse magnetoresistance, and Nernst effect. Also known as magnetogalvanic effect.

galvanometer [ENG] An instrument for indicating or measuring a small electric current by means of a mechanical motion derived from electromagnetic or electrodynamic forces produced by the current.

galvanometer constant [ELEC] Number by which a certain function of the reading of a galvanometer must be multiplied to obtain the current value in ordinary units.

galvanometer recorder [ENG ACOUS] A sound recorder in which the audio signal voltage is applied to a coil suspended in a magnetic field; the resulting movements of the coil cause a tiny attached mirror to move a reflected light beam back and forth across a slit in front of a moving photographic film.

galvanometer shunt [ELEC] Resistor connected in parallel with a galvanometer to increase its range under certain conditions; it allows only a known fraction of the current to pass through the galvanometer.

galvanostat [ELEC] A device to deliver constant current from a high-voltage battery.

galvanotaxis [BIOL] Movement of a free-living organism in response to an electrical stimulus.

galvanotropism [BIOL] Response of an organism to electrical stimulation.

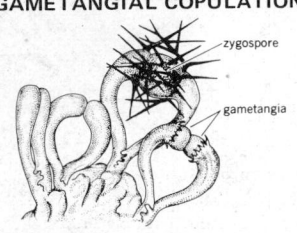

GAMETANGIAL COPULATION

zygospore

gametangia

Gametangial copulation, a sexual mechanism in fungi.

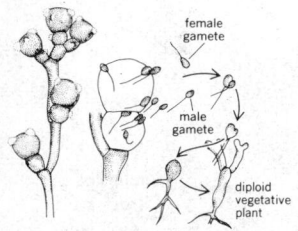

GAMETIC COPULATION

female gamete

male gamete

diploid vegetative plant

Gametic copulation, a sexual mechanism in fungi.

galvo *See* metal fume fever.

gambir [MATER] The yellowish extract from the twigs and leaves of the Malayan wood vine *Uncaria gambir* (Rubiaceae); used for tanning and dyeing and as an astringent. Also known as pale catechu; terra japonica; white cutch.

gambrel roof [ARCH] A roof consisting of two sections on either side of a horizontal peak, an upper section having a slight slope and a lower section having a steep slope.

game [MATH] A mathematical model expressing a contest between two players under specified rules.

game bird [BIOL] A bird that is legal quarry for hunters.

gamene *See* madder.

gametangial copulation [MYCOL] Direct fusion of certain fungal gametangia without differentiation of the gametes.

gamete [BIOL] A mature germ cell.

game theory [MATH] The mathematical study of games or abstract models of conflict situations from the viewpoint of determining an optimal policy or strategy. Also known as theory of games.

gametic copulation [MYCOL] The fusion of pairs of differentiated, uninucleate sexual cells or gametes formed in specialized gametangia.

gametocyte [HISTOL] An undifferentiated cell from which gametes are produced.

gametogenesis [BIOL] The formation of gametes, or reproductive cells such as ova or sperm.

gametophore [BOT] A branch that bears gametangia.

gametophyte [BOT] **1.** The haploid generation producing gametes in plants exhibiting metagenesis. **2.** An individual plant of this generation.

gamma [CHEM] The gamma position (the third carbon atom in an aliphatic carbon chain) on a chemical compound. [ELECTROMAG] A unit of magnetic field strength, equal to 10 microersteds, or 0.00001 oersted. [GRAPHICS] A numerical indication of the degree of contrast in a television or photographic image; equal to the slope of the straight-line portion of the H and D curve for the emulsion or screen. [MECH] A unit of mass equal to 10^{-6} gram or 10^{-9} kilogram.

gamma-absorption gage *See* gamma gage.

gamma acid [ORG CHEM] $C_{10}H_5NH_2OHSO_3H$ White crystals, slightly soluble in water; an intermediate in dyestuff manufacture. Also known as 2-amino-8-naphthol-6-sulfonic acid; 7-amino-1-naphthol-3-sulfonic acid; 2,5-naphthylamine sulfonic acid; 3-sulfonic acid; 6-sulfonic acid.

gamma counter [ENG] A device for detecting gamma radiation, primarily through the detection of fast electrons produced by the gamma rays; it either yields information about integrated intensity within a time interval or detects each photon separately.

gamma cross section [NUC PHYS] The cross section for absorption or scattering of gamma rays by a nucleus or atom.

gamma decay *See* gamma emission.

gamma distribution [STAT] A normal distribution whose frequency function involves a gamma function.

gamma emission [NUC PHYS] A quantum transition between two energy levels of a nucleus in which a gamma ray is emitted. Also known as gamma decay.

gamma flux density [NUC PHYS] The number of gamma rays passing through a unit area in a unit time.

gamma function [MATH] The complex function given by the integral with respect to t from 0 to ∞ of $e^{-t}t^{z-1}$; this function helps determine the general solution of Gauss' hypergeometric equation.

gamma gage [NUCLEO] A penetration-type thickness gage that measures the thickness or density of a sample by measuring its absorption of gamma rays. Also known as gamma-absorption gage.

gamma globulin [IMMUNOL] Any of the serum proteins with antibody activity. Also known as immune globulin; immunoglobulin (Ig).

gamma heating [NUCLEO] Heating resulting from absorption of gamma-ray energy by a material.

gamma iron [MET] Iron having a face-centered cubic lattice structure, stable between 910 and 1400°C.

gamma irradiation [NUCLEO] Exposure of a material to gamma rays.

gamma logging [ENG] Obtaining, by means of a gamma-ray

probe, a record of the intensities of gamma rays emitted by the rock strata penetrated by a borehole.

gamma matrix *See* Dirac matrix.

gamma radiation [NUCLEO] Radiation of gamma rays.

gamma radiography [NUCLEO] Radiography by means of gamma rays.

gamma ray [NUC PHYS] A high-energy photon, especially as emitted by a nucleus in a transition between two energy levels.

gamma-ray altimeter [ENG] An altimeter, used at altitudes under several hundred feet, that measures the photon backscatter from the earth resulting from the transmission of photons to earth from a cobalt-60 gamma source in the plane.

gamma-ray astronomy [ASTRON] The study of gamma rays from extraterrestrial sources, especially gamma-ray bursts.

gamma-ray bursts [ASTRON] Intense blasts of soft gamma rays of unknown origin, which range in duration from a tenth of a second to tens of seconds and occur several times a year from sources widely distributed over the sky.

gamma-ray detector [ENG] An instrument used on ships to identify and measure abnormal concentrations of gamma rays in the oceans.

gamma-ray laser [PHYS] A hypothetical device which would generate coherent radiation in the range 0.005–0.5 nm by inducing isomeric radiative transitions between isomeric nuclear states. Also known as graser.

gamma-ray level indicator [ENG] A level indicator in which the rising level of the liquid or other material reduces the amount of radiation passing from a gamma-ray source through the container to a Geiger counter or other radiation detector.

gamma-ray probe [ENG] A gamma-ray counter built into a watertight case small enough to be lowered into a borehole.

gamma-ray scattering *See* Compton scattering.

gamma-ray source [NUCLEO] A quantity of radioactive material that emits gamma radiation and is in a form convenient for radiology.

gamma-ray spectrometry [NUCLEO] 1. Determination of the energy distribution of gamma rays emitted by nuclei. Also known as gamma-ray spectroscopy. 2. In particular, a variation of neutron activation analysis in which the induced radiation from the sample is gamma rays instead of neutrons.

gamma-ray spectroscopy *See* gamma-ray spectrometry.

gamma-ray tracking [ENG] Use of three tracking stations, located at the three corners of a triangle centered on a missile about to be launched, to obtain accurate azimuthal tracking of a cobalt-60 gamma source in the tail.

gamma-ray well logging [ENG] Measurement of gamma-ray intensity versus depth down the wellbore; used to identify rock strata, their position, and their thicknesses.

Gammaridea [INV ZOO] The scuds or sand hoppers, a suborder of amphipod crustaceans; individuals are usually compressed laterally, are poor walkers, and lack a carapace.

gamma scanning [NUCLEO] The scanning of a fuel rod in a nuclear reactor for gamma activity by moving the rod past a slit in a lead block; photons emerging from the slit are detected by a scintillation spectrometer and recorded as a function of rod position.

gamma structure [SOLID STATE] A Hume-Rothery designation for structurally analogous phases or intermetallic phases having 21 valence electrons to 13 atoms, analogous to the γ-brass structure.

gamma taxonomy [SYST] The third stage in the history of animal systematics, involving renewed interest in the species, with growth of the genetic approach and widespread recognition of the significance of variation.

gamma transition *See* glass transition.

gammil [CHEM] A unit of concentration, equal to a concentration of 1 milligram of solute in 1 liter of solvent. Also known as micril; microgammil.

gamodeme [ECOL] An isolated breeding community.

gamogony [INV ZOO] Spore formation by multiple fission in sporozoans. [ZOO] Sexual reproduction.

gamont [INV ZOO] The gametocyte of sporozoans.

gamopetalous [BOT] Having petals united at their edges. Also known as sympetalous.

gamophobia [PSYCH] An abnormal fear of marriage.

gamophyllous [BOT] Having the leaves of the perianth united.

gamosepalous [BOT] Having sepals united at their edges. Also known as synsepalous.

Gamow barrier [NUC PHYS] The potential barrier which retards the escape of alpha particles from the nucleus according to the Gamow-Condon-Gurney theory.

Gamow-Condon-Gurney theory [NUC PHYS] An early quantum-mechanical theory of alpha-particle decay according to which the alpha particle penetrates a potential barrier near the surface of the nucleus by a tunneling process.

Gamow-Teller interaction [NUC PHYS] Interaction between a nucleon source current and a lepton field which has an axial vector or tensor form.

Gamow-Teller selection rules [NUC PHYS] Selection rules for beta decay caused by the Gamow-Teller interaction; that is, in an allowed transition there is no parity change of the nuclear state, and the spin of the nucleus can either remain unchanged or change by ± 1; transitions from spin 0 to spin 0 are excluded, however.

Gampsonychidae [PALEON] A family of extinct crustaceans in the order Palaeocaridacea.

gang capacitor [ELEC] A combination of two or more variable capacitors mounted on a common shaft to permit adjustment by a single control.

gang chart [IND ENG] A multiple-activity process chart used for groups of men on materials-handling operations.

gang drill [MECH ENG] A set of drills operated together in the same machine; used in rock drilling.

ganged control [ELECTR] Controls of two or more circuits mounted on a common shaft to permit simultaneous control of the circuits by turning a single knob.

gangliated cord [ANAT] One of the two main trunks of the sympathetic nervous system, one trunk running along each side of the spinal column.

ganglioma [MED] A form of ganglioneuroma in which neuronal and glial elements appear in about equal proportions.

ganglion [ANAT] A group of nerve cell bodies, usually located outside the brain and spinal cord.

ganglioneuroma [MED] A tumor composed of sympathetic ganglion cells and sheathed nerve fibers.

ganglionitis [MED] Inflammation of a ganglion.

ganglioside [BIOCHEM] One of a group of glycosphingolipids found in neuronal surface membranes and spleen; they contain an *N*-acyl fatty acid derivative of sphingosine linked to a carbohydrate (glucose or galactose); they also contain *N*-acetylglucosamine or *N*-acetylgalactosamine, and *N*-acetylneuraminic acid.

gang milling [ENG] Rolling of material by means of a composite machine with numerous cutting blades.

gangosa [MED] Destructive lesions of the nose and hard palate, sometimes more extensive, considered to be the tertiary stage of yaws.

gangplank [NAV ARCH] A long, narrow, movable bridge or plank from a ship to a pier or to another ship alongside.

gangplow [AGR] A plow with two or more cutters that turn parallel furrows.

gang-punch [ADP] To punch identical or constant information into all of a group of punched cards.

gangrene [MED] A form of tissue death usually occurring in an extremity due to insufficient blood supply.

gangrenous stomatitis *See* noma.

gang saw [MECH ENG] A steel frame in which thin, parallel saws are arranged to operate simultaneously in cutting logs.

gang switch [ELEC] A combination of two or more switches mounted on a common shaft to permit operation by a single control. Also known as deck switch.

gangue [GEOL] The valueless rock or aggregates of minerals in an ore.

gangway [MIN ENG] 1. A principal underground haulage road. 2. A passageway into or out of an underground mine. [NAV ARCH] An opening in the rail or bulwarks of a ship through which one can enter or leave it.

ganister [MATER] A fine mixture of quartz and fireclay used to line certain furnaces for metallurgical processes. [PETR] A fine, hard quartzose sandstone; used to make refractory silica brick to line furnace reactors.

ganoid scale [VERT ZOO] A structure having several layers of

GANOID SCALE

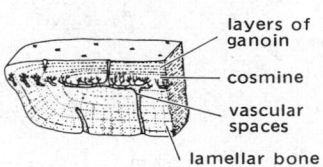

layers of ganoin

cosmine

vascular spaces

lamellar bone

Ganoid scale, a type of primitive vertebrate scale. *(From A. S. Romer, The Vertebrate Body, Saunders, 1962)*

enamellike material (ganoin) on the upper surface and laminated bone below.

ganoin [VERT ZOO] The enamellike covering of a ganoid scale.

ganomalite [MINERAL] $(Ca_2)Pb_3Si_3O_{11}$ A colorless to gray silicate of lead with calcium crystallizing in the tetragonal system.

ganophyllite [MINERAL] $(Na,K)(Mn,Fe,Al)_5(Si,Al)_6O_{15}(OH)_5 \cdot 2H_2O$ A brown, prismatic crystalline or foliated mineral composed of a hydrous silicate of manganese and aluminum.

gantlet [CIV ENG] A stretch of overlapping railroad track, with one rail of one track being between the two rails of another track; used over narrow bridges and passes.

gantry [ENG] A frame erected on side supports so as to span an area and support and hoist machinery and heavy materials.

gantry crane [MECH ENG] A bridgelike hoisting machine having fixed supports or arranged for running along tracks on ground level.

Gantt chart [IND ENG] In production planning and control, a type of bar chart depicting the work planned and done in relation to time; each division of space represents both a time interval and the amount of work to be done during that interval.

Gantt task and bonus plan [IND ENG] A wage incentive plan in which high task efficiency is maintained by providing a percentage bonus as a reward for production in excess of standard.

Ganymede [ASTRON] A satellite of Jupiter orbiting at a mean distance of 1,071,000 kilometers. Also known as Jupiter III.

gap [ADP] A uniformly magnetized area in a magnetic storage device (tape, disk), used to indicate the end of an area containing information. [COMMUN] A region not adequately covered by the main lobes of a radar antenna. [ELEC] The spacing between two electric contacts. [ELECTROMAG] A break in a closed magnetic circuit, containing only air or filled with a nonmagnetic material. [GEOGR] Any sharp, deep notch in a mountain ridge or between hills. [MET] An opening at the point of closest approach between faces of members in a weld joint.

gap coding [COMMUN] A process for conveying information by inserting gaps or periods of nontransmission in a system that normally transmits continuously.

gape [ANAT] The margin to margin distance between open jaws. [INV ZOO] The space between the margins of a closed mollusk valve.

gap factor [ELECTR] Ratio of the maximum energy gained in volts to the maximum gap voltage in a tube employing electron accelerating gaps, that is, a traveling-wave tube.

gap-filler radar [ENG] Radar used to fill gaps in radar coverage of other radar.

gap filling [ELECTROMAG] Electrical or mechanical rearrangement of an antenna array, or the use of a supplementary array, to produce lobes where gaps previously occurred.

gap-framepress [MECH ENG] A punch press whose frame is open at bed level so that wide work or strip work can be inserted.

gap-graded aggregate [MATER] Aggregate in which certain size particles are entirely or substantially absent.

gap lathe [MECH ENG] An engine lathe with a sliding bed providing enough space for turning large-diameter work.

gapless tape [ADP] A magnetic tape upon which raw data is recorded in a continuous manner; the data are streamed onto the tape without the word gaps; the data still may contain signs and end-of-record marks in the gapless form.

gap markers [ORD] In land mine warfare, markers used to identify a minefield gap; gap markers at the entrance to and exit from the gap are referenced to a landmark or intermediate marker.

gapped tape [ADP] A magnetic tape upon which blocked data has been recorded; it contains all of the flag bits and format to be read directly into a computer for immediate use.

gap scatter [ADP] The deviation from the exact distance required between read/write heads and the magnetized surface.

gar [VERT ZOO] The common name for about seven species

of bony fishes in the order Semionotiformes having a slim form, an elongate snout, and close-set ganoid scales.

Garamycin [MICROBIO] Trade name for the antibiotic gentamicin.

Garand rifle [ORD] A name for the .30-caliber M1 rifle, a weapon used by the U.S. Army up through 1957.

garbage See hash.

garbage in, garbage out [ADP] A phrase often stressed during introductory courses in computer utilization as a reminder that, regardless of the correctness of the logic built into the program, no answer can be valid if the input is erroneous. Abbreviated GIGO.

garbage pitch [MATER] Dark-brown to black pitch material obtained as a by-product residue from the burning of garbage; properties are analogous to complex hydrocarbons; used to make paints, varnishes, tarred paper, and waterproofing compound.

garbin [METEOROL] A sea breeze; in southwest France it refers to a southwesterly sea breeze which sets in about 9 A.M., reaches its maximum toward 2 P.M., and ceases about 5 P.M.

garble [COMMUN] To alter a message intentionally or unintentionally so that it is difficult to understand.

garboard strake [NAV ARCH] The strake of shell plating adjacent to the keel; this row of plates acts in conjunction with the keel, and the plates are made heavier than the other bottom plates.

Garbutt rod [PETRO ENG] A device used to pull the standing valve out of a tubing-type oil-well sucker-rod pump.

Gardner crusher [MIN ENG] A swing-and-hammer crusher; the U-shaped hammers are thrown by a revolving shaft against the feed and a heavy anvil inside the housing.

Gardner's syndrome [MED] A hereditary disorder transmitted as an autosomal dominant; manifested in childhood by multiple neoplasms, including bony and mesenteric tumors, fatty and fibrous skin, and intestinal polyps.

gargoylism See Hurler's syndrome.

garigue [ECOL] A low, open scrubland restricted to limestone sites in the Mediterranean area; characterized by small evergreen shrubs and low trees.

garland [MIN ENG] A channel fixed around a shaft in order to catch the water draining down the walls and conduct it to a lower level. Also known as water curb; water garland; water ring.

garlic [BOT] *Allium sativum.* A perennial plant of the order Liliales grown for its pungent, edible bulbs.

garlic oil [MATER] An essential oil obtained from steam distillation of garlic; contains chiefly a mixture of terpenes, with organic sulfides also present.

garner [AGR] **1.** A building or section of a building in which grain is stored. **2.** A bin for weighing grain.

garnet [MINERAL] A generic name for a group of mineral silicates that are isometric in crystallization and have the general chemical formula $A_3B_2(SiO_4)_3$, where A is Fe^{2+}, Mn^{2+}, Mg, or Ca, and B is Al, Fe^{3+}, Cr^{3+}, or Ti^{3+}; used as a gemstone and as an abrasive.

garnet hinge [DES ENG] A hinge with a vertical bar and horizontal strap.

garnet maser [ELECTR] A name incorrectly applied to a ferromagnetic amplifier.

garnet paper [MATER] Paper with a layer of crushed garnet on one side; used as an abrasive or polisher.

garnett [TEXT] **1.** A machine for removing foreign materials from fiber before it is carded. **2.** A machine for converting textile waste to fiber. **3.** The waste products of a garnett machine.

garnetting [TEXT] The reduction of textile waste materials to fiber.

garnierite [MINERAL] $(Ni,Mg)_3Si_2O_5(OH)_4$ An apple-green or pale-green, monoclinic serpentine; a gemstone and an ore of nickel. Also known as nepuite; noumeite.

garret [BUILD] The part of a house just under the roof.

garronite [MINERAL] $Na_2Ca_5Al_{12}Si_{20}O_{64} \cdot 27H_2O$ A zeolite mineral belonging to the phillipsite group; cyrstallizes in the tetragonal system.

garter spring [DES ENG] A closed ring formed of helically wound wire.

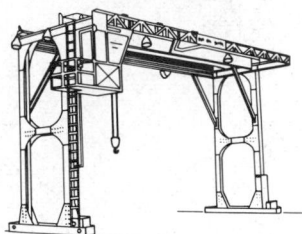
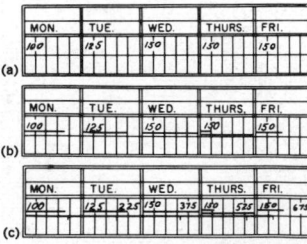

Gartner's duct [ANAT] The remnant of the embryonic Wolffian duct in the adult female mammal.

garúa [METEOROL] A dense fog or drizzle from low stratus clouds on the west coast of South America, creating a raw, cold atmosphere that may last for weeks in winter, and supplying a limited amount of moisture to the area. Also known as camanchaca.

gas [MATER] *See* gasoline. [ORD] To expose to a war gas. [PHYS] A phase of matter in which the substance expands readily to fill any containing vessel; characterized by relatively low density.

gas absorption operation [CHEM ENG] The recovery of solute gases present in gaseous mixtures of noncondensables; this recovery is generally achieved by contacting the gas stream with a liquid that offers specific or selective solubility for the solute gas to be recovered, or with an adsorbent (for example, synthetic or natural zeolite) that accepts only specific molecule sizes or shapes.

gas-activated battery [ELEC] A reserve battery which is activated by introducing a gas which reacts with a material between the electrodes of the battery to form an electrolyte.

gas adsorption [PHYS CHEM] The concentration of a gas upon the surface of a solid substance by attractive forces between the surface and the gas molecules.

gas alarm [MIN ENG] A signal system which warns mine workers of dangerous concentration of firedamp.

gas amplification [NUCLEO] The ratio of the charge collected to the charge liberated by the initial ionizing event in a radiation-counter tube.

gas analysis [ANALY CHEM] Analysis of the constituents or properties of a gas (either pure or mixed); composition can be measured by chemical adsorption, combustion, electrochemical cells, indicator papers, chromatography, mass spectroscopy, and so on; properties analyzed for include heating value, molecular weight, density, and viscosity.

gas anchor [PETRO ENG] A downhole gas separator used to reduce gas-in-oil froth before the pump to increase pump efficiency.

gas and mist sampler [MIN ENG] An instrument for automatic collection of one sample per hour of airborne contaminants such as sulfur dioxide or ammonia.

gas bearing [MECH ENG] A journal or thrust bearing lubricated with gas. Also known as gas-lubricated bearing.

gas black [CHEM] Fine particles of carbon formed by partial combustion or thermal decomposition of natural gas; used to reinforce rubber products such as tires. Also known as carbon black; channel black.

gas bomb [ORD] A bomb designed to produce casualties among personnel and to contaminate an area; a burster charge splits the bomb case and disperses the gas filler over the area.

gas brazing *See* gas-flame brazing.

gas buoy [NAV] A metal buoy having a gas light.

gas burner [ENG] A hole or a group of holes through which a combustible gas or gas-air mixture flows and burns.

gas cap [GEOPHYS] The gas immediately in front of a meteoroid as it travels through the atmosphere. [PETRO ENG] Gas occurring above liquid hydrocarbons in a reservoir under such trap conditions as the presence of water which prevents downward migration or the abutment of an impermeable formation against the reservoir.

gas capacitor [ELEC] A capacitor consisting of two or more electrodes separated by a gas, other than air, that serves as a dielectric.

gas-cap drive [PETRO ENG] Driving liquid hydrocarbons through a porous reservoir and toward well holes by utilizing the pressure of gas overlying the liquid pool.

gas-cap expansion [PETRO ENG] Process of reservoir-liquids displacement by the natural expansion of the reservoir gas cap to fill the voids vacated by recovered liquids.

gas-cap injection *See* external gas injection.

gas-cap reservoir [PETRO ENG] Two-phase reservoir in which a free area of gas (a gas cap) is underlain by an oil or liquid phase.

gas carburizing [MET] Surface hardening by heating a metal in gas of high carbon content in order to introduce carbon into the surface layers.

gas cell [ELEC] Cell in which the action depends on the absorption of gases by the electrodes.

gas-cell frequency standard [ATOM PHYS] An atomic frequency standard in which the frequency-determining element is a gas cell containing rubidium, cesium, or sodium vapor.

gas centrifuge process [NUCLEO] A method of isotope separation in which a mixture of isotopes in the gaseous state is spun at high speeds in a centrifuge, and centrifugal forces cause a concentration of heavy isotopes near the walls and light isotopes near the center.

gas check [ORD] Device in a gun that prevents escape of gas through the breech.

gas chromatograph [ANALY CHEM] The instrument used in gas chromatography to detect volatile compounds present; also used to determine certain physical properties such as distribution or partition coefficients and adsorption isotherms, and as a preparative technique for isolating pure components or certain fractions from complex mixtures.

gas chromatography [ANALY CHEM] A separation technique involving passage of a gaseous moving phase through a column containing a fixed adsorbent phase; it is used principally as a quantitative analytical technique for volatile compounds.

gas cleaning [ENG] Removing ingredients, pollutants, or contaminants from domestic and industrial gases.

gas-compression cycle [MECH ENG] A refrigeration cycle in which hot, compressed gas is cooled in a heat exchanger, then passes into a gas expander which provides an exhaust stream of cold gas to another heat exchanger that handles the sensible-heat refrigeration effect and exhausts the gas to the compressor.

gas compressor [MECH ENG] A machine that increases the pressure of a gas or vapor by increasing the gas density and delivering the fluid against the connected system resistance.

gas-condensate liquid [ORG CHEM] A hydrocarbon, such as propane, butane, and pentane, obtained as condensate when wet natural gas is compressed or refrigerated.

gas-condensate reservoir [GEOL] Hydrocarbon reservoir in which conditions of temperature and pressure have resulted in the condensation of the heavier hydrocarbon constituents from the reservoir gas.

gas-condensate well [PETRO ENG] A well producing hydrocarbons from a gas-condensate reservoir.

gas coning [PETRO ENG] The tendency of gas in a gas-drive reservoir to push oil downward in an inverse cone contour toward the casing perforations; at the extreme of coning, gas, not oil, will be produced from the well.

gas constant [THERMO] The constant of proportionality appearing in the equation of state of an ideal gas, equal to the pressure of the gas times its molar volume divided by its temperature. Also known as gas-law constant.

gas-cooled reactor [NUCLEO] A nuclear reactor in which a gas, such as air, carbon dioxide, or helium, is used as a coolant.

gas counter [NUCLEO] A counter in which the radioactive sample is prepared in the form of a gas and introduced into the counter tube.

gas current [ELECTR] A positive-ion current produced by collisions between electrons and residual gas molecules in an electron tube. Also known as ionization current.

gas cutting [MET] Cutting metal with the heat of an oxyacetylene flame.

gas cyaniding *See* carbonitriding.

gas cycle [THERMO] A sequence in which a gaseous fluid undergoes a series of thermodynamic phases, ultimately returning to its original state.

gas cylinder [MECH ENG] The chamber in which a piston moves in a positive displacement engine or compressor.

gas dehydrator [CHEM ENG] A device or system to remove moisture vapor from a gas stream, usually incorporates desiccant-type packed towers.

gas depletion drive *See* internal gas drive.

gas detector [MIN ENG] A device which indicates the existence of firedamp or other combustible or noxious gas in a mine.

gas-deviation factor *See* compressibility factor.

gas discharge [ELECTR] Conduction of electricity in a gas,

GAS CHROMATOGRAPH

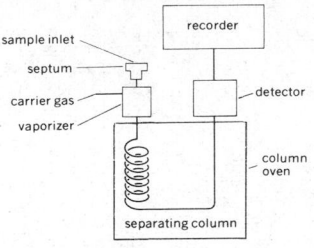

Diagram of a gas chromatograph.

due to movements of ions produced by collisions between electrons and gas molecules.

gas-discharge lamp *See* discharge lamp.

gas-discharge laser [OPTICS] A gas laser in which optical pumping is caused by nonequilibrium processes in a gas discharge.

gas-drive reservoir [PETRO ENG] An oil reservoir in which gas (either natural or reinjected) provides the driving force to sweep liquids through the formation and into the wellbore.

gas dynamics [PHYS] The study of the motion of gases, and of its causes, which takes into account thermal effects generated by the motion.

gas embolus [MED] An embolus composed of a gas resulting from trauma or other causes.

gas emission [MIN ENG] The release of gas from the strata into the mine workings.

gas engine [MECH ENG] An internal combustion engine that uses gaseous fuel.

gaseous conduction analyzer [ENG] A device to detect organic vapors in air by measuring the change in current that flows between a heated platinum anode and a concentric platinum cathode.

gaseous diffusion [CHEM ENG] **1.** Pressure-induced free-molecular transfer of gas through microporous barriers as in the process of making fissionable fuel. **2.** Selective solubility diffusion of gas through nonporous polymers by absorption and solution of the gas in the polymer matrix.

gaseous diffusion plant [NUCLEO] A facility where the gaseous diffusion process is used in the separation of uranium isotopes in order to provide fissionable fuel for nuclear power plants and weapons.

gaseous diffusion process [NUCLEO] A method of separating isotopes in which an isotopic mixture of gases is allowed to diffuse through a porous wall; the lighter molecules pass through the porous wall more readily than the heavier molecules.

gaseous fuel [MATER] A combustible gas that can be burned in a furnace or an engine.

gaseous nebulae [ASTRON] Clouds of gas, such as the Network Nebula in Cygnus, that are members of the Milky Way galactic system and are small compared with its overall dimensions.

gas explosion [MIN ENG] An explosion of firedamp in a coal mine; coal dust apparently does not play a significant part.

gas field [PETRO ENG] An area underlain with little or no interruption by one or more reservoirs of commercially valuable gas.

gas-filled cable [ELEC] A coaxial or other cable containing gas under pressure to serve as insulation and keep out moisture.

gas filled diode [ELECTR] A gas tube which is a diode, such as a cold-cathode rectifier or phanotron.

gas-filled porosity [GEOL] A reservoir formation in which the pore space is filled by gas instead of liquid hydrocarbons.

gas-filled radiation counter [NUCLEO] A gas tube used to detect radiation by means of gas ionization.

gas-filled rectifier *See* cold-cathode rectifier.

gas-filled thermometer [ENG] A thermometer which uses a gas (usually nitrogen or hydrogen), that approximately follows the ideal gas law.

gas-filled triode [ELECTR] A gas tube which has a grid or other control element, such as a thyratron or ignitron.

gas filter [CHEM ENG] A device used to remove liquid or solid particles from a flowing gas stream.

gas-flame brazing [MET] A brazing process for which the heat is supplied by a gas flame. Also known as gas brazing.

gas-flow counter tube [NUCLEO] A radiation-counter tube in which an appropriate atmosphere is maintained by a flow of gas through the tube. Also known as flow counter; gas-flow radiation counter.

gas-flow radiation counter *See* gas-flow counter tube.

gas focusing [ELECTR] A method of concentrating an electron beam by utilizing the residual gas in a tube; beam electrons ionize the gas molecules, forming a core of positive ions along the path of the beam which attracts beam electrons and thereby makes the beam more compact. Also known as ionic focusing.

gas furnace [ENG] An enclosure in which a gaseous fuel is burned.

gas gangrene [MED] A localized, but rapidly spreading, necrotizing bacterial wound infection characterized by edema, gas production, and discoloration; caused by several species of *Clostridium*.

gas generator [CHEM] A device used to generate gases in the laboratory. [CHEM ENG] A chemical plant for producing gas from coal, for example, water gas. [MECH ENG] An apparatus that supplies a high-pressure gas flow to drive compressors, airscrews, and other machines.

gas gland [VERT ZOO] A structure inside the swim bladder of many teleosts which secretes gas into the bladder.

gas grenade [ORD] Popular name for a chemical grenade designed to release a war gas, usually tear gas or other irritants.

gas gun [ORD] Automatic gun operated solely by gas, utilizing a portion of the gas pressure from firing a bullet to act on some form of piston-and-cylinder arrangement.

gas heater [MECH ENG] A unit heater designed to supply heat by forced convection, using gas as a heat source.

gash fracture [GEOL] Open gashes that are formed diagonally to a fault or fault zone.

gas holder [ENG] Gas storage container with vertically free top section that moves up or down to adjust to the volume of gas held.

gash vein [GEOL] A mineralized fissure that extends a short distance vertically.

gasification [CHEM ENG] Any chemical or heat process used to convert a substance to a gas; coal is converted by the Hygas process to a gaseous fuel. [MIN ENG] A method for exploiting poor-quality coal and thin coal seams by burning the coal in place to produce combustible gas which can be burned to generate power or processed into chemicals and fuels. Also known as underground gasification.

gas ignition [MIN ENG] The setting on fire of an accumulation of firedamp in a coal mine.

gas injection [MECH ENG] Injection of gaseous fuel into the cylinder of an internal combustion engine at the appropriate part of the cycle. [PETRO ENG] The injection of gas into a reservoir to maintain formation pressure and to drive liquid hydrocarbons toward the wellbores.

gas-injection well [PETRO ENG] A well hole in a reservoir into which pressurized gas is injected to maintain formation pressure or to drive liquid hydrocarbons toward other well holes.

gas ionization [ELECTR] Removal of the planetary electrons from the atoms of gas filling an electron tube, so that the resulting ions participate in current flow through the tube.

gasket [ENG] A packing made of deformable material, usually in the form of a sheet or ring, used to make a pressure-tight joint between stationary parts. Also known as static seal.

gas kinematics [FL MECH] The motion of a gas considered by itself, without regard for the causes of motion.

gas laser [OPTICS] A laser whose active material is gas in a quartz or glass tube with a Brewster angle window at each end; it is usually pumped by a discharge, but sometimes by other mechanisms, such as chemical reactions.

gas law [THERMO] Any law relating the pressure, volume, and temperature of a gas.

gas-law constant *See* gas constant.

gas lift [CHEM ENG] Solids movement operation in which an upward-flowing gas stream in a closed conduit or vessel is used to lift and move powdered or granular solid material.

gas-liquid chromatography [ANALY CHEM] A form of gas chromatography in which the fixed phase (column packing) is a liquid solvent distributed on an inert solid support. Abbreviated GLC. Also known as gas-liquid partition chromatography.

gas-liquid partition chromatography *See* gas-liquid chromatography.

gas logging [PETRO ENG] Hot-wire-detector or gas-chromatographic analysis and record of gas contained in the mud stream and cuttings for a well being drilled; a common way of detecting subsurface oil and gas shows.

gas-lubricated bearing *See* gas bearing.

gas magnification [ELECTR] Increase in current through a phototube due to ionization of the gas in the tube.

gas making [CHEM ENG] Making water gas or air gas by the action of steam and air upon hot coke.

gasman [MIN ENG] An underground official who examines the mine for firedamp and has charge of its removal.

gas manometer [ENG] A gage for determining the difference in pressure of two gases, usually by measuring the difference in height of liquid columns in the two sides of a U-tube.

gas maser [PHYS] A maser in which the microwave electromagnetic radiation interacts with the molecules of a gas such as ammonia; used chiefly in highly stable oscillator applications, as in atomic clocks.

gas mask [ENG] A device to protect the eyes and respiratory tract from noxious gases, vapors, and aerosols, by removing contamination with a filter and a bed of adsorbent material.

gas mechanics [FL MECH] The action of forces on gases.

gas meter [ENG] An instrument for measuring and recording the amount of gas flow through a pipe.

gas munition [ORD] A munition such as a bomb, projectile, pot, candle, or spray tank containing a war gas and a means of release.

gasogene [MATER] A fuel gas formed by incomplete combustion of charcoal; a European development as a substitute for gasoline. Also spelled gazogene.

gas oil [MATER] A petroleum distillate boiling within the general range 450-800°F (232-426°C); usually includes kerosine, diesel fuel, heating oils, and light fuel oils.

gas-oil ratio [PETRO ENG] Approximation of oil-reservoir composition, expressed in cubic feet of gas per barrel of liquid at 14.7 psia and 60°F (15.6°C).

gas-oil separator [PETRO ENG] An oil-field stock tank or series of tanks in which wellhead pressure is reduced so that the dissolved gas associated with reservoir oil is flashed off or separated as a separate phase. Also known as gas separator; oil-field separator; oil-gas separator; oil separator; separator.

gasoline [MATER] A fuel for internal combustion engines consisting essentially of volatile flammable liquid hydrocarbons; derived from crude petroleum by processes such as distillation reforming, polymerization, catalytic cracking, and alkylation; the common name is gas. Also known as petrol.

gasoline alkylate [MATER] A gasoline component usually formed by union of an olefin with isobutane by a refinery process known as alkylation; the product has high octane value and is blended with motor and aviation gasoline to improve antiknock value.

gasoline engine [MECH ENG] An internal combustion engine that uses a mixture of air and gasoline vapor as a fuel.

gasoline gel *See* gelatinized gasoline.

gasoline locomotive [MIN ENG] A mine locomotive which uses a gaseous fuel, is comparable to the steam locomotive in radius of travel, and has a speed of 3-12 miles per hour (5-19 kilometers per hour).

gasoline pool [MATER] A concept that considers gasolines of different qualities as a single group for the purpose of blending to meet final product specifications.

gasoline pump [MECH ENG] A device that pumps and measures the gasoline supplied to a motor vehicle, as at a filling station.

gasometer [ENG] A piece of equipment that holds and measures gas; may be used in analytical chemistry to measure the quantity of gas evolved in a reaction.

gasometric method [ANALY CHEM] An analytical technique for gases; the gas may be measured by instrumental methods or through chemical reactions with specific reagents.

gas packing [IND ENG] Packing a material such as food in an atmosphere consisting of an oxygen-free gas.

gaspar [MATER] A mixture of finely ground glass and quartz; a feldspar substitute in some applications and a hard-rubber filler.

gaspeite [MINERAL] NaCO₃ An anhydrous normal carbonate mineral with calcite structure.

gas phototube [ELECTR] A phototube into which a quantity of gas has been introduced after evacuation, usually to increase its sensitivity.

gas pliers [DES ENG] Pliers for gripping round objects such as pipes, tubes, and circular rods.

gas pocket [GEOL] A gas-filled cavity in rocks, especially

above an oil pocket. [MET] A cavity in a metal which contains trapped gases.

gas port [ORD] An opening for the passage of gas in the cylinder of a gas-operated automatic weapon to admit some of the propellant gases from the gun barrel.

gas-pressure maintenance [PETRO ENG] The maintenance of oil-reservoir gas pressure, usually by gas injection, to increase hydrocarbon recovery and to improve reservoir production characteristics.

gas producer [CHEM ENG] A device for complete gasification of coal by utilizing simultaneously the air and water-gas reactions.

gas reservoir [GEOL] An accumulation of natural gas found with or near accumulations of crude oil in the earth's crust.

gas rig [MIN ENG] A borehole drill, either rotary or churn type, driven by a combustion-type engine energized by a combustible liquid, such as gasoline, or a combustible gas, such as bottle gas.

gas scattering [ELECTR] The scattering of electrons or other particles in a beam by residual gas in the vacuum system.

gas scrubbing [CHEM ENG] Removal of gaseous or liquid impurities from a gas by the action of a liquid; the gas is contacted with the liquid which removes the impurities by dissolving or by chemical combination.

gas seal [ENG] A seal which prevents gas from leaking to or from a machine along a shaft.

gas separator *See* gas-oil separator.

Gasserian ganglion [ANAT] A group of nerve cells of the sensory root of the trigeminal nerve. Also known as semilunar ganglion.

gas-shielded arc welding [MET] Use of a gas atmosphere to shield the molten metal from air in arc welding.

gassiness [ELECTR] Presence of unwanted gas in a vacuum tube, usually in relatively small amounts, caused by the leakage from outside or evolution from the inside walls or elements of the tube.

gassing [ENG] **1.** Absorption of gas by a material. **2.** Formation of gas pockets in a material. **3.** Evolution of gas from a material during a process or procedure. [TEXT] *See* singeing.

gas slippage [FL MECH] Phenomenon of gas bypassing liquids that occurs when the diameter of capillary openings approaches the mean free path of the gas; occurs not only in capillary tubing, but in porous oil-reservoir formations.

gas-solid chromatography [ANALY CHEM] A form of gas chromatography in which the moving phase is a gas and the stationary phase is a surface-active sorbent (charcoal, silica gel, or activated alumina). Abbreviated GSC.

gas solubility [PHYS CHEM] The extent that a gas dissolves in a liquid to produce a homogeneous system.

gas-solubility factor [PETRO ENG] The number of standard cubic feet of gas liberated under specified gas-oil separator conditions that are in solution in one barrel of stock-tank oil at reservoir temperature and pressure.

gas spurt [GEOL] An accumulation of organic matter on certain strata caused by escaping gas.

gas sterilization [MICROBIO] Sterilization of heat and liquid-labile materials by means of gaseous agents, such as formaldehyde, ethylene oxide, and β-propiolactone.

gas stimulation [PETRO ENG] The detonation of a nuclear explosive in the strata of a natural-gas field to make the gas flow more freely.

gas storage [FOOD ENG] Storage in a low-oxygen atmosphere to delay ripening.

gassy [MIN ENG] A coal mine rating by the U.S. Bureau of Mines, applicable when an ignition occurs or if a methane content exceeding 0.25% can be detected; work must be halted if the methane exceeds 1.5% in a return airway.

gassy tube [ELECTR] A vacuum tube that has not been fully evacuated or has lost part of its vacuum due to release of gas by the electrode structure during use, so that enough gas is present to impair operating characteristics appreciably. Also known as soft tube.

gas tank [ENG] A tank for storing gas or gasoline.

Gasteromycetes [MYCOL] A group of basidiomycetous fungi in the subclass Homobasidiomycetidae with enclosed basidia and with basidiospores borne symmetrically on long sterigmata and not forcibly discharged.

Gasterophilidae [INV ZOO] The horse bots, a family of myo-

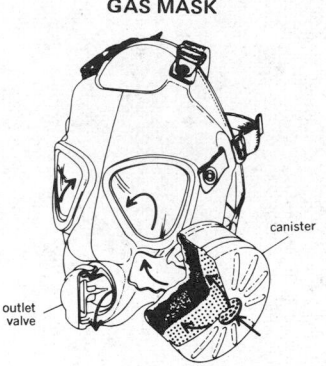

GAS MASK

canister

outlet
valve

A gas mask with arrows showing airflow direction.

darian cyclorrhaphous dipteran insects in the subsection Calypteratae, including individuals that cause myiasis in horses.

Gasterophilus [INV ZOO] A large genus of botflies in the family Gasterophilidae.

Gasterosteidae [VERT ZOO] The sticklebacks, a family of actinopterygian fishes in the order Gasterosteiformes.

Gasterosteiformes [VERT ZOO] An order of actinopterygian fishes characterized by a ductless swim bladder, a pelvic fin that is abdominal to subthoracic in position, and an elongate snout.

Gasteruptiidae [INV ZOO] A family of hymenopteran insects in the superfamily Proctotrupoidea.

gas tetrode *See* tetrode thyratron.

gas thermometer [ENG] A device to measure temperature by measuring the pressure exerted by a definite amount of gas enclosed in a constant volume; the gas (preferably hydrogen or helium) is enclosed in a glass or fused-quartz bulb connected to a mercury manometer.

gas thermometry [ENG] Measurement of temperatures with a gas thermometer; used with helium down to about 1K.

gas thermostatic switch [ELEC] A thermostatic switch in which heat causes the pressure of gas in a sealed metal bellows to increase, thereby moving the bellows and closing the contacts of a switch.

gas tracer [MIN ENG] Dust clouds, chemical smoke, or gaseous or radioactive tracers used to detect slowly moving air currents in a mine.

gastralium [INV ZOO] A microsclere located just beneath the inner cell layer of hexactinellid sponges. [VERT ZOO] One of the riblike structures in the abdomen of certain reptiles.

gas trap [CIV ENG] A bend or chamber in a drain or sewer pipe that prevents sewer gas from escaping.

gas-treating system [CHEM ENG] A process system to remove nonhydrocarbon impurities (such as water vapor, hydrogen sulfide, or carbon dioxide) from wellhead gas.

gastrectomy [MED] Surgical removal of all or part of the stomach.

gastric acid [BIOCHEM] Hydrochloric acid secreted by parietal cells in the fundus of the stomach.

gastric cecum [INV ZOO] One of the elongated pouchlike projections of the upper end of the stomach in insects.

gastric enzyme [BIOCHEM] Any digestive enzyme secreted by cells lining the stomach.

gastric filament [INV ZOO] In scyphozoans, a row of filaments on the surface of the gastric cavity which function to kill or paralyze live prey taken into the stomach. Also known as phacella.

gastric gland [ANAT] Any of the glands in the wall of the stomach that secrete components of the gastric juice.

gastric hypothermia [MED] Cooling of the upper digestive tract; useful in the management of bleeding disorders.

gastric juice [PHYSIO] The digestive fluid secreted by gastric glands; contains gastric acid and enzymes.

gastric mill [INV ZOO] A grinding apparatus consisting of calcareous or chitinous plates in the pharynx or stomach of certain invertebrates.

gastric ostium [INV ZOO] The opening into the gastric pouch in scyphozoans.

gastric pouch [INV ZOO] One of the pouchlike diversions of a scyphozoan stomach.

gastric shield [INV ZOO] A thickening of the stomach wall in some mollusks for mixing the contents.

gastric ulcer [MED] An ulcer of the mucous membrane of the stomach.

gastrin [BIOCHEM] A polypeptide hormone secreted by the pyloric mucosa which stimulates the pancreas to release pancreatic fluid and the stomach to release gastric acid.

gastritis [MED] Inflammation of the stomach.

gastroanastomosis [MED] The surgical formation of a communication between the two pouches of the stomach. Also known as gastrogastrostomy.

gastroblast [INV ZOO] A feeding zooid of a tunicate colony.

gastrocnemius [ANAT] A large muscle of the posterior aspect of the leg, arising by two heads from the posterior surfaces of the lateral and medial condyles of the femur, and inserted with the soleus muscle into the calcaneal tendon, and through this into the back of the calcaneus.

gastrocoele *See* archenteron.

gastrodermis [INV ZOO] The cellular lining of the digestive cavity of certain invertebrates.

gastroduodenitis [MED] Inflammation of the stomach and duodenum.

gastroenteritis [MED] Inflammation of the mucosa of the stomach and intestine.

gastroenterology [MED] The branch of medicine concerned with study of the stomach and intestines.

gastroenterostomy [MED] Surgical formation of a connection between the stomach and small intestine.

gastroepiploic artery [ANAT] Either of two arteries arising from the gastroduodenal and splenic arteries respectively and forming an anastomosis along the greater curvature of the stomach.

gastrogastrostomy *See* gastroanastomosis.

gastrointestinal hormone [BIOCHEM] Any hormone secreted by the gastrointestinal system.

gastrointestinal system [ANAT] The portion of the digestive system including the stomach, intestine, and all accessory organs.

gastrointestinal tract [ANAT] The stomach and intestine.

gastrojejunostomy [MED] Surgical establishment of an anastomosis between the jejunum and the anterior or posterior wall of the stomach.

gastrolith [VERT ZOO] A pebble swallowed by certain animals and retained in the gizzard or stomach, where it serves to grind food.

gastrolysis [MED] The breaking up of adhesions between the stomach and adjacent organs.

Gastromyzontidae [VERT ZOO] A small family of actinopterygian fishes of the suborder Cyprinoidei found in southeastern Asia.

gastropexy [MED] The fixation of a prolapsed stomach in its normal position by suturing it to the abdominal wall or other structure.

gastroplication [MED] An operation for relief of chronic dilation of the stomach by suturing a large horizontal fold in the stomach wall.

Gastropoda [INV ZOO] A large, morphologically diverse class of the phylum Mollusca, containing the snails, slugs, limpets, and conchs.

gastropore [INV ZOO] A pore containing a gastrozooid in hydrozoan corals.

gastroptosis [MED] Prolapse or downward displacement of the stomach.

gastroscope [MED] A hollow, tubular instrument used to examine the inside of the stomach by passage through the mouth and esophagus.

gastrosplenic ligament [ANAT] The fold of peritoneum passing from the stomach to the spleen. Also known as gastrosplenic omentum.

gastrosplenic omentum *See* gastrosplenic ligament.

gastrostome [INV ZOO] The opening of a gastropore.

gastrostomy [MED] The establishment of a fistulous opening into the stomach, with an external opening in the skin; used for artificial feeding.

gastrostyle [INV ZOO] A spiculated projection that extends into the gastrozooid from the base of the gastropore.

Gastrotricha [INV ZOO] A group of microscopic, pseudocoelomate animals considered either to be a class of the Aschelminthes or to constitute a separate phylum.

gastrozooid [INV ZOO] A nutritive polyp of colonial coelenterates, characterized by having tentacles and a mouth.

gastrula [EMBRYO] The stage of development in animals in which the endoderm is formed and invagination of the blastula has occurred.

gastrulation [EMBRYO] The process by which the endoderm is formed during development.

gas tube [ELECTR] An electron tube into which a small amount of gas or vapor is admitted after the tube has been evacuated; ionization of gas molecules during operation greatly increases current flow.

gas-tube boiler *See* waste-heat boiler.

gas-tungsten arc welding [MET] Arc welding by means of an electric arc between a single tungsten electrode and the work, with a shield of inert gas, such as argon or helium, around the electrode.

GASTROTRICHA

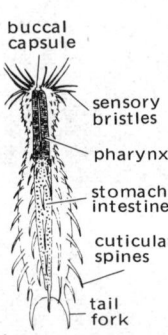

buccal
capsule

sensory
bristles

pharynx

stomach-
intestine

cuticular
spines

tail
fork

Chaetonotus, a freshwater
gastrotrich.

gas turbine [MECH ENG] A heat engine that converts energy of fuel into work by using compressed, hot gas as the working medium and that usually delivers its mechanical output through a rotating shaft. Also known as combustion turbine.

gas-turbine nozzle [MECH ENG] The component of a gas turbine in which the hot, high-pressure gas expands and accelerates to high velocity.

gas vacuole [BIOL] A membrane-bound, gas-filled cavity in some algae and protozoans; thought to control buoyancy.

gas vacuum breakdown [ELECTR] Ionization of residual gas in a vacuum, causing reverse conduction in an electron tube.

gas valve [ENG] An exhaust valve, held shut by rubber springs, used to discharge gas from the extreme top of a balloon.

gas vent [ENG] A pipe or hole that allows gas to pass off.

gas viscosity [FL MECH] The internal fluid function of a gas.

gas welding [MET] A welding process in which metals are joined by the heat of an oxyacetylene flame.

gas well [PETRO ENG] A well drilled for extraction of natural gas from a gas reservoir.

gas zone [GEOL] A rock formation containing gas under a pressure large enough to force the gas out if tapped from the surface.

gat [NAV] A natural or artificial passage or channel extending inland through shoals or steep banks.

GAT *See* Greenwich apparent time.

gate [CIV ENG] A movable barrier across an opening in a large barrier, a fence, or a wall. [ELECTR] **1.** A circuit having an output and a multiplicity of inputs and so designed that the output is energized only when a certain combination of pulses is present at the inputs. **2.** A circuit in which one signal, generally a square wave, serves to switch another signal on and off. **3.** One of the electrodes in a field-effect transistor. **4.** An output element of a cryotron. **5.** To control the passage of a pulse or signal. **6.** In radar, an electric waveform which is applied to the control point of a circuit to alter the mode of operation of the circuit at the time when the waveform is applied. Also known as gating waveform. [ENG] **1.** A device, such as a valve or door, for controlling the passage of materials through a pipe, channel, or other passageway. **2.** A device for positioning the film in a camera, printer, or projector. [GRAPHICS] The area or component in which the film is held at a fixed relationship to a lens. [MET] The opening in a casting mold through which molten metal enters the cavity. Also known as in-gate. [NAV] The position on the extension of the axis of a runway in use above which an aircraft heading toward that runway is required to pass at a time assigned by proper control authority. [NUCLEO] A movable barrier of shielding material used for closing a hole in a nuclear reactor. [ORD] A metal part in the rear of the cylinder of old-pattern revolvers that was turned out to expose the cylinder for loading.

gate conveyor [MIN ENG] A conveyor that carries coal from one source or face only, that is, from a single-unit or double-unit face.

gated-beam tube [ELECTR] A pentode electron tube having special electrodes that form a sheet-shaped beam of electrons; this beam may be deflected away from the anode by a relatively small voltage applied to a control electrode, thus giving extremely sharp cutoff of anode current.

gated sweep [ELECTR] Sweep in which the duration as well as the starting time is controlled to exclude undesired echoes from the indicator screen.

gate generator [ELECTR] A circuit used to generate gate pulses; in one form it consists of a multivibrator having one stable and one unstable position.

gate interlock [MIN ENG] A system that prevents movement of shaft conveyances or transmission of action signals until all shaft gates are closed.

gate multivibrator [ELECTR] Rectangular-wave generator designed to produce a single positive or negative gate voltage when triggered and then to become inactive until the next trigger pulse.

gate pulse [ELECTR] A pulse that triggers a gate circuit so it will pass a signal.

gate road bunker [MIN ENG] An appliance for coal storage from the face conveyors during peaks of production or during a stoppage of the outbye transport.

Gates crusher [MECH ENG] A gyratory crusher which has a cone or mantle that is moved eccentrically by the lower bearing sleeve.

gate turnoff [ELECTR] A *pnpn* switching device comparable to a silicon-controlled rectifier, but having a more complex gate structure that permits easy and fast turnoff as well as turn-on from its gate input terminal, at frequencies up to 100 kilohertz.

gate-turnoff silicon-controlled rectifier [ELECTR] A silicon-controlled rectifier that can be turned off by applying a current to its gate; used largely for direct-current switching, because turnoff can be achieved in a fraction of a microsecond.

gate valve [DES ENG] A valve with a disk-shaped closing element that fits tightly over an opening through which water passes.

gate winding [ELECTR] A winding used in a magnetic amplifier to produce on-off action of load current.

gather [GRAPHICS] In bookbinding, to arrange in sequence the folded sheets or signatures of a book. [MIN ENG] **1.** To assemble loaded cars from several production points and deliver them to main haulage for transport to the surface or pit bottom. **2.** To drive a heading through disturbed or faulty ground so as to meet the seam of coal at a convenient level or point on the opposite side.

gathering area [PETRO ENG] The area, usually down the regional dip from a hydrocarbon trap, from which the oil or gas may have migrated updip into the trap.

gathering arm loader [MIN ENG] A machine for loading loose rock or coal; has a tractor-mounted chassis and carries a chain conveyor whose front end is built into a wedge-shaped blade.

gathering conveyor [MIN ENG] Any conveyor which gathers coal from other conveyors and delivers it either into mine cars or onto another conveyor.

gathering iron [ENG] A rod used to collect molten glass for glassblowing.

gathering locomotive *See* gathering motor.

gathering mine locomotive *See* gathering motor.

gathering motor [MIN ENG] A lightweight type of electric locomotive used to haul loaded cars from the working places to the main haulage road and to replace them with empties. Also known as electric gathering locomotive; gathering locomotive; gathering mine locomotive.

gathering motorman [MIN ENG] In bituminous coal mining, one who operates a mine locomotive to haul loaded mine cars from working places to sidings.

gathering mule [MIN ENG] The mule used to collect the loaded cars from the separate working places and to return empties.

gathering pallet [HOROL] A finger that moves the rack in a striking timepiece for each blow struck.

gathering pump [MIN ENG] A portable or semiportable pump that is required for removing water encountered while opening a new mine, for extending headings or entries in an operating mine, for pump rooms or rib sections lying in the dip, for collecting water from local pools, or for sinking a shaft.

gathering ring [ENG] A clay ring placed on molten glass to collect impurities and thus permit high-quality glass to be taken from the center.

gathering system [PETRO ENG] A pipeline system used to gather gas or oil production from a number of separate wells, bringing the combined production to a central storage, pipelining, or processing terminal.

gather way [NAV] To attain headway.

gather write [ADP] An operation that creates a single output record from data items gathered from nonconsecutive locations in main memory.

gating [ELECTR] The process of selecting those portions of a wave that exist during one or more selected time intervals or that have magnitudes between selected limits.

gating waveform *See* gate.

Gatling gun [ORD] Type of machine gun using multibarrels which fire in rotation; it permits very rapid fire because it reduces the problem of overheating in a single barrel.

Gatterman-Koch synthesis [ORG CHEM] A synthesis of alde-

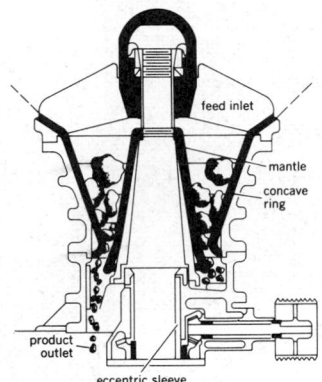

GATES CRUSHER

feed inlet
mantle
concave ring
product outlet
eccentric sleeve

Gyratory crusher of Gates type. *(Allis Chalmers Co.)*

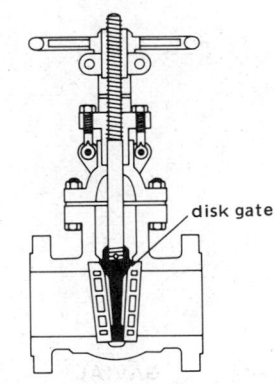

GATE VALVE

disk gate

Gate valve with disk gate; rising threaded stem shows when valve is open.

hydes; aldehydes form when an aromatic hydrocarbon is heated in the presence of hydrogen chloride, certain metallic chloride catalysts, and either carbon monoxide or hydrogen cyanide.

Gatterman reaction [ORG CHEM] **1.** Reaction of a phenol or phenol ester, and hydrogen chloride or hydrogen cyanide, in the presence of a metallic chloride such as aluminum chloride to form, after hydrolysis, an aldehyde. **2.** Reaction of an aqueous ethanolic solution of diazonium salts with precipitated copper powder or other reducing agent to form diaryl compounds.

Gaucher's cells [PATH] Abnormal macrophages associated with Gaucher's disease.

Gaucher's disease [MED] A rare chronic, probably hereditary disease in which cells loaded with cerebrosides become localized in reticuloendothelial tissue and eventually cause tissue destruction; manifestations include enlargement of the spleen, bronzing of the skin, and anemia. Also known as cerebroside lipoidosis; familial splenic anemia.

gauge *See* gage.

gauging *See* gaging.

Gaultheria [BOT] A genus of upright or creeping evergreen shrubs (Ericaceae).

gaultheria oil *See* methyl salicylate.

gauss [ELECTROMAG] Unit of magnetic induction in the electromagnetic and Gaussian systems of units, equal to 1 maxwell per square centimeter, or 10^{-4} weber per square meter. Also known as abtesla (abt).

Gauss-Codazzi equations [MATH] Equations dealing with the components of the fundamental tensor and Riemann-Christoffel tensor of a surface.

Gauss' error curve *See* normal distribution.

Gauss eyepiece [OPTICS] A Ramsden eyepiece which has a thin glass plate between the two lenses, making an angle of 45° with the optical axis; used to set a telescope perpendicular to a plane reflecting surface.

Gauss formulas [MATH] Formulas dealing with the sine and cosine of angles in a spherical triangle.

Gauss' hypergeometric equation [MATH] The differential equation, commonly arising in many physical contexts, $x(1 - x)y'' + [c - (a + b + 1)x]y' - aby = 0$.

Gaussian complex integers [MATH] Complex numbers whose real and imaginary parts are both integers.

Gaussian constant [ASTRON] The acceleration caused by the attraction of the sun at the mean distance of the earth from the sun.

Gaussian curvature [MATH] The invariant of a surface specified by Gauss' theorem.

Gaussian curve [STAT] The bell-shaped curve corresponding to a population which has a normal distribution. Also known as normal curve.

Gaussian distribution *See* normal distribution.

Gaussian error [NAV] Deviation of a magnetic compass due to transient magnetism which remains in a vessel's structure for short periods after the inducing force has been removed, usually appearing after a vessel has been on the same heading for a considerable time.

Gaussian noise [COMMUN] Noise that has a frequency distribution which follows the Gaussian curve. [MATH] *See* Wiener process.

Gaussian noise generator [ELECTR] A signal generator that produces a random noise signal whose frequency components have a Gaussian distribution centered on a predetermined frequency value.

Gaussian optics [OPTICS] An approximation which describes rays which are very close to the axis of an optical system and are nearly parallel to this axis, so that only the linear terms of Taylor series for the distance of a point from the axis or the angle which a ray makes with the axis need be considered. Also known as first-order theory.

Gaussian pulse [PHYS] A pulse for which the graph of intensity as a function of time is a Gaussian curve.

Gaussian reduction [MATH] A procedure of simplification of the rows of a matrix which is based upon the notion of solving a system of simultaneous equations. Also known as Gauss-Jordan elimination.

Gaussian system [ELECTROMAG] A combination of the electrostatic and electromagnetic systems of units (esu and emu), in which electrostatic quantities are expressed in esu and magnetic and electromagnetic quantities in emu, with appropriate use of the conversion constant c (the speed of light) between the two systems. Also known as Gaussian units.

Gaussian units *See* Gaussian system.

Gauss image point [OPTICS] A point through which pass all paraxial rays from a specified point source in an optical system.

Gauss-Jordan elimination *See* Gaussian reduction.

Gauss' law of flux [ELEC] The law that the total electric flux which passes out from a closed surface equals (in rationalized units) the total charge within the surface.

Gauss' law of the arithmetic mean [MATH] The law that a harmonic function can attain its maximum value only on the boundary of its domain of definition, unless it is a constant.

Gauss-Legendre rule [MATH] An approximation technique of definite integrals by a finite series which uses the zeros and derivatives of the Legendre polynomials.

Gauss lens system *See* Celor lens system.

Gauss' mean value theorem [MATH] The value of a harmonic function at a point in a planar region is equal to its integral about a circle centered at the point.

gauss meter [ENG] A magnetometer whose scale is graduated in gauss or kilogauss, and usually measures only the intensity, and not the direction, of the magnetic field.

Gauss objective lens *See* Celor lens system.

Gauss point *See* cardinal point.

Gauss positions [ELECTROMAG] The Gauss A and B positions; that is, a point on the axis of a bar magnet is in Gauss A position, and a point on the magnetic equator of the magnet is in Gauss B position, with respect to the magnet.

Gauss' principle of least constraint [MECH] The principle that the motion of a system of interconnected material points subjected to any influence is such as to minimize the constraint on the system; here the constraint, during an infinitesimal period of time, is the sum over the points of the product of the mass of the point times the square of its deviation from the position it would have occupied at the end of the time period if it had not been connected to other points.

Gauss-Seidel method *See* Seidel method.

Gauss test [MATH] In an infinite series with general term a_n, if $a_{n+1}/a_n = 1 - (x/n) - [f(n)/n^\lambda]$ where x and λ are greater than 1, and $f(n)$ is a particular integer function, then the series converges.

Gauss' theorem [MATH] **1.** The assertion, under certain light restrictions, that the volume integral through a volume V of the divergence of a vector function is equal to the surface integral of the exterior normal component of the vector function over the boundary surface of V. Also known as divergence theorem. **2.** At a point on a surface the product of the principal curvatures is an invariant of the surface, called the Gaussian curvature.

gauze [MATER] **1.** A sheer, loosely woven textile fabric similar to cheesecloth; used for surgical dressings. **2.** A plastic or wire mesh.

gavage [MED] The administration of nourishment through a stomach tube.

gavial [VERT ZOO] The name for two species of reptiles composing the family Gavialidae.

Gavialidae [VERT ZOO] The gavials, a family of reptiles in the order Crocodilia distinguished by an extremely long, slender snout with an enlarged tip.

Gaviidae [VERT ZOO] The single, monogeneric family of the order Gaviiformes.

Gaviiformes [VERT ZOO] The loons, a monofamilial order of diving birds characterized by webbed feet, compressed, bladelike tarsi, and a heavy, pointed bill.

Gay-Lussac acid [MATER] Product of a Gay-Lussac tower in the chamber process for sulfuric acid manufacture; a mixture of sulfuric acid and nitrogen oxides.

Gay-Lussac's law *See* Charles' law; combining-volumes principle.

Gay-Lussac tower [CHEM ENG] A component part in the chamber process for sulfuric acid production that absorbs nitrogen oxides to form nitrous vitriol.

gaylussite [MINERAL] $Na_2Ca(CO_3)_2 \cdot 5H_2O$ A translucent,

GAVIAL

Indian gavial, fierce looking but not dangerous to man.

yellowish-white hydrous carbonate mineral, with a vitreous luster, crystallizing in the monoclinic system; found in dry lakes.

gazogene *See* gasogene.

gc *See* gigahertz.

g-cal *See* calorie.

GCA radar *See* ground-controlled approach radar.

gcd *See* greatest common divisor.

GCI radar *See* ground-controlled intercept radar.

g-cm *See* gram-centimeter.

Gd *See* gadolinium.

GDCH *See* α-dichlorohydrin.

G display [ELECTR] A rectangular radar display in which horizontal and vertical aiming errors are indicated by horizontal and vertical displacement, respectively, and range is indicated by the length of wings appearing on the blip, with length increasing as range decreases.

G drift [NAV] A drift component in gyros and accelerometers which is proportional to the nongravitational acceleration and which is caused by torques generated by mass unbalance.

G² drift [NAV] A drift component in gyros and accelerometers which is proportional to the square of the nongravitational acceleration and which is generated by the anisoelasticity of the rotor support.

Ge *See* germanium.

geanticline [GEOL] A broad land uplift; refers to the land mass from which sediments in a geosyncline are derived.

gear [DES ENG] A toothed machine element used to transmit motion between rotating shafts when the center distance of the shafts is not too large. [MECH ENG] **1.** A mechanism performing a specific function in a machine. **2.** An adjustment device of the transmission in a motor vehicle which determines mechanical advantage, relative speed, and direction of travel.

gearbox *See* transmission.

gear cutter [MECH ENG] A machine or tool for cutting teeth in a gear.

gear cutting [MECH ENG] The cutting or forming of a uniform series of toothlike projections on the surface of a workpiece.

gear down [MECH ENG] To arrange gears so the driven part rotates at a slower speed than the driving part.

gear drive [MECH ENG] Transmission of motion or torque from one shaft to another by means of direct contact between toothed wheels.

geared turbine [MECH ENG] A turbine connected to a set of reduction gears.

gear forming [MECH ENG] A method of gear cutting in which the desired tooth shape is produced by a tool whose cutting profile matches the tooth form.

gear generating [MECH ENG] A method of gear cutting in which the tooth is produced by the conjugate or total cutting action of the tool plus the rotation of the workpiece.

gear grinding [MECH ENG] A gear-cutting method in which gears are shaped by formed grinding wheels and by generation; primarily a finishing operation.

gear hobber [MECH ENG] A machine that mills gear teeth; the rotational speed of the hob has a precise relationship to that of the work.

gearing [MECH ENG] A set of gear wheels.

gearing chain [MECH ENG] A continuous chain used to transmit motion from one toothed wheel, or sprocket, to another.

gearksutite [MINERAL] $CaAl(OH)F_4 \cdot H_2O$ A clayey mineral composed of hydrous calcium aluminum fluoride, occurring with cryolite.

gearless traction [MECH ENG] Direct drive, without reduction gears.

gear level [MECH ENG] To arrange gears so that the driven part and driving part turn at the same speed.

gear loading [MECH ENG] The power transmitted or the contact force per unit length of a gear.

gear meter [ENG] A type of positive-displacement fluid quantity meter in which the rotating elements are two meshing gear wheels.

gearmotor [MECH ENG] A motor combined with a set of speed-reducing gears.

gear oil [MATER] A lubricating oil for use in transmissions, most types of differential gears, and gears in gear boxes.

gear pump [MECH ENG] A rotary pump in which two meshing gear wheels contrarotate so that the fluid is entrained on one side and discharged on the other.

gear ratio [MECH ENG] The ratio of the angular speed of the driving member of a gear train or similar mechanism to that of the driven member; specifically, the number of revolutions made by the engine per revolution of the rear wheels of an automobile.

gear shaper [MECH ENG] A machine that makes gear teeth by means of a reciprocating cutter that rotates slowly with the work.

gear-shaving machine [MECH ENG] A finishing machine that removes excess metal from machined gears by the axial sliding motion of a straight-rack cutter or a circular gear cutter.

gearshift [MECH ENG] A device for engaging and disengaging gears.

gear teeth [DES ENG] Projections on the circumference or face of a wheel which engage with complementary projections on another wheel to transmit force and motion.

gear train [MECH ENG] A combination of two or more gears used to transmit motion between two rotating shafts or between a shaft and a slide.

gear up [MECH ENG] To arrange gears so that the driven part rotates faster than the driving part.

gear wheel [MECH ENG] A wheel that meshes gear teeth with another part.

gebli *See* ghibli.

Gecarcinidae [INV ZOO] The true land crabs, a family of decapod crustaceans belonging to the Brachygnatha.

gecko [VERT ZOO] The common name for more than 300 species of arboreal and nocturnal reptiles composing the family Gekkonidae.

Geco sampler [MIN ENG] Straight-line cutter designed to traverse a falling stream of ore or pulp at regular intervals, so as to divert a representative sample to a holding vessel.

gedanite [MINERAL] A brittle, wine-yellow variety of amber containing little succinic acid; found on the shore of the Baltic Sea.

Gedanken experiment [PHYS] German for "thought" experiment; a hypothetical experiment which is possible in principle and is analyzed (but not performed) to test some hypothesis.

gedrite [MINERAL] An aluminous variety of the mineral anthophyllite.

Gee [NAV] An electronic navigational system that establishes hyperbolic lines of position similar to those produced by loran, but employing a radio frequency of 22–30 and 40–85 megahertz, whereas standard loran operates in the 1700–2000-megahertz band.

Gee chart [NAV] A chart showing the hyperbolic lines of position which are produced by the Gee navigation system.

geepound *See* slug.

geg [METEOROL] A desert dust whirl of China and Tibet.

Gegenbauer polynomials [MATH] A family of polynomials solving a special case of the Gauss hypergeometric equation.

gegenschein [ASTRON] A round or elongated, faint, ill-defined spot of light in the sky at a point 180° from the sun. Also known as counterglow; zodiacal counterglow.

gehlenite [MINERAL] Ca_2Al_2SiO A mineral of the melitite group that crystallizes in the tetragonal crystal system and is isomorphous with akermanite; a green, resinous material found with spinel.

Geiger-Briggs rule [NUCLEO] The rule that the range of an alpha ray in dry air, at 15°C and 1 atmosphere, above 5 centimeters, is proportional to its initial velocity raised to the 3.26 power; an improvement on the Geiger formula.

Geiger counter *See* Geiger-Müller counter.

Geiger counter tube *See* Geiger-Müller tube.

Geiger formula [NUCLEO] A formula which states that the range of an alpha particle in dry air, at 15°C and 1 atmosphere, is proportional to the cube of its initial velocity.

Geiger-Müller counter [NUCLEO] **1.** A radiation counter that uses a Geiger-Müller tube in appropriate circuits to detect and count ionizing particles; each particle crossing the tube produces ionization of gas in the tube which is roughly independent of the particle's nature and energy, resulting in

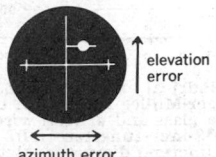

G DISPLAY

Type-G radar display, a three-dimensional display for use in an airplane cockpit.

GEAR CUTTER

Gear blank and cutter as used on a hypoid generator. Cutter has a series of cutting blades. (*Gleason Works*)

GEAR DRIVE

Bevel gears, a type of gear drive.

GECKO

Banded gecko, which is brown with broad yellow bands, and is economically important as a destroyer of insects.

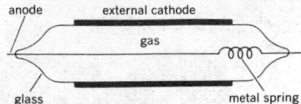

Cylindrical external-cathode Geiger-Müller tube, with thin soda glass and central wire of 0.003-inch-tungsten (0.08-millimeter-) diameter. Metal spring keeps central wire taut.

GEMINI

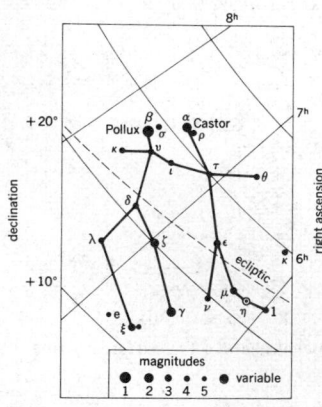

Line pattern of the constellation Gemini. Grid lines represent coordinates of the sky. Apparent brightness, or magnitude, of stars is shown by sizes of dots graded by appropriate numbers as indicated.

a uniform discharge across the tube. Abbreviated GM counter. Also known as Geiger counter. **2.** *See* Geiger-Müller tube.

Geiger-Müller counter tube *See* Geiger-Müller tube.

Geiger-Müller probe [ENG] A Geiger-Müller counter in a watertight container, lowered into a borehole to log the intensity of the gamma rays emitted by radioactive substances in traversed rock. Also known as electronic logger; Geiger probe.

Geiger-Müller tube [NUCLEO] A radiation-counter tube operated in the Geiger region; it usually consists of a gas-filled cylindrical metal chamber containing a fine-wire anode at its axis. Also known as Geiger counter tube; Geiger-Müller counter; Geiger-Müller counter tube.

Geiger-Nutall rule [NUC PHYS] The rule that the logarithm of the decay constant of an alpha emitter is linearly related to the logarithm of the range of the alpha particles emitted by it.

Geiger probe *See* Geiger-Müller probe.

geikielite [MINERAL] $MgTiO_3$ A bluish-black or brownish-black mineral that crystallizes in the rhombohedral system and occurs in the form of rolled pebbles; it is isomorphous with ilmenite.

Geissler pump [ENG] A type of air pump that uses the principle of the Torricellian vacuum, and in which the vacuum is produced by the flow of mercury back and forth between a vertically adjustable and a fixed reservoir.

Geissler tube [ELECTR] An experimental discharge tube with two electrodes at opposite ends, used to demonstrate and study the luminous effects of electric discharges through various gases at low pressures.

geitonogamy [BOT] Pollination and fertilization of one flower by another on the same plant.

geking [OCEANOGR] Obtaining measurements of ocean movements with a geomagnetic electrokinetograph (GEK).

Gekkonidae [VERT ZOO] The geckos, a family of small lizards in the order Squamata distinguished by a flattened body, a long sensitive tongue, and adhesive pads on the toes of many species.

gel [CHEM] A two-phase colloidal system consisting of a solid and a liquid in more solid form than a sol.

Gelamite [MATER] Trademark for a semigelatin high explosive used in underground mining, quarrying, construction, and general blasting.

Gelastocoridae [INV ZOO] The toad bugs, a family of tropical and subtropical hemipteran insects in the subdivision Hydrocorisae.

gelatin [MATER] *See* gelatin dynamite. [ORG CHEM] A protein derived from the skin, white connective tissue, and bones of animals; used as a food and in photography, the plastics industry, metallurgy, and pharmaceuticals.

gelatinase [BIOCHEM] An enzyme, found in some yeasts and molds, that liquefies gelatin.

gelatin duplicating [GRAPHICS] A gelatin process in which the item to be reproduced is typed or drawn on a hard-surfaced master with inks containing aniline dyes; the master is then placed face down on a gelatin surface, and the dye is transferred to the gelatin within seconds; when blank paper is pressed against the gelatin, a small amount of the dye is released and a copy made.

gelatin dynamite [MATER] A high explosive consisting mainly of a jellylike mass of nitroglycerin, with sodium nitrate, meal, collodion cotton, and sodium carbonate; commonly used by drillers to shatter boulders encountered in driving pipe through overburden, especially in water-filled or saturated ground. Also known as gelatin; gelignite; nitrogelatin.

gelatin extra [MATER] An explosive in which some of the nitroglycerin is replaced with ammonium nitrate.

gelatinized gasoline [MATER] Gasoline treated with a thickening agent; used in napalm bombs and flamethrowers. Also known as gasoline gel; jellied gasoline; thickened fuel.

gelatinizing agent [MATER] In manufacture of propellants, a material which softens the nitrocellulose, permitting the mixture to be processed and formed. Also known as gelling agent.

gelatin liquefaction [MICROBIO] Reduction of a gelatin culture medium to the liquid state by enzymes produced by

bacteria in a stab culture; used in identifying bacteria.

gelatin model *See* electrolytic model.

gelatinobromide [MATER] A preparation of gelatin silver bromide that is light-sensitive and used in photography.

gelatin process [GRAPHICS] A process for reproducing a direct image by putting the image (writing, typed matter, or drawn images) on a master paper using a special process, pressing it into a gelatin pad, then transferring it from the gelatin to duplicator paper.

gelation [CHEM] **1.** The act or process of freezing. **2.** Formation of a gel from a sol.

gelation model [PETRO ENG] Electrolytic analog of a reservoir; used to investigate the areal sweep movement of water from multiple injection wells; operates by movement of copper ammonium and zinc ammonium ions through the gelatin in a flat tray that simulates the reservoir.

gel cement [MATER] Cement containing a small percentage of bentonite, which makes the mixture more homogeneous, increases the water-cement ratio, and reduces loss of water to the formation.

gel coat [MATER] A resin applied to the surface of a mold and gelled prior to placing plastic material in the mold in position for production; the gel coat becomes an integral part of the finished laminate and improves surface appearance.

Gelechiidae [INV ZOO] A large family of minute to small moths in the lepidopteran superfamily Tineoidea, generally having forewings and trapezoidal hindwings.

gel electrophoresis [CHEM] Electrophoresis performed in silica gel, a porous, inert medium.

gelignite *See* gelatin dynamite.

gelisol *See* frozen ground.

gelling agent *See* gelatinizing agent.

Gell-Man–Nishijima scheme [PARTIC PHYS] A classification of elementary particles according to hypercharge, total isotopic spin, and its third component.

Gell-Mann–Okubo mass formula [PARTIC PHYS] A formula, based on SU(3) symmetry, giving the masses of the members of a unitary multiplet of mesons or baryons in terms of their total isotopic spin, hypercharge, and coefficients characteristic of the multiplet considered.

Gell-Mann relation [PARTIC PHYS] A relation, derived from the Gell-Mann–Okubo mass formula, between the masses of the pi, eta, and K mesons of a meson octet.

gel mineral *See* mineraloid.

Gelocidae [PALEON] A family of extinct pecoran ruminants in the superfamily Traguloidea.

gelose *See* ulmin.

gel paint [MATER] Paint formulation made thixotropic by the reaction of a small amount of polyamide resin with an alkyd resin vehicle.

gel permeation chromatography [ANALY CHEM] Analysis by chromatography in which the stationary phase consists of beads of porous polymeric material such as a cross-linked dextran carbohydrate derivative sold under the trade name Sephadex; the moving phase is a liquid.

gel point [PHYS CHEM] Stage at which a liquid begins to exhibit elastic properties and increased viscosity.

gem [MINERAL] A natural or artificially produced mineral or other material that has sufficient beauty and durability for use as a personal adornment.

GEM *See* air-cushion vehicle.

gem cutting [LAP] The polishing of rough gem materials into faceted or rounded forms for use in jewelry.

geminate [BIOL] Growing in pairs or couples.

Gemini [ASTRON] A northern constellation; right ascension 7 hours, declination 20°N. Also known as Twins.

Geminids [ASTRON] A meteor shower that reaches maximum about December 13.

geminiflorous [BOT] Having flowers in pairs.

gemma [BOT] A small, multicellular, asexual reproductive body of some liverworts and mosses.

gemmation *See* budding.

gemmho [ELEC] A unit of conductance, equal to the conductance of a substance which has a resistance of 10^6 ohms, or to 10^{-6} mho.

gemmiform [BOT] Resembling a gemma or bud.

gemmiparous [BIOL] Producing a bud or reproducing by a bud.

gem mounting [LAP] Setting gems and gemstones in rings or other jewelry pieces made of precious metals.

gemmule [ANAT] A minute dendritic process functioning as a synaptic contact point. [BIOL] Any bud formed by gemmation. [INV ZOO] A cystlike, asexual reproductive structure of many Porifera that germinates when proper environmental conditions exist; it is a protective, overwintering structure which germinates the following spring.

Gemolite [OPTICS] A binocular magnifier with dark-field illumination, used to distinguish natural from synthetic gem materials.

gemology [MINERAL] The science concerned with the identification, grading, evaluation, fashioning, and other aspects of gemstones.

Gempylidae [VERT ZOO] The snake mackerels, a family of the suborder Scombroidei comprising compressed, elongate, or eel-shaped spiny-rayed fishes with caniniform teeth.

gem stick [LAP] A stick on which a gem is mounted before it is cut.

gemstone [GEOL] A mineral or petrified organic matter suitable for use in jewelry.

Gemuendinoidei [PALEON] A suborder of extinct raylike placoderm fishes in the order Rhenanida.

gena [ANAT] Cheek, or side of the head.

gender identity [PSYCH] The sum of those aspects of personal appearance and behavior culturally attributed to masculinity or femininity.

gending [METEOROL] A local dry wind in the northern plains of Java that resembles the foehn, caused by a wind crossing the mountains near the south coast and pushing between the volcanoes.

gene [GEN] The basic unit of inheritance.

gene action [GEN] The functioning of a gene in determining the phenotype of an individual.

genecology [BIOL] The study of species and their genetic subdivisions, their place in nature, and the genetic and ecological factors controlling speciation.

gene conversion [GEN] A situation in which gametocytes of an individual that is heterozygous for a pair of alleles undergo meiosis, and the gametes produced are in a 3:1 ratio rather than the expected 2:2 ratio, implying that one allele was converted to the other.

gene flow [GEN] The passage and establishment of genes characteristic of a breeding population into the gene complex of another population through hybridization and backcrossing.

gene frequency [GEN] The number of occurrences of a specific gene within a population.

gene penetrance *See* penetrance.

general address [COMMUN] Group of characters included in the heading of a message that causes the message to be routed to all addresses included in the general address category.

general anesthesia [MED] Loss of sensation with loss of consciousness, produced by administration of anesthetic drugs.

general anesthetic [PHARM] An agent that produces general anesthesia.

general arrangement plans [NAV ARCH] Plans showing the quarters, spaces, and compartments of a ship.

general broadcast [COMMUN] The radio broadcast, to all U.S. Navy ships and merchant ships, in which storm warnings, area forecasts, map analyses, surface weather reports, and upper air, upper wind, and aircraft reports are given.

general chart [NAV] A nautical chart intended for offshore coastwise navigation, using scales ranging from about 1:100,000 to 1:600,000, which are smaller than those of a coast chart, but larger than those of a sailing chart.

general circulation [METEOROL] The complete statistical description of atmospheric motions over the earth. Also known as planetary circulation.

general depot [ORD] Large supply establishment for receiving, storing, and issuing supplies for more than one technical service.

general formula [CHEM] A formula that can apply not only to one specific compound but to a series of related compounds; for example, the general formula for an aldehyde RCHO, where R is hydrogen in formaldehyde (the simplest aldehyde) and is a hydrocarbon radical for other aldehydes in the series such as CH_3 for acetaldehyde and C_2H_5 for proprionaldehyde.

generalized binomial trials model [STAT] A product model in which the nth factor model has two simple events with probabilities p_n and $q_n = 1 - p_n$. Also known as Poisson binomial trials model.

generalized coordinates [MECH] A set of variables used to specify the position and orientation of a system, in principle defined in terms of cartesian coordinates of the system's particles and of the time in some convenient manner; the number of such coordinates equals the number of degrees of freedom of the system.

generalized function *See* distribution.

generalized hydrostatic equation [GEOPHYS] The vertical component of the vector equation of motion in natural coordinates when the acceleration of gravity is replaced by the virtual gravity; for most purposes it is identical to the hydrostatic equation.

generalized momentum *See* conjugate momentum.

generalized routine [ADP] A routine which can process a wide variety of jobs; for example, a generalized sort routine which will sort in ascending or descending order on any number of fields whether alphabetic or numeric, or both, and whether binary coded decimals or pure binaries.

generalized transmission function [GEOPHYS] In atmospheric-radiation theory, a set of values, variable with wavelength, each one of which represents an average transmission coefficient for a small wavelength interval and for a specified optical path through the absorbing gas in question.

generalized velocity [MECH] The derivative with respect to time of one of the generalized coordinates of a particle. Also known as Lagrangian generalized velocity.

general manager [IND ENG] The person of general authority who performs all reasonable tasks in conducting the usual and customary business of the principal head or owner.

general paresis [MED] An inflammatory and degenerative disease of the brain caused by infection with *Treponema pallidum*. Also known as syphilitic meningoencephalitis.

general precession [ASTRON] The resultant motion of the components causing precession of the equinoxes westward along the ecliptic at the rate of about 50″.3 per year.

general program [ADP] A computer program designed to solve a specific type of problem when values of appropriate parameters are supplied.

general purpose automatic test system [ELECTR] Modular, computer-type, automatic electronic checkout system capable of finding faults in electronic equipment at the system, subsystem, line replaceable unit, module, and piece part levels.

general-purpose bomb [ORD] An item designed to be dropped from an aircraft to destroy or reduce the utility of a target by explosive effect.

general-purpose computer [ADP] A device that manipulates data without detailed, step-by step control by human hand and is designed to be used for many different types of problems.

general-purpose vehicle [ORD] Motor vehicle designated to be used interchangeably for movement of personnel, supplies, ammunition, or equipment and for towing artillery carriages, trailers, or semitrailers; used without modification to body or chassis to satisfy general automotive transport needs.

general relativistic collapse [RELAT] Process in which a star undergoing gravitational collapse cannot release its kinetic energy to the outside universe, but continues to collapse into a general relativistic singularity of infinite density.

general relativity [RELAT] The theory of Einstein which generalizes special relativity to noninertial frames of reference and incorporates gravitation, and in which events take place in a curved space.

general research [SCI TECH] Pure or applied research which is not directly concerned with an applied research program.

general reserve artillery [ORD] Consists of all artillery units not integral to divisions and corps; a pool of reserve artillery

GEMOLITE

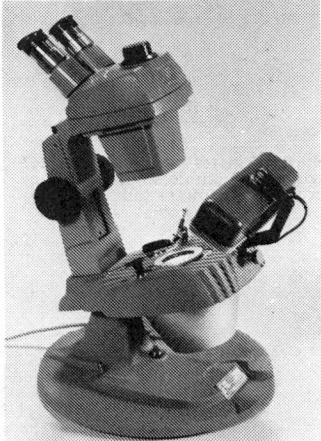

The Gemolite, a binocular magnifier with dark-field illuminator. (*Gemological Institute of America*)

GENERAL RELATIVISTIC COLLAPSE

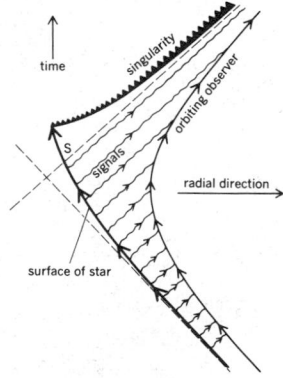

Space-time diagram for the collapse of a star past its Schwarzschild radius (*S*) and into a general relativistic singularity of infinite density.

available to the field forces commander for assignment or attachment to subordinate commands as required by combat conditions.

general routine [ADP] In computers, a routine, or program, applicable to a class of problems; it provides instructions for solving a specific problem when appropriate parameters are supplied.

general supplies [ORD] Intraservice classification applied to ordnance, quartermaster, and transportation supplies; ordnance general supplies include all ordnance supplies, with the exception of ammunition, required for the maintenance of an organization.

general-support artillery [ORD] Artillery which executes the fire directed by the commandeer of the unit to which it organically belongs or is attached; it fires in support of a specific subordinate unit.

general term [MATH] The general term of a sequence or series is an expression subscripted by an integer which determines any desired entry.

generate [ADP] To create a particular program by selecting parts of a general-program skeleton (or outline) and specializing these parts into a cohesive entity.

generating area See fetch.

generating flow [FL MECH] For a liquid allowed to flow smoothly into a duct, the flow while the boundary layer, which starts at the entrance and grows until it fills the duct, is growing.

generating function [MATH] A function $g(x,y)$ corresponding to a family of orthogonal polynomials $f_0(x), f_1(x), \ldots$, where a Taylor series expansion of $g(x,y)$ in powers of y will have the polynomial $f_n(x)$ as the coefficient for the term y^n.

generating plant See generating station.

generating routine See generator.

generating station [MECH ENG] A stationary plant containing apparatus for large-scale conversion of some form of energy (such as hydraulic, steam, chemical, or nuclear energy) into electrical energy. Also known as generating plant.

generation [ADP] **1.** Any one of three groups used to historically classify computers according to their electronic hardware components, logical organization and software, or programming techniques; computers are thus known as first-, second-, or third-generation; a particular computer may possess characteristics of all generations simultaneously. **2.** One of a family of data sets, related to one another in that each is a modification of the next most recent data set. [BIOL] A group of organisms having a common parent or parents and comprising a single level in line of descent.

generation rate [ELECTR] In a semiconductor, the time rate of creation of electron-hole pairs.

generation time [MICROBIO] The time interval required for a bacterial cell to divide or for the population to double. [NUCLEO] The mean time required for a neutron arising from a fission to produce a new fission.

generator [ADP] A program that produces specific programs as directed by input parameters. Also known as generating routine. [ELEC] A machine that converts mechanical energy into electrical energy; in its commonest form, a large number of conductors are mounted on an armature that is rotated in a magnetic field produced by field coils. Also known as dynamo; electric generator. [ELECTR] **1.** A vacuum-tube oscillator or any other nonrotating device that generates an alternating voltage at a desired frequency when energized with direct-current power or low-frequency alternating-current power. **2.** A circuit that generates a desired repetitive or nonrepetitive waveform, such as a pulse generator. [MATH] One of the set of elements of an algebraic system such as a group, ring, or module which determine all other elements when all admissible operations are performed upon them.

generator field control [ELEC] Method of regulating the output voltage of a generator by controlling the voltage which excites the field of the generator.

generator resistance [ELEC] The resistance of the current source in a network; usually much smaller than the load but taken into account in some network calculations.

generatrix [MATH] The straight line generating a ruled surface.

gene redundancy [GEN] The presence of many copies of one gene within a cell.

GENET

Spotted genet (*Genetta genetta*), which is distinguished from the civet by its longer tail.

genesis rocks [GEOL] Rocks that have retained their character from nearly 4.6×10^9 years ago, when planets were still occulting out of the cloud of dust and gas referred to as the solar nebula; examples are meteorites and asteroids.

gene suppression [GEN] The development of a normal phenotype in a mutant individual or cell due to a second mutation either in the same gene or in a different gene.

genet [VERT ZOO] The common name for nine species of small, arboreal African carnivores in the family Viverridae.

genetic code [MOL BIO] The genetic information in the nucleotide sequences in deoxyribonucleic acid represented by a four-letter alphabet that makes up a vocabulary of 64 three-nucleotide sequences, or codons; a sequence of such codons (averaging about 100 codons) constructs a message for a polypeptide chain.

genetic drift [GEN] The random fluctuation of gene frequencies from generation to generation that occurs in small populations.

genetic fingerprinting [MOL BIO] Identification of chemical entities in animal tissues as indicative of the presence of specific genes.

genetic homeostasis [GEN] The tendency of Mendelian populations to maintain a constant genetic composition.

genetic load [GEN] The abnormalities, deformities, and deaths produced in every generation by defective genetic material carried in the gene pool of the human race.

genetic map [GEN] A graphic presentation of the linear arrangement of genes on a chromosome; gene positions are determined by percentages of recombination in linkage experiments. Also known as chromosome map.

genetic material [GEN] The ultramicroscopic particles or genes, first defined by H. J. Muller, the influences of which permeate the cell and play a fundamental role in determining the nature of all cell substances, cell structures, and cell effects; the genes have properties of self-propagation and variation.

genetics [BIOL] The science that is concerned with the study of biological inheritance.

Geneva stop [HOROL] A device that transmits power only from the middle portion of a watch mainspring to provide an even force.

Geneva system [ORG CHEM] An international system of nomenclature by which organic compounds can be named in a logical way.

genial See mental.

genic hybrid sterility [GEN] Sterility resulting from the interaction of genes in a hybrid to cause disturbances of sex-cell formation or meiosis.

geniculate [SCI TECH] Bent abruptly at an angle, as a bent knee.

geniculate body [ANAT] Any of the four oval, flattened prominences on the posterior inferior aspect of the thalamus; functions as the synaptic center for fibers leading to the cerebral cortex.

geniculate ganglion [ANAT] A mass of sensory and sympathetic nerve cells located along the facial nerve.

geniculum [ANAT] **1.** A small, kneelike, anatomical structure. **2.** A sharp bend in any small organ.

genioglossus [ANAT] An extrinsic muscle of the tongue, arising from the superior mental spine of the mandible.

Geniohyidae [PALEON] A family of extinct ungulate mammals in the order Hyracoidea; all members were medium to large-sized animals with long snouts.

genistin [ORG CHEM] $C_{21}H_{20}O_{10}$ A pale-yellow glucoside derived from soybean meal, crystallizes from 80% methanol solution, melting point 256°C, soluble in hot 80% ethanol, hot 80% methanol, and hot acetone. Also known as 7-D-glucoside.

genital atrium [ZOO] A common chamber receiving openings of male, female, and accessory organs.

genital coelom [INV ZOO] In mollusks, the lumina of the gonads.

genital cord [EMBRYO] A mesenchymal shelf bridging the coeloms in mammalian embryos, produced by fusion of the caudal part of the urogenital folds; fuses with the urinary bladder in the male, and is the primordium of the broad ligament and the uterine walls in the female. [INV ZOO] Strands of cells located in the genital canal which are primordial sex cells in crinoids. Also known as genital rachis.

genitalia [ANAT] The organs of reproduction, especially those which are external.

genital orifice *See* genital pore.

genital pore [INV ZOO] A small opening on the side of the head in some gastropods through which the penis is protruded. Also known as genital orifice.

genital rachis *See* genital cord.

genital recess [INV ZOO] A depression between the calyx surface and anal cone in entoprocts which serves as a brood chamber.

genital ridge [EMBRYO] A medial ridge or fold on the ventromedial surface of the mesonephros in the embryo, produced by growth of the peritoneum; the primordium of the gonads and their ligaments.

genital scale [INV ZOO] Any of the small calcareous plates in ophiuroids associated with the buccal shields.

genital segment *See* gonosomite.

genital shield [INV ZOO] In ophiuroids, a support of a bursal slit in the arms located near the arm base.

genital stolon [INV ZOO] Part of the axial complex in ophiuroids.

genital sucker [INV ZOO] In some trematodes, a suckerlike structure surrounding the gonopore.

genital tract [ANAT] The ducts of the reproductive system.

genital tube [INV ZOO] A blood lacuna in crinoids, connected with the subtegminal plexus and suspended in the genital canal.

genitourinary system *See* urogenital system.

Genoa cyclone [METEOROL] A cyclone, or low, which appears to have formed or developed in the vicinity of the Gulf of Genoa. Also known as Genoa low.

Genoa low *See* Genoa cyclone.

genome [GEN] **1.** The genetic endowment of a species. **2.** The haploid set of chromosomes.

genophobia [PSYCH] An abnormal fear of sex.

genotype [GEN] The genetic constitution of an organism, usually in respect to one gene or a few genes relevant in a particular context. [SYST] The type species of a genus.

gentamicin [MICROBIOL] A broad-spectrum antibiotic produced by a species of *Micromonospora*.

Gentianaceae [BOT] A family of dicotyledonous herbaceous plants in the order Gentianales distinguished by lacking stipules and having parietal placentation.

Gentianales [BOT] A family of dicotyledonous plants in the subclass Asteridae having well-developed internal phloem and opposite, simple, mostly entire leaves.

gentianic acid *See* gentisic acid.

gentisic acid [ORG CHEM] $C_7H_6O_4$ A crystalline compound that forms monoclinic prisms from a water solution, sublimes at 200°C, melts at 250°C, and is soluble in water, alcohol, ether, sodium, and salt; used in medicine. Also known as gentianic acid; hydroquinone carboxylic acid; 5-hydroxysalicylic acid.

gentle breeze [METEOROL] In the Beaufort wind scale, a wind whose speed is from 7 to 10 knots (13–19 kilometers per hour).

gentnerite [MINERAL] $Cu_8Fe_3Cr_{11}S_{18}$ A sulfide mineral known only in meteorites.

genu *See* knee.

genus [MATH] An integer associated to a surface which measures the number of holes in the surface. [SYST] A taxonomic category that includes groups of closely related species; the principal subdivision of a family.

genu valgum [MED] Inward or medial curving of the knee; knock-knee.

geo [GEOGR] A narrow coastal inlet bordered by steep cliffs. Also spelled gio.

geoacoustics [ACOUS] Study of the acoustic properties of rock, mainly to study possible use of the rock system as a carrier of seismic signals in a communications system.

geobotanical prospecting [GEOL] The use of the distribution, appearance, and growth anomalies of plants in locating ore deposits.

geobotany [BOT] The study of plants as related to their geologic environment.

geocentric [ASTRON] Relative to the earth as a center; that is measured from the center of the earth.

geocentric coordinates [ASTRON] Coordinates that define the position of a point with respect to the center of the earth; can be either cartesian (x, y, and z) or spherical (latitude, longitude, and radial distance). Also known as geocentric coordinate system; geocentric position.

geocentric coordinate system *See* geocentric coordinates.

geocentric latitude [ASTRON] The latitude of a celestial body from the center of the earth. [GEOD] Of a position on the earth's surface, the angle between a line to the center of the earth and the plane of the equator.

geocentric longitude [ASTRON] The celestial longitude of the position of a body projected on the celestial sphere when the body is viewed from the center of the earth.

geocentric parallax [ASTRON] The difference in the apparent direction or position of a celestial body, measured in seconds of arc, as determined from the center of the earth and from a point on its surface; this varies with the body's altitude and distance from the earth. Also known as diurnal parallax.

geocentric position *See* geocentric coordinates.

geocentric vertical [GEOD] The direction of the radius vector drawn from the center of the earth through the location of the observer. Also known as geometric vertical.

geocentric zenith [ASTRON] The point where a line from the center of the earth through a point on its surface meets the celestial sphere.

geocerite [MINERAL] A white, waxy mineral composed of carbon, oxygen, and hydrogen, occurring in brown coal.

geochemical anomaly [GEOCHEM] Above-average concentration of a chemical element in a sample of rock, soil, vegetation, stream, or sediment; indicative of nearby mineral deposit.

geochemical cycle [GEOCHEM] During geologic changes, the sequence of stages in the migration of elements between the lithosphere, hydrosphere, and atmosphere.

geochemical evolution [GEOCHEM] **1.** A change in any constituent of a rock beyond that amount present in the parent rock. **2.** A change in chemical composition of a major segment of the earth during geologic time, as the oceans.

geochemical prospecting [ENG] The use of geochemical and biogeochemical principles and data in the search for economic deposits of minerals, petroleum, and natural gases.

geochemical well logging [ENG] Well logging dependent on geochemical analysis of the data.

geochemistry [GEOL] The study of the chemical composition of the various phases of the earth and the physical and chemical processes which have produced the observed distribution of the elements and nuclides in these phases.

geochronology [GEOL] **1.** The dating of the events in the earth's history. **2.** A system of dating developed for the purposes of study of the earth's history.

geochronometry [GEOL] The study of the absolute age of the rocks of the earth based on the radioactive decay of isotopes, such as U^{238}, U^{235}, Th^{232}, Rb^{87}, K^{40}, and C^{14}, present in minerals and rocks.

Geocorisae [INV ZOO] A subdivision of hemipteran insects containing those land bugs with conspicuous antennae and an ejaculatory bulb in the male.

geocosmogony [GEOL] The study of the origin of the earth.

geocronite [MINERAL] $Pb_5(Sb,As)_2S_3$ A mineral composed of lead-gray lead antimony arsenic sulfide.

geode [GEOL] A roughly spheroidal, hollow body lined inside with inward-projecting, small crystals; found frequently in limestone beds but may occur in shale.

geodesic [MATH] A curve joining two points in a Riemannian manifold which has minimum length.

geodesic coordinates [RELAT] Coordinates in the neighborhood of a point P such that the gradient of the metric tensor is zero at P.

geodesic dome [ARCH] A dome constructed of many light, straight structural elements in tension, arranged in a framework of triangles to reduce stress and weight.

geodesic line [MATH] The shortest line between two points on a mathematically derived surface.

geodesic motion [RELAT] Motion of a particle along a geodesic path in the four dimensional space-time continuum; according to general relativity, this is the motion which occurs in the absence of nongravitational forces.

geodesy [GEOPHYS] A subdivision of geophysics which in-

Selenium indicators used in prospecting for uranium.
(a) *Astragalus pattersoni*.
(b) *A. preussi*.

GEODE

Geode, lined with quartz crystals, Keokuk, Iowa. (*Brooks Museum, University of Virginia*)

cludes determination of the size and shape of the earth, the earth's gravitational field, and the location of points fixed to the earth's crust in an earth-referred coordinate system.

geodetic astronomy [GEOD] The branch of geodesy which utilizes astronomic observations to extract geodetic information.

geodetic coordinates [GEOD] The quantities latitude, longitude, and elevation which define the position of a point on the surface of the earth with respect to the reference spheroid.

geodetic datum [GEOD] A datum consisting of five quantities: the latitude and longitude of an initial point, the azimuth of a line from this point, and two constants necessary to define the terrestrial spheroid.

geodetic-distance meter *See* Geodimeter.

geodetic equator [GEOD] The great circle midway between the poles of revolution of the earth, connecting points of 0° geodetic latitude.

geodetic gravimetry [GEOD] Worldwide relative measurements of gravitational acceleration used in geodetic studies of the earth.

geodetic latitude [GEOD] Angular distance between the plane of the equator and a normal to the spheroid; a geodetic latitude differs from the corresponding astronomical latitude by the amount of the meridional component of station error. Also known as geographic latitude; topographical latitude.

geodetic line [GEOD] The shortest line between any two points on the surface of the spheroid.

geodetic longitude [GEOD] The angle between the plane of the reference meridian and the plane through the polar axis and the normal to the spheroid; a geodetic longitude differs from the corresponding astronomical longitude by the amount of the prime-vertical component of station error divided by the cosine of the latitude. Also known as geographic longitude.

geodetic meridian [GEOD] A line on a spheroid connecting points of equal geodetic longitude. Also known as geographic meridian.

geodetic parallel [GEOD] A line connecting points of equal geodetic latitude. Also known as geographic parallel.

geodetic position [GEOD] 1. A point on the earth, the coordinates of which have been determined by triangulation from an initial station, whose location has been established as a result of astronomical observations, the coordinates depending upon the reference spheroid used. 2. A point on the earth, defined in terms of geodetic latitude and longitude.

geodetic satellite [AERO ENG] An artificial earth satellite used to obtain data for geodetic triangulation calculations.

geodetic survey [ENG] A survey in which the figure and size of the earth are considered; it is applicable for large areas and long lines and is used for the precise location of basic points suitable for controlling other surveys.

Geodimeter [ENG] A trade name for electronic-optical distance-measuring equipment. Also known as geodetic-distance meter.

geodynamic height *See* dynamic height.

geoeconomy [GEOGR] The study of economic conditions that are influenced by geographic factors.

geoelectricity *See* terrestrial electricity.

geognosy [GEOL] The science dealing with the solid body of the earth as a whole, occurrences of minerals and rocks, and the origin of these and their relations.

geographical botany *See* plant geography.

geographical coordinates [GEOGR] Spherical coordinates, designating both astronomical and geodetic coordinates, defining a point on the surface of the earth, usually latitude and longitude. Also known as terrestrial coordinates.

geographical cycle *See* geomorphic cycle.

geographical mile [MECH] The length of 1 minute of arc of the Equator, or 6087.08 feet (1855.34 meters), which approximates the length of the nautical mile.

geographical plot [NAV] A plot of the movements of one or more craft relative to the surface of the earth.

geographical position [ASTRON] That point on the earth at which a given celestial body is in the zenith at a specified time. [GEOGR] Any position on the earth defined by means of its geographical coordinates, either astronomical or geodetic.

geographic latitude [GEOGR] A general term applying to either astronomic or geodetic latitude.

geographic longitude [GEOGR] A general term applying to either astronomical or geodetic latitude.

geographic meridian *See* geodetic meridian.

geographic number [NAV] The number assigned to an aid to navigation for identification purposes in accordance with the lateral system of numbering.

geographic parallel *See* geodetic parallel.

geographic position [GEOGR] The position of a point on the surface of the earth expressed in terms of geographical coordinates either geodetic or astronomical.

geographic range [GEOD] The extreme distance at which an object or light can be seen when limited only by the curvature of the earth and the heights of the object and the observer.

geographic search [NAV] An orderly arrangement of course lines in an area in which the area is defined in relation to one or more geographical points on the earth.

geographic sector search [NAV] An orderly arrangement of course lines consisting of three legs, the turning points being at equal distances along radial lines from a fixed point, the first leg being along the first radial, the second leg along the straight line connecting the equidistant points on the two radials, and the third leg along the course line intersecting a fixed or moving base at the time of return.

geographic speciation [EVOL] Evolution of two or more species from a single species following geographic isolation.

geographic square search [NAV] An orderly arrangement of course lines (a search) consisting of a series of lines of increasing length, each change of course being 90° in the same direction (right or left), so that the pattern of the search is an expanding square relative to a geographic point. Also known as fixed square search.

geographic vertical [GEOD] A line perpendicular to the surface of the geoid; it is the direction in which the force of gravity acts. [MAP] The direction of a line normal to the surface of the geoid. Also known as map vertical.

geography [SCI TECH] The study of all aspects of the earth's surface, comprising its natural and political divisions, the differentiation of areas, and, sometimes people in relationship to the environment.

geohydrology [HYD] The science dealing with underground water, often referred to as hydrogeology.

geoid [GEOD] The figure of the earth considered as a sea-level surface extended continuously over the entire earth's surface.

geoidal horizon [ASTRON] That circle of the celestial sphere formed by the intersection of the celestial sphere and a plane tangent to the sea-level surface of the earth at the zenith-nadir line.

geoisotherm [GEOPHYS] The locus of points of equal temperature in the interior of the earth; a line in two dimensions or a surface in three dimensions. Also known as geotherm; isogeotherm.

geolith *See* rock-stratigraphic unit.

geologic age [GEOL] 1. Any great time period in the earth's history marked by special phases of physical conditions or organic development. 2. A formal geologic unit of time that corresponds to a stage. 3. An informal geologic time unit that corresponds to any stratigraphic unit.

geological oceanography [GEOL] The study of the floors and margins of the oceans, including descriptions of topography, composition of bottom materials, interaction of sediments and rocks with air and sea water, the effects of movements in the mantle on the sea floor, and action of wave energy in the submarine crust of the earth. Also known as marine geology; submarine geology.

geological survey [GEOL] 1. An organization making geological surveys and studies. 2. A systematic geologic mapping of a terrain.

geological transportation [GEOL] Shifting of material by the action of moving water, ice, or air.

geologic climate *See* paleoclimate.

geologic column [GEOL] 1. The vertical sequence of strata of various ages found in an area or region. 2. The geologic time scale as represented by rocks.

geologic erosion *See* normal erosion.

geologic log [GEOL] A graphic presentation of the lithologic or stratigraphic units or both traversed by a borehole; used in petroleum and mining engineering as well as geological surveys.

geologic map [GEOL] A representation of the geologic surface or subsurface features by means of signs and symbols and with an indicated means of orientation; includes nature and distribution of rock units, and the occurrence of structural features, mineral deposits, and fossil localities.

geologic noise [GEOPHYS] Disturbances in observed data caused by random inhomogeneities in surface and near-surface material.

geologic province [GEOL] An area in which geologic history has been the same.

geologic section [GEOL] Any succession of rock units found at the surface or below ground in an area. Also known as section.

geologic structure [GEOL] The total structural features in an area.

geologic thermometer *See* geothermometer.

geologic thermometry *See* geothermometry.

geologic time [GEOL] The period of time covered by historical geology, from the end of the formation of the earth as a separate planet to the beginning of written history.

geologic time scale [GEOL] The relative age of various geologic periods and the absolute time intervals.

geologist [GEOL] An individual who specializes in the geological sciences.

geograph [ENG] A device that records the penetration rate of a bit during the drilling of a well.

geology [SCI TECH] The study or science of the earth, its history, and its life as recorded in the rocks; includes the study of geologic features of an area, such as the geometry of rock formations, weathering and erosion, and sedimentation.

geomagnetic coordinates [GEOPHYS] A system of spherical coordinates based on the best fit of a centered dipole to the actual magnetic field of the earth.

geomagnetic cutoff [GEOPHYS] The minimum energy of a cosmic-ray particle able to reach the top of the atmosphere at a particular geomagnetic latitude.

geomagnetic dipole [GEOPHYS] The magnetic dipole caused by the earth's magnetic field.

geomagnetic electrokinetograph [ENG] An instrument that can be suspended from the side of a ship to measure the direction and speed of ocean currents while the ship is under way by measuring the voltage induced in the moving conductive seawater by the magnetic field of the earth.

geomagnetic equator [GEOPHYS] That terrestrial great circle which is 90° from the geomagnetic poles.

geomagnetic field [GEOPHYS] The earth's magnetic field.

geomagnetic field reversal [GEOPHYS] Reversed magnetization in sedimentary and igneous rock, that is, polarized opposite to the mean geomagnetic field.

geomagnetic latitude [GEOPHYS] The magnetic latitude that a location would have if the field of the earth were to be replaced by a dipole field closely approximating it.

geomagnetic longitude [GEOPHYS] Longitude that is determined around the geomagnetic axis instead of around the rotation axis of the earth.

geomagnetic meridian [GEOPHYS] A circle passing around the earth and through the geomagnetic poles.

geomagnetic noise [COMMUN] Interference in radio communications arising from terrestrial magnetism. [GEOPHYS] Unwanted frequencies caused by fluctuations in the geomagnetic field of the earth.

geomagnetic pole [GEOPHYS] Either of two antipodal points marking the intersection of the earth's surface with the extended axis of a powerful bar magnet assumed to be located at the center of the earth and having a field approximating the actual magnetic field of the earth.

geomagnetic reversal [GEOPHYS] Reversed magnetization of the earth's geomagnetic dipole.

geomagnetic secular variation *See* secular variation.

geomagnetic storm [GEOPHYS] A large disturbance of the earth's magnetic field.

geomagnetic variation [GEOPHYS] Temporal changes in the geomagnetic field, both long-term (secular) and short-term (transient).

geomagnetism [GEOPHYS] **1.** The magnetism of the earth. Also known as terrestrial magnetism. **2.** The branch of science that deals with the earth's magnetism.

geometrical acoustics *See* ray acoustics.

geometrical dip [GEOD] The vertical angle, at the eye of an observer, between the horizontal and a straight line tangent to the surface of the earth.

geometrical factor [NAV] In loran and similar navigation systems, the ratio of the linear distance between a point on a loran line of position (hyperbola) and the nearest point on an adjacent line of the same family of lines to the interval between the corresponding time difference.

geometrical horizon [GEOD] The intersection of the celestial sphere and an infinite number of straight lines radiating from the eye of the observer and tangent to the earth's surface.

geometrical isomerism [PHYS CHEM] The phenomenon in which isomers contain atoms attached to each other in the same order and with the same bonds but with different spatial, or geometrical, relationships; the explicit geometry imposed upon a molecule by, say, a double bond between carbon atoms makes possible the existence of these isomers.

geometrical optics [OPTICS] The geometry of paths of light rays and their imagery through optical systems.

geometrical pitch [AERO ENG] The distance a component of an airplane propeller would move forward in one complete turn of the propeller if the path it was moving along was a helix that had an angle equal to an angle between a plane perpendicular to the axis of the propeller and the chord of the component.

geometrical similarity [FL MECH] Property of two fluid flows for which a simple alteration of scales of length and velocity transforms one into the other.

geometric average *See* geometric mean.

geometric construction [ENG] Construction that employs only straightedge and compasses or is carried out by drawing only straight lines and circles.

geometric distribution [STAT] A discrete probability distribution whose probability function is given by $p(x) = p(1-p)^{x-1}$ for x any positive integer, $p(x) = 0$ otherwise, when $0 \leq p \leq 1$; the mean is $1/p$.

geometric mean [MATH] The geometric mean of n given quantities is the nth root of their product. Also known as geometric average.

geometric number theory [MATH] The branch of number theory studying relationships among numbers by examining the geometric properties of ordered pair sets of such numbers.

geometric programming [SYS ENG] A nonlinear programming technique in which the relative contribution of each of the component costs is first determined; only then are the variables in the component costs determined.

geometric progression [MATH] A sequence which has the form $a, ar, ar^2, ar^3, \ldots$.

geometric series [MATH] An infinite series of the form $a + ar + ar^2 + ar^3 + \ldots$.

geometric vertical *See* geocentric vertical.

Geometridae [INV ZOO] A large family of lepidopteran insects in the superfamily Geometroidea that have slender bodies and relatively broad wings; includes measuring worms, loopers, and cankerworms.

geometrodynamics [RELAT] A theory involving only geometry which attempts to combine gravitational and electromagnetic theory; characterized by a multiply connected space-time manifold containing structures, descriptively called wormholes, associated with electric charge.

Geometroidea [INV ZOO] A superfamily of lepidopteran insects in the suborder Heteroneura comprising small to large moths with reduced maxillary palpi and tympanal organs at the base of the abdomen.

geometry [MATH] The qualitative study of shape and size. [NUCLEO] The physical arrangement of a device or assembly, such as a neutron counter.

geomorphic cycle [GEOL] The cycle of change in the surface configuration of the earth. Also known as cycle of erosion; geographical cycle.

geomorphology [GEOL] The study of the origin of secondary topographic features which are carved by erosion in the primary elements and built up of the erosional debris.

Geomyidae [VERT ZOO] The pocket gophers, a family of

rodents characterized by fur-lined cheek pouches which open outward, a stout body with short legs, and a broad, blunt head.

geonavigation [NAV] Plotting a course by computation based on information obtained by reference to other places on earth.

geopetal [PETR] Pertaining to the top-to-bottom relations in rocks at the time of formation.

geopetal fabric [PETR] The internal structure of a rock indicating the original orientation of the top-to-bottom strata.

geophagous [ZOO] Feeding on soil, as certain worms.

Geophilomorpha [INV ZOO] An order of centipedes in the class Chilopoda including specialized forms that are blind, epimorphic, and dorsoventrally flattened.

geophilous [ECOL] Living or growing in or on the ground.

geophone [ELECTR] A transducer, used in seismic work, that responds to motion of the ground at a location on or below the surface of the earth.

geophysical engineering [ENG] A branch of engineering that applies scientific methods for locating mineral deposits.

geophysical prospecting [ENG] Application of quantitative concepts and principles of physics and mathematics in geologic explorations to discover the character of and mineral resources in underground rocks in the upper portions of the earth's crust.

geophysicist [GEOPHYS] An individual who specializes in geophysics.

geophysics [GEOL] The physics of the earth and its environment, that is, earth, air, and (by extension) space.

geophyte [ECOL] A perennial plant that is deeply embedded in the soil substrata.

geopotential [PHYS] The potential energy of a unit mass relative to sea level, numerically equal to the work that would be done in lifting the unit mass from sea level to the height at which the mass is located, against the force of gravity.

geopotential height [GEOPHYS] The height of a given point in the atmosphere in units proportional to the potential energy of unit mass (geopotential) at this height, relative to sea level.

geopotential number [GEOPHYS] The numerical value C' that is assigned to a given geopotential surface when expressed in geopotential units (1 gpu = 1 meter × 1 kilogalileo).

geopotential surface [GEOPHYS] A surface of constant geopotential, that is, a surface along which a parcel of air could move without undergoing any changes in its potential energy. Also known as equigeopotential surface; level surface.

geopotential thickness [GEOPHYS] The difference in the geopotential height of two constant-pressure surfaces in the atmosphere, proportional to the appropriately defined mean air temperature between the two surfaces.

geopotential topography [GEOPHYS] The topography of any surface as represented by lines of equal geopotential; these lines are the contours of intersection between the actual surface and the level surfaces (which everywhere are normal to the direction of the force of gravity), and are spaced at equal intervals of dynamic height. Also known as absolute geopotential topography.

geopotential unit [GEOPHYS] A unit of gravitational potential used in describing the earth's gravitational field; it is equal to the difference in gravitational potential of two points separated by a distance of 1 meter when the gravitational field has a strength of 10 meters per second squared and is directed along the line joining the points. Abbreviated gpu.

georef See world geographic reference system.

georef grid [MAP] The grid system used on U.S. Air Force aeronautical charts for identifying the location of any point or area in the world.

Georges Banks [GEOL] An elevation beneath the sea east of Cape Cod, Mass.

georgiadesite [MINERAL] $Pb_3(AsO_4)Cl_3$ A white or brownish-yellow mineral composed of lead chloroarsenate, occurring in orthorhombic crystals.

georgiaite [GEOL] Any of a group of North American tektites, 134 million years of age, found in Georgia.

Georyssidae [INV ZOO] The minute mud-loving beetles, a family of coleopteran insects belonging to the Polyphaga.

geosere [GEOL] A series of ecological climax communities

following each other in geologic time and changing in response to changing climate and physical conditions.

Geosiridaceae [BOT] A monotypic family of monocotyledonous plants in the order Orchidales characterized by regular flowers with three stamens that are opposite the sepals.

geosphere [GEOL] **1.** The solid mass of earth, as distinct from the atmosphere and hydrosphere. **2.** The lithosphere, hydrosphere, and atmosphere combined.

Geospizinae [VERT ZOO] Darwin finches, a subfamily of perching birds in the family Fringillidae.

geostatic pressure See ground pressure.

geostationary satellite [AERO ENG] A satellite that orbits the earth from west to east at such a speed as to remain fixed over a given place on the earth's equator at approximately 35,900 kilometers altitude; makes one revolution in 24 hours, synchronous with the earth's rotation. Also known as synchronous satellite.

geostrophic [GEOPHYS] Pertaining to deflecting force resulting from the earth's rotation.

geostrophic approximation [GEOPHYS] The assumption that the geostrophic current can represent the actual horizontal current. Also known as geostrophic assumption. [METEOROL] A supposition that the geostrophic wind can represent the horizontal wind.

geostrophic assumption See geostrophic approximation.

geostrophic current [GEOPHYS] A current defined by assuming the existence of an exact balance between the horizontal pressure gradient force and the Coriolis force.

geostrophic departure [METEOROL] A vector representing the difference between the real wind and the geostrophic wind. Also known as ageostrophic wind; geostrophic deviation.

geostrophic deviation See geostrophic departure.

geostrophic distance [METEOROL] The distance (in degrees latitude) along a constant-pressure surface over which the change in height (in feet) is equal to the geostrophic wind speed (in knots).

geostrophic equation [GEOPHYS] An equation, used to compute geostrophic current speed, which represents a balance between the horizontal pressure gradient force and the Coriolis force.

geostrophic equilibrium [GEOPHYS] A state of motion of a nonviscous fluid in which the horizontal Coriolis force exactly balances the horizontal pressure force at all points of the field so described.

geostrophic flow [GEOPHYS] A form of gradient flow where the Coriolis force exactly balances the horizontal pressure force.

geostrophic flux [METEOROL] The transport of an atmospheric property by means of the geostrophic wind.

geostrophic vorticity [METEOROL] The vorticity of the geostrophic wind.

geostrophic wind [METEOROL] That horizontal wind velocity for which the Coriolis acceleration exactly balances the horizontal pressure force.

geostrophic-wind level [METEOROL] The lowest level at which the wind becomes geostrophic in the theory of the Ekman spiral. Also known as gradient-wind level.

geostrophic-wind scale [METEOROL] A graphical device used for the determination of the speed of the geostrophic wind from the isobar or contour-line spacing on a synoptic chart; it is a nomogram representing solutions of the geostrophic-wind equation.

geostrophy [GEOPHYS] Condition in which geostrophic assumption holds or geostrophic balance exists.

geosynclinal couple See orthogeosyncline.

geosynclinal cycle See tectonic cycle.

geosyncline [GEOL] A part of the crust of the earth that sank deeply through time.

geotaxis [PHYSIO] Movement of a free-living organism in response to the stimulus of gravity.

geotechnics [CIV ENG] The application of scientific methods and engineering principles to civil engineering problems through acquiring, interpreting, and using knowledge of materials of the crust of the earth.

geotechnology [ENG] Application of the methods of engi-

GEOPHILOMORPHA

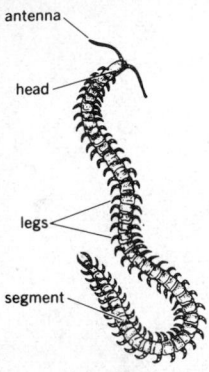

An epimorphic chilopod, an unidentified species 40 millimeters long. (*From R. E. Snodgrass, A Textbook of Arthropod Anatomy, Cornell University Press, 1952*)

GEOSYNCLINE

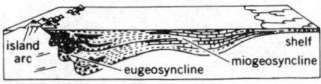

Diagrammatic section of Cordilleran geosyncline in southeastern Alaska and British Columbia at the close of the Permian. Volcanic deposits are indicated in black. (*After A. J. Eardley, J. Geol., 55:319–342, 1947*)

neering and science to exploitation of natural resources.

geotectogene *See* tectogene.

geotectonic cycle *See* orogenic cycle.

geotectonics *See* tectonics.

geotherm *See* geoisotherm.

geothermal [GEOPHYS] Pertaining to heat within the earth.

geothermal gradient [GEOPHYS] The change in temperature with depth of the earth.

geothermal prospecting [ENG] Exploration for sources of geothermal energy.

geothermal well logging [ENG] Measurement of the change in temperature of the earth by means of well logging.

geothermometer [ENG] A thermometer constructed to measure temperatures in boreholes or deep-sea deposits. [GEOL] A mineral that yields information about the temperature range within which it was formed. Also known as geologic thermometer.

geothermometry [GEOL] Measurement of the temperatures at which geologic processes occur or occurred. Also known as geologic thermometry.

geotropism [BOT] Response of a plant to the force of gravity.

Gephyrea [INV ZOO] A class of burrowing worms in the phylum Annelida.

gephyrocercal [VERT ZOO] Having the dorsal and anal fins coming together smoothly at the aborted end of the vertebral column of a fish's tail.

Geraniaceae [BOT] A family of dicotyledonous plants in the order Geraniales in which the fruit is beaked, styles are usually united, and the leaves have stipules.

geranial *See* citral.

geranialdehyde *See* citral.

Geraniales [BOT] An order of dicotyledonous plants in the subclass Rosidae comprising herbs or soft shrubs with a superior ovary and with compound or deeply cleft leaves.

geraniol [ORG CHEM] $(CH_3)_2CCH(CH_2)_2C(CH_3)$-$CHCH_2OH$ A colorless to pale-yellow liquid, an alcohol and a terpene, boiling point 230°C; soluble in alcohol and ether, insoluble in water; used in perfumery and flavoring.

Geranium [BOT] A genus of plants in the family Geraniaceae characterized by regular flowers, and glands alternating with the petals.

geranium oil [MATER] A pale-yellow or green liquid distilled from the herb of several Pelargonium species; chief known constituents are citronellol and geraniol, both used in perfumery and as flavoring agents.

geranyl [ORG CHEM] $C_{10}H_{17}$ The radical from geraniol, $(CH_3)_2:CHCH_2CH_2 \cdot CHCH_3:CH \cdot CH_2OH.$

Gerardiidae [INV ZOO] A family of anthozoans in the order Zoanthidea.

Gerard reagent [CHEM] The quaternary ammonium compounds, acethydrazide-pyridinium chloride and trimethyl-acethydrazide ammonium chloride; used to separate aldehydes and ketones from oily or fatty natural materials and to extract sex hormones from urine.

gerber beam [CIV ENG] A long, straight beam that functions essentially as a cantilevered beam by the insertion of two hinges in alternate spans.

Gerber test [FOOD ENG] A volumetric measurement of the quantity of fat in a sample of milk; the European counterpart of the Babcock test.

gerbil [VERT ZOO] The common name for about 100 species of African and Asian rodents composing the subfamily Gerbillinae.

Gerbillinae [VERT ZOO] The gerbils, a subfamily of rodents in the family Muridae characterized by hindlegs that are longer than the front ones, and a long, slightly haired, usually tufted tail.

gerhardtite [MINERAL] $Cu_2(NO_3)(OH)_3$ An emerald-green mineral composed of basic copper nitrate.

Gerhardt's test [PATH] A test for acetoacetic acid in urine.

geriatrics [MED] The study of the biological and physical changes and the diseases of old age.

germ [BIOL] A primary source, especially one from which growth and development are expected. [MICROBIO] General designation for a microorganism.

German cupellation [MET] A refining method using a large reverberatory furnace with a fixed bed and a movable roof;

the bullion to be cupelled is all charged at once, and the silver is not refined in the same furnace where the cupellation is carried on.

germane [INORG CHEM] **1.** A hydride of germanium whose general formula is Ge_nH_{2n+2}. **2.** The compound GeH_4, a hydride of germanium, a colorless gas that is combustible in air and burns with a blue flame.

germanide [INORG CHEM] A compound of an alkaline earth or alkali metal with germanium; an example is magnesium germanide, Mg_2Ge; the germanides are reactive with water.

germanite [MINERAL] $Cu_3(Ge,Ga,Fe)(S,As)_4$ Reddish-gray mineral occurring in massive form; an important source of germanium.

germanium [CHEM] A brittle, water-insoluble, silvery-gray metallic element in the carbon family, symbol Ge, atomic number 32, atomic weight 72.59, melting at 959°C. [MET] A rare metal used in semiconductors, alloys, and glass.

germanium diode [ELECTR] A semiconductor diode that uses a germanium crystal pellet as the rectifying element. Also known as germanium rectifier.

germanium halide [INORG CHEM] A dihalide or tetrahalide of fluorine, chlorine, bromine, or iodine with germanium.

germanium oxide [INORG CHEM] The monoxide GeO or dioxide GeO_2; a study of GeO indicates it exists in polymeric form; GeO_2 is a white powder, soluble in alkalies; used in special glass and in medicine.

germanium rectifier *See* germanium diode.

germanium transistor [ELECTR] A transistor in which the semiconductor material is germanium, to which electric contacts are made.

German measles *See* rubella.

German R unit [NUCLEO] A unit of radiation dose rate due to x-rays, equal to approximately 2.5 Solomon R units, or approximately 1.5 roentgens per second. Also known as R unit.

German silver *See* nickel silver.

German tubbing [MIN ENG] A form of tubbing, with internal flanges and bolts, for lining circular shafts sunk through heavily watered strata.

germarium [INV ZOO] The egg-producing portion of an ovary and the sperm-producing portion of a testis in Platyhelminthes and Rotifera.

germ ball [INV ZOO] A group of cells in digenetic trematode miracidial larvae which are embryos.

germ cell [BIOL] An egg or sperm or one of their antecedent cells.

germfree animal [MICROBIO] An animal having no demonstrable, viable microorganisms living in intimate association with it.

germfree isolator [MICROBIO] An apparatus that provides a mechanical barrier surrounding the area in which germfree vertebrates and accessory equipment are housed.

germicide [MATER] An agent that destroys germs.

germinal epithelium [EMBRYO] The region of the dorsal coelomic epithelium lying between the dorsal mesentery and the mesonephros.

germinal vesicle [CYTOL] The enlarged nucleus of the primary oocyte before reduction divisions are complete.

germination [BOT] The beginning or the process of development of a spore or seed. [MET] *See* grain growth.

germ layer [EMBRYO] One of the primitive cell layers which appear in the early animal embryo and from which the embryo body is constructed.

germ-layer theory [EMBRYO] The theory that three primary germ layers, ectoderm, mesoderm, and endoderm, are established in the early embryo and all organs and structures are derived from a specific germ layer.

germovitellarium [INV ZOO] A sex gland which differentiates into a yolk-producing or egg-producing region.

germ theory [MED] The theory that contagious and infectious diseases are caused by microorganisms.

geroderma [MED] The skin of old age, showing atrophy, loss of fat, and loss of elasticity.

gerontology [PHYSIO] The scientific study of aging processes in biological systems, particularly in humans.

Gerrhosauridae [VERT ZOO] A small family of lizards in the suborder Sauria confined to Africa and Madagascar.

GERANIALES

A common eastern United States species of geranium *(Geranium maculatum)*, which is characteristic of the order Geraniales. *(Courtesy of A. W. Ambler, from National Audubon Society)*

GERBIL

A gerbil, with long tail for balance when it hops.

GERMANIUM

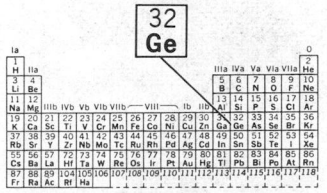

Periodic table of the chemical elements showing the position of germanium.

GEYSER

Old Faithful, Yellowstone Park, Wyoming. *(National Park Service, U.S. Department of the Interior)*

Gerridae [INV ZOO] The water striders, a family of hemipteran insects in the subdivision Amphibicorisae having long middle and hind legs and a median scent gland opening on the metasternum.

Gerroidea [INV ZOO] The single superfamily of the hemipteran subdivision Amphibicorisae; all members have conspicuous antennae and hydrofuge hairs covering the body.

gersdorffite [MINERAL] NiAsS A silver-white to steel-gray mineral, crystallizing in the isometric system; resembles cobaltite and may contain some iron and cobalt. Also known as nickel glance.

Gerstner wave [FL MECH] A rotational gravity wave of finite amplitude.

Gesneriaceae [BOT] A family of dicotyledonous plants in the order Scrophulariales characterized by parietal placentation, mostly opposite or whorled leaves, and a well-developed embryo.

gesso [MATER] A material made from chalk and gelatin or casein glue; painted on panels to furnish a surface for tempera work or for polymer-based paints.

gestalt [METEOROL] A complex of weather elements occurring in a familiar form, and though not necessarily referring to basic hydrodynamical or thermodynamical quantities, may persist for an appreciable length of time and is often considered to be an entity in itself.

Gestalt psychology [PSYCH] A school of psychology that views and examines the person as a whole.

gestate [EMBRYO] To carry the young in the uterus from conception to delivery.

gestation period [EMBRYO] The period in mammals from fertilization to birth.

get [IND ENG] A combination of two or more of the elemental motions of search, select, grasp, transport empty, and transport loaded; applied to time-motion studies.

getter [PHYS CHEM] 1. A substance, such as thallium, that binds gases on its surface and is used to maintain a high vacuum in a vacuum tube. 2. A special metal alloy that is placed in a vacuum tube during manufacture and vaporized after the tube has been evacuated; when the vaporized metal condenses, it absorbs residual gases. Also known as degasser.

getter-ion pump [ENG] A high-vacuum pump that employs chemically active metal layers which are continuously or intermittently deposited on the wall of the pump, and which chemisorb active gases while inert gases are "cleaned up" by ionizing them in an electric discharge and drawing the positive ions to the wall, where the neutralized ions are buried by fresh deposits of metal. Also known as sputter-ion pump.

GeV *See* giga electron volt.

gewel hinge [DES ENG] A hinge consisting of a hook inserted in a loop.

geyser [HYD] A natural spring or fountain which discharges a column of water or steam into the air at more or less regular intervals.

geyserite *See* siliceous sinter.

gf *See* gram-force.

g factor *See* Landé g factor.

g force [PHYS] A force such that a body subjected to it would have the acceleration of gravity at sea level; used as a unit of measurement for bodies undergoing the stress of acceleration.

GH *See* growth hormone.

GHA *See* Greenwich hour angle.

gharbi [METEOROL] A fresh westerly wind of oceanic origin in Morocco.

gharra [METEOROL] Hard squalls from the northeast in Libya and Africa that are sudden and frequent, and are accompanied by heavy rain and thunder.

ghatti gum [MATER] A water-soluble gum from the tree *Anogeissus latifolia,* forming a viscous glue in water; used as an emulsifier.

ghedda wax [MATER] A beeswax that is obtained from African and Indian bees.

ghibli [METEOROL] A hot, dust-bearing, desert wind in North Africa, similar to the foehn. Also known as chibli; gebli; gibleh; gibli; kibli.

Ghon complex [MED] The combination of a focus of subpleural tuberculosis with associated hilar and mediastinal lymph node tuberculosis.

ghost [ORD] In passive detection, one of the intersection

points of lines of position which do not represent actual targets but are only crossover points of multiple plotted lines of position from two or more detection stations. [PETR] The discernible outline of the shape of a former crystal or of another rock structure that has been partly obliterated and has as its boundaries inclusions, bubbles, or other foreign matter. Also known as phantom.

ghost crystal *See* phantom crystal.

ghost image [ELECTR] 1. An undesired duplicate image at the right of the desired image on a television receiver; due to multipath effect, wherein a reflected signal traveling over a longer path arrives slightly later than the desired signal. 2. *See* ghost pulse.

ghost lines *See* ferrite banding.

ghost mode [ELECTROMAG] Waveguide mode having a trapped field associated with an imperfection in the wall of the waveguide; a ghost mode can cause trouble in a waveguide operating close to the cutoff frequency of a propagation mode.

ghost pulse [ELECTR] An unwanted signal appearing on the screen of a radar indicator and caused by echoes which have a basic repetition frequency differing from that of the desired signals. Also known as ghost image; ghost signal.

ghost signal [ELECTR] 1. The reflection-path signal that produces a ghost image on a television receiver. Also known as echo. 2. *See* ghost pulse.

ghost spot [PL PATH] A disease of tomato characterized by small white rings on the fruit.

ghost structure *See* ferrite banding.

GHz *See* gigahertz.

giant *See* hydraulic monitor.

giant cell arteritis [MED] Inflammation of the arteries, particularly the carotid branches, characterized by the appearance of multinucleate giant cells in the exudate. Also known as temporal arteritis.

giant-cell leukemia *See* megakaryocytic leukemia.

giant granite *See* pegmatite.

giant planets [ASTRON] The planets Jupiter, Saturn, Uranus, and Neptune.

giant powder [MATER] A blasting powder made of nitroglycerin, sodium nitrate, sulfur, and rosin, sometimes with kieselguhr.

giant pulse laser *See* Q-switched laser.

giant star [ASTRON] One of a class of stars that is 20 or 30 or more times larger than the sun and over 100 times more luminous.

Giaque-Debye method *See* adiabatic demagnetization.

Giaque's temperature scale [THERMO] The internationally accepted scale of absolute temperature, in which the triple point of water is defined to have a temperature of 273.16K.

Giardia [INV ZOO] A genus of zooflagellates that inhabit the intestine of numerous vertebrates, and may cause diarrhea in humans.

giardiasis [MED] Presence of the protozoon *Giardia lamblia* in the human small intestine.

gib [ENG] A removable plate designed to hold other parts in place or act as a bearing or wear surface. [MIN ENG] 1. A temporary support at the face to prevent coal from falling before the cut is complete, either by hand or by machine. 2. A prop put in the holing of a seam while being undercut. 3. A piece of metal often used in the same hole with a wedge-shaped key for holding pieces together.

gibberellic acid [BIOCHEM] $C_{18}H_{22}O_6$ A crystalline acid occurring in plants that is similar to the gibberellins in its growth-promoting effects.

gibberellin [BIOCHEM] Any member of a family of naturally derived compounds which have a gibbane skeleton and a broad spectrum of biological activity but are noted as plant growth regulators.

gibberish *See* hash.

gibbon [VERT ZOO] The common name for seven species of large, tailless primates belonging to the genus *Hylobates*; the face and ears are hairless, and the arms are longer than the legs.

gibbous [MATH] Bounded by convex curves.

gibbous moon [ASTRON] The shape of the moon's visible surface when the sun is illuminating more than half of the side facing the earth.

GIBBON

Hylobates lar, which is a typical gibbon with extremely long arms and opposable thumb and big toe, and is found in Sumatra and southern Asia.

gibbs [PHYS] A unit of amount of adsorption, equal to a surface concentration of 10^{-6} mole per square meter.

Gibbs adsorption equation [PHYS CHEM] A formula for a system involving a solvent and a solute, according to which there is an excess surface concentration of solute if the solute decreases the surface tension, and a deficient surface concentration of solute if the solute increases the surface tension.

Gibbs adsorption isotherm [PHYS CHEM] An equation for the surface pressure of surface monolayers,

$$\phi = RT \int_0^p \Gamma d(\ln p)$$

where ϕ is surface pressure, T is absolute temperature, R is the gas constant, Γ is the number of molecules adsorbed per gram per unit surface area, and p is the pressure of the gas.

Gibbs apparatus [ENG] A compressed-oxygen breathing apparatus used in the United States.

Gibbs diaphragm cell [CHEM ENG] A type of electrolytic diaphragm cell for chlorine production, with graphite electrodes and a cylindrical shape.

Gibbs-Donnan equilibrium *See* Donnan equilibrium.

Gibbs-Duhem equation [PHYS CHEM] A relation that imposes a condition on the composition variation of the set of chemical potentials of a system of two or more components,

$$SdT - VdP + \sum_{i=1}^{r} n_i d\mu_i = 0$$

where S is entropy, T absolute temperature, P pressure, n_i the number of moles of the ith component, and μ_i is the chemical potential of the ith component. Also known as Duhem's equation.

Gibbs elasticity [PHYS] The elasticity of a film of liquid, equal to twice the product of the surface area and the derivative of the surface tension with respect to surface area.

Gibbs free energy [THERMO] The thermodynamic function $G = H - TS$, where H is enthalpy, T absolute temperature, and S entropy; Also known as free energy; free enthalpy; Gibbs function.

Gibbs function *See* Gibbs free energy.

Gibbs-Helmholtz equation [PHYS CHEM] An expression for the influence of temperature upon the equilibrium constant of a chemical reaction, $(d\ln K^0/dT)_P = \Delta H^0/RT^2$, where K^0 is the equilibrium constant, ΔH^0 the standard heat of the reaction at the absolute temperature T, and R the gas constant.

gibbsite [MINERAL] $Al(OH)_3$ A white or tinted mineral, crystallizing in the monoclinic system; a principal constituent of bauxite. Also known as hydrargillite.

Gibbs paradox [STAT MECH] The paradox in which there is an increase in entropy when two separate volumes of gases of the same kind, at the same temperature and pressure, are mixed.

Gibbs phase rule [PHYS CHEM] A relation describing the nature of a heterogeneous chemical system at equilibrium, $F = C + 2 - P$, where F is the degrees of freedom, P the number of phases, and C the number of components. Also known as Gibbs rule.

Gibbs' phenomenon [MATH] A convergence phenomenon occurring when a function with a discontinuity is approximated by a finite number of terms from a Fourier series.

Gibbs-Poynting equation [PHYS CHEM] An expression relating the effect of the total applied pressure P upon the vapor pressure p of a liquid, $(dp/dP)_T = V_l/V_g$, where V_l and V_g are molar volumes of the liquid and vapor.

Gibbs rule *See* Gibbs phase rule.

Gibbs system [STAT MECH] **1.** A hypothetical replica of a physical system. **2.** A set of such replicas forming an ensemble.

gibleh *See* ghibli.

gibli *See* ghibli.

giboulee *See* galerne.

Gibraltar stone *See* onyx marble.

gid [VET MED] A chronic brain disease of sheep, less frequently of cattle, characterized by forced movements of circling or rolling, caused by the larval form of the tapeworm *Multiceps multiceps*.

Giegy-Hardisty process [CHEM ENG] The production of sebacic acid from castor oil or its acids by reaction of the acid at a high temperature with caustic alkali.

Giemsa stain [CHEM] A stain for hemopoietic tissue and hemoprotozoa consisting of a stock glycerol methanol solution of eosinates of Azure B and methylene blue with some excess of the basic dyes.

Giesler coal test [ENG] A plastometric method for estimating the coking properties of coals.

gifblaar poison *See* fluoroacetic acid.

giga- [SCI TECH] A prefix representing 10^9, which is 1,000,000,000, or a billion. Abbreviated G. Also known as kilomega- (deprecated usage).

gigabit [COMMUN] One billion bits, or 1,000,000,000 bits. Also known as billibit (deprecated usage).

gigacycle *See* gigahertz.

giga electron volt [PHYS] A unit of energy, used primarily in high-energy physics, equal to 10^9 electron volts or $(1.60210 \pm 0.00007) \times 10^{-10}$ joule. Abbreviated GeV.

gigahertz [COMMUN] Unit of frequency equal to 10^9 hertz. Abbreviated GHz. Also known as gigacycle (gc); kilomegacycle; kilomegahertz.

gigantism [MED] Abnormal largeness of the body due to hypersecretion of growth hormone.

Giganturoidei [VERT ZOO] A suborder of small, mesopelagic actinopterygian fishes in the order Cetomimiformes having large mouths and strong teeth.

gigawatt [ELEC] One billion watts, or 10^9 watts. Abbreviated GW.

gigging [TEXT] Passing a fabric across rollers equipped with teasels to produce a nap on the surface.

gig mill [TEXT] A textile mill employing rotary wire cylinders for napping.

GIGO *See* garbage in, garbage out.

gigohm [ELEC] One thousand megohms, or 10^9 ohms.

Gila monster [VERT ZOO] The common name for two species of reptiles in the genus *Heloderma* (Helodermatidae) distinguished by a rounded body that is covered with multicolored beaded tubercles, and a bifid protrusible tongue.

gilbert [ELECTROMAG] The unit of magnetomotive force in the electromagnetic system, equal to the magnetomotive force of a closed loop of one turn in which there is a current of $1/4\pi$ abamp.

Gilbert circuit [ELECTR] A circuit that compensates for nonlinearities and instabilities in a monolithic variable-transconductance circuit by using the logarithmic properties of diodes and transistors.

Gilbreth's micromotion study [IND ENG] A time and motion study based on the concept that all work is performed by using a relatively few basic operations in varying combinations and sequence; basic elements (therbligs) include grasp, search, move, reach, and hold.

gilding [GRAPHICS] Overlaying material with a thin layer of gold.

gilding metal [MET] A copper alloy (about 90% copper, 10% zinc) used to jacket small-arms bullets, to form detonator or primer cups, and to form rotating bands for artillery projectiles; it can be readily engraved by the lands as the projectile moves down the bore.

gill [MECH] **1.** A unit of volume used in the United States for the measurement of liquid substances, equal to 1/4 U.S. liquid pint, or to $1.1829411825 \times 10^{-4}$ cubic meter. **2.** A unit of volume used in the United Kingdom for the measurement of liquid substances, and occasionally of solid substances, equal to 1/4 U.K. pint, or to approximately 1.42065×10^{-4} cubic meter. [VERT ZOO] The respiratory organ of water-breathing animals. Also known as branchia.

gill cover [VERT ZOO] The fold of skin providing external protection for the gill apparatus of most fishes; it may be stiffened by bony plates and covered with scales.

Gillespie equilibrium still [ANALY CHEM] A recirculating equilibrium distillation apparatus used to establish azeotropic properties of liquid mixtures.

gillespite [MINERAL] $BaFeSi_4O_{10}$ A micalike mineral composed of barium and iron silicate.

Gilliland correlation [CHEM ENG] Approximation method for distillation-column calculations; correlates reflux ratio and number of plates for the column as functions of minimum reflux and minimum plates.

gill net [ENG] A net that entangles the gill covers of fish.

gill raker [VERT ZOO] One of the bony processes on the inside

GILA MONSTER

Gila monster (*Heloderma suspectum*), about 20 inches (50 centimeters) long.

GINKGOALES

Two *Ginkgo* trees.

GINSENG

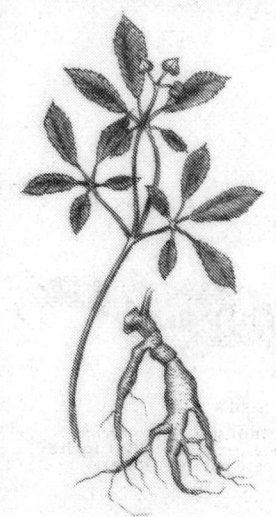

Panax quinquefolius, a ginseng, showing shoot and root.

GIRAFFE

The giraffe of Africa, with greater height in the forepart of the body because of the heavy muscular development at the base of the neck.

of the branchial arches of fishes which prevents the passage of solid substances through the branchial clefts.

Gilmour heat-exchange method [ENG] Thermal design method for heat exchangers by solution of five unique equations containing a minimum number of variables and involving tube-side, shell-side, tube-wall, and dirt resistance.

gilsonite [MINERAL] A variety of asphalt; it has black color, brilliant luster, brown streaks, and conchoidal fracture.

gimbal [ENG] **1.** A device with two mutually perpendicular and intersecting axes of rotation, thus giving free angular movement in two directions, on which an engine or other object may be mounted. **2.** In a gyro, a support which provides the spin axis with a degree of freedom. **3.** To move a reaction engine about on a gimbal so as to obtain pitching and yawing correction moments. **4.** To mount something on a gimbal.

gimbaled nozzle [MECH ENG] A nozzle supported on a gimbal.

gimbal freedom [ENG] Of a gyro, the maximum angular displacement about the output axis of a gimbal.

gimbaling error [NAV] That error introduced in a gyro compass by the tilting of its gimbal mounting system due to horizontal acceleration caused by motion of the vessel, such as rolling.

gimballess inertial navigation equipment See strapped-down inertial navigation equipment.

gimbal lock [ENG] A condition of a two-degree-of-freedom gyro wherein the alignment of the spin axis with an axis of freedom deprives the gyro of a degree-of-freedom and therefore its useful properties.

gimlet [DES ENG] A small tool consisting of a threaded tip, grooved shank, and a cross handle; used for boring holes in wood.

gimlet bit [DES ENG] A bit with a threaded point and spiral flute; used for drilling small holes in wood.

gimmick [ELEC] Length of twisted two-conductor cable, used as a variable capacitive load, in which the capacitance is varied by untwisting and separating the individual conductors.

gin [AGR] **1.** A machine used to separate cotton fiber from the seed and waste. **2.** To thus separate cotton fiber. [FOOD ENG] An alcoholic beverage made from distilled spirits flavored with an extract of the juniperberry or other flavoring botanicals. [MECH ENG] A hoisting machine in the form of a tripod with a windlass, pulleys, and ropes.

gin block [NAV ARCH] A block with a metal frame and a single large-diameter sheave, used to facilitate overhauling, and to hoist numerous packages, such as in cargo hoisting.

G indicator See G scope.

gingelly oil See sesame oil.

ginger [BOT] *Zingiber officinale*. An erect perennial herb of the family Zingiberaceae having thick, scaly branched rhizomes; a spice oleoresin is made by an organic solvent extraction of the ground dried rhizome.

ginger-grass oil [MATER] A type of citronella oil derived from sofia grass that contains about 50% geraniol; used in perfumery.

ginger oil [MATER] A thick, yellowish essential oil, soluble in most organic solvents and insoluble in water; distilled from dried ginger; main components are citral, borneol, and phellandrene; used as flavoring in liqueurs and soft drinks.

gingham [TEXT] Plain-weave cotton, or cotton-polyester blends, in checked or striped patterns or plain colors.

ginging [MIN ENG] **1.** Lining a shaft with masonry or brick. **2.** The brick or masonry of a shaft lining.

gingiva [ANAT] The mucous membrane surrounding the teeth sockets.

gingival crevice [ANAT] The space between the free margin of the gingiva and the surface of a tooth. Also known as gingival sulcus.

gingival sulcus See gingival crevice.

gingivectomy [MED] Excision of a portion of the gingiva.

gingivitis [MED] Inflammation of the gingiva.

gingivostomatitis [MED] An inflammation of the gingiva and oral mucosa.

ginglymoarthrodia [ANAT] A composite joint consisting of one hinged and one gliding element.

Ginglymodi [VERT ZOO] An equivalent name for Semionotiformes.

ginglymus [ANAT] A type of diarthrosis permitting motion only in one plane. Also known as hinge joint.

Ginkgoales [BOT] An order of gymnosperms composing the class Ginkgoopsida with one living species, the dioecious maidenhair tree (*Ginkgo biloba*).

Ginkgoatae See Ginkgoopsida.

Ginkgoopsida [BOT] A class of the subdivision Pinicae containing the single, monotypic order Ginkgoales.

Ginkgophyta [BOT] The equivalent name for Ginkgoopsida.

ginorite [MINERAL] $Ca_2B_{14}O_{23} \cdot 8H_2O$ A white monoclinic mineral composed of hydrous borate of calcium.

gin pit [MIN ENG] A shallow mine, the hoisting from which is done by a gin.

gin pole [MECH ENG] A hand-operated derrick which has a nearly vertical pole supported by guy ropes; the load is raised on a rope that passes through a pulley at the top and over a winch at the foot. Also known as guyed-mast derrick; pole derrick; standing derrick.

ginseng [BOT] The common name for plants of the genus *Panax*, a group of perennial herbs in the family Araliaceae; the aromatic root of the plant has been used medicinally in China.

gin tackle [MECH ENG] A tackle made for use with a gin.

Ginzburg-Landau theory [CRYO] A phenomenological theory of superconductivity which accounts for the coherence length; the ordered state of a superconductor is described by a complex order parameter which is similar to a Schrödinger wave function, but describes all the condensed superelectrons, rather than a single charged particle. Also known as Landau-Ginzburg theory.

Ginzburg-London superconductivity theory [SOLID STATE] A modification of the London superconductivity theory to take into account the boundary energy.

gio See geo.

giobertite See magnesite.

Giorgi system See meter-kilogram-second-ampere system.

gipsy head See gypsy head.

gipsy winch [MIN ENG] A small winch that may be attached to a post and operated by a rotary motion or the reciprocating action of a handle having a pair of pawls and a ratchet.

giraffe [VERT ZOO] *Giraffa camelopardalis*. An artiodactyl mammal in the family Giraffidae characterized by extreme elongation of the neck vertebrae, and two prominent horns on the head.

Giraffe See Camelopardalis.

Giraffidae [VERT ZOO] A family of pecoran ruminants in the superfamily Bovoidea including giraffe, okapi, and relatives.

Girbotal process [CHEM ENG] A regenerative absorption process to remove carbon dioxide, hydrogen sulfide, and other acid impurities from natural gas, using mono-, di-, or triethanolamine as the reagent.

girder [CIV ENG] A large beam made of metal or concrete, and sometimes of wood.

girdle [ANAT] Either of the ringlike groups of bones supporting the forelimbs (arms) and hindlimbs (legs) in vertebrates. [INV ZOO] **1.** Either of the hooplike bands constituting the sides of the two valves of a diatom. **2.** The peripheral portion of the mantle in chitons. [LAP] The periphery of a cut gemstone that is usually grasped by the setting or mounting. [PETR] With reference to a fabric diagram or equal-area projection net, a belt showing concentration of points which is approximately coincident with a great circle of the net and which represents orientation of the fabric elements.

girt [CIV ENG] **1.** A timber in the second-floor corner posts of a house to serve as a footing for roof rafters. **2.** A horizontal member to stiffen the framework of a building frame or trestle. [ENG] A brace member running horizontally between the legs of a drill tripod or derrick. [MIN ENG] In square-set timbering, a horizontal brace running parallel to the drift.

girth [NAV ARCH] The distance measured along a frame line from a waterline on one side around the hull to the corresponding point on the opposite side.

gismondite [MINERAL] $CaAl_2Si_2O_8 \cdot 4H_2O$ A light-colored mineral composed of hydrous calcium aluminum silicate, occurring in pyramidal crystals.

gitogenin [PHARM] $C_{27}H_{44}O_4$ A crystalline compound obtained from gitonin by heating with dilute hydrochloric acid, forms leaflets from benzene solutions, soluble in organic solvents such as chloroform, hot alcohol, and ether; used in medicine for treating coronary disease.

gitonin [ORG CHEM] The gitogenin tetraglycoside in *Digitalis purpurea* seed; resembles digitonin.

gitoxin [PHARM] $C_{41}H_{64}O_{14}$ A secondary glycoside derived from *Digitalis purpurea*, crystallizes in stout prisms from chloroform methanol solution, soluble in a mixture of chloroform and alcohol; used in medicine for coronary disease.

gitter cell [PATH] A compound granule cell that is characteristic of certain brain lesions.

gizzard [VERT ZOO] The muscular portion of the stomach of most birds where food is ground with the aid of ingested pebbles.

glabella [ANAT] The bony prominence on the frontal bone joining the supraorbital ridges.

glabello-occipital length [ANTHRO] The distance between the glabella and the opisthocranion.

glabrous [BIOL] Having a smooth surface; specifically, having the epidermis devoid of hair or down.

glacial [GEOL] Pertaining to an interval of geologic time which was marked by an equatorward advance of ice during an ice age; the opposite of interglacial; these intervals are variously called glacial periods, glacial epochs, glacial stages, and so on. [HYD] Pertaining to ice, especially in great masses such as sheets of land ice or glaciers.

glacial abrasion [GEOL] Alteration of portions of the earth's surface as a result of glacial flow.

glacial accretion [GEOL] Deposition of material as a result of glacial flow.

glacial acetic acid [ORG CHEM] CH_3COOH Pure acetic acid (containing less than 1% water); a clear, colorless, caustic hygroscopic liquid, boiling at 118°C, soluble in water, alcohol, and ether, and crystallizing readily; used as a solvent for oils and resins.

glacial advance [GEOL] **1.** Increase in the thickness and area of a glacier. **2.** A time period equal to that increase.

glacial anticyclone [METEOROL] A type of semipermanent anticyclone which overlies the ice caps of Greenland and Antarctica. Also known as glacial high.

glacial boulder [GEOL] A boulder moved to a point distant from its original site by a glacier.

glacial deposit [GEOL] Material carried to a point beyond its original location by a glacier.

glacial drift [GEOL] All rock material in transport by glacial ice, and all deposits predominantly of glacial origin made in the sea or in bodies of glacial meltwater, including rocks rafted by icebergs.

glacial epoch [GEOL] **1.** Any of the geologic epochs characterized by an ice age; thus, the Pleistocene epoch may be termed a glacial epoch. **2.** Generally, an interval of geologic time which was marked by a major equatorward advance of ice; the term has been applied to an entire ice age or (rarely) to the individual glacial stages which make up an ice age.

glacial erosion [GEOL] Movement of soil or rock from one point to another by the action of the moving ice of a glacier. Also known as ice erosion.

glacial flour See rock flour.

glacial flow See glacier flow.

glacial geology [GEOL] The study of land features resulting from glaciation.

glacial high See glacial anticyclone.

glacial ice [HYD] Ice that is flowing or that exhibits evidence of having flowed.

glacial lake [GEOL] A lake that exists because of the effects of the glacial period.

glacial lobe [HYD] A tonguelike projection from a continental glacier's main mass.

glacial maximum [GEOL] The time or position of the greatest extent of any glaciation; most frequently applied to the greatest equatorward advance of Pleistocene glaciation.

glacial mill See moulin.

glacial outwash See outwash.

glacial period [GEOL] **1.** Any of the geologic periods which embraced an ice age; for example, the Quaternary period may be called a glacial period. **2.** Generally, an interval of geologic time which was marked by a major equatorward advance of ice.

glacial plucking See plucking.

glacial retreat [GEOL] A condition occurring when backward melting at the front of a glacier takes place at a rate exceeding forward motion.

glacial scour [GEOL] Erosion resulting from glacial action, whereby the surface material is removed and the rock fragments carried by the glacier abrade, scratch, and polish the bedrock. Also known as scouring.

glacial striae [GEOL] Scratches, commonly parallel, on smooth rock surfaces due to glacial abrasion.

glacial till See till.

glacial varve See varve.

glaciated terrain [GEOL] A region that once bore great masses of glacial ice; a distinguishing feature is marks of glaciation.

glaciation [GEOL] Alteration of any part of the earth's surface by passage of a glacier, chiefly by glacial erosion or deposition. [METEOROL] The transformation of cloud particles from waterdrops to ice crystals, as in the upper portion of a cumulonimbus cloud.

glaciation limit [GEOPHYS] For a given locality, the lowest altitude at which glaciers can develop.

glacier [HYD] A mass of land ice, formed by the further recrystallization of firn, flowing slowly (at present or in the past) from an accumulation area to an area of ablation.

glacieret See snowdrift ice.

glacier flow [HYD] The motion that exists within a glacier's body. Also known as glacial flow.

glacier front [HYD] The leading edge of a glacier.

glacier ice [HYD] Any ice that is or was once a part of a glacier, consolidated from firn by further melting and refreezing and by static pressure; for example, an iceberg.

glacier mill See moulin.

glacier pothole See moulin.

glacier table [GEOL] A stone block supported by an ice pedestal above the surface of a glacier.

glacier well See moulin.

glacier wind [METEOROL] A shallow gravity wind along the icy surface of a glacier, caused by the temperature difference between the air in contact with the glacier and free air at the same altitude.

glaciofluvial [GEOL] Pertaining to streams fed by melting glaciers, or to the deposits and landforms produced by such streams.

glaciolacustrine [GEOL] Pertaining to lakes fed by melting glaciers, or to the deposits forming therein.

glaciology [GEOL] The study of existing or modern glaciers in their entirety.

glaçon [OCEANOGR] A piece of sea ice which is smaller than a medium-sized floe.

gladiate [BOT] Sword-shaped.

gladiolus See mesosternum.

Gladiolus [BOT] A genus of chiefly African plants in the family Iridaceae having erect, sword-shaped leaves and spikes of brightly colored irregular flowers.

Gladstone-Dale law [OPTICS] A law for the variation of the index of refraction n of a substance, according to which $n + 1$ is proportional to its density.

glance pitch [GEOL] A variety of asphaltite having brilliant conchoidal fracture, and resembling gilsonite but having higher specific gravity and percentage of fixed carbon.

glancing angle [PHYS] The angle between a surface and a beam of particles or radiation incident upon it; it is the complement of the angle of incidence.

gland [ANAT] A structure which produces a substance essential and vital to the existence of the organism. [ENG] **1.** A device for preventing leakage at a machine joint, as where a shaft emerges from a vessel containing a pressurized fluid. **2.** A movable part used in a stuffing box to compress the packing.

glanders [VET MED] A bacterial disease of equines caused by

GLADIOLUS

A specimen of the genus *Gladiolus* showing the erect sword-shaped leaves and spikes of flowers.

Actinobacillus mallei; involves the respiratory system, skin, and lymphatics. Also known as farcy.

glands of Brunner *See* Brunner's glands.

glands of Leydig [VERT ZOO] Unicellular, epidermal structures of urodele larvae and the adult *Necturus* that secrete a substance which digests the egg capsule and permits hatching.

glandular fever *See* infectious mononucleosis.

glandulomuscular [ANAT] Of or pertaining to glands and muscles.

glans [ANAT] The conical body forming the distal end of the clitoris or penis.

glare [OPTICS] **1.** Discomfort produced in an observer by one or more visible sources of light. Also known as discomfort glare. **2.** Visual disability caused by visible sources or areas of luminance which are in an observer's field of view but do not assist in viewing. Also known as disability glare. **3.** Dazzling brightness of the atmosphere, caused by excessive reflection and scattering of light by particles in the line of sight.

glare ice [HYD] Ice with a smooth, shiny surface.

Glareolidae [VERT ZOO] A family of birds in the order Charadriiformes including the ploverlike coursers and the swallowlike pratincoles.

glareous [ECOL] Growing in gravelly soil; refers specifically to plants.

glaserite *See* arcanite.

glass [MATER] A hard, amorphous, inorganic, usually transparent, brittle substance made by fusing silicates, sometimes borates and phosphates, with certain basic oxides and then rapidly cooling to prevent crystallization. [METEOROL] In nautical terminology, a contraction for "weather glass" (a mercury barometer).

glass armor [ORD] Any of several special purpose armors composed of glass or containing glass.

glassblowing [ENG] Shaping a mass of viscid glass by inflating it with air introduced through a tube.

glass-bonded mica [MATER] An insulating material made by compressing a mixture of powdered glass and powdered natural or synthetic mica at high temperatures.

glass brick [MATER] A hollow block of translucent glass with patterns molded on the faces; used in partitions.

glass capacitor [ELEC] A capacitor whose dielectric material is glass.

glass cutter [ENG] A tool equipped with a steel wheel or a diamond point used to cut glass.

glass dosimeter [NUCLEO] A dosimeter using as its radiation-sensing element a fluorod of special glass that fluoresces under ultraviolet light following gamma irradiation.

glassed steel [CHEM ENG] Process piping or vessels lined with glass; a glass-steel composite has structural strength of steel and corrosion resistance of glass.

glass electrode [PHYS CHEM] An electrode or half cell in which potential measurements are made through a glass membrane, which acts as a cation-exchange membrane; thus, the potential arises from phase-boundary and diffusion potentials which, depending on the composition of the glass, are logarithmic functions of the activity of the cations such as H^+, Na^+, or K^+ of the solutions in which the electrode is immersed.

Glasser's disease [VET MED] A generalized bacterial infection of swine caused by *Mycoplasma hyorhinis.*

glass fiber [MATER] A glass thread less than a thousandth of an inch thick, used loosely or in woven form as an acoustic, electrical, or thermal insulating material and as a reinforcing material in laminated plastics.

glass film plates [GRAPHICS] Film plates made of glass sheets coated with a sensitized emulsion; they have been largely replaced by sheet film, but are still used in some critical work because of their dimensional stability.

glass fission detector [NUCLEO] A piece of glass in which fission fragments, flying apart with high energy, can create narrow but continuous, submicroscopic trails of altered material, which can be seen in an ordinary microscope after the altered material has been dissolved by a chemical reagent.

glass furnace [ENG] A large, covered furnace or tank for melting large batches of glass, in which heat is supplied by a flame playing over the glass surface, and regenerative heating

of combustion air and gas is usually employed. Also known as glass tank.

glass guide [GRAPHICS] In microfilm photography, a transparent bar of optical glass employed to guide documents through the photographic field.

glass heat exchanger [ENG] Any heat exchanger in which glass replaces metal, such as shell-and-tube, cascade, double-pipe, bayonet, and coil exchangers.

glassine [MATER] A thin, dense, transparent, supercalendered paper from highly refined sulfite pulp, used for envelope windows, for sanitary wrapping, and as an insulating paper between layers of iron-core transformer windings.

glass insulator [MATER] An insulator for a power transmission line made of annealed or toughened (tempered) glass.

glassivation [ELECTR] Method of transistor passivation by a pyrolytic glass-deposition technique, whereby silicon semiconductor devices, complete with metal contact systems, are fully encapsulated in glass.

glassmakers' soap [MATER] A substance such as manganese dioxide added to glass to remove the green color created by iron salts.

glass paper [MATER] **1.** Paper with a layer of pulverized glass; used as an abrasive. **2.** Paper made of glass fibers.

glass-plate capacitor [ELEC] High-voltage capacitor in which the metal plates are separated by sheets of glass serving as the dielectric, with the complete assembly generally immersed in oil.

glass porphyry *See* virtophyre.

glass pot [ENG] A crucible used for making small amounts of glass.

glass resistor [ELEC] A glass tube with a helical carbon resistance element painted on it.

glass sand [MATER] High-quartz sand used in glassmaking; contains small amounts of aluminum oxide, iron oxide, calcium oxide, and magnesium oxide.

glass seal [ENG] An airtight seal made by molten glass.

glass sponge [INV ZOO] A siliceous sponge belonging to the class Hyalospongiae.

glass switch [ELECTR] An amorphous solid-state device used to control the flow of electric current. Also known as ovonic device.

glass tank *See* glass furnace.

glass textile [TEXT] Fabric woven from glass fibers; used for electrical insulation, as filter cloth, and in plastic laminates.

glass-to-metal seal [ELECTR] An airtight seal between glass and metal parts of an electron tube, made by fusing together a special glass and special metal alloy having nearly the same temperature coefficients of expansion.

glass transition [PHYS CHEM] The change in an amorphous region of a partially crystalline polymer from a viscous or rubbery condition to a hard and relatively brittle one; usually brought about by changing the temperature. Also known as gamma transition; glassy transition.

glass-tube manometer [ENG] A manometer for simple indication of difference of pressure, in contrast to the metallic-housed mercury manometer, used to record or control difference of pressure or fluid flow.

glassware [MATER] Articles, especially tableware, made of glass.

glass wool [MATER] A mass of glass fibers resembling wool and used as insulation, packing, and air filters.

glassy feldspar *See* sanidine.

glassy state *See* vitreous state.

glassy transition *See* glass transition.

glauberite [MINERAL] $Na_2Ca(SO_4)_2$ A brittle, gray-yellow monoclinic mineral having vitreous luster and saline taste.

Glauber's salt [INORG CHEM] $Na_2SO_4 \cdot 10H_2O$ Crystalline hydrated sodium sulfate; loses water when exposed to air; water soluble, alcohol insoluble; used in textile dyeing and medicine.

glaucocerinite [MINERAL] A mineral composed of a hydrous basic sulfate of copper, zinc, and aluminum.

glaucochroite [MINERAL] $CaMnSiO_4$ A bluish-green mineral that is related to monticellite, is composed of calcium manganese silicate, and occurs in prismatic crystals.

glaucodot [MINERAL] $(Co,Fe)AsS$ A grayish-white, metal-

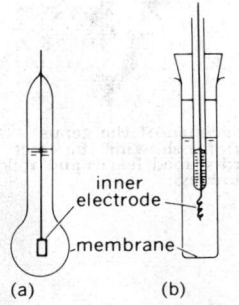

GLASS ELECTRODE

inner electrode

membrane

(a)   (b)

Two types of glass electrode: *(a)* bulb type; *(b)* flat-membrane type.

lic-looking mineral composed of cobalt iron sulfarsenide, occurring in orthorhombic crystals.

glaucoma [MED] A disease of the eye characterized by increased fluid pressure within the eyeball.

glauconite [MINERAL] $K_{1.5}(Fe,Mg,Al)_{4-6}(Si,Al)_8O_{20}(OH)_4$ A type of clay mineral; it is dioctahedral and occurs in flakes and as pigmentary material.

glauconitic sandstone [PETR] A quartz sandstone or an arkosic sandstone that has many glauconite grains.

glaucophane [MINERAL] $Na_2Mg_3Al_2Si_8$ A blue to black monoclinic sodium amphibole; blue to black coloration with marked pleochroism.

glaucophane schist [PETR] Metamorphic schist that contains glaucophane.

glaucous [BOT] Having a white or grayish powdery coating that gives a frosty appearance and rubs off easily.

glave *See* glaves.

glaves [METEOROL] A foehnlike wind of the Faroe Islands. Also known as glave; glavis.

glavis *See* glaves.

glaze [ENG] A glossy coating. Also known as enamel. [HYD] A coating of ice, generally clear and smooth but usually containing some air pockets, formed on exposed objects by the freezing of a film of supercooled water deposited by rain, drizzle, or fog, or possibly condensed from supercooled water vapor. Also known as glaze ice; glazed frost; verglas.

glazed frost *See* glaze.

glaze ice *See* glaze.

glaze stain [INORG CHEM] Colorant for ceramic glazes; made of a finely ground calcined oxide, such as of cobalt, copper, manganese, or iron.

glazier's point [ENG] A small piece of sheet metal, usually shaped like a triangle, used to hold a pane of glass in place.

glazing [ENG] **1.** Cutting and fitting panes of glass into frames. **2.** Smoothing the lead of a wiped pipe joint by passing a hot iron over it.

glazing compound [MATER] A caulking compound used to seal and support a glass pane in place.

glb *See* greatest lower bound.

GLC *See* gas-liquid chromatography.

Gleason bevel gear system [DES ENG] The standard for bevel gear designs in the United States; employs a basic pressure angle of 20° with long and short addenda for ratios other than 1:1 to avoid undercut pinions and to increase strength.

gleba [MYCOL] The central, sporogenous tissue of the sporophore in certain basidiomycetous fungi.

gleet [MED] The chronic stage of gonorrheal urethritis, characterized by a slight mucopurulent discharge.

glenoid [ANAT] A smooth, shallow, socketlike depression, particularly of the skeleton.

glenoid cavity [ANAT] The articular surface on the scapula for articulation with the head of the humerus.

glessite [GEOL] Fossil resin similar to amber.

gley [GEOL] A sticky subsurface layer of clay in some waterlogged soils.

glide [AERO ENG] Descent of an aircraft at a normal angle of attack, with little or no thrust.

glide angle *See* gliding angle.

glide bomb [ORD] A bomb fitted with airfoils to provide lift, released in the direction of a target by an airplane.

glide fold *See* shear fold.

glide path [AERO ENG] **1.** The flight path of an aeronautical vehicle in a glide, seen from the side. Also known as glide trajectory. **2.** The path used by an aircraft or spacecraft in a landing approach procedure.

glide path indicator [NAV] A system which provides signals for indicating vertical guidance of an aircraft along an inclined surface.

glide plane [CRYSTAL] A lattice plane in a crystal on which translation or twin gliding occurs. Also known as slip plane. [NAV] *See* glide slope.

glider [AERO ENG] A fixed-wing aircraft designed to glide, and sometimes to soar; usually does not have a power plant.

glide slope [AERO ENG] *See* gliding angle. [NAV] An inclined electromagnetic surface which is generated by instrument-landing approach facilities and which includes a glide

path supplying guidance in the vertical plane. Also known as glide plane.

glide slope facility [NAV] An instrument approach landing facility furnishing vertical guidance information to aircraft from its altitude down to the runway.

glide slope sector [NAV] A vertical sector containing that portion of the glide slope within which the cooperating aircraft equipment provides a quantitative measurement of deviations above or below the glide slope.

glide trajectory *See* glide path.

gliding angle [AERO ENG] The angle between the horizontal and the glide path of an aircraft. Also known as glide angle; glide slope.

gliding bacteria [MICROBIO] The descriptive term for members of the orders Beggiatoales and Myxobacterales; they are motile by means of creeping movements.

gliding joint *See* arthrodia.

gliding motility [MICROBIO] A means of bacterial self-propulsion by slow gliding or creeping movements on the surface of a substrate.

glime [HYD] An ice coating with a consistency intermediate between glaze and rime.

glimmer ice [HYD] Ice newly formed within the cracks or holes of old ice, or on the puddles on old ice.

G line [ELECTROMAG] A single dielectric-coated, round wire used for transmitting microwave energy.

glint [ELECTR] **1.** Pulse-to-pulse variation in amplitude of a reflected radar signal, owing to the reflection of the radar from a body that is rapidly changing its reflecting surface, for example, a spinning airplane propeller. **2.** The use of this effect to degrade tracking or seeking functions of an enemy weapons system.

glioma [MED] A malignant tumor derived from the supporting tissue of the central nervous system.

gliosis [MED] Proliferation of neuroglia in the brain or spinal cord, either as a replacement process or in response to a low-grade inflammation.

gliotoxin [MICROBIO] $C_{13}H_{14}O_4N_2S_2$ A heat-labile, bacteriostatic antibiotic produced by species of *Trichoderma* and *Cliocladium* and by *Aspergillus fumigatus*.

Gliridae [VERT ZOO] The dormice, a family of mammals in the order Rodentia.

glissile dislocation *See* Shockley partial dislocation.

Glisson's capsule [ANAT] The membranous sheet of collagenous and elastic fibers covering the liver.

glitter [MATER] A group of decorative materials consisting of flakes large enough so that each flake produces a plainly visible sparkle or reflection; incorporated into plastic during compounding. [OPTICS] The spots of light reflected from a point source by the surface of the sea or wave facets, that is, specular reflection.

Glivenko-Cantelli lemma [MATH] The empirical distribution functions of a random variable converge uniformly in probability to the distribution function of the random variable.

g load [PHYS] The numerical ratio of any applied force to the gravitational force at the earth's surface.

global radiation [GEOPHYS] The total of direct solar radiation and diffuse sky radiation received by a unit horizontal surface.

Globar lamp [ELEC] A lamp whose illuminating element is a silicon carbide rod which gives off blackbody radiation when heated.

globe [MAP] A sphere on the surface of which is a map of the world.

globe lightning *See* ball lightning.

globe valve [ENG] A device for regulating flow in a pipeline, consisting of a movable disk-type element and a stationary ring seat in a generally spherical body.

Globigerinacea [INV ZOO] A superfamily of foraminiferan protozoans in the suborder Rotaliina characterized by a radial calcite test with bilamellar septa and a large aperture.

globigerina ooze [GEOL] A pelagic sediment consisting of than 30% calcium carbonate in the form of foraminiferal tests of which *Globigerina* is the dominant genus.

globin [BIOCHEM] Any of a class of soluble histone proteins obtained from animal hemoglobins.

globin zinc insulin [PHARM] A preparation of insulin modi-

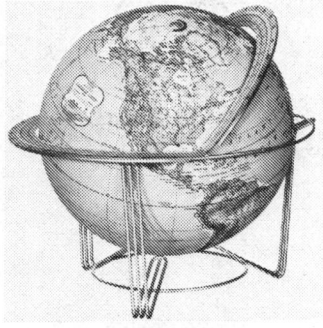

GLOBE

Globe with a hinged horizon ring. In the north polar area is a small disk which is separately rotated to show simultaneous time over the earth. *(Rand McNally)*

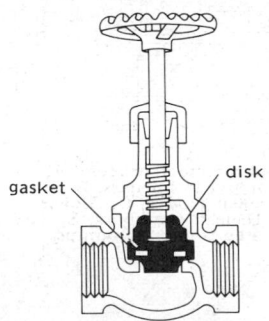

GLOBE VALVE

gasket — disk

A globe valve with the gasket in the disk.

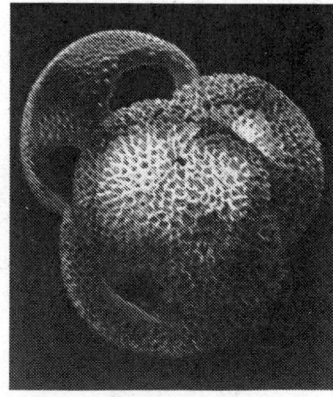

GLOBIGERINACEA

Scanning electron micrograph of *Globigerinoides* from the Holocene of the Caribbean Sea. *(R. B. MacAdam, Chevron Oil Field Research Co.)*

GLOSSIPHONIIDAE

Glossiphonia complanata, a small leech that sucks blood of aquatic invertebrates, especially snails.

GLOVER TOWER

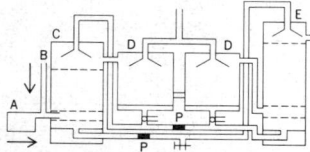

The lead chamber process. A, sulfur or pyrite burners; B, inlet for nitrogen oxides; C, Glover tower; D, lead chambers; E, Gay-Lussac tower; P, pumps.

GLOW-TUBE OSCILLATOR

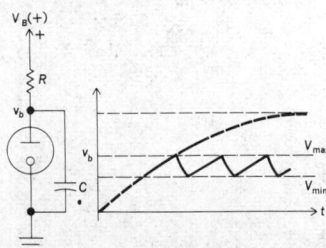

Circuit diagram of glow-tube oscillator, and fixed-amplitude periodic sawtooth waveform it generates. V_B = quiescent value of plate voltage; v_b = instantaneous value of varying component of plate voltage. Broken line on graph represents value that v_b would assume in absence of glow-discharge tube. When v_b reaches breakdown potential V_{max} of the tube, capacitance C discharges through tube and v_b is lowered until extinction potential V_{min} is reached, at which time cycle is repeated.

fied by the addition of globin (derived from the hemoglobin of beef blood) and zinc chloride; it has intermediate duration of action.

globoside [BIOCHEM] A glycoside of ceramide containing several sugar residues, but not neuraminic acid; obtained from human, sheep, and hog erythrocytes.

globular *See* spherulitic.

globular projection [MAP] A projection, in perspective, of a hemisphere upon a plane parallel to the base of the hemisphere.

globular protein [BIOCHEM] Any protein that is readily soluble in aqueous solvents.

globular star cluster [ASTRON] A group of many thousands of stars that are much closer to each other than the stars around the group and that are traveling through space together; a globular cluster has a slightly flattened spheroidal shape.

globule [ASTRON] A black volume of cosmic dust viewed against the brighter background of bright nebulae.

globulin [BIOCHEM] A heat-labile serum protein precipitated by 50% saturated ammonium sulfate and soluble in dilute salt solutions.

globulite [GEOL] A small, isotropic, globular of spherulelike crystallite; usually dark in color and found in glassy extrusive rocks.

globulomaxillary cyst [MED] A cyst in the alveolar process between the upper lateral incisor and canine teeth.

glochid *See* glochidium.

glochidium [BOT] A barbed hair. Also known as glochid. [INV ZOO] The larva of fresh-water mussels in the family Unionidae.

gloea [INV ZOO] An adhesive mucoid substance secreted by certain protozoans and other lower organisms.

glomerulonephritis [MED] Inflammation of the kidney, primarily involving the glomeruli.

glomerulosclerosis [MED] Fibrosis of the renal glomeruli.

glomerulus [ANAT] A tuft of capillary loops projecting into the lumen of a renal corpuscle.

glomus [ANAT] **1.** A fold of the mesothelium arising near the base of the mesentery in the pronephros and containing a ball of blood vessels. **2.** A prominent portion of the choroid plexus of the lateral ventricle of the brain.

glomus aorticum *See* paraaortic body.

glomus caroticum *See* carotid body.

gloom [METEOROL] The condition existing when daylight is very much reduced by dense cloud or smoke accumulation above the surface, the surface visibility not being materially reduced.

glory [OPTICS] A set of concentric, colored rings of light around the shadow cast by an observer or his head onto a cloud or fog bank.

glory hole [CIV ENG] A funnel-shaped, fixed-crest spillway. [ENG] A furnace for resoftening or fire polishing glass during working, or an entrance in such a furnace. [MIN ENG] An opening formed by the removal of soft or broken ore through an underground passage. [NUCLEO] *See* beam hole.

glory hole system *See* chute system.

gloss [OPTICS] The ratio of the light specularly reflected from a surface to the total light reflected.

glossa [INV ZOO] A tongue or tonguelike structure in insects, especially the median projection of the labium.

glossalgia [MED] Pain in the tongue.

glossate [INV ZOO] Having a glossa or tonguelike structure.

glossimeter [ENG] An instrument, often photoelectric, for measuring the ratio of the light reflected from a surface in a definite direction to the total light reflected in all directions. Also known as glossmeter.

Glossinidae [INV ZOO] The tsetse flies, a family of cyclorrhaphous dipteran insects in the section Pupipara.

Glossiphoniidae [INV ZOO] A family of small leeches with flattened bodies in the order Rhynchobdellae.

glossitis [MED] Inflammation of the tongue.

glossmeter *See* glossimeter.

gloss oil [MATER] Low-grade varnish composed of rosin dissolved in solvent naphtha.

glossopalatine nerve [ANAT] The intermediate branch of the facial nerve.

glossopharyngeal nerve [ANAT] The ninth cranial nerve in vertebrates; a paired mixed nerve that supplies autonomic innervation to the parotid gland and contains sensory fibers from the posterior one-third of the tongue and the anterior pharynx.

glossopterid flora [PALEOBOT] Permian and Triassic fossil ferns of the genus *Glossopteris*.

glossopyrosis [MED] Burning sensation of the tongue.

glossy [OPTICS] Property of a surface from which much more light is specularly reflected than is diffusely reflected.

glossy print [GRAPHICS] A photograph dried on a ferrotype plate or drum; the surface has a glazed appearance and is not easily scratched or soiled in handling.

glottis [ANAT] The opening between the margins of the vocal folds.

gloup [GEOL] An opening in the roof of a sea cave.

glove-and-stocking anesthesia [MED] Loss or diminution of sensation in the hands and feet, corresponding to the areas covered by gloves and stockings.

glove anesthesia [MED] Loss or diminution of sensation in the hands, corresponding to the area covered by gloves.

glove box [ENG] A sealed box with gloves attached and passing through openings into the box, so that workers can handle materials in the box; used to handle certain radioactive and biologically dangerous materials and to prevent contamination of materials and objects such as germfree rats or lunar rocks.

Glover tower [CHEM ENG] A tower in the lead chamber process for manufacturing sulfuric acid; in this tower the nitrogen oxide, sulfur dioxide, and air mixture is passed upward and sprayed with a sulfuric acid–nitrosyl sulfuric acid mixture.

glow discharge [ELECTR] A discharge of electricity through gas at relatively low pressure in an electron tube, characterized by several regions of diffuse, luminous glow and a voltage drop in the vicinity of the cathode that is much higher than the ionization voltage of the gas. Also known as cold-cathode discharge.

glow-discharge cold-cathode tube *See* glow-discharge tube.

glow-discharge microphone [ENG ACOUS] Microphone in which the action of sound waves on the current forming a glow discharge between two electrodes causes corresponding variations in the current.

glow-discharge tube [ELECTR] A gas tube that depends for its operation on the properties of a glow discharge. Also known as glow-discharge cold-cathode tube; glow tube.

glow-discharge voltage regulator [ELECTR] Gas tube that varies in resistance, depending on the value of the applied voltage; used for voltage regulation.

glowing cloud *See* nuée ardente.

glow lamp [ELECTR] A two-electrode electron tube containing a small quantity of an inert gas, in which light is produced by a negative glow close to the negative electrode when voltage is applied between the electrodes.

glow plug [MECH ENG] A small electric heater, located inside a cylinder of a diesel engine, that preheats the air and aids the engine in starting.

glow potential [ELECTR] The potential across a glow discharge, which is greater than the ionization potential and less than the sparking potential, and is relatively constant as the current is varied across an appreciable range.

glow tube *See* glow-discharge tube.

glow-tube oscillator [ELECTR] A circuit using a glow-discharge tube which functions as a simple relaxation oscillator, generating a fixed-amplitude periodic sawtooth waveform.

glucagon [BIOCHEM] The protein hormone secreted by α-cells of the pancreas which plays a role in carbohydrate metabolism. Also known as hyperglycemic factor; hyperglycemic glycogenolytic factor.

glucamine [BIOCHEM] $C_6H_{15}O_4N$ An amine formed by reduction of glucosylamine or of glucose oxime.

glucan [BIOCHEM] A polysaccharide composed of the hexose sugar D-glucose.

glucinium [CHEM] The former name for the element beryllium, coined because the salts of beryllium are sweet-tasting.

glucocerebroside [BIOCHEM] A glycoside of ceramide that contains glucose.

glucocorticoid [BIOCHEM] A corticoid that affects glucose metabolism; secreted principally by the adrenal cortex.

glucogenesis [BIOCHEM] Formation of glucose within the animal body from products of glycolysis.

glucokinase [BIOCHEM] An enzyme that catalyzes the phosphorylation of D-glucose to glucose-6-phosphate.

glucolipid [BIOCHEM] A glycolipid that yields glucose on hydrolysis.

glucomannan [BIOCHEM] A polysaccharide composed of D-glucose and D-mannose; a prominent component of coniferous trees.

gluconate [ORG CHEM] A salt of gluconic acid.

gluconeogenesis [BIOCHEM] Formation of glucose within the animal body from substances other than carbohydrates, particularly proteins and fats.

gluconic acid [ORG CHEM] $C_6H_{12}O_7$ A crystalline acid obtained from glucose by oxidation; used in cleaning metals.

gluconic acid sodium salt *See* sodium gluconate.

D-glucopyranose *See* glucose.

glucopyranoside [BIOCHEM] Any glucoside that contains a six-membered ring.

glucosamine [BIOCHEM] $C_6H_{13}O_5$ An amino sugar; the most abundant in nature, occurring in glycoproteins and chitin.

glucose [BIOCHEM] $C_6H_{12}O_6$ A monosaccharide; occurs free or combined and is the most common sugar. Also known as cerelose; D-glucopyranose.

glucose phosphate [BIOCHEM] A phosphoric derivative of glucose, as glucose-1-phosphate.

glucose-1-phosphate [BIOCHEM] $C_6H_{12}O_8P$ An ester of glucopyranose in which a phosphate group is attached to carbon atom 1; there are two types: α-D- and β-D-glucose-1-phosphates. Also known as Cori ester.

glucose-6-phosphate [BIOCHEM] $C_6H_{13}O_9P$ An ester of glucose with phosphate attached to carbon atom 6. Also known as Robisonester.

glucose-6-phosphate dehydrogenase [BIOCHEM] The mammalian enzyme that catalyzes the oxidation of glucose-6-phosphate by TPN+ (triphosphopyridine nucleotide).

glucose tolerance test [PATH] A test to measure the ability of the liver to convert glucose to glycogen.

glucosidase [BIOCHEM] An enzyme that hydrolyzes glucosides.

glucoside [BIOCHEM] One of a group of compounds containing the cyclic forms of glucose, in which the hydrogen of the hemiacetal hydroxyl has been replaced by an alkyl or aryl group.

7-D-glucoside *See* genistin.

glucosulfone sodium *See* sodium glucosulfone.

glucosyltransferase [BIOCHEM] An enzyme that catalyzes the glucosylation of hydroxymethyl cytosine; a constituent of bacteriophage deoxyribonucleic acid.

glucuronic acid [BIOCHEM] $C_6H_{10}O_7$ An acid resulting from oxidation of the CH_2OH radical of D-glucose to COOH; a component of many polysaccharides and certain vegetable gums. Also known as glycuronic acid.

glucuronidase [BIOCHEM] An enzyme that catalyzes hydrolysis of glucuronides. Also known as glycuronidase.

glucuronide [BIOCHEM] A compound resulting from the interaction of glucuronic acid with a phenol, an alcohol, or an acid containing a carboxyl group. Also known as glycuronide.

D-glucuronolactone [BIOCHEM] $C_6H_8O_6$ A water-soluble crystalline compound found in plant gums in polymers with other carbohydrates, and an important structural component of almost all fibrous and connective tissues in animals; used in medicine as an antiarthritic.

glue [MATER] A crude, impure, amber-colored form of commercial gelatin of unknown detailed composition produced by the hydrolysis of animal collagen; gelatinizes in aqueous solutions and dries to form a strong, adhesive layer.

glue cell *See* adhesive cell.

glued-laminated wood [MATER] A wooden member formed by assembling a set of boards or planks with glue so that the grain of all laminations is essentially parallel to the length of the member. Also known as glulam.

glue-joint ripsaw [MECH ENG] A heavy-gage ripsaw used on straight-line or self-feed rip machines; the cut is smooth enough to permit gluing of joints from the saw.

glue-line heating [ENG] Dielectric heating in which the electrodes are designed to give preferential heating to a thin film of glue or other relatively high-loss material located between layers of relatively low-loss material such as wood.

glug [MECH] A unit of mass, equal to the mass which is accelerated by 1 centimeter per second per second by a force of 1 gram-force, or to 980.665 grams.

glulam *See* glued-laminated wood.

glume [BOT] One of two bracts at the base of a spikelet of grass.

glumiferous [BOT] Bearing glumes.

Glumiflorae [BOT] An equivalent name for Cyperales.

gluside *See* saccharin.

glutamate [BIOCHEM] A salt or ester of glutamic acid.

glutamic acid [BIOCHEM] $C_5H_9O_4N$ A dicarboxylic amino acid of the α-ketoglutaric acid family occurring widely in proteins.

glutaminase [BIOCHEM] The enzyme which catalyzes the conversion of glutamine to glutamic acid and ammonia.

glutamine [BIOCHEM] $C_5H_{10}O_3N_2$ An amino acid; the monamide of glutamic acid; found in the juice of many plants and essential to the development of certain bacteria.

glutarate [BIOCHEM] The salt or ester of glutaric acid.

glutaric acid [BIOCHEM] $C_5H_5O_4$ A water-soluble, crystalline acid that occurs in green sugarbeets and in water extracts of crude wool.

glutathione [BIOCHEM] $C_{10}H_{17}O_6N_3S$ A widely distributed tripeptide that is important in plant and animal tissue oxidation reactions.

glutelin [BIOCHEM] A class of simple, heat-labile proteins occurring in seeds of cereals; soluble in dilute acids and alkalies.

gluten [BIOCHEM] 1. A mixture of proteins found in the seeds of cereals; gives dough elasticity and cohesiveness. 2. An albuminous element of animal tissue.

glutenin [BIOCHEM] A glutelin of wheat.

glutethimide [PHARM] $C_{13}H_{15}NO_2$ A minor or sedative antianxiety tranquilizer that acts as a central nervous system depressant.

gluteus maximus [ANAT] The largest and most superficial muscle of the buttocks.

gluteus medius [ANAT] The muscle of the buttocks lying between the gluteus maximus and gluteus minimus.

gluteus minimus [ANAT] The smallest and deepest muscle of the buttocks.

glutinant nematocyst [INV ZOO] A nematocyst characterized by an open, sticky tube used for anchoring the coelenterate when walking on its tentacles.

glutinous [BOT] Having a sticky surface.

glycemia [PHYSIO] The presence of glucose in the blood.

glyceraldehyde [BIOCHEM] $CH_2OHCOHCHO$ A colorless solid, isomeric with dehydroxyacetone; soluble in water and insoluble in organic solvents; an important intermediate in carbohydrate metabolism; used as a chemical intermediate in biochemical research and nutrition. Also known as 2,3-dihydroxypropanal; glyceric aldehyde.

glycerate [BIOCHEM] A salt or ester of glyceric acid.

glyceric acid [BIOCHEM] $C_3H_6O_4$ A hydroxy acid obtained by oxidation of glycerin.

glyceric aldehyde *See* glyceraldehyde.

Glyceridae [INV ZOO] A family of polychaete annelids belonging to the Errantia and characterized by an enormous eversible proboscis.

glyceride [BIOCHEM] An ester of glycerin and an organic acid radical; fats are glycerides of certain long-chain fatty acids.

glycerin *See* glycerol.

glycerinated vaccine virus *See* smallpox vaccine.

glycerokinase [BIOCHEM] An enzyme that catalyzes the phosphorylation of glycerol to glycerophosphate during microbial fermentation of propionic acid.

glycerol [ORG CHEM] $CH_2OHCHOHCH_2OH$ The simplest trihedric alcohol; when pure, it is a colorless, odorless, viscous liquid with a sweet taste; it is completely soluble in water and alcohol but only partially soluble in common solvents such as ether and ethyl acetate; used in manufacture

GLUCOSE-6-PHOSPHATE

Structural formula for glucose-6-phosphate.

GLUCOSIDE

α-D-Glucoside

β-D-Glucoside

Structural formulas of two forms of glucoside.

GLUTAMIC ACID

Structural formula of glutamic acid.

GLUTINANT NEMATOCYST

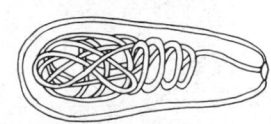

The glutinant nematocyst of *Hydra.* (*After Schulze, from T. I. Storer and R. L. Usinger, General Zoology, 3d ed., McGraw-Hill, 1957*)

of alkyd resins, explosives, antifreezes, medicines, inks, perfumes, cosmetics, soaps, and finishes. Also known as glycerin; glycyl alcohol.

glycerol dichlorohydrin See α-dichlorohydrin.

glycerol monoacetate [ORG CHEM] $CH_2OHCHOHCH_2O\text{-}OCCH_2$ A colorless oil, specific gravity 1.206 (at 20°C); soluble in water and ethyl alcohol; made by heating glycerol and strong acetic acid, distilling the weak acetic acid that is left, and repeating the process; used as a solvent and in the manufacture of explosives. Also known as acetin; monoacetin.

glycerophosphate [BIOCHEM] Any salt of glycerophosphoric acid.

glycerophosphoric acid [BIOCHEM] $C_3H_5(OH)_2OPO_3H_2$ Either of two pale-yellow, water-soluble, isomeric dibasic acids occurring in nature in combined form as cephalin and lecithin.

glyceryl [ORG CHEM] $OCH_2OCHOCH_2\equiv$ The radical group from glycerol, $(CH_2OH)_2CHOH$.

glyceryl diacetate See diacetin.

glyceryl triacetate See triacetin.

glyceryl tripalmitate See tripalmitin.

glyceryl tristearate See stearin.

glycide See glycidol.

glycidic acid [ORG CHEM] $C_2H_3O\cdot CO_2H$ A volatile liquid. Also known as epoxy-propionic acid.

glycidol [ORG CHEM] $C_3H_6O_2$ A colorless, liquid epoxide that boils at 162°C and is miscible with water; used in organic synthesis. Also known as epihydrin alcohol; 2,3-epoxy-1-propanol; glycide.

glycin [ORG CHEM] $C_8H_9NO_3$ A crystalline compound that forms shiny leaflets from water solution, melts at 245–247°C, and is soluble in alkalies and mineral acids; used as a photographic developer and in the analytical determination of iron, phosphorus, and silicon. Also known as para-hydroxyanilinoacetic acid; para-hydroxyphenylaminoacetic acid; photoglycine.

glycine [BIOCHEM] $C_2H_5O_2N$ A white, crystalline amino acid found as a constituent of many proteins. Also known as aminoacetic acid.

glyco- [ORG CHEM] Chemical prefix indicating sweetness, or relating to sugar or glycine.

glycocalyx [CYTOL] The outer component of a cell surface, outside the plasmalemma; usually contains strongly acidic sugars, hence it carries a negative electric charge.

glycocholic acid [BIOCHEM] $C_{26}H_{43}NO_6$ A bile obtained by the conjugation of cholic acid with glycine.

glycocyamine [BIOCHEM] $C_3H_7N_3O_2$ A product of interaction of aminocetic acid and arginine, which on transmethylation with methionine is converted to creatine. Also known as guanidine-acetic acid.

glycogen [BIOCHEM] A nonreducing, white, amorphous polysaccharide found as a reserve carbohydrate stored in muscle and liver cells of all higher animals, as well as in cells of lower animals.

glycogenesis [BIOCHEM] The metabolic formation of glycogen from glucose.

glycogenolysis [BIOCHEM] The metabolic breakdown of glycogen.

glycogenosis [MED] One of several inborn errors in the metabolism of glycogen, classified on the basis of the enzyme deficiency and clinical findings as von Gierke's disease, Pompe's disease, limit dextrinosis, amylopectinosis, McArdle's disease, or Hers' disease.

glycogen storage disease See von Gierke's disease.

glycogen synthetase [BIOCHEM] An enzyme that catalyzes the synthesis of the amylose chain of glycogen.

glycol [ORG CHEM] **1.** $C_nH_{2n}(OH)_2$ An organic chemical with two hydroxyl groups on an essentially aliphatic carbon chain. Also known as dihydroxy alcohol. **2.** $HOCH_2CH_2OH$ A colorless dihydroxy alcohol used as an antifreeze, in hydraulic fluids, and in the manufacture of dynamites and resins. Also known as ethlene glycol.

glycoldinitrate See ethylene nitrate.

glycol ester [ORG CHEM] Chemical compound composed of the reaction products of a glycol, $C_nH_{2n}(OH)_2$, and an organic acid; an example is ethylene glycol diacetate, the product of ethylene glycol and acetic acid.

glycol ether [ORG CHEM] A colorless liquid used as a solvent, in detergents, and as a diluent; a typical example is ethylene glycol diethyl ether, $C_2H_5OCH_2CH_2OC_2H_5$.

glycol ethylene ether See 1,4-dioxane.

glycolic acid [ORG CHEM] $CH_2OHCOOH$ Colorless, deliquescent leaflets, decomposing about 78°C; soluble in water, alcohol, and ether; used as a chemical intermediate in fabric dyeing. Also known as glycollic acid; hydroxyacetic acid.

glycolipid [BIOCHEM] Any of a class of complex lipids which contain carbohydrate residues.

glycollic acid See glycolic acid.

glycolysis [BIOCHEM] The enzymatic breakdown of glucose or other carbohydrate, with the formation of lactic acid or pyruvic acid and the release of energy in the form of adenosinetriphosphate.

glycolytic pathway [BIOCHEM] The principal series of phosphorylative reactions involved in pyruvic acid production in phosphorylative fermentations. Also known as Embden-Meyerhof pathway; hexose diphosphate pathway.

glycolyurea See hydantoin.

glyconeogenesis [BIOCHEM] The metabolic process of glycogen formation from noncarbohydrate precursors.

glycopeptide See glycoprotein.

glycophyte [BOT] A plant requiring more than 0.5% sodium chloride solution in the substratum.

glycoprotein [BIOCHEM] Any of a class of conjugated proteins containing both carbohydrate and protein units. Also known as glycopeptide.

glycosidase [BIOCHEM] An enzyme that hydrolyzes a glycoside.

glycoside [BIOCHEM] A compound that yields on hydrolysis a sugar (glucose, galactose) and an aglycone; many of the glycosides are therapeutically valuable.

glycosuria [MED] Presence of sugar in the urine.

glycotropic [BIOCHEM] Acting to antagonize the action of insulin.

glycuresis [PHYSIO] Excretion of sugar seen normally in urine.

glycuronic acid See glucuronic acid.

glycuronidase See glucuronidase.

glycuronide See glucuronide.

glycyl [ORG CHEM] NH_2CH_2COO- or $NHCH_2COO=$ The radical from glycine, NH_2CH_2COOH; found in peptides.

glycyl alcohol See glycerol.

glyoxal [ORG CHEM] $(CHO)_2$ Colorless, deliquescent powder or liquid with mild odor, melting point 15°C, boiling point 51°C; used to insolubilize starches, cellulosic materials, and proteins, in embalming fluids, for leather tanning, and for rayon shrinkproofing.

glyoxalase [BIOCHEM] An enzyme present in various body tissues which catalyzes the conversion of methylglyoxal into lactic acid.

glyoxalic acid [ORG CHEM] $CHOCOOH$ Colorless crystals that are soluble in water, forming glyoxylic acid. Also known as ethanol acid; oxaldehydic acid; oxoethanoic acid.

glyoxaline See imidazole.

glyoxylate cycle [BIOCHEM] A sequence of biochemical reactions related to respiration in germinating fatty seeds by which acetyl coenzyme A is converted to succinic acid and then to hexose.

glyoxylic acid [BIOCHEM] $CH(OH)_2COOH$ An aldehyde acid found in many plant and animal tissues, especially unripe fruit.

Glyphocyphidae [PALEON] A family of extinct echinoderms in the order Temnopleuroida comprising small forms with a sculptured test, perforate crenulate tubercles, and diademoid ambulacral plates.

glyptal resin [ORG CHEM] A phthalic anhydride glycerol made from an emulsion of an alkyd resin; used in lacquers and insulation.

Glyptocrinina [PALEON] A suborder of extinct crinoids in the order Monobathrida.

glyptolith See ventifact.

gm See gram.

GM counter See Geiger-Müller counter.

GLYCINE

Structural formula of glycine.

gmelinite [MINERAL] $(Na_2Ca)Al_2Si_4O_{12} \cdot 6H_2O$ Zeolite mineral that is colorless or lightly colored and crystallizes in the hexagonal system.

Gmelin's test [PATH] A qualitative test for the pigments in bile; test solution is mixed with nitric acid containing nitrous acid; reaction is positive if color appears at the acid-solution junction.

GMR Stirling thermal engine [MECH ENG] A hot-air engine, developed by the Philips Research Laboratory and General Motors Corporation, which operates on the Stirling cycle.

GMT *See* Greenwich mean time.

gnat [INV ZOO] The common name for a large variety of biting insects in the order Diptera.

gnathic index [ANTHRO] The ratio of the distance from the nasion to the basion to that from the basion to the alveolar point multiplied by 100.

Gnathiidea [INV ZOO] A suborder of isopod crustaceans characterized by a much reduced second thoracomere, short antennules and antennae, and a straight pleon.

gnathion [ANTHRO] The midpoint of the lower margin of the mandible in humans. [VERT ZOO] The most anterior point of the premaxillae on or near the middle line in certain lower mammals.

gnathite [INV ZOO] A mouth appendage in arthropods.

Gnathobdellae [INV ZOO] A suborder of leeches in the order Arhynchobdellae having jaws and a conspicuous posterior sucker; contains most of the important blood-sucking leeches of humans and other warm-blooded animals.

Gnathobelodontinae [PALEON] A subfamily of extinct elephantoid proboscideans containing the shovel-jawed forms of the family Gomphotheriidae.

gnathocephalon [INV ZOO] The part of the insect head lying behind the protocephalon; bears the maxillae and mandibles.

gnathochilarium [INV ZOO] The lower lip of certain arthropods; thought to be fused maxillae.

Gnathodontidae [PALEON] A family of extinct conodonts having platforms with large, cup-shaped attachment scars.

gnathopod [INV ZOO] Any of the crustacean paired thoracic appendages modified for manipulation of food but sometimes functioning in copulatory amplexion.

gnathopodite [INV ZOO] A segmental, modified appendage which serves as a jaw in arthropods.

gnathos [INV ZOO] A mid-ventral plate on the ninth tergum in lepidopterans.

gnathostegite [INV ZOO] One of a pair of broad plates formed from the outer maxillipeds of some crustaceans, which function to cover other mouthparts.

Gnathostomata [INV ZOO] A suborder of echinoderms in the order Echinoidea characterized by a rigid, exocyclic test and a lantern or jaw apparatus. [VERT ZOO] A group of the subphylum Vertebrata which possess jaws and usually have paired appendages.

Gnathostomidae [INV ZOO] A family of parasitic nematodes in the superfamily Spiruroidea; sometimes placed in the superfamily Physalopteroidea.

Gnathostomulida [INV ZOO] Microscopic marine worms of uncertain systematic relationship, mainly characterized by cuticular structures in the pharynx and a monociliated skin epithelium.

gnathothorax [INV ZOO] The thorax and part of the head bearing feeding organs in arthropods, regarded as a primary region of the body.

gnd *See* ground.

gneiss [PETR] A variety of rocks with a banded or coarsely foliated structure formed by regional metamorphism.

gneissic granodiorites [PETR] Granodiorite rocks with gneissic characteristics.

Gnetales [BOT] A monogeneric order of the subdivision Gneticae; most species are lianas with opposite, oval, entire-margined leaves.

Gnetatae *See* Gnetopsida.

Gneticae [BOT] A subdivision of the division Pinophyta characterized by vessels in the secondary wood, ovules with two integuments, opposite leaves, and an embryo with two cotyledons.

Gnetophyta [BOT] The equivalent name for Gnetopsida.

Gnetopsida [BOT] A class of gymnosperms comprising the subdivision Gneticae.

gnomic [ASTRON] Pertaining to the gnomon of a sundial.

gnomon [ENG] On a sundial, the inclined plate or pin that casts a shadow. [MATH] A geometric figure formed by removing from a parallelogram a similar parallelogram that contains one of its corners.

gnomonic chart [MAP] A chart on the gnomonic projection where great circles project as straight lines. Also known as great-circle chart.

gnomonic projection [MAP] A projection on a plane tangent to the surface of a sphere having the point of projection at the center of the sphere.

Gnostidae [INV ZOO] An equivalent name for Ptinidae.

gnotobiology [BIOL] That branch of biology dealing with known living forms; the study of higher organisms in the absence of all demonstrable, viable microorganisms except those known to be present.

gnotobiote [MICROBIO] An individual (host) living in intimate association with another known species (microorganism). 2. The known microorganism living on a host.

gnu [VERT ZOO] Any of several large African antelopes of the genera *Connochaetes* and *Gorgon* having a large oxlike head with horns that characteristically curve downward and outward and then up, with the bases forming a frontal shield in older individuals.

goaf [MIN ENG] 1. That part of a mine from which the coal has been worked away and the space more or less filled up. 2. The refuse or waste left in the mine. Also known as gob.

goat [VERT ZOO] The common name for a number of artiodactyl mammals in the genus *Capra*; closely related to sheep but differing in having a lighter build and hollow, swept-back, sometimes spiral or twisted horns.

gob *See* goaf.

gobi [GEOL] Sedimentary deposits in a synclinal basin.

Gobiatheriinae [PALEON] A subfamily of extinct herbivorous mammals in the family Uintatheriidae known from one late Eocene genus; characterized by extreme reduction of anterior dentition and by lack of horns.

Gobiesocidae [VERT ZOO] The single family of the order Gobiesociformes.

Gobiesociformes [VERT ZOO] The clingfishes, a monofamilial order of scaleless bony fishes equipped with a thoracic sucking disk which serves for attachment.

Gobiidae [VERT ZOO] A family of perciform fishes in the suborder Gobioidei characterized by pelvic fins united to form a sucking disk on the breast.

Gobioidei [VERT ZOO] The gobies, a suborder of morphologically diverse actinopterygian fishes in the order Perciformes; all lack a lateral line.

goblet cell [HISTOL] A unicellular, mucus-secreting intraepithelial gland that is distended on the free surface. Also known as chalice cell. [INV ZOO] Any of the unicellular choanocytes of the genus *Monosiga*.

gob stink [MIN ENG] 1. The odor from the burning coal given off by an underground fire. 2. The odor given off by the spontaneous heating of coal, not necessarily in the gob. Also known as stink.

godet [TEXT] A roller used in stretching synthetic fiber filaments.

go-devil [ENG] 1. A device inserted in a pipe or hole for purposes such as cleaning or for detonating an explosive. 2. A sled for moving logs or cultivating. 3. A large rake for gathering hay. 4. A small railroad car used for transporting workers and materials.

Goertler parameter [FL MECH] A dimensionless number used in studying boundary-layer flow on curved surfaces, equal to the Reynolds number, where the characteristic length is the boundary-layer momentum thickness, times the square root of this thickness, divided by the square root of the surface's radius of curvature.

goethite [MINERAL] FeO(OH) A yellow, red, or dark-brown mineral crystallizing in the orthorhombic system, although it is usually found in radiating fibrous aggregates; a common constituent of natural rust or limonite. Also known as xanthosiderite.

go gage [DES ENG] A test device that just fits a part if it has the proper dimensions (often used in pairs with a "no go" gage to establish maximum and minimum dimensions).

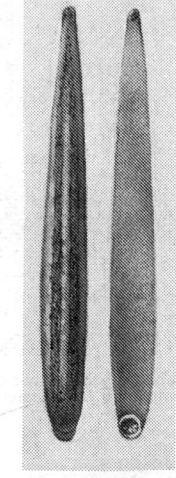

Haemopis grandis, dorsal and ventral view showing the conspicuous posterior sucker.

GNOMONIC PROJECTION

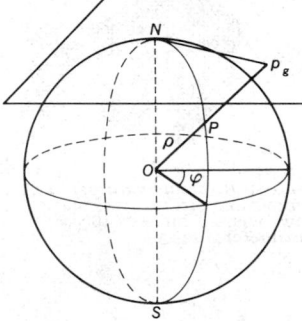

Gnomonic projection of P. Line OP cuts planes tangent to the sphere at N at point p_g. Coordinates φ and ρ define pole P.

GOETHITE

2.5 cm

Goethite, taken from Negaunee, Michigan. (*Specimen from Department of Geology, Bryn Mawr College*)

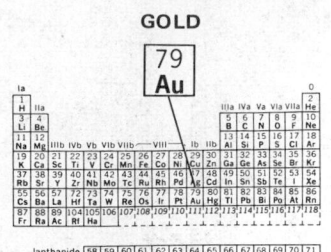

GOLD

79
Au

Periodic table of the chemical elements showing the position of gold.

GOLDFISH

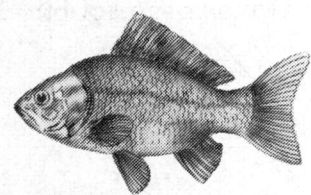

Goldfish (*Crassius auratus*), a common aquarium fish, may grow over 18 inches (46 centimeters) long.

GOLDHABER TRIANGLE

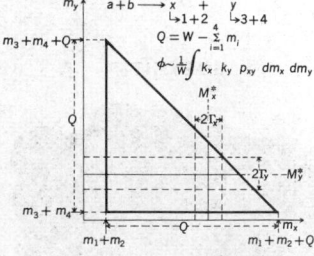

Goldhaber triangle for reaction between two particles, *a* and *b*, yielding four particles, 1, 2, 3, and 4, with masses m_1, m_2, m_3, and m_4. Coordinates are invariant masses m_x and m_y of intermediate-state quasi-particle composites *x* and *y* which decay into two particles each as indicated at top of figure. *Q* is total center of mass energy of *a* and *b* minus sum of m_1, m_2, m_3, and m_4.

goggles [ENG] Spectaclelike eye protectors having shields at the sides and short, projecting eye tubes.

going [CIV ENG] On a staircase, the distance between the faces of two successive risers.

going barrel [HOROL] A mainspring barrel with a toothed edge for driving the train; it is mounted on an arbor that is stationary except while the timepiece is being wound.

going train [HOROL] Gears that drive the hands in a striking timepiece.

goiter [MED] An enlargement of all or part of the thyroid gland; may be accompanied by a hormonal dysfunction.

Golay cell [ENG] A radiometer in which radiation absorbed in a gas chamber heats the gas, causing it to expand and deflect a diaphragm in accordance with the amount of radiation.

gold [CHEM] A chemical element, symbol Au, atomic number 79, atomic weight 196.967; soluble in aqua regia; melts at 1065°C. [MET] The native metallic element; a deep-yellow, very dense, soft, isometric metal, usually found alloyed with silver or copper; used in jewelry, dentistry, gilding, anodes, alloys, and solders.

gold-198 [NUC PHYS] The radioisotope of gold, with atomic number 198 and a half-life of 2.7 days; used in medical treatment of tumors by injecting it in colloidal form directly into tumor tissue.

gold alloy [MET] Any alloy containing gold.

Goldbach conjecture [MATH] The unestablished conjecture that every even number save the number 2 is the sum of two primes.

goldbeater's skin [MATER] The treated outside membrane of the large intestine of cattle; used between leaves of metal in goldbeating, and sometimes in hygrometers.

goldbeater's-skin hygrometer [ENG] A hygrometer using goldbeater's skin as the sensitive element; variations in the physical dimensions of the skin caused by its hygroscopic character indicate relative atmospheric humidity.

goldbeating [MET] The process of producing gold leaf.

Goldberg-Mohn friction [FL MECH] A force proportional to the velocity of a current and the density of the medium; used as a first approximation in estimating frictional effects in the atmosphere and the ocean.

Goldblatt unit [BIOL] A unit for the standardization of angiotonin.

gold bronze [MET] A powdered alloy of copper used to simulate gold, or an alloy of copper containing 3–5% aluminum.

gold chloride [INORG CHEM] $AuCl_3$ A red, soluble compound made by reaction of gold and chlorine or by reaction of $HAuCl_4$ with chlorine; decomposes by heat; soluble in water, alcohol, and ether; used in photography, plating, inks, medicine, and ceramics.

gold doping [ELECTR] A technique for controlling the lifetime of minority carriers in a transistor; gold is diffused into the base and collector regions to reduce storage time in transistor circuits.

golden algae [BOT] The common name for members of the class Chrysophyceae.

golden antimony sulfide *See* antimony pentasulfide.

golden-brown algae [BOT] The common name for members of the division Chrysophyta.

golden section [MATH] The division of a line so that the ratio of the whole line to the larger interval equals the ratio of the larger interval to the smaller. Also known as extreme and mean ratio.

golden section search [ADP] A dichotomizing search in which, in each step, the remaining items are divided as closely as possible according to the golden section.

gold-filled [MET] Covered by a layer of gold alloy.

goldfish [VERT ZOO] *Crassius auratus.* An orange cypriniform fish of the family Cyprinidae that can grow to over 18 inches; closely related to the carps.

gold foil [MET] A thin sheet of gold, thicker than gold leaf, formed by rolling or hammering.

Goldhaber triangle [PARTIC PHYS] A plot describing a high-energy reaction leading to four or more particles; its coordinates are the invariant masses of two intermediate-state quasi-particle composites, and its kinematical limits form a right-angle isosceles triangle.

gold hydroxide [INORG CHEM] $Au(OH)_3$ A yellow-brown, light-sensitive, water-insoluble powder; dissolves in most acids; easily reduced to metallic gold; used in medicine, porcelain, gold plating, and daguerreotypes.

gold leaf [MET] Gold beaten or rolled into extremely thin sheets or leaves (10^{-6} inch or 25 nanometers thick); leaves are stored in books (a book consists of 25 leaves), the paper of which is rubbed with chalk to keep the leaves from sticking.

gold-leaf electroscope [ELEC] An electroscope in which two narrow strips of gold foil or leaf suspended in a glass jar spread apart when charged; the angle between the strips is related to the charge.

gold metallurgy [MET] The science and technology of gold recovery and refining.

gold number [ANALY CHEM] A measure of the amount of protective colloid which must be added to a standard red gold sol mixed with sodium chloride solution to prevent the solution from causing the sol to coagulate, as manifested by a change in color from red to blue.

gold oxide [INORG CHEM] Au_2O_3 Water-insoluble, heat-decomposable, brownish-black powder; soluble in hydrochloric acid; used to gild, in medicine and porcelain, and for daguerreotypes. Also known as auric oxide; gold trioxide.

gold plate [MET] Gold which has been electroplated on a material in a thin layer of controlled thickness; used on electric contacts for corrosion resistance and solderability and on jewelry and oraments.

gold point [THERMO] The temperature of the freezing point of gold at a pressure of 1 standard atmosphere; used to define the International Practical Temperature Scale of 1968, on which it is assigned a value of 1337.58K or 1064.43°C.

goldschmidtine *See* stephanite.

goldschmidtite *See* sylvanite.

Goldschmidt process [MET] **1.** The thermite process of welding. **2.** A process by which dry chlorine is employed to remove tin from scrap tinplate.

Goldschmidt's law [SOLID STATE] The law that crystal structure is determined by the ratios of the numbers of the constituents, the ratios of their sizes, and their polarization properties.

Goldschmidt's mineralogical phase rule [GEOL] The rule that the probability of finding a system with degrees of freedom less than two is small under natural rock-forming conditions.

gold size [CHEM] A solution of white and red lead and yellow ocher in linseed oil; used to seal permanently microscopical preparations. [MATER] An adhesive used to fix gold leaf to a surface.

Gold slide [ENG] A slide rule used on British ships to compute barometric corrections and reduction of pressure to sea level; it includes the effects of temperature, latitude, index correction, and barometric height above sea level.

gold sodium thiosulfate [PHARM] $Na_3Au(S_2O_3)_2 \cdot 2H_2O$ A white crystalline compound that is freely soluble in water; used for treatment of rheumatoid arthritis and lupus erythematosus.

gold solder [MET] Solder composed of 60% gold, 20% silver, and 20% copper.

gold trioxide *See* gold oxide.

golfada [METEOROL] A heavy gale of the Mediterranean.

Golgi apparatus [CYTOL] A cellular organelle that is part of the cytoplasmic membrane system; it is composed of regions of stacked cisternae and it functions in secretory processes.

Golgi cell [ANAT] **1.** A nerve cell with long axons. **2.** A nerve cell with short axons that branch repeatedly and terminate near the cell body.

Golgi-Mazzoni's corpuscle [ANAT] A small sensory lamellar corpuscle located in the parietal pleura.

Golgi tendon organ [PHYSIO] Any of the kinesthetic receptors situated near the junction of muscle fibers and a tendon which act as muscle-tension recorders.

Gomberg-Bachmann-Hey reaction [ORG CHEM] Production of diaryl compounds by adding alkali to a mixture of a diazonium salt and a liquid aromatic hydrocarbon or a derivative.

Gomberg reaction [ORG CHEM] The production of free radicals by reaction of metals with triarylmethyl halides.

Gomphidae [INV ZOO] A family of dragonflies belonging to the Anisoptera.

gomphosis [ANAT] An immovable articulation, as that formed by the insertion of teeth into the bony sockets.

Gomphotheriidae [PALEON] A family of extinct proboscidean mammals in the suborder Elephantoidea consisting of species with shoveling or digging specializations of the lower tusks.

Gomphotheriinae [PALEON] A subfamily of extinct elephantoid proboscideans in the family Gomphotheriidae containing species with long jaws and bunomastodont teeth.

gon *See* grade.

gonad [ANAT] A primary sex gland; an ovary or a testis.

gonadal agenesis [MED] Failure of the gonad to develop, or retrogression of the gonad at very early stages. Also known as gonadal dysgenesis.

gonadal dysgenesis *See* gonadal agenesis.

gonadectomy [MED] Surgical removal of a gonad.

gonadotropic hormone [BIOCHEM] Either of two adenohypophyseal hormones, FSH (follicle-stimulating hormone) or ICSH (interstitial-cell-stimulating hormone), that act to stimulate the gonads.

gonadotropin [BIOCHEM] A substance that acts to stimulate the gonads.

gonapophysis [INV ZOO] A paired, modified appendage of the anal region in insects that functions in copulation, oviposition, or stinging.

gondola car [ENG] A flat-bottomed railroad car which has no top, fixed sides, and often removable ends, in which steel, rock, or heavy bulk commodities are transported.

Gondwanaland [GEOL] The ancient continent that is supposed to have fragmented and drifted apart to form eventually the present continents.

gong buoy [NAV] A buoy similar in construction to a bell buoy, but sounding a distinctive note because of the use of sets of gongs, each gong having a different tone.

Goniadidae [INV ZOO] A family of marine polychaete annelids belonging to the Errantia.

gonidium [BIOL] An asexual reproductive cell or group of cells arising in a special organ on or in a gametophyte.

gonimoblast [BOT] A filament arising from the fertilized carpogonium of most red algae.

goniometer [ELECTROMAG] An instrument for determining the direction of maximum response to a received radio signal, or selecting the direction of maximum radiation of a transmitted radio signal; consists of two fixed perpendicular coils, each attached to one of a pair of loop antennas which are also perpendicular, and a rotatable coil which bears the same space relationship to the coils as the direction of the signal to the antennas. [ENG] **1.** An instrument used to measure the angles between crystal faces. **2.** An instrument which uses x-ray diffraction to measure the angular positions of the axes of a crystal. **3.** Any instrument for measuring angles.

goniometric locator [ELECTROMAG] In radio detection finding, a rotating device that samples signals from orthogonal fixed antennas.

goniophotometer [OPTICS] A photometer designed to measure the intensity of light reflected from a surface at various angles.

gonioscope [MED] A special optical instrument for studying in detail the angle of the anterior chamber of the eye.

gonitis [MED] Inflammation of the knee joint.

gonnardite [MINERAL] $Na_2CaAl_4Si_6O_{20} \cdot 7H_2O$ Zeolite mineral occurring in fibrous, radiating spherules; specific gravity is 2.3.

gonococcal arthritis [MED] A blood-borne joint infection by *Neisseria gonorrhoeae* occurring as a manifestation of gonorrhea.

gonococcal epididymitis [MED] Inflammation of the epididymitis due to infection by *Neisseria gonorrhoeae*; a secondary manifestation of gonorrhea.

gonococcus *See* Neisseria gonorrhoeae.

Gonodactylidae [INV ZOO] A family of mantis shrimp in the order Stomatopoda.

gonodendrum [INV ZOO] A branched structure which bears clusters of gonophores.

go no-go detector [ENG] An instrument having only two operating states, such as a common fuse which is either intact or melted.

gonopalpon [INV ZOO] Tentaclelike, sensitive structures associated with cnidarian gonophores.

gonophore [BOT] An elongation of the receptacle extending between the stamens and corolla. [INV ZOO] Reproductive zooid of a hydroid colony.

gonopodium [VERT ZOO] Anal fin modified as a copulatory organ in certain fishes.

Gonorhynchiformes [VERT ZOO] A small order of soft-rayed teleost fishes having fusiform or moderately compressed bodies, single short dorsal and anal fins, a forked caudal fin, and weak toothless jaws.

gonorrhea [MED] A bacterial infection of man caused by the gonococcus (*Neisseria gonorrhoeae*) which invades the mucous membrane of the urogenital tract.

gonorrheal urethritis [MED] Inflammation of the urethra, particularly in males, as the result of gonorrhea.

gonorrheal vulvovaginitis [MED] Inflammation of the vulva and vagina as the result of gonorrhea.

gonosome [INV ZOO] Aggregate of gonophores in a hydroid colony.

gonosomite [INV ZOO] The ninth segment of the abdomen of the male insect. Also known as genital segment.

gonostome [INV ZOO] The part of the genital duct of a coelomate invertebrate known as the coelomic funnel. Also known as coelomostome.

gonostyle [INV ZOO] Gonapophysis of dipteran insects.

gonotome [EMBRYO] The part of an embryonic somite involved in gonad formation.

gonozooid [INV ZOO] A zooid of bryozoans and tunicates which produces gametes.

gonys [VERT ZOO] A ridge along the mid-ventral line of the lower mandible of certain birds.

Gooch crucible [ANALY CHEM] A ceramic crucible with a perforated base; in analysis it is used for filtration through asbestos or glass.

good geometry [NUC PHYS] An arrangement of source and detecting equipment such that little error is produced by the finite sizes of the source and the detector aperture.

Goodman duckbill loader [MIN ENG] A gathering and loading assembly for coal, consisting of a shovel trough with a shovel head fitting inside the feeder trough, an operating carrier which controls the connection between these troughs, a sliding shoe which moves to and fro on the floor of the seam, a swivel trough, and a pendulum jack; it loads the coal into the shaker conveyor pan column.

Goodman loader [MIN ENG] **1.** An electrohydraulic power shovel designed for loading coal where the seams are 6 feet (1.8 meters) or more in thickness. **2.** A loader designed for loading coal from thin seams; has a telescoping fan-shaped apron that extends from the entry of the room to the working face.

goodness of fit [STAT] The degree to which the observed frequencies of occurrence of events in an experiment correspond to the probabilities in a model of the experiment.

good oil *See* raffinate.

Goodpasture's syndrome [MED] A complex of symptoms associated with acute glomerulonephritis and pulmonary hemorrhage.

good seamanship [NAV] Any precaution which may be required by the ordinary practice of seamen.

googol [MATH] A name for 10 to the power 100.

googolplex [MATH] A name for 10 to the power googol.

goongarrite [MINERAL] $Pb_4Bi_2S_7$ A mineral composed of a sulfide of lead and bismuth.

goop [MATER] A compound in paste form containing finely divided magnesium, used as a constituent of certain incendiary bomb fillings.

goose [VERT ZOO] The common name for a number of

GONIADIDAE

prostomium

Multiannulate prostomium of *Goniada* in dorsal view.

GOOSE

Canada goose (*Branta canadensis*).

GOOSEBERRY

Gooseberry branch bearing leaves and fruits. (*USDA*)

GOPHER

Pocket gopher of North America.

GOVERNOR

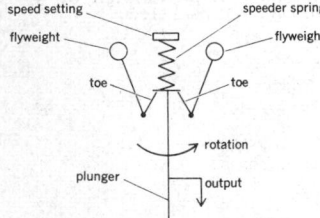

Ballhead of governor containing whirling flyweights driven at a speed proportional to the speed of the prime mover.

GRABEN

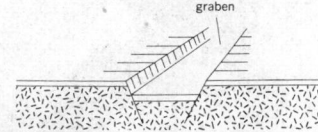

Diagram of simple graben. (*From A. K. Lobeck, Geomorphology, McGraw-Hill, 1939*)

waterfowl in the subfamily Anatinae; they are intermediate in size and features between ducks and swans.

gooseberry [BOT] The common name for about six species of thorny, spreading bushes of the genus *Ribes* in the order Rosales, producing small, acidic, edible fruit.

gooseberry stone *See* grossularite.

goosefoot oil *See* chenopodium oil.

gooseneck [DES ENG] **1.** A pipe, bar, or other device having a curved or bent shape resembling that of the neck of a goose. **2.** *See* water swivel. [NAV ARCH] An iron hook which joins a spar to a mast.

gooseneck barnacle [INV ZOO] Any stalked barnacle, especially of the genus *Lepas*.

gopher [VERT ZOO] The common name for North American rodents composing the family Geomyidae. Also known as pocket gopher.

gopher hole [ENG] Horizontal T-shaped opening made in rock in preparation for blasting. Also known as coyote hole. [MIN ENG] An irregular pitting hole made during prospecting.

gophering [MIN ENG] A method of breaking up a sandy medium-hard overburden where usual blastholes tend to cave in, by firing an explosive charge in each of a series of shallow holes; debris is cleared, and holes are made deeper for further charges, until they are deep enough to take enough explosives to break up the deposit.

gorceixite [MINERAL] $BaAl_3(PO_4)_2(OH)_5 \cdot H_2O$ A brown mineral composed of a hydrous basic phosphate of barium and aluminum.

Gordiidae [INV ZOO] A monogeneric family of worms in the order Gordioidea distinguished by a smooth cuticle.

Gordioidea [INV ZOO] An order of worms belonging to the Nematomorpha in which there is one ventral epidermal cord, a body cavity filled with mesenchymal tissue, and paired gonads.

gordioid larva [INV ZOO] The developmental stage of nematomorphs, free-living for a short time.

gordonite [MINERAL] $MgAl_2(PO_4)_2(OH)_2 \cdot 8H_2O$ A colorless mineral composed of a hydrous basic phosphate of magnesium and aluminum.

Gordon's formula [CIV ENG] An empirical formula which gives the collapsing load of a column in terms of its cross-sectional area, length, and least diameter.

gore [CIV ENG] A small triangular parcel of land.

gorge [ARCH] The entrance to a bastion. [GEOGR] A narrow passage between mountains or the walls of a canyon, especially one with steep, rocky walls. [OCEANOGR] A collection of solid matter obstructing a channel or a river, as an ice gorge.

gorge wind *See* canyon wind.

Gorgonacea [INV ZOO] The horny corals, an order of the coelenterate subclass Alcyonaria; colonies are fanlike or featherlike with branches spread radially or oppositely in one plane.

gorgonin [BIOCHEM] The protein, frequently containing iodine and bromine, composing the horny skeleton of members of the Gorgonacea; contains iodine and bromine.

Gorgonocephalidae [INV ZOO] A family of ophiuroid echinoderms in the order Phrynophiurida in which the individuals often have branched arms.

gorilla [VERT ZOO] *Gorilla gorilla.* An anthropoid ape, the largest living primate; the two African subspecies are the lowland gorilla and the mountain gorilla.

gorlic acid [ORG CHEM] $C_5H_7(C_{12}H_{22})COOH$ An unsaturated acid derived from sapucainha oil, obtained from the seeds of a tree in the Amazon Valley.

goslarite [MINERAL] $ZnSO_4 \cdot 7H_2O$ A white mineral composed of hydrous zinc sulfate.

gosling blast [METEOROL] A sudden squall of rain or sleet in England. Also known as gosling storm.

gosling storm *See* gosling blast.

gossan [GEOL] A rusty, ferruginous deposit filling the upper regions of mineral veins and overlying a sulfide deposit; formed by oxidation of pyrites. Also known as capping; gozzan; iron hat.

gossypose *See* raffinose.

Gotlandian [GEOL] A geologic time period recognized in Europe to include the Ordovician; it appears before the Devonian.

Goto pair [ELECTR] Two tunnel diodes connected in series in such a way that when one is in the forward conduction region, the other is in the reverse tunneling region; used in high-speed gate circuits.

gouge [DES ENG] A curved chisel for wood, bone, stone, and so on. [MIN ENG] A layer of soft material along the wall of a vein which favors the miner by enabling him, after gouging it out with a pick, to attack the solid vein from the side.

gout [MED] A condition of purine metabolism resulting in increased blood levels of uric acid with ultimate deposition as urates in soft tissues around joints.

gouy [PHYS CHEM] An electrokinetic unit equal to the product of the electrokinetic potential and the electric displacement divided by 4π times the polarization of the electrolyte.

Gouy balance [ANALY CHEM] Device for measurement of diamagnetic and paramagnetic susceptibilities of samples (solid, liquid, solution).

government frequency bands [COMMUN] Radio-frequency bands which are allotted to various departments and services of the federal government.

government specifications [ORD] Any description of the technical requirements for an ordnance material, item, or service, including the procedure by which it can be determined that the requirements have been met, according to government expectations.

governor [MECH ENG] A device, especially one actuated by the centrifugal force of whirling weights opposed by gravity or by springs, used to provide automatic control of speed or power of a prime mover.

gowk storm [METEOROL] In England, a storm or gale occurring at about the end of April or the beginning of May.

goyazite [MINERAL] $SrAl_3(PO_4)_2(OH)_5 \cdot H_2O$ A granular, yellowish-white mineral composed of a hydrous strontium aluminum phosphate.

gozzan *See* gossan.

G parity [PARTIC PHYS] The eigenvalue of a system under the operation of inversion in isotopic spin space; it is conserved by the strong interactions. Also known as isotopic parity.

GPI *See* ground point of intercept.

gpu *See* geopotential unit.

gr *See* grain.

Graafian follicle [HISTOL] The mature mammalian ovum with its surrounding epithelial cells.

grab [ENG] An instrument for extricating broken boring tools from a borehole.

grabbing crane [MECH ENG] An excavator made up of a crane carrying a large grab or bucket in the form of a pair of half scoops, hinged to dig into the earth as they are lifted.

grab bucket [MECH ENG] A bucket with hinged jaws or teeth that is hung from cables on a crane or excavator and is used to dig and pick up materials.

grab dredger [MECH ENG] Dredging equipment comprising a grab or grab bucket that is suspended from the jib head of a crane. Also known as grapple dredger.

graben [GEOL] A block of the earth's crust, generally with a length much greater than its width, that has dropped relative to the blocks on either side.

grabhook [DES ENG] A hook used for grabbing, as in lifting blocks of stone, in which case the hooks are used in pairs connected with a chain, and are so constructed that the tension of the chain causes them to adhere firmly to the rock.

grab sample [MIN ENG] A random mode of sampling; the samples may be taken from the pile broken in the process of mining, or from a truck or car of ore or coal.

graceful degradation [ADP] A programming technique to prevent catastrophic system failure by allowing the machine to operate, though in a degraded mode, despite failure or malfunction of several integral units or subsystems.

Gracilariidae [INV ZOO] A family of small moths in the superfamily Tineoidea; both pairs of wings are lanceolate and widely fringed.

gracilis [ANAT] A long slender muscle on the medial aspect of the thigh.

Gradall [MECH ENG] A hydraulic backhoe equipped with an extensible boom that excavates, backfills, and grades.

gradation period [GEOL] The time during which the base level of the sea remains in one position. Also known as base-leveling epoch.

grade [CIV ENG] **1.** To prepare a roadway or other land surface of uniform slope. **2.** A surface prepared for the support of rails, a road, or a conduit. [COMMUN] One of two types of television service, designated grade A and grade B, each having a specified signal strength, that of grade A being several times as large as B. [ENG] The degree of strength of a high explosive. [GEOL] The slope of the bed of a stream, or of a surface over which water flows, upon which the current can just transport its load without either eroding or depositing. [MATH] A unit of plane angle, equal to 0.01 right angle, or $\pi/200$ radians, or 0.9°. Also known as gon. [MIN ENG] **1.** A classification of ore according to recoverable amount of a valuable metal. **2.** To sort and classify diamonds.

gradeability [MECH ENG] The performance of earthmovers on various inclines, measured in percent grade.

grade beam [CIV ENG] A reinforced concrete beam placed directly on the ground to provide the foundation for the superstructure.

grade crossing [CIV ENG] The intersection of roadways, railways, pedestrian walks, or combinations of these at grade.

graded [GEOL] Brought to or established at grade.

graded bedding [GEOL] A stratification in which each stratum displays a gradation in the size of grains from coarse below to fine above.

graded-junction transistor *See* rate-grown transistor.

graded periodicity technique [ELECTR] A technique for modifying the response of a surface acoustic wave filter by varying the spacing between successive electrodes of the interdigital transducer.

graded profile *See* profile of equilibrium.

graded stream [HYD] A stream in which, over a period of years, slope is adjusted to yield the velocity required for transportation of the load supplied from the drainage basin.

graded topocline [ECOL] A topocline having a wide range, or ranging into different kinds of environment, thus subjecting its members to differential selection so that divergence between local races may become sufficient to warrant creation of varietal, or even specific, names.

grade line [CIV ENG] A line or slope used as a longitudinal reference for a railroad or highway.

grade of coal [MIN ENG] A classification of coal based on the amount and nature of the ash and sulfur content.

grader [MECH ENG] A high-bodied, wheeled vehicle with a leveling blade mounted between the front and rear wheels; used for fine-grading relatively loose and level earth.

grade scale [GEOL] A continuous scale of particle sizes divided into a series of size classes.

grade separation [CIV ENG] A grade crossing employing an underpass and overpass.

grade slab [CIV ENG] A reinforced concrete slab placed directly on the ground to provide the foundation for the superstructure.

grade stake [CIV ENG] A stake used as an elevation reference.

gradient [MATH] A vector obtained from a real function $f(x_1, x_2, \ldots, x_n)$ whose components are the partial derivatives of f; this measures the maximum rate of change of f in a given direction. [NAV] A slope expressed in feet per mile as a ratio of the horizontal to the vertical distance; for example, 40:1 means 40 feet horizontally to 1 foot vertically.

gradient coupling [PARTIC PHYS] A hypothetical interaction of particles in which the interaction Hamilton depends explicitly on first derivatives of wave functions associated with the particles with respect to position and time.

gradient current [OCEANOGR] A current defined by assuming that the horizontal pressure gradient in the sea is balanced by the sum of the Coriolis and bottom frictional forces; at some distance from the bottom the effect of friction becomes negligible, and above this the gradient and geostrophic currents are equivalent. Also known as slope current.

gradient elution analysis [ANALY CHEM] A form of gas-liquid chromatography in which the eluting solvent is changed with time, either by gradually mixing a second solvent of greater eluting power with the first, less powerful solvent, or by a gradual change in pH or other property.

gradienter [ENG] An attachment placed on a surveyor's transit to measure angle of inclination in terms of the tangent of the angle.

gradient flow [METEOROL] Horizontal frictionless flow in which isobars and streamlines coincide, or equivalently, in which the tangential acceleration is everywhere zero; the balance of normal forces (pressure force, Coriolis force, centrifugal force) is then given by the gradient wind equation.

gradient method [MATH] A finite iterative procedure for solving a system of n equations in n unknowns.

gradient microphone [ENG ACOUS] A microphone whose electrical response corresponds to some function of the difference in pressure between two points in space.

gradient of equal traction [MIN ENG] The gradient at which the tractive force necessary to pull an empty tram inby (slightly uphill) is equal to that force required to pull a loaded tram outby.

gradient projection method [MATH] Computational method used in nonlinear programming when constraint functions are linear.

gradient tints [MAP] A series of color tints used on maps or charts to indicate relative heights or depths. Also known as elevation tints; hypsometric tints.

gradient wind [METEOROL] A wind for which Coriolis acceleration and the centripetal acceleration exactly balance the horizontal pressure force.

gradient-wind level *See* geostrophic-wind level.

grading [IND ENG] Segregating a product into a number of adjoining categories which often form a spectrum of quality. Also known as classification.

graduate [CHEM] A cylindrical vessel that is calibrated in fluid ounces or milliliters or both; used to measure the volume of liquids.

Graebe-Ullman reaction [ORG CHEM] **1.** Production of fluorenone by boiling 2-benzoylbenzenediazonium salts in dilute acid solution. **2.** Reaction of 2-aminodiphenylamines with nitrous acid to form a benzotriazole which on heating loses nitrogen to form a carbazole.

Graeffe's method [MATH] A method of solving algebraic equations by means of squaring the exponents and making appropriate substitutions.

Graetz number [THERMO] A dimensionless number used in the study of streamline flow, equal to the mass flow rate of a fluid times its specific heat at constant pressure divided by the product of its thermal conductivity and a characteristic length. Also spelled Grätz number. Symbolized N_{Gz}.

graft [BIOL] **1.** To unite to form a graft. **2.** A piece of tissue transplanted from one individual to another or to a different place on the same individual. **3.** An individual resulting from the grafting of parts. [BOT] To unite a scion to an understock in such manner that the two grow together and continue development as a single plant without change in scion or stock.

graft copolymer [ORG CHEM] Any high polymer composed of two or more different polymeric entities chemically united.

graftonite [MINERAL] $(Fe,Mn,Ca)_3(PO_4)_2$ A salmon-pink mineral, crystallizing in the monoclinic system, and found as laminated intergrowths of triphylite; hardness is 5 on Mohs scale, and specific gravity is 3.7.

Graham-Cole test *See* cholecystography.

grahamite [GEOL] *See* mesosiderite. [MINERAL] A solid, jet-black hydrocarbon that occurs in veinlike masses; soluble in carbon disulfide and chloroform.

Grahm's law of diffusion [FL MECH] The law that the rate of diffusion of a gas is inversely proportional to the square root of its density.

grain [BOT] **1.** A rounded, granular prominence on the back of a sepal. **2.** *See* cereal. **3.** *See* drupelet. [GEOL] The particles or discrete crystals that make up a sediment or rock. [GRAPHICS] A small particle of metallic silver remaining in a photographic emulsion after developing and fixing; these grains together form the dark areas of a photographic image. [HYD] The particles which make up settled snow, firn, and glacier ice. [MATER] **1.** The appearance and texture of wood due to the arrangement of constituent fibers. **2.** The woodlike appearance or texture of a rock, metal, or other material. [MECH] A unit of mass in the United States and United Kingdom, common to the avoirdupois, apothecaries', and troy systems, equal to 1/7000 of a pound, or to 6.479891×10^{-5} kilogram. Abbreviated gr. [ORD] A single piece of solid propellant, regardless of size or shape, used in a gun or rocket; a rocket grain is often very large and shaped to fit its requirements.

grain alcohol *See* ethanol.

grain boundary [MET] Surface between individual grains in a metal.

grain direction [ADP] In character recognition, the arrangement of paper fibers in relation to a document's travel through a character reader.

grain drill [AGR] A drill used to sow small grains or seeds.

grain fineness number [MATER] Average grain size of a granular material.

grain growth [MET] Enlargement of grains in a metal, usually through heat treatment. [PETR] Enlargement of some individual crystals in a monomineralic rock, producing a coarser texture. Also known as germination.

graininess [GRAPHICS] A mottled effect in film caused by clumping of the silver particles.

grain leather [MATER] Leather made from the grain side of a skin.

grain size [GEOL] Average size of mineral particles composing a rock or sediment. [MET] Average size of grains in a metal expressed as average diameter, or grains per unit area or volume.

grain sorghum [BOT] *Sorghum bicolor*. A grass plant cultivated for its grain and to a lesser extent for forage. Also known as nonsaccharine sorghum.

grain spacing [DES ENG] Relative location of abrasive grains on the surface of a grinding wheel.

gram [MECH] The unit of mass in the centimeter-gram-second system of units, equal to 0.001 kilogram. Abbreviated g; gm.

gramagrass [BOT] Any grass of the genus *Bouteloua*; pasture grass.

gram-atomic weight [CHEM] The atomic weight of an element expressed in grams, that is, the atomic weight on a scale on which the atomic weight of carbon-12 isotope is taken as 12 exactly.

gram-calorie *See* calorie.

gram-centimeter [MECH] A unit of energy in the centimeter-gram-second gravitational system, equal to the work done by a force of magnitude 1 gram force when the point at which the force is applied is displaced 1 centimeter in the direction of the force. Abbreviated g-cm.

gram determinant [MATH] The gram determinant of vectors $v_1, \ldots, v_n$ from an inner product space is the determinant of the $n \times n$ matrix with the inner product of v_i and v_j as entry in the ith column and jth row; its vanishing is a necessary and sufficient condition for linear dependence.

gramenite *See* nontronite.

gram-equivalent weight [CHEM] The equivalent weight of an element or compound expressed in grams on a scale in which carbon-12 has an equivalent weight of 3 grams in those compounds in which its formal valence is 4.

gram-force [MECH] A unit of force in the centimeter-gram-second gravitational system, equal to the gravitational force on a 1-gram mass at a specified location. Abbreviated gf. Also known as fors; gram-weight; pond.

gramicidin [MICROBIO] A polypeptide antibacterial antibiotic produced by *Bacillus brevis*; active locally against gram-positive bacteria.

Graminales [BOT] The equivalent name for Cyperales.

Gramineae [BOT] The grasses, a family of monocotyledonous plants in the order Cyperales characterized by distichously arranged flowers on the axis of the spikelet.

graminicolous [ECOL] Living upon grass.

graminivorous [ZOO] Feeding on grasses.

graminoid [BOT] Of or resembling the grasses.

grammar [COMMUN] The set of rules governing the order of words in a communication, clause, or sentence.

gram molecular volume [CHEM] The volume occupied by a gram-molecular weight of a chemical in the gaseous state at 0°C and 760 millimeters of pressure.

gram-molecular weight [CHEM] The molecular weight of compound expressed in grams, that is, the molecular weight on a scale on which the atomic weight of carbon-12 isotope is taken as 12 exactly.

gram-negative [MICROBIO] Of bacteria, decolorizing and staining with the counterstain when treated with Gram's stain.

gram-negative diplococci [MICROBIO] Three bacteriologic genera composing the family Neisseriaceae: *Gemella, Veillonella,* and *Neisseria*.

gram-positive [MICROBIO] Of bacteria holding the color of the primary stain when treated with Gram's stain.

gram-rad [NUCLEO] A unit of integral absorbed dose of radiation, equal to 100 ergs.

gram-roentgen [NUCLEO] A unit of energy conversion, equal to a dose of 1 roentgen delivered to 1 gram of air.

Gram-Schmidt orthogonalization [MATH] A process for converting a set of linearly independent vectors into an orthogonal set spanning the same space.

Gram-Schmidt process [MATH] A process by which an orthogonal set of vectors is obtained from a linearly independent set of vectors in an inner product space.

Gram's stain [MICROBIO] A differential bacteriological stain; a fixed smear is stained with a slightly alkaline solution of basic dye, treated with a solution of iodine in potassium iodide, and then with a neutral decolorizing agent, and usually counterstained; bacteria stain either blue (gram-positive) or red (gram-negative).

gram-variable [MICROBIO] Pertaining to staining inconsistently with Gram's stain.

gram-weight *See* gram-force.

grana [CYTOL] A multilayered membrane unit formed by stacks of the lobes or branches of a chloroplast thylakoid.

Granby car [MIN ENG] An automatically dumped car for hand loading or power-shovel loading; a wheel attached to its side engages an inclined track at the dumping point.

Grand Banks [GEOGR] Banks off southeastern Newfoundland, important for cod fishing.

grand canonical ensemble [STAT MECH] A collection of systems of particles used to describe an individual system which is allowed to exchange both energy and particles with its environment.

grandfather [ADP] A data set that is two generations earlier than the data set under consideration.

grandfather cycle [ADP] The period during which records are kept but not used except to reconstruct other records which are accidentally lost.

grandite [MINERAL] A garnet that is intermediate in chemical composition between grossular and androdite.

grand mal [MED] A complete epileptic seizure involving sudden loss of consciousness and tonic convulsion of the skeletal musculature followed by clonic muscular spasms.

grandrelle yarn [TEXT] Yarn of blended or mixed colors made by twisting together two contrasting single yarns.

granite [PETR] A visibly crystalline plutonic rock with granular texture; composed of quartz and alkali feldspar with subordinate plagioclase and biotite and hornblende.

granite cloth [TEXT] A fabric having a hard finish with an irregular pebbled surface made by weaving tightly twisted wool or blended yarns.

granite-gneiss [PETR] A banded metamorphic rock derived from igneous or sedimentary rocks mineralogically equivalent to granite.

granite moss [BOT] The common name for a group of the class Bryatae represented by two Arctic genera and distinguished by longitudinal splitting of the mature capsule into four valves.

granite pegmatite *See* pegmatite.

granite porphyry *See* quartz porphyry.

granite wash [GEOL] Material eroded from granites and redeposited, forming a rock with the same major mineral constituents as the original rock.

granitic batholith [GEOL] A granitic shield mass intruded as the fusion of older formations.

granitic layer *See* sial.

granitic magma [PETR] A coarse-grained igneous rock.

granitization [PETR] A process whereby various types of rock may be converted to granite or closely related material.

granoblastic fabric [PETR] The texture of metamorphic

GRANA

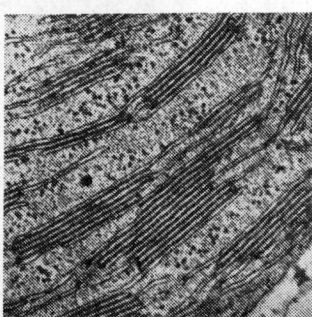

Part of oat leaf chloroplast at high magnification. Grana are seen in side view, revealing profiles of membranes. Thylakoids that interconnect grana are shown, with components of stroma, or ground substance, in which thylakoids lie. The dense granules are chloroplast ribosomes.

GRANITE MOSS

sporophytes

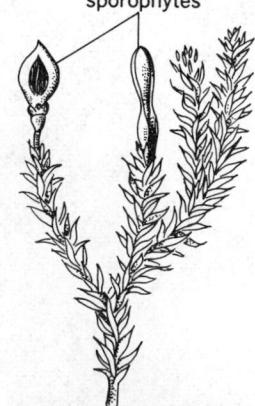

Granite moss (*Andreaea rupestris*), showing gametophyte with sporophytes. (*From G. M. Smith, Cryptogamic Botany, vol. 2, 2d ed., McGraw-Hill, 1955*)

rocks composed of equidimensional elements formed during recrystallization.

granodiorite [PETR] A visibly crystalline plutonic rock composed chiefly of sodic plagioclase, alkali feldspar, quartz, and subordinate dark-colored minerals.

granogabbro [PETR] Plutonic rock composed of quartz, basic plagioclase, potash-feldspar, and at least one ferromagnesian mineral; intermediate between a granite and a gabbro, and in a strict sense, a granodiorite with more than 50% boric plagioclase.

granophyre [PETR] A quartz porphyry or fine-grained porphyritic granite.

Grantiidae [INV ZOO] A family of calcareous sponges in the order Sycettida.

granular [SCI TECH] Having a grainy texture.

granular-bed separator [ENG] Vessel or chamber in which a bed of granular material is used to remove dust from a dust-laden gas as it passes through the bed.

granular fracture [MET] Grain-like or crystalline surface appearance of a broken metal.

granular gland [PHYSIO] A gland that produces and secretes a granular material.

granular ice [HYD] Ice composed of many tiny, opaque, white or milky pellets or grains frozen together and presenting a rough surface; this is the type of ice deposited as rime and compacted as névé.

granularity [PETR] The feature of rock texture relating to the size of the constituent grains or crystals.

granular leukocyte *See* granulocyte.

granular powder [MET] Equidimensional metal particles that are not spherical.

granular snow *See* snow grains.

granulate [CHEM] To form or crystallize into grains, granules, or small masses.

granulated metal [MET] Small pellets produced by pouring molten metal through a screen or similar device and chilling the droppings in water.

granulation [ASTRON] The small "rice grain" markings on the sun's photosphere. Also known as photospheric granulation. [MED] **1.** Tiny red granules made of capillary loops in the base of an ulcer. **2.** Process of granular tissue formation around a focus of inflammation. [PL PATH] Dry, tasteless condition of citrus fruit due to hardening of the juice sacs when fruit is left on trees too late in the season. [SCI TECH] The state or process of reducing a material to grains or small particles.

granulator [FOOD ENG] A revolving cylinder in which sugar is dried and granulated.

granule [GEOL] A somewhat rounded rock fragment ranging in diameter from 2 to 4 millimeters; larger than a coarse sand grain and smaller than a pebble.

granulite [PETR] **1.** Granite that contains muscovite. **2.** A relatively coarse, granuloblastic rock formed at the high temperatures and pressures of the granulite facies.

granulite facies [PETR] A group of gneissic rocks characterized by a granoblastic fabric and formed by regional dynamo-thermal metamorphism at temperatures above 650°C and pressures of 3000–12,000 bars.

granuloblastosis [VET MED] An avian leukosis characterized by the presence of excessive numbers of immature granulocytes in the blood of affected birds.

granulocyte [HISTOL] A leukocyte containing granules in the cytoplasm. Also known as granular leukocyte; polymorph; polymorphonuclear leukocyte.

granulocytic leukemia [MED] A blood disease involving neoplastic transformation of granulocytes, principally the neutrophilic series. Also known as myelogenous leukemia; myeloid leukemia.

granulocytopenia [MED] A deficiency of granulocytes in circulating blood. Also known as granulopenia.

granulocytosis [MED] An increase in the number of granulocytes in the circulation.

granuloma [MED] A discrete nodular lesion of inflammatory tissue in which granulation is significant.

granuloma inguinale [MED] An infectious, chronic, destructive granulomatous lesion of humans most frequently local-

ized in the genital and inguinal regions; caused by Donovan bodies (*Donovania granulomatis*).

granuloma pyogenicum [MED] A hemangioma with superimposed inflammation on the skin or other epithelial surfaces.

granulomatosis [MED] Any disease characterized by multiple granulomas.

granulometry [PETR] Measurement of grain sizes of sedimentary rock.

granulopenia *See* granulocytopenia.

Granuloreticulosia [INV ZOO] A subclass of the protozoan class Rhizopodea characterized by reticulopodia which often fuse into networks.

granulosis [INV ZOO] A virus disease of lepidopteran larvae characterized by the accumulation of small granular inclusion bodies (capsules) in the infected cells.

Granville wilt [PL PATH] A bacterial wilt of tobacco caused by *Pseudomonas solanacearum*.

grape [BOT] The common name for plants of the genus *Vitis* characterized by climbing stems with cylindrical-tapering tendrils and polygamodioecious flowers; grown for the edible, pulpy berries.

grapefruit [BOT] *Citrus paradisi.* An evergreen tree with a well-rounded top cultivated for its edible fruit, a large, globose citrus fruit characterized by a yellow rind and white, pink, or red pulp.

grapefruit oil [MATER] Pale-yellow, volatile liquid with grapefruit aroma; soluble in oils, insoluble in glycerin; derived from fresh rind of *Citrus paradisi*; used in flavors and toiletries. Also known as oil of grapefruit; oil of shaddock.

grapefruit seed oil [MATER] Reddish-brown oil expressed from grapefruit seeds; has nutlike aroma and bitter taste; solidifies at −10°C; used as lubricant for textile fibers and leather.

grapestone [GEOL] A cluster of sand-size grains, such as calcareous pellets, held together by incipient cementation shortly after deposition; the outer surface is lumpy, resembling a bunch of grapes.

grape sugar *See* dextrose.

grapevine drainage *See* trellis drainage.

grapevine stopper [NAV ARCH] A device that allows temporary grip on a large-diameter cable at high tension; consists of a webbed sleeve that contracts under tension.

graph [MATH] **1.** The planar object, formed from points and line segments between them, used in the study of circuits and networks. **2.** The graph of a function f is the set of all ordered pairs $[x, f(x)]$, where x is in the domain of f. **3.** *See* graphical representation.

graph component [MATH] A particular type of maximal connected subgraph of a graph.

graphechon [ELECTR] A storage tube having two electron guns, one for writing and the other for reading and simultaneous erasing, on opposite sides of the storage medium, which consists of an insulator or semiconductor deposited on a thin substratum of metal supported by a fine mesh.

grapheme [COMMUN] A pictorial representation of a semanteme, such as X-reference for cross-reference.

graph follower *See* curve follower.

graphical analysis [MATH] The study of interdependent phenomena by analyzing graphical representations.

graphical design [ELECTR] Methods of obtaining operating data for an electron tube or semiconductor circuit by using graphs which plot the relationship between two variables, such as plate voltage and grid voltage, while another variable, such as plate current, is held constant.

graphical representation [MATH] The plot of the points in the plane which constitute the graph of a given real function or a pictorial diagram depicting interdependence of variables. Also known as graph.

graphical statics [MECH] A method of determining forces acting on a rigid body in equilibrium, in which forces are represented on a diagram by straight lines whose lengths are proportional to the magnitudes of the forces.

graphical symbol [ELEC] A true symbol, rather than a coarse picture, representing an element in an electrical diagram.

graphic arts [GRAPHICS] Those methods of applied arts used to form a visual end product that conveys information or is a

GRANULATION

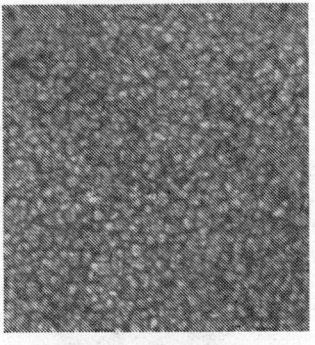

Large-scale photograph of photospheric granulation in white light taken from an altitude of 80,000 feet (24.38 kilometers) above sea level. The length of this section is about 55,000 kilometers on the sun. *(Princeton Observatory)*

GRAPE

Foliage, tendrils, fruit, and stem characteristics of the Concord grape (*Vitis labruscana*).

GRAPEFRUIT

Foliage, flowers, and fruit of grapefruit.

GRAPHITE

├─ 2.5 cm ─┤

An earthy aggregate of graphite. *(American Museum of Natural History Specimen)*

GRAPHITE ANODE

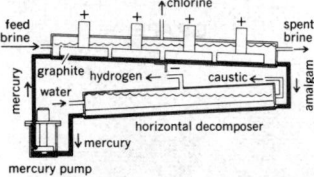

Diagram of a mercury cell for producing chlorine, showing the graphite anode.

GRASSHOPPER

0.5 cm

Melanophus mexicanus, a grasshopper. *(From Illinois Natural History Survey)*

GRASSHOPPER LINKAGE

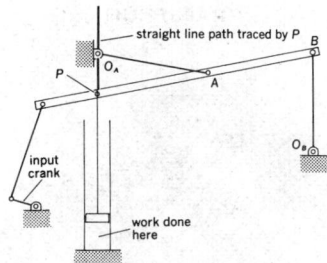

Grasshopper linkage used in the early steam engine. As input crank rotates, the cranks AO_A and BO_B pivot about O_A and O_B respectively, and point P moves approximately in a straight line. *(Georgia Institute of Technology, Theoretical Analysis of Four-Bar Mechanisms, Final Report, NSF Grant GP-2748)*

decoration; methods include drawing, painting, photography, all types of printing, and bookmaking.

graphic display [ELECTR] The display of data in graphical form on the screen of a cathode-ray tube.

graphic granite [PETR] A distinct type of pegmatite in which quartz and orthoclase crystals grew together along a parallel axis. Also known as Hebraic granite; runite.

graphic methods [GRAPHICS] Pencil-and-paper methods that employ the geometry of a plane to express mathematical relationships and to carry out mathematical operations in analog form.

graphic panel [CONT SYS] A master control panel which indicates the status of equipment and operations in a system, and their relationships.

graphic recording instrument [ENG] An instrument that makes a graphic record of one or more quantities as a function of another variable, usually time.

graphics [COMMUN] 1. In communications systems, an information mode in which a graphic system is used to reproduce intelligence; a variation of facsimile. 2. Nonvoice analog information devices and modes such as facsimile, photographics, and television. [SCI TECH] 1. The graphic media. 2. The art of drawing a three-dimensional object on a two-dimensional surface according to mathematical rules of projection.

graphic scale [MAP] A graduated line that indicates the length of miles or kilometers as they appear on a map; the line has the advantage of remaining true after the map is enlarged or reduced in reproduction. Also known as bar scale.

graphic tellurium See sylvanite.

graphic texture [GEOL] A pattern of rocks that is similar to cuneiform characters.

Graphidaceae [BOT] A family of mosses formerly grouped with lichenized Hysteriales but now included in the order Lecanorales; individuals have true paraphyses.

Graphiolaceae [MYCOL] A family of parasitic fungi in the order Ustilaginales in which teleutospores are produced in a cuplike fruiting body.

graphite [MINERAL] A mineral consisting of a low-pressure allotropic form of carbon; it is soft, black, and lustrous and has a greasy feeling; it occurs naturally in hexagonal crystals or massive or can be synthesized from petroleum coke; hardness is 1–2 on Mohs scale, and specific gravity is 2.09–2.23; used in pencils, crucibles, lubricants, paints, and polishes. Also known as black lead; plumbago.

graphite anode [CHEM ENG] One of the electrodes of graphite used in a mercury cell to produce chlorine by electrolysis. [ELECTR] 1. The rod of graphite which is inserted into the mercury-pool cathode of an ignitron to start current flow. 2. The collector of electrons in a beam power tube or other high-current tube.

graphite grease [MATER] A lubricating grease that contains 2–10% amorphous graphite; used for bearings, especially in damp places.

graphite-moderated reactor [NUCLEO] A nuclear reactor in which graphite is the principal moderating material.

graphite oil [MATER] A deflocculated suspension of graphite in oil; used as a lubricant.

graphite resistor [ELEC] A resistor made of carbon for resistance heating and suitable for any temperature that can be used.

graphitic carbon [MET] Carbon in iron or steel present in the form of graphite.

graphitic corrosion [MET] Corrosion of gray cast iron in which the metallic iron constituent is converted into corrosion products which cement together the residual graphite.

graphitic steel [MET] Alloy steel containing graphitic carbon.

graphitization [ORG CHEM] The formation of graphitelike material from organic compounds.

graphitizing [MET] Annealing cast iron to convert all or some of the combined carbon to graphitic carbon.

graph theory [MATH] 1. The mathematical study of the structure of graphs and networks. 2. The body of techniques used in graphing functions in the plane.

grapnel [DES ENG] An implement with claws used to recover a lost core, drill fittings, and junk from a borehole or for other

grappling operations. Also known as grapple. [NAV ARCH] An anchor with four or five hooks used for dragging the bottom or for anchorage.

grappier cement [MATER] A cement composed of finely ground lumps of leftover underburned and overburned slaked lime.

grapple See grapnel.

grapple dredger See grab dredger.

grapple hook [DES ENG] An iron hook used on the end of a rope to snag lines, to hold one ship alongside another, or as a fishing tool. Also known as grappling iron.

grappling iron See grapple hook.

Grapsidae [INV ZOO] The square-backed crabs, a family of decapod crustaceans in the section Brachyura.

graptolite shale [GEOL] Shale containing an abundance of extinct colonial marine organisms known as graptolites.

Graptolithina [PALEON] A class of extinct colonial animals believed to be related to the class Pterobranchia of the Hemichordata.

Graptoloidea [PALEON] An order of extinct animals in the class Graptolithina including branched, planktonic forms described from black shales.

Graptozoa [PALEON] The equivalent name for Graptolithina.

graser See gamma-ray laser.

Grashof formula [FL MECH] A formula, $m = 0.0165 A_2 p_1^{0.97}$, used to express the discharge m of saturated steam, where A_2 is the area of the orifice in square inches, and p_1 is reservoir pressure in pounds per square inch.

Grashof number [FL MECH] A dimensionless number used in the study of free convection of a fluid caused by a hot body, equal to the product of the fluid's coefficient of thermal expansion, the temperature difference between the hot body and the fluid, the cube of a typical dimension of the body and the square of the fluid's density, divided by the square of the fluid's dynamic viscosity. Also known as free convection number.

grasp [IND ENG] A basic element (therblig) in time-motion study; a useful element that accomplishes work.

grass [BOT] The common name for all members of the family Gramineae; moncotyledonous plants having leaves which consist of a sheath which fits around the stem like a split tube, and a long, narrow blade. [ELECTR] Clutter due to circuit noise in a radar receiver, seen on an A scope as a pattern resembling a cross section of turf. Also known as hash.

grass crops [AGR] Plants cultivated as forage and grain to be consumed by domestic livestock.

grasserie [INV ZOO] A polyhedrosis disease of silkworms characterized by spotty yellowing of the skin and internal liquefaction. Also known as jaundice.

grasshopper [INV ZOO] The common name for a number of plant-eating orthopteran insects composing the subfamily Saltatoria; individuals have hindlegs adapted for jumping, and mouthparts adapted for biting and chewing.

grasshopper fuse [ELEC] Small fuse incorporating a spring which, upon release by the fusing wire, connects an auxiliary circuit to operate an alarm.

grasshopper linkage [MECH ENG] A straight-line mechanism used in some early steam engines.

grassland [ECOL] Any area of herbaceous terrestrial vegetation dominated by grasses and graminoid species.

grassland climate See subhumid climate.

Grassmannian See Grassman manifold.

Grassmann manifold [MATH] The differentiable manifold whose points are all k-dimensional planes passing through the origin in n-dimensional euclidean space. Also known as Grassmannian.

Grassmann's laws [ANALY CHEM] Seven laws of color identification and mixing that form the basis of modern analytical colorimetry.

grass minimum [METEOROL] The minimum temperature shown by a minimum thermometer exposed in an open situation with its bulb on the level of the tops of the grass blades of short turf.

grass-roots deposit [MIN ENG] A deposit that is discovered in surface croppings, is easily exploited, and can pay for its own development while in progress.

grass-roots mining [MIN ENG] Also known as mining on a

shoestring. **1.** Inadequately financed mining operation, with catch-as-catch-can practices. **2.** Mining from surface down to bedrock.

grass-roots plant [CHEM ENG] A complete plant erected on a virgin site.

grass sickness [VET MED] A disease of horses occurring mainly in Scotland; thought to be caused by a virus similar to the one that causes poliomyelitis in humans.

grass temperature [METEOROL] The temperature registered by a thermometer with its bulb at the level of the tops of the blades of grass in short turf.

grass tetany [VET MED] A magnesium-deficiency disease of cows.

grate [ENG] A support for burning solid fuels; usually made of closely spaced bars to hold the burning fuel, while allowing combustion air to rise up to the fuel from beneath, and ashes to fall away from the burning fuel.

graticule [MAP] A network of lines representing the earth's parallels of latitude and meridians of longitude on a map, chart, or plotting sheet. [OPTICS] A scale at the focal plane of an optical instrument to aid in the measurement of objects.

grating [ELECTROMAG] **1.** An arrangement of fine, parallel wires used in waveguides to pass only a certain type of wave. **2.** An arrangement of crossed metal ribs or wires that acts as a reflector for a microwave antenna and offers minimum wind resistance. [SPECT] *See* diffraction grating.

grating constant [OPTICS] The distance between consecutive diffraction centers of a ultrasonic wave which is producing a light diffraction spectrum. [SPECT] The distance between consecutive grooves of a diffraction grating.

grating spectrograph [SPECT] A grating spectroscope provided with a photographic camera or other device for recording the spectrum.

grating spectroscope [SPECT] A spectroscope which employs a transmission or reflection grating to disperse light, and usually also has a slit, a mirror or lenses to collimate the light sent through the slit and to focus the light dispersed by the grating into spectrum lines, and an eyepiece for viewing the spectrum.

gratonite [MINERAL] $Pb_9As_4S_{15}$ A mineral composed of lead arsenic sulfide, occurring in rhombohedral crystals.

Grätz number *See* Graetz number.

Gratz rectifier [ELECTR] Three-phase, full-wave rectifying circuit using six rectifiers connected in a bridge circuit.

graupel *See* snow pellets.

grav *See* G.

gravel [GEOL] A loose or unconsolidated deposit of rounded pebbles, cobbles, or boulders.

gravel bank [GEOL] A natural mound or exposed face of gravel, particularly such a place from which gravel is dug.

gravel mine [MIN ENG] A mine extracting gold from sand or gravel. Also known as placer mine.

gravel pump [MECH ENG] A centrifugal pump with renewable impellers and lining, used to pump a mixture of gravel and water.

gravel stop [BUILD] Metal flashing placed at the edge of a roof to prevent gravel from falling off.

Grave's disease *See* hyperthyroidism.

graveyard *See* burial ground.

graveyard shift [IND ENG] The shift of workers that begins at or around midnight; the last shift of the day.

gravid [ZOO] **1.** Of the uterus, containing a fetus. **2.** Pertaining to female animals when carrying young or eggs.

Gravigrada [VERT ZOO] The sloths, a group of herbivorous xenarthran mammals in the order Edentata; members are completely hairy and have five upper and four lower prismatic teeth without enamel.

gravimeter [ENG] A highly sensitive weighing device used for relative measurement of the force of gravity by detecting small weight differences of a constant mass at different points on the earth. Also known as gravity meter.

gravimetric absorption method [ANALY CHEM] A method of measuring the moisture content of a gas in which a known volume of gas is passed through a suitable desiccant, such as phosphorus pentoxide or silica gel, and the change in weight of the desiccant is observed.

gravimetric analysis [ANALY CHEM] That branch of quantita-

tive analytical chemistry in which a desired constituent is converted, usually by precipitation or combustion, to a pure compound or element, of definite known composition, and is weighed; in a few cases a compound or element is formed which does not contain the constituent but bears a definite mathematical relationship to it.

gravimetric geodesy [GEOD] The science that utilizes measurements and characteristics of the earth's gravity field, as well as theories regarding this field, to deduce the shape of the earth and, in combination with arc measurements, the earth's size.

gravimetry [ENG] Measurement of gravitational force.

graving dock [CIV ENG] A form of dry dock consisting of an artificial basin fitted with a gate or caisson, into which a vessel can be floated and the water pumped out to expose the vessel's bottom.

gravitation [PHYS] The mutual attraction between all masses in the universe. Also known as gravitational attraction.

gravitational acceleration [PHYS] The acceleration imparted to a body by the attraction of the earth; approximately equal to 980.7 cm/sec², or 32.2 ft/sec².

gravitational astronomy *See* celestial mechanics.

gravitational attraction *See* gravitation.

gravitational clustering [ASTRON] A theory that attributes the hierarchy structure of the universe to growth of density fluctuations in a statistically uniform and isotropic universe.

gravitational collapse [ASTRON] The implosion of a star or other astronomical body from an initial size to a size hundreds or thousands of times smaller.

gravitational constant [MECH] The constant of proportionality in Newton's law of gravitation, equal to the gravitational force between any two particles times the square of the distance between them, divided by the product of their masses. Also known as constant of gravitation.

gravitational displacement [MECH] The gravitational field strength times the gravitational constant. Also known as gravitational flux density.

gravitational energy *See* gravitational potential energy.

gravitational field [MECH] The field in a region in space in which a test particle would experience a gravitational force; quantitatively, the gravitational force per unit mass on the particle at a particular point.

gravitational-field theory [RELAT] A theory in which gravity is treated as a field, as opposed to a theory in which the force acts instantaneously at a distance.

gravitational flux density *See* gravitational displacement.

gravitational force [MECH] The force on a particle due to its gravitational attraction to other particles.

gravitational gradient guidance [NAV] A system for maintaining the desired orientation of a space vehicle in the vicinity of the earth (or any large celestial body) which utilizes the differences in the pull of gravity between points on the craft having different distances from the earth.

gravitational instability [MECH] Instability of a dynamic system in which gravity is the restoring force.

gravitational mass [PHYS] The mass of a particle as it determines the force it experiences in a gravitational field; equal to inertial mass according to the equivalence principle.

gravitational potential [MECH] The amount of work which must be done against gravitational forces to move a particle of unit mass to a specified position from a reference position, usually a point at infinity.

gravitational potential energy [MECH] The energy that a system of particles has by virtue of their positions, equal to the work that must be done against gravitational forces to assemble the particles from some reference configuration, such as mutually infinite separation. Also known as gravitational energy.

gravitational pressure *See* hydrostatic pressure.

gravitational radiation *See* gravitational wave.

gravitational radius *See* Schwarzschild radius.

gravitational red shift [RELAT] A displacement of spectral lines toward the red when the gravitational potential at the observer of the light is greater than at its source.

gravitational repulsion [PHYS] Hypothetical repulsion of matter and antimatter; however, experimental results indicate

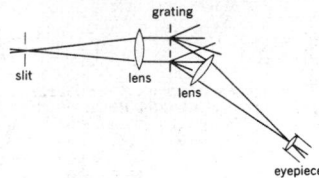

GRATING SPECTROSCOPE

Transmission grating spectroscope.

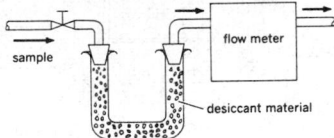

**GRAVIMETRIC
ABSORPTION METHOD**

Diagram of the gravimetric absorption method.

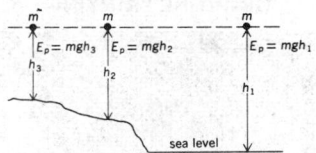

**GRAVITATIONAL
POTENTIAL ENERGY**

The gravitational potential energy E_p of a body of weight mg (m = mass, g = force of gravity) for different reference levels (h_1, h_2, h_3).

that matter and antimatter attract according to the same laws as matter and matter.

gravitational settling [GEOL] A movement of sediment resulting from gravitational forces.

gravitational sliding [GEOL] Extensive sliding of strata down a slope of an uplifted area. Also known as sliding.

gravitational systems of units [MECH] Systems in which length, force, and time are regarded as fundamental, and the unit of force is the gravitational force on a standard body at a specified location on the earth's surface.

gravitational tide [OCEANOGR] An atmospheric tide due to gravitational attraction of the sun and moon.

gravitational wave [RELAT] A propagating gravitational field predicted by general relativity, which is produced by some change in the distribution of matter; it travels at the speed of light, exerting forces on masses in its path. Also known as gravitational radiation.

gravitometer See densimeter.

graviton [PHYS] A theoretically deduced particle postulated as the quantum of the gravitational field, having a rest mass and charge of zero and a spin of 2.

gravity [MECH] The gravitational attraction at the surface of a planet or other celestial body.

gravity anomaly [MECH] A correction which must be added to a theoretical model describing gravity on the surface of the earth, resulting in the main from surface irregularities.

gravity bar [MIN ENG] A 5-foot length of heavy half-round rod forming the link between the wedge-oriented coupling and the drill-rod swivel coupling on an assembled Thompson retrievable borehole-deflecting wedge.

gravity bed [ENG] A moving body of solids in which particles (granules, pellets, beads, or briquets) flow downward by gravity through a vessel, while process fluid flows upward; the moving-bed technique is used in blast and shaft furnaces, petroleum catalytic cracking, pellet dryers, and coolers.

gravity cell [PHYS CHEM] An electrolytic cell in which two ionic solutions are separated by means of gravity.

gravity chute [ENG] A gravity conveyor in the form of an inclined plane, trough, or framework that depends on sliding friction to control the rate of descent.

gravity classification [MIN ENG] The grading of ores, or the separation of waste from coal, by the differences in specific gravities of the substances.

gravity-collapse structure See collapse structure.

gravity concentration [ENG] 1. Any of various methods for separating a mixture of particles, such as minerals, based on the differences in density of the various species and on the resistance to relative motion exerted upon the particles by the fluid or semifluid medium in which separation takes place. 2. The separation of liquid-liquid dispersions based on settling out of the dense phase by gravity.

gravity conveyor [ENG] Any unpowered conveyor such as a gravity chute or a roller conveyor, which uses the force of gravity to move materials over a downward path.

gravity corer [ENG] Any type of corer that achieves bottom penetration solely as a result of gravitational force acting upon its mass.

gravity dam [CIV ENG] A dam which depends on its weight for stability.

gravity drainage [HYD] Withdrawal of water from strata as a result of gravitational forces.

gravity fault See normal fault.

gravity feed [ENG] Movement of materials from one location to another using the force of gravity.

gravity flow [HYD] A form of glacier movement in which the flow of the ice results from the downslope gravitational component in an ice mass resting on a sloping floor.

gravity-flow gathering system [PETRO ENG] The use of gravity (downhill flow) through pipelines to transport and collect liquid at a central location; used for gathering of waste water from waterflooding operations for treatment prior to reuse or disposal.

gravity-gradient attitude control [AERO ENG] A device that regulates automatically attitude or orientation of an aircraft or spacecraft by responding to changes in gravity acting on the craft.

gravity haulage [MIN ENG] A type of haulage system in which the set of full cars is lowered at the end of a rope, and gravity force pulls up the empty cars, the rope being passed around a sheave at the top of the incline. Also known as self-acting incline.

gravity incline [MIN ENG] An opening made in the direction of, and along the same gradient as, the dip of the deposit.

gravity map [GEOPHYS] A map of gravitational variations in an area displaying gravitational highs and lows.

gravity meter [ENG] 1. U-tube-manometer type of device for direct reading of solution specific gravities in semimicro quantities. 2. An electrical device for measuring variations in gravitation through different geologic formations; used in mineral exploration. 3. See gravimeter.

gravity pendulum See pendulum.

gravity prospecting [ENG] Identifying and mapping the distribution of rock masses of different specific gravity by means of a gravity meter.

gravity railroad [ENG] A cable railroad in which cars descend a slope by gravity and are hauled back up the slope by a stationary engine, or there may be two tracks with cars so connected that cars going down may help to raise the cars going up and thus conserve energy.

gravity segregation [ENG] Tendency of immiscible liquids or multicomponent granular mixtures to separate into distinct layers in accordance with their respective densities.

gravity separation [ENG] Separation of immiscible phases (gas-solid, liquid-solid, liquid-liquid, solid-solid) by allowing the denser phase to settle out under the influence of gravity; used in ore dressing and various industrial chemical processes.

gravity settling chamber [ENG] Chamber or vessel in which the velocity of heavy particles (solids or liquids) in a fluid stream is reduced to allow them to settle downward by gravity, as in the case of a dust-laden gas stream.

gravity spring [HYD] A spring that issues under the influence of gravity, not internal pressure.

gravity stamp [MIN ENG] Unit in a stamp battery which directs a heavy falling weight onto a die on which rock is crushed.

gravity station [ENG] The site of installation of gravimeters.

gravity survey [ENG] The measurement of the differences in gravity force at two or more points.

gravity tide [GEOPHYS] Cyclic motion of the earth's surface caused by interaction of gravitational forces of the moon, sun, and earth.

gravity vector [MECH] The force of gravity per unit mass at a given point. Symbolized **g**.

gravity wall [CIV ENG] A retaining wall which is kept upright by the force of its own weight.

gravity wave [FL MECH] 1. A wave at a gas-liquid interface which depends primarily upon gravitational forces, surface tension and viscosity being of secondary importance. 2. A wave in a fluid medium in which restoring forces are provided primarily by buoyancy (that is, gravity) rather than by compression.

gravity wheel conveyor [MECH ENG] A downward-sloping conveyor trough with closely spaced axle-mounted wheel units on which flat-bottomed containers or objects are conveyed from point to point by gravity pull.

gravity wind [METEOROL] A wind (or component thereof) directed down the slope of an incline and caused by greater air density near the slope than at the same levels some distance horizontally from the slope. Also known as drainage wind; katabatic wind.

gravity yard See hump yard.

gravure coating See engraved-roll coating.

gravure printing [GRAPHICS] A process that prints from sunken or depressed surfaces or cups on a plate, in contrast to the raised printing surfaces of letterpress; the depth (and the area) of the depressed areas varies, thus yielding more or less ink on the paper.

Grawitz's tumor See renal-cell carcinoma.

gray antimony See jamesonite.

gray blight [PL PATH] A fungus disease of tea caused by *Pestalotia* (*Pestalozzia*) *theae*, which invades the tissues and causes the formation of black dots on the leaves.

graybody [THERMO] An energy radiator which has a black-

GRAVITY CHUTE

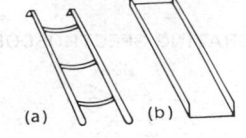

(a)   (b)

Two types of gravity chutes.
(a) Barrel skid. *(b)* Plain chute.

GRAVITY CONVEYOR

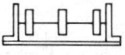

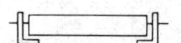

Wheels and roller used in some types of gravity conveyors.

GRAVURE PRINTING

Sheet-fed gravure printing surface.

body energy distribution, reduced by a constant factor, throughout the radiation spectrum or within a certain wavelength interval. Also known as nonselective radiator.

gray casting [MET] A casting of gray iron.

Gray clay treating [CHEM ENG] A fixed-bed, vapor-phase treating process used to polymerize selectively unsaturated gum-forming constituents (diolefins); a fixed bed is used of 30- to 60-mesh fuller's earth.

Gray code [COMMUN] A modified binary code in which sequential numbers are represented by expressions that differ only in one bit, to minimize errors. Also known as reflective code.

gray copper ore *See* tetrahedrite.

gray filter *See* neutral-density filter.

gray hematite *See* specularite.

gray iron [MET] Pig or cast iron in which the carbon not contained in pearlite is present in the form of graphitic carbon.

gray leaf spot [PL PATH] A fungus disease of tomatoes caused by *Stemphylium solani* and characterized by water-soaked brown spots on the leaves that become gray with age.

Grayloc tubing joint [PETRO ENG] Special wellbore-tubing joint that has greater leak resistance and strength than standard API tubing joints.

gray manganese ore *See* manganite.

gray matter [HISTOL] The part of the central nervous system composed of nerve cell bodies, their dendrites, and the proximal and terminal unmyelinated portions of axons.

gray mold [PL PATH] Any fungus disease characterized by a gray surface appearance of the affected part.

gray scab [PL PATH] A fungus disease of willow caused by *Sphaceloma murrayae* and characterized by irregular raised leaf spots having grayish-white centers and dark-brown margins.

gray scale [OPTICS] A series of achromatic tones having varying proportions of white and black, to give a full range of grays between white and black; a gray scale is usually divided into 10 steps.

gray sour *See* lime sour.

gray speck [PL PATH] A manganese-deficiency disease of oats characterized by light-green to grayish leaf spots, and later by the buff or light-brown discoloration of the blades.

graywacke [PETR] An argillaceous sandstone characterized by an abundance of unstable mineral and rock fragments and a fine-grained clay matrix binding the larger, sand-size detrital fragments.

graywall [PL PATH] A disease of tomatoes thought to be caused by excess sunlight and characterized by translucent grayish-brown streaks or blotches on the fruit and browning of the vascular strands.

graze [ORD] **1.** To pass close to the surface, as a shot that follows a path nearly parallel to the ground and low enough to strike a standing person. **2.** Burst of a projectile at the instant of impact with the ground. Also known as graze burst. **3.** In time fire, a burst on impact with the ground or other material object on a level with or below the target. [VERT ZOO] To feed by browsing on, cropping, and eating grass.

graze burst *See* graze.

grazing angle [PHYS] A very small glancing angle.

grazing fire [ORD] Small-arms fire which is approximately parallel to the ground and does not rise above the height of a standing person.

grazing incidence [PHYS] Incidence at a small glancing angle.

grease [MATER] **1.** Rendered, inedible animal fat that is soft at room temperature and is obtained from lard, tallow, bone, raw animal fat, and other waste products. **2.** A lubricant in the form of a solid to semisolid dispersion of a thickening agent in a fluid lubricant, such as petroleum oil thickened with metallic soap.

grease cup [ENG] A receptacle used to apply a solid or semifluid lubricant to a bearing; the receptacle is packed with grease and the cap forces the grease to the bearing.

greased-deck concentration [MIN ENG] A separation process based on selective adhesion of certain grains (diamonds) to quasi-solid grease.

grease gun [ENG] A small hand-operated device that pumps grease under pressure into bearings.

grease ice [HYD] A kind of slush with a greasy appearance, formed from the congelation of ice crystals in the early stages of freezing. Also known as ice fat; lard ice.

greasepaint [MATER] Makeup made of melted tallow or grease used by theatrical performers.

grease seal [ENG] **1.** Type of seal used on floating pistons of some hydropneumatic recoil systems to prevent leakage past the piston of gas or oil; also used in cylinders of some hydropneumatic equilibrators. **2.** Seal used to retain grease in a case or housing, as on an axle shaft.

grease spot [PL PATH] A fungus disease of turf grasses caused by *Pythium aphanidermatum* and characterized by spots that have a greasy border of blackened leaves and intermingled cottony mycelia. Also known as spot blight.

grease table [MIN ENG] An apparatus for concentrating minerals, such as diamonds, which adhere to grease; usually a shaking table coated with grease or wax over which an aqueous pulp is flowed.

grease trap [CIV ENG] A trap in a drain or waste pipe to stop grease from entering a sewer system.

greasewood [BOT] Any plant of the genus *Sarcobatus*, especially *S. vermiculatus*, which is a low shrub that grows in alkali soils of the western United States.

greasy quartz *See* milky quartz.

Great Basin high [METEOROL] A high-pressure system centered over the Great Basin of the western United States; it is a frequent feature of the surface chart in the winter season.

great circle [GEOD] A circle, or near circle, described on the earth's surface by a plane passing through the center of the earth. [MATH] The circle on the Riemann sphere produced by a plane passing through the center of the sphere.

great-circle bearing [NAV] The initial direction along a great circle through two terrestrial points, expressed as angular distance from a reference direction.

great-circle chart *See* gnomonic chart.

great-circle course [NAV] Course along the direction of the great circle through the departure point and the destination; expressed as the angular distance from a reference direction, usually north to the direction of the great circle; the angle varies from point to point along the great circle.

great-circle direction [GEOD] Horizontal direction of a great circle, expressed as angular distance from a reference direction.

great-circle distance [GEOD] The length of the shorter arc of the great circle joining two points.

great-circle sailing [NAV] Any method of solving the various problems involving courses, distance, and so on, as they are relating to a great-circle track.

great-circle track [NAV] The track of a craft following a great circle, or the approximate great circle which a craft intends to follow.

great diurnal range [OCEANOGR] The difference in height between mean higher high water and mean lower low water. Also known as diurnal range.

Greater Dog *See* Canis Major.

greater ebb [OCEANOGR] The stronger of two ebb currents occurring during a tidal day.

greater flood [OCEANOGR] The stronger of two flood currents occurring during a tidal day.

greater omentum [ANAT] A fold of peritoneum that is attached to the greater curvature of the stomach and hangs down over the intestine and fuses with the mesocolon.

greatest common divisor [MATH] The greatest common divisor of integers $n_1, n_2, \ldots, n_k$ is the largest of all integers that divide each n_i. Abbreviated gcd. Also known as highest common factor (hcf).

greatest elongation [ASTRON] The maximum angular distance of a body of the solar system from the sun, as observed from the earth.

greatest lower bound [MATH] The greatest lower bound of a set of numbers S is the largest number among the lower bounds of S. Abbreviated glb. Also known as infimum.

great galago *See* thick-tailed bushbaby.

Great Ice Age [GEOL] The Pleistocene epoch.

Great Nebula of Orion *See* Orion Nebula.

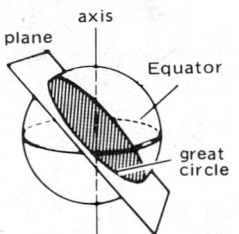

GREAT CIRCLE

axis
plane
Equator
great circle

Diagram of a great circle described by a plane through the center of the earth.

great soil group [GEOL] A group of soils having common internal soil characteristics; a subdivision of a soil order.

great tropic range [OCEANOGR] The difference in height between tropic higher high water and tropic lower low water.

great year [ASTRON] The period of one complete cycle of the equinoxes around the ecliptic, about 25,800 years. Also known as platonic year.

grebe [VERT ZOO] The common name for members of the family Podicipedidae; these birds have legs set far posteriorly, compressed bladelike tarsi, individually broadened and lobed toes, and a rudimentary tail.

greco [METEOROL] An Italian name for the northeast wind.

Greco-Latin square [STAT] An arrangement of combinations of two sets of letters (one set Greek, the other Roman) in a square array, in such a way that no letter occurs more than once in the array.

Greeffiellidae [INV ZOO] A superfamily of free-living nematodes in the order Desmoscolecoidea.

green [MET] Pertaining to an unsintered powder. [OPTICS] The hue evoked in an average observer by monochromatic radiation having a wavelength in the approximate range from 492 to 577 nanometers; however, the same sensation can be produced in a variety of other ways.

green algae [BOT] The common name for members of the plant division Chlorophyta.

green belt *See* frostless zone.

Greenburg-Smith impinger [MIN ENG] A dust-sampling apparatus based on the principle of impingement of the dust-carrying air at high velocity upon a wetted glass surface; also involves bubbling the air through a liquid medium.

green copperas *See* ferrous sulfate.

green flash [ASTRON] A brilliant green coloration of the upper limb of the sun occasionally observed just as the sun's apparent disk is about to sink below a distant clear horizon. Also known as blue flash; blue-green flame; green segment; green sun.

green gland *See* antennal gland.

green glass [MATER] Glass given a blue-green hue by substituting cupric oxide for the chromium compound used in ordinary glass.

green gold [MET] A greenish alloy of gold obtained by using silver, silver and cadmium, or silver and copper as the alloying metal.

greenhouse [BOT] Glass-enclosed, climate-controlled structure in which young or out-of-season plants are cultivated and protected.

greenhouse effect [METEOROL] The effect of the earth's atmosphere in trapping heat from the sun; the atmosphere acts like a greenhouse.

Greenland anticyclone [METEOROL] The glacial anticyclone which is supposed to overlie Greenland; analogous to the Antarctic anticyclone.

Greenland spar *See* cryolite.

green laser [OPTICS] A gas laser using mercury and argon to generate a green line at 5225 angstroms, corresponding to the wavelength that is most readily transmitted through seawater.

green lead ore *See* pyromorphite.

green lumber [MATER] Freshly sawed lumber, before drying.

green manure [AGR] Herbaceous plant material plowed into the soil while still green.

green mold [MYCOL] Any fungus, especially *Penicillium* and *Aspergillus* species, that is green or has green spores.

green mud [GEOL] 1. A fine-grained, greenish terrigenous mud or oceanic ooze found near the edge of a continental shelf at depths of 300–7500 feet (90–2300 meters). 2. A deep-sea terrigenous deposit characterized by the presence of a considerable proportion of glauconite and calcium carbonate.

green nickel oxide *See* nickel oxide.

green oats [AGR] Oat crops before separating and grading.

greenockite [MINERAL] CdS A green or orange mineral that crystallizes in the hexagonal system; occurs as an earthy encrustation and is dimorphous with hawleyite. Also known as cadmium blende; cadmium ocher; xanthochroite.

green oil [MATER] In the Scottish shale-oil industry, once-run crude shale oil.

green roof [MIN ENG] A mine roof which has not broken down or which shows no sign of taking weight.

green rosette [PL PATH] A virus disease of the peanut characterized by bunching and yellowing of the leaves with severe stunting of the plant.

green rot [PL PATH] A decay of fallen deciduous trees in which the wood is colored a malachite green by the fungus *Peziza aeruginosa*.

green salt *See* uranium tetrafluoride.

greensand [GEOL] A greenish sand consisting principally of grains of glauconite and found between the low-water mark and the inner mud line. [PETR] Sandstone composed of greensand with little or no cement.

greenschist [PETR] A schistose metamorphic rock with abundant chlorite, epidote, or actinolite present, giving it a green color.

greenschist facies [PETR] Any schistose rock containing an abundance of green minerals and produced under conditions of low to intermediate temperatures (300–500°C) and low to moderate hydrostatic pressures (3000–8000 bars).

Green's dyadic [MATH] A vector operator which plays a role analogous to a Green's function in a partial differential equation expressed in terms of vectors.

green segment *See* green flash.

Green's function [MATH] A function, associated with a given boundary value problem, which appears as an integrand for an integral representation of the solution to the problem.

Green's identities [MATH] Formulas, obtained from Green's theorem, which relate the volume integral of a function and its gradient to a surface integral of the function and its partial derivatives.

green sky [METEOROL] A greenish tinge to part of the sky, supposed by seamen to herald wind or rain, or in some cases, a tropical cyclone.

green smut [PL PATH] A fungus disease of rice characterized by enlarged grains covered with a green powder consisting of conidia, and caused by *Ustilaginoidea virens*. Also known as false smut.

green snow [HYD] A snow surface that has attained a greenish tint as a result of the growth within it of certain microscopic algae.

Green's theorem [MATH] Determines when the integral along a line of the sum of functions $P(x,y)$ and $Q(x,y)$ is equal to a surface integral of the partial derivatives of P and Q.

greenstick fracture [MED] An incomplete fracture of a long bone, seen in children; the bone is bent but splintered only on the convex side.

greenstone [MINERAL] *See* nephrite. [PETR] Any altered basic igneous rock which is green due to the presence of chlorite, hornblende, or epidote.

greenstone belts [GEOL] Oceanic and island arclike sequences that are similar to, and run to the south and north of, the Swaziland System.

greenstone schist [PETR] Greenstone with a foliated structure.

green strength [MET] The mechanical strength which a compacted powder must have in order to withstand mechanical operations to which it is subjected after pressing and before sintering, without damaging its fine details and sharp edges.

green sulfur bacteria [MICROBIO] A physiologic group of green photosynthetic bacteria of the Chloraceae that are capable of using H_2S and other inorganic electron donors.

green sun *See* green flash.

green vitriol *See* ferrous sulfate.

greenware [MATER] Ceramic ware that has not yet been fired.

Greenwell formula [MIN ENG] A formula used for calculating the thickness of tubing and involving the required thickness of tubing in feet, the vertical depth in feet, the diameter of the shaft in feet, and an allowance for possible flaws or corrosion.

Greenwich apparent noon [ASTRON] Local apparent noon at the Greenwich meridian; twelve o'clock Greenwich apparent time, or the instant the apparent sun is over the upper branch of the Greenwich meridian.

Greenwich apparent time [ASTRON] Local apparent time at the Greenwich meridian. Abbreviated GAT.

Greenwich civil time *See* Greenwich mean time.

Greenwich hour angle [ASTRON] Angular distance west of the Greenwich celestial meridian; the arc of the celestial equator, or the angle at the celestial pole, between the upper branch of the Greenwich celestial meridian and the hour circle of a point on the celestial sphere, measured westward from the Greenwich celestial meridian through 360°. Abbreviated GHA.

Greenwich interval [ASTRON] An interval based on the moon's transit of the Greenwich celestial meridian, as distinguished from a local interval based on the moon's transit of the local celestial meridian.

Greenwich lunar time [ASTRON] Local lunar time at the Greenwich meridian; the arc of the celestial equator, or the angle at the celestial pole, between the lower branch of the Greenwich celestial meridian and the hour circle of the moon, measured westward from the lower branch of the Greenwich celestial meridian through 24 hours.

Greenwich mean noon [ASTRON] Local mean noon at the Greenwich meridian, twelve o'clock Greenwich mean time, or the instant the mean sun is over the upper branch of the Greenwich meridian.

Greenwich mean time [ASTRON] Mean solar time at the meridian of Greenwich. Abbreviated GMT. Also known as Greenwich civil time; universal time; Z time; zulu time.

Greenwich meridian [GEOD] The meridian passing through Greenwich, England, and serving as the reference for Greenwich time; it also serves as the origin of measurement of longitude.

Greenwich sidereal time [ASTRON] Local sidereal time at the Greenwich meridian. Abbreviated GST.

gregale [METEOROL] The Maltese and best-known variant of a term for a strong northeast wind in the central and western Mediterranean and adjacent European land areas; it occurs either with high pressure over central Europe or the Balkans and low pressure over Libya, when it may continue for up to 5 days, or with the passage of a depression to the south or southeast, when it lasts only 1 or 2 days; it is most frequent in winter.

Gregarinia [INV ZOO] A subclass of the protozoan class Telosporea occurring principally as extracellular parasites of invertebrates.

Gregorian calendar [ASTRON] The calendar used for civil purposes throughout the world, replacing the Julian calendar and closely adjusted to the tropical year.

Gregorian telescope [OPTICS] A reflecting telescope having a paraboloidal mirror with a hole in the center and a small secondary (concave ellipsoidal) mirror placed beyond the focus of the primary mirror; light is reflected to the secondary mirror and back to an eyepiece at the hole; the telescope produces an erect image but a small field of view.

Gregory formula [MATH] A formula used in the numerical evaluation of integrals derived from the Newton formula.

greisen [PETR] A pneumatolytically altered granite consisting of mainly quartz and a light-green mica.

grenade [ORD] A small explosive or chemical missile, originally designed to be thrown by hand, but now also designed to be projected from special grenade launchers, usually fitted to rifles or carbines.

grenade cartridge extractor [ORD] An extractor designed for pulling an empty cartridge case or unfired cartridge out of the chamber of a grenade launcher.

grenade launcher [ORD] A device designed to hold a grenade in position for firing but does not itself propel the missile.

grenade net [ORD] A net of chicken wire or some similar material placed over a trench, for example, as a protection against grenades.

grenade pit [ORD] A pit, usually at the bottom of an incline, to catch hand grenades and limit the effects of their explosion.

grenadine [TEXT] A gauzelike fabric made on a jacquard loom.

grenatite *See* leucite; staurolite.

Grenville orogeny [GEOL] A Precambrian mountain-forming epoch.

grenz ray [NUCLEO] An x-ray produced at the long-wavelength end of the x-ray spectrum, involving wavelengths of the order of 1 to 10 angstroms, by using special x-ray tubes that operate at voltages from only 5000 to 15,000 volts.

gressorial [VERT ZOO] Adapted for walking, as certain birds' feet.

grid [ADP] In optical character recognition, a system of two groups of parallel lines, perpendicular to each other, used to measure or specify character images. [DES ENG] A network of equally spaced lines forming squares, used for determining permissible locations of holes on a printed circuit board or a chassis. [ELEC] **1.** A metal plate with holes or ridges, used in a storage cell or battery as a conductor and a support for the active material. **2.** Any systematic network, such as of telephone lines or power lines. [ELECTR] An electrode located between the cathode and anode of an electron tube, which has one or more openings through which electrons or ions can pass, and serves to control the flow of electrons from cathode to anode. [MAP] A system of uniformly spaced perpendicular lines and horizontal lines running north and south, and east and west on a map, chart, or aerial photograph; used in locating points. [MED] *See* Potter-Bucky grid. [MIN ENG] Imaginary line used to divide the surface of an area when following a checkerboard placement of boreholes.

grid amplitude [NAV] Amplitude relative to grid east or west.

grid-anode transconductance *See* transconductance.

grid azimuth [NAV] In grid navigation, the angle in the plane of projection measured clockwise between a straight line and the central meridian of a plane-rectangular coordinate system.

grid battery *See* C battery.

grid bearing [NAV] In grid navigation, the angle in the plane of the projection between a line and a north-south grid line.

grid bias [ELECTR] The direct-current voltage applied between the control grid and cathode of an electron tube to establish the desired operating point. Also known as bias; C bias; direct grid bias.

grid-bias cell *See* bias cell.

grid blocking [ELECTR] **1.** Method of keying a circuit by applying negative grid bias several times cutoff value to the grid of a tube during key-up conditions; when the key is down, the blocking bias is removed and normal current flows through the keyed circuit. **2.** Blocking of capacitance-coupled stages in an amplifier caused by the accumulation of charge on the coupling capacitors due to grid current passed during the reception of excessive signals.

grid blocking capacitor *See* grid capacitor.

grid cap [ELECTR] A top-cap terminal for the control grid of an electron tube.

grid capacitor [ELECTR] A small capacitor used in the grid circuit of an electron tube to pass signal current while blocking the direct-current anode voltage of the preceding stage. Also known as grid blocking capacitor; grid condenser.

grid cathode capacitance [ELECTR] Capacitance between the grid and the cathode in a vacuum tube.

grid characteristic [ELECTR] Relationship of grid current to grid voltage of a vacuum tube.

grid circuit [ELECTR] The circuit connected between the grid and cathode of an electron tube.

grid condenser *See* grid capacitor.

grid conductance *See* electrode conductance.

grid control [ELECTR] Control of anode current of an electron tube by variation (control) of the control grid potential with respect to the cathode of the tube.

grid-controlled mercury-arc rectifier [ELECTR] A mercury-arc rectifier in which one or more electrodes are employed exclusively to control the starting of the discharge. Also known as grid-controlled rectifier.

grid-controlled rectifier *See* grid-controlled mercury-arc rectifier.

grid control tube [ELECTR] Mercury-vapor-filled thermionic vacuum tube with an external grid control.

grid convergence *See* grid declination.

grid course [NAV] In grid navigation, the horizontal direction expressed as the angular distance from grid north.

grid current [ELECTR] Electron flow to a positive grid in an electron tube.

grid declination [NAV] In grid navigation, the angular difference in direction between grid north and true north; it is measured east or west from true north. Also known as declination of grid north; grid convergence.

grid-dip meter [ELECTR] A multiple-range electron-tube oscillator incorporating a meter in the grid circuit to indicate grid current; the meter reading dips (reads lower grid current) when an external resonant circuit is tuned to the oscillator frequency. Also known as grid-dip oscillator.

grid-dip oscillator   *See* grid-dip meter.

grid direction   [NAV] Horizontal direction expressed as angular distance from grid north, measured from grid north, clockwise through 360°.

grid drive   [ELECTR] A signal applied to the grid of a transmitting tube.

grid driving power   [ELECTR] Average product of the instantaneous value of the grid current and of the alternating component of the grid voltage over a complete cycle; this comprises the power supplied to the biasing device and to the grid.

grid element   [ELEC] A sinuous resistor used to heat a furnace, made of heavy wire, strap, or casting and suspended from refractory or stainless supports built into the furnace walls, floor, and roof.

grid equator   [NAV] A line perpendicular to a prime meridian, at the origin; for the usual orientation in polar regions, the grid equator is the 90°W–90°E meridian forming the basic grid parallel, from which grid latitude is measured.

grid-glow tube   [ELECTR] A glow-discharge tube in which one or more control electrodes initiate but do not limit the anode current except under certain operating conditions.

grid heading   [NAV] In grid navigation, the heading relative to grid north.

gridiron pendulum   [HOROL] A compensation pendulum in which the unequal thermal expansion of two metals is utilized to keep the length constant.

gridistor   [ELECTR] Field-effect transistor which uses the principle of centripetal striction and has a multichannel structure, combining advantages of both field effect transistors and minority carrier injection transistors.

grid latitude   [NAV] In grid navigation, the angular distance from a grid equator.

grid leak   [ELECTR] A resistor used in the grid circuit of an electron tube to provide a discharge path for the grid capacitor and for charges built up on the control grid.

grid-leak detector   [ELECTR] A detector in which the desired audio-frequency voltage is developed across a grid leak and grid capacitor by the flow of modulated radio-frequency current; the circuit provides square-law detection on weak signals and linear detection on strong signals, along with amplification of the af signal.

grid limiter   [ELECTR] Limiter circuit which operates by limiting positive grid voltages by means of a large ohmic value resistor; as the exciting signal moves in a positive direction with respect to the cathode, current through the resistor causes an *IR* drop which holds the grid voltage essentially at cathode potential; during negative excursions no current flows in the grid circuit, so no voltage drop occurs across the resistor.

grid locking   [ELECTR] Defect of tube operation in which the grid potential becomes continuously positive due to excessive grid emission.

grid longitude   [NAV] In grid navigation, the angular distance between a prime grid and any designated grid meridian.

grid magnetic angle   [NAV] In grid navigation, the angular difference in direction between grid north and magnetic north; it is measured east or west from grid north.

grid meridian   [NAV] One of the grid lines extending in a grid north-south direction.

grid metal   [MET] An alloy of lead with 5–12% antimony and sometimes with a smaller amount of tin; used for grids in storage batteries.

grid modulation   [ELECTR] Modulation produced by feeding the modulating signal to the control-grid circuit of any electron tube in which the carrier is present.

grid navigation   [NAV] A navigation system related to a reference grid instead of the true north for the measurement of angles.

grid nephoscope   [ENG] A nephoscope constructed of a grid work of bars mounted horizontally on the end of a vertical column and rotating freely about the vertical axis; the observer rotates the grid and adjusts the position until some feature of the cloud appears to move along the major axis of

the grid; the azimuth angle at which the grid is set is taken as the direction of the cloud motion.

grid neutralization   [ELECTR] Method of amplifier neutralization in which a portion of the grid-cathode alternating-current voltage is shifted 180° and applied to the plate-cathode circuit through a neutralizing capacitor.

grid north   [NAV] In grid navigation, the northerly or zero direction indicated by the grid datum of directional reference.

grid parallel   [NAV] A line parallel to a grid equator, connecting all points of equal grid latitude.

grid-plate capacitance   [ELECTR] Direct capacitance between the grid and the plate in a vacuum tube.

grid-plate transconductance   *See* transconductance.

grid-pool tube   [ELECTR] An electron tube having a mercury-pool cathode, one or more anodes, and a control electrode or grid that controls the start of current flow in each cycle; the excitron and ignitron are examples.

grid prime vertical   [NAV] The vertical circle through the grid east and west points of the horizon.

grid pulse modulation   [ELECTR] Modulation produced in an amplifier or oscillator by applying one or more pulses to a grid circuit.

grid pulsing   [ELECTR] Circuit arrangement of a radio-frequency oscillator in which the grid of the oscillator is biased so negatively that no oscillation takes place even when full plate voltage is applied; pulsing is accomplished by removing this negative bias through the application of a positive pulse on the grid.

grid-rectification meter   [ENG] A type of vacuum-tube voltmeter in which the grid and cathode of a tube act as a diode rectifier, and the rectified grid voltage, amplified by the tube, operates a meter in the plate circuit.

grid resistor   [ELECTR] A general term used to denote any resistor in the grid circuit.

grid return   [ELECTR] External conducting path for the return grid current to the cathode.

grid rhumb line   [NAV] A line making the same oblique angle with all grid meridians; grid parallels and meridians may be considered special cases of the grid rhumb line.

grid spectrometer   [SPECT] A grating spectrometer in which a large increase in light flux without loss of resolution is achieved by replacing entrance and exit slits with grids consisting of opaque and transparent areas, patterned to have large transmittance only when the entrance grid image coincides with that of the exit grid.

grid suppressor   [ELECTR] Resistor of low ohmic value inserted in the grid circuit of a radio-frequency amplifier to prevent low-frequency parasitic oscillations.

grid swing   [ELECTR] Total variation in grid-cathode voltage from the positive peak to the negative peak of the applied signal voltage.

grid track   [NAV] The direction of the track relative to grid north.

grid transformer   [ELECTR] Transformer to supply an alternating voltage to a grid circuit or circuits.

grid-type level detector   [ELECTR] A detector using a vacuum tube with input applied to a grid.

grid variation   [NAV] In grid navigation, the angular difference between grid north and magnetic north.

grid voltage   [ELECTR] The voltage between a grid and the cathode of an electron tube.

Griebe-Schiebe method   [SOLID STATE] A method of observing the piezoelectric behavior of small crystals, in which the crystals are placed between two electrodes connected to the resonant circuit of an oscillator, and tuning of the resonant circuit results in jumps in the oscillator frequency which produce clicks in headphones or a loudspeaker attached to the plate circuit of the oscillator.

Griebhard's rings   [ELEC] A method of producing lines of constant color on a copper sheet, coinciding with the equipotential lines of an electric field.

grief stem   *See* kelly.

Griess reagent   [ANALY CHEM] A reagent used to test for nitrous acid; it is a solution of sulfanilic acid, α-naphthylamine and acetic acid in water.

Griffith crack   [MET] Any small flaw in a metal theorized as creating a low order of fracture strength.

griffithite [MINERAL] A micalike mineral containing magnesium, iron, calcium, and aluminosilicate.

Griffiths method [THERMO] A method of measuring the mechanical equivalent of heat in which the temperature rise of a known mass of water is compared with the electrical energy needed to produce this rise.

Griffith's white *See* lithopone.

Grifulvin [MICROBIO] A trade name for griseofulvin.

Grignard reaction [ORG CHEM] A reaction between an alkyl or aryl halide and magnesium metal in a suitable solvent, usually absolute ether, to form an organometallic halide.

Grignard reagent [ORG CHEM] RMgX The organometallic halide formed in the Grignard reaction; an example is C_2H_5MgCl; it is useful in organic synthesis.

Grignard synthesis [ORG CHEM] Use of the Grignard reagent in any one of a vast number of reactions, usually condensations; typical syntheses involve formation of a hydrocarbon, acid, ketone, or secondary or tertiary alcohol.

grillage [CIV ENG] A footing that consists of two or more tiers of closely spaced structural steel beams resting on a concrete block, each tier being at right angles to the one below.

grille [ENG ACOUS] An arrangement of wood, metal, or plastic bars placed across the front of a loudspeaker in a cabinet for decorative and protective purposes.

grille cloth [ENG ACOUS] A loosely woven cloth stretched across the front of a loudspeaker to keep out dust and provide protection without appreciably impeding sound waves.

Grimmiales [BOT] An order of mosses commonly growing upon rock in dense tufts or cushions and having hygroscopic, costate, usually lanceolate leaves arranged in many rows on the stem.

grindability [MATER] Relative ease with which a material can be ground.

grindability index [MATER] A numerical indication of the capacity of a material to be ground.

grinder [MECH ENG] Any device or machine that grinds, such as a pulverizer or a grinding wheel.

grinding [ELECTR] 1. A mechanical operation performed on silicon substrates of semiconductors to provide a smooth surface for epitaxial deposition or diffusion of impurities. 2. A mechanical operation performed on quartz crystals to alter their physical size and hence their resonant frequencies. [MECH ENG] 1. Reducing a material to relatively small particles. 2. Removing material from a workpiece with a grinding wheel. [MIN ENG] The act or process of continuing to drill after the bit or core barrel is blocked, thereby crushing and destroying any core that might have been produced.

grinding aid [ENG] An additive to the charge in a ball mill or rod mill to accelerate the grinding process.

grinding burn [MECH ENG] Overheating a localized area of the work in grinding operations.

grinding cracks [MATER] Cracks in a workpiece resulting from grinding.

grinding fluid [MATER] Cutting fluid used in metal-grinding operations.

grinding medium [ENG] Any material including balls and rods, used in a grinding mill.

grinding mill [LAP] A lathe designed for lapidary work. [MECH ENG] A machine consisting of a rotating cylindrical drum, that reduces the size of particles of ore or other materials fed into it; three main types are ball, rod, and tube mills.

grinding pebbles [ENG] Pebbles, of chert or quartz, used for grinding in mills, where contamination with iron has to be avoided.

grinding ratio [MECH ENG] Ratio of the volume of ground material removed from the workpiece to the volume removed from the grinding wheel.

grinding relief [MET] A groove at the edge of a metal surface which permits overhang of the corner of the grinding wheel.

grinding sensitivity [MATER] Susceptibility of a material to the formation of grinding cracks.

grinding stress [MECH] Residual tensile or compressive stress, or a combination of both, on the surface of a material due to grinding.

grinding-type resin [ORG CHEM] Vinyl or other resin that

requires grinding before dispersal into plastisols or organosols.

grinding wheel [DES ENG] A wheel or disk having an abrasive material such as alumina or silicon carbide bonded to the surface.

grindle *See* bowfin.

grindstone [ENG] A stone disk on a revolving axle, used for grinding, smoothing, and shaping.

grip [ORD] One of a pair of wooden or plastic items designed to be attached by threaded fasteners to the two sides of the frame of a weapon, such as a revolver or bayonet; it is shaped to fit the hand and to provide a formed surface to hold the weapon.

griphite [MINERAL] $(Na,Al,Ca,Fe)_6Mn_4(PO_4)_5(OH)_4$ Mineral composed of a basic phosphate of sodium, calcium, iron, aluminum, and manganese.

grip safety [ORD] Safety mechanism that prevents a gun from being fired unless the stock is firmly grasped while the trigger is pulled; used mainly on automatic pistols.

griquaite [PETR] A hypabyssal rock that contains garnet and diopside, and sometimes olivine or phlogopite, and is found in kimberlite pipes and dikes.

Grisactin [MICROBIO] A trade name for griseofulvin.

grisein [MICROBIO] $C_{40}H_{61}O_{20}N_{10}SFe$ A red, crystalline, water-soluble antibiotic produced by strains of *Streptomyces griseus*.

griseofulvin [MICROBIO] $C_{17}H_{17}O_6Cl$ A colorless, crystalline antifungal antibiotic produced by several species of *Penicillium*. Also known as curling factor.

griseolutein [MICROBIO] Either of two fractions, A or B, of broad-spectrum antibiotics produced by *Streptomyces griseoluteus*; more active against gram-positive than gram-negative microorganisms.

griseomycin [MICROBIO] A white, crystalline antibiotic produced by an actinomycete resembling *Streptomyces griseolus*.

grissaille [GRAPHICS] 1. A technique of painting to imitate a bas-relief and done in shades of gray. 2. All methods of painting in which full modeling is done in black and white or other contrasting tones, and then finished by the application of transparent glazes.

grist [AGR] Grain for grinding or the products and by-products of grinding. [FOOD ENG] Ground malt for brewing.

gristmill [FOOD ENG] A mill that grinds grains.

grit [GEOL] 1. A hard, sharp granule, as of sand. 2. A coarse sand. [MATER] An abrasive material composed of angular grains. [PETR] A sandstone composed of angular grains of different sizes.

grit blasting [ENG] Surface treatment in which steel grit, sand, or other abrasive material is blown against an object to produce a roughened surface or to remove dirt, rust, and scale. Also known as sandblasting.

grit chamber [CIV ENG] A chamber designed to remove sand, gravel, or other heavy solids that have subsiding velocities or specific gravities substantially greater than those of the organic solids in waste water.

grit size [DES ENG] Size of the abrasive particles on a grinding wheel.

grizzly [ENG] 1. A coarse screen used for rough sizing and separation of ore, gravel, or soil. 2. A grating to protect chutes, manways, in mines, or to prevent debris from entering a water inlet.

grizzly bear [VERT ZOO] The common name for a number of species of large carnivorous mammals in the genus *Ursus*, family Ursidae.

grizzly chute [MIN ENG] A chute equipped with grizzlies which separate fine from coarse material as it passes through the chute.

grizzly crusher [MECH ENG] A machine with a series of parallel rods or bars for crushing rock and sorting particles by size.

grizzly worker [MIN ENG] In metal mining, a laborer who works underground at a grizzly.

Groeberiidae [PALEON] A family of extinct rodentlike marsupials.

grog [FOOD ENG] Liquor diluted with water and served hot. [MATER] Fired refractory material that is used in the manufacture of products which must withstand extreme heat.

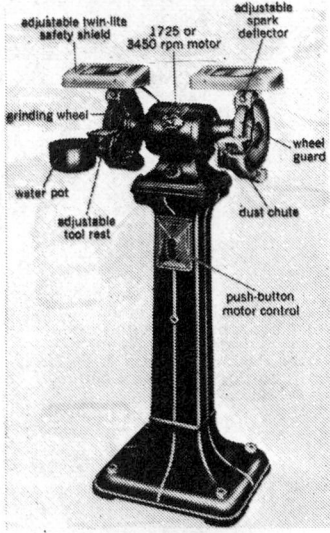

GRINDING WHEEL

Power-driven grinding wheels used on handheld workpieces. (*Rockwell International Corp.*)

GRISEOFULVIN

Structural formula for griseofulvin.

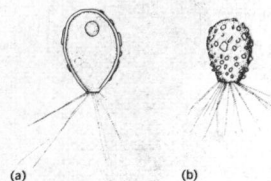

GROMIIDA

Representative Gromiida.
(a) Plagiophrys parvipunctata.
(b) Pseudodifflugia fluva
(after Penard). (From R. P. Hall,
Protozoology, Prentice-Hall, 1953)

groin [ANAT] Depression between the abdomen and the thigh. [ARCH] The projecting edge at the intersection of two vaults. [CIV ENG] A barrier built out from a seashore or riverbank to protect the land from erosion and sand movements, among other functions. Also known as groyne; jetty; spur dike; wing dam.

Gromiida [INV ZOO] An order of protozoans in the subclass Filosia; the test, which is chitinous in some species and thin and somewhat flexible in others, is reinforced with sand grains or siliceous particles.

grommet [ENG] **1.** A metal washer or eyelet. **2.** A piece of fiber soaked in a packing material and used under bolt and nut heads to preserve tightness. [ORD] Device made of rope, plastic, rubber, or metal to protect the rotating band of projectiles.

grommet nut [DES ENG] A blind nut with a round head; used with a screw to attach a hinge to a door.

groove [DES ENG] A long, narrow channel in a surface. [GEOL] Glaciated marks of large size on rock. [ORD] One of the spiral depressions in the rifling of a gun to impart a spinning motion to a projectile which stabilizes it in flight.

groove casts [GEOL] Rounded or sharp, crested, rectilinear ridges that are a few millimeters high and a few centimeters long; found on the undersurfaces of sandstone layers lying on mudstone.

grooved drum [DES ENG] Drum with a grooved surface to support and guide a rope.

groove diameter [ORD] One of the two diameters of a rifled barrel; it is measured between two grooves that are diametrically opposed.

groove face [MIN ENG] The portion of a surface of a member that is included in a groove.

groove sample [MIN ENG] A sample of coal or ore obtained by cutting appropriate grooves along or across the road exposures. Also known as channel sample.

groove weld [MET] A weld in the groove between a pair of members.

grooving saw [MECH ENG] A circular saw for cutting grooves.

grosgrain [TEXT] A ribbed material, usually with a dull finish, made by using cotton in the filling.

Grosh's law [ADP] The law that the processing power of a computer is proportional to the square of its cost.

gross anatomy [ANAT] Anatomy that deals with the naked-eye appearance of tissues.

gross area [BUILD] Sum of the areas of all stories included within the outside face of the exterior walls of a building.

gross-austausch [METEOROL] The exchange of air mass properties and the associated momentum and energy transports produced on a worldwide scale by the migratory large-scale disturbances of middle latitudes.

Grossberg unit [BIOL] A unit for the standardization of lipase.

gross errors [STAT] Errors that occur when a measurement process is subject occasionally to being very far off.

gross index [ADP] The first of two indexes consulted to gain access to a record.

gross information content [COMMUN] Measure of the total information, redundant or otherwise, contained in a message; expressed as the number of bits, nits, or Hartleys required to transmit the message with specified accuracy over a noiseless medium without coding.

gross porosity [MET] Gas pockets or pores of undesirable size and quantity in a casting or a weld metal.

gross recoverable value [MIN ENG] The part of the total metal recovered from an ore multiplied by the price.

gross stress reaction [PSYCH] A transient personality disorder in which, under conditions of great or unusual stress, a normal person utilizes neurotic mechanisms to deal with danger.

gross ton *See* ton.

gross tonnage [NAV ARCH] The total volume of the interior of a ship, measured in tons (units of 100 cubic feet).

grossular *See* grossularite.

grossularite [MINERAL] $Ca_3Al_2(SiO_4)_3$ The colorless or green, yellow, brown, or red end member of the garnet group,

GROSSULARITE

Grossularite, a subspecies of
garnet, with formula
$Ca_3Al_2(SiO_4)_3$. *(American Museum*
of Natural History Specimen)

often occurring in contact-metamorphosis impure limestones. Also known as gooseberry stone; grossular.

gross unit value [MIN ENG] The weight of metal per long or short ton as determined by assay or analysis, multiplied by the market price of the metal.

gross weight [IND ENG] The weight of a vehicle or container when it is loaded with goods. Abbreviated gr wt.

grothite *See* sphene.

Grotthus' chain theory [PHYS CHEM] An early theory used to explain the conductivity of an electrolyte, in which it was assumed that the cathode and anode attract hydrogen and oxygen respectively, and the molecules of the electrolyte are stretched out in chains between the electrodes, with decomposition occurring in molecules closest to the electrodes.

ground [AERO ENG] To forbid (an aircraft or individual) to fly, usually for a relatively short time. [ELEC] **1.** A conducting path, intentional or accidental, between an electric circuit or equipment and the earth, or some conducting body serving in place of the earth. Abbreviated gnd. Also known as earth (British usage); earth connection. **2.** To connect electrical equipment to the earth or to some conducting body which serves in place of the earth. [GEOL] **1.** Any rock or rock material. **2.** A mineralized deposit. **3.** Rock in which a mineral deposit occurs. [NAV] To touch bottom or run aground; in a serious grounding, the vessel is said to strand.

ground absorption [ELECTROMAG] Loss of energy in transmission of radio waves, due to dissipation in the ground.

ground-air [ORD] **1.** Of missiles, signals, and so on, from the ground to the air. **2.** Of a military force, made up of army or marine and air force units. **3.** Of actions, conducted by both ground and air units.

ground area [BUILD] The area of a building at ground level.

ground-based navigation aid [NAV] That portion of a navigation system which is located on the ground and emits signals or receives them from crafts or vehicles; these signals provide the navigation information.

ground cable [ELEC] A heavy cable connected to earth for the purpose of grounding electric equipment.

ground chain [NAV ARCH] A chain that keeps the anchor, when it is being pulled up from the ocean mooring, away from the side of the ship; it is attached along the first length of the anchor chain.

ground check [ENG] **1.** A procedure followed prior to the release of a radiosonde in order to obtain the temperature and humidity corrections for the radiosonde system. **2.** Any instrumental check prior to the ground launch of an airborne experiment. Also known as base-line check.

ground-check chamber [ENG] A chamber that is used to check the sensing elements of radiosonde equipment and that houses sources of heat and water vapor plus instruments for measuring temperature, humidity, and pressure, and in which air circulation is maintained by a motor-driven fan.

ground circuit [ELEC] A telephone or telegraph circuit part of which passes through the ground.

ground clutter [ELECTROMAG] Clutter on a ground or airborne radar due to reflection of signals from the ground or objects on the ground. Also known as ground flutter; ground return; land return; terrain echoes.

ground coal [MIN ENG] The bottom of a coal seam.

ground conductivity [ELEC] The effective conductivity of the ground, used in calculating the attenuation of radio waves.

ground control [CIV ENG] Supervision or direction of all airport surface traffic, except an aircraft landing or taking off.

ground-controlled approach [NAV] Technique or procedures by which a ground controller directs an aircraft to reach a point from which a landing may be made; the controller utilizes information from various sensors, including radar.

ground-controlled approach minimums [NAV] The minimum ceiling or visibility under which an aircraft may be landed with the use of ground-controlled approach (GCA); the pilot must be qualified to use GCA (and the aircraft must be equipped with the necessary radio equipment) before GCA minimums are applicable to the particular flight; at most airports GCA minimums are less than other minimums because of the double check on the "glide path" position of the aircraft.

ground-controlled approach radar [ENG] A ground radar

system providing information by which aircraft approaches may be directed by radio communications. Abbreviated GCA radar.

ground-controlled interception [ORD] In air defense, an interception in which the interceptor weapon is vectored to the target by instructions transmitted from the ground.

ground-controlled intercept radar [ENG] A radar system by means of which a controller may direct an interception of another aircraft. Abbreviated GCI radar.

ground controller [ENG] Aircraft controller stationed on the ground; a generic term, applied to the controller in ground-controlled approach, ground-controlled interception, and so on.

ground cover [BOT] Prostrate or low plants that cover the ground instead of grass. [FOR] All forest plants except trees.

ground current *See* earth current.

ground data equipment [ENG] Any device located on the ground that aids in obtaining space-position or tracking data (including computation function); reads out data telemetry, video, and so on, from payload instrumentation, or is capable of transmitting command and control signals to a satellite or space vehicle.

ground-derived navigation data [NAV] That data used in navigation and obtained by systems or parts of systems located on the ground or sea.

ground detector [ELEC] An instrument or equipment used for indicating the presence of a ground on an ungrounded system. Also known as ground indicator.

ground dielectric constant [ELEC] Dielectric constant of the earth at a given location.

ground discharge *See* cloud-to-ground discharge.

ground distance [NAV] The great-circle distance between two ground positions, as contrasted with slant distance or slant range, the straight-line distance between two points. Also known as ground range.

grounded-anode amplifier *See* cathode follower.

grounded-base amplifier [ELECTR] An amplifier that uses a transistor in a grounded-base connection.

grounded-base connection [ELECTR] A transistor circuit in which the base electrode is common to both the input and output circuits; the base need not be directly connected to circuit ground. Also known as common-base connection.

grounded-cathode amplifier [ELECTR] Electron-tube amplifier with a cathode at ground potential at the operating frequency, with input applied between control grid and ground, and with the output load connected between plate and ground.

grounded-collector connection [ELECTR] A transistor circuit in which the collector electrode is common to both the input and output circuits; the collector need not be directly connected to circuit ground. Also known as common-collector connection.

grounded-emitter amplifier [ELECTR] An amplifier that uses a transistor in a grounded-emitter connection.

grounded-emitter connection [ELECTR] A transistor circuit in which the emitter electrode is common to both the input and output circuits; the emitter need not be directly connected to circuit ground. Also known as common-emitter connection.

grounded-gate amplifier [ELECTR] Amplifier that uses thin-film transistors in which the gate electrode is connected to ground; the input signal is fed to the source electrode and the output is obtained from the drain electrode.

grounded-grid amplifier [ELECTR] An electron-tube amplifier circuit in which the control grid is at ground potential at the operating frequency; the input signal is applied between cathode and ground, and the output load is connected between anode and ground.

grounded-grid-triode circuit [ELECTR] Circuit in which the input signal is applied to the cathode and the output is taken from the plate; the grid is at radio-frequency ground and serves as a screen between the input and output circuits.

grounded-grid-triode mixer [ELECTR] Triode in which the grid forms part of a grounded electrostatic screen between the anode and cathode, and is used as a mixer for centimeter wavelengths.

grounded ice *See* stranded ice.

grounded-plate amplifier *See* cathode follower.

grounded system [ELEC] Any conducting apparatus connected to ground. Also known as earthed system.

ground effect [AERO ENG] Increase in the lift of an aircraft operating close to the ground caused by reaction between high-velocity downwash from its wing or rotor and the ground. [COMMUN] The effect of ground conditions on radio communications.

ground-effect machine *See* air-cushion vehicle.

ground electrode [ELEC] A conductor buried in the ground, used to maintain conductors connected to it at ground potential and dissipate current conducted to it into the earth, or to provide a return path for electric current in a direct-current power transmission system.

ground environment [ENG] **1.** Environment that surrounds and affects a system or piece of equipment that operates on the ground. **2.** System or part of a system, as of a guidance system, that functions on the ground; the aggregate of equipment, conditions, facilities, and personnel that go to make up a system, or part of a system, functioning on the ground. [ORD] In air defense, the portion of the air defense system that provides for the detection, surveillance, and control of airborne objects; normally these are ground-based facilities, but also include overwater facilities, such as picket vessels, and airborne early warning and control aircraft.

ground equalizer inductors [ELECTROMAG] Coils, having relatively low inductance, inserted in the circuit to one or more of the grounding points of an antenna to distribute the current to the various points in any desired manner.

ground fallout plot [NUCLEO] Time lines plotted within a radioactive fallout plot that approximate the distances to which local fallout will have spread at the end of succeeding hours; these lines are determined by using the rate of fall of the radioactive material and the mean wind vector (fallout wind) in the layer of air through which the material is falling.

ground fault [ELEC] Accidental grounding of a conductor.

ground fire [ORD] Fire, such as antiaircraft fire, that originates from the ground.

ground flutter *See* ground clutter.

ground fog [METEOROL] A fog that hides less than 0.6 of the sky and does not extend to the base of any clouds that may lie above it.

ground frost [METEOROL] In British usage, a freezing condition injurious to vegetation, which is considered to have occurred when a minimum thermometer exposed to the sky at a point just above a grass surface records a temperature (grass temperature) of 30.4°F (−0.9°C) or below.

ground glass [OPTICS] A sheet of matte-surfaced glass on the back of a view camera or process camera so that the image of the subject can be focused on it; it is exactly in the film plane.

ground handling equipment *See* ground support equipment.

groundhog *See* barney.

ground ice [HYD] A body of clear ice in frozen ground, most commonly found in more or less permanently frozen ground (permafrost), and may be of sufficient age to be termed fossil ice. Also known as stone ice; subsoil ice; subterranean ice; underground ice.

ground ice mound [GEOL] A frost mound containing bodies of ice. Also known as ice mound.

ground indicator *See* ground detector.

grounding [ELEC] Intentional electrical connection to a reference conducting plane, which may be earth, but which more generally consists of a specific array of interconnected electrical conductors referred to as the grounding conductor.

grounding conductor [ELEC] An array of interconnected electric conductors at a uniform potential, to which electrical connections are made for the purpose of grounding.

grounding electrode [ELEC] Conductor embedded in the earth, used for maintaining ground potential on conductors connected to it, and for dissipating into the earth, current conducted to it.

grounding reactor [ELEC] A reactor sometimes used in a grounded alternating-current system which joins a conductor or neutral point to ground and serves to limit ground current in case of a fault. Also known as earthing reactor (British usage).

grounding receptacle [ELEC] A receptacle which has an

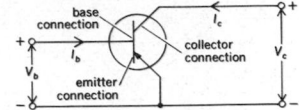

GROUNDED-BASE CONNECTION

Schematic diagram of an alloy junction transistor showing a grounded-base connection.

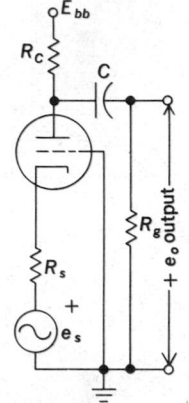

GROUNDED-GRID AMPLIFIER

Schematic diagram of a grounded-grid amplifier.

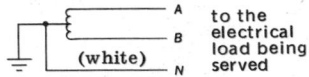

GROUNDING

Grounding used in a single-phase three-wire 240/120-volt service; *A* and *B* conduct the current; *N* is the grounding wire.

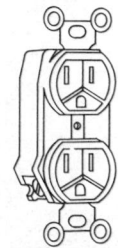

GROUNDING RECEPTACLE

Grounding-type 15-amp receptacle.

extra contact that accepts the third round or U-shaped prong of a grounding attachment plug and is connected internally to a supporting strap, providing a ground both through the outlet box and the grounding conductor, armor, or raceway of the wiring system.

grounding transformer [ELEC] Transformer intended primarily for the purpose of providing a neutral point for grounding purposes.

ground instrumentation *See* spacecraft ground instrumentation.

ground inversion *See* surface inversion.

ground junction *See* grown semiconductor junction.

ground layer *See* surface boundary layer.

ground lead *See* work lead.

ground log [NAV] A device for determining the course and speed made good over the ground in shallow water, consisting of a lead or weight attached to a line and thrown overboard to the bottom; the course being made good is indicated by the direction the line tends, and the speed by the amount of line paid out in unit time.

ground loop [COMMUN] Return currents or magnetic fields from relatively high-powered circuits or components which generate unwanted noisy signals in the common return of relatively low-level signal circuits. [ELEC] Potentially detrimental loop formed when two or more points in an electric system that are nominally at ground potential are connected by a conducting path.

ground magnetic survey [ENG] A determination of the magnetic field at the surface of the earth by means of ground-based instruments.

groundman [ENG] A person employed in digging or excavating. [MIN ENG] *See* mucker.

groundmass *See* matrix.

ground mine [ORD] An underwater mine with negative buoyancy, intended to rest on the bottom of relatively shallow water.

ground moraine [GEOL] Rock material carried and deposited in the base of a glacier. Also known as bottom moraine; subglacial moraine.

ground noise [ENG ACOUS] The residual system noise in the absence of the signal in recording and reproducing; usually caused by inhomogeneity in the recording and reproducing media, but may also include tube noise and noise generated in resistive elements in the amplifier system. [GEOPHYS] In seismic exploration, disturbance of the ground due to some cause other than the shot.

ground outlet [ELEC] Outlet equipped with a receptacle of the polarity type having, in addition to the current-carrying contacts, one grounded contact which can be used for the connection of an equipment-grounding conductor.

ground plane antenna [ELECTROMAG] Vertical antenna combined with a grounded horizontal disk, turnstile element, or similar ground plane simulation; such antennas may be mounted several wavelengths above the ground, and provide a low radiation angle.

ground plate [ELEC] A plate of conducting material embedded in the ground to act as a ground electrode.

ground plot *See* dead-reckoning plot.

ground point of intercept [NAV] In terminal approach procedures, a point in the vertical plane on the runway center line by which it is assumed that the straight-line extension of the slide slope intercepts the runway approach surface base line. Abbreviated as GPI.

ground position [NAV] A position on the spherical or spheroidal earth, particularly such a position vertically below an aircraft.

ground-position indicator [NAV] An airborne computing mechanism that provides a continuous indication of position through data derived from instruments which give heading, airspeed, elapsed time, and drift.

ground potential [ELEC] Zero potential with respect to the ground or earth.

ground pressure [GEOPHYS] The pressure to which a rock formation is subjected by the weight of the superimposed rock and rock material or by diastrophic forces created by movements in the rocks forming the earth's crust. Also known as geostatic pressure; lithostatic pressure; rock pressure.

ground protection [ELEC] Protection provided a circuit by a device which opens the circuit when a fault to ground occurs.

ground range *See* ground distance.

ground recharge [ELEC] The flow of electrons from the ground, in reference to lightning effects.

ground-referenced navigation data [NAV] Navigation data which are referenced to points on the earth's surface.

ground-reflected wave [ELECTROMAG] Component of the ground wave that is reflected from the ground.

ground resistance [ELEC] Opposition of the earth to the flow of current through it; its value depends on the nature and moisture content of the soil, on the material, composition, and nature of connections to the earth, and on the electrolytic action present.

ground return [ELEC] Use of the earth as the return path for a transmission line. [ELECTROMAG] **1.** An echo received from the ground by an airborne radar set. **2.** *See* ground clutter.

ground-return circuit [ELEC] Circuit which has a conductor (or two or more in parallel) between two points and which is completed through the ground or earth.

ground rod [ELEC] A rod that is driven into the earth to provide a good ground connection.

groundscatter propagation [COMMUN] Multihop ionospheric radio propagation along other than the great-circle path between transmitting and receiving stations; radiation from the transmitter is first reflected back to earth from the ionosphere, then scattered in many directions from the earth's surface.

ground sluice [MIN ENG] A channel through which gold-bearing earth is passed in placer mining.

ground speed [NAV] Speed of an aircraft relative to the surface of the earth.

ground state [QUANT MECH] The stationary state of lowest energy of a particle or a system of particles.

ground strafing [ORD] Attack upon ground troops by low-flying aircraft using bombs, machine gun, and cannon.

ground streamer [METEOROL] An upward advancing column of high-ion density which rises from a point on the surface of the earth toward which a stepped leader descends at the start of a lightning discharge.

ground substance *See* matrix.

ground support equipment [AERO ENG] That equipment on the ground, including all implements, tools, and devices (mobile or fixed), required to inspect, test, adjust, calibrate, appraise, gage, measure, repair, overhaul, assemble, disassemble, transport, safeguard, record, store, or otherwise function in support of a rocket, space vehicle, or the like, either in the research and development phase or in an operational phase, or in support of the guidance system used with the missile, vehicle, or the like. Abbreviated GSE. Also known as ground handling equipment.

ground surveillance radar [ENG] A surveillance radar operated at a fixed point on the earth's surface for observation and control of the position of aircraft or other vehicles in the vicinity.

ground swell [OCEANOGR] Swell as it passes through shallow water, characterized by a marked increase in height in water shallower than one-tenth wavelength.

ground system [ELECTROMAG] The portion of an antenna that is closely associated with an extensive conducting surface, which may be the earth itself.

ground tackle [NAV ARCH] The anchoring equipment on a ship.

ground-to-cloud discharge [GEOPHYS] A lightning discharge in which the original streamer processes start upward from an object located on the ground.

ground trace [ENG] The theoretical mark traced upon the surface of the earth by a flying object, missile, or satellite as it passes over the surface, the mark being made vertically from the object making the trace.

ground visibility [METEOROL] In aviation terminology, the horizontal visibility observed at the ground, that is, surface visibility or control-tower visibility.

groundwater [HYD] All subsurface water, especially that part that is in the zone of saturation.

groundwater decrement *See* groundwater discharge.

groundwater depletion curve [HYD] A recession curve of streamflow, so adjusted that the slope of the curve represents the runoff (depletion rate) of the groundwater; it is formed by the observed hydrograph during prolonged periods of no precipitation. Also known as groundwater recession.

groundwater discharge [HYD] **1.** Water released from the zone of saturation. **2.** Release of such water. Also known as decrement; groundwater decrement; phreatic-water discharge.

groundwater flow [HYD] That portion of the precipitation that has been absorbed by the ground and has become part of the groundwater.

groundwater increment *See* recharge.

groundwater level [HYD] **1.** The level below which the rocks and subsoil are full of water. **2.** *See* water table.

groundwater recession *See* groundwater depletion curve.

groundwater recharge *See* recharge.

groundwater replenishment *See* recharge.

groundwater surface *See* water table.

groundwater table *See* water table.

ground wave [GEOPHYS] A radio wave that is propagated along the earth and is ordinarily affected by the presence of the ground and the troposphere; includes all components of a radio wave over the earth except ionospheric and tropospheric waves. Also known as surface wave. [ORD] One of the waves formed in the ground by an explosion.

ground ways [CIV ENG] Supports, usually made of heavy timbers, which are placed on the ground on either side of the keel of a ship under construction, providing a track for launching, and supporting the sliding ways. Also known as standing ways.

ground wire [CIV ENG] A small-gage, high-strength steel wire used to establish line and grade for air-blown mortar or concrete. Also known as alignment wire; screed wire. [ELEC] A conductor used to connect electric equipment to a ground rod or other grounded object.

groundwood pulp *See* mechanical pulp.

ground zero [ORD] The point on the surface of the earth (including water) at which, above which, or below which an atomic detonation has actually occurred. Also known as surface zero.

group [COMMUN] A subdivision containing a number of voice channels, either within a supergroup or separately, normally comprised of up to 12 voice channels occupying the frequency band 60–108 kilohertz; each voice channel may be multiplexed for teletypewriter operation, if required; the number of voice channels which may be simultaneously multiplexed for teletypewriter operation will vary according to equipment design. [MATH] A set G with an associative binary operation where $g_1 \cdot g_2$ always exists and is an element of G; each g has an inverse element g^{-1}, and G contains an identity element.

group A kits [ELECTR] Normally those items of electronic equipment which may be permanently or semipermanently installed in an aircraft for supporting, securing, or interconnecting the components and controls of the equipment, and which will not in any manner compromise the security classification of the equipment.

group B kits [ELECTR] Normally, the operating or operable component of the electronic equipment in an aircraft which, when installed on or in connection with group A parts, constitute the complete operable equipment.

group bus [ELEC] A scheme of electrical connections for a generating station in which more than two feeder lines are supplied by two bus-selector circuit breakers which lead to a main bus and an auxiliary bus.

group busy tone [COMMUN] High tone connected to the jack sleeves of an outgoing trunk group as an indication that all trunks in the group are busy.

group diffusion method [NUCLEO] An approximation used in studying the diffusion of neutrons in a nuclear reactor, in which the range of neutron energies, from source energy to thermal energy, is divided into a finite number of intervals, or groups, and the neutrons in each group are assumed to diffuse with no loss of energy, until they have undergone the average number of collisions needed to reduce their energy to that of the next lower group.

group dynamics [PSYCH] A branch of social psychology which studies problems involving the structure of a group.

grouped-frequency operation [COMMUN] Use of different frequency bands for channels in opposite directions in a two-wire carrier system.

group flashing light [NAV] A light showing groups of flashes at regular intervals, the duration of light being less than that of darkness.

group frequency [ELECTROMAG] Frequency corresponding to group velocity of propagated waves in a transmission line or waveguide.

group incentive [IND ENG] Any wage incentive applied to more than one employee who is engaged in group work characterized by interdependent relationship between operations with consequent physical proximity and unification of interest.

group-indicate [ADP] To print indicative information from only the first record of a group.

grouping [COMMUN] Periodic error in the spacing of recorded lines in a facsimile system.

grouping circuits [COMMUN] Circuits used to interconnect two or more switchboard positions together, so that one operator may handle the several switchboard positions from one operator's set.

grouping of records [ADP] Placing records together in a group to either conserve storage space or reduce access time.

group mark [ADP] A character signaling the beginning or end of a group of data.

group modulation [COMMUN] Process by which a number of channels, already separately modulated to a specific frequency range, are again modulated to shift the group to another range.

group number [OCEANOGR] The first two numbers in the argument number in A. T. Doodson's scheme for predicting tides.

group occulting light [NAV] A light having groups of eclipses at regular intervals; the duration of the light period is equal to or greater than that of the dark period.

group of stars [ASTRON] A number of stars moving in the same direction with the same speed.

groupoid [MATH] A set having a binary relation everywhere defined.

group psychotherapy [PSYCH] Therapy given to a group of people by a therapist relying on the group effect on the individual and his interactions with the group.

group theory [MATH] The study of the structure of groups which especially deals with the classification of finite groups.

group therapy [PSYCH] A technique of psychotherapy, consisting of a group of persons who discuss personal problems under the guidance of a therapist.

group velocity [PHYS] The velocity of the envelope of a group of interfering waves having slightly different frequencies and phase velocities.

grouse [VERT ZOO] Any of a number of game birds in the family Tetraonidae having a plump body and strong, feathered legs.

grouser [ENG] A temporary pile or a heavy, iron-shod pole driven into the bottom of a stream to hold a drilling or dredging boat or other floating object in position. Also known as spud.

grout [MATER] **1.** A fluid mixture of cement and water, or a mixture of cement, sand, and water. **2.** Waste material of all sizes obtained in quarrying stone.

grout curtain [ENG] A row of vertically drilled holes filled with grout under pressure to form the cutoff wall under a dam, or to form a barrier around an excavation through which water cannot seep or flow.

grout hole [ENG] **1.** One of the holes in a grout curtain. **2.** Any hole into which grout is forced under pressure to consolidate the surrounding earth or rock.

grouting [ENG] The act or process of applying grout or of injecting grout into grout holes or crevices of a rock.

grout injector [ENG] A machine that mixes the dry ingredients for a grout with water and injects it, under pressure, into a grout hole.

GROUSE

The ruffed grouse *(Bonasa umbellus)*, a popular game bird of North America.

GROWTH CURVE

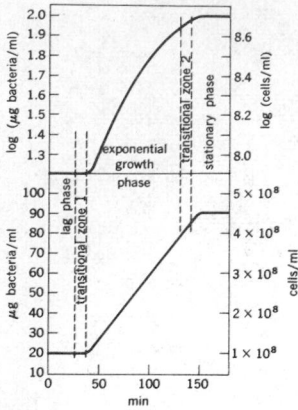

Growth curve of a typical bacterium. Lag phase = 25 minutes, doubling time = 45 minutes, and the specific growth rate = 0.154 min⁻¹.

GROWTH SPIRAL

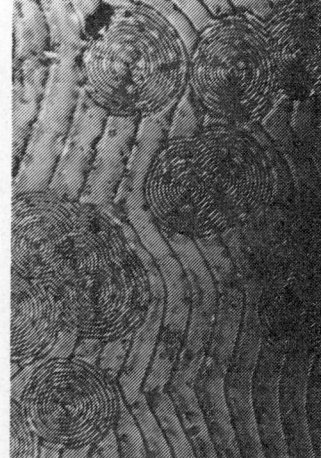

Crystal-growth spirals on the surface of a silicon carbide crystal magnified 250 diameters. Each spiral step originates in a screw dislocation defect. From the side the surface looks like an ascending ramp wound around a flat cone. *(General Electric)*

GRYLLIDAE

Grass cricket *(Nemobius fasciatus)*. *(From Illinois Natural History Survey)*

groutite [MINERAL] $HMnO_2$ A mineral of the diaspore group, composed of manganese, hydrogen, and oxygen; it is polymorphous with manganite.

grout pipe [ENG] A pipe that transports grout under pressure for injection into a grout hole or a rock formation.

grove cell [ELEC] Primary cell, having a platinum electrode in an electrolyte of nitric acid within a porous cup, outside of which is a zinc electrode in an electrolyte of sulfuric acid; it normally operates on a closed circuit.

Grove's synthesis [ORG CHEM] Production of alkyl chlorides by passing hydrochloric acid into an alcohol in the presence of anhydrous zinc chloride.

grower's year [CLIMATOL] In Great Britain, the 12-month period starting November 6, and referring to the cycle of seasonal change of weather.

growing season [AGR] The period of the year when climatic conditions are favorable for plant growth, common to a place or an area.

growl [GEOPHYS] Noise heard when strata are subjected to great pressure.

growler [ELEC] An electromagnetic device consisting essentially of two field poles arranged as in a motor, used for locating short-circuited coils in the armature of a generator or motor and for magnetizing or demagnetizing objects; a growling noise indicates a short-circuited coil. [OCEANOGR] A small piece of floating sea ice, usually a fragment of an iceberg or floeberg; it floats low in the water, and its surface often is heavily pitted; it often appears greenish in color.

grown-diffused transistor [ELECTR] A junction transistor in which the final junctions are formed by diffusion of impurities near a grown junction.

grown junction [ELECTR] A junction produced by changing the types and amounts of donor and acceptor impurities that are added during the growth of a semiconductor crystal from a melt.

grown-junction photocell [ELECTR] A photodiode consisting of a bar of semiconductor material having a *pn* junction at right angles to its length and an ohmic contact at each end of the bar.

grown-junction transistor [ELECTR] A junction transistor in which different impurities are placed in the melt in sequence as the silicon or germanium seed crystal is slowly withdrawn, to produce the alternate *pn* and *np* junctions.

grown semiconductor junction [ELECTR] Junction produced during growth of the crystal from a melt. Also known as ground junction.

growth [MED] Any abnormal, localized increase in cells, such as a tumor. [PHYSIO] Increase in the quantity of metabolically active protoplasm, accompanied by an increase in cell number or cell size, or both.

growth curve [MICROBIO] A graphic representation of the growth of a bacterial population in which the log of the number of bacteria or the actual number of bacteria is plotted against time. [NUCLEO] A curve showing how some quantity associated with a radioactive transformation or induced nuclear reaction increases with time.

growth factor [AERO ENG] The additional weight of fuel and structural material required by the addition of 1 pound of payload to the original payload. [PHYSIO] Any factor, genetic or extrinsic, which affects growth.

growth form [ECOL] The habit of a plant determined by its appearance of branching and periodicity.

growth hormone [BIOCHEM] **1.** A polypeptide hormone secreted by the anterior pituitary which promotes an increase in body size. Abbreviated GH. **2.** Any hormone that regulates growth in plants and animals.

growth rate [MICROBIO] Increase in the number of bacteria in a population per unit time.

growth regulator [BIOCHEM] A synthetic substance that produces the effect of a naturally occurring hormone in stimulating plant growth; an example is dichlorophenoxyacetic acid.

growth spiral [CRYSTAL] A structure on a crystal surface, observed after growth, consisting of a growth step winding downward and outward in an Archimedean spiral which may be distorted by the crystal structure.

growth step [CRYSTAL] A ledge on a crystal surface, one or

more lattice spacings high, where crystal growth can take place.

groyne *See* groin.

GR-S rubber [ORG CHEM] Former designation for general-purpose synthetic rubbers formed by copolymerization of emulsions of styrene and butadiene; used in tires and other rubber products; previously also known as Buna-S, currently known as SBR (styrene-butadiene rubber).

grubbing [CIV ENG] Clearing stumps and roots.

grub screw [DES ENG] A headless screw with a slot at one end to receive a screwdriver.

grubstake [MIN ENG] In the United States, supplies or money furnished to a mining prospector for a share in his discoveries.

grubstake contract [MIN ENG] An agreement between two or more persons to locate mines upon the public domain by their joint aid, effort, labor, or expense, with each to acquire by virtue of the act of location such an interest in the mine as agreed upon in the contract.

gruenlingite [MINERAL] Bi_4TeS_3 A mineral composed of sulfide and telluride of bismuth.

Gruidae [VERT ZOO] The cranes, a family of large, tall, cosmopolitan wading birds in the order Gruiformes.

Gruiformes [VERT ZOO] A heterogeneous order of generally cosmopolitan birds including the rails, coots, limpkins, button quails, sun grebes, and cranes.

Grüneisen constant [SOLID STATE] Three times the bulk modulus of a solid times its linear expansion coefficient, divided by its specific heat per unit volume; it is reasonably constant for most cubic crystals. Also known as Grüneisen gamma.

Grüneisen gamma *See* Grüneisen constant.

Grüneisen relation [SOLID STATE] The relation stating that the electrical resistivity of a very pure metal is proportional to a mathematical function which depends on the ratio of the temperature to a characteristic temperature.

grunerite [MINERAL] $(Mg,Fe)_7Si_8O_{22}(OH)_2$ Variety of amphibole; forms monoclinic crystals.

grus *See* gruss.

Grus [ASTRON] A constellation, right ascension 22 hours, declination 45°S. Also known as Crane.

gruss [GEOL] A loose accumulation of fragmental products formed from the weathering of granite. Also spelled grus.

gr wt *See* gross weight.

Gryllidae [INV ZOO] The true crickets, a family of orthopteran insects in which individuals are dark-colored and chunky with long antennae and long, cylindrical ovipositors.

Grylloblattidae [INV ZOO] A monogeneric family of crickets in the order Orthoptera; members are small, slender, wingless insects with hindlegs not adapted for jumping.

Gryllotalpidae [INV ZOO] A family of North American insects in the order Orthoptera which live in sand or mud; they eat the roots of seedlings growing in moist, light soils.

GSC *See* gas-solid chromatography.

G scan *See* G scope.

G scope [ELECTR] A cathode-ray scope on which a single signal appears as a spot on which wings grow as the distance to the target is decreased, with bearing error as the horizontal coordinate and elevation angle error as the vertical coordinate. Also known as G indicator; G scan.

GSE *See* ground support equipment.

g service [COMMUN] A Federal Aviation Administration service pertaining to aural and visual monitoring of radio aids to air navigation and of the landlines and radio communications systems, to detect faulty operation.

GST *See* Greenwich sidereal time.

G star [ASTRON] A star of spectral type G; many metallic lines are seen in the spectra, with hydrogen and potassium being strong; G stars are yellow stars, with surface temperatures of 5500–4200K for giants, 6000–5000K for dwarfs.

G string *See* field waveguide.

g suit [ENG] A suit that exerts pressure on the abdomen and lower parts of the body to prevent or retard the collection of blood below the chest under positive acceleration. Also known as anti-g suit.

guaiacol [ORG CHEM] $C_6H_4(OH)OCH_3$ A colorless, crystalline compound, soluble in water; used as a reagent to determine the presence of such substances as lignin, narceine, and nitrous acid. Also known as *ortho*-hydroxyanisole; 1-hydroxy-2-methoxybenzene; methylpyrocatechine.

guanajuatite [MINERAL] Bi_2Se_3 Bluish-gray mineral composed of bismuth selenide, occurring in crystals or masses.

guanidine [BIOCHEM] CH_5N_2 Aminomethanamidine, a product of protein metabolism found in urine. Also known as carbamidine; iminourea.

guanidine-acetic acid *See* glycocyamine.

guanine [BIOCHEM] $C_5H_5ON_5$ A purine base; occurs naturally as a fundamental component of nucleic acids.

guano [MATER] Phosphate- and nitrogen-rich, partially decomposed excrement of seabirds; used as a fertilizer.

guanophore *See* iridocyte.

guanosine [BIOCHEM] $C_{10}H_{13}O_5N_5$ Guanine riboside, a nucleoside composed of guanine and ribose. Also known as vernine.

guanylic acid [BIOCHEM] A nucleotide composed of guanine, a pentose sugar, and phosphoric acid and formed during the hydrolysis of nucleic acid.

guar [BOT] *Cyanopsis psoralioides*. A legume cultivated for forage and for its seed.

guard [ENG] A shield or other fixture designed to protect against injury. [MIN ENG] A support in front of a roll train to guide the bar into the groove.

guard ammunition [ORD] Ammunition specifically designed for use by guards, usually containing a reduced propelling charge.

guard arm [ELEC] 1. Crossarm placed across and in line with a cable to prevent damage to the cable. 2. Crossarm located over wires to prevent foreign wires from falling into them.

guard band [ELECTR] A narrow frequency band provided between adjacent channels in certain portions of the radio spectrum to prevent interference between stations.

guard cell [BOT] Either of two specialized cells surrounding each stoma in the epidermis of plants; functions in regulating stoma size.

guard circle [DES ENG] The closed loop at the end of a record groove.

guard-electrode system [PETRO ENG] System of extra electrodes used during electrical logging of reservoir formations to confine the surveying current from the measuring electrode to a generally horizontal path.

guarding [ELEC] A method of eliminating surface-leakage effects from measurements of electrical resistance which employs a low-resistance conductor in the vicinity of one of the terminals or a portion of the measuring circuit.

guard lock [CIV ENG] *See* entrance lock. [ENG] An auxiliary lock that must be opened before the key can be turned in a main lock.

guard magnet [MIN ENG] A magnet employed in a crushing system to remove or arrest tramp iron ahead of the machinery.

guardrail [CIV ENG] 1. A handrail. 2. A rail made of posts and a metal strip used on a road as a divider between lines of traffic in opposite directions or used as a safety barrier on curves. 3. A rail fixed close to the outside of the inner rail on railway curves to hold the inner wheels of a railway car on the rail. Also known as check rail; safety rail; slide rail.

guard relay [ELEC] Used in the linefinder circuit to make sure that only one linefinder can be connected to any line circuit when two or more line relays are operated simultaneously.

guard ring [ELEC] A ring-shaped auxiliary electrode surrounding one of the plates of a parallel-plate capacitor to reduce edge effects. [ELECTR] A ring-shaped auxiliary electrode used in an electron tube or other device to modify the electric field or reduce insulator leakage; in a counter tube or ionization chamber a guard ring may also serve to define the sensitive volume.

guard screen *See* oversize control screen.

guard shield [ELECTR] Internal floating shield that surrounds the entire input section of an amplifier; effective shielding is achieved only when the absolute potential of the guard is stabilized with respect to the incoming signal.

guar gum [MATER] A mucilage formed from seeds of the guar plant; light-gray powder dispersible in water; used as a

thickening agent in paper, foods, pharmaceuticals, and cosmetics.

Guarnieri body [PATH] Eosinophilic cytoplasmic inclusion bodies found in the epidermal cells of patients with smallpox or chickenpox.

guava [BOT] *Psidium guajava*. A shrub or low tree of tropical America belonging to the family Myrtaceae; produces an edible, aromatic, sweet, juicy berry.

guayule [BOT] *Parthenium argentatum*. A subshrub of the family Compositae that is native to Mexico and the southwestern United States; it has been cultivated as a source of rubber.

guba [METEOROL] In New Guinea, a rain squall on the sea.

gubernaculum [ANAT] A guiding structure, as the fibrous cord extending from the fetal testes to the scrotal swellings. [INV ZOO] 1. A posterior flagellum of certain protozoans. 2. A sclerotized structure associated with the copulatory spicules of certain nematodes.

Gudden-Pohl effect [ELECTR] The momentary illumination produced when an electric field is applied to a phosphor previously excited by ultraviolet radiation.

Gudermannian [MATH] The function y of the variable x satisfying $\tan y = \sinh x$ or $\sin y = \tanh x$; written gd$x$.

gudgeon [ENG] 1. A pivot. 2. A pin for fastening stone blocks. [NAV ARCH] Metal fittings on the sternpost of a boat or on the rudderpost of a ship on which the rudder is hung; the gudgeon forms the pivot point.

gudmundite [MINERAL] FeSbS A silver-white to steel-gray orthorhombic mineral composed of a sulfide and antimonide of iron.

Guerbet reaction [ORG CHEM] A condensation of alcohols at high temperatures through the action of sodium alkoxides.

guerrilla warfare [ORD] Operations carried on by independent or semi-independent forces in the rear of the enemy; these operations usually are conducted by irregular forces acting either separately from, or in conjunction with, regular forces but may at times be conducted entirely with regular troops.

guess-warp [NAV ARCH] A line used in conjunction with a Jacob's ladder on the thimble of which boat crews reeve their bowlines.

guest element *See* trace element.

Guest unit [BIOL] A unit for the standardization of plasmin.

Guggenheim process [CIV ENG] A method of chemical precipitation which employs ferric chloride and aeration to prepare sludge for filtration.

Guiana Current [OCEANOGR] A current flowing northwestward along the northeastern coast of South America.

guidance [NAV] The process of directing the movements of any vehicle, especially an aeronautical vehicle or space vehicle, with particular reference to the selection of a course or flight path.

guidance site [ENG] Specific location of high-order geodetic accuracy containing equipment and structures necessary to provide guidance services or a given launch rate; it may be an integrated part of a launch site, or it may be a remote facility.

guidance station equipment [ENG] The ground-based portion of the missile guidance system necessary to provide guidance during missile flight; it specifically includes the tracking radar, the rate measuring equipment, the data link equipment, and the computer, test, and maintenance equipment integral to these items.

guidance system [AERO ENG] The control devices used in guidance of an aircraft or spacecraft. [NAV] Apparatus for generating and detecting the path along which a vehicle or craft is guided, often remotely and automatically.

guidance tape [ADP] A magnetic tape or punched paper tape that is placed in a missile or its computer to program desired events during flight.

guide *See* waveguide.

guide bearing [MECH ENG] A plain bearing used to guide a machine element in its lengthwise motion, usually without rotation of the element.

guide bracket [MIN ENG] A steel bracket fixed to a bunton to secure rigid guides in a shaft.

guide coupling [MIN ENG] A short coupling with a projecting

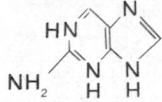

reamer guide or pup to which is attached a reaming bit, which it couples to a reaming barrel.

guided bend test [MET] A bend test in which the specimen is bent to a predetermined shape.

guided bomb [ORD] An aerial bomb that is guided in range or azimuth, or both, during its drop.

guided missile [ORD] An unmanned self-propelled vehicle, with or without a warhead, which is designed to move in a trajectory or flight path all or partially above the earth's surface and whose trajectory or course, while in flight, is capable of being directed by remote control, by homing systems, or by inertial or programmed guidance from within; excludes drones, torpedoes, and rockets and other vehicles whose trajectory or course cannot be controlled while in flight.

guided-missile control [ORD] The guidance or direction exercised over a guided missile during its flight to a target.

guided missile exercise head [ORD] An item designed to simulate a guided missile warhead, with or without telemetering devices.

guided-missile ship [NAV ARCH] A warship equipped with guided-missile launchers, and sometimes with a gun battery, long-range sonar, and antisubmarine warfare weapons. Also known as missile ship.

guided-missile submarine [NAV ARCH] A submarine designed to have an additional capability to launch guided-missile attacks from surface condition.

guided propagation [COMMUN] Type of radio-wave propagation in which radiated rays are bent excessively by refraction in the lower layers of the atmosphere; this bending creates an effect much as if a duct or waveguide has been formed in the atmosphere to guide part of the radiated energy over distances far beyond the normal range. Also known as trapping.

guided rocket [ORD] A guided missile having rocket propulsion.

guided wave [ELECTROMAG] A wave whose energy is concentrated near a boundary or between substantially parallel boundaries separating materials of different properties and whose direction of propagation is effectively parallel to these boundaries; waveguides transmit guided waves.

guide frame [MIN ENG] A frame held rigidly in place by roof jacks or timbers, with provisions for attaching a shaker conveyor pan line to the movable portion of the frame; prevents jumping or side movement of the pan line.

guide holes [ADP] One or more rows of holes prepunched around the margins of hand-sorted cards.

guide idler [MECH ENG] An idler roll with its supporting structure mounted on a conveyor frame to guide the belt in a defined horizontal path, usually by contact with the edge of the belt.

guide mill [MET] A hand rolling mill with a series of stands and guides at the entrance to the rolls.

guide number [GRAPHICS] A number that relates the output of a light source (such as a flashbulb) to the sensitivity of a particular film; when the number is divided by the distance in feet to the subject, it gives the T stop at which the lens should be set.

guide pin [ENG] A pin used to line up a tool or die with the work.

guide post [CIV ENG] A post along a road that bears direction signs or guide boards.

guide rail [CIV ENG] A track or rail that serves to guide movement, as of a sliding door, window, or similar element.

guide ring [MIN ENG] A longitudinally grooved, annular ring made almost full borehole size, which is fitted to an extension coupling between the core barrel and the first drill rod.

guide rod [MIN ENG] A heavy drill rod coupled to and having the same diameter as a core barrel on which it is used; gives additional rigidity to the core barrel and helps to prevent deflection of the borehole. Also known as core barrel rod; oversize rod.

guides [MECH ENG] **1.** Pulleys to lead a driving belt or rope in a new direction or to keep it from leaving its desired direction. **2.** Tracks that support and determine the path of a skip bucket and skip bucket bail. **3.** Tracks guiding the chain or buckets of a bucket elevator. **4.** The runway paralleling the path of

the conveyor which limits the conveyor or parts of a conveyor to movement in a defined path. [MIN ENG] **1.** Steel, wood, or steel-wire rope conductors in a mine shaft to guide the movement of the cages. **2.** Timber, rope, or metal tracks in a hoisting shaft, which are engaged by shoes on the cage or skip so as to steady it in transit. **3.** The holes in a crossbeam through which the stems of the stamps in a stamp mill rise and fall.

guide wavelength [ELECTROMAG] Wavelength of electromagnetic energy conducted in a waveguide; guide wavelength for all air-filled guides is always longer than the corresponding free-space wavelength.

guiding center [ELECTROMAG] A slowly moving point about which a charged particle rapidly revolves; this is used in an approximation for the motion of a charged particle in slowly varying electric and magnetic fields.

guiding telescope [OPTICS] A telescope that is mounted so that it remains parallel to a photographic telescope and is used by a person observing through it to supplement the clock motion in keeping the image of a celestial body motionless on a photographic plate.

guildite [MINERAL] $(Cu,Fe)_3 (Fe,Al)_4 (SO_4)_7 (OH)_4 \cdot 15H_3O$ A dark-brown mineral composed of a basic hydrated sulfate of copper, iron, and aluminum.

Guillain-Barré syndrome *See* Landry-Guillain-Barré syndrome.

Guillemin effect [ELECTROMAG] The tendency of a bent magnetorestrictive rod to straighten in a magnetic field parallel to its length.

Guillemin line [ELECTR] A network or artificial line used in high-level pulse modulation to generate a nearly square pulse, with steep rise and fall; used in radar sets to control pulse width.

guillotine [GRAPHICS] A heavy steel knife used to cut printing plates made of copper, zinc, or magnesium and to trim excess metal from electrotype or stereotype casts.

Guinea Current [OCEANOGR] A current flowing eastward along the southern coast of northwestern Africa into the Gulf of Guinea.

guinea fowl [VERT ZOO] The common name for plump African game birds composing the family Numididae; individuals have few feathers on the head and neck, but may have a crest of feathers and various fleshy appendages.

guinea pig [VERT ZOO] The common name for several species of wild and domestic hystricomorph rodents in the genus *Cavia*, family Caviidae; individuals are stocky, short-eared, short-legged, and nearly tailless.

guinea worm [INV ZOO] *Dracunculus medinensis*. A parasitic nematode that infects the subcutaneous tissues of humans and other mammals.

Guinier-Preston zones [MET] The initially formed zones of a precipitate as it comes out of solid solution.

guitarfish [VERT ZOO] The common name for fishes composing the family Rhinobatidae.

guitermanite [MINERAL] $Pb_{10}As_6S_{19}$ A bluish-gray mineral composed of lead, arsenic, and sulfur, occurring in compact masses.

Gukhman number [THERMO] A dimensionless number used in studying convective heat transfer in evaporation, equal to $(t_0 - t_m)/T_0$, where t_0 is the temperature of a hot gas stream, t_m is the temperature of a moist surface over which it is flowing, and T_0 is the absolute temperature of the gas stream. Symbolized Gu; N_{Gu}.

gular [ANAT] Of, pertaining to, or situated in the gula or upper throat. [VERT ZOO] A horny shield on the plastron of turtles.

gulch [GEOGR] A gulley, sometimes occupied by a torrential stream.

gulch claim [MIN ENG] A claim laid upon and along the bed of an unnavigable stream winding through a canyon, with precipitous, nonmineral, and uncultivable bands wherein have accumulated placer deposits.

Guldberg and Waage law *See* mass action law.

Guldberg-Waage group [CHEM ENG] A dimensionless number used in studying chemical reactions in blast furnaces; it is given by an equation relating volumes of reacting gases and reacting products. Symbolized N_{GW}.

GUINEA FOWL

The common guinea fowl (*Numida meleagris*), with red wattles and pearl-green plumage.

GUINEA PIG

The guinea pig, with short legs and large angular claws.

gulf [GEOGR] **1.** An abyss or chasm. **2.** A large extension of the sea partially enclosed by land.

Gulf HDS process [CHEM ENG] A fixed-bed catalytic hydrocracking process to convert heavy petroleum stocks to lower-boiling distillates with accompanying desulfurization; the feed may be crude oils or residuals.

Gulfinning [CHEM ENG] A catalytic hydrogen-treating process used to remove sulfur from cracked and straight-run distillates and fuel oils, also improving carbon residue, color, and general stability and effecting a slight increase in gravity.

Gulf Stream [OCEANOGR] A relatively warm, well-defined, swift, relatively narrow, northward-flowing ocean current which originates north of Grand Bahama Island where the Florida Current and the Antilles Current meet, and which eventually becomes the eastward-flowing North Atlantic Current.

Gulf Stream Countercurrent [OCEANOGR] **1.** A surface current opposite to the Gulf Stream, one current component on the Sargasso Sea side and the other component much weaker, on the inshore side. **2.** A predicted, but as yet unobserved, large current deep under the Gulf Stream but opposite to it.

Gulf Stream eddy [OCEANOGR] A cutoff meander of the Gulf Stream.

Gulf Stream front [OCEANOGR] The pronounced horizontal temperature gradient that defines a cross section of the Gulf Stream.

Gulf Stream meander [OCEANOGR] One of the changeable, winding bends in the Gulf Stream; such bends intensify as the Gulf Stream merges into North Atlantic Drift and break up into detached eddies at times, at about 40°N.

Gulf Stream system [OCEANOGR] The Florida Current, Gulf Stream, and North Atlantic Current, collectively.

gulfweed [BOT] Brown algae of the genus *Sargassum*.

gull [VERT ZOO] The common name for a number of long-winged swimming birds in the family Laridae having a stout build, a thick, somewhat hooked bill, a short tail, and webbed feet.

gullet [ANAT] *See* esophagus. [INV ZOO] A canal between the cytostome and reservoir that functions in food intake in ciliates.

gully [GEOGR] A narrow ravine.

gully erosion [GEOL] Erosion of soil by running water.

gully-squall [METEOROL] A nautical term for a violent squall of wind from mountain ravines on the Pacific side of Central America.

gulp [ADP] A series of bytes considered as a unit.

gum accroides *See* acaroid resin.

gum ammoniac *See* ammoniac.

gum arabic [MATER] A water-soluble gum obtained from acacia trees in Africa and Australia; produced commercially as a white powder; used in the manufacture of inks and adhesives, in textile finishing, and as the principal binder in watercolor and gouache. Also known as acacia gum; gum Kordofan; gum Senegal.

gum benzoin *See* benzoin gum.

gumbo [BOT] *See* okra. [GEOL] A soil that forms a sticky mud when wet.

gumboti [GEOL] Deoxidized, leached clay that contains siliceous stones.

gum dammar *See* dammar.

gum Kordofan *See* gum arabic.

gumme [PATH] A mass of rubberlike necrotic tissue found in any of various organs and tissues in tertiary syphilis.

gummed paper [MATER] **1.** A variety of colored, patterned papers with adhesive on one side for easy application. **2.** Coated or uncoated book paper, gummed on one side, used for stickers, labels, stamps, seals, and tapes.

gummite [MINERAL] Any of various yellow, orange, red, or brown secondary minerals containing hydrous oxides of uranium, thorium, and lead. Also known as uranium ocher.

gummosis [PL PATH] Production of gummy exudates in diseased plants as a result of cell degeneration.

gum resin [MATER] A group of oleoresinous substances obtained from plants; mixtures of true gums and resins less soluble in alcohol than natural resins; examples are rubber, gutta-percha, gamboge, myrrh, and olibanum; used to make certain pharmaceuticals.

gum Senegal *See* gum arabic.

gum test [CHEM ENG] A standard ASTM test to determine the amount of gums in gasolines.

gum thus *See* olibanum.

gum vein [BOT] Local accumulation of resin occurring as a wide streak in certain hardwoods.

gun [ORD] A piece of ordnance, consisting essentially of a tube or barrel, for throwing projectiles by force, usually the force of an explosive but sometimes that of compressed gas, a spring, or so on.

gunbarrel [CHEM ENG] An atmospheric vessel used for treatment of waterflood waste water.

gunboat [NAV ARCH] A small, moderate-speed, heavily armed vessel for general patrol and escort duty; usually unarmored and of less than 2000 tons.

gun-bomb-rocket sight [ORD] A computing sight used in a fighter aircraft in which different data may be set for use in aiming gunfire, bombs, or rockets.

gun book [ORD] Log that records the history of the operations and inspections of a particular gun.

gun bore line [ORD] The extended bore axis of a gun.

gun breech [ORD] The mass of metal at the rear end of a cannon, from the front slope to the rear face, exclusive of the breech mechanism; in this metal are formed the seat for the breech mechanism, the powder chamber, and the slopes connecting the latter with the rifled portion of the bore.

gun burner [ENG] A burner which sprays liquid fuel into a furnace for combustion.

gun camera [OPTICS] **1.** A camera used in gunnery training that records the image of the target at which it is aimed on a strip of motion-picture film. **2.** A camera synchronized to a gun to record the results of firing, with the film revealing if the camera was correctly sighted on the target.

gun carriage [ORD] Mobile or fixed support for a gun; sometimes includes the elevating and traversing mechanisms.

gun charger [ORD] A mechanism on a gun that operates to retract the breech mechanism or bolt to the rear and to insert a charge into the chamber.

guncotton [MATER] Any of various nitrocellulose explosives of high nitration (13.35–13.4% nitrogen) made by treating cotton with nitric and sulfuric acids; used principally in the manufacture of single-base and double-base propellants.

gundeck [NAV ARCH] A deck on old-time warships on which the ship's guns were carried.

gun deflection board *See* deflection board.

gun-directing radar [ORD] Radar used to direct antiaircraft artillery or similar fire.

gun displacement [ORD] **1.** Distance from a gun to the directing point or the base piece of a battery. **2.** Movement of a gun to a new firing position.

gun emplacement [ORD] Firing location of a gun together with necessary installations, such as camouflage, and ammunition supply.

gunfire [ORD] Use of artillery, rifles, and small arms as distinguished from the use of bayonets, swords, torpedoes, and bombs.

gun fore-end [ORD] A wood or plastic piece that is usually designed with a semicircular groove to fit under the barrel of a gun and is attached by metal fastening devices; it is shaped to fit the hand and is used to steady the weapon during firing.

gun group [ORD] Major parts of a gun, considered as a unit distinct from its mount.

gun handguard [ORD] A wood or plastic piece that is usually designed with a semicircular groove to fit over the top of the barrel of a carbine or rifle and is attached by metal fastening devices; it is shaped to fit the hand and to protect it during firing; it excludes the gun fore-end.

gun hoist [ORD] A device placed near the breech of a gun for lifting propellant and projectiles.

Gunite [MATER] Trade name for a mixture of sand and cement, applied pneumatically with a pressure gun; acts as a sealing agent to prevent erosion by air and moisture.

gun jack [ORD] A jack for forcing a tube of a gun out of battery.

gunk *See* rod dope.

gun launcher [ORD] A gun adapted to launching guided missiles or rockets.

GURNEY-MOTT THEORY

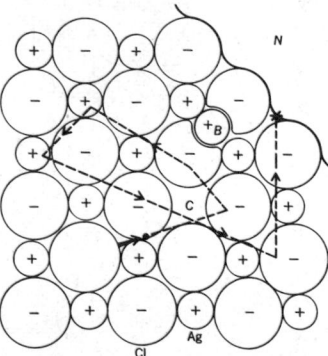

Schematic representation of Gurney-Mott theory. Broken line shows path of photoelectron which is eventually trapped in vicinity of a silver speck *N*. The interstitial silver ion *B* has come up to neutralize the charge of the electron and thus to add to the silver speck. *(After J. R. Haynes and W. Schockley)*

gun-laying radar [ENG] Radar equipment specifically designed to determine range, azimuth, and elevation of a target and sometimes also to automatically aim and fire antiaircraft artillery or other guns.

gun line [ORD] For a single aircraft gun, the axis of the bore extended; for multiple guns, it is the mean gun line.

gunlock [ORD] Fastening device used to prevent a weapon from being fired.

gunmetal [MET] **1.** Bronze composed of copper and tin in proportions of 9:1, formerly used to make cannons. **2.** Any metal or alloy from which guns are made. **3.** Any metal or alloy treated to give the appearance of black, tarnished copper-alloy gunmetal.

gun mount [ORD] An item designed to support a gun.

Gunn diode *See* Gunn oscillator.

Gunn effect [ELECTR] Development of a rapidly fluctuating current in a small block of a semiconductor (perhaps *n*-type gallium arsenide) when a constant voltage above a critical value is applied to contacts on opposite faces.

gunnel *See* gunwale.

Gunneraceae [BOT] A family of dicotyledonous terrestrial herbs in the order Haloragales, distinguished by two to four styles, a unilocular bitegmic ovule, large inflorescences with no petals, and drupaceous fruit.

gunner's quadrant [ENG] A mechanical device having scales graduated in mils, with fine micrometer adjustments and leveling or cross-leveling vials; it is a separate, unattached instrument for hand placement on a reference surface.

gunnery [ORD] The art or practice of using machine guns or cannon either on the ground or in the air.

Gunn oscillator [ELECTR] A microwave oscillator utilizing the Gunn effect. Also known as Gunn diode.

gun parallax [ORD] The difference in azimuth between the line from the directing point to the target and the line from the gun to the target.

gun pendulum [ENG] A device used to determine the initial velocity of a projectile fired from a gun in which the gun is mounted as a pendulum and its excursion upon firing is measured.

gunpowder [MATER] A black or brown explosive mixture of potassium nitrate, charcoal, and sulfur; originally, it was made in powder form, now generally in grains of various sizes.

gunpowder paper [MATER] Paper with an explosive on it that is rolled up for use in loading.

gun pressure [ORD] Pressure within a gun tube or barrel, as used in design practices.

gun reaction [MECH] The force exerted on the gun mount by the rearward movement of the gun resulting from the forward motion of the projectile and hot gases. Also known as recoil.

gunsight [ORD] A mechanical or optical device for aiming a firearm or for placing a gun or rocket launcher in position; based on the principle that two points (the observer's eye and a suitable mark in the instrument) in fixed relation to each other may be brought in line with a third (the target).

gunsight computer [ORD] The component of a computing gunsight that calculates for variables in gunnery and computes a prediction angle.

gun silencer [ORD] An item specifically designed to silence the explosive report caused by the discharge of cartridges by small arms weapon; it incorporates integral chambers or baffles which allow the gases to expand gradually. Also known as silencer.

gun slide [ORD] Portion of a gun which rests on the cradle guides.

gunstock [ORD] The wooden stock in which the barrel and mechanism of a gun are fixed.

gun stoppage [ORD] A condition which prevents a gun from being fired, usually in automatic weapons as the result of a cartridge not feeding properly from the magazine to the breech of the gun.

Gunter's chain [ENG] A chain 66 feet (20.1168 meters) long, consisting of 100 steel links, each 7.92 inches (20.1168 centimeters) long, joined by rings, which is used as the unit of length for surveying public lands in the United States. Also known as chain.

gun turret [ORD] Dome-shaped or cylindrical armored structure containing one or more guns, located on forts, warships, airplanes, or tanks.

gun-type burner [ENG] An oil burner that uses a nozzle to atomize the fuel.

gunwale [NAV ARCH] The upper edge of the side of a boat. Also spelled gunnel.

Günz [GEOL] A European stage of geologic time, in the Pleistocene (above Astian of Pliocene, below Mindel); it is the first stage of glaciation of the Pleistocene in the Alps.

Günzberg reagent [ANALY CHEM] A solution of 2 grams of vanillin and 4 grams of phloroglucinol in 80 milliliters of 95% alcohol; used as a test reagent for determining free hydrochloric acid in gastric juice.

Günz-Mindel [GEOL] The first interglacial stage of the Pleistocene in the Alps, between Günz and Mindel glacial stages.

Gurevich effect [SOLID STATE] An effect observed in electric conductors in which phonon-electron collisions are important, in the presence of a temperature gradient, in which phonons carrying a thermal current tend to drag the electrons with them from hot to cold.

Gurney formulas [ORD] A series of formulas proposed by R. W. Gurney for determining initial fragment velocity as a function of the type of explosive and the ratio of explosive charge to metal weight; each formula corresponds to a particular shape of container.

Gurney-Mott theory [CHEM] A theory of the photographic process that proposes a two-stage mechanism; in the first stage, a light quantum is absorbed at a point within the silver halide gelatin, releasing a mobile electron and a positive hole; these mobile defects diffuse to trapping sites (sensitivity centers) within the volume or on the surface of the grain; in the second stage, trapped (negatively charged) electron is neutralized by an interstitial (positively charged) silver ion, which combines with the electron to form a silver atom; the silver atom is capable of trapping a second electron, after which the process repeats itself, causing the silver speck to grow.

gusset [MIN ENG] A V-shaped cut in the face of a heading.

gusset plate [CIV ENG] A rectangular or triangular steel plate that connects members of a truss.

gust [METEOROL] A sudden, brief increase in the speed of the wind; it is of a more transient character than a squall and is followed by a lull or slackening in the wind speed.

gustation [PHYSIO] The act or the sensation of tasting.

gustatoreceptor [ANAT] A taste bud. [PHYSIO] Any sense organ that functions as a receptor for the sense of taste.

gust-gradient distance [AERO ENG] The horizontal distance along an aircraft flight path from the "edge" of a gust to the point at which the gust reaches its maximum speed.

gustiness [METEOROL] A quality of airflow characterized by gusts.

gustiness components [METEOROL] **1.** The ratios, to the mean wind speed, of the average magnitudes of the component fluctuations of the wind along three mutually perpendicular axes. **2.** The ratios of the root-mean-squares of the eddy velocities to the mean wind speed. Also known as intensity of turbulence.

gustiness factor [METEOROL] A measure of the intensity of wind gusts; it is the ratio of the total range of wind speeds between gusts and the intermediate periods of lighter wind to the mean wind speed, averaged over both gusts and lulls.

gust load [MECH] The wind load on an antenna due to gusts.

gustsonde [ENG] An instrument dropped from high altitude by a stable parachute, to measure the vertical component of turbulence aloft; consists of an accelerometer and radio telemetering equipment.

gust tunnel [AERO ENG] A type of wind tunnel that has an enclosed space and is used to test the effect of gusts on an airplane model in free flight to determine how atmospheric gusts affect the flight of an airplane.

gut [ANAT] The intestine. [EMBRYO] The embryonic digestive tube.

Guthrie test [PATH] A screening test for the detection of phenylketonuria in which the inhibition of growth of a strain of *Bacillus subtilis* by a phenylalanine analog is reversed by L-phenylalanine, as found in elevated concentration in the plasma of patients with phenylketonuria.

guti weather [METEOROL] In Rhodesia, a dense stratocumulus overcast, frequently with drizzle, occurring mainly in early summer, and associated with easterly winds that invade the interior, bringing in cool and stable maritime air when an anticyclone moves eastward south of Africa.

gutta-balata *See* balata.

gutta-percha [MATER] A leathery, thermoplastic substance consisting of gutta hydrocarbon with some resin obtained from the latex of certain Malaysian sapotaceous trees; used as insulation for submarine cables, and in golf balls and other products.

guttation [BOT] The discharge of water from a plant surface, especially from a hydathode.

gutter [BUILD] A trough along the edge of the eaves of a building to carry off rainwater. [CIV ENG] A shallow trench provided beside a canal, bordering a highway, or elsewhere, for surface drainage. [MIN ENG] A drainage trench cut along the side of a mine shaft to conduct the water back into a lodge or sump.

guttering [ENG] A process of quarrying stone in which channels, several inches wide, are cut by hand tools, and the stone block is detached from the bed by pinch bars. [MIN ENG] The process of cutting gutters in a mine shaft.

Guttiferae [BOT] A family of dicotyledonous plants in the order Theales characterized by extipulate leaves and conspicuous secretory canals or cavities in all organs.

guttra [METEOROL] In Iran, sudden squalls in May.

Guttulinaceae [MICROBIO] A family of microorganisms in the Acrasiales characterized by simple fruiting structures with only slightly differentiated component cells containing little or no cellulose.

Gutzeit test [ANALY CHEM] A test for arsenic; zinc and dilute sulfuric acid are added to the substance, which is then covered with a filter paper moistened with mercuric chloride solution; a yellow spot forms on the paper if arsenic is in the sample.

Gutzkow's process [MET] A modification of the sulfuric acid parting process for bullion containing large amounts of copper; a large excess of acid is used, and the silver sulfate is then reduced with charcoal, or, in the original process, ferrous sulfate.

guxen [METEOROL] A cold wind of the Alps in Switzerland.

guy [ENG] A rope or wire securing a pole, derrick, or similar temporary structure in a vertical position.

guy derrick [MECH ENG] A derrick having a vertical pole supported by guy ropes to which a boom is attached by rope or cable suspension at the top and by a pivot at the foot.

guyot [GEOL] A seamount, usually deeper than 100 fathoms (180 meters), having a smooth platform top. Also known as tablemount.

guzzle [METEOROL] In the Shetland Islands, an angry blast of wind, dry and parching.

G value [NUCLEO] The number of molecules produced or destroyed for each 100 electron volts absorbed by a substance from ionizing radiation.

GW *See* gigawatt.

Gymnarchidae [VERT ZOO] A monotypic family of electrogenic fishes in the order Osteoglossiformes in which individuals lack pelvic, anal, and caudal fins.

Gymnarthridae [PALEON] A family of extinct lepospondylous amphibians that have a skull with only a single bone representing the tabular and temporal elements of the primitive skull roof.

gymnite *See* deweylite.

Gymnoascaceae [MYCOL] A family of ascomycetous fungi in the order Eurotiales including dermatophytes and forms that grow on dung, soil, and feathers.

Gymnoblastea [INV ZOO] A suborder of coelenterates in the order Hydroida comprising hydroids without protective cups around the hydranths and gonozooids.

gymnoblastic [INV ZOO] Having naked medusa buds, referring to anthomedusan hydroids.

gymnocarpous [BOT] Having the hymenium uncovered on the surface of the thallus or fruiting body of lichens or fungi.

gymnocephalous cercaria [INV ZOO] A type of digenetic trematode larva.

Gymnocerata [INV ZOO] An equivalent name for Hydrocorisae.

Gymnocodiaceae [PALEOBOT] A family of fossil red algae.

Gymnodinia [INV ZOO] A suborder of flagellate protozoans in the order Dinoflagellida that are naked or have thin pellicles.

gymnogynous [BOT] Having a naked ovary.

Gymnolaemata [INV ZOO] A class of ectoproct bryozoans possessing lophophores which are circular in basal outline and zooecia which are short, wide, and vaselike or boxlike.

Gymnonoti [VERT ZOO] An equivalent name for Cypriniformes.

Gymnophiona [VERT ZOO] An equivalent name for Apoda.

Gymnopleura [INV ZOO] A subsection of brachyuran decapod crustaceans including the primitive burrowing crabs with trapezoidal or elongate carapaces, the first pereiopods subchelate, and some or all of the remaining pereiopods flattened and expanded.

gymnosperm [BOT] The common name for members of the division Pinophyta; seed plants having naked ovules at the time of pollination.

Gymnospermae [BOT] The equivalent name for Pinophyta.

Gymnostomatida [INV ZOO] An order of the protozoan subclass Holotrichia containing the most primitive ciliates, distinguished by the lack of ciliature in the oral area.

Gymnotidae [VERT ZOO] The single family of the suborder Gymnotoidei; eel-shaped fishes having numerous vertebrae, and anus located far forward, and lacking pelvic and developed dorsal fins.

Gymnotoidei [VERT ZOO] A monofamilial suborder of actinopterygian fishes in the order Cypriniformes.

gynaecandrous [BOT] Having staminate and pistillate flowers on the same spike.

gynander [BIOL] A mosaic individual composed of diploid female portions derived from both parents and haploid male portions derived from an extra egg or sperm nucleus.

gynandry [PHYSIO] A form of pseudohermaphroditism in which the external sexual characteristics are partly or wholly of the male aspect, but internal female genitalia are present. Also known as female pseudohermaphroditism; virilism.

gynecology [MED] The branch of the medical science dealing with diseases of women, particularly those affecting the sex organs.

gynephobia [PSYCH] An abnormal fear of women.

gynobase [BOT] A gynoecium-bearing elongation of the receptacle in certain plants.

gynodioecious [BOT] Dioecious but with some perfect flowers on a plant bearing pistillate flowers.

gynoecium [BOT] The aggregate of carpels in a flower.

gynogenesis [EMBRYO] Development of a fertilized egg through the action of the egg nucleus, without participation of the sperm nucleus.

gynomerogony [EMBRYO] Development of a fragment of a fertilized egg containing the haploid egg nucleus.

gynomonoecious [BOT] Having complete and pistillate flowers on the same plant.

gynophore [BOT] 1. A stalk that bears the gynoecium. 2. An elongation of the receptacle between pistil and stamens.

gynostemium [BOT] The column composed of the united gynoecia and androecium.

gyotaku [GRAPHICS] The art of Japanese fish printing, a modern application of stone-rubbing technique.

gypsite [GEOL] A variety of gypsum consisting of dirt and sand; found as an efflorescent deposit in arid regions, overlying gypsum. Also known as gypsum earth.

gypsophilous [ECOL] Flourishing on a gypsum-rich substratum.

gypsum [MINERAL] $CaSO_4 \cdot 2H_2O$ A mineral, the commonest sulfate mineral; crystals are monoclinic, clear, white to gray, yellowish, or brownish in color, with well-developed cleavages; luster is subvitreous to pearly, hardness is 2 on Mohs scale, and specific gravity is 2.3; it is calcined at 190–200°C to produce plaster of paris.

gypsum cement *See* gypsum plaster.

gypsum earth *See* gypsite.

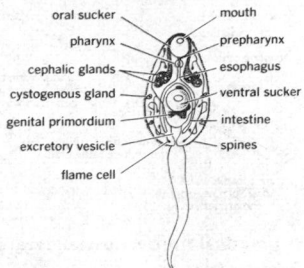

GYMNOCEPHALOUS CERCARIA

oral sucker — mouth
pharynx — prepharynx
cephalic glands — esophagus
cystogenous gland — ventral sucker
genital primordium — intestine
excretory vesicle — spines
flame cell

Anatomy of gymnocephalous cercaria. *(From R. M. Cable, An Illustrated Laboratory Manual of Parasitology, Burgess, 1958)*

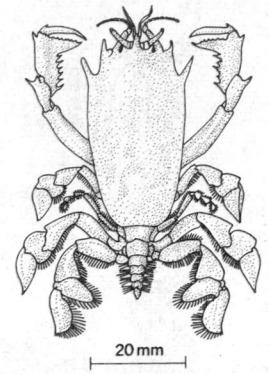

GYMNOPLEURA

20 mm

Gymnopleuran crab, *Raninoides louisianensis. (Smithsonian Institution)*

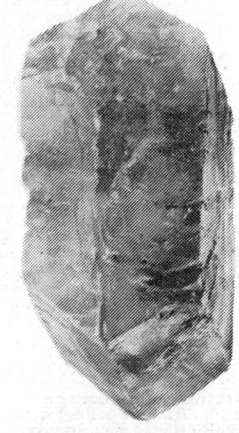

GYPSUM

A gypsum crystal found in Oxfordshire, England. *(American Museum of Natural History Specimen)*

GYRATOR

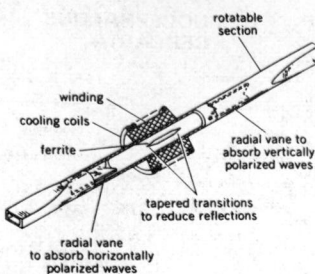

winding
cooling coils
ferrite
rotatable section
radial vane to absorb vertically polarized waves
tapered transitions to reduce reflections
radial vane to absorb horizontally polarized waves

A practical nonreciprocal gyrator.

GYRODACTYLOIDEA

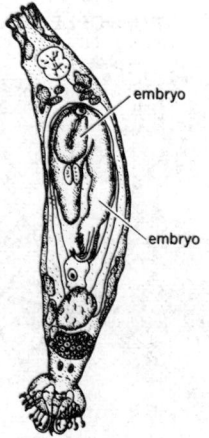

embryo
embryo

Gyrodactylus funduli Hargis from the longnose killfish (*Fundulus similis*).

GYROSCOPE

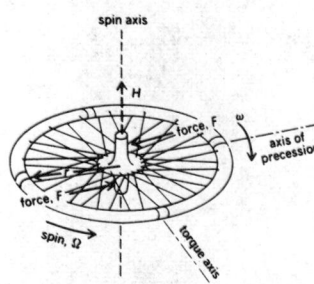

spin axis
H
force, F
ω
axis of precession
force, F
spin, Ω
torque axis

Illustration of gyroscope principle. Bicycle wheel with high spin velocity Ω has angular momentum $H = mr^2\Omega$, where m and r are mass and radius of wheel. Torque T resulting from force F produces precession with small angular velocity $\omega = T/H$ about axis perpendicular to both spin axis and torque axis.

---

gypsum lath [MATER] Lath consisting of a core of set gypsum surfaced with paper that is treated to receive plaster.

gypsum plank [MATER] A structural precast unit consisting of gypsum core reinforced with welded galvanized steel mesh and bounded on all four edges with a tongue-and-groove steel form; used as the roof deck of steel-frame buildings, and sometimes for the floor system.

gypsum plaster [MATER] Plaster made principally from gypsum. Also known as gypsum cement.

gypsum wallboard [MATER] Wallboard consisting of a core of set gypsum surfaced with paper or other fibrous material suitable to receive paint or paper.

gypsy [NAV ARCH] A small auxiliary drum fitted to one or both ends of a winch or windlass.

gypsy head [NAV ARCH] A small auxiliary drum at the end of a windlass or capstan, used to handle lines. Also spelled gipsy head.

gypsy moth [INV ZOO] *Porthetria dispar.* A large lepidopteran insect of the family Lymantriidae that was accidentally imported into New England from Europe in the late 19th century; larvae are economically important as pests of deciduous trees.

Gyracanthididae [PALEON] A family of extinct acanthodian fishes in the suborder Diplacanthoidei.

gyration tensor [SOLID STATE] A tensor characteristic of an optically active crystal, whose product with a unit vector in the direction of propagation of a light ray gives the gyration vector.

gyration vector [OPTICS] For light propagating in an optically active medium, a vector whose cross product with the time derivative of the electric displacement vector gives a negative contribution to the electric field.

gyrator [ELECTROMAG] A waveguide component that uses a ferrite section to give zero phase shift for one direction of propagation and 180° phase shift for the other direction; in other words, it causes a reversal of signal polarity for one direction of propagation but not for the other direction. Also known as microwave gyrator.

gyratory breaker *See* gyratory crusher.

gyratory crusher [MECH ENG] A primary breaking machine in the form of two cones, an outer fixed cone and a solid inner erect cone mounted on an eccentric bearing. Also known as gyratory breaker.

gyratory screen [MECH ENG] Boxlike machine with a series of horizontal screens nested in a vertical stack with downward-decreasing mesh-opening sizes; near-circular motion causes undersized material to sift down through each screen in succession.

gyre [OCEANOGR] A closed circulatory system, but larger than a whirlpool or eddy.

Gyrinidae [INV ZOO] The whirligig beetles, a family of large coleopteran insects in the suborder Adephaga.

Gyrinocheilidae [VERT ZOO] A monogeneric family of cypriniform fishes in the suborder Cyprinoidei.

gyro *See* gyroscope.

gyrocompass [NAV] A north-seeking form of gyroscope used as a vehicle's or craft's directional reference. Also known as gyroscopic compass.

gyrocompass alignment [NAV] Automatic alignment of a gyrocompass with the celestial meridian.

gyrocompassing [NAV] In inertial navigation equipment, a process of self-alignment in azimuth accomplished without any external equipment, consisting basically of using the inertial platform's accelerometers to slave it to the local gravity vector, and employing one of the gyros to seek true north by sensing the rotation of the earth in the manner of a gyro compass.

Gyrocotylidae [INV ZOO] An order of tapeworms of the subclass Cestodaria; species are intestinal parasites of chimaeroid fishes and are characterized by an anterior eversible proboscis and a posterior ruffled adhesive organ.

Gyrocotyloidea [INV ZOO] A class of trematode worms according to some systems of classification.

Gyrodactyloidea [INV ZOO] A superfamily of ectoparasitic trematodes in the subclass Monogenea; the posterior holdfast is solid and is armed with central anchors and marginal hooks.

gyro error [NAV] The error in the reading of the gyro compass, expressed in degrees east or west to indicate the direction in which the axis of the compass is offset from the north.

gyro flux-gate compass [NAV] A compass in which a flux gate, horizontally stabilized by a gyroscope, senses the horizontal component of the earth's magnetic field; being fixed with respect to the aircraft, the compass reacts to each change in heading with a change in current.

gyrofrequency *See* cyclotron frequency.

gyrogonite [PALEOBOT] A minute, ovoid body that is the residue of the calcareous encrustation about the female sex organs of a fossil stonewort.

gyro graph [NAV] A graph for recording and indicating drift of a directional gyro; usually a graph of drift from the desired heading versus time.

gyro log [NAV] **1.** A written record of directional gyro drift. **2.** A written record of the performance of a gyro compass.

gyromagnetic compass [NAV] A magnetic compass in which gyroscopic stabilization is used to indicate direction.

gyromagnetic coupler [ELECTR] A coupler in which a single-crystal yig (yttrium iron garnet) resonator provides coupling at the required low signal levels between two crossed stripline resonant circuits.

gyromagnetic effect [ELECTROMAG] The rotation induced in a body by a change in its magnetization, or the magnetization resulting from a rotation.

gyromagnetic radius *See* Larmor radius.

gyromagnetic ratio [PHYS] **1.** The ratio of the magnetic dipole moment to the angular momentum for a classical, atomic, or nuclear system. **2.** Occasionally, the reciprocal of the quantity in the first definition.

gyromagnetics [ELECTROMAG] The study of the relation between the angular momentum and the magnetization of a substance as exhibited in the gyromagnetic effect.

gyropendulum [MECH ENG] A gravity pendulum attached to a rapidly spinning gyro wheel.

Gyropidae [INV ZOO] A family of biting lice in the order Mallophaga; members are ectoparasites of South American rodents.

Gyropilot [ENG] A trade name for a type of automatic pilot used in guiding a missile.

gyroplane [AERO ENG] A rotorcraft whose rotors are not power-driven.

gyrorepeater [ENG] That part of a remote indicating gyro compass system which repeats at a distance the indications of the master gyro compass system.

gyroscope [ENG] An instrument that maintains an angular reference direction by virtue of a rapidly spinning, heavy mass; all applications of the gyroscope depend on a special form of Newton's second law, which states that a massive, rapidly spinning body rigidly resists being disturbed and tends to react to a disturbing torque by precessing (rotating slowly) in a direction at right angles to the direction of torque. Also known as gyro.

gyroscopic-clinograph method [ENG] A method used in borehole surveying which measures time, temperature, and temperature on 16-millimeter film while a gyroscope maintains the casing on a fixed bearing.

gyroscopic compass *See* gyrocompass.

gyroscopic couple [MECH ENG] The turning moment which opposes any change of the inclination of the axis of rotation of a gyroscope.

gyroscopic drift [NAV] The horizontal rotation of the spin axis of a gyroscope about the drift axis.

gyroscopic horizon [NAV] A gyroscopic instrument that indicates the lateral and longitudinal attitude of an aircraft by simulating the natural horizon.

gyroscopic precession [MECH] The turning of the axis of spin of a gyroscope as a result of an external torque acting on the gyroscope; the axis always turns toward the direction of the torque.

gyroscopics [MECH] The branch of mechanics concerned with gyroscopes and their use in stabilization and control of ships, aircraft, projectiles, and other objects.

gyrosextant [NAV] A sextant provided with a gyroscope for the purpose of determining the horizontal.

gyrostabilizer [ENG] A gyroscope used to stabilize ships and airplanes.

gyrotron [ELECTR] A device that detects motion of a system by measuring the phase distortion that occurs when a vibrating tuning fork is moved.

gyro wheel [MECH ENG] The rapidly spinning wheel in a gyroscope, which resists being disturbed.

gyrus [ANAT] One of the convolutions (ridges) on the surface of the cerebrum.

gyttja [GEOL] A fresh-water anaerobic mud containing an abundance of organic matter; capable of supporting aerobic life.

h *See* Planck's constant.

H *See* henry; hydrogen.

H [MED] A symbol used in electrocardiography for the longitudinal axis of the heart as projected on the frontal plane.

ha *See* hectare.

HAA *See* height above airport.

Haanel depth rule [GEOPHYS] A rule for estimating the depth of a magnetic body, provided the body may be considered magnetically equivalent to a single pole; the depth of the pole is then equal to the horizontal distance from the point of maximum vertical magnetic intensity to the points where the intensity is one-third of the maximum value.

haar [METEOROL] A wet sea fog or very fine drizzle which drifts in from the sea in coastal districts of eastern Scotland and northeastern England; it occurs most frequently in summer.

Haar measure [MATH] A measure on the Borel subsets of a locally compact topological group whose value on a Borel subset U is unchanged if every member of U is multiplied by a fixed element of the group.

Haase system [MIN ENG] Shaft sinking in loose ground or quicksand by piles in the form of iron tubes connected by webs; downward movement is facilitated by water forced down the tubes to wash away loose material beneath their points.

habenula [ANAT] **1.** Stalk of the pineal body. **2.** A ribbonlike structure.

habenular commissure [ANAT] The commissure connecting the habenular ganglia in the roof of the diencephalon.

habenular ganglia [ANAT] Olfactory centers anterior to the pineal body.

habenular nucleus [ANAT] Either of a pair of nerve centers that are located at the base of the pineal body on either side and serve as an olfactory correlation center.

Haber-Bosch process [CHEM ENG] Early nitrogen-fixation process for production of ammonia from hydrogen and nitrogen, catalyzed by iron; now replaced by more efficient ammonia synthesis processes. Also known as Haber process.

Haber process *See* Haber-Bosch process.

habit [CRYSTAL] *See* crystal habit. [PSYCH] A repetitious behavior pattern.

habitat [ECOL] The part of the physical environment in which a plant or animal lives.

habit plane [CRYSTAL] The crystallographic plane or system of planes along which certain phenomena such as twinning occur.

habitual abortion [MED] Recurring, successive spontaneous abortion.

habituation [MED] **1.** A condition of tolerance to the effects of a drug or a poison, acquired by its continued use; marked by a psychic craving for it when the drug is withdrawn. **2.** Mild drug addiction in which withdrawal symptoms are not severe.

habitus [BIOL] General appearance or constitution of an organism.

haboob [METEOROL] A strong wind and sandstorm or duststorm in the northern and central Sudan, especially around Khartum, where the average number is about 24 haboobs a year.

habutai [TEXT] A plain-weave Japanese silk, soft and lightweight.

hachure [GEOL] A short line at some angle to contours, used to denote slopes of the ground in geologic maps.

H acid [ORG CHEM] $H_2NC_{10}H_4(OH)(SO_3H)_2$ A gray powder or crystalline substance that is soluble in water, ether, and alcohol; used as a dye intermediate. Also known as 8-amino-1-naphthol-3,6-disulfonic acid.

hackberry [BOT] **1.** *Celtis occidentalis.* A tree of the eastern United States characterized by corky or warty bark, and by alternate, long-pointed serrate leaves unequal at the base;

produces small, sweet, edible drupaceous fruit. **2.** Any of several other trees of the genus *Celtis.*

hacking [LAP] A system of grooves in a lap that hold diamond powder for cutting and polishing gems.

hackling [TEXT] Passing a comb through flax to clean and straighten the fibers.

hackmanite [MINERAL] A mineral of the sodalite family containing a small amount of sulfur; fluoresces orange or red in ultraviolet light.

hackmarack *See* tamarack.

hacksaw [ENG] A hand or power tool consisting of a fine-toothed blade held in tension in a bow-shaped frame; used for cutting metal, wood, and other hard materials.

hack watch [HOROL] A watch having a device for stopping the balance so that the hour, minute, and second hands may be reset precisely; used for timing observations of celestial bodies and regulating ship clocks.

hadal [OCEANOGR] Pertaining to the environment of the ocean trenches, over 6.5 kilometers in depth.

Hadamard's inequality [MATH] An inequality that gives an upper bound for the square of the determinant of a matrix in terms of the squares of the matrix entries.

haddock [VERT ZOO] *Melanogrammus aeglefinus.* A fish of the family Gadidae characterized by a black lateral line and a dark spot behind the gills.

hade [GEOL] **1.** The angle of inclination of a fault as measured from the vertical. **2.** The inclination angle of a vein or lode.

Hadfield manganese steel [MET] Austenitic steel (face-centered cubic structure) containing 11–14% manganese; resistant to shock and wear. Also known as austenitic manganese steel; manganese steel.

Hadley cell [METEOROL] A direct, thermally driven, and zonally symmetric circulation first proposed by George Hadley as an explanation for the trade winds; it consists of the equatorward movement of the trade winds between about latitude 30° and the equator in each hemisphere, with rising wind components near the equator, poleward flow aloft, and finally descending components at about latitude 30° again.

Hadromerina [INV ZOO] A suborder of sponges in the class Clavaxinellida having monactinal megascleres, usually with a terminal knob at one end.

hadromycosis [PL PATH] Any plant disease resulting from infestation of the xylem by a fungus.

hadron [PARTIC PHYS] An elementary particle which has strong interactions.

hadronic atom [ATOM PHYS] An atom consisting of a negatively charged, strongly interacting particle orbiting around an ordinary nucleus.

hadrosaur [PALEON] A duck-billed dinosaur.

Hadsel mill [MIN ENG] An early autogenous grinding mill in which comminution was caused by the fall of ore on ore that was rotating in a large-diameter horizontal cylinder.

Haeckel's law *See* recapitulation theory.

haematodocha [INV ZOO] A sac in the palpus of male spiders that fills with hemolymph and becomes distended during pairing.

Haemosporina [INV ZOO] A suborder of sporozoan protozoans in the subclass Coccidia; all are parasites of vertebrates, and human malarial parasites are included.

haerangium [MYCOL] The fruiting body of *Fugascus* and *Ceratostomella.*

HAF black *See* high-abrasion furnace black.

haff [GEOGR] A freshwater lagoon separated from the sea by a sandbar.

hafnium [CHEM] A metallic element, symbol Hf, atomic number 72, atomic weight 178.49; melting point 2000°C, boiling point above 5400°C.

hafnium carbide [INORG CHEM] HfC Gray powder, melt-

HACKBERRY

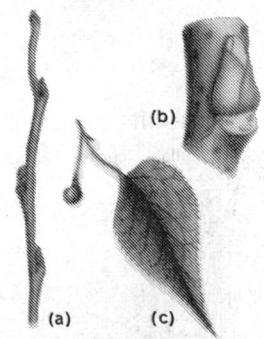

Common hackberry (*Celtis occidentalis*). (a) Twig. (b) Lateral bud. (c) Leaf.

HADROMERINA

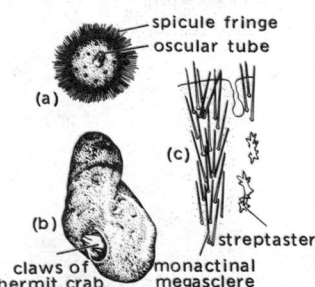

Representative hadromerine sponges. (a) *Radiella sol*, deep-sea species with peripheral fringe of spicules (*from Hyman, 1940, after Ridley and Dendy, 1887*). (b) *Suberites ficus* living on shell occupied by hermit crab (*from Hyman, 1940, after Muller, 1914*). (c) Monactinal megasclere arrangement of *Spirastrella* (*after Hyman, 1940*),

HAFNIUM

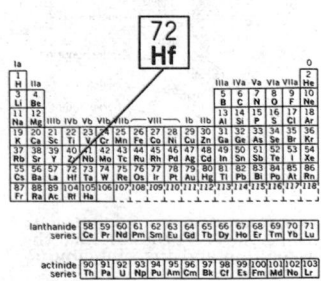

Periodic table of the chemical elements showing the position of hafnium.

ing at 3887°C; used in the control rods of nuclear reactors.

Haftplatte [INV ZOO] An adhesive plate or disk in some turbellarians; it is a glanduloepidermal organ.

Hageman factor *See* factor XII.

Hagen-Poiseuille law [FL MECH] In the case of laminar flow of fluid through a circular pipe, the loss of head due to fluid friction is 32 times the product of the fluid's viscosity, the pipe length, and the fluid velocity, divided by the product of the acceleration of gravity, the fluid density, and the square of the pipe diameter.

Hagen-Rubens relation [OPTICS] An equation for the reflectivity of a solid surface in terms of the frequency of radiation of the conductivity of the solid; it applies at wavelengths long enough that the product of the frequency and the relaxation time is much less than unity.

hagfish [VERT ZOO] The common name for the jawless fishes composing the order Myxinoidea.

Haggenmacher equation [CHEM] Equation to calculate latent heats of vaporizations of pure compounds by using critical conditions with Antoine constants.

H agglutinin [IMMUNOL] An antibody that is type-specific for the flagella of cells or microorganisms.

Hahn-Banach extension theorem [MATH] The theorem that every continuous linear functional defined on a subspace or linear manifold in a normed linear space X may be extended to a continuous linear functional defined on all of X.

Hahnemannism *See* homeopathy.

Haidinger fringes [OPTICS] Interference fringes produced by nearly normal incidence of light on thick, flat plates. Also known as constant-angle fringes; constant-deviation fringes.

haidingerite [MINERAL] $HCaAsO_4 \cdot H_2O$ A white mineral composed of hydrous calcium arsenate.

hail [METEOROL] Precipitation in the form of balls or irregular lumps of ice, always produced by convective clouds, nearly always cumulonimbus.

hail stage [METEOROL] The thermodynamic process of freezing of suspended water drops in adiabatically rising air with temperature below the freezing point, under the assumption that release of latent heat of fusion maintains constant temperature until all water is frozen.

hailstone [METEOROL] A single unit of hail, ranging in size from that of a pea to that of a grapefruit, or from less than $\frac{1}{4}$ inch (6 millimeters) to more than 5 inches (13 centimeters) diameter; may be spheroidal, conical, or generally irregular in shape.

hair [ZOO] **1.** A threadlike outgrowth of the epidermis of animals, especially a keratinized structure in mammalian skin. **2.** The hairy coat of a mammal, or of a part of the animal.

hair cell [HISTOL] The basic sensory unit of the inner ear of vertebrates; a columnar, polarized structure with specialized cilia on the free surface.

haircloth [TEXT] A stiff, wiry fabric made with a weft of hair, especially of horsehair or camel's hair, and a warp of cotton, linen, or wool.

hair copper *See* chalcotrichite.

hair cracks [MATER] Fine, random cracks in the surface of the top coat of paint or other coating material.

hair cycle [PHYSIO] The formation and growth of a new hair, followed by a resting stage, and ending with growth of another new hair from the same follicle.

hair felt [MATER] Felt made of cattle hair; used as insulation in buildings.

hair follicle [ANAT] An epithelial ingrowth of the dermis that surrounds a hair.

hair gland [ANAT] Sebaceous gland associated with hair follicles.

hair hygrometer [ENG] A hygrometer in which the sensing element is a bundle of human hair, which is held under slight tension by a spring and which expands and contracts with changes in the moisture of the surrounding air or gas.

hairline *See* air line.

hair pyrites *See* millerite.

hairspring *See* balance spring.

hairstone [GEOL] Quartz embedded with hairlike crystals of rutile, actinolite, or other mineral.

hair trigger [ORD] The trigger of a firearm when it is delicately balanced so as to fire easily.

hairworm [INV ZOO] The common name for about 80 species of worms composing the class Nematomorpha.

hairy tongue [MED] Hyperplasia of the papillae forming hairlike projections on the tongue.

Halacaridae [INV ZOO] A family of marine arachnids in the order Acarina.

halation [ELECTR] An area of glow surrounding a bright spot on a fluorescent screen, due to scattering by the phosphor or to multiple reflections at front and back surfaces of the glass faceplate. [OPTICS] A halo on a photographic image of a bright object caused by light reflected from the back of the film or plate.

halazone [ORG CHEM] $COOHC_6H_4SO_2NCl_2$ White crystals, with strong chlorine aroma; slightly soluble in water and chloroform; used as water disinfectant.

halcyon days [METEOROL] A period of fine weather.

Haldane's rule [GEN] The rule that if one sex in a first generation of hybrids is rare, absent, or sterile, then it is the heterogametic sex.

Halecomorphi [VERT ZOO] The equivalent name for Amiiformes.

Halecostomi [VERT ZOO] The equivalent name for Pholidophoriformes.

half-adder [ELECTR] A logic element which operates on two binary digits (but no carry digits) from a preceding stage, producing as output a sum digit and a carry digit.

half-adjust [ADP] A rounding process in which the least significant digit is dropped and, if the least significant digit is one-half or more of the number base, one is added to the next more significant digit and all carries are propagated.

half-and-half solder [MET] Solder composed of tin and lead in equal parts.

half-angle formulas [MATH] In trigonometry, formulas that express the trigonometric functions of half an angle in terms of trigonometric functions of the angle.

half-arc angle [METEOROL] The elevation angle of that point which a given observer regards as the bisector of the arc from his zenith to his horizon; a measure of the apparent degree of flattening of the dome of the sky.

half bat [MATER] One half of a brick, cut across the length.

half block [ADP] The unit of transfer between main storage and the buffer control unit; it consists of a column of 128 elements, each element 16 bytes long.

half-breadth plan [NAV ARCH] A plan of a ship showing outlines of horizontal sections or waterlines at various levels, from the main deck to the keel; only the left or right half is shown.

half-bridge [ELEC] A bridge having two power supplies, located in two of the bridge arms, to replace the single power supply of a conventional bridge.

half-cell [PHYS CHEM] A single electrode immersed in an electrolyte.

half-cell potential [PHYS CHEM] In electrochemical cells, the electrical potential developed by the overall cell reaction; can be considered, for calculation purposes, as the sum of the potential developed at the anode and the potential developed at the cathode, each being a half-cell.

half chronometer [HOROL] **1.** A watch in which the lever and chronometer escapements are combined. **2.** A watch adjusted for temperature.

half cock [ORD] Position of the hammer of a small arm weapon when it is held by the first cocking notch, with the trigger locked and the weapon relatively safe.

half column [ARCH] A column projecting from a wall by about half its diameter.

half-convergency *See* conversion angle.

half-course [MIN ENG] The drift or opening driven at an angle of about 45° to the strike and in the plane of the seam.

half cycle [ENG] The time interval corresponding to half a cycle, or 180°, at the operating frequency of a circuit or device.

half-cycle transmission [COMMUN] Data transmission and control system that uses synchronized sources of 60-hertz power at the transmitting and receiving ends; either of two receiver relays can be actuated by choosing the appropriate half-cycle polarity of the 60-hertz transmitter power supply.

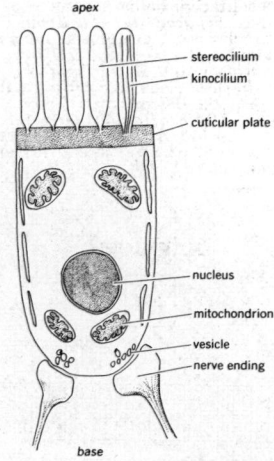

HAIR

medulla — cortex

cuticle

Longitudinal section of a hair shaft.

HAIR CELL

apex

stereocilium
kinocilium

cuticular plate

nucleus
mitochondrion
vesicle
nerve ending

base

Generalized diagram of hair cell showing the two types of cilia: the stereocilium and the kinocilium.

half-dog setscrew [DES ENG] A setscrew with a short, blunt point.

half-duplex circuit [COMMUN] A circuit designed for half-duplex operation. Abbreviated HDX.

half-duplex operation [COMMUN] Operation of a telegraph system in either direction over a single channel, but not in both directions simultaneously.

half-duplex repeater [ELECTR] Duplex telegraph repeater provided with interlocking arrangements which restrict the transmission of signals to one direction at a time.

half-hard [MET] A rolled-metal product of intermediate hardness or temper.

half-header [MIN ENG] A large cap piece; used by sawing a header in two and placing, generally, two timbers under the half header on one side of the haulage, with the end extending over the haulage.

half-life [CHEM] The time required for one-half of a given material to undergo chemical reactions. [NUCLEO] The average time interval required for one-half of any quantity of identical radioactive atoms to undergo radioactive decay. Also known as half-value period; radioactive half-life.

half line *See* ray.

half-loaded [ORD] In automatic arms, the condition wherein the belt or magazine is inserted and the receiver is charged, but without the first cartridge actually in the chamber.

half-moon [ASTRON] The moon as seen in the first quarter and the last, or third, quarter.

half nut [DES ENG] A nut split lengthwise so that it can be clamped around a screw.

half plane [MATH] The portion of a plane lying on one side of some line in the plane; in particular, all points of the complex plane either above or below the real axis.

half-power frequency [ELECTR] One of the two values of frequency, on the sides of an amplifier response curve, at which the voltage is $1/\sqrt{2}$ (70.7%) of a midband or other reference value. Also known as half-power point.

half-power point [ELECTR] 1. A point on the graph of some quantity in an antenna, network, or control system, versus frequency, distance, or some other variable at which the power is half that of a nearby point at which power is a maximum. 2. *See* half-power frequency.

half-pulse-repetition-rate delay [ELECTR] In the loran navigation system, an interval of time equal to half the pulse repetition rate of a pair of loran transmitting stations, introduced as a delay between transmission of the master and slave signals, to place the slave station signal on the B trace when the master station signal is mounted on the A trace pedestal.

half-round file [DES ENG] A file that is flat on one side and convex on the other.

half section [GRAPHICS] View of a section of a symmetrical workpiece or assembly on an engineering drawing, terminating at an axis of symmetry.

half set [MIN ENG] In mine timbering, one leg piece and a collar.

half-shift register [ELECTR] Logic circuit consisting of a gated input storage element, with or without an inverter.

half-side formulas [MATH] In trigonometry, formulas that express the tangents of one-half of each of the sides of a spherical triangle in terms of its angles.

half-silvered surface [OPTICS] A surface covered with metallic film of a thickness such that approximately half the light falling on it at normal incidence is reflected and half is transmitted.

half space [BUILD] A broad step between two half flights of a stair. [MATH] A space bounded only by an infinite plane.

half step *See* semitone.

half tap [ELEC] Bridge placed across conductors without disturbing their continuity.

half thickness [PHYS] The thickness of a sheet of material which reduces the intensity of a beam of radiation passing through it to one-half its initial value. Also known as half-value layer; half-value thickness.

half-through arch [CIV ENG] A bridge arch having the roadway running through it at an elevation midway between the base and the crown.

half tide [OCEANOGR] The condition when the tide is at the level between any given high tide and the following or preceding low tide. Also known as mean tide.

half-tide basin [CIV ENG] A lock of very large size and usually of irregular shape, the gates of which are kept open for several hours after high tide so that vessels may enter as long as there is sufficient depth over the sill; vessels remain in the half-tide basin until the ensuing flood tide, when they may pass through the gate to the inner harbor; if entry to the inner harbor is required before this time, water must be admitted to the half-tide basin from some external source.

half-tide level [OCEANOGR] The level midway between mean high water and mean low water.

half-timbered [BUILD] Pertaining to a timber frame building with brickwork, plaster, or wattle and daub filling the spaces between the timbers.

half time [NUCLEO] The time during which half the radioactive material resulting from a nuclear explosion remains in the atmosphere.

halftone [GRAPHICS] An engraving used in printing to reproduce photographs and drawings that contain continuous tones, that is, grays (middle tones or halftones) in addition to black and white; preparation involves photographing the artwork through a screen.

halftone characteristic [COMMUN] Fidelity of the recorded density shadings compared with the subject copy transmitted; the term may also be used to express the relationship between the facsimile signal and the subject copy or recorded copy.

halftone contact screen [GRAPHICS] A screen consisting of a regular pattern of dots that is used for making halftone negatives in a camera.

half-track [MECH ENG] 1. A chain-track drive system for a vehicle; consists of an endless metal belt on each side of the vehicle driven by one of two inside sprockets and running on bogie wheels; the revolving belt lays down on the ground a flexible track of cleated steel or hard-rubber plates; the front end of the vehicle is supported by a pair of wheels. 2. A motor vehicle equipped with half-tracks.

half-track tape reacorder *See* double-track tape recorder.

half-value layer *See* half thickness.

half-value period *See* half-life.

half-value thickness *See* half thickness.

half-wave [ELEC] Pertaining to half of one cycle of a wave. [ELECTROMAG] Having an electrical length of a half wavelength.

half-wave amplifier [ELECTR] A magnetic amplifier whose total induced voltage has a frequency equal to the power supply frequency.

half-wave antenna [ELECTROMAG] An antenna whose electrical length is half the wavelength being transmitted or received.

half-wave dipole *See* dipole antenna.

half-wavelength [ELECTROMAG] The distance corresponding to an electrical length of half a wavelength at the operating frequency of a transmission line, antenna element, or other device.

half-wave plate [OPTICS] A thin section of a doubly refracting crystal, of a thickness such that the ordinary and extraordinary components of normally incident light emerge from it with a phase difference corresponding to an odd number of half wavelengths.

half-wave rectification [ELECTR] Rectification in which current flows only during alternate half cycles.

half-wave rectifier [ELECTR] A rectifier that provides half-wave rectification.

half-wave transmission line [ELECTROMAG] Transmission line which has an electrical length equal to one-half the wavelength of the signal being transmitted or received.

half-wave vibrator [ELEC] A vibrator having only one pair of contacts; interrupts the flow of direct current through the primary of a power transformer, but does not reverse the current.

half-width [MATH] For a function which has a maximum and falls off rapidly on either side of the maximum, the difference between the two values of the independent variable for which the dependent variable has one-half its maximum value.

half-word I/O buffer [ADP] A buffer, the upper half being used to store the upper half of a word for both input and

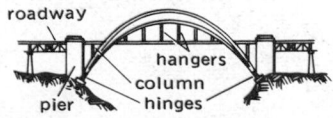

HALF-THROUGH ARCH

roadway / hangers / column / hinges / pier

Drawing of a half-through, two-hinged, crescent-rib arch. (*From G. A. Hool and W. S. Kinne, Movable and Long-span Steel Bridges, 2d ed., McGraw-Hill, 1943*)

HALFTONE

Greatly enlarged detail of photoengraved halftone plate, showing formation of dots of different sizes. (*American Museum of Photography*)

output characters, the lower half of the buffer being used for purposes such as the storage of constants.

halibut [VERT ZOO] Either of two large species of flatfishes in the genus *Hippoglossus;* commonly known as a right-eye flounder.

halibut liver oil [MATER] Fishy-tasting oil, pale yellow to dark red, with slightly fishy smell; soluble in alcohol, ether, chloroform, and carbon disulfide, insoluble in water; derived from halibut livers; used as medicine, as vitamin A and D source, and to dress leather.

Halichondrida [INV ZOO] A small order of sponges in the class Demospongiae with a skeleton of diactinal or monactinal, siliceous megascleres (or both), a skinlike dermis, and small amounts of spongin.

Halictidae [INV ZOO] The halictid and sweat bees, a family of hymenopteran insects in the superfamily Apoidea.

halide [CHEM] A compound of the type MX, where X is fluorine, chlorine, iodine, bromine, or astatine, and M is another element or organic radical.

Halimeda [BOT] A genus of small, bushy green algae in the family Codiaceae composed of thick, leaflike segments; important as a fossil and as a limestone builder.

Halimond tube [MIN ENG] A miniature pneumatic flotation cell used for examination of small ore samples under closely controllable conditions.

Haliotis [INV ZOO] A genus of gastropod mollusks commonly known as the abalones.

Haliplidae [INV ZOO] The crawling water beetles, a family of coleopteran insects in the suborder Adephaga.

halite [MINERAL] NaCl Native salt; an evaporite mineral occurring as isometric crystals or in massive, granular, or compact form. Also known as common salt; rock salt.

Halitheriinae [PALEON] A subfamily of extinct sirenian mammals in the family Dugongidae.

Hall accelerator [PL PHYS] A plasma accelerator based on the Hall effect.

Hall angle [ELECTROMAG] The electric field, resulting from the Hall effect, perpendicular to a current, divided by the electric field generating the current.

Hall coefficient [ELECTROMAG] A measure of the Hall effect, equal to the transverse electric field (Hall field) divided by the product of the current density and the magnetic induction. Also known as Hall constant.

Hall constant *See* Hall coefficient.

Hall cyclic thermal reforming [CHEM ENG] A gas-making process that uses component parts of carbureted-water gas apparatus to generate high-Btu gas from feedstocks ranging from naphtha to Bunker C.

Hall effect [ELECTROMAG] The development of a transverse electric field in a current-carrying conductor placed in a magnetic field; ordinarily the conductor is positioned so that the magnetic field is perpendicular to the direction of current flow and the electric field is perpendicular to both.

Hall-effect isolator [ELECTROMAG] An isolator that makes use of the Hall effect in a semiconductor plate mounted in a magnetic field, to provide greater loss in one direction of signal travel through a waveguide than in the other direction.

Hall-effect modulator [ELECTR] A Hall-effect multiplier used as a modulator to give an output voltage that is proportional to the product of two input voltages or currents.

Hall-effect multiplier [ELECTR] A multiplier based on the Hall effect, used in analog computers to solve such problems as finding the square root of the sum of the squares of three independent variables.

Haller's organ [INV ZOO] A chemoreceptor on the tarsus of certain ticks.

Hallett table [MIN ENG] A Wilfley-type concentrating table having the tops of the riffles in the same plane as the cleaning planes, and riffles inclined toward the waste water side.

Halley's Comet [ASTRON] A member of the solar system, with an orbit and a period of about 76 years; its nucleus has been estimated to be about 15 kilometers in radius; next due to appear in 1985–1986.

Hall generator [ELECTROMAG] A generator using the Hall effect to give an output voltage proportional to magnetic field strength.

halliard *See* halyard.

Hallinger shield [MIN ENG] A tunneling shield valuable for working in very soft ground; incorporates a mechanical excavator and does not entail the use of timbering to protect the miners.

Hall mobility [SOLID STATE] The product of conductivity and the Hall constant for a conductor or semiconductor; a measure of the mobility of the electrons or holes in a semiconductor.

halloysite [MINERAL] $Al_2Si_2O_5(OH)_4 \cdot 2H_2O$ Porcelainlike clay mineral whose composition is like that of kaolinite but contains more water and is structurally distinct; varieties are known as metahalloysites.

Hall process [MET] An electrolytic recovery process for aluminum employing a fused-bauxite (aluminum oxide), cryolite electrolyte.

hallucination [PSYCH] A perception without an appropriate stimulus.

hallucinogen [PHARM] A substance, such as LSD, that induces hallucinations.

hallucinogenic [PHARM] Of or pertaining to a hallucinogen. [PSYCH] Referring to any stimulus that creates the impression of experiencing a hallucination.

hallucinosis [PSYCH] The condition of being possessed by more or less persistent hallucinations, especially while fully conscious.

hallux [ANAT] The first digit of the hindlimb; the big toe of man.

hallux valgus [MED] A deformity of the great toe, in which the head of the first metatarsal deviates away from the second metatarsal, that is, toward the outside of the foot, and the phalanges are deviated toward the second toe, causing prominence of the metatarsophalangeal joint.

Hallwachs effect [PHYS] The ability of ultraviolet radiation to discharge a negatively charged body in a vacuum.

halmophagous [ZOO] Pertaining to organisms which infest and eat stalks or culms of plants.

halmyrolysis [GEOCHEM] Postdepositional chemical changes that occur while sediment is on the sea floor.

halo [ELECTR] An undesirable bright or dark ring surrounding an image on the fluorescent screen of a television cathode-ray tube; generally due to overloading or maladjustment of the camera tube. [GEOL] A ring or cresent surrounding an area of opposite sign; it is a diffusion of a high concentration of the sought mineral into surrounding ground or rock; it is encountered in mineral prospecting and in magnetic and geochemical surveys. [METEOROL] Any one of a large class of atmospheric optical phenomena which appear as colored or whitish rings and arcs about the sun or moon when seen through an ice crystal cloud or in a sky filled with falling ice crystals. [OPTICS] A ring around the photographic image of a bright source caused by light scattering in any one of a number of possible ways.

haloalkane [ORG CHEM] Halogenated aliphatic hydrocarbon.

halobacteria [MICROBIO] Rod-shaped bacteria which display extreme halophilism.

halo blight [PL PATH] A bacterial blight of beans and sometimes other legumes caused by *Pseudomonas phaseolicola* and characterized by water-soaked lesions surrounded by a yellow ring on the leaves, stems, and pods.

halocarbon plastic [ORG CHEM] Plastic made from halocarbon resins.

halocarbon resin [ORG CHEM] Resin made by the polymerization of monomers made up of halogenated hydrocarbons, such as tetrafluoroethylene, C_2F_4, and trifluorochloroethylene, C_2F_3Cl (the respective trade names are Teflon and Kel-F).

halocline [OCEANOGR] A well-defined vertical gradient of salinity in the oceans and seas.

halococci [MICROBIO] Coccoid bacteria which display extreme halophilism.

Halocypridacea [INV ZOO] A superfamily of ostracods in the suborder Myodocopa; individuals are straight-backed with a very thin, usually calcified carapace.

halo effect [IND ENG] A tendency when rating a person in regard to a specific trait to be influenced by a general impression or by another trait of the person.

haloform [ORG CHEM] CHX_3 A compound made by reaction of acetaldehyde or methyl ketones with NaOX, where X is a halogen; an example is iodoform, HCI_3, or bromoform, $HCBr_3$ or chloroform, $HCCl_3$.

haloform reaction [ORG CHEM] Halogenation of acetaldehyde or a methyl ketone in aqueous basic solution; the reaction is characteristic of compounds containing a CH_3CO group linked to a hydrogen or to another carbon.

halogen [CHEM] Any of the elements of the halogen family, consisting of fluorine, chlorine, bromine, iodine, and astatine.

halogenated hydrocarbon [ORG CHEM] One of a group of halogen derivatives of organic hydrogen- and carbon-containing compounds; the group includes monohalogen compounds (alkyl or aryl halides) and polyhalogen compounds that contain the same or different halogen atoms.

halogenation [ORG CHEM] A chemical process or reaction in which a halogen element is introduced into a substance, generally by the use of the element itself.

halogen counter [NUCLEO] A Geiger counter in which the self-quenching action is provided by a halogen gas, such as chlorine or bromine.

halogen mineral [MINERAL] Any of the naturally occurring compounds containing a halogen as the sole or principal anionic constituent.

halohydrin [ORG CHEM] A compound with the general formula $X-R-OH$ where X is a halide such as Cl^-; an example is chlorohydrin.

halonate [MYCOL] Pertaining to a spore surrounded by a colored circle. [PL PATH] A leaf spot surrounded by a halo.

halo of 22° [METEOROL] A halo phenomenon in the form of a prismatically colored circle of 22° angular radius around the sun or moon, exhibiting coloration from red on the inside to blue on the outside.

halo of 46° [METEOROL] A halo phenomenon in the form of a prismatically colored circle or incomplete arc thereof, centered on the sun or moon and having an angular radius of about 46°; the coloration is red on the inner edge to blue on the outer edge.

halo of 90° See Hevelian halo.

halo of Hevelius See Hevelian halo.

halophile [BIOL] An organism that requires high salt concentrations for growth and maintenance.

halophilism [BIOL] The phenomenon of demand for high salt concentrations for growth and maintenance.

halophyte [ECOL] A plant or microorganism that grows well in soils having a high salt content.

haloprogin [ORG CHEM] $C_9H_4OCl_3I$ An antimicrobial agent.

Haloragaceae [BOT] A family of dicotyledonous plants in the order Haloragales distinguished by an apical ovary of 2–4 loculi, small inflorescences, and small, alternate or opposite or whorled, exstipulate leaves.

Haloragales [BOT] An order of dicotyledonous plants in the subclass Rosidae containing herbs with perfect or often unisexual, more or less reduced flowers, and a minute or vestigial perianth.

Halosauridae [VERT ZOO] A family of mostly extinct deepsea teleost fishes in the order Notacanthiformes.

halosere [ECOL] The series of communities succeeding one another, from the pioneer stage to the climax, and commencing in salt water or on saline soil.

halothane [PHARM] $C_2HBrClF_3$ A colorless, nonflammable liquid used as a general anesthetic, by inhalation.

halotrichite [MINERAL] **1.** $FeAl_2(SO_4)_4 \cdot 22H_2O$ A mineral composed of hydrous sulfate of iron and aluminum. Also known as butter rock; feather alum; iron alum; mountain butter. **2.** Any sulfate mineral resembling halotrichite in structure and habit.

Halsey premium plan [IND ENG] A wage-incentive plan which sets a guaranteed daily rate to an employee and provides for predetermined compensation for superior performance.

Halsted's radical mastectomy [MED] Surgical removal of the breast, subcutaneous fat, muscle, lymph glands, and a wide area of skin for cancer.

haltere [INV ZOO] Either of a pair of capitate filaments representing rudimentary hindwings in Diptera. Also known as balancer.

halving [NAV] The process of adjusting magnetic compass correctors so as to remove half of the deviation on the opposite cardinal or adjacent intercardinal headings to those on which adjustment was originally made when all deviation was removed; this action is taken to equalize the error on opposite headings.

halyard [NAV ARCH] A rope or tackle used to raise or lower a flag, sail, or spar. Also spelled halliard.

Halysitidae [PALEON] A family of extinct Paleozoic corals of the order Tabulata.

Hamal [ASTRON] A second-magnitude star in the constellation Aries; the star α Ari.

Hamamelidaceae [BOT] A family of dicotyledonous trees or shrubs in the order Hamamelidales characterized by united carpels, alternate leaves, perfect or unisexual flowers, and free filaments.

Hamamelidae [BOT] A small subclass of plants in the class Magnoliopsida having strongly reduced, often unisexual flowers with poorly developed or no perianth.

Hamamelidales [BOT] A small order of dicotyledonous plants in the subclass Hamamelidae characterized by vessels in the wood and a gynoecium consisting either of separate carpels or of united carpels that open at maturity.

hamartoma [MED] An abnormal condition resulting in the formation of a mass of tissue of disproportionate size and distribution but composed of the normal tissue of the region.

hamartophobia [PSYCH] An abnormal fear of error or sin.

hamate [BIOL] Hook-shaped or hooked.

hambergite [MINERAL] Be_2BO_3OH A grayish-white or colorless mineral composed of beryllium borate and occurring as prismatic crystals; hardness is 7.5 on Mohs scale, and specific gravity is 2.35.

Hamel basis [MATH] For a normed space, a collection of vectors with every finite subset linearly independent, while any vector of the space is a linear combination of at most countably many vectors from this subset.

Hamilton-Cayley theorem See Cayley-Hamilton theorem.

Hamiltonian function [MECH] A function of the generalized coordinates and momenta of a system, equal in value to the sum over the coordinates of the product of the generalized momentum corresponding to the coordinate, and the coordinate's time derivative, minus the Lagrangian of the system; it is numerically equal to the total energy if the Lagrangian does not depend on time explicitly; the equations of motion of the system are determined by the functional dependence of the Hamiltonian on the generalized coordinates and momenta.

Hamiltonian operator See energy operator.

Hamilton-Jacobi equation [MATH] A particular partial differential equation useful in studying certain systems of ordinary equations arising in the calculus of variations, dynamics, and optics: $H(q_1, \ldots, q_n, \partial\phi/\partial q_1, \ldots, \partial\phi/\partial q_n, t) + \partial\phi/\partial t = 0$, where $q_1, \ldots, q_n$ are generalized coordinates, t is the time coordinate, H is the Hamiltonian function, and ϕ is a function that generates a transformation by means of which the generalized coordinates and momenta may be expressed in terms of new generalized coordinates and momenta which are constants of motion.

Hamilton-Jacobi theory [MATH] The study of the solutions of the Hamilton-Jacobi equation and the information they provide concerning solutions of the related systems of ordinary differential equations. [MECH] A theory that provides a means for discussing the motion of a dynamic system in terms of a single partial differential equation of the first order, the Hamilton-Jacobi equation.

Hamilton's equations of motion [MECH] A set of first-order, highly symmetrical equations describing the motion of a classical dynamical system, namely $\dot{q}_j = \partial H/\partial p_j$, $\dot{p}_j = -\partial H/\partial q_j$; here $q_j(j = 1, 2, \ldots)$ are generalized coordinates of the system, p_j is the momentum conjugate to q_j, and H is the Hamiltonian. Also known as canonical equations of motion.

Hamilton's principle [MECH] A variational principle which states that the path of a conservative system in configuration space between two configurations is such that the integral of the Lagrangian function over time is a minimum or maximum relative to nearby paths between the same end points and taking the same time.

hammer [DES ENG] **1.** A hand tool used for pounding and consisting of a solid metal head set crosswise on the end of a handle. **2.** An arm with a striking head for sounding a bell or gong. [MECH ENG] A power tool with a metal block or a drill for the head. [ORD] A metallic pivoted item in a firing mechanism designed to strike a firing pin or percussion cap and thus fire a gun.

hammer code *See* Hamming code.

hammer drill [MECH ENG] Any of three types of fast-cutting, compressed-air rock drills (drifter, sinker, and stoper) in which a hammer strikes rapid blows on a loosely held piston, and the bit remains against the rock in the bottom of the hole, rebounding slightly at each blow, but does not reciprocate.

hammer forging [MET] Forging by means of repeated blows of a hammer.

hammer gun [ORD] Gun with an outside, visible hammer.

hammerhead [DES ENG] The striking part of a hammer.

hammerhead crane [MECH ENG] A crane with a horizontal jib that is counterbalanced.

hammerless gun [ORD] Gun with a wholly enclosed hammer and firing mechanism.

hammer mill [MECH ENG] 1. A type of impact mill or crusher in which materials are reduced in size by hammers revolving rapidly in a vertical plane within a steel casing. Also known as beater mill. 2. A grinding machine which pulverizes feed and other products by several rows of thin hammers revolving at high speed.

hammer milling [MECH ENG] Crushing or fracturing materials in a hammer mill.

hammer pick [MIN ENG] A pneumatic hand-held machine used to break up the harder rocks in a mine; consists of a pick which is driven by a hammer set in a cylinder which receives compressed air.

hammer test [MET] An impact test conducted by dropping weights from increasing heights until a specified deflection of the weight is produced.

hammertoe [MED] A condition of the toe, usually the second, in which the proximal phalanx is extremely extended while the two distal phalanges are flexed.

hammer welding [MET] Forge welding by means of a hammer.

Hammett acidity function [CHEM] An expression for the acidity of a medium, defined as $h_0 = K_{BH^+}[BH^+]/[B]$, where K_{BH^+} is the dissociation constant of the acid form of the indicator, and $[BH^+]$ and $[B]$ are the concentrations of the protonated base and the unprotonated base respectively.

Hamming code [COMMUN] An error-correcting code used in data transmission. Also known as hammer code.

hamming distance *See* signal distance.

Hamoproteidae [INV ZOO] A family of parasitic protozoans in the suborder Haemosporina; only the gametocytes occur in blood cells.

hamper [NAV ARCH] Articles of outfit, especially spars, rigging, and so on, above the deck.

ham radio *See* amateur radio.

hamster [VERT ZOO] The common name for any of 14 species of rodents in the family Cricetidae characterized by scent glands in the flanks, large cheek pouches, and a short tail.

hamstring muscles [ANAT] The biceps femoris, semitendinosus, and semimembranosus collectively.

hamulus [VERT ZOO] A hooklike process, especially a small terminal hook on the barbicel of a feather.

hamus [BIOL] A hook or a curved process.

hancockite [MINERAL] A complex silicate mineral containing lead, calcium, strontium, and other minerals; it is isomorphous with epidote.

Hancock jig [MIN ENG] A moving-screen jig used to treat lead-zinc ores; the material is jigged in a tank of water, and the heavy layer settles through slots.

hand [ANAT] The terminal part of the upper extremity modified for grasping. [TEXT] The quality or feel of a fabric.

hand-and-foot counter *See* hand-and-foot monitor.

hand-and-foot monitor [NUCLEO] An instrument routinely used to monitor the hands and feet of atomic energy workers as they leave locations in which radioactive materials are handled. Also known as hand-and-foot counter.

hand arm *See* hand weapon.

hand auger [DES ENG] A hand tool resembling a large carpenters' bit or comprising a short cylindrical container with cutting lips attached to a rod; used to bore shallow holes in the

soil to obtain samples of it and other relatively unconsolidated near-surface materials.

handbarrow [ENG] A flat, rectangular frame with handles at both ends, carried by two persons to transport objects. Also known as barrow.

hand brake [MECH ENG] A manually operated brake.

hand breadth [ANTHRO] The distance between the outside projections of the distal ends of the second and fifth metacarpals of the right hand, with the fingers extended and together.

hand cable [MIN ENG] A flexible cable for electrical connection between a mining machine and a truck carrying a reel of portable cable. Also known as butt cable; head cable.

handcar [MECH ENG] A small, four-wheeled, hand-pumped car used on railroad tracks to transport men and equipment for construction or repair work; other cars for the same purpose are motor-operated.

hand drill [DES ENG] A small, portable drilling machine which is operated by hand.

handedness [PHYS] A division of objects, such as coordinate systems, screws, and circularly polarized light beams, into two classes (right and left), which distinguishes an object from a mirror image but not from a rotated object.

hand electric lamp [MIN ENG] In mining, a portable, battery operated hand lamp with a tungsten-filament light source that forms a self-contained unit.

hand feed [ENG] A drill machine in which the rate at which the bit is made to penetrate the rock is controlled by a hand-operated ratchet and lever or a hand-turned wheel meshing with a screw mechanism.

hand-foot-and-mouth disease [MED] An infectious disease of humans caused by a coxsackie virus and characterized by maculopapular and vesicular eruptions in the mouth and on the hands and feet.

hand forging [MET] Plastic deformation of a metal by manual force.

hand generator [ELEC] A manually cranked dynamo or alternator, usually used as the prime mover for emergency radio transmitters.

hand grenade [ORD] A grenade designed to be thrown by hand.

handguard [ORD] 1. A wooden part on a rifle to protect the shooter's hand from the hot barrel. 2. The part of a sword or dagger designed to protect the hand.

hand hammer drill [ENG] A hand-held rock drill.

handhole [ENG] A shallow access hole large enough for a hand to be inserted for maintenance and repair of machinery or equipment.

handie-talkie [ELECTR] Two-way radio communications unit small enough to be carried in the hand.

hand jig [MIN ENG] A moving-screen jig operated by hand and used to treat small batches of ore; the jig box is attached to a rocking beam and moved in a tank of water.

handle [MECH ENG] The arm connecting the bucket with the boom in a dipper shovel or hoe.

hand lead [ENG] A light sounding lead (7–14 pounds or 3–6 kilograms) usually having a line not more than 25 fathoms (46 meters) in length.

hand length [ANTHRO] The distance measured from the end of the small wrist bone at the base of the thumb to the tip of the middle finger of the right hand, palm turned up, with the fingers extended and together.

hand lens *See* simple microscope.

hand level [ENG] A hand-held surveyor's level, basically a telescope with a bubble tube attached so that the position of the bubble can be seen when looking through the telescope.

handling time [IND ENG] The time needed to transport parts or materials to or from a work area.

hand loader [MIN ENG] A miner who uses a shovel, rather than a machine to load coal.

hand-operating device [ORD] Mechanism on certain automatic firing weapons that permits the piece to be prepared by hand for firing.

handpicking [MIN ENG] Manual removal of a selected fraction of coarse run-of-mine ore, after washing and screening away waste.

handrail [ENG] A narrow rail to be grasped by a person for support.

HAMSTER

Hamster, *Cricetus oricetus*, body about 6 inches (15 centimeters) long, used as a pet and a subject in laboratory experiments.

hand-rammed [ORD] Indicating that the cartridge is intended to be rammed into the gun by hand rather than by power.

hand-reset [ELEC] Pertaining to a relay in which the contacts must be reset manually to their original positions when normal conditions are resumed.

hand rule [ELECTROMAG] The rule that, when grasping the conductor in the right hand with the thumb pointing in the direction of the current, the fingers will then point in the direction of the lines of flux.

hand sampling [MIN ENG] Using manual methods to detach and reduce in size representative samples of ore; one of the major methods in sampling small batches of ore, others being grab sampling, trench or channel sampling, fractional selection, coning and quartering, and pipe sampling.

handsaw [DES ENG] A saw operated by hand, with a backward and forward arm movement.

Hand-Schüller-Christian disease [MED] A childhood syndrome characterized by exophthalmos, diabetes insipidus, and softened or punched-out areas in the bones.

handset [DES ENG] A combination of a telephone-type receiver and transmitter, designed for holding in one hand.

handset bit [DES ENG] A bit in which the diamonds are manually set into holes that are drilled into a malleable-steel bit blank and shaped to fit the diamonds.

handshaking [COMMUN] The establishment of synchronization between sending and receiving equipment by means of exchange of specific character configurations.

hand-shoulder syndrome See shoulder-hand syndrome.

handspike [ORD] Handle attached to the trails of certain firearms for ease in handling.

hand sugar refractometer [ANALY CHEM] Portable device to read refractive indices of sugar solutions. Also known as proteinometer.

hand time [IND ENG] The time necessary to complete a manual element. Also known as manual time.

hand tool [ENG] Any implement used by hand.

hand tramming [MIN ENG] Pushing of mine cars by manpower; limited to small mines with low output.

hand truck [ENG] **1.** A manually operated, two-wheeled truck consisting of a rectangular frame with handles at the top and a plate at the bottom to slide under the load. **2.** Any of various small, manually operated, multiwheeled platform trucks for transporting materials.

hand type See foundry type.

hand viewer [OPTICS] An optical magnifying device small enough for hand use.

hand weapon [ORD] A weapon, such as a pistol, knife, or sword, used by one hand. Also known as hand arm.

hand winch [MECH ENG] A winch that is operated by hand.

hang [ORD] To lock the receiver or bolt of a gun in an open position.

hangar [CIV ENG] A building at an airport specially designed in height and width to enable aircraft to be stored or maintained in it.

hangar deck [NAV ARCH] A deck, below the flight deck of a carrier, where aircraft are parked and serviced.

hanger [GEOL] See hanging wall. [PETRO ENG] **1.** A device to seat in the bowl of a lowermost casing head to suspend the next-smaller casing string and form a seal between the two. Also known as casing hanger. **2.** A device to provide a seal between the tubing and the tubing head. Also known as tubing hanger.

hanger bolt [DES ENG] A bolt with a machine-screw thread on one end and a lag-screw thread on the other.

hangfire [ENG] Delay in the explosion of a charge.

hanging [GEOL] See hanging wall. [MET] Sticking or wedging of part of the charge in a blast furnace.

hanging bolt [MIN ENG] A bolt used to suspend wall plates in shaft construction.

hanging coal [MIN ENG] A portion of the coal seam which, by undercutting, has had its natural support removed.

hanging compass See inverted compass.

hanging-drop atomizer [MECH ENG] An atomizing device used in gravitational atomization; functions by quasi-static emission of a drop from a wetted surface. Also known as pendant atomizer.

hanging-drop preparation [MICROBIO] A technique used in microscopy in which a specimen is placed in a drop of a suitable fluid on a cover slip and the cover slip is inverted over a concavity on a slide.

hanging glacier [HYD] A glacier lying above a cliff or steep mountainside; as the glacier advances, calving can cause ice avalanches.

hanging load [MECH ENG] **1.** The weight that can be suspended on a hoist line or hook device in a drill tripod or derrick without causing the members of the derrick or tripod to buckle. **2.** The weight suspended or supported by a bearing.

hanging scaffold [CIV ENG] A movable platform suspended by ropes and pulleys; used by workers for above-ground building construction and maintenance.

hanging sets [MIN ENG] Timbers from which cribs are suspended in working through soft strata.

hanging sheave [MIN ENG] The grooved wheel or pulley which is suspended from the drill tripod clevis or from the roof or side of a haulage road, and over which the hoist line runs to minimize friction.

hanging side See hanging wall.

hanging valley [GEOL] A valley whose floor is higher than the level of the shore or other valley to which it leads.

hanging wall [GEOL] The rock mass above a fault plane, vein, lode, ore body, or other structure. Also known as hanger; hanging; hanging side.

hanging-wall drift [MIN ENG] A horizontal gallery that is driven in the hanging wall of a vein.

hangover [COMMUN] **1.** In television, overlapping and blurring of successive frames opposite to direction of subject motion, due to improper adjustment of transient response. **2.** In facsimile, distortion produced when the signal changes from maximum to minimum conditions at a slower rate than required, resulting in tailing on the lines in the recorded copy. [MED] Aftereffect following excessive intake of alcohol or certain drugs, such as barbiturates.

hang-up [ADP] A nonprogrammed stop in a computer routine caused by a human mistake or a computer malfunction. [MIN ENG] Blockage of the movement of ore by rock in an underground chute.

hangwire [ORD] A length of wire connecting the fuse assembly of an aerial flare or bomb to the structure of an aircraft; it removes the safety and arms the fuse at the beginning of the drop; it may also open a parachute or stabilizing sleeve.

hank [TEXT] A unit of measure of yarn or thread equal to 840 yards (768.096 meters) for cotton, and 560 yards (512.064 meters) for first-grade wool.

Hankel functions [MATH] The Bessel functions of the third kind, occurring frequently in physical studies.

Hankel transform [MATH] The Hankel transform of order m of a real function $F(s)$ given by the integral from 0 to ∞ of $f(t)tJ_m(st)dt$, where J_m denotes the mth-order Bessel function. Also known as Bessel transform; Fourier-Bessel transform.

hanksite [MINERAL] $Na_{22}K(SO_4)_9(CO_3)_2Cl$ A white or yellow mineral crystallizing in the hexagonal system; found in California.

hannayite [MINERAL] $Mg_3(NH_4)_2H_2(PO_4)_4 \cdot 8H_2O$ Mineral composed of hydrous acid ammonium magnesium phosphate; occurs as yellow crystals in guano.

Hansa yellow [ORG CHEM] Group of organic azo pigments with strong tinting power, but poor opacity in paints; used where nontoxicity is important.

Hansen's bacillus See Mycobacterium leprae.

Hansen's disease [MED] An infectious disease of humans thought to be caused by *Mycobacterium leprae*; common manifestations are cutaneous and neural lesions. Also known as leprosy.

Hanson unit [BIOL] A unit for the standardization of parathyroid hormone.

H antigen [MICROBIO] A general term for microbial flagellar antigens; former designation for species-specific flagellar antigens of *Salmonella*.

Hantzsch synthesis [ORG CHEM] The reaction whereby a pyrrole compound is formed when a β-ketoester, chloroacetone, and a primary amine condense.

HAND TRUCK

handles curved

nose

legs

straight frame

straight crossmember

An Eastern type of hand truck showing curved handles, straight frame, straight crossmember and other components.

Hanus solution [ANALY CHEM] Iodine monobromide in glacial acetic acid; used to determine iodine values in oils containing unsaturated organic compounds.

haplobiont [BOT] A plant that produces only sexual haploid individuals.

haplocaulescent [BOT] Having a simple axis with the reproductive organs on the principal axis.

Haplodoci [VERT ZOO] The equivalent name for Batrachoidiformes.

haploid [GEN] Having half of the diploid or full complement of chromosomes, as in mature gametes.

Haplolepidae [PALEON] A family of Carboniferous chondrostean fishes in the suborder Palaeoniscoidei having a reduced number of fin rays and a vertical jaw suspension.

Haplomi [VERT ZOO] An equivalent name for Salmoniformes.

haplomitosis [CYTOL] Type of primitive mitosis in which the nuclear granules form into threadlike masses rather than clearly differentiated chromosomes.

haplont [BOT] A plant with only haploid somatic cells; the zygote is diploid.

haplophase [BIOL] Haploid stage in the life cycle of an organism.

haplopore [PALEON] Any randomly distributed pore on the surface of fossil cystoid echinoderms.

Haplosclerida [INV ZOO] An order of sponges in the class Demospongiae including species with a skeleton made up of siliceous megascleres embedded in spongin fibers or spongin cement.

haplosis [CYTOL] Reduction of the chromosome number to half during meiosis.

Haplosporea [INV ZOO] A class of Protozoa in the subphylum Sporozoa distinguished by the production of spores lacking polar filaments.

Haplosporida [INV ZOO] An order of Protozoa in the class Haplosporea distinguished by the production of uninucleate spores that lack both polar capsules and filaments.

haplostele [BOT] A type of protostele with the core of xylem characterized by a smooth outline.

hapten [IMMUNOL] A simple substance that reacts like an antigen in vitro by combining with antibody; may function as an allergen when linked to proteinaceous substances of the tissue.

haptephobia [PSYCH] An abnormal fear of being touched.

hapteron [BOT] A disklike holdfast on the stem of certain algae.

haptochlamydeous [BOT] Having the sporophylls protected by rudimentary perianth leaves.

haptoglobin [BIOCHEM] An alpha globulin that constitutes 1–2% of normal blood serum; contains about 5% carbohydrate.

Haptophyceae [BOT] A class of the phylum Chrysophyta that contains the Coccolithophorida.

haptor [INV ZOO] The posterior organ of attachment in certain monogenetic trematodes characterized by multiple suckers and the presence of hooks.

haptotropism [BIOL] Movement of sessile organisms in response to contact, especially in plants.

harassing agent [ORD] A chemical agent that forces troops to wear gas masks and to cut down their efficiency; examples are an irritating gas or smoke.

harassing fire [ORD] Rifle or artillery fire designed to inflict losses, or to disturb the enemy troops by the threat of losses; the fire restricts movement and lowers troop morale.

harbor [GEOGR] Any body of water of sufficient depth for ships to enter and find shelter from storms or other natural phenomena. Also known as port.

harbor chart [NAV] A nautical chart for navigation and anchorage in small waterways and harbors.

harbor engineering [CIV ENG] Planning and design of facilities for ships to discharge or receive cargo and passengers.

harbor line [CIV ENG] The line beyond which wharves and other structures cannot be extended.

harbor reach [GEOGR] The stretch of a river or estuary which leads directly to the harbor.

hard [MATER] Quality of a material that is compact, solid, and difficult to deform.

hard acid [CHEM] A Lewis acid of low polarizability, small size, and high positive oxidation state; it does not have easily excitable outer electrons; some examples are H^+, Li^+, and Al^{3+}.

hard base [CHEM] A Lewis base (electron donor) that has high polarizability and low electronegativity, is easily oxidized, or possesses lowlying empty orbitals; some examples are H_2O, HO^-, OCH_3^-, and F^-. [ORD] A launching base that is protected against a nuclear explosion.

hard beach [CIV ENG] A portion of a beach especially prepared with a hard surface extending into the water, employed for the purpose of loading or unloading directly into or from landing ships or landing craft.

hardboard [MATER] A fiberboard formed to a density of 30–50 pounds per cubic foot (480–800 kilograms per cubic meter) and having one textured and one smooth face.

hard bottom [MIN ENG] A condition encountered in some opencut mines wherein the rock occasionally does not break down to grade because of an extra-hard streak of ground or because insufficient powder is used.

hard bronze [MET] A high-tensile-strength alloy containing 88% copper, 7% tin, 3% zinc, and 2% lead.

hard-burned brick [MATER] A brick that has been fired and sintered at high temperature.

hard chromium [MET] A thick coating of electrodeposited chromium on a base metal; increases wear resistance.

hard coal *See* anthracite.

hard-coal plough [MIN ENG] A plough-type cutter-loader consisting of rigid or swiveling kerfing bits which precut the coal in hard-coal seams.

hard copy [ADP] Human-readable typewritten or printed characters produced on paper at the same time that information is being keyboarded in a coded machine language, as when punching cards or paper tape.

hard cosmic ray [NUCLEO] A cosmic-radiation component that penetrates a moderate thickness of an absorber, such as 4 inches of lead.

hard data [SCI TECH] Data in the form of numbers or graphs, as opposed to qualitative information.

hard detergent [CHEM] A nonbiodegradable detergent.

hard-drawn wire [MET] Cold-drawn metal wire, usually of high tensile strength.

hardenability [MET] In a ferrous alloy, the property that determines the depth and distribution of hardness induced by quenching from elevated temperatures.

hardened links [COMMUN] Transmission links that require special construction or installation to assure a high probability of survival under nuclear attack.

hardened site [ORD] An underground missile-launching site, control center, or other facility that can withstand in varying degrees the effects of an enemy nuclear blast.

hardened steel [MET] Steel hardened by quenching from high temperatures.

hardener [MET] A master alloy added to a melt to control hardness. [ORG CHEM] Compound reacted with a resin polymer to harden it, such as the amines or anhydrides that react with epoxides to cure or harden them into plastic materials. Also known as curing agent.

hardener bath [GRAPHICS] A fixing bath containing chemicals such as formalin and chrome alum to toughen the emulsion of a film.

hardening [MET] **1.** Imparting hardness to carbon steel by abrupt cooling (quenching) through a critical temperature range. **2.** Heat-treating an age-hardening or precipitation-hardening alloy at intermediate temperatures.

Harderian gland [VERT ZOO] An accessory lacrimal gland associated with lower eyelid structures in all vertebrates except land mammals.

hard-face [MET] To apply a layer of hard, abrasion-resistant metal to a less resistant metal part by plating, welding, spraying, or other techniques. Also known as hard-surface.

hard fiber [BOT] A heavily lignified leaf fiber used in making cordage, twine, and textiles.

hard-fiber [MATER] Indicating vulcanization with zinc chloride; used of paper or boards.

hard freeze [HYD] A freeze in which seasonal vegetation is destroyed, the ground surface is frozen solid underfoot, and

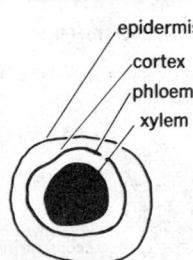

HAPLOSTELE

epidermis
cortex
phloem
xylem

Cross section of a haplostele.

heavy ice is formed on small water surfaces such as puddles and water containers.

hard frost *See* black frost.

hard glass [MATER] A potash-lime glass with a high silica content, used for making brilliant glassware. Also known as Bohemian glass.

hard grease [MATER] A lubricating grease that flows at a temperature of about 90°C.

hard ground [MIN ENG] Ground that is difficult to work.

Hardgrove grindability index [ENG] The relative grindability of ores and minerals in comparison with standard coal, chosen as 100 grindability, as determined by a miniature ball-ring pulverizer. Also known as Hardgrove number.

Hardgrove number *See* Hardgrove grindability index.

hard hat [ENG] A safety hat usually having a metal crown; used by construction workers and miners.

hardhead [MET] A hard white deposit formed during tin refining by liquation.

Hardinge feeder-weigher [MECH ENG] A pivoted, short belt conveyor which controls the rate of material flow from a hopper by weight per cubic foot.

Hardinge mill [MECH ENG] A tricone type of ball mill; the cones become steeper from the feed end toward the discharge end.

Hardinge thickener [ENG] A machine for removing the maximum amount of liquid from a mixture of liquid and finally divided solids by allowing the solids to settle out on the bottom as sludge while the liquid overflows at the top.

hard iron [MET] Iron or steel which is not readily magnetized by induction, but which retains a high percentage of the magnetism acquired.

hard lac [MATER] Solvent-extracted shellac. Also known as hard-lac resin.

hard-lac resin *See* hard lac.

hard-laid [DES ENG] Pertaining to rope with strands twisted at a 45° angle.

hard landing [AERO ENG] A landing made without deceleration, as by impact on the moon.

hard lead [MET] Lead alloy with reduced malleability due to the presence of impurities, usually antimony.

hard-limiting [COMMUN] Limiting condition for which limiting takes place for at least 20 decibels into the signal noise.

hard magnetic material [MET] A metal having a high coercive force which gives a large magnetic hysteresis.

hardness [CHEM] The amount of calcium carbonate dissolved in water, usually expressed as parts of calcium carbonate per million parts of water. [ELECTROMAG] That quality which determines the penetrating ability of x-rays; the shorter the wavelength, the harder and more penetrating the rays. [ENG] Property of an installation, facility, transmission link, or equipment that will prevent an unacceptable level of damage. [MATER] Resistance of a metal or other material to indentation, scratching, abrasion, or cutting.

hardness number [ENG] A number representing the relative hardness of a mineral, metal, or other material as determined by any of more than 30 different hardness tests.

hardness test [ANALY CHEM] Test to determine the calcium and magnesium content of water. [ENG] A test to determine the relative hardness of a metal, mineral, or other material according to one of several scales, such as Brinell, Mohs, or Shore.

hard pad [VET MED] A disease of dogs, probably associated with the canine distemper virus; often characterized by encephalitis and hardening of the foot pads.

hard palate [ANAT] The anterior portion of the roof of the mouth formed by paired palatine processes of the maxillary bones and by the horizontal part of each palate bone.

hardpan [GEOL] A hard, impervious layer of soil clay cemented by insoluble materials. Also known as caliche.

hard paste porcelain *See* porcelain.

hard porcelain [MATER] A ceramic material having good resistance to thermal shock.

hard radiation [PHYS] Radiation whose particles or photons have a high energy and, as a result, readily penetrate all kinds of materials, including metals.

hard rime [HYD] Opaque, granular masses of rime deposited chiefly on vertical surfaces by a dense super-cooled fog; it is

more compact and amorphous than soft rime, and may build out into the wind as glazed cones or feathers.

hard rock [GEOL] Rock which needs drilling and blasting for removal.

hard-rock driller [MIN ENG] A worker who operates a drill in a mine where the rocks are generally igneous or metamorphosed and considered hard, such as rocks in which coal and salt are generally found.

hard-rock mine [MIN ENG] A mine located in hard rock, especially a mine difficult to drill, blast, and square up.

hard-rock miner [MIN ENG] A worker competent to mine in hard rock, usually an expert miner.

hard-rock tunnel boring [MIN ENG] A tunneling method utilizing a mole to cut out 7-foot-diameter (2.1-meter-diameter) drifts in hard rock at an average rate of 5 feet (1.5 meters) per hour.

hard rot [PL PATH] **1.** Any plant disease characterized by lesions with hard surfaces. **2.** A fungus disease of gladiolus caused by *Septoria gladioli* which produces hard-surfaced lesions on the leaves and corms.

hard rubber [MATER] Rubber that has been vulcanized at high temperatures and pressures to give hardness; used as an electrical insulating material and in tool handles. Also known as ebonite.

hard silk [TEXT] Silk that has not been degummed, so that the finished fabric contains sericin, or silk gum.

hard site [ORD] Underground missile-launching site, control center, or other facility that can withstand in varying degrees the effects of an enemy nuclear blast.

hard solder [MET] *See* brazing alloy.

hard-solder *See* braze.

hardstand [CIV ENG] **1.** A paved or stabilized area where vehicles or aircraft are parked. **2.** Open ground area having a prepared surface and used for storage of material.

hard superconductor [CRYO] A superconductor that requires a strong magnetic field, over 1000 oersteds, to destroy superconductivity; niobium and vanadium are examples.

hard-surface [CIV ENG] To treat a ground surface in order to prevent muddiness. [MET] *See* hard-face.

hard tube *See* high-vacuum tube.

hardware [ADP] The physical, tangible, and permanent components of a computer or a data-processing system. [ENG] Items made of metal, such as tools, fittings, fasteners, and appliances. [ORD] Metal military items for use in combat.

hardware check *See* machine check.

hardware control [ADP] The control of, and communications between, the various parts of a computer system.

hardware monitor [ADP] A system used to evaluate the performance of computer hardware; it collects information such as central processing unit usage from voltage level sensors that are attached to the circuitry and measure the length of time or the number of times various signals occur, and displays this information or stores it on a medium that is then fed into a special data-reduction program.

hard water [CHEM] Water that contains certain salts, such as those of calcium or magnesium, which form insoluble deposits in boilers and form precipitates with soap.

Hardwick conveyor loader head [MIN ENG] A dust collector for belt conveyors used at the loading station; a scraper chain runs at the bottom of a coal hopper and collects underbelt fines.

hard-wire [ELEC] To connect electric components with solid, metallic wires as opposed to radio links, and the like.

hard-wire telemetry *See* wire-link telemetry.

hardwood [MATER] Dense, close-grained wood of an angiospermous tree, such as oak, walnut, cherry, and maple.

hardwood bearing [MECH ENG] A fluid-film bearing made of lignum vitae which has a natural gum, or of hard maple which is impregnated with oil, grease, or wax.

hardwood forest [ECOL] **1.** An ecosystem having deciduous trees as the dominant form of vegetation. **2.** An ecosystem consisting principally of trees that yield hardwood.

hard x-ray [ELECTR] An x-ray having high penetrating power.

Hardy plankton indicator [ENG] Metal-shrouded net sampler designed to collect specimens of plankton during normal passage of a ship.

HARELIP

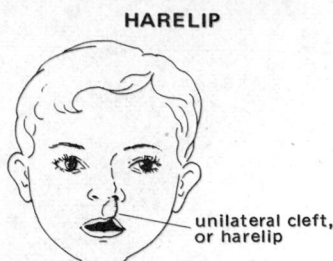

Drawing of a harelip. (From L. B.
Arey, Developmental Anatomy,
7th ed., Saunders, 1965)

HARMONIC SPEED CHANGER

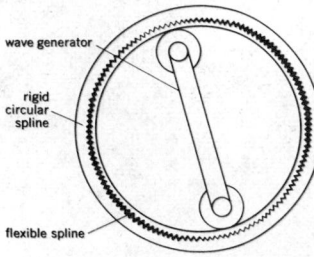

Rotary-to-rotary harmonic speed
changer that effects a speed
reduction of 1:66.

Hardy-Schulz rule [PHYS CHEM] An increase in the charge of ions results in a large increase in their flocculating power.

hardystonite [MINERAL] $Ca_2ZnSi_2O_7$ A white mineral composed of zinc calcium silicate.

hare [VERT ZOO] The common name for a number of lagomorphs in the family Leporidae; they differ from rabbits in being larger with longer ears, legs, and tails.

Hare See Lepus.

harelip [MED] A congenital defect, sometimes hereditary, marked by an abnormal cleft between the upper lip and the base of the nose. Also known as cleft lip.

Haring cell [PHYS CHEM] An electrolytic cell with four electrodes used to measure electrolyte resistance and polarization of electrodes.

Harker-Kasper inequalities [SOLID STATE] Inequalities used in the analysis of crystal structure by x-ray diffraction which relate the structure factors and help to determine their phase factors.

Harkin's rule [PHYS] An empirical rule for the calculation of the nuclear abundances of an element's isotopes stating that isotopes with an odd mass number are less abundant than their even-mass-number neighbors.

Harlechian [GEOL] A European stage of geologic time: Lower Cambrian.

harmatan See harmattan.

harmattan [METEOROL] A dry, dust-bearing wind from the northeast or east which blows in West Africa especially from late November until mid-March; it originates in the Sahara as a desert wind and extends southward to about 5°N in January and 18°N in July. Also spelled harmatan; harmetan; hermitan.

harmetan See harmattan.

harmful interference [COMMUN] Radiation or induction which endangers the functioning of a radionavigation broadcasting service or of a safety broadcasting service, or obstructs or repeatedly interrupts a radio service operating in accordance with the appropriate regulations.

harmless-depth theory [MIN ENG] Formerly, hypothesis that there was a certain depth below which mining could be carried on without risk of damage to the surface.

harmonic [ACOUS] One of a series of sounds, each of which has a frequency which is an integral multiple of some fundamental frequency. [MATH] A solution of Laplace's equation which is separable in a specified coordinate system. [PHYS] A sinusoidal component of a periodic wave, having a frequency that is an integral multiple of the fundamental frequency. Also known as harmonic component.

harmonica bug [ELECTR] A surreptitious interception technique applied to telephone lines; the target instrument is modified so that a tuned relay bypasses the switch hook and ringing circuit when a 500-hertz tone is received; this tone was originally generated by use of a harmonica.

harmonic analysis [MATH] A study of functions by attempting to represent them as infinite series or integrals which involve functions from some particular well-understood family; it subsumes studying a function via its Fourier series. [PHYS] Any method of identifying and evaluating the harmonics that make up a complex waveform of sound pressure, voltage, current, or some other varying quantity.

harmonic analyzer [ELECTR] An instrument that measures the strength of each harmonic in a complex wave. Also known as harmonic wave analyzer.

harmonic antenna [ELECTROMAG] An antenna whose electrical length is an integral multiple of a half-wavelength at the operating frequency of the transmitter or receiver.

harmonic attenuation [ELECTR] Attenuation of an undesired harmonic component in the output of a transmitter.

harmonic component See harmonic.

harmonic conjugates [MATH] **1.** Two points, P_3 and P_4, that are collinear with two given points, P_1 and P_2, such that P_3 lies in the line segment $P_1 P_2$ while P_4 lies outside it, and, if $x_1, x_2, x_3,$ and x_4 are the abscissas of the points, $(x_3 - x_1) / (x_3 - x_2) = - (x_4 - x_1) / (x_4 - x_2)$. **2.** A pair of harmonic functions, u and v, such that $u + iv$ is an analytic function, or, equivalently, u and v satisfy the Cauchy-Riemann equations.

harmonic content [PHYS] The components remaining after the fundamental frequency has been removed from a complex wave.

harmonic conversion transducer [ELECTR] A conversion transducer of which the useful output frequency is a multiple or a submultiple of the input frequency.

harmonic decline [PETRO ENG] One of three types of decline in oil or gas production rate (the others are constant-percentage and hyperbolic), in which the nominal decline in production rate per unit of time expressed as a fraction of the production rate is proportional to the production rate itself.

harmonic detector [ELECTR] Voltmeter circuit so arranged as to measure only a particular harmonic of the fundamental frequency.

harmonic distortion [ELECTR] Nonlinear distortion in which undesired harmonics of a sinusoidal input signal are generated because of circuit nonlinearity.

harmonic division [MATH] The division of a line segment externally and internally in the same ratio; that is, the division of a line segment by the harmonic conjugates of its end points.

harmonic fields [ELECTROMAG] The sinusoidal Fourier components of a magnetic or other field confined to a finite region of space; their half-wavelengths are integral divisors of the length of the space in which the field is confined.

harmonic filter [ELECTR] A filter that is tuned to suppress an undesired harmonic in a circuit.

harmonic folding [GEOL] Folding in the earth's surface, with no sharp changes with depth in the form of the folds.

harmonic frequency [PHYS] An integral multiple of the fundamental frequency of a periodic wave.

harmonic function [MATH] **1.** A function of two real variables which is a solution of Laplace's equation in two variables. **2.** A function of three real variables which is a solution of Laplace's equation in three variables.

harmonic generator [ELECTR] A generator operated under conditions such that it generates strong harmonics along with the fundamental frequency.

harmonic interference [COMMUN] Interference due to the presence of harmonics in the output of a radio station.

harmonic loss [ELECTROMAG] Energy loss in a generator due to space harmonics of the magnetomotive force produced by armature current, especially losses resulting from the fifth and seventh harmonics.

harmonic mean [MATH] For n positive numbers $x_1, x_2, \ldots, x_n$ their harmonic mean is the number $n/(1/x_1 + 1/x_2 + \ldots + 1/x_n)$.

harmonic motion [MECH] A periodic motion that is a sinusoidal function of time, that is, motion along a line given by the equation $x = a \cos(kt + \theta)$, where t is the time parameter, and a, k, and θ are constants. Also known as harmonic vibration; simple harmonic motion (SHM).

harmonic oscillator [ELECTR] See sinusoidal oscillator. [MECH] Any physical system that is bound to a position of stable equilibrium by a restoring force or torque proportional to the linear or angular displacement from this position. [PHYS] Anything which has equations of motion that are the same as the system in the mechanics definition. Also known as simple oscillator.

harmonic producer [ELECTR] Tuning-fork controlled oscillator device capable of producing odd and even harmonics of the fundamental tuning-fork frequency; used to provide carrier frequencies for broad-band carrier systems.

harmonic progression [MATH] A sequence of numbers whose reciprocals form an arithmetic progression.

harmonic selective ringing [COMMUN] Selective ringing which employs currents of several frequencies and ringers, each tuned mechanically or electrically to the frequency of one of the ringing currents, so that only the desired ringer responds.

harmonic speed changer [MECH ENG] A mechanical-drive system used to transmit rotary, linear, or angular motion at high ratios and with positive motion.

harmonic synthesizer [MECH] A machine which combines elementary harmonic constituents into a single periodic function; a tide-predicting machine is an example.

harmonic telephone ringer [ELECTR] Telephone ringer which responds only to alternating current within a very narrow frequency band.

harmonic tide plane See Indian spring low water.

harmonic vibration See harmonic motion.

harmonic vibration-rotation band [SPECT] A vibration-rotation band of a molecule in which the harmonic oscillator approximation holds for the vibrational levels, so that the vibrational levels are equally spaced.

harmonic wave [PHYS] A transverse waveform obtained by mapping onto a time base the periodic up and down excursions of simple harmonic motion.

harmonic wave analyzer *See* harmonic analyzer.

harmonize [ORD] **1.** To align the sights of a gun so that the curving path of the projectile will meet the straight line of sight at the target. **2.** To adjust or align the gunsight, guns, rocket launchers, or gun camera of a fighter aircraft so that an accurate aim is obtained at a given range, or so that the guns will produce a desired pattern of fire.

harmotome [MINERAL] $(K,Ba)(Al,Si)_2(Si_6O_{16}) \cdot 6H_2O$ A zeolite mineral with ion-exchange properties that forms cruciform twin crystals. Also known as cross-stone.

Harnack's first convergence theorem [MATH] The theorem that if a sequence of functions harmonic in a common domain of three-dimensional space and continuous on the boundary of the domain converges uniformly on the boundary, then it converges uniformly in the domain to a function which is itself harmonic; the sequence of any partial derivative of the functions in the original sequence converges uniformly to the corresponding partial derivative of the limit function in every closed subregion of the domain.

Harnack's second convergence theorem [MATH] The theorem that if a sequence of functions is harmonic in a common domain of three-dimensional space and their values are monotonically decreasing at any point in the domain, then convergence of the sequence at any point in the domain implies uniform convergence of the sequence in every closed subregion of the domain to a function which is itself harmonic.

harness [AERO ENG] **1.** Straps arranged to hold an occupant of a spacecraft or aircraft in his seat. **2.** Straps worn by a parachutist or used to suspend a load from a parachute. [ELEC] Wire and cables so arranged and tied together that they may be inserted and connected, or may be removed after disconnection, as a unit. [TEXT] One of two or more frames on a loom which are raised to separate the warp from the filler yarns to allow the shuttle to pass between them.

Harpacticoida [INV ZOO] An order of minute copepod crustaceans of variable form, but generally being linear and more or less cylindrical.

harpago [INV ZOO] Part of the clasper on the copulatory organ of certain male insects.

Harpidae [INV ZOO] A family of marine gastropod mollusks in the order Neogastropoda.

harpoon [DES ENG] A barbed spear used to catch whales.

harpoon log [ENG] A log which consists essentially of a rotator and distance registering device combined in a single unit, and towed through the water; it has been largely replaced by the taffrail log; the two types of logs are similar except that the registering device of the taffrail log is located at the taffrail and only the rotator is in the water.

Harris flow [ELECTR] Electron flow in a cylindrical beam in which a radial electric field is used to overcome space charge divergence.

Harris process [MET] A method for refining lead in which the liquid bullion is sprayed through molten caustic soda and molten sodium nitrate; arsenic, antimony, and tin are oxidized, converted into sodium salts, and skimmed from the bath.

harrow [AGR] An implement that is pulled over plowed soil to break clods, level the surface, and destroy weeds.

harrowing [AGR] Cultivation of the soil with a harrow.

harstigite [MINERAL] $Be_2Ca_3Si_3O_{11}$ A mineral composed of silicate of beryllium and calcium.

Hartford loop [MECH ENG] A condensate return arrangement for low-pressure, steam-heating systems featuring a steady water line in the boiler.

hartite [GEOL] A white, crystalline, fossil resin that is found in lignites. Also known as bombiccite; branchite; hofmannite; josen.

hartley [COMMUN] A unit of information content, equal to the designation of 1 of 10 possible and equally likely values or states of anything used to store or convey information.

Hartley formula [COMMUN] The relationship expressing that

as the time function is made narrower, the frequency spectrum must become broader.

Hartley oscillator [ELECTR] A vacuum-tube oscillator in which the parallel-tuned tank circuit is connected between grid and anode; the tank coil has an intermediate tap at cathode potential, so the grid-cathode portion of the coil provides the necessary feedback voltage.

Hartley principle [COMMUN] The principle that the total number of bits of information that can be transmitted over a channel in a given time is proportional to the product of channel bandwidth and transmission time.

Hartmann diaphragm [ANALY CHEM] Comparison device for positive-element-identification readings from emission spectra.

Hartmann dispersion formula [OPTICS] A semiempirical formula relating the index of refraction n and wavelengths λ; $n = n_0 + a/(\lambda - \lambda_0)$, where n_0, a, and λ_0 are empirical constants.

Hartmann flow [PL PHYS] The steady flow of an electrically conducting fluid between two parallel plates when there is a uniform applied magnetic field normal to the plates.

Hartmann lines *See* Lüders lines.

Hartmann number [PL PHYS] A dimensionless number which gives a measure of the relative importance of drag forces resulting from magnetic induction and viscous forces in Hartmann flow, and determines the velocity profile for such flow.

Hartmann test [OPTICS] A test for telescope mirrors in which the mirror is covered with a screen with regularly spaced holes, and a photographic plate is placed near the focus; for a perfect mirror, this results in regularly spaced dots on the plate. [SPECT] A test for spectrometers in which light is passed through different parts of the entrance slit; any resulting changes of the spectrum indicate a fault in the instrument.

Hartman's solution [ANALY CHEM] Solution of thymol, ethyl alcohol, and sulfuric ether; used for selective dentin analysis.

Hartman unit [BIOL] A unit for the standardization of adrenal cortical hormone.

Hart-Park virus [VIROL] A ribonucleic acid–containing animal virus of the rhabdovirus group.

hartree [ATOM PHYS] A unit of energy used in studies of atomic spectra and structure, equal (in centimeter-gram-second units) to $4\pi^2 me^4/h^2$, where e and m are the charge and mass of the electron, and h is Planck's constant; equal to approximately 27.21 electron volts or 4.360×10^{-18} joule.

Hartree equation [ELECTR] An equation which gives the lowest anode voltage at which it is theoretically possible to maintain oscillation in the different modes of a magnetron.

Hartree-Fock approximation [QUANT MECH] A refinement of the Hartree method in which one uses determinants of single-particle wave functions rather than products, thereby introducing exchange terms into the Hamiltonian.

Hartree method [QUANT MECH] An iterative variational method of finding an approximate wave function for a system of many electrons, in which one attempts to find a product of single-particle wave functions, each one of which is a solution of the Schrödinger equation with the field deduced from the charge density distribution due to all the other electrons. Also known as self-consistent field method.

Hartree units [ATOM PHYS] A system of units in which the unit of angular momentum is Planck's constant divided by 2π, the unit of mass is the mass of the electron, and the unit of charge is the charge of the electron. Also known as atomic units.

hartshorn oil *See* bone oil.

harvester [AGR] A machine used to reap field crops.

harvester-thresher [AGR] A machine that combines the harvesting and threshing of grain crops.

harvesting [AGR] The gathering of mature field crops.

harvest moon [ASTRON] A full moon that is seen nearest the autumnal equinox.

harzburgite [PETR] A peridotite consisting principally of olivine and orthopyroxene.

Harz jig [MIN ENG] A device used to separate coal and foreign matter which gives pulsion intermittently with suction.

Hasche process [CHEM ENG] A thermal reforming process for hydrocarbon fuels; it is a noncatalytic regenerative

---

HARPACTICOIDA

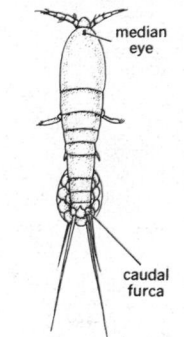

Harpacticus chelifer, a typical harpacticoid.

HARTLEY OSCILLATOR

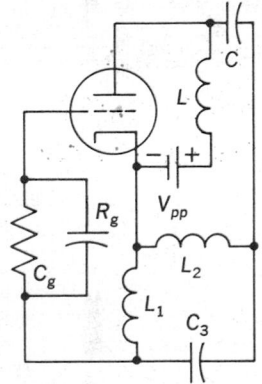

Circuit diagram for a Hartley oscillator.

HARVESTER

Self-propelled forage harvester for corn. *(Sperry New Holland, Division of Sperry Rand Corp.)*

method in which a mixture of hydrocarbon gas or vapor and air is passed through a regenerative mass that is progressively hotter in the direction of the gas flow; partial combustion occurs, liberating heat to crack the remaining hydrocarbons in a combustion zone.

Hasenclever turntable [MIN ENG] A turntable that is made to rotate by the friction between the positively driven pulley, the car, and the table; used as an alternative to the shunt-back or the traverser for changing the direction of mine cars or tubs, either on the surface or underground.

hash [ADP] Data which are obviously meaningless, caused by human mistakes or computer malfunction. Also known as garbage; gibberish. [ELEC] Electric noise produced by the contacts of a vibrator or by the brushes of a generator or motor. [ELECTR] *See* grass.

Hashimoto's disease *See* struma lymphomatosa.

Hashimoto's struma *See* struma lymphomatosa.

hashish [PHARM] A narcotic drug derived from the plant *Cannabis sativa;* can be smoked, chewed, or drunk.

hash total [ADP] A sum obtained by adding together numbers having different meanings; the sole purpose is to ensure that the correct number of data have been read by the computer.

Hassal's body *See* thymic corpuscle.

hastate [BIOL] Shaped like an arrowhead with divergent barbs.

haster [METEOROL] In England, a violent rain storm.

hastingsite [MINERAL] $NaCa_2(Fe,Mg)_5Al_2Si_6O_{22}(OH)_2$ A mineral of the amphibole group crystallizing in the monoclinic system and composed chiefly of sodium, calcium, and iron, but usually with some potassium and magnesium.

hasty mine field [ORD] Field of mines quickly laid as a protection against an enemy attack; when practicable, it is laid in a definite pattern, as is a deliberate field, but measurements are approximate rather than exact.

hat [COMMUN] To arrange a fixed quantity of symbols or groups of symbols in an entirely haphazard sequence, as if they had been drawn from a hat.

HAT *See* height above touch-down.

hatch [ENG] A door or opening, especially on an airplane, spacecraft, or ship.

hatch battens [NAV ARCH] Flat bars used to fasten and make tight the edges of a tarpaulin covering a hatch.

hatch beam [ENG] A heavy, portable beam which supports a hatch cover.

hatch carlings [NAV ARCH] Fore and aft girders running under the coamings of hatches, to which the partial or half deck beams are attached.

hatch coaming [NAV ARCH] A raised frame around a hatch; it forms a support for the hatch cover and strengthens the edges of the opening.

hatch cover [ENG] A steel or wooden cover for a hatch.

hatch end beam [NAV ARCH] The deck beam at the fore and aft end of a hatch.

hatchet [DES ENG] A small ax with a short handle and a hammerhead in addition to the cutting edge.

hatchettine *See* hatchettite.

hatchettite [MINERAL] $C_{38}H_{78}$ A yellow-white mineral paraffin wax, melting at 55–65°C in the natural state and 79°C in the pure state; occurs in masses in ironstone nodules or in cavities in limestone. Also known as adipocerite; adipocire; hatchettine; mineral tallow; mountain tallow; naphthine.

hatchettolite *See* ellsworthite.

hatching [GRAPHICS] Parallel lines drawn on sections of plans for buildings or machines to distinguish between different materials.

hatchite [MINERAL] A lead-gray mineral composed of sulfide of lead and arsenic; occurs in triclinic crystals.

hatted code [COMMUN] Randomized code consisting of an encoding section; the plain text groups are arranged in alphabetical or other significant order, accompanied by their code groups arranged in a nonalphabetical or random order.

H attenuator *See* H network.

haud [METEOROL] In Scotland, a squall.

hauerite [MINERAL] MnS_2 A reddish-brown or brownish-black mineral composed of native manganese sulfide; occurs massive or in octahedral or pyritohedral crystals.

haughtonite [PETR] A black variety of biotite that is rich in iron.

haul [ENG] A single tow of a net or dredge. [NAV] To change the course of a vessel so as to bring the wind farther forward.

haulage [MIN ENG] The movement, in cars or otherwise, of men, supplies, ore, and waste, underground and on the surface.

haulage conveyor [MIN ENG] A conveyor used to transport material between the gathering conveyor and the outside.

haulage curve [MIN ENG] A bend in a haulage road in any direction.

haulage drum [MIN ENG] A cylinder on which steel haulage rope is coiled.

haulage level [MIN ENG] An underground level, either along and inside the ore body or closely parallel to it, and usually in the footwall, in which mineral is loaded into trams and moved out to the hoisting shaft.

haulage stage [MIN ENG] A mine roadway along which a load is moved by one form of haulage without coupling or uncoupling of cars.

haulageway [MIN ENG] The gangway, entry, or tunnel through which loaded or empty mine cars are hauled by animal or mechanical power.

haul-cycle time [MIN ENG] The time required for the scraper to haul a load to the dumping area and to return to its position in the loading area.

haul road [MIN ENG] A road built to carry heavily loaded trucks at a good speed; the grade is limited and usually kept to less than 17% of climb.

Hausdorff maximal principle [MATH] The principle that every partially ordered set has a linearly ordered subset S which is maximal in the sense that S is not a proper subset of another linearly ordered subset.

Hausdorff space [MATH] A topological space where each pair of distinct points can be enclosed in disjoint open neighborhoods.

hausmannite [MINERAL] Mn_3O_4 Brownish-black, opaque mineral composed of manganese tetroxide.

haustellum [INV ZOO] A proboscis modified for sucking.

haustorium [BOT] 1. An outgrowth of certain parasitic plants which serves to absorb food from the host. 2. Food-absorbing cell of the embryo sac in nonparasitic plants.

haustrum [ANAT] An outpocketing or pouch of the colon.

Hauterivian [GEOL] A European stage of geologic time, in the Lower Cretaceous, above Valanginian and below Barremian.

Haüy law [CRYSTAL] The law that for a given crystal there is a set of ratios such that the ratios of the intercepts of any crystal plane on the crystal axes are rational fractions of these ratios.

haüyne [MINERAL] $(Na,Ca)_{4-8}(Al_6Si_6O_{24})(SO_4,S)_{1-2}$ An isometric silicate mineral of the sodalite group occurring as grains embedded in various igneous rocks; hardness is 5.5–6 on Mohs scale, and specific gravity is 2.4–2.5. Also known as haüynite.

haüynite *See* haüyne.

Hauzeur furnace [MET] A double furnace for the distillation of zinc wherein waste heat from one set of retorts is utilized for heating the second set.

hav *See* haversine.

Havelock's law [OPTICS] The law that in a substance displaying the Kerr effect, $n_p - n = 2(n_s - n)$, where n is the index of refraction in the absence of an electric field, and n_p and n_s are the indices of refraction of light whose magnetic vector is parallel and perpendicular to the applied electric field.

haven [NAV] A place of safety for vessels; it is accessible in all states of weather and tides.

Haverhill fever [MED] An acute bacterial infection caused by *Streptobacillus moniliformis,* usually acquired by rat bite, and characterized by acute onset, intermittent fever, erythematous rash, and polyarthritis. Also known as streptobacillary fever.

Haversian canal [HISTOL] The central, longitudinal channel of an osteon containing blood vessels and connective tissue.

Haversian lamella [HISTOL] One of the concentric layers of bone composing a Haversian system.

haversine [MATH] The haversine of an angle A is half of the versine of A, or is $\frac{1}{2}(1 - \cos A)$. Abbreviated hav.

havgul *See* havgull.

havgula *See* havgull.

havgull [METEOROL] Cold, damp wind blowing from the sea during summer in Scotland and Norway. Also known as havgul; havgula.

HA virus *See* hemadsorption virus.

hawk [ENG] A board with a handle underneath used by a workman to hold mortar. [VERT ZOO] Any of the various smaller diurnal birds of prey in the family Accipitridae; some species are used for hunting hare and partridge in India and other parts of Asia.

Hawk [ORD] A U.S. Army surface-to-air guided missile having a range of over 15 miles and a speed of over Mach 2; it is guided by radio command to attack low-flying enemy airplanes, and may also have radar homing.

hawse [NAV ARCH] **1.** The area in the bow of the ship where the hawsepipes are located. **2.** The distance between the bow of a ship and the anchors.

hawse bolster [NAV ARCH] A steel ring or fairing at the end of or around a hawsepipe to aid the motion of a cable through the pipe, prevent its chafing, and guide the anchor into stowed position. Also known as anchor bolster.

hawsepipe [NAV ARCH] A pipe, made of heavy cast iron or steel, through which the anchor chain runs; placed in the ship's bow on each side of the stem, or in some cases also at the stern when a stern anchor is used.

hawser [NAV ARCH] A large rope or cable, usually over 5 inches in diameter, generally used to tow or moor a ship or secure it at a dock.

hay [AGR] Forage plants cut and dried for animal feed.

Hay bridge [ELEC] A four-arm alternating-current bridge used to measure inductance in terms of capacitance, resistance, and frequency; bridge balance depends on frequency.

Hayden process [MET] A method of electrolytic copper refining; anodes of unrefined copper are suspended in an acid electrolyte, and one side of each then acts as an anode and the other as a cathode.

haydite [MATER] Expanded shale, slate, or clay characterized by low unit weight and satisfactory structural properties; used as an aggregate to produce lightweight structural concrete.

hay fever [MED] An allergic disorder of the nasal membranes and related structures due to sensitization by certain plant pollens. Also known as allergic rhinitis.

haymaker [AGR] A machine for curing hay.

Hay's test [PATH] A test for bile salts; the salts lower the surface tension of water, and therefore a light powder such as flowers of sulfur will not float in a solution containing a high concentration of the salts.

Hayward grab bucket [MECH ENG] A clamshell type of grab bucket used for handling coal, sand, gravel, and other flowable materials.

Hayward orange peel [MECH ENG] A grab bucket that operates like the clamshell type but has four blades pivoted to close.

hazard [IND ENG] Any risk to which a worker is subject as a direct result (in whole or in part) of his being employed.

hazard beacon *See* obstruction beacon.

haze [METEOROL] Fine dust or salt particles dispersed through a portion of the atmosphere; the particles are so small that they cannot be felt, or individually seen with the naked eye, but they diminish horizontal visibility and give the atmosphere a characteristic opalescent apparance that subdues all colors. [OPTICS] The degree of cloudiness in a solution, cured plastic material, or coating material.

haze droplet [METEOROL] Any small liquid droplet contributing to an atmospheric haze condition.

haze factor [METEOROL] The ratio of the luminance of a mist or fog through which an object is viewed to the luminance of the object.

haze horizon [METEOROL] The top of a haze layer which is confined by a low-level temperature inversion and has the appearance of the horizon when viewed from above against the sky.

haze layer [METEOROL] A layer of haze in the atmosphere, usually bounded at the top by a temperature inversion and frequently extending downward to the ground.

haze level *See* haze line.

haze line [METEOROL] The boundary surface in the atmo-

sphere between a haze layer and the relatively clean, transparent air above the top of a haze layer. Also known as haze level.

hazelnut *See* filbert.

hazel sandstone [GEOL] An arkosic, iron-bearing redbed sandstone from the Precambrian found in western Texas.

hazemeter *See* transmissometer.

h-bar [QUANT MECH] A fundamental constant equal to $h/2\pi$, where h is Planck's constant. Symbolized $\hbar$. Also known as Dirac h; h-line.

H beacon [NAV] Nondirectional homing beacon which has a power output of 50 to 2000 watts.

H beam [CIV ENG] A beam similar to the I beam but with longer flanges.

H bend *See* H-plane bend.

H bit [DES ENG] A core bit manufactured and used in Canada having inside and outside diameters of 2.875 and 3.875 inches (73.025 and 98.425 millimeters), respectively; the matching reaming shell has an outside diameter of 3.906 inches (99.2124 millimeters).

H bomb *See* hydrogen bomb.

H-B virus [VIROL] A subgroup-A picornavirus associated with diseases in rodents.

H carrier system [COMMUN] Low-frequency carrier system which provides one carrier channel, using frequencies up to about 10 kilohertz, by means of effective four-wire transmission on a single open-wire pair.

hcf *See* greatest common divisor.

HCG *See* human chorionic gonadotropin.

hcp structure *See* hexagonal close-packed structure.

H display [NAV] In radar, a B display modified to include indication of angle of elevation.

HDX *See* half-duplex circuit.

He *See* helium.

head [ADP] A device that reads, records, or erases data on a storage medium such as a drum or tape; examples are a small electromagnet or a sensing or punching device. [ANAT] The region of the body consisting of the skull, its contents, and related structures. [ELECTR] The photoelectric unit that converts the sound track on motion picture film into corresponding audio signals in a motion picture projector. [ENG ACOUS] *See* cutter. [FL MECH] *See* pressure head. [GEOGR] *See* headland.

headache [MED] A deep form of pain, with a characteristic aching quality, localized in the head.

headblock [MIN ENG] **1.** A stop at the head of a slope or shaft to keep cars from going down the shaft or slope. **2.** A cap piece.

headboard [MIN ENG] **1.** A wooden wedge placed against the hanging wall, and against which one end of the stull is jammed. **2.** A board in the roof of a heading, contacting the earth above and supported by a headtree on each side.

head breadth [ANTHRO] Greatest horizontal breadth of the head above the ear openings, at whichever level is found, with moderate pressure applied on the caliper points.

head bulb [INV ZOO] A structure armed with spines behind the lips of spiruroid nematodes.

head cable *See* hand cable.

head circumference [ANTHRO] Maximum of three measurements taken above the eyebrows.

header [BUILD] A framing beam positioned between trimmers and supported at each end by a tail beam. [CIV ENG] Brick or stone laid in a wall with its narrow end facing the wall. [COMMUN] The first section of a message, which contains information such as the addressee, routing, and origination time. [ELEC] A mounting plate through which the insulated terminals or leads are brought out from a hermetically sealed relay, transformer, transistor, tube, or other device. [ENG] A pipe, conduit, or chamber which distributes fluid from a series of smaller pipes or conduits; an example is a manifold. [MECH ENG] A machine used for gathering or upsetting materials; used for screw, rivet, and bolt heads. [MIN ENG] **1.** An entry-boring machine that bores the entire section of the entry in one operation. **2.** A rock that heads off or delays progress. **3.** A blasthole at or above the head.

header bond [CIV ENG] A masonry bond consisting of header courses exclusively.

HAWK

Rough-legged hawk *(Buteo lagopus).*

header card [ADP] A card that contains supplemental information related to the data on the succeeding cards.

header course [CIV ENG] A masonry course of bricks laid as headers.

head erosion *See* headward erosion.

header record [ADP] Computer input record containing common, constant, or identifying information for records that follow.

header-type boiler *See* straight-tube boiler.

head fold [EMBRYO] A ventral fold formed by rapid growth of the head of the embryo over the embryonic disk, resulting in the formation of the foregut accompanied by anteroposterior reversal of the anterior part of the embryonic disk.

headframe [MIN ENG] **1.** The frame at the top of a shaft, on which is mounted the hoisting pulley. Also known as gallows; gallows frame; headstock; hoist frame. **2.** The shaft frame, sheaves, hoisting arrangements, dumping gear, and connected works at the top of a shaft. Also known as headgear.

head gap [ADP] The space between the read/write head and the recording medium, such as a disk in a computer.

head gate [CIV ENG] **1.** A gate on the upstream side of a lock or conduit. **2.** A gate at the starting point of an irrigation ditch.

headgear *See* headframe.

head height [ANTHRO] Average perpendicular distance between the tragion point and the mid-longitudinal line on top of the head, measured on both sides, when the head is positioned so that the line between the tragion point and the bottom of the bony orbit is horizontal.

headhouse [MIN ENG] **1.** A timber framing located at the top of a shaft and receiving the shaft guides that carry the cage or elevator. **2.** A structure that houses the headframe.

heading [NAV] **1.** The horizontal direction in which a ship actually points or heads at any instant, expressed in angular units from a reference direction, usually from 0° at the reference direction clockwise through 360°. **2.** In air navigation, the horizontal direction in which an aircraft points or heads, that is the direction of the longitudinal axis, measured as in the first definition.

heading-and-bench mining [MIN ENG] A stoping method used in thicker ore where it is customary first to take out a slice or heading 7 or 8 feet high directly under the top of the ore, and then to bench or stope down the ore between the bottom of the heading and the bottom of the ore or floor of the level.

heading angle [NAV] Heading measured from 0° at the reference direction clockwise or counterclockwise through 90 or 180° it is labeled with the reference direction as a prefix and the direction of measurement from the reference as a suffix; for example, heading angle S 107° E is 107° east of south, or heading 073°.

heading blasting *See* coyote blasting.

heading line [NAV] The line extending in the direction of a heading.

heading–overhand bench method [MIN ENG] A tunneling method in which the heading is the lower part of the section and is driven at least a round or two in advance of the upper part (bench), which is taken out by overhand excavating. Also known as inverted heading and bench method.

heading side *See* footwall.

heading-upward plan position indicator [ELECTR] A plan position indicator in which the heading of the craft appears at the top of the indicator at all times.

heading wall *See* footwall.

headland [GEOGR] **1.** A high, steep-faced promontory extending into the sea. Also known as head; mull. **2.** High ground surrounding a body of water.

head length [ANTHRO] The distance measured between glabella and opisthocranion, with application of moderate pressure.

headline [MIN ENG] In dredging, the line which is anchored ahead of the dredge pond and holds the dredge up to its digging front.

head loss [FL MECH] The drop in the sum of pressure head, velocity head, and potential head between two points along the path of a flowing fluid, due to causes such as fluid friction.

headman [MIN ENG] **1.** A person who brings coal to the tramway from the workings. **2.** One who engages or disengages grips on mine cars at the top of a haulage slope.

head mast [MIN ENG] The tower carrying the working lines of a cable excavator.

head meter [ENG] A flowmeter that is dependent upon change of pressure head to operate.

head motion [MECH ENG] The vibrator on a reciprocating table concentrator which imparts motion to the deck.

head organ [INV ZOO] One of the bulbous structures in the prohaptor of monogenetic trematodes which are openings for adhesive glands.

head-per-track [ADP] An arrangement having one read/write head for each magnetized track on a disk or drum to eliminate the need to move a single head from track to track.

headphone [ENG ACOUS] An electroacoustic transducer designed to be held against an ear by a clamp passing over the head, for private listening to the audio output of a communications, radio, or television receiver or other source of audio-frequency signals. Also known as phone.

head process [EMBRYO] The notochord or notochordal plate formed as an axial outgrowth of the primitive node.

head pulley [MECH ENG] The pulley at the discharge end of a conveyor belt; may be either an idler or a drive pulley.

head-pulley-drive conveyor [MECH ENG] A conveyor having the belt driven by the head pulley without a snub pulley.

headroom [MIN ENG] **1.** Distance between the drill platform and the bottom of the sheave wheel. **2.** Height between the floor and the roof in a mine opening.

headrope [MIN ENG] In rope haulage, that rope used to pull the loaded transportation device toward the discharge point.

heads [MIN ENG] **1.** Material removed from the ore in the treatment plant and containing the valuable metallic constituents. **2.** The feed to a concentrating system in ore dressing.

head section [COMMUN] *See* end section. [ENG] That part of belt conveyor which consists of a drive pulley, a head pulley which may or may not be a drive pulley, belt idlers if included, and the necessary framing.

headset [ENG ACOUS] A single headphone or a pair of headphones, with a clamping strap or wires holding them in position.

head shaft [MECH ENG] The shaft driven by a chain and mounted at the delivery end of a chain conveyor; it serves as the mount for a sprocket which drives the drag chain.

head sheave [MIN ENG] Pulley in the headgear of a winding shaft over which the hoisting rope runs.

head shield [INV ZOO] A conspicuous structure arching over the lips of certain nematodes.

headsill [BUILD] A horizontal beam at the top of the frame of a door or window.

head smut [PL PATH] A fungus disease of corn and sorghum caused by *Sphacelotheca reiliana* which destroys the head of the plant.

headspace [ORD] The distance between the face of the locked bolt or breechblock of a gun and some specified point in the chamber; in guns using rimless bottlenecked cartridges, the space between the boltface and a specified point on the shoulder of the cartridge.

headstock [MECH ENG] **1.** The device on a lathe for carrying the revolving spindle. **2.** The movable head of certain measuring machines. **3.** The device on a cylindrical grinding machine for rotating the work. [MIN ENG] *See* headframe.

head tide [NAV] A tidal current setting in a direction approximately opposite to the heading of a vessel.

headtree [MIN ENG] The horizontal timber placed at each side of a rectangular heading to support the headboard.

head value [MIN ENG] Assay value of the feed to a concentrating system.

headwall [CIV ENG] A retaining wall at the outlet of a drain or culvert. [GEOL] The steep cliff at the back of a cirque.

headward erosion [GEOL] Erosion caused by water flowing at the head of a valley. Also known as head erosion; headwater erosion.

headwater erosion *See* headward erosion.

headwaters [HYD] The source and upstream waters of a stream.

headway [MIN ENG] *See* cross heading. [NAV] Motion in a forward direction.

headworks [CIV ENG] Any device or structure at the head or diversion point of a waterway.

health [MED] A state of dynamic equilibrium between an organism and its environment in which all functions of mind and body are normal.

health physics [NUCLEO] The protection of personnel from harmful effects of ionizing radiation by such means as routine radiation surveys, area and personnel monitoring, and protective equipment and procedures.

heap clouds *See* clouds with vertical development.

heaped capacity [MIN ENG] In scraper loading, the volume of heaped material that a scraper will hold.

heap leaching [MET] A process used for the recovery of copper from weathered ore and material from mine dumps; material is laid to a thickness of 20 feet in alternately fine and coarse beds and treated with water at intervals during which oxidation occurs; liquor that runs off is treated with scrap iron to precipitate copper.

heap roasting [MIN ENG] A process in which ore with a high sulfur content is roasted by the combustion of the sulfur.

heap sampling [MIN ENG] Method of ore sampling in which the material is shoveled into a conical heap which is then flattened with a spade and shoveled into four equal heaps, two of which are retained, crushed, mixed, and formed into another, smaller cone; the process is repeated until the required small sample is produced.

hearing [PHYSIO] The general perceptual behavior and the specific responses made in relation to sound stimuli.

hearing aid [ENG ACOUS] A miniature, portable sound amplifier for persons with impaired hearing, consisting of a microphone, audio amplifier, earphone, and battery.

heart [ANAT] The hollow muscular pumping organ of the cardiovascular system in vertebrates.

heartbeat [PHYSIO] Pulsation of the heart coincident with ventricular systole.

heart block [MED] The cardiac condition resulting from defective transmission of impulses from atrium to ventricle.

heart bond [CIV ENG] A masonry bond in which two header stones meet in the middle of the wall, their joint being covered by another stone; no headers stretch across the wall.

heartburn [MED] A burning sensation emanating from the esophagus below the sternum.

heart cam [HOROL] A heart-shaped cam used in stopwatches and chronographs to return the recording hand to zero.

heart failure *See* cardiac failure.

heart-failure cells [PATH] Macrophages containing hemosiderin granules found in the lung in certain heart disorders.

hearth [BUILD] **1.** The floor of a fireplace or brick oven. **2.** The projection in front of a fireplace, made of brick, stone, or cement. [MET] The floor of a reverberatory, open-hearth, cupola, or blast furnace; it is made of refractory material able to support the charge and to collect the molten products.

hearth furnace [MET] A furnace designed to heat the charge, resting on the hearth, by passing hot gases over it.

hearth roasting [MIN ENG] A process in which ore or concentrate enters at the top of a multiple hearth roaster and drops from hearth until it is discharged at the bottom.

heart-lung machine [MED] A machine through which blood is shunted to maintain circulation during heart surgery.

heart rate [PHYSIO] The number of heartbeats per minute.

heartrot [PL PATH] **1.** A rot involving disintegration of the heartwood of a tree. **2.** A fungus disease of beets and rutabagas caused by *Mycosphaerella tabifica* which results in decay of the central tissues of the plant. **3.** A boron-deficiency disease of sugarbeets that causes rot.

heart valve [ANAT] Flaps of tissue that prevent reflux of blood from the ventricles to the atria or from the pulmonary arteries or aorta to the atria.

heartwater disease [VET MED] A septicemic infectious disease of cattle, sheep, and goats in Africa caused by the rickettsial microorganism *Cowdria ruminantium*.

heartwood [BOT] Xylem of an angiosperm.

heart worm [INV ZOO] *Dirofilaria immitis*. A filarial nematode parasitic on dogs and other carnivores.

heat [THERMO] Energy in transit due to a temperature difference between the source from which the energy is coming and a sink toward which the energy is going; other types of energy in transit are called work.

heat-activated battery *See* thermal battery.

heat-affected zone [MET] The zone within a base metal that undergoes structural changes but does not melt during welding, cutting, or brazing.

heat balance [GEOPHYS] The equilibrium which exists on the average between the radiation received by the earth and atmosphere from the sun and that emitted by the earth and atmosphere. [MET] The calculation used in fluidization roasting so that the addition or removal of heat can be controlled to maintain the optimum temperature in the reacting vessel. [THERMO] The equilibrium which is known to exist when all sources of heat gain and loss for a given region or body are accounted for.

heat barrier *See* thermal barrier.

heat budget [GEOPHYS] Amount of heat needed to raise a lake's water from the winter temperature to the maximum summer temperature. [THERMO] The statement of the total inflow and outflow of heat for a planet, spacecraft, biological organism, or other entity.

heat capacity [THERMO] The quantity of heat required to raise a system one degree in temperature in a specified way, usually at constant pressure or constant volume. Also known as thermal capacity.

heat check [MET] Parallel surface cracks forming a pattern on the surface of a metal as a result of thermal fatigue.

heat coil [ELEC] Protective device which uses a mechanical element which is allowed to move when the fusible substance that holds it in place is heated above a predetermined temperature by current in the circuit.

heat conductivity *See* thermal conductivity.

heat content *See* enthalpy.

heat convection [THERMO] The transfer of thermal energy by actual physical movement from one location to another of a substance in which thermal energy is stored. Also known as thermal convection.

heat cramps [MED] Painful voluntary-muscle spasm and cramps following strenuous exercise, usually in persons in good physical condition, due to loss of sodium chloride and water from excessive sweating.

heat cycle *See* thermodynamic cycle.

heat dissipation *See* heat loss.

heat distortion point [ENG] The temperature at which a standard test bar (ASTM test) deflects 0.010 inch (0.254 millimeter) under a load of either 66 or 264 psi (4.55 × 10⁵ or 18.20 × 10⁵ newtons per square meter), as specified.

heat dump *See* heat sink.

heat energy *See* internal energy.

heat engine [MECH ENG] A machine that converts heat into work (mechanical energy). [THERMO] A thermodynamic system which undergoes a cyclic process during which a positive amount of work is done by the system; some heat flows into the system and a smaller amount flows out in each cycle.

heat equation [THERMO] A parabolic second-order differential equation for the temperature of a substance in a region where no heat source exists: $\partial t/\partial\tau = (k/\rho c)(\partial^2 t/\partial x^2 + \partial^2 t/\partial y^2 + \partial^2 t/\partial z^2)$, where x, y, and z are space coordinates, τ is the time, $t(x,y,z,\tau)$ is the temperature, k is the thermal conductivity of the body, ρ is its density, and c is its specific heat; this equation is fundamental to the study of heat flow in bodies.

heat equator [METEOROL] **1.** The line which circumscribes the earth and connects all points of highest mean annual temperature for their longitudes. **2.** The parallel of latitude of 10°N, which has the highest mean temperature of any latitude. Also known as thermal equator.

heater [ELECTR] An electric heating element for supplying heat to an indirectly heated cathode in an electron tube. Also known as electron-tube heater. [ENG] A contrivance designed to give off heat.

heater oil *See* heating oil.

heater-type cathode *See* indirectly heated cathode.

heat exchange [CHEM ENG] A unit operation based on heat transfer which functions in the heating and cooling of fluids with or without phase change.

heat exchanger [ENG] Any device, such as an automobile radiator, that transfers heat from one fluid to another or to the environment. Also known as exchanger.

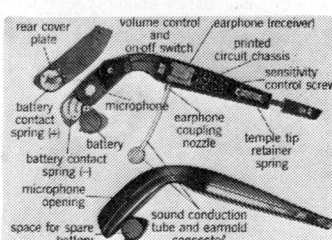

HEARING AID

Eyeglass hearing aid. *(Zenith Radio Corp.)*

heat exhaustion [MED] A heat-exposure syndrome characterized by weakness, vertigo, headache, nausea, and peripheral vascular collapse, usually precipitated by physical exertion in a hot environment.

heat filter [OPTICS] Special glass in condenser lens systems to keep heat from film.

heat flow [THERMO] Heat thought of as energy flowing from one substance to another; quantitatively, the amount of heat transferred in a unit time. Also known as heat transmission.

heat flow equation [THERMO] The relationship stating that in the absence of thermal sources or sinks, the rate of change of temperature of a substance at a fixed point due to heat conduction, times the heat capacity of the substance per unit volume, divided by the thermal conductivity, is equal to the Laplacian of the temperature.

heat flux [THERMO] The amount of heat transferred across a surface of unit area in a unit time. Also known as thermal flux.

heath *See* temperate and cold scrub.

heather [BOT] *Calluna vulgaris.* An evergreen heath of northern and alpine regions distinguished by racemes of small purple-pink flowers.

heating degree-day [METEOROL] A form of degree-day used as an indication of fuel consumption; in United States usage, one heating degree-day is given for each degree that the daily mean temperature departs below the base of 65°F (where the Celsius scale is used, the base is usually 19°C).

heating element [ELEC] The part of a heating appliance in which electrical energy is transformed into heat.

heating fuel *See* heating oil.

heating load [CIV ENG] The quantity of heat per unit time that must be provided to maintain the temperature in a building at a given level.

heating oil [MATER] No. 2 fuel oil; used in domestic heating units. Also known as heater oil; heating fuel.

heating plant [CIV ENG] The whole system for heating an enclosed space. Also known as heating system.

heating system *See* heating plant.

heating value *See* heat of combustion.

heat insulator [MATER] A substance having relatively low heat conductivity.

heat lamp [ELEC] An infrared lamp used for brooders in farming, for drying paint or ink, for keeping food warm, and for therapeutic and other applications requiring heat with or without some visible light.

heat lightning [GEOPHYS] Nontechnically, the luminosity observed from ordinary lightning too far away for its thunder to be heard.

heat loss [PHYS] Energy or power transmitted out of a system in the form of heat. Also known as heat dissipation.

heat low *See* thermal low.

heat of ablation [THERMO] A measure of the effective heat capacity of an ablating material, numerically the heating rate input divided by the mass loss rate which results from ablation.

heat of activation [PHYS CHEM] The increase in enthalpy when a substance is transformed from a less active to a more reactive form at constant pressure.

heat of adsorption [THERMO] The increase in enthalpy when 1 mole of a substance is adsorbed upon another at constant pressure.

heat of aggregation [THERMO] The increase in enthalpy when an aggregate of matter, such as a crystal, is formed at constant pressure.

heat of association [PHYS CHEM] Increase in enthalpy accompanying the formation of 1 mole of a coordination compound from its constituent molecules or other particles at constant pressure.

heat of combustion [PHYS CHEM] The amount of heat released in the oxidation of 1 mole of a substance at constant pressure, or constant volume. Also known as heat value; heating value.

heat of compression [THERMO] Heat generated when air is compressed.

heat of condensation [THERMO] The increase in enthalpy accompanying the conversion of 1 mole of vapor into liquid at constant pressure and temperature.

heat of cooling [THERMO] Increase in enthalpy during cooling of a system at constant pressure, resulting from an internal change such as an allotropic transformation.

heat of crystallization [THERMO] The increase in enthalpy when 1 mole of a substance is transformed into its crystalline state at constant pressure.

heat of decomposition [PHYS CHEM] The change in enthalpy accompanying the decomposition of 1 mole of a compound into its elements at constant pressure.

heat of dilution [PHYS CHEM] **1.** The increase in enthalpy accompanying the addition of a specified amount of solvent to a solution at constant pressure. Also known as integral heat of dilution; total heat of dilution. **2.** The increase in enthalpy when an infinitesimal amount of solvent is added to a solution at constant pressure. Also known as differential heat of dilution.

heat of dissociation [PHYS CHEM] The increase in enthalpy at constant pressure, when molecules break apart or valence linkages rupture.

heat of emission [ELECTR] Additional heat energy that must be supplied to an electron-emitting surface to maintain it at a constant temperature.

heat of evaporation *See* heat of vaporization.

heat of formation [PHYS CHEM] The increase in enthalpy resulting from the formation of 1 mole of a substance from its elements at constant pressure.

heat of fusion [THERMO] The increase in enthalpy accompanying the conversion of 1 mole, or a unit mass, of a solid to a liquid at its melting point at constant pressure and temperature. Also known as latent heat of fusion.

heat of hydration [PHYS CHEM] The increase in enthalpy accompanying the formation of 1 mole of a hydrate from the anhydrous form of the compound and from water at constant pressure.

heat of ionization [PHYS CHEM] The increase in enthalpy when 1 mole of a substance is completely ionized at constant pressure.

heat of linkage [PHYS CHEM] The bond energy of a particular type of valence linkage between atoms in a molecule, as determined by the energy required to dissociate all bonds of the type in 1 mole of the compound divided by the number of such bonds in a compound.

heat of mixing [THERMO] The difference between the enthalpy of a mixture and the sum of the enthalpies of its components at the same pressure and temperature.

heat of reaction [PHYS CHEM] **1.** The negative of the change in enthalpy accompanying a chemical reaction at constant pressure. **2.** The negative of the change in internal energy accompanying a chemical reaction at constant volume.

heat of solidification [THERMO] The increase in enthalpy when 1 mole of a solid is formed from a liquid or, less commonly, a gas at constant pressure and temperature.

heat of solution [PHYS CHEM] The enthalpy of a solution minus the sum of the enthalpies of its components. Also known as integral heat of solution; total heat of solution.

heat of sublimation [THERMO] The increase in enthalpy accompanying the conversion of 1 mole, or unit mass, of a solid to a vapor at constant pressure and temperature. Also known as latent heat of sublimation.

heat of transformation [THERMO] The increase in enthalpy of a substance when it undergoes some phase change at constant pressure and temperature.

heat of vaporization [THERMO] The quantity of energy required to evaporate 1 mole, or a unit mass, of a liquid, at constant pressure and temperature. Also known as enthalpy of vaporization; heat of evaporation; latent heat of vaporization.

heat of wetting [THERMO] **1.** The heat of adsorption of water on a substance. **2.** The additional heat required, above the heat of vaporization of free water, to evaporate water from a substance in which it has been absorbed.

heat pipe [NUCLEO] A heat-transfer device used to cool the collecting electrode of a low-voltage nuclear battery, consisting of an empty cylinder which absorbs heat at one end by vaporization of a liquid and releases heat at the other end by condensation of the vapor.

heat pump [MECH ENG] A device which transfers heat from a cooler reservoir to a hotter one, expending mechanical energy in the process, especially when the main purpose is to heat the hot reservoir rather than refrigerate the cold one.

heat quantity [THERMO] A measured amount of heat; units are the small calorie, normal calorie, mean calorie, and large calorie.

heat radiation [THERMO] The energy radiated by solids, liquids, and gases in the form of electromagnetic waves as a result of their temperature. Also known as thermal radiation.

heat rash *See* miliaria.

heat rate [MECH ENG] An expression of the conversion efficiency of a thermal power plant or engine, as heat input per unit of work output; for example, Btu/kWhr.

heat reactor [NUCLEO] A nuclear reactor designed primarily to supply heat for industrial purposes.

heat release [THERMO] The quantity of heat released by a furnace or other heating mechanism per second, divided by its volume.

heat resistance *See* thermal resistance.

heat-resistant alloy [MET] An oxidation-resistant alloy.

heat-resistant glass [MATER] Glass, such as borosilicate glass, that is heat-treated or leached to remove alkali so that it withstands high heat and sudden cooling without shattering.

heat run [ELEC] A series of temperature measurements made on an electric device during operating tests under various conditions.

heat seal [ENG] A union between two thermoplastic surfaces by application of heat and pressure to the joint.

heatseeker [ORD] A guided missile incorporating an infrared device for homing on heat-radiating machines or installations, such as an aircraft engine or a blast furnace.

heat shield [MATER] Any protective layer that gives protection from heat; used on the front of a reentry capsule.

heat-shrinkable tubing [MATER] A type of plastic tubing that can be heated and shrink-fitted over terminals and other objects of varying sizes and shapes, for insulating and other purposes.

heat sink [AERO ENG] **1.** A type of protective device capable of absorbing heat and used as a heat shield. **2.** In nuclear propulsion, any thermodynamic device, such as a radiator or condenser, that is designed to absorb the excess heat energy of the working fluid. Also known as heat dump. [ELEC] A mass of metal that is added to a device for the purpose of absorbing and dissipating heat; used with power transistors and many types of metallic rectifiers. Also known as dissipator. [THERMO] Any (gas, solid, or liquid) region where heat is absorbed.

heat-sink cooling [ENG] Cooling a body or system by allowing heat to be absorbed from it by another body.

heat source [THERMO] Any device or natural body that supplies heat.

heat sterilization [ENG] An act of destroying all forms of life on and in bacteriological media, foods, hospital supplies, and other materials by means of moist or dry heat.

heat storage [OCEANOGR] The tendency of the ocean to act as a heat reservoir; results in smaller daily and annual variations in temperature over the sea.

heat stress index [PHYSIO] Relation of the amount of evaporation or perspiration required for particular job conditions as related to the maximum evaporative capacity of an average person. Abbreviated HSI.

heatstroke [MED] A heat-exposure syndrome characterized by hyperpyrexia and prostration due to diminution or cessation of sweating, occurring most commonly in persons with underlying disease.

heat thunderstorm [METEOROL] In popular terminology, a thunderstorm of the air mass type which develops near the end of a hot, humid summer day.

heat time [MET] Duration of a single current impulse in pulsation welding.

heat tinting [MET] Oxidation of a polished metal surface by heating to reveal the microstructure.

heat transfer [THERMO] The movement of heat from one body to another (gas, liquid, solid, or combinations thereof) by means of radiation, convection, or conduction.

heat-transfer coefficient [THERMO] The amount of heat which passes through a unit area of a medium or system in a unit time when the temperature difference between the boundaries of the system is 1 degree.

heat-transfer oil [MATER] An oil used to transport heat or cold between two areas of process-equipment surface, and especially compounded to avoid heat degradation in the temperature range of application.

heat transmission *See* heat flow.

heat transport [THERMO] Process by which heat is carried past a fixed point or across a fixed plane, as in a warm current.

heat-treatable alloy [MET] An alloy that can be hardened by thermal treatment.

heat-treating film [MET] An oxide coating formed on a metal surface by heat treating.

heat treatment [MET] Heating and cooling a metal or alloy to obtain desired properties or conditions.

heat value *See* heat of combustion.

heat wave [ELECTROMAG] Infrared radiation, much higher in frequency than radio waves. [METEOROL] A period of abnormally and uncomfortably hot and usually humid weather; the condition must prevail at least 1 day to be a heat wave, but conventionally the term is reserved for periods of several days to several weeks. Also known as hot wave; warm wave.

heave [GEOL] The horizontal component of the slip, measured at right angles to the strike of the fault. [MIN ENG] A rising of the floor of a mine caused by its being too soft to resist the weight on the pillars. [OCEANOGR] The motion imparted to a floating body by wave action.

heavenly body *See* celestial body.

heaves [VET MED] Chronic emphysema in horses marked by labored breathing due to overdistension of the alveoli. Also known as broken wind.

heave to [NAV] To bring a ship into such a position that there is no headway.

heavier-than-air craft [AERO ENG] Any aircraft weighing more than the air it displaces.

heaving [NAV ARCH] Vertical motion of a ship, as distinguished from pitching.

Heaviside calculus [MATH] A type of operational calculus that is used to completely analyze a linear dynamical system which represents some vibrating physical system.

Heaviside layer *See* E layer.

Heaviside-Lorentz system [ELECTROMAG] A system of electrical units which is the same as the Gaussian system except that the units of charge and current are smaller by a factor of $1/\sqrt{4\pi}$, and those of electric and magnetic field are larger by a factor by $\sqrt{4\pi}$. Also known as Lorentz-Heaviside system.

Heaviside's expansion theorem [MATH] A theorem providing an infinite series representation for the inverse Laplace transforms of functions of a particular type.

Heaviside unit function [MATH] The real function $f(x)$ whose value is 0 if x is negative and whose value is 1 otherwise.

heavy acid *See* phosphotungstic acid.

heavy alloy [MET] A tungsten-nickel alloy produced by pressing and sintering the metallic powders; used for screens for x-ray tubes and radioactivity units and for contact surfaces of circuit breakers.

heavy antiaircraft artillery [ORD] Conventional antiaircraft artillery pieces larger than 90-millimeter, the weight of which in a trailed mount is greater than 40,000 pounds (18,000 kilograms).

heavy artillery [ORD] Artillery other than antiaircraft artillery; consists of howitzers and longer-barreled cannon not classified as medium artillery.

heavy bombardment [ORD] A bombardment of great intensity, especially one with large aerial bombs or other missiles.

heavy bomber [AERO ENG] Any large bomber considered to be relatively heavy, such as, in 1955, a bomber having a gross weight, including bomb load, of 250,000 pounds (113,000 kilograms) or more, as the B-36 and the B-52.

heavy concrete [MATER] Concrete in which some or all rock aggregate is replaced by metal aggregate.

heavy cruiser [NAV ARCH] A warship designed to operate with strike, antisubmarine-warfare, or amphibious forces against air and surface threats.

heavy drop [ORD] An airdrop in which heavy articles, such as trucks or artillery pieces, are dropped by parachute.

heavy-duty [ENG] Designed to withstand excessive strain.

heavy-duty car [MECH ENG] A railway motorcar weighing

more than 1400 pounds (635 kilograms), propelled by an engine of 12–30 horsepower (8900–22,400 watts), and designed for hauling heavy equipment and for hump-yard service.

heavy-duty oil [MATER] Lubricating oil with good oxidation stability and corrosion-preventive and detergent-dispersant characterisitics; used in high-speed diesel and gasoline engines under heavy-duty service conditions.

heavy-duty tool block See open-side tool block.

heavy ends [MATER] The highest boiling portion of a petroleum fraction.

heavy field artillery [ORD] Field artillery of the largest calibers, such as the 155-millimeter gun, the 6-inch (152.4-millimeter) gun, the 8-inch (203.2 millimeter) howitzer, and the 240-millimeter howitzer.

heavy force fit [DES ENG] A fit for heavy steel parts or shrink fits in medium sections.

heavy ground [MIN ENG] Dangerous hanging wall requiring vigilance against possible rock fall.

heavy howitzer [ORD] A complete projectile-firing weapon, with a medium muzzle velocity and a curved trajectory; the bore diameter is larger than 200 millimeters.

heavy hydrogen [NUC PHYS] Hydrogen consisting of isotopes whose mass number is greater than one, namely deuterium or tritium.

heavy ion See large ion.

heavy-ion linac See heavy-ion linear accelerator.

heavy-ion linear accelerator [NUCLEO] A linear accelerator which produces a beam of heavy particles of high intensity and sharp energy; used to produce transuranic elements and short-lived isotopes, and to study nuclear reactions, nuclear spectroscopy, and the absorption of heavy ions in matter. Also known as heavy-ion linac; hilac.

heavy liquid [MATER] Any of a group of heavy organic liquids, inorganic solutions, and fused salts used for determination of specific gravity of mineral particles or for separation of minerals having lower and higher specific gravities than the liquids; examples are methylene iodide and bromoform.

heavy-liquid bubble chamber [NUCLEO] A bubble chamber which contains deuterium or an organic liquid such as propane or Freon.

heavy-liquid separation [MIN ENG] A laboratory technique for separating ore particles by allowing them to settle through, or float above, a fluid of intermediate density.

heavy machine gun [ORD] **1.** Any machine gun of relatively heavy weight, including caliber-.30 water-cooled machine guns and caliber-.50 machine guns. **2.** Any aircraft machine gun above caliber .50.

heavy-media separation [MIN ENG] A series of processes for the concentration of ore developed at one time, but now used in coal cleaning; uses suspensions of magnetic materials such as magnetite.

heavy metal [MET] A metal whose specific gravity is approximately 5.0 or higher.

heavy mineral [MINERAL] A mineral with a density above 2.9, which is the density of bromoform, the liquid used to separate the heavy from the light minerals.

heavy-mineral prospecting [MIN ENG] Locating the source of an economic mineral by determining the relative amounts of the mineral in stream sediments and tracing the drainage upstream.

heavy naphtha [MATER] A dark amber to red liquid that is a mixture of xylene and higher homologs; it is flammable; used as a solvent for asphalt and in production of coumarone resins. Also known as high-flash naphtha.

heavy oil [MATER] The high-boiling, relatively viscous fractions of petroleum or coal tar oils.

heavy oxygen See oxygen-18.

heavy particle See baryon.

heavy resin oil [MATER] Reddish-brown, high-boiling, heavy coal tar oils.

heavy section car [MECH ENG] A railway motorcar weighing 1200–1400 pounds (544–635 kilograms) and propelled by an 8–12 horsepower (6000–8900 watts) engine.

heavy tank [ORD] A full-track combat tank with a weight of 56–85 tons.

heavy water [INORG CHEM] A compound of hydrogen and oxygen containing a higher proportion of the hydrogen

isotope deuterium than does naturally occurring water. Also known as deuterium oxide.

heavy-water reactor [NUCLEO] A nuclear reactor in which heavy water serves as moderator and sometimes also as coolant.

heavy weapon [ORD] Any weapon such as a howitzer, mortar, heavy machine gun, and recoilless rifle that is usually part of infantry equipment.

heazelwoodite [MINERAL] Ni_3S_2 A meteorite mineral consisting of a sulfide of nickel.

hebephrenia [PSYCH] A type of schizophrenia marked by disorganized thinking, mannerisms, and regressive caricaturing that is seen in some adolescents, such as silliness, unpredictable giggling, and posturing.

Heberden-Rosenbach node See Heberden's node.

Heberden's arthritis [MED] Degenerative joint disease of the terminal joints of the fingers, producing enlargement (Heberden's nodes) and flexion deformities.

Heberden's node [MED] Nodose deformity of the fingers in degenerative joint disease. Also known as Heberden-Rosenbach node.

Hebraic granite See graphic granite.

Hebrovellidae [INV ZOO] A family of hemipteran insects in the subdivision Amphibicorisae.

hecatolite See moonstone.

hectare [MECH] A unit of area in the metric system equal to 100 ares or 10,000 square meters. Abbreviated ha.

hecto- [SCI TECH] A prefix representing 10^2 or 100.

hectocotylus [INV ZOO] A specialized appendage of male cephalopods adapted for the transference of sperm.

hectogram [MECH] A unit of mass equal to 100 grams. Abbreviated hg.

hectoliter [MECH] A metric unit of volume equal to 100 liters or to 0.1 cubic meter. Abbreviated hl.

hectometer [MECH] A unit of length equal to 100 meters. Abbreviated hm.

hectometric wave [COMMUN] A radio wave between the wavelength limits of 100 and 1000 meters, corresponding to the frequency range of 3000 to 300 kilohertz.

hectorite [MINERAL] $(Mg,Li)_3Si_4O_{10}(OH)_2$ A trioctohedral clay mineral of the montmorillonite group composed of a hydrous silicate of magnesium and lithium.

heddle [TEXT] **1.** A twisted wire with an eye, attached to the harness to guide the warp threads. **2.** One of the sets of parallel cords or wires composing the harness of a loom.

hedenbergite [MINERAL] $CaFeSi_2O_6$ A black mineral consisting of calcium-iron pyroxene and occurring at the contacts of limestone with granitic masses.

hedgehog [ORD] **1.** A portable obstacle, made of crossed poles laced with barbed wire, in the general shape of an hourglass. **2.** A beach obstacle, usually made of steel bars or channel iron, imbedded in concrete and used to interfere with beach landings. **3.** A concentration of troops securely entrenched or fortified, with arms and defenses facing all directions. [VERT ZOO] The common name for members of the insectivorous family Erinaceidae characterized by spines on their back and sides.

hedgehog round [ORD] A small, mortarlike, antisubmarine projectile.

hedleyite [MINERAL] A mineral composed of an alloy of bismuth and tellurium.

hedonic gland [VERT ZOO] One of the mucus-secreting scent glands in many urodeles; functions in courtship.

hedonism [PSYCH] The doctrine that every act is motivated by the desire for pleasure or the aversion from pain and unpleasantness.

hedonophobia [PSYCH] An abnormal fear of pleasure.

hedreocraton [GEOL] A craton that influenced later continental development.

Hedwigiaceae [BOT] A family of mosses in the order Isobryales.

hedyphane [MINERAL] $(Ca,Pb)_5Cl(AsO_4)_3$ Yellowish-white mineral composed of lead and calcium arsenate and chloride; occurs in monoclinic crystals.

HEED See high-energy electron diffraction.

heel [MECH ENG] See heel block. [NAV] Of a ship, to incline or to be inclined to one side. [ORD] Upper corner of the butt of a rifle stock held in firing position.

HEDGEHOG

Hedgehog (*Erinaceus europaeus*), with well-developed ears and eyes, pointed snout, short legs, and spines.

heel block [MECH ENG] A block or plate that is usually fixed on the die shoe to minimize deflection of a punch or cam. Also known as heel.

heeling adjuster [ENG] A dip needle with a sliding weight that can be moved along one of its arms to balance the magnetic force; used to determine the correct position of a heeling magnet. Also known as heeling error instrument; vertical force instrument.

heeling error [NAV] The change in the indication of a magnetic compass when a craft heels, due to the change in the position of the magnetic influences on the craft relative to the earth's magnetic field and to the compass.

heeling error instrument *See* heeling adjuster.

heeling magnet [ENG] A permanent magnet placed vertically in a tube under the center of a marine magnetic compass, to correct for heeling error.

heel of a shot [ENG] **1.** In blasting, the front or face of a shot farthest from the charge. **2.** The distance between the mouth of the drill hole and the corner of the nearest free face. **3.** That portion of a drill hole which is filled with the tamping. [MIN ENG] That portion of the coal to be fractured which is outside the powder.

heel post [CIV ENG] A post to which are secured the hinges of a gate or door.

Hefner candle [OPTICS] A luminous intensity standard, formerly used in Germany, equal to 0.9 international candle; produced by a Hefner lamp burning under standard conditions. Abbreviated HK. Also known as Hefnerkerze.

Hefnerkerze *See* Hefner candle.

Hefner lamp [CHEM] A flame lamp that burns amyl acetate.

Hegeler furnace [MET] A muffle furnace having seven tiers of hearths; lower hearths are heated by gas burned in flues beneath them.

Hehner number [ANALY CHEM] Weight percent of water-insoluble fatty acids in fats and oils.

Heidelberg capsule [ELECTR] A radio pill for telemetering pH values of gastric acidity.

Heidelberg man [PALEON] An early type of European fossil man known from an isolated lower jaw; considered a variant of *Homo erectus* or an early stock of Neanderthal man.

height [MATH] The perpendicular distance between horizontal lines or planes passing through the top and bottom of an object. [ORD] The vertical distance to a target from a horizontal plane passing through the location of the height computer.

height above airport [NAV] In air operations, the height of the minimum descent altitude above the published airport elevation. Abbreviated HAA.

height above touch-down [NAV] In air operations, the height of the minimum descent altitude above the highest elevation in the touch-down zone. Abbreviated HAT.

height-change chart [METEOROL] A chart indicating the change in height of a constant-pressure surface over a specified previous time interval; comparable to a pressure-change chart.

height-change line [METEOROL] A line of equal change in height of a constant-pressure surface over a specified previous interval of time; the lines drawn on a height-change chart. Also known as contour-change line; isallohypse.

height control [ELECTR] The television receiver control that adjusts picture height.

height equivalent of theoretical plate [CHEM ENG] In a packed fractionating column, a height of packing that makes a separation equivalent to that of a theoretical plate; used in sorption and distillation calculations. Abbreviated HETP.

height finder [ENG] A radar equipment, used to determine height of aerial targets.

height finding [ENG] Dermination of the height of an airborne object.

height-finding radar [ENG] A radar set that measures and determines the height of an airborne object.

height gage [ENG] A gage used to measure heights by either a micrometer or a vernier scale. [GRAPHICS] A C-shaped device for measuring foot-to-face height of printing type or mounted printing plates.

height gain [ELECTR] A radio-wave interference phenomenon which results in a more or less periodic signal strength variation with height; this specifically refers to interference between direct and surface-reflected waves; maxima or minima in these height-gain curves occur at those elevations at which the direct and reflected waves are exactly in phase or out of phase respectively.

height input [ELECTR] Radar height information on target received by a computer from height finders and relayed via ground-to-ground data link or telephone.

height of burst [ORD] Vertical distance from the ground, or target, to the point of burst.

height-of-eye correction [NAV] A correction applied to sextant altitude to compensate for the dip of the horizon. Also known as dip correction.

height of instrument [ENG] **1.** In survey leveling, the vertical height of the line of collimation of the instrument over the station above which it is centered, or above a specified datum level. **2.** In spirit leveling, the vertical distance from datum to line of sight of the instrument. **3.** In stadia leveling the height of center of transit above the station stake. **4.** In differential leveling, the elevation of the line of sight of the telescope when the instrument is leveled.

height of tide [OCEANOGR] Vertical distance from the chart datum to the level of the water at any time; it is positive if the water level is higher than the chart datum.

height of transfer unit [CHEM ENG] A dimensionless parameter used to calculate countercurrent sorption tower operations; it is proportional to the apparent resident time of the fluid. Abbreviated HTU.

height overlap coverage [ELECTR] Height-finder coverage within which there is an area of overlapping coverage from adjacent height finders or other radar stations.

height pattern [METEOROL] The general geometric characteristics of the distribution of height of a constant-pressure surface as shown by contour lines on a constant-pressure chart. Also known as baric topography; isobaric topography; pressure topography.

height-position indicator [ELECTR] Radar display which shows simultaneously angular elevation, slant range, and height of objects detected in the vertical sight plane.

height-range indicator [ELECTR] **1.** Radar display which shows an echo as a bright spot on a rectangular field, slant range being indicated along the X axis, height above the horizontal plane being indicated (on a magnified scale) along the Y axis, and height above the earth being shown by a cursor. **2.** Cathode-ray tube from which altitude and range measurements of flightborne objects may be viewed.

heiligenschein [OPTICS] A diffuse white ring surrounding the shadow cast by the observer's head upon a dew-covered lawn when the solar elevation is low and, therefore, the distance from observer to shadow is great.

Heine-Borel theorem [MATH] The theorem that the only compact subsets of the real line are those which are closed and bounded.

Heine-Medin disease *See* poliomyelitis.

Heinz bodies [PATH] Refractile spots seen in erythrocytes in hemolytic anemia that may represent denatured globulin.

Heisenberg algebra [QUANT MECH] The Lie algebra formed by the operators of position and momentum.

Heisenberg equation of motion [QUANT MECH] An equation which gives the rate of change of an operator corresponding to a physical quantity in the Heisenberg picture.

Heisenberg exchange coupling [SOLID STATE] The exchange forces between electrons in neighboring atoms which give rise to ferromagnetism in the Heisenberg theory.

Heisenberg force [NUC PHYS] A force between two nucleons derivable from a potential with an operator which exchanges both the positions and the spins of the particles.

Heisenberg picture [QUANT MECH] A mode of description of a system in which dynamic states are represented by stationary vectors and physical quantities are represented by operators which evolve in the course of time. Also known as Heisenberg representation.

Heisenberg representation *See* Heisenberg picture.

Heisenberg theory of ferromagnetism [SOLID STATE] A theory in which exchange forces between electrons in neighboring atoms are shown to depend on relative orientations of electron spins, and ferromagnetism is explained by the assumption that parallel spins are favored so that all the spins in a lattice have a tendency to point in the same direction.

HEIDELBERG MAN

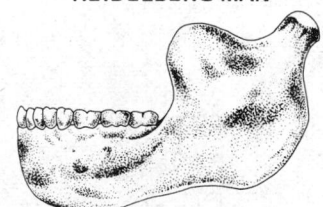

Reconstruction of the Heidelberg jaw. (*From M. F. Ashley Montagu, An Introduction to Physical Anthropology, 2d ed., Charles C. Thomas, 1951*)

HELICAL GEAR

A helical gear. *(Fellows Corp.)*

HELICAL SPRING

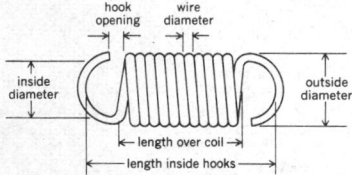

A helical spring wound tight to extend under axial tension.

HELICOID CYME

Drawing of a helicoid cyme.

Heisenberg uncertainty principle *See* uncertainty principle.

Heisenberg uncertainty relation *See* uncertainty relation.

Heising modulation *See* constant-current modulation.

Heitler-London covalence theory [PHYS CHEM] A calculation of the binding energy and the distance between the atoms of a diatomic hydrogen molecule, which assumes that the two electrons are in atomic orbitals about each of the nuclei, and then combines these orbitals into a symmetric or antisymmetric function.

hekistotherm [ECOL] Plant adapted for conditions of minimal heat; can withstand long dark periods.

HeLa cells [PATH] Human cancer cells maintained in tissue culture since 1953, originally excised from the cervical carcinoma of a patient named Helen Lane.

Helaletidae [PALEON] A family of extinct perissodactyl mammals in the superfamily Tapiroidea.

Helanca [TEXT] Trade name of nylon or polyester yarn manufactured by the Heberlein Patent Corporation; filaments have minuscule curls which are heat-set to produce stretch characteristics.

Helcionellacea [PALEON] A superfamily of extinct gastropod mollusks in the order Aspidobranchia.

helcosol *See* bismuth pyrogallate.

Helderbergian [GEOL] A North American stage of geologic time, in the lower Lower Devonian.

held in common [MIN ENG] Pertaining to a claim whereof there is more than one owner.

Heleidae [INV ZOO] The biting midges, a family of orthorrhaphous dipteran insects in the series Nematocera.

Helferich unit [BIOL] A unit for the standardization of phosphatase.

heliacal rising [ASTRON] The rising of a celestial body at the same time or just before that of the sun.

Heliarc welding *See* inert gas-shielded arc welding.

Heliasteridae [INV ZOO] A family of echinoderms in the subclass Asteroidea lacking pentameral symmetry but structurally resembling common asteroids.

helical [MATH] Pertaining to a cylindrical spiral, for example, a screw thread.

helical angle [MECH] In the study of torsion, the angular displacement of a longitudinal element, originally straight on the surface of an untwisted bar, which becomes helical after twisting.

helical antenna [ELECTROMAG] An antenna having the form of a helix. Also known as helix antenna.

helical conveyor [MECH ENG] A conveyor for the transport of bulk materials which consists of a horizontal shaft with helical paddles or ribbons rotating inside a stationary tube.

helical-fin section [CHEM ENG] Helical-shaped, extended-surface addition for the external surfaces of process-fluid tubes to increase heat-exchange efficiency; used for gas heating and cooling and in fuel oil residuum exchangers.

helical-flow turbine [MECH ENG] A steam turbine in which the steam is directed tangentially and radially inward by nozzles against buckets milled in the wheel rim; the steam flows in a helical path, reentering the buckets one or more times. Also known as tangential helical-flow turbine.

helical gear [MECH ENG] Gear wheels running on parallel axes, with teeth twisted oblique to the gear axis.

helical line [ELECTROMAG] A transmission line with a helical inner conductor.

helical milling [MECH ENG] Milling in which the work is simultaneously rotated and translated.

helical potentiometer [ELEC] A multiturn precision potentiometer in which a number of complete turns of the control knob are required to move the contact arm from one end of the helically wound resistance element to the other end.

helical resonator [ELECTROMAG] A cavity resonator with a helical inner conductor.

helical scanning [COMMUN] A method of facsimile scanning in which a single-turn helix rotates against a stationary bar to give horizontal movement of an elemental area. [ENG] A method of radar scanning in which the antenna beam rotates continuously about the vertical axis while the elevation angle changes slowly from horizontal to vertical, so that a point on the radar beam describes a distorted helix.

helical spline broach [MECH ENG] A broach used to produce internal helical splines having a straight-sided or involute form.

helical spring [DES ENG] A bar or wire of uniform cross section wound into a helix.

helical steel support [MIN ENG] A continuous, screw-shaped steel joist lining used for staple shafts.

helical traveling-wave tube *See* helix tube.

Helicinidae [INV ZOO] A family of gastropod mollusks in the order Archeogastropoda containing tropical terrestrial snails.

helicity [QUANT MECH] The component of the spin of a particle along its momentum.

helicoid [INV ZOO] Of a gastropod shell, shaped like a flat coil or flattened spiral. [MATH] A surface generated by a curve which is rotated about a straight line and also is translated in the direction of the line at a rate that is a constant multiple of its rate of rotation.

helicoid cyme [BOT] A type of determinate inflorescence having a coiled cluster, with flowers on only one side of the axis.

Helicon gear [MECH ENG] A gear, manufactured by the Illinois Tool Works, which couples skew shafts and in which the pinion is not tapered.

Heliconiaceae [BOT] A family of monocotyledonous plants in the order Zingiberales characterized by perfect flowers with a solitary ovule in each locule, schizocarpic fruit, and capitate stigma.

Helicoplacoidea [PALEON] A class of free-living, spindle- or pear-shaped, plated echinozoans known only from the Lower Cambrian of California.

helicopter [AERO ENG] An aircraft fitted to sustain itself by motor-driven horizontal rotating blades (rotors) that accelerate the air downward, providing a reactive lift force, or accelerate the air at an angle to the vertical, providing lift and thrust.

helicopter yarding [FOR] A technique that uses a helicopter to bring logs from a forest to a wide level spot along a road where the logs can be loaded onto trucks for hauling to a sawmill; used where steep terrain and unstable ground makes it impossible to build roads without damaging watersheds.

Helicosporae [MYCOL] A spore group of the Fungi Imperfecti characterized by spirally coiled, septate spores.

helicospore [INV ZOO] Mature spore of the Helicosporida characterized by a peripheral spiral filament.

Helicosporida [INV ZOO] An order of protozoans in the class Myxosporidea characterized by production of spores with a relatively thick, single intrasporal filament and three uninucleate sporoplasms.

helicotrema [ANAT] The opening at the apex of the cochlea through which the scala tympani and the scala vestibuli communicate with each other.

Heligmosomidae [INV ZOO] A family of parasitic roundworms belonging to the Strongyloidea.

helimagnet [SOLID STATE] A metal, alloy, or salt that possesses helimagnetism.

helimagnetism [SOLID STATE] A property possessed by some metals, alloys, and salts of transition elements or rare earths, in which the atomic magnetic moments, at sufficiently low temperatures, are arranged in ferromagnetic planes, the direction of the magnetism varying in a uniform way from plane to plane.

heliocentric [ASTRON] Relative to the sun as a center.

heliocentric coordinates [ASTRON] A coordinate system relative to the sun as a center.

heliocentric latitude [ASTRON] Sun-centered coordinate of angular distance perpendicular to the ecliptic plane.

heliocentric longitude [ASTRON] The angular distance east or west from a given point on the sun's equator.

heliocentric orbit [ASTRON] An orbit relative to the sun as a center.

heliocentric parallax *See* annual parallax.

Heliodinidae [INV ZOO] A family of lepidopteran insects in the suborder Heteroneura.

heliogram [COMMUN] A message transmitted on a heliograph.

heliograph [COMMUN] An instrument for sending telegraphic messages by reflecting the sun's rays from a mirror. [ENG] An instrument that records the duration of sunshine and gives a qualitative measure of its amount by action of sun's rays on blueprint paper.

heliographic latitude [ASTRON] On the sun, angular distance north or south of its equator.

heliographic longitude [ASTRON] On the sun, angular distance east or west from given point on the equator of the sun.

heliolite *See* sunstone.

Heliolitidae [PALEON] A family of extinct corals in the order Tabulata.

heliometer [OPTICS] A split-lens telescope used to measure the sun's diameter as well as small distances between stars or other celestial bodies.

heliophilous [ECOL] Attracted by and adapted for a high intensity of sunlight.

heliophobe [MED] An individual who is extremely sensitive to the sun's rays. [PSYCH] One who has an abnormal fear of the sun's rays.

heliophobia [PSYCH] An abnormal fear of exposure to the sun's rays.

heliophyte [ECOL] A plant that thrives in full sunlight.

Heliornithidae [VERT ZOO] The lobed-toed sun grebes, a family of pantropical birds in the order Gruiformes.

helioscope [OPTICS] A telescope for observing the sun that protects the observer's eyes from the sun's glare.

heliosphere [GEOPHYS] The region in the ionosphere where helium ions are predominant (sometimes there may be no region in which helium ions dominate).

heliostat [ENG] A clock-driven instrument mounting which automatically and continuously points in the direction of the sun; it is used with a pyrheliometer when continuous direct solar radiation measurements are required.

heliotaxis [BIOL] Orientation movement of an organism in response to the stimulus of sunlight.

heliotrope [BOT] A plant whose flower or stem turns toward the sun. [ENG] An instrument that reflects the sun's rays over long distances; used in geodetic surveys. [MINERAL] *See* bloodstone.

heliotropic wind [METEOROL] A subtle, diurnal component of the wind velocity leading to a diurnal shift of the wind or turning of the wind with the sun, produced by the east-to-west progression of daytime surface heating.

heliotropism [BIOL] Growth or orientation movement of a sessile organism or part, such as a plant, in response to the stimulus of sunlight.

heliotype *See* photogelatin printing plate.

heliox [MATER] A mixture of helium and a few percent of oxygen used for breathing during deep dives.

Heliozoia [INV ZOO] A subclass of the protozoan class Actinopodea; individuals lack a central capsule and have either axopodia or filopodia.

heliozooid [BIOL] Ameboid, but with distinct filamentous pseudopodia.

helipad [CIV ENG] The launch and landing area of a heliport.

heliport [CIV ENG] A place built for helicopter takeoffs and landings.

helitron [ELECTR] An electrostatically focused, low-noise backward-wave oscillator; the microwave output signal frequency can be swept rapidly over a wide range by varying the voltage applied between the cathode and the associated radio-frequency circuit.

helium [CHEM] A gaseous chemical element, symbol He, atomic number 2, and atomic weight 4.0026; one of the noble gases in group 0 of the periodic table.

helium I [CRYO] The phase of liquid helium-4 which is stable at temperatures above the lambda point (about 2.2 K) and has the properties of a normal liquid, except low density.

helium II [CRYO] The phase of liquid helium-4 which is stable at temperatures between absolute zero and the lambda point (about 2.2 K), and has many remarkable properties such as vanishing viscosity, extremely high heat conductivity, and the fountain effect.

helium-3 [NUC PHYS] The isotope of helium with mass number 3, constituting approximately 1.3 parts per million of naturally occurring helium.

helium-4 [NUC PHYS] The isotope of helium with mass number 4, constituting nearly all naturally occurring helium.

helium liquefier [CRYO] Any one of several machines which liquefy helium by causing it to undergo adiabatic expansion and to do external work.

helium magnetometer [PHYS] A device for measuring magnetic fields by observing the Zeeman effect in the lowest triplet level of helium atoms subjected to the field.

helium-3 maser [PHYS] A gas maser in which the gas used is helium-3.

helium-neon laser [OPTICS] A laser using a combination of helium and neon gases.

helium-oxygen diving [ENG] Diving operations employing a breathing mixture of helium and oxygen.

helium refrigerator [MECH ENG] A refrigerator which uses liquid helium to cool substances to temperatures of 4K or less.

helium spectrometer [SPECT] A small mass spectrometer used to detect the presence of helium in a vacuum system; for leak detection, a jet of helium is applied to suspected leaks in the outer surface of the system.

helium stars [ASTRON] The class B stars.

helix [ELEC] A spread-out, single-layer coil of wire, either wound around a supporting cylinder or made of stiff enough wire to be self-supporting. [MATH] A curve traced on a cylindrical or conical surface where all points of the surface are cut at the same angle; this screwlike winding curve is parametrically given by equations $x = a \cos \theta$, $y = a \sin \theta$, $z = b\theta$, where θ is the parameter and a and b are constants.

Helix [INV ZOO] A genus of pulmonate land mollusks including many of the edible snails; individuals have a coiled shell with a low conical spire.

helix angle [MATH] The constant angle between the tangent to a helix and a generator of the cylinder upon which the helix lies.

helix antenna *See* helical antenna.

helix tube [ELECTR] A traveling-wave tube in which the electromagnetic wave travels along a wire wound in a spiral about the path of the beam, so that the wave travels along the tube at a velocity approximately equal to the beam velocity. Also known as helical traveling-wave tube.

hellandite [MINERAL] Mineral composed of silicate of metals in the cerium group with aluminum, iron, manganese, and calcium.

hellbender [VERT ZOO] *Cryptobranchus alleganiensis.* A large amphibian of the order Urodela which is the most primitive of the living salamanders, retaining some larval characteristics.

Heller's test [PATH] A test for albumin in urine; presence of albumin is indicated by formation of a white ring at the junction of the solution and a concentrate solution of nitric acid.

Hellespontus [ASTRON] A surface region of the planet Mars between the regions Hellas and Noachis.

Hell-Volhard-Zelinsky reaction [ORG CHEM] Preparation of an ester or α-halo substituted acid (chloro or bromo) by reacting the halogen on the acid in the presence of phosphorus or phosphorus halide, and then followed by hydrolysis or alcoholysis of the haloacyl halide resulting.

helm [NAV ARCH] **1.** The tiller or wheel controlling a ship's rudder. **2.** The entire apparatus for steering a ship.

Helmert's formula [GEOPHYS] A formula for the acceleration due to gravity in terms of the latitude and the altitude above sea level.

helmet [ENG] A globe-shaped head covering made of copper and supplied with air pumped through a hose; attached to the breastplate of a diving suit for deep-sea diving.

Helmholtz coils [ELECTROMAG] A pair of flat, circular coils having equal numbers of turns and equal diameters, arranged with a common axis, and connected in series; used to obtain a magnetic field more nearly uniform than that of a single coil.

Helmholtz double layer [PHYS] An electrical double layer of positive and negative charges one molecule thick which occurs at a surface where two bodies of different materials are in contact, or at the surface of a metal or other substance

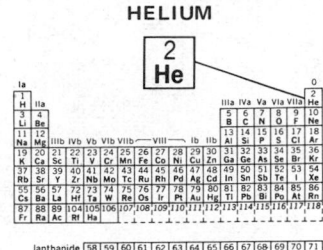

HELIUM

Periodic table of the chemical elements showing the position of helium.

HELLBENDER

The hellbender, with a maximum length of 18 inches (46 centimeters).

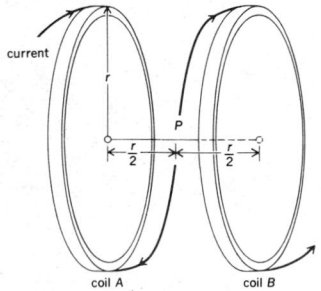

HELMHOLTZ COILS

Helmholtz coils showing the optimum arrangement in which the distance between the two coils is equal to the diameter r of one of the coils. P = point of nearly uniform field strength. *(From L. B. Loeb, Fundamentals of Electricity and Magnetism, 3d ed., Wiley, 1947)*

capable of existing in solution as ions and immersed in a dissociating solvent.

Helmholtz equation [MATH] A partial differential equation obtained by setting the Laplacian of a function equal to the function multiplied by a negative constant. [OPTICS] An equation which relates the linear and angular magnifications of a spherical refracting interface. Also known as Lagrange-Helmholtz equation. [PHYS CHEM] The relationship stating that the emf (electromotive force) of a reversible electrolytic cell equals the work equivalent of the chemical reaction when charge passes through the cell plus the product of the temperature and the derivative of the emf with respect to temperature.

Helmholtz flow [FL MECH] Flow with free streamlines or vortex sheets.

Helmholtz free energy *See* free energy.

Helmholtz function *See* free energy.

Helmholtz instability [FL MECH] The hydrodynamic instability arising from a shear, or discontinuity, in current speed at the interface between two fluids in two-dimensional motion; the perturbation gains kinetic energy at the expense of that of the basic currents. Also known as shearing instability.

Helmholtz-Keteller formula [OPTICS] A dispersion formula in which the difference between the square of the index of refraction and unity is set equal to a sum of terms each of which is associated with a resonant wavelength of the medium.

Helmholtz potential *See* free energy.

Helmholtz resonator [ENG ACOUS] An enclosure having a small opening consisting of a straight tube of such dimensions that the enclosure resonates at a single frequency determined by the geometry of the resonator.

Helmholtz's theorem [ELEC] *See* Thévenin's theorem. [FL MECH] The theorem that in the isentropic flow of a nonviscous fluid which is not subject to body forces, individual vortices always consist of the same fluid particles. [MATH] The theorem determining a general class of vector fields as being everywhere expressable as the sum of an irrotational vector with a divergence-free vector.

Helmholtz theory *See* Young-Helmholtz theory.

Helmholtz wave [FL MECH] An unstable wave in a system of two homogeneous fluids with a velocity discontinuity at the interface.

helminth [INV ZOO] Any parasitic worm.

helminthemesis [MED] Vomiting of worms.

helminthiasis [MED] Any disease caused by the presence of parasitic worms in the body.

helminthic abscess [MED] An abscess caused by worms.

helminthogogue [PHARM] An anthelminthic.

helminthoid [BIOL] Resembling a helminth.

helminthologist [BIOL] One who studies helminths.

helminthophobia [PSYCH] An abnormal fear of worms or becoming infested with worms.

helminthosporin [BIOCHEM] $C_{15}H_{10}O_5$ A maroon, crystalline pigment formed by certain fungi growing on a sugar substrate.

Helminthosporium [MYCOL] A genus of parasitic fungi of the family Dematiaceae having conidiophores which are more or less irregular or bent and bear conidia successively on new growing tips.

Helminthosporium leaf spot [PL PATH] Any leaf spot, commonly of grasses, caused by *Helminthosporium* species.

helm roof [ARCH] A steeply pitched roof with four faces rising from gables to a point.

helm wind [METEOROL] A strong, cold northeasterly wind blowing down into the Eden valley from the western slope of the Crossfell Range in northern England.

Helobiae [BOT] The equivalent name for Helobiales.

Helobiales [BOT] An order embracing most of the Alismatidae in certain systems of classification.

Heloderma [VERT ZOO] The single genus in the reptilian family Helodermatidae; contains the only known poisonous lizards, the Gila monster (*H. suspectum*) and the beaded lizard (*H. horridum*).

Helodermatidae [VERT ZOO] A family of lizards in the suborder Sauria.

Helodidae [INV ZOO] The marsh beetles, a family of coleopteran insects in the superfamily Dascilloidea.

Helodontidae [PALEON] A family of extinct ratfishes conditionally placed in the order Bradyodonti.

Helomyzidae [INV ZOO] The sun flies, a family of myodarian cyclorrhaphous dipteran insects in the subsection Acalypteratae.

helophyte [ECOL] A marsh plant; buds overwinter underwater.

helophytia [ECOL] Differences in ecological control by fluctuations in water level such as in marshes.

Heloridae [INV ZOO] A family of hymenopteran insects in the superfamily Proctotrupoidea.

Helotiales [MYCOL] An order of fungi in the class Ascomycetes.

Helotidae [INV ZOO] The metallic sap beetles, a family of coleopteran insects in the superfamily Cucujoidea.

helotism [ECOL] Symbiosis in which one organism is a slave to the other, as between certain species of ants.

Helotrephidae [INV ZOO] A family of true aquatic, tropical hemipteran insects in the subdivision Hydrocorisae.

helper grade [MIN ENG] A grade on which helper engines are required to assist road locomotives. Also known as pusher grade.

helper set [MIN ENG] A set of timbers to reinforce the normal set of timbers in a mine.

helper virus [VIROL] A virus that, by its infection of a cell, enables a defective virus to multiply by supplying one or more functions that the defective virus lacks.

help-yourself system [IND ENG] A tool-crib system for temporary issue of tools employed in small shops; employees have access to tools in the crib and help themselves.

helve hammer [MET] A belt-driven trip hammer with the hammer face or swage carried on the end of a beam; used for welding, forging, plating, drawing, and other metal-working operations.

helvine *See* helvite.

helvite [MINERAL] (Mn,Fe,Zn)$_4$Be$_3$(SiO$_4$)$_3$S A silicate mineral isomorphous with danalite and genthelvite. Also known as helvine.

hem-, hema-, hemo-, haem- [HISTOL] Combining form for blood.

hemacytometer *See* hemocytometer.

hemadsorption virus [VIROL] A descriptive term for myxoviruses that agglutinate red blood cells and cause the cells to adsorb to each other. Abbreviated HA virus.

hemafibrite [MINERAL] Mn$_3$(AsO$_4$)(OH)$_3$·H$_2$O A brownish to garnet-red mineral composed of basic manganese arsenate.

hemagglutination [IMMUNOL] Agglutination of red blood cells.

hemagglutination-inhibition test [IMMUNOL] A test to identify a virus antigen or to quantitate an antibody by adding virus-specific antibody to a mixture of agglutinating virus and red blood cells.

hemagglutinin [IMMUNOL] An erythrocyte-agglutinating antibody.

hemal arch [ANAT] 1. A ventral loop on the body of vertebrate caudal vertebrae surrounding the blood vessels. 2. In man, the ventral vertebral process formed by the centrum together with the ribs.

hemal ring [INV ZOO] A vessel in certain echinoderms, variously located, associated with the coelom and axial gland.

hemal sinus [INV ZOO] The two principal lacunae along the digestive tube in certain echinoderms.

hemal tuft [INV ZOO] Series of fine vessels in echioderms arising from the axial gland.

hemangioendothelioma [MED] A malignant tumor composed of anoplastic endothelial cells. Also known as hemangiosarcoma.

hemangioma [MED] A tumor composed of blood vessels. Also known as capillary angioma.

hemangiopericytoma [MED] A tumor composed of endothelium-lined tubes or cords of cells surrounded by spherical cells with supporting reticulin network.

hemangiosarcoma *See* hemangioendothelioma.

HELMHOLTZ RESONATOR

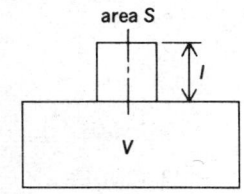

Schematic drawing of a Helmholtz resonator. Here l = length of straight tube, S = cross-sectional area of tube, V = closed volume. Resonant frequency is $f_0 = 1/2\pi \sqrt{c^2 S/l_e V}$ where c = speed of sound and l_e = effective length of tube, approximately $l + 0.8\sqrt{S}$.

hemapodium [INV ZOO] The dorsal lobe of a parapodium.

hemarthrosis [MED] Passage of blood into a joint.

hematein [BIOCHEM] $C_{16}H_{12}O_6$ A brownish stain and chemical indicator obtained by oxidation of hematoxylin.

hematemesis [MED] Vomiting of blood.

hematidrosis [MED] The appearance of blood or blood products in sweat gland secretions.

hematin [ORG CHEM] $C_{34}H_{33}O_5N_4Fe$ The hydroxide of ferriheme derived from oxidized heme.

hematite [MINERAL] Fe_2O_3 An iron mineral crystallizing in the rhombohedral system; the most important ore of iron, it is dimorphous with maghemite, occurs in black metallic-looking crystals, in reniform masses or fibrous aggregates, or in reddish earthy forms. Also known as bloodstone; red hematite; red iron ore; red ocher; rhombohedral iron ore.

hematoblast [HISTOL] An immature erythrocyte.

hematocele [MED] Collection of blood in a body part.

hematochrome [BIOCHEM] A red pigment occurring in green algae, especially when plants are exposed to intense light on subaerial habitats.

hematocrit [PATH] The volume, after centrifugation, occupied by the cellular elements of blood, in relation to the total volume.

hematogenous [PHYSIO] **1.** Pertaining to the production of blood or of its fractions. **2.** Carried by way of the bloodstream. **3.** Originating in blood.

hematoidin crystals [PATH] Yellow to brown crystals in the feces following gastrointestinal hemorrhage.

hematolite [MINERAL] $(Mn,Mg)_4Al(AsO_4)(OH)_8$ A brownish-red mineral composed of aluminum manganese arsenate; occurs in rhombohedral crystals.

hematologist [MED] A specialist in the study of blood.

hematology [MED] The science of the blood, its nature, functions, and diseases.

hematoma [MED] A localized mass of blood in tissue; usually it clots and becomes encapsulated by connective tissue.

hematometra [MED] Accumulation of blood in the uterus.

hematomyelia [MED] Hemorrhage into the spinal cord.

hematopathology *See* hemopathology.

hematophagous [ZOO] Feeding on blood.

hematophanite [MINERAL] $Pb_5Fe_4O_{10}(Cl,OH)_2$ A mineral composed of oxychloride lead and iron.

hematopoiesis [PHYSIO] The process by which the cellular elements of the blood are formed. Also known as hemopoiesis.

hematopoietic system *See* reticuloendothelial system.

hematopoietic tissue [HISTOL] Blood-forming tissue, consisting of reticular fibers and cells. Also known as hemopoietic tissue.

hematopoietin [BIOCHEM] A substance which is produced by the juxtaglomerular apparatus in the kidney and controls the rate of red cell production. Also known as hemopoietin.

hematoporphyrin [BIOCHEM] $C_{34}H_{38}O_6N$ Iron-free heme, a porphyrin obtained by treating hemoglobin with sulfuric acid in vitro. Also known as hemoporphyrin.

hematorrhachis [MED] Hemorrhage into the spinal meninges, producing irritative phenomena.

hematosalpinx [MED] Accumulation of blood in a Fallopian tube. Also known as hemosalpinx.

hematoxylin [ORG CHEM] $C_{16}H_{14}O_6$ A colorless, crystalline compound occurring in hematoxylon; upon oxidation, it is converted to hematein which forms deeply colored lakes with various metals; used as a stain in microscopy.

hematoxylon [BOT] The heartwood of *Haematoxylon campechianum.* Also known as logwood.

hematuria [MED] A pathological condition in which the urine contains blood.

heme [BIOCHEM] $C_{34}H_{32}O_4N_4Fe$ An iron-protoporphyrin complex associated with each polypeptide unit of hemoglobin.

hemeralopia [MED] Day blindness.

hemerythrin *See* hemoerythrin.

hemi- [BIOL] **1.** Prefix for half. **2.** Prefix denoting one side of the body.

hemiacetal [ORG CHEM] A class of compounds that have the grouping $>C(OH)-(OR)$ and that result from the reaction of an aldehyde and alcohol.

hemianesthesia [MED] Loss of sensation on one side of the body.

hemianopsia [MED] Bilateral or unilateral blindness in one-half of the field of vision.

Hemiascomycetes [MYCOL] The equivalent name for Hemiascomycetidae.

Hemiascomycetidae [MYCOL] A subclass of fungi in the class Ascomycetes.

hemiazygous vein [ANAT] A vein on the left side of the vertebral column which drains blood from the left ascending lumbar vein to the azygos vein.

hemiballismus [MED] Sudden, violent, spasmodic movements involving particularly the proximal portions of the extremities of one side of the body; caused by a destructive lesion of the contralateral subthalamic nucleus or its neighboring structures or pathways.

Hemibasidiomycetes [MYCOL] The equivalent name for Heterobasidiomycetidae.

hemicellulose [BIOCHEM] $(C_6H_{10}O_5)_n$ A type of polysaccharide found in plant cell walls in association with cellulose and lignin; it is soluble in and extractable by dilute alkaline solutions. Also known as hexosan.

hemicephaly [MED] Congenital absence of the cerebrum.

Hemichordata [SYST] A group of marine animals categorized as either a phylum of deuterostomes or a subphylum of chordates; includes the Enteropneusta, Pterobranchia, and Graptolithina.

Hemicidaridae [PALEON] A family of extinct Echinacea in the order Hemicidaroida distinguished by a stirodont lantern, and ambulacra abruptly widened at the ambitus.

Hemicidaroida [PALEON] An order of extinct echinoderms in the superorder Echinacea characterized by one very large tubercle on each interambulacral plate.

hemic murmur [MED] Blowing or rasping sound heard in the heart or vessels, usually in association with systole, in abnormal conditions of increased velocity of blood flow.

hemicolectomy [MED] Surgical removal of a portion of the colon.

hemicryptophyte [ECOL] A plant having buds at the soil surface and protected by scales, snow, or litter.

hemicrystalline *See* hypocrystalline.

hemicycle [MATH] A curve in the form of a semicircle.

hemicydic [BOT] Of flowers, having the floral leaves arranged partly in whorls and partly in spirals.

hemidiaphragm [ANAT] Lateral half of a diaphragm. [MED] Diaphragm with normal muscle development only on one side.

Hemidiscosa [INV ZOO] An order of sponges in the subclass Amphidiscophora distinguished by birotulates that are hemidiscs with asymmetrical ends.

hemihedral symmetry [CRYSTAL] The possession by a crystal of only half of the elements of symmetry which are possible in the crystal system to which it belongs.

hemiholohedral [CRYSTAL] Of hemihedral form but with half the octants having the full number of planes.

Hemileucinae [INV ZOO] A subfamily of lepidopteran insects in the family Saturnidae consisting of the buck moths and relatives.

hemimellitic acid [ORG CHEM] $C_6H_3(COOH)_3$ A compound crystallizing in colorless needles; melting point 196°C; slightly soluble in water. Also known as benzene-1,2,3-tricarboxylic acid.

Hemimetabola [INV ZOO] A division of the insect subclass Pterygota; members are characterized by hemimetabolous metamorphosis.

hemimetabolous metamorphosis [INV ZOO] An incomplete metamorphosis; gills are present in aquatic larvae, or naiads.

hemimorphic crystal [CRYSTAL] A crystal with no transverse plane of symmetry and no center of symmetry; composed of forms belonging to only one end of the axis of symmetry.

hemimorphite [MINERAL] $Zn_4Si_2O_7(OH)_2 \cdot H_2O$ A white, colorless, pale-green, blue, or yellow mineral having an orthorhombic crystal structure; an ore of zinc. Also known as calamine; electric calamine; galmei.

hemin [BIOCHEM] $C_{34}H_{32}O_4N_4FeCl$ The crystalline salt of ferriheme, containing iron in the ferric state.

hemiparasite [ECOL] A parasite capable of a saprophytic

HEMATITE

Crystals of hematite from Saint Gothard, Switzerland. (*American Museum of Natural History Specimen*)

— 3.6 cm —

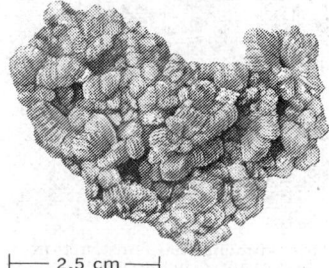

HEMIMORPHITE

Crystals forming coxcomb masses, Granby, Missouri. (*Specimen from Department of Geology, Bryn Mawr College*)

— 2.5 cm —

HEMLOCK

Branch and cone of Eastern hemlock (*Tsuga canadensis*).

HEMOCHORIAL PLACENTA

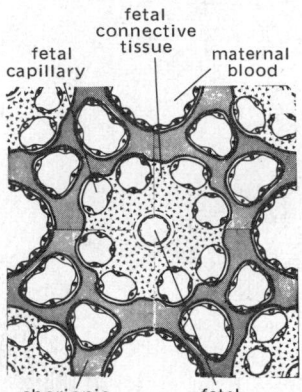

Diagram of a hemochorial placenta found in rodents, many insectivores, and tarsiers.

HEMOGLOBIN

Three-dimensional construction representing a hemoglobin molecule bound with two molecules of oxygen (O_2). (From A. F. Cullis et al., Structures of haemoglobin IX: A three-dimensional Fourier synthesis at 5.5 resolution: Description of the structure, Proc. Roy. Soc. London Ser. A, 265:161, 1962)

existence, especially certain parasitic plants containing some chlorophyll. Also known as semiparasite.

hemiparesis [MED] Muscle weakness on one side of the body.

hemipelagic [ECOL] Of the biogeographic environment of the hemipelagic region with both neritic and pelagic qualities.

hemipelagic region [OCEANOGR] The region of the ocean extending from the edge of a shelf to the pelagic environment; roughly corresponds to the bathyal zone, in which the bottom is 200 to 1000 meters below the surface.

hemipelagic sediment [GEOL] Deposits containing terrestrial material and the remains of pelagic organisms, found in the ocean depths.

hemipenis [VERT ZOO] Either of a pair of nonerectile, evertible sacs that lie on the floor of the cloaca in snakes and lizards; used as intromittent organs.

Hemipeplidae [INV ZOO] An equivalent name for Cucujidae.

hemiplegia [MED] Unilateral paralysis of the body.

hemiprism [CRYSTAL] A pinacoid that cuts two crystallographic axes.

Hemiprocnidae [VERT ZOO] The crested swifts, a family comprising three species of perching birds found only in southeastern Asia.

Hemiptera [INV ZOO] The true bugs, an order of the class Insecta characterized by forewings differentiated into a basal area and a membranous apical region.

Hemisphaeriales [MYCOL] A group of ascomycetous fungi characterized by the wall of the fruit body being a stroma.

hemisphere [GEOGR] A half of the earth divided into north and south sections by the equator, or into an east section containing Europe, Asia, and Africa, and a west section containing the Americas. [MATH] One of the two pieces of a sphere divided by a great circle.

hemispherical candlepower [OPTICS] Luminous intensity of a hemispherical light source.

hemispherical pyrheliometer [ENG] An instrument for measuring the total solar energy from the sun and sky striking a horizontal surface, in which a thermopile measures the temperature difference between white and black portions of a thermally insulated horizontal target within a partially evacuated transparent sphere or hemisphere.

hemispheric wave number *See* angular wave number.

hemispheroid [MATH] One of the halves into which a spheroid is divided by a plane of symmetry.

hemithorax [ANAT] One side of the chest.

hemitropic [CRYSTAL] Pertaining to a twinned structure in which, if one part were rotated 180°, the two parts would be parallel.

Hemizonida [PALEON] A Paleozoic order of echinoderms of the subclass Asteroidea having an ambulacral groove that is well defined by adambulacral ossicles, but with restricted or undeveloped marginal plates.

hemizygous [GEN] Pertaining to the condition or state of having a gene present in a single dose; for instance, in the X chromosome of male mammals.

hemlock [BOT] The common name for members of the genus *Tsuga* in the pine family characterized by two white lines beneath the flattened, needlelike leaves.

hemlock oil *See* spruce oil.

hemming [MECH ENG] Forming of an edge by bending the metal back on itself.

hemobilirubin [BIOCHEM] Bilirubin in normal blood serum before passage through the liver.

hemoblast *See* hemocytoblast.

hemochorial placenta [EMBRYO] A type of placenta having the maternal blood in direct contact with the chorionic trophoblast. Also known as labyrinthine placenta.

hemochromatosis [MED] A disorder of iron metabolism characterized by excessive accumulation of iron in the liver and other tissues and by development of severe cirrhosis.

hemocoel [INV ZOO] An expanded portion of the blood system in arthropods that replaces a portion of the coelom.

hemoconcentration [MED] An increase in the concentration of blood cells resulting from the loss of plasma or water from the bloodstream.

hemoconia [BIOCHEM] Round or dumbbell-shaped, refractile, colorless particles found in blood plasma.

hemoconiosis [MED] Condition of having an abnormal amount of hemoconia in the blood.

hemocyanin [BIOCHEM] A blue respiratory pigment found only in mollusks and in arthropods other than insects.

hemocyte [INV ZOO] A cellular element of blood, especially in invertebrates.

hemocytoblast [HISTOL] A pluripotential blast cell thought to be capable of giving rise to all other blood cells. Also known as hemoblast; stem cell.

hemocytolysis [PHYSIO] The dissolution of blood cells.

hemocytometer [PATH] A specifically designed, ruled and calibrated glass slide used with a microscope to count red and white blood cells. Also spelled hemacytometer.

hemodichorial placenta [EMBRYO] A placenta with a double trophoblastic layer.

hemodynamics [PHYSIO] A branch of physiology concerned with circulatory movements of the blood and the forces involved in circulation.

hemoendothelial placenta [EMBRYO] A placenta having the endothelium of vessels of chorionic villi in direct contact with the maternal blood.

hemoerythrin [BIOCHEM] A red respiratory pigment found in a few annelid and sipunculid worms and in the brachiopod *Lingula*. Also known as hemerythrin.

hemoflagellate [INV ZOO] A parasitic, flagellate protozoan that lives in the blood of the host.

hemoglobin [BIOCHEM] The iron-containing, oxygen-carrying molecule of the red blood cells of vertebrates comprising four polypeptide subunits in a heme group.

hemoglobin A [BIOCHEM] The type of hemoglobin found in normal adults, which moves as a single component in an electrophoretic field, is rapidly denatured by highly alkaline solutions, and contains two titratable sulfhydryl groups per molecule.

hemoglobin C [PATH] A slow-moving abnormal hemoglobin associated with intraerythrocytic crystal formation, target cells, and chronic hemolytic anemia.

hemoglobin E [PATH] An abnormal hemoglobin found in people of Southeast Asia, migrating slightly faster than hemoglobin C; in the homozygous form it causes a mild hemolytic anemia with normochromic target cells.

hemoglobinemia [MED] The presence of hemoglobin in the blood plasma.

hemoglobin H [PATH] An abnormal hemoglobin migrating more rapidly than normal hemoglobin on electrophoresis, and usually associated with thalassemia.

hemoglobin M [PATH] An abnormal hemoglobin associated with hereditary methemoglobinemia, differing from normal hemoglobin in its electrophoretic mobility by the starch-block method.

hemoglobinopathy [MED] Any blood dyscrasia resulting from the genetically determined alteration of the chemical nature of hemoglobin.

hemoglobin S *See* sickle-cell hemoglobin.

hemoglobinuria [MED] A pathological condition in which the urine contains hemoglobin.

hemoglobinuric nephrosis *See* lower nephron nephrosis.

hemogram [PATH] **1.** Erythrocyte and leukocyte count per cubic millimeter of blood plus the differential leukocyte count and hemoglobin level in grams per 100 milliliters of blood. **2.** The differential leukocyte count.

hemohistioblast [HISTOL] The hypothetical reticuloendothelial cell from which all the cells of the blood are eventually differentiated.

hemolymph [INV ZOO] The circulating fluid of the open circulatory systems of many invertebrates.

hemolysin [IMMUNOL] A substance that lyses erythrocytes.

hemolysis [PHYSIO] The lysis, or destruction, of erythrocytes with the release of hemoglobin.

hemolytic anemia [MED] A decrease in the blood concentration of hemoglobin and the number of erythrocytes, due to the inability of the mature erythrocytes to survive in the circulating blood.

hemolytic disease of newborn *See* erythroblastosis fetalis.

hemolytic jaundice [MED] Accumulation of bile pigments in the plasma as a result of excessive hemolysis.

hemomonochorial placenta [EMBRYO] A placenta with a single trophoblastic layer.

hemoparasite [INV ZOO] A parasitic animal that lives in the blood of a vertebrate.

hemopathology [MED] A branch of medicine dealing with blood diseases. Also known as hematopathology.

hemopathy [MED] Any disease of the blood.

hemopericardium [MED] The presence of blood or bloody effusion in the pericardial sac.

hemoperitoneum [MED] An effusion of blood in the peritoneal cavity.

hemopexin [BIOCHEM] A heme-binding protein in human plasma that may be a regulator of heme and drug metabolism, and a distributor of heme.

hemophilia [MED] A rare, hereditary blood disorder marked by a tendency toward bleeding and hemorrhages due to a deficiency of factor VIII.

hemophilic bacteria [MICROBIO] Bacteria of the genera *Hemophilus, Bordetella,* and *Moraxella;* all are small, gram-negative, nonmotile, parasitic rods, dependent upon blood factors for growth.

hemophilioid disease [MED] Any hemophilialike disease; it is the same as hemophilia clinically, but caused by a deficiency of factors IX, X, and XII.

Hemophilus [MICROBIO] A genus of hemophilic bacteria in the family Brucellaceae requiring hemin and nicotinamide nucleoside for growth.

Hemophilus aegyptius [MICROBIO] A nonhemolytic hemophilic bacterium that causes pinkeye. Also known as Koch-Weeks bacillus.

Hemophilus ducreyi [MICROBIO] A hemolytic hemophilic bacterium that causes chancroid in man. Also known as Ducrey bacillus.

Hemophilus influenzae [MICROBIO] A hemophilic bacterium that causes serious infections of man, especially infants, including influenza, purulent meningitis, and pneumonia.

Hemophilus parainfluenzae [MICROBIO] A nonpathogenic hemophilic bacterium found in the upper respiratory tract of man.

Hemophilus pertussis [MICROBIO] A pathogenic hemophilic bacterium that causes whooping cough. Also known as Bordet-Gengou bacillus.

hemophobia [PSYCH] An abnormal fear of the sight of blood.

hemopoiesis *See* hematopoiesis.

hemopoietic tissue *See* hematopoietic tissue.

hemopoietin *See* hematopoietin.

hemoporphyrin *See* hematoporphyrin.

Hemoprotein [BIOCHEM] Trademark for a mixture of protein fractions of digested ox blood used in nonspecific protein therapy.

hemoptysis [MED] Discharge of blood from the larynx, trachea, bronchi, or lungs.

hemorrhage [MED] The escape of blood from the vascular system.

hemorrhagic diathesis [MED] Any condition marked by abnormal bleeding tendency.

hemorrhagic fever virus [VIROL] Any of several arboviruses causing acute infectious human diseases characterized by fever, prostration, vomiting, and hemorrhage.

hemorrhagic measles [MED] A grave variety of measles with a hemorrhagic eruption and severe constitutional symptoms. Also known as black measles.

hemorrhagic pericarditis [MED] Inflammation of the pericardium accompanied by the hemorrhagic appearance of the exudate imparted by the presence of red cells.

hemorrhagic pleuritis [MED] Inflammation of the pleura characterized by the presence of bloody fluid in the pleural cavity.

hemorrhagic septicemia [VET MED] An infectious bacterial disease of fowl, rabbit, buffalo, and other animals caused by *Pasteurella mulfocida.* Also known as pasteurellosis.

hemorrhagic unit [BIOL] A unit for the standardization of snake venom.

hemorrhoid [MED] A varicosity of the external hemorrhoidal veins, causing painful swelling in the anal region.

hemorrhoidectomy [MED] Surgical removal of hemorrhoids.

hemosalpinx *See* hematosalpinx.

hemosiderin [BIOCHEM] An iron-containing glycoprotein found in most tissues and especially in liver.

hemosiderosis [PHYSIO] Deposition of hemosiderin in body tissues without tissue damage, reflecting an increase in body iron stores.

hemostasis [MED] **1.** The arrest of a flow of blood or hemorrhage. **2.** The stopping or slowing of circulation.

hemostat [MED] An instrument to compress a bleeding vessel.

hemostatic [MED] An agent that arrests or checks bleeding, especially by shortening clotting time.

hemothorax [MED] Accumulation of blood in the pleural cavity.

hemotrichorial placenta [EMBRYO] A placenta with a triple trophoblastic layer.

hemotrophe [BIOCHEM] The nutritive substance supplied via the placenta to embryos of viviparous animals.

hemp [BOT] *Cannabis sativa.* An Asiatic annual herb cultivated for its tough bast fiber, used for making twines, linenlike fabrics, and canvases; the plant is the source of the drug marijuana.

hemp-core cable *See* standard wire rope.

hempseed oil [MATER] A fatty oil, light green or brownish yellow when dry, obtained from hempseed; used in soft soap, paints, and varnishes, and in Asia in foods.

hen [VERT ZOO] The female of several bird species, especially gallinaceous species.

henbane [BOT] *Hyoscyamus niger.* A poisonous herb containing the toxic alkaloids hyoscyamine and hyoscine; extracts have properties similar to belladonna.

hendecane *See* undecane.

hendecanoic acid *See* undecanoic acid.

hendecyl *See* undecyl.

Henderson equation for pH [PHYS CHEM] An equation for the pH of an acid during its neutralization: $pH = pK_a + \log [salt]/[acid]$ where pK_a is the logarithm to base 10 of the reciprocal of the dissociation constant of the acid; the equation is found to be useful for the pH range 4–10, providing the solutions are not too dilute.

Henderson process [MET] The treatment of copper sulfide ores by roasting with salt to form chlorides, which are then leached out and precipitated.

heneicosane [ORG CHEM] $C_{21}H_{44}$ Saturated hydrocarbon of the methane series; the crystals melt at 40°C and boil at 215°C (at 15 mm Hg).

henequen [MATER] A hard plant fiber, obtained from the leaves of the American agave (*Agave fourcroydes*) and other agave species; used to make rope, twine, and cord.

Hengstebeck approximation [CHEM ENG] A method of calculation to estimate the distribution of non-key components in distillation column products.

Henicocephalidae [INV ZOO] A family of hemopteran insects of uncertain affinities.

Henle's loop *See* loop of Henle.

henna [BOT] *Lawsonia inermis.* An Old World plant having small opposite leaves and axillary panicles of white flowers; a reddish-brown dye extracted from the leaves is used in hair dyes. Also known as Egyptian henna.

henry [ELECTROMAG] The mks unit of self and mutual inductance, equal to the self-inductance of a circuit or the mutual inductance between two circuits if there is an induced electromotive force of 1 volt when the current is changing at the rate of 1 ampere per second. Abbreviated H.

Henry's law [PHYS CHEM] The law that at sufficiently high dilution in a liquid solution, the fugacity of a nondissociating solute becomes proportional to its concentration.

Henry unit [BIOL] A unit for the standardization of thyroid extracts.

Hensen's node [EMBRYO] Thickening formed by a group of cells at the anterior end of the primitive streak in vertebrate gastrulas.

hentriacontane [ORG CHEM] $C_{31}H_{64}$ A hydrocarbon; a crystalline material melting at 68°C and boiling at 302°C (at 15 mm Hg); derived from roots of *Oenanthe crocata* and found in beeswax.

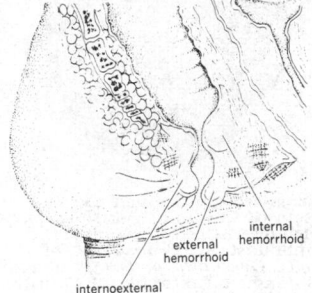

HEMORRHOID

internal hemorrhoid

external hemorrhoid

internoexternal hemorrhoid

Types of hemorrhoids. *(After Pennington, from W. A. N. Dorland, ed., American Illustrated Medical Dictionary, 19th ed., Saunders, 1942)*

HEMP

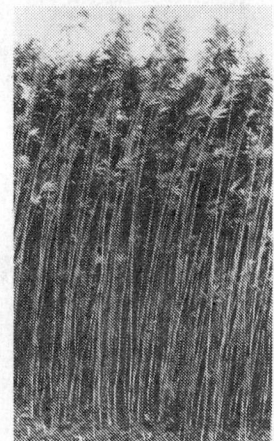

Standing crop of hemp at maturity; plants range from 6 to 8 feet (1.8 to 2.4 meters) and have stems about the diameter of a pencil. *(Courtesy of H. L. Dean, University of Iowa)*

HENBANE

Drawing of henbane. *(Adapted from Webster's New International Dictionary, 2d ed., Merriam, 1959)*

HERBACEOUS DICOTYLEDON

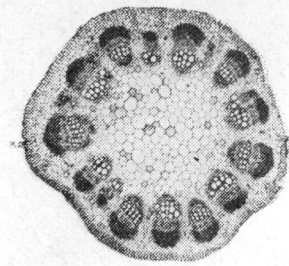

Cross section through the stem of the sunflower showing characteristics of ectophloic siphonostele. *(From J. B. Hill, L. O. Overholts, and H. W. Popp, Botany: A Textbook for Colleges, 2d ed., McGraw-Hill, 1950)*

HERBACEOUS MONOCOTYLEDON

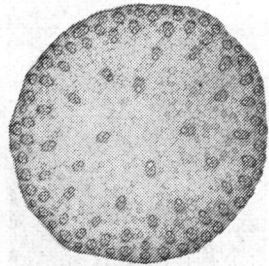

Photomicrograph of transverse section through corn stem. *(From J. B. Hill, L. O. Overholts, and H. W. Popp, Botany: A Textbook for Colleges, 2d ed., McGraw-Hill, 1950)*

HERBARIUM

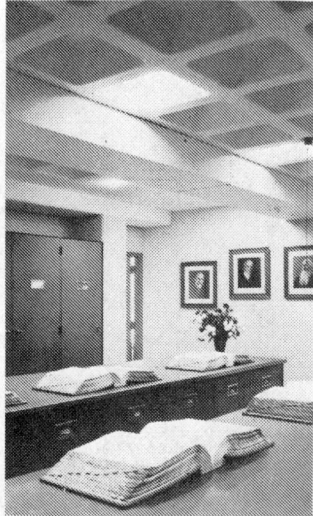

Interior of herbarium. *(Gray Herbarium, Harvard)*

hepar calcies   *See* calcium sulfide.

heparin   [BIOCHEM] An acid mucopolysaccharide acting as an antithrombin, antithromboplastin, and antiplatelet factor to prolong the clotting time of whole blood; occurs in a variety of tissues, most abundantly in liver.

hepar lobatum   [MED] The liver in syphilitic cirrhosis, having a nodular lobulated appearance.

hepar sulfuris   *See* potassium sulfide.

HEPAT charge   *See* high-explosive plastic antitank charge.

hepatectomy   [MED] Surgical removal of the liver or a part of it.

Hepaticae   [BOT] The equivalent name for Marchantiatae.

hepatic artery   [ANAT] A branch of the celiac artery that carries blood to the stomach, pancreas, great omentum, liver, and gallbladder.

hepatic cecum   [INV ZOO] A hollow outpocketing of the foregut of *Branchiostoma*; receives veins from the intestine.

hepatic coma   [MED] Unconscious state associated with advanced liver disease.

hepatic duct   [ANAT] The common duct draining the liver. Also known as common hepatic duct.

hepatic duct system   [ANAT] The biliary tract including the hepatic ducts, gallbladder, cystic duct, and common bile duct.

hepatic encephalopathy   [MED] Behavioral, psychological, and neurological changes associated with advanced liver disease.

hepatic glycogenosis   *See* von Gierke's disease.

hepatic plexus   [ANAT] Nerve network accompanying the hepatic artery to the liver.

hepatic portal system   [ANAT] A system of veins in vertebrates which collect blood from the digestive tract and spleen and pass it through capillaries in the liver.

hepatic vein   [ANAT] A blood vessel that drains blood from the liver into the inferior vena cava.

hepatitis   [MED] Inflammation of the liver; commonly of viral origin but also occurring in association with syphilis, typhoid fever, malaria, toxemias, and parasitic infestations.

hepatitis virus   [VIROL] Any of several viruses causing hepatitis in man and lower mammals.

hepatization   [PATH] The conversion of tissue into a liverlike substance, as of the lungs during the exudative stage of pneumonia.

hepatolenticular degeneration   *See* Wilson's disease.

hepatoma   [MED] A usually malignant neoplasm arising from parenchymal cells of the liver.

hepatomegaly   [MED] Enlargement of the liver.

hepatopancreas   [INV ZOO] A gland in crustaceans and certain other invertebrates that combines the digestive functions of the liver and pancreas of vertebrates.

hepatorenal syndrome   [MED] A complex of syndromes due to hepatic and renal failure, including hyperpyrexia, oliguria, and coma. Also known as Heyd's syndrome.

hepatoscopy   [MED] Inspection of the liver, as by laparotomy or peritoneoscopy.

hepatotoxin   [PHARM] An agent capable of damaging the liver.

Hepialidae   [INV ZOO] A family of lepidopteran insects in the superfamily Hepialoidea.

Hepialoidea   [INV ZOO] A superfamily of lepidopteran insects in the suborder Homoneura including medium- to large-sized moths which possess rudimentary mouthparts.

Hepplewhite-Gray lamp   [MIN ENG] A lamp which drew its air from the top down through four tubular pillars into the base, where the air fed the flame through a gauze ring; the outlet was through a metal chimney closed by a gauze disk.

HEP projectile   *See* high-explosive plastic projectile.

Hepsogastridae   [INV ZOO] A family of parasitic insects in the order Mallophaga.

heptachlor   [ORG CHEM] $C_{10}H_7Cl_7$ An insecticide; a white to tan, waxy solid; insoluble in water, soluble in alcohol and xylene; melts at 95–96°C.

heptacosane   [ORG CHEM] $C_{27}H_{56}$ A hydrocarbon; water-insoluble crystals melting at 60°C and boiling at 27 C (at 15 mm Hg); soluble in alcohol; found in beeswax.

heptad   [SCI TECH] A group of seven.

heptadecane   [ORG CHEM] $C_{17}H_{36}$ A hydrocarbon; water-insoluble, alcohol-soluble solid melting at 23°C and boiling at

303°C; used as a chemical intermediate. Also known as dioctylmethane.

***n*-heptadecanoic acid**   [ORG CHEM] $CH_3(CH_2)_{15}COOH$ A fatty acid that is saturated; soluble in ether and alcohol, insoluble in water; colorless crystals melt at 61°C. Also known as margaric acid.

heptadecanol   [ORG CHEM] $C_{17}H_{35}OH$ An alcohol; colorless liquid boiling at 309°C; slightly soluble in water; used as a chemical intermediate, as a perfume fixative, in cosmetics and soaps, and to manufacture surfactants.

heptagon   [MATH] A seven-sided polygon.

heptakaidecagon   [MATH] A polygon with 17 sides.

heptaldehyde   [ORG CHEM] $C_6H_{13}CHO$ An aldehyde; ether-soluble, colorless oil with fruity aroma; slightly soluble in water; boils at 153°C; used as a chemical intermediate and for perfumes and pharmaceuticals. Also known as heptanal.

heptanal   *See* heptaldehyde.

heptane   [ORG CHEM] $CH_3(CH_2)_5CH_3$ A hydrocarbon; water-insoluble, flammable, colorless liquid boiling at 98°C; soluble in alcohol, chloroform, and ether; used as an anesthetic, solvent, and chemical intermediate, and in standard octane-rating tests.

1,7-heptanedicarboxylic acid   *See* azelaic acid.

heptanedioic acid   *See* pimelic acid.

heptanoic acid   [ORG CHEM] $CH_3(CH_2)_5COOH$ Clear oil boiling at 223°C; soluble in alcohol and ether, insoluble in water; used as a chemical intermediate. Also known as enanthic acid; oenanthic acid.

1-heptanol   [ORG CHEM] $C_7H_{15}OH$ An alcohol; a fragrant, colorless liquid boiling at 174°C; soluble in water, ether, or alcohol; used as a chemical intermediate, as a solvent, and in cosmetics. Also known as heptyl alcohol.

2-heptanol   *See* methyl amyl carbinol.

3-heptanol   [ORG CHEM] $CH_3CH_2CH(OH)C_4H_9$ An alcohol; a liquid boiling at 156°C; used as a coating, solvent, and diluent, as a chemical intermediate, and as a flotation frother.

2-heptanone   *See* methyl *n*-amyl ketone.

3-heptanone   *See* ethyl butyl ketone.

4-heptanone   [ORG CHEM] $(CH_3CH_2CH_2)_2CO$ A colorless liquid that is stable and has a pleasant odor; boils at approximately 98°C; used to put nitrocellulose and raw and blown oils into solution, and used in lacquers and as a flavoring in foods. Also known as butyrone; dipropyl ketone.

heptene   [ORG CHEM] $C_{17}H_{14}$ A liquid that is a mixture of isomers; boils at 189.5°C; used as an additive in lubricants, as a catalyst, and as a surface active agent. Also known as heptylene.

heptode   [ELECTR] A seven-electrode electron tube containing an anode, a cathode, a control electrode, and four additional electrodes that are ordinarily grids.

heptose   [BIOCHEM] Any member of the group of monosaccharides containing seven carbon atoms.

heptoxide   [CHEM] An oxide whose molecule contains seven atoms of oxygen.

heptulose   [BIOCHEM] The generic term for a ketose formed from a seven-carbon monosaccharide.

heptyl   [ORG CHEM] $CH_3(CH_2)_6-$ The radical from heptane, $CH_3(CH_2)_5CH_3$.

***n*-heptylacetylene**   *See* nonyne.

heptyl alcohol   *See* 1-heptanol.

heptylene   *See* heptene.

herb   [BOT] **1.** A seed plant that lacks a persistent, woody stem aboveground and dies at the end of the season. **2.** An aromatic plant or plant part used medicinally or for food flavoring.

herbaceous   [BOT] **1.** Resembling or pertaining to a herb. **2.** Pertaining to a stem with little or no woody tissue.

herbaceous dicotyledon   [BOT] A type of dicotyledon in which the primary vascular cylinder forms an ectophloic siphonostele with widely separated vascular strands.

herbaceous monocotyledon   [BOT] A type of monocotyledon with a vascular system composed of widely spaced strands arranged in one of four ways.

herbarium   [BOT] A collection of plant specimens, pressed and mounted on paper or placed in liquid preservatives, and systematically arranged.

Herbert cloudburst test   [MET] A hardness test in which a shower of steel balls, dropped from a predetermined height, dulls the surface of a hardened part in proportion to its softness and thus reveals defective areas.

Herbert pendulum method [MET] Hardness testing in which a 1-millimeter steel or jewel ball resting on the surface to be tested acts as the fulcrum for a 4-kilogram compound pendulum of 10-second period; the swinging of the pendulum causes a rolling indentation in the material, and several hardness factors, such as work hardenability, are determined.

herbicide [MATER] A chemical agent that destroys or inhibits plant growth.

herbicolous [ECOL] Living on herbs.

herbivore [VERT ZOO] An animal that eats only vegetation.

Herbst corpuscle [VERT ZOO] A cutaneous sense organ found in the mucous membrane of the tongue of the duck.

Hercoal F [MATER] A high explosive, used in mines.

Hercogel [MATER] A high explosive, used in mines.

Hercomite [MATER] A high-count, ammonia-type dynamite, used in mines.

herbularc lining [MIN ENG] A German method of lining mine roadways subjected to heavy pressures; a closed circular arch of wedge-shaped precast concrete blocks made in two sizes are erected so that alternate blocks offer their wedge action in opposite direction, the larger blocks toward the center of the roadway and the smaller outward.

Hercules [ASTRON] A constellation with no stars brighter than third magnitude; right ascension 17 hours, declination 30° north.

Hercules coal powder 2 [MATER] A high explosive, used in mines.

Hercules powder [MATER] A weak dynamite, based on nitroglycerin and a semiactive carrying dope.

Hercules stone See lodestone.

Hercules trap [ANALY CHEM] Water-measuring liquid trap used in aquametry when the material collected is heavier than water.

Hercules X-1 [ASTROPHYS] A source of x-rays that pulses with a period of 1.237 seconds, and is eclipsed for 6 of every 42 hours, associated with a variable star, designated HZ Herculis, that also has a period of 42 hours and faint 1.237-second pulsations; believed to be a binary star whose invisible member is a rotating neutron star. Abbreviated Her X-1.

Herculon [TEXT] Trade name of an olefin fiber manufactured from polypropylene by Hercules, Inc.

Hercynian geosyncline [GEOL] A principal area of geosynclinal sediment accumulation in Devonian time; found in south-central and southern Europe and northern Africa.

Hercynian orogeny See Variscan orogeny.

hercynite [MINERAL] $(Fe, Mg)Al_2O_4$ A black mineral of the spinel group; crystallizes in the isometric system. Also known as ferrospinel; iron spinel.

herd [VERT ZOO] A number of one kind of wild, semidomesticated, or domesticated animals grouped or kept together under human control.

herderite [MINERAL] $CaBe(PO_4)(F,OH)$ A colorless to pale-yellow or greenish-white mineral consisting of phosphate and fluoride of calcium and beryllium; hardness is 7.5–8 on Mohs scale, and specific gravity is 3.92.

herd instinct [PSYCH] Psychic need for identification with a group.

hereditary [GEN] Of or pertaining to heredity or inheritance.

hereditary deforming chondrodysplasia See multiple hereditary exostoses.

hereditary disease [MED] A genetically determined illness transmitted from parent to child.

hereditary hemorrhagic telangiectasia [MED] An inherited disease characterized by dilatation of groups of capillaries and a tendency to hemorrhage. Also known as Osler-Rendu-Weber disease.

hereditary hypophosphatemic rickets [MED] A sex-linked syndrome involving defective bone growth and decreased concentrations of phosphates in the serum.

hereditary mechanics [MECH] A field of mechanics in which quantities, such as stress, depend not only on other quantities, such as strain, at the same instant but also on integrals involving the values of such quantities at previous times.

hereditary nephritis [MED] A familial disease characterized by recurrent attacks of interstitial inflammation of the kidneys and discharge of blood in the urine.

hereditary spherocytosis [MED] A chronic congenital disorder of the erythrocytopoietic system characterized by a preponderance of spherical erythrocytes, increased osmotic fragility, hemolytic anemia, and splenomegaly.

hereditary spinal ataxia See Friedrich's ataxia.

heredity [GEN] The sum of genetic endowment obtained from the parents.

heredofamilial [MED] Referring to a disease or disorder having a familial pattern of occurrence and thought to be hereditary.

Hering theory [PHYSIO] A theory of color vision which assumes that three qualitatively different processes are present in the visual system, and that each of the three is capable of responding in two opposite ways.

Hermann-Mauguin symbols [CRYSTAL] Symbols representing the 32 symmetry classes, consisting of series of numbers giving the multiplicity of symmetry axes in descending order, with other symbols indicating inversion axes and mirror planes.

hermaphrodite [BIOL] An individual (animal, plant, or higher vertebrate) exhibiting hermaphroditism. [PHYSIO] An abnormal condition, especially in humans and other higher vertebrates, in which both male and female reproductive organs are present in the individual.

hermaphrodite caliper [DES ENG] A layout tool having one leg pointed and the other like that of an inside caliper; used to locate the center of irregularly shaped stock or to lay out a line parallel to an edge.

hermaphroditic See monoecious.

hermatype See hermatypic coral.

hermatypic coral [INV ZOO] Reef-building coral characterized by the presence of symbiotic algae within their endodermal tissue. Also known as hermatype.

Hermes [ASTRON] One of the asteroids of the Eros group.

hermetic seal [ENG] An airtight seal.

hermit crab [INV ZOO] The common name for a number of marine decapod crustaceans of the families Paguridae and Parapaguridae; all lack right-sided appendages and have a large, soft, coiled abdomen.

Hermite polynomials [MATH] A family of orthogonal polynomials which arise as solutions to Hermite's differential equation, a particular case of the hypergeometric differential equation.

Hermite's differential equation [MATH] A particular case of the hypergeometric equation; it has the form $w'' - 2zw' + 2nw = 0$, where n is an integer.

Hermitian conjugate of a matrix [MATH] The transpose of the complex conjugate of a matrix. Also known as adjoint of a matrix; associate matrix.

Hermitian conjugate operator See adjoint operator.

Hermitian form [MATH] A polynomial in n real or complex variables where the matrix constructed from its coefficients is Hermitian.

Hermitian inner product See inner product.

Hermitian kernel [MATH] A kernel $K(x,t)$ of an integral transformation or integral equation is Hermitian if $K(x,t)$ equals its adjoint kernel, $K^*(t,x)$.

Hermitian matrix [MATH] A matrix which equals its conjugate transpose matrix, that is, is self-adjoint.

Hermitian operator [MATH] A linear operator A on vectors in a Hilbert space, such that if x and y are in the range of A then the inner products (Ax,y) and (x,AY) are equal.

Hermitian scalar product See inner product.

hernia [MED] Abnormal protrusion of an organ or other body part through its containing wall. Also called rupture.

hernial sac [MED] A pouch of peritoneum containing a herniated organ or other body part.

herniated disk [MED] An intervertebral disk in which the pulpy center has pushed through the fibrocartilage. Also known as slipped disk.

herniation of the nucleus pulposus [MED] Occurrence of a herniated disk.

herniorrhaphy [MED] Operation for repair and suturing of a hernia.

HERMAPHRODITE CALIPER

Hermaphrodite caliper. *(From R. J. Sweeney, Measurement Techniques in Mechanical Engineering, Wiley, 1953)*

herniotomy [MED] An operation for the relief of irreducible hernia, by cutting through the neck of the sac.

heroin [PHARM] $C_{21}H_{23}O_5N$ A white, crystalline powder made from morphine; the hydrochloride compound is used as a sedative and narcotic.

heron [VERT ZOO] The common name for wading birds composing the family Ardeidae characterized by long legs and neck, a long tapered bill, large wings, and soft plumage.

herpangia [MED] A mild viral disease of humans caused by a coxsackie virus and characterized by fever, anorexia, and grayish papules surrounded by a red areola in the mouth.

herpes [MED] An acute inflammation of the skin or mucous membranes, characterized by the development of groups of vesicles on an inflammatory base.

herpes simplex [MED] An acute vesicular eruption of the skin or mucous membranes caused by a virus, commonly seen as cold sores or fever blisters.

herpes simplex encephalitis *See* herpetic encephalitis.

herpes simplex virus [VIROL] Either of two types of subgroup A herpesviruses that are specific for humans; given the binomial designation *Herpesvirus hominis*.

herpesvirus [VIROL] A major group of deoxyribonucleic acid–containing animal viruses, distinguished by a cubic capsid, enveloped virion, and affinity for the host nucleus as a site of maturation.

herpes zoster [MED] A systemic virus infection affecting spinal nerve roots, characterized by vesicular eruptions distributed along the course of a cutaneous nerve. Also known as shingles; zoster.

herpetic encephalitis [MED] A type of meningoencephalitis characterized by large intranuclear inclusion bodies in the brain. Also known as herpes simplex encephalitis.

herpetic stomatitis [MED] Inflammation of the soft tissues of the mouth characterized by fever blisters.

herpetic tonsillitis [MED] Acute inflammation of the tonsils characterized by fever and vesicles caused by a herpesvirus.

herpolhode [MECH] The curve traced out on the invariable plane by the point of contact between the plane and the inertia ellipsoid of a rotating rigid body not subject to external torque.

Herreshoff furnace [MET] **1.** A rectangular-shaft blast furnace for smelting copper ore. **2.** A mechanical, cylindrical, multiple-deck muffle furnace of the McDougall type.

herring [VERT ZOO] The common name for fishes composing the family Clupeidae; fins are soft-rayed and have no supporting spines, there are usually four gill clefts, and scales are on the body but absent on the head.

Herring body [HISTOL] Any of the distinct colloid masses in the vertebrate pituitary gland, possibly representing greatly dilated endings of nerve fibers.

herringbone gear [MECH ENG] The equivalent of two helical gears of opposite hand placed side by side.

herringbone pattern [ELECTR] An interference pattern sometimes seen on television receiver screens, consisting of a horizontal band of closely spaced V- or S-shaped lines.

herringbone stoping [MIN ENG] Method used in flattish Rand stope panels 500–1000 feet (150–300 meters) long for breaking and moving the ore; the stope is divided into 20-foot (6-meter) panels, and a different gang works each panel; a tramming system delivers the cut rock to a central scraper system.

herringbone timbering [MIN ENG] A method of timber support in a roadway with a weak roof and strong sides, using neither arms nor side uprights; the crossbar is notched into the sides and supported at its center by a bar under it and parallel with the roadway; the bar is supported by struts notched into the sides at about half height.

herring oil [MATER] Pale-yellow to dark-red liquid obtained from herring; soluble in carbon disulfide, chloroform, and ether, insoluble in water; used in soaps, in leather dressing, and as machinery lubricant.

Herschel effect [GRAPHICS] **1.** The reaction whereby a photographic image on silver chloride is destroyed by exposure to red light. Also known as visual Herschel effect. **2.** The reaction whereby a latent photographic image in gelatin emulsion which is not dye-sensitized is destroyed by exposure to red light. Also known as latent Herschel effect.

Herschel-type venturi tube [ENG] A type of venturi tube in which the converging and diverging sections are cones, the throat section is relatively short, the diverging cone is long, and the pressures preceding the inlet cone and in the throat are transferred through multiple openings into annular openings, called piezometer rings.

hertz [PHYS] Unit of frequency; a periodic oscillation has a frequency of *n* hertz if in 1 second it goes through *n* cycles. Also known as cycle per second. Abbreviated cps; Hz.

Hertz antenna [ELECTROMAG] An ungrounded half-wave antenna.

Hertz effect [ELECTR] Increase in the length of a spark induced across a spark gap when the gap is irradiated with ultraviolet light.

Hertzian oscillator [ELECTROMAG] **1.** A generator of electric dipole radiation; consists of two capacitors joined by a conducting rod having a small spark gap; an oscillatory discharge occurs when the two halves of the oscillator are raised to a sufficiently high potential difference. **2.** A dumbbell-shaped conductor in which electrons oscillate from one end to the other, producing electric dipole radiation.

Hertzian wave *See* electric wave.

Hertz's law [MECH] A law which gives the radius of contact between a sphere of elastic material and a surface in terms of the sphere's radius, the normal force exerted on the sphere, and Young's modulus for the material of the sphere.

Hertzsprung-Russell diagram [ASTRON] A plot showing the relation between the luminosity and surface temperature of stars; other related quantities frequently used in plotting this diagram are the absolute magnitude for luminosity, and spectral type or color index for the surface temperatures. Abbreviated H-R diagram.

Hertz vector *See* polarization potential.

hervidero *See* mud volcano.

Her X-1 *See* Hercules X-1.

Hesionidae [INV ZOO] A family of small polychaete worms belonging to the Errantia.

hesperidium [BOT] A modified berry, with few seeds, a leathery rind, and membranous extensions of the endocarp dividing the pulp into chambers; an example is the orange.

Hesperiidae [INV ZOO] The single family of the superfamily Hesperioidea comprising butterflies known as skippers because of their rapid, erratic flight.

Hesperioidea [INV ZOO] A monofamilial superfamily of lepidopteran insects in the suborder Heteroneura including heavy-bodied, mostly diurnal insects with clubbed antennae that are bent, curved, or reflexed at the tip.

Hesperornithidae [PALEON] A family of extinct North American birds in the order Hesperornithiformes.

Hesperornithiformes [PALEON] An order of ancient extinct birds; individuals were large, flightless, aquatic diving birds with the shoulder girdle and wings much reduced and the legs specialized for strong swimming.

Hesperus [ASTRON] Greek name for the planet Venus as an evening star.

Hesser's variation [ADP] A variation of a Kiviat graph in which all variables are arranged so that their plots approach the circumference of the graph as the system being evaluated approaches saturation, and the scales on the various axes may not cover the full 0–100% range, or may be in units other than percent.

hessian *See* burlap.

Hessian [MATH] For a function $f(x_1, \ldots, x_n)$ of n real variables, the real-valued function of $(x_1, \ldots, x_n)$ given by the determinant of the matrix with entry $\partial^2 f/\partial x_i \partial x_j$ in the ith row and jth column; used for analyzing critical points.

hessite [MINERAL] Ag_2Te A lead-gray sectile mineral crystallizing in the isometric system; usually massive and often auriferous.

Hess's law [PHYS CHEM] The law that the evolved or absorbed heat in a chemical reaction is the same whether the reaction takes one step or several steps. Also known as the law of constant heat summation.

hetaerolite [MINERAL] $ZnMn_2O_4$ A black mineral consisting of zinc-manganese oxide found with chalcophanite.

Heteractinida [PALEON] A group of Paleozoic sponges with calcareous spicules; probably related to the Calcarea.

Heterakidae [INV ZOO] A group of nematodes assigned either to the suborder Oxyurina or the suborder Ascaridina.

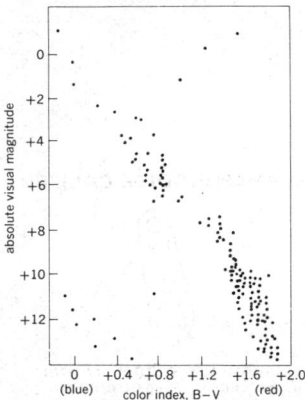
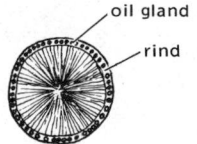

heterandrous [BOT] Having stamens differing from each other in length or form.

heterauxesis *See* allometry.

hetero- [CHEM] Chemical prefix meaning different; for example, a heterocyclic compound is one in which the ring is made of more than one kind of atom.

heteroagglutinin [IMMUNOL] An antibody in normal blood serum capable of agglutinating foreign particles and erythrocytes of other species.

heteroauxin [BIOCHEM] $C_{10}H_9O_2N$ A plant growth hormone with an indole skeleton.

heteroazeotrope [CHEM] Liquid mixture that is not completely miscible in all proportions in the liquid phase, yet does not form an azeotrope. Also known as heterogeneous zeotrope.

Heterobasidiomycetidae [MYCOL] A class of fungi in which the basidium either is branched or is divided into cells by cross walls.

heteroblastic [EMBRYO] Arising from different tissues or germ layers, in referring to similar organs in different species. [PETR] Pertaining to rocks in which the essential constituents are of two distinct orders of magnitude of size.

Heterocapsina [BOT] An order of green algae in the class Xanthophyceae. [INV ZOO] A suborder of yellow-green to green flagellate protozoans in the order Heterochlorida.

heterocarpous [BOT] Producing two distinct types of fruit.

Heterocera [INV ZOO] A formerly recognized suborder of Lepidoptera including all forms without clubbed antennae.

heterocercal [VERT ZOO] Pertaining to the caudal fin of certain fishes and indicating that the upper lobe is larger, with the vertebral column terminating in this lobe.

Heteroceridae [INV ZOO] The variegated mud-loving beetles, a family of coleopteran insects in the superfamily Dryopoidea.

Heterocheilidae [INV ZOO] A family of parasitic roundworms in the superfamily Ascaridoidea.

heterochlamydeous [BOT] Having the perianth differentiated into a distinct calyx and a corolla.

Heterochlorida [INV ZOO] An order of yellow-green to green flagellate oraganisms of the class Phytamastigophorea.

heterochromatin [CYTOL] Specialized chromosome material which remains tightly coiled even in the nondividing nucleus and stains darkly in interphase.

heterochromia [PHYSIO] A condition in which the two irises of an individual have different colors, or in which one iris has two colors.

heterochronism [EMBRYO] Deviation from the normal sequence of organ formation; a factor in evolution.

heterococcolith [BIOL] A coccolith with crystals arranged into boat, trumpet, or basket shapes.

heterocoelous [ANAT] Pertaining to vertebrae with centra having saddle-shaped articulations.

Heterocorallia [PALEON] An extinct small, monofamilial order of fossil corals with elongate skeletons; found in calcareous shales and in limestones.

Heterocotylea [INV ZOO] The equivalent name for Monogenea.

heterocyclic compound [ORG CHEM] Compound in which the ring structure is a combination of more than one kind of atom; for example, pyridine, C_5H_5N.

heterocyst [BOT] Clear, thick-walled cell occurring at intervals along the filament of certain blue-green algae.

heterodactylous [VERT ZOO] Having the first two toes turned backward.

Heterodera [INV ZOO] The cyst nematodes, a genus of phytoparasitic worms that live in the internal root systems of many plants.

heterodesmic [CRYSTAL] Pertaining to those atoms bonded in more than one way in crystals.

heterodont [ANAT] Having teeth that are variable in shape and differentiated into incisors, canines, and molars. [INV ZOO] In bivalves, having two types of teeth on one valve which fit into depressions on the other valve.

Heterodonta [INV ZOO] An order of bivalve mollusks in some systems of classification; hinge teeth are few in number and variable in form.

Heterodontoidea [VERT ZOO] A suborder of sharks in the order Selachii represented by the single living genus *Heterodontus*.

heterodyne [ELECTR] To mix two alternating-current signals of different frequencies in a nonlinear device for the purpose of producing two new frequencies, the sum of and difference between the two original frequencies.

heterodyne analyzer [ENG ACOUS] A type of constant-band-width analyzer in which the electric signal from a microphone beats with the signal from an oscillator, and one of the side bands produced by this modulation is then passed through a fixed filter and detected.

heterodyne conversion transducer *See* converter.

heterodyne detector [ELECTR] A detector in which an unmodulated carrier frequency is combined with the signal of a local oscillator having a slightly different frequency, to provide an audio-frequency beat signal that can be heard with a loudspeaker or headphones; used chiefly for code reception.

heterodyne frequency [COMMUN] Either of the two new frequencies resulting from heterodyne action between the two input frequencies of a heterodyne detector.

heterodyne frequency meter [ELECTR] A frequency meter in which a known frequency, which may be adjustable or fixed, is heterodyned with an unknown frequency to produce a zero beat or an audio-frequency signal whose value is measured by other means. Also known as heterodyne wavemeter.

heterodyne interference *See* heterodyne whistle.

heterodyne measurement [ELECTR] A measurement carried out by a type of harmonic analyzer which employs a highly selective filter, at a frequency well above the highest frequency to be measured, and a heterodyning oscillator.

heterodyne modulator *See* mixer.

heterodyne oscillator [ELECTR] 1. A separate variable-frequency oscillator used to produce the second frequency required in a heterodyne detector for code reception. 2. *See* beat-frequency oscillator.

heterodyne reception [ELECTR] Radio reception in which the incoming radio-frequency signal is combined with a locally generated rf signal of different frequency, followed by detection. Also known as beat reception.

heterodyne repeater [ELECTR] A radio repeater in which the received radio signals are converted to an intermediate frequency, amplified, and reconverted to a new frequency band for transmission over the next repeater section.

heterodyne wavemeter *See* heterodyne frequency meter.

heterodyne whistle [COMMUN] A steady, high-pitched audio tone heard in an ordinary amplitude-modulation radio receiver under certain conditions when two signals that differ slightly in carrier frequency enter the receiver and heterodyne to produce an audio beat. Also known as heterodyne interference.

heteroecious [BIOL] Pertaining to forms which pass through different stages of a life cycle in different hosts.

heteroerotism [PSYCH] Sexual desire directed away from one's self or sex.

heterogamete [BIOL] A gamete that differs in size, appearance, structure, or sex chromosome content from the gamete of the opposite sex. Also known as anisogamete.

heterogametic sex [GEN] That sex of some species in which the two sex chromosomes are different in gene content or size and which therefore produces two or more different kinds of gametes.

heterogamety [GEN] The production of different kinds of gametes by one sex of a species.

heterogamous [BIOL] Of or pertaining to heterogamy.

heterogamy [BIOL] 1. Alternation of a true sexual generation with a parthenogenetic generation. 2. Sexual reproduction by fusion of unlike gametes. Also known as anisogamy. [BOT] Condition of producing two kinds of flowers.

heterogeneity [BIOL] The condition or state of being different in kind or nature.

heterogeneous [CHEM] Pertaining to a mixture of phases such as liquid-vapor, or liquid-vapor-solid. [MATH] Pertaining to quantities having different degrees or dimensions. [SCI TECH] Composed of dissimilar or nonuniform constituents.

heterogeneous catalysis [CHEM] Catalysis occurring at a phase boundary, usually a solid-fluid interface.

heterogeneous chemical reaction [CHEM] Chemical reac-

HETEROAUXIN

Structural formula for heteroauxin.

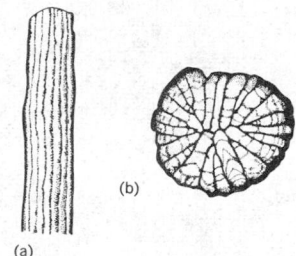

HETEROCOCCOLITH

A heterococcolith whose crystals are arranged in a basket shape. (*A. McIntyre, Lamont-Doherty Geological Observatory of Columbia University*)

HETEROCORALLIA

(b)

(a)

Heterocorallian skeleton. (*a*) External view. (*b*) Transverse thin section. (*From R. C. Moore, ed., Treatise on Invertebrate Paleontology, pt. F, Geological Society of America, University of Kansas Press, 1956*)

tion system in which the reactants are of different phases; for example, gas with liquid, liquid with solid, or a solid catalyst with liquid or gaseous reactants.

heterogeneous fluid [FL MECH] A fluid within which the density varies from point to point; for most purposes the atmosphere must be treated as heterogeneous, particularly with regard to the decrease of density with height.

heterogeneous radiation [PHYS] Radiation having a number of different frequencies, different particles, or different particle energies.

heterogeneous reactor [NUCLEO] A nuclear reactor in which fissionable material and moderator are arranged in a regular pattern of discrete bodies with dimensions such that a nonhomogeneous medium is presented to neutrons.

heterogeneous reservoir [GEOL] Formation with two or more noncommunicating sand members, each possibly with different specific- and relative-permeability characteristics.

Heterogeneratae [BOT] A class of brown algae distinguished by a heteromorphic alteration of generations.

heterogenesis [BIOL] Alternation of generations in a complete life cycle, especially the alternation of a dioecious generation with one or more parthenogenetic generations.

heterogenetic antigen *See* heterophile antigen.

heterogenite [MINERAL] CoO(OH) A black cobalt mineral, sometimes with some copper and iron, found in mammillary masses. Also known as stainierite.

heterogenous [BIOL] Not originating within the body of the organism.

heterogenous vaccine [IMMUNOL] A vaccine derived from a source other than the patient.

Heterognathi [VERT ZOO] An equivalent name for Cypriniformes.

heterogony [BIOL] **1.** *See* allometry. **2.** Alteration of generations in a complete life cycle, especially of a dioecious and hermaphroditic generation. [BOT] Having heteromorphic perfect flowers with respect to lengths of the stamens or styles.

heterograft [IMMUNOL] A tissue or organ obtained from an animal of one species and transplanted to the body of an animal of another species. Also known as heterologous graft.

heterohemolysis [IMMUNOL] Hemolytic amboceptor against the erythrocytes of a species different from that used to obtain the amboceptor.

heterojunction [ELECTR] The boundary between two different semiconductor materials, usually with a negligible discontinuity in the crystal structure.

heterokaryon [MYCOL] A bi- or multinucleate cell having genetically different kinds of nuclei.

heterokaryosis [MYCOL] The condition of a bi- or multinucleate cell having nuclei of genetically different kinds.

heterokont [BIOL] An individual, especially among certain algae, having unequal flagella.

heterolactic fermentation [MICROBIO] A type of lactic acid fermentation by which small yields of lactic acid are produced and much of the sugar is converted to carbon dioxide and other products.

heterolalia [PSYCH] Saying one thing and meaning another as by an unconscious mechanism or in motor aphasia.

heterolateral [ANAT] Of, pertaining to, or located on the opposite side.

heterolecithal [CYTOL] Of an egg, having the yolk distributed unevenly throughout the cytoplasm.

heterologous graft *See* heterograft.

heterologous stimulus [PHYSIO] A form of energy capable of exciting any sensory receptor or form of nervous tissue.

heterologous tumor [MED] A neoplasm composed of tissues that differ from those of the organ at the site of the tumor.

heterolytic cleavage [ORG CHEM] Breaking of a covalent bond involving carbon to produce two oppositely charged fragments.

heteromedusoid [INV ZOO] A styloid type of sessile gonophore.

Heteromera [INV ZOO] The equivalent name for Tenebrionoidea.

heteromerous [BOT] Of a flower, having one or more whorls made up of a different number of members than the remaining whorls.

heterometaplasia [MED] Change in the character of an autograft.

Heteromi [VERT ZOO] An equivalent name for Notacanthiformes.

heteromixis [MYCOL] In Fungi, sexual reproduction which involves the fusion of genetically different nuclei, each from a different thallus.

heteromorphic [CYTOL] Having synoptic or sex chromosomes that differ in size or form. [MED] Differing from the normal in size or morphology. [ZOO] Having a different form at each stage of the life history.

heteromorphosis [BIOL] Regeneration of an organ or part that differs from the original structure at the site. [EMBRYO] Formation of an organ at an abnormal site. Also known as homeosis.

Heteromyidae [VERT ZOO] A family of the mammalian order Rodentia containing the North American kangaroo mice and the pocket mice.

Heteromyinae [VERT ZOO] The spiny pocket mice, a subfamily of the rodent family Heteromyidae.

Heteromyota [INV ZOO] A monospecific order of wormlike animals in the phylum Echiurida.

Heterodontidae [VERT ZOO] The Port Jackson sharks, a family of aberrant modern elasmobranchs in the suborder Heterodontoidea.

Heteronemertini [INV ZOO] An order of the class Anopla; individuals have a middorsal blood vessel and a body wall composed of three muscular layers.

Heteroneura [INV ZOO] A suborder of Lepidoptera; individuals are characterized by fore- and hindwings that differ in shape and venation and by sucking mouthparts.

heteropelmous [VERT ZOO] Having bifid flexor tendons of the toes.

heterophagic vacuole *See* food vacuole.

heterophile agglutination test [PATH] A test for the presence of heterophile antibodies in the serum produced in infectious mononucleosis; agglutination of sheep red cells is a positive test. Also known as heterophile antibody test; Paul-Bunnell test.

heterophile antibody [IMMUNOL] Substance that will react with heterophile antigen; found in the serum of patients with infectious mononucleosis.

heterophile antibody test *See* heterophile agglutination test.

heterophile antigen [IMMUNOL] A substance that occurs in unrelated species of animals but has similar serologic properties among them. Also known as heterogenetic antigen.

heterophile leukocyte [HISTOL] A neutrophile of vertebrates other than humans.

heterophyiasis [MED] Presence of the minute intestinal fluke *Heterophyes heterophyes* in the small intestine of humans.

Heterophyllidae [PALEON] The single family of the extinct coelenterate order Heterocorallia.

heterophyllous [BOT] Having more than one form of foliage leaves on the same plant or stem.

heterophyte [BOT] A plant that depends upon living or dead plants or their products for food materials.

Heteropiidae [INV ZOO] A family of calcareous sponges in the order Sycettida.

heteroplasia [MED] **1.** The presence of a tissue in an abnormal location. **2.** A process whereby tissues are displaced to or developed in locations foreign to their normal habitats.

heteroploidy [GEN] The condition of a chromosome complement in which one or more chromosomes, or parts of chromosomes, are present in number different from the numbers of the rest.

heteropoly acid [INORG CHEM] Complex acids of metals, whose specific gravity is greater than 4, with phosphoric acid; an example is phosphomolybdic acid.

heteropoly compound [INORG CHEM] Polymeric compounds of molybdates with anhydrides of other elements such as phosphorus; the yellow precipitate $(NH_4)_3[P(Mo_3O_{10})_4]$ is such a compound.

Heteroporidae [INV ZOO] A family of trepostomatous-like bryozoans in the order Cyclostomata.

Heteroptera [INV ZOO] The equivalent name for Hemiptera.

heteropycnosis [CYTOL] Differential condensation of certain chromosomes, such as sex chromosomes, or chromosome parts.

heterosexuality [PSYCH] Having sexual feeling toward members of the opposite sex.

heterosis [GEN] The increase in size, yield, and performance found in some hybrids, especially of inbred parents. Also known as hybrid vigor.

heterosite [MINERAL] A mineral composed of phosphate of iron and manganese; it is isomorphous with purpurite.

Heterosomata [VERT ZOO] The equivalent name for Pleuronectiformes.

Heterosoricinae [PALEON] A subfamily of extinct insectivores in the family Soricidae distinguished by a short jaw and hedgehoglike teeth.

heterosphere [METEOROL] The upper portion of a two-part division of the atmosphere (the lower portion is the homosphere) according to the general homogeneity of atmospheric composition; characterized by variation in composition, and in mean molecular weight of constituent gases; starts at 80–100 kilometers above the earth and therefore closely coincides with the ionosphere and the thermosphere.

Heterospionidae [INV ZOO] A monogeneric family of spioniform worms found in shallow and abyssal depths of the Atlantic and Pacific oceans.

heterospory [BOT] Development of more than one type of spores, especially relating to the microspores and megaspores in ferns and seed plants.

heterostatic [ELEC] Pertaining to the measurement of one electrostatic potential by means of a different potential.

heterostemony [BOT] Presence of two or more different types of stamens in the same flower.

Heterostraci [PALEON] An extinct group of ostracoderms, or armored, jawless vertebrates; armor consisted of bone lacking cavities for bone cells.

heterostyly [BOT] Condition or state of flowers having unequal styles.

heterosuggestion [PSYCH] A suggestion from a source outside an individual's mind.

Heterotardigrada [INV ZOO] An order of the tardigrades exhibiting wide morphologic variations.

heterothallic [BOT] Pertaining to a mycelium with genetically incompatible hyphae, therefore requiring different hyphae to form a zygospore; refers to fungi and some algae.

heterotopia [ECOL] An abnormal habitat. [MED] Displacement of an organ or other body part from its natural position.

heterotopic [MED] Occurring in an abnormal anatomic location.

heterotopic epithelium [MED] Intestinal epithelium and goblet cells occurring in the stomach.

heterotopic pregnancy [MED] Double pregnancy with one fetus within and the other outside the uterus.

heterotopic transplantation [BIOL] A graft transplanted to an abnormal anatomical location on the host.

Heterotrichida [INV ZOO] A large order of large ciliates in the protozoan subclass Spirotrichia; buccal ciliature is well developed and some species are pigmented.

Heterotrichina [INV ZOO] A suborder of the protozoan order Heterotrichida.

heterotroph [BIOL] An organism that obtains nourishment from the ingestion and breakdown of organic matter.

heterotropia See strabismus.

heteroxenous [BIOL] Requiring more than one host to complete a life cycle.

heterozooid [INV ZOO] Any of the specialized, nonfeeding zooids in a bryozoan colony.

heterozygote [GEN] An individual that has different alleles at one or more loci and therefore produces gametes of two or more different kinds.

heterozygous [GEN] Of or pertaining to a heterozygote.

HETP See height equivalent of theoretical plate.

Hettangian [GEOL] A stage of Lower Jurassic geologic time.

heulandite [MINERAL] $CaAl_2Si_6O_{16} \cdot 5H_2O$ A zeolite mineral that crystallizes in the monoclinic system; often occurs as foliated masses or in crystal form in cavities of decomposed basic igneous rocks.

Heulinger equations [PHYS] Equations which relate the values of various quantities, such as the Hall coefficient, thermoelectric power, electrical resistivity, and thermal conductivity, in isothermal and adiabatic thermoelectric and thermomagnetic effects.

heuristic method [MATH] A method of solving a problem in which one tries each of several approaches or methods and evaluates progress toward a solution after each attempt.

heuristic program [ADP] A program in which a computer tries each of several methods of solving a problem and judges whether the program is closer to solution after each attempt. Also known as heuristic routine.

heuristic routine See heuristic program.

heuristics [PSYCH] The study of the mental processes involved in problem solving.

Heusler alloy [MET] Any of a group of ferromagnetic nonferrous alloys typically composed of 18–25% manganese, 10–25% aluminum, and the balance copper.

hevea rubber [MATER] Rubber made from latex obtained from the rubber tree (*Hevea brasiliensis*); used for electrical insulation.

Hevelian halo [METEOROL] A faint, white halo with an angular radius of 90°, centered on the sun or moon, and only occasionally seen; it is a member of the class of haloes reported but not yet fully explained. Also known as halo of Hevelius; halo of 90°.

Hevelius's parhelia [METEOROL] Bright spots infrequently observed on the parhelic circle halfway around from the sun to the anthelion; these two brighter areas on the parhelic circle are probably a result of superposition of luminosity of the parhelic circle and the Hevelian halo.

hewettite [MINERAL] $CaV_6O_{16} \cdot 9H_2O$ A deep-red mineral composed of hydrated calcium vanadate; found in silky orthorhombic crystal aggregates in Colorado, Utah, and Peru.

hexacanth [INV ZOO] Having six hooks; refers specifically to the embryo of certain tapeworms.

hexachlorobutadiene [ORG CHEM] $Cl_2C:CClCCl:CCl_2$ A colorless liquid with mild aroma, boiling at 210–220°C; soluble in alcohol and ether, insoluble in water; used as solvent, heat-transfer liquid, and hydraulic fluid.

hexachloroethane [ORG CHEM] Cl_3CCCl_3 Colorless crystals with a camphorlike odor, melting point 185°C, toxic; used in organic synthesis, as a retarding agent in fermentation, and as a rubber accelerator. Also known as carbon trichloride; perchloroethane.

hexachlorophene [ORG CHEM] $(C_6HCl_3OH)_2CH_2$ A white powder melting at 161°C; soluble in alcohol, ether, acetone, and chloroform, insoluble in water; bacteriostat used in antiseptic soaps, cosmetics, and dermatologicals.

hexachloropropylene [ORG CHEM] $CCl_3CCl:CCl_2$ Water-white liquid boiling at 210°C, soluble in alcohol, ether, and chlorinated solvents, insoluble in water; used as a solvent, plasticizer, and hydraulic fluid.

hexacontane [ORG CHEM] $C_{60}H_{122}$ Solid, saturated hydrocarbon of the methane series; melts at 101°C.

Hexacorallia [INV ZOO] The equivalent name for Zoantharia.

hexacosane [ORG CHEM] $C_{26}H_{54}$ Saturated hydrocarbon of the methane series; colorless crystals melting at 57°C.

hexacosanoic acid See cerotic acid.

hexactin [INV ZOO] A spicule, especially in Porifera, having six equal rays at right angles to each other.

Hexactinellida [INV ZOO] A class of the phylum Porifera which includes sponges with a skeleton made up basically of hexactinal siliceous spicules.

Hexactinosa [INV ZOO] An order of sponges in the subclass Hexasterophora; parenchymal megascleres form a rigid framework and consist of simple hexactins.

n-hexadecane [ORG CHEM] $C_{16}H_{34}$ A colorless, solid hydrocarbon, melting point 20°C; a standard reference fuel in determining the ignition quality (cetane number) of diesel fuels. Also known as cetane.

1-hexadecanol See cetyl alcohol.

1-hexadecene [ORG CHEM] $CH_3(CH_2)_{13}CH:CH_2$ A colorless liquid made by treating cetyl alcohol with phosphorous pentoxide; boils at 274°C; soluble in organic solvents such as alcohol, ether, and petroleum; used as an intermediate in organic synthesis. Also known as cetene; cetylene; α-hexadecylene.

cis-9-hexadecenoic acid See palmitoleic acid.

hexadecimal [MATH] Pertaining to a number system using the base 16. Also known as sexadecimal.

hexadecimal notation [ADP] A notation in the scale of 16, using decimal digits 0 to 9 and six more digits that are sometimes represented by A, B, C, D, E, and F.

HETEROTRICHIDA

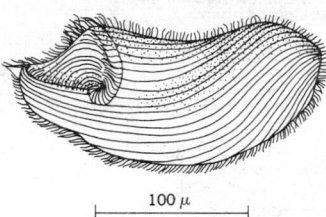

Climacostomum, an example of a heterotrich.

HEULANDITE

heulandite

2.5 cm

calcite

Crystals with calcite and quartz, Paterson, New Jersey. *(Specimen from Department of Geology, Bryn Mawr College)*

HEXAGONAL
CLOSE-PACKED STRUCTURE

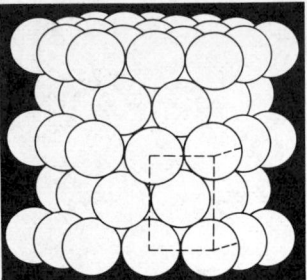

Hexagonal close packing of spheres that simulates the arrangement of the atoms in a crystal that is a hexagonal close-packed structure.

HEXAGONAL-HEAD BOLT

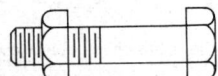

A drawing of a heavy-weight standard bolt that has a hexagonal head.

HEXAGONAL NUT

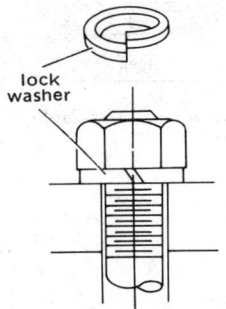

lock washer

A hexagonal nut used with a lock washer.

HEXASTEROPHORA

A representative hexasterophoran, *Polylophus*. (*After Shulze, 1887*)

hexadecimal number system [MATH] A digital system based on powers of 16, as compared with the use of powers of 10 in the decimal number system. Also known as sexadecimal number system.

α-hexadecylene *See* 1-hexadecene.

hexadiene [ORG CHEM] C_6H_{10} A group of unsaturated hydrocarbons with two double bonds; some members of the group are 1,4-hexadiene, 1,5-hexadiene, and 2,4-hexadiene.

2,4-hexadienoic acid *See* sorbic acid.

hexagon [MATH] A six-sided polygon.

hexagonal close-packed structure [CRYSTAL] Close-packed crystal structure characterized by the regular alternation of two layers; the atoms in each layer lie at the vertices of a series of equilateral triangles, and the atoms in one layer lie directly above the centers of the triangles in neighboring layers. Abbreviated hcp structure.

hexagonal column [METEOROL] One of the many forms in which ice crystals are found in the atmosphere; this crystal habit is characterized by hexagonal cross-section in a plane perpendicular to the long direction (principal axis, optic axis, or *c*-axis) of the columns; it differs from that found in hexagonal platelets only in that environmental conditions have favored growth along the principal axis rather than perpendicular to that axis.

hexagonal-head bolt [DES ENG] A standard wrench head bolt with a hexagonal head.

hexagonal lattice [CRYSTAL] A Bravais lattice whose unit cells are right prisms with hexagonal bases and whose lattice points are located at the vertices of the unit cell and at the centers of the bases.

hexagonal nut [DES ENG] A plain nut in hexagon form.

hexagonal platelet [METEOROL] A small ice crystal of the hexagonal tabular form; the distance across the crystal from one side of the hexagon to the opposite side may be as large as about 1 millimeter, and the thickness perpendicular to this dimension is of the order of one-tenth as great; this crystal form is usually formed at temperatures of −10 to −20°C by sublimation; at higher temperatures the apices of the hexagon grow out and develop dendritic forms.

hexagonal system [CRYSTAL] A crystal system that has three equal axes intersecting at 120° and lying in one plane; a fourth, unequal axis is perpendicular to the other three.

hexahedrite [GEOL] An iron meteorite composed of single crystals or aggregates of kamacite, usually containing 4–6% nickel in the metal phase.

hexahedron [MATH] A polyhedron with six faces.

hexahydric alcohol [ORG CHEM] A member of the mannitol-sorbitol-dulcitol sugar group; isomer of $C_6H_8(OH)_6$.

hexahydrite [MINERAL] $MgSO_4 \cdot 6H_2O$ A white or greenish-white monoclinic mineral composed of hydrous magnesium sulfate.

hexahydrocresol *See* methyl cyclohexanol.

hexahydrocymene *See* menthane.

hexahydromethyl phenol *See* methyl cyclohexanol.

hexahydrothymol *See* menthol.

hexahydrotoluene *See* methyl cyclohexane.

***n*-hexaldehyde** [ORG CHEM] $CH_3(CH_2)_4CHO$ Colorless liquid with sharp aroma, boiling at 128.6°C; used as an intermediate for plasticizers, dyes, insecticides, resins, and rubber chemicals. Also known as caproic aldehyde.

hexamethonium chloride [PHARM] One of a homologous series of polymethylene bis (trimethylammonium) ions, of the general formula $(CH_3)_3N^+(CH_2)_nN^+(CH_3)_3$, in which *n* is 6; possesses potent ganglion-blocking action, effecting reduction in blood pressure; used clinically as a salt, commonly bromide or iodide.

hexamethylene *See* cyclohexane.

hexamethylenediamine [ORG CHEM] $H_2N(CH_2)_6NH_2$ Colorless solid boiling at 205°C; slightly soluble in water, alcohol, and ether; used to make nylon and other high polymers.

hexamethylene tetramine *See* cystamine.

Hexanchidae [VERT ZOO] The six- and seven-gill sharks, a group of aberrant modern elasmobranchs in the suborder Notidanoidea.

hexane [ORG CHEM] C_6H_{14} Water-insoluble, toxic, flammable, colorless liquid with faint aroma; forms include: *n*-hexane, a straight-chain compound boiling at 68.7°C and used as a solvent, paint diluent, alcohol denaturant, and polymerization-reaction medium; isohexane, a mixture of hexane isomers boiling at 54–61°C and used as a solvent and freezing-point depressant; and neohexane.

hexanedinitrile *See* adiponitrile.

hexanedioic acid *See* adipic acid.

hexanitrodiphenyl amine [ORG CHEM] $(NO_2)_3C_6H_2NH-C_6H_2(NO_2)_3$ Explosive, yellow solid melting at 238–244°C; insoluble in water, ether, alcohol, or benzene; soluble in alkalies and acetic and nitric acids; used as an explosive and in potassium analysis.

hexanitromannite *See* mannitol hexanitrate.

hexanoic acid *See* caproic acid.

1-hexanol *See* hexyl alcohol.

hexapetalous [BOT] Having or being a perianth comprising six petaloid divisions.

hexaphenylethane [ORG CHEM] $(C_6H_5)_3CC(C_6H_5)_3$ The dimer of triphenylmethyl radical.

Hexapoda [INV ZOO] An equivalent name for Insecta.

hexaster [INV ZOO] A type of hexactin with branching rays that form star-shaped figures.

Hexasterophora [INV ZOO] A subclass of sponges of the class Hexactinellida in which parenchymal microscleres are typically hexasters.

hexenbesen *See* witches'-broom disease.

1-hexene [ORG CHEM] $CH_3(CH_2)_3HC:CH_2$ Colorless, olefinic hydrocarbon boiling at 64°C; soluble in alcohol, acetone, ether, and hydrocarbons, insoluble in water; used as a chemical intermediate and for resins, drugs, and insecticides. Also known as hexylene.

hexobarbital [PHARM] $C_{12}H_{16}N_2O_3$ A sedative and hypnotic of short duration of action; used also, as the sodium derivative, to induce surgical anesthesia.

hexoctahedron [CRYSTAL] A cubic crystal form that has 48 equal triangular faces, each of which cuts the three crystallographic axes at different distances.

hexode [ELECTR] A six-electrode electron tube containing an anode, a cathode, a control electrode, and three additional electrodes that are ordinarily grids.

***n*-hexoic acid** *See* caproic acid.

hexokinase [BIOCHEM] Any enzyme that catalyzes the phosphorylation of hexoses.

hexone *See* methyl isobutyl ketone.

hexosamine [BIOCHEM] A primary amine derived from a hexose by replacing the hydroxyl with an amine group.

hexosan *See* hemicellulose.

hexose [BIOCHEM] Any monosaccharide that contains six carbon atoms in the molecule.

hexose diphosphate pathway *See* glycolytic pathway.

hexose monophosphate cycle [BIOCHEM] A pathway for carbohydrate metabolism in which one molecule of hexose monophosphate is completely oxidized.

hexose phosphate [BIOCHEM] Any one of the phosphoric acid esters of a hexose, notably glucose, formed during the metabolism of carbohydrates by living organisms.

hextetrahedron [CRYSTAL] A 24-faced form of crystal in the tetrahedral group of the isometric system.

hexulose [BIOCHEM] A ketose made from a six-carbon-chain monosaccharide.

***n*-hexyl acetate** [ORG CHEM] $CH_3COOC_6H_{13}$ Colorless liquid boiling at 169°C; soluble in alcohol and ether, insoluble in water; used as a solvent for resins and cellulosic esters.

hexylacetic acid *See* caprylic acid.

hexylacetylene *See* octyne.

hexyl alcohol [ORG CHEM] $CH_3(CH_2)_4CH_2OH$ Colorless liquid boiling at 156°C; soluble in alcohol and ether, slightly soluble in water; used as a chemical intermediate for pharmaceuticals, perfume esters, and antiseptics. Also known as amyl carbinol; 1-hexanol.

hexylamine [ORG CHEM] $CH_3(CH_2)_5NH_2$ Poisonous, water-white liquid with amine aroma; boils at 129°C; a ptomaine base from the autolysis of protoplasm. Also known as aminohexane.

hexylene *See* 1-hexene.

hexylene glycol [ORG CHEM] $C_6H_{14}O_2$ Water-miscible, colorless liquid boiling at 198°C; used in hydraulic brake fluids, in printing inks, and in textile processing.

***n*-hexyl ether** [ORG CHEM] $C_6H_{13}OC_6H_{13}$ Faintly colored liquid with a characteristic odor, only slightly water-soluble; used in solvent extraction and in the manufacture of collodion and various cellulosic products.

n-hexylic acid See caproic acid.

hexylresorcinol [ORG CHEM] $C_6H_{13}C_6H_3(OH)_2$ Sharp-tasting, white to yellowish crystals melting at 64°C; slightly soluble in water, soluble in glycerin, vegetable oils, and organic solvents; used in medicine.

1-hexyne [ORG CHEM] C_4H_9CCH A colorless, water-white liquid, either n-butylacetylene, boiling at 71.5°C, or methyl-propylacetylene, boiling at 84°C.

Heyd's syndrome See hepatorenal syndrome.

Heyn's reagent [MET] Double chloride of copper and ammonia; used in microanalysis of carbon steels.

Hf See hafnium.

HF See high frequency.

HF akylation [CHEM ENG] Petroleum refinery alkylation process in which olefins (C_3, C_4, C_5) are reacted with isobutane in the presence of hydrofluoric acid catalyst.

HFIR See high-flux isotope reactor.

Hfr See high-frequency recombination.

hfs See hyperfine structure.

HFS See type II superconductor.

hg See hectogram.

Hg See mercury.

HH beacon [NAV] Nondirectional radio homing beacon which has a power output of 2000 watts or greater.

H hour [ORD] The hour at which some specified operation will begin, as an attack, amphibious assault, or movement.

hiascent [BIOL] Gaping.

hiatus [ANAT] A space or passage through an organ. [GEOL] A gap in a rock sequence due to a lack of deposition of a bed or to erosion of beds.

hiatus hernia [MED] Hernia through the esophageal hiatus, usually of a portion of the stomach.

hibernaculum [BIOL] A winter shelter for plants or dormant animals. [BOT] A winter bud or other winter plant part. [INV ZOO] A winter resting bud produced by a few fresh-water bryozoans which grows into a new colony in the spring.

hibernal [METEOROL] Of or pertaining to winter.

hibernation [PHYSIO] **1.** Condition of dormancy and torpor found in cold-blooded vertebrates and invertebrates. **2.** See deep hibernation.

Hibernian orogeny See Erian orogeny.

hickey [ELEC] A threaded coupling for attaching an electrical fixture to an outlet box, used when wires from the fixture come out of the end of a stem in the fixture, rather than through an opening in the side of the stem.

hickory [BOT] The common name for species of the genus *Carya* in the order Fagales; tall deciduous tree with pinnately compound leaves, solid pith, and terminal, scaly winter buds.

hiddenite [MINERAL] A transparent green or yellowish-green spodumene mineral containing chromium and valued as a gem.

hide [MATER] A raw or dressed pelt, especially from a large, adult animal such as a cow. [VERT ZOO] Outer covering of an animal.

hide glue [MATER] A strong, light-brown glue made from animal hides.

hidradenitis [MED] Inflammation of sweat glands.

hidradenoma papilliferum [MED] Benign tumor of sweat glands, usually on the vulva or perineum.

hidrosis [MED] Abnormally profuse sweat. [PHYSIO] The formation and excretion of sweat.

hiemal climate [CLIMATOL] Climate pertaining to winter.

hieratite [MINERAL] K_2SiF_6 A grayish mineral composed of potassium fluosilicate; occurs as deposits in volcanic holes.

hierophobia [PSYCH] An abnormal fear of sacred objects.

hi-fi See high fidelity.

Hi-Fix [NAV] A Decca radio navigation system utilizing lightweight, portable stations, and used in either of two configurations; in the hyperbolic configuration, the relative phases of continuous-wave signals from the master and slave stations ashore, as measured at the receiver at the mobile station, provide hyperbolic lines of position; in the range configuration, the master station is located at the mobile receiver; phase comparison of continuous-wave signals from one shore station and the master provides range to the shore station.

Higbie model [CHEM ENG] Mass-transfer theory for packed absorption towers, stating that liquid flows across each packing piece in laminar flow and is mixed with other liquids meeting it at the points of discontinuity between packing elements.

high [METEOROL] An area of high pressure, referring to a maximum of atmospheric pressure in two dimensions (closed isobars) in the synoptic surface chart, or a maximum of height (closed contours) in the constant-pressure chart; since a high is, on the synoptic chart, always associated with anticyclonic circulation, the term is used interchangeably with anticyclone.

high-abrasion furnace black [MATER] A variety of carbon black made by burning oil in a deficiency of air and then quenching to stop the reaction short of equilibrium; particles are 26–30 nanometers in diameter; used in tires and mechanical rubber goods. Abbreviated HAF black.

high-alloy steel [MET] Steel containing large percentages of elements other than carbon.

high aloft See upper-level anticyclone.

high-altitude bombing [ORD] Bombing, especially horizontal bombing, from a high altitude, such as over 15,000 feet (4500 meters).

high-altitude disease See mountain sickness.

high-altitude erythremia See mountain sickness.

high-altitude method [NAV] The establishing of a circular line of position from the observation of the altitude of a celestial body; the geographical position and zenith distance of the body are used in constructing this line of position.

high-altitude radio altimeter See radar altimeter.

high-altitude station [METEOROL] A weather observing station at a sufficiently high elevation to be nonrepresentative of conditions near sea level; 6500 feet (about 2000 meters) has been given as a reasonable lower limit.

high-alumina brick [MATER] Refractory brick made from raw materials rich in alumina, such as diaspore and bauxite; when well fired, they contain a large amount of mullite; used for unusually severe temperature or load conditions.

high-alumina cement See aluminous cement.

high-angle fault [GEOL] A fault with a dip greater than 45°.

high-angle fire [ORD] **1.** Fire delivered at elevations greater than the elevation of maximum range. **2.** Fire whose range decreases as the angle of elevation is increased; typical of mortar and howitzer fire.

high-angle gun [ORD] A cannon, such as an antiaircraft cannon, capable of firing at a high angle of elevation.

high-angle strafing [ORD] Strafing from an airplane at a comparatively high dive angle; for example, approximately 45°.

high band [COMMUN] The television band extending from 174 to 216 megahertz, which includes channels 7 to 13.

high boiler [MATER] A solvent added to lacquer thinner to slow the rate of evaporation; boiling point 150–200°C.

high boost See high-frequency compensation.

high brass [MET] The most common commercial wrought brass containing about 35% zinc and 65% copper.

high-burst ranging [ORD] Adjustment of gunfire by observation of airbursts.

high-capacity projectile [ORD] A projectile with thin walls and high explosive loading, for use where no special penetrative qualities are required.

high-carbon chromium [MET] Chromium containing at least 86% chromium, 8–11% carbon, and a maximum of 0.5% each of iron and silicon.

high-carbon steel [MET] A cast or forged steel containing more than 0.5% carbon.

high clouds [METEOROL] Types of clouds whose mean lower level is above 20,000 feet (6100 meters); principal clouds in this group are cirrus, cirrocumulus, and cirrostratus.

high contrast [GRAPHICS] That area where the degree of difference between black and white approaches the maximum.

high-current rectifier [ELECTR] A solid-state device, gas tube, or vacuum tube used to convert alternating to direct current for powering low-impedance loads.

high-current switch [ELEC] A switch used to redirect heavy current flow; usually has a make-before-break feature to prevent excessive arcing.

HICKORY

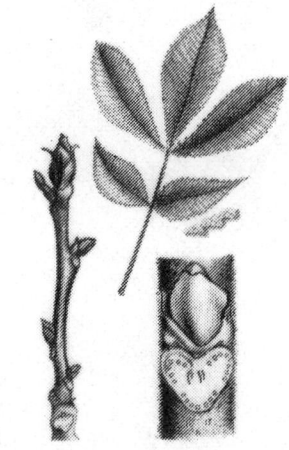

Twig, bud, and leaf of the shagbark hickory, *Carya ovata*.

high definition [COMMUN] Television or facsimile equivalent of high fidelity, in which the reproduced image contains such a large number of accurately reproduced elements that picture details approximate those of the original scene.

high enema [MED] An enema injected into the colon.

high-energy astrophysics [ASTROPHYS] A science concerned with studies of acceleration of charged particles to high energies in space, cosmic rays, radio galaxies, pulsars, and quasi-stellar sources.

high-energy bond [PHYS CHEM] Any chemical bond yielding a decrease in free energy of at least 5 kilocalories per mole.

high-energy electron diffraction [PHYS] Diffraction of electrons with high energies, usually in the range 30,000–70,000 electron volts, mainly to study the structure of atoms and molecules in gases and liquids. Abbreviated HEED.

high-energy fuel [MATER] Fuel that upon combustion provides greater energy than that from conventional carbonaceous fuels; specifically, a hydroboron.

high-energy particle [PARTIC PHYS] An elementary particle having an energy of hundreds of Mev (million electron volts) or more.

high-energy physics *See* particle physics.

high-energy scattering [PARTIC PHYS] Collisions of particles with energies of hundreds of Mev (million electron volts) or more, sufficient to produce new particles.

high-epithermal neutron range [NUCLEO] The neutron energy range of 1000 to 100,000 electron volts.

higher high water [OCEANOGR] The higher of two high tides occurring during a tidal day.

higher low water [OCEANOGR] The higher of two low tides occurring during a tidal day.

higher mode [ELECTROMAG] A waveguide mode whose frequency is higher than the lowest one.

higher pair [MECH ENG] A link in a mechanism in which the mating parts have surface (instead of line or point) contact.

higher plane curve [MATH] Any algebraic curve whose degree exceeds 2.

highest common factor *See* greatest common divisor.

high etch *See* dry-relief offset.

high-expansion alloy [MET] An alloy possessing a high coefficient of expansion.

high-expansion foam [MATER] Noncombustible foam made from ammonium lauryl sulfate; used in underground mine fire fighting.

high explosive [MATER] An explosive with a nitroglycerin base requiring a detonator; the explosion is violent and practically instantaneous.

high-explosive bomb [ORD] **1.** Any aerial bomb charged with a high explosive. **2.** Any bomb chiefly dependent upon its explosion or blast effect to create damage.

high-explosive plastic [MATER] High-explosive substance or mixture which, within normal ranges of atmospheric temperature, is capable of being molded into desired shapes. Also known as plastic explosive.

high-explosive plastic antitank charge [ORD] A shaped charge coupled with a high-explosive plastic charge, intended to produce jet penetration followed by a detonation of the plastic charge. Abbreviated HEPAT charge.

high-explosive plastic projectile [ORD] A thin-walled projectile, filled with plastic explosive; designed to squash against an armored target before detonation, and to defeat the armor by producing spalls which are detached with considerable velocity from the back of the target plate. Abbreviated HEP projectile.

high-explosive projectile [ORD] Projectile with a bursting charge of high explosive.

high fidelity [ENG ACOUS] Audio reproduction that closely approximates the sound of the original performance. Also known as hi-fi.

high-field superconductor *See* type II superconductor.

high-flash naphtha *See* heavy naphtha.

high-flux isotope reactor [NUCLEO] A thermal research reactor at the Oak Ridge National Laboratory, used mainly in the production of elements with atomic numbers greater than that of plutonium. Abbreviated HFIR.

high foehn [METEOROL] The occurrence of warm, dry air above the level of the general surface, accompanied by clear skies, resembling foehn conditions; it is due to subsiding air in an anticyclone, above a cold surface layer; in such circumstances the mountain peaks may be warmer than the lowlands. Also known as free foehn.

high fog [METEOROL] The frequent fog on the slopes of the coastal mountains of California, especially applied when the fog overtops the range and extends as stratus over the leeward valleys.

high frequency [COMMUN] Federal Communications Commission designation for the band from 3 to 30 megahertz in the radio spectrum. Abbreviated HF.

high-frequency carrier telegraphy [COMMUN] Form of carrier telegraphy in which the carrier currents have their frequencies above the range transmitted over a voice-frequency telephone channel.

high-frequency compensation [ELECTR] Increasing the amplification at high frequencies with respect to that at low and middle frequencies in a given band, such as in a video band or an audio band. Also known as high boost.

high-frequency furnace [ENG] An induction furnace in which the heat is generated within the charge, within the walls of the containing crucible, or within both, by currents induced by high-frequency magnetic flux produced by a surrounding coil. Also known as coreless-type induction furnace; high-frequency heater.

high-frequency heater *See* high-frequency furnace.

high-frequency heating *See* electronic heating.

high-frequency propagation [COMMUN] Propagation of radio waves in the high-frequency band, which depends entirely on reflection from the ionosphere.

high-frequency recombination [MICROBIO] A bacterial cell type, especially *Escherichia coli,* having an integrated F factor and characterized by a high frequency of recombination. Abbreviated Hfr.

high-frequency resistance [ELEC] The total resistance offered by a device in an alternating-current circuit, including the direct-current resistance and the resistance due to eddy current, hysteresis, dielectric, and corona losses. Also known as alternating-current resistance; effective resistance; radio-frequency resistance.

high-frequency titration [ANALY CHEM] A conductimetric titration in which two electrodes are mounted on the outside of the beaker or vessel containing the solution to be analyzed and an alternating current source in the megahertz range is used to measure the course of a titration.

high-frequency transformer [ELECTR] A transformer which matches impedances and transmits a frequency band in the carrier (or higher) frequency ranges.

high-frequency triode [ELECTR] A triode designed for operation at high frequency, having small spacings between the grid and the cathode and anode, large emission and power densities, and low active and inactive capacitances.

high-frequency voltmeter [ELECTR] A voltmeter designed to measure currents alternating at high frequencies.

high-frequency welding [MET] Resistance welding in which the heat is produced by the current flow induced by a high-frequency electromagnetic field. Also known as radio-frequency welding.

high-front shovel [MECH ENG] A power shovel with a dipper stick mounted high on the boom for stripping and overburden removal.

high-grade [MIN ENG] To steal or pilfer ore or gold from a mine or miner.

high-grade dynamite [MATER] Dynamite of 40% strength or over.

high hat [ENG] A very low tripod head resembling a formal top hat in shape.

high heat [THERMO] Heat absorbed by the cooling medium in a calorimeter when products of combustion are cooled to the initial atmospheric (ambient) temperature.

high-heat cement [MATER] A type of cement which releases a large amount of heat during curing.

high-helix drill [DES ENG] A two-flute twist drill with a helix angle of 35–40°; used for drilling deep holes in metals, such as aluminum, copper, hard brass, and soft steel. Also known as fast-spiral drill.

high-impedance voltmeter [ELEC] A voltage-measuring device with a high-impedance input to reduce load on the unit under test; a vacuum-tube voltmeter is one type.

high index [METEOROL] A relatively high value of the zonal index which, in middle latitudes, indicates a relatively strong westerly component of wind flow and the characteristic weather features attending such motion; a synoptic circulation pattern of this type is commonly called a high-index situation.

high-intensity atomizer [MECH ENG] A type of atomizer used in electrostatic atomization, based on stress sufficient to overcome tensile strength of the liquid.

high-K capacitor [ELEC] A capacitor whose dielectric material is a ferroelectric having a high dielectric constant, up to about 6000.

high key [GRAPHICS] A negative or print that has an overall light effect, with very few dark tones or shadows.

high-key lighting [GRAPHICS] In photography, lighting that produces tones that fall mostly between white and gray, with very few dark-gray or black tones.

highland climate *See* mountain climate.

highland glacier [HYD] A semicontinuous ice cap or glacier that covers the highest or central portion of a mountainous area and partly reflects irregularities of the land surface lying beneath it. Also known as highland ice.

highland halftone [GRAPHICS] A halftone with elimination of the residual dot formation in the pure highlights, done either by spotting the highlight areas on the negative with opaque or by etching the printing plate.

highland ice *See* highland glacier.

high-lead bronze [MET] Bronze containing high percentages of lead to give a soft matrix alloy.

high-level anticyclone *See* upper-level anticyclone.

high-level cyclone *See* upper-level cyclone.

high-level modulation [COMMUN] Modulation produced at a point in a system where the power level approximates that at the output of the system.

high-level ridge *See* upper-level ridge.

high-level thunderstorm [METEOROL] Generally, a thunderstorm based at a comparatively high altitude in the atmosphere, roughly 8000 feet (2400 meters) or higher.

high-level trough *See* upper-level trough.

high-lift truck [MECH ENG] A forklift truck with a fixed or telescoping mast to permit high elevation of a load.

highlights [ELECTR] Bright areas in a television image. [GRAPHICS] The bright parts of a subject that show up as the densest part of the negative and the lightest or whitest part of a print.

high-low bias test [ELECTR] A routine maintenance procedure that tests equipment over and under normal operating conditions in order to detect defective units.

high-modulus furnace black [MATER] A variety of carbon black made by burning oil in a deficiency of air; particle diameter is 49-60 nanometers; production is now negligible, but it was used in tire carcasses. Abbreviated HMF black.

highmoor bog [ECOL] A bog whose surface is covered by sphagnum mosses and is not dependent upon the water table.

high-mu tube [ELECTR] A tube having a very high amplification factor.

high-octane [MATER] Having an octane number in the middle or high 90s and therefore having good antiknock properties.

high-order [ADP] Pertaining to a digit location in a numeral, the leftmost digit being the highest-order digit.

high-pass filter [ELECTR] A filter that transmits all frequencies above a given cutoff frequency and substantially attenuates all others.

high plain [GEOGR] A large area of level land situated above sea level.

high polymer [ORG CHEM] A large molecule (of molecular weight greater than 10,000) usually composed of repeat units of low-molecular-weight species; for example, ethylene or propylene.

high-positive indicator [ADP] A component in some computers whose status is "on" if the number tested is positive and nonzero.

high-potting [ELEC] Testing with a high voltage, generally on a production line.

high-precision shoran *See* hiran.

high-pressure chemistry [PHYS CHEM] The study of chemical reactions and phenomena that occur at pressures exceed-

ing 10,000 bars (a bar is nearly equivalent to a kilogram per square centimeter), mainly concerned with the properties of the solid state.

high-pressure cloud chamber [NUCLEO] A cloud chamber in which the gas is maintained at high pressure to reduce the range of high-energy particles and thereby increase the probability of observing events.

high-pressure gage glass [ENG] A gage glass consisting of a metal tube with thick glass windows.

high-pressure gas injection [PETRO ENG] Oil reservoir pressure maintenance by injection of gas at pressures higher than those used in conventional equilibrium gas drives.

high-pressure laminate [MATER] A plastic-substrate laminate molded and cured at pressures of, commonly, 1200-2000 psi (8-14 $\times$ 10^6 newtons per square meter).

high-pressure mercury-vapor lamp [ELECTR] A discharge tube containing an inert gas and a small quantity of liquid mercury; the initial glow discharge through the gas heats and vaporizes the mercury, after which the discharge through mercury vapor produces an intensely brilliant light.

high-pressure phenomena [PHYS] Natural conditions and processes occurring at high pressures, and their duplication in the laboratory.

high-pressure physics [PHYS] The study of the effects of high pressure on the properties of matter.

high-pressure process [CHEM ENG] A chemical process operating at elevated pressure; for example, phenol manufacture at 330 atmospheres (1 atmosphere = 1.01325 $\times$ 10^5 newtons per square meter), ethylene polymerization at 2000 atm, ammonia synthesis at 100-1000 atm, and synthetic-diamond manufacture up to 100,000 atm.

high-pressure separator [PETRO ENG] A horizontal vessel through which a low-temperature, high-pressure gas stream is fed, and in which free liquids separate out from the gas.

high-pressure torch [ENG] A type of torch in which both acetylene and oxygen are delivered to the mixing chamber under pressure.

high-pressure well [PETRO ENG] A well with a shut-in wellhead pressure of more than 2000 psia (1.4 $\times$ 10^7 newtons per square meter, absolute).

high Q [ELECTR] A characteristic wherein a component has a high ratio of reactance to effective resistance, so that its Q factor is high.

high-Q cavity [ELECTROMAG] A cavity resonator which has a large Q factor, and thus has a small energy loss. Also known as high-Q resonator.

high-Q resonator *See* high-Q cavity.

high quartz [MINERAL] Quartz that was formed at high temperatures.

high-rank coal [GEOL] Coal consisting of less than 4% moisture when air-dried, or more than 84% carbon.

high-rank graywacke *See* feldspathic graywacke.

high relief [GRAPHICS] Sculpture projecting from the background block to the extent of at least half of the natural depth of the modeled forms.

high-residual-phosphorus copper [MET] Deoxidized copper having reduced conductivity due to the presence of residual phosphorus, usually less than 0.1%.

high-resistance voltmeter [ELEC] A voltmeter having a resistance considerably higher than 1000 ohms per volt, so that it draws little current from the circuit in which a measurement is made.

high-resolution radar [ENG] A radar system which can discriminate between two close targets.

high-side capacitance coupling [ELECTR] Taking the output of an oscillator or amplifier from a point of high potential, using a capacitor to block direct current flow.

high-solvency naphtha [MATER] Any of the petroleum-based solvents that boil in the naphtha range (95-650°F; 35-343°C) and have a high aromatic-chemical content to give high-solvency power for nitrocelluloses, dry paints, and certain resins.

high-speed carry [ADP] A technique in parallel addition to speed up the propagation of carries.

high-speed data acquisition system [ADP] A system which collects and transmits data rapidly to a monitoring and controlling center.

high-speed excitation system [ELEC] Excitation system capable of changing its voltage rapidly in response to a change in the excited generator field circuit.

high-speed machine [MECH ENG] A diamond drill capable of rotating a drill string at a minimum of 2500 revolutions per minute, as contrasted with the normal maximum speed of 1600–1800 revolutions per minute attained by the average diamond drill.

high-speed oscilloscope [ELECTR] An oscilloscope with a very fast sweep, capable of observing signals with rise times or periods on the order of nanoseconds.

high-speed photography [GRAPHICS] Photography to record movement or events that occur too quickly to be observed by usual visual or photographic means; motion pictures may be shot at high speeds (50–500 frames per second) and projected at normal rates, so that the action of the subject is slowed to a point where it can be observed; or a series of individual still photos may be produced.

high-speed printer [ADP] A printer which can function at a high rate, relative to the state of the art; 600 lines per minute is considered high speed. Abbreviated HSP.

high-speed reader [ADP] The fastest input device existing at a particular time in the state of the technology.

high-speed relay [ELECTR] A relay specifically designed for short operate time, short release time, or both.

high-speed steel [MET] An alloy steel that remains hard and tough at red heat.

high-speed storage *See* rapid storage.

high-strength alloy *See* high-tensile alloy.

high-strength steel *See* high-tensile steel.

high-temperature alloy [MET] An alloy suitable for use at temperatures of 500°C and above.

high-temperature cement [MATER] A cement that resists fusing, softening, or spalling at elevated temperatures; used to bond refractory materials.

high-temperature chemistry [PHYS CHEM] The study of chemical phenomena occurring above about 500K.

high-temperature coke [MATER] Coke produced at temperatures of 900–1150°C; used mainly for metallurgical purposes.

high-temperature fuel cell [ELEC] A fuel cell which operates at temperatures above about 550°C, can use inexpensive hydrocarbon fuels, and usually uses a molten salt as an electrolyte.

high-temperature gas-cooled reactor [NUCLEO] A prototype gas-cooled reactor in which the coolant is pressurized helium gas with an inlet temperature of about 325°C and an outlet temperature of about 750°C, and the fuel consists of fully enriched uranium and thorium. Abbreviated HTGR.

high-temperature material [MATER] A material with high-temperature capability, including the superalloys, refractory alloys, and ceramics; used in structures such as spacecraft subjected to extreme thermal environments.

high-temperature phenomena [PHYS] Phenomena occurring at temperatures above about 500K.

high-temperature reactor [NUCLEO] A power reactor in which the temperature is high enough for efficient generation of mechanical power.

high-temperature water boiler [MECH ENG] A boiler which provides hot water, under pressure, for space heating of large areas.

high-tensile alloy [MET] An alloy having a high tensile strength. Also known as high-strength alloy.

high-tensile bolt [ENG] A bolt that is adjusted to a carefully controlled tension by means of a calibrated torsion wrench.

high-tensile steel [MET] Low-alloy steel having a yield strength range of 50,000–100,000 pounds per square inch (3.4×10^8 to 6.9×10^8 newtons per square meter). Also known as high-strength steel.

high-tension detonator [ENG] A detonator requiring an electric potential of about 50 volts for firing.

high-tension separation *See* electrostatic separation.

high-test chain [ENG] Chain made from heat-treatable plain-carbon steel, usually with a carbon content of 0.15–0.20%; used for load binding, tie-downs, and other applications where failure would be costly.

high tide [OCEANOGR] The maximum height reached by a rising tide. Also known as high water.

high vacuum [PHYS] A vacuum with a pressure between 1×10^{-3} and 1×10^{-6} mm Hg.

high-vacuum insulation [CHEM ENG] High vacuum between the walls of double-wall vessels to serve as thermal insulation at ultralow (cryogenic) temperatures, such as in Dewar vessels.

high-vacuum rectifier [ELECTR] Vacuum-tube rectifier in which conduction is entirely by electrons emitted from the cathode.

high-vacuum switching tube [ELECTR] A microwave transmit-receive (TR) tube of the high-vacuum variety, as contrasted with gas-tube or semiconductor devices.

high-vacuum tube [ELECTR] Electron tube evacuated to such a degree that its electrical characteristics are essentially unaffected by gaseous ionization. Also known as hard tube.

high-velocity stars [ASTRON] Those stars moving across the galactic track along which the majority of the stars execute their galactic rotation, thus exhibiting high velocity with respect to the sun, low velocity with respect to the galactic center.

high-volatile bituminous coal [GEOL] A bituminous coal composed of more than 31% volatile matter.

high-voltage electron microscope [ELECTR] An electron microscope whose accelerating voltage is on the order of 10^6 volts, as compared with 40–100 kilovolts for an ordinary electron microscope; it has the advantages of increased specimen penetration, reduced specimen damage, better theoretical resolution, and more efficient dark-field operation.

high-voltage insulation [ELEC] Electrical insulation designed to prevent breakdown in a circuit operating at high voltages.

highwall [MIN ENG] The unexcavated face of exposed overburden and coal or ore in an opencast mine or the face or bank of the uphill side of a contour strip-mine excavation.

high water *See* high tide.

high-water full and change [GEOPHYS] The average interval of time between the transit (upper or lower) of the full or new moon and the next high water at a place. Also known as common establishment; vulgar establishment.

high-water inequality [OCEANOGR] The difference between the heights of the two high tides during a tidal day.

high-water line [OCEANOGR] The intersection of the plane of mean high water with the shore.

high-water lunitidal interval [GEOPHYS] The interval of time between the transit (upper or lower) of the moon and the next high water at a place.

high-water platform *See* wave-cut bench.

high-water quadrature [OCEANOGR] The average high-water interval when the moon is at quadrature.

high-water springs *See* mean high-water springs.

high-water stand [OCEANOGR] The condition at high tide when there is no change in the height of the water.

highway [CIV ENG] A public road where traffic has the right to pass and to which owners of adjacent property have access.

highway engineering [CIV ENG] A branch of civil engineering dealing with highway planning, location, design, and maintenance.

Hikojima serotype [IMMUNOL] An immunologically distinct group of *Vibrio* somatic O antigens.

hilac *See* heavy-ion linear accelerator.

Hilbert cube [MATH] The topological space which is the cartesian product of a countable number of copies of *I*, the unit interval.

Hilbert-Schmidt theory [MATH] A body of theorems which investigates the kernel of an integral equation via its eigenfunctions, and then applies these functions to help determine solutions of the equation.

Hilbert space [MATH] A Banach space which also is an inner-product space with the inner product of a vector with itself being the same as the square of the norm of the vector.

Hilbert transform [MATH] The transform of a function $f(x)$ realized by taking the integral of $f(x)[1 + \cot(y-x)/2]dx$.

Hilda group [ASTRON] A group of asteroids whose mean motion is close to that of Jupiter.

Hildebrand function [THERMO] The heat of vaporization of a compound as a function of the molal concentration of the vapor; it is nearly the same for many compounds.

hilgardite [MINERAL] $Ca_8(B_6O_{11})_3Cl_4 \cdot H_2O$ Colorless mineral composed of hydrous borate and chloride of calcium; occurs as monoclinic domatic crystals.

hill [GEOGR] A land surface feature characterized by strong relief; it is a prominence smaller than a mountain.

hill-and-dale recording *See* vertical recording.

Hill bandwidth [ELECTR] The difference between the upper and lower frequencies at which the gain of an amplifier is 3 decibels less than its maximum value.

hill-climbing [MATH] Any numerical procedure for finding the maximum or maxima of a function. [MECH ENG] Adjustment, either continuous or periodic, of a self-regulating system to achieve optimum performance.

hill creep [GEOL] Slow gravity movement of rock and soil waste down a steep hillside. Also known as hillside creep.

hillebrandite [MINERAL] $Ca_2SiO_3(OH)_2$ A white mineral composed of hydrous calcium silicate; occurs in masses.

Hill reaction [ORD CHEM] Production of substituted phenylacetic acids by the oxidation of the corresponding alkylbenzene by potassium permanganate in the presence of acetic acid.

hillside creep *See* hill creep.

hillside quarry [MIN ENG] A quarry cut along a hillside.

Hiltner-Hall effect [ASTRON] The polarization of the light received from distant stars; this effect is thought to take place in interstellar space.

Hilt's law [GEOL] The law that in a small area the deeper coals are of higher rank than those above them.

hilum [ANAT] *See* hilus. [BOT] Scar on a seed marking the point of detachment from the funiculus.

hilus [ANAT] An opening or recess in an organ, usually for passage of a vessel or duct. Also known as hilum.

Himantandraceae [BOT] A family of dicotyledonous plants in the order Magnoliales characterized by several, uniovulate carpels and laminar stamens.

himantioid [MYCOL] Pertaining to a mycelium arranged in spreading fanlike cords.

Himantopterinae [INV ZOO] A subfamily of lepidopteran insects in the family Zygaenidae including small, brightly colored moths with narrow hindwings, ribbonlike tails, and long hairs covering the body and wings.

hindbrain *See* rhombencephalon.

hindered contraction [MET] Thermal contraction of a casting that is hindered locally due to the particular geometry.

hindered settling [MIN ENG] Settling of particles in a thick suspension in water through which their fall is hindered by rising water.

hindered-settling ratio [MIN ENG] The ratio of the specific gravity of a mineral to that of the suspension of ore raised to a power between one-half and unity.

hindgut [EMBRYO] The caudal portion of the embryonic digestive tube in vertebrates.

H indicator *See* H scope.

Hindley screw [DES ENG] An endless screw or worm of hourglass shape that fits a part of the circumference of a worm wheel so as to increase the bearing area and thus diminish wear. Also known as hourglass screw; hourglass worm.

hindrance factor *See* drag factor.

hinge [DES ENG] A pair of metal leaves forming a jointed device on which a swinging part turns.

hinged arch [CIV ENG] A structure that can rotate at its supports or in the center or at both places.

hinged bar [MIN ENG] A steel extension bar placed in contact with the mine roof perpendicular to the longwall face and supported by yielding steel props. Also known as link bar.

hinge fault [GEOL] A fault whose movement is an angular or rotational one on a side of an axis that is normal to the fault plane.

hinge joint *See* ginglymus.

hinge line [GEOL] 1. The line separating the region in which a beach has been thrust upward from that in which it is horizontal. 2. A line in the plane of a hinge fault separating the part of a fault along which thrust or reverse movement occurred from that having normal movement.

hinge moment [AERO ENG] The tendency of an aerodynamic force to produce motion about the hinge line of a control surface.

hinge plate [INV ZOO] 1. In bivalve mollusks, the portion of

a valve that supports the hinge teeth. 2. The socket-bearing part of the dorsal valve in brachiopods.

hinge tooth [INV ZOO] A projection on a valve of a bivalve mollusk near the hinge line.

Hinsberg test [ANALY CHEM] A test to distinguish between primary and secondary amines; it involves reaction of an amine with benzene disulforyl chloride in alkaline solution; primary amines give sulfonamides that are soluble in basic solution; secondary amines give insoluble derivatives; tertiary amines do not react with the reagent.

hinsdalite [MINERAL] $(Pb,Sr)Al_3(PO_4)(SO_4)(OH)_6$ A dark-gray or greenish rhombohedral mineral composed of basic lead and strontium aluminum sulfate and phosphate; occurs in coarse crystals and masses.

hinterland [GEOL] 1. The region behind the coastal district. 2. The terrain on the back of a folded mountain chain. 3. The moving block which forces geosynclinal sediments toward the foreland.

Hiodontidae [VERT ZOO] A family of tropical, fresh-water actinopterygian fishes in the order Osteoglossiformes containing the mooneyes of North America.

hip [ANAT] 1. The region of the junction of thigh and trunk. 2. The hip joint, formed by articulation of the femur and hipbone.

hip bone [ANAT] A large broad bone consisting of three parts, the ilium, ischium, and pubis; makes up a lateral half of the pelvis in mammals. Also known as innominate.

Hipernik [MET] A magnetic alloy consisting of 50% iron and 50% nickel.

hip joint [CIV ENG] The junction of an inclined head post and the top chord of a truss.

hi pot [ELEC] High potential voltage applied across a conductor to test the insulation or applied to an etched circuit to burn out tenuous conducting paths that might later fail in service.

Hippidea [INV ZOO] A group of decapod crustaceans belonging to the Anomura and including cylindrical or squarish burrowing crustaceans in which the abdomen is symmetrical and bent under the thorax.

Hippoboscidae [INV ZOO] The louse flies, a family of cyclorrhaphous dipteran insects in the section Pupipara.

hippocampal sulcus [ANAT] A fissure on the brain situated between the para hippocampal gyrus and the fimbria hippocampi. Also known as dentate fissure.

hippocampus [ANAT] A ridge that extends over the floor of the descending horn of each lateral ventricle of the brain.

Hippocampus [VERT ZOO] A genus of marine fishes in the order Gasterosteiformes which contains the sea horses.

Hippocrateaceae [BOT] A family of dicotyledonous plants in the order Celastrales distinguished by an extrastaminal disk, mostly opposite leaves, seeds without endosperm, and a well-developed latex system.

hippocrepiform [BIOL] Horseshoe-shaped.

Hippoglossidae [VERT ZOO] A family of actinopterygian fishes in the order Pleuronectiformes composed of the flounders and plaice.

Hippomorpha [VERT ZOO] A suborder of the mammalian order Perissodactyla containing horses, zebras, and related forms.

Hippopotamidae [VERT ZOO] The hippopotamuses, a family of palaeodont mammals in the superfamily Anthracotherioidea.

hippopotamus [VERT ZOO] The common name for two species of artiodactyl ungulates composing the family Hippopotamidae.

Hipposideridae [VERT ZOO] The Old World leaf-nosed bats, a family of mammals in the order Chiroptera.

hippuric acid [ORG CHEM] $C_6H_5CONHCH_2\cdot COOH$ Colorless crystals melting at 188°C; soluble in hot water, alcohol, and ether; used in medicine and as a chemical intermediate.

hip roof [ARCH] A roof having four slopes; the two shorter ends are triangular.

hiran [NAV] A modification of the shoran system; special operating techniques permit distance measurements of an accuracy comparable to first-order triangulation. Derived from high-precision shoran.

Hirschback method [MIN ENG] A method for draining fire-

HILUM

The structure of a mature kidney bean showing the hilum.

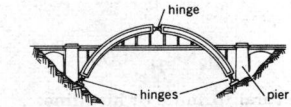

HINGED ARCH

hinge

hinges   pier

A three-hinged bridge arch, which has the advantage that it is statically determinate, and no stresses result from temperature, shrinkage, or rib-shortening effects.

HIPPOPOTAMUS

The great African hippopotamus, more at home in water than on land because of its size.

HIRUDINEA

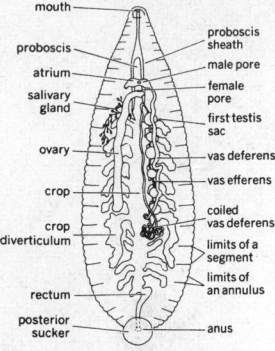

General structure of a leech. Male reproductive system is shown on the right, the female on the left. *(From K. H. Mann, A key to the British freshwater leeches, Freshwater Biol. Assoc. Sci. Publ., 14:3–21, 1954)*

Labels: mouth, proboscis, atrium, salivary gland, ovary, crop, crop diverticulum, rectum, posterior sucker; proboscis sheath, male pore, female pore, first testis sac, vas deferens, vas efferens, coiled vas deferens, limits of a segment, limits of an annulus, anus

HISTIDINE

Structural formula of histidine.

HITTORF METHOD

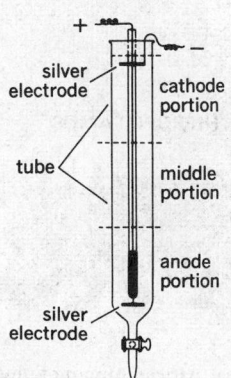

Labels: silver electrode, tube, silver electrode; cathode portion, middle portion, anode portion

The cell used in the Hittorf method.

damp from coal seams by means of superjacent entries located 80–138 feet (24–42 meters) above the seams and supplemented by boreholes drilled at right angles to the entry walls. Also known as superjacent roadway system.

Hirschsprung's disease [MED] A disease caused by absence of the myenteric ganglion cells in a segment of rectum or distal colon, resulting in spasm of the affected part and dilation of the bowel proximal to the defect.

hirsute [BIOL] Shaggy; hairy.

hirsutism [MED] An abnormal condition characterized by growth of hair in unusual places and in unusual amounts.

Hirudinea [INV ZOO] A class of parasitic or predatory annelid worms commonly known as leeches; all have 34 body segments and terminal suckers for attachment and locomotion.

Hirudinidae [VERT ZOO] The swallows, a family of passeriform birds in the suborder Oscines.

hisingerite [MINERAL] $Fe_2^{3+}Si_2O_5(OH)_4 \cdot 2H_2O$ A black, amorphous mineral composed of hydrous ferric silicate; an iron ore.

hispid [BIOL] Having a covering of bristles or minute spines.

hispidulous [BIOL] Hispid to a minute degree.

hiss [COMMUN] Random noise in the audio-frequency range, similar to prolonged sibilant sounds.

histamine [BIOCHEM] $C_5H_9N_3$ An amine derivative of histadine which is widely distributed in human tissues.

Histeridae [INV ZOO] The clown beetles, a large family of coleopteran insects in the superfamily Histeroidea.

Histeroidea [INV ZOO] A superfamily of coleopteran insects in the suborder Polyphaga.

histidase [BIOCHEM] An enzyme found in the liver of higher vertebrates that catalyzes the deamination of histidine to urocanic acid.

histidine [BIOCHEM] $C_6H_9O_2N_3$ A crystalline basic amino acid present in large amounts in hemoglobin and resulting from the hydrolysis of most proteins.

histidinemia [MED] An asymptomatic, hereditary metabolic disorder involving a deficiency of histidase with high blood and urine levels of histidine, urocanic acid, and sometimes alanine.

histiocyte *See* macrophage.

histiocytoma [MED] 1. Benign tumor composed of histiocytes. 2. Dermatofibroma.

histiocytosis [MED] Abnormal proliferation of histiocytes, especially in hematopoietic tissues.

Histioteuthidae [INV ZOO] A family of cephalopod mollusks containing several species of squids.

histochemistry [BIOCHEM] A science that deals with the distribution and activities of chemical components in tissues.

histocompatibility [IMMUNOL] The capacity to accept or reject a tissue graft.

histodifferentiation [EMBRYO] Differentiation of cell groups into tissues.

histogen [BOT] A clearly delimited region or primary tissue from which the specific parts of a plant organ are thought to be produced.

histogenesis [EMBRYO] The developmental process by which the definite cells and tissues which make up the body of an organism arise from embryonic cells.

histogram [STAT] A graphical representation of a distribution function by means of rectangles whose widths represent intervals into which the range of observed values is divided and whose heights represent the number of observations occurring in each interval.

histoincompatibility [IMMUNOL] The condition in which a recipient rejects a tissue graft.

histologist [ANAT] An individual who specializes in histology.

histology [ANAT] The study of the structure and chemical composition of animal and plant tissues as related to their function.

histolysis [PATH] Disintegration of organic tissue.

histomycosis [MED] Infection of deep tissues by a fungus.

histone [BIOCHEM] Any of the strong, soluble basic proteins of cell nuclei that are precipitated by ammonium hydroxide.

histopathology [PATH] A branch of pathology that deals with tissue changes associated with disease.

histophysiology [PHYSIO] The science of tissue functions.

Histoplasma [MYCOL] A genus of parasitic fungi.

Histoplasma capsulatum [MYCOL] The parasitic fungus that causes histoplasmosis in man.

histoplasmin [PHARM] A standardized liquid concentrate of soluble growth factors developed by the fungus *Histoplasma capsulatum.*

histoplasmin test [IMMUNOL] Skin test for hypersensitivity reaction to *Histoplasma capsulatum* products in the diagnosis of histoplasmosis.

histoplasmoma [MED] A tumorlike swelling caused by an inflammatory reaction to *Histoplasma capsulatum.*

histoplasmosis [MED] An infectious fungus disease of the lungs of man caused by *Histoplasma capsulatum.*

historadiography [BIOPHYS] A technique for taking x-ray pictures of cells, tissues, or sometimes whole small organisms.

historical climate [CLIMATOL] A climate of the historical period (the past 7000 years).

historical geology [GEOL] A branch of geology concerned with the systematic study of bedded rocks and their relations in time and the study of fossils and their locations in a sequence of bedded rocks.

Histosol [GEOL] An order of wet soils consisting mostly of organic matter, popularly called peats and mucks.

histotome [BIOL] A microtome used to cut tissue sections for microscopic examination.

Histriobdellidae [INV ZOO] A small family of errantian polychaete worms that live as ectoparasites on crayfishes.

histrionism [PSYCH] Dramatic affect, associated with some psychoneuroses and psychoses.

hit [ADP] The obtaining of a correct answer in a mechanical information-retrieval system. [ELEC] A momentary electrical disturbance on a transmission line. [ORD] 1. A blow or impact on a target by a bullet, bomb, or other projectile. 2. An instance of striking something with a bomb or the like.

hitch [GEOL] 1. A fault of strata common in coal measures, accompanied by displacement. 2. A minor dislocation of a vein or stratum not exceeding in extent the thickness of the vein or stratum. [MIN ENG] 1. A step cut in the rock face to hold timber support in an underground working. 2. A hole cut in side rock solid enough to hold the cap of a set of timbers, permitting the leg to be dispensed with.

hitcher [MIN ENG] 1. The worker who runs trams into or out of the cages, gives the signals, and attends at the shaft when workers are riding in the cage. 2. A worker at the bottom of a haulage slope or plane who engages the clips or grips by means of which mine cars are attached to a hoisting cable or chain. Also known as hitcher-on.

hitcher-on *See* hitcher.

hitch timbering [MIN ENG] Installing timbers in hitches either cut or drilled in the rock.

Hitenso [MET] Trade name for a high-tensile-strength bronze of high conductivity.

hit-on-the-fly system [ADP] A printer in a computer system where either the print roller or the paper is in continuous motion.

hit probability [ORD] The probability of hits being made on a target out of a given number of projectiles directed at the target.

Hittorf dark space *See* cathode dark space.

Hittorf method [PHYS CHEM] A procedure for determining transference numbers in which one measures changes in the composition of the solution near the cathode and near the anode of an electrolytic cell, due to passage of a known amount of electricity.

Hittorf principle [ELECTR] The principle that a discharge between electrodes in a gas at a given pressure does not necessarily occur between the closest points of the electrodes if the distance between these points lies to the left of the minimum on a graph of spark potential versus distance. Also known as short-path principle.

hive *See* beehive.

hives *See* urticaria.

hjelmite [MINERAL] A black mineral containing yttrium, iron, manganese, uranium, calcium, columbium, tantalum, tin, and tungsten oxide; often occurs with crystal structure disrupted by radiation.

HK *See* Hefner candle.

hl *See* hectoliter.

h-line *See* h-bar.

hm *See* hectometer.

H mode *See* transverse electric mode.

H Monel [MET] A Monel containing 3.2% silicon available in cast form; it is harder and stronger than Monel.

H network [ELECTR] An attenuation network composed of five branches and having the form of the letter H. Also known as H attenuator; H pad.

Ho *See* holmium.

hoar crystal [HYD] An individual ice crystal in a deposit of hoarfrost; always grows by sublimation.

hoarding behavior [VERT ZOO] The carrying of food to the home nest for storage, in quantities exceeding daily need.

hoarfrost [HYD] A deposit of interlocking ice crystals formed by direct sublimation on objects. Also known as white frost.

hoarhound *See* marrubium.

hoarse [MED] Having a harsh, discordant voice, caused by an abnormal condition of the larynx or throat.

hoary [BOT] Having grayish or whitish color, referring to leaves.

hob [MECH ENG] A rotary cutting tool with its teeth arranged along a helical thread; used for generating gear teeth.

hobber *See* hobbing machine.

hobbing [MECH ENG] Cutting evenly spaced forms, such as gear teeth, on the periphery of cylindrical workpieces.

hobbing machine [MECH ENG] A machine for cutting gear teeth in gear blanks or for cutting worm, spur, or helical gears. Also known as hobber.

hobbing steel [MET] A high-speed steel used to make gear teeth cutters.

hobnail [DES ENG] A short, large-headed, sharp-pointed nail; used to attach soles to heavy shoes.

hobo connection [ENG] A parallel electrical connection used in blasting.

hod [CIV ENG] A tray fitted with a handle by which it can be carried on the shoulder for transporting bricks or mortar.

Hodgkin's disease [MED] A disease characterized by a neoplastic proliferation of atypical histiocytes in one or several lymph nodes. Also known as lymphogranulomatosis.

hodgkinsonite [MINERAL] $MnZnSiO_5 \cdot H_2O$ A pink to reddish-brown mineral composed of hydrous zinc manganese silicate; occurs as crystals.

Hodgson number [CHEM ENG] Method of predicting the metering error during pulsating gas flow when a surge tank is located between the pulsation source (pump or compressor) and the meter (orifice, nozzle, or venturi).

hodograph [PHYS] The curve traced out in the course of time by the tip of a vector representing some physical quantity.

hodoscope [NUCLEO] An array of small Geiger counters, scintillation counters, or other radiation counters used in tracing paths of high-energy particles in experiments with particle accelerators or in cosmic rays. [OPTICS] *See* conoscope.

Hodotermitidae [INV ZOO] A family of lower (primitive) termites in the order Isoptera.

hoe [DES ENG] An implement consisting of a long handle with a thin, flat, straight-edged blade attached transversely to the end; used for cultivating and weeding.

hoegbomite [MINERAL] $Mg(Al,Fe,Ti)_4O_7$ A black mineral composed of an oxide of magnesium, aluminum, iron, and titanium. Also spelled högbomite.

Hoepfner process *See* Hopfner process.

hoernesite [MINERAL] $Mg_3As_2O_8 \cdot H_2O$ A white, monoclinic mineral composed of hydrous magnesium arsenate; occurs as gypsumlike crystals.

hoe scraper [MIN ENG] A scraper-loader consisting of a box-sided hoe pulled by cables and used in mining to gather and transport severed rock.

hoe shovel [MECH ENG] A revolving shovel with a pull-type bucket rigidly attached to a stick hinged on the end of a live boom.

Hofbauer cell [HISTOL] A large, possibly phagocytic cell found in chorionic villi.

Hofmann amine separation [ORG CHEM] A technique to separate a mixture of primary, secondary, and tertiary amines; they are heated with ethyl oxalate; there is no reaction with tertiary amines, primary amines form a di-

amide, and the secondary amines form a monoamide; when the reaction mixture is distilled, the mixture is separated into components.

Hofmann degradation [ORG CHEM] The action of bromine and an alkali on an amide so that it is converted into a primary amine with one less carbon atom.

Hofmann exhaustive methylation reaction [ORG CHEM] The thermal decomposition of quaternary ammonium hydroxide compounds to yield an olefin and water; an exception is tetramethylammonium hydroxide, which decomposes to give an alcohol.

hofmannite *See* hartite.

Hofmann mustard-oil reaction [ORG CHEM] Preparation of alkylisothiocyanates by heating together a primary amine, mercuric chloride, and carbon disulfide.

Hofmann reaction [ORG CHEM] A reaction in which amides are degraded by treatment with bromine and alkali (caustic soda) to amines containing one less carbon; used commercially in the production of nylon.

Hofmann rearrangement [ORG CHEM] A chemical rearrangement of the hydrohalides of *N*-alkylanilines upon heating to give aminoalkyl benzenes.

Hofmeister series [CHEM] An arrangement of anions or cations in order of decreasing ability to produce coagulation when their salts are added to lyophilic sols. Also known as lyotopic series.

hog [AGR] A domestic swine.

hogback [GEOL] Alternate ridges and ravines in certain areas of mountains, caused by erosive action of mountain torrents.

högbomite *See* hoegbomite.

hog cholera [VET MED] A fatal infectious virus disease of swine characterized by fever, diarrhea, and inflammation and ulceration of the intestine; secondary infection by *Salmonella cholerae suis* is common. Also known as African swine fever.

hog frame [NAV ARCH] A rigid framework of beams, usually extending the length of a ship and above the deck, used to prevent hogging and increase longitudinal stiffness.

hogged fuel [MATER] Sawmill refuse that has been fed through a disintegrator, or hog, by which the various sizes and forms are reduced to a practically uniform size of chips or shreds.

hogging [ENG] Mechanical chipping of wood waste for fuel. [NAV ARCH] Sagging of the bow and stern of a ship with respect to the amidships section.

hog gum [MATER] A sticky gum, an exudate from various species of *Sterculia* trees, whose chief constituent is galactan; the gum is dried and marketed either as a powder or as white flakes; used in the food, cosmetic, and textile industry.

hoghorn antenna *See* horn antenna.

hohlraum *See* blackbody.

Hohmann orbit [AERO ENG] A minimum-energy-transfer orbit.

Hohmann trajectory [AERO ENG] The minimum-energy trajectory between two planetary orbits, utilizing only two propulsive impulses.

hoist [MECH ENG] **1.** To move or lift something by a rope-and-pulley device. **2.** A power unit for a hoisting machine, designed to lift from a position directly above the load and therefore mounted to facilitate mobile service.

hoist back-out switch [MECH ENG] A protective switch that permits hoist operation only in the reverse direction in case of overwind.

hoist cable [MECH ENG] A fiber rope, wire rope, or chain by means of which force is exerted on the sheaves and pulleys of a hoisting machine.

hoist frame *See* headframe.

hoist hook [DES ENG] A swivel hook attached to the end of a hoist cable for securing a load.

hoisting [MECH ENG] **1.** Raising a load, especially by means of tackle. **2.** Either of two power-shovel operations: the raising or lowering of the boom, or the lifting or dropping of the dipper stick in relation to the boom.

hoisting compartment [MIN ENG] The section of a shaft used for hoisting the mined mineral to the surface.

hoisting cycle [MIN ENG] The periods of acceleration, uniform speed, retardation, and rest; the deeper the shaft, the

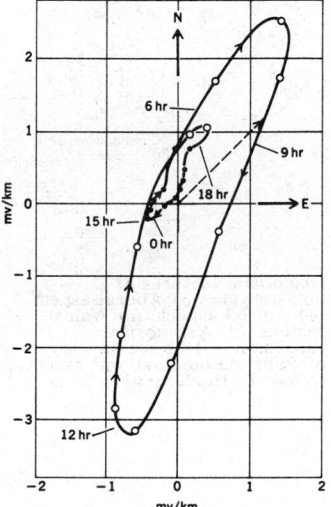

HODOGRAPH

Hodograph representing the potential-gradient vector of the average diurnal variation of earth currents at Tucson, Arizona, for 10 quiet days each month for 1932 to 1942 inclusive.

greater is the ratio of the time of full-speed hoisting to the whole cycle.

hoisting drum *See* drum.

hoisting machine [MECH ENG] A mechanism for raising and lowering material with intermittent motion while holding the material freely suspended.

hoisting power [MECH ENG] The capacity of the hoisting mechanism on a hoisting machine.

hoistman [ENG] One who operates steam or electric hoisting machinery to lower and raise cages, skips, or instruments into a mine or an oil or gas well. Also known as hoist operator; winch operator.

hoist operator *See* hoistman.

hoist overspeed device [MECH ENG] A device used to prevent a hoist from operating at speeds greater than predetermined values by activating an emergency brake when the predetermined speed is exceeded.

hoist overwind device [MECH ENG] A device which can activate an emergency brake when a hoisted load travels beyond a predetermined point into a danger zone.

hoist slack-brake switch [MECH ENG] A device that automatically cuts off power to the hoist motor and sets the brake if the links in the brake rigging require tightening or if the brakes require relining.

hoist tower [CIV ENG] A temporary shaft of scaffolding used to hoist materials for building construction.

hoja blanca [PL PATH] A major virus disease of rice in Cuba and Venezuela.

holarctic zoogeographic region [ECOL] A major unit of the earth's surface extending from the North Pole to 30–45°N latitude and characterized by faunal homogeneity.

Holasteridae [INV ZOO] A family of exocyclic Euechinoidea in the order Holasteroida; individuals are oval or heart-shaped, with fully developed pore pairs.

Holasteroida [INV ZOO] An order of exocyclic Euechinoidea in which the apical system is elongated along the anteroposterior axis and teeth occur only in juvenile stages.

holcodont [VERT ZOO] Having the teeth fixed in a long, continuous groove.

hold [ADP] To retain information in a computer storage device for further use after it has been initially utilized. [ELECTR] To maintain storage elements at equilibrium voltages in a charge storage tube by electron bombardment. [ENG] The interior of a ship or plane, especially the cargo compartment. [IND ENG] A therblig, or basic operation, in time-and-motion study in which the hand or other body member maintains an object in a fixed position and location.

hold back [GRAPHICS] In printing of photos, to shade certain portions of the image while printing.

holdback [MECH ENG] A brake on an inclined-belt conveyor system which is automatically activated in the event of power failure, thus preventing the loaded belt from running downward.

holdbeam [NAV ARCH] Any one of a tier of beams which go from side to side, spanning the hold from frame to frame, and upon which a deck is not attached.

hold circuit [ELECTR] A circuit in a sampled-data control system that converts the series of impulses, generated by the sampler, into a rectangular function, in order to smooth the signal to the motor or plant.

hold control [ELECTR] A manual control that changes the frequency of the horizontal or vertical sweep oscillator in a television receiver, so that the frequency more nearly corresponds to that of the incoming synchronizing pulses.

holddown [MET] A device that holds the outer portion of a metal sheet in place during deep-drawing operations in order to keep it from becoming wrinkled. [PETRO ENG] A device to anchor an oil well rod pump in its position.

holdenite [MINERAL] A red, orthorhombic mineral composed of basic manganese zinc arsenate with a small amount of calcium, magnesium, and iron.

holder [ELEC] A device that mechanically and electrically accommodates one or more crystals, fuses, or other components in such a way that they can readily be inserted or removed.

Hölder condition [MATH] A function $f(x)$ satisfies the Hölder condition in a neighborhood of a point x_0 if $|f(x) - f(x_0)| \leq c|(x - x_0)|^n$, where c and n are constants.

Hölder's inequality [MATH] A generalization of the Schwarz inequality; for real functions it says that $|\int f(x)g(x)\,dx| \leq (\int |f(x)|^p\,dx)^{1/p}(\int |g(x)|^q\,dx)^{1/q}$ where $1/p + 1/q = 1$.

holdfast [BOT] **1.** A suckerlike base which attaches the thallus of certain algae to the support. **2.** A disklike terminal structure on the tendrils of various plants used for attachment to a flat surface. [INV ZOO] An organ by which parasites such as tapeworms attach themselves to the host.

holding anode [ELECTR] A small auxiliary anode used in a mercury-pool rectifier to keep a cathode spot energized during the intervals when the main-anode current is zero.

holding beam [ELECTR] A diffused beam of electrons used to regenerate the charges stored on the dielectric surface of a cathode-ray storage tube.

holding coil [ELECTR] A separate relay coil that is energized by contacts which close when a relay pulls in, to hold the relay in its energized position after the orginal operating circuit is opened.

holding furnace [MET] A heated reservoir to hold molten metal preparatory to casting.

holding ground [NAV] The bottom ground of an anchorage; the term is usually preceded by an adjective to indicate the quality of the holding power of the material constituting the bottom.

holding magnet *See* lifting magnet.

holding point [NAV] An identifiable point designated by an air-traffic control center as the position in the vicinity of which an aircraft is instructed to remain.

holding time [COMMUN] Period of time a trunk or circuit is in use on a call, including operator's time in connecting and subscriber's or user's conversation time.

hold lamp [ELEC] Indicating lamp which remains lighted while a telephone connection is being held.

hold-over command [ADP] A command punched at the end of each card to cause machines to treat the several cards as if they were one continuous record.

hold time [MET] In resistance welding, the time that pressure is applied to the electrodes after the welding current is cut off.

holdup [CHEM ENG] **1.** Volume of material held or contained in a process vessel or line. **2.** Liquid held up (suspended) in a vertical process vessel or line by rising gas or vapor streams.

hole [SOLID STATE] A vacant electron energy state near the top of an energy band in a solid; behaves as though it were a positively charged particle. Also known as electron hole.

hole conduction [ELECTR] Conduction occurring in a semiconductor when electrons move into holes under the influence of an applied voltage and thereby create new holes.

Holectypidae [PALEON] A family of extinct exocyclic Euechinoidea in the order Holectypoida; individuals are hemispherical.

Holectypoida [INV ZOO] An order of exocyclic Euechinoidea with keeled, flanged teeth, with distinct genital plates, and with the ambulacra narrower than the interambulacra on the adoral side.

hole deviation [ENG] The change in the course or direction that a borehole follows.

hole director [MIN ENG] A steel framework used in underground tunneling to set the angle at which holes for a blasting round are to be drilled.

hole injection [ELECTR] The production of holes in an n-type semiconductor when voltage is applied to a sharp metal point in contact with the surface of the material.

hole layout [MIN ENG] In quarrying, an arrangement of vertical and horizontal holes.

hole mobility [ELECTR] A measure of the ability of a hole to travel readily through a semiconductor, equal to the average drift velocity of holes divided by the electric field.

hole saw *See* crown saw.

hole site [ADP] The area on a punch card where a hole may be punched.

hole-through [MIN ENG] The meeting of two approaching tunnel heads.

hole trap [ELECTR] A semiconductor impurity capable of releasing electrons to the conduction or valence bands, equivalent to trapping a hole.

holing [MIN ENG] **1.** The working of a lower part of a bed of

HOLASTEROIDA

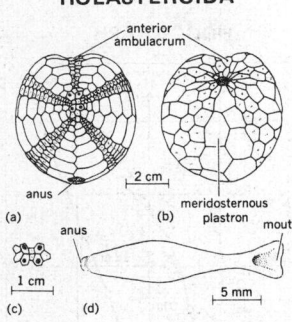

Diagnostic features of holasteroids. *(a)* Aboral aspect. *(b)* Adoral aspect. *(c)* Apical system. *(d)* *Echinosigra paradoxa,* a deep-sea species of Pourtalesiidae, aboral aspect. *(After T. Mortensen)*

coal to bring down the upper mass. **2.** The final act of connecting two workings underground. **3.** The meeting of two mine roadways driven to intersect. Also known as thirling.

holism [BIOL] The view that the whole of a complex system, such as a cell or organism, is functionally greater than the sum of its parts. Also known as organicism.

holistic masks [ADP] In character recognition, that set of characters which resides within a character reader and theoretically represents the exact replicas of all possible input characters.

hollander [MECH ENG] An elongate tube with a central mid-feather and a cylindrical beater roll; formerly used for stock preparation in paper manufacture.

Holland formula [ENG] A formula used to calculate the height of a plume formed by pollutants emitted from a stack in terms of the diameter of the stack exit, the exit velocity and heat emission rate of the stack, and the mean wind speed.

hollandite [MINERAL] $Ba(Mn^{2+},Mn^{4+})_8O_{16}$ A silvery-gray to black mineral composed of manganate of barium and manganese; occurs as crystals.

Hollerith code [ADP] A code used to represent letters, numbers, or special symbols to be punched in standard 80-column punch cards.

Hollerith string [ADP] A sequence of characters preceded by an H and a character count in FORTRAN, as 4HSTOP.

Hollinacea [PALEON] A dimorphic superfamily of extinct ostracods in the suborder Beyrichicopina including forms with sulci, lobation, and some form of velar structure.

Hollinidae [PALEON] An extinct family of ostracods in the superfamily Hollinacea distinguished by having a bulbous third lobe on the valve.

hollow [SCI TECH] **1.** Having a concave surface. **2.** Having an interior cavity.

hollow block *See* hollow tile.

hollow cathode [ELECTR] A cathode which is hollow and closed at one end in a discharge tube filled with inert gas, designed so that radiation is emitted from the cathode glow inside the cathode.

hollow drill [DES ENG] A drill rod or stem having an axial hole for the passage of water or compressed air to remove cuttings from a drill hole. Also known as hollow rod; hollow stem.

hollow ended [NAV ARCH] Pertaining to when the extremities of the waterlines in the neighborhood of the designed load line are concave to the surrounding water, and when the sectional area curve at the ends is fine, indicating relatively small displacement in these locations.

hollow gravity dam [CIV ENG] A fixed gravity dam, usually of reinforced concrete, constructed of inclined slabs or arched sections supported by transverse buttresses.

hollow mill [MECH ENG] A milling cutter with three or more cutting edges that revolve around the cylindrical workpiece.

hollow-pipe waveguide [ELECTROMAG] A waveguide consisting of a hollow metal pipe; electromagnetic waves are transmitted through the interior and electric currents flow on the inner surfaces.

hollow-plunger pump [MIN ENG] A pump for mining and quarrying in gritty and muddy water.

hollow reamer [ENG] A tool or bit used to correct the curvature in a crooked borehole.

hollow rod *See* hollow drill.

hollow-rod churn drill [MECH ENG] A churn drill with hollow rods instead of steel wire rope.

hollow-rod drilling [ENG] A modification of wash boring in which a check valve is introduced at the bit so that the churning action may be also used to pump the cuttings up the drill rods.

hollow shafting [MECH ENG] Shafting made from hollowed-out rods or hollow tubing to minimize weight, allow internal support, or permit other shafting to operate through the interior.

hollow stalk [PL PATH] Any plant disease characterized by deterioration of the pith in the stalk.

hollow stem *See* hollow drill.

hollow tile [MATER] A hollow building block of concrete or burnt clay used for making partitions, exterior walls, or

suspended floors or roofs. Also known as hollow block.

hollow wall [BUILD] A masonry wall provided with an air space between the inner and outer wythes.

holly [BOT] The common name for the trees and shrubs composing the genus *Ilex*; distinguished by spiny leaves and small berries.

hollywood lignumvitae *See* lignumvitae.

Holman Airleg [MIN ENG] A drill support consisting of a cylinder in which the piston is actuated by compressed air controlled by a twist-grip valve.

Holman counterbalanced drill rig [MIN ENG] A drill rig consisting of a rail-track carriage with a counterbalanced boom 10 feet (3.0 meters) long.

Holman dust extractor [MIN ENG] A dust-trapping system in which the dust and chippings from percussive drilling operations are drawn back through the hollow drill rod and along a hose to a metal container with filter elements.

Holman stamp [MIN ENG] A crushing stamp raised by a crank and accelerated in its fall by compressed air.

Holme mud sampler [ENG] A scooplike device which can be lowered by cable to the ocean floor to collect sediment samples.

holmium [CHEM] A rare-earth element belonging to the yttrium subgroup, symbol Ho, atomic number 67, atomic weight 164.93, melting point 1400–1525°C.

holmquisite [MINERAL] $(Na,K,Ca)Li(Mg,Fe)_3Al_2Si_8O_{22}$-$(OH)_2$ A bluish-black, orthorhombic mineral composed of alkali and silicate of iron, magnesium, lithium, and aluminum.

holoaxial [CRYSTAL] Having all possible axes of symmetry.

Holobasidiomycetes [MYCOL] An equivalent name for Homobasidiomycetidae.

holoblastic [EMBRYO] Pertaining to eggs that undergo total cleavage due to the absence of a mass of yolk.

holobranch [VERT ZOO] A gill with a row of filaments on each side of the branchial arch.

holocarpic [BOT] **1.** Having the entire thallus developed into a fruiting body or sporangium. **2.** Lacking rhizoids and haustoria.

holocellulose [BIOCHEM] The total polysaccharide fraction of wood, straw, and so on, that is composed of cellulose and all of the hemicelluloses and that is obtained when the extractives and the lignin are removed from the natural material.

Holocene *See* Recent.

Holocentridae [VERT ZOO] A family of nocturnal beryciform fishes found in shallow tropical and subtropical reefs; contains the squirrelfishes and soldierfishes.

Holocephali [VERT ZOO] The chimaeras, a subclass of the Chondrichthyes, distinguished by four pairs of gills and gill arches, an erectile dorsal fin and spine, and naked skin.

holoclastic [PETR] Being or belonging to ordinary (sedimentary) clastic rock.

holococcolith [BIOL] A coccolith with simple rhombic or hexagonal crystals arranged like a mosaic.

holocoenosis [ECOL] The nature of the action of the environment on living organisms.

holocrine gland [PHYSIO] A structure whose cells undergo dissolution and are entirely extruded, together with the secretory product.

holocrystalline [PETR] Pertaining to igneous rocks that are entirely crystallized minerals, without glass.

holoenzyme [BIOCHEM] A complex, fully active enzyme, containing an apoenzyme and a coenzyme.

hologamy [BIOL] Condition of having gametes similar in size and form to somatic cells. [BOT] Condition of having the whole thallus develop into a gametangium.

hologony [INV ZOO] Condition of having the germinal area extend the full length of a gonad; refers specifically to certain nematodes.

hologram [OPTICS] The special photographic plate used in holography; when this negative is developed and illuminated from behind by a coherent gas-laser beam, it produces a three-dimensional image in space. Also known as hologram interferometer.

hologram interferometer *See* hologram.

holography [PHYS] A technique for recording, and later re-

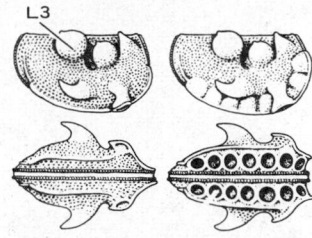
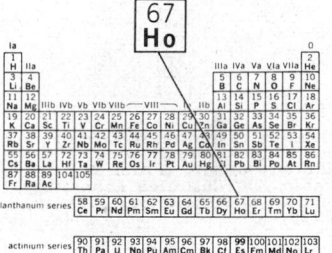
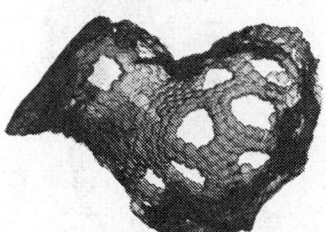
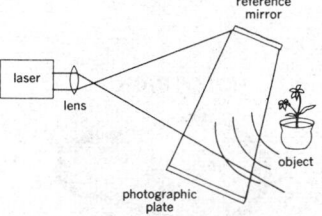

constructing, the amplitude and phase distributions of a wave disturbance; widely used as a method of three-dimensional optical image formation, and also with acoustical and radio waves; in optical image formation, the technique is accomplished by recording on a photographic plate the pattern of interference between coherent light reflected from the object of interest, and light that comes directly from the same source or is reflected from a mirror.

holohedral [CRYSTAL] Pertaining to a crystal structure having the highest symmetry in each crystal class. Also known as holosymmetric; holosystemic.

holohedron [CRYSTAL] A crystal form of the holohedral class, having all the faces needed for complete symmetry.

holohyaline [PETR] Pertaining to an entirely glassy rock.

Holometabola [INV ZOO] A division of the insect subclass Pterygota whose members undergo holometabolous metamorphosis during development.

holometabolous metamorphosis [INV ZOO] Complete metamorphosis, during which there are four stages; the egg, larva, pupa, and imago or adult.

holomictic lake [HYD] A lake whose water circulates completely from top to bottom.

holomorphic function *See* analytic function.

holomorphosis [BIOL] Complete regeneration of a lost body structure.

holomyarian [INV ZOO] Having zones of muscle layers but no muscle cells; refers specifically to certain nematodes.

holonephros [VERT ZOO] Type of kidney having one nephron beside each somite along the entire length of the coelom; seen in larvae of myxinoid cyclostomes.

holonomic constraints [MECH] An integrable set of differential equations which describe the restrictions on the motion of a system; a function relating several variables, in the form $f(x_1, \ldots, x_n) = 0$, in optimization or physical problems.

holonomic system [MECH] A system in which the constraints are such that the original coordinates can be expressed in terms of independent coordinates and possibly also the time.

holophotal [OPTICS] 1. Pertaining to a holophote. 2. Reflecting all the light from a source in one direction.

holophote [OPTICS] An optical system consisting of lenses or reflectors that collect a large amount of the light from a source (such as the lamp of a lighthouse) and send it in a desired direction.

holophyte [BIOL] An organism that obtains food in the manner of a green plant, that is, by synthesis of organic substances from inorganic substances using the energy of light.

holoplankton [ZOO] Organisms that live their complete life cycle in the floating state.

Holoptychidae [PALEON] A family of extinct lobefin fishes in the order Osteolepiformes.

holorhinal [VERT ZOO] Among birds, having a rounded anterior margin on the nasal bones.

Holostei [VERT ZOO] An infraclass of fishes in the subclass Actinopterygii descended from the Chondrostei and ancestral to the Teleostei.

holostome [INV ZOO] A type of adult digenetic trematode having a portion of the ventral surface modified as a complex adhesive organ.

holosymmetric *See* holohedral.

holosystemic *See* holohedral.

Holothuriidae [VERT ZOO] A family of aspidochirotacean echinoderms in the order Aspidochirotida possessing tentacular ampullae and only the left gonad.

Holothuroidea [INV ZOO] The sea cucumbers, a class of the subphylum Echinozoa characterized by a cylindrical body and smooth, leathery skin.

Holothyridae [INV ZOO] The single family of the acarine suborder Holothyrina.

Holothyrina [INV ZOO] A suborder of mites in the order Acarina which are large and hemispherical with a deep-brown, smooth, heavily sclerotized cuticle.

Holotrichia [INV ZOO] A major subclass of the protozoan class Ciliatea; body ciliation is uniform with cilia arranged in longitudinal rows.

holotype [SYST] A nomenclatural type for the single specimen designated as "the type" by the orginal author at the time of publication of the original description.

holozoic [ZOO] Obtaining food in the manner of most animals, by ingesting complex organic matter.

holster [ORD] A pocket-type device with a single compartment designed to be worn on a belt or shoulder harness, which may be furnished with it, and used to carry a pistol, revolver, or the like.

Holtz machine *See* Toepler-Holtz machine.

Holuridae [PALEON] A group of extinct chondrostean fishes in the suborder Palaeoniscoidei distinguished in having lepidotrichia of all fins articulated but not bifurcated, fins without fulcra, and the tail not cleft.

holystone [MATER] A soft sandstone that is used in scrubbing a ship's deck.

Homacodontidae [PALEON] A family of extinct palaeodont mammals in the superfamily Dichobunoidea.

Homalopteridae [VERT ZOO] A small family of cypriniform fishes in the suborder Cyprinoidei.

Homalorhagae [INV ZOO] A suborder of the class Kinorhyncha having a single dorsal plate covering the neck and three ventral plates on the third zonite.

Homalozoa [INV ZOO] A subphylum of echinoderms characterized by the complete absence of radial symmetry.

Homan's sign [MED] Pain in the calf and popliteal area on passive dorsiflexion of the foot, indicating deep venous thrombosis of the calf. Also known as dorsiflexion sign.

Homaridae [INV ZOO] A family of marine decapod crustaceans containing the lobsters.

Homarus [INV ZOO] A genus of the family Homaridae comprising most species of lobsters.

homatropine [ORG CHEM] $C_{16}H_{21}O_3N$ An alkaloid that causes pupil dilation and paralysis of accommodation.

homaxial [BIOL] Having all axes equal.

home [ELEC] To return to the starting position, as in a stepping relay or turning motor. [NAV] To navigate toward a point by maintaining constant some navigational parameter other than altitude. [ORD] To travel to a target by guidance of heat radiation, radar echoes, radio waves, sound waves, laser beams, or other phenomena reflected from or originating in the target, as in homing missiles and homing torpedoes.

home address [ADP] A technique used to identify each disk track uniquely by means of a 9-byte record immediately following the index marker; the record contains a flag (good or defective track), cylinder number, head number, cyclic check, and bit count appendage.

home loop [COMMUN] A telephone pair terminating at the local central office. Also known as local loop.

homeoblastic [PETR] Of a metamorphic crystalloblastic texture, having constituent minerals of approximately the same size.

homeomorphic spaces [MATH] Two topological spaces with a homeomorphism existing between them; intuitively one can be obtained from the other by stretching, twisting, or shrinking.

homeomorphism [MATH] A continuous map between topological spaces which is one-to-one, onto, and its inverse function is continuous. Also known as bicontinuous function; topological mapping.

home-on-jam [ELECTR] A feature that permits radar to track a jamming source in angle.

homeopathy [MED] A system of medicine expounded by Samuel Hahnemann that treats disease by administering to the patient small doses of drugs which produce the signs and symptoms of the disease in a healthy person. Also known as Hahnemannism.

homeosis *See* heteromorphosis.

homeostasis [BIOL] In higher animals, the maintenance of an internal constancy and an independence of the environment.

homeotherm [VERT ZOO] An animal (mammal or bird) which routinely regulates and maintains body temperature.

homeothermia [PHYSIO] The condition of being warm-blooded.

homer [NAV] A ground-based direction-finding station that utilizes radio transmissions from aircraft to determine their bearing, then guides the aircraft toward the station by voice communications. [ORD] *See* target seeker.

HOLOSTOME

Diagram of ventral surface of adult holostome showing adhesive organs. *(From R. M. Cable, An Illustrated Laboratory Manual of Parasitology, Burgess, 1940)*

HOLOTHUROIDEA

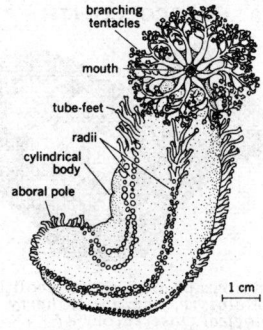

branching tentacles
mouth
tube-feet
radii
cylindrical body
aboral pole

1 cm

Cucumaria, a representative holothurian.

HOLOTRICHIA

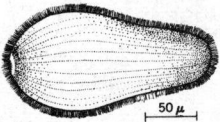

50 μ

Prorodon, a primitive holotrich.

home record [ADP] The first record in the chaining method of file organization.

home signal [CIV ENG] A signal at the beginning of a block of railroad track that indicates whether the block is clear.

hometaxial-base transistor [ELECTR] Transistor manufactured by a single-diffusion process to form both emitter and collector junctions in a uniformly doped silicon slice; the resulting homogeneously doped base region is free from accelerating fields in the axial (collector-to-emitter) direction, which could cause undesirable high current flow and destroy the transistor.

homeward bound [NAV] Of a vessel or person, headed for the home port or country.

homichlophobia [PSYCH] An abnormal fear of fog.

homilite [MINERAL] $(Ca_2(Fe,Mg)B_2Si_2O_{10}$ A black or blackish brown mineral composed of iron calcium borosilicate.

homing adapter [NAV] A device which, when used with an aircraft radio receiver, produces aural or visual signals which indicate the direction of a transmitting radio station with respect to the heading of the aircraft.

homing antenna [ELECTROMAG] A directional antenna array used in flying directly to a target that is emitting or reflecting radio or radar waves.

homing beacon [NAV] A radio beacon, either airborne or on the ground, toward which an aircraft can fly if equipped with a radio compass or homing adapter. Also known as radio homing beacon.

homing device [ELECTR] A control device that automatically starts in the correct direction of motion or rotation to achieve a desired change, as in a remote-control tuning motor for a television receiver. [ENG] A device incorporated in a guided missile or the like to home it on a target. [NAV] A transmitter, receiver, or adapter used for homing aircraft or used by aircraft for homing purposes.

homing guidance [ENG] A guidance system in which a missile directs itself to a target by means of a self-contained mechanism that reacts to a particular characteristic of the target.

homing range [NAV] The maximum distance from a target or homing beacon at which a homing device is effective.

homing station [NAV] A station at which a beacon emits signals that may be used for homing.

homing torpedo [ORD] A torpedo having homing guidance, designed for homing on a surface vessel or a submerged submarine.

homing transponder [NAV] A small acoustic transponder used in navigation of a submersible vehicle, which can be carried by the vehicle and quickly dropped when an area of interest is reached; a single transponder and a dead reckoning system are used.

Hominidae [VERT ZOO] A family of primates in the superfamily Hominoidea containing one living species, *Homo sapiens*.

Hominoidea [VERT ZOO] A superfamily of the order Primates comprising apes and humans.

homo- [ORG CHEM] **1.** Indicating the homolog of a compound differing in formula from the latter by an increase of one CH_2 group. **2.** Indicating a homopolymer made up of a single type of monomer, such as polyethylene from ethylene. [SCI TECH] Prefix indicating the same or similar.

Homo [VERT ZOO] The genus of human beings, including modern humans and many extinct species.

Homobasidiomycetidae [MYCOL] A subclass of basidiomycetous fungi in which the basidium is not divided by cross walls.

homobront *See* isobront.

homocentric [OPTICS] Pertaining to rays which have the same focal point, or which are parallel. Also known as stigmatic.

homocercal [VERT ZOO] Pertaining to the caudal fin of certain fishes which has almost equal lobes, with the vertebral column terminating near the middle of the base.

homochlamydeous [BIOL] Having all members of the perianth similar or not differentiated into calyx or corolla.

homochromy [ZOO] A form of protective coloration whereby the individual blends into the background.

homocline [GEOL] Any rock unit in which the strata exhibit the same dip.

homocyclic compound [ORG CHEM] A ring compound that has one type of atom in its structure; an example is benzene.

homocysteine [BIOCHEM] $C_4H_9O_2NS$ An amino acid formed in animals by demethylation of methionine.

homocystinuria [MED] A hereditary metabolic disorder in which homocysteine appears in the urine because cystathionine synthetase activity is absent; there is also malpositioning of the lens, and mental retardation.

homodesmic [CRYSTAL] Of a crystal, having atoms bonded in a single way.

homodont [VERT ZOO] Having all teeth similar in form; characteristic of nonmammalian vertebrates.

homodynamic [INV ZOO] Developing through continuous successive generations without a diapause; applied to insects.

homodyne reception [ELECTR] A system of radio reception for suppressed-carrier systems of radiotelephony, in which the receiver generates a voltage having the original carrier frequency and combines it with the incoming signal. Also known as zero-beat reception.

homoecious [BIOL] Having one host for all stages of the life cycle.

homoeomerous [BOT] Having algae distributed uniformly throughout the thallus of a lichen.

Homo erectus [PALEON] A type of fossil human from the Pleistocene of Java and China representing a specialized side branch in human evolution.

homoerotism [PSYCH] Sexual desire directed toward a member of the same sex; usually sublimated and not expressed.

homofermentative lactobacilli [MICROBIO] Bacteria that produce a single end product, lactic acid, from fermentation of carbohydrates.

homogametic sex [GEN] The sex of a species in which the paired sex chromosomes are of equal size and which therefore produces homogametes.

homogamety [GEN] The production of homogametes by one sex of a species.

homogamous [BIOL] Of or pertaining to homogamy.

homogamy [BIOL] Inbreeding due to isolation. [BOT] Condition of having all flowers alike.

homogenate [BIOL] A tissue that has been finely divided and mixed.

homogeneity [PHYS] Quality of a substance whose properties are independent of position. [STAT] Equality of the distribution functions of several populations.

homogeneous [CHEM] Pertaining to a substance having uniform composition or structure. [MATH] Pertaining to a group of mathematical symbols of uniform dimensions or degree. [SCI TECH] Uniform in structure or composition.

homogeneous atmosphere [METEOROL] A hypothetical atmosphere in which the density is constant with height.

homogeneous catalysis [CHEM] Catalysis occurring within a single phase, usually a gas or liquid.

homogeneous chemical reaction [CHEM] Chemical reaction system in which all constitutents (reactants and catalyst) are of the same phase.

homogeneous coordinates [MATH] To a point in the plane with cartesian coordinates (x,y) there corresponds the homogeneous coordinates (x_1,x_2,x_3), where $x_1/x_3 = x$, $x_2/x_3 = y$; any polynomial equation in cartesian coordinates becomes homogeneous if a change into these coordinates is made.

homogeneous differential equation [MATH] A differential equation where every scalar multiple of a solution is also a solution.

homogeneous equation [MATH] An equation that can be rewritten into the form having zero on one side of the equal sign and a homogeneous function of all the variables on the other side.

homogeneous function [MATH] A real function $f(x_1, x_2, \ldots, x_n)$ is homogeneous of degree r if $f(ax_1, ax_2, \ldots, ax_n) = a^r f(x_1, x_2, \ldots, x_n)$ for every real number a.

homogeneous integral equation [MATH] An integral equation where every scalar multiple of a solution is also a solution.

homogeneous polynomial [MATH] A polynomial all of

whose terms have the same total degree; equivalently it is a homogenous function of the variables involved.

homogeneous radiation [PHYS] Radiation having an extremely narrow band of frequencies, or a beam of monoenergetic particles of a single type, so that all components of the radiation are alike.

homogeneous reactor [NUCLEO] A nuclear reactor in which fissionable material and moderator (if used) are intimately mixed to form an effectively homogeneous medium for neutrons.

homogeneous space [MATH] A topological space having a group of transformations acting upon it, that is, a transformation group, where for any two points x and y some transformation from the group will send x to y.

homogenize [MET] To hold metal at a high temperature long enough to eliminate by diffusion any chemical segregation of the components.

homogenizer [MECH ENG] A machine that blends or emulsifies a substance by forcing it through fine openings against a hard surface.

homogentisase [BIOCHEM] The enzyme that catalyzes the conversion of homogentisic acid to fumaryl acetoacetic acid.

homogentisic acid [BIOCHEM] $C_8H_8O_4$ An intermediate product in the metabolism of phenylalanine and tyrosine; found in excess in persons with phenylketonuria and alkaptonuria.

homogony [BOT] Condition of having one type of flower, with stamens and pistil of uniform length.

homograft [BIOL] Graft from a donor transplanted to a genetically dissimilar recipient of the same species. Also known as allograft.

homograft rejection [IMMUNOL] An immunologic process by which an individual destroys and casts off a tissue transplanted from a donor of the same species.

homoiochlamydeous [BOT] Having perianth leaves alike, not differentiated into sepals and petals.

homoiogenetic [EMBRYO] Of a determined part of an embryo, capable of inducing formation of a similar part when grafted into an undetermined field.

Homoistela [PALEON] A class of extinct echinoderms in the subphylum Homalozoa.

homokaryon [MYCOL] A bi- or multinucleate cell having nuclei all of the same kind.

homokaryosis [MYCOL] The condition of a bi- or multinucleate cell having nuclei all of the same kind.

homolecithal [CYTOL] Referring to eggs having small amounts of evenly distributed yolk. Also known as isolecithal.

homological algebra [MATH] The study of the structure of modules, particularly by means of exact sequences; it has application to the study of a topological space via its homology groups.

homologous [BIOL] Pertaining to a structural relation between parts of different organisms due to evolutionary development from the same or a corresponding part, such as the wing of a bird and the pectoral fin of a fish.

homologous serum jaundice [MED] A type of hepatitis caused by a filtrable virus that exists in the blood plasma and may be passed to another person through blood transfusion.

homologous stimulus [PHYSIO] A form of energy to which a specific sensory receptor is most sensitive.

homologous transformation [ASTRON] A mathematical transformation in the study of stellar models.

homologous tumor [MED] A neoplasm composed of tissue identical with those of the organ at the site of the tumor.

homology [ORG CHEM] That state, in a series of organic compounds that differ from each other by a CH_2 such as the methane series C_nH_{2n+2}, in which there is a similarity between the compounds in the series and a graded change of their properties.

homology group [MATH] Associated to a topological space X, one of a sequence of Abelian groups $H_n(X)$ that reflect how n-dimensional simplicial complexes can be used to fill up X and also help determine the presence of n-dimensional holes appearing in X. Also known as Betti group.

homology theory [MATH] Theory attempting to compare topological spaces and investigate their structures by determining the algebraic nature and interrelationships appearing in the various homology groups.

homolysis [CHEM] Symmetrical breaking of a covalent electron bond; for example, $A:B = A \cdot + B \cdot$.

homometric pair [CRYSTAL] A pair of crystal structures whose x-ray diffraction patterns are identical.

homomorphism [BOT] Having perfect flowers consisting of only one type. [MATH] A function between two algebraic systems of the same type which preserves the algebraic operations.

homomorphs [CHEM] Chemical molecules that are similar in size and shape, but not necessarily having any other characteristics in common.

Homoneura [INV ZOO] A suborder of the Lepidoptera with mandibulate mouthparts, and fore- and hindwings that are similar in shape and venation.

homonomous hemianopsia [MED] Partial blindness affecting the inner half of one field of vision or the outer half of the other; caused by optic nerve lesions posterior to the chiasma.

homopause [GEOPHYS] The level of transition between the homosphere and the heterosphere; it lies about 80 to 90 kilometers above the earth.

homopetalous [BOT] Having all petals identical.

homoplasy [BIOL] Correspondence between organs or structures in different organisms acquired as a result of evolutionary convergence or of parallel evolution.

homopolar [ELEC] **1.** Electrically symmetrical. **2.** Having equal distribution of charge.

homopolar bond [PHYS CHEM] A covalent bond whose total dipole moment is zero.

homopolar generator [ELECTR] A direct-current generator in which the poles presented to the armature are all of the same polarity, so that the voltage generated in active conductors has the same polarity at all times; a pure direct current is thus produced, without commutation. Also known as acyclic machine; homopolar machine; unipolar machine.

homopolar machine See homopolar generator.

homopolymer [ORG CHEM] A polymer formed from a single monomer; an example is polyethylene, formed by polymerization of ethylene.

Homoptera [INV ZOO] An order of the class Insecta including a large number of sucking insects of diverse forms.

Homo sapiens [VERT ZOO] Modern human species; a large, erect, omnivorous terrestrial biped of the primate family Hominidae.

homoscedastic [STAT] **1.** Pertaining to two or more distributions whose variances are equal. **2.** Pertaining to a variate in a bivariate distribution whose variance is the same for all values of the other variate.

Homosclerophorida [INV ZOO] An order of primitive sponges of the class Demospongiae with a skeleton consisting of equirayed, tetraxonid, siliceous spicules.

homoserine [BIOCHEM] $C_4H_9O_3N$ An amino acid formed as an intermediate product in animals in the metabolic breakdown of cystathionine to cysteine.

homosexual [BIOL] Of, pertaining to, or being the same sex. [PSYCH] **1.** Of, pertaining to, or exhibiting homosexuality. **2.** One who practices homosexuality.

homosexuality [PSYCH] **1.** State of being sexually attracted to members of the same sex. **2.** A form of homoerotism involving sexual interest without genital expression.

homosexual panic [PSYCH] An acute syndrome that comes as a climax of prolonged tension from unconscious homosexual conflicts or sometimes bisexual tendencies.

homosphere [METEOROL] The lower portion of a two-part division of the atmosphere (the upper portion is the heterosphere) according to the general homogeneity of atmospheric composition; the region in which there is no gross change in atmospheric composition, that is, all of the atmosphere from the earth's surface to about 80–100 kilometers.

homospory [BOT] Production of only one kind of asexual spore.

homothallic [MYCOL] Having genetically compatible hyphae and therefore forming zygospores from two branches of the same mycelium.

homotopy [MATH] Between two mappings of the same topological spaces, a continuous function representing how, in a step-by-step fashion, the image of one mapping can be continuously deformed onto the image of the other.

homotopy groups [MATH] Associated to a topological space X, the groups appearing for each positive integer n, which reflect the number of different ways (up to homotopy) than an n-dimensional sphere may be mapped to X.

homotopy theory [MATH] The study of the topological structure of a space by examining the algebraic properties of its various homotopy groups.

homotropous [BOT] Having the radicle directed toward the hilum.

homotype [SYST] A taxonomic type for a specimen which has been compared with the holotype by another than the author of the species and determined by him to be conspecific with it.

homotypy [BIOL] Protective device based on resemblance of shape to the background.

homozeotrope [CHEM] Mixture in which the liquid components are miscible in all proportions in the liquid phase, and may be separated by ordinary distillation.

homozygote [GEN] An individual that has identical alleles at one or more loci and therefore produces gametes which are all identical.

homozygous [GEN] Of or pertaining to a homozygote.

hone [MATER] A fine-grit stone that is used for sharpening a cutting tool. [MECH ENG] A machine for honing that consists of a holding device containing several oblong stones arranged in a circular pattern.

Honest John [ORD] The 762-millimeter rocket system, a surface-to-surface tactical missile employing solid propellant, developed by the Army; it is released from a rail-type launcher at an elevation which can be varied to obtain the desired point of impact; it can carry either a conventional high-explosive warhead or a nuclear warhead.

honey [FOOD ENG] The sweet, viscous secretion composed principally of levulose and dextrose that is deposited in the honeycomb by the honeybee.

honeybee [INV ZOO] *Apis mellifera.* The bee kept for the commercial production of honey; a member of the dipterous family Apidae.

honeycomb [INV ZOO] A mass of wax cells in the form of hexagonal prisms constructed by honeybees for their brood and honey.

honeycomb coil [ELECTROMAG] A coil wound in a crisscross manner to reduce distributed capacitance. Also known as duolateral coil; lattice-wound coil.

honeycomb coral [PALEON] The common name for members of the extinct order Tabulata; has prismatic sections arranged like the cells of a honeycomb.

honeycombing [MATER] **1.** Internal fiber separation in drying timber. **2.** Local roughness and weakening on the face of a concrete wall due to·segregation of the concrete, with the result that there is little sand to fill in between the stone aggregate.

honeycomb lung [MED] **1.** Condition of the lung in emphysema. **2.** A lung containing small pus-filled cavities.

honeycomb radiator [MECH ENG] A heat-exchange device utilizing many small cells, shaped like a bees' comb, for cooling circulating water in an automobile.

honeycomb weave *See* waffle weave.

honeydew [INV. ZOO] The viscous secretion deposited on leaves by many aphids and scale insects; an attractant for ants.

Honey Dew melon [BOT] A variety of muskmelon (*Cucumis melo*) belonging to the Violales; fruit is large, oval, smooth, and creamy yellow to ivory, without surface markings.

honey tube [INV ZOO] Either of a pair of cornicles on the dorsal aspect of one abdominal segment in certain aphids.

Honigmann process [MIN ENG] A method of shaft sinking through water-bearing sand; the shaft is bored in stages, increasing in size from the pilot hole, about 4 feet in diameter, to the final size.

honing [MECH ENG] The process of removing a relatively small amount of material from a cylindrical surface by means of abrasive stones to obtain a desired finish or extremely close dimensional tolerance.

hood [DES ENG] An opaque shield placed above or around the screen of a cathode-ray tube to eliminate extraneous light. [ENG] **1.** Close-fitting, rubber head covering that leaves the

face exposed; used in scuba diving. **2.** A protective covering, usually providing special ventilation to carry away objectionable fumes, dusts, and gases, in which dangerous chemical, biological, or radioactive materials can be safely handled.

hoodmold [ARCH] A molding projecting over the hood of an arch.

hoof [VERT ZOO] **1.** Horny covering for terminal portions of the digits of ungulate mammals. **2.** A hoofed foot, as of a horse.

hoof-and-mouth disease *See* foot-and-mouth disease.

hoof oil *See* neat's-foot oil.

hook [DES ENG] A piece of hard material, especially metal, formed into a curve for catching, holding, or pulling something. [ELECTR] A circuit phenomenon occurring in four-zone transistors, wherein hole or electron conduction can occur in opposite directions to produce voltage drops that encourage other types of conduction. [GEOGR] The end of a spit of land that is turned toward shore. Also known as recurved spit.

hook-and-eye hinge [DES ENG] A hinge consisting of a hook (usually attached to a gate post) over which an eye (usually attached to the gate) is placed.

hook bolt [DES ENG] A bolt with a hook or L band at one end and threads at the other to fit a nut.

hook collector transistor [ELECTR] A transistor in which there are four layers of alternating n- and p-type semiconductor material and the two interior layers are thin compared to the diffusion length. Also known as hook transistor; pn hook transistor.

Hookean deformation [MECH] Deformation of a substance which is proportional to the force applied to it.

Hookean solid [MECH] An ideal solid which obeys Hooke's law exactly for all values of stress, however large.

hooked spit *See* recurved spit.

Hooke number *See* Cauchy number.

hooker [MIN ENG] A worker who detaches empty downcoming buckets and hook-loads buckets or cans onto the hoisting rope.

Hooker diaphragm cell [CHEM ENG] A device used in industry for the electrolysis of brine (sodium chloride) to make chlorine and caustic soda (sodium hydroxide) or caustic potash (potassium hydroxide); saturated purified brine fed around the anode passes through the diaphragm to the cathode; chlorine is formed at the anode and hydrogen released at the cathode, leaving sodium hydroxide and residual sodium chloride in the cell liquor; the diaphragm prevents the products from mixing.

Hookeriales [BOT] An order of the mosses with irregularly branched stems and leaves that appear to be in one plane.

Hooker process [MET] A forming process in which pierced slugs or cups are punched through a die to produce tubing and other shapes.

Hooke's joint [MECH ENG] A simple universal joint; consists of two yokes attached to their respective shafts and connected by means of a spider. Also known as Cardan joint.

Hooke's law [MECH] The law that the stress of a solid is directly proportional to the strain applied to it.

hook gage [ENG] An instrument used to measure changes in the level of the water in an evaporation pan; it consists of a pointed metal hook, mounted in the vertical, whose position with respect to its supporting member may be adjusted by means of a micrometer arrangement; the gage is placed on the still well, and a measurement is taken when the point of the hook just breaks above the surface of the water.

hook tender [MIN ENG] In bituminous coal mining, a worker who attaches the hook of a hoisting cable to the link of a trip of cars to be hauled up or lowered down an incline. Also known as rope cutter.

hook transistor *See* hook collector transistor.

hookup [ELEC] An arrangement of circuits and apparatus for a particular purpose.

hookup wire [ELECR] Tinned and insulated solid or stranded soft-drawn copper wire used in making circuit connections.

hook-wall packer [PETRO ENG] Fluid-proof seal between the outside of oil well tubing and the inside of the casing; hooks hold it in place.

hookworm [INV ZOO] The common name for parasitic roundworms composing the family Ancylostomidae.

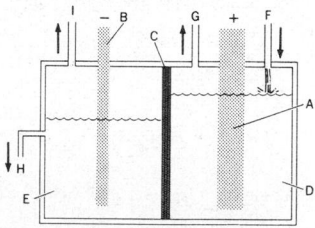

HOOKER DIAPHRAGM CELL

Diagram of diaphragm cell for making chlorine and caustic soda. A = graphite anode, B = iron screen cathode, C = asbestos diaphragm, D = anode compartment for brine and chlorine, E = cathode compartment for NaOHNaCl cell liquor, F = brine inlet, G = chlorine outlet, H = cell liquor (caustic soda) outlet, I = hydrogen outlet.

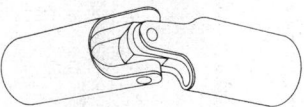

HOOKE'S JOINT

A drawing of a Hooke's joint. (*From C. W. Ham, E. J. Crane, and W. L. Rogers, Mechanics of Machinery, McGraw-Hill, 1958*)

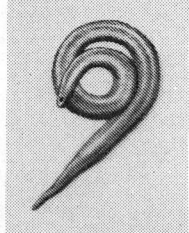

HOOKWORM

A typical hookworm.

HOP

Hop, female inflorescences. *(USDA)*

HOPHORNBEAM

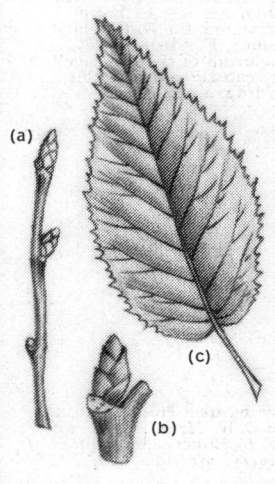

American hophornbeam *(Ostrya virginiana). (a)* Twig. *(b)* Terminal bud. *(c)* Leaf.

hookworm disease [MED] Microcytic hypochromic anemia in man produced by the nematodes *Necator americanus* or *Ancylostoma duodenale* in the the intestine.

hook wrench [DES ENG] A wrench with a hook for turning a nut or bolt.

hoop [CIV ENG] A ring-shaped binder placed around the main reinforcement in a reinforced concrete column.

hooped column [CIV ENG] A column of reinforced concrete with hoops around the main reinforcements.

Hooper jig [MIN ENG] A pneumatic jig, used when water is scarce or the ore must be kept dry, to concentrate values from sands.

hop [BOT] *Humulus lupulus.* A dioecious liana of the order Urticales distinguished by herbaceous vines produced from a perennial crown; the inflorescence, a catkin, of the female plant is used commercially for beer production. [COMMUN] A single reflection of a radio wave from the ionosphere back to the earth in traveling from one point to another.

Hopcalite [MATER] Trade name for a mixture of manganese dioxide with other oxides used as a catalyst for removing carbon monoxide from air and for detecting carbon monoxide in gas analysis.

hopeite [MINERAL] $Zn_3(PO_4)_2 \cdot 4H_2O$ A gray, orthorhombic mineral composed of hydrous phosphate of zinc; specific gravity is 2.76–2.85; dimorphous with parahopeite.

Hopfner process [MET] A process for the recovery of copper from its sulfide ores by leaching with a solution of cuprous chloride in sodium or calcium chloride and electrolyzing the resulting solution in tanks protected by diaphragms. Also spelled Hoepfner process.

hophornbeam [BOT] Any tree of the genus *Ostrya* in the birch family recognized by its very scaly bark and the fruit which closely resembles that of the hopvine.

Hopkins-Cole reaction [ANALY CHEM] The appearance of a violet ring when concentrated sulfuric acid is added to a mixture that includes a protein and glyoxylic acid; however, gelatin and zein do not show the reaction.

Hopkinson's coefficient [ELECTROMAG] The average magnetic flux per turn of an induction coil divided by the average flux per turn of another coil linked with it.

Hoplocarida [INV ZOO] A superorder of the class Crustacea with the single order Stomatopoda.

Hoplonemertini [INV ZOO] An order of unsegmented, ribbonlike worms in the class Enopla; all species have an armed proboscis.

Hoplophoridae [INV ZOO] A family of prawns containing numerous bathypelagic representatives.

hopper [ADP] *See* card hopper. [ENG] A funnel-shaped receptacle with an opening at the top for loading and a discharge opening at the bottom for bulk-delivering material such as grain or coal.

hopperburn [PL PATH] A disease of potato and peanut plants caused by a leafhopper which secretes a toxic substance on the leaves, causing browning and shriveling.

hopper car [ENG] A freight car with a permanent roof and a hinged floor sloping to one or more hoppers for discharging contents by gravity.

hoppit [MIN ENG] A large bucket, usually up to about 80 cubic feet (2.3 cubic meters), used in shaft sinking for hoisting men, rock, materials, and tools. Also known as bowk; kibble; sinking bucket.

hops oil [MATER] Greenish-yellow essential oil with strong aroma; soluble in alcohol, ether, and chloroform, insoluble in water; distilled from strobiles of the hop (*Humulus lupulus*); main components are humulene, geraniol, and terpenes; used to aromatize beer and tobacco.

horadiam drilling [MIN ENG] The drilling of a number of horizontal boreholes radiating outward from a common center. Also known as horizontal-ring drilling.

hordeolum [MED] A furuncular inflammation of the connective tissue of the eyelids near a hair follicle. Also known as sty.

Hordeum [BOT] A genus of the order Cyperales containing all species of barley.

horehound *See* marrubium.

horizon [ASTRON] The apparent boundary line between sky and earth or sea. Also known as apparent horizon. [GEOL]

1. The surface separating two beds. 2. One of the layers, each of which is a few inches to a foot thick, that make up a soil.

horizon bar [NAV] The gyro-stabilized line or bar in an artificial-horizon flight instrument.

horizon distance [GEOD] The distance, at any given azimuth, to the point on the earth's surface constituting the horizon for some specified observer.

horizon glass [NAV] That glass of a marine sextant, through which the horizon is observed.

horizon lights [NAV] Lights arranged to provide a ground reference for pilots taking off.

horizon mining [MIN ENG] A system of mining suitable for inclined, and perhaps faulted, coal seams; main stone headings are driven, at predetermined levels, from the winding shaft to intersect the seams to be developed.

horizon mirror [NAV] On a marine sextant, the mirror part of the horizon glass; the term is sometimes used loosely to refer to the horizon glass.

horizon prism [NAV] A prism which can be inserted in the optical path of an instrument, such as a bubble sextant, to permit observation of the visible horizon.

horizon scanner *See* horizon tracker.

horizon sensor [ENG] A passive infrared device that detects the thermal discontinuity between the earth and space; used in establishing a stable vertical reference for control of the attitude or orientation of a missile or satellite in space.

horizon system of coordinates [ASTRON] A set of celestial coordinates based on the celestial horizon as the primary great circle.

horizontal [SCI TECH] Being in a plane perpendicular to the gravitational field, that is, perpendicular to a plumb line, at a given point on the earth's surface.

horizontal auger [MECH ENG] A rotary drill, usually powered by a gasoline engine, for making horizontal blasting holes in quarries and opencast pits.

horizontal base-line method [ORD] Method of locating targets or other points by intersection from two observation stations located at opposite ends of a base line.

horizontal blanking [ELECTR] Blanking of a television picture tube during the horizontal retrace.

horizontal blanking pulse [ELECTR] The rectangular pulse that forms the pedestal of the composite television signal between active horizontal lines and causes the beam current of the picture tube to be cut off during retrace. Also known as line-frequency blanking pulse.

horizontal bombing [ORD] Bombing from an airplane in horizontal flight, as distinguished from dive bombing, glide bombing, toss bombing, and the like. Also known as level bombing.

horizontal borer [MIN ENG] A machine that makes holes 2–6 inches (5–15 centimeters) in diameter, used for drilling at opencut coal mines.

horizontal boring machine [MECH ENG] A boring machine adapted for work not conveniently revolved, for milling, slotting, drilling, tapping, boring, and reaming long holes and for making interchangeable parts that must be produced without jigs and fixtures.

horizontal broaching machine [MECH ENG] A pull-type broaching machine having the broach mounted on the horizontal plane.

horizontal centering control [ELECTR] The centering control provided in a television receiver or cathode-ray oscilloscope to shift the position of the entire image horizontally in either direction on the screen.

horizontal chromatography [ANALY CHEM] Paper chromatography in which the chromatogram is horizontal instead of vertical.

horizontal convergence control [ELECTR] The control that adjusts the amplitude of the horizontal dynamic convergence voltage in a color television receiver.

horizontal crosscut *See* horizontal drive.

horizontal crusher [MECH ENG] Rotary size reducer in which the crushing cone is supported on a horizontal shaft; needs less headroom than vertical models.

horizontal danger angle [NAV] The maximum or minimum angle between two points on a chart, as observed from a craft, indicating the limit of safe approach to an off-lying danger. Also known as danger angle.

horizontal definition *See* horizontal resolution.

horizontal deflection electrode [ELECTR] One of a pair of electrodes that move the electron beam horizontally from side to side on the fluorescent screen of a cathode-ray tube employing electrostatic deflection.

horizontal deflection oscillator [ELECTR] The oscillator that produces, under control of the horizontal synchronizing signals, the sawtooth voltage waveform that is amplified to feed the horizontal deflection coils on the picture tube of a television receiver. Also known as horizontal oscillator.

horizontal displacement *See* strike slip.

horizontal drilling machine [MECH ENG] A drilling machine in which the drill bits extend in a horizontal direction.

horizontal drive [MIN ENG] An opening with a small inclination directed toward the shaft to drain the water and facilitate hauling of full cars to the shaft. Also known as horizontal crosscut.

horizontal drive control [ELECTR] The control in a television receiver, usually at the rear, that adjusts the output of the horizontal oscillator. Also known as drive control.

horizontal earth rate [NAV] The rate at which the spin axis of a gyroscope must be tilted about the horizontal axis to remain parallel to the earth's surface to compensate for the effect of earth rate; horizontal earth rate is maximum at the equator, zero at the poles, and varies as the cosine of the latitude.

horizontal engine [MECH ENG] An engine with horizontal stroke.

horizontal error [ORD] The error in range, deflection, or radius which a weapon may be expected to exceed as often as not.

horizontal field-strength diagram [ELECTROMAG] Representation of the field strength at a constant distance from an antenna and in a horizontal plane; unless otherwise specified, this plane is that passing through the antenna.

horizontal flyback [ELECTR] Flyback in which the electron beam of a television picture tube returns from the end of one scanning line to the beginning of the next line. Also known as horizontal retrace.

horizontal fold *See* nonplunging fold.

horizontal force instrument [ENG] An instrument used to make a comparison between the intensity of the horizontal component of the earth's magnetic field and the magnetic field at the compass location on board a craft; basically, it consists of a magnetized needle pivoted in a horizontal plane, as a dry-card compass; it settles in some position which indicates the direction of the resultant magnetic field; if the needle is started swinging, it damps down with a certain period of oscillation dependent upon the strength of the magnetic field. Also known as horizontal vibrating needle.

horizontal frequency *See* line frequency.

horizontal hold control [ELECTR] The hold control that changes the free-running period of the horizontal deflection oscillator in a television receiver, so that the picture remains steady in the horizontal direction.

horizontal intensity [GEOPHYS] The strength of the horizontal component of the earth's magnetic field.

horizontal intensity variometer [ENG] Essentially a declination variometer with a larger, stiffer fiber than in the standard model; there is enough torsion in the fiber to cause the magnet to turn 90° out of the magnetic meridian; the magnet is aligned with the magnetic prime vertical to within 0.5° so it does not respond appreciably to changes in declination. Also known as H variometer.

horizontal lathe [MECH ENG] A horizontally mounted lathe with which longitudinal and radial movements are applied to a workpiece that rotates.

horizontal linearity control [ELECTR] A linearity control that permits narrowing or expanding of the width of the left half of a television receiver image, to give linearity in the horizontal direction so that circular objects appear as true circles.

horizontal line frequency *See* line frequency.

horizontal magnetometer [ENG] A measuring instrument for ascertaining changes in the horizontal component of the magnetic field intensity.

horizontal milling machine [MECH ENG] A knee-type milling machine with a horizontal spindle and a swiveling table for cutting helices.

horizontal obstacle sonar [NAV] A constant-transmission, frequency-modulated sonar used in underwater navigation to detect objects ahead of a submersible vehicle and to determine the bearing and range of specially designed transponders. Abbreviated HOS.

horizontal oscillator *See* horizontal deflection oscillator.

horizontal output stage [ELECTR] The television receiver stage that feeds the horizontal deflection coils of the picture tube through the horizontal output transformer; may also include a part of the second-anode power supply for the picture tube.

horizontal output transformer [ELECTR] A transformer used in a television receiver to provide the horizontal deflection voltage, the high voltage for the second-anode power supply of the picture tube, and the filament voltage for the high-voltage rectifier tube. Also known as flyback transformer; horizontal sweep transformer.

horizontal parallax *See* absolute stereoscopic parallax.

horizontal parity check *See* longitudinal parity check.

horizontal pendulum [MECH] A pendulum that moves in a horizontal plane, such as a compass needle turning on its pivot.

horizontal plane [ANAT] A transverse plane at right angles to the longitudinal axis of the body.

horizontal polarization [COMMUN] Transmission of radio waves so that the electric lines of force are horizontal, while the magnetic lines of force are vertical.

horizontal-position welding [MET] **1.** Making a fillet weld on the upper side of the intersection of a vertical surface and a horizontal surface. **2.** Making a horizontal groove weld on a vertical surface.

horizontal pressure force [GEOPHYS] The horizontal pressure gradient per unit mass, $-\alpha\nabla_H p$, where α is the specific volume, p the pressure, and ∇_H the horizontal component of the del-operator; this force acts normal to the horizontal isobars toward lower pressure; it is one of the three important forces appearing in the horizontal equations of motion, the others being the Coriolis force and friction.

horizontal projection [GRAPHICS] A drawing of a structure as it would appear if projected on a horizontal plane.

horizontal range [ORD] The distance measured horizontally between a gun and its target; specifically in antiaircraft gunnery, the distance between a gun and a spot on the ground directly beneath the target.

horizontal resolution [ELECTR] The number of individual picture elements or dots that can be distinguished in a horizontal scanning line of a television or facsimile image. Also known as horizontal definition.

horizontal retort [MET] An intermittent unit made from a siliceous fireclay and formerly used for zinc smelting.

horizontal retort process [MET] A zinc-smelting process that employs vast, honeycomblike batteries of fireclay or silicon-carbide retorts set horizontally in a gas- or coal-fired furnace. Also known as Belgian retort process.

horizontal retrace *See* horizontal flyback.

horizontal-ring drilling *See* horadiam drilling.

horizontal-rolled-position welding [MET] Top-side welding of a butt joint connecting two horizontal pieces of rotating pipe.

horizontal scanning [ENG] In radar scanning, rotating the antenna in azimuth around the horizon or in a sector. Also known as searching lighting.

horizontal scanning frequency [ELECTR] The number of horizontal lines scanned by the electron beam in a television receiver in 1 second.

horizontal screen [MECH ENG] Shaking screen with horizontal plates.

horizontal separation *See* strike slip.

horizontal separator [PETRO ENG] Horizontal tank used to separate free oil well gas from liquid hydrocarbons.

horizontal silo [AGR] A trench or bunker for storing silage.

horizontal sweep [ELECTR] The sweep of the electron beam from left to right across the screen of a cathode-ray tube.

horizontal sweep transformer *See* horizontal output transformer.

horizontal synchronizing pulse [ELECTR] The rectangular pulse transmitted at the end of each line in a television system,

HORIZONTAL OBSTACLE SONAR

Configuration of horizontal obstacle sonar and configurations of vertical obstacle and altitude depth sonars.

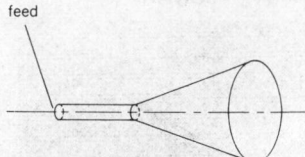

HORN ANTENNA

feed

Horn antenna, a type of direct-aperture antenna.

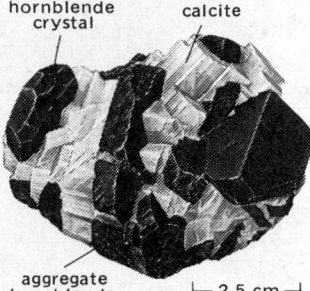

HORNBLENDE

hornblende crystal

calcite

aggregate hornblende

⊢ 2.5 cm ⊣

Hornblende crystals and aggregate with calcite from Franklin, New Jersey. (Specimen from Department of Geology, Bryn Mawr College)

HORNED TOAD

Horned toad (Phrynosoma), a lizard, uses its tongue to catch insects while on the run.

HORSEHEAD NEBULA

Horsehead Nebula in Orion (IC 434, Bernard 33), which was photographed in red light with a 200-inch (508-centimeter) telescope. (Hale Observatories)

to keep the receiver in line-by-line synchronism with the transmitter. Also known as line synchronizing pulse.

horizontal system [ADP] A programming system in which instructions are written horizontally, that is, across the page.

horizontal-tube evaporator [MECH ENG] A horizontally mounted tube-and-shell type of liquid evaporator, used most often for preparation of boiler feedwater.

horizontal vee [ELECTROMAG] An antenna consisting of two linear radiators in the form of the letter V, lying in a horizontal plane.

horizontal vibrating needle See horizontal force instrument.

horizon tracker [NAV] A device for establishing a vertical reference in a navigation system by precisely tracking the visible horizon. Also known as horizon scanner.

hormesis [BIOL] Providing stimulus by nontoxic amounts of a toxic agent.

hormogonium [BOT] Portion of a filament between heterocysts in certain algae; detaches as a reproductive body.

hormone [BIOCHEM] 1. A chemical messenger produced by endocrine glands and secreted directly into the bloodstream to exert a specific effect on a distant part of the body. 2. An organic compound that is synthesized in minute quantities in one part of a plant and translocated to another part, where it influences a specific physiological process.

horn [ELECTROMAG] See horn antenna. [ENG ACOUS] A tube whose cross-sectional area increases from one end to the other, used to radiate or receive sound waves and to intensify and direct them. Also known as acoustic horn. [MET] Holding arm for the electrode of a resistance spot-welding machine.

horn angle [MATH] A geometric figure formed by two tangent plane curves that lie on the same side of their mutual tangent line in the neighborhood of the point of tangency.

horn antenna [ELECTROMAG] A microwave antenna produced by flaring out the end of a circular or rectangular waveguide into the shape of a horn, for radiating radio waves directly into space. Also known as electromagnetic horn; flare (British usage); hoghorn antenna (British usage); horn; horn radiator.

horn arrester [ELEC] A lightning arrester in which the spark gap has thick wire horns that spread outward and upward; the arc forms at the narrowest bottom part of the gap, travels upward, and extinguishes itself when it reaches the widest part of the gap. Also known as horn lightning arrester.

hornbeam [BOT] Any tree of the genus Carpinus in the birch family distinguished by doubly serrate leaves and by small, pointed, angular winter buds with scales in four rows.

hornblende [MINERAL] A general name given to the monoclinic calcium amphiboles that form an extensive solid-solution series between the various metals in the generalized formula $(Ca,Na)_2(Mg,Fe,Al)_5(Al,Si)_8O_{22}(OH,F)_2$.

hornblendite [PETR] A plutonic rock consisting mainly of hornblende.

horn buoy [NAV] A buoy provided with a horn.

horned dinosaur [PALEON] Common name for extinct reptiles of the suborder Ceratopsia.

horned liverwort [BOT] The common name for bryophytes of the class Anthocerotae. Also known as hornwort.

horned toad [VERT ZOO] The common name for any of the lizards of the genus Phrynosoma; a reptile that resembles a toad but is less bulky.

horned-toad dinosaur [PALEON] The common name for extinct reptiles composing the suborder Ankylosauria.

horn equation [ACOUS] A second-order partial differential equation for the velocity potential as a function of time and of distance along an acoustic horn.

Horner's method [MATH] A technique for approximating the real roots of an algebraic equation; a root is located between consecutive integers, then a successive search is performed.

Horner's syndrome [MED] A complex of symptoms due to unilateral destruction of the cervical sympathetics.

hornet [INV ZOO] The common name for a number of large wasps in the hymenopteran family Vespidae.

hornfels [PETR] A common name for a class of metamorphic rocks produced by contact metamorphism and characterized by equidimensional grains without preferred orientation.

hornfels facies [PETR] Rock formed at depths in the earth's

crust not exceeding 10 kilometers at temperatures of 250–800°C; includes albite-epidote hornfels facies, pyroxene-hornfels facies, and hornblende-hornfels facies.

horn gap [ELEC] Type of spark gap which is provided with divergent electrodes.

horn-gap switch [ELEC] Form of air switch provided with arcing horns.

horn lead See phosgenite.

horn lightning arrester See horn arrester.

horn loudspeaker [ENG ACOUS] A loudspeaker in which the radiating element is coupled to the air or another medium by means of a horn.

horn quicksilver See calomel.

horn radiator See horn antenna.

horn silver See cerargyrite.

horn socket [DES ENG] A cone-shaped fishing tool especially designed to recover lost collared drill rods, drill pipe, or tools in bored wells.

horn spacing [MET] Unobstructed work space in a resistance-welding machine between horns at right angles to the throat depth.

horn spoon [MIN ENG] A troughlike section cut from a cow horn and scraped thin; used for washing auriferous gravel and pulp when exacting tests are to be performed.

hornwort See horned liverwort.

horny coral [INV ZOO] The common name for coelenterate members of the order Gorgonacea.

horography [HOROL] The making of watches and clocks.

Horologium [ASTRON] A constellation with right ascension 3 hours, declination 60° south. Also known as Clock.

horology [SCI TECH] The science of measuring time and the technology of constructing instruments for this measurement.

horse [GEOL] A large rock caught along a fault. [MIN ENG] See horseback. [VERT ZOO] Equus caballus. A herbivorous mammal in the family Equidae; the feet are characterized by a single functional digit.

horseback [GEOL] A low and sharp ridge of sand, gravel, or rock. [MIN ENG] 1. Shale or sandstone occurring in a channel that was cut by flowing water in a coal seam. Also known as cutout; horse; roll; swell; symon fault; washout. 2. To move or raise a heavy piece of machinery or timber by using a pinch bar as a lever. Also known as pinch.

horse chestnut [BOT] Aesculus hippocastanum. An ornamental buckeye tree in the order Sapindales, usually with seven leaflets per leaf and resinous buds.

horsehair blight [PL PATH] A fungus disease of tea and certain other tropical plants caused by Marasmius equicrinis and characterized by black festoons of mycelia hanging from the branches.

horsehead [MIN ENG] Timbers or steel joists used to support planks in tunneling through loose ground.

Horsehead Nebula [ASTRON] A cloud of obscuring particles between the earth and a gaseous emission nebula in the constellation Orion.

horse latitudes [METEOROL] The belt of latitudes over the oceans at approximately 30–35°N and S where winds are predominantly calm or very light and weather is hot and dry.

horsepower [MECH] The unit of power in the British engineering system, equal to 550 foot-pounds per second, approximately 745.7 watts. Abbreviated hp.

horsepox [VET MED] A disease of horses such as pseudotuberculosis, contagious pustular stomatitis, or a vesicular exanthema.

horseradish [BOT] Armoracia rusticana. A perennial crucifer belonging to the order Capparales and grown for its pungent roots, used as a condiment.

horse serum [IMMUNOL] Immune serum obtained from the blood of horses.

horseshoe bend See oxbow.

horseshoe crab [INV ZOO] The common name for arthropods composing the subclass Xiphosurida, especially the subgroup Limulida.

horseshoe kidney [MED] Congenital fusion of two kidneys at one pole.

horseshoe lake See oxbow lake.

horseshoe magnet [ELECTROMAG] A permanent magnet or electromagnet in which the core is horseshoe-shaped or U-shaped, to bring the two poles near each other.

horsetail [BOT] The common name for plants of the genus *Equisetum* composing the order Equisetales.

horsetail ore [GEOL] Ore occurring in fractures which diverge from a larger fracture.

horsfordite [MINERAL] Cu_5Sb A silver-white mineral composed of copper-antimony alloy.

horst [GEOL] **1.** A block of the earth's crust uplifted along faults relative to the rocks on either side. **2.** A mass of the earth's crust limited by faults and standing in relief. **3.** One of the older mountain masses limiting the Alps on the west and north. **4.** A knobby ledge of limestone beneath a thin soil mantle.

horticultural crop [AGR] Any food-producing plant.

horticulture [BOT] The art and science of growing plants.

Horton number [HYD] A dimensionless number that is formed by the product of runoff intensity and erosion proportionality factor; expresses the relative intensity of erosion on the slopes of a drainage basin.

hortonolite [MINERAL] $(Fe,Mg,Mn)_2SiO_4$ A dark mineral composed of silicate of iron, magnesium, and manganese; a member of the olivine series.

Hortvet sublimator [ANALY CHEM] Device for the determination of the condensation temperature (sublimation point) of sublimed solids.

HOS *See* horizontal obstacle sonar.

hose [DES ENG] Flexible tube used for conveying fluids.

hose clamp [DES ENG] Band or brace to attach the raw end of a hose to a water outlet.

hose coupling [DES ENG] Device to interconnect two or more pieces of hose.

hose fitting [DES ENG] Any attachment or accessory item for a hose.

Hoskold formula [MIN ENG] A two-rate valuation formula formerly used to determine present value of mining properties or shares, with redemption of invested capital.

hospital information system [ADP] The collection, evaluation or verification, storage, and retrieval of information about a patient.

hospital ship [NAV ARCH] An unarmed ship equipped as a hospital, in particular, one assigned only to assist sick and wounded men in time of war and protected from attack by international law.

host [BIOL] **1.** An organism on or in which a parasite lives. **2.** The dominant partner of a symbiotic or commensal pair.

host computer [ADP] The computer upon which depends a specialized computer handling the input/output functions in a real-time system.

hostile identification [PSYCH] The assumption by an individual, particularly a child, of socially undesirable characteristics of an older person important to the child so as to gain some special recognition from that person.

hostility [PSYCH] The feeling or display of anger, antagonism, or resistance toward an individual or group.

host rock [GEOL] Rock which serves as a host for other rocks or for mineral deposits.

hot [ELEC] *See* energized. [NUCLEO] Being highly radioactive.

hot acid [PETRO ENG] The use of hot hydrochloric acid (200–300°F; 93–149°C) for oil well acidizing where wellbore scale is slow-dissolving and hard to remove.

hot-air engine [MECH ENG] A heat engine in which air or other gases, such as hydrogen, helium, or nitrogen, are used as the working fluid, operating on cycles such as the Stirling or Ericsson.

hot-air furnace [MECH ENG] An encased heating unit providing warm air to ducts for circulation by gravity convection or by fans.

hot-air soldering [MET] Soldering with a narrow blast of air whose temperature is closely controlled at the value required for soldering individual joints on printed circuit boards.

hot-air sterilization [ENG] A method of sterilization using dry heat for glassware and other heat-resistant materials which need to be dry after treatment; temperatures of 160–165°C are generated for at least 2 hours.

hot atom [NUCLEO] An atom that has high internal or kinetic energy as a result of a nuclear process such as beta decay or neutron capture.

hotbed [MET] An area where hot-rolled metal is placed to cool. Also known as cooling table.

hot belt [CLIMATOL] The belt around the earth within which the annual mean temperature exceeds 20°C.

hot-blast stove [MET] A retracting device for preheating the incoming air in an iron blast furnace by using heat from the burning gases of the furnace.

hot-bulb [MECH ENG] Pertaining to an ignition method used in semidiesel engines in which the fuel mixture is ignited in a separate chamber kept above the ignition temperature by the heat of compression.

hot carrier [ELECTR] A carrier, which may be either an electron or a hole, that has relatively high energy with respect to the carriers normally found in majority-carrier devices such as thin-film transistors.

hot-carrier diode [ELECTR] A semiconductor diode in which hot carriers (either electrons or holes) are emitted from a semiconductor layer into the metal base.

hot cathode [ELECTR] A cathode in which electron or ion emission is produced by heat. Also known as thermionic cathode.

hot-cathode gas-filled tube *See* thyratron.

hot-cathode tube *See* thermionic tube.

hot cell *See* cave.

Hotchkiss drive [MECH ENG] An automobile rear suspension designed to take torque reactions through longitudinal leaf springs.

Hotchkiss superdip [ENG] A sensitive dip needle consisting of a freely rotating magnetic needle about a horizontal axis and a nonmagnetic bar with a counterweight at the end which is attached to the pivot point of the needle.

hot-die steel [MET] A high-temperature, shock-resistant alloy steel used in forging machines.

hot dipping [MET] Coating metal components by immersion in a molten metal bath, such as tin or zinc.

hot-draw [ENG] To draw a material while it is hot.

hot electron [ELECTR] An electron that is in excess of the thermal equilibrium number and, for metals, has an energy greater than the Fermi level; for semiconductors, the energy must be a definite amount above that of the edge of the conduction band.

hot-electron triode [ELECTR] Solid-state, evaporated thin-film structure directly equivalent to a vacuum triode.

hot extrusion [MET] The process of extruding metal at very high temperatures.

hot-filament ionization gage [ELECTR] An ionization gage in which electrons emitted by an incandescent filament, and attracted toward a positively charged grid electrode, collide with gas molecules to produce ions which are then attracted to a negatively charged electrode; the ion current is a measure of the number of gas molecules.

hot flash [PHYSIO] A sudden transitory sensation of heat, often involving the whole body, due to cessation of ovarian function; a symptom of the climacteric.

hot forming [MET] Shaping operations performed at temperatures above the recrystallization temperature of the metal.

hot-gas welding [ENG] Joining of thermoplastic materials by softening first with a jet of hot air, then joining at the softened points.

hot hole [ELECTR] A hole that can move at much greater velocity than normal holes in a semiconductor.

hothouse [ENG] A greenhouse heated to grow plants out of season.

hot junction [ELECTR] The heated junction of a thermocouple.

hot laboratory [NUCLEO] A laboratory designed for research with radioactive materials that have such high strengths that special handling precautions are required.

hot line [COMMUN] Direct circuit between two points, available for immediate use without patching or switching.

hot patching [ENG] Repair of a hot refractory lining in a furnace, usually by spraying with a refractory slurry.

hot press forge [MET] A press in which metal parts are formed by forcing hot metal into dies under high pressure.

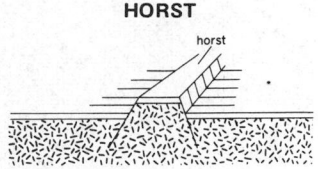

HORST

A simple horst with associated faults. *(From A. K. Lobeck, Geomorphology, McGraw-Hill, 1939)*

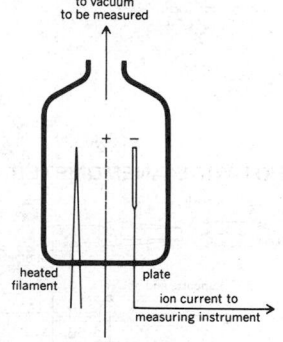

**HOT-FILAMENT
IONIZATION GAGE**

Diagram of hot-filament ionization gage.

hot pressing [ENG] Forming a metal-powder compact or a ceramic shape by applying pressure and heat simultaneously at temperatures high enough for sintering to occur.

hot-press printing *See* hot stamping.

hot-quenching [MET] Quenching from high temperatures into a medium of lower but still high temperature.

hot rolling [MET] Rolling of metal bars or sheets when hot.

hot saw [MECH ENG] A power saw used to cut hot metal.

hot shortness [MET] Brittleness, usually of steel or wrought iron, when the metal is hot, due to a high sulfur content.

hotshot wind tunnel [AERO ENG] A wind tunnel in which electrical energy is discharged into a pressurized arc chamber, increasing the temperature and pressure in the arc chamber so that a diaphragm separating the arc chamber from an evacuated chamber is ruptured, and the heated gas from the arc chamber is then accelerated in a conical nozzle to provide flows with mach numbers of 10 to 27 for durations of 10 to 100 milliseconds.

hot-solder coating [ENG] The application of a protective finish to a printed circuit board by dip soldering in a solder bath.

hot spot [CHEM ENG] An area or point within a reaction system at which the temperature is appreciably higher than in the bulk of the reactor; usually locates the reaction front. [FOR] A forest region where fires occur at frequent intervals. [GRAPHICS] A region of excessive illumination on a photo. [MOL BIO] A site in a gene at which there is an unusually high frequency of mutation. [NUCLEO] **1.** A surface area of higher than average radioactivity. **2.** A part of a reactor fuel surface element that has become overheated. [PHYS] A localized region with temperature higher than the surroundings.

hot spring [HYD] A thermal spring whose water temperature is above 98°F (37°C).

hot stamping [GRAPHICS] A method of printing in which heated type or stamping dies are pressed against a thin leaf of gold or other metal, or against pigment upon a surface such as paper, board, plastic, or leather; the thin leaf contains a sizing agent that is set by heat; this system is sometimes used for making movie titles and artwork for slides. Also known as hot-press printing.

hot-swage [MET] To reduce the cross section of a hot metal tube or rod.

hot-water heating [MECH ENG] A heating system for a building in which the heat-conveying medium is hot water and the heat-emitting means are radiators, convectors, or panel coils. Also known as hydronic heating.

hot wave *See* heat wave.

hot well [MECH ENG] A chamber for collecting condensate, as in a steam condenser serving an engine or turbine.

hot wind [METEOROL] General term for winds characterized by intense heat and low relative humidity, such as summertime desert winds or an extreme foehn.

hot-wire ammeter [ENG] An ammeter which measures alternating or direct current by sending it through a fine wire, causing the wire to heat and to expand or sag, deflecting a pointer. Also known as thermal ammeter.

hot-wire anemometer [ENG] An anemometer used in research on air turbulence and boundary layers; the resistance of an electrically heated fine wire placed in a gas stream is altered by cooling by an amount which depends on the fluid velocity.

hot-wire instrument [ENG] An instrument that depends for its operation on the expansion by heat of a wire carrying a current.

hot-wire microphone [ENG ACOUS] A velocity microphone that depends for its operation on the change in resistance of a hot wire as the wire is cooled by varying particle velocities in a sound wave.

hot working [MET] Plastic deformation of a metal at a rate and temperature such that strain hardening cannot occur.

Houben-Hoesch synthesis [ORG CHEM] Condensation of cyanides with polyhydric phenols in the presence of hydrogen chloride and zinc chloride to yield phenolic ketones.

Houdriflow catalytic cracking [CHEM ENG] A continuous, moving-bed petroleum-refinery catalytic-cracking process for producing high-octane gasoline from heavier petroleum frac-

tions; uses an integrated reactor-regenerator vessel; the catalyst is transported upward through the unit gas lift.

Houdriforming [CHEM ENG] A continuous petroleum-refinery catalytic-reforming process for producing high-octane aromatic concentrates from low-octane naphthas.

Houdry butane dehydrogenation [CHEM ENG] A catalytic process for dehydrogenating light hydrocarbons from crude oil to their corresponding mono- or diolefins; chromia-alumina catalysts with inert material are used in pellet form.

Houdry fixed-bed catalytic cracking [CHEM ENG] A cyclic, regenerable process for cracking of petroleum distillates to produce high-octane gasoline from higher-boiling petroleum fractions; synthetic or natural bead catalysts of activated hydrosilicate of alumina may be used. Also known as Houdry process.

Houdry hydrocracking [CHEM ENG] A catalytic process combining cracking and desulfurization of crude petroleum oil in the presence of hydrogen; catalysts may be nickel oxide or nickel sulfide on silica alumina, and cobalt molybdate on alumina.

Houdry process *See* Houdry fixed-bed catalytic cracking.

Hound Dog [ORD] A U.S. Air Force air-to-surface guided missile having a range of several hundred miles, capable of carrying a nuclear warhead, and intended for launching from long-range bombers of the Strategic Air Command.

hour [MECH] A unit of time equal to 3600 seconds. Abbreviated hr.

hour angle [ASTRON] Angular distance west of a celestial meridian or hour circle; the arc of the celestial equator, or the angle at the celestial pole, between the upper branch of a celestial meridian or hour circle and the hour circle of a celestial body or the vernal equinox, measured westward through 360°.

hour angle difference *See* meridian angle difference.

hour circle [ASTRON] An imaginary great circle passing through the celestial poles on the celestial sphere above which declination is measured. Also known as circle of declination; circle of right ascension.

hourglass [HOROL] An instrument for measuring time, consisting of two somewhat funnel-shaped vessels joined at their narrow points by a thin, hollow neck; one hour is required for a given quantity of sand to fall from the upper to the lower vessel.

hourglass screw *See* Hindley screw.

hourglass stomach [MED] A stomach with an equatorial constriction usually caused by formation of scar tissue around an ulcer.

hourglass worm *See* Hindley screw.

hour hand [HOROL] The pointer that indicates the hour on a timepiece.

hour-out line *See* time front.

hour wheel [HOROL] The wheel carrying the hour hand of a timepiece.

houseboat [NAV ARCH] A barge-type craft with cabins on its deck, used as a dwelling and for leisurely cruising.

house drain [CIV ENG] Horizontal drain in a basement receiving waste from stacks.

housefly [INV ZOO] *Musca domestica.* A dipteran insect with lapping mouthparts commonly found near human habitations; a vector in the transmission of many disease pathogens.

housekeeping [ADP] Those operations or routines which do not contribute directly to the solution of a computer program, but rather to the organization of the program.

house physician [MED] A physician employed by a hospital.

house sewer [CIV ENG] Connection between house drain and public sewer.

housing [ENG] A case or enclosure to cover and protect a structure or a mechanical device.

Houskeeper seal [ENG] A vacuumtight seal made between copper and glass by bringing the copper to a flexible feather edge before fusing it to the glass; the copper then flexes as the glass shrinks during cooling.

hovercraft *See* air-cushion vehicle.

howardite [GEOL] An achondritic stony meteorite composed chiefly of calcic plagioclase and orthopyroxene.

Howell-Bunger valve *See* cone valve.

Howell-Jolly bodies [PATH] Small, round basophilic inclu-

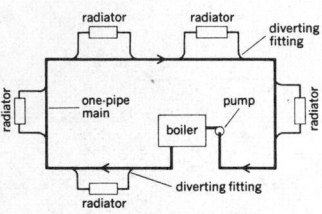

HOT-WATER HEATING

One-pipe hot-water heating system.

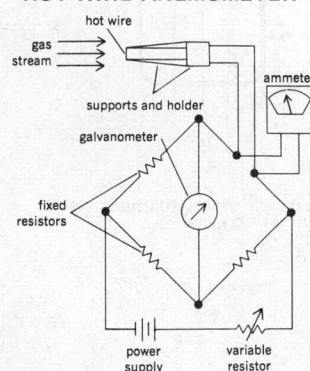

HOT-WIRE ANEMOMETER

Constant-resistance hot-wire anemometer.

sions of nuclear material in erythrocytes of splenectomized persons.

Howell's unit [BIOL] A unit for the standardization of heparin.

Howe truss [CIV ENG] A truss for spans up to 80 feet (24 meters) having both vertical and diagonal members; made of steel or timber or both.

howitzer [ORD] A complete projectile-firing weapon with a bore diameter greater than 30 millimeters; length of the bore is usually 25 to 30 caliber, and maximum angle of velocity about 65°; used to deliver curved fire with projectiles of lower muzzle velocities than those from the gun.

howl [ENG ACOUS] Undesirable prolonged sound produced by a radio receiver or audio-frequency amplifier system because of either electric or acoustic feedback.

howler [COMMUN] In telephone practice, an associated unit by which the test desk operator may connect a high tone of varying loudness to a subscriber's line to call the subscriber's attention to the fact that his receiver is off the hook. [ELECTR] An audio device used to warn a radar operator that signals are appearing on a radar screen.

howlite [MINERAL] $Ca_2Bi_5SiO_9(OH)_5$ A white mineral occurring in nodular or earthy form.

howl repeater [COMMUN] Condition in telephone repeater operation where more energy is returned than sent, resulting in an oscillation being set up on the circuit.

Howship's lacuna [HISTOL] Minute depressions in the surface of a bone undergoing resorption.

Hoyer method of prestressing *See* pretensioning.

hp *See* horsepower.

H pad *See* H network.

h parameter [ELECTR] One of a set of four transistor equivalent-circuit parameters that conveniently specify transistor performance for small voltages and currents in a particular circuit. Also known as hybrid parameter.

H pile [CIV ENG] A steel pile that is H-shaped in section.

H plane [ELECTROMAG] The plane of an antenna in which lies the magnetic field vector of linearly polarized radiation.

H-plane bend [ELECTROMAG] A rectangular waveguide bend in which the longitudinal axis of the waveguide remains in a plane parallel to the plane of the magnetic field vector throughout the bend. Also known as H bend.

H-plane T junction [ELECTROMAG] Waveguide T junction in which the change in structure occurs in the plane of the magnetic field. Also known as shunt T junction.

hr *See* hour.

H-R diagram *See* Hertzsprung-Russell diagram.

H I region [ASTRON] A region of interstellar space where neutral hydrogen is present.

H II region [ASTRON] A region of interstellar space occupied by gas that is largely atomic hydrogen and mostly ionized.

H rod [DES ENG] A drill rod having an outside diameter of $3\frac{1}{2}$ inches.

H scan *See* H scope.

H scope [NAV] A modified form of B scope on which the signal appears as two dots, the slope of the line joining them indicating elevation angle; the horizontal coordinate indicates bearing angle and the vertical coordinate range, with respect to the left-hand dot. Also known as H indicator; H scan.

HSI *See* heat stress index.

HSP *See* high-speed printer.

H substance [BIOCHEM] An agent similar to histamine and believed to play a role in local blood vessel response in tissue damage.

H system [NAV] A radar navigation system that uses two ground radar beacons, in conjunction with airborne equipment which gives the direction and distance to each beacon.

HTGR *See* high-temperature gas-cooled reactor.

H-theorem of Boltzmann *See* Boltzmann H-theorem.

HTU *See* height of transfer unit.

hub [ADP] An electric socket in a plugboard into which one may insert or connect leads or may plug wires. [DES ENG] **1.** The cylindrical central part of a wheel, propeller, or fan. **2.** A piece in a lock that is turned by the knob spindle, causing the bolt to move. **3.** A short coupling that joins plumbing pipes.

[MET] A steel punch used in making a working die for a coin or medal.

Hubbard Glacier [GEOL] A valley glacier which reaches tidewater from a source area of Mount St. Elias of Alaska and the Yukon.

Hubbard tank [MED] A large, specially designed tank in which a patient may be immersed for various therapeutic underwater exercises.

hubbing [MET] Forcing a male die into a blank to form a female die.

hubble [ASTRON] A unit of astronomical distance equal to 10^9 light-years or 9.4605×10^{24} meters.

Hubble constant [ASTROPHYS] The rate at which the velocity of recession of the galaxies increases with distance; the value is about 30 kilometers per second per million light-years (or 3.2×10^{-18} sec^{-1}) with an uncertainty of about ± 15 km/sec.

Hubble effect *See* red shift.

Hubble's Variable Nebula [ASTRON] A variable-brightness nebula associated with variable stars and fan-shaped in appearance.

hub cannon [ORD] A cannon mounted through a propeller hub of an aircraft.

hubcap [DES ENG] A metal cap fastened or clamped to the end of an axle, as on motor vehicles.

Huber's reagent [ANALY CHEM] Aqueous solution of ammonium molybdate and potassium ferrocyanide used as a reagent to detect free mineral acid.

Hubl's reagent [ANALY CHEM] Solution of iodine and mercuric chloride in alcohol used to determine the iodine content of oils and fats.

Hübner rhomb [OPTICS] A glass rhombohedron used in photometry to compare two illuminated surfaces.

huckleberry [BOT] The common name for shrubs of the genus *Gaylussacia* in the family Ericaceae distinguished by an ovary with 10 locules and 10 ovules; the dark-blue berries are edible.

Hudsonian life zone [ECOL] A zone comprising the climate and biotic communities of the northern portions of North American coniferous forests and the peaks of high mountains.

hue [OPTICS] The name of a color, such as red, yellow, green, blue, or purple, as perceived subjectively.

huebnerite [MINERAL] $MnWO_4$ A brownish-red to black manganese member of the wolframite series, occurring in short, monoclinic, prismatic crystals; isomorphous with ferberite.

hue control [ELECTR] A control that varies the phase of the chrominance signals with respect to that of the burst signal in a color television receiver, in order to change the hues in the image. Also known as phase control.

Huggenberger tensometer [ENG] A type of extensometer having a short gage length (10 to 20 millimeters) and employing a compound lever system that gives a magnification of about 1200.

Huggins-Talalay unit [BIOL] A unit for the standardization of phosphatase.

Hugoniot function [PHYS] A function specifying the locus of states which are possible immediately after the passage of a shock front; gives the state's pressure as a function of its specific volume.

hühnerkobelite [MINERAL] $(Na,Ca)(Fe,Mn)_2(PO_4)_2$ A mineral composed of phosphate of sodium, calcium, iron, and manganese; it is isomorphous with varulite.

Huhner's test [PATH] An examination of seminal fluid obtained from the vaginal fornix and cervical canal after a specified interval following coitus; used in fertility studies to evaluate spermatozoal survival and activity in the lower female genital tract.

hull [BOT] The outer, usually hard, covering of a fruit or seed. [FOOD ENG] **1.** To remove husks from fruits and seeds, as from ears of corn, nuts, or peas. **2.** To remove the shell of a crustacean or mollusk, as an oyster. [NAV ARCH] The body or shell of a ship. [ORD] **1.** The outer casing of a rocket, guided missile, or the like. **2.** Massive armored body of a tank, exclusive of tracks, motor, turret, and armament.

Hull cell [PHYS CHEM] An electrodeposition cell that operates within a simultaneous range of known current densities.

hulless buckwheat *See* tartary buckwheat.

HUBBARD GLACIER

Oblique air photograph of a 4-mile- (6-kilometer-) wide tidewater terminus of the Hubbard Glacier, a prototypical valley glacier. *(Photograph by H. B. Washburn, August, 1938)*

HUEBNERITE

├─ 2.5 cm ─┤

Radiating groups of huebnerite crystals in quartz vein, Silverton, Colorado. *(Specimen from Department of Geology, Bryn Mawr College)*

HUMIDISTAT

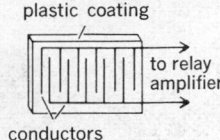

plastic coating

to relay
amplifier

conductors

Electronic humidistat.
(Honeywell Inc.)

HUMMINGBIRD

Hummingbird, capable of
hovering in flight.

hulsite [MINERAL] $(Fe^{2+},Mg)_2(Fe^{3+},Sn)(BO_3)O_2$ A black mineral composed of iron calcium magnesium tin borate.

hum [ELEC] A sound produced by an iron core of a transformer due to loose laminations or magnetostrictive effects; the frequency of the sound is twice the power line frequency. [ELECTR] An electrical disturbance occurring at the power supply frequency or its harmonics, usually 60 or 120 hertz in the United States.

human biogeography [ECOL] The science concerned with the distribution of human populations on the earth.

human chorionic gonadotropin [BIOCHEM] A gonadotropic and luteotropic hormone secreted by the chorionic vesicle. Abbreviated HCG. Also known as chorionic gonadotropin.

human community [ECOL] That portion of a human ecosystem composed of human beings and associated plant and animal species.

human ecology [ECOL] The branch of ecology that considers the relations of individual persons and of human communities with their particular environment.

human-factors engineering [ENG] The area of knowledge dealing with the capabilities and limitations of human performance in relation to design of machines, jobs, and other modifications of the human's physical environment.

human genetics [GEN] The study of human heredity and variation, including the physical basis, mechanisms of inheritance, inheritance patterns, and population analysis.

human measles immune serum [IMMUNOL] Serum from the blood of a person who has recovered from measles.

human threadworm *See* pinworm.

hum bar [ELECTR] A dark horizontal band extending across a television picture due to excessive hum in the video signal applied to the input of the picture tube.

Humble gage [PETRO ENG] Device to measure oil well bottom-hole pressure; a piston acts through a stuffing box against a helical spring in tension.

Humble relation [PETRO ENG] Equation used by oil companies to estimate porosity of the oil-bearing formation from measurements made with a contact resistivity device, such as a microlog.

Humboldt Current *See* Peru Current.

Humboldt Glacier [HYD] The largest Arctic iceberg, at latitude 79°, with a seaward front extending 65 miles (105 kilometers).

humboldtine [MINERAL] $FeC_2O_4 \cdot 2H_2O$ A mineral composed of hydrous ferrous oxalate. Also known as humboldtite; oxalite.

humboldtite *See* humboldtine.

Humboldt jig [MIN ENG] An ore jig with a movable screen.

hum-bucking coil [ENG ACOUS] A coil wound on the field coil of an excited-field loudspeaker and connected in series opposition with the voice coil, so that hum voltage induced in the voice coil is canceled by that induced in the hum-bucking coil.

humectant [CHEM] A substance which absorbs or retains moisture; examples are glycerol, propylene glycol, and sorbitol; used in preparing confectioneries and dried fruit.

humeral [ANAT] Of or pertaining to the humerus or the shoulder region.

humeroglandular effector system [PHYSIO] Glandular effector system in which the activating agent is a blood-borne chemical.

humeromuscular effector system [PHYSIO] Muscular effector system in which the activating agent is a blood-borne chemical.

Hume-Rothery compound *See* electron compound.

Hume-Rothery rule [SOLID STATE] The rule that the ratio of the number of valence electrons to the number of atoms in a given phase of an electron compound depends only on the phase, and not on the elements making up the compounds.

humerus [ANAT] The proximal bone of the forelimb in vertebrates; the bone of the upper arm in humans, articulating with the glenoid fossa and the radius and ulna.

humic acid [ORG CHEM] Any of various complex organic acids obtained from humus; insoluble in acids and organic solvents.

humic coal [GEOL] A coal whose attritus is composed mainly of transparent humic degradation material.

humicolous [ECOL] Of or pertaining to plant species inhabiting medium-dry ground.

humid climate [CLIMATOL] A climate whose typical vegetation is forest. Also known as forest climate.

humidification [ENG] The process of increasing the water vapor content of a gas.

humidifier [MECH ENG] An apparatus for supplying moisture to the air and for maintaining desired humidity conditions.

humidistat [ENG] An instrument that measures and controls relative humidity. Also known as hydrostat.

humidity [METEOROL] Atmospheric water vapor content, expressed in any of several measures, such as relative humidity.

humidity capacitor [ELECTR] A device for measuring ambient relative humidity by sensing a change in capacitance.

humidity coefficient [METEOROL] A measure of the precipitation effectiveness of a region; it recognizes the exponential relationship of temperature versus plant growth and is expressed as humidity coefficient $= P/(1.07)^t$, where P is the precipitation in centimeters, and t is the mean temperature in degrees Celsius for the period in question; the denominator approximately doubles with each 10°C rise in temperature.

humidity element [ENG] The transducer of any hygrometer, that is, that part of a hygrometer that quantitatively senses atmospheric water vapor.

humidity index [CLIMATOL] An index of the degree of water surplus over water need at any given station; it is calculated as humidity index $= 100s/n$, where s (the water surplus) is the sum of the monthly differences between precipitation and potential evapotranspiration for those months when the normal precipitation exceeds the latter, and where n (the water need) is the sum of monthly potential evapotranspiration for those months of surplus.

humidity indicator [INORG CHEM] Cobalt salt (for example, cobaltous chloride) that changes color as the surrounding humidity changes; changes from pink when hydrated, to greenish-blue when anhydrous.

humidity mixing ratio [METEOROL] The amount of water vapor mixed with one unit mass of dry air, usually expressed as grams of water vapor per kilogram of air.

humidity province [CLIMATOL] A region in which the precipitation effectiveness of its climate produces a definite type of biological consequence, in particular the climatic climax formations of vegetation (rain forest, tundra, and the like).

humidity strip [ENG] The humidity transducing element in a Diamond-Hinman radiosonde; it consists of a flat plastic strip bounded by electrodes on two sides and coated with a hygroscopic chemical compound such as lithium chloride; the electrical resistance of this coating is a function of the amount of moisture absorbed from the atmosphere and the temperature of the strip. Also known as electrolytic strip.

humidity test [MET] Corrosion test in which a specimen is exposed to an environment of controlled humidity and temperature.

humid transition life zone [ECOL] A zone comprising the climate and biotic communities of the northwest moist coniferous forest of the north-central United States.

humification [GEOL] Formation of humus.

humifuse [BIOL] Spread over the ground surface.

humin [GEOL] *See* ulmin. [ORG CHEM] An insoluble pigment formed in the acid hydrolysis of a protein that contains tryptophan.

Humiriaceae [BOT] A family of dicotyledonous plants in the order Linales characterized by exappendiculate petals, usually five petals, flowers with an intrastaminal disk, and leaves lacking stipules.

humite [MINERAL] 1. A humic coal mineral. 2. A series of magnesium neosilicate minerals closely related in crystal structure and chemical composition.

humivore [ECOL] An organism that feeds on humus.

hummer screen [MIN ENG] An electrically vibrated ore screen for sizing moderately small material.

hummingbird [VERT ZOO] The common name for members of the family Trochilidae; fast-flying, short-legged, weak-footed insectivorous birds with a tubular, pointed bill and a fringed tongue.

hummocked ice [OCEANOGR] Pressure ice, characterized by haphazardly arranged mounds or hillocks; it has less definite form, and show the effects of greater pressure, than either rafted ice or tented ice, but in fact may develop from either of those.

hum modulation [ELECTR] Modulation of a radio-frequency signal or detected audio-frequency signal by hum; heard in a radio receiver only when a station is tuned in.

humodurite *See* translucent attritus.

humogelite *See* ulmin.

humor [PHYSIO] A fluid or semifluid part of the body.

Humpage gears [MECH ENG] A train of bevel gears used for speed reduction.

humpback *See* kyphosis.

Humphrey gas pump [MECH ENG] A combined internal combustion engine and pump in which the metal piston has been replaced by a column of water.

Humphrey's spiral [MIN ENG] An ore concentrator consisting of a stationary spiral trough through which pulp gravitates; heavy particles stay on the inside and lighter ones climb to the outside.

Humphries equation [THERMO] An equation which gives the ratio of specific heats at constant pressure and constant volume in moist air as a function of water vapor pressure.

hump yard [CIV ENG] A switch yard in a railway system that has a hump or steep incline down which freight cars can coast to prescheduled locations. Also known as gravity yard.

humus [GEOL] The amorphous, ordinarily dark-colored, colloidal matter in soil; a complex of the fractions of organic matter of plant, animal, and microbial origin that are most resistant to decomposition.

hundred-percent rectangle [ORD] A rectangle whose length is eight probable errors in range, and whose breadth is eight probable errors in direction; its center is the center of dispersion; it is the area in which all shots (except wild shots) will fall when fired with the same data under identical conditions. Also known as rectangle of dispersion.

hundredweight [SCI TECH] Abbreviated cwt. **1.** A unit of weight, in common use in the United States, equal to 100 pounds or the weight of 45.359237 kilograms. Also known as cental; centner; kintal; quintal; short hundredweight. **2.** A unit of weight in common use in the United Kingdom equal to 112 pounds or to the weight of 50.80234544 kilograms. Also known as long hundredweight. **3.** A unit of weight in troy measure equal to 100 troy pounds or the weight of 37.32417216 kilograms.

Hund rules [ATOM PHYS] Two rules giving the order in energy of atomic states formed by equivalent electrons: of the terms given by equivalent electrons, the ones with greatest multiplicity have the least energy, and of these the one with greatest orbital angular momentum is lowest; the state of a multiplet with lowest energy is that in which the total angular momentum is the least possible, if the shell is less than half-filled, and the greatest possible, if more than half filled.

Hundsdieke reaction [ORG CHEM] Production of an alkyl halide by boiling a silver carboxylate with an equivalent weight of bromine in carbon tetrachloride.

hunger [PSYCH] The need for food and the physiological and psychological mechanisms regulating food intake.

hungry joint *See* starved joint.

hung shot [ENG] A shot whose explosion is delayed after detonation or ignition.

Hunner's ulcer [MED] A chronic ulcer of the urinary bladder, frequently in association with interstitial cystitis. Also known as elusive ulcer.

hunt [AERO ENG] **1.** Of an aircraft or rocket, to weave about its flight path, as if seeking a new direction or another angle of attack; specifically, to yaw back and forth. **2.** Of a control surface, to rotate up and down or back and forth without being detected by the pilot.

Hunt and Douglas process [MET] Smelting process involving the roasting of matte carrying copper, lead, gold, and silver to form copper sulfate and oxide (but not silver sulfate); this product is leached with sulfuric acid for copper; the resulting solution is treated with calcium chloride by passing sulfur dioxide through it; the cuprous chloride is then reduced to cuprous oxide by milk of lime, (regenerating calcium chloride), and the cuprous oxide is smelted.

Hunt continuous filter [MIN ENG] A continuous-vacuum filter consisting of a horizontally revolving, annular filter bed on which pulp is washed and then vacuum-dried.

hunter's moon [ASTRON] The full moon next following the harvest moon.

hunting [CONT SYS] Undesirable oscillation of an automatic control system, wherein the controlled variable swings on both sides of the desired value. [ELECTR] Operation of a selector in moving from terminal to terminal until one is found which is idle.

hunting circuit *See* lockout circuit.

Hunting Dogs *See* Canes Venatici.

Huntington-Heberlein process [MIN ENG] A sink-float process employing a galena medium, which is recovered by froth flotation.

Huntington's chorea [MED] A rare hereditary disease of the basal ganglia and cerebral cortex resulting in choreiform (dancelike) movements, intellectual deterioration, and psychosis.

hunting tooth [DES ENG] An extra tooth on the larger of two gear wheels so that the total number of teeth will not be an integral multiple of the number on the smaller wheel.

huntite [MINERAL] $CaMg_3(CO_3)_4$ A white mineral consisting of calcium magnesium carbonate.

hurdle sheet [MIN ENG] A brattice-cloth screen across a roadway below a roof cavity or at the ripping lip to divert air current upward, thus diluting and removing firedamp.

Hurler's syndrome [MED] Mucopolysaccharoidosis I, a hereditary condition transmitted as an autosomal recessive in which there is excessive chondroitin sulfate B and heparin sulfate in the urine and tissues, and which is marked clinically by a complex of symptoms including grotesque skeletal and facial deformities, skin and cardiac changes, clouding of the cornea, and mental deficiency. Also known as gargoylism; lipochondrodystrophy.

Huronian [GEOL] The lower system of the restricted Proterozoic.

hurricane [METEOROL] A tropical cyclone of great intensity; any wind reaching a speed of more than 73 miles per hour (117 kilometers per hour) is said to have hurricane force.

hurricane air stemmer [MIN ENG] A mechanical device for rapidly tamping shotholes; consists of a funnel connected by a T piece to the charge tube, with a connection to a compressed-air column; sand is put in the funnel and injected into the shothole as the charge tube is withdrawn.

hurricane band *See* spiral band.

hurricane beacon [ENG] An air-launched balloon designed to be released in the eye of a tropical cyclone, to float within the eye at predetermined levels, and to transmit radio signals.

hurricane deck *See* awning deck; promenade deck.

hurricane-force wind [METEOROL] In the Beaufort wind scale, a wind whose speed is 64 knots (117 kilometers per hour) or higher.

hurricane lamp [ENG] An oil lamp with a glass chimney and perforated lid to protect the flame, or a candle with a glass chimney.

hurricane monitoring buoy [METEOROL] A free-floating automatic weather station designed as an expendable instrument in connection with hurricane and typhoon monitoring and forecasting services.

hurricane radar band *See* spiral band.

hurricane surge *See* hurricane wave.

hurricane tide *See* hurricane wave.

hurricane tracking [ENG] Recording of the movement of individual hurricanes by means of airplane sightings and satellite photography.

hurricane warning [METEOROL] A warning of impending winds of hurricane force; for maritime interests, the storm-warning signals for this condition are two square red flags with black centers by day, and a white lantern between two red lanterns by night.

hurricane watch [METEOROL] An announcement for a specific area that hurricane conditions pose a threat; residents are cautioned to take stock of their preparedness needs but, otherwise, are advised to continue normal activities.

hurricane wave [OCEANOGR] As experienced on islands and along a shore, a sudden rise in the level of the sea associated

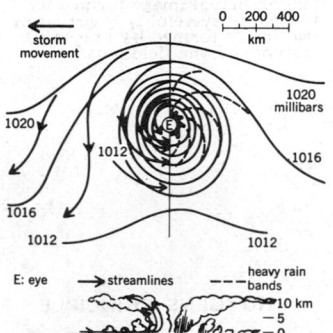

HURRICANE

Simplified model diagram of circulation of a Northern Hemisphere hurricane. Isobars (in millibars) have been omitted near the cyclone center. On the left are streamlines of airflow; on the right are regions of heavy rain. Below is a vertical section through the center, showing clouds and vertical circulation.

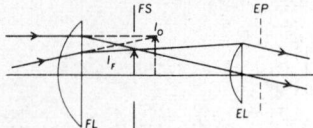

HUYGENS EYEPIECE

Arrangement of lenses in Huygens eyepiece. FL = field lens; EL = eye lens; FS = field stop; EP = exit pupil or eye joint; I_O = real or virtual image formed by preceding system; I_f = virtual or real image formed by preceding system and the field lens.

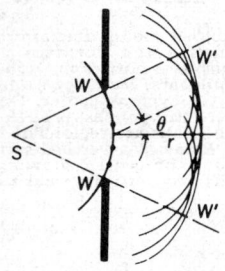

HUYGENS' PRINCIPLE

The construction for a spherical wave. WW = wave originating at S; $W'W'$ = envelope of secondary waves; θ = angle between the normal to the original wavefront and any point on the secondary wave; r = radius representing the distance wave would travel in time t.

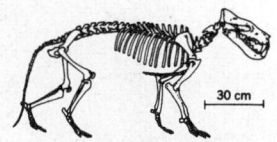

HYAENODONTIDAE

The deltatheridian *Hyaenodon*, original about 4 feet (1.2 meters) long. *(from Scott). (From A. S. Romer, Vertebrate Paleontology, 3d ed., University of Chicago Press, 1966)*

with a hurricane. Also known as hurricane surge; hurricane tide.

hurricane wind [METEOROL] In general, the severe wind of an intense tropical cyclone (hurricane or typhoon); the term has no further technical connotation, but is easily confused with the strictly defined hurricane-force wind. Also known as typhoon wind.

Hurst formula [PETRO ENG] Relationship used in reservoir material-balance analysis; interrelates field pressure and production data at a number of different times.

Hurst method [PETRO ENG] A calculation method for the bottom-hole static pressure of a well; uses graphical extrapolation of pressure buildup over a short period of time.

Hurthle cells [PATH] Enlarged epithelial cells of the thyroid follicles containing acidophilic cytoplasm, seen most frequently in adenomas.

Hurtley's test [PATH] A test for the presence of acetoacetic acid in urine: take 10 cubic centimeters (cm^3) of a urine specimen and add 2.5 cm^3 of concentrated hydrochloric acid and 1 cm^3 of a freshly prepared 1% solution of sodium nitrite, then shake and let stand for 2 minutes; add 15 cm^3 of 0.880 ammonia and 5 cm^3 of 10% ferrous sulfate, then shake and pour into a large boiling tube, allowing it to stand; a violet or purple color confirms the presence of acetoacetic acid.

Hurwitz polynomial [MATH] A polynomial whose zeros all have negative real parts.

husk [BOT] The outer coat of certain seeds, particularly if it is a dry, membranous structure.

hutch [MIN ENG] The bottom compartment of a jig used in ore dressing.

hutchinsonite [MINERAL] $(Pb,Tl)_2(Cu,Ag)As_5S_{10}$ Red mineral composed of sulfide of lead, copper, and arsenic, with varying amounts of thallium and silver, occurring in small orthorhombic crystals.

Hutchinson's freckle *See* melanotic freckle.

Hutchinson's teeth [MED] Deformity of permanent incisor teeth associated with congenital syphilis; the crown of the incisor is wider in the cervical portion than at the incisal edge, and the incisal edge has a characteristic crescent-shaped notch.

hutch product [MIN ENG] Fine, heavy materials that pass through the screen of a jig and collect in the hutch.

Huttig equation [THERMO] An equation which states that the ratio of the volume of gas adsorbed on the surface of a nonporous solid at a given pressure and temperature to the volume of gas required to cover the surface completely with a unimolecular layer equals $(1 + r) c^r/(1 + c^r)$, where r is the ratio of the equilibrium gas pressure to the saturated vapor pressure of the adsorbate at the temperature of adsorption, and c is the product of a constant and the exponential of $(q - q_l)/RT$, where q is the heat of adsorption into a first layer molecule, q_l is the heat of liquefaction of the adsorbate, T is the temperature, and R is the gas constant.

huttonite [MINERAL] $ThSiO_4$ A colorless to pale-green monoclinic mineral composed of silicate of thorium; it is dimorphous with thorite.

Huwood loader [MIN ENG] A machine consisting of a number of horizontal rotating flight bars working near the floor of the seam which push prepared coal up a ramp onto a low, bottom-loaded coveyor belt.

Huxley's anastomosis [HISTOL] Polyhedral cells forming the middle layer of a hair root sheath.

Huygens' approximation [MATH] The length of a small circular arc is approximately $\frac{1}{3}(8c' - c)$, where c is the chord of the arc and c' is the chord of half the arc.

Huygens eyepiece [OPTICS] An eyepiece in which there are two plano-convex lenses, and the plane sides of both lenses face the eye.

Huygens' principle [OPTICS] The principle that each point on a light wavefront may be regarded as a source of secondary waves, the envelope of these secondary waves determining the position of the wavefront at a later time.

Huygens wavelet [OPTICS] A secondary wave as used in Huygens' principle.

HVAP *See* hypervelocity armor-piercing.

H vector [ELECTROMAG] A vector that is the magnetic field. For a plane wave in free space, it is perpendicular to the E

vector and to the direction of propagation.

H wave *See* transverse electric wave.

hyacinth *See* zircon.

Hyaenidae [VERT ZOO] A family of catlike carnivores in the superfamily Feloidea including the hyenas and aardwolf.

Hyaenodontidae [PALEON] A family of extinct carnivorous mammals in the order Deltatheridia.

Hyalellidae [INV ZOO] A family of amphipod crustaceans in the suborder Gammaridea.

hyaline [BIOCHEM] A clear, homogeneous, structureless material found in the matrix of cartilage, vitreous body, mucin, and glycogen. [GEOL] Transparent and resembling glass.

hyaline cartilage [HISTOL] A translucent connective tissue comprising about two-thirds clear, homogeneous matrix with few or no collagen fibrils.

hyaline cast [PATH] A clear, structureless mass of proteinaceous material found in the urine in association with certain kidney diseases.

hyaline degeneration [PATH] Degenerative change involving tissues and cells so that they become clear, structureless, and homogeneous.

hyaline membrane [HISTOL] **1.** A basement membrane. **2.** A membrane of a hair follicle between the inner fibrous layer and the outer root sheath.

hyaline membrane disease [MED] A disease occurring during the first few days of neonatal life, characterized by respiratory distress due to formation of a hyalinelike membrane within the alveoli.

hyaline test [INV ZOO] A translucent wall or shell of certain foraminiferans composed of layers of calcite interspersed with separating membranes.

hyalinization [PATH] Replacement of the normal components of cells or tissues by a hyaline material.

hyalinocrystalline [PETR] Of porphyritic rock texture, having the phenocrysts lying in a glassy ground mass.

hyalite [MINERAL] A colorless, clear or translucent variety of opal occurring as globular concretions or botryoidal crusts in cavities or cracks of rocks. Also known as Müller's glass; water opal.

hyalobasalt *See* tachylite.

hyaloclastite [GEOL] A tufflike deposit formed by the flowing of basalt under water and ice and its consequent fragmentation. Also known as aquagene tuff.

Hyalodictyae [MYCOL] A subdivision of the spore group Dictyosporae characterized by hyaline spores.

Hyalodidymae [MYCOL] A subdivision of the spore group Didymosporae characterized by hyaline spores.

hyalography [GRAPHICS] The art of engraving on glass, with diamond or emery cutting tools or with hydrofluoric acid or other etching solutions.

Hyalohelicosporae [MYCOL] A subdivision of the spore group Helicosporae characterized by hyaline spores.

hyaloid membrane [ANAT] The limiting membrane surrounding the vitreous body of the eyeball, and forming the suspensory ligament.

hyaloophitic [PETR] Of the texture of igneous rocks, being composed principally of a glassy ground mass with little interstitial texture.

hyalophane [MINERAL] $BaAl_2Si_2O_8$ A colorless feldspar mineral crystallizing in the monoclinic system; isomorphous with adularia. Also known as baryta feldspar.

hyalophobia [PSYCH] An abnormal fear of glass.

Hyalophragmiae [MYCOL] A subdivision of the spore group Phragmosporae characterized by hyaline spores.

hyaloplasm [CYTOL] The optically clear, viscous to gelatinous ground substance of cytoplasm in which formed bodies are suspended.

hyalopsite *See* obsidian.

Hyaloscolecosporae [MYCOL] A subdivision of the spore group Scalecosporae characterized by hyaline spores.

Hyalospongia [PALEON] A class of extinct glass sponges, equivalent to the living Hexactinellida, having siliceous spicules made of opaline silica.

Hyalosporae [MYCOL] A subdivision of the spore group Amerosporae characterized by hyaline spores.

Hyalostaurosporae [MYCOL] A subdivision of the spore group Staurosporae characterized by hyaline spores.

hyalotekite [MINERAL] (Pb,Ca,Ba)$_4$BSi$_6$O$_{17}$(OH,F) A white gray mineral composed of borosilicate and fluoride of lead, barium, and calcium, occurring in crystalline masses.

hyaluronate [BIOCHEM] A salt or ester of hyaluronic acid.

hyaluronate lyase See hyaluronidase.

hyaluronic acid [BIOCHEM] A polysaccharide found as an integral part of the gellike substance of animal connective tissue.

hyaluronidase [BIOCHEM] Any one of a family of enzymes which catalyze the breakdown of hyaluronic acid. Also known as hyaluronate lyase; spreading factor.

Hybinette process [MET] A process used for refining of crude nickel anodes; anodes are placed in asphalt-lined, reinforced concrete tanks and dissolved electrochemically so that impurities such as copper and iron pass into solution while pure nickel electrolyte is continuously added.

Hybodontoidea [PALEON] An ancient suborder of extinct fossil sharks in the order Selachii.

hybrid [GEN] The offspring of genetically dissimilar parents. [PETR] Pertaining to a rock formed by the assimilation of two magmas.

hybrid balance [ELEC] Loss between two conjugate sides of a hybrid set less the same loss when one of the other sides is open or shorted.

hybrid circuit [ELEC] A circuit in which two or more basically different types of components, such as tubes and transistors, performing similar functions are used together.

hybrid coil See hybrid transformer.

hybrid computer [ADP] A computer designed to handle both analog and digital data. Also known as analog-digital computer.

hybrid electromagnetic wave [ELECTROMAG] Wave which has both transverse and longitudinal components of displacement.

hybrid enzyme [BIOCHEM] A form of polymeric enzyme occurring in heterozygous individuals that shows a hybrid molecular form made up of subunits differing in one or more amino acids.

hybrid hardware control [ADP] The control of and communication between the various parts of a hybrid computer.

hybrid input/output [ADP] The routines required to handle inputs to and outputs from a computer system comprising digital and analog computers.

hybrid integrated circuit [ELECTR] A circuit in which one or more discrete components are used in combination with integrated-circuit construction.

hybridization [GEN] The act or process of producing hybrids.

hybridized orbital [PHYS CHEM] A molecular orbital which is a linear combination of two or more orbitals of comparable energy (such as $2s$ and $2p$ orbitals), is concentrated along a certain direction in space, and participates in formation of a directed valence bond.

hybrid junction [ELECTR] A transformer, resistor, or waveguide circuit or device that has four pairs of terminals so arranged that a signal entering at one terminal pair divides and emerges from the two adjacent terminal pairs, but is unable to reach the opposite terminal pair. Also known as bridge hybrid.

hybrid merogony [EMBRYO] The fertilization of cytoplasmic fragments of the egg of one species by the sperm of a related species.

hybrid microcircuit [ELECTR] Microcircuit in which thinfilm, thick-film, or diffusion techniques are combined with separately attached semiconductor chips to form the circuit.

hybrid molecule [BIOCHEM] A single molecule, usually protein, peculiar to heterozygotes and containing two structurally different polypeptide chains determined by two different alleles.

hybrid network [COMMUN] Nonhomogeneous communications network required to operate with signals of dissimilar characteristics (such as analog and digital modes).

hybrid parameter See h parameter.

hybrid problem analysis [ADP] The determination of the parts of a problem best suited for the digital computer.

hybrid programming [ADP] Hybrid system routines that handle timing, function generation, and simulation.

hybrid propellant [MATER] A propellant using a combination of liquid and solid materials to provide propulsion energy and working fluid.

hybrid propulsion [AERO ENG] Propulsion utilizing energy released by a liquid propellant with a solid propellant in the same rocket engine.

hybrid repeater See hybrid transformer.

hybrid rocket [AERO ENG] A rocket with an engine utilizing a liquid propellant with a solid propellant in the same rocket engine.

hybrid set [ELEC] Two or more transformers interconnected to form a hybrid junction. Also known as transformer hybrid.

hybrid simulation [ADP] The use of a hybrid computer for purposes of simulation.

hybrid sterility [GEN] Inability to form functional gametes in a hybrid due to disturbances in sex-cell development or in meiosis, caused by incompatible genetic constitution.

hybrid system checkout [ADP] The static check of a hybrid system and of the digital program and analog wiring required to solve a problem.

hybrid tee [ELECTROMAG] A microwave hybrid junction composed of an E-H tee with internal matching elements; it is reflectionless for a wave propagating into the junction from any arm when the other three arms are match-terminated. Also known as magic tee.

hybrid thin-film circuit [ELECTR] Microcircuit formed by attaching discrete components and semiconductor devices to networks of passive components and conductors that have been vacuum-deposited on glazed ceramic, sapphire, or glass substrates.

hybrid transformer [ELEC] A single transformer that performs the essential functions of a hybrid set. Also known as bridge transformer; hybrid coil; hybrid repeater.

hybrid vigor See heterosis.

hybrid wave function [QUANT MECH] A linear combination of wave functions of one problem used as an approximation to the wave function in another problem; for example, a linear combination of atomic orbitals used to represent a molecular bond.

hydantoin [ORG CHEM] C$_3$N$_2$O$_2$H A white, crystalline compound, melting point 220°C; used as an intermediate in certain pharmaceutical manufacturing and as a textile softener and lubricant. Also known as glycolyurea.

hydathode [BOT] An opening of the epidermis of higher plants specialized for exudation of water.

hydatid [MED] 1. A cyst formed in tissues due to growth of the larval stage of *Echinococcus granulosus.* 2. A cystic remnant of an embryonal structure.

hydatid disease See echinococcosis.

hydatid of Morgagni See appendix testis.

hydatidosis See echinococcosis.

hydatiform mole [MED] A benign placental tumor formed as a cystic growth of the chorionic villi. Also known as hydatiform tumor.

hydatiform tumor See hydatiform mole.

hydatogenesis [GEOL] Crystallization and deposition of minerals from aqueous solutions.

hydatosis [MED] Multiple hydatid cysts.

hydnocarpic acid [ORG CHEM] C$_{16}$H$_{28}$O$_2$ A nonedible fat and oil isolated from chaulmoogra oil, forming white crystals that melt at 60°C; used to treat Hansen's disease. Also known as cyclopentenylundecylic acid.

hydra [INV ZOO] Any species of *Hydra* or related genera, consisting of a simple, tubular body with a mouth at one end surrounded by tentacles, and a foot at the other end for attachment.

Hydra [ASTRON] A large constellation of the Southern Hemisphere, right ascension 10 hours, declination 20° south. Also known as Water Monster. [INV ZOO] A common genus of coelenterates in the suborder Anthomedusae.

Hydrachnellae [INV ZOO] A family of generally fresh-water predacious mites in the suborder Trombidiformes, including some parasitic forms.

hydracrylic acid [ORG CHEM] CH$_2$OH·CH$_2$COOH An oily liquid that is an isomer of lactic acid and that breaks down on heating to acrylic acid. Also known as 3-hydroxypropanoic acid; β-hydroxypropionic acid.

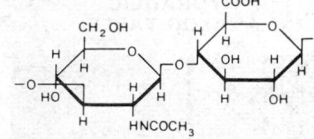

HYALURONIC ACID

Repeating unit in hyaluronic acid molecule.

HYBRID COMPUTER

A hybrid computer console with the special feature of double patch bays: one bay for the analog circuits and one bay for the digital logic circuits. *(Applied Dynamics Computer Systems, Division of Reliance Electric Co.)*

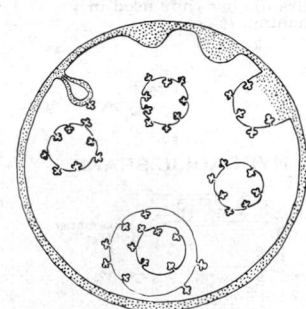

HYDATID

A drawing of the hydatid cyst showing large numbers of attached and free-floating brood capsules, each with one or more scoleces.

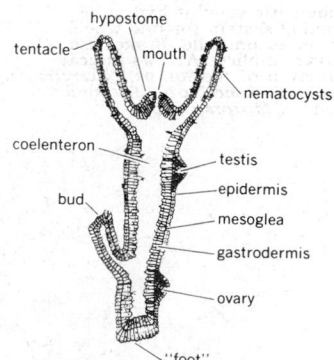

HYDRA

Longitudinal section of *Hydra.* *(From T. I. Storer and R. L. Usinger, General Zoology, 4th ed., McGraw-Hill, 1965)*

HYDRAULIC ANALOG TABLE

Shadowgraph of airfoil in hydraulic analog table.

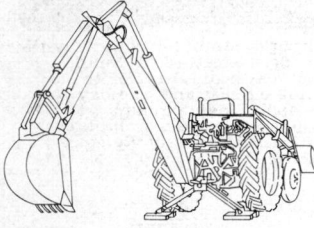

HYDRAULIC BACKHOE

Hydraulic backhoe used in trenching. *(Case)*

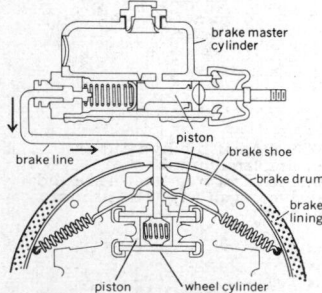

HYDRAULIC BRAKE

brake master cylinder

piston

brake line

brake shoe

brake drum

brake lining

piston   wheel cylinder

Schematic view of hydraulic braking system for one wheel on an automobile. Brakes are shown applied. Arrows indicate direction of motion of hydraulic fluid. *(Pontiac Motor Division, General Motors Corp.)*

Hydraenidae [INV ZOO] The equivalent name for Limnebiidae.

hydragogue [MED] Causing the discharge of watery fluid, especially from the bowel.

hydralazine [PHARM] $C_8H_8N_4$ An antihypertensive drug; used as the hydrochloride salt.

Hydra-Matic [MECH ENG] A trade name for an automatic transmission that made use of a fluid coupling rather than a torque converter; two separate planetary systems gave the desired gear reductions.

hydramnios *See* polyhydramnios.

hydranencephaly [MED] A congenital anomaly in which there are vestiges of a cerebellum, occipital lobes, and basal nuclei; frontal and parietal lobes are replaced by a cyst; and the neurocranium is undeveloped.

hydrant *See* fire hydrant.

hydranth [INV ZOO] Nutritive individual in a hydroid colony.

hydrargillite *See* gibbsite.

hydrargyrism *See* mercurialism.

hydrargyrum *See* mercury.

hydrarthrosis [MED] An accumulation of fluid in a joint.

hydrase [BIOCHEM] An enzyme that catalyzes removal or addition of water to a substrate without hydrolyzing it.

hydrastine [ORG CHEM] $C_{21}H_{21}NO_6$ An alkaloid isolated from species of the family Ranunculaceae and from *Hydrastis canadensis;* orthorhombic prisms crystallize from alcohol solution, melting point 132°C; highly soluble in acetone and benzene, soluble in chloroform, less soluble in ether and alcohol.

hydrastinine [ORG CHEM] $C_{11}H_{13}O_3N$ A compound formed by the decomposition of hydrastine; crystallizes as needles from petroleum-ether solution, soluble in organic solvents such as alcohol, chloroform, and ether; used in medicine as a stimulant in coronary disease and as a hemostatic in uterine hemorrhage.

hydrate [CHEM] A form of a solid compound which has water in the form of H_2O molecules associated with it; for example, anhydrous copper sulfate is a white solid with the formula $CuSO_4$, but when crystallized from water a blue crystalline solid with formula $CuSO_4 \cdot 5H_2O$ results, and the water molecules are an integral part of the crystal.

hydrate aluminum oxide *See* alumina trihydrate.

hydrated alumina *See* alumina trihydrate.

hydrated cellulose *See* hydrocellulose.

hydrated chloral *See* chloral hydrate.

hydrated halloysite *See* endellite.

hydrated lime *See* calcium hydroxide.

hydrated manganic hydroxide *See* manganic hydroxide.

hydrated mercurous nitrate [INORG CHEM] $Hg_2(NO_3)_2 \cdot 2H_2O$ Poisonous, light-sensitive crystals, soluble in warm water, decomposes at 70°C; used as an analytical reagent and in cosmetics and medicine.

hydrated silica *See* silicic acid.

hydrate inhibitor [CHEM] A material (such as alcohol or glycol) added to a gas stream to prevent the formation and freezing of gas hydrates in low-temperature systems.

hydration [CHEM] The incorporation of molecular water into a complex molecule with the molecules or units of another species; the complex may be held together by relatively weak forces or may exist as a definite compound.

hydraucone [DES ENG] A conical, spreading type of draft tube used on hydraulic turbine installations.

hydraulic [ENG] Operated or effected by the action of water or other fluid of low viscosity.

hydraulic accumulator [MECH ENG] A hydraulic flywheel that stores potential energy by accumulating a quantity of pressurized hydraulic fluid in a suitable enclosed vessel.

hydraulic actuator [MECH ENG] A cylinder or fluid motor that converts hydraulic power into useful mechanical work; mechanical motion produced may be linear, rotary, or oscillatory.

hydraulic air compressor [MECH ENG] A device in which water falling down a pipe entrains air which is released at the bottom under compression to do useful work.

hydraulic amplifier [CONT SYS] A device which increases the power of a signal in a hydraulic servomechanism or other system through the use of fixed and variable orifices. Also known as hydraulic intensifier.

hydraulic analog table [FL MECH] An experimental facility based on the hydraulic analogy; the water flows over a smooth horizontal surface and is bounded by vertical walls geometrically similar to the boundaries of the corresponding compressible gas flow; flow patterns are easily observed, and boundary changes may be made rapidly and inexpensively during exploratory studies.

hydraulic analogy [FL MECH] The analogy between the flow of a shallow liquid and the flow of a compressible gas; various phenomena such as shock waves occur in both systems; the analogy requires neglect of vertical accelerations in the liquid, and restrictions on the ratio of specific heats for the gas.

hydraulic backhoe [MECH ENG] A backhoe operated by a hydraulic mechanism.

hydraulic blasting [MIN ENG] Fracturing coal by means of a hydraulic cartridge.

hydraulic bottom-hole pump [PETRO ENG] Liquid power-operated oil production pump; the liquid is oil, piped under pressure to the bottom of the well to operate the engine that drives the pump.

hydraulic brake [MECH ENG] A brake in which the retarding force is applied through the action of a hydraulic press.

hydraulic cartridge [MIN ENG] A device used in mining to split coal or rock and having 8–12 small hydraulic rams in the sides of a steel cylinder.

hydraulic cement [MATER] Cement that hardens underwater.

hydraulic chock [MIN ENG] A steel face-support structure consisting of one to four hydraulic legs mounted in a steel frame with a large head and base plate.

hydraulic chuck *See* automatic chuck.

hydraulic circuit [MECH ENG] A circuit whose operation is analogous to that of an electric circuit except that electric currents are replaced by currents of water or other fluids, as in a hydraulic control.

hydraulic circulating system [MIN ENG] A method used to drill a borehole wherein water or a mud-laden liquid is circulated through the drill string.

hydraulic classification [ENG] Classification of particles in a tank by specific gravity, utilizing the action of rising water currents.

hydraulic classifier [MECH ENG] A classifier in which particles are sorted by specific gravity in a stream of hydraulic water that rises at a controlled rate; heavier particles gravitate down and are discharged at the bottom, while lighter ones are carried up and out. Also known as hydrosizer.

hydraulic clutch *See* fluid drive.

hydraulic computer [ADP] A computer in which electric current and gates are replaced by fluid and valves.

hydraulic conductivity *See* permeability coefficient.

hydraulic conveyor [MECH ENG] A system for handling material, such as ash from a coal-fired furnace; refuse is flushed from a hopper or slag tank to a grinder which discharges to a pump for conveying to a disposal area or a dewatering bin.

hydraulic coupling *See* fluid coupling.

hydraulic current [OCEANOGR] A current in a channel, due to a difference in the water level at the two ends.

hydraulic cylinder [MECH ENG] The cylindrical chamber of a positive displacement pump.

hydraulic discharge [HYD] The direct discharge of groundwater from the zone of saturation upon the land or into a body of surface water.

hydraulic dredge [MECH ENG] A dredge consisting of a large suction pipe which is mounted on a hull and supported and moved about by a boom, a mechanical agitator or cutter head which churns up earth in front of the pipe, and centrifugal pumps mounted on a dredge which suck up water and loose solids.

hydraulic drill [MECH ENG] A rotary drill powered by hydrodynamic means and used to make shot-firing holes in coal or rock, or to make a well hole.

hydraulic drive [MECH ENG] A mechanism transmitting motion from one shaft to another, the velocity ratio of the shafts being controlled by hydrostatic or hydrodynamic means.

hydraulic ejector [ENG] A pipe for removing excavated material from a pneumatic caisson.

hydraulic elevator [MECH ENG] An elevator operated by water pressure. Also known as hydraulic lift.

hydraulic engineering [CIV ENG] A branch of civil engineering concerned with the design, erection, and construction of sewage disposal plants, waterworks, dams, water-operated power plants, and such.

hydraulic excavation See hydraulicking.

hydraulic excavator digger [MECH ENG] An excavation machine which employs hydraulic pistons to actuate mechanical digging elements.

hydraulic extraction See hydraulicking.

hydraulic filling [MIN ENG] The use of water to wash waste material into stopes in order to prevent failure of rock walls and subsidence.

hydraulic fluid [MATER] A low-viscosity fluid used in operating a hydraulic mechanism.

hydraulic flume transport [MIN ENG] The transport of coal, pulp, or minerals in water flowing in semicircular or rectangular channels.

hydraulic fracturing [PETRO ENG] A method in which sand-water mixtures are forced into underground wells under pressure; the pressure splits the petroleum-bearing sandstone, thereby allowing the oil to move toward the wells more freely.

hydraulic friction [FL MECH] Resistance to flow which is exerted on the surface of contact between a stream and its conduit and which induces a loss of energy.

hydraulic giant See hydraulic monitor.

hydraulic grade line [FL MECH] **1.** In a closed channel, a line joining the elevations that water would reach under atmospheric pressure. **2.** The free water surface in an open channel.

hydraulic gradient [FL MECH] With regard to an aquifer, the rate of change of pressure head per unit of distance of flow at a given point and in a given direction. [HYD] The slope of the hydraulic grade line of a stream.

hydraulic intensifier See hydraulic amplifier.

hydraulic jack [MECH ENG] A jack in which force is applied through the mechanism of a hydraulic press.

hydraulic jetting [ENG] Use of high-pressure water forced through nozzles to clean tube interiors and exteriors in heat exchangers and boilers.

hydraulic jump [FL MECH] A steady-state, finite-amplitude disturbance in a channel, in which water passes turbulently from a region of (uniform) low depth and high velocity to a region of (uniform) high depth and low velocity; when applied to hydraulic jumps, the usual hydraulic formulas governing the relations of velocity and depth do not conserve energy.

hydraulicking [MIN ENG] Excavating alluvial or other mineral deposits by means of high-pressure water jets. Also known as hydraulic excavation; hydraulic extraction; hydro-extraction.

hydraulic lift See hydraulic elevator.

hydraulic lime [MATER] A hydraulic cementitious product that is produced by burning of hydraulic limestone.

hydraulic limestone [MATER] Limestone, containing silica and alumina, which produces lime that hardens in water.

hydraulic loading [MIN ENG] The flushing of coal or other material broken down by jets of water along the mine floor and into flumes.

hydraulic locomotive [MIN ENG] A diesel locomotive in which traction wheels are driven by hydraulic motors powered by a hydraulic system on the unit; used in mine haulage.

hydraulic loss [FL MECH] The loss in fluid power due to flow friction within the system.

hydraulic machine [MECH ENG] A machine powered by a motor activated by the confined flow of a stream of liquid, such as oil or water under pressure.

hydraulic mine [MIN ENG] A placer mine worked by means of a water stream directed against a bank.

hydraulic monitor [MIN ENG] A device for directing a high-pressure jet of water in hydraulicking; essentially, a swivel-mounted, counterweighted nozzle attached to a tripod or other type of stand and so designed that one worker can easily control and direct the vertical and lateral movements of the nozzle. Also known as giant; hydraulic giant; monitor.

hydraulic motor [MECH ENG] A motor activated by water or other liquid under pressure.

hydraulic nozzle [MECH ENG] An atomizing device in which fluid pressure is converted into fluid velocity.

hydraulic packer holddown [PETRO ENG] A pressure-actuated anchor located below a production packer (the seal between tubing and casing) to prevent well pressure from forcing the packer upward in the casing.

hydraulic packing [ENG] Packing material that resists the effects of water even under high pressure.

hydraulic power oil [MATER] Well production oil from which corrosive and abrasive impurities have been removed; used to power downhole hydraulic pumps.

hydraulic power system [MECH ENG] A power transmission system comprising machinery and auxiliary components which function to generate, transmit, control, and utilize hydraulic energy.

hydraulic press [MECH ENG] A combination of a large and a small cylinder connected by a pipe and filled with a fluid so that the fluid pressure created by a small force acting on the small-cylinder piston will result in a large force on the large piston. Also known as hydrostatic press.

hydraulic profile [HYD] A vertical section of an aquifer's potentiometric surface.

hydraulic prop [MIN ENG] A supporting device consisting of two telescoping steel cylinders extended by hydraulic pressure provided by a built-in hand pump.

hydraulic pump See hydraulic ram.

hydraulic radius [FL MECH] The ratio of the cross-sectional area of a conduit in which a fluid is flowing to the inner perimeter of the conduit.

hydraulic ram [MECH ENG] A device for forcing running water to a higher level by using the kinetic energy of flow; the flow of water in the supply pipeline is periodically stopped so that a small portion of water is lifted by the velocity head of a larger portion. Also known as hydraulic pump.

hydraulic ratio [GEOL] The weight of a heavy mineral multiplied by 100 and divided by the weight of a hydraulically equivalent light mineral.

hydraulic rope-geared elevator [MECH ENG] An elevator hoisted by a system of ropes and sheaves attached to a piston in a hydraulic cylinder.

hydraulics [FL MECH] The branch of science and technology concerned with the mechanics of fluids, especially liquids.

hydraulic scale [MECH ENG] An industrial scale in which the load applied to the load-cell piston is converted to hydraulic pressure.

hydraulic separation [MECH ENG] Mechanical classification using a hydraulic classifier.

hydraulic shovel [MECH ENG] A revolving shovel in which hydraulic rams or motors are substituted for drums and cables.

hydraulic sprayer [MECH ENG] A machine that sprays large quantities of insecticide or fungicide on crops.

hydraulic stacker [MECH ENG] A tiering machine whose carriage is raised or lowered by a hydraulic cylinder.

hydraulic swivel head [MECH ENG] In a drill machine, a swivel head equipped with hydraulically actuated cylinders and pistons to exert pressure on and move the drill rod string longitudinally.

hydraulic telemetry [COMMUN] A form of mechanical telemetry in which signals are transmitted in the form of sound or other waves through water or some other liquid.

hydraulic transport [ENG] Movement of material by water.

hydraulic turbine [MECH ENG] A machine which converts the energy of an elevated water supply into mechanical energy of a rotating shaft.

hydrazide [INORG CHEM] An acyl hydrazine; a compound of the formula $R-\overset{O}{\overset{\|}{C}}-NH-NH_2$ where R may be an alkyl group.

hydrazine [INORG CHEM] H_2NNH_2 A colorless, hygroscopic liquid, boiling point 114°C, with an ammonialike

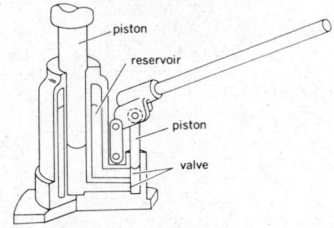

HYDRAULIC JACK

Hydraulic jack showing working parts.

HYDRAULIC NOZZLE

A simple jet, a type of hydraulic nozzle.

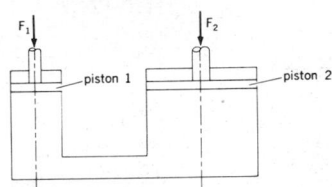

HYDRAULIC PRESS

Principle of hydraulic press. F_1 and F_2 are forces applied to pistons 1 and 2.

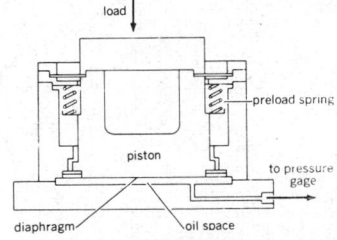

HYDRAULIC SCALE

Hydraulic scale, in which load cell is used. (*From D. M. Considine, ed., Process Instruments and Controls Handbook, McGraw-Hill, 1957*)

odor; it is reducing, decomposable, basic, and bifunctional; used as a rocket fuel, in corrosion inhibition in boilers, and in the synthesis of biologically active materials, explosives, antioxidants, and photographic chemicals.

hydrazine hydrate [ORG CHEM] $H_2NNH_2OH_2O$ A colorless, fuming liquid that boils at 119.4°C; used as a component in jet fuels and as an intermediate in organic synthesis.

hydrazinobenzene *See* phenylhydrazine.

hydrazobenzene [ORG CHEM] $C_{12}H_{12}N_2$ A colorless, crystalline compound, melts at 132°C, slightly soluble in water, soluble in alcohol; used as an intermediate in the synthesis of benzidine. Also known as *N,N'*-diphenylhydrazine.

hydrazoic acid [INORG CHEM] NHN:N Explosive liquid, a strong protoplasmic poison boiling at 37°C. Also known as azoimide; diazoimide; hydronitric acid.

hydrazone [ORG CHEM] A compound containing the grouping $-NH\cdot N:C-$, and obtained from a condensation reaction involving hydrazines with aldehydes or ketones; has been used as an exotic fuel.

hydremia [MED] An excessive amount of water in the blood; disproportionate increase in plasma volume as compared with red blood cell volume.

hydric [ECOL] Characterized by or thriving in abundance of moisture.

hydride [INORG CHEM] A compound containing hydrogen and another element; examples are H_2S, which is a hydride although it may be properly called hydrogen sulfide, and lithium hydride, LiH.

hydride descaling [MET] Removing surface deposits of oxides from a metal by immersion in molten alkali that contains hydrides.

hydriodic acid [INORG CHEM] A yellow liquid that is a water solution of the gas hydrogen iodide; a solution of 59% hydrogen iodide produces a liquid that is constant-boiling; it is a strong acid used in organic synthesis and as a reagent in analytical chemistry.

hydriodic acid gas *See* hydrogen iodide.

hydroacoustics *See* underwater acoustics.

hydrobasaluminite [MINERAL] $Al_4(SO_4)(OH)_{10}\cdot 36H_2O$ Mineral composed of a hydrous sulfate and hydroxide of aluminum.

Hydrobatidae [VERT ZOO] The storm petrels, a family of oceanic birds in the order Procellariiformes.

hydrobenzoin [ORG CHEM] $C_{14}H_{14}O_2$ A colorless, crystalline compound formed by action of sodium amalgam on benzaldehyde, melts at 136°C, and is slightly soluble in water. Also known as benzyleneglycol; diphenyldihydroxyethane.

hydrobiotite [MINERAL] A light-green, trioctahedral clay mineral of mixed layers of biotite and vermiculite.

hydroboracite [MINERAL] $CaMgB_6O_{11}\cdot 6H_2O$ A white mineral composed of hydrous calcium magnesium borate, occurring in fibrous and foliated masses.

hydroboration [ORG CHEM] The process of producing organoboranes by the addition of a compound with a B-H bond to an unsaturated hydrocarbon; for example, the reaction of diborane ion with a carbonyl compound.

hydrobromic acid [INORG CHEM] HBr A solution of hydrogen bromide in water, usually 40%; a clear, colorless liquid; used in medicine, analytical chemistry, and synthesis of organic compounds.

hydrocalumite [MINERAL] $Ca_2Al(OH)_7\cdot 3H_2O$ A colorless to light-green mineral composed of a hydrous hydroxide of calcium and aluminum.

hydrocarbon [ORG CHEM] One of a very large group of chemical compounds composed only of carbon and hydrogen; the largest source of hydrocarbons is from petroleum crude oil.

hydrocarbon blending value [ENG] Octane number rating for a 20% blend of a hydrocarbon with a 60:40 mixture of isooctane:*n*-heptane, which has been recalculated for a hypothetical 100% concentration of the tested hydrocarbon.

hydrocarbon-mud log [PETRO ENG] Record of oil, gas, or cuttings released into mud during rock drilling; used to detect the presence of hydrocarbon-bearing strata.

hydrocarbon pore volume [PETRO ENG] The pore volume in a reservoir formation available to hydrocarbon intrusion.

hydrocarbon resins [ORG CHEM] Brittle or gummy materials prepared by the polymerization of several unsaturated constituents of coal tar, rosin, or petroleum; they are inexpensive and find uses in rubber and asphalt formulations and in coating and caulking compositions.

hydrocarbon stabilization [PETRO ENG] The stepwise pressure reduction of a well stream to allow the release of dissolved gases until the liquid is stable at storage-tank conditions.

hydrocaulus [INV ZOO] Branched, upright stem of a hydroid colony.

hydrocele [MED] Accumulation of fluid in the membranes surrounding the testis.

hydrocellulose [MATER] A gelatinous mass formed from the reaction of cellulose with water either by grinding cellulose and mixing with water, or by using strong salt solutions, acids, or alkalies; used in the manufacture of artifical fibers such as rayon, mercerized cotton, paper, and vulcanized fiber. Also known as hydrated cellulose.

hydrocephaly [MED] Increased volume of cerebrospinal fluid in the skull.

hydrocerussite [MINERAL] $Pb_3(OH)_2(CO_3)_2$ A colorless mineral composed of basic lead carbonate, occurring as crystals in thin hexagonal plates.

Hydrocharitaceae [BOT] The single family of the order Hydrocharitales, characterized by an inferior, compound ovary with laminar placentation.

Hydrocharitales [BOT] A monofamilial order of aquatic monocotyledonous plants in the subclass Alismatidae.

hydrochinone *See* hydroquinone.

hydrochloric acid [INORG CHEM] HCl A solution of hydrogen chloride gas in water; a poisonous, pungent liquid forming a constant-boiling mixture at 20% concentration in water; widely used as a reagent, in organic synthesis, in acidizing oil wells, ore reduction, food processing, and metal cleaning and pickling. Also known as muriatic acid.

hydrochloric ether *See* ethyl chloride.

hydrochlorthiazide [PHARM] $C_7H_8ClN_3O_4S_2$ An orally effective diuretic and antihypertensive drug.

hydrocholeresis [MED] Choleresis characterized by an increase of water output, or of a bile relatively low in specific gravity, viscosity, and content of total solids.

hydrochory [BIOL] Dispersal of disseminules by water.

hydrocinnamic acid [ORG CHEM] $C_6H_5CH_2CH_2COOH$ A compound whose crystals have a floral odor (hyacinth-rose) and melt at 46°C; used in perfumes and flavoring. Also known as 3-phenylpropionic acid.

hydrocladium [INV ZOO] Branchlet of a hydrocaulus.

hydrocoele [INV ZOO] 1. Water vascular system in Echinodermata. 2. Embryonic precursor of the system.

Hydrocol process [CHEM ENG] A catalytic process based on Fischer-Tropsch synthesis for production of gasoline and other hydrocarbon products from a mixture of hydrogen and carbon monoxide derived from natural gas, coal, or petroleum fractions.

hydrocooling [FOOD ENG] Removing heat from freshly harvested fruits and vegetables by immersion in ice water.

Hydrocorallina [INV ZOO] An order in some systems of classification set up to include the coelenterate groups Milleporina and Stylasterina.

Hydrocorisae [INV ZOO] A subdivision of the Hemiptera containing water bugs with concealed antennae and without a bulbus ejaculatorius in the male.

hydrocortisone [BIOCHEM] $C_{21}H_{30}O_5$ The generic name for 17-hydroxycorticosterone; an adrenocortical steroid occurring naturally and prepared synthetically; its effects are similar to cortisone, but it is more active. Also known as cortisol.

hydrocrackate [ORG CHEM] The product of a hydrocracker.

hydrocracker [CHEM ENG] A high-pressure processing unit that cracks long hydrocarbon molecules under a high-hydrogen-content atmosphere.

hydrocracking [CHEM ENG] A catalytic, high-pressure petroleum refinery process that is flexible enough to produce either high-octane gasoline or aviation jet fuel; the two main reactions are the adding of hydrogen to petroleum-derived molecules too massive and complex for gasoline and then the cracking of them to the required fuels; the catalyst is an acidic solid and a hydrogenating metal component.

hydrocyanic acid [INORG CHEM] HCN A highly toxic liquid that has the odor of bitter almonds and boils at 25.6°C; used to manufacture cyanide salts, acrylonitrile, and dyes, and as a fumigant in agriculture. Also known as formonitrile; hydrogen cyanide; prussic acid.

hydrocyanic ether See ethyl cyanide.

hydrocyanite See chalcocyanite.

hydrocyclone [MECH ENG] A cyclone separator in which granular solids are removed from a stream of water and classified by centrifugal force.

hydrocystoma [MED] A group of clear vesicles, usually located around the eyes, composed of cystic sweat glands.

Hydrodamalinae [VERT ZOO] A monogeneric subfamily of sirenian mammals in the family Dugongidae.

hydrodesulfurization [CHEM ENG] A catalytic process in which the petroleum feedstock is reacted with hydrogen to reduce the sulfur content in the oil.

hydrodynamic equations [FL MECH] Three equations which express the net acceleration of a unit water particle as the sum of the partial accelerations due to pressure gradient force, frictional force, earth's deflecting force, gravitational force, and other factors.

hydrodynamic oscillator [ENG ACOUS] A transducer for generating sound waves in fluids, in which a continuous flow through an orifice is modulated by a reciprocating valve system controlled by acoustic feedback.

hydrodynamic pressure [FL MECH] The difference between the pressure of a fluid and the hydrostatic pressure; this concept is useful chiefly in problems of the steady flow of an incompressible fluid in which the hydrostatic pressure is constant for a given elevation (as when the fluid is bounded above by a rigid plate), so that the external force field (gravity) may be eliminated from the problem.

hydrodynamics [FL MECH] The study of the motion of a fluid and of the interactions of the fluid with its boundaries, especially in the incompressible inviscid case.

hydroecium [INV ZOO] The closed, funnel-shaped tube at the upper end of coelenterates belonging to the Siphonophora.

hydroelasticity [FL MECH] **1.** Theory of elasticity of a fluid. **2.** The interaction between the flow of water or other liquid and the elastic behavior of a body immersed in it.

hydroelectric generator [MECH ENG] An electric rotating machine that transforms mechanical power from a hydraulic turbine or water wheel into electric power.

hydroelectricity [ELEC] Electric power produced by hydroelectric generators.

hydroextraction See hydraulicking.

hydrofining [CHEM ENG] A fixed-bed catalytic process to desulfurize and hydrogenate a wide range of charge stocks, from gases through waxes; the catalyst comprises cobalt oxide and molybdenum oxide on an extruded alumina support and may be regenerated in place by air and steam or flue gas.

Hydroflo [MATER] Trade name for a free-flowing, trinitrotoluene-base explosive.

hydrofluoric acid [INORG CHEM] An aqueous solution of hydrogen fluoride, HF; colorless, fuming, poisonous liquid; extremely corrosive, it is a weak acid as compared to hydrochloric acid, but will attack glass and other silica materials; used to polish, frost, and etch glass, to pickle copper, brass, and alloy steels, to clean stone and brick, to acidize oil wells, and to dissolve ores.

hydrofluorosilicic acid See fluorosilicic acid.

hydrofoil [NAV ARCH] **1.** A flat or airfoil-shaped plate attached to a ship to stabilize it in roll. **2.** Foils attached by struts to the bottom of a hydrofoil boat to lift the hull out of the water as its speed is increased.

hydrofoil boat [NAV ARCH] A marine craft of generally conventional hull form which is supported above the water, when the craft is traveling at high speed, by dynamic lift produced by winglike surfaces (hydrofoils) below the hull.

hydroforming [CHEM ENG] A petroleum-refinery process in which naphthas are passed over a catalyst at elevated temperatures and moderate pressures in the presence of added hydrogen or hydrogen-containing gases, to form high-octane BTX aromatics for motor fuels or chemical manufacture.

hydroformylation [CHEM ENG] The reaction of adding hydrogen and the −CHO group to the carbon atoms across a double bond to yield oxygenated derivatives; an example is in the oxo process where the term hydroformylation applies to those reactions brought about by treating olefins with a mixture of hydrogen and carbon monoxide in the presence of a cobalt catalyst.

hydrofuge [ZOO] Of a structure, shedding water, as the hair on certain animals.

hydrogarnets [MINERAL] A group of minerals having the general formula $A''_3B''_2(SiO_4)_{3-x}(OH)_{4x}$; isomorphous with certain garnets.

hydrogasification [CHEM ENG] A technique to manufacture synthetic pipeline gas from coal; pulverized coal is reacted with hot, raw, hydrogen-rich gas containing a substantial amount of steam at 1000 psig (6.9×10^6 newtons per square meter, gage) to form methane.

hydrogel [CHEM] The formation of a colloid in which the disperse phase (colloid) has combined with the continuous phase (water) to produce a viscous jellylike product; for example, coagulated silicic acid.

hydrogen [CHEM] The first chemical element, symbol H, in the periodic table, atomic number 1, atomic weight 1.00797; under ordinary conditions it is a colorless, odorless, tasteless gas composed of diatomic molecules, H_2; used in manufacture of ammonia and methanol, for hydrofining, for desulfurization of petroleum products, and to reduce metallic oxide ores.

hydrogen 21-centimeter line [SPECT] A spectral line of microwave radiations with a wavelength of 21 centimeters, emitted by neutral hydrogen during a hyperfine transition.

hydrogenase [BIOCHEM] Enzyme that catalyzes the oxidation of hydrogen.

hydrogenated oil [ORG CHEM] Unsaturated liquid vegetable oil that has had hydrogen catalytically added so as to convert the oil to a hydrogen-saturated solid.

hydrogenation [CHEM ENG] Saturation of diolefin impurities in gasolines to form a stable product. [ORG CHEM] Catalytic reaction of hydrogen with other compounds, usually unsaturated; for example, unsaturated cottonseed oil is hydrogenated to form solid fats.

hydrogen bacteria [MICROBIO] Bacteria capable of obtaining energy from the oxidation of molecular hydrogen.

hydrogen blistering [MET] Cracks or blisters caused when atomic hydrogen penetrates steel via submicroscopic discontinuities or voids and becomes molecular hydrogen and develops internal pressures.

hydrogen bomb [ORD] A device in which heavy hydrogen nuclei, under intense heat and pressure, undergo an uncontrolled, self-sustaining fusion reaction to produce an explosion. Also known as H bomb.

hydrogen bond [PHYS CHEM] A type of bond formed when a hydrogen atom bonded to atom A in one molecule makes an additional bond to atom B either in the same or another molecule; the strongest hydrogen bonds are formed when A and B are highly electronegative atoms, such as fluorine, oxygen, or nitrogen.

hydrogen brazing [MET] Brazing in an atmosphere rich in hydrogen.

hydrogen bromide [INORG CHEM] HBr A hazardous, toxic gas used as a chemical intermediate and as an alkylation catalyst; forms hydrobromic acid in aqueous solution.

hydrogen burning [ASTROPHYS] Thermonuclear reactions occurring in the cores of main-sequence stars, in which nuclei of hydrogen fuse to form helium nuclei.

hydrogen chloride [INORG CHEM] HCl A fuming, highly toxic, colorless gas soluble in water, alcohol, and ether; used in the production of vinyl chloride and alkyl chloride, and in polymerization, isomerization, and other reactions.

hydrogen cyanide See hydrocyanic acid.

hydrogen cyanide laser [OPTICS] A gas laser using hydrogen cyanide, which emits infrared radiation at wavelengths of 311 and 337 micrometers.

hydrogen damage [MET] Corrosion, common in boilers, caused by diffusion of hydrogen through steel reacting with carbon to form methane, which builds up local stresses at the

HYDROFOIL BOAT

Hydrofoil boat *Denison*, with surface-piercing foils, designed by Grumman Aircraft Engineering Corp.

HYDROGEN

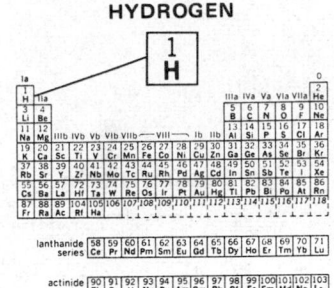

Periodic table of the chemical elements showing the position of hydrogen.

interfaces between grains, forming voids that ultimately produce failure.

hydrogen-discharge lamp [ELECTR] A discharge lamp containing hydrogen and used as a source of ultraviolet radiation.

hydrogen disulfide *See* hydrogen sulfide.

hydrogen electrode [PHYS CHEM] A noble metal (such as platinum) of large surface area covered with hydrogen gas in a solution of hydrogen ion saturated with hydrogen gas; metal is used in a foil form and is welded to a wire sealed in the bottom of a hollow glass tube, which is partially filled with mercury; used as a standard electrode with a potential of zero to measure hydrogen ion activity.

hydrogen embrittlement *See* acid brittleness.

hydrogen equivalent [CHEM] The number of replaceable hydrogen atoms or hydroxyl groups in a molecule of an acid or a base.

hydrogen fluoride [INORG CHEM] HF The hydride of fluoride; anhydrous HF is a mobile, colorless, liquid that fumes in air, melts at $-83°C$, boils at $19.8°C$; used to make fluorine-containing refrigerants (such as Freon) and organic fluorocarbon compounds, as a catalyst in alkylate gasoline manufacture, as a fluorinating agent, and in preparation of hydrofluoric acid.

hydrogenic rock *See* aqueous rock.

hydrogen iodide [INORG CHEM] HI A water-soluble, colorless gas that may be used in organic synthesis and as a reagent. Also known as hydriodic acid gas.

hydrogen ion *See* hydronium ion.

hydrogen ion concentration [CHEM] The normality of a solution with respect to hydrogen ions, H^+; it is related to acidity measurements in most cases by $pH = \log 1/2 [1/(H^+)]$, where (H^+) is the hydrogen ion concentration in gram equivalents per liter of solution.

hydrogen ion exponent [CHEM] A way of expressing pH; namely, $pH = -\log c_H$, where $c_H =$ hydrogen ion concentration.

hydrogen line [SPECT] A spectral line emitted by neutral hydrogen having a frequency of 1420 megahertz and a wavelength of 21 centimeters; radiation from this line is used in radio astronomy to study the amount and velocity of hydrogen in the galaxy.

hydrogen loss [MET] Loss of weight by a compact or a metal powder when heated in a hydrogen atmosphere; used as a measure of oxygen content of the sample.

hydrogen maser [PHYS] A maser in which hydrogen gas is the basis for providing an output signal with a high degree of stability and spectral purity.

hydrogenolysis [CHEM] A reaction in which hydrogen gas causes a chemical change that is similar to the role of water in hydrolysis.

Hydrogenomonas [MICROBIO] A genus of gram-negative, rod-shaped bacteria in the family Methanomonadaceae characterized by the ability to grow on inorganic substrates and to derive energy from the oxidation of hydrogen.

hydrogenous [CHEM] Of, pertaining to, or containing hydrogen.

hydrogen overvoltage [MET] An overvoltage occurring at an electrode as a result of the liberation of hydrogen gas.

hydrogen oxide *See* water.

hydrogen peroxide [INORG CHEM] H_2O_2 Unstable, colorless, heavy liquid boiling at 158°C; soluble in water and alcohol; used as a bleach, chemical intermediate, rocket fuel, and antiseptic. Also known as peroxide.

hydrogen phosphide *See* phosphine.

hydrogen-reduced powder [MET] Metal powder produced by hydrogen-reduction of a metal, metallic compound, or surface-contaminated metal particles.

hydrogen selenide [INORG CHEM] H_2Se A toxic, colorless gas, soluble in water, carbon disulfide, and phosgene; used to make metallic selenides and organoselenium compounds and in the preparation of semiconductor materials.

hydrogen sulfide [INORG CHEM] H_2S Flammable, toxic, colorless gas with offensive odor, boiling at $-60°C$; soluble in water and alcohol; used as an analytical reagent, as a sulfur source, and for purification of hydrochloric and sulfuric acids. Also known as hydrogen disulfide.

hydrogen tellurate *See* telluric acid.

hydrogen thyratron [ELECTR] A thyratron containing hydrogen instead of mercury vapor to give freedom from effects of changes in ambient temperature; used in radar pulse circuits and stroboscopic photography.

hydrogeochemistry [GEOCHEM] The study of the chemical characteristics of ground and surface waters as related to areal and regional geology.

hydrogeology [HYD] The science dealing with the occurrence of surface and ground water, its utilization, and its functions in modifying the earth, primarily by erosion and deposition.

hydrograph [HYD] A graphical representation of stage, flow, velocity, or other characteristics of water at a given point as a function of time.

hydrographical theorem of Knudsen *See* Knudsen's hydrographical theorem.

hydrographic chart [MAP] A map designed from data obtained by hydrographic surveys for purposes of navigation.

hydrographic cruise [OCEANOGR] Exploration of a body of water for hydrographic surveys.

hydrographic sextant [ENG] A surveying sextant similar to those used for celestial navigation but smaller and lighter, constructed so that the maximum angle that can be read is slightly greater than that on the navigating sextant; usually the angles can be read only to the nearest minute by means of a vernier; it is fitted with a telescope with a large object glass and field of view. Also known as sounding sextant; surveying sextant.

hydrographic sonar [ENG] An echo sounder used in mapping ocean bottoms.

hydrographic survey [OCEANOGR] Survey of a water area with particular reference to tidal currents, submarine relief, and any adjacent land.

hydrographic table [OCEANOGR] Tabular arrangement of data relating sea-water density to salinity, temperature, and pressure.

hydrography [GEOGR] Science which deals with the measurement and description of the physical features of the oceans, lakes, rivers, and their adjoining coastal areas, with particular reference to their control and utilization. [NAV] Measurement of the tides and currents as an aid to navigation.

hydrohalite [MINERAL] $Na_2Cl \cdot 2H_2O$ A mineral composed of hydrated sodium chloride, formed only from salty water cooled below 0°C.

hydrohalloysite *See* endellite.

hydroid [INV ZOO] **1.** The polyp form of a hydrozoan coelenterate. Also known as hydroid polyp; hydropolyp. **2.** Any member of the Hydroida.

Hydroida [INV ZOO] An order of coelenterates in the class Hydrozoa including usually colonial forms with a well-developed polyp stage.

hydroid polyp *See* hydroid.

hydroiodic ether *See* ethyl iodide.

hydrokaolin *See* endellite.

hydrokinematics [FL MECH] The study of the motion of a liquid apart from the cause of motion.

hydrokinetics [FL MECH] The study of the forces produced by a liquid as a consequence of its motion.

hydrolaccolith [GEOL] A frost mound, 0.1–6 meters in height, having a core of ice and resembling a laccolith in section. Also known as cryolaccolith.

HYDROLANT [NAV] An urgent notice of dangers to navigation in the Atlantic Ocean, originated by the U.S. Naval Oceanographic Office and disseminated for the immediate safeguarding of shipping.

hydrolase [BIOCHEM] Any of a class of enzymes which catalyze the hydrolysis of proteins, nucleic acids, starch, fats, phosphate esters, and other macromolecular substances.

hydrolith [PETR] **1.** A chemically precipitated aqueous rock, such as rock salt. **2.** A rock that is free of organic material.

hydrologic accounting [HYD] A systematic summary of the terms (inflow, outflow, and storage) of the storage equation as applied to the computation of soil-moisture changes, ground-water changes, and so forth; an evaluation of the hydrologic balance of an area. Also known as basin accounting; water budget.

hydrologic cycle [HYD] The complete cycle through which

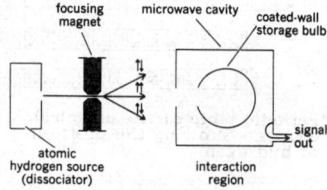

HYDROGEN MASER

focusing magnet
microwave cavity
coated-wall storage bulb
signal out
atomic hydrogen source (dissociator)
interaction region

Basic elements of the hydrogen maser. The hydrogen atoms in the upper hyperfine level are focused into the coated bulb, and the energy coupled out through a loop and coaxial cable.

water passes, from the oceans, through the atmosphere, to the land, and back to the ocean. Also known as water cycle.

hydrologist [HYD] An individual who specializes in hydrology.

hydrology [GEOPHYS] The science that treats the occurrence, circulation, distribution, and properties of the waters of the earth, and their reaction with the environment.

hydrolysis [CHEM] **1.** Decomposition or alteration of a chemical substance by water. **2.** In aqueous solutions of electrolytes, the reactions of cations with water to produce a weak base or of anions to produce a weak acid.

hydrolytic enzyme [BIOCHEM] A catalyst that acts like a hydrolase.

hydrolytic process [CHEM] A reaction of both organic and inorganic chemistry wherein water effects a double decomposition with another compound, hydrogen going to one compound and hydroxyl to another.

hydrolyzate [GEOL] A sediment characterized by elements such as aluminum, potassium, or sodium which are readily hydrolyzed.

hydromagnesite [MINERAL] $Mg_4(OH)_2(CO_3)_3 \cdot 3H_2O$ White, earthy mineral crystallizing in the monoclinic system and found in small crystals, amorphous masses, or chalky crusts.

hydromagnetic instability See magnetohydrodynamic instability.

hydromagnetics See magnetohydrodynamics.

hydromagnetic stability See magnetohydrodynamic stability.

hydromagnetic wave See magnetohydrodynamic wave.

hydromechanics [FL MECH] The study of liquids, traditionally water, as a medium for the transmission of forces.

hydrometallurgy [MET] Treatment of metals and metal-containing materials by wet processes.

hydrometamorphism [GEOL] Alteration of rocks by material carried in solution by water without the influence of high temperature or pressure.

hydrometeor [HYD] **1.** Any product of condensation or sublimation of atmospheric water vapor, whether formed in the free atmosphere or at the earth's surface. **2.** Any water particles blown by the wind from the earth's surface.

hydrometeorology [METEOROL] That part of meteorology of direct concern to hydrologic problems, particularly to flood control, hydroelectric power, irrigation, and similar fields of engineering and water resources.

hydrometer [ENG] A direct-reading instrument for indicating the density, specific gravity, or some similar characteristic of liquids.

Hydrometridae [INV ZOO] The marsh treaders, a family of hemipteran insects in the subdivision Amphibicorisae.

hydrometry [FL MECH] The science and technology of measuring specific gravities, particularly of liquids.

hydromica [GEOL] Any of several varieties of muscovite, especially illite, which are less elastic than mica, have a pearly luster, and sometimes contain less potash and more water than muscovite. Also known as hydrous mica.

hydronephrosis [MED] Accumulation of urine in and distension of the renal pelvis and calyces due to obstructed outflow.

hydronic heating See hot-water heating.

hydronic radiation [COMMUN] A form of electromagnetic radiation used for communication underwater.

hydronitric acid See hydrazoic acid.

hydronium ion [INORG CHEM] H_3O^+ A proton combined with a molecule of water; found in pure water and in all aqueous solutions. Also known as hydrogen ion; oxonium ion.

HYDROPAC [NAV] An urgent notice of dangers to navigation in the Pacific Ocean, originated by the U.S. Naval Oceanographic Office and disseminated for the immediate safeguarding of shipping and life at sea.

hydropathy [MED] The system of internal and external use of water in attempting to cure disease.

hydropericardium [MED] Accumulation of serous fluid in the pericardial sac.

Hydrophiidae [VERT ZOO] A family of proglyphodont snakes in the suborder Serpentes found in Indian-Pacific oceans.

hydrophilic [CHEM] Having an affinity for, attracting, adsorbing, or absorbing water.

Hydrophilidae [INV ZOO] The water scavenger beetles, a large family of coleopteran insects in the superfamily Hydrophiloidea.

Hydrophiloidea [INV ZOO] A superfamily of coleopteran insects in the suborder Polyphaga.

hydrophilous [ECOL] Inhabiting moist places.

hydrophobia [MED] See rabies. [PSYCH] An abnormal fear of water.

hydrophobic [CHEM] Lacking an affinity for, repelling, or failing to adsorb or absorb water. [MED] Of, pertaining to, or suffering from hydrophobia.

hydrophobophobia [PSYCH] An abnormal fear of hydrophobia.

hydrophone [ENG ACOUS] A device which receives underwater sound waves and converts them to electric waves.

hydrophone array [COMMUN] A group of two or more hydrophones which feed into a common receiver.

hydrophone noise [ELEC] Any unwanted disturbance in the electric waves delivered by a hydrophone.

hydrophone response [ELEC] The electric waves delivered by a hydrophone in response to waterborne sound waves.

hydrophotometer [OPTICS] An instrument for measuring the attenuation coefficient of collimated light in sea water, in which light from a collimated source is directed through a column of sea water and is measured by a photocell or other electronic device at the other end of the column.

Hydrophyllaceae [BOT] A family of dicotyledonous plants in the order Polemoniales distinguished by two carpels, parietal placentation, and generally imbricate corolla lobes in the bud.

hydrophyllium [INV ZOO] A transparent body partly covering the spore sacs of siphonophoran coelenterates.

hydrophyte [BOT] **1.** A plant that grows in a moist habitat. **2.** A plant requiring large amounts of water for growth. Also known as hygrophyte.

hydroplane [NAV ARCH] **1.** A boat which when operated at high speed planes on the surface of the water; the bottom of such a craft is normally a prismatic surface. **2.** See diving rudder.

hydroplanula [INV ZOO] A coelenterate larval stage between the planula and actinula stages.

hydropneumatic [ENG] Operated by both water and air power.

hydropneumatic recoil system [MECH ENG] A recoil mechanism that absorbs the energy of recoil by the forcing of oil through orifices and returns the gun to battery by compressed gas.

hydropolyp See hydroid.

hydroponics [BOT] Growing of plants in a nutrient solution with the mechanical support of an inert medium such as sand.

hydropore [INV ZOO] In certain asteroids and echinoids, an opening on the aboral surface of a canal which extends from the ring canal in one of the interradii.

hydroquinol See hydroquinone.

hydroquinone [ORG CHEM] $C_6H_4(OH)_2$ White crystals melting at 170°C and boiling at 285°C; soluble in alcohol, ether, and water; used in photographic dye chemicals, in medicine, as an antioxidant and inhibitor, and in paints, varnishes, and motor fuels and oils. Also known as hydrochinone; hydroquinol; quinol.

hydroquinone carboxylic acid See gentisic acid.

hydroquinone monopentyl ether See para-pentyloxyphenol.

hydrorhiza [INV ZOO] Rootlike structure of a hydroid colony.

hydrosalpinx [MED] A distension of a fallopian tube with fluid.

Hydroscaphidae [INV ZOO] The skiff beetles, a small family of coleopteran insects in the suborder Myxophaga.

hydroscope [OPTICS] An instrument designed to observe objects an appreciable distance below the surface of water, consisting of a series of mirrors enclosed in a steel tube.

hydroseparator [MECH ENG] A separator in which solids in suspension are agitated by hydraulic pressure or stirring devices.

hydrosere [ECOL] Community in which the pioneer plants invade open water, eventually forming some kind of soil such as peat or muck.

HYDROMETER

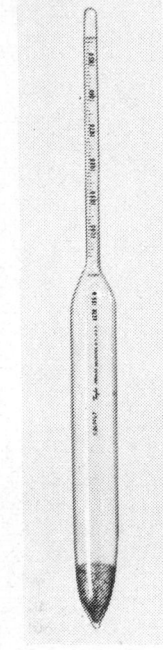

A plain hydrometer. *(Taylor Instrument Co.)*

HYDROPHILIDAE

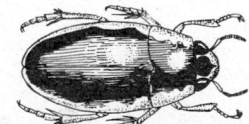

A drawing of a water scavenger beetle. *(From T. I. Storer and R. L. Usinger, General Zoology, 3d ed., McGraw-Hill, 1957)*

HYDROQUINONE

Structural formula of hydroquinone.

hydrosizer *See* hydraulic classifier.

hydroskeleton [INV ZOO] Water contained within the coelenteron and serving a skeletal function in most coelenterate polyps.

hydrosol [CHEM] A colloidal system in which the dispersion medium is water, and the dispersed phase may be a solid, a gas, or another liquid.

hydrospace detection [ORD] Detection of underwater targets, as with sonar.

hydrosphere [HYD] The water portion of the earth as distinguished from the solid part (lithosphere) and from the gaseous outer envelope (atmosphere).

hydrospire [INV ZOO] Either of a pair of flattened tubes composing part of the respiratory system in blastoids.

hydrospore [INV ZOO] Opening into the hydrocoele on the right side in echinoderm larvae.

hydrostat *See* humidistat.

hydrostatic approximation [METEOROL] The assumption that the atmosphere is in hydrostatic equilibrium.

hydrostatic assumption [GEOPHYS] **1.** The assumption that the pressure of seawater increases by 1 atmosphere over approximately 10 meters of depth, the exact value depending on the water density and the local acceleration of gravity. **2.** Specifically, the assumption that fluid is not undergoing vertical accelerations, hence the vertical component of the passive gradient force per unit mass is equal to g, the local acceleration due to gravity.

hydrostatic bearing [MECH ENG] A sleeve bearing in which high-pressure oil is pumped into the area between the shaft and the bearing so that the shaft is raised and supported by an oil film.

hydrostatic equation [PHYS] The form assumed by the vertical component of the vector equation of motion when all Coriolis force, earth curvature, frictional, and vertical acceleration terms are considered negligible compared with those involving the vertical pressure force and the force of gravity.

hydrostatic equilibrium [PHYS] The state of a fluid whose surfaces of constant mass (or density) coincide and are horizontal throughout; complete balance exists between the force of gravity and the pressure force; the relation between the pressure and the geometic height is given by the hydrostatic equation.

hydrostatic forging [MET] Forging a metal part by using pressure supplied by a liquid.

hydrostatic fuse [ORD] A fuse used with a depth bomb to cause an underwater explosion at a predetermined depth; initiation is caused by the pressure of the water as the depth bomb sinks.

hydrostatic modulus *See* bulk modulus of elasticity.

hydrostatic press *See* hydraulic press.

hydrostatic pressing [ENG] Compacting ceramic or metal powders by packing them in a rubber bag which is subjected to hydrostatic press in a cylinder.

hydrostatic pressure [FL MECH] **1.** The pressure at a point in a fluid at rest due to the weight of the fluid above it. Also known as gravitational pressure. **2.** The negative of the stress normal to a surface in a fluid.

hydrostatic roller conveyor [MECH ENG] A portion of a roller conveyor that has rolls weighted with liquid to control the speed of the moving objects.

hydrostatics [FL MECH] The study of liquids at rest and the forces exerted on them or by them.

hydrostatic stability *See* static stability.

hydrostatic strength [MECH] The ability of a body to withstand hydrostatic stress.

hydrostatic stress [MECH] The condition in which there are equal compressive stresses or equal tensile stresses in all directions, and no shear stresses on any plane.

hydrostatic test [ENG] Test of strength and leak-resistance of a vessel, pipe, or other hollow equipment by internal pressurization with a test liquid.

hydrostatic weighing [FL MECH] A method of determining the density of a sample in which the sample is weighed in air, and then weighed in a liquid of known density; the volume of the sample is equal to the loss of weight in the liquid divided by the density of the liquid.

hydrosulfide [CHEM] A compound that has the SH— radical; for example, sulfhydrates, sulfhydryls, thioalcohols, thiols, sulfur alcohols, and mercaptans.

hydrotalcite [MINERAL] $Mg_6Al_2(OH)_{16}(CO_3) \cdot 4H_2O$ Pearly-white mineral composed of hydrous aluminum and magnesium hydroxide and carbonate.

hydrotheca [INV ZOO] Cup-shaped portion of the perisarc in some coelenterates that serves to hold and protect a withdrawn hydranth.

hydrotherapy [MED] Treatment of disease by external application of water.

hydrothermal [GEOL] Of or pertaining to heated water, to its action, or to the products of such action.

hydrothermal alteration [GEOL] Rock or mineral phase changes that are caused by the interaction of hydrothermal liquids and wall rock.

hydrothermal crystal growth [CHEM ENG] Formation of simple crystals of quartz at elevated temperatures and pressures in an autoclave with an alkaline solution.

hydrothermal deposit [GEOL] A mineral deposit precipitated from a hot, aqueous solution.

hydrothermal synthesis [GEOL] Mineral synthesis in the presence of heated water.

hydrothorax [MED] Collection of serous fluid in the pleural spaces.

hydrotreating [CHEM ENG] Oil refinery catalytic process in which hydrogen is contacted with petroleum intermediate or product streams to remove impurities, such as oxygen, sulfur, nitrogen, or unsaturated hydrocarbons.

hydrotroilite [MINERAL] $FeS \cdot NH_2O$ A black, finely divided colloidal material reported in many muds and clays; thought to be formed by bacteria on bottoms of marine basins.

hydrotrope [CHEM] Compound with the ability to increase the solubilities of certain slightly soluble organic compounds.

hydrotropism [BIOL] Orientation involving growth or movement of a sessile organism or part, especially plant roots, in response to the presence of water.

hydrotungstite [MINERAL] $H_2WO_4 \cdot H_2O$ A mineral composed of hydrous tungstic acid.

hydroureter [MED] Accumulation of urine in and distension of the ureter due to obstructed outflow.

hydrous [CHEM] Indicating the presence of an indefinite amount of water. [MINERAL] Indicating a definite proportion of combined water.

hydrous mica *See* hydromica.

Hydrox [MIN ENG] An explosive device, used in some English coal mines, consisting of a steel tube that has a thin steel shearing disk and is filled with a charge of ammonium chloride and sodium nitrate; when the charge is ignited, the powder is gasified and the disk is sheared.

hydroxamic acid [ORG CHEM] An organic compound that contains the group $-C(=O)NHOH$.

hydroxide [CHEM] Compound containing the OH^- group; the hydroxides of metals are usually bases and those of nonmetals are usually acids; a hydroxide can be organic or inorganic.

hydroximino *See* nitroso.

hydroxy- [ORG CHEM] Chemical prefix indicating the OH^- group in an organic compound, such as hydroxybenzene for phenol, C_6H_5OH; the use of just oxy- for the prefix is incorrect. Also spelled hydroxyl-.

hydroxyacetic acid *See* glycolic acid.

hydroxy acid [ORG CHEM] Any organic acid, with an OH^- group, such as hydroxyacetic acid.

hydroxyamphetamine [PHARM] C_9H_3ON A sympathetic amine used as the hydrobromide salt orally as a drug and locally as a mydriatic and nasal decongestant.

2-hydroxy-3-amylene-α-naphthoquinone *See* lapachol.

para-**hydroxyanilinoacetic acid** *See* glycin.

ortho-**hydroxyanisole** *See* guaiacol.

5-hydroxybarbituric acid *See* dialuric acid.

ortho-**hydroxybenzaldehyde** *See* salicylaldehyde.

ortho-**hydroxybenzamide** *See* salicylamide.

hydroxybenzoic acid [ORG CHEM] $C_7H_6O_3$ Any one of three crystalline derivatives of benzoic acid: ortho, meta, and para forms; the ester of the para compound is used as a bacteriostatic agent.

ortho-**hydroxybenzoic acid** *See* salicylic acid.

***para*-hydroxybenzoic acid** [ORG CHEM] $C_6H_4(OH)COOH \cdot 2H_2O$ Colorless crystals melting at 210°C; soluble in alcohol, water, and ether; used as a chemical intermediate and for synthetic drugs.

***ortho*-hydroxybenzyl alcohol** *See* salicyl alcohol.

3-hydroxybutanal *See* aldol.

3-hydroxy-2-butanone *See* acetoin.

β-hydroxybutyric dehydrogenase [BIOCHEM] The enzyme that catalyzes the conversion of L-β-hydroxybutyric acid to acetoacetic acid by dehydrogenation.

2-hydroxycamphane *See* borneol.

hydroxychloroquine [PHARM] $C_{18}H_{26}ClN_3O$ A drug used as the sulfate salt for the treatment of malaria, lupus erythematosus, and rheumatoid arthritis.

hydroxycholine *See* muscarine.

1-hydroxy-2-cyanoethane *See* ethylene cyanohydrin.

2-hydroxy-*para*-cymene *See* carvacrol.

hydroxydimethylbenzene *See* xylenol.

***para*-hydroxydiphenyl** *See* phenylphenol.

β-hydroxyethylamine *See* ethanolamine.

hydroxyethylcellulose [MATER] A white powder made from cellulose, used for textile finishes and as a thickener for water-base paints.

1-hydroxy fenchane *See* fenchyl alcohol.

hydroxyine [PHARM] $C_{21}H_{27}ClN_2O_2$ A tranquilizer, also possessing antiemetic and antihistaminic effects; used as the hydrochloride salt.

hydroxyl- *See* hydroxy-.

hydroxylamine [INORG CHEM] NH_2OH A colorless, crystalline compound produced commercially by acid hydrolysis of nitroparaffins, decomposes on heating, melts at 33°C; used in organic synthesis and as a reducing agent.

***ortho*-hydroxylaniline** [ORG CHEM] $C_6H_4NH_2OH$ White crystals that turn brownish upon standing for some time; melts at 172–173°C, and will sublime upon more heating; soluble in cold water and benzene; used as a dye for hair and furs, and as a dye intermediate. Also known as *ortho*-aminophenol.

hydroxylapatite [MINERAL] $Ca_5(PO_4)_3OH$ A rare form of the apatite group that crystallizes in the hexagonal system.

hydroxylase [BIOCHEM] Any of several enzymes that catalyze certain hydroxylation reactions involving atomic oxygen.

hydroxylation reaction [ORG CHEM] One of several types of reactions used to introduce one or more hydroxyl groups into organic compounds; an oxidation reaction as opposed to hydrolysis.

hydroxylherderite [MINERAL] $CaBe(PO_4)(OH)$ A monoclinic mineral composed of a phosphate and hydroxide of calcium and beryllium; isomorphous with herderite.

3-hydroxymenthane *See* menthol.

1-hydroxy-2-methoxybenzene *See* guaiacol.

2-hydroxy methyl furan *See* furfuryl alcohol.

1-hydroxynaphthalene *See* α-naphthol.

2-hydroxynaphthalene *See* β-naphthol.

3-hydroxy-2-naphthoic acid *See* β-hydroxynaphthoic acid.

β-hydroxynaphthoic acid [ORG CHEM] $C_{10}H_6OHCOOH$ A yellow solid that is soluble in ether and alcohol and melts at about 218°C; used as a dye and a pigment. Also known as 3-hydroxy-2-naphthoic acid; 3-naphthol-2-carboxylic acid; β-oxynaphthoic acid.

***cis*-12-hydroxyoctadec-9-enoic acid** *See* ricinoleic acid.

12-hydroxyoleic acid *See* ricinoleic acid.

2-hydroxy-2-phenyl acetophenone *See* benzoin.

***para*-hydroxyphenylaminoacetic acid** *See* glycin.

hydroxyproline [BIOCHEM] $C_5H_9O_3N$ An amino acid that is essentially limited to structural proteins of the collagen type.

2-hydroxy-1,2,3-propane-tricarboxylic acid *See* isocitric acid.

2-hydroxypropanoic acid *See* lactic acid.

3-hydroxypropanoic acid *See* hydracrylic acid.

α-hydroxypropanoic acid *See* lactic acid.

1-hydroxy-2-propanone *See* acetol.

β-hydroxypropionic acid *See* hydracrylic acid.

α-hydroxypropionitrile *See* lactonitrile.

8-hydroxyquinoline [ORG CHEM] C_9H_6NOH White crystals or powder that darken on exposure to light, slightly soluble in water, soluble in benzene, melting at 73–75°C; used in preparing fungicides and in the separation of metals by acting as a precipitating agent. Also known as oxine; oxyquinoline; 8-quinolinol.

4-hydroxysalicylic acid *See* β-resorcylic acid.

5-hydroxysalicylic acid *See* gentisic acid.

5-hydroxytryptamine *See* serotonin.

8-hydroxyxanthine *See* uric acid.

hydrozincite [MINERAL] $Zn_5(OH)_5(CO_3)_2$ A white, grayish, or yellowish mineral composed of basic zinc carbonate, occurring as masses or crusts.

Hydrozoa [INV ZOO] A class of the phylum Coelenterata which includes the fresh-water hydras, the marine hydroids, many small jellyfish, a few corals, and the Portuguese man-of-war.

Hydrus [ASTRON] A southern constellation, right ascension 2 hours, declination 75°S. Also known as Water Snake.

hyena [VERT ZOO] An African carnivore represented by three species of the family Hyaenidae that resemble dogs but are more closely related to cats.

Hyeniales [PALEOBOT] An order of Devonian plants characterized by small, dichotomously forked leaves borne in whorls.

Hyeniatae *See* Hyeniopsida.

Hyeniopsida [PALEOBOT] An extinct class of the division Equisetophyta.

hyetal coefficient *See* pluviometric coefficient.

hyetal equator [CLIMATOL] A line (or transition zone) which encircles the earth (north of the geographical equator) and lies between two belts that typify the annual time distribution of rainfall in the lower latitudes of each hemisphere; a form of meteorological equator.

hyetal region [CLIMATOL] A region in which the amount and seasonal variation of rainfall are of a given type.

hyetograph [CLIMATOL] A map or chart displaying temporal or areal distribution of precipitation.

hyetography [CLIMATOL] The study of the annual variation and geographic distribution of precipitation.

hyetology [METEOROL] The science which treats of the origin, structure, and various other features of all the forms of precipitation.

hygiene [MED] The science that deals with the principles and practices of good health.

Hygrobiidae [INV ZOO] The squeaker beetles, a small family of coleopteran insects in the suborder Adephaga.

hygrodeik [ENG] A form of psychrometer with wet-bulb and dry-bulb thermometers mounted on opposite edges of a specially designed graph of the psychrometric tables, arranged so that the intersections of two curves determined by the wet-bulb and dry-bulb readings yield the relative humidity, dew-point, and absolute humidity.

hygrogram [ENG] The record made by a hygrograph.

hygrograph [ENG] A recording hygrometer.

hygrokinematics [METEOROL] The descriptive study of the motion of water substances in the atmosphere.

hygrology [METEOROL] The study which deals with the water vapor content (humidity) of the atmosphere.

hygroma [MED] A congenital disorder in which a lymph-filled cystic cavity is formed from distended lymphatics.

hygrometer [ENG] An instrument for giving a direct indication of the amount of moisture in the air or other gas, the indication usually being in terms of relative humidity as a percentage which the moisture present bears to the maximum amount of moisture that could be present at the location temperature without condensation taking place.

hygrometry [METEOROL] The study which treats of the measurement of the humidity of the atmosphere and other gases.

hygromycin [MICROBIO] $C_{25}H_{33}O_{12}N$ A weakly acidic, soluble antibiotic with a fairly broad spectrum, produced by a strain of *Streptomyces hygroscopicus*.

hygrophobia [PSYCH] An abnormal fear of liquids or moisture.

hygrophyte *See* hydrophyte.

hygroscopic [CHEM] **1.** Possessing a marked ability to accelerate the condensation of water vapor; applied to condensation nuclei composed of salts which yield aqueous solutions of a very low equilibrium vapor pressure compared with that of pure water at the same temperature. **2.** Pertaining to a

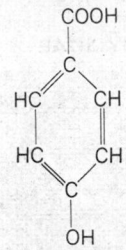

Structural formula of *para*-hydroxybenzoic acid.

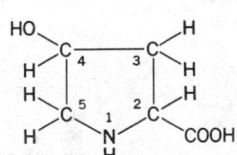

Structural formula of hydroxyproline.

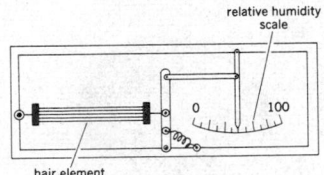

Hygrometer which uses hair as the sensing element. *(From D. M. Considine, ed., Process Instruments and Controls Handbook, McGraw-Hill, 1957)*

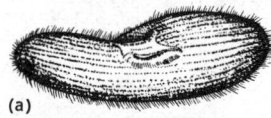

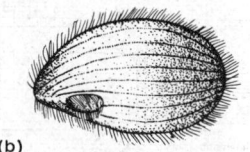

substance, whose physical characteristics are appreciably altered by effects of water vapor.

hygroscopic coefficient [HYD] The percentage of water that a soil will absorb and hold in equilibrium in a saturated atmosphere.

hygroscopic depression [CHEM] The measure of a desiccant's capacity to take on water.

hygroscopicity [CHEM] The tendency of a material to absorb water from the atmosphere.

hygroscopic water [HYD] The component of soil water that is held adsorbed on the surface of soil particles and is not available to vegetation.

hygrothermograph [ENG] An instrument for recording temperature and humidity on a single chart.

hyl *See* metric-technical unit of mass.

hylaea *See* tropical rainforest.

Hylidae [VERT ZOO] The tree frogs, a large amphibian family in the suborder Procoela; many are adapted to arboreal life, having expanded digital disks.

Hylleraas coordinates [ATOM PHYS] Coordinates for two particles used in studying the helium atom; they comprise the distance between the two particles, the sum of the distances of the particles from the origin, and the difference of the distances of the particles from the origin.

Hylobatidae [VERT ZOO] A family of anthropoid primates in the superfamily Hominoidea including the gibbon and the siamang of southeastern Asia.

hylophagous [ZOO] Feeding on wood, as termites.

hylophobia [PSYCH] An abnormal fear of forests.

hylotomous [ZOO] Cutting wood, as wood-boring insects.

hymen [ANAT] A mucous membrane partly closing off the vaginal orifice. Also known as maidenhead.

hymenium [MYCOL] The outer, sporebearing layer of certain fungi or their fruiting bodies.

hymenolepiasis [MED] Intestinal infection by tapeworms of the genus *Hymenolepis*.

Hymenolepis [INV ZOO] A genus of tapeworms parasitic in humans, birds, and mammals.

Hymenolepis nana [INV ZOO] The dwarf tapeworm that infests humans.

Hymenomycetes [MYCOL] A group of the Homobasidiomycetidae including forms such as mushrooms and pore fungi in which basidia are formed in an exposed layer (hymenium) and basidiospores are borne asymmetrically on slender stalks.

hymenophore [MYCOL] Portion of a sporophore that bears the hymenium.

hymenopodium [MYCOL] **1.** Tissue beneath the hymenium in certain fungi. **2.** A genus of the Moniliales.

Hymenoptera [INV ZOO] A large order of insects including ants, wasps, bees, sawflies, and related forms; head, thorax and abdomen are clearly differentiated; wings, when present, and legs are attached to the thorax.

Hymenostomatida [INV ZOO] An order of ciliated protozoans in the subclass Holotrichia having fairly uniform ciliation and a definite buccal ciliature.

hymenotomy [MED] Surgical incision of the hymen.

Hymu [MET] A magnetic alloy consisting of about 80% nickel and 20% iron.

Hynobiidae [VERT ZOO] A family of salamanders in the suborder Cryptobranchoidea.

hyobranchium [VERT ZOO] A Y-shaped bone supporting the tongue and tongue muscles in a snake.

Hyocephalidae [INV ZOO] A monospecific family of hemipteran insects in the superfamily Pentatomorpha.

hyoglossus [ANAT] An extrinsic muscle of the tongue arising from the hyoid bone.

hyoid [ANAT] **1.** A bone or complex of bones at the base of the tongue supporting the tongue and its muscles. **2.** Of or pertaining to structures derived from the hyoid arch.

hyoid arch [EMBRYO] Either of the second pair of pharyngeal segments or gill arches in vertebrate embryos.

hyoid tooth [VERT ZOO] One of a number of teeth on the tongue of fishes.

hyomandibular [VERT ZOO] The upper portion of the hyoid arch in fishes.

hyomandibular cleft [EMBRYO] The space between the hyoid arch and the mandibular arch in the vertebrate embryo.

hyomandibular pouch [EMBRYO] A portion of the endodermal lining of the pharyngeal cavity which separates the paired hyoid and mandibular arches in vertebrate embryos.

Hyopssodontidae [PALEON] A family of extinct mammalian herbivores in the order Condylarthra.

hyoscyamine [ORG CHEM] $C_{17}H_{23}O_3N$ A white, crystalline alkaloid isolated from henbane, belladonna, and other plants of the family Solanaceae, which is freely soluble in alcohol and dilute acids; used in medicine as an anticholinergic.

hyostylic [VERT ZOO] Having the jaws and cranium connected by the hyomandibular, as certain fishes.

hypabyssal rock [PETR] Those igneous rocks that rose from great depths as magmas but solidified as minor intrusions before reaching the surface.

hypacusia [MED] Impairment of hearing.

hypalgesia [PHYSIO] Diminished sensitivity to pain.

hypandrium [INV ZOO] A plate covering the genitalia on the ninth abdominal segment of certain male insects.

hypanthium [BOT] Expanded receptacle margin to which the sepals, petals, and stamens are attached in some flowers.

hypantrum [VERT ZOO] In reptiles, a notch on the anterior portion of the neural arch that articulates with the hyposphene.

hypautomorphic *See* hypidiomorphic.

hypaxial musculature [ANAT] The ventral portion of the axial musculature of vertebrates including subvertebral flank and ventral abdominal muscle groups.

hypengyophobia [PSYCH] An abnormal fear of responsibility.

hyperacid [PHYSIO] Containing more than the normal concentration of acid in the gastric juice.

hyperacoustic zone [GEOPHYS] The region in the upper atmosphere, between 100 and 160 kilometers, where the distance between the rarefied air molecules roughly equals the wavelength of sound, so that sound is transmitted with less volume than at lower levels.

hyperactivity [PHYSIO] Excessive or pathologic activity.

hyperadrenalism [MED] Hypersecretion of adrenal hormones marked by increased basal metabolism, decreased sugar tolerance, and glycosuria. Also known as hypercorticism.

hyperadrenocorticism [MED] Hypersecretion of adrenocortical hormones resulting in Cushing's syndrome, or virilism.

hyperaldosteronism [MED] Hypersecretion of aldosterone by the adrenal cortex.

hyperalgesia [PHYSIO] Increased or heightened sensitivity to pain stimulation.

hyperbaric [MED] Pertaining to an anesthetic solution with a specific gravity greater than that of the cerebrospinal fluid.

hyperbaric chamber [ENG] A specially equipped pressure vessel used in medicine and physiological research to administer oxygen at elevated pressures.

hyperbaric medicine [PHARM] Any agent having a specific gravity greater than that of spinal fluid; used for spinal anesthesia.

hyperbaric oxygenation [MED] The administration of oxygen, under greater than atmospheric pressure, by placing the patient in a room or chamber especially designed for the purpose.

hyperbilirubinemia [MED] **1.** Excessive amounts of bilirubin in the blood. **2.** A severe, prolonged physiologic jaundice.

hyperbola [MATH] The plane curve obtained by intersecting a circular cone of two nappes with a plane parallel to the axis of the cone.

hyperbolic amplitude [COMMUN] Excursion of a signal measured along hyperbolic rather than cartesian coordinates.

hyperbolic antenna [ELECTROMAG] A radiator whose reflector in cross section describes a half hyperbola.

hyperbolic cosecant [MATH] A function whose value is equal to the reciprocal of the value of the hyperbolic sine. Abbreviated csch.

hyperbolic cosine [MATH] A function whose value at the complex number z is one-half the sum of the exponential of z and the exponential of $-z$. Abbreviated cosh.

hyperbolic cotangent [MATH] A function whose value is equal to the value of the hyperbolic cosine divided by the value of the hyperbolic sine. Abbreviated coth.

hyperbolic decline [PETRO ENG] One of three types of decline in oil or gas production rate (the others are constant-percentage and harmonic decline).

hyperbolic differential equation [MATH] A general type of real second-order partial differential equation which includes the wave equation, and having the general form

$$\sum_{i,j=1}^{n} a_{ij} \frac{\partial^2 f}{\partial x_i \partial x_j} + F\left(x_1, \ldots, x_n, f, \frac{\partial f}{\partial x_1}, \ldots, \frac{\partial f}{\partial x_n}\right) = 0$$

where the quadratic form

$$\sum_{i,j}^{n} a_{ij} y_i y_j$$

can be reduced to a sum of n squares, not all of the same sign, by a linear transformation.

hyperbolic distance [ELECTROMAG] A function of pairs of points within a unit circle, where the interior of this circle is a conformal or projective representation of a hyperbolic space used in transmission line theory and waveguide analysis.

hyperbolic Dovap [NAV] System using four or more Dovap stations with a common reference signal that is not coherent with the interrogation signal.

hyperbolic fix [NAV] A fix established by means of hyperbolic lines of position.

hyperbolic flareout [AERO ENG] A flareout obtained by changing the glide slope from a straight line to a hyperbolic curve at an appropriate distance from touchdown at an airport.

hyperbolic functions [MATH] The real or complex functions sinh (x), cosh (x), tanh (x), coth (x), sech (x), csch (x); they are related to the hyperbola in somewhat the same fashion as the trigonometric functions are related to the circle, and have properties analogous to those of the trigonometric functions.

hyperbolic geometry *See* Lobachevski geometry.

hyperbolic guidance [ORD] Missile guidance in which the difference in the arrival times of radio signals transmitted simultaneously from two ground stations is used to control the position of the missile.

hyperbolic horn [ENG] Horn whose equivalent cross-sectional radius increases according to a hyperbolic law.

hyperbolic line of position [NAV] A line of position in the shape of a hyperbola, determined by measuring the difference in the phase or time of transit of radiations from fixed points; measurement may be made through the use of radio, sound, or light.

hyperbolic logarithm *See* logarithm.

hyperbolic navigation [NAV] Navigation by maintaining constant the indication of two parameters; the parameters can have any reasonable ratio to each other.

hyperbolic paraboloid [MATH] A surface which can be so situated that sections parallel to one coordinate plane are parabolas while those parallel to the other plane are hyperbolas.

hyperbolic point [FL MECH] A singular point in a streamline field which constitutes the intersection of a convergence line and a divergence line; it is analogous to a col in the field of a single-valued scalar quantity. Also known as neutral point. [MATH] A point on a surface where the Gaussian curvature is strictly negative.

hyperbolic secant [MATH] A function whose value is equal to the reciprocal of the value of the hyperbolic cosine. Abbreviated sech.

hyperbolic sine [MATH] A function whose value at the complex number z is one-half the difference between the exponential of z and the exponential of $-z$. Abbreviated sinh.

hyperbolic space [MATH] A space described by hyperbolic rather than cartesian coordinates.

hyperbolic spiral [MATH] A plane curve for which the radius vector is inversely proportional to the polar angle. Also known as reciprocal spiral.

hyperbolic sweep generator [ELECTR] A sweep generator that generates a waveform resembling a hyperbola.

hyperbolic tangent [MATH] A function whose value is equal to the value of the hyperbolic sine divided by the value of the hyperbolic cosine. Abbreviated tanh.

hyperbolic trajectory [AERO ENG] A trajectory entered by a spacecraft when its velocity exceeds the escape velocity of a planet, satellite, or star.

hyperbolic waveform [ELECTR] A waveform which is an approximate hyperbola.

hyperboloid of revolution [MATH] A surface generated by rotating a hyperboloid about one of its axes.

hyperboloids [MATH] Quadric surfaces given by equations of the form $(x^2/a^2) \pm (y^2/b^2) - (z^2/c^2) = 1$; in certain cases they can be realized by spinning the pieces of a hyperbola about an appropriate axis.

hyperbrachycephalic [ANTHRO] Having a round or broad head with a cephalic index of more than 85.

hyperbrachycranial [ANTHRO] Having a round or broad head with a cranial index of 85 to 90.

hyperbrachyskelic [ANTHRO] Having the legs 75% of the length of the trunk with a skelic index below 75.

hypercalcemia [MED] Excessive amounts of calcium in the blood. Also known as calcemia.

hypercapnia [MED] Excessive amount of carbon dioxide in the blood.

hyperchamaerrhine [ANTHRO] Having a short broad nose with a nasal index of 58 or over.

hypercharge [PARTIC PHYS] A quantum number conserved by strong interactions, equal to twice the average of the charges of the members of an isospin multiplet.

hyperchlorhydria [MED] Excessive secretion of hydrochloric acid in the stomach.

hypercholesteremia [MED] Elevated cholesterol levels in the blood.

hyperchromatic [BIOL] Staining more intensely than normal.

hyperchromatism [PATH] 1. Excessive pigment formation in the skin. 2. A condition in which cells or parts of cells stain more intensely than is normal.

hyperchromic [PATH] Pertaining to increased hemoglobin content in erythrocytes due to increased cell thickness, not increased hemoglobin concentration.

hyperchromic anemia [MED] Any of several blood disorders in which erythrocytes show an increase in hemoglobin and a reduction in number.

hypercoagulability [MED] Coagulation of blood more rapidly than normal.

hypercomplex number *See* quaternion.

hypercomplex system *See* algebra.

hyperconjugation [PHYS CHEM] An arrangement of bonds in a molecule that is similar to conjugation in its formulation and manifestations, but the effects are weaker; it occurs when a CH_2 or CH_3 group (or in general, an AR_2 or AR_3 group where A may be any polyvalent atom and R any atom or radical) is adjacent to a multiple bond or to a group containing an atom with a lone π-electron, π-electron pair or quartet, or π-electron vacancy; it can be sacrificial (relatively weak) or isovalent (stronger).

hypercoracoid [VERT ZOO] The upper of two bones at the base of the pectoral fin in teleosts.

hypercorticism *See* hyperadrenalism.

hyperdisk [ADP] A mass-storage technique which uses a large-capacity storage and a disk for overflow.

hyperdolichocephalic [ANTHRO] Having a long narrow head with a cephalic index of less than 70.

hyperdolichocranial [ANTHRO] Having a long narrow skull with a cranial index of 65 to 70.

hyperdynamic ileus *See* spastic ileus.

hyperemesis [MED] Excessive vomiting.

hyperemesis gravedorium [MED] Pernicious vomiting in pregnancy.

hyperemia [MED] An excess of blood within an organ or tissue caused by blood vessel dilation or impaired drainage, especially of the skin.

hyperergia [IMMUNOL] An altered state of reactivity to antigenic materials, in which the response is more marked than usual; one form of allery or pathergy.

hyperesthesia [PHYSIO] Increased sensitivity or sensation.

hypereuryene [ANTHRO] Having a high wide forehead with an upper facial index less than 45.

hypereutectic alloy [MET] Any binary alloy whose composition lies to the right of the eutectic on an equilibrium diagram and which contains some eutectic structure.

hypereutectoid steel [MET] Steel containing more than 0.8% carbon.

hyperfine structure [SPECT] A splitting of spectral lines due to the spin of the atomic nucleus or to the occurrence of a mixture of isotopes in the element. Abbreviated hfs.

hyperfocal distance [OPTICS] The distance from the camera lens to the nearest object in acceptable focus when the lens is focused on infinity.

hyperforming [CHEM ENG] A catalytic, petroleum-refinery hydrogenation process to improve naphtha octane number by removal of sulfur and nitrogen compounds; the catalyst is cobalt molybdate on a silica-alumina base.

hyperfrequency waves [ELECTROMAG] Microwaves having wavelengths in the range from 1 centimeter to 1 meter.

hypergammaglobulinemia [MED] Increased blood levels of gamma globulin, usually associated with hepatic disease.

hypergene *See* supergene.

hypergeometric differential equation *See* Gauss' hypergeometric equation.

hypergeometric distribution [STAT] The distribution of the number D of special items in a random sample of size s drawn from a population of size N that contains r of the special items:

$$P(D = d) = \binom{r}{d}\binom{N-r}{s-d} \Big/ \binom{N}{s}.$$

hypergeometric function [MATH] A function which is a solution to the hypergeometric equation and obtained as an infinite series expansion.

hypergeometric series [MATH] A particular infinite series which in certain cases is a solution to the hypergeometric equation, and having the form:

$$1 + \frac{ab}{c}z + \frac{1}{2!}\frac{a(a+1)b(b+1)}{c(c+1)}z^2 + \cdots$$

hyperglobulinemia [MED] Increased blood levels of globulin.

hyperglycemia [MED] Excessive amounts of sugar in the blood.

hyperglycemic factor *See* glucagon.

hyperglycemic glycogenolytic factor *See* glucagon.

hyperglycinemia [MED] A hereditary metabolic disorder of males in which blood levels of glycine are excessive, resulting in vomiting, dehydration, osteoporosis, and mental retardation.

hypergolic [CHEM] Capable of igniting spontaneously upon contact.

hypergolic fuel [MATER] A combination of fuel and oxidizer that ignite spontaneously on contact, such as methanol and hydrogen peroxide; used as rocket propellant.

hyperhidrosis [MED] Excessive sweating, which may be localized or generalized, chronic or acute, and often accumulating in visible drops on the skin. Also known as ephidrosis; polyhidrosis; sudatoria.

Hyperiidea [INV ZOO] A suborder of amphipod crustaceans distinguished by large eyes which cover nearly the entire head.

hyperimmune antibody [IMMUNOL] An antibody having the characteristics of a blocking antibody.

hyperimmune serum [IMMUNOL] An antiserum that provides a very high degree of immunity due to a high antibody titer.

hyperinsulinism [MED] Condition caused by abnormally high levels of insulin in the blood.

Hyperion [ASTRON] A satellite of Saturn approximately 300 miles (480 kilometers) in diameter.

hyperkalemia *See* hyperpotassemia.

hyperkeratosis [MED] **1.** Hypertrophy of the cornea. **2.** Hypertrophy of the horny layer of the skin.

hyperkinesia [MED] Excessive and usually uncontrollable muscle movement.

hyperleptene [ANTHRO] Having a high narrow forehead with an upper facial index of 60 or more.

hyperleptoprosopic [ANTHRO] Having a long narrow face with a facial index of 93 or more on the living and of 95 or more on the skull.

hyperleptorrhine [ANTHRO] Having a long narrow nose with a nasal index of 40 to 55.

hyperlipemia [MED] Excessive amounts of fat in the blood.

hypermakroskelic [ANTHRO] Having long legs in proportion to the length of the trunk with a skelic index of 100 or more.

Hypermastigida [INV ZOO] An order of the multiflagellate protozoans in the class Zoomastigophorea; all inhabit the alimentary canal of termites, cockroaches, and woodroaches.

hypermenorrhea *See* menorrhagia.

hypermetabolism [MED] Any state in which there is an increase in basal metabolic rate.

hypermetamorphism [INV ZOO] Type of embryological development in certain insects in which one or more stages have been interpolated between the full-grown larva and the adult.

hypermetropia [MED] A defect of vision resulting from too short an eyeball so that unaccommodated rays focus behind the retina. Also known as farsightedness; hyperopia.

hypermotility [MED] Increased motility, as of the stomach or intestines.

hypernatremia [MED] Excessive amounts of sodium in the blood.

hypernephroma *See* renal-cell carcinoma.

Hyperoartii [VERT ZOO] A superorder in the subclass Monorhina distinguished by the single median dorsal nasal opening leading into a blind hypophyseal sac.

hyperoid axle [MECH ENG] A type of rear-axle drive gear set which generally carries the pinion 1.5–2 inches (38–51 millimeters) or more below the centerline of the gear.

hyperon [PARTIC PHYS] **1.** An elementary particle which has baryon number $B = +1$, that is, which can be transformed into a nucleon and some number of mesons or lighter particles, and which has nonzero strangeness number. **2.** A hyperon (as in the first definition) which is semistable (the lifetime is much longer than 10^{-22} second).

Σ⁻-hyperonic atom [ATOM PHYS] An atom consisting of a negatively charged sigma hyperon orbiting around an ordinary nucleus.

hyperopia *See* hypermetropia.

hyperorthognathous [ANTHRO] Having a face that is flat in profile with a facial angle of 93° or more.

hyperosmia [MED] An abnormally acute sense of smell.

hyperostosis [MED] Hypertrophy of bony tissue.

Hyperotreti [VERT ZOO] A suborder in the subclass Monorhina distinguished by the nasal opening which is located at the tip of the snout and communicates with the pharynx by a long duct.

hyperoxemia [MED] Extreme acidity of the blood.

hyperparasite [ECOL] An organism that is parasitic on other parasites.

hyperparathyroidism [MED] Condition caused by increased functioning of the parathyroid glands.

hyperpathia [MED] An exaggerated or excessive perception of or response to any stimulus as being disagreeable or painful.

hyperperistalsis [MED] An increase in the rate and depth of the peristaltic waves.

hyperphagia *See* bulimia.

hyperphosphaturia [MED] An excess of phosphates in the urine. Also known as phosphaturia.

hyperpigmentation [MED] Increased pigmentation.

hyperpituitarism [MED] Any abnormal condition resulting from overactivity of the anterior pituitary.

hyperplane [MATH] A hyperplane is an $(n - 1)$-dimensional subspace of an n-dimensional vector space.

hyperplasia [MED] Increase in cell number causing an increase in the size of a tissue or organ.

hyperplatycnemic [ANTHRO] Having a shinbone that is laterally flattened with a platycnemic index below 55.

hyperploid [GEN] Having one or more chromosomes or parts of chromosomes in excess of the haploid number, or of a whole multiple of the haploid number.

hyperploidy [GEN] The condition or state of being hyperploid.

hyperpnea [MED] Increase in depth and rate of respiration.

hyperpotassemia [MED] Excessive amounts of potassium in the blood. Also known as hyperkalemia.

hyperprognathous [ANTHRO] Having prominent jaws with a facial profile angle less than 70°.

hyperproteinemia [MED] Excessive protein levels in the blood.

hyperpycnal inflow [HYD] A denser inflow that occurs when a sediment-laden fluid flows down the side of a basin and along the bottom as a turbidity current.

hyperpyrexia [MED] Extremely high fever.

hyperreflexia [MED] A condition of abnormally increased reflex action.

hyperresonance [MED] Exaggeration of normal resonance on percussion of the chest; heard chiefly in pulmonary emphysema and pneumothorax.

hypersaline [GEOL] Geologic material with high salinity.

hypersensitivity [IMMUNOL] The state of being abnormally sensitive, especially to allergens; responsible for allergic reactions.

hypersensitization [IMMUNOL] The process of producing hypersensitivity.

hypersensor [ELECTR] Single-component, resettable circuit breaker which operates as a majority-carrier tunneling device, and is used for overcurrent or overvoltage protection of integrated circuits.

Hypersil [MET] A magnetic alloy of iron and about 3.5% silicon.

hypersomnia [MED] Excessive sleepiness.

hypersonic [ACOUS] Pertaining to frequencies above 500 megahertz. [FL MECH] Pertaining to hypersonic speeds, or air currents moving at hypersonic speeds.

hypersonic flight [AERO ENG] Flight at speeds well above the local velocity of sound; by convention, hypersonic regime starts at about five times the speed of sound and extends upward indefinitely.

hypersonic flow [FL MECH] Flow of a fluid over a body at hypersonic speeds, and in which shock waves start at a finite distance from the surface of the body.

hypersonic glider [AERO ENG] An unpowered vehicle, specifically a reentry vehicle, designed to fly at hypersonic speeds.

hypersonic inlet [FL MECH] An entrance or orifice for admission of fluids at hypersonic speeds.

hypersonic nozzle [FL MECH] A supersonic nozzle designed to accelerate a fluid to hypersonic speeds.

hypersonics [ACOUS] Production and utilization of sound waves of frequencies above 500 megahertz.

hypersonic speed [FL MECH] A speed of an object greater than about five times the speed of sound in the fluid through which the object is moving.

hypersonic wind tunnel [ENG] A wind tunnel in which air flows at speeds roughly in the range from 5 to 15 times the speed of sound.

hypersorption [CHEM ENG] Process with recirculating bed of activated-carbon adsorbent for continuous recovery of ethylene from methane and other low-molecular-weight gases.

hypersplenism [MED] Condition caused by abnormal spleen activity.

hyperstereoscopy [MAP] Stereoscopic viewing of a map in which the scale (usually vertical) along the line of sight is exaggerated in comparison with the scale perpendicular to the line of sight. Also known as appearance ratio.

hypersthene [MINERAL] $(Mg,Fe)SiO_3$ A grayish, greenish, black, or dark-brown rock-forming mineral of the orthopyroxene group, with bronzelike luster on the cleavage surface.

hypersthenfels *See* norite.

hypertape control unit *See* tape control unit.

hypertape drive *See* cartridge tape drive.

hypertelorism [ANAT] An unusually large distance between paired body parts or organs.

hypertely [EVOL] An extreme overdevelopment of an organ or body part during evolution that is disadvantageous to the organism. [ZOO] An extreme degree of imitative coloration, beyond the aspect of utility.

hypertensin *See* angiotensin.

hypertension [MED] Abnormal elevation of blood pressure, generally regarded to be levels of 165 systolic and 95 diastolic.

hyperthecosis [PATH] Abnormal thickening of the inner layer of the Graafian follicle with increased leutein formation.

hyperthermia [PHYSIO] A condition of elevated body temperature.

hyperthyroidism [MED] The constellation of signs and symptoms caused by excessive thyroid hormone in the blood, either from exaggerated functional activity of the thyroid gland or from excessive administration of thyroid hormone, and manifested by thyroid enlargement, emaciation, sweating, tachycardia, exophthalmos, and tremor. Also known as exophthalmic goiter; Grave's disease; thyrotoxicosis; toxic goiter.

hyperthyrotropinism [MED] Excessive thyrotropic hormone secretion by the adenohypophysis.

hypertonia [MED] Abnormal increase in muscle tonicity.

hypertonic [PHYSIO] 1. Excessive or above normal in tone or tension, as a muscle. 2. Having an osmotic pressure greater than that of physiologic salt solution or of any other solution taken as a standard.

hypertonic bladder [MED] Hypertonia of the urinary bladder.

hypertonic contracture [MED] Prolonged muscular spasms in spastic paralysis.

Hypertragulidae [PALEON] A family of extinct chevrotainlike pecoran ruminants in the superfamily Traguloidea.

hypertrophic arthritis *See* degenerative joint disease.

hypertrophic gastritis [MED] Chronic inflammation of the stomach with hypertrophy of the mucosa and rugae.

hypertrophy [PATH] Increase in cell size causing an increase in the size of an organ or tissue.

hyperuricemia [MED] Abnormally high level of uric acid in the blood. Also known as lithemia.

hypervelocity [MECH] 1. Muzzle velocity of an artillery projectile of 3500 feet per second (1067 meters per second) or more. 2. Muzzle velocity of a small-arms projectile of 5000 feet per second (1524 meters per second) or more. 3. Muzzle velocity of a tank-cannon projectile in excess of 3350 feet per second (1021 meters per second).

hypervelocity armor-piercing [ORD] Designating a type of artillery projectile consisting of a core of extremely hard, high-density material, such as tungsten carbide, contained within a lightweight carrier called a sabot. Abbreviated HVAP.

hypervelocity wind tunnel [ENG] A wind tunnel in which higher airspeeds and temperatures can be attained than in a hypersonic wind tunnel.

hyperventilation [MED] Increase in air intake or of the rate or depth of respiration.

hypervisor [ADP] A control program enabling two operating systems to share a common computing system.

hypervitaminosis [MED] Condition caused by intake of toxic amounts of a vitamin.

hypesthesia [MED] Reduced or subnormal tactile sensibility.

hypha [MYCOL] One of the filaments composing the mycelium of a fungus.

hyphidium [MYCOL] A sterile hymenial structure of hyphal origin.

Hyphochytriales [MYCOL] An order of aquatic fungi in the class Phycomycetes having a saclike to limited hyphal thallus and zoospores with two flagella.

Hyphochytridiomycetes [MYCOL] A class of the true fungi; usually grouped with other classes under the general term Phycomycetes.

hyphoid [MYCOL] Hyphalike.

Hyphomicrobiaceae [MICROBIO] A family of bacteria in the order Hyphomicrobiales; cells occur in free-floating groups with individual cells attached to each other by a slender filament.

Hyphomicrobiales [MICROBIO] An order of bacteria in the class Schizomycetes containing forms that multiply by budding.

Hyphomicrobium [MICROBIO] A genus of nonpigmented, gram-negative, aerobic bacteria in the family Hyphomicrobiaceae; cells usually occur in groups attached by slender filaments.

hyphopodium [MYCOL] Hypha with a haustorium in certain ectoparasitic fungi.

hypidiomorphic [PETR] Of the texture of igneous rocks, having the crystals bounded partly by the crystal faces characteristic of the mineral species. Also known as hypautomorphic; subidiomorphic.

hypnagogic hallucination [PSYCH] Mental images occurring normally while falling asleep.

Hypnineae [BOT] A suborder of the Hypnobryales character-

HYPHOMICROBIACEAE

Electron micrograph of *Rhodomicrobium vannielii*, pink to orange in color and photosynthetic.

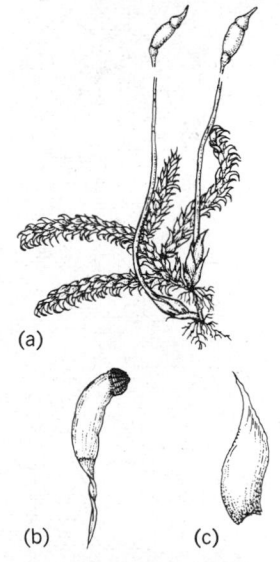

HYPNINEAE

Hypnum reptile. **(a)** Portion of plant, with stems shortened. **(b)** Urn and peristome. **(c)** Faintly bicostate leaf. *(From W. H. Welch, Mosses of Indiana, Indiana Department of Conservation, 1957)*

ized by complanate, glossy plants with ecostate or costate leaves and paraphyllia rarely present.

hypnoanalysis [PSYCH] Technique used in psychotherapy combining hypnosis with psychoanalysis.

Hypnobryales [BOT] An order of mosses composed of procumbent and pleurocumbent plants with usually symmetrical leaves arranged in more than two rows.

hypnophobia [PSYCH] An abnormal fear of sleeping.

hypnosis [PSYCH] An altered state of consciousness in which the individual is more susceptible to suggestion and in which regressive behavior may spontaneously occur.

hypnotherapy [MED] Treatment of disease by means of hypnotism.

hypnotic [PHARM] A drug which induces sleep. Also known as somnificant; soporific.

hypnotism [PSYCH] The practice or study of inducing hypnosis.

hypnotize [PSYCH] To induce a state of hypnosis.

hypnotoxin [BIOCHEM] A supposed hormone produced by brain tissue and inducing sleep.

hypo [GRAPHICS] In photography, the common fixing agent sodium thiosulfate, which formerly was incorrectly called hyposulfate of soda. [INORG CHEM] *See* sodium thiosulfate. [PSYCH] Informal term for a hypochondriac or hypochondria.

hypoadrenalism *See* hypoadrenia.

hypoadrenia [MED] Reduced functioning of the adrenal glands. Also known as hypoadrenalism.

hypoadrenocorticism [MED] Lowered or subnormal adrenal cortex activity.

hypoalbuminemia [MED] Abnormally low levels of albumin in the blood.

hypoallergenic [PHARM] Having a low tendency to induce allergic reactions; used particularly for formulated dermatologic preparations.

hypobaric [MED] Pertaining to an anesthetic solution of specific gravity lower than the cerebrospinal fluid. [PHYS] Having less weight or pressure.

hypobasal [BOT] Located posterior to the basal wall.

hypoblast *See* endoderm.

hypobranchial musculature [ANAT] The ventral musculature in vertebrates extending from the pectoral girdle forward to the hyoid arch, chin, and tongue.

hypocalcemia [MED] Condition in which there are reduced levels of calcium in the blood.

hypocalcification [MED] Reduction of normal amounts of mineral salts in calcified tissue.

hypocalciuria [MED] Decreased excretion of calcium in the urine.

hypocapnia [MED] Reduced or subnormal blood levels of carbon dioxide.

hypochil [BOT] Lower portion of the lip in orchids. Also known as hypochillium.

Hypochilidae [INV ZOO] A family of true spiders in the order Araneida.

hypochillium *See* hypochil.

Hypochilomorphae [INV ZOO] A monofamilial suborder of spiders in the order Araneida.

hypochloremia [MED] Reduction in the amount of blood chlorides.

hypochlorhydria [MED] Reduction in the hydrochloric acid content of gastric juice.

hypochlorite [INORG CHEM] ClO_3^- A negative ion derived from hypochlorous acid, $HClO$; the ion is an oxidizing agent and a constituent of bleaching agents.

hypochlorite sweetening [CHEM ENG] A petroleum refinery process to oxidize gasoline mercaptans by agitation with an aqueous, alkaline hypochlorite solution.

hypochlorization [MED] Reduction of the dietary intake of sodium chloride.

hypochlorous acid [INORG CHEM] $HOCl$ Weak, unstable acid existing in solution only; its salts (such as calcium hypochlorite) are used as bleaching agents.

hypochnoid [MYCOL] Having generally compacted hyphae.

hypocholesterolemia [MED] Subnormal levels of serum cholesterol.

hypochondriac [PSYCH] A person affected with hypochondriasis.

hypochondriac region [ANAT] The upper, lateral abdominal region just below the ribs on each side of the body.

hypochondriasis [PSYCH] A chronic condition in which the patient is morbidly concerned with his own health and believes himself suffering from grave bodily diseases.

hypochromia [PATH] Lack of complete saturation of the erythrocyte stroma with hemoglobin, as judged by pallor of the unstained or stained erythrocytes when examined microscopically.

hypochromic microcytic anemia [MED] An anemia associated with erythrocytes of reduced size and hemoglobin content.

hypocleidium [ANAT] The median, ventral bone between clavicles. [VERT ZOO] Median process on the wishbone of birds.

hypocone [ANAT] The posterior inner cusp of an upper molar. [INV ZOO] Region of a dinoflagellate posterior to the girdle.

Hypocopridae [INV ZOO] A small family of coleopteran insects in the superfamily Cucujoidea.

hypocoracoid [VERT ZOO] The lower of two bones at the base of the pectoral fin in teleosts.

hypocotyl [BOT] The portion of the embryonic plant axis below the cotyledon.

hypocrateriform [BIOL] Saucer-shaped.

Hypocreales [MYCOL] An order of fungi belonging to the Ascomycetes and including several entomophilic fungi.

hypocrystalline [PETR] Pertaining to the texture of igneous rock characterized by crystalline components in an amorphous groundmass. Also known as hemicrystalline; hypohyaline; merocrystalline; miocrystalline; semicrystalline.

hypocycloid [MATH] The curve composed of four identical arcs which is traced in the plane as a given point fixed on a circle moves while this circle rolls along the inside of another circle.

Hypodermatidae [INV ZOO] The warble flies, a family of myodarian cyclorrhaphous dipteran insects in the subsection Calypteratae.

hypodermic needle [MED] A hollow needle, with a slanted open point, used for subcutaneous and intramuscular injections of fluid.

hypodermic syringe *See* syringe.

hypodermis [BOT] The outermost cell layer of the cortex of plants. Also known as exodermis. [INV ZOO] The layer of cells that underlies and secretes the cuticle in arthropods and some other invertebrates.

hypodermoclysis [MED] Subcutaneous injections of large quantities of fluid for therapeutic purposes.

hypo eliminator [GRAPHICS] A solution used to facilitate the removal of fixer solutions during the wash bath.

hypoergia [IMMUNOL] A state of less than normal reactivity to antigenic materials, in which the response is less marked than usual; one form of allergy or pathergy.

hypoeutectic alloy [MET] Any binary alloy whose composition lies to the left of the eutectic.

hypoeutectoid steel [MET] Steel containing less than 0.8% carbon.

hypofibrinogenemia [MED] A decrease in plasma fibrogen level.

hypogammaglobulinemia [MED] Reduced blood levels of gamma globulin.

hypogene [GEOL] 1. Of minerals or ores, formed by ascending waters. 2. Of geologic processes, originating within or below the crust of the earth.

hypogeous [BIOL] Living or maturing below the surface of the ground.

hypoglossal nerve [ANAT] The twelfth cranial nerve; a paired motor nerve in tetrapod vertebrates innervating tongue muscles; corresponds to the hypobranchial nerve in fishes.

hypoglossal nucleus [ANAT] A long nerve nucleus throughout most of the length of the medulla oblongata; cells give rise to the hypoglossal nerve fibers.

hypoglycemia [MED] Condition caused by low levels of sugar in the blood.

hypogonadism [MED] Reduced hormonal secretion by the testes or ovaries.

hypogynium [BOT] Structure that supports the ovary in plants such as sedges.

hypogynous [BOT] Having all flower parts attached to the receptacle below the pistil and free from it.

hypohidrosis [MED] Deficient perspiration.

hypohyaline *See* hypocrystalline.

hypoid gear [MECH ENG] Gear wheels connecting nonparallel, nonintersecting shafts, usually at right angles.

hypoid generator [MECH ENG] A gear-cutting machine for making hypoid gears.

hypoiodous acid [INORG CHEM] HIO A very weak unstable acid that occurs as the result of the weak hydrolysis of iodine in water.

hypokinesia [MED] Subnormal muscular movements.

hypokinetic syndrome [MED] General decrease in motor functions due to a form of minimal brain dysfunction.

hypolimnion [HYD] The lower level of water in a stratified lake, characterized by a uniform temperature that is generally cooler than that of other strata in the lake.

hypomenorrhea [MED] A deficient amount of menstrual flow at the regular period.

hypomere [EMBRYO] The lateral or lower mesodermal plate zone in vertebrate embryos. [INV ZOO] The basal portion of certain sponges that contain no flagellated chambers.

hypometabolism [MED] Metabolism below the normal rate.

hypomorph [GEN] An allele having an effect similar to the normal allele, but being less active.

hypomotility [MED] Decreased motility, especially of the gastrointestinal tract.

hyponasty [BOT] A nastic movement involving inward and upward bending of a plant part.

hyponatremia [MED] Subnormal or reduced blood sodium levels.

hyponeural sinus [INV ZOO] Tubular portion of the coelom containing hemal vessels and motor nerves in certain echinoderms.

hyponychium [HISTOL] The thickened stratum corneum of the epidermis, which lies under the free edge of the nail.

hypoovarianism [MED] Decrease in ovarian endocrine activity.

hypoparathyroidism [MED] Condition caused by insufficient functioning of the parathyroid gland.

hypophalangism [MED] Congenital absence of one or more phalanges in a finger or toe.

hypopharynx [ANAT] *See* laryngopharynx. [INV ZOO] A sensory, tonguelike structure on the floor of the mouth of many insects; sometimes modified for piercing.

hypophosphatasia [MED] 1. Alkaline phosphatase deficiency. 2. A hereditary metabolic disorder characterized by subnormal amounts of alkaline phosphatase in the tissues.

hypophrenia [PSYCH] Mental retardation.

hypophyseal cachexia *See* Simmonds' disease.

hypophyseal duct tumor [MED] Any tumor derived from epithelial remnants of Rathke's pouch.

hypophysectomy [MED] Surgical removal of the pituitary gland.

hypophysis [ANAT] A small rounded endocrine gland which lies in the sella turcica of the sphenoid bone and is attached to the floor of the third ventricle of the brain in all craniate vertebrates. Also known as pituitary gland.

hypopituitarism [MED] Condition caused by insufficient secretion of pituitary hormones, especially of the adenohypophysis. Also known as panhypopituitarism.

hypopituitary cachexia *See* Simmonds' disease.

hypoplankton [BIOL] Forms of marine life whose swimming ability lies somewhere between that of the plankton and the nekton; includes some mysids, amphipods, and cumacids.

hypoplasia [MED] Failure of a tissue or organ to achieve complete development.

hypoplastic dwarf [MED] A normally proportioned individual of subnormal size.

hypoplastron [VERT ZOO] Either of the third pair of lateral bony plates in the plastron of most turtles.

hypopleura [INV ZOO] Sclerite above and in front of the hind coxa in Diptera.

hypoploid [GEN] Having one or more less chromosomes, or parts of chromosomes, than a whole multiple of the haploid number.

hypoploidy [GEN] The condition or state of being hypoploid.

hypoproliferative anemia [MED] Decreased concentration of hemoglobin and number of red blood cells due to subnormal numbers of erythrocyte primordial cells in relation to the stimulus of anemia.

hypoproteinemia [MED] Abnormally low levels of protein in the blood.

hypoprothrombinemia [MED] Deficiency of prothrombin in the blood.

hypopus [INV ZOO] The resting larval stage of certain mites.

hypopycnal inflow [HYD] Flowing water of lower density than the body of water into which it flows.

hypopygium [INV ZOO] A modified ninth abdominal segment together with the copulatory apparatus in Diptera.

hypopyon [MED] A collection of pus in the anterior chamber of the eye.

hyporeactive [MED] Characterized by decreased responsiveness to stimuli.

hyporeflexia [MED] A condition in which reflexes are below normal, due to a variety of causes.

hyposensitivity [MED] Condition marked by diminished sensitivity to stimuli.

hyposensitization *See* desensitization.

hypospadias [MED] 1. Congenital anomaly in which the urethra opens on the ventral surface of the penis or in the perineum. 2. Congenital anomaly in which the urethra opens into the vagina.

hypospermatogenesis [MED] Decreased sperm production.

hypostatic [GEN] Subject to being suppressed, as a gene that can be suppressed by a nonallelic gene.

hyposthenia [MED] Weakness; subnormal strength.

hyposthenuria [MED] The secretion of urine of low specific gravity.

hypostome [INV ZOO] 1. Projection surrounding the oral aperture in many coelenterate polyps. 2. Anteroventral part of the head in Diptera. 3. Median ventral mouthpart in ticks. 4. Raised area on the posterior oral margin in crustaceans.

hypostracum [INV ZOO] The innermost layer of the cuticle of ticks lying above the hypodermis.

hyposulculus [INV ZOO] A groove of the siphonoglyph below the pharynx in anthozoans.

hyposynergia [MED] Defective coordination.

hypotarsus [VERT ZOO] A process on the tarsometatarsal bone in birds.

hypotelorism [MED] Decrease in distance between two organs or body parts.

hypotension [MED] Abnormally low blood pressure, commonly considered to be levels below 100 diastolic and 40 systolic.

hypotenuse [MATH] On a right triangle, the side opposite the right angle.

hypo test [GRAPHICS] A method of checking the washing efficiency of processed film or paper.

hypothalamic center [ANAT] Any of the neural centers which regulate autonomic functions.

hypothalamoneurohypophyseal system [PHYSIO] The hormones and neurosecretory structures involved in the endocrine activity of the adenohypophysis, neurohypophysis, and hypothalamus.

hypothalamoneurohypophyseal tract [ANAT] A bundle of nerve fibers connecting the supraoptic and paraventricular neurons of the hypothalamus with the infundibular stem and neurohypophysis.

hypothalamus [ANAT] The floor of the third brain ventricle; site of production of several substances that act on the adenohypophysis.

hypotheca [INV ZOO] 1. The lower valve of a diatom frustule. 2. Covering on the hypocone in dinoflagellates.

hypothenar [ANAT] Of or pertaining to the prominent portion of the palm above the base of the little finger.

hypothermal [GEOL] Referring to the high-temperature (300–500°C) environment of hypothermal deposits.

hypothermal deposit [MINERAL] Mineral deposit formed at great depths and high (300–500°C) temperatures.

hypothermia [PHYSIO] Condition of reduced body temperature in homeotherms.

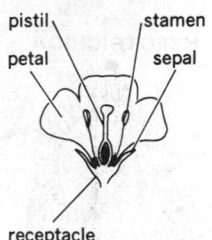

pistil · stamen
petal · sepal
receptacle

Flower arrangement and parts on receptacle in hypogynous flowers.

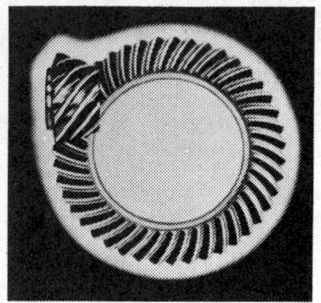

Picture of hypoid gear. *(Gleason Works)*

HYPOTRICHIDA

50 μ

Euplotes, an example of
Hypotrichida.

HYPOXANTHINE

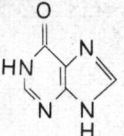

Structural formula of
hypoxanthine.

HYSTERESIS LOOP

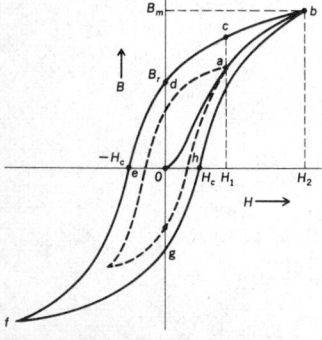

Hysteresis loop. H = magnetizing
force; B = flux density; Oab =
normal magnetization curve; H_c =
coercive force; B_r = retentivity;
B_m = maximum value of flux
density; $bdefghb$ = the hysteresis
loop.

hypothesis [SCI TECH] **1.** A proposition which is assumed to be true in proving another proposition. **2.** A proposition which is thought to be true because its consequences are found to be true. [STAT] A statement which specifies a population or distribution, and whose truth can be tested by sample evidence.

hypothesis testing [STAT] The branch of statistics which considers the problem of choosing between two actions on the basis of the observed value of a random variable whose distribution depends on a parameter, the value of which would indicate the correct action.

hypothetical parallax *See* dynamic parallax.

hypothyroidism [MED] Condition caused by deficient secretion of the thyroid hormone.

hypotonia [MED] Decrease of normal tonicity or tension, especially diminution of intraocular pressure or of muscle tone.

hypotonic [PHYSIO] **1.** Pertaining to subnormal muscle strength or tension. **2.** Referring to a solution with a lower osmotic pressure than physiological saline.

Hypotrichida [INV ZOO] An order of highly specialized protozoans in the subclass Spirotrichia characterized by cirri on the ventral surface and a lack of ciliature on the dorsal surface.

hypotrochoid [MATH] A curve traced by a point rigidly attached to a circle at a point other than the center when the circle rolls without slipping on the inside of a fixed circle.

hypotype [SYST] A specimen of a species, which, though not a member of the original type series, is known from a published description or listing.

hypovitaminosis [MED] Condition due to deficiency of an essential vitamin.

hypovolemia [MED] Low blood volume.

hypovolemic shock [MED] Shock caused by reduced blood volume which may be due to loss of blood or plasma as in burns, the crush syndrome, perforating gastrointestinal wounds, or other trauma. Also known as wound shock.

hypoxanthine [BIOCHEM] $C_5H_4ON_4$ An intermediate product derived from adenine in the hydrolysis of nucleic acid.

hypoxanthylic acid *See* inosinic acid.

hypoxemia *See* hypoxia.

hypoxia [MED] Oxygen deficiency; any state wherein a physiologically inadequate amount of oxygen is available to or is utilized by tissue, without respect to cause or degree. Also known as hypoxemia.

hypoxic encephalopathy [MED] Brain damage syndrome caused by hypoxia.

hypozygal [INV ZOO] In comatulids, the proximal member of adjacent brachials in an articulation.

hypsicephalic [ANTHRO] Having a high forehead with a length-height index of 62.6 or more.

hypsiconch [ANTHRO] Having high orbits with an orbital index of 89 or more.

hypsicranial [ANTHRO] Having a high skull with a length-height index of 75 or more.

hypsidolichocephalic [ANTHRO] Having a head that is high and narrow, high and long, or high, long, and narrow.

hypsistenocephalic [ANTHRO] Having a very high and narrow head.

hypsodont [VERT ZOO] Of teeth, having crowns that are high or deep and roots that are short.

hypsographic map [MAP] A chart showing topographic relief in reference to a given datum, usually sea level.

hypsography [GEOGR] The science of measuring or describing elevations of the earth's surface with reference to a given datum, usually sea level.

hypsometer [ENG] **1.** An instrument for measuring atmospheric pressure to ascertain elevations by determining the boiling point of liquids. **2.** Any of several instruments for determining tree heights by triangulation.

hypsometric [ENG] Pertaining to hypsometry.

hypsometric formula [GEOPHYS] A formula, based on the hydrostatic equation, for either determining the geopotential difference or thickness between any two pressure levels, or for reducing the pressure observed at a given level to that at some other level.

hypsometric map [MAP] In topographic surveying, a map giving elevations by contours, or sometimes by means of shading, tinting, or batching.

hypsometric tinting [MAP] A technique of showing relief on maps and charts by coloring, in different shades, those parts which lie between different levels. Also known as altitude tints.

hypsometry [ENG] The measuring of elevation with reference to sea level.

hypsophobia [PSYCH] An abnormal fear of being at a great height.

hypural [VERT ZOO] Of or pertaining to the bony structure formed by fusion of the hemal spines of the last few vertebrae in most teleost fishes.

Hyracodontidae [PALEON] The running rhinoceroses, an extinct family of perissodactyl mammals in the superfamily Rhinoceratoidea.

Hyracoidea [VERT ZOO] An order of ungulate mammals represented only by the conies of Africa, Arabia, and Syria.

hyster- [MED] A combining form that denotes either a relation to or a connection with the uterus, or hysteria.

hysterectomy [MED] Surgical removal of all or part of the uterus.

hysteresimeter [ENG] A device for measuring hysteresis.

hysteresis [ELECTR] An oscillator effect wherein a given value of an operating parameter may result in multiple values of output power or frequency. [ELECTROMAG] *See* magnetic hysteresis. [NUCLEO] A temporary change in the counting-rate-voltage characteristic of a radiation counter tube, caused by its previous operation. [PHYS] The dependence of the state of a system on its previous history, generally in the form of a lagging of a physical effect behind its cause.

hysteresis clutch [MECH ENG] A clutch in which torque is produced by attraction between induced poles in a magnetized iron ring and the control field.

hysteresis coefficient [PHYS] A constant, characteristic of a particular material, in a formula for hysteresis loss.

hysteresis damping [MECH] Damping of a vibration due to energy lost through mechanical hysteresis.

hysteresis error [PHYS] The maximum separation due to hysteresis between upscale-going and downscale-going indications of a measured variable.

hysteresis loop [PHYS] The closed curve followed by a material displaying hysteresis (such as a ferromagnet or ferroelectric) on a graph of a driven variable (such as magnetic flux density or electric polarization) versus the driving variable (such as magnetic field or electric field).

hysteresis loss [PHYS] The energy converted to heat in a material because of magnetic or other hysteresis, accompanying cyclic variation of the magnetic field or other driving variable.

hysteresis motor [ELEC] A synchronous motor without salient poles and without direct-current excitation which utilizes the hysteresis and eddy-current losses induced in its hardened-steel rotor to produce rotor torque.

hysteria [PSYCH] A type of neurosis characterized by extreme emotionalism involving disorders of somatic and psychological functions; the conversion type is associated with neuromuscular and sensory symptoms such as paralysis, tremors, seizures, or blindness, whereas the dissociative displays disorders of consciousness such as amnesia, somnolence, and multiple personalities.

hysteriaceous [MYCOL] Of, belonging to, or resembling the Hysteriales.

Hysteriales [BOT] An order of lichens in the class Ascolichenes including those species with an ascolocular structure.

hysterical anesthesia [MED] Loss of cutaneous pain sensation accompanying hysteria.

hysterical paralysis [MED] Muscle weakness or paralysis without loss of reflex activity, in which no organic nerve lesion can be demonstrated, but which is due to psychogenic factors. Also known as functional paralysis.

hysterics [PSYCH] **1.** Attack of hysteria. **2.** Extreme display of emotions.

hysterochroic [MYCOL] Having fruiting bodies which discolor progressively from base to apex with age.

hysterogram [MED] A roentgenogram with opacification of the cavity of the uterus by the injection of contrast medium.

hysterography [MED] Roentgenologic examination of the uterus after the introduction of a contrast medium.

hysteromania *See* nymphomania.

hystero-oophorectomy [MED] Surgical removal of the uterus and ovaries.

hysteropexy [MED] Fixation of the uterus by a surgical operation to correct displacement.

hysterorrhaphy [MED] The closure of a uterine incision by suture.

hysterorrhexis [MED] Rupture of the uterus.

hysterosalpingectomy [MED] Surgical removal of the uterus and oviducts.

hysterosalpingography [MED] Roentgenographic examination of the uterus and oviducts after injection of a radiopaque substance.

hysterosalpingo-oophorectomy [MED] The excision of the uterus, oviducts, and ovaries.

hysteroscope [MED] A uterine speculum with a reflector.

hysterosoma [INV ZOO] A body division of an acarid mite composed of the metapodosoma and opisthosoma.

hysterotomy [MED] **1.** Incision into the uterine wall. **2.** Cesarean section.

Hystrichospherida [PALEON] A group of protistan microfossils.

Hystricidae [VERT ZOO] The Old World porcupines, a family of Rodentia ranging from southern Europe to Africa and eastern Asia and into the Philippines.

Hystricomorpha [VERT ZOO] A superorder of the class Rodentia.

Hz *See* hertz.

H zone [HISTOL] The central portion of an A band in a sarcomere; characterized by the presence of myosin filaments.

HYSTRICHOSPHERIDA

75 μ

A drawing of a hystrichospherid.

I *See* iodine.

IA *See* international angstrom.

IAA *See* indoleacetic acid.

IAC *See* international analysis code.

ianthinite [MINERAL] $2UO_2 \cdot 7H_2O$ A violet mineral composed of hydrous uranium dioxide, occurring as orthorhombic crystals.

Iapetus [ASTRON] A satellite of Saturn with diameter of about 1000 miles.

IAS *See* indicated airspeed.

iatrogenic [MED] Effected by a physician, usually referring to psychosomatic illness induced by his words, actions, or treatments upon the patient.

IBA *See* indolbutyric acid.

Iballidae [INV ZOO] A small family of hymenopteran insects in the superfamily Cynipoidea.

I band [HISTOL] The band on either side of a Z line; encompasses portions of two adjacent sarcomeres and is characterized by the presence of actin filaments.

I beam [CIV ENG] A rolled iron or steel joist having an I section, with short flanges.

Ibe wind [METEOROL] A local strong wind which blows through the Dzungarian Gate (western China), a gap in the mountain ridge separating the depression of Lakes Balkash and Ala Kul from that of Lake Ebi Nor; the wind resembles the foehn and brings a sudden rise of temperature, in winter from about -15 to about $30°F$ (-20 to $-1°C$).

ibis [VERT ZOO] The common name for wading birds making up the family Threskiornithidae and distinguished by a long, slender, downward-curving bill.

I blood group [IMMUNOL] The erythrocyte antigens defined by reactions with anti-I and anti-i antibodies, which occur both in acquired hemolytic anemia and naturally in normal persons of the rare phenotype i.

ibogaine [ORG CHEM] $C_{26}H_{32}O_2N_2$ An alkaloid isolated from the stems and leaves of the shrub *Tabernanthe iboga*, crystallizing from absolute ethanol as prismatic needles, melting at 152-153°C, soluble in ethanol, ether, and chloroform; used in medicine.

IC *See* integrated circuit.

Icacinaceae [BOT] A family of dicotyledonous plants in the order Celastrales characterized by haplostemonous flowers, pendulous ovules, stipules wanting or vestigial, a polypetalous corolla, valvate petals, and usually one (sometimes three) locules.

Icarus [ASTRON] An asteroid with an orbit of highest known eccentricity and smallest perihelion and distance of all the asteroids.

ICBM *See* intercontinental ballistic missile.

ice [PHYS CHEM] **1.** The dense substance formed by the freezing of water to the solid state; has a melting point of 32°F (0°C) and commonly occurs in the form of hexagonal crystals. **2.** A layer or mass of frozen water.

ice accretion [HYD] The process by which a layer of ice builds up on solid objects which are exposed to freezing precipitation or to supercooled fog or cloud droplets.

ice-accretion indicator [ENG] An instrument used to detect the occurrence of freezing precipitation, usually consisting of a strip of sheet aluminum about 1½ inches (4 centimeters) wide, and is exposed horizontally, face up, in the free air a few meters above the ground.

ice age [GEOL] A major interval of geologic time during which extensive ice sheets (continental glaciers) formed over many parts of the world.

ice anchor [NAV ARCH] An anchor, usually with only one fluke, used for securing a vessel to ice.

ice apron [CIV ENG] A wedge-shaped structure which protects a bridge pier from floating ice. [HYD] **1.** The snow and ice attached to the walls of a cirque. **2.** The ice that is flowing

from an ice sheet over the edge of a plateau. **3.** A piedmont glacier's lobe. **4.** Ice that adheres to a wall of a valley below a hanging glacier.

ice atlas [NAV] A publication containing a series of ice charts showing geographic distribution of ice, usually by seasons or months; used in navigation.

ice band [HYD] A layer of ice in firn or snow.

ice barrier [HYD] The periphery of the Antarctic ice sheet; or used generally for any ice dam.

ice bay [OCEANOGR] A baylike recess in the edge of a large ice floe or ice shelf. Also known as ice bight.

ice belt [OCEANOGR] A band of fragments of sea ice in otherwise open water. Also known as ice strip.

iceberg [OCEANOGR] A large mass of detached land ice floating in the sea or stranded in shallow water.

ice bight *See* ice bay.

ice blink [METEOROL] A relatively bright, usually yellowish-white glare on the underside of a cloud layer, produced by light reflected from an ice-covered surface such as pack ice; used in polar regions with reference to the sky map; ice blink is not as bright as snow blink, but much brighter than water sky or land sky.

icebound [NAV] Surrounded so closely by ice as to be incapable of proceeding.

icebreaker [NAV ARCH] A vessel designed for operating in and breaking heavy ice, having a special-shaped, reinforced bow, protected propellers, and powerful engines.

ice bridge [OCEANOGR] Surface river ice of sufficient thickness to impede or prevent navigation.

ice buoy [ENG] A sturdy buoy, usually a metal spar, used to replace a more easily damaged buoy during a period when heavy ice is anticipated.

ice calorimeter *See* Bunsen ice calorimeter.

ice calving *See* calving.

ice canopy *See* pack ice.

ice cap [HYD] **1.** A perennial cover of ice and snow in the shape of a dome or plate on the summit area of a mountain through which the mountain peaks emerge. **2.** A perennial cover of ice and snow on a flat land mass such as an Arctic island.

ice-cap climate *See* perpetual frost climate.

ice cascade *See* icefall.

ice cave [GEOL] A cave that is cool enough to hold ice through all or most of the warm season. [HYD] A cave in ice such as a glacier formed by a stream of melted water.

ice chart [NAV] A chart showing prevalence of ice, usually with reference to navigable waters.

ice clearing *See* polyn'ya.

ice color *See* azoic dye.

ice concentration [OCEANOGR] In sea ice reporting, the ratio of the areal extent of ice present and the total areal extent of ice and water. Also known as ice cover.

ice-contact delta [GEOL] A delta formed by a stream flowing between a valley slope and the margin of glacial ice. Also known as delta moraine; morainal delta.

ice cover *See* ice concentration.

ice crust [HYD] A type of snow crust; a layer of ice, thicker than a film crust, upon a snow surface, formed by the freezing of meltwater or rainwater which has flowed onto it.

ice crystal [PHYS CHEM] Any one of a number of macroscopic crystalline forms in which ice appears, including hexagonal columns, hexagonal platelets, dendritic crystals, ice needles, and combinations of these forms; although the crystal lattice of ice is hexagonal in its symmetry, varying conditions of temperature and vapor pressure can lead to growth of crystalline forms in which the simple hexagonal pattern is almost undiscernible.

ice-crystal cloud [METEOROL] A cloud consisting entirely of ice crystals, such as cirrus (in this sense distinguished from

ICEBERG

Arctic iceberg, eroded to form a valley or drydock type; grotesque shapes are common to the glacially produced icebergs of the North.

water clouds and mixed clouds), and having a diffuse and fibrous appearance quite different from that typical of water droplet clouds.

ice-crystal fog *See* ice fog.

ice-crystal haze [METEOROL] A type of very light ice fog composed only of ice crystals and at times observable to altitudes as great as 20,000 feet (6100 meters), and usually associated with precipitation of ice crystals.

ice-crystal theory *See* Bergeron-Findeisen theory.

ice day [CLIMATOL] A day on which the maximum air temperature in a thermometer shelter does not rise above 32°F (0°C), and ice on the surface of water does not thaw.

ice desert [CLIMATOL] Any polar area permanently covered by ice and snow, with no vegetation other than occasional red snow or green snow.

iced firn [HYD] A mixture of glacier ice and firn; firn permeated with meltwater and then refrozen. Also known as firn ice.

ice erosion [GEOL] 1. Erosion due to freezing of water in rock fractures. 2. *See* glacial erosion.

icefall [HYD] That portion of a glacier where a sudden steepening of descent causes a chaotic breaking up of the ice. Also known as ice cascade.

ice fat *See* grease ice.

ice feathers [HYD] A type of hoarfrost formed on the windward side of terrestrial objects and on aircraft flying from cold to warm air layers. Also known as frost feathers.

ice field [HYD] A mass of land ice resting on a mountain region and covering all but the highest peaks. [OCEANOGR] A flat sheet of sea ice that is more than 5 miles (8 kilometers) across.

ice floe *See* floe.

ice flowers [HYD] 1. Formations of ice crystals on the surface of a quiet, slowly freezing body of water. 2. Delicate tufts of hoarfrost that occasionally form in great abundance on an ice or snow surface. Also known as frost flowers. 3. Frost crystals resembling a flower, formed on salt nuclei on the surface of sea ice as a result of rapid freezing of sea water. Also known as salt flowers.

ice fog [METEOROL] A type of fog composed of suspended particles of ice, partly ice crystals 20–100 micrometers in diameter but chiefly, especially when dense, droxtals 12–20 micrometers in diameter; occurs at very low temperatures and usually in clear, calm weather in high latitudes. Also known as frost flakes; frost fog; frozen fog; ice-crystal fog; pogonip; rime fog.

ice foot [OCEANOGR] Sea ice firmly frozen to a polar coast at the high-tide line and unaffected by tide; this fast ice is formed by the freezing of seawater during ebb tide, and of spray, and it is separated from the floating sea ice by a tide crack.

ice-free [HYD] 1. Referring to a harbor, river, estuary, and so on, when there is not sufficient ice present to interfere with navigation. 2. Descriptive of a water surface completely free of ice.

ice fringe [HYD] An ice deposit on plant surfaces, not of hoarfrost from atmospheric water vapor, but of moisture exuded from the stems of plants and appearing as frosted fringes or ribbons. Also known as ice ribbon. [OCEANOGR] A belt of sea ice extending a short distance from the shore.

ice gland [HYD] A column of ice in the granular snow at the top of a glacier.

ice gruel [HYD] A type of slush formed by the irregular freezing together of ice crystals.

ice island [OCEANOGR] A large tabular fragment of shelf ice found in the Arctic Ocean and having an irregular surface, thickness of 15–50 meters, and an area between a few thousand square meters and 500 square kilometers or more.

ice island iceberg [OCEANOGR] An iceberg having a conical or dome-shaped summit, often mistaken by mariners for ice-covered islands.

ice jam [HYD] 1. An accumulation of broken river ice caught in a narrow channel, frequently producing local floods during a spring breakup. 2. Fields of lake or sea ice thawed loose from the shores in early spring, and blown against the shore, sometimes exerting great pressure.

ice-laid drift *See* till.

Iceland agate *See* obsidian.

Iceland crystal *See* Iceland spar.

Iceland disease *See* epidemic neuromyasthenia.

Icelandic low [METEOROL] 1. The low-pressure center located near Iceland (mainly between Iceland and southern Greenland) on mean charts of sea-level pressure. 2. On a synoptic chart, any low centered near Iceland.

Iceland spar [MINERAL] A pure, transparent form of calcite found particularly in Iceland; easily cleaved to form rhombohedral crystals that are doubly refracting. Also known as Iceland crystal.

ice layer [HYD] An ice crust covered with new snow; when exposed at a glacier front or in crevasses, the ice layers viewed in cross section are termed ice bands.

ice load [ENG] The weight of glaze deposited on an overhead wire in a power supply system; standard safety codes require allowance for ½-inch (12.7 millimeters) radial thickness in heavy loading districts and ¼-inch (6.35 millimeters) in medium.

ice mantle *See* ice sheet.

ice mine [ORD] A waterproof mine placed in or under the ice, detonated by a pressure device on the surface to break up river or lake ice.

ice needle [PHYS CHEM] A long, thin ice crystal whose cross section perpendicular to its long dimension is typically hexagonal. Also called ice spicule.

ice nucleus [METEOROL] Any particle which may act as a nucleus in formation of ice crystals in the atmosphere.

ice pack *See* pack ice.

ice pellets [METEOROL] A type of precipitation consisting of transparent or translucent pellets of ice 5 millimeters or less in diameter; may be spherical, irregular, or (rarely) conical in shape.

ice pick [DES ENG] A hand tool for chipping ice.

ice pillar [HYD] A column of glacial ice covered with stones or debris which tend to protect the ice from melting.

ice point [PHYS CHEM] The true freezing point of water; the temperature at which a mixture of air-saturated pure water and pure ice may exist in equilibrium at a pressure of 1 standard atmosphere.

ice pole [GEOGR] The approximate center of the most consolidated portion of the arctic pack ice, near 83 or 84°N and 160°W. Also known as pole of inaccessibility.

ice push [GEOL] Lateral pressure that is caused by expansion of shoreward-moving ice on a lake or a bay of the sea and that follows a rise in temperature. Also known as ice shove; ice thrust.

ice ribbon *See* ice fringe.

ice rind [HYD] A thin but hard layer of sea ice, river ice, or lake ice, which is either a new encrustation upon old ice or a single layer of ice usually found in bays and fiords, where fresh water freezes on top of slightly colder sea water.

ice run [HYD] The initial stage in the spring or summer breakup of river ice, being an exceedingly rapid process, seldom taking more than 1 day.

ice sheet [HYD] A thick glacier, more than 50,000 square kilometers in area, forming a cover of ice and snow that is continuous over a land surface and moving outward in all directions. Also known as ice mantle.

ice shelf [OCEANOGR] A thick sheet of ice with a fairly level or undulating surface, formed along a polar coast and in shallow bays and inlets, fastened to the shore along one side but mostly afloat and nourished by annual accumulation of snow and by the seaward extension of land glaciers.

ice shove *See* ice push.

ice spar *See* sanidine.

ice spicule *See* ice needle.

ice splinters [PHYS CHEM] Minute, electrically charged fragments of ice which have been observed under laboratory conditions to be torn away from dendritic crystals or spatial aggregates exposed to moving air.

ice stone *See* cryolite.

ice storm [METEOROL] A storm characterized by a fall of freezing precipitation, forming a glaze on terrestrial objects that creates many hazards. Also known as silver storm.

ice stream [HYD] A current of ice flowing in an ice sheet or ice cap; usually moves toward an ocean or to an ice shelf.

ice strip *See* ice belt.

ice thrust   *See* ice push.

ice tongs   [DES ENG] Tongs for handling cubes or blocks of ice.

ice tongue   [HYD] Any narrow extension of a glacier or ice shelf, such as a projection floating in the sea or an outlet glacier of an ice cap.

ich   [VET MED] A dermatitis of fresh-water fishes caused by the invasion of the skin by the ciliated protozoan *Ichthyophthirius multifiliis*.

ichneumon   [INV ZOO] The common name for members of the family Ichneumonidae.

Ichneumonidae   [INV ZOO] The ichneumon flies, a large family of cosmopolitan, parasitic wasps in the superfamily Ichneumonoidea.

Ichneumonoidea   [INV ZOO] A superfamily of hymenopteran insects; members are parasites of other insects.

ichnofossil   *See* trace fossil.

ichnography   [GRAPHICS] **1.** The art of making decorative drawings by use of compass and rule, in a sense revived in Op art (optical art) techniques. **2.** The art of tracing plans and illustrations.

ichor   [GEOL] A fluid rich in mineralizers. [MED] An acrid, watery or blood-tinged discharge from an ulcer or wound.

ichthammol   [PHARM] A viscid fluid obtained by the destructive distillation of certain bituminous schists, followed by sulfonation of the distillate and neutralization with ammonia; used as a weak antiseptic and stimulant in skin diseases, and occasionally as an expectorant. Also known as ammonium bithiolicum; ammonium ichthosulfonate; ammonium sulfoichthyolate.

ichthyism   [MED] Food poisoning caused by eating spoiled fish.

Ichthyobdellidae   [INV ZOO] A family of leeches in the order Rhynchobdellae distinguished by cylindrical bodies with conspicuous, powerful suckers.

ichthyocolla   *See* isinglass.

Ichthyodectidae   [PALEON] A family of Cretaceous marine osteoglossiform fishes.

ichthyol   [PHARM] A trademark for ichthammol.

ichthyology   [VERT ZOO] A branch of vertebrate zoology that deals with the study of fishes.

ichthyophagous   [ZOO] Subsisting on a diet of fish.

ichthyophobia   [PSYCH] An abnormal fear of fish.

Ichthyopterygia   [PALEON] A subclass of extinct Mesozoic reptiles composed of predatory fish-finned and sea-swimming forms with short necks and a porpoiselike body.

Ichthyornis   [PALEON] The type genus of Ichthyornithidae.

Ichthyornithes   [PALEON] A superorder of fossil birds of the order Ichthyornithiformes according to some systems of classification.

Ichthyornithidae   [PALEON] A family of extinct birds in the order Ichthyornithiformes.

Ichthyornithiformes   [PALEON] An order of ancient fossil birds including strong flying species from the Upper Cretaceous that possessed all skeletal characteristics of modern birds.

Ichthyosauria   [PALEON] The only order of the reptilian subclass Ichthyopterygia, comprising the extinct predacious fish-lizards; all were adapted to a sea life in having tail flukes, paddles, and dorsal fins.

ichthyosis   [MED] A congenital skin disease characterized by dryness and scales, especially on the extensor surfaces of the extremities.

ichthyosis congenita   [MED] A severe form of ichthyosis characterized by cracked, thickened skin and mucous membranes.

ichthyosis simplex   [MED] A common, childhood form of ichthyosis characterized by large, papery scales on the skin.

Ichthyostegalia   [PALEON] An extinct Devonian order of labyrinthodont amphibians, the oldest known representatives of the class.

Ichthyotomidae   [INV ZOO] A monotypic family of errantian annelids in the superfamily Eunicea.

icicle   [HYD] Ice shaped like a narrow cone, hanging point downward from a roof, fence, or other sheltered or heated source from which water flows and freezes in below-freezing air.

icing   [HYD] **1.** Any deposit or coating of ice on an object,

caused by the impingement and freezing of liquid (usually supercooled) hydrometeors. **2.** A mass or sheet of ice formed on the ground surface during the winter by successive freezing of sheets of water that may seep from the ground, from a river, or from a spring. Also known as flood icing; flooding ice.

icing level   [METEOROL] The lowest level in the atmosphere at which an aircraft in flight does, or could, encounter aircraft icing conditions over a given locality.

icing-rate meter   [ENG] An instrument for the measurement of the rate of ice accretion on an unheated body.

ICL   *See* lifting condensation level.

ICNI   *See* integrated communications-navigation-identification.

iconic imagery   [PSYCH] Fleeting images due perhaps to continuing activity of sensory organs following stimulation.

iconocenter   [ELECTROMAG] The image of the reflection coefficient of a matched load as plotted on an Argand diagram.

iconometer   [OPTICS] **1.** An instrument used to find the size of an object of known distance or the distance of an object of known size by measurement of the image of it produced by a lens whose focal length is known. **2.** A direct viewfinder with a metal frame.

iconoscope   [ELECTR] A television camera tube in which a beam of high-velocity electrons scans a photoemissive mosaic that is capable of storing an electric charge pattern corresponding to an optical image focused on the mosaic. Also known as storage camera; storage-type camera tube.

icosahedral virus   [VIROL] A virion in the form of an icosahedron.

icosahedron   [MATH] A 20-sided polyhedron.

icositetrahedron   *See* trapezohedron.

Icosteidae   [VERT ZOO] The ragfishes, a family of perciform fishes in the suborder Stromateoidei found in high seas.

icotype   [SYST] A typical, accurately identified specimen of a species, but not the basis for a published description.

ICSH   *See* luteinizing hormone.

ICS system   *See* intercarrier sound system.

ICT   *See* International Critical Tables.

icteric   [MED] Pertaining to or characterized by jaundice.

Icteridae   [VERT ZOO] The troupials, a family of New World perching birds in the suborder Oscines.

icteroanemia   [VET MED] A disease of swine characterized by jaundice, anemia, and erythrocytolysis.

icterogenic   [MED] Causing icterus, or jaundice.

icterohematuria   [VET MED] A disease of sheep caused by the protozoan *Babesia ovis* and characterized by hemolysis of erythrocytes accompanied by jaundice.

icterohemorrhagic fever   *See* Weil's disease.

icterus   *See* jaundice.

icterus gravis   [MED] Acute yellow atrophy of the liver marked by jaundice and nervous system dysfunctions.

icterus gravis neonatorum   [MED] Icterus gravis in the newborn caused by physiologic jaundice, erythroblastosis fetalis, and severe jaundice.

icterus index   [PATH] Measure of serum bilirubin levels by comparing the yellow blood serum from a jaundiced patient with the colors of standard potassium dichromate solutions.

Ictidosauria   [PALEON] An extinct order of mammallike reptiles in the subclass Synapsida including small carnivorous and herbivorous terrestrial forms.

ICW   *See* interrupted continuous wave.

id   [PSYCH] The primitive, psychic energy source of the unconscious.

ID   *See* inside diameter.

ID$_{50}$ *See* infective dose 50.

iddingsite   [MINERAL] A reddish-brown mixture of silicates, forming patches in basic igneous rocks.

idea   [PSYCH] **1.** A mental impression or thought. **2.** An experience or thought not directly due to an external sensory stimulation.

ideal   [MATH] A subset I of a ring R where $x - y$ is in I for every x,y in I and either rx is in I for every r in R and x in I or xr is in I for every r in R and x in I; in the first case I is called a left ideal, and in the second a right ideal; an ideal is two-sided if it is both a left and a right ideal.

ideal aerodynamics   [FL MECH] A branch of aerodynamics that deals with simplifying assumptions that help explain

ICHNEUMON

Long-tailed ichneumon (*Megarhyssa lunator*), about 6 inches (15 centimeters) long of which more than 4 inches (10 centimeters) is "tail."

ICHTHYOSAURIA

A Jurassic ichthyosaur, much reduced. (*Simplified from E. von Stromer, as used by A. S. Romer, Vertebrate Paleontology, 3d ed., University of Chicago Press, 1966*)

some airflow problems and provide approximate answers. Also known as ideal fluid dynamics.

ideal bunching [ELECTR] Theoretical condition in which the bunching of electrons in a velocity-modulated tube would give a single infinitely large current peak during each cycle.

ideal dielectric [ELEC] Dielectric in which all the energy required to establish an electric field in the dielectric is returned to the source when the field is removed. Also known as perfect dielectric.

ideal exhaust velocity [FL MECH] The theoretical maximum velocity, relative to the nozzle, of the gas flow as it passes from a given nozzle inlet temperature and pressure to a given ambient pressure, when the combustion gas has a given mean molecular weight.

ideal flow [FL MECH] **1.** Fluid flow which is incompressible, two-dimensional, irrotational, steady, and nonviscous. **2.** See inviscid flow.

ideal fluid [FL MECH] **1.** A fluid which has ideal flow. **2.** See inviscid fluid.

ideal fluid dynamics See ideal aerodynamics.

ideal gas [THERMO] Also known as perfect gas. **1.** A gas whose molecules are infinitely small and exert no force on each other. **2.** A gas that obeys Boyle's law (the product of the pressure and volume is constant at constant temperature) and Joule's law (the internal energy is a function of the temperature alone).

ideal gas law [THERMO] The equation of state of an ideal gas which is a good approximation to real gases at sufficiently high temperatures and low pressures; that is, $PV = RT$, where P is the pressure, V is the volume per mole of gas, T is the temperature, and R is the gas constant.

ideal grain [ORD] Propellant granulation which produces the maximum obtainable velocity of a projectile without exceeding the permissible pressure at any point along the bore.

ideal line [MATH] The collection of all ideal points, each corresponding to a given family of parallel lines.

ideal network [ELECTR] An interconnection of lumped, constant electrical quantities analyzed without consideration of noise and distributed parameters that would exist in actual settings.

ideal point [MATH] In projective geometry, all lines parallel to a given line are hypothesized to meet at a point at infinity, called an ideal point.

ideal productivity index [PETRO ENG] Theoretical straight-line relationship between oil production from a reservoir and the resultant pressure drop within that reservoir.

ideal radiator See blackbody.

ideal rocket [AERO ENG] A rocket motor or rocket engine that would have a velocity equal to the velocity of its jet gases.

ideal solution [CHEM] A solution that conforms to Raoult's law over all ranges of temperature and concentration and shows no internal energy change on mixing and no attractive force between components.

ideal theory [MATH] The branch of algebra studying the properties of ideals.

ideal transducer [ELEC] Hypothetical passive transducer which transfers the maximum possible power from the source to the load.

ideal transformer [ELEC] A hypothetical transformer that neither stores nor dissipates energy, has unity coefficient of coupling, and has pure inductances of infinitely great value.

ideas of influence [PSYCH] A clinical manifestation of certain psychotic disorders in which the patients may believe that their thoughts are read, that their limbs move without their consent, or that they are under the control of someone else or some external force or influence.

ideas of reference [PSYCH] A symptom complex in which, through the mechanism of projection, an individual incorrectly believes himself or herself to be the direct object of casual remarks or incidents or of external events.

ideation [PSYCH] Conceptualization of an idea.

ideational apraxia [MED] Inability to perform meaningful motor functions or to use objects properly due to mental confusion caused by diffuse brain disease.

idem factor [MATH] The dyadic $I = ii + jj + kk$ such that scalar multiplication of I by any vector yields that vector.

I demodulator [ELECTR] Stage of a color television receiver which combines the chrominance signal with the color oscillator output to restore the I signal.

idempotent [MATH] **1.** An element x of an algebraic system satisfying the equation $x^2 = x$. **2.** An algebraic system in which every element x satisfies $x^2 = x$.

idempotent law [MATH] A law which states that every element x of an algebraic system satisfies $x^2 = x$.

idempotent matrix [MATH] A matrix E satisfying the equation $E^2 = E$.

identical twins See monozygotic twins.

identifiable flying object [SCI TECH] A reported event or sighting identifiable as such; a thing as a balloon, meteor, planet, aircraft, or mirage reflection. Abbreviated IFO.

identification [CONT SYS] The procedures for deducing a system's transfer function from its response to a step-function input or to an impulse. [PSYCH] **1.** The tendency of children to model their behavior after that of one or more selected adults. **2.** A defense mechanism in which a person likens himself or herself to someone else.

identification division [ADP] The section of a program, written in the COBOL language, which contains the name of the program and the name of the programmer.

identification, friend or foe [ENG] A system using pulsed radio transmissions to which equipment carried by friendly forces automatically responds, by emitting a pulse code, thereby identifying themselves from enemy forces; a method of determining the friendly or unfriendly character of aircraft, ships, and army units by other aircraft, ships, or ground force units. Abbreviated IFF.

identifier [ADP] A symbol whose purpose is to specify a body of data.

identifier word [ADP] A full-length computer word associated with a search function.

identity [MATH] An equation satisfied for all possible choices of values for the variables involved.

identity crisis [PSYCH] The critical period in emotional maturation and personality development, occurring usually during adolescence, which involves the reworking and abandonment of childhood identifications and the integration of new personal and social identifications.

identity element [MATH] The unique element e of a group where $g \cdot e = e \cdot g$ for every element g of the group. Also known as identity of a group.

identity function [MATH] The function of a set to itself which assigns to each element the same element.

identity matrix [MATH] The square matrix all of whose entries are zero save along the principal diagonal where they all are 1.

identity of a group See identity element.

ideogenous See syngenetic.

ideokinetic apraxia See ideomotor apraxia.

ideomotor [PHYSIO] **1.** Pertaining to involuntary movement resulting from or accompanying some mental activity, as moving the lips while reading. **2.** Pertaining to both ideation and motor activity.

ideomotor apraxia [MED] A nervous system disorder caused by lesions of the cerebral cortex in which simple single acts can be performed but not a sequence of associated acts. Also known as ideokinetic apraxia.

ideotype [SYST] A specimen identified as belonging to a specific taxon but collected from other than the type locality.

ID grinding [MET] The grinding of the inner diameter of a piece of tubing or piping.

idioblast [BOT] A plant cell that differs markedly in shape or function from neighboring cells within the same tissue.

idiochromatic [MINERAL] Having characteristic color, usually applied to minerals.

idiochromatin [CYTOL] The portion of the nuclear chromatin thought to function as the physical carrier of genes.

idiocy [PSYCH] The lowest grade of mental deficiency; the individual's mental age is less than 3 years.

idioglossia [PSYCH] Speech or other vocalizations unique to an individual and generally incomprehensible to others; may be normal, as during childhood development, or representative of a pathological process.

idiomorphic See automorphic.

idiomuscular [PHYSIO] Pertaining to any phenomenon occurring in a muscle which is independent of outside stimuli.

idiopathic arthritis *See* Reiter's syndrome.

idiopathic colitis [MED] Any form of colitis for which the causative agent is not identified.

idiopathic eunuchoidism [MED] A primary type of eunuchoidism in which there is no mammary gland enlargement and testes remain at a prepubertal stage of development.

idiopathic familial jaundice [MED] A familial form of obstructive jaundice of unknown cause in which there is decreased ability to excrete conjugated bilirubin into the bile ducts.

idiopathic hypercholesterolemia [MED] A genetic derangement of fat metabolism characterized by high blood, cell, and plasma levels of cholesterol.

idiopathic megacolon [MED] Hypertrophy and dilation of the colon.

idiopathic pulmonary hemisiderosis [MED] A disease of unknown etiology characterized by recurrent hemorrhaging from pulmonary capillaries.

idiopathic steatorrhea *See* celiac syndrome.

idiopathic thrombocytopenic purpura [MED] Thrombocytopenic purpura of unknown causes.

idiopathic ulcerative colitis [MED] A form of ulcerative colitis of unknown cause.

idiopathy [MED] **1.** A primary disease; one not a result of any other disease, but of spontaneous origin. **2.** Disease for which no cause is known.

idiosome [CYTOL] **1.** A hypothetical unit of a cell, such as the region of modified cytoplasm surrounding the centriole or centrosome. **2.** A sex chromosome.

Idiostolidae [INV ZOO] A small family of hemipteran insects in the superfamily Pentatomorpha.

idiosyncrasy [MED] A peculiarity of constitution that makes an individual react differently from most persons, to drugs, diet, treatment, or other situations. [PSYCH] Any special or peculiar characteristic or temperament by which a person differs from other persons.

idiot [PSYCH] A person afflicted with idiocy and requiring custodial or protective care.

I display [ELECTR] A radarscope display in which a target appears as a complete circle when the radar antenna is correctly pointed at it, the radius of the circle being proportional to target distance; when the antenna is not aimed at the target, the circle reduces to a circle segment.

idle [MECH ENG] To run without a load.

idle current [ELEC] Reactive volt-amperes.

idler frequency [ELECTR] Of a parametric device, a sum or difference frequency generated within the parametric device other than the input, output, or pump frequencies which require specific circuit consideration to achieve the desired device performance; it is called an idler frequency since, in conventional parametric amplifiers, it is more or less a useless by-product of the parametric process.

idler gear [MECH ENG] A gear situated between a driving gear and a driven gear to transfer motion, without any change of direction or of gear ratio.

idler pulley [MECH ENG] A pulley used to guide and tighten the belt or chain of a conveyor system.

idler wheel [MECH ENG] **1.** A wheel used to transmit motion or to guide and support something. **2.** A roller with a rubber surface used to transfer power by frictional means in a sound-recording or sound-reproducing system.

idle time [ADP] The time during which a piece of hardware in good operating condition is unused. [IND ENG] A period of time during a regular work cycle when a worker is not active because of waiting for materials or instruction. Also known as waiting time.

idle trunk lamp [ELEC] Signal lamp associated with an outgoing trunk to indicate that the trunk is not busy.

idling jet [MECH ENG] A carburetor part that introduces gasoline during minimum load or speed of the engine.

idling system [MECH ENG] A system to obtain adequate metering forces at low airspeeds and small throttle openings in an automobile carburetor in the idling position.

idocrase *See* vesuvianite.

Idoteidae [INV ZOO] A family of isopod crustaceans in the

suborder Valvifera having a flattened body and seven pairs of similar walking legs.

IDP *See* integrated data processing.

id reaction [MED] Papular and vesicular eruptions on the skin, occurring suddenly following exacerbation of foci of some cutaneous fungus infections.

idrialite [MINERAL] A mineral composed of crystalline hydrocarbon, $C_{22}H_{14}$.

i-f *See* intermediate frequency.

i-f amplifier *See* intermediate-frequency amplifier.

IF canceler [ELECTR] In radar, a moving-target indicator canceler that operates at intermediate frequencies.

IFF *See* identification, friend or foe.

IFO *See* identifiable flying object.

IFR *See* instrument flight rules.

IFR terminal minimums [METEOROL] The operational weather limits concerned with minimum conditions of ceiling and visibility at an airport under which aircraft may legally approach and land under instrument flight rules; these minimums frequently are in the form of a sliding scale, and also vary with aircraft type, pilot experience, and from airport to airport.

IFR weather *See* instrument weather.

i-f transformer *See* intermediate-frequency transformer.

Ig *See* gamma globulin.

Igewsky's solution [MET] Etchant used to prepare carbon steels for microscopic analysis; consists of 5% picric acid in absolute alcohol.

igneous [PETR] Pertaining to rocks which have congealed from a molten mass.

igneous complex [PETR] An assemblage of igneous rocks that are intimately associated and roughly contemporaneous.

igneous facies [PETR] A part of an igneous rock differing in structure, texture, or composition from the main mass.

igneous meteor [GEOPHYS] A visible electric discharge in the atmosphere; lightning is the most common and important type, but types of corona discharge are also included.

igneous mineral [MINERAL] Mineral material forming igneous rock.

igneous petrology [PETR] The study of igneous rocks, their occurrence, composition, and origin.

igneous province *See* petrographic province.

ignimbrite [PETR] A silicic volcanic rock that forms thick, compact, lavalike sheets over a wide area of New Zealand. Also known as flood tuff.

ignite [CHEM] To start a fuel burning.

igniter [ENG] **1.** A device for igniting a fuel mixture. **2.** A charge, as of black powder, to facilitate ignition of a propelling or bursting charge.

igniter cord [ENG] A cord which passes an intense flame along its length at a uniform rate to light safety fuses in succession.

igniter pad [ORD] A black powder charge in the form of a thin pad of cartridge cloth attached to separate-loading propelling charge to facilitate complete and uniform ignition.

igniter train [ORD] Step-by-step arrangement of charges in pyrotechnic munitions by which the initial fire from the primer is transmitted and intensified until it reaches and sets off the main charge. Also known as burning train.

igniting fuse [ORD] A fuse designed to initiate its main munition by an igniting action, as compared to the detonating action of a detonating fuse, and suitable only for munitions using a main charge of low explosive or other readily ignitable material.

igniting primer [ORD] An auxiliary primer that carries the fire from a primer to the propelling charge in certain subcaliber tubes.

ignition [CHEM] The process of starting a fuel mixture burning, or the means for such a process.

ignition charge [MATER] A small quantity of explosive, usually composed of black powder, that facilitates the firing of the main charge.

ignition coil [ELECTROMAG] A coil in an ignition system which stores energy in a magnetic field relatively slowly and releases it suddenly to ignite a fuel mixture.

ignition delay *See* ignition lag.

ignition interference [COMMUN] Radio interference due to

IDOTEIDAE

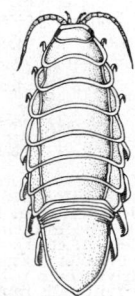

5 mm

Idotea neglecta, showing flattened body and seven pairs of legs. *(From G. O. Sars, An Account of the Crustacea of Norway, vol. 2, 1899)*

IGNITRON

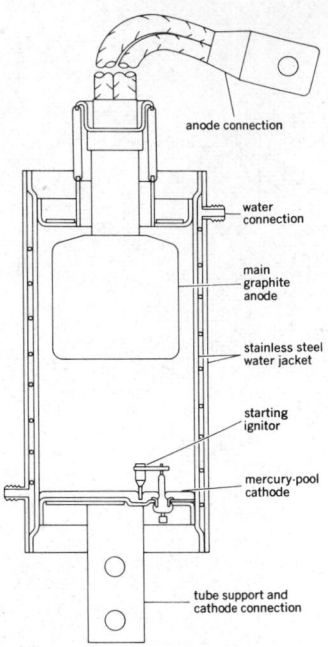

anode connection

water connection

main graphite anode

stainless steel water jacket

starting ignitor

mercury-pool cathode

tube support and cathode connection

Diagram of a typical sealed ignitron for welding service. *(General Electric Co.)*

IGNITION SYSTEM

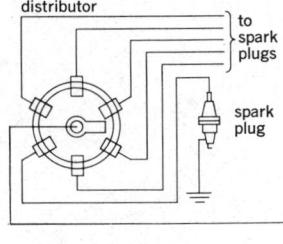

distributor

to spark plugs

spark plug

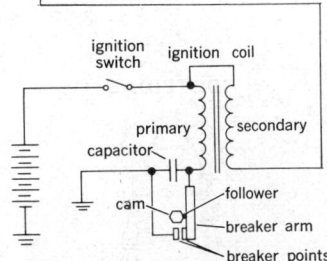

ignition switch

ignition coil

primary

secondary

capacitor

follower

cam

breaker arm

breaker points

Diagram of battery ignition system. When switch and breaker parts are closed current flows through primary, inducing magnetic field in core of ignition coil. When breaker points open, as engine driven cam rotates, primary current falls rapidly and magnetic field collapses, inducing very high voltage (10,000–20,000 V) in secondary, which is led to spark plugs, in proper sequence, by distributor. *(Modified from A. R. Rogowski, Elements of Internal-combustion Engines, McGraw-Hill, 1953)*

the spark discharges in an automotive or other ignition system.

ignition lag [MECH ENG] In the internal combustion engine, the time interval between the passage of the spark and the inflammation of the air-fuel mixture. Also known as ignition delay.

ignition point *See* ignition temperature.

ignition system [MECH ENG] The system in an internal combustion engine that initiates the chemical reaction between fuel and air in the cylinder charge by producing a spark.

ignition temperature [CHEM] The lowest temperature at which combustion begins and continues in a substance when it is heated in air. Also known as autogenous ignition temperature; ignition point.

ignitor [ELECTR] 1. An electrode used to initiate and sustain the discharge in a switching tube. Also known as keep-alive electrode (deprecated). 2. A pencil-shaped electrode, made of carborundum or some other conducting material that is not wetted by mercury, partly immersed in the mercury-pool cathode of an ignitron and used to initiate conduction at the desired point in each alternating-current cycle.

ignitron [ELECTR] A single-anode pool tube in which an ignitor electrode is employed to initiate the cathode spot on the surface of the mercury pool before each conducting period.

ignitron contactor [ELECTR] A circuit containing an ignitron and control contacts that serves as a heavy-duty switch in the primary of a resistance-welding transformer.

iguana [VERT ZOO] The common name for a number of species of herbivorous, arboreal reptiles in the family Iguanidae characterized by a dorsal crest of soft spines and a dewlap; there are only two species of true iguanas.

iguanid [VERT ZOO] The common name for members of the reptilian family Iguanidae.

Iguanidae [VERT ZOO] A family of reptiles in the order Squamata having teeth fixed to the inner edge of the jaws, a nonretractile tongue, a compressed body, five clawed toes, and a long but rarely prehensile tail.

IGY *See* International Geophysical Year.

I-head cylinder [MECH ENG] The internal combustion engine construction having both inlet and exhaust valves located in the cylinder head.

ihleite *See* copiapite.

ihp *See* indicated horsepower.

Ihrigizing [MET] Trade name for forming of a corrosion-resistant silicon alloy layer by heating a metal in a gaseous silicon chloride.

I indicator *See* I scope.

ijolite [PETR] A plutonic rock of nepheline and 30–60% mafic materials, generally sodic pyroxene, with accessory apatite, sphene, calcite, and titaniferous garnet.

ilang-ilang oil [MATER] An oil made synthetically or derived from the flowers of the tree *Canangium odorata*, containing pinene, geraniol, and linalol. Also known as cananga oil; ylang-ylang oil.

ileitis [MED] Inflammation of the ileum.

ileocecal valve [ANAT] A muscular structure at the junction of the ileum and cecum which prevents reflex of the cecal contents.

ileocolic artery [ANAT] A branch of the superior mesenteric artery that supplies blood to the terminal part of the ileum and the beginning of the colon.

ileocolic intussusception [MED] Slipping of the ileum through the ileocecal valve into the colon.

ileocolitis [MED] Inflammation of both the ileum and the colon.

ileocolostomy [MED] Surgical formation of a bypass channel between the ileum and the colon.

ileocutaneous [MED] Pertaining to a joining of the ileum and the skin resulting from a fistulous connection.

ileostomy [MED] Surgical formation of an artificial anus through the abdominal wall into the ileum.

ilesite [MINERAL] $(Mn,Zn,Fe)SO_4 \cdot 4H_2O$ A green mineral composed of hydrous manganese zinc iron sulfate.

ileum [ANAT] The last portion of the small intestine, extending from the jejunum to the large intestine.

ileus [MED] Acute intestinal obstruction of neurogenic origin.

ILF *See* infralow frequency.

Ilgner flywheel [MIN ENG] A flywheel in the Ward-Leonard control of winding engines in mine hoists.

Ilgner system [MIN ENG] A variation of the Ward-Leonard speed control system; the Ilgner flywheel, carried on the motor generator shaft, smooths out peak loads.

iliac artery [ANAT] Either of the two large arteries arising by bifurcation of the abdominal aorta and supplying blood to the lower trunk and legs (or hind limbs in quadrupeds). Also known as common iliac artery.

iliac crest height [ANTHRO] A measure of the vertical distance from the top of the iliac crest to the floor while the subject stands.

iliac fascia [ANAT] The fascia covering the pelvic surface of the iliacus muscle.

iliac index [ANTHRO] The ratio multiplied by 100 of the distance between the iliac spines and the distance between the lower margin of the acetabulum and the topmost crest of the ilium.

iliac region *See* inguinal region.

iliacus [ANAT] The portion of the iliopsoas muscle arising from the iliac fossa and sacrum.

iliac vein [ANAT] Any of the three veins on each side of the body which correspond to and accompany the iliac artery.

iliocostalis [ANAT] The lateral portion of the erector spinal muscle that extends the vertebral column and assists in lateral movements of the trunk.

iliofemoral ligament [ANAT] A strong band of dense fibrous tissue extending from the anterior inferior iliac spine to the lesser trochanter and the intertrochanteric line. Also known as Y ligament.

iliolumbar ligament [ANAT] A fibrous band that radiates laterally from the transverse processes of the fourth and fifth lumbar vertebrae and attaches to the pelvis by two main bands.

iliopsoas [ANAT] The combined iliacus and psoas muscles.

iliotibial tract [ANAT] A thickened portion of the fascia lata extending from the lateral condyle of the tibia to the iliac crest.

ilium [ANAT] Either of a pair of bones forming the superior portion of the pelvis bone in vertebrates.

Ilkovič equation [ANALY CHEM] Mathematical relationship between diffusion current, diffusion coefficient, and active-substance concentration; used for polarographic analysis calculations.

illegal character [ADP] A character or combination of bits that is not accepted as a valid representation by a computer or by a specific routine; commonly detected and used as an indication of a machine malfunction.

illepé fat *See* mowrah fat.

Illinoian [GEOL] The third glaciation of the Pleistocene in North America, between the Yarmouth and Sangamon interglacial stages.

illite [MINERAL] A group of gray, green, or yellowish-brown micalike clay minerals found in argillaceous sediments; intermediate in composition and structure between montmorillonite and muscovite.

illness [MED] 1. The state of being sick. 2. A sickness, disease, or disorder.

illuminance [OPTICS] The density of the luminous flux on a surface. Also known as illumination; luminous flux density.

illuminating [GRAPHICS] The hand decoration of books, as done in medieval times, with drawings and miniature paintings, or with embellishments, usually in red, blue, and gold, added to initial letters and borders.

illuminating gas [MATER] Flammable mixture of gases suitable for illuminating purposes; contains hydrogen, methane, ethane, carbon monoxide, and some nitrogen and oxygen.

illuminating grenade [ORD] Hand grenade or rifle grenade designed to provide illumination by a burning action; used also as a trip flare or as an incendiary device.

illuminating oil [MATER] An oil such as kerosine suitable for burning to provide illumination.

illuminating projectile [ORD] Projectile, with a time fuse, that releases a parachute flare at any desired height; used for lighting up an area. Also known as star shell.

illumination [ELECTROMAG] **1.** The geometric distribution of power reaching various parts of a dish reflector in an antenna system. **2.** The power distribution to elements of an antenna array. [OPTICS] **1.** The science of the application of lighting. **2.** *See* illuminance.

illumination climate [METEOROL] Also known as light climate. **1.** The worldwide distribution of natural light from the sun and sky (direct solar radiation plus diffuse sky radiation) as received on a horizontal surface. **2.** The character of total illumination at any given place.

illumination control [ELECTR] A photoelectric control that turns on lights when outdoor illumination decreases below a predetermined level.

illumination design [ENG] Design of sources of lighting and of systems which distribute light in order to effect a comfortable and satisfactory environment for seeing.

illumination distribution [OPTICS] The manner in which light is dispersed on a surface.

illuminometer [OPTICS] A portable photometer which is used in the field or outside the laboratory and yields results of lower accuracy than a laboratory photometer.

illusion [PSYCH] A false interpretation of a real sensation; a perception that misinterprets the object perceived.

illustration board [GRAPHICS] A board made by mounting good drawing paper on a stiff backing, usually a filled pulpboard; surfaces vary from smooth to rough.

illuvial [GEOL] Pertaining to a region or material characterized by the accumulation of soil by the illuviation of another zone or material.

illuvial horizon *See* B horizon.

illuviation [GEOL] The deposition of colloids, soluble salts, and small mineral particles in an underlying layer of soil.

illuvium [GEOL] Material leached by chemical or other processes from one soil horizon and deposited in another.

ilmenite [MINERAL] $FeTiO_3$ An iron-black, opaque, rhombohedral mineral that is the principal ore of titanium. Also known as mohsite; titanic iron ore.

Ilotycin [MICROBIO] A trade name for erythromycin.

ILS *See* instrument landing system.

ilsemannite [MINERAL] A black, blue-black, or blue mineral composed of hydrous molybdenum oxide or perhaps sulfate, occurring in earthy massive form.

ILS reference point [NAV] A point that is on the centerline of a runway served by an ILS and that is designated as the optimum point of contact for the landing aircraft. Derived from instrument landing system reference point.

image [ACOUS] An acoustic counterpart of an extended source of sound; for each point of the source there is a corresponding point in the image to which the sound from the source point is focused by an acoustic lens, mirror, or other acoustic system, or from which it appears to emanate. Also known as acoustic image. [ADP] A copy of the information contained in a punched card (or other data record) recorded on a different data medium. [COMMUN] **1.** One of two groups of side bands generated in the process of modulation; the unused group is referred to as the unwanted image. **2.** The scene reproduced by a television or facsimile receiver. [ELEC] *See* electric image. [ELECTROMAG] The input reflection coefficient corresponding to the reflection coefficient of a specified load when the load is placed on one side of a waveguide junction and a slotted line is placed on the other. [OPTICS] An optical counterpart of a self-luminous or illuminated object formed by the light rays that traverse an optical system; each point of the object has a corresponding point in the image from which rays diverge or appear to diverge. [PHYS] Any reproduction of an object produced by means of focusing light, sound, electron radiation, or other emanations coming from the object or reflected by the object. [PSYCH] A representation of a sensory experience, occurring in the brain.

image antenna [ELECTROMAG] A fictitious electrical counterpart of an actual antenna, acting mathematically as if it existed in the ground directly under the real antenna and served as the direct source of the wave that is reflected from the ground by the actual antenna.

image attenuation constant [ELECTR] The real part of the image transfer constant.

image card [ADP] A representation in storage of the holes punched in a card, in such a manner that the holes are represented by one binary digit, and the unpunched spaces are represented by the other binary digit.

image converter [ELECTR] *See* image tube. [OPTICS] A converter that uses a fiber optic bundle to change the form of an image, for more convenient recording and display or for the coding of secret messages.

image converter camera [ELECTR] A camera consisting of an image tube and an optical system which focuses the image produced on the phosphorescent screen of the tube onto photographic film.

image dissection photography [ELECTR] A method of high-speed photography in which an image is split in any one of various ways into interlaced space and time elements which can be unscrambled or played back through the system either to be viewed or to give a master negative.

image dissector [ADP] In optical character recognition, a device that optically examines an input character for the purpose of breaking it down into its prescribed elements.

image dissector tube [ELECTR] A television camera tube in which an electron image produced by a photoemitting surface is focused in the plane of the defining aperture and is scanned past that aperture. Also known as Farnsworth image dissector tube.

image effect [ELECTROMAG] Effect produced on the field of an antenna due to the presence of the earth; electromagnetic waves are reflected from the earth's surface, and these reflections often are accounted for by an image antenna at an equal distance below the earth's surface.

image force [ELEC] The electrostatic force on a charge in the neighborhood of a conductor, which may be thought of as the attraction to the charge's electric image.

image frequency [ELECTR] An undesired carrier frequency that differs from the frequency to which a superheterodyne receiver is tuned by twice the intermediate frequency.

image iconoscope [ELECTR] A camera tube in which an optical image is projected on a semitransparent photocathode, and the resulting electron image emitted from the other side of the photocathode is focused on a separate storage target; the target is scanned on the same side by a high-velocity electron beam, neutralizing the elemental charges in sequence to produce the camera output signal at the target. Also known as superemitron camera (British usage).

image impedance [ELECTR] One of the impedances that, when connected to the input and output of a transducer, will make the impedances in both directions equal at the input terminals and at the output terminals.

image intensifier *See* light amplifier.

image interference [COMMUN] Interference occurring in a superheterodyne receiver when a station broadcasting on the image frequency is received along with the desired station.

image isocon [ELECTR] A television camera tube which is similar to the image orthicon but whose return beam consists of scanning beam electrons that are scattered by positive stored charges on the target.

image load [ELECTR] Load parameters reflected back to the source by line discontinuities.

image orthicon [ELECTR] A television camera tube in which an electron image is produced by a photoemitting surface and focused on one side of a separate storage tube that is scanned on its opposite side by a beam of low-velocity electrons; electrons that are reflected from the storage tube, after positive stored charges are neutralized by the scanning beam, form a return beam which is amplified by an electron multiplier.

image parameter design [ELECTR] A method of filter design using image impedance and image transfer functions as the fundamental network functions.

image parameter filter [ELECTR] A filter constructed by image parameter design.

image phase constant [ELECTR] The imaginary part of the image transfer constant.

image plane [OPTICS] The plane in which an image produced by an optical system is formed; if the object plane is

ILMENITE

├─2.5 cm─┤

Ilmenite crystals from Arendal, Norway. *(Specimen from Department of Geology, Bryn Mawr College)*

perpendicular to the optical axis, the image plane will ordinarily also be perpendicular to the axis.

image ratio [ELECTR] In a heterodyne receiver, the ratio of the image frequency signal input at the antenna to the desired signal input for identical outputs.

image reject mixer [ELECTR] Combination of two balanced mixers and associated hybrid circuits designed to separate the image channel from the signal channels normally present in a conventional mixer; the arrangement gives image rejection up to 30 decibels without the use of filters.

image response [ELECTR] The response of a superheterodyne receiver to an undesired signal at its image frequency.

image spacing [GRAPHICS] In microfilm technology, the area between the trailing edge of one image and the leading edge of the next.

image transfer constant [ELECTR] One-half the natural logarithm of the complex ratio of the steady-state apparent power entering and leaving a network terminated in its image impedance.

image tube [ELECTR] An electron tube that reproduces on its fluorescent screen an image of the optical image or other irradiation pattern incident on its photosensitive surface. Also known as electron image tube; image converter.

imaginal disk [INV ZOO] Any of the thickened areas within the sac of the body wall in holometabolous insects which give rise to specific organs in the adult.

imaginary axis [MATH] All complex numbers $x + iy$ where $x = 0$; the vertical coordinate axis for the complex plane.

imaginary number [MATH] A complex number of the form $a + bi$, with b not equal to zero, where a and b are real numbers, and $i = \sqrt{-1}$; some writers require also that $a = 0$. Also known as imaginary quantity.

imaginary part [MATH] For a complex number $x + iy$ the imaginary part is the real number y.

imaginary quantity *See* imaginary number.

imaging [PHYS] The formation of images of objects.

imaging radar [ENG] Radar carried on aircraft which forms images of the terrain.

imago [INV ZOO] The sexually mature, usually winged stage of insect development. [PSYCH] An unconscious mental picture, usually idealized, of a parent or loved person important in the early development of an individual and carried into adulthood.

imbecile [PSYCH] A person of middle-grade mental deficiency; the individual's mental age is between 3 and 7 years.

imbed *See* embed.

imbedding [MATH] A homeomorphism of one topological space into another.

imbibition [PHYS CHEM] Absorption of liquid by a solid or a semisolid material.

imbricate [BIOL] Having overlapping edges, such as scales, or the petals of a flower.

imbricate structure [GEOL] **1.** A sedimentary structure characterized by shingling of pebbles all inclined in the same direction with the upper edge of each leaning downstream or toward the sea. Also known as shingle structure. **2.** Tabular masses that overlap one another and are inclined in the same direction. Also known as schuppen structure; shingle-block structure.

imbrication [GEOL] Formation of an imbricate structure. Also known as shingling.

imerinite [MINERAL] $Na_2(Mg,Fe)_6Si_8O_{22}(O,OH)_2$ A colorless to blue mineral composed of a basic silicate of sodium, iron, and magnesium, occurring as acicular crystals.

Imhoff cone [CIV ENG] A graduated glass vessel for measuring settled solids in testing the composition of sewage.

Imhoff tank [CIV ENG] A sewage treatment tank in which digestion and settlement take place in separate compartments, one below the other.

imidazole [ORG CHEM] $C_3H_4N_2$ One of a group of organic heterocyclic compounds containing a five-membered diunsaturated ring with two nonadjacent nitrogen atoms as part of the ring; the particular compound imidazole is a member of the group. Also known as 1,3-diazole; glyoxaline; iminazole.

imidazoletrione *See* parabanic acid.

imidazolyl [ORG CHEM] $C_3H_3N_2$ The radical from imidazole.

imide [ORG CHEM] **1.** A compound derived from acid anhy-

drides by replacing the oxygen (O) with the =NH group. **2.** A compound that has either the =NH group or a secondary amine in which R is an acyl radical, as R_2NH.

iminazole *See* imidazole.

imine [ORG CHEM] A class of compounds that are the product of condensation reactions of aldehydes or ketones with ammonia or amines; they have the NH radical attached to the carbon with the double bond, as $R - HC = NH$; an example is benzaldimine.

imino acid [ORG CHEM] Organic acid in which the =NH group is attached to one or two carbons; for example, acetic acid, $NH(CH_2COOH)_2$.

imino compound [ORG CHEM] A compound that has the =NH radical attached to one or two carbon atoms.

imino nitrogen [ORG CHEM] Nitrogen combined with hydrogen in the imino group. Also known as ammonia nitrogen.

iminourea *See* guanidine.

imipramine [PHARM] $C_{19}H_{24}N_2$ A water-soluble compound crystallizing from acetone solution, melting point 175°C; used in medicine as an antidepressant.

imitative deception [ELECTR] Introduction of electromagnetic radiations into enemy channels which imitate their own emissions, in order to mislead them.

immarginate [BIOL] Lacking a clearly defined margin.

immature soil [GEOL] A soil in which erosion has exceeded the rate at which the soil develops downward.

immediate-access [ADP] **1.** Pertaining to an access time which is relatively brief, or to a relatively fast transfer of information. **2.** Pertaining to a device which is directly connected with another device.

immediate address [ADP] The value of an operand contained in the address part of an instruction and used as data by this instruction.

immediate hypersensitivity [IMMUNOL] A type of hypersensitivity in which the response rapidly occurs following exposure of a sensitized individual to the antigen.

immersion [ASTRON] The disappearance of a celestial body either by passing behind another or passing into another's shadow. [MATH] A mapping f of a topological space X into a topological space Y such that for every $x \epsilon X$ there exists a neighborhood N of x, such that f is a homeomorphism of N onto $f(N)$. [SCI TECH] Placement into or within a fluid, usually water.

immersion cleaning [MET] Removing surface dirt from metal by dipping into a cleaning liquid.

immersion coating [ENG] Applying material to the surface of a metal or ceramic by dipping into a liquid.

immersion electron microscope [ELECTR] An emission electron microscope in which the specimen is a flat conducting surface which may be heated, illuminated, or bombarded by high-velocity electrons or ions so as to emit low-velocity thermionic, photo-, or secondary electrons; these are accelerated to a high velocity in an immersion objective or cathode lens and imaged as in a transmission electron microscope.

immersion electrostatic lens *See* bipotential electrostatic lens.

immersion foot [MED] A serious and disabling condition of the feet due to prolonged immersion in seawater at 60°F (15.6°C) or lower, but not at freezing temperature.

immersion heater [ELEC] An electric device for heating a liquid by direct immersion in the liquid.

immersion lens *See* immersion objective.

immersion objective [OPTICS] A high-power microscope objective designed to work with the space between the objective and the cover glass over the object filled with an oil whose index of refraction is nearly the same as that of the objective and the cover glass, in order to reduce reflection losses and increase the index of refraction of the object space. Also known as immersion lens.

immersion plating [MET] Applying an adherent layer of more-noble metal to the surface of a metal object by dipping in a solution of more-noble metal ions; a replacement reaction. Also known as dip plating; metal replacement.

immersion proof [ORD] An item of equipment when ready for field transport can be submerged for 2 hours in salt or fresh water to a covering depth of 3 feet (91 centimeters), and be capable of operating at normal effectiveness immediately after being removed from the water.

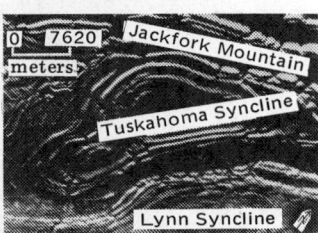

Geologic structures in the Ouachita Mountains of Oklahoma mapped by imaging radar. The aircraft was flying NW–SE above the top of this area. The dark areas are radar shadow since the energy transmitted at low angle does not pick up terrain information on the southwest side of the ridges. *(NASA)*

IMIDAZOLE

$$4 \quad HC \underline{\qquad} N \quad 3$$
$$5 \quad HC \qquad CH \quad 2$$
$$N \quad 1$$
$$H$$

Structural formula of imidazole; the positions on the five-membered ring are shown by arabic numbers.

immersion refractometer [OPTICS] Device to measure refractive indices by immersing the prism portion in the sample being checked. Also known as dipping refractometer.

immersion sampling [ANALY CHEM] Collection of a liquid sample for laboratory or other analysis by immersing a container in the liquid and filling it.

immiscible [CHEM] Pertaining to liquids that will not mix with each other.

immittance [ELEC] A term used to denote both impedance and admittance, as commonly applied to transmission lines, networks, and certain types of measuring instruments.

immittance bridge [ELECTROMAG] A modification of an admittance bridge which compares the output current of a four-terminal device with admittance standards in a T configuration in order to measure transfer admittance by a null method.

immobilize [MED] To render motionless or to fix in place, as by splints or surgery.

immune [IMMUNOL] **1.** Safe from attack; protected against a disease by an innate or an acquired immunity. **2.** Pertaining to or conferring immunity.

immune body *See* antibody.

immune globulin *See* gamma globulin.

immune hemolysin [IMMUNOL] A substance formed in blood in response to an injection of erythrocytes from another species.

immune horse serum [IMMUNOL] Serum obtained from the blood of an immunized horse.

immune lysin [IMMUNOL] An antibody that will disrupt a particular type of cell in the presence of complement and cofactors, such as magnesium or calcium ions.

immune opsonin [IMMUNOL] A substance produced in blood in response to an infection or to inoculation with dead cells of the infecting species of bacteria.

immune precipitation [IMMUNOL] A method of isolating a protein from mixtures by using a specific antibody as the precipitating agent.

immune protein [IMMUNOL] Any antibody.

immune serum [IMMUNOL] Blood serum obtained from an immunized individual and carrying antibodies.

immunity [IMMUNOL] The condition of a living organism whereby it resists and overcomes infection or disease.

immunization [IMMUNOL] Rendering an organism immune to a specific communicable disease.

immunization therapy [MED] The use of vaccines or antiserums to produce immunity against a specific disease.

immunochemistry [IMMUNOL] A branch of science dealing with the chemical changes associated with immunity factors.

immunodiffusion [IMMUNOL] A serological procedure in which antigen and antibody solutions are permitted to diffuse toward each other through a gel matrix; interaction is manifested by a precipitin line for each system.

immunoelectrophoresis [IMMUNOL] A serological procedure in which the components of an antigen are separated by electrophoretic migration and then made visible by immunodiffusion of specific antibodies.

immunofluorescence [IMMUNOL] Fluorescence as the result of, or identifying, an immune response; a specifically stained antigen fluoresces in ultraviolet light and can thus be easily identified with a homologous antigen.

immunogen [IMMUNOL] A substance which stimulates production of specific antibody or of cellular immunity, and which can react with these products.

immunogenetics [MED] A branch of immunology dealing with the relationships between immunity and genetic factors in disease.

immunogenic [IMMUNOL] Producing immunity.

immunoglobulin *See* gamma globulin.

immunogranulomatous disease [MED] A condition in which a deviation from the standard immune mechanisms is considered to be associated with widespread granulomatosis.

immunological paralysis *See* acquired immunological tolerance.

immunologic suppression [IMMUNOL] The use of x-irradiation, chemicals, corticosteroid hormones, or antilymphocyte antisera to suppress antibody production, particularly in graft transplants.

immunologic tolerance [IMMUNOL] **1.** A condition in which an animal will accept a homograft without rejection. **2.** A

state of specific unresponsiveness to an antigen or antigens in adult life as a consequence of exposure to the antigen in utero or in the neonatal period.

immunologist [IMMUNOL] A person who specializes in immunology.

immunology [BIOL] A branch of biological science concerned with the native or acquired resistance of higher animal forms and humans to infection with microorganisms.

immunopathology [MED] The study of various human and animal diseases in which humoral and cellular immune factors seem important in causing pathological damage to cells, tissues, and the host.

immunosuppression [IMMUNOL] Suppression of an immune response by the use of drugs or radiation.

immunosuppressive [PHARM] Any drug or agent used to suppress antibody production.

immunotherapy [MED] **1.** *See* serotherapy. **2.** Therapy utilizing immunosuppressives.

impact [MECH] A forceful collision between two bodies which is sufficient to cause an appreciable change in the momentum of the system on which it acts. Also known as impulsive force.

impact avalanche and transit time diode *See* IMPATT diode.

impact bar [ENG] Specimen used to test the relative susceptibility of a plastic material to fracture by shock.

impact breaker [MECH ENG] A device that utilizes the energy from falling stones in addition to power from massive impellers for complete breaking up of stone. Also known as double impeller breaker.

impact cast *See* prod cast.

impact crusher [MECH ENG] A machine for crushing large chunks of solid materials by sharp blows imposed by rotating hammers, or steel plates or bars; some crushers accept lumps as large as 28 inches (about 70 centimeters) in diameter, reducing them to ¼ inch (6 millimeters) and smaller.

impact energy [MECH] The energy necessary to fracture a material. Also known as impact strength.

impact excitation [ELEC] Starting of damped oscillations in a radio circuit by a sudden surge, such as that produced by a spark discharge.

impact extrusion [MET] A cold extrusion process for producing tubular components by striking a slug of the metal, which has been placed in the cavity of the die, with a punch moving at high velocity.

impact force *See* set forward force.

impact forging [MET] Plastic deformation of a metal using an impactive force.

impact fuse [ORD] A fuse that detonates upon striking an object. Also known as contact fuse.

impact grinding [MECH ENG] A technique used to break up particles by direct fall of crushing bodies on them.

impaction [MED] **1.** The state of being lodged and retained in a body part. **2.** Confinement of a tooth in the jaw so that its eruption is prevented. **3.** A condition in which one fragment of bone is firmly driven into another fragment so that neither can move against the other.

impactite [GEOL] Glassy fused rock or meteor fragments resulting from heat of impact of a meteor on the earth.

impact law [PHYS] The relationship of fluid density, particle density, and fluid viscosity in the settling velocity of large particles in a given liquid: settling velocity is directly proportional to the square root of the particle diameter.

impact load [ENG] A force delivered by a blow, as opposed to a force applied gradually and maintained over a long period.

impact loss [FL MECH] Loss of head in a flowing stream due to the impact of water particles upon themselves or some bounding surface.

impact mark *See* prod mark.

impact microphone [ENG ACOUS] An instrument that picks up the vibration of an object impinging upon another, used especially on space probes to record the impact of small meteoroids.

impact mill [MECH ENG] A unit that reduces the size of rocks and minerals by the action of rotating blades projecting the material against steel plates.

impactometer *See* impactor.

impactor [ENG] A general term for instruments which sample atmospheric suspensoids by impaction; such instruments consist of a housing which constrains the air flow past a sensitized sampling plate. Also known as impactometer. [MECH ENG] A machine or part whose operating principle is striking blows. [MIN ENG] A rotary hammermill which crushes ore by impacting it against crushing plates or elements.

impact parameter [NUC PHYS] In a nuclear collision, the perpendicular distance from the target nucleus to the initial line of motion of the incident particle.

impact predictor [AERO ENG] A device which takes information from a trajectory measuring system and continuously computes the point (in real time) at which the rocket will strike the earth.

impact pressure [MECH] The sum of dynamic pressure and static head.

impact roll [MECH ENG] An idler roll protected by a covering of a resilient material from the shock of the loading of material onto a conveyor belt, so as to reduce the damage to the belt.

impact screen [MECH ENG] A screen designed to swing or rock forward when loaded and to stop abruptly by coming in contact with a stop.

impact strength *See* impact energy.

impact stress [MECH] Force per unit area imposed on a material by a suddenly applied force.

impact test [ENG] Determination of the degree of resistance of a material to breaking by impact, under bending, tension, and torsion loads; the energy absorbed is measured in breaking the material by a single blow.

impact tube *See* pitot tube.

impact velocity [MECH] The velocity of a projectile or missile at the instant of impact. Also known as striking velocity.

impact wrench [MECH ENG] A compressed-air or electrically operated wrench that gives a rapid succession of sudden torques.

imparipinnate *See* odd-pinnate.

impasto [GRAPHICS] **1.** The thick, heavy application of oil paint to a canvas, often with a palette knife; impasto sections stand out in considerable relief. **2.** The thick application of polymer or other paint to any surface.

IMPATT diode [ELECTR] A *p-n* junction diode that has a depletion region adjacent to the junction, through which electrons and holes can drift, and is biased beyond the avalanche breakdown voltage. Derived from impact avalanche and transit time diode.

impedance [ELEC] *See* electrical impedance. [PHYS] **1.** The ratio of a sinusoidally varying quantity to a second quantity which measures the response of a physical system to the first, both being considered in complex notation; examples are electrical impedance, acoustic impedance, and mechanical impedance. Also known as complex impedance. **2.** The ratio of the greatest magnitude of a sinusoidally varying quantity to the greatest magnitude of a second quantity which measures the response of a physical system to the first; equal to the magnitude of the quantity in the first definition.

impedance-admittance matrix [ELECTR] A four-element matrix used to describe analytically a transistor in terms of impedances or admittances.

impedance bridge [ELEC] A device similar to a Wheatstone bridge, used to compare impedances which may contain inductance, capacitance, and resistance.

impedance coil [ELEC] A coil of wire designed to provide impedance in an electric circuit.

impedance compensator [ELEC] Electric network designed to be associated with another network or a line with the purpose of giving the impedance of the combination a desired characteristic with frequency over a desired frequency range.

impedance component [ELEC] **1.** Resistance or reactance. **2.** A device such as a resistor, inductor, or capacitor designed to provide impedance in an electric circuit.

impedance coupling [ELEC] Coupling of two signal circuits with an impedance.

impedance irregularities [ELEC] Discontinuities or abrupt changes which result from junctions between unlike sections of a transmission line or irregularities on a line.

impedance match [ELEC] The condition in which the exter-

nal impedance of a connected load is equal to the internal impedance of the source or to the surge impedance of a transmission line, thereby giving maximum transfer of energy from source to load, minimum reflection, and minimum distortion.

impedance meter *See* electrical impedance meter.

impedometer [ELECTROMAG] An instrument used to measure impedances in waveguides.

impeller [MECH ENG] The rotating member of a turbine, blower, fan, axial or centrifugal pump, or mixing apparatus. Also known as rotor.

impeller pump [MECH ENG] Any pump using a mechanical agency to provide continuous power to move liquids.

Impennes [VERT ZOO] A superorder of birds for the order Sphenisciformes in some systems of classification.

imperative statement [ADP] A statement in a symbolic program which is translated into actual machine-language instructions by the assembly routine.

imperfect flower [BOT] A flower lacking either stamens or carpels.

imperfect gas *See* real gas.

imperforate [BIOL] Lacking a normal opening. [GRAPHICS] In printing, no perforations between repeated designs, for example, stamps or labels, which are to be used separately.

imperforate anus [MED] A congenital malformation in which the large intestine ends blindly.

imperial gallon *See* gallon.

imperial pint *See* pint.

imperial red [INORG CHEM] Any of the red varieties of ferric oxide used as pigment.

imperial smelting process [MET] A pyrometallurgical process which treats a complex concentrated feed to a single furnace to recover zinc, copper, lead, cadmium, silver, gold, and other metals in one pass. Abbreviated ISP.

impermeable [SCI TECH] Not permitting water or other fluid to pass through. Also known as impervious.

impervious *See* impermeable.

impervious carbon [MATER] Carbon compressed with bituminous binder, then carbonized by sintering to produce a dense, impervious material; used as brick to line chemical process and storage vessels.

impetigo [MED] An acute, contagious, inflammatory skin disease caused by streptococcal or staphylococcal infections and characterized by vesicular or pustular lesions.

impingement [ENG] Removal of liquid droplets from a flowing gas or vapor stream by causing it to collide with a baffle plate at high velocity, so that the droplets fall away from the stream. Also known as liquid knockout.

impingement attack [MET] Accelerated corrosive attack on a metal by moving liquids, resulting usually from erosion of a protective surface layer.

impinger [ENG] A device used to sample dust in the air that draws in a measured volume of dusty air and directs it through a jet to impact on a wetted glass plate; the dust particles adhering to the plate are counted.

implant [MED] **1.** A quantity of radioactive material in a suitable container, intended to be embedded in a tissue or tumor for therapeutic purposes. **2.** A tissue graft placed in depth in the body.

implantation [MED] **1.** Placement of a tissue transplant in depth in the body. **2.** Placement in the body of a device for mechanical repair, such as for a ventral hernia or a fracture. **3.** Embedding of an embryo into the endometrium.

implexed [INV ZOO] In insects, having the integument infolded for muscle attachment.

implication [MATH] The logical relation between two statements p and q, usually expressed as "if p then q."

implicit function [MATH] A function defined by an equation $f(x,y) = 0$, when x is considered as an independent variable and y, called an implicit function of x, as a dependent variable.

implicit function theorem [MATH] A theorem that gives conditions under which an equation in variables x and y may be solved so as to express y directly as a function of x; it states that if $F(x,y)$ and $\partial F(x,y)/\partial y$ are continuous in a neighborhood of the point (x_0, y_0) and if $F(x,y) = 0$ and $\partial F(x,y)/\partial y \neq 0$, then there is a number $\varepsilon > 0$ such that there is one and only

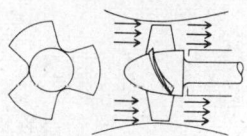

IMPELLER

Axial-flow, propeller-type impeller used for high-speed centrifugal pumps.

one function $f(x)$ that is continuous and satisfies $F[x, f(x)] = 0$ for $|x - x_0| < \varepsilon$, and satisfies $f(x_0) = y_0$.

implosion [CHEM] The sudden reduction of pressure by chemical reaction or change of state which causes an inrushing of the surrounding medium. [PHYS] A bursting inward, as in the inward collapse of an evacuated container (such as the glass envelope of a cathode-ray tube) or the compression of fissionable material by ordinary explosives in a nuclear weapon.

implosion weapon [ORD] A nuclear weapon in which a quantity of fissionable material, less than a critical mass in its untriggered configuration, has its volume suddenly reduced by compression with ordinary explosives, so that it becomes supercritical and a nuclear explosion can occur.

imposition [GRAPHICS] The pattern of arranging pages for a signature of a book so that the pages will be in sequence when folding occurs.

impotence [MED] **1.** Inability in the male to perform the sexual act. **2.** Lack of sexual vigor.

impound [CIV ENG] To collect water for irrigation, flood control, or similar purpose.

impounding reservoir [CIV ENG] A reservoir with outlets controlled by gates that release stored surface water as needed in a dry season; may also store water for domestic or industrial use or for flood control. Also known as storage reservoir.

impregnate [ENG] To force a liquid substance into the spaces of a porous solid in order to change its properties, as the impregnation of turquoise gems with plastic to improve color and durability, the impregnation of porous tungsten with a molten barium compound to manufacture a dispenser cathode, or the impregnation of wood with creosote to preserve its integrity against water damage. [MED] To fertilize or cause to become pregnant.

impregnated bit [DES ENG] A sintered, powder-metal matrix bit with fragmented bort or whole diamonds of selected screen sizes uniformly distributed throughout the entire crown section.

impregnated timber [MATER] Timber which has been made flame-resistant, fungi-resistant, or insect-proof by forcing into it under vacuum or pressure a flame retardant or a fungal or insect poison.

impressed voltage [ELEC] Voltage applied to a circuit or device.

impression [GEOL] A form left on a soft soil surface by plant parts; the soil hardens and usually the imprint is a concave feature. [GRAPHICS] **1.** A print made from an engraved plate. **2.** A press run or printing of a book.

impression cylinder [GRAPHICS] A cylinder onto which an inked image is pressed, so that this image can be transferred to paper in the offset duplicating process.

imprinter [GRAPHICS] Any device for entering markings onto a form, including, but not limited to, printing presses, typewriters, pressure imprinting devices such as those used with credit cards and address plates, pencils, pens, cash registers, adding machines, and bookkeeping machines.

imprinting [PSYCH] The very rapid development of a response or learning pattern to a stimulus at an early and usually critical period of development; particularly characteristic of some species of birds.

improper fraction [MATH] **1.** In arithmetic, the quotient of two integers in which the numerator is greater than or equal to the denominator. **2.** In algebra, the quotient of two polynomials in which the degree of the numerator is greater than or equal to that of the denominator.

improper integral [MATH] Any integral in which either the integrand becomes unbounded on the domain of integration, or the domain of integration is itself unbounded.

improvement factor See noise improvement factor.

improvement threshold [COMMUN] The value of carrier-to-noise ratio below which the signal-to-noise ratio decreases more rapidly than the carrier-to-noise ratio.

improvised grenade [ORD] Any nonstandard type of grenade which is prepared by the user, for example, frangible grenades and fragmentation grenades composed of nails, cartridge cases, or other fragments taped to the sides of a trinitrotoluene block, with suitable detonating device.

improvised mine [ORD] A mine manufactured of available materials because standard mines are either unavailable or are incapable of producing the desired result.

impsonite [GEOL] A black, asphaltic pyrobitumen with a high fixed-carbon content derived from the metamorphosis of petroleum.

impulse [MECH] The integral of a force over an interval of time. [PHYS] A pulse which lasts for so short a time that its duration can be thought of as infinitesimal. [PSYCH] A sudden psychogenic urge to act.

impulse excitation See shock excitation.

impulse face [HOROL] Lifting surface of a club tooth on an escape wheel, or the surface of a pallet stone engaged by such a wheel.

impulse function [MATH] An idealized or generalized function defined not by its values but by its behavior under integration, such as the (Dirac) delta function.

impulse generator [ELEC] An apparatus which produces very short surges of high-voltage or high-current power by discharging capacitors in parallel or in series. Also known as pulse generator.

impulse modulation [CONT SYS] Modulation of a signal in which it is replaced by a series of impulses, equally spaced in time, whose strengths (integrals over time) are proportional to the amplitude of the signal at the time of the impulse.

impulse movement [HOROL] Clock movement in which the hands are driven by electromagnet impulses from the pendulum.

impulse noise See pulse noise.

impulse pallet [HOROL] Pallet stone in a chronometer balance, driven by impulses from the escape wheel.

impulse period See pulse period.

impulse response [CONT SYS] The response of a system to an impulse which differs from zero for an infinitesimal time, but whose integral over time is unity; this impulse may be represented mathematically by a Dirac delta function.

impulse sealing [ENG] Heat-sealing of plastic materials by applying a pulse of intense thermal energy to the sealing area for a very short time, followed immediately by cooling.

impulse separator [ELECTR] In a television receiver, the circuit that separates the horizontal synchronizing impulses in the received signal from the vertical synchronizing impulses.

impulse signaling [COMMUN] Conveying information by means of on-off conditions transmitted down a line or over free space.

impulse strength [ELEC] Voltage breakdown of insulation under voltage surges on the order of microseconds in duration.

impulse tachometer [ENG] A tachometer in which each rotation of a shaft generates an electric pulse and the time rate of pulses is then measured; classified as capacitory-current, inductory, or interrupted direct-current tachometer.

impulse train [CONT SYS] An input consisting of an infinite series of unit impulses, equally separated in time.

impulse transmission [COMMUN] Form of signaling which employs impulses of either or both polarities for transmission to indicate the occurrence of transitions in the signals; used principally to reduce the effects of low-frequency interference; the impulses are generally formed by suppressing the low-frequency components, including direct current, of the signals.

impulse turbine [MECH ENG] A prime mover in which fluid under pressure enters a stationary nozzle where its pressure (potential) energy is converted to velocity (kinetic) energy and absorbed by the rotor.

impulse-type telemeter [COMMUN] A telemeter that employs electric impulses as the translating means.

impulsive force See impact.

impulsive sound equation [ACOUS] An equation which states that the total sound energy produced by a short burst of sound in a room is an exponentially decreasing function of time, whose decay constant depends on the speed of sound, the sound absorption coefficient, and the volume and surface area of the room.

impunctate [BIOL] Lacking pores.

impurity [SCI TECH] An undesirable foreign material in a

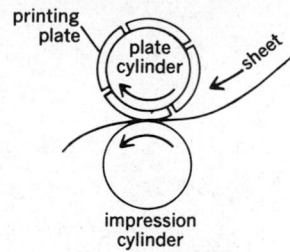

IMPRESSION CYLINDER

The plate and impression cylinders in a rotary press.

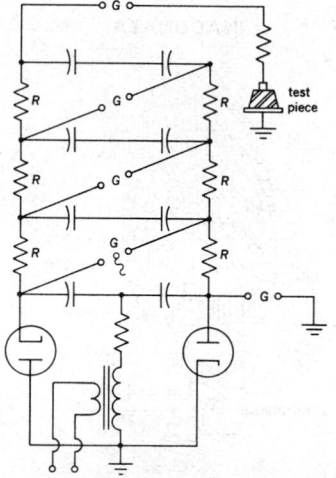

IMPULSE GENERATOR

Typical four-stage Marx impulse generator circuit. R = resistor, G = spark gap.

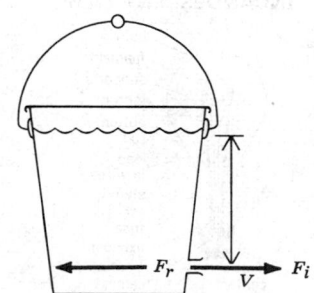

IMPULSE TURBINE

Bucket used to demonstrate the impulse principle basic to many impulse turbines. Water escapes through a nozzle near the base illustrating the impulse F_i and reaction F_r forces of the jet issuing with a velocity V. Vertical arrow indicates distance from nozzle to water level.

pure substance. [SOLID STATE] A substance that, when diffused into semiconductor metal in small amounts, either provides free electrons to the metal or accepts electrons from it.

impurity scattering [SOLID STATE] Scattering of electrons by holes or phonons in the crystal.

impurity semiconductor [SOLID STATE] A semiconductor whose properties are due to impurity levels produced by foreign atoms.

IMVIC test [MICROBIO] A group of four cultural tests used to differentiate genera of bacteria in the family Enterobacteriaceae and to distinguish them from other bacteria; tests are indole, methyl red, Voges-Proskauer, and citrate.

In See indium.

in. See inch.

inactive front [METEOROL] A front, or portion thereof, that produces very little cloudiness and no precipitation, as opposed to an active front. Also known as passive front.

inactive tartaric acid See racemic acid.

inadequate personality [PSYCH] An individual showing no obvious mental or physical defect, but characterized by inappropriate or inadequate response to intellectual, emotional, and physical demand, and whose behavior pattern shows inadaptability, ineptitude, poor judgment, lack of stamina, and social incompatibility.

Inadunata [PALEON] An extinct subclass of stalked Paleozoic Crinozoa characterized by branched or simple arms that were free and in no way incorporated into the calyx.

inadunate [BIOL] Not united. [INV ZOO] In crinoids, having the arms free from the calyx.

in-and-out bond [CIV ENG] Masonry bond composed of vertically alternating stretchers and headers.

inanimate [MED] Lacking consciousness or life. [SCI TECH] A lifeless object.

inanition [MED] The exhausted, pathologic condition resulting from starvation.

inaperturate [BIOL] Lacking apertures.

inarching [BOT] A kind of repair grafting in which two plants growing on their own roots are grafted together and one plant is severed from its roots after the graft union is established.

Inarticulata [INV ZOO] A class of the phylum Brachiopoda; valves are typically not articulated and are held together only by soft tissue of the living animal.

inboard [ENG] Toward or close to the longitudinal axis of a ship or aircraft.

inboard profile [NAV ARCH] A plan having a longitudinal section through the center of a ship, showing all dimensions, positions of bulkheads, location of spaces, machinery, and fittings.

inbond [CIV ENG] Pertaining to bricks or stones laid as headers across a wall.

inborn [BIOL] Of or pertaining to a congenital or hereditary characteristic.

inbreeding [GEN] Breeding of closely related individuals; self-fertilization, as in some plants, is the most extreme form.

inby [MIN ENG] Away from the shaft or mine entrance and therefore toward the working face.

incandescence [OPTICS] The emission of visible radiation by a hot body.

incandescent lamp [ELEC] An electric lamp that produces light when a metallic filament is heated white-hot in a vacuum by passing an electric current through the filament. Also known as filament lamp; light bulb.

incandescent readout [ELECTR] A readout in which each character is formed by energizing an appropriate combination of seven bar-shaped incandescent lamps.

incarbonization See coalification.

incarcerated hernia [MED] A hernia in which the intestinal loop is permanently trapped in the hernia sac.

incendiary [ORD] Ammunition equipped so that an incendiary effect at the target occurs.

incendiary bomb [ORD] A bomb designed to be dropped from an aircraft to destroy or reduce the utility of a target by the effects of combustion.

incendiary file destroyer [ORD] An incendiary device designed for use in destroying combustible file material.

incendiary grenade [ORD] Hand grenade designed to be

filled with incendiary materials, or used primarily for incendiary purposes.

incendiary rocket [ORD] A rocket with a warhead designed to produce an incendiary effect at the target.

incentive wage system See wage incentive plan.

Inceptisol [GEOL] A soil order in which materials other than carbonates or amorphous silica are altered or removed from the pedogenic horizon.

incertae sedis [SYST] Placed in an uncertain taxonomic position.

inch [MECH] A unit of length in common use in the United States and Great Britain, equal to $\frac{1}{12}$ foot or 2.54 centimeters. Abbreviated in.

inching See jogging.

inch of mercury [MECH] The pressure exerted by a 1-inch-high (2.54-centimeter-high) column of mercury that has a density of 13.5951 grams per cubic centimeter when the acceleration of gravity has the standard value of 9.80665 m/sec^2 or approximately 32.17398 ft/sec^2; equal to 3386.388640341 newtons per square meter; used as a unit in the measurement of atmospheric pressure.

incidence angle See angle of incidence.

incidence matrix [MATH] In a graph the p by q matrix (b_{ij}) for which $b_{ij} = 1$ if the ith vertex is an end point of the jth edge, and $b_{ij} = 0$ otherwise.

incidence plane See plane of incidence.

incidental element See irregular element.

incident field intensity [ELECTROMAG] Field strength of a sky wave without including the effects of earth reflections at the receiving location.

incident light [OPTICS] The direct light that falls on a surface.

incident power [ELEC] Product of the outgoing current and voltage, from a transmitter, traveling down a transmission line to the antenna.

incident wave [ELECTR] A current or voltage wave that is traveling through a transmission line in the direction from source to load. [PHYS] A wave that impinges on a discontinuity, particle, or body, or on a medium having different propagation characteristics.

incineration [CHEM] The process of burning a material so that only ashes remain.

incinerator [ENG] A furnace or other container in which materials are burned.

Incirrata [INV ZOO] A suborder of cephalopod mollusks in the order Octopoda.

incised [BIOL] Having a deeply and irregularly notched margin. [MED] Made by cutting, as a wound.

incised meander [GEOL] A deep, tortuous valley cut by a meandering stream that was rejuvenated.

incision [MED] A cut or wound of the body tissue, as an abdominal incision or a vertical or oblique incision.

incisional hernia [MED] Abnormal protrusion of an organ through an operative or accidental incision. Also known as postoperative hernia; posttraumatic hernia.

incisive canal [ANAT] The bifurcated bony passage from the floor of the nasal cavity to the incisive fossa.

incisive foramen [ANAT] One of the two to four openings of the incisive canal on the floor of the incisive fossa.

incisive fossa [ANAT] 1. A bony pit behind the upper incisors into which the incisive canals open. 2. A depression on the maxilla at the origin of the depressor muscle of the nose. 3. A depression of the mandible at the origin of the mentalis muscle.

incisor [ANAT] A tooth specialized for cutting, especially those in front of the canines on the upper jaw of mammals.

inclination [GEOPHYS] In magnetic inclination, the dip angle of the earth's magnetic field. Also known as magnetic dip. [MATH] 1. The inclination of a line in a plane is the angle made with the positive x axis. 2. The inclination of a line in space with respect to a plane is the smaller angle the line makes with its orthogonal projection in the plane. 3. The inclination of a plane with respect to a given plane is the smaller of the dihedral angles which it makes with the given plane.

inclination of planetary orbits [ASTRON] The angle between the plane of the orbit and the plane of the ecliptic, which is the plane of the earth's orbit.

INADUNATA

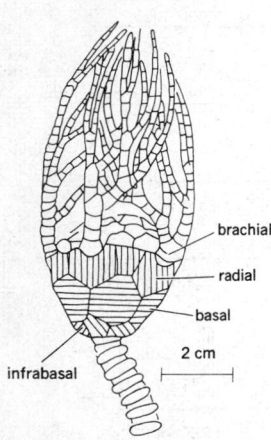

brachial

radial

basal

2 cm

infrabasal

Sketch of *Cyathocrinus*, a dicyclic inadunate crinoid, showing the main plates.

INCANDESCENT LAMP

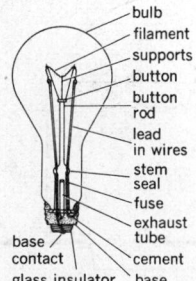

bulb
filament
supports
button
button rod
lead in wires
stem
seal
fuse
exhaust tube
base
contact
glass insulator
cement
base

Parts of an incandescent lamp. (*General Electric Co.*)

inclination of the wind [METEOROL] The angle between the direction of the wind and the isobars.

incline [SCI TECH] An upward- or downward-sloping surface.

inclined bedding [GEOL] A type of bedding in which the strata dip in the direction of current flow.

inclined cableway [MECH ENG] A monocable arrangement in which the track cable has a slope sufficiently steep to allow the carrier to run down under its own weight.

inclined contact [GEOL] A contact plane of gas or oil with water underlying, in which the plane slopes or is inclined.

inclined drilling [ENG] The drilling of blastholes at an angle with the vertical.

inclined extinction [OPTICS] Extinction in which the vibration directions are inclined to a crystal axis or direction of cleavage. Also known as oblique extinction.

inclined orbit [AERO ENG] A satellite orbit which is inclined with respect to the earth's equator.

inclined plane [MECH] A plane surface at an angle to some force or reference line.

inclined skip hoist [MIN ENG] A skip hoist that operates on steeply inclined rails placed on a mine pit slope or wall.

inclined-tube manometer [ENG] A glass-tube manometer with the leg inclined from the vertical to extend the scale for more minute readings.

incline man [MIN ENG] In anthracite coal mining, bituminous coal mining, and metal mining, a laborer who controls the movement of cars on a self-acting incline.

incline shaft [MIN ENG] A shaft which has been dug at an angle to the vertical to follow the depth of the lode.

inclining experiment [NAV ARCH] An experimental determination of the weight of a ship and of the position of its center of gravity, in which a known weight, already aboard, is moved a measured distance perpendicular to the ship's centerline plane, and the resulting angle of list is measured.

inclinometer [ENG] **1.** An instrument that measures the attitude of an aircraft with respect to the horizontal. **2.** An instrument for measuring the angle between the earth's magnetic field vector and the horizontal plane. **3.** An apparatus used to ascertain the direction of the magnetic field of the earth with reference to the plane of the horizon.

inclusion [CRYSTAL] **1.** A crystal or fragment of a crystal found in another crystal. **2.** A small cavity filled with gas or liquid in a crystal. [CYTOL] A visible product of cellular metabolism within the protoplasm. [PETR] A fragment of older rock enclosed in an igneous rock.

inclusion blennorrhea *See* inclusion conjunctivitis.

inclusion body [VIROL] Any of the abnormal structures appearing within the cell nucleus or the cytoplasm during the course of virus multiplication.

inclusion complex [CHEM] Crystalline mixture in which the molecules of one component are contained within the crystal lattice of another component. Also known as inclusion compound.

inclusion compound *See* inclusion complex.

inclusion conjunctivitis [MED] An acute inflammation of the conjunctiva with pus formation caused by a virus and identified from epithelial-cell inclusion bodies in conjunctival scrapings. Also known as inclusion blennorrhea; paratrachoma; swimmer's conjunctivitis; swimming pool conjunctivitis.

inclusion cyst [MED] A cyst formed by the implantation of epithelial tissue into another structure.

inclusion encephalitis [MED] A chronic inflammation of the brain in which large inclusion bodies occur in the nuclei of oligodendria and sometimes in nerve cells.

inclusive or *See* or.

incoalation *See* coalification.

incoherence [MED] Lack of coherence, relevance, or continuity of ideas or language.

incoherent light [OPTICS] Electromagnetic radiant energy not all of the same phase, and possibly also consisting of various wavelengths.

incoherent scattering [PHYS] Scattering of particles or photons in which the scattering elements act independently of one another, so that there are no definite phase relationships among the different parts of the scattered beam.

incoherent waves [PHYS] Waves having no fixed phase relationship.

incoming first selector [ELEC] Connects incoming calls from outlying dial offices to local second selectors.

incoming selector [ELEC] Selector associated with trunk circuits from another central office.

incommensurable numbers [MATH] Two numbers whose ratio is irrational.

incompatible equations [MATH] Two or more equations that are not satisfied by any set of values for the variables appearing. Also known as inconsistent equations.

incompetence [MED] **1.** Insufficiency or inadequacy in performing natural functions. **2.** In forensic medicine, inability to function within the law, as the incompetence of an individual to drive when under the influence of alcohol.

incompetent bed [GEOL] A bed not combining sufficient firmness and flexibility to transmit a thrust and to lift a load by bending.

incompetent rock [ENG] Soft or fragmented rock in which an opening, such as a borehole or an underground working place, cannot be maintained unless artificially supported by casing, cementing, or timbering.

incomplete abortion [MED] Expulsion of only part of the product of conception, with some of the membranes or placenta remaining in the uterus.

incomplete beta function [MATH] The function $\beta_x(p,q)$ defined by

$$\beta_x(p,q) = \int_0^x t^{p-1}(1-t)^{q-1}\,dt$$

where $0 \leq x \leq 1$, $p > 0$, and $q > 0$.

incomplete flower [BOT] A flower lacking one or more modified leaves, such as petals, sepals, pistils, or stamens.

incomplete fusion *See* quasifission.

incomplete gamma function [MATH] Either of the functions $\gamma(a,x)$ and $\Gamma(a,x)$ defined by

$$\gamma(a,x) = \int_0^x t^{a-1} e^{-t}\,dt, \text{ and}$$

$$\Gamma(a,x) = \int_x^\infty t^{a-1} e^{-t}\,dt,$$

where $0 \leq x \leq \infty$ and $a > 0$.

incomplete lubrication [MECH ENG] Lubrication that takes place when the load on the rubbing surfaces is carried partly by a fluid viscous film and partly by areas of boundary lubrication; friction is intermediate between that of fluid and boundary lubrication.

incompressibility [MECH] Quality of a substance which maintains its original volume under increased pressure.

incompressibility condition [FL MECH] The condition prevailing when dp/dt, the time rate of change of the density of a fluid, is zero; this is a valid assumption for most problems in dynamic oceanography.

incompressible flow [FL MECH] Fluid motion without any change in density.

incompressible fluid [FL MECH] A fluid which is not reduced in volume by an increase in pressure.

Inconel [MET] Trade name for a nickel-base alloy containing 16% chromium and 7% iron and characterized by marked resistance to aqueous corrosion and by high-temperature-oxidation resistance. Also known as Alloy 600.

incongruous [GEOL] Of a drag fold, having an axis and axial surface not parallel to the axis and axial surface of the main fold to which it is related.

inconsistent equations *See* incompatible equations.

incontinence [MED] Inability to control the natural evacuations, as the feces or the urine; specifically, involuntary evacuation from organic causes.

incorporate [ADP] To place in storage.

incrassate [BIOL] State of being swollen or thickened.

increaser [ENG] An adapter for connecting a small-diameter pipe to a larger-diameter pipe.

increasing function *See* monotone nondecreasing function.

increment [HYD] *See* recharge. [MATH] A change in the argument or values of a function, usually restricted to be a small positive or negative quantity. [SCI TECH] A small change in the value of a variable.

incremental computer [ADP] A special-purpose computer designed to process changes in variables as well as absolute values; for instance, a digital differential analyzer.

incremental digital recorder [ADP] Magnetic tape recorder

INCLINED PLANE

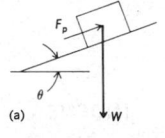

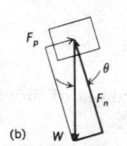

Weight resting on an inclined plane *(a)* with principal forces applied, and *(b)* their resolution into normal force. θ is angle of inclination of plane, W is weight of body, F_p is force parallel to the surface, F_n is force normal to the surface.

INCLINED-TUBE MANOMETER

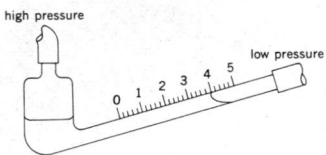

Drawing of inclined-tube manometer.

in which the tape advances across the recording head step by step, as in a punched-paper-tape recorder; used for recording an irregular flow of data economically and reliably.

incremental dump tape [ADP] A safety technique used in time-sharing which consists in copying all files (created or modified by a user during a day) on a magnetic tape; in case of system failure, the file storage can then be reconstructed. Also known as failsafe tape.

incremental frequency shift [COMMUN] Method of superimposing incremental intelligence on another intelligence by shifting the center frequency of an oscillator a predetermined amount.

incremental hysteresis loss [ELECTROMAG] Hysteresis loss when a magnetic material is subjected to a pulsating magnetizing force.

incremental induction [ELECTROMAG] The quantity lying between the highest and lowest value of a magnetic induction at a point in a polarized material, when subjected to a small cycle of magnetization.

incremental mode [ADP] The plotting of a curve on a cathode-ray tube by illuminating a fixed number of points at a time.

incremental permeability [ELECTROMAG] The ratio of a small cyclic change in magnetic induction to the corresponding cyclic change in magnetizing force when the average magnetic induction is greater than zero.

incretion [PHYSIO] An internal secretion.

incubation [CHEM] Maintenance of chemical mixtures at specified temperatures for varying time periods to study chemical reactions, such as enzyme activity. [MED] The phase of an infectious disease process between infection by the pathogen and appearance of symptoms. [VERT ZOO] The act or process of brooding or incubating.

incubation period [MED] The period of time required for the development of symptoms of a disease after infection, or of altered reactivity after exposure to an allergen. [VERT ZOO] The brooding period required to bring an egg to hatching.

incubator [AGR] A device for the artificial hatching of eggs. [MED] A small chamber with controlled oxygen, temperature, and humidity for newborn infants requiring special care. [MICROBIO] A laboratory cabinet with controlled temperature for the cultivation of bacteria, or for facilitating biologic tests.

incubator oil [MATER] Special grade of long-burning petroleum heating oil used to heat farm incubators.

incubatory carrier [MED] A person infected with a certain microorganism but in such an early stage of disease that clinical manifestations are not apparent.

incubous [BOT] The juxtaposition of leaves such that the anterior margins of older leaves overlap the posterior margins of younger leaves.

incudate [BIOL] Of, pertaining to, or having an incus.

incumbent [BIOL] Lying on or down. [GEOL] Lying above, said of a stratum that is superimposed or overlies another stratum.

incunabula printing See cradle printing.

incurrent canal [INV ZOO] A canal through which water enters a sponge.

incurrent siphon See inhalant siphon.

Incurvariidae [INV ZOO] A family of lepidopteran insects in the superfamily Incurvarioidea; includes yucca moths and relatives.

Incurvarioidea [INV ZOO] A monofamilial superfamily of lepidopteran insects in the suborder Heteroneura having wings covered with microscopic spines, a single genital opening in the female, and venation that is almost complete.

incus [ANAT] The middle one of three ossicles in the middle ear. Also known as anvil. [METEOROL] A supplementary cloud feature peculiar to cumulonimbus capillatus; the spreading of the upper portion of cumulonimbus when this part takes the form of an anvil with a fibrous or smooth aspect. Also known as anvil; thunderhead.

indamine [ORG CHEM] HN:C_6H_4:N·$C_6H_4NH_2$ An unstable dye obtained by the reaction of *para*-phenylenediamine and aniline. Also known as phenylene blue.

indan [ORG CHEM] $C_6H_4(CH_2)_3$ Colorless liquid boiling at

INDENE

Structural formula of indene.

177°C; soluble in alcohol and ether, insoluble in water; derived from coal tar.

indanthrene See indanthrone.

indanthrone [ORG CHEM] $C_{28}H_{14}N_2O_4$ A blue pigment or vat dye soluble in dilute base solutions; used in cotton dyeing and as a pigment in paints and enamels. Also known as 6,15-dihydro-5,9,14,18-anthrazinetetrone; indanthrene.

indeciduate placenta [EMBRYO] A placenta having the maternal and fetal elements associated but not fused.

indefinite ceiling [METEOROL] After United States weather observing practice, the ceiling classification applied when the reported ceiling value represents the vertical visibility upward into surface-based, atmospheric phenomena (except precipitation), such as fog, blowing snow, and all of the lithometeors. Formerly known as ragged ceiling.

indefinite integral [MATH] An indefinite integral of a function $f(x)$ is a function $F(x)$ whose derivative equals $f(x)$. Also known as antiderivative; integral.

indehiscent [BOT] **1.** Remaining closed at maturity, as certain fruits. **2.** Not splitting along regular lines.

indelible ink [MATER] An ink that cannot be removed, for example, India ink.

indene [ORG CHEM] C_9H_8 A colorless, liquid, polynuclear hydrocarbon; boils at 181°C and freezes at −2°C; derived from coal tar distillates; copolymers with benzofuran have been manufactured on a small scale for use in coatings and floor coverings.

indent [SCI TECH] To form a depression by forcing inward.

indentation hardness [MET] The resistance of a metal surface to indention when subjected to pressure by a hard pointed or rounded tool. Also known as penetration hardness.

indented bolt [DES ENG] A type of anchor bolt that has indentations to hold better in cemented grout.

independent assortment [GEN] The random assortment of the alleles at two or more loci on different chromosome pairs or far apart on the same chromosome pair which occurs at meiosis; first discovered by G. Mendel.

independent axioms [MATH] A list of axioms such that no axiom can be deduced as a theorem from the others.

independent chuck [DES ENG] A chuck for holding work by means of four jaws, each of which is moved independently of the others.

independent contractor [ENG] One who exercises independent control over the mode and method of operations to produce the results demanded by the contract.

independent equations [MATH] A system of equations such that no one of them is necessarily satisfied by a solution to the rest.

independent events [STAT] Two events in probability such that the occurrence of one of them does not affect the probability of the occurrence of the other.

independent footing [CIV ENG] A footing that supports a concentrated load, such as a single column.

independent functions [MATH] A set of functions such that knowledge of the values obtained by all but one of them at a point is insufficient to determine the value of the remaining function.

independent line of sighting [ORD] A system for laying a gun, whereby the angle of site and the angle of elevation (range) mechanisms work independently of each other.

independent migration law [ANALY CHEM] The law that each ion in a conductiometric titration contributes a definite amount to the total conductance, irrespective of the nature of the other ions in the electrolyte.

independent random variables [STAT] The discrete random variables $X_1, X_2, \ldots, X_n$ are independent if for arbitrary values $x_1, x_2, \ldots, x_n$ of the variables the probability that $X_1 = x_1$ and $X_2 = x_2$, etc., is equal to the product of the probabilities that $X_i = x_i$ for $i = 1, 2, \ldots, n$; random variables which are unrelated.

independent recoil system [ORD] A recoil mechanism for artillery that has an independent recuperator, that is, the recuperator is entirely independent of the recoil brake in the recoil mechanism.

independent-sideband modulation [COMMUN] Modulation in which the radio-frequency carrier is reduced or eliminated and two channels of information are transmitted, one on an upper and one on a lower sideband. Abbreviated ISB modulation.

independent-sideband receiver [ELECTR] A radio receiver designed for the reception of independent-sideband modulation, having provisions for restoring the carrier.

independent-sideband transmitter [ELECTR] A transmitter which produces independent-sideband modulated signals.

independent suspension [MECH ENG] In automobiles, a system of springs and guide links by which wheels are mounted independently on the chassis.

independent variable [MATH] In an equation $y = f(x)$, the input variable x. Also known as argument.

independent wire-rope core [DES ENG] A core of steel in a wire rope made in accordance with the best practice and design, either bright (uncoated) galvanized or drawn galvanized wire.

inderborite [MINERAL] $CaMgB_6O_{11} \cdot 11H_2O$ A monoclinic mineral composed of hydrous calcium and magnesium borate.

inderite [MINERAL] $Mg_2B_6O_{11} \cdot 15H_2O$ A hydrated borate mineral.

indeterminancy principle See uncertainty principle.

indeterminate cleavage [EMBRYO] Cleavage in which all the early cells have the same potencies with respect to development of the entire zygote.

indeterminate equations [MATH] A set of equations possessing an infinite number of solutions.

indeterminate forms [MATH] Products, quotients, differences, or powers of functions which are undefined when the argument of the function has a certain value, because one or both of the functions are zero or infinite; however, the limit of the product, quotient, and so on as the argument approaches this value is well defined.

indeterminate growth [BOT] Growth of a plant in which the axis is not limited by development of a reproductive structure, and therefore growth continues indefinitely.

indeterminate truss [CIV ENG] A truss having redundant bars.

index [ADP] **1.** A list of record surrogates arranged in order of some attribute expressible in machine-orderable form. **2.** To produce a machine-orderable set of record surrogates, as in indexing a book. **3.** To compute a machine location by indirection, as is done by index registers. **4.** The portion of a computer instruction which indicates what index register (if any) is to be used to modify the address of an instruction. [MATH] **1.** Unity of a logarithmic scale, as the C scale of a slide rule. **2.** A subscript or superscript used to indicate a specific element of a set or sequence. [PHYS] A numerical quantity, usually dimensionless, denoting the magnitude of some physical effect, such as the refractive index.

index arm [NAV] On a marine sextant, a slender bar carrying the index; the bar pivots at the center of curvature of the arc of the sextant and carries the index and the vernier or micrometer.

index array [ADP] A group of registers corresponding one-for-one with the buffer registers, each register containing the main storage address of the data contained in the corresponding buffer register.

index bed See key bed.

index catalog [ASTRON] A supplement to the New General Catalog of nebulae.

index center [MECH ENG] One of two machine-tool centers used to hold work and to rotate it by a fixed amount.

index chart [NAV] In marine operations, an outline chart showing the limits and identifying designations of navigational charts, volumes of sailing directions, and so on.

index cycle [METEOROL] A roughly cyclic variation in the zonal index.

indexed address [ADP] An address which is modified, generally by means of index registers, before or during execution of a computer instruction.

indexed sequential organization [ADP] A sequence of records arranged in collating sequence used with direct-access devices.

index ellipsoid [OPTICS] An ellipsoid whose three perpendicular axes are proportional in length to the principal values of the index of refraction of light in an anisotropic medium

and point in the direction of the corresponding electric vector. Also known as ellipsoid of wave normals; indicatrix; optical indicatrix; polarizability ellipsoid; reciprocal ellipsoid.

index error [ENG] An error caused by the misalignment of the vernier and the graduated circle (arc) of an instrument.

index forest [FOR] A forest reaching the highest average in a given locality for density, volume, and increment.

index fossil [PALEON] The ancient remains and traces of an organism that lived during a particular geologic time period and that geologically date the containing rocks.

index glass See index mirror.

index head [MECH ENG] A headstock that can be affixed to the table of a milling machine, planer, or shaper; work may be mounted on it by a chuck or centers, for indexing.

indexing fixture [MECH ENG] A fixture that changes position with regular steplike movements.

index line See isopleth.

index liquid [OPTICS] A liquid whose index of refraction is known, used to find the index of refraction of powdered substances with a microscope.

index marker [ADP] The beginning (and end) of each track in a disk, which is recognized by a special sensing device within the disk mechanism.

index mineral [PETR] A mineral whose first appearance in passing from low to higher grades of metamorphism indicates the outer limit of a zone.

index mirror [NAV] The mirror attached to the index arm of a marine sextant. Also known as index glass.

index number [STAT] A number indicating change in magnitude, as of cost or of volume of production, as compared with the magnitude at a specified time, usually taken as 100; for example, if production volume in 1970 was two times as much as the volume in 1950 (taken as 100), its index number is 200.

index of absorption See absorption index.

index of precision [STAT] The constant h in the normal curve $y = K \exp[-h^2(x-u)^2]$; a large value of h indicates a high precision, or small standard deviation.

index of a radical [MATH] The number above and to the left of the radical sign indicating the root to be extracted.

index of aridity [CLIMATOL] A measure of the precipitation effectiveness or aridity of a region, given by the following relationship: index of aridity $= P/(T+10)$, where P is the annual precipitation in centimeters, and T the annual mean temperature in degrees centigrade.

index of a subgroup [MATH] The quotient of the order of the group by the order of a subgroup.

index of cooperation [COMMUN] In rectilinear scanning or recording, the product of the total length of a scanning or recording line by the number of scanning or recording lines per unit length.

index of modulation See modulation factor.

index of refraction [OPTICS] The ratio of the phase velocity of light in a vacuum to that in a specified medium. Also known as refractive index; refracture index.

index plane [GEOL] A surface used as a reference point in determining geological structure.

index plate [DES ENG] A plate with circular graduations or holes arranged in circles, each circle with different spacing; used for indexing on machines.

index prism [NAV] A sextant prism which can be rotated to any angle corresponding to altitudes between established limits; the bubble or pendulum sextant counterpart of the index mirror of a marine sextant.

index ratio [ELECTROMAG] The ratio of the radius of a conductor used in induction heating to its skin depth at the frequency used.

index register [ADP] A hardware element which holds a number that can be added to (or, in some cases, subtracted from) the address portion of a computer instruction to form an effective address. Also known as base register; B box; B line; B register; B store; modifier register.

index stock [GRAPHICS] A stiff paper receptive to writing ink and to printing; available in smooth and antique finishes and used for index records and business cards.

index thermometer [ENG] A thermometer in which steel index particles are carried by mercury in the capillary and

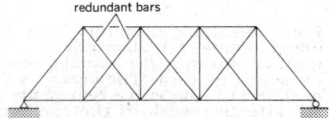

INDETERMINATE TRUSS

redundant bars

Redundant bars in an indeterminate truss.

adhere to the capillary wall in the high and low positions, thus indicating minimum and maximum inertial scales.

index word *See* modifier.

india ink [MATER] A permanent black ink made of lampblack and blue binder; some varieties are waterproof. Also known as Chinese ink; sumi ink.

indialite [MINERAL] $Mg_2Al_4Si_5O_{18}$ A hexagonal cordierite mineral; it is isotypic with beryl.

Indian *See* Indus.

indianaite [MINERAL] A white porcelainlike clay mineral; a variety of halloysite found in Indiana.

Indiana limestone *See* spergenite.

Indian balsam *See* Peru balsam.

Indian grass oil *See* palmarosa oil.

Indian gum [MATER] Any of the gums, such as ghatti gum and sterculia gum, with mucilage consistency from trees in the forests in India and Ceylon.

Indian Ocean [GEOGR] The smallest and geologically the most youthful of the three oceans, whose surface area is 75,900,000 square kilometers; it is bounded on the north by India, Pakistan, and Iran; on the east by the Malay Peninsula; on the south by Antarctica; and on the west by the Arabian peninsula and Africa.

Indian red [MATER] Iron-oxide-base, maroon pigment; used to polish gold, silver, and other metals. Also known as iron saffron.

Indian spring low water [OCEANOGR] An arbitrary tidal datum approximating the level of the mean of the lower low waters at spring time, first used in waters surrounding India. Also known as harmonic tide plane; Indian tide plane.

Indian summer [CLIMATOL] A period, in mid- or late autumn, of abnormally warm weather, generally clear skies, sunny but hazy days, and cool nights; in New England, at least one killing frost and preferably a substantial period of normally cool weather must precede this warm spell in order for it to be considered a true Indian summer; it does not occur every year, and in some years there may be two or three Indian summers; the term is most often heard in the northeastern United States, but its usage extends throughout English-speaking countries.

Indian tide plane *See* Indian spring low water.

Indian tragacanth *See* karaya gum.

Indian yellow [MATER] A yellow pigment which may be aureolin, made of cobalt and potassium nitrates; or puree, the impure basic magnesium salt of euxanthic acid; or the synthetic dye primuline.

indican [BIOCHEM] C_8H_6NOSOK The potassium salt of indoxylsulfate found in urine as a result of bacterial action on tryptophan in the bowel. [ORG CHEM] $C_{14}H_{17}O_6N$ A glucoside of indoxyl occurring in the indigo plant; on hydrolysis indican gives rise to indoxyl, which is oxidized to indigo by air.

indicated airspeed [AERO ENG] The airspeed as shown by a differential-pressure airspeed indicator, uncorrected for instrument and installation errors; a simple computation for altitude and temperature converts indicated airspeed to true airspeed. Abbreviated IAS.

indicated air temperature [METEOROL] The uncorrected reading from the free air temperature gage. Also known as outside air temperature.

indicated altitude [AERO ENG] The uncorrected reading of a barometric altimeter.

indicated horsepower [MECH ENG] The horsepower delivered by an engine as calculated from the average pressure of the working fluid in the cylinders and the displacement. Abbreviated ihp.

indicated ore [MIN ENG] An ore whose tonnage and grade are determined partly by a combination of specific measurements, samples, production data, and partly by a projection based on reasonable geological evidence.

indicating gage [ENG] A gage consisting essentially of a case and mounting, a spindle carrying the contact point, an amplifying mechanism, a pointer, and a graduated dial; used to amplify and measure the displacement of a movable contact point.

indicator [ADP] A device announcing an error or failure. [ELECTR] A cathode-ray tube or other device that presents

information transmitted or relayed from some other source, as from a radar receiver. [ENG] An instrument for obtaining a diagram of the pressure-volume changes in a running positive-displacement engine, compressor, or pump cylinder during the working cycle.

indicator card [ENG] A chart on which an indicator diagram is produced by an instrument called an engine indicator which traces the real-performance cycle diagram as the machine is running.

indicator diagram [ENG] A pressure-volume diagram representing and measuring the work done by or on a fluid while performing the work cycle in a reciprocating engine, pump, or compressor cylinder.

indicator element [ELECTR] A component whose variability under conditions of manufacture or use is likely to cause the greatest variation in some measurable parameter.

indicator function *See* characteristic function.

indicator gate [ELECTR] Rectangular voltage waveform which is applied to the grid or cathode circuit of an indicator cathode-ray tube to sensitize or desensitize it during a desired portion of the operating cycle.

indicator lamp [ELEC] A neon lamp whose on-off condition is used to convey information.

indicator medium [MICROBIO] A usually solid culture medium containing substances capable of undergoing a color change in the vicinity of a colony which has effected a particular chemical change, such as fermenting a certain sugar.

indicator plant [BOT] A plant used in geobotanical prospecting as an indicator of a certain geological phenomenon.

indicator tube [ELECTR] An electron-beam tube in which useful information is conveyed by the variation in cross section of the beam at a luminescent target.

indicator unit [ENG] An instrument which detects the presence of an electrical quantity without necessarily measuring it.

indicatrix *See* index ellipsoid.

indicolite [MINERAL] An indigo-blue variety of tourmaline that is used as a gemstone. Also known as indigolite.

indifferent equilibrium *See* neutral equilibrium; neutral stability.

indifferent stability *See* neutral stability.

indigenous [SCI TECH] Existing and having originated naturally in a particular region or environment.

indigenous coal *See* autochthonous coal.

indigo [ORG CHEM] **1.** A blue dye extracted from species of the *Indigofera* bush. **2.** *See* indigo blue.

indigo blue [ORG CHEM] $C_{16}H_{10}O_2N_2$ A component of the dye indigo, crystallizing as dark-blue rhomboids that break down at 30°C, that are soluble in hot aniline and hot chloroform, and that are also made synthetically; used as a reagent and a dye. Also known as indigo.

indigoid dye [ORG CHEM] Any of the vat dyes with $C_{16}H_{10}O_2N_2$ (indigo) or $C_{16}H_8S_2O_2$ (thioindigo) groupings; used to dye cotton and rayon, sometimes silk.

indigolite *See* indicolite.

indigo red [ORG CHEM] $C_{16}H_{10}O_2N_2$ A red isomer of indigo obtained in the manufacture of indigo. Also known as indirubin.

indirect address [ADP] An address in a computer instruction that indicates a location where the address of the referenced operand is to be found. Also known as multilevel address.

indirect addressing [ADP] A programming device whereby the address part of an instruction is not the address of the operand but rather the location in storage where the address of the operand may be found.

indirect-arc furnace [ENG] A refractory-lined furnace in which the burden is heated indirectly by the radiant heat from an electric arc.

indirect cell [METEOROL] A closed circulation in a vertical plane in which the rising motion occurs at lower potential temperature than the descending motion, thus forming an energy sink.

indirect Coombs test *See* Rh blocking test.

indirect cost [IND ENG] A cost that is not readily indentifiable with or chargeable to a specific product or service.

indirect cycle [NUCLEO] A nuclear reactor cycle in which a

INDICATOR DIAGRAM

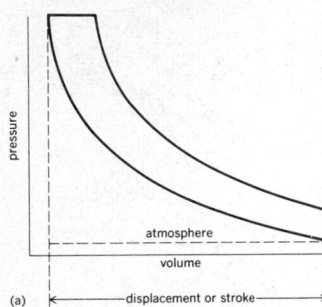

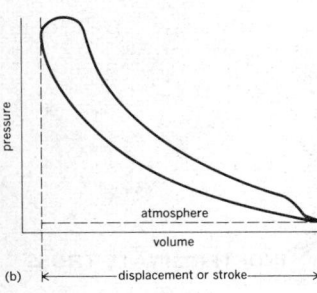

Pressure-volume diagrams for a diesel engine. *(a)* Theoretical ideal conditions; *(b)* actual two-cycle engine card. The areas outlined by the curve represent the effective work of the cycle.

heat exchanger transfers heat from the reactor coolant to a second fluid, which then drives a prime mover.

indirect developing test   *See* Rh blocking test.

indirect extrusion   [MET] An extrusion process in which the billet remains stationary while a hollow die stand forces the die back into the cylinder.

indirect fire   [ORD] Gunfire delivered at a target that cannot be seen from the gun position or firing ship.

indirect hernia   [MED] A form of inguinal hernia that passes out of the abdomen through the inguinal canal.

indirect labor   [IND ENG] Labor not directly engaged in the actual production of the product or performance of a service.

indirect laying   [ORD] Aiming a gun either by sighting at a fixed object called the aiming point instead of at the target or by using a means of pointing other than a sight, such as a gun director, when the target cannot be seen from the gun position.

indirectly heated cathode   [ELECTR] A cathode to which heat is supplied by an independent heater element in a thermionic tube; this cathode has the same potential on its entire surface, whereas the potential along a directly heated filament varies from one end to the other. Also known as equipotential cathode; heater-type cathode; unipotential cathode.

indirect material   [IND ENG] Any material used in the manufacture of a product which does not itself become a part of the product and whose cost is indirect.

indirect stratification   *See* secondary stratification.

indirect vision   *See* peripheral vision.

indirect wave   [PHYS] Any radio wave which arrives by an indirect path, having undergone an abrupt change of direction by refraction or reflection.

indirubin   *See* indigo red.

indium   [CHEM] A metallic element, symbol In, atomic number 49, atomic weight 114.82; soluble in acids; melts at 156°C, boils at 1450°C. [MET] A ductile, silver-white, shiny metal that resists tarnishing and is used in precious-metal alloys for jewelry and dentistry, in glass-sealing alloys, lubricants, and bearing metals, and as an atomic-pile neutron indicator.

indium antimonide   [INORG CHEM] InSb Crystals that melt at 535°C; an intermetallic compound having semiconductor properties and the highest room-temperature electron mobility of any known material; used in Hall-effect and magnetoresistive devices and as an infrared detector.

indium arsenide   [INORG CHEM] InAs Metallic crystals that melt at 943°C; an intermetallic compound having semiconductor properties; used in Hall-effect devices.

indium chloride   [INORG CHEM] $InCl_3$ Hygroscopic white powder, soluble in water and alcohol.

indium phosphide   [INORG CHEM] InP A metallic mass that is brittle and melts at 1070°C; an intermetallic compound having semiconductor properties.

indium sulfate   [INORG CHEM] $In_2(SO_4)_3$ Deliquescent, water-soluble, grayish powder; decomposes when heated.

individual line   [COMMUN] Subscriber line arranged to serve only one main station, although additional stations may be connected to the line as extensions; an individual line is not arranged for discriminatory ringing with respect to the stations on that line.

individual psychology   [PSYCH] A system of psychology in which traits of an individual are compared in terms of striving for superiority and then restated in the form of a composite of this single tendency.

indogen   [ORG CHEM] The radical $C_6H_4(NH)COC=$; it occurs, for example, in the molecule indigo.

indogenide   [ORG CHEM] A compound containing the radical $C_6H_4(NH)\cdot CO \cdot C=$ from indogen.

indole   Also known as 2,3-benzopyrrole. [BIOCHEM] $C_6H_4(CHNH)CH$ A decomposition product of tryptophan formed in the intestine during putrefaction and by certain cultures of bacteria. [ORG CHEM] Carcinogenic, white to yellowish scales with unpleasant aroma; soluble in alcohol, ether, hot water, and fixed oils; melt at 52°C; used as a chemical reagent and in perfumery and medicine.

indoleacetic acid   [BIOCHEM] $C_{10}H_9O_2N$ A decomposition product of tryptophan produced by bacteria and occurring in urine and feces; used as a hormone to promote plant growth. Abbreviated IAA.

indolebutyric acid   [ORG CHEM] $C_{12}H_{13}O_2N$ A crystalline acid similar to indoleacetic acid in auxin activity. Abbreviated IBA.

indolent   [MED] **1.** Of or relating to a slow-growing, nonpainful neoplasm. **2.** Slowness in the process of healing.

indole test   [MICROBIO] A test for the production of indole from tryptophan by microorganisms; a solution of *para*-dimethylaminobenzaldehyde, amyl alcohol, and hydrochloric acid added to the incubated culture of bacteria shows a red color in the alcoholic layer if indole is present.

Indo-Pacific faunal region   [ECOL] A marine littoral faunal region extending eastward from the east coast of Africa, passing north of Australia and south of Japan, and ending in the east Pacific south of Alaska.

indoxyl   [ORG CHEM] $(C_8H_6N)OH$ A yellow crystalline glycoside, used as an intermediate in the manufacture of indigo.

Indriidae   [VERT ZOO] A family of Madagascan prosimians containing wholly arboreal vertical clingers and leapers.

induced angle of attack   [AERO ENG] The downward vertical angle between the horizontal and the velocity (relative to the wing of an aircraft) of the airstream passing over the wing.

induced anisotropy   [SOLID STATE] A type of uniaxial anisotropy in a magnetic material produced by annealing the magnetic material in a magnetic field.

induced capacity   *See* absolute permeability.

induced current   [ELECTROMAG] A current produced in a conductor by a time-varying magnetic field, as in induction heating.

induced dipole   [ELEC] An electric dipole produced by application of an electric field.

induced draft   [MECH ENG] A mechanical draft produced by suction stream jets or fans at the point where air or gases leave a unit.

induced-draft cooling tower   [MECH ENG] A structure for cooling water by circulating air where the load is on the suction side of the fan.

induced drag   [FL MECH] That part of the drag caused by the downflow or downwash of the airstream passing over the wing of an aircraft, equal to the lift times the tangent of the induced angle of attack.

induced electromotive force   [ELECTROMAG] An electromotive force resulting from the motion of a conductor through a magnetic field, or from a change in the magnetic flux that threads a conductor.

induced fission   [NUC PHYS] Fission which takes place only when a nucleus is bombarded with neutrons, gamma rays, or other carriers of energy.

induced magnetism   [ELECTROMAG] The magnetism acquired by magnetic material while it is in a magnetic field.

induced magnetization   [GEOPHYS] That component of a rock's magnetization which is proportional to, and has the same direction as, the ambient magnetic field.

induced moment   [ELEC] The average electric dipole moment per molecule which is produced by the action of an electric field on a dielectric substance.

induced potential   *See* induced voltage.

induced precession   *See* real precession.

induced radioactivity   [NUCLEO] Radioactivity created by bombarding a substance with radiation. Also known as artificial radioactivity.

induced voltage   [ELECTROMAG] A voltage produced by electromagnetic or electrostatic induction. Also known as induced potential.

inducer   [EMBRYO] The cell group that functions as the acting system in embryonic induction by controlling the mode of development of the reacting system. Also known as inductor.

inducible enzyme   [BIOCHEM] A bacterial enzyme formed only in response to a specific inducer.

inductance   [ELECTROMAG] **1.** That property of an electric circuit or of two neighboring circuits whereby an electromotive force is generated (by the process of electromagnetic induction) in one circuit by a change of current in itself or in the other. **2.** Quantitatively, the ratio of the emf (electromotive force) to the rate of change of the current. **3.** *See* coil.

inductance bridge   [ELECTROMAG] **1.** A device, similar to a Wheatstone bridge, for comparing inductances. **2.** A four-coil alternating-current bridge circuit used for transmitting a

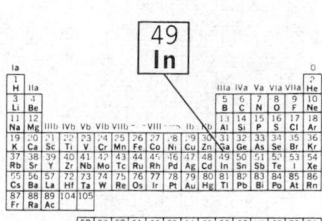

INDIUM

Periodic table of the chemical elements showing the position of indium.

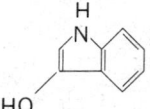

INDOXYL

Structural formula of indoxyl.

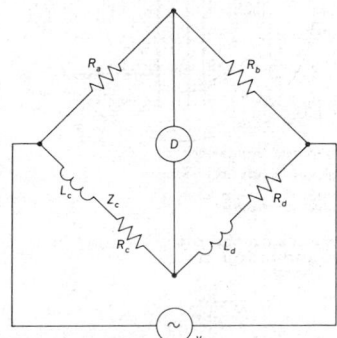

INDUCTANCE BRIDGE

Circuit diagram for a general inductance bridge (def. 1); R_{a-d} are resistors; L_{a-d} are inductances; Z_c is impedance; V is source of ac voltage; and D is a null detector.

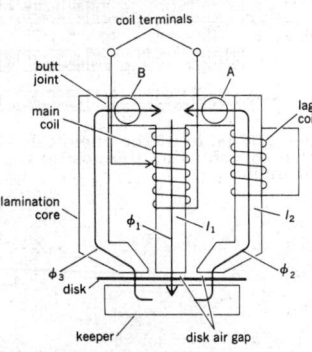

INDUCTION DISK RELAY

Schematic of the induction disk type of relay. I_1 = current in main coil, I_2 = current in lag coil, ϕ_1 = flux in main coil, ϕ_2 = flux in lag coil, ϕ_3 = differential induction flux, A = flux field about ϕ_3, B = flux field about ϕ_2.

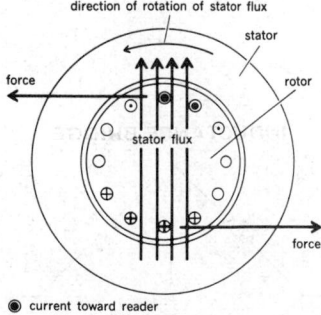

INDUCTION MOTOR

- ⊙ current toward reader
- ⊕ current away from reader

weight of ⊙ or + indicates magnitude of current

Forces on the rotor winding in an induction motor.

mechanical movement to a remote location over a three-wire circuit; half of the bridge is at each location.

inductance coil *See* coil.

inductance measurement [ELECTROMAG] The determination of the self-inductance of a circuit or the mutual inductance of two circuits.

inductance meter [ELECTROMAG] A device which measures the self-inductance of a circuit or the mutual inductance of two circuits.

inductance standards [ELECTROMAG] Two equal, multilayer coils, wound on toroidal cores of nonmagnetic materials, connected in series and located so that their interactions with external fields tend to cancel one another.

induction [ELEC] *See* electrostatic induction. [ELECTROMAG] *See* electromagnetic induction. [EMBRYO] *See* embryonic induction. [MED] The period from administration of the anesthetic to loss of consciousness by the patient. [SCI TECH] The act or process of causing.

induction accelerator *See* betatron.

induction brazing [MET] A brazing process in which coalescence is produced by heat generated within the work by an induced electric current.

induction burner [ENG] Fuel-air burner into which the fuel is fed under pressure to entrain needed air into the combustion nozzle area.

induction charging [ELEC] Production of electric charge on a body by means of electrostatic induction.

induction coil [ELECTROMAG] A device for producing high-voltage alternating current or high-voltage pulses from low-voltage direct current, in which interruption of direct current in a primary coil, containing relatively few turns of wire, induces a high voltage in a secondary coil, containing many turns of wire wound over the primary.

induction disk relay [ELECTROMAG] A unit widely used in regulating and protective relays, in which alternating current applied to a coil produces torque to rotate a disk.

induction-electrical survey [ENG] Study of subterranean formations by combined induction and electrical logging.

induction field [ELECTROMAG] A component of an electromagnetic field associated with an alternating current in a loop, coil, or antenna which carries energy alternately away from and back into the source, with no net loss, and which is responsible for self-inductance in a coil or mutual inductance with neighboring coils.

induction frequency converter [ELEC] Slip-ring induction machine which is driven by an external source of mechanical power and whose primary circuits are connected to a source of electric energy having a fixed frequency; the secondary circuits deliver energy at a frequency proportional to the relative speed of the primary magnetic field and the secondary member.

induction furnace [ENG] An electric furnace in which heat is produced in a metal charge by electromagnetic induction.

induction generator [ELEC] A nonsynchronous alternating-current generator whose construction is identical to that of an ac motor, and which is driven above synchronous speed by external sources of mechanical power.

induction hardening [MET] A quench-hardening technique in which the required elevated temperature is obtained by electromagnetic induction.

induction inclinometer *See* earth inductor.

induction instrument [ENG] Meter that depends for its operation on the reaction between magnetic flux set up by current in fixed windings, and other currents set up by electromagnetic induction in conducting parts of the moving system.

induction log [ENG] An electric log of the conductivity of rock with depth obtained by lowering into an uncased borehole a generating coil that induces eddy currents on the rocks and these are detected by a receiver coil.

induction loudspeaker [ENG ACOUS] Loudspeaker in which the current which reacts with the steady magnetic field is induced in the moving member.

induction machine [ELEC] An asynchronous alternating-current machine, such as an induction motor or induction generator, in which the windings of two electric circuits rotate with respect to each other and power is transferred

from one circuit to the other by electromagnetic induction.

induction melting [MET] Converting a solid metal to the molten state in an induction furnace.

induction method [GEOPHYS] In studies of the radioactivity of the atmosphere, a technique for estimating the concentration of the radioactive gases by exposing a negatively charged wire to the air and then using an ionization chamber to count the activity of the radioactive deposit formed on the wire.

induction motor [ELEC] An alternating-current motor in which a primary winding on one member (usually the stator) is connected to the power source, and a secondary winding on the other member (usually the rotor) carries only current induced by the magnetic field of the primary.

induction noise [ACOUS] The noise caused by the periodic inrush of intake air into the cylinder of an automobile engine or an air compressor as the intake valve opens and the piston moves on the intake stroke.

induction period [PHYS CHEM] A time of acceleration of a chemical reaction from zero to a maximum rate.

induction problems [ELECTROMAG] The effects of potentials and currents induced in conductors of a telephone system by paralleling power facilities or power lines.

induction pump [MECH ENG] Any pump operated by electromagnetic induction.

induction salinometer [ENG] A device for measuring salinity by taking voltage readings of the current in seawater.

induction silencer [ENG] A device for reducing engine induction noise, which consists essentially of a low-pass acoustic filter with the inertance of the air-entrance tube and the acoustic compliance of the annular and central volumes providing acoustic filtering elements.

induction soldering [MET] A soldering process in which the metals are heated by an induced electric current.

induction valve *See* inlet valve.

induction voltage regulator [ELECTROMAG] A type of transformer having a primary winding connected in parallel with a circuit and a secondary winding in series with the circuit; the relative positions of the primary and secondary windings are changed to vary the voltage or phase relations in the circuit.

induction watthour meter [ELEC] A watthour meter used with alternating current; the energy taken by a circuit over a period of time is proportional to the rotation in that period of a light aluminum disk, in which a driving torque is developed by the joint action of the alternating magnetic flux produced by the potential circuit and by the load current.

induction welding [MET] A process of welding by means of heat generated within the work by induced electric currents.

inductive charge [ELEC] The charge that exists on an object as a result of its being near another charged object.

inductive circuit [ELEC] A circuit containing a higher value of inductive reactance than capacitive reactance.

inductive coordination [ELECTROMAG] Measures to reduce induction problems.

inductive coupler [ELEC] A mutual inductance that provides electrical coupling between two circuits; used in radio equipment.

inductive coupling [ELEC] Coupling of two circuits by means of the mutual inductance provided by a transformer. Also known as transformer coupling.

inductive divider [ELECTROMAG] A device for incorporating a desired fraction of an inductance into a circuit, usually consisting of an autotransformer with an intermediate tap.

inductive feedback [ELECTR] **1.** Transfer of energy from the plate circuit to the grid circuit of a vacuum tube by means of induction. **2.** Transfer of energy from the output circuit to the input circuit of an amplifying device through an inductor, or by means of inductive coupling.

inductive filter [ELECTR] A low-pass filter used for smoothing the direct-current output voltage of a rectifier; consists of one or more sections in series, each section consisting of an inductor on one of the pair of conductors in series with a capacitor between the conductors. Also known as LC filter.

inductive grounding [ELEC] Use of grounding connections containing an inductance in order to reduce the magnitude of short-circuit currents created by line-to-ground faults.

inductive interference [COMMUN] Effect arising from the

characteristics and inductive relations of electric supply and communications systems of such character and magnitude as would prevent the communications circuits from rendering service satisfactorily and economically if methods of inductive coordination were not applied.

inductive line pair [COMMUN] A telephone line displaying induction whose effects are of consequence, as in crosstalk; opposed to twisted pair.

inductive load [ELEC] A load that is predominantly inductive, so that the alternating load current lags behind the alternating voltage of the load. Also known as lagging load.

inductive neutralization [ELECTR] Neutralizing an amplifier whereby the feedback susceptance due to an interelement capacitance is canceled by the equal and opposite susceptance of an inductor. Also known as coil neutralization; shunt neutralization.

inductive post [ELECTROMAG] Metal post or screw extending across a waveguide parallel to the E field, to add inductive susceptance in parallel with the waveguide for tuning or matching purposes.

inductive pressure transducer [ELECTROMAG] A type of pressure transducer in which changes in pressure cause a bourdon tube or other pressure-sensing element to move a magnetic core, and this results in a change in inductance of one or more windings of a coil surrounding the core.

inductive reactance [ELEC] Reactance due to the inductance of a coil or circuit.

inductive relay [ELECTROMAG] A relay displaying inductance, as opposed to one wound to be noninductive.

inductive spacing [ELECTROMAG] Spacing of parallel transmission lines so that there is transfer of energy by mutual inductance.

inductive surge [ELECTROMAG] A surge in voltage caused by sudden interruption of current in an inductive circuit.

inductive susceptance [ELEC] In a circuit containing almost no resistance, the part of the susceptance due to inductance.

inductive waveform [ELEC] A graph or trace of the effect of current buildup across an inductive network; proportional to the exponential of the product of a negative constant and the time.

inductive window [ELECTROMAG] Conducting diaphragm extending into a waveguide from one or both sidewalls of the waveguide, to give the effect of an inductive susceptance in parallel with the waveguide.

inductometer [ELECTROMAG] A coil of wire of known inductance; the inductance may be fixed as in the case of primary standards, adjustable by means of switches, or continuously variable by means of a movable-coil construction.

inductor *See* coil; inducer.

inductor alternator [ELEC] A synchronous generator in which the field winding is fixed in magnetic position relative to the armature conductors.

inductor generator [ELEC] An alternating-current generator in which all the windings are fixed, and the flux linkages are varied by rotating an appropriately toothed ferromagnetic rotor; sometimes used for generating high power at frequencies up to several thousand hertz for induction heating.

inductor microphone [ENG ACOUS] Moving-conductor microphone in which the moving element is in the form of a straight-line conductor.

inductor tachometer [ENG] A type of impulse tachometer in which the rotating member, consisting of a magnetic material, causes the magnetic flux threading a circuit containing a magnet and a pickup coil to rise and fall, producing pulses in the circuit which are rectified for a permanent-magnet, movable-coil instrument.

inductura [INV ZOO] A layer of lamellar shell material along the inner lip of the aperture in gastropods.

indumentum [BOT] A covering, such as one that is woolly. [MYCOL] A covering of hairs. [VERT ZOO] The plumage covering a bird.

induplicate [BOT] Having the edges turned or rolled inward without twisting or overlapping; applied to the leaves of a bud.

induration [BIOL] The process of hardening, especially by increasing the fibrous elements. [GEOL] **1.** The hardening of a rock material by the application of heat or pressure or by the introduction of a cementing material. **2.** A hardened mass formed by such processes. [MED] Hardening of a tissue or organ due to an accumulation of blood, inflammation, or neoplastic growth.

Indus [ASTRON] A constellation, right ascension 21 hours, declination 55° south. Also known as Indian.

indusium [ANAT] A covering membrane such as the amnion. [BOT] An epidermal outgrowth covering the sori in many ferns. [MYCOL] The annulus of certain fungi.

industrial alcohol [ORG CHEM] Ethyl alcohol that has been denatured by acetates, ketones, gasoline, or other additives to make it unfit for beverage purposes.

industrial anthropometry [IND ENG] Application of the knowledge of physical anthropology to the design and construction of equipment for human use, such as automobiles.

industrial car [IND ENG] Any of various narrow-gage railcars used for indoor or outdoor handling of bulk and package materials.

industrial climatology [CLIMATOL] A type of applied climatology which studies the effect of climate and weather on industry's operations; the goal is to provide industry with a sound statistical basis for all administrative and operational decisions which involve a weather factor.

industrial cost control [IND ENG] A specific system or procedure used to keep manufacturing costs in line. Also known as cost control.

industrial crop [AGR] Any crop that provides materials for industrial processes and products such as soybeans, cotton (lint and seed), flax, and tobacco.

industrial diamond [MINERAL] Diamond that is too hard or too radial-grained to be used for jewel cutting.

industrial engineering [ENG] A branch of engineering concerned with the design, improvement, and installation of integrated systems of people, materials, and equipment. Also known as management engineering.

industrial frequency bands [COMMUN] The radio-frequency bands allocated in the United States for land mobile communications of private industries other than transportation.

industrial geography [GEOGR] A branch of geography that deals with location, raw materials, products, and distribution, as influenced by geography.

industrial glass [MATER] Any glass molded into shapes for product parts, for example, lime glass and lead glass.

industrial jewel [MINERAL] A hard stone, such as ruby or sapphire, used for bearings and impulse pins in instruments and for recording needles.

industrial meteorology [METEOROL] The application of meteorological information and techniques to industrial problems.

industrial microbiology [MICROBIO] The study, utilization, and manipulation of those microorganisms capable of economically producing desirable substances or changes in substances, and the control of undesirable microorganisms.

industrial microorganism [MICROBIO] Any microorganism utilized for industrial microbiology.

industrial mobilization [IND ENG] Transformation of industry and other productive facilities and contributory services from their peacetime activities to the fulfillment of the munitions program necessary to support a military effort.

industrial psychology [PSYCH] Psychology applied to problems in industry, dealing chiefly with the selection, efficiency, and mental health of personnel.

industrial railway [IND ENG] **1.** A usually short feeder line that is either owned or controlled and wholly operated by an industrial firm. **2.** Narrow-gage rail lines used on construction jobs or around industrial plants.

industrial revolution [IND ENG] A widespread change in industrial or production methods, toward production by machine and away from manual labor.

industrial security [IND ENG] The portion of internal security which refers to the protection of industrial installations, resources, utilities, materials, and classified information essential to protection from loss or damage.

industrial television [COMMUN] Closed-circuit television used for remote viewing of industrial processes and operations. Abbreviated ITV.

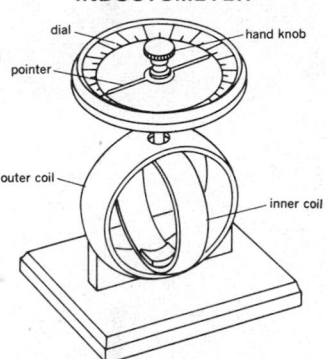

INDUCTOMETER

A variable inductometer showing the movable inner coil.

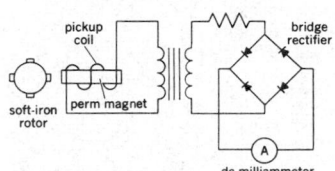

INDUCTOR TACHOMETER

Circuit for an inductor form of tachometer.

industrial waste [ENG] Worthless materials remaining from industrial operations.

industrial yeast [MICROBIO] Any yeast used for the production of fermented foods and beverages, for baking, or for the production of vitamins, proteins, alcohol, glycerol, and enzymes.

inelastic [MECH] Not capable of sustaining a deformation without permanent change in size or shape.

inelastic buckling [MECH] Sudden increase of deflection or twist in a column when compressive stress reaches the elastic limit but before elastic buckling develops.

inelastic collision [MECH] A collision in which the total kinetic energy of the colliding particles is not the same after the collision as before it.

inelastic cross section [PHYS] The cross section for an inelastic collision.

inelastic scattering [PHYS] Scattering that results from inelastic collisions.

inelastic stress [MECH] A force acting on a solid which produces a deformation such that the original shape and size of the solid are not restored after removal of the force.

inequality [MATH] A statement that one quantity is less than, less than or equal to, greater than, or greater than or equal to another quantity.

inequality of Clausius *See* Clausius inequality.

inequilateral [BIOL] Having the two sides or ends unequal, as the ends of a bivalve mollusk on either side of a line from umbo to gape.

inermous [BIOL] Lacking mechanisms for defense or offense, especially spines.

inert [SCI TECH] Lacking an activity, reactivity, or effect.

inert atmosphere [CHEM ENG] A nonreactive gas atmosphere, such as nitrogen, carbon dioxide, or helium; used to blanket reactive liquids in storage, to purge process lines and vessels of reactive gases and liquids, and to cover a reaction mix in a partially filled vessel.

inert gas *See* noble gas.

inert-gas cutting [MET] Cutting of metal while inert gas flows around the cutting area to prevent oxidation.

inert gas-shielded arc welding [MET] An arc-welding process in which the weld area is shielded by an inert gas to prevent oxidation. Also known as Heliarc welding.

inertia [MECH] That property of matter which manifests itself as a resistance to any change in the momentum of a body. [MED] Sluggishness, especially of muscular activity.

inertia currents [OCEANOGR] Currents resulting after the cessation of wind in a generating area or after the water movement has left the generating area; circular currents with a period of one-half pendulum day.

inertia ellipsoid [MECH] An ellipsoid used in describing the motion of a rigid body; it is fixed in the body, and the distance from its center to its surface in any direction is inversely proportional to the square root of the moment of inertia about the corresponding axis. Also known as Poinsot ellipsoid.

inertia governor [MECH ENG] A speed-control device utilizing suspended masses that respond to speed changes by reason of their inertia.

inertial circle [METEOROL] A loop in the path of an air parcel in inertial flow, which is approximately circular if the latitudinal displacement is small. Also known as circle of inertia. [OCEANOGR] The circle described by inertial motion in a body of ocean water and having a radius $R = C/f$, where C is the particle velocity in a given direction and f is the Coriolis parameter.

inertial coordinate system *See* inertial reference frame.

inertial flow [FL MECH] Flow in which no external forces are exerted on a fluid. [GEOPHYS] Frictionless flow in a geopotential surface in which there is no pressure gradient; the centrifugal and Coriolis accelerations must therefore be equal and opposite, and the constant inertial wind speed V_i is given by $V_i = fR$, where f is the Coriolis parameter and R the radius of curvature of the path.

inertial force [MECH] The fictitious force acting on a body as a result of using a noninertial frame of reference; examples are the centrifugal and Coriolis forces that appear in rotating coordinate systems. Also known as effective force.

inertial guidance [NAV] **1.** Guidance by means of accelerations measured and integrated within the craft. **2.** Guidance by the use of an inertial navigation system.

inertial instability [FL MECH] **1.** Generally, instability in which the only form of energy transferred between the steady state and the disturbance in the fluid is kinetic energy. **2.** The hydrodynamic instability arising in a rotating fluid mass when the velocity distribution is such that the kinetic energy of a disturbance grows at the expense of kinetic energy of the rotation. Also known as dynamic instability.

inertial mass [MECH] The mass of an object as determined by Newton's second law, in contrast to the mass as determined by the proportionality to the gravitational force.

inertial navigator [NAV] A device which determines position by automatic dead reckoning; this operation is performed by the double integration of the outputs of accelerometers stabilized with respect to inertial space.

inertial platform [NAV] In an inertial navigator a platform that maintains sensing instruments in a precise known angular orientation in space.

inertial reference frame [MECH] A coordinate system in which a body moves with constant velocity as long as no force is acting on it. Also known as inertial coordinate system.

inertial size *See* aerodynamic size.

inertial space [NAV] A coordinate system or frame of reference defined with respect to the stars whose apparent positions relative to surrounding stars appear to be fixed or unvarying for long periods of time.

inertial theory [OCEANOGR] The theory associated with the motion of an ocean current under the influences of inertia and the Coriolis force, which cause it to take a circular path.

inertia of energy [RELAT] The principle that the inertial properties of matter both determine and are determined by its total energy content.

inertia period [OCEANOGR] The time required for a given particle to complete an inertia circle.

inertia starter [MECH ENG] A device utilizing inertial principles to start the rotator of an internal combustion engine.

inertia switch [ELEC] A switch that is actuated by an abrupt change in the velocity of the item on which it is mounted.

inertia tensor [MECH] A tensor associated with a rigid body whose product with the body's rotation vector yields the body's angular momentum.

inertia wave [FL MECH] **1.** Any wave motion in which no form of energy other than kinetic energy is present; in this general sense, Helmholtz waves, barotropic disturbances, Rossby waves, and so forth, are inertia waves. **2.** More restrictedly, a wave motion in which the source of kinetic energy of the disturbance is the rotation of the fluid about some given axis; in the atmosphere a westerly wind system is such a source, the inertia waves here being, in general, stable.

inertinite [GEOL] A carbon-rich maceral group, which includes micrinite, sclerotinite, fusinite, and semifusinite.

inert primer [ENG] A cylinder which enshrouds a detonator but does not interfere with the detonation of the explosive charge.

inesite [MINERAL] $Ca_2Mn_7Si_{10}O_{28}(OH)_2 \cdot 5H_2O$ A pale-red mineral composed of hydrous manganese calcium silicate, occurring in small prismatic crystals or massive.

inevitable abortion [MED] An abortion that has progressed to a stage where termination of the pregnancy cannot be avoided.

inextensional deformation [MECH] A bending of a surface that leaves unchanged the length of any line drawn on the surface and the curvature of the surface at each point.

in extremis [MED] At the point of death.

inface *See* scarp slope.

infall process [ASTROPHYS] A process in which gas falls upon a very compact object such as a neutron star or black hole, reaching a high velocity and forming a hot plasma; postulated as a model for x-ray sources such as Centaurus X-1 and Hercules X-1.

infancy [GEOL] The initial (youthful) or very early stage of the cycle of erosion characterized by smooth, nearly level erosional surfaces dissected by narrow stream gorges, numerous depressions filled by marshy lakes and ponds, and shallow streams. Also known as topographic infancy.

infant [ANTHRO] **1.** A baby; child under 2 years of age. **2.** An individual under legal age.

infantile amaurotic familial idiocy *See* Tay-Sachs disease.

infantile autism [PSYCH] A disorder of children characterized by extreme withdrawal and introspection without regard for reality.

infantile celiac disease [MED] Celiac syndrome of infants and young children.

infantile cortical hyperostosis [MED] A condition occurring during the first 3 months of life in which there is fever and painful swelling of the soft tissue of the lower jaw, characterized by periosteal proliferation of the mandible.

infantile diarrhea [MED] An acute gastrointestinal disease in infants resulting from damage of the intestinal mucosa by an infectious organism.

infantile eczema [MED] An allergic inflammation of the skin in young children, usually due to common antigens such as food or inhalants.

infantile genitalia [ANAT] The genital organs of an infant. [MED] Underdevelopment of the adult genitals.

infantile neuroaxonal dystrophy [MED] A familial disease of the central nervous system occurring early in life and characterized by axonal swellings, arrested development, atrophy of the optic nerves, and eventual blindness.

infantile paralysis *See* poliomyelitis.

infantile scurvy [MED] Acute scurvy of infants and young children characterized by periosteal hemorrhage and swelling, especially of long bones. Also known as Cheadle's disease; Moeller-Barlow disease.

infantile sexuality [PSYCH] An infant's or child's capacity for and enjoyment of activities and experiences that are essentially sexual.

infantile spasm [MED] A type of seizure seen in infants and young children, characterized by a sudden, brief, massive myoclonic jerk.

infantilism [MED] Persistence of physical, behavioral, or mental infantile characteristics into childhood, adolescence, or adult life.

infarct [MED] Localized death of tissue caused by obstructed inflow of arterial blood. Also known as infarction.

infarction [MED] **1.** Condition or process leading to the development of an infarct. **2.** *See* infarct.

infauna [ZOO] Aquatic animals which live in the bottom sediment of a body of water.

infect [MED] To cause an infection, as by contamination with or invasion by a pathogen. [MICROBIO] To cause a phage infection of bacteria.

infection [MED] **1.** Invasion of the body by a pathogenic organism, with or without disease manifestation. **2.** Pathologic condition resulting from invasion of a pathogen.

infectious [MED] Caused by infection.

infectious abortion *See* contagious abortion.

infectious anemia [VET MED] A virus disease of horses and mules characterized by intermittent fever, weakness, jaundice, and hemorrhages of mucous membranes; it is often fatal. Also known as swamp fever.

infectious arthritis [MED] An inflammatory joint disease caused by microbial invasion of the articular tissue.

infectious bronchitis [VET MED] A highly contagious respiratory viral disease of chickens.

infectious canine hepatitis [VET MED] An acute inflammatory liver disease of dogs caused by a virus.

infectious chlorosis [PL PATH] A virus disease of plants characterized by yellowing of the green parts.

infectious conjunctivitis [MED] Conjunctivitis due to invasion by a microorganism.

infectious disease [MED] Any disease caused by invasion by a pathogen which subsequently grows and multiplies in the body.

infectious drug resistance [MICROBIO] A type of drug resistance that is transmissible from one bacterium to another by infectivelike agents referred to as resistance factors.

infectious endocarditis [MED] Inflammation of the endocardium due to an infectious microorganism.

infectious hepatitis [MED] An acute infectious virus disease of the liver associated with hepatic inflammation and characterized by fever, liver enlargement, and jaundice. Also known as catarrhal jaundice; epidemic hepatitis; epidemic jaundice; virus hepatitis.

infectious laryngotracheitis [VET MED] A highly contagious respiratory disease of viral etiology affecting chickens.

infectious mononucleosis [MED] A disorder of unknown etiology characterized by irregular fever, pathology of lymph nodes, lymphocytosis, and high serum levels of heterophil antibodies against sheep erythrocytes. Also known as acute benign lymphoblastosis; glandular fever; kissing disease; lymphocytic angina; monocytic angina; Pfeiffer's disease.

infectious myocarditis [MED] Inflammation of the myocardium due to an infectious microorganism.

infectious myxomatosis [VET MED] An infectious virus disease of rabbits characterized by myxomatous lesions.

infectious papillomatosis [VET MED] A virus disease of cattle characterized by the appearance of warts on the body.

infectious rhinitis [MED] Inflammation of the nasal mucous membrane due to an infectious microorganism.

infectious uroarthritis *See* Reiter's syndrome.

infective dose 50 [MICROBIO] The dose of microorganisms required to cause infection in 50% of the experimental animals; a special case of the median effective dose. Abbreviated ID_{50}. Also known as median infective dose.

inferential flow meter [ENG] A flow meter in which the flow is determined by measurement of a phenomenon associated with the flow, such as a drop in static pressure at a restriction in a pipe, or the rotation of an impeller or rotor, rather than measurement of the actual mass flow.

inferential liquid-level meter [ENG] A liquid-level meter in which the level of a liquid is determined by measurement of some phenomenon associated with this level, such as the buoyancy of a solid partly immersed in the liquid, the pressure at a certain level, the conductance of the liquid, or its absorption of gamma radiation, rather than by direct measurement.

inferior [BIOL] The lower of two structures.

inferior alveolar artery [ANAT] A branch of the internal maxillary artery supplying the mucous membrane of the mouth and teeth of the lower jaw.

inferior alveolar nerve [ANAT] A branch of the mandibular nerve that innervates the teeth of the lower jaw.

inferior cerebellar peduncle [ANAT] A large bundle of nerve fibers running from the medulla oblongata to the cerebellum. Also known as restiform body.

inferior colliculus [ANAT] One of the posterior pair of rounded eminences arising from the dorsal portion of the mesencephalon.

inferior conjunction [ASTRON] A type of configuration in which two celestial bodies have their least apparent separation; the smaller body is nearer the observer than the larger body, about which it orbits; for example, Venus is closest to the earth at its inferior conjunction.

inferior ganglion [ANAT] **1.** The lower sensory ganglion in the glossopharyngeal nerve. **2.** The lower sensory ganglion on the vagus.

inferior hypogastric plexus [ANAT] A network of nerves in the pelvic fascia containing autonomic nerve elements.

inferiority [MED] An organic or psychic state or condition of being inferior or less adequate.

inferiority complex [PSYCH] Repressed unconsious fears and feelings of physical or social inadequacy or both, which may result in excessive anxiety, inability to function, or actual failure.

inferior mediastinal syndrome *See* inferior vena cava syndrome.

inferior mesenteric ganglion [ANAT] A sympathetic ganglion within the inferior mesenteric plexus at the origin of the inferior mesenteric artery.

inferior mirage [OPTICS] A spurious image of an object formed below the true position of that object by abnormal refraction conditions along the line of sight; one of the most common types of mirage, and the opposite of a superior mirage.

inferior planet [ASTRON] A planet that circles the sun in an orbit that is smaller than the earth's.

inferior temporal gyrus [ANAT] A convolution on the temporal lobe of the cerebral hemispheres lying below the middle temporal sulcus and extending to the inferior sulcus.

inferior vena cava [ANAT] A large vein which drains blood from the iliac veins, lower extremities, and abdomen into the right atrium.

inferior vena cava syndrome [MED] Edema and venous distention of the abdomen and legs due to obstruction of the inferior vena cava. Also known as inferior mediastinal syndrome.

inferior vermis [ANAT] The inferior portion of the vermis of the brain.

inferior vestibular nucleus [ANAT] The terminal nucleus for the spinal vestibular nerve tract.

infernal machine [ORD] Disguised or cleverly concealed explosive device, usually intended for sabotage.

inferred ore [MIN ENG] An ore whose estimate of tonnage and grade is based largely on knowledge of the deposit's geological character and to a lesser degree on samples and other data.

infertility [MED] Involuntary reduction in reproductive ability.

infest [MED] To live on or within the host's body.

infestation [MED] The state or condition of having animal parasites on or in the body.

infiltrating lipoma *See* liposarcoma.

infiltration [GEOL] Deposition of mineral matter among the pores or grains of a rock by permeation of water carrying the matter in solution. [HYD] Movement of water through the soil surface into the ground. [MET] **1.** Filling the pores of a metal powder compact with metal having a lower melting point. **2.** Movement of molten metal into the pores of a fiber or foam metal.

infiltration capacity [HYD] The maximum rate at which water enters the soil or other porous material in a given condition.

infiltration gallery [CIV ENG] A large, horizontal underground conduit of porous material or with openings on the sides for collecting percolating water by infiltration.

infiltration vein [GEOL] Vein deposited in rock by percolating water.

infimum *See* greatest lower bound.

infinite [MATH] Larger than any fixed number.

infinite aquifer [HYD] The portion of a formation that contains water, and for which the exterior boundary is at an effectively infinite distance from the oil reservoir.

infinite baffle [ENG ACOUS] A loudspeaker baffle which prevents interaction between the front and back radiation of the loudspeaker.

infinite integral [MATH] An integral at least one of whose limits of integration is infinite.

infinite multiplication factor [NUCLEO] The multiplication factor of a theoretical system from which there is no leakage of neutrons, that is, a reactor of infinite size.

infinite reservoir [PETRO ENG] In reservoir unsteady-state liquid-diffusion calculations, a reservoir in which the outer boundary is considered to be effectively at an infinite distance from the inner boundary at the well of an aquifer.

infinite series [MATH] An indicated sum of an infinite sequence of quantities, written $a_1 + a_2 + a_3 + \ldots$, or

$$\sum_{k=1}^{\infty} a_k.$$

infinite set [MATH] A set with more elements than any fixed integer; such a set can be put into a one to one correspondence with a proper subset of itself.

infinitesimal generator [MATH] A closed linear operator defined relative to some semigroup of operators and which uniquely determines that semigroup.

infinity [ADP] Any number larger than the maximum number that a computer is able to store in any register. [MATH] The concept of a value larger than any finite value.

infinity method [OPTICS] Method of adjusting two lines of sight to make them parallel; lines are adjusted on an object at great distance, for example, a star.

infinity transmitter [ELECTR] A way of tapping a telephone; the telephone instrument is so modified that an interception device can be actuated from a distant source without the caller's becoming aware.

infix operation [ADP] An operation carried out within an operation, as the addition of a and b prior to the multiplication by c or division by d in the operation $(a+b)c/d$.

inflammability *See* flammability.

inflammation [MED] Local tissue response to injury characterized by redness, swelling, pain, and heat.

inflammatory carcinoma [MED] A carcinoma, usually of the breast, associated with inflammation and characterized by rapid metastasis.

inflammatory tissue [MED] Tissue characterized by exudation or cell proliferation caused by trauma.

inflatable packer [PETRO ENG] A packer (downhole pressure seal between tubing and casing) set and held in place by an element that is inflated with hydraulic pressure.

inflated [BIOL] **1.** Distended, applied to a hollow structure. **2.** Open and enlarged. [ENG] Filled or distended with air or gas.

inflected [BOT] Curved or bent sharply inward, downward, or toward the axis. Also known as inflexed.

inflected arch *See* inverted arch.

inflection point *See* point of inflection.

inflexed *See* inflected.

inflorescence [BOT] **1.** A flower cluster. **2.** The arrangement of flowers on a plant.

influence factor *See* telephone influence factor.

influence function [PETRO ENG] Mathematical statement of the influence on pressure and production of each oil reservoir pool in a multipool aquifer.

influence fuse *See* proximity fuse.

influence line [MECH] A graph of the shear, stress, bending moment, or other effect of a movable load on a structural member versus the position of the load.

influence mine [ORD] A mine that is detonated by methods that are different than target contact.

influent [ECOL] An organism that disturbs the ecological balance of a community. [SCI TECH] An input stream of a fluid, as water into a reservoir, or liquid into a process vessel.

influent stream [HYD] A stream that contributes water to the zone of saturation of groundwater and develops bank storage. Also known as losing stream.

influenza [MED] An acute virus disease of the respiratory system characterized by headache, muscle pain, fever, and prostration.

influenzal meningitis [MED] Inflammation of the meninges caused by *Hemophilus influenzae*.

influenzal pneumonia [MED] Pneumonia resulting from infection by *Hemophilus influenzae*.

influenza vaccine [IMMUNOL] A vaccine prepared from formaldehyde-attenuated mixtures of strains of influenza virus.

influenza virus [VIROL] Any of three immunological types, designated A, B, and C, belonging to the myxovirus group which cause influenza.

influx *See* mouth.

information [COMMUN] Data which has been recorded, classified, organized, related, or interpreted within a framework so that meaning emerges.

information bit [COMMUN] Bit that is generated by the data source but is not used by the data-transmission system.

information center [COMMUN] Center designed specifically for storing, processing, and retrieving information for dissemination at regular intervals, on demand or selectively, according to express needs of users.

information content [COMMUN] A numerical measure of the information generated in selecting a specific symbol (or message), equal to the negative of the logarithm of the probability of the symbol (or message) selected. Also known as negentropy.

information flow [ADP] The graphic representation of data collection, data processing, and report distribution throughout an organization.

information function of a partition [MATH] If ξ is a finite partition of a probability space, the information function of ξ is a step function whose sets of constancy are the elements of ξ and whose value on an element of ξ is the negative of the logarithm of the probability of this element.

information interchange [COMMUN] The exchange of information in code between machines.

information link *See* data link.

information process analysis chart *See* form process chart.

information processing [ADP] **1.** The manipulation of data so that new data (implicit in the original) appear in a useful form. **2.** *See* data processing.

information Processing Language *See* IPL.

information rate [COMMUN] The information content generated per symbol or per second by an information source.

information requirements [ADP] Actual or anticipated questions which may be posed to an information retrieval system.

information retrieval [ADP] The technique and process of searching, recovering, and interpreting information from large amounts of stored data.

information selection systems [ADP] A class of information processing systems which carry out a sequence of operations necessary to locate in storage one or more items assumed to have certain specified characteristics and to retrieve such items directly or indirectly, in whole or in part.

information source [COMMUN] A system which produces messages by making successive selections from a group of symbols.

information system [COMMUN] Any means for communicating knowledge from one person to another, such as by simple verbal communication, punched-card systems, optical coincidence systems based on coordinate indexing, and completely computerized methods of storing, searching, and retrieving of information.

information theory [COMMUN] A branch of communications theory which is devoted to problems in coding, and which provides criteria for comparing different communications systems on the basis of signaling rate, using a numerical measure of the amount of information gained when the content of a message is learned. [MATH] The branch of probability theory concerned with the likelihood of the transmission of messages, accurate to within specified limits, when the bits of information composing the message are subject to possible distortion.

information unit [COMMUN] A unit of information content, equal to a bit, nit, or hartley, according to whether logarithms are taken to base 2, e, or 10.

information word *See* data word.

infrabasal [BIOL] Inferior to a basal structure.

infrabranchial [VERT ZOO] Situated below the gills.

Infracambrian *See* Eocambrian.

infracentral [ANAT] Located below the centrum.

infracerebral gland [INV ZOO] A structure lying ventral to the brain in annelids which is thought to produce a hormone that inhibits maturation of the gametes.

infraciliature [INV ZOO] The neuromotor apparatus, silverline system, or neuroneme system of ciliates.

infraclass [SYST] A subdivision of a subclass; equivalent to a superorder.

infraclavicle [VERT ZOO] A bony element of the shoulder girdle located below the cleithrum in some ganoid and crossopterygian fishes.

infradyne receiver [ELECTR] A superheterodyne receiver in which the intermediate frequency is higher than the signal frequency, so as to obtain high selectivity.

infrafoliar [BOT] Located below the leaves.

infraglenoid [ANAT] Below the glenoid cavity of the scapula.

infraglenoid tubercle [ANAT] A rough impression below the glenoid cavity, from which the long head of the triceps muscle arises.

infralateral tangent arcs [METEOROL] Two oblique, colored arcs, convex toward the sun and tangent to the halo of 46° at points below the altitude of the sun, produced by refraction (90° effective prism angle) in hexagonal columnar ice crystals whose principal axes are horizontal but randomly directed in azimuth; if the sun's elevation exceeds about 68°, the arcs cannot appear.

infralow frequency [COMMUN] A designation for the band from 0.3 to 3 kilohertz in the radio spectrum. Abbreviated ILF.

infraorbital [ANAT] Located beneath the orbit.

infrared [ELECTROMAG] Pertaining to infrared radiation.

infrared absorption [ELECTROMAG] The taking up of energy from infrared radiation by a medium through which the radiation is passing.

infrared astronomy [ASTROPHYS] The study of electromagnetic radiation in the spectrum between 0.75 and 1000 micrometers emanating from astronomical sources.

infrared beacon [NAV] A source of infrared radiation used to establish a geographical reference point, the bearings of which may be determined.

infrared binoculars [OPTICS] An instrument for viewing an enlarged infrared image with both eyes; it has two infrared telescopes whose lens systems are similar to those of ordinary binoculars.

infrared bolometer [ELECTR] A bolometer adapted to detecting infrared radiation, as opposed to microwave radiation.

infrared brazing [MET] A brazing process in which coalescence is produced by heat generated by infrared radiation.

infrared catastrophe [QUANT MECH] The logarithmic divergence in the cross section (which one would expect to be finite) for the emission of low-energy photons in bremsstrahlung and in the double Compton effect, according to quantum electrodynamics; the difficulty is resolved by taking radiative corrections to elastic scattering into account. Also known as infrared problem.

infrared communications set [ELECTR] Components required to operate a two-way electronic system using infrared radiation to carry intelligence.

infrared detector [ELECTR] A device responding to infrared radiation, used in detecting fires, or overheating in machinery, planes, vehicles, and people, and in controlling temperature-sensitive industrial processes.

infrared dome *See* irdome.

infrared emission [PHYS] The act of emitting infrared waves.

infrared filter [OPTICS] A substance or device which is highly transparent to infrared radiation at certain wavelengths while absorbing other types of electromagnetic radiation.

infrared galaxy [ASTRON] A galaxy or quasar whose nucleus emits enormous amounts of infrared radiation, in some cases more than 1000 times the output of the entire Milky Way Galaxy at all wavelengths.

infrared heating [ENG] Heating by means of infrared radiation.

infrared homing [ENG] Homing in which the target is tracked by means of its emitted infrared radiation.

infrared image converter [ELECTR] A device for converting an invisible infrared image into a visible image, consisting of an infrared-sensitive, semitransparent photocathode on one end of an evacuated envelope and a phosphor screen on the other, with an electrostatic lens system between the two. Also known as infrared image tube.

infrared image tube *See* infrared image converter.

infrared jamming [ELECTR] An attempt to confuse heat-seeking missiles by emissions which overload their inputs or misdirect them.

infrared lamp [ELEC] An incandescent lamp which operates at reduced voltage with a filament temperature of 4000°F (2200°C) so that it radiates electromagnetic energy primarily in the infrared region.

infrared laser [PHYS] A laser which emits infrared radiation, especially in the near- and intermediate-infrared regions.

infrared mapping [MAP] Mapping in which a sensitive infrared detector is mounted on an infrared scanner, and the resulting thermal image is translated into varying shades of gray on photographic film, to give a line-pattern image much like that seen on a television screen.

infrared maser [PHYS] A laser which emits infrared radiation, especially in the far-infrared region, or which is pumped with radiation at infrared frequencies and emits radiation at millimeter wavelengths.

infrared microscope [OPTICS] A type of reflecting microscope which uses radiation of wavelengths greater than 700 nanometers and is used to reveal detail in materials that are opaque to light, such as molybdenum, wood, corals, and many red-dyed materials.

infrared optical material [ELECTROMAG] A material which is transparent to infrared radiation.

infrared phosphor [SOLID STATE] A phosphor which, when exposed to infrared radiation during or even after decay of luminescence resulting from its usual or dominant activator,

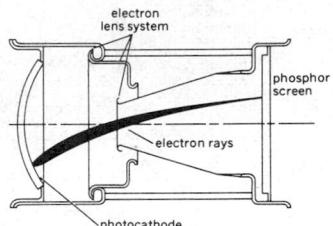

**INFRARED
IMAGE CONVERTER**

Components of the infrared image converter.

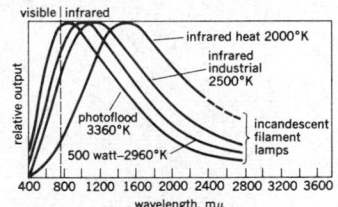

INFRARED LAMP

Spectral distribution of energy from various types of infrared lamps. *(From Illuminating Engineering Society, IES Lighting Handbook, 4th rev. ed., 1968)*

emits light having the same spectrum as that of the dominant activator; sulfide and selenide phosphors are the most important examples.

infrared photoconductor [ELECTR] A conductor whose conductivity increases when it is exposed to infrared radiation.

infrared problem *See* infrared catastrophe.

infrared radiation [ELECTROMAG] Electromagnetic radiation whose wavelengths lie in the range from 0.75 or 0.8 micrometer (the long-wavelength limit of visible red light) to 1000 micrometers (the shortest microwaves).

infrared receiver [ELECTR] A device that intercepts or demodulates infrared radiation that may carry intelligence. Also known as nancy receiver.

infrared scanner [ELECTR] An infrared detector mounted on a motor-driven platform which causes it to scan a field of view line by line, much as in television.

infrared searchlight [OPTICS] A device for illuminating a scene with infrared radiation so that it may be viewed through an infrared image converter tube, although it is invisible to the unaided eye.

infrared spectrometer [SPECT] Device used to identify and measure the concentrations of heteroatomic compounds in gases, in many nonaqueous liquids, and in some solids by arc or spark excitation and subsequent measurement of the electromagnetic emissions in the wavelength range of 0.78 to 300 micrometers.

infrared spectrophotometry [SPECT] Spectrophotometry in the infrared region, usually for the purpose of chemical analysis through measurement of absorption spectra associated with rotational and vibrational energy levels of molecules.

infrared spectroscopy [SPECT] The study of the properties of material systems by means of their interaction with infrared radiation; ordinarily the radiation is dispersed into a spectrum after passing through the material.

infrared spectrum [ELECTROMAG] **1.** The range of wavelengths of infrared radiation. **2.** A display or graph of the intensity of infrared radiation emitted or absorbed by a material as a function of wavelength or some related parameter.

infrared star [ASTROPHYS] A star that emits a large amount of radiant energy in the infrared portion of the electromagnetic spectrum.

infrared telescope [OPTICS] An instrument that converts an invisible infrared image into a visible image and enlarges this image, consisting of an infrared image converter tube, an objective lens for imaging the scene to be viewed onto the photocathode of the tube, and an ocular for viewing the phosphor screen of the tube.

infrared thermistor [ELECTR] A thermistor used to measure the power of infrared radiation.

infrared transmitter [ELECTR] A transmitter that emits energy in the infrared spectrum; may be modulated with intelligence signals.

infrared vidicon [ELECTR] A vidicon whose photoconductor surface is sensitive to infrared radiation.

infrared window [GEOPHYS] A frequency region in the infrared where there is good transmission of electromagnetic radiation through the atmosphere.

infrasonic [ACOUS] Pertaining to signals, equipment, or phenomena involving frequencies below the range of human hearing, hence below about 15 hertz. Also known as subsonic (deprecated usage).

infraspinous [ANAT] Below the spine of the scapula.

infraspinous fossa [ANAT] The recess on the posterior surface of the scapula occupied by the infraspinatus muscle.

infratemporal [ANAT] Situated below the temporal fossa.

infratemporal fossa [ANAT] An irregular space situated below and medial to the zygomatic arch, behind the maxilla and medial to the upper part of the ramus of the mandible.

infructescence [BOT] An inflorescence's fruiting stage.

infundibular canal [INV ZOO] A pathway from the mantle cavity through the funnel for water in cephalopods.

infundibular ganglion [INV ZOO] A branch of pedal ganglion which supplies the funnel in cephalopods.

infundibular process [ANAT] The distal portion of the neural lobe of the pituitary.

infundibulum [ANAT] **1.** A funnel-shaped passage or part. **2.** The stalk of the neurohypophysis.

infusion [CHEM] The aqueous solution of a soluble constituent of a substance as the result of the substance's steeping in the solvent for a period of time. [MED] The slow injection of a solution into a vein or into subcutaneous or other tissue of the body.

infusorial earth [GEOL] Formerly, and incorrectly, a soft rock or an earthy substance composed of siliceous remains of diatoms.

infusoriform larva [INV ZOO] The final larval stage, arising from germ cells within the infusorigen, in the life cycle of dicyemid mesozoans.

infusorigen [INV ZOO] An individual that gives rise to the infusoriform larva in dicyemid mesozoans.

in-gate *See* gate.

Ingersoll-Rand jumbo columns [MIN ENG] Columns held in place by the pressure of air-operated pistons against the roof; drills are attached to movable arms mounted on the columns.

ingesta [BIOL] Food and other substances taken into an animal body.

ingestion [BIOL] The act or process of taking food and other substances into the animal body.

Ingolfiellidea [INV ZOO] A suborder of amphipod crustaceans in which both abdomen and maxilliped are well developed and the head often bears a separate ocular lobe lacking eyes.

ingot [MET] **1.** A solid metal casting suitable for remelting or working. **2.** A bar of gold or silver.

ingot iron [MET] Relatively pure iron produced in an open-hearth furnace.

ingrain color *See* azoic dye.

ingress [ASTRON] The entrance of the moon into the shadow of the earth in an eclipse, of a planet into the disk of the sun, or of a satellite (or its shadow) onto the disk of the parent planet. [SCI TECH] The act of entering, as of air into the lungs or a liquid into an orifice.

ingrown [MED] Of a hair or nail, grown inward so that the normally free end is embedded in or under the skin.

ingrown meander [GEOL] A meander of a stream with an undercut bank on one side and a gentle slope on the other.

inguinal canal [ANAT] A short, narrow passage between the abdominal ring and the inguinal ring in which lies the spermatic cord in males and the round ligament in females.

inguinal fold [EMBRYO] A fold of embryonic tissue on the urogenital ridge in which the gubernaculum testis develops.

inguinal gland [ANAT] Any of the superficial lymphatic glands in the groin.

inguinal hernia [MED] Protrusion of the abdominal viscera through the inguinal canal.

inguinal ligament [ANAT] The thickened lower portion of the aponeurosis of the external oblique muscle extending from the anterior superior spine of the ileum to the tubercle of the pubis and the pectineal line. Also known as Poupart's ligament.

inguinal region [ANAT] The abdominal region occurring on each side of the body as a depression between the abdomen and the thigh. Also known as iliac region.

inhabited building distance [ENG] The minimum distance permitted between an ammunition or explosive location and any building used for habitation or where people are accustomed to assemble, except operating buildings or magazines.

inhalant canal [INV ZOO] The incurrent canal in sponges and mollusks.

inhalant siphon [INV ZOO] A channel for water intake in the mantle of bivalve mollusks. Also known as incurrent siphon.

inhalation [PHYSIO] The process of breathing in.

inhalator [MED] A device for facilitating the inhalation of a gas or spray, as for providing oxygen or oxygen-carbon dioxide mixtures for respiration in resuscitation.

inhaler [MED] **1.** A device containing a solid medication through which air is drawn into the air passages. **2.** An atomizer containing a liquid medication. [MIN ENG] An apparatus, of different forms, for permitting the supply of fresh air to a miner.

inhaul cable [MECH ENG] In a cable excavator, the line that pulls the bucket to dig and bring in soil. Also known as digging line.

inherent bursts [MIN ENG] Rock bursts that occur in development.

inherent damping [MECH ENG] A method of vibration damping which makes use of the mechanical hysteresis of such materials as rubber, felt, and cork.

inheritance [GEN] **1.** The acquisition of characteristics by transmission of germ plasm from ancestor to descendant. **2.** The sum total of characteristics dependent upon the constitution of the fertilized ovum.

inherited error [ADP] The error existing in the data supplied at the beginning of a step in a step-by-step calculation as executed by a program.

inhibit-gate [ELECTR] Gate circuit whose output is energized only when certain signals are present and other signals are not present at the inputs.

inhibiting antibody [IMMUNOL] A substance sometimes produced in the blood of immunized persons which is thought to prevent the expected antigen-reagin reaction.

inhibiting input [ELECTR] A gate input which, if in its prescribed state, prevents any output which might otherwise occur.

inhibiting signal [ELECTR] A signal, which when entered into a specific circuit will prevent the circuit from exercising its normal function; for example, an inhibit signal fed into an AND gate will prevent the gate from yielding an output when all normal input signals are present.

inhibition [PSYCH] An unconscious mechanism for restraining an impulse by means of an opposing impulse. [SCI TECH] The act of repressing or restraining a physical or chemical action.

inhibition-action balance [PSYCH] Relative balance maintained in every individual between his experiencing of emotional feelings and his outward behavior in response to them.

inhibition index [BIOCHEM] The amount of antimetabolite that will overcome the biological effect of a unit weight of metabolite.

inhibitor [CHEM] A substance which is capable of stopping or retarding a chemical reaction; to be technically useful, it must be effective in low concentration.

inhibitor sweetening [CHEM ENG] Petroleum-refinery treating process to sweeten gasoline (convert mercaptans to disulfides) of low mercaptan content; uses a phenylenediamine inhibitor, air, and caustic.

inhibit pulse [ELECTR] A drive pulse that tends to prevent flux reversal of a magnetic cell by certain specified drive pulses.

inhour [NUCLEO] A unit of reactivity of a reactor; 1 inhour is the reactivity that will give the reactor a period of 1 hour. Derived from inverse hour.

inhour equation [NUCLEO] An equation relating the reactivity of a nuclear reactor to the parameters of the delayed-neutron emitters and the neutron lifetime of the reactor.

in-house system *See* in-plant system.

Iniomi [VERT ZOO] An equivalent name for Salmoniformes.

inion [ANTHRO] The external occipital protuberance of the skull.

initial aiming point [ORD] Point on which a gun is sighted to establish a reference line from which direction angles for targets are measured; from this reference line, other aiming points that give the direction of the targets are measured off.

initial boiling point [CHEM ENG] According to ASTM petroleum-analysis distillation procedures, the recorded temperature when the first drop of distilled vapor is liquefied and falls from the end of the condenser.

initial condition [METEOROL] A prescription of the state of a dynamical system at some specified time; for all subsequent times the equations of motion and boundary conditions determine the state of the system; the appropriate synoptic weather charts, for example, constitute a (discrete) set of initial conditions for a forecast; in many contexts, initial conditions are considered as boundary conditions in the dimension of time.

initial detention *See* surface storage.

initial dip *See* primary dip.

initial free space [MECH] In interior ballistics, the portion of the effective chamber capacity not displaced by propellant.

initial great-circle course [NAV] The direction, at the point of departure, of the great circle through that point and the destination, expressed as the angular distance from a reference direction, usually north, to that part of the great circle extending toward the destination. Also known as initial great-circle direction.

initial great-circle direction *See* initial great-circle course.

initial heading [NAV] The aircraft heading at the beginning of a rating period while using gyro steering.

initial inverse voltage [ELECTR] Of a rectifier tube, the peak inverse anode voltage immediately following the conducting period.

initialize [ADP] **1.** To set counters, switches, and addresses to zero or other starting values at the beginning of, or at prescribed points in, a computer routine. **2.** To begin an operation, and more specifically, to adjust the environment to the required starting configuration.

initial lead [ORD] The amount a gun is pointed in front of, above, or below a moving target when opening fire; this amount allows for the distance the target will travel while the projectile is in flight.

initial lock mechanism [ORD] Device for preventing inadvertent motion of stroking member in a cartridge-actuated device prior to firing.

initial mass [AERO ENG] The mass of a rocket missile at the beginning of its flight.

initial nuclear radiation [NUCLEO] Radiation emitted from the fireball of a nuclear explosive during the first minute (an arbitrary time interval) after detonation.

initial saturation [PETRO ENG] A reservoir's initial relative content (saturation) of water, oil, and gas.

initial set [MATER] The onset of hardening after water has been added to concrete, cement, or plaster.

initial shot start pressure [MECH] In interior ballistics, the pressure required to start the motion of the projectile from its initial loaded position; in fixed ammunition, it includes pressure required to separate projectile and cartridge case and to start engraving the rotating band.

initial surge voltage [ELEC] A spike of voltage experienced when a noncompensated load is first connected to a generator.

initial-value problem [FL MECH] A dynamical problem whose solution determines the state of a system at all times subsequent to a given time at which the state of the system is specified by given initial conditions; the initial-value problem is contrasted with the steady-state problem, in which the state of the system remains unchanged in time. Also known as transient problem. [MATH] An nth order ordinary or partial differential equation in which the solution and its first $(n-1)$ derivatives are required to take on specified values at a particular value of a given independent variable.

initial-value theorem [MATH] The theorem that, if a function $f(t)$ and its first derivative have Laplace transforms, and if $g(s)$ is the Laplace transform of $f(t)$, and if the limit of $sg(s)$ as s approaches infinity exists, then this limit equals the limit of $f(t)$ as t approaches zero.

initial velocity [PHYS] The velocity of anything at the beginning of a specific phase of its motion.

initial yaw [MECH] The yaw of a projectile the instant it leaves the muzzle of a gun.

initiating agent [MATER] An explosive material which has the necessary sensitivity to heat, friction, or percussion to make it suitable for use as the initial element in an explosive train.

initiation [ORD] **1.** As applied to an explosive item, the beginning of the deflagration or detonation of the explosive. **2.** The first action in a fuse which occurs as a direct result of the action of the functioning medium. **3.** In a time fuse, the starting of the action which is terminated in the functioning of the fused munition.

initiator [CHEM] The substance or molecule (other than reactant) that initiates a chain reaction, as in polymerization; an example is acetyl peroxide. [ORD] A device used as the first element of an explosive train, such as a detonator or squib, which upon receipt of the proper mechanical or electrical impulse produces a burning or detonating action; it generally contains a small quantity of a sensitive explosive.

injected [PETR] Pertaining to intrusive igneous rock or other mobile rock that has erupted through rock walls to neighboring older rocks.

injected gas [PETRO ENG] Gas that has been pumped into an oil-producing reservoir to provide a gas-drive for increased oil production.

injected hole [MIN ENG] A borehole into which a cement slurry or grout has been forced by high-pressure pumps and allowed to harden.

injection [AERO ENG] The process of placing a spacecraft into a specific trajectory, such as an earth orbit or an encounter trajectory to Mars. [GEOL] Also known as intrusion; sedimentary injection. **1.** A process by which sedimentary material is forced under abnormal pressure into a preexisting rock or deposit. **2.** A structure formed by an injection process. [MECH ENG] The introduction of fuel, fuel and air, fuel and oxidizer, water, or other substance into an engine induction system or combustion chamber. [MED] **1.** Introduction of a fluid into the skin, vessels, muscle, subcutaneous tissue, or any cavity of the body. **2.** The substance injected. [MIN ENG] The introduction under pressure of a liquid or plastic material into cracks, cavities, or pores in a rock formation.

injection blow molding [ENG] Plastics molding process in which a hollow-plastic tube is formed by injection molding.

injection carburetor [MECH ENG] A carburetor in which fuel is delivered under pressure into a heated part of the engine intake system. Also known as pressure carburetor.

injection electroluminescence [ELECTR] Radiation resulting from recombination of minority charge carriers injected in a *pn* or *pin* junction that is biased in the forward direction. Also known as Lossev effect; recombination electroluminescence.

injection fluid [PETRO ENG] Gas or water, depending on the nature of the reservoir and its fluid content, for injection into the formation to increase hydrocarbon production.

injection-fluid front [PETRO ENG] The moving interfacial contact between an injected fluid (gas or water) and the natural fluid content of the reservoir formation.

injection gas-fluid ratio [PETRO ENG] The ratio of gas injected into a reservoir formation to the fluid hydrocarbons produced by the resultant gas lift.

injection gneiss [PETR] A composite rock with banding entirely or partly caused by layer-by-layer injection of granitic magma into rock layers.

injection grid [ELECTR] Grid introduced into a vacuum tube in such a way that it exercises control over the electron stream without causing interaction between the screen grid and control grid.

injection laser [OPTICS] A laser in which a forward-biased gallium arsenide diode converts direct-current input power directly into coherent light, without optical pumping.

injection luminescent diode [ELECTR] Gallium arsenide diode, operating in either the laser or the noncoherent mode, that can be used as a visible or near-infrared light source for triggering such devices as light-activated switches.

injection mold [ENG] A plastics mold into which the material to be formed is introduced from an exterior heating cylinder.

injection molding [ENG] Molding metal, plastic, or nonplastic ceramic shapes by injecting a measured quantity of the molten material into dies.

injection pressure [PETRO ENG] Pressure of fluid injected into oil formations for waterflood (water) or pressure maintenance (gas).

injection pump [MECH ENG] A pump that forces a measured amount of fuel through a fuel line and atomizing nozzle in the combustion chamber of an internal combustion engine.

injection ram [ENG] In injection molding, the ram that applies pressure to the feed plunger in the process of either injection or transfer molding.

injection temperature [NAV ARCH] The temperature of the sea water as measured at the sea-water intakes in the engine room of a vessel.

injection well [HYD] *See* recharge well. [PETRO ENG] In secondary recovery of petroleum, a well in which a fluid such as gas or water is injected to provide supplemental energy to

drive the oil remaining in the reservoir to the vicinity of production wells.

injection-well plugging [PETRO ENG] Plugging of the sand face of an injection well because of lubricant or corrosion-product carryover from surface lines or well equipment.

injectivity index [PETRO ENG] The number of barrels per day of gross liquid pumped into an injection well per psi pressure differential between the mean injection pressure and the mean formation pressure.

injectivity test [PETRO ENG] A test series of reservoir water injection rates at different pressures to predict the performance of an injection well.

injector [ELECTR] An electrode through which charge carriers (holes or electrons) are forced to enter the high-field region in a spacistor. [MECH ENG] **1.** An apparatus containing a nozzle in an actuating fluid which is accelerated and thus entrains a second fluid, so delivering the mixture against a pressure in excess of the actuating fluid. **2.** A plug with a valved nozzle through which fuel is metered to the combustion chambers in diesel- or full-injection engines. **3.** A jet through which feedwater is injected into a boiler, or fuel is injected into a combustion chamber.

injector torch *See* low-pressure torch.

injury [MED] **1.** A structural or functional stress or trauma that induces a pathologic process. **2.** Damage resulting from the stress.

injury current *See* injury potential.

injury potential [PHYSIO] The potential difference observed between the injured and the noninjured regions of an injured tissue or cell. Also known as demarcation potential; injury current.

ink [MATER] A dispersion of a pigment or a solution of a dye in a carrier vehicle, yielding a fluid, paste, or powder to be applied to and dried on a substrate; writing, marking, drawing, and printing inks are applied by several methods to paper, metal, plastic, wood, glass, fabric, or other substrate.

ink bleed [ADP] In character recognition, the capillary extension of ink beyond the original edges of a printed or handwritten character.

ink disease [PL PATH] A fungus disease of the chestnut in Europe caused by *Phytophthora cambivora* which produces black cankers and a black exudate in the trunk. Also known as black canker.

ink knife [GRAPHICS] An instrument resembling a large spatula, with a handle and a blade with square or rounded end; used for handling paste inks. Also known as ink slice.

ink-mist recording *See* ink-vapor recording.

inkometer [ENG] An instrument for measuring adhesion of liquids by rotating drums in contact with the liquid.

ink recorder [GRAPHICS] A recorder that employs an ink-filled pen or capillary tube to produce the graphic record.

ink sac [INV ZOO] An organ attached to the rectum in many cephalopods which secretes and ejects an inky fluid.

ink slice *See* ink knife.

ink smudge [ADP] In character recognition, the overflow of ink beyond the original edges of a printed or handwritten character.

ink squeezeout [ADP] In character recognition, the overflow of ink from the stroke centerline to the edges of a printed or handwritten character.

ink-vapor recording [GRAPHICS] A type of recording in which vaporized ink particles are directly deposited on the record sheet. Also known as ink-mist recording.

inland [GEOGR] Interior land, not bordered by the sea.

inland ice [HYD] Ice composing the inner portion of a continental glacier or large ice sheet; applied particularly to Greenland ice.

inland rules of the road [NAV] Rules to be followed by all vessels while navigating upon certain inland waters of the United States.

inland sea *See* epicontinental sea.

inland water [GEOGR] **1.** A lake, river, or other body of water wholly within the boundaries of a state. **2.** An interior body of water not bordered by the sea.

inlay [GRAPHICS] A picture or ornament made by inserting a material such as metal into a space in metal, stone, or wood;

INKOMETER

The inkometer, used to measure ink tack.

the material (such as wire) may be burnished, heated, or fused.

inlet [ENG] An entrance or orifice for the admission of fluid. [GEOGR] **1.** A short, narrow waterway connecting a bay or lagoon with the sea. **2.** A recess or bay in the shore of a body of water. **3.** A waterway flowing into a larger body of water.

inlet of the pelvis [ANAT] The space within the brim of the pelvis.

inlet valve [MECH ENG] The valve through which a fluid is drawn into the cylinder of a positive-displacement engine, pump, or compressor. Also known as induction valve.

inlier [GEOL] A circular or elliptical area of older rocks surrounded by strata that are younger.

in line [ENG] **1.** Over the center of a borehole and parallel with its long axis. **2.** Of a drill motor, mounted so that its drive shaft and the drive rod in the drill swivel head are parallel, or mounted so that the shaft driving the drill-swivel-head bevel gear and the drill-motor drive shaft are centered in a direct line and parallel with each other. **3.** Having similar units mounted together in a line.

in-line assembly machine [IND ENG] An assembly machine that inserts components into a wiring board one at a time as the board is moved from station to station by a conveyor or other transport mechanism.

in-line coding [ADP] Any group of instructions within the main body of a program.

in-line engine [MECH ENG] A multiple-cylinder engine with cylinders aligned in a row.

in-line equipment [ENG] **1.** A sequence of equipment or processing items mounted along the same vertical or horizontal plane. **2.** Equipment mounted within a process line, such as an in-line pump, pressure-drop flowmeter, or nozzle mixer.

in-line gun-type picture tube [ELECTR] A color television picture tube in which three electron guns that provide electron beams to excite phosphors on the screen are arranged horizontally in a line, rather than in the triangular formation of a conventional delta-gun tube.

in-line linkage [MECH ENG] A power-steering linkage which has the control valve and actuator combined in a single assembly.

in-line procedure [ADP] A short body of coding or instruction which accomplishes some purpose.

in-line processing [ADP] The processing of data in random order, not subject to preliminary editing or sorting.

in-line subroutine [ADP] A subroutine which is an integral part of a program.

in-line tuning [ELECTR] Method of tuning the intermediate-frequency strip of a superheterodyne receiver in which all the intermediate-frequency amplifier stages are made resonant to the same frequency.

innage [ENG] The volume or the measured height of liquid introduced into a tank or container.

innate [BIOL] Pertaining to a natural or inborn character dependent on genetic constitution. [BOT] Positioned at the apex of a supporting structure. [MYCOL] Embedded in, especially of an organ such as the fruiting body embedded in the thallus of some fungi.

inner automorphism [MATH] An automorphism h of a group where $h(g) = g_0^{-1} \cdot g \cdot g_0$, for every g in the group with g_0 some fixed group element.

inner barrel See inner tube.

inner bottom [NAV ARCH] A watertight plating laid over the frames and longitudinals, forming the inner layer of the double bottom of the hull.

inner cell mass [EMBRYO] The cells at the animal pole of a blastocyst which give rise to the embryo and certain extraembryonic membranes.

inner ear [ANAT] The part of the vertebrate ear concerned with labyrinthine sense and sound reception; consists generally of a bony and a membranous labyrinth, made up of the vestibular apparatus, three semicircular canals, and the cochlea. Also known as internal ear.

inner function [MATH] A continuous open mapping of a topological space X into a topological space Y where the inverse image of each point in Y is zero dimensional.

inner harbor [GEOGR] The part of a harbor more remote from the sea, as contrasted with the outer harbor; this expression is

normally used only in a harbor that is clearly divided into parts, by a narrow passageway or artificial structure; the inner harbor generally has additional protection and is often the principal berthing area.

inner keel [NAV ARCH] The inner plate of a double, flat plate keel.

inner mantle See lower mantle.

inner marker [NAV] A 75-megahertz marker beacon normally used with the instrument landing system (ILS) to indicate that the aircraft is over the boundary of the airport.

inner planet [ASTRON] Any of the four planets (Mercury, Venus, Earth, and Mars) in the solar system whose orbits are closest to the sun.

inner product [MATH] **1.** A scalar valued function of pairs of vectors from a vector space, denoted by (x,y) where x and y are vectors, and with the properties that (x,x) is always positive and is zero only if $x = 0$, that $(ax + by, z) = a(x,z) + b(y,z)$ for any scalars a and b, and that $(x,y) = (y,x)$ if the scalars are real numbers, $(x,y) = \overline{(y,x)}$ if the scalars are complex numbers. Also known as Hermitian inner product; Hermitian scalar product. **2.** The inner product of vectors $(x_1, \ldots, x_n)$ and $(y_1, \ldots, y_n)$ from n-dimensional euclidean space is the sum of $x_i y_i$ as i ranges from 1 to n. Also known as dot product; scalar product.

inner product of functions [MATH] For two functions f and g of a real or complex variable, this is defined to be the integral of $f(x)\overline{g(x)}dx$, where $\overline{g(x)}$ denotes the conjugate of $g(x)$.

inner product of tensors [MATH] The inner product of two tensors is the contracted tensor obtained from their product by means of pairing contravariant indices of one with covariant indices of the other.

inner product space [MATH] A vector space that has an inner product defined on it.

inner quantum number [ATOM PHYS] A quantum number J which gives an atom's total angular momentum, excluding the nuclear spin.

inner strake [NAV ARCH] The inner part of an in and out system of shell plating; the strakes adjacent to the molded frame line.

inner tube [ENG] A rubber tube used inside a pneumatic tire casing to hold air under pressure. [MIN ENG] The inside tube which acts as the core container of a double-tube core barrel; used to obtain core samples for analysis of an ore formation. Also known as inner barrel.

inner-tube extension See lifter case.

innervation [ANAT] The distribution of nerves to a part. [PHYSIO] The amount of nerve stimulation received by a part.

innominate See hip bone.

innominate artery [ANAT] The first artery branching from the aortic arch; distributes blood to the head, neck, shoulder, and arm on the right side of the body.

inoculation [BIOL] Introduction of a disease agent into an animal or plant to produce a mild form of disease and render the individual immune. [MET] Treating a molten material with another material before casting in order to nucleate crystals. [MICROBIO] Introduction of microorganisms onto or into a culture medium.

inoculum [MICROBIO] A small amount of substance containing bacteria from a pure culture which is used to start a new culture or to infect an experimental animal.

inoperculate [BIOL] Lacking an operculum.

inorganic [INORG CHEM] Pertaining to or composed of chemical compounds that do not contain carbon as the principal element (excepting carbonates, cyanides, and cyanates), that is, matter other than plant or animal.

inorganic acid [INORG CHEM] A compound composed of hydrogen and a nonmetal element or radical; examples are hydrochloric acid, HCl, sulfuric acid, H_2SO_4, and carbonic acid, H_2CO_3.

inorganic chemistry [CHEM] The study of chemical reactions and properties of all the elements and their compounds, with the exception of hydrocarbons, and usually including carbides, oxides of carbon, metallic carbonates, carbon-sulfur compounds, and carbon-nitrogen compounds.

inorganic chert [PETR] Chert derived from siliceous colloids precipitated from silica-saturated waters.

inorganic peroxide [INORG CHEM] Inorganic compound con-

IN-LINE LINKAGE

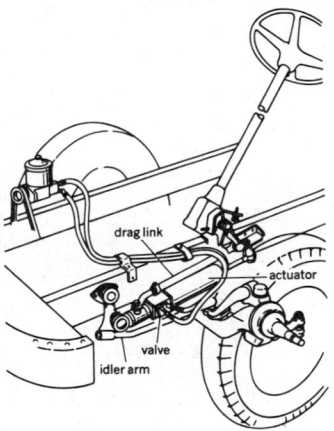

In-line linkage power steering.

taining an element at its highest state of oxidation (such as perchloric acid, $HClO_4$), or having the peroxy group, $-O-O-$ (such as perchromic acid, $H_3CrO_8 \cdot 2H_2O$).

inorganic pigment [INORG CHEM] A natural or synthetic metal oxide, sulfide, or other salt used as a coloring agent for paints, plastics, and inks.

inorganic polymer [INORG CHEM] Large molecules, usually linear or branched chains with atoms other than carbon in their backbone; an example is glass, an inorganic polymer made up of rings and chains of repeating silicate units.

inosculation *See* anastomosis.

inosilicate [GEOL] A class or structural type of silicate in which the SiO_4 tetrahedrons are linked together by the sharing of oxygens to form linear chains of indefinite length.

inosine [BIOCHEM] $C_{10}H_{12}N_4O_5$ A compound occurring in muscle; a hydrolysis product of inosinic acid.

inosinic acid [BIOCHEM] $C_{10}H_{13}N_4O_8P$ A nucleotide constituent of muscle, formed by deamination of adenylic acid; on hydrolysis it yields hypoxanthine and D-ribose-5-phosphoric acid. Also known as hypoxanthylic acid.

inositol [ORG CHEM] $C_6H_6(OH)_6 \cdot 2H_2O$ A water-soluble alcohol often grouped with the vitamins; there are nine stereoisomers of hexahydroxycyclohexane, and the only one of biological importance is optically inactive *meso*-inositol, comprising white crystals, widely distributed in animals and plants; it serves as a growth factor for animals and microorganisms.

in phase [PHYS] Having waveforms that are of the same frequency and that pass through corresponding values at the same instant.

in-phase component [ELEC] The component of the phasor representing an alternating current which is parallel to the phasor representing voltage.

in-phase rejection *See* common-mode rejection.

in-phase signal *See* common-mode signal.

in-pile [NUCLEO] A term used to designate experiments or equipment inside a reactor.

in-pile loop [NUCLEO] An experiment inserted directly in a nuclear reactor (pile) incorporating a closed circuit (loop) of fluid usually for cooling purposes.

in-place stress field *See* ambient stress field.

in-plant system [COMMUN] Communications system wholly confined to a single manufacturing establishment. Also known as in-house system.

input [ADP] The information that is delivered to a dataprocessing device from the external world, the process of delivering this data, or the equipment that performs this process. [ELECTR] **1.** The power or signal fed into an electrical or electronic device. **2.** The terminals to which the power or signal is applied. [SCI TECH] Those resources and other environmental factors converted by a system.

input admittance [ELEC] The admittance measured across the input terminals of a four-terminal network with the output terminals short-circuited.

input area [ADP] A section of internal storage reserved for storage of data or instructions received from an input unit such as cards or tape. Also known as input block; input storage.

input block [ADP] **1.** A block of data read or transferred into a computer. **2.** *See* input area.

input capacitance [ELECTR] The short-circuited transfer capacitance that exists between the input terminals and all other terminals of an electron tube (except the output terminal) connected together.

input data [ADP] Data employed as input.

input equipment [ADP] **1.** The equipment used for transferring data and instructions into an automatic data-processing system. **2.** The equipment by which an operator transcribes original data and instructions to a medium that may be used in an automatic data-processing system.

input gap [ELECTR] An interaction gap used to initiate a variation in an electron stream; in a velocity-modulated tube it is in the buncher resonator.

input impedance [ELEC] The impedance across the input terminals of a four-terminal network when the output terminals are short-circuited.

input-limited [ADP] Pertaining to a system or operation whose speed or efficiency depends mainly on the speed of

INOSITOL

Structural formula for inositol.

input into the machine rather than the speed of the machine itself.

input magazine [ADP] A part of a card-handling device which supplies the cards to the processing portion of the machine. Also known as magazine.

input/output [ADP] Pertaining to all equipment and activity that transfers information into or out of a computer. Abbreviated I/O.

input/output adapter [ADP] A circuitry which allows input/output devices to be attached directly to the central processing unit.

input/output bound [ADP] Pertaining to a system or condition in which the time for input and output operation exceeds other operations. Also known as input/output limited.

input/output channel [ADP] The physical link connecting the computer to an input device or to an output device.

input/output controller [ADP] An independent processor which provides the data paths between input and output devices and main memory.

input/output control system [ADP] A set of flexible routines that supervise the input and output operations of a computer at the detailed machine-language level. Abbreviated IOCS.

input/output control unit [ADP] The piece of hardware which controls the operation of one or more of a type of devices such as tape drives or disk drives; this unit is frequently an integral part of the input/output device itself.

input/output device [ADP] A unit that accepts new data, sends it into the computer for processing, receives the results, and translates them into a usable medium.

input/output interrupt [ADP] A technique by which the central processor needs only initiate an input/output operation and then handle other matters, while other units within the system carry out the rest of the operation.

input/output library [ADP] A set of programs which take over the job from the programmer of creating the required instructions to access the various peripheral devices. Also known as input/output routines.

input/output limited *See* input/output bound.

input/output register [ADP] Computer register that provides the transfer of information from inputs to the central computer, or from it to output equipment.

input/output relation [SYS ENG] The relation between two vectors whose components are the inputs (excitations, stimuli) of a system and the outputs (responses) respectively.

input/output routines *See* input/output library.

input/output wedge [ADP] The characteristic shape of a Kiviat graph of a system which is approaching complete input/output boundedness.

input register [ADP] A register that accepts input information from a computer at one speed and supplies the information to the central processing unit at another speed, usually much greater.

input resistance *See* transistor input resistance.

input resonator *See* buncher resonator.

input routine [ADP] A routine which controls the loading and reading of programs, data, and other routines into a computer for storage or immediate use. Also known as loading routine.

input storage *See* input area.

inquartation [MET] A step in bullion assay that uses nitric acid to dissolve silver from associated gold. Also known as quartation.

inquiline [ZOO] An animal that inhabits the nest of another species.

inquiry [ADP] A request for the retrieval of a particular item or set of items from storage.

inquiry display terminal [ADP] A cathode-ray-tube terminal which allows the user to query the computer through a keyboard, the answer appearing on the screen.

inquiry station [ADP] A remote terminal from which an inquiry may be sent to a computer over wire lines.

inquiry unit [ADP] Any terminal which enables a user to query a computer and get a hard-copy answer.

inradius [MATH] The radius of a circle or sphere inscribed in a given geometric figure.

insanity [PSYCH] **1.** Any mental disorder. **2.** In forensic psychiatry, a mental disorder which prevents one from managing one's affairs, impairs one's ability to distinguish right from wrong, or renders one harmful to oneself or others.

inscribed [MATH] A polygon is inscribed in a circle or some curve if every vertex of the polygon lies on the circle or curve.

insect [INV ZOO] **1.** A member of the Insecta. **2.** An invertebrate that resembles an insect, such as a spider, mite, or centipede.

Insecta [INV ZOO] A class of the Arthropoda typically having a segmented body with an external, chitinous covering, a pair of compound eyes, a pair of antennae, three pairs of mouthparts, and two pairs of wings.

insect attractant [MATER] A chemical agent, usually associated with an insect's sexual drive, which may be used to attract pests to poisoned bait or for insect surveys.

insect control [ECOL] Regulation of insect populations by biological or chemical means.

insecticide [MATER] A chemical agent that destroys insects.

Insectivora [VERT ZOO] An order of mammals including hedgehogs, shrews, moles, and other forms, most of which have spines.

insectivorous [BIOL] Feeding on a diet of insects.

insectivorous plant [BOT] A plant that captures and digests insects as a source of nutrients by using specialized leaves. Also known as carnivorous plant.

insect pathology [INV ZOO] A biological discipline embracing the general principles of pathology as applied to insects.

insect physiology [INV ZOO] The study of the functional properties of insect tissues and organs.

inselberg [GEOL] A large, steep-sided residual hill, knob, or mountain, generally rocky and bare, rising abruptly from an extensive, nearly level lowland erosion surface in arid or semiarid regions. Also known as island mountain.

insemination [BIOL] The planting of seed. [PHYSIO] **1.** The introduction of sperm into the vagina. **2.** Impregnation.

insensitive time *See* dead time.

insequent stream [HYD] A stream that has developed on the present surface, but not consequent upon it, and seemingly not controlled or adjusted by the rock structure and surface features.

insert [MET] **1.** The part of a die or mold that can be removed. **2.** A part, usually metal, which is placed in a mold and appears as an integral part of the final casting.

insert bit [DES ENG] A bit into which inset cutting points of various preshaped pieces of hard metal (usually a sintered tungsten carbide–cobalt powder alloy) are brazed or hand-peened into slots or holes cut or drilled into a blank bit. Also known as slug bit.

inserted [BIOL] United or attached to the supporting structure by natural growth.

inserted-tooth cutter [DES ENG] A milling cutter in which the teeth can be replaced.

insert film [GRAPHICS] A microfilm strip cut in lengths to fit a jacket channel length.

insertion gain [ELECTR] The ratio of the power delivered to a part of the system following insertion of an amplifier, to the power delivered to that same part before insertion of the amplifier; usually expressed in decibels.

insertion loss [ELECTR] The loss in load power due to the insertion of a component or device at some point in a transmission system; generally expressed as the ratio in decibels of the power received at the load before insertion of the apparatus, to the power received at the load after insertion.

insertion switch [ADP] Process by which information is inserted into the computer by an operator who manually operates switches.

insert pump *See* rod pump.

inshore [GEOGR] **1.** Located near the shore. **2.** Indicating a shoreward position.

inshore current [OCEANOGR] The horizontal movement of water inside the surf zone, including longshore and rip currents.

inshore zone [GEOL] The zone of variable width extending from the shoreline at low tide through the breaker zone.

inside caliper [DES ENG] A caliper that has two legs with feet that turn outward; used to measure inside dimensions, as the diameter of a hole.

inside diameter [DES ENG] The length of a line which passes through the center of a hollow cylindrical or spherical object, and whose end points lie on the inner surface of the object. Abbreviated ID.

inside face [DES ENG] That part of the bit crown nearest to or parallel with the inside wall of an annular or coring bit.

inside gage [DES ENG] The inside diameter of a bit as measured between the cutting points, such as between inset diamonds on the inside-wall surface of a core bit.

inside micrometer [DES ENG] A micrometer caliper with the points turned outward for measuring the internal dimensions of an object.

inside work *See* internal work.

insight therapy [PSYCH] Treatment of a personality disorder by attempting to uncover the deep causes of the individual's problem and to help eliminate defense mechanisms.

in situ [SCI TECH] In the original location.

in situ combustion [PETRO ENG] A method of driving high-viscosity, low-gravity ore otherwise unrecoverable from a formation by setting fire to the oil sand and thereby heating the oil in the horizon to increase its mobility by reducing its viscosity.

in situ foaming [ENG] Depositing of the ingredients of a foamable plastic onto the location where foaming is to take place; for example, in situ foam insulation on equipment or walls.

insol *See* insoluble.

insolation [ASTRON] **1.** Exposure of an object to the sun. **2.** Solar energy received, often expressed as a rate of energy per unit horizontal surface.

insoluble [CHEM] Incapable of being dissolved in another material; usually refers to solid-liquid or liquid-liquid systems. Abbreviated insol.

insoluble anode [CHEM] An anode that resists dissolution during electrolysis.

insoluble residue [GEOL] Material remaining after a geological specimen is dissolved in hydrochloric or acetic acid.

insomnia [MED] Sleeplessness; disturbed sleep; prolonged inability to sleep.

insomniac [MED] A person who is susceptible to insomnia.

inspect [IND ENG] To examine an object to determine whether it conforms to standards; may employ sight, hearing, touch, odor, or taste.

inspection [IND ENG] The critical examination of a product to determine its conformance to applicable quality standards or specifications.

inspector [MIN ENG] One employed to make examinations of and to report upon mines and surface plants relative to compliance with mining laws, rules and regulations, and safety methods.

inspiration [PHYSIO] The drawing in of the breath.

inspirometer [MED] An instrument for measuring the amount of air inspired.

inspissation [GEOCHEM] Thickening of an oil deposit by evaporation or oxidation, resulting, for example, after long exposure in pitch or gum formation.

instability [CONT SYS] A condition of a control system in which excessive positive feedback causes persistent, unwanted oscillations in the output of the system. [PHYS] A property of the steady state of a system such that certain disturbances or perturbations introduced into the steady state will increase in magnitude, the maximum perturbation amplitude always remaining larger than the initial amplitude.

instability line [METEOROL] Any nonfrontal line or band of convective activity in the atmosphere; this is the general term and includes the developing, mature, and dissipating stages; however, when the mature stage consists of a line of active thunderstorms, it is properly termed a squall line; therefore, in practice, instability line often refers only to the less active phases.

installation [ENG] Procedures for setting up equipment for use or service.

installation kit [ORD] A collection of items, which are of a supplementary nature to a major component or equipment; the items within the collection are used to establish and install an accessory or equipment to an operational condition or to the component or equipment.

installed capacity [ELEC] The maximum runoff of a hydro-

INSIDE CALIPER

Inside caliper. *(From R. J. Sweeney, Measurement Techniques in Mechanical Engineering, Wiley, 1953)*

electric facility that can be constantly maintained and utilized by equipment.

instantaneous automatic gain control [ELECTR] Portion of a radar system that automatically adjusts the gain of an amplifier for each pulse to obtain a substantially constant output-pulse peak amplitude with different input-pulse peak amplitudes; the circuit is fast enough to act during the time a pulse is passing through the amplifier.

instantaneous axis [MECH] The axis about which a rigid body is carrying out a pure rotation at a given instant in time.

instantaneous carrying current [ELEC] The maximum value of current which a switch, circuit breaker, or similar apparatus can carry instantaneously.

instantaneous center [MECH] A point about which a rigid body is rotating at a given instant in time. Also known as instant center.

instantaneous companding [ELECTR] Companding in which the effective gain variations are made in response to instantaneous values of the signal wave.

instantaneous condition [PHYS] The condition of a system at a particular instant in time.

instantaneous cut [ENG] A cut that is set off by instantaneous detonators to be certain that all charges in the cut go off at the same time; the drilling and ignition are carried out so that all the holes break smaller top angles.

instantaneous detonator [ENG] A type of detonator that does not have a delay period between the passage of the electric current through the detonator and its explosion.

instantaneous effects [COMMUN] Impairment of telephone or telegraph transmission caused by instantaneous changes in phase or amplitude of the wave in a transmission line.

instantaneous frequency [COMMUN] The time rate of change, divided by 2π, of a phase angle whose sine is proportional to the amplitude of a frequency-modulated wave.

instantaneous fuse [ENG] A fuse with an ignition rate of several thousand feet per minute; an example is PETN.

instantaneous power [ELEC] The product of the instantaneous voltage and the instantaneous current for a circuit or component.

instantaneous readout [COMMUN] Readout by a radio transmitter instantaneous with the computation of data to be transmitted.

instantaneous recording [ENG ACOUS] A recording intended for direct reproduction without further processing.

instantaneous recovery [MECH] The immediate reduction in the strain of a solid when a stress is removed or reduced, in contrast to creep recovery.

instantaneous sample [COMMUN] One of a sequence of instantaneous values of a wave taken at regular intervals.

instantaneous strain [MECH] The immediate deformation of a solid upon initial application of a stress, in contrast to creep strain.

instantaneous value [PHYS] The value of a sinusoidal or otherwise varying quantity at a particular instant.

instant center *See* instantaneous center.

instantizing [FOOD ENG] Redrying a wet agglomerate of nonfat dry milk powder to render the product more easily reconstituted.

instant-on switch [ELECTR] A switch that applies a reduced filament voltage to all tubes in a television receiver continuously, so the picture appears almost instantaneously after the set is turned on.

instant replay *See* video replay.

instar [INV ZOO] A stage between molts in the life of arthropods, especially insects.

instep [ANAT] The arch on the medial side of the foot.

instinct [PSYCH] A primary tendency or inborn drive, as toward life, sexual reproduction, and death. [ZOO] A precise form of behavior in which there is an invariable association of a particular series of responses with specific stimuli; an unconditioned compound reflex.

instinctive behavior [ZOO] Any species-typical pattern of responses not clearly acquired through training.

instinctual [PSYCH] Pertaining to an emotional, impulsive, and generally unreasoned behavior or mental process which is a function of the id. [ZOO] Of or pertaining to instincts.

instruction [ADP] A pattern of digits which signifies to a computer that a particular operation is to be performed and which may also indicate the operands (or the locations of operands) to be operated on.

instruction area [ADP] A section of storage used for storing program instructions.

instruction card [IND ENG] A written description of the standard method used by a worker, to guide his activities.

instruction code [ADP] That part of an instruction which distinguishes it from all other instructions and specifies the action to be performed.

instruction constant [ADP] A dummy instruction of the type $K = 1$, where K is irrelevant to the program.

instruction counter [ADP] A counter that indicates the location of the next computer instruction to be interpreted. Also known as location counter; program counter; sequence counter.

instruction deck [ADP] Set of cards punched to contain a symbolic coded program to be read into a computer.

instruction format [ADP] Any rule which assigns various functions to the various digits of an instruction.

instruction length [ADP] The number of bits or bytes (eight bits per byte) which defines an instruction.

instruction modification [ADP] A change, carried out by the program, in an instruction so that, upon being repeated, this instruction will perform a different operation.

instruction register [ADP] A hardware element that receives and holds an instruction as it is extracted from memory; the register either contains or is connected to circuits that interpret the instruction (or discover its meaning). Also known as current-instruction register.

instruction repertory [ADP] **1.** The set of instructions which a computing or data-processing system is capable of performing. **2.** The set of instructions which an automatic coding system assembles.

instruction time [ADP] The time required to carry out an instruction having a specified number of addresses in a particular computer.

instruction transfer [ADP] An instruction which transfers control to one or another subprogram, depending upon the value of some operation.

instruction word [ADP] A computer word containing an instruction rather than data. Also known as coding line.

instrument [ENG] A device for measuring and sometimes also recording and controlling the value of a quantity under observation.

instrumental analysis [ENG] The use of an instrument to measure a component, to detect the completion of a quantitative reaction, or to detect a change in the properties of a system.

instrumental conditioning *See* operant conditioning.

instrument approach chart [NAV] An aeronautical chart designed for use under instrument flight conditions, for making instrument approach and letdown to contact flight conditions in the vicinity of an aerodrome.

instrument approach procedure [NAV] A series of predetermined maneuvers for the orderly transfer of an aircraft under instrument flight conditions from the initial approach to a landing, or to a point from which a landing can be made visually.

instrument approach system [NAV] An aircraft navigation system that furnishes guidance in the vertical and horizontal planes to aircraft during descent from an initial-approach altitude to a point near the landing area; completion of a landing requires guidance to touchdown by visual or other means.

instrumentation [ENG] Designing, manufacturing, and utilizing physical instruments or instrument systems for detection, observation, measurement, automatic control, automatic computation, communication, or data processing.

instrument correction [ENG] A correction of measurements made on a unit under test for either inaccuracy of the instrument or eroding effect of the instrument.

instrument flight [NAV] A flight in which the navigation of the aircraft is controlled solely by reference to instruments.

instrument flight rules [NAV] Regulations governing flying when weather conditions are below the minimum for visual flight rules. Abbreviated IFR.

instrument housing [ENG] A case or enclosure to cover and protect an instrument.

instrument landing [NAV] A landing made through the use of a system of electronic beacons and radar.

instrument landing system [NAV] A system of radio navigation which provides lateral and vertical guidance, as well as other navigational parameters required by a pilot in a low approach or a landing. Abbreviated ILS.

instrument landing system localizer [NAV] System of horizontal guidance embodied in the instrument landing system which indicates the horizontal deviation of the aircraft from its optimum path of descent along the axis of the runway.

instrument landing system reference point *See* ILS reference point.

instrument multiplier [ELEC] A highly accurate resistor used in series with a voltmeter to extend its voltage range. Also known as voltage multiplier; voltage-range multiplier.

instrument oil [MATER] Special grade of lubricating oil that has been refined to have oxidation resistance and gum resistance, that has compatibility with electrical insulation, and that prevents tarnish or oxidation of contacted metal surfaces; used to lubricate instruments and other intricate mechanisms.

instrument panel [ENG] A panel or board containing indicating meters.

instrument reading time [ENG] The time, after a change in a measured quantity, which it takes for the indication of an instrument to come and remain within a specified percentage of its final value.

instrument resistor [ELEC] A high-accuracy, four-terminal resistor used to bypass the major portion of currents around the low-current elements of an instrument, such as a direct-current ammeter.

instrument shelter [ENG] A boxlike structure designed to protect certain meteorological instruments from exposure to direct sunshine, precipitation, and condensation, while providing adequate ventilation. Also known as thermometer screen; thermometer shelter; thermoscreen.

instrument shunt [ELEC] A resistor designed to be connected in parallel with an ammeter to extend its current range.

instrument system [ENG] A system which integrates one or more instruments with auxiliary or associated devices for detection, observation, measurement, automatic control, automatic computation, communication, or data processing.

instrument transformer [ELEC] A transformer that transfers primary current, voltage, or phase values to the secondary circuit with sufficient accuracy to permit connecting an instrument to the secondary rather than the primary; used so only low currents or low voltages are brought to the instrument.

instrument-type relay [ELEC] A relay constructed like a meter, with one adjustable contact mounted on the scale and the other contact mounted on the pointer. Also known as contact-making meter.

instrument weather [METEOROL] Route or terminal weather conditions of sufficiently low visibility to require the operation of aircraft under instrument flight rules (IFR). Also known as IFR weather.

insulated [ELEC] Separated from other conducting surfaces by a nonconducting material.

insulated conductor [ELEC] A conductor surrounded by insulation to prevent current leakage or short circuits. Also known as insulated wire.

insulated-gate field-effect transistor *See* metal oxide semiconductor field-effect transistor.

insulated-substrate monolithic circuit [ELECTR] Integrated circuit which may be either an all-diffused device or a compatible structure so constructed that the components within the silicon substrate are insulated from one another by a layer of silicon dioxide, instead of reverse-biased *pn* junctions used for isolation in other techniques.

insulated wire *See* insulated conductor.

insulating board [MATER] Any board used in a wall or ceiling to provide insulation.

insulating compound [MATER] A liquid, at low temperatures, which is poured into joint boxes and allowed to solidify; as a poor conductor of heat and electricity, it provides good insulation.

insulating concrete [MATER] Concrete with insulating properties, often made with asbestos fibers and in the form of blocks, corrugated slabs, or sheathing.

insulating oil [MATER] A chlorinated hydrocarbon, such as trichlorobenzene, mixed with fluorinated hydrocarbons, whose high dielectric strength and high flash point allow it to be used in switches, circuit breakers, and transformers as an insulator and cooling medium. Also known as electrical oil.

insulating paper [MATER] A standard material for insulating electrical equipment, usually consisting of bond or kraft paper coated with black or yellow insulating varnish on both sides. Also known as electrical insulating paper; varnish paper.

insulating strength [ELEC] Measure of the ability of an insulating material to withstand electric stress without breakdown; it is defined as the voltage per unit thickness necessary to initiate a disruptive discharge; usually measured in volts per centimeter.

insulating tape [MATER] Tape impregnated with insulating material, and usually adhesive; used to cover joints in insulated wires or cables. Also known as electrical tape.

insulation [BUILD] Material used in walls, ceilings, and floors to retard the passage of heat and sound. [ELEC] A material having high electrical resistivity and therefore suitable for separating adjacent conductors in an electric circuit or preventing possible future contact between conductors. Also known as electrical insulation.

insulation coordination [ELEC] Steps taken to ensure that electric equipment is not damaged by overvoltages and that flashovers are localized in regions where no damage results from them.

insulation porcelain [MATER] Any of the various insulating materials consisting of molded silica, molded steatite, or specially compounded ceramics, often containing zirconia or beryllia. Also known as electrical porcelain.

insulation protection [ELEC] Use of devices to protect insulators of power transmission lines from damage by heavy arcs.

insulation resistance [ELEC] The electrical resistance between two conductors separated by an insulating material.

insulation sampler [ENG] A device for collecting deep water which prevents any significant conduction of heat from the water sample so that it maintains its original temperature as it is hauled to the surface.

insulator [ELEC] A device having high electrical resistance and used for supporting or separating conductors to prevent undesired flow of current from them to other objects. Also known as electrical insulator. [MATER] A material that is a poor conductor of heat, sound, or electricity. [SOLID STATE] A substance in which the normal energy band is full and is separated from the first excitation band by a forbidden band that can be penetrated only by an electron having an energy of several electron volts, sufficient to disrupt the substance.

insulator arc-over [ELEC] Discharge of power current in the form of an arc, following a surface discharge over an insulator.

insulator arrangement [ELECTROMAG] The placement of insulators on a transmission mast.

insulin [BIOCHEM] A protein hormone produced by the beta cells of the islets of Langerhans which participates in carbohydrate and fat metabolism.

insulinase [BIOCHEM] An enzyme produced by the liver which is able to inactivate insulin.

insulinoma *See* islet-cell tumor.

insulin shock [MED] Clinical manifestation of hypoglycemia due to excess amounts of insulin in the blood.

insulin shock therapy [MED] Administration of large doses of insulin to induce hypoglycemic comas, followed by administration of glucose, in the treatment of certain psychotic disorders.

insuloma *See* islet-cell tumor.

intaglio [LAP] A type of carved gemstone in which the figure is engraved on the surface of the stone rather than left in relief by cutting away the background, as in a cameo.

intaglio plate [GRAPHICS] A metal surface into which the printing elements are formed in intaglio printing.

intaglio printing [GRAPHICS] A printing method in which the printing elements are all below the plate surface, having been cut, scratched, engraved, or etched into the metal to form ink-

INSTRUMENT TRANSFORMER

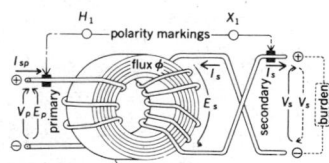

magnetic flux circuit ferromagnetic core

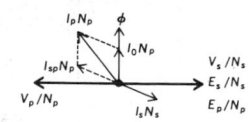

A simple instrument transformer and its phasor diagram. Core-loss components and impedance drops have been omitted from the diagram. Subscripts *p* and *s* = primary and secondary windings respectively; *E* = induced voltage; *V* = applied voltage; *N* = number of turns; *I* = current. (*From I. F. Kinnard, Applied Electrical Measurements, Wiley, 1956*)

retaining grooves or cups; surplus ink on the surface must be wiped or scraped off after each inking and before each printing impression.

intake [ENG] **1.** An entrance for air, water, fuel, or other fluid, or the amount of such fluid taken in. **2.** A main passage for air in a mine. [HYD] *See* recharge.

intake chamber [CIV ENG] A large chamber that gradually narrows to an intake tunnel; designed to avoid undesirable water currents.

intake gate [CIV ENG] A movable partition for opening or closing a water intake opening.

intake manifold [MECH ENG] A system of pipes which feeds fuel to the various cylinders of a multicylinder internal combustion engine.

intake stroke [MECH ENG] The fluid admission phase or travel of a reciprocating piston and cylinder mechanism as, for example, in an engine, pump, or compressor.

intarsia [GRAPHICS] Decorative designs of inlaid wood in a background of wood; often used in furniture making. Also known as tarsia.

integer [MATH] Any positive or negative counting number or zero.

integer constant [ADP] A constant that uses the values 0, 1, . . . , 9 with no decimal point in FORTRAN.

integer programming [SYS ENG] A series of procedures used in operations research to find maxima or minima of a function subject to one or more constraints, including one which requires that the values of some or all of the variables be whole numbers.

integer spin [QUANT MECH] Property of a particle whose spin angular momentum is a whole number times Planck's constant divided by 2π; bosons have this property; in contrast, fermions have half-integer spin.

integer variable [ADP] A variable in FORTRAN whose first character is normally I, J, K, L, M, or N.

integral [MATH] **1.** A solution of a differential equation is sometimes called an integral of the equation. **2.** *See* definite Riemann integral; indefinite integral.

integral absorbed dose *See* integral dose.

integral action [CONT SYS] A control action in which the rate of change of the correcting force is proportional to the deviation.

integral calculus [MATH] The study of integration and its applications to finding areas, volumes, or solutions of differential equations.

integral compensation [CONT SYS] Use of a compensator whose output changes at a rate proportional to its input.

integral control [CONT SYS] Use of a control system in which the control signal changes at a rate proportional to the error signal.

integral discriminator [ELECTR] A circuit which accepts only pulses greater than a certain minimum height.

integral domain [MATH] A commutative ring with identity where the product of nonzero elements is never zero.

integral dose [NUCLEO] The total energy imparted to an irradiated body by an ionizing radiation; usually expressed in gram-rads or gram-roentgens. Also known as integral absorbed dose; volume dose.

integral equation [MATH] An equation where the unknown function occurs under an integral sign.

integral function [MATH] A function taking on integer values.

integral-furnace boiler [MECH ENG] A type of steam boiler which incorporates furnace water-cooling in the circulatory system.

integral heat of dilution *See* heat of dilution.

integral heat of solution *See* heat of solution.

integral-joint casing [PETRO ENG] Oil well casing lengths on whose ends the connection joints are integrally formed.

integral-mode controller [CONT SYS] A controller which produces a control signal proportional to the integral of the error signal.

integral network [CONT SYS] A compensating network which produces high gain at low input frequencies and low gain at high frequencies, and is therefore useful in achieving low steady-state errors. Also known as lagging network; lag network.

integral operator [MATH] A rule for transforming one func-

tion into another function by means of an integral; this often is in context a linear transformation on some vector space of functions.

integral procedure decomposition temperature [PHYS CHEM] Decomposition temperatures derived from graphical integration of the thermogravimetric analysis of a polymer.

integral test [MATH] If $f(x)$ is a function that is positive and decreasing for positive x, then the infinite series with nth term $f(n)$ and the integral of $f(x)$ from 1 to ∞ are either both convergent (finite) or both infinite.

integral transform *See* integral transformation.

integral transformation [MATH] A transform of a function $F(x)$ given by the function

$$f(y) = \int_a^b K(x,y)F(x)\,dx$$

where $K(x,y)$ is some function. Also known as integral transform.

integral waterproofing [ENG] Waterproofing concrete by adding the waterproofing material to the cement or to the mixing water.

integrand [MATH] The function which is being integrated in a given integral.

integrated circuit [ELECTR] An interconnected array of active and passive elements integrated with a single semiconductor substrate or deposited on the substrate by a continuous series of compatible processes, and capable of performing at least one complete electronic circuit function. Abbreviated IC. Also known as integrated semiconductor; monolithic circuit.

integrated communications-navigation-identification [NAV] The concept of coordinating the electronic units of a system to improve the efficiency of providing communications, navigation, and identification for civilian and military aircraft. Abbreviated ICNI.

integrated communications system [COMMUN] Communications system on either a unilateral or joint basis in which a message can be filed at any communications center in that system and be delivered to the addressee by any other appropriate communications center in that system without reprocessing enroute.

integrated console [COMMUN] Computer control console that is capable of controlling the operation of the switching center equipment of an integrated communications system.

integrated data processing [ADP] Data processing that has been organized and carried out as a whole, so that intermediate outputs may serve as inputs for subsequent processing with no human copying required. Abbreviated IDP.

integrated data retrieval system [ADP] A section of a data-processing system that provides facilities for simultaneous operation of several video-data interrogations in a single line and performs required communications with the rest of the system; it provides storage and retrieval of both data subsystems and files and standard formats for data representation.

integrated drainage [HYD] Drainage resulting after folding and faulting of a surface under arid conditions; the streams by working headward have joined basins across intervening mountains or ridges.

integrated electronics [ELECTR] A generic term for that portion of electronic art and technology in which the interdependence of material, device, circuit, and system-design consideration is especially significant; more specifically, that portion of the art dealing with integrated circuits.

integrated fire control system [ORD] A system which combines target acquisition and tracking data computation, and weapon laying and firing, primarily using electronic means assisted by electromechanical devices.

integrated information processing [ADP] System of computers and peripheral systems arranged and coordinated to work concurrently or independently on different problems at the same time.

integrated information system [COMMUN] An expansion of a basic information system achieved through system design of an improved or broader capability by functionally or technically relating two or more information systems, or by incorporating a portion of the functional or technical elements of one information system into another.

integrated neutron flux [NUCLEO] A measure of radiation

INTEGRATED CIRCUIT

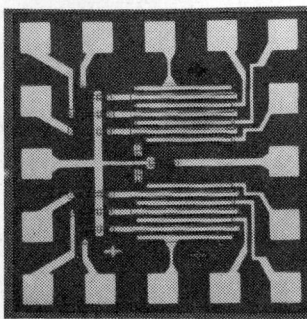

Photomicrograph of a simple MOS (metal oxide semiconductor) integrated circuit, a three-input logic gate circuit.

exposure, equal to the product of the number of free neutrons per unit volume, the average speed of neutrons, and the exposure time.

integrated optics [OPTICS] A thin-film device containing tiny lenses, prisms, and switches to transmit very thin laser beams, and serving the same purposes as the manipulation of electrons in thin-film devices of integrated electronics.

integrated semiconductor *See* integrated circuit.

integrated train [MIN ENG] A long string of cars, permanently coupled together, that shuttles endlessly between one mine and one generating plant, not even stopping to load and unload, since rotary couplers permit each car to be flipped over and dumped as the train moves slowly across a trestle.

integrating accelerometer [ENG] A device whose output signals are proportional to the velocity of the vehicle or to the distance traveled (depending on the number of integrations) instead of acceleration.

integrating amplifier [ELECTR] An operational amplifier with a shunt capacitor such that mathematically the waveform at the output is the integral (usually over time) of the input.

integrating circuit *See* integrating network.

integrating detector [ELECTR] A frequency-modulation detector in which a frequency-modulated wave is converted to an intermediate-frequency pulse-rate modulated wave, from which the original modulating signal can be recovered by use of an integrator.

integrating factor [MATH] A factor which when multiplied into a differential equation makes the portion involving derivatives an exact differential.

integrating filter [ELECTR] A filter in which successive pulses of applied voltage cause cumulative buildup of charge and voltage on an output capacitor.

integrating gyroscope [ENG] A gyroscope that senses the rate of angular displacement and measures and transmits the time integral of this rate.

integrating network [ELECTR] A circuit or network whose output waveform is the time integral of its input waveform. Also known as integrating circuit; integrator.

integrating-sphere photometer [OPTICS] An instrument for measuring the total luminous flux of a lamp or luminaire; the source is placed inside a sphere whose inside surface has a diffusely reflecting white finish, and the light reflected from this surface onto a window is measured by an ordinary photometer. Also known as sphere photometer.

integration by parts [MATH] A technique used to find the integral of the product of two functions by means of an identity involving another simpler integral; for functions of several variables this is tantamount to using Stokes' theorem or the divergence theorem.

integrator [ELECTR] **1.** A computer device that approximates the mathematical process of integration. **2.** *See* integrating network.

integrodifferential equation [MATH] An equation relating a function, its derivatives, and its integrals.

integument [ANAT] An outer covering, especially the skin, together with its various derivatives.

integumentary musculature [VERT ZOO] Superficial skeletal muscles which are spread out beneath the skin and are inserted into it in some terrestrial vertebrates.

integumentary pattern [ANAT] Any of the features of the skin and its derivatives that are arranged in designs, such as scales, epidermal ridges, feathers coloration, or hair.

integumentary system [ANAT] A system encompassing the integument and its derivatives.

intelligence [COMMUN] Data, information, or messages that are to be transmitted. [PSYCH] **1.** The intellect or astuteness of the mind. **2.** Ability to recognize and understand qualities and attributes of the physical world and of mankind. **3.** Ability to solve problems and engage in abstract thought processes.

intelligence quotient [PSYCH] The numerical designation for intelligence expressed as a ratio of an individual's performance on a standardized test to the average performance according to age. Abbreviated IQ.

intelligence test [PSYCH] A series of standardized tasks or problems presented to an individual to measure his innate capacity to think, conceive, or reason; examples are the

Stanford-Binet test and the Wechsler-Bellevue intelligence test.

intelligibility [COMMUN] The percentage of speech units understood correctly by a listener in a communications system; customarily used for regular messages where the context aids the listener, in distinction to articulation. Also known as speech intelligibility.

intelligible crosstalk [COMMUN] Crosstalk which is sufficiently understandable under pertinent circuit and room noise conditions that meaningful information can be obtained by more sensitive listeners.

intensifier [GRAPHICS] In photography, a means used to strengthen the image of either a negative or a positive; usually, metal is added to the silver image, and the increase in density is proportional to the existing density; thus when a negative is intensified, the highlights are always intensified more than the shadows. [PETRO ENG] Hydrofluoric acid added to hydrochloric acid for oil well acidizing; the fluoride destroys silica films that are insoluble by hydrochloric acid.

intensifier electrode [ELECTR] An electrode used to increase the velocity of electrons in a beam near the end of their trajectory, after deflection of the beam. Also known as post-accelerating electrode; post-deflection accelerating electrode.

intensifier image orthicon [ELECTR] An image orthicon combined with an image intensifier that amplifies the electron stream originating at the photocathode before it strikes the target.

intensity [PHYS] **1.** The strength or amount of a quantity, as of electric field, current, magnetization, radiation, or radioactivity. **2.** The power transmitted by a light or sound wave across a unit area perpendicular to the wave.

intensity control *See* brightness control.

intensity level [PHYS] The logarithm of the ratio of two intensities, powers or energies, usually expressed in decibels.

intensity modulation [ELECTR] Modulation of electron beam intensity in a cathode-ray tube in accordance with the magnitude of the received signal.

intensity of magnetization *See* intrinsic induction.

intensity of turbulence *See* gustiness components.

intensive properties [CHEM] Properties independent of the quantity or shape of the substance under consideration; for example, temperature, pressure, or composition.

intention tremor [MED] A clinical manifestation of certain diseases of the nervous system characterized by involuntary trembling of the limbs brought on by voluntary movements, and ceasing on rest.

interaction [FL MECH] With respect to wave components, the nonlinear action by which properties of fluid flow (such as momentum, energy, vorticity), are transferred from one portion of the wave spectrum to another, or viewed in another manner, between eddies of different size-scales. [STAT] The phenomenon which causes the response to applying two treatments not to be the simple sum of the responses to each treatment.

interaction picture [QUANT MECH] A mode of description of a system in which the time dependence is carried partly by the operators and partly by the state vectors, the time dependence of the state vectors being due entirely to that part of the Hamiltonian arising from interactions between particles. Also known as interaction representation.

interaction representation *See* interaction picture.

interaction space [ELECTR] A region of an electron tube in which electrons interact with an alternating electromagnetic field.

interactive graphical input [ADP] Information which is delivered to a computer by using hand-held devices, such as writing styli used with electronic tablets and light-pens used with cathode-ray tube displays, to sketch a problem description in an on-line interactive mode in which the computer acts as a drafting assistant with unusual powers, such as converting rough freehand motions of a pen or stylus to accurate picture elements.

interambulacrum [INV ZOO] In echinoderms, an area between two ambulacra.

interarticular [ANAT] Situated between articulating surfaces.

interatrial [ANAT] Located between the atria of the heart.

interatrial septal defect [MED] A congenital malformation of the septum between the atria of the heart.

INTEGRATING-SPHERE PHOTOMETER

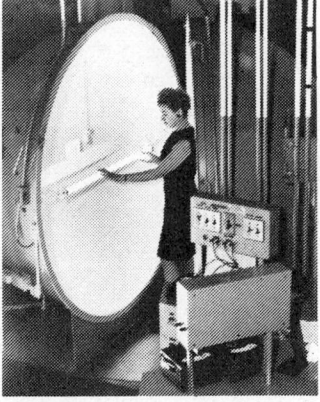

Uhlbricht sphere for measuring luminous flux and efficiency. (*Westinghouse Electric Corp.*)

INTEGUMENT

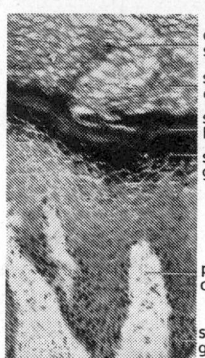

duct of sweat gland
stratum corneum
stratum lucidum
stratum granulosum
papilla of corium
stratum germinativum

The characteristic strata of thick skin of the human finger as seen in cross section at high magnification. (*From J. F. Nonidez and W. F. Windle, Textbook of Histology, McGraw-Hill, 1949*)

INTENSIFIER IMAGE ORTHICON

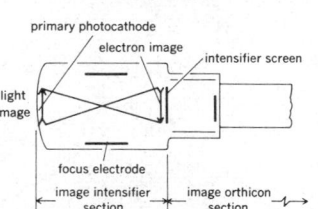

primary photocathode
electron image
intensifier screen
light image
focus electrode
image intensifier section
image orthicon section

Components of an intensifier image orthicon.

INTERAMBULACRUM

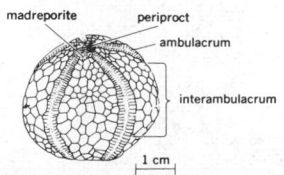

madreporite
periproct
ambulacrum
interambulacrum
1 cm

Morphological features of *Aulechinus grayae*, an echinocystitoid example from the Upper Ordovician of Scotland.

interatrial septum See atrial septum.

interaxillary [BOT] Located within or between the axils of leaves.

interbase current [ELECTR] The current that flows from one base connection of a junction tetrode transistor to the other, through the base region.

interbedded [GEOL] Having beds lying between other beds with different characteristics.

intercalary [BOT] Referring to growth occurring between the apex and leaf. [SCI TECH] Inserted between two original components.

intercalary day [ASTRON] A day inserted or introduced among others in a calendar, as February 29 during leap years.

intercalary meristem [BOT] A meristem that is forming between regions of permanent or mature meristem.

intercalated disc [HISTOL] A dense region at the junction of cellular units in cardiac muscle.

intercalated nucleus [ANAT] A nucleus of the medulla oblongata in the central gray matter of the ventricular floor located between the hypoglossal nucleus and the dorsal motor nucleus of the vagus.

intercalation [GEOL] A layer located between layers of different character.

intercapillary [ANAT] Located between capillaries.

intercapillary glomerulosclerosis [PATH] Nodular eosinophilic hyalin deposits on the periphery of glomeruli in individuals with diabetes. Also known as diabetic glomerulosclerosis; Kimmelstiel-Wilson disease.

intercardinal heading [NAV] A heading in the direction of any of the intercardinal points.

intercardinal point [GEOD] Any of the four directions midway between the cardinal points, that is, northeast, southeast, southwest, and northwest. Also known as quadrantal point.

intercardinal rolling error [NAV] Quadrantal error of a gyro compass.

intercarpal [ANAT] Located between the carpal bones.

intercarrier channel [COMMUN] A carrier telegraph channel in the available frequency spectrum between carrier telephone channels.

intercarrier noise suppression [ELECTR] Means of suppressing the noise resulting from increased gain when a high-gain receiver with automatic volume control is tuned between stations; the suppression circuit automatically blocks the audio-frequency input of the receiver when no signal exists at the second detector. Also known as interstation noise suppression.

intercarrier sound system [ELECTR] A television receiver arrangement in which the television picture carrier and the associated sound carrier are amplified together by the video intermediate-frequency amplifier and passed through the second detector, to give the conventional video signal plus a frequency-modulated sound signal whose center frequency is the 4.5 megahertz difference between the two carrier frequencies. Abbreviated ICS system.

intercavernous sinuses [ANAT] Venous sinuses located on the median line of the dura mater, connecting the cavernous sinuses of each side.

intercellular [HISTOL] Of or pertaining to the region between cells.

intercellular cement [HISTOL] A substance bonding epithelial cells together.

intercellular plexus [HISTOL] A network of neuronal processes surrounding the cells in a sympathetic ganglion.

intercellular space [HISTOL] A space between adjacent cells.

intercellular substance [HISTOL] Tissue component that lies between cells.

intercentrum [VERT ZOO] A type of crescentic intervertebral structure between successive centra in certain reptilian and mammalian tails.

intercept [MAP] See altitude difference. [MATH] The point where a straight line crosses one of the axes of a cartesian coordinate system.

intercepting [COMMUN] In telephone practice, routing of a call placed to a disconnected or nonexisting telephone number, to an operator, or to a machine answering device.

intercepting sewer [CIV ENG] A sewer that receives flow from transverse sewers and conducts the water to a treatment plant or disposal point.

interception [COMMUN] Tapping or tuning in to a telephone or radio message not intended for the listener. [HYD] 1. The process by which precipitation is caught and retained on vegetation or structures and subsequently evaporated without reaching the ground. 2. That part of the precipitation intercepted by vegetation. [METEOROL] 1. The loss of sunshine, a part of which may be intercepted by hills, trees, or tall buildings. 2. The depletion of part of the solar spectrum by atmospheric gases and suspensoids; this commonly refers to the absorption of ultraviolet radiation by ozone and dust. [ORD] Meeting or interrupting the course of a moving vessel, aircraft, or missile.

intercept method [MET] A method for determining grain size or the quantity of a phase in a microstructure by measuring the number of grains or phase particles per unit length intersected by straight lines. [NAV] See St. Hilaire method.

interceptometer [ENG] A rain gage which is placed under trees or in foliage to determine the rainfall in that location; by comparing this catch with that from a rain gage set in the open, the amount of rainfall which has been intercepted by foliage is found.

interceptor [AERO ENG] A manned aircraft utilized for the identification or engagement of airborne objects.

intercepts [CRYSTAL] Distances cut off a crystal's reference axis by planes.

intercept station [COMMUN] Provides service for subscribers whereby calls to disconnected stations or dead lines are either routed to an intercept operator for explanation, or the calling party receives a distinctive tone that informs him that he has made such a call.

intercept tape [COMMUN] A tape used for temporary storage of messages for trunk channels and tributary stations that are having equipment or circuit trouble.

intercept trunk [COMMUN] Trunk to which a call for a vacant number, a changed number, or a line out of order is connected for action by an operator.

interchange [CIV ENG] A junction of two or more highways at a number of separate levels so that traffic can pass from one highway to another without the crossing at grade of traffic streams. [ELEC] The current flowing into or out of a power system which is interconnected with one or more other power systems.

interchangeability [ENG] The ability to replace the components, parts, or equipment of one manufacturer with those of another, without losing function or suitability.

interchangeable lens [OPTICS] A lens which can be used in place of another, generally of different magnification.

interchange coefficient See exchange coefficient.

interchannel crosstalk [COMMUN] Crosstalk between channels in a multiplex system.

interclavicle [VERT ZOO] A membrane bone in front of the sternum and between the clavicles in monotremes and most reptiles.

intercloud discharge See cloud-to-cloud discharge.

intercolumniation [ARCH] Distance between columns, measured between the bottoms of shafts, just above the apophyge, and expressed in terms of the lower diameter of the column.

intercom See intercommunicating system.

intercommunicating porosity [MET] The type of porosity in a sintered metal powder compact that allows fluid to pass from pore to pore.

intercommunicating system Also known as intercom. [COMMUN] 1. A telephone system providing direct communication between telephones on the same premises. 2. A two-way communication system having a microphone and loudspeaker at each station and providing communication within a limited area.

intercommunication [PETRO ENG] Flow interconnection between the reservoir areas being drained by adjacent wells.

intercondenser [MECH ENG] A condenser between stages of a multistage steam jet pump.

interconnection [ELEC] A link between power systems enabling them to draw on one another's reserves in time of need and to take advantage of energy cost differentials resulting from such factors as load diversity, seasonal conditions, time-zone differences, and shared investment in larger generating units.

intercontinental ballistic missile [ORD] A missile flying a ballistic trajectory after guided powered flight, usually over ranges in excess of 4000 miles (6500 kilometers). Abbreviated ICBM.

intercontinental sea [GEOGR] A large body of salt water extending between two continents.

interconversion [COMMUN] Changing the representation of information from one code to another, as from six-bit to ASCII.

intercooler [MECH ENG] A heat exchanger for cooling fluid between stages of a multistage compressor with consequent saving in power.

intercostal [ANAT] Situated or occurring between adjacent ribs. [NAV ARCH] Situated or fitted between adjacent members of a ship's frame.

intercostal floor [NAV ARCH] A ship's floor constructed of a range of plates fitted between and clipped to longitudinal or side keelsons.

intercostal muscles [ANAT] Voluntary muscles between adjacent ribs.

intercostal nerve [ANAT] Any of the branches of the thoracic nerves in the intercostal spaces.

intercourse _See_ coitus.

intercrescence [BIOL] A growing together of tissues.

intercrystalline corrosion [MET] Localized attack occurring along the crystal boundaries of a metal or alloy. Also known as intergranular corrosion.

interdendritic attack _See_ interdendritic corrosion.

interdendritic corrosion [MET] Preferential corrosion of the metal immediately surrounding dendrites in unworked or slightly worked alloys caused by composition gradients. Also known as interdendritic attack.

interdiction [ORD] The prevention or destruction of, or interference with, enemy movements, communications, and lines of communication, as by gunfire, shelling, or bombing.

interdiction bombing [ORD] Bombing done for purposes of interdiction.

interdiction fire [ORD] Gunfire delivered in the process of interdiction; for example, firing on a specified place, as a rail yard, assembly area, crossroad, or the like, to prevent effective use of that area.

interdiffusion [PHYS CHEM] The self-mixing of two fluids, initially separated by a diaphragm.

interdigital magnetron [ELECTR] Magnetron having axial anode segments around the cathode, alternate segments being connected together at one end, remaining segments connected together at the opposite end.

interelectrode capacitance [ELECTR] The capacitance between one electrode of an electron tube and the next electrode on the anode side. Also known as direct interelectrode capacitance.

interelectrode transit time [ELECTR] Time required for an electron to traverse the distance between the two electrodes.

interface [ADP] The place (or places) at which two different systems (or subsystems) meet and interact with each other. [GEOPHYS] _See_ seismic discontinuity. [PHYS CHEM] The boundary between any two phases; among the three phases (gas, liquid, and solid), there are five types of interfaces: gas-liquid, gas-solid, liquid-liquid, liquid-solid, and solid-solid.

interface connection _See_ feedthrough.

interface control module [ADP] Relocatable modularized compiler allowing for efficient operation and easy maintenance.

interface mixing [PHYS CHEM] The mixing of two immiscible or partially miscible liquids at the plane of contact (interface).

interface resistance [THERMO] 1. Impairment of heat flow caused by the imperfect contact between two materials at an interface. 2. Quantitatively, the temperature difference across the interface divided by the heat flux through it.

interfacial angle [CRYSTAL] The angle between two crystal faces.

interfacial energy [PHYS] The free energy of the surfaces at an interface, resulting from differences in the tendencies of each phase to attract its own molecules; equal to the surface tension. Also known as surface energy.

interfacial force _See_ interfacial tension.

interfacial polarization [ELEC] _See_ space-charge polarization. [OPTICS] Polarization of light by reflection from the surface of a dielectric at Brewster's angle.

interfacial tension [PHYS] A kind of surface tension, occurring at the interface between two liquids. Also known as interfacial force.

interfacility transfer trunk [COMMUN] Trunk interconnecting switching centers of two different facilities.

interference [COMMUN] Any undesired energy that tends to interfere with the reception of desired signals. Also known as electrical interference; radio interference. [PHYS] The variation with distance or time of the amplitude of a wave which results from the superposition (algebraic or vector addition) of two or more waves having the same, or nearly the same, frequency. Also known as wave interference.

interference analyzer [ELECTR] An instrument that discloses the frequency and amplitude of unwanted input.

interference blanker [ELECTR] Device that permits simultaneous operation of two or more pieces of radio or radar equipment without confusion of intelligence, or that suppresses undesired signals when used with a single receiver.

interference colors [OPTICS] Colors formed by interference of a beam of light passed through a thin section of a mineral placed in a polarizing microscope.

interference fading [COMMUN] Fading of the signal produced by different wave components traveling slightly different paths in arriving at the receiver.

interference figure [OPTICS] A pattern of light and dark areas observed with a conoscope when a birefringent crystal is placed in a convergent beam of linearly polarized light.

interference filter [ELECTR] 1. A filter used to attenuate artificial interference signals entering a receiver through its power line. 2. A filter used to attenuate unwanted carrier-frequency signals in the tuned circuits of a receiver. [OPTICS] An optical filter in which the wavelengths that are not transmitted are removed by interference phenomena rather than by absorption or scattering.

interference fringes [OPTICS] A series of light and dark bands produced by interference of light waves.

interference microscope [OPTICS] A microscope used for visualizing and measuring differences in phase or optical paths in transparent or reflecting specimens; it differs from the phase contrast microscope in that the incident and diffracted waves are not separated, but interference is produced between the transmitted wave and another wave which originates from the same source.

interference pattern [ELECTR] Pattern produced on a radarscope by interference signals. [PHYS] Resulting space distribution of pressure, particle density, particle velocity, energy density, or energy flux when progressive waves of the same frequency and kind are superimposed.

interference phenomenon [VIROL] Inhibition by a virus of the simultaneous infection of host cells by some other virus.

interference prediction [ELECTR] Process of estimating the interference level of a particular equipment as a function of its future electromagnetic environment.

interference reduction [ELECTR] Reduction of interference from such causes as power lines and equipment, radio transmitters, and lightning, usually through the use of electric filters. Also known as interference suppression.

interference region [COMMUN] That region in space in which interference between wave trains occurs; in microwave propagation, it refers to the region bounded by the ray path and the surface of the earth which is above the radio horizon.

interference rejection [ELECTR] Use of a filter to reject (to bypass to ground) unwanted input.

interference ripple mark [GEOL] A pattern resulting from two sets of symmetrical ripples formed by waves crossing at right angles.

interference source suppression [ELECTR] Techniques applied at or near the source to reduce its emission of undesired signals.

interference spectrum [ELECTR] Frequency distribution of the jamming interference in the propagation medium external to the receiver. [SPECT] A spectrum that results from interference of light, as in a very thin film.

interference suppression _See_ interference reduction.

interference test [PETRO ENG] Test of pressure interrelationships (interference) between wells serving the same formation.

interference time [IND ENG] Idle machine time occurring

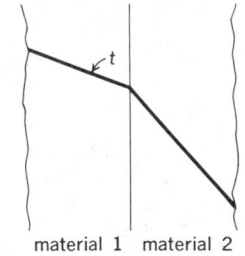

INTERFACE RESISTANCE

material 1   material 2
(a)

interface Δt_i

material 1   material 2
(b)

Distribution of temperature t through composite wall. _(a)_ With perfect interface contact. _(b)_ For typical actual surfaces. Δt_i is temperature difference across interface. Interface resistance equals Δt_i divided by heat flux.

INTERFERENCE FRINGES

Interference fringes formed with Fresnel biprism and mercury-arc light source.

when a machine operator, assigned to two or more semiautomatic machines, is unable to service a machine requiring attention.

interference wave [COMMUN] A radio wave reflected by the lower atmosphere which produces an interference pattern when combined with the direct wave.

interferogram [GRAPHICS] A photograph of a shock wave or other fluid flow, produced with the aid of an interferometer, in which the displacement of interference fringes is proportional to the density change across the fluid.

interferometer [OPTICS] An instrument in which light from a source is split into two or more beams, which are subsequently reunited after traveling over different paths and display interference.

interferometer systems [ELECTR] Method of determining the position of a target in azimuth by using an interferometer to compare the phases of signals at the output terminals of a pair of antennas receiving a common signal from a distant source.

interferometry [OPTICS] The design and use of optical interometers; uses include precise measurement of wavelength, measurement of very small distances and thicknesses, study of hyperfine structure of spectral lines, precise measurement of indices of refraction, and determination of separations of binary stars and diameters of very large stars.

interferon [BIOCHEM] A protein produced by intact animal cells when infected with viruses; acts to inhibit viral reproduction and to induce resistance in host cells.

interferonogen [VIROL] A preparation made of inactivated virus particles used as an inoculant to stimulate formation of interferon.

interfix [ADP] A technique for describing relationships of key words in an item or document in a way which prevents crosstalk from causing false retrievals when very specific entries are made.

interflow [HYD] The water, derived from precipitation, that infiltrates the soil surface and then moves laterally through the upper layers of soil above the water table until it reaches a stream channel or returns to the surface at some point downslope from its point of infiltration.

interfoliaceous [BOT] Between a pair of leaves, such as between those which are opposite or verticillate.

intergalactic matter [ASTRON] The material between the galaxies.

intergelisol *See* pereletok.

intergenic crossing-over [MOL BIO] Recombination between distinct genes or cistrons.

intergenic suppression [GEN] The restoration of a suppressed function or character by a second mutation that is located in a different gene than the original or first mutation.

interglacial [GEOL] Pertaining to or formed during a period of geologic time between two successive glacial epochs or between two glacial stages.

intergranular [SCI TECH] Occurring between grains.

intergranular corrosion *See* intercrystalline corrosion.

intergranular fracture [MET] Propagation of a crack along the grain boundaries of a metal or alloy.

intergrowth [MINERAL] A state of interlocking of different mineral crystals because of simultaneous cyrstallization.

interhalogen [INORG CHEM] Any of the compounds formed from the elements of the halogen family that react with each other to form a series of binary compounds; for example, iodine monofluoride.

interionic attraction [PHYS CHEM] The coulomb attraction between ions of opposite sign in a solution.

interior angle [MATH] **1.** An angle between two adjacent sides of a polygon that lies within the polygon. **2.** For a line (called the transversal) that intersects two other lines, an angle between the transversal and one of the two lines that lies within the space between the two lines.

interior ballistics [MECH] The science concerned with the combustion of powder, development of pressure, and movement of a projectile in the bore of a gun.

interior distribution [ELEC] Distribution of electric power within a building or plant.

interior label [ADP] A label attached to the data that it identifies.

interior of a set [MATH] The set of all interior points of a set in a topological space.

interior point [MATH] A point in a subset of a topological space where some open neighborhood of the point lies completely within the subset.

interkinesis *See* interphase.

interlabium [INV ZOO] A small lobe situated between the lips in certain nematodes.

interlace [ADP] To assign successive memory location numbers to physically separated locations on a storage tape or magnetic drum of a computer, usually to reduce access time.

interlaced scanning [ELECTR] A scanning process in which the distance from center to center of successively scanned lines is two or more times the nominal line width, so that adjacent lines belong to different fields. Also known as line interlace.

interlace operation [ADP] System of computer operation where data can be read out or copied into memory without interfering with the other activities of the computer.

interlacing arches [ARCH] Arches with intersecting curved moldings that appear to be interlaced.

interlaminated [SCI TECH] **1.** Having insertions between laminae. **2.** Arranged in alternate laminae.

interleave [ADP] To alternate parts of one sequence with parts of one or more other sequences in a cyclic fashion such that each sequence retains its identity.

interleaved windings [ELEC] An arrangement of winding coils around a transformer core in which the coils are wound in the form of a disk, with a group of disks for the low-voltage windings stacked alternately with a group of disks for the high-voltage windings.

interlobate moraine *See* intermediate moraine.

interlock [ENG] A switch or other device that prevents activation of a piece of equipment when a protective door is open or some other hazard exists.

interlock relay [ELEC] A relay composed of two or more coils, each with its own armature and associated contacts, so arranged that movement of one armature or the energizing of its coil is dependent on the position of the other armature.

interlock switch [ELEC] A switch designed for mounting on a door, drawer, or cover so that it opens automatically when the door or other part is opened.

intermediary metabolism [BIOCHEM] Intermediate steps in the chemical synthesis and breakdown of foodstuffs within body cells.

intermediate [CHEM] A precursor to a desired product; ethylene is an intermediate for polyethylene, and ethane is an intermediate for ethylene. [GRAPHICS] That print which is used as a master for further reproduction.

intermediate annealing [MET] Softening of a metal by heat treatment at one or more stages during cold working and before final treatment.

intermediate distributing frame [ELEC] Frame in a local telephone central office, the primary purpose of which is to cross-connect the subscriber line multiple to the subscriber line circuit; in a private exchange, the intermediate distributing frame is for similar purposes.

intermediate flux [MET] A flux consisting of organic halide compounds whose residues are decomposed by the heat of soldering; fluxing action approaches that of corrosive flux.

intermediate frequency [ELECTR] The frequency produced by combining the received signal with that of the local oscillator in a superheterodyne receiver. Abbreviated i-f.

intermediate-frequency amplifier [ELECTR] The section of a superheterodyne receiver that amplifies signals after they have been converted to the fixed intermediate-frequency value by the frequency converter. Abbreviated i-f amplifier.

intermediate-frequency jamming [ELECTR] Form of continuous wave jamming that is accomplished by transmitting two continuous wave signals separated by a frequency equal to the center frequency of the radar receiver intermediate-frequency amplifier.

intermediate-frequency response ratio [ELECTR] In a superheterodyne receiver, the ratio of the intermediate-frequency signal input at the antenna to the desired signal input for identical outputs. Also known as intermediate-interference ratio.

intermediate-frequency signal [ELECTR] A modulated or continuous-wave signal whose frequency is the intermediate-

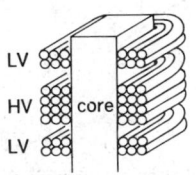

INTERLEAVED WINDINGS

LV
HV — core
LV

Interleaved windings arrangement. LV = low voltage, HV = high voltage.

frequency value of a superheterodyne receiver and is produced by frequency conversion before demodulation.

intermediate-frequency stage [ELECTR] One of the stages in the intermediate-frequency amplifier of a superheterodyne receiver.

intermediate-frequency strip [ELECTR] A receiver subassembly consisting of the intermediate-frequency amplifier stages, installed or replaced as a unit.

intermediate-frequency transformer [ELECTR] The transformer used at the input and output of each intermediate-frequency amplifier stage in a superheterodyne receiver for coupling purposes and to provide selectivity. Abbreviated i-f transformer.

intermediate ganglion [ANAT] Any of certain small groups of nerve cells found along communicating branches of spinal nerves.

intermediate haulage See relay haulage.

intermediate haulage conveyor [MIN ENG] A type of conveyor, usually 500 to 3000 feet (150 to 900 meters) in length, that transports material between the gathering conveyor and the main haulage conveyor.

intermediate horizon [ELECTROMAG] Screening object (hill, mountain, ridge, building, and so on) similar to the radar horizon, but not the most distant.

intermediate host [BIOL] The host in which a parasite multiplies asexually.

intermediate-infrared radiation [ELECTROMAG] Infrared radiation having a wavelength between about 2.5 micrometers and about 50 micrometers; this range includes most molecular vibrations.

intermediate-interference ratio See intermediate-frequency response ratio.

intermediate ion [METEOROL] An atmospheric ion of size and mobility intermediate between the small ion and the large ion.

intermediate layer See sima.

intermediate lobe [ANAT] The intermediate portion of the adenohypophysis.

intermediate memory storage [ADP] An electronic device for holding working figures temporarily until needed and for releasing final figures to the output.

intermediate moraine [GEOL] A type of lateral moraine formed at the junction of two adjacent glacial lobes. Also known as interlobate moraine.

intermediate neutron [NUCLEO] A neutron having energy in a range from about 100 to 100,000 electron volts.

intermediate phase [MET] In an alloy system a distinct phase whose composition ranges do not extend to any of the pure constituents of the system.

intermediate-range ballistic missile [ORD] A missile flying a ballistic trajectory after guided powered flight and having a range of about 200 to 1500 miles (300 to 2500 kilometers). Abbreviated IRBM.

intermediate reactor [NUCLEO] A reactor in which the chain reaction is sustained mainly by intermediate neutrons.

intermediate repeater [ELECTR] Repeater for use in a trunk or line at a point other than an end.

intermediate state [CRYO] A state of partial superconductivity that occurs when a magnetic field of approximate strength is applied to a superconducting material below its critical temperature. [QUANT MECH] A state through which a system may pass during transition from an initial state to a final state.

intermediate storage [ADP] The portion of the computer storage facilities that usually stores information in the processing stage.

intermediate trunk distributing frame [ELEC] Frame which mounts terminal blocks for connecting linefinders and first selectors.

intermediate value theorem [MATH] If $f(x)$ is a continuous real-valued function on the interval from a to b, then for any y between $f(a)$ and $f(b)$ there is an x between a and b with $f(x) = y$.

intermediate vector boson [PARTIC PHYS] One of the hypothetical particles with spin quantum number 1 and negative parity, which would interact with weak currents and mediate the weak interactions in the same way that photons interact with electromagnetic currents and mediate the electromagnetic interactions.

intermedin [BIOCHEM] A hormonal substance produced by the intermediate portion of the hypophysis of certain animal species which influences pigmentation; similar to melanocyte-stimulating hormone in humans.

intermembranous ossification [HISTOL] Ossification within connective tissue with no prior formation of cartilage.

intermeningeal [ANAT] Between any two of the three meninges covering the brain and spinal cord.

intermenstrual [PHYSIO] Between periods of menstruation.

intermenstrual flow See metrorrhagia.

intermetallic compound See electron compound.

intermetameric [ANAT] Between adjacent metameres.

intermetatarsal [ANAT] Between adjacent bones of the metatarsus.

intermitotic [CYTOL] Of or pertaining to a stage of the cell cycle between two successive mitoses.

intermittent current [ELEC] A unidirectional current that flows and ceases to flow at irregular or regular intervals. [OCEANOGR] A unidirectional current interrupted at intervals.

intermittent defect [ENG] A defect that is not continuously present.

intermittent-duty rating [ENG] An output rating based on operation of a device for specified intervals of time rather than continuous duty. Also known as intermittent rating.

intermittent gas lift [PETRO ENG] A gas-drive oil reservoir that is valved and timed for intermittent activity.

intermittent light [NAV] In marine operations, a light having equal periods of light and darkness.

intermittent operation [ENG] Condition in which a device operates normally for a time, then becomes defective for a time, with the process repeating itself at regular or irregular intervals.

intermittent quick flashing light [NAV] In marine operations, a light showing flashes for several seconds, followed by an equal period of darkness.

intermittent rating See intermittent-duty rating.

intermittent scanning [ELECTR] Scans of an antenna beam at irregular intervals to increase difficulty of detection by intercept receivers.

intermittent spring [HYD] A spring that ceases flow after a long dry spell but flows again after heavy rains.

intermittent stream [HYD] A stream which carries water a considerable portion of the time, but which ceases to flow occasionally or seasonally because bed seepage and evapotranspiration exceed the available water supply.

intermittent weld [MET] A weld in which the continuity is broken by recurring unwelded spaces.

intermodulation [ELECTR] Modulation of the components of a complex wave by each other, producing new waves whose frequencies are equal to the sums and differences of integral multiples of the component frequencies of the original complex wave.

intermodulation distortion [ELECTR] Nonlinear distortion characterized by the appearance of output frequencies equal to the sums and differences of integral multiples of the input frequency components; harmonic components also present in the output are usually not included as part of the intermodulation distortion.

intermodulation interference [ELECTR] Interference that occurs when the signals from two undesired stations differ by exactly the intermediate-frequency value of a superheterodyne receiver, and both signals are able to pass through the preselector due to poor selectivity.

intermolecular force [PHYS CHEM] The force between two molecules; it is that negative gradient of the potential energy between the interacting molecules, if energy is a function of the distance between the centers of the molecules.

intermontane [GEOL] Located between or surrounded by mountains.

intermontane glacier [GEOL] A glacier that is formed by the confluence of several valley glaciers and occupies a trough between separate ranges of mountains.

intermontane trough [GEOL] 1. A subsiding area in an island arc of the ocean, lying between the stable elements of a region. 2. A basinlike area between mountains.

intermural [ANAT] Between the walls of an organ.

intermuscular [ANAT] Between muscles.

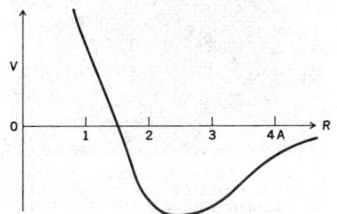

INTERMOLECULAR FORCE

General characteristics of intermolecular force curve. V = potential energy between two interacting molecules; R = distance between centers of two molecules; A = angstroms.

INTERNAL GEAR

A helical internal gear and pinion.
(Fellows Corp.)

INTERNAL VIBRATOR

Appearance of concrete after
internal vibrator has consolidated
it; vibrator appears on the right.

intermuscular hernia *See* interstitial hernia.

intermuscular septum [ANAT] A connective-tissue partition between muscles.

internal acoustic meatus [ANAT] An opening in the hard portion of the temporal bone for passage of the facial and acoustic nerves and internal auditory vessels.

internal ankle height [ANTHRO] The measure of the vertical distance from the lower end of the tibia to the floor.

internal arithmetic [ADP] Arithmetic operations carried out in a computer's arithmetic unit within the central processing unit.

internal brake [MECH ENG] A friction brake in which an internal shoe follows the inner surface of the rotating brake drum, wedging itself between the drum and the point at which it is anchored; used in motor vehicles.

internal capsule [ANAT] A layer of nerve fibers on the outer side of the thalamus and caudate nucleus, which it separates from the lenticular nucleus.

internal carotid [ANAT] A main division of the common carotid artery, distributing blood through three sets of branches to the cerebrum, eye, forehead, nose, internal ear, trigeminal nerve, dura mater, and hypophysis.

internal carotid nerve [ANAT] A sympathetic nerve which forms networks of branches around the internal carotid artery and its branches.

internal cast *See* steinkern.

internal combustion engine [MECH ENG] A prime mover in which the fuel is burned within the engine and the products of combustion serve as the thermodynamic fluid, as with gasoline and diesel engines.

internal conversion [NUC PHYS] A nuclear de-excitation process in which energy is transmitted directly from an excited nucleus to an orbital electron, causing ejection of that electron from the atom.

internal conversion coefficient *See* conversion coefficient.

internal diffusion [CHEM ENG] The diffusion of liquid or gaseous reactants to the innermost pore depths of an adsorbent-base catalyst, necessary for full catalytic effect.

internal drift current [OCEANOGR] Motion in an underlying layer of water caused by shearing stresses and friction created by current in a top layer that has different density.

internal ear *See* inner ear.

internal elastic membrane [HISTOL] A sheet of elastin found between the tunica intima and the tunica media in medium- and small-caliber arteries.

internal energy [THERMO] A characteristic property of the state of a thermodynamic system, introduced in the first law of thermodynamics; it includes intrinsic energies of individual molecules, kinetic energies of internal motions, and contributions from interactions between molecules, but excludes the potential or kinetic energy of the system as a whole; it is sometimes erroneously referred to as heat energy.

internal fertilization [PHYSIO] Fertilization of the egg within the body of the female.

internal fistula [ANAT] A fistula which has no opening through the skin.

internal floating-head exchanger [MECH ENG] Tube-and-shell heat exchanger in which the tube sheet (support for tubes) at one end of the tube bundle is free to move.

internal friction [FL MECH] *See* viscosity. [MECH] Conversion of mechanical strain energy to heat within a material subjected to fluctuating stress.

internal gas drive [PETRO ENG] A primary oil recovery process in which oil is displaced from the reservoir by the expansion of the gas originally dissolved in the liquid. Also known as dissolved-gas drive; gas depletion drive; solution gas drive.

internal gear [DES ENG] An annular gear having teeth on the inner surface of its rim.

internal granular layer [HISTOL] The fourth layer of the cerebral cortex.

internal grinder [MECH ENG] A machine designed for grinding the surfaces of holes.

internal hemorrhage [MED] Bleeding within a body cavity or organ that is concealed from an observer.

internal hernia [MED] A hernia of intraabdominal contents occurring within the abdominal cavity.

internal hydrocephaly *See* obstructive hydrocephaly.

internal iliac artery [ANAT] The medial terminal division of the common iliac artery.

internalization [PSYCH] A mental mechanism operating outside of and beyond conscious awareness by which certain external attributes, attitudes, or standards are taken within oneself.

internal loss *See* loss.

internally stored program [ADP] A sequence of instructions, stored inside the computer in the same storage facilities as the computer data, as opposed to external storage on tape, disk, drum, or cards.

internal memory *See* internal storage.

internal mix atomizer [MECH ENG] A type of pneumatic atomizer in which gas and liquid are mixed prior to the gas expansion through the nozzle.

internal oxidation [MET] The subsurface oxidation of components of an alloy due to oxygen diffusion into the metal.

internal phase *See* disperse phase.

internal pressure *See* intrinsic pressure.

internal reflectance spectroscopy *See* attenuated total reflectance.

internal resistance [ELEC] The resistance within a voltage source, such as an electric cell or generator.

internal secretion [PHYSIO] A secreted substance that is absorbed directly into the blood.

internal standard [SPECT] The principal line in spectrum analysis by the logarithmic sector method, a quantitative spectroscopy procedure.

internal storage [ADP] The total memory or storage that is accessible automatically to a computer without human intervention. Also known as internal memory.

internal storage capacity [ADP] The quantity of data that can be retained simultaneously in internal storage.

internal stress [MECH] A stress system within a solid that is not dependent on external forces. Also known as residual stress.

internal thread [DES ENG] A screw thread cut on the inner surface of a hollow cylinder.

internal vibrator [MECH ENG] A vibrating device which is drawn vertically through placed concrete to achieve proper consolidation.

internal wave [FL MECH] A wave motion of a stably stratified fluid in which the maximum vertical motion takes place below the surface of the fluid.

internal work [IND ENG] Manual work done by a machine operator while the machine is automatically operating. Also known as fill-up work; inside work.

international ampere [ELEC] The current that, when flowing through a solution of silver nitrate in water, deposits silver at a rate of 0.001118 gram per second; it has been superseded by the ampere as a unit of current, and is equal to approximately 0.999850 ampere.

international analysis code [METEOROL] An internationally recognized code for communicating details of synoptic chart analyses. Abbreviated IAC.

international angstrom [PHYS] A unit of length, equal to 1/6438.4696 of the wavelength of the red cadmium line in dry air at standard atmospheric pressure, at a temperature of 15°C containing 0.03% by volume of carbon dioxide; equal to 1.0000002 angstroms. Abbreviated IA.

international broadcasting [COMMUN] Radio broadcasting for public entertainment between different countries, on frequency bands between 5950 and 21, 750 kilohertz, assigned by international agreement.

international cable code *See* Morse cable code.

international call sign [COMMUN] Call sign assigned according to the provisions of the International Telecommunication Union to identify a radio station; the nationality of the radio station is identified by the first or the first two characters.

international candle [OPTICS] A unit of luminous intensity, now replaced by the candela; as defined in the United States, it was a specified fraction of the average luminous intensity radiated in a horizontal direction by a group of 45 carbon-filament lamps preserved at the National Bureau of Standards when the lamps were operated at a specified voltage. Also known as standard candle.

international code signal [COMMUN] Code adopted by many nations for international communications; it uses combinations of letters in lieu of words, phrases, and sentences; the letters are transmitted by the hoisting of international alphabet flags or by transmitting their dot and dash equivalents in the international Morse code. Also known as international signal code.

international control frequency bands [COMMUN] Radio-frequency bands assigned in the United States to links between stations used for international communication and their associated control centers.

international control station [COMMUN] Fixed station in the fixed public control service associated directly with the international fixed public radio communications service.

International Critical Tables [PHYS] A 7-volume series of tables of numerical data in physics, chemistry, and technology, published in 1926–30, prepared by experts who gave the "best" value which could be derived from all the data available at the time. Abbreviated ICT.

international date line [ASTRON] A jagged arbitrary line, roughly equal to the 180° meridian, where a date change occurs: if the line is crossed from east to west a day is skipped, if from west to east the same day is repeated.

international ellipsoid of reference [GEOD] The reference ellipsoid, based upon the Hayford spheroid, the semimajor axis of which is 6,378,388 meters; the flattening or ellipticity equals 1/297; by computation the semiminor axis is 6,356,-911.946 meters. Also known as international spheroid.

international fixed public radio communications service [COMMUN] Fixed service, the stations of which are open to public correspondence; this service is intended to provide radio communications between the United States and its territories and foreign or overseas points.

International Geophysical Year [GEOPHYS] An internationally accepted period, extending from July 1957 through December 1958, for concentrated and coordinated geophysical exploration, primarily of the solar and terrestrial atmospheres. Abbreviated IGY.

international henry [ELECTROMAG] A unit of electrical inductance which has been superseded by the henry, and is equal to 1.00049 henry. Also known as quadrant; secohm.

International Ice Patrol [OCEANOGR] An organization established in 1914 to protect shipping by providing iceberg warnings.

international index numbers [METEOROL] A system of designating meteorological observing stations by number, established and administered by the World Meteorological Organization; under this scheme, specified areas of the world are divided into blocks, each bearing a two-number designator; stations within each block have an additional unique three-number designator, the numbers generally increasing from east to west and from south to north.

international Morse code See continental code.

international nautical mile [NAV] A unit of length equal to 1852 meters.

international ohm [ELEC] A unit of resistance, equal to that of a column of mercury of uniform cross section that has a length of 160.3 centimeters and a mass of 14.4521 grams at the temperature of melting ice; it has been superseded by the ohm, and is equal to 1.00049 ohms.

International Polar Year [METEOROL] The years 1882 and 1932, during which participating nations undertook increased observations of geophysical phenomena in polar (mostly arctic) regions; the observations were largely meteorological, but included such as auroral and magnetic studies.

international practical temperature scale [THERMO] Temperature scale based on six points; the water triple point, the boiling points of oxygen, water, sulfur, and the solidification points of silver and gold; designated as °C, degrees Celsius, or t_{int}.

International Quiet Sun Year [GEOPHYS] An international cooperative effort, similar to the International Geophysical Year and extending through 1964 and 1965, to study the sun and its terrestrial and planetary effects during the minimum of the 11-year cycle of solar activity. Abbreviated IQSY. Also known as the International Year of the Quiet Sun.

international radio silence [COMMUN] Three-minute periods of radio silence, on the frequency of 500 kilohertz only, commencing 15 and 45 minutes after each hour, during which all marine radio stations must listen on that frequency for distress signals of ships and aircraft.

international rules of the road [NAV] In marine operations, international regulations for preventing collisions at sea.

international signal code See international code signal.

international spheroid See international ellipsoid of reference.

international standard annealed copper [MET] An annealed pure copper having a resistivity of 1.7241 microhm-centimeter at 20°C, which is taken as 100% conductivity.

international synoptic code [METEOROL] A synoptic code approved by the World Meteorological Organization in which the observable meteorological elements are encoded and transmitted in words of five numerical digits length.

international system of electrical units [ELEC] System of electrical units based on agreed fundamental units for the ohm, ampere, centimeter, and second, in use between 1893 and 1947, inclusive; in 1948, the Giorgi, or meter-kilogram-second-absolute system, was adopted for international use.

International System of Units [PHYS] A system of physical units in which the fundamental quantities are length, time, mass, electric current, temperature, luminous intensity, and amount of substance, and the corresponding units are the meter, second, kilogram, ampere, kelvin, candela, and mole; it has been given official status and recommended for universal use by the General Conference on Weights and Measures. Also known as (in French) as Système International d'Unites. Abbreviated SI (in all languages).

international table British thermal unit See British thermal unit.

international table calorie See calorie.

international telecommunications service [COMMUN] Telecommunications service between offices or stations of different states, or between mobile stations which are not in the same state, or are subject to different states.

international telegraph alphabet See CCIT2 code.

International Telegraphic Consultative Committee code 2 See CCIT2 code.

international thread [DES ENG] A standardized metric system in which the pitch and diameter of the thread are related, with the thread having a rounded root and flat crest.

international unit [BIOL] A quantity of a vitamin, hormone, antibiotic, or other biological that produces a specific internationally accepted biological effect.

international volt [ELEC] A unit of potential difference or electromotive force, equal to 1/1.01858 of the electromotive force of a Weston cell at 20°C; it has been superseded by the volt, and is equal to 1.00034 volts.

International Year of the Quiet Sun See International Quiet Sun Year.

internode [BIOL] The interval between two nodes, as on a stem or along a nerve fiber.

internuclear [SCI TECH] Located between nuclei.

internuclear distance [PHYS CHEM] The distance between two nuclei in a molecule.

interoceptor [PHYSIO] A sense receptor located in visceral organs and yielding diffuse sensations.

interocular diameter [ANTHRO] The measure of the distance between the internal canthi.

interoffice trunk [COMMUN] A direct trunk between local central offices in the same exchange.

interorbital [ANAT] Between the orbits of the eyes.

interparietal [ANAT] Between the parietal bones.

interparietal hernia See interstitial hernia.

interpass temperature [MET] In a multiple-pass weld, the lowest temperature of the deposited weld metal before the next run is started.

interpenetration twin [CRYSTAL] Two or more individual crystals so twinned that they appear to have grown through one another. Also known as penetration twin.

interphase [CYTOL] Also known as interkinesis. **1.** The period between succeeding mitotic divisions. **2.** The period between the first and second meiotic divisions in those organisms where nuclei are reconstituted at the end of the first division.

interphase transformer [ELECTR] Autotransformer or a set of mutually coupled reactors used in conjunction with three-phase rectifier transformers to modify current relations in the

INTERNATIONAL DATE LINE

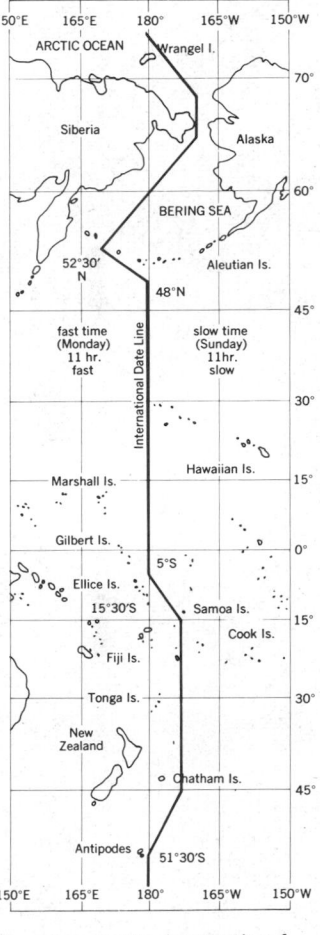

Map showing the international date line.

rectifier system to increase the number of anodes of different phase relations which carry current at any instant.

interphone [COMMUN] An intercommunication system using headphones and microphones for communication between adjoining or nearby studios or offices, or between crew locations on an aircraft, vessel, or tank or other vehicle. Also known as talk-back circuit.

interplanetary dust [ASTRON] Dust particles between the planets.

interplanetary flight [AERO ENG] Flight to the moon or to a planet.

interplanetary magnetic field [ASTROPHYS] The magnetic field between the planets.

interplanetary medium [ASTRON] That part of space containing electromagnetic radiation, dust, gas, and plasma between the planets.

interplanetary probe [AERO ENG] An instrumented spacecraft that flies close to or lands upon a planet.

interplanetary space [ASTRON] The region that extends beyond near-space away from earth to the other planets in the solar system.

interplanetary spacecraft [AERO ENG] A spacecraft designed for flight between planets.

interplanetary transfer orbit [AERO ENG] An elliptical trajectory tangent to the orbits of both the departure planet and the target planet.

interpluvial [GEOL] Pertaining to an episode or period of geologic time that was dryer than the pluvial period occurring before or after it.

interpolation [MATH] A process used to estimate an intermediate value of one (dependent) variable which is a function of a second (independent) variable when values of the dependent variable corresponding to several discrete values of the independent variable are known.

interpole *See* commutating pole.

interposition trunk [COMMUN] Trunk which connects two positions of a large switchboard so that a line on one position can be connected to a line on another position.

interpret [ADP] To print on a punched card the infomation punched in that card.

interpreter [ADP] **1.** A program that translates and executes each source program statement before proceeding to the next one. Also known as interpretive routine. **2.** A machine that senses a punched card and prints the punched information on that card. Also known as punched-card interpreter.

interpretive code *See* interpretive language.

interpretive language [ADP] A computer programming language in which each instruction is immediately translated and acted upon by the computer, as opposed to a compiler which decodes a whole program before a single instruction can be executed. Also known as interpretive code; pseudocode.

interproglottid gland [INV ZOO] Any of a number of cell clusters or glands arranged transversely along the posterior margin of the proglottids of certain tapeworms.

interpterygoid [ZOO] A space between palatal plates in certain chordates.

interpulmonary [ANAT] Located between the lungs.

interpupillary diameter [ANTHRO] A measure of the distance between the centers of the pupils, as the subject looks straight ahead.

interquartile range [STAT] The distance between the top of the lower quartile and the bottom of the upper quartile of a distribution.

interradial canal [INV ZOO] Any of the radially arranged gastrovascular canals in certain jellyfishes and ctenophores.

interradius [INV ZOO] The area between two adjacent arms in echinoderms.

interray [INV ZOO] A division of the radiate body of echinoderms.

interrecord gap *See* record gap.

interrenal [ANAT] Located between the kidneys.

interrogation [COMMUN] The transmission of a radio-frequency pulse, or combination of pulses, intended to trigger a transponder or group of transponders, a racon system, or an IFF system, in order to elicit an electromagnetic reply. Also known as challenging signal.

interrogation suppressed time delay [COMMUN] Overall fixed time delay between transmission of an interrogation and reception of the reply to this interrogation at zero distance.

interrogator [ELECTR] **1.** A radar transmitter which sends out a pulse that triggers a transponder; usually combined in a single unit with a responsor, which receives the reply from a transponder and produces an output suitable for actuating a display of some navigational parameter. Also known as challenger; interrogator-transmitter. **2.** *See* interrogator-responsor.

interrogator-responsor [ELECTR] A transmitter and receiver combined, used for sending out pulses to interrogate a radar beacon and for receiving and displaying the resulting replies. Also known as interrogator.

interrogator-transmitter *See* interrogator.

interrupt [ADP] **1.** To stop a running program in such a way that it can be resumed at a later time, and in the meanwhile permit some other action to be performed. **2.** The action of such a stoppage.

interrupted continuous wave [COMMUN] A continuous wave that is interrupted at a constant audio-frequency rate high enough to give several interruptions for each keyed code dot. Abbreviated ICW.

interrupted current [ELEC] A current produced by opening and closing at regular intervals a circuit that would otherwise carry a steady current or one that varied continuously with time.

interrupted dc tachometer [ENG] A type of impulse tachometer in which the frequency of pulses generated by the interrupted direct current of an ignition-circuit primary of an internal combustion engine is used to measure the speed of the engine.

interrupted fire [ORD] Automatic fire delivered in short series of bursts.

interrupted quick flashing light [NAV] In marine operations, a light showing quick flashes for several seconds, followed by a period of darkness.

interrupted screw [DES ENG] A screw with longitudinal grooves cut into the thread, and which locks quickly when inserted into a similar mating part.

interrupter [ELEC] An electric, electronic, or mechanical device that periodically interrupts the flow of a direct current so as to produce pulses. [ORD] A barrier in a fuse which prevents transmission of an explosive effect to some element beyond the interrupter.

interrupter gear [ORD] A synchronizing gear for machine guns, so called because it interrupts the firing mechanism of the gun or guns to allow a propeller blade to pass the muzzle.

interrupter vibrator [ELEC] A mechanical device used to change direct current to alternating current.

interrupting capacity [ELEC] Maximum power in the arc that a circuit breaker or fuse can successfully interrupt without restrike or violent failure; rated in volt-amperes for alternating-current circuits and watts for direct-current circuits.

interrupt mask [ADP] A technique of suppressing certain interrupts and allowing the control program to handle these masked interrupts at a later time.

interrupt signal [ADP] A control signal which requests the immediate attention of the central processing unit.

interrupt system [ADP] The means of interrupting a program and proceeding with it at a later time; this is accomplished by saving the contents of the program counter and other specific registers, storing them in reserved areas, starting the new instruction sequence, and upon completion, reloading the program counter and registers to return to the original program, and reenabling the interrupt.

interrupt trap [ADP] A program-controlled technique which either recognizes or ignores an interrupt, depending upon a switch setting.

interscapular [ANAT] Between the shoulders or shoulder blades.

intersect [ENG] To find a position by the triangulation method. [SCI TECH] To pass through or across.

intersection [CIV ENG] **1.** A point of junction or crossing of two or more roadways. **2.** A surveying method in which a plane table is used alternately at each end of a measured baseline. [MATH] The intersection of two sets is the set consisting of all elements common to both of the two sets.

intersection angle [CIV ENG] The angle of deflection at the intersection point between the straights of a railway or highway curve.

intersection point [CIV ENG] That point where two straights or tangents to a railway or road curve would meet if extended.

intersegmental [BIOL] Situated between or involving segments. [EMBRYO] Situated between the primordial segments of the embryo.

intersegmental reflex [PHYSIO] An unconditioned reflex arc connecting input and output by means of afferent pathways in the dorsal spinal roots and efferent pathways in the ventral spinal roots.

intersertal [PETR] Referring to the texture of a porphyritic igneous rock in which the groundmass forms a small proportion of the rock, filling the interstices between unoriented feldspar laths.

intersex [PHYSIO] An individual who is intermediate in sexual constitution between male and female.

interspace [ANAT] An interval between the ribs or the fibers or lobules of a tissue or organ. [BUILD] An air space. [PHYS] An interval of space or time.

interspersion [ECOL] **1.** An intermingling of different organisms within a community. **2.** The level or degree of intermingling of one kind of organism with others in the community.

interspinal [ANAT] Situated between or connecting spinous processes: interspinous.

interstadial [GEOL] Pertaining to a period during a glacial stage in which the ice retreated temporarily.

interstage transformer [ELECTR] A transformer used to provide coupling between two stages.

interstation noise suppression *See* intercarrier noise suppression.

interstellar [ASTRON] Between the stars.

interstellar communication [COMMUN] Hypothetical communication between beings in different stellar systems.

interstellar lines [ASTRON] Dark, narrow lines in the spectra of stars, caused by absorption of radiation by a gaseous medium in space.

interstellar matter [ASTRON] The gaseous and dust material throughout the galaxy.

interstellar probe [AERO ENG] An instrumentated spacecraft propelled beyond the solar system to obtain specific information about interstellar environment.

interstellar space [ASTRON] The space between the stars and other celestial bodies.

interstellar travel [AERO ENG] Space flight between stars.

intersternite [INV ZOO] An intersegmental plate on the ventral surface of the abdomen in insects.

interstice [SOLID STATE] A space or volume between atoms of a lattice, or between groups of atoms or grains of a solid structure.

interstitial [CRYSTAL] A crystal defect in which an atom occupies a position between the regular lattice positions of a crystal. [SCI TECH] Of, pertaining to, or situated in a space between two things.

interstitial atom [CRYSTAL] A displaced atom which is forced into a nonequilibrium site within a crystal lattice.

interstitial cell [HISTOL] A cell that is not peculiar to or characteristic of a particular organ or tissue but which comprises fibrous tissue binding other cells and tissue elements; examples are neuroglial cells and Leydig cells.

interstitial cell–stimulating hormone *See* luteinizing hormone.

interstitial-cell tumor [MED] A benign tumor of the testes composed of Leydig cells. Also known as interstitioma; Leydig-cell tumor.

interstitial emphysema [MED] Escape of air from the alveoli into the interstices of the lung, commonly due to trauma or violent cough.

interstitial endometrosis [MED] The presence of endometrial tissue in the form of the stroma thoughout the myometrium. Also known as endolymphatic stromomyosis; fibromyosis; parathelioma; stromal endometriosis; stromal myosis; stromatosis.

interstitial gland [HISTOL] **1.** Groups of Leydig cells which secrete angiogens. **2.** Groups of epithelioid cells in the ovarian medulla of some lower animals.

interstitial hepatitis [MED] Pathologic deterioration and death of parenchymal cells in the liver associated with infiltration of lymphocytes and monocytes in the portal canals. Also known as acute nonsuppurative hepatitis; nonspecific hepatitis.

interstitial hernia [MED] Protrusion of the intestine between the muscular planes of the abdominal wall. Also known as intermuscular hernia; interparietal hernia.

interstitial implants [MED] Solid or encapsulated radiation sources, made in the form of seeds, wires, or other shapes, to be inserted directly into tissue that is to be irradiated.

interstitial inflammation [MED] Inflammation of the interstitial tissues of an organ.

interstitial keratitis [MED] Inflammation of the cornea in which the iris is almost completely hidden by the diffuse haziness of the corneal tissue.

interstitial lamella [HISTOL] Any of the layers of bone between adjacent Haversian systems.

interstitial myocarditis [MED] Inflammation of the myocardium accompanied by cellular infiltration of interstitial tissues. Also known as Fiedler's myocarditis.

interstitial nephritis *See* pyelonephritis.

interstitial plasma-cell pneumonia *See* Pneumocystis carinii pneumonia.

interstitial pneumonia [MED] Inflammation of the lungs, particularly the stroma, including peribronchial tissue and interalveolar septa.

interstitial radiation [MED] Radiation of tissues by implantation of a radioactive source material.

interstitial water [HYD] Subsurface water contained in pore spaces between the grains of rock and sediments.

interstitial water saturation [HYD] The water content of a subterranean reservoir formation.

interstitioma *See* interstitial-cell tumor.

intersymbol interference [COMMUN] In a transmission system, extraneous energy from the signal in one or more keying intervals which tends to interfere with the reception of the signal in another keying interval, or the disturbance which results.

intertergite [INV ZOO] One of the small plates between the tergites of certain insects.

interterminal switching [CIV ENG] Moving railroad cars from one line to another within a switching area.

intertidal zone [OCEANOGR] The part of the littoral zone above low-tide mark.

intertoll trunk [COMMUN] A trunk between toll offices in different telephone exchanges.

intertropical convergence zone [METEOROL] The axis, or a portion thereof, of the broad trade-wind current of the tropics; this axis is the dividing line between the southeast trades and the northeast trades (of the Southern and Northern hemispheres, respectively). Also known as equatorial convergence zone; meteorological equator.

intertropical front [METEOROL] The interface or transition zone occurring within the equatorial trough between the Northern and Southern hemispheres. Also known as equatorial front; tropical front.

intertubercular sulcus [ANAT] A deep groove on the anterior surface of the upper end of the humerus, separating the greater and lesser tubercles; contains the tendon of the long head of the biceps brachii muscle. Also known as bicipital groove.

interurban [GEOGR] Connecting or extending between urban areas.

interval [ACOUS] The spacing in pitch or frequency between two sounds; the frequency interval is the ratio of the frequencies or the logarithm of this ratio. [MATH] A set of numbers which consists of those numbers that are greater than one fixed number and less than another, and that may also include one or both of the end numbers. [PHYS] The time separating two events, or the distance between two objects. [RELAT] **1.** In special relativity, the Lorentz invariant quantity $c^2(\Delta t)^2 - (\Delta x)^2 - (\Delta y)^2 - (\Delta z)^2$, where c is the speed of light, Δt is the difference in the time coordinates of two specified events, and Δx, Δy, and Δz are the differences in their x, y, and z coordinates, respectively. **2.** In general relativity, a generalization of this concept, namely the sum over the indices μ and

ν of $g_{\mu\nu}dx^\mu dx^\nu$, where dx^μ and dx^ν are the differences in the x^μ and x^ν coordinates of two specified neighboring events, and $g_{\mu\nu}$ is an element of the metric tensor.

intervallum [INV ZOO] The space between the walls of pleosponges.

intervalometer [ORD] An electrical device used in bombing, by which data is preset in order to drop a desired number of bombs at a constant predetermined interval.

interval timer [ENG] A device which operates a set of contacts during a preset time interval and, at the end of the interval, returns the contacts to their normal positions. Also known as timer.

intervascular [ANAT] Located between or surrounded by blood vessels.

interventricular foramen [ANAT] Either one of the two foramens that connect the third ventricle of the brain with each lateral ventricle. Also known as foramen of Monro.

interventricular septal defect [MED] A congenital malformation of the septum between the ventricles of the heart.

interventricular septum [ANAT] The muscular wall between the heart ventricles. Also known as ventricular septum.

intervertebral [ANAT] Being or located between the vertebrae.

intervillous spaces [HISTOL] Spaces in the placenta which communicate with the maternal blood vessels.

intestinal crura [INV ZOO] The main intestinal branches in certain trematodes.

intestinal digestion [PHYSIO] Conversion of food to an assimilable form by the action of intestinal juices.

intestinal dyspepsia [MED] Disturbed digestion due to diminished secretion of intestinal juices or lack of tonus in the wall of the intestine.

intestinal hormone [BIOCHEM] Either of two hormones, secretin and cholecystokinin, secreted by the intestine.

intestinal juice [PHYSIO] An alkaline fluid composed of the combined secretions of all intestinal glands.

intestinal lipodystrophy See Whipple's disease.

intestinal villi [ANAT] Fingerlike projections of the small intestine, composed of a core of vascular tissue covered by epithelium and containing smooth muscle cells and an efferent lacteal end capillary.

intestine [ANAT] The tubular portion of the vertebrate digestive tract, usually between the stomach and the cloaca or anus.

in-the-seam mining [MIN ENG] The usual method of mining characterized by the driving of development shafts into the coal seam.

in the wind [NAV] In the direction from which the wind is blowing; windward; used particularly in reference to a heading or to the position of an object.

intima [HISTOL] The innermost coat of a blood vessel. Also known as tunica intima.

intorsion [BIOL] Inward rotation of a structure about a fixed point or axis.

into the solid [MIN ENG] Of a shot, going into the coal beyond the point to which the coal can be broken by the blast. Also known as on the solid.

intoxication [MED] **1.** Poisoning. **2.** The state produced by overindulgence in alcohol.

intraabdominal [ANAT] Occurring or being within the cavity of the abdomen.

intraatrial heart block [MED] A type of heart block which shows a broad, notched P wave of longer than normal duration on the electrocardiographic record.

intracartilaginous ossification See endochondral ossification.

intracavernous aneurysm [MED] A dilation of the wall of the internal carotid artery within the cavernous sinus.

intracellular [CYTOL] Within a cell.

intracellular canaliculi [CYTOL] A system of minute canals within certain gland cells which are thought to drain the glandular secretions.

intracellular digestion [PHYSIO] Digestion which takes place within the cytoplasm of the organism, as in many unicellular protozoans.

intracellular enzyme [BIOCHEM] An enzyme that remains active only within the cell in which it is formed. Also known as organized ferment.

intracellular symbiosis [CYTOL] Existence of a self-duplicating unit within the cytoplasm of a cell, such as a kappa particle in *Paramecium*, which seems to be an infectious agent and may influence cell metabolism.

intracervical [ANAT] Located within the cervix of the uterus.

intracistron complementation [GEN] The process whereby two different mutant alleles, each of which determines in homozygotes an inactive enzyme, determine the formation of the active enzyme when present in the same nucleus.

intraclast [GEOL] A fragment of limestone formed by erosion within a basin of deposition and redeposited there to form a new sediment.

intracloud discharge See cloud discharge.

Intracoastal Waterway [NAV] An inside protected route extending through New Jersey; from Norfolk, Virginia to Key West, Florida; across Florida from St. Lucie Inlet to Fort Myers, Charlotte Harbor, Tampa Bay, and Tarpon Springs; and from Carabelle, Florida, to Brownsville, Texas.

Intracoastal Waterway charts [NAV] Charts of scale 1:40,000, for navigating the Intracoastal Waterway, and designed especially for small-boat and yacht operators.

intracortical [ANAT] Occurring or located within the cortex.

intracranial [ANAT] Within the cranium.

intracranial aneurysm [MED] Dilation of a cerebral artery.

intracranial angiography [MED] Roentgenography of the blood vessels within the cranial cavity following the intravascular injection of a radiopaque material.

intracratonic basin See autogeosyncline.

intracrystalline See transcrystalline.

intracytoplasmic [CYTOL] Being or occurring within the cytoplasm of a cell.

intradermal [ANAT] Within the skin.

intradermal nevus [MED] A lesion containing melanocytes and located principally or completely within the derma.

intrados [ARCH] The inner curved surface of an arch.

intraductal [ANAT] Within a duct.

intradural [ANAT] Within the dura mater.

intraembryonic [EMBRYO] Within the embryo.

intraepidermal [ANAT] Within the epidermis.

intraepidermal epithelioma [MED] Carcinoma in situ, of either the squamous-cell or basal-cell type.

intraepithelial [ANAT] Within the epithelium.

intraesophageal [ANAT] Within the esophagus.

intrafascicular [BOT] Located or occurring within a vascular bundle.

intraformational breccia [PETR] A rock resulting from cracking and desiccation-shrinking of a mud after withdrawal of water followed by almost contemporaneous sedimentation.

intraformational conglomerate [GEOL] **1.** A conglomerate in which clasts and the matrix are contemporaneous in origin. **2.** A conglomerate formed in the midst of a geologic formation.

intrafusal fiber [HISTOL] Any of the striated muscle fibers contained in a muscle spindle.

intragenic [MOL BIO] Within a gene, in referring to certain events.

intragenic recombination [MOL BIO] Recombination occurring between the mutons of one cistron.

intragenic suppression [GEN] The restoration of a suppressed function or character as a consequence of a second mutation located within the same gene as the original or first mutation.

intrahepatic [ANAT] Within the liver.

intrajugular process [ANAT] **1.** A small, curved process on some occipital bones which partially or completely divides the jugular foramen into lateral and medial parts. **2.** A small process on the hard portion of the temporal bone which completely or partly separates the jugular foramen into medial and lateral parts.

intralaminar nuclei [ANAT] A diffuse group of nuclei located in the internal medullary lamina of the thalamus.

intraline distance [ENG] The minimum distance permitted between any two buildings within an explosives operating line; to protect buildings from propagation of explosions due to blast effect.

intraluminal [ANAT] Within the lumen of a structure.

intramarginal [BIOL] Within a margin.

intramedullary [ANAT] **1.** Within the tissues of the spinal cord or medulla oblongata. **2.** Within the bone marrow. **3.** Within the adrenal medulla.

intramembranous [HISTOL] Formed or occurring within a membrane.

intramembranous ossification [HISTOL] Formation of bone tissue directly from connective tissue without a preliminary cartilage stage.

intramural [ANAT] Within the substance of the walls of an organ.

intramuscular [ANAT] Lying within or going into the substance of a muscle.

intraocular [ANAT] Occurring within the globe of the eye.

intraocular pressure [PHYSIO] The hydrostatic pressure within the eyeball.

intraparietal [ANAT] **1.** Within the wall of an organ or cavity. **2.** Within the parietal region of the cerebrum. **3.** Within the body wall.

intrapartum [MED] Occurring during parturition.

intraperitoneal [ANAT] **1.** Within the peritoneum. **2.** Within the peritoneal cavity.

intrapetiolar [BOT] **1.** Enclosed by the base of the petiole. **2.** Between the petiole and the stem.

intrapulmonic [ANAT] Being or occurring within the lungs.

intraspecific [BIOL] Being within or occurring among the members of the same species.

intraspinal block [MED] Anesthesia of the spinal column by injection of an anesthetic into the spinal canal.

intrastratal solution [GEOCHEM] A chemical attrition of the constituents of a rock after deposition.

intratelluric [GEOL] **1.** Pertaining to a phenocryst that is formed earlier than its matrix. **2.** Pertaining to a period in which igneous rocks crystallized prior to their eruption. **3.** Located, formed, or originating at great depths within the earth.

intrathecal [ANAT] Within the subarachnoid space.

intrathoracic [ANAT] Within the thoracic cavity.

intratracheal [ANAT] Being or occurring within the trachea.

intrauterine [ANAT] Being or occurring within the uterus.

intravaginal [ANAT] **1.** Being or occurring within the vagina. **2.** Located within a tendon sheath. [BOT] Located within a sheath, referring to branches of grass.

intravenous [ANAT] Located within, or going into, the veins.

intraventricular heart block [MED] Prolongation of the process of ventricular excitation, measured by the QRS complex of the electrocardiogram. Also known as arborization block; parietal block; peri-infarction block.

intravesical [ANAT] Within the urinary bladder.

intravital [BIOL] Occurring while the cell or organism is alive.

intravital stain [BIOL] A nontoxic dye injected into the body to selectively stain certain cells or tissues.

intrazonal soil [GEOL] A group of soils with well-developed characteristics that reflect the dominant influence of some local factor of relief, parent material, or age over the usual effect of vegetation and climate.

intrinsic asthma [MED] Asthma caused by a respiratory tract infection.

intrinsic-barrier diode [ELECTR] A *pin* diode, in which a thin region of intrinsic material separates the *p*-type region and the *n*-type region.

intrinsic-barrier transistor [ELECTR] A *pnip* or *npin* transistor, in which a thin region of intrinsic material separates the base and collector.

intrinsic conductivity [SOLID STATE] The conductivity of a semiconductor or metal in which impurities and structural defects are absent or have a very low concentration.

intrinsic contact potential difference [ELEC] True potential difference between two perfectly clean metals in contact.

intrinsic equations of a curve [MATH] The equations describing the radius of curvature and torsion of a curve as a function of arc length; these equations determine the curve up to its position in space.

intrinsic factor [BIOCHEM] A substance, produced by the stomach, which combines with the extrinsic factor (vitamin B_{12}) in food to yield an antianemic principle; lack of the intrinsic factor is believed to be a cause of pernicious anemia. Also known as Castle's intrinsic factor.

intrinsic flux density *See* intrinsic induction.

intrinsic geometry of a surface [MATH] The description of the intrinsic properties of a surface.

intrinsic induction [ELECTROMAG] The vector difference between the magnetic flux density at a given point and the magnetic flux density which would exist there, for the same magnetic field strength, if the point were in a vacuum. Symbol B_i. Also known as intensity of magnetization; intrinsic flux density; magnetic polarization.

intrinsic luminosity [ASTROPHYS] The total amount of radiation emitted by a star over a specified range of wavelengths.

intrinsic mobility [SOLID STATE] The mobility of the electrons in an intrinsic semiconductor.

intrinsic nerve supply [ANAT] The nerves contained entirely within an organ or structure.

intrinsic parity [PARTIC PHYS] A quantum number, equal to $+1$ or -1, which is assigned to particles so that the product of the intrinsic parities of the particles composing a system times the parity of the system's wave function yields the total parity.

intrinsic photoemission [SOLID STATE] Photoemission which can occur in an ideally pure and perfect crystal, in contrast to other types of photoemission which are associated with crystal defects.

intrinsic pressure [PHYS] Pressure in a fluid resulting from inward forces on molecules near the fluid surface, caused by attraction between molecules. Also known as internal pressure.

intrinsic properties of a surface [MATH] A property of a surface which can be described without reference to the surrounding space.

intrinsic semiconductor [SOLID STATE] A semiconductor in which the concentration of charge carriers is characteristic of the material itself rather than of the content of impurities and structural defects of the crystal.

intrinsic temperature range [SOLID STATE] In a semiconductor, the temperature range in which its electrical properties are essentially not modified by impurities or imperfections within the crystal.

intrinsic variable star [ASTRON] A star that is variable not because of an eclipse.

intrinsic viscosity [PHYS CHEM] The ratio of a solution's specific viscosity to the concentration of the solute, extrapolated to zero concentration. Also known as limiting viscosity number.

introductory column [MIN ENG] The highest and first column that is inserted in casing a borehole.

introfaction [CHEM] Change in fluidity and specific wetting properties (for impregnation acceleration) of an impregnating compound, caused by an introfier (impregnation accelerator).

introgressive hybridization [GEN] The spreading of genes of a species into the gene complex of another due to hybridization between numerically dissimilar populations, the extensive backcrossing preventing formation of a single stable population.

introitus [ANAT] An opening or entryway, especially the opening into the vagina.

introjection [PSYCH] The symbolic absorption into and toward oneself of concepts and feelings generated toward another person or object; motivates irrational behavior toward oneself.

intromission [ZOO] The act or process of inserting one body into another, specifically, of the penis into the vagina.

intromittent [ZOO] Adapted for intromission; applied to a copulatory organ.

introrse [BIOL] Turned inward or toward the axis.

introspective diplopia *See* physiologic diplopia.

introversion [MED] The act or process of turning in upon itself, as a hollow organ. [PSYCH] Preoccupation with the self associated with diminished interest in external events.

introvert [PSYCH] An individual whose interests are self-directed, and not directed toward the outside world. [ZOO] **1.**

A structure capable of introversion. **2.** To turn inward.

intrusion [GEOL] **1.** The prócess of emplacement of magma in preexisting rock. Also known as injection; invasion; irruption. **2.** A large-scale sedimentary injection. Also known as sedimentary intrusion. **3.** Any rock mass formed by an intrusive process.

intrusion grouting [ENG] A method of placing concrete by intruding the mortar component in position and then converting it into concrete as it is introduced into voids.

intrusive [PETR] Pertaining to material forced while still in a fluid state into cracks or between layers of rock.

intubation [MED] The introduction of a tube into a hollow organ to keep it open, especially into the larynx to ensure the passage of air.

intumescent agent [MATER] A water-repellent coating that swells and snuffs out a fire when it is heated.

intussusception [MED] Passing of a portion of a structure into another part of the same structure, such as the invagination of a part of the intestine.

inulase [BIOCHEM] An enzyme produced by certain molds that catalyzes the conversion of inulin to levulose.

inulin [BIOCHEM] A polysaccharide made up of polymerized fructofuranose units.

inunction [MED] Act of applying an oil or fatty material, especially rubbing an ointment into the skin as a therapeutic measure.

inundation [HYD] Flooding, by the rise and spread of water, of a land surface that is not normally submerged.

in utero [EMBRYO] Within the uterus, referring to the fetus.

in vacuo [PHYS] In a vacuum.

invaded zone [PETRO ENG] Transitional downhole area between the area invaded completely by drilling mud and uncontaminated bulk of the reservoir.

invagination [EMBRYO] The enfolding of a part of the wall of the blastula to form a gastrula. [PHYSIO] **1.** The act of ensheathing or becoming ensheathed. **2.** The process of burrowing or enfolding to form a hollow space within a previously solid structure, as the invagination of the nasal mucosa within a bone of the skull to form a paranasal sinus.

Invar [MET] A low-expansion alloy consisting of 36% nickel, 0.35% manganese, and the balance iron with carbon and other elements at a minimum; used for watch parts and the measuring guides of accurate instruments.

invariable plane [MECH] A plane which is perpendicular to the angular momentum vector of a rotating rigid body not subject to external torque, and which is always tangent to its inertia ellipsoid.

invariance [MATH] *See* invariant property. [PHYS] The property of a physical quantity or physical law of being unchanged by certain transformations or operations, such as reflection of spatial coordinates, time reversal, charge conjugation, rotations, or Lorentz transformations. Also known as symmetry.

invariance principle [PHYS] Any principle which states that a physical quantity or physical law possesses invariance under certain transformations. Also known as symmetry law; symmetry principle. [RELAT] In general relativity, the principle that the laws of motion are the same in all frames of reference, whether accelerated or not.

invariant property [MATH] A mathematical property of some space unchanged after the application of any member from some given family of transformations. Also known as invariance.

invariant subgroup *See* normal subgroup.

invasion [GEOL] **1.** The movement of one material into a porous reservoir area that has been occupied by another material. **2.** *See* intrusion; transgression. [MED] The phase of an infectious disease during which the pathogen multiplies and is distributed; precedes signs and symptoms. **2.** The process by which microorganisms enter the body.

invasion efficiency [PETRO ENG] Completeness of invasion of a reservoir formation by a fluid.

inventory control [IND ENG] Systematic management of the balance on hand of inventory items, involving the supply, storage, distribution, and recording of items.

inverna [METEOROL] A southeast wind of Lake Maggiore, Italy.

inverse beta decay [NUC PHYS] A reaction providing evidence for the existence of the neutrino, in which an antineutrino (or neutrino) collides with a proton (or neutron) to produce a neutron (or proton) and a positron (or electron).

inverse cam [MECH ENG] A cam that acts as a follower instead of a driver.

inverse Compton effect [QUANT MECH] A process in which relativistic particles give up some of their energy to long-wavelength radiation, converting it to shorter-wavelength radiation.

inverse current [ELECTR] The current resulting from an inverse voltage in a contact rectifier.

inverse cylindrical orthomorphic projection *See* transverse Mercator projection.

inverse electrode current [ELECTR] Current flowing through an electrode in the direction opposite to that for which the tube is designed.

inverse element [MATH] In a group G the inverse of an element g is the unique element g^{-1} such that $g \cdot g^{-1} = g^{-1} \cdot g = e$, where $\cdot$ denotes the group operation and e is the identity element.

inverse feedback *See* negative feedback.

inverse function [MATH] An inverse function for a function f is a function g whose domain is the range of f and whose range is the domain of f with the property that both f composed with g and g composed with f give the identity function.

inverse function theorem [MATH] If f is a continuously differentiable function of euclidean n-space to itself and at a point x_0 the matrix with the entry $\partial f_i / \partial x_j^{(x_0)}$ in the ith row and jth column is nonsingular, then there is a continuously differentiable function $g(y)$ defined in a neighborhood of $f(x_0)$ which is an inverse function for $f(x)$ at all points near x_0.

inverse hour *See* inhour.

inverse latitude *See* transverse latitude.

inverse limiter [ELECTR] A transducer, the output of which is constant for input of instantaneous values within a specified range and a linear or other prescribed function of the input for inputs above and below that range.

inverse logarithm of a number *See* antilogarithm of a number.

inverse matrix [MATH] The inverse of a nonsingular matrix A is the matrix A^{-1} where $A \cdot A^{-1} = A^{-1} \cdot A = I$, the identity matrix.

inverse Mercator chart *See* transverse Mercator chart.

inverse Mercator projection *See* transverse Mercator projection.

inverse network [ELEC] Two 2-terminal networks are said to be inverse when the product of their impedances is independent of frequency within the range of interest.

inverse neutral telegraph transmission [COMMUN] Form of transmission in which marking signals are zero current intervals and spacing signals are current pulses of either polarity.

inverse of a number [MATH] The additive inverse of a real or complex number a is the number which when added to a gives 0; the multiplicative inverse of a is the number which when multiplied with a gives 1.

inverse operator [MATH] The inverse of an operator L is the operator which is the inverse function of L.

inverse parallel *See* transverse parallel.

inverse peak voltage [ELECTR] **1.** The peak value of the voltage that exists across a rectifier tube or x-ray tube during the half cycle in which current does not flow. **2.** The maximum instantaneous voltage value that a rectifier tube or x-ray tube can withstand in the inverse direction (with anode negative) without breaking down and becoming conductive.

inverse points [MATH] A pair of points lying on a diameter of a circle or sphere such that the product of the distances of the points from the center equals the square of the radius.

inverse ranks [STAT] Ranking responses to treatments from largest response to smallest response.

inverse rhumb line *See* transverse rhumb line.

inverse segregation [MET] Segregation in a cast metal in which an excess of lower-melting metal occurs in the earlier-freezing portions because liquid metal enters cavities developed in the earlier-solidified metal.

inverse-square law [PHYS] Any law in which a physical

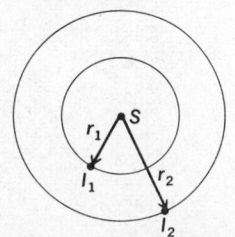

INVERSE-SQUARE LAW

Point source S emitting energy of intensity I. The inverse-square law states that $I_1 / I_2 = r_2^2 / r_1^2$.

quantity varies with distance from a source inversely as the square of that distance.

inverse Stark effect [ATOM PHYS] The Stark effect as observed with absorption lines, in contrast to emission lines.

inverse voltage [ELECTR] The voltage that exists across a rectifier tube or x-ray tube during the half cycle in which the anode is negative and current does not normally flow.

inversion [CHEM] Change of a compound into an isomeric form. [COMMUN] The process of scrambling speech for secrecy by beating the voice signal with a fixed, higher audio frequency and using only the difference frequencies. [CRYSTAL] A change from one crystal polymorph to another. Also known as transformation. [GEN] A type of chromosomal rearrangement in which two breaks take place in a chromosome and the fragment between breaks rotates 180° before rejoining. [GEOL] 1. Development of inverted relief through which anticlines are transformed into valleys and synclines are changed into mountains. 2. The occupancy by a lava flow of a ravine or valley that occurred in the side of a volcano. [MECH ENG] The conversion of basic four-bar linkages to special motion linkages, such as parallelogram linkage, slider-crank mechanism, and slow-motion mechanism by successively holding fast, as ground link, members of a specific linkage (as drag link). [MED] The act or process of turning inward or upside down. [METEOROL] A departure from the usual decrease or increase with altitude of the value of an atmospheric property, most commonly temperature. [PHYS] The simultaneous reflection of all three directions in space, so that each coordinate is replaced by the negative of itself. Also known as space inversion.

inversion axis See rotation-inversion axis.

inversion layer [METEOROL] The atmosphere layer through which an inversion occurs.

inversion spectrum [SPECT] Lines in the microwave spectra of certain molecules (such as ammonia) which result from the quantum mechanical analog of an oscillation of the molecule between two configurations which are mirror images of each other.

inversion symmetry [PHYS] The principle that the laws of physics are unchanged by the operation of inversion; it is violated by the weak interactions.

invert [CIV ENG] The floor or bottom of a conduit.

invertase See saccharase.

Invertebrata [INV ZOO] A division of the animal kingdom including all animals which lack a spinal column; has no taxonomic status.

invertebrate [INV ZOO] An animal lacking a backbone and internal skeleton.

invertebrate pathology [INV ZOO] All studies having to do with the principles of pathology as applied to invertebrates.

invertebrate zoology [ZOO] A branch of biology that deals with the study of Invertebrata.

inverted See overturned.

inverted amplifier [ELECTR] A two-tube amplifier in which the control grids are grounded and the input signal is applied between the cathodes; the grid then serves as a shield between the input and output circuits.

inverted arch [CIV ENG] An arch with the crown downward, below the line of the springings; commonly used in tunnels and foundations. Also known as inflected arch.

inverted compass [NAV] A marine magnetic compass designed and installed for observation from below the compass card; frequently used as a telltale compass. Also known as hanging compass; overhead compass.

inverted engine [MECH ENG] An engine in which the cylinders are below the crankshaft.

inverted file [ADP] 1. A file, or method of file organization, in which labels indicating the locations of all documents of a given type are placed in a single record. 2. A file whose usual order has been inverted.

inverted heading and bench method See heading-overhand bench method.

inverted image [OPTICS] An image in which up and down, as well as left and right, are interchanged; that is, an image that results from rotating the object 180° about a line from the object to the observer; such images are formed by most astronomical telescopes. Also known as reversed image.

inverted L antenna [ELECTROMAG] An antenna consisting of one or more horizontal wires to which a connection is made by means of a vertical wire at one end.

inverted microscope [OPTICS] A microscope in which the body of the microscope, including the objective and the ocular, are below the stage, the illumination for transmitted light is above the stage, and with opaque materials, the vertical illuminator is used under the stage near the objective.

inverted plunge [GEOL] A plunge of a fold whose inclination has been carried past the vertical, so that the plunge is less than 90° in the direction opposite from the original attitude; younger rocks plunge beneath the older rocks.

inverted siphon [CIV ENG] A pressure pipeline crossing a depression or passing under a highway; sometimes called a sag line from its U-shape.

inverted vee [ELECTROMAG] 1. A directional antenna consisting of a conductor which has the form of an inverted V, and which is fed at one end and connected to ground through an appropriate termination at the other. 2. A center-fed horizontal dipole antenna whose arms have ends bent downward 45°.

inverter [ELEC] A device for converting direct current into alternating current; it may be electromechanical, as in a vibrator or synchronous inverter, or electronic, as in a thyratron inverter circuit. Also known as dc-to-ac converter; dc-to-ac inverter. [ELECTR] See phase inverter.

inverter circuit See NOT circuit.

invertin See saccharase.

inverting amplifier [ELECTR] Amplifier whose output polarity is reversed as compared to its input; such an amplifier obtains its negative feedback by a connection from output to input, and with high gain is widely used as an operational amplifier.

inverting parametric device [ELECTR] Parametric device whose operation depends essentially upon three frequencies, a harmonic of the pump frequency and two signal frequencies, of which the higher signal frequency is the difference between the pump harmonic and the lower signal frequency.

inverting telescope [OPTICS] A telescope that inverts the usual telescopic image, allowing the object to be seen right side up.

invert level [ENG] The level of the lowest portion at any given section of a liquid-carrying conduit, such as a drain or a sewer, and which determines the hydraulic gradient available for moving the contained liquid.

invert sugar [FOOD ENG] A mixture consisting of 50% glucose and 50% fructose, obtained by hydrolysis of sucrose, which absorbs water readily; used in the food industry.

investing bone See dermal bone.

investment casting [MET] A casting method designed to achieve high dimensional accuracy for small castings by making a mold of refractory slurry, which sets at room temperature, surrounding a wax pattern which is then melted out to leave a mold without joints.

investment compound [MATER] A mixture containing a refractory filler, a binder, and a liquid vehicle which is used to make molds for investment casting.

inviscid flow [FL MECH] Flow of an inviscid fluid. Also known as frictionless flow; ideal flow; nonviscous flow.

inviscid fluid [FL MECH] A fluid which has no viscosity; it therefore can support no shearing stress, and flows without energy dissipation. Also known as ideal fluid; nonviscous fluid; perfect fluid.

invisible image [PHYS] An image having a form which cannot be perceived by unaided vision, such as a latent image on a photographic emulsion.

invisible writing ink [GRAPHICS] Ink that remains invisible until the color is brought out by the application of heat or chemical; made with sal ammoniac or salts of metals.

in vitro [BIOL] Pertaining to a biological reaction taking place in an artificial apparatus.

in vivo [BIOL] Pertaining to a biological reaction taking place in a living cell or organism.

involucrate [BOT] Having an involucre.

involucre [BOT] Bracts forming one or more whorls at the base of an inflorescence or fruit in certain plants.

involucrum [ANAT] 1. The covering of a part. 2. New bone laid down by periosteum around a sequestrum in osteomyelitis.

involuntary fibers *See* smooth muscle fibers.

involuntary muscle [PHYSIO] Muscle not under the control of the will; usually consists of smooth muscle tissue and lies within a vescus, or is associated with skin.

involute [BIOL] Being coiled, curled, or rolled in at the edge. [MATH] A curve produced by any point of a perfectly flexible inextensible thread that is kept taut as it is wound upon or unwound from another curve.

involute gear tooth [DES ENG] A gear tooth whose profile is established by an involute curve outward from the base circle.

involute spline [DES ENG] A spline having the same general form as involute gear teeth, except that the teeth are one-half the depth and the pressure angle is 30°.

involute spline broach [MECH ENG] A broach that cuts multiple keys in the form of internal or external involute gear teeth.

involution [BIOL] A turning or rolling in. [EMBRYO] Gastrulation by ingrowth of blastomeres around the dorsal lip. [MED] 1. The retrogressive change to their normal condition that organs undergo after fulfilling their functional purposes, as the uterus after pregnancy. 2. The period of regression or the processes of decline or decay which occur in the human constitution after middle life.

involutional melancholia *See* involutional psychosis.

involutional psychosis [PSYCH] A prolonged psychotic reaction occurring in late middle life, characterized by depression and paranoid ideas. Also known as involutional melancholia.

involucel [BOT] A secondary involucre.

involucellate [BOT] Possessing an involucel.

inward bound [NAV] Heading toward the land or up a harbor, away from the open sea.

inward-outward dialing system [COMMUN] Dialing system whereby calls within the local exchange area may be dialed directly to and from base private branch exchange telephone stations without the assistance of the base private branch exchange operator; CENTREX, a service offered by some telephone companies, is a form of inward-outward dialing.

inyoite [MINERAL] $Ca_2B_6O_{11} \cdot 13H_2O$ A colorless, monoclinic mineral consisting of a hydrous calcium borate; hardness is 2 on Mohs scale, and specific gravity is 2.

Io [ASTRON] A satellite of Jupiter; its diameter is 2300 miles (3700 kilometers). [NUC PHYS] *See* ionium.

I/O *See* input/output.

IOCS *See* input/output control system.

iodargyrite [MINERAL] AgI A yellowish or greenish hexagonal mineral composed of native silver iodide, usually occurring in thin plates. Also known as iodyrite.

iodate [INORG CHEM] A salt of iodic acid containing the IO_3^- radical; sodium and potassium iodates are the most important salts and are used in medicine.

iodcyanin *See* cyanine dye.

iodic acid [INORG CHEM] HIO_3 Water-soluble, moderately strong acid; colorless or white powder or crystals; decomposes at 110°C; used in analytical chemistry and medicine.

iodic acid anhydride *See* iodine pentoxide.

iodide [CHEM] 1. A compound which contains the iodine atom in the −1 oxidation state and which may be considered to be derived from hydriodic acid (HI); examples are KI and NaI. 2. A compound of iodine, such as CH_3CH_2I, in which the iodine has combined with a more electropositive group.

iodide process [MET] A refining process in which a metal, such as titanium or zirconium, is combined with iodine vapor and then the iodide volatilized and decomposed at high temperatures to yield a pure solid metal.

iodine [CHEM] A nonmetallic halogen element, symbol I, atomic number 53, atomic weight 126.9044; melts at 114°C, boils at 184°C; the poisonous, corrosive, dark plates or granules are readily sublimed; insoluble in water, soluble in common solvents; used as germicide and antiseptic, in dyes, tinctures, and pharmaceuticals, in engraving lithography, and as a catalyst and analytical reagent.

iodine-131 [NUC PHYS] A radioactive, artificial isotope of iodine, mass number 131; its half-life is 8 days with beta and gamma radiation; used in medical and industrial radioactive tracer work; moderately radiotoxic.

iodine bisulfide *See* sulfur iodine.

iodine cyanide [INORG CHEM] ICN Poisonous, colorless needles with pungent aroma and acrid taste; melts at 147°C; soluble in water, alcohol, and ether; used in taxidermy as a preservative. Also known as cyanogen iodide.

iodine disulfide *See* sulfur iodine.

iodine number [ANALY CHEM] A measure of the iodine absorbed in a given time by a chemically unsaturated material, such as a vegetable oil or a rubber; used to measure the unsaturation of a compound or mixture. Also known as iodine value.

iodine pentoxide [INORG CHEM] I_2O_5 White crystals, decomposing at 275°C, very soluble in water, insoluble in absolute alcohol, ether, and chloroform; used as an oxidizing agent to oxidize carbon monoxide to dioxide at ordinary temperatures, and in organic synthesis. Also known as iodic acid anhydride.

iodine solution *See* tincture of iodine.

iodine test [ANALY CHEM] Placing a few drops of potassium iodide solution on a sample to detect the presence of starch; test is positive if sample turns blue.

iodine tincture *See* tincture of iodine.

iodine value *See* iodine number.

iodized oil [MATER] A thick, viscous, oily liquid with a garliclike odor that is an iodine addition product of vegetable oil, containing about 40% organically combined iodine; used in medicine as a radiopaque medium for radiography.

iodized salt [FOOD ENG] Common table salt which has been treated with iodide to provide iodine as nutritional supplement.

iodoacetic acid [ORG CHEM] CH_2ICOOH White or colorless crystals that are soluble in water and alcohol, and melt at 82–83°C; used in biological research for its inhibitive effect on enzymes.

iodoalkane [ORG CHEM] An alkane hydrocarbon in which an iodine atom replaces one or more hydrogen atoms in the molecule; an example is iodomethane, CH_3I, better known as methyl iodide.

iodobromite [MINERAL] Ag(Br,Cl,I) An isometric mineral composed of chloride, iodide, and bromide of silver; it is isomorphous with cerargyrite and bromyrite.

iodoeasin *See* easin.

iodoethane *See* ethyl iodide.

iodoform [ORG CHEM] CHI_3 A yellow, hexagonal solid; melting point 119°C; soluble in chloroform, ether, and water; has weak bactericidal qualities and is used in ointments for minor skin diseases. Also known as triiodomethane.

iodohydrocarbon [ORG CHEM] A hydrocarbon in which an iodine atom replaces one or more hydrogen atoms in the molecule, as in an alkane, aromatic, or olefin.

iodomethane *See* methyl iodide.

iodometry [ANALY CHEM] An application of iodine chemistry to oxidation-reduction titrations for the quantitative analysis in certain chemical compounds, in which iodine is used as a reductant and the iodine freed in the associated reaction is titrated, usually in neutral or slightly acid mediums with a standard solution of a reductant such as sodium thiosulfate or sodium arsenite; examples of chemicals analyzed are copper-(III), gold(VI), arsenic(V), antimony(V), chlorine, and bromine.

iodophor [CHEM] Any compound that is a carrier of iodine.

iodopsin [BIOCHEM] The visual pigment found in the retinal cones, consisting of retinene, combined with photopsin.

iodosobenzene [ORG CHEM] C_6H_5IO A yellowish-white amorphous solid that explodes at 200°C, soluble in hot water and alcohol; a strong oxidizing agent.

iodoxybenzene [ORG CHEM] $C_6H_5IO_2$ Clear white crystals that explode at 227–228°C, slightly soluble in water, insoluble in chloroform, acetone, and benzene; a strong oxidizing agent.

iodyrite *See* iodargyrite.

ion [CHEM] An isolated electron or positron or an atom or molecule which by loss or gain of one or more electrons has acquired a net electric charge.

ion accelerator [NUCLEO] A linear accelerator in which ions are accelerated by an electric field in a standing-wave pattern that is set up in a resonant cavity by external oscillators or amplifiers.

ion-acoustic wave [PL PHYS] A longitudinal compression

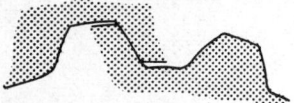

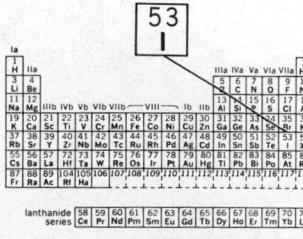

wave in the ion density of a plasma which can occur at high electron temperatures and low frequencies, caused by a combination of ion inertia and electron pressure.

ion atmosphere   See ion cloud.

ion-beam scanning   [ELECTR] The process of analyzing the mass spectrum of an ion beam in a mass spectrometer either by changing the electric or magnetic fields of the mass spectrometer or by moving a probe.

ion chamber   See ionization chamber.

ion cloud   [GEOPHYS] An inhomogeneity or patch of unusually great ion density in one of the regular regions of the ionosphere; such patches occur quite often in the E region. [PHYS CHEM] A slight preponderance of negative ions around a positive ion in an electrolyte, and vice versa, according to the Debye-Hückel theory. Also known as ion atmosphere.

ion column   [GEOPHYS] The trail of ionized gases in the trajectory of a meteoroid entering the upper atmosphere; a part of the composite phenomenon known as a meteor. Also known as meteor trail.

ion concentration   See ion density.

ion counter   See ionization counter.

ion current   [PHYS] The electric current resulting from motion of ions.

ion density   [PHYS] The number of ions per unit volume. Also known as ion concentration.

ion detector   [ANALY CHEM] Device for detection of presence or concentration of liquid solution ions, such as with a pH meter or by conductimetric techniques.

ion emission   [PHYS] The ejection of ions from the surface of a substance into the surrounding space.

ion engine   [AERO ENG] An engine which provides thrust by expelling accelerated or high velocity ions; ion engines using energy provided by nuclear reactors are proposed for space vehicles.

ion exchange   [PHYS CHEM] A chemical reaction in which mobile hydrated ions of a solid are exchanged, equivalent for equivalent, for ions of like charge in solution; the solid has an open, fishnetlike structure, and the mobile ions neutralize the charged, or potentially charged, groups attached to the solid matrix; the solid matrix is termed the ion exchanger.

ion-exchange chromatography   [ANALY CHEM] A chromatographic procedure in which the stationary phase consists of ion-exchange resins which may be acidic or basic.

ion-exchange electrolyte cell   [ELEC] Fuel cell which operates on hydrogen and oxygen in the air, similar to the standard hydrogen-oxygen fuel cell with the exception that the liquid electrolyte is replaced by an ion-exchange membrane; operation is at atmospheric pressure and room temperature.

ion-exchange resin   [MATER] A synthetic resin that can combine or exchange ions with a solution; such a resin produces the exchange of sodium for calcium ions in the softening of hard water.

ion exclusion   [CHEM] Ion-exchange resin system in which the mobile ions in the resin-gel phase electrically neutralize the immobilized charged functional groups attached to the resin, thus preventing penetration of solvent electrolyte into the resin-gel phase; used in separations where electrolyte is to be excluded from the resin, but not nonpolar materials, as the separation of salt from nonpolar glycerin.

ion-exclusion chromatography   [ANALY CHEM] Chromatography in which the adsorbent material is saturated with the same mobile ions (cationic or anionic) as are present in the sample-carrying eluent (solvent), thus repelling the similar sample ions.

ion fractionation   [CHEM ENG] Separation of cations or anions from an ionic solution by use of a membrane permeable to the desired ion; equipment includes electrodialyzers and ion-fractionation stills.

ion gage   See ionization gage.

ion gun   See ion source.

ionic bond   [PHYS CHEM] A type of chemical bonding in which one or more electrons are transferred completely from one atom to another, thus converting the neutral atoms into electrically charged ions; these ions are approximately spherical and attract one another because of their opposite charge. Also known as electrovalent bond.

ionic charge   [PHYS] **1.** The total charge of an ion. **2.** The

charge of an electron; the charge of any ion is equal to this electron charge in magnitude, or is an integral multiple of it.

ionic column   [ARCH] A column characterized by the spiral ornament of its head.

ionic conductance   [PHYS CHEM] The contribution of a given type of ion to the total equivalent conductance in the limit of infinite dilution.

ionic crystal   [CRYSTAL] A crystal in which the lattice-site occupants are charged ions held together primarily by their electrostatic interaction.

ionic equilibrium   [PHYS CHEM] The condition in which the rate of dissociation of nonionized molecules is equal to the rate of combination of the ions.

ionic equivalent conductance   [PHYS CHEM] The contribution made by each ion species of a salt toward an electrolyte's equiviconductance.

ionic focusing   See gas focusing.

ionic gel   [CHEM] A gel with ionic groups attached to the structure of the gel; the groups can not diffuse out into the surrounding solution.

ionic-heated cathode   [ELECTR] Hot cathode heated primarily by ionic bombardment of the emitting surface.

ionic lattice   [CRYSTAL] The lattice of an ionic crystal.

ionic membrane   [CHEM ENG] Semipermeable membrane that conducts electricity; the application of an electric field to the membrane achieves an electrophoretic movement of ions through the membrane; used in electrodialysis.

ionic mobility   [PHYS] The ratio of the average drift velocity of an ion in a liquid or gas to the electric field.

ionic polymerization   [ORG CHEM] Polymerization that proceeds via ionic intermediates (carbonium ions or carbanions) than through neutral species (olefins or acetylenes).

ionic radii   [PHYS CHEM] Radii which can be assigned to ions because the rapid variation of their repulsive interaction with distance makes them repel like hard spheres; these radii determine the dimensions of ionic crystals.

ionic ratios   [OCEANOGR] The ratios by weight of major constituents of seawater to the chloride ion content; for example, $SO_4/Cl = 0.1396$, $Ca/Cl = 0.02150$, $Mg/Cl = 0.06694$.

ionic strength   [PHYS CHEM] A measure of the average electrostatic interactions among ions in an electrolyte; it is equal to one-half the sum of the terms obtained by multiplying the molality of each ion by its valence squared.

ion irradiation   [PHYS] Bombardment of a substance by high-velocity ions.

ionium   [NUC PHYS] A naturally occurring radioisotope, symbol Io, of thorium, atomic weight 230.

ionization   [CHEM] A process by which a neutral atom or molecule loses or gains electrons, thereby acquiring a net charge and becoming an ion; occurs as the result of the dissociation of the atoms of a molecule in solution ($NaCl \rightarrow Na^+ + Cl^-$) or of a gas in an electric field ($H_2 \rightarrow 2H^+$).

ionization arc-over   [ELEC] **1.** Arcing across terminals or contacts due to ionization of the adjacent air or gas. **2.** Arcing across satellite antenna terminals as the satellite passes through the ionized regions of the ionosphere.

ionization chamber   [NUCLEO] A particle detector which measures the ionization produced in the gas filling the chamber by the fast-moving charged particles as they pass through. Also known as ion chamber.

ionization constant   [PHYS CHEM] Analog of the dissociation constant, where $k = [H^+][A^-]/[HA]$; used for the application of the law of mass action to ionization; in the equation HA represents the acid, such as acetic acid.

ionization counter   [NUCLEO] An ionization chamber in which there is no internal amplification by gas multiplication; used for counting ionizing particles. Also known as ion counter.

ionization cross section   [PHYS] The cross section for a particle or photon to undergo a collision with an atom, thus removing or adding one or more electrons to the atom.

ionization current   See gas current.

ionization degree   [PHYS CHEM] The proportion of potential ionization that has taken place for an ionizable material in a solution or reaction mixture.

ionization density   [ELECTR] The density of ions in a gas.

ionization energy   [ATOM PHYS] The amount of energy needed to remove an electron from a given kind of atom or molecule to an infinite distance; usually expressed in electron volts, and numerically equal to the ionization potential in volts.

ionization gage [ELECTR] An instrument for measuring low gas densities by ionizing the gas and measuring the ion current. Also known as ion gage; ionization vacuum gage.

ionization potential [ATOM PHYS] The energy per unit charge needed to remove an electron from a given kind of atom or molecule to an infinite distance; usually expressed in volts. Also known as ion potential.

ionization radiation *See* ionizing radiation.

ionization source *See* ion source.

ionization spectrometer *See* Bragg spectrometer.

ionization time [ELECTR] Of a gas tube, the time interval between the initiation of conditions for and the establishment of conduction at some stated value of tube voltage drop.

ionization vacuum gage *See* ionization gage.

ionized atom [CHEM] An atom with an excess or deficiency of electrons, so that it has a net charge.

ionized gas [PHYS] A gas, some of whose atoms or molecules have undergone ionization.

ionized layers [GEOPHYS] Layers of increased ionization within the ionosphere produced by cosmic radiation; responsible for absorption and reflection of radio waves and important in connection with communications and tracking of satellites and other space vehicles.

ionizing event [PHYS] Any occurrence in which an ion or group of ions is produced; for example, by passage of charged particles through matter.

ionizing radiation [NUCLEO] **1.** Particles or photons that have sufficient energy to produce ionization directly in their passage through a substance. Also known as ionization radiation. **2.** Particles that are capable of nuclear interactions in which sufficient energy is released to produce ionization.

ion kinetic energy spectrometry [SPECT] A spectrometric technique that uses a beam of ions of high kinetic energy passing through a field-free reaction chamber from which ionic products are collected and energy analyzed; it is a generalization of metastable ion studies in which both unimolecular and bimolecular reactions are considered.

ion mean life [PHYS CHEM] The average time between the ionization of an atom or molecule and its recombination with one or more electrons, or its loss of excess electrons.

ion microprobe *See* secondary ion mass spectrometer.

ion microprobe mass spectrometer [ENG] A type of secondary ion mass spectrometer in which primary ions are focused on a spot 1–2 micrometers in diameter, mass-charge separation of secondary ions is carried out by a double focusing mass spectrometer or spectrograph, and a magnified image of elemental or isotopic distributions on the sample surface is produced using synchronous scanning of the primary ion beam and an oscilloscope.

ion microscope *See* field-ion microscope.

ion migration [ELEC] Movement of ions produced in an electrolyte, semiconductor, and so on, by the application of an electric potential between electrodes.

ionogram [ENG] A record produced by an ionosonde, that is, a graph of the virtual height of the ionosphere plotted against frequency.

ionography [ANALY CHEM] A type of electrochromatography involving migration of ions.

ionomer [ORG CHEM] Polymer with covalent bonds between the elements of the chain, and ionic bonds between the chains.

ionophone [ENG ACOUS] A high-frequency loudspeaker in which the audio-frequency signal modulates the radio-frequency supply to an arc maintained in a quartz tube, and the resulting modulated wave acts directly on ionized air to create sound waves.

ionophore [BIOCHEM] Any of a class of compounds, generally cyclic, having the ability to carry ions across lipid barriers due to the property of cation selectivity; examples are valinomycin and nonactin.

ionosonde [ENG] A radar system for determining the vertical height at which the ionosphere reflects signals back to earth at various frequencies; a pulsed vertical beam is swept periodically through a frequency range from 0.5 to 20 megahertz, and the variation of echo return time with frequency is photographically recorded.

ionosphere [GEOPHYS] That part of the earth's upper atmosphere which is sufficiently ionized by solar ultraviolet radiation so that the concentration of free electrons affects the propagation of radio waves; its base is at about 70 or 80 kilometers and it extends to an indefinite height.

ionospheric disturbance [GEOPHYS] A temporal variation in electron concentration in the ionosphere that is caused by solar activity and that makes the heights of the ionosphere layers go beyond the normal limits for a location, date, and time of day.

ionospheric D scatter meteor burst [GEOPHYS] Phenomenon affecting ionospheric scatter communications resulting from the penetration of meteors through the D region of the ionospheric layer.

ionospheric error [COMMUN] Variation in the character of the ionospheric transmission path or paths used by the radio waves of electronic navigation systems which if not compensated will produce an error in the information generated by the system.

ionospheric propagation [COMMUN] Propagation of radio waves over long distances by reflection from the ionosphere, useful at frequencies up to about 25 megahertz.

ionospheric recorder [ELECTR] A radio device for determining the distribution of virtual height with frequency, and the critical frequencies of the various layers of the ionosphere.

ionospheric scatter [COMMUN] A form of scatter propagation in which radio waves are scattered by the lower E layer of the ionosphere to permit communication over distances from 600 to 1400 miles (1000 to 2250 kilometers) when using the frequency range of about 25 to 100 megahertz.

ionospheric storm [GEOPHYS] A turbulence in the F region of the ionosphere, usually due to a sudden burst of radiation from the sun; it is accompanied by a decrease in the density of ionization and an increase in the virtual height of the region.

ionospheric wave *See* sky wave.

ion pair [NUCLEO] A positive ion and an equal-charge negative ion, usually an electron, that are produced by the action of radiation on a neutral atom or molecule.

ion-permeable membrane [MATER] A film or sheet of a substance which is preferentially permeable to some species or types of ions.

ion potential *See* ionization potential.

ion probe *See* secondary ion mass spectrometer.

ion propulsion [AERO ENG] Vehicular motion caused by reaction from the high-speed discharge of a beam of electrically equally charged minute particles ejected behind the vehicle.

ion pump [ELECTR] A vacuum pump in which gas molecules are first ionized by electrons that have been generated by a high voltage and are spiraling in a high-intensity magnetic field, and the molecules are then attracted to a cathode, or propelled by electrodes into an auxiliary pump or an ion trap.

ion retardation [CHEM ENG] Sorbent extraction of strong electrolytes with an anion-exchange resin in which a cationic monomer has been polymerized, or vice versa.

ion source [ELECTR] A device in which gas ions are produced, focused, accelerated, and emitted as a narrow beam. Also known as ion gun; ionization source.

ion spot [ELECTR] Of a cathode-ray tube screen, an area of localized deterioration of luminescence caused by bombardment with negative ions.

iontophoresis [MED] A medical treatment used to drive positive or negative ions into a tissue, in which two electrodes are placed in contact with tissue, one of the electrodes being a pad of absorbent material soaked with a solution of the material to be administered, and a voltage is applied between the electrodes.

ion trap [ELECTR] **1.** An arrangement whereby ions in the electron beam of a cathode-ray tube are prevented from bombarding the screen and producing an ion spot, usually employing a magnet to bend the electron beam so that it passes through the tiny aperture of the electron gun, while the heavier ions are less affected by the magnetic field and are trapped inside the gun. **2.** A metal electrode, usually of titanium, into which ions in an ion pump are absorbed.

iophobia [PSYCH] An abnormal fear of poison.

Iospilidae [INV ZOO] A small family of pelagic polychaetes assigned to the Errantia.

Iowan glaciation [GEOL] The earliest substage of the Wisconsin glacial stage; occurred more than 30,000 years ago.

ipecac [BOT] Any of several low, perennial, tropical South American shrubs or half shrubs in the genus *Cephaelis* of the family Rubiaceae; the dried rhizome and root, containing emetine, cephaeline, and other alkaloids, is used as an emetic and expectorant.

Ipidae [INV ZOO] The equivalent name for Scolytidae.

IPL [ADP] Collective term for a series of list-processing languages developed principally by Newell, Simon, and Shaw. Derived from Information Processing Language.

IPL-V [ADP] The fifth IPL language; a good language for manipulating tree structures.

ipsonite [GEOL] The final stage of weathered asphalt; a black, infusible substance, only slightly soluble in carbon disulfide, containing 50-80% fixed carbon and very little oxygen.

IQ See intelligence quotient.

IQSY See International Quiet Sun Year.

Ir See iridium.

IRBM See intermediate-range ballistic missile.

irdome [OPTICS] A dome used to protect an infrared detector and its optical elements, generally made from quartz, silicon, germanium, sapphire, calcium aluminate, or other material having high transparency to infrared radiation. Derived from infrared dome.

IR drop See resistance drop.

IRG See record gap.

Iridaceae [BOT] A family of monocotyledonous herbs in the order Liliales distinguished by three stamens and an inferior ovary.

iridectomy [MED] Surgical removal of part of the iris.

iridescence [OPTICS] 1. A rainbow color effect exhibited in various bodies as a result of interference in a thin film (as of soap bubbles or mother of pearl) or of diffraction of light reflected from a ribbed surface (as of the plumage of some birds).

iridescent clouds [METEOROL] Ice-crystal clouds which exhibit brilliant spots or borders of colors, usually red and green, observed up to about 30° from the sun.

iridic chloride [INORG CHEM] $IrCl_4$ Hygroscopic brownish-black mass, soluble in water and alcohol; used to analyze for nitric acid, HNO_3, and in analytical microscopic work. Also known as iridium chloride; iridium tetrachloride.

iridium [CHEM] A metallic element, symbol Ir, atomic number 77, atomic weight 192.2, in the platinum group; insoluble in acids, melting at 2454°C. [MET] A silver-white, brittle, hard metal used in jewelry, electric contacts, electrodes, resistance wires, and pen tips.

iridium-192 [NUC PHYS] Radioactive isotope of iridium with a 75-day half-life; β and γ radiation; used in cancer treatment and for radiography of light metal castings.

iridium chloride See iridic chloride.

iridium tetrachloride See iridic chloride.

iridocyclitis [MED] Inflammation of the iris and the ciliary body.

iridocyclochoroiditis [MED] Inflammation of the iris, the ciliary body, and the choroid. Also known as uveitis.

iridocyte [HISTOL] A specialized cell in the integument of certain animal species which is filled with iridescent crystals of guanine and a variety of lipophores. Also known as guanophore; iridophore.

iridophore See iridocyte.

iris [ANAT] A pigmented diaphragm perforated centrally by an adjustable pupil which regulates the amount of light reaching the retina in vertebrate eyes. [BOT] Any plant of the genus *Iris*, the type genus of the family Iridaceae, characterized by linear or sword-shaped leaves, erect stalks, and bright-colored flowers with the three inner perianth segments erect and the outer perianth segments drooping. [ELECTROMAG] A conducting plate mounted across a waveguide to introduce impedance; when only a single mode can be supported, an iris acts substantially as a shunt admittance and may be used for matching the waveguide impedance to that of a load. Also known as diaphragm; waveguide window. [OPTICS] A circular mechanical device, whose diameter can be varied continuously, which controls the amount of light reaching the film of a camera.

irisation [METEOROL] The coloration exhibited by iridescent clouds and at times along the borders of lenticular clouds.

Irish linen [TEXT] Thin linen fabric woven of Irish flax.

Irish potato See potato.

Irish Sea [GEOGR] A marginal sea of the Atlantic Ocean between Ireland and England, approximately 53°N latitude and 5°W longitude.

iritis [MED] Inflammation of the iris.

I²R loss See copper loss.

IRM See isothermal remanent magnetization.

Irminger Current [OCEANOGR] An ocean current that is one of the terminal branches of the Gulf Stream system, flowing west off the southern coast of Iceland.

iron [CHEM] A silvery-white metallic element, symbol Fe, atomic number 26, atomic weight 55.847, melting at 1530°C. [MET] A heavy, magnetic, malleable and ductile metal occurring in meteorites and combined in a wide range of ores and most igneous rocks; it is one of the most widely used metals, and plays a role in biological processes.

iron-55 [NUC PHYS] Radioactive isotope of iron, symbol Fe^{55}, with a 2.91-year half-life; highly toxic.

iron-59 [NUC PHYS] Radioactive isotope of iron, symbol Fe^{59}, 46.3-day half-life; β and γ radiation; highly toxic; used to study metallic welds, corrosion mechanisms, engine wear, and bodily functions.

iron acetate See ferrous acetate.

iron acetate liquor [MATER] Black liquor containing 5-5.5% iron and sometimes copperas or tannin; results from pyroligneous acid attack on iron filings; used for dyeing and printing with logwood, and as a mordant for alizarine and nitroso dyes. Also known as black liquor; black mordant; iron liquor.

Iron Age [ARCHEO] Period characterized by production and widespread use of iron, starting approximately 1000 B.C.

iron alloy [MET] An alloy having iron as the principal component.

iron alum See halotrichite.

iron ammonium sulfate See ferric ammonium sulfate; ferrous ammonium sulfate.

iron and steel sheet piling [MIN ENG] A technique that uses iron and steel piling instead of wood to drive a shaft through loose, wet ground near the surface.

Ironarc process [MET] An ultra-high temperature smelting process employing plasma chemistry to process refractory materials, such as zirconia.

iron arsenate See ferrous arsenate.

iron bacteria [MICROBIO] The common name for bacteria capable of oxidizing ferrous iron to the ferric state.

iron-binding protein [BIOCHEM] A serum protein, such as hemoglobin, for the transport of iron ions.

iron black [CHEM] Fine black antimony powder used to give a polished-steel look to papier-mâché and plaster of Paris; made by reaction of zinc with acid solution of an antimony salt and precipitation of black antimony powder.

iron blue [INORG CHEM] Ferric ferrocyanide used as blue pigment by the paint industry for permanent body and trim paints; also used in blue ink, in paper dyeing, and as a fertilizer ingredient.

iron bromide See ferric bromide.

iron carbide See cementite.

iron carbonyl See iron pentacarbonyl.

iron castings [MET] Shapes cast in molds from iron.

iron cement [MATER] A mixture of small iron pieces with ammonium chloride, used to join iron or steel surfaces.

iron chloride See ferric chloride; ferrous chloride.

iron citrate See ferric citrate.

iron-Constantan [MET] Dual-metal combination for thermocouple junctions, used for temperature measurement in oxidizing or reducing atmospheres.

iron core [ELECTROMAG] A core made of solid or laminated iron, or some other magnetic material which may contain very little iron.

iron-core choke See iron-core coil.

iron-core coil [ELECTROMAG] A coil in which solid or laminated iron or other magnetic material forms part or all of the magnetic circuit linking its winding. Also known as iron-core choke; magnet coil.

iron-core transformer [ELECTROMAG] A transformer in which laminations of iron or other magnetic material make

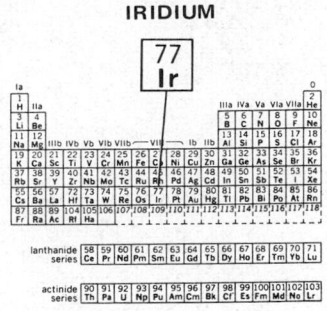

IRIDIUM

Periodic table of the chemical elements showing the position of iridium.

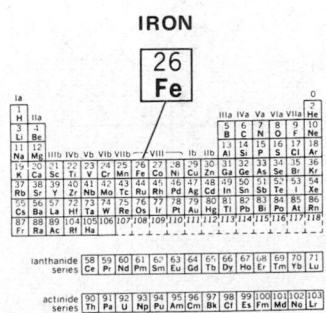

IRIS

The iris, an ornamental flower.

IRON

Periodic table of the chemical elements showing the position of iron.

up part or all of the path for magnetic lines of force that link the transformer windings.

iron-deficiency anemia [MED] Hypochromic microcytic anemia due to excessive loss, deficient intake, or poor absorption of iron. Also known as nutritional hypochromic anemia.

iron dichloride See ferrous chloride.

iron-dust core [ELECTROMAG] A core made by mixing finely powdered magnetic material with an insulating binder and molding under pressure to form a rod-shaped core that can be moved into or out of a coil or transformer to vary the inductance or degree of coupling for tuning purposes.

irone [ORG CHEM] $C_{14}H_{22}O$ A colorless liquid terpene; a component of essential oil from the orrisroot; used in perfumes.

iron ferrocyanide See ferric ferrocyanide.

iron fluoride See ferric fluoride.

iron formation [GEOL] Sedimentary, low-grade iron ore bodies consisting mainly of chert or fine-grained quartz and ferric oxide segregated in bands or sheets irregularly mingled.

iron foundry [MET] A building in which iron castings are made.

iron glance See specularite.

iron hat See gossan.

iron hydroxide See ferric hydroxide.

ironing [MET] Reducing the wall thickness of a deep-drawn object by reducing the clearance between punch and die.

iron liquor See iron acetate liquor.

iron loss See core loss.

iron metabolism [BIOCHEM] The chemical and physiological processes involved in absorption of iron from the intestine and in its role in erythrocytes.

iron metavanadate See ferric vanadate.

iron meteorite [ASTRON] A type of meteorite that consists mainly of iron and nickel and is several times heavier than any ordinary rock.

iron mica See lepidomelane.

iron monoxide See ferrous oxide.

iron-nickel alloy [MET] An iron alloy containing 20–80% nickel; it has high permeability and low hysteresis losses at low flux densities and is more readily rolled into thin laminations than silicon steels.

iron nitrate See ferric nitrate.

iron nonacarbonyl [INORG CHEM] $Fe_2(CO)_9$ Orange-yellow crystals that break down at 100°C to yield tetracarbonyl, slightly soluble in alcohol and acetone, almost insoluble in water, ether, and benzene. Also known as di-iron enneacarbonyl.

iron ore [GEOL] Rocks or deposits containing compounds from which iron can be extracted.

iron-ore cement [MATER] A cement in which iron ore is used instead of clay or shale.

iron oxalate See ferrous oxalate.

iron oxide [INORG CHEM] Any of the hydrated, synthetic, or natural oxides of iron: ferrous oxide, ferric oxide, ferriferous oxide.

iron pentacarbonyl [INORG CHEM] $Fe(CO)_5$ An oily liquid that decomposes upon exposure to light, soluble in most organic solvents; used as a source of a pure iron catalyst and for magnet cores. Also known as iron carbonyl.

iron phosphate See ferric phosphate.

iron-porphyrin protein [BIOCHEM] Any protein containing iron and porphyrin; examples are hemoglobin, the cytochromes, and catalase.

iron pyrite See pyrite.

iron red [MATER] Any of the pigments made from red varieties of ferric oxide.

iron resinate See ferric resinate.

iron-retention agent [MATER] Complexing agent that ties dissolved iron into complex ions to prevent reprecipitation and wellbore plugging after well acidizing.

iron saffron See Indian red.

iron scurf [MATER] Glazing material used to color blue bricks; a mixture of stone and iron particles from grinding of gun barrels.

ironshot [MINERAL] Pertaining to a mineral with streaks or spots of iron or iron ore.

iron sight [ORD] Any metallic gunsight, such as a blade sight, as distinguished from an optical or computing sight.

iron spar See siderite.

iron spinel See hercynite.

iron stearate See ferric stearate.

iron-stony meteorite See stony-iron meteorite.

iron sulfate See ferric sulfate; ferrous sulfate.

iron sulfide See ferrous sulfide.

iron tetracarbonyl [INORG CHEM] $Fe_3(CO)_{12}$ Dark-green lustrous crystals that break down at 140–150°C; soluble in organic solvents. Also known as tri-iron dodecacarbonyl.

iron whiskers [MET] Single-crystal pure iron filaments or fibers.

iron winds [METEOROL] Northeasterly winds of Central America, prevalent during February and March, and blowing steadily for several days at a time.

ironwood [BOT] Any of a number of hardwood trees in the United States, including the American hornbeam, the buckwheat, and the eastern hophornbeam.

irradiance See radiant flux density.

irradiation [BIOPHYS] Subjection of a biological system to sound waves of sufficient intensity to modify their structure or function. [ENG] The exposure of a material, object, or patient to x-rays, gamma rays, ultraviolet rays, or other ionizing radiation.

irradiation cataract [MED] A cataract that develops slowly following prolonged or intense irradiation, as by radium or roentgen rays. Also known as cyclotron cataract; radiation cataract.

irradiation correction [NAV] A correction to the readings of sextant altitudes made necessary by irradiation that is caused by the apparent enlargement of the bright surface of a celestial body against the darker background of the sky.

irradiation cystitis [MED] Inflammation of the urinary bladder following radiation therapy of pelvic organs.

irrational equation [MATH] An equation having an unknown raised to some fractional power. Also known as radical equation.

irrational number [MATH] A number which is not the quotient of two integers.

irreducible equation [MATH] An equation that is equivalent to one formed by setting an irreducible polynomial equal to zero.

irreducible function See irreducible polynomial.

irreducible polynomial [MATH] A polynomial is irreducible over a field K if it cannot be written as the product of two polynomials of lesser degree whose coefficients come from K. Also known as irreducible function.

irreducible representation of a group [MATH] A representation of a group as a family of linear operators of a vector space V where there is no proper closed subspace of V invariant under these operators.

irreducible saturation [PETRO ENG] In a permeable reservoir, that condition in which the nonwetting phase saturation is so large that the wetting phase can be reduced no more.

irreducible tensor [MATH] A tensor that cannot be written as the inner product of two tensors of lower degree.

irregular [BOT] Lacking symmetry, as of a flower having petals unlike in size or shape.

irregular cleavage [EMBRYO] Division of a zygote into random masses of cells, as in certain coelenterates.

irregular connective tissue [HISTOL] A loose or dense connective tissue with fibers irregularly woven and irregularly distributed; collagen is the dominant fiber type.

irregular crystal [METEOROL] A snow particle, sometimes covered by a coating of rime, composed of small crystals randomly grown together; generally, component crystals are so small that the crystalline form of the particle can be seen only through a magnifying glass or microscope.

irregular element [IND ENG] An element whose frequency of occurrence is irregular but predictable. Also known as incidental element.

irregular galaxy [ASTRON] A galaxy which shows no definite order or shape, except that of a general flattened appearance.

Irregularia [INV ZOO] An artificial assemblage of echinoderms in which the anus and periproct lie outside the apical system, the ambulacral plates remain simple, the primary radioles are hollow, and the rigid test shows some degree of bilateral symmetry.

irregular iceberg See pinnacled iceberg.

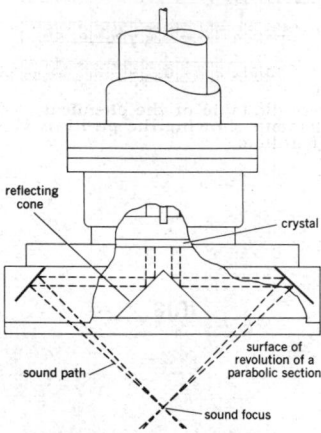

IRRADIATION

Reflector-type focusing irradiator, a device for producing irradiation by sound waves in the megahertz range. The sound waves emanating from the circular crystal plate impinge on a 90° included angle cone, from which surface they are directed to a surface of revolution generated by a section of a parabola. Focusing is obtained by reflection of the waves from this surface.

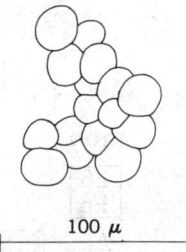

IRREGULAR CLEAVAGE

100 μ

Pattern of cells in irregular cleavage.

irregular variable star [ASTRON] A star with no fixed period.

irrespirable atmosphere [MIN ENG] Atmosphere in a coal mine requiring men to wear breathing apparatus because of poisonous gas or insufficient oxygen as a result of an explosion from firedamp or coal dust, or mine fires.

irreversible energy loss [THERMO] Energy transformation process in which the resultant condition lacks the driving potential needed to reverse the process; the measure of this loss is expressed by the entropy increase of the system.

irreversible process [THERMO] A process which cannot be reversed by an infinitesimal change in external conditions.

irreversible thermodynamics *See* nonequilibrium thermodynamics.

irrigation [CIV ENG] Artificial application of water to arable land for agricultural use. [MED] Therapeutic washing out by means of a continuous stream of water.

irrigation canal [CIV ENG] An artificial open channel for transporting water for crop irrigation.

irrigation pipe [CIV ENG] A conduit of connected pipes for transporting water for crop irrigation.

irritability [PHYSIO] 1. A condition or quality of being excitable; the power of responding to a stimulus. 2. A condition of abnormal excitability of an organism, organ, or part, when it reacts excessively to a slight stimulation.

irritable colon [MED] Any of several disturbed colonic functions associated with anxiety or emotional stress. Also known as adaptive colitis; mucous colitis; spastic colon; unstable colon.

irritant gas [MATER] A nonlethal gas, causing irritation of the skin and flow of tears; any one of the family of tear gases used for training and riot control.

irrotational flow [FL MECH] Fluid flow in which the curl of the velocity function is zero everywhere, so that the circulation of the velocity about any closed curve vanishes. Also known as acyclic motion; irrotational motion.

irrotational motion *See* irrotational flow.

irrotational strain [GEOL] Strain in which the orientation of the axes of strain does not change. Also known as nonrotational strain.

irrotational vector field [MATH] A vector field whose curl is identically zero; every such field is the gradient of a scalar function. Also known as lamellar vector field.

irrotational wave *See* compressional wave.

irruption *See* intrusion.

isallobar [METEOROL] A line of equal change in atmospheric pressure during a specified time interval; an isopleth of pressure tendency; a common form is drawn for the three-hourly local pressure tendencies on a synoptic surface chart.

isallobaric [METEOROL] Of equal or constant pressure change; this may refer either to the distribution of equal pressure tendency in space or to the constancy of pressure tendency with time.

isallobaric high *See* pressure-rise center.

isallobaric low *See* pressure-fall center.

isallobaric maximum *See* pressure-rise center.

isallobaric minimum *See* pressure-fall center.

isallobaric wind [METEOROL] The wind velocity whose Coriolis force exactly balances a locally accelerating geostrophic wind. Also known as Brunt-Douglas isallobaric wind.

isallohypse *See* height-change line.

isallohypsic wind [METEOROL] An isallobaric wind, using height tendency in a constant-pressure surface instead of pressure tendency in a constant-height surface.

isallotherm [METEOROL] A line connecting points of equal change in temperature within a given time period.

isanabat [METEOROL] A line drawn through points of equal vertical component of wind velocity; positive values indicate upward motion, negative values indicate downward motion.

isanakatabar [METEOROL] A line on a chart of equal atmospheric-pressure range during a specified time interval.

isanomal *See* isanomalous line.

isanomalous line [METEOROL] A line drawn through geographical points having equal anomaly of some meteorological quantity. Also known as isanomal.

isano oil [MATER] A pale-yellow, viscous drying oil obtained from the nut of an African tree; used as a varnish oil.

isanthous [BOT] Having regular flowers.

isarithm *See* isopleth.

isatin [ORG CHEM] $C_6H_5NO_2$ An indole substituted with oxygen at carbon position 2 and 3; crystallizes as red needles that are soluble in hot water; used in dye manufacture.

isaurore *See* isochasm.

ISB modulation *See* independent-sideband modulation.

I scan *See* I scope.

ischemia [MED] Localized tissue anemia as a result of obstruction of the blood supply or to vasoconstriction.

ischemic necrosis [MED] Local tissue death due to impaired blood supply.

ischemic neuropathy [MED] Nerve lesions characterized by numbness, tingling, and pain with loss of sensory and motor functions of the parts involved, due to obstruction of the blood supply to the nerves.

ischemic paralysis [MED] Impaired motor function due to obstructed circulation to the area.

ischemic tubulorrhexis *See* lower nephron nephrosis.

ischiopodite [INV ZOO] The segment nearest the basipodite of walking legs in certain crustaceans. Also known as ischium.

ischiorectal region [ANAT] The region between the ischium and the rectum.

ischium [ANAT] Either of a pair of bones forming the dorsoposterior portion of the vertebrate pelvis; the inferior part of the pelvis in humans upon which the body rests in sitting. [INV ZOO] *See* ischiopodite.

ischium-pubis index [ANAT] The ratio (length of pubis × 100/length of ischium) by which the sex of an adult pelvis may usually be determined; the index is greater than 90 in females, and less than 90 in males.

Ischnacanthidae [PALEON] The single family of the acanthodian order Ischnacanthiformes.

Ischnacanthiformes [PALEON] A monofamilial order of extinct fishes of the order Acanthodii; members were slender, lightly armored predators with sharp teeth, deeply inserted fin spines, and two dorsal fins.

I scope [ELECTR] A cathode-ray scope on which a single signal appears as a circular segment whose radius is proportional to the range and whose circular length is inversely proportional to the error of aiming the antenna, true aim resulting in a complete circle; the position of the arc, relative to the center, indicates the position of the target relative to the beam axis. Also known as broken circle indicator; I indicator; I scan.

Isectolophidae [PALEO] A family of extinct ceratomorph mammals in the superfamily Tapiroidea.

isenthalpic expansion [THERMO] Expansion which takes place without any change in enthalpy.

isentrope [THERMO] A line of equal or constant entropy.

isentropic [THERMO] Having constant entropy; at constant entropy.

isentropic chart [METEOROL] A constant-entropy chart; a synoptic chart presenting the distribution of meteorological elements in the atmosphere on a surface of constant potential temperature (equivalent to an isentropic surface); it usually contains the plotted data and analysis of such elements as pressure (or height), wind, temperature, and moisture at that surface.

isentropic compression [THERMO] Compression which occurs without any change in entropy.

isentropic condensation level *See* lifting condensation level.

isentropic expansion [THERMO] Expansion which occurs without any change in entropy.

isentropic flow [THERMO] Fluid flow in which the entropy of any part of the fluid does not change as that part is carried along with the fluid.

isentropic map [GEOL] A map indicating constant entropy function for facies.

isentropic mixing [METEOROL] Any atmospheric mixing process which occurs within an isentropic surface; the fact that many atmospheric motions are reversible adiabatic processes renders this type of mixing important, and exchange coefficients have been computed therefor.

isentropic process [THERMO] A change that takes place without any increase or decrease in entropy, such as a process which is both reversible and adiabatic.

isentropic surface [METEOROL] A surface in space in which potential temperature is everywhere equal.

ISATIN

Structural formula of isatin.

isentropic thickness chart [METEOROL] A thickness chart of an atmospheric layer bounded by two selected isentropic surfaces (surfaces of constant potential temperature); the thickness of such a layer is directly proportional to the static instability of that layer; hence, these charts have been called instability charts. Also known as thick-thin chart.

isentropic weight chart [METEOROL] A chart of atmospheric pressure difference between two selected isentropic surfaces (surfaces of constant potential temperature); the greater the pressure difference the greater the weight of the air column separating the two surfaces.

isethionic acid [ORG CHEM] $CH_2OH \cdot CH \cdot SO_2OH$ A water-soluble liquid, boiling at 100°C; used in the manufacture of detergents. Also known as ethylenehydrinsulfonic acid; oxyethylsulfonic acid.

ishkyldite [MINERAL] $Mg_{15}Si_{11}O_{27}(OH)_{20}$ A mineral composed of a basic silicate of magnesium.

I signal [ELECTR] The in-phase component of the chrominance signal in color television, having a bandwidth of 0 to 1.5 megahertz, and consisting of $+0.74(R - Y)$ and $-0.27(B - Y)$, where Y is the luminance signal, R is the red camera signal, and B is the blue camera signal.

Ising coupling [SOLID STATE] A model of coupling between two atoms in a lattice, used to study ferromagnetism, in which the spin component of each atom along some axis is taken to be $+1$ or -1, and the energy of interaction is proportional to the negative of the product of the spin components along this axis.

isinglass [MATER] A gelatin made from the dried swim bladders of sturgeon and other fishes; used in glues, cements, and printing inks. Also known as fish gelatin; fish glue; ichthyocolla.

Ising model [SOLID STATE] A crude model of a ferromagnetic material or an analogous system, used to study phase transitions, in which atoms in a one-, two-, or three-dimensional lattice interact via Ising coupling between nearest neighbors, and the spin components of the atoms are coupled to a uniform magnetic field.

island [GEOGR] A tract of land smaller than a continent and surrounded by water; normally in an ocean, sea, lake, or stream.

island arc [GEOGR] A group of islands usually with a curving archlike pattern, generally convex toward the open ocean, having a deep trench or trough on the convex side and usually enclosing a deep basin on the concave side.

island mountain *See* inselberg.

island of Langerhans *See* islet of Langerhans.

island of Reil [ANAT] The insula of the cerebral hemisphere.

islet-cell carcinoma [MED] A metastatic tumor of pancreatic cells of the islet of Langerhans in the pancreas.

islet-cell tumor [MED] A benign tumor of the pancreatic islet cells. Also known as insulinoma; insuloma; Langerhansian adenoma.

islet of Langerhans [HISTOL] A mass of cell cords in the pancreas that is of an endocrine nature, secreting insulin and a minor hormone like lipocaic. Also known as island of Langerhans; islet of the pancreas.

islet of the pancreas *See* islet of Langerhans.

iso- [CHEM] A prefix indicating an isomer of an element in which there is a difference in the nucleus when compared to the most prevalent form of the element. [ORG CHEM] A prefix indicating a single branching at the end of the carbon chain.

isoagglutinin [IMMUNOL] An agglutinin which acts upon the red blood cells of members of the same species. Also known as isohemagglutinin.

isoalloxazine mononucleotide *See* riboflavin 5'-phosphate.

isoamyl acetate *See* amyl acetate.

isoamyl alcohol *See* isobutyl carbinol.

isoamyl benzoate [ORG CHEM] $C_6H_5COOC_5H_{11}$ Colorless liquid with fruity aroma; boils at 260°C: soluble in alcohol, insoluble in water; used in flavors and perfumes. Also known as amyl benzoate.

isoamyl butyrate [ORG CHEM] $C_5H_{11}COOC_3H_7$ A water-white liquid boiling at 150–180°C; soluble in alcohol and ether; used as a solvent and plasticizer for cellulose acetate and in flavor extracts.

isoamyl chloride [ORG CHEM] $C_5H_{11}Cl$ Water-insoluble,

colorless liquid boiling at 100°C; it can be any one of several compounds, such as 1-chloro-3-methylbutane, $(CH_3)_2$-$CH(CH_2)_2Cl$, or mixtures thereof; used as a solvent, in inks, for soil fumigation, and as a chemical intermediate.

isoamyl isovalerate *See* isoamyl valerate.

isoamyl nitrite *See* amyl nitrite.

isoamyl salicylate *See* amyl salicylate.

isoamyl valerate [ORG CHEM] $C_4H_9CO_2C_5H_{11}$ Clear liquid with apple aroma; boils at 204°C; soluble in alcohol and ether, insoluble in water; used in medicine and fruit flavors. Also known as apple essence; apple oil; isoamyl isovalerate.

isoantibody [IMMUNOL] An antibody, found only in some members of a species, which acts upon cells or cell components of other members of the same species.

isoantigen [IMMUNOL] An antigen in an individual capable of stimulating production of a specific antibody in another member of the same species.

isobar [METEOROL] A line drawn through all points of equal atmospheric pressure along a given reference surface, such as a constant-height surface (notably mean sea level on surface charts), an isentropic surface, or the vertical plan of a synoptic cross section. [NUC PHYS] One of two or more nuclides having the same number of nucleons in their nuclei but differing in their atomic numbers and chemical properties.

isobaric [METEOROL] Of equal or constant pressure, with respect to either space or time.

isobaric chart *See* constant-pressure chart.

isobaric contour chart *See* constant-pressure chart.

isobaric divergence [METEOROL] The horizontal divergence in a constant-pressure surface; expressed in a system of coordinates with pressure as an independent variable.

isobaric equivalent temperature *See* equivalent temperature.

isobaric map [METEOROL] A map depicting points in the atmosphere of equal barometric pressure.

isobaric spin *See* isotopic spin.

isobaric surface [METEOROL] A surface on which the pressure is uniform. Also known as constant-pressure surface.

isobaric topography *See* height pattern.

isobaric vorticity [METEOROL] Relative vorticity in a constant-pressure surface; that is, expressed in a system of coordinates with pressure as an independent variable.

isobath [OCEANOGR] A contour line connecting points of equal water depths on a chart. Also known as depth contour; depth curve; fathom curve.

isobathytherm [OCEANOGR] A line or surface showing the depth in oceans or lakes at which points have the same temperatures.

isobiochore [ECOL] A boundary line on a map connecting world environments that have similar floral and faunal constituents.

isobits [ADP] Binary digits having the same value.

isobront [METEOROL] A line drawn through geographical points at which a given phase of thunderstorm activity occurred simultaneously. Also known as homobront.

Isobryales [BOT] An order of mosses in which the plants are slender to robust and up to 90 centimeters in length.

isobutane [ORG CHEM] $(CH_3)_2CHCH_3$ A colorless, stable gas, noncorrosive to metals, nonreactive with water; boils at -11.7°C; used as a chemical intermediate, refrigerant, and fuel. Also known as 2-methyl propane.

isobutanol *See* isobutyl alcohol.

isobutene *See* isobutylene.

isobutyl [ORG CHEM] The radical $(CH_3)_2CHCH_2-$, occurring, for example, in isobutanol (isobutyl alcohol), $(CH_3)_2CHCH_2OH$.

isobutyl acetate [ORG CHEM] $C_4H_9OOCCH_3$ Colorless liquid with fruitlike aroma; soluble in alcohols, ether, and hydrocarbons, insoluble in water; boils at 116°C; used as a solvent for lacquer and nitrocellulose.

isobutyl alcohol [ORG CHEM] $(CH_3)_2CHCH_2OH$ A colorless liquid that is a by-product of the synthetic production of methanol, boils at 107°C; soluble in water, ether, and alcohol; used as a solvent in paints and lacquers, in organic synthesis, and in resin coatings. Also known as isobutanol; isopropylcarbinol; 2-methyl-1-propanol.

isobutyl aldehyde [ORG CHEM] $(CH_3)_2CHCHO$ Colorless,

transparent liquid with pungent aroma; soluble in alcohol, insoluble in water; boils at 64°C; used as a chemical intermediate. Also known as isobutyraldehyde.

isobutyl carbinol [ORG CHEM] $(CH_3)_2CH(CH_2)_2OH$ Colorless liquid with pungent taste and disagreeable aroma; soluble in alcohol and ether, slightly soluble in water; boils at 132°C; used as a chemical intermediate and solvent, and in pharmaceutical products and medicines. Also known as isoamyl alcohol; 3-methyl-1-butanol; primary isoamyl alcohol.

isobutylene [ORG CHEM] $(CH_3)_2CCH_2$ Flammable, colorless, volatile liquid boiling at −7°C; easily polymerized; used in gasolines, as a chemical intermediate, and to make butyl rubber. Also known as isobutene; 2-methylpropene.

isobutyraldehyde See isobutyl aldehyde.

isobutyric acid [ORG CHEM] $(CH_3)_2CHCOOH$ Colorless liquid boiling at 154°C; soluble in water, alcohol, and ether; used as a chemical intermediate and disinfectant, in flavor and perfume bases, and for leather treating.

isobutyryl [ORG CHEM] $(CH_3)_2C \cdot CHO$ The radical group from isobutyric acid, $(CH_3)_2CHCOOH$.

isocarb [GEOCHEM] A line on a map that connects points of equal content of fixed carbon in coal.

isocarpic [BOT] Having the same number of carpels and perianth divisions.

isoceraunic [METEOROL] Indicating or having equal frequency or intensity of thunderstorm activity. Also spelled isokeraunic.

isoceraunic line [METEOROL] A line drawn through geographical points at which some phenomenon connected with thunderstorms has the same frequency or intensity; used for lines of equal frequency of lightning discharges.

isocercal [VERT ZOO] Of the tail fin of a fish, having the upper and lower lobes symmetrical and the vertebral column gradually tapering.

isochasm [GEOPHYS] A line connecting points on the earth's surface at which the aurora is observed with equal frequency. Also known as isaurore.

isochela [INV ZOO] 1. A chela having two equally developed parts. 2. A chelate spicule with both ends identical.

isochemical metamorphism [PETR] Theoretically, a metamorphism involving no great change in its chemical composition. Also known as treptomorphism.

isochemical series [PETR] A series of rocks with identical chemical compositions.

isochor See isochore.

isochore [PHYS] A graph that shows the variation of one quantity with another; for example, the variation of pressure with temperature, when the volume of the substance is held constant. Also known as isochor; isometric.

isochoric [PHYS] Taking place without change in volume. Also known as isovolumic.

isochromatic [OPTICS] 1. Pertaining to a variation of certain quantities related to light (such as density of the medium through which the light is passing, index of refraction), in which the color or wavelength of the light is held constant. 2. Pertaining to lines connecting points of the same color.

isochromatic fringe pattern [OPTICS] A pattern of bands, each of uniform color, observed when a plate is placed in a polariscope and subjected to stress, making it birefringent.

isochromosome [CYTOL] An abnormal chromosome with a medial centromere and identical arms formed as a result of transverse, rather than longitudinal, splitting of the centromere.

isochronal test [PETRO ENG] Short-time back-pressure test for low-permeability reservoirs that otherwise require excessively long times for pressure stabilization when wells are shut in.

isochrone [PHYS] A line on a chart connecting all points having the same time of occurrence of particular phenomena or of a particular value of a quantity.

isochronism [MECH] The property of having a uniform rate of operation or periodicity, for example, of a pendulum or watch balance.

isochronon [HOROL] A clock designed to keep very accurate time.

isochronous [PHYS] Having a fixed frequency or period.

isochronous circuits [ELEC] Circuits having the same resonant frequency.

isochronous governor [MECH ENG] A governor that keeps the speed of a prime mover constant at all loads. Also known as astatic governor.

isocitric acid [BIOCHEM] $HOOCCH_2CH(COOH)CH(OH)-COOH$ An isomer of citric acid that is involved in the Krebs tricarboxylic acid cycle in bacteria and plants. Also known as 2-hydroxy-1,2,3-propane-tricarboxylic acid.

isoclasite [MINERAL] $Ca_2(PO_4)(OH) \cdot 2H_2O$ A white mineral composed of a basic hydrous calcium phosphate; occurring in small crystals or columnar forms.

isoclinal See isoclinic line.

isoclinal chart [GEOPHYS] A chart showing isoclinic lines. Also known as isoclinic chart.

isocline [GEOL] A fold of strata so tightly compressed that parts on each side dip in the same direction.

isoclinic chart See isoclinal chart.

isoclinic line [GEOPHYS] A line connecting points on the earth's surface which have the same magnetic dip. Also known as isoclinal. [SOLID STATE] A line joining points in a plate at which the principal stresses have parallel directions.

isoconcentration [CHEM ENG] Constant concentration values.

isoconcentration map [CHEM ENG] Map or diagram or a liquid or gas system's concentration with respect to a single component of the system, shown by constant-concentration contour lines.

isocrackate [MATER] The liquid products of the process of isocracking.

isocracking [CHEM ENG] A hydrocracking process for conversion of hydrocarbons into more valuable, lower-boiling products; operates at relatively low temperatures and pressures in the presence of hydrogen and a catalyst.

Isocrinida [INV ZOO] An order of stalked articulate echinoderms with nodal rings of cirri.

isocyanate [ORG CHEM] 1. One of a group of neutral derivatives of primary amines; its formula is $R-N=C=O$, where R may be an alkyl or aryl group; an example is 2,4-toluene diisocyanate. 2. Any compound containing the $-N=L=O$ radical.

isocyanate resin [ORG CHEM] A linear alkyd resin lengthened by reaction with isocyanates, then treated with a glycol or diamine to cross-link the molecular chain; the product has good abrasion resistance.

isocyanic acid [ORG CHEM] $HN=C=O$ One of two forms of cyanic acid; a gas used as an intermediate in the preparation of polyurethane and other resins.

isocyanine [ORG CHEM] Any one of a series of dyes whose structure has two heterocyclic or quinoline rings connected by an odd number chain of carbon atoms containing conjugated double bonds; for example, cyanine blue.

isocyanuric acid See fulminuric acid.

isodiametric [BIOL] Having equal diameters or dimensions.

isodont [VERT ZOO] 1. Having all teeth alike. 2. Of a snake, having the maxillary teeth of equal length.

isodose curve [NUCLEO] A curve, drawn on a chart of an object, connecting points receiving equal doses of radiation.

isodrosotherm [METEOROL] An isogram of dew-point temperature.

isodulcitol See rhamnose.

isodynamic [MECH] Pertaining to equality of two or more forces or to constancy of a force.

isodynamic line [GEOPHYS] One of the lines on a map of a magnetic field that connect points having equal strengths of the earth's field.

isoelectric [ELEC] Pertaining to a constant electric potential.

isoelectric point [PHYS CHEM] The pH value of the dispersion medium of a colloidal suspension at which the colloidal particles do not move in an electric field.

isoelectric precipitation [CHEM] Precipitation of materials at the isoelectric point (the pH at which the net charge on a molecule in solution is zero); proteins coagulate best at this point.

isoelectronic [ATOM PHYS] Pertaining to atoms having the same number of electrons outside the nucleus of the atom.

isoelectronic sequence [SPECT] A set of spectra produced by different chemical elements ionized so that their atoms or ions contain the same number of electrons.

isoenzyme [BIOCHEM] Any of the electrophoretically distinct

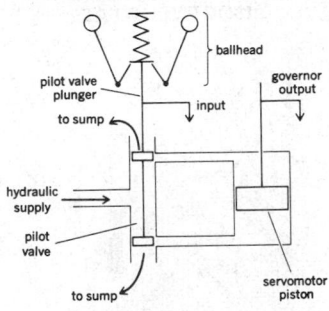
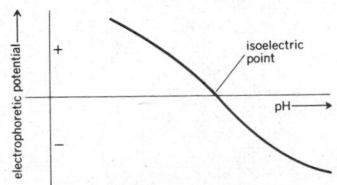

forms of an enzyme, representing different polymeric states but having the same function. Also known as isozyme.

Isoetaceae [BOT] The single family assigned to the order Isoetales in some systems of classification.

Isoetales [BOT] A monotypic order of the class Isoetopsida containing the single genus *Isoetes*, characterized by long, narrow leaves with a spoonlike base, spirally arranged on an underground cormlike structure.

Isoetatae See Isoetopsida.

Isoetopsida [BOT] A class of the division Lycopodiophyta; members are heterosporous and have a distinctive appendage, the ligule, on the upper side of the leaf near the base.

isoeugenol [ORG CHEM] $C_{10}H_{12}O_2$ An oily liquid prepared from eugenol by heating, slightly soluble in water; used in the manufacture of vanillin.

isofacies map [GEOL] A stratigraphic map showing the distribution of one or more facies within a particular stratigraphic unit.

isoflavone [BIOCHEM] $C_{15}H_{10}O_2$ A colorless, crystalline ketone, occurring in many plants, generally in the form of a hydroxy derivative.

isofootcandle See isolux.

isoforming [CHEM ENG] A petroleum refinery process in which olefinic naphtha is contacted with an alumina catalyst at high temperature and low pressure to produce isomers of higher octane number.

isofronts-preiso code [METEOROL] A code in which data on isobars and fronts at sea level (or earth's surface) are encoded and transmitted; a modified form of the international analysis code.

isogamete [BIOL] A reproductive cell that is morphologically similar in both male and female and cannot be distinguished on form alone.

isogamy [BIOL] Sexual reproduction by union of gametes or individuals of similar form or structure.

Isogeneratae [BOT] A class of brown algae distinguished by having an isomorphic alternation of generations.

isogeotherm See geoisotherm.

isogonic chart [GEOPHYS] A chart showing isogonics.

isogonic line [GEOPHYS] A line connecting points of equal magnetic variation.

isogony [BIOL] Growth of parts at such a rate as to maintain relative size differences.

isogor [PETRO ENG] Constant gas-oil ratio.

isogor map [PETRO ENG] Oil reservoir contour-line map that shows constant gas-oil ratios.

isograd [GEOL] A line on a map joining those rocks comprising the same metamorphic grade.

isogradient [METEOROL] A line connecting points having the same horizontal gradient of atmospheric pressure, temperature, and so on.

isogram [METEOROL] A line, on a given reference surface, drawn through all points where a given quantity has the same numerical value; the reference surface can be any coordinate plane functionally related to the given quantity (this includes physically defined surfaces in space).

isograph [ELECTR] An electronic calculator that ascertains both real and imaginary roots for algebraic equations. [GRAPHICS] An instrument combining the functions of a protractor and a set square and comprising two short straight-edges connected by a large circular joint with an angular-degree scale.

isogriv [NAV] A line drawn on a map or chart joining points of equal grivation.

isogriv chart [NAV] A chart showing isogrivs.

isogyre [OPTICS] A dark band in an interference figure located at those points that correspond to directions of transmission through the crystal plate in which the polarization of the incident light is not affected by passing through the plate.

isohaline [OCEANOGR] 1. Of equal or constant salinity. 2. A line on a chart connecting all points of equal salinity.

isoheight See contour line.

isohel [METEOROL] A line drawn through geographical points having the same duration of sunshine (or other function of solar radiation) during any specified time period.

isohemagglutinin See isoagglutinin.

isohemolysin [IMMUNOL] A hemolysin produced by an indi-

vidual injected with erythrocytes from another individual of the same species.

isohemolysis [IMMUNOL] Hemolysis induced by the action of an isohemolysin.

isohexane [ORG CHEM] C_6H_{14} A liquid mixture of isomeric hydrocarbons, flammable and explosive, insoluble in water, soluble in most organic solvents, boils at 54–61°C; used as a solvent, freezing-point depressant, and chemical intermediate.

isohume [GEOL] A line of a map or chart connecting points of equal moisture content in a coal bed. [METEOROL] A line drawn through points of equal humidity on a given surface; an isopleth of humidity; the humidity measures used may be the relative humidity or the actual moisture content (specific humidity or mixing ratio).

isohydric [CHEM] Referring to a set of solutions with the same hydrogen ion concentration and not affecting the conductivity of each of the various solutions on mixing.

isohyet [METEOROL] A line drawn through geographic points recording equal amounts of precipitation for a specified period or for a particular storm.

isohypse See contour line.

isohypsic chart See constant-height chart.

isohypsic surface See constant-height surface.

isoimmunization [IMMUNOL] Immunization of an individual by the introduction of antigens from another individual of the same species.

Iso-Kel process [CHEM ENG] A proprietary petroleum isomerization process for the manufacture of high-octane fuel components from petroleum feedstocks.

isokeraunic See isoceraunic.

isokinetic See isotach.

isokinetic sampling [ENG] Any technique for collecting airborne particulate matter in which the collector is so designed that the airstream entering it has a velocity equal to that of the air passing around and outside the collector.

isolate [CHEM ENG] To separate two portions of a process system by means of valving or line blanks; used as safety measure during maintenance or repair, or to redirect process flows.

isolated camera [ELECTR] 1. A television camera that views a particular portion of a scene of action and produces a tape which can then be used either immediately for instant replay or for video replay at a later time. 2. The technique of video replay involving such a camera.

isolated footing [CIV ENG] A concrete slab or block under an individual load or column.

isolated point [MATH] A point is isolated from a subset of a topological space if it can be enclosed in an open neighborhood including no other points of the subset. Also known as acnode.

isolated set [MATH] A set consisting entirely of isolated points.

isolating switch [ELEC] A switch intended for isolating an electric circuit from the source of power; it has no interrupting rating and is intended to be operated only after the circuit has been opened by some other means.

isolation [CHEM] Separation of a pure chemical substance from a compound or mixture; as in distillation, precipitation, or absorption. [MED] Separation of an individual with a communicable disease from other, healthy individuals. [MICROBIO] Separation of an individual or strain from a natural, mixed population. [PHYSIO] Separation of a tissue, organ, system, or other part of the body for purposes of study. [PSYCH] Dissociation of a memory or thought from the emotions or feelings associated with it.

isolation amplifier [ELECTR] An amplifier used to minimize the effects of a following circuit on the preceding circuit.

isolation network [ELEC] A network inserted in a circuit or transmission line to prevent interaction between circuits on each side of the insertion point.

isolation transformer [ELEC] A transformer inserted in a system to separate one section of the system from undesired influences of other sections.

isolator [ELECTR] A passive attenuator in which the loss in one direction is much greater than that in the opposite direction; a ferrite isolator for waveguides is an example.

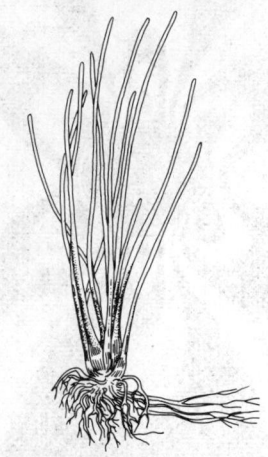
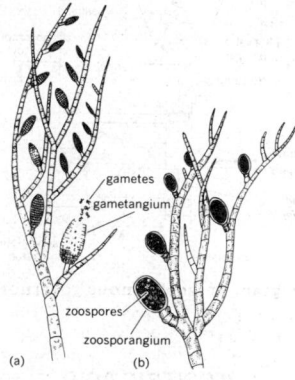

[ENG] Any device that absorbs vibration or noise, or prevents its transmission.

isolead curve [ORD] A curved line, on a chart or diagram, used to show how far ahead of a moving target a weapon must be aimed to allow for the time the projectile takes to reach the target; the isolead curve connects points of equal lead on the chart or diagram.

isolecithal *See* homolecithal.

isoleucine [BIOCHEM] $C_6H_{13}O_2$ An essential monocarboxylic amino acid occurring in most dietary proteins.

isolith [ELECTR] Integrated circuit of components formed on a single silicon slice, but with the various components interconnected by beam leads and with circuit parts isolated by removal of the silicon between them. [GEOL] A line on a contour-type map that denotes the aggregate thickness of a single lithology in a stratigraphic succession composed of one or more lithologies.

isolith map [GEOL] A contour-line map depicting the thickness of an exclusive lithology.

isolux [OPTICS] A curve or surface connecting points at which light intensity is the same. Also known as isofootcandle; isophot.

isomagnetic [GEOPHYS] Of or pertaining to lines connecting points of equality in some magnetic element.

Isomate process [CHEM ENG] A proprietary petroleum isomerization process that takes place in the liquid phase for the manufacture of high-octane components from petroleum feedstocks.

isomer [CHEM] One of two or more chemical substances having the same elementary percentage composition and molecular weight but differing in structure, and therefore in properties; there are many ways in which such structural differences occur; one example is provided by the compounds *n*-butane, $CH_3(CH_2)_2CH_3$, and isobutane, $CH_3CH(CH_3)_2$. [NUC PHYS] One of two or more nuclides having the same mass number and atomic number, but existing for measurable times in different quantum states with different energies and radioactive properties.

isomerase [BIOCHEM] An enzyme that catalyzes isomerization reactions.

Isomerate process [CHEM ENG] A proprietary petroleum isomerization process that takes place in the vapor phase for the manufacture of high-octane fuel components from petroleum feedstocks.

isomeric shift [PHYS CHEM] Shift in the Mössbauer resonance caused by the effect of the valence of the atom on the interaction of the electron density at the nucleus with the nuclear charge. Also known as chemical shift.

isomeric transition [NUC PHYS] A radioactive transition from one nuclear isomer to another of lower energy.

isomerism [BIOL] The condition of having two or more comparable parts made up of identical numbers of similar segments. [CHEM] The phenomenon whereby certain chemical compounds have structures that are different although the compounds possess the same elemental composition. [NUC PHYS] The occurrence of nuclear isomers.

isomerization [CHEM] A process whereby a compound is changed into an isomer; for example, conversion of butane into isobutane.

isomerous [BIOL] Characterized by isomerism.

Isometopidae [INV ZOO] A family of hemipteran insects in the superfamily Cimicimorpha.

isometric *See* isochore.

isometric drawing [GRAPHICS] A method of nonperspective pictorial drawing in which the object being drawn is turned so that three mutually perpendicular edges are equally foreshortened.

isometric particle [VIROL] A plant virus particle that appears at first sight to be spherical when viewed in the electron microscope, but which is actually an icosahedron, possessing 20 sides.

isometric process [THERMO] A constant-volume, frictionless thermodynamic process in which the system is confined by mechanically rigid boundaries.

isometric projection *See* axonometric projection.

isometric system [CRYSTAL] The crystal system in which the

forms are referred to three equal, mutually perpendicular axes. Also known as cubic system.

isometry [MATH] A mapping f from a metric space X to a metric space Y where the distance between any two points of X equals the distance between their images under f in Y.

isomolecule *See* nonlinear molecule.

isomorphism [MATH] A one to one function of an algebraic structure (for example, group, ring, module, vector space) onto another of the same type, preserving all algebraic relations; its inverse function behaves likewise. [PHYS CHEM] A condition present when an ion at high dilution is incorporated by mixed crystal formation into a precipitate, even though such formation would not be predicted on the basis of crystallographic and ionic radii; an example is coprecipitation of lead with potassium chloride. [SCI TECH] The quality or state of being identical or similar in form, shape, or structure, such as between organisms resulting from evolutionary convergence, or crystalline forms of similar composition.

isoneph [METEOROL] A line drawn through all points on a map having the same amount of cloudiness.

isoniazid [PHARM] $C_6H_7N_3O$ A drug used as a tuberculostatic. Also known as isonicotinic acid hydrazide.

isonicotinic acid [ORG CHEM] $C_6H_5NO_2$ White platelets or powder, slightly soluble in water, sublimes at 260°C; used in the manufacture of isonicotinic acid hydrazide, an antitubercular agent. Also known as pyridine-4-carboxylic acid.

isonicotinic acid hydrazide *See* isoniazid.

isooctane [ORG CHEM] $(CH_3)_2CHCH_2C(CH_3)_3$ Flammable, colorless liquid boiling at 99°C; slightly soluble in alcohol and ether, insoluble in water; used in motor fuels and as a chemical intermediate. Also known as 2,2,4-trimethylpentane.

isooctyl alcohol [ORG CHEM] $C_7H_{15}CH_2OH$ Mixture of isomers from oxo-process synthesis; boils at 182–195°C; used as a chemical intermediate, resin solvent, emulsifier, and antifoaming agent.

isopach map [GEOL] Map of the areal extent and thickness variation of a stratigraphic unit; used in geological exploration for oil and for underground structural analysis.

isopachous line [GEOL] One of the lines drawn on a map to indicate equal thickness.

isoparaffin [ORG CHEM] A branched-chain version of a straight-chain (normal) saturated hydrocarbon; for example, isooctane, or 2,2,4-trimethyl pentane, $(CH_3)_3C_5H_9$, is the branched-chain version of *n*-octane, $CH_3(CH_2)_6CH_3$.

isopathic principle [PSYCH] The rule according to which the cause cures the effect, as when a feeling of guilt is relieved by an exhibition of guilt, namely hate.

isopectic [CLIMATOL] A line on a map connecting points at which ice begins to form at the same time of winter.

isopentane [ORG CHEM] $CH_3CHCH_3CH_2CH_3$ Flammable, colorless liquid with pleasant aroma; boils at 28°C; soluble in oils, ether, and hydrocarbons, insoluble in water; used as a solvent and chemical intermediate. Also known as 2-methylbutane.

isoperimetrical figures *See* isoperimetric figures.

isoperimetric figures [MATH] Figures whose perimeters are equal. Also known as isoperimetrical figures.

isoperimetric line [MAP] A line on a map projection that indicates no variation from exact scale.

isoperimetric problem [MATH] In the calculus of variations this problem deals with finding a closed curve in the plane which encloses the greatest area given its length as fixed.

isoperm [PETRO ENG] One of the lines of equal (constant) permeability plotted on a reservoir map.

isophene [BIOL] A line on a chart connecting those places within a given region where a particular biological phenomenon (as the flowering of a certain plant) occurs at the same time.

isophenomenal [METEOROL] Having the same amount of any atmospheric phenomenon.

isophorone [ORG CHEM] $COCHC(CH_3)CH_2C(CH_3)_2CH_2$ A water-white liquid boiling at 215°C; used as a solvent for lacquers and polyvinyl and nitrocellulose resins.

isophot *See* isolux.

isophotometer [OPTICS] A direct-recording photometer that automatically scans and measures optical density of all points in a film transparency or plate, and plots the measured density values in a quantitative two-dimensional isodensity tracing of the scanned areas.

isophthalic acid [ORG CHEM] $C_6H_4(COOH)_2$ Colorless crystals subliming at 345°C; slightly soluble in water, soluble in alcohol and acetic acid, and insoluble in benzene; used as an intermediate for polyester and polyurethane resins, and as a plasticizer. Also known as *meta*-phthalic acid.

isophyllous [BOT] Having foliage leaves of similar form on a plant or stem.

isopiestic [PHYS] Denoting equal or constant pressure.

isopiestic line [HYD] A line indicating on a map the piezometric surface of an aquifer.

isopleth [MATH] The straight line which cuts the three scales of a nomograph at values satisfying some equation. Also known as index line. [METEOROL] **1.** A line of equal or constant value of a given quantity with respect to either space or time. Also known as isogram. **2.** More specifically, a line drawn through points on a graph at which a given quantity has the same numerical value (or occurs with the same frequency) as a function of the two coordinate variables. Also known as isarithm.

Iso-Plus Houdriforming [CHEM ENG] A naphtha-reforming process used in petroleum refineries; a Houdriformer catalytic reforming unit is operated at moderate severity, and aromatics are recovered from the product, with raffinate reformed separately or recycled.

isopluvial [METEOROL] A line on a map drawn through geographical points having the same pluvial index.

Isopoda [INV ZOO] An order of malacostracan crustaceans characterized by a cephalon bearing one pair of maxillipeds in addition to the antennae, mandibles, and maxillae.

isopolymolybdate [INORG CHEM] A class of compounds formed by the acidification of a molybdate solution, or in some cases by heating normal molybdates.

isopolytungstate [INORG CHEM] A compound formed by the condensation of tungstate compounds, usually classified into metatungstates, such as $Na_6W_{12}O_{40} \cdot xH_2O$, and paratungstates, such as $Na_{10}W_{12}O_{41} \cdot xH_2O$.

isopor [GEOPHYS] An imaginary line connecting points on the earth's surface having the same annual change in a magnetic element.

isopotential level *See* potentiometric surface.

isopotential map [PETRO ENG] A contour-line map to show the initial or calculated daily rate of oil well production in a multiwell field.

isoprecipitin [IMMUNOL] A precipitin effective only against the serum of individuals of the same species from which it is derived.

ISOPRENE

Structural formula of isoprene.

isoprene [ORG CHEM] C_5H_8 A conjugated diolefin; a mobile, colorless liquid having a boiling point of 34.1°C; insoluble in water, soluble in alcohol and ether; polymerizes readily to form dimers and high-molecular-weight elastomer resins.

isopropanol *See* isopropyl alcohol.

isopropyl [ORG CHEM] The radical $(CH_3)_2CH$, from isopropane; an example of its occurrence is in isopropyl alcohol, $(CH_3)_2CHOH$.

isopropyl alcohol [ORG CHEM] $(CH_3)_2CHOH$ A colorless liquid that boils at 82.4°C; soluble in water, ether, and ethanol; used in manufacturing of acetone and its derivatives, of glycerol, and as a solvent. Also known as isopropanol; 2-propanol; *sec*-propyl alcohol.

para-**isopropylaniline** *See* cumidine.

isopropylarterenol *See* isoproterenol.

isopropylbenzene *See* cumene.

isopropylcarbinol *See* isobutyl alcohol.

isopropyl-*meta*-cresol *See* thymol.

isopropyl-*ortho*-cresol *See* carvacrol.

isopropyl ether [ORG CHEM] $(CH_3)_2CHOCH(CH_3)_2$ Water-soluble, flammable, colorless liquid with etherlike aroma; boils at 68°C; used as a solvent and extractant, in paint and varnish removers, and in spotting formulas. Also known as diisopropyl ether.

4-isopropyl-1-methylcyclohexane *See* menthane.

7-isopropyl-1-methyl phenanthrene *See* retene.

para-**isopropyltoluene** *See* cymene.

isoproterenol [PHARM] $C_{11}H_{17}NO_3$ A sympathomimetic amine used as a bronchodilator. Also known as isopropylarterenol.

Isoptera [INV ZOO] An order of Insecta containing morphologically primitive forms characterized by gradual metamorphosis, lack of true larval and pupal stages, biting and prognathous mouthparts, two pairs of subequal wings, and the abdomen joined broadly to the thorax.

isopulegol [ORG CHEM] $C_{10}H_{17}OH$ An alcohol derived from terpene as a water-white liquid that has a mintlike odor; used in making perfumes.

isopulse system [COMMUN] In adaptive communications, a pulse coding system wherein the number of information pulses transmitted is indicated by special inserted pulses.

isopycnic [METEOROL] A line on a chart connecting all points of equal or constant density. [PHYS] Of equal or constant density, with respect to either space or time.

isopycnic level [METEOROL] Specifically, a level surface in the atmosphere, at about 8 kilometers altitude, where the air density is approximately constant in space and time; this level corresponds to the maximum upper-tropospheric interdiurnal pressure variation.

isopygous [INV ZOO] Having a pygidium and a cephalon of equal size, as in certain trilobites.

isoquinoline [ORG CHEM] $C_6H_4CHNCHCH$ Colorless liquid boiling at 243°C; soluble in most organic solvents and dilute mineral acids, insoluble in water; derived from coal tar or made synthetically; used to make dyes, insecticides, pharmaceuticals, and rubber accelerators, and as a chemical intermediate.

isosafrole [ORG CHEM] $C_{10}H_{10}O_2$ A liquid with the odor of anise that is obtained from safrole, and that boils at 253°C; used to make perfumes and flavors. Also known as 1,2-methylenedioxy-4-propenylbenzene.

isosceles triangle [MATH] A triangle with two sides of equal length.

isoseismal [GEOPHYS] Pertaining to points having equal intensity of earthquake shock, or to a line on a map of the earth's surface connecting such points.

isoshear [METEOROL] A line on a chart of equal magnitude of vertical wind shear.

isospin *See* isotopic spin.

isospin multiplet [PARTIC PHYS] A collection of elementary particles which have approximately the same mass and the same quantum numbers except for charge, but have a sequence of charge values, $Y/2 - I$, $Y/2 - I + 1, \ldots$, $Y/2 + I$ times the proton charge, where Y is an integer known as the hypercharge, and I is an integer or half-integer known as the isospin; examples are the pions ($Y = 0$, $I = 1$) and the nucleons ($Y = 1$, $I = \frac{1}{2}$). Also known as charge multiplet; particle multiplet.

Isospondyli [VERT ZOO] A former equivalent name for Clupeiformes.

isospore [BIOL] A spore that does not display sexual dimorphism.

isostasy [GEOPHYS] A theory of the condition of approximate equilibrium in the outer part of the earth, such that the gravitational effect of masses extending above the surface of the geoid in continental areas is approximately counterbalanced by a deficiency of density in the material beneath those masses, while deficiency of density in ocean waters is counterbalanced by an excess in density of the material under the oceans.

isostatic adjustment *See* isostatic compensation.

isostatic anomaly [GEOPHYS] A gravity anomaly based on a generalized hypothesis that the gravitational effect of masses above sea level is approximately compensated by a density deficiency of the subsurface materials.

isostatic compensation [GEOL] The process in which lateral transport at the surface of the earth by erosion or deposition is compensated by lateral movements in a subcrustal layer. Also known as isostatic adjustment; isostatic correction.

isostatic correction *See* isostatic compensation.

isostatics [MECH] In photoelasticity studies of stress analyses, those curves, the tangents to which represent the progressive change in principal-plane directions. Also known as stress trajectories.

isostemonous [BOT] Having the number of stamens of a flower equal to the number of perianth divisions.

isosteric [CHEM] Referring to similar electronic arrangements in chemical compounds. [PHYS] Of equal or constant specific volume with respect to either time or space.

isosterism [PHYS CHEM] A similarity in the physical properties of ions, compounds, or elements, as a result of electron arrangements that are identical or similar.

isostructural [CRYSTAL] Pertaining to crystalline materials that have corresponding atomic positions, and have a considerable tendency for ionic substitution.

isosulfur map [PETRO ENG] A contour-line map to show the percentage of sulfur in underground crude oil.

isotach [METEOROL] A line in a given surface connecting points with equal wind speed. Also known as isokinetic; isovel.

isotach chart [METEOROL] A synoptic chart showing the distribution of wind by means of isotachs.

isotactic [ORG CHEM] Designating crystalline polymers in which substituents in the asymmetric carbon atoms have the same (rather than random) configuration in relation to the main chain.

isothere [CLIMATOL] A line on a map connecting points having the same mean summer temperature.

isotherm [GEOPHYS] A line on a chart connecting all points of equal or constant temperature. [THERMO] A curve or formula showing the relationship between two variables, such as pressure and volume, when the temperature is held constant. Also known as isothermal.

isothermal [THERMO] **1.** Having constant temperature; at constant temperature. **2.** See isotherm.

isothermal annealing [MET] Transformation of an austenitic steel to ferrite and pearlite at constant temperature. Also known as isothermal transformation.

isothermal atmosphere [METEOROL] An atmosphere in hydrostatic equilibrium, in which the temperature is constant with height and the pressure decreases exponentially upward. Also known as exponential atmosphere.

isothermal calorimeter [THERMO] A calorimeter in which the heat received by a reservoir, containing a liquid in equilibrium with its solid at the melting point or with its vapor at the boiling point, is determined by the change in volume of the liquid.

isothermal chart [GEOPHYS] A map showing the distribution of air temperature (or sometimes sea-surface or soil temperature) over a portion of the earth or at some level in the atmosphere; places of equal temperature are connected by lines called isotherms.

isothermal compression [THERMO] Compression at constant temperature.

isothermal equilibrium [METEOROL] The state of an atmosphere at rest, uninfluenced by any external agency, in which the conduction of heat from one part to another has produced, after a sufficient length of time, a uniform temperature throughout its entire mass. Also known as conductive equilibrium. [THERMO] The condition in which two or more systems are at the same temperature, so that no heat flows between them.

isothermal expansion [THERMO] Expansion of a substance while its temperature is held constant.

isothermal flow [THERMO] Flow of a gas in which its temperature does not change.

isothermal layer [METEOROL] The approximately isothermal region of the atmosphere immediately above the tropopause. [THERMO] A layer of fluid, all points of which have the same temperature.

isothermal magnetization [THERMO] Magnetization of a substance held at constant temperature; used in combination with adiabatic demagnetization to produce temperatures close to absolute zero.

isothermal process [THERMO] Any constant-temperature process, such as expansion or compression of a gas, accompanied by heat addition or removal from the system at a rate just adequate to maintain the constant temperature.

isothermal remanent magnetization [GEOPHYS] Remanent magnetization that has been acquired under normal temperature conditions in a short period of time as a result of the application of a magnetic field. Abbreviated IRM.

isothermal transformation [MET] See isothermal annealing. [THERMO] Any transformation of a substance which takes place at a constant temperature.

isothermal treatment [MET] Heat treatment of metals at constant temperature.

isothermobath [OCEANOGR] A line connecting points having the same temperature in a diagram of a vertical section of the ocean.

isotherm ribbon [METEOROL] A zone of crowded isotherms on a synoptic upper-level chart; the temperature gradient is many times greater than normally encountered in the atmosphere.

isothiocyanate [ORG CHEM] A compound of the type $R-N=C=S$, where R may be an alkyl or aryl group; an example is mustard oil. Also known as sulfocarbimide.

isotimic [METEOROL] Pertaining to a quantity which has equal value in space at a particular time.

isotimic line [METEOROL] On a given reference surface in space, a line connecting points of equal value of some quantity; most of the lines drawn in the analysis of synoptic charts are isotimic lines.

isotimic surface [METEOROL] A surface in space on which the value of a given quantity is everywhere equal; isotimic surfaces are the common reference surfaces for synoptic charts, principally constant-pressure surfaces and constant-height surfaces.

isotone [NUC PHYS] One of several nuclides having the same number of neutrons in their nuclei but differing in the number of protons.

isotonic [PHYSIO] **1.** Having uniform tension, as the fibers of a contracted muscle. **2.** Of a solution, having the same osmotic pressure as the fluid phase of a cell or tissue.

isotonic sodium chloride solution See normal saline.

isotope [NUC PHYS] One of two or more atoms having the same atomic number but different mass number.

isotope abundance [NUC PHYS] The ratio of the number of atoms of a particular isotope in a sample of an element to the number of atoms of a specified isotope, or to the total number of atoms of the element.

isotope dilution [NUCLEO] The introduction of a radioisotope into stable isotopes of an element in order to make volume, mass, and age measurements of the element.

isotope-dilution analysis [ANALY CHEM] Variation on paper-chromatography analysis; a labeled radioisotope of the same type as the one being quantitated is added to the solution, then quantitatively analyzed afterward via radioactivity measurement.

isotope effect [PHYS CHEM] The effect of difference of mass between isotopes of the same element on nonnuclear physical and chemical properties, such as the rate of reaction or position of equilibrium, of chemical reactions involving the isotopes. [SOLID STATE] Variation of the transition temperatures of the isotopes of a superconducting element in inverse proportion to the square root of the atomic mass.

isotope exchange [NUCLEO] **1.** Exchange of places by two atoms, but different isotopes, of the same element in two different molecules, or in different locations of the same molecule. **2.** The transfer of isotopically tagged atoms from one chemical form or valence state to another, without net chemical reaction.

isotope exchange reaction [CHEM] A chemical reaction in which interchange of the atoms of a given element between two or more chemical forms of the element occurs, the atoms in one form being isotopically labeled so as to distinguish them from atoms in the other form.

isotope farm [BOT] A carbon-14 growth chamber, or greenhouse, arranged as a closed system in which plants can be grown in a carbon-14 dioxide ($CO_2{}^{14}$) atmosphere and thus become labeled with C^{14}; isotope farms also can be used with other materials, such as heavy water (D_2O), phosphorus-35 (P^{35}), and so forth, to produce biochemically labeled compounds.

isotope fractionation [NUCLEO] Natural or artificial alteration of the isotopic composition of an element via processes of diffusion, evaporation, and chemical exchange, utilizing small differences in physical and chemical properties of isotopes. Also known as isotopic fractionation.

isotope separation [NUCLEO] The physical separation of different stable isotopes of an element from one another.

isotope shift [SPECT] A displacement in the spectral lines due to the different isotopes of an element.

isotopic age determination *See* radiometric dating.

isotopic carrier *See* carrier.

isotopic chronometer [NUCLEO] A method of determining the age of geological, archeological, or other samples by measuring the amount of a particular radioisotope and of its daughter isotope in a sample.

isotopic element [NUC PHYS] An element which has more than one naturally occurring isotope.

isotopic enrichment [NUCLEO] The process by which the relative abundances of the isotopes of a given element are altered in a batch, thus producing a form of the element enriched in a particular isotope.

isotopic fractionation *See* isotope fractionation.

isotopic indicator *See* isotopic tracer.

isotopic irradiation [NUCLEO] The subjection of a material to radiation from radioactive isotopes for therapeutic or other purposes.

isotopic label *See* isotopic tracer.

isotopic molecule [NUCLEO] A molecule in which the nucleus of one of the atoms is a special isotope.

isotopic number *See* neutron excess.

isotopic parity *See* G parity.

isotopic spin [NUC PHYS] A quantum-mechanical variable, resembling the angular momentum vector in algebraic structure whose third component distinguished between members of groups of elementary particles, such as the nucleons, which apparently behave in the same way with respect to strong nuclear forces, but have different charges. Also known as isobaric spin; isospin; i-spin.

isotopic tracer [CHEM] An isotope of an element, either radioactive or stable, a small amount of which may be incorporated into a sample material (the carrier) in order to follow the course of that element through a chemical, biological, or physical process, and also follow the larger sample. Also known as isotopic indicator; isotopic label; label; tag.

isotron [NUCLEO] A device for sorting isotopes of an element in which ions are accelerated to a fixed energy in a strong electric field, and a radio-frequency field then selects ions according to their velocity, which is inversely proportional to the square root of their mass.

isotropic [BIOL] Having a tendency for equal growth in all directions. [CYTOL] An ovum lacking any predetermined axis. [PHYS] Having identical properties in all directions.

isotropic antenna *See* unipole.

isotropic dielectric [ELEC] A dielectric whose polarization always has a direction that is parallel to the applied electric field, and a magnitude which does not depend on the direction of the electric field.

isotropic fabric [PETR] A random orientation in space of the elements that compose a rock.

isotropic fluid [FL MECH] A fluid whose properties are not dependent on the direction along which they are measured.

isotropic flux [PHYS] Radiation, or a flow of particles or matter, which reaches a location from all directions with equal intensity.

isotropic gain of an antenna *See* absolute gain of an antenna.

isotropic material [PHYS] A material whose properties are not dependent on the direction along which they are measured.

isotropic noise [ELECTROMAG] Random noise radiation which reaches a location from all directions with equal intensity.

isotropic plasma [PL PHYS] A plasma whose properties, such as pressure, are not dependent on the direction along which they are measured.

isotropic radiation [ELECTROMAG] Radiation which is emitted by a source in all directions with equal intensity, or which reaches a location from all directions with equal intensity.

isotropic radiator [PHYS] An energy source that radiates uniformly in all directions.

isotropic turbulence [FL MECH] Turbulence whose proper-

ties, especially statistical correlations, do not depend on direction.

isotropic universe [ASTRON] A universe postulated to have the same properties when viewed from all directions.

isotropy [PHYS] The quality of a property which does not depend on the direction along which it is measured, or of a medium or entity whose properties do not depend on the direction along which they are measured.

isotypes [GEN] A series of antigens, for example, blood types, common to all members of a species but differentiating classes and subclasses within the species.

isotypic [CRYSTAL] Pertaining to a crystalline substance whose chemical formula is analogous to, and whose structure is like, that of another specified compound.

isovalent conjugation [PHYS CHEM] An arrangement of bonds in a conjugated molecule such that alternative structures with an equal number of bonds can be written; an example occurs in benzene.

isovalent hyperconjugation [PHYS CHEM] An arrangement of bonds in a hyperconjugated molecule such that the number of bonds is the same in the two resonance structures but the second structure is energetically less favorable than the first structure; examples are $H_3 \equiv C - C^+ H_2$ and $H_3 \equiv C - CH_2$.

isovaleric acid [ORG CHEM] $(CH_3)_2CHCH_2COOH$ Colorless liquid with disagreeable taste and aroma; boils at 176°C; soluble in alcohol and ether; found in valeriana, hop, tobacco, and other plants; used in flavors, perfumes, and medicines.

isovel *See* isotach.

isovolumic *See* isochoric.

isozyme *See* isoenzyme.

ISP *See* imperial smelting process.

i-spin *See* isotopic spin.

isthmus [BIOL] A passage or constricted part connecting two parts of an organ. [GEOGR] A narrow strip of land having water on both sides and connecting two large land masses.

Istiophoridae [VERT ZOO] The billfishes, a family of oceanic perciform fishes in the suborder Scombroidei.

Isuridae [VERT ZOO] The mackerel sharks, a family of pelagic, predacious galeoids distinguished by a heavy body, nearly symmetrical tail, and sharp, awllike teeth.

itacolumite [PETR] A fine-grained, thin-bedded sandstone or a schistose quartzite that contains mica, chlorite, and talc and that exhibits flexibility when split into slabs. Also known as articulite.

itaconic acid [ORG CHEM] $CH_2:C(COOH)CH_2COOH$ A colorless crystalline compound that decomposes at 165°C, prepared by fermentation with *Aspergillus terreus;* used as an intermediate in organic synthesis and in resins and plasticizers. Also known as methylene succinic acid.

itatartaric acid [ORG CHEM] $C_5H_8O_6$ A compound produced experimentally by fermentation; formed as a minor product, 5.8% of total acidity produced, of an itaconic-acid producing strain of *Aspergillus niger.*

IT calorie *See* calorie.

itch [PHYSIO] An irritating cutaneous sensation allied to pain.

item [ADP] A set of adjacent digits, bits, or characters which is treated as a unit and conveys a single unit of information.

item advance [ADP] A technique of efficiently grouping records to optimize the overlap of read, write, and compute times.

item design [ADP] The specification of what fields make up an item, the order in which the fields are to be recorded, and the number of characters to be allocated to each field.

item size [ADP] The length of an item expressed in characters, words, or blocks.

iterated integral [MATH] An integral over an area or volume designated to be performed by successive integrals over line segments.

iteration *See* iterative method.

iteration process [ADP] The process of repeating a sequence of instructions with minor modifications between successive repetitions.

iterations per second [ADP] In computers, the number of approximations per second in iterative division; the number of times an operational cycle can be repeated in 1 second.

iterative array [ADP] In a computer, an array of a large

ITACONIC ACID

Structural formula of itaconic acid.

number of interconnected identical processing modules, used with appropriate driver and control circuits to permit a large number of simultaneous parallel operations.

iterative division [ADP] In computers, a method of dividing by use of the operations of addition, subtraction, and multiplication; a quotient of specified precision is obtained by a series of successively better approximations.

iterative filter [ELECTR] Four-terminal filter that provides iterative impedance.

iterative impedance [ELECTR] Impedance that, when connected to one pair of terminals of a four-terminal transducer, will cause the same impedance to appear between the other two terminals.

iterative method [MATH] Any process of successive approximation used in such problems as numerical solution of algebraic equations, differential equations, or the interpolation of the values of a function. Also known as iteration.

iterative process [MATH] A process for calculating a desired result by means of a repeated cycle of operations, which comes closer and closer to the desired result; for example, the arithmetical square root of N may be approximated by an iterative process using additions, subtractions, and divisions only.

Ithomiinae [INV ZOO] The glossy-wings, a subfamily of weak-flying lepidopteran insects having on the wings broad, transparent areas in which the scales are reduced to short hairs.

Itonididae [INV ZOO] The gall midges, a family of orthorrhaphous dipteran insects in the series Nematocera; most are plant pests.

Itô's formula *See* stochastic chain rule.

Itô's integral *See* stochastic integral.

ITV *See* industrial television.

Ivanov reagent [ORG CHEM] A reagent that is similar to a Grignard reagent, and that is formed by reacting an arylacetic acid or its sodium salt with isopropyl magnesium halide.

ivory [MATER] The ivory-white material composing the tusks and teeth of the elephant; specific gravity is 1.87; takes a high polish and is used for ornamental parts and art objects, and formerly for piano keys.

ivory black [MATER] Animal black made by calcining ivory; used as a pigment.

ivory board [MATER] A highly finished cardboard that is clay-coated on both sides; used for art printing and menu cards.

ivory point [ENG] A small pointer extending downward from the top of the cistern of a Fortin barometer; the level of the mercury in the cistern is adjusted so that it just comes in contact with the end of the pointer, thus setting the zero of the barometer scale.

Ixion [PHYS] An experimental magnetic-mirror device used for research on controlled fusion; involves study of plasma rotation in a magnetic-mirror confinement system using crossed electric and magnetic fields.

Ixodides [INV ZOO] The ticks, a suborder of the Acarina distinguished by spiracles behind the third or fourth pair of legs.

Izod test [MET] An impact test in which a falling pendulum strikes a fixed, usually notched specimen with 120 foot-pounds (163 joules) of energy at a velocity of 11.5 feet (3.5 meters) per second; the height of the pendulum swing after striking is a measure of the energy absorbed and thus indicates impact strength.

J-1 [PARTIC PHYS] A neutral particle which has a mass of 3.105 giga electron volts and probably spin quantum number 1 and negative parity; it has a lifetime on the order of 10^{-19} second, at least 10^3 times longer than expected of a particle decaying through the strong interactions.

J-2 [PARTIC PHYS] A neutral particle which has a mass of 3.695 giga electron volts and probably spin quantum number 1 and negative parity; like the J-1, it has an anomalously long lifetime.

jaagsiekte [VET MED] A contagious disease of sheep, sometimes of goats and guinea pigs, resembling the more benign and diffuse forms of bronchiolar carcinoma in humans.

Jablochkoff candle [ELECTR] An early type of arc lamp in which carbons were placed side by side and separated by plaster of Paris.

Jacanidae [VERT ZOO] The jacanas or lily-trotters, the single family of the avian superfamily Jacanoidea.

Jacanoidea [VERT ZOO] A monofamilial superfamily of colorful marsh birds distinguished by greatly elongated toes and claws, long legs, a short tail, and a straight bill.

jacaranda *See* carob wood.

J acid *See* 2-amino-5-naphthol-7-sulfonic acid.

jacinth *See* zircon.

jack [ELEC] A connecting device into which a plug can be inserted to make circuit connections; may also have contacts that open or close to perform switching functions when the plug is inserted or removed. [MECH ENG] A portable device for lifting heavy loads through a short distance, operated by a lever, a screw, or a hydraulic press. [MINERAL] *See* sphalerite. [TEXT] **1.** A frame in lace-manufacturing equipment that has horizontal bars to support fixed vertical wires, against which bobbins containing the yarn can freely revolve. **2.** An oscillating lever that raises the harness of a dobby loom.

jackal [VERT ZOO] **1.** *Canis aureus.* A wild dog found in southeastern Europe, southern Asia, and northern Africa. **2.** Any of various similar Old World wild dogs; they resemble wolves but are smaller and more yellowish.

jackbit [DES ENG] A drilling bit used to provide the cutting end in rock drilling; the bit is detachable and either screws on or is taper-fitted to a length of drill steel. Also known as ripbit.

jack chain [DES ENG] **1.** A chain made of light wire, with links arranged in figure-eights with loops at right angles. **2.** A toothed endless chain for moving logs.

jacket [NUCLEO] A thin container for one or more fuel slugs, used to prevent the fuel from escaping into the coolant of a reactor. Also known as can; cartridge. [ORD] **1.** Cylinder of steel covering and strengthening the breech end of a gun or howitzer tube. **2.** The water jacket on some machine guns.

jacket crown [MED] An artificial crown of a tooth consisting of a covering of porcelain or resin.

jacket face [GRAPHICS] The index-readable side of a microfilm.

jacket gage [GRAPHICS] Acetate thickness measurement of microfilm in thousandths of an inch.

jackhammer [MECH ENG] A hand-held rock drill operated by compressed air.

jacking [TEXT] A process in spinning to provide extra twist or draft to the roving in a mule.

jack ladder [ENG] A V-shaped trough holding a toothed endless chain, and used to move logs from pond to sawmill. [NAV ARCH] *See* Jacob's ladder.

jackleg [ENG] A supporting bar used with a jackhammer.

jack line [PETRO ENG] A steel cable or rod that connects the arms of the central pumping engine to the two or more wells that are being pumped.

jack plane [DES ENG] A general-purpose bench plane measuring over 1 foot (30 centimeters) in length.

jack post [MIN ENG] Timber used where a coal seam is separated by a rock band and one bench is loaded out before the other.

jack rafter [BUILD] A short, secondary, or simulated rafter.

jackscrew [MECH ENG] **1.** A jack operated by a screw mechanism. Also known as screw jack. **2.** The screw of such a jack.

jackshaft [MECH ENG] A countershaft, especially when used as an auxiliary shaft between two other shafts.

Jacksonian epilepsy [MED] Recurrent Jacksonian seizures.

Jacksonian seizure [MED] A focal seizure originating in one part of the motor or sensorimotor cortex and manifested usually by spasmodic contractions of or crawling or burning sensations of the skin; may become generalized, leading to loss of consciousness.

jack spool [TEXT] A large spool used to hold wool strands or yarn.

jack staff [NAV ARCH] A staff or flagpole attached to the bow of a ship, or the end of a spar projecting from the bow.

jackstay [NAV ARCH] A rope, rod, or pipe rove through the eyebolts fitted on a yard or mast for the purpose of attaching sails to the yard or mast.

jack timber [MIN ENG] A timber such as a rafter that is shorter than others with which it is used.

jack truss [BUILD] A minor truss in a hip roof where the roof has a reduced section.

Jacobian [MATH] The Jacobian of functions $f_i(x_1, x_2, \ldots, x_n)$, $i = 1, 2, \ldots, n$, of real variables x_i is the determinant of the matrix whose ith row lists all the first-order partial derivatives of the function $f_i(x_1, x_2, \ldots, x_n)$. Also known as Jacobian determinant.

Jacobian determinant *See* Jacobian.

Jacobian elliptic function [MATH] For m a real number between 0 and 1, and u a real number, let ϕ be that number such that

$$\int_0^\phi d\theta / (1 - m \sin^2 \theta)^{1/2} = u;$$

the 12 Jacobian elliptic functions of u with parameter m are sn $(u|m) = \sin \phi$, cn $(u|m) = \cos \phi$, dn $(u|m) = (1 - m \sin^2\phi)^{1/2}$, the reciprocals of these three functions, and the quotients of any two of them.

Jacobian matrix [MATH] The matrix used to form the Jacobian.

Jacobi condition [MATH] In the calculus of variations, a differential equation used to study the extremals in a variational problem.

Jacobi polynomials [MATH] Polynomials that are constructed from the hypergeometric function and satisfy the differential equation $(1-x^2)y'' + [\beta-\alpha-(\alpha+\beta+2)x]y' + n(\alpha + \beta + n + 1)y = 0$, where n is an integer and α and β are constants greater than -1; in certain cases these generate the Legendre and Chebycheff polynomials.

Jacobi's transformations [MATH] Transformations of Jacobian elliptic functions to other Jacobian elliptic functions given by change of parameter and variable.

Jacobshavn Glacier [HYD] A glacier on the west coast of Greenland at latitude 68°N; it is the most productive glacier in the Northern Hemisphere, calving about 1400 icebergs yearly.

jacobsite [MINERAL] $MnFe_2O_4$ A black magnetic mineral composed of an oxide of manganese and iron; a member of the magnetite series.

Jacob's ladder [NAV ARCH] A portable ladder, having rope, chain, or wire sides, and wooden or iron rungs, slung over the ship's side for temporary use. Also known as jack ladder.

Jacobsoniidae [INV ZOO] The false snout beetles, a small family of coleopteran insects in the superfamily Dermestoidea.

Jacobson's cartilage *See* vomeronasal cartilage.

JACKSCREW

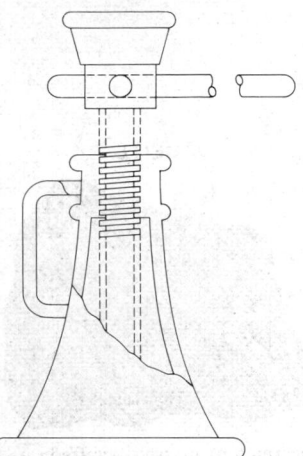

Jackscrew used to raise heavy objects.

JACQUARD

Weaving linen on a jacquard damask loom.

JADEITE

├─2.5 cm─┤

Specimen of massive jadeite in its dark-green variety chloromelonite, from Susa in Sardinia. *(Specimen from Department of Geology, Bryn Mawr College)*

Jacobson's organ [VERT ZOO] An olfactory canal in the nasal mucosa which ends in a blind pouch; it is highly developed in reptiles and vestigial in humans.

Jacob's staff *See* cross-staff.

Jacobs taper [DES ENG] A machine tool used for mounting drill chucks in drilling machines.

jacquard [TEXT] **1.** A loom or knitting machine for weaving figured fabrics and whose apparatus is controlled by punched cards. **2.** A fabric of jacquard weave.

Jacquemart's reagent [ANALY CHEM] Analytical reagent used to test for ethyl alcohol; consists of an aqueous solution of mercuric nitrate and nitric acid.

Jadassohn's nevus *See* nevus sebaceus.

jade [MINERAL] A hard, compact, dark-green or greenish-white gemstone composed of either jadeite or nephrite. Also known as jadestone.

jadeite [MINERAL] $NaAl(SiO_3)_2$ A clinopyroxene mineral occurring as green, fibrous monoclinic crystals; the most valuable variety of jade.

jadestone *See* jade.

Jaeger method [FL MECH] A method of determining surface tension of a liquid in which one measures the pressure required to cause air to flow from a capillary tube immersed in the liquid.

Jaeger-Steinwehr method [THERMO] A refinement of the Griffiths method for determining the mechanical equivalent of heat, in which a large mass of water, efficiently stirred, is used, the temperature rise of the water is small, and the temperature of the surroundings is carefully controlled.

jaff [ORD] A combination of electronic and chaff jamming.

jag bolt [DES ENG] An anchor bolt with barbs on a flaring shank.

jaguar [VERT ZOO] *Felis onca.* A large, wild cat indigenous to Central and South America; it is distinguished by a buff-colored coat with black spots, and has a relatively large head and short legs.

Jahn-Teller effect [PHYS CHEM] The effect whereby, except for linear molecules, degenerate orbital states in molecules are unstable.

jalap [MATER] An orange or reddish solid or a yellowish to brown powder with acrid taste and slight odor; the dried tuberous root of a Mexican plant (*Exogonium purga*), or the drug prepared from it; used as a cathartic in medicine.

jam [ADP] In punched-card equipment, a feed malfunction causing blockage of passages with crumpled cards.

Jamaica bayberry *See* bayberry.

jamb [BUILD] The vertical member on the side of an opening, as a door or window.

jamb brick [MATER] A brick with one rounded corner; used to provide a rounded edge on wall openings.

jamb liner [BUILD] A small strip of wood applied to the edge of a window jamb to increase its width for use in thicker walls.

James concentrator [MIN ENG] A concentration table whose deck is divided into two sections; one section contains riffles for the coarse material, and the other section is smooth to allow settling of the fine particles which will not settle on a riffled surface.

jamesonite [MINERAL] $Pb_4FeSb_6S_{14}$ A lead-gray to gray-black mineral that crystallizes in the orthorhombic system, occurs in acicular crystals with fibrous or featherlike forms, and has a metallic luster. Also known as feather ore; gray antimony.

Jamin effect [FL MECH] Resistance to flow of a column of liquid divided by air bubbles in a capillary tube, even when subjected to a substantial pressure difference between the ends of the tube.

Jamin refractometer [OPTICS] An instrument for measuring the index of refraction of a gas in which two light beams from a common source are each passed through an evacuated tube and recombined, and the displacement of interference fringes is noted as gas is slowly admitted into one of the tubes.

jammer [ELECTR] A transmitter used in jamming of radio or radar transmissions. Also known as electronic jammer.

jammer finder [ELECTR] Radar which attempts to obtain the range of the target by training a highly directional pencil beam on a jamming source. Also known as burn-through.

jamming [ELECTR] Radiation or reradiation of electromagnetic waves so as to impair the usefulness of a specific segment of the radio spectrum that is being used by the enemy for communication or radar. Also known as active jamming; electronic jamming.

jam nut *See* locknut.

Janecke coordinates [CHEM ENG] Use of a rectangular or Ponchon-type diagram to plot the solvent content of liquid-liquid equilibrium phases; used for solvent-extraction design calculations.

J antenna [ELECTROMAG] Antenna having a configuration resembling a J, consisting of a half-wave antenna end-fed by a parallel-wire quarter-wave section.

January thaw [CLIMATOL] A period of mild weather popularly supposed to recur each year in late January in New England and other parts of the northeastern United States.

Janus system [NAV] A technique used in Doppler navigator design which utilizes radar or sound beams directed forward and astern for computation of ground-speed components; by heterodyning the returned pulses, the Doppler shift can be determined independently of the transmitted frequency.

japan [MATER] A glossy, black baking paint or varnish that consists primarily of a hard asphalt base.

Japan Current *See* Kuroshio Current.

Japanese encephalitis [MED] A human viral infection epidemic in Japan, transmitted by the common house mosquito (*Culex pipiens*) and characterized by severe inflammation of the brain.

Japanese peppermint oil [MATER] An oil distilled from *Mentha arvensis*, grown in Japan, Brazil, and the United States; the oil has less odor than peppermint oil; used for the production of menthol.

japanning [MET] The finishing of metal objects with japan.

Japan paper [MATER] A special paper made with an irregular mottled effect on the surface; used for greeting cards and other decorative applications.

Japan tallow *See* Japan wax.

Japan wax [MATER] A pale-yellow wax with rancid aroma obtained from the berries of sumac; soluble in benzene and naphtha, insoluble in water; melts at 53°C; used in wax products, polishes, and soaps, and as a beeswax substitute. Also known as Japan tallow; sumac wax.

Japygidae [INV ZOO] A family of wingless insects in the order Diplura with forcepslike anal appendages; members attack and devour small soil arthropods.

jar [ELEC] A unit of capacitance equal to 1000 statfarads, or approximately 1.11265×10^{-9} farad; it is approximately equal to the capacitance of a Leyden jar; this unit is now obsolete.

jar coupling *See* jars.

jardiniere glaze [MATER] A type of unfritted soft and hard lead glaze; contains oxides of lead, zinc, potassium, calcium, aluminum, and silicon.

jarlite [MINERAL] $NaSr_3Al_3F_{16}$ A colorless to brownish mineral composed of aluminofluoride of sodium and strontium.

jarosite [MINERAL] $KFe_3(SO_4)_2(OH)_6$ An ocher-yellow or brown alunite mineral having rhombohedral crystal structure. Also known as utahite.

jarosite process [MET] A zinc electrometallurgical process in which ferric ions are precipitated from feed solutions in the form of jarosite, a hydrous sulfate of iron and potassium or sodium.

jars [PETRO ENG] A series of links in the drill string to connect drill cables to the drill bit; sets up the uneven motion on the upstroke that helps free the string of tools. Also known as jar coupling.

jasmine oil [MATER] A colorless fragrant essential oil from flowers of a jasmine, as *Jasminum officinale* or *J. grandiflorum*; the oil is extracted from the flowers by enfleurage and is used in perfumery.

jasmone [ORG CHEM] $C_{11}H_{16}O$ A liquid ketone found in jasmine oil and other essential oils from plants.

jaspagate *See* agate jasper.

jasper [PETR] A dense, opaque to slightly translucent crypto-crystalline quartz containing iron oxide impurities; characteristically red. Also known as jasperite; jasperoid; jaspis.

jasperite *See* jasper.

jasperoid *See* jasper.

jaspilite [PETR] A compact siliceous rock resembling jasper and containing iron oxides in bands.

jaspis *See* jasper.

jaspoid *See* tachylite.

JATO engine [AERO ENG] Derived from jet-assisted-takeoff engine. **1.** An auxiliary jet-producing unit or units, usually rockets, for additional thrust. **2.** A JATO bottle or unit; the complete auxiliary power system used for assisted takeoff.

jauch *See* jauk.

jauk [METEOROL] A local name for the foehn in the Klagenfurt basin of Austria; it may come from the south, but is developed as a north foehn. Also spelled jauch.

jaundice [INV ZOO] *See* grasserie. [MED] Yellow coloration of the skin, mucous membranes, and secretions resulting from hyperbile-rubinemia. Also known as icterus.

jaundice of newborn [MED] Jaundice in infants during the first few days after birth, due to various causes.

Java black rot [PL PATH] A fungus disease of stored sweet potatoes caused by *Diplodia tubericola*; the inside of the root becomes black and brittle.

Java cotton *See* kapok.

Java man [PALEON] An overspecialized, apelike form of *Homo sapiens* from the middle Pleistocene having a small brain capacity, low cranial vault, and massive browridges.

jaw [ANAT] Either of two bones forming the skeleton of the mouth of vertebrates: the upper jaw or maxilla, and the lower jaw or mandible. [ENG] A notched part that permits a railroad-car axle box to move vertically. [GEOL] The side of a narrow passage such as a gorge.

jawbreaker *See* jaw crusher.

jaw clutch [MECH ENG] A clutch that provides positive connection of one shaft with another by means of interlocking faces; may be square or spiral; the most common type of positive clutch.

jaw crusher [MECH ENG] A machine for breaking rock between two steel jaws, one fixed and the other swinging. Also known as jawbreaker.

jawless vertebrate [VERT ZOO] The common name for members of the Agnatha.

J bolt [DES ENG] A J-shaped bolt, threaded on the long leg of the J.

J box *See* junction box.

J-carrier system [COMMUN] Broad-band carrier system, providing 12 telephone channels, which uses frequencies up to about 140 kilohertz by means of effective four-wire transmission on a single open-wire pair.

J display [ELECTR] A modified radarscope A display in which the time base is a circle; the target signal appears as an outward radial deflection from the time base.

Jeans viscosity equation [THERMO] An equation which states that the viscosity of a gas is proportional to the temperature raised to a constant power, which is different for different gases.

jeep [MECH ENG] A one-quarter-ton, four-wheel-drive utility vehicle in wide use in all United States military services.

jeffersonite [MINERAL] $Ca(Mn,Zn,Fe)Si_2O_6$ A dark-green or greenish-black mineral composed of pyroxene.

Jeffrey crusher [MIN ENG] A crusher to break soft minerals, such as limestone. Also known as whizzer mill.

Jeffrey diaphragm jig [MIN ENG] A plunger-type jig with the plunger beneath the screen.

Jeffrey molveyor [MIN ENG] A string of short conveyors on driven wheels connected together to run alongside a heading or room conveyor; used to keep a continuous miner in operation at all times.

Jeffrey single-roll crusher [MIN ENG] A simple type of crusher for coal, with a drum to which are bolted toothed segments designed to grip the coal, thus forcing it down into the crushing opening.

Jeffrey swing-hammer crusher [MIN ENG] A crusher with swing arms on a revolving shaft for crushing coal, ore, or other material against the iron casing of the crusher; a screen at the bottom allows sufficiently fine pieces to pass through.

Jeffrey-Traylor vibrating feeder [MIN ENG] A feed chute vibrated electromagnetically in a direction oblique to its surface; rate of movement of rock depends on amplitude and frequency of vibration.

Jeffrey-Traylor vibrating screen [MIN ENG] A vibrating screen whose action results from an oscillating armature and a stationary coil.

jejunitis [MED] Inflammation of the jejunum.

jejunostomy [MED] The making of an artificial opening through the abdominal wall into the jejunum.

jejunum [ANAT] The middle portion of the small intestine, extending between the duodenum and the ileum.

jellied gasoline *See* gelatinized gasoline.

jelly *See* ulmin.

jellyfish [INV ZOO] Any of various free-swimming marine coelenterates belonging to the Hydrozoa or Scyphozoa and having a bell- or bowl-shaped body. Also known as medusa.

jelly fungus [MYCOL] The common name for many members of the Heterobasidiomycetidae, especially the orders Tremallales and Dacromycetales, distinguished by a jellylike appearance or consistency.

jelutong *See* pontianak gum.

Jennerian vaccine *See* smallpox vaccine.

Jenner's stain *See* May-Grünwald stain.

jenny [VERT ZOO] **1.** A female animal, as a jenny wren. **2.** A female donkey.

Jensen's inequality [MATH] **1.** A general inequality satisfied by a convex function:

$$f\left(\sum_{i=1}^{n} a_i x_i\right) \leq \sum_{i=1}^{n} a_i f(x_i)$$

where the x_i are any numbers in the region where f is convex and the a_i are nonegative numbers whose sum is equal to 1. **2.** If $a_1, a_2, \ldots, a_n$ are positive numbers and $s > t > 0$, then $(a_1{}^s + a_2{}^s + \ldots + a_n{}^s)^{1/s}$ is less than or equal to $(a_1{}^t + a_2{}^t + \ldots + a_n{}^t)^{1/t}$.

Jensen's sarcoma [VET MED] A transmissible malignant tumor originally observed in a rat inoculated with acid-fast bacteria from a cow with pseudotuberculous enteritis.

Jeppel's oil *See* bone oil.

jerboa [VERT ZOO] The common name for 25 species of rodents composing the family Dipodidae; all are adapted for jumping, having extremely long hindlegs and feet.

jeremejevite [MINERAL] $AlBO_3$ A colorless or yellowish mineral composed of aluminum borate that occurs in hexagonal crystals.

Jeremiassen crystallizer [CHEM ENG] Device used to grow solid crystals in a supersaturated liquid solution and to separate them from it.

jerk [MECH] **1.** The rate of change of acceleration; it is the third derivative of position with respect to time. **2.** A unit of rate of change of acceleration, equal to 1 foot (30.48 centimeters) per second squared per second.

jerkinhead [ARCH] Section of a roof hipped for only part of its height, forming a truncated gable on the wall below.

jerk pump [MECH ENG] A pump that supplies a precise amount of fuel to the fuel injection valve of an internal combustion engine at the time the valve opens; used for fuel injection.

jerry can [ORD] A 5-gallon (19-liter), flat-sided, narrow can adapted from a German-made can, easily stacked and transported, and adapted by special openings for discharging fuel.

jersey [TEXT] A knitted wool, cotton, polyester, rayon, or other fabric with a slight rib on one side.

jet [FL MECH] A strong, well-defined stream of compressible fluid, either gas or liquid, issuing from an orifice or nozzle or moving in a contracted duct.

jet aircraft [AERO ENG] An aircraft with a jet engine or engines.

jet coal [GEOL] A hard, lustrous, pure black variety of lignite, occurring in isolated masses in bituminous shale; thought to be derived from waterlogged driftwood. Also known as black amber.

jet compressor [MECH ENG] A device, utilizing an actuating nozzle and a combining tube, for the pumping of a compressible fluid.

jet condenser [MECH ENG] A direct-contact steam condenser utilizing the aspirating effect of a jet for the removal of noncondensables.

jet drilling [MECH ENG] A drilling method that utilizes a chopping bit, with a water jet run on a string of hollow drill rods, to chop through soils and wash the cuttings to the surface. Also known as wash boring.

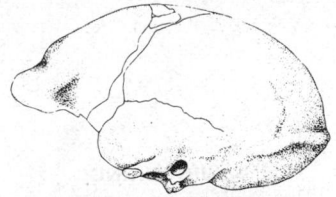

JAVA MAN

Lateral view of the cranium of *Homo erectus II*, one of the first specimens of Java man. (*Carnegie Institution of Washington, as used in M. F. Ashley Montagu, An Introduction to Physical Anthropology, 2d ed., Charles C. Thomas, 1951*)

J DISPLAY

J type of radar display showing the signals as radial pips.

JERBOA

Jerboa, with body 3–6 inches (7–15 centimeters) long and tail up to 8 inches (20 centimeters) long.

JET CONDENSER

A low-level jet condenser. (*C. H. Wheeler Manufacturing Co.*)

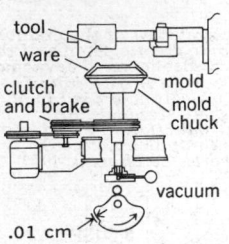

JIGGERING

tool
ware
clutch
and brake
mold
mold
chuck
vacuum
.01 cm

The jiggering tool and cam-raised
jiggering apparatus used for
plastic forming of clay. *(From
W. D. Kingery, ed., Ceramic
Fabrication Processes, Technology
Press, MIT, and Wiley, 1958)*

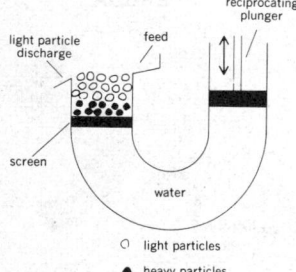

JIGGING

reciprocating
plunger
light particle
discharge
feed
screen
water
○ light particles
● heavy particles

The principle of the jigging
method for beneficiation of ore.

jet-effect wind [METEOROL] A wind which is increased in speed through the channeling of air by some mountainous configuration, such as a narrow mountain pass or canyon.

jet engine [AERO ENG] An aircraft engine that derives all or most of its thrust by reaction to its ejection of combustion products (or heated air) in a jet and that obtains oxygen from the atmosphere for the combustion of its fuel. [MECH ENG] Any engine that ejects a jet or stream of gas or fluid, obtaining all or most of its thrust by reaction to the ejection.

jet-flame drill [MIN ENG] A mining drill that utilizes a high-velocity flame to spall out a hole.

jet flap [AERO ENG] A sheet of fluid discharged at high speed close to the trailing edge of a wing so as to induce lift over the whole wing.

jet fuel [MATER] Special grade of kerosine with a flash point of 125°F (52°C), used for jet aircraft; may have methane or naphthene added to produce a 110°F (43°C) flash point, for military aircraft.

jet hole [ENG] A borehole drilled by use of a directed, forceful stream of fluid or air.

jet mill *See* fluid-energy mill.

jet mixer [MECH ENG] A type of flow mixer or line mixer, depending on impingement of one liquid on the other to produce mixing.

jet molding [ENG] Molding method in which most of the heat is applied to the material to be molded as it passes through a nozzle or jet, rather than in a conventional heating cylinder.

jet nozzle [DES ENG] A nozzle, usually specially shaped, for producing a jet, such as the exhaust nozzle on a jet or rocket engine.

jet-piercing drill *See* fusion-piercing drill.

jet propulsion [AERO ENG] The propulsion of a rocket or other craft by means of a jet engine. [ENG] Propulsion by means of a jet of fluid.

jet pump [MECH ENG] A pump in which an accelerating jet entrains a second fluid to deliver it at elevated pressure.

jetsam [ENG] Articles that sink when thrown overboard, particularly those jettisoned for the purpose of lightening a vessel in distress.

jet spinning [ENG] Production of plastic fibers in which a directed blast or jet of hot gas pulls the molten polymer from a die lip; similar to melt spinning.

jet stream [AERO ENG] The stream of gas or fluid expelled by any reaction device, in particular the stream of combustion products expelled from a jet engine, rocket engine, or rocket motor. [METEOROL] A relatively narrow, fast-moving wind current flanked by more slowly moving currents; observed principally in the zone of prevailing westerlies above the lower troposphere, and in most cases reaching maximum intensity with regard to speed and concentration near the tropopause.

jetting [CIV ENG] A method of driving piles or well points into sand by using a jet of water to break the soil. [ENG] During molding of plastics, the turbulent flow of molten resin from an undersized gate or thin section into a thicker mold section, as opposed to laminar, progressive flow.

jetting tool [PETRO ENG] Downhole device that jets a high-pressure, sand-laden fluid stream to clean out wellbore holes, to disintegrate perforating pipe, and to perform other operations.

jettison [ENG] The throwing overboard of objects, especially to lighten a craft in distress.

jetty *See* groin.

Jevons effect [METEOROL] The effect upon the measurement of rainfall caused by the presence of the rain gage; in 1861 W. S. Jevons pointed out that the rain gage causes a disturbance in airflow past it, and this carries part of the rain past the gage which would normally be captured.

jewel [ENG] **1.** A bearing usually made of synthetic corundum and used in precision timekeeping devices, gyros, and other instruments. **2.** A bearing lining of soft metal, used in railroad cars, for example.

jeweler's rouge *See* ferric oxide.

jewelry alloy [MET] Any ductile, malleable alloy, usually bronze, of good corrosion resistance, used as a base metal in jewelry.

jezekite *See* morinite.

J-factor [THERMO] A dimensionless equation used for the calculation of free convection heat transmission through fluid films.

J function [GEOPHYS] A dimensionless mathematical relationship to correlate capillary pressure data of similar geologic formations.

jib [NAV ARCH] A triangular sail bent to a foremast stay.

jib boom [MECH ENG] An extension that is hinged to the upper end of a crane boom. [NAV ARCH] A spar used as an extension of the bowsprit on sailing ships.

jib crane [MECH ENG] Any of various cranes having a projecting arm (jib).

jib end [MIN ENG] The delivery end in conveyor systems in which a jib is fitted to deliver the load in advance of and remote from the drive.

jig [MECH ENG] A device used to position and hold parts for machining operations and to guide the cutting tool. [MIN ENG] A vibrating device in which coal is cleaned and ore is concentrated in water.

jig back [MECH ENG] An aerial ropeway with a pair of containers that move in opposite directions and are loaded or stopped alternately at opposite stations but do not pass around the terminals. Also known as reversible tramway; to-and-fro ropeway.

jig borer [MECH ENG] A machine tool resembling a vertical milling machine designed for locating and drilling holes in jigs.

jigger *See* jigging conveyor.

jigger boss [MIN ENG] A first-line supervisor in some western United States mines.

jiggering [ENG] A mechanization of the ceramic-forming operation consisting of molding the outside of a piece by throwing plastic clay on a plaster of paris mold, placing the mold and clay on a rotating head, and forming the inner surface by forcing a template or jigger knife against the clay; method used in mass-producing dinnerware.

jigging [MIN ENG] A gravity method which separates mineral from gangue particles by utilizing an effective difference in settling rate through a periodically dilated bed.

jigging conveyor [MIN ENG] A series of steel troughs suspended from the roof of the stope or laid on rollers on its floor, and given reciprocating motion mechanically, to move mineral. Also known as chute conveyor; jigger; pan conveyor.

jig grinder [MECH ENG] A precision grinding machine used to locate and grind holes to size, especially in hardened steels and carbides.

jigsaw [MECH ENG] A tool with a narrow blade suitable for cutting intricate curves and lines.

jig washer [MIN ENG] A coal or mineral washer for relatively coarse material; the broken ore is placed on a screen and pulsed vertically with water; the heavy portion passes through the screen and the light portion goes over the sides.

jim crow [DES ENG] A device with a heavy buttress screw thread used for bending rails by hand.

jimsonweed [BOT] *Datura stramonium*. A tall, poisonous annual weed having large white or violet trumpet-shaped flowers and globose prickly fruits. Also known as apple of Peru.

J indicator *See* J scope.

jitter [COMMUN] In facsimile, distortion in the received copy caused by momentary errors in synchronism between the scanner and recorder mechanisms; does not include slow errors in synchronism due to instability of the frequency standards used in the facsimile transmitter and recorder. [ELECTR] Small, rapid variations in a waveform due to mechanical vibrations, fluctuations in supply voltages, control-system instability, and other causes.

jittered pulse recurrence frequency [COMMUN] Random variation of the pulse repetition period; provides a discrimination capability against repeater-type jammers.

j-j coupling [ATOM PHYS] A process for building up many-electron wave functions; the spin and orbital functions of each particle are combined to form eigenfunctions of the particle's total angular momentum, and then the wave functions of all the particles are combined to form eigenfunctions of the total angular momentum of the system; this coupling is used when the spin-orbit interaction is strong compared to the electrostatic interaction.

J-K flip-flop [ELECTR] A storage stage consisting only of transistors and resistors connected as flip-flops between input and output gates, and working with charge-storage transistors; gives a definite output even when both inputs are 1.

JND *See* just-noticeable difference.

joaquinite [MINERAL] $NaBa_2Ce_2Fe(Ti,Nb)_2Si_8O_{26}(OH,F)_2$ A honey-yellow mineral composed of sodium iron titanium silicate, occurring in orthorhombic crystals.

job [ADP] A unit of work to be done by the computer; it is a single entity from the standpoint of computer installation management, but may consist of one or more job steps. [IND ENG] **1.** The combination of duties, skills, knowledge, and responsibilities assigned to an individual employee. **2.** A work order.

jobber's reamer [DES ENG] A machine reamer that is solid with straight or helical flutes and taper shanks.

job breakdown [IND ENG] Separation of an operation into elements. Also known as operation breakdown.

job characteristic *See* job factor.

job class [IND ENG] A group of jobs involving a similar type of work, difficulty of performance, or range of pay. Also known as job family; job grade; labor grade.

job classification [IND ENG] Designating job classes on the basis of job factors or level of pay, or on the basis of job evaluation.

job control block [ADP] A group of data containing the execution-control data and the job identification when the job is initiated as a unit of work to the operating system.

job control statement [ADP] Any of the statements used to direct an operating system in its functioning, as contrasted to data, programs, or other information needed to process a job but not intended directly for the operating system itself. Also known as control statement.

job description [IND ENG] A detailed description of the essential activities required to perform a task.

job evaluation [IND ENG] Orderly qualitative appraisal of each job or position in an establishment either by a point system for the specific job characteristics or by comparison of job factors; used for establishing a job hierarchy and wage plans.

job factor [IND ENG] An essential job element which provides a basis for selecting and training employees and establishing the wage plan for the job. Also known as job characteristic.

job family *See* job class.

job grade *See* job class.

job library [ADP] A partitioned data set, or a concatenation of partitioned data sets, used as the primary source of object programs (load modules) for a particular job, and more generally, as a source of runnable programs from which all or most of the programs for a given job will be selected.

Jo block *See* Johansson block.

job plan [IND ENG] The organized approach to production management involving formal, step-by-step procedures.

job processing control [ADP] The section of the control program responsible for initiating operations, assigning facilities, and proceeding from one job to the next.

job stacking [ADP] The presentation of jobs to a computer system, each job followed by another.

job step [ADP] A unit of work from the viewpoint of the user.

jochwinde [METEOROL] The mountain-gap wind of the Tauern Pass in the Alps.

jogging [ELEC] Quickly repeated opening and closing of a circuit to produce small movements of the driven machine. Also known as inching.

joggle [DES ENG] **1.** A flangelike offset on a flat piece of metal. **2.** A projection or notch on a sheet of building material to prevent protrusion. **3.** A dowel for joining blocks of masonry. [MIN ENG] Trusses or sets of timbers joined for taking pressure at right angles.

joggled frame [NAV ARCH] A frame offset to fit the laps of the shell plating.

joggled plating [NAV ARCH] A plating construction used in riveted steel ships in which one or both edges of a plate are offset with respect to the edge of an adjacent one; it became essentially obsolete with the advent of all-welded ships.

jog method [NAV] Following one or more doglegs to avoid a direct approach, as when it is desired to delay arrival at a destination.

Johann crystal geometry [CRYSTAL] The focusing shape of a diffracting crystal for x-ray dispersion used in electron-probe microanalysis; less stringent than Johannson crystal geometry.

johannsenite [MINERAL] $CaMnSi_2O_6$ A clove-brown, grayish, or greenish clinopyroxene mineral composed of a silicate of calcium and manganese; a member of the pyroxene group.

Johannson crystal geometry [CRYSTAL] The full-focusing shape of a diffracting crystal for x-ray dispersion used in electron-probe microanalyzers; more stringent than Johann crystal geometry.

Johansson block [DES ENG] A type of gage block ground to an accuracy of at least 1/100,000 inch (0.25 micrometer). Also known as Jo block.

Johne's bacillus *See* Mycobacterium paratuberculosis.

Johne's disease [VET MED] A chronic inflammation of the intestinal tract of sheep, cattle, and deer caused by *Mycobacterium paratuberculosis*.

Johnson and Lark-Horowitz formula [SOLID STATE] A formula according to which the resistivity of a metal or degenerate semiconductor resulting from impurities which scatter the electrons is proportional to the cube root of the density of impurities.

Johnson concentrator [MIN ENG] A device used to separate heavy particles such as metallic gold from auriferous pulp; composed of a shell in the shape of a cylinder that is lined with rubber grooves parallel to the inclined axis.

Johnson noise *See* thermal noise.

johnstrupite [MINERAL] A mineral composed of a complex silicate of cerium and other metals, approximately $(Ca,Na)_3$-$(Ce,Ti,Zr)(SiO_4)_2F$; occurs in prismatic crystals.

joiner plans [NAV ARCH] Plans showing the arrangement of quarters and living spaces on a ship.

joint [ANAT] A contact surface between two individual bones. Also known as articulation. [ELEC] A juncture of two wires or other conductive paths for current. [ENG] The surface at which two or more mechanical or structural components are united. [GEOL] A fracture that traverses a rock and does not show any discernible displacement of one side of the fracture relative to the other.

joint bar [CIV ENG] A rigid steel member used in pairs to join, hold, and align rail ends.

joint body *See* joint mouse.

joint capsule [ANAT] A sheet of fibrous connective tissue enclosing a synovial joint.

joint communications [COMMUN] Common use of communications facilities by two or more services of the same nation.

joint distribution of two random variables [STAT] The distribution which gives the probability that $Z = z$ and $W = w$ for all values z and w of the random variables Z and W respectively.

joint drag *See* kink band.

joint efficiency [MET] A numerical value expressed as the ratio of the strength of a riveted, welded, or brazed joint to the strength of the parent metal.

jointer [ENG] **1.** Any tool used to prepare, make, or simulate joints, such as a plane for smoothing wood surfaces prior to joining them, or a hand tool for inscribing grooves in fresh cement. **2.** A file for making sawteeth the same height. **3.** An attachment to a plow that covers discarded material. **4.** A worker who makes joints, particularly a construction worker who cuts stone to proper fit.

jointer gage [DES ENG] An attachment to a bench vise that holds a board at any angle desired for planing.

jointing [CIV ENG] Caulking of masonry joints. [GEOL] A condition of rock characterized by joints.

joint marginal distribution [STAT] The distribution obtained by summing the joint distribution of three random variables over all possible values of one of these variables.

joint mouse [PATH] A small, loose body within a synovial joint, frequently calcified, derived from synovial membrane, organized fibrin fragments of articular cartilage, or arthritic osteophytes. Also known as joint body.

joint penetration [MET] The distance extended into a weld joint by the weld metal or fusion zone.

joint pole [ELEC] Pole used in common by two or more utility companies.

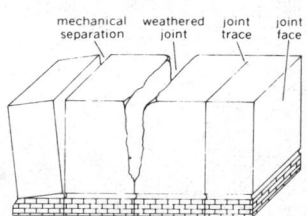

JOINT

mechanical separation   weathered joint   joint trace   joint face

Aspects of systematic joint, a type of geological joint.

joint set [GEOL] A group of parallel joints in a geologic formation.

joint system [GEOL] Two or more joint sets.

joint vein [GEOL] A small vein in a joint.

joist [CIV ENG] A steel or wood beam providing direct support for a floor.

joist anchor *See* wall anchor.

Jolly balance [ENG] A spring balance used to measure specific gravity of mineral specimens by weighing a specimen when in the air and when immersed in a liquid of known density.

jolt-and-jumble tests [ORD] A standardized program of tests intended to simulate the shocks to which various components of ammunition are subjected in transportation and handling.

jolt molding [ENG] A process for shaping refractory blocks in which a mold containing prepared batch is jolted mechanically to consolidate the material.

Jominy end quench test *See* Jominy test.

Jominy test [MET] A hardenability test in which a steel bar is heated to the desired austenitizing temperature and quench-hardened at one end and then measured for hardness along its length, beginning at the quenched end. Also known as Jominy end quench test.

Jonathan freckle [PL PATH] A storage disease of apples characterized by small circular discolorations in the skin.

Jonathan spot [PL PATH] A disease of apples characterized by circular depressed necrotic areas around the lenticels.

Jones reductor [CHEM] A device used to chemically reduce solutions, such as ferric salt solutions, consisting of a vertical tube containing granular zinc into which the solution is poured.

Jones riffle [MIN ENG] An apparatus used to reduce the size of a sample to a desired weight; consists of a hopper which passes samples to a series of open-bottom pockets, each of which divides the sample into two equal parts, and the next pass of each part gives a quarter of the original sample, and so on until the desired sample is obtained.

Jones splitter [MIN ENG] A device used to reduce the volume of a sample, consisting of a belled, rectangular container, the bottom of which is fitted with a series of narrow slots or alternating chutes designed to cast material in equal quantities to opposite sides of the device.

Jones sucker rod [PETRO ENG] Connecting rod between the subsurface pump and the lifting or pumping device on the surface; serves to lift oil out of the cased hole.

jonquil oil [MATER] A colorless oil obtained from flowers of jonquil (*Narcissus jonquilla*); used in perfumes. Also known as narcissus oil.

Joppeicidae [INV ZOO] A monospecific family of hemipteran insects included in the Pentatomorpha; found in the Mediterranean regions.

joran *See* juran.

jordan [MECH ENG] A machine or engine used to refine paper pulp, consisting of a rotating cone, with cutters, that fits inside another cone, also with cutters.

Jordan algebra [MATH] A nonassociative algebra over a field in which products satisfy the Jordan identity $(xy)x^2 = x(yx^2)$.

Jordan curve [MATH] A simple closed curve in the plane, that is, a curve that is closed, connected, and does not cross itself.

Jordan curve theorem [MATH] The theorem that in the plane every simple closed curve separates the plane into two parts.

Jordan-Hölder theorem [MATH] The theorem that for a group any two composition series have the same number of subgroups listed, and both series produce the same quotient groups.

jordanite [MINERAL] $(Pb,Tl)_{13}As_7S_{23}$ A lead-gray mineral composed of lead arsenic sulfide, occurring as monoclinic crystals.

jordanon *See* microspecies.

Jordan's rule [EVOL] The rule that organisms which are closely related tend to occupy adjacent rather than identical or distant ranges. [VERT ZOO] The rule that fishes in areas of low temperatures tend to have more vertebrae than those in warmer waters.

Jordan sunshine recorder [ENG] A sunshine recorder in which the time scale is supplied by the motion of the sun; it consists of two opaque metal semicylinders mounted with their curved surfaces facing each other; each of the semicylinders has a short narrow slit in its flat side; sunlight entering one of the slits falls on light-sensitive paper (blueprint paper) which lines the curved side of the semicylinder.

Jordan-Wigner commutation rules [QUANT MECH] Rules obtained by replacing commutators of creation and destruction operators by anticommutators; applicable to fermion fields in a quantized field theory.

joseite [MINERAL] $Bi_3Te(Si,S)$ A mineral composed of telluride of bismuth containing sulfur and selenium.

josen *See* hartite.

josephinite [MINERAL] A mineral consisting of an alloy of iron and nickel; occurs naturally in stream gravel.

Josephson current [CRYO] The current across a Josephson junction in the absence of voltage across the junction, resulting from the Josephson effect.

Josephson effect [CRYO] The tunneling of electron pairs through a thin insulating barrier between two superconducting materials. Also known as Josephson tunneling.

Josephson equation [CRYO] An equation according to which the Josephson current is a sinusoidally varying function of the applied magnetic field.

Josephson junction [CRYO] A thin insulator separating two superconducting materials; it displays the Josephson effect.

Josephson tunneling *See* Josephson effect.

JOSS [ADP] A time-sharing language, designed for concurrent use by a number of people, each at his own console typewriter, and for programs of moderate size and running time.

joule [MECH] The unit of energy or work in the meter-kilogram-second system of units, equal to the work done by a force of magnitude of 1 newton when the point at which the force is applied is displaced 1 meter in the direction of the force. Also known as newton-meter of energy.

Joule-Clausius velocity [STAT MECH] A quantity used in the description of the kinetic behavior of a gas, equal to the square root of the ratio of the pressure of the gas to one-third of its density.

Joule cycle *See* Brayton cycle.

Joule effect [PHYS] **1.** The heating effect produced by the flow of current through a resistance. **2.** A change in the length of a ferromagnetic substance which occurs parallel to an applied magnetic field. Also known as Joule magnetostriction; longitudinal magnetorestriction.

Joule equivalent [THERMO] The numerical relation between quantities of mechanical energy and heat; the present accepted value is 1 fifteen-degrees calorie equals 4.1855 ± 0.0005 joules. Also known as mechanical equivalent of heat.

Joule heat [ELEC] The heat which is evolved when current flows through a medium having electrical resistance, as given by Joule's law.

Joule-Kelvin effect *See* Joule-Thomson effect.

Joule magnetorestriction *See* Joule effect.

Joule's law [ELEC] The law that when electricity flows through a substance, the rate of evolution of heat in watts equals the resistance of the substance in ohms times the square of the current in amperes. [THERMO] The law that at constant temperature the internal energy of a gas tends to a finite limit, independent of volume, as the pressure tends to zero.

Joule-Thomson effect [THERMO] A change of temperature in a gas undergoing Joule-Thomson expansion. Also known as Joule-Kelvin effect.

Joule-Thomson expansion [THERMO] The adiabatic, irreversible expansion of a fluid flowing through a porous plug or partially opened valve. Also known as Joule-Thomson process.

Joule-Thomson process *See* Joule-Thomsom expansion.

Joule-Thomson valve [CRYO] A valve through which a gas is allowed to expand adiabatically, resulting in lowering of its temperature; used in production of liquid hydrogen and helium.

journal bearing [MECH ENG] An antifriction device in which a cylindrical shaft, called the journal, is supported in a stationary part, called the bearing.

journal box [ENG] A metal housing for a journal bearing.

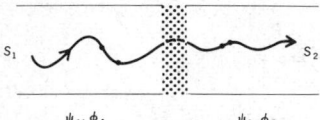

journal oil [MATER] A special grade of lubricating oil for use on journal bearings.

JOVIAL [ADP] A procedure-oriented language derived from ALGOL, commonly used in programming command and control procedures.

Jovian planet [ASTRON] Any of the four major planets (Jupiter, Saturn, Uranus, and Neptune) that are at a greater distance from the sun than the terrestrial planets (Mercury, Venus, Earth, and Mars).

Jovian Van Allen belts [ASTROPHYS] The extended belts of high-energy charged particles that are trapped in Jupiter's magnetic field and cause the microwave non-thermal emission of radio waves observed in the band from about 3 to 70 centimeters.

Joy double-ended miner [MIN ENG] A cutter-loader for continuous mining on a longwall face.

Joy extensible conveyor [MIN ENG] A type of belt conveyor consisting of a head section and a tail section, each mounted on crawler tracks and independently driven; used between a loader or continuous miner and the main transport.

Joy extensible steel band [MIN ENG] A hydraulically driven system linking a continuous miner to the main transport; the steel band is coiled on the drivehead.

Joy loader [MIN ENG] A loading machine which uses mechanical arms to collect coal or ore onto an apron that is pushed onto the broken material.

Joy longwall loading machine [MIN ENG] A modified Joy loader comprising a hydraulically elevated loading head fitted with mechanical gathering arms.

Joy microdyne [MIN ENG] A dust collector that wets and traps dust pulled into it and releases the dust as a slurry to be removed by pumps; used at the return end of tunnels or hard headings.

Joy miner [MIN ENG] A continuous miner weighing about 15 tons (13,600 kilograms) and made up of a turntable, a ripper bar, and a discharge boom conveyor; used mainly in coal headings and in extraction of coal pillars.

Joy-Sullivan hydrodrill rig [MIN ENG] A drill rig set on a jib or boom which can be moved to and locked in any position by hydraulic power controlled from the drill carriage.

Joy transloader [MIN ENG] A rubber-tired, self-propelled machine for loading, transporting, and dumping.

Joy walking miner [MIN ENG] A continuous miner designed to make thin seams; a walking mechanism is used instead of caterpillar tracks for moving the miner.

JP-1 [MATER] A kerosinelike petroleum product; the first jet aircraft fuel.

JP-4 [MATER] Jet engine test fuel made up of 35% light petroleum distillates and 65% gasoline distillates.

JP-5 [MATER] Military jet engine fuel made of specially refined kerosine with specified flash and freezing points.

J scan *See* J scope.

J scope [ELECTR] A modification of an A scope in which the trace appears as a circular range scale near the circumference of the cathode-ray tube face, the signal appearing as a radial deflection of the range scale; no bearing indication is given. Also known as J indicator; J scan.

Judson powder [MATER] An explosive containing sodium nitrate, coal, sulfur, and some nitroglycerin.

jugal [ANAT] Pertaining to the zygomatic bone. [VERT ZOO] In lower vertebrates, a bone lying below the orbit of the eye.

Jugatae [INV ZOO] The equivalent name for Homoneura.

jugate [BIOL] Structures which are joined together.

Juglandaceae [BOT] A family of dicotyledonous plants in the order Juglandales having unisexual flowers, a solitary basal ovule in a unilocular inferior ovary, and pinnately compound, exstipulate leaves.

Juglandales [BOT] An order of dicotyledonous plants in the subclass Hamamelidae distinguished by compound leaves; includes hickory, walnut, and butternut.

jugular [ANAT] Of or pertaining to the region of the neck above the clavicle.

jugular compression [PATH] A test for a spinal subarachnoid block by noting the rate of rise and fall of the spinal fluid pressure following compression and release of the jugular veins.

jugular foramen [ANAT] An opening in the cranium formed by the jugular notches of the occipital and temporal bones for passage of an internal jugular vein, the ninth, tenth, and eleventh cranial nerves, and the inferior petrosal sinus.

jugular foramen syndrome [MED] Injury to the jugular foramen, usually due to a basilar skull fracture, with resultant paralysis of ninth, tenth, and eleventh cranial nerves.

jugular process [ANAT] A rough process external to the condyle of the occipital bone.

jugular vein [ANAT] The vein in the neck which drains the brain, face, and neck into the innominate.

jugum [BOT] One pair of opposite leaflets of a pinnate leaf. [INV ZOO] **1.** The most posterior and basal portion of the wing of an insect. **2.** A crossbar connecting the two arms of the brachidium in some brachiopods.

Julian calendar [ASTRON] A calendar replaced by the Gregorian calendar; the Julian year was 365.25 days, the fraction allowed for the extra day every fourth year (leap year). There are 12 months, each 30 or 31 days except for February which has 28 days or in leap year 29.

Julian day [ASTRON] The number of each day, as reckoned consecutively since the beginning of the present Julian period on January 1, 4713 B.C.; it is used primarily by astronomers to avoid confusion due to the use of different calendars at different times and places; the Julian day begins at noon, 12 hours later than the corresponding civil day.

Julian ephemeris century [ASTRON] The unit of ephemeris time (ET) in Simon Newcomb's formulas which relate the orbital position of the earth to ephemeris time; the Julian ephemeris century is subdivided into 36,525 days, and 1 ephemeris day = 86,400 ephemeris seconds.

julienite [MINERAL] $Na_2Co(SCN)_4 \cdot 8H_2O$ A blue mineral composed of a hydrous thiocyanate of sodium and cobalt, and occurring as small needlelike crystals.

jumbo *See* drill carriage.

jump [ADP] A transfer of control which terminates one sequence of instructions and begins another sequence at a different location. Also known as branch; transfer.

jump discontinuity [MATH] A point *a* where for a real-valued function $f(x)$ the limit on the left of $f(x)$ as *x* approaches *a* and the limit on the right both exist but are distinct.

jumper [ELEC] A short length of conductor used to make a connection between two points or terminals in a circuit or to provide a path around a break in a circuit.

jump fire [FOR] A fire carried ahead of a forest fire by wind-borne burning material.

jump function [MATH] A function used to represent a sampled data sequence arising in the numerical study of linear difference equations.

jumping a claim [MIN ENG] **1.** Taking possession of a mining claim which has been abandoned. **2.** Taking possession of a mining claim that is liable to forfeiture because the requirements of the law are unfulfilled. **3.** Taking possession of a mining claim by stealth, fraud, or force.

jumping trace routine [ADP] A trace routine which is primarily concerned with providing a record of jump instructions in order to show the sequence of program steps that the computer followed.

jump phenomenon [CONT SYS] A phenomenon occurring in a nonlinear system subjected to a sinusoidal input at constant frequency, in which the value of the amplitude of the forced oscillation can jump upward or downward as the input amplitude is varied through either of two fixed values, and the graph of the forced amplitude versus the input amplitude follows a hysteresis loop.

jump resonance [CONT SYS] A jump discontinuity occurring in the frequency response of a nonlinear closed-loop control system with saturation in the loop.

Juncaceae [BOT] A family of monocotyledonous plants in the order Juncales characterized by an inflorescence of diverse sorts, vascular bundles with abaxial phloem, and cells without silica bodies.

Juncales [BOT] An order of monocotyledonous plants in the subclass Commelinidae marked by reduced flowers and capsular fruits with one too many anatropous ovules per carpel.

junction [CIV ENG] A point of intersection of roads or highways, especially where one terminates. [ELEC] *See* major

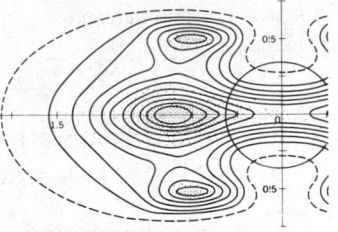

JOVIAN VAN ALLEN BELTS

Contours of radio emission from Jupiter originating from one-half of the Van Allen belt, and extending far beyond the planet's disk.

JUNCTION DETECTOR

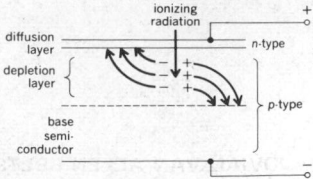

Silicon junction detector.

JUNCTION-GATE FIELD-EFFECT TRANSISTOR

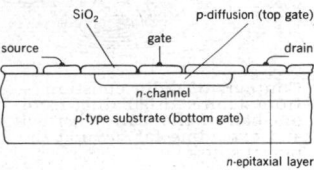

Components of a junction-gate field-effect transistor.

JUPITER

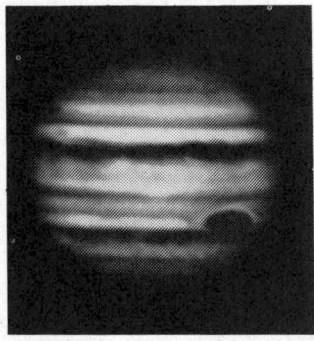

Telescopic appearance of Jupiter showing bands and red spot. *(Hale Observatories)*

JURASSIC

PRECAMBRIAN		
CAMBRIAN		
ORDOVICIAN		
SILURIAN	PALEOZOIC	
DEVONIAN		
Mississippian	CARBON-IFEROUS	
Pennsylvanian		
PERMIAN		
TRIASSIC		
JURASSIC	MESOZOIC	
CRETACEOUS		
TERTIARY	CENOZOIC	
QUATERNARY		

Chart showing the geologic periods and eras, with the Jurassic in the Mesozoic.

node. [ELECTR] A region of transition between two different semiconducting regions in a semiconductor device, such as a *pn* junction, or between a metal and a semiconductor. [ELECTROMAG] A fitting used to join a branch waveguide at an angle to a main waveguide, as in a tee junction. Also known as waveguide junction.

junctional nevus [MED] A skin lesion containing nevus cells at the junction of the epidermis and dermis.

junction battery [NUCLEO] A nuclear-type battery in which a radioactive material, such as strontium-90, irradiates a *pn* silicon junction.

junction box [ENG] A protective enclosure into which wires or cables are led and connected to form joints. Also known as J box.

junction buoy [NAV] A buoy marking the junction of two channels or two parts of a channel when proceeding seaward.

junction detector [NUCLEO] A reverse-biased semiconductor junction functioning as a solid ionization chamber to produce an electric output pulse whose amplitude is linearly proportional to the energy deposited in the junction depletion layer by the incident ionizing radiation.

junction diode [ELECTR] A semiconductor diode in which the rectifying characteristics occur at an alloy, diffused, electrochemical, or grown junction between *n*-type and *p*-type semiconductor materials. Also known as junction rectifier.

junction-gate field-effect transistor [ELECTR] A field-effect transistor in which there is normally a channel of relatively low-conductivity semiconductor joining the source and drain, and this channel is reduced and eventually cut off by junction depletion regions, reducing the conductivity, when a voltage is applied between the gate electrodes. Also known as depletion-mode field-effect transistor.

junction laser [OPTICS] A laser in which a junction in a semiconductor serves as the source of the coherent laser beam.

junction loss [COMMUN] In telephone circuits, that part of the repetition equivalent assignable to interaction effects arising at trunk terminals.

junction phenomena [ELECTR] Phenomena which occur at the boundary between two semiconductor materials, or a semiconductor and a metal, such as the existence of an electrostatic potential in the absence of current flow, and large injection currents which may arise when external voltages are applied across the junction in one direction.

junction point *See* branch point.

junction pole [ELEC] Pole at the end of a transposition section of an open-wire line or the pole common to two adjacent transposition sections.

junction rectifier *See* junction diode.

junction station [ELECTR] Microwave relay station that joins a microwave radio leg or legs to the main or through route.

junction streamer [GEOPHYS] The streamer process by which negative charge centers at successively higher altitudes in a thundercloud are believed to be "tapped" for discharge by lightning.

junction transistor [ELECTR] A transistor in which emitter and collector barriers are formed between semiconductor regions of opposite conductivity type.

junction transposition [ELEC] Transposition located at the junction pole between two transposition sections of an open-wire line.

junctor [ELEC] In crossbar systems, a circuit extending between frames of a switching unit and terminating in a switching device on each frame.

June solstice [ASTRON] Summer solstice in the Northern Hemisphere.

Jungermanniales [BOT] The leafy liverworts, an order of bryophytes in the class Marchantiatae characterized by chlorophyll-containing, ribbonlike or leaflike bodies and an undifferentiated thallus.

Jungian psychology *See* analytic psychology.

jungle [ECOL] An impenetrable thicket of second-growth vegetation replacing tropical rain forest that has been disturbed; lower growth layers are dense.

jungle yellow fever [MED] A form of yellow fever endemic in forested areas of Brazil.

junior [OPTICS] A 1000- or 2000-watt Fresnel spotlight.

juniper berry oil [MATER] Colorless oil that darkens and thickens in air; bitter taste, turpentinelike aroma; derived from the dried ripe fruit of the common juniper (*Juniperus communis*), used in medicine, gins, and liqueurs, and in veterinary practice. Also known as juniper oil.

juniperic acid [ORG CHEM] $C_{16}H_{32}O_3$ A crystalline hydroxy acid that melts at 95°C, obtained from waxy exudations from conifers.

juniper oil *See* juniper berry oil.

juniper tar oil *See* cade oil.

juniper wood oil [MATER] Oil made by diluting juniper berry oil with turpentine oil or by distilling turpentine oil over juniper wood; used as an external medicine and in veterinary practice.

Junkers engine [MECH ENG] A double-opposed-piston, two-cycle internal combustion engine with intake and exhaust ports at opposite ends of the cylinder.

junk wind [METEOROL] A south or southeast monsoon wind, favorable for the sailing of junks; the wind is known in Thailand, China, and Japan.

Juno [ASTRON] An asteroid whose disk has an estimated diameter of 120 miles (190 kilometers); the asteroids Ceres, Pallas, and Vesta together with Juno are sometimes called the Big Four.

junta [METEOROL] A wind blowing through Andes Mountain passes, sometimes reaching hurricane force.

Jupiter [ASTRON] The largest planet in the solar system, and the fifth in order of distance from the sun; semimajor axis = 485×10^6 miles (780×10^6 kilometers); sidereal revolution period = 11.86 years; mean orbital velocity = 8.2 miles per second (13.2 kilometers per second); inclination of orbital plane to ecliptic = 1.03; equatorial diameter = 88,700 miles (142,700 kilometers); polar diameter = 82,800 miles (133,300 kilometers); mass = about 318.4 (earth = 1).

Jupiter III *See* Ganymede.

Jupiter IV *See* Callisto.

Jura *See* Jurassic.

juran [METEOROL] A wind blowing from the Jura Mountains in Switzerland from the northwest toward Lake Geneva; it is a cold and snowy wind and may be very turbulent, especially in spring. Also spelled joran.

Jurassic [GEOL] Also known as Jura. **1.** The second period of the Mesozoic era of geologic time. **2.** The corresponding system of rocks.

Jurin rule [FL MECH] The rule that a height to which a liquid rises in a capillary tube is twice the liquid's surface tension times the cosine of its contact angle with the capillary, divided by the product of the liquid's weight density and the internal radius of the tube.

jurupaite [MINERAL] $(Ca,Mg)_2(Si_2O_5)(OH)_2$ A mineral composed of hydrous calcium magnesium silicate.

jury rig [MIN ENG] Any makeshift or temporary device, rig, or piece of equipment.

jury rudder [NAV ARCH] A temporary device used to steer a boat when the rudder is out of commission.

justification [GRAPHICS] In type composition, the adjustment of spacing in each line of type so that all lines are filled out to the same desired length.

justify [ADP] To shift data so that they assume a particular position relative to one or more reference points, lines, or marks in a storage medium.

justifying space [GRAPHICS] In type composition, a variable space used between words so that each line of type will attain a desired length.

justifying typewriter [GRAPHICS] A typewriter that is designed to type test material with regular right-hand margins or left-hand margins or both, especially for reproduction by offset lithography; the process usually requires two typings of the text because correct word spacing cannot be determined until the first typing is done; some typewriters in combination with a computer properly programmed can justify in one typing.

just-noticeable difference [PSYCH] A subjective scale used in psychophysical tests, defined by C. Fechner as the subjective unit, and the absolute threshold as the zero point of the subjective scale; for example, the subjective intensity of a particular brightness of light would be specified when it was given as 100 just-noticeable differences above the absolute threshold. Abbreviated JND. Also known as difference limen; difference threshold.

just scale [ACOUS] A diatonic scale rendered in the just tuning system.

just ton *See* ton.

just tuning [ACOUS] A tuning system generated by octave rearrangements of the notes of three consecutive triads, each having the frequency ratio 4:5:6, with the highest note of one triad serving as the lowest note of the next.

jute [BOT] Either of two Asiatic species of tall, slender, half-shrubby annual plants, *Corchorus capsularis* and *C. olitorius*, in the family Malvaceae, useful for their fiber.

jute board [MATER] A fiberboard made of jute fiber.

jute paper [MATER] A strong paper composed principally of jute fiber.

juvenile cell *See* metamyelocyte.

juvenile hormone *See* neotenin.

juvenile melanoma [MED] A benign compound nevus in young people; resembles a malignant melanoma histologically.

juvenile water *See* magmatic water.

Jynginae [VERT ZOO] The wrynecks, a family of Old World birds in the order Piciformes; a subfamily of the Picidae in some systems of classification.

JUTE

Morphological features of
Corchorus capsularis.

K

k *See* kilo-.

K *See* potassium.

K-A age [GEOL] The radioactive age of a rock determined from the ratio of potassium-40 (K^{40}) to argon-40 (A^{40}) present in the rock.

kaavie [METEOROL] In Scotland, a heavy fall of snow.

kachchan [METEOROL] A hot, dry west or southwest wind of foehn type in the lee of the Sri Lanka (Ceylon) hills during the southwest monsoon in June and July; it is well developed at Batticaloa on the east coast, where it is strong enough to overcome the sea breeze and bring maximum temperatures of nearly 100°F (38°C).

K acid [ORG CHEM] $C_{10}H_4NH_2OH(SO_3H)_2$ An acid derived from naphthylamine trisulfonic acid; used in dye manufacture. Also known as 1-amino-8-naphthol-4,6-disulfonic acid; 8-amino-1-naphthol-3,5-disulfonic acid.

kadaya gum *See* karaya gum.

K-A decay [NUC PHYS] Radioactive decay of potassium-40 (K^{40}) to argon-40 (A^{40}), as the nucleus of potassium captures an orbital electron and then decays to argon-40; the ratio of K^{40} to A^{40} is used to determine the age of rock (K-A age).

kaempferol [BIOCHEM] A flavonoid with a structure similar to that of quercetin but with only one hydroxyl in the B ring; acts as an enzyme cofactor and causes growth inhibition in plants.

Kahler's disease *See* multiple myeloma.

Kahn flocculation test [PATH] A macroscopic precipitin test for identification of the antibody resulting from syphilitic infection, by using an antigen prepared from normal beef heart.

kainite [MINERAL] $MgSO_4 \cdot KCl \cdot 3H_2O$ A white, gray, pink, or black monoclinic mineral, occurring in irregular granular masses; used as a fertilizer and as a source of potassium and magnesium compounds.

kainosite [MINERAL] $Ca_2(Ce,Y)_2(SiO_4)_3CO_3 \cdot H_2O$ A yellowish-brown mineral composed of a hydrous silicate and carbonate of calcium, cerium, and yttrium.

Kaiserling's method [BIOL] A method for preserving organ specimens and retaining their color by fixing in a solution of formalin, water, potassium nitrate, and potassium acetate, immersing in ethyl alcohol to restore color, and preserving in a solution of glycerin, aqueous arsenious acid, water, potassium acetate, and thymol.

kaki [BOT] *Diospyros kaki.* The Japanese persimmon; it provides a type of ebony wood that is black with gray, yellow, and brown streaks, has a close, even grain, and is very hard.

kakidrosis [MED] Secretion of sweat having a disagreeable odor.

kakorrhaphiophobia [PSYCH] Abnormal fear of failure.

kala azar [MED] Visceral leishmaniasis due to the protozoan *Leishmania donovani,* transmitted by certain sandflies (*Phlebotomus*); characterized by chronic, irregular fever, enlargement of the spleen and liver, emaciation, anemia, and leukopenia.

kal Baisakhi [METEOROL] A short-lived dusty squall at the onset of the southwest monsoon (April-June) in Bengal.

kale [BOT] Either of two biennial crucifers, *Brassica oleracea* var. *acephala* and · *B. fimbriata,* in the order Capparales, grown for the nutritious curled green leaves.

kaleidoscope [OPTICS] An optical toy consisting of a tube containing two plane mirrors placed at an angle of 60° and mounted so that a symmetrical pattern produced by multiple reflection is observed through a peephole at one end when objects (such as pieces of colored glass) at the other end are suitably illuminated.

kalema [OCEANOGR] A very heavy surf breaking on the Guinea coast of Africa during the winter.

kalfax [GRAPHICS] An emulsion sensitive to ultraviolet light coated on a mylar base, processed by heat.

kaliborite [MINERAL] $HKMg_2B_{12}O_{21} \cdot 9H_2O$ A colorless to white mineral composed of a hydrous borate of potassium and magnesium.

kalinite [MINERAL] $KAl(SO_4)_2 \cdot 11H_2O$ A birefringent mineral of the alum group composed of a hydrous sulfate of potassium and aluminum, occurring in fibrous form. Also known as potash alum.

kaliophilite [MINERAL] $KAlSiO_4$ A rare hexagonal tectosilicate mineral found in volcanic rocks; high in potassium and low in silica, it is dimorphous with kalsilite. Also known as facellite; phacellite.

kalium *See* potassium.

kalkowskite [MINERAL] $Fe_2Ti_3O_9$ A rare, brownish or black mineral composed of an oxide of iron and titanium, usually with small amounts of rare-earth elements, niobium, and tantalum.

kallidin I *See* bradykinin.

kallitype [GRAPHICS] An early photographic process using paper sensitized with ferrous oxalate and a silver salt, and a developer containing borax and Rochelle salt.

Kalman filter [CONT SYS] A linear system in which the mean squared error between the desired output and the actual output is minimized when the input is a random signal generated by white noise.

Kalotermitidae [INV ZOO] A family of relatively primitive, lower termites in the order Isoptera.

kalsilite [MINERAL] $KAlSiO_4$ A rare mineral from volcanic rocks in southwestern Uganda; the crystal system is hexagonal; kalsilite is dimorphous with kaliophilite and sometimes contains sodium.

kalsomine *See* calcimine.

kalunite [MINERAL] The naturally occurring form of alum.

Kaluza theory [RELAT] An attempted unified field theory in which the four-dimensional world that one observes is taken to be a projection of a five-dimensional continuum.

Kalvar process [GRAPHICS] A procedure for producing microfilms in which exposure of the film causes photosensitive compounds to release nitrogen gas; upon development of the film by heat, the pressure of the gas creates air sacs in the film to scatter light during projection for reading.

kamacite [MINERAL] A mineral composed of a nickel-iron alloy and comprising with taenite the bulk of most iron meteorites.

kame [GEOL] A low, long, steep-sided mound of glacial drift, commonly stratified sand and gravel, deposited as an alluvial fan or delta at the terminal margin of a melting glacier.

Kamptozoa [INV ZOO] An equivalent name for Entoprocta.

kanamycin [MICROBIO] $C_{18}H_{36}O_{11}N_4$ A water-soluble, basic antibiotic produced by strains of *Streptomyces kanamyceticus;* the sulfate salt is effective in infections caused by gramnegative bacteria.

kangaroo [VERT ZOO] Any of various Australian marsupials in the family Macropodidae generally characterized by a long, thick tail that is used as a balancing organ, short forelimbs, and enlarged hindlegs adapted for jumping.

Kansan glaciation [GEOL] The second glaciation of the Pleistocene epoch in North America; began about 400,000 years ago, after the Aftonian and before the Yarmouth interglacials.

Kansasii disease [MED] A mycobacterial tuberculosislike infection caused by *Mycobacterium kansasii,* an orange-yellow acid-fast bacterium.

kansite *See* mackinawite.

Kantrax [MICROBIO] A trade name for the antibiotic kanamycin.

kaolin [MINERAL] Any of a group of clay minerals, including kaolinite, nacrite, dickite, and anauxite, with a two-layer

KALE

Kale (*Brassica oleracea* var. *acephala*), cultivar Vates. *(Joseph Harris Co., Rochester, N.Y.)*

KANAMYCIN

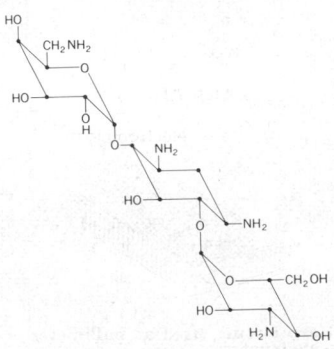

Structural formula of kanamycin A; the major component (95%) of commercial kanamycin sulfate USP.

KANGAROO

The red kangaroo at rest, with its weight carried by a "tripod" consisting of hindlegs and tail.

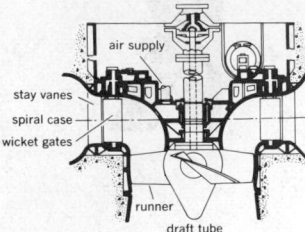

KAPLAN TURBINE

Cross section of a Kaplan type of hydraulic turbine installation.

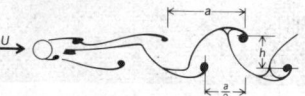

KÁRMÁN VORTEX STREET

Kármán vortex street showing double row of line vortices. Here U = stream speed, h = perpendicular distance between the two lines of vortices, and a = distance between successive vortices on the same line.

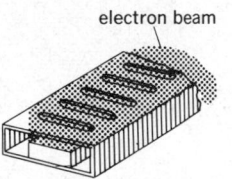

KARP CIRCUIT

Karp circuit, used at millimeter wavelengths.

KARST

Sinkholes in Meade County, Kansas. Depth to the underground water is about 50 feet (15 meters). (*U.S. Geological Survey*)

crystal in which silicon-oxygen and aluminum-hydroxyl sheets alternate; approximate composition is $Al_2O_3 \cdot 2SiO_2 \cdot 2H_2O$. [PETR] A soft, nonplastic white rock composed principally of kaolin minerals. Also known as bolus alba; white clay.

kaolinite [MINERAL] $Al_2Si_2O_5(OH)_4$ The principal mineral of the kaolin group of clay minerals; a white, gray, or yellowish high-alumina mineral consisting of sheets of tetrahedrally coordinated silicon linked by an oxygen shared with octahedrally coordinated aluminum.

kaolinization [GEOL] The forming of kaolin by the weathering of aluminum silicate minerals or other clay minerals.

kaon See K meson.

kaonic atom [ATOM PHYS] An atom consisting of a negatively charged kaon orbiting around an ordinary nucleus.

Kapetyn selected areas [ASTRON] Certain areas in the Milky Way Galaxy that the astronomer J. C. Kapetyn suggested be studied intensively in order to determine the structure of the galaxy. Also known as selected areas.

Kapetyn's star [ASTRON] A star 13.0 light-years from the solar system, absolute magnitude 11.2, spectrum type MO; has a large proper motion.

Kapitza expander [CHEM ENG] Reciprocating-piston gas expander used for helium liquefaction; relies on close fit rather than packing or rings on the pistons.

Kaplan turbine [MECH ENG] A propeller-type hydraulic turbine in which the positions of the runner blades and the wicket gates are adjustable for load change with sustained efficiency.

kapoc oil See kapok oil.

kapok [BOT] Silky fibers that surround the seeds of the kapok or ceiba tree. Also known as ceiba; Java cotton; silk cotton.

kapok oil [MATER] Yellow-green oil with pleasant aroma and taste; soluble in alcohol, ether, and chloroform; derived by pressing seeds of the kapok tree; used in edible oils and soap stock. Also spelled kapoc oil.

kapok tree [BOT] *Ceiba pentandra*. A tree of the family Bombacaceae which produces pods containing seeds covered with silk cotton. Also known as silk cotton tree.

kappa particle [CYTOL] A self-duplicating nucleoprotein particle found in various strains of *Paramecium* and thought to function as an infectious agent; classed as an intracellular symbiont, occupying a position between the viruses and the bacteria and organelles.

karaburan [METEOROL] A violent northeast wind of Central Asia occurring during spring and summer; it carries clouds of dust (which darken the sky) instead of snow. Also known as black buran; black storm.

karajol [METEOROL] On the Bulgarian coast, a west wind which usually follows rain and persists 1–3 days.

karat [MET] A unit for designating the fineness of gold in an alloy; represents a twenty-fourth part; thus, 18-karat gold is 18/24 or 75% pure.

karaya gum [MATER] Exudation from the Indian tree (*Sterculia urens*); white to dark-colored tears; used in pharmaceuticals, textiles, and foods, and as tragacanth gum substitute. Also known as Indian tragacanth; kadaya gum.

karema [METEOROL] A violent east wind on Lake Tanganyika in Africa.

karif [METEOROL] A strong southwest wind on the southern shore of the Gulf of Aden, especially at Berbera, Somaliland, during the southwest monsoon.

Karl Fischer reagent [ANALY CHEM] A solution of 8 moles pyridine to 2 moles sulfur dioxide, with the addition of about 15 moles methanol and then 1 mole iodine; used to determine trace quantities of water by titration.

Karl Fischer technique [ANALY CHEM] A method of determining trace quantities of water by titration; the Karl Fischer reagent is added in small increments to a glass flask containing the sample until the color changes from yellow to brown or a change in potential is observed at the end point.

Kármán constant [FL MECH] A dimensionless number formed from the velocity of turbulent flow parallel to a plane wall, the distance from the wall, the shear stress, and the density of the fluid; for a wide range of flow patterns it has a constant value.

Kármán vortex street [FL MECH] A double row of line vortices in a fluid which, under certain conditions, is shed in the wake of cylindrical bodies when the relative fluid velocity is perpendicular to the axis of the cylinder.

Karnaugh map [ELECTR] A truth table that has been rearranged to show a geometrical pattern of functional relationships for gating configurations; with this map, essential gating requirements can be recognized in their simplest form.

Karnian See Carnian.

karoo See karroo.

Karp circuit [ELECTR] A slow-wave circuit used at millimeter wavelengths for backward-wave oscillators.

Karrer method [CHEM ENG] An industrial method for the chemical synthesis of riboflavin.

karroo [GEOGR] A dry, broad, level, elevated area found especially in southern Africa, often rising to considerable elevations in terrace formations; does not support vegetation in the dry season but supports grass during the wet season. Also spelled karoo.

Karroo System [GEOL] Glaciated strata formed in Permian times in southern Africa.

karrusel [HOROL] A revolving escapement that reduces position errors.

karst [GEOL] A topography formed over limestone, dolomite, or gypsum and characterized by sinkholes, caves, and underground drainage.

karstbora [METEOROL] The bora of the Yugoslavian coast.

karst plain [GEOL] A plain on which karst features are developed.

Karumiidae [INV ZOO] The termitelike beetles, a small family of coleopteran insects in the superfamily Cantharoidea distinguished by having a tenth tergum.

karyocyte See normoblast.

karyogamy [CYTOL] Fusion of gametic nuclei, as in fertilization.

karyokinesis [CYTOL] Nuclear division characteristic of mitosis.

karyolymph [CYTOL] The clear material composing the ground substance of a cell nucleus.

karyolysis [CYTOL] Dissolution of a cell nucleus.

karyomastigont [INV ZOO] Pertaining to members of the protozoan order Oxymonadida; individuals can be uni- or multinucleate, and unattached forms give rise to two pairs of flagella.

karyoplasm See nucleoplasm.

karyoplasmic ratio See nucleocytoplasmic ratio.

karyorrhexis [CYTOL] Fragmentation of a nucleus with scattering of the pieces in the cytoplasm.

karyotype [CYTOL] The normal diploid or haploid complement of chromosomes, with respect to size, form, and number, characteristic of an individual, species, genus, or other grouping.

kasolite [MINERAL] $Pb(UO_2)SiO_4 \cdot H_2O$ Yellow-ocher mineral composed of a hydrous lead uranium silicate, occurring in monoclinic crystals.

katabaric See katallobaric.

katabatic wind See gravity wind.

katafront [METEOROL] A front (usually a cold front) at which warm air descends the frontal surface (except, presumably, in the lowest layers).

katallobaric [METEOROL] Of or pertaining to a decrease in atmospheric pressure. Also known as katabaric.

katallobaric center See pressure-fall center.

katamorphism [GEOL] A process of metamorphism occurring at or near the earth's surface and resulting in production of simpler minerals from complex minerals through oxidation, hydration, solution, and other processes. Also spelled catamorphism.

Kata thermometer [ENG] An alcohol thermometer used to measure low velocities in air circulation, by heating the large bulb of the thermometer above 100°F (38°C) and noting the time it takes to cool from 100 to 95°F (38 to 35°C) or some other interval above ambient temperature, the time interval being a measure of the air current at that location.

katchung oil See peanut oil.

Kater's reversible pendulum [MECH] A gravity pendulum designed to measure the acceleration of gravity and consisting of a body with two knife-edge supports on opposite sides of the center of mass.

katharometer [ENG] An instrument for detecting the presence of small quantities of gases in air by measuring the resulting change in thermal conductivity of the air. Also known as thermal conductivity cell.

Kathlaniidae [INV ZOO] A family of nematodes assigned to the Ascaridina by some authorities and to the Oxyurina by others.

katoptric system [OPTICS] An optical system such that, when the object is displaced in a direction parallel to the axis, the image is displaced in the opposite direction (in contrast to a dioptric system). Also known as contracurrent system.

Kauertz engine [MECH ENG] A type of cat-and-mouse rotary engine in which the pistons are vanes which are sections of a right circular cylinder; two pistons are attached to one rotor so that they rotate with constant angular velocity, while the other two pistons are controlled by a gear-and-crank mechanism, so that angular velocity varies.

kauri-butanol value [ANALY CHEM] The measure of milliliters of paint or varnish petroleum thinner needed to cause cloudiness in a solution of kauri gum in butyl alcohol.

kauri gum [MATER] Hard copal resins from kauri pine (*Agathis australis*); used in lacquers and varnishes.

kaus [METEOROL] A moderate to gale-force southeasterly wind in the Persian Gulf, accompanied by gloomy weather, rain, and squalls; it is most frequent between December and April. Also known as cowshee; sharki.

kavaburd *See* cavaburd.

kaver *See* caver.

kay *See* key.

kayser [SPECT] A unit of reciprocal length, especially wave number, equal to the reciprocal of 1 centimeter. Also known as rydberg.

Kayser-Fleischer ring [PATH] A ring of golden-brown or brownish-green pigment behind the limbic border of the cornea, due to the deposition of copper.

Kazanian [GEOL] A European stage of geologic time: Upper Permian (above Kungurian, below Tatarian).

K band [COMMUN] A band of radio frequencies extending from 10,900 to 36,000 megahertz, corresponding to wavelengths of 2.75 to 0.834 centimeters. [SOLID STATE] An optical absorption band which appears together with an *F*-band and has a lower intensity and shorter wavelength than the latter.

K bentonite *See* potassium bentonite.

kc *See* kilohertz.

KC-97 *See* Stratofreighter.

KC-135 *See* Stratotanker.

kcal *See* kilocalorie.

K capture [NUC PHYS] A type of beta interaction in which a nucleus captures an electron from the *K* shell of atomic electrons (the shell nearest the nucleus) and emits a neutrino.

K-carrier system [COMMUN] Carrier system providing 12 telephone channels with a bandwidth up to 60 kilohertz, either on a four-wire cable system or on microwave (line-of-sight) and trophospheric scatter radio systems.

K corona [ASTRON] The inner portion of the sun's corona, having a continuous spectrum caused by electron scattering.

K damage [ORD] **1.** Combat damage sufficient to cause a vehicle to be destroyed. **2.** Combat damage such that an aircraft will fall out of control immediately after the damage occurs.

K display [ELECTR] A modified radarscope A display in which a target appears as a pair of vertical deflections instead of as a single deflection; when the radar antenna is correctly pointed at the target in azimuth, the deflections are of equal height; when the antenna is not correctly pointed, the difference in pulse heights is an indication of direction and magnitude of azimuth pointing error.

kedge [NAV] To move, as a vessel, by carrying out an anchor, letting it go, and hauling the ship up to the anchor.

kedge anchor [NAV ARCH] A light anchor that is used to warp or kedge a ship.

keel [NAV ARCH] A steel beam or timber, or a series of steel beams and plates or timbers joined together, extending along the center of the bottom of a ship from stem to stern and often projecting below the bottom, to which the frames and hull plating are attached. [VERT ZOO] The median ridge on the breastbone in certain birds. Also known as carina.

Keel *See* Carina.

keel block [CIV ENG] docking block used to support a ship's keel. [MET] A simple shape from which a test casting is made in the form of a large head, which is removed and discarded, with a keel on the bottom.

keel condenser [NAV ARCH] Pipes, made of material that will not induce electrolysis with other parts of a ship, placed outside the ship near the keel for condensing steam.

keel molding [ARCH] A molding with two ogees, the central fillet of which projects like a keel.

keelson [NAV ARCH] A structure of timbers or steel beams which are bolted to the top of a keel to increase its strength. Also spelled kelson.

Keene's cement [MATER] An anhydrous calcined gypsum mixed with an accelerator; used as a hard-finish plaster.

keep-alive circuit [ELECTR] A circuit used with a transmit-receive (TR) tube or anti-TR tube to produce residual ionization for the purpose of reducing the initiation time of the main discharge.

keep-alive electrode *See* ignitor.

keeper [ELECTROMAG] A bar of iron or steel placed across the poles of a permanent magnet to complete the magnetic circuit when the magnet is not in use, to avoid the self-demagnetizing effect of leakage lines. Also known as magnet keeper.

keeps *See* folding boards.

Keesom relationship [PHYS CHEM] An equation for the potential energy associated with the interaction of the dipole moments of two polar molecules.

Keewatin [GEOL] A division of the Archeozoic rocks of the Canadian Shield.

kefir [FOOD ENG] A low-alcohol-content beverage prepared from cow's milk treated with *Lactobacillus casei* and then fermented by the action of *Saccharomyces kefir* on kefir grain.

Keflin [MICROBIO] The trade name for cephalothin.

keg buoy [NAV] A buoy consisting of a keg to which is attached a small pole with a flag, used by fishermen to mark the position of a trawl line.

kehoeite [MINERAL] An amorphous mineral composed of a basic hydrous calcium aluminum zinc phosphate, occurring massive.

Keilor skull [PALEON] An Australian fossil type specimen of *Homo sapiens* from the Pleistocene.

Keldysh theory [ATOM PHYS] A theory of multiphoton ionization, in which an atom is ionized by rapid absorption of a sufficient number of photons; it predicts that the ionization rate depends primarily upon the ratio of the mean binding electric field to the peak strength of the incident electromagnetic field, and upon the ratio of the binding energy to the energy of photons in the field.

K electron [ATOM PHYS] An electron in the *K* shell.

Kel-F [ORG CHEM] Trademark of M. W. Kellogg Company for their fluorocarbon products based on polychlorotrifluoroethylene resin.

Kell blood group system [IMMUNOL] A family of antigens found in erythrocytes and designated K, k, Kp^a, Kp^b, and Ku; antibodies to the K antigen, which occurs in about 10% of the population of England, have been associated with hemolytic transfusion reactions and with hemolytic disease.

kellering [MECH ENG] Three-dimensional machining of a contoured surface by tracer-milling the die block or punch; the cutter path is controlled by a tracer that follows the contours on a die model.

Kellner eyepiece [OPTICS] A Ramsden eyepiece with an achromatic eye lens.

Kellogg equation [THERMO] An equation of state for a gas, of the form

$$p = RT\rho + \sum_{n=2}^{\infty} (b_n T - a_n - C_n/T^2)\rho^n$$

where p is the pressure, T the absolute temperature, ρ the density, R the gas constant, and a_n, b_n, and c_n are constants.

kelly [PETRO ENG] In rotary drilling, the rod attached to the top of the drill column. Also known as grief stem.

Kelly ball test [ENG] A test for the consistency of concrete using the penetration of a half sphere; a 1-inch penetration by the Kelly ball corresponds to about 2 inches of slump.

keloid [MED] A firm, elevated fibrous formation of tissue at the site of a scar.

keloid acne [MED] Acnelike infection of hair follicles, especially on the nape of the neck, resulting in hard, white or reddish keloids and scarring. Also known as folliculitis keloidalis.

kelp [BOT] The common name for brown seaweed belonging to the Laminariales and Fucales.

kelsher [METEOROL] In England, a heavy fall of rain.

kelson *See* keelson.

kelvin [ELEC] A name formerly given to the kilowatt-hour. Also known as thermal volt. [THERMO] A unit of absolute temperature equal to 1/273.16 of the absolute temperature of the triple point of water. Symbolized K. Formerly known as degree Kelvin.

Kelvin absolute temperature scale [THERMO] A temperature scale in which the ratio of the temperatures of two reservoirs is equal to the ratio of the amount of heat absorbed from one of them by a heat engine operating in a Carnot cycle to the amount of heat rejected by this engine to the other reservoir; the temperature of the triple point of water is defined as 273.16K. Also known as Kelvin temperature scale.

Kelvin balance [ELECTROMAG] An ammeter in which the force between two coils in series that carry the current to be measured, one coil being attached to one arm of a balance, is balanced against a known weight at the other end of the balance arm.

Kelvin body [MECH] An ideal body whose shearing (tangential) stress is the sum of a term proportional to its deformation and a term proportional to the rate of change of its deformation with time.

Kelvin bridge [ELEC] A specialized version of the Wheatstone bridge network designed to eliminate, or greatly reduce, the effect of lead and contact resistance, and thus permit accurate measurement of low resistance. Also known as double bridge; Kelvin network; Thomson bridge.

Kelvin equation [THERMO] An equation giving the increase in vapor pressure of a substance which accompanies an increase in curvature of its surface; the equation describes the greater rate of evaporation of a small liquid droplet as compared to that of a larger one, and the greater solubility of small solid particles as compared to that of larger particles.

Kelvin guard-ring capacitor [ELEC] A capacitor with parallel circular plates, one of which has a guard ring separated from the plate by a narrow gap; it is used as a standard, whose capacitance can be accurately calculated from its dimensions.

Kelvin-Helmholtz contraction [ASTROPHYS] A contraction of a star once it is formed and before it is hot enough to ignite its hydrogen; the contraction converts gravitational potential energy into heat, some of which is radiated, with the remainder used to raise the internal temperature of the star.

Kelvin network *See* Kelvin bridge.

Kelvin relations *See* Thomson relations.

Kelvin scale [THERMO] The basic scale used for temperature definition; the triple point of water (comprising ice, liquid, and vapor) is defined as 273.16K; given two reservoirs, a reversible heat engine is built operating in a cycle between them, and the ratio of their temperatures is defined to be equal to the ratio of the heats transferred.

Kelvin's circulation theorem [FL MECH] The theorem that, if the external forces acting on an inviscid fluid are conservative and if the fluid density is a function of the pressure only, then the circulation along a closed curve which moves with the fluid does not change with time.

Kelvin skin effect *See* skin effect.

Kelvin's minimum-energy theorem [FL MECH] The theorem that the irrotational motion of an incompressible, inviscid fluid occupying a simply connected region has less kinetic energy than any other fluid motion consistent with the boundary condition of zero relative velocity normal to the boundaries of the region.

Kelvin's statement of the second law of thermodynamics [THERMO] The statement that it is *not* possible that, at the end of a cycle of changes, heat has been extracted from a reservoir and an equal amount of work has been produced without producing some other effect.

Kelvin temperature scale [THERMO] **1.** An International Practical Temperature Scale which agrees with the Kelvin absolute temperature scale within the limits of experimental determination. **2.** *See* Kelvin absolute temperature scale.

Kelvin wave [OCEANOGR] A type of wave progression in relatively confined water bodies where because of Coriolis force the wave is higher to the right of direction of advance (in the Northern Hemisphere).

kelyphite *See* corona.

kelyphytic rime [PETR] A peripheral zone of pyroxene or amphibole developed around olivine in some igneous rocks. Also known as kelyphytic border.

kempite [MINERAL] $Mn_2(OH)_3Cl$ An emerald-green orthorhombic mineral composed of a basic manganese oxychloride, occurring in small crystals.

Kendall effect [COMMUN] A spurious pattern or other distortion in a facsimile record caused by unwanted modulation products arising from the transmission of a carrier signal; occurs principally when the width of one side band is greater than half the facsimile carrier frequency.

Kennard packet [QUANT MECH] A wave packet for which the product of the root-mean-square deviations of position and momentum from their respective mean values is as small as possible, being equal to Planck's constant divided by 4π.

Kennedy key [DES ENG] A square taper key fitted into a keyway of square section and driven from opposite ends of the hub.

Kennelly-Heaviside layer *See* E layer.

kenophobia [PSYCH] An abnormal fear of large, empty spaces.

Kenoran orogeny *See* Algoman orogeny.

kenotron [ELEC] A high-vacuum diode designed to serve as a rectifier in appliances requiring high voltage and low current.

kenozooecium [INV ZOO] The outer, nonliving, hardened portion of a kenozooid.

kenozooid [INV ZOO] A type of bryozoan heterozooid possessing a slender tubular or boxlike chamber, completely enclosed and lacking an aperture.

kentrolite [MINERAL] $Pb_2Mn_2Si_2O_9$ A dark reddish-brown mineral composed of a lead manganese silicate.

Kentucky coffee tree [BOT] *Gymnocladus diocai.* An extremely tall, dioecious tree of the order Rosales readily recognized when in fruit by its leguminous pods containing heavy seeds, once used as a coffee substitute.

Kenyapithecus [PALEON] An early member of Hominidae from the Miocene.

kep interlock [MIN ENG] A system designed to prevent the lowering of a shaft conveyance before all keps are fully withdrawn, and to indicate the position of the keps.

Keplerian ellipse *See* Keplerian orbit.

Keplerian motion [ASTRON] Orbital movement of a body about another that is not disturbed by the presence of a third celestial body.

Keplerian orbit [ASTRON] An elliptical orbit of a celestial body about another, the latter at a focus of the ellipse. Also known as Keplerian ellipse.

Keplerian telescope [OPTICS] A telescope that forms a real intermediate image in the focal plane and can be used for introducing a reticle or a scale into the focal plane.

Kepler's equations [ASTRON] The mathematical relationship between two different systems of angular measurements of the position of a body in an ellipse.

Kepler's laws [ASTRON] Three laws, determined by Johannes Kepler, that describe the motions of planets in their orbits: the orbits of the planets are ellipses with the sun at a common focus; the line joining a planet and the sun sweeps over equal areas during equal intervals of time; the squares of the periods of revolution of any two planets are proportional to the cubes of their mean distances from the sun.

Kepler's nova [ASTRON] A nova that appeared in October 1604 and was visible until March 1606.

keps *See* folding boards.

kerabitumen *See* kerogen.

keratectomy [MED] Surgical removal of a portion of the cornea.

keratin [BIOCHEM] Any of various albuminoids characteristic of epidermal derivatives, such as nails and feathers, which are insoluble in protein solvents, have a high sulfur content, and generally contain cystine and arginine as the predominating amino acids.

keratinized tissue [HISTOL] Any tissue with a high keratin content, such as the epidermis or its derivatives.

KELVIN BRIDGE

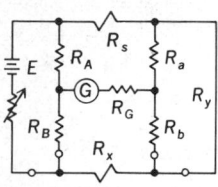

Circuit of the Kelvin bridge. E = battery; R_A, R_B = main ratio resistors; R_a, R_b = auxiliary ratio; R_x = unknown; R_s = standard; R_y = a heavy copper yoke of low resistance connected between the unknown and standard resistors; R_G = resistance in series with galvanometer G.

KENTUCKY COFFEE TREE

Branch of Kentucky coffee tree (*Gymnocladus dioica*).

keratinocyte [HISTOL] A specialized epidermal cell that synthesizes keratin.

keratinous degeneration [CYTOL] The occurrence of keratin granules in the cytoplasm of a cell, other than a keratinocyte.

keratitis [MED] Inflammation of the cornea.

keratitis rosacea [MED] The occurrence of small, sterile infiltrates at the periphery of the cornea.

keratoconjunctivitis [MED] Concurrent inflammation of the cornea and the conjunctiva. Also known as shipyard eye.

keratohyalin [HISTOL] Granules in the stratum granulosum of keratinized stratified squamous epithelium which become keratin.

keratomalacia [MED] Degeneration of the cornea characterized by infiltration and keratinization of the epithelium, eventually leading to thinning and perforation of the cornea; generally occurs in vitamin-A deficiency.

keratophyre [PETR] Any dike rock or salic lava that is characterized by the presence of albite or albite oligoclase, chlorite, epidote, and calcite.

keratoplasty [MED] A plastic operation on the cornea, especially the transplantation of a portion of the cornea.

keratosis [MED] Any disease of the skin characterized by an overgrowth of the cornified epithelium.

kerf [ENG] A cut made in wood, metal, or other material by a saw or cutting torch. [MIN ENG] A narrow, deep cut made in the face of coal to facilitate mining.

Kerguelen faunal region [ECOL] A marine littoral faunal region comprising a large area surrounding Kerguelen Island in the southern Indian Ocean.

kerma [NUCLEO] The kinetic energy imparted to charged particles in a unit mass of material by uncharged particles such as neutrons; it may be expressed as joules per kilogram or ergs per gram.

kermesite [MINERAL] Sb_2S_2O A cherry-red mineral occurring as tufts of capillary crystals, and formed from an alteration of stibnite. Also known as antimony blende; purple blende; pyrostibite; red antimony.

Kern counter *See* dust counter.

kernel [ATOM PHYS] An atom that has been stripped of its valence electrons, or a positively charged nucleus lacking the outermost orbital electrons. [BOT] **1.** The inner portion of a seed. **2.** A whole grain or seed of a cereal plant, such as corn or barley. [MATH] *See* null space.

kernel blight [PL PATH] Any of several fungus diseases of barley caused chiefly by *Gibberella zeae, Helminthosporium sativum,* and *Alternaria* species shriveling and discoloring the grain.

kernel ice [HYD] In aircraft icing, an extreme form of rime ice, that is, very irregular, opaque, and of low density; it forms at temperatures of $-15°C$ and lower.

kernel of a homomorphism [MATH] For a homomorphism *h* from a group *G* to a group *H*, this consists of all elements of *G* which *h* sends to the identity element of *H*.

kernel of a linear transformation [MATH] All those vectors which a linear transformation maps to the zero vector.

kernel of an integral equation [MATH] For Fredholm and Volterra type of equations, this is the function $K(x,t)$.

kernel of an integral transform [MATH] The function $K(x,t)$ in the transformation which sends the function $f(x)$ to the function $\int K(x,t)f(t)dt = F(x)$.

kernel spot [PL PATH] A fungus disease of the pecan kernel caused by *Coniothyrium caryogenum* and characterized by dull-brown roundish spots.

kernicterus [PATH] Deposition of bilirubin in the gray matter of the brain and spinal cord, especially the basal ganglia, accompanied by nerve cell degeneration.

Kernig's sign [MED] In meningeal irritation, with the patient lying face up and the thigh flexed at the hip, the pain and spasm of the hamstring muscles when an attempt is made to completely extend the leg at the knee.

kernite [MINERAL] $Na_2B_4O_7 \cdot 4H_2O$ A colorless to white hydrous borate mineral crystallizing in the monoclinic system and having vitreous luster; an important source of boron. Also known as rasorite.

kerogen [GEOL] The complex, fossilized organic material present in sedimentary rocks, especially in shales; converted to petroleum products by distillation. Also known as kerabitumen; petrologen.

kerosine [MATER] A refined petroleum fraction used as a fuel for heating and cooking, jet engines, lamps, and weed burning and as a base for insecticides; specific gravity is about 0.8; components are mostly paraffinic and naphthenic hydrocarbons in the C_{10} to C_{14} range. Also known as lamp oil.

kerosine distillate [MATER] The distilled cut in the 150–300°C range from petroleum or shale oil; used as lamp, stove, or illuminating oil, as a solvent, and as a component of jet aircraft fuels. Also known as burning oil.

kerosine propellant [MATER] A propellant consisting of highly refined, low-aromatic liquid propellant distillate with a gravity range not exceeding three degrees American Petroleum Institute gravity at 60°F (15.6°C); may contain additives.

kerosine shale *See* torbanite.

Kerr cell [OPTICS] A glass cell containing a dielectric liquid that exhibits the Kerr effect, such as nitrobenzene, in which is inserted the two plates of a capacitor, used to observe the Kerr effect on light passing through the cell.

Kerr constant [OPTICS] A measure of the strength of the Kerr effect in a substance, equal to the difference between the extraordinary and ordinary indices of refraction divided by the product of the light's wavelength and the square of the electric field.

Kerr effect *See* electrooptical Kerr effect.

Kerr magnetooptical effect *See* magnetooptic Kerr effect.

kersantite [PETR] Dark dike rocks consisting mostly of biotite, plagioclase, and augite.

ketal [ORG CHEM] **1.** Former term for the $=CO$ group, as in dimethyl ketal (acetone). **2.** Any of the ketone acetates from condensation of alkyl orthoformates with ketones in the presence of alcohols.

ketene [ORG CHEM] C_2H_2O A colorless, toxic, highly reactive gas, with disagreeable taste; boils at -56°C; soluble in ether and acetone, and decomposes in water and alcohol; used as an acetylating agent in organic synthesis.

ketene lamp [CHEM ENG] An electrically heated Chromel filament by the means of which acetone is hydrolyzed to produce ketene.

ketimide [ORG CHEM] A compound that is represented by $R_2:C:NX$, where X is an acyl radical.

ketimine [ORG CHEM] An organic compound that contains the divalent group $>C=NH$; a Schiff base is an example.

keto- [ORG CHEM] Organic chemical prefix for the keto or carbonyl group, $C:O$, as in a ketone.

keto acid [ORG CHEM] A compound that is both an acid and a ketone; an example is β-acetoacetic acid.

ketoacidosis *See* ketosis.

ketoadipic acid [BIOCHEM] $C_6H_8O_5$ An intermediate product in the metabolism of lysine to glutaric acid.

ketogenesis [BIOCHEM] Production of ketone bodies.

ketogenic hormone [BIOCHEM] A factor originally derived from crude anterior hypophysis extract which stimulated fatty-acid metabolism; now known as a combination of adrenocorticotropin and the growth hormone. Also known as fat-metabolizing hormone.

ketogenic substance [BIOCHEM] Any foodstuff which provides a source of ketone bodies.

ketoglutaric acid [BIOCHEM] $C_5H_6O_5$ A dibasic keto acid occurring as an intermediate product in carbohydrate and protein metabolism.

ketohexose [BIOCHEM] Any monsaccharide composed of a six-carbon chain and containing one ketone group.

ketolase [BIOCHEM] A type of enzyme that catalyzes cleavage of carbohydrates at the carbonyl carbon position.

ketolysis [BIOCHEM] Dissolution of ketone bodies.

ketone [ORG CHEM] One of a class of chemical compounds of the general formula $RR'CO$, where R and R' are alkyl, aryl, or heterocyclic radicals; the groups R and R' may be the same or different, or incorporated into a ring; the ketones, acetone, and methyl ethyl ketone are used as solvents, and ketones in general are important intermediates in the synthesis of organic compounds.

ketone body [BIOCHEM] Any of various ketones which increase in blood and urine in certain conditions, such as

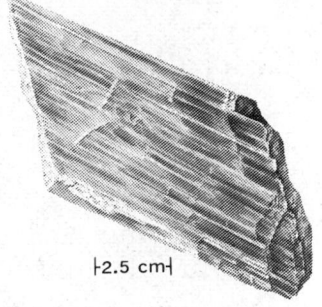

KERNITE

⊢2.5 cm⊣

Kernite crystal, Kern County, California. *(Specimen from Department of Geology, Bryn Mawr College)*

KETENE

Structural formula of ketene.

KETONE

Structural formula of a ketone.

diabetic acidosis, starvation, and pregnancy. Also known as acetone body.

ketonemia See acetonemia.

ketonuria [MED] Presence of ketone bodies in the urine.

ketose [BIOCHEM] A carbohydrate that has a ketone group.

ketosis [MED] Excess amounts of ketones in the body, especially associated with diabetes mellitus. Also known as ketoacidosis.

ketosteroid [BIOCHEM] One of a group of neutral steroids possessing keto substitution, which produces a characteristic red color with m-dinitrobenzene in an alkaline solution; these compounds are principally metabolites of adrenal cortical and gonadal steroids.

γ-ketovaleric acid See levulinic acid.

kettle [GEOL] **1.** A bowl-shaped depression with steep sides in glacial drift deposits that is formed by the melting of glacier ice left behind by the retreating glacier and buried in the drift. Also known as kettle basin; kettle hole. **2.** See pothole.

kettle basin See kettle.

kettle hole See kettle.

kettleman [IND ENG] A worker who operates a kettle, such as in a brewery, lead refinery, or paint factory.

kettle reboiler [CHEM ENG] Tube-and-shell heat exchange device in which liquid is vaporized on the shell side from heat transferred from hot liquid flowing through the tubes; dome space allows liquid-vapor separation above the tube bundle.

ket vector [QUANT MECH] A vector in Hilbert space specifying the state of a system (opposed to bra vector); represented by the symbol "$|>$," with a letter or one or more indices inserted to distinguish it from other vectors.

Keuper [GEOL] A European stage of geologic time, especially in Germany; Upper Triassic.

keV See kilo electron volt.

Kew barometer [ENG] A type of cistern barometer; no adjustment is made for the variation of the level of mercury in the cistern as pressure changes occur; rather, a uniformly contracting scale is used to determine the effective height of the mercury column.

Keweenawan [GEOL] The younger of two Precambrian time systems that constitute the Proterozoic period in Michigan and Wisconsin.

key [ADP] A data item that serves to uniquely identify a data record. [BUILD] **1.** Plastering that is forced between laths to secure the rest of the plaster in place. **2.** The roughening on a surface to be glued or plastered to increase adhesiveness. [CIV ENG] A projecting portion that serves to prevent movement of parts at a construction joint. [DES ENG] **1.** An instrument that is inserted into a lock to operate the bolt. **2.** A device used to move in some manner in order to secure or tighten. **3.** One of the levers of a keyboard. **4.** See machine key. [ELEC] **1.** A hand-operated switch used for transmitting code signals. Also known as signaling key. **2.** A special lever-type switch used for opening or closing a circuit only as long as the handle is depressed. Also known as switching key. [ENG] The pieces of core causing a block in a core barrel, the removal of which allows the rest of the core in the barrel to slide out. [GEOL] A cay, especially one of the islets off the south of Florida. Also spelled kay. [SYST] An arrangement of the distinguishing features of a taxonomic group to serve as a guide for establishing relationships and names of unidentified members of the group.

key bed [GEOL] Also known as index bed; key horizon; marker bed. **1.** A stratum or body of strata that has distinctive characteristics so that it can be easily identified. **2.** A bed whose top or bottom is employed as a datum in the drawing of structure contour maps.

keyboard [ENG] A set of keys or control levers having a systematic arrangement and used to operate a machine or other piece of equipment such as a typewriter, typesetter, processing unit of a computer, or piano.

keyboard entry [ADP] A piece of information fed manually into a computing system by means of a set of keys, such as a typewriter.

keyboard inquiry [ADP] A question asked a computer concerning the status of a program being run, or concerning the value achieved by a specific variable, by means of a console typewriter.

KEYED CLAMP CIRCUIT

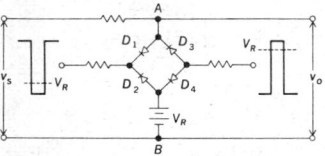

Four-diode bidirectional keyed clamp circuit. D = diode; V_R = reference voltage; v_S = voltage when switch is in series with reference voltage; A, B = terminals.

keyboardless typesetter [ADP] An automatic typesetting machine that has no keyboard and is operated by perforated tape at a speed of 12–15 lines per minute; the text is punched on tape at separate keyboard machines.

keyboard lockout [ADP] An arrangement for preventing transmission from a particular keyboard while other transmissions are taking place on the same circuit.

keyboard perforator [ENG] A typewriterlike device that prepares punched paper tape for communications or computing equipment.

keyboard send/receive [ELECTR] A manual teleprinter that can transmit or receive. Abbreviated KSR. Also known as keyboard teleprinter.

keyboard teleprinter See keyboard send/receive.

key cabinet [ELECTR] A case, installed on a customer's premises, to permit different lines to the control office to be connected to various telephone stations; it has signals to indicate originating calls and busy lines.

key click [COMMUN] Transient signal sometimes produced when a radiotelegraph sending key is opened or closed; it is heard in a loudspeaker or headphone as a click or chirp.

key cut [MIN ENG] In strip mine operations, the section excavated adjacent to the new highwall.

key day See control day.

key-driven calculator [ADP] A mechanical desk calculator in which numeric selection keys, arranged in columns, are coupled directly to accumulator dials on which the results are visible.

keyed clamp [ELECTR] Clamping circuit in which the time of clamping is determined by a control signal.

keyed clamp circuit [ELECTR] A clamp circuit in which the time of clamping is controlled by separate voltage or current sources, rather than by the signal itself. Also known as synchronous clamp circuit.

keyer [ELECTR] Device which changes the output of a transmitter from one condition to another according to the intelligence to be transmitted.

keyer adapter [ELECTR] Device which detects a modulated signal and produces the modulating frequency as a direct-current signal of varying amplitude.

Keyes equation [THERMO] An equation of state of a gas which is designed to correct the van der Waals' equation for the effect of surrounding molecules on the term representing the volume of a molecule.

keyhole [DES ENG] A hole or slot for receiving a key. [ORD] Of a bullet, to strike a target after tumbling in flight so that the long axis of the bullet and the line of flight are not the same; usually caused by failure of the bullet to receive sufficient spin from the rifling in the barrel.

keyhole saw [DES ENG] A fine compass saw with a blade 11–16 inches (28–41 centimeters) long.

keyhole specimen [MET] A metal specimen containing a keyhole shaped notch and used in certain impact tests.

key horizon See key bed.

keying [CIV ENG] Establishing a mechanical bond in a construction joint. [ELEC] The forming of signals, such as for telegraph transmission, by modulating a direct-current or other carrier between discrete values of some characteristic.

keying frequency [COMMUN] In facsimile, the maximum number of times a second that a black-line signal occurs when scanning the subject copy.

keying interval [COMMUN] In a periodically keyed transmission system, one of the set of intervals starting from a change in state and equal in length to the shortest time between changes of state.

keying wave See marking wave.

key job [IND ENG] A job that has been evaluated and is considered representative of similar jobs in the same labor market and is used as a benchmark to evaluate the similar jobs and to establish non-key-job wages.

key joint [CIV ENG] A mortar joint with a concave pointing.

keyless ringing [COMMUN] Form of machine ringing on a manual telephone switchboard which is started automatically by the insertion of the calling plug into the jack of the called line.

key plate [GRAPHICS] A plate used in printing two-color stamps that prints the central design of the stamp.

key pulse [COMMUN] System of signaling where numbered keys are depressed instead of using a dial.

key punch [ADP] A keyboard-actuated device that punches holes in a card. Also known as card punch.

key seat See keyway.

keyseater [MECH ENG] A machine for milling beds or grooves in mechanical parts which receive keys.

keyshelf [COMMUN] Horizontal shelf of a manual telephone switchboard on which are mounted the keys by which the operator switches one or more of the switchboard circuits.

keystone [ARCH] Wedge-shaped stone at the crown of an arch. [MATER] Small crushed stone used as filler for the large aggregate in bituminous bound roads.

keystone distortion [COMMUN] Distortion produced by scanning in a rectilinear manner, with constant-amplitude sawtooth waves, a plane target area which is not normal to the average direction of the beam.

keystoning [ELECTR] Producing a keystone-shaped (wider at the top than at the bottom, or vice versa) scanning pattern because the electron beam in the television camera tube is at an angle with the principal axis of the tube.

key telephone system [COMMUN] A telephone system consisting of phones with several keys, connecting cables, and relay switching apparatus, which does not need a special operator to handle incoming or outgoing calls and which generally permits users to select one of several possible lines and to hold calls.

key telephone unit [COMMUN] A small mounting plate with relays which performs pickup and hold switching functions in a key telephone system.

key verify [ADP] The use of a key punch verifier to ascertain that the data punched on a card corresponds to the data of the original document.

keyway [DES ENG] 1. An opening in a lock for passage of a flat metal key. 2. The pocket in the driven element to provide a driving surface for the key. 3. A groove or channel for a key in any mechanical part. Also known as key seat.

key-word-in-context index [ADP] A computer-generated listing of titles of documents, produced on a line printer, with the key words lined up vertically in a fixed position within the title and arranged in alphabetical order. Abbreviated KWIC index.

K factor [GRAPHICS] See base-altitude ratio. [NUCLEO] A measure of the energy of the gamma rays produced by a particular type of emitter; it is the gamma-ray dose rate in roentgens per hour at a distance of 1 centimeter from a source having a radioactive disintegration rate of 1 millicurie (3.7×10^7 disintegrations per second). [ORD] In artillery ground fire, a factor to be applied to the actual range to a point in order to determine the range which must be fired to hit that point; it is the result of registration and the solution of a meteorological message, and is expressed as plus or minus so many yards per thousand units of actual range.

K feldspar See potassium feldspar.

kg See kilogram; kilogram force.

kg-cal See kilocalorie.

kgf-m See meter-kilogram.

K gun [ORD] A type of U.S. Navy gun in the shape of a K for firing depth charges.

kg-wt See kilogram force.

khamsin [METEOROL] A dry, dusty, and generally hot desert wind in Egypt and over the Red Sea; it is generally southerly or southeasterly, occurring in front of depressions moving eastward across North Africa or the southeastern Mediterranean.

khelin See khellin.

khellin [PHARM] $C_{14}H_{12}O_5$ A synthetic compound that crystallizes from methanol solution, has a bitter taste, melts at 154–155°C, and is more soluble in water than in organic solvents; used in medicine as an antispasmodic, a coronary vasodilator, and a bronchodilator. Also spelled chellin; khelin. Also known as visammin.

khibinite See mosandrite.

kHz See kilohertz.

kibble See hoppit.

kibli See ghibli.

kick [ORD] Violent backward movement of a gun after being fired, caused by the rearward force of the propellant gases acting on the gun.

kickback [MECH ENG] A backward thrust, such as the backward starting of an internal combustion engine as it is cranked, or the reverse push of a piece of work as it is fed to a rotary saw.

kickdown [MECH ENG] 1. Shifting to lower gear in an automotive vehicle. 2. The device for shifting.

kick over [MECH ENG] To start firing; applied to internal combustion engines.

kickpipe [BUILD] A short pipe protecting an electrical cable at the point where it emerges from a floor.

kickplate [BUILD] A plate used on the bottom of doors and cabinets or on the risers of steps to protect them from shoe marks.

Kick's law [ENG] The law that the energy needed to crush a solid material to a specified fraction of its original size is the same, regardless of the original size of the feed material.

kick-sorter See pulse-height analyzer.

kick wheel [ENG] A potter's wheel worked by a foot pedal.

Kidd blood group system [IMMUNOL] The erythrocyte antigens defined by reactions to anti-Jk^a antibodies, originally found in the mother (Mrs. Kidd) of the erythroblastotic infant, and to anti-Jk^b antibodies.

kidney [ANAT] Either of a pair of organs involved with the elimination of water and waste products from the body of vertebrates; in humans they are bean-shaped, about 5 inches (12.7 centimeters) long, and are located in the posterior part of the abdomen behind the peritoneum.

kidney joint [ELECTROMAG] Flexible joint, or air-gap coupling, used in the waveguide of certain radars and located near the transmitting-receiving position.

kidney ore [MINERAL] A form of hematite found in compact masses, concretions, or nodules that are kidney-shaped.

kidney stone See nephrite.

Kienböck's disease See osteochondrosis.

kier [TEXT] A tank in which unfinished cotton is boiled under pressure to remove foreign materials such as sizing or wax.

kieselguhr See diatomaceous earth.

kieserite [MINERAL] $MgSO_4 \cdot H_2O$ A white mineral that crystallizes in the monoclinic system, is composed of hydrous magnesium sulfate, and occurs in saline residues.

Kikuchi lines [CRYSTAL] A pattern consisting of pairs of white and dark parallel lines, obtained when an electron beam is scattered (diffracted) by a crystalline solid; the pattern gives information on the structure of the crystal.

Kiliani reaction [ORG CHEM] A method of synthesizing a higher aldose from a lower aldose; monosaccharides, such as aldehydes and ketones, react with hydrogen cyanide to form cyanohydrins, which are hydrolyzed to hydroxy acids, converted to lactones, and reduced to aldoses with sodium amalgams.

Kilkenny coal See anthracite.

kill [MATER] To treat in such a way as to destroy undesirable properties; for example, neutralization of an acid by the addition of an alkali. [MET] To add a strong deoxidizer, such as silicon or aluminum, to molten steel in order to stop the reaction between carbon and oxygen forming gaseous carbon monoxide and dioxide during solidification.

killed steel [MET] Thoroughly deoxidized steel, for example, by additions of aluminum or silicon, in which the reaction between carbon and oxygen during solidification is suppressed.

killed vaccine [IMMUNOL] A suspension of killed microorganisms used as antigens to produce immunity.

killer circuit [ELECTR] Vacuum tube or tubes and associated circuits in which are generated the blanking pulses used to temporarily disable a radar set.

killer pulse [ELECTR] Blanking pulse generated by a killer circuit.

killer stage See color killer circuit.

killer whale [VERT ZOO] Orcinus orca. A predatory, cosmopolitan cetacean mammal, about 30 feet (9 meters) long, found only in cold waters.

kill probability [ORD] The probability that, given a hit, a single projectile or missile will kill the target against which it is fired.

kiln [ENG] A heated enclosure used for drying, burning, or firing materials such as ore or ceramics.

kilo- [SCI TECH] A prefix representing 10^3 or 1000. Abbreviated k.

kiloampere [ELEC] A metric unit of current flow equal to 1000 amperes. Abbreviated kA.

kilobar [MECH] A unit of pressure equal to 1000 bars. Abbreviated kb.

kilocalorie [THERMO] A unit of heat energy equal to 1000 calories. Abbreviated kcal. Also known as kilogram-calorie (kg-cal); large calorie (Cal).

kilocycle *See* kilohertz.

kilo electron volt [PHYS] A unit of energy, equal to 1000 electron volts. Abbreviated keV.

kilogram [MECH] **1.** The unit of mass in the meter-kilogram-second system, equal to the mass of the international prototype kilogram stored at Sèvres, France. Abbreviated kg. **2.** *See* kilogram force.

kilogram-calorie *See* kilocalorie.

kilogram-equivalent weight [CHEM] A unit of mass 1000 times the gram-equivalent weight.

kilogram force [MECH] A unit of force equal to the weight of a 1-kilogram mass at a point on the earth's surface where the acceleration of gravity is 9.80665 meters/sec². Abbreviated kgf. Also known as kilogram (kg); kilogram weight (kg-wt).

kilogram-meter *See* meter-kilogram.

kilogram weight *See* kilogram force.

kilohertz [PHYS] A unit of frequency equal to 1000 hertz. Abbreviated kHz. Also known as kilocycle (kc).

kilohm [ELEC] A unit of electrical resistance equal to 1000 ohms. Abbreviated K; kohm.

kilojoule [PHYS] A unit of energy or work equal to 1000 joules. Abbreviated kJ.

kiloliter [MECH] A unit of volume equal to 1000 liters or to 1 cubic meter. Abbreviated kl.

kilomega- *See* giga-.

kilomegacycle *See* gigahertz.

kilomegahertz *See* gigahertz.

kilometer [MECH] A unit of length equal to 1000 meters. Abbreviated km.

kilometric wave [COMMUN] British term for electromagnetic waves having wavelengths between 1000 and 10,000 meters.

kiloparsec [ASTRON] A distance of 1000 parsecs (3260 light-years).

kiloton [PHYS] A unit used in specifying the yield of a fission or fusion bomb, equal to the explosive power of 1000 metric tons of trinitrotoluene (TNT). Abbreviated kt.

kilovar [ELEC] A unit of 1000 volt-amperes reactive. Abbreviated kvar.

kilovolt [ELEC] A unit of potential difference equal to 1000 volts. Abbreviated kV.

kilovolt-ampere [ELEC] A unit of apparent power in an alternating-current circuit, equal to 1000 volt-amperes. Abbreviated kVA.

kilovoltmeter [ELEC] A voltmeter which measures potential differences on the order of several kilovolts.

kilovolts peak [ELECTR] The peak voltage applied to an x-ray tube, expressed in kilovolts. Abbreviated kVp.

kilowatt [PHYS] A unit of power equal to 1000 watts. Abbreviated kW.

kilowatt-hour [ELEC] A unit of energy or work equal to 1000 watt-hours. Abbreviated kW-hr. Also known as Board of Trade Unit.

Kimberley reefs [GEOL] Gold-bearing reefs in southern Africa that lie above the Main reef and Bird reef groups. Also known as battery reefs.

kimberlite [PETR] A form of mica periodite that is formed mainly of phenocrysts, olivine, phlogopite, and subordinate melilite with minor amounts of pyroxene, apatite, perovskite, and opaque oxides.

Kimmelstiel-Wilson disease *See* intercapillary glomerulosclerosis.

Kimmeridgian [GEOL] A European stage of geologic time; middle Upper Jurassic, above Oxfordian, below Portlandian.

kimzeyite [MINERAL] $Ca_3(Zr,Ti)_2(Al,Si)_3O_{12}$ A mineral of the garnet group.

kinase [BIOCHEM] Any enzyme that catalyzes phosphorylation reactions.

Kind-Chaudron process [MIN ENG] A technique used to sink a large-diameter deep shaft; a pilot shaft of smaller diameter is first dug, then enlarged until the full diameter is reached; a lining with a moss box at the bottom is forced into place when water is found.

Kinderhookian [GEOL] Lower Mississippian geologic time, above the Chautauquan of Devonian, below Osagian.

K indicator *See* K scope.

kinematic boundary condition [FL MECH] The condition that the component of fluid velocity perpendicular to a solid boundary must vanish on the boundary itself; when the boundary is a fluid surface, the condition applies to the vector difference of velocities across the interface.

kinematic fluidity [FL MECH] The reciprocal of the kinematic viscosity.

kinematics [MECH] The study of the motion of a system of material particles without reference to the forces which act on the system.

kinematic similarity [FL MECH] A relationship between fluid-flow systems in which corresponding fluid velocities and velocity gradients are in the same ratios at corresponding locations.

kinematic viscosity [FL MECH] The absolute viscosity of a fluid divided by its density. Also known as coefficient of kinematic viscosity.

Kinepanorama [GRAPHICS] A Soviet motion picture process similar to Cinerama, using standard 35-millimeter film, a frame size that is six perforations high and 1 inch (25 millimeters) wide, three lens sets, three camera mechanisms, and three strips of film to take the picture, and three projectors to show it.

kineplasty [MED] An amputation of a limb in which tendons are arranged in the stump to permit their use in moving parts of the prosthetic appliance. Also spelled cineplasty.

kinescope *See* picture tube.

kinescope recording [COMMUN] A motion picture film made by photographing images on the face of the picture tube in a television monitor or receiver, to permit repeating the same television program later and at different stations. Also known as television recording.

kinesiatrics [MED] The treatment of disease by systematic active or passive movements. Also known as kinesitherapy; kinetotherapy.

kinesis [PHYSIO] The general term for physical movement, including that induced by stimulation, for example, light.

kinesitherapy *See* kinesiatrics.

kinesthesis [PHYSIO] The system of sensitivity present in the muscles and their attachments.

kinesthetic apraxia *See* motor apraxia.

kinetic [SCI TECH] Pertaining to or producing motion.

kinetic art [GRAPHICS] The use of material objects in motion to produce an artistic effect.

kinetic energy [MECH] The energy which a body possesses because of its motion; in classical mechanics, equal to one-half of the body's mass times the square of its speed.

kinetic energy ammunition [ORD] Ammunition designed to inflict damage to fortifications, armored vehicles, or ships by reason of the kinetic energy of the missile upon impact; the damage may consist of shattering, spalling, or piercing; the missile may be solid, or may contain an explosive charge, intended to function after penetration.

kinetic friction [MECH] The friction between two surfaces which are sliding over each other.

kinetic lead [ORD] The correction or allowance made for the relative motion of a target when computing the lead angle in gunnery.

kinetic momentum [MECH] The momentum which a particle possesses because of its motion; in classical mechanics, equal to the particle's mass times its velocity.

kinetic potential *See* Lagrangian.

kinetics [MECH] The dynamics of material bodies.

kinetic stress [STAT MECH] A stress which arises, in a theory taking the motions of individual molecules into account, from the existence of a velocity distribution of molecules, an example is the pressure of an ideal gas.

kinetic theory [STAT MECH] A theory which attempts to explain the behavior of physical systems on the assumption that they are composed of large numbers of atoms or molecules in vigorous motion; it is further assumed that energy and momentum are conserved in collisions of these particles, and that statistical methods can be applied to deduce the particles' average behavior. Also known as molecular theory.

kinetin [BIOCHEM] $C_{10}H_9ON_5$ A cytokinin formed in many plants which has a stimulating effect on cell division.

kinetochore *See* centromere.

kinetoplast [CYTOL] A genetically autonomous, membrane-bound organelle associated with the basal body at the base of flagella in certain flagellates, such as trypanosomes. Also known as parabasal body.

Kinetoplastida [INV ZOO] An order of colorless protozoans in the class Zoomastigophorea having pliable bodies and possessing one or two flagella in some stage of their life.

kinetosome *See* basal body.

kinetotherapy *See* kinesiatrics.

kingdom [SYST] One of the primary divisions that include all living organisms: most authorities recognize two, the animal kingdom and the plant kingdom, while others recognize three or more, such as Protista, Plantae, Animalia, and Mychota.

kingfisher [VERT ZOO] The common name for members of the avian family Alcedinidae; most are tropical Old World species characterized by short legs, long bills, bright plumage, and short wings.

kingpin [MECH ENG] The pin for articulation between an automobile stub axle and an axle-beam or steering head. Also known as swivel pin.

king post [NAV ARCH] A short, strong post for supporting a cargo boom on a cargo ship. Also known as derrick post; samson post.

king post truss [BUILD] A wooden roof truss having two principal rafters held by a horizontal tie beam, a king post upright between tie beam and ridge, and usually two struts to the rafters from a thickening at the king post foot.

king's blue *See* cobalt blue.

kingston valve [NAV ARCH] A sea valve so arranged that the pressure of the sea forces the valve on its seat or closes it, thus differing from most valves which are so arranged that the pressure is in the direction of opening of the valve.

kinic acid *See* quinic acid.

kinin [PHARM] Any of several pharmacologically active polypeptides that act as hypotensives, contracting isolated smooth muscles and increasing capillary permeability; an example is bradykinin.

kink [ENG] A tightened loop in a wire rope resulting in permanent deformation and damage to the wire.

kink band [GEOL] A deformation band in a single crystal or in foliated rocks in which the orientation is changed due to slipping on several parallel slip planes. Also known as joint drag; knick band; knick zone.

kink instability [PL PHYS] A type of hydromagnetic instability in which the ionized gas and its magnetic confining field tend to form a loop or kink, which then grows steadily larger. Also known as sausage instability.

kinocilium [CYTOL] A type of cilium containing one central pair of microfibrils and nine peripheral pairs; they extend from the apex of hair cells in all vertebrate ears except mammals.

kinomere *See* centromere.

kinoplasm [CYTOL] The substance of the protoplasm that is thought to form astral rays and spindle fibers.

Kinorhyncha [INV ZOO] A class of the phylum Aschelminthes consisting of superficially segmented microscopic marine animals lacking external ciliation.

Kinosternidae [VERT ZOO] The mud and musk turtles, a family of chelonian reptiles in the suborder Cryptodira found in North, Central, and South America.

kintal *See* hundredweight.

kinzigite [PETR] A coarse-grained metamorphic rock that is formed principally of garnet and biotite, with K feldspar, quartz, mica, cordierite, and sillimanite.

kip [MECH] A 1000-pound (453.6-kilogram) load.

Kirchhoff's current law [ELEC] The law that at any given instant the sum of the instantaneous values of all the currents flowing toward a point is equal to the sum of instantaneous values of all the currents flowing away from the point. Also known as Kirchhoff's first law.

Kirchhoff's equations [THERMO] Equations which state that the partial derivative of the change of enthalpy (or of internal energy) during a reaction, with respect to temperature, at constant pressure (or volume) equals the change in heat capacity at constant pressure (or volume).

Kirchhoff's first law *See* Kirchhoff's current law.

Kirchhoff's laws [ELEC] Either of the two fundamental laws dealing with the relation of currents at a junction and voltages around closed loops in an electric network; they comprise Kirchhoff's current law and Kirchhoff's voltage law. [THERMO] The law that the ratio of the emissivity of a heat radiator to the absorptivity of the same radiator is the same for all bodies, depending on frequency and temperature alone, and is equal to the emissivity of a blackbody. Also known as Kirchhoff's principle.

Kirchhoff's principle *See* Kirchhoff's law.

Kirchhoff's second law *See* Kirchhoff's voltage law.

Kirchhoff's voltage law [ELEC] The law that at each instant of time the algebraic sum of the voltage rises around a closed loop in a network is equal to the algebraic sum of the voltage drops, both being taken in the same direction around the loop. Also known as Kirchhoff's second law.

Kirchhoff theory [OPTICS] A theory of diffraction of light which gives a mathematical formulation of Huygens' principle, based on the wave equation and Green's theorem, and enables quantitative determination of the amplitude and phase at any point to a very close approximation.

Kirkbyacea [PALEON] A monomorphic superfamily of extinct ostracods in the suborder Beyrichicopina, all of which are reticulate.

Kirkbyidae [PALEON] A family of extinct ostracods in the superfamily Kirkbyacea in which the pit is reduced and lies below the middle of the valve.

Kirkendall effect [MET] The phenomenon whereby a marker placed at the interface between an alloy and a metal moves toward the alloy region when the temperature of the system is raised to the point where diffusion can occur.

Kirkwood-Brinkley's theory [MECH] In terminal ballistics, a theory formulating the scaling laws from which the effect of blast at high altitudes may be inferred, based upon observed results at ground level.

Kirkwood gaps [ASTRON] Regions in the main zone of asteroids where almost no asteroids are found.

kirovite [MINERAL] $(Fe,Mg)SO_4 \cdot 7H_2O$ A mineral composed of a hydrous sulfate of iron and magnesium; it is isomorphous with malanterite and pisanite.

Kirsten propeller [NAV ARCH] A vertical-axis propeller whose vertical blades are interlocked by gears so that each one makes half a revolution about its axis for each revolution of the whole propeller.

Kiruna method [MIN ENG] A borehole-inclination survey method whereby the electrolytic deposition of copper from a solution is used to make a mark on the inside of a metal container.

kish [MET] Free graphite that floats to the surface of molten hypereutectic cast iron as it cools.

kissing disease *See* infectious mononucleosis.

kiss-roll coating [ENG] Procedure for coating a substrate web in which the coating roll carries a metered film of coating material; part of the film transfers to the web, part remains on the roll.

Kistiakowsky-Fishtine equation [PHYS CHEM] An equation to calculate latent heats of vaporization of pure compounds; useful when vapor pressure and critical data are not available.

kite observation [METEOROL] An atmospheric sounding by means of instruments carried aloft by a kite.

kitol [ORG CHEM] $C_{40}H_{60}O_2$ One of the provitamins of vitamin A derived from whale liver oil; crystallizes from methanol solution.

Kiviat graph [ADP] A circular diagram used in computer performance evaluation, in which variables are plotted on axes of the circle with 0% at the center of the circle and 100% at the circumference, and variables which are "good" and "bad" as they approach 100% are plotted on alternate axes.

kiwi [VERT ZOO] The common name for three species of nocturnal ratites of New Zealand composing the family Apterygidae; all have small eyes, vestigial wings, a long slender bill, and short powerful legs.

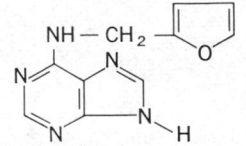

KINETIN

Structural formula for kinetin.

KINGFISHER

The belted kingfisher with white underparts, a blue-gray back, and a band across the chest.

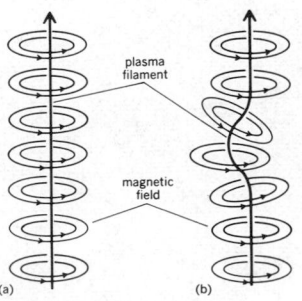

KINK INSTABILITY

Plasma-magnetic-field configuration for cylindrical pinch. *(a)* Equilibrium configuration of plasma filament and magnetic field generated by axial current flow through plasma. *(b)* Onset of kink instability.

Kiwi nuclear reactor [NUCLEO] One of several test reactors for the nuclear engine for a rocket vehicle.

kJ *See* kilojoule.

Kjeldahl method [ANALY CHEM] Quantitative analysis of organic compounds to determine nitrogen content by interaction with concentrated sulfuric acid; ammonia is distilled from the NH_4SO_4 formed.

KK damage [ORD] Combat damage of such extent or nature that an aircraft will disintegrate immediately after the damage occurs.

kl *See* kiloliter.

klaprothite [MINERAL] $Cu_6Bi_4S_9$ A gray mineral composed of copper bismuth sulfide.

klaxon [ENG ACOUS] A diaphragm horn sometimes operated by hand.

klebelsbergite [MINERAL] A mineral composed of basic antimony sulfate, occurring between crystals of stibnite.

Klebsiella [MICROBIO] A genus of nonmotile, gram-negative, rod-shaped bacteria in the family Enterobacteriaceae; species are human pathogens.

Klebsiella pneumoniae [MICROBIO] An encapsulated pathogenic bacterium that causes severe pneumonitis in humans. Formerly known as Friedlander's bacillus; pneumobacillus.

Klebs-Loeffler bacillus *See* Corynebacterium diphtheriae.

Klein bottle [MATH] The nonorientable surface having only one side with no inside or outside; it resembles a bottle pulled into itself.

Klein-Gordon equation [QUANT MECH] A wave equation describing a spinless particle which is consistent with the special theory of relativity. Also known as Schrödinger-Klein-Gordon equation.

kleinite [MINERAL] A yellow to orange mineral composed of a basic oxide, sulfate, and chloride of mercury and ammonium.

Klein-Nishina formula [QUANT MECH] A formula, based on the Dirac electron theory without radiative correction, for the differential cross section for scattering of a photon by an unbound electron.

Klein paradox [QUANT MECH] The paradox whereby, according to the Dirac electron theory, an electron can penetrate into a potential barrier which is greater than twice the rest energy of the electron (about 1 MeV) by making a transition from a positive energy state to a negative energy state, provided the potential change occurs over a distance on the order of a Compton wavelength or less.

Klein-Rydberg method [PHYS CHEM] A method for determining the potential energy function of the distance between the nuclei of a diatomic molecule from the molecule's vibrational and rotational levels.

Kleinschmidt printer [ADP] A page printer capable of receiving ASCII (American Standard Code for Information Interchange) or Baudot code from transmission circuits exceeding 300 bits per second.

Klein's hypothesis [ASTRON] A theory of the overall structure of the universe that regards the visible universe as part of a large but finite astronomical system called a metagalaxy, which may itself belong to a much larger bounded system.

Klein's reagent [CHEM] Saturated solution of borotungstate; used to separate minerals by specific gravity.

klendusity [BOT] The tendency of a plant to resist disease due to a protective covering, such as a thick cuticle, that prevents inoculation.

kleptomania [PSYCH] An obsessive desire to steal; stolen objects are usually petty and useless, being of symbolic value only.

kleptophobia [PSYCH] **1.** An abnormal fear of thieves or of being robbed. **2.** An abnormal fear of becoming a kleptomaniac.

Klieglight [ELEC] A trademark name of Kliegl Brothers Company that at one time referred to a large arc light used in motion picture photography; since 1932, it refers to an ellipsoidal spotlight with an incandescent lamp.

Klinefelter's syndrome [MED] A complex of symptoms associated with hypogonadism in males as an accompaniment of an anomaly of the sex chromosomes; somatic cells are found to have a Y chromosome and more than one X chromosome.

Kline flocculation test [PATH] A microscopic precipitin test for identification of the antibody resulting from syphilitic infection.

Klinkenberg correction [PETRO ENG] Mathematical conversion of laboratory air-permeability measurements (made on formation material) into equivalent liquid-permeability values.

klinotaxis [BIOL] Positive orientation movement of a motile organism induced by a stimulus.

klint [GEOL] An exhumed coral reef or bioherm that is more resistant to the processes of erosion than the rocks that enclose it so that the core remains in relief as hills and ridges.

klintite [GEOL] The dense, hard dolomite composing a klint; gives to the core a strength and resistance to erosion.

klippe [GEOL] A block of rock that is separated from underlying rocks by a fault that usually has a gentle dip.

Klischograph [GRAPHICS] A type of electronic engraver that produces printing plates for letterpress.

Kloedenellacea [PALEON] A dimorphic superfamily of extinct ostracods in the suborder Kloedenellocopina having the posterior part of one dimorph longer and more inflated than the other dimorph.

Kloedenellocopina [PALEON] A suborder of extinct ostracods in the order Paleocopa characterized by a relatively straight dorsal border with a gently curved or nearly straight ventral border.

kloof wind [METEOROL] A cold southwest wind of Simons Bay, South Africa.

K/L ratio [NUC PHYS] The ratio of the number of internal conversion electrons emitted from the K shell of an atom during de-excitation of a nucleus to the number of such electrons emitted from the L shell.

kludge [ADP] A poorly designed data-processing system composed of ill-fitting mismatched components.

klydonograph [ENG] A device attached to electric power lines for estimating certain electrical characteristics of lightning by means of the figures produced on photographic film by the lightning-produced surge carried over the lines; the size of the figure is a function of the potential and polarity of the lightning discharge.

klystron [ELECTR] An evacuated electron-beam tube in which an initial velocity modulation imparted to electrons in the beam results subsequently in density modulation of the beam; used as an amplifier in the microwave region or as an oscillator.

klystron generator [ELECTR] Klystron tube used as a generator, with its second cavity or catcher directly feeding waves into a waveguide.

klystron oscillator *See* velocity-modulated oscillator.

klystron repeater [ELECTR] Klystron tube operated as an amplifier and inserted directly in a waveguide in such a way that incoming waves velocity-modulate the electron stream emitted from a heated cathode; a second cavity converts the energy of the electron clusters into waves of the original type but of greatly increased amplitude and feeds them into the outgoing guide.

km *See* kilometer.

K meson [PARTIC PHYS] **1.** Collective name for four pseudoscalar mesons, having masses of about 495 MeV (million electron volts) and decaying via weak interactions: K^+, K^-, K_S^0, and K_L^0; they consist of two isotopic spin doublets, the (K^+, K^0) doublet and its antiparticle doublet $(K^-, \bar{K}^0)$, having hypercharge or strangeness of $+1$ and -1 respectively, where K^0 and $\bar{K}^0$ are certain combinations of K_L^0 and K_S^0 states. Also known as kaon. **2.** Collective name for any meson resonance belonging to an isotopic doublet with hypercharge $+1$ or -1, denoted $K_{JP}(m)$ or $\bar{K}_{JP}(m)$ respectively, where m is the mass, and J and P are the spin and parity.

K Monel [MET] A nonmagnetic, age-hardenable alloy of nickel (28–34%) and copper and 2.75% aluminum that can be heat-treated after finishing to produce a material that is both corrosion-resistant and extra strong.

kneaded eraser [MATER] An eraser made of unvulcanized rubber whose shape can be readily changed by the user for removing pencil marks from paper.

kneader [FOOD ENG] Mixer for doughy materials in which the dough is repeatedly pulled out, folded back on itself, and pushed down to join the separate layers into a homogeneous mixture.

knebelite [MINERAL] $(Fe,Mn)_2SiO_4$ A mineral composed of an iron manganese silicate.

knee [ANAT] **1.** The articulation between the femur and the tibia in humans. Also known as genu. **2.** The corresponding articulation in the hindlimb of a quadrupedal vertebrate.

knee brace [BUILD] A stiffener between a column and a supported truss or beam to provide greater rigidity in a building frame under transverse loads.

kneecap *See* patella.

kneecap height [ANTHRO] A measure of the vertical distance taken from the floor at the base of the heel to the top of the muscle mass near the end of the thigh bone as the subject sits with both feet on the floor.

knee frequency *See* break frequency.

knee pad [ENG] A protective cushion, usually made of sponge rubber, that can be strapped to a workman's knee.

knee rafter [BUILD] A brace placed diagonally between a principal rafter and a tie beam.

knee tool [MECH ENG] A tool holder with a shape resembling a knee, such as the holder for simultaneous cutting and interval operations on a screw machine or turret lathe.

knee wall [BUILD] A partition that forms a side wall or supports roof rafters under a pitched roof.

Kneriidae [VERT ZOO] A small family of tropical African fresh-water fishes in the order Gonorynchiformes.

knick *See* knickpoint.

knick band *See* kink band.

knickpoint [GEOL] A point of sharp change of slope, especially in the longitudinal profile of a stream or of its valley. Also known as break; knick; nick; nickpoint; rejuvenation head; rock step.

knick zone *See* kink band.

knife [DES ENG] A sharp-edged blade for cutting.

knife coating [ENG] Procedure for coating a continuous-web substrate in which coating thickness is controlled by the distance between the substrate and a movable knife or bar.

knife-edge [DES ENG] A sharp narrow edge resembling that of a knife, such as the fulcrum for a lever arm in a measuring instrument.

knife-edge bearing [MECH ENG] A balance beam or lever arm fulcrum in the form of a hardened steel wedge; used to minimize friction.

knife-edge cam follower [DES ENG] A cam follower having a sharp narrow edge or point like that of a knife; useful in developing cam profile relationships.

knife-edge refraction [ELECTROMAG] Radio propagation effect in which the atmospheric attenuation of a signal is reduced when the signal passes over and is diffracted by a sharp obstacle such as a mountain ridge.

knife file [DES ENG] A tapered file with a thin triangular cross section resembling that of a knife.

knife harrow [AGR] A type of harrow that consists of a frame holding a number of knives which scrape and partly invert the soil surface to smooth it and destroy small weeds.

knife switch [ELEC] An electric switch consisting of a metal blade hinged at one end to a stationary jaw, so that the blade can be pushed over to make contact between spring clips.

knik wind [METEOROL] Local name for a strong southeast wind in the vicinity of Palmer in the Matanuska Valley of Alaska; it blows most frequently in the winter, although it may occur at any time of year.

knitting [TEXT] Making a fabric by interlocking loops of yarn by means of needles or wires.

knob [DES ENG] A component that is placed on a control shaft to facilitate manual rotation of the shaft; sometimes has a pointer or markings to indicate shaft position.

knob-and-tube wiring [ELEC] An electric wiring method used for light and power circuits that uses open insulated wiring on solid insulators; now obsolete and illegal in most countries.

knocker *See* shell knocker.

knock intensity [ENG] The intensity of knock (detonation) recorded when testing a motor gasoline for octane or knock rating.

knockmeter [ENG] A fuels-testing device used to measure the output of the detonation meter used in ASTM knock-test ratings of motor fuels.

knock-off [MECH ENG] **1.** The automatic stopping of a machine when it is operating improperly. **2.** The device that causes automatic stopping.

knock-off bit *See* detachable bit.

knockout [ENG] A partially cutout piece in metal or plastic that can be forced out when a hole is needed.

knockout vessel [CHEM ENG] A vessel, drum, or trap used to remove fluid droplets from flowing gases.

knock rating [ENG] Rating of gasolines according to knocking tendency.

knock suppressor [MATER] A material added to motor fuel to retard or prevent detonation and resultant knock in reciprocating internal combustion engines; an example is tetraethyllead.

Knoevenagel reaction [ORG CHEM] The condensation of aldehydes with compounds containing an activated methylene $(=CH_2)$ group.

knoll [GEOL] A mound rising less than 1000 meters from the sea floor. Also known as sea knoll.

Knoop hardness [MET] The relative microhardness of a material, such as metal, determined by the Knoop indentation test.

Knoop indentation test [MET] A diamond pyramid hardness test employing the Knoop indenter; hardness is determined by the depth to which the Knoop indenter penetrates.

Knoop indenter [MET] A diamond indenter which has a rhombic base with diagonals in a 1:7 ratio and included apical angles of 130° and 172°30′; used in the Knoop indentation test.

knopite [MINERAL] A cerium-bearing variety of perovskite.

Knorr synthesis [ORG CHEM] A condensation reaction carried out in either glacial acetic acid or an aqueous alkali in which an α-aminoketone combines with an α-carbonyl compound to form a pyrrole; possibly the most versatile pyrrole synthesis.

knot [MATER] A scar on lumber marking a place where a branch grew out of the tree trunk. [MATH] In the general case, a knot consists of an embedding of an n-dimensional sphere in an $(n+2)$-dimensional sphere; classically, it is an interlaced closed curve, homeomorphic to a circle. [PHYS] A speed unit of 1 nautical mile (1.852 kilometers) per hour, equal to approximately 0.51444 meters per second.

knot theory [MATH] The topological and algebraic study of knots emphasizing their classification and how one may be continuously deformed into another.

Knox and Oxborne furnace [MIN ENG] A continuously working shaft furnace for roasting quicksilver ores, having the fireplace built in the masonry at one side; the fuel is wood.

knoxvillite *See* copiapite.

knuckle [MIN ENG] The place on an incline where there is a sudden change in grade.

knuckle joint [DES ENG] A hinge joint between two rods in which an eye on one piece fits between two flat projections with eyes on the other piece and is retained by a round pin.

knuckle joint press [MECH ENG] A short-stroke press in which the slide is actuated by a crank attached to a knuckle joint hinge.

knuckle man [MIN ENG] A worker who connects mine cars to and disconnects them from cables and also couples cars into trains.

knuckle pin [DES ENG] The pin of a knuckle joint.

knuckle post [MECH ENG] A post which acts as the pivot for the steering knuckle in an automobile.

Knudsen cell [PHYS CHEM] A vessel used to measure very low vapor pressures by measuring the mass of vapor which escapes when the vessel contains a liquid in equilibrium with its vapor.

Knudsen flow *See* free molecule flow.

Knudsen gage [ENG] An instrument for measuring very low pressures, which measures the force of a gas on a cold plate beside which there is an electrically heated plate.

Knudsen-Langmuir equation [CHEM ENG] Relationship of molecular distillation rate to vapor saturation pressure, solution temperature, and molecular weight during evaporation and no-recycle condensation.

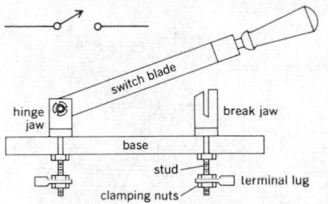

KOALA

The koala, a slow-moving arboreal animal with opposable digits and a rudimentary tail.

KOEHLER LAMP

Picture of the Koehler lamp. *(U.S. Bureau of Mines)*

KOHLRABI

Kohlrabi *(Brassica caulorapa),* cultivar Early White Vienna. *(Joseph Harris Co., Rochester, N.Y.)*

Knudsen number [FL MECH] The ratio of the mean free path length of the molecules of a fluid to a characteristic length; used to describe the flow of low-density gases.

Knudsen reversing water bottle [ENG] A type of frameless reversing bottle for collecting water samples; carries reversing thermometers.

Knudsen's tables [OCEANOGR] Hydrographical tables published by Martin Knudsen in 1901 to facilitate the computation of results of seawater chlorinity titrations and hydrometer temperature readings, and their conversion to salinity and density.

Knudsen vacuum gage [ENG] Device to measure negative gas pressures; a rotatable vane is moved by the pressure of heated molecules, proportionally to the concentration of molecules in the system.

knurl [ENG] To provide a surface, usually a metal, with small ridges or knobs to ensure a firm grip or as a decorative feature.

koala [VERT ZOO] *Phascolarctos cinereus.* An arboreal marsupial mammal of the family Phalangeridae having large hairy ears, gray fir, and two clawed toes opposing three others on each limb.

kobellite [MINERAL] $Pb_2(Bi,Sb)_2S_5$ A blackish-gray mineral composed of antimony bismuth lead sulfide.

Kochab [ASTRON] The brighter of the two stars called the Guardian of the Pole in the constellation Ursa Minor.

Koch freezing process [MIN ENG] A process used to sink a shaft through a formation such as clay that will not sustain a shaft; magnesium chloride cooled to about -30°C is circulated through pipes sunk in the ground until the ground is frozen.

Koch's postulates [MICROBIO] A set of laws elucidated by Robert Koch: the microorganism identified as the etiologic agent must be present in every case of the disease; the etiologic agent must be isolated and cultivated in pure culture; the organism must produce the disease when inoculated in pure culture into susceptible animals; the organism must be present in the animal after recovery from the disease. Also known as law of specificity of bacteria.

Kodel [TEXT] Trade name of a polyester made by Eastman Chemical Products.

koechlinite [MINERAL] Bi_2MoO_6 A greenish-yellow orthorhombic mineral composed of a bismuth molybdate.

Koehler lamp [MIN ENG] A naptha-burning flame safety lamp for use in gaseous mines.

koembang [METEOROL] A dry foehnlike wind from the southeast or south in Cheribon and Tegal in Java, caused by the east monsoon which develops a jet effect in passing through the gaps in the mountain ranges and descends on the leeward side.

koenenite [MINERAL] $Mg_5Al_2(OH)_{12}Cl_4$ A very soft mineral composed of a basic magnesium aluminum chloride.

Koepe hoist *See* Koepe winder.

Koepe shear [MIN ENG] A wheel used in place of a winding drum in the Koepe winder; made up of a cast steel hub with steel arms and a welded rim.

Koepe winder [MIN ENG] A hoisting system in which the winding drum is replaced by large wheels or sheaves over which passes an endless rope. Also known as Koepe hoist.

Koepe winder brake [MIN ENG] A device that works directly on the Koepe shear to slow or stop the hoist; can be applied by the engineman's brake lever or by safety devices.

koettigite [MINERAL] $Zn_3(AsO_4)_2 \cdot 8H_2O$ A carmine mineral composed of a hydrated zinc arsenate.

Kohler illumination [OPTICS] A method of illumination for the optical microscope used with coiled filaments or other sources of irregular form or brightness; an image of the filament large enough to fill the iris opening is focused on the condenser which is focused so that the image of the iris diaphragm on the lamp is in focus with the specimen, and the lamp iris is opened only enough to fill the field of view; the iris of the microscope is opened only enough to illuminate the back aperture of the objective; no ground glass is used.

Kohler's disease *See* osteochondrosis.

kohlrabi [BOT] A biennial crucifer, designated *Brassica caulorapa* and *B. oleracea* var. *caulo-rapa*, of the order Capparales grown for its edible turniplike, enlarged stem.

Kohlrausch method [PHYS CHEM] A method of measuring the electrolytic conductance of a solution using a Wheatstone bridge.

kohm *See* kilohm.

koilonychia [MED] Spoon-shaped deformity of the fingernails, which may be familial or associated with a disease, such as iron-deficiency anemia. Also known as spoon nail.

koilorachic [MED] Having the lumbar spinal region concave ventrally.

koktaite [MINERAL] $(NH_4)_2Ca(SO_4)_2 \cdot H_2O$ A mineral composed of a hydrous calcium ammonium sulfate.

kolbeckite [MINERAL] A blue to gray mineral composed of a hydrous beryllium aluminum calcium silicate and phosphate.

Kolbe hydrocarbon synthesis [ORG CHEM] The production of an alkane by the electrolysis of a water-soluble salt of a carboxylic acid.

Kolbe-Schmitt synthesis [ORG CHEM] The reaction of carbon dioxide with sodium phenoxide at 125°C to give salicylic acid.

Kollsman window [AERO ENG] A small window on the dial face of an aircraft pressure altimeter in which the altimeter setting in inches of mercury is indicated.

Kolmer test [PATH] A complement-fixation test for syphilis and other diseases.

Kolmogoroff inertial subrange [OCEANOGR] The middle portion of the turbulence spectrum, between the low-wave-number (long-wave) part and the high-wave-number (short-wave) part.

Kolmogorov consistency conditions [MATH] For each finite subset F of the real numbers or integers, let P_F denote a probability measure defined on the Borel subsets of the cartesian product of $k(F)$ copies of the real line indexed by elements in F, where $k(F)$ denotes the number of elements in F; the family $\{P_F\}$ of measures satisfy the Kolmogorov consistency conditions if given any two finite sets F_1 and F_2 with F_1 contained in F_2, the restriction of P_{F_2} to those sets which are independent of the coordinates in F_2 which are not in F_1 coincides with P_{F_1}.

Kolmogorov inequalities [MATH] For each integer K let X_k be a random variable with finite variance σ_k and suppose $\{X_k\}$ is an independent sequence which is uniformly bounded by some constant c; then for every $\varepsilon > 0$, and integer n,

$$1 - (\varepsilon + 2c)^2 / \sum_{k=1}^{n} \sigma_k^2 \leq \text{Prob} \{ \max_{k \leq n} |S_k + ES_k| \geq \varepsilon \} \text{ and}$$

$$\frac{1}{\varepsilon^2} \sum_{k=1}^{n} \sigma_k^2 \geq \text{Prob} \{ \max_{k \leq n} |S_k + ES_k| \geq \varepsilon \};$$

here $S_k = \sum_{i=1}^{k} X_i$ and ES_k denotes the expected value of S_k.

Kolmogorov-Sinai invariant [MATH] An isomorphism invariant of measure-preserving transformations; if T is a measure-preserving transformation on a probability space, the Kolmogorov-Sinai invariant is the least upper bound of the set of entropies of T given each finite partition of the probability space. Also known as entropy of a transformation.

Komodo dragon [VERT ZOO] *Varanus komodoensis.* A predatory reptile of the family Varanidae found only on the island of Komodo; it is the largest living lizard and may grow to 10 feet.

kona [METEOROL] A stormy, rain-bringing wind from the southwest or south-southwest in Hawaii; it blows about five times a year on the southwest slopes, which are in the lee of the prevailing northeast trade winds.

kona cyclone [METEOROL] A slow-moving extensive cyclone which forms in subtropical latitudes during the winter season. Also known as kona storm.

kona storm *See* kona cyclone.

Kondo alloy [MET] A dilute alloy of a magnetic material in a nonmagnetic host which exhibits the Kondo effect.

Kondo effect [MET] The large anomalous increase in the resistance of certain dilute alloys of magnetic materials in nonmagnetic hosts as the temperature is lowered.

Kondo temperature [MET] The temperature below which the Kondo effect predominates for a specified magnetic impurity and host material.

kongsbergite [MINERAL] A silver-rich variety of a native amalgam composed of silver (95%) and mercury (5%).

konig [OPTICS] The X tristimulus value.

Konig's theorem [MATH] In combinatorial theory, the theorem that the minimum number m of lines is the same as the maximum number m of independent points.

konimeter [ENG] An air-sampling device used to measure dust as in a cement mill or a mine; a measured volume of air drawn through a jet impacts on a glycerin-jelly-coated glass surface; the particles are counted with a microscope.

koninckite [MINERAL] $FePO_4 \cdot 3H_2O$ A yellow mineral composed of a hydrous ferric phosphate.

koniscope [ENG] An instrument which indicates the presence of dust particles in the atmosphere. Also spelled coniscope.

Konowaloff rule [PHYS CHEM] An empirical rule which states that in the vapor over a liquid mixture there is a higher proportion of that component which, when added to the liquid, raises its vapor pressure, than of other components.

Koonungidae [INV ZOO] A family of Australian crustaceans in the order Anaspidacea with sessile eyes and the first thoracic limb modified for digging.

kopfring [ORD] A metal ring which is attached to the nose of a bomb to reduce its penetration in earth or water.

Koplik's sign [MED] Small red spots surrounded by white areas seen in the mucous membrane of the mouth in the prodromal stage of measles. Also known as Koplik's spots.

Koplik's spots See Koplik's sign.

kopophobia [PSYCH] An abnormal fear of fatigue or exhaustion.

Köppen-Supan line [METEOROL] The isotherm connecting places which have a mean temperature of 10°C (50°F) for the warmest month of the year.

koppite [MINERAL] Mineral composed of a form of pyrochlore containing cerium, iron, and potassium.

Kopp's law [PHYS CHEM] The law that for solids the molal heat capacity of a compound at room temperature and pressure approximately equals the sum of heat capacities of the elements in the compound.

Korfmann arch saver [MIN ENG] A machine that uses a controlled hydraulic system to withdraw steel arches.

Korfmann power loader [MIN ENG] A cutter-loader that is able to cart and load in both directions; its components are four drilling heads and one cutter chain surrounding them.

kornelite [MINERAL] $Fe_2(SO_4)_3 \cdot 7H_2O$ A colorless to brown mineral composed of hydrous ferric sulfate.

Korner's method [ORG CHEM] A method for determining the absolute position of substituents for positional isomers in benzene by the experimental production of positional isomers from a given disubstituted benzene.

kornerupine [MINERAL] $(Mg,Fe,Al)_{20}(Si,B)_9O_{43}$ A colorless, yellow, brown, or sea-green mineral composed of magnesium iron borosilicate.

Koroseal [ORG CHEM] Trade name of plasticized vinyl resins made by the B.F. Goodrich Company.

Korsakoff's neurosis See Korsakoff's syndrome.

Korsakoff's psychosis See Korsakoff's syndrome.

Korsakoff's syndrome [PSYCH] A form of amnesic-confabulatory syndrome characterized by confusion, loss of memory, retrograde amnesia with compensatory confabulation, and polyneuritis; seen in chronic alcoholism and other causes of vitamin B deficiency. Also known as Korsakoff's neurosis; Korsakoff's psychosis.

Korshun method [ANALY CHEM] Microdetermination of carbon and hydrogen in organic compounds; the sample is prepyrolyzed (cracked) in a shortage of oxygen, then oxidized in an excess of oxygen.

Kort nozzle [NAV ARCH] A cylindrical tube enclosing a screw propeller on a ship to decrease the tip losses and increase the efficiency; the walls of the tube are streamlined where clear of the ship's hull; adopted in many river towboats and recently in large tankers.

kossava [METEOROL] A cold, very squally wind descending from the east or southeast in the region of the Danube "Iron Gate" through the Carpathians, continuing westward over Belgrade, then spreading northward to the Rumanian and Hungarian borderlands and southward as far as Nish.

Kossel-Sommerfeld law [SPECT] The law that the arc spectra of the atom and ions belonging to an isoelectronic sequence resemble each other, especially in their multiplet structure.

kotoite [MINERAL] $Mg_3(BO_3)_2$ An orthorhombic borate mineral; it is isostructural with jimboite.

Kourbatoff's reagents [MET] Etching agents used for microanalysis of carbon steels; there are four different formulations, three with nitric acid, one with hydrochloric acid.

Kovar [MET] Trademark of Westinghouse Electric Corporation for an iron-nickel-cobalt alloy used in making metal-to-glass seals.

Kovat's retention indexes [ANALY CHEM] Procedure to identify compounds in gas chromatography; the behavior of a compound is indicated by its position on a scale of normal alkane values (for example, methane = 100, ethane = 200).

Kozeny-Carmen equation [FL MECH] Equation for streamline flow of fluids through a powdered bed.

Kozeny's equation [PETRO ENG] Mathematical relationship of flow network permeability to capillary pore dimensions; used for reservoir calculations.

Kr See krypton.

kraft paper [MATER] A strong paper or cardboard made from sulfate-process wood pulp; unbleached varieties are used for wrapping paper and shipping cartons.

kraft process See sulfate pulping.

kraft pulping See sulfate pulping.

Krakatao winds [METEOROL] A layer of easterly winds over the tropics at an altitude of about 18 to 24 kilometers, which tops the mid-tropospheric westerlies (the antitrades), is at least 6 kilometers deep, and is based at about 2 kilometers above the tropopause.

Kramers-Kronig relation [OPTICS] A relation between the real and imaginary parts of the index of refraction of a substance, based on the causality principle and Cauchy's theorem.

Kramer's rule [MATH] A rule that provides a method to compute the determinant of a matrix by cofactors. Also known as Kramer's theorem.

Kramer's theorem [MATH] See Kramer's rule. [SOLID STATE] The theorem that the states of a system consisting of an odd number of electrons in an external electrostatic field are at least twofold degenerate.

Krassowski ellipsoid of 1938 [GEOD] The reference ellipsoid of which the semimajor axis is 6,378,245 meters and the flattening or ellipticity equals 1/298.3.

K ratio [AERO ENG] The ratio of propellant surface to nozzle throat area.

kraurosis [MED] A progressive, sclerosing, shriveling process of the skin, due to glandular atrophy.

Krause rolling mill [MET] A type of rolling mill in which the rolls translate as well as rotate, accomplishing a high reduction of thickness for the single passage of a metal sheet.

Krause's corpuscle [ANAT] One of the spheroid nerve-end organs resembling lamellar corpuscles, but having a more delicate capsule; found especially in the conjunctiva, the mucosa of the tongue, and the external genitalia; they are believed to be cold receptors. Also known as end bulb of Krause.

krausite [MINERAL] $KFe(SO_4)_2 \cdot H_2O$ Yellowish-green mineral composed of hydrous potassium iron sulfate.

Krebs cycle [BIOCHEM] A sequence of enzymatic reactions involving oxidation of a two-carbon acetyl unit to carbon dioxide and water to provide energy for storage in the form of high-energy phosphate bonds. Also known as citric acid cycle; tricarboxylic acid cycle.

Krebs-Henseleit cycle [BIOCHEM] A cyclic reaction pathway involving the breakdown of arginine to urea in the presence of arginase.

Krebspest [INV ZOO] A fatal fungus disease of crayfish caused by *Aphanomyces mystaci*.

Krein-Milman theorem [MATH] The theorem that in a locally convex topological vector space, any compact convex set K is identical with the intersection of all convex sets containing the extreme points of K.

kremastic water See vadose water.

kremersite [MINERAL] $[NH_4,K]_2FeCl_5 \cdot H_2O$ A red mineral composed of hydrous potassium ammonium iron chloride, occurring in octahedral crystals.

Kremser formula [CHEM ENG] Equation for calculating distillation-column material balances and equilibrium, assuming the ideal distribution law, that is, the concentrations in the two phases (vapor and liquid) are proportional to each other.

krennerite [MINERAL] $AuTe_2$ A silver-white to pale-yellow mineral composed of gold telluride and often containing silver. Also known as white tellurium.

kribergite [MINERAL] $Al_5(PO_4)_3(SO_4)(OH)_4 \cdot 2H_2O$ White, chalklike mineral composed of hydrous basic aluminum sulfate and phosphate.

krill [INV ZOO] A name applied to planktonic crustaceans that constitute the diet of many whales, particularly whalebone whales.

krohnkite [MINERAL] $Na_2Cu(SO_4)_2 \cdot 2H_2O$ An azure-blue monoclinic mineral composed of hydrous copper sodium sulfate, occurring in massive form.

krokidolite *See* crocidolite.

Kroll process [MET] A reduction process for the production principally of titanium metal sponge from titanium tetrachloride.

Kronecker delta [MATH] The function or symbol δ_{ij} dependent upon the subscripts i and j which are usually integers; its value is 1 if $i = j$ and 0 if $i \neq j$.

Kronecker product [MATH] Given two different representations of the same group, their Kronecker product is a representation of the group constructed by taking direct products of matrices from the respective representations.

Kronig-Penney model [SOLID STATE] An idealized one-dimensional model of a crystal in which the potential energy of an electron is an infinite sequence of periodically spaced square wells.

Krukenberg's tumor [MED] Bilateral carcinoma of the ovaries; originally described as primary, but now denoting a metastatic form usually of gastric origin.

krummholz [ECOL] Stunted alpine forest vegetation. Also known as elfinwood.

Krupp ball mill [MIN ENG] An ore pulverizer in which the grinding is done by chilled iron or steel balls of various sizes moving against each other and the die ring, composed of five perforated spiral plates, each of which overlaps the next; material is discharged through a cylindrical screen.

kryolithionite [MINERAL] $Na_3Li_3(AlF_6)_2$ Variety of spodumene found in Greenland; has a crystal structure resembling that of garnet.

kryptoclimate *See* cryptoclimate.

kryptoclimatology *See* cryptoclimatology.

krypton [CHEM] A colorless, inert gaseous element, symbol Kr, atomic number 36, atomic weight 83.80; it is odorless and tasteless; used to fill luminescent electric tubes.

krypton-86 [NUC PHYS] An isotope of krypton, atomic mass 86; used in measurement of the standard meter.

krypton lamp [ELEC] An arc lamp filled with krypton; one type pierces fog for 1000 feet (300 meters) or more and is used to light airplane runways at night.

K scan *See* K scope.

K scope [ELECTR] A modified form of A scope on which one signal appears as two pips, the relative amplitudes of which indicate the error of aiming the antenna. Also known as K indicator; K scan.

K shell [ATOM PHYS] The innermost shell of electrons surrounding the atomic nucleus, having electrons characterized by the principal quantum number 1.

K-14 sight [ORD] A gyroscopic computing gunsight employing a mechanical range-control system.

K-18 sight [ORD] A gyroscopic computing gunsight employing an electrical range-control system.

k-space *See* wave-vector space.

KSR *See* keyboard send/receive.

K transfer [ORD] In artillery ground fire, the shift of fire from one point to another in the transfer limits of the piece, the actual range being corrected by application of the *K*.

K truss [BUILD] A building truss in the form of a K due to the orientation of the vertical member and two oblique members in each panel.

Kubelka-Munk model [OPTICS] A widely used theoretical model of reflectance; the model supposes that some light passing through a homogeneous sample is scattered and absorbed so that the light is attenuated in both directions.

kudzu [BOT] Any of various perennial vine legumes of the genus *Pueraria* in the order Rosales cultivated principally as a forage crop.

Kuehneosauridae [PALEON] The gliding lizards, a family of Upper Triassic reptiles in the order Squamata including the earliest known aerial vertebrates.

kukersite [GEOL] An organic sediment rich in remains of the alga *Gloexapsamorpha prisca*; found in the Ordovician of Estonia.

Kullenberg piston corer [MECH ENG] A piston-operated coring device used to obtain 2-inch-diameter (5-centimeter-diameter) core samples.

kumiss [FOOD ENG] A low-alcohol-content beverage made of fermented mare's or cow's milk by the action of *Lactobacillus casei, Streptococcus lactis,* and yeasts.

kumquat [BOT] A citrus shrub or tree of the genus *Fortunella* in the order Sapindales grown for its small, flame- to orange-colored edible fruit having three to five locules filled with an acid pulp, and a sweet, pulpy rind.

Kundt effect *See* Faraday effect.

Kundt rule [SPECT] The rule that the optical absorption bands of a solution are displaced toward the red when its refractive index increases because of changes in composition or other causes.

Kundt's constant [OPTICS] A measure of the strength of the Faraday effect in a material, equal to the ratio of the Faraday rotation to the product of the path length and the magnetization of the material; it depends only on the temperature for any magnetic material.

Kundt tube [ACOUS] A tube used to measure the speed of sound; it is filled with air or other gas and contains a light powder which becomes lumped at nodes, giving the length of standing waves generated in the tube.

Kungurian [GEOL] A European stage of geologic time; Middle Permian, above Artinskian, below Kazanian.

kunzite [MINERAL] A pinkish gem variety of spodumene.

Kupffer cell [HISTOL] One of the fixed macrophages lining the hepatic sinusoids.

Kuratowski's lemma [MATH] Each linearly ordered subset of a partially ordered set is contained in a maximal linearly ordered subset.

Kurie plot [NUC PHYS] Graph used in studying beta decay, in which the square root of the number of beta particles whose momenta (or energy) lie within a certain narrow range, divided by a function worked out by Fermi, is plotted against beta-particle energy; it is a straight line for allowed transitions and some forbidden transitions, in accord with the Fermi beta-decay theory. Also known as Fermi plot.

kurnakovite [MINERAL] $Mg_2B_6O_{11} \cdot 13H_2O$ A white mineral composed of hydrous magnesium borate.

Kuron [TEXT] Trade name of a stretch material manufactured by Uniroyal, Inc.

Kuroshio Countercurrent [OCEANOGR] A component of the Kuroshio system flowing south and southwest between latitudes 155° and 160°E about 70 kilometers from the coast of Japan on the right-hand side of the Kuroshio Current.

Kuroshio Current [OCEANOGR] A fast ocean current (2–4 knots) flowing northeastward from Taiwan to the Ryukyu Islands and close to the coast of Japan to about 150°E. Also known as Japan Current.

Kuroshio extension [OCEANOGR] A general term for the warm, eastward-transitional flow that connects the Kuroshio and the North Pacific currents.

Kuroshio system [OCEANOGR] A system of ocean currents which includes part of the North Equatorial Current, the Tsushima Current, the Kuroshio Current, and the Kuroshio extension.

Kurrol's salt [INORG CHEM] $NaPO_3(IV)$ A crystalline high-temperature form of sodium phosphate made by seeding a melt at 550°C.

Kurtoidei [VERT ZOO] A monogeneric suborder of perciform fishes having a unique ossification that encloses the upper part of the swim bladder, and an occipital hook in the male for holding eggs during brooding.

KRONIG-PENNEY MODEL

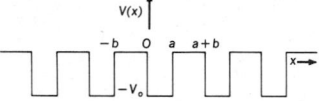

Potential energy $V(x)$ which is assumed for the one-dimensional Kronig-Penney model. Here x is directed distance from origin O. Square wells of depth V_o and width a are separated by distance b.

KRYPTON

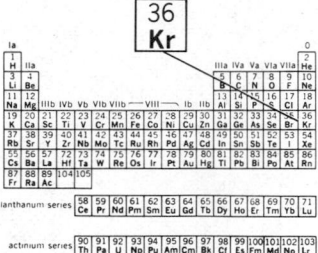

Periodic table of the chemical elements showing the position of krypton.

KUMQUAT

Fruit cluster and foliage of Nagami kumquat (*Fortunella margarita*). (*J. Horace McFarland Co.*)

kurtorachic [MED] Having the lumbar spinal region concave dorsally.

kurtosis [STAT] The extent to which a frequency distribution is concentrated about the mean or peaked; it is sometimes defined as the ratio of the fourth moment of the distribution to the square of the second moment.

kutnahorite [MINERAL] $Ca(Mn,Mg,Fe)(CO_3)_2$ A rare carbonate of calcium and manganese, found with some magnesium and iron substituting for manganese; forms rhombohedral crystals and is isomorphous with dolomite.

Kutorginida [PALEON] An order of extinct brachiopod mollusks that is unplaced taxonomically.

Kutta-Joukowski equation [FL MECH] An equation which states that the lift force exerted on a body by an ideal fluid, per unit length of body perpendicular to the flow, is equal to the product of the mass density of the fluid, the linear velocity of the fluid relative to the body, and the fluid circulation. Also known as Kutta-Joukowski theorem.

Kutta-Joukowski theorem See Kutta-Joukowski equation.

kV See kilovolt.

kVA See kilovolt-ampere.

kvar See kilovar.

K virus [VIROL] A group 2 papovavirus affecting rats and mice.

kVp See kilovolts peak.

kW See kilowatt.

kwashiorkor [MED] A nutritional deficiency disease in infants and young children, mainly in the tropics, caused primarily by a diet low in proteins and rich in carbohydrates. Also known as nutritional dystrophy.

KWIC index See key-word-in-context index.

kyanite [MINERAL] Al_2SiO_5 A blue or light-green neosilicate mineral; crystallizes in the triclinic system, and luster is vitreous to pearly; occurs in long, thin bladed crystals and crystalline aggregates. Also known as cyanite; disthene; sappare.

Kyasanur Forest virus [VIROL] A group B arbovirus recognized as an agent that causes hemorrhagic fever.

kymograph [MED] A device for recording internal body movements by making tracings with a stylus on a revolving smoked drum.

kymography [MED] A recording the movements of internal organs by a kymograph.

kynurenic acid [BIOCHEM] $C_{10}H_7O_3N$ A product of tryptophan metabolism found in the urine of mammals.

kynurenine [BIOCHEM] $C_{10}H_{12}O_3N_2$ An intermediate product of tryptophan metabolism occurring in the urine of mammals.

kyphos [MED] The anteroposterior hump in the spine occurring in kyphoscoliosis.

kyphoscoliosis [MED] Lateral curvature of the spine accompanied by rotation of the vertebrae.

kyphosis [MED] Angular curvature of the spine, usually in the thoracic region. Also known as humpback; hunchback.

kyrohydratic point [OCEANOGR] The temperature at which a particular salt crystallizes in brine which is trapped by frozen seawater, the eutectic temperature of that salt.

kytoon [AERO ENG] A captive balloon used to maintain meteorological equipment aloft at approximately a constant height; it is streamlined, and combines the aerodynamic properties of a balloon and a kite.

KYANITE

⊢29 mm⊣

Kyanite on paragonite schist, St. Gotthard, Switzerland. (*American Museum of Natural History Specimen*)

l *See* liter.

L *See* lambert.

La *See* lanthanum.

Labarraque's solution [MATER] Aqueous solution of 4–6% sodium hypochlorite and 4–6% sodium chloride with sodium hydroxide or sodium carbonate stabilizer; used as disinfectant.

labbé [METEOROL] An infrequent, moderate to strong southwest wind that occurs only in March in Provence (southeastern France), bringing mild, humid, and very cloudy or rainy weather, while on the coast it raises a rough sea.

labdanum oil [MATER] An essential oil, golden yellow with ambergris aroma; soluble in alcohol, chloroform, and ether; derived from gum resin of various rockroses, such as *Cistus ladaniferus;* used in perfumes. Also known as ladanum oil.

label [ADP] A data item that serves to identify a data record (much in the same way as a key is used), or a symbolic name used in a program to mark the location of a particular instruction or routine. [NUCLEO] *See* isotopic tracer.

labeled cargo [IND ENG] Cargo of a dangerous nature, such as explosives and flammable or corrosive liquids, which is designated by different-colored labels to indicate the requirements for special handling and storage.

labellate [BIOL] Having a labellum.

labellum [BOT] The median membrane of the corolla of an orchid often differing in size and morphology from the other two petals. [INV ZOO] **1.** A prolongation of the labrum in certain beetles and true bugs. **2.** In Diptera, either of a pair of sensitive fleshy lobes consisting of the expanded end of the labium.

label paper [GRAPHICS] Any of the various papers used for can and bottle labels; made for both offset and letterpress printing, they have a smooth side for printing and a rough side to receive an adhesive.

label record [ADP] A tape record containing information concerning the file on that tape, such as format, record length, and block size.

labial gland [ANAT] Any of the small, tubular mucous and serous glands underneath the mucous membrane of mammalian lips. [INV ZOO] A salivary gland, or modification thereof, opening at the base of the labium in certain insects.

labial palp [INV ZOO] **1.** Either of a pair of fleshy appendages on either side of the mouth of certain bivalve mollusks. **2.** A jointed appendage attached to the labium of certain insects.

labial papilla [INV ZOO] Any of the sensory bristles around the mouth of many nematodes; they are jointed projections of the cuticle.

Labiatae [BOT] A large family of dicotyledonous plants in the order Lamiales; members are typically aromatic and usually herbaceous or merely shrubby.

labiate [ANAT] Having liplike margins that are thick and fleshy. [BIOL] Having lips. [BOT] Having the limb of a tubular calyx or corolla divided into two unequal overlapping parts.

labile [PSYCH] Unstable in mood. [SCI TECH] Also known as metastable. **1.** Readily changed, as by heat, oxidation, or other processes. **2.** Moving from place to place.

labile factor *See* proaccelerin.

labile oscillator [ELECTR] An oscillator whose frequency is controlled from a remote location by wire or radio.

lability [PSYCH] Very rapid fluctuations in intensity and modality of emotions; seen in the affective reaction or in certain organic brain disorders.

labite [MINERAL] $MgSi_3O_6(OH)_2 \cdot H_2O$ A mineral composed of hydrous basic silicate of magnesium.

labium [BIOL] **1.** A liplike structure. **2.** The lower lip, as of a labiate corolla or of an insect.

labium majus [ANAT] Either of the two outer folds surrounding the vulva in the female.

labium minus [ANAT] Either of the two inner folds, at the inner surfaces of the labia majora, surrounding the vulva in the female.

laboratory [SCI TECH] A place for experimental study.

laboratory coordinate system [MECH] A reference frame attached to the laboratory of the observer, in contrast to the center-of-mass system.

laboratory pack film [GRAPHICS] Unexposed film which is wound on a plain core (no flange), usually in 100-foot (30-meter) lengths.

labor cost [IND ENG] That part of the cost of goods and services attributable to wages, especially for direct labor.

labor grade *See* job class.

labor relations [IND ENG] The management function that deals with a company's work force; usually the term is restricted to relations with organized labor.

Laboulbeniales [MYCOL] An order of ascomycetous fungi made up of species that live primarily on the external surfaces of insects.

La Bour centrifugal pump [MIN ENG] A self-priming centrifugal pump with a trap which ensures sufficent water for the pump to function, and a separator to remove the entrained air in the water.

Labrador Current [OCEANOGR] A current that flows southward from Baffin Bay, through the Davis Strait, and southwestward along the Labrador and Newfoundland coasts.

labradorite [MINERAL] A gray, blue, green, or brown plagioclase feldspar with composition ranging from $Ab_{50}An_{50}$ to $Ab_{30}An_{70}$, where $Ab = NaAlSi_3O_8$ and $An = CaAl_2Si_2O_8$; in the course of formation when the natural material cools, the feldspar sometimes exhibits a variously colored luster. Also known as Labrador spar.

Labrador spar *See* labradorite.

Labridae [VERT ZOO] The wrasses, a family of perciform fishes in the suborder Percoidei.

labrum [INV ZOO] **1.** The upper lip of certain arthropods, lying in front of or above the mandibles. **2.** The outer edge of a gastropod shell.

labyrinth [ANAT] **1.** Any body structure full of intricate cavities and canals. **2.** The inner ear. [ENG ACOUS] A loudspeaker enclosure having air chambers at the rear that absorb rearward-radiated acoustic energy, to prevent it from interfering with the desired forward-radiated energy.

labyrinthine placenta *See* hemochorial placenta.

labyrinthine reflex [PHYSIO] The involuntary response to stimulation of the vestibular apparatus in the inner ear.

labyrinthine syndrome *See* Ménière's syndrome.

labyrinthitis [MED] Inflammation of the labyrinth of the inner ear.

Labyrinthodontia [PALEON] A subclass of fossil amphibians descended from crossopterygian fishes, ancestral to reptiles, and antecedent to at least part of other amphibian types.

labyrinth seal [ENG] A minimum-leakage seal that offers resistance to fluid flow while providing radial or axial clearance; a labyrinth of circumferential knives or touch points provides for successive expansion of the fluid being piped; used for gas pipes, steam engines, and turbines.

Labyrinthulia [INV ZOO] A subclass of the protozoan class Rhizopoda containing mostly marine, ovoid to spindle-shaped, uninucleate organisms that secrete a network of filaments (slime tubes) along which they glide.

Labyrinthulida [INV ZOO] The single order of the protozoan subclass Labyrinthulia.

lac [MATER] A resinous material secreted by some insects that live on the sap of certain trees, principally in India; used in the manufacture of shellac.

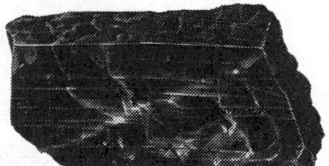
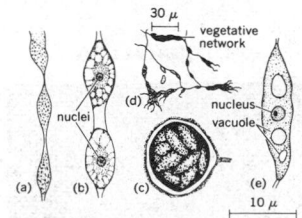

LACTAM

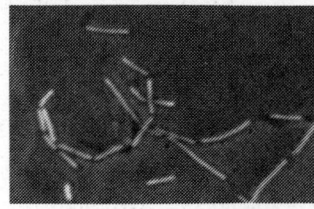

Equation showing conversion of γ-aminobutyric acid to γ-butyrolactam.

LACTIC ACID

(a)

(b)

Structural formulas of (a) dextro form and (b) levo form of lactic acid.

LACTIM

Equation showing tautomeric equilibrium between the lactam and lactim forms of isatin.

LACTOBACILLEAE

Photomicrograph showing morphology of *Lactobacillus brevis*, tribe Lactobacilleae.

Lacaille's constellations [ASTRON] The 14 southern constellations identified by N. L. de Lacaille in 1763: Antlia, Callum, Circinus, Crux, Fornax, Horologium, Mensa, Microscopium, Norma, Octans, Pictor, Reticulum, Sculptor, and Telescopium.

laccal [BIOCHEM] $C_{17}H_{31}C_6H_3(OH)_2$ A phenol compound which is found in the sap of lacquer trees, and which can be isolated in crystalline form.

laccase [BIOCHEM] Any of a class of plant oxidases which catalyze the oxidation of phenols.

laccate [BIOL] Having a lacquered appearance.

Lacciferinae [INV ZOO] A subfamily of scale insects in the superfamily Coccoidea in which the male lacks compound eyes, the abdomen is without spiracles in all stages, and the apical abdominal segments of nymphs and females do not form a pygidium.

laccolith [GEOL] A body of igneous rock intruding into sedimentary rocks so that the overlying strata have been notably lifted by the force of intrusion.

lace [ADP] To punch all the holes in some area of a punch card, such as a card row or card column. [TEXT] A patterned, openwork fabric made by hand with needles or hooks, or by machinery.

lacerate [MED] To inflict a wound by tearing.

lacerated [BIOL] Having a deeply and irregularly incised margin or apex.

laceration [MED] A wound made by tearing.

Lacerta [ASTRON] A small northern constellation lying between Cygnus and Andromeda, and adjoining the northern boundary of Pegasus. Also known as Lizard.

Lacertidae [VERT ZOO] A family of reptiles in the suborder Sauria, including all typical lizards, characterized by movable eyelids, a fused lower jaw, homodont dentition, and epidermal scales.

lachesne [ORG CHEM] $C_{20}H_{26}ClNO_3$ A compound that crystallizes from a solution of ethanol and acetone, and whose melting point is 213°C; used in ophthalmology. Also known as chloride benzilate.

lacing [ADP] Extra multiple punching in a card column to signify the end of a specific card run; the term is derived from the lacework appearance of the card. [CIV ENG] **1.** A lightweight metallic piece that is fixed diagonally to two channels or four angle sections, forming a composite strut. **2.** A course of brick, stone, or tiles in a wall of rubble to give strength. **3.** A course of upright bricks forming a bond between two or more arch rings. **4.** Distribution steel in a slab of reinforced concrete. **5.** A light timber fastened to pairs of struts or walings in the timbering of excavations (including mines).

laciniate [BIOL] **1.** Having a fringed border. **2.** Narrowly and deeply incised to form irregular lobes, which may be pointed.

lacmus *See* litmus.

lacquer [MATER] A material which contains a substantial quantity of a cellulose derivative, most commonly nitrocellulose but sometimes a cellulose ester, such as cellulose acetate or cellulose butyrate, or a cellulose ether such as ethyl cellulose; used to give a glossy finish, especially on brass and other bright metals.

lacquer diluent [MATER] An organic liquid with no solvent power added to lacquer formulations to reduce viscosity and to adjust flow or other properties.

lacquer tree *See* varnish tree.

lacrimal [ANAT] Pertaining to tears, tear ducts, or tear-secreting organs.

lacrimal apparatus [ANAT] The functional and structural mechanisms for secreting and draining tears; includes the lacrimal gland, lake, puncta, canaliculi, sac, and nasolacrimal duct.

lacrimal bone [ANAT] A small bone located in the anterior medial wall of the orbit, articulating with the frontal, ethmoid, maxilla, and inferior nasal concha.

lacrimal canal *See* nasolacrimal canal.

lacrimal canaliculus [ANAT] A small tube lined with stratified squamous epithelium which runs vertically a short distance from the punctum of each eyelid and then turns horizontally in the lacrimal part of the lid margin to the lacrimal sac. Also known as lacrimal duct.

lacrimal duct *See* lacrimal canaliculus.

lacrimal gland [ANAT] A compound tubuloalveolar gland that secretes tears. Also known as tear gland.

lacrimal sac [ANAT] The dilation at the upper end of the nasolacrimal duct within the medial canthus of the eye. Also known as dacryocyst.

lacrimation [PHYSIO] **1.** Normal secretion of tears. **2.** Excessive secretion of tears, as in weeping.

lacrimator *See* tear gas.

lacroixite [MINERAL] A pale yellowish-green mineral composed of basic phosphate of aluminum, calcium, manganese, and sodium (often with fluorine), occurring as crystals.

LACT *See* lease automatic custody transfer.

lactalbumin [BIOCHEM] A simple protein contained in milk which resembles serum albumin and is of high nutritional quality.

lactam [ORG CHEM] An internal (cyclic) amide formed by heating gamma (γ) and delta (δ) amino acids; thus γ-aminobutyric acid readily forms γ-butyrolactam lactam (pyrrolidone); many lactams have physiological activity.

lactase [BIOCHEM] An enzyme that catalyzes the hydrolysis of lactose to dextrose and galactose.

lactase deficiency syndrome [MED] Diarrhea induced by ingestion of a lactose-containing food such as milk, secondary to a congenital or acquired deficiency of lactase.

lactate [ORG CHEM] A salt or ester of lactic acid in which the acidic hydrogen of the carboxyl group has been replaced by a metal or an organic radical. [PHYSIO] To secrete milk.

lactate dehydrogenase [BIOCHEM] A zinc-containing enzyme which catalyzes the oxidation of several α-hydroxy acids to corresponding α-keto acids.

lactation [PHYSIO] Secretion of milk by the mammary glands.

lacteal [ANAT] One of the intestinal lymphatics that absorb chyle. [PHYSIO] Pertaining to or resembling milk.

lactescent [BIOL] Having a milky appearance. [PHYSIO] Secreting milk or a milklike substance.

lactic acid [BIOCHEM] $C_3H_6O_3$ A hygroscopic α-hydroxy acid, occurring in three optically isomeric forms: L form, in blood and muscle tissue as a product of glucose and glycogen metabolism; D form, obtained by fermentation of sucrose; and DL form, a racemic mixture present in foods prepared by bacterial fermentation, and also made synthetically. Also known as 2-hydroxypropanoic acid; α-hydroxypropanoic acid.

lactic dehydrogenase [BIOCHEM] An enzyme that catalyzes the dehydrogenation of L-lactic acid to pyruvic acid. Abbreviated LDH.

lactic dehydrogenase virus [VIROL] A virus of the rubella group which infects mice.

lactide [ORG CHEM] A cyclic, intermolecular, double ester formed from α-hydroxy acids; most lactides are relatively low melting solids and are easily hydrolyzed by base to form salts of the parent acid, such as sodium lactate.

lactim [ORG CHEM] A tautomeric enol form of a lactam with which it forms an equilibrium whenever the lactam nitrogen carries a free hydrogen.

lactin *See* lactose.

lactivorous [ZOO] Feeding on milk.

Lactobacillaceae [MICROBIO] A family of sugar-fermenting bacteria in the order Eubacteriales including both spherical and rod-shaped forms.

Lactobacilleae [MICROBIO] A tribe of rod-shaped bacteria in the family Lactobacillaceae.

Lactobacillus [MICROBIO] The lactic acid bacteria, a genus of nonmotile gram-positive bacteria in the family Lactobacillaceae; they produce lactic acid from certain carbohydrates.

lactoferrin [BIOCHEM] An iron-binding protein found in milk, saliva, tears, and intestinal and respiratory secretions that interferes with the iron metabolism of bacteria; in conjunction with antibodies, it plays an important role in resistance to certain infectious diseases.

lactoflavin *See* riboflavin.

lactogenic hormone *See* prolactin.

lactoglobulin [BIOCHEM] A crystalline protein fraction of milk, which is soluble in half-saturated ammonium sulfate solution and insoluble in pure water.

lactometer [ENG] A hydrometer used to measure the specific gravity of milk.

lactonase [BIOCHEM] The enzyme that catalyzes the hydrolysis of 6-phosphoglucono-Δ-lactone to 6-phosphogluconic acid in the pentose phosphate pathway.

lactone [ORG CHEM] An internal cyclic mono ester formed by gamma (γ) or delta (δ) hydroxy acids spontaneously; thus γ-hydroxybutyric acid forms γ-butyrolactone.

lactonitrile [ORG CHEM] $CH_3CHOHCN$ A straw-colored liquid boiling at 183°C; soluble in water; insoluble in carbon disulfide and petroleum ether; used as a solvent, and as a chemical intermediate in making esters of lactic acid. Also known as acetaldehyde cyanohydrin; α-hydroxypropionitrile.

lactophenene See para-lactophenetide.

para-lactophenetide [PHARM] $C_{11}H_{15}NO_3$ A water-soluble compound, crystallizing from ethyl acetate and hexane solution, and melting at 117–118°C; used in medicine as an analgesic and antipyretic. Also known as fenolactine; lactophenene; para-lactophenetidide; lactyl-para-phenetiden.

para-lactophenetidide See para-lactophenetide.

lactoprene [MATER] Any of several synthetic rubbers that have good resistance to hydrocarbon oil, ozone, oxygen, and other weather elements excepting cold, and that are polymers or copolymers of an acrylic acid ester.

lactose [BIOCHEM] $C_{12}H_{22}O_{11}$ A disaccharide composed of D-glucose and D-galactose which occurs in milk. Also known as lactin; milk sugar.

lactosuria [MED] Presence of lactose in the urine.

lactron [TEXT] A thread made from natural rubber latex by extrusion.

lactyl-para-phenetiden See para-lactophenetide.

lacuna [BIOL] A small space or depression. [HISTOL] A cavity in the matrix of bone or cartilage which is occupied by the cell body.

lacunaris See lacunosus.

lacunar system [INV ZOO] A series of intercommunicating spaces branching from two longitudinal vessels in the hypodermis of many acanthocephalins.

lacunosus [METEOROL] A cloud variety characterized more by the appearance of the spaces between the cloud elements than by the elements themselves, the gaps being generally rounded, often with fringed edges, and the overall appearance being that of a honeycomb or net; it is the negative of clouds composed of separate rounded elements. Formerly known as lacunaris.

lacustrine [GEOL] Belonging to or produced by lakes.

lacustrine sediments [GEOL] Sediments that are deposited in lakes.

lacustrine soil [GEOL] Soil that is uniform in texture but variable in chemical composition and that has been formed by deposits in lakes which have become extinct.

Lacydonidae [INV ZOO] A benthic family of pelagic errantian polychaetes.

ladanum oil See labdanum oil.

ladar [OPTICS] A missile-tracking system that uses a visible light beam in place of a microwave radar beam to obtain measurements of speed, altitude, direction, and range of missiles. Derived from laser detecting and ranging. Also known as colidar; laser radar.

ladder [ENG] A structure, often portable, for climbing up and down; consists of two parallel sides joined by a series of crosspieces that serve as footrests.

ladder attenuator [ELECTR] A type of ladder network designed to introduce a desired, adjustable loss when working between two resistive impedances, one of which has a shunt arm that may be connected to any of various switch points along the ladder.

ladder-bucket dredge See bucket-ladder dredge.

ladder ditcher See ladder trencher.

ladder dredge See bucket-ladder dredge.

ladder drilling [MECH ENG] An arrangement of retractable drills with pneumatic powered legs mounted on banks of steel ladders connected to a holding frame; used in large-scale rock tunneling, with the advantage that many drills can be worked at the same time by a small labor force.

ladder fire [ORD] A method of establishing a gun range to a target by firing a succession of salvos with established differences in elevation.

ladder network [ELECTR] A network composed of a sequence of H, L, T, or pi networks connected in tandem. Also known as series-shunt network.

ladder road See ladderway.

ladder track [CIV ENG] A main track that joins successive body tracks in a railroad yard.

ladder trencher [MECH ENG] A machine that digs trenches by means of a bucket-ladder excavator. Also known as ladder ditcher.

ladderway [MIN ENG] Also known as ladder road. **1.** Mine shaft between two main levels, equipped with ladders. **2.** The particular shaft, or compartment of a shaft, containing ladders.

laddic [ELECTR] Multiaperture magnetic structure resembling a ladder, used to perform logic functions; operation is based on a flux change in the shortest available path when adjacent rungs of the ladder are initially magnetized with opposite polarity.

Ladenburg f value See oscillator strength.

Ladinian [GEOL] A European stage of geologic time: upper Middle Triassic (above Anisian, below Carnian).

ladle [DES ENG] A deep-bowled spoon with a long handle for dipping up, transporting, and pouring liquids. [MET] A receptacle for transporting and pouring molten metal.

Laemobothridae [INV ZOO] A family of lice in the order Mallophaga including parasites of aquatic birds, especially geese and coots.

Laennec's cirrhosis See portal cirrhosis.

Lafarge cement [MATER] A cement made of plaster of paris, lime, and marble powder; used in mortar for marble and limestone pieces because it is nonstaining.

Lafond's Tables [METEOROL] A set of tables and associated information for correcting reversing thermometers and computing dynamic height anomalies, compiled by E. C. Lafond and published by the U.S. Navy Hydrographic Office.

lag [ELECTR] A persistence of the electric charge image in a camera tube for a small number of frames. [PHYS] **1.** The difference in time between two events or values considered together. **2.** See lag angle.

lagan [ENG] A heavy object thrown overboard and buoyed to mark its location for future recovery.

lag angle [PHYS] The negative of phase difference between a sinusoidally varying quantity and a reference quantity which varies sinusoidally at the same frequency, when this phase difference is negative. Also known as angle of lag; lag.

lag bolt See coach screw.

Lagenidiales [MYCOL] An order of aquatic fungi belonging to the class Phycomycetales characterized by a saclike to limited hyphal thallus and zoospores having two flagella.

lageniform [BIOL] Flask-shaped.

lager beer [FOOD ENG] A variety of beer produced by bottom fermentation and allowed to age from 6 weeks to 6 months; it is usually dry, light in color, and well carbonated.

lag fault [GEOL] A minor low-angle thrust fault occurring within an overthrust; it develops when one part of the mass is thrust farther than an adjacent higher or lower part.

lagging [CIV ENG] **1.** Horizontal wooden strips fastened across an arch under construction to transfer weight to the centering form. **2.** Wooden members positioned vertically to prevent cave-ins in earthworking.

lagging load See inductive load.

lagging network See integral network.

lag gravel [GEOL] Residual accumulations of particles that are coarser than the material that has blown away.

lag-lead network See lead-lag network.

lag network See integral network.

Lagomorpha [VERT ZOO] The order of mammals including rabbits, hares, and pikas; differentiated from rodents by two pairs of upper incisors covered by enamel, vertical or transverse jaw motion, three upper and two lower premolars, fused tibia and fibula, and a spiral valve in the cecum.

lagoon [GEOGR] **1.** A shallow sound, pond, or lake generally near but separated from or communicating with the open sea. **2.** A shallow fresh-water pond or lake generally near or communicating with a larger body of fresh water.

LACTONE

Structural formula of γ-butyrolactone.

LACTOSE

The α-D anomer of lactose, the usual form in which lactose is obtained.

LACUNAR SYSTEM

Lacunar system of Moniliformis, dorsal view, showing regular circular branches. (After Meyer in L. H. Hyman, The Invertebrates, vol. 3, McGraw-Hill, 1951)

LAGOON NEBULA

Lagoon Nebula NGC 6523 and star cluster NGC 6530; dark grains and globules are caused by solid grains in the neighborhood of the nebula. (*Curtis Schmidt Telescope, University of Michigan*)

Lagoon Nebula [ASTRON] A patchy, luminous gaseous nebula that appears to be surrounded by a much larger region of cold, neutral hydrogen.

lag phase [MICROBIO] The period of physiological activity and diminished cell division following the addition of inoculum of bacteria to a new culture medium.

Lagrange bracket [MECH] Given two functions of coordinates and momenta in a system, their Lagrange bracket is an expression measuring how coordinates and momenta change jointly with respect to the two functions.

Lagrange function *See* Lagrangian.

Lagrange-Hamilton theory [MECH] The formalized study of continuous systems in terms of field variables where a Lagrangian density function and Hamiltonian density function are introduced to produce equations of motion.

Lagrange-Helmholtz equation *See* Helmholtz equation.

Lagrange's equations [MECH] Equations of motion of a mechanical system for which a classical (non-quantum-mechanical) description is suitable, and which relate the kinetic energy of the system to the generalized coordinates, the generalized forces, and the time. Also known as Lagrangian equations of motion.

Lagrange's interpolation formula [MATH] A formula by use of which a polynomial $f(x)$ of degree less than or equal to n can be completely determined provided that one knows $n + 1$ distinct values of it.

Lagrange's theorem [MATH] In a group of finite order, the order of any subgroup must divide the order of the entire group.

Lagrange stream function [FL MECH] A scalar function of position used to describe steady, incompressible two-dimensional flow; constant values of this function give the streamlines, and the rate of flow between a pair of streamlines is equal to the difference between the values of this function on the streamlines. Also known as current function; stream function.

Lagrangian [MECH] **1.** The difference between the kinetic energy and the potential energy of a system of particles, expressed as a function of generalized coordinates and velocities from which Lagrange's equations can be derived. Also known as kinetic potential; Lagrange function. **2.** For a dynamical system of fields, a function which plays the same role as the Lagrangian of a system of particles; its integral over a time interval is a maximum or a minimum with respect to infinitesimal variations of the fields, provided the initial and final fields are held fixed.

Lagrangian current measurement [OCEANOGR] Observation of the speed direction of an ocean current by means of a device, such as a parachute drogue, which follows the water movement.

Lagrangian density [MECH] For a dynamical system of fields or continuous media, a function of the fields, of their time and space derivatives, and the coordinates and time, whose integral over space is the Lagrangian.

Lagrangian equations of motion *See* Lagrange's equations.

Lagrangian function [MECH] The function which measures the difference between the kinetic and potential energy of a dynamical system.

Lagrangian generalized velocity *See* generalized velocity.

Lagrangian method [FL MECH] A method of studying fluid motion and the mechanics of deformable bodies in which one considers volume elements which are carried along with the fluid or body, and across whose boundaries material does not flow; in contrast to Euler method.

Lagrangian multipliers [MATH] A technique whereby potential extrema of functions of several variables are obtained.

Lagrangian points [ASTRON] In planetary orbits, two positions in which the motion of a body of negligible mass (such as an asteroid) is stable under the gravitational influence of two other bodies (such as the sun and Jupiter), one of which is moving about the other in an approximately circular orbit; the two positions are located on the orbit, 60° ahead of or behind the orbiting body, so that the three bodies form an equilateral triangle.

Lagriidae [INV ZOO] The long-jointed bark beetles, a family of coleopteran insects in the superfamily Tenebrionoidea.

lag screw *See* coach screw.

lag time [ELEC] The time between the application of current and rupture of the circuit within the detonator.

Laguerre polynomials [MATH] A sequence of orthogonal polynomials which solves a linear homogeneous second-order differential equation.

Laguerre's differential equation [MATH] The equation $xy'' + (1 - x)y' + \alpha y = 0$, where α is a constant.

Lagynacea [INV ZOO] A superfamily of foraminiferan protozoans in the suborder Allogromiina having a free or attached test that has a membranous to tectinous wall and a single, ovoid, tubular, or irregular chamber.

lahar [GEOL] **1.** A mudflow or landslide of pyroclastic material occurring on the flank of a volcano. **2.** The deposit of mud or land so formed.

laid fabric [TEXT] Fabric in which the warp threads are bonded in a material, such as latex, there being no weft threads.

laitance [MATER] Weak material, consisting principally of lime, that is formed on the surface of concrete, especially when excess water is mixed with the cement.

lake [HYD] An inland body of water, small to moderately large, with its surface water exposed to the atmosphere. [MATER] Any of a large group of dyes that have been combined with or adsorbed by salts of calcium, barium, chromium, aluminum, phosphotungstic acid, or phosphomolybdic acid; used for textile dyeing. Also known as color lake.

lake breeze [METEOROL] A wind, similar in origin to the sea breeze but generally weaker, blowing from the surface of a large lake onto the shores during the afternoon; it is caused by the difference in surface temperature of land and water, as in the land and sea breeze system.

lake copper [MET] A pure type of copper produced from ores taken from the Lake Superior region; has high conductivity.

lake effect [METEOROL] Generally, the effect of any lake in modifying the weather about its shore and for some distance downwind; in the United States, this term is applied specifically to the region about the Great Lakes.

lake ore *See* bog iron ore.

lake peat [GEOL] A sedimentary peat formed near lakes.

lake plain [GEOL] One of the surfaces of the earth that represent former lake bottoms; these featureless surfaces are formed by deposition of sediments carried into the lake by streams.

Lalande cell [ELEC] A type of wet cell that uses a zinc anode and cupric oxide cathode cast as flat plates or hollow cylinders, and an electrolyte of sodium hydroxide in aqueous solution (caustic soda).

laliophobia [PSYCH] An abnormal fear of talking or stuttering.

lalopathy [MED] Any disorder of speech or disturbance of language.

laloplegia [MED] Inability to speak, caused by paralysis of the muscles concerned in speech, except those of the tongue.

Lamarckism [EVOL] The theory that organic evolution takes place through the inheritance of modifications caused by the environment, and by the effects of use and disuse of organs.

lamb [VERT ZOO] A young sheep.

Lambast [ORG CHEM] Trade name for 2,4-bis(3-methoxy propylamino)-6-mercapto-s-triazine, a selective postemergence herbicide, soluble in organic solvents, available as an emulsifiable concentrate.

lamb-blasts *See* lambing storm.

lambda [MECH] A unit of volume equal to 10^{-6} liter or 10^{-9} cubic meter.

lambda dispatch [IND ENG] The solution of the problem of finding the most economical use of generators to supply a given quantity of electric power, using the method of Lagrange multipliers, which are symbolized λ.

lambda hyperon [PARTIC PHYS] **1.** A quasi-stable baryon, forming an isotopic singlet, having zero charge and hypercharge, a spin of $\frac{1}{2}$, positive parity and mass of 1115.5 MeV (million electron volts). Designated Λ. Also known as lambda particle. **2.** Any baryon resonance having zero hypercharge and total isotopic spin; designated $\Lambda_p(m)$, where m is the mass of the baryon in MeV, and J and P are its spin and parity (if known).

lambda particle *See* lambda hyperon.

lambda point [CRYO] The temperature (2.1780 K), at atmospheric pressure, at which the transformation between the liquids helium I and helium II takes place; a special case of the thermodynamics definition. [THERMO] A temperature at which the specific heat of a substance has a sharply peaked maximum, observed in many second-order transitions.

lambda sulfur [CHEM] One of the two components of plastic (or gamma) sulfur; soluble in carbon disulfide.

lamb dysentery [VET MED] A bacterial infection and inflammation of the intestinal tract of lambs caused by *Clostridium perfringens*, chiefly along the English-Scottish border.

lambert [OPTICS] A unit of luminance (photometric brightness) that is equal to $1/\pi$ candela per square centimeter, or to the uniform luminance of a perfectly diffusing surface emitting or reflecting light at the rate of 1 lumen per square centimeter. Abbreviated L.

Lambert bearing [NAV] A bearing as measured on a Lambert conformal chart or plotting sheet; approximates a great-circle bearing.

Lambert-Beer law *See* Bouguer-Lambert-Beer law.

Lambert conformal chart [MAP] A chart on the Lambert conformal projection.

Lambert conformal projection [MAP] A conformal conic projection with two standard parallels, or a conformal conic map projection in which the surface of a sphere or spheroid, such as the earth, is conceived as developed on a cone which intersects the sphere or spheroid at two standard parallels; the cone is then spread out to form a plane which is the map.

Lambert course [NAV] Course as measured on a Lambert conformal chart or plotting sheet.

Lambert's law [OPTICS] 1. The law that the illumination of a surface by a light ray varies as the cosine of the angle of incidence between the normal to the surface and the incident ray. 2. The law that the luminous intensity in a given direction radiated or reflected by a perfectly diffusing plane surface varies as the cosine of the angle between that direction and the normal to the surface. 3. *See* Bouguer-Lambert law.

Lambert surface [THERMO] An ideal, perfectly diffusing surface for which the intensity of reflected radiation is independent of direction.

lambing storm [METEOROL] A slight fall of snow in the spring in England. Also known as lamb-blasts; lamb-showers; lamb-storm.

Lamb shift [ATOM PHYS] A small shift in the energy levels of a hydrogen atom, and of hydrogenlike ions, from those predicted by the Dirac electron theory, in accord with principles of quantum electrodynamics.

lamb-showers *See* lambing storm.

lamb-storm *See* lambing storm.

Lamb wave [ACOUS] *See* plate wave. [ELECTROMAG] Electromagnetic wave propagated over the surface of a solid whose thickness is comparable to the wavelength of the wave.

lamé [TEXT] A fabric, usually of silk, rayon, or polyester, ornamented with flat metal threads.

Lamé constants [MECH] Two constants which relate stress to strain in an isotropic, elastic material.

lamella [ANAT] A thin scale or plate.

lamella arch [CIV ENG] An arch consisting basically of a series of intersecting skewed arches made up of relatively short straight members; two members are bolted, riveted, or welded to a third piece at its center.

lamellar bone [HISTOL] Any bone with a microscopic structure consisting of thin layers or plates.

lamellar chloroplast [CYTOL] A type of chloroplast in which the layered structure extends more or less uniformly through the whole chloroplast body.

lamella roof [BUILD] A large span vault built of members connected in a diamond pattern.

lamellar vector field *See* irrotational vector field.

Lamellibranchiata [INV ZOO] An equivalent name for Bivalvia.

Lamellisabellidae [INV ZOO] A family of marine animals in the order Thecanephria.

Lamé's equations [MATH] A general collection of second-order differential equations which have five regular singularities.

Lamé's relations [MATH] Six independent relations which

when satisfied by the covariant metric tensor of a three-dimensional space provide necessary and sufficient conditions for the space to be euclidean.

Lamiaceae [BOT] An equivalent name for Labiatae.

Lamiales [BOT] An order of dicotyledonous plants in the subclass Asteridae marked by its characteristic gynoecium, consisting of usually two biovulate carpels, with each carpel divided between the ovules by a false partition, or with the two halves of the carpel seemingly wholly separate.

lamina [ANAT] A thin sheet or layer of tissue; a scalelike structure. [GEOL] A thin, clearly differentiated layer of sedimentary rock or sediment, usually less than 1 centimeter thick.

lamina cribrosa [ANAT] 1. The portion of the sclera which is perforated for the passage of the optic nerve. 2. The fascia covering the saphenous opening in the thigh. 3. The anterior or posterior perforated space of the brain. 4. The perforated plates of bone through which pass branches of the cochlear part of the vestibulocochlear nerve.

laminal placentation [BOT] Condition in which the ovules occur on the inner surface of the carpels.

laminar [SCI TECH] 1. Arranged in thin layers. 2. Pertaining to viscous streamline flow without turbulence.

laminar boundary layer [FL MECH] A thin layer over the surface of a body immersed in a fluid, in which the fluid velocity relative to the surface increases rapidly with distance from the surface and the flow is laminar.

laminar flow [FL MECH] Streamline flow of an incompressible, viscous Newtonian fluid; all particles of the fluid move in distinct and separate lines.

Laminariophyceae [BOT] A class of algae belonging to the division Phaeophyta.

laminar sublayer [FL MECH] The laminar boundary layer underlying a turbulent boundary layer.

laminar wing [AERO ENG] A low-drag wing in which the distribution of thickness along the chord is so selected as to maintain laminar flow over as much of the wing surface as possible.

laminate [MATER] A sheet of material made of several different bonded layers.

laminated contact [ELEC] Switch contact made up of a number of laminations, each making individual contact with the opposite conducting surface.

laminated core [ELECTROMAG] An iron core for a coil transformer, armature, or other electromagnetic device, built up from laminations stamped from sheet iron or steel and more or less insulated from each other by surface oxides and sometimes also by application of varnish.

laminated glass *See* nonshattering glass.

laminated metal [MET] A sheet or bar of composite metal composed of two or more bonded layers.

laminated plastic [MATER] A thin sheet made of superposed layers of plastic bonded or impregnated with resin or compressed under heat.

laminated spring [DES ENG] A flat or curved spring made of thin superimposed plates and forming a cantilever or beam of uniform strength.

laminated wood [MATER] Board or timber composed of layers of wood glued together with the grains parallel.

lamina terminalis [ANAT] The layer of gray matter in the brain connecting the optic chiasma and the anterior commissure where the latter becomes continuous with the rostral lamina.

lamination [MED] An operation in embryotomy in which the skull is cut in slices. [SCI TECH] Arrangement in layers.

laminectomy [MED] Surgical removal of the lateral portion of the neural arch from one or more vertebrae.

laminography *See* sectional radiography.

Lami's theorem [MECH] When three forces act on a particle in equilibrium, the magnitude of each is proportional to the sine of the angle between the other two.

lamp [ENG] A device that produces light, such as an electric lamp.

lampadite [MINERAL] A mineral composed chiefly of hydrous manganese oxide with as much as 18% copper oxide and often cobalt oxide.

lamp bank [ELEC] A number of incandescent lamps con-

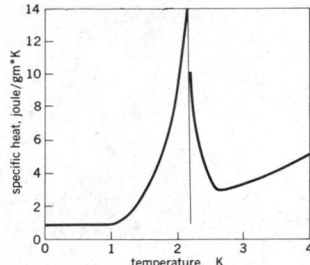

Specific heat of liquid helium computed as a function of temperature. Helium I is the high-temperature liquid; helium II occurs below the lambda point.

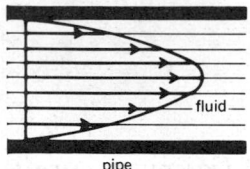

Laminar flow in a circular pipe. In this case the velocity adjacent to the wall is zero and increases to a maximum in the center of the pipe.

nected in parallel or series to serve as a resistance load for full-load tests of electric equipment.

lampblack [MATER] A grayish-black amorphous, practically pure form of carbon made by burning oil, coal tar, resin, or other carbonaceous substance in an insufficient supply of air; used in making paints, lead pencils, metal polishes, electric brush carbons, crayons, and carbon papers.

lampbrush chromosome [CYTOL] An exceptionally large chromosome characterized by fine lateral projections which are associated with active ribonucleic acid and protein synthesis.

lamp cabin *See* lamp room.

lamp-charging rack [MIN ENG] Mine-lamp-charging racks which allow miners to store lamp units for recharging after daily use.

lamp cord [ELEC] Two twisted or parallel insulated wires, usually no. 18 or no. 20, used chiefly for connecting electric equipment to wall outlets.

lampholder [ELEC] A device designed to connect an electric lamp to a circuit and to support it mechanically.

lamphouse [ENG] The light housing in a motion picture projector, located behind the projector head ordinarily consisting of a carbon arc lamp operating on direct current at about 60 volts, a concave reflector behind the arc which collects the light and concentrates it on the film, and cooling devices. **2.** A box with a small hole containing an electric lamp and a concave mirror behind it, used as a concentrated source of light in a microscope, photographic enlarger, or other instrument.

lamping [MIN ENG] Use of a portable ultraviolet lamp to reveal fluorescent minerals in prospecting.

lampman [MIN ENG] A person responsible for maintaining and servicing miners' lamps.

lamp oil *See* kerosine.

lamprey [VERT ZOO] The common name for all members of the order Petromyzonida.

Lampridiformes [VERT ZOO] An order of teleost fishes characterized by a compressed, often ribbonlike body, fins composed of soft rays, a ductless swim bladder, and protractile maxillae among other distinguishing features.

lamprobolite *See* basaltic hornblende.

lamp room [MIN ENG] A room or building at the surface of a mine for charging, servicing, and issuing all cap, hand, and flame safety lamps. Also known as lamp cabin; lamp station.

lamprophyllite [MINERAL] $Na_2SrTiSi_2O_8$ A mineral composed of titanium strontium sodium silicate.

lamprophyre [PETR] Any of a group of igneous rocks characterized by a porphyritic texture in which abundant, large crystals of dark-colored minerals appear set in a not visibly crystalline matrix.

lampshade paper [MATER] Paper that is translucent and either flame-resistant or flame-retardant; often made of wood pulp, vegetable parchment, or laminated glassine.

lamp station *See* lamp room.

Lampyridae [INV ZOO] The firefly beetles, a large cosmopolitan family of coleopteran insects in the superfamily Cantharoidea.

lanac [NAV] A proprietary navigation system that depends on secondary radar to aid airplanes to avoid midair collisions and maintain a specified altitude prior to landing. Derived from laminar, air navigation, and collision.

lanarkite [MINERAL] Pb_2OSO_4 A white, greenish, or gray monoclinic mineral consisting of basic lead sulfate, with specific gravity of 6.92; formed by action of heat and air on galena.

lanatoside [BIOCHEM] Any of three natural glycosides from the leaves of *Digitalis lanata*; on hydrolysis with acid, it yields one molecule of D-glucose, three molecules of digitoxose, and one molecule of acetic acid; all three glycosides are cardioactive.

Lancashire boiler [MECH ENG] A cylindrical steam boiler consisting of two longitudinal furnace tubes which have internal grates at the front.

lance [MED] To cut or open, as with a lancet. [MET] To cut into but not through the piece of work.

Lancefield groups [MICROBIO] Antigenically determined categories for classification of β-hemolytic streptococci.

lancelet [ZOO] The common name for members of the subphylum Cephalochordata.

lanceolate [BIOL] Shaped like the head of a lance.

Lanceolidae [INV ZOO] A family of bathypelagic amphipod crustaceans in the suborder Hyperiidea.

lancet [MED] A sharp-pointed, double-edged cutting instrument used to make small incisions.

lancet window [ARCH] A narrow window with a sharply pointed top.

Lanchester balancer [MECH ENG] A device for balancing four-cylinder engines; consists of two meshed gears with eccentric masses, driven by the crankshaft.

land [AERO ENG] Of an aircraft, to alight on land or a ship deck. [ELECTR] *See* terminal area. [ENG] **1.** In plastics molding equipment, the horizontal bearing surface of a semipositive or flash mold to allow excess material to escape; or the bearing surface along the top of the screw flight in a screw extruder; or the surface of an extrusion die that is parallel to the direction of melt flow. **2.** The surface between successive grooves of a diffraction grating or phonograph record. [GEOGR] The portion of the earth's surface that stands above sea level. [ORD] One of the raised ridges in the bore of a rifled gun barrel.

land accretion [CIV ENG] Gaining land in a wet area, such as a marsh or by the sea, by planting maritime plants to encourage silt deposition or by dumping dredged materials in the area. Also known as land reclamation.

land and sea breeze [METEOROL] The complete cycle of diurnal local winds occurring on seacoasts due to differences in surface temperature of land and sea; the land breeze component of the system blows from land to sea, and the sea breeze blows from sea to land.

Landau damping [PL PHYS] Damping of a plasma oscillation wave which occurs in situations where the particles of the plasma are able to increase their average energy at the expense of the wave, and thus to damp it out, even in cases where the dissipative effects of collisions are unimportant.

Landau fluctuations [NUCLEO] Variations in the losses of energy of different particles in a thin detector, resulting from random variations in the number of collisions and in the energies lost in each collision of the particle.

Landau-Ginzburg theory *See* Ginzburg-Landau theory.

Landau levels [SOLID STATE] Energy levels of conduction electrons which occur in a metal subjected to a magnetic field at very low temperatures and which are quantized because of the quantization of the electron motion perpendicular to the field.

landblink [METEOROL] A yellowish glow observed over snow-covered land in the polar regions.

land breeze [METEOROL] A coastal breeze blowing from land to sea, caused by the temperature difference when the sea surface is warmer than the adjacent land; therefore, the land breeze usually blows by night and alternates with a sea breeze which blows in the opposite direction by day.

land bridge [GEOGR] A strip of land linking two landmasses, often subject to temporary submergence, but permitting intermittent migration of organisms.

land drainage [CIV ENG] The removal of water from land to improve the soil as a medium for plant growth and a surface for land management operations.

land effect *See* coastal refraction.

Landé g factor [ATOM PHYS] Also known as g factor. **1.** The negative ratio of the magnetic moment of an electron or atom, in units of the Bohr magneton, to its angular momentum, in units of Planck's constant divided by 2π. **2.** The ratio of the difference in energy between two energy levels which differ only in magnetic quantum number to the product of the Bohr magneton, the applied magnetic field, and the difference between the magnetic quantum numbers of the levels; identical to the first definition for free atoms. Also known as Landé splitting factor; spectroscopic splitting factor. [NUC PHYS] The ratio of the magnetic moment of a nucleon, in units of the nuclear magneton, to its angular momentum in units of Planck's constant divided by 2π.

Landé interval rule [ATOM PHYS] The rule that when the spin-orbit interaction is weak enough to be treated as a perturbation, an energy level having definite spin angular

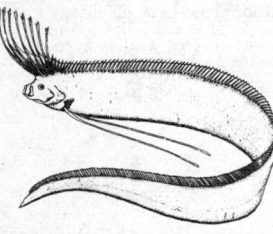

LAMPRIDIFORMES

Oarfish (*Regalecus glesne*), an example of Lampridiformes; length to over 20 feet (6 meters). (*After D. S. Jordan and B. W. Evermann, The Fishes of North and Middle America, U.S. Nat. Mus. Bull. no. 47, 1900*)

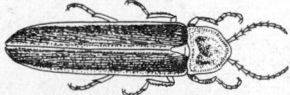

LAMPYRIDAE

Drawing of a firefly beetle; there are 11109 species. (*From T. I. Storer and R. L. Usinger, General Zoology, 3d ed., McGraw-Hill 1957*)

momentum and orbital angular momentum is split into levels of differing total angular momentum, so that the interval between successive levels is proportional to the larger of their total angular momentum values.

Landenian [GEOL] A European stage of geologic time: upper Paleocene (above Montian, below Ypresian of Eocene).

Landé Γ-permanence rule [ATOM PHYS] The rule that the sum of the shifts of energy levels produced by the spin-orbit interaction, over a series of states having the same spin and orbital angular momentum quantum numbers (or the same total angular momentum quantum numbers for individual electrons) but different total angular momenta, and having the same total magnetic quantum number, is independent of the strength of an applied magnetic field.

lander [MIN ENG] **1.** A worker at one of the levels of a mine shaft to unload rock and to load drilling and blasting supplies to be lowered. **2.** In the quarry industry, one who guides, steadies, and loads trucks or railroad cars with the blocks of stone hoisted from the quarry floor. Also known as top hooker. **3.** In metal mining, one who cleans skips by directing a blast of compressed air into them through a hose, records number of loaded skips hoisted to surface, and loads railroad cars with ore from bins. **4.** In coal mining, one who works with shaft sinking crew at the top of the shaft or at a level immediately above shaft bottom, dumping rock into mine cars from a bucket in which it is raised. Also known as bucket dumper; landing tender; top lander. **5.** The worker who receives the loaded bucket or tub at the mouth of the shaft. Also known as banksman.

Landé splitting factor *See* Landé g factor.

landfall [NAV] **1.** The first sighting of land when approaching from seaward; by extension, the term is sometimes used to refer to the first contact with land by any means, as by radar. **2.** A navigational procedure in which an aircraft, after a relatively long overwater flight, turns on to a line of position passing through its destination and follows the line to the destination.

landfast ice *See* fast ice.

landfill [CIV ENG] Disposal of solid waste by burying in layers of earth in low ground.

landform [GEOGR] All the physical, recognizable, naturally formed features of land, having a characteristic shape; includes major forms such as a plain, mountain, or plateau, and minor forms such as a hill, valley, or alluvial fan.

land hemisphere [GEOGR] The half of the globe, with its pole located at 47.25°N 2.5°W, in which most of the earth's land area is concentrated.

Landholt fringe [OPTICS] A black fringe that crosses the darkened field which is produced when a brilliant source of light is viewed through two Nicol prisms oriented with their principal axes at right angles to one another.

land ice [HYD] Any part of the earth's seasonal or perennial ice cover which has formed over land as the result, principally, of the freezing of precipitation.

landing [CIV ENG] A place where boats receive or discharge passengers, freight, and so on. [MIN ENG] **1.** Level stage in a shaft at which cages are loaded and discharged. **2.** The top or bottom of a slope, shaft, or inclined plane. [NAV] The termination of an aircraft's flight or of a ship's voyage.

landing aid [NAV] A lamp, searchlight, radio beacon, radar device, communicating device, or any system of such devices for aiding aircraft in an approach and landing. Also known as landing system.

landing area [AERO ENG] An area intended primarily for landing and takeoff of aircraft.

landing beacon [NAV] A beacon transmitting a beam to guide aircraft in making a landing.

landing chart [NAV] An aeronautical chart showing obstructions in the immediate vicinity of an aerodrome and the layout of the runways or landing area for use in landing and taxiing.

landing circle [AERO ENG] The approximately circular path flown by an airplane to get into the landing pattern; used particularly with naval aircraft landing on an aircraft carrier.

landing compass [NAV] A compass taken ashore so as to be unaffected by deviation; if reciprocal bearings of the landing compass and the magnetic compass on board a ship are observed, the deviation of the magnetic compass can be determined.

landing craft [NAV ARCH] A combat vessel employed in amphibious warfare to transport mobile equipment, amphibious vehicles, tanks, general cargo, and personnel, and to discharge such directly on to the beach; the vessel ranges from 36 to 135 feet (11 to 41 meters) in length and differs from landing ships in that it is not designed for long transoceanic voyages.

landing direction indicator [NAV] A device which indicates visually to pilots of aircraft the direction designated for landing or takeoff.

landing flap [AERO ENG] A movable airfoil-shaped structure located aft of the rear beam or spar of the wing; extends about two-thirds of the span of the wing and functions to substantially increase the lift, permitting lower takeoff and landing speeds.

landing flare [NAV] A pyrotechnic flare sometimes dropped from an aircraft for illumination during a night landing.

landing gear [AERO ENG] Those components of an aircraft or spacecraft that support and provide mobility for the craft on land, water, or other surface.

landing light [AERO ENG] One of the floodlights mounted on the leading edge of the wing and below the nose of the fuselage to enable an airplane to land at night.

landing load [AERO ENG] The load on an aircraft's wings produced during landing; depends on descent velocity and landing attitude.

landing ship [NAV ARCH] Large-type assault ship generally over 200 feet (60 meters) long which is designed for long sea voyages and for rapid unloading over or onto a beach.

landing stage [CIV ENG] A platform, usually floating and attached to the shore, for the discharge and embarkation of passengers, freight, and so on.

landing strip [AERO ENG] A portion of the landing area prepared for the landing and takeoff of aircraft in a particular direction; it may include one or more runways. Also known as air strip.

landing system *See* landing aid.

landing tee *See* wind tee.

landing tender *See* lander.

landline [ELEC] A communications cable on or under the earth's surface, in contrast to a submarine cable.

landlocked [GEOGR] Pertaining to a harbor which is surrounded or almost completely surrounded by land.

landmark [ENG] Any fixed natural or artificial monument or object used to designate a land boundary. [NAV] A conspicuous natural or artificial object near or on land, other than an established aid to navigation, and observable by eye or radar; used in the piloting type of navigation.

landmark beacon [NAV] A light beacon, flashing white and red, marking the location of a natural or man-made landmark or a point on the Federal airways.

land measure [MECH] **1.** Units of area used in measuring land. **2.** Any system for measuring land.

land mile *See* mile.

land mine [ORD] A container filled with high explosives or chemicals, placed on the ground or lightly covered, and fitted with a fuse or a firing device or both.

land mobile service [COMMUN] Mobile service between base stations and mobile stations, or between land mobile stations.

land mobile station [COMMUN] Mobile station in the land mobile service, capable of surface movement within the geographical limits of a country or continent.

land navigation [NAV] Navigation of vehicles across land or ice, generally used in connection with the crossing of a region devoid of roads or landmarks, so that methods similar to those employed in air or marine navigation must be used.

land pebble *See* land pebble phosphate.

land pebble phosphate [GEOL] A pebble phosphate in a clay or sand bed below the ground surface; a small amount of uranium is often present and is recovered as a by-product; used as a source of phosphate fertilizer. Also known as land pebble; land rock; matrix rock.

land plaster [MATER] Finely ground gypsum, used as a fertilizer and as a corrective for soil with excess sodium and potassium carbonates.

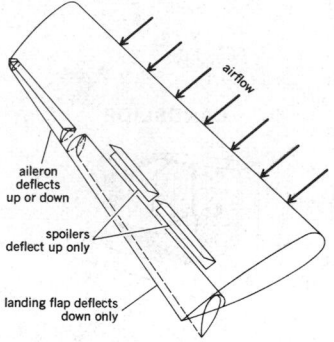

land reclamation *See* land accretion.

land return *See* ground clutter.

land rock *See* land pebble phosphate.

Landry-Guillain-Barré syndrome [MED] A diffuse motor-neuron paresis, rapid in onset, and usually ascending and symmetrical in distribution, with proximal involvement greater than distal, and motor deficits greater than sensory. Also known as Guillain-Barré syndrome; Landry's paralysis.

Landry's paralysis *See* Landry-Guillain-Barré syndrome.

landscape [GEOGR] The distinct association of landforms that can be seen in a single view.

landscape agate [MINERAL] A type of chalcedony that is translucent and contains inclusions which give it an appearance reminiscent of familiar natural scenes. Also known as fortification agate.

landscape architecture [CIV ENG] The art of arranging and fitting land for human use and enjoyment.

landscape engineer [CIV ENG] A person who applies engineering principles and methods to planning, design, and construction of natural scenery arrangements on a tract of land.

land sky [METEOROL] The relatively dark appearance of the underside of a cloud layer when it is over land that is not snow-covered, used largely in polar regions with reference to the sky map; it is brighter than water sky, but much darker than iceblink or snowblink.

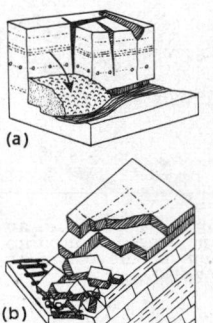

(a)

(b)

Two principal types of landslide: *(a)* debris fall and *(b)* rockslide.

landslide [GEOL] The perceptible downward sliding or falling of a relatively dry mass of earth, rock, or combination of the two under the influence of gravity. Also known as landslip.

landslip *See* landslide.

land station [COMMUN] Station in the mobile service not intended for operation while in motion.

land surveyor [CIV ENG] A specialist who measures land and its natural features and any man-made features such as buildings or roads for drawing to scale as plans or maps.

land tie [CIV ENG] A rod or chain connecting an outside structure such as a retaining wall to a buried anchor plate.

land transportation frequency bands [COMMUN] A group of radio-frequency bands between 25 megahertz and 30,000 megahertz allocated for use by taxicabs, railroads, buses, and trucks.

land transportation radio services [COMMUN] Any service of radio communications operated by and for the sole use of certain land transportation carriers, the radio transmitting facilities of which are defined as fixed, land, or mobile stations.

land-use classes [CIV ENG] Categories into which land areas can be grouped according to present or potential economic use.

land-use map [MAP] A map showing land-use classes as well as other earth surface features such as roads, manufacturing plants, and harbors.

lane [CIV ENG] An established route, as an air lane, shipping lane, or highway traffic lane. [NAV] One of the sections of the coverage area of a pair of Decca stations in which any phase relationship may be measured.

Lane's law [ASTROPHYS] For the contraction of a star that is assumed to be a sphere of perfect gas, the law that the temperature of the perfect-gas sphere is inversely proportional to its radius.

langbanite [MINERAL] An ironblack hexagonal mineral composed of silicate and oxide of manganese, iron, and antimony, occurring in prismatic crystals.

langbeinite [MINERAL] $K_2Mg_2(SO_4)_3$ Colorless, yellowish, reddish, or greenish hexagonal mineral with vitreous luster, found in salt deposits; used in the fertilizer industry as a source of potassium sulfate.

Langerhans cell [HISTOL] **1.** A type of cytotrophoblast in the human chorionic vesicle which is thought to secrete chorionic gonadotropin. **2.** A highly branched dendritic cell of the mammalian epidermis showing a lobulated nucleus and a diagnostic organelle resembling a tennis racket.

Langerhansian adenoma *See* islet-cell tumor.

Lange's nerve [INV ZOO] One of the paired cords of nervous tissue lying in the wall of the radial perihemal canal of asteroids.

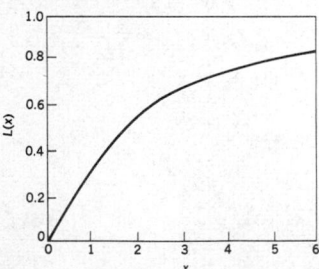

Plot of Langevin function.

Drawing of lang lay method of winding wire rope.

Langevin-Debye formula [STAT MECH] A formula for the polarizability of a dielectric material or the paramagnetic susceptibility of a magnetic material, in which these quantities are the sum of a temperature-independent contribution and a contribution arising from the partial orientation of permanent electric or magnetic dipole moments which varies inversely with the temperature. Also known as Langevin-Debye law.

Langevin-Debye law *See* Langevin-Debye formula.

Langevin function [ELECTROMAG] A mathematical function, $L(x)$, which occurs in the expressions for the paramagnetic susceptibility of a classical (non-quantum-mechanical) collection of magnetic dipoles, and for the polarizability of molecules having a permanent electric dipole moment; given by $L(x) = \coth x - 1/x$.

Langevin ion *See* large ion.

Langevin ion-mobility theories [ELECTR] Two theories developed to calculate the mobility of ions in gases; the first assumes that atoms and ions interact through a hard-sphere collision and have a constant mean free path, while the second assumes that there is an attraction between atoms and ions arising from the polarization of the atom in the ion's field, in addition to hard-sphere repulsion for close distances of approach.

Langevin ion-recombination theory [ELECTR] A theory predicting the rate of recombination of negative with positive ions in an ionized gas on the assumption that ions of opposite sign approach one another under the influence of mutual attraction, and that their relative velocities are determined by ion mobilities; applicable at high pressures, above 1 or 2 atmospheres.

Langevin theory of diamagnetism [ELECTROMAG] A theory based on the idea that diamagnetism results from electronic currents caused by Larmor precession of electrons inside atoms.

Langevin theory of paramagnetism [ELECTROMAG] A theory which treats a substance as a classical (non-quantum-mechanical) collection of permanent magnetic dipoles with no interactions between them, having a Boltzmann distribution with respect to energy of interaction with an applied field.

langite [MINERAL] A blue to green mineral composed of basic hydrous copper sulfate.

lang lay [DES ENG] A wire rope lay in which the wires of each strand are twisted in the same direction as the strands.

langley [PHYS] A unit of energy per unit area commonly employed in radiation theory; equal to 1 gram-calorie per square centimeter.

Langmuir-Child equation *See* Child's law.

Langmuir dark space [ELECTR] A nonluminous region surrounding a negatively charged probe inserted in the positive column of a glow discharge.

Langmuir isotherm equation [PHYS CHEM] An equation, useful chiefly for gaseous systems, for the amount of material adsorbed on a surface as a function of pressure, while the temperature is held constant, assuming that a single layer of molecules is adsorbed; it is $f = ap/(1 + ap)$, where f is the fraction of surface covered, p is the pressure, and a is a constant.

Langmuir plasma frequency [PL PHYS] The frequency of nonpropagating oscillations in a plasma; in rationalized mks units, it is $(ne^2/\varepsilon_0 m)^{1/2}$, where e and m are the charge and mass of the oscillating electrons or ions, n is their number density, and ε_0 is the permittivity of empty space. Also known as plasma frequency.

Langmuir probe [PL PHYS] A device for measuring the temperature and electron density of a plasma, consisting of an electrode in contact with the plasma whose potential is varied while the resulting collection currents are measured.

Lang's vesicle [INV ZOO] A seminal bursa in many polyclad flatworms.

language [ADP] The set of words and rules used to construct sentences with which to express and process information for handling by computers and associated equipment.

language converter [ADP] A device which translates a form of data (such as that on microfilm) into another form of data (such as that on magnetic tape).

language theory [MATH] A branch of automata theory which attempts to formulate the grammar of a language in mathematical terms; it has been applied to automatic language translation and to the construction of higher-level programming languages and systems such as the propositional calculus, nerve networks, sequential machines, and programming schemes.

language translator [ADP] **1.** Any assembler or compiler that accepts human-readable statements and produces equivalent outputs in a form closer to machine language. **2.** A program designed to convert one computer language to equivalent statements in another computer language, perhaps to be executed on a different computer. **3.** A routine that performs or assists in the performance of natural language translations, such as Russian to English, or Chinese to Russian.

Languriidae [INV ZOO] The lizard beetles, a cosmopolitan family of coleopteran insects in the superfamily Cucujoidea.

Langweiler charge *See* traveling charge.

Laniatores [INV ZOO] A suborder of arachnids in the order Phalangida having flattened, often colorful bodies and found chiefly in tropical areas.

lanolin [MATER] The hydrous sheep's-wool wax (primarily cholesterol esters of higher fatty acids) derived as a by-product from the preparation of raw wool for the spinner; used as a base for emollients in cosmetics and shampoos.

lanosterol [BIOCHEM] $C_{30}H_{50}O$ An unsaturated sterol occurring in wool fat and yeast.

lansan [METEOROL] A strong southeast trade wind of the New Hebrides and East Indies.

lansfordite [MINERAL] $MgCO_3 \cdot 5H_2O$ A mineral composed of hydrous basic carbonate of magnesium when extracted from the earth, changing to nesquehovite after exposure to the air.

Lanston Monotype [GRAPHICS] A kind of composing machine that casts type in single characters in justified lines of predetermined length.

L antenna [ELECTROMAG] An antenna that consists of an elevated horizontal wire having a vertical down-lead connected at one end.

lantern [ENG] A portable lamp.

lantern clock [HOROL] **1.** A wall-mounted clock with weights and pendulum outside the case. **2.** A 17th-century clock characterized by a dome and open fretwork.

lantern fish [VERT ZOO] The common name for the deep-sea teleost fishes composing the family Myctophidae and distinguished by luminous glands that are widely distributed upon the body surface.

lantern pinion [DES ENG] A pinion with bars (between parallel disks) instead of teeth.

lantern ring [DES ENG] A ring or sleeve around a rotating shaft; an opening in the ring provides for forced feeding of oil or grease to bearing surfaces; particularly effective for pumps handling liquids.

lantern slide [GRAPHICS] A positive transparent picture on glass or film which can be viewed by projection onto reflecting screens, by hand viewers, or by devices in which the picture is projected onto the rear of a diffusing screen.

lanthana *See* lanthanum oxide.

lanthanide contraction [ATOM PHYS] A phenomenon encountered in the rare-earth elements; the radii of the atoms of the members of the series decrease slightly as the atomic numbers increase; starting with element 58 in the periodic table, the balancing electron fills in an inner incomplete $4f$ shell as the charge on the nucleus increases.

lanthanide series [CHEM] Rare-earth elements of atomic numbers 57 through 71; their chemical properties are similar to those of lanthanum, atomic number 57.

lanthanite [MINERAL] $(La,Ce)_2(CO_3)_3 \cdot 8H_2O$ A colorless, white, pink, or yellow mineral composed of hydrous lanthanum carbonate, occurring in crystals or in earthy form.

lanthanum [CHEM] A chemical element, symbol La, atomic number 57, atomic weight 138.91; it is the second most abundant element in the rare-earth group. [MET] A white, soft, malleable metal; tarnishes in moist air; a major component of misch metal.

lanthanum nitrate [INORG CHEM] $La(NO_3)_3 \cdot 6H_2O$ Hygroscopic white crystals melting at 40°C; soluble in alcohol and water; used as an antiseptic and in gas mantles.

lanthanum oxide [INORG CHEM] La_2O_3 A white powder melting at about 2000°C; soluble in acid, insoluble in water; used to replace lime in calcium lights and in optical glass. Also known as lanthana; lanthanum sesquioxide; lanthanum trioxide.

lanthanum sesquioxide *See* lanthanum oxide.

lanthanum sulfate [INORG CHEM] $La_2(SO_4)_3 \cdot 9H_2O$ White crystals; slightly soluble in water, soluble in alcohol; used for atomic weight determinations for lanthanum.

lanthanum trioxide *See* lanthanum oxide.

Lanthonotidae [VERT ZOO] A family of lizards (Sauria) belonging to the Anguimorpha line; restricted to North Borneo.

lanugo [ANAT] A downy covering of hair, especially that seen on the fetus or persisting on the adult body.

lanyard [NAV ARCH] A small rope or line used to fasten something in ships, especially one passing through deadeyes and used to extend shrouds or stays.

lap [CIV ENG] The length by which a reinforcing bar must overlap the bar it will replace. [MATER] An abrasive material used for lapping. [MET] A defect caused by folding and then rolling or forging a hot metal fin or corner onto a surface without welding. Also known as fold.

lapachoic acid *See* lapachol.

lapachol [BIOCHEM] $C_{15}H_{14}O_3$ A yellow crystalline compound obtained from lapacho, a hardwood in Argentina and Paraguay. Also known as 2-hydroxy-3-amylene-α-naphthoquinone; lapachoic acid; targusic acid.

laparoscopy [MED] A method of visually examining the peritoneal cavity by means of a long slender endoscope equipped with sheath, obturator, biopsy forceps, a sphygmomanometer bulb and tubing, scissors, and a syringe; the endoscope is introduced into the peritoneal cavity through a small incision in the abdominal wall. Also known as peritoneoscopy.

laparotomy [MED] A surgical incision through the abdominal wall into the abdominal cavity.

lap dissolve [ELECTR] Changeover from one television scene to another so that the new picture appears gradually as the previous picture simultaneously disappears. [GRAPHICS] *See* dissolve.

lapel microphone [ENG ACOUS] A small microphone that can be attached to a lapel or pocket on the clothing of the user, to permit free movement while speaking.

lapidary [SCI TECH] **1.** Of or pertaining to precious stones. **2.** The art of cutting precious stones. **3.** A person skilled in such art.

lapidicolous [ECOL] Living under a stone.

lapilli [GEOL] Pyroclasts that range from 1 to 64 millimeters in diameter.

lapilli-tuff [GEOL] A pyroclastic deposit that is indurated and consists of lapilli in a fine tuff matrix.

lapis lazuli [PETR] An azure-blue, violet-blue, or greenish-blue, translucent to opaque crystalline rock used as a semiprecious stone; composed chiefly of lazurite and calcite with some haüyne, sodalite, and other minerals. Also known as lazuli.

lap joint [ENG] A simple joint between two members made by overlapping the ends and fastening them together with bolts, rivets, or welding.

Laplace irrotational motion [FL MECH] Irrotational flow of an inviscid, incompressible fluid.

Laplace law *See* Ampère law.

Laplace operator [MATH] The linear operator defined on differentiable functions which gives for each function the sum of all its nonmixed second partial derivatives. Also known as Laplacian.

Laplace's equation [MATH] The partial differential equation which states that the sum of all the nonmixed second partial derivatives equals 0; the potential functions of many physical systems satisfy this equation.

Laplace's expansion [MATH] An expansion by means of which the determinant of a matrix may be computed in terms of the determinants of all possible smaller square matrices contained in the original.

Laplace's measure of dispersion [STAT] The expected value of the absolute value of the difference between a random variable and its mean.

Laplace transform [MATH] For a function $f(x)$ its Laplace

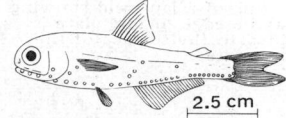

LANTERN FISH

Drawing of the lantern fish (*Myctophym punctatum*); luminous glands are shown as small circles.

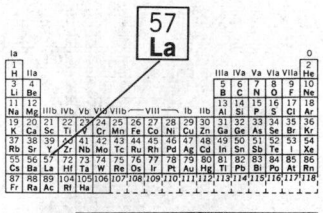

LANTHANUM

Periodic table of the chemical elements showing the position of lanthanum.

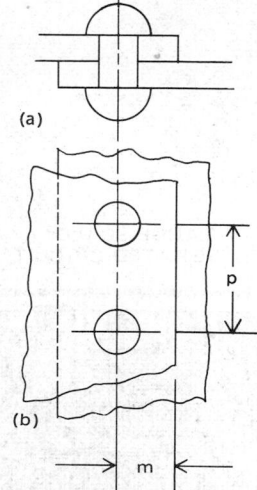

LAP JOINT

An example of a lap joint involving plates and rivets. (a) Side view (b) top view; p = the distance between adjacent rivets along the line through the centers of the rivets parallel to the edge of the plate; m = margin which is the distance between the line through the center of the rivets and the edge of the plate.

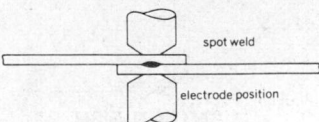

Drawing of a lap weld showing how the edge of one plate is welded to the surface of the other plate.

Cone and needles of Western larch (*Larix occidentalis*).

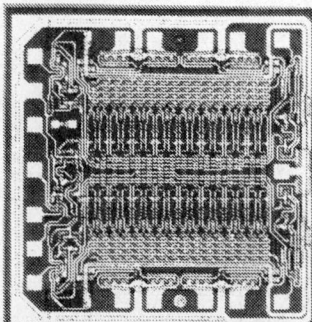

Monolithic silicon 128-bit read-only memory unit. (*Courtesy of Motorola Semiconductor Products Inc.*)

transform is the function $F(y)$ defined as the integral over x from 0 to ∞ of the function $e^{-yx}f(x)$.

Laplacian *See* Laplace operator.

Laplacian speed of sound [FL MECH] The phase speed of a sound wave in a compressible fluid under the assumption that the expansions and compressions are adiabatic.

Laporte selection rule [ATOM PHYS] The rule that an electric dipole transition can occur only between states of opposite parity.

lappet [TEXT] **1.** An attachment for a loom, used to insert floating warp threads into a fabric. **2.** A lightweight patterned fabric such as dotted swiss made with such an attachment. [ZOO] A lobe or flaplike projection, such as on the margin of a jellyfish or the wattle of a bird.

lapping [ELECTR] Moving a quartz, semiconductor, or other crystal slab over a flat plate on which a liquid abrasive has been poured, to obtain a flat polished surface or to reduce the thickness a carefully controlled amount. [MET] Polishing with a material such as cloth, lead, plastic, wood, iron, or copper having fine abrasive particles incorporated or rubbed into the surface.

lap-rivet [MET] To rivet a lap joint.

lapse line [METEOROL] A curve showing the variation of temperature with height in the free air.

lapse rate [METEOROL] **1.** The rate of decrease of temperature in the atmosphere with height. **2.** Sometimes, the rate of change of any meteorological element with height.

lap siding [BUILD] Beveled boards used for siding that are similar to clapboards but longer and wider. [CIV ENG] Two railroad sidings, the turnout of one overlapping that of the other.

lapstrake [NAV ARCH] A method of hull construction where each continuous band of hull planking (strake) is lapped on the outside of the one beneath.

lap weld [MET] A welded lap joint.

lap winding [ELEC] A two-layer winding in which each coil is connected in series to the adjacent coil.

Laramic orogeny *See* Laramidian orogeny.

Laramide orogeny *See* Laramidian orogeny.

Laramide revolution *See* Laramidian orogeny.

Laramidian orogeny [GEOL] An orogenic era typically developed in the eastern Rocky Mountains; phases extended from Late Cretaceous until the end of the Paleocene. Also known as Laramic orogeny; Laramide orogeny; Laramide revolution.

Laray viscometer [ENG] An instrument designed to measure viscosity and other properties of ink.

larch [BOT] The common name for members of the pine genus *Larix*, with deciduous needles and short, spurlike branches which annually bear a crown of needles.

larch canker [PL PATH] A destructive fungus disease of larch and sometimes pine and fir caused by *Dasyscypha willkommii* and characterized by flat, depressed cankers on the twigs and branches.

lard [FOOD ENG] A solid fat prepared by rendering the fatty tissue from hogs.

larderillite [MINERAL] $(NH_4)B_5O_8 \cdot 2H_2O$ A white mineral composed of hydrous ammonium borate, occurring as a crystalline powder.

lard ice *See* grease ice.

lardite *See* agalmatolite.

lard oil [MATER] Yellowish to colorless oil with characteristic aroma and bland taste; melts at $-2°C$; soluble in carbon disulfide, ether, benzene, and chloroform; main components are olein and glycerides of solid fatty acids; used as a lubricant, wool oil, and illuminant, and in soap manufacture.

large calorie *See* kilocalorie.

large dyne *See* newton.

large intestine *See* colon.

large ion [METEOROL] An ion created by a small ion attaching to an Aitken nucleus; it is characterized by relatively large mass and low mobility. Also known as heavy ion; Langevin ion; slow ion.

Large Magellanic Cloud [ASTRON] An irregular cloud of stars in the constellation Doradus; it is 150,000 light-years away and nearly 30,000 light-years in diameter. Abbreviated LMC.

large nuclei [OCEANOGR] Particles of concentrated seawater

or crystalline salt in the marine atmosphere having radii larger than 10^{-5} centimeter.

large scale [MAP] A scale of sufficient size to permit the plotting of much detail with exactness. [METEOROL] A scale such that the curvature of the earth may not be considered negligible; this scale is applicable to the high tropospheric long-wave patterns, with four or five waves around the hemisphere in the middle latitudes.

large-scale convection [METEOROL] Organized vertical motion on a larger scale than atmospheric free convection associated with cumulus clouds; the patterns of vertical motion in hurricanes or in migratory cyclones are examples of such convection.

large-scale integrated circuit [ELECTR] A very complex integrated circuit, which may contain a thousand or more individual devices, such as basic logic gates and transistors, placed on a single semiconductor chip. Abbreviated LSI circuit. Also known as chip circuit; multiple-function chip.

Largidae [INV ZOO] A family of hemipteran insects in the superfamily Pentatomorpha.

Laridae [VERT ZOO] A family of birds in the order Charadriiformes composed of the gulls and terns.

Larinae [VERT ZOO] A subfamily of birds in the family Laridae containing the gulls and characterized by a thick, slightly hooked beak, a square tail, and a stout white body, with shades of gray on the back and the upper wing surface.

Larmor formula [ELECTROMAG] The rate at which energy is radiated by a nonrelativistic, accelerated charge is $2q^2a^2/3c^3$, where q is the particle's charge in esu (electrostatic units), a is its acceleration, and c is the speed of light.

Larmor frequency [ELECTROMAG] The angular frequency of the Larmor precession, equal in esu (electrostatic units) to the negative of a particle's charge times the magnetic induction divided by the product of twice the particle's mass and the speed of light.

Larmor orbit [ELECTROMAG] The motion of a charged particle in a uniform magnetic field, which is a superposition of uniform circular motion in a plane perpendicular to the field, and uniform motion parallel to the field.

Larmor precession [ELECTROMAG] A common rotation superposed upon the motion of a system of charged particles, all having the same ration of charge to mass, by a magnetic field.

Larmor radius [ELECTROMAG] For a charged particle moving transversely in a uniform magnetic field, the radius of curvature of the projection of its path on a plane perpendicular to the field. Also known as gyromagnetic radius.

Larmor's theorem [ELECTROMAG] The theorem that for a system of charged particles, all having the same ratio of charge to mass, moving in a central field of force, the motion in a uniform magnetic induction B is, to first order in B, the same as a possible motion in the absence of B except for the superposition of a common precession of angular frequency equal to the Larmor frequency.

larnite [MINERAL] β-Ca_2SiO_4 A gray mineral that is a metastable monoclinic phase of calcium orthosilicate, stable from 520 to 670°C. Also known as belite.

larry [MIN ENG] **1.** A car with a hopper bottom and adjustable chutes for feeding coke ovens. Also known as lorry. **2.** *See* barney.

larsenite [MINERAL] $PbZnSiO_4$ A colorless or white mineral composed of lead zinc silicate, occurring in orthorhombic crystals.

Larsen's pile [MIN ENG] A collection of hollow cylinders that increases resistance against bending and crumpling; useful for sinking a shaft in sand or gravel.

Larsen's spiles [MIN ENG] Steel sheet made in various forms to resist bending; used in place of wooden spiles in timbering a weak roof.

Larson-Miller parameter [MECH] The effects of time and temperature on creep which is defined empirically as $P = T(C + \log t) \times 10^{-3}$, where T = test temperature in degrees Rankine (degrees Fahrenheit + 460) and t = test time in hours; the constant C depends upon the material but is frequently taken to be 20.

larva [INV ZOO] An independent, immature, often vermiform stage that develops from the fertilized egg and must usually undergo a series of form and size changes before assuming characteristic features of the parent.

Larvacea [INV ZOO] A class of the subphylum Tunicata consisting of minute planktonic animals in which the tail, with dorsal nerve cord and notochord, persists throughout life.

Larvaevoridae [INV ZOO] The tachina flies, a large family of dipteran insects in the suborder Cyclorrhapha distinguished by a thick covering of bristles on the body; most are parasites of arthropods.

larva migrans [INV ZOO] Fly larva, *Hypoderma* or *Gastrophilus*, that produces a creeping eruption in the dermis. [MED] Infestation of the dermis by various burrowing nematode larvae, producing a creeping eruption that may become contaminated with bacteria.

larvicide [MATER] A pesticide used to kill larva.

larvikite [PETR] An alkali syenite consisting of cryptoperthite or anorthoclase in rhombic crystals; used as an ornamental building material.

larviporous [INV ZOO] Feeding on larva, referring especially to insects.

laryngeal pouch [VERT ZOO] A lateral saclike expansion of the cavity of the larynx that is greatly developed in certain monkeys.

laryngectomy [MED] Surgical removal of all or part of the larynx.

laryngismus stridulus [MED] 1. Spasmodic croup. 2. The laryngeal spasm sometimes seen in hypocalcemic states.

laryngitis [MED] Inflammation of the larynx.

laryngology [MED] The science of anatomy, physiology, and diseases of the larynx.

laryngopharynx [ANAT] The lower portion of the pharynx, lying adjacent to the larynx. Also known as hypopharynx.

laryngophone [ENG ACOUS] A microphone designed to be placed against the throat of a speaker, to pick up voice vibrations directly without responding to background noise.

laryngoscope [MED] A tubular instrument, combining a light system and a telescopic system, used in the visualization of the interior larynx and adaptable for diagnostic, therapeutic, and surgical procedures.

laryngotracheal groove [EMBRYO] A channel in the floor of the pharynx serving as the anlage of the respiratory system.

laryngotracheitis [MED] Inflammation of the larynx and trachea.

laryngotracheobronchitis [MED] Acute inflammation of the mucosa of the larynx, trachea, and bronchi.

laryngotracheobronchitis virus *See* croup-associated virus.

larynx [ANAT] The complex of cartilages and related structures at the opening of the trachea into the pharynx in vertebrates; functions in protecting the entrance of the trachea, and in phonation in higher forms.

Lasater's bubble-point pressure correction [PETRO ENG] Relation of the gas-oil ratio in a high-pressure oil reservoir to the bubble-point-pressure factor.

laser [OPTICS] A device that uses the maser principle of amplification of electromagnetic waves by stimulated emission of radiation and operates in the optical or infrared regions. Derived from light amplification by stimulated emission of radiation. Also known as optical maser.

laser altimeter [NAV] An altimeter in which a laser beam, modulated by radio frequencies, is directed downward from an aircraft, and the laser light reflected from the terrain is picked up by a telescope system, sensed by a photomultiplier, and phase-compared with the transmitted signal to obtain the round-trip propagation time; the height above ground equals one-half the product of the propagation time and the speed of light.

laser amplifier [ELECTR] A laser which is used to increase the output of another laser. Also known as light amplifier.

laser anemometer [ENG] An anemometer in which the wind being measured passes through two perpendicular light beams, and the resulting change in velocity of one or both beams is measured.

laser beam [OPTICS] A narrow beam of coherent, powerful, and nearly monochromatic electromagnetic radiation emitted by a laser.

laser camera [OPTICS] An airborne camera system for night photography in which a laser beam is split into two beams; one beam, which is almost invisible, scans the ground, while the second beam is modulated by a detector of light reflected from the ground area being scanned, and is in turn swept back and forth over a moving film by the same scanner.

laser ceilometer [ENG] A ceilometer in which the time taken by a light pulse from a ground laser to travel straight up to a cloud ceiling and be reflected to a receiving photomultiplier is measured and converted into a cathode-ray display that indicates cloud-base height.

laser detecting and ranging *See* ladar.

laser diode *See* diode laser.

laser drill [OPTICS] A drill in which concentrated light from a ruby laser generates intense heat for drilling holes as small as 0.0001 inch (0.00025 centimeter) in diameter in tungsten, gemstones, and other hard materials.

laser earthquake alarm [ENG] An early-warning system proposed for earthquakes, involving the use of two lasers with beams at right angles, positioned across a known geologic fault for continuous monitoring of distance across the fault.

laser extensometer [OPTICS] A device which uses interference of laser beams to measure small changes in distance; it can operate between points as much as 1 kilometer apart, and has been used to measure effects produced by earth tides.

laser flash tube [ELECTR] A high-power, air-cooled or water-cooled xenon flash tube designed to produce high-intensity flashes for pumping applications.

laser-induced fusion [NUCLEO] A proposed method of producing controlled nuclear fusion, in which small pellets of frozen deuterium and tritium are bombarded with short, intense bursts of radiation from one or more lasers, producing a plasma and raising it to sufficiently high temperatures and pressures for fusion to occur.

laser infrared radar *See* lidar.

laser interferometer [OPTICS] An interferometer which uses a laser as a light source; because of the monochromaticity and high intrinsic brilliance of laser light, it can operate with path differences in the interfering beams of hundreds of meters, in contrast to a maximum of about 20 centimeters for classical interferometers.

laser photocoagulator [MED] A laser combined with an ophthalmoscope for directing bursts of coherent light through a human eye to burn selected points on a detached retina; subsequent healing of the burns causes scars that weld the retina back into position.

laser radar *See* ladar.

laser range finder [OPTICS] A portable range finder using a battery-powered ruby laser in combination with an optical telescope to aim a laser beam and a photomultiplier for picking up the laser beam reflected from the target.

laser spectroscopy [SPECT] A branch of spectroscopy in which a laser is used as an intense, monochromatic light source; in particular, it includes saturation spectroscopy, as well as the application of laser sources to Raman spectroscopy and other techniques.

laser tracking [ENG] Determination of the range and direction of a target by echoed coherent light.

laser velocimeter *See* diffraction velocimeter.

laser welding [MET] Micro-spot welding with a laser beam.

lashing [ENG] A rope, chain, or wire used for binding, fastening, or wrapping. [MIN ENG] Planks nailed inside of frames or sets in a shaft to keep them in place. Also known as listing.

lashing chain [MIN ENG] A short chain to attach tubs to an overrope in endless rope haulage by wrapping the chain around the rope.

lash-up [ENG] A model or test sample of equipment required in the testing of a new concept or idea which is in the embryo stage.

lasing [OPTICS] Generation of visible or infrared light waves having very nearly a single frequency by pumping or exciting electrons into high-energy states in a laser.

Lasiocampidae [INV ZOO] The tent caterpillars and lappet moths, a family of cosmopolitan (except New Zealand) lepidopteran insects in the suborder Heteroneura.

lasso cell *See* adhesive cell.

Lastex [TEXT] Trade name of an elastic yarn covered with cotton, nylon, or other fiber, manufactured by Uniroyal, Inc.

last in, first out [IND ENG] A method of determining the inventory costs by transferring the costs of material to the

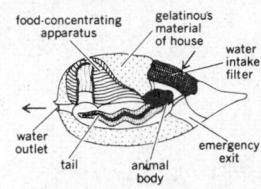

LARVACEA

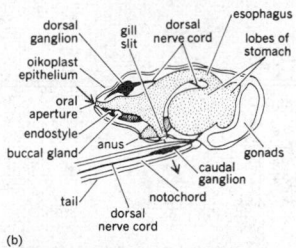

Oikopleura dioica, a larvacean. *(a)* In its house, a gelatinous tunic with strainers to filter food. *(b)* Body and anterior tail region. Arrows indicate the direction of water flow. *(Modified after Körner)*

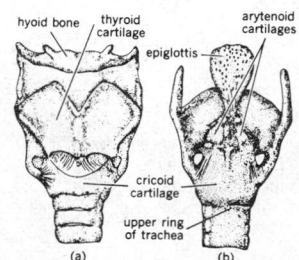

LARYNX

Human laryngeal cartilages and ligaments. *(a)* Front view. *(b)* Back view. *(After Sappey, from J. Symington, ed., Quain's Elements of Anatomy, vol. 2, pt. 2, Longmans, Green, 1914)*

LASER

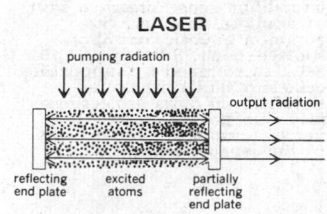

Structure of a parallel-plate laser.

product in reverse chronological order. Abbreviated LIFO.

last quarter [ASTRON] The phase of the moon at western quadrature, half of the illuminated hemisphere being visible from the earth; has the characteristic half-moon shape.

latch [ENG] Any of various closing devices on a door that fit into a hook, notch, or cavity in the frame. [MIN ENG] To make an underground survey with a dial and chain, or to mark out upon the surface, with the same instruments, the position of the workings underneath.

latch bolt [DES ENG] A self-acting spring bolt with a beveled head.

latching relay [ELEC] A relay having contacts that lock in either the energized or the de-energized position or both, until reset manually or electrically. Also known as latch-in relay; locking relay; lock-up relay.

latch-in relay See latching relay.

late blight [PL PATH] A fungus blight disease in which symptoms do not appear until late in the growing season and vary for different species.

latency [ADP] The waiting time between the order to read/write some information from/to a specified place and the beginning of the data-read/write operation. [MED] The stage of an infectious disease, other than the incubation period, in which there are no clinical signs or symptoms. [PHYSIO] The period between the introduction of and the response to a stimulus. [PSYCH] The phase between the Oedipal period and adolescence, characterized by an apparent cessation of psychosexual development.

latent bud [BOT] An axillary bud whose development is inhibited, sometimes for many years, due to the influence of apical and other buds. Also known as dormant bud.

latent defect [IND ENG] A flaw or other imperfection in any article which is discovered after delivery; usually, latent defects are inherent weaknesses which normally are not detected by examination or routine tests, but which are present at time of manufacture and are aggravated by use.

latent heat [THERMO] The amount of heat absorbed or evolved by 1 mole, or a unit mass, of a substance during a change of state (such as fusion, sublimation or vaporization) at constant temperature and pressure.

latent heat of fusion See heat of fusion.

latent heat of sublimation See heat of sublimation.

latent heat of vaporization See heat of vaporization.

latent Herschel effect See Herschel effect.

latent homosexuality [PSYCH] Homosexual tendencies present in the unconscious but not felt or expressed overtly.

latent image [GRAPHICS] An invisible image produced by the physical or chemical effects of light on the individual crystals (usually silver halide) of photographic emulsions; the development process makes the image visible, in the negative.

latent instability [METEOROL] The state of that portion of a conditionally unstable air column lying above the level of free convection; latent instability is released only if an initial impulse on a parcel gives it sufficient kinetic energy to carry it through the layer below the level of free convection, within which the environment is warmer than the parcel.

latent load [MECH ENG] Cooling required to remove unwanted moisture from an air-conditioned space.

latent period [MED] Any stage of an infectious disease in which there are no clinical signs of symptoms of the infection. [PHYSIO] The period between the introduction of a stimulus and the response to it. [VIROL] The initial period of phage growth after infection during which time virus nucleic acid is manufactured by the host cell.

latent-virus infection [MED] A chronic, inapparent virus infection in which a virus-host equilibrium is established.

laterad [ANAT] Toward the lateral aspect.

lateral [ANAT] At, pertaining to, or in the direction of the side; on either side of the medial vertical plane. [MIN ENG] **1.** In horizon mining, a hard heading branching off a horizon along the strike of the seams. **2.** A horizontal mine working.

lateral aberration [OPTICS] **1.** The distance from the axis of an optical system at which a ray intersects a plane perpendicular to the axis through the focus of paraxial rays. **2.** The difference between the reciprocals of the image distances for paraxial and rim rays. **3.** For chromatic aberration, the

difference in sizes of the images of an object for two different colors.

lateral acceleration [AERO ENG] The component of the linear acceleration of an aircraft or missile along its lateral, or Y, axis.

lateral accretion [GEOL] The digging away of material at the outer bank of a meandering stream and the simultaneous building up to the water level by deposition of material brought there by pushing and rolling along the stream bottom.

lateral area [MATH] The area of a surface with the bases (if any) excluded.

lateral axis [NAV ARCH] The athwartship line approximately through the center of gravity of a craft, around which it pitches.

lateral bud [BOT] Any bud that develops on the side of a stem.

lateral compliance [ENG ACOUS] That characteristic of a stylus based on the force required to move it from side to side as it follows the grooves of a phonograph record.

lateral cone See adventive cone.

lateral controller [AERO ENG] A primary flight control mechanism, generally a part of the longitudinal controller, which controls the ailerons; often resembles an automobile steering wheel but may be a control column.

lateral deflection angle [ORD] The horizontal angle representing the difference between the azimuth of the target at the instant of firing and the azimuth at which the gun must be pointed in order to hit the target; it is the algebraic sum of the principal lateral deflection angle and the lateral pointing correction.

lateral deflection setting [ORD] The setting on the lateral deflection scale of the sighting mechanism of the gun, corresponding to the lateral deflection angle.

lateral deviation [ORD] Horizontal distance between the point of impact or burst and the gun-target line.

lateral displacement [NAV] The number of nautical miles an aircraft has been displaced, perpendicular to its effective air path, determined by a pressure pattern formula. Also known as lateral drift; pressure pattern displacement.

lateral drift See lateral displacement.

lateral erosion [GEOL] The action of a stream in undermining a bank on one side of its channel so that material falls into the stream and disintegrates; simultaneously, the stream shifts toward the bank that is being undercut.

lateral extensometer [ENG] An instrument used in photoelastic studies of the stresses on a plate; it measures the change in the thickness of the plate resulting from the stress at various points.

lateral face [MATH] The lateral face for a prism or pyramid is any edge or face which is not part of a base.

lateral flow spillway See side-channel spillway.

lateral hermaphroditism [MED] A form of human hermaphroditism in which there is an ovary on one side and a testis on the other.

lateral jump [ORD] The horizontal angle between the plane of fire and the plane of departure.

lateral lemniscus [ANAT] The secondary auditory pathway arising in the cochlear nuclei and terminating in the inferior colliculus and medial geniculate body.

lateral line [INV ZOO] A longitudinal lateral line along the sides of certain oligochaetes consisting of cell bodies of the layer of circular muscle. [VERT ZOO] A line along the sides of the body of most fishes, often distinguished by differently colored scales, which marks the lateral line organ.

lateral-line organ [VERT ZOO] A small, pear-shaped sense organ in the skin of many fishes and amphibians that is sensitive to pressure changes in the surrounding water.

lateral-line system [VERT ZOO] The complex of lateral-line end organs and nerves in skin on the sides of many fishes and amphibians.

lateral magnification [OPTICS] The ratio of some linear dimension, perpendicular to the optical axis, of an image formed by an optical system, to the corresponding linear dimension of the object. Also known as magnification.

lateral meristem [BOT] Strips or cylinders of dividing cells located parallel to the long axis of the organ in which they

LATERAL-LINE ORGAN

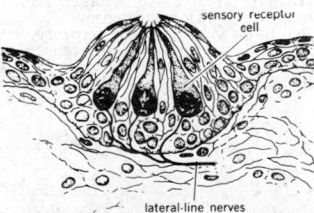

sensory receptor cell

lateral-line nerves

Schematic drawing of a lateral-line sense organ in skin of adult salamander, the common aquatic vermilion spotted newt. Sensory receptor cells, surrounded by supporting cells and skin epithelium, terminate in hairs projecting into skin pores. Lateral-line nerves terminate around bases of the sensory cells.

occur; the lateral meristem functions to increase the diameter of the organ.

lateral mirage [OPTICS] A very rare type of mirage in which the apparent position of an object appears displaced to one side of its true position.

lateral moraine [GEOL] Drift material, usually thin, that was deposited by a glacier in a valley after the glacier melted.

lateral observation [ORD] Observation of gunfire from a point considerably to the right or left of the line of fire.

lateral parity check [ADP] The number of one bits counted across the width of the magnetic tape; this number plus a one or a zero must always be odd (or even), depending upon the manufacturer.

lateral planation [GEOL] Reduction in land in interstream areas in a plane parallel to the stream profile; the reduction is caused by lateral movement of the stream against its banks.

lateral pointing correction [ORD] That part of the lateral deflection angle due to causes other than the travel of the target, such as wind, drift, and lateral adjustment correction.

lateral quadrupole [ACOUS] A sound source resulting from a variation of a component of the velocity of matter in a direction that is perpendicular to the velocity component. [ELECTROMAG] An electric or magnetic quadrupole which produces a field equivalent to that of two equal and opposite electric or magnetic dipoles separated by a small distance perpendicular to the direction of the dipoles.

lateral recording [ENG ACOUS] A type of disk recording in which the groove modulation is parallel to the surface of the recording medium so that the cutting stylus moves from side to side during recording.

lateral root [BOT] A root branch arising from the main axis.

lateral sclerosis See amyotrophic lateral sclerosis; primary lateral sclerosis.

lateral secretion [GEOL] A supposed phenomenon whereby a lode's or vein's mineral content is derived from the adjacent wall rock.

lateral separation [NAV] In air-traffic control, the separation between aircraft on parallel course lines.

lateral sewer [CIV ENG] A sewer discharging into a branch or other sewer and having no tributary sewer.

lateral support [CIV ENG] Horizontal propping applied to a column, wall, or pier across its smallest dimension.

lateral system [NAV] In marine operations, a system of buoyage in which the shape, color, and number distinction are assigned in accordance with the buoy's location in respect to navigable waters; when used to mark a channel, the buoys are assigned colors to indicate the side they mark and numbers to indicate their sequence along the channel; the lateral system is used in the United States.

lateral thrombus See mural thrombus.

lateral ventricle [ANAT] The cavity of a cerebral hemisphere; communicates with the third ventricle by way of the interventricular foramen.

laterite [GEOL] Weathered material composed principally of the oxides of iron, aluminum, titanium, and manganese; laterite ranges from soft, earthy, porous soil to hard, dense rock.

lateritic soil [GEOL] **1.** Soil containing laterite. **2.** Any reddish soil developed from weathering. Also known as latosol.

laterization [GEOL] Those conditions of weathering that lead to removal of silica and alkalies, resulting in a soil or rock with high concentrations of iron and aluminum oxides (laterite).

laterlog [ENG] A downhole resistivity measurement method wherein electric current is forced to flow radially through the formation in a sheet of predetermined thickness; used to measure the resistivity in hard-rock reservoirs as a method of determining subterranean structural features.

latex [MATER] **1.** Milky colloid in which natural or synthetic rubber or plastic is suspended in water. **2.** An elastomer product made from latex.

latex cement [MATER] A highly adhesive solvent solution of latex.

latex paint [MATER] A paint consisting of a water suspension or emulsion of latex combined with pigments and additives such as binders and suspending agents. Also known as latex water paint.

latex water paint See latex paint.

lath [CIV ENG] **1.** A narrow strip of wood used in making a level base, as for plaster or tiles, or in constructing a light framework, as a trellis. **2.** A sheet of material used as a base for plaster.

lath brick [MATER] A long, narrow brick.

lath crib See lath frame.

lath door-set See lath frame.

lathe [MECH ENG] A machine for shaping a workpiece by gripping it in a holding device and rotating it under power against a suitable cutting tool for turning, boring, facing, or threading.

lath frame [MIN ENG] A weak construction of laths, surrounding a main crib, the space between being for the insertion of piles. Also known as lath crib; lath door-set.

Lathridiidae [INV ZOO] The minute brown scavenger beetles, a large cosmopolitan family of coleopteran insects in the superfamily Cucujoidea.

lathyrism [MED] Poisoning produced by ingestion of vetch (*Lathyrus*) and characterized by spastic paraplegia and decreased connective-tissue tensile strength. Also known as neurolathyrism.

laticifer [BOT] A latex duct found in the mid-cortex of certain plants.

latiferous [BOT] Containing or secreting latex.

Latimeridae [VERT ZOO] A family of deep-sea lobefin fishes (Coelacanthiformes) known from a single living species, *Latimeria chalumnae*.

Latin square [MATH] An $n \times n$ square array of n different symbols, each symbol appearing once in each row and once in each column; these symbols prove useful in ordering the observations of an experiment.

latite [PETR] A not visibly crystalline rock of volcanic origin composed chiefly of sodic plagioclase and alkali feldspar with subordinate quantities of dark-colored minerals in a finely crystalline to glassy groundmass.

latitude [GEOD] Angular distance from a primary great circle or plane, as on the celestial sphere or the earth.

latitude effect [GEOPHYS] The variation of a quantity with latitude; applied particularly to the increase in cosmic-ray intensity with increasing magnetic latitude.

latitude factor [NAV] The change in latitude along a celestial line of position per 1-minute change in longitude.

latitude line [NAV] A line of position extending in an approximate east-west direction.

latosol See lateritic soil.

latrappite [MINERAL] $(Ca,Na)(Nb,Ti,Fe)O_3$ A variety of the mineral perovskite.

latrine [ENG] A toilet facility, either fixed or of a portable nature, such as is maintained underground for use by miners.

latrine cleaner [MIN ENG] A laborer who brings toilet cars in a mine to the surface on a cage and flushes the contents into a sewer. Also called sanitary nipper.

latten [MET] A thin metal sheet, particularly of brass or similar alloy, hot-rolled steel, or tin-covered iron, used for ornamental purposes.

lattice [CIV ENG] A network of crisscrossed strips of metal or wood. [CRYSTAL] A regular periodic arrangement of points in three-dimensional space; it consists of all those points P for which the vector from a given fixed point to P has the form $n_1 \mathbf{a} + n_2 \mathbf{b} + n_3 \mathbf{c}$, where n_1, n_2, and n_3 are integers, and $\mathbf{a}$, $\mathbf{b}$, and $\mathbf{c}$ are fixed, linearly independent vectors. Also known as periodic lattice; space lattice. [MATH] A partially ordered set in which each pair of elements has both a greatest lower bound and least upper bound. [NAV] A pattern formed by two or more families of intersecting lines of position, such as the hyperbolic lines of position from two or more loran stations. [NUCLEO] An orderly array or pattern of nuclear fuel elements and moderator in a reactor or critical assembly.

lattice constant [CRYSTAL] A parameter defining the unit cell of a crystal lattice, that is, the length of one of the edges of the cell or an angle between edges. Also known as lattice parameter.

lattice defect See crystal defect.

lattice drainage pattern See rectangular drainage pattern.

lattice dynamics [SOLID STATE] The study of the thermal

LATIMERIDAE

Living coelacanth, *Latimeria chalumnae*; 5 feet (1.5 meters). (*After P. P. Grasse, ed., Traite de Zoologie, tome 13, fasc. 3, 1958*)

vibrations of a crystal lattice. Also known as crystal dynamics.

lattice energy [SOLID STATE] The energy required to separate ions in an ionic crystal an infinite distance from each other.

lattice filter [ELECTR] An electric filter consisting of a lattice network whose branches have L-C parallel-resonant circuits shunted by quartz crystals.

lattice girder [CIV ENG] An open girder, beam, or column built from members joined and braced by intersecting diagonal bars. Also known as open-web girder.

lattice network [ELEC] A network that is composed of four branches connected in series to form a mesh; two nonadjacent junction points serve as input terminals, and the remaining two junction points serve as output terminals.

lattice parameter See lattice constant.

lattice polarization [SOLID STATE] Electric polarization of a solid due to displacement of ions from equilibrium positions in the lattice.

lattice reactor [NUCLEO] A heterogeneous nuclear reactor in which both fuel and moderator are in the form of long rods.

lattice scattering [SOLID STATE] Scattering of electrons by collisions with vibrating atoms in a crystal lattice, reducing the mobility of charge carriers in the crystal and thereby affecting its conductivity.

lattice truss [CIV ENG] A truss that resembles latticework because of diagonal placement of members connecting the upper and lower chords.

lattice vibration [SOLID STATE] A periodic oscillation of the atoms in a crystal lattice about their equilibrium positions.

lattice wave [SOLID STATE] A disturbance propagated through a crystal lattice in which atoms oscillate about their equilibrium positions.

lattice winding [ELEC] A winding made of lattice coils and used for electric machines.

lattice-wound coil See honeycomb coil.

lattissimus dorsi [ANAT] The widest muscle of the back; a broad, flat muscle of the lower back that adducts and extends the humerus, is used to pull the body upward in climbing, and is an accessory muscle of respiration.

Lattorfian See Tongrian.

latus rectum [MATH] The length of a chord through the focus and perpendicular to the axis of symmetry in a conic section.

lauan [MATER] Any of several light-yellow to reddish-brown, moderately close-grained, and stringy timbers having moderate strength and durability, obtained from several trees of the Philippines, for example, Philippine mahogany.

laubmannite [MINERAL] $Fe_3Fe_6(PO_4)_4(OH)_2$ Mineral composed of basic ferrous iron phosphate and ferric iron phosphate.

laudanidine [ORG CHEM] $C_{20}H_{25}NO_4$ An optically active alkaloid found in opium that crystallizes as prisms from an alcohol solution, and melts at 185°C. Also known as l-laudanine; tritopine.

laudanine [ORG CHEM] $C_{20}H_{25}NO_4$ An optically inactive alkaloid derived from alkaline mother liquors from morphine extraction; it crystallizes in orthorhombic prisms from alcohol and chloroform; the prisms melt at 167°C, and are soluble in hot alcohol, benzene, and chloroform. Also known as dl-laudanine.

dl-laudanine See laudanine.

l-laudanine See laudanidine.

laudanosine [ORG CHEM] $C_{21}H_{27}NO_4$ An alkaloid that is the methyl ether of laudanine; the optically inactive form crystallizes from dilute alcohol and melts at about 115°C; the levorotatory active form crystallizes from light petroleum solution and melts at 89°C.

Laue camera [CRYSTAL] The apparatus used in the Laue method; the x-ray beam usually enters through a hole in the x-ray film, which records beams bent through an angle of nearly 180° by the crystal; less commonly, the film is placed beyond the crystal.

Laue condition [CRYSTAL] **1.** The condition for a vector to lie in a Laue plane: its scalar product with a specified vector in the reciprocal lattice must be one-half of the scalar product of the latter vector with itself. **2.** See Laue equations.

Laue equations [CRYSTAL] Three equations which must be satisfied for an x-ray beam of specified wavelength to be

diffracted through a specified angle by a crystal; they state that the scaler products of each of the crystallographic axial vectors with the difference between unit vectors in the directions of the incident and scattered beams, are integral multiples of the wavelength. Also known as Laue condition.

Laue method [CRYSTAL] A method of studying crystalline structures by x-ray diffraction, in which a finely collimated beam of polychromatic x-rays falls on a single crystal whose orientation can be set as desired, and diffracted beams are recorded on a photographic film.

Laue pattern [CRYSTAL] The characteristic photographic record obtained in the Laue method.

Laue plane [CRYSTAL] A plane which is the perpendicular bisector of a vector in the reciprocal lattice; such planes form the boundaries of Brillouin zones.

Laue theory [CRYSTAL] A theory of diffraction of x-rays by crystals, based on the Laue equations.

laughing gas See nitrous oxide.

Laugiidae [PALEON] A family of Mesozoic fishes in the order Coelacanthiformes.

laumonite See laumontite.

laumontite [MINERAL] $CaAl_2Si_4O_{12} \cdot 4H_2O$ A white zeolite mineral crystallizing in the monoclinic system; loses water on exposure to air, eventually becoming opaque and crumbling. Also known as laumonite; lomonite; lomontite.

launch [AERO ENG] **1.** To send off a rocket vehicle under its own rocket power, as in the case of guided aircraft rockets, artillery rockets, and space vehicles. **2.** To send off a missile or aircraft by means of a catapult or by means of inertial force, as in the release of a bomb from a flying aircraft. **3.** To give a space probe an added boost for flight into space just before separation from its launch vehicle.

launch complex [AERO ENG] The composite of facilities and support equipment needed to assemble, check out, and launch a rocket vehicle.

launcher [ORD] **1.** A device designed to support and launch a rocket or rocket shell. **2.** A device on a rifle for firing a grenade.

launching [CIV ENG] The act or process of floating a ship after only hull construction is completed; in some cases ships are not launched until after all construction is completed. [ELECTROMAG] The process of transferring energy from a coaxial cable or transmission line to a waveguide.

launching angle [AERO ENG] The angle between the horizontal plane and the longitudinal axis of a rocket or missile at the time of launching.

launching cradle [CIV ENG] A framework made of wood to support a vessel during launching from sliding ways.

launching ramp [AERO ENG] A ramp used for launching an aircraft or missile into the air.

launching site [AERO ENG] **1.** A site from which launching is done. **2.** The platform, ramp, rack, or other installation at such a site.

launching tube [ORD] A tube used to guide a rocket missile or other projectile during launching.

launching ways [CIV ENG] Two (or more) sets of long, heavy timbers arranged longitudinally under the bottom of a ship during building and launching, with one set on each side, and sloping toward the water; the lower set, or ground ways, remain stationary and support the upper set, or sliding ways, which carry the weight of the ship after the shores and keel blocks are removed.

launch pad [AERO ENG] The load-bearing base or platform from which a rocket vehicle is launched. Also known as pad.

launch vehicle [AERO ENG] A rocket or other vehicle used to launch a probe, satellite, or the like. Also known as booster.

launch window [AERO ENG] The time period during which a spacecraft or missile must be launched in order to achieve a desired encounter, rendezvous, or impact.

launder [ENG] An inclined channel or trough for the conveyance of a liquid, such as for water in mining and construction engineering or for molten metal.

launder screen [MIN ENG] A screen used in a launder for the sizing and dewatering of small sizes of anthracite.

launder separation process [MIN ENG] A hydraulic process for separating heavy gravity product from the lighter product that flows above it; a stream of fluid carries material down a

LAUE CAMERA

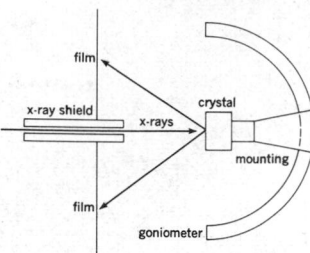

Schematic diagram of back-reflection Laue camera. Orientation of single crystal can be set as desired by system of goniometer circles. Diffracted beams, bent through angles of nearly 180°, are recorded on film.

LAUNCH COMPLEX

Aerial view of the Titan 3 complex at Cape Canaveral, Florida.

channel that has draws to separate the heavy material from the light material.

launder washer [MIN ENG] A type of coal washer in which the coal is separated from the refuse by stratification due to hindered settling while being carried in aqueous suspension through the launder.

laundry blue [MATER] Dye-containing solution used to give a blue tint to laundry-yellowed white cottons and linens; usually contains Prussian blue.

lauoho o pele *See* Pele's hair.

Lauraceae [BOT] The laurel family of the order Magnoliales distinguished by definite stamens in series of three, a single pistil, and the lack of petals.

lauraldehyde *See* lauryl aldehyde.

Laurasia [GEOL] A continent theorized to have existed in the Northern Hemisphere; supposedly it broke up to form the present northern continents about the end of the Pennsylvania period.

Lauratonematidae [INV ZOO] A family of marine nematodes of the superfamily Enoploidea; many females possess a cloaca.

laurel forest *See* temperate rainforest.

laurel wax *See* bayberry wax.

laurence [OPTICS] A shimmering seen over a hot surface on a calm, cloudless day, caused by the unequal refraction of light by innumerable convective air columns of different temperatures and densities.

Laurence-Moon-Biedl syndrome [MED] A hereditary endocrine disorder of the pituitary or other hypothalamic structures transmitted as a dominant mutant gene and characterized principally by girdle-type obesity, mental retardation, and hypogenitalism.

Laurent expansion [MATH] An infinite series in which an analytic function $f(z)$ defined on an annulus about the point z_0 may be expanded, with nth term $a_n(z - z_0)^n$, n ranging from $-\infty$ to ∞, and $a_n = 1/2\pi i$ times the integral of $f(t)/(t - z_0)^{n+1}$ along a simple closed curve interior to the annulus. Also known as Laurent series.

Laurent half-shade plate [OPTICS] A device used to determine the direction of polarization of plane polarized light; it consists of a quartz plate of special thickness that covers half of the plane polarized beam, followed by a plane polarization analyzer.

Laurentian Plateau *See* Laurentian Shield.

Laurentian Shield [GEOL] A Precambrian plateau extending over half of Canada from Labrador southwest along Hudson Bay and northwest to the Arctic Ocean. Also known as Canadian Shield; Laurentian Plateau.

Laurentide ice sheet [HYD] A major recurring glacier that at its maximum completely covered North America east of the Rockies from the Arctic Ocean to a line passing through the vicinity of New York, Cincinnati, St. Louis, Kansas City, and the Dakotas.

Laurent series *See* Laurent expansion.

Laurer's canal [INV ZOO] An accessory canal connecting the female reproductive tract to the outside in Digenoidea and Aspidobothroidea.

lauric acid [ORG CHEM] $CH_3(CH_2)_{10}COOH$ A fatty acid melting at 44°C, boiling at 225°C (100 mm Hg); colorless needles soluble in alcohol and ether, insoluble in water; found as the glyceride in vegetable fats, such as coconut and laurel oils; used for wetting agents, in cosmetics, soaps, resins, and insecticides, and as a chemical intermediate. Also known as dodecanoic acid.

lauric aldehyde *See* lauryl aldehyde.

laurionite [MINERAL] $Pb(OH)Cl$ A colorless mineral composed of basic lead chloride, occurring in prismatic crystals; it is dimorphous with paralaurionite.

laurisilva *See* temperate rainforest.

laurite [MINERAL] RuS_2 A black mineral composed of ruthenium sulfide (often with osmium), occurring as small crystals or grains.

Lauritsen electroscope [ELEC] A rugged and sensitive electroscope in which a metallized quartz fiber is the sensitive element.

lauryl alcohol [ORG CHEM] $CH_3(CH_2)_{11}OH$ A colorless solid which is obtained from coconut oil fatty acids, has a floral odor, and boils at 259°C; used in detergents, lubricating oils, and pharmaceuticals. Also known as alcohol C-12; *n*-dodecanol; dodecyl alcohol.

lauryl aldehyde [ORG CHEM] $CH_3(CH_2)_{10}CHO$ A constituent of an essential oil from the silver fir; a colorless solid or a liquid, with a floral odor, that is soluble in 90% alcohol; used in perfumes. Also known as aldehyde C-12 lauric; lauraldehyde; lauric aldehyde.

lauryl mercaptan [ORG CHEM] $C_{12}H_{25}SH$ Pale-yellow or water-white liquid with mild odor; insoluble in water, soluble in organic solvents; used to manufacture plastics, pharmaceuticals, insecticides, fungicides, and elastomers.

Lauson engine [ENG] Single-cylinder engine used in screening tests prior to the L-series lube oil tests (such as L-1 or L-2 tests).

lautal [MET] A hard aluminum alloy with small percentages of copper and silicon and traces of iron, manganese, or magnesium

lautarite [MINERAL] $Ca(IO_3)_2$ A monoclinic mineral composed of calcium iodate that occurs in prismatic crystals.

lauter tub [FOOD ENG] A tank with a perforated false bottom in which the extract (wort) is separated from the insoluble grain residue during mashing of barley malt.

lautite [MINERAL] $CuAsS$ A mineral composed of copper sulfide and copper arsenide.

Lauxaniidae [INV ZOO] A family of myodarian cyclorrhaphous dipteran insects in the subsection Acalypteratae; larvae are leaf miners.

lava [GEOL] **1.** Molten extrusive material that reaches the earth's surface through volcanic vents and fissures. **2.** The rock mass formed by consolidation of molten rock issuing from volcanic vents and fissures, consisting chiefly of magnesium silicate; used for insulators.

lava blisters [GEOL] Small, steep-sided swellings that are hollow and raised on the surfaces of some basaltic lava flows; formed by gas bubbles pushing up the lava's viscous surface.

lava cone [GEOL] A volcanic cone that was formed of lava flows.

lava dome *See* shield volcano.

lava field [GEOL] A wide area of lava flow; it is commonly several square kilometers in area and forms along the base of a large compound volcano or on the flanks of shield volcanoes.

lava flow [GEOL] **1.** A lateral, surficial stream of molten lava issuing from a volcanic cone or from a fissure. **2.** The solidified mass of rock formed when a lava stream congeals.

lava fountain [GEOL] A jetlike eruption of lava that issues vertically from a volcanic vent or fissure. Also known as fire fountain.

lava lake [GEOL] A lake of lava that is molten and fluid; usually contained within a summit volcanic crater or in a pit crater on the flanks of a shield volcano.

Laval nozzle *See* de Laval nozzle.

lava plateau [GEOL] An elevated tableland or flat-topped highland that is several hundreds to several thousands of square kilometers in area; underlain by a thick succession of lava flows.

lava tube [GEOL] A long, tubular opening under the crust of solidified lava.

lavender oil [MATER] Colorless to yellow or green-yellow essential oil with sweet aroma and bitter taste; distilled from fresh flowers of several species of lavender (*Lavandula*); main components are linalool, linalyl acetate, geraniol, cumarin, furfurol, and borneol; used in perfumery, and in medicine as a stimulant.

lavenite [MINERAL] $(Na,Ca)_3Zr(Si_2O_7)(O,OH,F)_2$ A mineral composed of complex silicate, occurring in prismatic crystals.

Laves phases [MET] Alloy phases which have the general formula AB_2, and the crystal structures of either $MgCu_2$ (cubic) or $MgZn_2$ or $MgNi_2$ (both hexagonal).

law [SCI TECH] A regularity which applies to all members of a broad class of phenomena.

lawn [TEXT] A sheer cotton or cotton and polyester fabric made of combed or carded yarn.

lawnmower [ELECTR] Type of radio-frequency preamplifier used with radar receivers. [ENG] A helix-type recorder

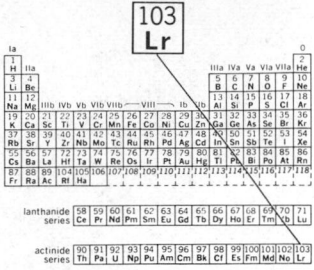

LAWRENCIUM

Periodic table of the chemical elements showing the position of lawrencium.

mechanism. [MECH ENG] A machine for cutting grass on lawns.

law of action and reaction *See* Newton's third law.

law of constant angles [CRYSTAL] The law that the angles between the faces of a crystal remain constant as the crystal grows.

law of constant heat summation *See* Hess's law.

law of corresponding states [CHEM] The law that when, for two substances, any two ratios of pressure, temperature, or volume to their respective critical properties are equal, the third ratio must equal the other two.

law of cosines [MATH] Given a triangle with angles A, B, and C and sides a, b, c opposite these angles respectively: $a^2 = b^2 + c^2 - 2bc \cos A$.

law of definite composition *See* law of definite proportion.

law of definite proportion [CHEM] The law that a given chemical compound always contains the same elements in the same fixed proportion by weight. Also known as law of definite composition.

law of electric charges [ELEC] The law that like charges repel, and unlike charges attract.

law of electromagnetic induction *See* Faraday's law of electromagnetic induction.

law of electrostatic attraction *See* Coulomb's law.

law of exponents [MATH] Any one of the laws $a^m a^n = a^{m+n}$, $a^m/a^n = a^{m-n}$, $(a^m)^n = a^{mn}$, $(ab)^n = a^n b^n$, $(a/b)^n = a^n/b^n$; these laws are valid when m and n are any integers, or when a and b are positive and m and n are any real numbers. Also known as exponential law.

law of extralateral rights *See* apex law.

law of gravitation *See* Newton's law of gravitation.

law of large numbers [STAT] The law that if, in a collection of independent identical experiments, $N(B)$ represents the number of occurrences of an event B in n trials, and p is the probability that B occurs at any given trial, then for large enough n it is unlikely that $N(B)/n$ differs from p by very much. Also known as Bernoulli theorem.

law of magnetism [ELECTROMAG] The law that like poles repel, and unlike poles attract.

law of mass action [CHEM] The law stating that the rate at which a chemical reaction proceeds is directly proportional to the molecular concentrations of the reacting compounds.

law of minimum [BIOL] The law that those essential elements for which the ratio of supply to demand (A/N) reaches a minimum will be the first to be removed from the environment by life processes; it was proposed by J. Von Liebig, who recognized phosphorus, nitrogen, and potassium as minimum in the soil; in the ocean the corresponding elements are phosphorus, nitrogen, and silicon. Also known as Liebig's law of minimum.

law of parallel solenoids [PHYS] The law that under stationary conditions isopycnals and isobars must be parallel at all levels, and isobars and isopycnals at one level must be parallel to those at all other levels.

law of partial pressures *See* Dalton's law.

law of rational intercepts *See* Miller law.

law of reflection *See* reflection law.

law of signs [MATH] The product or quotient of two numbers is positive if the numbers have the same sign, negative if they have opposite signs.

law of sines [MATH] Given a triangle with angles A, B, and C and sides a, b, c opposite these angles respectively: $\sin A/a = \sin B/b = \sin C/c$.

law of specificity of bacteria *See* Koch's postulate.

law of storms [METEOROL] Historically, the general statement of the manner in which the winds of a cyclone rotate about the cyclone's center, and the way that the entire disturbance moves over the earth's surface.

law of superposition [GEOL] The law that strata underlying other strata must be the older if there has been neither overthrust nor inversion.

law of tangents [MATH] Given a triangle with angles A, B, and C and sides a, b, c opposite these angles respectively: $(a - b)/(a + b) = [\tan \frac{1}{2}(A - B)]/[\tan \frac{1}{2}(A + B)]$.

Lawrence tube *See* chromatron.

lawrencite [MINERAL] $(Fe,Ni)Cl_2$ A brown or green mineral composed of ferrous chloride and found as an abundant accessory mineral in iron meteorites.

lawrencium [CHEM] A chemical element, symbol Lr, atomic number 103; two isotopes have been discovered, mass number 257 or 258 and mass number 256.

laws of refraction *See* Snell laws of refraction.

Lawson criterion [PL PHYS] The requirement for the energy produced by fusion in a plasma to exceed that required to produce the confined plasma; it states that for a mixture of deuterium and tritium in the temperature range from 1×10^8 to 5×10^8 degrees Celsius, the product of the ionic density and the confinement time must be about 10^{14} seconds per cubic centimeter.

lawsonite [MINERAL] $CaAl_2(Si_2O_7)(OH)_2 \cdot H_2O$ A colorless or grayish-blue mineral crystallizing in the orthorhombic system; found in gneisses and schists.

laxative [PHARM] An agent that stimulates bowel movement and relieves constipation.

lay [DES ENG] The direction, length, or angle of twist of the strands in a rope or cable. [MET] The direction of the prevailing surface pattern on a piece of metal after grinding, cutting, lapping, or other processing. [MIN ENG] A share of profit.

lay-by [MIN ENG] Siding in single-track underground tramming road.

layer [GEOL] A tabular body of rock, ice, sediment, or soil lying parallel to the supporting surface and distinctly limited above and below. [GEOPHYS] One of several strata of ionized air, some of which exist only during the daytime, occurring at altitudes between 50 and 400 kilometers; the layers reflect radio waves at certain frequencies and partially absorb others. [MET] The stratum of weld metal consisting of one or more passes and lying parallel to the welding surface.

layer capacitance *See* cathode interface capacitance.

layer depth [OCEANOGR] 1. The thickness of the mixed layer in an ocean. 2. The depth to the top of the thermocline.

layer impedance *See* cathode interface impedance.

layering [BOT] A propagation method by which root formation is induced on a branch or a shoot attached to the parent stem by covering the part with soil. [ECOL] A stratum of plant forms in a community, such as mosses, shrubs, or trees in a bog area.

layering of firedamp [MIN ENG] The formation of a layer of firedamp at the roof of a mine working and above the ventilating air current.

layer loading [MIN ENG] A procedure whereby the coal is placed in the railroad cars in horizontal layers.

layer of no motion [OCEANOGR] A layer, assumed to be at rest, at some depth in the ocean.

layer silicate *See* phyllosilicate.

layer winding [ELEC] Coil-winding method in which adjacent turns are laid evenly and side by side along the length of the coil form; any number of additional layers may be wound over the first, usually with sheets of insulating material between the layers.

lay off [ENG] The process of fairing a ship's lines or an airplane's in a mold loft in order to make molds and templates for structural units. [NAV] The act of steering a ship away from the shore, a pier, or another ship.

layout [GRAPHICS] A design drawing or graphical statement of the overall form of a component, system, or device, which is usually prepared during innovative stages of a design.

lay-up [ENG] Production of reinforced plastics by positioning the reinforcing material (such as glass fabric) in the mold prior to impregnation with resin.

lazaret [NAV ARCH] A space between decks above the afterpeak.

lazuli *See* lapis lazuli.

lazulite [MINERAL] $(Mg,Fe)Al_2(OH)_2(PO_4)_2$ A violet-blue or azure-blue mineral with vitreous luster; composed of basic aluminum phosphate and occurring in small masses or monoclinic crystals; hardness is 5-6 on Mohs scale, and specific gravity is 3.06-3.12. Also known as berkeyite; blue spar; false lapis.

lazurite [MINERAL] $(Na,Ca)_8(Al,Si)_{12}O_{24}(S,SO_4)$ A blue or violet-blue feldspathoid mineral crystallizing in the isometric system; the chief mineral constituent of lapis lazuli.

LAZURITE

├── 12.7 cm ──┤

Lazurite from Afghanistan. *(Specimen from Department of Geology, Bryn Mawr College)*

lazy H antenna [ELECTROMAG] An antenna array in which two or more dipoles are stacked one above the other to obtain greater directivity.

lazy jack [ENG] A device that accommodates changes in length of a pipeline or similar structure through the motion of two linked bell cranks.

lb *See* pound.

L band [COMMUN] A band of radio frequencies extending from 390 to 1550 megahertz, corresponding to wavelengths of 76.9 to 19.37 centimeters.

lb ap *See* pound.

lb apoth *See* pound.

lbf *See* pound.

lbf-ft *See* foot-pound.

lb t *See* pound.

lb tr *See* pound.

L capture [NUC PHYS] A type of generalized beta interaction in which a nucleus captures an electron from the *L* shell of atomic electrons (the shell second closest to the nucleus).

L carrier system [COMMUN] Telephone carrier system used on coaxial cable systems, and microwave (line of sight), and tropospheric scatter radio systems; it occupies a frequency band from 68 kilohertz to over 8 megahertz.

LC filter *See* inductive filter.

LCL *See* less-than-carload; lifting condensation level.

LC ratio [ELEC] The inductance of a circuit in henrys divided by capacitance in farads.

LD$_{50}$ *See* lethal dose 50.

LDA localizer [NAV] A facility of comparable utility and accuracy to the standard ILS (instrument landing system) localizer, but which is not part of a full ILS and may not be in line with the runway.

LDH *See* lactic dehydrogenase.

L display [ELECTR] A radarscope display in which the target appears as two horizontal pulses or blips, one extending to the right and one to the left from a central vertical time base; when the radar antenna is correctly aimed in azimuth at the target, both blips are of equal amplitude; when not correctly aimed, the relative blip amplitudes indicate the pointing error; the position of the signal along the baseline indicates target distance; the display may be rotated 90° when used for elevation aiming instead of azimuth aiming.

L/D ratio [ENG] Length to diameter ratio, a frequently used engineering relationship.

leaching [CHEM ENG] The dissolving, by a liquid solvent, of soluble material from its mixture with an insoluble solid; leaching is an industrial separation operation based on mass transfer; examples are the washing of a soluble salt from the surface of an insoluble precipitate, and the extraction of sugar from sugarbeets. [GEOCHEM] The separation or dissolving out of soluble constituents from a rock or ore body by percolation of water. [MIN ENG] Dissolving soluble minerals or metals out of the ore, as by the use of percolating solutions such as cyanide or chlorine solutions, acids, or water. Also known as lixiviation.

leach ion-exchange flotation process [MET] A method of extraction developed for treatment of copper ores not amenable to direct flotation; the metal is dissolved by leaching, for example, with sulfuric acid, in the presence of an ion-exchange resin; the resin recaptures the dissolved metal and is then recovered in a mineralized froth by the flotation process.

leach material [MIN ENG] Material sufficiently mineralized to be economically recoverable by leaching.

leach pile [MIN ENG] Mineralized materials stacked so as to permit wanted minerals to be effectively and selectively dissolved by leaching.

lead [CHEM] A chemical element, symbol Pb, atomic number 82, atomic weight 207.19. [DES ENG] The distance a screw will advance or move into a nut in one complete turn. [ELEC] A wire used to connect two points in a circuit. [ENG] A mass of lead attached to a line, as used for sounding at sea. [GEOL] A small, narrow passage in a cave. [MET] A soft, heavy metal with a silvery-bluish color; when freshly cut it is malleable and ductile; occurs naturally, mostly in combination; used principally in alloys in pipes, cable sheaths, type metal, and shields against radioactivity. [ORD] **1.** The action of aiming ahead of a moving target with a gun, bomb, rocket, or torpedo so as to hit the target, including whatever action is necessary to correct for deflection. **2.** The lead angle. **3.** The distance between the moving target and the point at which the gun or missile is aimed. **4.** The number of diameters for one complete turn of the rifling. **5.** An explosive train component which consists of a column of high explosive, usually small in diameter, used to transmit detonation from one detonating component to a succeeding high explosive component; it is generally used to transmit the detonation from a detonator to a booster charge. [PHYS] *See* lead angle.

lead-208 [NUC PHYS] Lead isotope, atomic mass number of 208, formed by the radioactive decay of thorium.

lead acetate [ORG CHEM] $Pb(C_2H_3O_2)_2 \cdot 3H_2O$ Poisonous, water-soluble white crystals decomposing at 280°C; loses water at 75°C; used in hair dyes, medicines, and textile mordants, for waterproofing, for manufacture of varnishes and pigments, and as an analytical reagent. Also known as sugar of lead.

lead-acid battery [ELEC] A storage battery in which the electrodes are grids of lead containing lead oxides that change in composition during charging and discharging, and the electrolyte is dilute sulfuric acid.

lead angle [DES ENG] The angle that the tangent to a helix makes with the plane normal to the axis of the helix. [MET] The angle at the point of welding between an electrode and a line perpendicular to the weld axis. [ORD] **1.** The angle between the line of sight to a moving target and the line of sight to a point ahead of the target. **2.** A dropping angle. [PHYS] The phase difference between a sinusoidally varying quantity and a reference quantity which varies sinusoidally at the same frequency, when this phase difference is positive. Also known as angle of lead; lead; phase lead.

lead antimonite [INORG CHEM] $Pb_3(SbO_4)_2$ Poisonous, water-insoluble orange-yellow powder; used as a paint pigment and to stain glass and ceramics. Also known as antimony yellow; Naples yellow.

lead arsenate [INORG CHEM] $Pb_3(AsO_4)_2$ Poisonous, water-insoluble white crystals; soluble in nitric acid; used as an insecticide.

lead azide [INORG CHEM] $Pb(N_3)_2$ Unstable, colorless needles that explode at 350°C; lead azide is shipped submerged in water to reduce sensitivity; used as a detonator for high explosives.

lead-base babbitt [MET] Alloy of 10-15% antimony, 2-10% tin, up to 0.2% copper, sometimes with arsenic, the remainder being lead; used as a bearing metal; a variation used in diesel engine and railway bearings contains alkaline-earth metals. Also known as white-metal bearing alloy.

lead borate [INORG CHEM] $Pb(BO_2)_2 \cdot H_2O$ Poisonous, water-insoluble white powder; soluble in dilute nitric acid; used as varnish and paint drier, for galvanoplastic work, in lead glass, and in waterproofing paints.

lead bromide [INORG CHEM] $PbBr_2$ An alcohol-insoluble white powder melting at 373°C, boiling at 916°C; slightly soluble in hot water.

lead bronze [MET] An alloy of 60-70% copper, up to 2% nickel, and up to 15% tin with the balance lead; used as a bearing metal.

lead bullet [ORD] In small-arms ammunition, a bullet composed of lead or of a composition with a high percentage of lead.

lead burning *See* lead welding.

lead carbonate [INORG CHEM] $PbCO_3$ Poisonous, acid-soluble white crystals decomposing at 315°C; insoluble in alcohol and water; used as a paint pigment.

lead-chamber process [CHEM ENG] A process for the preparation of impure or dilute (60-78%) sulfuric acid; sulfur dioxide is oxidized by moist air with nitrogen oxide catalysts in a series of lead-lined chambers, the Gay-Lussac tower and the Glover tower; used primarily in the manufacture of fertilizer.

lead chart [ORD] A chart or table giving the leads necessary to strike a moving target under various conditions, such as the distance to the target and its speed and direction of travel.

lead chloride [INORG CHEM] $PbCl_2$ Poisonous white crystals melting at 498°C, boiling at 950°C; slightly soluble in hot water, insoluble in alcohol and cold water; used to make lead

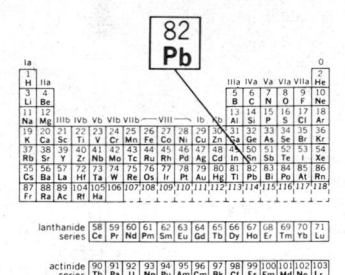

LEAD

Periodic table of the chemical elements showing the position of lead.

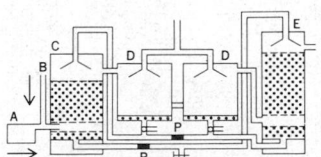

LEAD-CHAMBER PROCESS

Schematic drawing of the lead-chamber process for manufacture of sulfuric acid, with concentration of 60–78%. A, sulfur or pyrite burners; B, inlet for nitrogen oxides; C, Glover tower; D, lead chamber; E, Gay-Lussac tower; P, pumps.

salts and lead chromate pigments and as an analytical reagent.

lead chromate [INORG CHEM] PbCrO₄ Poisonous, water-insoluble yellow crystals melting at 844°C; soluble in acids; used as a paint pigment.

lead computer [ORD] A device for calculating leads to hit a moving target, often placed in the sights or mount of a machine gun that automatically gives the proper lead when the target is tracked by the gun.

lead curve [CIV ENG] The curve in a railroad turnout between the switch and the frog.

lead cyanide [INORG CHEM] Pb(CN)₂ Poisonous white to yellow powder; slightly soluble in water, decomposed by acids; used in metallurgy.

lead dioxide [INORG CHEM] PbO₂ Poisonous brown crystals that decompose when heated; insoluble in water and alcohol, soluble in glacial acetic acid; used as an oxidizing agent, in electrodes, batteries, matches, and explosives, as a textile mordant, in dye manufacture, and as an analytical reagent. Also known as anhydrous plumbic acid; brown lead oxide; lead peroxide.

leaded alloy [MET] An alloy, especially of brass, bronze, or steel, to which lead is added to improve machinability and mechanical properties.

leaded gasoline [MATER] Motor gasoline into which a small amount of TEL (tetraethyllead) has been added to increase octane number or rating.

leaded zinc oxide [MATER] A mixture of zinc oxide and basic lead sulfate; used in paints.

lead encephalopathy [MED] Degeneration of the neurons of the brain accompanied by cerebral edema, due to lead poisoning.

leader [ADP] A record which precedes a group of detail records, giving information about the group not present in the detail records; for example, "beginning of batch 17." [ENG] The unrecorded length of magnetic tape that enables the operator to thread the tape through the drive and onto the take-up reel without losing data or recorded music, speech, or such. [GEOPHYS] The streamer which initiates the first phase of each stroke of a lightning discharge; it is a channel of very high ion density which propagates through the air by the continual establishment of an electron avalanche ahead of its tip. Also known as leader streamer. [GRAPHICS] A short piece of blank film at each end of a strip of photographic film or reel of motion picture film, used to thread the film through the mechanism of a camera or projector.

leader cable [NAV] A cable used as a navigation aid, in which the path to be followed is defined by the magnetic field produced by current flowing through the cable.

leader label [ADP] A record appearing at the beginning of a magnetic tape to uniquely identify the tape as one required by the system.

leader streamer *See* leader.

leader stroke [GEOPHYS] The entire set of events associated with the propagation of any leader between cloud and ground in a lightning discharge.

lead fluosilicate *See* lead hexafluorosilicate.

lead foil [MET] A foil of lead or of lead alloy containing, for example, 10–12% tin and 1% copper.

lead formate [ORG CHEM] Pb(CHO₂)₂ Poisonous, water-soluble brownish-white crystals that decompose at 190°C; used as an analytical reagent.

lead glance *See* galena.

lead glass [MATER] Glass into which lead oxide is incorporated to give high refractive index, optical dispersion, and surface brilliance; used in optical glass.

lead halide [INORG CHEM] PbX₂, where X is a halogen (such as F, Br, Cl, or I).

lead hexafluorosilicate [INORG CHEM] PbSiF₆·2H₂O Poisonous, colorless, water-soluble crystals; used in the electrolytic method for refining lead. Also known as lead fluosilicate; lead silicofluoride.

leadhillite [MINERAL] Pb₄(SO₄)(CO₃)₂(OH)₂ A yellowish or greenish- or grayish-white monoclinic mineral consisting of basic sulfate and carbonate of lead; dimorphous with susanite.

lead-I-lead junction [SOLID STATE] A Josephson junction consisting of two pieces of lead separated by a thin insulating barrier of lead oxide. Abbreviated Pb-I-Pb junction.

lead-in [ELEC] A single wire used to connect a single-terminal outdoor antenna to a receiver or transmitter. Also known as down-lead.

lead-in-air indicator [MIN ENG] An instrument that utilizes reagents to measure the concentration of lead in the air.

leading [GRAPHICS] Space inserted between lines of type to open them up vertically.

leading current [ELEC] An alternating current that reaches its maximum value up to 90° ahead of the voltage that produces it.

leading edge [ADP] The edge of a punched card or document which enters a machine first. [AERO ENG] The front edge of an airfoil or wing. [DES ENG] The surfaces or inset cutting points on a bit that face in the same direction as the rotation of the bit. [PHYS] The major portion of the rise of a pulse.

leading edge slat [AERO ENG] A small airfoil attached to the leading edge of a wing of an aircraft that automatically improves airflow at large angles of attack.

leading fire [ORD] Fire delivered ahead of a moving target to allow for its motion.

leading heading [MIN ENG] **1.** The heading of a pair of parallel headings that is a short distance in front of the other; used to drain a mining area. **2.** A heading dug into the solid coal before the advance of the general face.

leading light [NAV] A light or lights arranged to indicate the path to be followed.

leading load [ELEC] Load that is predominately capacitive, so that its current leads the voltage applied to the load.

leading phase [ELEC] In three-phase power measurement, the phase whose voltage is leading upon that of one of the other phases by 120°.

lead-in groove [DES ENG] A blank spiral groove at the outside edge of a disk recording, generally of a pitch much greater than that of the recorded grooves, provided to bring the pickup stylus quickly to the first recorded groove. Also known as lead-in spiral.

leading stone *See* lodestone.

leading truck [MECH ENG] A swiveling frame with wheels under the front end of a locomotive.

lead-in insulator [ELEC] A tubular insulator inserted in a hole drilled through a wall, through which the lead-in wire can be brought into a building.

lead-in spiral *See* lead-in groove.

lead iodide [INORG CHEM] PbI₂ Poisonous, water- and alcohol-insoluble golden-yellow crystals melting at 402°C, boiling at 954°C; used in photography, medicine, printing, mosaic gold, and bronzing.

lead joint [ENG] A pipe joint made by caulking with lead wool or molten lead.

lead-lag network [CONT SYS] Compensating network which combines the characteristics of the lag and lead networks, and in which the phase of a sinusoidal response lags a sinusoidal input at low frequencies and leads it at high frequencies. Also known as lag-lead network.

lead line *See* sounding line.

lead lining [ENG] Lead sheeting used to line the inside surfaces of liquid-storage vessels and process equipment to prevent corrosion.

lead marcasite *See* sphalerite.

lead metallurgy [MET] The science and technology of lead.

lead metasilicate *See* lead silicate.

lead molybdate [INORG CHEM] PbMoO₄ Poisonous, acid-soluble yellow powder; insoluble in water and alcohol; used in pigments and as an analytical reagent.

lead monoxide [INORG CHEM] PbO Yellow, tetragonal crystals that melt at 888°C and are soluble in alkalies and acids; used in storage batteries, ceramics, pigments, and paints. Also known as litharge; plumbous oxide; yellow lead oxide.

lead naphthenate [MATER] Soft, combustible, alcohol-soluble, transparent, resinous material, melting about 100°C; made from addition of lead salt to solution of sodium naphthenate; used as a paint and varnish drier, wood preservative, catalyst, insecticide, and lubricating oil additive.

lead network   *See* derivative network.

lead nitrate   [INORG CHEM] $Pb(NO_3)_2$ Strongly oxidizing, poisonous, water- and alcohol-soluble white crystals that decompose at 205–223°C; used as a textile mordant, paint pigment, and photographic sensitizer and in medicines, matches, explosives, tanning, and engraving.

lead ocher   *See* massicot.

lead oleate   [ORG CHEM] $Pb(C_{18}H_{33}O_2)_2$ Poisonous, water-insoluble, white, ointmentlike material; soluble in alcohol, benzene, and ether; used in varnishes, lacquers, and high-pressure lubricants, and as a paint drier.

lead orthoplumbate   *See* lead tetroxide.

lead-out groove   [DES ENG] A blank spiral groove at the end of a disk recording, generally of a pitch much greater than that of the recorded grooves, connected to either the locked or eccentric groove. Also known as throw-out spiral.

lead-over groove   [DES ENG] A groove cut between separate selections or sections on a disk recording to transfer the pickup stylus from one cut to the next. Also known as cross-over spiral.

lead oxide red   *See* lead tetroxide.

lead palsy   *See* lead polyneuropathy.

lead peroxide   *See* lead dioxide.

lead phosphate   [INORG CHEM] Pb_3PO_4 A poisonous, white powder that melts at 1014°C; soluble in nitric acid and in fixed alkali hydroxide; used as a stabilizer in plastics.

lead pigments   [CHEM] Chemical compounds of lead used in paints to give color; examples are white lead; basic lead carbonate; lead carbonate; lead thiosulfate; lead sulfide; basic lead sulfate (sublimed white lead); silicate white lead; basic lead silicate; lead chromate; basic lead chromate; lead oxychloride; and lead oxide (monoxide and dioxide).

lead poisoning   [MED] Poisoning due to ingestion or absorption of lead over a prolonged period of time; characterized by colic, brain disease, anemia, and inflammation of peripheral nerves.

lead polyneuropathy   [MED] A distal polyneuropathy, affecting mainly the wrist and hand, seen principally in adults with chronic lead poisoning; characterized by weakness, paresthesias, pain, and glove-and-stocking anesthesia. Also known as lead palsy.

lead rail   [CIV ENG] In an ordinary rail switch, the turnout rail lying between the rails of the main track.

lead resinate   [ORG CHEM] $Pb(C_{20}H_{29}O_2)_2$ Poisonous, insoluble, brown, lustrous, translucent lumps; used as a paint and varnish drier and for textile waterproofing.

lead screw   [MECH ENG] A threaded shaft used to convert rotation to longitudinal motion; in a lathe it moves the tool carriage when cutting threads; in a disk recorder it guides the cutter at a desired rate across the surface of an ungrooved disk.

lead silicate   [INORG CHEM] $PbSiO_3$ Toxic, insoluble white crystals; used in ceramics, paints, and enamels, and to fireproof fabrics. Also known as lead metasilicate.

lead silicofluoride   *See* lead hexafluorosilicate.

lead-silver babbitt   [MET] Alloy of lead with 10–15% antimony, 2.5–5.1% silicon, up to 5% tin, and up to 0.2% copper; used as a bearing metal.

lead-soap lubricant   [MATER] Hard, high-melting-point, extreme-pressure lubricant made of lead salts saponified with fats.

lead sodium hyposulfate   *See* lead sodium thiosulfate.

lead sodium thiosulfate   [INORG CHEM] $Na_4Pb(S_2O_3)_3$ Poisonous, small, white, heavy crystals that are soluble in thiosulfate solutions; used in the manufacture of matches. Also known as lead sodium hyposulfate; sodium lead hyposulfate; sodium lead thiosulfate.

lead solder   [MET] Solder composed of a lead alloy.

lead stearate   [ORG CHEM] $Pb(C_{18}H_{35}O_2)_2$ Poisonous white powder; soluble in alcohol and ether, insoluble in water; used as a lacquer and varnish drier and in high-pressure lubricants.

lead sulfate   [INORG CHEM] $PbSO_4$ Poisonous white crystals melting at 1170°C; slightly soluble in hot water, insoluble in alcohol; used in storage batteries and as a paint pigment.

lead sulfide   [INORG CHEM] PbS Blue, metallic, cubic crystals that melt at 1120°C, derived from the mineral galena or

by reacting hydrogen sulfide gas with a solution of lead nitrate; used in semiconductors and ceramics. Also known as plumbous sulfide.

lead sulfide cell   [ELECTR] A cell used to detect infrared radiation; either its generated voltage or its change of resistance may be used as a measure of the intensity of the radiation.

lead susceptibility   [CHEM ENG] The increase in octane number of gasoline imparted by the addition of a specified amount of TEL (tetraethyllead).

lead telluride   [INORG CHEM] PbTe A crystalline solid that is very toxic if inhaled or ingested; melts at 902°C; used as a semiconductor and photoconductor in the form of single crystals.

lead tetraacetate   [ORG CHEM] $Pb(CH_3COO)_4$ Crystals that are faintly pink or colorless; melts at 175°C; used as an oxidizing agent in organic chemistry, cleaving 1,2-diols to form aldehydes or ketones.

lead tetroxide   [INORG CHEM] Pb_3O_4 A poisonous, bright-red powder, soluble in excess glacial acetic acid and dilute hydrochloric acid; used in medicine, in cement for special applications, in manufacture of colorless glass, and in ship paint. Also known as lead orthoplumbate; lead oxide red; red lead.

lead thiocyanate   [INORG CHEM] $Pb(SCN)_2$ Yellow, monoclinic crystals, soluble in potassium thiocyanate and slightly soluble in water; used in the powder mixture that primes small arm cartridges, in dyes, and in safety matches.

lead time   [IND ENG] The time allowed or required to initiate and develop a piece of equipment that must be ready for use at a given time.

lead titanate   [INORG CHEM] $PbTiO_3$ A water-insoluble, pale-yellow solid; used as coloring matter in paints.

lead track   [CIV ENG] A distance measured along a straight railroad track from a switch to a frog.

lead tungstate   [INORG CHEM] $PbWO_4$ A yellowish powder, melting at 1130°C; insoluble in water, soluble in acid; used as a pigment. Also known as lead wolframate.

lead vanadate   [INORG CHEM] $Pb(VO_3)_2$ A water-insoluble, yellow powder; used as a pigment and for the preparation of other vanadium compounds.

lead welding   [MET] Welding of lead by fusion. Incorrectly known as lead burning.

lead wire   [ENG] One of the heavy wires connecting a firing switch with the cap wires.

lead wolframate   *See* lead tungstate.

lead wool   [MATER] A coarse lead fiber used to caulk pipe joints.

leaf   [BOT] A modified aerial appendage which develops from a plant stem at a node, usually contains chlorophyll, and is the principal organ in which photosynthesis and transpiration occur.

leaf blight   [PL PATH] Any of various blight diseases which cause browning, death, and falling of the leaves.

leaf blotch   [PL PATH] A plant disease characterized by discolored areas in the leaves with indistinct or diffuse margins.

leaf bud   [BOT] A bud that produces a leafy shoot.

leaf cast   [PL PATH] Any of several diseases of conifers characterized by falling of the needles.

leaf curl   [PL PATH] A fungus or viral disease of plants marked by the curling of leaves.

leaf cushion   [BOT] The small part of the thickened leaf base that remains after abscission in various conifers, and also in some extinct plants.

leaf drop   [PL PATH] Premature falling of leaves, associated with disease.

leaf fiber   [BOT] A long, multiple-celled fiber extracted from the leaves of many plants that is used for cordage, such as sisal for binder, and abaca for manila hemp.

leafhopper   [INV ZOO] The common name for members of the homopteran family Cicadellidae.

leaflet   [BOT] 1. A division of a compound leaf. 2. A small or young foliage leaf.

leaflet projectile   [ORD] A projectile designed for or adapted to usage as a carrier for leaflets.

leaf miner   [INV ZOO] Any of the larvae of various insects which burrow into and eat the parenchyma of leaves.

leaf mold [GEOL] A soil layer or compost consisting principally of decayed vegetable matter.

leaf mottle [PL PATH] A fungus disease characterized by chlorotic mottling of the leaves; for example, caused by *Verticillium dahliae* in sun-flower.

leaf-nosed [VERT ZOO] Having a leaflike membrane on the nose, as certain bats.

leaf roll [PL PATH] Any of several virus diseases characterized by upward or inward rolling of the leaf margins.

leaf rot [PL PATH] Any plant disease characterized by breakdown of leaf tissues; for example, caused by *Pellicularia koleroga* in coffee.

leaf rust [PL PATH] Any rust disease primarily affecting leaves; common in coffee, alfalfa, and wheat, barley, and other cereals.

leaf scald [PL PATH] A bacterial disease of sugarcane caused by *Bacterium albilineans* which invades the vascular tissues, causing creamy or grayish streaking and withering of the leaves.

leaf scar [BOT] A mark on a stem, formed by secretion of suberin and a gumlike substance, showing where a leaf has abscised.

leaf scorch [BOT] Any of several disorders and fungus diseases marked by a burned appearance of the leaves; for example, caused by the fungus *Diplocarpon earliana* in strawberry.

leaf sight [ORD] Rear sight for small arms, hinged so that it can be raised for aiming or lowered to keep it from being broken when not in use, and containing a peep sight that can be moved up and down to make adjustments for range.

leaf spot [PL PATH] Any of various diseases or disorders characterized by the appearance of well-defined discolored spots on the leaves.

leaf spring [DES ENG] A beam of cantilever design, firmly anchored at one end and with a large deflection under a load. Also known as flat spring.

leaf stripe [PL PATH] Any of various plant diseases characterized by striped discolorations on the foliage.

league [MECH] A unit of length equal to 3 miles or 4828.032 meters.

leak [PL PATH] A watery rot of fruits and vegetables caused by various fungi, such as *Rhizopus nigricans* in strawberry.

leakage [ENG] Undesired and gradual escape or entry of a quantity, such as loss of neutrons by diffusion from the core of a nuclear reactor, escape of electromagnetic radiation through joints in shielding, flow of electricity over or through an insulating material, and flow of magnetic lines of force beyond the working region. [MIN ENG] An unintentional diversion of ventilation air from its designed path.

leakage coefficient *See* leakage factor.

leakage conductance [ELEC] The conductance of the path over which leakage current flows; it is normally a low value.

leakage current [ELEC] **1.** Undesirable flow of current through or over the surface of an insulating material or insulator. **2.** The flow of direct current through a poor dielectric in a capacitor. [ELECTR] The alternating current that passes through a rectifier without being rectified.

leakage factor [ELECTROMAG] The total magnetic flux in an electric rotating machine or transformer divided by the useful flux that passes through the armature or secondary winding. Also known as leakage coefficient.

leakage flux [ELECTROMAG] Magnetic lines of force that go beyond their intended path and do not serve their intended purpose. [NUCLEO] The number of neutrons which pass outward through a unit area at the surface of a reactor core in a unit time, and are not reflected back into the core.

leakage halo [GEOCHEM] The dispersion of elements along channels and paths followed by mineralizing solutions leading into and away from the central focus of mineralization.

leakage indicator [ELEC] An instrument used to measure or detect current leakage from an electric system to earth. Also known as earth detector.

leakage inductance [ELECTROMAG] Self-inductance due to leakage flux in a transformer.

leakage intake system [MIN ENG] A ventilation circuit with two adjacent intake roadways leading to the coal face.

leakage radiation [ELECTROMAG] In a radio transmitting system, radiation from anything other than the intended radiating system.

leakage reactance [ELECTROMAG] Inductive reactance due to leakage flux that links only the primary winding of a transformer.

leakage resistance [ELEC] The resistance of the path over which leakage current flows; it is normally high.

leak detector [ENG] An instrument used for finding small holes or cracks in the walls of a vessel; the helium mass spectrometer is an example.

leaky [ELEC] Pertaining to a condition in which the leakage resistance has dropped so much below its normal value that excessive leakage current flows; usually applied to a capacitor.

leaky mutant gene [GEN] An allele with reduced activity relative to that of the normal allele.

leaky-wave antenna [ELECTROMAG] A wide-band microwave antenna that radiates a narrow beam whose direction varies with frequency; it is fundamentally a perforated waveguide, thin enough to permit flush mounting for aircraft and missile radar applications.

lean [MATER] **1.** Of concrete or mortar, containing little or insufficient cement. **2.** Of clay, deficient in plasticity. **3.** Of coal, having little or no volatile matter. **4.** Of lime, containing impurities. **5.** Of fuel mixture, expecially for internal combustion engines, being low in combustible component. **6.** Of ore, being low-grade.

lean fuel mixture *See* lean mixture.

lean gas [MATER] **1.** In natural-gas absorption processes (such as natural-gasoline recovery from natural gas), gas from which desired liquid components have been removed. **2.** Natural gas poor in butane-and-heavier liquids.

leaning wheel grader [CIV ENG] A grader with skewed wheels to help cut or spread the soil.

lean mixture [MECH ENG] A fuel-air mixture containing a low percentage of fuel and a high percentage of air, as compared with a normal or rich mixture. Also known as lean fuel mixture.

lean oil [MATER] Absorbent oil from which absorbed gas has been stripped; an example is absorber oil in a natural-gasoline plant from which absorbed liquids (ethane, propane, butane) have been removed.

lean-to [BUILD] A single-pitched roof whose summit is supported by the wall of a higher structure.

leapfrog system [MIN ENG] A system used in mining coal on a longwall face; self-advancing supports are used, with alternate supports advancing as each web of coal is removed.

leapfrog test [ADP] A computer test using a special program that performs a series of arithmetical or logical operations on one group of storage locations, transfers itself to another group, checks the correctness of the transfer, then begins the series of operations again; eventually, all storage positions will have been tested.

leap year [ASTRON] A year with 366, and not 365, days.

learning [PSYCH] The gathering, processing, storage, and recall of information received through the senses.

learning curve [PSYCH] Graphical representation of the relationship between acquisition of knowledge or skill, and the amount of practice or trials.

learning machine [ADP] A machine that is capable of improving its future actions as a result of analysis and appraisal of past actions.

lease [IND ENG] **1.** Contract between landowner and another granting the latter the right to use the land, usually upon payment of an agreed rental, bonus, or royalty. **2.** A piece of land that is leased. [TEXT] A means of keeping warp threads in position during weaving and beaming by passing them alternately over and under a set of rods.

lease automatic custody transfer [PETRO ENG] Automatic unattended system to receive and record oil produced from a drilling lease, then to transfer the contents to a pipeline. Abbreviated LACT.

leased facility [COMMUN] A collection of communication lines dedicated to a particular service; sometimes the lines have a predetermined path through system switching equipment.

LEAF SPRING

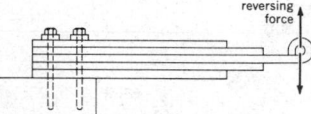

reversing
force

Drawing of a leaf spring showing
response to perpendicular force
in either direction.

lease-distribution system [PETRO ENG] Any of the electrical distribution systems serving oil-field pump motors and other electrical equipment.

lease rod [TEXT] A rod used to position warp threads during weaving.

lease tank [PETRO ENG] An oil-field storage tank that stores oil flowing from designated wells.

least-action principle *See* principle of least action.

least common denominator [MATH] The least common multiple of the denominators of a collection of fractions.

least common multiple [MATH] The least common multiple of a set of quantities (for example, numbers or polynomials) is the smallest quantity divisible by each of them.

least-energy principle [MECH] The principle that the potential energy of a system in stable equilibrium is a minimum relative to that of nearby configurations.

least significant character [ADP] The character in the rightmost position in a number or word.

least squares method [STAT] A technique of fitting a curve close to some given points which minimizes the sum of the squares of the deviations of the given points from the curve.

least-time principle *See* Fermat's principle.

least upper bound [MATH] The least upper bound of a subset A of a set S with ordering $<$ is the smallest element of S which is greater than or equal to every element of A. Also known as supremum.

least-work theory [MECH] A theory of statically indeterminate structures based on the fact that when a stress is applied to such a structure the individual parts of it are deflected so that the energy stored in the elastic members is minimized.

leather [MATER] Dressed hide or skin of an animal.

leather rot [PL PATH] A hard rot of strawberry caused by the fungus *Phytophthora cactorum*.

leaven [FOOD ENG] **1.** A substance (such as yeast) used to produce fermentation. **2.** A substance such as yeast or baking powder used to produce a gas to lighten dough.

Lebesgue integral [MATH] The generalization of Riemann integration of real valued functions, which allows for integration over more complicated sets, existence of the integral even though the function has many points of discontinuity, and convergence properties which are not valid for Riemann integrals.

Lebesgue measure [MATH] A measure defined on subsets of euclidean space which expresses how one may approximate a set by coverings consisting of intervals.

Lebesgue number [MATH] The Lebesgue number of an open cover of a compact metric space X is a positive real number so that any subset of X whose diameter is less than this number must be completely contained in a member of the cover.

Lebesgue-Stieltjes integral [MATH] A Lebesgue integral of the form

$$\int_b^a f(x)\, d\phi(x)$$

where ϕ is of bounded variation; if $\phi(x) = x$, it reduces to the Lebesgue integral of $f(x)$; if $\phi(x)$ is differentiable, it reduces to the Lebesgue integral of $f(x)\phi'(x)$.

Lecanicephaloidea [INV ZOO] An order of tapeworms of the subclass Cestoda distinguished by having the scolex divided into two portions; all species are intestinal parasites of elasmobranch fishes.

Lecanoraceae [BOT] A temperate and boreal family of lichens in the order Lecanorales characterized by a crustose thallus and a distinct thalloid rim on the apothecia.

Lecanorales [BOT] An order of the Ascolichenes having open, discoid apothecia with a typical hymenium and hypothecium.

lechatelierite [MINERAL] A natural silica glass, occurring in fulgurites and impact craters and formed by the melting of quartz sand at high temperatures generated by lightning or by the impact of a meteorite.

Le Chatelier's principle [PHYS] The principle that when an external force is applied to a system at equilibrium, the system adjusts so as to minimize the effect of the applied force.

Lecher line *See* Lecher wires.

Lecher wires [ELECTROMAG] Two parallel wires that are several wavelengths long and a small fraction of a wavelength apart, used to measure the wavelength of a microwave source that is connected to one end of the wires; a shorting bar which slides along the wires is used to determine the position of standing-wave nodes. Also known as Lecher line; Lecher wire wavemeter.

Lecher wire wavemeter *See* Lecher wires.

Lecideaceae [BOT] A temperate and boreal family of lichens in the order Lecanorales; members lack a thalloid rim around the apothecia.

lecithin [BIOCHEM] Any of a group of phospholipids having the general composition $CH_2OR_1 \cdot CHOR_2 \cdot CH_2OPO_2OHR_3$, in which R_1 and R_2 are fatty acids and R_3 is choline, and with emulsifying, wetting, and antioxidant properties. [MATER] **1.** A mixture of phosphatides and oil obtained by drying the separate gums from the degumming of soybean oil; consists of the phosphatides (lecithin), cephalin, other fatlike phosphorus-containing compounds, and 30–35% entrained soybean oil; may be treated to produce more refined grades; used in foods, cosmetics, and paints. Also known as commercial lecithin; crude lecithin; soybean lecithin; soy lecithin. **2.** A waxy mixture of phosphatides obtained by refining commercial lecithin to remove the soybean oil and other materials; used in pharmaceuticals. Also known as refined lecithin.

lecithinase [BIOCHEM] An enzyme that catalyzes the breakdown of a lecithin into its constituents.

lecithinase A [BIOCHEM] An enzyme that catalyzes the removal of only one fatty acid from lecithin, yielding lipolecithin.

lecithinase C [BIOCHEM] An enzyme that catalyzes the removal of the nitrogenous base of lecithin to produce the base and a phosphatidic acid.

lecithinase D [BIOCHEM] An enzyme that catalyzes the removal of the phosphorylated base from lecithins, producing α-β-diglyceride.

Leclanche cell [ELEC] The common dry cell, which is a primary cell having a carbon positive electrode and a zinc negative electrode in an electrolyte of sal ammoniac and a depolarizer.

lecontite [MINERAL] $Na(NH_4,K)SO_4 \cdot 2H_2O$ Colorless mineral composed of a hydrous sodium potassium ammonium sulfate; found in bat guano.

le cri du chat syndrome [MED] A complex of congenital malformations resulting from a deletion in chromosome 4 or 5 and characterized by mental retardation and the production of a catlike cry.

lectotype [SYST] A specimen selected as the type of a species or subspecies if the type was not designated by the author of the classification.

Lecythidaceae [BOT] The single family of the order Lecythidales.

Lecythidales [BOT] A monofamilial order of dicotyledonous tropical woody plants in the subclass Dilleniidae; distinguished by entire leaves, valvate sepals, separate petals, numerous centrifugal stamens, and a syncarpous, inferior ovary with axile placentation.

ledeburite [MET] The eutectic of the iron-carbon system, the constituents being cementite and austenite at high temperatures; cooling decomposes the austenite to ferrite and cementite.

Lederberg technique [MICROBIO] A method for rapid isolation of individual bacterial cells for demonstrating the spontaneous origin of bacterial mutants.

ledge [BUILD] A horizontal timber on the back of a batten door or on a framed and braced door. [ENG] **1.** A raised edge or molding. **2.** A narrow shelf projecting from the side of a vertical structure. **3.** A horizontal timber that supports the putlogs of scaffolding. [MET] Ingate for a foundry mold.

ledger [CIV ENG] A main horizontal member of formwork, supported on uprights and supporting the soffit of the formwork.

ledger balance [COMMUN] Facility used with message switching equipment to check the number of addresses received with the number of addresses transmitted, to ensure that no messages are lost within the center.

ledger paper [GRAPHICS] A heavy white paper made from

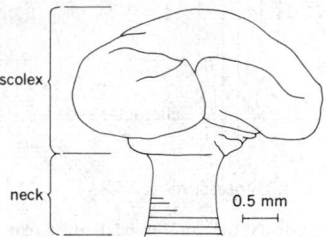

LECANICEPHALOIDEA

scolex

neck

0.5 mm

Anterior end of a lecanicephaloid tapeworm showing scolex and neck.

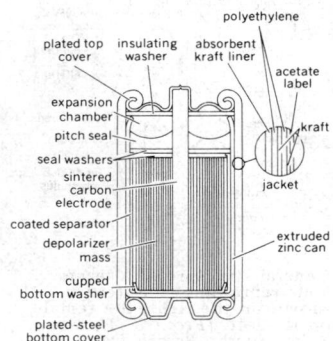

LECLANCHE CELL

polyethylene
plated top cover
insulating washer
absorbent kraft liner
acetate label
expansion chamber
pitch seal
kraft
seal washers
sintered carbon electrode
jacket
coated separator
depolarizer mass
extruded zinc can
cupped bottom washer
plated-steel bottom cover

Modern Leclanche dry cell. *(Bright Star Industries Inc.)*

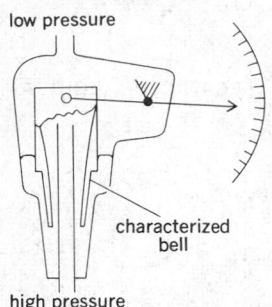
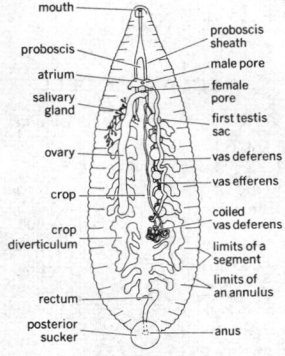
rag pulp, chemical wood pulp, or a mixture of both and well suited for ink and wash drawings; it is a strong, durable, and permanent paper, and the best varieties are fully opaque.

Ledian [GEOL] Lower upper Eocene geologic time. Also known as Auversian.

Ledoux bell meter [ENG] A type of manometer used to measure the difference in pressure between two points generated by any one of several types of flow measurement devices such as a pitot tube; it is equipped with a shaped plug which makes the reading of the meter directly proportional to the flow rate.

Leduc current [ELEC] An asymmetrical alternating current obtained from, or similar to that obtained from, the secondary winding of an induction coil; used in electrobiology.

Leduc effect *See* Righi-Leduc effect.

Leduc law *See* Amagat-Leduc rule.

lee [SCI TECH] The side of an object, such as an island or a ship, away from the direction in which the wind is coming, and sheltered from wind or waves.

Leeaceae [BOT] A family of dicotyledonous plants in the order Rhamnales distinguished by solitary ovules in each locule, simple to compound leaves, a small embryo, and hypogynous flowers.

leech [INV ZOO] The common name for members of the annelid class Hirudinea.

LEED *See* low-energy electron diffraction.

lee dune [GEOL] A dune formed to the leeward of a source of sand or of an obstacle.

lee eddies [FL MECH] The small, irregular motions or eddies produced immediately in the rear of an obstacle in a turbulent fluid.

leek [BOT] *Allium porrum*. A biennial herb known only by cultivation; grown for its mildly pungent succulent leaves and thick cylindrical stalk.

Lee-Norse miner [MIN ENG] A continuous miner for driving headings in medium or thick coal seams.

lee tide *See* leeward tidal current.

lee trough *See* dynamic trough.

leeward [SCI TECH] 1. Situated away from the wind. 2. On the lee side.

leeward tidal current [OCEANOGR] A tidal current setting in the same direction as that in which the wind is blowing. Also known as lee tide; leeward tide.

leeward tide *See* leeward tidal current.

lee wave [FL MECH] Any wave disturbance which is caused by, and is therefore stationary with respect to, some barrier in the fluid flow.

leeway [NAV] The off-course, leeward motion of a vessel due to wind or current; may be expressed as distance, speed, or angular difference between course steered and course through the water.

LE factor [PATH] A substance in body fluids, especially the blood, of patients with systemic lupus erythematosus, and sometimes in other diseases. Derived from lupus erythematosus factor.

left bank [GEOGR] The bank of a stream or river on the left of an observer when he is facing in the direction of flow, or downstream.

left-continuous function [MATH] A function $f(x)$ of a real variable is left-continuous at a point c if $f(x)$ approaches $f(c)$ as x approaches c from the left, that is, $x < c$ only.

left-hand [DES ENG] Of drilling and cutting tools, screw threads, and other threaded devices, designed to rotate clockwise or cut to the left.

left-hand derivative [MATH] The limit of the difference quotient $[f(x) - f(c)]/[x - c]$ as x approaches c from the left, that is, $x < c$ only.

left-handed [CRYSTAL] Having a crystal structure with a mirror-image relationship to a right-handed structure. [DES ENG] *See* left-laid.

left-handed coordinate system [MATH] 1. A three-dimensional rectangular coordinate system such that when the thumb of the left hand extends in the positive direction of the first (or x) axis, the fingers fold in the direction in which the second (or y) axis could be rotated about the first axis to coincide with the third (or z) axis. 2. A coordinate system of a Riemannian space which has negative scalar density function.

left-hand limit *See* limit on the left.

left-hand polarization [ELECTROMAG] In elementary-particle discussions, circular or elliptical polarization of an electromagnetic wave in which the electric field vector at a fixed point in space rotates about the direction of propagation; in the left-hand sense about the direction of propagation; in optics, the opposite convention is used; in facing the source of the beam, the electric vector is observed to rotate counterclockwise.

left-hand rule [ELECTROMAG] 1. For a current-carrying wire, the rule that if the fingers of the left hand are placed around the wire so that the thumb points in the direction of electron flow, the fingers will be pointing in the direction of the magnetic field produced by the wire. 2. For a current-carrying wire in a magnetic field, such as a wire on the armature of a motor, the rule that if the thumb, first, and second fingers of the left hand are extended at right angles to one another, with the first finger representing the direction of magnetic lines of force and the second finger representing the direction of current flow, the thumb will be pointing in the direction of force on the wire. Also known as Fleming's rule.

left-hand taper [ELEC] A taper in which there is greater resistance in the counterclockwise half of the operating range of a rheostat or potentiometer (looking from the shaft end) than in the clockwise half.

left-justify [GRAPHICS] To align characters horizontally so that the leftmost character of a string is in a specified position.

left-laid [DES ENG] The lay of a wire or fiber rope or cable in which the individual wires or fibers in the strands are twisted to the right and the strands to the left. Also known as left-handed; regular-lay left twist.

left lateral fault [GEOL] A fault in which movement is such that an observer walking toward the fault along an index plane (a bed, vein, or dike) would turn to the left to find the other part of the displaced index plane. Also known as sinistral fault.

left rudder [NAV] The operation of moving the rudder, and consequently the bow of the ship, to port.

left ventricular thrust *See* apex impulse.

leg [ANAT] The lower extremity of a human limb, between the knee and the ankle. [ENG] 1. Anything that functionally or structurally resembles an animal leg. 2. One of the branches of a forked or jointed object. 3. One of the main upright members of a drill derrick or tripod. [GEOPHYS] A single cycle of more or less periodic motion in a wave train on a seismogram. [MECH ENG] The case that encloses the vertical part of the belt carrying the buckets within a grain elevator. [MIN ENG] 1. In mine timbering, a prop or upright member of a set or frame. 2. A stone that has to be wedged out from beneath a larger one. [NAV] 1. One part of a craft's track, consisting of a single course line. 2. A track identified by an aid to navigation. [ZOO] An appendage or limb used for support and locomotion.

legena [VERT ZOO] An appendage of the sacculus containing sensory areas in the inner ear of tetrapods; termed the cochlea in humans.

legend [GRAPHICS] A title or explanation on a chart, diagram, or other illustration.

Legendre contact transformation *See* Legendre transformation.

Legendre equation [MATH] A second-order linear homogeneous differential equation related to the hypergeometric differential equation: $(1 - x^2)y'' - 2xy' + n(n + 1)y = 0$.

Legendre polynomials [MATH] A collection of orthogonal polynomials which provide solutions to the Legendre equation.

Legendre transformation [MATH] A mathematical procedure in which one replaces a function of several variables with a new function which depends on partial derivatives of the original function with respect to some of the original independent variables. Also known as Legendre contact transformation.

leg of fillet weld [MET] The distance between the original intersecting surfaces and the toe of the fillet.

legrandite [MINERAL] $Zn_{14}(OH)(AsO_4)_9 \cdot 12H_2O$ A yellow to nearly colorless mineral composed of basic hydrous zinc arsenate.

legume [BOT] A dry, dehiscent fruit derived from a single simple pistil; common examples are alfalfa, beans, peanuts, and vetch.

legume forage [AGR] Any plant of the legume family (Leguminosae) used for livestock feed, grazing, hay, or silage.

Leguminosae [BOT] The legume family of the plant order Rosales characterized by stipulate, compound leaves, 10 or more stamens, and a single carpel; many members harbor symbiotic nitrogen-fixing bacteria in their roots.

leg wire [ENG] One of the two wires forming a part of an electric blasting cap or squib.

Lehigh jig [MIN ENG] A plunger-type jig in which check valves open on the upstroke of the plunger, makeup water is introduced with the feed, the screen plate is at two levels, and the bottom of the discharge end is hinged; used to wash anthracite.

lehite [MINERAL] $(Na,K)_2Ca_5Al_8(PO_4)_8(OH)_{12}\cdot 6H_2O$ White mineral composed of hydrous basic calcium aluminum phosphate.

Lehmann process [MIN ENG] A process for treating coal by disintegration and separation of the petrographic constituents.

Leibnitz's rule [MATH] A formula to compute the nth derivative of the product of two functions f and g:

$$d^n(f\cdot g)/dx^n = \sum_{k=0}^{n} \binom{n}{k} d^{n-k} f/dx^{n-k}\cdot d^k g/dx^k$$

where $\binom{n}{k} = n!/(n-k)!\,k!$

Leibnitz's test [MATH] If the sequence of positive numbers a_n approaches zero monotonically, then the series

$$\sum_{n=1}^{\infty} (-1)^n a_n \text{ is convergent.}$$

Leidenfrost point [THERMO] The lowest temperature at which a hot body submerged in a pool of boiling water is completely blanketed by a vapor film; there is a minimum in the heat flux from the body to the water at this temperature.

Leidenfrost's phenomenon [THERMO] A phenomenon in which a liquid dropped on a surface that is above a critical temperature becomes insulated from the surface by a layer of vapor, and does not wet the surface as a result.

leifite [MINERAL] $Na_2AlSi_4O_{10}F$ A colorless mineral composed of fluoride and silicate of sodium and aluminum.

leightonite [MINERAL] $K_2Ca_2Cu(SO_4)_4\cdot 2H_2O$ A pale-blue mineral composed of hydrous sulfate of copper, calcium, and potassium.

Leiodidae [INV ZOO] The round carrion beetles, a cosmopolitan family of coleopteran insects in the superfamily Staphylinoidea; commonly found under decaying bark.

leiomyofibroma [MED] A benign tumor composed of smooth muscle cells and fibrocytes.

leiomyoma [MED] A benign tumor composed of smooth muscle cells.

leiomyosarcoma [MED] A malignant tumor composed of anaplastic smooth muscle cells.

leiosporous [MYCOL] Having smooth spores.

Leishman-Donovan bodies [PATH] Small, oval protozoans lacking flagella and undulating membranes, found within macrophages of the skin, liver, and spleen in leishmanial infections such as kala-azar and mucocutaneous leishmaniasis.

Leishmania [INV ZOO] A genus of flagellated protozoan parasites that are the etiologic agents of several diseases of humans, such as leishmaniasis.

Leishmania donovani [INV ZOO] The protozoan parasite that causes kala-azar.

Leishmania infantum [INV ZOO] The protozoan parasite that causes infantile leishmaniasis.

leishmaniasis [MED] Any of several infections caused by *Leishmania* species.

Leitneriales [BOT] A monofamilial order of flowering plants in the subclass Hamamelidae; members are simple-leaved, dioecious shrubs with flowers in catkins, and have a superior, pseudomonomerous ovary with a single ovule.

Lelapiidae [INV ZOO] A family of calcaronean sponges in the order Sycettida characterized by a rigid skeleton composed of tracts or bundles of modified triradiates.

L electron [ATOM PHYS] An electron in the L shell.

LEM *See* lunar excursion module.

lemma [BOT] Either of the upper pair of bracts enclosing the flower of a grass spikelet. [MATH] A mathematical fact germane to the proof of some theorem.

lemma of duBois-Reymond [MATH] A continuous function $f(x)$ is constant in the interval (a,b) if for certain functions g whose integral over (a,b) is zero, the integral over (a,b) of f times g is zero.

lemming [VERT ZOO] The common name for the small burrowing rodents composing the subfamily Microtinae.

Lemnaceae [BOT] The duckweeds, a family of monocotyledonous plants in the order Arales; members are small, free-floating, thalloid aquatics with much reduced flowers that form a miniature spadix.

lemniscate of Bernoulli [MATH] The locus of points (x,y) in the plane satisfying the equation $(x^2 + y^2)^2 = a^2(x^2 - y^2)$ or, in polar coordinates (r,θ), the equation $r^2 = a^2 \cos 2\theta$.

lemniscus [ANAT] A secondary sensory pathway of the central nervous system, usually terminating in the thalamus.

lemon [BOT] *Citrus limon.* A small evergreen tree belonging to the order Sapindales cultivated for its acid citrus fruit which is a modified berry called a hesperidium.

lemongrass oil [MATER] An essential oil with the odor of lemon, distilled from either of two lemongrasses (*Cymbopogon citratus* or *C. flexuosus*) in the East Indies; contains citral, citronellol, and geraniol; used as a perfume and as a flavoring.

lemon oil [MATER] A yellow essential oil squeezed from lemon rind; it is high in citral, limonene, terpinol, and citronellol; used in soap, perfumes, and flavors.

lemur [VERT ZOO] The common name for members of the primate family Lemuridae; characterized by long tails, foxlike faces, and scent glands on the shoulder region and wrists.

Lemuridae [VERT ZOO] A family of prosimian primates of Madagascar belonging to the Lemuroidea; all members are arboreal forest dwellers.

Lemuroidea [VERT ZOO] A suborder or superfamily of Primates including the lemurs, tarsiers, and lorises, or sometimes simply the lemurs.

Lenard rays [ELECTR] Cathode rays produced in air by a Lenard tube.

Lenard tube [ELECTR] An early experimental electron-beam tube that had a thin glass or metallic foil window at the end opposite the cathode, through which the electron beam could pass into the atmosphere.

length [MECH] Extension in space.

length between perpendiculars [NAV ARCH] The summer load waterline as measured from the fore side of the stern post or to the center of the rudder stock.

length block [ADP] The total number of records, words, or characters contained in one block.

lengthened dipole [ELECTROMAG] An antenna element with lumped inductance to compensate an end loss.

lengthening joint [ENG] A joint between two members running in the same direction.

lengthening reaction [PHYSIO] Sudden inhibition of the stretch reflex when extensor muscles are subjected to an excessive degree of stretching by forceful flexion of a limb.

length-height index [ANTHRO] The ratio of the basion-bregma height of the skull to its greatest length multiplied by 100.

length of a curve [MATH] A curve represented by $x = x(t)$, $y = y(t)$ for $t_1 \leq t \leq t_2$, with $x(t_1) = x_1$, $x(t_2) = x_2$, $y(t_1) = y_1$, $y(t_2) = y_2$, has length from (x_1,y_1) to (x_2,y_2) given by the integral from t_1 to t_2 of the function $\sqrt{(dx/dt)^2 + (dy/dt)^2}$.

length of barrel *See* length of bore.

length of bore [ORD] The length of the bore, measured from the rear face of the tube or barrel to the muzzle, expressed in inches or calibers. Also known as length of barrel; length of cannon; length of tube.

length of cannon *See* length of bore.

length of lay [DES ENG] The distance measured along a line parallel to the axis of the rope in which the strand makes one complete turn about the axis of the rope, or the wires make a complete turn about the axis of the strand.

length of record [CLIMATOL] The period during which obser-

A lemming, a small burrowing rodent with short legs and stout claws adapted for digging.

Fruit and branch of lemon tree (*Citrus limon*).

Ring-tailed lemur, measuring 4 feet (1 meter), including tail.

LENTICEL

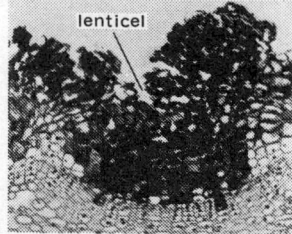

A lenticel as seen in a cross section of the stem periderm of *Sambucus nigra*.

LEO

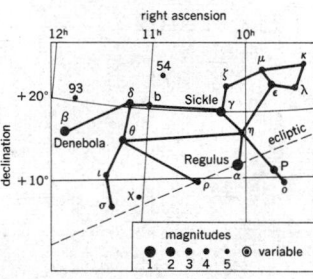

Line pattern of the constellation Leo. The grid lines represent the coordinates of the sky. Right ascension (E–W) is in hours, and declination (N–S) in degrees, corresponding to the longitude and latitude of the earth. The apparent brightness, or magnitude, of the stars is shown by the sizes of the dots, which are graded by appropriate numbers as indicated.

LEPIDOBLASTIC

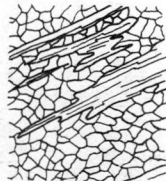

The fabric of lepidoblastic rock seen under the microscope. (*From T. F. W. Barth, Theoretical Petrology, Wiley, 1952*)

LEPIDOLITE

├─2.5 cm─┤

Group of lepidolite crystals found in Zinnwald, Czechoslovakia. (*Specimen from Department of Geology, Bryn Mawr College*)

vations have been maintained at a meteorological station, and which serves as the frame of reference for climatic data at that station.

length of shot [ENG] The depth of the shothole, in which powder is placed, or the size of the block of coal or rock to be loosened by a single blast, measured parallel with the hole. [MIN ENG] In open-pit mining, the distance from the first drill hole to the last drill hole along the bank.

length of tube *See* length of bore.

length on waterline [NAV ARCH] The length along the designed waterline from the forepart of the stem to the afterpart of the stern.

length overall [NAV ARCH] The total length of a ship's hull from the foremost to the aftermost points.

lenitic *See* lentic.

Lennard-Jones potential [PHYS CHEM] A semiempirical approximation to the potential of the force between two molecules, given by $V = (A/R^{12})-(B/R^6)$, where R is the distance between the centers of the molecules, and A and B are constants.

leno [TEXT] A mesh-weave cotton or cotton-and-polyester fabric made by twisting two warp yarns around each other.

lens [ANAT] A transparent, encapsulated, nearly spherical structure located behind the pupil of vertebrate eyes, and in the complex eyes of many invertebrates, that focuses light rays on the retina. Also known as crystalline lens. [COMMUN] A dielectric or metallic structure that is transparent to radio waves and can bend them to produce a desired radiation pattern; used with antennas for radar and microwave relay systems. [ELECTR] *See* electron lens. [ELECTROMAG] *See* magnetic lens. [GEOL] A geologic deposit that is thick in the middle and converges toward the edges, resembling a convex lens. [MATER] *See* acoustic lens. [OPTICS] A curved piece of ground and polished or molded material, usually glass, used for the refraction of light, its two surfaces having the same axis; or two or more such surfaces cemented together. Also known as optical lens.

lens antenna [ELECTROMAG] A microwave antenna in which a dielectric lens is placed in front of the dipole or horn radiator to concentrate the radiated energy into a narrow beam or to focus received energy on the receiving dipole or horn.

lens coating [OPTICS] A transparent substance coated on an optical surface to derive maximum light transmission.

lens element [OPTICS] A separate component lens of a multielement lens.

lens equation [OPTICS] Any equation which relates the distance of a point object from some well-defined reference point in an optical system to the distance of its image from a similar point.

lens placode [EMBRYO] The ectodermal anlage of the lens of the eye; its formation is induced by the presence of the underlying optic vesicle.

lens shim [OPTICS] Thin piece of material used to position and focus a lens.

lens tissue [MATER] Specially prepared paperlike material for cleaning lenses.

lentic [ECOL] Of or pertaining to still waters such as lakes, reservoirs, ponds, and bogs. Also spelled lenitic.

lenticel [BOT] A loose-structured opening in the periderm beneath the stomata in the stem of many woody plants that facilitates gas transport.

lenticle [GEOL] A bed or rock stratum or body that is lens-shaped.

lenticular [OPTICS] Of or pertaining to a lens. [SCI TECH] Having the shape of a lentil or double convex lens.

lenticular cloud *See* lenticularis.

lenticular film process [GRAPHICS] An additive color-photography process in which the camera lens carries a three-color banded filter, the film is embossed on the back with minute lenses facing the camera lens, development by reversal produces a photograph broken up into minute elements each of which is an image of the filter, and projection through a similar filter gives a color reproduction.

lenticularis [METEOROL] A cloud species, the elements of which have the form of more or less isolated, generally smooth lenses; the outlines are sharp. Also known as lenticular cloud.

lenticule [GRAPHICS] One of the minute lenses embossed on the back of a color photographic film in the lenticular film process.

lentiginose [ANAT] Of or pertaining to pigment spots in the skin; freckled.

lentil [BOT] *Lens esculenta.* A semivivy annual legume having pinnately compound, vetchlike leaves; cultivated for its thin, lens-shaped, edible seed. [GEOL] **1.** A rock body that is lens-shaped and enclosed in a stratum of different material. **2.** A rock stratigraphic unit that is a subdivision of a formation and has limited geographic extent; it thins out in all directions.

lentor *See* stoke.

Lenz's law [ELECTROMAG] The law that whenever there is an induced electromotive force (emf) in a conductor, it is always in such a direction that the current it would produce would oppose the change which causes the induced emf.

leo [MECH] A unit of acceleration, equal to 10 meters per second per second; it has rarely been employed.

Leo [ASTRON] A northern constellation, right ascension 11 hours, declination 15° north. Also known as Lion.

Leo Minor [ASTRON] A northern constellation, right ascension 10 hours, declination 35° north. Also known as Lesser Lion.

Leonardian [GEOL] A North American provincial series: Lower Permian (above Wolfcampian, below Guadalupian).

Leon firedamp tester [MIN ENG] A device, based on a form of Wheatstone bridge, that is used to detect firedamp.

Leonids [ASTRON] A meteor shower, the radiant of which lies in the constellation Leo; it is visible between November 10 and 15.

leonite [MINERAL] $K_2Mg(SO_4)_2 \cdot 4H_2O$ A colorless, white, or yellowish mineral composed of hydrous magnesium potassium sulfate, occurring in monoclinic crystals.

leopard [VERT ZOO] *Felis pardus.* A species of wildcat in the family Felidae found in Africa and Asia; the coat is characteristically buff-colored with black spots.

leopoldite *See* sylvite.

Leotichidae [INV ZOO] A small Oriental family of hemipteran insects in the superfamily Leptopodoidea.

Lepadomorpha [INV ZOO] A suborder of barnacles in the order Thoracica having a peduncle and a capitulum which is usually protected by calcareous plates.

leper [MED] A person afflicted with Hansen's disease.

Leperditicopida [PALEON] An order of extinct ostracods characterized by very thick, straight-backed valves which show unique muscle scars and other markings.

Leperditillacea [PALEON] A superfamily of extinct paleocopan ostracods in the suborder Kloedenellocopina including the unisulcate, nondimorphic forms.

Lepiceridae [INV ZOO] Horn's beetle, a family of Central American coleopteran insects composed of two species.

lepidine [ORG CHEM] $C_9H_6NCH_3$ An alkaloid derived as an oily liquid from cinchona bark; boils at 266°C; soluble in ether, benzene, and alcohol; used in organic synthesis. Also known as cincholepidine; γ-methylquinoline.

lepidoblastic [PETR] Of the texture of a metamorphic rock, having a fabric of minerals characterized as flaky or scaly, such as mica.

Lepidocentroida [INV ZOO] The name applied to a polyphyletic assemblage of echinoids now regarded as members of the Echinocystitoida and Echinothurioida.

lepidocrocite [MINERAL] $\alpha FeO(OH)$ A ruby- or blood-red mineral crystallizing in the orthorhombic system; it is associated with limonite in iron ores and is a component of meteorites.

Lepidodendrales [PALEOBOT] The giant club mosses, an order of extinct lycopods (Lycopodiopsida) consisting primarily of arborescent forms characterized by dichotomous branching, small amounts of secondary vascular tissue, and heterospory.

lepidolite [MINERAL] $K(Li,Al)_3(Si,Al)_4O_{10}(F,OH)_2$ A rose-colored mineral of the mica group crystallizing in the monoclinic system. Also known as lithionite; lithium mica.

lepidomelane [MINERAL] A black variety of biotite that is characterized by the presence of large amounts of ferric iron. Also known as iron mica.

lepidophyllous [BOT] Having scaly leaves.

Lepidoptera [INV ZOO] A large order of scaly-winged insects, including the butterflies, skippers, and moths; adults are characterized by two pairs of membranous wings and sucking mouthparts, featuring a prominent, coiled proboscis.

Lepidorsirenidae [VERT ZOO] A family of slender, obligate air-breathing, eellike fishes in the order Dipteriformes having small thin scales, slender ribbonlike paired fins, and paired ventral lungs.

Lepidosaphinae [INV ZOO] A family of homopteran insects in the superfamily Coccoidea having dark-colored, noncircular scales.

Lepidosauria [VERT ZOO] A subclass of reptiles in which the skull structure is characterized by two temporal openings on each side which have reduced bony arcades, and by the lack of an antorbital opening in front of the orbit.

Lepidotrichidae [INV ZOO] A family of wingless insects in the order Thysanura.

Lepismatidae [INV ZOO] A family of silverfish in the order Thysanura characterized by small or missing compound eyes.

Lepisostei [VERT ZOO] An equivalent name for Semionotiformes.

Lepisosteidae [VERT ZOO] A family of fishes in the order Semionotiformes.

Lepisosteiformes [VERT ZOO] An equivalent name for Semionotiformes.

lepisphere [PETR] A microspherical aggregate of platy, blade-shaped crystals of opal-CT.

Lepley-Hull hypothesis [PSYCH] An explanation of the empirical phenomenon in serial learning which postulates both direct and remote associations among items, with the former excitatory and the latter inhibitory.

Leporidae [VERT ZOO] A family of mammals in the order Lagomorpha including the rabbits and hares.

Lepospondyli [PALEON] A subclass of extinct amphibians including all forms in which the vertebral centra are formed by ossification directly around the notochord.

lepospondylous [VERT ZOO] Having the notochord enclosed by cylindrical vertebrae shaped like an hourglass in longitudinal section.

leproma [MED] The cutaneous nodular lesion of leprosy.

lepromatous leprosy [MED] A severely debilitating form of Hansen's disease characterized by the presence of multiple nodular lesions (lepromata) on the skin.

lepromin [IMMUNOL] An emulsion of ground lepromata containing the leprosy bacillus; used for intradermal skin tests in Hansen's disease.

leprophobia [PSYCH] An abnormal fear of Hansen's disease.

leprosarium [MED] An institution for the treatment of lepers.

leprosy *See* Hansen's disease.

Leptaleinae [INV ZOO] A subfamily of the Formicidae including largely arboreal ant forms which inhabit plants in tropical and subtropical regions.

Leptictidae [PALEON] A family of extinct North American insectivoran mammals belonging to the Proteutheria which ranged from the Cretaceous to middle Oligocene.

leptine [ANTHRO] Having a high, a narrow, or a high and narrow forehead with an upper facial index of 55–60 on the skull and of 53–57 on the living person.

Leptinidae [INV ZOO] The mammal nest beetles, a small European and North American family of coleopteran insects in the superfamily Staphylinoidea.

leptite [PETR] A quartz-feldspathic metamorphic rock that is fine-grained with little or no foliation; formed by regional metamorphism of the highest grade.

Leptocardii [ZOO] The equivalent name for Cephalochordata.

leptocephalous larva [VERT ZOO] The marine larva of the fresh-water European eel *Anguilla vulgaris.*

leptocephaly [ANTHRO] Abnormal narrowness and tallness of the head or skull.

leptocercal [VERT ZOO] Of the tail of a fish, tapering to a long, slender point.

Leptochoeridae [PALEON] An extinct family of palaeodont artiodactyl mammals in the superfamily Dichobunoidea.

Leptodactylidae [VERT ZOO] A large family of frogs in the suborder Procoela found principally in the American tropics and Australia.

leptodactylous [VERT ZOO] Having slender toes, as certain birds.

Leptodiridae [INV ZOO] The small carrion beetles, a cosmopolitan family of coleopteran insects in the superfamily Staphylinoidea.

leptogeosyncline [GEOL] A deep oceanic trough that has not been filled with sedimentation and is associated with volcanism.

leptokurtic distribution [STAT] A distribution in which the ratio of the fourth moment to the square of the second moment is greater than 3, which is the value for a normal distribution; it appears to be more heavily concentrated about the mean, or more peaked, than a normal distribution.

Leptolepidae [PALEON] An extinct family of fishes in the order Leptolepiformes representing the first teleosts as defined on the basis of the advanced structure of the caudal skeleton.

Leptolepiformes [PALEON] An extinct order of small, ray-finned teleost fishes characterized by a relatively strong, ossified axial skeleton, thin cycloid scales, and a preopercle with an elongated dorsal portion.

Leptomedusae [INV ZOO] A suborder of hydrozoan coelenterates in the order Hydroida characterized by the presence of a hydrotheca.

leptomeninges [ANAT] The pia mater and arachnoid considered together.

leptomeningitis [MED] Inflammation of the leptomeninges of the brain or the spinal cord.

Leptomitales [MYCOL] An order of aquatic Phycomycetes characterized by a hyphal thallus, or basal rhizoids and terminal hyphae, and zoospores with two flagella.

lepton [PARTIC PHYS] A fermion having a mass smaller than the proton mass; leptons interact with electromagnetic and gravitational fields, but beyond this they interact only through weak interactions.

lepton conservation [PARTIC PHYS] The principle that the number of electrons and *e*-neutrinos minus the number of positrons and *e*-antineutrinos is unchanged in any interaction; similarly, the number of negatively charged muons and μ-neutrinos minus the number of positively charged muons and μ-antineutrinos is unchanged.

leptonic decay [PARTIC PHYS] Decay of an elementary particle in which at least some of the products are leptons.

leptophyll [ECOL] A growth-form class of plants having a leaf surface area of 25 square millimeters or less; common in alpine and desert habitats.

Leptopodidae [INV ZOO] A tropical and subtropical family of hemipteran insects in the superfamily Leptopodoidea distinguished by the spiny body and appendages.

Leptopodoidea [INV ZOO] A superfamily of hemipteran insects in the subdivision Geocorisae.

leptorrhine [ANTHRO] Having a long, narrow nose with a nasal index of less than 47 on the skull or of less than 70 on the living person.

Leptosomatidae [VERT ZOO] The cuckoo rollers, a family of Madagascan birds in the order Coraciiformes composed of a single species distinguished by the downy covering on the newly hatched young.

Leptospira [MICROBIO] A genus of aerobic spirochetes in the family Treponemataceae characterized by twisted filaments with hooked or recurved extremities.

Leptospira icterohemorrhagiae [MICROBIO] The organism that causes spirochetal jaundice in humans.

leptospirosis [MED] Infection with spirochetes of the genus *Leptospira.*

leptospirosis icterohemorrhagia *See* Weil's disease.

leptostaphyline [ANTHRO] Having a high, narrow palate with a palatal index of less than 80 on the skull.

Leptostraca [INV ZOO] A primitive group of crustaceans considered as one of a series of Malacostraca distinguished by an additional abdominal somite that lacks appendages, and a telson bearing two movable articulated prongs.

Leptostromataceae [MYCOL] A family of fungi of the order Sphaeropsidales; pycnidia are black and shield-shaped, circu-

LEPTOLEPIDAE

Leptolepis dubia (Blainville) restoration, scales omitted. (*From A. S. Woodward, Catalogue of the Fossil Fishes in the British Museum (Natural History), pt. 3, 1895*)

LEPTOSTROMATACEAE

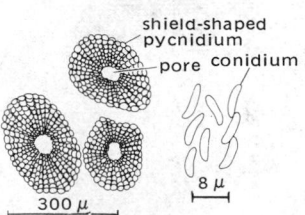

Pycnidia and conidia of *Leptothyrium vulgare.*

LERNAEIDAE

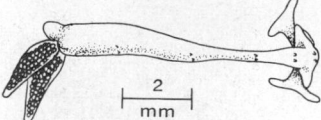

Lernaea cyprinacea L., the female, showing the branched processes of the head that anchor the parasite deep in the flesh of the host.

LETTUCE

Head lettuce (*Lactuca sativa*), cultivar Fulton, of the plant order Asterales. (*J. Harris Co., Rochester, N.Y.*)

LEUCALTIDAE

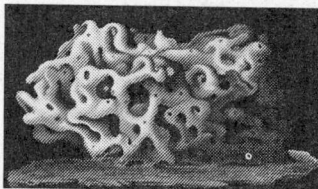

Leucaltis clathria showing the small chambers. (*After Polejaeff, 1883*)

LEUCINE

$$H_3C \quad CH_3$$
$$CH$$
$$CH_2$$
$$C$$
$$H_2N \quad COOH$$
$$H$$

Structural formula of leucine.

LEUCITE

|← 2.5 cm →|

Leucite crystals from Wiesenthal, Saxony. (*Specimen from Department of Geology, Bryn Mawr College*)

lar or oblong, and slightly asymmetrical; included are some fruit-tree pathogens.

leptotene [CYTOL] The first stage of meiotic prophase, when the chromosomes appear as thin threads having well-defined chromomeres.

Leptotyphlopidae [VERT ZOO] A family of small, harmless, burrowing circumtropical snakes (Serpentes) in the order Squamata; teeth are present only on the lower jaw and are few in number.

Lepus [ASTRON] A southern constellation, right ascension 5.5 hours, declination 20°S. Also known as Hare. [VERT ZOO] The type genus of the family Leporidae, comprising the typical hares.

Lernaeidae [INV ZOO] A family of copepod crustaceans in the suborder Caligoida; all are fixed ectoparasites, that is, they penetrate the skin of fresh-water fish.

Lernaeopodidae [INV ZOO] A family of ectoparasitic crustaceans belonging to the Lernaeopodoida; individuals are attached to the walls of the fishes' gill chambers by modified second maxillae.

Lernaeopodoida [INV ZOO] The fish maggots, a group of ectoparasitic crustaceans characterized by a modified postembryonic development reduced to two or three stages, a free-swimming larva, and the lack of external signs of physical maturity in adults.

LES *See* Lincoln experimental satellite.

lesbian [PSYCH] 1. Pertaining to female homosexuality. 2. A female homosexual.

Lesch-Nyhan syndrome [MED] A hereditary disease of male children, transmitted as an X-linked recessive, characterized by hyperuricemia, deficiency of hypoxanthine-guanine phosphoribosyl transferase, mental retardation, spastic cerebral palsy, choreathetosis, and self-mutilating biting.

lesion [BIOL] A structural or functional alteration due to injury or disease.

Leskeineae [BOT] A suborder of mosses in the order Hypnobryales; plants are not complanate, paraphyllia are frequently present, leaves are costate, and alar cells are not generally differentiated.

lespedeza [BOT] Any of various legumes of the genus *Lespedeza* having trifoliate leaves, small purple pea-shaped blossoms, and one seed per pod.

lesser circulation *See* pulmonary circulation.

Lesser Dog *See* Canis Minor.

lesser ebb [OCEANOGR] The weaker of two ebb currents occurring during a tidal day.

lesser flood [OCEANOGR] The weaker of two flood currents occurring during a tidal day.

Lesser Lion *See* Leo Minor.

lesser omentum [ANAT] A fold of the peritoneum extending from the lesser curvature of the stomach to the transverse hepatic fissure.

less-than-carload [IND ENG] Too light to fill a freight car and therefore not eligible for carload rate. Abbreviated LCL.

leste [METEOROL] Spanish nautical term for east wind, specifically the hot, dry, dusty easterly or southeasterly wind which blows from the Atlantic coast of Morocco out to Madeira and the Canary Islands; it is a form of sirocco, occurring in front of depressions advancing eastward.

Lestidae [INV ZOO] A family of odonatan insects belonging to the Zygoptera; distinguished by the wings being held in a V position while at rest.

Lestoniidae [INV ZOO] A monospecific family of hemipteran insects in the superfamily Pentatomorpha found only in Australia.

LET *See* stopping power.

letdown [AERO ENG] Gradual and orderly reduction in altitude, particularly in preparation for landing.

lethal area [ORD] In terminal ballistics, a figure of merit, having the dimension of area, which permits prediction of the number of casualties a missile may be expected to produce when employed under specified conditions; an equation states the relationship between the lethal area and the numerous factors affecting its numerical value.

lethal dose 50 [PHARM] The dose of a substance which is fatal to 50% of the test animals. Abbreviated LD_{50}. Also known as median lethal dose.

lethal gene [GEN] A gene mutation that causes premature death in heterozygotes if dominant, and in homozygotes if recessive. Also known as lethal mutation.

lethal mutation *See* lethal gene.

lethal radius [ORD] The distance from point of burst or ground zero at which a projectile, missile, or the like will probably destroy a target or kill persons.

lethargy [MED] A morbid condition of drowsiness or stupor; mental torpor. [NUCLEO] A measure of the energy which has been lost by a neutron, equal to the natural logarithm of the ratio of the initial energy of a neutron to its energy at any given point in the slowing-down process.

letovicite [MINERAL] $(NH_3)_3H(SO_4)_2$ A mineral composed of acid ammonium sulfate.

letter [COMMUN] A character used in an alphabet generally representing one or more sounds of a spoken language.

letter code [ADP] A Baudot code function which cancels errors by causing the receiving terminal to print nothing.

Letterer-Siwe disease [MED] A fatal disease of infants and young children, of unknown etiology, characterized by hyperplasia of the reticuloendothelial system without lipid storage.

letterpress [GRAPHICS] A method of printing by impressing paper on a raised inked surface; the oldest printing process, it employs type or plates cast or engraved in relief on materials such as metal, wood, or rubber. Also known as relief press.

letterset [GRAPHICS] A method of printing that uses a blanket for transferring an image from plate to paper; unlike offset lithography, it uses a relief plate and requires no dampening system. Also known as dry offset.

letter spacing [GRAPHICS] Space inserted between letters of a word to open it up horizontally.

letters patent *See* patent.

letters shift [COMMUN] Abbreviated LTRS. 1. A movement of a teletypewriter carriage which permits printing of alphabetic characters in an appropriate, generally linear sequence. 2. The control which actuates this movement.

lettuce [BOT] *Lactuca sativa.* An annual plant of the order Asterales cultivated for its succulent leaves; common varieties are head lettuce, leaf or curled lettuce, romaine lettuce, and iceberg lettuce.

Leucaltidae [INV ZOO] A family of calcinean sponges in the order Leucettida having numerous small, interstitial, flagellated chambers.

Leucascidae [INV ZOO] A family of calcinean sponges in the order Leucettida having a radiate arrangement of flagellated chambers.

Leucettida [INV ZOO] An order of calcareous sponges in the subclass Calcinea having a leuconoid structure and a distinct dermal membrane or cortex.

leucine [BIOCHEM] $C_6H_{13}O_2N$ A monocarboxylic essential amino acid obtained by hydrolysis of protein-containing substances such as milk.

leucine amino peptidase [BIOCHEM] An enzyme that acts on peptides to catalyze the release of terminal amino acids, especially leucine residues, having free α-amino groups.

leucite [MINERAL] $KAlSi_2O_6$ A white or gray rock-forming mineral belonging to the feldspathoid group; at ordinary temperatures the mineral exists as aggregates of trapezohedral crystals with glassy fracture; hardness is 5.5–6.0 on Mohs scale, and specific gravity is 2.45–2.50. Also known as amphigene; grenatite; vesuvian; Vesuvian garnet; white garnet.

leucite phonolite [PETR] An extrusive rock composed of alkali feldspar, mafic minerals, and leucite.

leucitite [PETR] A fine-grained or porphyritic extrusive rock or hypabyssal igneous rock composed mostly of pyroxene and leucite.

leucitohedron *See* trapezohedron.

leucochalcite *See* olivenite.

leucocratic [PETR] Light-colored as applied to igneous rock containing 0–50% dark-colored minerals.

Leucodontineae [BOT] A family of mosses in the order Isobryales with foliated branches, often bearing catkins.

leucoline *See* quinoline.

leucon [INV ZOO] A type of sponge having the choanocytes restricted to flagellated chambers inserted between the incurrent and excurrent canals, and a reduced or absent paragastric cavity.

leucophanite [MINERAL] $(Na,Ca)_2BeSi_7(O,F,OH)_7$ Greenish mineral composed of beryllium sodium calcium silicate containing fluorine and occurring in glassy, tabular crystals.

leucophore [HISTOL] A white reflecting chromatophore.

leucophosphite [MINERAL] $K_2Fe_4(PO_4)_4(OH)_2 \cdot 9H_2O$ White mineral composed of hydrous basic phosphate of potassium and iron.

leucoplast [BOT] A nonpigmented plastid; capable of developing into a chromoplast.

leucopyrite See loellingite.

Leucosiidae [INV ZOO] The purse crabs, a family of true crabs belonging to the Oxystomata.

leucosin [BIOCHEM] A simple protein of the albumin type found in wheat and other cereals.

Leucosoleniida [INV ZOO] An order of calcareous sponges in the subclass Calcaronea characterized by an asconoid structure and the lack of a true dermal membrane or cortex.

leucosphenite [MINERAL] $Na_4BaTi_2Si_{10}O_{27}$ A white mineral composed of sodium barium silicotitanate and occurring as wedge-shaped crystals.

Leucospidae [INV ZOO] A small family of hymenopteran insects in the superfamily Chalcidoidea distinguished by a longitudinal fold in the forewings.

leucosporous [MYCOL] Having white spores.

Leucothoidae [INV ZOO] A family of amphipod crustaceans in the suborder Gammaridea including semiparasitic and commensal species.

Leucotrichaceae [MICROBIO] A family of short, cylindrical bacteria in the order Beggiatoales occurring as long, unbranched, nonmotile trichomes attached to solid substrates.

leucovorin [PHARM] Folinic acid used as a calcium salt to counteract the toxic effects of folic acid antagonists and for treatment of megaloblastic anemias.

leucoxene [MINERAL] A mineral composed of rutile with some anatase or sphene; occurs in igneous rocks, usually as an alteration product of ilmenite.

leukemia [MED] Any of several diseases of the hemopoietic system characterized by uncontrolled leukocyte proliferation. Also known as leukocythemia.

leukemia virus See leukovirus.

leukemoid [MED] Similar to leukemia, that is, characterized by the presence of immature leukocytes in the blood, but of a different etiology.

leukocidin [BIOCHEM] A toxic substance released by certain bacteria which destroys leukocytes.

leukocyte [HISTOL] A colorless, ameboid blood cell having a nucleus and granular or nongranular cytoplasm. Also known as white blood cell; white corpuscle.

leukocythemia See leukemia.

leukocytolysin [BIOCHEM] A leukocyte-desintegrating lysin.

leukocytopenia See leukopenia.

leukocytopoiesis [PHYSIO] Formation of leukocytes.

leukocytosis [MED] Elevation of the leukocyte count to values above the normal limit.

leukoderma [MED] Defective skin pigmentation, especially the congenital absence of pigment in patches or bands.

leukodystrophy [MED] A condition thought to result from an inborn error of metabolism and characterized by progressive degeneration of the white matter of the cerebrum, or by defective buildup of myelin.

leukoencephalitis [MED] Inflammation of the white matter of the cerebrum.

leukoerythroblastic anemia See myelophthisic anemia.

leukoerythroblastosis See myelophthisic anemia.

leukol See quinoline.

leukolymphosarcoma See leukosarcoma.

leukoma [MED] A large and dense opacity of the cornea as a result of an ulcer, wound, or inflammation, which presents an appearance of ground glass.

leukonychia [MED] Whitish discoloration or spotting of the fingernails.

leukopenia [MED] A reduction in the leukocyte count to values below the normal limit. Also known as leukocytopenia.

leukoplakia [MED] Formation of thickened white patches on mucous membranes, particularly of the mouth and vulva.

leukorrhea [MED] A whitish, mucopurulent discharge from the female genital canal.

leukosarcoma [MED] Lymphosarcoma accompanied by a small number of anaplastic lymphoid cells in the peripheral blood. Also known as leukolymphosarcoma; sarcoleukemia.

leukosis [MED] An excess of white blood cells.

leukotomy See lobotomy.

leukovirus [VIROL] A major group of animal viruses including those causing leukemia in birds, mice, and rats. Also known as leukemia virus.

leuneburgite [MINERAL] $Mg_3B_2(PO_4)_2(OH)_6 \cdot 5H_2O$ A colorless mineral consisting of a hydrous basic phosphate of magnesium and boron.

leurocristine See vincristine.

levan [BIOCHEM] $(C_6H_{10}O_5)_n$ A polysaccharide consisting of repeating units of D-fructose and produced by a range of microorganisms, such as *Bacillus mesentericus*.

levante [METEOROL] The Spanish and most widely used term for an east or northeast wind occurring along the coast and inland from southern France to the Straits of Gibraltar; it is moderate or fresh, mild, very humid, and rainy, and occurs with a depression over the western Mediterranean Sea.

levantera [METEOROL] A persistent east wind of the Adriatic, usually accompanied by cloudy weather.

Levantine Basin [OCEANOGR] A basin in the Mediterranean Ocean between Asia Minor and Egypt.

levator [MED] An instrument used for raising a depressed portion of the skull. [PHYSIO] Any muscle that raises or elevates a part.

leveche [METEOROL] A warm wind in Spain, either a foehn or a hot southerly wind in advance of a low-pressure area moving from the Sahara Desert.

levee [CIV ENG] 1. A dike for confining a stream. 2. A pier along a river. [GEOL] 1. An embankment bordering one or both sides of a sea channel or the low-gradient seaward part of a canyon or valley. 2. A low ridge sometimes deposited by a stream on its sides.

level [ACOUS] Volume of sound. [ADP] 1. The status of a data item in COBOL language indicating whether this item includes additional items. 2. See channel. [CIV ENG] 1. A surveying instrument with a telescope and bubble tube used to take level sights over various distances, commonly 100 feet (30 meters). 2. To make the earth surface horizontal. [COMMUN] A specified position on an amplitude scale applied to a signal waveform, such as reference white level and reference black level in a standard television signal. [DES ENG] A device consisting of a bubble tube that is used to find a horizontal line or plane. Also known as spirit level. [ELEC] A single bank of contacts, as on a stepping relay. [ELECTR] 1. The difference between a quantity and an arbitrarily specified reference quantity, usually expressed as the logarithm of the ratio of the quantities. 2. A charge value that can be stored in a given storage element of a charge storage tube and distinguished in the output from other charge values. [MIN ENG] 1. Mine workings that are at the same elevation. 2. A gutter for the water to run in. [QUANT MECH] See energy level.

level above threshold [PHYSIO] Also known as sensation level. 1. The pressure level of a sound in decibels above its threshold of audibility for the individual observer. 2. In general, the level of any psychophysical stimulus, such as light, above its threshold of perception.

level bombing [ORD] Releasing bombs in level flight.

level compensator [ELECTR] 1. Automatic transmission-regulating feature or device used to minimize the effects of variations in amplitude of the received signal. 2. Automatic gain control device used in the receiving equipment of a telegraph circuit.

level crosscut [MIN ENG] A horizontal crosscut.

level drive [MIN ENG] A horizontal shaft which allows access to the length of a deposit and forms the basis for the splitting of the deposit into levels.

leveled elemental time See normal elemental time.

leveled time See normal time.

leveler [ENG] A back scraper, drag, or other form of device for smoothing land.

level fold See nonplunging fold.

LEUCOSIIDAE

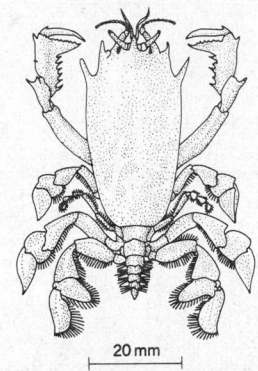

Drawing of a purse crab, *Persephona punctata.* (Smithsonian Institution)

level indicator [ENG] An instrument that indicates liquid level. [ENG ACOUS] An indicator that shows the audio voltage level at which a recording is being made; may be a volume-unit meter, neon lamp, or cathode-ray tuning indicator.

leveling [ENG] Adjusting any device, such as a launcher, gun mount, or sighting equipment, so that all horizontal or vertical angles will be measured in the true horizontal and vertical planes. [IND ENG] A method of performance rating which seeks to rate the principal factors that cause the speed of motions rather than speed itself; it considers that the level at which the operator works is influenced by effort and skill. [MET] Flattening rolled sheet by evening out irregularities, using a roller or tensile straining. [MIN ENG] Measurement of rises and falls, heights, and contour lines.

leveling action [MET] The property exhibited by a plating solution in making the coating smoother than the base metal.

leveling screw [ENG] An adjusting screw used to bring an instrument into level.

level measurement [MECH] The determination of the linear vertical distance between a reference point or datum plane and the surface of a liquid or the top of a pile of divided solid.

level of burst [ORD] Point at which a projectile bursts.

level off [AERO ENG] To bring an aircraft to level flight after an ascent or descent.

level-off position [AERO ENG] That position over which a craft ends an ascent or descent and begins relatively horizontal motion.

level of free convection [METEOROL] The level at which a parcel of air lifted dry and adiabatically until saturated, and lifted saturated and adiabatically thereafter, would first become warmer than its surroundings in a conditionally unstable atmosphere. Abbreviated LFC.

level of nondivergence [METEOROL] A level in the atmosphere throughout which the horizontal velocity divergence is zero; although in some meteorological situation there may be several such surfaces (not necessarily level), the level of nondivergence usually considered is that mid-tropospheric surface which separates the major regions of horizontal convergence and divergence associated with the typical vertical structure of the migratory cyclonic-scale weather systems.

level of saturation *See* water table.

level of significance of a test [STAT] The probability of false rejection of the null hypothesis. Also known as significance level.

level point *See* point of fall.

level scheme *See* energy-level diagram.

level surface [ENG] A surface which is perpendicular to the plumb line at every point. [GEOPHYS] *See* geopotential surface.

level width [QUANT MECH] A measure of the spread in energy of an unstable state, equal to the difference between the energies at which intensity of emission or absorption of photons or particles, or the cross section for a reaction, is one-half its maximum value.

Levenstein process [CHEM ENG] A process for the manufacture of mustard gas from ethene, $CH_2=CH_2$, and sulfur chloride, S_2Cl_2.

lever [ENG] A rigid bar, pivoted about a fixed point (fulcrum), used to multiply force or motion; used for raising, prying, or dislodging an object.

leverage [MECH] The multiplication of force or motion achieved by a lever.

lever escapement [HOROL] A clock movement in which the balance wheel is connected to the escapement by a lever attached to a roller; the wheel swings through a much larger angle than does a pendulum.

leveret [VERT ZOO] A young hare.

Leverett function [FL MECH] A dimensionless number used in studying two-phase flow in porous mediums, equal to $(\xi/e)^{1/2}(p/\sigma)$, where ξ is the permeability of a medium (as defined by Darcy's law), e is the medium's porosity, σ is the surface tension between two liquids flowing through it, and p is the capillary pressure.

lever shears [DES ENG] A shears in which the input force at the handles is related to the output force at the cutting edges

LEVER ESCAPEMENT

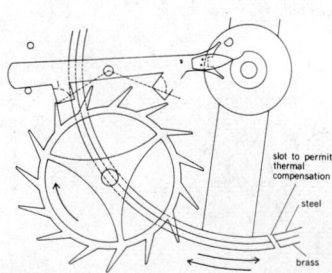

slot to permit thermal compensation

steel

brass

The lever escapement, used in good spring clocks and watches, is detached from the balance wheel during most of a cycle, thus promoting accurate timekeeping.

by the principle of the lever. Also known as alligator shears; crocodile shears.

lever switch [ELEC] A switch having a lever-shaped operating handle.

Levi-Civita symbol [MATH] A symbol $\varepsilon_{i,j,\ldots,s}$ where $i,j,\ldots,s$ are n indices, each running from 1 to n; the symbol equals zero if any two indices are identical, and 1 or -1 otherwise, depending on whether $i,j,\ldots,s$ form an even or an odd permutation of $1, 2, \ldots, n$.

levigate [BOT] *See* glabrous. [ENG] **1.** To separate fine powder from coarser material by suspending the fine material in a liquid. Also known as elutriation. **2.** To grind a moist solid to a fine powder.

levigated abrasive [MATER] A fine abrasive powder for final burnishing of metals or for metallographic polishing; the powder is usually processed to make it chemically neutral.

levitation [MIN ENG] In froth flotation, raising of particles in a froth to the surface of the pulp, to facilitate separation of selected minerals in the froth.

levitation heating [MET] Providing heat through high-frequency magnetic fields; employed in levitation melting.

levitation melting [MET] Melting metal out of contact with a supporting material by using the induced current provided by a high-frequency surrounding magnetic field to suspend the melt.

levo form [PHYS CHEM] An optical isomer which induces levorotation in a beam of plane polarized light.

levophobia [PSYCH] An abnormal fear of objects to the left of the body.

levorotation [OPTICS] Rotation of the plane of polarization of plane polarized light in a counterclockwise direction, as seen by an observer facing in the direction of light propagation. Also known as levulorotation.

levotropic cleavage [EMBRYO] Spiral cleavage with the cells displaced counterclockwise.

levulinic acid [ORG CHEM] $CH_3COCH_2CH_2COOH$ Crystalline compound forming plates or leaflets that melt at 37°C; freely soluble in alcohol, ether, and chloroform; used in the manufacture of pharmaceuticals, plastics, rubber, and synthetic fibers. Also known as acetylpropionic acid; γ-ketovaleric acid; 4-oxypentanoic acid.

levulorotation *See* levorotation.

levulose [BIOCHEM] Levorotatory D-fructose.

levulose tolerance test [PATH] A liver function test based on the observation that blood sugar increases in cases of hepatic disease following oral administration of levulose.

levyine *See* levynite.

levyite *See* levynite.

levyne *See* levynite.

levynite [MINERAL] $NaCa_3Al_7Si_{11}O_{36} \cdot 15H_2O$ A white or light-colored mineral of the zeolite group, composed of hydrous silicate of aluminum, sodium, and calcium, and occurring in rhombohedral crystals. Also known as levyine; levyite; levyne.

lewis [DES ENG] A device for hoisting heavy stones; employs a dovetailed tenon that fits into a mortise in the stone.

Lewis acid [CHEM] A substance that can accept an electron pair from a base; thus, $AlCl_3$, BF_3, and SO_3 are acids.

Lewis base [CHEM] A substance that can donate an electron pair; examples are the hydroxide ion, OH^-, and ammonia, NH_3.

Lewis blood group system [IMMUNOL] An antigen, designated by Le^a, first recognized in a Mrs. Lewis, occurring in about 22% of the population, detected by anti-Le^a antibodies; primarily composed of soluble antigens of serum and body fluids like saliva, with secondary absorption by erythrocytes.

lewis bolt [DES ENG] A bolt with an enlarged, tapered head that is inserted into masonry or stone and fixed with lead; used as a foundation bolt.

Lewis gun [ORD] A gas-operated machine gun with a horizontal drum magazine, manufactured in several modifications; now obsolete.

lewis hole [MIN ENG] A series of two or more holes drilled as closely together as possible, but then connected by knocking out the thin partition between them, thus forming one wide hole having its greatest diameter in a plane with the desired rift.

lewisite [MINERAL] $(Ca,Fe,Na)_2$ A titanian romeite mineral. [ORG CHEM] $C_2H_2AsCl_3$ An oily liquid, colorless to brown or violet; forms a toxic gas, used in World War I.

Lewis-Matheson method [CHEM ENG] Trial-and-error calculation method for the design of multicomponent distillation columns, or for the determination of the separating ability of an existing column.

Lewis number [PHYS] **1.** A dimensionless number used in studies of combined heat and mass transfer, equal to the thermal diffusivity divided by the diffusion coefficient. Symbolized Le; N_{Le}. **2.** Sometimes, the reciprocal of this quantity.

lewis pin [MIN ENG] A pin used for attachment to a key block; it is placed in a shallow drill hole with a wedge at either side.

lewistonite [MINERAL] $(Ca,K,Na)_5(PO_4)_3(OH)$ White mineral composed of basic calcium potassium sodium phosphate.

Lexell's Comet [ASTRON] A small comet that approached to within 2,000,000 miles (3,200,000 kilometers) of earth in 1770; it has not been seen since.

lexicographic order [MATH] Given sets A and B with a common ordering $<$, one defines an ordering between all sequences (finite or infinite) of elements of A and of elements of B by $(a_1,a_2,\ldots) < (b_1,b_2,\ldots)$ if either $a_i = b_i$ for every i, or $a_n < b_n$, where n is the first place in which they differ; this is the way words are ordered in a dictionary.

Leyden jar [ELEC] An early type of capacitor, consisting simply of metal foil sheets on the inner and outer surfaces of a glass jar.

Leydig cell [HISTOL] One of the interstitial cells of the testes; thought to produce androgen.

Leydig-cell tumor See interstitial-cell tumor.

LF See low-frequency.

LFC See level of free convection.

LF loran See low-frequency loran.

LGV See lymphogranuloma venereum.

L-head engine [MECH ENG] A type of four-stroke cycle internal combustion engine having both inlet and exhaust valves on one side of the engine block which are operated by pushrods actuated by a single camshaft.

lherzolite [PETR] Peridotite composed principally of olivine with orthopyroxene and clinopyroxene.

L'Hospital's rule [MATH] A rule useful in evaluating indeterminate forms: if both the functions $f(x)$ and $g(x)$ and all their derivatives up to order $(n-1)$ vanish at $x = a$, but the nth derivatives both do not vanish or both become infinite at $x = a$, then $\lim_{x \to a} f(x)/g(x) = f^{(n)}(a)/g^{(n)}(a)$, $f^{(n)}$ denoting the nth derivative.

l'Huilier's equation [MATH] An equation used in the solution of a spherical triangle, involving tangents of various functions of its angles and sides.

Li See lithium.

liana [BOT] A woody or herbaceous climbing plant with roots in the ground.

Liapunov function See Lyapunov function.

Lias See Liassic.

Liassic [GEOL] The Lower Jurassic period of geologic time. Also known as Lias.

Libby counter See black carbon counter.

libeccio [METEOROL] Italian name for a southwest wind, used in northern Corsica for the west or southwest wind which blows throughout the year, especially in winter when it is often stormy.

Libellulidae [INV ZOO] A large family of odonatan insects belonging to the Anisoptera and distinguished by a notch on the posterior margin of the eyes and a foot-shaped anal loop in the hindwing.

Liberty-Gel [MATER] Gelatinous permissible explosive; used in mining.

libethenite [MINERAL] $Cu_2(PO_4)OH$ An olive-green mineral composed of basic copper sulfate, occurring as small prismatic crystals or in masses.

libido [PSYCH] **1.** Sexual desire. **2.** The sum total of all instinctual forces; psychic energy or drive usually associated with the sexual instinct.

Libman-Sacks endocarditis [MED] Inflammation of, accompanied by the presence of watery vegetations on, the endocardium; complicates systemic lupus erythematosus.

Libra [ASTRON] A southern constellation, right ascension 15 hours, declination 15° south. Also known as Balance.

librarian [ADP] The program which maintains and makes available all programs and routines composing the operating system.

library [ADP] **1.** A computerized facility containing a collection of organized information used for reference. **2.** An organized collection of computer programs together with the associated program listings, documentation, users' directions, decks, and tapes.

library routine [ADP] A computer program that is part of some program library.

libration [PHYS] Any oscillatory rotational motion, such as that of the moon, or of a molecule in a solid which does not have enough energy to make full rotations.

libration in latitude See lunar libration.

Libytheidae [INV ZOO] The snout butterflies, a family of cosmopolitan lepidopteran insects in the suborder Heteroneura distinguished by long labial palps; represented in North America by a single species.

licanic oil [MATER] A fatty acid found in drying oils that are in paint formulations.

Liceaceae [MYCOL] A family of plasmodial slime molds in the order Liceales.

Liceales [MYCOL] An order of plasmodial slime molds in the subclass Myxogastromycetidae.

licensed material [NUCLEO] Source material, special nuclear material, or by-product material received, possessed, used, or transferred under a general or special license issued by the Atomic Energy Commission.

lichen [BOT] The common name for members of the Lichenes.

Lichenberger figures See Lichtenberg figures.

lichen blue See litmus.

Lichenes [BOT] A group of organisms consisting of fungi and algae growing together symbiotically.

Lichenes Imperfecti [BOT] A class of the Lichenes containing species with no known method of sexual reproduction.

lichenification [MED] The process whereby the skin becomes leathery and hardened; often the result of chronic pruritis and the irritation produced by scratching or rubbing eruptions.

lichenology [BOT] The study of lichens.

lichenophagous [ZOO] Animals which feed on lichens.

Lichnophorina [INV ZOO] A suborder of ciliophoran protozoans belonging to the Heterotrichida.

Lichtenberg figures [ELEC] Patterns produced on a photographic emulsion, or in fine powder spread over the surface of a solid dielectric, by an electric discharge produced by a high transient voltage. Also known as Lichenberger figures.

licorice [BOT] *Glycyrrhiza glabra.* A perennial herb of the legume family (Leguminosae) cultivated for its roots, which when dried provide a product used as a flavoring in medicine, candy, and tobacco and in the manufacture of shoe polish.

lidar [OPTICS] An instrument in which a ruby laser generates intense infrared pulses in beam widths as small as 30 seconds of arc; beam reflections and scattering effects of clouds, smog layers, and some atmospheric discontinuities are measured by radar techniques; it can also be used for tracking weather balloons, smoke puffs, and rocket trails. Derived from laser infrared radar.

lidocaine [ORG CHEM] $C_{14}H_{22}N_2O$ A crystalline compound, used as a local anesthetic. Also known as lignocaine.

Lie algebra [MATH] The algebra of vector fields on a manifold with additive operation given by pointwise sum and multiplication by the Lie bracket.

liebigite [MINERAL] $Ca_2U(CO_3)_4 \cdot 10H_2O$ An apple- or yellow-green mineral composed of hydrous uranium calcium carbonate; occurs as a coating or concretion in rock.

Liebig's law of the minimum See law of minimum.

Liebmann effect [OPTICS] The effect whereby it is more difficult to visually distinguish contrasting forms when they have the same luminance and different chromaticities than when they have different luminances and the same chromaticity.

Lie bracket [MATH] Given vector fields X, Y on a manifold M, their Lie bracket is the vector field whose value is the difference between the values of XY and YX.

lie detector [ENG] An instrument that indicates or records

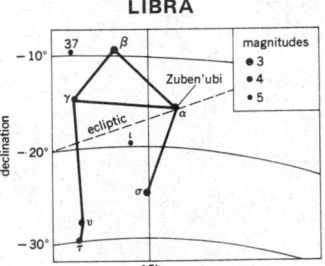

LIBRA

Line pattern of the constellation Libra. The grid lines represent the coordinates of the sky. The apparent brightness, or magnitude, of the stars is shown by the sizes of the dots, graded by appropriate numbers.

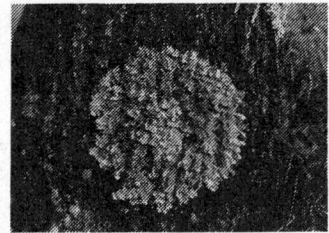

LICHENES

A picture of the foliose (leaflike) type of plant body of the *Parmelia* species. (*From H. J. Fuller and O. Tippo, College Botany, rev. ed., Holt, 1954*)

one or more functional variables of a person's body while the person undergoes the emotional stress associated with a lie. Also known as polygraph; psychintegroammeter.

Lie group [MATH] A topological group which is also a differentiable manifold in such a way that the group operations are themselves analytic functions.

Lienard-Wiechert potentials [ELECTROMAG] The retarded and advanced electromagnetic scaler and vector potentials produced by a moving point charge, expressed in terms of the (retarded or advanced) position and velocity of the charge.

Liesegang banding [GEOL] Colored or compositional rings or bands in a fluid-saturated rock due to rhythmic precipitation. Also known as Liesegang rings.

Liesegang rings *See* Liesegang banding.

lifeboat [NAV ARCH] A small boat hoisted on davits or carried on one of the upper decks of a vessel, which can be quickly lowered into the water in case of an emergency.

life cycle [BIOL] The functional and morphological stages through which an organism passes between two successive primary stages.

life expectancy [BIOL] The expected number of years an organism will live based on statistical probability. [ENG] The predicted useful service life of an item of equipment.

life form [ECOL] The form characteristically taken by a plant at maturity.

life of mine [MIN ENG] The time in which, through the employment of the available capital, the ore reserves—or such reasonable extension of the ore reserves as conservative geological analysis may justify—will be extracted.

life preserver [ENG] A buoyant device used to prevent drowning by supporting a person in the water.

life raft [NAV ARCH] A very buoyant raft designed to be used by people forced into the water; made of a metal tube covered with wood or canvas, of balsa wood, or of rubber which may be automatically inflated.

life-saving station *See* Coast Guard station.

life support system [ENG] A system providing atmospheric control and monitoring, such as a breathing mixture supply system, air purification and filtering system, or carbon dioxide removal system; used in oceanographic submersibles and spacecraft.

life test [CHEM ENG] In petroleum testing, an ASTM oxidation test made on inhibited steam-turbine oils to determine their stability under oxidizing conditions. [ENG] A test in which a device is operated under conditions that simulate a normal lifetime of use, to obtain an estimate of service life.

lifetime *See* mean life.

life zone [ECOL] A portion of the earth's land area having a generally uniform climate and soil, and a biota showing a high degree of uniformity in species composition and adaptation.

LIFO *See* last in first out.

lift [FL MECH] *See* aerodynamic lift. [MECH ENG] *See* elevator. [MIN ENG] 1. The vertical height traveled by a cage in a shaft. 2. The distance between the first level and the surface or between any two levels. 3. Any of the various gangways from which coal is raised at a slope colliery.

lift bridge [CIV ENG] A drawbridge whose movable spans are raised vertically.

lift coefficient [AERO ENG] The quantity $C_L = 2L/\rho V^2 S$, where L is the lift of a whole airplane wing, ρ is the mass density of the air, V is the free-stream velocity, and S is the wing area; this is also applicable to other airfoils.

lift-drag ratio [AERO ENG] The lift of an aerodynamic form, such as an airplane wing, divided by the drag.

lifter [MIN ENG] A shothole drilled near the floor when tunneling, and fired subsequent to the cut and relief holes.

lifter case [MIN ENG] The sleeve or tubular part attached to the lower end of the inner tube of M-design core barrels and some other types of core barrels, in which is fitted a core lifter. Also known as core-catcher case; core-gripper case; core-lifter case; core-spring case; inner-tube extension; ring-lifter case; spring-lifter case.

lifter flight [DES ENG] Spaced plates or projections on the inside surfaces of cylindrical rotating equipment (such as rotary dryers) to lift and shower the solid particles through the gas-drying stream during their passage through the dryer cylinder.

LIFTING REENTRY

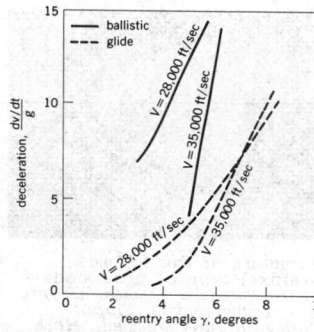

Maximum deceleration (in units of acceleration of gravity, g) versus initial reentry angle, γ, for glide and ballistic vehicles and various initial reentry speeds, V.

lifter roof [ENG] Gas storage tank in which the roof is raised by the incoming gas as the tank fills.

lift fan [AERO ENG] A special turbofan engine used primarily for lift in VTOL/STOL aircraft and often mounted in a wing with vertical thrust axis.

lift fire [ORD] 1. Command to advance the range of fire by elevating the muzzle of a weapon. 2. Command to cease or suspend fire.

lifting [MATH] Given a fiber bundle $(\tilde{X},B,p)$ and a continuous map of a topological space $\tilde{Y}$ to B, $g:\tilde{Y}\to B$, lifting entails finding a continuous map $\bar{g}:\tilde{Y}\to\tilde{X}$ such that the function g is the composition $p \circ \bar{g}$.

lifting block [MECH ENG] A combination of pulleys and ropes which allows heavy weights to be lifted with least effort.

lifting condensation level [METEOROL] The level at which a parcel of moist air lifted dry adiabatically would become saturated. Abbreviated LCL. Also known as isentropic condensation level (ICL).

lifting dog [ENG] 1. A component part of the overshot assembly that grasps and lifts the inner tube or a wire-line core barrel. 2. A clawlike hook for grasping cylindrical objects, such as drill rods or casing, while raising and lowering them.

lifting guard [MIN ENG] Fencing placed around the mouth of a shaft and lifted out of the way by the ascending cage.

lifting magnet [ELECTROMAG] A type of electromagnet in which a material to be held or moved is initially placed in contact with the magnet, in contrast to a traction magnet. Also known as holding magnet. [ENG] A large circular, rectangular, or specially shaped magnet used for handling pig iron, scrap iron, castings, billets, rails, and other magnetic materials.

lifting reentry [AERO ENG] A reentry into the atmosphere by a space vehicle where aerodynamic lift is used, allowing a more gradual descent, greater accuracy in landing at a predetermined spot; it can accommodate greater errors in the guidance system and greater temperature control.

lifting reentry vehicle [AERO ENG] A space vehicle designed to utilize aerodynamic lift upon entering the atmosphere.

lift-off [AERO ENG] The action of a rocket vehicle as it leaves its launch pad in a vertical ascent.

lift pump [MECH ENG] A pump for lifting fluid to the pump's own level.

lift-slab construction [CIV ENG] Pouring reinforced concrete roof and floor slabs at ground level, then lifting them into position after hardening.

lift truck [MECH ENG] A small hand- or power-operated dolly equipped with a platform or forklift.

lift valve [MECH ENG] A valve that moves perpendicularly to the plane of the valve seat.

ligament [HISTOL] A flexible, dense white fibrous connective tissue joining, and sometimes encapsulating, the articular surfaces of bones.

ligamentum nuchae *See* nuchal ligament.

ligand [CHEM] The molecule, ion, or group bound to the central atom in a chelate or a coordination compound; an example is the ammonia molecules in $[Co(NH_3)_6]^{3+}$.

ligase [BIOCHEM] An enzyme that catalyzes the union of two molecules, involving the participation of a nucleoside triphosphate which is converted to a nucleoside diphosphate or monophosphate. Also known as synthetase.

ligation [MED] Surgical tying of vessels or ducts with a ligature.

ligature [MED] A cord or thread used for tying vessels and ducts.

light [OPTICS] Electromagnetic radiation with wavelengths capable of causing the sensation of vision, ranging approximately from 4000 (extreme violet) to 7700 angstroms (extreme red). Also known as light radiation; visible radiation. 2. More generally, electromagnetic radiation of any wavelength; thus, the term is sometimes applied to infrared and ultraviolet radiation.

light absorption [OPTICS] The process in which energy of light radiation is transferred to a medium through which it is passing.

light-activated silicon-controlled rectifier [ELECTR] A silicon-controlled rectifier having a glass window for incident light that takes the place of, or adds to the action of, an electric gate current in providing switching action. Also known as light-activated switch.

light-activated switch *See* light-activated silicon-controlled rectifier.

light adaptation [PHYSIO] The disappearance of dark adaptation; the chemical processes by which the eyes, after exposure to a dim environment, become accustomed to bright illumination, which initially is perceived as quite intense and uncomfortable.

light amplification by stimulated emission of radiation *See* laser.

light amplifier [ELECTR] **1.** Any electronic device which, when actuated by a light image, reproduces a similar image of enhanced brightness, and which is capable of operating at very low light levels without introducing spurious brightness variations (noise) into the reproduced image. Also known as image intensifier. **2.** *See* laser amplifier.

light antiaircraft artillery [ORD] Conventional antiaircraft artillery pieces, usually under 90-millimeter, the weight of which in a trailed mount, including on-carriage fire control, does not exceed 20,000 pounds (9000 kilograms); self-propelled versions are rated in the same category as the trailed version.

light artillery [ORD] All guns and howitzers of 105-millimeter caliber (4.13-inch bore) or smaller.

light beacon [NAV] A beacon from which a light is displayed for directing ships.

light-beam galvanometer *See* d'Arsonval galvanometer.

light-beam oscillograph [ELECTROMAG] An oscillograph in which a beam of light, focused to a point by a lens, is reflected from a tiny mirror attached to the moving coil of a galvanometer onto a photographic film moving at constant speed.

light-beam pickup [ENG ACOUS] A phonograph pickup in which a beam of light is a coupling element of the transducer.

light blasting [ENG] Loosening of shallow or small outcrops of rock and breaking boulders by explosives.

light bomber [AERO ENG] Any bomber with a gross weight of less than 100,000 pounds (45,000 kilograms), including bombs; for example, the A-20 and A-26 bombers in World War II.

light bulb *See* incandescent lamp.

light carrier injection [ELECTR] A method of introducing the carrier in a facsimile system by periodic variation of the scanner light beam, the average amplitude of which is varied by the density changes of the subject copy. Also known as light modulation.

light-case bomb [ORD] A type of general-purpose bomb having a thin, light metal casing and designed to accomplish damage by blast primarily.

light chopper [ELECTR] A rotating fan or other mechanical device used to interrupt a light beam that is aimed at a phototube, to permit alternating-current amplification of the phototube output and to make its output independent of strong, steady ambient illumination.

light climate *See* illumination climate.

light cruiser [NAV ARCH] A vessel whose armaments consist primarily of 6-inch guns.

light curve [ASTROPHYS] A graph showing the variations in brightness of a celestial object; the stellar magnitude is usually shown on the vertical axis, and time is the horizontal coordinate.

light-distribution photometer [OPTICS] A device which measures the luminous intensity of a light source in various directions; the light source is fixed, and a mirror system is rotated about an axis passing through the centers of the light source and a photocell so that the light emitted by the source in any direction perpendicular to this axis is reflected to the photocell. Also known as distribution photometer.

light-drawn [MET] Cold-worked very slightly; for copper or copper alloy tubing, drawing entails between 10 and 25% reduction in area.

lighted buoy [NAV] In marine operations, a buoy with a light having definite characteristics for detection and identification during darkness.

light-emitting diode [ELECTR] A semiconductor diode, generally made from gallium arsenide, that can serve as a light source when voltage is applied continuously or in pulses.

light ends [MATER] The lower-boiling components of a mixture of hydrocarbons, such as those evaporated or distilled off easily in comparison to the bulk of the mixture; for hydrocarbon mixtures, usually considered to be butane and lighter.

lighter [NAV ARCH] A barge used to load and unload ships not lying at piers, or to move cargo around a harbor.

lighterage [IND ENG] **1.** Loading or unloading ships by means of a lighter. **2.** The fee charged for this operation.

lighter-than-air craft [AERO ENG] An aircraft, such as a dirigible, that weighs less than the air it displaces.

light filter *See* color filter.

light float [NAV] A boatlike structure (smaller than a light vessel and usually unmanned) used instead of a light buoy in waters where strong streams or currents are experienced, or when a greater elevation than that of a light buoy is necessary.

light fog [GRAPHICS] A graying of the image produced by exposure to white light.

light freeze [METEOROL] The condition when the surface temperature of the air drops to below the freezing point of water for a short time period, so that only the tenderest plants and vines are adversely affected.

light frost [HYD] A thin and more or less patchy deposit of hoarfrost on surface objects and vegetation.

light gun [ELECTR] A light pen mounted in a gun-type housing.

light homing [ORD] Guidance of a missile to its target by using light which the target emits.

lighthouse [NAV] A structure equipped with a powerful light to aid in navigation.

lighthouse tube *See* disk-seal tube.

light howitzer [ORD] A complete projectile-firing weapon with a medium muzzle velocity and a curved trajectory, a bore diameter is over 30 millimeters, through 125 millimeters.

light hydrogen *See* protium.

lighting branch circuit [ELEC] A circuit that supplies power to outlets for lighting fixtures only.

light-inspection car [MECH ENG] A railway motorcar weighing 400–600 pounds (180–270 kilograms) and having a capacity of 650–800 pounds (295–360 kilograms).

light intensity *See* luminous intensity.

light ion *See* small ion.

light list [NAV] A publication tabulating navigational lights, with their locations, candlepowers, and other characteristics, to assist in their identification, and details of any accompanying fog signals.

light list number [NAV] In marine operations, a number assigned each aid to navigation (except unlighted buoys), published in the light list for identification purposes.

light machine gun [ORD] **1.** Any aircraft machine gun of .30 caliber or smaller. **2.** In U.S. Army usage, any lightweight machine gun, including the .30-caliber air-cooled machine gun and excluding fully automatic rifles, submachine guns, and machine pistols. **3.** In specific nonofficial context, any fully automatic rifle or any machine gun having total weight, with mount, between 20–30 pounds (9–14 kilograms) approximately; for example, the Browning automatic rifle and the Lewis machine gun.

light measurement *See* photometry.

light metal [MET] A metal or alloy of low density, especially aluminum and magnesium alloys.

light meter [ENG] A small, portable device for measuring illumination; often calibrated to give photographic exposures.

light microscope *See* optical microscope.

light microsecond [ELECTROMAG] Distance a light wave travels in free space in one-millionth of a second.

light mineral [MINERAL] **1.** A rock with minerals that have a specific gravity lower than a standard, usually 2.85. **2.** A light-colored mineral.

light modulation *See* light carrier injection.

light modulator [ELECTR] The combination of a source of light, an appropriate optical system, and a means for varying the resulting light beam to produce an optical sound track on motion picture film.

light negative [ELECTR] Having negative photoconductivity, hence decreasing in conductivity (increasing in resistance) under the action of light.

LIGHT-BEAM OSCILLOGRAPH

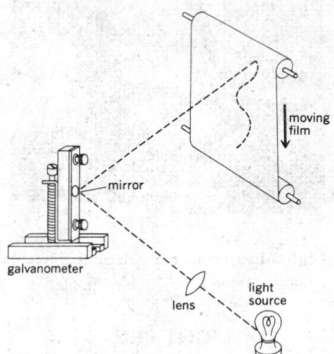

Essential components of a light-beam oscillograph. *(From I. F. Kinnard, Applied Electrical Measurements, Wiley, 1956)*

LIGHT-DISTRIBUTION PHOTOMETER

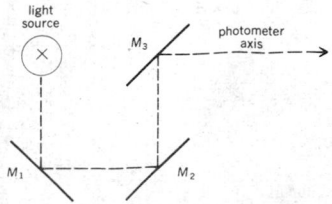

Mirror system in a light-distribution photometer. M_1, M_2, M_3 are mirrors.

LIGHTNING ARRESTER

Installation of three single-pole station lightning arresters rated 276 kilovolts for lightning and surge protection of large 345-kilovolt power transformer. *(General Electric Co.)*

LIGHTNING ROD

Lightning rods on a house.

LIGHT PEN

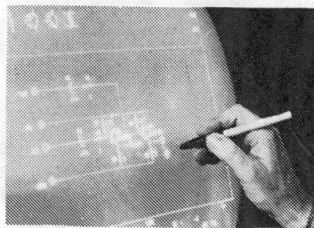

Electronic light pen is used by designer on a cathode-ray tube to instruct computer in the drafting of a schematic. *(Lockheed)*

lightning [GEOPHYS] The large spark produced by an abrupt discontinuous discharge of electricity through the air, resulting most often from the creation and separation of electric charge in cumulonimbus clouds.

lightning arrester [ELEC] A protective device designed primarily for connection between a conductor of an electrical system and ground to limit the magnitude of transient overvoltages on equipment. Also known as arrester.

lightning channel [GEOPHYS] The irregular path through the air along which a lightning discharge occurs.

lightning conductor [ELEC] A conductor designed to carry the current of a lightning discharge from a lightning rod to ground.

lightning discharge [GEOPHYS] The series of electrical processes by which charge is transferred within the atmosphere along a channel of high ion density between electric charge centers of opposite sign.

lightning flash [GEOPHYS] In atmospheric electricity, the total observed luminous phenomenon accompanying a lightning discharge.

lightning generator [ELEC] A high-voltage power supply used to generate surge voltages resembling lightning, for testing insulators and other high-voltage components.

lightning protection [ELEC] Means, such as lightning rods and lightning arresters, of protecting electrical systems, buildings, and other property from lightning.

lightning recorder See sferics receiver.

lightning rod [ELEC] A metallic rod set up on an exposed elevation of a structure and connected to a low-resistance ground to intercept lightning discharges and to provide a direct conducting path to ground for them.

lightning stroke [GEOPHYS] Any one of a series of repeated discharges comprising a single lightning discharge (or lightning flash); specifically, in the case of the cloud-to-ground discharge, a leader plus its subsequent return streamer.

lightning switch [ELEC] A manually operated switch used to connect a radio antenna to ground during electrical storms, rather than to the radio receiver.

light-of-the-night-sky See airglow.

light-operated switch [ELECTR] A switch operated by a beam or pulse of light, such as a light-activated silicon-controlled rectifier.

light panel See electroluminescent panel.

light pen [ELECTR] A tiny photocell or photomultiplier, mounted with or without fiber or plastic light pipe in a pen-shaped housing; it is held against a cathode-ray screen to make measurements from the screen or to change the nature of the display.

light pillar See sun pillar.

light pipe [OPTICS] A solid, transparent plastic rod that transmits light from one end to the other even when bent.

light-positive [ELECTR] Having positive photoconductivity; selenium ordinarily has this property.

light projector [OPTICS] A device designed to produce controlled beams of light that can be projected over considerable distances.

light quantum See photon.

light radiation See light.

light ratio [ASTROPHYS] A number (2.512) that expresses the ratio of a star's light to that of another star that is one magnitude fainter or brighter.

light ray [OPTICS] A beam of light having a small cross section.

light-red silver ore See proustite.

light reflex [PHYSIO] The postural orientation response of certain aquatic forms stimulated by the source of light; receptors may be on the ventral or dorsal surface.

light relay See photoelectric relay.

light-ruby silver See proustite.

light scattering [OPTICS] The process in which energy is removed from a beam of light radiation and reemitted without appreciable change in wavelength.

light section car [MECH ENG] A railway motorcar weighing 750–900 pounds (340–408 kilograms) and propelled by 4-6-horsepower (3000–4500-watt) engines.

light sector [NAV] **1.** A sector in which a navigational light has a distinctive color differing from that of adjoining sectors. **2.** A sector in which a navigational light is visible.

light-sensitive [ELECTR] Having photoconductive, photoemissive, or photovoltaic characteristics. Also known as photosensitive.

light-sensitive cell See photodetector.

light-sensitive detector See photodetector.

light-sensitive tube See phototube.

light sensor See photodetector.

lightship [NAV] A distinctively marked vessel anchored or moored at a charted point, to serve as an aid to navigation.

light source [OPTICS] A lamp used to supply radiant energy, as for an optical microscope, projector, or photoelectric control system.

light stability [ADP] In optical character recognition, the ability of an image to retain its spectral appearance when exposed to radiant energy.

light station [NAV] A group of buildings including a lighthouse and additional buildings housing personnel, fog signal, radio-beacon, and any other equipment associated with the lighthouse.

light transmission [OPTICS] The process in which light travels through a medium without being absorbed or scattered.

light valve [ELECTR] A device whose light transmission can be made to vary in accordance with an externally applied electrical quantity, such as voltage, current, electric field, or magnetic field, or an electron beam.

light water [NUCLEO] Water in which both the hydrogen atoms in each molecule are of the isotope protium.

light waterline [NAV ARCH] The waterline of a vessel without cargo.

light-water reactor [NUCLEO] A nuclear reactor that uses ordinary water as moderator, in contrast to heavy water.

light watt [OPTICS] A unit of luminous power equal to the luminous power of light of a single wavelength λ whose radiant power is $1/V_\lambda$ watts, where V_λ is the value of the luminosity function at λ.

lightweight aggregate [MATER] A lightweight inert material, such as foamed slag, vermiculite, clinker, and perlite, used in unreinforced concrete for making structures of low weight and high insulation.

lightweight concrete [MATER] A type of concrete made with lightweight aggregate.

light-year [ASTROPHYS] A unit of measurement of astronomical distance; it is the distance light travels in one sidereal year and is equivalent to 9.461×10^{12} kilometers or 5.879×10^{12} miles.

Ligiidae [INV ZOO] A family of primitive terrestrial isopods in the suborder Oniscoidea.

ligneous [BIOL] Of, pertaining to, or resembling wood.

lignify [BOT] To convert cell wall constituents into wood or woody tissue by chemical and physical changes.

lignin [BIOCHEM] A substance that together with cellulose forms the woody cell walls of plants and cements them together.

lignite [GEOL] Coal of relatively recent origin, intermediate between peat and bituminous coal; often contains patterns from the wood from which it formed.

lignocaine See lidocaine.

lignocellulose [BIOCHEM] Any of a group of substances in woody plant cells consisting of cellulose and lignin.

lignosa [BOT] Woody vegetation.

lignumvitae [BOT] *Guaiacum sanctum.* A medium-sized evergreen tree in the order Sapindales that yields a resin or gum known as gum guaiac or resin of guaiac. Also known as hollywood lignumvitae.

ligroin See petroleum ether.

ligulate [BOT] **1.** Strap-shaped. **2.** Having ligules.

ligule [BOT] **1.** A small outgrowth in the axis of the leaves in Selaginellales. **2.** A thin outgrowth of a foliage leaf or leaf sheath. [INV ZOO] A small lobe on the parapodium of certain polychaetes.

likelihood [MATH] The likelihood of a sample of independent values of $x_1, x_2, \ldots, x_n$, with $f(x)$ the probability function, is the product $f(x_1) \circ f(x_2) \circ \ldots \circ f(x_n)$.

likelihood ratio [STAT] The probability of a random drawing of a specified sample from a population, assuming a given hypothesis about the parameters of the population, divided by the probability of a random drawing of the same sample, assuming that the parameters of the population are such that this probability is maximized.

Liliaceae [BOT] A family of monocotyledonous plants in the order Liliales distinguished by six stamens and an inferior ovary.

Liliales [BOT] An order of monocotyledonous plants in the subclass Liliidae having the typical characteristics of the subclass.

Liliatae See Liliopsida.

Liliidae [BOT] A subclass of the Liliopsida; all plants are syncarpous and have a six-membered perianth, with all members petaloid.

Liliopsida [BOT] The monocotyledons, making up a class of the Magnoliophyta; characterized generally by a single cotyledon, parallel-veined leaves, and stems and roots lacking a well-defined pith and cortex.

lillianite [MINERAL] $Pb_3Bi_2S_6$ A steel-gray mineral composed of lead bismuth sulfide.

Lilly controller [MECH ENG] A device on steam and electric winding engines that protects against overspeed, overwind, and other incidents injurious to workers and the engine.

lily [BOT] **1.** Any of the perennial bulbous herbs with showy unscented flowers constituting the genus *Lilium*. **2.** Any of various other plants having similar flowers.

lily-pad ice See pancake ice.

Limacidae [INV ZOO] A family of gastropod mollusks containing the slugs.

limacon [MATH] The locus of points of the plane which in polar coordinates (r,θ) satisfy the equation $r = a \cos \theta + b$. Also known as Pascal's limacon.

limb [ANAT] An extremity or appendage used for locomotion or prehension, such as an arm or a leg. [ASTRON] The circular outer edge of a celestial body; the half with the greater altitude is called the upper limb, and the half with the lesser altitude, the lower limb. [BOT] A large primary tree branch. [DES ENG] **1.** The graduated margin of an arc or circle in an instrument for measuring angles, as that part of a marine sextant carrying the altitude scale. **2.** The graduated staff of a leveling rod. [GEOL] One of the two sections of an anticline or syncline on either side of the axis. Also known as flank.

limbate [BIOL] Having a part of one color bordered with a different color.

limb bud [EMBRYO] A mound-shaped lateral proliferation of the embryonic trunk; the anlage of a limb.

limb darkening [ASTROPHYS] An observed darkening near the surface of the sun's limb as compared to its brighter center.

limber [ORD] A two-wheeled vehicle designed primarily to support the trail section of a gun carriage while in transit.

limber board [NAV ARCH] One of several movable planks for covering the bilge-water passages on either side of the keelson. Also known as bilge board.

limber hole [NAV ARCH] A drainage hole running through the bottom of the frame of ship.

limbic system [ANAT] The inner edge of the cerebral cortex in the medial and temporal regions of the cerebral hemispheres.

limb-kinetic apraxia See motor apraxia.

limburgite [PETR] A dark, glass-rich igneous rock with abundant large crystals of olivine and pyroxene and with little or no feldspar.

limbus [BIOL] A border clearly defined by its color or structure, as the margin of a bivalve shell or of the cornea of the eye.

lime [BOT] *Citrus aurantifolia*. A tropical tree with elliptic oblong leaves cultivated for its acid citrus fruit which is a hesperidium.

lime-and-cement mortar [MATER] Mortar made of mortar cement, lime putty or hydrated lime, and sand in proportions, by volume, normally of one part cement, one or two lime, and five or six sand; suited for all kinds of masonry.

lime glass [MATER] A type of glass containing a high proportion of lime; used in many commercial glass products, such as bottles.

lime grease [MATER] A type of grease that emulsifies less readily than one made with a soda base and therefore is used in an environment where water may occur.

lime kiln [CHEM ENG] Furnace-type apparatus, usually a long, tilted cylinder that is slowly rotated, used to heat calcium carbonate, $CaCO_3$ above 900°C to produce lime.

limelight [ENG] A light source once used in spotlights; it consisted of a block of lime heated to incandescence by means of an oxyhydrogen flame torch.

lime mortar [MATER] A mixture of hydrated lime, sand, and water having a compressive strength up to 400 pounds per square inch (2.8×10^6 newtons per square meter); used for interior non-load-bearing walls in buildings.

lime nitrogen [INORG CHEM] Trade name for calcium cyanamide.

lime oil [MATER] **1.** An edible essential oil squeezed from the rind of lime and other citrus fruit, whose components are limonene and citral; used in flavorings and perfumes. **2.** The distilled essential oil from citron.

lime-pan playa [GEOL] A playa with a smooth, hard surface composed of calcium carbonate.

lime putty [MATER] A puttylike cement made from lime slaked in water.

lime sour [TEXT] The first treatment with a weak acid solution given to cotton goods in the finishing process. Also known as gray sour.

limestone [PETR] **1.** A sedimentary rock composed dominantly (more than 95%) of calcium carbonate, principally in the form of calcite; examples include chalk and travertine. **2.** Any rock containing 80% or more of calcium carbonate or magnesium carbonate.

limestone log [ENG] A log that employs an electrical resistivity element in the form of four symmetrically arranged current electrodes to give accurate readings in borehole surveying of hard formations.

limestone pebble conglomerate [GEOL] A well-sorted conglomerate composed of limestone pebbles resulting from special conditions involving rapid mechanical erosion and short transport distances.

lime water [PHARM] An alkaline aqueous solution of calcium hydroxide; used in medicine as an antacid.

limicolous [ECOL] Living in mud.

liminal contrast See threshold contrast.

liming [AGR] Treating the soil with lime. [CHEM ENG] Soaking hides and skins in milk of lime and causing them to swell, to facilitate the removal of hair.

limit cycle of a differential equation [MATH] A closed trajectory C in the plane (corresponding to a periodic solution of the equation) where every point of C has a neighborhood so that every trajectory through it spirals toward C.

limit dimensioning method [DES ENG] Method of dimensioning and tolerancing wherein the maximum and minimum permissible values for a dimension are stated specifically to indicate the size or location of the element in question.

limited-access highway See expressway.

limited-pressure cycle See mixed cycle.

limited-rotation hydraulic actuator [MECH ENG] A type of hydraulic actuator that produces limited reciprocating rotary force and motion; used for lifting, lowering, opening, closing, indexing, and transferring movements; examples are the piston-rack actuator, single-vane actuator, and double-vane actuator.

limited signal [ELECTR] Radar signal that is intentionally limited in amplitude by the dynamic range of the radar system.

limited traverse [ORD] Restricted movement of a gun to right or left, caused by a mechanical device or by a natural obstacle.

limiter [ELECTR] An electronic circuit used to prevent the amplitude of an electronic waveform from exceeding a specified level while preserving the shape of the waveform at amplitudes less than the specified level. Also known as amplitude limiter; amplitude-limiting circuit; automatic peak limiter; clipper; clipping circuit; limiter circuit; peak limiter.

limiter circuit See limiter.

limit governor [MECH ENG] A mechanical governor that takes over control from the main governor to shut the machine down when speed reaches a predetermined excess above the allowable rated. Also known as topping governor.

LILIACEAE

Carolina lily (*Lilium michauxii*), a characteristic member of the family Liliaceae in the order Liliales. The six-membered perianth, with all members petaloid, is typical of the family, order, and subclass (Liliidae). (*Courtesy of John H. Gerard, from National Audubon Society*)

LIME

Tahiti (Persian) lime (*Citrus aurantifolia*).

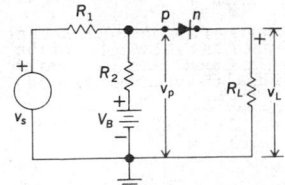

LIMITER

Circuit diagram of series diode clipping.

LIMONITE

├2.5 cm┤

Limonite in stalactitic masses, which are found in the Tintic District of Utah. *(Specimen from Department of Geology, Bryn Mawr College)*

LIMPET

Two views of the rough keyhole limpet *(Diodora aspera)* of the Pacific Coast, maximum length 2 inches (5 centimeters).

limit inferior [MATH] **1.** The limit inferior of a sequence whose nth term is a_n is the limit as N approaches infinity of the greatest lower bound of the terms a_n for which n is greater than N; denoted by $\lim\inf_{n\to\infty} a_n$ or $\underline{\lim}_{n\to\infty} a_n$. **2.** The limit inferior of a function f at a point c is the limit as ε approaches zero of the greatest lower bound of $f(x)$ for $|x - c| < \varepsilon$ and $x \neq c$; denoted by $\lim\inf_{x\to c} f(x)$ or $\underline{\lim}_{x\to c} f(x)$.

limiting [ELECTR] A desired or undesired amplitude-limiting action performed on a signal by a limiter. Also known as clipping.

limiting current density [PHYS CHEM] The maximum current density to achieve a desired electrode reaction before hydrogen or other extraneous ions are discharged simultaneously.

limiting viscosity number *See* intrinsic viscosity.

limit lines [IND ENG] Lines on a chart designating specification limits.

limit-load design *See* ultimate-load design.

limit of a function [MATH] A function $f(x)$ has limit L as x tends to c if given any positive number ε (no matter how small) there is a positive number δ such that if x is in the domain of f, x is not c, and $|x - c| < \delta$, then $|f(x) - L| < \varepsilon$; written $\lim_{x\to c} f(x) = L$.

limit of a sequence [MATH] A sequence $\{a_n : n = 1, 2, \ldots\}$ has limit L if given a positive number ε (no matter how small), there is a positive integer N such that for all integers n greater than N, $|a_n - L| < \varepsilon$.

limit of fire [ORD] **1.** Any angular limit, established for safety purposes, for firing at a towed aerial target. **2.** The boundary of an area within which gunfire is placed.

limit of resolution [OPTICS] The minimum distance or angular separation between two point objects which allows them to be resolved according to the Rayleigh criterion.

limit of the atmosphere [GEOPHYS] The level at which the atmospheric density becomes the same as the density of interplanetary space, which is usually taken to be about one particle per cubic centimeter.

limit on the left [MATH] The limit on the left of the function f at a point c is the limit of f at c which would be obtained if only values of x less than c were taken into account; more precisely, it is the number L which has the property that for any positive number ε, there is a positive number δ so that if x is the domain of f and $0 < c - x < \delta$ then $|f(x) - L| < \varepsilon$; denoted by $\lim_{x\to c^-} f(x) = L$, or $f(c^-) = L$. Also known as left-hand limit.

limit on the right [MATH] The limit on the right of the function $f(x)$ at a point c is the limit of f at c which would be obtained if only values of x greater than c were taken into account; more precisely, it is the number L which has the property that for any positive number ε there is a positive number δ so that if x is in the domain of f and $0 < x - c < \delta$, then $|f(x) - L| < \varepsilon$; denoted by $\lim_{x\to c^+} f(x) = L$ or $f(c^+) = L$. Also known as right-hand limit.

limit point *See* cluster point.

limit priority [ADP] An upper bound to the dispatching priority that a task can assign to itself or any of its subtasks.

limit ratio [ELECTR] Ratio of peak value to limited value, or comparison of such ratios.

limits [DES ENG] In dimensioning, the maximum and minimum values prescribed for a specific dimension; the limits may be of size if the dimension concerned is a size dimension, or they may be of location if the dimension concerned is a location dimension.

limits of integration [MATH] The end points of the interval over which a function is being integrated.

limit stop [ORD] An arm or part used to limit angular motions, as of a gun turret.

limit superior [MATH] **1.** The limit superior of a sequence whose nth term is a_n is the limit as N approaches infinity of the least upper bound of the terms a_n for which n is greater than N; denoted by $\lim\sup_{n\to\infty} a_n$ or $\overline{\lim}_{n\to\infty} a_n$. **2.** The limit superior of a function f at a point c is the limit as ε approaches zero of the least upper bound of $f(x)$ for $|x - c| < \varepsilon$ and $x \neq c$; denoted by $\lim\sup_{x\to c} f(x)$ or $\overline{\lim}_{x\to c} f(x)$.

limit switch [ELEC] A switch designed to cut off power automatically at or near the limit of travel of a moving object controlled by electrical means.

limit velocity [MECH] In armor and projectile testing, the lowest possible velocity at which any one of the complete penetrations is obtained; since the limit velocity is difficult to obtain, a more easily obtainable value, designated as the ballistic limit, is usually employed.

limivorous [ZOO] Feeding on mud, as certain annelids, for the organic matter it contains.

Limnebiidae [INV ZOO] The minute moss beetles, a family of coleopteran insects in the superfamily Hydrophiloidea.

limnetic [ECOL] Of, pertaining to, or inhabiting the pelagic region of a body of fresh water.

Limnichidae [INV ZOO] The minute false water beetles, a cosmopolitan family of coleopteran insects in the superfamily Dryopoidea.

limnimeter [ENG] A type of tide gage for measuring lake level variations.

limnite *See* bog iron ore.

Limnocharitaceae [BOT] A family of monocotyledonous plants in the order Alismatales characterized by schizogenous secretory canals, multiaperturate pollen, several or many ovules, and a horseshoe-shaped embryo.

limnograph [ENG] A recording made on a limnimeter.

limnology [ECOL] The science of the life and conditions for life in lakes, ponds, and streams.

Limnomedusae [INV ZOO] A suborder of hydrozoan coelenterates in the order Hydroida characterized by naked hydroids.

limnoplankton [BIOL] Plankton found in fresh water, especially in lakes.

Limnoriidae [INV ZOO] The gribbles, a family of isopod crustaceans in the suborder Flabellifera that burrow into submerged marine timbers.

limonene [ORG CHEM] $C_{10}H_{16}$ A terpene with a lemon odor that is optically active and is found in oils from citrus fruits and in oils from peppermint and spearmint; a colorless, water-insoluble liquid that boils at 176°C.

limoniform [BOT] Lemon-shaped.

limonite [MINERAL] A group of brown or yellowish-brown, amorphous, naturally occurring ferric oxides of variable composition; commonly formed secondary material by oxidation of iron-bearing minerals; a minor ore of iron. Also known as brown hematite; brown iron ore.

limpet [INV ZOO] Any of several species of marine gastropod mollusks composing the families Patellidae and Acmaeidae which have a conical and tentlike shell with ridges extending from the apex to the border.

Limulacea [INV ZOO] A group of horseshoe crabs belonging to the Limulida.

Limulida [INV ZOO] A subgroup of Xiphosurida including all living members of the subclass.

Limulodidae [INV ZOO] The horseshoe crab beetles, a family of coleopteran insects in the superfamily Staphylinoidea.

Limulus [INV ZOO] The horseshoe crab; the type genus of the Limulacea.

linac *See* linear accelerator.

Linaceae [BOT] A family of herbaceous or shrubby dicotyledonous plants in the order Linales characterized by mostly capsular fruit, stipulate leaves, and exappendiculate petals.

Linales [BOT] An order of dicotyledonous plants in the subclass Orsidae containing simple-leaved herbs or woody plants with hypogynous, regular, syncarpous flowers having five to many stamens which are connate at the base.

linalool [ORG CHEM] $(CH_3)_2C{:}CH(CH_2)_2CCH_3OHCH{:}CH_2$ A terpene that is a colorless liquid, has a bergamot odor, boils at 195–196°C, and is found in many essential oils, particularly bergamot and rosewood; used as a flavoring agent and in perfumes. Also known as coriandrol; 3,7-dimethyl-1,6-octadiene-3-ol.

linalyl acetate [ORG CHEM] $(CH_3)_2C{:}CH(CH_2)_2CCH_3(OCOCH_3)CH{:}CH_2$ The acetic acid ester of linalool, a colorless oily liquid with a bergamot odor that boils at 108–110°C; used in perfumes and as a flavoring agent.

linarite [MINERAL] $PbCu(SO_4)(OH)_2$ A deep-blue mineral composed of basic lead copper sulfate and occurring as monoclinic crystals.

Lincocin [MICROBIO] A trade name for the antibiotic lincomycin.

Lincoln experimental satellite [AERO ENG] A series of military communication satellites initiated in 1965; carried out successful X band experiments. Abbreviated LES.

lincomycin [MICROBIO] $C_{18}H_{34}O_6N_2S \cdot HCl$ Monobasic crystalline antibiotic, produced by *Streptomyces lincolnensis*, that is active as lincomycin hydrochloride mainly toward gram-positive microorganisms.

lindackerite [MINERAL] $Cu_6Ni_3(AsO_4)_4(SO_4)(OH)_4$ A light-green or apple-green mineral composed of hydrous basic sulfate and arsenate of nickel and copper; occurs in tabular crystals or massive.

Lindeck potentiometer [ELEC] A potentiometer in which an unknown potential difference is balanced against a known potential difference derived from a fixed resistance carrying a variable current; the converse of most potentiometers.

Linde copper sweetening [CHEM ENG] A petroleum-refinery process to treat gasolines and distillates with a slurry of clay and cupric chloride to remove mercaptans.

Linde drill See fusion-piercing drill.

Lindelof space [MATH] A topological space where if a family of open sets covers the space, then a countable number of these sets also covers the space.

Lindemann electrometer [ELEC] A variant of the quadrant electrometer, designed for portability and insensitivity to changes in position, in which the quadrants are two sets of plates about 6 millimeters apart, mounted on insulating quartz pillars; a needle rotates about a taut silvered quartz suspension toward the oppositely charged plates when voltage is applied to it, and its movement is observed through a microscope.

Lindemann glass [MATER] A lithium borate–beryllium oxide glass having no element higher in atomic number than oxygen; used as window material for low-voltage x-ray tubes because it will pass x-rays of extremely long wavelength, such as Grenz rays.

linden See basswood.

Linde's rule [SOLID STATE] The rule that the increase in electrical resistivity of a monovalent metal produced by a substitutional impurity per atomic percent impurity is equal to $a + b(v - 1)^2$, where a and b are constants for a given solvent metal and a given row of the periodic table for the impurity, and v is the valence of the impurity.

lindgrenite [MINERAL] $Cu_3(MoO_4)_2(OH)_2$ A green mineral composed of basic copper molybdate.

L indicator See L scope.

lindstromite [MINERAL] $PbCuBi_3S_6$ A lead-gray to tin-white mineral composed of bismuth copper lead sulfide.

line [BOT] A unit of length, equal to 1/12 inch, or approximately 2.117 millimeters; it is most frequently used by botanists in describing the size of plants. [ELEC] See power line; transmission line. [ELECTR] **1.** The path covered by the electron beam of a television picture tube in one sweep from left to right across the screen. **2.** One horizontal scanning element in a facsimile system. **3.** See trace. [MATH] The set of points $(x_1, \ldots, x_n)$ in euclidean space, each of whose coordinates is a linear function of a single parameter t; $x_i = f_i(t)$. [SPECT] See spectral line.

linea alba [ANAT] A tendinous ridge extending in the median line of the abdomen from the pubis to the tiphoid process and formed by the blending of aponeuroses of the oblique and transverse muscles of the abdomen.

lineage [GEN] Descent from a common progenitor.

lineage structure [CRYSTAL] An imperfection structure characterizing a crystal, parts of which have slight differences in orientation.

lineament [ASTRON] A prominent linear feature on the lunar surface. [GEOL] A straight or gently curved, lengthy topographic feature expressed as depressions or lines of depressions. Also known as linear. [GRAPHICS] A structurally controlled line on an aerial photograph; applied to lines representing beds, veins, faults, rock boundaries, and such.

line-and-staff organization [IND ENG] A form of organization structure which combines functional subunits with staff officers in line functions.

line and trunk group [COMMUN] A group consisting of four-wire circuits, incoming private automatic branch exchange trunks, and intertoll trunk groups.

linear [CONT SYS] Having an output that varies in direct proportion to the input. [GEOL] See lineament. [SCI TECH] **1.** Of or relating to a line. **2.** Having a single dimension.

linear accelerator [NUCLEO] A particle accelerator which accelerates electrons, protons, or heavy ions in a straight line by the action of alternating voltages. Also known as linac.

linear actuator [MECH ENG] A device that converts some kind of power, such as hydraulic or electric power, into linear motion.

linear algebra [MATH] The study of vector spaces and linear transformations.

linear algebraic equation [MATH] An equation in some algebraic system where the unknowns occur linearly, that is, to the first power.

linear amplifier [ELECTR] An amplifier in which changes in output current are directly proportional to changes in applied input voltage.

linear array [ELECTROMAG] An antenna array in which the dipole or other half-wave elements are arranged end to end on the same straight line. Also known as collinear array.

linear birefringence [OPTICS] Birefringence effects which are proportional to applied stresses.

linear burning rate [ORD] The distance normal to any burning surface of the propellant grain burned through in unit time.

linear circuit See linear network.

linear cleavage [GEOL] The property of metamorphic rocks of breaking into long planar fragments.

linear combination [MATH] A linear combination of vectors $\mathbf{v}_1, \ldots, \mathbf{v}_n$ in a vector space is any expression of the form $a_1\mathbf{v}_1 + a_2\mathbf{v}_2 + \ldots + a_n\mathbf{v}_n$, where the a_i are scalars.

linear comparator [ELECTR] A comparator circuit which operates on continuous, or nondiscrete, waveforms. Also known as continuous comparator.

linear computing element [ELEC] A linear circuit in an analog computer.

linear conductor antenna [ELECTROMAG] An antenna consisting of one or more wires which all lie along a straight line.

linear control [ELEC] Rheostat or potentiometer having uniform distribution of graduated resistance along the entire length of its resistance element.

linear control system [CONT SYS] A linear system whose inputs are forced to change in a desired manner as time progresses.

linear detection [ELECTR] Detection in which the output voltage is substantially proportional, over the useful range of the detecting device, to the voltage of the input wave.

linear differential equation [MATH] A differential equation in which all derivatives occur linearly, and all coefficients are functions of the independent variable.

linear distortion [ELECTR] Amplitude distortion in which the output signal envelope is not proportional to the input signal envelope and no alien frequencies are involved.

linear electrical constants of a uniform line [ELEC] Series resistance, series inductance, shunt conductance, and shunt capacitance per unit length of line.

linear electrical parameters See transmission-line parameters.

linear energy transfer See stopping power.

linear equation [MATH] A linear equation in the variables $x_1, \ldots, x_n$, and y is any equation of the form $a_1x_1 + a_2x_2 + \ldots + a_nx_n = y$.

linear expansion [PHYS] Expansion of a body in one direction.

linear feedback control [CONT SYS] Feedback control in a linear system.

linear flow structure See platy flow structure.

linear fractional transformations See Möbius transformations.

linear function See linear transformation.

linear functional [MATH] A linear transformation from a vector space to its scalar field.

linear independence [MATH] The property of a set of vectors $\mathbf{v}_1, \ldots, \mathbf{v}_n$ in a vector space where if $a_1\mathbf{v}_1 + a_2\mathbf{v}_2 + \ldots + a_n\mathbf{v}_n = 0$, then all the scalars $a_i = 0$.

LINDEMANN ELECTROMETER

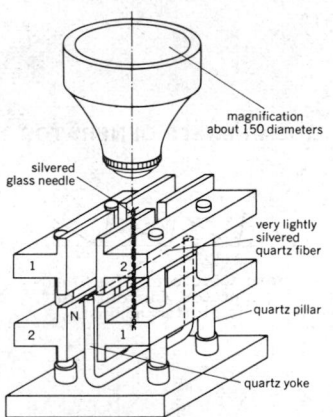

Lindemann electrometer. Here 1 and 2 indicate the quadrants. *(From F. A. Laws, Electrical Measurements, 2d ed., McGraw-Hill, 1938)*

LINEAR ACTUATOR

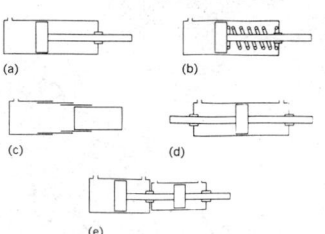

Hydraulic cylinder types of linear actuators. *(a)* Single-acting, external return. *(b)* Single-acting, spring return. *(c)* Telescopic. *(d)* Double-acting double-ended rod. *(e)* Tandem cylinders.

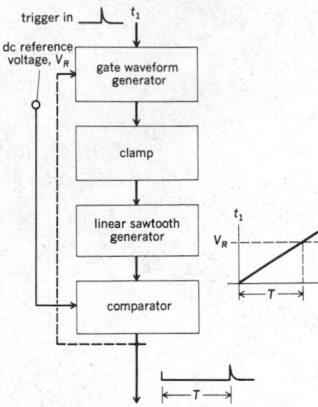

LINEAR-SWEEP DELAY CIRCUIT

Elements of linear-sweep delay circuit. T = delay time; V = reference voltage; t_1 = time.

LINEAR-SWEEP GENERATOR

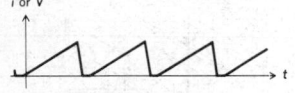

Sawtooth waveform of a linear-sweep generator. Current i or voltage v is plotted against time t.

linear inequalities [MATH] A collection of relations among variables x_i, where at least one relation has the form $\Sigma_i a_i x_i \geq 0$.

linear interpolation [MATH] A process to find a value of a function between two known values under the assumption that the three plotted points lie on a straight line.

linearity [MATH] The property whereby a mathematical system is well behaved (in the context of the given system) with regard to addition and scalar multiplication. [PHYS] The relationship that exists between two quantities when a change in one of them produces a directly proportional change in the other.

linearity control [ELECTR] A cathode-ray-tube control which varies the distribution of scanning speed throughout the trace interval. Also known as distribution control.

linearization [CONT SYS] 1. The modification of a system so that its outputs are approximately linear functions of its inputs, in order to facilitate analysis of the system. 2. The mathematical approximation of a nonlinear system, whose departures from linearity are small, by a linear system corresponding to small changes in the variables about their average values.

linear light [NAV] In marine operations, a luminous signal having perceptible length, as contrasted with a point light, which does not have perceptible length.

linear-logarithmic intermediate-frequency amplifier [ELECTR] Amplifier used to avoid overload or saturation as a protection against jamming in a radar receiver.

linearly independent quantities [MATH] Quantities which do not jointly satisfy a homogeneous linear equation unless all coefficients are zero.

linear magnetic amplifier [ELECTR] A magnetic amplifier employing negative feedback to make its output load voltage a linear function of signal current.

linear manifold [MATH] A subset of a vector space which is itself a vector space with the induced operations of addition and scalar multiplication.

linear meter [ENG] A meter in which the deflection of the pointer is proportional to the quantity measured.

linear modulation [COMMUN] Modulation in which the amplitude of the modulation envelope (or the deviation from the resting frequency) is directly proportional to the amplitude of the intelligence signal at all modulation frequencies.

linear molecule [PHYS CHEM] A molecule whose atoms are arranged so that the bond angle between each is 180°; an example is carbon dioxide, CO_2.

linear momentum See momentum.

linear motion See rectilinear motion.

linear network [ELEC] A network in which the parameters of resistance, inductance, and capacitance are constant with respect to current or voltage, and in which the voltage or current of sources is independent of or directly proportional to other voltages and currents, or their derivatives, in the network. Also known as linear circuit.

linear operator See linear transformation.

linear order [MATH] Any order $<$ on a set S with the property that for any two elements a and b in S either $a < b$ or $b < a$.

linear parallax See absolute stereoscopic parallax.

linear-phase [ELECTR] Pertaining to a filter or other network whose image phase constant is a linear function of frequency.

linear polarization [OPTICS] Polarization of an electromagnetic wave in which the electric vector at a fixed point in space remains pointing in a fixed direction, although varying in magnitude. Also known as plane polarization.

linear polymer [ORG CHEM] A polymer whose molecule is arranged in a chainlike fashion with few branches or bridges between the chains.

linear power amplifier [ELECTR] A power amplifier in which the signal output voltage is directly proportional to the signal input voltage.

linear programming [MATH] The study of maximizing or minimizing a linear function $f(x_1, \ldots, x_n)$ subject to given constraints which are linear inequalities involving the variables x_i.

linear rectifier [ELECTR] A rectifier, the output current of

voltage of which contains a wave having a form identical with that of the envelope of an impressed signal wave.

linear regression [STAT] The straight line running among the points of a scatter diagram about which the amount of scatter is smallest, as defined, for example, by the least squares method.

linear scanning [ENG] Radar beam which moves with constant angular velocity through the scanning sector, which may be a complete 360°.

linear space See vector space.

linear speed method [ORD] Method of calculating artillery firing data in which the future position of a moving target is determined by finding the direction of flight and the ground speed of the target; by multiplying the ground speed by the time of flight of the projectile, the future position is determined.

linear Stark effect [ATOM PHYS] A splitting of spectral lines of hydrogenlike atoms placed in an electric field; each energy level of principal quantum number n is split into $2n-1$ equidistant levels of separation proportional to the field strength.

linear stopping power See stopping power.

linear strain [MECH] The ratio of the change in the length of a body to its initial length. Also known as longitudinal strain.

linear sweep [ELECTR] A cathode-ray sweep in which the beam moves at constant velocity from one side of the screen to the other, then suddenly snaps back to the starting side.

linear-sweep delay circuit [ELECTR] A widely used form of linear time-delay circuit in which the input signal initiates action by a linear sawtooth generator, such as the bootstrap or Miller integrator, whose output is then compared with a calibrated direct-current reference voltage level.

linear-sweep generator [ELECTR] An electronic circuit that provides a voltage or current that is a linear function of time; the waveform is usually recurrent at uniform periods of time.

linear system [CONT SYS] A system in which the outputs are components of a vector which is equal to the value of a linear operator applied to a vector whose components are the inputs. [MATH] A system where all the interrelationships among the quantities involved are expressed by linear equations which may be algebraic, differential, or integral.

linear taper [ELEC] A taper that gives the same change in resistance per degree of rotation over the entire range of a potentiometer.

linear time base [ELECTR] A time base that makes the electron beam of a cathode-ray tube move at a constant speed along the horizontal time scale.

linear transducer [ELECTR] A transducer for which the pertinent measures of all the waves concerned are linearly related.

linear transformation [MATH] A function T defined in a vector space E and having its values in another vector space over the same field, such that if f and g are vectors in E, and c is a scalar, then $T(f+g) = Tf + Tg$ and $T(cf) = c(Tf)$. Also known as linear function; linear operator.

linear velocity See velocity.

lineation [GEOL] Any linear structure on or within a rock; examples are ripple marks and flow lines.

line balance [ELEC] 1. Degree of electrical similarity of the two conductors of a transmission line. 2. Matching impedance, equaling the impedance of the line at all frequencies, that is used to terminate a two-wire line.

line-balance converter See balun.

line block See line cut.

line blow [METEOROL] A strong wind on the equator side of an anticyclone, probably so called because there is little shifting of wind direction during the blow, as contrasted with the marked shifting which occurs with a cyclonic windstorm.

line brattice [MIN ENG] A partition in an opening to divide it into intake and return airways.

line building out network [ELEC] An impedance-matching network.

linecasting machine [GRAPHICS] A composing machine which assembles a line of type, casts it, and distributes the matrices to a magazine.

line characteristic distortion [COMMUN] A kind of teletypewriter transmission distortion caused when the lengths of the received signal impulses are affected by the presence of changing current transitions in wire circuits.

line circuit [ELEC] 1. Relay equipment associated with each station connected to a dial or manual switchboard. 2. A circuit to interconnect an individual telephone and a channel terminal.

line clinometer [ENG] A clinometer designed to be inserted between rods at any point in a string of drill rods.

line code [ADP] The single instruction required to solve a specific type of problem on a special-purpose computer.

line conductor [ELEC] A metal used as a conductor in a power line; the most frequently used conductors are copper and aluminum.

line-controlled blocking oscillator [ELECTR] A circuit formed by combining a monostable blocking oscillator with an open-circuit transmission line in the regenerative circuit; it is capable of generating pulses with large amounts of power.

line conversion [GRAPHICS] A picture obtained by photographing continuous-tone pictures with line film, and no screen, thus reducing all the values in the original picture to black or white.

line cord [ELEC] A two-wire cord terminating in a two-prong plug at one end and connected permanently to a radio receiver or other appliance at the other end; used to make connections to a source of power. Also known as power cord.

line-cord resistor [ELEC] An asbestos-enclosed wire-wound resistor incorporated in a line cord along with the two regular wires.

line cut [GRAPHICS] A relief printing plate made by photographing a design and then transferring the negative onto a zinc or copper plate that is then developed, with the lines that will form the printing surface being protected and the rest of the plate etched down; used exclusively for the reproduction of materials executed in black (or a color) and white, with no intermediate shades of gray (or tones). Also known as line block; line engraving; line etching; line plate.

line defect *See* dislocation.

line displacement [ASTROPHYS] Widening or shifting of spectral lines of celestial objects arising from several causes, such as gas under high pressure.

line drilling [MIN ENG] The combined methods of drilling and broaching for the primary cut in quarrying; deep, closely spaced holes are drilled in a straight line by means of a reciprocating drill, and webs between holes are removed by a drill or a flat broaching tool.

line drop [ELEC] The voltage drop existing between two points on a power line or transmission line, due to the impedance of the line.

line-drop compensator [ELEC] A device that restores the voltage lost when electricity is transmitted along a wire.

line drop signal [COMMUN] Signal associated with a subscriber line on a manual switchboard.

line engraving *See* line cut.

line equalizer [ELEC] An equalizer containing inductance or capacitance, inserted in a transmission line to modify the frequency response of the line.

line etching *See* line cut.

line facility [COMMUN] A transmission line in a communication system, together with amplifiers spaced at regular intervals to offset attenuation in the line.

line fault [ELEC] A defect, such as an open circuit, short circuit, or ground, in an electric line for transmission or distribution of power or of speech, music, or other content.

line feed [ADP] 1. Signal that causes a printer to feed the paper up a discrete number of lines. 2. Rate at which paper is fed through a printer.

line fill [COMMUN] Ratio of the number of connected main telephone stations on a line to the nominal main station capacity of that line.

line filter [ELEC] 1. A filter inserted between a power line and a receiver, transmitter, or other unit of electric equipment to prevent passage of noise signals through the power line in either direction. Also known as power-line filter. 2. A filter inserted in a transmission line or high-voltage power line for carrier communication purposes.

line filter balance [COMMUN] Network designed to maintain phantom group balance when one side of the group is equipped with a carrier system.

line finder [ADP] A device that automatically advances the platen of a line printer or typewriter. [COMMUN] A switching device that automatically locates an idle telephone or telegraph circuit going to the desired destination.

line-finder shelf [COMMUN] Usually, 20 line finders with the equipment required for connecting any of the associated calling telephones to a selector or connector which will receive the dial pulses from the calling telephone.

line-finder switch [COMMUN] In telephony, an automatic switch for seizing selector apparatus which provides dial tone to the calling party.

line flux [ELECTROMAG] A local inductive field of a telephone or power line.

line-formula method [ORG CHEM] A system of notation for hydrocarbons showing the chemical elements, functional groups, and ring systems in linear form; an example is acetone, CH_3COCH_3.

line frequency [ELECTR] The number of times per second that the scanning spot sweeps across the screen in a horizontal direction in a television system. Also known as horizontal frequency; horizontal line frequency.

line-frequency blanking pulse *See* horizontal blanking pulse.

line functions [IND ENG] Organizational functions having direct authority and responsibility.

line gale *See* equinoctial storm.

line graph [MATH] A graph in which successive points representing the value of a variable at selected values of the independent variable are connected by straight lines.

line hydrophone [ENG ACOUS] A directional hydrophone consisting of one straight-line element, an array of suitably phased elements mounted in line, or the acoustic equivalent of such an array.

Lineidae [INV ZOO] A family of the Heteronemertini.

line impedance [ELECTROMAG] The impedance measured across the terminals of a transmission line.

line influence [ELECTROMAG] The effect of a local inductive field around a telephone line.

line integral [MATH] 1. For a curve in a vector space defined by $\mathbf{x} = \mathbf{x}(t)$, and a vector function $\mathbf{V}$ defined on this curve, the line integral of $\mathbf{V}$ along the curve is the integral over t of the scalar product of $\mathbf{V}[\mathbf{x}(t)]$ and $d\mathbf{x}/dt$; this is written $\int \mathbf{V} \cdot d\mathbf{x}$. 2. For a curve defined by $x = x(t)$, $y = y(t)$, and a scalar function f depending on x and y, the line integral of f along the curve is the integral over t of $f[x(t),y(t)]\sqrt{(dx/dt)^2 + (dy/dt)^2}$; this is written $\int f ds$, where $ds = \sqrt{(dx)^2 + (dy)^2}$ is an infinitesimal element of length along the curve. 3. For a curve in the complex plane defined by $z = z(t)$, and a function f depending on z, the line integral of f along the curve is the integral over t of $f[z(t)] (dz/dt)$; this is written $\int f dz$.

line interlace *See* interlaced scanning.

line lengthener [ELECTROMAG] Device for altering the electrical length of a waveguide or transmission line without altering other electrical characteristics, or the physical length.

line level [COMMUN] Signal level in decibels at a particular position on a transmission line.

line link [COMMUN] A frame on which several hundred telephone lines appear in a crossbar switching system. Abbreviated LL.

line location [ELEC] The location of power and communications lines when two or more such lines run along the same route; they should either be used jointly, or located with respect to each other so as to avoid unnecessary crossings, conflicts, and inductive exposures.

line loop [COMMUN] Portion of a telephone circuit that includes a user's telephone set and the pair of wires that connect it with the distributing frame of a central office.

line loop resistance [ELEC] Metallic resistance of the line wires that extend from an individual telephone set to the dial central office.

line loss [ELEC] Total of the various energy losses occurring in a transmission line.

line lubricator *See* line oiler.

line map *See* planimetric map.

line microphone [ENG ACOUS] A highly directional microphone consisting of a single straight-line element or an array

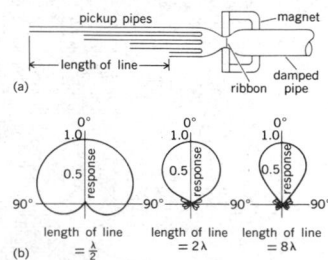

LINE MICROPHONE

Line microphone. (a) Sectional view showing open-ended pipes. (b) Directivity patterns. The maximum voltage response is arbitrarily chosen as unity.

of small parallel tubes of different lengths, with one end of each abutting a microphone element. Also known as machine-gun microphone.

line misregistration [ADP] In character recognition, the improper appearance of a line of characters, on site in a character reader, with respect to a real or imaginary horizontal line.

line mixer *See* flow mixer.

linen [TEXT] A cloth made from flax fibers, noted for its strength, weavability, durability, and minimum discharge of lint.

line noise [COMMUN] Noise originating in a transmission line from such causes as poor joints and inductive interference from power lines.

line of action [MECH ENG] The locus of contact points as gear teeth profiles go through mesh.

line of aim [ORD] A line from a person's eye, as that of a gunner or bombardier, through a sight, along which aim is taken.

line of apsides [ASTRON] **1.** The line connecting the two points of an orbit that are nearest and farthest from the center of attraction, as the perigee and apogee of the moon or the perihelion and aphelion of a planet. **2.** The length of this line.

line of balance [IND ENG] A production planning system that schedules key events leading to completion of an assembly on the basis of the delivery date for the completed system. Abbreviated LOB.

line of collimation [OPTICS] In a surveying telescope, the imaginary line through the optical center of the object glass and the cross-hair intersection in the diaphragm.

line of departure [MECH] **1.** The direction of a projectile at the instant it clears the muzzle of the gun. **2.** The direction of a bomb or rocket at the instant of launching. [NAV] The initial position of a scouting line, from which scouts proceed on their prescribed courses for search.

line of electrostatic induction [ELEC] A unit of electric flux equal to the electric flux associated with a charge of 1 statcoulomb.

line of elevation [ORD] The prolongation of the bore of a gun when the piece is set to fire.

line of fall [MECH] The line tangent to the ballistic trajectory at the level point.

line of flight [MECH] The line of movement, or the intended line of movement, of an aircraft, guided missile, or projectile in the air.

line of flux *See* line of force.

line of force [PHYS] An imaginary line in a field of force (such as an electric, magnetic, or gravitational field) whose tangent at any point gives the direction of the field at that point; the lines are spaced so that the number through a unit area perpendicular to the field represents the intensity of the field. Also known as flux line; line of flux.

line of impact [MECH] A line tangent to the trajectory of a missile at the point of impact.

line of magnetic induction *See* maxwell.

line of observation [ORD] A line from a position finder to a target at the exact time of a recorded observation.

line of position [NAV] A line indicating a series of possible positions of a craft, determined by observation or measurement. Also known as position line.

line of retirement [NAV] The position of a scouting line when it has reached its outer limit, and a search to the rear is initiated.

line of return [NAV] The final position of the scouting line, where individual scouts leave their stations and return to their bases.

line of sight [ELECTROMAG] The straight line for a transmitting radar antenna in the direction of the beam. [SCI TECH] A straight, unobstructed path or line between two points, as between an observer's eye and a target.

line-of-sight velocity *See* radial velocity.

line of soundings [NAV] A series of soundings obtained by a vessel when it is underway.

line of strike *See* strike.

line of thrust [MECH] Locus of the points through which the resultant forces pass in an arch or retaining wall.

line of tunnel [ENG] The width marked by the exterior lines or sides of a tunnel.

line oiler [MECH ENG] An apparatus inserted in a line conducting air or steam to an air- or steam-activated machine that feeds small controllable amounts of lubricating oil into the air or steam. Also known as air-line lubricator; line lubricator.

Lineolaceae [MICROBIO] A family of bacteria in some systems of classification that includes coenocytic members (*Lineola*) of the Caryophanales.

lineolate [BIOL] Marked with fine lines.

line pad [ELECTR] Pad inserted between a program amplifier and a transmission line, to isolate the amplifier from impedance variations of the line.

line parameters *See* transmission-line parameters.

line plate *See* line cut.

line printer [ADP] A device that prints an entire line in a single operation, without necessarily printing one character at a time.

line printing [ADP] The printing of an entire line of characters as a unit.

line profile [ASTROPHYS] A curve that indicates the internal variation in intensity of a spectral line of a celestial body.

line pulsing [ELECTR] Method of pulsing a transmitter in which an artificial line is charged over a relatively long period of time and then discharged through the transmitter tubes in a short interval determined by the line characteristic.

liner [DES ENG] A replaceable tubular sleeve inside a hydraulic or pump-pressure cylinder in which the piston travels. [ENG] A string of casing in a borehole. [MET] **1.** The cylindrical chamber that holds the billet for extrusion. **2.** The slab of coating metal that is placed on the core alloy and is subsequently rolled down to form a clad composite. [MIN ENG] **1.** A foot piece for uprights in timber sets. **2.** Timber supports erected to reinforce existing sets which are beginning to collapse due to heavy strata pressure. **3.** A bar put up between two other bars to assist in carrying the roof. **4.** Replaceable facings inside a grinding mill. [NAV ARCH] A merchant vessel engaged in regular, usually high-speed service.

line radiation [ELECTROMAG] Electromagnetic radiation from a power line caused mainly by corona pulses; gives rise to radio interference.

liner bushing [DES ENG] A bushing, provided with or without a head, that is permanently installed in a jig to receive the renewable wearing bushings. Also known as master bushing.

line reflection [COMMUN] Reflection of a signal at the end of a transmission line, at the junction of two or more lines, or at a substation.

line relay [ELEC] Relay which is controlled over a subscriber line or trunkline.

liner plate cofferdam [CIV ENG] A cofferdam made from steel plates about 16 inches (41 centimeters) high and 3 feet (91 centimeters) long, and corrugated for added stiffness.

lines [NAV ARCH] The outline of a ship, either as projected onto one of three perpendicular planes or as viewed visually. Also known as ship's lines.

line segment [MATH] A connected piece of a line.

line shafting [MECH ENG] One or more pieces of assembled shafting to transmit power from a central source to individual machines.

line side [ELEC] Terminal connections to an external or outstation source, such as data terminal connections to a communications circuit connecting to another data terminal.

line skew [ADP] In character recognition, a form of line misregistration, when the string of characters to be recognized appears in a uniformly slanted condition with respect to a real or imaginary baseline.

linesman [ENG] **1.** A worker who sets up and repairs communication and power lines. **2.** An assistant to a surveyor.

line source [OPTICS] An idealized source of light consisting of an infinitely long line from which light is emitted with uniform intensity.

line space lever [MECH ENG] A lever on a typewriter used to move the carriage to a new line.

line spectrum [SPECT] **1.** A spectrum of radiation in which the quantity being studied, such as frequency or energy, takes on discrete values. **2.** Conventionally, the spectra of atoms, ions, and certain molecules in the gaseous phase at low pressures; distinguished from band spectra of molecules, which consist of a pattern of closely spaced spectral lines which could not be resolved by early spectroscopes.

line speed [COMMUN] Maximum rate at which signals may be transmitted over a given channel, usually in bauds or bits per second.

lines per minute [ADP] A measure of the speed of the printer. Abbreviated LPM.

line squall [METEOROL] A squall that occurs along a squall line.

line-stabilized oscillator [ELECTR] Oscillator in which a section of line is used as a sharply selective circuit element for the purpose of controlling the frequency.

line storm *See* equinoctial storm.

line strength [ATOM PHYS] The intensity of a spectrum line.

line stretcher [ELECTROMAG] Section of waveguide or rigid coaxial line whose physical length is variable to provide impedance matching.

line switching [COMMUN] A telephone switching system in which a switch attached to a subscriber line connects an originating call to an idle part of the switching apparatus. [ELECTR] Connecting or disconnecting the line voltage from a piece of electronic equipment.

line switching concentrator [COMMUN] Switching center used between a group of users and the switching center to reduce the number of trunks and increase efficiency of switching equipment usage.

line synchronizing pulse *See* horizontal synchronizing pulse.

line timbers [MIN ENG] Timbers placed along the sides of the track of a working place in rows according to a predetermined plan.

line-to-ground fault [ELEC] A defect in a power or communications line in which faulty insulation allows the conductor to make contact with the earth.

line transducer [ELECTR] A special type of electret transducer consisting essentially of a coaxial cable with polarized dielectric, and with the center conductor and shield serving as electrodes; mechanical excitation resulting in a deformation of the shield at any point along the length of the cable produces an electrical output signal.

line transformer [ELEC] Transformer connecting a transmission line to terminal equipment; used for isolation, line balance, impedance matching, or additional circuit connections.

line trap [ELEC] A filter consisting of a series inductance shunted by a tuning capacitor, inserted in series with the power or telephone line for a carrier-current system to minimize the effects of variations in line attenuation and reduce carrier energy loss.

line tuning [ELEC] Adjustment of the frequency of carrier current of a communication system to tune out the reactance of a capacitor with suitable inductance.

line unit [ELECTR] Electric control device used to send, receive, and control the impulses of a teletypewriter.

line up [MIN ENG] **1.** A command signifying that the drill runner wants the hoisting cable attached to the drill stem, threaded through the sheave wheel, or wound on the hoist drum. **2.** To reposition a drill so that the drill stem is centered over the parallel to a newly collared drill hole.

line-use ratio [COMMUN] As applied to facsimile broadcasting, the ratio of the available line to the total length of scanning line.

line voltage [ELEC] The voltage provided by a power line at the point of use.

line-voltage regulator [ELEC] A regulator that counteracts variations in power-line voltage, so as to provide an essentially constant voltage for the connected load.

line vortex [FL MECH] A type of fluid motion in which fluid flows approximately in circles about a line, at speeds inversely proportional to the distance from the line, so that there is an infinite concentration of vorticity on the line, and vorticity vanishes elsewhere.

lingual artery [ANAT] An artery originating in the external carotid and supplying the tongue.

lingual gland [ANAT] A serous, mucous, or mucoserous gland lying deep in the mucous membrane of the mammalian tongue.

lingual nerve [ANAT] A branch of the mandibular nerve having somatic sensory components and innervating the

mucosa of the floor of the mouth and the anterior two-thirds of the tongue.

lingual tonsil [ANAT] An aggregation of lymphoid tissue composed of 35–100 separate tonsillar units occupying the posterior part of the tongue surface.

Linguatulida [INV ZOO] The equivalent name for Pentastomida.

Linguatuloidea [INV ZOO] A suborder of pentastomid arthropods in the order Porocephalida; characterized by an elongate, ventrally flattened, annulate, posteriorly attenuated body, simple hooks on the adult, and binate hooks in the larvae.

linguistics [COMMUN] The study of human speech in its various aspects, especially units of language, phonetics, syntax, accent, semantics, and grammar.

lingula [ANAT] A tongue-shaped organ, structure, or part thereof.

Lingulacea [INV ZOO] A superfamily of inarticulate brachiopods in the order Lingulida characterized by an elongate, biconvex calcium phosphate shell, with the majority having a pedicle.

lingulate [BIOL] Tongue- or strap-shaped.

Lingulida [INV ZOO] An order of inarticulate brachiopods represented by two living genera, *Lingula* and *Glottidia*.

linguloid ripple mark *See* linguoid ripple mark.

linguoid current ripple *See* linguoid ripple mark.

linguoid ripple mark [GEOL] An aqueous current ripple mark with tonguelike projections which are formed by action of a current of water and which point into the current. Also known as cuspate ripple mark; linguloid ripple mark; linguoid current ripple.

liniment [PHARM] A heat-generating liquid that is thinner than ointment and is applied to the skin with friction.

lining [MATER] A material used to protect inner surfaces, as of tunnels, pipes, or process equipment.

lining bar [DES ENG] A crowbar with a pinch, wedge, or diamond point at its working end.

lining sight [MIN ENG] An instrument consisting of a plate with a slot in the middle, and the means of suspending it; used with a plumbline for directing the courses of underground drifts or headings.

linin net [CYTOL] The reticulum composed of chromatinic or oxyphilic substances in a cell nucleus.

linishing *See* belt grinding.

link [CIV ENG] A standardized part of a surveyor's chain, being 7.92 inches (20.1168 centimeters) in the Gunter's chain and 1 foot (30.48 centimeters) in the engineer's chain. [COMMUN] General term used to indicate the existence of communications facilities between two points. [DES ENG] **1.** One of the rings of a chain. **2.** A connecting piece in the moving parts of a machine.

linkage [ADP] In programming, coding that connects two separately coded routines. [ELECTROMAG] *See* flux linkage. [GEN] Failure of nonallelic genes to recombine at random as a result of their being located within the same chromosome or chromosome fragment.

linkage editor [ADP] A service routine that converts the output of assemblers and compilers into a form that can be loaded and executed.

linkage group [GEN] The genes located on a single chromosome.

link bar *See* hinged bar.

link chute adapter [ORD] A unit attached to a gun to allow a link ejection chute to be fastened to the gun, and to lead ejected links to the chute.

link circuit [ELECTROMAG] Closed loop used for coupling purposes; it generally consists of two coils, each having a few turns of wire, connected by a twisted pair of wires or by other means, with each coil placed over, near, or in one of the two coils that are to be coupled.

link control message [COMMUN] **1.** Message sent over a link of a network to condition the link to handle transmissions in a prearranged manner. **2.** Message used only between a pair of terminals for the conditioning of the link for digital system control.

link coupling [ELECTROMAG] Modification of inductive coupling where the two coils are connected together by a short

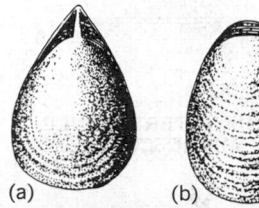

LINGULACEA

(a)   (b)

Internal molds of Cambrian lingulacean *Lingulella.* (a) Pedicle valve. (b) Brachial valve. (*From C. D. Walcott, Cambrian Brachiopoda, USGS Monogr. no. 51, 1912*)

length of transmission line, with each coil inductively coupled to the coil of a separate tuned circuit.

linked ammunition [ORD] Cartridges fastened side by side with metal links, forming a belt for ready feed to a machine gun.

link ejection chute [ORD] A chute or passage attached to a machine gun, through which links are thrown or conveyed to a desired point after being separated from the cartridges; it may be either fixed or flexible.

link encryption [COMMUN] The application of on-line crypto-operation to the individual links of relay systems so that all messages passing over the link are encrypted in their entirety.

linker-delinker [ORD] A machine designed to assemble or disassemble a metallic disintegrating linked belt for ammunition.

Linke scale [METEOROL] A type of cyanometer; used to measure the blueness of the sky; it is simply a set of eight cards of different standardized shades of blue, numbered (evenly) 2 to 16; the odd numbers are used by the observer if the sky color lies between any of the given shades. Also known as blue-sky scale.

link group [COMMUN] A collection of links that employ the same multiplex terminal equipment.

link-loading machine [ORD] Machine that quickly loads ammunition into interlocking metal links, which in turn form an ammunition belt for certain types of automatic weapons.

link stretch [ORD] The change in the center to center distance of the individual rounds of belted ammunition as the load is applied.

lin-log amplifier [ELECTR] Automatic gain control amplifier that operates in a linear manner for low-amplitude input signals, but responds in a logarithmic manner to high-amplitude input signals.

linnaeite [MINERAL] $(Co,Ni)_3S_4$ A steel-gray mineral with a coppery-red tarnish, occurring in isometric crystals; an ore of cobalt. Also known as cobalt pyrites; linneite.

linneite *See* linnaeite.

Linnik interference microscope [OPTICS] A type of interference microscope used for studying the surface structure of reflecting specimens; light from a source is divided by a semireflecting mirror into two beams, one of which is focused through an objective onto the specimen surface, the other onto a comparison surface; after reflection from the respective surfaces, the beams are reunited by the mirror.

Linofilm typesetter [GRAPHICS] A photographic typesetting machine consisting of a keyboard, photographic unit, corrector, and composer; the keyboard produces a perforated paper tape containing information for operating the photographic unit, which produces right-reading positive type on film or on photographic paper; the corrector affixes correct lines in position automatically; and the composer produces a made-up page.

linoleic acid [BIOCHEM] $C_{17}H_{31}COOH$ A yellow unsaturated fatty acid, boiling at 229°C (14 mm Hg), occurring as a glyceride in drying oils; obtained from linseed, safflower, and tall oils; a principal fatty acid in plants, and considered essential in animal nutrition; used in medicine, feeds, paints, and margarine. Also known as linolic acid; 9,12-octadecadienoic acid.

linolenate [BIOCHEM] A salt or ester of linolenic acid.

linolenic acid [BIOCHEM] $C_{17}H_{29}COOH$ One of the principal unsaturated fatty acids in plants and an essential fatty acid in animal nutrition; a colorless liquid that boils at 230°C (17 mm Hg), soluble in many organic solvents; used in medicine and drying oils. Also known as 9,12,15-octadecatrienoic acid.

linoleum [MATER] A floor covering made by applying a mixture of gelled linseed oil, pigments, fillers, and other materials to a burlap backing, and curing to produce a hard, resilient sheet.

linolic acid *See* linoleic acid.

Linotron [GRAPHICS] A high-speed cathode-ray photocomposition machine, designed to take computer magnetic tapes directly and expose negatives at speeds from 600 to over 1000 characters per second, depending on type size.

Linotype [GRAPHICS] A typesetting machine in which the type molds of letters are arranged in lines; solid slugs, or lines

of type, are cast; to make a correction, the entire line must be reset and recast.

linseed cake [MATER] The residue formed during pressing of commercial linseed oil; used for cattle feed and fertilizer.

linseed oil [MATER] A product made from the seeds of the flax plant by crushing and pressing either with or without heat; formulated in various grades and with various drying agents and used as a vehicle in oil paints and as a component of oil varnishes.

lintel [BUILD] A horizontal member over an opening, such as a door or window, usually carrying the wall load.

linter [MECH ENG] A machine for removing fuzz linters from ginned cottonseed.

linters [BOT] Short residual fibers that adhere to ginned cottonseed; used for making fabrics that do not require long fibers.

linuron [ORG CHEM] A powder, 3-(3,4-dichlorophenyl)-1-methoxy-1-methylurea, soluble in organic solvents; used as a selective pre- or postemergence herbicide.

LIOCS [ADP] Set of routines handling buffering, blocking, label checking, and overlap of input/output with processing. Derived from logical input/output control system.

lion [VERT ZOO] *Felis leo.* A large carnivorous mammal of the family Felidae distinguished by a tawny coat and blackish tufted tail, with a heavy blackish or dark-brown mane in the male.

Lion *See* Leo.

Liopteridae [INV ZOO] A small family of hymenopteran insects in the superfamily Cynipoidea.

Liouville equation [STAT MECH] An equation which states that the density of points representing an ensemble of systems in phase space which are in the neighborhood of some given system does not change with time.

Liouville-Neumann series [MATH] An infinite series of functions constructed from the given functions in the Fredholm equation which under certain conditions provides a solution. Also known as Neumann series.

Liouville's theorem [MATH] Every function of a complex variable which is bounded and analytic in the entire complex plane must be constant.

lip [ANAT] A fleshy fold above and below the entrance to the mouth of mammals. [MED] The margin of an open wound. [SCI TECH] The edge of a hollow cavity or container.

Lipalian [GEOL] A hypothetical geologic period that supposedly antedated the Cambrian.

Liparidae [INV ZOO] The equivalent name for Lymantriidae.

lipase [BIOCHEM] An enzyme that catalyzes the hydrolysis of fats or the breakdown of lipoproteins.

lipemia [MED] The presence of a fine emulsion of fatty substance in the blood. Also known as lipidemia; lipoidemia.

Liphistiidae [INV ZOO] A family of spiders in the suborder Liphistiomorphae in which the abdomen shows evidence of true segmentation by the presence of tergal and sternal plates.

Liphistiomorphae [INV ZOO] A suborder of arachnids in the order Araneida containing families with a primitively segmented abdomen.

lipid [BIOCHEM] One of a class of compounds which contain long-chain aliphatic hydrocarbons and their derivatives, such as fatty acids, alcohols, amines, amino alcohols, and aldehydes; includes waxes, fats, and derived compounds. Also known as lipin; lipoid.

lipidemia *See* lipemia.

lipid histiocytosis [MED] **1.** Any collection of histiocytes containing lipids. **2.** *See* Niemann-Pick disease.

lipid metabolism [BIOCHEM] The physiologic and metabolic processes involved in the assimilation of dietary lipids and the synthesis and degradation of lipids.

lipid nephrosis [MED] A chronic kidney disease of children associated with thickening of the basement membranes of glomeruli and characterized by edema, presence of protein in the urine, and abnormally high blood levels of albumin and cholesterol.

lipidosis [MED] The generalized deposition of fat or fatty substances in reticuloendothelial cells. Also known as lipoidosis.

lipid pneumonia [MED] **1.** Pneumonia resulting from aspiration of oily substances, such as nose drops. **2.** Deposition of

LINNIK INTERFERENCE MICROSCOPE

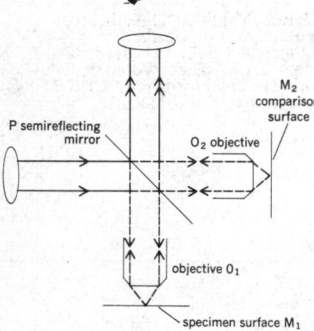

The Linnik interference microscope for reflecting specimens.

LINOTYPE

Linotype, a kind of keyboard typesetter that produces type in the form of slugs, or solid lines.

lipids in tissues of chronically inflamed lungs. Also known as lipoid pneumonia.

lipid proteinosis [MED] A hereditary disorder characterized by extracellular deposits of phospholipid-protein conjugate involving various areas of the body, including the skin and air passages.

lipid storage disease [MED] Any of various rare diseases characterized by the accumulation of large histiocytes containing lipids throughout reticuloendothelial tissues; examples are Goucher's disease, Niemann-Pick disease, and amaurotic familial idiocy.

lipin [BIOCHEM] 1. A compound lipid, such as a cerebroside. 2. See lipid.

lipoblastoma See liposarcoma.

lipochondrodystrophy See Hurler's syndrome.

lipochrome [BIOCHEM] Any of various fat-soluble pigments, such as carotenoid, occurring in natural fats. Also known as chromolipid.

lipodystrophy [MED] A disturbance of fat metabolism in which the subcutaneous fat disappears over some regions of the body, but is unaffected in others.

lipofuscin [BIOCHEM] Any of a group of lipid pigments found in cardiac and smooth muscle cells, in macrophages, and in parenchyma and interstitial cells; differential reactions include sudanophilia, Nile blue staining, fatty acid, glycol, and ethylene.

lipogranuloma [MED] A small mass of fatty tissue associated with granulomatous inflammation.

lipoic acid [BIOCHEM] $C_8H_{14}O_2S_2$ A compound which participates in the enzymatic oxidative decarboxylation of α-keto acids in a stage between thiamine pyrophosphate and coenzyme A.

lipoid [BIOCHEM] 1. A fatlike substance. 2. See lipid.

lipoidemia See lipemia.

lipoidosis See lipidosis.

lipoid pneumonia See lipid pneumonia.

lipoma [MED] A benign tumor composed of fat cells.

lipomatosis [MED] 1. Multiple lipomas. 2. Obesity.

lipomelanotic reticulosis [MED] A form of lymph node hyperplasia characterized by preservation of the architectural structure, inflammatory exudate, and hyperplasia of the reticulum cells which show phagocytosis of hemosiderin, melanin, and occasionally fat. Also known as dermatopathic lymphadenitis.

Lipomycetoideae [MICROBIO] A subfamily of oxidative yeasts in the family Saccharomycetaceae characterized by budding cells and a saclike appendage which develops into an ascus.

lipomyxoma See liposarcoma.

lipophore [HISTOL] A chromatophore which contains lipochrome.

lipopolysaccharide [BIOCHEM] Any of a class of conjugated polysaccharides consisting of a polysaccharide combined with a lipid.

lipoprotein [BIOCHEM] Any of a class of conjugated proteins consisting of a protein combined with a lipid.

liposarcoma [MED] A sarcoma originating in adipose tissue. Also known as embryonal-cell lipoma; fetal fat-cell lipoma; infiltrating lipoma; lipoblastoma; lipomyxoma; myxolipoma; myxoma lipomatodes.

Lipostraca [PALEON] An order of the subclass Branchiopoda erected to include the single fossil species *Lepidocaris rhyniensis*.

lipotropic [BIOCHEM] Having an affinity for lipid compounds. [PHARM] Having a preventive or curative effect on the deposition of excessive fat in abnormal sites.

lipotropic hormone [BIOCHEM] Any hormone having lipolytic activity on adipose tissue.

Lipotyphla [VERT ZOO] A group of insectivoran mammals composed of insectivores which lack an intestinal cecum and in which the stapedial artery is the major blood supply to the brain.

lipoxidase [BIOCHEM] An enzyme catalyzing the oxidation of the double bonds of an unsaturated fatty acid.

Lippich prism [OPTICS] A Nicol prism which is placed in the eyepiece of a polarimeter, covering half the field of view, to identify the character of polarized light emerging from the instrument.

Lippman fringes [OPTICS] Interference fringes in standing electromagnetic waves generated when light is reflected by a mercury coating at the back of a special fine-grained photographic emulsion; originally used in color photography.

Lipschitz condition [MATH] A function f satisfies such a condition at a point b if $|f(x) - f(b)| \leq K|x - b|$, with K a constant, for all x in some neighborhood of b.

lip-sync [COMMUN] Synchronization of sound and motion picture so that facial movements of speech coincide with the sounds.

liptinite See exinite.

liq pt See pint.

liquation [MET] 1. Separation of fusible metals from less fusible ones by applying heat. 2. The partial melting of an alloy.

liquefaction [PHYS] A change in the phase of a substance to the liquid state; usually, a change from the gaseous to the liquid state, especially of a substance which is a gas at normal pressure and temperature.

liquefied gas [MATER] A gaseous compound or mixture converted to the liquid phase by cooling or compression; examples are liquefied petroleum gas (LPG), liquefied natural gas (LNG), liquid oxygen, and liquid ammonia.

liquefied natural gas [MATER] A product of natural gas which consists primarily of methanes; its critical temperature is about $-100°F$ $(-73°C)$, and thus it must be liquefied by cooling to cryogenic temperatures and must be well insulated to be held in the liquid state; used as a domestic fuel. Abbreviated LNG.

liquefied petroleum gas [MATER] A product of petroleum gases; principally propane and butane, it must be stored under pressure to keep it in a liquid state; it is often stored in metal cylinders (bottled gas) and used as fuel for tractors, trucks, and buses, and as a domestic cooking or heating fuel in rural areas. Abbreviated LPG.

liquefier [ENG] Equipment or system used to liquefy gases; usually employs a combination of compression, heat exchange, and expansion operations.

liqueur [FOOD ENG] An alcoholic beverage prepared by combining a spirit, usually brandy, with certain flavorings and sugar.

liquid [PHYS] A state of matter intermediate between that of crystalline substances and gases in which a substance has the capacity to flow under extremely small shear stresses and conforms to the shape of a confining vessel, but is relatively incompressible, lacks the capacity to expand without limit, and can possess a free surface.

liquid air [PHYS] Air in the liquid state obtained as a faintly bluish, transparent, mobile, intensely cold liquid by compressing purified air and cooling it to a temperature below the boiling points of its principal components, nitrogen and oxygen; used chiefly as a refrigerant.

liquid asphalt See residual oil.

liquid blast cleaning [MET] Cleaning metal surfaces with a suspension of abrasive in water accelerated to high velocities by compressed air, or by a centrifugal wheel.

liquid blocking [PETRO ENG] The blocking or plugging of the sand around an injection-well borehole, usually caused by lubricant carryover from compressors.

liquid bright gold [MATER] Any of several gold compounds applied to ceramics in the form of varnish which is dried and heated to redness, decomposing the compound and leaving a thin film of gold firmly attached to the underlying ceramic; used in decorating china and for the production of printed electrical circuits on ceramics.

liquid-bubble tracer [FL MECH] A method of observing the motion of a liquid by following tiny particles of an immiscible liquid of the same density as the moving liquid.

liquid carburizing [MET] Surface hardening of steel by immersion into a molten bath consisting of cyanides and other salts, for example, at 1600-1750°F (850-950°C).

liquid chromatography [ANALY CHEM] A form of chromatography employing a liquid as the moving phase and a solid or a liquid on a solid support as the stationary phase; techniques include column chromatography, gel permeation chromatography, and partition chromatography.

liquid-column gage See U-tube manometer.

liquid compass [ENG] A compass in a bowl filled with liquid.

LIQUID HELIUM

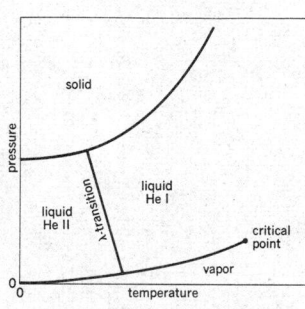

Phase diagram for He⁴ (not to scale). There are two liquid phases, helium I and helium II; the transition between them is called the λ-transition. Solid phase is never in equilibrium with vapor phase, and occurs only at elevated pressures. Critical point has temperature of 5.2K and pressure of 2.26 atmospheres.

LIQUID PISTON ROTARY COMPRESSOR

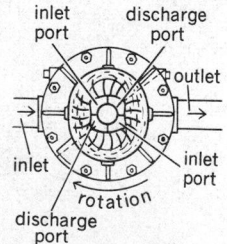

Schematic of the compressor showing its components.

liquid-cooled dissipator   *See* cold plate.

liquid-cooled engine   [MECH ENG] An internal combustion engine with a jacket cooling system in which liquid, usually water, is circulated to maintain acceptable operating temperatures of machine parts.

liquid cooling   [ENG] Use of circulating liquid to cool process equipment and hermetically sealed components such as transistors.

liquid crystal   [PHYS CHEM] A liquid which is not isotropic; it is birefringent and exhibits interference patterns in polarized light; this behavior results from the orientation of molecules parallel to each other in large clusters.

liquid-dielectric capacitor   [ELEC] A capacitor in which the plate assemblies are mounted in a tank filled with a suitable oil or liquid dielectric.

liquid dioxide   *See* nitrogen dioxide.

liquid-drop model of nucleus   [NUC PHYS] A model of the nucleus in which it is compared to a drop of incompressible liquid, and the nucleons are analogous to molecules in the liquid; used to study binding energies, fission, collective motion, decay, and reactions. Also known as drop model of nucleus.

liquid extraction   *See* solvent extraction.

liquid-filled porosity   [GEOL] The condition in porous rock or sand formations in which pore spaces contain fresh or salt water, liquid petroleum, pressure-liquefied butane or propane, or tar.

liquid filter   [CHEM ENG] A device for the removal of solids or coalesced droplets out of a liquid stream by use of a filter medium, such as a screen, cartridge, or granular bed.

liquid flow   [FL MECH] The flow or movement of materials in the liquid phase.

liquid fluorine   [CRYO] Cold, liquefied fluorine gas; used as a cryogenic propellant.

liquid fuel   [MATER] A rocket fuel which is liquid under the conditions in which it is utilized in the rocket. Also known as liquid propellant.

liquid fuse unit   [ELEC] Fuse unit in which the fuse link is immersed in a liquid, or provision is made for drawing the arc into the liquid when the fuse link melts.

liquid gas   [PHYS] A gas in the liquid state.

liquid glass   *See* sodium silicate.

liquid helium   [CRYO] The state of helium which exists at atmospheric pressure at temperatures below −268.95°C (4.2K), and for temperatures near absolute zero at pressures up to about 25 atmospheres; has two phases, helium I and helium II.

liquid holdup   [FL MECH] A condition in two-phase flow through a vertical pipe; when gas flows at a greater linear velocity than the liquid, slippage takes place and liquid holdup occurs.

liquid honing   *See* vapor blasting.

liquid hydrocarbon   [ORG CHEM] A hydrocarbon that has been converted from a gas to a liquid by pressure or by reduction in temperature; usually limited to butanes, propane, ethane, and methane.

liquid hydrogen   [CRYO] Hydrogen that exists as a liquid at atmospheric pressure, at −252.7°C (20.46K); used for high-impulse rocket fuels.

liquid-hydrogen bubble chamber   [NUCLEO] A bubble chamber in which the active liquid is hydrogen; particularly useful in research on elementary particles produced in high-energy interactions, because the hydrogen provides a dense target of protons.

liquid-in-glass thermometer   [ENG] A thermometer in which the thermally sensitive element is a liquid contained in a graduated glass envelope; the indication of such a thermometer depends upon the difference between the coefficients of thermal expansion of the liquid and the glass; mercury and alcohol are liquids commonly used in meteorological thermometers.

liquid-in-metal thermometer   [ENG] A thermometer in which the thermally sensitive element is a liquid contained in a metal envelope, frequently in the form of a Bourdon tube.

liquid insulator   [MATER] A liquid with a resistivity greater than about 10^{14} ohm-centimeters, such as a petroleum oil, silicone oil, or halogenated aromatic hydrocarbon.

liquid junction emf   [PHYS CHEM] The emf (electromotive force) generated at the area of contact between the salt bridge and the test solution in a pH cell electrode.

liquid knockout   *See* impingement.

liquid laser   [OPTICS] A laser whose active material is dissolved in a liquid contained in a transparent cylindrical shell; rare-earth ions in suitable dissolved molecules and organic dye solutions are used.

liquid level control   [ENG] Regulation of the linear vertical distance between the surface of a liquid and some reference point.

liquid limit   [GEOL] The moisture content boundary that exists between the plastic and semiliquid states of a sediment.

liquid-liquid chemical reaction   [CHEM] Chemical reaction in which the reactants, two or more, are liquids.

liquid-liquid extraction   [CHEM ENG] The removal of a soluble component from a liquid mixture by contact with a second liquid, immiscible with the carrier liquid in which the component is preferentially soluble.

liquid measure   [MECH] A system of units used to measure the volumes of liquid substances in the United States; the units are the fluid dram, fluid ounce, gill, pint, quart, and gallon.

liquid-metal fuel cell   [ELEC] A fuel cell that uses molten potassium and bismuth as reactants and a molten salt electrolyte; has very high power output, but a relatively short life.

liquid-metal MHD generator   [ELEC] A system for generating electric power in which the kinetic energy of a flowing, molten metal is converted to electric energy by magnetohydrodynamic (MHD) interaction.

liquid-metal nuclear fuel   [NUCLEO] A nuclear fuel consisting of a solution of uranium or plutonium in a molten metal such as bismuth.

liquid methane   [CRYO] Methane that has been cooled to at least −161°C; used for cryogenic applications and for tank-ship transport of methane.

liquid nitrogen   [CRYO] Nitrogen that exists as a liquid at atmospheric pressure, at −195°C (77.4K); used in research work, cryogenics, and cryosurgery.

liquid oxygen   [CRYO] Oxygen that exists as a liquid at 59 atmospheres, at −113°C (160.2K); a pale-blue, transparent, mobile liquid.

liquid-oxygen explosive   [MATER] Sawdust or other carbonaceous material formed into a cartridge and dipped into liquid oxygen, to use in blasting. Abbreviated LOX.

liquid penetrant test   [ENG] A penetrant method of nondestructive testing used to locate defects open to the surface of nonporous materials; penetrating liquid is applied to the surface, and after 1–30 minutes excess liquid is removed, and a developer is applied to draw the penetrant out of defects, thus showing their location, shape, and size.

liquid petrolatum   *See* white mineral oil.

liquid-phase hydrogenation   [CHEM ENG] Hydrogen reaction with liquid-phase hydrogenatable material, such as unsaturated aliphatic or aromatic hydrocarbons.

liquid pint   *See* pint.

liquid piston rotary compressor   [MECH ENG] A rotary compressor in which a multiblade rotor revolves in a casing partly filled with liquid, for example, water.

liquid poison   [NUCLEO] A neutron-absorbing liquid that can be injected quickly into the cooling system of a nuclear reactor by explosive-actuated valves; used for automatic or manual scramming to shut down a reactor.

liquid propellant   *See* liquid fuel.

liquid rheostat   [ELECTR] A variable-resistance type of voltage regulator in which the variable-resistance element is liquid, usually water; carbon electrodes are raised or lowered in the liquid to change resistance ratings and control voltage flow.

liquid rosin   *See* tall oil.

liquid scintillation detector   [NUCLEO] Scintillation counter in which the sensitive material is a liquid, such as p-terphenyl dissolved in toluene, placed in a glass or metal container.

liquid seal   [CHEM ENG] 1. The depth of liquid above an opening from which gas or vapor issues, as for a riser in a distillation-column tray. 2. Product drawoff in which a depth of liquid prevents the outflow of gas or vapor.

liquid semiconductor   [ELECTR] An amorphous material in solid or liquid state that possesses the properties of varying resistance induced by charge carrier injection.

liquid-solid chemical reaction [CHEM] Chemical reaction in which at least one of the reactants is a liquid, and another of the reactants is a solid.

liquid-solid equilibrium *See* solid-liquid equilibrium.

liquid-sorbent dehumidifier [MECH ENG] A sorbent type of dehumidifier consisting of a main circulating fan, sorbent-air contactor, sorbent pump, and reactivator; dehumidification and reactivation are continuous operations, with a small part of the sorbent constantly bled off from the main circulating system and reactivated to the concentration required for the desired effluent dew point.

liquid sulfur dioxide–benzene process [CHEM ENG] A petroleum-refinery process using a mixed solvent (SO_2 and benzene) to dewax lubricating oils or improve their viscosity indices.

liquidus line [THERMO] For a two-component system, a curve on a graph of temperature versus concentration which connects temperatures at which fusion is completed as the temperature is raised.

liquid-vapor chemical reaction [CHEM] Chemical reaction in which at least one of the reactants is a liquid, and another of the reactants is a vapor.

liquid-vapor equilibrium [PHYS CHEM] The equilibrium relationship between the liquid and its vapor phase for a partially vaporized compound or mixture at specified conditions of pressure and temperature; for mixtures, it is expressed by $K = x/y$, where K is the equilibrium constant, x the mole fraction of a key component in the vapor, and y the mole fraction of the same key component in the liquid. Also known as vapor-liquid equilibrium.

liquid-water content *See* water content.

liquor [CHEM ENG] **1.** Supernatant liquid decanted from a liquid-solids mixture in which the solids have settled. **2.** Liquid overflow from a liquid-liquid extraction unit. [FOOD ENG] **1.** Sugarcane sap before it is crystallized into sugar. **2.** A strong distilled alcoholic beverage. [PHARM] A solution of a medicinal substance in water.

liquor finish [MET] A bright, smooth finish on wet-drawn wire achieved by using fermented-grain mash liquor as a lubricant.

lirella [BOT] A long, narrow apothecium with a medial longitudinal furrow, occurring in certain lichens.

Liriopeidae [INV ZOO] The phantom craneflies, a family of dipteran insects in the suborder Orthorrhapha distinguished by black and white banded legs.

liroconite [MINERAL] $Cu_2Al(AsO_4)(OH)_4 \cdot 4H_2O$ A light-blue or yellowish-green mineral composed of basic hydrous aluminum copper arsenate, occurring in monoclinic crystals.

liskeardite [MINERAL] $(Al,Fe)_3(AsO_4)(OH)_6 \cdot 5H_2O$ A soft, white mineral composed of basic hydrous aluminum iron arsenate.

lisle [TEXT] Fine-quality, tightly twisted, long-staple cotton yarn with a sleek surface produced by passing it near a gas flame to remove fuzz.

LISP [ADP] An interpretive language developed for the manipulation of symbolic strings of recursive data; can also be used to manipulate mathematical and arithmetic logic. Derived from list processing language.

Lissajous figure [PHYS] The path of a particle moving in a plane when the components of its position along two perpendicular axes each undergo simple harmonic motions and the ratio of their frequencies is a rational number.

Lissamphibia [VERT ZOO] A subclass of Amphibia including all living amphibians; distinguished by pedicellate teeth and an operculum-plectrum complex of the middle ear.

list [ADP] **1.** A last-in, first-out storage organization, usually implemented by software, but sometimes implemented by hardware. **2.** In FORTRAN, a set of data items to be read or written. [ENG] To lean to one side, or deviate from the vertical.

listening station [ENG] A radio or radar receiving station that is continuously manned for various purposes, such as for radio direction finding or for gaining information about enemy electronic devices.

Listeria [MICROBIO] A genus of motile, gram-negative, non-

sporeforming rod-shaped bacteria of the family Corynebacteriaceae.

listeriosis [MED] A bacterial disease of humans and some animals caused by *Listeria monocytogenes* (Corynebacteriaceae); occurs primarily as meningitis or granulomatosis infantiseptica in humans, and takes many forms, such as meningoencephalitis, distemperlike disease, or generalized infection, in animals.

listing *See* lashing.

Listomatic camera [GRAPHICS] A machine used in photocopying which photographs data on tabulating cards and prints it in columnar form on roll film; the film negative is then used for printing by photolithography.

list processing [ADP] A programming technique in which list structures are used to organize memory.

list processing language *See* LISP.

list structure [ADP] A set of data items, connected together because each element contains the address of a successor element (and sometimes of a predecessor element).

litchi *See* lychee.

liter [MECH] A unit of volume or capacity, equal to 1 decimeter cubed, or 0.001 cubic meter, or 1000 cubic centimeters. Abbreviated l.

literal constant [MATH] A letter denoting a constant.

literal expression [MATH] An expression or equation in which the constants are represented by letters.

literal notation [MATH] The use of letters to denote numbers, known or unknown.

liter-atmosphere [PHYS] A unit of energy equal to the work done on a piston by a fluid at a pressure of 1 standard atmosphere when the piston sweeps out a volume of 1 liter; equal to 101.325 joules.

lithamide *See* lithium amide.

litharge *See* lead monoxide.

litharge-glycerin cement [MATER] Mixture of glycerin, water, and litharge (lead monoxide) to give, when cured, an acid-resistant cement.

lithemia *See* hyperuricemia.

lithian muscovite [MINERAL] A form of the mineral lepidolite containing 3–4% lithium oxide and having a modified two-layer monoclinic muscovite structure.

lithiasis [MED] The formation of calculi in the body.

lithic [PETR] Pertaining to stone.

lithic graywacke [PETR] A low-grade graywacke, that is, containing an abundance of unstable materials, especially a sandstone containing less than 75% quartz and chert, 15–75% detrital clay matrix, and more rock fragments than feldspar grains.

lithic sandstone [PETR] A sandstone that contains more rock fragments than feldspar grains.

lithic tuff [GEOL] **1.** A tuff that is mostly crystalline rock fragments. **2.** An indurated volcanic ash deposit whose fragments are composed of previously formed rocks that first solidified in the volcanic vent and were then blown out.

lithifaction *See* lithification.

lithification [GEOL] **1.** Conversion of a newly deposited sediment into an indurated rock. Also known as lithifaction. **2.** Compositional change of coal to bituminous shale or other rock.

lithionite *See* lepidolite.

lithiophilite [MINERAL] $Li(Mn,Fe)PO_4$ A salmon-pink or clove-brown mineral crystallizing in the orthorhombic system; isomorphous with triphylite.

lithiophorite [MINERAL] $(Al,Li)MnO_2(OH)_2$ Mineral composed of basic manganese aluminum lithium oxide.

Lithistida [PALEON] An order of fossil sponges in the class Demospongia having a reticulate skeleton composed of irregular and knobby siliceous spicules.

lithium [CHEM] A chemical element, symbol Li, atomic number 3, atomic weight 6.939; an alkali metal.

lithium aluminum hydride [INORG CHEM] $LiAlH_2$ A compound made by the reaction of lithium hydride and aluminum chloride; a powerful reducing agent for specific linkages in complex molecules; used in organic synthesis.

lithium amide [INORG CHEM] $LiNH_2$ A compound crystallizing in the cubic form, and melting at 380–400°C; used in organic synthesis. Also known as lithamide.

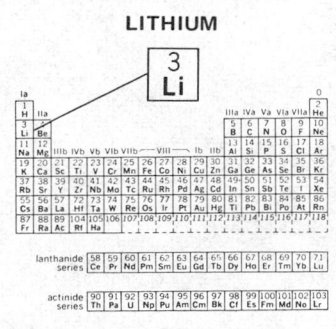

LITHIUM

Periodic table of the chemical elements showing the position of lithium.

LITHIUM CELL

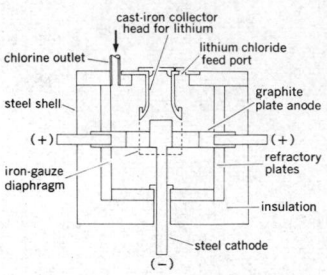

Cross section of lithium cell.

lithium bromide [INORG CHEM] LiBr·H₂O A white, deliquescent, granular powder with a bitter taste, melting at 547°C; soluble in alcohol and glycol; used to add moisture to air-conditioning systems and as a sedative and hypnotic in medicine.

lithium carbonate [INORG CHEM] Li₂CO₃ A colorless, crystalline compound that melts at 700°C and has slight solubility in water; used in ceramic industries in the manufacture of powdered glass for porcelain enamel formulation.

lithium cell [CHEM] An electrolytic cell for the production of metallic lithium. [ELEC] A primary cell for producing electrical energy using lithium metal for one electrode immersed in usually an organic electrolyte.

lithium chloride [INORG CHEM] LiCl·2H₂O A colorless, water-soluble compound, forming octahedral crystals and melting at 614°C; used to form concentrated brine in commercial air-conditioning systems and as a pyrotechnic in welding and brazing fluxes.

lithium citrate [ORG CHEM] Li₃C₆H₅O₇·4H₂O White powder that decomposes when heated; slightly soluble in alcohol; soluble in water; used in beverages and pharmaceuticals.

lithium-drifted germanium crystal [ELECTR] A high-resolution junction detector, used especially for more penetrating gamma-radiation and higher-energy electrons, produced by drifting lithium ions through a germanium crystal to produce an intrinsic region where impurity-based carrier generation centers are deactivated, sandwiched between a *p* layer and an *n* layer.

lithium fluoride [INORG CHEM] LiF Poisonous, white powder melting at 870°C, boiling at 1670°C; insoluble in alcohol, slightly soluble in water, and soluble in acids; used as a heat-exchange medium, as a welding and soldering flux, in ceramics, and as crystals in infrared instruments.

lithium grease [MATER] Heat-stable, water-resistant lubricating grease with lithium salts of higher fatty acids (or lithium soaps of fatty glycerides) as a base; used for low-temperature service in aircraft.

lithium halide [INORG CHEM] A binary compound of lithium, LiX, where X is a halide; examples are lithium chloride, LiCl, and lithium fluoride, LiF.

lithium hydride [INORG CHEM] LiH Flammable, brittle, white, translucent crystals; decomposes in water; insoluble in ether, benzene, and toluene; used as a hydrogen source and desiccant, and to prepare lithium amide and double hydrides.

lithium hydroxide [INORG CHEM] LiOH; LiOH·H₂O Colorless crystals; used as a storage-battery electrolyte, as a carbon dioxide absorbent, and in lubricating greases and ceramics.

lithium iodide [INORG CHEM] LiI; LiI·3H₂O White, water- and alcohol-soluble crystals; LiI melts at 446°C; LiI·3H₂O loses water at 72°C; used in medicine, photography, and mineral waters.

lithium mica *See* lepidolite.

lithium molybdate [INORG CHEM] Li₂MoO₄ Water-soluble white crystals melting at 705°C; used as a catalytic cracking (petroleum) catalyst and as a mill additive for steel.

lithium nitrate [INORG CHEM] LiNO₃ Water- and alcohol-soluble colorless powder melting at 261°C; used as a heat-exchange medium and in ceramics, pyrotechnics, salt baths, and refrigeration systems.

lithium perchlorate [INORG CHEM] LiClO₄·3H₂O A compound with high oxygen content (60% available oxygen), used as a source of oxygen in rockets and missiles.

lithium tetraborate [INORG CHEM] Li₂B₄O₇·5H₂O White crystals that lose water at 200°C; insoluble in alcohol, soluble in water; used in ceramics.

lithium titanate [INORG CHEM] Li₂TiO₃ A water-insoluble white powder with strong fluxing ability when used in titanium-containing enamels; also used as a mill additive in vitreous and semivitreous glazes.

Lithobiomorpha [INV ZOO] An order of chilopods in the subclass Pleurostigmophora; members are anamorphic and have 15 leg-bearing trunk segments, and when eyes are present, they are ocellar.

lithoclase [GEOL] A naturally produced rock fracture.

lithocyst [BOT] Epidermal plant cell in which cytoliths are formed. [INV ZOO] One of the minute sacs containing lith-ites in many invertebrates; thought to function in audition and orientation.

lithocyte [INV ZOO] A special cell in anthomedusae containing a statolith.

Lithodidae [INV ZOO] The king crabs, a family of anomuran decapods in the superfamily Paguridea distinguished by reduced last pereiopods and by the asymmetrical arrangement of the abdominal plates in the female.

lithodomous [ZOO] Burrowing in rock.

lithofacies [GEOL] A subdivision of a specified stratigraphic unit distinguished on the basis of lithologic features.

lithofacies map [GEOL] The facies map of an area based on lithologic characters; shows areal variation in all aspects of the lithology of a stratigraphic unit.

lithogenesis [PATH] The process of formation of calculi or stones. [PETR] The branch of science dealing with the formation of rocks, especially the formation of sedimentary rocks.

lithogeochemical survey [GEOCHEM] A geochemical survey that involves the sampling of rocks.

lithograph [GRAPHICS] Originally, a reproduction of a writing sample or a drawing made from a litho stone onto which the writing or drawing had been drawn with a greasy ink or crayon; now, a reproduction from litho metal plates produced by photolithography and run on an offset press.

lithographic film [GRAPHICS] Orthochromatic film used in the production of lithographic plates.

lithographic limestone [GEOL] A dense, compact, fine-grained crystalline limestone having a pale creamy-yellow or grayish color. Also known as lithographic stone; litho stone.

lithographic plate [GRAPHICS] A metal plate, usually having little porosity, on which an image is produced for lithographic printing.

lithographic stone *See* lithographic limestone.

lithographic texture [GEOL] The texture of certain calcareous sedimentary rocks characterized by grain size of less than 1/256 millimeter and having a smooth appearance.

lithography [GRAPHICS] A printing process in which a design is sketched with an oily ink or a litho crayon on a flat, smooth stone; in printing, the entire surface of the stone is wetted, and the design areas repel the water, but accept a greasy ink; a clean impression is then made by pressing a sheet of paper against the surface of the stone and running the whole through a press.

lithologic unit *See* rock-stratigraphic unit.

lithology [GEOL] The description of the physical character of a rock as determined by eye or with a low-power magnifier, and based on color, structures, mineralogic components, and grain size.

lithol red [MATER] Any of various pigments derived from combination of β-naphthol and Tobias acid; available as sodium, barium, and calcium toners and lakes; used in outside, drum, and toy enamels.

lithometeor [METEOROL] The general term for dry atmospheric suspensoids, including dust, haze, smoke, and sand.

lithopedion [MED] A retained fetus that has become calcified.

lithophagous [ZOO] Feeding on stone, as certain mollusks.

lithophile [GEOCHEM] Pertaining to elements that have become concentrated in the silicate phase of meteorites or the slag crust of the earth.

lithophone *See* lithopone.

lithophysa [GEOL] A large spherulitic hollow or bubble in glassy basalts and certain rhyolites. Also known as stone bubble.

lithophyte [ECOL] A plant that grows on rock.

lithopone [MATER] A white pigment produced as a filtered, heated, quenched precipitate from reaction of barium sulfide and zinc sulfide; used as a pigment for paint, ink, filled leather, paper, linoleum, oilcloth, and cosmetics. Also known as Charlton white; Griffith's white; lithophone; Orr's white; zinc baryta white; zinc sulfide white.

lithosere [ECOL] A succession of plant communities that originate on rock.

lithosiderite *See* stony-iron meteorite.

lithosol [GEOL] A group of shallow soils lacking well-defined horizons and composed of imperfectly weathered fragments of rock.

lithosphere [GEOL] **1.** The rigid outer crust of rock on the earth about 50 miles (80 kilometers) thick, above the asthenosphere. Also known as oxysphere. **2.** Since the development of plate tectonics theory, a term referring to the rigid, upper 100 kilometers of the crust and upper mantle, above the asthenosphere.

lithostatic pressure *See* ground pressure.

litho stone *See* lithographic limestone.

lithostratic unit *See* rock-stratigraphic unit.

lithostratigraphic unit *See* rock-stratigraphic unit.

lithostyle [INV ZOO] A static organ in Narcomedusae. Also known as tentaculocyst.

lithotomy [MED] Surgical removal of a calculus.

lithotope [GEOL] **1.** The environment under which a sediment is deposited. **2.** An area of uniform sedimentation.

lithotype [GEOL] A macroscopic band in humic coals, analyzed on the basis of physical characteristics rather than botanical origin.

lithuria [MED] A condition marked by excess of uric (lithic) acid or its salts in the urine.

litmus [MATER] Blue, water-soluble powder from various lichens, especially *Variolaria lecanora* and *V. rocella*; turns red in solutions at pH 4.5, and blue at pH 8.3; used as an acid-base indicator. Also known as lacmus; lichen blue.

litmus paper [MATER] White, unsized paper saturated by litmus in water; used as a pH indicator.

Litopterna [PALEON] An order of hoofed, herbivorous mammals confined to the Cenozoic of South America; characterized by a skull without expansion of the temporal or squamosal sinuses, a postorbital bar, primitive dentition, and feet that were three-toed or reduced to a single digit.

lit-par-lit [GEOL] Pertaining to the penetration of bedded, schistose, or other foliate rocks by innumerable narrow sheets and tongues of granitic rock.

Little Bear *See* Ursa Minor.

little brother [METEOROL] A subsidiary tropical cyclone that sometimes follows a more severe disturbance.

little cherry disease [PL PATH] A virus disease of sweet cherries characterized by small, angular pointed fruits which retain the bright red color of immaturity and never reach mature size.

Little Dipper *See* Ursa Minor.

little-drop technique [BIOL] A method for isolating single cells in which a drop of a cellular suspension containing a single cell, as determined by microscopic examination, is transferred with a capillary pipet to an appropriate culture medium.

Little Fox *See* Vulpecula.

little giant [MIN ENG] A jointed iron nozzle used in hydraulic mining.

Little Horse *See* Equuleus.

Little Ice Age [GEOL] A period of expansion of mountain glaciers, marked by climatic deterioration, that began about 5500 years ago and extended to as late as A.D. 1550–1850 in some regions, as the Alps, Norway, Iceland, and Alaska.

Little John [ORD] Name applied to a U.S. Army rocket system (318-millimeter) consisting of a surface-to-surface tactical missile, similar to but smaller than Honest John, and employing solid fuel.

little leaf [PL PATH] Any of various plant diseases and disorders characterized by chlorotic, underdeveloped, and sometimes distorted leaves.

little peach disease [PL PATH] A virus disease of the peach tree in which the fruit is dwarfed and delayed in ripening, the leaves yellow, and the tree dies.

Little Ruler *See* Regulus.

Little's disease [MED] Spastic diplegia of infants which is characterized by spasticity of the lower extremities; involves degenerative and atrophic cerebral changes as well as congenital malformation.

littoral current [OCEANOGR] A current, caused by wave action, that sets parallel to the shore; usually in the nearshore region within the breaker zone. Also known as alongshore current; longshore current.

littoral drift [GEOL] Materials moved by waves and currents of the littoral zone. Also known as longshore drift.

littoral sediments [GEOL] Deposits of littoral drift.

littoral transport [GEOL] The movement of littoral drift.

littoral zone [ECOL] Of or pertaining to the biogeographic zone between the high- and low-water marks.

Littorinacea [PALEON] An extinct superfamily of gastropod mollusks in the order Prosobranchia.

Littorinidae [INV ZOO] The periwinkles, a family of marine gastropod mollusks in the order Pectinibranchia distinguished by their spiral, globular shells.

Littrow grating spectrograph [SPECT] A spectrograph having a plane grating at an angle to the axis of the instrument, and a lens in front of the grating which both collimates and focuses the light.

Littrow mounting [SPECT] The arrangement of the grating and other components of a Littrow grating spectrograph, which is analogous to that of a Littrow quartz spectrograph.

Littrow prism [OPTICS] A prism having angles of 30, 60, and 90°, silvered on the side opposite the 60° angle; a lens used with it can serve both as a telescope and as a collimator.

Littrow quartz spectrograph [SPECT] A spectrograph in which dispersion is accomplished by a Littrow quartz prism with a rear reflecting surface that reverses the light; a lens in front of the prism acts as both collimator and focusing lens.

lituate [BOT] Having a forked member or part with the ends turned slightly outward, as in certain fungi.

Lituolacea [INV ZOO] A superfamily of benthic marine foraminiferans in the suborder Textulariina having a multilocular, rectilinear, enrolled or uncoiled test with a simple to labyrinthic wall.

lituus [MATH] The trumpet-shaped plane curve whose points in polar coordinates (r, θ) satisfy the equation $r^2 = a/\theta$.

litz wire [ELEC] Wire consisting of a number of separately insulated strands woven together so each strand successively takes up all possible positions in the cross section of the entire conductor, to reduce skin effect and thereby reduce radio-frequency resistance.

live [COMMUN] Being broadcast directly at the time of production, instead of from recorded or filmed program material. [ELEC] *See* energized.

live ammunition [ORD] Ammunition containing explosives or active chemicals, as distinguished from inert or drill ammunition.

live axle [MECH ENG] An axle to which wheels are rigidly fixed.

live end [ACOUS] The end of a radio studio that gives almost complete reflection of sound waves.

live load [MECH] A moving load or a load of variable force acting upon a structure, in addition to its own weight.

liver [ANAT] A large vascular gland in the body of vertebrates, consisting of a continuous parenchymal mass covered by a capsule; secretes bile, manufactures certain blood proteins and enzymes, and removes toxins from the systemic circulation. [MATER] Intermediate layer of dark-colored, oily material formed by hydrolyzation of acid sludge from sulfuric acid treatment of petroleum oil; insoluble in weak acid and oil.

liver failure [MED] Severe functional disability of the liver marked clinically by a variety of signs and symptoms, including jaundice, coma, and abnormal blood levels of such things as ammonia, bilirubin, and alkaline phosphatase.

liver fluke [INV ZOO] Any trematode, especially *Clonorchis sinensis*, that lodges in the biliary passages within the liver.

live-roller conveyor [MECH ENG] Conveying machine which moves objects over a series of rollers by the application of power to all or some of the rollers.

live room [ACOUS] A room having a minimum of sound-absorbing material.

liver phosphorylase [BIOCHEM] An enzyme that catalyzes the breakdown of liver glycogen to glucose-1-phosphate.

liverwort [BOT] The common name for members of the Marchantiatae.

live-virus vaccine [IMMUNOL] A suspension of attenuated live viruses injected to produce immunity.

living fossil [BIOL] A living species belonging to an ancient stock otherwise known only as fossils.

Livingstone sphere [ENG] A clay atmometer in the form of a sphere; evaporation indicated by this instrument is supposed to be somewhat representative of that from plant growth.

LITTROW GRATING SPECTROGRAPH

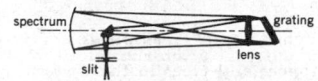

Littrow mounting of a plane grating.

LITTROW QUARTZ SPECTROGRAPH

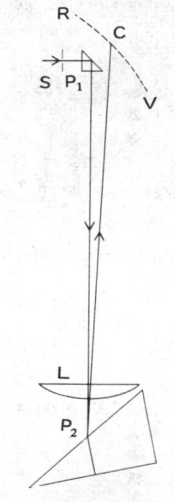

Littrow quartz spectrograph, typical arrangement for double-beam recording instrument. S, slit; P_1, totally reflecting quartz prism; L, autocollimating quartz lens; P_2, Littrow quartz prism; C, camera; RV, red to violet spectrum.

LIVERWORT

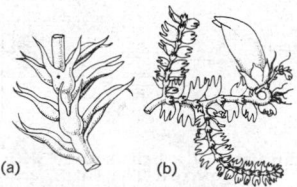

Leafy liverworts of the Jungermanniales. (a) *Herberta*, showing three ranks of equal bifid leaves (*after Muller*). (b) *Lepidozia*, ventral aspect, showing reduced ventral leaves (*after A. Lorenz*). (*From E. W. Sinnot and K. S. Wilson, Botany: Principles and Problems, 5th ed., McGraw-Hill, 1955*)

LIZARD

Fence lizard (*Sceloporus undulatus*) often runs along fences and trees in the eastern United States.

LLAMA

A South American llama.

LLOYD'S MIRROR INTERFERENCE

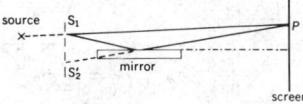

Splitting of a light source with Lloyd's mirror. The slit S_1 and its virtual image S'_2 constitute the double source. Part of the light falls directly on the screen at P, and part is reflected at grazing incidence from a plane mirror.

livingstonite [MINERAL] $HgSb_4S_7$ A lead-gray mineral with red streak and metallic luster; a source of mercury.

livor mortis [PATH] The reddish-blue discoloration of the cadaver that occurs in the dependent portions of the body due to gradual gravitational flow of unclotted blood.

livre [MECH] A unit of mass, used in France, equal to 0.5 kilogram.

lixiviate [CHEM ENG] To extract a soluble component from a solid mixture by washing or percolation processes.

lizard [VERT ZOO] Any reptile of the suborder Sauria.

Lizard *See* Lacerta.

lizard-hipped dinosaur [PALEON] The name applied to members of the Saurichia because of the comparatively unspecialized three-pronged pelvis.

Ljungström heater [MECH ENG] Continuous, regenerative, heat-transfer air heater (recuperator) made of slow-moving rotors packed with closely spaced metal plates or wires with a housing to confine the hot and cold gases to opposite sides.

Ljungström steam turbine [MECH ENG] A radial outward-flow turbine having two opposed rotation rotors.

LK virus [VIROL] A type of equine herpesvirus.

LL *See* line link.

llama [VERT ZOO] Any of three species of South American artiodactyl mammals of the genus *Lama* in the camel family; differs from the camel in being smaller and lacking a hump.

Llandeilian [GEOL] Upper Middle Ordovician geologic time.

Llandoverian [GEOL] Lower Silurian geologic time.

llano [ECOL] A savannah of Spanish America and the southwestern United States generally having few trees.

Llanvirnian [GEOL] Lower Middle Ordovician geologic time.

llebetjado [METEOROL] In northeastern Spain, a hot, squally wind descending from the Pyrenees and lasting for a few hours.

llerzolite [PETR] A form of periodotite containing olivine and monoclinic and orthorhombic pyroxenes.

LLL circuit *See* low-level logic circuit.

Lloyd's mirror interference [OPTICS] The interference pattern produced when part of the light from a slit falls directly on a screen, and part is reflected from a mirror whose surface makes a small angle with the incident beam.

lm *See* lumen.

LM *See* lunar excursion model.

L/M [NUC PHYS] The ratio of the number of internal conversion electrons emitted from the L shell in the de-excitation of a nucleus to the number of such electrons emitted from the M shell.

LMC *See* Large Magellanic Cloud.

L meson [PARTIC PHYS] A K-meson resonance having a mass of 1781 ± 14 MeV (million electron volts), and a width of 70 MeV; its decay modes are dominantly $K\pi\pi$, with some $K\omega$ (5%).

lm-hr *See* lumen-hour.

lm-sec *See* lumen-second.

lm/w *See* lumen per watt.

L network [ELECTR] A network composed of two branches in series, with the free ends connected to one pair of terminals; the junction point and one free end are connected to another pair of terminals.

LNG *See* liquefied natural gas.

LNG ship [NAV ARCH] A specially designed, insulated tanker for shipping liquefied natural gas.

loach [VERT ZOO] The common name for fishes composing the family Cobitidae; most are small and many are eel-shaped.

load [ADP] 1. To place data into an internal register under program control. 2. To place a program from external storage into central memory under operator (or program) control, particularly when loading the first program into an otherwise empty computer. 3. An instruction, or operator control button, which causes the computer to initiate the load action. [ELEC] 1. A device that consumes electric power. 2. The amount of electric power that is drawn from a power line, generator, or other power source. 3. The material to be heated by an induction heater or dielectric heater. Also known as work. [ELECTR] The device that receives the useful signal output of an amplifier, oscillator, or other signal source. [ENG] To place ammunition in a gun, bombs on an

airplane, explosives in a missile or borehole, fuel in a fuel tank, cargo or passengers into a vehicle, and the like. [MECH] 1. The weight that is supported by a structure. 2. Mechanical force that is applied to a body. 3. The burden placed on any machine, measured by units such as horsepower, kilowatts, or tons. [MIN ENG] Unit of weight of ore used in the South African diamond mines; equal to 1600 pounds (725 kilograms); the equivalent of about 16 cubic feet of broken ore.

load-and-carry equipment [MECH ENG] Earthmoving equipment designed to load and transport material.

load-and-go [ADP] An operating technique with no stops between the loading and execution phases of a program; may include assembling or compiling.

load-bearing tile [MATER] A tile with the capacity to support superimposed loads.

load-break switch [ELEC] An electric switch in a circuit with several hundred thousand volts, designed to carry a large amount of current without overheating the open position, having enough insulation to isolate the circuit in closed position, and equipped with arc interrupters to interrupt the load current.

load cast [GEOL] An irregularity at the base of an overlying stratum, usually sandstone, that projects into an underlying stratum, usually shale or clay.

load cell [ELEC] A device which measures large pressures by applying the pressure to a piezoelectric crystal and measuring the voltage across the crystal; the cell plus a recording mechanism constitutes a strain gage.

load characteristic [ELECTR] Relation between the instantaneous values of a pair of variables such as an electrode voltage and an electrode current, when all direct electrode supply voltages are maintained constant. Also known as dynamic characteristic.

load chart [IND ENG] A graph showing the amount of work still to be performed by a factory producing unit such as a machine or assembly group.

load circuit [ELECTR] Complete circuit required to transform power from a source such as an electron tube to a load.

load circuit efficiency [ELECTR] Ratio between useful power delivered by the load circuit to the load and the load circuit power input.

load compensation [CONT SYS] Compensation in which the compensator acts on the output signal after it has generated feedback signals. Also known as load stabilization.

load controller [MIN ENG] A device to control the load and prevent spillage on a gathering conveyor receiving coal or mineral from several loading points or subsidiary conveyors; it is a simplified weightometer.

load diagram [CIV ENG] A diagram showing the distribution and intensity of loads on a structure.

load divider [ELEC] Unit for distributing power to various units.

loaded line [ELEC] Wire line in which loading coils have been inserted at regular intervals to reduce attenuation and phase lag at the frequencies within the band used.

loaded motional impedance *See* motional impedance.

loaded Q [ELEC] The Q factor of an impedance which is connected or coupled under working conditions. Also known as working Q. [ELECTROMAG] The Q factor of a specific mode of resonance of a microwave tube or resonant cavity when there is external coupling to that mode.

loader [MECH ENG] A machine such as a mechanical shovel used for loading bulk materials. [ORD] Mechanical device which loads guns with cartridges.

load factor [ELEC] The ratio of average electric load to peak load, usually calculated over a 1-hour period. [MECH] The ratio of load to the maximum rated load.

load impedance [ELECTR] The complex impedance presented to a transducer by its load.

loading [CHEM ENG] Condition of vapor overcapacity in a liquid-vapor-contact tower, in which rising vapor lifts or holds falling liquid. [ELEC] The addition of inductance to a transmission line to improve its transmission characteristics throughout a given frequency band. Also known as electrical loading. [ENG] 1. Buildup on a cutting tool of the material removed in cutting. 2. Filling the pores of a grinding wheel with material removed in the grinding process. [ENG ACOUS]

Placing material at the front or rear of a loudspeaker to change its acoustic impedance and thereby alter its radiation. [MET] Filling of a die cavity with powdered metal. [NUCLEO] Placing fuel in a nuclear reactor.

loading angle [ORD] Angle of elevation specified for loading a particular weapon with its ammunition.

loading coil [ELECTROMAG] **1.** An iron-core coil connected into a telephone line or cable at regular intervals to lessen the effect of line capacitance and reduce distortion. Also known as Pupin coil; telephone loading coil. **2.** A coil inserted in series with a radio antenna to increase its electrical length and thereby lower the resonant frequency.

loading density [ENG] The number of pounds of explosive per foot length of drill hole. [ORD] A term applied specifically to explosive charges of projectiles, bombs, warheads, and so on; it is the quantity of explosive per unit volume, usually expressed as grams per cubic centimeter.

loading disk [ELECTROMAG] Circular metal piece mounted at the top of a vertical antenna to increase its natural wavelength.

loading head [MECH ENG] The part of a loader which gathers the bulk materials.

loading pan [MIN ENG] A box or scoop into which broken rock is shoveled in a sinking shaft while the hoppit is traveling in the shaft.

loading program [ADP] Program used to load other programs into computer memory. Also known as bootstrap program.

loading rack [ENG] The shelter and associated equipment for the withdrawal of liquid petroleum or a chemical product from a storage tank and loading it into a railroad tank car or tank truck.

loading routine *See* input routine.

loading station [MECH ENG] A device which receives material and puts it on a conveyor; may be one or more plates or a hopper.

loading tray [ORD] **1.** Trough-shaped carrier on which heavy projectiles are placed so that they can be more easily and safely slipped into the breech of a gun. **2.** Hollowed slide which guides the projectiles into the breech of some types of automatic weapons.

loading weight [ENG] Weight of a powder put into a container.

load isolator [ELECTROMAG] Waveguide or coaxial device that provides a good energy path from a signal source to a load, but provides a poor energy path for reflections from a mismatched load back to the signal source.

load line [ELECTR] A straight line drawn across a series of tube or transistor characteristic curves to show how output signal current will change with input signal voltage when a specified load resistance is used. [NAV ARCH] A line, painted or cut on the outside of a ship, which marks the maximum waterline when the ship is loaded with the greatest cargo which it can carry safely.

load loss [ELEC] The sum of the copper loss of a transformer, due to resistance in the windings, plus the eddy current loss in the winding, plus the stray loss.

load metamorphism *See* static metamorphism.

load module [ADP] A program in a form suitable for loading into memory and executing.

load point [ADP] Preset point on a magnetic tape from which reading or writing will start.

load stabilization *See* load compensation.

loadstone *See* lodestone.

load waterline [NAV ARCH] The waterline of a fully loaded vessel.

load water plane [NAV ARCH] The water plane of a fully loaded vessel.

loam [GEOL] Soil mixture of sand, silt, clay, and humus. [MET] Molding material consisting of sand, silt, and clay used over backup material for producing massive castings, usually of iron or steel.

LOB *See* line of balance.

Lobachevski geometry [MATH] A system of planar geometry in which the euclidean parallel postulate fails; any point p not on a line L has at least two lines through it parallel to L. Also known as Bolyai geometry; hyperbolic geometry.

lobar pneumonia [MED] An acute febrile disease involving one or more lobes of the lung, usually following pneumococcal infection.

lobar sclerosis [MED] Neuroglial proliferation accompanied by atrophy of a cerebral lobe leading to mental and neurological deficits; most common in infants and children who have suffered prolonged hypoxia.

Lobata [INV ZOO] An order of the Ctenophora in which the body is helmet-shaped.

lobate [BIOL] Having lobes. [VERT ZOO] Of a fish, having the skin of the fin extend onto the bases of the fin rays.

lobate rill mark [GEOL] A flute cast formed by current action.

lobe [BIOL] A rounded projection on an organ or body part. [DES ENG] A projection on a cam wheel or a noncircular gear wheel. [ELECTROMAG] A part of the radiation pattern of a directional antenna representing an area of stronger radio-signal transmission. Also known as radiation lobe. [HYD] A curved projection on the margin of a continental ice sheet.

lobectomy [MED] Surgical removal of a lobe of an organ, particularly of a lung.

lobed impeller meter [ENG] A type of positive displacement meter in which a fluid stream is separated into discrete quantities by rotating, meshing impellers driven by interlocking gears.

lobefin fish [VERT ZOO] The common name for members composing the subclass Crossopterygii.

lobe-half-power width [ELECTROMAG] In a plane containing the direction of the maximum energy of a lobe, the angle between the two directions in that plane about the maximum in which the radiation intensity is one-half the maximum value of the lobe.

lobe penetration [ELECTROMAG] Penetration of the radar coverage of a station which is not limited by pulse repetition frequency, scope limitations, or the screening angle at the azimuth of penetration.

lobe switching *See* beam switching.

lobing [ELECTROMAG] Formation of maxima and minima at various angles of the vertical plane antenna pattern by the reflection of energy from the surface surrounding the radar antenna; these reflections reinforce the main beam at some angles and detract from it at other angles, producing fingers of energy.

loblolly pine [BOT] *Pinus taeda.* A hard yellow pine of the central and southeastern United States having a reddish-brown fissured bark, needles in groups of three, and a full bushy top.

lobopodia [INV ZOO] Broad, thick pseudopodia.

Lobosia [INV ZOO] A subclass of the protozoan class Rhizopodea generally characterized by lobopodia.

lobotomy [MED] An operative section of the fibers between the frontal lobes of the brain. Also known as leukotomy; prefrontal lobotomy.

Lobry de Bruyn–Ekenstein transformation [ORG CHEM] The change in which an aldose sugar treated with dilute alkali results in a mixture of an epimeric pair and 2-keto-hexose due to the production of enolic forms in the presence of hydroxyl ions, followed by a rearrangement.

lobster [INV ZOO] The common name for several bottom-dwelling decapod crustaceans making up the family Homaridae which are commercially important as a food item.

lobular pneumonia *See* bronchopneumonia.

lobule [BIOL] **1.** A small lobe. **2.** A division of a lobe.

local action [ELEC] **1.** Internal losses of a battery caused by chemical reactions producing local currents between different parts of a plate. **2.** Quantitatively, the percentage loss per month in the capacity of a battery on open circuit, or the amount of current needed to keep the battery fully charged. [MET] Electrochemical corrosion resulting from the action of local cells.

local anesthetic [PHARM] A drug which induces loss of sensation only in the region to which it is applied.

local angular momentum [METEOROL] Angular momentum about an arbitrarily located vertical axis which is fixed with respect to the earth.

local apparent noon [ASTRON] Twelve o'clock local apparent time, or the instant the apparent sun is over the upper branch of the local meridian.

LOAD LINE

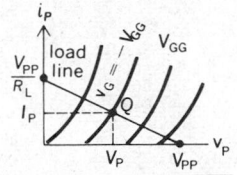

Load line drawn across characteristic curves giving plate current, i_P, as function of plate voltage, v_P, for various values of grid supply voltage, V_{GG}. V_{PP} = plate supply voltage, R_L = load resistance, v_G = grid voltage. Quiescent point, Q, determines quiescent plate current I_P and quiescent plate voltage, V_P.

LOBATA

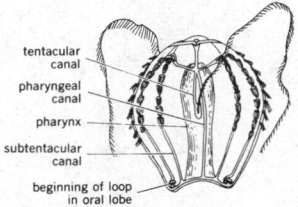

Bolinopsis mikado.

LOBED IMPELLER METER

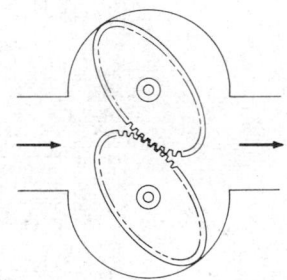

Meshing impellers in the lobed impeller meter. Arrows show direction of fluid flow.

LOBSTER

The lobster, showing the stout pincers on the front legs which are used to crush prey.

local apparent time [ASTRON] The arc of the celestial equator, or the angle at the celestial pole, between the lower branch of the local celestial meridian and the hour circle of the apparent or true sun, measured westward from the lower branch of the local celestial meridian through 24 hours.

local attraction *See* local magnetic disturbance.

local base level *See* temporary base level.

local battery [ELEC] Battery that actuates the telegraphic station recording instruments, as distinguished from the battery furnishing current to the line.

local battery telephone set [ELECTR] Telephone set for which the transmitter current is supplied from a battery, or other current supply circuit, individual to the telephone set; the signaling current may be supplied from a local hand generator or from a centralized power source.

local buckling [MECH] Buckling of thin elements of a column section in a series of waves or wrinkles.

local cable [COMMUN] Handmade cable form for terminations of circuits at the attendant's switchboard, at unit equipment, and other locations, where wiring is run inside the section or unit.

local cell [ELEC] A galvanic cell resulting from differences in potential between adjacent areas on the surface of a metal immersed in an electrolyte.

local central office [COMMUN] A telephone central office, which terminates subscriber lines and makes connections with other central offices, usually equipped to serve 10,000 main telephones of its immediate community.

local change [OCEANOGR] The time rate of change of a scalar quantity (such as temperature, salinity, pressure, or oxygen content) in a fixed locality.

local circuit [COMMUN] Circuit to a main or auxiliary circuit which can be made available at any station or patched from point to point through one or more stations.

local civil time [ASTRON] United States terminology during 1925–1952 for local mean time.

local cluster of stars *See* local star system.

local coefficient [MATH] By using fiber bundles where the fiber is a group, one may generalize cohomology theory for spaces; one uses such bundles as the algebraic base for such a theory and calls the bundle a system of local coefficients.

local coefficient of heat transfer [THERMO] The heat transfer coefficient at a particular point on a surface, equal to the amount of heat transferred to an infinitesimal area of the surface at the point by a fluid passing over it, divided by the product of this area and the difference between the temperatures of the surface and the fluid.

local control [COMMUN] System or method of radio-transmitter control whereby the control functions are performed directly at the transmitter.

local coordinate system [MATH] The coordinate system about a point which is induced when the global space is locally euclidean.

local derivative [FL MECH] The rate of change of a quantity with respect to time at a fixed point of a fluid, $\partial f/\partial t$; it is related to the individual derivative df/dt through the expression $\partial f/\partial t = df/dt - V \cdot \nabla f$, where f is a thermodynamic property $f(x,y,z,t)$ of the fluid, V the vector velocity of the fluid, and ∇ the del operator.

local exchange *See* exchange.

local extra observation [METEOROL] An aviation weather observation taken at specified intervals, usually every 15 minutes, when there are impending aircraft operations and when weather conditions are below certain operational weather limits; the observation includes ceiling, sky condition, visibility, atmospheric phenomena, and pertinent remarks.

local first selector [COMMUN] The second portion of a line that connects to a calling line, through a line primary switch, to a local second selector and special service second selector, and returns a dial tone to the calling subscribers.

local forecast [METEOROL] Generally, any weather forecast of conditions over a relatively limited area, such as a city or airport.

local group [ASTRON] A group of at least 20 known galaxies in the vicinity of the sun; the Andromeda Spiral is the largest of the group, and the Milky Way Galaxy is the second largest.

local hour angle [ASTRON] Angular distance west of the local celestial meridian.

local immunity [IMMUNOL] Immunity localized in a specific tissue or region of the body.

local inflow [HYD] The water that enters a stream between two stream-gaging stations.

localization [ADP] Imposing some physical order upon a set of objects, so that a given object has a greater probability of being in some particular regions of space than in others.

localized state [QUANT MECH] A state of motion in which an electron may be found anywhere within a region of a material of linear extent smaller than that of the material.

localizer [NAV] A directional radio beacon to provide aircraft with signals for lateral guidance with respect to the runway centerline.

local level [NAV] The plane normal to the local vertical.

local line [COMMUN] A telephone line terminating at the local central office.

local loop *See* home loop.

local lunar time [ASTRON] The arc of the celestial equator, or the angle at the celestial pole, between the lower branch of the local celestial meridian and the hour circle of the moon, measured westward from the lower branch of the local celestial meridian through 24 hours; local hour angle of the moon, expressed in time units, plus 12 hours; local lunar time at the Greenwich meridian is called Greenwich lunar time.

locally arcwise connected topological space [MATH] A topological space in which every point has an arcwise connected neighborhood, that is, an open set any two points of which can be joined by an arc.

locally compact topological space [MATH] A topological space in which every point lies in a compact neighborhood.

locally connected topological space [MATH] A topological space in which every point has a connected neighborhood.

locally euclidean topological space [MATH] A topological space in which every point has a neighborhood which is homeomorphic to a euclidean space.

locally one to one [MATH] A function is locally one to one if it is one to one in some neighborhood of each point.

local Mach number [AERO ENG] The Mach number of an isolated section of an airplane or its airframe.

local magnetic disturbance [GEOPHYS] An anomaly of the magnetic field of the earth, extending over a relatively small area, due to local magnetic influences. Also known as local attraction.

local maximum [MATH] A local maximum of a function f is a value $f(c)$ of f where $f(x) \leq f(c)$ for all x in some neighborhood of c; if $f(c)$ is a local maximum, f is said to have a local maximum at c.

local mean noon [ASTRON] Twelve o'clock local mean time, or the instant the mean sun is over the upper branch of the local meridian; local mean noon at the Greenwich meridian is called Greenwich mean noon.

local mean time [ASTRON] The arc of the celestial equator, or the angle at the celestial pole, between the lower branch of the local celestial meridian and the hour circle of the mean sun, measured westward from the lower branch of the local celestial meridian through 24 hours.

local meridian [ASTRON] The meridian through any particular position which serves as the reference for local time.

local minimum [MATH] A local minimum of a function f is a value $f(c)$ of f where $f(x) \geq f(c)$ for all x in some neighborhood of c; if $f(c)$ is a local minimum, f is said to have a local minimum at c.

local noon [ASTRON] Noon at the local meridian.

Local Notice to Mariners [NAV] A locally disseminated issue of *Notice to Mariners* affecting a specific area.

local oscillator [ELECTR] The oscillator in a superheterodyne receiver, whose output is mixed with the incoming modulated radio-frequency carrier signal in the mixer to give the frequency conversions needed to produce the intermediate-frequency signal.

local oscillator injection [ELECTR] Adjustment used to vary the magnitude of the local oscillator signal that is coupled into the mixer.

local oscillator radiation [ELECTR] Radiation of the fundamental or harmonics of the local oscillator of a superheterodyne receiver.

local peat [GEOL] Peat formed by groundwater. Also known as basin peat.

local procurement [ORD] **1.** Procurement of supplies or equipment in the continental United States by other than a centralized purchasing office, such as purchase by an installation of supplies and equipment for use of that installation. **2.** Procurement of supplies or equipment for its own use in an area outside the United States by a United States military command located in that area.

local quasi-F martingale [MATH] A stochastic process $\{X_t\}$ such that the process obtained from $\{X_t\}$ by stopping it when it reaches n or $-n$ is a quasi-F martingale for each integer n.

local second selector [COMMUN] Interconnects a local first selector to a connector switch which is controlled and directed by impulses received from the local first selector.

local side [COMMUN] Terminal connections to an internal or in-station source such as data terminal connections to input or output devices.

local sidereal noon [ASTRON] Zero hour local sidereal time, or the instant the vernal equinox is over the upper branch of the local meridian; local sidereal noon at the Greenwich meridian is called Greenwich sidereal noon.

local sidereal time [ASTRON] The arc of the celestial equator, or the angle at the celestial pole which is between the upper branch of the local celestial meridian and the hour circle of the vernal equinox.

local solution [MATH] A function which solves a system of equations only in a neighborhood of some point.

local star cloud *See* local star system.

local star system [ASTRON] The group of stars of which the sun is a member. Also known as local cluster of stars; local star cloud.

local storm [METEOROL] A storm of mesometeorological scale; thus, thunderstorms, squalls, and tornadoes are often put in this category.

local time [ASTRON] **1.** Time based upon the local meridian as reference, as contrasted with that based upon a zone meridian, or the meridian of Greenwich. **2.** Any time kept locally.

local trunk [COMMUN] Trunk between local and long-distance switchboards, or between local and private branch exchange switchboards.

local vertical [NAV] A reference direction determined through the use of a plumb-bob or level.

local winds [METEOROL] Winds which, over a small area, differ from those which would be appropriate to the general pressure distribution, or which possess some other peculiarity.

locate [MIN ENG] To mark out the boundaries of a mining claim and establish the right of possession.

locate mode [ADP] A method of communicating with an input/output control system (IOCS), in which the address of the data involved, but not the data themselves, is transferred between the IOCS routine and the program.

locating [MECH ENG] A function of tooling operations accomplished by designing and constructing the tooling device so as to bring together the proper contact points or surfaces between the workpiece and the tooling.

locating hole [MECH ENG] A hole used to position the part in relation to a cutting tool or to other parts and gage points.

location [ADP] Any place in which data may be stored; usually expressed as a number.

location counter *See* instruction counter.

location damages [MIN ENG] Compensation by an operator to the surface owner for injury to the surface or to growing crops, resulting from the drilling of a well.

location dimension [DES ENG] A dimension which specifies the position or distance relationship of one feature of an object with respect to another.

location fit [DES ENG] The characteristic wherein mechanical sizes of mating parts are such that, when assembled, the parts are accurately positioned in relation to each other.

location notice [MIN ENG] A written sign placed prominently on a claim, showing the locator's name and describing the claim's extent and boundaries.

location plan [MIN ENG] A scale map of the projected mine development indicating, among other things, proposed shafts and works in relation to existing surface features.

location work [MIN ENG] Labor required by law to be done on mining claims within 60 days of location, in order to establish ownership.

locator [ENG] A radar or other device designed to detect and locate airborne aircraft.

locellus [BOT] **1.** In some legumes, a secondary compartment of a unilocular ovary that is formed by a false partition. **2.** One of the two cavities of a pollen sac.

lochia [MED] The discharge from the uterus and vagina during the first few weeks after labor.

lociation [ECOL] One of the subunits of a faciation, distinguished by the relative abundance of a dominant species.

lock [CIV ENG] A chamber with gates on both ends connecting two sections of a canal or other waterway, to raise or lower the water level in each section. [DES ENG] A fastening device in which a releasable bolt is secured. [ELECTR] To fasten onto and automatically follow a target by means of a radar beam. [MET] A condition in forging in which the flash line is in more than one plane. [ORD] **1.** Position of a safety mechanism which prevents a weapon from being fired. **2.** Fastening device used to secure against accidental movement, as on a control surface. **3.** To secure or make safe, as to set the safety on a weapon.

lockalloy [MET] A beryllium-base alloy composed of 62% beryllium and 38% aluminum; used as a structural aerospace alloy because of low density and high (47,000 pounds per square inch or 3.2×10^8 newtons per square meter) yield strength.

lock bolt [ENG] **1.** The bolt of a lock. **2.** A bolt equipped with a locking collar instead of a nut. **3.** A bolt for adjusting and securing parts of a machine.

lock chamber [CIV ENG] A compartment between lock gates in a canal.

locked-coil rope [DES ENG] A completely smooth wire rope that resists wear, made of specially formed wires arranged in concentric layers about a central wire core. Also known as locked-wire rope.

locked groove [DES ENG] A blank and continuous groove placed at the end of the modulated grooves on a disk recording to prevent further travel of the pickup. Also known as concentric groove.

locked oscillator [ELECTR] A sine-wave oscillator whose frequency can be locked by an external signal to the control frequency divided by an integer.

locked-oscillator detector [ELECTR] A frequency-modulation detector in which a local oscillator follows, or is locked to, the input frequency; the phase difference between local oscillator and input signal is proportional to the frequency deviation, and an output voltage is generated proportional to the phase difference.

locked-rotor current [ELEC] The current drawn by a stalled electric motor.

locked-wire rope *See* locked-coil rope.

lock frame [ORD] A mechanical unit in certain firearms, used to assist in unlocking the bolt from the barrel after recoil has started.

lock gate [CIV ENG] A movable barrier separating the water in an upper or lower section of waterway from that in the lock chamber.

lock-in [ELECTR] Shifting and automatic holding of one or both of the frequencies of two oscillating systems which are coupled together, so that the two frequencies have the ratio of two integral numbers.

locking [ELECTR] Controlling the frequency of an oscillator by means of an applied signal of constant frequency. [ENG] Automatic following of a target by a radar antenna.

locking fastener [DES ENG] A fastening used to prevent loosening of a threaded fastener in service, for example, a seating lock, spring stop nut, interference wedge, blind, or quick release.

locking lugs [ORD] Metal projections on the bolt of a small-arms weapon which cam into recesses cut in the side of the receiver to lock the weapon prior to firing.

locking relay *See* latching relay.

lockjaw *See* tetanus.

lock joint [DES ENG] A joint made by interlocking the joined elements, with or without other fastening.

LOCOMOTIVE

Electric locomotive rated at 5000 horsepower (3.7×10^6 watts). *(General Electric Co.)*

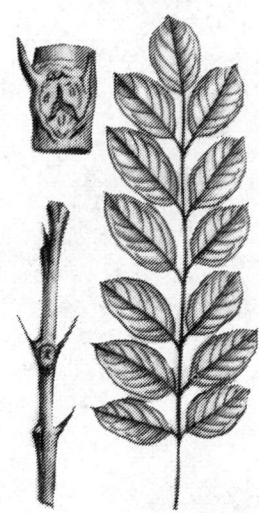

LOCUST

Leaf scar, branch, and leaves of the black locust (*Robinia pseudoacacia*).

LOCUSTIDAE

Drawing of the grasshopper (*Melanoplus mexicanus*).

locknut [DES ENG] **1.** A nut screwed down firmly against another or against a washer to prevent loosening. Also known as jam nut. **2.** A nut that is self-locking when tightened. **3.** A nut fitted to the end of a pipe to secure it and prevent leakage.

lock-on [ELECTR] **1.** The procedure wherein a target-seeking system (such as some types of radars) is continuously and automatically following a target in one or more coordinates (for example, range, bearing, elevation). **2.** The instant at which radar begins to track a target automatically.

lockout [ADP] In computer communications, the inability of a remote terminal to achieve entry to a computer system until project programmer number, processing authority code, and password have been validated against computer-stored lists. [COMMUN] **1.** In a telephone circuit controlled by two voice-operated devices, the inability of one or both subscribers to get through, because of either excessive local circuit noise or continuous speech from one or both subscribers. Also known as receiver lockout system. **2.** In mobile communications, an arrangement of control circuits whereby only one receiver can feed the system at one time to avoid distortion. Also known as receiver lockout system.

lockout circuit [ELECTR] A switching circuit which responds to concurrent inputs from a number of external circuits by responding to one, and only one, of these circuits at any time. Also known as finding circuit; hunting circuit.

lock primer [ORD] Primer intended to be placed by hand in the firing lock of a gun; it is used for initiation of the ignition of bag charges.

lockset [ENG] **1.** A complete lock including the lock mechanism, keys, plates, and other parts. **2.** A jig or template for making cuts in a door for holding a lock.

lock-up relay *See* latching relay.

lock washer [DES ENG] A solid or split washer placed underneath a nut or screw to prevent loosening by exerting pressure.

loco disease [VET MED] Poisoning in livestock resulting from ingestion of selenium-containing plants (loco weed); characterized by atrophy, delirium, convulsions, and stupor, often terminating in death.

locomotion [SCI TECH] Progressive movement, as of an animal or a vehicle.

locomotive [MECH ENG] A self-propelling machine with flanged wheels, for moving loads on railroad tracks; utilizes fuel (for steam or internal combustion engines), compressed air, or electric energy.

locomotive crane [MECH ENG] A crane mounted on a railroad flatcar or a special chassis with flanged wheels. Also known as rail crane.

locomotive gradient [MIN ENG] The gradient set by law for a locomotive haulage; maximum is 1 in 15, but the limit for practical purposes is 1 in 25.

locomotive haulage [MIN ENG] The use of locomotive-hauled mine cars to carry coal ore, workers, and materials in a mine.

locomotor ataxia *See* tabes dorsalis.

locomotor system [ZOO] Appendages and associated parts, such as muscles, joints, and bones, concerned with motor activities and locomotion of the animal body.

loco weed [BOT] Any species of *Astragalus* containing selenium taken up from the soil.

loculate [BIOL] Having, or divided into, loculi.

loculus [BIOL] A small cavity or chamber.

locus [GEN] The fixed position of a gene in a chromosome. [MATH] A collection of points in a euclidean space whose coordinates satisfy one or more algebraic conditions.

locus of an equation [MATH] A collection of points, all of which satisfy a single equation.

locust [BOT] Either of two species of commercially important trees, black locust (*Robinia pseudoacacia*) and honey locust (*Gladitsia triacanthos*), in the family Leguminosae. [INV ZOO] The common name for various migratory grasshoppers of the family Locustidae.

Locustidae [INV ZOO] A family of insects in the order Orthoptera; antennae are usually less than half the body length, hindlegs are adapted for jumping, and the ovipositor is multipartite.

lodar [NAV] A direction finder used to determine the direc-

tion of arrival of loran signals, free of night effect, by observing the separately distinguishable ground and sky-wave loran signals on a cathode-ray oscilloscope and positioning a loop antenna to obtain a null indication of the component selected to be most suitable. Also known as lorad.

lode [GEOL] A fissure in consolidated rock filled with mineral; usually applied to metalliferous deposits.

lode claim [MIN ENG] That portion of a vein or lode, and of the adjoining surface, which has been acquired by a compliance with the law, both Federal and state.

lodestone [MINERAL] The naturally occurring magnetic iron oxide, or magnetite, possessing polarity, and attracting iron objects to itself. Also known as Hercules stone; leading stone; loadstone.

lodos [METEOROL] A southerly wind on the Black Sea coast of Bulgaria.

lodranite [GEOL] A stony iron meteorite composed of bronzite and olivine within a fine network of nickel-iron.

Loeffler's syndrome [MED] Extensive infiltration of the lung by eosinophils, and eosinophilia of the peripheral circulation. Also known as eosinophilic pneumonitis.

loellingite [MINERAL] $FeAs_2$ A silver-white to steel-gray mineral composed of iron arsenide with some cobalt, nickel, antimony, and sulfur; isomorphous with arsenopyrite; a source of arsenic. Also known as leucopyrite; löllingite.

loess [GEOL] An essentially unconsolidated, unstratified calcareous silt; commonly it is homogeneous, permeable, and buff to gray in color, and contains calcareous concretions and fossils.

loess kindchen [GEOL] An irregular or spheroidal nodule of calcium carbonate that is found in loess.

loewite [MINERAL] $Na_4Mg_2(SO_4)_4 \cdot 5H_2O$ A white to pale-yellow mineral composed of hydrous sulfate of sodium and magnesium.

lofar [NAV] A submarine detection system using autocorrelation techniques for long-range analysis of patterned sound picked up at the low-frequency end of the sound spectrum by underwater hydrophones of the Caesar submarine detection system. Derived from low-frequency acquisition and ranging.

Lofco car feeder [MIN ENG] A carrying chain running between the rails that controls mine cars at loading points, for marshaling trains and for loading cars into cages and tipplers.

loft [BUILD] **1.** An upper part of a building. **2.** A work area in a factory or warehouse.

loft bombing [ORD] A method of aerial bombing in which the delivery plane approaches the target at a very low altitude, makes a definite pull-up at a given point, releases bomb at predetermined point during the pull-up, and tosses the missile on the target.

loft building [BUILD] A building with a large open floor area.

log [ADP] A record of computer operating runs, including tapes used, control settings, halts, and other pertinent data. [COMMUN] A written record of radio and television station operating data, required by law. [ENG] The record of, or the act or process of recording, events or the type and characteristics of the rock penetrated in drilling a borehole as evidenced by the cuttings, core recovered, or information obtained from electronic devices. [MATER] Unshaped timber either rough or squared. [NAV] **1.** An instrument for measuring the speed or distance or both traveled by a vessel. **2.** A written record of the movements of a craft, with regard to courses, speeds, positions, and other information of interest to navigators, and of important happenings aboard the craft. **3.** A written record of specific related information, such as that concerning performance of an instrument.

Loganiaceae [BOT] A family of mostly woody dicotyledonous plants in the order Gentianales; members lack a latex system and have fully united carpels and axile placentation.

Logan slabbing machine [MIN ENG] A machine that has three cutting chains; two are horizontal—one at the base of the coal seam, the other at a distance from the floor; the third is mounted vertically and shears off the coal at the back of the cut; a short conveyor transfers the coal to the face conveyor.

logarithm [MATH] **1.** The real-valued function log u defined by log $u = v$ if $e^v = u$, e^v denoting the exponential function. Also known as hyperbolic logarithm; Naperian logarithm;

natural logarithm. **2.** An analog in complex variables relative to the function e^z.

logarithmic amplifier [ELECTR] An amplifier whose output signal is a logarithmic function of the input signal.

logarithmic coordinate paper [MATH] Paper ruled with two sets of mutually perpendicular, parallel lines spaced according to the logarithms of consecutive numbers, rather than the numbers themselves.

logarithmic coordinates [MATH] In the plane, logarithmic coordinates are defined by two coordinate axes, each marked with a scale where the distance between two points is the difference of the logarithms of the two numbers.

logarithmic curve [MATH] A curve whose equation in cartesian coordinates is $y = \log ax$, where a is greater than 1.

logarithmic decrement [PHYS] The natural logarithm of the ratio of the amplitude of one oscillation to that of the next which has the same polarity, when no external forces are applied to maintain the oscillation.

logarithmic derivative [MATH] The logarithmic derivative of a function $f(z)$ of a real (complex) variable is the ratio $f'(z)/f(z)$, that is, the derivative of $\log f(z)$.

logarithmic differentiation [MATH] A technique often helpful in computing the derivatives of a differentiable function $f(x)$; set $g(x) = \log f(x)$ where $f(x) \neq 0$, then $g'(x) = f'(x)/f(x)$, and if there is some other way to find $g'(x)$, then one also finds $f'(x)$.

logarithmic distribution [STAT] A frequency distribution whose value at any integer $n = 1,2, \ldots$ is $\lambda^n/(-n)\log(1-\lambda)$, where λ is fixed.

logarithmic equation [MATH] An equation which involves a logarithmic function of some variable.

logarithmic fast time constant [ELECTR] Constant false alarm rate scheme which has a logarithmic intermediate-frequency amplifier followed by a fast time constant circuit.

logarithmic growth *See* exponential growth.

logarithmic multiplier [ELECTR] A multiplier in which each variable is applied to a logarithmic function generator, and the outputs are added together and applied to an exponential function generator, to obtain an output proportional to the product of two inputs.

logarithmic potential [PHYS] A potential function that is proportional to the logarithm of some coordinate; for example, a straight, electrically charged cylinder of circular cross section and effectively infinite length gives rise to an electrostatic potential that is the sum of a constant and a term proportional to the logarithm of the distance from the cylinder's axis.

logarithmic profile of velocity [FL MECH] The mean velocity parallel to a boundary of a fluid in turbulent motion as a function of distance from the boundary, on the assumption that the shearing stress is independent of distance from the boundary, and the mixing length is proportional either to the distance from the boundary or to the ratio of the first derivative of the profile of velocity itself to the second derivative.

logarithmic scale [MATH] A scale in which the distances that numbers are at from a reference point are proportional to their logarithms.

logarithmic spiral [MATH] The spiral plane curve whose points in polar coordinates (r,θ) satisfy the equation $\log r = a\theta$. Also known as equiangular spiral.

logarithmic transformation [STAT] The replacement of a variate y with a new variate $z = \log y$ or $z = \log(y + c)$, where c is a constant; this operation is often performed when the resulting distribution is normal, or if the resulting relationship with another variable is linear.

logarithmic velocity profile [METEOROL] The theoretical variation of the mean wind speed with height in the surface boundary layer under certain assumptions.

logbook [ADP] A bound volume in which operating data of a computer is noted. [NAV] A book in which all affairs and events of navigational importance of a ship are recorded, such as speed and ship's progress.

logger [ENG] A recorder that automatically scans measured quantities at specified times and records, or logs, their values on a chart.

loggia [ARCH] A roofed open arcade on the side of a building.

logging [ENG] Continuous recording versus depth of some characteristic datum of the formations penetrated by a drill hole; for example, resistivity, spontaneous potential, conductivity, fluid content, radioactivity, or density. [FOR] The cutting and removal of the woody stem portions of forest trees.

logical addition [MATH] The additive binary operation of a Boolean algebra.

logical comparison [ADP] The operation of comparing two items in a computer and producing a one output if they are equal or alike, and a zero output if not alike.

logical connectives [MATH] Symbols which link mathematical statements; these symbols represent the terms "and," "or," "implication," and "negation."

logical construction [ADP] A simple logical property that determines the type of characters which a particular code represents; for example, the first two bits can tell whether a character is numeric or alphabetic.

logical decision [ADP] The ability to select one of many paths, depending upon intermediate programming data.

logical expression [ADP] Two arithmetic expressions connected by a relational operator indicating whether an expression is greater than, equal to, or less than the other, or connected by a logical variable, logical constant (true or false), or logical operator.

logical flow chart [ADP] A detailed graphic solution in terms of the logical operations required to solve a problem.

logical gate *See* switching gate.

logical instruction [ADP] A digital computer instruction which forms a logical combination (on a bit-by-bit basis) of its operands and leaves the result in a known location.

logical multiplication [MATH] The multiplicative binary operation of a Boolean algebra.

logical operation [ADP] A nonarithmetical operation in a computer, such as comparing, selecting, making references, matching, sorting, and merging, where logical yes-or-no quantities are involved.

logical record [ADP] A group of adjacent, logically related data items.

logical shift [ADP] A shift operation that treats the operand as a set of bits, not as a signed numeric value or character representation.

logical sum [ADP] A computer addition in which the result is 1 when either one or both input variables is 1, and the result is 0 when the input variables are both 0.

logical symbol [ADP] A graphical symbol used to represent a logic element.

logic-arithmetic unit *See* arithmetical unit.

logic card [ELECTR] A small fiber chassis on which resistors, capacitors, transistors, magnetic cores, and diodes are mounted and interconnected in such a way as to perform some computer function; computers employing this type of construction may be repaired by removing the faulty card and replacing it with a new card.

logic circuit [ADP] A computer circuit that provides the action of a logic function or logic operation.

logic design [ADP] The design of a computer at the level which considers the operation of each functional block and the relationships between the functional blocks.

logic diagram [ADP] A graphical representation of the logic design or a portion thereof; displays the existence of functional elements and the paths by which they interact with one another.

logic element [ADP] A hardware circuit that performs a simple, predefined transformation on its input and presents the resulting signal as its output. Occasionally known as functor.

logic section *See* arithmetical unit.

logic switch [ELECTR] A diode matrix or other switching arrangement that is capable of directing an input signal to one of several outputs.

logistic curve [STAT] **1.** A type of growth curve, representing the size of a population y as a function of time t: $y = k/(1 + e^{-kbt})$, where k and b are positive constants. Also known as Pearl-Reed curve. **2.** More generally, a curve representing a function of the form $y = k/(1 + e^{cf(t)})$, where c is a constant and $f(t)$ is some function of time.

log line *See* current line.

log-mean temperature difference [THERMO] The log-mean temperature difference $T_{LM} = (T_2 - T_1)/\ln T_2/T_1$, where T_2 and T_1 are the absolute (K or °R) temperatures of the two extremes being averaged; used in heat transfer calculations in which one fluid is cooled or heated by a second held separate by pipes or process vessel walls.

logo *See* logotype.

logomania [MED] Logorrhea so excessive as to be a form of a manic state; new words may be invented to keep up the garrulity.

logoplegia [MED] Loss of ability to articulate, usually due to paralysis of the speech organs.

logorrhea [MED] Excessive, usually rapid, incoherent, and uncontrollable talkativeness.

logotype [GRAPHICS] A single slug or line of type cast in one piece, that carries one or more words such as the name of a firm or a product; a special style of type may be used, and the logotype may include a trademark or other art work with the type. Also known as logo. [SYST] The selection or designation of a genotype after the generic name was published.

log-periodic antenna [ELECTROMAG] A broad-band antenna which consists of a sheet of metal with two wedge-shaped cutouts, each with teeth cut into its radii along circular arcs; characteristics are repeated at a number of frequencies that are equally spaced on a logarithmic scale.

log rule [FOR] A table showing expected log output in board feet or other units.

log volume [FOR] The cubic volume of a log computed inside the bark as determined by any of several formulas; parameters are the cross-sectional areas of log midpoint, large end and small end of log, and log length.

logwood *See* hematoxylon.

loiasis [MED] A filariasis of tropical Africa, caused by the filaria *Loa,* and characterized by diurnal periodicity of microfilariae in the blood, and transient cutaneous swelling caused by migrating adult worms.

loktal base [ELECTR] A special base for small vacuum tubes, so designed that it locks the tube firmly in a corresponding special eight-pin loktal socket; the tube pins are sealed directly into the glass envelope.

löllingite *See* loellingite.

lolly ice [OCEANOGR] Saltwater frazil, a heavy concentration of which is called sludge.

lombarde [METEOROL] An easterly wind (from Lombardy) that predominates along the French-Italian frontier, and comes from the High Alps; in winter it is violent and forms snow drifts in the mountain valleys; in the plains it is gentle and very dry.

loment [BOT] A dry, indehiscent single-celled fruit that is formed from a single superior ovary; splits transversely in numerous segments at maturity.

lomonite *See* laumontite.

Lomonosov ridge [GEOL] An undersea ridge which subdivides the Arctic Basin, extending from Ellesmere Land to the New Siberian Islands.

lomontite *See* laumontite.

Lonchaeidae [INV ZOO] A family of minute myodarian cyclorrhaphous dipteran insects in the subsection Acalypteratae.

London dispersion force *See* Van der Waals force.

London equations [SOLID STATE] Equations for the time derivative and the curl of the current in a superconductor in terms of the electric and magnetic field vectors respectively, derived in the London superconductivity theory.

London penetration depth [SOLID STATE] A measure of the depth which electric and magnetic fields can penetrate beneath the surface of a superconductor from which they are otherwise excluded, according to the London superconductivity theory.

London superconductivity theory [SOLID STATE] An extension of the two-fluid model of superconductivity, in which it is assumed that superfluid electrons behave as if the only force acting on them arises from applied electric fields, and that the curl of the superfluid current vanishes in the absence of a magnetic field.

London superfluidity theory [CRYO] A theory, based on the fact that helium-4 obeys Bose-Einstein statistics, in which helium-4 is treated as an ideal Bose-Einstein gas, and its superfluid component is equated with the finite fraction of the atoms of such a gas which are in the ground state at very low temperatures.

long-base-line system [COMMUN] System in which the distance separating ground stations approximates the distance to the target being tracked.

long bone [ANAT] A bone in which the length markedly exceeds the width, as the femur or the humerus.

long clay [MATER] A clay used in ceramics that has a high degree of plasticity.

long column [CIV ENG] A column so slender that bending is the primary deformation, generally having a slenderness ratio greater than 120–150.

long-conductor antenna *See* long-wire antenna.

long-day plant [BOT] A plant that flowers in response to a long photoperiod.

long discharge [ELEC] **1.** A capacitor or other electrical charge accumulator which takes a long time to leak off. **2.** A gaseous electrical discharge in which the length of the discharge channel is very long compared with its diameter; lightning discharges are natural examples of long discharges. Also known as long spark.

long-distance loop [COMMUN] Line from a subscriber's station directly to a long-distance switchboard.

long-distance navigation [NAV] Navigation performed in an area where aids to navigation are spaced more than 200 miles (322 kilometers) apart, exclusive of the short-range, the approach and landing zone, and the airport zone.

long-distance xerography [COMMUN] A facsimile system that uses a cathode-ray scanner at the microwave transmitting terminal; at the receiving terminal, a lens projects the received cathode-ray image onto the selenium-coated drum of a xerographic copying machine.

longeron [AERO ENG] A principal longitudinal member of the structural framework of a fuselage, nacelle, or empennage boom.

long-haul carrier system [COMMUN] An intercity telephone communication system in which a frequency-division multiplexed signal modulates a subcarrier.

long-haul radio [COMMUN] A microwave radio system for transmitting telephone and telegraph signals over distances on the order of 4000 miles (6500 kilometers) or more on line-of-sight paths between a series of repeaters which demodulate the signal to intermediate frequency.

long hole [MIN ENG] An underground borehole and blasthole exceeding 10 feet (3 meters) in depth or requiring the use of two or more lengths of drill steel or rods coupled together to attain the desired depth.

long-hole drill [MIN ENG] A rotary- or a percussive-type drill used to drill long holes.

long-hole jetting [MIN ENG] A hydraulic mining system consisting essentially of drilling a hole down the pitch of the vein, replacing the drilling head with a jet cutting head, and then retracting the drill column with the jets in operation to remove the coal.

long hundredweight *See* hundredweight.

Longhurst-Hardy plankton sampler [ENG] A nonquantitative metal-shrouded net for trapping plankton.

longicollous [BIOL] Having a long beak or neck.

Longidorinae [INV ZOO] A subfamily of nematodes belonging to the Dorylaimoidea including economically important plant parasites.

Longipennes [VERT ZOO] An equivalent name for Charadriiformes.

longitude [GEOD] Angular distance, along the Equator, between the meridian passing through a position and, usually, the meridian of Greenwich.

longitude factor [NAV] The change in longitude along a celestial line of position per 1-minute change in latitude.

longitude line [NAV] A line of position extending in a north-south direction generally.

longitude method [NAV] In celestial navigation, the establishing of a line of position from the observation of the altitude of a celestial body by assuming a latitude (or longitude), and calculating the longitude (or latitude) through which the line of position passes, and the azimuth; the line of position is drawn through the point thus found, perpendicular to the azimuth.

longitude signal [NAV] A radio or telegraphic time signal transmitted by a station of known longitude and received at stations whose longitude can then be calculated from differences in local time.

longitudinal [SCI TECH] Pertaining to the lengthwise dimension.

longitudinal aberration [OPTICS] **1.** The distance along the optical axis from the focus of paraxial rays to the point where rays coming from the outer edges of its lens or reflecting surface intersect this axis. **2.** In chromatic aberration, the distance along the optical axis between the foci of two standard colors.

longitudinal acceleration [MECH] The component of the linear acceleration of an aircraft, missile, or particle parallel to its longitudinal, or X, axis.

longitudinal baffle [CHEM ENG] Baffle sheets or plates within a process vessel (such as a heat exchanger) that are parallel to the long dimension of the vessel; used to direct fluid flow in the desired flow pattern.

longitudinal bulkhead [NAV ARCH] A partition wall running fore and aft, made of planking or plating.

longitudinal center of gravity [NAV ARCH] That point at which the combined weight of all the items that constitute a ship's weight are considered to be concentrated; usually stated as either aft or forward of the middle perpendicular or the midship frame.

longitudinal circuit [ELEC] Circuit formed by one telephone wire (or by two or more telephone wires in parallel) with return through the earth or through any other conductors except those which are taken with the original wire or wires to form a metallic telephone circuit.

longitudinal coefficient [NAV ARCH] The ratio of the volume of water displaced by a ship to the product of its length on the waterline and the area of its midship section below the water plane. Also known as prismatic coefficient.

longitudinal controller [AERO ENG] A primary flight control mechanism which controls pitch attitude; located in the cockpit, this may be a control column or a side stick.

longitudinal current [ELEC] Current which flows in the same direction in the two wires of a parallel pair using the earth or other conductors for a return path.

longitudinal direction [MET] The principal direction of flow in a plastically deformed metal.

longitudinal dune [GEOL] A type of linear dune ridge that extends parallel to the direction of the dominant dune-building winds.

longitudinal fault [GEOL] A fault parallel to the trend of the surrounding structure.

longitudinal flow reactor [CHEM ENG] Theoretical reactor system in which there is no longitudinal mixing (back mixing) of reactants and products as they flow through the reactor, but in which there is complete radial (side-to-side) mixing.

longitudinal frame [NAV ARCH] Any of the frames of a ship running fore and aft.

longitudinal framed ship [NAV ARCH] A ship constructed of widely spaced, transverse frames, which support a number of small fore and aft frames.

longitudinal magnetoresistance [ELECTROMAG] The change of electrical resistance produced in a current-carrying metal or semiconductor upon application of a magnetic field parallel to the current flow.

longitudinal magnetorestriction *See* Joule effect.

longitudinal parity [COMMUN] Parity associated with bits recorded on one track in a data block, to indicate whether the number of recorded bits in the block is even or odd.

longitudinal parity check [COMMUN] The count for even or odd parity of all the bits in a message as a precaution against transmission error. Also known as horizontal parity check.

longitudinal quadrupole [ACOUS] A sound source resulting from a variation of a component of the velocity of matter in a direction that is parallel to the velocity component. [ELECTROMAG] An electric or magnetic quadrupole which produces a field equivalent to that of two equal and opposite electric or magnetic dipoles separated by a small distance

parallel to the direction of the dipoles. Also known as axial quadrupole.

longitudinal redundancy check [COMMUN] A method of checking for errors, in which data are arranged in blocks according to some rule, and the correctness of each character in the block is determined according to the rule. Abbreviated LRC.

longitudinal section [SCI TECH] A section taken through the lengthwise dimension of a structure, organism, or other object.

longitudinal separation [NAV] In air-traffic control, the separation of aircraft along a longitudinal line.

longitudinal stability [ENG] The ability of a ship or aircraft to recover a horizontal position after a vertical motion of its ends about a horizontal axis perpendicular to the centerline.

longitudinal strain *See* linear strain.

longitudinal stream *See* subsequent stream.

longitudinal vibration [MECH] A continuing periodic change in the displacement of elements of a rod-shaped object in the direction of the long axis of the rod.

longitudinal wave [PHYS] A wave in which the direction of some vector characteristic of the wave, for example, the displacement of particles of the transmitting medium, is along the direction of propagation.

long-life items [ORD] Ordnance items which have an estimated average service life of 5 years or more.

long-line current [ELEC] A current that flows through the earth from an anodic to a cathodic area and returns along an underground pipe or other metal structure, often over a considerable distance and as the result of concentration cell action.

long-line effect [ELECTR] An effect occurring when an oscillator is coupled to a transmission line with a bad mismatch; two or more frequencies may then be equally suitable for oscillation, and the oscillator jumps from one of these frequencies to another as its load changes.

long-lines engineering [COMMUN] Engineering performed to develop, modernize, or expand long-haul, point-to-point communications facilities using radio, microwave, or wire circuits.

long-nose pliers [DES ENG] Small pincer with long, tapered jaws.

long oil [MATER] Varnish containing a large percentage of oil.

long-period tide [OCEANOGR] A tide or tidal current constituent with a period which is independent of the rotation of the earth but which depends upon the orbital movement of the moon or of the earth.

long period variable [ASTRON] A variable star with a period from about 100 to more than 600 days.

long-persistence screen [ELECTR] A fluorescent screen containing phosphorescent compounds that increase the decay time, so a pattern may be seen for several seconds after it is produced by the electron beam.

long-pillar work [MIN ENG] A coal-winning technique used in underground mining in which large pillars of coal are left as the face is advanced; finally all the pillars are removed together.

long-playing record [ENG ACOUS] A 10- or 12-inch (25.4- or 30.48-centimeter) phonograph record that operates at a speed of $33\frac{1}{3}$ rpm (revolutions per minute) and has closely spaced grooves, to give playing times up to about 30 minutes for one 12-inch side. Also known as LP record; microgroove record.

long primer [ORD] Primer with relatively long body, designed to provide central ignition for a propelling charge.

long-range chart [NAV] An aeronautical chart of small scale, covering a large area, designed for long flights in which celestial or electronic navigation is expected to be used principally.

long-range forecast [METEOROL] A forecast for a period greater than 5 days (or a week) in advance. Also known as extended-range forecast.

long-range materiel requirements [ORD] Ordnance items required by operational and organizational concepts established for a period 5 to 10 years hence.

long-range navigation zone [NAV] **1.** A zone in which navigational aids furnish service at distances greater than 200 miles (322 kilometers). **2.** *See* loran.

long-range order [SOLID STATE] A tendency for some property of atoms in a lattice (such as spin orientation or type of atom) to follow a pattern which is repeated every few unit cells.

long residuum [MATER] Residue from crude-oil distillation in a petroleum refinery when a relatively small proportion of the feed is distilled overhead.

long run frequency [STAT] The ratio of the number of occurrences of an event in a large number of trials to the number of trials.

Long's coefficient [PATH] The number 2.6, multiplied by the last two figures of the urine specific gravity, determined at 25°C, to derive the number of grams of solids per liter of urine.

longshore bar [GEOL] A ridge of sand, gravel, or mud built on the seashore by waves and currents, generally parallel to the shore and submerged by high tides. Also known as offshore bar.

longshore current *See* littoral current.

longshore drift *See* littoral drift.

longshore trough [GEOL] A long, wide, shallow depression of the sea floor parallel to the shore.

long span [ENG] Span of open wire exceeding 250 feet (76 meters) in length.

long-span steel framing [BUILD] Framing system used when there is a greater clear distance between supports than can be spanned with rolled beams; girders, simple trusses, arches, rigid frames, and cantilever suspension spans are used in this system.

long spark *See* long discharge.

long-staple cotton [TEXT] A grade of cotton whose staple is $1\frac{1}{8}$ inch (29 millimeters) or over in length.

long-time burning oil [MATER] Carefully refined kerosine used in railway semaphore signal lamps; heavy ends are removed so that the oil will burn without charring the wick.

long tom [MIN ENG] A trough, longer than a rocker, for washing gold-bearing earth.

Long Tom [ORD] Popular name for 155-millimeter self-propelled gun.

long ton *See* ton.

long-tube vertical evaporator [CHEM ENG] A liquid evaporator in which the material is force-fed into the bottom of a bundle of long, vertical tubes; hot liquid on the outsides of the tubes transfers heat to the rising liquid feed, causing partial evaporation.

longwall coal cutter [MIN ENG] Compact machine driven by electricity or compressed air which cuts into the coal on relatively long faces, with its jib at right angles to its body.

longwall peak stoping [MIN ENG] An underland stoping method in which the rapid advance of the face goes on, and by working the faces at a 60° angle to the strike, the peak travels down the dip at twice the face advance rate.

longwall pillar working [MIN ENG] A technique used to extract coal pillars left behind in long-pillar working.

longwall retreating [MIN ENG] A longwall working system in which all the roadways are in the solid coal seam and the waste areas are left behind; developing headings are driven close to the boundary or limit, and the coal is taken out by the longwall retreating toward the shaft.

longwall system [MIN ENG] A method of mining in which the faces are advanced from the shaft toward the boundary, and the roof is allowed to cave in behind the miners as work progresses.

long wave [COMMUN] An electromagnetic wave having a wavelength longer than the longest broadcast-band wavelength of about 545 meters, corresponding to frequencies below about 550 kilohertz. [METEOROL] With regard to atmospheric circulation, a wave in the major belt of westerlies which is characterized by large length (thousands of kilometers) and significant amplitude; the wavelength is typically longer than that of the rapidly moving individual cyclonic and anticyclonic disturbances of the lower troposphere. Also known as major wave; planetary wave.

long-wave radio [COMMUN] A radio which can receive fre-

quencies below the lowest broadcast frequency of 550 kilohertz.

long-wire antenna [ELECTROMAG] An antenna whose length is a number of times greater than its operating wavelength, so as to give a directional radiation pattern. Also known as long-conductor antenna.

long-wool [TEXT] A straight wool with a shine or gloss and a staple length of 7 to 20 inches (18 to 51 centimeters).

lonsdaleite [MINERAL] A mineral composed of a form of carbon; found in meteorites.

loo [METEOROL] A hot wind from the west in India.

look angle [AERO ENG] The elevation and azimuth at which a particular satellite is predicted to be found at a specified time. [ENG] The solid angle in which an instrument operates effectively, generally used to describe radars, optical instruments, and space radiation detectors.

look box [CHEM ENG] Box with glass windows built into distillation-column rundown lines (or other flow lines) so that the stream of condensate from the condenser can be watched.

lookout station [ENG] A structure or place on shore at which personnel keep watch of events at sea or along the shore.

lookout tower [ENG] In marine operations, any tower surmounted by a small house in which a watch is habitually kept, as distinguished from an observation tower in which no watch is kept.

look-through [ELECTR] **1.** When jamming, a technique whereby the jamming emission is interrupted irregularly for extremely short periods to allow monitoring of the victim signal during jamming operations. **2.** When being jammed, the technique of observing or monitoring a desired signal during interruptions in the jamming signal.

look-up [ADP] An operation or process in which a table of stored values is scanned (or searched) until a value equal to (or sometimes, greater than) a specified value is found.

look-up table [ADP] A stored matrix of data for reference purpose.

loom [METEOROL] The glow of light below the horizon produced by greater-than-normal refraction in the lower atmosphere; it occurs when the air density decreases more rapidly with height than in the normal atmosphere. [TEXT] A machine on which fabrics are woven.

Loomis-Wood diagram [SPECT] A graph used to assign lines in a molecular spectrum to the various branches of rotational bands when these branches overlap, in which the difference between observed wave numbers and wave numbers extrapolated from a few lines that apparently belong to one branch are plotted against arbitrary running numbers for that branch.

loon [VERT ZOO] The common name for birds composing the family Gaviidae, all of which are fish-eating diving birds.

loop [ADP] A sequence of computer instructions which are executed repeatedly, but usually with address modifications changing the operands of each iteration, until a terminating condition is satisfied. [AERO ENG] A flight maneuver in which an airplane flies a circular path in an approximately vertical plane, with the lateral axis of the airplane remaining horizontal, that is, an inside loop. [ELEC] **1.** A closed path or circuit over which a signal can circulate, as in a feedback control system. **2.** Commercially, the portion of a connection from central office to subscriber in a telephone system. [ELECTROMAG] *See* coupling loop; loop antenna. [ENG] **1.** A reel of motion picture film or magnetic tape whose ends are spliced together, so that it may be played repeatedly without interruption. **2.** A closed circuit of pipe in which materials and components may be placed to test them under different conditions of temperature, irradiation, and so forth. **3.** *See* mesh. [PHYS] **1.** A closed curve on a graph, such as a hysteresis loop. **2.** *See* antinode.

loop antenna [ELECTROMAG] A directional-type antenna consisting of one or more complete turns of a conductor, usually tuned to resonance by a variable capacitor connected to the terminals of the loop. Also known as loop.

loop body [ADP] The set of statements to be performed iteratively with the range of a loop.

loop checking [COMMUN] Sending signals from the central office to test the integrity of local loops.

loop circuit [COMMUN] Common communications circuit shared by more than two parties; when applied to a teletype-

LOOP ANTENNA

Shape of loop antenna.

869

writer operation, all machines print all data entered on the loop.

loop control *See* photoelectric loop control.

loop coupling [ELECTROMAG] A method of transferring energy between a waveguide and an external circuit, by inserting a conducting loop into the waveguide, oriented so that electric lines of flux pass through it.

loop dialing [COMMUN] Return-path method of dialing in which the dial pulses are sent out over one side of the interconnecting line or trunk and are returned over the other side; limited to short-haul traffic.

looped pipeline [PETRO ENG] A pipeline that is paralleled (looped) by a second pipeline, both of which serve the same liquid or gas source and destination.

loop filter [ELECTR] A low-pass filter, which may be a simple *RC* filter or may include an amplifier, and which passes the original modulating frequencies but removes the carrier-frequency components and harmonics from a frequency-modulated signal in a locked-oscillator detector.

loop gain [CONT SYS] The ratio of the magnitude of the primary feedback signal in a feedback control system to the magnitude of the actuating signal. [ELECTR] Total usable power gain of a carrier terminal or two-wire repeater; maximum usable gain is determined by, and may not exceed, the losses in the closed path.

loop head [ADP] The first instruction of a loop, which contains the mode of execution, induction variable, and indexing parameters.

looping mill [MET] An arrangement of mills such that a hot bar discharged from one mill is fed into a second mill in the opposite direction.

loop lake *See* oxbow lake.

loop-mile [ELEC] Length of wire in a mile of two-wire line.

Loop Nebula [ASTRON] A large, bright gaseous nebula in the Large Magellanic Cloud; its diameter is about 260 light-years. Also known as 30 Doradus; Tarantula.

loop of Henle [ANAT] The U-shaped portion of a renal tubule formed by a descending and an ascending tubule. Also known as Henle's loop.

loop pattern [ANAT] A type of fingerprint pattern in which one or more of the ridges enter on either side of the impression, recurve, touch, or pass an imaginary line from the delta to the core and terminate on the entering side.

loop pulsing [COMMUN] Regular, momentary interruptions of the direct-current path at the sending end of a transmission line. Also known as dial pulsing.

loop rating [HYD] A rating curve that has higher values of discharge for a certain stage when the river is rising than it does when the river is falling; thus, the curve (stage versus discharge) describes a loop with each rise and fall of the river.

loop ratio *See* loop transfer function.

loop seal [CHEM ENG] Antivapor seal for liquid drawoffs from process or storage vessels; liquid drawoff is made to flow through an immersed loop or beneath an obstruction, thus sealing off vapor flow.

loopstick antenna *See* ferrite-rod antenna.

loop test [ELEC] A telephone or telegraph line test that is made by connecting a faulty line to good lines in such a way as to form a loop in which measurements can be made to determine the position of the fault.

loop transfer function [CONT SYS] For a feedback control system, the ratio of the Laplace transform of the primary feedback signal to the Laplace transform of the actuating signal. Also known as loop ratio.

loop transmission [COMMUN] A series connection of subscribers in a communications system, as in the ARPA computer network or McCullough burglar alarm loops.

loop transmittance [CONT SYS] 1. The transmittance between the source and sink created by the splitting of a specified node in a signal flow graph. 2. The transmittance between the source and sink created by the splitting of a node which has been inserted in a specified branch of a signal flow graph in such a way that the transmittance of the branch is unchanged.

loop tunnel [ENG] A tunnel which is looped or folded back on itself to gain grade in a tunnel location.

loop-type pit bottom [MIN ENG] An arrangement at the pit bottom in which loaded cars are fed to the cage on one side

only, and the empties are returned by a loop roadway to the same place.

loop-type radio range [NAV] An A-N radio range employing two loop antennas placed at right angles to one another.

loopway [MIN ENG] A double-track loop in a main single-track haulage plane at which mine cars may pass.

loose connective tissue [HISTOL] A type of irregularly arranged connective tissue having a relatively large amount of ground substance.

loose coupling [ELEC] Coupling of a degree less than the critical coupling.

loose-detail mold [ENG] A plastics mold with parts that come out with the molded piece.

loose fit [DES ENG] A fit with enough clearance to allow free play of the joined members.

loose ground [MIN ENG] Broken, fragmented, or loosely cemented bedrock material that tends to slough from sidewalls into a borehole and must be supported, as with timber sets. Also known as broken ground.

loose-joint butt [DES ENG] A knuckle hinge in which the pin on one half slides easily into a slot on the other half.

loose kernel smut [PL PATH] A type of kernel smut disease distinguished by the ruptured spore-containing gall.

Looser-Milkman syndrome *See* Milkman's syndrome.

loose round [ORD] Defective cartridge in which the bullet is loose in the cartridge case.

Lopadorrhynchidae [INV ZOO] A small family of pelagic polychaete annelids belonging to the Errantia.

loparite [MINERAL] (Ce,Na,Ca)₂(Ti,Nb)₂O₆ A brown to black mineral; a variety of perovskite containing alkalies and cerium.

lopezite [MINERAL] K₂Cr₂O₇ An orange-red mineral composed of potassium dichromate.

Lophialetidae [PALEON] A family of extinct perissodactyl mammals in the superfamily Tapiroidea.

Lophiiformes [VERT ZOO] A modified order of actinopterygian fishes distinguished by the reduction of the first dorsal fin to a few flexible rays, the first of which is on the head and bears a terminal bulb; includes anglerfish and allies.

lophine [ORG CHEM] C₂₁H₁₆O₂ A colorless, crystalline, water-insoluble compound that melts at 275°C; used as an indicator in fluorescent neutralization tests.

Lophiodontidae [PALEON] An extinct family of perissodactyl mammals in the superfamily Tapiroidea.

lophocercous [VERT ZOO] Having a ridgelike caudal fin that lacks rays.

lophocyte [INV ZOO] A specialized cell of uncertain function beneath the dermal membrane of certain Demospongiae which bears a process terminating in a tuft of fibrils.

lophodont [VERT ZOO] Having molar teeth whose grinding surfaces have transverse ridges.

Lophogastrida [INV ZOO] A suborder of free-swimming marine crustaceans in the order Mysidacea characterized by imperfect fusion of the sixth and seventh abdominal somites, seven pairs of gills and brood lamellae, and natatory, biramous pleopods.

lophophore [INV ZOO] A food-gathering organ consisting of a fleshy basal ridge or lobe, from which numerous ciliated tentacles arise; found in Bryozoa, Phoronida, and Brachiopoda.

lophotrichous [CYTOL] Having a polar tuft of flagella.

lopolith [GEOL] A large, floored intrusive body that is sunken centrally into the shape of a basin due to sagging of the underlying country rock.

lopping shears [DES ENG] Long-handled shears used for pruning branches.

lorac [NAV] A radio navigation system which utilizes phase comparison techniques to provide hyperbolic lines of position; the system permits an unlimited number of users. Derived from long-range accuracy.

lorad *See* lodar.

loran [NAV] The designation of a family of radio navigation systems by which hyperbolic lines of position are determined by measuring the difference in the times of reception of synchronized pulse signals from two or more fixed transmitters. Derived from long-range navigation.

loran A [NAV] A medium-frequency radio navigation system

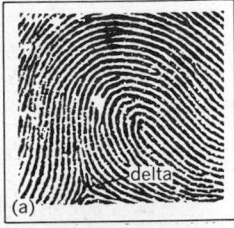

Loop fingerprint patterns. .
(a) Twelve-count ulna loop.
(b) Radial loop. (Federal Bureau of Investigation)

LOPADORRHYNCHIDAE

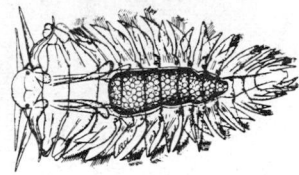

Lopadorrhynchus, dorsal view.
(After McIntosh)

LOPHOCYTE

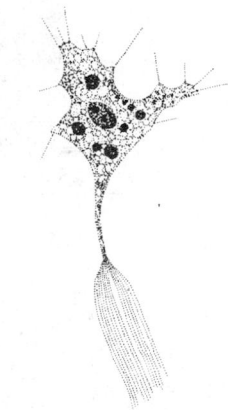

Lophocyte of a freshwater sponge showing terminal tuft of fibrils. (After Ankel and Wintermann-Kilian)

LOPHOGASTRIDA

Lophogaster typicus, adult male.

by which hyperbolic lines of position are determined by measuring the difference in the times of reception of synchronized pulse signals from two fixed transmitters. Also known as standard loran.

loran C [NAV] A low-frequency radio navigation system by which hyperbolic lines of position are determined by measuring the difference in the times of reception of synchronized pulse signals from two fixed transmitters; as compared to loran A, time difference measurements are increased in accuracy through utilizing phase comparison techniques in addition to relatively coarse matches of pulse envelopes of received signals within the loran C receiver.

loran chain [NAV] **1.** A system or combination of three or more loran A stations, forming two or more pairs of stations for loran A navigation. **2.** A system or combination of a master loran C station and two or more loran C slave stations, forming two or more pairs of stations for loran C navigation.

loran chart [NAV] A chart showing loran lines of position along with a limited amount of topographic detail.

loran D [NAV] Tactical loran system that uses the coordinate converter of low-frequency loran C and can operate in conjunction with inertial systems on aircraft, independently of ground facilities and without radiating radio-frequency energy that could reveal the aircraft's location.

lorandite [MINERAL] $TlAsS_2$ A cochineal- to carmine-red or dark lead-gray mineral composed of thallium sulfarsenide, occurring in monoclinic form.

loran fix [NAV] A fix established by means of loran lines of position.

loran line [NAV] A line of position on a loran chart; each line is the locus of points whose distances from two fixed stations differ by a constant amount.

loran rate [NAV] **1.** The frequency channel and pulse repetition rate by which a pair of loran stations is identified. **2.** By extension, the term refers to a pair of transmitting stations, their signals, and resulting lines of position.

loranskite [MINERAL] $(Y,Ce,Ca,Zr)TaO_4$ A black mineral composed of an oxide of yttrium, cerium, calcium, tantalum, and zirconium.

loran tables [NAV] A series of tables containing tabular data for constructing loran hyperbolic lines of position.

Loranthaceae [BOT] A family of dicotyledonous plants in the order Santalales in which the ovules have no integument and are embedded in a large, central placenta.

lorate [BIOL] Strap-shaped.

lordosis [MED] Exaggerated forward curvature of the lumbar spine.

Lorentz-Boltzmann equation [STAT MECH] An approximation to the Boltzmann transport equation for states that are near equilibrium, which shows that the Maxwell-Boltzmann distribution applies at equilibrium.

Lorentz conductivity theory *See* classical conductivity theory.

Lorentz contraction *See* FitzGerald-Lorentz contraction.

Lorentz electron [ELECTROMAG] A model of the electron as a damped harmonic oscillator; used to explain the variation of the real and imaginary parts of the index of refraction of a substance with frequency.

Lorentz equation [ELECTROMAG] The equation of motion for a charged particle, which sets the rate of change of its momentum equal to the Lorentz force.

Lorentz-FitzGerald contraction *See* FitzGerald-Lorentz contraction.

Lorentz force [ELECTROMAG] The force on a charged particle moving in electric and magnetic fields, equal to the particle's charge times the sum of the electric field and the cross product of particle's velocity with the magnetic flux density.

Lorentz force density [ELECTROMAG] The force per unit volume on a charge density and current density, assuming that these densities arise from large numbers of charged particles experiencing a Lorentz force.

Lorentz four-vector *See* four-vector.

Lorentz frame [RELAT] Any of the family of inertial coordinate systems, with three space coordinates and one time coordinate, used in the special theory of relativity; each frame is in uniform motion with respect to all the other Lorentz frames, and the interval between any two events is the same in all frames.

Lorentz gage [ELECTROMAG] Any gage in which the sum of the divergence of the vector potential and the partial derivative of the scalar potential divided by the speed of light (in Gaussian units) vanishes identically; it is always possible to find a gage satisfying this condition.

Lorentz gas [ELECTR] A model of completely ionized gas in which ions are assumed to be stationary and interactions between electrons are neglected.

Lorentz group [MATH] The group of all Lorentz transformations of euclidean four-space with composition as the operation.

Lorentz-Heaviside system *See* Heaviside-Lorentz system.

Lorentz invariance [RELAT] The property, possessed by the laws of physics and of certain physical quantities, of being the same in any Lorentz frame, and thus unchanged by a Lorentz transformation.

Lorentz line-splitting theory [ATOM PHYS] A theory predicting that when a light source is placed in a strong magnetic field, its spectral lines are each split into three components, one of them retaining the zero-field frequency, and the other two shifted upward and downward in frequency by the Larmor frequency (the normal Zeeman effect).

Lorentz local field [ELEC] In a theory of electric polarization, the average electric field due to the polarization at a molecular site that is calculated under the assumption that the field due to polarization by molecules inside a small sphere centered at the site may be neglected. Also known as Mosotti field.

Lorentz-Lorenz equation [OPTICS] The equation that results from replacing the relative dielectric constant with the square of the index of refraction in the Clausius-Mosotti equation.

Lorentz-Lorenz molar refraction *See* molar refraction.

Lorentz matrix [RELAT] A matrix whose product with a vector whose components are the space and time coordinates of an event yields a vector whose components are new coordinates derived from the original ones by a Lorentz transformation.

Lorentz number [PL PHYS] The ratio of the velocity of a fluid to the velocity of light. Symbolized N_{Lo}. [SOLID STATE] The thermal conductivity of a metal divided by the product of its temperature and its electrical conductivity, according to the Wiedemann-Franz law.

Lorentz polarization factor [OPTICS] A geometric factor in the equation for the intensity of x-rays or other radiation diffracted through a given angle by a crystalline substance.

Lorentz relation *See* Wiedemann-Franz law.

Lorentz theory of light sources [ATOM PHYS] A theory according to which light is emitted by vibrations of electrons, which are damped harmonic oscillators attached to atoms.

Lorentz transformation [MATH] Any linear transformation of euclidean four space which preserves the quadratic form $q(x,y,z,t) = t^2 - x^2 - y^2 - z^2$. [RELAT] Any of the family of mathematical transformations used in the special theory of relativity to relate the space and time variables of different Lorentz frames.

Lorentz unit [SPECT] A unit of reciprocal length used to measure the difference, in wave numbers, between a (zero field) spectrum line and its Zeeman components; equal to $eH/4\pi mc^2$, where H is the magnetic field strength, c is the speed of light, and e and m are the charge and mass of the electron respectively (gaussian units).

Lorenz curve [STAT] A graph for showing the concentration of ownership of economic quantities such as wealth and income; it is formed by plotting the cumulative distribution of the amount of the variable concerned against the cumulative frequency distribution of the individuals possessing the amount.

lorettoite [MINERAL] $Pb_7O_6Cl_2$ A honey-yellow to brownish-yellow mineral composed of lead oxychloride.

L organisms [MICROBIO] Pleomorphic forms of bacteria occurring spontaneously, or favored by agents such as penicillin, which lack cell walls and grow in minute colonies; transition may be reversible under certain conditions.

lorhumb line [NAV] A line along which the rates of change of the values of two families of hyperbolas are constants.

lorica [INV ZOO] A hard shell or case in certain invertebrates, as in many rotifers and protozoans; functions as an exoskeleton.

Loricaridae [VERT ZOO] A family of catfishes in the suborder Siluroidei found in the Andes.

Loricata [INV ZOO] An equivalent name for Polyplacophora.

loricate [INV ZOO] Of, pertaining to, or having a lorica.

loris [VERT ZOO] Either of two slow-moving, nocturnal, arboreal primates included in the family Lorisidae: the slender loris (*Loris tardigradus*) and the slow loris (*Nycticebus coucang*).

Lorisidae [VERT ZOO] A family of prosimian primates comprising the lorises of Asia and the galagos of Africa.

lorry *See* larry.

Loschmidt number [PHYS] The number of molecules in 1 cubic centimeter of an ideal gas at 1 atmosphere pressure and 0°C, equal to approximately 2.687×10^{19}.

lose returns *See* lost circulation.

lose water *See* lost circulation.

losing stream *See* influent stream.

loss [COMMUN] *See* transmission loss. [ENG] Power that is dissipated in a device or system without doing useful work. Also known as internal loss.

loss angle [ELECTROMAG] A measure of the power loss in an inductor or a capacitor, equal to the amount by which the angle between the phasors denoting voltage and current across the inductor or capacitor differs from 90°.

loss cone [PL PHYS] A cone in the velocity space of particles in a plasma confined by magnetic mirrors; particles with velocities in the cone are not trapped by the mirrors and are lost out of the system.

loss-cone instability [PL PHYS] An instability in a plasma confined between magnetic mirrors.

loss current [ELEC] The current which passes through a capacitor as a result of the conductivity of the dielectric and results in power loss in the capacitor. [ELECTROMAG] The component of the current across an inductor which is in phase with the voltage (in phasor notation) and is associated with power losses in the inductor.

losser circuit [ELEC] Resonant circuit having sufficient high-frequency resistance to prevent sustained oscillation at the resonant frequency.

losser effect [ELECTR] Radiation resulting from recombinations of charge carriers injected in a *pn* or *pin* junction biased in the forward direction.

loss evaluation [ELEC] A method of achieving an economic balance between buyer and seller in adding material to a transformer design to get lower losses, in which one calculates a value in dollars per kilowatt for load loss and for no-load loss.

Lossev effect *See* injection electroluminescence.

loss factor [ELEC] The power factor of a material multiplied by its dielectric constant; determines the amount of heat generated in a material.

loss function [MATH] In decision theory, the function, dependent upon the decision and the true underlying distributions, which expresses the loss produced in taking the decision.

loss-in-weight feeder [MECH ENG] A device to apportion the output of granulated or powdered solids at a constant rate from a feed hopper; weight-measured decrease in hopper content actuates further opening of the discharge chute to compensate for flow loss as the hopper overburden decreases; used in the chemical, fertilizer, and plastics industries.

lossless junction [ELECTROMAG] A waveguide junction in which all the power incident on the junction is reflected from it.

lossless material [PHYS] An ideal material that dissipates none of the energy of electromagnetic or acoustic waves passing through it.

loss modulation *See* absorption modulation.

loss of head [FL MECH] Energy decrease between two points in a hydraulic system due to such causes as friction, bends, obstructions, or expansions.

loss-of-head gage [ENG] A gage on a rapid sand filter, which indicates loss of head for a filtering operation.

lossy attenuator [ELECTROMAG] In waveguide technique, a length of waveguide deliberately introducing a transmission loss by the use of some dissipative material.

lossy line [ELEC] **1.** Cable used in test measurements which has a large attenuation per unit length. **2.** Transmission line designed to have a high degree of attenuation.

lossy material [PHYS] A material that dissipates energy of electromagnetic or acoustic energy passing through it.

lost circulation [PETRO ENG] A condition that occurs when the drilling fluid escapes into crevices or porous sidewalls of a borehole and does not return to the collar of the drill hole. Also called lose returns; lose water; lost returns; lost water.

lost motion [MECH ENG] The delay between the movement of a driver and the movement of a follower.

lost returns *See* lost circulation.

lost water *See* lost circulation.

lost-wax process [MET] A method used in investment casting in which a wax pattern between a two-layered mold is removed by melting and replaced with molten metal; used for casting bronze statues and in jewelry casting.

lot [CIV ENG] A piece of land with fixed boundaries. [IND ENG] A quantity of material, such as propellant, the units of which were manufactured under identical conditions.

lotic [ECOL] Of or pertaining to swiftly moving waters.

lot number [IND ENG] Identification number assigned to a particular quantity or lot of material from a single manufacturer.

lot plot method [IND ENG] A variables acceptance sampling plan based on the frequency plot of a random sample of 50 items taken from a lot.

lotrite *See* pumpellyite.

lot tolerance percent defective [IND ENG] The percent of defectives in a lot which is considered bad and should be rejected for some specified fraction, usually 90%, of the time.

loudness [ACOUS] The magnitude of the physiological sensation produced by a sound, which varies directly with the physical intensity of sound but also depends on frequency of sound and waveform.

loudness control [ENG ACOUS] A combination volume and tone control that boosts bass frequencies when the control is set for low volume, to compensate automatically for the reduced response of the ear to low frequencies at low volume levels. Also known as compensated volume control.

loudness level [ACOUS] The level of a sound, in phons, equal to the sound pressure level in decibels, relative to 0.0002 microbar, of a pure 1000-hertz tone that is judged to be equally loud by listeners.

loudness unit [ACOUS] A unit of loudness equal to the loudness of a sound having a loudness level of 0 phon; the loudness unit has been replaced by the sone.

loudspeaker [ENG ACOUS] A device that converts electrical signal energy into acoustical energy, which it radiates into a bounded space, such as a room, or into outdoor space. Also known as speaker.

loudspeaker dividing network *See* crossover network.

loudspeaker voice coil *See* voice coil.

loughlinite [MINERAL] $Na_2Mg_3Si_6O_{16} \cdot 8H_2O$ A pearly-white mineral that resembles asbestos, consisting of a hydrous silicate of sodium and magnesium.

louping ill [VET MED] A virus disease of sheep, similar to encephalomyelitis, transmitted by the tick *Ixodes racinus*. Also known as ovine encephalomyelitis; trembling ill.

louping-ill virus [VIROL] A group B arbovirus that is infectious in sheep, monkeys, mice, horses, and cattle.

louse [INV ZOO] The common name for the apterous ectoparasites composing the orders Anoplura and Mallophaga.

louver [BUILD] An opening in a wall or ceiling with slanted or sloping slats to allow sun and ventilation and exclude rain; may be fixed or adjustable, and may be at the opening of a ventilating duct. [ENG] Any arrangement of fixed or adjustable slatlike openings to provide ventilation. [ENG ACOUS] An arrangement of concentric or parallel slats or equivalent grille members used to conceal and protect a loudspeaker while allowing sound waves to pass.

lovchorrite *See* mosandrite.

Love wave [GEOPHYS] A horizontal dispersive surface wave, multireflected between internal boundaries of an elastic body, applied chiefly in the study of seismic waves in the earth's crust.

Lovibond tintometer [OPTICS] A colorimeter which com-

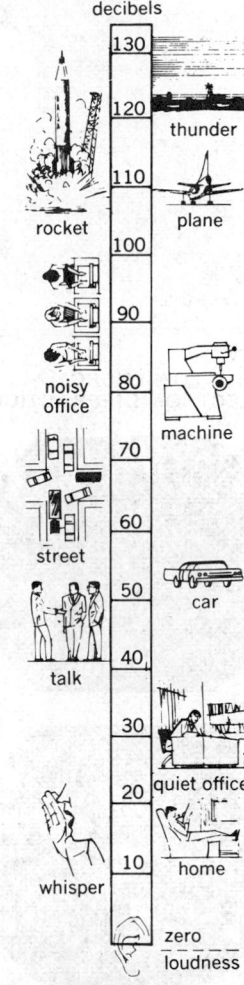

LOUDNESS

Loudness of sound is expressed in decibel units; shown is the decibel scale of common sounds.

LOUSE

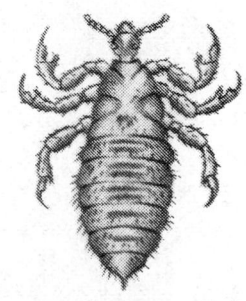

The head louse of humans, a pale-gray-brown insect with black margins and a maximum length of $1/10$ inch (0.25 centimeter).

pares a solution or object under examination with a series of slides of each of three colors.

lovozerite [MINERAL] $(Na,K)_2(Mn,Ca)ZrSi_6O_{16}\cdot3H_2O$ Mineral composed of hydrous silicate of sodium, potassium, manganese, calcium, and zirconium.

low *See* depression.

low-alloy steel [MET] A hardenable carbon steel generally containing not more than about 1% carbon and one or more of the following alloyed components: < (less than) 2% manganese, < 4% nickel, < 2% chromium, < 0.6% molybdenum, and < 0.2% vanadium.

low aloft *See* upper-level cyclone.

low-altitude bombing [ORD] Horizontal bombing at altitudes between about 900 and 8000 feet (0.27 and 2.4 kilometers). Also known as low-level bombing.

low-angle bombing [ORD] Bombing from an airplane at a slight dive angle.

low-angle fault [GEOL] A fault that dips at an angle less than 45°.

low-angle fire [ORD] Gunfire delivered at angles of elevation below the elevation that corresponds to the maximum range of the piece, so that ranges increase with increases in angles of elevation.

low-angle thrust *See* overthrust.

low-approach system [NAV] A means for furnishing guidance in the vertical and horizontal planes to aircraft during descent from an initial approach altitude to a point near the ground.

low band [COMMUN] The band that includes television channels 2 to 6, extending from 54 to 88 megahertz.

low boiler [MATER] A fast-evaporating solvent used in lacquer thinner to give a rapid initial set; boiling point is 70–100°C.

low-boiling butene-2 *See* butene-2.

low brass [MET] Brass containing 20% zinc, 80% copper.

low-carbon steel [MET] Steel containing 0.15% or less of carbon.

low clouds [METEOROL] Types of clouds, the mean level of which is between the surface and 6500 feet (1980 meters); the principal clouds in this group are stratocumulus, stratus, and nimbostratus.

low-definition television [COMMUN] Television that involves less than about 200 scanning lines per complete image.

low-density dynamite [MATER] Any dynamite containing up to 80% ammonium nitrate as the principal explosive ingredient.

low-energy electron diffraction [SOLID STATE] A technique for studying the atomic structure of single crystal surfaces, in which electrons of uniform energy in the approximate range 5–500 electron volts are scattered from a surface, and those scattered electrons that have lost no energy are selected and accelerated to a fluorescent screen where the diffraction pattern from the surface can be observed. Abbreviated LEED.

low-energy np scattering [NUC PHYS] An elastic collision of a neutron, having an energy from less than 1 electron volt to 10,000,000 electron volts, with a proton (usually the nucleus of a hydrogen atom).

low-energy physics [PHYS] That part of physics which studies microscopic phenomena involving energies of several million electron volts or less, such as the arrangement of electrons in an atom or a solid, and the arrangement of protons and neutrons within the atomic nucleus, and the nature of forces between these particles.

low-energy pp scattering [NUC PHYS] An elastic collision of a proton, having an energy of less than 10,000,000 electron volts, with another proton (usually the nucleus of a hydrogen atom).

Lowenhertz thread [DES ENG] A screw thread that differs from U.S. Standard form in that the angle between the flanks measured on an axial plane is 53°8′; height equals 0.75 times the pitch, and width of flats at top and bottom equals 0.125 times the pitch.

lower atmosphere [METEOROL] That part of the atmosphere in which most weather phenomena occur (that is, the troposphere and lower stratosphere); in other contexts, the term implies the lower troposphere.

LOW-ENERGY ELECTRON DIFFRACTION

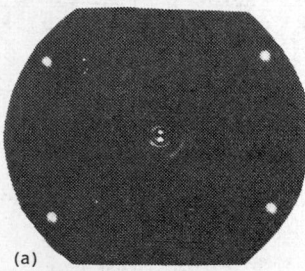

(a)

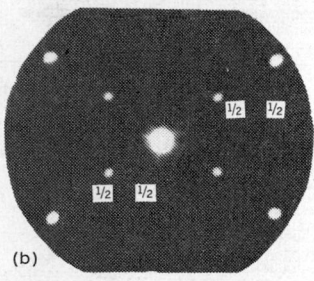

(b)

Low-energy electron diffraction patterns from a (110) tungsten surface. *(a)* Clean, 54 volts. *(b)* Half monolayer of oxygen atoms, 53 volts. Fractions refer to Miller indices of superstructure produced by oxygen absorption. *(From L. H. Germer and J. W. May, Diffraction study of oxygen absorption on a (110) tungsten face, Surface Sci., 4:452, 1966)*

Lower Austral life zone [ECOL] A term used by C. H. Merriam to describe the southern portion of the Austral life zone, characterized by accumulated temperatures of 18,000°F (10,000°C).

lower bound [MATH] A lower bound of a subset A of a set S is a point of S which is smaller than every element of A.

lower branch [ASTRON] That half of a meridian or celestial meridian from pole to pole which passes through the antipode or nadir of a place.

Lower Cambrian [GEOL] The earliest epoch of the Cambrian period of geologic time, ending about 540,000,000 years ago.

lower chord [CIV ENG] The bottom member of a truss.

Lower Cretaceous [GEOL] The earliest epoch of the Cretaceous period of geologic time, extending from about 140- to 120,000,000 years ago.

Lower Devonian [GEOL] The earliest epoch of the Devonian period of geologic time, extending from about 400- to 385,000,000 years ago.

lower half-power frequency [ELECTR] The frequency on an amplifier response curve which is smaller than the frequency for peak response and at which the output voltage is $1/\sqrt{2}$ of its midband or other reference value.

lower heating value *See* low heat value.

lower high water [OCEANOGR] The lower of two high tides occurring during a tidal day.

Lower Jurassic [GEOL] The earliest epoch of the Jurassic period of geologic time, extending from about 185- to 170,000,000 years ago.

lower limb [ASTRON] That half of the outer edge of a celestial body having the least altitude.

lower low water [OCEANOGR] The lower of two low tides occurring during a tidal day.

lower mantle [GEOL] The portion of the mantle below a depth of about 1000 kilometer. Also known as inner mantle; mesosphere; pallasite shell.

Lower Mississippian [GEOL] The earliest epoch of the Mississippian period of geologic time, beginning about 350,000,-000 years ago.

lower motor neuron [ANAT] An efferent neuron which has its body located in the anterior gray column of the spinal cord or in the brainstem nuclei, and its axon passing by way of a peripheral nerve to skeletal muscle. Also known as final common pathway.

lower motor neuron disease [MED] An injury to any part of a lower motor neuron, characterized by flaccid paralysis of the muscle, diminished or absent reflexes, and progressive atrophy of the muscle.

lower nephron nephrosis [MED] Retrogressive kidney changes following traumatic injury and other conditions producing shock; sometimes accompanied by distal and collecting tubule necrosis. Also known as acute tubular necrosis; crush kidney; hemoglobinuric nephrosis; ischemic tubulorrhexis.

Lower Ordovician [GEOL] The earliest epoch of the Ordovician period of geologic time, extending from about 490- to 460,000,000 years ago.

lower pair [MECH ENG] A link in a mechanism in which the mating parts have surface (instead of line or point) contact.

Lower Pennsylvanian [GEOL] The earliest epoch of the Pennsylvanian period of geologic time, beginning about 310,000,-000 years ago.

Lower Permian [GEOL] The earliest epoch of the Permian period of geologic time, extending from about 275- to 260,000,000 years ago.

lower pitch limit [ACOUS] Minimum frequency, for a sinusoidal sound wave, that will produce a pitch sensation.

lower plate *See* footwall.

lower punch [MET] In powder metallurgy, the portion of the die forming the bottom of the die cavity.

lower rib height [ANTHRO] The measure of the vertical distance taken from the lower edge of the last front-attached rib to the floor as the subject stands.

lower-side-band upconverter [ELECTR] Parametric amplifier in which the frequency, power, impedance, and gain considerations are the same as for the nondegenerate amplifier; here, however, the output is taken at the difference frequency, or the lower side band, rather than the signal-input frequency.

Lower Silurian [GEOL] The earliest epoch of the Silurian period of geologic time, beginning about 420,000,000 years ago.

Lower Sonoran life zone [ECOL] A term used by C. H. Merriam to describe the climate and biotic communities of subtropical deserts and thorn savannas in the southwestern United States.

lower transit [ASTRON] Transit across the lower branch of the celestial meridian.

Lower Triassic [GEOL] The earliest epoch of the Triassic period of geologic time, extending from about 230- to 215,000,000 years ago.

lower yield point [MET] In annealed carbon steels, the lowest value of stress after the initial dropoff and before the load begins to rise continuously.

lowest required radiating power [COMMUN] The smallest power output of an antenna which will suffice to maintain a specified grade of broadcast service. Abbreviated LRRP.

lowest useful high frequency [COMMUN] The lowest high frequency that is effective at a specified time for ionospheric propagation of radio waves between two specified points. Abbreviated LUHF.

low-expansion alloy [MET] An alloy whose dimensions do not vary appreciably with temperature.

low explosive [MATER] An explosive which when used in its normal manner deflagrates or burns rather than detonates; used when shattering must be prevented.

low-freezing dynamite [MATER] A dynamite designed to work under freezing conditions; part of the nitroglycerin of the straight dynamite is replaced by nitrated sugar, nitrotoluene, nitrated polymerized glycerin, or ethylene glycol dinitrate.

low-frequency [COMMUN] A Federal Communications Commission designation for the band from 30 to 300 kilohertz in the radio spectrum. Abbreviated LF.

low-frequency acquisition and ranging *See* lofar.

low-frequency antenna [ELECTROMAG] An antenna designed to transmit or receive radiation at frequencies of less than about 300 kilohertz.

low-frequency compensation [ELECTR] Compensation that serves to extend the frequency range of a broad-band amplifier to lower frequencies.

low-frequency current [ELEC] An alternating current having a frequency of less than about 300 kilohertz.

low-frequency cutoff [ELECTR] A frequency below which the gain of a system or device decreases rapidly.

low-frequency gain [ELECTR] The gain of the voltage amplifier at frequencies less than those frequencies at which this gain is close to its maximum value.

low-frequency impedance corrector [ELEC] Electric network designed to be connected to a basic network, or to a basic network and a building-out network, so that the combination will simulate, at low frequencies, the sending-end impedance, including dissipation, of a line.

low-frequency induction furnace [ENG] An induction furnace in which current flow at the commercial power-line frequency is induced in the charge to be heated.

low-frequency loran [NAV] A modification of standard loran, which operates in the low-frequency range of approximately 100 to 200 kilohertz to increase range over land and during daytime, and which matches cycles rather than envelopes of pulses to obtain a more accurate fix. Abbreviated LF loran. Also known as cycle-matching loran.

low-frequency padder [ELECTR] In a superheterodyne receiver, a small adjustable capacitor connected in series with the oscillator tuning coil and adjusted during alignment to obtain correct calibration of the circuit at the low-frequency end of the tuning range.

low-frequency propagation [ELECTROMAG] Propagation of radio waves at frequencies between 30 and 300 kilohertz.

low-frequency spectrum [SPECT] Spectrum of atoms and molecules in the microwave region, arising from such causes as the coupling of electronic and nuclear angular momenta, and the Lamb shift.

low-frequency transconductance [ELECTR] The change in the plate current of a vacuum tube divided by the change in the control-grid voltage that produces it, at frequencies small

enough for these two quantities to be considered in phase.

low-frequency tube [ELECTR] An electron tube operated at frequencies small enough so that the transit time of an electron between electrodes is much smaller than the period of oscillation of the voltage.

low-grade [MATER] An arbitrary designation of dynamites of less strength than 40%. [MIN ENG] Pertaining to ores that have a relatively low content of minerals. Also known as coarse; lean. [SCI TECH] Being of inferior quality.

low-heat cement [MATER] A chemically altered portland cement with a low initial heat liberation.

low heat value [THERMO] The heat value of a combustion process assuming that none of the water vapor resulting from the process is condensed out, so that its latent heat is not available. Also known as lower heating value; net heating value.

low-helix drill [DES ENG] A two-flute twist drill with a lower helix angle than a conventional drill. Also known as slow-spiral drill.

low-hydrogen electrode [MET] A covered electrode used in arc welding that provides an atmosphere low in hydrogen.

low-impedance measurement [ELECTR] The measurement of an impedance which is small enough to necessitate use of indirect methods.

low-impedance switching tube [ELECTR] A gas tube which has a static impedance on the order of 10,000 ohms, but zero or negative dynamic impedance, and therefore can be used as a relay and transmits information with negligible loss as well.

low index [METEROL] A relatively low value of the zonal index which, in middle latitudes, indicates a relatively weak westerly component of wind flow (usually implying stronger north-south motion), and the characteristic weather attending such motion; a circulation pattern of this type is commonly called a low-index situation.

low-intensity atomizer [MECH ENG] A type of electrostatic atomizer operating on the principle that atomization is the result of Rayleigh instability, in which the presence of charge in the surface counteracts surface tension.

low-key photograph [GRAPHICS] A photograph which has mostly dark tones and very few light or white areas.

low-level bombing *See* low-altitude bombing.

low-level condenser [MECH ENG] A direct-contact water-cooled steam condenser that uses a pump to remove liquid from a vacuum space.

low-level counting [NUCLEO] The measurement of very small amounts of radioactivity, such as that generated by long-lived natural radioactive isotopes, and isotopes produced by cosmic rays and nuclear explosions.

low-level language [ADP] A computer language consisting of mnemonics that directly correspond to machine language instructions; for example, an assembler that converts the interpreted code of a higher-level language to machine language.

low-level logic circuit [ELECTR] A modification of a diode-transistor logic circuit in which a resistor and capacitor in parallel are replaced by a diode, with the result that a relatively small voltage swing is required at the base of the transistor to switch it on or off. Abbreviated LLL circuit.

low-level modulation [ELECTR] Modulation produced at a point in a system where the power level is low compared with the power level at the output of the system.

low-lift truck [MECH ENG] A hand or powered lift truck that raises the load sufficiently to make it mobile.

low-loss [ELEC] Having a small dissipation of electric or electromagnetic power.

low-melting glass [MATER] Glass to which selenium, thallium, arsenic, or sulfur is added to give melting points of 260-660°F (127-349°C).

low-melting solder *See* soft solder.

low-moor bog [GEOL] A bog that is at or slightly below the ground water table.

low-moor peat [GEOL] Peat found in low-moor bogs or swamps and containing little or no sphagnum. Also known as fen peat.

low-noise amplifier [ELECTR] An amplifier having very low background noise when the desired signal is weak or absent;

LOW-LEVEL LOGIC CIRCUIT

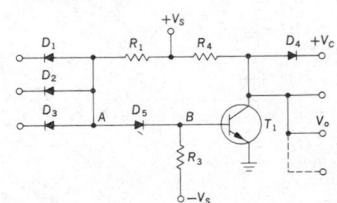

Circuit diagram of low-level logic circuit. D_1–D_5 = diodes; R_1–R_4 = resistors; T = transistor; $+V_S$ = positive supply voltage; B = common node.

field-effect transistors are used in audio preamplifiers for this purpose.

low-noise preamplifier [ELECTR] A low-noise amplifier placed in a system prior to the main amplifier, sometimes close to the source; used to establish a satisfactory noise figure at an early point in the system.

low-order [ADP] Pertaining to the digit which contributes the smallest amount to the value of a numeral, or to its position, or to the rightmost position of a word.

low-order burst [ORD] Functioning of a projectile or bomb in which the explosive fails to attain a high-order detonation.

low-pass band-pass transformation *See* frequency transformation.

low-pass filter [ELEC] A filter that transmits alternating currents below a given cutoff frequency and substantially attenuates all other currents.

low population zone [ENG] An area of low population density sometimes required around a nuclear installation; the number and density of residents is of concern in providing, with reasonable probability, that effective protection measures can be taken if a serious accident should occur.

low-pressure area [MECH ENG] The point in a bearing where the pressure is the least and the area or space for a lubricant is the greatest.

low-pressure fluid flow [FL MECH] Flow of fluids below atmospheric pressures, particularly gases and vapors following ideal gas laws, in pipes, fittings, and other common configurations.

low-pressure laminate [MATER] A plastic laminate molded and cured at pressures in general of 400 psi (approximately 27 atmospheres or 2.8×10^6 newtons per square meter).

low-pressure torch [ENG] A type of torch in which acetylene enters a mixing chamber, where it meets a jet of high-pressure oxygen; the amount of acetylene drawn into the flame is controlled by the velocity of this oxygen jet. Also known as injector torch.

low-pressure well [PETRO ENG] An oil or gas well with a shut-in wellhead pressure of less than 2000 psia (1.38×10^7 newtons per square meter).

low-Q filter [ELECTR] A filter in which the energy dissipated in each cycle is a fairly large fraction of the energy stored in the filter.

low quartz [MINERAL] Quartz that has been formed below 573°C; the tetrahedral crystal structure is less symmetrically arranged than a quartz formed at a higher temperature.

low-rank graywacke [PETR] A graywacke that is nonfeldspathic.

low-rank metamorphism [GEOL] A metamorphic process that occurs under conditions of low to moderate pressure and temperature.

low-reactance grounding [ELEC] Use of grounding connections with a moderate amount of inductance to effect a moderate reduction in the short-circuit current created by a line-to-ground fault.

low-reflection film [OPTICS] A transparent film covering a glass surface, designed so that a small proportion of the light incident will be reflected and a correspondingly large proportion transmitted into the glass.

low relief [GRAPHICS] Sculpture having only a slight projection. Also known as bas-relief.

low-residual-phosphorus copper [MET] Deoxidized copper with a 0.004–0.012% residual phosphorus content.

Lowry process [ENG] A system for wood preservation which uses atmospheric pressure at the start and then introduces preservative into the wood in a vacuum.

low-shaft furnace [MET] A blast furnace having a short shaft; used to produce pig iron, ferroalloys, alumina, and other products from low-grade ores using low-grade fuel.

low-speed wind tunnel [ENG] A wind tunnel that has a speed up to 300 miles (480 kilometers) per hour and the essential features of most wind tunnels.

low-temperature carbonization [CHEM ENG] Low-temperature destructive distillation of coal to produce liq d products.

low-temperature coke [MATER] Coke produced at mperatures of 500–750°C, used chiefly for house heating, particularly in England. Also known as char.

low-temperature hygrometry [METEOROL] The study that deals with the measurement of water vapor at low temperatures; the techniques used differ from those of conventional hygrometry because of the extremely small amounts of moisture present at low temperatures and the difficulties imposed by the increase of the time constants of the standard instruments when operated at these temperatures.

low-temperature physics [CRYO] A study of the properties of gross matter at low temperature's especially at temperatures so low that the quantum character of the substance becomes observable in effects such as superconductivity, superfluid liquid helium, magnetic cooling, and nuclear orientation.

low-temperature production [CRYO] Production of temperatures from about 80K down to about 10^{-6}K by techniques such as isentropic expansion of gases, refrigeration cycles, and adiabatic demagnetization.

low-temperature separation [CHEM ENG] Liquid condensate recovery from wet gases at temperatures of 20 to $-20°F$ (-6.7 to $-28.9°C$), the temperature range at which the gas-oil separator operates.

low-temperature thermometry [CRYO] The assignment of numbers on the Kelvin absolute temperature scale to achievable and reproducible low-temperature states, and the choice and calibration of suitable instruments for the practical measurement of low temperatures, such as thermocouples, and resistance, vapor-pressure, gas, and magnetic thermometers.

low tide *See* low water.

low-tide terrace [GEOL] A flat area of a beach adjacent to the low-water line.

low velocity [MECH] Muzzle velocity of an artillery projectile of 2499 feet (762 meters) per second or less.

low-velocity drop [ORD] The act or process of delivering personnel, supplies, or equipment from aircraft in flight, utilizing sufficient parachute retardation to prevent injury or damage upon ground impact.

low-velocity layer [GEOPHYS] A layer in the solid earth in which seismic wave velocity is lower than the layers immediately below or above.

low-volatile coal [GEOL] A coal that is nonagglomerating, has 78% to less than 86% fixed carbon, and 14% to less than 22% volatile matter.

low voltage [ELEC] **1.** Voltage which is small enough to be regarded as safe for indoor use, usually 120 volts in the United States. **2.** Voltage which is less than that needed for normal operation; a result of low voltage may be burnout of electric motors due to loss of electromotive force.

low-voltage relay [COMMUN] A relay that responds to the drop in voltage (increase in current) when a telephone line becomes active; used to activate bugging equipment.

low-voltage winding [ELECTROMAG] The coil of wire wound around the core of a power transformer which has the smaller number of turns, and therefore the lower voltage.

low water [OCEANOGR] The lowest limit of the surface water level reached by the lowering tide. Also known as low tide.

low-water inequality [OCEANOGR] The difference between the heights of two successive low tides.

low-water interval *See* low-water lunitidal interval.

low-water lunitidal interval [GEOPHYS] For a specific location, the interval of time between the transit (upper or lower) of the moon and the next low water. Also known as low-water interval.

low-water neaps *See* mean low-water neaps.

low-water springs *See* mean low-water springs.

LOX *See* liquid-oxygen explosive.

loxodont [VERT ZOO] Having molar teeth with shallow hollows between the ridges.

loxodrome *See* rhumb line.

loxodromic spiral [MATH] A curve on a surface of revolution which cuts the meridians at a constant angle other than 90°.

loxolophodont [VERT ZOO] Having crests on the molar teeth that connect three of the tubercles and with the fourth or posterior inner tubercle being rudimentary or absent.

Loxonematacea [PALEON] An extinct superfamily of gastropod mollusks in the order Prosobranchia.

lozenge file [DES ENG] A small file with four sides and a lozenge-shaped cross section; used in forming dies.

L pad [ENG ACOUS] A volume control having essentially the same impedance at all settings.

LPF process [MIN ENG] Recovery of metals from tailings by a sequence of leaching, precipitation, and flotation.

LPG *See* liquefied petroleum gas.

LPM *See* lines per minute.

LP record *See* long-playing record.

Lr *See* lawrencium.

LRC *See* longitudinal redundancy check.

LRRP *See* lowest required radiating power.

L scan *See* L scope.

L scope [ELECTR] A cathode-ray scope on which a trace appears as a vertical or horizontal range scale, the signals appearing as left and right horizontal (or up and down vertical) deflections as echoes are received by two antennas, the left and right (or up and down) deflections being proportional to the strength of the echoes received by the two antennas. Also known as L indicator; L scan.

LS coupling *See* Russell-Saunders coupling.

LSD *See* dock landing ship; lysergic acid diethylamide.

LSD-25 *See* lysergic acid diethylamide.

L shell [ATOM PHYS] The second shell of electrons surrounding the nucleus of an atom, having electrons whose principal quantum number is 2.

LSI circuit *See* large-scale integrated circuit.

L-1 test [ENG] A 480-hour engine test in a single-cylinder Caterpillar diesel engine to determine the detergency of heavy-duty lubricating oils.

L-2 test [ENG] An engine test made in a single-cylinder Caterpillar diesel engine to determine the oiliness of an engine oil. Also known as scoring test.

L-3 test [ENG] An engine test in a four-cylinder Caterpillar engine to determine stability of crankcase oil at high temperatures and under severe operating conditions.

L-4 test [ENG] An engine test in a six-cylinder spark-ignition Chevrolet engine to evaluate crankcase oil oxidation stability, bearing corrosion, and engine deposits.

L-5 test [ENG] An engine test in a General Motors diesel engine to determine detergency, corrosiveness, ring sticking, and oxidation stability properties of lubricating oils.

LTPD *See* lot tolerance percent defective.

LTRS *See* letters shift.

Lu *See* lutetium.

lubber line *See* lubber's line.

lubber's line [NAV] A reference line on any direction-indicating instrument, marking the reading which coincides with the heading. Also known as lubber line; lubber's point.

lubber's line error [NAV] In a magnetic compass, the angular difference between the heading as indicated by a lubber's line, and the actual heading; this error is caused by faulty calibration.

lubber's point *See* lubber's line.

lube cut [MATER] The distilled fraction of crude oil with suitable boiling range and viscosity to yield a lubricating oil when it is completely refined. Also known as lube-oil distillate; lube stock.

lube oil *See* lubricating oil.

lube-oil distillate *See* lube cut.

lube stock *See* lube cut.

lubricant [MATER] A substance used to reduce friction between parts or objects in relative motion.

lubricant additive [MATER] Any material added to lubricants (greases or oils) to give the product special properties, such as resistance to extremes of pressure, cold, or heat, improved viscosity, and detergency.

lubricated gasoline [MATER] A motor gasoline into which a lubricant has been added.

lubricating grease [MATER] A solid or semisolid lubricant consisting of a thickening agent (soap or other additives) in a fluid lubricant (usually petroleum lubricating oil).

lubricating oil [MATER] Selected fractions of refined petroleum or other oils (with or without additives) used to lessen friction between moving surfaces. Also known as lube oil.

lubrication action [MATER] The ability of the lubricant to maintain a fluid film between solid surfaces and to prevent their physical contact.

lubricator [ENG] A device for applying a lubricant.

lubricity [MATER] The ability of a material to lubricate.

Lucanidae [INV ZOO] The stag beetles, a cosmopolitan family of coleopteran insects in the superfamily Scarabaeoidea.

lucca oil *See* olive oil.

lucerne *See* alfalfa.

luciferase [BIOCHEM] An enzyme that catalyzes the oxidation of luciferin.

luciferin [BIOCHEM] A species-specific pigment in many luminous organisms that emits heatless light when combined with oxygen.

Luciocephalidae [VERT ZOO] A family of fresh-water fishes in the suborder Anabantoidei.

Lucite [MATER] Trademark of DuPont for its transparent acrylic resins.

Luckiesh-Moss visibility meter [ENG] A type of photometer that consists of two variable-density filters (one for each eye) that are adjusted so that an object seen through them is just barely discernible; the reduction in visibility produced by the filters is read on a scale of relative visibility related to a standard task.

Lüders' lines [MET] Surface markings on a metal caused by flow of the material strained beyond its elastic limit. Also known as deformation bands; Hartmann lines; Lüders' bands; Piobert lines; stretcher strains.

ludlamite [MINERAL] $(Fe,Mg,Mn)_3(PO_4)_2 \cdot 4H_2O$ A green mineral crystallizing in the monoclinic system and occurring in small, transparent crystals.

Ludlovian [GEOL] A European stage of geologic time; Upper Silurian, below Gedinnian of Devonian, above Wenlockian.

ludwigite [MINERAL] $(Mg,Fe)_2FeBO_5$ Blackish-green mineral that crystallizes in the monoclinic system and occurs in fibrous masses; isomorphous with ronsenite.

Ludwig's angina [MED] Acute streptococcal cellulitis of the floor of the mouth.

lueneburgite [MINERAL] $Mg_3B_2(OH)_6(PO_4)_2 \cdot 6H_2O$ A colorless mineral composed of hydrous basic phosphate of magnesium and boron.

lueshite [MINERAL] $NaNbO_3$ An orthorhombic mineral having perovskite-type structure; it is dimorphous with natroniobite.

lug [DES ENG] A projection or head on a metal part to serve as a cap, handle, support, or fitting connection.

luganot [METEOROL] A strong south or south-southeast wind of Lake Garda, Italy.

lug bolt [DES ENG] **1.** A bolt with a flat extension or hook instead of a head. **2.** A bolt designed for securing a lug.

lug brick [MATER] A brick with lugs for spacing adjacent bricks.

Lugol solution [CHEM] A solution of 5 grams of iodine and 10 grams of potassium iodide per 100 milliliters of water; used in medicine.

LUHF *See* lowest useful high frequency.

Luidiidae [INV ZOO] A family of echinoderms in the suborder Paxillosina.

Luisian [GEOL] A North American stage of geologic time; Miocene (above Relizian, below Mohnian).

lum *See* trolley.

lumbago [MED] Backache in the lumbar or lumbosacral region. [VET MED] *See* azoturia.

lumbang oil [MATER] Colorless or yellow liquid with pleasant aroma and bland taste; soluble in alcohol, ether, chloroform, and carbon disulfide; expressed from candlenut; used as an illuminant and wood preservative, and in paints, calking, and soap manufacture. Also known as candlenut oil.

lumbar artery [ANAT] Any of the four or five pairs of branches of the abdominal aorta opposite the lumbar region of the spine; supplies blood to loin muscles, skin on the sides of the abdomen, and the spinal cord.

lumbar nerve [ANAT] Any of five pairs of nerves arising from lumbar segments of the spinal cord; characterized by motor, visceral sensory, somatic sensory, and sympathetic components; they innervate the skin and deep muscles of the lower back and the lumbar plexus.

lumbar vertebrae [ANAT] Those vertebrae located between

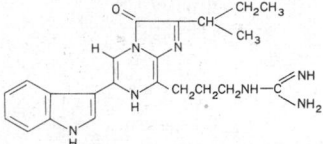

LUCIFERIN

Structural formula of *Cypridina* luciferin.

LUCKIESH-MOSS VISIBILITY METER

Photograph of Luckiesh-Moss visibility meter. (*General Electric Co.*)

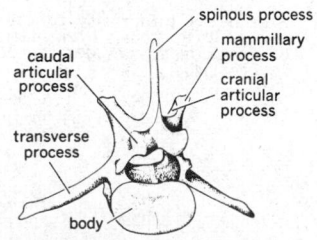

LUMBAR VERTEBRAE

Fifth lumbar vertebra of the dog, the caudal lateral aspect showing the body or centrum. (*From M. E. Miller, G. C. Christensen, and H. E. Evans, Anatomy of the Dog, Saunders, 1964*)

the lowest ribs and the pelvic girdle in all vertebrates.
lumber [MATER] Logs that have been sawed and prepared for market.

lumberg [OPTICS] A unit of luminous energy equal to the luminous energy corresponding to a radiant energy of $1/K$ ergs, where K is the luminous efficiency in lumens per watt. Formerly known as lumerg.

lumbodorsal fascia [ANAT] The sheath of the erector spinae muscle alone, or the sheaths of the erector spinae and the quadratus lumborum muscles.

lumbosacral plexus [ANAT] A network formed by the anterior branches of lumbar, sacral, and coccygeal nerves which for descriptive purposes are divided into the lumbar, sacral, and pudendal plexuses.

Lumbricidae [INV ZOO] A family of annelid worms in the order Oligochaeta; includes the earthworm.

Lumbriclymeninae [INV ZOO] A subfamily of mud-swallowing sedentary worms in the family Maldanidae.

Lumbriculidae [INV ZOO] A family of aquatic annelids in the order Oligochaeta.

Lumbricus [INV ZOO] A genus of earthworms recognized as the type genus of the family Lumbricidae.

Lumbrineridae [INV ZOO] A family of errantian polychaetes in the superfamily Eunicea.

lumen [OPTICS] The unit of luminous flux, equal to the luminous flux emitted within a unit solid angle (1 steradian) from a point source having a uniform intensity of 1 candela, or to the luminous flux received on a unit surface, all points of which are at a unit distance from such a source. Abbreviated lm. [SCI TECH] The space within a tube.

lumen-hour [OPTICS] A unit of quantity of light (luminous energy), equal to the quantity of light radiated or received for a period of 1 hour by a flux of 1 lumen. Abbreviated lm-hr.

lumen per watt [OPTICS] The unit of luminosity factor and of luminous efficacy. Abbreviated lm/w.

lumen-second [OPTICS] A unit of quantity of light (luminous energy), equal to the quantity of light radiated or received for a period of 1 second by a flux of 1 lumen. Abbreviated lm-sec.

lumerg *See* lumberg.

luminaire [ELEC] An electric lighting fixture, wall bracket, portable lamp, or other complete lighting unit designed to contain one or more electric lighting sources and associated reflectors, refractors, housing, and such support for those items as necessary.

luminance [OPTICS] The ratio of the luminous intensity in a given direction of an infinitesimal element of a surface containing the point under consideration, to the orthogonally projected area of the element on a plane perpendicular to the given direction. Also known as brightness.

luminance carrier *See* picture carrier.

luminance channel [COMMUN] A path intended primarily for the luminance signal in a color television system.

luminance factor [OPTICS] The ratio of the luminance of a body when illuminated and observed under certain conditions to that of a perfect diffuser under the same conditions.

luminance primary [COMMUN] One of the three transmission primaries whose amount determines the luminance of a color in a color television system.

luminance signal [COMMUN] The color television signal that is intended to have exclusive control of the luminance of the picture. Also known as Y signal.

luminescence [PHYS] Light emission that cannot be attributed merely to the temperature of the emitting body, but results from such causes as chemical reactions at ordinary temperatures, electron bombardment, electromagnetic radiation, and electric fields.

luminescent [PHYS] Capable of exhibiting luminescence.

luminescent cell *See* electroluminescent panel.

luminescent center [SOLID STATE] A point-lattice defect in a transparent crystal that exhibits luminescence.

luminescent dye [MATER] A dye that is made luminous by excitation with an outside energy source; used in luminous paint.

luminescent screen [ELECTR] The screen in a cathode-ray tube, which becomes luminous when bombarded by an electron beam and maintains its luminosity for an appreciable time.

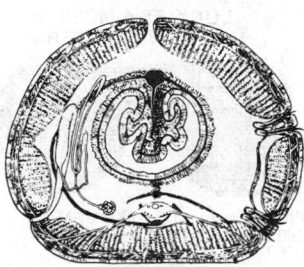

LUMBRICUS

Cross section of the earthworm. *(Lumbricus terrestris). (From T. I. Storer, General Zoology, 3d ed., McGraw-Hill, 1957)*

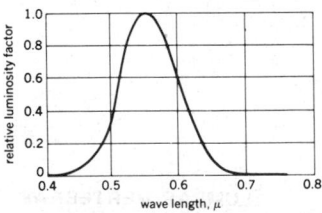

LUMINOSITY FUNCTION

Human-eye luminosity function. *(From W. B. Boast, Illumination Engineering, 2d ed., McGraw-Hill, 1953)*

luminol [ORG CHEM] $NH_2 \cdot C_6H_3 \cdot (CO \cdot NH)_2$ A white, water-soluble, crystalline compound that melts at 320°C; used in an alkaline solution for analytical testing in chemistry. Also known as 3-aminophthalic acid cyclic hydrazide.

luminophor [PHYS] A luminescent material that converts part of the absorbed primary energy into emitted luminescent radiation. Also known as fluophor; fluor; phosphor.

luminosity *See* luminosity factor.

luminosity classes [ASTRON] A classification of stars in an orderly sequence according to their absolute brightness.

luminosity curve *See* luminosity function.

luminosity factor [OPTICS] The ratio of luminous flux in lumens emitted by a source at a particular wavelength to the corresponding radiant flux in watts at the same wavelength; thus this is a measure of the visual sensitivity of the eye. Also known as luminosity.

luminosity function [ASTRON] The functional relationship between stellar magnitude and the number and distribution of stars of each magnitude interval. Also known as relative luminosity factor. [OPTICS] A standard measure of the response of an eye to monochromatic light at various wavelengths; the function is normalized to unity at its maximum value. Also known as luminosity curve; spectral luminous efficiency; visibility function.

luminous cloud *See* sheet lightning.

luminous coefficient [OPTICS] A measure of the fraction of the radiant power of a light source which contributes to its luminous properties, equal to the average of the luminosity function at various wavelengths, weighted according to the spectral intensity of the source. Also known as luminous efficiency.

luminous efficacy [OPTICS] **1.** The ratio of the total luminous flux in lumens emitted by a light source over all wavelengths to the total radiant flux in watts. Also known as luminous efficiency. **2.** The ratio of the total luminous flux emitted by a light source to the power input of the source; expressed in lumens per watt.

luminous efficiency *See* luminous coefficient; luminous efficacy.

luminous emittance [OPTICS] The emittance of visible radiation weighted to take into account the different response of the human eye to different wavelengths of light; in photometry, luminous emittance is always used as a property of a self-luminous source, and therefore should be distinguished from luminance.

luminous energy [OPTICS] The total radiant energy emitted by a source, evaluated according to its capacity to produce visual sensation; measured in lumen-hours or lumen-seconds.

luminous flux [OPTICS] The time rate of flow of radiant energy, evaluated according to its capacity to produce visual sensations; measured in lumens.

luminous flux density *See* illuminance.

luminous intensity [OPTICS] The luminous flux incident on a small surface which lies in a specified direction from a light source and is normal to this direction, divided by the solid angle (in steradians) which the surface subtends at the source of light. Also known as light intensity.

luminous meteor [METEOROL] According to United States weather observing practice, any one of a number of atmospheric phenomena which appear as luminous patterns in the sky, including halos, coronas, rainbows, aurorae, and their many variations, but excluding lightning (an igneous meteor or electrometeor).

luminous nebula [ASTRON] A nebula made bright by radiation from stars in the vicinity.

luminous paint [MATER] A type of paint in which luminous pigments are used.

luminous pigment [MATER] A pigment that absorbs light energy and radiates visible light when exposed to ultraviolet light; made of phosphors such as strontium, zinc, and cadmium sulfides.

luminous range [NAV] The distance at which a marine light may be seen in clear weather, expressed in nautical miles.

luminous sensitivity of phototube [ELECTR] Quotient of the anode current by the incident luminous flux.

luminous time ratio [NAV] Of a navigational light, the ratio of the length of a flash to the period of rotation.

luminous visibility diagram [NAV] A diagram by which the luminous ranges, as given in light lists, may be adjusted to various conditions of visibility.

Lummer-Brodhun sight box [OPTICS] A device, having a series of prisms, for viewing simultaneously the two sides of a white diffuse plaster screen illuminated by light sources whose luminous intensities are being compared.

Lummer-Gehrcke plate [OPTICS] An interferometer consisting of a glass or quartz plate with parallel surfaces and sizable thickness in which multiple reflections take place.

lump coal [MIN ENG] Bituminous coal that passes through a 6-inch (15-centimeter) round mesh in initial screening.

lumped constant [ELEC] A single constant that is electrically equivalent to the total of that type of distributed constant existing in a coil or circuit.

lumped-constant network [ELEC] An analytical tool in which distributed constants (inductance, capacitance, and resistance) are represented as hypothetical components.

lumped discontinuity [ELECTROMAG] An analytical tool in the study of microwave circuits in which the effective values of inductance, capacitance, and resistance representing a discontinuity in a waveguide are shown as discrete components of equivalent value.

lumped element [ELECTROMAG] A section of a transmission line designed so that electric or magnetic energy is concentrated in it at specified frequencies, and inductance or capacitance may therefore be regarded as concentrated in it, rather than distributed over the length of the line.

lumped impedance [ELECTROMAG] An impedance concentrated in a single component rather than distributed throughout the length of a transmission line.

lumpy jaw *See* actinomycosis.

Luna [ASTRON] A name for the moon.

Luna program [AERO ENG] A series of Soviet space probes launched for flight missions to the moon.

lunar appulse [ASTRON] An eclipse of the moon in which the penumbral shadow of the earth falls on the moon. Also known as penumbral eclipse.

lunar atmosphere [ASTROPHYS] The volatile elements postulated to have been present on the moon's surface at one time.

lunar atmospheric tide [METEOROL] An atmospheric tide due to the gravitational attraction of the moon; the only detectable components are the 12-lunar-hour or semidiurnal component, as in the oceanic tides, and two others of very nearly the same period; the amplitude of this atmospheric tide is so small that it is detected only by careful statistical analysis of a long record.

lunar crater [ASTRON] A crater on the moon's surface.

lunar crust [ASTRON] The outer layer of the moon.

lunar day [ASTRON] The time interval between two successive crossings of the meridian by the moon.

lunar dust [ASTRON] Small particles adhering to the moon's surface.

lunar eclipse [ASTRON] Obscuration of the full moon when it passes through the shadow of the earth.

lunar ephemeris [ASTRON] A computed list of positions the moon will occupy in the sky on certain dates.

lunar excursion module [AERO ENG] A manned spacecraft designed to be carried on top of the Apollo service module and having its own power plant for making a manned landing on the moon and a return from the moon to the orbiting Apollo spacecraft. Abbreviated LEM. Also known as lunar module (LM).

lunar flight [AERO ENG] Flight by a spacecraft to the moon.

lunar geology *See* selenology.

lunar inequality [ASTRON] Variation in the moon's motion in its orbit, due to attraction by other bodies of the solar system. [GEOPHYS] A minute fluctuation of a magnetic needle from its mean position, caused by the moon.

lunar interval [ASTRON] The difference in time between the transit of the moon over the Greenwich meridian and a local meridian; the lunar interval equals the difference between the Greenwich and local intervals of a tide or current phase.

lunar libration [ASTRON] **1.** The effect wherein the face of the moon appears to swing east and west about 8° from its central position each month. Also known as apparent libration in longitude. **2.** The state wherein the inclination of the moon's

polar axis allows an observer on earth to see about 59% of the moon's surface. Also known as libration in latitude. **3.** The small oscillation with which the moon rocks back and forth about its mean rotation rate. Also known as physical libration of the moon.

lunar magnetic field [ASTROPHYS] The magnetic field of the moon.

lunar mass [ASTROPHYS] The mass of the moon.

lunar meteoroid [ASTRON] A meteoric particle before it strikes the moon.

lunar module *See* lunar excursion module.

lunar month [ASTRON] The period of revolution of the moon about the earth, especially a synodical month.

lunar mountain [ASTRON] A mountain on the moon.

lunar node [ASTRON] A node of the moon's orbit.

lunar nodule [ASTRON] A rock nodule found on the moon.

lunar noon [ASTRON] The instant at which the sun is over the upper branch of any meridian of the moon.

lunar orbit [AERO ENG] Orbit of a spacecraft around the moon.

lunar polarization [ASTROPHYS] Polarization of light by the moon's surface.

lunar pole [ASTRON] A pole of the moon.

lunar probe [AERO ENG] Any space probe launched for flight missions to the moon.

lunar rainbow *See* moonbow.

lunar rock [ASTRON] Rock found on the moon.

lunar satellite [AERO ENG] A satellite making one or more revolutions about the moon.

lunar spacecraft [AERO ENG] A spacecraft designed for flight to the moon.

lunar tide [OCEANOGR] The portion of a tide produced by forces of the moon.

lunar time [ASTRON] **1.** Time based upon the rotation of the earth relative to the moon; it may be designated as local or Greenwich, as the local or Greenwich meridian is used as the reference. **2.** Time on the moon.

lunar topology [ASTRON] Topology of the moon.

lunar year [ASTRON] A time interval comprising 12 lunar (synodic) months.

lunate [BIOL] Crescent-shaped

lunate bar [GEOL] A crescent-shaped bar of sand that is frequently found off the entrance to a harbor.

lunation [ASTRON] The time period between two successive new moons.

Lundegardh vaporizer [ANALY CHEM] A device used for emission flame photometry in which a compressed air aspirator vaporizes the solution within a chamber; smaller droplets are carried into the fuel-gas stream and to the burner orifice where the solvent is evaporated, dissociated, and optically excited.

lune [MATH] A section of a plane bounded by two circular arcs, or of a sphere bounded by two great circles.

Luneberg lens [ELECTROMAG] A type of antenna consisting of a dielectric sphere whose index of refraction varies with distance from the center of the sphere so that a beam of parallel rays falling on the lens is focused at a point on the lens surface diametrically opposite from the direction of incidence, and, conversely, energy emanating from a point on the surface is focused into a plane wave.

lunette [GEOL] A broad, low crescentic mound of windblown fine silt and clay. [ORD] Towing ring in the trial plate or tongue of a towed vehicle, such as a gun carriage or trailer, used for attaching the towed vehicle to the prime mover or towing vehicle.

lung [ANAT] Either of the paired air-filled sacs, usually in the anterior or anteroventral part of the trunk of most tetrapods, which function as organs of respiration.

lung bud [EMBRYO] A primary outgrowth of the embryonic trachea; the anlage of a primary bronchus and all its branches.

lungfish [VERT ZOO] The common name for members of the Dipnoi; all have lungs that arise from a ventral connection with the gut.

lung-governed breathing apparatus [ENG] A breathing apparatus in which the oxygen supplied to the wearer is governed by the wearer's demand.

lunisolar precession [ASTROPHYS] Precession of the earth's

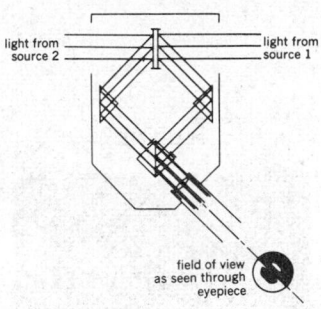

**LUMMER-BRODHUN
SIGHT BOX**

light from source 2   light from source 1

field of view as seen through eyepiece

Lummer-Brodhun contrast sight box.

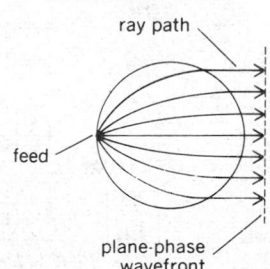

LUNEBERG LENS

ray path

feed

plane-phase wavefront

Luneberg lens with dielectric sphere between feed and plane-phase wavefront.

equinox caused by the gravitational attraction of the sun and moon.

lunitidal interval [OCEANOGR] The period between the moon's upper or lower transit over a specified meridian and a specified phase of the tidal current following the transit.

lunk [COMMUN] Access line that terminates in an automatic dial exchange where it functions as an access line for subscribers, but functions as a trunk for the automatic dial exchange equipment.

lunule [BIOL] A crescent-shaped organ, structure, or mark.

lupine [BOT] A leguminous plant of the genus *Lupinus* with an upright stem, leaves divided into several digitate leaflets, and terminal racemes of pea-shaped blossoms.

lupinidine *See* sparteine.

Lupus [ASTRON] A southern constellation lying between Centaurus and Scorpius. Also known as Wolf.

lupus erythematosus [MED] An acute or subacute febrile collagen disease characterized by a butterfly-shaped rash over the cheeks and perilingual erythema.

lupus erythematosus factor *See* LE factor.

lupus vulgaris [MED] True tuberculosis of the skin; a slow-developing, scarring, and deforming disease, often asymptomatic, frequently involving the face, and occurring in a wide variety of appearances.

Lurex [TEXT] Trade name for a yarn with an aluminum base and a plastic coating made by the Dow Badische Company.

lusec [PHYS] A unit used for the measurement of power of evacuation of a vacuum pump, equal to the power associated with a leak rate of 1 liter per second at a pressure of 1 millitorr, or to approximately 1.33322×10^{-4} watt.

Lusin's theorem [MATH] Given a measurable function f which is finite almost everywhere in a euclidean space, then for every number $\varepsilon > 0$ there is a continuous function g which agrees with f, except on a set of measure less than ε.

Lusitanian [GEOL] Lower Jurassic geologic time.

luster [OPTICS] The appearance of a surface dependent on reflected light; types include metallic, vitreous, resinous, adamantine, silky, pearly, greasy, dull, and earthy; applied to minerals, textiles, and many other materials.

lustering [TEXT] A series of processes used to treat cottons to improve their sales appeal; the processes are mercerizing, frictioning, and filling.

lusterless paint [MATER] Paint which absorbs light rays so that no shine or polish appears on its surface; used extensively on U.S. Army vehicles.

luster mottlings [GEOL] The spotted, shimmering appearance of certain rocks caused by reflection of light from cleavage faces of crystals that contain small inclusions of other minerals.

lutaceous [GEOL] Claylike.

lute [MATER] A substance, such as cement or clay, for packing a joint or coating a porous surface to produce imperviousness to gas or liquid.

lutecite [GEOL] A fibrous, chalcedonylike quartz with optical anomalies that have led to its being considered a distinct species.

lutein [BIOCHEM] **1.** A dried, powdered preparation of corpus luteum. **2.** *See* xanthophyll.

luteinization [PHYSIO] Acquisition of characteristics of cells of the corpus luteum by ovarian follicle cells following ovulation.

luteinizing hormone [BIOCHEM] A glycoprotein hormone secreted by the adenohypophysis of vertebrates which stimulates hormone production by interstitial cells of gonads. Also known as interstitial-cell-stimulating hormone (ICSH).

luteoma [MED] A tumor of the ovary composed of cells resembling those of the corpus luteum.

luteotropic hormone *See* prolactin.

lutetium [CHEM] A chemical element, symbol Lu, atomic number 71, atomic weight 174.97; a very rare metal and the heaviest member of the rare-earth group.

Lutheran blood group [IMMUNOL] The erythrocyte antigens defined by reactions with an antibody designated anti-Lu^a, initially detected in the serum of a multiply transfused patient with lupus erythematosus, who developed antibodies against erythrocytes of a donor named Lutheran, and by anti-Lu^b.

lutite [GEOL] A consolidated rock or sediment formed principally of clay or clay-sized particles.

LUPINE

A field crop of lupine near maturity.

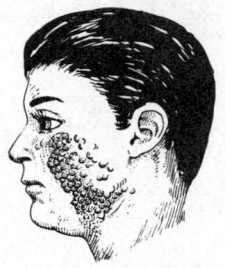

LUPUS ERYTHEMATOSUS

Manifestation of lupus erythematosus. *(From W. A. N. Dorland, American Illustrated Medical Dictionary, 24th ed., Saunders, 1965)*

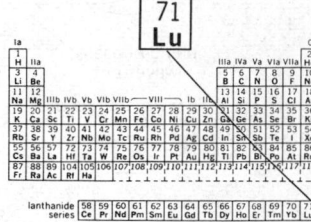

LUTETIUM

Periodic table of the chemical elements showing the position of lutetium.

Lutjanidae [VERT ZOO] The snappers, a family of perciform fishes in the suborder Percoidei.

lux [OPTICS] A unit of illumination, equal to the illumination on a surface 1 square meter in area on which there is a luminous flux of 1 lumen uniformly distributed, or the illumination on a surface all points of which are at a distance of 1 meter from a uniform point source of 1 candela. Abbreviated lx. Also known as meter-candle.

Luxemburg effect [COMMUN] Cross modulation between two radio signals during their passage through the ionosphere, due to the nonlinearity of the propagation characteristics of free charges in space.

luxon *See* troland.

luzonite *See* enargite.

L wave [GEOPHYS] A phase designation for an earthquake wave that is a surface wave, without respect to type.

lx *See* lux.

Lyapunov function [MATH] A function of a vector and of time which is positive-definite and has a negative-definite derivative with respect to time for nonzero vectors, is identically zero for the zero vector, and approaches infinity as the norm of the vector approaches infinity; used in determining the stability of control systems. Also spelled Liapunov function.

lyase [BIOCHEM] An enzyme that catalyzes the nonhydrolytic cleavage of its substrate with the formation of a double bond; examples are decarboxylases.

Lycaenidae [INV ZOO] A family of heteroneuran lepidopteran insects in the superfamily Papilionoidea including blue, gossamer, hairstreak, copper, and metalmark butterflies.

Lycaeninae [INV ZOO] A subfamily of the Lycaenidae distinguished by functional prothoracic legs in the male.

lychee [BOT] A tree of the genus *Litchi* in the family Sapindaceae, especially *L. chinensis* which is cultivated for its edible fruit, a one-seeded berry distinguished by the thin, leathery, rough pericarp that is red in most varieties. Also spelled litchi.

lychnisc [INV ZOO] A hexactin in which the central part of the spicule is surrounded by a system of 12 struts.

Lychniscosa [INV ZOO] An order of sponges in the subclass Hexasterophora in which parenchymal megascleres form a rigid framework and are all or in part lychniscs.

lycine *See* betaine.

lycopene [BIOCHEM] $C_{40}H_{50}$ A red, crystalline hydrocarbon that is the coloring matter of certain fruits, as tomatoes; it is isomeric with carotene.

lycoperdonosis [MED] A respiratory disease caused by inhalation of spores from the puffball mushroom, *Lycoperdon*.

lycophore larva [INV ZOO] A larva of certain cestodes characterized by cilia, large frontal glands, and 10 hooks. Also known as decanth larva.

Lycopodiales [BOT] The type order of Lycopodiopsida.

Lycopodiatae *See* Lycopodiopsida.

Lycopodineae [BOT] The equivalent name for Lycopodiopsida.

Lycopodiophyta [BOT] A division of the subkingdom Embryobionta characterized by a dominant independent sporophyte, dichotomously branching roots and stems, a single vascular bundle, and small, simple, spirally arranged leaves.

Lycopodiopsida [BOT] The lycopods, the type class of Lycophodiophyta.

lycopodium [MATER] A yellow powder prepared from the spores of *Lycopodium clavatum*; used as a desiccant and absorbent.

Lycopsida [BOT] Former subphylum of the Embryophyta now designated as the division Lycopodiophyta.

Lycoriidae [INV ZOO] A family of small, dark-winged dipteran insects in the suborder Orthorrhapha.

Lycosidae [INV ZOO] A family of hunting spiders in the suborder Dipneumonomorphae that actively pursue their prey.

Lycoteuthidae [INV ZOO] A family of squids.

Lycra [TEXT] Trade name of an elastic fiber made by Du Pont.

Lyctidae [INV ZOO] The large-winged beetles, a large cosmopolitan family of coleopteran insects in the superfamily Bostrichoidea.

Lyddane-Sachs-Teller relation [SOLID STATE] For an infinite ionic crystal, the relation $\varepsilon(0)/\varepsilon(\infty) = \omega_L{}^2/\omega_T{}^2$, where $\varepsilon(0)$ is the crystal's static dielectric constant, $\varepsilon(\infty)$ is the dielectric constant at a frequency at which electronic polarizability is effective but ionic polarizability is not, ω_L is the frequency of longitudinal optical phonons with zero wave vectors, and ω_T is the frequency of transverse optical phonons with large wave vector.

lyddite [MATER] An explosive composed chiefly of picric acid.

Lydian stone *See* basanite.

lydite *See* basanite.

lye [INORG CHEM] **1.** A solution of potassium hydroxide or sodium hydroxide used as a strong alkaline solution in industry. **2.** The alkaline solution that is obtained from the leaching of wood ashes.

Lygaeidae [INV ZOO] The lygaeid bugs, a large family of phytophagous hemipteran insects in the superfamily Lygaeoidea.

Lygaeoidea [INV ZOO] A superfamily of pentatomorphan insects having four-segmented antennae and ocelli.

Lyginopteridaceae [PALEOBOT] An extinct family of the Lyginopteridales including monostelic pteridosperms having one or two vascular traces entering the base of the petiole.

Lyginopteridales [PALEOBOT] An order of the Pteridospermae.

Lyginopteridatae [PALEOBOT] The equivalent name for Pteridospermae.

Lyman-alpha radiation [SPECT] Radiation emitted by hydrogen associated with the spectral line in the Lyman series whose wavelength is 1215 angstrom units.

Lyman band [SPECT] A band in the ultraviolet spectrum of molecular hydrogen, extending from 125 to 161 nanometers.

Lyman continuum [SPECT] A continuous range of wavelengths (or wave numbers or frequencies) in the spectrum of hydrogen at wavelengths less than the Lyman limit, resulting from transitions between the ground state of hydrogen and states in which the single electron is freed from the atom.

Lyman ghost [SPECT] A false line observed in a spectroscope as a result of a combination of periodicities in the ruling.

Lyman limit [SPECT] The lower limit of wavelengths of spectral lines in the Lyman series (912 angstrom units), or the corresponding upper limit in frequency, energy of quanta, or wave number (equal to the Rydberg constant for hydrogen).

Lyman series [SPECT] A group of lines in the ultraviolet spectrum of hydrogen covering the wavelengths of 1215–912 angstrom units.

Lymantriidae [INV ZOO] The tussock moths, a family of heteroneuran lepidopteran insects in the superfamily Noctuoidea; the antennae of males is broadly pectinate and there is a tuft of hairs on the end of the female abdomen.

Lymexylonidae [INV ZOO] The ship timber beetles composing the single family of the coleopteran superfamily Lymexylonoidea.

Lymexylonoidea [INV ZOO] A monofamilial superfamily of wood-boring coleopteran insects in the suborder Polyphaga characterized by a short neck and serrate antennae.

lymph [HISTOL] The colorless fluid which circulates through the vessels of the lymphatic system.

lymphadenitis [MED] Inflammation of lymph nodes.

lymphadenoid goiter *See* struma lymphomatosa.

lymphadenoma [MED] A tumorlike enlargement of a lymph node.

lymphadenomatosis [MED] A malignant neoplasm of lymphoid tissue.

lymphadenopathy [MED] Enlargement or disease of lymph nodes.

lymphadenosis [MED] Neoplasia or hyperplasia of lymph nodes.

lymphagogue [PHARM] An agent that stimulates lymph flow.

lymphangiectasis [MED] Dilation in the wall of a lymphatic vessel.

lymphangiectomy [MED] Surgical removal of a pathologic lymphatic channel, as for cancer.

lymphangioendothelial sarcoma *See* lymphangiosarcoma.

lymphangioendothelioma [MED] A tumor composed of aggregations of lymphatic vessels, between which are large mononuclear cells presumed to be endothelial cells.

lymphangiofibroma [MED] A benign tumor composed of lymphangiomatous and fibromatous elements.

lymphangioma [MED] An abnormal mass of lymphatic vessels.

lymphangiosarcoma [MED] A sarcoma whose parenchymal cells form vascular channels resembling lymphatics. Also known as lymphangioendothelial sarcoma.

lymphangitis [MED] Inflammation of lymphatic vessels.

lymphatic system [ANAT] A system of vessels and nodes conveying lymph in the vertebrate body, beginning with capillaries in tissue spaces and eventually forming the thoracic ducts which empty into the subclavian veins.

lymphatic tissue [HISTOL] Tissue consisting of networks of lymphocytes and reticular and collagenous fibers. Also known as lymphoid tissue.

lymphedema [MED] Edema resulting from lymph vessel obstruction.

lymph gland *See* lymph node.

lymph heart [VERT ZOO] A muscular expansion of a lymphatic vessel which contracts, driving lymph to the veins, as in amphibians.

lymph node [ANAT] An aggregation of lymphoid tissue surrounded by a fibrous capsule; found along the course of lymphatic vessels. Also known as lymph gland.

lymphoblast [HISTOL] Precursor of a lymphocyte.

lymphoblastic leukemia *See* acute lymphocytic leukemia.

lymphoblastic reticulosarcoma [MED] A sarcoma of reticulum cells.

lymphoblastoma [MED] A malignant neoplasm of lymphoid tissue composed of lymphoblasts.

lymphoblastosis [MED] An excessive number of lymphoblasts in peripheral blood; occasionally found in tissues.

lymphocyte [HISTOL] An agranular leukocyte formed primarily in lymphoid tissue; occurs as the principal cell type of lymph and composes 20–30% of the blood leukocytes.

lymphocytic angina *See* infectious mononucleosis.

lymphocytic choriomeningitis [MED] An acute viral meningitis caused by a specific virus endemic in mice; characterized clinically by rapid onset of symptoms of meningeal irritation, pleocytosis and often a rise in protein in the cerebrospinal fluid, and a short, benign course with recovery.

lymphocytic leukemia [MED] A type of leukemia in which lymphocytic cells predominate.

lymphocytic lymphoma [MED] A malignant neoplasm of lymphoid tissue composed predominantly of lymphocytic cells.

lymphocytic sarcoma *See* lymphosarcoma.

lymphocytoma [MED] A malignant neoplasm of lymphoid tissue composed principally of cells which resemble mature lymphocytes.

lymphocytopenia [MED] Reduction of the absolute number of lymphocytes per unit volume of peripheral blood. Also known as lymphopenia.

lymphocytosis [MED] An abnormally high lymphocyte count in the blood.

lymphoepithelioma [MED] A squamous-cell carcinoma of the nasopharynx whose parenchymal cells resemble elements of the reticuloendothelial system.

lymphogranuloma inguinale *See* lymphogranuloma venereum.

lymphogranulomatosis *See* Hodgkin's disease.

lymphogranuloma venereum [MED] A systemic infectious venereal disease caused by a microorganism belonging to the PLT-Bedsonia group, characterized by enlargement of inguinal lymph nodes and genital ulceration. Abbreviated LGV. Also known as lymphogranuloma inguinale; lymphopathia venereum; venereal bubo.

lymphoid cell [HISTOL] A mononucleocyte that resembles a leukocyte.

lymphoid hemoblast of Pappenheim *See* pronormoblast.

lymphoid tissue *See* lymphatic tissue.

lymphoma [MED] Any neoplasm, usually malignant, of the lymphoid tissues.

lymphomatosis [MED] Multiple malignant lymphomas.

lymphopathia venereum *See* lymphogranuloma venereum.

lymphopenia *See* lymphocytopenia.

lymphopoiesis [PHYSIO] The production of lymph.

lymphosarcoma [MED] A malignant lymphoma composed of

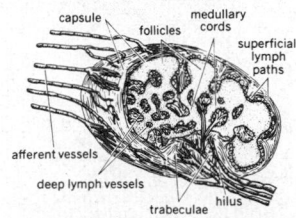

LYMPH NODE

Diagram of a lymph node. *(After C. Toldt, An Atlas of Human Anatomy for Students and Physicians, vol. 1, 2d ed., Macmillan, 1941)*

LYNX

Lynx canadensis, with tufted ears, large broad feet, long legs, and a short tail.

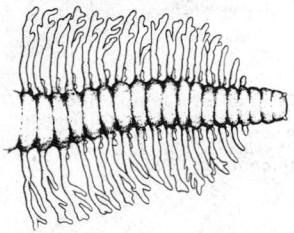

LYSARETIDAE

Anterior end of *Iphitime* in dorsal view.

LYSINE

$$NH_2$$
$$|$$
$$CH_2$$
$$|$$
$$CH_2$$
$$|$$
$$CH_2$$
$$|$$
$$CH_2$$
$$|$$
$$C$$
$$H_2N \quad | \quad COOH$$
$$H$$

Structural formula of lysine.

anaplastic lymphoid cells resembling lymphocytes or lymphoblasts, according to the degree of differentiation. Also known as lymphocytic sarcoma.

lymph sinus [ANAT] One of the tracts of diffuse lymphatic tissue between the cords and nodules, and between the septa and capsule of a lymph node.

lymph vessel [ANAT] A tubular passage for conveying lymph.

lynx [VERT ZOO] Any of several wildcats of the genus *Lynx* having long legs, short tails, and usually tufted ears; differs from other felids in having 28 instead of 30 teeth.

Lyomeri [VERT ZOO] The equivalent name for Saccopharyngiformes.

Lyon hypothesis [GEN] The assumption that mammalian females are X-chromosome mosaics as a result of the inactivation of one X chromosome in some embryonic cells and the other in the rest.

Lyon phenomenon [GEN] In the normal human female, the rendering of two X chromosomes inactive, or at least largely so, at an early stage in embryogenesis.

lyophilic [CHEM] Referring to a substance which will readily go into colloidal suspension in a liquid.

lyophilization [CHEM ENG] Rapid freezing of a material, especially biological specimens for preservation, at a very low temperature followed by rapid dehydration by sublimation in a high vacuum.

lyophobic [CHEM] Referring to a substance in a colloidal state that has a tendency to repel liquids.

Lyopomi [VERT ZOO] An equivalent name for Notacanthiformes.

lyotopic series *See* Hofmeister series.

Lyra [ASTRON] A northern constellation; right ascension 19 hours, declination 40° north; its first-magnitude star, Vega, is a navigational star and the most brilliant star in this part of the sky.

α Lyrae *See* Vega.

Lyrids [ASTRON] An important meteor shower occurring about April 22; it is regular and predictable, but not heavy, the hourly rate usually being about 7–10.

Lysaretidae [INV ZOO] A family of errantian polychaete worms in the superfamily Eunicea.

lyse [CYTOL] To undergo lysis.

lysergic acid [ORG CHEM] $C_{16}H_{16}N_2O_2$ A compound that crystallizes in the form of hexagonal plates that melt and decompose at 240°C; derived from ergot alkaloids; used as a psychotomimetic agent.

lysergic acid diethylamide [ORG CHEM] $C_{15}H_{15}N_2CON-(C_2H_5)_2$ A psychotomimetic drug synthesized from compounds derived from ergot. Abbreviated LSD; LSD-25.

Lysianassidae [INV ZOO] A family of pelagic amphipod crustaceans in the suborder Gammaridea.

lysigenous [BIOL] Of or pertaining to the space formed following lysis of cells.

lysimeter [ENG] An instrument for measuring the water percolating through soils and determining the materials dissolved by the water.

lysin [IMMUNOL] A substance, particularly antibodies, capable of lysing a cell.

lysine [BIOCHEM] $C_6H_{14}O_2N_2$ An essential, basic amino acid obtained from many proteins by hydrolysis.

Lysiosquillidae [INV ZOO] A family of crustaceans in the order Stomatopoda.

lysis [CYTOL] Dissolution of a cell or tissue by the action of a lysin. [MED] **1.** Gradual decline in the manifestations of a disease, especially an infectious disease. Also known as defervescence. **2.** Gradual fall of fever.

lysogeny [MICROBIO] Lysis of bacteria, with the liberation of bacteriophage particles.

lysosome [CYTOL] A specialized cell organelle surrounded by a single membrane and containing a mixture of hydrolytic (digestive) enzymes.

lysozyme [BIOCHEM] An enzyme present in certain body secretions, principally tears, which acts to hydrolyze certain bacterial cell walls.

lyssacine [INV ZOO] An early stage of the skeletal network in hexactinellid sponges.

Lyssacinosa [INV ZOO] An order of sponges in the subclass Hexasterophora in which parenchymal megascleres are typically free and unconnected but are sometimes secondarily united.

lyssophobia [PSYCH] **1.** Abnormal fear of hydrophobia. **2.** Abnormal fear of becoming insane.

lytic infection [MICROBIO] Penetration of a host cell by lytic phage.

lytic phage [VIROL] Any phage that cause host cells to lyse.

lytic reaction [CYTOL] A reaction that leads to lysis of a cell.

m *See* meter; milli-.

M *See* mega-; molarity.

M$_s$ [MET] In a carbon steel, the temperature at which martensite formation begins during cooling of austenite.

mA *See* milliampere.

maar [GEOL] A volcanic crater that was created by violent explosion but not accompanied by igneous extrusion; frequently, it is filled by a small circular lake.

maarad [TEXT] A type of Egyptian cotton cultivated from American pima seeds.

macadam [CIV ENG] Uniformly graded stones consolidated by rolling to form a road surface; may be bound with water or cement, or coated with tar or bitumen.

macadamia nut [BOT] The hard-shelled seed obtained from the fruit of a tropical evergreen tree, *Macadamia ternifolia*.

macaluba *See* mud volcano.

macanilla oil [MATER] An oil obtained from the nuts of the palm *Guilielma garipaes* of Venezuela and Central America.

macaque [VERT ZOO] The common name for 12 species of Old World monkeys composing the genus *Macaca*, including the Barbary ape and the rhesus monkey.

Macarthur and Forest cyanide process [MET] A process for recovering gold by leaching the pulped gold ore with a solution of 0.2–0.8% potassium cyanide and next with water; the gold is then obtained by precipitation on zinc or aluminum or by electrolysis.

Macassar oil [MATER] **1.** A fat obtained from seeds of kusam, used in hair dressing, cooking, and illumination. **2.** Any of several fatty oils or oily preparations that have similar properties and are used in hair preparations.

macaw [VERT ZOO] The common name for large South and Central American parrots of the genus *Ara* and related genera; individuals are brilliantly colored with a long tail, a hooked bill, and a naked area around the eyes.

Macbeth illuminometer [OPTICS] A type of portable visual photometer in which the light to be measured is balanced by a Lummer-Brodhun sight box against a comparison lamp, whose apparent brightness can be varied by moving it along a tube; a control box supplies a calibrated current to the comparison lamp, and calibrated optical filters can be placed in the light paths to correct for color differences between the comparison and measured sources and to extend the range of the instrument.

MacCullagh's formula [PHYS] A formula for the potential due to a distribution of mass or charge at an external point: the potential V at a point P resulting from a distribution of mass or positive charge centered about a point O is $V = kM/r + (k/2r^3)(A + B + C − 3I) + O(1/r^4)$, where r is the distance from O to P, k is the gravitational or electrostatic constant, M is the total mass or charge, A, B, and C are the principal moments about O, I is the moment about OP, and $O(1/r^4)$ is a quantity that falls off at least as rapidly as $1/r^4$.

MacDonald functions *See* modified Hankel functions.

mace [FOOD ENG] Spice made from the covering of the nutmeg.

Mace [MATER] Trade name for a tear gas containing chloroacetophenone; causes severe dermatitis and can cause permanent damage to the eyes.

macedonite [MINERAL] PbTiO$_3$ A mineral composed of an oxide of lead and titanium. [PETR] A basaltic rock that contains orthoclase, sodic plagioclase, biotite, olivine, and rare pyriboles.

mace oil [MATER] An essential oil obtained by distillation from mace and containing pinene and dipentene; used in flavoring.

maceral [GEOL] The microscopic organic constituents found in coal.

macgovernite [MINERAL] Mn$_5$(AsO$_3$)SiO$_3$(OH)$_2$ A mineral composed of basic manganese arsenite and silicate. Also spelled mcgovernite.

Machaeridea [INV ZOO] A class of homolozoan echinoderms in older systems of classification.

Mach angle [FL MECH] The vertex half angle of the Mach cone generated by a body in supersonic flight.

Mach cone [FL MECH] **1.** The cone-shaped shock wave theoretically emanating from an infinitesimally small particle moving at supersonic speed through a fluid medium; it is the locus of the Mach lines. **2.** The cone-shaped shock wave generated by a sharp-pointed body, as at the nose of a high-speed aircraft.

machete [DES ENG] A knife with a broad blade 2 to 3 feet (60 to 90 centimeters) long.

Mach front *See* Mach stem.

Machilidae [INV ZOO] A family of insects belonging to the Thysanura having large compound eyes and ocelli and a monocondylous mandible of the scraping type.

machinability [MET] **1.** The ability of a metal to be machined. **2.** The difficulty or ease with which a metal can be machined.

machinability index [MET] A numerical value that designates the degree of difficulty or ease with which a particular material can be machined; originally based on turning B1112 steel at 180 feet/minute with a high-speed tool for an index of 100%; with replacement of high-speed steels with carbides in turning operations, it has been found that the machinability index of a given material changes with the type of operation and the tool material.

machinable *See* machine-sensible.

machinable carbide [MET] Titanium carbide in a matrix of Ferro-Tic C tool steel.

Mach indicator *See* Machmeter.

machine [ADP] **1.** A mechanical, electric, or electronic device, such as a computer, tabulator, sorter, or collator. **2.** A simplified, abstract model of an internally programmed computer, such as a Turing machine. [MECH ENG] A combination of rigid or resistant bodies having definite motions and capable of performing useful work.

machine address [ADP] The actual and unique internal designation of the location at which an instruction or datum is to be stored or from which it is to be retrieved.

machine attention time [IND ENG] Time during which a machine operator must observe the machine's functioning and be available for immediate servicing, while not actually operating or servicing the machine.

machine available time [ADP] The time during which a computer has its power turned on, is not undergoing maintenance, and is thought to be operating properly.

machine bolt [DES ENG] A heavy-weight bolt with a square, hexagonal, or flat head used in the automotive, aircraft, and machinery fields.

machine check [ADP] A check that tests whether the parts of equipment are functioning properly. Also known as hardware check.

machine-check indicator [ADP] A protective device which turns on when certain conditions arise within the computer; the computer can be programmed to stop or to run a separate correction routine or to ignore the condition.

machine codes [ADP] The absolute numbers assigned by the manufacturer to the various parts of the computer.

machine controlled time [IND ENG] The time necessary for a machine to complete the automatic portion of a work cycle. Also known as independent machine time; machine element; machine time.

machine cut [MIN ENG] A groove or slot made horizontally or vertically in a coal seam by a coal cutter, as a step to shot firing.

machine cycle [ADP] **1.** The shortest period of time at the end

MACADAMIA NUT

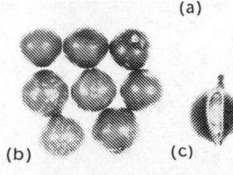

Macadamia integrifolia. (a) Mature nuts. (b) Nuts without husks. (c) Nuts in husk showing method of dehiscence. (From R. A. Jaynes, ed., Handbook of North American Nut Trees, Humphrey Press, 1969)

MACH CONE

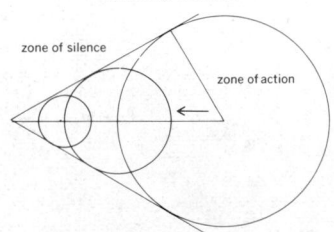

Generation of Mach cone (def. 1) by particle moving through fluid at supersonic speed. Arrow indicates direction of particle's motion. Cone is tangent to spherical surfaces of disturbances generated by particle at successive times, and separates zones of action and silence.

MACHINE BOLT

A hexagon-head machine bolt with nut. *(Reynolds Metals Co.)*

of which a series of events in the operation of a computer is repeated. **2.** The series of events itself.

machine design [DES ENG] Application of science and invention to the development, specification, and construction of machines.

machine drill [MECH ENG] Any mechanically driven diamond, rotary, or percussive drill.

machine element [DES ENG] Any of the elementary mechanical parts, such as gears, bearings, fasteners, screws, pipes, springs, and bolts used as essentially standardized components for most devices, apparatus, and machinery. [IND ENG] *See* machine controlled time.

machine error [ADP] A deviation from correctness in computer-processed data, caused by equipment failure.

machine file [DES ENG] A file that can be clamped in the chuck of a power-driven machine.

machine forging [MET] Forging operations performed in and by certain machines.

machine gun [ORD] **1.** A weapon that automatically fires small-arms ammunition, caliber .60 or 15.24 millimeters or under, and is capable of sustained rapid fire. **2.** To riddle a target with machine gun fire.

machine-gun microphone *See* line microphone.

machine-hour [IND ENG] A unit representing the operation of one machine for 1 hour; used in the determination of costs and economics.

machine idle time [IND ENG] Time during a work cycle when a machine is idle, awaiting completion of manual work.

machine-independent [ADP] Referring to programs and procedures which function in essentially the same manner regardless of the machine on which they are carried out.

machine instruction [ADP] A set of digits, binary bits, or characters that a computer can recognize and act upon, and that, when interpreted or decoded, indicates the action to be performed and which operand is to be involved in the action.

machine instruction statement [ADP] A statement consisting usually of a tag, an operating code, and one or more addresses.

machine interruption [ADP] A halt in computer operations followed by the beginning of a diagnosis procedure, as a result of an error detection.

machine key [DES ENG] A piece inserted between a shaft and a hub to prevent relative rotation. Also known as key.

machine language [ADP] The set of instructions available to a particular digital computer, and by extension the format of a computer program in its final form, capable of being executed by a computer.

machine language code [ADP] A set of instructions appearing as combinations of binary digits.

machine learning [ADP] The process or technique by which a device modifies its own behavior as the result of its past experience and performance.

machine loading [IND ENG] **1.** Feeding work into a machine. **2.** Planning the amount of use of a unit of equipment during a given time period.

machine logic [ADP] The structure of a computer, the operation it performs, and the type and form of data used internally.

machine mining [MIN ENG] The use of power machines and equipment in the excavation and extraction of ore or coal.

machine oil [MATER] Medium-density lubricating oil used for machine parts.

machine operator [ADP] The person who manipulates the computer controls, brings up and closes down the computer, and can override a number of computer decisions.

machine-oriented programming system [ADP] A system written in assembly language (or macro code) directly oriented toward the computer's internal language.

machine-paced operation [IND ENG] The proportion of an operation cycle during which the machine controls the speed of work progress.

machine pistol [ORD] A pistol capable of fully automatic fire.

machine processible form [ADP] Any input medium such as a punch card, paper tape, or magnetic tape.

machine-readable *See* machine-sensible.

machine-recognizable *See* machine-sensible.

machine ringing [COMMUN] In a telephone system, ringing which is started either mechanically or by an operator, after which it continues automatically until the call is answered or abandoned.

machine run *See* run.

machinery [MECH ENG] A group of parts or machines arranged to perform a useful function.

machine screw [DES ENG] A blunt-ended screw with a standardized thread and a head that may be flat, round, fillister, or oval, and may be slotted, or constructed for wrenching; used to fasten machine parts together.

machine-sensible [ADP] Capable of being read or sensed by a device, usually by one designed and built specifically for this task. Also known as machinable; machine-readable; machine-recognizable; mechanized.

machine-sensible information [ADP] Information in a form which can be read by a specified machine.

machine setting *See* mechanical setting.

machine shop [ENG] A workshop in which work, metal or other material, is machined to specified size and assembled.

machine steel [MET] Plain carbon steel with a 0.2–0.3% carbon content.

machine switching system *See* automatic exchange.

machine taper [MECH ENG] A taper that provides a connection between a tool, arbor, or center and its mating part to ensure and maintain accurate alignment between the parts; permits easy separation of parts.

machine time *See* machine controlled time.

machine tool [MECH ENG] A stationary power-driven machine for the shaping, cutting, turning, boring, drilling, grinding, or polishing of solid parts, especially metals.

machine-tool control [ADP] The computer control of a machine tool for a specific job by means of a special programming language.

machine translation *See* mechanical translation.

machine welding [MET] Welding with a machine under the control and observation of an operator; may be loaded and unloaded either manually or mechanically.

machine word [ADP] The fundamental unit of information in a word-organized digital computer, consisting of a fixed number of binary bits, decimal digits, characters, or bytes.

machining [MET] Performing various cutting or grinding operations on a piece of work.

machining stress [MET] Residual stress in the work caused by machining.

machinist's file [DES ENG] A type of double-cut file that removes metal fast and is used for rough metal filing.

Mach line [FL MECH] **1.** A line representing a Mach wave. **2.** *See* Mach wave.

Machmeter [ENG] An instrument that measures and indicates speed relative to the speed of sound, that is, indicates the Mach number. Also known as Mach indicator.

Mach number [FL MECH] The ratio of the speed of a body or of a point on a body with respect to the surrounding air or other fluid, or the ratio of the speed of a fluid, to the speed of sound in the medium. Symbolized Ma; N_{Ma}. Also known as relative Mach number.

machopolyp *See* machozooid.

machozooid [INV ZOO] A defensive polyp equipped with stinging organs in certain hydroid colonies. Also known as machopolyp.

Mach principle [RELAT] The principle that the motion of a particle is only meaningful when referred to the rest of the matter in the universe; this motion is thus determined by the distribution of this matter and is not an intrinsic property of an absolute space.

Mach reflection [FL MECH] The reflection of a shock wave from a rigid wall in which the shock strength of the reflected wave and the angle of reflection both have the smaller of the two values which are theoretically possible.

Mach refractometer *See* Mach-Zehnder interferometer.

Mach stem [FL MECH] A shock wave or front formed above the surface of the earth by the fusion of direct and reflected shock waves resulting from an airburst bomb. Also known as Mach front.

Mach wave [FL MECH] Also known as Mach line. **1.** A shock wave theoretically occurring along a common line of intersection of all the pressure disturbances emanating from an

MACHINE SCREW

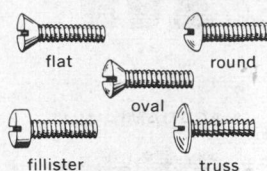

flat   round

oval

fillister   truss

Types of machine screws.
(Reynolds Metals Co.)

infinitesimally small particle moving at supersonic speed through a fluid medium, with such a wave considered to exert no changes in the condition of the fluid passing through it. **2.** A very weak shock wave appearing, for example, at the nose of a very sharp body, where the fluid undergoes no substantial change in direction.

Mach-Zehnder interferometer [OPTICS] A variation of the Michelson interferometer used mainly in measuring the spatial variation of the index of refraction of a gas; the device has two semitransparent mirrors and two wholly reflecting mirrors at alternate corners of a rectangle, and half the beam travels along each side of the rectangle. Also known as Mach refractometer.

M acid [ORG CHEM] $NH_2C_{10}H_5(OH)SO_3H$ A sulfonic acid formed by alkaline fusion of a disulfonic acid of α-naphthylamine; used as a dye intermediate. Also known as 5-amino-1-naphthol-3-sulfonic acid.

mackayite [MINERAL] $FeTe_2O_5(OH)$ A green mineral composed of basic iron tellurite.

mackerel [VERT ZOO] The common name for perciform fishes composing the subfamily Scombroidei of the family Scombridae, characterized by a long slender body, pointed head, and large mouth.

mackerel shark [VERT ZOO] The common name for isurid galeoid elasmobranchs making up the family Isuridae; heavy-bodied fish with sharp-edged, awllike teeth and a nearly symmetrical tail.

mackerel sky [METEOROL] A sky with considerable cirrocumulus or small-element altocumulus clouds, resembling the scales on a mackerel.

mackinawite [MINERAL] $(Fe,Ni)S$ A tetragonal mineral occurring as a corrosion product in iron pipes. Also known as kansite.

Mack's cement [MATER] Cement made of dehydrated gypsum with a small amount of calcined sodium sulfate and potassium sulfate.

Macky effect [METEOROL] The reduction of the effective dielectric strength of air when waterdrops are present.

Mac-Lane system [MIN ENG] A means of conveying dirt to the top of a heap; an inclined rail track goes from the loading station at the bottom of the heap to an extending frame at the top; the rope that hauls the tubs with dirt up the rail track passes around a return sheave in the extending frame; a tub is tipped over at the top by a gear and the dirt discharged.

Maclaurin-Cauchy test *See* Cauchy's test for convergence.

Maclaurin expansion [MATH] The power series representation of a function arising from Maclaurin's theorem.

Maclaurin series [MATH] The power series in the Maclaurin expansion.

Maclaurin's theorem [MATH] The theorem giving conditions when a function, which is infinitely differentiable, may be represented in a neighborhood of the origin as an infinite series with nth term $(1/n!) \cdot f^{(n)}(0) \cdot x^n$, where $f^{(n)}$ denotes the nth derivative.

macle *See* chiastolite.

Macleod equation [FL MECH] An equation which states that the fourth root of the surface tension of a liquid is proportional to the difference between the densities of the liquid and of its vapor.

MacMichael degree [FL MECH] An arbitrary unit used in measuring viscosity with a type of Couette viscometer; its size depends on the stiffness of the suspension of the inner cylinder of the viscometer.

Macquer's salt *See* potassium arsenate.

macrandrous [BOT] Having both antheridia and oogonia on the same plant; used especially for certain green algae.

Macraucheniidae [PALEON] A family of extinct herbivorous mammals in the order Litopterna; members were proportioned much as camels are, and eventually lost the vertebral arterial canal of the cervical vertebrae.

macrencephaly [MED] The condition of having an abnormally large brain.

Macristiidae [VERT ZOO] A family of oceanic teleostean fishes assigned by some zoologists to the order Ctenothrissiformes.

macro- [SCI TECH] Prefix meaning large.

macroanalysis [ANALY CHEM] Qualitative or quantitative analysis of chemicals that are in quantities of the order of grams.

macroanalytical balance [ENG] A relatively large type of analytical balance that can weigh loads of up to 200 grams to the nearest 0.1 milligram.

macroassembler [ADP] A program made up of one or more sequences of assembly language statements, each sequence represented by a symbolic name.

macroassembly program [ADP] A set of assembly languages for the IBM 7090 and 7040 series of computers, and the assemblers for these languages; the assemblers operate under the IBSYS systems and prepare relocatable or absolute binary output in computer language. Abbreviated MAP.

macroblast of Naegeli *See* pronormoblast.

macroblepharia [MED] The condition of having abnormally large eyelids.

macrobrachia [MED] The condition of having excessive arm development.

macrocephalus [MED] An individual with an abnormally large head.

macrocephaly [MED] The condition of having an abnormally large head.

macrocheiria [MED] The condition of having abnormal hand enlargement.

macroclastic [PETR] Rock that is composed of fragments that are visible without magnification.

macroclimate [CLIMATOL] The climate of a large geographic region.

macrocode [ADP] A coding and programming language that assembles groups of computer instructions into single instructions.

macroconsumer [ECOL] A large consumer which ingests other organisms or particulate organic matter. Also known as biophage.

macrocrania [MED] The condition of having abnormally large skull size compared with face size.

macrocrystalline [PETR] **1.** Pertaining to the texture of holocrystalline rock in which the constituents are visible without magnification. **2.** Pertaining to the texture of a rock with grains or crystals greater than 0.75 millimeter in diameter in recrystallized sediment.

macrocyclic [MYCOL] Of a rust fungus, having binuclear spores as well as teliospores and sporidia, or having a life cycle that is long or complex. [ORG CHEM] An organic molecule with a large ring structure usually containing over 15 atoms.

Macrocypracea [INV ZOO] A superfamily of marine ostracods in the suborder Podocopa having all thoracic legs different from each other, greatly reduced furcae, and long, thin Zenker's organs.

macrocyst [MED] A large cyst, visible to the naked eye.

macrocyte [HISTOL] An erythrocyte whose diameter or mean corpuscular volume exceed that of the mean normal by more than two standard deviations. Also known as macronormocyte.

macrocytic anemia [MED] A form of anemia characterized by the presence of macrocytes in the blood.

macrocytosis [MED] Presence of macrocytes in the blood.

macrodactyly [MED] The condition of having abnormally large fingers or toes.

Macrodasyoidea [INV ZOO] An order of wormlike invertebrates of the class Gastrotricha characterized by distinctive, cylindrical adhesive tubes in the cuticle which are moved by delicate muscle strands.

macrodome [CRYSTAL] Dome of a crystal in which planes are parallel to the longer lateral axis.

macrodontia [MED] The condition of having abnormally large teeth.

macroetching *See* deep-etching.

macroevolution [EVOL] The larger course of evolution by which the categories of animal and plant classification above the species level have been evolved from each other and have differentiated into the forms within each.

macrofacies [GEOL] A collection of sedimentary facies that are related genetically.

macrofauna [ECOL] **1.** Widely distributed fauna. **2.** Fauna of a macrohabitat. [ZOO] Animals visible to the naked eye.

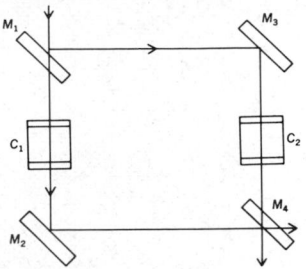

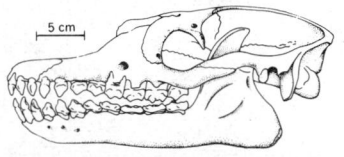

macroflora [BOT] Plants visible to the naked eye. [ECOL] **1.** Widely distributed flora. **2.** Flora of a macrohabitat.

macrofollicular adenoma [MED] **1.** A benign tumor of the thyroid with enlarged follicles. **2.** A type of malignant lymphoma.

macrofossil [PALEON] A fossil large enough to be observed with the naked eye.

macrogamete [BIOL] The larger, usually female gamete produced by a heterogamous organism.

macrogeneration [ADP] The creation of many machine instructions from one macroword.

macroglia [HISTOL] That portion of the neuroglia composed of astrocytes.

macroglobulin [BIOCHEM] Any gamma globulin with a sedimentation constant of 195.

macroglobulinemia [MED] **1.** Abnormal increase in macroglobulins in the blood. **2.** A disease characterized by proliferation of lymphocytes and plasmocytes and abnormally high macroglobulin blood levels.

macroglossia [MED] Enlargement of the tongue.

macrograph [GRAPHICS] A photograph or representation of an object that may be about 10 times natural size, as large as the natural object, or slightly smaller.

macrogyria [MED] The condition of having congenitally enlarged brain convolutions.

macrohabitat [ECOL] An extensive habitat presenting considerable variation of the environment, containing a variety of ecological niches, and supporting a large number and variety of complex flora and fauna.

macroinstruction [ADP] An instruction in a higher-level language which is equivalent to a specific set of one or more ordinary instructions in the same language.

Macrolepidoptera [INV ZOO] A former division of Lepidoptera that included the larger moths and butterflies.

macrolibrary [ADP] A collection of prewritten specialized but unparticularized routines (or sets of statements) which reside in mass storage.

macrolide antibiotic [MICROBIO] A basic antibiotic characterized by a macrocyclic ring structure.

macrolymphocyte [HISTOL] A large lymphocyte.

macromastia [MED] The condition of having abnormally enlarged breasts.

macromelia [MED] The condition of having abnormally large arms or legs.

macromere [EMBRYO] Any of the large blastomeres composing the vegetative hemisphere of telolecithal morulas and blastulas.

macrometeorology [METEOROL] The study of the largest-scale aspects of the atmosphere, such as the general circulation, and weather types.

macrometer [OPTICS] Instrument that has two mirrors and a focusing telescope with which the ranges of distant objects can be found.

macromolecular [ORG CHEM] Composed of or characterized by large molecules.

macromolecule [ORG CHEM] A large molecule in which there is a large number of one or several relatively simple structural units, each consisting of several atoms bonded together.

macronormocyte *See* macrocyte.

macronucleus [INV ZOO] A large, densely staining nucleus of most ciliated protozoans, believed to influence nutritional activities of the cell.

macronutrient [BIOCHEM] An element, such as potassium and nitrogen, essential in large quantities for plant growth.

macroparameter [ADP] The character in a macro operand which will complete an open subroutine created by the macroinstruction.

macrophage [HISTOL] A large phagocyte of the reticuloendothelial system. Also known as histiocyte.

macrophagy [BIOL] Feeding on large particulate matter.

macrophotography [GRAPHICS] The photography of a subject so that the final image is either unmagnified or magnified to no more than 10 times the object.

macrophreate [INV ZOO] A comatulid with a large, deep cavity in the calyx.

macrophyllous [BOT] Having large or long leaves.

macrophyte [ECOL] A macroscopic plant, especially one in an aquatic habitat.

macropinacoid *See* front pinacoid.

Macropodidae [VERT ZOO] The kangaroos, a family of Australian herbivorous mammals in the order Marsupialia.

macropodous [BOT] **1.** Having a large or long hypocotyl. **2.** Having a long stem or stalk.

macropore [GEOL] A pore in soil of a large enough size so that water is not held in it by capillary attraction.

macroprogramming [ADP] The process of writing machine procedure statements in terms of macroinstructions.

macroprosopus [MED] An individual with an abnormally large face.

macropsia [MED] A disturbance of vision in which objects seem larger than they are. Also known as megalopia.

macropterous [ZOO] Having large or long wings or fins.

macrorheology [MECH] A branch of rheology in which materials are treated as homogeneous or quasi-homogeneous, and processes are treated as isothermal.

Macroscelidea [VERT ZOO] A monofamilial order of mammals containing the elephant shrews and their allies.

Macroscelididae [VERT ZOO] The single, African family of the mammalian order Macroscelidea.

macroscopic [SCI TECH] Large enough to be observed by the naked eye.

macroscopic anisotropy [ENG] Phenomenon in electrical downhole logging wherein electric current flows more easily along sedimentary strata beds than perpendicular to them.

macroscopic property [NUCLEO] A nuclear reactor property that can be treated independently of other factors. [THERMO] *See* thermodynamic property.

macroscopic stress [MET] Residual stress in a metal in a distance comparable to the gage length of strain measurement specimens and therefore detectable by x-ray or dissection techniques. Also known as macrostress.

macroscopic theory [PHYS] A theory concerning only phenomena observable with the naked eye or with an ordinary light microscope, and not with the behavior of atoms, molecules, or their constituents which may underlie these phenomena.

macroscopy [SCI TECH] The study or observation of objects visible to the unaided eye.

macroseptum [INV ZOO] A primary septum in certain anthozoans.

macroskeleton [ADP] A definition of a macroinstruction in a precise but content-free way, which can be particularized by a processor as directed by macroinstruction parameters. Also known as model.

macrosonics [ACOUS] The technology of sound at signal amplitudes so large that linear approximations are not valid, as in the use of ultrasonics for cleaning or drilling.

macrosporangium [BOT] A spore case in which macrospores are produced. Also known as megasporangium.

macrospore [BOT] The larger of two spore types produced by heterosporous plants; the female gamete. Also known as megaspore. [INV ZOO] The larger gamete produced by certain radiolarians; the female gamete.

Macrostomida [INV ZOO] An order of rhabdocoels having a simple pharynx, paired protonephridia, and a single pair of longitudinal nerves.

Macrostomidae [INV ZOO] A family of rhabdocoels in the order Macrostomida; members are broad and flattened in shape and have paired sex organs.

macrostress *See* macroscopic stress.

macrostructure [MET] Structure of an etched metal visible to the naked eye or at magnifications up to 10 diameters.

macrostylous [BOT] **1.** Having long styles. **2.** Having long styles and short stamens.

macrosystem [ADP] A language in which words represent a number of machine instructions.

Macrotermitinae [INV ZOO] A subfamily of termites in the family Termitidae.

macrothermophyte *See* megathermophyte.

macrotome [ENG] A device for making large anatomical sections.

Macrouridae [VERT ZOO] The grenadiers, a family of actinopterygian fishes in the order Gadiformes in which the body tapers to a point, and the dorsal, caudal, and anal fins are continuous.

Macroveliidae [INV ZOO] A family of hemipteran insects in the subdivision Amphibicorisae.

Macrura [INV ZOO] A group of decapod crustaceans in the suborder Reptantia including eryonids, spiny lobsters, and true lobsters; the abdomen is extended and bears a well-developed tail fan.

macrurous [ZOO] Having a long tail.

macula [ANAT] Any anatomical structure having the form of a spot or strain.

macula densa [MED] A thickening of the epithelium of the ascending limb of the loop of Henle.

macula lutea [ANAT] A yellow spot on the retina; the area of maximum visual acuity, being made up almost entirely of retinal cones.

maculate [BOT] Marked with speckles or spots.

maculopapule [MED] A small, circumscribed, discolored elevation of the skin; a macule and papule combined.

maculose [GEOL] Of a group of contact-metamorphosed rocks or their structures, having spotted or knotted character.

MAD *See* magnetic anomaly detector.

madarosis [MED] Loss of the eyelashes or eyebrows.

madder [MATER] The root of the madder plant (*Rubia tinctorium*), pulverized and used as source of glucosides to produce alizarin by fermentation. Also known as gamene.

madder lake [MATER] Bluish-red, transparent pigment produced from alizarin red; used to make stains and inks, and as a component of artists' oil colors.

Madelung constant [SOLID STATE] A dimensionless constant which determines the electrostatic energy of a three-dimensional periodic crystal lattice consisting of a large number of positive and negative point charges when the number and magnitude of the charges and the nearest-neighbor distance between them is specified.

madia oil [MATER] An oil that is made by crushing the seeds of the melosa plant; used as a substitute for olive oil.

madistor [ELECTR] A cryogenic semiconductor device in which injection plasma can be steered or controlled by transverse magnetic fields, to give the action of a switch.

Madreporaria [INV ZOO] The equivalent name for Scleractinia.

madreporite [INV ZOO] A delicately perforated sieve plate at the distal end of the stone canal in echinoderms.

Madsen impedance meter [ENG] An instrument for measuring the acoustic impedance of normal and deaf ears, based on the principle of the Wheatstone bridge.

MADT *See* microalloy diffused transistor.

madura foot *See* mycetoma.

maduromycosis *See* mycetoma.

maelstrom [OCEANOGR] A powerful and often destructive water current caused by the combined effects of high, wind-generated waves and a strong, opposing tidal current.

Maerz and Paul color system [ANALY CHEM] A dictionary of 7056 different colors in the form of two-impression screen-plate printing on semiglossy paper.

Maestrichtian [GEOL] A European stage of geologic time: Upper Cretaceous (above Menevian, below Fastiniogian).

maestro [METEOROL] A northwesterly wind with fine weather which blows, especially in summer, in the Adriatic, most frequently on the western shore; it is also found on the coasts of Corsica and Sardinia.

mafic mineral [MINERAL] 1. A mineral that is composed predominantly of the ferromagnesian rock-forming silicates. 2. In general any dark mineral.

mafura tallow [MATER] A bitter-tasting, heavy vegetable fat obtained from the nuts of the mafura tree (*Trichilia emetica*); used for soaps, candles, and ointments.

MAG *See* maximum available gain.

magamp *See* magnetic amplifier.

magazine [ADP] 1. A holder of microfilm or magnetic recording media strips. 2. *See* input magazine. [ENG] 1. A storage area for explosives. 2. A building, compartment, or structure constructed and located for the storage of explosives or ammunition. [ORD] That part of a gun or firearm that holds ammunition ready for chambering.

magazine filler [ORD] An item which attaches to the rear top

portion of a rifle magazine to retain and position a cartridge clip to facilitate loading.

magazine safety [ORD] A mechanism for an automatic pistol, to make firing impossible unless the weapon contains the magazine.

Magellanic Clouds [ASTRON] Two irregular clouds of stars that are the nearest galaxies to the galactic system; both the Large and Small Magellanic Clouds are identified as Irregular in the classification of E. P. Hubble.

Magelonidae [INV ZOO] A monogeneric family of spioniform annelid worms belonging to Sedentaria.

magenstrasse [ANAT] Gastric canal.

magenta *See* fuchsin.

Maggi-Righi-Leduc effect [PHYS] A phenomenon in which the thermal conductivity of a conductor changes when it is placed in a magnetic field.

maggot [INV ZOO] Larva of a dipterous insect.

maggot therapy [MED] Implantation of sterile cultivated maggots of the bluebottle fly into wounds in the treatment of chronic soft tissue infections and chronic osteomyelitis.

maghemite [MINERAL] γ-Fe_2O_3 A mineral form of iron oxide that is strongly magnetic and a member of the magnetite series.

magic eye *See* cathode-ray tuning indicator.

magic numbers [NUC PHYS] The integers 8, 20, 28, 50, 82, 126; nuclei in which the number of protons, neutrons, or both is magic have a stability and binding energy which is greater than average, and have other special properties.

magic square [MATH] A square array of integers where the sum of the entries of each row, each column, and each diagonal is the same.

magic tee *See* hybrid tee.

magister of sulfur [CHEM] Amorphous sulfur produced by acid precipitation from solutions of hyposulfites or polysulfides.

magma [GEOL] The molten rock material from which igneous rocks are formed.

magma chamber [GEOL] A larger reservoir in the crust of the earth that is occupied by a body of magma.

magma province *See* petrographic province.

magmatic differentiation [PETR] 1. The process by which the different types of igneous rocks are derived from a single parent magma. 2. The process by which ores are formed by solidification from magma. Also known as magmatic segregation.

magmatic rock [PETR] A rock derived from magma.

magmatic segregation *See* magmatic differentiation.

magmatic stoping [GEOL] A process of igneous intrusion in which magma gradually works its way upward by breaking off and engulfing blocks of the country rock.

magmatic water [HYD] Water derived from or existing in molten igneous rock or magma. Also known as juvenile water.

magmatism [PETR] The formation of igneous rock from magma.

magn [ELECTROMAG] A unit of absolute permeability equal to 1 henry per meter; it has been proposed by the Soviet Union but has not won general acceptance.

magnafacies [GEOL] A major, continuous belt of deposits that is homogeneous in lithologic and paleontologic characteristics and that extends obliquely across time planes or through several time-stratigraphic units.

magnaflux method [MET] Trade name for magnetic particle test.

Magnamycin [MICROBIO] A trade name for the antibiotic carbomycin.

Magne-gage [MET] Trade name for a magnetic instrument used to measure thickness of nickel coatings on nonmagnetic metals such as brass, nonmagnetic coatings such as copper on steel, and magnetic coatings such as nickel on steel.

magnesia [INORG CHEM] Magnesium oxide that is processed for a particular purpose.

magnesia brick [MATER] A type of refractory brick composed of magnesium oxide with about 15% of other oxides. Also known as magnesite brick.

magnesia cement *See* magnesium oxychloride cement.

magnesia magma *See* milk of magnesia.

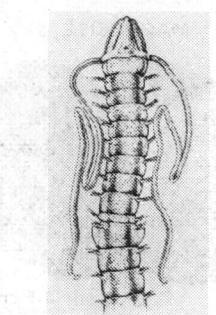

MAGELONIDAE

Anterior end of a species of *Magelona* in dorsal view.

MAGNESITE

Magnesite crystal in chlorite schist; the sample is from Tyrol, Austria. (Specimen from Department of Geology, Bryn Mawr College)

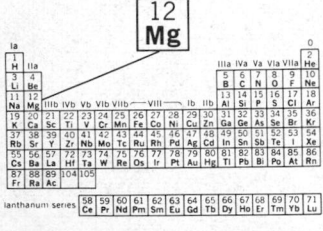

MAGNESIUM

Periodic table of the chemical elements showing the position of the element magnesium.

magnesia mixture [ANALY CHEM] Reagent used to analyze for phosphorus; consists of the filtered liquor from an aqueous mixture of ammonium chloride, magnesium sulfate, and ammonia.

magnesian calcite [MINERAL] $(Ca,Mg)CO_3$ A variety of calcite consisting of randomly substituted magnesium carbonate in a disordered calcite lattice. Also known as magnesium calcite.

magnesian limestone [PETR] Limestone with at least 90% calcite, a maximum of 10% dolomite, an approximate magnesium oxide equivalent of 1.1–2.1%, and an approximate magnesium carbonate equivalent of 2.3–4.4%.

magnesian marble [PETR] A type of magnesian limestone that has been metamorphosed; contains some dolomite. Also known as dolomitic marble.

magnesia refractory [MATER] Heat- and corrosion-resistant material made of magnesium oxide; used in cement or brick form to line high-temperature process vessels or furnaces.

Magnesil [MET] A magnetic alloy used in making cores for magnetic amplifiers, similar to Hypersod and Selectron.

magnesiochromite [MINERAL] $MgCr_2O_4$ A mineral of the spinel group composed of magnesium chromium oxide; it is isomorphous with chromite. Also known as magnochromite.

magnesiocopiapite [MINERAL] $MgFe_4(SO_4)_6(OH)_2 \cdot 20H_2O$ A mineral of the copiapite group composed of hydrous basic magnesium and iron sulfate; it is isomorphous with copiapite and cuprocopiapite.

magnesioferrite [MINERAL] $(Mg,Fe)Fe_2O_4$ Black, strongly magnetic mineral of the magnetite series in the spinel group. Also known as magnoferrite.

magnesite [MINERAL] $MgCO_3$ The mineral form of magnesium carbonate, usually massive and white, with hexagonal symmetry; specific gravity is 3, and hardness is 4 on Mohs scale. Also known as giobertite.

magnesite brick *See* magnesia brick.

magnesite wheel [ENG] A grinding wheel made with magnesium oxychloride as the bonding agent.

magnesium [CHEM] A metallic element, symbol Mg, atomic number 12, atomic weight 24.312. [MET] A silvery-white, lightweight, malleable, ductile metal, used in metallurgical and chemical processes, photography, pyrotechny, and light alloys.

magnesium acetate [ORG CHEM] $Mg(OOCCH_3)_2 \cdot 4H_2O$ or $Mg(OOCCH_3)_2$ A compound, forming colorless crystals that are soluble in water, and melt at 80°C; used in textile printing, in medicine as an antiseptic, and as a deodorant.

magnesium anode [ELEC] Bar of magnesium buried in the earth, connected to an underground cable to prevent cable corrosion due to electrolysis.

magnesium arsenate [INORG CHEM] $Mg_3(AsO_4)_2 \cdot xH_2O$ A white, poisonous, water-insoluble powder used as an insecticide.

magnesium benzoate [ORG CHEM] $Mg(C_7H_5O_2)_2 \cdot 3H_2O$ A crystalline white powder melting at 200°C; soluble in alcohol and hot water; used in medicine.

magnesium bomb [ORD] 1. An incendiary bomb in which the burning agent is magnesium. 2. A magnesium flare for use from aircraft.

magnesium borate [INORG CHEM] $3MgO \cdot B_2O_3$ Crystals that are white or colorless and transparent; soluble in alcohol and acids, slightly soluble in water; used as a fungicide, antiseptic, and preservative.

magnesium bromate [INORG CHEM] $Mg(BrO_3)_2 \cdot 6H_2O$ A white crystalline compound, insoluble in alcohol, soluble in water; a fire hazard; used as an analytical reagent.

magnesium bromide [INORG CHEM] $MgBr_2 \cdot 6H_2O$ Deliquescent, colorless, bitter-tasting crystals, melting at 172°C; soluble in alcohol, slightly soluble in alcohol; used in medicine and in the synthesis of organic chemicals.

magnesium calcite *See* magnesian calcite.

magnesium carbonate [INORG CHEM] $MgCO_3$ A water-insoluble, white powder, decomposing at about 350°C; used as a refractory material.

magnesium cell [ELEC] A primary cell in which the negative electrode is made of magnesium or one of its alloys.

magnesium chlorate [INORG CHEM] $Mg(ClO_3)_2 \cdot 6H_2O$ A white powder, bitter-tasting and hygroscopic; slightly soluble in alcohols, soluble in water; used in medicine.

magnesium chloride [INORG CHEM] $MgCl_2 \cdot 6H_2O$ Deliquescent white crystals; soluble in water and alcohol; used in disinfectants and fire extinguishers, and in ceramics, textiles, and paper manufacture.

magnesium–copper sulfide rectifier [ELECTR] Dry-disk rectifier consisting of magnesium in contact with copper sulfide.

magnesium dust [MET] Magnesium metal powder; flammable; used in photographic flash lights and pyrotechnics.

magnesium flare [ORD] A flare using magnesium as the illuminating agent.

magnesium fluoride [INORG CHEM] MgF_2 White, fluorescent crystals; insoluble in water and alcohol, soluble in nitric acid; melts at 1263°C; used in ceramics and glass. Also known as magnesium flux.

magnesium fluosilicate [INORG CHEM] $MgSiF_6 \cdot 6H_2O$ Water-soluble, efflorescent white crystals; used in ceramics, in mothproofing and waterproofing, and as a concrete hardener. Also known as magnesium silicofluoride.

magnesium flux *See* magnesium fluoride.

magnesium formate [ORG CHEM] $Mg(CHO_2)_2 \cdot 2H_2O$ Colorless, water-soluble crystals; insoluble in alcohol and ether; used in analytical chemistry and medicine.

magnesium gluconate [ORG CHEM] $Mg(C_6H_{11}O_7)_2 \cdot 2H_2O$ An odorless, tasteless, water-soluble powder; used in medicine.

magnesium halide [INORG CHEM] A compound formed from the metal magnesium and any of the halide elements; an example is magnesium bromide.

magnesium hydrate *See* magnesium hydroxide.

magnesium hydride [INORG CHEM] MgH_2 A hydride compound formed from the metal magnesium; it decomposes violently in water, and in a vacuum at about 280°C.

magnesium hydroxide [INORG CHEM] $Mg(OH)_2$ A white powder, very slightly soluble in water, decomposing at 350°C; used as an intermediate in extraction of magnesium metal, and as a reagent in the sulfite wood pulp process. Also known as magnesium hydrate.

magnesium hyposulfite *See* magnesium thiosulfate.

magnesium iodide [INORG CHEM] $MgI_2 \cdot 8H_2O$ Crystalline powder, white and deliquescent, discoloring in air; soluble in water, alcohol, and ether; used in medicine.

magnesium–iron mica *See* biotite.

magnesium lactate [ORG CHEM] $Mg(C_3H_5O_3)_2 \cdot 3H_2O$ Bitter-tasting, water-soluble white crystals; slightly soluble in alcohol; used in medicine.

magnesium lime [MATER] Lime containing more than 20% magnesium oxide; slakes more slowly, evolves less heat, expands less, sets more rapidly, and produces higher-strength mortars than does high-calcium quicklime.

magnesium–manganese dioxide cell [ELEC] Type of electrochemical (dry) cell battery in which the active elements are magnesium and manganese dioxide.

magnesium methoxide [ORG CHEM] $(CH_3O)_2Mg$ Colorless crystals that decompose when heated; used as a catalyst, dielectric coating, and cross-linking agent, and to form gels. Also known as magnesium methylate.

magnesium methylate *See* magnesium methoxide.

magnesium nitrate [INORG CHEM] $Mg(NO_3)_2 \cdot 6H_2O$ Deliquescent white crystals; soluble in alcohol and water; a fire hazard; used as an oxidizing material in pyrotechnics.

magnesium oleate [ORG CHEM] $Mg(C_{18}H_{33}O_2)_2$ Water-insoluble, yellowish mass; soluble in hydrocarbons, alcohol, and ether; used as a plasticizer lubricant and emulsifying agent, and in varnish driers and dry-cleaning solutions.

magnesium oxide [INORG CHEM] MgO A white powder that (depending on the method of preparation) may be light and fluffy, or dense; melting point 2800°C; insoluble in acids, slightly soluble in water; used in making refractories, and in cosmetics, pharmaceuticals, insulation, and medicine.

magnesium oxychloride cement [MATER] Cement made by adding a magnesium chloride solution to magnesia; used for interior flooring. Also known as magnesia cement.

magnesium perchlorate [INORG CHEM] $Mg(ClO_4)_2 \cdot 6H_2O$ White, deliquescent crystals; soluble in water and alcohol;

explosive when in contact with reducing materials; used as a drying agent for gases.

magnesium peroxide [INORG CHEM] MgO_2 A tasteless, odorless white powder; soluble in dilute acids, insoluble in water; a fire hazard; used as a bleaching and oxidizing agent, and in medicine.

magnesium phosphate [INORG CHEM] A compound with three forms: monobasic, $MgH_4(PO_4)_2 \cdot 2H_2O$, used in medicine and wood fireproofing; dibasic, $MgHPO_4 \cdot 3H_2O$, used in medicine and as a plastics stabilizer; tribasic, $Mg_3(PO_4)_2 \cdot 8H_2O$, used in dentifrices, as an adsorbent, and in pharmaceuticals.

magnesium salicylate [ORG CHEM] $Mg(C_7H_5O)_3 \cdot 4H_2O$ Efflorescent colorless crystals; soluble in water and alcohol; used in medicine.

magnesium silicate [INORG CHEM] $3MgSiO_3 \cdot 5H_2O$ White, water-insoluble powder, containing variable proportions of water of hydration; used as a filler for rubber and in medicine.

magnesium silicofluoride See magnesium fluosilicate.

magnesium–silver chloride cell [ELEC] A reserve primary cell that is activated by adding water; active elements are magnesium and silver chloride.

magnesium stearate [ORG CHEM] $Mg(C_{18}H_{35}O_2)_2$ Tasteless, odorless white powder; soluble in hot alcohol, insoluble in water; melts at 89°C; used in paints and medicine, and as a plastics stabilizer and lubricant. Also known as dolomol.

magnesium sulfite [INORG CHEM] $MgSO_3 \cdot 6H_2O$ A white, crystalline powder; insoluble in alcohol, slightly soluble in water; used in medicine and paper pulp.

magnesium thiosulfate [INORG CHEM] $MgS_2O_3 \cdot 6H_2O$ Colorless crystals that lose water at 170°C; used in medicine. Also known as magnesium hyposulfite.

magnesium trisilicate [INORG CHEM] $Mg_2Si_3O_8 \cdot 5H_2O$ A white, odorless, tasteless powder; insoluble in water and alcohol; used as an industrial odor absorbent and in medicine.

magnesium tungstate [INORG CHEM] $MgWO_4$ White crystals, insoluble in alcohol and water, soluble in acid; used in luminescent paint and for fluorescent x-ray screens.

magnesyn [ELEC] A portion of a repeater unit; a two-pole permanently magnetized rotor within a three-phase two-pole delta-connected stator which carries the indicating pointer and is free to rotate in any direction.

magnet [ELECTROMAG] A piece of ferromagnetic or ferrimagnetic material whose domains are sufficiently aligned so that it produces a net magnetic field outside itself and can experience a net torque when placed in an external magnetic field.

magnet alloy [MET] An alloy such as Alnico or Alcomax having strong magnetic properties; used in making permanent magnets.

magnet coil See iron-core coil.

magnet grate [MIN ENG] A series of magnetized bars used to trap and remove tramp iron from a flow of pulverized or granulated dry solids passing over the grate; used to protect crushing or grinding equipment.

magnetic [ELECTROMAG] Pertaining to magnetism or a magnet.

magnetically hard alloy [MET] A ferromagnetic alloy that can be permanently magnetized after the removal of an externally applied magnetic field.

magnetically soft alloy [MET] A ferromagnetic alloy which is capable of being magnetized upon application of an external magnetic field, but which returns to a nonmagnetic condition when the field is removed.

magnetic amplifier [ELECTR] A device that employs saturable reactors to modulate the flow of alternating-current electric power to a load in response to a lower-energy-level direct-current input signal. Abbreviated magamp. Also known as transductor.

magnetic amplitude [NAV] Angular distance of a celestial body north or south of the prime vertical circle and relative to magnetic east or west.

magnetic analysis inspection [MET] A nondestructive inspection method to determine the presence of variations in magnetic flux in ferromagnetic materials of constant cross section caused by defects, variations in hardness, discontinuities, or other irregularities.

magnetic anisotropy [ELECTROMAG] The dependence of the magnetic properties of some materials on direction.

magnetic annealing [MET] Annealing and cooling in a strong magnetic field.

magnetic annual change [GEOPHYS] The amount of secular change in the earth's magnetic field which occurs in 1 year. Also known as annual magnetic change.

magnetic annual variation [GEOPHYS] The small, systematic temporal variation in the earth's magnetic field which occurs after the trend for secular change has been removed from the average monthly values. Also known as annual magnetic variation.

magnetic anomaly detector [ELECTROMAG] A sensitive magnetometer carried at the end of a boom projecting from the tail of a patrol plane which can detect very small changes in the earth's magnetic field caused by a ferrous object, such as a submerged submarine; used to pinpoint a submarine's location, for effective deployment of suitable weapons. Abbreviated MAD.

magnetic azimuth [NAV] Azimuth relative to magnetic north.

magnetic bay [GEOPHYS] A small magnetic disturbance whose magnetograph resembles an indentation of a coastline; on earth, magnetic bays occur mainly in the polar regions and have a duration of a few hours.

magnetic bearing [NAV] Bearing relative to magnetic north, with the compass bearing corrected for deviation.

magnetic bias [ELECTROMAG] A steady magnetic field applied to the magnetic circuit of a relay or other magnetic device.

magnetic blowout [ELECTROMAG] **1.** A permanent magnet or electromagnet used to produce a magnetic field that lengthens the arc between opening contacts of a switch or circuit breaker, thereby helping to extinguish the arc. **2.** See blowout.

magnetic bottle [PL PHYS] A magnetic field used to confine or contain a plasma in controlled fusion experiments.

magnetic brake [MECH ENG] A friction brake under the control of an electromagnet.

magnetic card [ADP] A card with a magnetic surface on which data can be stored by selective magnetization.

magnetic cell [ELECTR] One unit of a magnetic memory, capable of storing one bit of information as a zero state or a one state.

magnetic character [ADP] A character printed with magnetic ink, as on bank checks, for reading by machines as well as by humans.

magnetic character figure See C index.

magnetic character sorter [ADP] A device that reads documents printed with magnetic ink; all data read are stored, and records are sorted on any required field. Also known as magnetic document sorter-reader.

magnetic charge See magnetic grenade.

magnetic chart [NAV] A chart showing the distribution of one of the magnetic elements, as by isogonic lines, or of its secular change.

magnetic chuck [MECH ENG] A chuck in which the workpiece is held by magnetic force.

magnetic circuit [ELECTROMAG] A group of magnetic flux lines each forming a closed path, especially when this circuit is regarded as analogous to an electric circuit because of the similarity of its magnetic field equations to direct-current circuit equations.

magnetic clutch See magnetic fluid clutch; magnetic friction clutch.

magnetic coercive force See coercive force.

magnetic compass [NAV] A compass depending for its directive force upon the attraction of the horizontal component of the earth's magnetic field for a magnetized needle or sensing element free to turn with a minimum of friction in any horizontal direction.

magnetic compass table See deviation table.

magnetic confinement [PL PHYS] The containment of a plasma within a region of space by the forces of magnetic fields on the charged particles in the gas.

magnetic cooling See adiabatic demagnetization.

magnetic core Also known as core. [ELECTR] A configura-

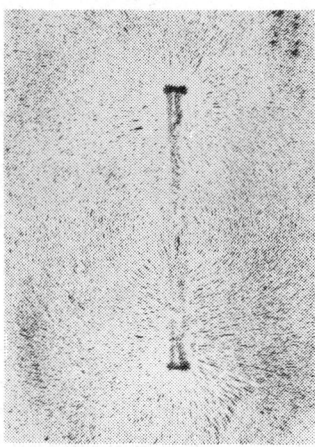

MAGNET

Photograph of the iron-filing map of the magnetic field of a permanent bar magnet. Note that the magnetic flux lines can be traced by the lines of iron filings, which act like tiny compass needles.

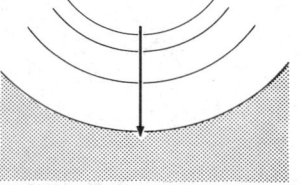

MAGNETIC CONFINEMENT

Stable magnetic confinement by a convex magnetic field. The lines of force are convex to the plasma; arrow indicates decrease of magnetic field strength.

tion of magnetic material, usually a mixture of iron oxide or ferrite particles mixed with a binding agent and formed into a tiny doughnutlike shape, that is placed in a spatial relationship to current-carrying conductors, and is used to maintain a magnetic polarization for the purpose of storing data, or for its nonlinear properties as a logic element. Also known as memory core. [ELECTROMAG] A quantity of ferrous material placed in a coil or transformer to provide a better path than air for magnetic flux, thereby increasing the inductance of the coil and increasing the coupling between the windings of a transformer.

magnetic core multiplexer [ADP] A device which channels many bit inputs into a single output.

magnetic core storage [ADP] A computer storage system in which each of thousands of magnetic cores stores one bit of information; current pulses are sent through wires threading through the cores to record or read out data. Also known as core memory; core storage.

magnetic coupling [ELECTROMAG] For a pair of particles or systems, the effect of the magnetic field created by one system on the magnetic moment or angular momentum of the other.

magnetic course [NAV] Course relative to magnetic north, with the compass course corrected for deviation.

magnetic Curie temperature [SOLID STATE] The temperature below which a magnetic material exhibits ferromagnetism, and above which ferromagnetism is destroyed and the material is paramagnetic.

magnetic cutter [ENG ACOUS] A cutter in which the mechanical displacements of the recording stylus are produced by the action of magnetic fields.

magnetic daily variation See magnetic diurnal variation.

magnetic damping [ELECTROMAG] Damping of a mechanical motion by means of the reaction between a magnetic field and the current generated by the motion of a coil through the magnetic field.

magnetic declination [GEOPHYS] The deviation of a magnetized compass needle from geographic north; it is reckoned positive to the east and negative to the west.

magnetic deflection [ELECTR] Deflection of an electron beam by the action of a magnetic field, as in a television picture tube.

magnetic delay line [ELECTR] Delay line, used for the storage of data in a computer, consisting essentially of a metallic medium along which the velocity of the propagation of magnetic energy is small compared to the speed of light; storage is accomplished by the recirculation of wave patterns containing information, usually in binary form.

magnetic deviation [NAV] The angle between the magnetic meridian and the axis of a compass card, expressed in degress east or west to indicate in which direction the end of the compass card is offset from magnetic north.

magnetic diffusivity [ELECTROMAG] A measure of the tendency of a magnetic field to diffuse through a conducting medium at rest; it is equal to the partial derivative of the magnetic field strength wth respect to time divided by the Laplacian of the magnetic field, or to the reciprocal of $4\pi\mu\sigma$, where μ is the magnetic permeability and σ is the conductivity in electromagnetic units.

magnetic dip See inclination.

magnetic dipole [ELECTROMAG] An object, such as a permanent magnet, current loop, or particle with angular momentum, which experiences a torque in a magnetic field, and itself gives rise to a magnetic field, as if it consisted of two magnetic poles of oposite sign separated by a small distance.

magnetic dipole antenna [ELECTROMAG] Simple loop antenna capable of radiating an electromagnetic wave in response to a circulation of electric current in the loop.

magnetic dipole density See magnetization.

magnetic dipole moment [ELECTROMAG] A vector associated with a magnet, current loop, particle, or such, whose cross product with the magnetic induction (or alternatively, the magnetic field strength) of a magnetic field is equal to the torque exerted on the system by the field. Also known as dipole moment; magnetic moment.

magnetic direction [GEOD] Horizontal direction expressed as angular distance from magnetic north.

magnetic disk See disk.

magnetic displacement See magnetic induction.

magnetic diurnal variation [GEOPHYS] Oscillations of the earth's magnetic field which have a periodicity of about a day and which depend to a close approximation only on local time and geographic latitude. Also known as magnetic daily variation.

magnetic document sorter-reader See magnetic character sorter.

magnetic domain See ferromagnetic domain.

magnetic double refraction [OPTICS] The double refraction of light passing through certain substances when the substance is placed in a transverse magnetic field.

magnetic drag dynamometer See eddy-current brake.

magnetic drum See drum.

magnetic drum receiving equipment [ELECTR] Radar developed for detection of targets beyond line of sight using ionospheric reflection and very low power.

magnetic drum storage See drum.

magnetic earphone [ENG ACOUS] An earphone in which variations in electric current produce variations in a magnetic field, causing motion of a diaphragm.

magnetic element [ENG] That part of an instrument producing or influenced by magnetism. [GEOPHYS] Magnetic declination, dip, or intensity at any location on the surface of the earth.

magnetic equator [GEOPHYS] That line on the surface of the earth connecting all points at which the magnetic dip is zero. Also known as aclinic line; geomagnetic equator.

magnetic field [ELECTROMAG] 1. One of the elementary fields in nature; it is found in the vicinity of a magnetic body or current-carrying medium and, along with electric field, in a light wave; charges moving through a magnetic field experience the Lorentz force. 2. See magnetic field strength.

magnetic field intensity See magnetic field strength.

magnetic field strength [ELECTROMAG] An auxiliary vector field, used in describing magnetic phenomena, whose curl, in the case of static charges and currents, equals (in meter-kilogram-second units) the free current density vector, independent of the magnetic permeability of the material. Also known as magnetic field; magnetic field intensity; magnetic force; magnetic intensity; magnetizing force.

magnetic film See magnetic thin film.

magnetic filter [CHEM ENG] Filtration device in which the filter screen is magnetized to trap and remove fine iron from liquids or liquid suspensions being filtered.

magnetic firing circuit [ELECTR] A type of firing circuit in which the capacitor is discharged through the igniter by saturating a reactor, which is connected in series with the capacitor; often used in ignitron rectifiers to obtain longer life and greater reliability than is possible with thyratron firing tubes.

magnetic flaw detector [ELECTROMAG] A flaw detector in which a ferrous object is magnetized with an electromagnet or permanent magnet and sprayed with magnetic particles or a solution containing fine suspended magnetic particles which outline surface or near-surface flaws.

magnetic fluid [MATER] A mixture of iron particles in oil or other liquid; viscosity increases sharply in a strong magnetic field.

magnetic fluid clutch [MECH ENG] A friction clutch that is engaged by magnetizing a liquid suspension of powdered iron located between pole pieces mounted on the input and output shafts. Also known as magnetic clutch.

magnetic flux [ELECTROMAG] 1. The integral over a specified surface of the component of magnetic induction perpendicular to the surface. 2. See magnetic lines of force.

magnetic flux density See magnetic induction.

magnetic focusing [ELECTROMAG] Focusing a beam of electrons or other charged particles by using the action of a magnetic field.

magnetic force See magnetic field strength.

magnetic force parameter [PL PHYS] A dimensionless number used in magnetofluid dynamics, equal to the product of the square of the magnetic permeability, the square of the magnetic field strength, the electrical conductivity, and a characteristic length, divided by the product of the mass density and the fluid velocity. Symbolized N.

magnetic-force welding [MET] A welding process in which the mechanical force is exerted by a magnetic field.

magnetic forming [MET] The forming of metal into desired shapes by using strong magnetic fields, produced by charging a large capacitor bank and then discharging it into an induction coil in less than 10^{-6} second, to push the metal against a forming die.

magnetic friction clutch [MECH ENG] A friction clutch in which the pressure between the friction surfaces is produced by magnetic attraction. Also known as magnetic clutch.

magnetic gap [ELECTROMAG] The space between a magnet's pole faces.

magnetic grenade [ORD] Small explosive charge with attached magnets, designed to be thrown or placed against tanks; the magnets hold the explosive in place until detonated by a time fuse. Also known as magnetic charge.

magnetic groups *See* Shubnikov groups.

magnetic hardness comparator [ENG] A device for checking the hardness of steel parts by placing a unit of known proper hardness within an induction coil; the unit to be tested is then placed within a similar induction coil, and the behavior of the induction coils compared; if the standard and test units have the same magnetic properties, the hardness of the two units is considered to be the same.

magnetic head [ELECTR] The electromagnet used for reading, recording, or erasing signals on a magnetic disk, drum, or tape. Also known as magnetic read/write head.

magnetic heading [NAV] Heading relative to magnetic north, with the compass heading corrected for deviation.

magnetic hysteresis [ELECTROMAG] Lagging of changes in the magnetization of a substance behind changes in the magnetic field as the magnetic field is varied. Also known as hysteresis.

magnetic induction [ELECTROMAG] A vector quantity that is used as a quantitative measure of magnetic field; the force on a charged particle moving in the field is equal to the particle's charge times the cross product of the particle's velocity with the magnetic induction (mks units). Also known as magnetic displacement; magnetic flux density; magnetic vector.

magnetic induction gyroscope [ENG] A gyroscope without moving parts, in which alternating- and direct-current magnetic fields act on water doped with salts which exhibit nuclear paramagnetism.

magnetic ink [MATER] Ink containing magnetic particles to permit reading of printed characters by a magnetic character reader as well as by humans.

magnetic-ink character recognition [ADP] That branch of character recognition which involves the sensing of magnetic-ink characters for the purpose of determining the character's most probable identity. Abbreviated MICR.

magnetic inspection oil [MET] A light petroleum oil, such as kerosine or naphtha, to which has been added fine ferromagnetic particles (usually colored black or red for contrast) to form an inspection penetrant; when the penetrant is applied to a metal surface being inspected, the ferrous particles accumulate in any surface cracks by magnetic attraction, thereby permitting the cracks to be discernible.

magnetic inspection paste [MET] A paste containing ferromagnetic particles designed to be added to a light distilled petroleum oil, such as kerosine or naphtha, to form an inspection penetrant; when the inspection penetrant is applied to a metal, the ferrous particles accumulate in any surface cracks by magnetic attraction, thereby permitting the cracks to be discerned.

magnetic inspection powder [MET] A dry powder containing ferromagnetic particles colored gray, black, or red for contrast, designed to be dusted on metal parts being inspected by a magnetic inspection machine; the ferrous powder accumulates in any surface cracks (flaws) by magnetic attraction, thereby permitting the cracks to be readily discerned; if the ferrous particles are fluorescent, surface cracks will be brilliantly illuminated under black light.

magnetic intensity *See* magnetic field strength.

magnetic iron ore *See* magnetite.

magnetic latitude [GEOPHYS] Angular distance north or south of the magnetic equator.

magnetic leakage [ELECTROMAG] Passage of magnetic flux outside the path along which it can do useful work.

magnetic lens [ELECTROMAG] A magnetic field with axial symmetry, capable of converging beams of charged particles of uniform velocity and of forming images of objects placed in the path of such beams; the field may be produced by solenoids, electromagnets, or permanent magnets. Also known as lens.

magnetic lines of flux *See* magnetic lines of force.

magnetic lines of force [ELECTROMAG] Lines used to represent the magnetic induction in a magnetic field, selected so that they are parallel to the magnetic induction at each point, and so that the number of lines per unit area of a surface perpendicular to the induction is equal to the induction. Also known as magnetic flux; magnetic lines of flux.

magnetic local anomaly [GEOPHYS] A localized departure of the geomagnetic field from its average over the surrounding area.

magnetic loudspeaker [ENG ACOUS] Loudspeaker in which acoustic waves are produced by mechanical forces resulting from magnetic reactions. Also known as magnetic speaker.

magnetic Mach number [PL PHYS] A dimensionless number equal to the ratio of the velocity of a fluid to the velocity of Alfven waves in the fluid. Symbolized M_{Ma}.

magnetic material [ELECTROMAG] A material exhibiting ferromagnetism.

magnetic memory *See* magnetic storage.

magnetic-memory plate [ELECTR] Magnetic memory consisting of a ferrite plate having a grid of small holes through which the read-in and read-out wires are threaded; printed wiring may be applied directly to the plate in place of conventionally threaded wires, permitting mass production of plates having a high storage capacity.

magnetic meridian [GEOPHYS] A line which is at any point in the direction of horizontal magnetic force of the earth; a compass needle without deviation lies in the magnetic meridian.

magnetic microphone [ENG ACOUS] A microphone consisting of a diaphragm acted upon by sound waves and connected to an armature which varies the reluctance in a magnetic field surrounded by a coil. Also known as reluctance microphone; variable-reluctance microphone.

magnetic mine [ORD] An underwater mine intended to be detonated when the hull of a passing vessel causes a change in the magnetic field at the mine.

magnetic mine detector [ORD] An electrical device for the detection and indication of metallic mines buried in the earth or on the surface of land or water.

magnetic mirror [PL PHYS] A magnetic field used in controlled-fusion experiments to reflect charged particles into the central region of a magnetic bottle; reflection occurs in the region where the magnetic field increases abruptly in strength.

magnetic modulator [ELECTR] A modulator in which a magnetic amplifier serves as the modulating element for impressing an intelligence signal on a carrier.

magnetic moment *See* magnetic dipole moment.

magnetic monopole [ELECTROMAG] A hypothetical particle carrying magnetic charge; it would be a source for magnetic field in the same way that a charged particle is a source for electric field. Also known as monopole.

magnetic multipole [ELECTROMAG] One of a series of types of static or oscillating distributions of magnetization, which is a magnetic multipole of order 2; the electric and magnetic fields produced by a magnetic multipole of order 2^n are equivalent to those of two magnetic multipoles of order 2^{n-1} of equal strength but opposite sign, separated from each other by a short distance.

magnetic multipole field [ELECTROMAG] The electric and magnetic fields generated by a static or oscillating magnetic multipole.

magnetic needle [ELECTROMAG] 1. A bar magnet or collection of bar magnets which is hung so as to show the direction of the magnetic field. 2. In particular, a slender bar magnet, pointed at both ends, that is pivoted or freely suspended in a magnetic compass.

magnetic north [GEOPHYS] At any point on the earth's sur-

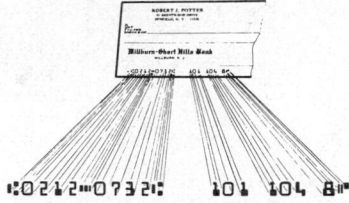

MAGNETIC-INK CHARACTER RECOGNITION

E-13B font in magnetic ink on the bottom of a typical bank check to permit electronic processing.

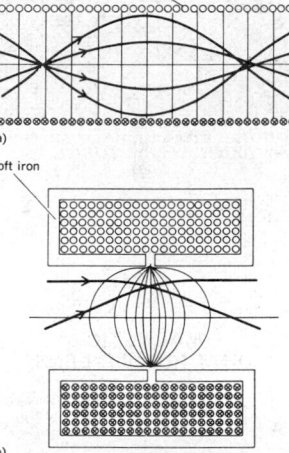

MAGNETIC LENS

Two types of magnetic lens. (a) Uniform magnetic field. (b) Short magnetic lens formed at gap in soft-iron casing about coil. Arrows show direction of movement of a beam of charged particles. (*From E. G. Ramberg and G. A. Morton, J. Appl. Phys., vol. 10, 1939, and J. Hillier and E. G. Ramberg, J. Appl. Phys., vol. 18, 1947*)

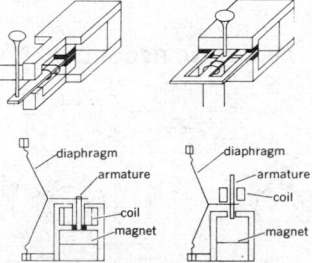

MAGNETIC MICROPHONE

Two magnetic microphones of the balanced-armature type, perspective and sectional views. (*From H. F. Olson, Acoustical Engineering, Van Nostrand, 1957*)

face, the horizontal direction of the earth's magnetic lines of force (direction of a magnetic meridian) toward the north magnetic pole; a particular direction indicated by the needle of a magnetic compass.

magnetic nuclear resonance *See* nuclear magnetic resonance.

magnetic number [PL PHYS] A dimensionless number used in magnetofluid dynamics, equal to the square root of the magnetic force parameter. Symbolized R_M.

magnetic observatory [GEOPHYS] A geophysical measuring station employing some form of magnetometer to measure the intensity of the earth's magnetic field.

magnetic octupole moment [ELECTROMAG] A quantity characterizing a distribution of magnetization; obtained by integrating the product of the divergence of the magnetization, the third power of the distance from the origin, and a spherical harmonic Y^*_{3m} over the magnetization distribution.

magnetic oscillograph [ELECTROMAG] An instrument that records a trace measuring one component of the earth's magnetic field.

magnetic Oseen number [PL PHYS] A dimensionless number used in magnetofluid dynamics, equal to $\frac{1}{2}(1 - N_{AL}^2)R_M$, where N_{AL} is the Alfvén number, and R_M is the magnetic number. Symbolized k.

magnetic-particle test [MET] A nondestructive test to determine the existence and extent of macrodefects such as cracks in ferromagnetic materials; discontinuities in the material create variations of magnetic field which are outlined by fine magnetic particles.

magnetic pendulum [ELECTROMAG] A bar magnet which is hung by a thread or balanced on a pivot so that it oscillates in a horizontal plane when disturbed and released in a magnetic field having a horizontal component.

magnetic permeability *See* permeability.

magnetic pickup *See* variable-reluctance pickup.

magnetic pinch *See* pinch effect.

magnetic polarization *See* intrinsic induction.

magnetic pole [ELECTROMAG] **1.** One of two regions located at the ends of a magnet that generate and respond to magnetic fields in much the same way that electric charges generate and respond to electric fields. Also known as pole. **2.** A particle which generates and responds to magnetic fields in exactly the same way that electric charges generate and respond to electric fields; the particle probably does not have physical reality, but it is often convenient to imagine that a magnetic dipole consists of two magnetic poles of opposite sign, separated by a small distance. [GEOPHYS] In geomagnetism, either of the two points on the earth's surface where the magnetic meridians converge, that is, where the magnetic field is vertical. Also known as dip pole.

magnetic pole strength [ELECTROMAG] The magnitude of a (fictional) magnetic pole, equal to the force exerted on the pole divided by the magnetic induction (or, alternatively, by the magnetic field strength). Also known as pole strength.

magnetic potential *See* magnetic scalar potential.

magnetic potentiometer [ENG] Instrument that measures magnetic potential differences.

magnetic pressure [PL PHYS] A function, proportional to the square of the magnetic induction, such that the force exerted by a magnetic field on an electrically conducting fluid (excluding the force associated with curvature of magnetic flux lines) is the same as the force that would be exerted by a hydrostatic pressure equal to this function.

magnetic pressure transducer [ENG] A type of pressure transducer in which a change of pressure is converted into a change of magnetic reluctance or inductance when one part of a magnetic circuit is moved by a pressure-sensitive element, such as a bourdon tube, bellows, or diaphragm.

magnetic prime vertical [GEOPHYS] The vertical circle through the magnetic east and west points of the horizon.

magnetic printing [ELECTR] The permanent and usually undesired transfer of a recorded signal from one section of a magnetic recording medium to another when these sections are brought together, as on a reel of tape. Also known as crosstalk; magnetic transfer.

magnetic probe [ELECTROMAG] A small coil inserted in a magnetic field to measure changes in field strength.

magnetic profile [GEOPHYS] A profile of a geologic structure showing magnetic anomalies.

magnetic prospecting [ENG] Carrying out airborne or ground surveys of variations in the earth's magnetic field, using a magnetometer or other equipment, to locate magnetic deposits of iron, nickel, or titanium, or nonmagnetic deposits which either contain magnetic gangue minerals or are associated with magnetic structures.

magnetic pulley [ENG] Magnetized pulley device for a conveyor belt; removes tramp iron from dry products being moved by the belt.

magnetic pumping [ELECTROMAG] A method of moving a conducting liquid by applying a magnetic field which varies with time. [PL PHYS] A method of heating a plasma to a high ion temperature by applying an oscillating electromagnetic field.

magnetic quadrupole lens [ELECTROMAG] A magnetic field generated by four magnetic poles of alternating sign arranged in a circle; used to focus beams of charged particles in devices such as electron microscopes and particle accelerators.

magnetic quantum number [ATOM PHYS] The eigenvalue of the component of an angular momentum operator in a specified direction, such as that of an applied magnetic field, in units of Planck's constant divided by 2π.

magnetic read/write head *See* magnetic head.

magnetic recorder [ELECTR] An instrument that records information, generally in the form of audio-frequency or digital signals, on magnetic tape or magnetic wire as magnetic variations in the medium.

magnetic recording [ELECTR] Recording by means of a signal-controlled magnetic field.

magnetic refrigerator [CRYO] A device for keeping substances cooled to about 0.2K, in which a working substance consisting of a paramagnetic salt undergoes a cycle of processes which approximates a Carnot cycle between a high-temperature reservoir consisting of a liquid-helium bath at 1.2K and a low-temperature reservoir consisting of the substance to be cooled, and isentropic cooling of the working substance is accomplished by demagnetization.

magnetic relaxation [PHYS] The approach of a magnetic system to an equilibrium or steady-state condition, over a period of time.

magnetic reluctance *See* reluctance.

magnetic reluctivity *See* reluctivity.

magnetic reproducer [ELECTR] An instrument which moves a magnetic recording medium, such as a tape, wire, or disk, past an electromagnetic transducer that converts magnetic signals on the medium into electric signals.

magnetic reproducing [ELECTR] The conversion of information on magnetic tape or magnetic wire, which was originally produced by electric signals, back into electric signals.

magnetic resonance [PHYS] A phenomenon exhibited by the magnetic spin systems of certain atoms whereby the spin systems absorb energy at specific (resonant) frequencies when subjected to magnetic fields alternating at frequencies which are in synchronism with natural frequencies of the system. Also known as spin resonance.

magnetic Reynolds number [PL PHYS] A dimensionless number used to compare the transport of magnetic lines of force in a conducting fluid to the leakage of such lines from the fluid, equal to a characteristic length of the fluid times the fluid velocity, divided by the magnetic diffusivity. Symbolized R_M.

magnetic rigidity [ELECTROMAG] A measure of the momentum of a particle moving perpendicular to a magnetic field, equal to the magnetic induction times the particle's radius of curvature. [PL PHYS] The existence of restoring forces which resist displacements of a conducting fluid when a magnetic field is present.

magnetic rotation *See* Faraday effect.

magnetic rubber [MATER] Synthetic rubber to which magnetic metal powder is added; produced in sheets or strips.

magnetics [ELECTROMAG] The study of magnetic phenomena, comprising magnetostatics and electromagnetism.

magnetic saturation [ELECTROMAG] The condition in which, after a magnetic field strength becomes sufficiently large, further increase in the magnetic field strength produces no

---

MAGNETIC PROSPECTING

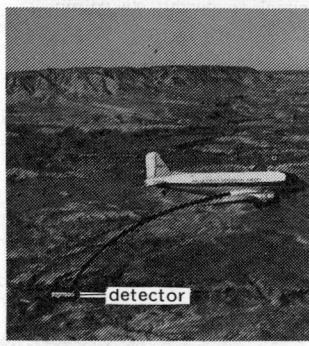

Airborne magnetometer survey over rough terrain. *(USGS)*

MAGNETIC QUADRUPOLE LENS

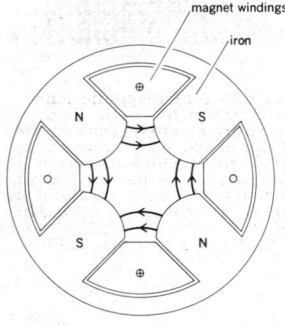

Section normal to the lens axis of magnetic quadrupole lens.

MAGNETIC RECORDER

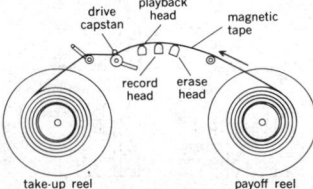

Elements of a typical magnetic tape recording and reproducing system. *(After H. F. Olson, Acoustical Engineering, Van Nostrand, 1957)*

additional magnetization in a magnetic material. Also known as saturation.

magnetic scalar potential [ELECTROMAG] The work which must be done against a magnetic field to bring a magnetic pole of unit strength from a reference point (usually at infinity) to the point in question. Also known as magnetic potential.

magnetic scanning [SPECT] The magnetic field sorting of ions into their respective spectrums for analysis by mass spectroscopy; accomplished by varying the magnetic field strength while the electrostatic field is held constant.

magnetic scattering [PHYS] Scattering of neutrons as a result of the interaction of the magnetic moment of the neutron with the magnetic moments of atoms or other particles.

magnetic secular change [GEOPHYS] The gradual variation in the value of a magnetic element which occurs over a period of years.

magnetic separator [ENG] A machine for separating magnetic from less magnetic or nonmagnetic materials by using strong magnetic fields; used for example, in tramp iron removal, or concentration and purification.

magnetic shell [ELECTROMAG] Two layers of magnetic charge of opposite sign, separated by an infinitesimal distance.

magnetic shielding See magnetostatic shielding.

magnetic shift register [ADP] A shift register in which the pattern of settings of a row of magnetic cores is shifted one step along the row by each new input pulse; diodes in the coupling loops between cores prevent backward flow of information.

magnetic shunt [ELECTROMAG] Piece of iron, usually adjustable as to position, used to divert a portion of the magnetic lines of force passing through an air gap in an instrument or other device.

magnetic sound track [ENG ACOUS] A magnetic tape, attached to a motion picture film, on which a sound recording is made.

magnetic spark chamber [NUCLEO] A spark chamber in a magnetic field up to 20,000 gauss, in which the sign of the charge and the momentum of charged particles · can be measured by measuring the curvature of their trajectories.

magnetic speaker See magnetic loudspeaker.

magnetic spectrograph [NUCLEO] A magnetic spectrometer that provides a permanent record of the distribution of intensity versus momentum of a beam of charged particles.

magnetic spectrometer [NUCLEO] A device for measuring the momentum of charged particles, or their distribution of intensity versus momentum, by passing the particles through a magnetic field which bends their paths in proportion to their momentum.

magnetic station [GEOPHYS] A facility equipped with instruments for measuring local variations in the earth's magnetic field.

magnetic stepping motor See stepper motor.

magnetic storage [ADP] A device utilizing magnetic properties of materials to store data; may be roughly divided into two categories, moving (drum, disk, tape) and static (core, thin film). Also known as magnetic memory.

magnetic storm [GEOPHYS] A worldwide disturbance of the earth's magnetic field; frequently characterized by a sudden onset, in which the magnetic field undergoes marked changes in the course of an hour or less, followed by a very gradual return to normalcy, which may take several days.

magnetic strain energy [SOLID STATE] The potential energy of a magnetic domain, subject to both a tensile stress and a magnetic field, associated with the domain's magnetostriction expansion.

magnetic stratigraphy See paleomagnetic stratigraphy.

magnetic stress [PL PHYS] The force which acts across a surface in a conducting fluid because of curving or stretching of magnetic flux lines.

magnetic stress tensor [PL PHYS] A second-rank tensor, proportional to the dyad product of the magnetic induction with itself, whose divergence gives that part of the force of a magnetic field on a unit volume of conducting fluid which is due to curvature or stretching of magnetic flux lines.

magnetic striped ledger [ADP] A ledger sheet used on a special typing device which stores the coded data on a

magnetic strip on the sheet while typing out the data on the sheet; the magnetic strip can be read directly by a special reader linked to a computer.

magnetic survey [GEOPHYS] **1.** Magnetometer map of variations in the earth's total magnetic field; used in petroleum exploration to determine basement-rock depths and geologic anomalies. **2.** Measurement of a component of the geomagnetic field at different locations.

magnetic susceptibility [ELECTROMAG] The ratio of the magnetization of a material to the magnetic field strength; it is a tensor when these two quantities are not parallel; otherwise it is a simple number. Also known as susceptibility.

magnetic tape [ELECTR] A plastic, paper, or metal tape that is coated or impregnated with magnetizable iron oxide particles; used in magnetic recording.

magnetic-tape core [ELECTR] Toroidal core formed by winding a strip of thin magnetic-core material around a form.

magnetic tape file operation [ADP] All the jobs related to creating, sorting, inputting, and maintenance of magnetic tapes in a magnetic tape environment.

magnetic tape librarian [ADP] Routine which provides a computer the means to automatically run a sequence of programs.

magnetic tape master file [ADP] A magnetic tape consisting of a set of related elements such as is found in a payroll, an inventory, or an accounts receivable; a master file is, as a rule, periodically updated.

magnetic tape parity [ADP] A check performed on the data bits on a tape; usually an odd (or even) condition is expected and the occurrence of the wrong parity indicates the presence of an error.

magnetic tape reader [ELECTR] A computer device that is capable of reading information recorded on magnetic tape by transforming this information into electric pulses.

magnetic tape station [ADP] On-line device that provides write, read, and erase data on magnetic tape to permit high-speed storage of data.

magnetic tape storage [ADP] Storage of binary information on magnetic tape, generally on 5 to 10 tracks, with up to several thousand bits per inch (more than a thousand bits per centimeter) on each track.

magnetic tape switching unit [ADP] A device which permits the computer operator to bring into play any number of tape drives as required by the system.

magnetic tape terminal [ADP] Device which converts pulses in series to pulses in parallel while checking for bit parity prior to the entry in buffer storage.

magnetic tape unit [ADP] A computer unit that usually consists of a tape transport, reading and recording heads, and associated electric and electronic equipment.

magnetic temporal variation [GEOPHYS] Any change in the earth's magnetic field which is a function of time.

magnetic test coil See exploring coil.

magnetic thermometer [SOLID STATE] A sample of a paramagnetic salt whose magnetic susceptibility is measured and whose temperature is then calculated from the inverse relationship between the two quantities; useful at temperatures below about 1K.

magnetic thin film [SOLID STATE] A sheet or cylinder of magnetic material less than 5 micrometers thick, usually possessing uniaxial magnetic anisotropy; used mainly in computer storage and logic elements. Also known as ferromagnetic film; magnetic film.

magnetic track [NAV] The direction of the track relative to magnetic north.

magnetic transducer [ELECTROMAG] A device for transforming mechanical into electrical energy, which consists of a magnetic field including a variable-reluctance path and a coil surrounding all or a part of this path, so that variation in reluctance leads to a variation in the magnetic flux through the coil and a corresponding induced emf (electromotive force).

magnetic transfer See magnetic printing.

magnetic variation [GEOPHYS] Small changes in the earth's magnetic field in time and space.

magnetic vector See magnetic induction.

magnetic vector potential See vector potential.

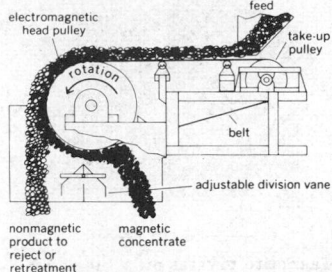

MAGNETIC SEPARATOR

Belt-type magnetic cobber. (Stearns Magnetics Inc.)

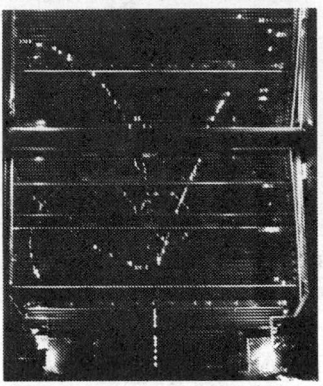

MAGNETIC SPARK CHAMBER

A photograph taken through a pair of cylindrical field lenses of an event in a narrow-gap spark-chamber system in a magnetic field. (From A. Roberts, Spark chambers, Encyclopaedic Dictionary of Physics, Pergamon Press, 1962–1964)

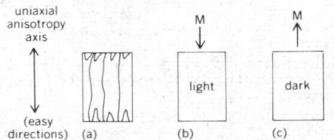

MAGNETIC THIN FILM

Typical domain patterns in a thin film viewed by use of the Kerr magnetooptic technique. (a) Magnetic film demagnetized. (b) Magnetization pointing downward. (c) Magnetization pointing upward.

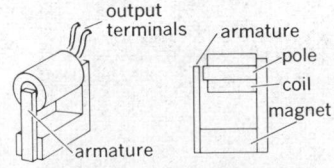

MAGNETIC TRANSDUCER

Perspective and sectional view of a single-pole and armature magnetic transducer.

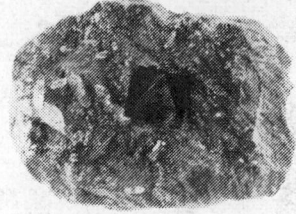

MAGNETITE

Magnetite crystal on a block of mica schist with small quartz crystals on left. (*American Museum of Natural History Specimen*)

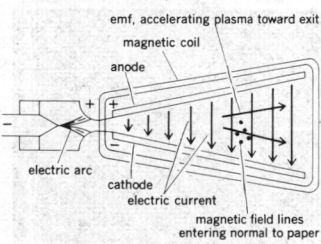

MAGNETOHYDRODYNAMIC ARCJET

Sectional view through a magnetohydrodynamic arc engine showing the various components.

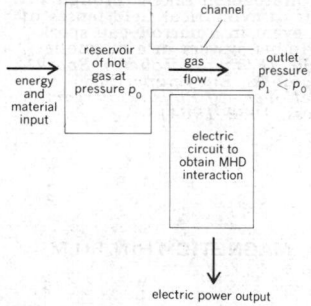

MAGNETOHYDRODYNAMIC GENERATOR

Schematic drawing showing how the generator produces electric power.

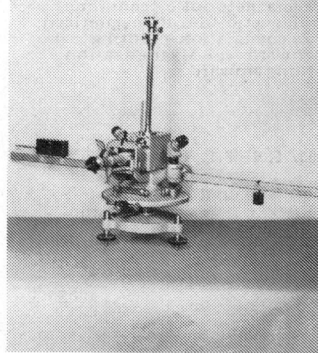

MAGNETOMETER

Observatory-type magnetometer, capable of accuracy of 3–4 γ. (*U.S. Coast and Geodetic Survey*)

magnetic viscosity [PL PHYS] The effect, possessed by a magnetic field in the absence of sizable mechanical forces or electric fields, of damping motions of a conducting fluid perpendicular to the field similar to ordinary viscosity.

magnetic wave [SOLID STATE] The spread of magnetization from a small portion of a substance where an abrupt change in the magnetic field has taken place.

magnetic wind direction [METEOROL] The direction, with respect to magnetic north, from which the wind is blowing; distinguished from true wind direction.

magnetic wire [MATER] A wire made from magnetic material suitable for magnetic recording.

magnetism [PHYS] Phenomena involving magnetic fields and their effects upon materials.

magnetite [MINERAL] An opaque iron-black and streak-black isometric mineral and member of the spinel structure type, usually occurring in octahedrals or in granular to massive form; hardness is 6 on Mohs scale, and specific gravity is 5.20. Also known as magnetic iron ore; octahedral iron ore.

magnetization [ELECTROMAG] 1. The property and in particular, the extent of being magnetized; quantitatively, the magnetic moment per unit volume of a substance. Also known as magnetic dipole density; magnetization intensity. 2. The process of magnetizing a magnetic material.

magnetization curve *See* B-H curve; normal magnetization curve.

magnetization intensity *See* magnetization.

magnetizing current [ELEC] The current that flows through the primary winding of a power transformer when no loads are connected to the secondary winding; this current established the magnetic field in the core and furnished energy for the no-load power losses in the core. Also known as exciting current.

magnetizing force *See* magnetic field strength.

magnet keeper *See* keeper

magneto [ELEC] An alternating-current generator that uses one or more permanent magnets to produce its magnetic field; frequently used as a source of ignition energy on tractor, marine, industrial, and aviation engines. Also known as magnetoelectric generator.

magnetoacoustics [PHYS] The study of the effects of magnetic fields on acoustical phenomena, such as various oscillations in the attenuation of ultrasonic sound waves by a crystal placed in a magnetic field at a very low temperature, as the magnetic field strength or sound frequency is varied.

magnetoaerodynamics [PL PHYS] Study of the properties and characteristics of, and the forces exerted by, highly ionized air and other gases; applied principally to study of reentering ballistic missiles and spacecraft.

magneto anemometer [ENG] A cup anemometer with its shaft mechanically coupled to a magnet; both the frequency and amplitude of the voltage generated are proportional to the wind speed, and may be indicated or recorded by suitable electrical instruments.

magnetocaloric effect [THERMO] The reversible change of temperature accompanying the change of magnetization of a ferromagnetic material.

magnetochemistry [PHYS CHEM] A branch of chemistry which studies the interrelationship between the bulk magnetic properties of a substance and its atomic and molecular structure.

magnetoelastic coupling [SOLID STATE] The interaction between the magnetization and the strain of a magnetic material.

magnetoelasticity [SOLID STATE] Phenomenon in which an elastic strain alters the magnetization of a ferromagnetic material.

magnetoelectric generator *See* magneto.

magnetoelectricity [ELECTROMAG] Magnetic techniques for generating voltages, such as in an ordinary generator. [SOLID STATE] The appearance of an electric field in certain substances, such as chromic oxide (Cr_2O_3), when they are subjected to a static magnetic field.

magnetoencephalography [MED] A method to detect the brain's electrical activity with a superconducting magnet.

magnetofluid [PHYS CHEM] A Newtonian or shear-thinning fluid whose flow properties become viscoplastic when it is modulated by a magnetic field.

magnetofluid dynamics [PHYS] 1. The study of the motion of an electrically conducting metal, such as mercury, in the presence of electric and magnetic fields. 2. *See* magnetohydrodynamics.

magnetogalvanic effect *See* galvanomagnetic effect.

magnetogas dynamics [PL PHYS] The science of motion in a plasma under the influence of mechanical, electric, and magnetic forces.

magnetograph [ELECTROMAG] A set of three variometers attached to a suitable recording unit, which records the components of the magnetic field vector in each of three perpendicular directions.

magnetohydrodynamic arcjet [AERO ENG] An electromagnetic propulsion system utilizing a plasma that is heated in an electric arc and then adiabatically expanded through a nozzle and further accelerated by a crossed electric and magnetic field.

magnetohydrodynamic generator [ELEC] A system for generating electric power in which the kinetic energy of a flowing conducting fluid is converted to electric energy by a magnetohydrodynamic interaction. Abbreviated MHD generator.

magnetohydrodynamic instability [PL PHYS] An instability of a plasma in which the plasma expands while moving into a region of weaker magnetic field, until it is expelled from the field. Also known as hydromagnetic instability.

magnetohydrodynamics [PHYS] The study of the dynamics or motion of an electrically conducting fluid, such as an ionized gas or liquid metal, interacting with a magnetic field. Abbreviated MHD. Also known as hydromagnetics; magnetofluid dynamics.

magnetohydrodynamic stability [PL PHYS] The condition of a plasma in which fluctuations in density, pressure, velocity, or the distribution of particles in phase space, die out rather than increase. Also known as hydromagnetic stability.

magnetohydrodynamic turbulence [PL PHYS] Motion of a plasma in which velocities and pressures fluctuate irregularly.

magnetohydrodynamic wave [PHYS] Wave motion in an electrically conducting fluid, such as plasma or liquid metal, in a strong magnetic field at a frequency much less than that of the ion cyclotron frequency. Also known as hydromagnetic wave.

magneto ignition system [ELECTROMAG] An ignition system in which the voltage required to cause a flow of current in the primary winding of the ignition coil is generated by a set of permanent magnets, instead of being supplied by a battery.

magnetoionic duct [GEOPHYS] Duct along the geomagnetic lines of force which exhibits waveguide characteristics for radio-wave propagation between conjugate points on the earth's surface.

magnetoionic theory [GEOPHYS] The theory of the combined effect of the earth's magnetic field and atmospheric ionization on the propagation of electromagnetic waves.

magnetoionic wave component [GEOPHYS] Either of the two elliptically polarized wave components into which a linearly polarized wave incident on the ionosphere is separated because of the earth's magnetic field.

magnetomechanical factor [PHYS] The gyromagnetic ratio of an atom or substance (magnetic dipole moment divided by angular momentum) divided by the quantity $e/2mc$, where e and m are the charge (in esu, or electrostatic units) and mass of the electron respectively, and c is the speed of light. Also known as g factor.

magnetomechanics [PHYS] The study of the effects which the magnetization of a material and its strain have on each other.

magnetometer [ENG] An instrument for measuring the magnitude and sometimes also the direction of a magnetic field, such as the earth's magnetic field.

magnetomotive force [ELECTROMAG] The work that would be required to carry a magnetic pole of unit strength once around a magnetic circuit. Abbreviated mmf.

magneton [PHYS] A unit of magnetic moment used for atomic, molecular, or nuclear magnets, such as the Bohr magneton, Weiss magneton, or nuclear magneton.

magneton number [PHYS] The ratio of the magnetic moment

per atom, ion, or molecule of a paramagnetic or ferromagnetic material to the Bohr magneton.

magnetooptical shutter [OPTICS] A device in which light passes through crossed Nicol prisms and a glass cell containing a liquid displaying the Faraday effect between the prisms; light can pass through the system only when a magnetic field is applied to the cell at an angle of 45° to the polarization planes of both prisms.

magnetooptical switch [ADP] A thin-film modulator which acts on a laser beam by polarization, causing the beam to emerge from the output prism at a different angle.

magnetooptic Kerr effect [OPTICS] Changes produced in the optical properties of a reflecting surface of a ferromagnetic substance when the substance is magnetized; this applies especially to the elliptical polarization of reflected light, when the ordinary rules of metallic reflection would give only plane polarized light. Also known as Kerr magnetooptical effect.

magnetooptics [OPTICS] The study of the effect of a magnetic field on light passing through a substance in the field.

magnetopause [GEOPHYS] A boundary that marks the transition from the earth's magnetosphere to the interplanetary medium.

magnetoplasmadynamics [ELECTROMAG] The generation of electric current by shooting a beam of ionized gas through a magnetic field, to give the same effect as moving copper bars near a magnet.

magnetoplumbite [MINERAL] $(PbMn)_2Fe_6O_{11}$ Black mineral consisting of a ferric oxide of plumbite and manganese, and occurring in acute metallic hexagonal crystals.

magnetoresistance [ELECTROMAG] The change in electrical resistance produced in a current-carrying conductor or semiconductor on application of a magnetic field.

magnetoresistivity [ELECTROMAG] The change in resistivity produced in a current-carrying conductor or semiconductor on application of a magnetic field.

magnetoresistor [ELECTR] Magnetic field–controlled variable resistor.

magnetosheath [GEOPHYS] The relatively thin region between the earth's magnetopause and the shock front in the solar wind.

magnetosphere [GEOPHYS] The region of the earth in which the geomagnetic field plays a dominant part in controlling the physical processes that take place; it is usually considered to begin at an altitude of about 100 kilometers and to extend outward to a distant boundary that marks the beginning of interplanetary space.

magnetospheric plasma [GEOPHYS] A low-energy plasma with particle energies less than a few electron volts that permeates the entire region of the earth's magnetosphere.

magnetospheric ring current [GEOPHYS] A belt of charged particles around the earth whose perturbations give rise to ionospheric storms.

magnetostatic [ELECTROMAG] Pertaining to magnetic properties that do not depend upon the motion of magnetic fields.

magnetostatic mode [SOLID STATE] A spin wave in a magnetic material whose wavelength is greater than about one-tenth the size of the sample.

magnetostatics [ELECTROMAG] The study of magnetic fields that remain constant with time. Also known as static magnetism.

magnetostatic shielding [ELECTROMAG] The use of an enclosure made of a high-permeability magnetic material to prevent a static magnetic field outside the enclosure from reaching objects inside it, or to confine a magnetic field within the enclosure. Also known as magnetic shielding.

magnetostriction [ELECTROMAG] The dependence of the state of strain (dimensions) of a ferromagnetic sample on the direction and extent of its magnetization.

magnetostriction transducer [ELECTROMAG] A transducer used with sonar equipment to change an alternating current to sound energy at the same frequency and to form the sound energy into a beam; its operation depends on the interaction between the magnetization and the deformation of a material having magnetostrictive properties.

magnetostrictive delay line [ELECTROMAG] A delay line made of nickel or other magnetostrictive material, in which the amount of delay is determined by a shock wave traveling through the length of the line at the speed of sound.

magnetostrictive filter [ELECTR] Filter network which uses the magnetostrictive phenomena to form high-pass, low-pass, band-pass, or band-elimination filters; the impedance characteristic is the inverse of that of a crystal.

magnetostrictive loudspeaker [ENG ACOUS] Loudspeaker in which the mechanical forces result from the deformation of a material having magnetostrictive properties.

magnetostrictive microphone [ENG ACOUS] Microphone which depends for its operation on the generation of an electromotive force by the deformation of a material having magnetostrictive properties.

magnetostrictive oscillator [ELECTR] An oscillator whose frequency is controlled by a magnetostrictive element.

magnetostrictive resonator [SOLID STATE] Ferromagnetic rod so designed that it can be excited magnetically into resonant vibration at one or more definite and known frequencies.

magnetostrictor [ELECTROMAG] A device for converting electric oscillations to mechanical oscillations by employing the property of magnetostriction.

magnetoswitchboard exchange [COMMUN] A manual exchange at which the subscribers and operators call and clear by means of magnetoelectric generators.

magneto telephone set [ELEC] Local battery telephone set in which current for signaling by the telephone station is supplied from a local hand generator, usually a magneto.

magnet power [ELECTROMAG] The electric power supplied to the coils of an electromagnet.

magnetron [ELECTR] One of a family of crossed-field microwave tubes, wherein electrons, generated from a heated cathode, move under the combined force of a radial electric field and an axial magnetic field in such a way as to produce microwave radiation in the frequency range 1–40 gigahertz; a pulsed microwave radiation source for radar, and continuous source for microwave cooking.

magnetron beam-switching tube *See* beam-switching tube.

magnetron oscillator [ELECTR] Oscillator circuit employing a magnetron tube.

magnetron pulling [ELECTR] Frequency shift of a magnetron caused by factors which vary the standing waves or the standing-wave ratio on the radio-frequency lines.

magnetron pushing [ELECTR] Frequency shift of a magnetron caused by faulty operation of the modulator.

magnetron vacuum gage [ELECTR] A vacuum gage that is essentially a magnetron operated beyond cutoff in the vacuum being measured.

magnet wire [ELEC] The insulated copper or aluminum wire used in the coils of all types of electromagnetic machines and devices.

magnification [OPTICS] 1. A measure of the effectiveness of an optical system in enlarging or reducing an image; the magnification may be lateral, longitudinal, or angular. 2. *See* lateral magnification.

magnifier *See* simple microscope.

magnifying glass [OPTICS] 1. Any device that uses a simple lens which enlarges the object being viewed. 2. *See* simple microscope.

magnifying power [OPTICS] The ratio of the tangent of the angle subtended at the eye by an image formed by an optical system, to the tangent of the angle subtended at the eye by the corresponding object at a distance for convenient viewing.

magnistor [ELECTR] A device that utilizes the effects of magnetic fields on injection plasmas in semiconductors such as indium antimonide.

magnitude [ASTRON] The relative luminance of a celestial body; the smaller (algebraically) the number indicating magnitude, the more luminous the body. Also known as stellar magnitude.

magnitude method [ORD] Method of adjusting gunfire for range when the amount and direction of the deviation are known.

magnitude of a complex number *See* absolute value of a complex number.

magnitude of a real number *See* absolute value of a real number.

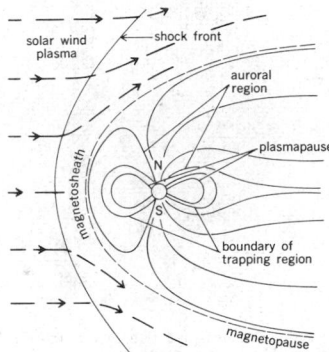

MAGNETOSHEATH

Configuration of the magnetosphere in the plane containing the sun-earth line and the geomagnetic axis; position of magnetosheath is shown.

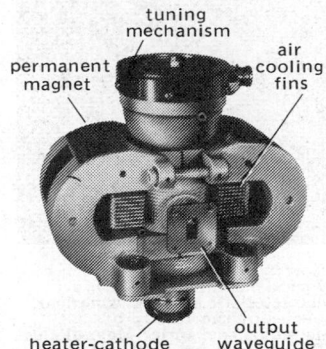

MAGNETRON

Coaxial cavity magnetron with horseshoe-shaped magnets. *(From J. W. Gewartowski and H. A. Watson, Principles of Electron Tubes, Van Nostrand, 1965)*

magnitude of a vector   *See* absolute value of a vector.

magnitude ratio   [ASTRON] The ratio (2.512) of relative brightness of two celestial bodies differing in magnitude by 1.0.

magnitude system   [ASTRON] A system for designating the relative brightness of stars when photography is used; emulsions of different color sensitivities, used with color filters, permit measurements of starlight of different wavelengths with corresponding determination of magnitude at these wavelengths.

magno   [MET] An alloy of 95.5% nickel and 4.5% manganese, used in the manufacture of incandescent lamps and radio tubes.

magnocellular   [CYTOL] Having large cell bodies; said of various nuclei of the central nervous system.

magnochromite   *See* magnesiochromite.

magnoferrite   *See* magnesioferrite.

MAGNOLIA

Leaf and twig of the cucumber tree, *Magnolia acuminata*.

Magnolia   [BOT] A genus of trees, the type genus of the Magnoliaceae, with large, chiefly white, yellow, or pinkish flowers, and simple, entire, usually large evergreen or deciduous alternate leaves.

Magnoliaceae   [BOT] A family of dicotyledonous plants of the order Magnoliales characterized by hypogynous flowers with few to numerous stamens, stipulate leaves, and uniaperturate pollen.

Magnoliales   [BOT] The type order of the subclass Magnoliidae; members are woody plants distinguished by the presence of spherical ethereal oil cells and by a well-developed perianth of separate tepals.

Magnoliatae   *See* Magnoliopsida.

Magnoliidae   [BOT] A primitive subclass of flowering plants in the class Magnoliopsida generally having a well-developed perianth, numerous centripetal stamens, and bitegmic, crassinucellate ovules.

Magnoliophyta   [BOT] The angiosperms, a division of vascular seed plants having the ovules enclosed in an ovary and well-developed vessels in the xylem.

Magnoliopsida   [BOT] The dicotyledons, a class of flowering plants in the division Magnoliophyta generally characterized by having two cotyledons and net-veined leaves, with vascular bundles borne in a ring enclosing a pith.

magnon   [SOLID STATE] A quasiparticle which is introduced to describe small departures from complete ordering of electronic spins in ferro-, ferri-, antiferro-, and helimagnetic arrays. Also known as quantized spin wave.

magnophorite   [MINERAL] $NaKCaMg_5Si_8O_{23}OH$ A monoclinic mineral composed of a basic silicate of sodium, potassium, calcium, and magnesium; member of the amphibole group.

magnum   [ANAT] Large, as in foramen magnum.

Magnus effect   [FL MECH] A force on a rotating cylinder in a fluid flowing perpendicular to the axis of the cylinder; the force is perpendicular to both flow direction and cylinder axis. Also known as Magnus force.

Magnus force   *See* Magnus effect.

Magnus moment   [FL MECH] A torque associated with the Magnus effect, such as moments about the pitch and yaw axes of a missile or aircraft due to rotation about the roll axis.

mag-slip   *See* synchro.

maguey   [MATER] A fiber obtained from the agave (*Agave cantala*); maguey fibers are white, stiff, brilliant, and light in weight, and are used chiefly for binder twine.

mahlstick   [GRAPHICS] A stick held in the palette hand of a painter and used as a support for his painting hand; it is a light rod of wood often with a leather-covered ball at one end. Also known as maulstick; rest stick.

mahogany   [BOT] Any of several tropical trees in the family Meliaceae of the Geraniales.   [MATER] The hard wood of these trees, especially the red or yellow-brown wood of the West Indies mahogany tree (*Swietenia mahogani*).

mahogany acid   [MATER] A dark-colored mixture of sulfonic acid derived from petroleum; the salts are used as emulsifying agents and in lubricants.

mahogany soap   [MATER] The sodium salt of crude or refined petroleum sulfonic acids, used as flotation agents and to increase the oil absorption of mineral pigments in paint.

mahubarana fat   [MATER] A pale-yellow solid oil, melting

MAGNOLIACEAE

Flower of the tulip tree (*Liriodendron tulipifera*), a characteristic eastern American species of the family of Magnoliaceae, order Magnoliales. (*Photograph by F. W. Westlake, from National Audubon Society*)

point 40–44°C, obtained from seeds of trees of the genus *Boldoa*; used for soaps and candles.

maidenhead   *See* hymen.

maieusiophobia   [PSYCH] An abnormal fear of childbirth.

main   [CIV ENG] A duct or pipe that supplies or drains ancillary branches.   [ELEC] 1. One of the conductors extending from the service switch, generator bus, or converter bus to the main distribution center in interior wiring. 2. *See* power transmission line.

main-and-tail haulage   [MIN ENG] A single-track haulage system that is operated by a haulage engine with two drums, each with a separate rope.

main bang   [ELECTR] Transmitted pulse, within a radar system.

main clock   *See* master clock.

main crosscut   [MIN ENG] A crosscut that traverses the entire mining field and penetrates all deposits.

main deck   [NAV ARCH] 1. The uppermost complete deck of a naval vessel extending over its entire length and width. 2. The uppermost deck of a merchant ship for which it is possible to close all openings securely.

main distributing frame   [ELEC] Frame which terminates the permanent outside lines entering the central office building on one side and the subscriber-line multiple cabling, trunk multiple cabling, and so on, used for associating an outside line with any desired terminal on the other side; it usually carries the control-office protective devices, and functions as a test point between line and office. Also known as main frame.

Maindroniidae   [INV ZOO] A family of wingless insects belonging to the Thysanura proper.

main entry   [MIN ENG] The principal entry or set of entries driven through the coalbed from which cross entries, room entries, or rooms are turned.

main exciter   [ELEC] Exciter which supplies energy for the field excitation of a principal electric machine.

main fans   [MIN ENG] Fans that produce the general ventilating current of the mine, being of large capacity and permanently installed.

main firing   [ENG] The firing of a round of shots by means of current supplied by a transformer fed from a main power supply.

main frame   *See* central processing unit; main distributing frame.

main haulage   [MIN ENG] The section of the haulage system which moves the coal from the secondary or intermediate haulage system to the shaft or mine opening.

main haulage conveyor   [MIN ENG] A conveyor used to transport material in the main haulage section of the mine, between the intermediate haulage conveyor and a car-loading point or the outside.

main instruction buffer   [ADP] A section of storage in the instruction unit, 16 bytes in length, used to hold prefetched instructions.

main-line locomotive   [MIN ENG] A large, high-powered locomotive which hauls trains of cars over the main haulage system.

main lobe   *See* major lobe.

mainmast   [NAV ARCH] The principal mast of a sailing ship.

main memory   *See* main storage.

main path   [ADP] The principal branch of a routine followed by a computer in the process of carrying out the routine.

main return   [MIN ENG] The main return airway of a mine.

main rope   *See* pull rope.

main-rope haulage system   [MIN ENG] A system of haulage for hauling loaded trains of tubs or cars up, or lowering them down, a comparatively steep gradient which is not steep enough, in the latter case, for a self-acting incline.

mainsail   [NAV ARCH] The principal sail of a sailing vessel, carried on the mainmast.

main sequence   [ASTRON] The band in the spectrum luminosity diagram which has the great majority of stars; their energy derives from core burning of hydrogen into helium.

main sequence star   [ASTRON] 1. Any of those stars in the smooth curve termed the main sequence in a Hertzsprung-Russell diagram. 2. *See* dwarf star.

main shaft   [MECH ENG] The line of shafting receiving its power from the engine or motor and transmitting power to other parts.

main station [COMMUN] Telephone station with a distinct call number designation, directly connected to a central office.

main storage [ADP] A digital computer's principal working storage, from which instructions can be executed or operands fetched for data manipulation. Also known as main memory.

main stream [HYD] The principal or largest stream of a given area or drainage system. Also known as master stream; trunk stream.

main stroke *See* return streamer.

main sweep [ELECTR] On certain fire-control radar, the longest range scale available.

maintenance [IND ENG] The upkeep of industrial facilities and equipment.

maintenance engineering [IND ENG] The function of providing policy guidance for maintenance activities, and of exercising technical and management review of maintenance programs.

maintenance kit [ENG] A collection of items not all having the same basic name, which are of a supplementary nature to a major component or equipment; the items within the collection may provide replacement parts and facilitate such functions as inspection, test repair, or preventive types of maintenance, for the specific purpose of restoring and improving the operational status of a component or equipment comparable to its original capacity and efficiency.

maintenance vehicle [ENG] Vehicle used for carrying parts, equipment, and personnel for maintenance or evacuation of vehicles.

main thermocline [OCEANOGR] A thermocline that is deep enough in the ocean to be unaffected by seasonal temperature changes in the atmosphere. Also known as permanent thermocline.

maize [BOT] *Zea mays.* Indian corn, a tall cereal grass characterized by large ears.

Majac mill [MIN ENG] A mill for dry-grinding mica by means of fluid energy; consists of a chamber which contains two horizontal, directly opposing jets and into which mica is fed continuously from a screw conveyor.

Majidae [INV ZOO] The spider, or decorator, crabs, a family of decapod crustaceans included in the Brachyura; members are slow-moving animals that often conceal themselves by attaching seaweed and sessile animals to their carapace.

Majorana force [NUC PHYS] A force between two nucleons postulated to explain various phenomena, which can be derived from a potential containing an operator which exchanges the nucleons' positions but not their spins.

Majorana neutrino [PARTIC PHYS] A particle described by a wave function that satisfies the Dirac equation with mass equal to zero, and that is self–charge-conjugate.

major arc [MATH] The longer of the two arcs produced by a secant of a circle.

major assembly [ENG] A self-contained unit of individual identity; a completed assembly of component parts ready for operation, but utilized as a portion of, and intended for further installation in, an end item or major item.

major axis [MATH] The longer of the two axes with respect to which an ellipse is symmetric.

major combination [ENG] A single composite unit of mechanical equipment inherently complete for independent use and consisting of one or more major items; as issued, it is complete in respect to both equipment and spare parts, including items furnished by services other than the issuing service; for example, a tank, complete with armament, equipment, and spare parts.

major cycle [ADP] The time interval between successive appearances of a given storage position in a serial-access computer storage.

major defect [IND ENG] Defect which causes serious malfunctioning of a product.

major diameter [DES ENG] The largest diameter of a screw thread, measured at the crest for an external (male) thread and at the root for an internal (female) thread.

major diatonic scale [ACOUS] A diatonic scale in which the relative sizes of the sequence of intervals are approximately 2,2,1,2,2,2,1.

major fog signal [NAV] A sound signal which has a normal range of reception in excess of 2 miles (3.2 kilometers), to aid watercraft in avoiding obstacles.

major fold [GEOL] A large-scale fold with which minor folds are usually associated.

major hysteria [PSYCH] A conversion reaction manifested in movements that suggest a generalized convulsion.

major item [ORD] An end item, a group of end items individually classified by the responsible technical service, or an assembled group of items as procured or issued for a specific tactical role, excluding combinations required to complete the assigned tactical mission.

majority [MATH] A logic operator having the property that if P, Q, R are statements, then the function (P, Q, R, . . .) is true if more than half the statements are true, or false if half or less are true.

majority carrier [ELECTR] The type of carrier, that is, electron or hole, that constitutes more than half the carriers in a semiconductor.

majority emitter [ELECTR] Of a transistor, an electrode from which a flow of minority carriers enters the interelectrode region.

major light [NAV] A light of high candlepower and reliability exhibited from a fixed structure ashore or on a marine site (except range lights).

major lobe [ELECTROMAG] Antenna lobe indicating the direction of maximum radiation or reception. Also known as main lobe.

major mine disaster [MIN ENG] As defined by the U.S. Bureau of Mines, any accident that results in the death of five or more persons.

major node [ELEC] A point in an electrical network at which three or more elements are connected together. Also known as junction.

major operation [MED] An extensive, difficult, and potentially dangerous surgical procedure, usually requiring general anesthesia.

major planet [ASTRON] Any of the four planets that are larger than earth: Jupiter, Saturn, Neptune, and Uranus.

major relay station [ELECTR] Tape relay station which has two or more trunk circuits connected thereto to provide an alternate route or to meet command requirements.

major repair [ENG] Repair work on items of material or equipment that need complete overhaul or substantial replacement of parts, or that require special tools.

major trough [METEOROL] A long-wave trough in the large-scale pressure pattern of the upper troposphere.

major wave *See* long wave.

make [ELEC] Closing of relay, key, or other contact.

make-break operation [COMMUN] A circuit operation in which there is a cessation of current flow as a pulse transmission occurs.

make contact [ELEC] Contact of a device which closes a circuit upon the operation of the device (normally open).

makeready [GRAPHICS] The careful leveling of relief printing plates on the bed of the press so that they yield the best possible impression.

make the land [NAV] To sight and approach or reach land from seaward.

makeup water [CHEM ENG] Water feed needed to replace that which is lost by evaporation or leakage in a closed-circuit, recycle operation.

make way [NAV] To progress through the water.

makroskelic [ANTHRO] Being long-legged relative to the trunk length, with a skelic index of 95–100.

Maksutov system [OPTICS] A catadioptric telescope optical system capable of covering a large field (60° and more); used to survey large areas of the sky.

mal- [SCI TECH] A combining form meaning bad, wrong, irregular, abnormal, inadequate.

Malachiidae [INV ZOO] An equivalent name for Malyridae.

malachite [MINERAL] $Cu_2(OH)_2(CO_3)$ A bright-green monoclinic mineral consisting of a basic carbonate of copper and usually occurring in massive forms or in bundles of

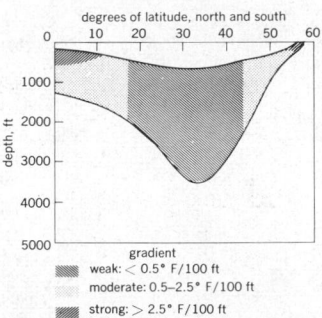

MAIN THERMOCLINE

Diagram showing variations with latitude in average depth, thickness and temperature gradient of the main thermocline.

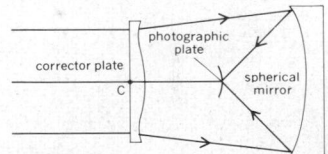

MAKSUTOV SYSTEM

Optics of Maksutov system; *C* is the center of curvature of the mirror.

MALACHITE

Rounded masses of malachite crystals. *(Specimen from Department of Geology, Bryn Mawr College)*

radiating fibers; specific gravity is 4.05, and hardness is 3.5–4 on Mohs scale.

malacia [MED] Abnormal softening of tissues of an organ or other body structure.

Malacobothridia [INV ZOO] A subclass of worms in the class Trematoda; members typically have one or two soft, flexible suckers and are endoparasitic in vertebrates and invertebrates.

Malacocotylea [INV ZOO] The equivalent name for Digenea.

malacology [INV ZOO] The study of mollusks.

malacoplakia [MED] The accumulation of modified histiocytes (malacoplakia cells) to produce soft, pale, elevated plaques, usually in the urinary bladder of middle-aged women.

malacoplakia cell [PATH] A large histiocyte, occasionally multinucleate and containing Michaelis-Gutmann calcospherules in the cytoplasm; seen in malacoplakia.

Malacopoda [INV ZOO] A subphylum of invertebrates in the phylum Oncopoda.

Malacopterygii [VERT ZOO] An equivalent name for Clupeiformes in older classifications.

Malacostraca [INV ZOO] A large, diversified subclass of Crustacea including shrimps, lobsters, crabs, sow bugs, and their allies; generally characterized by having a maximum of 19 pairs of appendages and trunk limbs which are sharply differentiated into thoracic and abdominal series.

maladie du sommeil *See* African sleeping sickness.

maladjustment [PSYCH] Failure to conform or inadequate conformity due to the inability or a lack of motivation to change one's feelings or attitudes to adjust to the demands of the environment.

malady [MED] A disorder, disease, or illness.

malaise [MED] A general state of ill-being or the feeling of poor health.

Malapteruridae [VERT ZOO] A family of African catfishes in the suborder Siluroidei.

malar [ANAT] Of or pertaining to the cheek or to the zygomatic bone.

malar bone *See* zygomatic bone.

malaria [MED] A group of human febrile diseases with a chronic relapsing course caused by hemosporidian blood parasites of the genus *Plasmodium,* transmitted by the bite of the *Anopheles* mosquito.

malaria pigment [PATH] Dark-brown, amorphous, microcrystalline and birefringent pigment found in parasitized erythrocytes, especially with malarial parasites, and in littoral phagocytes of spleen, liver, and bone marrow.

malar stripe [VERT ZOO] **1.** The area extending from the corner of the mouth backward and down in birds. **2.** Area on side of throat below the base of the lower mandible.

malassimilation [MED] Faulty or inadequate assimilation.

malathion [ORG CHEM] $C_{10}H_{19}O_6PS_2$ A yellow liquid, slightly soluble in water; malathion is the generic name for *S*-[1,2-bis(ethoxycarbonyl)ethyl] *O,O*-dimethylphosphorodithioate; used as an insecticide.

Malayan filariasis [MED] Filariasis of man caused by *Brugia malayi,* transmitted by *Mansonia* and *Anopheles* mosquitoes.

malchite [PETR] A fine-grained lamprophyre with small, rare phenocrysts or hornblende, labradorite, and sometimes biotite embedded in a matrix of hornblende, andesine, and some quartz.

Malcidae [INV ZOO] A small family of Ethiopian and Oriental hemipteran insects in the superfamily Pentatomorpha.

Malcodermata [INV ZOO] The equivalent name for Cantharoidea.

Maldanidae [INV ZOO] The bamboo worms, a family of mudswallowing annelids belonging to the Sedentaria.

Maldaninae [INV ZOO] A subfamily of the Maldanidae distinguished by cephalic and anal plaques with the anal aperture located dorsally.

mal de pinto *See* pinta.

male [BOT] A flower lacking pistils. [ZOO] **1.** Of or pertaining to the sex that produces spermatozoa. **2.** An individual of this sex.

maleate [ORG CHEM] An ester or salt of maleic acid.

male climacteric [PHYSIO] A condition presumably due to loss of testicular function, associated with an elevated urinary excretion of gonadotropins and symptoms of loss of sexual desire and potency, hot flashes, and vasomotor instability.

male heterogamety [GEN] Having paired sex chromosomes of unequal sizes in the male.

male homogamety [GEN] Having the paired sex chromosomes of equal size in the male.

maleic acid [ORG CHEM] HOOCCH:CHCOOH A colorless, crystalline dibasic acid; soluble in water, acetone, and alcohol; melting point 130–131°C; used in textile processing, and as an oil and fat preservative.

maleic anhydride [ORG CHEM] $C_4H_2O_3$ Colorless crystals, soluble in acetone, hydrolyzing in water; used to form polyester resins. Also known as 2,5-furandione.

maleic hydrazide [ORG CHEM] $C_4N_2H_4O_2$ Solid material, decomposing at 260°C; slightly soluble in alcohol and water; used as a weed killer and growth inhibitor.

male pseudohermaphroditism *See* androgyny.

male sterility [PHYSIO] The condition in which male gametes are absent, deficient in number, or nonfunctional.

male Turner's syndrome *See* Ullrich-Turner syndrome.

malformation [MED] A deformity of a part of the body resulting from abnormal development.

malfunction [SCI TECH] Failure to function normally.

malfunction routine [ADP] A program used in troubleshooting.

malic acid [BIOCHEM] COOH·CH₂·CHOH·COOH Hydroxysuccinic acid: a dibasic hydroxy acid existing in two optically active isomers and a racemic form; found in apples and many other fruits.

malic dehydrogenase [BIOCHEM] An enzyme in the Krebs cycle that catalyzes the conversion of L-malic acid to oxaloacetic acid.

malignant [MED] **1.** Endangering the life or health of an individual. **2.** Of or pertaining to the growth and proliferation of certain neoplasms which terminate in death if not checked by treatment.

malignant adenoma [MED] A neoplasm which resembles an adenoma cytologically and gives rise to metastases.

malignant catarrh [VET MED] A catarrhal fever of cattle caused by a virus and characterized by acute inflammation and edema of the respiratory and digestive systems.

malignant disease [MED] Any disease that endangers the life of an individual over a short period of time, especially cancer.

malignant edema [VET MED] Inflammatory edema associated with certain infections, especially an acute wound infection in wild and domestic animals.

malignant embolus [MED] A blood-borne mass of malignant cells which have become dislodged from the parent neoplasm.

malignant glaucoma [MED] A form of glaucoma associated with severe pain and rapidly leading to blindness.

malignant hypertension [MED] A severe form of hypertension with a rapid course leading to progressive cardiac and renal vascular disease. Also known as accelerated hypertension.

malignant malaria *See* falciparum malaria.

malignant pustule [MED] The commonest form of anthrax in man, resulting from contamination of the skin; characterized by a necrotic pustule surrounded by an area of edema and vesicles containing yellow fluid. Also known as cutaneous anthrax.

malignant rhabdomyoma *See* rhabdomyosarcoma.

malinger [MED] To pretend or exaggerate illness in order to avoid responsibilities.

malladrite [MINERAL] Na₂SiF₆ A hexagonal mineral composed of sodium fluosilicate, occurring as small crystals in volcanic holes in Vesuvius.

mallardite [MINERAL] MnSO₄·7H₂O A pale-rose, monoclinic mineral composed of hydrous manganese sulfate.

malleable [MET] Capable of undergoing plastic deformation without rupture; a property characteristic of metals.

malleable brass *See* Muntz metal.

malleable iron [MET] White cast iron which has been rendered malleable by heat treatment.

malleableize [MET] To render a material malleable, such as by heat-treating white cast iron.

malleable pig iron [MET] A grade of pig iron suitable for production of white cast iron from which malleable iron is made.

malleate trophus [INV ZOO] A type of crushing masticatory apparatus in rotifers that are incidentally predatory, such as brachionids.

mallee See tropical scrub.

malleolus [ANAT] A projection on the distal end of the tibia and fibula at the ankle.

malleoramate trophus [INV ZOO] An intermediate type of rotiferan masticatory apparatus having a looped manubrium and teeth on the incus (comprising the fulcrum and rami); developed for grinding.

mallet [DES ENG] An implement with a barrel-shaped head made of wood, rubber, or other soft material; used for driving another tool, such as a chisel, or for striking a surface without causing damage.

malleus [ANAT] The outermost, hammer-shaped ossicle of the middle ear; attaches to the tympanic membrane and articulates with the incus.

Mallophaga [INV ZOO] Biting lice, a comparatively small order of wingless insects characterized by five-segmented antennae, distinctly developed mandibles, one or two terminal claws on each leg, and a prothorax developed as a distinct segment.

Mallory bodies [PATH] Oval acidophilic hyalin inclusion bodies seen in the cytoplasm of hepatic cells in Laennec's cirrhosis.

Mallory-Weiss syndrome [MED] Painless vomiting of blood secondary to lacerations of the distal esophagus and esophagogastric junction, usually a result of prolonged violent vomiting, coughing, or hiccuping.

Malm [GEOL] The Upper Jurassic geologic series, above Dogger and below Cretaceous.

malmstone [MATER] A name applied to chert when it is used in building and paving.

malnutrition [MED] Defective nutrition due to inadequate intake of nutrients or to their faulty digestion, assimilation, or metabolism.

malocclusion [MED] Faulty occlusion of the teeth.

malodorant See odorant.

malonic acid [ORG CHEM] $CH_2(COOH)_2$ A white, crystalline dicarboxylic acid, melting at 132–134°C; used to manufacture pharmaceuticals. Also known as methanedicarbonic acid; methanedicarboxylic acid.

malonic ester See ethyl malonate.

malonyl [ORG CHEM] $CH_2(COO)_2$ A bivalent radical formed from malonic acid.

malonyl urea See barbituric acid.

Malotte's metal [MET] A fusible alloy composed of 46% bismuth, 20% lead, and 34% tin; melts at 96–123°C.

Malpighiaceae [BOT] A family of dicotyledonous plants in the order Polygalales distinguished by having three carpels, several fertile stamens, five petals that are commonly fringed or toothed, and indehiscent fruit.

Malpighian corpuscle [ANAT] 1. A lymph nodule of the spleen. 2. See renal corpuscle.

Malpighian layer [HISTOL] The germinative layer of the epidermis.

Malpighian pyramid See renal pyramid.

Malpighian tubule [INV ZOO] Any of the blind tubes that open into the posterior portion of the gut in most insects and certain other arthropods and excrete matter or secrete substances such as silk.

malposition [MED] Abnormal position of an organ or other body structure, or of the fetus.

malpractice [MED] Improper or injurious medical or surgical treatment, through carelessness, ignorance, or intent.

malpresentation [MED] Abnormal position of the child at birth, making normal delivery difficult or impossible.

malt [FOOD ENG] A nutrient material made from grain, commonly barley, which has been soaked, allowed to germinate, and dried.

Malta fever See brucellosis.

maltase [BIOCHEM] An enzyme that catalyzes the conversion of maltose to dextrose.

malt beverage [FOOD ENG] Any of various fermented alcoholic beverages, including beer, ale, stout, and porter; barley malt is the principal ingredient.

Malter effect [SOLID STATE] A phenomenon in which a metal with a nonconducting surface film has a large coefficient of secondary electron emission; this is particularly notable in aluminum whose surface has been oxidized and then coated with cesium oxide.

Malthusianism [BIOL] The theory that population increases more rapidly than the food supply unless held in check by epidemics, wars, or similar phenomena.

malting [FOOD ENG] Converting the starches of a distillery mash to fermentable sugar by the enzymes of the malt.

maltobiose See maltose.

maltose [BIOCHEM] $C_{12}H_{22}O_{11}$ A crystalline disaccharide that is a product of the enzymatic hydrolysis of starch, dextrin, and glycogen; does not appear to exist free in nature. Also known as maltobiose; malt sugar.

maltosuria [MED] Presence of maltose in the urine.

malt sugar See maltose.

malt whiskey [FOOD ENG] Whiskey produced in a pot still from malted barley.

malunion [MED] Faulty union of the pieces of a fractured bone.

Malus See Pyxis.

Malus cosine-squared law [OPTICS] The law that if a beam of plane polarized light passes through a Nicol prism, the intensity of light emerging from the prism is proportional to the square of the cosine between the plane of polarization of the incident light and the plane of polarization of the prism.

Malus' law of rays [OPTICS] The law that an orthotomic system of rays is still orthotomic after the rays have been reflected and refracted any number of times.

Malvaceae [BOT] A family of herbaceous dicotyledons in the order Malvales characterized by imbricate or contorted petals, mostly unilocular anthers, and minutely spiny, multiporate pollen.

Malvales [BOT] An order of flowering plants in the subclass Dilleniidae having hypogynous flowers with valvate calyx, mostly separate petals, numerous centrifugal stamens, and a syncarpous pistil.

malysite [MINERAL] $FeCl_3$ A halogen mineral deposited by sublimation; found most commonly at Mount Vesuvius, Italy.

mamarron oil [MATER] A cream-colored fat high in lauric acid and similar to coconut oil in characteristics and odor; obtained from a species of *Attalea* palm.

mamelon [BIOL] Any dome-shaped protrusion or elevation.

mamm-, mammo- [ANAT] A combining form meaning breast.

mamma [ANAT] A milk-secreting organ characterizing all mammals.

mamma aberrans [MED] A supernumerary breast.

mammal [VERT ZOO] A member of Mammalia.

Mammalia [VERT ZOO] A large class of warm-blooded vertebrates containing animals characterized by mammary glands, a body covering of hair, three ossicles in the middle ear, a muscular diaphragm separating the thoracic and abdominal cavities, red blood cells without nuclei, and embryonic development in the allantois and amnion.

mammary [ANAT] Of or pertaining to the mamma, or breast.

mammary gland [PHYSIO] A highly modified sebaceous gland that secretes milk; a unique anatomical feature of mammals.

mammary lymphatic plexus [ANAT] A network of anastomosing lymphatic vessels in the walls of the ducts and between the lobules of the mamma; also functions to drain skin, areola, and nipple.

mammary region [ANAT] The space on the anterior surface of the chest between a line drawn through the lower border of the third rib and one drawn through the upper border of the xiphoid cartilage.

mammary ridge [EMBRYO] An ectodermal thickening forming a longitudinal elevation on the chest between the limbs from which the mammary glands develop.

mammary-stimulating hormone [BIOCHEM] 1. Estrogen and progesterone considered together as the hormones which induce proliferation of the mammary ductile and acinous elements respectively. 2. See prolactin.

mammectomy See mastectomy.

mammillary [ANAT] 1. Of or pertaining to the nipple. 2.

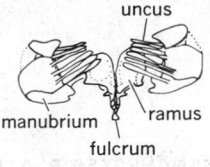
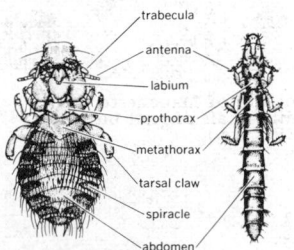
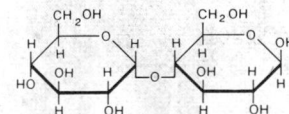

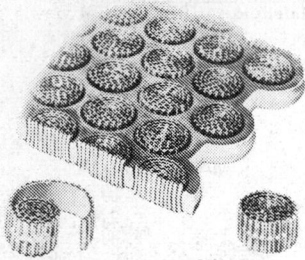

MANCHESTER PLATE

Section of Manchester plate with detail of lead button. *(ESB Inc.)*

MANDRILL

The mandrill *(Mandrillus sphinx)*.

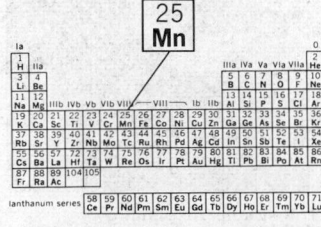

MANGANESE

Periodic table of the chemical elements showing the position of manganese.

Breast- or nipple-shaped. [MINERAL] Of or pertaining to an aggregate of crystals in the form of a rounded mass.

mammillary body [ANAT] Either of two small, spherical masses of gray matter at the base of the brain in the space between the hypophysis and oculomotor nerve, which receive and relay olfactory impulses.

mammillary line [ANAT] A vertical line passing through the center of the nipple.

mammillary process [ANAT] One of the tubercles on the posterior part of the superior articular processes of the lumbar vertebrae.

mammillary structure *See* pillow structure.

mammillitis [MED] Inflammation of the nipple.

mammogen *See* prolactin.

mammogenic hormone [BIOCHEM] **1.** Any hormone that stimulates or induces development of the mammary gland. **2.** *See* prolactin.

mammography [MED] Radiographic examination of the breast, performed with or without injection of the glandular ducts with a contrast medium.

mammoplasia [PHYSIO] Development of breast tissue.

mammoplasty [MED] Plastic surgery performed to alter the shape of the breast.

mammoth [PALEON] Any of various large Pleistocene elephants having long, upcurved tusks and a heavy coat of hair.

mammotropin *See* prolactin.

Mammutinae [PALEON] A subfamily of extinct proboscidean mammals in the family Mastodontidae.

man [ANTHRO] **1.** *Homo sapiens.* A member of the human race. **2.** An adult human male.

management engineering *See* industrial engineering.

management game [IND ENG] A training exercise in which prospective decision makers act out managerial decision-making roles in a simulated environment. Also known as business game; operational game.

management information system [COMMUN] A communication system in which data are recorded and processed to form the basis for decisions by top management of an organization.

manandonite [MINERAL] $Li_4Al_{14}B_4Si_6O_{29}(OH)_{24}$ A white mineral composed of basic borosilicate of lithium and aluminum.

manasseite [MINERAL] $MgAl_2(OH)_{16}(CO_3)\cdot4H_2O$ A hexagonal mineral composed of basic hydrous magnesium and aluminum carbonate; it is dimorphous with hydrotalcite.

man cage [MIN ENG] A special cage for raising and lowering workers in a mine shaft.

man car [MIN ENG] A kind of car for transporting miners up and down the steeply inclined shafts of some mines.

Manchester plate [ELEC] A storage battery consisting of a heavy alloy grid with circular openings into which are pressed pure lead buttons that are made from lead tape by crimping and rolling to develop a large surface area and are coated with lead peroxide, PbO_2.

mandarin [BOT] A large and variable group of citrus fruits in the species *Citrus reticulata* and some of its hybrids; many varieties of the trees are compact with willowy twigs and small, narrow, pointed leaves; includes tangerines, King oranges, Temple oranges, and tangelos.

mandarin oil [MATER] Golden-yellow or olive-green essential oil with refreshing aroma, obtained from the peel of mandarin oranges; chief constituents are limonene and esters; used in medicine and as flavoring. Also known as tangerine oil.

mandatory layer [METEOROL] A layer of the atmosphere between two consecutive (or any two) specified mandatory levels.

mandatory level [METEOROL] One of several constant-pressure levels in the atmosphere for which a complete evaluation of data from upper-air observations is required. Also known as mandatory surface.

mandatory surface *See* mandatory level.

mandelic acid [ORG CHEM] $C_6H_5CHOHCOOH$ A white, crystalline compound, melting at 117–119°C, darkening upon exposure to light; used in organic synthesis. Also known as amygdalic acid; benzoglycolic acid; phenylglycolic acid; α-phenylhydroxyacetic acid.

Mandelstam plane [PARTIC PHYS] A method of plotting en-

ergy versus scattering angle of three reactions, each having two particles both before and after scattering, which can be derived from each other by the crossing principle; the three reactions are on an equal footing, and poles in the scattering amplitude representing exchanged particles lie along straight lines.

Mandelstam representation [PARTIC PHYS] For a reaction in which there are two particles both before and after scattering: an expression, containing several integrals, for a function related to the scattering amplitude; the arguments of the function are the center-of-mass energy and scattering angle, extended to complex values; the function is conjectured to be analytic in these variables except for certain cuts and to have values along these cuts which give the scattering amplitude of the reaction, and of the two reactions derivable from it by the crossing principle.

mandible [ANAT] **1.** The bone of the lower jaw. **2.** The lower jaw. [INV ZOO] Any of various mouthparts in many invertebrates designed to hold or bite into food.

mandibular arch [EMBRYO] The first visceral arch in vertebrates.

mandibular cartilage [EMBRYO] The bar of cartilage supporting the mandibular arch.

mandibular fossa [ANAT] The depression in the temporal bone that articulates with the mandibular condyle.

mandibular gland *See* submandibular gland.

mandibular nerve [ANAT] A mixed nerve branch of the trigeminal nerve; innervates various structures of the lower jaw and face.

Mandibulata [INV ZOO] A subphylum of Arthropoda; members possess a pair of mandibles which characterize the group.

mandrel [MECH ENG] A shaft inserted through a hole in a component to support the work during machining. [MET] A metal bar serving as a core around which other metals are cast, forged, or extruded, forming a true central hole.

mandrel hanger [PETRO ENG] A device used to provide a liquid- or gas-tight seal (blowout preventer) between oil-well tubing and the tubing head.

mandrel press [MECH ENG] A press for driving mandrels into holes.

mandrill [VERT ZOO] *Mandrillus sphinx.* An Old World cercopithecoid monkey found in West-central Africa and characterized by large red callosities near the ischium and by blue ridges on each side of the nose in males.

maneb [ORG CHEM] $Mn[SSCH(CH_2)_2NHCSS]$ A generic term for manganese ethylene-1,2-bisdithiocarbamate; irritating to eyes, nose, skin, and throat; used as a fungicide.

maneuverability [NAV] The rate at which the orientation of a craft can be changed.

maneuvering board [NAV] A polar coordinate plotting sheet devised to facilitate solution of problems involving relative movement of naval vessels.

maneuvering craft [NAV] A craft whose movements are defined relative to a given craft called the reference craft; it may be a maneuvering ship, maneuvering aircraft, or maneuvering vehicle.

manganate [INORG CHEM] **1.** Salts that have manganese in the anion. **2.** In particular, a salt of manganic acid formed by fusion of manganese dioxide with an alkali.

manganese [CHEM] A metallic element, symbol Mn, atomic weight 54.938, atomic number 25; a transition element whose properties fall between those of chromium and iron. [MET] A hard, brittle, grayish-white metal used chiefly in making steel.

manganese acetate [ORG CHEM] $Mn(C_2H_3O_2)_2\cdot4H_2O$ A pale-red crystalline compound melting at 80°C; soluble in water and alcohol; used in textile dyeing, as a catalyst, and for leather tanning.

manganese-aluminum [MET] A hardener alloy employed for making additions of manganese to aluminum alloys such as Duralumin; typical composition is 25% manganese, 75% aluminum.

manganese binoxide *See* manganese dioxide.

manganese black *See* manganese dioxide.

manganese borate [INORG CHEM] MnB_4O_7 Water-insoluble, reddish-white powder; used as a varnish and oil drier.

manganese-boron [MET] Manganese alloyed with boron; used as an ingredient for hardening and deoxidizing bronze.

manganese brass [MET] A brass containing about 70% copper, 29% zinc, and 1.3% manganese.

manganese bromide *See* manganous bromide.

manganese bronze [MET] A type of brass or bronze containing about 59% copper, 39% zinc, 1.5% iron, 1% tin, and 0.1% manganese; another composition by the same name contains about 66% copper, 23% zinc, 3% iron, 4.5% aluminum, and 3.7% manganese.

manganese carbonate [INORG CHEM] $MnCO_3$ Rose-colored crystals found in nature as rhodocrosite; soluble in dilute acids, insoluble in water; used in medicine, in fertilizer, and as a paint pigment.

manganese citrate [ORG CHEM] $Mn_3(C_6H_5O_7)_2$ A white powder, water-insoluble in the presence of sodium citrate; used in medicine.

manganese dioxide [INORG CHEM] MnO_2 A black, crystalline, water-insoluble compound, decomposing to manganese sesquioxide, Mn_2O_3, and oxygen when heated to 535°C; used as a depolarizer in certain dry-cell batteries, as a catalyst, and in dyeing of textiles. Also known as battery manganese; manganese binoxide; manganese black; manganese peroxide.

manganese epidote *See* piemontite.

manganese fluoride *See* manganous fluoride.

manganese gluconate [ORG CHEM] $Mn(C_6H_{11}O_7)_2 \cdot 2H_2O$ A pinkish powder, insoluble in benzene and alcohol, soluble in water; used in medicine, in vitamin tablets, and as a feed additive and dietary supplement.

manganese halide [INORG CHEM] Compound of manganese with a halide, such as chlorine, bromine, fluorine, or iodine.

manganese heptoxide [INORG CHEM] Mn_2O_7 A compound formed as an explosive dark-green oil by the action of concentrated sulfuric acid on permanganate compounds.

manganese hydrogen phosphate *See* acid manganous phosphate.

manganese hydroxide *See* manganous hydroxide.

manganese hypophosphite [INORG CHEM] $Mn(H_2PO_2)_2 \cdot H_2O$ Odorless, tasteless pink crystals which explode if heated with oxidants; used in medicine.

manganese iodide *See* manganous iodide.

manganese lactate [ORG CHEM] $Mn(C_3H_5O_3)_2 \cdot 3H_2O$ Pale-red crystals; insoluble in water and alcohol; used in medicine.

manganese linoleate [ORG CHEM] $Mn(C_{18}H_{31}O_2)_2$ A dark-brown mass, soluble in linseed oil; used in pharmaceutical preparations and as a varnish and paint drier.

manganese monoxide *See* manganese oxide.

manganese naphthenate [ORG CHEM] Hard brown resinous mass, soluble in mineral spirits; melts at 135°C; contains 6% manganese in commercial solutions; used as a paint and varnish drier.

manganese nodule [GEOL] Small, irregular black to brown concretions consisting chiefly of manganese salts and manganese oxide minerals; formed in oceans as a result of pelagic sedimentation or precipitation.

manganese oleate [ORG CHEM] $Mn(C_{18}H_{33}O_2)_2$ Granular brown mass, soluble in oleic acid and ether, insoluble in water; used in medicine and as a varnish drier.

manganese oxalate [ORG CHEM] $MnC_2O_4 \cdot 2H_2O$ A white crystalline compound, soluble in dilute acids, only slightly soluble in water; used as a paint and varnish drier.

manganese oxide [INORG CHEM] MnO Green powder, soluble in acids, insoluble in water; melts at 1650°C; used in medicine, in textile printing, as a catalyst, in ceramics, and in dry batteries. Also known as manganese monoxide; manganous oxide.

manganese peroxide *See* manganese dioxide.

manganese resinate [ORG CHEM] $Mn(C_{20}H_{29}O_2)_2$ Water-insoluble mass, flesh-colored or brownish black; used as a varnish and oil drier.

manganese silicate *See* manganous silicate.

manganese-silicon [MET] An alloy that contains 73-78% silicon, 20-25% manganese, a maximum of 1.5% iron, and a maximum of 0.25% carbon; used for adding manganese and silicon to metals.

manganese steel *See* Hadfield manganese steel.

manganese sulfate *See* manganous sulfate.

manganese sulfide *See* manganous sulfide.

manganese-titanium [MET] An alloy usually composed of 38% manganese, 29% titanium, 8% aluminum, 3% silicon, 22% iron, and no carbon; used as a deoxidizer for high-grade steels and for nonferrous alloys.

manganic fluoride [INORG CHEM] MnF_3 Poisonous red crystals, decomposed by heat and water; used as a fluorinating agent.

manganic hydroxide [INORG CHEM] $Mn(OH)_3$ A brown powder that rapidly loses water to form $MnO(OH)$; used in ceramics and as a fabric pigment. Also known as hydrated manganic hydroxide.

manganic oxide [INORG CHEM] Mn_2O_3 Hard black powder, insoluble in water, soluble in cold hydrochloric acid, hot nitric acid, and sulfuric acid; occurs in nature as manganite. Also known as manganese sesquioxide.

Manganin [MET] Trade name for an alloy containing 80-85% copper, 12-15% manganese, and 2-4% nickel; used in making precision wire-wound resistors because of its low temperature coefficient of resistance.

manganite [MINERAL] $MnO(OH)$ A brilliant steel-gray or black polymorphous mineral; crystallizes in the orthorhombic system. Also known as gray manganese ore.

manganolangbeinite [MINERAL] $K_2Mn_2(SO_4)_3$ A rose-red, isometric mineral composed of potassium manganese sulfate; occurs in lava on Vesuvius.

manganosite [MINERAL] MnO An emerald-green isometric mineral occurring in small octahedrons that blacken on exposure; hardness is 5-6 on Mohs scale, and specific gravity is 5.18.

manganous bromide [INORG CHEM] $MnBr_2 \cdot 4H_2O$ Water-soluble, deliquescent red crystals. Also known as manganese bromide.

manganous chloride [INORG CHEM] $MnCl_2 \cdot 4H_2O$ Water-soluble, deliquescent rose-colored crystals melting at 88°C; used as a catalyst and in paints, dyeing, and pharmaceutical preparations.

manganous fluoride [INORG CHEM] MnF_2 Reddish powder, insoluble in water, soluble in acid. Also known as manganese fluoride.

manganous hydroxide [INORG CHEM] $Mn(OH)_2$ Heat-decomposable white-pink crystals; insoluble in water and alkali, soluble in acids; occurs in nature as pyrochroite. Also known as manganese hydroxide.

manganous iodide [INORG CHEM] $MnI_2 \cdot 4H_2O$ Water-soluble, deliquescent yellowish-brown crystals. Also known as manganese iodide.

manganous silicate [MINERAL] $MnSiO_3$ Water-insoluble red crystals or yellowish-red powder; occurs in nature as rhodonite. Also known as manganese silicate.

manganous sulfate [INORG CHEM] $MnSO_4 \cdot 4H_2O$ Water-soluble, translucent, efflorescent rose-red prisms; melts at 30°C; used in medicine, textile printing, and ceramics, as a fungicide and fertilizer, and in paint manufacture. Also known as manganese sulfate.

manganous sulfide [INORG CHEM] MnS An almost water-insoluble powder that decomposes on heating; used as a pigment and as an additive in making steel. Also known as manganese sulfide.

manganous sulfite [INORG CHEM] $MnSO_3$ Grayish-black or brownish-red powder, soluble in sulfur dioxide, insoluble in water.

mange [VET MED] Infestation of the skin of mammals by certain mites (Sarcoptoidea) which burrow into the epidermis; characterized by multiple lesions accompanied by severe itching.

Manger *See* Praesepe.

Mangin mirror [OPTICS] A negative meniscus lens whose shallower surface is silvered to act as a spherical mirror while the other surface corrects for spherical aberration of the reflecting surface; used in searchlights and aircraft gunsights.

mangle gearing [MECH ENG] Gearing for producing reciprocating motion; a pinion rotating in a single direction drives a rack with teeth at the ends and on both sides.

mango [BOT] *Mangifera indica.* A large evergreen tree of the sumac family (Anacardiaceae), native to southeastern Asia,

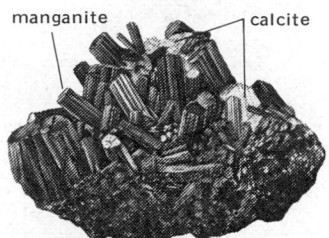

MANGANITE

Crystals of manganite with calcite from Ilfeld, Harz, Germany. *(Specimen from Department of Geology, Bryn Mawr College)*

MANGO

Fruit on a branch of the mango tree.

but now cultivated in Africa, tropical America, Florida, and California for its edible fruit, a thick-skinned, yellowish-red, fleshy drupe.

mangrove [BOT] A tropical tree or shrub of the genus *Rhizophora* characterized by an extensive, impenetrable system of prop roots which contribute to land building. [MATER] Liquid derived from the mangrove tree *Rhizophora mucronata* and used in the leather tanning industry.

mangrove swamp [ECOL] A tropical or subtropical marine swamp distinguished by the abundance of low to tall trees, especially mangrove trees.

Manhattan Project [ENG] A United States project lasting from August 1942 to August 1946, which developed the atomic-energy program, with special reference to the atomic bomb.

manhead *See* manhole.

manhole [ENG] An access hole to a tank or boiler, to underground passages, or in a deck or bulkhead of a ship; usually covered with a cast iron or steel plate. Also known as manhead.

manhole coaming [NAV ARCH] The raised framework or stiffening around the edge of a manhole in the deck of a ship to strengthen the opening and provide a support for the cover.

man-hour [IND ENG] A unit of measure representing one person working for one hour.

mania [PSYCH] Excessive enthusiasm or excitement; a violent desire or passion; manifestation of a psychotic disorder.

-mania [PSYCH] A combining form denoting obsession, abnormal preoccupation, or compulsion.

manic-depressive psychosis [PSYCH] A severe disturbance of affect characterized by extreme and pathological elation alternating with severe dejection, both of which may last for months or years.

Manidae [VERT ZOO] The pangolins, a family of mammals comprising the order Pholidota.

manifest content [PSYCH] Any idea, feeling, or action considered to be the conscious expression of repressed motives or desires, particularly the remembered content of a dream or fantasy which conceals and distorts the unconscious meaning.

manifest covariance [RELAT] Property of an expression composed of Lorentz invariant numbers and operators, four-vectors, and tensors in such a way that its Lorentz covariance is immediately obvious.

manifest stimulus [PSYCH] The obvious or external basis for anxiety, fear, or dread, such as the immediate or apparent cause for a phobic reaction.

manifold [ENG] The branch pipe arrangement which connects the valve parts of a multicylinder engine to a single carburetor or to a muffler. [MATH] A topological space which is locally euclidean; there are four types: topological, piecewise linear, differentiable, and complex, depending on whether the local coordinate systems are obtained from continuous, piecewise linear, differentiable, or complex analytic functions of those in euclidean space; intuitively, a surface.

manifolding [ENG] The gathering of multiple-line fluid inputs into a single intake chamber (intake manifold), or the division of a single fluid supply into several outlet streams (distribution manifold).

manifold of states [ATOM PHYS] A set of states sufficient to form a representation of an operator or a Lie group of operators.

manifold pressure [MECH ENG] The pressure in the intake manifold of an internal combustion engine.

manihot *See* cassava.

manikin [ENG] A correctly proportioned doll-like figure that is jointed and will assume any human position and hold it; useful in art to draw a human figure in action, or in medicine to show the relations of organs by means of movable parts. [MED] **1.** A model of a term fetus; used in the teaching of obstetrics. **2.** A model of an adult human used to teach first aid and basic nursing skills.

Manila copal *See* Manila resin.

Manila hemp *See* abaca.

Manila paper [MATER] Yellowish paper or bristol board; the term originally referred to paper manufactured from Manila hemp.

MANNED SPACECRAFT

The Gemini manned spacecraft. *(NASA)*

Manila resin [MATER] A type of resin extracted from trees of the genus *Agathis* in the Philippines that is soluble in ethyl and methyl alcohol, insoluble in water; used in printing ink, varnishes, paints, and linoleum. Also known as Manila copal.

manioc *See* cassava.

manipulated variable [ADP] Variable whose value is being altered to bring a change in some condition.

manipulation [MED] Skillful use of the hands in moving body parts, as reducing a dislocation, or changing the position of a fetus.

manipulators [ENG] Mechanical devices used for safe handling of dangerous materials of any kind, especially radioactive materials; frequently, they are remotely operated from behind a protective shield.

manjak [MATER] A variety of grahamite, manjak is the blackest of the asphalts; used for insulation and varnishes.

mankato stone [PETR] A variety of limestone containing more than 49% calcium carbonate, with about 4.5% alumina and some silica.

manketti oil [MATER] A light-yellow, viscous varnish oil, obtained from the nuts of the African tree *Ricinodendron rautonemii*.

man-machine chart [IND ENG] A two-column, multiple-activity process chart listing the steps performed by an operator and the operations performed by a machine and showing the corresponding idle times for each.

man-machine system [ENG] A system in which the functions of the worker and the machine are interrelated and necessary for the operation of the system.

man-made fiber [TEXT] A textile fiber or filament manufactured from chemical substances of natural, inorganic, or synthetic origin.

man-made interference [COMMUN] Any electromagnetic interference due to the operation of electrical or electronic equipment, but particularly harmonic or spurious signals from radio-frequency devices, as opposed to noise.

man-made noise [COMMUN] Noise in communications channels produced by sources which are not extraterrestrial or associated with atmospheric disturbances, such as electric motors, electric switching gear, motorcar ignition, high-tension-line leakage, diathermy, and industrial-heating generators.

manna [MATER] The concrete, yellowish, saccharine exudation of the flowering ash (*Fraxinus ornus*); contains mannitol, sugar, mucilage, and resin and has been used as a mild laxative.

mannan [BIOCHEM] Any of a group of polysaccharides composed chiefly or entirely of D-mannose units.

manna sugar *See* mannitol.

manned orbiting laboratory [AERO ENG] An earth-orbiting satellite containing instrumentation and personnel for continuous measurement and surveillance of the earth, its atmosphere, and space. Abbreviated MOL.

manned spacecraft [AERO ENG] A vehicle capable of sustaining a person above the terrestrial atmosphere.

Mannesmann mill [MET] A mill consisting of two rolls mounted with their axes slightly inclined.

Mannesmann process [MET] A process for making seamless tubing by forcing a billet between the rolls of a Mannesmann mill so as to pierce the center, and then forcing the metal over a mandrel to form the central bore.

Mannich condensation reaction *See* Mannich reaction.

Mannich reaction [ORG CHEM] Condensation of a primary or secondary amine or ammonia (usually as the hydrochloride) with formaldehyde and a compound containing at least one reactive hydrogen atom, for example, acetophenone. Also known as Mannich condensation reaction.

Manning equation [FL MECH] An equation used to compute the velocity of uniform flow in an open channel.

mannite *See* mannitol.

mannitol [ORG CHEM] $C_6H_8(OH)_6$ A straight-chain alcohol with six hydroxyl groups; a white, water-soluble, crystalline powder; used in medicine and as a dietary supplement. Also known as manna sugar; mannite.

mannitol hexanitrate [ORG CHEM] $C_6H_8(ONO_2)_6$ Explosive colorless crystals; soluble in alcohol, acetone, and ether, insoluble in water; melts at 112°C; used in explosives and medicine. Also known as hexanitromannite (HNM); nitromannite; nitromannitol.

mannose [BIOCHEM] $C_6H_{12}O_6$ A fermentable monosaccharide obtained from manna.

Mann-Whitney test [STAT] A modification of the Wilcoxon two-sample test.

manocryometer [THERMO] An instrument for measuring the change of a substance's melting point with change in pressure; the height of a mercury column in a U-shaped capillary supported by an equilibrium between liquid and solid in an adjoining bulb is measured, and the whole apparatus is in a thermostat.

manometer [ENG] A double-leg liquid-column gage used to measure the difference between two fluid pressures.

manometry [ENG] The use of manometers to measure gas and vapor pressures.

manostat [ENG] Fluid-filled, upside-down manometer-type device used to control pressures within an enclosure, as for laboratory analytical distillation systems.

man process chart [IND ENG] A graphic representation of the activities performed or to be performed by a worker. Abbreviated MPC.

mansard roof [ARCH] A roof with two slopes on all sides, the lower slope being steeper than the upper one.

M-A-N scavenging system [MECH ENG] A system for removing used oil and waste gases from a cylinder of an internal combustion engine in which the exhaust ports are located above the intake ports on the same side of the cylinder, so that gases circulate in a loop, leaving a dead spot in the center of the loop.

mansfieldite [MINERAL] $Al(AsO_4) \cdot 2H_2O$ A white to pale-gray orthorhombic mineral composed of hydrous aluminum arsenate; it is isomorphous with scorodite.

mansonelliasis [MED] A parasitic infection of humans by the filarioid nematode *Mansonella ozzardi*.

Mantidae [INV ZOO] A family of predacious orthopteran insects characterized by a long, slender prothorax bearing a pair of large, grasping legs, and a freely moving head with large eyes.

mantis [INV ZOO] The common name for insects comprising the family Mantidae.

mantissa [ADP] A fixed point number composed of the most significant digits of a given floating point number. Also known as fixed-point part; floating-point coefficient. [MATH] The positive decimal part of a common logarithm.

mantle [ANAT] Collectively, the convolutions, corpus callosum, and fornix of the brain. [BIOL] An enveloping layer, as the external body wall lining the shell of many invertebrates, or the external meristematic layers in a stem apex. [GEOL] The intermediate shell zone of the earth below the crust and above the core (to a depth of 3480 kilometers). [MET] That part of the outer wall and casing of a blast furnace located above the hearth. [VERT ZOO] The back and wing plumage of a bird if distinguished from the rest of the plumage by a uniform color.

mantle cavity [INV ZOO] The space between mantle and body proper in bivalve mollusks.

mantled gneiss dome [GEOL] A dome in metamorphic terrains that has a remobilized core of gneiss surrounded by a concordant sheath of the basal part of the overlying metamorphic sequence.

mantle lobe [INV ZOO] Either of the flaps on the dorsal and ventral sides of the mantle in bivalve mollusks.

mantle rock *See* regolith.

mantlet [ORD] Protective shield or armor, as in front of a gun, or attached to the front of a tank.

manto [GEOL] A sedimentary or igneous ore body occurring in flat-lying depositional layers.

Mantodea [INV ZOO] An order equivalent to the family Mantidae in some systems of classification.

Mantoux test [IMMUNOL] An intradermal test for tuberculin sensitivity, that is, for past or present infection with tubercle bacilli.

man trip [MIN ENG] A trip made by mine cars and locomotives to take workers, rather than coal, to and from the working places.

manual casing hanger [PETRO ENG] A device in the bowl of the lowermost (or intermediate) casing head to suspend the next smaller casing string and to provide a seal between the suspended casing and the casing-head bowl.

manual central office [COMMUN] Central office of a manual telephone system.

manual control [CONT SYS] Controls arranged for their operation by hand.

manual direction finder *See* manual radio direction finder.

manual element [IND ENG] An element completed by hand.

manual exchange [COMMUN] Any exchange where calls are completed by an operator.

manual input [ADP] The entry of data by hand into a device at the time of processing.

manually controlled work *See* effort-controlled cycle.

manual operation [ADP] Any processing operation performed by hand.

manual radio direction finder [NAV] A radio direction finder which requires manual rotation of an antenna or goniometer, in contrast with an automatic radio direction finder, which indicates automatically and continuously the great-circle direction of the transmitter to which it is tuned. Also known as manual direction finder.

manual rate-aided tracking [ELECTR] Radar circuit which tracks individual targets by computing the velocity from position fixes inserted manually into the circuitry.

manual ringing [COMMUN] Ringing which is started by the manual operation of a key and continues only while the key is held in operation.

manual spinning [MET] A sheet-metal forming process that forms the material over a rotating mandrel with little or no change in the thickness of the original blank. Also known as conventional spinning.

manual switchboard [ELEC] Telephone switchboard in which the connections are made manually, by plugs and jacks, or by keys.

manual switching [ELECTR] Method by which manual connection is made between two or more teletypewriter circuits.

manual system [ADP] A system involving data processing which does not make use of stored-program computing equipment; by this somewhat arbitrary definition, systems using other types of tabulating equipment, such as the card-programmed calculator, are considered to be manual.

manual telephone set [ELECTR] Telephone set not equipped with a dial.

manual telephone system [COMMUN] A telephone system in which connections between customers are ordinarily established manually by telephone operators in accordance with orders given verbally by calling parties.

manual time *See* hand time.

manual tracking [ENG] System of tracking a target in which all the power required is supplied manually through the tracking handwheels.

manual welding [MET] A welding method in which the operator manually guides an electrode, clamped in a hand-held electrode holder.

manubrium [ANAT] **1.** The triangular cephalic portion of the sternum in humans and certain other mammals. **2.** The median anterior portion of the sternum in birds. **3.** The process of the malleus. [BOT] A cylindrical cell that projects inward from the middle of each shield composing the antheridium in stoneworts. [INV ZOO] The elevation bearing the mouth in hydrozoan polyps.

manufactured gas [MATER] A gaseous fuel that is manufactured from soft coal or from various petroleum products; the gas mixture is composed of producer gas or carbureted water gas.

manufacturer's part number [IND ENG] Identification number of symbol assigned by the manufacturer to a part, subassembly, or assembly.

manure [MATER] Animal excreta collected from stables and barnyards with or without litter; used to enrich the soil.

manure salts [INORG CHEM] Potash salts that have a high proportion of chloride and 20–30% potash; used in fertilizers.

manus [ANAT] The hand of a human or the forefoot of a quadruped. [INV ZOO] The proximal enlargement of the propodus of the chela of arthropods.

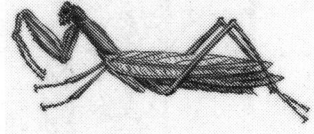

MANTIS

The praying mantis *(Stagomantis carolina)*. *(Illinois Natural History Survey)*

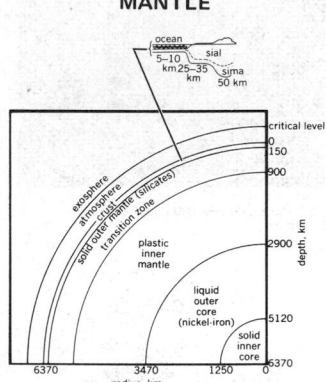

MANTLE

The principal layers of the earth.

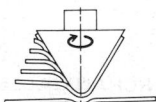

MANUAL SPINNING

Drawing of manual-spinning method of sheet-metal forming, showing the rotating mandrel.

MAPLE

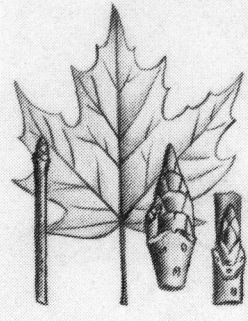

Twig, leaf, and terminal and axillary buds of the sugar maple (*Acer saccharum*).

MAQUIS

Typical low-shrub maquis, Turkey.

MARCASITE

⊢— 2.5 cm —⊣

Typical cockscomb groups of marcasite crystals found in galena-sphalerite ore, Carterville, Missouri. *(Specimen from Department of Geology, Bryn Mawr College)*

MARCHANTIALES

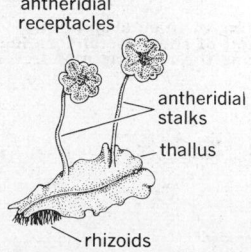

antheridial receptacles — antheridial stalks — thallus — rhizoids

Marchantia, a genus in Marchantiales; example of male gametophyte (antheridial plant). *(From H. J. Fuller and O. Tippo, College Botany, Holt, 1949)*

manus valga [MED] Clubhand with deviation of the ulna.

manway [MIN ENG] A compartment for ladders, pipes, and timber chutes. Also known as ladderway.

many-body force [PHYS] A force exerted on a particle, in the presence of two or more other particles, which differs from the vector sum of the forces which would be exerted on it if each of the other particles were present alone.

many-body problem [MECH] The problem of predicting the motions of three or more objects obeying Newton's laws of motion and attracting each other according to Newton's law of gravitation. Also known as *n*-body problem.

many-body theory [PHYS] A scheme for calculating physical quantities for systems with large numbers of particles, without finding details of each particle's motion, often at temperatures close to absolute zero.

manyplies *See* omasum.

map [ADP] **1.** An output produced by an assembler, compiler, linkage editor, or relocatable loader which indicates the (absolute or relocatable) locations of such elements as programs, subroutines, variables, or arrays. **2.** By extension, an index of the storage allocation on a magnetic disk or drum. [GRAPHICS] A representation, usually on a plane surface, of all or part of the surface of the earth, celestial sphere, or other area; shows relative size and position, according to a given projection, of the physical features represented and such other information as may be applicable to the purpose intended. [MATH] *See* mapping.

MAP *See* macroassembly program.

map chart [MAP] A representation of a land-sea area, using the characteristics of a map to represent the land area and the characteristics of a chart to represent the sea area, with such special characteristics as to make the map chart most useful in military operations, particularly amphibious operations.

maple [BOT] Any of various broad-leaved, deciduous trees of the genus *Acer* in the order Sapindales characterized by simple, opposite, usually palmately lobed leaves and a fruit consisting of two long-winged samaras. [MATER] The hard, light-colored, close-grained wood, especially from sugar maple (*A. saccharum*).

maple sugar [FOOD ENG] Sugar obtained from boiled maple syrup.

maple syrup [FOOD ENG] Syrup made from the sap of certain maple trees, particularly the sugar maple.

maple syrup urine disease [MED] A hereditary metabolic disorder caused by deficiency of branched-chain keto acid decarboxylase; characterized by the maple-syrup-like odor of urine. Also known as branched-chain ketoaciduria.

map parallel *See* axis of homology.

mapping [GRAPHICS] Preparation of a map or engaging in a mapping operation. [MATH] **1.** Any function or multiple-valued relation. Also known as map. **2.** In topology, a continuous function.

mapping radar [NAV] Radar carried on an aircraft as an aid to navigation, which displays on a cathode-ray tube imagery of the ground in the vicinity of the aircraft.

map plotting [METEOROL] The process of transcribing weather information onto maps, diagrams, and so on; it usually refers specifically to decoding synoptic reports and entering those data in conventional station-model form on synoptic charts. Also known as map spotting.

map projection [MAP] A representation or method of representing all or part of the surface of a sphere or spheroid, such as the earth, upon a plane surface. Also known as projection.

map range [ORD] The range from an artillery piece to any point as scaled or computed from a map.

map reading [MAP] Interpretation of the symbols, lines, abbreviations, and terms appearing on maps. [NAV] The determination of position by identification of landmarks with their representations on a map or chart.

map scale [MAP] The ratio between a distance on a map and the corresponding distance on the earth, often represented as 1:80,000 (natural scale) or 30 miles (48.27 kilometers) to an inch.

map spotting *See* map plotting.

map symbol [MAP] A character, letter, or similar graphic representation used on a map to indicate some object, characteristic, and so on.

map vertical *See* geographic vertical.

maquis [ECOL] A type of vegetation composed of shrubs, or scrub, usually not exceeding 3 meters in height, the majority having small, hard, leathery, often spiny or needlelike drought-resistant leaves and occurring in areas with a Mediterranean climate.

maraging steel [MET] High-strength, low-carbon iron-nickel alloy in which a martensitic structure is formed on cooling; contains 7–6% nickel, 0–11% cobalt, 0–5% molybdenum, and small percentages of titanium, aluminum, and columbium; hardening is accomplished by heating the quenched alloy at 400–500°C.

Marangoni effect [CHEM ENG] The effect that a disturbance of the liquid-liquid interface (due to interfacial tension) has on mass transfer in a liquid-liquid extraction system.

Marantaceae [BOT] A family of monocotyledonous plants in the order Zingiberales characterized by one functional stamen with a single functional pollen sac, solitary ovules in each locule, and mostly arillate seeds.

marantic [MED] **1.** Of or pertaining to marasmus. **2.** Of or pertaining to slowed circulation.

marantic endocarditis [MED] Nonbacterial thrombic endocarditis, usually associated with neoplasm or other debilitating disease.

marasmus [MED] Chronic severe wasting of the tissues of the body, particularly in children, due to malnutrition.

Marattiaceae [BOT] A family of ferns coextensive with the order Marattiales.

Marattiales [BOT] An ancient order of ferns having massive eusporangiate sporangia in sori on the lower side of the circinate leaves.

marble [PETR] **1.** Metamorphic rock composed of recrystallized calcite or dolomite. **2.** Commercially, any limestone or dolomite taking polish.

marble bone disease *See* osteopetrosis.

marble dust *See* marble flour.

marble flour [MATER] Finely divided marble chips; used as a filler or abrasive in hand soaps and for casting. Also known as marble dust.

marble paper [GRAPHICS] A decorated paper, with a coloration that resembles marble, used as end leaves in blank books and often in printed books.

marble shot [PETRO ENG] An explosive shot in open-hole well completions in which glass marbles are packed around the explosive in the wellbore; the marbles become projectiles that help break up the formation.

Marburg virus [VIROL] A large virus transmitted to humans by the grivet monkey (*Cercopithecus aethiops*).

marcasite [MINERAL] FeS_2 A pale bronze-yellow to nearly white mineral, crystallizing in the orthorhombic system; hardness is 6–6.5 on Mohs scale, and specific gravity is 4.89.

marcescent [BOT] Withering without falling off.

Marcgraviaceae [BOT] A family of dicotyledonous shrubs or vines in the order Theales having exstipulate leaves with scanty or no endosperm, two integuments, and highly modified bracts.

march [METEOROL] The variation of any meteorological element throughout a specific unit of time, such as a day, month, or year; as the daily march of temperature, the complete cycle of temperature during 24 hours.

Marchantiales [BOT] The thallose liverworts, an order of the class Marchantiopsida having a flat body composed of several distinct tissue layers, smooth-walled and tuberculate-walled rhizoids, and male and female sex organs borne on stalks on separate plants.

Marchantiatae *See* Marchantiopsida.

Marchantiopsida [BOT] The liverworts, a class of lower green plants; the plant body is usually a thin, prostrate thallus with rhizoids on the lower surface.

March equinox *See* vernal equinox.

Marconaflo slurry transport [MIN ENG] A system which recovers solids from ore or tailings piles and mixes them with water to produce a transportable slurry in pipelines.

Marconi antenna [ELECTROMAG] Antenna system of which the ground is an essential part, as distinguished from a Hertz antenna.

Marcq–St.-Hilaire method [NAV] A mathematical process of deriving from an observation the information needed for establishing a line of position.

Marcy mill [MIN ENG] A ball mill with a vertical grate diaphragm placed near the discharge end; screens for sizing the material are located between the diaphragm and the end of the tube.

mare [ASTRON] **1.** One of the large, dark, flat areas on the lunar surface. **2.** One of the less well-defined areas on Mars. [VERT ZOO] A mature female horse or other equine.

marekanite [GEOL] Rounded to subangular obsidian bodies that occur in masses of perlite.

mare's tail *See* precipitation trajectory.

Marfan's syndrome [MED] A hereditary connective-tissue disorder transmitted as an autosomal dominant; manifested by skeletal and ocular changes and by congenital heart disease.

margaric acid *See* n-heptadecanoic acid.

margarine [FOOD ENG] A plastic food fat product composed of processed vegetable oils or animal fats or both, cultured milk, salt, and emulsifiers.

margarine oil [FOOD ENG] Edible vegetable oil used to make oleomargarine.

margarite [GEOL] A string of beadlike globulites; commonly found in glassy igneous rocks. [MINERAL] $CaAl_2(Al_2Si_2)$-$O_{10}(OH)_2$ A pink, reddish, or yellow, brittle mica mineral.

Margaritiferidae [INV ZOO] A family of gastropod mollusks with nacreous shells that provide an important source of commercial pearls.

Margarodinae [INV ZOO] A subfamily of homopteran insects in the superfamily Coccoidea in which abdominal spiracles are present in all stages of development.

margarosanite [MINERAL] $PbCa_2(SiO_3)_3$ A colorless or snow-white triclinic mineral composed of lead calcium silicate, occurring in lamellar masses.

margin [GEOGR] The boundary around a body of water. [GRAPHICS] The blank area at the vertical and horizontal edges of a printed page. [SCI TECH] An outside limit.

marginal blight [PL PATH] A bacterial disease of lettuce caused by *Pseudomonas marginalis*, characterized by brownish marginal discoloration of the foliage.

marginal checking [ELECTR] A preventive-maintenance procedure in which certain operating conditions, such as supply voltage or frequency, are varied about their normal values in order to detect and locate incipient defective units.

marginal chlorosis [PL PATH] A virus disease characterized by yellowing or blanching of leaf margins; common disease of peanut plants.

marginal cost [IND ENG] The extra cost incurred for an extra unit of output.

marginal escarpment [GEOL] A seaward slope of a marginal plateau with a gradient of 1:10 or more.

marginal fissure [GEOL] A magma-filled fracture bordering an igneous intrusion.

marginal moraine *See* terminal moraine.

marginal placentation [BOT] Arrangement of ovules near the margins of carpels.

marginal plain *See* outwash plain.

marginal plateau [GEOL] A relatively flat shelf adjacent to a continent and similar topographically to, but deeper than, a continental shelf.

marginal probability [STAT] Probability expressed by the two conditional probability distributions which arise from the joint distribution of two random variables.

marginal product [IND ENG] The extra unit of output obtained by one extra unit of some factor, all other factors being held constant.

marginal relay [ELEC] Relay with a small margin between its nonoperative current value (maximum current applicable without operation) and its operative value (minimum current that operates the relay).

marginal revenue [IND ENG] The extra revenue achieved by selling an extra unit of output.

marginal salt pan [GEOL] A natural, coastal salt pan.

marginal sea [GEOGR] A semiclosed sea adjacent to a continent and connected with the ocean at the water surface.

marginal sinus [ANAT] **1.** One of the small, bilateral sinuses of the dura mater which skirt the edge of the foramen magnum, usually uniting posteriorly to form the occipital sinus. **2.** *See* terminal sinus. [EMBRYO] An enlarged venous sinus incompletely encircling the margin of the placenta.

marginal test [ELECTR] A test of electronic equipment in which conditions are varied until failures occur or faults can be detected, allowing measurement of permissible operating margins.

marginal thrust [GEOL] One of a series of faults bordering an igneous intrusion and crossing both the intrusion and the wall rock. Also known as marginal upthrust.

marginal ulcer [MED] A peptic ulcer of the jejunum on the efferent margin of a gastrojejunostomy.

marginal upthrust *See* marginal thrust.

marginate [BOT] Having a distinct margin or border.

margin line [NAV ARCH] A line drawn parallel to the bulkhead deck of a ship at the side and 76 millimeters (3 inches) below the upper surface of that deck.

margin of safety [DES ENG] A design criterion, usually the ratio between the load that would cause failure of a member or structure and the load that is imposed upon it in service.

margin plate [NAV ARCH] The plate forming the sides of the inner bottom tank of a ship.

margin-punched card [ADP] A card which is punched only along the edges, the remaining area being used for printed or written data.

margosa oil *See* neem oil.

Margoulis number *See* Stanton number.

Margules equation *See* Witte-Margules equation.

marialite [MINERAL] $3NaAlSi_3O_8 \cdot NaCl$ A scapolite mineral that is isomorphous with meronite.

maricolous [ECOL] Living in the sea.

mariculture [AGR] The cultivation of marine organisms, plant and animal, for purposes of human consumption.

Marie's ataxia [MED] A hereditary ataxia combining features of cerebellar, posterior column, and pyramidal tract lesions, with onset after age 20, normal or exaggerated deep tendon reflexes, and frequently optic atrophy and oculomotor palsies, but no clubfeet or scoliosis.

Marie's disease [MED] Rheumatic spondylitis involving the spine only, or invading the shoulders and hips.

Marie-Strümpell disease *See* rheumatoid spondylitis.

Marietta miner [MIN ENG] Trade name for a continuous miner mounted on caterpillar treads; has two cutter arms and two cutter chains at the working face, and a conveyor system to carry the broken coal to be loaded onto cars.

marigram [OCEANOGR] A graphic record of the rising and falling movements of the tide expressed as a curve.

marigraph [ENG] A self-registering gage that records the heights of the tides.

marihuana *See* marijuana.

marijuana [BOT] The Spanish name for the dried leaves and flowering tops of the hemp plant (*Cannabis sativa*), which have narcotic ingredients and are smoked in cigarettes. Also spelled marihuana.

marina [CIV ENG] A harbor facility for small boats, yachts, and so on, where supplies, repairs, and various services are available.

marine [OCEANOGR] Pertaining to the sea.

marine abrasion [GEOL] Erosion of the ocean floor by sediment moved by ocean waves. Also known as wave erosion.

marine alidade [NAV] An instrument used aboard ship in conjunction with, but not including, a pair of binoculars to indicate the relative bearing of a target or object; it is usually fastened to a solid mounting with the zero degree position parallel to the centerline of the ship.

marine biocycle [ECOL] A major division of the biosphere composed of all biochores of the sea.

marine biology [BIOL] A branch of biology that deals with those living organisms which inhabit the sea.

marine boiler [NAV ARCH] A steam boiler designed to meet requirements of ship operation in propulsion, running of auxiliary machinery, generation of electricity, and heating.

marine cave *See* sea cave.

marine chart *See* nautical chart.

marine climate [CLIMATOL] A regional climate which is under the predominant influence of the sea, that is, a climate characterized by oceanity; the antithesis of a continental climate. Also known as maritime climate; oceanic climate.

marine-cut terrace [GEOL] A terrace or platform cut by wave

MARGINAL PLACENTATION

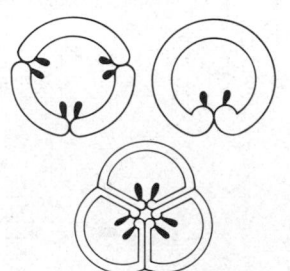

Types of marginal placentation; ovules in black.

MARIJUANA

Branch of the marijuana (*Cannabis sativa*) plant.

erosion of marine origin. Also known as wave-cut terrace.

marine distiller [NAV ARCH] A machine that distills fresh water from sea water on a ship; evaporation must be carried out at temperatures below 60°C to prevent the sea water from forming a hard, insoluble scale, and the boiling liquid must therefore be kept below atmospheric pressure. Also known as marine evaporator.

marine engine [NAV ARCH] An engine that propels a water-borne vessel.

marine engineering [ENG] The design, construction, installation, operation, and maintenance of main power plants, as well as the associated auxiliary machinery and equipment, for the propulsion of ships.

marine evaporator *See* marine distiller.

marine forecast [METEOROL] A forecast, for a specified oceanic or coastal area, of weather elements of particular interest to maritime transportation, including wind, visibility, the general state of the weather, and storm warnings.

marine geology *See* geological oceanography.

marine glue [MATER] An adhesive that is insoluble in water; usually made of rubber and shellac, sometimes with resins.

marine light [NAV] A luminous or lighted aid to navigation intended primarily for marine navigation.

marine littoral faunal region [ECOL] A geographically determined division of that portion of the zoosphere composed of marine animals.

marine meteorology [METEOROL] That part of meteorology which deals mainly with the study of oceanic areas, including island and coastal regions; in particular, it serves the practical needs of surface and air navigation over the oceans.

marine microbiology [MICROBIO] The study of the microorganisms living in the sea.

marine navigation [NAV] The process of directing the movements of watercraft from one point to another; the process, always present in some form when a vessel is under way and not drifting, varies with the type of craft, its mission, and its area of operation.

marine propeller [NAV ARCH] A component of a ship-propulsion power plant which converts engine torque force into propulsive force or thrust, thus overcoming the hull resistance of a moving ship by creating a sternward accelerated column of water.

marine radio beacon [NAV] A radio-beacon station which produces service primarily for the guidance of ships.

marine railway [CIV ENG] A type of dry dock consisting of a cradle of wood or steel with rollers on which the ship may be hauled out of the water along a fixed inclined track leading up the bank of a waterway.

marine rainbow [OPTICS] A rainbow seen in ocean spray. Also known as sea rainbow.

Mariner program [AERO ENG] A United States program, begun in 1962, to send a series of unmanned, solar-powered spacecraft to the vicinity of Venus and Mars, to carry out observations of cosmic rays and solar wind in interplanetary space, to investigate interactions of the solar wind with planets, and to make atmospheric and surface measurements of planets, including photographic scanning of the surface of Mars.

marine salina [GEOGR] A body of salt water found along an arid coast and separated from the sea by a sand or gravel barrier.

Marinesco-Sjögren-Garland syndrome [MED] A hereditary, congenital form of cerebral ataxia transmitted as an autosomal recessive; characterized by mental retardation, cataracts, minor skeletal anomalies, and hypertension.

marine sextant [NAV] A sextant designed primarily for ship navigation.

marine stack *See* stack.

marine swamp [GEOGR] An area of low, salty, or brackish water found along the shore and characterized by abundant grasses, mangrove trees, and similar vegetation. Also known as paralic swamp.

marine terminal [CIV ENG] That part of a port or harbor with facilities for docking, cargo-handling, and storage.

marine terrace [GEOL] A seacoast terrace formed by the merging of a wave-built terrace and a wave-cut platform. Also known as sea terrace; shore terrace.

MARINE RAILWAY

Bow-end view of a marine railway which has been pulled out of water.

marine traffic [NAV] Traffic on the waterways.

marine transgression *See* transgression.

marine weather observation [METEOROL] The weather as observed from a ship at sea, usually taken in accordance with procedures specified by the World Meteorological Organization.

Mariotte's law *See* Boyle's law.

maritime air [METEOROL] A type of air whose characteristics are developed over an extensive water surface and which, therefore, has the basic maritime quality of high moisture content in at least its lower levels.

maritime climate *See* marine climate.

maritime frequency bands [COMMUN] In the United States, a collection of radio frequencies allocated for communication between coast stations and ships or between ships.

maritime law [NAV] Law that concerns navigation and commerce on the oceans and other navigable bodies of water.

maritime mobile service [COMMUN] A mobile service between coast stations and ship stations, or between ship stations, in which survival craft stations may also participate.

maritime polar air [METEOROL] Polar air initially possessing similar properties to those of continental polar air, but in passing over warmer water it becomes unstable with a higher moisture content.

maritime position [NAV] The location of a seaport or other point along a coast.

maritime tropical air [METEOROL] The principal type of tropical air, produced over the tropical and subtropical seas; it is very warm and humid, and is frequently carried poleward on the western flanks of the subtropical highs.

marjoram [BOT] Any of several perennial plants of the genera *Origanum* and *Majorana* in the mint family, Labiatae; the leaves are used as a food seasoning.

marjoram oil [MATER] A colorless essential liquid whose chief components are terpenes, obtained from marjoram plants of the genus *Origanum*; used as a perfume in soaps, and in flavorings.

mark [ADP] A distinguishing feature used to signal some particular location or condition. [COMMUN] The closed-circuit condition in telegraphic communication, during which the signal actuates the printer; the opposite of space. [NAV] **1.** A charted conspicuous object, structure, or light serving as an indicator for guidance or warning to craft; a beacon; it may be a day-beacon or sea-mark depending upon its location, or a day-mark or lighted beacon depending upon its period of usefulness. **2.** Fathoms marked on a lead line. [ORD] A designation followed by a serial number, used to identify models of military equipment.

mark detection [ADP] That class of character recognition systems which employs coded documents, in the form of boxes or windows, in order to convey intended information by means of pencil or ink marks made in specific boxes.

marked anatomy *See* pathologic anatomy.

marker [ORD] A sign or signal for marking a location on land or water; frequently contains pyrotechnics.

marker beacon [NAV] A low-power radio beacon transmitting a signal to designate a small area, as an aid to navigation.

marker bed [GEOL] **1.** A stratified unit with distinctive characteristics making it an easily recognized geologic horizon. **2.** A rock layer which accounts for a characteristic portion of a seismic refraction time-distance curve. **3.** *See* key bed.

marker buoy [NAV] **1.** A temporary buoy used in surveying to mark a location of particular interest, such as a shoal or reef. **2.** *See* station buoy.

marker pulse [COMMUN] A pulse used for synchronization between transmitter and receiver in a time-division multiplex system.

market analysis [IND ENG] The collection and evaluation of data concerned with the past, present, or future attributes of potential consumers for a product or service.

mark-hold [COMMUN] The transmission of a steady mark to indicate that there is no traffic over a telegraph channel; the upper marking frequency of a duplex channel (2225 hertz) is used to disable echo suppressors which may interfere with data communications.

Mark-Houwink equation [PHYS CHEM] The relationship between intrinsic viscosity and molecular weight for homogeneous linear polymers.

marking and spacing intervals [COMMUN] Intervals of closed and open conditions in transmission circuits.

marking bias [COMMUN] Bias distortion that lengthens the marking impulse.

marking current [ELEC] Magnitude and polarity of current in the line when the receiving mechanism is in the operating position.

marking-end distortion [COMMUN] End distortion that lengthens the marking impulse.

marking pulse [ELEC] In a teletypewriter, the signal interval during which time the teletypewriter selector unit is operated.

marking wave [ELEC] In telegraphic communications, that portion of the emission during which the active portions of the code character are being transmitted. Also known as keying wave.

Markov chain [MATH] A Markov process whose state space is finite or countably infinite.

Markov inequality [STAT] If x is a random variable with probability P and expectation E, then $P(|x| \geq a) \leq E(|x|^n/a^n)$.

Markovnikoff's rule [ORG CHEM] In an addition reaction, the additive molecule RH adds as H and R, with the R going to the carbon atom with the lesser number of hydrogen atoms bonded to it.

Markov process [MATH] A process in which the future state is completely determined by the present state. [STAT] A stochastic process which assumes that in a series of random events the outcome or probability of each event depends only on the immediately preceding event.

mark reading [ADP] In character recognition, that form of mark detection which employs a photoelectric device to locate and convey intended information; the information appears as special marks on sites (windows) within the document coding area.

mark sensing [ADP] In character recognition, that form of mark detection which depends on the conductivity of graphite pencil marks to locate and convey intended information; the information appears as special marks on sites (windows) within the document coding area.

mark-to-space transition [COMMUN] The process of switching from a mark to a space.

marl [GEOL] A deposit of crumbling earthy material composed principally of clay with magnesium and calcium carbonate; used as a fertilizer for lime-deficient soils.

marline [NAV ARCH] A tarred two-stranded, left-handed hemp about 1/8 inch (3 millimeters) in diameter; used for neat seizings.

marline spike [NAV ARCH] A tapered metal tool used to separate the strands of rope in splicing, and as a lever in marling and seizing.

marlite See marlstone.

marlstone [PETR] 1. A consolidated rock that has about the same composition as marl; considered to be an earthy or impure argillaceous limestone. Also known as marlite. 2. A hard ferruginous rock of the Middle Lias in England.

marly [GEOL] Pertaining to, containing, or resembling marl.

marmatite [MINERAL] A dark-brown to black mineral composed of iron-bearing sphalerite. Also known as christophite.

marmolite [MINERAL] A pale-green serpentine mineral, occurring in thin laminations; a variety of chrysotile.

marmoset [VERT ZOO] Any of 10 species of South American primates belonging to the family Callithricidae; individuals are primitive in that they have claws rather than nails and a nonprehensile tail.

marmot [VERT ZOO] Any of several species of stout-bodied, short-legged burrowing rodents of the genus *Marmota* in the squirrel family Sciuridae.

marrite [MINERAL] PbAgAsS$_3$ A monoclinic mineral, occurring as small crystals in Valais, Switzerland.

marrubium [BOT] *Marrubium vulgari.* An aromatic plant of the mint family, Labiatae; leaves have a bitter taste and are used as a tonic and anthelmintic. Also known as hoarhound; horehound.

Mars [ASTRON] The planet fourth in distance from the sun; it is visible to the naked eye as a bright red star, except for short periods when it is near its conjunction with the sun; its diameter is about 6700 kilometers.

Marsden chart [METEOROL] A system for showing the distribution of meteorological data on a chart, especially over the oceans; using a Mercator map projection, the world between 80°N and 70°S latitudes is divided into Marsden "squares," each of 10° latitude by 10° longitude and systematically numbered to indicate position; each square may be divided into quarter squares, or into 100 one-degree subsquares numbered from 00 to 99 to give the position to the nearest degree.

Marseilles fever See fièvre boutonneuse.

Marseilles soap [MATER] A castile soap to which is added olive oil and soda.

marsh [ECOL] A transitional land-water area, covered at least part of the time by estuarine or coastal waters, and characterized by aquatic and grasslike vegetation, especially without peatlike accumulation.

Marsh-Berzelius test See Marsh test.

marsh gas [GEOCHEM] Combustible gas, consisting chiefly of methane, produced as a result of decay of vegetation in stagnant water.

marshite [MINERAL] CuI A reddish, oil-brown isometric mineral composed of cuprous iodide and occurring as crystals; hardness is 2.5 on Mohs scale, and specific gravity is 5.6.

marsh ore See bog iron ore.

Marsh test [ANALY CHEM] A test for the presence of arsenic in a compound; the substance to be tested is mixed with granular zinc, and dilute hydrochloric acid is added to the mixture; gaseous arsine forms, which decomposes to a black deposit of arsenic, when the gas is passed through a heated glass tube. Also known as Marsh-Berzelius test.

Marsileales [BOT] A small monofamilial order of heterosporous, leptosporangiate ferns (Polypodiophyta); leaves arise on long stalks from the rhizome, and sporangia are enclosed in modified folded leaves or leaf segments called sporocarps.

Marsipobranchii [VERT ZOO] An equivalent name for Cyclostomata.

Mars pigments [INORG CHEM] A group of five pigments produced when milk of lime is added to a ferrous sulfate solution, and the precipitate is calcined; color is controlled by calcination temperature to give yellow, orange, brown, red, or violet.

Mars probe [AERO ENG] A United States unmanned spacecraft intended to be sent to the vicinity of the planet Mars, such as in the Mariner or Viking programs.

marsupial [VERT ZOO] 1. A member of the Marsupialia. 2. Having a marsupium. 3. Of, pertaining to, or constituting a marsupium.

marsupial anteater [VERT ZOO] *Myrmecobius fasciatus.* An anteater belonging to the Marsupialia. Also known as banded anteater.

marsupial frog [VERT ZOO] Any of various South American tree frogs which carry the eggs in a pouch on the back.

Marsupialia [VERT ZOO] The single order of the mammalian infraclass Metatheria, characterized by the presence of a marsupium in the female.

marsupialization [MED] Surgical evacuation of pancreatic, hydatid, and other cysts when complete removal and closure are not possible.

marsupial mole [VERT ZOO] *Notoryctes typhlops.* A marsupial of Australia that resembles the euterian mole.

Marsupicarnivora [VERT ZOO] An order proposed to include the polydactylous and polyprotodont carnivorous superfamilies of Marsupialia.

marsupium [VERT ZOO] A fold of skin that forms a pouch enclosing the mammary glands on the abdomen of most marsupials.

martempering [MET] Quenching austenitized steel to a temperature just above, or in the upper part of, the martensite range, holding it at this point until the temperature is equalized throughout, and then cooling in air to room temperature.

marten [VERT ZOO] Any of seven species of carnivores of the genus *Martes* in the family Mustelidae which resemble the weasel but are larger and of a semiarboreal habit.

martensite [MET] A metastable transitional structure formed

MARMOSET

The common marmoset, about 9 inches (23 centimeters) long with a 12-inch (30-centimeter) tail, an inhabitant of the mouth of the Amazon River.

MARMOT

Though short-legged and heavy, the marmot runs, jumps, and climbs with speed and agility.

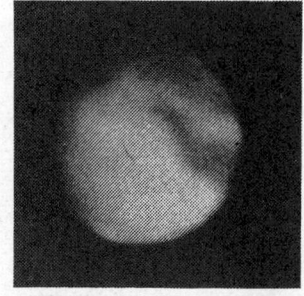

MARS

Photograph of Mars through yellow filter at Lick Observatory showing surface details.

MARTEN

Though scarce, the American marten is important as predator on mice and other injurious rodents.

by a shear process during a phase transformation, characterized by an acicular or needlelike pattern; in carbon steel it is a hard, supersaturated solid solution of carbon in a body-centered tetragonal lattice of iron.

martensite range [MET] The temperature interval between the temperature (M_s) at which formation of martensite initiates and the temperature (M_f) at which the formation is complete.

martensitic stainless steel [MET] A hard, quenched magnetic martensitic steel containing principally 11–18% chromium and 0.1–1.2% carbon.

martensitic steel [MET] Quenched carbon steel composed chiefly of martensite.

martensitic structure [MET] Of, pertaining to, or having the structure of martensite, that is, an interstitial, supersaturated solid solution of carbon in iron having a body-centered tetragonal lattice; the microstructure is characterized by an acicular or needlelike pattern.

martensitic transformation [MET] A phase transformation which occurs in some metals, resulting in the formation of martensite.

Martens wedge [OPTICS] A wedge-shaped piece of quartz used to rotate the plane of polarization of linearly polarized light.

martingale [STAT] A sequence of random variables $x_1, x_2, \ldots,$ where the conditional expected value of x_{n+1} given $x_1, x_2, \ldots, x_n$, equals x_n.

Martin's cement [MATER] A gypsum cement made with potassium carbonate instead of alum.

martite [MINERAL] Hematite occurring in iron-black octahedral crystals pseudomorphous after magnetite.

martonite [MATER] A poison gas composed of 20% chloroacetone and 80% bromoacetone; acts as a powerful lacrimator.

Marvin sunshine recorder [ENG] A sunshine recorder in which the time scale is supplied by a chronograph, and consisting of two bulbs (one of which is blackened) that communicate through a glass tube of small diameter, which is partially filled with mercury and contains two electrical contacts; when the instrument is exposed to sunshine, the air in the blackened bulb is warmed more than that in the clear bulb; the warmed air expands and forces the mercury through the connecting tube to a point where the electrical contacts are shorted by the mercury; this completes the electrical circuit to the pen on the chronograph.

Marx circuit [ELEC] An electric circuit used in an impulse generator in which capacitors are charged in parallel through charging resistors, and then connected in series and discharged through the test piece by the simultaneous sparkover of spark gaps.

Marx effect [SOLID STATE] The effect wherein the energy of photoelectrons emitted from an illuminated surface is decreased when the surface is simultaneously illuminated by light of lower frequency than that causing the emission.

mA s *See* milliampere-second.

mascagnite [MINERAL] $(NH_4)_2SO_4$ A yellowish-gray mineral found in guano, near burning coal beds, or as lava incrustation; specific gravity is 1.77; hardness is 2–2.5 on Mohs scale.

mascaret *See* bore.

mascon [GEOL] A large, high-density mass concentration below a ringed mare on the surface of the moon.

masculine [BIOL] Having an appearance or qualities distinctive for a male.

masculine pelvis [ANAT] A female pelvis similar to the normal male pelvis in having a deeper cavity and more conical shape. Also known as android pelvis.

masculine protest [PSYCH] The struggle to dominate, exhibited primarily by women but to some extent also by men, with the desire to escape identification with the feminine role.

masculinize [PHYSIO] To cause a female or a sexually immature animal to take on male secondary sex characteristics.

masculinoma [MED] Adrenocorticoid adenoma of the ovary.

maser [PHYS] A device for coherent amplification or generation of electromagnetic waves in which an ensemble of atoms or molecules, raised to an unstable energy state, is stimulated by an electromagnetic wave to radiate excess energy at the same frequency and phase as the stimulating wave. Derived

from microwave amplification by stimulated emission of radiation. Also known as paramagnetic amplifier.

maser amplifier [ELECTR] A maser which is used to increase the power produced by another maser.

mash [FOOD ENG] **1.** Mixture of grain and other ingredients fermented to produce whiskey. **2.** Malted barley or other grain mixed with water to prepare wort for brewing operations.

mash seam weld [MET] A seam weld at a lap joint in which the overall lap thickness is reduced plastically to the approximate thickness of one of the lapped parts.

mask [ADP] A pattern of characters used to control the retention or elimination of portions of another pattern of characters. Also known as extractor. [DES ENG] A frame used in front of a television picture tube to conceal the rounded edges of the screen. [ELECTR] A thin sheet of metal or other material containing an open pattern, used to shield selected portions of a semiconductor or other surface during a deposition process.

mask face [MED] An expressionless face seen in certain degenerative and inflammatory diseases of the basal ganglia and the extrapyramidal system; voluntary movements are near normal while involuntary movements are infrequent.

masking [ACOUS] The amount by which the threshold of audibility of a sound is raised by the presence of another sound; the unit customarily used is the decibel. Also known as audio masking; aural masking. [ADP] **1.** Replacing specific characters in one register by corresponding characters in another register. **2.** Extracting certain characters from a string of characters. [ELECTR] **1.** Using a covering or coating on a semiconductor surface to provide a masked area for selective deposition or etching. **2.** A programmed procedure for eliminating radar coverage in areas where such transmissions may be of use to the enemy for navigation purposes, by weakening the beam in appropriate directions or by use of additional transmitters on the same frequency at suitable sites to interfere with homing; also used to suppress the beam in areas where it would interfere with television reception.

mask matching [ADP] In character recognition, a method employed in character property detection in which a correlation or match is attempted between a specimen character and each of a set of masks representing the characters to be recognized.

mask register [ADP] Filter which determines the parts of a word which are to be tested.

mask word [ADP] A word modifier used in a logical AND operation.

masochism [PSYCH] Pleasure derived from experiencing physical or psychological pain.

Masonite [MATER] A trademark for fiberboards made by the Masonite Corporation; they are produced by reducing wood chips to basic cellulose fibers under high steam pressure; the resulting pulp, containing long fibers and the natural lignin adhesive of the wood, is pressed into boards.

masonry [CIV ENG] A construction of stone or similar materials such as concrete or brick.

masonry cement [MATER] A blended cement, made by combining either natural or portland cements with fattening materials such as hydrated lime and, sometimes, with air-entraining mixtures; used in the mortar of brick and block masonry.

masonry dam [CIV ENG] A dam constructed of stone or concrete blocks set in mortar.

masonry drill [DES ENG] A drill tipped with cemented carbide for drilling in concrete or masonry.

masonry nail [DES ENG] Spiral-fluted nail designed to be driven into mortar joints in masonry.

mason's hydrated lime [MATER] Any hydrated lime suitable for use in mortars, base-coat plasters, and concrete.

Mason's theorem [CONT SYS] A formula for the overall transmittance of a signal flow graph in terms of transmittances of various paths in the graph.

mass [MECH] A quantitative measure of a body's resistance to being accelerated; equal to the inverse of the ratio of the body's acceleration to the acceleration of a standard mass under otherwise identical conditions.

mass absorption coefficient [PHYS] The linear absorption coefficient divided by the density of the medium.

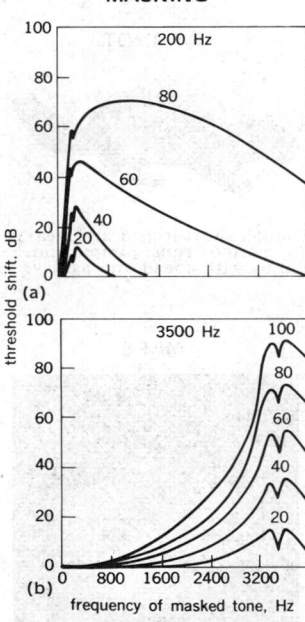

MASKING

threshold shift, dB

(a) 200 Hz — 80, 60, 40, 20

(b) 3500 Hz — 100, 80, 60, 40, 20

frequency of masked tone, Hz

Masking of pure tones by pure tones at masking frequencies of (a) 200 hertz, (b) 3500 hertz. The number above each curve is the level in decibels above the threshold of audibility of each masking tone. (*From H. Fletcher, Speech and Hearing in Communication, 2d ed., Van Nostrand, 1953*)

mass action law [PHYS CHEM] The law that the rate of a chemical reaction for a uniform system at constant temperature is proportional to the concentrations of the substances reacting. Also known as Guldberg and Waage law.

massage [MED] The act of rubbing, kneading, or stroking the superficial parts of the body with the hand or with an instrument, for therapeutic purposes.

mass-analyzed ion kinetic energy spectrometry [SPECT] A type of ion kinetic energy spectrometry in which the ionic products undergo mass analysis followed by energy analysis. Abbreviated MIKES.

mass attraction vertical [GEOPHYS] The vertical which is a function only of the distribution of mass and is unaffected by forces resulting from the motions of the earth.

mass burning rate [ORD] Rate of consumption of propellant charge, usually expressed in pounds per second.

mass communication [COMMUN] Communication which is directed to or reaches an appreciable fraction of the population.

mass concrete [CIV ENG] Concrete set without structural reinforcement.

mass-data multiprocessing [ADP] The basic concept of time sharing, with many inquiry stations to a central location capable of on-line data retrieval.

mass defect [NUC PHYS] The difference between the mass of an atom and the sum of the masses of its individual components in the free (unbound) state.

mass-distance [ENG] The mass carried by a vehicle multiplied by the distance it travels.

mass divergence [FL MECH] The divergence of the momentum field, a measure of the rate of net flux of mass out of a unit volume of a system; in symbols, $\nabla \cdot \rho \mathbf{V}$, where ρ is the fluid density, $\mathbf{V}$ the velocity vector, and ∇ the del operator.

massecuite [FOOD ENG] Sugar industry term for sugar-molasses mixture prior to the removal of the molasses.

mass-energy conservation [RELAT] The principle that energy cannot be created or destroyed; however, one form of energy is that which a particle has because of its rest mass, equal to this mass times the square of the speed of light.

mass-energy relation [RELAT] The relation whereby the total energy content of a body is equal to its inertial mass times the square of the speed of light.

Massenfilter *See* quadrupole spectrometer.

masseter [ANAT] The masticatory muscle, arising from the zygomatic arch and inserted into the lower jaw.

mass extinction *See* faunal extinction.

Massey formula [ATOM PHYS] A formula for the probability that an excited atom approaching the surface of a metal will emit secondary electrons.

mass flow [FL MECH] The mass of a fluid in motion which crosses a given area in a unit time.

mass formula [NUC PHYS] An equation giving the atomic mass of a nuclide as a function of its atomic number and mass number.

mass-haul curve [CIV ENG] A curve showing the quantity of excavation in a cutting which is available for fill.

massicot [MINERAL] PbO A yellow, orthorhombic mineral consisting of lead monoxide; found in the western and southern United States. Also known as lead ocher.

Massieu function [THERMO] The negative of the Helmholtz free energy divided by the temperature.

massif [GEOL] A massive block of rock within an erogenic belt, generally more rigid than the surrounding rocks, and commonly composed of crystalline basement or younger plutons.

massive [GEOL] Of a mineral deposit, having a large concentration of ore in one place. [MINERAL] Of a mineral, lacking an internal structure. [PALEON] Of corallum, composed of closely packed corallites. [PETR] **1.** Of a competent rock, being homogeneous, isotropic, and elastically perfect. **2.** Of a metamorphic rock, having constituents which do not show parallel orientation and are not arranged in layers. **3.** Of igneous rocks, being homogeneous over wide areas and lacking layering, foliation, cleavage, or similar features.

mass law of sound insulation [CIV ENG] The rule stating that sound insulation for a single wall is determined almost wholly by its weight per unit area; doubling the weight of the partition increases the insulation by 5 decibels.

mass-luminosity relation [ASTROPHYS] A relation between stellar magnitudes and mass of the stars; when the absolute magnitudes of stars are plotted versus the logarithms of their masses, the points fall closely along a smooth curve.

mass-memory unit [ADP] Drum or disk memory that provides rapid access bulk storage for messages that are awaiting availability of outgoing channels.

mass movement [GEOL] Movement of a portion of the land surface as a unit.

mass number [NUC PHYS] The sum of the numbers of protons and neutrons in the nucleus of an atom or nuclide. Also known as nuclear number; nucleon number.

mass operator [QUANT MECH] An operator which is added to the Lagrangian in a quantized field theory in order to eliminate certain infinite quantities, and whose sum with the mechanical mass gives the observed mass.

mass ratio [AERO ENG] The ratio of the mass of the propellant charge of the rocket to the total mass of the rocket when charged with the propellant.

mass reflex [PHYSIO] A spread of reflexes suggesting lack of control by higher cortical centers; seen in normal newborns, in persons under the influence of drugs or in severe emotional states, and in encephalopathy or high spinal cord transections.

mass renormalization [QUANT MECH] The mathematical operation of adding the mass which a particle possesses because of its self interaction, to its mechanical mass in order to obtain its measured mass.

mass spectrograph [ENG] A mass spectroscope in which the ions fall on a photographic plate which after development shows the distribution of particle masses.

mass spectrometer [ENG] A mass spectroscope in which a slit moves across the paths of particles with various masses, and an electrical detector behind it records the intensity distribution of masses.

mass spectrometry [ANALY CHEM] An analytical technique for identification of chemical structures, determination of mixtures, and quantitative elemental analysis, based on application of the mass spectrometer.

mass spectroscope [ENG] An instrument used for determining the masses of atoms or molecules, in which a beam of ions is sent through a combination of electric and magnetic fields so arranged that the ions are deflected according to their masses.

mass spectrum [PARTIC PHYS] A plot of masses of elementary particles, including unstable states. Also known as particle spectrum. [PHYS] A display, record, or plot of the distribution in mass, or in mass-to-charge ratio, of ionized atoms, molecules, or molecular fragments.

mass stopping power [NUCLEO] The decrease, per unit surface density traversed, in kinetic energy of an ionizing particle passing through matter; equal to the linear energy transfer (energy loss per unit path length) divided by the density of the material.

mass storage [ADP] A computer storage with large capacity, especially one whose contents are directly accessible to a computer's central processing unit.

mass-storage executive capability [ADP] The ability of the executive to relieve operators from handling cards, tapes, and the like and achieve a more efficient operation.

mass-storage system [ADP] A computer system containing a large number of storage devices, with one of these devices containing the master file of the operating system, routines, and library routines.

mass susceptibility [PHYS CHEM] Magnetic susceptibility of a compound per gram. Also known as specific susceptibility.

mass-to-charge ratio [ANALY CHEM] In analysis by mass spectroscopy, the measurement of the sample mass as a ratio to its ionic charge.

mass-transfer rate [PHYS] The measurement of the movement of matter as a function of time.

mass transport [FL MECH] **1.** Carrying of loose materials in a moving medium such as water or air. **2.** The movement of fluid, especially water, from one place to another.

mass units [MECH] Units of measurement having to do with masses of materials, such as pounds or grams.

mass velocity [FL MECH] The weight flow rate of a fluid

MASS SPECTROMETER

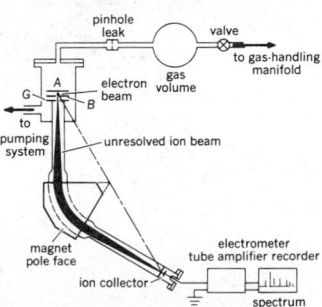

Schematic diagram of mass spectrometer tube. Electric field caused by potential difference of several volts between plates *A* and *B* draws ions through slit in *B*. Ions are further accelerated by potential difference of hundreds or thousands of volts between *B* and *G*.

divided by the cross-sectional area of the enclosing chamber or conduit; for example, 1b/hr ft².

mass wasting [GEOL] Dislodgement and downslope transport of loose rock and soil material under the direct influence of gravitational body stresses.

mast [ENG] **1.** A vertical metal pole serving as an antenna or antenna support. **2.** A slender vertical pole which must be held in position by guy lines. **3.** A drill, derrick, or tripod mounted on a drill unit, which can be raised to operating position by mechanical means. **4.** A single pole, used as a drill derrick, supported in its upright or operating position by guys. [NAV ARCH] A long wooden or metal pole or spar, usually vertical, on the deck or keel of a ship, to support other spars which in turn support or are attached to sails, as well as derricks.

mast-, masto- [ANAT] A combining form denoting breast; denoting mastoid.

Mastacembeloidei [VERT ZOO] The spiny eels, a suborder of perciform fishes that are eellike in shape and have the pectoral girdle suspended from the vertebral column.

mastalgia [MED] Pain in the breast.

mastatrophy [MED] Atrophy of the breast.

mastax [INV ZOO] The muscular pharynx in rotifers.

mast cell [HISTOL] A connective-tissue cell with numerous large, basophilic, metachromatic granules in the cytoplasm.

mast-cell disease *See* mastocytosis.

mastectomy [MED] Surgical removal of the breast. Also known as mammectomy.

master *See* master phonograph record; master station.

master alloy [MET] An alloy of selected elements that can be added to a charge of molten metal to provide a desired composition or texture or to deoxidize the material. Also known as foundry alloy.

masterbatch [MATER] A plastic, rubber, or elastomer mixture in which there is a high additives concentration, such as rubber with carbon black, or plastic with color pigment; used to proportion additives accurately into large bulks of plastic, rubber, or elastomer.

master bushing *See* liner bushing.

master card [ADP] A computer card that contains information about a group of computer cards, and is usually the first card of this group.

master clock [ADP] The electronic or electric source of standard timing signals, often called clock pulses, required for sequencing the operation of a computer. Also known as main clock; master synchronizer; master timer.

master compass [NAV] That part of a remote-indicating compass system which determines direction for transmission to various repeaters.

master control [ADP] A computer program, oriented toward applications, which carries out the highest level of control in a hierarchy of programs, routines, and subroutines. [COMMUN] The control console that contains the main program controls for a radio or television transmitter or network.

master control interrupt [ADP] A signal which causes the master control program to take over control of a computer system.

master cylinder [MECH ENG] The container for the fluid and the piston, forming part of a device such as a hydraulic brake or clutch.

master data [ADP] A set of data which are rarely changed, or changed in a known and constant manner.

master equation [ATOM PHYS] An equation which determines the rate of change of the population of an energy level in terms of the populations of other levels and transition probabilities.

master file [ADP] **1.** A computer file containing relatively permanent information, usually updated periodically, such as subscriber records or payroll data other than time worked. **2.** A computer file that is used as an authoritative source of data in carrying out a particular job on the computer.

master gage [DES ENG] A locating device with fixed hole locations or part positions; locates in three dimensions and generally occupies the same space as the part it represents.

master gain [ELECTR] Control of overall gain of an amplifying system as opposed to varying the gain of several individual inputs.

master group [COMMUN] In carrier telephony, ten supergroups (600 voice channels) multiplexed together and treated as a unit.

master gyro compass [NAV] A gyro compass for controlling one or more remote indicators, called gyro repeaters.

master instruction tape [ADP] A computer magnetic tape on which all programs for a system of runs are recorded.

master layout [DES ENG] A permanent template record laid out in reference planes and used as a standard of reference in the development and coordination of other templates.

master mechanic [ENG] The supervisor, as at the mine, in charge of the maintenance and installation of equipment.

master multivibrator [ELECTR] Master oscillator using a multivibrator unit.

master oscillator [ELECTR] An oscillator that establishes the carrier frequency of the output of an amplifier or transmitter.

master-oscillator power amplifier [ELECTR] Transmitter using an oscillator followed by one or more stages of radio-frequency amplification.

master phonograph record [ENG ACOUS] The negative metal counterpart of a disk recording, produced by electroforming as one step in the production of phonograph records. Also known as master.

master plan position indicator [ELECTR] In a radar system, a plan position indicator which controls remote indicators or repeaters.

master program file [ADP] The tape record of all programs for a system of runs.

master record [ADP] The basic updated record which will be used for the next run.

master routine *See* executive routine.

master/slave manipulator [NUCLEO] A pair of mechanical hands used to handle radioactive materials by remote control by an operator behind a protective shield.

master/slave mode [ADP] The feature ensuring the protection of each program when more than one program resides in memory.

master/slave system [ADP] A system of interlinked computers under the control of one computer (master computer).

master station [NAV] In a radio navigation system such as loran, the controlling station of two or more synchronized transmitting stations. Also known as master.

master stream *See* main stream.

Master's two-step test *See* two-step test.

master switch [ELEC] **1.** Switch that dominates the operation contactors, relays, or other magnetically operated devices. **2.** Switch electrically ahead of a number of individual switches.

master synchronization pulse [COMMUN] In telemetry, a pulse distinguished from other telemetering pulses by amplitude and duration, used to indicate the end of a sequence of pulses.

master synchronizer *See* master clock.

master system tape [ADP] A monitor program centralizing the control of program operation by loading and executing any program on a system tape.

master timer *See* master clock.

masthead bombing [ORD] Very low bombing, for example, against shipping.

mastic [MATER] **1.** A glasslike, brittle, yellow to greenish yellow resinous exudation of the mastic tree (*Pistacia lentiscus*); used in medicine, condiments, adhesive, incense, and lacquer. Also known as mastiche; mastix; pistachia galls. **2.** Mixture of finely powdered rock and asphaltic material used for highway construction.

masticate [PHYSIO] To chew.

masticatory [PHARM] A medicine to be chewed but not swallowed.

mastiche *See* mastic.

mastic oil [MATER] Colorless essential oil with balsamic odor; the chief constituents are pinenes; used in medicine.

Mastigamoebidae [INV ZOO] A family of ameboid protozoans possessing one or two flagella, belonging to the order Rhizomastigida.

Mastigophora [INV ZOO] A superclass of the Protozoa characterized by possession of flagella.

mastitis [MED] Inflammation of the breast.

mastix *See* mastic.

MASTACEMBELOIDEI

Spiny eel (*Mastacembelus circumcinctus*); length to 7 inches (18 centimeters). (*After H. M. Smith, The Fresh-Water Fishes of Siam or Thailand, U.S. Nt. Mus. Bull. no. 188, 1945*)

MASTIGOPHORA

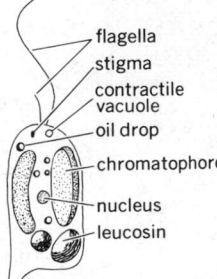

flagella
stigma
contractile
vacuole
oil drop
chromatophore
nucleus
leucosin

Ochromonas ludibunda, a holophytic and holozoic mastigophoran.

mastocytoma [VET MED] A local proliferation of mast cells forming a tumorous nodule; seen most frequently in dogs, but occasionally noted in humans.

mastocytosis [MED] Excessive mast cell proliferation. Also known as mast-cell disease.

mastodon [PALEON] A member of the Mastodontidae, especially the genus *Mammut*.

Mastodontidae [PALEON] An extinct family of elephantoid proboscideans that had low-crowned teeth with simple ridges and without cement.

mastodynia [MED] A type of cystic hyperplasia of the breast marked by an increase of connective tissue in the breast, without a proportionate increase in glandular epithelium.

mastoid [ANAT] **1.** Breast-shaped. **2.** The portion of the temporal bone where the mastoid process is located.

mastoid air cell *See* mastoid cell.

mastoid antrum [ANAT] An air-filled space between the upper portion of the middle ear and the mastoid cells.

mastoid canaliculus [ANAT] A small canal opening just above the stylomastoid foramen; gives passage to the auricular branch of the vagus nerve.

mastoid cell [ANAT] One of the compartments in the mastoid portion of the temporal bone, connected with the mastoid antrum and lined with a mucous membrane. Also known as mastoid air cell; mastoid sinus.

mastoid foramen [ANAT] A small opening behind the mastoid process.

mastoid fossa [ANAT] The depression behind the suprameatal spine on the lateral surface of the temporal bone.

mastoiditis [MED] Inflammation of the mastoid cells.

mastoid process [ANAT] A nipple-shaped, inferior projection of the mastoid portion of the temporal bone.

mastoid sinus *See* mastoid cell.

mastopathy [MED] Any disease or pain of the mammary gland. Also known as mazopathy.

mastopexy [MED] Surgical fixation of a pendulous breast. Also known as mazopexy.

mastoplasia [MED] Hypertrophy of breast tissue. Also known as mastoplastia.

mastoplastia *See* mastoplasia.

mastoplasty [MED] Plastic surgery on the breast.

mastorrhagia [MED] Hemorrhage from the breast.

mastoscirrhus [MED] Hardening of breast tissue, usually indicative of cancer.

Mastotermitidae [INV ZOO] A family of lower termites in the order Isoptera with a single living species, in Australia.

mastotomy [MED] Incision of the breast.

mast step [NAV ARCH] A wooden or steel foundation supporting a mast.

mat [CIV ENG] **1.** A steel or concrete footing under a post. **2.** Mesh reinforcement in a concrete slab. **3.** A heavy steel-mesh blanket to suppress rock fragments during blasting. [MATER] Randomly distributed felt or glass fibers used in reinforced-plastics lay-up molding. [MIN ENG] An accumulation of broken mine timbers, rock, earth, and other debris coincident with the caving system of mining.

Matador [ORD] Name applied to a U.S. Air Force surface-to-surface tactical missile, controlled electronically from the ground; powered by a jet engine, rocket-boosted, and attaining subsonic speed; it has a range of several hundred miles, can carry a conventional or atomic warhead, and has a combination radar and inertial guidance system; it is transported to the launching site on a transport trailer.

matamorphic differentiation [PETR] Processes by which different mineral assemblages develop in some sequence from an initially uniform parent rock.

Matanuska wind [METEOROL] A strong, gusty, northeast wind which occasionally occurs during the winter in the vicinity of Palmer, Alaska.

match [ADP] A data-processing operation similar to a merge, except that instead of producing a sequence of items made up from the input sequences, the sequences are matched against each other on the basis of some key. [ENG] **1.** A charge of gunpowder put in a paper several inches long and used for igniting explosives. **2.** A short flammable piece of wood, paper, or other material tipped with a combustible mixture that bursts into flame through friction. [IMMUNOL] To select

blood donors whose erythrocytes are compatible with those of the recipient.

matched filter [ADP] In character recognition, a method employed in character property detection in which a vertical projection of the input character produces an analog waveform which is then compared to a set of stored waveforms for the purpose of determining the character's identity. [ELECTR] A filter with the property that, when the input consists of noise in addition to a specified desired signal, the signal-to-noise ratio is the maximum which can be obtained in any linear filter.

matched impedance [ELEC] An impedance of a load which is equal to the impedance of a generator, so that maximum power is delivered to the load.

matched load [ELECTR] A load having the impedance value that results in maximum absorption of energy from the signal source.

matched-metal molding [ENG] Forming of reinforced-plastic articles between two close-fitting metal molds mounted in a hydraulic press.

matched pairs [STAT] The design of an experiment for paired comparison in which the assignment of subjects to treatment or control is not completely at random, but the randomization is restricted to occur separately within each pair.

matched pulse intercepting [COMMUN] System for intercepting calls on party lines in a terminal-per-line office; operates on a ground pulse which is matched in time with the intercepted station's particular ringing frequency.

matched transmission line [ELEC] Transmission line terminated with a load equivalent to its characteristic impedance.

matching [ELEC] Connecting two circuits or parts together with a coupling device in such a way that the maximum transfer of energy occurs between the two circuits, and the impedance of either circuit will be terminated in its image. [NAV] The bringing of two or more signals or indications into suitable position or condition preliminary to making a measurement, as on a loran indicator or a sky compass.

matching diaphragm [ELECTROMAG] Diaphragm consisting of a slit in a thin sheet of metal, placed transversely across a waveguide for matching purposes; the orientation of the slit with respect to the long dimension of the waveguide determines whether the diaphragm acts as a capacitive or inductive reactance.

matching distribution [STAT] The distribution of number of matches obtained if N tickets labeled 1 to N are drawn at random one at a time and laid in a row, and a match is counted when a ticket's label matches its position.

matching impedance [ELEC] Impedance value that must be connected to the terminals of a signal-voltage source for proper matching.

matching section [ELECTROMAG] A section of transmission line, a quarter or half wavelength long, inserted between a transmission line and a load to obtain impedance matching.

matching stub [ELECTROMAG] Device placed on a radio-frequency transmission line which varies the impedance of the line; the impedance of the line can be adjusted in this manner.

match plate [MET] A plate on which metal-casting patterns are mounted or formed as an integral part, to facilitate the molding operation.

mate [BIOL] **1.** To pair for breeding. **2.** To copulate.

material balance [CHEM ENG] A calculation to inventory material inputs versus outputs in a process system.

material particle [MECH] An object which has rest-mass and an observable position in space, but has no geometrical extension, being confined to a single point. Also known as particle.

materials control [IND ENG] Inventory control of materials involved in manufacturing or assembly.

materials handling [ENG] The loading, moving, and unloading of materials.

materials science [ENG] The study of the nature, behavior, and use of materials applied to science and technology.

materials testing reactor [NUCLEO] A nuclear reactor designed mainly for studying the behavior of materials and equipment subjected to large fluxes of neutrons and other radiation.

materia medica [PHARM] **1.** The science that treats of the sources, properties, and preparation of medicinal substances. **2.** The materials from which medicinal substances are prepared. **3.** A treatise on the subject.

materiel [ORD] **1.** Things of all kinds required for the equipment, maintenance, operation, and support of military activities, both combat and noncombat. **2.** In a restricted sense, those things used in combat or logistic support operations, such as weapons, motor vehicles, or special-purpose clothing, as distinguished from items of ordinary use, such as uniforms, food, or medicine.

maternal [BIOL] Of, pertaining to, or related to a mother.

maternal effect [GEN] Determination of characters of the progeny by the maternal parent; mediated by the genetic constitution of the mother.

maternal impressions [PSYCH] The congenital developmental effects formerly thought to be produced upon the fetus in the uterus by mental impressions of a vivid character received by the mother during pregnancy.

maternal inheritance [GEN] The acquisition of characters transmitted through the cytoplasm of the egg.

maternal mortality rate [MED] The number of deaths reported as due to puerperal causes in a calendar year per 100 live births reported in the same year and place.

maternal placenta [EMBRYO] The outer placental layer, developed from the decidua basalis.

maternity [BIOL] **1.** Motherhood. **2.** The state of being pregnant.

mathematical analysis *See* analysis.

mathematical biology [BIOL] A discipline that encompasses all applications of mathematics, computer technology, and quantitative theorizing to biological systems, and the underlying processes within the systems.

mathematical biophysics [BIOL] A discipline which attempts to utilize mathematics to explain biophysical processes.

mathematical check [ADP] A programmed computer check of a sequence of operations, using the mathematical properties of that sequence.

mathematical climate [CLIMATOL] An elementary generalization of the earth's climatic pattern, based entirely on the annual cycle of the sun's inclination; this early climatic classification recognized three basic latitudinal zones (the summerless, intermediate, and winterless), which are now known as the Frigid, Temperate, and Torrid Zones, and which are bounded by the Arctic and Antarctic Circles and the Tropics of Cancer and Capricorn.

mathematical forecasting *See* numerical forecasting.

mathematical function program [ADP] A set of routinely used mathematical functions, such as square root, which are efficiently coded and called for by special symbols.

mathematical geography [GEOGR] The branch of geography that deals with the features and processes of the earth, and their representations on maps and charts.

mathematical induction [MATH] A general method of proof of statements concerning a positive integral variable: if a statement is proven true for $x = 1$, and if it is proven that, if the statement is true for $x = 1, \ldots, n$, then it is true for $x = n + 1$, then it follows that the statement is true for any integer. Also known as complete induction.

mathematical logic [MATH] The study of mathematical theories from the viewpoint of model theory, recursive function theory, proof theory, and set theory.

mathematical model [MATH] **1.** A mathematical representation of a process, device, or concept by means of a number of variables which are defined to represent the inputs, outputs, and internal states of the device or process, and a set of equations and inequalities describing the interaction of these variables. **2.** A mathematical theory or system together with its axioms.

mathematical physics [PHYS] The study of the mathematical systems which represent physical phenomena; particular areas are, for example, quantum and statistical mechanics and field theory.

mathematical probability [MATH] The ratio of the number of mutually exclusive, equally likely outcomes of interest to the

total number of such outcomes when the total is exhaustive. Also known as a priori probability.

mathematical programming *See* optimization theory.

mathematical subroutine [ADP] A computer subroutine in which a well-defined mathematical function, such as exponential, logarithm, or sine, relates the output to the input.

mathematical table [MATH] A listing of the values of a function of one or several variables at a series of values of the arguments, usually equally spaced.

mathematics [SCI TECH] The deductive study of shape, quantity, and dependence; the two main areas are applied mathematics and pure mathematics, the former arising from the study of physical phenomena, the latter the intrinsic study of mathematical structures.

Matheson joint [DES ENG] A wrought-pipe joint made by enlarging the end of one pipe length to receive the male end of the next length.

Mathieu equation [MATH] A differential equation of the form $y'' + (a + b \cos 2x)y = 0$, whose solution depends on periodic functions.

Mathieu functions [MATH] Any solution of the Mathieu equation which is periodic and an even or odd function.

matico [BOT] *Piper angustifolium.* An aromatic wild pepper found in tropical America whose leaves are rich in volatile oil, gums, and tannins; leaves were used medicinally as a stimulant and hemostatic.

matico oil [MATER] An alcohol-soluble, yellowish-brown volatile oil distilled from the leaves or flowers of matico; chief constituents are asarone and methyl eugenol.

matinal [METEOROL] The morning winds, that is, an east wind.

mating [BIOL] The meeting of individuals for sexual reproduction.

matlockite [MINERAL] PbFCl A mineral consisting of lead chloride and fluoride.

m-atm *See* meter-atmosphere.

mat packs [MIN ENG] Timbers laid side by side into a solid mass 2 by 2½ feet (61 by 76 centimeters) square and 4–6 inches (10–15 centimeters) thick, kept together by wires through holes drilled through the edges of the mass, carried into a mine and built up to make effective roof supports.

Matricaria [BOT] A genus of weedy herbs having a strong odor and white and yellow disk flowers; chamomile oil is obtained from certain species.

matrix [ADP] A latticework of input and output leads with logic elements connected at some of their intersections. [ELECTR] **1.** The section of a color television transmitter that transforms the red, green, and blue camera signals into color-difference signals and combines them with the chrominance subcarrier. Also known as color coder; color encoder; encoder. **2.** The section of a color television receiver that transforms the color-difference signals into the red, green, and blue signals needed to drive the color picture tube. Also known as color decoder; decoder. [ENG] A recessed mold in which something is formed or cast. [HISTOL] **1.** The intercellular substance of a tissue. Also known as ground substance. **2.** The epithelial tissue from which a toenail or fingernail develops. [MATER] A binding agent used to make an agglomerate mass. [MATH] A rectangular array of numbers or scalars from a vector space. [MET] **1.** The principal component of an alloy. **2.** The precisely shaped form used as the cathode in electroforming. [MYCOL] The substrate on or in which fungus grows. [PETR] The continuous, fine-grained material in which large grains of a sediment or sedimentary rock are embedded. Also known as groundmass.

matrix algebra [MATH] An algebra whose elements are matrices and whose operations are addition and multiplication of matrices.

matrix algebra tableau [ADP] The current matrix at the end of an iteration while running a linear program.

matrix calculus [MATH] The treatment of matrices whose entries are functions as functions in their own right with a corresponding theory of differentiation; this has application to the study of multidimensional derivatives of functions of several variables.

matrix effects [ANALY CHEM] The enhancement or suppression of minor element spectral lines from metallic oxides during emission spectroscopy by the matrix element (such as graphite) used to hold the sample.

matrix element [MATH] One of the set of numbers which form a matrix. [QUANT MECH] The scalar product of a member of a complete, orthogonal set of vectors, representing states, with a vector which results from applying a specified operator to another member of this set.

matrix game [MATH] A game involving two persons, which gives rise to a matrix representing the amount received by the two players.

matrix mechanics [QUANT MECH] The theory of quantum mechanics developed by using the Heisenberg picture and representing operators by their matrix elements between eigenfunctions of the Hamiltonian operator; Heisenberg's original formulation of quantum mechanics.

matrix of a linear transformation [MATH] A unique matrix A, such that for a specified linear transformation L from one vector space to another, and for specified finite bases in each space, L applied to a vector is equal to A times that vector.

matrix porosity [GEOL] Core-sample porosity determined from a small sample of the core, in contrast to total porosity, where the whole core is used.

matrix printing [ADP] High-speed printing in which characterlike configurations of dots are printed through the proper selection of wire ends from a matrix of wire ends. Also known as stylus printing; wire printing.

matrix rock *See* land pebble phosphate.

matrix spectrophotometry [SPECT] Spectrophotometric analysis in which the specimen is irradiated in sequence at more than one wavelength, with the visible spectrum evaluated for the energy leaving for each wavelength of irradiation.

matrix theory [MATH] The algebraic study of matrices and their use in evaluating linear processes.

matrix velocity [GEOPHYS] The velocity of sound through a formation's rock matrix during an acoustic-velocity log.

matromycin *See* oleandomycin.

matte [MATER] Dull, as applied to appearance of a surface. [MET] An impure metallic sulfide mixture produced by smelting the sulfide ores of such metals as copper, lead, or nickel.

matte dip [MET] An etching solution which reacts with the surface of a metal, giving a dull finish.

matter [PHYS] The substance composing bodies perceptible to the senses; includes any entity possessing mass when at rest.

matte smelting [MET] Smelting of copper-bearing materials in a reverberatory furnace.

Matteuci effect [PHYS] A phenomenon in which an electric potential difference appears between the ends of a ferromagnet that is twisted in a magnetic field.

Matthias' rules [SOLID STATE] Several empirical rules giving the dependence of the transition temperatures of superconducting metals and alloys on the position of the metals in the periodic table and in the composition of the alloys.

Matthiessen's rule [SOLID STATE] An empirical rule which states that the total resistivity of a crystalline metallic specimen is the sum of the resistivity due to thermal agitation of the metal ions of the lattice and the resistivity due to imperfections in the crystal.

mattock [DES ENG] A tool with the combined features of an adz, an ax, and a pick.

mattress array *See* billboard array.

maturation [BIOL] **1.** The process of coming to full development. **2.** The final series of changes in the growth and formation of germ cells.

mature [BIOL] **1.** Fully grown and developed. **2.** Ripe. [FOOD ENG] Having attained the final state of processing, as certain wines. [GEOL] **1.** Pertaining to a topography or region, and to its landforms, having undergone maximum development and accentuation of form. **2.** Pertaining to the third stage of textural maturity of a clastic sediment. [PSYCH] Having the emotional qualities of a well-adjusted adult.

mature soil *See* zonal soil.

maturing *See* curing.

maturity [GEOL] **1.** The second stage of the erosion cycle in the topographic development of a landscape or region characterized by numerous and closely spaced mature streams, reduction of level surfaces to slopes, large well-defined drainage systems, and the absence of swamps or lakes on the uplands. Also known as topographic maturity. **2.** A stage in the development of a shore or coast that begins with the attainment of a profile of equilibrium. **3.** The extent to which the texture and composition of a clastic sediment approach the ultimate end product. **4.** The stage of stream development at which maximum vigor and efficiency has been reached.

maturity index [GEOL] A measure of the progress of a clastic sediment in the direction of chemical or mineralogic stability; for example, a high ratio of quartz + cherts/feldspar + rock fragments indicates a highly mature sediment.

maucherite [MINERAL] $Ni_{11}As_8$ A reddish silver-white mineral composed of nickel arsenide.

Mauler [ORD] Name given to a U.S. Army developed surface-to-air, lightweight, highly mobile, air defense missile for use in forward areas; it has infrared guidance.

maulstick *See* mahlstick.

Maunoir's hydrocele [MED] A congenital lymphatic cyst of the neck.

Maupertius' principle [MECH] The principle of least action is sufficient to determine the motion of a mechanical system.

Mauriac syndrome [MED] A complex of symptoms associated with diabetes mellitus in children including retarded growth, obesity, and enlargement of the liver, probably related to inadequate control of the condition.

Mauritius hemp [BOT] A hard fiber obtained from the leaves of the cabuya, grown on the island of Mauritius; not a true hemp.

mavar *See* parametric amplifier.

max *See* maximum.

max-flow min-cut theorem [IND ENG] In the analysis of networks, the concept that for any network with a single source and sink, the maximum feasible flow from source to sink is equal to the minimum cut value for any of the cuts of the network.

maxilla [ANAT] **1.** The upper jawbone. **2.** The upper jaw. [INV ZOO] Either of the first two pairs of mouthparts posterior to the mandibles in certain arthropods.

maxillary air sinus *See* maxillary sinus.

maxillary antrum *See* maxillary sinus.

maxillary arch *See* palatomaxillary arch.

maxillary artery [ANAT] A branch of the external carotid artery which supplies the deep structures of the face (internal maxillary) and the side of the face and nose (external maxillary).

maxillary hiatus [ANAT] An opening in the maxilla connecting the nasal cavity with the maxillary sinus.

maxillary nerve [ANAT] A somatic sensory branch of the trigeminal nerve; innervates the meninges, the skin of the upper portion of the face, upper teeth, and mucosa of the nose, palate, and cheeks.

maxillary process of the embryo [EMBRYO] An outgrowth of the dorsal part of the mandibular arch that forms the lateral part of the upper lip, the upper cheek region, and the upper jaw except the premaxilla.

maxillary sinus [ANAT] A paranasal air cavity in the body of the maxilla. Also known as maxillary air sinus; maxillary antrum.

maxilliped [INV ZOO] One of the three pairs of crustacean appendages immediately posterior to the maxillae.

maxillo-alveolar index [ANTHRO] Ratio of the breadth to the length of the alveolar arch, multiplied by 100.

maximal flow [IND ENG] Maximum total flow from the source to the sink in a connected network.

maximal member [MATH] In a partially ordered set a maximal member is one for which no other element follows it in the ordering.

maximax criterion [MATH] In decision theory, one of several possible prescriptions for making a decision under conditions of uncertainty; it prescribes the strategy which will maximize the maximum possible profit.

maxim criterion [MATH] One of several prescriptions for making a decision under conditions of uncertainty; it prescribes the strategy which will maximize the minimum profit. Also known as maximin criterion.

maximin criterion *See* maxim criterion.

maximizing a function [MATH] Finding the largest value assumed by a function.

maximum [MATH] The maximum of a real-valued function is the greatest value it assumes. Abbreviated max.

maximum angle of inclination [MECH ENG] The maximum angle at which a conveyor may be inclined and still deliver an amount of bulk material within a given time.

maximum available gain [ELECTR] The theoretical maximum power gain available in a transistor stage; it is seldom achieved in practical circuits because it can be approached only when feedback is negligible. Abbreviated MAG.

maximum average power output [ELECTR] In television, the maximum of radio-frequency output power that can occur under any combination of signals transmitted, averaged over the longest repetitive modulation cycle.

maximum belt slope [MECH ENG] A slope beyond which the material on the belt of a conveyor tends to roll downhill.

maximum belt tension [MECH ENG] The total of the starting and operating tensions in a conveyor.

maximum breathing capacity [PHYSIO] The greatest volume of air an individual can breathe voluntarily in 10–30 seconds; expressed as liters per minute. Abbreviated MBC.

maximum credible accident [NUCLEO] The most serious nuclear reactor accident that can be hypothesized from an adverse combination of equipment malfunction, operating errors, and other reasonable foreseen causes.

maximum depression [ORD] The maximum vertical angle below the horizontal plane at which a piece of artillery can be laid and still deliver effective fire.

maximum ebb [OCEANOGR] The greatest speed of an ebb current.

maximum effective range [ORD] The greatest distance at which a weapon may be expected to fire accurately to inflict casualties or damage.

maximum elevation [ORD] The greatest vertical angle at which a gun or launcher can be laid; usually limited by the mechanical structure of the piece.

maximum flood [OCEANOGR] The greatest speed of a flood current.

maximum gradability [MECH ENG] Steepest slope a vehicle can negotiate in low gear; usually expressed in percentage of slope, namely, the ratio between the vertical rise and the horizontal distance traveled; sometimes expressed by the angle between the slope and the horizontal.

maximum keying frequency [ELECTR] In facsimile, the frequency in hertz that is numerically equal to the spot speed divided by twice the horizontal dimension of the spot.

maximum likelihood method [STAT] A technique in statistics where the likelihood distribution is so maximized as to produce an estimate to the random variables involved.

maximum-minimum principle See min-max theorem.

maximum modulating frequency [ELECTR] Highest picture frequency required for a facsimile transmission system; the maximum modulating frequency and the maximum keying frequency are not necessarily equal.

maximum-modulus theorem [MATH] For a complex analytic function in a closed bounded simply connected region its modulus assumes its maximum value on the boundary of the region.

maximum operating frequency [ADP] The highest rate at which the modules perform iteratively and reliably.

maximum ordinate [MECH] Difference in altitude between the origin and highest point of the trajectory of a projectile.

maximum permissible concentration [MED] The maximum quantity/unit volume of radioactive material in air, water, and foodstuffs that is not considered an undue risk to human health.

maximum permissible dose [MED] The dose of ionizing radiation that a person may receive in his lifetime without appreciable bodily injury.

maximum producible oil index [PETRO ENG] An approximation of the maximum amount of oil per bulk formation volume that is producible with water drive.

maximum retention time [ELECTR] Maximum time between writing into and reading an acceptable output from a storage element of a charge storage tube.

maximum signal level [ELECTR] In an amplitude-modulated facsimile system, the level corresponding to copy black or copy white, whichever has the highest amplitude.

maximum subsidence [GEOL] The maximum amount of subsidence in a basin.

maximum thermometer [ENG] A thermometer that registers the maximum temperature attained during an interval of time.

maximum thrust [ORD] The highest thrust recorded on the thrust-time trace in missile testing.

maximum unambiguous range [ELECTROMAG] The range beyond which the echo from a pulsed radar signal returns after generation of the next pulse, and can thus be mistaken as a short-range echo of the next cycle.

maximum undistorted power output [ELECTR] Of a transducer, the maximum power delivered under specified conditions with a total harmonic output not exceeding a specified percentage.

maximum usable frequency [COMMUN] The upper limit of the frequencies that can be used at a specified time for point-to-point radio transmission involving propagation by reflection from the regular ionized layers of the ionosphere. Abbreviated MUF.

maximum-wind and shear chart [METEOROL] A synoptic chart on which are plotted the altitudes of the maximum wind speed, the maximum wind velocity (wind direction optional), plus the velocity of the wind at mandatory levels both above and below the level of maximum wind. Also known as max-wind and shear chart.

maximum-wind level [METEOROL] The height at which the maximum wind speed occurs, determined in a winds-aloft observation. Also known as max-wind level.

maximum-wind topography [METEOROL] The topography of the surface of maximum wind speed. Also known as max-wind topography.

maximum working area [IND ENG] That portion of the working area that is readily accessible to the hands of a worker when in his normal operating position.

maximum zonal westerlies [METEOROL] The average west-to-east component of wind over the continuous 20° belt of latitude in which this average is a maximum; it is usually found, in the winter season, in the vicinity of 40–60° north latitude.

maxoplasia [MED] Degenerative disease of the breast.

maxwell [ELECTROMAG] A centimeter-gram-second electromagnetic unit of magnetic flux, equal to the magnetic flux which produces an electromotive force of 1 abvolt in a circuit of one turn linking the flux, as the flux is reduced to zero in 1 second at a uniform rate. Abbreviated Mx. Also known as abweber (abWb); line of magnetic induction.

Maxwell body See Maxwell liquid.

Maxwell-Boltzmann density function See Maxwell-Boltzmann distribution.

Maxwell-Boltzmann distribution [STAT MECH] Any function giving the probability (or some function proportional to it) that a molecule of a gas in thermal equilibrium will have values of certain variables within given infinitesimal ranges, assuming that the gas molecules obey classical mechanics, and possibly making other assumptions; examples are the Maxwell distribution and the Boltzmann distribution. Also known as Maxwell-Boltzmann density function.

Maxwell-Boltzmann equation See Boltzmann transport equation.

Maxwell-Boltzmann statistics [STAT MECH] The classical statistics of identical particles, as opposed to the Bose-Einstein or Fermi-Dirac statistics. Also known as Boltzmann statistics.

Maxwell bridge [ELEC] A four-arm alternating-current bridge used to measure inductance (or capacitance) in terms of resistance and capacitance (or inductance); bridge balance is independent of frequency. Also known as Maxwell-Wien bridge; Wien-Maxwell bridge.

Maxwell distribution [STAT MECH] A function giving the number of molecules of a gas in thermal equilibrium whose velocities lie within a given, infinitesimal range of values, assuming that the molecules obey classical mechanics, and do not interact. Also known as Maxwellian distribution.

Maxwell effect [OPTICS] Double refraction of a viscous liquid having anisotropic molecules, which results from components of the velocity gradient perpendicular to the fluid velocity itself.

Maxwell equal area rule [THERMO] At temperatures for which the theoretical isothermal of a substance, on a graph of pressure against volume, has a portion with positive slope (as occurs in a substance with liquid and gas phases obeying the Van der Waals equation), a horizontal line drawn at the equilibrium vapor pressure and connecting two parts of the isothermal with negative slope has the property that the area between the horizontal and the part of the isothermal above it is equal to the area between the horizontal and the part of the isothermal below it.

Maxwell equations *See* Maxwell field equations.

Maxwell field equations [ELECTROMAG] Four differential equations which relate the electric and magnetic fields to electric charges and currents, and form the basis of the theory of electromagnetic waves. Also known as electromagnetic field equations; Maxwell equations.

Maxwellian distribution *See* Maxwell distribution.

Maxwellian distribution law [STAT MECH] Equation relating the statistical distribution of speeds and energies of molecules of a pure gas at a uniform temperature where there are no convection currents.

Maxwellian equilibrium [STAT MECH] Thermal equilibrium of a gas, or of some group of particles, in which the velocity distribution of the particles is the Maxwell distribution corresponding to the temperature of the object with which they are in equilibrium.

Maxwellian gas [STAT MECH] A gas whose molecules have the Maxwell distribution of velocities.

Maxwell liquid [FL MECH] A liquid whose rate of deformation is the sum of a term proportional to the shearing stress acting on it and a term proportional to the rate of change of this stress. Also known as Maxwell body.

Maxwell primaries [OPTICS] The primary colors in a system of colorimetry devised by J. C. Maxwell; they are cyan, green, and magenta.

Maxwell relation [ELECTROMAG] According to Maxwell's electromagnetic theory, that relation wherein the dielectric constant of a substance equals the square of its index of refraction. [THERMO] One of four equations for a system in thermal equilibrium, each of which equates two partial derivatives, involving the pressure, volume, temperature, and entropy of the system.

Maxwell's coefficient of diffusion [FL MECH] A number in an equation for the difference between mean velocities of two gases which are allowed to mix, which determines the contribution to this quantity of the concentration gradient.

Maxwell's demon *See* demon of Maxwell.

Maxwell's displacement current *See* displacement current.

Maxwell's electromagnetic theory [ELECTROMAG] A mathematical theory of electric and magnetic fields which predicts the propagation of electromagnetic radiation, and is valid for electromagnetic phenomena where effects on an atomic scale can be neglected.

Maxwell's law [ELECTROMAG] A movable portion of a circuit will always move in such a direction as to give maximum magnetic flux linkages through the circuit.

Maxwell's stress tensor [ELECTROMAG] A second-rank tensor whose product with a unit vector normal to a surface gives the force per unit area transmitted across the surface by an electromagnetic field.

Maxwell's theorem [MECH] If a load applied at one point A of an elastic structure results in a given deflection at another point B, then the same load applied at B will result in the same deflection at A.

Maxwell's theory of light [OPTICS] An application of Maxwell's electromagnetic theory in which light is treated as a propagating electromagnetic wave.

Maxwell triangle [OPTICS] Color-matching chromaticity values plotted on an x,y diagram. Also known as x,y chromaticity diagram.

Maxwell-Wagner mechanism [ELEC] A capacitor consisting of two parallel metal plates with two layers of material between them, one with vanishing conductivity, the other with finite conductivity and vanishing electric susceptibility.

Maxwell-Wien bridge *See* Maxwell bridge.

max-wind and shear chart *See* maximum-wind and shear chart.

max-wind level *See* maximum-wind level.

max-wind topography *See* maximum wind topography.

mayer [THERMO] A unit of heat capacity equal to the heat capacity of a substance whose temperature is raised 1° centigrade by 1 joule.

Mayer condensation theory [STAT MECH] A theory of the condensation and critical state of a system of chemically saturated molecules, in which the system is assumed to consist of independent clusters of molecules.

mayfly [INV ZOO] The common name for insects composing the order Ephemeroptera.

May-Grünwald stain [MATER] A saturated solution of methylene blue eosinate in methyl alcohol; used to stain blood. Also known as Jenner's stain.

mayonnaise [FOOD ENG] A semisolid emulsion prepared by whipping a mixture of raw eggs or egg yolks, vegetable oil, vinegar or lemon juice, salt, and condiments; used as a dressing.

maz-, mazo- [EMBRYO] A combining form denoting placenta.

mazaedium [BOT] The fruiting body of certain lichens, with the spores lying in a powdery mass in the capitulum. [MYCOL] A slimy layer on the hymenial surface of some ascomycetous fungi.

maze [PSYCH] A network of paths, blind alleys, and compartments; used in intelligence tests and in experimental psychology for developing learning curves.

mazopathy [MED] **1.** Disease of the placenta. **2.** *See* mastopathy.

mazopexy *See* mastopexy.

Mazzoni's corpuscle [ANAT] A specialized encapsulated sensory nerve end organ on a tendon in series with muscle fibers.

mb *See* millibar; millibarn.

MBC *See* maximum breathing capacity.

MBE *See* molecular beam epitaxy.

MBT *See* mercaptobenzothiazole.

mc *See* millihertz.

Mc *See* megahertz.

McArdle's syndrome [MED] A hereditary metabolic disorder caused by deficiency of muscle phosphorylase, with abnormal glycogen deposition in skeletal muscle leading to muscle fiber destruction. Also known as myophosphorylase deficiency glycogenosis.

M-C asphalt *See* medium-curing asphalt.

McBurney's incision [MED] A short diagonal incision in the lower right quadrant in which the muscle fibers are separated rather than cut; used for appendectomy.

McBurney's point [ANAT] A point halfway between the umbilicus and the anterior superior iliac spine; a point of extreme tenderness in appendicitis.

McCaa breathing device [MIN ENG] A self-contained, compressed-oxygen breathing apparatus designed by the U.S. Bureau of Mines for mine rescue work; it is carried on the back in a protective aluminum cover.

McCabe-Thiele diagram [CHEM ENG] Graphical method for calculation of the number of theoretical plates or contacting stages required for a given binary distillation operation.

mcd *See* millicurie-destroyed.

M center [SOLID STATE] A color center consisting of an F center combined with two ion vacancies.

McGinty [MIN ENG] Three sheaves over which a rope is passed so as to take a course somewhat like that of the letter M.

mcgovernite *See* macgovernite.

McLeod gage [FL MECH] A type of instrument used to measure vacuum by measuring the height of a column of mercury supported by the gas whose pressure is to be measured, when this gas is trapped and compressed into a capillary tube.

McLuckie gas detector [MIN ENG] A portable nonautomatic means of analyzing air that can be used underground or aboveground if the air sample is brought out of the mine in small rubber bladders.

McMath telescope [OPTICS] A unique 60-inch (1.778-meter) solar telescope at Kitt Peak, Arizona, that has an unconven-

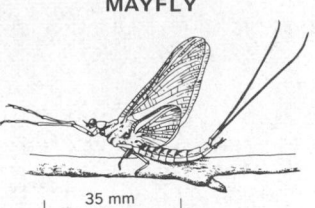

MAYFLY

Mayfly, adult male, *Hexagenia* species, in usual resting position. *(From A. H. Morgan, Field Book of Ponds and Streams, Putnam, 1930)*

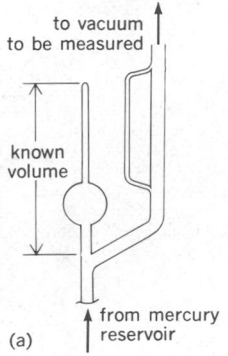
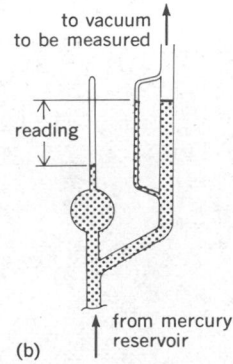

McLEOD GAGE

McLeod gage in *(a)* filling (charging) position, and *(b)* measuring position.

McMATH TELESCOPE

McMath 60-inch (1.524-meter) solard telescope, Kitt Peak, Arizona.

tional configuration; the sun's light is reflected from an 80-inch (2.032-meter) mirror into a long, fixed tube.

McNally-Carpenter centrifuge [MIN ENG] A machine that removes water from fine coal; it has a cone-shaped rotating vertical element, into the top of which the wet feed is put by gravity; a distributing disk forces the material onto the screen or basket lining the cone; as the material approaches the cone bottom, the drying action increases until the dry coal is discharged from the bottom.

McNally-Norton jig [MIN ENG] A device used to clean raw coal by carrying it to a wash box and using water whose level rises and falls as a result of air pulsations; the incoming coal is suspended by the pulsating water while the heavier refuse sinks to the bottom; the coal then spills over into the next wash box where the process is repeated, and the clean coal is discharged by the dewatering screens.

McNally tube [ELECTR] Reflex klystron tube, the frequency of which may be electrically controlled over a wide range; used as a local oscillator.

McNally-Vissac dryer [MIN ENG] A coal dryer that operates on the convection from the heavy forced draft from a coal-fired furnace; it consists of an inclined reciprocating screen over which coal moves; moisture is removed by the hot air from the furnace passing down through the bed of coal.

M-component hypergammaglobulinemia [MED] A form of hypergammaglobulinemia characterized by a single, prominent, more or less narrow band occurring from the slow gamma to the fast alpha-1 region of the electrophoretic strip.

M contour [CONT SYS] A line on a Nyquist diagram connecting points having the same magnitude of the primary feedback ratio.

McQuaid-Ehn test [MET] A test for determining the grain-size characteristics of a steel in which a sample is carburized for 8 hours at 1700°F (927°C) and cooled slowly; the high-carbon case on slow cooling will reject cementite at the austenite grain boundaries and, by polishing and etching, the grains will clearly be seen under a microscope.

md *See* millidarcy.

Md *See* mendelevium.

M damage [ORD] Damage causing immobilization of a combat vehicle.

M-derived filter [ELECTR] A filter consisting of a series of T or pi sections whose impedances are matched at all frequencies, even though the sections may have different resonant frequencies.

M-design bit [DES ENG] A long-shank, box-threaded core bit made to fit M-design core barrels.

M-design core barrel [DES ENG] A double-tube core barrel in which a 2½°-taper core lifter is carried inside a short tubular sleeve coupled to the bottom end of the inner tube, and the sleeve extends downward inside the bit shank to within a very short distance behind the face of the core bit.

MDF bearing indicator [NAV] A manual direction finder (MDF) indicator operated manually and used with a radio direction finder to indicate the relative, magnetic, or true bearing (or reciprocal) of a transmitter.

M display [ELECTR] A modified radarscope A display in which target distance is determined by moving an adjustable pedestal signal along the baseline until it coincides with the horizontal position of the target deflection.

MEA *See* ethanolamine; minimum enroute instrument altitude.

meaconing [ELECTROMAG] A system for receiving electromagnetic signals and rebroadcasting them with the same frequency so as, for instance, to confuse navigation; a confusion reflector, such as chaff, is an example.

meadow [ECOL] A vegetation zone which is a low grassland, dense and continuous, variously interspersed with forbs but few if any shrubs. Also known as pelouse; Wiesen. [ENG] Range of air-fuel ratio within which smooth combustion may be had.

meadow ore *See* bog iron ore.

mealybug [INV ZOO] Any of various scale insects of the family Pseudococcidae which have a powdery substance covering the dorsal surface; all are serious plant pests.

mean [MATH] The average or expected value.

mean-average boiling point [CHEM ENG] Pseudo boiling point for a hydrocarbon mixture; calculated from the ASTM distillation curve's volumetric average boiling point.

mean British thermal unit *See* British thermal unit.

mean calorie [THERMO] One-hundredth of the heat needed to raise 1 gram of water from 0 to 100°C.

mean camber line [AERO ENG] A line on a cross section of a wing of an aircraft which is equidistant from the upper and lower surfaces of the wing.

mean carrier frequency [ELECTR] Average carrier frequency of a transmitter corresponding to the resting frequency in a frequency-modulated system.

mean chart [METEOROL] Any chart on which isopleths of the mean value of a given meteorological element are drawn. Also known as mean map.

mean chord [AERO ENG] That chord of an airfoil that is equal to the sum of all the airfoil's chord lengths divided by the number of chord lengths; a mean chord is equal to the area divided by the span.

mean depth [HYD] Average water depth in a stream channel or conduit computed by dividing the cross-sectional area by the surface width.

meander [OCEANOGR] A deviation of the flow pattern of a current.

meander bar *See* point bar.

meander belt [GEOL] The zone along the floor of a valley across which a meandering stream periodically shifts its channel.

meander core [GEOL] A hill encircled by a stream meander. Also known as rock island.

meandering stream [HYD] A stream having a pattern of successive meanders. Also known as snaking stream.

meander niche [GEOL] A conical or crescentic opening in the wall of a cave formed by downward and lateral stream erosion.

meander plain [GEOL] A plain built by the meandering process, or a plain of lateral accretion.

meander scar [GEOL] A crescentic, concave mark on the face of a bluff or valley wall formed by a meandering stream.

meander spur [GEOL] An undercut projection of high land that extends into the concave part of, and is enclosed by, a meander.

mean deviation *See* average deviation.

mean difference [STAT] The average of the absolute values of the $n(n-1)/2$ differences between pairs of elements in a statistical distribution that has n elements.

mean diurnal high-water inequality [OCEANOGR] Half the average difference between the heights of the two high waters of each tidal day over a 19-year period; it is obtained by subtracting the mean of all high waters from the mean of the higher high waters.

mean diurnal low-water inequality [OCEANOGR] Half the average difference between the heights of the two low waters of each tidal day over a 19-year period; it is obtained by subtracting the mean of all lower low waters from the mean of the low waters.

mean draft [NAV ARCH] The mean of the drafts at the bow and the stern; for a vessel with a straight keel, the draft at the midpoint of the waterline length.

mean effective pressure [MECH ENG] A term commonly used in the evaluation for positive displacement machinery performance which expresses the average net pressure difference in pounds per square inch on the two sides of the piston in engines, pumps, and compressors. Abbreviated mep; mp. Also known as mean pressure.

mean free path [ACOUS] For sound waves in an enclosure, the average distance sound travels between successive reflections in the enclosure. [PHYS] The average distance traveled between two similar events, such as elastic collisions of molecules in a gas, of electrons or phonons in a crystal, or of neutrons in a moderator.

mean height of burst [ORD] Average of the heights of bursts of a group of shots fired with the same firing data.

mean higher high water [OCEANOGR] The average height of higher high waters at a place over a 19-year period.

mean high water [OCEANOGR] The average height of all high waters recorded at a given place over a 19-year period.

mean high-water lunitidal interval [OCEANOGR] The average

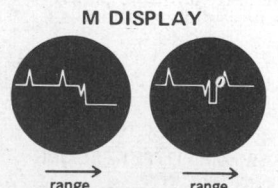

M DISPLAY

range   range

M type of radar display: type A with range step or range notch; when pip is aligned with step or notch, range can be read from dial or counter.

interval of time between the transit (upper or lower) of the moon and the next high water at a place. Also known as corrected establishment.

mean high-water neaps [OCEANOGR] The average height of the high waters of neap tides. Also known as neap high water.

mean high-water springs [OCEANOGR] The average height of the high waters of spring tides. Also known as high-water springs; spring high water.

mean latitude [GEOD] Half the arithmetical sum of the latitudes of two places on the same side of the equator; mean latitude is labeled N or S to indicate whether it is north or south of the equator.

mean life [PHYS] The average time during which a system, such as an atom, nucleus, or elementary particle, exists in a specified form; for a radionuclide or an excited state of an atom or nucleus, it is the reciprocal of the decay constant. Also known as average life; lifetime.

mean lower low water [OCEANOGR] The average height of the lower low waters at a place over a 19-year period.

mean lower low-water springs [OCEANOGR] The average height of lower low-water springs at a place.

mean low water [OCEANOGR] The average height of all low waters recorded at a given place over a 19-year period.

mean low-water lunitidal interval [OCEANOGR] The average interval of time between the transit (upper or lower) of the moon and the next low water at a place.

mean low-water neaps [OCEANOGR] The average height of the low water at neap tides. Also known as low-water neaps; neap low water.

mean low-water springs [OCEANOGR] The average height of the low waters of spring tides; this level is used as a tidal datum in some areas. Also known as low-water springs; spring low water.

mean map *See* mean chart.

mean neap range *See* neap range.

mean neap rise [OCEANOGR] The height of mean high-water neaps above the chart datum.

mean noon [ASTRON] Twelve o'clock mean time, or the instant the mean sun is over the upper branch of the meridian; it may be either local or Greenwich, depending upon the reference meridian.

mean normal stress [MECH] In a system stressed multiaxially, the algebraic mean of the three principal stresses.

mean point of impact [ORD] The point which is at the geometrical center of all the points of impact of the several shots of a salvo, excluding wild shots; when firing time-fused projectiles, the point of detonation of such a projectile is considered to be the point of impact.

mean point of impact error [ORD] Distance between the mean point of impact and the target.

mean power of a radio transmitter [ELECTR] Power supplied to the antenna transmission line by a transmitter during normal operation, averaged over a time sufficiently long compared with the period of the lowest frequency encountered in the modulation; a time of $1/10$ second during which the mean power is greatest will be selected normally.

mean pressure *See* mean effective pressure.

mean proportional [MATH] For two numbers a and b, a number x, such that $x/a = b/x$.

mean range [OCEANOGR] The difference in the height between mean high water and mean low water. [ORD] Average distance reached by a group of shots fired with the same firing data. [SCI TECH] The average difference in the extreme values of a variable quantity.

mean rank method [STAT] A method of handling data which has the same observed frequency occurring at two or more consecutive ranks; it consists of assigning the average of the ranks as the rank for the common frequency.

mean rise interval [OCEANOGR] The average interval of time between the transit (upper or lower) of the moon and the middle of the period of rise of the tide at a place; it may be either local or Greenwich, depending on the transit to which it is referred, but the local interval is assumed unless otherwise specified.

mean rise of tide [OCEANOGR] The height of mean high water above the chart datum.

mean river level [HYD] The average height of the surface of a river at any point for all stages of the tide over a 19-year period.

mean sea level [OCEANOGR] The average sea surface level for all stages of the tide over a 19-year period, usually determined from hourly height readings from a fixed reference level.

mean sidereal time [ASTRON] Sidereal time adjusted for nutation, to eliminate slight irregularities in the rate.

mean solar day [ASTRON] The duration of one rotation of the earth on its axis, with respect to the mean sun; the length of the mean solar day is 24 hours of mean solar time or $24^h 03^m 56.555^s$ of mean sidereal time.

mean solar second [ASTRON] A unit equal to $1/86,400$ of a mean solar day.

mean solar time [ASTRON] Time that has the mean solar second as its unit, and is based on the mean sun's motion.

mean specific heat [THERMO] The average over a specified range of temperature of the specific heat of a substance.

mean spring range *See* spring range.

mean spring rise [OCEANOGR] The height of mean high-water springs above the chart datum.

mean square deviation [STAT] A measure of the extent to which a collection $v_1, v_2, \ldots, v_n$ of numbers is unequal; it is given by the expression $(1/n)[(v_1 - \bar{v})^2 + \ldots + (v_n - \bar{v})^2]$, where $\bar{v}$ is the mean of the numbers.

mean-square-error criterion [CONT SYS] Evaluation of the performance of a control system by calculating the square root of the average over time of the square of the difference between the actual output and the output that is desired.

mean stress [MECH] The algebraic mean of the maximum and minimum values of a periodically varying stress.

mean sun [ASTRON] A fictitious sun conceived to move eastward along the celestial equator at a rate that provides a uniform measure of time equal to the average apparent time; used as a reference for reckoning time, such as mean time or zone time.

mean temperature [METEOROL] The average temperature of the air as indicated by a properly exposed thermometer during a given time period, usually a day, month, or year.

mean temperature difference [CHEM ENG] In heat exchange calculations, a pseudo average temperature difference between the warmer and colder fluids at inlet and outlet conditions.

Meantes [VERT ZOO] The mud eels, a small suborder of the Urodela including three species of aquatic eellike salamanders with only anterior limbs.

mean tide *See* half tide.

mean tide level [OCEANOGR] The tide level halfway between mean high water and mean low water.

mean time [ASTRON] Time based on the rotation of the earth relative to the mean sun.

meantime screw [HOROL] A screw in a watch balance, used to regulate precisely the speed of oscillation.

mean time to failure [ENG] A measure of reliability of a piece of equipment, giving the average time before the first failure.

mean time to repair [ENG] A measure of reliability of a piece of repairable equipment, giving the average time between repairs.

mean trajectory [MECH] The trajectory of a missile that passes through the center of impact or center of burst.

mean value [MATH] The arithmetic mean or average of a collection of quantities.

mean value theorem *See* first law of the mean.

mean velocity [PHYS] The average value of the velocities of a group of particles, such as the molecules of a gas.

mean water level [OCEANOGR] The average surface level of a body of water.

measles [MED] An acute, highly infectious viral disease with cough, fever, and maculopapular rash; the appearance of Koplik spots on the oral mucous membranes marks the onset. Also known as rubeola.

measles encephalitis [MED] Acute disseminated encephalitis following measles.

measles immune globulin [IMMUNOL] Sterile human globulin used to provide passive immunization against measles.

measles virus vaccine [IMMUNOL] A suspension of live attenuated or inactivated measles virus used for active immunization against measles.

measly [FOOD ENG] Of meat, containing larval tapeworms. [MED] Infected with measles.

measurable function [MATH] **1.** A real valued function f defined on a measurable space X, where for every real number a all those points x in X for which $f(x) \geq a$ form a measurable set. **2.** A function on a measurable space to a measurable space such that the inverse image of a measurable set is a measurable set.

measurable set [MATH] A member of the sigma-algebra of subsets of a measurable space.

measurable space [MATH] A set together with a sigma-algebra of subsets of this set.

measurand transmitter [COMMUN] A telemetry transmitter that transmits a signal modulated according to the values of the quantity being measured.

measure [MATH] A nonnegative real valued function m defined on a sigma-algebra of subsets of a set S whose value is zero on the empty set, and whose value on a countable union of disjoint sets is the sum of its values on each set.

measured daywork [IND ENG] Work done for an hourly wage on which specific productivity levels have been determined but which provides no incentive pay.

measured drilling depth [ENG] The apparent depth of a borehole as measured along its longitudinal axis.

measured mile [CIV ENG] The distance of 1 mile (1609.344 meters), the units of which have been accurately measured and marked. [NAV] A length of 1 nautical mile (1852 meters), the limits of which have been accurately measured and are indicated by ranges ashore; used by vessels to calibrate logs, engine revolution counters, and such, and to determine speed.

measured ore *See* developed reserves.

measured service [COMMUN] Telephone service for which charge is made according to the measured amount of usage.

measurement [SCI TECH] The process of determining the value of some quantity in terms of a standard unit.

measurement ton *See* ton.

measure space [MATH] A set together with a sigma-algebra of subsets of the set and a measure defined on this sigma-algebra.

measure zero [MATH] **1.** A set has measure zero if it is measurable and the measure of it is zero. **2.** A subset of euclidean n-dimensional space which has the property that for any positive number ε there is a covering of the set by n-dimensional rectangles such that the sum of the volumes of the rectangles is less than ε.

measuring chute [MIN ENG] An ore bin or coal bin installed adjacent to the shaft bottom in skip winding and having a capacity equal to that of the skip used; ensures rapid, correct loading of skips without spillage. Also known as measuring pocket.

measuring day [MIN ENG] The day when work is measured and recorded for assessing the wages.

measuring pocket *See* measuring chute.

meatal plate [EMBRYO] A mass of ectodermal cells on the bottom of the branchial groove in a 2-month embryo.

meatotomy [MED] Incision into and enlargement of a meatus.

meatus [ANAT] A natural opening or passage in the body.

mechanical [ENG] Of, pertaining to, or concerned with machinery or tools. [GRAPHICS] A finished copy that usually contains hand lettering, type proofs, and art especially positioned and mounted so that a photochemical reproduction can be made on a letterpress, offset, or other printing plate. Also known as keyline layout; paste-up.

mechanical advantage [MECH ENG] The ratio of the force produced by a machine such as a lever or pulley to the force applied to it.

mechanical analog computer [ADP] A machine aid to computation in which variables are represented as continuously variable displacements or motions of mechanical elements, such as gears and shafts.

mechanical analysis [MECH ENG] Mechanical separation of soil, sediment, or rock by sieving, screening, or other means to determine particle-size distribution.

mechanical area [BUILD] The areas in a building that include equipment rooms, shafts, stacks, tunnels, and closets used for heating, ventilating, air conditioning, piping, communication, hoisting, conveying, and electrical services.

mechanical bearing cursor *See* bearing cursor.

mechanical birefringence [OPTICS] A change in the double refraction of a solid material when it is subjected to stress. Also known as stress birefringence.

mechanical classification [MECH ENG] A sorting operation in which mixtures of particles of mixed sizes, and often of different specific gravities, are separated into fractions by the action of a stream of fluid, usually water.

mechanical classifier [MECH ENG] Any of various machines that are commonly used to classify mixtures of particles of different sizes, and sometimes of different specific gravities; the Dorr classifier is an example.

mechanical comparator [ENG] A contact comparator in which movement is amplified usually by a rack, pinion, and pointer or by a parallelogram arrangement.

mechanical computer [ADP] A machine such as a Charles Babbage's analytical engine.

mechanical damping [ENG ACOUS] Mechanical resistance which is generally associated with the moving parts of an electromechanically transducer such as a cutter or a reproducer.

mechanical dialer *See* automatic dialer.

mechanical differential analyzer [ADP] An analog computer using interconnected surfaces to solve differential equations, such as the device developed by Vannevar Bush at Massachusetts Institute of Technology.

mechanical draft [MECH ENG] A draft that depends upon the use of fans or other mechanical devices; may be induced or forced.

mechanical-draft cooling tower [MECH ENG] Cooling tower that depends upon fans for introduction and circulation of its air supply.

mechanical drawing [GRAPHICS] Drawing with the aid of instruments.

mechanical dysmenorrhea [MED] Painful menstruation due to mechanical obstruction of the discharge of menstrual fluids. Also known as obstructive dysmenorrhea.

mechanical efficiency [MECH ENG] In an engine, the ratio of brake horsepower to indicated horsepower.

mechanical engineering [ENG] The branch of engineering that deals with the generation, transmission, and utilization of heat and mechanical power and with the production of tools, machines, and their products.

mechanical equation of state [MET] An equation that expresses the relation of stress, strain, strain rate, and temperature for a metal.

mechanical equivalent of heat [THERMO] The amount of mechanical energy equivalent to a unit of heat.

mechanical equivalent of light [OPTICS] The ratio of the radiant power emitted by a monochromatic light source whose wavelength is that at which the sensitivity of phototopic vision is greatest (about 555 nanometers), to its luminous flux measured in lumens.

mechanical erosion *See* corrasion.

mechanical expression *See* expression.

mechanical filter [ELECTR] Filter, used in intermediate-frequency amplifiers of highly selective superheterodyne receivers, consisting of shaped metal rods that act as coupled mechanical resonators when used with piezoelectric input and output transducers. Also known as mechanical wave filter. [PETRO ENG] Granule-packed steel shell used to filter suspended floc or undissolved solids out of treated waterflood water; granules can be graded sand and gravel, anthracite coal, graphitic ore, or aluminum-oxide plates with granular filter medium.

mechanical flotation cell [MIN ENG] A device that separates minerals from ore water pulp; it consists of a cell in which the pulp is kept mixed and moving by an impeller at the bottom

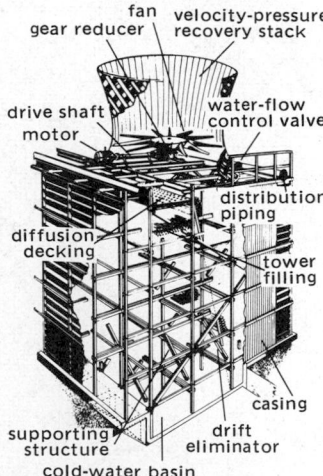

MECHANICAL-DRAFT COOLING TOWER

fan
gear reducer
velocity-pressure recovery stack
drive shaft motor
water-flow control valve
distribution piping
diffusion decking
tower filling
casing
supporting structure
drift eliminator
cold-water basin

An induced-draft cross-flow cooling tower showing component parts. (*N. P. Green and Associates*)

of the cell; the impeller pulls air down the standpipe and disperses it as bubbles through the pulp; the floatable minerals concentrate in the froth above, and the pulp is removed by a scraper.

mechanical hygrometer [ENG] A hygrometer in which an organic material, most commonly a bundle of human hair, which expands and contracts with changes in the moisture in the surrounding air or gas is held under slight tension by a spring, and a mechanical linkage actuates a pointer.

mechanical hysteresis [MECH] The dependence of the strain of a material not only on the instantaneous value of the stress but also on the previous history of the stress; for example, the elongation is less at a given value of tension when the tension is increasing than when it is decreasing.

mechanical impedance [MECH] The complex ratio of a phasor representing a sinusoidally varying force applied to a system to a phasor representing the velocity of a point in the system.

mechanical instability See absolute instability.

mechanical integrator [ADP] A mechanical device which draws the graph of the integral of a function when a tracing point is passed over a graph of the function.

mechanical jamming See passive jamming.

mechanical lift dock [CIV ENG] A type of dry dock or marine elevator in which a vessel, after being placed on the keel and bilge blocks in the dock, is bodily lifted clear of the water so that work may be performed on the underwater body.

mechanical linkage [MECH ENG] A set of rigid bodies, called links, joined together at pivots by means of pins or equivalent devices.

mechanical loader [MECH ENG] A power machine for loading mineral, coal, or dirt.

mechanical mass [QUANT MECH] The part of a particle's mass which is supposed to exist in the absence of any interaction of the particle with itself through a field.

mechanical metallurgy [MET] The science and technology of the behavior of metals relating to mechanical forces imposed on them; includes rolling, extruding, deep drawing, bending, and other processes.

mechanical mucking [MECH ENG] Loading of dirt or stone in tunnels or mines by machines.

mechanical mule [ORD] Popular name for a lightweight, low-silhouette United States infantry light weapons carrier, powered by an opposed-cylinder engine.

mechanical ohm [MECH] A unit of mechanical resistance, reactance, and impedance, equal to a force of 1 dyne divided by a velocity of 1 centimeter per second.

mechanical oil valve [PETRO ENG] A float-operated liquid-level control valve used to control liquid flow out of oil-lease gas-oil separator tank systems.

mechanical oscillograph See direct-writing recorder.

mechanical patent [ENG] A patent granted for an inventive improvement in a process, manufacture, or machine.

mechanical plating [MET] Deposition of one metal on another by a cold-peening process, such as tumbling.

mechanical plotting board See coordinate plotter.

mechanical press [MECH ENG] A press whose slide is operated by mechanical means.

mechanical property [MECH] A property that involves a relationship between stress and strain or a reaction to an applied force.

mechanical puddling See vibration puddling.

mechanical pulp [MATER] Wood pulp produced by grinding and soaking the wood fibers. Also known as groundwood pulp.

mechanical pulping [MECH ENG] Mechanical, rather than chemical, recovery of cellulose fibers from wood; unpurified, finely ground wood is made into newsprint, cheap Manila papers, and tissues.

mechanical pump [MECH ENG] A pump through which fluid is conveyed by direct contact with a moving part of the pumping machinery.

mechanical reactance [MECH] The imaginary part of mechanical impedance.

mechanical rectifier [ELEC] A rectifier in which rectification is accomplished by mechanical action, as in a synchronous vibrator.

mechanical refrigeration [MECH ENG] The removal of heat by utilizing a refrigerant subjected to cycles of refrigerating thermodynamics and employing a mechanical compressor.

mechanical replacement [ADP] The replacement of one piece of hardware by another piece of hardware at the instigation of the manufacturer.

mechanical resistance See resistance.

mechanical rotational impedance See rotational impedance.

mechanical rotational reactance See rotational reactance.

mechanical rotational resistance See rotational resistance.

mechanical scale [ENG] A weighing device that incorporates a number of levers with precisely located fulcrums to permit heavy objects to be balanced with counterweights or counterpoises.

mechanical scanner [ADP] In optical character recognition, a device that projects an input character into a rotating disk, on the periphery of which is a series of small, uniformly spaced apertures; as the disk rotates, a photocell collects the light passing through the apertures.

mechanical seal [MECH ENG] Mechanical assembly that forms a leakproof seal between flat, rotating surfaces to prevent high-pressure leakage.

mechanical separation [MECH ENG] A group of industrial operations by means of which particles of solid or drops of liquid are removed from a gas or liquid, or are separated into individual fractions, or both, by gravity separation (settling), centrifugal action, and filtration.

mechanical setting [MECH ENG] Producing bits by setting diamonds in a bit mold into which a cast or powder metal is placed, thus embedding the diamonds and forming the bit crown; opposed to hand setting. Also known as cast setting; machine setting; sinter setting.

mechanical shovel [MECH ENG] A loader limited to level or slightly graded drivages; when full, the shovel is swung over the machine, and the load is discharged into containers or vehicles behind.

mechanical splice [ENG] A splice made to terminate wire rope by pressing one or more metal sleeves over the rope junction.

mechanical spring See spring.

mechanical stage [ENG] A stage on a microscope provided with a mechanical device for positioning or changing the position of a slide.

mechanical stepping motor [ELEC] A device in which a voltage pulse through a solenoid coil causes reciprocating motion by a solenoid plunger, and this is transformed into rotary motion through a definite angle by ratchet-and-pawl mechanisms or other mechanical linkages.

mechanical stoker See automatic stoker.

mechanical telemetry [COMMUN] Telemetry by mechanical linkages through shafts and gear trains over distances of a few feet, or by propagation of pressure or acoustic waves through fluid media over distances of several hundred feet.

mechanical tilt [ELECTR] 1. Vertical tilt of the mechanical axis of a radar antenna. 2. The angle indicated by the tilt indicator dial.

mechanical torque converter [MECH ENG] A torque converter, such as a pair of gears, that transmits power with only incidental losses.

mechanical translation [ADP] Automatic translation of one language into another by means of a computer or other machine that contains a dictionary look-up in its memory, along with the programs needed to make logical choices from synonyms, supply missing words, and rearrange word order as required for the new language. Also known as machine translation.

mechanical twin [MET] A twin formed in a metal crystal by plastic deformation, involving shear of the lattice.

mechanical units [MECH] Units of length, time, and mass, and of physical quantities derivable from them.

mechanical vibration [MECH] A motion, usually unintentional and often undesirable, of parts of machines and structures.

mechanical wave filter See mechanical filter.

mechanical yielding prop [MIN ENG] A steel prop in which yield is controlled by friction between two sliding surfaces or telescopic tubes. Also known as friction yielding prop.

MECHANICAL LIFT DOCK

An electromechanical platform-lift dry dock, capacity 4800 tons (4350 metric tons). A submarine is on the multiwheel cradle. A tow-unit is moving into position to pull the vessel ashore. (*Pearlson Engineering Co.*)

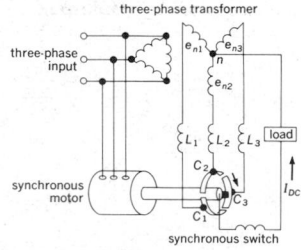

MECHANICAL RECTIFIER

Circuit of mechanical rectifier used on three-phase system; e_{n_1}, e_{n_2}, and e_{n_3} are transformers. L_1, L_2, and L_3 are nonlinear reactors which provide good commutation at contacts C_1, C_2, and C_3, and limit short circuit current.

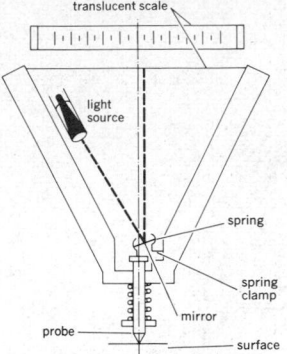

MECHANOOPTICAL VIBROMETER

translucent scale

light source

spring

spring clamp

mirror

probe

surface

Diagram of mechanooptical vibrometer. *(General Electric Co.)*

MECOPTERA

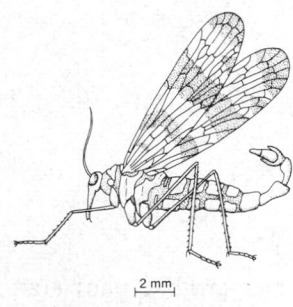

2 mm

Male scorpion fly *(Panorpa).*

mechanics [PHYS] **1.** In the original sense, the study of the behavior of physical systems under the action of forces. **2.** More broadly, the branch of physics which seeks to formulate general rules for predicting the behavior of a physical system under the influence of any type of interaction with its environment.

mechanism [MECH ENG] That part of a machine which contains two or more pieces so arranged that the motion of one compels the motion of the others.

mechanize [MECH ENG] **1.** To substitute machinery for human or animal labor. **2.** To produce or reproduce by machine. [ORD] To equip a military force with armed and armored vehicles.

mechanized *See* machine-sensible.

mechanized dew-point meter *See* dew-point recorder.

mechanized gun [ORD] Gun mounted on, and fired from, a motor vehicle; usually an artillery gun mounted on an armored, wheeled or tracklaying vehicle.

mechanocaloric effect [CRYO] An effect resulting from the fact that a temperature gradient in helium II is invariably accompanied by a pressure gradient, and conversely; examples are the fountain effect, and the heating of liquid helium left behind in a container when part of it leaks out through a small orifice.

mechanooptical vibrometer [ENG] A vibrometer in which the motion given to a probe by a surface whose vibration amplitude is to be measured is used to rock a mirror; a light beam reflected from the mirror and focused onto a scale provides an indication of the vibration amplitude.

mechanoreceptor [PHYSIO] A receptor that provides the organism with information about such mechanical changes in the environment as movement, tension, and pressure.

Meckel's cartilage [EMBRYO] The cartilaginous axis of the mandibular arch in the embryo and fetus.

Meckel's diverticulum [EMBRYO] The persistent blind end of the yolk stalk forming a tube connected with the lower ileum.

meconin [ORG CHEM] $C_{10}H_{10}O_4$ A neutral principle of opium; white crystals, soluble in hot water and alcohol and melting at 102–103°C. Also known as opianyl.

meconium [EMBRYO] A greenish mass of mucous, desquamated epithelial cells, lanugo, and vernix caseosa that collects in the fetal intestine, becoming the first fecal discharge of the newborn.

meconium ileus [MED] Intestinal obstruction in the newborn with cystic fibrosis due to trypsin deficiency.

Mecoptera [INV ZOO] The scorpion flies, a small order of insects; adults are distinguished by the peculiar prolongation of the head into a beak, which bears chewing mouthparts.

mecystasis [PHYSIO] Increase in muscle length with maintenance of the original degree of tension.

media [HISTOL] The middle, muscular layer in the wall of a vein, artery, or lymph vessel.

media conversion [ADP] The transfer of data from one storage type (such as punched cards) to another storage type (such as magnetic tape).

media conversion buffer [ADP] Large storage area, such as a drum, on which data may be stored at low speed during nonexecution time, to be later transferred at high speed into core memory during execution time.

mediad [ANAT] Toward the median line or plane of the body or of a part of the body.

medial [ANAT] **1.** Being internal as opposed to external (lateral). **2.** Toward the midline of the body. [SCI TECH] Located in the middle.

medial arteriosclerosis [MED] Calcification of the tunica media of small and medium-sized muscular arteries. Also known as medial calcinosis; Mönckeberg's arteriosclerosis.

medial calcinosis *See* medial arteriosclerosis.

medial lemniscus [ANAT] A lemniscus arising in the nucleus gracilis and nucleus cuneatus of the brain, crossing immediately as internal arcuate fibers, and terminating in the posterolateral ventral nucleus of the thalamus.

medial moraine [GEOL] **1.** An elongate moraine carried in or upon the middle of a glacier and parallel to its sides. **2.** A moraine formed by glacial abrasion of a rocky protuberance near the middle of a glacier.

medial necrosis [MED] Death of cells in the tunica media of arteries. Also known as medionecrosis.

medial necrosis of aorta [MED] Death of muscle cells and degeneration of elastic fibers in the tunica media of the aorta.

media migration [CHEM ENG] Carryover of fibers or other filter material by liquid effluent from a filter unit.

median [MATH] Any line in a triangle which joins a vertex to the midpoint of the opposite side. [SCI TECH] Located in the middle. [STAT] An average of a series of quantities or values; specifically, the quantity or value of that item which is so positioned in the series, when arranged in order of numerical quantity or value, that there are an equal number of items of greater magnitude and lesser magnitude.

median effective dose *See* effective dose 50.

median infective dose *See* infective dose 50.

median lethal dose *See* lethal dose 50.

median lethal time [MICROBIO] The period of time required for 50% of a large group of organisms to die following a specific dose of an injurious agent, such as a drug or radiation.

median mass [GEOL] A less disturbed structural block in the middle of an orogenic belt, bordered on both sides by orogenic structure, thrust away from it. Also known as betwixt mountains; Zwischengebirge.

median maxillary cyst [MED] Cystic dilation of embryonal inclusions in the incisive fossa or between the roots of the central incisors. Also known as nasopalatine cyst.

median nasal process [EMBRYO] The region below the frontonasal sulcus between the olfactory sacs; forms the bridge and mobile septum of the nose and various parts of the upper jaw and lip.

median nerve test [MED] A test for loss of function of the median nerve by having the patient abduct the thumb at right angles to the palm with fingertips in contact and forming a pyramid.

median particle diameter [GEOL] The middlemost particle diameter of a rock or sediment, larger than 50% of the diameter in the distribution and smaller than the other 50%.

median strip [CIV ENG] A paved or planted section dividing a highway into lanes according to direction of travel.

mediastinitis [MED] Inflammation of the mediastinum.

mediastinum [ANAT] **1.** A partition separating adjacent parts. **2.** The space in the middle of the chest between the two pleurae.

medical bacteriology [MED] A branch of medical microbiology that deals with the study of bacteria which affect human health, especially those which produce disease.

medical climatology [MED] The study of the relation between climate and disease.

medical electronics [ELECTR] A branch of electronics in which electronic instruments and equipment are used for such medical applications as diagnosis, therapy, research, anesthesia control, cardiac control, and surgery.

medical entomology [MED] The study of insects that are vectors for diseases and parasitic infestations in humans and domestic animals.

medical ethics [MED] Principles and moral values of proper medical conduct.

medical examiner [MED] A professionally qualified physician duly authorized and charged by a governmental unit to determine facts concerning causes of death, particularly deaths not occurring under natural circumstances, and to testify thereto in courts of law.

medical frequency bands [COMMUN] A collection of radio frequency bands allocated to medical equipment in the United States.

medical geography [MED] The study of the relation between geographic factors and disease.

medical history [MED] An account of a patient's past and present state of health obtained from the patient or relatives.

medical microbiology [MED] The study of microorganisms which affect human health.

medical mycology [MED] A branch of medical microbiology that deals with fungi that are pathogenic to humans.

medical parasitology [MED] A branch of medical microbiology which deals with the relationship between humans and those animals which live in or on them.

medical protozoology [MED] A branch of medical microbiology that deals with the study of Protozoa which are parasites of humans.

medical radiography [MED] The use of x-rays to produce photographic images for visualizing internal anatomy as an aid in diagnosis.

medication [MED] **1.** A medicinal substance. **2.** Treatment by or administration of a medicine.

medicinal [MED] Of, pertaining to, or having the nature of medicine.

medicinal oil [MATER] A highly refined, colorless, tasteless and odorless petroleum oil used medicinally as an internal lubricant and for the manufacture of salves and ointments. Also known as mineral oil.

medicine [MED] **1.** Any agent administered for the treatment of disease. **2.** The science and art of treating and healing.

medina quartzite [MINERAL] A variety of quartz containing 97.8% silica; melting point is about 1700°C.

medionecrosis [MED] Necrosis occurring in the tunica media of an artery.

mediterranean *See* mesogeosyncline.

Mediterranean anemia *See* thalassemia.

Mediterranean climate [CLIMATOL] A type of climate characterized by hot, dry, sunny summers and a winter rainy season; basically, this is the opposite of a monsoon climate. Also known as etesian climate.

Mediterranean faunal region [ECOL] A marine littoral faunal region including that offshore portion of the Atlantic Ocean from northern France to near the Equator.

Mediterranean fever *See* brucellosis.

mediterranean sea [GEOGR] A deep epicontinental sea that is connected with the ocean by a narrow channel.

Mediterranean Sea [GEOGR] A sea that lies between Europe, Asia Minor, and Africa and is completely landlocked except for the Strait of Gibraltar, the Bosporus, and the Suez Canal; total water area is 2,501,000 square kilometers.

medium [ADP] The material, or configuration thereof, on which data are recorded; usually not applied to disk, drum, or core, but to storable, removable media, such as paper tape, cards, and magnetic tape. [CHEM ENG] **1.** The carrier in which a chemical reaction takes place. **2.** Material of controlled pore size used to remove foreign particles or liquid droplets from fluid carriers. [PHYS] That entity in which objects exist and phenomena take place; examples are free space and various fluids and solids.

medium-altitude bombing [ORD] Horizontal bombing with the height of release at an altitude between 8000 and 15,000 feet (2400 and 4600 meters).

medium antiaircraft artillery [ORD] Conventional antiaircraft artillery pieces, 90 millimeters or larger, the weight of which in a trailed mount, excluding on-carriage fire control, does not exceed 40,000 pounds (18,000 kilograms).

medium artillery [ORD] Artillery which includes guns of caliber greater than 105 millimeters but less than 155 mm, and howitzers of caliber greater than 105 mm but not greater than 155 mm.

medium boiler [MATER] A solvent, intermediate in volatility between high and low boilers, which is added to lacquer thinner and influences flow and freedom from an orange peel condition during drying; boiling point is 115–145°C.

medium carbon steel [MET] Steel containing 0.15–0.30% carbon.

medium-curing asphalt [MATER] Liquid product composed of asphalt cement and a kerosine-type diluent. Also known as M-C asphalt.

medium frequency [COMMUN] A Federal Communications Commission designation for the band from 300 to 3000 kilohertz in the radio spectrum. Abbreviated mf.

medium-frequency propagation [COMMUN] Radio propagation at broadcast frequencies where skip is not an important factor.

medium-frequency tube [ELECTR] An electron tube operated at frequencies between 300 and 3000 kilohertz, at which the transit time of an electron between electrodes is much smaller than the period of oscillation of the voltage.

medium howitzer [ORD] A complete projectile-firing weapon, with a medium muzzle velocity and a curved trajectory; the bore diameter is from 126 through 200 millimeters.

medium-range forecast [METEOROL] A forecast for a period extending from about 2 to 5 days or a week in advance; there are no absolute limits to the period embraced by the definition.

medium-scale integration [ELECTR] Solid-state integrated circuits having hundreds of elements per square inch. Abbreviated MSI.

medium-volatile bituminous coal [GEOL] Bituminous coal consisting of 23–31% volatile matter.

medius [ANAT] The middle finger.

medulla [ANAT] **1.** The central part of certain organs and structures such as the adrenal glands and hair. **2.** Marrow, as of bone or the spinal cord. **3.** *See* medulla oblongata. [BOT] **1.** Pith. **2.** The central spongy portion of some fungi.

medulla oblongata [ANAT] The somewhat pyramidal, caudal portion of the vertebrate brain extending from the pons to the spinal cord. Also known as medulla.

medullary carcinoma [MED] A form of poorly differentiated adenocarcinoma, usually of the breast, grossly well circumscribed, gray-pink, and firm. Also known as encephaloid carcinoma.

medullary cord [ANAT] Dense lymphatic tissue separated by sinuses in the medulla of a lymph node. [EMBRYO] A primary invagination of the germinal epithelium of the embryonic gonad that differentiates into rete testis and seminiferous tubules or into rete ovarii.

medulloblastoma [MED] A malignant neoplasm of the brain with a tendency to metastasize in the meninges.

medulloepithelioma [MED] **1.** A locally invasive tumor of the eye arising from the ciliary epithelium of the iris. **2.** *See* ependymoma.

Medullosaceae [PALEOBOT] A family of seed ferns; these extinct plants all have large spirally arranged petioles with numerous vascular bundles.

medusa *See* jellyfish.

meerschaum *See* sepiolite.

mega- [SCI TECH] A prefix representing 10^6, or one million. Abbreviated M.

megabit [ADP] One million binary bits.

megacanthopore [INV ZOO] A large prominent tube, commonly projecting as a spine in a mature region of a bryozoan colony.

Megachilidae [INV ZOO] The leaf-cutting bees, a family of hymenopteran insects in the superfamily Apoidea.

Megachiroptera [VERT ZOO] The fruit bats, a group of Chiroptera restricted to the Old World; most species lack a tail, but when present it is free of the interfemoral membrane.

megacine [MICROBIO] The bacteriocidin produced by *Bacillus megaterium*.

megacolon [MED] Hypertrophy and dilation of the colon associated with prolonged constipation.

megacryst [PETR] Any crystal or grain in an igneous or metamorphic rock that is significantly larger than the surrounding matrix.

megacycle *See* megahertz.

megacyclothem [GEOL] A cycle of or combination of related cyclothems.

Megadermatidae [VERT ZOO] The false vampires, a family of tailless bats with large ears and a nose leaf; found in Africa, Australia, and the Malay Archipelago.

mega-electron-volt *See* million electron volts.

mega-electron-volt–curie [NUCLEO] A unit of radioactive power, equal to the power generated by 1 curie emitting an average energy of 10^6 electron volts per disintegration; equal to approximately 5.92777×10^{-3} watt. Abbreviated MeV Ci.

megahertz [PHYS] Unit of frequency, equal to 1,000,000 hertz. Abbreviated MHz. Also known as megacycle (Mc).

megakaryocyte [HISTOL] A giant bone-marrow cell characterized by a large, irregularly lobulated nucleus; precursor to blood platelets.

megakaryocytopenia *See* megakaryophthisis.

megakaryophthisis [MED] A scarcity of megakaryocytes in the bone marrow. Also known as megakaryocytopenia.

megaloblast [PATH] A large nucleated erythroblast appearing in bone marrow in vitamin B_{12} or folic acid deficiency.

megaloblastic anemia [MED] Anemia characterized by the occurrence of megaloblasts in the bone marrow and blood.

MEGACHILIDAE

Drawing of a leaf-cutting beetle. (*From T. I. Storer and R. L. Usinger, General Zoology, 3d ed., McGraw-Hill, 1957*)

MEGOHMMETER

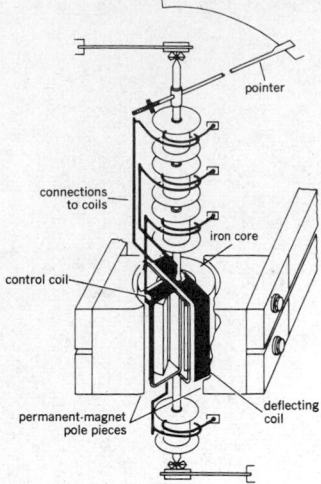

Coils and magnets of a megohmmeter. *(J. G. Biddle Co.)*

MEISSNER'S CORPUSCLE

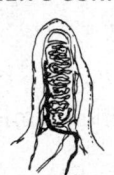

Drawing of Meissner's corpuscle. *(From F. A. Geldard, Human Senses, Wiley, 1953)*

megaloblast of Sabin *See* pronormoblast.

megalocardia [MED] Abnormal enlargement of the heart.

megalocephaly [MED] The condition of having a head whose maximum fronto-occipital circumference is greater than two standard deviations above the mean for age and sex.

Megalodontoidea [INV ZOO] A superfamily of hymenopteran insects in the suborder Symphyta.

megalomania [PSYCH] The delusion of greatness and omnipotence characterizing certain psychotic reactions.

Megalomycteroidei [VERT ZOO] The mosaic-scaled fishes, a monofamilial suborder of the Cetomimiformes; members are rare species of small, elongate deep-sea fishes with degenerate eyes and irregularly disposed scales.

megalopia *See* macropsia.

megalops larva [INV ZOO] A preimago stage of certain crabs having prominent eyes and chelae.

Megaloptera [INV ZOO] A suborder included in the order Neuroptera by some authorities.

Megalopygidae [INV ZOO] The flannel moths, a small family of lepidopteran insects in the suborder Heteroneura.

megalosphere [INV ZOO] The initial, large-chambered shell of sexual individuals of certain dimorphic species of Foraminifera.

megaloureter [MED] Abnormal enlargement of a ureter.

-megaly [MED] A combining form denoting abnormal enlargement.

Megamerinidae [INV ZOO] A family of myodarian cyclorrhaphous dipteran insects in the subsection Acalypteratae.

megaparsec [ASTRON] A unit equal to 1,000,000 parsecs.

megaphyllous [BOT] Having large leaves or leaflike extensions.

Megapodiidae [VERT ZOO] The mound birds and brush turkeys, a family of birds in the order Galliformes; distinguished by their method of incubating eggs in mounds of dirt or in decomposing vegetation.

megarectum [MED] Abnormal enlargement of the rectum.

megaripple [GEOL] A large sand wave.

megasclere [INV ZOO] A large sclerite.

megasecond [MECH] A unit of time, equal to 1,000,000 seconds. Abbreviated Msec.

megasporangium *See* macrosporangium.

megaspore *See* macrospore.

megasporophyll [BOT] A leaf bearing megasporangia.

megatectonics [GEOL] The tectonics of the very large structural features of the earth.

megathermophyte [ECOL] A plant that requires great heat and abundant moisture for normal growth. Also known as macrothermophyte.

Megathyminae [INV ZOO] The giant skippers, a subfamily of lepidopteran insects in the family Hesperiidae.

megaton [PHYS] The energy released by 1,000,000 metric tons of chemical high explosive calculated at a rate of 1000 calories per gram, or a total of 4.18×10^{15} joules; used principally in expressing the energy released by a nuclear bomb. Abbreviated MT.

megaton weapon [ORD] A nuclear fission or fusion bomb capable of exploding with megaton energy.

megatron *See* disk-seal tube.

megavolt [ELEC] A unit of potential difference or emf (electromotive force), equal to 1,000,000 volts. Abbreviated MV.

megawatt [MECH] A unit of power, equal to 1,000,000 watts. Abbreviated MW.

megawatt-day per ton [NUCLEO] A unit used for expressing the burnup of fuel in a reactor; specifically, the number of megawatt-days of heat output per metric ton of fuel in the reactor.

megawatt electric [NUCLEO] Unit of the electric power of a nuclear reactor, as opposed to thermal power. Abbreviated MW(E).

megawatt thermal [NUCLEO] Unit of the thermal power of a nuclear reactor, as opposed to electric power. Abbreviated MW(Th).

megawatt year of electricity [ELEC] A unit of electric energy, equal to the energy delivered by a power of 1,000,000 watts over a period of 1 tropical year, or to 3.1557×10^{13} joules. Abbreviated MWYE.

megohm [ELEC] A unit of resistance, equal to 1,000,000 ohms.

megohmmeter [ELEC] A high-range, permanent-magnet, moving-coil, direct-reading ohmmeter, commonly used as a portable instrument for measuring the high resistance of electrical materials of the order of 20,000 megohms at 1000 volts.

Mehlis' gland [INV ZOO] One of the large unicellular glands around the ootype of flatworms.

Meibomian cyst *See* chalazion.

Meibomian gland *See* tarsal gland.

meibomianitis [MED] Inflammation of the tarsal glands.

Meig's syndrome [MED] A complex of symptoms associated with ovarian fibroma including abnormal accumulation of serous fluid in the pleural and peritoneal cavities.

Meijer transform [MATH] The Meijer transform of a function $f(x)$ is the function $F(y)$ defined as the integral from 0 to ∞ of $\sqrt{xy}\, K_\nu(xy) f(x) dx$, where K_ν is a modified Bessel function.

Meinertellidae [INV ZOO] A family of wingless insects belonging to the Microcoryphia.

Meinzer unit *See* permeability coefficient.

meiocyte [CYTOL] A cell undergoing meiotic division.

meionite [MINERAL] $3CaAl_2Si_2O_8 \cdot CaCO_3$ A scapolite mineral composed of calcium aluminosilicate and calcium carbonate; it is isomorphous with marialite.

meiosis [CYTOL] A type of cell division occurring in diploid or polyploid tissues that results in a reduction in chromosome number, usually by half.

Meissner effect [SOLID STATE] The expulsion of magnetic flux from the interior of a piece of superconducting material as the material undergoes the transition to the superconducting phase. Also known as flux jumping; Meissner-Ochsenfeld effect.

Meissner-Ochsenfeld effect *See* Meissner effect.

Meissner oscillator [ELECTR] Electron-tube oscillator in which the grid and plate circuits are inductively coupled through an independent tank circuit which determines the frequency.

Meissner's corpuscle [ANAT] An ovoid, encapsulated cutaneous sense organ presumed to function in touch sensation in hairless portions of the skin.

Meissner's plexus *See* submucous plexus.

MEK *See* methyl ethyl ketone.

mel [ACOUS] A unit of pitch, equal to one-thousandth of the pitch of a simple tone whose frequency is 1000 hertz and whose loudness is 40 decibels above a listener's threshold.

mel-, melo- [SCI TECH] A combining form denoting dark or black; denoting or pertaining to melanin.

melamine [ORG CHEM] $C_3H_6N_4$ A white crystalline compound that is slightly soluble in water, melts at 354°C and is a cyclic trimer of cyanamide; used to make melamine resins and in tanning of leather. Also known as cyanurtriamide; 2,4,6-triamino-*s*-triazine.

melamine resin [MATER] An amino resin made from formaldehyde and melamine; it is used as a molding compound with fillers added to it; it may also be used for laminating.

Melamphaidae [VERT ZOO] A family of bathypelagic fishes in the order Beryciformes.

Melampsoraceae [MYCOL] A family of parasitic fungi in the order Uredinales in which the teleutospores are laterally united to form crusts or columns.

melancholia [PSYCH] A disordered mental condition of psychotic proportion characterized by severe depression.

Melanconiaceae [MYCOL] The single family of the order Melanconiales.

Melanconiales [MYCOL] An order of the class Fungi Imperfecti including many plant pathogens commonly causing anthracnose; characterized by densely aggregated cnidophores on an acervulus.

Melandryidae [INV ZOO] The false darkling beetles, a family of coleopteran insects in the superfamily Tenebrionoidea.

Melanesia [GEOGR] A group of islands in the Pacific Ocean northeast of Australia.

melanic *See* melanocratic.

melaniline *See* diphenylguanidine.

melanin [BIOCHEM] Any of a group of brown or black pigments occurring in plants and animals.

melanoblast [HISTOL] 1. Precursor cell of melanocytes and melanophores. 2. An immature pigment cell in certain vertebrates. 3. A mature cell that elaborates melanin.

melanoblastoma [MED] A malignant tumor composed principally of melanoblasts.

melanocarcinoma [MED] A malignant melanoma derived from epithelial tissue.

melanocerite [MINERAL] $(Ca,Ce,Y)_8(BO_3)(SiO_4)_4(F,OH)_4$ A brown or black rhombohedral mineral composed of complex silicate, borate, fluoride, tantalate, or other anion of cerium, yttrium, calcium, and other metals; occurs as crystals.

melanocratic [GEOL] Dark-colored, referring to igneous rock containing at least 50–60% mafic minerals. Also known as chromocratic; melanic.

melanocyte [HISTOL] A cell containing dark pigments.

melanocyte-stimulating hormone [BIOCHEM] A protein substance secreted by the intermediate lobe of the pituitary of man which causes dispersion of pigment granules in the skin; similar to intermedins in other vertebrates. Abbreviated MSH. Also known as melanophore-dilating principle; melanophore hormone.

melanocytoma [MED] A benign tumor composed principally of melanocytes.

melanocytosis [MED] An excessive number of melanocytes.

melanoderma [MED] Abnormal darkening of the skin.

melanogen [BIOCHEM] A colorless precursor of melanin.

melanogenesis [BIOCHEM] The formation of melanin.

melanoglosia [MED] Any blackening of the tongue associated with certain disorders and diseases in animal and man.

melanoid [MED] Dark-colored; resembling melanin.

melanoma [MED] 1. A malignant tumor composed of anaplastic melanocytes. 2. A benign or malignant tumor composed of melanocytes.

melanomatosis [MED] 1. Widespread distribution of melanoma. 2. Diffuse melanotic pigmentation of the meninges.

melanophage [HISTOL] A phagocytic cell which engulfs and contains melanin.

melanophlogite [MINERAL] A mineral composed chiefly of silicon dioxide and containing some carbon and sulfur.

melanophore [HISTOL] A type of chromatophore containing melanin.

melanophore-dilating principle See melanocyte-stimulating hormone.

melanophore hormone See melanocyte-stimulating hormone.

melanoprotein [BIOCHEM] A conjugated protein in which melanin is the associated chromagen.

melanosarcoma [MED] A malignant melanoma.

melanose [PL PATH] 1. A fungus disease of grapevine caused by *Septoria ampelina*; leaves are infected and fall off. 2. A fungus disease of citrus trees and fruits caused by *Diaporthe citri*, characterized by hard, brown, usually gummy elevations on the rind, twigs, and leaves.

melanosis coli [PATH] Melanotic pigmentation of the mucosa of the colon in large numbers of minute foci.

melanosis iridis [PATH] Abnormal melanotic pigmentation of the iris.

melanosome [CYTOL] An organelle which contains melanin and in which tyrosinase activity is not demonstrable.

melanotekite [MINERAL] $Pb_2Fe_2Si_2O_9$ A black or dark-gray mineral composed of lead iron silicate.

melanotic cancer [MED] A neoplasm producing excessive quantities of melanin.

melanotic freckle [MED] An unevenly pigmented macule that sometimes develops on the skin, usually on the face, of an individual beyond middle life and enlarges progressively. Also known as Hutchinson's freckle.

melanovandite [MINERAL] $Ca_2V_{10}O_{25}$ A black mineral composed of a complex oxide of calcium and vanadium.

melanterite [MINERAL] $FeSO_4 \cdot 7H_2O$ A green mineral occurring mainly in fibrous or concretionary masses, or in short, monoclinic, prismatic crystals; hardness is 2 on Mohs scale, and specific gravity is 1.90.

melanuria [MED] The presence of black pigment in the urine.

melaphyre [PETR] Altered basalt, especially of Carboniferous and Permian age.

Melasidae [INV ZOO] The equivalent name for Eucnemidae.

Melastomataceae [BOT] A large family of dicotyledonous plants in the order Myrtales characterized by an inferior ovary, axile placentation, up to twice as many stamens as petals (or sepals), anthers opening by terminal pores, and leaves with prominent, subparallel longitudinal ribs.

Meleagrididae [VERT ZOO] The turkeys, a family of birds in the order Galliformes characterized by a bare head and neck.

M electron [ATOM PHYS] An electron whose principal quantum number is 3.

melena [MED] The discharge of stools colored black by altered blood.

Meliaceae [BOT] A family of dicotyledonous plants in the order Sapindales characterized by mostly exstipulate, alternate leaves, stamens mostly connate by their filaments, and syncarpous flowers.

melilite [MINERAL] A sorosilicate mineral group of complex composition $[(Na, Ca)_2(Mg, Al)(Si, Al)_2O_7]$ crystallizing in the tetragonal system; luster is vitreous to resinous, and color is white, yellow, greenish, reddish, or brown; hardness is 5 on Mohs scale, and specific gravity varies from 2.95 to 3.04.

melilitite [PETR] An extrusive rock that is generally olivine-free and composed of more than 90% mafic mineral such as melilite and augite, with minor amounts of feldspathoids and sometimes plagioclase.

Melinae [VERT ZOO] The badgers, a subfamily of carnivorous mammals in the family Mustelidae.

Melinninae [INV ZOO] A subfamily of sedentary annelids belonging to the Ampharetidae which have a conspicuous dorsal membrane, with or without dorsal spines.

melioidosis [VET MED] An endemic bacterial disease, primarily of rodents but occasionally communicable to humans, caused by *Pseudomonas pseudomallei* and characterized by infectious granulomas.

Meliolaceae [MYCOL] The sooty molds, a family of ascomycetous fungi in the order Erysiphales, with dark mycelia and conidia.

meliphane See meliphanite.

meliphanite [MINERAL] $(Ca,Na)_2Be(Si,Al)_2(O,OH,F)_7$ A yellow, red, or black mineral composed of sodium calcium beryllium fluosilicate. Also known as meliphane.

melissic acid [ORG CHEM] $CH_3(CH_2)_{29}COOH$ Fatty acid found in beeswax; soluble in benzene and hot alcohol; melts at 90°C; used in biochemical research.

melissophobia [PSYCH] An abnormal fear of bees.

melitose See raffinose.

melitriose See raffinose.

Melittidae [INV ZOO] A family of hymenopteran insects in the superfamily Apoidea.

melituria [MED] The presence of sugar in the urine.

Mellin transform [MATH] The transform $F(s)$ of a function $f(t)$ defined as the integral over t from 0 to ∞ of $f(t)t^{s-1}$.

mellitate [ORG CHEM] An ester or salt of mellitic acid.

mellite [MINERAL] $Al_2[C_6(COO)_6] \cdot 18H_2O$ A honey-colored mineral with resinous luster composed of the hydrous aluminum salt of mellitic acid, occurring as nodules in brown coal; it is in part a product of vegetable decomposition.

mellitic acid [ORG CHEM] $C_6(COOH)_6$ A water-soluble compound forming colorless needles that melt at 287°C.

melodeon [ELECTR] Broadband panoramic receiver used for countermeasures reception; all types of received electromagnetic radiation are presented as vertical pips on a frequency-calibrated cathode-ray indicator screen.

Meloidae [INV ZOO] The blister beetles, a large cosmopolitan family of coleopteran insects in the superfamily Meloidea; characterized by soft, flexible elytra and the strongly vesicant properties of the body fluids.

Meloidea [INV ZOO] A superfamily of coleopteran insects in the suborder Polyphaga.

melon [BOT] Either of two soft-fleshed edible fruits, muskmelon or watermelon, or varieties of these.

melonite [MINERAL] $NiTe_2$ A reddish-white mineral composed of nickel telluride.

melt [CHEM] 1. To change a solid to a liquid by the application of heat. 2. A melted material. [MET] A charge of molten metal.

meltback transistor [ELECTR] A junction transistor in which the junction is made by melting a properly doped semiconductor and allowing it to solidify again.

Fibrous crust of melanterite in a sample from Kremnitz, Hungary. *(Specimen from Department of Geology, Bryn Mawr College)*

├── 2.5 cm ──┤

MELILITE

├─ 2.5 cm ─┤

Melilite crystals from Vesuvius, Italy. *(Specimen from Department of Geology, Bryn Mawr College)*

MELINAE

The American badger (*Taxidea taxus*).

MELOIDAE

Blister beetle. *(From T. I. Storer and R. L. Usinger, General Zoology, 3d ed., McGraw-Hill, 1957)*

melter [ENG] A chamber used for melting.

melt extractor [ENG] A device used to feed an injection mold, separating molten feed material from partially molten pellets.

melt fracture [MECH] Melt flow instability through a die during plastics molding, leading to helicular, rippled surface irregularities on the finished product.

melt index [ENG] Number of grams of thermoplastic resin at 190°C that can be forced through a 0.0825-inch (2.0955-millimeter) orifice in 10 minutes by a 2160-gram force.

melting *See* fusion.

melting furnace [ENG] A furnace in which the frit for glass is melted.

melting level [METEOROL] The altitude at which ice crystals and snowflakes melt as they descend through the atmosphere.

melting point [THERMO] 1. The temperature at which a solid of a pure substance changes to a liquid. Abbreviated mp. 2. For a solution of two or more components, the temperature at which the first trace of liquid appears as the solution is heated.

melting rate [MET] In electric arc welding, the weight or length of electrode melted in a specified unit of time. Also known as burn-off rate; melt-off rate.

melt loading [ORD] Process of melting solid explosive by heat and pouring into bombs, projectiles, and the like to solidify. Also known as cast loading.

melt-off rate *See* melting rate.

melt spinning [TEXT] A process by which nylon, polyester, or glass is melted to allow it to be extruded into fibers through a spinneret.

meltwater [HYD] Water derived from melting ice or snow, especially glacier ice.

Melusinidae [INV ZOO] A family of orthorrhaphous dipteran insects in the series Nematocera.

Melyridae [INV ZOO] The soft-winged flower beetles, a large family of cosmopolitan coleopteran insects in the superfamily Cleroidea.

member [CIV ENG] A structural unit such as a wall, column, beam, or tie, or a combination of any of these. [GEOL] A rock stratigraphic unit of subordinate rank comprising a specially developed part of a varied formation. [MATH] An element of a set.

Membracidae [INV ZOO] The treehoppers, a family of homopteran insects included in the series Auchenorrhyncha having a pronotum that extends backward over the abdomen, and a vertical upper portion of the head.

membrane [CHEM ENG] 1. The medium through which the fluid stream is passed for purposes of filtration. 2. The ion-exchange medium used in dialysis, diffusion, osmosis and reverse osmosis, and electrophoresis. [HISTOL] A thin layer of tissue surrounding a part of the body, separating adjacent cavities, lining cavities, or connecting adjacent structures.

membrane bone *See* dermal bone.

membrane curing *See* membrane waterproofing.

membrane potential [PHYSIO] A potential difference across a living cell membrane.

membrane waterproofing [CIV ENG] Curing concrete, especially in pavements, by spraying a liquid material over the surface to form a solid, impervious layer which holds the mixing water in the concrete. Also known as membrane curing.

membranous glomerulonephritis [MED] A type of glomerulonephritis characterized by thickening of the basement membrane due to deposition of electron-dense material.

membranous labyrinth [ANAT] The membranous portion of the inner ear of vertebrates.

membranous pregnancy [MED] Gestation in which there has been a rupture of the amniotic sac and the fetus is in direct contact with the wall of the uterus.

membranous urethra [ANAT] The part of the urethra between the two facial layers of the urogenital diaphragm.

memex [ADP] A hypothetical machine described by Vannevar Bush, which would store written records so that they would be available almost instantly by merely pushing the right button for the information desired.

memistor [ELEC] Nonmagnetic memory device consisting of a resistive substrate in an electrolyte; when used in an adaptive system, a direct-current signal removes copper from

an anode and deposits it on the substrate, thus lowering the resistance of the substrate; reversal of the current reverses the process, raising the resistance of the substrate.

memomotion study [IND ENG] A technique of work measurement and methods analysis using a motion picture camera operated at less than normal camera speed. Also known as camera study.

memory [ADP] Any apparatus in which data may be stored and from which the same data may be retrieved; especially, the internal, high-speed, large-capacity working storage of a computer, as opposed to external devices. Also known as computer memory. [PSYCH] The recollection of past events or sensations, or the performance of previously learned skills without practice.

memory address register [ADP] A special register containing the address of a word currently required.

memory buffer register [ADP] A special register in which a word is stored as it is read from memory or just prior to being written into memory.

memory capacity *See* storage capacity.

memory cell [ADP] A single storage element of a memory, together with associated circuits for storing and reading out one bit of information.

memory core *See* magnetic core.

memory cycle *See* cycle time.

memory dump *See* storage dump.

memory dump routine [ADP] A debugging routine which produces a listing of a consecutive section of memory, either numbers or instructions, at selected points in a program.

memory element [ADP] Any component part of core memory.

memory fill *See* storage fill.

memory guard [ADP] Built-in safety devices which prevent a program or a programmer from accessing certain memory areas reserved for the central processor. Also known as memory protect.

memory lockout register [ADP] A special register containing the limiting addresses of an area in memory which may not be accessed by the program.

memory map [ADP] The list of variables, constants, identifiers, and their memory locations when a FORTRAN program is being run. Also known as memory map list.

memory map list *See* memory map.

memory overlay [ADP] The efficient use of memory space by allowing for repeated use of the same areas of internal storage during the different stages of a program; for instance, when a subroutine is no longer required, another routine can replace all or part of it.

memory power [ADP] A relative characteristic pertaining to differences in access time speeds in different parts of memory; for instance, access time from the buffer may be a tenth of the access time from core.

memory print *See* storage dump.

memory printout [ADP] A listing of the contents of memory.

memory protect *See* memory guard.

memory protection *See* storage protection.

memory-reference instruction [ADP] A type of instruction usually requiring two machine cycles, one to fetch the instruction, the other to fetch the data at an address (part of the instruction itself) and to execute the instruction.

memory register *See* storage register.

memory search routine [ADP] A debugging routine which has as an essential feature the scanning of memory in order to locate specified instructions.

memory-segmentation control [ADP] Address-computing logic to address words in memory with dynamic allocation and protection of memory segments assigned to different users.

memory storage [ADP] The sum total of the computer's storage facilities, that is, core, drum, disk, cards, and paper tape.

memory trace [PHYSIO] *See* engram. [PSYCH] An experience intentionally forgotten but not fully repressed, which may result in the development of a neurotic conflict.

memory tube *See* storage tube.

memory twist *See* false twist.

memotron [ELECTR] An electrical-visual storage tube which

MEMBRACIDAE

Membracids on stems. *(a)* Adult. *(b)* Nymphs. *(Courtesy of C. H. Hanson)*

is capable of bistable visual-signal display, controllable in duration from a few milliseconds to infinity, and which is suited to specialized oscillography.

men-, meno- [PHYSIO] A combining form denoting menses.

menacme [PHYSIO] The period of a woman's life during which menstruation persists.

menadione [ORG CHEM] $C_{11}H_8O_2$ Yellow crystals, soluble in alcohol, benzene, and vegetable oils, insoluble in water; affected by sunlight; used in medicine and fungicides.

menarche [PHYSIO] The onset of menstruation.

mendelevium [CHEM] Synthetic radioactive element, symbol Md, with atomic number 101; made by bombarding lighter elements with light nuclei accelerated in cyclotrons.

Mendelian genetics [GEN] Scientific study of heredity as related to or in accordance with Mendel's laws.

Mendelian population [GEN] A group of individuals who interbreed according to a certain system of mating; the total genic content of the individuals of the group is called gene pool.

Mendelian ratio [GEN] The ratio of occurrence of various phenotypes in any cross involving Mendelian characters.

Mendelism [GEN] The basic laws of inheritance as formulated by Mendel.

Mendel's laws [GEN] Two basic principles of genetics formulated by Mendel: the law of segregation and law of independent assortment.

mendip [GEOL] **1.** A buried hill that is exposed as an inlier. **2.** A coastal-plain hill that was originally an offshore island.

mendipite [MINERAL] $Pb_3Cl_2O_2$ A white orthorhombic mineral consisting of an oxide and chloride of lead.

mendozite [MINERAL] $NaAl(SO_4)_2 \cdot 11H_2O$ A monoclinic mineral of the alum group composed of hydrous sodium aluminum sulfate.

Menelaus' theorem [MATH] If ABC is a triangle and PQR is a straight line that cuts AB, CA, and the extension of BC at P, Q, and R respectively, then $(AP/PB)(CQ/QA)(BR/CR) = 1$.

Menetrier's disease [MED] Benign, diffuse hypertrophic gastritis; symptoms include vomiting, diarrhea, weight loss, and excessive secretion of mucus.

Menger's theorem [MATH] A theorem in graph theory which states that if G is a connected graph and A and B are disjoint sets of points of G, then the minimum number of points whose deletion separates A and B is equal to the maximum number of disjoint paths between A and B.

menhaden oil [MATER] A combustible drying oil that is soluble in benzene or ether, and is derived by cooking or pressing the menhaden fish.

Ménière's syndrome [MED] A disease of the inner ear characterized by deafness, vertigo, and tinnitus; possibly an allergic process. Also known as labyrinthine syndrome.

meninges [ANAT] The membranes that cover the brain and spinal cord; there are three in mammals and one or two in submammalian forms.

meninginococcemia [MED] **1.** The presence of meningococci in the blood. **2.** A clinical disorder consisting of fever, skin hemorrhages, varying degrees of shock, and meningococci in the blood.

meningioma [MED] A localized tumor composed of meningeal cells, involving the meninges and other central nervous system structures. Also known as meningothelioma.

meningism [MED] A condition in which signs and symptoms suggest meningitis, but clinical evidence for the disease is absent. Also known as meningismus.

meningismus *See* meningism.

meningitis [MED] Inflammation of the meninges of the brain and spinal cord, caused by viral, bacterial, and protozoan agents.

meningitophobia [PSYCH] An abnormal fear of meningitis.

meningocele [MED] Hernia of the meninges through a defect in the skull or vertebral column, forming a cyst filled with cerebrospinal fluid.

meningococcal meningitis [MED] Inflammation of the meninges caused by the bacterium *Neisseria meningitidis* (meningococcus).

meningococcus [MICROBIO] Common name for *Neisseria meningitidis.*

meningoencephalitis [MED] Inflammation of the brain and its meninges.

meningoencephalocele [MED] A protrusion of the brain and its membranes through a defect in the skull.

meningoencephalomyelitis [MED] Combined inflammation of the meninges, brain, and spinal cord.

meningoencephalopathy [MED] Disease of the brain and its meninges.

meningomyelitis [MED] Inflammation of the spinal cord and its membranes.

meningomyocele [MED] Hernia of the spinal cord and its meninges through a defect in the vertebral column.

meningothelioma *See* meningioma.

meningothelium [HISTOL] Epithelium of the arachnoid which envelops the brain.

meningovascular [MED] Involving both the meninges and the cerebral blood vessels.

meningovascular syphilis [MED] Syphilis of the central nervous system involving the formation of gummas of the leptomeninges and endarteritis of cerebral vessels.

meninx [ANAT] Any one of the three membranes covering the brain and spinal cord.

meninx primitiva [VERT ZOO] The single membrane covering the brain and spinal cord of certain submammalian vertebrates.

menisc-, menisco- [SCI TECH] A combining form denoting crescentic, sickle-shaped, semilunar; denoting meniscus, semilunar cartilage.

meniscectomy [MED] Surgical removal of a meniscus or semilunar cartilage.

meniscitis [MED] Inflammation of the semilunar cartilages.

Meniscotheriidae [PALEON] A family of extinct mammals of the order Condylarthra possessing selenodont teeth and molarized premolars.

meniscus [ANAT] A crescent-shaped body, especially an interarticular cartilage. [FL MECH] The free surface of a liquid which is near the walls of a vessel and which is curved because of surface tension.

meniscus lens [OPTICS] A lens with one convex surface and one concave surface.

Menispermaceae [BOT] A family of dicotyledonous woody vines in the order Ranunculales distinguished by mostly alternate, simple leaves, unisexual flowers, and a dioecious habit.

menometrorrhagia [MED] Excessive uterine bleeding during menstruation, plus irregular uterine bleeding at other times.

menopause [PHYSIO] The natural physiologic cessation of menstruation, usually occurring in the last half of the fourth decade. Also known as climacteric.

menoplania [MED] Bleeding during menstruation from a part of the body other than the uterus.

Menoponidae [INV ZOO] A family of biting lice (Mallophaga) adapted to life only upon domestic and sea birds.

menorrhagia [MED] Excessive bleeding during menstruation. Also known as hypermenorrhea.

menorrhalgia [MED] Pelvic pain occurring at the menstrual period.

menorrhea [PHYSIO] The normal flow of the menses. [MED] Excessive menstruation.

menostasis [MED] Suppression of the menstrual flow.

menostaxis [MED] Prolonged menstruation.

menses *See* menstruation.

menstrual age [EMBRYO] The age of an embryo or fetus calculated from the first day of the mother's last normal menstruation preceding pregnancy.

menstrual cycle [PHYSIO] The periodic series of changes associated with menstruation and the intermenstrual cycle; menstrual bleeding indicates onset of the cycle.

menstrual period [PHYSIO] The time of menstruation.

menstruate [PHYSIO] To discharge the products of menstruation.

menstruation [PHYSIO] The periodic discharge of sanguineous fluid and sloughing of the uterine lining in women from puberty to the menopause. Also known as menses.

menstruum [MATER] A solvent, commonly one that extracts certain principles from entire plant or animal tissues.

mensuration [MATH] The measurement of geometric quanti-

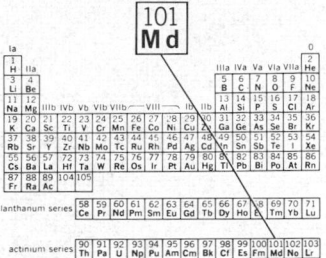

MENDELEVIUM

Periodic table of the chemical elements showing the position of mendelevium.

MENISCUS LENS

(a)   (b)

Two types of meniscus lens. (a) Positive. (b) Negative. (From F. A. Jenkins and H. E. White, Fundamentals of Optics, 3d ed., McGraw-Hill, 1957)

ties; for example, length, area, and volume. [SCI TECH] The act or process of measuring.

mental [ANAT] Pertaining to the chin. Also known as genial. [PSYCH] **1.** Pertaining to the mind, psyche, or inner self. **2.** Pertaining to the intellectual or cognitive functions. **3.** Imaginary or unreal, as when a pain is said to be purely mental.

mental aberration [PSYCH] A departure from normal mental function.

mental adjustment [PSYCH] The act or process by which an individual adapts his attitudes, traits, or feelings to the social environment.

mental age [PSYCH] The degree of mental development of an individual in terms of the chronological age of the average individual of equivalent mental ability; specifically, a score derived from intelligence tests.

mental deficiency [PSYCH] A condition characterized by intellectual retardation, social inadequacy, and persistent dependency.

mental health [PSYCH] A relatively enduring state of being in which an individual has effected an integration of his instinctual drives in a way that is reasonably satisfying to himself as reflected in his zest for living and his feeling of self-realization.

mental hygiene [PSYCH] That branch of hygiene dealing with the preservation of mental and emotional health.

mental illness [PSYCH] Any form of mental aberration; usually refers to a chronic or prolonged disorder in which there are wide deviations from the normal.

mental retardation [PSYCH] An abnormal slowness of mental function and behavior patterns relative to age and development.

mental telepathy [PSYCH] A form of extrasensory perception in which one person is aware of an external event through direct sensory perception, and another person, not in the same place, also becomes aware of the event but not through direct sensory perception.

mentation [PSYCH] Mental activity.

menthacamphor *See* menthol.

Menthaceae [BOT] An equivalent name for Labiatae.

menthane [ORG CHEM] $C_{10}H_{20}$ A colorless, water-insoluble liquid hydrocarbon; used in organic synthesis. Also known as hexahydrocymene; 4-isopropyl-1-methylcyclohexane; menthonaphthene; terpane.

para-menthan-3-ol *See* menthol.

menthene [ORG CHEM] $C_{10}H_{18}$ A colorless, water-insoluble, liquid hydrocarbon; used in organic synthesis.

menthol [ORG CHEM] $CH_3C_6H_9(C_3H_7)OH$ An alcohol-soluble, white crystalline compound that may exist in levo form or a mixture of dextro and levo isomers; used in medicines and perfumes, and as a flavoring agent. Also known as hexahydrothymol; 3-hydroxymenthane; menthacamphor; para-menthan-3-ol; methylhydroxyisopropylcyclohexane; peppermint camphor.

menthonaphthene *See* menthane.

menthone [ORG CHEM] $C_{10}H_{18}O$ Oily, colorless ketonic liquid with slight peppermint odor; slightly soluble in water, soluble in organic solvents.

menthyl [ORG CHEM] $C_{10}H_{19}$ A univalent radical that is derived from menthol by removal of the hydroxyl group.

menton-philtrum [ANTHRO] A measure of the distance from the midpoint of the lower edge of the chin to the midpoint of the philtrum or vertical groove of the upper lip.

menton-supramentale [ANTHRO] The measurement of the distance taken from the midpoint of the lower edge of the chin to the supramentale, or the angle between the chin and the lower lip.

mentum [ANAT] The chin. [BOT] A projection formed by union of the sepals at the base of the column in some orchids. [INV ZOO] **1.** A projection between the mouth and foot in certain gastropods. **2.** The median or basal portion of the labium in insects.

Menurae [VERT ZOO] A small suborder of suboscine perching birds restricted to Australia, including the lyrebirds and scrubbirds.

Menuridae [VERT ZOO] The lyrebirds, a family of birds in the suborder Menurae notable for their vocal mimicry.

mep *See* mean effective pressure.

meperidine hydrochloride [ORG CHEM] $C_{15}H_{21}O_2N \cdot HCl$ A white, odorless crystalline compound, melting at 186–189°C; soluble in water and alcohol; used in medicine.

mephenesin [PHARM] $C_{10}H_{14}O_3$ The compound 3-o-(tolyloxy)-1,2-propanediol, used as a skeletal muscle relaxant.

mephentermine sulfate [ORG CHEM] $(C_{11}H_{17}N)_2 \cdot H_2SO_4 \cdot 2H_2O$ White odorless crystals; slightly soluble in alcohol, soluble in water; used in medicine.

mepivacaine [PHARM] $C_{15}H_{22}N_2O$ The compound dl-1-methyl-2′6,′-pipecoloxylidide, a local anesthetic; used as the hydrochloride salt.

méplat [ORD] The flat nose formed by truncation of the ogival portion of a projectile or point fuse.

meprobromate [PHARM] $C_9H_{18}N_2O_4$ The compound 2-methyl-2-n-propyl-1,3-propanediol dicarbamate, a tranquilizer with anticonvulsant, muscle relaxant, and sedative actions.

mer-, mero- [SCI TECH] A combining form meaning part or partial.

-mer [ORG CHEM] A combining form denoting the repeating structure unit of any high polymer.

meralluride [PHARM] $C_9H_{16}HgN_2O_6$ A diuretic consisting of succinamic acid and theophylline, in approximately molecular proportions, administered as the sodium derivative.

Meramecian [GEOL] A North American provincial series of geologic time: Upper Mississippian (above Osagian, below Chesterian).

meraspis [PALEON] Advanced larva of a trilobite; stage in which the pygidium begins to form.

merbromin [ORG CHEM] $C_{20}H_8O_6Na_2Br_2Hg$ A green crystalline powder that gives a deep-red solution in water; used as an antiseptic. [PHARM] The disodium salt of 2′,7′,-dibromo-4′-(hydroxymercuri)-fluorescein, an organomercurial antibacterial agent applied topically.

Mercalli scale [GEOPHYS] A 12-point scale for classifying the magnitude of an earthquake.

mercapt-, mercapto- [CHEM] A combining form denoting the presence of the thiol (SH) group.

mercaptal [ORG CHEM] A group of organosulfur compounds that contain the group $=C(SR)_2$.

mercaptan [ORG CHEM] A group of organosulfur compounds that are derivatives of hydrogen sulfide in the same way that alcohols are derivatives of water; have a characteristically disagreeable odor, and are found with other sulfur compounds in crude petroleum; an example is methyl mercaptan. Also known as thiol.

mercaptoacetic acid *See* thioglycollic acid.

mercaptobenzothiazole [ORG CHEM] C_7H_5NS A yellow powder, melting at 164–174°C; used in rubber as a vulcanization accelerator with stearic acid. Also known as MBT.

mercapto compound *See* sulfhydryl compound.

mercaptoethanol [ORG CHEM] $HSCH_2CH_2OH$ Mobile liquid, water-white; soluble in water, benzene, ether, and most organic solvents; boils at 157°C; used as a solvent, chemical intermediate, and reducing agent.

mercaptol [ORG CHEM] A compound formed by combining a mercaptal and a ketone.

2-mercaptoproprionic acid *See* thiolactic acid.

Mercator bearing *See* rhumb bearing.

Mercator chart [MAP] A chart on the Mercator projection, commonly used for marine navigation. Also known as equatorial cylindrical orthomorphic chart.

Mercator course *See* rhumb-line course.

Mercator direction [NAV] Horizontal direction of a rhumb line, expressed as angular distance from a reference direction. Also known as rhumb direction.

Mercator projection [MAP] A conformal cylindrical map projection in which the surface of a sphere or spheroid, such as the earth, is conceived as developed on a cylinder tangent along the Equator; meridians appear as equally spaced vertical lines, and parallels as horizontal lines drawn farther apart as the latitude increases, such that the correct relationship between latitude and longitude scales at any point is maintained.

Mercator sailing [NAV] A method of solving the various problems involving course, distance, difference of latitude, difference of longitude, and departure by considering them in the relation in which they are plotted on a Mercator chart.

Mercer engine [MECH ENG] A revolving-block engine in which two opposing pistons operate in a single cylinder with two rollers attached to each piston; intake ports are uncovered when the pistons are closest together, and exhaust ports are uncovered when they are farthest apart.

mercerization [TEXT] A technique used to increase luster, dye absorptivity, and strength in cotton and linen goods; the cloth is put into a heated solution of caustic soda at a controlled temperature, then washed, neutralized, and rinsed.

mercerizing assistant [MATER] A wetting agent, such as cresylic acid and derivatives or oils, used to increase the penetration of textile mercerization baths.

merchant mill [MET] A mill, consisting of a group of stands of three rolls each, used to roll rounds, flats, or squares of smaller dimensions than could be rolled on a bar mill.

merchant ship [NAV ARCH] A power-driven ship employed in commercial transport on the oceans and large inland bodies of water such as the Great Lakes.

mercurial horn ore *See* calomel.

mercurialism [MED] Chronic type of mercury poisoning. Also known as hydrargyrism.

mercurial nephrosis [MED] Nephrosis caused by poisoning with mercury bichloride.

mercurial tremor [MED] A fine muscular tremor observed in persons with mercurialism or poisoning by other heavy metals.

mercuric [INORG CHEM] The mercury ion with a 2+ oxidation state, for example $Hg(NO_3)_2$.

mercuric acetate [ORG CHEM] $Hg(C_2H_3O_2)_2$ Poisonous, light-sensitive white crystals; soluble in alcohol and water; used in medicine and as a catalyst in organic synthesis. Also known as mercury acetate.

mercuric arsenate [INORG CHEM] $HgHAsO_4$ A poisonous yellow powder; soluble in hydrochloric acid, insoluble in water; used in antifouling and waterproof paints and in medicine. Also known as mercury arsenate; mercury arseniate.

mercuric barium iodide *See* barium mercuric iodide.

mercuric benzoate [ORG CHEM] $Hg(C_7H_5O_2)_2 \cdot H_2O$ Poisonous white crystals, sensitive to light, melting at 165°C; slightly soluble in alcohol and water; used in medicine. Also known as mercury benzoate.

mercuric bromide [INORG CHEM] $HgBr_2$ Poisonous white crystals, sensitive to light, melting at 235°C; soluble in alcohol and ether; used in medicine. Also known as mercury bromide.

mercuric chloride [INORG CHEM] $HgCl_2$ An extremely toxic compound that forms white, rhombic crystals which sublime at 300°C and are soluble in alcohol or benzene; used for the manufacture of other mercuric compounds, as a fungicide, and in medicine and photography. Also known as bichloride of mercury; corrosive sublimate.

mercuric cyanide [INORG CHEM] $Hg(CN)_2$ Poisonous, colorless, transparent crystals that darken in light, decompose when heated; soluble in water and alcohol; used in photography, medicine, and germicidal soaps. Also known as mercury cyanide.

mercuric fluoride [INORG CHEM] HgF_2 Poisonous, transparent crystals that decompose when heated; moderately soluble in alcohol and water; used to synthesize organic fluorides.

mercuric iodide [INORG CHEM] HgI_2 Poisonous red crystals that turn yellow when heated to 150°C; soluble in boiling alcohol; used in medicine and in Nessler's and Mayer's reagents.

mercuric lactate [ORG CHEM] $Hg(C_3H_5O_3)_2$ A poisonous white powder that decomposes when heated; soluble in water; used in medicine.

mercuric nitrate [INORG CHEM] $Hg(NO_3)_2 \cdot H_2O$ Poisonous, colorless crystals that decompose when heated; soluble in water and nitric acid, insoluble in alcohol; a fire hazard; used in medicine, in nitrating organic aromatics, and in felt manufacture. Also known as mercury nitrate; mercury pernitrate.

mercuric oleate [ORG CHEM] $Hg(C_{18}H_{33}O_2)_2$ A poisonous yellowish-to-red liquid or solid mass; insoluble in water; used

in medicine and antifouling paints, and as an antiseptic. Also known as mercury oleate.

mercuric oxide [INORG CHEM] HgO A compound of mercury that exists in two forms, red mercuric oxide and yellow mercuric oxide; the red form decomposes upon heating, is insoluble in water, and is used in pigments and paints, and in ceramics; the yellow form is insoluble in water, decomposes upon heating, and is used in medicine. Also known as mercury oxide; red precipitate; yellow precipitate.

mercuric phosphate [INORG CHEM] $Hg_3(PO_4)_2$ Poisonous yellowish or white powder; insoluble in alcohol and water, soluble in acids; used in medicine. Also known as mercury phosphate; neutral, normal, or tertiary mercuric phosphate; trimercuric orthophosphate.

mercuric salicylate [ORG CHEM] $Hg(C_7H_5O_3)_2$ Poisonous, white powder; odorless and tasteless; almost insoluble in water and alcohol; variable composition; used in medicine. Also known as salicylated mercury.

mercuric stearate [ORG CHEM] $Hg(C_{17}H_{35}CO_2)_2$ Poisonous yellow powder; soluble in fatty acids, slightly soluble in alcohol; used as a germicide and in medicine. Also known as mercury stearate.

mercuric sulfide [INORG CHEM] HgS **1.** The black variety is a poisonous powder; insoluble in water, alcohol, and nitric acid, soluble in sodium sulfide solution; sublimes at 583°C; used as a pigment. Also known as black mercury sulfide; ethiops mineral. **2.** The red variety is a poisonous powder; insoluble in water and alcohol; sublimes at 446°C; used as a medicine and pigment. Also known as Chinese vermilion; quicksilver vermilion; red mercury sulfide; vermilion.

mercuric sulfocyanate *See* mercuric thiocyanate.

mercuric thiocyanate [INORG CHEM] $Hg(SCN)_2$ Poisonous white powder; soluble in alcohol, slightly soluble in water; decomposes when heated; used in photography. Also known as mercuric sulfocyanide; mercury sulfocyanate; mercury thiocyanate.

Mercurochrome [PHARM] A trademark for merbromin.

mercurous [INORG CHEM] Referring to mercury with a 1+ valence; for example, mercurous chloride, Hg_2Cl_2, where the mercury is covalently bonded, as $Cl-Hg-Hg-Cl$.

mercurous acetate [ORG CHEM] $HgC_2H_3O_2$ Poisonous colorless plates or scales; decomposed by boiling water and by light; soluble in dilute nitric acids, slightly soluble in water. Also known as mercury acetate; mercury protoacetate.

mercurous bromide [INORG CHEM] $HgBr$ Poisonous white powder, crystals, or fibrous mass; odorless and tasteless; darkens in light; soluble in hot sulfuric acid and fuming nitric acid, insoluble in alcohol and ether; used in medicine. Also known as mercury bromide.

mercurous chlorate [INORG CHEM] $Hg_2(ClO_3)_2$ Poisonous white crystals that decompose at 250°C; soluble in alcohol and water; explodes in contact with combustible substances. Also known as mercury chlorate.

mercurous chloride [INORG CHEM] Hg_2Cl_2 Odorless, nonpoisonous white crystals that darken in light; insoluble in water, alcohol, and ether; melts at 302°C; used in medicine and pyrotechnics. Also known as mercury monochloride; mercury protochloride; mild mercury chloride.

mercurous chromate [INORG CHEM] Hg_2CrO_4 Red powder with variable composition; decomposes when heated; soluble in nitric acid, insoluble in water and alcohol; used to color ceramics green. Also known as mercury chromate.

mercurous iodide [INORG CHEM] Hg_2I_2 Odorless, tasteless, poisonous yellow powder; darkens when heated; insoluble in water, alcohol, and ether; sublimes at 140°C; used as external medicine. Also known as mercury protoiodide.

mercurous oxide [INORG CHEM] Hg_2O A poisonous black powder; insoluble in water, soluble in acids; decomposes at 100°C.

mercurous phosphate [INORG CHEM] Hg_3PO_4 Light-sensitive white powder with variable composition; insoluble in alcohol and water, soluble in nitric acids; used in medicine. Also known as mercury phosphate; neutral, normal, or tertiary mercurous phosphate; trimercurous orthophosphate.

mercurous sulfate [INORG CHEM] Hg_2SO_4 Poisonous yellow-to-white powder; soluble in hot sulfuric acid or dilute

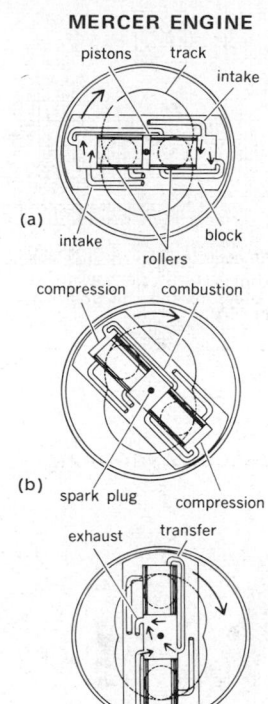

MERCER ENGINE

Mercer engine. *(a)* Pistons are closest together and intake ports are open admitting fresh charge. *(b)* Pistons separate as combustion takes place, compressing gases behind pistons, forcing roller to move outward, and rotating the entire engine block at the same time. *(c)* Pistons are farthest apart, exhaust ports are opened, gases are purged, and compressed fresh charge is transferred to region between pistons.

MERCURY

80
Hg

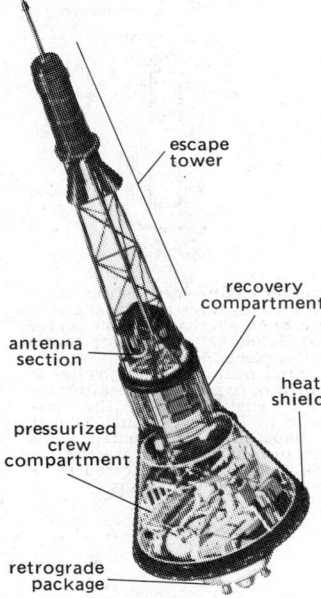

Periodic table of the chemical elements showing the position of mercury.

MERCURY PROGRAM

Cutaway view of a Mercury spacecraft. *(NASA)*

- escape tower
- recovery compartment
- antenna section
- heat shield
- pressurized crew compartment
- retrograde package

MERCURY-VAPOR LAMP

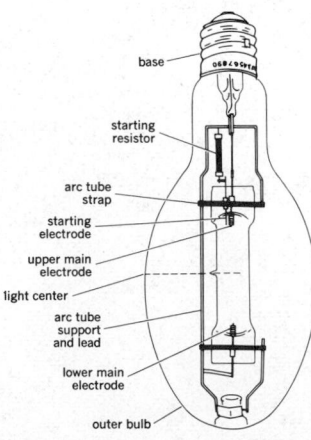

Parts of the mercury-vapor lamp.

- base
- starting resistor
- arc tube strap
- starting electrode
- upper main electrode
- light center
- arc tube support and lead
- lower main electrode
- outer bulb

---

nitric acid, insoluble in water; used as a catalyst and in laboratory batteries.

mercury [CHEM] A metallic element, symbol Hg, atomic number 80, atomic weight 200.59, existing at room temperature as a silvery, heavy liquid. Also known as quicksilver.

Mercury [ASTRON] The planet nearest to the sun; it is visible to the naked eye shortly after sunset or before sunrise when it is nearest to its greatest angular distance from the sun.

mercury acetate *See* mercuric acetate; mercurous acetate.

mercury arc [ELECTR] An electric discharge through ionized mercury vapor, giving off a brilliant bluish-green light containing strong ultraviolet radiation.

mercury-arc rectifier [ELECTR] A gas-filled rectifier tube in which the gas is mercury vapor; small sizes use a heated cathode, while larger sizes rated up to 8000 kilowatts and higher use a mercury-pool cathode. Also known as mercury rectifier; mercury-vapor rectifier.

mercury arsenate *See* mercuric arsenate.

mercury arseniate *See* mercuric arsenate.

mercury ballistic [NAV] A system of reservoirs and connecting tubes containing mercury used with a type of nonpendulous gyro compass; the action of gravity on this system provides the torques and resultant precessions required to convert the gyroscope into a compass.

mercury barometer [ENG] An instrument which determines atmospheric pressure by measuring the height of a column of mercury which the atmosphere will support; the mercury is in a glass tube closed at one end and placed, open end down, in a well of mercury. Also known as Torricellian barometer.

mercury benzoate *See* mercuric benzoate.

mercury bromide *See* mercuric bromide; mercurous bromide.

mercury-cathode cell [CHEM ENG] Electrolytic cell used to manufacture chlorine and caustic soda from sodium chloride brine; includes Castner and DeNora cells.

mercury cell [ELEC] A primary dry cell that delivers an essentially constant output voltage throughout its useful life by means of a chemical reaction between zinc and mercury oxide; widely used in hearing aids. Also known as mercury oxide cell.

mercury chlorate *See* mercurous chlorate.

mercury chromate *See* mercurous chromate.

mercury cosmetic *See* ammoniated mercury.

mercury cyanide *See* mercuric cyanide.

mercury delay line [ELECTR] An acoustic delay line in which mercury is the medium for sound transmission. Also known as mercury memory; mercury storage.

mercury lamp *See* mercury-vapor lamp.

mercury manometer [ENG] A manometer in which the instrument fluid is mercury; used to record or control difference of pressure or fluid flow.

mercury memory *See* mercury delay line.

mercury monochloride *See* mercurous chloride.

mercury naphthenate [ORG CHEM] Poisonous dark-amber liquid; soluble in mineral oils; used in gasoline antiknock compounds and as a paint antimildew promoter.

mercury nitrate *See* mercuric nitrate.

mercury oleate *See* mercuric oleate.

mercury oxide *See* mercuric oxide.

mercury oxide cell *See* mercury cell.

mercury pernitrate *See* mercuric nitrate.

mercury phosphate *See* mercuric phosphate; mercurous phosphate.

mercury-pool cathode [ELECTR] A cathode of a gas tube consisting of a pool of mercury; an arc spot on the pool emits electrons.

mercury-pool rectifier *See* pool-cathode mercury arc rectifier.

Mercury program [AERO ENG] First United States program to use manned spacecraft, carried out in 1961–1963; the craft carried one man.

mercury protoacetate *See* mercurous acetate.

mercury protochloride *See* mercurous chloride.

mercury protoiodide *See* mercurous iodide.

mercury stearate *See* mercuric stearate.

mercury storage *See* mercury delay line.

mercury switch [ELEC] A switch that is closed by making a

---

large globule of mercury move up to the contacts and bridge them; the mercury is usually moved by tilting the entire switch.

mercury tank [ELECTR] A container of mercury, with pairs of transducers at opposite ends, used in a mercury delay line.

mercury thermometer [ENG] A liquid-in-glass thermometer or a liquid-in-metal thermometer using mercury as the liquid.

mercury thiocyanate *See* mercuric thiocyanate.

mercury tube *See* mercury-vapor tube; pool tube.

mercury-vapor lamp [ELECTR] A lamp in which light is produced by an electric arc between two electrodes in an ionized mercury-vapor atmosphere; it gives off a bluish-green light rich in ultraviolet radiation. Also known as mercury lamp.

mercury-vapor rectifier *See* mercury-arc rectifier.

mercury-vapor tube [ELECTR] A gas tube in which the active gas is mercury vapor. Also known as mercury tube.

mercury-wetted reed switch [ELEC] A reed switch containing a pool of mercury at one end and normally operated vertically; the contacts on the reeds are covered with a mercury film by capillary action; each operation of the switch renews this mercury film contact, thereby increasing the operating life of the switch many times.

merganser [VERT ZOO] Any of several species of diving water fowl composing a distinct subfamily of Anatidae and characterized by a serrate bill adapted for catching fish.

merge [ADP] To create an ordered set of data by combining properly the contents of two or more sets of data, each originally ordered in the same manner as the output data set. Also known as mesh.

merge sort [ADP] To produce a single sequence of items ordered according to some rule, from two or more previously ordered or unordered sequences, without changing the items in size, structure, or total number; although more than one pass may be required for a complete sort, items are selected during each pass on the basis of the entire key.

merging routine [ADP] A program that creates a single sequence of items, ordered according to some rule, out of two or more sequences of items, each sequence ordered according to the same rule.

Merian's formula [OCEANOGR] A formula for the period of a seiche, $T = (1/n)(2L/\sqrt{gd})$, where n is the number of nodes, L is the horizontal dimension of the basin measured in the direction of wave motion, g is the acceleration of gravity, and d is the depth of the water.

mericarp [BOT] An individual, one-seeded carpel of a schizocarp.

meridian [GEOD] A north-south reference line, particularly a great circle through the geographical poles of the earth.

meridian altitude [ASTRON] The altitude of a celestial body when it is on the celestial meridian of the observer, bearing 000° or 180° true.

meridian angle [ASTRON] Angular distance east or west of the local celestial meridian; the arc of the celestial equator, or the angle at the celestial pole, between the upper branch of the local celestial meridian and the hour circle of a celestial body, measured eastward or westward from the local celestial meridian through 180°, and labeled E or W to indicate the direction of measurement. Also called hour angle.

meridian angle difference [ASTRON] The difference between two meridian angles, particularly between the meridian angle of a celestial body and the value used as an argument for entering into a table. Also called hour angle difference.

meridian observation [ASTRON] Measurement of the altitude of a celestial body on the celestial meridian of the observer, or the altitude so measured.

meridian passage [ASTRON] The passage of a celestial body across an observer's meridian.

meridian sailing [NAV] Following a true course of 000° or 180°, or sailing along a meridian; under these conditions the dead-reckoning latitude is assumed to change 1 minute for each mile run, and the dead-reckoning longitude is assumed to remain unchanged.

meridian transit *See* transit circle.

meridional [GEOL] Pertaining to longitudinal movements or directions, that is, northerly or southerly.

meridional cell [GEOPHYS] A very large-scale convection

circulation in the atmosphere or ocean which takes place in a meridional plane, with northward and southward currents in opposite branches of the cell, and upward and downward motion in the equatorward and poleward ends of the cell.

meridional circulation [METEOROL] An atmospheric circulation in a vertical plane oriented along a meridian; it consists, therefore, of the vertical and the meridional (north or south) components of motion only. [OCEANOGR] The exchange of water masses between northern and southern oceanic regions.

meridional difference [MAP] The difference between the meridional parts of any two given parallels of latitude; this difference is found by subtraction if the two parallels are on the same side of the equator, and by addition if on opposite sides. Also known as difference of meridional parts.

meridional flow [METEOROL] A type of atmospheric circulation pattern in which the meridional (north and south) component of motion is unusually pronounced; the accompanying zonal component is usually weaker than normal. [OCEANOGR] Current moving along a meridian.

meridional focus *See* primary focus.

meridional front [METEOROL] A front in the South Pacific separating successive migratory subtropical anticyclones; such fronts are essentially in the form of great arcs with meridians of longitudes as chords; they have the character of cold fronts.

meridional index [METEOROL] A measure of the component of air motion along meridians, averaged, without regard to sign, around a given latitude circle.

meridional parts [MAP] The length of the arc of a meridian between the equator and a given parallel on a mercator chart, expressed in units of 1 minute of longitude at the equator.

meridional ray [OPTICS] A ray that lies within a plane which also contains the axis of an optical system.

meridional wind [METEOROL] The wind or wind component along the local meridian, as distinguished from the zonal wind.

Meridosternata [INV ZOO] A suborder of echinoderms including various deep-sea forms of sea urchins.

merinthophobia [PSYCH] An abnormal fear of being tied up.

merismite [PETR] A type of chorismite in which penetration of the diverse units is irregular.

meristem [BOT] Formative plant tissue composed of undifferentiated cells capable of dividing and giving rise to other meristematic cells as well as to specialized cell types; found in growth areas.

meristic [BIOL] Pertaining to a change in number or in geometric relation of parts of an organism. [ZOO] Of, pertaining to, or divided into segments.

merit [ELECTR] A performance rating that governs the choice of a device for a particular application; it must be qualified to indicate type of rating, as in gain-bandwidth merit or signal-to-noise merit.

merit pay plan [IND ENG] Work performed for a set hourly wage that varies from one pay period to another as a function of the worker's productivity, but never declines below a guaranteed minimum wage.

Merkel's corpuscles [ANAT] Touch receptors consisting of flattened platelets at the tips of certain cutaneous nerves.

Mermithidae [INV ZOO] A family of filiform nematodes in the superfamily Mermithoidea; only juveniles are parasitic.

Mermithoidea [INV ZOO] A superfamily of nematodes composed of two families, both of which are invertebrate parasites.

meroblastic [EMBRYO] Of or pertaining to an ovum that undergoes incomplete cleavage due to large amounts of yolk.

merocrine [PHYSIO] Pertaining to glands in which the secretory cells undergo cytological changes without loss of cytoplasm during secretion.

merocrystalline *See* hypocrystalline.

merogony [EMBRYO] The normal or abnormal development of a part of an egg following cutting, shaking, or centrifugation of the egg before or after fertilization.

merohedral [CRYSTAL] Of a crystal class in a system, having a general form with only one-half, one-fourth, or one-eighth the number of equivalent faces of the corresponding form in the holohedral class of the same system. Also known as merosymmetric.

meromictic [HYD] Of or pertaining to a lake whose water is permanently stratified and therefore does not circulate completely throughout the basin at any time during the year.

meromorphic function [MATH] A function of complex variables which is analytic in its domain of definition save at a finite number of points which are poles.

meromyarian [INV ZOO] Having few muscle cells in each quadrant as seen in cross section; applied especially to nematodes.

meromyosin [BIOCHEM] Protein fragments of a myosin molecule, produced by enzymatic digestion.

Meropidae [VERT ZOO] The bee-eaters, a family of brightly colored Old World birds in the order Coraciiformes.

meroplankton [BIOL] Plankton composed of floating developmental stages (that is, eggs and larvae) of the benthos and nekton organisms. Also known as temporary plankton.

Merostomata [INV ZOO] A class of primitive arthropods of the subphylum Chelicerata distinguished by their aquatic mode of life and the possession of abdominal appendages which bear respiratory organs; only three living species are known.

merosymmetric *See* merohedral.

-merous [BIOL] Combining form that denotes having such parts or so many parts.

Merozoa [INV ZOO] The equivalent name for Cestoda.

merozoite [INV ZOO] An ameboid trophozoite in some sporozoans produced from a schizont by schizogony.

Merrick Weightometer [MECH ENG] An instrument which measures the weight of material being carried by a belt conveyor, in which the weight of the load on a portion of the conveyor is balanced, using a system of levers, by the buoyancy of a cylindrical steel float partly immersed in a bath of mercury.

merrihueite [MINERAL] $(K,Na)_2(Fe,Mg)_5Si_{12}O_{30}$ A silicate mineral found only in meteorites.

Merrill-Crowe process [MET] Removal of gold from cyanide solution by deoxygenation followed by precipitation on zinc dust, the work being completed by filtration to give the resultant auriferous gold slimes.

merrillite [MINERAL] $Ca_3(PO_4)_2$ Colorless phosphate mineral found only in meteorites.

Mersenne's law [MECH] The fundamental frequency of a vibrating string is proportional to the square root of the tension and inversely proportional both to the length and the square root of the mass per unit length.

Mersey yellow coal *See* tasmanite.

merthiolate [PHARM] A trademark for thimerosal.

merwinite [MINERAL] $Ca_3MgSi_2O_8$ A rare colorless or pale-green neosilicate mineral crystallizing in the monoclinic system; occurs in granular aggregates showing polysynthetic twinning; hardness is 6 on Mohs scale, and specific gravity is 3.15

merycism *See* rumination.

Merycoidodontidae [PALEON] A family of extinct tylopod ruminants in the superfamily Merycoidodontoidea.

Merycoidodontoidea [PALEON] A superfamily of extinct ruminant mammals in the infraorder Tylopoda which were exceptionally successful in North America.

merzlota *See* frozen ground.

mes-, meso- [SCI TECH] A combining form denoting mid-, middle, medial; medium, moderate, intermediate; mesentery; mesodermal.

mesa [GEOGR] A broad, isolated, flat-topped hill bounded by a steep cliff or slope on at least one side; represents an erosion remnant.

mesa-butte [GEOGR] A butte formed as the result of erosion and reduction of a mesa.

Mesacanthidae [PALEON] An extinct family of primitive acanthodian fishes in the order Acanthodiformes distinguished by a pair of small intermediate spines, large scales, superficially placed fin spines, and a short branchial region.

mesaconic acid [ORG CHEM] $C_5H_6O_4$ An unsaturated dibasic acid, an isomer of citraconic acid, that melts at 202°C. Also known as methyl fumaric acid.

mesa device [ELECTR] Any device produced by diffusing the

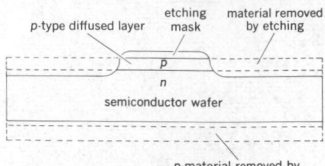

MESA DIODE

p-type diffused layer | etching mask | material removed by etching

semiconductor wafer

p-material removed by lapping and etching

Mesa structure of a high-speed diffused silicon diode.

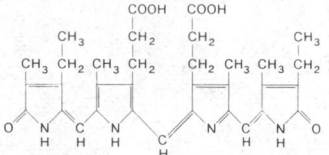

MESOBILIVERDIN

Structural formula for mesobiliverdin.

surface of a germanium or silicon wafer and then etching down all but selected areas, which then appear as physical plateaus or mesas.

mesa diode [ELECTR] A diode produced by diffusing the entire surface of a large germanium or silicon wafer and then delineating the individual diode areas by a photoresist-controlled etch that removes the entire diffused area except the island or mesa at each junction site.

mesa plain [GEOGR] A flat-topped summit of a hilly mountain.

mesappendix [ANAT] The mesentery of the vermiform appendix.

mesarch [BOT] Having metaxylem on both sides of the protoxylem in a siphonostele. [ECOL] Originating in a mesic environment.

mesarteritis [MED] Inflammation of the tunica media of an artery.

mesa transistor [ELECTR] A transistor in which a germanium or silicon wafer is etched down in steps so the base and emitter regions appear as physical plateaus above the collector region.

mescal buttons [BOT] The dried tops from the cactus *Lophophora williamsii*; capable of producing inebriation and hallucinations.

mescaline [ORG CHEM] $C_{11}H_{17}NO_3$ The alkaloid 3,4,5-trimethoxyphenethylamine, found in mescal buttons; produces unusual psychic effects and visual hallucinations.

mesectoderm [EMBRYO] The portion of the mesenchyme arising from ectoderm.

mesencephalon [EMBRYO] The middle portion of the embryonic vertebrate brain; gives rise to the cerebral peduncles and the tectum. Also known as midbrain.

mesenchymal cell [HISTOL] An undifferentiated cell found in mesenchyme and capable of differentiating into various specialized connective tissues.

mesenchymal epithelium [HISTOL] A layer of squamous epithelial cells lining subdural, subarachnoid, and perilymphatic spaces, and the chambers of the eyeball.

mesenchymal hyalin [PATH] A form of hyalin which results from degeneration or necrosis of nonepithelial tissue, usually of muscle, as in Zenker's hyaline necrosis, or of blood vessels.

mesenchymal tissue [EMBRYO] Undifferentiated tissue composed of branching cells embedded in a coagulable fluid matrix.

mesenchyme [EMBRYO] That part of the mesoderm from which all connective tissues, blood vessels, blood, lymphatic system proper, and the heart are derived.

mesenchymoma [MED] A tumor composed of cells resembling those of embryonic mesenchyme, or of mesenchyme with its derivatives.

mesendoderm [EMBRYO] Embryonic tissue which differentiates into mesoderm and endoderm.

mesenteric [ANAT] Of or pertaining to the mesentery.

mesenteric artery [ANAT] Either of two main arterial branches arising from the abdominal aorta: the inferior, supplying the descending colon and the rectum, and the superior, supplying the small intestine, the cecum, and the ascending and transverse colon.

mesenteric lymphadenitis [MED] Inflammation of the lymph nodes in the mesentery.

mesenteron [EMBRYO] See midgut. [INV ZOO] Central gastric cavity in an actinozoan.

mesentery [ANAT] A fold of the peritoneum that connects the intestine with the posterior abdominal wall.

mesentoderm [EMBRYO] **1.** The entodermal portion of the mesoderm. **2.** Undifferentiated tissue from which entoderm and mesoderm are derived. **3.** That part of the mesoderm which gives rise to certain structures of the digestive tract.

mesethmoid [ANAT] A bone or cartilage in the center of the ethmoid region of the vertebrate skull; usually constitutes the greater portion of the nasal septum.

mesh [ADP] See merge. [DES ENG] A size of screen or of particles passed by it in terms of the number of openings per linear inch. Also known as mesh size. [ELEC] A set of branches forming a closed path in a network so that if any one branch is omitted from the set, the remaining branches of the set do not form a closed path. Also known as loop. [MECH

ENG] Engagement or working contact of teeth of gears or of a gear and a rack. [MIN ENG] **1.** A closed path traversed through the network in ventilation surveys. **2.** The size of diamonds as determined by sieves.

mesh connection See delta connection.

mesh impedance [ELEC] The ratio of the voltage to the current in a mesh when all other meshes are open. Also known as self-impedance.

mesh size See mesh.

mesh weld [MET] A seam weld in which the finished weld is only slightly thicker than the sheets, and the lap disappears.

mesic [ECOL] **1.** Of or pertaining to a habitat characterized by a moderate amount of water. **2.** Of or pertaining to a mesophyte.

mesic atom [PARTIC PHYS] An atom in which one of the electrons is replaced by a negative muon or meson orbiting close to or within the nucleus. Also known as mesonic atom.

mesic molecule [PARTIC PHYS] A molecule in which one of the electrons is replaced by a negative muon or meson orbiting close to or within one of the nuclei. Also known as mesonic molecule.

mesmerism [PSYCH] Hypnotism induced by animal magnetism, a supposed force passing from operator to man.

meso- [CHEM] A prefix meaning intermediate or middle, as in denoting inactive optical isomers, the form of intermediate inorganic acid, the middle position in cyclic organic compounds, or a ring system with middle ring positions.

mesoappendix [ANAT] The mesentery of the vermiform appendix.

mesobenthos [OCEANOGR] Of or pertaining to the sea bottom at depths of 180–900 meters (100–500 fathoms).

mesobilirubin [BIOCHEM] $C_{33}H_{40}O_6N_4$ Yellow, crystalline by-product of bilirubin reduction.

mesobilirubinogen [BIOCHEM] $C_{33}H_{44}O_6N_4$ Colorless, crystalline by-product of bilirubin reduction; may be converted to urobilin, stercobilinogen, or stercobilin.

mesobiliverdin [BIOCHEM] $C_{28}H_{38}O_6N_4$ A structural isomer of phycoerythrin and phycocyanobilin released by certain biliprotein by treatment with alkali.

mesoblast [EMBRYO] Undifferentiated mesoderm of the middle layer of the embryo.

mesoblastema See mesoderm.

mesocardium [ANAT] Epicardium covering the blood vessels which enter and leave the heart. [EMBRYO] The mesentery supporting the embryonic heart.

mesocarp [BOT] The middle layer of the pericarp.

mesocercaria [INV ZOO] The developmental stage in the second intermediate host of *Alaria*, a digenetic trematode.

mesoclimate [CLIMATOL] **1.** The climate of small areas of the earth's surface which may not be representative of the general climate of the district. **2.** A climate characterized by moderate temperatures, that is, in the range 20–30°C. Also known as mesothermal climate.

mesoclimatology [CLIMATOL] The study of mesoclimates.

mesocolon [ANAT] The part of the mesentery that is attached to the colon.

mesoconch [ANTHRO] Having moderately rounded orbits with an orbital index of 83 to 89.

mesocranial [ANTHRO] Having a medium-sized skull with a cranial index of 75–80.

mesocratic [PETR] Of igneous rock, being intermediate in color between leucocratic and melanocratic due to equal amounts of light and dark constituents.

mesocrystalline [PETR] Of a crystalline rock, containing crystals whose diameters are intermediate between microcrystalline and macrocrystalline rock.

mesoderm [EMBRYO] The third germ layer, lying between the ectoderm and endoderm; gives rise to the connective tissues, muscles, urogenital system, vascular system, and the epithelial lining of the coelom. Also known as mesoblastema.

mesodermal tumor [MED] A tumor composed of cells normally derived from the mesoderm.

mesogaster [ANAT] The mesentery of the stomach.

Mesogastropoda [INV ZOO] The equivalent name for Pectinibranchia.

mesogeoscyncline [GEOL] A geosyncline between two continents. Also known as mediterranean.

mesoglea [INV ZOO] The gelatinous layer between the ectoderm and endoderm in coelenterates and certain sponges.

mesognathic [ANTHRO] Designating a condition of the upper jaw in which it has a mild degree of anterior projection with respect to the profile of the facial skeleton, when the skull is oriented on the Frankfort horizontal plane; having a gnathic index of 98.0 to 102.9.

Mesohippus [PALEON] An early ancestor of the modern horse; occurred during the Oligocene.

mesoinositol See myoinositol.

mesolamella [INV ZOO] A thin gelatinous membrane between the epidermis and gastrodermis in hydrozoans.

mesolite [MINERAL] $Na_2Ca_2Al_6Si_9O_{30}\cdot8H_2O$ Zeolite mineral composed of hydrous sodium calcium aluminosilicate, usually found in white or colorless tufts of acicular crystals; used as cation exchangers or molecular sieves.

mesomere [EMBRYO] The muscle-plate region between the epimere and hypomere in vertebrates.

mesometeorology [METEOROL] That portion of the science of meteorology concerned with the study of atmospheric phenomena on a scale larger than that of micrometeorology, but smaller than the cyclonic scale.

mesometrium [ANAT] The part of the broad ligament attached directly to the uterus.

mesomorph [PSYCH] A somatotype characterized by an athletic physique.

meson [PARTIC PHYS] Any elementary (noncomposite) particle with strong nuclear interactions and baryon number equal to zero.

meson capture [PARTIC PHYS] Process in which an atomic nucleus acquires a negative muon or meson which circles it in a tightly bound orbit until it decays.

mesonephric duct [EMBRYO] The efferent duct of the mesonephros. Also known as Wolffian duct.

mesonephric fold See mesonephric ridge.

mesonephric ridge [EMBRYO] A fold of the dorsal wall of the coelom lateral to the mesentery formed by development of the mesonephros. Also known as mesonephric fold.

mesonephroma [MED] Any of several benign or malignant tumors of the genital tract thought to be derived from mesonephros. Also known as teratoid adenocystoma.

mesonephroma ovarii [MED] A malignant ovarian mesonephroma.

mesonephros [EMBRYO] One of the middle of three pairs of embryonic renal structures in vertebrates; persists in adult fish and is replaced by the metanephros in higher forms.

mesonic atom See mesic atom.

mesonic molecule See mesic molecule.

mesonic x-ray [PARTIC PHYS] An x-ray emitted by a mesic atom when the muon or meson makes a transition from one bound state to another.

meson resonance [PARTIC PHYS] Any elementary particle with a baryon number of zero which decays through strong interactions, and therefore has an extremely short lifetime on the order of 10^{-23} second.

Mesonychidae [PALEON] A family of extinct mammals of the order Condylarthra.

mesopause [METEOROL] The top of the mesosphere; corresponds to the level of minimum temperature at 80 to 95 kilometers.

mesopeak [METEOROL] The temperature maximum at about 50 kilometers in the mesosphere.

mesophile [BIOL] An organism, as certain bacteria, that grows at moderate temperature.

mesophily [ECOL] Physiological response of organisms living in environments with moderate temperatures and a fairly high, constant amount of moisture.

mesophyll [BOT] Parenchymatous tissue between the upper and lower epidermal layers in foliage leaves.

mesophyte [ECOL] A plant requiring moderate amounts of moisture for optimum growth.

mesopore [PALEON] A tube paralleling the autopore or chamber in fossil bryozoans.

mesopterygium [VERT ZOO] The middle one of three basal cartilages in the pectoral fin of sharks and rays.

mesoptic vision [PHYSIO] Vision in which the human eye's spectral sensitivity is changing from the photoptic state to the scotoptic state.

mesorchium [EMBRYO] The mesentery that supports the embryonic testis in vertebrates.

mesorrhine [ANTHRO] Having a nose of moderate size: nasal index is 47–51 on the skull and 70–85 on the living person.

mesosalpinx [ANAT] The portion of the broad ligament forming the mesentery of the uterine tube.

Mesosauria [PALEON] An order of extinct aquatic reptiles known from a single genus, Mesosaurus, characterized by a long snout, numerous slender teeth, small forelimbs, and webbed hindfeet.

mesosere [ECOL] A sere originating in a mesic habitat and characterized by mesophytes.

mesosiderite [GEOL] A stony-iron meteorite containing about equal amounts of silicates and nickel-iron, with considerable troilite. Also known as grahamite.

mesosoma [INV ZOO] **1.** The anterior portion of the abdomen in certain arthropods. **2.** The middle of the body of some invertebrates, especially when the phylogenetic segmentation pattern cannot be determined.

mesosome [MICROBIO] An extension of the cell membrane within a bacterial cell; possibly involved in cross-wall formation, cell division, and the attachment of daughter chromosomes following deoxyribonucleic acid replication.

mesosphere [GEOL] See lower mantle. [METEOROL] The atmospheric shell between about 45–55 kilometers and 80–95 kilometers, extending from the top of the stratosphere to the mesopause; characterized by a temperature that generally decreases with altitude.

mesosternum [ANAT] The middle portion of the sternum in vertebrates. Also known as gladiolus. [INV ZOO] The ventral portion of the mesothorax in insects.

Mesostigmata [INV ZOO] The mites, a suborder of the Acarina characterized by a single pair of breathing pores (stigmata) located laterally in the middle of the idiosoma between the second and third, or third and fourth, legs.

Mesosuchia [PALEON] A suborder of extinct crocodiles of the Late Jurassic and Early Cretaceous.

Mesotaeniaceae [BOT] The saccoderm desmids, a family of fresh-water algae in the order Conjugales; cells are oval, cylindrical, or rectangular and have simple, undecorated walls in one piece.

Mesotardigrada [INV ZOO] An order of tardigrades which combines certain echiniscoidean features with eutardigradan characters.

mesotheca [INV ZOO] The middle lamina of bifoliate bryozoan colonies.

mesothelioma [MED] A primary benign or malignant tumor composed of cells resembling the mesothelium.

mesothelium [ANAT] The simple squamous-cell epithelium lining the pleural, pericardial, peritoneal, and scrotal cavities. [EMBRYO] The lining of the wall of the primitive body cavity situated between the somatopleure and splanchnopleure.

mesotherm [ECOL] A plant that grows successfully at moderate temperatures.

mesothermal [MINERAL] Of a hydrothermal mineral deposit, formed at great depth at temperatures of 200–300°C.

mesothermal climate See mesoclimate.

mesothorax [INV ZOO] The middle of three somites composing the thorax in insects.

mesotil [GEOL] A semiplastic or semifriable derivative of chemically weathered till; forms beneath a partially drained area.

mesovarium [ANAT] A fold of the peritoneum that connects the ovary with the broad ligament.

Mesoveliidae [INV ZOO] The water treaders, a small family of hemipteran insects in the subdivision Amphibicorisae having well-developed ocelli.

mesoxyalyurea See alloxan.

Mesozoa [INV ZOO] A division of the animal kingdom sometimes ranked intermediate between the Protozoa and the Metazoa; composed of two orders of small parasitic, wormlike organisms.

Mesozoic [GEOL] A geologic era from the end of the Paleozoic to the beginning of the Cenozoic; commonly referred to as the Age of Reptiles.

MESOSAURIA

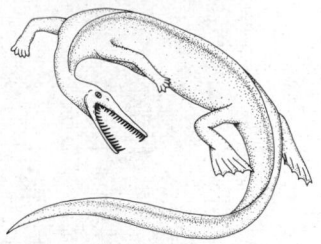

Restoration of *Mesosaurus*. (After McGregor)

MESOZOA

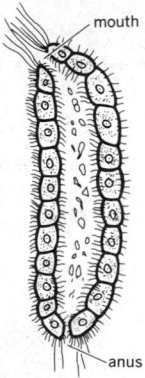

Salinella, in longitudinal section. (After Frenzel, 1892)

MESOZOIC

PRECAMBRIAN		
CAMBRIAN		
ORDOVICIAN		
SILURIAN		PALEOZOIC
DEVONIAN		
Mississippian	CARBONIFEROUS	
Pennsylvanian		
PERMIAN		
TRIASSIC		
JURASSIC		MESOZOIC
CRETACEOUS		
TERTIARY		CENOZOIC
QUATERNARY		

Chart showing the position of the Mesozoic era in relation to the other eras and to the periods of geologic time.

mesozone [PETR] The intermediate depth zone of metamorphism in metamorphic rock characterized by moderate temperatures (300–500°C), hydrostatic pressure, and shearing stress.

mesozooid [INV ZOO] A type of bryozoan heterozooid that produces slender tubes (mesozooecia or mesopores), internally subdivided by many closely spaced diaphragms, that open as tiny polygonal apertures.

mesquite [BOT] Any plant of the genus *Prosopis*, especially *P. juliflora*, a spiny tree or shrub bearing sugar-rich pods; an important livestock feed.

message [ADP] An arbitrary amount of information with beginning and end defined or implied: usually, it originates in one place and is intended to be transmitted to another place. [COMMUN] A series of words or symbols, transmitted with the intention of conveying information.

message accounting [COMMUN] Use of equipment to make records of telephone calls for billing purposes.

message authentication [COMMUN] Security measure designed to establish the authenticity of a message by means of an authenticator within the transmission derived from certain predetermined elements of the message itself.

message center [COMMUN] A communications agency charged with the responsibility for acceptance, preparation for transmission, transmission, receipt and delivery of messages.

message display console [ADP] A cathode-ray tube on which is displayed information requested by the user.

message exchange [ADP] A device which acts as a buffer between a communication line and a computer and carries out communication functions.

message indicator [COMMUN] Element placed within a message to serve as a guide to the selection or derivation and application of the correct key to facilitate the prompt decryption of the message.

message interpolation [COMMUN] Data message insertion during intersyllable periods or speech pauses on a busy voice channel without breaking down the voice connection or noticeably affecting the voice transmission.

message keying element [COMMUN] That part of the key which changes with every message.

message queuing [ADP] The stacking of messages according to some priority rule as the messages await processing.

message reference block [COMMUN] A set of signals denoting the beginning or end of a message.

message registration [COMMUN] A method for counting the number of completed charged calls which originate from a particular telephone line, making one scoring for each local call and more than one scoring for calls between zones.

message routing [COMMUN] Selection of the communication path over which a message is sent.

message switching [COMMUN] A system in which data transmitted between stations on different circuits within a network are routed through central points.

messenger [ENG] A small, cylindrical metal weight that is attached around an oceanographic wire and sent down to activate the tripping mechanism on various oceanographic devices. [NAV ARCH] A light line used to haul in a larger line or hawser.

messenger cable [COMMUN] A cable made of stranded steel which supports aerial cables between poles.

messenger ribonucleic acid [BIOCHEM] A linear sequence of nucleotides which is transcribed from and complementary to a single strand of deoxyribonucleic acid and which carries the information for protein synthesis to the ribosomes. Abbreviated m-RNA.

Messier number [ASTRON] A number by which star clusters and nebulae are listed in Messier's catalog; for example, the Andromeda Galaxy is M 31.

Messier's catalog [ASTRON] A listing of 103 star clusters and nebulae compiled in 1784.

meta- [ORG CHEM] A prefix for benzene-ring compounds when two side chains are connected to carbon atoms with an unsubstituted carbon atom between them.

metaanthracite [GEOL] Anthracite coal containing at least 98% fixed carbon.

metabentonite [GEOL] Altered bentonite, formed by compaction or metamorphism; it swells very little and lacks the usual high colloidal properties of bentonite.

metabiosis [ECOL] An ecological association in which one organism precedes and prepares a suitable environment for a second organism.

metabolic [PHYSIO] Of or pertaining to metabolism.

metabolic disorder [MED] Any disorder that involves an alteration in the normal metabolism of carbohydrates, lipids, proteins, water, and nucleic acids; evidenced by various syndromes and diseases.

metabolism [PHYSIO] The physical and chemical processes by which foodstuffs are synthesized into complex elements (assimilation, anabolism), complex substances are transformed into simple ones (disassimilation, catabolism), and energy is made available for use by an organism.

metabolite [BIOCHEM] A product of intermediary metabolism.

metabolize [PHYSIO] To transform by metabolism; to subject to metabolism.

metacarpus [ANAT] The portion of a hand or forefoot between the carpus and the phalanges.

metacenter [FL MECH] The intersection of a vertical line through the center of buoyancy of a floating body, slightly displaced from its equilibrium position, with a line connecting the center of gravity and the equilibrium center of buoyancy; the floating body is stable if the metacenter lies above the center of gravity.

metacentric [CYTOL] Having the centromere near the middle of the chromosome.

metacentric diagram [NAV ARCH] A curve indicating the height of metacenter (generally above base) for all drafts to which the vessel may be loaded.

metacentric height [NAV ARCH] The vertical distance between a ship's center of gravity and its metacenter, for transverse or longitudinal inclinations, as specified.

metacercaria [INV ZOO] Encysted cercaria of digenetic trematodes; the infective form.

metacestode [INV ZOO] Encysted larva of a tapeworm; occurs in the intermediate host.

Metachlamydeae [BOT] An artificial group of flowering plants, division Magnoliophyta, recognized in the Englerian system of classification; consists of families of dicotyledons in which petals are characteristically fused, forming a sympetalous corolla.

metachromasia [CHEM] 1. The property exhibited by certain pure dyestuffs, chiefly basic dyes, of coloring certain tissue elements in a different color, usually of a shorter wavelength absorption maximum, than most other tissue elements. 2. The assumption of different colors or shades by different substances when stained by the same dye. Also known as metachromatism.

metachromatic granules [CYTOL] Granules which assume a color different from that of the dye used to stain them.

metachromatic leukodystrophy [MED] A hereditary degenerative disease transmitted as an autosomal recessive, due to sulfatase A deficiency, with excess accumulation of sulfated lipids responsible for metachromasia in various tissues. Abbreviated MLD. Also known as sulfatide lipidosis.

metachromatic stain [MATER] A stain which changes apparent color when absorbed by certain cell constituents.

metachromatism *See* metachromasia.

metachrosis [VERT ZOO] The ability of some animals to change color by the expansion and contraction of chromatophores.

metacinnabar [MINERAL] HgS A black isometric mineral that represents an ore of mercury. Also known as metacinnabarite.

metacinnabarite *See* metacinnabar.

metacneme [INV ZOO] A secondary mesentery in many zoantharians.

metacone [VERT ZOO] 1. The posterior of three cusps of primitive upper molars. 2. The posteroexternal cusp of an upper molar in higher vertebrates, especially mammals.

metaconid [VERT ZOO] The posteroexternal cusp of a lower molar in mammals; corresponds with the metacone.

Metacopina [PALEON] An extinct suborder of ostracods in the order Podocopida.

metacryst [PETR] A large crystal, such as garnet, formed in metamorphic rock by recrystallization. Also known as metacrystal.

metacrystal *See* metacryst.

metadyne [ELECTR] A type of rotating magnetic amplifier having more than one brush per pole, used for voltage regulation or transformation.

metaformaldehyde *See* sym-trioxane.

metagalaxy [ASTRON] The total assemblage of recognized galaxies; essentially this represents the entire material universe.

metagenesis [BIOL] The phenomenon in which one generation of certain plants and animals reproduces asexually, followed by a sexually reproducing generation. Also known as alternation of generations.

metagranulocyte *See* metamyelocyte.

metahalloysite [GEOL] A term used in Europe for the less hydrous form of halloysite. Also known as halloysite in the United States.

metaharmosis *See* metharmosis.

metahydrate sodium carbonate [INORG CHEM] $Na_2CO_3 \cdot H_2O$ Water-soluble, white crystals with an alkaline taste, loses water at 109°C, melts at 851°C; used in medicine, photography, and water pH control, and as a food additive. Also known as crystal carbonate; soda crystals.

metakaryocyte *See* normoblast.

metal [MATER] An opaque crystalline material usually of high strength with good electrical and thermal conductivities, ductility, and reflectivity; properties are related to the structure, in which the positively charged ions are bonded through a field of free electrons which surrounds them forming a close-packed structure.

metal ague *See* metal fume fever.

metal-air battery *See* air depolarized battery.

metal alkyl [ORG CHEM] One of the family of organometallic compounds, a combination of an alkyl organic radical with a metal atom or atoms.

metal antenna [ELECTROMAG] An antenna which has a relatively small metal surface, in contrast to a slot antenna.

metal arc cutting [MET] Cutting metal with the heat of an arc between a metal electrode and the base metal.

metal arc welding [MET] Arc welding using covered metal electrodes.

metal ceramic *See* cermet.

metal coating [MET] A thin film of metal bonded to a base material in order to add specific surface properties, such as corrosion or oxidation resistance, color, wear resistance, or optical characteristics.

metaldehyde [ORG CHEM] $(CH_3CHO)_n$ White acetaldehyde-polymer prisms; soluble in organic solvents, insoluble in water; used as a pesticide or fuel.

metal detector [ELECTR] An electronic device for detecting concealed metal objects, such as guns, knives, or buried pipelines, generally by radiating a high-frequency electromagnetic field and detecting the change produced in that field by the ferrous or nonferrous metal object being sought. Also known as electronic locator; metal locator; radio metal locator.

metal distribution ratio [MET] In plating operations, the ratio of the thickness of metal deposited on a near portion of a cathode to that deposited on a far portion of the cathode.

metal dye [MATER] Any of the special dyes, such as alizarin cyanin RR or alizarin green S, used to color oxidized surfaces of aluminum or steel.

metal-film resistor [ELEC] A resistor in which the resistive element is a thin film of metal or alloy, deposited on an insulating substrate of an integrated circuit.

metal-foil paper [MATER] Paper backed with metal foils, manufactured in a number of vivid colors.

metal forming [MET] Any manufacturing process by which parts of components are fabricated by shaping or molding a piece of metal stock.

metal fouling [ORD] Deposits of metal that collect in the bore of a gun, coming from the jackets or rotating bands of projectiles.

metal fume fever [MED] A febrile influenzalike occupational disorder following the inhalation of finely divided particles and fumes of metallic oxides. Also known as brass chills; brass founder's ague; galvo; Monday fever; metal ague; polymer fume fever; spelter shakes; teflon shakes; zinc chills.

metaliding [MET] A process of depositing a metal as an alloy on a substrate from a fused complex metal salt.

metalimnion *See* thermocline.

metallaophobia [PSYCH] An abnormal fear of metal or metallic objects.

metal lath [ENG] A mesh of metal used to provide a base for plaster.

metal leaf [MET] Metal sheet, thinner than foil, formed by beating.

metallic [OPTICS] Having a brilliant mineral luster characteristic of metals. [SCI TECH] Pertaining to metals.

metallic bond [PHYS CHEM] The type of chemical bond that is present in all metals, and may be thought of as resulting from a sea of valence electrons which are free to move throughout the metal lattice.

metallic circuit [ELEC] Wire circuit of which the ground or earth forms no part.

metallic corrosion [MET] Destruction of a metal by dissolution, oxidation, or other chemical reaction of the metal with its environment.

metallic-disk rectifier *See* metallic rectifier.

metallic electrode arc lamp [ELEC] A type of arc lamp in which light is produced by luminescent vapor introduced into the arc by evaporation from the cathode; the anode is solid copper, and the cathode is formed of magnetic iron oxide with titanium as the light-producing element and other chemicals to control steadiness and vaporization.

metallic element [CHEM] An element generally distinguished (from a nonmetallic one) by its luster, electrical conductivity, malleability, and ability to form positive ions.

metallic insulator [ELECTROMAG] Section of transmission line used as a mechanical support device; the section is an odd number of quarter-wavelengths long at the frequency of interest, and the input impedance becomes high enough so that the section effectively acts as an insulator.

metallic mortar [MATER] Mortar made with ceramic oxide binders and containing a high percentage of lead powder; mixed with water to form plasters or for casting sections and blocks; used for x-ray and nuclear installation shielding.

metallic nuclear fuel [NUCLEO] A fissionable isotope of a metallic element, or an alloy containing such an isotope, used as the energy source for a nuclear reactor.

metallic paint [MATER] **1.** Paint used for covering metal surfaces; the pigment is commonly iron oxide. **2.** Paint with a metal pigment.

metallic paper [MATER] **1.** A paper coated with zinc white, or with clay and other materials; the surface can be marked on with metal points (of silver, aluminum, or gold, for example), but it cannot be erased. **2.** Paper coated with finely flaked metal.

metallic pigment [MATER] Thin, opaque aluminum or copper alloy flakes that are incorporated into plastic masses to produce metallike effects.

metallic rectifier [ELECTR] A rectifier consisting of one or more disks of metal under pressure-contact with semiconductor coatings or layers, such as a copper oxide, selenium, or silicon rectifier. Also known as contact rectifier; dry-disk rectifier; dry-plate rectifier; metallic-disk rectifier; semiconductor rectifier.

metallic soap [ORG CHEM] A salt of stearic, oleic, palmitic, lauric, or erucic acid with a heavy metal such as cobalt or copper; used as a drier in paints and inks, in fungicides, decolorizing varnish, and waterproofing.

metalliferous [MINERAL] Pertaining to mineral deposits from which metals can be extracted.

metallize [ENG] To coat or impregnate a metal or nonmetal surface with a metal, as by metal spraying or by vacuum evaporation.

metallized capacitor [ELEC] A capacitor in which a film of metal is deposited directly on the dielectric to serve in place of a separate foil strip; has self-healing characteristics.

metallized-paper capacitor [ELEC] A modification of a paper capacitor in which metal foils are replaced by extremely thin films of metal deposited on the paper; if a breakdown occurs, these films burn away in the area of the breakdown.

metallized resistor [ELEC] A resistor made by depositing a thin film of high-resistance metal on the surface of a glass or ceramic rod or tube.

metallized slurry blasting [ENG] The breaking of rocks by using slurried explosive medium containing a powdered metal, such as powdered aluminum.

metallized wood [MATER] Wood impregnated with molten metal, filling the cells in the wood to increase hardness, compressive strength, and flexural strength; the wood becomes an electrical conductor lengthwise of the grain.

metal locator *See* metal detector.

metallogenic province [GEOL] A region characterized by a particular mineral assemblage, or by one or more specific types of mineralization. Also known as metallographic province.

metallograph [OPTICS] An optical microscope equipped with a camera for both visual observation and photography of the structure and constitution of a metal or alloy.

metallographic province *See* metallogenic province.

metallographic tests [MET] Tests to determine the structural composition of a metal as shown at low and high magnification and by x-ray diffraction methods; tests include macroexamination, microexamination, and x-ray diffraction studies.

metallography [MET] The study of the structure of metals and alloys by various methods, especially by the optical and the electron microscope, and by x-ray diffraction.

metalloid [CHEM] A nonmetallic element, such as carbon or nitrogen, which can combine with a metal to form an alloy.

metalloporphyrin [BIOCHEM] A compound, such as heme, consisting of a porphyrin combined with a metal such as iron, copper, silver, zinc, or magnesium.

metalloprotein [BIOCHEM] A protein enzyme containing a metallic atom as an inherent portion of its molecule.

metallurgical balance sheet [MET] Material balance of a metallurgical process.

metallurgical coke [MATER] Coke resulting from high-temperature retorting of suitable coal; a dense, crush-resistant fuel for use in shaft furnaces.

metallurgical dust [MET] A mixture of particles of elements and nonmetallic and metallic compounds.

metallurgical engineer [ENG] A person who specializes in metallurgical engineering.

metallurgical engineering [ENG] Application of the principles of metallurgy to the engineering sciences.

metallurgical fume [MET] A mixture of fine particles of elements and metallic and nonmetallic compounds either sublimed or condensed from the vapor state.

metallurgical microscope [ENG] A microscope used in the study of metals, usually optical.

metallurgy [SCI TECH] The science and technology of metals and alloys.

metal mining [MIN ENG] The industry that supplies the various metals and associated products.

metal oxide semiconductor device [ELECTR] A diode, capacitor, or other semiconductor device in which a metallic oxide such as silicon dioxide serves as an insulating layer.

metal oxide semiconductor field-effect transistor [ELECTR] A field-effect transistor having a gate that is insulated from the semiconductor substrate by a thin layer of silicon dioxide. Abbreviated MOS FET. Also known as insulated-gate field-effect transistor.

metal plating *See* plating.

metal pointing *See* pointing.

metal powder [MET] A finely divided metal or alloy.

metal replacement *See* immersion plating.

metal rolling *See* rolling.

metal-slitting saw [MECH ENG] A milling cutter similar to a circular saw blade but sometimes with side teeth as well as teeth around the circumference; used for deep slotting and sinking in cuts.

metal spinning *See* spinning.

metal spraying [ENG] Coating a surface with droplets of molten metal or alloy by using a compressed gas stream.

metal-to-metal tap [COMMUN] A tapping procedure in which actual contact is made with the target pair.

metamathematics [MATH] The study of the principles of deductive logic as they are used in mathematical logic.

metamer [ORG CHEM] One of two or more chemical compounds that exhibits isomerism with the others.

metamere [ZOO] One of the linearly arranged similar segments of the body of metameric animals. Also known as somite.

metamerism [ZOO] The condition of an animal body characterized by the repetition of similar segments (metameres), exhibited especially by arthropods, annelids, and vertebrates in early embryonic stages and in certain specialized adult structures. Also known as segmentation.

metamict [MINERAL] Of a radioactive mineral, exhibiting lattice disruption due to radiation damage while the original external morphology is retained.

metamorphic aureole *See* aureole.

metamorphic breccia [PETR] Breccia formed by metamorphism.

metamorphic facies [PETR] All rocks of any composition that have reached chemical equilibrium with respect to certain ranges of pressure and temperature during metamorphism, characterized by the stability of specific index minerals. Also known as densofacies.

metamorphic facies series [PETR] A group of metamorphic facies characteristic of an individual area, represented in a pressure-temperature diagram by a curve or group of curves illustrating the range of the different types of metamorphism and metamorphic facies.

metamorphic rock [PETR] A rock formed from preexisting solid rocks by mineralogical, structural, and chemical changes, in response to extreme changes in temperature, pressure, and shearing stress.

metamorphic rock reservoir [GEOL] Uncommon type of formation for oil reservoir; developed when secondary porosity results from fracturing or weathering.

metamorphic zone *See* aureole.

metamorphism [PETR] The mineralogical and structural changes of solid rock in response to environmental conditions at depth in the earth's crust.

metamorphosis [BIOL] 1. A structural transformation. 2. A marked structural change in an animal during postembryonic development. [MED] A degenerative change in tissue or organ structure.

metamyelocyte [HISTOL] A granulocytic cell intermediate in development between the myelocyte and granular leukocyte; characterized by a full complement of cytoplasmic granules and a bean-shaped nucleus. Also known as juvenile cell; metagranulocyte.

metanauplius [INV ZOO] A primitive larval stage of certain decapod crustaceans characterized by seven pairs of appendages; follows the nauplius stage.

metanephridium [INV ZOO] A type of nephridium consisting of a tubular structure lined with cilia which opens into the coelomic cavity.

metanephrine [BIOCHEM] An inactive metabolite of epinephrine (3-O-methylepinephrine) that is excreted in the urine; it is recovered and measured as a test for pheochromocytoma.

metanephros [EMBRYO] One of the posterior of three pairs of vertebrate renal structures; persists as the definitive or permanent kidney in adult reptiles, birds, and mammals.

metanillic acid [ORG CHEM] $C_6H_4(NH_2)SO_3H$ A water-soluble, crystalline compound, isomeric with sulfanilic acid; used in medicines and dyes. Also known as *meta*-aminobenzenesulfonic acid; *meta*-sulfanilic acid.

metanitricyte *See* normoblast.

metaphase [CYTOL] 1. The phase of mitosis during which centromeres are arranged on the equator of the spindle. 2. The phase of the first meiotic division when centromeric regions of homologous chromosomes come to lie equidistant on either side of the equator.

metaphyseal aclasis *See* multiple hereditary exostoses.

metaphysis *See* epiphyseal plate.

Metaphyta [BIOL] A kingdom set up to include mosses, ferns, and other plants in some systems of classification.

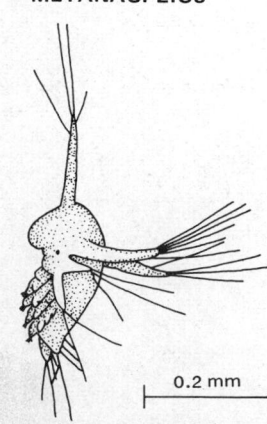

METANAUPLIUS

Lateral view of metanauplius larva of penaeid shrimp. (*Smithsonian Institution*)

0.2 mm

metaplasia [PATH] Transformation of one form of tissue to another.

metaplasm [CYTOL] The ergastic substance of protoplasm.

metapodium [ANAT] 1. The metatarsus in bipeds. 2. The metatarsus and metacarpus in quadrupeds. [INV ZOO] Posterior portion of the foot of a mollusk.

metapodosoma [INV ZOO] Portion of the body bearing the third and fourth pairs of legs in Acarina.

metapterygium [VERT ZOO] The posterior one of three basal cartilages in the pectoral fin of sharks and rays.

metaquartzite [MINERAL] A quartzite formed by metamorphic recrystallization.

metarheology [MECH] A branch of rheology whose approach is intermediate between those of macrorheology and microrheology; certain processes that are not isothermal are taken into consideration, such as kinetic elasticity, surface tension, and rate processes.

metaripple [GEOL] An asymmetrical sand ripple.

metarubricyte *See* normoblast.

metascope [ELECTR] An infrared receiver used for converting pulsed invisible infrared rays into visible signals for communication purposes; also used with an infrared source for reading maps in darkness.

metascutellum [INV ZOO] The scutellum of the metathorax in insects.

metasediment [GEOL] A sediment or sedimentary rock which shows evidence of metamorphism. [PETR] Metamorphic rock formed from sedimentary rock.

metasicula [INV ZOO] The succeeding part of the sicula or colonial tube of graptolites.

metasideronatrite [MINERAL] $Na_4Fe_2(SO_4)_4(OH)_2\cdot 3H_2O$ A yellow mineral composed of basic hydrous iron sodium sulfate.

metasilicate [MINERAL] A salt of the hypothetical metasilicic acid H_2SiO_3. Also known as bisilicate.

metasoma [INV ZOO] The posterior region of the body of certain invertebrates, a term used especially when the phylogenetic segmentation pattern cannot be identified.

metasomatic [PETR] Pertaining to the process or the result of metasomatism.

metasomatism [PETR] A variety of metamorphism in which one mineral or a mineral assemblage is replaced by another of different composition without melting.

metastable *See* labile.

metastable equilibrium [PHYS CHEM] A state of pseudo-equilibrium having higher free energy than the true equilibrium state.

metastable ion [ANALY CHEM] In mass spectroscopy, an ion formed by a secondary dissociation process in the analyzer tube (formed after the parent or initial ion has passed through the accelerating field).

metastable phase [PHYS CHEM] Existence of a substance as either a liquid, solid, or vapor under conditions in which it is normally unstable in that state.

metastable state [QUANT MECH] An excited stationary energy state whose lifetime is unusually long.

metastasis [MED] Transfer of the causal agent (cell or microorganism) of a disease from a primary focus to a distant one through the blood or lymphatic vessels. [PHYS] A transition of an electron or nucleon from one bound state to another in an atom or molecule, or the capture of an electron by a nucleus.

metastasize [MED] To be transferred by metastasis.

metastatic anemia *See* myelophthisic anemia.

metasternum [INV ZOO] The ventral portion of the metathorax in insects.

metastoma [INV ZOO] Median plate posterior to the mouth in certain crustaceans and related arthropods.

Metastrongylidae [INV ZOO] A family of roundworms belonging to the Strongyloidea; species are parasitic in sheep, cattle, horses, dogs, and other domestic animals.

metatarsal [ANAT] Of or pertaining to the metatarsus.

metatarsalgia [MED] Tenderness and burning pain in the metatarsal region.

metatarsus [ANAT] The part of a foot or hindfoot between the tarsus and the phalanges.

Metatheria [VERT ZOO] An infraclass of therian mammals including a single order, the Marsupialia; distinguished by a small braincase, a total of 50 teeth, the inflected angular process of the mandible, and a pair of marsupial bones articulating with the pelvis.

metathetical salts [CHEM] Salts that form a four-component, ternary equilibrium system in which there are four possible binary systems, resulting in two quadruple points.

metathorax [INV ZOO] Posterior segment of the thorax in insects.

metatitanic acid *See* titanic acid.

metatorbernite [MINERAL] $Cu(UO_2)_2(PO_4)_2\cdot 8H_2O$ A green secondary mineral composed of hydrous copper uranium phosphate; similar to torbernite, but with less water content.

metatroch [INV ZOO] A segmented larval form following the trochophore in annelids.

metavariscite [MINERAL] $AlPO_6\cdot 2H_2O$ A green monoclinic mineral composed of hydrous aluminum phosphate; it is isomorphous with phosphosiderite.

metavauxite [MINERAL] $FeAl_2(PO_4)_2(OH)_2\cdot 8H_2O$ A colorless mineral composed of hydrous basic phosphate of iron and aluminum; similar to vauxite, but with more water.

metaxylem [BOT] Primary xylem differentiated after and distinguished from protoxylem by thicker tracheids and vessels with pitted or reticulated walls.

Metazoa [ZOO] The multicellular animals that make up the major portion of the animal kingdom; cells are organized in layers or groups as specialized tissues or organ systems.

metazoea [INV ZOO] The last zoea of certain decapod crustaceans; metamorphoses into a megalopa.

metencephalon [EMBRYO] The cephalic portion of the rhombencephalon; gives rise to the cerebellum and pons.

meteor [ASTRON] The phenomena which accompany a body from space (a meteoroid) in its passage through the atmosphere, including the flash and streak of light and the ionized trail. [METEOROL] Anything in the air; hence, the science of meteorology.

Meteoriaceae [BOT] A family of mosses in the order Isobryales in which the calyptra is frequently hairy.

meteoric ionization [ASTROPHYS] Ionization resulting from collisional interactions of a meteoroid and its vaporization products with the air.

meteoric scatter [COMMUN] A form of scatter propagation in which meteor trails serve to scatter radio waves back to earth.

meteoric stone *See* stony meteorite.

meteoric water [HYD] Groundwater which originates in the atmosphere and reaches the zone of saturation by infiltration and percolation.

meteorite [GEOL] Any meteoroid that has fallen to the earth's surface.

meteorite crater [GEOL] An impact crater on the surface of the earth or of a celestial body caused by a meteorite; a characteristic feature on the earth is the upturned rim, which formed as the rocks rebounded following the impact.

meteorogram [ENG] A record obtained from a meteorograph. [METEOROL] A chart in which meteorological variables are plotted against time.

meteorograph [ENG] An instrument that measures and records meteorological data such as air pressure, temperature, and humidity.

meteoroid [ASTRON] Any solid object moving in interplanetary space that is smaller than a planet or asteroid but larger than a molecule.

meteorolite *See* stony meteorite.

meteorological [METEOROL] Of or pertaining to meteorology or weather.

meteorological balloon [ENG] A balloon, usually of high-quality neoprene, polyethylene, or Mylar, used to lift radiosondes to high altitudes.

meteorological chart [METEOROL] A weather map showing the spatial distribution, at an instant of time, of atmospheric highs and lows, rain clouds, and other phenomena.

meteorological check point in gunnery [ORD] Arbitrarily selected point for which meteorological corrections are determined as a time saving expedient; these corrections are applied to any target located within transfer limits of the meteorological check point.

meteorological correction [ORD] Adjustment made in the

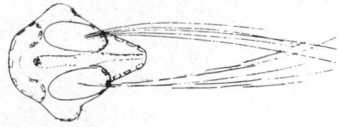

METATROCH

Metatroch of planktonic sabellarian larva, showing early segmentation and long swimming setae. *(From O. Hartman, Polychaetous annelids, Paraonidae, Magelonidae, Longosomidae, Ctenodrilidae and Sabellariidae, Allan Hancock Pacific Expedition, 10(3):311–389, 1944)*

METAZOEA

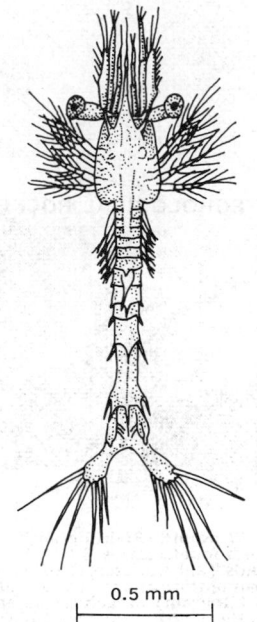

0.5 mm

Metazoea larva of a penaeid shrimp.

METEORITE CRATER

Canyon Diablo, Arizona, meteorite crater. The rocks immediately surrounding this impact crater are turned upward. Meteorites from the crater were found on the plains around the depression.

firing data of a gun or other weapon to allow for the effect of wind, air pressure, and so forth, on the flight of a projectile.

meteorological data [METEOROL] Facts pertaining to the atmosphere, especially wind, temperature, and air density.

meteorological datum plane [ORD] Reference plane assumed as a basis or starting point for atmospheric data furnished to artillery; its altitude is that of the meteorological station.

meteorological frequency bands [COMMUN] A collection of radio and microwave frequency bands allocated for use by radiosondes and ground-based radars used in weather forecasting in the United States.

meteorological instrumentation [ENG] Apparatus and equipment used to obtain quantitative information about the weather.

meteorological minima [METEOROL] Minimum values of meteorological elements prescribed for specific types of flight operation.

meteorological optics [OPTICS] A branch of atmospheric physics or physical meteorology in which optical phenomena occurring in the atmosphere are described and explained. Also known as atmospheric optics.

meteorological radar [ENG] Radar which is used to study the scattering of radar waves by various types of atmospheric phenomena, for making weather observations and forecasts.

meteorological range [METEOROL] An empirically consistant measure of the visual range of a target; a concept developed to eliminate from consideration the threshold contrast and adaptation luminance, both of which vary from observer to observer. Also known as standard visibility; standard visual range.

The 77-pound (35-kilogram) Arcas rocket carries 6.5 pounds (2.9 kilograms) of instrumentation and a 3-pound (1.4-kilogram) parachute to an altitude of 200,000 feet (60,000 meters), where payload is separated from the rocket motor. Meteorological measurements are obtained during the descent of the instrumentation on the parachute. (*Atlantic Research Corp.*)

meteorological rocket [ENG] Small rocket system used to extend observation of atmospheric character above feasible limits for balloon-borne observing and telemetering instruments. Also known as rocketsonde.

meteorological satellite [AERO ENG] Earth-orbiting spacecraft carrying a variety of instruments for measuring visible and invisible radiations from the earth and its atmosphere.

meteorological solenoid [METEOROL] A hypothetical tube formed in space by the intersection of a set of surfaces of constant pressure and a set of surfaces of constant specific volume of air. Also known as solenoid.

meteorological tide [OCEANOGR] A change in water level caused by local meteorological conditions, in contrast to an astronomical tide, caused by the attractions of the sun and moon.

meteorology [SCI TECH] The science concerned with the atmosphere and its phenomena; the meteorologist observes the atmosphere's temperature, density, winds, clouds, precipitation, and other characteristics and aims to account for its observed structure and evolution (weather, in part) in terms of external influence and the basic laws of physics.

meteor shower [ASTRON] A number of meteors with approximately parallel trajectories.

meteor stream [ASTRON] A group of meteoric bodies with nearly identical orbits.

meteor trail *See* ion column.

meter [MECH] The international standard unit of length, equal to 1,650,763.73 times the wavelength of the orange light emitted when a gas consisting of the pure krypton isotope of mass number 86 is excited in an electrical discharge. Abbreviated m. [ENG] A device for measuring the value of a quantity under observation; the term is usually applied to an indicating instrument alone.

meter-ampere [COMMUN] Measure of the strength of a radio transmitting station.

meter-atmosphere [PHYS] The depth of an equivalent atmosphere of a given gas, in meter-atmospheres, is equal to the depth in meters that the atmosphere would have if it were composed entirely of the gas in question and in the same amount as exists in the actual atmosphere, and had a uniform temperature and pressure of 0°C and 1 standard atmosphere. Abbreviated m-atm. Also known as atmo-meter.

meter-candle *See* lux.

meter factor [ENG] A factor used with a meter to correct for ambient conditions, for example, the factor for a fluid-flow

meter to compensate for such conditions as liquid temperature change and pressure shrinkage.

metering installation [PETR ENG] Oil-production receiving system that includes with the tank battery a metering separator, metering treater, or other type of meter used in conjunction with test separators or emulsion treaters.

metering pin *See* metering rod.

metering pump [CHEM ENG] Plunger-type pump designed to control accurately small-scale fluid-flow rates; used to inject small quantities of materials into continuous-flow liquid streams. Also known as proportioning pump.

metering rod [ENG] A device consisting of a long metallic pin of graduated diameters fitted to the main nozzle of a carburetor (on an internal combustion engine) or passage leading thereto in such a way that it measures or meters the amount of gasoline permitted to flow by it at various speeds. Also known as metering pin.

metering screw [MECH ENG] An extrusion-type screw feeder or conveyor section used to feed pulverized or doughy material at a constant rate.

metering separator [PETRO ENG] Oil-field process vessel that performs the dual functions of gas-oil separation and liquids metering.

meter-kilogram [MECH] **1.** A unit of energy or work in a meter-kilogram-second gravitational system, equal to the work done by a kilogram-force when the point at which the force is applied is displaced 1 meter in the direction of the force; equal to 9.80665 joules. Abbreviated m-kgf. Also known as meter kilogram-force. **2.** A unit of torque, equal to the torque produced by a kilogram-force acting at a perpendicular distance of 1 meter from the axis of rotation. Also known as kilogram-meter (kgf-m).

meter kilogram-force *See* meter-kilogram.

meter-kilogram-second-ampere system [PHYS] A system of electrical and mechanical units in which length, mass, time, and electric current are the fundamental quantities, and the units of these quantities are the meter, the kilogram, the second, and the ampere respectively. Abbreviated mksa system. Also known as Giorgi system; practical system.

meter-kilogram-second system [MECH] A metric system of units in which length, mass, and time are fundamental quantities, and the units of these quantities are the meter, the kilogram, and the second respectively. Abbreviated mks system.

meter oil [MATER] High-purity grade of oil used to lubricate the moving elements of meters.

meter-proving tank *See* calibrating tank.

meter run [ENG] The length of straight, unobstructed fluid-flow conduit preceding an orifice or venturi meter.

meter sensitivity [ENG] The accuracy with which a meter can measure a voltage, current, resistance, or other quantity.

meter sizing factor [FL MECH] A dimensionless number used in calculating the rate of flow of fluid through a pipe from the readings of a flowmeter that measures the drop in pressure when the fluid is forced to flow through a circular orifice; it is equal to $K(d/D)^2$, where K is the flow coefficient, d is the orifice bore diameter, and D is the internal diameter of the pipe.

meter-ton-second system [MECH] A modification of the meter-kilogram-second system in which the metric ton (1000 kilograms) replaces the kilogram as the unit of mass.

meter-type relay [ELEC] A relay that uses a meter movement having a contact-bearing pointer which moves toward or away from a fixed contact mounted on the meter scale.

meter wheel [ENG] A special block used to support the oceanographic wire paid out over the side of a ship; attached directly or connected by means of a speedometer cable to a gearbox which measures the length of wire.

metestrus [PHYSIO] The beginning of the luteal phase following estrus.

methacetin [ORG CHEM] $CH_3OC_6H_4NHCOCH_3$ A water-insoluble, white powder, melting at 127.1°C. Also known as acetanisidine; *para*-methoxyacetanilide.

methacrolein [ORG CHEM] $CH_2C(CH_3)CHO$ Liquid with 68°C boiling point; slightly soluble in water; used to make resins and copolymers.

methacrylate ester [ORG CHEM] $CH_2{:}C(CH_3)COOR$

Methacrylic acid ester in which R can be methyl, ethyl, isobutyl, or 50-50 *n*-butyl-isobutyl groups; used to make thermoplastic polymers or copolymers.

methacrylic acid [ORG CHEM] $CH_2C(CH_3)COOH$ Easily polymerized, colorless liquid melting at 15–16°C; soluble in water and most organic solvents; used to make water-soluble polymers and as a chemical intermediate.

methacrylonitrile [ORG CHEM] $CH_2:C(CH_3)CN$ Clear, colorless liquid boiling at 90°C; used to make solvent-resistant thermoplastic polymers and copolymers.

methadone [PHARM] $C_{21}H_{27}NO$ The compound 6-(di-methylamino)-4,4-diphenyl-3-heptanone, a narcotic analgesic, administered in the hydrochloride form for maintenance treatment of heroin addiction.

methallyl alcohol [ORG CHEM] $H_2C:C(CH_3)CH_2OH$ Flammable, toxic, water-soluble, colorless liquid boiling at 115°C; has pungent aroma; soluble in most organic solvents; used as a chemical intermediate. Also spelled methyl allyl alcohol.

methanamide *See* formamide.

methane [ORG CHEM] CH_4 A colorless, odorless, and tasteless gas, lighter than air and reacting violently with chlorine and bromine in sunlight, a chief component of natural gas; used as a source of methanol, acetylene, and carbon monoxide. Also known as methyl hydride.

methanedicarbonic acid *See* malonic acid.

methanedicarboxylic acid *See* malonic acid.

methane-disulfonic acid *See* methionic acid.

methane drainage *See* firedamp drainage.

methane indicator [MIN ENG] A portable analytical instrument that can determine the methane content in the mine air at the place where the sample is taken; air is brought into the instrument through an aspirator bulb and passed through a cartridge filter to remove moisture.

methane monitoring system [MIN ENG] A system that samples methane content in mine air continuously and feeds this information into an electrical device that cuts off power in each mining machine when the methane content rises above a predetermined level.

methane-oxidizing bacteria [MICROBIO] Bacteria that derive energy from oxidation of methane.

methanethiomethane *See* methyl sulfide.

Methanobacterium [MICROBIO] A genus of anaerobic, rod-shaped, gram-negative, chemoheterotrophic and chemoautotrophic bacteria in the family Spirillaceae which reduce carbon dioxide to methane.

Methanococcus [MICROBIO] A genus of anaerobic, spherical bacterial cells in the family Micrococcaceae which are chemoheterotrophic, fermenting various organic compounds, with the production of methane.

methanogenesis [BIOCHEM] The biosynthesis of the hydrocarbon methane; common in certain bacteria. Also known as bacterial methanogenesis.

methanoic acid *See* formic acid.

methanol *See* methyl alcohol.

Methanomonadaceae [MICROBIO] A family of bacteria in the suborder Pseudomonadineae; members are gram-negative rods able to use carbon monoxide (*Carboxydomonas*), methane (*Methanomonas*), and hydrogen (*Hydrogenomonas*) as their sole source of energy for growth.

Methanomonas [MICROBIO] A genus of obligate methane-methanol-utilizing bacteria in the family Methanomonadaceae.

metharmosis [GEOL] Changes that occur in a buried sediment after uplift or consolidation but before the onset of weathering. Also spelled metaharmosis.

methemoglobin *See* ferrihemoglobin.

methemoglobinemia [MED] The presence of methemoglobin in the blood.

methemoglobinuria [MED] The presence of methemoglobin in the urine.

methide [ORG CHEM] A binary compound consisting of methyl and, most commonly, a metal, such as sodium (sodium methide, $NaCH_3$).

methionic acid [ORG CHEM] $CH_2(SO_3H)_2$ An acid that exists as hygroscopic crystals; used in organic synthesis. Also known as methane-disulfonic acid.

methionine [BIOCHEM] $C_5H_{11}O_2NS$ An essential amino acid; furnishes both labile methyl groups and sulfur necessary for normal metabolism.

method of bisectors [NAV] As applied to celestial lines of position, the movement of each of three or four intersecting lines of position in equal amounts, in the same direction, toward or away from the celestial bodies, so as to bring them as nearly as possible to a common intersection; when there are more than four lines of position, the lines of position in the same general direction are combined to reduce the data to not more than four lines of position.

method of images [PETRO ENG] Method of calculating the interference between reservoirs by assuming a mirror image of one reservoir on the far side of a geologic fault.

method of joints [ENG] Determination of stresses for joints at which there are not more than two unknown forces by the methods of the stress polygon, resolution, or moments.

method of mixtures [THERMO] A method of determining the heat of fusion of a substance whose specific heat is known, in which a known amount of the solid is combined with a known amount of the liquid in a calorimeter, and the decrease in the liquid temperature during melting of the solid is measured.

methods design [IND ENG] Design for a new, more efficient method of job performance.

methods engineering [IND ENG] A technique used by management to improve working methods and reduce labor costs in all areas where human effort is required.

methods study [IND ENG] An analysis of the methods in use, of the means and potentials for their improvement, and of reducing costs.

methotrexate *See* amethopterin.

methoxide [ORG CHEM] A compound formed from a metal and the methoxy radical; an example is sodium methoxide. Also known as methylate.

methoxy- [ORG CHEM] OCH_3- A combining form indicating the oxygen-containing methane radical, found in many organic solvents, insecticides, and plasticizer intermediates.

para-**methoxyacetanilide** *See* methacetin.

4-methoxybenzaldehyde *See* anisaldehyde.

para-**methoxybenzaldehyde** *See* anisaldehyde.

methoxybenzene *See* anisole.

para-**methoxybenzyl alcohol** *See* anisic alcohol.

methoxychlor [ORG CHEM] $Cl_3CCH(C_6H_4OCH_3)_2$ White, water-insoluble crystals melting at 89°C; used as an insecticide. Also known as DMDT; methoxy DDT.

methoxy DDT *See* methoxychlor.

methoxyl [ORG CHEM] CH_3O- A radical which is univalent.

1-methoxy-4-propenyl benzene *See* anethole.

para-**methoxypropenylbenzene** *See* anethole.

methyl- [ORG CHEM] CH_3- A combining form indicating the organic radical derived from methane, CH_4, and present as an alkyl group in numerous organic compounds.

methyl abietate [ORG CHEM] $C_{19}H_{29}COOCH_3$ Colorless to yellow liquid boiling at 365°C; miscible with most organic solvents; used as a solvent and plasticizer for lacquers, varnishes, and coatings.

methyl acetate [ORG CHEM] $CH_3CO_2CH_3$ Flammable, colorless liquid with fragrant odor; boils at 54°C; partially soluble in water, miscible with hydrocarbon solvents; used as a solvent and extractant.

methylacetic acid *See* propionic acid.

methyl acetoacetate [ORG CHEM] $CH_3COCH_2CO_2CH_3$ Alcohol-soluble, colorless liquid boiling at 172°C; used as a chemical intermediate and as a solvent for cellulosics.

methyl acetone [MATER] Flammable, water-white liquid, a mixture of acetone, methanol, and methyl acetate in various proportions; miscible with water, oils, and hydrocarbons; used as a solvent.

methyl acetophenone [ORG CHEM] $CH_3C_6H_4COCH_3$ Fragrant (coumarin aroma), colorless or pale-yellow liquid, soluble in alcohol; used in perfumery. Also known as methyl tolyl ketone.

methylacetopyranone *See* dehydroacetic acid.

methyl acetylene *See* allylene.

methyl acrylate [ORG CHEM] $CH_2:CHCOOCH_3$ A readily polymerized, volatile, colorless liquid boiling at 80°C; slightly

METHIONINE

$$
\begin{array}{c}
CH_3 \\
| \\
S \\
| \\
CH_2 \\
| \\
CH_2 \\
| \\
H_2N - \overset{\displaystyle C}{\underset{\displaystyle H}{|}} - COOH
\end{array}
$$

Structural formula of methionine.

soluble in water; used as a chemical intermediate and in making polymers.

methylal [ORG CHEM] $CH_3OCH_2OCH_3$ Flammable, volatile, colorless liquid boiling at 42°C; soluble in ether, hydrocarbons, and alcohol, partially soluble in water; used as a solvent and chemical intermediate, and in perfumes, adhesives, coatings. Also known as dimethoxymethane; formal.

methyl alcohol [ORG CHEM] CH_3OH A colorless, toxic, flammable liquid, boiling at 64.5°C, miscible with water, ether, alcohol; used in manufacture of formaldehyde, chemical synthesis, antifreeze for autos, and as a solvent. Also known as methanol; wood alcohol.

methyl allyl alcohol *See* methallyl alcohol.

methyl allyl chloride [ORG CHEM] $CH_2:C(CH_3)CH_2Cl$ Volatile, flammable, colorless liquid boiling at 72°C; has disagreeable odor; used as an insecticide and fumigant, and for chemical synthesis.

methylamine [ORG CHEM] CH_3NH_2 A colorless gas that is highly toxic and flammable; used to prepare dyes, and as a chemical intermediate. Also known as aminomethane; monomethylamine.

methyl aminoacetic acid *See* sarcosine.

methyl *ortho*-aminobenzoate *See* methyl anthranilate.

methyl amyl acetate [ORG CHEM] $CH_3COOCH(CH_3)$ $CH_2CH(CH_3)_2$ Toxic, flammable, colorless liquid with mild, agreeable odor; boils at 146°C; used as nitrocellulose lacquer solvent. Also known as methyl isobutyl carbinol acetate.

methyl amyl alcohol [ORG CHEM] $(CH_3)_2CHCH_2CHOHCH_3$ Toxic, flammable, colorless liquid; boils at 132°C; miscible with water and most organic solvents; used as a solvent and as a chemical intermediate. Also known as methyl isobutyl carbinol (MIBC).

methyl amyl carbinol [ORG CHEM] $CH_3(CH_2)_4CHOHCH_3$ Colorless liquid with mild aroma; boils at 160°C; miscible with most organic liquids; used as an ore-flotation frothing agent and as a synthetic-resin solvent. Also known as 2-heptanol; heptyl alcohol.

methyl-*n*-amyl ketone [ORG CHEM] $CH_3(CH_2)_4COCH_3$ Stable, water-white liquid; miscible with organic lacquer solvents, slightly soluble in water; used as an inert reaction medium and as a solvent for nitrocellulose lacquers. Also known as 2-heptanone.

***N*-methylaniline** [ORG CHEM] $C_6H_5NH(CH_3)$ Oily liquid, colorless to reddish-brown; soluble in water and organic solvents; boils at 190°C; used as an acid acceptor, solvent, and chemical intermediate.

methyl anisole *See* methyl *para*-cresol.

methyl anthranilate [ORG CHEM] $H_2NC_6H_4CO_2CH_3$ A yellowish to colorless liquid, slightly soluble in water; used in flavoring and in perfumery. Also known as artificial neroli oil; methyl *ortho*-aminobenzoate.

methylate *See* methoxide.

methylbenzene *See* toluene.

methyl benzoate [ORG CHEM] $C_6H_25CO_2CH_3$ Colorless, fragrant liquid boiling at 199°C; slightly soluble in alcohol and water, soluble in ether; used in perfumery and as a solvent. Also known as niobe oil.

2-methylbenzoic acid *See* *ortho*-toluic acid.

3-methylbenzoic acid *See* *meta*-toluic acid.

4-methylbenzoic acid *See* *para*-toluic acid.

methylbenzoylecgonine *See* cocaine.

α-methyl bivinyl *See* pentadiene.

methyl blue [ORG CHEM] Dark-blue powder or dye; sodium triphenyl *para*-rosaniline sulfonate; used as a biological and bacteriological stain and as an antiseptic.

methyl borate *See* trimethyl borate.

methyl bromide [ORG CHEM] CH_3Br A toxic, colorless gas that forms a crystalline hydrate with cold water; used in synthesis of organic compounds, and as a fumigant. Also known as bromomethane.

2-methyl-1,3-butadiene *See* isoprene.

2-methylbutane *See* isopentane.

3-methyl-1-butanol *See* isobutyl carbinol.

methyl butene [ORG CHEM] C_5H_{10} Either of two colorless, flammable, volatile liquid isomers; soluble in alcohol, insoluble in water: 3-methyl-1-butene boils at 20°C, is used as a

chemical intermediate and in the manufacture of high-octane fuel, and is also known as isopropylethylene; 3-methyl-2-butene boils at 38°C, is used as an anesthetic and high-octane fuel and as a chemical intermediate and is also known as trimethylethylene.

methyl butyl ketone [ORG CHEM] $CH_3COC_4H_9$ A liquid boiling at 127°C; soluble in water, alcohol, and ether; used as a solvent. Also known as propylacetone.

methylbutynol [ORG CHEM] $HC:CCOH(CH_3)_2$ Water-miscible, colorless liquid boiling at 104°C; soluble in most organic solvents; used as a stabilizer for chlorinated organic compounds, as a solvent, and as a chemical intermediate.

methyl butyrate [ORG CHEM] $CH_3CH_2CH_2COOCH_3$ Liquid boiling at 102°C; used as a solvent for cellulosic materials.

methyl caproate [ORG CHEM] $CH_3(CH_2)_4COOCH_3$ Colorless liquid boiling at 150°C; soluble in alcohol and ether, insoluble in water; used as an intermediate to make caproic acid.

methyl caprylate [ORG CHEM] $CH_3(CH_2)_6COOCH_3$ Colorless liquid boiling at 193°C; soluble in ether and alcohol, insoluble in water; used as an intermediate to make caprylic acid.

methyl carbonate [ORG CHEM] $CO(OCH_3)_2$ Water-insoluble, colorless liquid boiling at 91°C; has pleasant odor; miscible with acids and alkalies; used as a chemical intermediate.

β-methylchalcone *See* dipnone.

methyl chavicol *See* estragole.

methyl chloride *See* chloromethane.

methyl chloroacetate [ORG CHEM] $ClHC_2COOCH_3$ Colorless liquid boiling at 131°C; miscible with ether and alcohol, slightly soluble in water; used as a solvent.

methyl chloroform *See* trichloroethane.

methyl cholanthrene [ORG CHEM] $C_{21}H_{16}$ A hydrocarbon that is carcinogenic; it is obtained by organic synthesis.

methyl cinnamate [ORG CHEM] $C_6H_5CH:CHCO_2CH_3$ A white crystalline compound with strawberry aroma; soluble in ether and alcohol, insoluble in water; boils at 260°C; used to flavor confectioneries and in perfumes.

methyl *para*-cresol [ORG CHEM] $CH_3C_6H_4OCH_3$ Colorless liquid with floral aroma; used in perfumery. Also known as *para*-cresyl methyl ether; methyl anisole.

methyl cyanide *See* acetonitrile.

methyl cyclohexane [ORG CHEM] C_7H_{14} Colorless liquid boiling at 101°C; used as a cellulosic solvent and as a chemical intermediate. Also known as hexahydrotoluene.

methyl cyclohexanol [ORG CHEM] $CH_3C_6H_{10}OH$ A toxic, colorless liquid with menthol aroma; a mixture of three isomers; used as a solvent for lacquer and cellulosics, as a lubricant antioxidant, and in detergents and textile soaps. Also known as hexahydrocresol; hexahydromethyl phenol.

methyl cyclohexanone [ORG CHEM] $CH_3C_5H_9CO$ A toxic, clear to pale-yellow liquid with acetonelike aroma; a mixture of cyclic ketones; used as a solvent and in lacquers.

methyl cyclopentane [ORG CHEM] $C_5H_9CH_3$ Flammable, colorless liquid boiling at 72°C; used as a chemical intermediate.

methyl diethanolamine [ORG CHEM] $CH_3N(C_2H_4OH)_2$ A colorless liquid miscible with water and benzene; has amine aroma; boils at 247°C; used as a chemical intermediate and as an acid-gas absorbent.

methyl dioxolane [ORG CHEM] $C_4H_7O_2$ Water-soluble, clear liquid boiling at 81°C; used as a solvent and extractant. Also known as 2-methyl-1,3-dioxolane.

2-methyl-1,3-dioxolane *See* methyl dioxolane.

methyl dipropylmethane *See* methyl heptane.

methylene [ORG CHEM] $-CH_2-$ A radical that contains a bivalent carbon.

methylene bromide [ORG CHEM] CH_2Br_2· Colorless, clear liquid boiling at 97°C; miscible with organic solvents, slightly soluble in water; used as a solvent and chemical intermediate. Also known as dibromomethane.

methylene chloride [ORG CHEM] CH_2Cl_2 A colorless liquid, practically nonflammable and nonexplosive; used as a refrigerant in centrifugal compressors, a solvent for organic materials, and a component in nonflammable paint-remover mixtures. Also known as carrene; dichloromethane.

methylenecyclopentadiene   *See* fulvene.

3,4-methylenedioxy-1-allyl benzene   *See* safrole.

1,2-methylenedioxy-4-propenylbenzene   *See* isosafrole.

methylene iodide   [ORG CHEM] CH_2I_2 Yellow liquid boiling at 180°C; soluble in ether and alcohol, insoluble in water; used as a chemical intermediate and to separate mineral mixtures. Also known as diiodomethane.

methylene oxide   *See* formaldehyde.

methylene succinic acid   *See* itaconic acid.

methyl ester   [ORG CHEM] An ester that forms methanol when hydrolyzed.

methyl ether   *See* dimethyl ether.

methyl ethylene   *See* propylene.

methyl ethylene glycol   *See* propylene glycol.

methyl ethyl ketone   [ORG CHEM] $CH_3COC_2H_5$ A water-soluble, colorless liquid that is miscible in oil; used as a solvent in vinyl films and nitrocellulose coatings, and as a reagent in organic synthesis. Also known as 2-butanone; ethyl methyl ketone; MEK.

methyl fumaric acid   *See* mesaconic acid.

2-methylfuran   [ORG CHEM] $C_4H_3OCH_3$ A colorless liquid with ether aroma; boils at 64°C; used as a chemical intermediate.

methyl furoate   [ORG CHEM] $C_4H_3OCO_2CH_3$ Colorless liquid that turns yellow in light; soluble in ether and alcohol, insoluble in water; used as a solvent and chemical intermediate.

methyl glucoside   [ORG CHEM] $C_7H_{14}O_6$ Odorless, water-soluble white crystals; used to make resins, drying oils, plasticizers, and surfactants.

methyl glycocoll   *See* sarcosine.

methyl glycol   *See* propylene glycol.

methyl heptane   [ORG CHEM] C_8H_{18} Either of two colorless, water-insoluble liquids, soluble in alcohol and ether, used as chemical intermediates: 2-methylheptane boils at 118°C, is flammable, and is also known as isooctane; 4-methylheptane boils at 122°C and is known as methyl dipropylmethane.

2-methylhexane   [ORG CHEM] C_7H_{16} Colorless liquid boiling at 90°C; insoluble in alcohol and water; used as a chemical intermediate. Also known as ethyl isobutylmethane.

methyl hydride   *See* methane.

methylhydroxyisopropylcyclohexane   *See* menthol.

methyl hydroxystearate   [ORG CHEM] $C_{19}H_{38}O_3$ A white, waxy material; slightly soluble in organic solvents, insoluble in water; used in cosmetics, inks, and adhesives.

3-methylindole   *See* skatole.

methyl iodide   [ORG CHEM] CH_3I Flammable colorless liquid that turns brown in light; boils at 42°C; soluble in ether and alcohol, insoluble in water; used as a chemical intermediate, in medicine, and in analytical chemistry. Also known as iodomethane.

methyl isobutyl carbinol   *See* methyl amyl alcohol.

methyl isobutyl carbinol acetate   *See* methyl amyl acetate.

methyl isobutyl ketone   [ORG CHEM] $(CH_3)_2CHCH_2COCH_3$ Flammable colorless liquid with pleasant aroma; boils at 116°C, miscible with most organic solvents; used as a solvent, extractant, and chemical intermediate. Also known as hexone.

methylisopropylphenanthrene   *See* retene.

2-methyl-5-isopropyl phenol   *See* carvacrol.

methyl lactate   [ORG CHEM] $CH_3CHCHCOOCH_3$ Liquid boiling at 145°C; miscible with water and most organic liquids; used as a solvent for lacquers, stains, and cellulosic materials.

methyl laurate   [ORG CHEM] $CH_3(CH_2)_{10}COOCH_3$ Water-insoluble, clear, colorless liquid boiling at 262°C; used as a chemical intermediate to make rust removers, and for leather treatment.

methyl maleic acid   *See* citraconic acid.

methyl mercaptan   [ORG CHEM] CH_3SH Colorless, toxic, flammable gas with unpleasant odor; boils at 6.2°C; insoluble in water, soluble in organic solvents; used as a chemical intermediate.

methyl methacrylate   [ORG CHEM] $CH_2C(CH_3)COOCH_3$ A flammable, colorless liquid, soluble in most organic solvents but insoluble in water; used as a monomer for polymethacrylate resins.

methylnaphthalene   [ORG CHEM] $C_{10}H_7CH_3$ A solid melting at 34°C; used in insecticides and organic synthesis.

methyl nitrate   [ORG CHEM] CH_3NO_3 Explosive liquid boiling at 60°C; slightly soluble in water, soluble in ether and alcohol; used as a rocket propellant.

***meta*-methylnitrobenzene**   *See meta*-nitrotoluene.

***ortho*-methylnitrobenzene**   *See ortho*-nitrotoluene.

***para*-methylnitrobenzene**   *See para*-nitrotoluene.

3-methyl-4-nitro-1-(*p*-nitrophenyl)-2-pyrazoline-5-one   *See* picolinic acid.

methyl oleate   [ORG CHEM] $C_{17}H_{33}COOCH_3$ Amber liquid with faint fatty odor; soluble in organic liquids, mineral spirits, and vegetable oil, insoluble in water; used as a plasticizer and softener.

methylol urea   [ORG CHEM] $H_2NCONHCH_2OH$ Water-soluble, colorless crystals melting at 111°C; used to treat textiles and wood, and in the manufacture of resins and adhesives.

3-methylpentane   [ORG CHEM] C_6H_{14} Flammable, colorless liquid; insoluble in water, soluble in alcohol; boils at 64°C; used as a chemical intermediate. Also known as diethylmethylmethane.

methylpentene polymer   [ORG CHEM] Thermoplastic material based on 4-methylpentene-1; has low gravity, excellent electrical properties, and 90% optical transmission.

methyl pentose   [ORG CHEM]   **1.** Any compound that is a methyl derivative of a five carbon sugar.   **2.** In particular, the compound $CH_3(CHOH)_4CHO$.

methyl phenyl acetate   [ORG CHEM] $C_6H_5CH_2COOCH_3$ A colorless liquid with honey odor; used to flavor tobacco and in perfumery.

methylphenyl ether   *See* anisole.

methylpicrylnitramine   *See* tetryl.

***N*-methylpiperazinyl-*N'*-propylphenothiazine**   *See* perazine.

2-methyl propane   *See* isobutane.

2-methylpropene   *See* isobutylene.

methyl propyl carbinol   [ORG CHEM] $CH_3CHOHC_3H_7$ Colorless liquid boiling at 119°C; miscible with ether and alcohol, slightly soluble in water; used as a pharmaceuticals intermediate and as a paint and lacquer solvent. Also known as 2-pentanol; sec-*n*-amyl alcohol.

methyl propyl ketone   *See* pentanone.

methylpyrocatechin   *See* guaiacol.

***N*-methyl-2-pyrrolidone**   [ORG CHEM] C_5H_9NO A liquid boiling at 202°C; miscible with water, castor oil, and organic solvents; used as a chemical intermediate and as a solvent for petroleum and resins, and in PVC spinning.

γ-methylquinoline   *See* lepidine.

methyl red test   [MICROBIO] A cultural test for the ability of bacteria to ferment carbohydrate to form acid; uses methyl red as the indicator.

methyl ricinoleate   [ORG CHEM] $C_{19}H_{36}O_3$ Clear, low-viscosity fluid used as a wetting agent, cutting oil additive, lubricant, and plasticizer.

methyl salicylate   [ORG CHEM] $C_6H_4OHCOOCH_3$ A colorless, yellow, or reddish liquid, slightly soluble in water, boiling at 222.2°C, with an odor of wintergreen; used in medicine and perfumery, and as a solvent for cellulose derivatives. Also known as betula oil; gaultheria oil; wintergreen oil.

methyl silicone   [ORG CHEM] $[(CH_3)_2SiO]_x$, $[C(CH_3)_2Si_2O_3]_y$, etc.   The common varieties of silicones with properties of oil, resin, or rubber, depending on molecular size and arrangement.

methyl stearate   [ORG CHEM] $C_{17}H_{35}COOCH_3$ Colorless crystals melting at 39°C; soluble in alcohol and ether, insoluble in water; used as an intermediate for stearic acid manufacture.

methyl styrene   *See* vinyltoluene.

α-methyl styrene   [ORG CHEM] $C_6H_5C(CH_3){:}CH_2$ Colorless, toxic, polymerizable liquid boiling at 165°C; used to produce polystyrene resins.

methyl sulfate   *See* dimethyl sulfate.

methyl sulfide   [ORG CHEM] $(CH_3)_2S$ Flammable, colorless liquid with disagreeable aroma; soluble in ether and alcohol, insoluble in water; boils at 38°C; used as a chemical intermediate. Also known as dimethyl sulfide; methanethiomethane.

methyl-1,2,5,6-tetrahydro-1-methylnicotinate *See* arecoline.

N-methyl-N,2,4,6-tetranitroaniline *See* tetryl.

methylthiouracil [PHARM] $C_5H_6N_2OS$ A crystalline compound; used as a medicine for overactivity of the thyroid. Also known as 6-methyl-2-thiouracil.

6-methyl-2-thiouracil *See* methylthiouracil.

methyl tolyl ketone *See* methyl acetophenone.

methyltrinitrobenzene *See* 2,4,6-trinitrotoluene.

metonic cycle [ASTRON] A time period of 235 lunar months, or 19 years 11 days; after this period the phases of the moon occur on the same days of the month.

metraterm [INV ZOO] The distal portion of the uterus in trematodes.

metria [MED] **1.** Any pathologic condition of the uterus. **2.** Any uterine inflammation occurring between childbirth and the 6 weeks following.

metric [MATH] A real valued "distance" function on a topological space X satisfying four rules: for x, y, and z in X, the distance from x to itself is zero; the distance from x to y is positive if x and y are different; the distance from x to y is the same as the distance from y to x; and the distance from x to y is less than or equal to the distance from x to z plus the distance from z to y (triangle inequality).

metricate [SCI TECH] To use the metric system in expressing all physical quantities.

metric carat *See* carat.

metric centner [MECH] **1.** A unit of mass equal to 50 kilograms. **2.** A unit of mass equal to 100 kilograms. Also known as quintal.

metric grain [MECH] A unit of mass, equal to 50 milligrams; used in commercial transactions in precious stones.

metric horsepower [PHYS] A unit of power, equal to 75 meter kilogram-force per second; equal to 753.49875 watts.

metric line *See* millimeter.

metric ounce *See* mounce.

metric slug *See* metric-technical unit of mass.

metric space [MATH] Any topological space which has a metric defined on it.

metric system [MECH] A system of units used in scientific work throughout the world and employed in general commercial transactions and engineering applications; its units of length, time, and mass are the meter, second, and kilogram respectively, or decimal multiples and submultiples thereof.

metric-technical unit of mass [MECH] A unit of mass, equal to the mass which per second per second is accelerated by 1 meter per second per second by a force of 1 kilogram-force; it is equal to 9.80665 kilograms. Abbreviated TME. Also known as hyl; metric slug.

metric tensor [MATH] A second rank tensor of a Riemannian space whose components are functions which help define magnitude and direction of vectors about a point.

metric thread gearing [DES ENG] Gears that may be interchanged in change-gear systems to provide feeds suitable for cutting metric and module threads.

metric ton *See* tonne.

metric waves [ELECTROMAG] Radio waves having wavelengths between 1 and 10 meters, corresponding to frequencies between 30 and 300 megahertz (the very-high-frequency band).

Metridiidae [INV ZOO] A family of zoantharian coelenterates in the order Actiniaria.

metriocranic [ANTHRO] Having a skull that is moderately high compared with its width, with a breadth-height index of 92 to 98.

metritis [MED] Inflammation of the uterus, usually involving both the endometrium and myometrium.

metrizable space [MATH] A topological space on which can be defined a metric whose topological structure is equivalent to the original one.

metrology [PHYS] The science of measurement.

metrorrhagia [MED] Uterine bleeding during the intermenstrual cycle. Also known as intermenstrual flow; polymenorrhea.

metrorrhea [MED] Any pathologic discharge from the uterus.

metrorrhexis [MED] Rupture of the uterine wall.

metrosalpingitis [MED] Inflammation of the uterus and oviducts.

metrostaxis [MED] Slight, chronic bleeding from the uterus.

Meusnier's theorem [MATH] A theorem stating that the curvature of a surface curve equals the curvature of the normal section through the tangent to the curve divided by the cosine of the angle between the plane of this normal section and the osculating plane of the curve.

MeV *See* million electron volts.

mevalonic acid [ORG CHEM] $HO_2C_5H_9COOH$ A dihydroxy acid used in organic synthesis. Also known as 3,5-dihydroxy-3-methyl valeric acid.

MeV Ci *See* mega-electron-volt–curie.

Mexican onyx *See* onyx marble.

Meyer atomic volume curve [ATOM PHYS] A graph of the atomic volumes of the elements versus their atomic numbers; it reveals a periodicity, with peaks at the alkali elements and valleys at the transition elements.

meyerhofferite [MINERAL] $Ca_2B_6O_{11} \cdot 7H_2O$ A colorless, hydrated borate mineral that crystallizes in the triclinic system.

Meyliidae [INV ZOO] A family of free-living nematodes in the superfamily Desmoscolecoidea.

Meziridae [INV ZOO] A family of hemipteran insects in the superfamily Aradoidea.

mezzograph [GRAPHICS] A halftone that has a grained surface rather than crossline screen dots.

mezzotint [GRAPHICS] An engraving process in which a copper or steel plate is first entirely roughened by rubbing Carborundum between it and another plate and by using a steel chisel with an edge set with minute teeth which rock into the plate at various angles; the rough-grain plate is burnished with a steel instrument to produce appropriate white areas of the design.

M_f [MET] In a carbon steel, the temperature at which martensite formation finishes during cooling of austenite.

mf *See* medium frequency.

mF *See* millifarad.

mg *See* milligram.

mG *See* milligauss.

Mg *See* magnesium.

mGal *See* milligal.

mg h *See* milligram-hour.

mH *See* millihenry.

MHD *See* magnetohydrodynamics.

MHD generator *See* magnetohydrodynamic generator.

mho [ELEC] A unit of conductance, admittance, and susceptance, equal to the conductance between two points of a conductor such that a potential difference of 1 volt between these points produces a current of 1 ampere; the conductance of a conductor in mhos is the reciprocal of its resistance in ohms. Also known as reciprocal ohm; siemens (S).

mHz *See* millihertz.

MHz *See* megahertz.

mi *See* mile.

Miacidae [PALEON] The single, extinct family of the carnivoran superfamily Miacoidea.

Miacoidea [PALEON] A monofamilial superfamily of extinct carnivoran mammals; a stem group thought to represent the progenitors of the earliest member of modern carnivoran families.

miargyrite [MINERAL] $AgSbS_2$ An iron-black to steel-gray mineral that crystallizes in the monoclinic system.

miarolithite [PETR] A chorismite type of igneous rock having miarolitic cavities or vestiges therof.

miarolitic [PETR] Of igneous rock, characterized by small irregular cavities into which well-formed crystals of the rock-forming mineral protrude.

MIBC *See* methyl amyl alcohol.

mica [MINERAL] A group of phyllosilicate minerals (with sheetlike structures) of general formula $(K,Na,Ca)(Mg,Fe,Li,Al)_{2-3}(Al,Si)_4O_{10}(OH,F)_2$ characterized by low hardness ($2-2\frac{1}{2}$) and perfect basal cleavage.

mica book [MINERAL] A crystal of mica, usually large and irregular, whose cleavage plates resemble the leaves of a book. Also known as book.

mica capacitor [ELEC] A capacitor whose dielectric consists of thin rectangular sheets of mica and whose electrodes are

MIACOIDEA

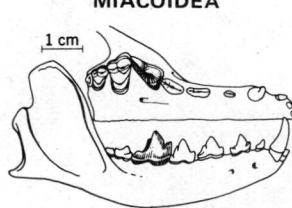

Right side of *Miacis* **jaw shown in ventral (upper) and lateral (lower) views.**

either thin sheets of metal foil stacked alternately with mica sheets, or thin deposits of silver applied to one surface of each mica sheet.

mica schist [PETR] A schist which is composed essentially of mica and quartz and whose characteristic foliation is mainly due to the parallel orientation of the mica flakes.

micelle [MOL BIO] A submicroscopic structural unit of protoplasm built up from polymeric molecules.

Michaelis constant [BIOCHEM] A constant K_m such that the initial rate of reaction V, produced by an enzyme when the substrate concentration is high enough to saturate the enzyme, is related to the rate of reaction v at a lower substrate concentration c by the formula $V = v (1 + K_m/c)$.

Michaelson actinograph [ENG] A pyrheliometer of the bimetallic type used to measure the intensity of direct solar radiation; the radiation is measured in terms of the angular deflection of a blackened bimetallic strip which is exposed to the direct solar beams.

Michel parameter [PARTIC PHYS] A number appearing in an equation for the momentum spectrum of muon decay, which depends on the nature of the weak interactions; the number is equal to 3/4 in any two-component neutrino theory before radiative corrections are taken into account.

Michelson interferometer [OPTICS] An interferometer in which light strikes a partially reflecting plate at an angle of 45°, the light beams reflected and transmitted by the plate are both reflected back to the plate by mirrors, and the beams are recombined at the plate, interfering constructively or destructively depending on the distances from the plate to the two mirrors.

Michelson-Morley experiment [OPTICS] An experiment which uses a Michelson interferometer to determine the difference between the speeds of light in two perpendicular directions.

Michelson stellar interferometer [OPTICS] An instrument for measuring angular diameters of astronomical objects, in which a system of mirrors directs two parallel beams of light into a telescope, and angular diameter is determined from the maximum distance between the beams at which interference fringes are observable.

Michigan cut [MIN ENG] A technique used to break off ore at a heading; a large hole or series of small holes at the center of the heading are drilled parallel to the tunnel direction but not charged with explosive; other holes are drilled in the heading and charged so that upon detonation they break out toward the uncharged holes.

Michigan tripod [MIN ENG] A support for a drilling outfit; consists of three debarked pine or fir timber poles about 25 feet (7.6 meters) long whose butt ends are about 12 inches (30 centimeters) in diameter; a sheave suspended from a clevis at the top of the tripod is aligned over the hoisting drum and the borehole; there is a minimum of 22 feet (6.7 meters) of headroom above the drill floor.

mickey-mouse [ADP] To play with something new, such as hardware, software, or a system, until a feel is gotten for it and the proper operating procedure is discovered, understood, and mastered.

MICR See magnetic-ink character recognition.

micracanthopore [INV ZOO] Small, minute tubes projecting from the surface of bryozoan colonies.

micrencephaly [MED] The condition of having an abnormally small brain. Also spelled microencephaly.

micril See gammil.

micrinite [PETR] An opaque granular variety of inertinite of medium hardness showing no plant-cell structure.

micrite [PETR] A semiopaque crystalline limestone matrix that consists of chemically precipitated calcite mud, whose crystals are generally 1–4 micrometers in diameter.

micro- [MATH] A prefix representing 10^{-6}, or one-millionth. [SCI TECH] **1.** A prefix indicating smallness, as in microwave. **2.** A prefix indicating extreme sensitivity, as in microradiometer and microphone.

microabscess [MED] A small abscess.

microaerophilic [MICROBIO] Pertaining to those microorganisms requiring free oxygen but in very low concentration for optimum growth.

microalloy diffused transistor [ELECTR] A microalloy transistor in which the semiconductor wafer is first subjected to gaseous diffusion to produce a nonuniform base region. Abbreviated MADT.

microalloy transistor [ELECTR] A transistor in which the emitter and collector electrodes are formed by etching depressions, then electroplating and alloying a thin film of the impurity metal to the semiconductor wafer, somewhat as in a surface-barrier transistor.

microammeter [ELEC] An ammeter whose scale is calibrated to indicate current values in microamperes.

microampere [ELEC] A unit of current equal to one-millionth of an ampere. Abbreviated μA.

microanalysis [ANALY CHEM] Identification and chemical analysis of material on a small scale so that specialized instruments such as the microscope are needed; the material analyzed may be on the scale of 1 microgram.

microanatomy [ANAT] Anatomical study of microscopic tissue structures.

microaneurysm [MED] Dilation of the wall of a capillary, characteristic of certain disease entities.

microangstrom [MECH] A unit of length equal to one-millionth of an angstrom, or 10^{-16} meter. Abbreviated μA.

microautoradiograph [GRAPHICS] An image which is produced by placing a specimen containing radioactive material (usually a radioactive tracer) in close contact with a photographic film and optically enlarging the developed image.

Microbacterium [MICROBIO] A genus of small, rod-shaped, gram-positive, nonsporulating, and nonmotile bacteria in the family Corynebacteraceae which are capable of fermenting the lactose in milk to acidic products, mostly lactic acid.

microbalance [ENG] A small, light type of analytical balance that can weigh loads of up to 0.1 gram to the nearest microgram.

microbar See barye.

microbarm [GEOPHYS] That portion of the record of a microbarograph between any two or a specified small number of the successive crossings of the average pressure level in the same direction; analogous to microseism.

microbarogram [ENG] The record or trace made by a microbarograph.

microbarograph [METEOROL] A type of aneroid barograph designed to record atmospheric pressure variations of very small magnitude.

microbe [MICROBIO] A microorganism, especially a bacterium of a pathogenic nature.

microbial insecticide [MICROBIO] Species-specific bacteria which are pathogenic for and used against injurious insects.

microbicide [MATER] An agent that kills microbes.

microbiology [BIOL] The science and study of microorganisms, including protozoans, algae, fungi, bacteria, viruses, and rickettsiae.

microbiophobia [PSYCH] An abnormal fear of microbes.

microbit [ADP] A unit of information equal to one-millionth of a bit.

microbreccia [GEOL] A poorly sorted sandstone containing large, angular sand particles in a fine silty or clayey matrix.

microcaliper log [PETRO ENG] A detailed and accurate record of drill-hole diameter; used to detect caved sections and to verify the presence of mud cake.

microcanonical ensemble [STAT MECH] A collection of systems describing a single isolated system of specified energy; its members are uniformly distributed over a part of phase space whose energies lie within an infinitesimal range.

microcapacitor [ELECTR] Any very small capacitor used in microelectronics, usually consisting of a thin film of dielectric material sandwiched between electrodes.

microcard [GRAPHICS] A type of microtext, consisting of photographic prints 7½ by 12½ centimeters in size prepared from 16- or 35-millimeter film, commonly at a reduction of 20 diameters.

microcentrum [CYTOL] The centrosome, or a group of centrosomes, functioning as the dynamic center of a cell.

microcephalus [MED] An individual with microcephaly.

microcephaly [MED] The condition of having an abnormally small head, with a circumference less than two standard deviations below the mean.

microceratous [INV ZOO] Having short antennae.

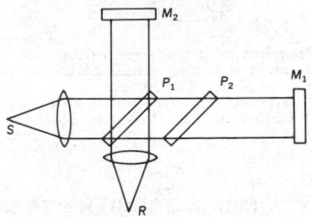

Michelson interferometer, S = narrow angle source; R = receiver; M_1, M_2 = mirrors; P_1 = 50% partially reflecting plate; P_2 = reflector plate which compensates for thickness of P_1. (From A. C. Hardy and F.H. Perrin, The Principles of Optics, McGraw-Hill, 1932)

**MICHELSON STELLAR
INTERFEROMETER**

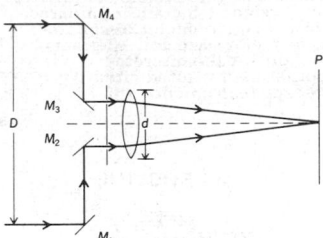

Michelson stellar interferometer, used for measuring the diameters of stars. M_2, M_3 = two fixed mirrors; M_1, M_4 = two mirrors that are movable so as to vary their separation D; d = diameter of lens; P = focal plane of lens.

MICROAUTORADIOGRAPH

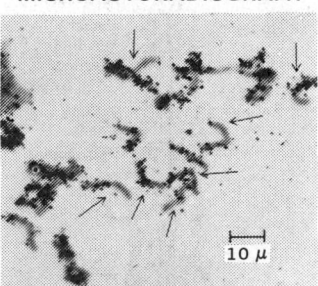

10 μ

Photomicrograph of an autoradiograph of Vicia faba chromosomes in metaphase labeled with tritiated thymidine. In the chromosome pairs indicated by the arrows, one chromosome is labeled but the other is not. (Courtesy of W. L. Hughes)

MICROCERCOUS CERCARIA

Drawing of a microcercous cercaria showing small tail. (*From R. M. Cable, An Illustrated Laboratory Manual of Parasitology, Burgess, 1958*)

MICROCIRCUITRY

Size comparison of solid-state circuit, at right, and discrete-component printed circuit. Each is a binary full-serial adder. Solid-state circuit occupies 0.02 cubic inch (0.33 cubic centimeter) and weighs 1.5 grams compared with 4-square-inch (26-square-centimeter) size and 42-gram weight for the discrete-component printed circuit. (*Texas Instruments, Inc.*)

MICROCLINE

5 cm

Microcline, variety amazon stone, from Pikes Peak, Colorado. (*American Museum of Natural History Specimen*)

MICROCOTYLOIDEA

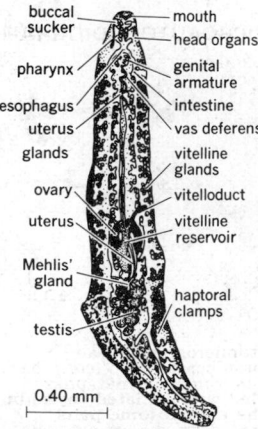

0.40 mm

Ventral view of *Heteraxinoides xanthophilis* (Hargis), an ectoparasite of the spot fish (*Leiostomus xanthurus*).

microcercous cercaria [INV ZOO] A cercaria with a very short broad tail.

microchemistry [BIOCHEM] The chemistry of individual cells and minute organisms. [CHEM] The study of chemical reactions, using small quantities of materials, frequently less than 1 milligram or 1 milliliter, and often requiring special small apparatus and microscopical observation.

Microchiroptera [VERT ZOO] A suborder of the mammalian order Chiroptera composed of the insectivorous bats.

microchronometer [HOROL] A spring-driven, fast-moving clock capable of indicating time intervals as small as 1/2000 of a minute; used as a timing device in micromotion studies.

microcircuitry [ELECTR] Electronic circuit structures that are orders of magnitude smaller and lighter than circuit structures produced by the most compact combinations of discrete components. Also known as microelectronic circuitry; microminiature circuitry.

microcirculation [PHYSIO] The flow of blood or lymph in the vessels of the microcirculatory system.

microcirculatory system [ANAT] Those vessels of the blood and lymphatic systems which are visible only with a microscope.

microclimate [CLIMATOL] The local, rather uniform climate of a specific place or habitat, compared with the climate of the entire area of which it is a part.

microclimatology [CLIMATOL] The study of a microclimate, including the study of profiles of temperature, moisture and wind in the lowest stratum of air, the effect of the vegetation and of shelterbelts, and the modifying effect of towns and buildings.

microcline [MINERAL] $KAlSi_3O_8$ A triclinic potassium-rich feldspar, usually containing minor amounts of sodium; may be clear, white, pale-yellow, brick-red, or green, and is generally characterized by crosshatch twinning.

microcneme [INV ZOO] Microsepta in certain anemones.

Micrococcaceae [MICROBIO] A family of spherical, gram-positive bacteria in the order Eubacteriales characterized by chemoorganotrophic energy metabolism.

Micrococcus [MICROBIO] A genus of bacteria of the family Micrococcaceae; cells are arranged in irregular masses, metabolism is oxidative, and some species form red, yellow, or orange pigments.

Micrococcus meningitidis *See* Neisseria meningitidis.

microcode [ADP] A code that employs microinstructions; not ordinarily used in programming.

microconsumer *See* decomposer.

microcoquina [PETR] A clastic limestone composed wholly or partially of cemented sand-size particles of shell detritus.

Microcotyloidea [INV ZOO] A superfamily of ectoparasitic trematodes in the subclass Monogenea.

microcoulomb [ELEC] A unit of electric charge equal to one-millionth of a coulomb. Abbreviated μC.

microcrack *See* microfissure.

microcrystalline [CRYSTAL] Composed of or containing crystals that are visible only under the microscope.

microcrystalline wax [MATER] A petroleum wax containing small, indistinct crystals, and having a higher molecular weight, melting point, and viscosity than paraffin wax; used in laminated paper and electrical coil coating.

Microcyprini [VERT ZOO] The equivalent name for Cyprinodontiformes.

microcyst [MED] A very small cyst.

microcyte [MED] A red blood cell whose diameter or mean corpuscular volume or both are more than two standard deviations below the normal mean. Also known as microerythrocyte.

microcythemia [MED] Blood characterized by the presence of small red blood cells.

microcytic anemia [MED] Any form of anemia in which small erythrocytes occur in the blood.

microcytosis [MED] A blood disorder characterized by a preponderance of microcytes.

microdactyly [MED] A condition of abnormal smallness of fingers or toes.

microdensitometer [SPECT] A high-sensitivity densitometer used in spectroscopy to detect spectrum lines too faint on a negative to be seen by the human eye.

microdissection [BIOL] Dissection under a microscope.

Microdomatacea [PALEON] An extinct superfamily of gastropod mollusks in the order Aspidobranchia.

microelectrolysis [PHYS CHEM] Electrolysis of small quantities of material.

microelectronic circuitry *See* microcircuitry.

microelectronics [ELECTR] The technology of constructing circuits and devices in extremely small packages by various techniques. Also known as microminiaturization; microsystem electronics.

microelectrophoresis [ANALY CHEM] Direct microscopic observation and measurement of the velocity of migration of ions or other charged bodies through a solution toward oppositely charged electrodes.

microelement [ELECTR] Resistor, capacitor, transistor, diode, inductor, transformer, or other electronic element or combination of elements mounted on a ceramic wafer 0.025 centimeter thick and about 0.75 centimeter square; individual microelements are stacked, interconnected, and potted to form micromodules.

microencephaly *See* micrencephaly.

microenvironment [ECOL] The specific environmental factors in a microhabitat.

microerythrocyte *See* microcyte.

microevolution [EVOL] 1. Evolutionary processes resulting from the accumulation of minor changes over a relatively short period of time; evolutionary changes due to gene mutation. 2. Evolution of species.

microfacies [PETR] The composition, features, or appearance of a rock or mineral in thin section under the microscope.

microfarad [ELEC] A unit of capacitance equal to one-millionth of a farad. Abbreviated μF.

microfibril [MOL BIOL] The submicroscopic unit of a microscopic cellular fiber.

microfiche [GRAPHICS] A microfilm card or sheet used in some information storage systems; consists of a film format about 4 by 6 inches (10 by 15 centimeters) containing microimages of type and other information, with a title heading large enough to be read by the unaided eye.

microfilaria [INV ZOO] Slender, motile prelarval forms of filarial nematodes measuring 150–300 micrometers in length; adult filaria are mammalian parasites.

microfilm [GRAPHICS] Greatly reduced film records of such things as books, newspapers, engineering drawings, reports, and manuscripts; copies are made on fine-grain film of 16, 35, 70, and 105-millimeter size, permitting easy storage and handling.

microfissure [MET] A crack of microscopic dimensions. Also known as microcrack.

microflora [BOT] Microscopic plants. [ECOL] The flora of a microhabitat.

microfluid [FL MECH] A fluid in which the effects of local motion of contained material particles on properties and behavior of the fluid are not disregarded.

microfluoroscope [ENG] A fluoroscope in which a very fine-grained fluorescent screen is optically enlarged.

microforge [ENG] In micromanipulation techniques, an optical-mechanical device for controlling the position of needles or pipets in the field of a low-power microscope by a simple micromanipulator.

microform [GRAPHICS] A miniature replica of data, such as microfiche or microfilm.

microfossil [PALEON] A small fossil which is studied and identified by means of the microscope.

microgamete [BIOL] The smaller, or male gamete produced by heterogametic species.

microgametocyte [BIOL] A cell that gives rise to microgametes.

microgammil *See* gammil.

microgamy [BIOL] Sexual reproduction by fusion of the small male and female gametes in certain protozoans and algae.

microgastria [MED] A condition of abnormal smallness of the stomach.

microgenesis [BIOL] Abnormally small development of a part.

microgenitalism [MED] Having abnormally small genitalia.

microglia [HISTOL] Small neuroglia cells of the central nervous system having long processes and exhibiting ameboid and phagocytic activity under certain pathologic conditions.

microglossia [MED] A condition of abnormal smallness of the tongue.

micrognathia [MED] A condition of abnormal smallness of the jaws, particularly the mandible.

microgram [MECH] A unit of mass equal to one-millionth of a gram. Abbreviated μg.

micrograph [ENG] An instrument for making very tiny writing or engraving. [GRAPHICS] A graphic reproduction of the surface of a prepared specimen at a magnification greater than 10 diameters.

micrographic [SCI TECH] Having graphic texture distinguishable only with the aid of a microscope.

microgroove record *See* long-playing record.

microgyria [MED] A condition of abnormal smallness of the gyri of the brain.

microhabitat [ECOL] A small, specialized, and effectively isolated location.

microhardness [MET] Hardness of microscopic areas of a metal or alloy.

microhm [ELEC] A unit of resistance, reactance, and impedance, equal to 10^{-6} ohm.

Microhylidae [VERT ZOO] A family of anuran amphibians in the suborder Diplasiocoela including many heavy-bodied forms with a pointed head and tiny mouth.

microhysteresis effect [SOLID STATE] Hysteresis that results from the motion of domain walls lagging behind an applied magnetic or elastic stress when these walls are held up by dislocations and other imperfections in the material.

microincineration [CHEM] Reduction of small quantities of organic substances to ash by application of heat.

microinfarct [MED] A very small infarct.

microinjection [CYTOL] Injection of cells with solutions by using a micropipet.

microinstruction [ADP] The smallest unit of action that a computer can perform.

microinterferometer [OPTICS] Functional combination of a microscope with an interferometer; used to study thin films, platings, or transparent coatings.

microinvasion [MED] Invasion by tumor, especially a squamous-cell carcinoma of the uterine cervix, a very short distance into the tissues beneath the point of origin.

microlaterolog [PETRO ENG] Modification of the downhole microlog in which extra electrodes focus electric current into a trumpet-shaped area; gives greater resistivity-measurement resolution than does the microlog.

microlayer [OCEANOGR] The thin zone beneath the surface of the ocean or any free water surface within which physical processes are modified by proximity to the air-water boundary.

Microlepidoptera [INV ZOO] A former division of Lepidoptera.

microlite [CRYSTAL] A microscopic crystal which polarizes light. Also known as microlith. [MINERAL] $(Na,Ca)_2$ $Ta_2O_6(O,OH,F)$ A pale-yellow, reddish, brown, or black isometric mineral composed of sodium calcium tantalum oxide with a small amount of fluorine; it is isomorphous with pyrochlore.

microlith [CRYSTAL] *See* microlite. [MED] A calculus of microscopic size.

microlithiasis [MED] The presence of numerous microliths.

microlithiasis alveolaris pulmonum [MED] A rare form of pulmonary calcification of unidentified etiology in which microliths, and larger osseous nodules, are found.

microlithology [PETR] Microscopic study of the characteristics of rocks.

microlitic [PETR] Of the texture of a porphyritic igneous rock, having a groundmass composed of an aggregate of microlites in a generally glassy base.

microlock [ELECTR] **1.** Satellite telemetry system that uses phase-lock techniques in the ground receiving equipment to achieve extreme sensitivity. **2.** A lock by a tracking station upon a minitrack radio transmitter. **3.** The system by which this lock is effected.

microlog [PETRO ENG] A drill-hole resistivity log recorded with electrodes mounted at short distances from each other in the face of a rubber-padded microresistivity sonde.

Micromalthidae [INV ZOO] A family of coleopteran insects in the superfamily Cantharoidea; the single species is the telephone pole beetle.

micromania [PSYCH] A delusional state in which the patient believes himself diminutive in size and mentally inferior.

micromanipulation [BIOL] The techniques and practice of microdissection, microvivisection, microisolation, and microinjection.

micromanipulator [ENG] A device for holding and moving fine instruments for the manipulation of microscopic specimens under a microscope.

micromanometer [ENG] Any manometer that is designed to measure very small pressure differences.

micromere [EMBRYO] A small blastomere of the upper or animal hemisphere in eggs that undergo uneven cleavage.

micrometeorite [ASTRON] A very small meteorite or meteoritic particle with a diameter generally less than a millimeter.

micrometeorite penetration [AERO ENG] Penetration of the thin outer shell (skin) of space vehicles by small particles traveling in space at high velocities.

micrometeoroid [ASTRON] A very small meteoroid with diameter generally less than a millimeter.

micrometeorology [METEOROL] That portion of the science of meteorology that deals with the observation and explanation of the smallest-scale physical and dynamic occurrences within the atmosphere; studies are confined to the surface boundary layer of the atmosphere, that is, from the earth's surface to an altitude where the effects of the immediate underlying surface upon air motion and composition become negligible.

micrometer [ENG] **1.** An instrument attached to a telescope or microscope for measuring small distances or angles. **2.** A caliper for making precise measurements; a spindle is moved by a screw thread so that it touches the object to be measured; the dimension can then be read on a scale. Also known as micrometer caliper. [MECH] A unit of length equal to one-millionth of a meter. Abbreviated μm. Also known as micron (μ).

micrometer caliper *See* micrometer.

micrometer of mercury *See* micron.

micromicro- *See* pico-.

micromicrofarad *See* picofarad.

micromicrosecond *See* picosecond.

micromicrowatt *See* picowatt.

microminiature circuitry *See* microcircuitry.

microminiaturization *See* microelectronics.

micromodule [ELECTR] Cube-shaped, plug-in, miniature circuit composed of potted microelements; each microelement can consist of a resistor, capacitor, transistor, or other element, or a combination of elements.

Micromonospora [MICROBIO] A genus of gram-positive, aerobic bacteria in the family Streptomycetaceae characterized by a well-developed, fine, nonseptate mycelium, and multiplication by means of conidia.

Micromonospora purpurea [MICROBIO] The bacterium that produces the antibiotic gentamycin.

micromotion study [IND ENG] Study of the fundamental elements of an operation by means of a motion picture camera operated at normal or faster than normal speed, and a timing device to measure element motion time.

micron [MECH] **1.** A unit of pressure equal to the pressure exerted by a column of mercury 1 micrometer high, having a density of 13.5951 grams per cubic centimeter, under the standard acceleration of gravity; equal to 0.133322387415 pascal; it differs from the millitorr by less than one part in seven million. Also known as micrometer of mercury. **2.** *See* micrometer.

micronekton [ECOL] Active pelagic crustaceans and other forms intermediate between thrusting nekton and feebler-swimming plankton.

micronized clay [MATER] A pure kaolin pulverized to a fineness of 400 to 800 mesh; used as a filler material in rubber.

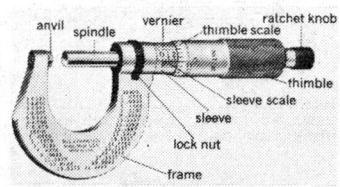

MICROMETER

Machinist's outside caliper with micrometer reading 0.250 inch (6.35 millimeters), showing component parts. (*L. S. Starrett Co.*)

micronized mica [MATER] Powdered mica of a fineness of 400 to 1000 mesh; used as a filler.

micronucleus [INV ZOO] The smaller, reproductive nucleus in multinucleate protozoans.

micronutrient [BIOCHEM] Trace elements and compounds required by living systems only in minute amounts.

microoperation [ADP] Any clock-timed step of an operation.

microorganism [MICROBIO] A microscopic organism, including bacteria, protozoans, yeast, viruses, and algae.

micropaleontology [PALEON] A branch of paleontology that deals with the study of microfossils.

micropane [GRAPHICS] A microphotograph on glass.

micropegmatite [PETR] Microcrystalline graphic granite.

microperthite [MINERAL] Perthite in which the lamellae are visible only under the microscope.

Micropezidae [INV ZOO] A family of myodarian cyclorrhaphous dipteran insects in the subsection Acalypteratae.

microphage [HISTOL] A small phagocyte, especially a neutrophil.

microphagy [BIOL] Feeding on minute organisms or particles.

microphakia [MED] Congenital condition of abnormal smallness of the crystalline lens.

microphone [ENG ACOUS] An electroacoustic device containing a transducer which is actuated by sound waves and delivers essentially equivalent electric waves.

microphone transducer [ENG ACOUS] A device which converts variation in the position or velocity of some body into corresponding variations of some electrical quantity, in a microphone.

microphonics [ELECTR] Noise caused by mechanical vibration of the elements of an electron tube, component, or system. Also known as microphonism.

microphonism *See* microphonics.

microphotograph [GRAPHICS] A microscopically small photograph, requiring magnification to be readable.

microphotometer [ENG] A photometer that provides highly accurate illumination measurements; in one form, the changes in illumination are picked up by a phototube and converted into current variations that are amplified by vacuum tubes.

microphthalmus [MED] A condition characterized by an abnormally small eyeball. Also known as nanophthalmus.

microphyllous [BOT] 1. Having small leaves. 2. Having leaves with a single, unbranched vein.

microphyric [PETR] Of the texture of an igneous rock, containing microscopic phenocrysts (longest dimension 0.2 millimeter). Also known as microporphyritic.

Microphysidae [INV ZOO] A palearctic family of hemipteran insects in the subfamily Cimicimorpha.

microphyte [ECOL] 1. A microscopic plant. 2. A dwarfed plant due to unfavorable environmental conditions.

micropipet [ENG] 1. A pipet with capacity of 0.5 milliliter or less, to measure small volumes of liquids with a high degree of accuracy; types include lambda, straight-bore, and Lang-Levy. 2. A fine-pointed pipette used for microinjection.

micropoikilitic [PETR] Of the texture of an igneous rock, having poikilitic character visible only under the microscope.

micropore [GEOL] A pore small enough to hold water against the pull of gravity and to retard water flow.

microporosity [MET] Extremely fine porosity, visible only with the aid of a microscope.

microporous barrier [CHEM ENG] A metallic or plastic membrane with micrometer-sized pores used for dialysis and other membrane-separation processes.

microporphyritic *See* microphyric.

microprobe [SPECT] An instrument for chemical microanalysis of a sample, in which a beam of electrons is focused on an area less than a micrometer in diameter, and the characteristic x-rays emitted as a result are dispersed and analyzed in a crystal spectrometer to provide a qualitative and quantitative evaluation of chemical composition.

microprobe spectrometry [SPECT] Microanalysis of a sample, using a microprobe.

microprocessor [ELECTR] An integrated circuit chip which incorporates the arithmetic and control functions of a general-purpose digital computer and has the ability to work with large amounts of external memory and large numbers of input/output chips, enabling it to supervise large operations such as the control of a factory automation line.

microprogram [ADP] A sequence of pseudocommands which will be translated by hardware into machine subcommands.

microprogrammable instruction [ADP] An instruction that does not refer to a core memory address and that can be microprogrammed, thus specifying various commands within one instruction.

microprogramming [ADP] Transformation of a computer instruction into a sequence of elementary steps (microinstructions) by which the computer hardware carries out the instruction.

micropsia [MED] A visual disturbance in which objects appear undersized.

Micropterygidae [INV ZOO] The single family of the lepidopteran superfamily Micropterygoidea; members are minute moths possessing toothed, functional mandibles and lacking a proboscis.

Micropterygoidea [INV ZOO] A monofamilial superfamily of lepidopteran insects in the suborder Homoneura.

micropulsation [GEOPHYS] A short-period geomagnetic variation in the range of about 0.2–600 seconds, typically exhibiting an oscillatory waveform.

micropump *See* electroosmotic driver.

micropycnometer [ENG] A small-volume pycnometer with a capacity from 0.25 to 1.6 milliliters; weighing precision is 1 part in 10,000, or better.

Micropygidae [INV ZOO] A family of echinoderms in the order Diadematoida that includes only one genus, *Micropyga*, which has noncrenulate tubercles and umbrellalike outer tube feet.

micropyle [BOT] A minute opening in the integument at the tip of an ovule through which the pollen tube commonly enters; persists in the seed as an opening or a scar between the hilum and point of radicle.

microradiogram [PHYS] A two-dimensional x-ray image of a sample, produced by one type of x-ray microscope used in microradiography; all levels of the sample object are imaged into essentially a single focal plane for subsequent microphotographic enlargement.

microradiograph [GRAPHICS] An enlarged radiographic image on photographic film produced either by increasing the distance from specimen to photographic plate to secure inherent enlargement of divergent x-ray beams, or by optical enlargement of a developed image.

microradiography [ANALY CHEM] Technique for the study of surfaces of solids by monochromatic-radiation (such as x-ray) contrast effects shown via projection or enlargement of a contact radiograph. [GRAPHICS] The radiography of small objects having details too fine to be seen by the unaided eye, with optical enlargement of the resulting negative.

microradiometer [ELECTR] A radiometer used for measuring weak radiant power, in which a thermopile is supported on and connected directly to the moving coil of a galvanometer. Also known as radiomicrometer.

micro-reciprocal-degree *See* mired.

microrefractometry [OPTICS] The measurement of refractive indices of microscopic objects; this is often done by immersing an object in a series of mediums of graded refractive index until one is found that makes the object invisible in a phase-contrast microscope.

microrelief [GEOGR] Irregularities of the land surface causing variations in elevation amounting to no more than a few feet.

microresistivity survey [PETRO ENG] General term for downhole resistivity surveys of oil-bearing formations; includes microlog and microlaterolog surveys.

microrheology [MECH] A branch of rheology in which the heterogeneous nature of dispersed systems is taken into account.

Microsauria [PALEON] An order of Carboniferous and early Permian lepospondylous amphibians.

microsclere [INV ZOO] A minute sclerite in Porifera.

microscope [OPTICS] An instrument through which minute objects are enlarged by means of a lens or lens system; principal types include optical, electron, and x-ray.

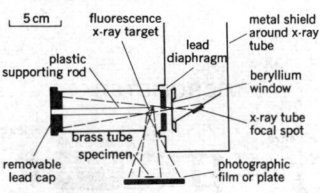
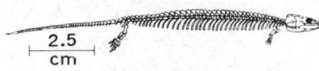
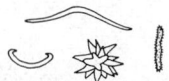

microscope stage [OPTICS] The platform on which specimens are placed for microscopic examination.

microscopic [OPTICS] *See* microscopical. [SCI TECH] Of extremely small size.

microscopical [OPTICS] Also known as microscopic. **1.** Of or pertaining to the microscope. **2.** Visible only under a microscope.

microscopical diagnosis [PATH] Identification of a disease by microscopic examination of specimens taken from the patient.

microscopic anisotropy [PETRO ENG] Phenomenon in electrical downhole logging wherein electric current flows most easily along the water-filled interstices, usually parallel to sedimentary bed strata.

microscopic reversibility [STAT MECH] A principle which requires that in a system at equilibrium any molecular process and its reverse take place at the same average rate. Also known as reversibility principle.

microscopic stress [MET] Residual stress ranging from compression to tension in a metal within a distance often comparable to the grain size. Also known as microstress.

microscopic theory [PHYS] A theory concerned with the interactions of atoms, molecules, or their constituents, involving distances on the order of 10^{-10} meter or less, which underlie observable phenomena.

microscopist [SCI TECH] An individual skilled in the use of the microscope.

microscopy [OPTICS] The interpretive application of microscope magnification to the study of materials that cannot be properly seen by the unaided eye.

microsecond [MECH] A unit of time equal to one-millionth of a second. Abbreviated μs.

microsegregation [MET] Segregation within a grain, crystal, or particle of microscopic size.

microseism [GEOPHYS] A weak, continuous, oscillatory motion in the earth having a period of 1–9 seconds and caused by a variety of agents, especially atmospheric agents; not related to an earthquake.

microseismic instrument [MIN ENG] An instrument for the study of roof strata and supports; it is inserted in holes, drilled at selected points, for listening to subaudible vibrations that precede rock failure.

microseptum [INV ZOO] An incomplete or imperfect mesentery in zoantharians.

microshrinkage [MET] A casting defect consisting of interdendritic voids, visible only at magnifications over 10 diameters.

microsome [CYTOL] **1.** A fragment of the endoplasmic reticulum. **2.** A minute granule of protoplasm.

microspec function [ADP] The set of microinstructions which performs a specific operation in one or more machine cycles.

microspecies [ECOL] A small, localized species population that is clearly differentiated from related forms. Also known as jordanon.

microspectrograph [SPECT] A microspectroscope provided with a photographic camera or other device for recording the spectrum.

microspectrophotometer [SPECT] A split-beam or double-beam spectrophotometer including a microscope for the localization of the object under study, and capable of carrying out spectral analyses within the dimensions of a single cell.

microspectroscope [SPECT] An instrument for analyzing the spectra of microscopic objects, such as living cells, in which light passing through the sample is focused by a compound microscope system, and both this light and the light which has passed through a reference sample are dispersed by a prism spectroscope, so that the spectra of both can be viewed simultaneously.

microspherulitic [PETR] Of the texture of an igneous rock, having spherulitic character visible only under the microscope.

Microsporaceae [BOT] A monogeneric family of green algae in the suborder Ulotrichineae; the chloroplast is a parietal network.

microsporangium [BOT] A sporangium bearing microspores.

microspore [BOT] The smaller spore of heterosporous plants; gives rise to the male gametophyte.

Microsporida [INV ZOO] The single order of the class Microsporidea.

Microsporidae [INV ZOO] The equivalent name for Sphaeriidae.

Microsporidea [INV ZOO] A class of Cnidospora characterized by the production of minute spores with a single intrasporal filament or one or two intracapsular filaments and a single sporoplasm; mainly intracellular parasites of arthropods and fishes.

microsporidiosis [VET MED] Infection with microsporidians.

microsporophyll [BOT] A sporophyll bearing microsporangia.

microstress *See* microscopic stress.

microstrip *See* strip transmission line.

microstructure [SCI TECH] The structure of an object, organism, or material as revealed by a microscope of a magnification over 10 times.

microstylolite [PETR] A stylolite in which the surface relief is less than 1 millimeter.

microsurgery [BIOL] Surgery on single cells by micromanipulation.

Microswitch [ELEC] A trade name for a small switch in which contact is made or broken by a slight motion.

microsystem electronics *See* microelectronics.

Microtatobiotes [BIOL] An artificial taxonomic category, comprising two unrelated groups of biological entities, the rickettsiae and the viruses.

microtectonics *See* structural petrology.

microtherm [ECOL] A plant requiring a mean annual temperature range of 0–14°C for optimum growth.

microthermal climate [CLIMATOL] A temperature province in both of C. W. Thornthwaite's climatic classifications, generally described as a "cool" or "cold winter" climate.

microthrowing power [PHYS CHEM] Relative ability of an electroplating solution to deposit metal in a small, shallow aperture or crevice not exceeding a few thousandths of an inch in dimensions.

Microtinae [VERT ZOO] A subfamily of rodents in the family Muridae that includes lemmings and muskrats.

microtome [ENG] An instrument for cutting thin sections of tissues or other materials for microscopical examination.

microtomy [BIOL] Cutting of thin sections of specimens with a microtome.

Microtragulidae [PALEON] A group of saltatorial caenolistoid marsupials that appeared late in the Cenozoic and paralleled the small kangaroos of Australia.

microtrichia [INV ZOO] Small hairs on the integument of various insects, especially on the wings.

microtron [NUCLEO] A type of circular particle accelerator for accelerating electrons to energies of several million electron volts, in which the time of successive revolutions of the particles increases by exactly one cycle of the accelerating radio-frequency voltage, so that synchronism is maintained.

microtubule [CYTOL] One of the thin rods or tubes contained by certain cytoplasmic components, such as in cilia or the mitotic spindle; they may occur free or in bundles in the cytoplasm, where they help to maintain cell shape.

microvillus [CYTOL] One of the filiform processes that form a brush border on the surfaces of certain specialized cells, such as intestinal epithelium.

microvitrain [GEOL] A coal lithotype; fine vitrainlike lenses or laminae in clarain.

microvolt [ELEC] A unit of potential difference equal to one-millionth of a volt. Abbreviated μV.

microvoltmeter [ELECTR] A voltmeter whose scale is calibrated to indicate voltage values in microvolts.

microvolts per meter [ELECTROMAG] Field strength of antenna which is the ratio of the antenna voltage in microvolts to the antenna length in meters, as measured at a given point.

microwatt [MECH] A unit of power equal to one-millionth of a watt. Abbreviated μW.

microwave [ELECTROMAG] An electromagnetic wave which has a wavelength between about 0.3 and 30 centimeters, corresponding to frequencies of 1–100 gigahertz; however, there are no sharp boundaries distinguishing microwaves from infrared and radio waves.

microwave acoustics [ACOUS] The production and study of elastic vibrations in materials at microwave frequencies, on

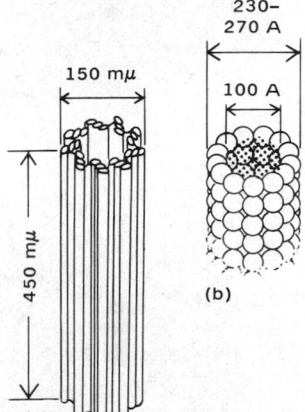

MICROTUBULE

Diagram of mature centriole showing (a) arrangement of the microtubules and (b) structure of one of the microtubules.

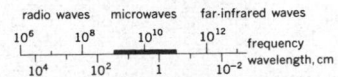

MICROWAVE

A portion of the electromagnetic spectrum showing the part occupied by microwaves.

the order of 10^9 to 10^{11} hertz, such as in single-crystal delay lines used in radar systems.

microwave amplification by stimulated emission of radiation *See* maser.

microwave amplifier [ELECTR] A device which increases the power of microwave radiation.

microwave antenna [ELECTROMAG] A combination of an open-end waveguide and a parabolic reflector or horn, used for receiving and transmitting microwave signal beams at microwave repeater stations.

microwave attenuator [ELECTROMAG] A device that causes the field intensity of microwaves in a waveguide to decrease by absorbing part of the incident power; usually consists of a piece of lossy material in the waveguide along the direction of the electric field vector.

microwave bridge [ELECTROMAG] A microwave circuit equivalent to an ordinary electrical bridge and used to measure impedance; consists of six waveguide sections arranged to form a multiple junction.

microwave cavity *See* cavity resonator.

microwave circuit [ELECTROMAG] Any particular grouping of physical elements, including waveguides, attenuators, phase changers, detectors, wavemeters, and various types of junctions, which are arranged or connected together to produce certain desired effects on the behavior of microwaves.

microwave circulator *See* circulator.

microwave communication [COMMUN] Transmission of messages using highly directional microwave beams, which are generally relayed by a series of microwave repeaters spaced up to 80 kilometers apart.

microwave detector [ELECTR] A device that can demonstrate the presence of a microwave by a specific effect that the wave produces, such as a bolometer, or a semiconductor crystal making a pinpoint contact with a tungsten wire.

microwave device [ELECTR] Any device capable of generating, amplifying, modifying, detecting, or measuring microwaves, or voltages having microwave frequencies.

microwave early warning [ENG] High-power, long-range radar with a number of indicators, giving high resolution, and with a large traffic-handling capacity; used for early warning of missiles.

microwave filter [ELECTROMAG] A device which passes microwaves of certain frequencies in a transmission line or waveguide while rejecting or absorbing other frequencies; consists of resonant cavity sections or other elements.

microwave frequency [PHYS] A frequency on the order of 10^9–10^{11} hertz.

microwave generator *See* microwave oscillator.

microwave gyrator *See* gyrator.

microwave link *See* microwave repeater.

microwave maser [PHYS] A maser which emits microwave radiation.

microwave network [COMMUN] A series of microwave repeaters, spaced up to 80 kilometers apart, which relay messages over long distances using highly directional microwave beams.

microwave optics [ELECTROMAG] The study of those properties of microwaves which are analogous to the properties of light waves in optics.

microwave oscillator [ELECTR] A type of electron tube or semiconductor device used for generating microwave radiation or voltage waveforms with microwave frequencies. Also known as microwave generator.

microwave radiometer *See* radiometer.

microwave receiver [ELECTR] Complete equipment that is needed to convert modulated microwaves into useful information.

microwave reflectometer [ELECTROMAG] A pair of single-detector couplers on opposite sides of a waveguide, one of which is positioned to monitor transmitted power, and the other to measure power reflected from a single discontinuity in the line.

microwave refractometer [ELECTROMAG] An instrument that measures the index of refraction of the atmosphere by measuring the travel time of microwave signals through each of two precision microwave transmission cavities, one of which is hermetically sealed to serve as a reference.

microwave relay *See* microwave repeater.

microwave repeater [COMMUN] A tower equipped with a receiver and transmitter for picking up, amplifying, and passing on in either direction the signals sent over a microwave network by highly directional microwave beams. Also known as microwave link; microwave relay.

microwave resonance cavity *See* cavity resonator.

microwave solid-state device [ELECTR] A semiconductor device for the generation or amplification of electromagnetic energy at microwave frequencies.

microwave spectrometer [SPECT] An instrument which makes a graphical record of the intensity of microwave radiation emitted or absorbed by a substance as a function of frequency, wavelength, or some related variable .

microwave spectroscope [SPECT] An instrument used to observe the intensity of microwave radiation emitted or absorbed by a substance as a function of frequency, wavelength, or some related variable.

microwave spectroscopy [SPECT] The methods and techniques of observing and the theory for interpreting the selective absorption and emission of microwaves at various frequencies by solids, liquids, and gases.

microwave spectrum [ELECTROMAG] The range of wavelengths or frequencies of electromagnetic radiation that are designated microwaves. [SPECT] A display, photograph, or plot of the intensity of microwave radiation emitted or absorbed by a substance as a function of frequency, wavelength, or some related variable.

microwave stripline *See* strip transmission line.

microwave transmission line [ELECTROMAG] A material structure forming a continuous path from one place to another and capable of directing the transmission of electromagnetic energy along this path.

microwave tube [ELECTR] A high-vacuum tube designed for operation in the frequency region from approximately 3000 to 300,000 megahertz.

microwave waveguide *See* waveguide.

microwave wavemeter [ELECTROMAG] Any device for measuring the free-space wavelengths (or frequencies) of microwaves; usually made of a cavity resonator whose dimensions can be varied until resonance with the microwaves is achieved.

microzoospermia [MED] A condition of abnormal smallness of sperm in the semen.

micrurgy [SCI TECH] The art and science of using minute tools in a magnified field.

mictic [BIOL] **1.** Requiring or produced by sexual reproduction. **2.** Of or pertaining to eggs which without fertilization develop into males and with fertilization develop into amictic females, as occurs in rotifers.

micturition *See* urination.

Midas [AERO ENG] A two-object trajectory-measuring system whereby two complete cotar antenna systems and two sets of receivers at each station, with the multiplexing done after phase comparison, are utilized in tracking more than one object at a time.

mid-Atlantic ridge [GEOL] The mid-oceanic ridge in the Atlantic.

midaxillary line [ANAT] A perpendicular line drawn downward from the apex of the axilla.

midbrain [ANAT] Those portions of the adult brain derived from the embryonic midbrain. [EMBRYO] The middle portion of the embryonic vertebrate brain. Also known as mesencephalon.

midclavicular line [ANAT] A vertical line parallel to and midway between the midsternal line and a vertical line drawn downward through the outer end of the clavicle.

midcourse correction [AERO ENG] A change in the course of a spacecraft some time between the end of the launching phase and some arbitrary point when terminal guidance begins.

middle body [NAV ARCH] That portion of the ship adjacent to the midship section.

Middle Cambrian [GEOL] The geologic epoch between Upper and Lower Cambrian, beginning approximately 540,000,000 years ago.

middle clouds [METEOROL] Types of clouds the mean level of which is between 6500 and 20,000 feet (1980 and 6100 meters);

the principal clouds in this group are altocumulus and altostratus.

Middle Cretaceous [GEOL] The geologic epoch between the Upper and Lower Cretaceous, beginning approximately 120,000,000 years ago.

Middle Devonian [GEOL] The geologic epoch between the Upper and Lower Devonian, beginning approximately 385,000,000 years ago.

middle ear [ANAT] The middle portion of the ear in higher vertebrates; in mammals it contains three ossicles and is separated from the external ear by the tympanic membrane and from the inner ear by the oval and round windows.

middle ground [NAV] A shoal with a channel on either side of it in a navigable part of a river, port, or harbor.

middle ground buoy [NAV] In maritime operations, one of the buoys placed at each end of a middle ground, that is, a shoal with channels on both sides.

Middle Jurassic [GEOL] The geologic epoch between the Upper and Lower Jurassic, beginning approximately 170,000,000 years ago.

middle latitude Also known as mid-latitude. [GEOGR] A point of latitude that is midway on a north-and-south line between two parallels. [NAV] The latitude at which the arc length of the parallel separating the meridians passing through two specific points is exactly equal to the departure in proceeding from one point to the other by middle-latitude sailing.

middle-latitude sailing [NAV] A method of converting departure into difference of longitude, or vice versa, when the true course is not 090° or 270° by assuming that such a course is steered at the middle latitude.

middle-latitude westerlies *See* westerlies.

middle lobe syndrome [MED] A complex of symptoms due to enlarged lymph nodes compressing the bronchus of the right middle lobe, causing atelectasis, bronchiectasis, or chronic pneumonitis of the lobe.

middle marker [NAV] That instrument-landing-system marker between the outer and inner markers.

Middle Mississippian [GEOL] The geologic epoch between the Upper and Lower Mississippian.

Middle Ordovician [GEOL] The geologic epoch between the Upper and Lower Ordovician, beginning approximately 460,000,000 years ago.

Middle Pennsylvania [GEOL] The geologic epoch between the Upper and Lower Pennsylvanian.

Middle Permian [GEOL] The geologic epoch between the Upper and Lower Permian, beginning approximately 260,000,000 years ago.

Middle Silurian [GEOL] The geologic epoch between the Upper and Lower Silurian.

middle-third rule [CIV ENG] The rule that no tension is developed in a wall or foundation if the resultant force lies within the middle third of the structure.

Middle Triassic [GEOL] The geologic epoch between the Upper and Lower Triassic, beginning approximately 215,000,000 years ago.

middle-ultraviolet lamp [ELECTR] A mercury-vapor lamp designed to produce radiation in the wavelength band from 2800 to 3200 angstrom units such as sunlamps and photochemical lamps.

middling [MIN ENG] An ore product intermediate in mineral content between a concentrate and a tailing.

middlings *See* sharps.

mid-extreme tide [OCEANOGR] A level midway between the extreme high water and extreme low water occurring at a place.

midfan [GEOL] The portion of an alluvial fan between the fanhead and the outer, lower margins.

mid-frequency gain [ELECTR] The maximum gain of an amplifier, when this gain depends on the frequency; for an *RC*-coupled voltage amplifier the gain is essentially equal to this value over a large range of frequencies.

midge [INV ZOO] Any of various dipteran insects, principally of the families Ceratopogonidae, Cecidomyiidae, and Chironomidae; many are biting forms and are vectors of parasites of man and other vertebrates.

midget [MED] An individual who is abnormally small, but otherwise normal.

midget impinger [MIN ENG] A dust-sampling impinger requiring only a 12-inch (30-centimeter) head of water for its operation.

midgut [EMBRYO] The middle portion of the digestive tube in vertebrate embryos. Also known as mesenteron. [INV ZOO] The mesodermal intermediate part of an invertebrate intestine.

mid-latitude *See* middle latitude.

mid-latitude westerlies *See* westerlies.

midnight sun [ASTRON] The sun when it is visible at midnight; occurs during the summer in high latitudes, poleward of the circle at which the latitude is approximately equal to the polar distance of the sun.

mid-ocean canyon *See* deep-sea channel.

mid-oceanic ridge [GEOL] A continuous, median, seismic mountain range on the floor of the ocean, extending through the North and South Atlantic oceans, the Indian Ocean, and the South Pacific Ocean; the topography is rugged, elevation is 1–3 kilometers (km), width is about 1500 km, and length is over 84,000 km. Also known as mid-ocean ridge; mid-ocean rise; oceanic ridge.

mid-ocean ridge *See* mid-oceanic ridge.

mid-ocean rift *See* rift valley.

mid-ocean rise *See* mid-oceanic ridge.

midpoint [MATH] The midpoint of a line segment is the point which separates the segment into two equal parts.

mid-range materiel requirements [ORD] Items required as soon as possible by current operational and organizational concepts, or required by new concepts to be implemented within the next 5 years.

midrib [BOT] The large central vein of a leaf.

midships [NAV ARCH] **1.** Halfway between a boat's or ship's stem and stern, at the design waterline. **2.** Halfway between the side of a boat's or ship's hull.

midship section [NAV ARCH] A drawing of the cross section of a ship halfway between the intersections of its stem and stern with the design waterline.

midship section coefficient [NAV ARCH] The ratio of the area of the midship section below the water plane to the product of its beam and its draft.

Mie-Grüneisen equation [THERMO] An equation of state particularly useful at high pressure, which states that the volume of a system times the difference between the pressure and the pressure at absolute zero equals the product of a number which depends only on the volume times the difference between the internal energy and the internal energy at absolute zero.

Mie scattering [OPTICS] The scattering of light by a sphere of dielectric material.

migma [GEOL] A mixture of solid rock materials and rock melt with mobility or potential mobility.

migmatite [PETR] A mixed rock exhibiting crystalline textures in which a truly metamorphic component is streaked and mixed with obviously once-molten material of a more or less granitic character.

migmatization [PETR] Formation of migmatite; involves either injection or in-place melting.

migraine [MED] Recurrent paroxysmal vascular headache, commonly having unilateral onset and often associated with nausea and vomiting.

migrant [ZOO] An animal that moves from one habitat to another.

migration [GEOL] **1.** Movement of a topographic feature from one place to another, especially movement of a dune by wind action. **2.** Movement of liquid or gaseous hydrocarbons from their source into reservoir rocks. [HYD] Slow, downstream movement of a system of meanders. [VERT ZOO] Periodic movement of animals to new areas or habitats.

migration area [NUCLEO] One-sixth the mean square distance that a neutron travels in a medium from its birth in fission until its absorption.

migration current [PHYS CHEM] Additional current produced by electrostatic attraction of cations to the surface of a dropping electrode; an unpredictable and undesirable effect to be avoided during analytical voltammetry.

migration length [NUCLEO] The square root of the migration area.

migratory [METEOROL] Commonly applied to pressure systems embedded in the westerlies and, therefore, moving in a general west-to-east direction.

migratory dune *See* wandering dune.

MIKES *See* mass-analyzed ion kinetic energy spectrometry.

Mikulicz's disease [MED] Enlargement of salivary and lacrimal glands from any of various causes.

mil [MATH] A unit of angular measure which, due to nonuniformity of usage, may have any one of three values: 0.001 radian or approximately 0.0572958°; 1/6400 of a full revolution or 0.5625°; 1/1000 of a right angle or 0.09°. [MECH] **1.** A unit of length, equal to 0.001 inch, or to 2.54×10^{-5} meter. Also known as milli-inch; thou. **2.** *See* milliliter.

milarite [MINERAL] $K_2Ca_4Be_4Al_2Si_{24}O_{62} \cdot H_2O$ A colorless to greenish, glassy, hexagonal mineral composed of a hydrous silicate of potassium, calcium, beryllium, and aluminum, occurring in crystals.

mild abrasive [MATER] An abrasive material, such as chalk or talc, having a hardness of 1–2 on Mohs scale; used in silver polishes and window cleaners.

mildew [MYCOL] **1.** A whitish growth on plants, organic matter, and other materials caused by a parasitic fungus. **2.** Any fungus producing such growth.

mild mental retardation [PSYCH] Subnormal general intellectual functioning in which the intelligence quotient is approximately 52–67.

mild mercury chloride *See* mercurous chloride.

mild steel [MET] Carbon steel containing 0.05–0.25% carbon.

mile [MECH] A unit of length in common use in the United States, equal to 5280 feet, or 1609.344 meters. Abbreviated mi. Also known as land mile; statute mile.

mileage chart [MAP] A chart showing distances between various points.

mileage number [NAV] For the Mississippi River system, a number assigned to aids to navigation which gives the distance in sailing miles along the river from a reference point to the aid; the number serves as a light list number for identification purposes.

milepost [CIV ENG] **1.** A post placed a mile away from a similar post. **2.** A post indicating mileage from a given point.

miles of relative movement [NAV] The distance, in miles, traveled relative to a reference point which is usually in motion.

miles on course [NAV] The actual or predicted distance, in miles, traveled on any given course.

mil formula [ORD] Mil relation used in gunnery; expressed by $n = W/R$, where n is the angular measurement in mils between two points, W is the lateral distance in yards between the points, and R is the mean distance to the points in thousands of yards; the mil relation is approximately true for angles less than 400 mils.

miliaria [MED] An acute inflammatory skin disease, the lesions consisting of vesicles and papules, which may be accompanied by a prickling or tingling sensation. Also known as heat rash; prickly heat.

Milichiidae [INV ZOO] A family of myodarian cyclorrhaphous dipteran insects in the subsection Acalypteratae.

milieu interieur [PHYSIO] The fundamental concept that the living organism exists in an aqueous internal environment which bathes all tissues and provides a medium for the elementary exchange of nutrients and waste.

milieu therapy [PSYCH] The treatment of mental disorder or maladjustment by making substantial changes in a patient's immediate life circumstances and environment in a way that will enhance the effectiveness of other forms of therapy. Also known as situation therapy.

Miliolacea [INV ZOO] A superfamily of marine or brackish foraminiferans in the suborder Miliolina characterized by an imperforate test wall of tiny, disordered calcite rhombs.

Miliolidae [INV ZOO] A family of foraminiferans in the superfamily Miliolacea.

Miliolina [INV ZOO] A suborder of the Foraminiferida characterized by a porcelaneous, imperforate calcite wall.

military aircraft [AERO ENG] Aircraft that are designed or modified for highly specialized use by the armed services of a nation.

military characteristics [ORD] Those characteristics of

equipment found desirable or necessary to the performance of a military mission, either combat or noncombat.

military department [ORD] In the United States, the Department of the Army, the Department of the Navy, or the Department of the Air Force.

military engineering [ENG] Science, art, and practice involved in design and construction of defensive and offensive military works as well as construction and maintenance of transportation systems.

military geology [ENG] The application of the earth sciences to such military concerns as terrain analysis, water supply, foundations, and construction of roads and airfields.

military grid [MAP] Two sets of parallel lines intersecting at right angles and forming squares; superimposed on maps or charts in an accurate and consistent manner to permit identification of ground locations with respect to other locations and the computation of direction and distance to other points.

military grid reference system [MAP] A system which uses a standard-scaled grid square, based on a point of origin on a map projection of the earth's surface in an accurate and consistent manner to permit either position referencing, or the computation of direction and distance between grid positions.

military map [MAP] A map containing a military grid.

military motor vehicle [ORD] Wheeled or tracklaying vehicle or combined wheeled and tracklaying vehicle, designed for transporting personnel or cargo for military purposes; it may be either propelled by a self-contained power unit or drawn by a vehicle containing a self-contained power unit.

military requirement [ORD] The statement of a recognized and approved need for a new item, a weapons system, or an assemblage for military use.

military research [ORD] Investigation or experimentation to discover and interpret new facts or to analyze existing facts with the aim of applying new or revised conclusions, theories, or laws to areas of interest to the military services.

military satellite [AERO ENG] An artificial earth satellite used for military purposes; the six mission categories are communication, navigation, geodesy, nuclear test detection, surveillance, and research and technology.

military specification [ORD] A procurement specification promulgated by the military agencies and used for the procurement of military supplies and equipment.

military standard test [ORD] Specifications approved by the United States Department of Defense, to ensure uniformity of test conditions.

military target [ORD] **1.** Any industrial plant, city, or other object, or any person, group of persons, or force marked as a target for destruction, damage, injury, or capture because of its direct or indirect use in the conduct or support of an enemy's military endeavor. **2.** In restricted usage, a military person, force, installation, or area marked as a target because of its use, or potential use, in direct military operations.

military technology [ENG] The technology needed to develop and support the armament used by the military.

military transport vehicle [ORD] Wheeled or tracked vehicle specifically designed for military purposes other than employment as a fighting vehicle, intended for transporting cargo, personnel, and equipment or for towing other vehicles over land and roads in close support of fighting vehicles and troops.

military-type vehicle [ORD] Vehicle designed primarily for military purposes to meet field requirements in connection with combat and tactical operations.

military vessel [NAV ARCH] A ship or boat designed primarily for use in war.

milk [CHEM] A suspension of certain metallic oxides, as milk of magnesia, iron, or bismuth. [PHYSIO] **1.** The whitish fluid secreted by the mammary gland for the nourishment of the young; composed of carbohydrates, proteins, fats, mineral salts, vitamins, and antibodies. **2.** Any whitish fluid in nature resembling milk, as coconut milk.

milk-alkali syndrome [MED] A complex of symptoms associated with prolonged excessive intake of milk and soluble alkali, including hypercalcemia, renal insufficiency, milk

alkalosis, conjunctivitis, and calcinosis. Also known as Burnett's syndrome; milk-drinker's syndrome.

milk-drinker's syndrome *See* milk-alkali syndrome.

milk factor [BIOCHEM] A filtrable, noncellular agent in the milk and tissues of certain strains of inbred mice; transmitted from the mother to the offspring by nursing. Also known as Bittner milk factor.

milk fever [MED] A fever occurring during the first six weeks after childbirth, believed to be caused by puerperal infection.

milk glass [MATER] A white, and sometimes colored, opaque glass.

milk intolerance [MED] Extreme sensitivity to milk due to allergy to milk protein or lactose deficiency; characterized by diarrhea, abdominal cramps, and vomiting.

milk leg *See* phlegmasia alba dolens.

milk line *See* mammary ridge.

Milkman's syndrome [MED] Decreased tubular reabsorption of phosphate, resulting in osteomalacia which gives a peculiar striped appearance (multiple pseudofractures) to the bones in roentgenograms. Also known as Looser-Milkman syndrome.

milk of magnesia [PHARM] A white suspension of magnesium hydroxide in water; used as a cathartic. Also known as magnesia magma.

milk sugar *See* lactose.

milk teeth *See* deciduous teeth.

milkweed [BOT] Any of several latex-secreting plants of the genus *Asclepias* in the family Asclepiadaceae.

milky disease [INV ZOO] A bacterial disease of Japanese beetle larvae or related grubs caused by *Bacillus papilliae* and *B. lentimorbus* that penetrate the intestine and sporulate in the body cavity; blood of the grub eventually turns milky white.

milky quartz [MINERAL] An opaque, milk-white variety of crystalline quartz, often with a greasy luster; milkiness is due to the presence of air-filled cavities. Also known as greasy quartz.

Milky Way [ASTRON] The faint band of light which encircles the sky and results from the combined light of the many stars near the plane of our galaxy.

Milky Way Galaxy [ASTRON] The large aggregation of stars and interstellar gas and dust of which the sun is a member.

milky weather *See* whiteout.

mill [FOOD ENG] **1.** A machine for grinding grain into flour. **2.** A building that houses milling machines. [IND ENG] **1.** A machine that manufactures paper, textiles, or other products by the continuous repetition of some simple process or action. **2.** A building that houses machinery for manufacturing processes. [MIN ENG] **1.** An excavation made in the country rock, by a crosscut from the workings on a vein, to obtain waste for filling; it is left without timber so that the roof can fall in and furnish the required rock. **2.** A passage connecting a stope or upper level with a lower level intended to be filled with broken ore that can then be drawn out at the bottom as desired for further transportation.

millboard [MATER] Hard, strong paperboard; used for furniture panels.

mill building [CIV ENG] A steel-frame building in which roof trusses span columns in the outside wall; originally, this type of building housed milling machinery, as for wood or metal, hence the name.

milled soap [MATER] Soap, such as toilet soap, made by adding color and perfume to soap chips and then passing the mixture through milling rollers and pressing in molds.

Milleporina [INV ZOO] An order of the class Hydrozoa known as the stinging corals; they resemble true corals because of a calcareous exoskeleton.

miller *See* milling machine.

Miller-Abbott tube [MED] A double-lumen rubber tube having a balloon at the end, inserted through the nasal passage and passed through the pylorus to locate and treat intestinal obstructions.

Miller bridge [ELECTR] Type of bridge circuit for measuring amplification factors of vacuum tubes.

Miller effect [ELECTR] The increase in the effective grid-cathode capacitance of a vacuum tube due to the charge induced electrostatically on the grid by the anode through the grid-anode capacitance.

Miller generator *See* bootstrap integrator.

Miller indices [CRYSTAL] Three integers identifying a type of crystal plane; the intercepts of a plane on the three crystallographic axes are expressed as fractions of the crystal parameters; the reciprocals of these fractions, reduced to integral proportions, are the Miller indices. Also known as crystal indices.

Miller integrator [ELECTR] A resistor-capacitor charging network having a high-gain amplifier paralleling the capacitor; used to produce a linear time-base voltage. Also known as Miller time-base.

millerite [MINERAL] NiS A brass to bronze-yellow mineral that crystallizes in the hexagonal system and usually contains trace amounts of cobalt, copper, and iron; hardness is 3–3.5 on Mohs scale, and specific gravity is 5.5; it generally occurs in fine crystals, chiefly as nodules in clay ironstone. Also known as capillary pyrites; hair pyrites; nickel pyrites.

Miller law [AGR] A law administered by the Food and Drug Administration that regulates the production and use of agricultural fungicides in the United States, and will not allow materials to leave poisonous residues on edible crops. [CRYSTAL] If the edges formed by the intersections of three faces of a crystal are taken as the three reference axes, then the three quantities formed by dividing the intercept of a fourth face with one of these axes by the intercept of a fifth face with the same axis are proportional to small whole numbers, rarely exceeding 6. Also known as law of rational intercepts.

Miller time-base *See* Miller integrator.

millet [BOT] A common name applied to at least five related members of the grass family grown for their edible seeds.

mill finish [MET] The characteristic surface finish on a rolled metal product. [TEXT] The slightly napped finish of a worsted fabric.

mill-head ore *See* run-of-mill.

milli- [MATH] A prefix representing 10^{-3}, or one-thousandth. Abbreviated m.

milliammeter [ELEC] An ammeter whose scale is calibrated to indicate current values in milliamperes.

milliampere [ELEC] A unit of current equal to one-thousandth of an ampere. Abbreviated mA.

milliampere-second [NUCLEO] A unit of radiation dose resulting from exposure to x-rays, equal to the dose produced by an electron beam, carrying a current of 1 milliampere, bombarding the target of an x-ray tube for 1 second. Abbreviated mA s.

millibar [MECH] A unit of pressure equal to one-thousandth of a bar. Abbreviated mb. Also known as vac.

millibarn [NUC PHYS] A unit of cross section equal to one-thousandth of a barn. Abbreviated mb.

millicurie-destroyed [NUCLEO] A unit of radiation dose, equal to the radiation emitted by a sample over a period during which its activity decreases by 1 millicurie; for radon-222 (for which this unit is most often used), it is approximately 133 milligram-hours. Abbreviated mcd.

millicurie-of-intensity-hour *See* sievert.

millicycle *See* millihertz.

millidarcy [PHYS] A unit of fluid permeability equal to one-thousandth of a darcy. Abbreviated md.

milliequivalent [CHEM] One-thousandth of a compound's or an element's equivalent weight.

millier *See* tonne.

millifarad [ELEC] A unit of capacitance equal to one-thousandth of a farad. Abbreviated mF.

milligal [MECH] A unit of acceleration commonly used in geodetic measurements, equal to 10^{-3} galileo, or 10^{-5} meter per second per second. Abbreviated mGal.

milligauss [ELECTROMAG] A unit of magnetic flux density equal to one-thousandth of a gauss. Abbreviated mG.

milligram [MECH] A unit of mass equal to one-thousandth of a gram. Abbreviated mg.

milligram-hour [NUCLEO] A unit of radiation dose, equal to the radiation emitted by a source with an equivalent radium content of 1 milligram for a period of 1 hour. Abbreviated mg h.

millihenry [ELECTROMAG] A unit of inductance equal to one-thousandth of a henry. Abbreviated mH.

millihertz [PHYS] A unit of frequency equal to one-thou-

A part of the Milky Way Galaxy. Dark line is caused by obscuring matter. *(Hale Observatories)*

MILLEPORINA

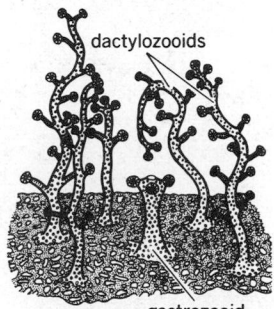

Polyps of Milleporina. *(From L. H. Hyman, The Invertebrates, vol. 1, McGraw-Hill, 1940)*

MILLER INDICES

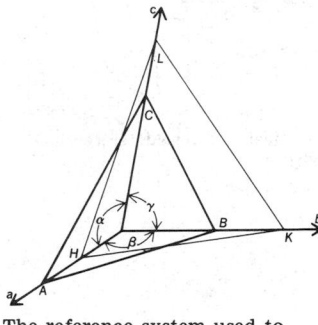

The reference system used to define Miller indices. *a*, *b* and *c* are chosen as crystallographic axes, and *ABC* as unit plane, defined by angles α, β and γ, and ratio *OA:OB:OC*. Miller indices of another plane, *HKL*, are integers proportional to *OA/OH*, *OB/OK* and *OC/OL*.

MILLERITE

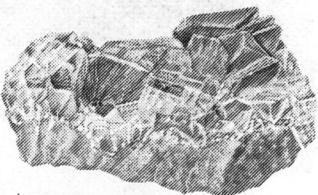

├─2.5 cm─┤

Millerite crystals on calcite from Keokuk, Iowa. *(Specimen from Department of Geology, Bryn Mawr College)*

sandth of a hertz. Abbreviated mHz. Also known as millicycle (mc).

millihg *See* millimeter of mercury.

milli-inch *See* mil.

milli-k [NUCLEO] A unit of reactivity; the reactivity of a reactor in milli-k is equal to 1000 ($k -$ 1), where k is the effective multiplication factor.

Millikan meter [ELECTR] An integrating ionization chamber in which a gold-leaf electroscope is charged a known amount and ionizing events reduce this charge, so that the resulting angle through which the gold leaf is repelled at any given time indicates the number of ionizing events that have occurred.

Millikan oil-drop experiment [ATOM PHYS] A method of determining the charge on an electron, in which one measures the terminal velocities of rise and fall of oil droplets in an electric field after the droplets have picked up charge from ionization in the surrounding gas produced by an x-ray beam.

milliliter [MECH] A unit of volume equal to 10^{-3} liter or 10^{-6} cubic meter. Abbreviated ml. Also known as mil.

milli-mass-unit [PHYS] 1/1000 of an atomic mass unit. Abbreviated mmu.

millimeter [MECH] A unit of length equal to one-thousandth of a meter. Abbreviated mm. Also known as metric line; strich.

millimeter of mercury [MECH] A unit of pressure, equal to the pressure exerted by a column of mercury 1 millimeter high with a density of 13.5951 grams per cubic centimeter under the standard acceleration of gravity; equal to 133.322387415 pascals; it differs from the torr by less than 1 part in 7,000,000. Abbreviated mmHg. Also known as millihg.

millimeter of water [MECH] A unit of pressure, equal to the pressure exerted by a column of water 1 millimeter high with a density of 1 gram per cubic centimeter under the standard acceleration of gravity; equal to 9.80665 pascals. Abbreviated mmH$_2$O.

millimeter wave [ELECTROMAG] An electromagnetic wave having a wavelength between 1 millimeter and 1 centimeter, corresponding to frequencies between 30 and 300 gigahertz. Also known as millimetric wave.

millimetric wave *See* millimeter wave.

milli-micro- *See* nano-.

millimicron [MECH] A unit of length equal to one-thousandth of a micron (or micrometer) or one-billionth of a meter. Abbreviated mμ.

milling [MECH ENG] Mechanical treatment of materials to produce a powder, to change the size or shape of metal powder particles, or to coat one powder mixture with another. [MIN ENG] A combination of open-cut and underground mining, wherein the ore is mined in open cut and handled underground.

milling cutter [DES ENG] A rotary tool-steel cutting tool with peripheral teeth, used in a milling machine to remove material from the workpiece through the relative motion of workpiece and cutter.

milling machine [MECH ENG] A machine for the removal of metal by feeding a workpiece through the periphery of a rotating circular cutter. Also known as miller.

milling planer [MECH ENG] A planer that uses a rotary cutter rather than single-point tools.

milling system *See* chute system.

Millington reverberation formula [ACOUS] A formula that states that the reverberation time of a chamber in seconds is 0.05 times its volume in cubic feet, divided by the sum over the surfaces of the chamber of the product of the surface's area in square feet by the natural logarithm of 1 minus its absorption coefficient.

milling width [MIN ENG] Width of lode designated for treatment in the mill, as calculated with regard to daily tonnage.

million electron volts [PHYS] A unit of energy commonly used in nuclear and particle physics, equal to the energy acquired by an electron in falling through a potential of 10^6 volts. Abbreviated MeV. Also known as mega-electron-volt.

millipede [INV ZOO] The common name for members of the arthropod class Diplopoda.

millirad [NUCLEO] A unit of absorbed ionizing radiation dose equal to one-thousandth of a rad. Abbreviated mrad.

milliroentgen [NUCLEO] A unit of radioactive dose of electromagnetic radiation equal to one-thousandth of a roentgen. Abbreviated mr.

millisecond [MECH] A unit of time equal to one-thousandth of a second. Abbreviated msec.

millisecond delay cap [ENG] A delay cap with an extremely short (20–500 thousandths of a second) interval between passing of current and explosion. Also known as short-delay detonator.

millisite [MINERAL] $(Na,K)CaAl_6(PO_4)_4(OH)_9 \cdot 3H_2O$ White mineral composed of a basic hydrous phosphate of sodium, potassium, calcium, and aluminum.

Millium [TEXT] Trade name of metal-insulated fabrics manufactured by Deering Milliken, Inc.

millivolt [ELEC] A unit of potential difference or emf equal to one-thousandth of a volt. Abbreviated mV.

millivoltmeter [ELEC] A voltmeter whose scale is calibrated to indicate voltage values in millivolts.

milliwatt [MECH] A unit of power equal to one-millionth of a watt. Abbreviated mW.

Millon's reagent [CHEM] Reagent used to test for proteins; made by dissolving mercury in nitric acid, diluting, then decanting the liquid from the precipitate.

mill ore [MIN ENG] An ore that must be given some preliminary treatment before a marketable grade or a grade suitable for further treatment can be obtained.

mill run [MIN ENG] **1.** A given quantity of ore tested for its quality by actual milling. **2.** The yield of such a test.

mill scale [MET] A surface layer of ferric oxide (Fe_3O_4) that forms on steel or iron during hot rolling.

Mills-Crowe process [MIN ENG] Method of regeneration of cyanide liquor from the gold leaching process; the barren solution is acidified, liberating hydrocyanic acid which is separated and reabsorbed in alkaline solution.

millsite [MIN ENG] A plot of ground suitable for the erection of a mill, or reduction works, to be used in connection with mining operations.

millstone *See* buhrstone.

millwright [ENG] **1.** A person who plans, builds, or sets up the machinery for a mill. **2.** A person who repairs milling machines.

Milne method [MATH] A technique which provides numerical solutions to ordinary differential equations.

Milorganite [MATER] Granular form of sewage-plant activated sludge; used as fertilizer.

Milroy's disease [MED] Familial chronic lymphedema of the lower extremities.

Mima mound [GEOGR] A term used in the northwestern United States for any of the numerous low, circular or oval domes composed of loose silt and soil material; probably built by pocket gophers.

Mimas [ASTRON] A satellite of Saturn orbiting at a mean distance of 186,000 kilometers.

mimetene *See* mimetite.

mimetesite *See* mimetite.

mimetic [CRYSTAL] Pertaining to a crystal that is twinned or malformed but whose crystal symmetry appears to be of a higher grade than it actually is. [PETR] Of a tectonite, having a deformation fabric, formed by mimetic crystallization, that reflects and is influenced by preexisting anisotropic structure. [ZOO] Pertaining to or exhibiting mimicry.

mimetic crystallization [PETR] Recrystallization or neomineralization in metamorphism which reproduces preexistent structures.

mimetite [MINERAL] $Pb_5(AsO_4)_3Cl$ A yellow to yellowish-brown mineral of the apatite group, commonly containing calcium or phosphate; a minor ore of lead. Also known as mimetene; mimetesite.

mimicry [ZOO] Assumption of color, form, or behavior patterns by one species of another species, for camouflage and protection.

Mimidae [VERT ZOO] The mockingbirds, a family of the Oscines in the order Passeriformes.

Mimosoideae [BOT] A subfamily of the legume family, Leguminosae; members are largely woody and tropical or subtropical with regular flowers and usually numerous stamens.

min *See* minim.

MILLING CUTTER

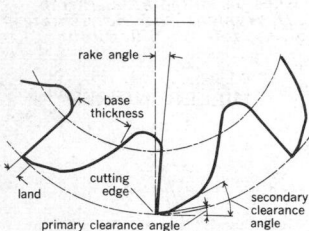

Typical milling cutter teeth.

MILLING MACHINE

Part of a horizontal milling machine. (*Brown and Sharpe Manufacturing Co.*)

minable [MIN ENG] Material that can be mined under present-day mining technology and economics.

mind [PSYCH] **1.** The sum total of the neural processes which receive, code, and interpret sensations, recall and correlate stored information, and act on it. **2.** The state of consciousness. **3.** The understanding, reasoning, and intellectual faculties and processes considered as a whole. **4.** The psyche, or the conscious, subconscious, and unconscious considered together.

Mindel glaciation [GEOL] The second glacial stage of the Pleistocene in the Alps.

Mindel-Riss interglacial [GEOL] The second interglacial stage of the Pleistocene in the Alps; follows the Mindel glaciation.

M indicator See M scope.

mine [MIN ENG] An opening or excavation in the earth for extracting minerals. [ORD] An encased explosive or chemical charge designed to be positioned so that it detonates when the target touches or moves near it or when it is fired by remote control; general types are land mines and underwater mines.

mine captain [MIN ENG] **1.** The director of work in a mine, with or without superior officials or subordinates. **2.** See mine superintendent.

mine car [MECH ENG] An industrial car, usually of the four-wheel type, with a low body; the door is at one end, pivoted at the top with a latch at the bottom used for hauling bulk materials.

mine characteristic [MIN ENG] The relation between pressure p and volume Q in the ventilation of a mine of resistance R, expressed as $p = RQ^2$.

mine characteristic curve [MIN ENG] A graph derived by plotting the static or total mine head, or both, against the quantity; used to solve problems in mine ventilation.

mine countermeasures [ORD] All methods for preventing or reducing damage or danger to ships, personnel, aircraft, and vehicles from mines.

mine development [MIN ENG] The operations involved in preparing a mine for ore extraction, including tunneling, sinking, crosscutting, drifting, and raising.

mine dust [MIN ENG] Dust from drilling, blasting, or handling rock.

mined volume [MIN ENG] A statistic used in mine subsidence, computed by multiplying the mined area by the mean thickness of the bed or of that part of the bed which has been dug out.

mine examiner See fire boss.

mine fan signal system [MIN ENG] A system which indicates by electric light or electric audible signal, or both, the slowing down or stopping of a mine ventilating fan.

mine field [ORD] An area in which mines have been placed.

mine field gap [ORD] A portion of a mine field in which no mines have been laid, of specified width to enable a friendly force to pass through the mine field in tactical formation; seldom less than 100 yards (91 meters) wide.

mine field lane [ORD] An unmined or demined route through any mine field, normally 8 yards (7.3 meters) wide and suitably marked; however, for lanes through enemy mine fields, the width depends on the method of breaching and the intended purpose.

mine fire truck [MIN ENG] A low-slung railcar designed to fight fires in mines; has a water supply and a pump to supply high pressure to the fire hoses.

mine foreman [MIN ENG] The worker charged with the general supervision of the underground workings of a mine and the persons employed therein.

Mine Gel [MATER] Brand name for a series of semigelatin dynamites.

mine hoist [MIN ENG] A device for raising and lowering ore, rock, or coal in a mine, and men and supplies.

mine inspector [MIN ENG] Generally, the state mine inspector, as contrasted to the Federal mine inspector; inspects mines to find fire and dust hazards and inspects the safety of working areas, electric circuits, and mine equipment.

mine jeep [MIN ENG] An electrically driven car for underground transportation of officials, inspectors, rescue workers, and repair, maintenance, and surveying crews.

minelite [MATER] An explosive made by mixing a chlorate compound with paraffin wax.

mine locomotive [MIN ENG] A low, heavy haulage engine designed for underground operation; usually propelled by electricity, gasoline, or compressed air.

mine-planting equipment [ORD] A group of items for uncrating, fusing, and magazine loading of antitank mines, and the mechanical emplacement of barrier-type mine fields.

mine props [MIN ENG] Sections of wood used for holding up pieces of rock in the roof of mines.

mine radio telephone system [MIN ENG] A communication system between the dispatcher and locomotive operators; radio impulses travel along the trolley wire and down the trolley pole to the radio telephone.

mineragraphy See ore microscopy.

mineral [GEOL] A naturally occurring substance with a characteristic chemical composition expressed by a chemical formula; may occur as individual crystals or may be disseminated in some other mineral or rock; most mineralogists include the requirements of inorganic origin and internal crystalline structure.

mineral black [MATER] Black pigment made from ground slate, shale, coke, coal, or slaty coal; used in inks, plastics, and coatings.

mineral charcoal See fusain.

mineral deposit [GEOL] A mass of naturally occurring mineral material, usually of economic value.

mineral dressing See beneficiation.

mineral dye [MATER] A natural dyestuff made from minerals; examples are ochre, chrome yellow, and Prussian blue.

mineral economics [MIN ENG] Study and application of the processes used in management and finance connected with the discovery, exploitation, and marketing of minerals.

mineral engineering See mining engineering.

mineral facies [PETR] Rocks of any origin whose components have been formed within certain temperature-pressure limits characterized by the stability of certain index minerals.

mineral fuel [MATER] A carbonaceous fuel mined or stripped from the earth, such as petroleum, coal, peat, shale oil, or tar sands.

mineralization [GEOL] **1.** The process of fossilization whereby inorganic materials replace the organic constituents of an organism. **2.** The introduction of minerals into a rock, resulting in a mineral deposit.

mineralize [GEOL] To convert to, or impregnate with mineral material; applied to processes of ore vein deposition and of fossilization.

mineralizer [GEOL] A gas or fluid dissolved in a magma that aids in the concentration and crystallization of ore minerals.

mineral jelly [MATER] A viscous, thick material derived from petroleum; used as a stabilizer in explosives.

mineral land [MIN ENG] Land which is worth more for mining than for agriculture or other use.

mineral lard oil [MATER] A mixture of refined mineral oil with lard oil, having a fatty content of 25–30%, and a flash point about 300°F (149°C).

mineral lease See mining lease.

mineral monument [MIN ENG] A permanent monument established in a mining district to provide for an accurate description of mining claims and their location.

mineralocorticoid [BIOCHEM] A steroid hormone secreted by the adrenal cortex that regulates mineral metabolism and, secondarily, fluid balance.

mineralogenetic epoch [GEOL] A geologic time period during which mineral deposits formed.

mineralogenetic province [GEOL] Geographic region where conditions were favorable for the concentration of useful minerals.

mineralogical phase rule [MINERAL] Any of several variations of the Gibbs phase rule, taking into account the number of degrees of freedom consumed by the fixing of physical-chemical variables in the natural environment; it assumes that temperature and pressure are fixed externally and that consequently the number of phases (minerals) in a system (rock) will not usually exceed the number of components.

mineralogist [MINERAL] A person who studies the occurrence, description, mode of formation, and uses of minerals.

mineralography   *See* ore microscopy.

mineralogy   [INORG CHEM] The science which concerns the study of natural inorganic substances called minerals.

mineraloid   [MINERAL] A naturally occurring, inorganic material that is amorphous and is therefore not considered to be a mineral. Also known as gel mineral.

mineral oil   *See* medicinal oil.

mineral processing   [MIN ENG] Procedures, such as dry and wet crushing and grinding of ore or other products containing minerals, to raise the concentration of the substance being mined.

mineral resources   [GEOL] Valuable mineral deposits of an area that are presently recoverable and may be so in the future; includes known ore bodies and potential ore.

mineral right   *See* mining right.

mineral rubber   [MATER] Asphaltine minerals (such as gilsonite and grahamite) and blown asphalts; used in compounding of rubber, coatings, and paints.

minerals beneficiation   *See* beneficiation.

mineral seal oil   [MATER] A petroleum distillate with a boiling point higher than that of kerosine; used as a solvent oil and illuminant.

mineral sequence   *See* paragenesis.

mineral soil   [GEOL] Soil composed of mineral or rock derivatives with little organic matter.

mineral spring   [HYD] A spring whose water has a definite taste due to a high mineral content.

mineral suite   [MINERAL] **1.** A group of associated minerals in one deposit. **2.** A representative group of minerals from a certain locality. **3.** A group of specimens showing variations, as in color or form, in a single mineral species.

mineral tallow   *See* hatchettite.

mineral water   [HYD] Water containing naturally or artificially supplied minerals or gases.

mineral wax   *See* ozocerite.

mineral wool   [MATER] A natural fiber of mineral origin resembling wool or glass fiber and formed by blowing air or steam through molten rock or slag; examples are asbestos and synthetics in the form of whiskers made from inorganic compounds; used for insulation and fireproofing, and as a filter medium.

mine rescue apparatus   [MIN ENG] Certain types of apparatus worn by workers and permitting them to perform in noxious or irrespirable atmospheres, such as during mine fires or following mine explosions.

mine rescue crew   [MIN ENG] A crew consisting usually of five to eight persons who are thoroughly trained in the use of mine rescue apparatus, which they wear in rescue or recovery work in a mine following an explosion or during a fire.

mine rescue lamp   [MIN ENG] A particular type of electric safety hand lamp used in rescue operations; equipped with a lens for concentrating or diffusing the light.

mine resistance   [MIN ENG] Resistance by a mine to the passage of an air current.

miner's friend   [MIN ENG] **1.** The Davy safety lamp. **2.** A steam engine once used to pump water from underground.

miner's hammer   [MIN ENG] A hammer used to break ore.

miner's hand lamp   [MIN ENG] A self-contained mine lamp with handle for carrying.

miner's helmet   [MIN ENG] A hat designed for miners to provide head protection and to hold the cap lamp.

miner's horn   [MIN ENG] A metal spoon or horn to collect ore particles in gold washing.

miner's inch   [MIN ENG] The quantity of water that will escape from an aperture 1-inch (2.54-centimeter) square through a 2-inch-thick (5.08-centimeter-thick) plank, with a steady flow of water standing 6 inches (15.24 centimeters) above the top of the escape aperture, the quantity so discharged amounting to 2274 cubic feet (64.39 cubic meters) in 24 hours.

miner's lamp   [MIN ENG] Any one of a variety of lamps used by a miner to furnish light, such as oil lamps, carbide lamps, flame safety lamps, and cap lamps.

miner's right   [MIN ENG] An annual permit from the government to occupy and work mineral land.

miners' rules   [MIN ENG] Rules and regulations proclaimed by the miners of any district, relating to the location of,

recording of, and the work necessary to hold possession of a mining claim.

miner's self-rescuer   [MIN ENG] A pocket gas mask effective against carbon monoxide; air passes through a cannister containing fused calcium chloride before entering the mouth.

mine run   [MIN ENG] The unscreened output of a mine. Also known as run-of-mine.

mine sample   [MIN ENG] Coal or mineral extracted at underground exposures for analysis.

mine signal system   [MIN ENG] Signal lights installed at individual track switches immediately indicating to the motorman whether or not he can safely proceed.

mine skips   [MIN ENG] Skips used to bring mined ore to the surface of a shaft.

mine superintendent   [MIN ENG] A mine manager or group manager. Also known as mine captain.

mine surveyor   [MIN ENG] The official who periodically surveys the mine workings and prepares plans for the manager.

minesweeper   [NAV ARCH] A ship specially designed and equipped to remove or destroy underwater mines. [ORD] Heavy road roller pushed in front of a tank, to destroy land mines by exploding them.

minesweeping   [ORD] Process of detecting and removing land mines or underwater mines.

mine track devices   [MIN ENG] Track devices to provide maximum safety for haulage trains in mines.

mine tractor   [MIN ENG] A trackless, self-propelled vehicle used to transport equipment and supplies.

minette   [PETR] A syenitic variety of lamprophyre composed principally of biotite phenocrysts in a matrix of orthoclase and biotite.

mine valuation   [MIN ENG] Properly weighing the financial considerations to place a present value on mineral reserves.

mine ventilating fan   [MIN ENG] A motor-driven disk, propeller, or wheel for blowing or exhausting air to provide ventilation of a mine; large units are used for stationary systems, while small portable types provide fresh air in inaccessible locations, such as dead ends.

mine ventilation system   [MIN ENG] A combination of connecting airways with air pressure sources and governing devices that are instrumental in making and controlling airflow.

mine water   [MIN ENG] Water pumped from mines.

miniature electron tube   [ELECTR] A small electron tube having no base, with tube electrode leads projecting through the glass bottom in positions corresponding to those of pins for either a seven-pin or nine-pin tube base.

miniaturization   [ELECTR] Reduction in the size and weight of a system, package, or component by using small parts arranged for maximum utilization of space.

minicomputer   [ADP] A small computer which in its basic configuration has at least 4096 words of memory, employs words between 8 and 16 bits in length.

minim   [MECH] A unit of volume in the apothecaries' measure; equals $\frac{1}{60}$ fluidram (approximately 0.061612 cubic centimeter) or about 1 drop (of water). Abbreviated min.

minimal brain dysfunction syndrome   [MED] A complex of learning and behavioral disabilities seen primarily in children of near-average or above-average intelligence exhibiting also deviations of function of the central nervous system.

minimal equation   [MATH] An algebraic equation whose zeros define a minimal surface.

minimal flight path   [NAV] The flight path between two points that provides the shortest possible time enroute; it may be planned to take advantage of winds and pressure systems, and is thus not necessarily the shortest air distance between the two points.

minimal-latency coding   *See* minimum-access coding.

minimal polynomial   [MATH] The polynomial of least degree which both divides the characteristic polynomial of a matrix and has the same roots.

minimal surface   [MATH] A surface whose mean curvature is identically zero.

mini-maxi regret   [CONT SYS] In decision theory, a criterion which selects that strategy which has the smallest maximum difference between its payoff and that of the best hindsight choice.

minimax technique   *See* min-max technique.

minimize [COMMUN] Condition when normal message and telephone traffic is drastically reduced so messages connected with an actual or simulated emergency will not be delayed.

minimum [MATH] The least value that a real valued function assumes.

minimum-access coding [ADP] Coding in such a way that a minimum time is required to transfer words to and from storage, for a computer in which this time depends on the location in storage. Also known as minimal-latency coding; minimum-delay coding; minimum latency coding.

minimum-access programming [ADP] The programming of a digital computer in such a way that minimum waiting time is required to obtain information out of the memory. Also known as forced programming; minimum-latency programming.

minimum-access routine *See* minimum-latency routine.

minimum-altitude bombing [ORD] Horizontal or glide bombing with the height of release at an altitude under 900 feet (274 meters); it includes masthead bombing (bombing at masthead level), which is sometimes erroneously referred to as skip bombing.

minimum bend radius [MET] The minimum radius through which a piece of metal can be bent to form a given angle without fracturing.

minimum-delay coding *See* minimum-access coding.

minimum detectable signal *See* threshold signal.

minimum deviation [OPTICS] For a prism, the smallest possible angle between the incident and refracted rays; this angle is realized when refraction is symmetrical.

minimum discernible signal [ELECTR] Receiver input power level that is just sufficient to produce a discernible signal in the receiver output; a receiver sensitivity test.

minimum distance [NAV] The shortest distance at which a navigational instrument or system will function satisfactorily.

minimum-distance code [COMMUN] A binary code in which the signal distance does not fall below a specified minimum value.

minimum ebb [OCEANOGR] The least speed of a current that runs continuously ebb.

minimum elevation [ORD] Lowest elevation of a weapon at which the projectile will safely clear an obstacle between the weapon and the target.

minimum enroute instrument altitude [NAV] In air operations, the altitude which is attained as the result of the height of an artificial or natural object or the limitation in the performance of navigation and communication facilities. Abbreviated MEA.

minimum firing current [ELEC] The limit below which firing will not occur in electric blasting caps.

minimum flight altitude [AERO ENG] The lowest altitude at which aircraft may safely operate.

minimum flood [OCEANOGR] The least speed of a current that runs continuously flood.

minimum frontal diameter [ANTHRO] The smallest distance measured between the temporal crests, with only moderate pressure.

minimum ionizing speed [ATOM PHYS] The smallest speed at which a charged particle passing through a gas can ionize an atom or molecule.

minimum-latency coding *See* minimum-access coding.

minimum-latency programming *See* minimum-access programming.

minimum-latency routine [ADP] A computer routine that is constructed so that the latency in serial-access storage is less than the random latency that would be expected if storage locations were chosen without regard for latency. Also known as minimum-access routine.

minimum lethal dose [PHARM] The amount of an injurious agent which is the average of the smallest dose that kills and the largest dose that fails to kill.

minimum-loss attenuator [ELECTR] A section linking two unequal resistive impedances which is designed to introduce the smallest attenuation possible. Also known as minimum-loss pad.

minimum-loss matching [ELECTR] Design of a network linking two resistive impedances so that it introduces a loss which is as small as possible.

minimum-loss pad *See* minimum-loss attenuator.

minimum metal condition [DES ENG] The condition corresponding to the removal of the greatest amount of material permissible in a machined part.

minimum-phase system [CONT SYS] A linear system for which the poles and zeros of the transfer function all have negative or zero real parts.

minimum range [ORD] **1.** Least range setting of a gun at which the projectile will clear an obstacle or friendly troops between the gun and the target. **2.** Shortest distance to which a gun can fire from a given position.

minimum reflux ratio [CHEM ENG] The smallest reflux ratio in a two-component liquid distillation system that will produce the desired overhead and bottom compositions.

minimum rent [MIN ENG] The right to work coal acquired by a mine owner by the payment of an annual rent and a royalty to the landowner (the coal or mineral owner). Also known as fixed rent.

minimum safe distance [ORD] In an atomic explosion, the total distance from desired ground zero (DGZ) to friendly positions required to ensure troops safety.

minimum signal level [ELECTR] In facsimile, level corresponding to the copy white or copy black signal, whichever is the lower.

minimum thermometer [ENG] A thermometer that automatically registers the lowest temperature attained during an interval of time.

minimum turning circle [ENG] The diameter of the circle described by the outermost projection of a vehicle when the vehicle is making its shortest possible turn.

minimum wetting rate [CHEM ENG] The smallest liquid-flow rate through a packed column that will thoroughly wet the column packing.

mining [MIN ENG] The technique and business of mineral discovery and exploitation.

mining camp [MIN ENG] A term loosely applied to any mining town.

mining claim [MIN ENG] That portion of the public mineral lands which a miner, for mining purposes, takes and holds in accordance with mining laws. Also known as claim.

mining effect [ORD] Violent upheaval or movement of earth and the destruction or damage resulting therefrom, generally caused by an explosion below the surface of the earth; mining effect may be contrasted with the blast effect produced by an explosion on or above the surface of the earth.

mining engineer [MIN ENG] One qualified by education, training, and experience in mining engineering.

mining engineering [ENG] Engineering concerned with the discovery, development, and exploitation of coal, ores, and minerals, as well as the cleaning, sizing, and dressing of the product. Also known as mineral engineering.

mining geology [MIN ENG] The study of the structure and occurrence of mineral deposits and the geologic aspects of mine planning.

mining ground [MIN ENG] Land from which a mineral substance is extracted by the process of mining.

mining hazard [MIN ENG] Any of the dangers unique to the winning and working of minerals and coal.

mining lease [MIN ENG] A contract to work a mine and extract mineral or other deposits from it under specified conditions. Also known as mineral lease.

mining machine truck [MIN ENG] A truck used to transport shortwall mining machines.

mining on a shoestring *See* grass-roots mining.

mining partnership [MIN ENG] The arrangement whereby two or more persons acquire a mining claim and actually engage in working it.

mining property [MIN ENG] Property valued for its mining possibilities.

mining retreating [MIN ENG] A process of mining by which the ore or coal is untouched until after all the gangways and such are driven, at which time the work of extraction begins at the boundary and progresses toward the shaft.

mining right [MIN ENG] A right to enter upon and occupy a specific piece of ground for the purpose of working it, either by underground excavations or open workings, to obtain the mineral ores which may be deposited therein. Also known as mineral right.

mining shield [MIN ENG] A canopy or cover for the protection of workers and machines at the face of a mechanized coal heading.

mining title [MIN ENG] A claim, exclusive prospecting license, right, concession, or lease.

mining town [MIN ENG] A town that has arisen next to a mine or mines.

mining width [MIN ENG] The minimum width needed to extract ore regardless of the actual width of ore-bearing rock.

minitrack [AERO ENG] A satellite tracking system consisting of a field of separate antennas and associated receiving equipment interconnected so as to form interferometers which track a transmitting beacon in the payload itself. [ELECTR] A subminiature radio transmitter capable of sending data over 4000 miles (6500 kilometers) on extremely low power.

minium [MINERAL] Pb_3O_4 A scarlet or orange-red mineral consisting of an oxide of lead; found in Wisconsin and the western United States. Also known as red lead.

mink [VERT ZOO] Any of three species of slender-bodied aquatic carnivorous mammals in the genus *Mustela* of the family Mustelidae.

Minkowski electrodynamics [ELECTROMAG] An electromagnetic theory, compatible with the special theory of relativity, which takes into account the presence of matter with electric and magnetic polarization.

Minkowski metric [RELAT] The metric tensor of the Minkowski world used in special relativity; it is a 4 × 4 matrix whose nonzero entries lie on the diagonal, with one entry (corresponding to the time coordinate) equal to 1, and three entries (corresponding to space coordinates) equal to −1; sometimes, the negative of this matrix is used.

Minkowski's inequality [MATH] **1.** An inequality involving powers of sums of sequences of real or complex numbers, a_k and b_k:

$$\left[\sum_{k=1}^{\infty} \left| a_k + b_k \right|^s \right]^{1/s} \leq \left[\sum_{k=1}^{\infty} \left| a_k \right|^s \right]^{1/s} + \left[\sum_{k=1}^{\infty} \left| b_k \right|^s \right]^{1/s}$$

provided $s \geq 1$. **2.** An inequality involving powers of integrals of real or complex functions, f and g, over an interval or region R:

$$\left[\int_R \left| f(x) + g(x) \right|^s dx \right]^{1/s} \leq \left[\int_R \left| f(x) \right|^s dx \right]^{1/s}$$
$$+ \left[\int_R \left| g(x) \right|^s dx \right]^{1/s}$$

provided $s \geq 1$ and the integrals involved exist.

Minkowski universe *See* Minkowski world.

Minkowski world [RELAT] Space-time as described by the four coordinates (x, y, z, ict), where i is the imaginary unit and c is the speed of light; Lorentz transformations of space-time are orthogonal tranformations of the Minkowski world. Also known as Minkowski universe.

min-max technique [MATH] A method of approximation of a function f by a function g from some class where the maximum of the modulus of $f - g$ is minimized over this class. Also known as Chebyshev (Tchebycheff) approximation; minimax technique.

min-max theorem [MATH] The theorem that provides information concerning the nth eigenvalue of a symmetric operator on an inner product space without necessitating knowledge of the other eigenvalues. Also known as maximum-minimum principle.

Minnesota multiphasic personality inventory [PSYCH] An empirical scale of an individual's personality based mainly on his own yes-or-no responses to a questionnaire of 550 items; included are special validating scales which measure the individual's test-taking attitude and degree of frankness. Abbreviated MMPI. Also known as multiphasic personality inventory (MPI).

Minnesota preschool scale [PSYCH] A verbal and nonverbal test designed to measure the learning ability of children from 18 months to 6 years of age.

minnow [VERT ZOO] The common name for any fresh-water fish composing the family Cyprinidae, order Cypriniformes.

Minofor Alloy [MET] A white alloy containing 68% tin, 18% antimony, 3% copper, and 10% zinc; properties are similar to those of Brittania metal.

minor [MATH] The minor of an entry of a matrix is the determinant of the matrix obtained by removing the row and column containing the entry. Also known as cofactor; complementary minor.

minor arc [MATH] The smaller of the two arcs on a circle produced by a secant.

minor axis [MATH] The smaller of the two axes of an ellipse.

minor bend [ELECTROMAG] Rectangular waveguide bent so that throughout the length of a bend a longitudinal axis of the guide lies in one plane which is parallel to the narrow side of the waveguide.

minor cycle [ADP] The time required for the transmission or transfer of one machine word, including the space between words, in a digital computer using serial transmission. Also known as word time.

minor defect [IND ENG] A defect which reduces the effectiveness of the product, without causing serious malfunctioning.

minor diameter [DES ENG] The diameter of a cylinder bounding the root of an external thread or the crest of an internal thread.

minor diatonic scale [ACOUS] A diatonic scale in which the relative sizes of the sequence of intervals are approximately 2,1,2,2,2,2,1.

minor fog signal [NAV] A sound signal of limited power generally having a normal range of 2 miles or less.

minority carrier [SOLID STATE] The type of carrier, electron, or hole that constitutes less than half the total number of carriers in a semiconductor.

minority emitter [ELECTR] Of a transistor, an electrode from which a flow of minority carriers enters the interelectrode region.

minor light [NAV] An automatic unwatched light on a fixed structure usually showing low to moderate candlepower.

minor lobe [ELECTROMAG] Any lobe except the major lobe of an antenna radiation pattern. Also known as secondary lobe; side lobe.

minor loop [CONT SYS] A portion of a feedback control system that consists of a continuous network containing both forward elements and feedback elements.

minor mine disaster [MIN ENG] **1.** A mine accident (not an explosion or fire) causing the death of less than five persons and considerable property damage. **2.** A mine explosion or fire causing injury to one or more persons or considerable property damage but no loss of life. **3.** A mine explosion or fire resulting in the death of less than five persons.

minor planet [ASTRON] **1.** Those planets smaller than the earth, specifically Mercury, Venus, Mars, and Pluto. **2.** *See* asteroid.

minor relay station [ELECTR] A tape relay station which has tape relay responsibility but does not provide an alternate route.

minor surgery [MED] Any superficial surgical procedure involving little hazard to the life of the patient and not requiring general anesthesia.

minor switch [ELEC] Single-motion stepping switch mounted atop the telephone connectors and most commonly used for the party-line selection.

minor trough [METEOROL] A pressure trough of smaller scale than a long-wave trough; it ordinarily moves rapidly and is associated with a migratory cyclonic disturbance in the lower troposphere.

minuano [METEOROL] A cold southwesterly wind of southern Brazil, occurring during the Southern Hemisphere winter (June to September).

minuend [MATH] The quantity from which another quantity is to be subtracted.

minus [MATH] *A* minus *B* means that the quantity *B* is to be subtracted from the quantity *A*.

minus angle *See* angle of depression.

minus-cement porosity [GEOL] The porosity that would characterize a sedimentary material if it contained no chemical cement.

minus sieve [MET] In powder metallurgy, the portion of a powder sample that passes through a specified standard sieve.

minus zone [ADP] The bit positions in a computer code that represent the algebraic minus sign.

minute [MATH] A unit of measurement of angle, equal to ⅟₆₀ of a degree. Symbolized ′. Also known as arcmin. [MECH] A unit of time, equal to 60 seconds.

MINK

The American mink *(Mustela vison)*, the largest mink. It is valued for its dark, soft, thick fur.

MINNOW

The common shiner *(Notropis cornutus)* which ranges eastward from the Rocky Mountains.

minute hand [HOROL] The long hand on a timepiece, indicating minutes.

Minuteman intercontinental ballistic missile [ORD] A surface-to-surface missile of the U.S. Air Force, having a range of over 5000 miles (8000 kilometers) and using solid fuel propellant to permit almost instant firing when needed.

minutes on leg [NAV] The interval of time, estimated or actual, required for an aircraft to fly a given distance at a given ground speed, usually along a single course line.

minute wheel [HOROL] A wheel in a dial train that is driven by the cannon pinion, and drives the hour wheel by means of a pinion.

minyulite [MINERAL] $KAl_2(PO_4)_2(OH,F)\cdot4H_2O$ A white mineral composed of hydrous basic potassium aluminum phosphate.

Miocene [GEOL] A geologic epoch of the Tertiary period, extending from the end of the Oligocene to the beginning of the Pliocene.

miocrystalline *See* hypocrystalline.

miogeosyncline [GEOL] The nonvolcanic portion of an orthogeosyncline, located adjacent to the craton.

Miosireninae [PALEON] A subfamily of extinct sirenian mammals in the family Dugongidae.

miosis [PHYSIO] Contraction of the pupil of the eye.

miotic [PHARM] **1.** Causing miosis. **2.** Any agent that causes miosis. [PHYSIO] Of or pertaining to miosis.

Mira [ASTRON] The first star recognized to be a periodic variable; has a period of 333–9 days and its spectrum changes from M5e at maximum to M9e at minimum. It is the prototype of long-period variable stars.

mirabilite [MINERAL] $Na_2SO_4\cdot10H_2O$ A yellow or white monoclinic mineral consisting of hydrous sodium sulfate, occurring as a deposit from saline lakes, playas, and springs, and as an efflorescence; the pure crystals are known as Glauber's salt.

miracidium [INV ZOO] The ciliated first larva of a digenetic trematode; forms a sporocyst after penetrating intermediate host tissues.

mirage [OPTICS] Any one of a variety of unusual images of distant objects seen as a result of the bending of light rays in the atmosphere during abnormal vertical distribution of air density.

mirage effect [COMMUN] Reception of radio waves at distances far beyond the normally expected range due to abnormal refraction caused by meteorological conditions such as abnormal vertical water-vapor and temperature gradients.

Miranda [ASTRON] A satellite of Uranus orbiting at a mean distance of 124,000 kilometers.

Mirapinnatoidei [VERT ZOO] A suborder of tiny oceanic fishes in the order Cetomimiformes.

Mira variables [ASTRON] A group of over 1300 stars having the same type of variability as the star Mira.

mire [GEOL] Wet spongy earth, as of a marsh, swamp, or bog.

mired [THERMO] A unit used to measure the reciprocal of color temperature, equal to the reciprocal of a color temperature of 10^6 kelvins. Derived from micro-reciprocal-degree.

Miridae [INV ZOO] The largest family of the Hemiptera; included in the Cimicomorpha, it contains herbivorous and predacious plant bugs which lack ocelli and have a cuneus and four-segmented antennae.

Miripinnati [VERT ZOO] The equivalent name for Marapinnatoidei.

mirror [OPTICS] A surface which specularly reflects a large fraction of incident light.

mirror coating [OPTICS] A thin film of highly reflective material spread over a correctly shaped glass surface to produce a mirror; aluminum is usually used in the visible region. Also known as reflective coating.

mirror galvanometer [ELEC] A galvanometer having a small mirror attached to the moving element, to permit use of a beam of light as an indicating pointer. Also known as reflecting galvanometer.

mirror glance *See* wehrlite.

mirror interference [OPTICS] Interference occurring between two beams, one or both of which are reflected from a mirror at a small angle.

mirror interferometer [ENG] An interferometer used in radio astronomy, in which the sea surface acts as a mirror to reflect radio waves up to a single antenna, where the reflected waves interfere with the waves arriving directly from the source. [OPTICS] Any interferometer which makes use of mirror interference.

mirror machine [PL PHYS] A device which confines plasma in a tube with magnetic mirrors at each end to prevent it from escaping.

mirror nephoscope [ENG] A nephoscope in which the motion of a cloud is observed by its reflection in a mirror. Also known as cloud mirror; reflecting nephoscope.

mirror nuclei [NUC PHYS] A pair of atomic nuclei, each of which would be transformed into the other by changing all its neutrons into protons, and vice versa.

mirror optics [OPTICS] The science and technology of mirrors which, by means of reflecting rays of light, either revert optical bundles or focus them to form images.

mirror reflection *See* specular reflection.

mirror stone *See* muscovite.

mirror plane of symmetry *See* plane of mirror symmetry.

miscarriage [MED] Expulsion of the fetus before it is viable.

miscegenation [ANTHRO] Intermarriage between different races.

misch metal [MET] An alloy consisting of a crude mixture of cerium, lanthanum, and other rare-earth metals obtained by electrolysis of the mixed chlorides of the metals dissolved in fused sodium chloride; used in making aluminum alloys, in some steels and irons, and in coating the cathodes of glow-type voltage regulator tubes.

miscibility [CHEM] The tendency or capacity of two or more liquids to form a uniform blend, that is, to dissolve in each other; degrees are total miscibility, partial miscibility, and immiscibility. Also known as solubility.

miscible [CHEM] Referring to liquids that are mutually soluble, that is, they will dissolve in each other.

miscible-phase displacement [PETRO ENG] Method of increasing reservoir oil recovery by displacement with an oil-miscible driving fluid, such as gas or liquefied petroleum gas.

miscible-slug process [PETRO ENG] A miscible-phase displacement in which reservoir oil recovery is increased by displacement with liquefied petroleum gas as the driving fluid.

misenite [MINERAL] $K_8H_6(SO_4)_7$ A white mineral composed of native acid potassium sulfate.

miser [PETRO ENG] A well-boring bit that is tubular with a valve at the bottom, and has a screw for forcing the earth upward. Also spelled mizer.

misfeed [ORD] Failure to supply ammunition properly, especially to a magazine-fed or belt-fed automatic gun.

misfire [CHEM] Failure of fuel or an explosive charge to ignite properly. [ELECTR] Failure to establish an arc between the main anode and the cathode of an ignitron or other mercury-arc rectifier during a scheduled conducting period.

misfit stream [HYD] A stream whose meanders are either too large or too small to have eroded the valley in which it flows.

mismatch [ELEC] The condition in which the impedance of a source does not match or equal the impedance of the connected load or transmission line. [MET] Failure to match forged surfaces formed in opposite dies.

mismatch factor *See* reflection factor.

mismatch loss [ELECTR] Loss of power delivered to a load as a result of failure to make an impedance match of a transmission line with its load or with its source.

mismatch slotted line [ELECTROMAG] A slotted line linking two waveguides which is not properly designed to minimize the power reflected or transmitted by it.

misogamy [PSYCH] A feeling of revulsion and repugnance toward marriage.

misogyny [PSYCH] Hatred of women.

misology [PSYCH] Unreasoning aversion to intellectual or literary matters, or to argument or speaking.

misoneism [PSYCH] Hatred of new circumstances or things, or change.

misopedia [PSYCH] Morbid hatred of all children, but especially of one's own.

mispickel *See* arsenopyrite.

misregistration [ADP] In character recognition, the improper state of appearance of a character, line, or document, on site in a character reader, with respect to a real or imaginary

**MINUTEMAN
INTERCONTINENTAL
BALLISTIC MISSILE**

A Minuteman ICBM on the launch pad with its service structure in place. *(Boeing Aircraft)*

MIOCENE

		PALEOZOIC										MESOZOIC			CENOZOIC	
PRECAMBRIAN							CARBON-IFEROUS									
		CAMBRIAN	ORDOVICIAN	SILURIAN	DEVONIAN	Mississippian	Pennsylvanian	PERMIAN	TRIASSIC	JURASSIC	CRETACEOUS	TERTIARY	QUATERNARY			

TERTIARY					QUATERNARY	
Paleocene	Eocene	Oligocene	Miocene	Pliocene	Pleistocene	Recent

Position of the Miocene epoch in relation to other epochs and to the periods and eras of geologic time.

MIRROR MACHINE

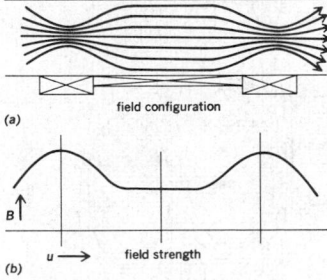

Mirror machine. *(a)* Longitudinal section. Lines represent magnetic lines of force. *(b)* Curve represents magnetic field strength B along path u in the tube. Regions where B has a large value are magnetic mirrors produced by coils and located at constrictions of field lines at each end; they confine plasma to central part.

MISSISSIPPIAN

PRECAMBRIAN		
CAMBRIAN		
ORDOVICIAN		
SILURIAN	PALEOZOIC	
DEVONIAN		
Mississippian	CARBON-IFEROUS	
Pennsylvanian		
PERMIAN		
TRIASSIC		
JURASSIC	MESOZOIC	
CRETACEOUS		
TERTIARY	CENOZOIC	
QUATERNARY		

Chart showing position of the Mississippian in geologic time.

MISTLETOE

Mistletoe (*Phoradendron flavescens*).

MITOCHONDRIA

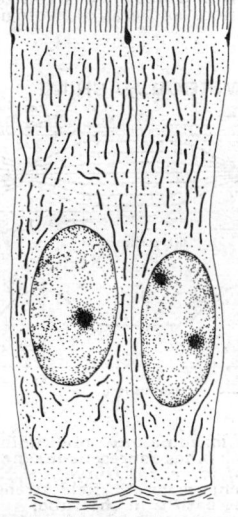

Drawing of intestinal epithelial cells showing numerous filamentous mitochondria concentrated in the apical cytoplasm.

horizontal baseline. [GRAPHICS] Improper alignment of register marks in matching overlays on a page or in matching the front of a type page to the back.

misrun [MET] Incompletely formed casting due to premature solidification of metal before the mold is filled.

missed abortion [MED] A condition in which a fetus weighing less than 500 grams dies and remains in the uterus for an extended period of time.

missed hole *See* failed hole.

missed round [ENG] A round in which all or part of the explosive has failed to detonate.

missense codon [GEN] A mutant codon that directs the incorporation of a different amino acid and results in the synthesis of a protein with a sequence in which one amino acid has been replaced by a different one; in some cases the mutant protein may be unstable or less active.

missense mutation [MOL BIO] A mutation that converts a codon coding for one amino acid to a codon coding for another amino acid.

missile [ORD] Any object that is, or is designed to be, thrown, dropped, projected, or propelled, for the purpose of making it strike a target; examples are guided missile and ballistic missile.

missile attitude [MECH] The position of a missile as determined by the inclination of its axes (roll, pitch, and yaw) in relation to another object, as to the earth.

missile checkout [ORD] Performance of procedures to determine whether all parts of the missile are apparently capable of performing their prescribed functions.

missile countermeasure [ORD] Any device or technique intended to impair the effectiveness of an enemy missile.

missile decoy [ORD] A vehicle used to simulate a missile in flight, attracting enemy radar and drawing fire so as to increase the odds for penetration by manned bombers or other weapon systems.

missile destruct system [ORD] That portion of a missile which, upon command or due to a failure, is capable of destroying the missile, generally for safety reasons.

missile guidance central [ORD] A collection of items specifically designed to provide target ranging facilities, and guidance and control equipment for directing the terminal flight phase of a guided missile.

missile impact predictor group [ORD] A group that provides facilities for determining the point of impact of a missile.

missile plume [ELECTROMAG] The region of electromagnetic and other disturbances that follow a missile during reentry and make the missile more readily detectable.

missile position tracking group [ORD] A group that provides facilities for determining the in-flight position of a missile.

missile ship *See* guided-missile ship.

missile site radar [ENG] Phased array radar located at a missile launch area to provide a guidance link with interceptor missiles enroute to their targets.

missile test [ORD] The test that includes missile checkout plus actual firing to determine whether the missile is capable of performing its operational function.

missing error [ADP] The result of calling for a subroutine not available in the library.

missing mass spectrometer [PARTIC PHYS] An apparatus which measures the momentum of the recoil protons in a reaction such as $\pi^- + p \rightarrow p + (MM)^-$, in order to determine the distribution of masses of the MM system, without any detailed observations on this system.

Mississippian [GEOL] A large division of late Paleozoic geologic time, after the Devonian and before the Pennsylvanian, named for a succession of highly fossiliferous marine strata consisting largely of limestones found along the Mississippi River between southeastern Iowa and southern Illinois; approximately equivalent to the European Lower Carboniferous.

Mississippi River–type buoy [NAV] A lightweight unlighted buoy of unique design developed for use on rivers, and now used in many inland areas.

Missnay-Schardin effect [ORD] The acceleration of a solid end plate (usually metal) from the face of an explosive charge under detonation, such that the end plate remains a solid and is usable as a missile.

Missourian [GEOL] A North American provincial series of

geologic time: lower Upper Pennsylvanian (above Desmoinesian, below Virgilian).

mist [FL MECH] Fine liquid droplets suspended in or falling through a moving or stationary gas atmosphere. [METEOROL] A hydrometeor consisting of an aggregate of microscopic and more or less hygroscopic water droplets suspended in the atmosphere; it produces, generally, a thin, grayish veil over the landscape; it reduces visibility to a lesser extent than fog; the relative humidity with mist is often less than 95%.

mistake [ADP] A human action producing an unintended result, in contrast to an error in a computer operation.

mistbow *See* fogbow.

mist droplet [METEOROL] A particle of mist, intermediate between a haze droplet and a fog drop.

mist extractor [ENG] A device that removes liquid mist or droplets from a gas stream via impingement, flow-direction change, velocity change, centrifugal force, filters, or coalescing packs.

mistletoe [BOT] 1. *Viscum album.* The true, Old World mistletoe having dichotomously branching stems, thick leathery leaves, and waxy-white berries. 2. Any of several species of green hemiparasitic plants of the family Loranthaceae.

mist projector [MIN ENG] An appliance to form a mist spray to allay dust and fume during blasting operations in a tunnel.

mistral [METEOROL] A north wind which blows down the Rhone Valley south of Valence, France, and into the Gulf of Lions. Strong, squally, cold, and dry, it is the combined result of the basic circulation, a fall wind, and jet-effect wind.

Mitchell's disease *See* erythromelalgia.

mite [INV ZOO] The common name for the acarine arthropods composing the diverse suborders Onychopalpida, Mesostigmata, Trombidiformes, and Sarcoptiformes.

miter bend [DES ENG] A pipe bend made by mitering (angle cutting) and joining pipe ends.

miter box [ENG] A troughlike device of metal or wood with vertical slots set at various angles in the upright sides, for guiding a handsaw in making a miter joint.

miter gate [CIV ENG] Either of a pair of canal lock gates that swing out from the side walls and meet at an angle pointing toward the upper level.

miter gear [DES ENG] A bevel gear whose bevels are in 1:1 ratio.

miter joint [DES ENG] A joint, usually perpendicular, in which the mating ends are beveled.

miter saw [DES ENG] A hollow-ground saw in diameters from 6 to 16 inches (15.24 to 40.64 centimeters), used for cutting off and mitering on light stock such as moldings and cabinet work.

miter valve [DES ENG] A valve in which a disk fits in a seat making a 45° angle with the axis of the valve.

miticide [MATER] An agent that kills mites. Also known as acaricide.

mitochondria [CYTOL] Minute cytoplasmic organelles in the form of spherical granules, short rods, or long filaments found in almost all living cells; submicroscopic structure consists of an external membrane system.

mitogenesis [CYTOL] 1. Induction of mitosis. 2. Formation as a result of mitosis.

mitogenetic [CYTOL] Inducing or stimulating mitosis.

mitomycin [MICROBIO] A complex of three antibiotics (mitomycin A, B, and C) produced by *Streptomyces caespitosus.*

mitosis [CYTOL] Nuclear division involving exact duplication and separation of the chromosome threads so that each of the two daughter nuclei carries a chromosome complement identical to that of the parent nucleus.

mitotic index [CYTOL] The number of cells undergoing mitosis per thousand cells.

mitotic inhibitor [CYTOL] A compound that inhibits mitosis.

mitotic poison [CYTOL] A compound that prevents or affects the completion of mitosis.

mitral commissurotomy [MED] Any of several surgical procedures performed to relieve mitral stenosis.

mitral stenosis [MED] Obstruction of the mitral valve, usually due to narrowing of the orifice.

mitral valve [ANAT] The atrioventricular valve on the left side of the heart.

mitriform [BIOL] Shaped like a miter.

Mitscherlich law of isomorphism [CHEM] Substances which

have similar chemical properties and crystalline forms usually have similar chemical formulas.

Mittag-Leffler's theorem   [MATH] A theorem that enables one to explicitly write down a formula for a meromorphic complex function with given poles.

mittelschmerz   [MED] Pain or discomfort in the lower abdomen in women occurring midway in the intermenstrual interval, thought to be secondary to the irritation of the pelvic peritoneum by fluid or blood escaping from the point of ovulation in the ovary.

mix crystal   See mixed crystal.

mixed acid   See nitrating acid.

mixed aphasia   [PSYCH] A combination of two or more forms of aphasia with impairment or loss of language function.

mixed arthritis   [MED] A combination of features of rheumatoid arthritis and degenerative joint disease seen in the same patient or the same joint.

mixed-based notation   [MATH] A computer number system in which a single base, such as 10 in the decimal system, is replaced by two number bases used alternately, such as 2 and 5.

mixed-base number   [MATH] A number in mixed-base notation. Also known as mixed-radix number.

mixed cloud   [METEOROL] A cloud containing both water drops and ice crystals, hence a cloud whose composition is intermediate between that of a water cloud and that of an ice-crystal cloud.

mixed cryoglobulin   [BIOCHEM] A cryoglobulin with a monoclonal component made of immunoglobulin belonging to two different classes, one of which is monoclonal.

mixed crystal   [CRYSTAL] A crystal whose lattice sites are occupied at random by different ions or molecules of two different compounds. Also known as mix crystal.

mixed current   [OCEANOGR] A type of tidal current characterized by a conspicuous difference in speed between the two flood currents or two ebb currents usually occurring each tidal day.

mixed cycle   [MECH ENG] An internal combustion engine cycle which combines the Otto cycle constant-volume combustion and the Diesel cycle constant-pressure combustion in high-speed compression-ignition engines. Also known as combination cycle; commercial Diesel cycle; limited-pressure cycle.

mixed decimal   [MATH] Any decimal plus an integer.

mixed flow   [CHEM ENG] Flow stream existing in two or more phases, such as gas, hydrocarbon, and water. Also known as mixed-phase flow.

mixed-flow fan   [MIN ENG] A mine fan in which the flow is both radial and axial.

mixed-flow impeller   [MECH ENG] An impeller for a pump or compressor which combines radial- and axial-flow principles.

mixed forest   [FOR] A forest consisting of two or more types of trees, with no more than 80% of the most common tree.

mixed gland   [PHYSIO] A gland that secretes more than one substance, especially a gland containing both mucous and serous components.

mixed highs   [COMMUN] In color television, a method of reproducing very fine picture detail by transmitting high-frequency components as part of luminance signals for achromatic reproduction in color pictures.

mixed indicator   [ANALY CHEM] Color-change indicator for acid-base titration end points in which a mixture of two indicator substances is used to give sharper end-point color changes.

mixed laterality   [PSYCH] The tendency, when there is a choice, to prefer to use parts of one side of the body for certain tasks and parts of the opposite side for others.

mixed layer   [OCEANOGR] The layer of water which is mixed through wave action or thermohaline convection. Also known as surface water.

mixed-layer mineral   [MINERAL] A mineral having an interstratified structure consisting of alternating layers of two different clays or of a clay and some other mineral.

mixed nerve   [PHYSIO] A nerve containing both sensory and motor components.

mixed nucleus   [METEOROL] A condensation nucleus of intermediate efficacy which, as a result of particle coagulation, contains both soluble hygroscopic matter and insoluble but wettable matter.

mixed number   [MATH] The sum of an integer and a fraction.

mixed oil   See nitrating acid.

mixed ore   [GEOL] Any ore with both oxidized and unoxidized minerals.

mixed-phase flow   See mixed flow.

mixed radix   [MATH] Pertaining to a numeration system using more than one radix, such as the biquinary system.

mixed-radix number   See mixed-base number.

mixed reflection   See spread reflection.

mixed salvo   [ORD] Series of shots in which some fall short of the target and some beyond it.

mixed tide   [OCEANOGR] A tide in which the presence of a diurnal wave is conspicuous by a large inequality in the heights of either the two high tides or the two low tides usually occurring each tidal day.

mixer   [ELECTR] **1.** A device having two or more inputs, usually adjustable, and a common output; used to combine separate audio or video signals linearly in desired proportions to produce an output signal. **2.** The stage in a superheterodyne receiver in which the incoming modulated radio-frequency signal is combined with the signal of a local r-f oscillator to produce a modulated intermediate-frequency signal. Also known as first detector; heterodyne modulator; mixer-first detector. [OPTICS] A nonlinear device in which two light beams are combined to form new beams having frequencies equal to the sum or the difference of the input wavelengths.

mixer-first detector   See mixer.

mixer-settler   [CHEM ENG] Solvent-extraction system with alternating or combined arrangement of mixers and settlers; used for chemicals extraction, lubricating-oil refining, and uranium oxide recovery. Also known as mixer-settler extractor.

mixer-settler extractor   See mixer-settler.

mixer tube   [ELECTR] A multigrid electron tube, used in a superheterodyne receiver, in which control voltages of different frequencies are impressed upon different control grids, and the nonlinear properties of the tube cause the generation of new frequencies equal to the sum and difference of the impressed frequencies.

mixing   [CHEM ENG] The intermingling of different materials (liquid, gas, solid) to produce a homogeneous mixture. [ELECTR] Combining two or more signals, such as the outputs of several microphones. [SCI TECH] The thorough intermingling of two or more different materials.

mixing chamber   [ENG] The space in a welding torch in which the gases are mixed.

mixing length   [PHYS] A mean length of travel, characteristic of a particular motion, over which an eddy maintains its identity; it is analogous to the mean free path of a molecule; physically, the idea implies that mixing occurs by discontinuous steps, that fluctuations which arise as eddies with different characteristics wander about, and that the mixing is done almost entirely by the small eddies.

mixing ratio   [METEOROL] In a system of moist air, the dimensionless ratio of the mass of water vapor to the mass of dry air; for many purposes, the mixing ratio may be approximated by the specific humidity.

mixing transformation   [MATH] A function of a measure space which moves the measurable sets in such a manner that, asymptotically as regards measure, any measurable set is distributed uniformly throughout the space.

mixing valve   [ENG] Multi-inlet valve used to mix two or more fluid intakes to give a mixed product of desired composition.

mixite   [MINERAL] $Cu_{11}Bi(AsO_4)_5(OH)_{10}\cdot 6H_2O$ A green to whitish mineral composed of a hydrous basic arsenate of copper and bismuth.

Mixodectidae   [PALEON] A family of extinct insectivores assigned to the Proteutheria; a superficially rodentlike group confined to the Paleocene of North America.

mixolimnion   [HYD] The upper layer of a meromictic lake, characterized by low density and free circulation; this layer is mixed by the wind.

mixotrophic   [BIOL] Obtaining nutrition by combining autotrophic and heterotrophic mechanisms.

mixtite *See* diamictite.

mixture [PHARM] A liquid medicine prepared by adding insoluble substances to a liquid medium, usually with a suspending agent. [SCI TECH] The product of mixing; components are not in a fixed proportion to each other.

mixture ratio [AERO ENG] The ratio of the weight of oxidizer used per unit of time to the weight of fuel used per unit of time.

mizer *See* miser.

mizzenmast [NAV ARCH] The third mast or the mast aft of the mainmast in a sailing ship.

mizzonite [MINERAL] A mineral of the scapolite group, composed of 54 to 57% silica. Also known as dipyre.

mksa system *See* meter-kilogram-second-ampere system.

mks system *See* meter-kilogram-second system.

ml *See* milliliter.

MLD *See* metachromatic leukodystrophy.

mm *See* millimeter.

M meter [ENG] A class of instruments which measure the liquid water content of the atmosphere.

mmf *See* magnetomotive force.

mmHg *See* millimeter of mercury.

mmH₂O *See* millimeter of water.

MMPI *See* Minnesota multiphasic personality inventory.

MMSCFD [CHEM ENG] Abbreviation for million standard cubic feet per day; usually refers to gas flow.

MMSCFH [CHEM ENG] Abbreviation for million standard cubic feet per hour; usually refers to gas flow.

MMSCFM [CHEM ENG] Abbreviation for million standard cubic feet per minute; usually refers to gas flow.

mmu *See* milli-mass-unit.

Mn *See* manganese.

mnemonic [PSYCH] **1.** Aiding or pertaining to memory. **2.** A device, such as combinations of letters, pictures, or words, to stimulate recall of the facts they represent.

mnemonic code [ADP] A programming code that is easy to remember because the codes resemble the original words, such as MPY for multiply and ACC for accumulator.

mnemonics [PSYCH] The science of the cultivation of memory functions using systematic methods.

Mnesarchaeidae [INV ZOO] A family of lepidopteran insects in the suborder Homoneura; members are confined to New Zealand.

Mo *See* molybdenum.

moat [GEOL] **1.** A ringlike depression around the base of a seamount. **2.** A valleylike depression around the inner side of a volcanic cone, between the rim and the lava dome. [HYD] **1.** A glacial channel in the form of a deep, wide trench. **2.** *See* oxbow lake.

mobile [GRAPHICS] A decorative three-dimensional art object constructed of metal, glass, wood, plastic, or other materials; it is mounted in a hanging position and is free to move in any of its planes.

mobile artillery [ORD] Artillery weapons designed for movement and ready conversion from traveling position to firing position; wheels or other suspension devices are not ordinarily removed in the firing position.

mobile belt [GEOL] A long, relatively narrow crustal region of tectonic acitivity.

mobile crane [MECH ENG] **1.** A cable-controlled crane mounted on crawlers or rubber-tired carriers. **2.** A hydraulic-powered crane with a telescoping boom mounted on truck-type carriers or as self-propelled models.

mobile digital computer [ADP] Large, mobile, fixed-point operation, one-address, parallel-mode type digital computer.

mobile drill [MIN ENG] A drill unit mounted on wheels or crawl-type tracks to facilitate moving.

mobile filling [MIN ENG] Filling which is supplemented only from above, and which sinks, filling the mined-out rooms.

mobile hoist [MECH ENG] A platform hoist mounted on a pair of pneumatic-tired road wheels, so it can be towed from one site to another.

mobile loader [MECH ENG] A self-propelling power machine for loading coal, mineral, or dirt.

mobile mount [ORD] Any weapons mount which is mobile, as contrasted to one which is fixed; as a mount for mobile artillery.

mobile radio [COMMUN] Radio communication in which the transmitter is installed in a vessel, vehicle, or airplane and can be operated while in motion.

mobile-relay station [COMMUN] Base station in which the base receiver automatically tunes on the base station transmitter and which retransmits all signals received by the base station receiver; used to extend the range of mobile units, and requires two frequencies for operation.

mobile station [COMMUN] Station in the mobile service intended to be used while in motion or during halts at unspecified points.

mobile systems equipment [ADP] Computers located on planes, ships, or vans.

Mobilina [INV ZOO] A suborder of ciliophoran protozoans in the order Peritrichida.

mobility [ELECTR] *See* drift mobility. [PHYS] Freedom of particles to move, either in random motion or under the influence of fields or forces.

mobility ratio [PETRO ENG] The ratio of the mobility of the driving fluid (such as water) to that of the driven fluid (such as gas) in a petroleum reservoir.

mobility tensor [PL PHYS] A second-rank tensor whose product with the electric field vector for a plane wave in a plasma gives a vector equal to the average velocity of electrons or ions; components of both vectors are in phasor notation.

mobilization [GEOL] Any process by which solid rock becomes sufficiently soft and plastic to permit it to flow or to permit geochemical migration of the mobile components.

Möbius band [MATH] The nonorientable surface obtained from a rectangular strip by twisting it once and then gluing the two ends.

Möbius function [MATH] The function μ of the positive integers where $\mu(1) = 1$, $\mu(n) = (-1)^r$ if n factors into r distinct primes, and $\mu(n) = 0$ otherwise; also, $\mu(n)$ is the sum of the primitive nth roots of unity.

Möbius resistor [ELEC] A nonreactive resistor made by placing strips of aluminum or other metallic tape on opposite sides of a length of dielectric ribbon, twisting the strip assembly half a turn, joining the ends of the metallic tape, then soldering leads to opposite surfaces of the resulting loop.

Möbius transformations [MATH] These are the most commonly used conformal mappings of the complex plane; their form is $f(z) = (az + b)/(cz + d)$ where the real numbers a, b, c, and d satisfy $ad - bc \neq 0$. Also known as linear fractional transformations.

Mobulidae [VERT ZOO] The devil rays, a family of batoids that are surface feeders and live mostly on plankton.

mocaya oil [MATER] Oil from kernels of the Paraguayan palm (*Acrocomia sclerocarpa*), found in South America and the West Indies; used in manufacture of tinplate, soaps, candles, and margarine.

mock fog [METEOROL] A simulation of true fog by atmospheric refraction.

mock lead *See* sphalerite.

mock moon *See* paraselene.

mock ore *See* sphalerite.

mock silver [MET] **1.** An aluminum alloy containing 5% copper and 10% tin, or 5% copper and 5% silver. **2.** A white brass containing 55% zinc and 45% copper.

mock sun *See* paranthelion; parhelion.

mock sun ring *See* parhelic circle.

mockup [ENG] A model, often full-sized, of a piece of equipment, or installation, so devised as to expose its parts for study, training, or testing.

modacrylic [TEXT] Of a synthetic fiber, composed of less than 85% and more than 35% by weight of acrylonitrile units.

mode [ADP] One of several alternative conditions or methods of operation of a device. [COMMUN] Form of the information in a communication such as literal language, digital data, and video. [ELECTROMAG] A form of propagation of guided waves that is characterized by a particular field pattern in a plane transverse to the direction of propagation. Also known as transmission mode. [PETR] The mineral composition of a rock, usually expressed as percentages of total weight or volume. [PHYS] A state of an oscillating system that corresponds to a particular field pattern and one of the possible resonant frequencies of the system. [STAT] The most frequently occurring member of a set of numbers.

mode converter *See* mode transducer.

mode eddies [OCEANOGR] Densely packed, irregularly oval high- and low-pressure centers roughly 400 kilometers in diameter in which current intensities are typically tenfold greater than the local means. Also known as mesoscale eddies.

mode filter [ELECTROMAG] A waveguide filter designed to separate waves of the same frequency but of different transmission modes.

mode jump [ELECTR] Change in mode of magnetron operation from one pulse to the next; each mode represents a different frequency and power level.

model [ADP] See macroskeleton. [SCI TECH] A mathematical or physical system, obeying certain specified conditions, whose behavior is used to understand a physical, biological, or social system to which it is analogous in some way.

model atmosphere [METEOROL] Any theoretical representation of the atmosphere, particularly of vertical temperature distribution.

model basin [ENG] A large basin or tank of water where scale models of ships can be tested. Also known as model tank; towing tank.

mode-locked laser [OPTICS] A laser designed so that several modes of oscillation with closely spaced wavelengths, in which the laser would normally oscillate, are synchronized so that a pulse of light, lasting for as little as a picosecond, is generated.

model reference system [CONT SYS] An ideal system whose response is agreed to be optimum; computer simulation in which both the model system and the actual system are subjected to the same stimulus is carried out, and parameters of the actual system are adjusted to minimize the difference in the outputs of the model and the actual system.

model symbol [ADP] The standard usage of geometrical figures, such as squares, circles, or triangles, to help illustrate the various working parts of a model: each symbol must, nevertheless, be footnoted for complete clarification.

model tank See model basin.

model theory [MATH] The general qualitative study of the structure of a mathematical theory.

modem [ELECTR] A carrier terminal panel containing a modulator and demodulator, some circuits of which may be in common. Derived from modulator-demodulator.

mode number [ELECTR] 1. The number of complete cycles during which an electron of average speed is in the drift space of a reflex klystron. 2. The number of radians of phase in the microwave field of a magnetron divided by 2π as one goes once around the anode.

mode of oscillation See mode of vibration.

mode of vibration [MECH] A characteristic manner in which a system which does not dissipate energy and whose motions are restricted by boundary conditions can oscillate, having a characteristic pattern of motion and one of a discrete set of frequencies. Also known as mode of oscillation.

moder See coder.

moderate breeze [METEOROL] In the Beaufort wind scale, a wind whose speed is from 11 to 16 knots (13 to 18 miles per hour or 20 to 30 kilometers per hour).

moderate gale [METEOROL] In the Beaufort wind scale, a wind whose speed is from 28 to 33 knots (32 to 38 miles per hour or 52 to 61 kilometers per hour).

moderator [NUCLEO] The material used in a nuclear reactor to moderate or slow down neutrons from the high velocities at which they are created in the fission process.

modern algebra [MATH] The study of algebraic systems such as groups, rings, modules, and fields.

mode shift [ELECTR] Change in mode of magnetron operation during a pulse.

mode skip [ELECTR] Failure of a magnetron to fire on each successive pulse.

mode switch [ADP] A preset control which affects the normal response of various components of a mechanical desk calculator. [ELECTR] A microwave control device, often consisting of a waveguide section of special cross section, which is used to change the mode of microwave power transmission in the waveguide.

mode transducer [ELECTR] Device for transforming an electromagnetic wave from one mode of propagation to another. Also known as mode converter; mode transformer.

mode transformer See mode transducer.

modification [ENG] A major or minor change in the design of an item, effected in order to correct a deficiency, to facilitate production, or to improve operational effectiveness. [MET] Treatment of molten aluminum alloys containing 8–13% silicon with small amounts of a sodium fluoride or sodium chloride mixture; improves mechanical properties. [SCI TECH] Any change brought about by external or internal factors.

modification kit [ENG] A collection of items not all having the same basic name which are employed individually or conjunctively to alter the design of a component or equipment.

modified asphalt [MATER] Asphalt modified by addition of a rosin ester or synthetic resin.

modified Bessel functions [MATH] The functions defined by $I_\nu(x) = \exp(-i\nu\pi/2) J_\nu(ix)$, where J_ν is the Bessel function of order ν, and x is real and positive.

modified constant-voltage charge [ELEC] Charging of a storage battery in which the voltage of the charging circuit is held substantially constant, but a fixed resistance is inserted in the battery circuit producing a rising voltage characteristic at the battery terminals as the charge progresses.

modified gunmetal [MET] Gunmetal containing about 2.5% lead; used for gears and bearings.

modified Hankel functions [MATH] The functions defined by $K_\nu(x) = (i\pi/2) \exp(i\nu\pi/2) H_\nu^{(1)}(ix)$, where $H_\nu^{(1)}$ is the first Hankel function of order ν, and x is real and positive. Also known as MacDonald functions.

modified index of refraction [METEOROL] An atmospheric index of refraction mathematically modified so that when its gradient is applied to energy propagation over a hypothetical flat earth, it is substantially equivalent to propagation over the true curved earth with the actual index of refraction. Also known as modified refractive index; refractive modulus.

modified Lambert conformal chart [MAP] A chart on the modified Lambert conformal projection. Also known as Ney's chart.

modified Lambert conformal projection [MAP] A modification of the Lambert conformal projection for use in polar regions, one of the standard parallels being at latitude 89°59′58″ and the other at latitude 71° or 74°, and the parallels being expanded slightly to form complete concentric circles. Also known as Ney's projection.

modified precision approach radar [NAV] A special precision radar approach landing procedure for high performance aircraft; radar guidance is provided to a landing flare point instead of a runway touchdown point.

modified rayon [TEXT] A woollike rayon fiber made with additives in the spinning solution.

modified refractive index See modified index of refraction.

modifier [ADP] A quantity used to alter the address of an operand in a computer, such as the cycle index. Also known as index word. [MATER] In flotation, any of the chemicals which increase the specific attraction between collector agents and particle surfaces or which increase the wettability of those surfaces.

modifier gene [GEN] A gene that alters the phenotypic expression of a nonallelic gene.

modifier register See index register.

modify [ADP] 1. To alter a portion of an instruction so its interpretation and execution will be other than normal; the modification may permanently change the instruction or leave it unchanged and affect only the current execution; the most frequent modification is that of the effective address through the use of index registers. 2. To alter a subroutine according to a defined parameter.

moding [ELECTR] Defect of magnetron oscillation in which it oscillates in one or more undesired modes.

modiolus [ANAT] The central axis of the cochlea.

modular circuit [ELECTR] Any type of circuit assembled to form rectangular or cubical blocks that perform one or more complete circuit functions.

modularity [ADP] The property of functional flexibility built into a computer system by assembling discrete units which can be easily joined to or arranged with other parts or units.

modular structure [BUILD] A building constructed of preassembled or presized units of standard sizes; uses a 4-inch

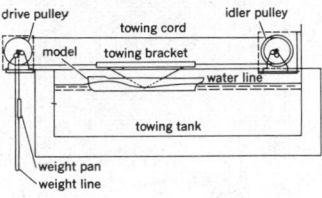

MODEL BASIN

Model basin with model towed by falling weight.

(10.16-centimeter) cubical module as a reference. [ELECTR] **1.** An assembly involving the use of integral multiples of a given length for the dimensions of electronic components and electronic equipment, as well as for spacings of holes in a chassis or printed wiring board. **2.** An assembly made from modules.

modulate [ELECTR] To vary the amplitude, frequency, or phase of a wave, or vary the velocity of the electrons in an electron beam in some characteristic manner.

modulated amplifier [ELECTR] Amplifier stage in a transmitter in which the modulating signal is introduced and modulates the carrier.

modulated carrier [COMMUN] Radio-frequency carrier wave whose amplitude phase or frequency has been varied according to the intelligence to be conveyed.

modulated continuous wave [COMMUN] Wave in which the carrier is modulated by a constant audio-frequency tone.

modulated Raman scattering [SPECT] Application of modulation spectroscopy to the study of Raman scattering; in particular, use of external perturbations to lower the symmetry of certain crystals and permit symmetry-forbidden modes, and the use of wavelength modulation to analyze second-order Raman spectra.

modulated stage [ELECTR] Radio-frequency stage to which the modulator is coupled and in which the continuous wave (carrier wave) is modulated according to the system of modulation and the characteristics of the modulating wave.

modulating electrode [ELECTR] Electrode to which a potential is applied to control the magnitude of the beam current.

modulating signal [COMMUN] Signal which causes a variation of some characteristics of a carrier.

modulation [COMMUN] The process or the result of the process by which some parameter of one wave is varied in accordance with some parameter of another wave.

modulation capability [ELECTR] Of an aural transmitter, the maximum percentage modulation that can be obtained without exceeding a given distortion figure.

modulation code [COMMUN] A code used to cause variations in a signal in accordance with a predetermined scheme; normally used to alter or modulate a carrier wave to transmit data.

modulation crest [COMMUN] The peak amplitude of an amplitude-modulated wave.

modulation envelope [COMMUN] A curve drawn through the peaks of a graph showing the waveform of a modulated signal; represents the waveform of the intelligence carried by the signal.

modulation factor [COMMUN] **1.** In general, the ratio of the peak variation in the modulation actually used in a transmitter to the maximum variation for which the transmitter was designed. **2.** In an amplitude-modulated wave, the ratio (usually expressed in percent) of the peak variation of the envelope from its reference value, to the reference value. Also known as index of modulation. **3.** In a frequency-modulated wave, the ratio of the actual frequency swing to the frequency swing required for 100% modulation.

modulation index [COMMUN] The ratio of the frequency deviation to the frequency of the modulating wave in a frequency-modulation system when using a sinusoidal modulating wave. Also known as ratio deviation.

modulation meter [ENG] Instrument for measuring the degree of modulation (modulation factor) of a modulated wave train, usually expressed in percent.

modulation rise [ELECTR] Increase of the modulation percentage caused by nonlinearity of any tuned amplifier, usually the last intermediate-frequency stage of a receiver.

modulation spectroscopy [SPECT] A branch of spectroscopy concerned with the measurement and interpretation of changes in transmission or reflection spectra induced (usually) by externally applied perturbation, such as temperature or pressure change, or an electric or magnetic field.

modulation transformer [ENG ACOUS] An audio-frequency transformer which matches impedances and transmits audio frequencies between one or more plates of an audio output stage and the grid or plate of a modulated amplifier.

modulation with a fixed reference [COMMUN] Phase modulation with a pilot carrier.

modulator [ELECTR] **1.** The transmitter stage that supplies the modulating signal to the modulated amplifier stage or that triggers the modulated amplifier stage to produce pulses at desired instants as in radar. **2.** A device that produces modulation by any means, such as by virtue of a nonlinear characteristic or by controlling some circuit quantity in accordance with the waveform of a modulating signal. **3.** One of the electrodes of a spacistor.

modulator crystal [OPTICS] Crystal which is used to modulate a polarized light beam by the use of the Pockel's effect; useful as a modulator in laser systems.

modulator-demodulator *See* modem.

modulator glow tube [ELECTR] Cold cathode recorder tube that is used for facsimile and sound-on-film recording; provides a modulated high-intensity point source of light.

module [ADP] **1.** A distinct and identifiable unit of computer program for such purposes as compiling, loading, and linkage editing. **2.** One memory bank and associated electronics in a computer. [AERO ENG] A self-contained unit which serves as a building block for the overall structure in space technology; usually designated by its primary function, such as command module or lunar landing module. [ELECTR] A packaged assembly of wired components, built in a standardized size and having standardized plug-in or solderable terminations. [ENG] A unit of size used as a basic component for standardizing the design and construction of buildings, building parts, and furniture. [MATH] A vector space in which the scalars are a ring rather than a field.

modulo [MATH] **1.** A group G modulo a subgroup H is the quotient group G/H of cosets of H in G. **2.** A technique of identifying elements in an algebraic structure in such a manner that the resulting collection of identified objects is the same type of structure.

modulo N [MATH] Two integers are said to be congruent modulo N (where N is some integer) if they have the same remainder when divided by N.

modulo N arithmetic [MATH] Calculations in which all integers are replaced by their remainders after division by N (where N is some fixed integer.)

modulo N check [ADP] A procedure for verification of the accuracy of a computation by repeating the steps in modulo N arithmetic and comparing the result with the original result (modulo N). Also known as residue check.

modulo-two adder [ADP] A logical circuit for adding one-digit binary numbers.

modulus of a complex number *See* absolute value of a complex number.

modulus of a congruence [MATH] A number a, such that two specified numbers b and c give the same remainder when divided by a; b and c are then said to be congruent, modulus a (or congruent, modulo a).

modulus of a logarithm [MATH] The modulus of a logarithm with a given base is the factor by which a logarithm with a second base must be multiplied to give the first logarithm.

modulus of continuity [MATH] For a real valued continuous function f, this is the function whose value at a real number r is the maximum of the modulus of $f(x) - f(y)$ where the modulus of $x - y$ is less than r; this function is useful in approximation theory.

modulus of distance [ASTRON] The quantity $m - M$, where M is the absolute magnitude of a given star and m is its apparent magnitude. Also known as distance modulus.

modulus of elasticity [MECH] The ratio of the increment of some specified form of stress to the increment of some specified form of strain, such as Young's modulus, the bulk modulus, or the shear modulus. Also known as coefficient of elasticity; elasticity modulus; elastic modulus.

modulus of elasticity in shear [MECH] A measure of a material's resistance to shearing stress, equal to the shearing stress divided by the resultant angle of deformation expressed in radians. Also known as coefficient of rigidity; modulus of rigidity; shear modulus; torsional modulus.

modulus of resilience [MECH] The maximum mechanical energy stored per unit volume of material when it is stressed to its elastic limit.

modulus of rigidity *See* modulus of elasticity in shear.

modulus of rupture in bending [MECH] The maximum stress per unit area that a specimen can withstand without breaking when it is bent, as calculated from the breaking load under the assumption that the specimen is elastic until rupture takes place.

modulus of rupture in torsion [MECH] The maximum stress per unit area that a specimen can withstand without breaking when its ends are twisted, as calculated from the breaking load under the assumption that the specimen is elastic until rupture takes place.

modulus of strain hardening *See* rate of strain hardening.

modulus of volume elasticity *See* bulk modulus of elasticity.

moellen *See* degras.

Moeller-Barlow disease *See* infantile scurvy.

Moeritheriidae [PALEON] The single family of the extinct order Moeritherioidea.

Moeritherioidea [PALEON] A suborder of extinct sirenian mammals considered as primitive proboscideans by some authorities and as a sirenian offshoot by others.

mofette [GEOL] A small opening emitting carbon dioxide in an area of late-stage volcanic activity.

mohair [TEXT] A sleek, lustrous fabric made of angora goat fibers.

Mohaupt effect [ORD] The effect of a metal liner introduced in a shaped charge to increase penetration.

mohavite *See* tincalconite.

mohm [MECH] A unit of mechanical mobility, equal to the reciprocal of 1 mechanical ohm.

Mohnian [GEOL] A North American stage of geologic time: Miocene (above Luisian, below Delmontian).

Moho *See* Mohorovičić discontinuity.

Mohole drilling [GEOL] Drilling aimed at penetration of the earth's crust, through the Mohorovičić discontinuity, to sample the mantle.

Mohorovičić discontinuity [GEOPHYS] A seismic discontinuity that separates the earth's crust from the subjacent mantle, inferred from travel time curves indicating that seismic waves undergo a sudden increase in velocity. Also known as Moho.

Mohr cubic centimeter [CHEM ENG] A unit of volume used in saccharimetry, equal to the volume of 1 gram of water at a specified temperature, usually 17.5°C, in which case, it is equal to 1.00238 cubic centimeters.

Mohr liter [MECH] A unit of volume, equal to 1000 Mohr cubic centimeters.

Mohr's circle [MECH] A graphical construction making it possible to determine the stresses in a cross section if the principal stresses are known.

Mohr's salt *See* ferrous ammonium sulfate.

Mohr titration [ANALY CHEM] Titration with silver nitrate to determine the concentration of chlorides in a solution; silver chromate precipitation is the end-point indicator.

mohsite *See* ilmenite.

Mohs scale [MINERAL] An empirical scale consisting of 10 minerals with reference to which the hardness of all other minerals is measured; it includes, from softest (designated 1) to hardest (10): talc, gypsum, calcite, fluorite, apatite, orthoclase, quartz, topaz, corundum, and diamond.

moiety [CHEM] A part or portion of a molecule, generally complex, having a characteristic chemical or pharmacological property.

moil [MIN ENG] A long steel wedge with a rounded point used for breaking up rocks in a mine.

moiré [COMMUN] In television, the spurious pattern in the reproduced picture resulting from interference beats between two sets of periodic structures in the image. [GRAPHICS] Undesirable patterns that occur when a halftone is made from a previously printed halftone or steel engraving; they are caused by the conflict between the dot arrangement produced by the halftone screen and the dots or lines of the original halftone or engraving; careful rotation of the halftone screen by the photographer or engraver may minimize moiré.

moiré effect [OPTICS] The effect whereby, when one family of curves is superposed on another family of curves so that the curves cross at angles of less than about 45°, a new family of curves appears which pass through intersections of the original curves.

moiré fringes [OPTICS] The bands which appear in the moiré effect.

moissanite [MINERAL] SiC A carbide mineral found in meteorites; identical with artificial carborundum.

moist adiabat *See* saturation adiabat.

moist-adiabatic lapse rate *See* saturation-adiabatic lapse rate.

moist air [METEOROL] **1.** In atmospheric thermodynamics, air that is a mixture of dry air and any amount of water vapor. **2.** Generally, air with a high relative humidity.

moist climate [CLIMATOL] In C. W. Thornthwaite's climatic classification, any type of climate in which the seasonal water surplus counteracts seasonal water deficiency; thus it has a moisture index greater than zero.

moist-heat sterilization [ENG] Sterilization with steam under pressure, as in an autoclave, pressure cooker, or retort; most bacteriological media are sterilized by autoclaving at 121°C, with 15 pounds pressure, for 20 minutes or more.

moisture [CLIMATOL] The quantity of precipitation or the precipitation effectiveness. [METEOROL] The water vapor content of the atmosphere, or the total water substance (gaseous, liquid, and solid) present in a given volume of air.

moisture adjustment [METEOROL] The adjustment of observed precipitation in a storm by the ratio of the estimated probable maximum precipitable water over the basin under study to the actual precipitable water calculated for the particular storm.

moisture content [MECH] The quantity of water in a mass of soil, sewage, sludge, or screenings; expressed in percentage by weight of water in the mass.

moisture factor [METEOROL] One of the simplest measures of precipitation effectiveness: moisture factor = P/T, where P is precipitation in centimeters and T is temperature degrees centigrade for the period in question.

moisture flux *See* eddy flux.

moisture inversion [METEOROL] An increase with height of the moisture content of the air; specifically, the layer through which this increase occurs, or the altitude at which the increase begins.

moisture-vapor transmission [FL MECH] The rate at which water vapor permeates a porous film (such as plastic or paper) or a wall.

Moko disease [PL PATH] Bacterial wilt of banana caused by *Pseudomonas solanacearum*.

MOL *See* manned orbiting laboratory.

molal average boiling point [PHYS CHEM] A pseudo boiling point for a mixture calculated as the summation of individual mole fraction–boiling point (in degrees Rankine) products.

molal heat capacity *See* molar heat capacity.

molality [CHEM] Concentration given as moles per 1000 grams of solvent.

molal quantity [CHEM] The number of moles (gram-molecular weights) present, expressed with weight in pounds, grams, or such units, numerically equal to the molecular weight; for example, pound-mole, gram-mole.

molal solution [CHEM] Concentration of a solution expressed in moles of solute divided by 1000 grams of solvent.

molal specific heat *See* molar specific heat.

molar [ANAT] **1.** A tooth adapted for grinding. **2.** Any of the three pairs of cheek teeth behind the premolars on each side of the jaws in man.

molar dispersion [OPTICS] In refractometry, the difference in molar refraction (refractive index) of a compound at two different light-beam wavelengths.

molar heat capacity [PHYS CHEM] The amount of heat required to raise 1 mole of a substance 1° in temperature. Also known as molal heat capacity; molecular heat capacity.

molarity [CHEM] Measure of the number of gram-molecular weights of a compound present (dissolved) in 1 liter of solution; it is indicated by M, preceded by a number to show solute concentration.

molar refraction [OPTICS] Equation for the refractive index of a compound modified by the compound's molecular weight and density. Also known as the Lorentz-Lorenz molar refraction.

molar solution [CHEM] Aqueous solution that contains 1 mole (gram-molecular weight) of solute in 1 liter of the solution.

molar specific heat [PHYS CHEM] The ratio of the amount of heat required to raise the temperature of 1 mole of a compound 1°, to the amount of heat required to raise the temperature of 1 mole of a reference substance, such as water,

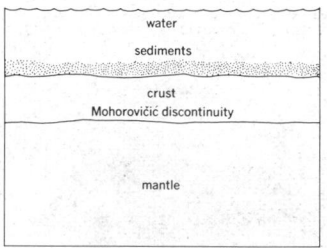

MOHOROVIČIĆ DISCONTINUITY

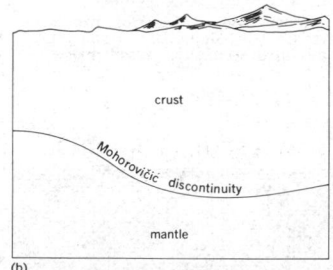

Mohorovičić discontinuity. *(a)* At average depths of about 10–12 kilometers beneath ocean basins. *(b)* At average subcontinental depths of 33–35 kilometers.

MOIRÉ EFFECT

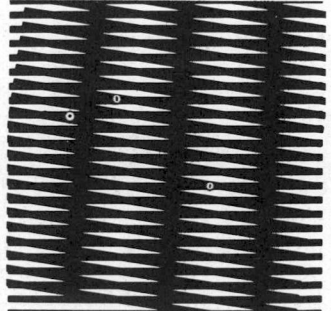

Two simple gratings crossed at a small angle.

MOLD LOFT

Interior of a mold loft. *(Newport News Shipbuilding photograph)*

MOLE

The European common mole, with thick black fur, a reduced tail, and spadelike forelimbs.

MOLECULAR BINDING

A model of a crystal surface showing different types of molecular binding sites. Molecule *A* resting on completed plane of molecules is more weakly bound than molecule *B* at ledge formed by incomplete plane. *(General Electric Co.)*

MOLECULAR DISTILLATION

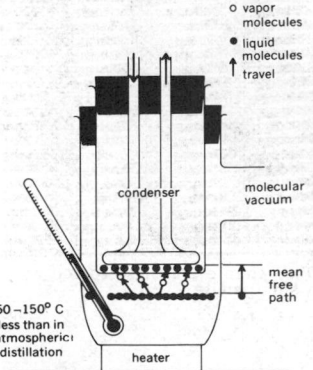

Schematic diagram illustrating distillation at molecular distillation conditions.

1° at a specified temperature. Also known as molal specific heat; molecular specific heat.

molar susceptibility [PHYS CHEM] Magnetic susceptibility of a compound per gram-mole of that compound.

molasse [GEOL] A paralic sedimentary facies consisting mainly of shale, subgraywacke sandstone, and conglomerate; it is more clastic and less rhythmic than the preceding flysch and is generally postorogenic.

molasses [FOOD ENG] A brown viscid syrup prepared from raw sugar during sugar manufacturing processes.

molasses/A.N. explosive [MATER] An explosive mixture consisting of ammonium nitrate mixed with molasses and water.

mold [ENG] **1.** A pattern or template used as a guide in construction. **2.** A cavity which imparts its form to a fluid or malleable substance. [ENG ACOUS] The metal part derived from the master by electroforming in reproducing disk recordings; has grooves similar to those of the recording. [GEOL] Soft, crumbling friable earth. [GRAPHICS] **1.** To form a plastic substance by placing it in a matrix or form. **2.** The form or matrix for shaping a plastic substance. [MYCOL] Any of various woolly fungus growths. [PALEON] An impression made in rock or earth material by an inner or outer surface of a fossil shell or other organic structure; a complete mold would be the hollow space.

moldability [MATER] The capability of being molded.

moldauite *See* moldavite.

moldavite [GEOL] A translucent, olive- to brownish-green or pale-green tektite from western Czechoslovakia, characterized by surface sculpturing due to solution etching. Also known as moldauite; pseudochrysolite; vitavite. [MINERAL] A variety of ozocerite from Moldavia.

moldboard plow [AGR] A plow equipped with a curved iron plate (moldboard) that lifts and turns the soil. Also known as turnplow.

molded breadth [NAV ARCH] The maximum breadth of a ship measured to the outside of the frame bar usually at the midship section.

molded capacitor [ELEC] Capacitor, usually mica, that has been encased in a molded plastic insulating material.

molded depth [NAV ARCH] The vertical distance from the base line to the molded line of the main deck at side measured at the midship section.

molded-fabric bearing [DES ENG] A bearing composed of laminations of cotton or other fabric impregnated with a phenolic resin and molded under heat and pressure.

molded lines [ENG] Full-size lines of a ship or airplane which are laid out in a mold loft. [NAV ARCH] Line drawings of a ship which represent the shape inside the plating or outside the planking on a wooden vessel.

mold efficiency [ENG] In a multimold blow-molding system, the percentage of the total turn-around time actually required for the forming, cooling, and ejection of the formed objects.

molding [ARCH] A continuous contour surface of rectangular or curved profile, used on a plane surface such as a wall to effect a transition or create a decorative effect by the play of shadow and light on it.

molding machine [MET] A machine that compacts sand around a pattern to form a mold.

molding press [MET] A press used to form compacts in powder metallurgy.

mold loft [ENG] A large building with a smooth wooden floor where full-size lines of a ship or airplane are laid down and templates are constructed from them to lay off the steel for cutting.

mold release *See* release agent.

mold steel [MET] Steel of tool-steel quality used to make molds for plastics; properties include uniform texture, good machinability with die-sinking tools, and lack of microscopic porosity.

mold wash [MET] An aqueous or alcoholic suspension or emulsion used to coat the surfaces of a mold cavity.

mole [CHEM] An amount of substance of a system which contains as many elementary units as there are atoms of carbon in 0.012 kilogram of the pure nuclide carbon-12; the elementary unit must be specified and may be an atom, molecule, ion, electron, photon, or even a specified group of such units. [CIV ENG] A breakwater or berthing facility,

extending from shore to deep water, with a core of stone or earth. [MECH ENG] A mechanical tunnel excavator. [MED] **1.** A mass formed in the uterus by the maldevelopment of all or part of the embryo or of the placenta and membranes. **2.** A fleshy, pigmented nevus. [VERT ZOO] Any of 19 species of insectivorous mammals composing the family Talpidae; the body is stout and cylindrical, with a short neck, small or vestigial eyes and ears, a long naked muzzle, and forelimbs adapted for digging.

molectronics *See* molecular electronics.

molecular adhesion [PHYS CHEM] A particular manifestation of intermolecular forces which causes solids or liquids to adhere to each other; usually used with reference to adhesion of two different materials, in contrast to cohesion.

molecular amplitude [ANALY CHEM] The difference between the molecular rotation at the extreme (peak or trough) value caused by the longer light wavelength and the molecular rotation at the extreme value caused by the shorter wavelength.

molecular association [PHYS CHEM] The formation of double molecules or polymolecules from a single species as a result of specific and moderately strong intermolecular forces.

molecular attraction [PHYS CHEM] A force which pulls molecules toward each other.

molecular beam [PHYS] A beam of neutral molecules whose directions of motion lie within a very small solid angle.

molecular-beam apparatus [PHYS] A device in which a molecular beam in a vacuum is subjected to magnetic fields, oscillating fields, or other influences, and a detector measures the resulting intensity of the beam at some location; used primarily in radio-frequency spectroscopy.

molecular beam epitaxy [SOLID STATE] A technique of growing single crystals in which beams of atoms or molecules are made to strike a single-crystalline substrate in a vacuum, giving rise to crystals whose crystallographic orientation is related to that of the substrate. Abbreviated MBE.

molecular binding [SOLID STATE] The force which holds a molecule at some site on the surface of a crystal.

molecular biology [BIOL] That part of biology which attempts to interpret biological events in terms of the physicochemical properties of molecules in a cell.

molecular circuit [ELECTR] A circuit in which the individual components are physically indistinguishable from each other.

molecular conductivity [PHYS CHEM] The conductivity of a volume of electrolyte containing 1 mole of dissolved substance.

molecular crystal [CRYSTAL] A solid consisting of a lattice array of molecules such as hydrogen, methane, or more complex organic compounds, bound by weak van der Waals forces, and therefore retaining much of their individuality.

molecular diamagnetism [PHYS CHEM] Diamagnetism of compounds, especially organic compounds whose susceptibilities can often be calculated from the atoms and chemical bonds of which they are composed.

molecular diameter [PHYS CHEM] The diameter of a molecule, assuming it to be spherical; has a numerical value of 10^{-8} centimeter multiplied by a factor dependent on the compound or element.

molecular diffusion [FL MECH] The transfer of mass between adjacent layers of fluid in laminar flow.

molecular dipole [PHYS CHEM] A molecule having an electric dipole moment, whether it is permanent or produced by an external field.

molecular distillation [CHEM] A process by which substances are distilled in high vacuum at the lowest possible temperature and with least damage to their composition.

molecular drag pump [ENG] A vacuum pump in which pumping is accomplished by imparting a high momentum to the gas molecules by impingement of a body rotating at very high speeds, as much as 16,000 revolutions per minute; such pumps achieve a vacuum as high as 10^{-6} torr.

molecular effusion [FL MECH] Mass-transfer flow mechanism of free-molecule transfer through pores or orifices.

molecular electronics [ELECTR] The branch of electronics that deals with the production of complex electronic circuits in microminiature form by producing semiconductor devices and circuit elements integrally while growing multizoned crystals in a furnace. Also known as molectronics.

(From molecular distillation diagram labels:) o vapor molecules · liquid molecules ↑ travel; condenser; molecular vacuum; mean free path; 50–150° C less than in atmospheric distillation; heater

molecular energy level [PHYS CHEM] One of the states of motion of nuclei and electrons in a molecule, having a definite energy, which is allowed by quantum mechanics.

molecular engineering [ELECTR] The use of solid-state techniques to build, in extremely small volumes, the components necessary to provide the functional requirements of overall equipments, which when handled in more conventional ways are vastly bulkier.

molecular field theory See Weiss theory.

molecular flow [FL MECH] Gas-flow phenomenon at low pressures or in small channels when the mean free path is of the same order of magnitude as the channel diameter; a gas molecule thus migrates along the channel independent of other gas molecules present.

molecular gas [CHEM] A gas composed of a single species, such as oxygen, chlorine, or neon.

molecular genetics [MOL BIO] The approach which deals with the physics and chemistry of the processes of inheritance.

molecular heat [THERMO] The heat capacity per mole of a substance.

molecular heat diffusion [THERMO] Transfer of heat through the motion of molecules.

molecular laser [OPTICS] Any gas laser in which the gas consists of molecules rather than atoms; such a laser can be operated on a large number of rotational-vibrational lines, and, at a sufficiently high pressure, these lines overlap and a wide gain region is obtained.

molecular magnet [PHYS CHEM] A molecule having a non-vanishing magnetic dipole moment, whether it is permanent or produced by an external field.

molecular optics [OPTICS] The study of the propagation of light and associated phenomena, such as refraction, absorption, and scattering, through collections of molecules in gases, liquids, and solids.

molecular orbital [PHYS CHEM] A wave function describing an electron in a molecule.

molecular paramagnetism [PHYS CHEM] Paramagnetism of molecules, such as oxygen, some other molecules, and a large number of organic compounds.

molecular pathology [PATH] The study of the bases and mechanisms of disease on a molecular or chemical level.

molecular physics [PHYS] The study of the behavior and structure of molecules, including the quantum-mechanical explanation of several kinds of chemical binding between atoms in a molecule, directed valence, the polarizability of molecules, the quantization of vibrational, rotational, and electronic motions of molecules, and the phenomena arising from intermolecular forces.

molecular polarizability [PHYS CHEM] The electric dipole moment induced in a molecule by an external electric field, divided by the magnitude of the field.

molecular pump [MECH ENG] A vacuum pump in which the molecules of the gas to be exhausted are carried away by the friction between them and a rapidly revolving disk or drum.

molecular relaxation [PHYS CHEM] Transition of a molecule from an excited energy level to another excited level of lower energy or to the ground state.

molecular rotation [OPTICS] In a solution of an optically active compound, the specific rotation (angular rotation of polarized light) multiplied by the compound's molecular weight.

molecular specific heat See molar specific heat.

molecular spectroscopy [SPECT] The production, measurement, and interpretation of molecular spectra.

molecular spectrum [SPECT] The intensity of electromagnetic radiation emitted or absorbed by a collection of molecules as a function of frequency, wave number, or some related quantity.

molecular still [CHEM] An apparatus used to conduct molecular distillation.

molecular stopping power [NUCLEO] For an ionizing particle passing through a compound, the particle's energy loss per molecule within a unit area normal to the particle's path; equal to the linear energy transfer (energy loss per unit path length) divided by the number of molecules per unit volume.

molecular structure [PHYS CHEM] The manner in which electrons and nuclei interact to form a molecule, as elucidated by quantum mechanics and a study of molecular spectra.

molecular theory See kinetic theory.

molecular vibration [PHYS CHEM] The theory that all atoms within a molecule are in continuous motion, vibrating at definite frequencies specific to the molecular structure as a whole as well as to groups of atoms within the molecule; the basis of spectroscopic analysis.

molecular volume [CHEM] The volume that is occupied by 1 mole (gram-molecular weight) of an element or compound; equals the molecular weight divided by the density.

molecular weight [CHEM] The sum of the atomic weights of all the atoms in a molecule.

molecular-weight distribution [ORG CHEM] Frequency of occurrence of the different molecular-weight chains in a homologous polymeric system.

molecule [CHEM] A group of atoms held together by chemical forces; the atoms in the molecule may be identical as in H_2, S_2, and S_8, or different as in H_2O and CO_2; a molecule is the smallest unit of matter which can exist by itself and retain all its chemical properties.

mole drain [CIV ENG] A subsurface channel for water drainage; formed by pulling a solid object, usually a solid cylinder having a wedge-shaped point at one end, through the soil at the proper slope and depth.

mole fraction [CHEM] The ratio of the number of moles of a substance in a mixture or solution to the total number of moles of all the components in the mixture or solution.

mole mining [MIN ENG] A method of working coal seams about 30 inches thick; a small continuous-miner type of machine is used, which is remote-controlled from the roadway, without associated supports.

mole percent [CHEM] Percentage calculation expressed in terms of moles rather than weight.

moleskin [TEXT] Rugged satin-weave cotton with a soft napped back having a one-warp, two-filling construction.

mole volume [CHEM] The volume, 22.414 liters, that is occupied by a mole (gram-molecular weight) of an ideal gas, measured at standard conditions. Also known as molal volume.

Molidae [VERT ZOO] A family of marine fishes, including some species of sunfishes, in the order Perciformes.

Møller scattering [QUANT MECH] Scattering of electrons by electrons.

Mollicutes [MICROBIO] A class containing the single order Mycoplasmatales; name refers to the absence of a true cell wall and the plasticity of the outer membrane.

Mollier diagram [THERMO] Graph of enthalpy versus entropy of a vapor on which isobars, isothermals, and lines of equal dryness are plotted.

Mollisol [GEOL] An order of soils having dark or very dark, friable, thick A horizons high in humus and bases such as calcium and magnesium; most have lighter-colored or browner B horizons that are less friable and about as thick as the A horizons; all but a few have paler C horizons, many of which are calcareous.

Moll thermopile [ENG] A thermopile used in some types of radiation instruments; alternate junctions of series-connected manganan-constantan thermocouples are embedded in a shielded nonconducting plate having a large heat capacity; the remaining junctions, which are blackened, are exposed directly to the radiation; the voltage developed by the thermocouple is proportional to the intensity of radiation.

Mollusca [INV ZOO] One of the divisions of phyla of the animal kingdom containing snails, slugs, octopuses, squids, clams, mussels, and oysters; characterized by a shell-secreting organ, the mantle, and a radula, a food-rasping organ located in the forward area of the mouth.

molluscicide [MATER] An agent that kills mollusks.

molluscum contagiosum [MED] A viral disease of the skin, characterized by one or more discrete, waxy, dome-shaped nodules with frequent umbilication.

mollusk [INV ZOO] Any member of the Mollusca.

Molossidae [VERT ZOO] The free-tailed bats, a family of tropical and subtropical insectivorous mammals in the order Chiroptera.

Molpadida [INV ZOO] An order of sea cucumbers belonging

MOLYBDENITE

Molybdenite on biotite mica from Edison, New Jersey. *(American Museum of Natural History Specimens)*

MOLYBDENUM

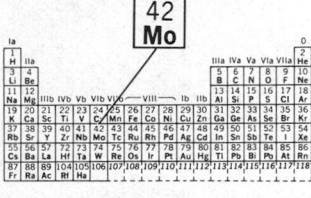

Periodic table of the chemical elements showing the position of molybdenum.

MONAURAL SOUND

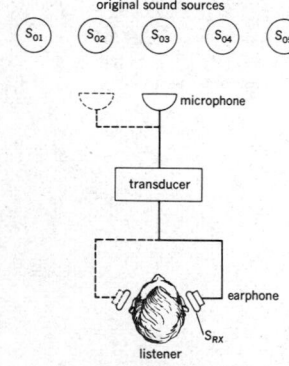

Monaural sound-reproducing system.

MONAZITE

2.5 cm

Monazite crystals from Minas Gerais, Brazil. *(Specimen from Department of Geology, Bryn Mawr College)*

---

to the Apodacea and characterized by a short, plump body bearing a taillike prolongation.

Molpadidae [INV ZOO] The single family of the echinoderm order Molpadida.

molt [PHYSIO] To shed an outer covering as part of a periodic process of growth.

molten-salt reactor [NUCLEO] A nuclear reactor in which the fissile and fertile material, in the form of fluoride salts, is dissolved in the coolant, which is a molten mixture of salts such as lithium fluoride and beryllium fluoride. Abbreviated MSR. Also known as fused-salt reactor.

molting hormone [BIOCHEM] Any of several hormones which activate molting in arthropods.

molybdate [INORG CHEM] A salt derived from a molybdic acid.

molybdate orange [MATER] Pigment that is a solid solution of lead chromate, molybdate, and sulfate; used in paints, inks, and plastics.

molybdenite [MINERAL] MoS_2 A metallic, lead-gray mineral that crystallizes in the hexagonal system and is commonly found in scales or foliated masses; hardness is 1.5 on Mohs scale, and specific gravity is 4.7; it is chief ore of molybdenum.

molybdenum [CHEM] A chemical element, symbol Mo, atomic number 42, and atomic weight 95.95. [MET] A silvery-gray metal used in iron-based alloys.

molybdenum cast iron [MET] Cast iron containing small amounts of molybdenum, added as ferromolybdenum or calcium molybdenum; increases strength, toughness, and wear resistance.

molybdenum dioxide [INORG CHEM] MoO_2 Lead-gray powder; insoluble in hydrochloric and hydrofluoric acids; used in pigment for textiles.

molybdenum disulfide [INORG CHEM] MoS_2 A black lustrous powder, melting at 1185°C, insoluble in water, soluble in aqua regia and concentrated sulfuric acid; used as a dry lubricant and an additive for greases and oils. Also known as molybdenum sulfide; molybdic sulfide.

molybdenum pentachloride [INORG CHEM] $MoCl_5$ Hygroscopic gray-black needles melting at 194°C; reacts with water and air; soluble in anhydrous organic solvents; used as a catalyst and as raw material to make molybdenum hexacarbonyl.

molybdenum sesquioxide [INORG CHEM] MoO_3 Water-insoluble, gray-black powder with slight solubility in acids; used as a catalyst and as a coating for metal articles.

molybdenum silicide [MET] A mixture of molybdenum, silicon, and iron in the proportion 60:30:10; used to introduce molybdenum into steel melts.

molybdenum steel [MET] **1.** A carbon steel containing usually less than 0.5% molybdenum to aid hardenability. **2.** A tool steel containing up to 10% molybdenum, up to 1.5% carbon, and varying amounts of chromium, vanadium tungsten, and sometimes cobalt.

molybdenum sulfide *See* molybdenum disulfide.

molybdic acid [INORG CHEM] Any acid derived from molybdenum trioxide, especially the simplest acid H_2MoO_4, obtained as white crystals.

molybdic ocher *See* molybdite.

molybdic sulfide *See* molybdenum disulfide.

molybdine *See* molybdite.

molybdite [MINERAL] MoO_3 A mineral, much of which is actually ferrimolybdite. Also known as molybdic ocher; molybdine.

molybdophyllite [MINERAL] $(Pb,Mg)_2SiO_4 \cdot H_2O$ A colorless, white, or pale-green mineral composed of a silicate of lead and magnesium.

moly-blacks [MATER] Lustrous, black, molybdenum-containing decorative coatings used to blacken zinc or zinc-base alloys.

molysite [MINERAL] $FeCl_3$ A brownish-red or yellow mineral composed of native ferric chloride, occurring in lava at Vesuvius.

molysmophobia [PSYCH] Abnormal fear of infection or contamination.

moment [MECH] Static moment of some quantity, except in the term "moment of inertia." [STAT] The nth moment of a distribution $f(x)$ about a point x_0 is the expected value of

---

$(x - x_0)^n$, that is, the integral of $(x - x_0)^n df(x)$, where $df(x)$ is the probability of some quantity's occurrence; the first moment is the mean of the distribution, while the variance may be found in terms of the first and second moments.

moment coefficient [AERO ENG] The coefficients used for moment are similar to coefficients of lift, drag, and thrust, and are likewise dimensionless; however, these must include a characteristic length, in addition to the area; the span is used for rolling or yawing moment, and the chord is used for pitching moment.

moment diagram [MECH] A graph of the bending moment at a section of a beam versus the distance of the section along the beam.

moment of force *See* torque.

moment of inertia [MECH] The sum of the products formed by multiplying the mass (or sometimes, the area) of each element of a figure by the square of its distance from a specified line. Also known as rotational inertia.

moment of momentum *See* angular momentum.

momentum [MECH] Also known as linear momentum; vector momentum. **1.** For a single nonrelativistic particle, the product of the mass and the velocity of a particle. **2.** For a single relativistic particle, $mv/(1 - v^2/c^2)^{1/2}$, where m is the rest-mass, v the velocity, and c the speed of light. **3.** For a system of particles, the vector sum of the momenta (as in the first or second definition) of the particles.

momentum conservation *See* conservation of momentum.

momentum thrust *See* thrust.

momentum-transport hypothesis [FL MECH] The hypothesis that the principle of conservation of momentum is valid in turbulent eddy transfer.

Momertz-Lentz system [MIN ENG] Placement of two winding engines alongside the top of the mine shaft using the shaft collar as a common foundation; results in practically vertical ropes and less rope oscillation.

Momotidae [VERT ZOO] The motmots, a family of colorful New World birds in the order Coraciiformes.

monactine [INV ZOO] A single-rayed spicule in the sponges.

monadelphous [BOT] Having the filaments of the stamens united into one set.

monadic operation [ADP] An operation on one operand, such as a negation.

Monadidae [INV ZOO] A family of flagellated protozoans in the order Kinetoplastida having two flagella of uneven length.

monadnock [GEOL] A remnant hill of resistant rock rising abruptly from the level of a peneplain; commonly represents an outcrop of rock that has withstood erosion. Also known as torso mountain.

monalbite [MINERAL] A modification of albite with monoclinic symmetry that is stable under equilibrium conditions at temperatures (about 1000°C) near the melting point.

monandrous [BOT] Having one stamen.

monatomic [CHEM] Composed of one atom.

monatomic gas [CHEM] A gas whose molecules have only one atom; the inert gases are examples.

monaural sound [ENG ACOUS] Sound produced by a system in which one or more microphones are connected to a single transducing channel which is coupled to one or two earphones worn by the listener.

monaxon [INV ZOO] A spicule formed by growth along a single axis.

monazite [MINERAL] A yellow or brown rare-earth phosphate monoclinic mineral with appreciable substitution of thorium for rare-earths and silicon for phosphorus; the principal ore of the rare earths and of thorium. Also known as cryptolite.

monchiquite [PETR] A lamprophyre composed of olivine, pyroxene, and usually mica or amphibole phenocrysts embedded in a glass or analcime groundmass.

Mönckeberg's arteriosclerosis *See* medial arteriosclerosis.

Monday fever *See* metal fume fever.

Mond process [MET] A process for extracting and purifying nickel whereby nickel carbonyl is first formed by reaction of the reduced metal with carbon monoxide, and then the nickel carbonyl is decomposed thermally, resulting in deposition of nickel.

Monel [MET] Trade name for a corrosion-resistant alloy containing 67% nickel and 30% copper; alloyed 3% aluminum increases strength; alloyed 3–4% silicon increases strength and nongalling characteristics.

Monel 402 [MET] An alloy of 58% nickel and 40% copper, used for its corrosion resistance, for example, for pickling equipment.

Monera [BIOL] A kingdom that includes the bacteria and blue-green algae in some classification schemes.

monestrous [PHYSIO] Having a single estrous cycle per year.

monetite [MINERAL] $CaHPO_4$ A yellowish-white mineral consisting of an acid calcium hydrogen phosphate, occurring in crystals.

Monge's disease See mountain sickness.

Monge's theorem [MATH] For three coplanar circles, and for radii of these circles which are parallel to each other, the three outer centers of similitude of the circles taken in pairs lie on a single straight line, and any two inner centers of similitude lie on a straight line with one of the outer centers.

Mongolian spot [MED] A focal bluish-gray discoloration of the skin of the lower back, also aberrantly on the face, present at birth and fading gradually.

mongolism See Down's syndrome.

mongoloid [ANTHRO] Characteristic of a Far Eastern racial stock and often American Indians; features yellow complexion, a broad flat face with small nose and prominent cheekbones, and eyes that have an epicanthal fold. [MED] Having physical characteristics associated with Down's syndrome.

mongoose [VERT ZOO] The common name for 39 species of carnivorous mammals which are members of the family Viveridae; they are plantigrade animals and have a long slender body, short legs, nonretractile claws, and scent glands.

Monhysterida [INV ZOO] An order of aquatic nematodes in the subclass Chromadoria.

Monhysteroidea [INV ZOO] A superfamily of free-living nematodes in the order Monhysterida characterized by single or paired outstretched ovaries, circular to cryptospiral amphids, and a stoma which is usually shallow and unarmed.

Monilia albicans See Candida albicans.

Moniliaceae [MYCOL] A family of fungi in the order Moniliales; sporophores are usually lacking, but when present they are aggregated into fascicles, and hyphae and spores are hyaline or brightly colored.

Moniliales [MYCOL] An order of fungi of the Fungi Imperfecti containing many plant pathogens; asexual spores are always formed free on the surface of the material on which the organism is living, and never occur in either pycnidia or acervuli.

monilial vaginitis [MED] Inflammation of the vagina caused by a fungus of the genus *Monilia*.

moniliasis See candidiasis.

moniliform [BIOL] Constructed with contractions and expansions at regular alternating intervals, giving the appearance of a string of beads.

monimolimnion [HYD] The dense bottom stratum of a meromictic lake; it is stagnant and does not mix with the water above.

monimolite [MINERAL] $(Pb,Ca)_3Sb_2O_8$ Yellowish to brownish or greenish mineral composed of lead calcium antimony oxide; it may contain ferrous iron.

monitor [ADP] To supervise a program, and check that it is operating correctly during its execution, usually by means of a diagnostic routine. [ENG] **1.** An instrument used to measure continuously or at intervals a condition that must be kept within prescribed limits, such as radioactivity at some point in a nuclear reactor, a variable quantity in an automatic process control system, the transmissions in a communication channel or bank, or the position of an aircraft in flight. **2.** To use meters or special techniques to measure such a condition. **3.** A person who watches a monitor. [VERT ZOO] Any of 27 carnivorous, voracious species of the reptilian family Varanidae characterized by a long, slender forked tongue and a dorsal covering of small, rounded scales containing pointed granules.

monitor board [COMMUN] A console at which a supervising telephone operator sits and from which she can intercept calls being handled by other operators.

monitor control dump [ADP] A memory dump routinely carried out by the system once a program has been run.

monitor display [ADP] The facility of stopping the central processing unit and displaying information of main storage and internal registers; after manual intervention, normal instruction execution can be initiated.

monitoring amplifier [ELECTR] A power amplifier used primarily for evaluation and supervision of a program.

monitoring key [ELECTR] Key which, when operated, makes it possible for an attendant or operator to listen on a telephone circuit without appreciably impairing transmission on the circuit.

monitor operating system [ADP] The control of the routines which achieves efficient use of all the hardware components.

monitor printer [ADP] Input-output device, capable of receiving coded signals from the computer, which automatically operates the keyboard to print a hard copy and, when desired, to punch paper tape. [COMMUN] A teleprinter used in a technical control facility or communications center for checking incoming teletypewriter signals.

monitor routine See executive routine.

monitor system See executive system.

monitor system components [ADP] The set of programs which form the complete operating system: the control program (task management, job management, and data management routines) and the processing programs.

monkey [MIN ENG] **1.** An appliance for mechanically gripping or releasing the rope in rope haulage. **2.** An airway in an anthracite mine. [VERT ZOO] Any of several species of frugivorous and carnivorous primates which compose the families Cercopithecidae and Cebidae in the suborder Anthropoidea; the face is typically flattened and hairless, all species are pentadactyl, and the mammary glands are always in the pectoral region.

monkey chatter See adjacent-channel interference.

monkey drift [MIN ENG] A small drift driven in for prospecting purposes, or a crosscut driven to an airway above the gangway.

monkey heading [MIN ENG] A narrow, low passage in the coal, providing refuge for miners while coal is blasted.

monkey ladder [MIN ENG] A ladder of saplings; the widely separated steps rest in the coal.

monkey winch [MIN ENG] A device for exerting a strong pull; consists of a framework containing a hand-operated drum, around which a steel rope 50 feet long is wound.

monkey wrench [DES ENG] A wrench having one jaw fixed and the other adjustable, both of which are perpendicular to a straight handle.

monk's cloth [TEXT] A coarse basket-weave fabric, usually made of thick cotton. Also known as friar's cloth.

monkshood See aconite.

mono- [CHEM] A prefix for chemical compounds to show a single radical; for example, monoglyceride, a glycol ester on which a single acid group is attached to the glycerol group.

monoacetate [ORG CHEM] A compound such as a salt or ester that contains one acetate group.

monoacetin See glycerol monoacetate.

monoacid [CHEM] Compound with a single acid group, such as hydrochloric acid, HCl, or 2-naphthol-7-sulfonic acid, $C_{10}H_6(OH)(SO_3H)$.

monoamine [ORG CHEM] An amine compound that has only one amino group.

monoamine oxidase [BIOCHEM] An enzyme that catalyzes oxidative deamination of monoamines.

monoamine oxidase inhibitor [PHARM] Any drug, such as isocarboxazid and tranylcypromine, that inhibits monoamine oxidase and therby leads to an accumulation of the amines on which the enzyme normally acts.

monoammonium tartrate See ammonium bitartrate.

monobasic [CHEM] Pertaining to an acid with one displaceable hydrogen atom, such as hydrochloric acid, HCl.

monobasic calcium phosphate See calcium phosphate.

monobasic sodium phosphate [INORG CHEM] NaH_2PO_4 White crystals that are slightly hygroscopic, soluble in water, insoluble in alcohol; used in baking powders and acid cleansers, and as a cattle-food supplement.

monoblast [HISTOL] A motile cell of the spleen and bone marrow from which monocytes are derived.

monoblastic leukemia See acute monocytic leukemia.

MONGOOSE

The golden-brown mongoose, a stout, weasellike animal which grows to a length of 15–18 inches (34–45 centimeters), excluding the tail.

MONKEY

The spider monkey (*Ateles*), with slender body, long prehensile tail, long muscular limbs, and no thumbs.

MONOCHROME SIGNAL

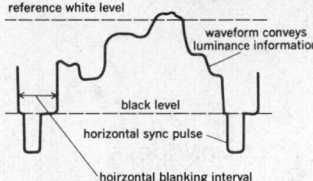

Waveform sketch of a normal monochrome signal, one of the major components of the color television signal.

MONOCOLPATE POLLEN

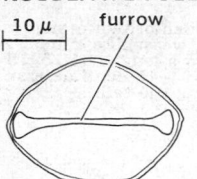

Monocolpate pollen of *Zamia floridana*.

MONOCYTE

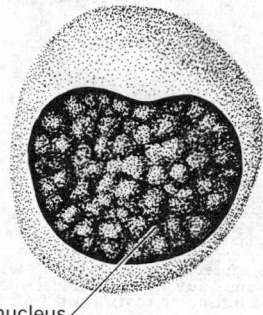

Diagrammatic representation of a monocyte.

MONODELLIDAE

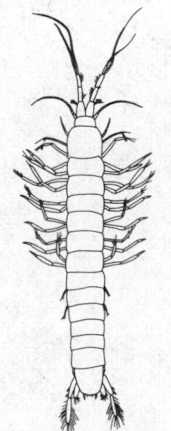

Drawing of a male *Monodella halophila* Karaman showing the seven pairs of appendages and the characteristic telson. *(After S. Karaman)*

Monoblepharidales [MYCOL] An order of aquatic fungi in the class Phycomycetes; distinguished by a mostly hyphal thallus and zoospores with one posterior flagellum.

monobloc projectile [ORD] Armor-piercing projectile which consists of one piece of steel, suitably heat-treated; may be provided with a false ogive to decrease air resistance.

Monobothrida [PALEON] An extinct order of monocyclic camerate crinoids.

monobrid circuit [ELECTR] Integrated circuit using a combination of monolithic and multichip techniques by means of which a number of monolithic circuits, or a monolithic device in combination with separate diffused or thin-film components, are interconnected in a single package.

monocable [MECH ENG] An aerial ropeway that uses one rope to both support and haul a load.

monocalcium phosphate *See* calcium phosphate.

monocardiogram *See* vectorcardiogram.

monocarpic [BOT] Bearing fruit once and then dying.

monocarpous [BOT] Having a single ovary.

Monoceros [ASTRON] A constellation, right ascension 7 hours, declination 5° south; it has mostly faint stars.

monocharge electret [ELECTR] A type of foil electret that carries electrical charge of the same sign on both surfaces.

monochlamydous [BOT] Referring to flowers having only one set of floral envelopes, that is, either a calyx or a corolla.

monochloroacetone *See* chloroacetone.

monochloroethane *See* ethyl chloride.

monochromasia [MED] Complete color blindness in which all colors appear as shades of gray. Also known as monochromatism.

monochromat [MED] An individual who suffers from total color blindness even at high light levels; such persons are typically deficient or lacking in cone receptors, so that their form vision is also poor.

monochromatic [OPTICS] Pertaining to the color of a surface which radiates light having an extremely small range of wavelengths. [PHYS] Consisting of electromagnetic radiation having an extremely small range of wavelengths, or particles having an extremely small range of energies.

monochromatic interference [OPTICS] Interference between beams coming from a source of monochromatic light.

monochromatic light [OPTICS] Light of one color, having wavelengths confined to an extremely narrow range.

monochromatic neutron beam [NUCLEO] A beam of neutrons whose energies are confined to an extremely narrow range of values.

monochromatic radiation [ELECTROMAG] Electromagnetic radiation having wavelengths confined to an extremely narrow range.

monochromatic temperature scale [THERMO] A temperature scale based upon the amount of power radiated from a blackbody at a single wavelength.

monochromatism *See* monochromasia.

monochromator [SPECT] A spectrograph in which a detector is replaced by a second slit, placed in the focal plane, to isolate a particular narrow band of wavelengths for refocusing on a detector or experimental object.

monochrome [OPTICS] Having only one chromaticity.

monochrome channel [ELECTR] In a color television system, any path which is intended to carry the monochrome signal; the monochrome channel may also carry other signals.

monochrome signal [ELECTR] **1.** A signal wave used for controlling luminance values in monochrome television. **2.** The portion of a signal wave that has major control of the luminance values in a color television system, regardless of whether the picture is displayed in color or in monochrome. Also known as M signal.

monochrome television [COMMUN] Television in which the final reproduced picture is monochrome, having only shades of gray between black and white. Also known as black-and-white television.

monoclimax [ECOL] A climax community controlled primarily by one factor, as climate.

monocline [GEOL] A stratigraphic unit that dips from the horizontal in one direction only, not as part of an anticline or syncline.

monoclinic [BOT] Having both stamens and pistils in the same flower.

monoclinic system [CRYSTAL] One of the six crystal systems characterized by a single, two-fold symmetry axis or a single symmetry plane.

monoclonal cryoglobulin [BIOCHEM] A cryoglobulin composed of immunoglobin with only one class or subclass of heavy and light chain.

monocolpate pollen [BOT] Pollen grains having a single furrow.

monocoque [AERO ENG] A type of construction, as of a rocket body, in which all or most of the stresses are carried by the skin.

monocord switchboard [ELEC] Local battery switchboard in which each telephone line terminates in a single jack and plug.

monocotyledon [BOT] Any plant of the class Liliopsida; all have a single cotyledon.

Monocotyledoneae [BOT] The equivalent name for Liliopsida.

monocrepid [INV ZOO] A desma formed by secondary deposits of silica on a monaxon.

monocular vision [MED] Sight with one eye.

Monocyathea [PALEON] A class of extinct parazoans in the phylum Archaeocyatha containing single-walled forms.

monocyte [HISTOL] A large (about 12 micrometers), agranular leukocyte with a relatively small, eccentric, oval or kidney-shaped nucleus.

monocytic angina *See* infectious mononucleosis.

monocytic leukemia [MED] A form of leukemia in which monocytic cells are predominant in the blood. Also known as myelomonocytic leukemia.

monocytoma [MED] A neoplasm composed principally of monocytes, usually anaplastic.

monocytopenia [MED] Reduction in the number of circulating monocytes per unit volume of blood to below the minimum normal levels.

monocytosis [MED] Increase in the number of circulating monocytes per unit volume of blood to above the maximum normal levels.

monodactylous [ZOO] Having a single digit or claw.

Monodellidae [INV ZOO] A monogeneric family of crustaceans in the order Thermosbaenacea distinguished by seven pairs of biramous pereiopods on thoracomeres 2–8, and by not having the telson united to the last pleonite.

monodelphic [VERT ZOO] **1.** Having a single genital tract, in the female. **2.** Having a single uterus.

monodispersity [ORG CHEM] Polymer system that is homogeneous in molecular weight, that is, it does not have a distribution of different molecular-weight chains within the total mass.

monodromy theorem [MATH] If a complex function is analytic at a point of a bounded simply connected domain and can be continued analytically along every curve from the point, then it represents a single-valued analytic function in the domain.

monoecious [BOT] **1.** Having both staminate and pistillate flowers on the same plant. **2.** Having archegonia and antheridia on different branches. [ZOO] Having male and female reproductive organs in the same individual. Also known as hermaphroditic.

Monoedidae [INV ZOO] An equivalent name for Colydiidae.

monoenergetic gamma rays [PHYS] A beam of gamma rays whose energies are confined to an extremely narrow range.

monoester [ORG CHEM] An ester that has only one ester group.

monoethanolamine *See* ethanolamine.

monofier [ELECTR] Complete master oscillator and power amplifier system in a single evacuated tube envelope; electrically, it is equivalent to a stable low-noise oscillator, an isolator, and a two- or three-cavity klystron amplifier.

monofilament [TEXT] A single, large, continuous filament (single-strand thread) of a natural or synthetic fiber.

monofuel propulsion [AERO ENG] Propulsion system which obtains its power from a single fuel; in rocket units, the fuel furnishes both oxygen supply and the hydrocarbon for combustion.

monogamy [ANTHRO] Marriage to only one person at a time.

Monogenea [INV ZOO] A diverse subclass of the Trematoda which are principally ectoparasites of fishes; individuals have enlarged anterior and posterior holdfasts with paired suckers anteriorly and opisthaptors posteriorly.

Monogenoidea [INV ZOO] A class of the Trematoda in some systems of classification; equivalent to the Monogenea of other systems.

monogeosyncline [GEOL] A primary geosyncline that is long, narrow, and deeply subsided; composed of the sediments of shallow water and situated along the inner margin of the borderlands.

monoglyceride [ORG CHEM] Any of the fatty-acid glycerol esters where only one acid group is attached to the glycerol group, for example, $RCOOCH_2CHOHCH_2OH$; examples are glycerol monostearate and monolaurate; used as emulsifiers in cosmetics and lubricants.

Monogonota [INV ZOO] An order of the class Rotifera, characterized by the presence of a single gonad in both males and females.

monogony [BIOL] Asexual reproduction.

monogynous [BOT] Having only one pistil. [VERT ZOO] **1.** Having only one female in a colony. **2.** Consorting with only one female.

monohybrid [GEN] A hybrid individual heterozygous for one gene or a single character.

monoid [MATH] A semigroup which has an identity element.

monoideism [PSYCH] A mental condition marked by the domination of a single idea; persistent and thorough preoccupation with one idea, but seldom an idea that is complete.

monolayer See monomolecular film.

monolith [MATER] A large concrete block.

monolithic circuit See integrated circuit.

monolithic filter [COMMUN] A device used to separate telephone communications sent simultaneously over the transmission line, consisting of a series of electrodes vacuum-deposited on a crystal plate so that the plated sections are resonant with ultrasonic sound waves, and the effect of the device is similar to that of an electric filter.

monolithic integrated circuit See integrated circuit.

monomer [CHEM] A simple molecule which is capable of combining with a number of like or unlike molecules to form a polymer; it is a repeating structure unit within a polymer.

monomethylamine See methylamine.

monomial [MATH] A polynomial of degree one.

monomineralic [PETR] Of a rock, composed entirely or principally of a single mineral.

Monommidae [INV ZOO] A family of coleopteran insects in the superfamily Tenebrionoidea.

monomolecular film [PHYS CHEM] A film one molecule thick. Also known as monolayer.

monomorphic [BIOL] Having or exhibiting only a single form.

mononuclear [CYTOL] Having only one nucleus.

mononucleosis [MED] Any of various conditions marked by an abnormal increase in monocytes in the peripheral blood.

monophagous [BIOL] Subsisting on a single kind of food. Also known as monotrophic.

Monophisthocotylea [INV ZOO] An order of the Monogenea in which the posthaptor is without discrete multiple suckers or clamps.

Monophlebinae [INV ZOO] A subfamily of the homopteran superfamily Coccoidea distinguished by a dorsal anus.

monophonic sound [ENG ACOUS] Sound produced by a system in which one or more microphones feed a single transducing channel which is coupled to one or more loudspeakers.

monophyodont [VERT ZOO] Having only one set of teeth throughout life.

monopinch [ELECTR] Antijam application of the monopulse technique where the error signal is used to provide discrimination against jamming signals.

Monopisthocotylea [INV ZOO] An order of trematode worms in the subclass Pectobothridia.

Monoplacophora [INV ZOO] A group of shell-bearing mollusks represented by few living forms; considered to be a sixth class of mollusks.

monoplegia [MED] Paralysis involving a single limb, muscle, or group of muscles.

monoploid [GEN] **1.** Having only one set of chromosomes. **2.** Having the haploid number of chromosomes.

monopodial [BOT] Stem branching in which there are lateral shoots on a primary axis.

monopodium [BOT] A primary axis that continues to grow while giving off successive lateral branches.

monopole See magnetic monopole.

monopole antenna [ELECTROMAG] An antenna, usually in the form of a vertical tube or helical whip, on which the current distribution forms a standing wave, and which acts as one part of a dipole whose other part is formed by its electrical image in the ground or in an effective ground plane. Also known as spike antenna.

monopropellant [MATER] A rocket propellant consisting of a single substance, especially a liquid, capable of creating rocket thrust without the addition of a second substance.

monopulse radar [ENG] Radar in which directional information is obtained with high precision by using a receiving antenna system having two or more partially overlapping lobes in the radiation patterns. Also known as monopulse tracking; simultaneous lobing; static split tracking.

monopulse tracking See monopulse radar.

Mono pump [MIN ENG] A pump designed and manufactured of special materials to meet mining conditions; consists of a rubber stator shaped like a double internal helix and a single helical rotor which travels in the stator with a slightly eccentric motion.

Monopylina [INV ZOO] A suborder of radiolarian protozoans in the order Oculosida in which pores lie at one pole of a single-layered capsule.

monopyroxene clinoaugite See clinopyroxene.

monorail [CIV ENG] A single rail used as a track; usually elevated, with cars straddling or hanging from it.

monorchid [ANAT] **1.** Having one testis. **2.** Having one testis descended into the scrotum.

Monorhina [VERT ZOO] The subclass of Agnatha that includes the jawless vertebrates with a single median nostril.

monorhinal [ANAT] Having only one nostril.

monosaccharide [BIOCHEM] A carbohydrate which cannot be hydrolyzed to a simpler carbohydrate; a polyhedric alcohol having reducing properties associated with an actual or potential aldehyde or ketone group; classified on the basis of the number of carbon atoms, as triose (3C), tetrose (4C), pentose (5C), and so on.

monoscope [ELECTR] A signal-generating electron-beam tube in which a picture signal is produced by scanning an electrode that has a predetermined pattern of secondary-emission response over its surface. Also known as monotron; phasmajector.

Monosigales [BOT] A botanical order equivalent to the Choanoflagellida in some systems of classification.

monosiphonous [BIOL] Having a single central tube, as in the thallus of certain filamentous algae or the hydrocaulus of some hydrozoans.

monosodium glutamate See sodium glutamate.

monospermous [BOT] Having or producing one seed.

monosporangium [BOT] A sporangium producing monospores.

monospore [BOT] A simple or undivided nonmotile asexual spore; produced by the diploid generation of some algae.

monostable [ELECTR] Having only one stable state.

monostable blocking oscillator [ELECTR] A blocking oscillator in which the electron tube or other active device carries no current unless positive voltage is applied to the grid. Also known as driven blocking oscillator.

monostable circuit [ELECTR] A circuit having only one stable condition, to which it returns in a predetermined time interval after being triggered.

monostable multivibrator [ELECTR] A multivibrator with one stable state and one unstable state; a trigger signal is required to drive the unit into the unstable state, where it remains for a predetermined time before returning to the stable state. Also known as one-shot multivibrator; single-shot multivibrator; start-stop multivibrator; univibrator.

monostat [ENG] Fluid-filled, upside-down manometer-type device used to control pressures within an enclosure, as for laboratory analytical distillation systems.

monostatic radar [ENG] Conventional radar, in which the transmitter and receiver are at the same location and share the same antenna; in contrast to bistatic radar.

MONOPLACOPHORA

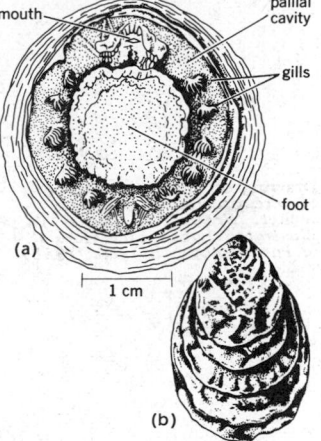

Living and fossil Monoplacophora. *(a) Neopilina galatheae* Lemche. *(b) Tryblidium reticulatum* Lindström. *(Adapted from R. C. Moore, ed., Treatise on Invertebrate Paleontology, pt. 1, 1957)*

MONOPOLE ANTENNA

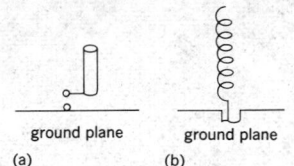

Types of monopole antenna with horizontal and vertical patterns. *(a)* Vertical tube. *(b)* Helical.

MONOSTABLE MULTIVIBRATOR

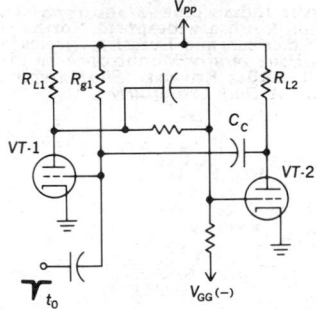

Circuit diagram of basic monostable multivibrator using vacuum triodes.

MONOSTOME

Drawing of monostome type of cercaria showing the one sucker. *(From R. M. Cable, An Illustrated Laboratory Manual of Parasitology, Burgess, 1958)*

MONOTROPACEAE

The Indian pipe *(Monotropa uniflora)*, a widespread North American and Eurasian species of the family Monotropaceae in the order Ericales. *(Photograph by Arthur Cronquist)*

MONSTRILLOIDA

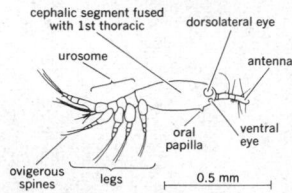

Monstrilla reticulata, adult female, shown in lateral view *(From C. C. Davis, 1949)*

monostome [INV ZOO] A cercaria having only one mouth or sucker.

Monostylifera [INV ZOO] A suborder of the Hoplonemertini characterized by a single stylet.

monoterpene [ORG CHEM] **1.** A class of terpenes with molecular formula $C_{10}H_{16}$; the members of the class contain two isoprene units. **2.** A derivative of a member of such a class.

Monotomidae [INV ZOO] The equivalent name for Rhizophagidae.

monotone [SCI TECH] A quantity which never increases (or which never decreases) as a function of some other quantity. Also known as monotonic.

monotone convergence theorem [MATH] If a monotone increasing sequence of nonnegative measurable functions of a measure space converges almost everywhere, then the integral of their limit is the limit of their integrals.

monotone function [MATH] A function which is either monotone nondecreasing or monotone nonincreasing. Also known as monotonic function.

monotone nondecreasing function [MATH] A function which never decreases, that is, if $x \leq y$ then $f(x) \leq f(y)$. Also known as increasing function; monotonically nondecreasing function.

monotone nonincreasing function [MATH] A function which never increases, that is, if $x \leq y$ then $f(x) \geq f(y)$. Also known as decreasing function; monotonically nonincreasing function.

monotonic *See* monotone.

monotonically nondecreasing function *See* monotone nondecreasing function.

monotonically nonincreasing function *See* monotone nonincreasing function.

monotonic function *See* monotone function.

Monotremata [VERT ZOO] The single order of the mammalian subclass Prototheria containing unusual mammallike reptiles, or quasi-mammals.

monotrichous [MICROBIO] Of bacteria, having an individual flagellum at one pole.

monotron [ELECTR] *See* monoscope. [MET] A machine for determining indentation hardness by measuring the load required to dent a specimen to a constant depth with a diamond 5/8 millimeter in diameter.

Monotropaceae [BOT] A family of dicotyledonous herbs or half shrubs in the order Ericales distinguished by a small, scarcely differentiated embryo without cotyledons, lack of chlorophyll, leaves reduced to scales, and anthers opening by longitudinal slits.

monotrophic [CRYSTAL] Of crystal pairs, having one of the pair always metastable with respect to the other. [ZOO] *See* monophagous.

monotropic [PHYS] Pertaining to an element which may exist in two or more forms, but in which one form is the stable modification at all temperatures and pressures.

monotropy coefficient [FL MECH] A coefficient ν related to the ratio of velocity coefficients, A_y/A_x, in an equation developed by P. Raethjen for the velocity profile in a fluid.

monotype [BIOL] A single type of organism that constitutes a species or genus. [GRAPHICS] A printing technique in which a picture is painted on a sheet of glass or metal; the picture is transferred to a sheet of paper by pressure; additional copies require that the subject be repainted on the plate.

Monotype [GRAPHICS] A trademarked name for a typesetting system in which each character is cast as a separate piece; corrections are made by hand from a case of type.

monotype metal [MET] A type metal typically composed of 76% lead, 16% antimony, and 8% tin, with good wear resistance and compressive strength.

monovalent [CHEM] A radical or atom whose valency is 1.

monoxide [CHEM] A compound that contains a single oxygen atom, such as carbon monoxide, CO.

monozygotic twins [BIOL] Twins which develop from a single fertilized ovum. Also known as identical twins.

mons [ANAT] An eminence.

monsoon [METEOROL] A large-scale wind system which predominates or strongly influences the climate of large regions, and in which the direction of the wind flow reverses from winter to summer; an example is the wind system over the Asian continent.

monsoon climate [CLIMATOL] The type of climate which is found in regions subject to monsoons.

monsoon current [OCEANOGR] A seasonal wind-driven current occurring in the northern part of the Indian Ocean.

monsoon fog [METEOROL] An advection type of fog occurring along a coast where monsoon winds are blowing, when the air has a high specific humidity and there is a large difference in the temperature of adjacent land and sea.

monsoon low [METEOROL] A seasonal low found over a continent in the summer and over the adjacent sea in the winter.

mons pubis [ANAT] The eminence of the lower anterior abdominal wall above the superior rami of the pubic bones.

monster [MED] A congenitally malformed fetus which is incapable of properly performing the vital functions, or which exhibits marked structural differences from the normal.

Monstrilloida [INV ZOO] A suborder or order of microscopic crustaceans in the subclass Copepoda; adults lack a second antenna and mouthparts, and the digestive tract is vestigial.

mons veneris [ANAT] The mons pubis of the female.

montage [GRAPHICS] In photography, a collection of photos pasted on a common background, or a composite picture made by printing two or more negatives on a single sheet of photographic paper.

montane [ECOL] Of, pertaining to, or being the biogeographic zone composed of moist, cool slopes below the timberline and having evergreen trees as the dominant lifeform.

montebrasite [MINERAL] $LiAlPO_4(OH)$ A mineral composed of basic lithium aluminum phosphate; it is isomorphous with amblygonite and natromontebrasite.

Monte Carlo method [STAT] A technique which obtains a probabilistic approximation to the solution of a problem by using statistical sampling techniques.

montgomeryite [MINERAL] $Ca_2Al_2(PO_4)_3(OH) \cdot 7H_2O$ A green to colorless mineral composed of hydrous basic calcium aluminum phosphate.

Montgomery's tubercles [ANAT] Elevations in the areola of the nipple due to apocrine sweat glands; most prominent during pregnancy and lactation.

month [ASTRON] **1.** The period of the revolution of the moon around the earth (sidereal month). **2.** The period of the phases of the moon (synodic month). **3.** The month of the calendar (calendar month).

Montian [GEOL] A European stage of geologic time: Paleocene (above Danian, below Thanetian).

Monticellidae [INV ZOO] A family of tapeworms in the order Proteocephaloidea, in which some or all of the organs are in the cortical mesenchyme; catfish parasites.

monticellite [MINERAL] $CaMgSiO_4$ A colorless or gray mineral of the olivine structure type; isomorphous with kirsch steinite.

monticulus [ANAT] The median dorsal portion of the cerebellum.

montmorillonite [MINERAL] **1.** A group name for all clay minerals with an expanding structure, except vermiculite. **2.** The high-alumina end member of the montmorillonite group; it is grayish, pale red, or blue and has some replacement of aluminum ion by magnesium ion. **3.** Any mineral of the montmorillonite group.

montroydite [MINERAL] HgO Natural mercury oxide mineral from Texas.

monzonite [PETR] A phaneritic (visibly crystalline) plutonic rock composed chiefly of sodic plagioclase and alkali feldspar, with subordinate amounts of dark-colored minerals, intermediate between syenite and dorite.

Moody formula [MECH ENG] A formula giving the efficiency e' of a field turbine, whose runner has diameter D', in terms of the efficiency e of a model turbine, whose runner has diameter D; $e' = 1 - (1-e)(D/D')^{1/5}$.

Moody friction factor [FL MECH] Modification of the friction factor-Reynolds number-fluid flow relationship into which a roughness factor has been incorporated.

moon [ASTRON] **1.** The natural satellite of the earth. **2.** A natural satellite of any planet.

moonbow [OPTICS] A rainbow formed by light from the

moon; the colors in a moonbow are usually very difficult to detect. Also known as lunar rainbow.

Mooney unit [CHEM ENG] An arbitrary unit used to measure the plasticity of raw, or unvulcanized rubber; the plasticity in Mooney units is equal to the torque, measured on an arbitrary scale, on a disk in a vessel that contains rubber at a temperature of 100°C and rotates at two revolutions per minute.

moon pillar [METEOROL] A halo consisting of a vertical shaft of light through the moon.

moonrise [ASTRON] The crossing of the visible horizon by the upper limb of the ascending moon.

moonset [ASTRON] The crossing of the visible horizon by the upper limb of the descending moon.

moon shot [AERO ENG] The launching of a rocket intended to travel to the vicinity of the moon.

moonstone [MINERAL] An alkali feldspar or cryptoperthite that is semitransparent to translucent and exhibits a bluish to milky-white, pearly, or opaline luster; used as a gemstone if flawless. Also known as hecatolite.

moor [ECOL] See bog. [ENG] Securing a ship or aircraft by attaching it to a fixed object or a mooring buoy with chains or lines, or with anchors or other devices.

moor coal [GEOL] A friable lignite or brown coal.

Moore code [COMMUN] A binary teleprinter code with seven binary digits for each letter.

moored mine [ORD] An underwater mine with positively buoyant mine case, held at a predetermined depth beneath the surface by a cable or chain mooring attached to an anchor that rests on the bottom.

mooreite [MINERAL] $(Mg,Zn,Mn)_8(SO_4)_4(OH)_{14} \cdot 4H_2O$ A glassy white mineral composed of hydrous basic magnesium zinc manganese sulfate.

Moore-Smith convergence [MATH] Convergence of a net.

mooring anchor [NAV ARCH] The second or additional anchor used to hold a ship at the mooring position.

mooring buoy [ENG] A buoy secured to the bottom by permanent moorings and provided with means for mooring a vessel by use of its anchor chain or mooring lines; in its usual form a mooring buoy is equipped with a ring.

mooring line [NAV ARCH] A rope or cable used for securing a ship to a pier or wharf.

mooring mast [AERO ENG] A mast or pole with fittings at the top to secure any lighter-than-air craft, such as a dirigible or blimp.

mooring winch [NAV ARCH] A hydraulic, electric, or steam machine on a ship used to haul in mooring lines when securing the ship to a pier or wharf.

moose [VERT ZOO] An even-toed ungulate of the genus *Alces* in the family Cervidae; characterized by spatulate antlers, long legs, a short tail, and a large head with prominent overhanging snout.

Mo-Permalloy [MET] A magnetic alloy having very high permeability and low saturation, consisting of 79% nickel, 17% iron, and 4% molybdenum.

mor See ectohumus.

Moraceae [BOT] A family of dicotyledonous woody plants in the order Urticales characterized by two styles or style branches, anthers inflexed in the bud, and secretion of a milky juice.

mora fat See mowrah fat.

morainal apron See outwash plain.

morainal delta See ice-contact delta.

morainal lake [HYD] A glacial lake filling a depression resulting from irregular deposition of drift in a terminal or ground moraine of a continental glacier.

morainal plain See outwash plain.

moraine [GEOL] An accumulation of glacial drift deposited chiefly by direct glacial action and possessing initial constructional form independent of the floor beneath it.

moraine bar [GEOL] A terminal moraine serving as a bar, rising out of deep water at some distance from the shore.

moraine kame [GEOL] One of a group of kames characterized by the same topography, constitution, and position as a terminal moraine.

moral idiocy [PSYCH] Inability to understand moral principles and values and to act in accordance with them, appar-

ently without impairment of the reasoning and intellectual faculties.

moral masochism [PSYCH] Masochism in which there is a need for punishment arising from unconscious sexual desires and reactivation of the Oedipus complex, characterized by self-destructive acts or the provocation of punishment from authority figures.

morass ore See bog iron ore.

moravite [MINERAL] $Fe_2(N,Fe)_4Si_7O_{20}(OH)_4$ A black mineral of the chlorite group, composed of basic iron aluminum silicate, occurring as fine scales.

Morax-Axenfeld bacillus See Moraxella lacunata.

Morax-Axenfeld conjunctivitis [MED] Chronic bacterial conjunctivitis caused by the *Morax-Axenfeld* bacillus.

Moraxella [MICROBIO] A genus of strictly aerobic bacteria in the family Brucellaceae; members are short, plump, rod-shaped cells that are chemoorganotrophic; most species are pathogens of man and other mammals.

Moraxella lacunata [MICROBIO] A short, nonmotile, gram-negative, rod-shaped bacterium that causes conjunctivitis in man. Also known as Morax-Axenfeld bacillus.

morbidity [MED] **1.** The quantity or state of being diseased. **2.** The conditions inducing disease. **3.** The ratio of the number of sick individuals to the total population of a community.

mordant [CHEM] An agent, such as alum, phenol, or aniline, that fixes dyes to tissues, cells, textiles, and other materials by combining with the dye to form an insoluble compound. Also known as dye mordant.

mordant dye [MATER] Textile dye that requires a mordant (third substance) to bind the dye onto the fiber.

mordanting assistant [MATER] A chemical used with textile dye mordants to cause decomposition of the mordant and uniform deposition on the fibers; examples are sulfuric, oxalic, and lactic acids.

mordant rouge [MATER] Aluminum acetate-acetic acid solution used in dyeing and calico printing. Also known as red acetate; red liquor.

Mordellidae [INV ZOO] The tumbling flower beetles, a family of coleopteran insects in the superfamily Meloidea.

mordenite [MINERAL] $(Ca,Na_2,K_2)_4Al_8Si_{40}O_{96} \cdot 28H_2O$ A zeolite mineral crystallizing in the orthorhombic system and found in minute crystals or fibrous concretions. Also known as arduinite; ashtonite; flokite; ptilolite.

morencite See nontronite.

morenosite [MINERAL] $NiSO_4 \cdot 7H_2O$ An apple-green or light-green mineral composed of hydrous nickel sulfate, occurring in crystals or fibrous crusts. Also known as nickel vitriol.

Morera's theorem [MATH] If a function of a complex variable is continuous in a simply connected domain D, and if the integral of the function about every simply connected curve in D vanishes, then the function is analytic in D.

mores [ECOL] Groups of organisms preferring the same physical environment and having the same reproductive season.

Morgan equation [THERMO] A modification of the Ramsey-Shields equation, in which the expression for the molar surface energy is set equal to a quadratic function of the temperature rather than to a linear one.

morganite See vorobyevite.

morgue [MED] A place where dead bodies are held pending identification and disposition.

moribund [BIOL] **1.** In a dying or deathlike state. **2.** In a state of suspended life functions; dormant.

Moridae [VERT ZOO] A family of actinopterygian fishes in the order Gadiformes.

morin [ORG CHEM] $C_{15}H_{10}O_7 \cdot 2H_2O$ Colorless needles soluble in boiling alcohol, slightly soluble in water; used as a mordant dye and analytical reagent.

Morinae [VERT ZOO] The deep-sea cods, a subfamily of the Moridae.

morinite [MINERAL] $Na_2Ca_3Al_3H(PO_4)_4F_6 \cdot 8H_2O$ A mineral composed of hydrous acid phosphate of sodium, calcium, and aluminum. Also known as jezekite.

Morisette expansion reamer [MIN ENG] A reaming device with three tapered lugs or cutters designed so that drilling pressure necessary to penetrate rock with a noncoring pilot bit forces the diamond-faced cutters of the reamer to expand

MOOSE

The male moose, noted for its "Roman" nose, overhanging upper lip, beard, and spatulate antlers.

outward, thereby enlarging the pilot hole sufficiently to allow the casing to follow the reamer as drilling progresses.

Mormyridae [VERT ZOO] A large family of electrogenic fishes belonging to the Osteoglossiformes; African river and lake fishes characterized by small eyes, a slim caudal peduncle, and approximately equal dorsal and anal fins in most.

Mormyriformes [VERT ZOO] Formerly an order of fishes which are now assigned to the Osteoglossiformes.

morning glory spillway *See* shaft spillway.

morning gun *See* reveille gun.

morning sickness [MED] Morning nausea associated with early pregnancy.

morning star [ASTRON] A misnomer given to a planet visible to the naked eye, when it rises before the sun.

morning twilight [ASTRON] The period of time between darkness and sunrise.

moron [PSYCH] A mentally defective individual with a mental age between 7 and 12 years; a child with an IQ between 50 and 69.

Moro reflex [PHYSIO] The startle reflex observed in normal infants from birth through the first few months, consisting in abduction and extension of all extremities, followed by flexion and abduction of the extremities.

morphallaxis [PHYSIO] Regeneration whereby one part is transformed into another by reorganization of tissue fragments rather than by cell proliferation.

Morphinae [INV ZOO] A subfamily of large tropical butterflies in the family Nymphalidae.

morphine [PHARM] $C_{17}H_{19}NO_3 \cdot H_2O$ A white crystalline narcotic powder, melting point 254°C, an alkaloid obtained from opium; used in medicine in the form of a hydrochloride or sulfate salt.

morphine methyl bromide *See* morphosan.

morphinism [MED] **1.** The condition caused by the habitual use of morphine. **2.** The morphine habit.

morphism [MATH] The class of elements which together with objects form a category; in most cases, morphisms are functions which preserve some structure on a set.

morphogenesis [EMBRYO] The transformation involved in the growth and differentiation of cells and tissue. Also known as topogenesis.

morphogenetic region [GEOL] A region in which, under certain climatic conditions, the predominant geomorphic processes will contribute regional characteristics to the landscape that contrast with those of other regions formed under different climatic conditions.

morpholine [ORG CHEM] C_4H_8ONH A hygroscopic liquid, soluble in water; used as a solvent and rubber accelerator. Also known as tetrahydro-1,4-oxazine.

morphology [BIOL] A branch of biology that deals with structure and form of an organism at any stage of its life history.

morphosan [ORG CHEM] $C_{17}H_{19}NO_3 \cdot CH_3Br$ A solid morphine derivative without morphine's disagreeable aftereffects; used in medicine. Also known as morphine methyl bromide.

morphotropism [CRYSTAL] Similarity of structure, axial ratios, and angles between faces of one or more zones in crystalline substances whose formulas can be derived one from another by substitution.

Morquio's syndrome [MED] A hereditary disease transmitted as an autosomal recessive and characterized by large quantities of keratosulfate in urine, dwarfism, and a typical facies with broad mouth, prominent maxilla, short nose, and widely spaced teeth. Also known as Brailsford-Morquio syndrome; familial osteochondrodystrophy; mucopolysaccharidosis IV.

Morse cable code [COMMUN] A code used chiefly in submarine cable telegraphy, in which positive and negative current impulses of equal length represent dots and dashes, and a space is represented by the absence of current. Also known as cable code; International cable code.

Morse code [COMMUN] **1.** A telegraph code for manual operating, consisting of short (dot) and long (dash) signals and various-length spaces; now used only for wire telegraphy. Also known as American Morse code. **2.** Collective term for Morse code (American Morse code) and continental code (International Morse code).

Morse equation [PHYS CHEM] An equation according to

which the potential energy of a diatomic molecule in a given electronic state is given by a Morse potential.

Morse potential [PHYS CHEM] An approximate potential associated with the distance r between the nuclei of a diatomic molecule in a given electronic state; it is $V(r) = D\{1 - \exp[-a(r-r_e)]\}^2$, where r_e is the equilibrium distance, D is the dissociation energy, and a is a constant.

Morse taper reamer [DES ENG] A machine reamer with a taper shank.

Morse theory [MATH] The study of differentiable mappings of differentiable manifolds, which by examining critical points shows how manifolds can be constructed from one another.

mortality rate [MED] For a given period of time, the ratio of the number of deaths occurring per 1000 population. Also known as death rate.

mortar [MATER] A mixture of cement, lime, and sand used for laying bricks or masonry. [ORD] A complete projectile-firing weapon, with rifled or smooth bore, characterized by a shorter barrel, lower velocity, shorter range, and higher angle of fire than a howitzer or a gun; most present-day mortars are muzzle-loaded and of simple construction for lightness and mobility. [SCI TECH] A bowl-shaped vessel made of hard material in which solids are crushed by hand with a pestle.

mortar structure [PETR] A cataclastic structure produced by dynamic metamorphism of crystalline rocks and characterized by a mica-free aggregate of finely crushed grains of quartz and feldspar filling the interstices between or forming borders on the edges of larger, rounded relicts. Also known as cataclastic structure; murbruk structure; porphyroclastic structure.

mortise [ENG] A groove or slot in a timber for holding a tenon.

mortise and tenon [DES ENG] A type of joint, principally used for wood, in which a hole, slot, or groove (mortise) in one member is fitted with a projection (tenon) from the second member.

mortising machine [MECH ENG] A machine employing an auger and a chisel to produce a square or rectangular mortise in wood.

mortlake *See* oxbow lake.

morula [EMBRYO] A solid mass of blastomeres formed by cleavage of the eggs of many animals; precedes the blastula or gastrula, depending on the type of egg. [INV ZOO] A cluster of immature male gametes in which differentiation occurs outside the gonad; common in certain annelids.

Moruloidea [INV ZOO] The only class of the phylum Mesozoa; embryonic development in the organisms proceeds as far as the morula or stereoblastula stage.

morvan [GEOL] The area where two peneplains intersect. Also known as skiou.

mosaic [BIOL] An organism or part made up of tissues or cells exhibiting mosaicism. [ELECTR] A light-sensitive surface used in television camera tubes, consisting of a thin mica sheet coated on one side with a large number of tiny photosensitive silver-cesium globules, insulated from each other. [EMBRYO] An egg in which the cytoplasm of early cleavage cells is of the type which determines its later fate. [PETR] **1.** Pertaining to a granoblastic texture in a rock formed by dynamic metamorphism in which the boundaries between individual grains are straight or slightly curved. Also known as cyclopean. **2.** Pertaining to a texture in a crystalline sedimentary rock in which contacts at grain boundaries are more or less regular. [SCI TECH] A surface pattern made by the assembly and arrangement of many small pieces.

mosaic gold *See* stannic sulfide.

mosaicism [GEN] The coexistence in an individual of somatic cells of genetically different types; it is caused by gene or chromosome mutations, especially nondisjunction, after fertilization, by double fertilization, or by fusion of embryos.

mosaic structure [CRYSTAL] In crystals, a substructure in which neighboring regions are oriented slightly differently.

mosandrite [MINERAL] A reddish-brown or yellowish-brown mineral composed of a silicate of sodium, calcium, titanium, zirconium, and cerium. Also known as khibinite; lovchorrite; rinkite; rinkolite.

mosasaur [PALEON] Any reptile of the genus *Mosasaurus*; large, aquatic, fish-eating lizards from the Cretaceous which

---

MORSE CODE

```
A  ● ▬
B  ▬ ● ● ●
C  ● ●   ● 
D  ▬ ● ●
E  ●
F  ● ▬ ●
G  ▬ ▬ ●
H  ● ● ● ●
I  ● ●
J  ● ▬ ▬ ▬
K  ▬ ● ▬
L  ▬▬▬▬
M  ▬ ▬
N  ▬ ●
O  ●   ●
P  ● ● ● ●
Q  ● ● ▬ ●
R  ●   ● ●
S  ● ● ●
T  ▬
U  ● ● ▬
V  ● ● ● ▬
W  ● ▬ ▬
X  ● ▬ ● ●
Y  ● ●   ● ●
Z  ● ● ●   ●
```

C, O, R, Y, Z and & are composed of dots and spaces
T is a short dash
L is a longer dash

Morse code for the alphabet.

are related to the monitors but had paddle-shaped limbs.

Moschcowitz's disease *See* thrombotic thrombocytopenic purpura.

moscovite *See* muscovite.

Moseley's law [SPECT] The law that the square-root of the frequency of an x-ray spectral line belonging to a particular series is proportional to the difference between the atomic number and a constant which depends only on the series.

mosesite [MINERAL] $Hg_2N(SO_4,MoO_4)\cdot H_2O$ Mineral composed of a hydrous nitride of mercury and various anions.

MOS FET *See* metal oxide semiconductor field-effect transistor.

Mosotti field [ELEC] The electric field which would exist at the location of a molecule in a dielectric if one removed all the molecules in a sphere centered at the molecule in question without changing the polarization of molecules outside this sphere, the radius of the sphere being large enough to view the dielectric outside the sphere as a continuum.

mosquito [INV ZOO] Any member of the dipterous subfamily Culicinae; a slender fragile insect, with long legs, a long slender abdomen, and narrow wings.

mosquito boat *See* motor torpedo boat.

moss [BOT] Any plant of the class Bryatae, occurring in nearly all damp habitats except the ocean.

moss agate [MINERAL] A milky or almost transparent chalcedony containing dark inclusions in a dendritic pattern.

Mössbauer effect [NUC PHYS] The emission and absorption of gamma rays by certain nuclei, bound in crystals, without loss of energy through nuclear recoil, with the result that radiation emitted by one such nucleus can be absorbed by another.

Mössbauer spectroscopy [SPECT] The study of Mössbauer spectra, for example, for nuclear hyperfine structure, chemical shifts, and chemical analysis.

Mössbauer spectrum [SPECT] A plot of the absorption, by nuclei bound in a crystal lattice, of gamma rays emitted by similar nuclei in a second crystal, as a function of the relative velocity of the two crystals.

moss forest *See* temperate rainforest.

mossite [MINERAL] $Fe(Nb,Ta)_2O_6$ A mineral composed of an iron tantalum oxide; it is isomorphous with tapiolite.

mossy zinc [MET] Zinc granules made by pouring molten zinc into water.

most probable position [NAV] 1. The computed position of an aircraft determined by comparing a dead-reckoning position and a loran position determined for the same period of time, in which relative weights are given to the estimated probable errors of each. 2. That position of a craft judged to be most accurate when an element of doubt exists as to the true position; it may be a fix, estimated position, dead-reckoning position, no-wind position, or some other position as determined from the squares of the estimated probable errors of each.

most significant bit [ADP] The left-most bit in a word. Abbreviated msb.

Motacillidae [VERT ZOO] The pipits, a family of passeriform birds in the suborder Oscines.

moth [INV ZOO] Any of various nocturnal or crepuscular insects belonging to the lepidopteran suborder Heteroneura; typically they differ from butterflies in having the antennae feathery and rarely clubbed, a stouter body, less brilliant coloration, and proportionately smaller wings.

mothball [ORD] Placing military equipment into a state of long storage.

mother [ENG ACOUS] A mold derived by electroforming from a master; used to produce the stampers from which disk records are molded in large quantities. Also known as metal positive.

mother liquor *See* discharge liquor.

mother lode [GEOL] A main unit of mineralized matter that may not have economic value but to which workable veins are related.

mother map *See* base map.

mother-of-pearl [INV ZOO] The pearly iridescent internal layer of the shell of various pearl-bearing bivalve mollusks.

mother-of-pearl clouds *See* nacreous clouds.

mother rock *See* source rock.

motile [BIOL] Capable of spontaneous movement. [PSYCH]

An individual characterized by motor-type mental imagery which takes the form of inner feelings of action.

motion [MECH] A continuous change of position of a body.

motional impedance [ELECTR] Of a transducer, the complex remainder after the blocked impedance has been subtracted from the loaded impedance. Also known as loaded motional impedance.

motion analysis [IND ENG] Detailed study of the motions used in a work task or at a given work area.

motion cycle [IND ENG] The complete sequence of motions and activities required to complete one work cycle.

motion economy [IND ENG] Simplification and reduction of body motions to simplify and reduce work content.

motion picture [GRAPHICS] 1. A sequence of filmed images viewed in rapid succession so that the illusion of continuity and motion is created. 2. The complete thematically related content of such images. Also known as cinema; movie; moving picture.

motion picture camera [OPTICS] A camera capable of capturing action by taking a series of still pictures at regular brief intervals on a lengthy strip of film.

motion picture pickup [ELECTR] Use of a television camera to pick up scenes directly from motion picture film.

motion picture projector [ENG] An optical and mechanical device capable of flashing pictures taken by a motion picture camera on a viewing screen at the same frequency the action was photographed, thus producing an image that appears to move.

motion register [ADP] The register which controls the go/stop, forward/reverse motion of a tape drive.

motion sickness [MED] A complex of symptoms, including nausea, vertigo, and vomiting, occurring as the result of random multidirectional accelerations of a vehicle.

motivated forgetting [PSYCH] Forgetting, such as by repression, activated by the needs of the individual.

motivation [PSYCH] The comparatively spontaneous drive, force, or incentive, which partly determines the direction and strength of the response of a higher organism to a given situation; it arises out of the internal state of the organism.

motive column [MIN ENG] The ventilating pressure in a mine in units of feet of air column; the height of a column of air whose density is the same as the air in the downcast shaft, which exerts a pressure equal to the ventilating pressure.

motoneuron *See* motor neuron.

motor [ELEC] A machine that converts electric energy into mechanical energy by utilizing forces produced by magnetic fields on current-carrying conductors. Also known as electric motor. [PHYSIO] 1. That which causes action or movement. 2. Pertaining to efferent nerves which innervate muscles and glands.

motor alexia [MED] Inability to read aloud, while comprehension of the written word is preserved.

motor aphasia [MED] A form of aphasia in which the patient knows what he wishes to say but is unable to get the words out, and is able to perceive and comprehend both spoken and written language but is unable to repeat what he sees or hears; due principally to a brain lesion.

motor apraxia [MED] Inability to carry out, on command, a complex or skilled movement, though the purpose thereof is clear to the patient. Also known as kinesthetic apraxia; limb-kinetic apraxia.

motor area [ANAT] The ascending frontal gyrus containing nerve centers for voluntary movement; characterized by the presence of Betz cells. Also known as Broadman's area 4; motor cortex; pyramidal area.

motor ataxia [MED] Inability to coordinate the muscles, which becomes apparent only on body movement.

motor benzol [MATER] Grade of benzene used in fuels for internal combustion engines.

motorboat [NAV ARCH] A boat propelled by an internal combustion engine or an electric motor. Also known as autoboat.

motorboating [ELECTR] Undesired oscillation in an amplifying system or transducer, usually of a pulse type, occurring at a subaudio or low-audio frequency.

motor branch circuit [ELEC] A branch circuit that terminates at a motor; it must have conductors with current-carrying capacity at least 125% of the motor full-load current rating,

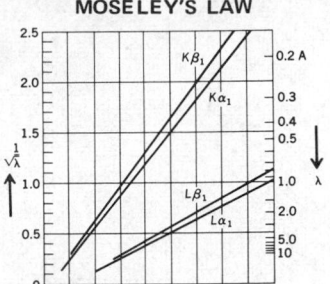

MOSELEY'S LAW

Plot of Moseley's law showing dependence of characteristic x-ray line wavelengths on atomic number. (*Philips Tech. Rev., vol. 17, no. 10, 1956*)

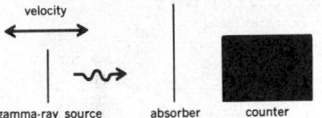

MÖSSBAUER EFFECT

velocity

gamma-ray source   absorber   counter

Experiment demonstrating Mossbauer effect. Source is moved toward stationary absorber at various known velocities to Doppler-shift the gamma-ray energy received by absorber and study absorption by absorber of gamma rays as a function of energy.

MOTH

The luna moth (*Tropaea luna*).

MOTOR

N

force   direction of current

direction of field

S

Relative directions of field flux, current, and force in an electric motor.

and overcurrent protection capable of carrying the starting current of the motor.

motor cell [BOT] *See* bulliform cell. [PHYSIO] An efferent nerve cell in the anterior horn of the spinal cord.

motor control *See* electronic motor control.

motor-converter [ELEC] Induction motor and a synchronous converter with their rotors mounted on the same shaft and with their rotor windings connected in series; such converter operates synchronously at a speed corresponding to the sum of the numbers of poles of the two machines.

motor cortex *See* motor area.

motorcycle [MECH ENG] An automotive vehicle, essentially a motorized bicycle, with two tandem and sometimes three rubber wheels.

motor effect [ELECTROMAG] The mutually repulsive force exerted by neighboring conductors that carry current in opposite directions.

motor element [ENG ACOUS] That portion of an electroacoustic receiver which receives energy from the electric system and converts it into mechanical energy.

motor end plate [ANAT] A specialized area beneath the sarcolemma where functional contact is made between motor nerve fibers and muscle fibers.

motor-generator set [ELEC] A motor and one or more generators that are coupled mechanically for use in changing one power-source voltage to other desired voltages or frequencies.

motor grader *See* autopatrol.

motor method [ENG] ASTM motor octane test to determine the knock rating of fuels for use in spark-ignition engines.

motor nerve [PHYSIO] A nerve composed wholly or principally of motor fibers.

motor neuron [PHYSIO] An efferent nerve cell. Also known as motoneuron.

motor noise [ACOUS] The noisy sound made by an electric motor.

motor reducer [MECH ENG] Speed-reduction power transmission equipment in which the reducing gears are integral with drive motors.

motorship [NAV ARCH] A ship propelled by a motor which can travel on the seas; in particular, one propelled by an internal combustion engine such as a diesel engine.

motor speech area [ANAT] The cortical area located in the triangular and opercular portions of the inferior frontal gyrus; in right-handed people it is more developed on the left side.

motor system [PHYSIO] Any portion of the nervous system that regulates and controls the contractile activity of muscle and the secretory activity of glands.

motor torpedo boat [NAV ARCH] A motor boat 60 to 100 feet (18 to 30 meters) long and capable of speeds of about 60 knots (110 kilometers per hour), armed with two to four torpedo tubes, machine and antiaircraft guns, and depth charges. Also known as mosquito boat; PT boat.

motortruck [MECH ENG] An automotive vehicle which is used to transport freight.

motor unit [ANAT] The axon of an anterior horn cell, or the motor fiber of a cranial nerve, together with the striated muscle fibers innervated by its terminal branches.

motor vehicle [MECH ENG] Any automotive vehicle that does not run on rails, and generally having rubber tires.

mottled [GEOL] Of a soil, irregularly marked with spots of different colors. [PETR] Of a sedimentary rock, marked with spots of various colors.

mottled iron [MET] A cast iron showing gray areas that contain graphite, perlite, and sometimes ferrite, and white areas containing primarily cementite.

mottle-leaf [PL PATH] 1. A virus disease characterized by chlorotic mottling and wrinkling of leaves. 2. A zinc-deficiency disease characterized by partial chlorosis of the leaves and stunting of the plant.

mottramite [MINERAL] $(Cu,Zn)Pb(VO_4)(OH)$ A mineral composed of a basic lead copper zinc vanadate; it is isomorphous with descloizite. Also known as cuprodescloizite; psittacinite.

Mott scattering [QUANT MECH] The scattering of identical particles due to a Coulomb force.

moulage [GRAPHICS] 1. The technique of making a mold. 2.

A material used to make molds; it is extremely delicate and can be used on human hair or skin, or on antiques or other objects that are too fragile to be molded in rubber, gelatin, or plaster; it is applied warm with brushes or palette knives and can be reused.

moulin [HYD] A shaft or hole in the ice of a glacier which is roughly cylindrical and nearly vertical, formed by swirling meltwater pouring down from the surface. Also known as glacial mill; glacier mill; glacier pothole; glacier well; pothole.

mounce [MECH] A unit of mass, equal to 25 grams. Also known as metric ounce.

mound [GEOL] 1. A low, isolated, rounded natural hill, usually of earth. Also known as tuft. 2. A structure built by fossil colonial organisms.

mount [ELECTROMAG] The flange or other means by which a switching tube, or tube and cavity, is connected to a waveguide. [ENG] 1. Structure supporting any apparatus, as a gun, searchlight, telescope, or surveying instrument. 2. To fasten an apparatus in position, as a gun on its support. [ORD] To equip; to put into operation; to go into operation, as to mount an offensive.

mountain [GEOGR] A feature of the earth's surface that rises high above the base and has generally steep slopes and a relatively small summit area.

mountain and valley winds [METEOROL] A system of diurnal winds along the axis of a valley, blowing uphill and upvalley by day, and downhill and downvalley by night; they prevail mostly in calm, clear weather.

mountain blue [INORG CHEM] $2CuCO_3 \cdot Cu(OH)_2$ Ground azurite used as a paint pigment. Also known as copper blue.

mountain breeze [METEOROL] A breeze that blows down a mountain slope due to the gravitational flow of cooled air.

mountain brown ore [GEOL] Name used in Virginia for limonite or brown iron ore.

mountain butter *See* halotrichite.

mountain chain *See* mountain system.

mountain climate [CLIMATOL] Very generally, the climate of relatively high elevations; mountain climates are distinguished by the departure of their characteristics from those of surrounding lowlands, and the one common basis for this distinction is that of atmospheric rarefaction; aside from this, great variety is introduced by differences in latitude, elevation, and exposure to the sun; thus, there exists no single, clearly defined, mountain climate. Also known as highland climate.

mountain cork [MINERAL] 1. A white or gray variety of asbestos composed of thick, interwoven fibers and having a corklike weight and texture. Also known as rock cork. 2. A fibrous clay mineral, such as sepiolite.

mountain crystal *See* rock crystal.

mountain effect [ELECTROMAG] The effect of rough terrain on radio-wave propagation, causing reflections that produce errors in radio direction-finder indications.

mountain-gap wind [METEOROL] A local wind blowing through a gap between mountains.

mountain glacier *See* alpine glacier.

mountain lion *See* puma.

mountain mahogany *See* obsidian.

mountain pediment [GEOL] A plain of combined erosion and transportation at the base of and surrounding a desert mountain range; at a distance it has the appearance of a broad triangular mass.

mountain range [GEOGR] A succession of mountains or narrowly spaced mountain ridges closely related in position, direction, and geologic features.

mountain sickness [MED] A disease occurring in persons living at high altitudes when homeostatic adjustments to the lowered atmospheric oxygen tension fail or develop disproportionately. Also known as high-altitude disease; high-altitude erythremia; Monge's disease; seroche.

mountain slope [GEOGR] The inclined surface that forms a mountainside.

mountain soap *See* saponite.

mountain system [GEOGR] A group of mountain ranges tied together by common geological features. Also known as mountain chain.

mountain tallow *See* hatchettite.

mountain tick fever *See* Colorado tick fever.

Mount Rose snow sampler [ENG] A particular pattern of snow sampler having an internal diameter of 1.485 inches (3.7727 centimeters), so that each inch of water in the sample weighs 1 ounce (28.349 grams).

mouse [VERT ZOO] Any of various rodents which are members of the families Muridae, Heteromyidae, Cricetidae, and Zapodidae; characterized by a pointed snout, short ears, and an elongated body with a long, slender, sparsely haired tail.

mousebane *See* aconite.

mouse deer *See* chevrotain.

mousse de chêne *See* oakmoss resin.

mouth [ANAT] The oral or buccal cavity and its related structures. [ENG ACOUS] The end of a horn that has the larger cross-sectional area. [GEOGR] **1.** The place where one body of water discharges into another. Also known as influx. **2.** The entrance or exit of a geomorphic feature, such as of a cave or valley. [MIN ENG] **1.** The end of a shaft, adit, drift, entry, or tunnel emerging at the surface. **2.** The collar of a borehole. [SCI TECH] Something resembling a mouth, that is, a place where one thing enters another or an opening at the receiving end of a container or enclosure.

mouth breadth [ANTHRO] The measure of the distance from one to the other corner of the mouth in a natural relaxed position.

mouth-to-mouth resuscitation [MED] A method of artificial respiration in which the rescuer's mouth is placed over the victim's mouth and air is blown forcefully into the victim's lungs every few seconds to inflate them.

movable bridge [CIV ENG] A bridge which can be raised or turned to allow passage of vessels on a waterway.

movable contact [ELEC] The relay contact that is mechanically displaced to engage or disengage one or more stationary contacts. Also known as armature contact.

movable-point crossing [CIV ENG] A small-angle rail crossing with two center frogs, each of which consists essentially of a knuckle rail and two opposed movable center points.

movable propeller blade [NAV ARCH] A propeller blade cast separate from the boss or hub and attached to the boss by bolts.

move mode [ADP] A method of communicating between an operating program and an input/output control system in which the data records to be read or written are actually moved into and out of program-designated memory areas; in contrast to locate mode.

movie *See* motion picture.

Movieola [ENG] A mechanical aid with which the operator of a motion picture projector can start and stop a film sequence for editing, reviewing, or examining in detail.

moving bed [CHEM ENG] Granulated solids in a process vessel that are circulated (moved) either mechanically or by gravity flow; used in catalytic and absorption processes.

moving-bed catalytic cracking [CHEM ENG] Petroleum refining process for cracking (breaking) of long hydrocarbon molecules by use of heat, pressure, and a granular cracking catalyst that is continuously cycled between the reactor vessel and the catalyst regenerator.

moving-boundary electrophoresis [ANALY CHEM] A U-tube variation of electrophoresis analysis that uses buffered solution so that all ions of a given species move at the same rate to maintain a sharp, moving front (boundary).

moving cluster [ASTRON] **1.** A star cluster with common motions. **2.** An open star cluster near the sun such that measurements may be made of the individual proper motions of the stars.

moving-coil instrument [ELEC] Any instrument in which current is sent through one or more coils suspended or pivoted in a magnetic field, and the motion of the coils is used to measure either the current in the coils or the strength of the field.

moving-coil loudspeaker *See* dynamic loudspeaker.

moving-coil meter [ELEC] A meter in which a pivoted coil is the moving element.

moving-coil microphone *See* dynamic microphone.

moving-coil pickup *See* dynamic pickup.

moving-coil voltmeter [ENG] A voltmeter in which the current, produced when the voltage to be measured is applied across a known resistance, is sent through coils pivoted in the magnetic field of permanent magnets, and the resulting torque on the coils is balanced by control springs so that the deflection of a pointer attached to the coils is proportional to the current.

moving-coil wattmeter *See* electrodynamic wattmeter.

moving-conductor loudspeaker [ENG ACOUS] A loudspeaker in which the mechanical forces result from reactions between a steady magnetic field and the magnetic field produced by current flow through a moving conductor.

moving constraint [MECH] A constraint that changes with time, as in the case of a system on a moving platform.

moving-iron meter [ENG] A meter that depends on current in one or more fixed coils acting on one or more pieces of soft iron, at least one of which is movable.

moving-iron voltmeter [ENG] A voltmeter in which a field coil is connected to the voltage to be measured through a series resistor; current in the coil causes two vanes, one fixed and one attached to the shaft carrying the pointer, to be similarly magnetized; the resulting torque on the shaft is balanced by control springs.

moving load [MECH] A load that can move, such as vehicles or pedestrians.

moving-magnet voltmeter [ENG] A voltmeter in which a permanent magnet aligns itself with the resultant magnetic field produced by the current in a field coil and another permanent control magnet.

moving map display [NAV] An air navigation device which displays the aircraft's navigational position on moving film of an aeronautical chart; the positional information may be obtained from various systems such as Tacan, Doppler, inertial, and so forth; such display systems may indicate course and distance to destination, track over ground, and so on.

moving picture *See* motion picture.

moving sidewalk [CIV ENG] A sidewalk constructed on the principle of an endless belt, on which pedestrians are moved.

moving submarine haven [NAV] A haven surrounding submarines in transit, extending 50 miles (80 kilometers) ahead, 100 miles (161 kilometers) astern, and 15 miles (24 kilometers) on each side of the estimated position of the submarine along the stated track.

moving surface ship haven [NAV] A haven which will normally be a circle with a specified radius centered on the estimated position of the ship or the guide of a group of ships.

moving-target indicator [ELECTR] A device that limits the display of radar information primarily to moving targets; signals due to reflections from stationary objects are canceled by a memory circuit. Abbreviated MTI.

mowa fat *See* mowrah fat.

mowrah fat [MATER] A vegetable fat extracted from the seeds of *Bassia longifola;* used to make soap. Also known as illepé fat; mora fat; mowa fat.

Mozambique Current [OCEANOGR] The portion of the South Equatorial Current that turns and flows along the coast of Africa in the Mozambique Channel, forming one of the western boundary currents in the Indian Ocean.

mp *See* mean effective pressure; melting point.

MPC *See* man process chart.

MPI *See* Minnesota multiphasic personality inventory.

MPK *See* pentanone.

MQ register [ADP] Temporary-storage register whose contents can be transferred to or from, or swapped with, the accumulator.

mr *See* milliroentgen.

mrad *See* millirad.

M region [ASTROPHYS] Any of the areas on the surface of the sun that are theoretically responsible for magnetic disturbances on the earth.

m-RNA *See* messenger ribonucleic acid.

msb *See* most significant bit.

M scan *See* M scope.

MSCFD [CHEM ENG] Abbreviation for thousand standard cubic feet per day; usually refers to gas flow.

MSCFH [CHEM ENG] Abbreviation for thousand standard cubic feet per hour; usually refers to gas flow.

MOVING-IRON VOLTMETER

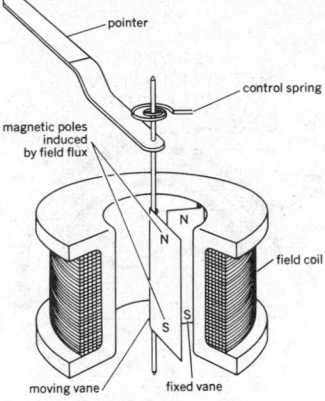

A representative form of the moving-iron voltmeter. *(General Electric Co.)*

MOVING-MAGNET VOLTMETER

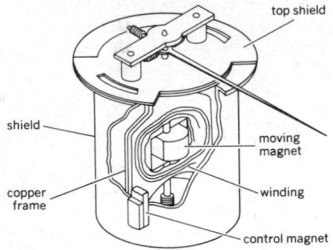

Moving-magnet direct-current voltmeter. *(General Electric Co.)*

MOVING-TARGET INDICATOR

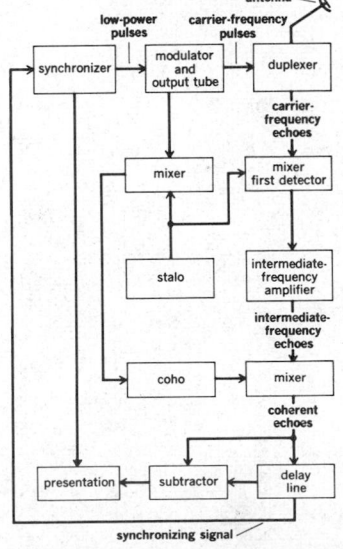

Diagram of a radar system with moving-target indicator capability. Here stalo = stable local oscillator, coho = coherent oscillator.

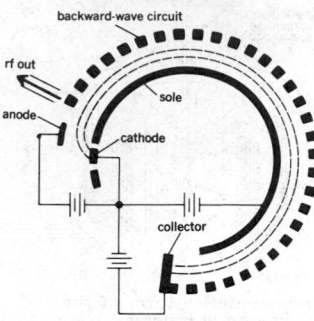

M-TYPE BACKWARD-WAVE OSCILLATOR

backward-wave circuit

rf out

anode

sole

cathode

collector

M-type backward-wave oscillator with the beam focused by a static magnetic field directed into the plane of the page. Here rf out = radio frequency energy out.

MSCFM [CHEM ENG] Abbreviation for thousand standard cubic feet per minute; usually refers to gas flow.

M scope [ELECTR] A modified form of A scope on which part of the time base is slightly displaced in a vertical direction by insertion of an adjustable step which serves as a range marker. Also known as M indicator; M scan.

msec *See* millisecond.

Msec *See* megasecond.

MSH *See* melanocyte-stimulating hormone.

M shell [ATOM PHYS] The third layer of electrons about the nucleus of an atom, having electrons characterized by the principal quantum number 3.

MSI *See* medium-scale integration.

M signal *See* monochrome signal.

MSR *See* molten salt reactor.

M star [ASTRON] A spectral classification for a star whose spectrum is characterized by the presence of titanium oxide bands; M stars have surface temperatures of 3000K for giants and 3400K for dwarfs.

M synchronization [ENG] A linking arrangement between a camera lens and the flashbulb unit to allow a 15-millisecond delay of the shutter so that the bulb burns to its brightest point before the shutter opens.

MT *See* megaton.

MTI *See* moving-target indicator.

M-type backward-wave oscillator [ELECTR] A backward-wave oscillator in which focusing and interaction are through magnetic fields, as in a magnetron. Also known as M-type carcinotron; type-M carcinotron.

M-type carcinotron *See* M-type backward-wave oscillator.

muc-, muci-, muco- [ZOO] A combining form denoting pertaining to mucus, mucin, mucosa.

Mucedinaceae [MYCOL] The equivalent name for Moniliaceae.

mucigen [BIOCHEM] A substance from which mucin is derived; contained in mucus-secreting epithelial cells.

mucilage [MATER] 1. A sticky material employed as an adhesive. 2. A gummy material derived from plants.

mucin [BIOCHEM] A glycoprotein constituent of mucus and various other secretions of man and lower animals.

mucinosis [MED] Accumulations of materials containing mucin or mucinous substances in the skin; sometimes accompanied by papule and nodule formation.

mucinous cyst *See* mucous cyst.

muck [CIV ENG] Rock or earth removed during excavation. [GEOL] Dark, finely divided, well-decomposed, organic matter intermixed with a high percentage of mineral matter, usually silt, forming a surface deposit in some poorly drained areas. [MIN ENG] *See* waste rock.

mucker [MIN ENG] A worker who loads broken mineral into trams or pushes them from stope chute to shaft. Also known as groundman.

mucking [ENG] Clearing and loading broken rock and other excavated materials, as in tunnels or mines.

mucocutaneous [ANAT] Pertaining to a mucous membrane and the skin, and to the line where these join.

mucoid [BIOCHEM] Any of various glycoproteins, similar to mucins but differing in solubilities and precipitation properties and found in cartilage, in the crystalline lens, and in white of egg. 2. Resembling mucus. [MICROBIO] Pertaining to large colonies of bacteria characterized by being moist and sticky.

mucolytic [BIOCHEM] Effecting the dissolution, liquefaction, or dispersion of mucus and mucopolysaccharides.

mucopolysaccharide [BIOCHEM] Any of a group of polysaccharides containing an amino sugar and uronic acid; a constituent of mucoproteins, glycoproteins, and blood-group substances.

mucopolysaccharidosis [MED] Any of several inborn metabolic disorders involving mucopolysaccharides; the six types are MPS I, Hurler's syndrome; MPS II, Hunter's syndrome; MPS III, Sanfilippo's syndrome; MPS IV, Morquio's syndrome; MPS V, Scheil's syndrome; and MPS VI, Maroteaux-Lamy's syndrome.

mucoprotein [BIOCHEM] Any of a group of glycoproteins containing a sugar, usually chondroitinsulfuric or mucoitinsulfuric acid, combined with amino acids or polypeptides.

mucopurulent [MED] Containing mucus and pus.

Mucorales [MYCOL] An order of terrestrial fungi in the class Phycomycetes, characterized by a hyphal thallus and nonmotile sporangiospores, or conidiospores.

mucormycosis [MED] An acute, usually fulminating fungus infection of man caused by several genera of Mucorales, including *Absidia, Rhizopus,* and *Mucor.*

mucosa [HISTOL] A mucous membrane.

mucosanguineous [MED] Containing mucus and blood.

mucous [PHYSIO] Of or pertaining to mucus; secreting mucus.

mucous cell [PHYSIO] A mucus-secreting cell.

mucous colitis *See* irritable colon.

mucous connective tissue [HISTOL] A type of loose connective tissue in which the ground substance is especially prominent and soft; occurs in the umbilical cord.

mucous cyst [MED] A retention cyst of a gland, containing a secretion rich in mucin. Also known as mucinous cyst.

mucous degeneration [MED] Any retrogressive change associated with abnormal production of mucus.

mucous epithelium [EMBRYO] The epidermis of an embryo, excluding the epitrichium. [HISTOL] The germinative layer of a stratified squamous epithelium.

mucous gland [PHYSIO] A gland that secretes mucus.

mucous membrane [HISTOL] The type of membrane lining cavities and canals which have communication with air; it is kept moist by glandular secretions. Also known as tunica mucosa.

mucoviscidosis *See* cystic fibrosis.

mucro [BIOL] An abrupt, sharp terminal tip or process.

mucronate [BIOL] Terminated abruptly by a sharp terminal tip or process.

mucus [PHYSIO] A viscid fluid secreted by mucous glands, consisting of mucin, water, inorganic salts, epithelial cells, and leukocytes, held in suspension.

mud [ENG] *See* slime. [GEOL] An unindurated mixture of clay and silt with water; it is slimy with a consistency varying from that of a semifluid to that of a soft and plastic sediment. [PETR] The silt plus clay portion of a sedimentary rock.

mud auger [DES ENG] A diamond-point bit with the wings of the point twisted in a shallow augerlike spiral. Also known as clay bit; diamond-point bit; mud bit.

mud ball [GEOL] A rounded mass of mud or mudstone up to 20 centimeters in diameter in a sedimentary rock. Also known as chalazoidite; tuff ball.

mud berth [CIV ENG] A berth where a vessel rests on the bottom at low water.

mud bit *See* mud auger.

mud blasting [ENG] The detonation of sticks of explosive stuck on the side of a boulder with a mud covering, so that little of the explosive energy is used in breaking the boulder.

mud cake [ENG] A caked layer of clay adhering to the walls of a well or borehole, formed where the water in the drilling mud filtered into a porous formation during rotary drilling.

mud-cake resistivity [PETRO ENG] Resistivity of drilling mud cake pressed from a sample of the mud; important in mud log interpretation.

mudcap [ENG] A quantity of wet mud, wet earth, or sand used to cover a charge of dynamite or other high explosive fired in contact with the surface of a rock in mud blasting.

mud cone [GEOL] A cone of sulfurous mud built around the opening of a mud volcano or mud geyser, with slopes as steep as 40° and diameters ranging upward to several hundred yards. Also known as puff cone.

mud crack [GEOL] An irregular fracture formed by shrinkage of clay, silt, or mud under the drying effects of atmospheric conditions at the surface. Also known as desiccation crack; sun crack.

mud crack polygon *See* mud polygon.

mud drilling [MIN ENG] Drilling operations in which a mud-laden circulation fluid is used. Also known as mud flush drilling.

mudfish *See* bowfin.

mud flat [GEOL] A relatively level, sandy or muddy coastal strip along a shore or around an island; may be alternately covered and uncovered by the tide or may be covered by shallow water. Also known as flat.

mudflow [GEOL] A flowing mass of fine-grained earth material having a high degree of fluidity during movement.

mud flush drilling *See* mud drilling.

mud flush test [MIN ENG] A test carried out at the boring site to determine whether the mud solution is of the correct viscosity and density.

mud log [PETRO ENG] A continuous record of changes in oil or gas contents of circulating drilling mud while drilling a well.

mudlump [GEOL] A diapiric sedimentary structure consisting of clay or silt and forming an island in deltaic areas; produced by the loading action of rapidly deposited delta front sands upon lighter-weight prodelta clays.

mud pilot [NAV] A person who pilots a vessel by visually observing changes in the color of the water as the depth of the water increases or decreases.

mud pit *See* slushpit.

mud polygon [GEOL] A nonsorted polygon whose center lacks vegetation but whose peripheral fissures contain peat and plants. Also known as mud crack polygon.

mud puppy [VERT ZOO] Any of several American salamanders of the genera *Necturus* and *Proteus* making up the family Proteidae; distinguished by having both lungs and gills as an adult.

mud-removal acid [PETRO ENG] Mixture of hydrochloric and hydrofluoric acids with inhibitors, surfactants, and demulsifiers; used to dissolve drilling-mud clays away from the drillhole face.

mudsill [CIV ENG] The lowest sill of a structure, usually embedded in the earth.

mudstone [GEOL] An indurated equivalent of mud in the form of a blocky or massive, fine-grained sedimentary rock containing approximately equal proportions of silt and clay; lacks the fine lamination or fissility of shale.

mud sump [CHEM ENG] Upstream area in a process vessel where, because of a velocity drop, entrained solids drop out and are collected in a sump.

mud volcano [GEOL] A conical accumulation of variable admixtures of sand and rock fragments, the whole resulting from eruption of wet mud and impelled upward by fluid or gas pressure. Also known as hervidero; macaluba.

Mueller matrices [OPTICS] Matrix operators in a calculus used to treat polarized light; in this calculus, the light vector is split into four components one of which is the intensity of the light, and unpolarized light can be treated directly.

MUF *See* maximum usable frequency.

mu factor [ELECTR] Ratio of the change in one electrode voltage to the change in another electrode voltage under the conditions that a specified current remains unchanged and that all other electrode voltages are maintained constant; a measure of the relative effect of the voltages on two electrodes upon the current in the circuit of any specified electrode.

muffle furnace [ENG] A furnace with an externally heated chamber; the walls of which radiantly heat the contents of the chamber.

muffler [ENG] A device to deaden the noise produced by escaping gases or vapors.

mugearite [PETR] A dark-colored, fine-grained igneous rock in which the chief feldspar is oligoclase, plus orthoclase and olivine with some apatite and opaque oxides; originates by differentiation and volcanic crystallization of the primary magma.

muggy [METEOROL] Colloquially descriptive of warm and especially humid weather.

Mugilidae [VERT ZOO] The mullets, a family of perciform fishes in the suborder Mugiloidei.

Mugiloidei [VERT ZOO] A suborder of fishes in the order Perciformes; individuals are rather elongate, terete fishes with a short spinous dorsal fin that is well separated from the soft dorsal fin.

mulberry [BOT] Any of various trees of the genus *Morus* (family Moraceae), characterized by milky sap and simple, often lobed alternate leaves.

mulch [MATER] A mixture of manure and compost, used as a fertilizer.

mule [MIN ENG] A small car or truck used to push mine cars up a slope or inclined plane. [VERT ZOO] The sterile hybrid offspring of the male ass and the mare, or female horse.

mule skinner [MIN ENG] A mule driver.

mull [ENG] To mix thoroughly or grind. [GEOGR] *See* headland. [GEOL] Granular forest humus that is incorporated with mineral matter. [TEXT] A thin, sheer cotton or cotton and polyester fabric.

Müllerian duct *See* paramesonephric duct.

Müllerian duct cyst [MED] A congenital cyst arising from vestiges of the Müllerian ducts.

Müllerian mimicry [ZOO] Mimicry between two aposometic species.

Müller's glass *See* hyalite.

Müller's larva [INV ZOO] The ciliated larva characteristic of various members of the Polycladida; resembles a modified ctenophore.

mullite [MINERAL] $Al_6Si_2O_{13}$ An orthorhombic mineral consisting of an aluminum silicate that is resistant to corrosion and heat; used as a refractory. Also known as porcelainite.

mullock *See* waste rock.

mull technique [SPECT] Method for obtaining infrared spectra of materials in the solid state; material to be scanned is first pulverized, then mulled with mineral oil.

multi- [SCI TECH] A prefix meaning many.

multiaddress [ADP] Referring to an instruction that has more than one address part.

multianode tube [ELECTR] Electron tube having two or more main anodes and a single cathode.

multiaperture reluctance switch [ELECTR] Two-aperture ferrite storage core which may be used to provide a nondestructive readout computer memory.

multiaspect [ADP] Pertaining to searches or systems which permit more than one aspect, or facet, of information to be used in combination, one with the other to effect identifying or selecting operations.

multicavity klystron [ELECTR] A klystron in which there is at least one cavity between the input and output cavities, each of which remodulates the beam so that electrons are more closely bunched.

multicavity magnetron [ELECTR] A magnetron in which the circuit includes a plurality of cavities, generally cut into the solid cylindrical anode so that the mouths of the cavities face the central cathode.

multicellular [BIOL] Consisting of many cells.

multicellular horn [ELECTROMAG] A cluster of horn antennas having mouths that lie in a common surface and that are fed from openings spaced one wavelength apart in one face of a common waveguide. [ENG ACOUS] A combination of individual horn loudspeakers having individual driver units or joined in groups to a common driver unit. Also known as cellular horn.

multichannel analyzer *See* pulse-height analyzer.

multichannel communication [COMMUN] Communication in which there are two or more communication channels over the same path, such as a communication cable, or a radio transmitter which can broadcast on two different frequencies, either individually or simultaneously.

multichannel field-effect transistor [ELECTR] A field-effect transistor in which appropriate voltages are applied to the gate to control the space within the current flow channels.

multichannel loading [COMMUN] Behavior of a multichannel communications system with all channels active.

multichannel telephone system [COMMUN] A telephone system in which two or more communications channels are carried over a single telephone cable or radio link.

multichip circuit [ELECTR] Microcircuit in which discrete, miniature, active electronic elements (transistor or diode chips) and thin-film or diffused passive components or component clusters are interconnected by thermocompression bonds, alloying, soldering, welding, chemical deposition, or metallization.

multicipital [BIOL] Having many heads or branches arising from one point.

multicollector electron tube [ELECTR] An electron tube in which electrons travel to more than one electrode.

MUD PUPPY

Necturus maculosus, the most common species of mud puppy found in the eastern United States and Canada.

MUD VOLCANO

Mud volcanoes near Douglaston, New York. Photograph taken at low tide, September 12, 1903. *(USGS)*

MULTIPATH SYSTEM

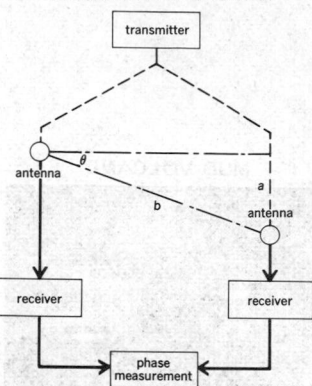

Multipath system. Transmission paths (dashed lines) are assumed parallel for long distances. Here b = base line between the two antennas. Path from transmitter to one antenna is larger by length a than path to second antenna, causing phase difference $n\theta$ between signals received at two receivers, where n is a positive integer.

multicompletion well   *See* multiple-completion well.

multicomponent distillation   [CHEM ENG] The distillation separation of a single liquid feed stream containing three or more components into a single overhead product and a single bottoms product.

multicomputer system   [ADP] A system consisting of more than one computer, usually under the supervision of a master computer, in which smaller computers handle input/output and routine jobs while the large computer carries out the more complex computations.

multicoupler   [ELECTR] A device for connecting several receivers to one antenna and properly matching the impedances of the receivers to the antenna.

multicycle   [GEOL] Pertaining to a landscape or landform produced by more than one cycle of erosion.

multideck cage   [MIN ENG] A cage with two or more compartments or platforms to hold the mine cars.

multideck clarifiers   [ENG] Extraction units which remove pollutants from recycled plant waste water.

multideck screen   [MIN ENG] A screen with two or more superimposed screening surfaces mounted within a common frame.

multideck sinking platform   [MIN ENG] A sinking platform of several decks so that various shaft-sinking operations may be performed simultaneously.

multideck table   [MIN ENG] A type of shaking table that is double-decked; while each deck is fed and discharged independently, one mechanism vibrates both.

multidimensional derivative   [MATH] The generalized derivative of a function of several variables which is usually represented as a matrix involving the various partial derivatives of the function.

multidrop line   [COMMUN] A telephone pair which terminates at several locations.

multielectrode tube   [ELECTR] Electron tube containing more than three electrodes associated with a single electron stream.

multielement array   [ELECTROMAG] An antenna array having a large number of antennas.

multielement parasitic array   [ELECTROMAG] Antennas consisting of an array of driven dipoles and parasitic elements, arranged to produce a beam of high directivity.

multielement vacuum tube   [ELECTR] A vacuum tube which has one or more grids in addition to the cathode and plate electrodes.

multifid   [BIOL] Divided into many lobes.

multifilament   [TEXT] A yarn made of two or more monofilaments, natural or synthetic.

multifunction array radar   [ENG] Electronic scanning radar which will perform target detection and identification, tracking, discrimination, and some interceptor missile tracking on a large number of targets simultaneously and as a single unit.

multifuse igniter   [ENG] A black powder cartridge that allows several fuses to be fired at the same time by lighting a single fuse.

multiglandular   [ANAT] Of or pertaining to several glands.

multigrid tube   [ELECT] An electron tube having two or more grids between cathode and anode, as a tetrode or pentode.

multigroup diffusion   [NUCLEO] Diffusion of neutrons in a material as it is regarded in the multigroup model.

multigroup model   [NUCLEO] A model for the behavior of neutrons in a material in which they are grouped into several energy ranges, taking into account differences in spatial behavior of neutrons in the various groups, and transfer of neutrons between groups.

multigun tube   [ELECTR] A cathode-ray tube having more than one electron gun.

multijob operation   [ADP] The concurrent or interleaved execution of job steps from more than one job.

multilayer bit   [DES ENG] A bit set with diamonds arranged in successive layers beneath the surface of the crown.

multilevel address   *See* indirect address.

multilevel transmission   [COMMUN] Transmission of digital information in which three or more levels of voltage are recognized as meaningful, as 0,1,2 instead of simply 0,1.

multilinear algebra   [MATH] The study of functions of several variables which are linear relative to each variable.

multilinear form   [MATH] A multilinear form of degree n is a polynomial expression which is linear in each of n variables.

Multillidae   [INV ZOO] An economically important family of Hymenoptera; includes the cow killer, a parasite of bumblebee pupae.

multilocular   [BIOL] Having many small chambers or vesicles.

multimeter   *See* volt-ohm-milliammeter.

multinomial   [MATH] An algebraic expression which involves the sum of at least two terms.

multinomial distribution   [MATH] The joint distribution of the set of random variables which are the number of occurrences of the possible outcomes in a sequence of multinomial trials.

multinomial trials   [STAT] Unrelated trials with more than two possible outcomes the probabilities of which do not change from trial to trial.

multipactor   [ELECTR] A high-power, high-speed microwave switching device in which a thin electron cloud is driven back and fourth between two parallel plane surfaces in a vacuum by a radio-frequency electric field.

multipass sort   [ADP] Computer program designed to sort more data than can be contained within the internal storage of a computer; intermediate storage, such as disk, tape, or drum, is required.

multipath   *See* multipath transmission.

multipath cancellation   [COMMUN] Occurrence of essentially complete cancellation of radio signals because of the relative amplitude and phase differences of the components arriving over the separate paths.

multipath system   [NAV] An electronic navigation system that measures differences in or otherwise compares transmission times of radio signals.

multipath transmission   [ELECTROMAG] The propagation phenomenon that results in signals reaching a radio receiving antenna by two or more paths, causing distortion in radio and ghost images in television. Also known as multipath.

multiphase flow   [CHEM ENG] Mixture of two or more distinct phases (such as oil, water, and gas) flowing through a closed conduit.

multiphasic personality inventory   *See* Minnesota multiphasic personality inventory.

multiple   [ELEC] 1. Group of terminals arranged to make a circuit or group of circuits accessible at a number of points at any one of which connection can be made. 2. To connect in parallel. 3. *See* parallel.

multiple-access computer   [ADP] A computer system whose facilities can be made available to a number of users at essentially the same time, normally through terminals, which are often physically far removed from the central computer and which typically communicate with it over telephone lines.

multiple accumulating registers   [ADP] Special registers capable of handling factors larger than one computer word in length.

multiple-activity process chart   [IND ENG] A chart showing the coordinated synchronous or simultaneous activities of a work system comprising one or more machines or men; each machine or man is shown in a separate, parallel column indicating its or his activities as related to the other parts of the work system.

multiple-address code   [ADP] A computer instruction code in which more than one address or storage location is specified; the instruction may give the locations of the operands, the destination of the result, and the location of the next instruction.

multiple-address computer   [ADP] A computer whose instruction contains more than one address, for example, an operation code and three addresses A, B, C, such that the content of A is multiplied by the content of B and the product stored in location C.

multiple-address instruction   [ADP] An instruction which has more than one address in a computer; the addresses give locations of other instructions, or of data or instructions that are to be operated upon.

multiple-anomaly syndrome   [MED] Any syndrome associated with numerous congenital abnormalities.

multiple appearance   [ELEC] Jack arrangement in telephone switchboards whereby a single-line circuit appears before two or more operators.

multiple-arch dam [CIV ENG] A dam composed of a series of arches inclined at about 45° and carried on parallel buttresses or piers.

multiple-beam antenna [ELECTROMAG] An antenna or antenna array which radiates several beams in different directions.

multiple-beam interference [OPTICS] Interference which arises when part of a beam is reflected several times back and forth between a pair of strongly reflecting surfaces before being reflected or transmitted from the pair.

multiple-beam interferometer [OPTICS] An interferometer in which a beam is reflected several times back and forth between a pair of parallel plane surfaces; examples are the Fizeau interferometer and the Fabry-Perot interferometer.

multiple call transmission [COMMUN] Routing and transmitting to two or more stations are done by the operator switching the message, cross-office, to the multiple address processing unit for routing line segregation.

multiple cancellous exostoses *See* multiple hereditary exostoses.

multiple cartilaginous exostoses *See* multiple hereditary exostoses.

multiple cartridges [CHEM ENG] Filter medium made up of two or more filter cartridges, either fastened end to end or arranged side by side (in series or parallel flow respectively).

multiple colloid goiter *See* adenomatous goiter.

multiple-completion packer [PETRO ENG] A device used to provide a seal between the outside surfaces of the two or more parallel tubing strings in a multiple-completion well, and the inside surface of the common wellbore casing.

multiple-completion well [PETRO ENG] Oil well in which there is production from more than one oil-bearing zone (different depths) with parallel tubing strings within a single wellbore casing string. Also known as multicompletion well.

multiple computer operation [ADP] The utilization of any one computer of a group of computers by means of linkages provided by multiplexor channels, all computers being linked through their channels or files.

multiple-contact switch *See* selector switch.

multiple-current hypothesis [OCEANOGR] The proposal that the Gulf Stream, instead of being composed of a single tortuous current, actually consists of many quasipermanent currents, countercurrents, and eddies.

multiple decay *See* branching.

multiple discharge *See* composite flash.

multiple disintegration *See* branching.

multiple-effect evaporation [CHEM ENG] Series-operation energy economizer system in which heat from the steam generated (evaporated liquid) in the first stage is used to evaporate additional liquid in the second stage (by reducing system pressure), and so on, up to 10 or more effects; commonly used in the pulp and paper industry.

multiple-entry system [MIN ENG] A system of access or development openings generally in bituminous coal mines involving more than one pair of parallel entries, one for haulage and fresh-air intake and the other for return air.

multiple epidermis [BOT] Epidermis that is several layers thick, occurring in many species of *Ficus, Begonia,* and *Peperomia.*

multiple fabric [TEXT] Material made of two or more fabrics bound together as they are woven.

multiple-factor incentive plan [IND ENG] A wage incentive plan based on productivity and other factors such as yield, material usage, and reduction of scrap.

multiple fault *See* step fault.

multiple firing [ENG] Electrically firing with delay blasting caps in a number of holes at one time.

multiple-function chip *See* large-scale integrated circuit.

multiple gun [ORD] Any group of guns mounted in one position and fired as a unit.

multiple hereditary exostoses [MED] An inherited form of exostosis, revealing itself at several sites in childhood or adolescence. Also known as diaphyseal aclasis; hereditary deforming chondrodysplasia; metaphyseal aclasis; multiple cancellous exostoses; multiple cartilaginous exostoses.

multiple hydraulic pump [PETRO ENG] Oil-well pump arrangement by which a single pump can be used in alternating operation to lift oil from two producing zones within a single multiple completion well.

multiple-impulse welding [MET] Spot, upset, or projection welding in which more than one current impulse is generated during a single machine cycle. Also known as pulsation welding.

multiple integral [MATH] An integral of a function of several variables which involves integrating the function successively over a prescribed range for each variable.

multiple jacks [ELEC] Series of jacks with tip, ring, and sleeve, respectively connected in parallel, and appearing in different panels of the face equipment of a telephone exchange.

multiple lamp holder [ELEC] A device that can be inserted in a lamp holder to act as two or more lamp holders. Also known as current tap.

multiple-loop system [CONT SYS] A system whose block diagram has at least two closed paths, along each of which all arrows point in the same direction.

multiple modulation [COMMUN] A succession of modulating processes in which the modulated wave from one process becomes the modulating wave for the next. Also known as compound modulation.

multiple module access [ADP] Device which establishes priorities in storage access in a multiple computer environment.

multiple myeloma [MED] A primary bone malignancy characterized by diffuse osteoporosis, anemia, hyperglobulinemia, and other clinical features. Also known as Kahler's disease.

multiple neuritis *See* polyneuritis.

multiple neurofibroma *See* neurofibromatosis.

multiple neurofibromatosis *See* neurofibromatosis.

multiple openings [MIN ENG] Any series of underground openings separated by rib pillars or connected at frequent intervals to form a system of rooms and pillars.

multiple-parallel-tubing string [PETRO ENG] Two or more parallel and closely packed oil-well tubing strings used in multiple-completion wells. Also known as multiple-tubing string.

multiple-pass weld [MET] A weld made by depositing filler metal with two or more passes in succession.

multiple personality [PSYCH] A personality capable of dissociation into several or many other personalities at the same time, whereby the delusion is entertained that the one person is many separate persons; a symptom in schizophrenic patients.

multiple piece rate plan [IND ENG] A wage incentive plan wherein increasingly higher unit pay rates are given to the worker as his productivity increases.

multiple precision arithmetic [ADP] Method of increasing the precision of a result by increasing the length of the number to encompass two or more computer words in length.

multiple pregnancy [MED] Being pregnant with more than one fetus.

multiple programming [ADP] The execution of two or more operations simultaneously.

multiple punching [ADP] The punching of two or more holes in a column of a punch card.

multiple-purpose tester *See* volt-ohm-milliammeter.

multiple reflection [GEOPHYS] A seismic wave which has more than one reflection. Also known as repeated reflection; secondary reflection. [OPTICS] Reflection of light back and forth several times between a pair of strongly reflecting surfaces.

multiple reflection echoes [ELECTROMAG] Radar echoes returned from a real target by reflection from some object in the radar beam; such echoes appear at a false bearing and false range.

multiple resonance [ELEC] Two or more resonances at different frequencies in a circuit consisting of two or more coupled circuits which are resonant at slightly different frequencies. [QUANT MECH] Two or more resonances at slightly different energies, resulting from the splitting of a single resonance by an interaction that is relatively weak.

multiple root [MATH] A polynomial $f(x)$ has c as a multiple root if $(x - c)^n$ is a factor for some $n > 1$. Also known as repeated root.

multiple-row blasting [ENG] The drilling, charging, and firing of rows of vertical boreholes.

multiple sampling [IND ENG] A plan for quality control in which a given number of samples from a group are inspected, and the group is either accepted, resampled, or rejected, depending on the number of failures found in the samples.

multiple scattering [PHYS] Process in which a particle undergoes a large number of collisions, and the total change in its momentum is the sum of the many small changes occurring during individual collisions.

multiple sclerosis [MED] A degenerative disease of the nervous system of unknown cause in which there is demyelination followed by gliosis.

multiple series [ENG] A method of wiring a large group of blasting charges by connecting small groups in series and connecting these series in parallel. Also known as parallel series.

multiple serositis *See* polyserositis.

multiple shooting [ENG] The firing of an entire face at one time by means of connecting shot holes in a single series and shooting all holes at the same instant.

multiple-shot survey [PETRO ENG] The determining and recording of drill-hole direction.

multiple-slide press [MECH ENG] A press with individual adjustable slides built into the main slide or connected independently to the main shaft.

multiple spot welding [MET] Spot welding in which several spots are welded during a single machine cycle.

multiple-stage rocket *See* multistage rocket.

multiple-stage separator [PETRO ENG] Oilwell gas-oil separator in which wellhead pressure is reduced in several stages, with the flashing off of gas at each pressure reduction.

multiple star [ASTRON] A system of three or more stars which appear to the naked eye as a single star.

multiple-strand conveyor [MECH ENG] A conveyor with two or more spaced strands of chain, belts, or cords as the supporting or propelling medium.

multiple switchboard [ELEC] Manual telephone switchboard in which each subscriber line is attached to two or more jacks, to be within reach of several operators.

multiplet [QUANT MECH] A collection of relatively closely spaced energy levels which result from the splitting of a single energy level by an interaction which is relatively weak; examples are spin-orbit multiplets and isospin multiplets. [SPECT] A collection of relatively closely spaced spectral lines resulting from transitions to or from the members of a multiplet (as in the quantum-mechanics definition).

multiple target generator [ELECTR] An electronic countermeasures device that produces several false responses in a hostile radar set.

multiplet intensity rules [SPECT] Rules for the relative intensities of spectral lines in a spin-orbit multiplet, stating that the sum of the intensities of all lines which start from a common initial level, or end on a common final level, is proportional to $2J + 1$, where J is the total angular momentum of the initial level or final level respectively.

multiple tropopause [GEOPHYS] A frequent condition in which the tropopause appears not as a continuous single "surface" of discontinuity between the troposphere and the stratosphere, but as a series of quasi-horizontal "leaves," which are partly overlapping in steplike arrangement.

multiple-tubing string *See* multiple-parallel-tubing string.

multiple-tuned antenna [ELECTROMAG] Low-frequency antenna having a horizontal section with a multiplicity of tuned vertical sections.

multiple twin quad [ELEC] Quad cable in which the four conductors are arranged in two twisted pairs, and the two pairs twisted together.

multiple-unit semiconductor device [ELECTR] Semiconductor device having two or more seats of electrodes associated with independent carrier streams.

multiple-unit steerable antenna *See* musa.

multiple-unit tube *See* multiunit tube.

multiple-valued [MATH] A relation between sets is multiple-valued if it associates to an element of one more than one

element from the other; sometimes functions are allowed to be multiple-valued.

multiplex [ENG] Stereoscopic device to project aerial photographs onto surfaces so that the images may be viewed in three dimensions by using anaglyphic spectacles; used to prepare topographic maps.

multiplexer [ELECTR] A device for combining two or more signals, as for multiplex, or for creating the composite color video signal from its components in color television. Also spelled multiplexor.

multiplex mode [ADP] The utilization of differences in operating speeds between a computer and transmission lines; the multiplexor channel scans each line in sequence, and any transmitted pulse on a line is assembled in an area reserved for this line; consequently, a number of users can be handled by the computer simultaneously. Also known as multiplexor channel operation.

multiplex operation [COMMUN] Simultaneous transmission of two or more messages in either or both directions over a carrier channel.

multiplexor *See* multiplexer.

multiplexor channel operation *See* multiplex mode.

multiplexor terminal unit [ADP] Device which permits a large number of data transmission lines to access a single computer.

multiplex transmission [COMMUN] The simultaneous transmission of two or more programs or signals over a single radio-frequency channel, such as by time division, frequency division, or phase division.

multiple x-y recorder [ENG] Recorder that plots a number of independent charts simultaneously, each showing the relation of two variables, neither of which is time.

multiple zone [PETRO ENG] Two or more discrete oil or gas reservoirs in the same geographic area, but at different depths.

multiplicand [MATH] If a number x is to be multiplied by a number y, then x is called the multiplicand.

multiplication [ELECTR] An increase in current flow through a semiconductor due to increased carrier activity. [MATH] Any algebraic operation analogous to multiplication of real numbers. [NUCLEO] The ratio of neutron flux in a subcritical reactor to that supplied by a neutron source; it is the factor by which, in effect, the reactor multiplies the source strength. Also known as neutron multiplication factor.

multiplication constant *See* multiplication factor.

multiplication factor [NUCLEO] The ratio of the number of neutrons present in a reactor in any one neutron generation to that in the immediately preceding generation. Also known as multiplication constant; neutron multiplication factor.

multiplication time [ADP] The time required for a computer to perform a multiplication; for a binary number it will be equal to the total of all the addition times and all the shift times involved in the multiplication.

multiplicity [MATH] A root of a polynomial $f(x)$ has multiplicity n if $(x - a)^n$ is a factor of $f(x)$ and n is the largest possible integer for which this is true. [PHYS] In a system having Russell-Saunders coupling, the quantity $2S + 1$, where S is the total spin quantum number.

multiplier [ELEC] A resistor used in series with a voltmeter to increase the voltage range. Also known as multiplier resistor. [ELECTR] 1. A device that has two or more inputs and an output that is a representation of the product of the quantities represented by the input signals; voltages are the quantities commonly multiplied. 2. *See* electron multiplier. 3. *See* frequency multiplier. [MATH] If a number x is to be multiplied by a number y, then y is called the multiplier.

multiplier field [ADP] The area reserved for a multiplication, equal to the length of multiplier plus multiplicand plus one character.

multiplier phototube [ELECTR] A phototube with one or more dynodes between its photocathode and the output electrode; the electron stream from the photocathode is reflected off each dynode in turn, with secondary emission adding electrons to the stream at each reflection. Also known as electron-multiplier phototube; photoelectric electron-multiplier tube; photomultiplier; photomultiplier tube.

multiplier-quotient register [ADP] A register equal to two

MULTIPLIER PHOTOTUBE

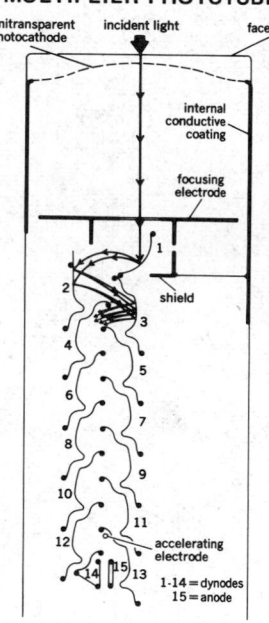

Typical multiplier phototube construction.

1-14 = dynodes
15 = anode

words in length in which the quotient is developed and in which the multiplier is entered for multiplication.

multiplier resistor   *See* multiplier.

multiplier traveling-wave photodiode   [ELECTR] Photodiode in which the construction of a traveling-wave tube is combined with that of a multiplier phototube to give increased sensitivity.

multiplier tube   [ELECTR] Vacuum tube using secondary emission from a number of electrodes in sequence to obtain increased output current; the electron stream is reflected, in turn, from one electrode of the multiplier to the next.

multipling   [COMMUN] Use of multidrop lines to provide for changes in telephone service patterns or requirements; unused terminals afford convenient access to wiretappers.

multiply connected region   [MATH] An open set in the plane which has holes in it.

multiply defined symbol   [ADP] Common assembler or compiler error printout indicating that a label has been used more than once.

multipoint line   [COMMUN] A line which is shared by two or more different tributary stations.

multipole   [ELECTROMAG] One of a series of types of static or oscillating distributions of charge or magnetization; namely, an electric multipole or a magnetic multipole.

multipole fields   [ELECTROMAG] The electric and magnetic fields generated by static or oscillating electric or magnetic multipoles.

multipole radiation   [PHYS] **1.** Electromagnetic radiation which has characteristics equivalent to those of radiation generated by an oscillating electric or magnetic multipole, and is made up of photons of well-defined angular momentum and parity. **2.** Internal conversion electrons, or positron-electron pairs having similar characteristics, emitted from an atom when the nucleus makes a transition between two energy states.

multipole transition   [PHYS] A transition between two energy states of an atom or nucleus in which a quantum of multipole radiation is emitted or absorbed.

multiport memory   [ADP] A memory shared by many processors to communicate among themselves.

multiprecision arithmetic   [ADP] A form of arithmetic similar to double precision arithmetic except that two or more words may be used to represent each number.

multiprocessing   [ADP] Carrying out of two or more sequences of instructions at the same time in a computer.

multiprocessing system   *See* multiprocessor.

multiprocessor   [ADP] A data-processing system that can carry out more than one program, or more than one arithmetic operation, at the same time. Also known as multiprocessing system.

multiprocessor interleaving   [ADP] Technique used to speed up processing time; by splitting banks of memory each with x microseconds access time and accessing each one in sequence $1/n$-th of a cycle later, a reference to memory can be had every x/n microseconds; this speed is achieved at the cost of hardware complexity.

multiprogramming   [ADP] The interleaved execution of two or more programs by a computer, in which the central processing unit executes a few instructions from each program in succession.

multiprogramming executive control   [ADP] Control program structure required to handle multiprogramming with either a fixed or a variable number of tasks.

multipurpose projectile   [ORD] A projectile designed so that the type of payload can be changed; this is accomplished by using prepared loads in canister form and providing a removable base plug to permit change of canister; thus a canister containing colored smoke mixture can be replaced by, for instance, one containing leaflets.

multirope friction winder   [MECH ENG] A winding system in which the drive to the winding ropes is the frictional resistance between the ropes and the driving sheaves.

multisegment magnetron   [ELECTR] Magnetron with an anode divided into more than two segments, usually by slots parallel to its axis.

multistable circuit   [ELECTR] A circuit having two or more stable operating conditions.

multistage   [ENG] Functioning or occurring in separate steps.

multistage amplifier   *See* cascade amplifier.

multistage compressor   [MECH ENG] A machine for compressing a gaseous fluid in a sequence of stages, with or without intercooling between stages.

multistage pump   [MECH ENG] A pump in which the head is developed by multiple impellers operating in series.

multistage queuing   [IND ENG] A situation involving two or more sequential stages in a process, each of which involves waiting in line.

multistage rocket   [AERO ENG] A vehicle having two or more rocket units, each unit firing after the one in back of it has exhausted its propellant; normally, each unit, or stage, is jettisoned after completing its firing. Also known as multiple-stage rocket; step rocket.

multistatic radar   [ENG] Radar in which successive antenna lobes are sequentially engaged to provide a tracking capability without physical movement of the antenna.

multistation   [COMMUN] Pertaining to a network in which each station can communicate with each of the other stations.

multistator watt-hour meter   [ELEC] An induction type of watt-hour meter in which several stators exert a torque on the rotor.

multistrip coupler   [ELECTR] A series of parallel metallic strips placed on a surface acoustic wave filter between identical apodized interdigital transducers; it converts the spatially nonuniform surface acoustic wave generated by one transducer into a spatially uniform wave received at the other transducer, and helps to reject spurious bulk acoustic modes.

multitask operation   [ADP] A sophisticated form of multijob operation in a computer which allows a single copy of a program module to be used for more than one task.

multitrack operation   [ADP] The selection of the next read/write head in a cylinder, usually indicated by bit zero of the operation code in the channel command word.

multitrack recording system   [ENG] Recording system which provides two or more recording paths on a medium, which may carry either related or unrelated recordings in common time relationship.

Multituberculata   [PALEON] The single order of the nominally mammalian suborder Allotheria; multituberculates had enlarged incisors, the coracoid bones were fused to the scapula, and the lower jaw consisted of the dentary bone alone.

multituberculate   [VERT ZOO] Of teeth, having several or many simple conical cusps.

multiturn potentiometer   [ELEC] A precision wire-wound potentiometer in which the resistance element is formed into a helix, generally having from 2 to 10 turns.

multiunit tube   [ELECTR] Electron tube containing within one glass or metal envelope, two or more groups of electrodes, each associated with separate electron streams. Also known as multiple-unit tube.

multivalent   *See* polyvalent.

multivariate analysis   [STAT] The study of random variables which are multidimensional.

multivator   [ELEC] Type of dc-to-dc up-converter. [ELECTR] An automatic device for analyzing a number of dust samples that might be collected by spacecraft on the moon, Mars, and other planets, to detect the presence of microscopic organisms with a multiplier phototube that measures the fluorescence given off.

multivibrator   [ELECTR] A relaxation oscillator using two tubes, transistors, or other electron devices, with the output of each coupled to the input of the other through resistance-capacitance elements or other elements to obtain in-phase feedback voltage.

multiwell gas-lift system   [PETRO ENG] An installation to allow gas-lift production from the various tubing strings involved in a multiple completion well.

mu meson   *See* muon.

Mumetal   [MET] An alloy of high magnetic permeability, containing 14% iron, 5% copper, 1.5% chromium, and the balance nickel.

mummification   [MED] **1.** Drying of a part of the body into a hard mass. **2.** Dry gangrene.

mumps   [MED] An acute contagious viral disease characterized chiefly by painful enlargement of a parotid gland.

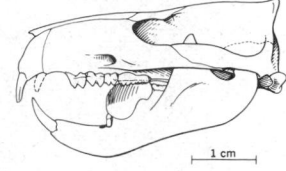

MULTITUBERCULATA

Skull and jaw of *Ptilodus*, an early Tertiary multituberculate. *(After G. G. Simpson, 1937)*

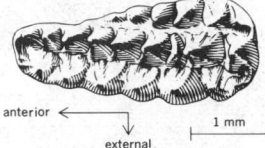

MULTITUBERCULATE

Occlusal view of a first upper molar of a multituberculate mammal.

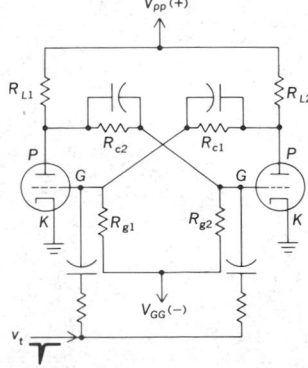

MULTIVIBRATOR

Circuit diagram of a bistable multivibrator.

mumps orchitis [MED] Inflammation of the testis due to the mumps virus.

Munchausen syndrome [PSYCH] A personality disorder in which the patient describes dramatic but false symptoms or simulates acute illness, happily undergoing examinations, hospitalization, and diagnostic and therapeutic manipulations, and upon discovery of the real nature of his case often leaves without notice and moves on to another hospital.

mundic *See* pyrite.

municipal engineering [CIV ENG] Branch of engineering dealing with the form and functions of urban areas.

munition [ORD] **1.** In a broad sense, any and all supplies and equipment required to conduct offensive or defensive war, including war machines, ammunition, transport, fuel, food, and clothing, but excluding personnel and excluding supplies and equipment for purposes other than for direct military operations. **2.** In a restricted sense, ordnance.

Munsell chroma *See* chroma.

Munsell color system [OPTICS] A system for designating colors which employs three perceptually uniform scales (Munsell hue, Munsell value, Munsell chroma) defined in terms of daylight reflectance.

Munsell hue [OPTICS] The dimension of the Munsell system of color that determines whether a color is blue, green, yellow, red, purple, or the like, without regard to its lightness or saturation.

Munsell value [OPTICS] The dimension, in the Munsell system of object-color specificiation, that indicates the apparent luminous transmittance or reflectance of the object on a scale having approximately equal perceptual steps under the usual conditions of observation.

muntins *See* sash bars.

Muntz metal [MET] A 60/40 type of brass composed of 58–61% copper, up to 1% lead, and remainder zinc. Also known as malleable brass.

muon [PARTIC PHYS] Collective name for two semistable elementary particles with positive and negative charge, designated μ^+ and μ^- respectively, which are leptons and have a spin of ½ and a mass of approximately 105.7 MeV. Also known as mu meson.

muonic atom [PARTIC PHYS] An atom in which an electron is replaced by a negatively charged muon orbiting close to or within the nucleus.

muonium [PARTIC PHYS] An atom consisting of an electron bound to a positively charged muon by their mutual Coulomb attraction, just as an electron is bound to a proton in the hydrogen atom.

mural thrombus [MED] A thrombus attached to the wall of a blood vessel or mural endocardium. Also known as lateral thrombus.

muramic acid [BIOCHEM] An organic acid found in the mucopeptide (murein) in the cell walls of bacteria and blue-green algae.

muramidase [BIOCHEM] Lysozyme when acting as an enzyme on the hydrolysis of the muramic acid-containing mucopeptide in the cell walls of some bacteria.

murbruk structure *See* mortar structure.

Murchisoniacea [PALEON] An extinct superfamily of gastropod mollusks in the order Prosobranchia.

murein [BIOCHEM] The peptidoglycan of bacterial cell walls.

muriatic acid *See* hydrochloric acid.

muriatic ether *See* ethyl chloride.

Muricacea [INV ZOO] A superfamily of gastropod mollusks in the order Prosobranchia.

muricate [ZOO] Covered with sharp, hard points.

Muricidae [INV ZOO] A family of predatory gastropod mollusks in the order Neogastropoda; contains the rock snails.

Muridae [VERT ZOO] A large diverse family of relatively small cosmopolitan rodents; distinguished from closely related forms by the absence of cheek pouches.

muriform [BIOL] **1.** Resembling the arrangement of courses in a brick wall, especially having both horizontal and vertical septa. **2.** Pertaining to or resembling a rat or mouse.

Murinae [VERT ZOO] A subfamily of the Muridae which contains such forms as the striped mouse, house mouse, harvest mouse, and field mouse.

murine plague [VET MED] Infection of the rat by the bacte-

rium *Pasteurella pestis;* transmitted from rat to rat and from rat to man by a flea.

murine typhus [MED] A relatively mild, acute, febrile illness of worldwide distribution caused by *Rickettsia mooseri,* transmitted from rats to man by the flea and characterized by headache, macular rash, and myalgia. Also known as endemic typhus; flea-borne typhus; rat typhus; shop typhus; urban typhus.

murmur [MED] A blowing or roaring heart sound heard through the wall of the chest.

muromontite [MINERAL] $Be_2FeY_2(SiO_4)_3$ A mineral composed of yttrium iron beryllium silicate.

Murphree efficiency [CHEM ENG] In a plate-distillation column, the ratio of the actual change in vapor composition when the vapor passes through the liquid on a tray (plate) to the composition change of the vapor if it were in vapor-liquid equilibrium with the tray liquid.

Murray code [COMMUN] A binary code with five binary digits per letter which was developed to be used with a typewriter-like device which would punch holes in paper tape, and is now the basis of the widely used CCIT 2 code.

Murray loop test [ELEC] A method of localizing a fault in a cable by replacing two arms of a Wheatstone bridge with a loop formed by the cable under test and a good cable connected to the far end of the defective cable.

Murray Valley encephalitis [MED] An acute inflammation of the brain and spinal cord caused by a virus; confined to Australia and New Guinea. Also known as Australian X disease.

murumuru oil [MATER] An oil obtained from nuts of the palm *Attalea orbignya,* containing 40% lauric acid, 35% myristic acid, and some palmitic, stearic, linoleic, and oleic acids.

musa [ELECTROMAG] An electrically steerable receiving antenna whose directional pattern can be rotated by varying the phases of the contributions of the individual units. Derived from multiple-unit steerable antenna.

Musaceae [BOT] A family of monocotyledonous plants in the order Zingiberales characterized by five functional stamens, unisexual flowers, spirally arranged leaves and bracts, and fleshy, indehiscent fruit.

Musca [ASTRON] A southern constellation, right ascension 12 hours, declination 70°S. Also known as Fly.

muscarine [ORG CHEM] $C_8H_{19}NO_3$ A quaternary ammonium compound, the toxic ingredient of certain mushrooms, as *Amanita muscaria.* Also known as hydroxycholine.

muscarinism [MED] Poisoning due to ingestion of certain mushrooms.

Muschelkalk [GEOL] A European stage of geologic time equivalent to the Middle Triassic, above Bunter and below Keuper.

Musci *See* Bryopsida.

Muscicapidae [VERT ZOO] A family of passeriform birds assigned to the Oscines; includes the Old World flycatchers or fantails.

Muscidae [INV ZOO] A family of myodarian cyclorrhaphous dipteran insects in the subsection Calypteratae; includes the houseflies, stable flies, and allies.

muscle [ANAT] A contractile organ composed of muscle tissue that changes in length and effects movement when stimulated. [HISTOL] A tissue composed of cells containing contractile fibers; three types are smooth, cardiac, and skeletal.

muscle fiber [HISTOL] The contractile cell or unit of which muscle is composed.

muscle hemoglobin *See* myoglobin.

muscovite [MINERAL] $KAl_2(AlSi_3)O_{10}(OH)_2$ One of the mica group of minerals, occurring in some granites and abundant in pegmatites; it is colorless, whitish, or pale brown, and the crystals are tabular sheets with prominent base and hexagonal or rhomboid outline; hardness is 2–2.5 on Mohs scale, and specific gravity is 2.7–3.1. Also known as common mica; mirror stone; moscovite; Muscovy glass; potash mica; white mica.

Muscovy glass *See* muscovite.

muscul-, musculo- [ZOO] A combining form denoting muscle, muscular.

muscular atrophy [MED] Degenerative reduction of muscle

MURRAY LOOP TEST

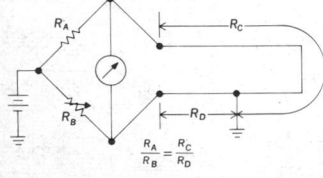

Circuit diagram for a Murray loop for location of ground fault.

size, especially skeletal muscles, due to a lesion involving either the cell body or axon of the lower motor neuron.

muscular dystrophy [MED] A group of diseases characterized by degeneration of or injury to individual muscle cells, not primarily involving the nerve supply; the most common form is Duchenne-Greisinger disease.

muscularis externa [HISTOL] The layer of the digestive tube consisting of smooth muscles.

muscularis mucosae [HISTOL] Thin, deep layer of smooth muscle in some mucous membranes, as in the digestive tract.

muscular system [ANAT] The muscle cells, tissues, and organs that effect movement in all vertebrates.

musculoaponeurotic [HISTOL] Composed of muscle and of fibrous connective tissue in the form of a membrane.

musculocutaneous [ANAT] Of or pertaining to muscles and skin.

musculocutaneous nerve [ANAT] A branch of the brachial plexus with both motor and somatic sensory components; innervates flexor muscles of the upper arm, and skin of the lateral aspect of the forearm.

musculoskeletal system [ANAT] The muscular and skeletal elements of vertebrates, considered as a functional unit.

mushroom [MYCOL] **1.** A fungus belonging to the basidiomycetous order Agaricales. **2.** The fruiting body (basidiocarp) of such a fungus.

mushroom anchor [NAV ARCH] A large anchor having the approximate shape of a mushroom, capable of grasping the ground whichever way it falls; used mainly where the bottom is sandy or muddy for permanent moorings of light ships and moorings of modern submarines.

musical acoustics [ACOUS] That part of acoustics which is relevant to the composition, performance, and appreciation of music, including the physical characteristics of sounds that may be heard as music, laws governing the action, design, and construction of musical instruments, and the effects of musical sounds upon listeners.

musical quality *See* timbre.

Musidoridae [INV ZOO] A family of orthorrhaphous dipteran insects in the series Brachycera distinguished by spear-shaped wings.

musk [PHYSIO] Any of various strong-smelling substances obtained from the musk glands of musk deer or similar animals; used in the form of a tincture as a fixative for perfume.

musk ambrette [ORG CHEM] $C_{12}H_{16}N_2O_5$ White to yellow powder with heavy musky aroma; soluble in various oils and phthalates, insoluble in water; congeals at 83°C; used as a perfume fixative. Also known as 2,6-dinitro-3-methoxy-4-tert-butyltoluene.

Muskat equation [PETRO ENG] Equation used to calculate oil reservoir permeability from the pressure buildup curve recorded when a producing well is shut in after a flow test.

musk bag *See* musk gland.

muskeg [ECOL] A peat bog or tussock meadow, with variably woody vegetation.

musk gland [VERT ZOO] A large preputial scent gland of the musk deer and various other animals, including skunk and musk-ox. Also known as musk bag.

musk ketone [ORG CHEM] $C_{14}H_{18}N_2O_5$ White to yellow crystals with sweet musk aroma; soluble in various oils and phthalates, insoluble in water; used as a perfume fixative. Also known as 3,5-dinitro-2,6-dimethyl-4-tert-butylacetophenone.

muskmelon [BOT] *Cucumis melo.* The edible, fleshy, globular to long-tapered fruit of a trailing annual plant of the order Violales; surface is uniform to broadly sutured to wrinkled, and smooth to heavily netted, and flesh is pale green to orange; varieties include cantaloupe, Honey Dew, Casaba, and Persian melons.

musk-ox [VERT ZOO] *Ovibos moschatus.* An even-toed ungulate which is a member of the mammalian family Bovidae; a heavy-set animal with a shag pilage, splayed feet, and flattened horns set low on the head.

muskrat [VERT ZOO] *Ondatra zibethica.* The largest member of the rodent subfamily Microtinae; essentially a water rat with a laterally flattened, long, naked tail, a broad blunt head with short ears, and short limbs.

musk xylene *See* musk xylol.

musk xylol [ORG CHEM] $(NO_2)_3C_6(CH_3)_2C(CH_3)_3$ White to yellow crystals with powerful musk aroma; soluble in various oils and phthalates, insoluble in water; congeals at 105°C; used as a perfume fixative. Also known as musk xylene; 2,4,6-trinitro-1,3-dimethyl-5-tert-butylbenzene.

muslin [TEXT] A thin to coarse-textured plain-weave cotton.

Musophagidae [VERT ZOO] The turacos, an African family of birds of uncertain affinities usually included in the order Cuculiformes; resemble the cuckoos anatomically but have two unique pigments, turacin and turacoverdin.

mustard [BOT] Any of several annual crucifers belonging to the genus *Brassica* of the order Capparales; leaves are lyrately lobed, flowers are yellow, and pods have linear beaks; the mustards are cultivated for their pungent seed and edible foliage, and the seeds of *B. niger* are used as a condiment, prepared as a powder, paste, or oil.

mustard gas [ORG CHEM] $HS(CH_2ClCH_2)_2S$ An oil, density 1.28, boiling point 215°C; used in chemical warfare. Also known as dichlorodiethylsulfide.

mustard oil *See* allyl isothiocyanate.

Mustilidae [VERT ZOO] A large, diverse family of low-slung, long-bodied carnivorous mammals including minks, weasels, and badgers; distinguished by having only one molar in each upper jaw, and two at the most in the lower jaw.

mutagen [GEN] An agent that raises the frequency of mutation above the spontaneous rate.

mutant [GEN] An individual bearing an allele that has undergone mutation and is expressed in the phenotype.

mutarotation [CHEM] A change in the optical rotation of light that takes place in the solutions of freshly prepared sugars.

mutase [BIOCHEM] An enzyme able to catalyze a dismutation or a molecular rearrangement.

mutation [GEN] An abrupt change in the genotype of an organism, not resulting from recombination; genetic material may undergo qualitative or quantitative alteration, or rearrangement.

Mutillidae [INV ZOO] The velvet ants, a family of hymenopteran insects in the superfamily Scolioidea.

muting circuit [ELECTR] **1.** Circuit which cuts off the output of a receiver when no radio-frequency carrier greater than a predetermined intensity is reaching the first detector. **2.** Circuit for making a receiver insensitive during operation of its associated transmitter.

muting switch [ELEC] **1.** A switch used in connection with automatic tuning systems to silence the receiver while tuning from one station to another. **2.** A switch used to ground the output of a phonograph pickup automatically while a record changer is in its change cycle.

mutual capacitance [ELEC] The accumulation of charge on the surfaces of conductors of each of two circuits per unit of potential difference between the circuits.

mutual conductance *See* transconductance.

mutual exclusion rule [PHYS CHEM] The rule that if a molecule has a center of symmetry, then no transition is allowed in both its Raman scattering and infrared emission (and absorption), but only in one or the other.

mutual impedance [ELEC] For two meshes of a network carrying alternating current, the ratio of the complex voltage in one mesh to the complex current in the other, when all meshes besides the latter one carry no current.

mutual inductance [ELECTROMAG] Property of two neighboring circuits, equal to the ratio of the electromotive force induced in one circuit to the rate of change of current in the other circuit.

mutual induction [ELECTROMAG] The generation of a voltage in one circuit by a varying current in another.

mutual interference [COMMUN] Man-made interference from two or more electrical or electronic systems which affects these systems on a reciprocal basis.

mutualism [ECOL] Mutual interactions between two species that are beneficial to both species.

mutuality of phases [CHEM] The rule that if two phases, with respect to a reaction, are in equilibrium with a third phase at a certain temperature, then they are in equilibrium with respect to each other at that temperature.

muzzle [ORD] The end of the barrel of a gun from which the

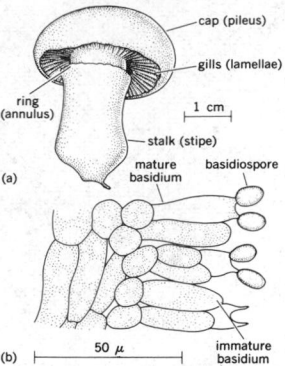

MUSHROOM

Agaricus bisporus, the common cultivated mushroom. *(a)* Basidiocarp. *(b)* Cross section of gill, showing basidia in various stages of development.

MUSK-OX

The musk-ox *(Ovibos moschatus)*; flattened horns, implanted low in the skull, are used defensively.

MUSKRAT

The muskrat *(Ondatra zibethica).*

MUTILLIDAE

Drawing of a velvet ant. *(From T. I. Storer and R. L. Usinger, General Zoology, 3d ed., McGraw-Hill, 1957)*

bullet or projectile emerges. [VERT ZOO] The snout of an animal, as a dog or horse.

muzzle energy [ORD] Kinetic energy of the projectile as it emerges from the muzzle; it is a measure of the power of the weapon.

mV *See* millivolt.

MV *See* megavolt.

mW *See* milliwatt.

MW *See* megawatt.

MW(E) *See* megawatt electric.

MW(Th) *See* megawatt thermal.

MWYE *See* megawatt year of electricity.

Mx *See* maxwell.

myalgia [MED] Pain in the muscles.

myasthenia [MED] Muscular weakness.

myasthenia gravis [MED] A muscle disorder of unknown etiology characterized by varying degrees of weakness and excessive fatigability of voluntary muscle.

myasthenia reaction [MED] The electromyographic reaction observed in myasthenia gravis in which there is a gradual loss of intensity and duration for the tetanic contraction, and a gradual diminution in amplitude and frequency of motor unit discharges until the muscle is fatigued.

myasthenic crisis [MED] Profound myasthenia and respiratory paralysis associated with myasthenia gravis.

myatonia [MED] Lack of muscle tone.

Mycelia Sterilia [MYCOL] An order of fungi of the class Fungi Imperfecti distinguished by the lack of spores; certain members are plant pathogens.

mycelium [MYCOL] A mass of fungal filaments (hyphae) which may be septate and compose the vegetative body of a fungus.

Mycetaeidae [INV ZOO] The equivalent name for Endomychidae.

mycetocyte [INV ZOO] **1.** One of the cells clustered together to form a mycetome. **2.** An individual cell functioning like a mycetome.

mycetoma [MED] A chronic fungus or bacterial infection, usually of the feet, resulting in swelling. Also known as madura foot; maduromycosis.

mycetome [INV ZOO] One of the specialized structures in the body of certain insects for holding endosymbionts.

Mycetophagidae [INV ZOO] The hairy fungus beetles, a cosmopolitan family of coleopteran insects in the superfamily Cucujoidea.

Mycetozoa [BIOL] A zoological designation for organisms that exhibit both plant and animal characters during their life history (Myxomycetes); equivalent to the botanical Myxomycophyta.

Mycetozoia [INV ZOO] A subclass of the protozoan class Rhizopodea.

Mycobacteriaceae [MICROBIO] A family of nonsporeforming, nonmotile, aerobic gram-positive bacteria belonging to the order Actinomycetales.

mycobacterial disease [MED] Any disease caused by species of *Mycobacterium*.

Mycobacterium [MICROBIO] A genus of acid-fast, rod-shaped bacteria in the family Mycobacteriaceae.

Mycobacterium leprae [MICROBIO] Bacterium that causes leprosy. Also known as Hansen's bacillus.

Mycobacterium tuberculosis [MICROBIO] Bacterium that causes tuberculosis; a slow-growing, drug-sensitive species. Also known as tubercle bacillus.

mycobactin [BIOCHEM] Any compound produced by some strains, and required for growth by other strains, of *Mycobacteria*.

mycology [BOT] The branch of botany that deals with the study of fungi.

mycomycin [MICROBIO] $C_{13}H_{10}O_2$ An antibiotic produced by *Nocardia acidophilus* and a species of *Actinomyces*; characterized as a highly unsaturated aliphatic acid that shows strong activity against *Mycobacterium tuberculosis*.

mycophagous [ZOO] Feeding on fungi.

Mycophiformes [VERT ZOO] An equivalent name for Salmoniformes.

Mycoplasma [MICROBIO] The single genus of the Mycoplasmatales.

Mycoplasma pneumoniae [MICROBIO] A bacterial agent of pneumonia; formerly it was thought to be a virus and was known as Eaton agent.

Mycoplasmataceae [MICROBIO] The single family of the Mycoplasmatales.

Mycoplasmatales [MICROBIO] An order of the class Mollicutes; organisms are gram-negative, generally nonmotile, nonsporing bacteria which lack a true cell wall.

mycorrhiza [BOT] A mutual association in which the mycelium of a fungus invades the roots of a seed plant.

mycosis [MED] An infection with or a disease caused by a fungus.

mycostatin [MICROBIO] Trade name for the antibiotic mystatin.

Mycota [MYCOL] An equivalent name for Eumycetes.

mycotic stomatitis *See* thrush.

Myctophidae [VERT ZOO] The lantern fishes, a family of deep-sea forms of the suborder Myctophoidei.

Myctophoidei [VERT ZOO] A large suborder of marine salmoniform fishes characterized by having the upper jaw bordered only by premaxillae, and lacking a mesocoracoid arch in the pectoral girdle.

Mydaidae [INV ZOO] The mydas flies, a family of orthorrhaphous dipteran insects in the series Brachycera.

mydriasis [MED] Prolonged dilation of the pupil of the eye.

mydriatic [PHARM] An agent which produces dilation of the pupil, such as eucatropine hydrochloride.

myel-, myelo- [ANAT] A combining form for bone marrow, spinal cord.

myelencephalon [EMBRYO] The caudal portion of the hindbrain; gives rise to the medulla oblongata.

myelin [BIOCHEM] A soft, white fatty substance that forms a sheath around certain nerve fibers.

myelin sheath [HISTOL] An investing cover of myelin around the axis cylinder of certain nerve fibers.

myelitis [MED] **1.** Inflammation of the spinal cord. **2.** Inflammation of the bone marrow.

myeloblast [HISTOL] The youngest precursor cell for blood granulocytes, having a nucleus with finely granular chromatin and nucleoli and intensely basophilic cytoplasm.

myeloblastemia [MED] The presence of myeloblasts in the peripheral circulation.

myeloblastic [HISTOL] Of, pertaining to, or characterized by the presence of myeloblasts.

myeloblastic leukemia *See* acute granulocytic leukemia.

myeloblastoma [MED] A malignant tumor composed of myeloblasts.

myeloblastosis [MED] Diffuse proliferation of myeloblasts, with involvement of blood, bone marrow, and other tissues and organs.

myelocele [ANAT] The canal of the spinal cord. [MED] Spina bifida, with protrusion of the spinal cord.

myelocyte [HISTOL] A motile precursor cell of blood granulocytes found in bone marrow.

myelocytoma [MED] A malignant plasmacytoma.

myelocytosis [MED] The presence of myelocytes in the blood.

myelodysplasia [MED] Abnormal spinal cord development, especially the lumbosacral portion.

myeloencephalitis [MED] Inflammation of the brain and spinal cord.

myelofibrosis [PATH] Growth of white, fibrous connective tissue in the bone marrow.

myelogenous leukemia *See* granulocytic leukemia.

myelogram [MED] Roentgenogram of the spinal cord, made by myelography. [PATH] Differential cell study of material extracted from bone marrow.

myelography [MED] Roentgenographic visualization of the subarachnoid space, after the injection of air or an opaque medium.

myeloid [ANAT] **1.** Of or pertaining to bone marrow. **2.** Of or pertaining to the spinal cord.

myeloid leukemia *See* granulocytic leukemia.

myeloid metaplasia [MED] The occurrence of hemopoietic tissue in abnormal places in the body.

myeloid myeloma [MED] A malignant plasmacytoma.

myeloid reaction [MED] Increased numbers of granulocytes

MYCOPLASMA

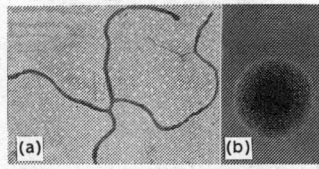

Mycoplasma. (a) Electron micrograph of *M. mycoides* var. *mycoides* from a 24-hour broth culture. *(b) M. hominis* colony on solid nutrient medium, 5 days old.

MYCORRHIZA

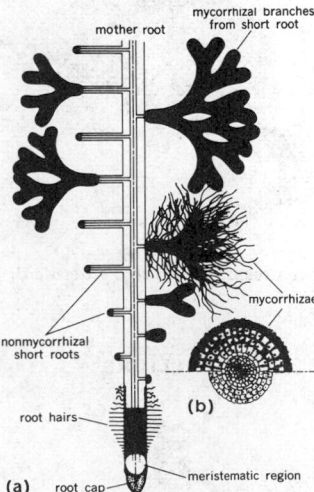

Development of mycorrhizae on a pine root. Solid areas indicate absorbing surfaces. *(a)* The main axis, a mother root. *(b)* Cross section, representing a mycorrhizal root. *(From P. J. Kramer, Plant and Soil Water Relationships, McGraw-Hill, 1949)*

in the bone marrow and peripheral circulation, often with the appearance of immature granulocytes in the blood.

myeloid tissue [HISTOL] Red bone marrow attached to argyrophile fibers which form wide meshes containing scattered fat cells, erythroblasts, myelocytes, and mature myeloid elements.

myeloma [MED] A primary tumor of the bone marrow composed of any of the bone marrow cell types.

myelomalacia [MED] Softening of the spinal cord.

myelomeningitis [MED] Inflammation of the spinal cord and its meninges.

myelomeningocele [MED] Spina bifida with protrusion of the spinal meninges.

myelomonocyte [HISTOL] 1. A monocyte developing in bone marrow. 2. A blood cell intermediate between monocytes and granulocytes.

myelomonocytic leukemia *See* monocytic leukemia.

myelopathic anemia *See* myelophthisic anemia.

myelophthisic anemia [MED] An anemia associated with space-occupying disorders of the bone marrow. Also known as leukoerythroblastic anemia; leukoerythroblastosis; metastatic anemia; myelopathic anemia; myelosclerotic anemia; osteosclerotic anemia.

myelophthisis [MED] 1. Loss of bone marrow. 2. Atrophy of the spinal cord.

myeloplegia [MED] Spinal paralysis.

myelopoiesis [PHYSIO] The process by which blood cells form in the bone marrow.

myelosclerosis [MED] 1. Multiple sclerosis of the spinal cord. 2. Hardening of the bone marrow.

myelosclerotic anemia *See* myelophthisic anemia.

myenteric [HISTOL] Of or pertaining to the muscular coat of the intestine.

myenteric plexus [ANAT] A network of nerves between the circular and longitudinal layers of the muscular coat of the digestive tract. Also known as Auerbach's plexus.

myenteron [HISTOL] The muscular coat of the intestine.

Mygalomorphae [INV ZOO] A suborder of spiders (Araneida) including American tarantulas, trap-door spiders, and purse-web spiders; the tarantulas may attain a leg span of 10 inches (25 centimeters).

myiasis [MED] Infestation of vertebrates by the larvae, or maggots, of flies.

Myklestad method [AERO ENG] A method of determining the mode shapes and frequencies of the lateral bending modes of space vehicles, taking into account secondary effects of shear and rotary inertia, in which one imagines masses to be concentrated at a finite number of points along the beam, with elastic properties remaining constant between consecutive mass points.

Mylabridae [INV ZOO] The equivalent name for Bruchidae.

Mylar [MATER] Trademark of Du Pont for a polyester film widely used for insulating purposes and as a backing for magnetic tape.

Mylar capacitor [ELEC] A capacitor that uses Mylar film as a dielectric between rolled strips of metal foil.

Myliogatidae [VERT ZOO] The eagle rays, a family of batoids which may reach a length of 15 feet (4.6 meters).

mylonite [PETR] A hard, coherent, often glassy-looking rock that has suffered extreme mechanical deformation and granulation but has remained chemically unaltered; appearance is flinty, banded, or streaked, but the nature of the parent rock is easily recognized.

mylonite gneiss [PETR] A metamorphic rock intermediate in character between mylonite and schist.

mylonitic structure [PETR] A structure characteristic of mylonites, produced by extreme microbrecciation and shearing which gives the appearance of a flow structure.

mylonitization [GEOL] Rock deformation produced by intense microbrecciation without appreciable chemical alteration of granulated materials.

Mymaridae [INV ZOO] The fairy flies, a family of hymenopteran insects in the superfamily Chalcidoidea.

myoblast [EMBRYO] A precursor cell of a muscle fiber.

myocardial infarct [MED] An infarct in heart muscle.

myocardiopathy [MED] Disease of the myocardium. Also known as cardiomyopathy.

myocarditis [MED] Inflammation of the myocardium.

myocardium [HISTOL] The muscular tissue of the heart wall.

myocardosis [MED] Any noninflammatory disease of the myocardium.

myoclonic epilepsy [MED] Recurrent irregular, arrhythmic clonic muscle spasms, usually occurring more frequently in the morning or on going to sleep and often associated with other types of seizures.

myoclonic status [MED] Continual clonic spasms lasting an hour or more.

myoclonus [MED] 1. Clonic muscle spasm. 2. Any disorder characterized by scattered, irregular, arrhythmic muscle spasms.

myocoel [EMBRYO] Portion of the coelom enclosed in a myotome.

myocomma [HISTOL] A ligamentous connection between successive myomeres. Also known as myoseptum.

myocyte [HISTOL] 1. A contractile cell. 2. A muscle cell.

Myodaria [INV ZOO] A section of the Schizophora series of cyclorrhaphous dipterans; in this group adult antennae consist of three segments, and all families except the Conopidae have the second cubitus and the second anal veins united for almost their entire length.

Myodocopa [INV ZOO] A suborder of the order Myodocopida; includes exclusively marine ostracods distinguished by possession of a heart.

Myodocopida [INV ZOO] An order of the subclass Ostracoda.

Myodopina [INV ZOO] The equivalent name for Myodocopa.

myodystrophy [MED] Muscle degeneration.

myoelastic fiber [HISTOL] An elastic fiber associated with the smooth muscles in bronchi and bronchioles.

myoelectric potential [PHYSIO] The electrical potential created by muscle action.

myofascitis [MED] Muscular pain of obscure nature and origin in the lower back.

myofibril [CYTOL] A contractile fibril in a muscle cell. [INV ZOO] *See* myoneme.

myofilament [CYTOL] The structural unit of muscle proteins in a muscle cell.

myofrisk [INV ZOO] A contractile structure surrounding the spines of certain radiolarians.

myoglobin [BIOCHEM] A hemoglobinlike iron-containing protein pigment occurring in muscle fibers. Also known as muscle hemoglobin; myohemoglobin.

myoglobinuria [MED] The presence of myoglobin in the urine.

myohematin [BIOCHEM] A cytochrome respiratory enzyme allied to hematin.

myohemoglobin *See* myoglobin.

myoinositol [BIOCHEM] The commonest isomer of inositol. Also known as mesionositol.

myokinase [BIOCHEM] An enzyme that catalyzes the reversible transfer of phosphate groups in adenosinediphosphate; occurs in muscle and other tissues.

myolipoma [MED] A benign tumor composed of adipose and smooth muscle cells.

myology [MED] The study of muscles in both the normal and diseased states.

myoma [MED] 1. A benign uterine tumor composed principally of smooth muscle cells. 2. Any neoplasm originating in muscle.

myomalacia [MED] Degeneration, with softening, of muscle tissue.

myomere [EMBRYO] A muscle segment differentiated from the myotome, which divides to form the epimere and hypomere.

myometritis [MED] Inflammation of the myometrium.

myometrium [HISTOL] The muscular tissue of the uterus.

Myomorpha [VERT ZOO] A suborder of rodents recognized in some systems of classification.

myoneme [INV ZOO] A contractile fibril in a protozoan. Also known as myofibril.

myoneural junction [ANAT] The point of junction of a motor nerve with the muscle which it innervates. Also known as neuromuscular junction.

myopathia *See* myopathy.

myopathic facies [MED] An expressionless face with sunken

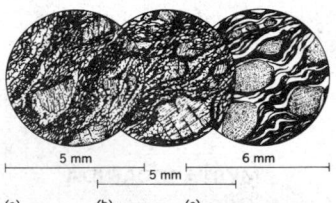

MYLONITE

5 mm   5 mm   6 mm

(a)   (b)   (c)

Mylonite types: *(a)* porphyroclasts mylonite; *(b)* granite mylonite; *(c)* mylonite augen gneiss. *(From H. Williams, F. J. Turner and C. M. Gilbert, Petrography: An Introduction to the Study of Rocks in Thin Sections, Freeman, 1954)*

cheeks and a drooping lower lip characteristic of patients with myopathies, especially myotonic dystrophy.

myopathy [MED] Any disease of the muscles. Also known as myopathia.

myopericarditis [MED] A combination of myocarditis and pericarditis.

myophagia [PATH] The invasion of degenerated muscle sarcoplasm by histiocytes.

myophosphorylase deficiency glycogenosis See McArdle's syndrome.

myopia [MED] A condition in which the focal image is formed in front of the retina of the eye. Also known as nearsightedness.

myoplasty [MED] Plastic surgery performed on muscle tissue.

Myopsida [INV ZOO] A natural assemblage of cephalopod mollusks considered as a suborder in the order Teuthoida according to some systems of classification, and a group of the Decapoda according to other systems; the eye is covered by the skin of the head in all species.

myopsychopathy [MED] Any disease of the muscles associated with mental retardation or loss of intellect.

myorhythmia [MED] Muscle tremor with a rate of 2–4 per second, and irregular intervals between cycles.

myosarcoma [MED] A sarcoma derived from muscle.

myoseptum See myocomma.

myosin [BIOCHEM] A muscle protein, comprising up to 50% of the total muscle proteins; combines with actin to form actomycin.

myositis [MED] Inflammation of muscle. Also known as fibromyositis.

myositis ossificans [MED] Muscle inflammation with bone formation in muscle, tendons, or ligaments.

myostatic reflex See stretch reflex.

myosynovitis [MED] Inflammation of synovial membranes and surrounding musculature.

myotasis [PHYSIO] Stretching of a muscle.

myotome [ANAT] A group of muscles innervated by a single spinal nerve. [EMBRYO] The muscle plate that differentiates into myomeres. [ENG] An instrument used to divide a muscle, particularly through its belly.

myotonia [MED] Tonic muscular spasm occurring after injury or infection.

myotonia congenita intermittens See paramyotonia congenita.

myotonic [MED] Of, pertaining to, or characterized by myotonia.

myotonic dystrophy [MED] A hereditary disease, transmitted as an autosomal dominant, characterized by lack of normal relaxation of muscles after contraction, slowly progressive muscular weakness and atrophy, especially of the face and neck, cataract formation, early baldness, gonadal atrophy, abnormal glucose tolerance curve, and, frequently, mental deficiency.

myria- [SCI TECH] A prefix representing 10^4 or 10,000.

myriametric waves [ELECTROMAG] Electromagnetic waves having wavelengths between 10 and 100 kilometers, corresponding to the very low frequency band.

Myriangiales [MYCOL] An order of parasitic fungi of the class Ascomycetes which produce asci at various levels in uniascal locules within stromata.

Myriapoda [INV ZOO] Informal designation for those mandibulate arthropods having two body tagmata, one pair of antennae, and more than three pairs of adult ambulatory appendages.

Myricaceae [BOT] The single family of the plant order Myricales.

Myricales [BOT] An order of dicotyledonous plants in the subclass Hamamelidae, marked by its simple, resinous-dotted, aromatic leaves, and a unilocular ovary with two styles and a single ovule.

Myrientomata [INV ZOO] The equivalent name for the Protura.

myringitis [MED] Inflammation of the tympanic membrane.

Myriotrochidae [INV ZOO] A family of holothurian echinoderms in the order Apodida, distinguished by eight or more spokes in each wheel-shaped spicule.

myristic acid [ORG CHEM] $CH_3(CH_2)_{12}COOH$ Oily white crystals melting at 58°C; soluble in ether and alcohol, insoluble in water; used to synthesize flavor and perfume esters, and in soaps and cosmetics.

myristica oil See nutmeg oil.

myristyl alcohol [ORG CHEM] $C_{14}H_{29}OH$ Liquid boiling at 264°C; soluble in ether and alcohol, insoluble in water; used as a chemical intermediate, plasticizer, and perfume fixative. Also known as 1-tetradecanol.

Myrmecophagidae [VERT ZOO] A small family of arboreal anteaters in the order Edentata.

myrmecophagous [ZOO] Feeding on ants.

myrmecophile [ECOL] An organism, usually a beetle, that habitually inhabits the nest of ants.

myrmecophyte [ECOL] A plant that houses and benefits from the habitation of ants.

myrmekite [PETR] Intergrowth of plagioclase feldspar and vermicular quartz in an igneous rock.

myrmekitic [PETR] 1. Pertaining to the texture of an igneous rock marked by intergrowths of feldspar and vermicular quartz. 2. Having characteristic properties of myrmekite.

Myrmeleontidae [INV ZOO] The ant lions, a family of insects in the order Neuroptera; larvae are commonly known as doodlebugs.

Myrmicinae [INV ZOO] A large diverse subfamily of ants (Formicidae); some members are inquilines and have no worker caste.

myrrh [MATER] A gum resin of species of myrrh (*Commiphora*); partially soluble in water, alcohol, and ether; used in dentifrices, perfumery, and pharmaceuticals.

Myrsinaceae [BOT] A family of mostly woody dicotyledonous plants in the order Primulales characterized by flowers without staminodes, a schizogenous secretory system, and gland-dotted leaves.

Myrtaceae [BOT] A family of dicotyledonous plants in the order Myrtales characterized by an inferior ovary, numerous stamens, anthers usually opening by slits, and fruit in the form of a berry, drupe, or capsule.

Myrtales [BOT] An order of dicotyledonous plants in the subclass Rosidae characterized by opposite, simple, entire leaves and perigynous to epigynous flowers with a compound pistil.

myrtle oil [MATER] Light-yellow liquid distilled from the flowers and leaves of the European myrtle (*Myrtus communis*); aromatic aroma; formerly used in medicine; now used for flavors and as perfume fixative.

myrtle wax See bayberry wax.

Mysida [INV ZOO] A suborder of the crustacean order Mysidacea characterized by fusion of the sixth and seventh abdominal somites in the adult, lack of gills, and other specializations.

Mysidacea [INV ZOO] An order of free-swimming Crustacea included in the division Pericarida; adult consists of 19 somites, each bearing one pair of functionally modified, biramous appendages, and the carapace envelops most of the thorax and is fused dorsally with up to four of the anterior thoracic segments.

mysis [INV ZOO] A larva of certain higher crustaceans, characterized by biramous thoracic appendages.

mysophobia [PSYCH] An abnormal fear of contamination or of dirt.

Mystacinidae [VERT ZOO] A monospecific family of insectivorous bats (Chiroptera) containing the New Zealand short-tailed bat; hindlegs and body are stout, and fur is thick.

Mystacocarida [INV ZOO] An order of primitive Crustacea; the body is wormlike and the cephalothorax bears first and second antennae, mandibles, and first and second maxillae.

Mysticeti [VERT ZOO] The whalebone whales, a suborder of the mammalian order Cetacea, distinguished by horny filter plates of suspended from the upper jaws.

mythophobia [PSYCH] An abnormal fear of making an incorrect statement.

Mytilacea [INV ZOO] A suborder of bivalve mollusks in the order Filibranchia.

Mytilidae [INV ZOO] A family of mussels in the bivalve order Anisomyaria.

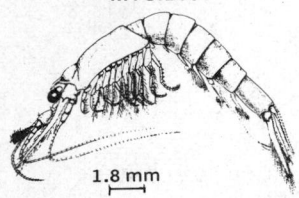
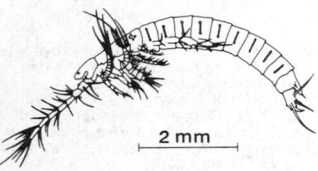

myx-, myxo- [ZOO] A combining form denoting mucus, mucous, mucin, mucinous.

myxadenitis [MED] Inflammation of mucous glands.

myxadenoma [MED] An adenoma of a mucous gland.

myxameba [BIOL] An independent ameboid cell of the vegetative phase of Acrasiales.

myxedema [MED] A condition caused by hypothyroidism characterized by a subnormal basal metabolic rate, dry coarse hair, loss of hair, mental dullness, anemia, and slowed reflexes.

Myxicolinae [INV ZOO] A subfamily of sedentary polychaete annelids in the family Sabellidae.

Myxiniformes [VERT ZOO] The equivalent name for the Myxinoidea.

Myxinoidea [VERT ZOO] The hagfishes, an order of eellike, jawless vertebrates (Agnatha) distinguished by having the nasal opening at the tip of the snout and leading to the pharynx, with barbels around the mouth and 6–15 pairs of gill pouches.

myxoadenoma [MED] An adenoma of a mucous gland.

Myxobacterales [MICROBIO] The slime bacteria, an order of bacteria of the class Schizomycetes; vegetative cells are flexible, slender motile rods that tend to swarm and arrange themselves in parallel rows prior to aggregating to form a fruiting body. Also known as fruiting myxobacteria.

myxochondrofibrosarcoma [MED] A sarcoma composed of anaplastic myxoid, chondroid, and fibrous cells.

Myxococcaceae [MICROBIO] A family of slime bacteria (Myxobacterales) which have spherical microcysts surrounded by a thick wall.

myxofibroma of nerve sheath *See* neurofibroma.

Myxogastromycetidae [MYCOL] A large subclass of plasmodial slime molds (Myxomycetes).

myxolipoma *See* liposarcoma.

myxoma [MED] A benign tumor composed of mucinous connective tissue.

myxoma lipomatodes *See* liposarcoma.

myxomatosis [VET MED] A virus disease of rabbits producing fever, skin lesions resembling myxomas, and mucoid swelling of mucous membranes.

Myxomycetes [BIOL] Plasmodial (acellular or true) slime molds, a class of microorganisms of the division Mycota; they are on the borderline of the plant and animal kingdoms and have a noncellular, multinucleate, jellylike, creeping, assimilative stage (the plasmodium) which alternates with a myxameba stage.

Myxomycophyta [BOT] An order of microorganisms, equivalent to the Mycetozoia of zoological classification.

Myxophaga [INV ZOO] A suborder of the Coleoptera.

Myxophyceae [BOT] An equivalent name for the Cyanophyceae.

myxosarcoma [MED] A sarcoma whose parenchyma is composed of anaplastic myxoid cells.

Myxosporida [INV ZOO] An order of the protozoan class Myxosporidea characterized by the production of spores with one or more valves and polar capsules, and by possession of a single sporoplasm.

Myxosporidea [INV ZOO] A class of the protozoan subphylum Cnidospora; members are parasitic in some fish, a few amphibians, and certain invertebrates.

myxovirus [VIROL] A group of ribonucleic-acid animal viruses characterized by hemagglutination and hemadsorption; includes influenza and fowl plague viruses and the paramyxoviruses.

Myzopodidae [VERT ZOO] A monospecific order of insectivorous bats (Chiroptera) containing the Old World disk-winged bat of Madagascar; characterized by long ears and by a vestigial thumb with a monostalked sucking disk.

myzorhynchus [INV ZOO] An apical sucker on the scolex of certain tapeworms.

Myzostomaria [INV ZOO] An aberrant group of Polychaeta; most are greatly depressed, broad, and very small, and true segmentation is delayed or absent in the adult; all are parasites of echinoderms.

Myzostomidae [INV ZOO] A monogeneric family of the Myzostomaria.

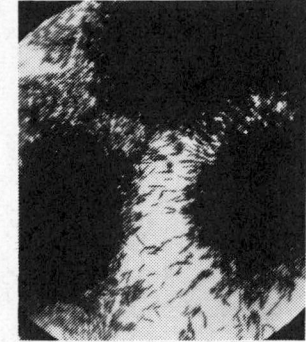

MYXOBACTERALES

Myxococcus fulvus, vegetative cells. (*After N. A. Woods, from A. T. Henrici and E. J. Ordal, The Biology of Bacteria, 3d ed., Heath, 1948*)

MYXOSPORIDEA

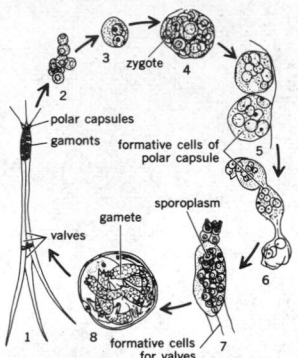

The life cycle of *Triactinomyxon legeri*, a representative member of the Actinomyxida showing: 1, mature spore; 2, liberated gamonts; 3, gamonts pairing; 4, zygote formation; 5, spore formation from zygote; 6–7, later stages in spore formation. (*After MacKinnon and Adam, 1924*)

MYZOSTOMARIA

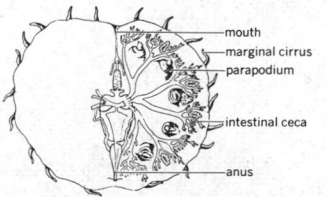

Myzostomum (Myzostomidae) in ventral view, with organ systems indicated. (*From R. R. von Stummer-Traunfels, Myzostomida, Kukenthal's Handbuch der Zoologie, Bd. 3, Lief. 2, Bogen 9–17 pp. 132–210, 1927*)

n- [ORG CHEM] Chemical prefix for "normal" (straight-carbon-chain) hydrocarbon compounds.

N *See* newton; normality.

Na *See* sodium.

N.A. *See* numerical aperture.

NAA *See* naphthaleneacetic acid.

nabam [ORG CHEM] $NaSSCNHCH_2CH_2NHCSSNa$ Water-soluble, colorless crystals that will irritate skin and eyes; used as a pesticide and pesticides intermediate. Also known as disodium ethylene-bis-dithiocarbamate.

Nabidae [INV ZOO] The damsel bugs, a family of hemipteran insects in the superfamily Cimicimorpha.

nabla *See* del operator.

Nabothian cyst [MED] Cystic distention of the Nabothian glands of the uterine cervix.

Nabothian glands [ANAT] Mucous glands of the uterine cervix.

nacelle [AERO ENG] A separate streamlined enclosure on an airplane for sheltering or housing something, as the crew or an engine.

nacre [INV ZOO] An iridescent inner layer of many mollusk shells.

nacreous [OPTICS] Having an iridescent luster resembling that of mother-of-pearl. Also known as pearly.

nacreous clouds [METEOROL] Clouds of unknown composition, whose form resembles that of cirrus or altocumulus lenticularis, and which show very strong irisation similar to that of mother-of-pearl, especially when the sun is several degrees below the horizon; they occur at heights of about 20 or 30 kilometers. Also known as mother-of-pearl clouds.

nacrite [MINERAL] $Al_2Si_2O_5(OH)_4$ A crystallized clay mineral of the kaolinite group; structurally distinct in being the most closely stacked in the *c*-axis direction.

NAD *See* diphosphopyridine nucleotide.

nadir [ASTRON] That point on the celestial sphere vertically below the observer, or 180° from the zenith.

Naegeli-type leukemia [MED] A type of monocytic leukemia in which the leukoyctes resemble cells of the granulocytic series.

naftalan *See* naphthalan.

nagatelite [MINERAL] Black mineral composed of phosphosilicate of an aluminum, rare-earth elements, calcium, and iron; occurs in tabular masses.

nagyagite [MINERAL] $Pb_5Au(Te,Sb)_4S_{5-8}$ A lead-gray mineral consisting of a sulfide of lead, gold, tellurium, and antimony. Also known as black tellurium; tellurium glance.

nahcolite [MINERAL] $NaHCO_3$ A white, monoclinic mineral consisting of natural sodium bicarbonate.

nail [ANAT] The horny epidermal derivative covering the dorsal aspect of the terminal phalanx of each finger and toe. [DES ENG] A slender, usually pointed fastener with a head, designed for insertion by impact. [ENG] To drive nails in a manner that will position and hold two or more members, usually of wood, in a desired relationship. [MED] A metallic rod with one blunt end and one sharp end, used surgically to anchor bone fragments.

nailhead [DES ENG] Flat protuberance at the end of a nail opposite the point.

nailheaded molding [ARCH] A molding consisting of a series of short protuberances resembling nailheads.

nailhead spot [PL PATH] A fungus rot of tomato caused by *Alternaria tomato* and marked by small brown to black sunken spots on the fruit.

nail-patella syndrome [GEN] A genetic disorder inherited as an autosomal dominant that is part of a linkage group with ABO blood group genes; characterized by defects in the nails and abnormalities of the elbows and other bones, skin, eyes, and kidneys.

Najadaceae [BOT] A family of monocotyledonous, submerged aquatic plants in the order Najadales distinguished by branching stems and opposite or whorled leaves.

Najadales [BOT] An order of aquatic and semiaquatic flowering plants in the subclass Alismatidae; the perianth, when present, is not differentiated into sepals and petals, and the flowers are usually not individually subtended by bracts.

naked bud [BOT] A bud covered only by rudimentary foliage leaves.

naked light [MIN ENG] Open flame, such as a match or a burning cigarette, that is a fire risk in mines.

naked-light mine [MIN ENG] A coal mine that is nongassy, where naked lights can be used by miners.

nakhlite [GEOL] An achondritic stony meteorite composed of an aggregate of diopside and olivine.

Namanereinae [INV ZOO] A subfamily of largely fresh-water errantian annelids in the family Nereidae.

Namurian [GEOL] A European stage of geologic time; divided into a lower stage (Lower Carboniferous or Upper Mississippian) and an upper stage (Upper Carboniferous or Lower Pennsylvanian).

nancy receiver *See* infrared receiver.

NAND [ADP] A logic operator having the characteristic that if P, Q, R, . . . are statements, then the NAND of P, Q, R, . . . is true if at least one statement is false, false if all statements are true. Derived from NOT-AND. Also known as sheffer stroke.

NAND circuit [ELECTR] A logic circuit whose output signal is a logical 1 if any of its inputs is a logical 0, and whose output signal is a logical 0 if all of its inputs are logical 1.

nanism [MED] Dwarfed stature due to arrested development.

nannoplankton [BIOL] Minute plankton; the smallest plankton, including algae, bacteria, and protozoans.

nano- [BIOL] A prefix meaning dwarfed. [MATH] A prefix representing 10^{-9}, which is 0.000000001 or one-billionth of the unit adjoined. Also known as milli-micro- (deprecated usage).

nanocephalus [MED] A fetus with an undersized head.

nanogram [MECH] One-billionth (10^{-9}) of a gram. Abbreviated ng.

nanometer [MECH] A unit of length equal to one-billionth of a meter, or 10^{-9} meter. Also known as nanon.

nanon *See* nanometer.

nanophanerophyte [ECOL] A shrub not exceeding 2 meters in height.

nanophthalmus *See* microphthalmus.

nanosecond [MECH] A unit of time equal to one-billionth of a second, or 10^{-9} second.

nanozooid [INV ZOO] Dwarf zooid; bryozoan heterozooid possessing only a single tentacle.

Nansen bottle [ENG] A bottlelike water-sampling device with valves at both ends that is lowered into the water by wire; at the desired depth it is activated by a messenger which strikes the reversing mechanism and inverts the bottle, closing the valves and trapping the water sample inside. Also known as Petterson-Nansen water bottle; reversing water bottle.

Nansen cast [OCEANOGR] A series of Nansen-bottle water samples and associated temperature observations resulting from one release of a messenger.

nantokite *See* cuprous chloride.

nap [TEXT] Fuzzy fibers on the surface of a fabric; produced by a finishing process called raising.

napalm [MATER] **1.** Aluminum soap in powder form, used to gelatinize oil or gasoline for use in napalm bombs or flame throwers. **2.** The resultant gelatinized substance.

napalm bomb [ORD] A bomb filled with napalm; primarily an antipersonnel weapon.

NAKED BUD

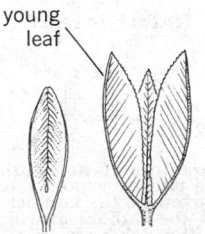

young leaf

Closed and open naked buds of the hobblebush.

NAPHTHALENE

Structural formula of naphthalene showing two benzenoid rings fused together; the numbers identify the carbons at which substitution takes place.

NAPHTHYLAMINE

(a)

(b)

Structural formulas for (a) α-naphthylamine; (b) β-naphthylamine.

nape [ANAT] The back of the neck.

Naperian logarithm *See* logarithm.

napex [ANAT] That portion of the scalp just below the occipital protuberance.

naphtha [MATER] **1.** Petroleum fraction with volatility between gasoline and kerosine; used as a gasoline ingredient, solvent for paints and rubber, and cleaning solvent. **2.** Aromatic solvent from coal tar, either solvent naphtha or heavy naphtha.

naphthacene [ORG CHEM] $C_{18}H_{12}$ A hydrocarbon molecule that may be considered to be four benzene rings fused together; it is explosive when shocked; used in organic synthesis. Also known as rubene; tetracene.

naphthalan [MATER] Soft, greenish-black mass distilled from Armenian naphtha; soluble in ether and hydrocarbons, insoluble in water; melts at 70°C; used in medicine. Also known as naftalan.

naphthalene [ORG CHEM] $C_{10}H_8$ White, volatile crystals with coal tar aroma; insoluble in water, soluble in organic solvents; structurally it is represented as two benzenoid rings fused together; boiling point 218°C, melting point 80.1°C; used for moth repellents, fungicides, lubricants, and resins, and as a solvent. Also known as naphthalin; tar camphor.

naphthaleneacetic acid [ORG CHEM] $C_{10}H_7CH_2COOH$ White, odorless crystals, melting at 132–135°C; soluble in organic solvents, slightly soluble in water; used as an agricultural spray. Abbreviated NAA. Also known as 1-naphthylacetic acid.

naphthalene-1,5-disulfonic acid [ORG CHEM] $C_{10}H_6(SO_3H)_2$ White crystals, decomposing when heated; used to make dyes. Also known as Armstrong's acid.

naphthalic acid *See* phthalic acid.

naphthalin *See* naphthalene.

naphthene [ORG CHEM] Any of the cycloparaffin derivatives of cyclopentane (C_5H_{10}) or cyclohexane (C_6H_{12}) found in crude petroleum.

naphthenic acid [ORG CHEM] Any of the derivatives of cyclopentane, cyclohexane, cycloheptane, or other naphthenic homologs derived from petroleum; molecular weights 180 to 350; soluble in organic solvents and hydrocarbons, slightly soluble in water; used as a paint drier and wood preservative, and in metals production.

naphthenic crude [MATER] Crude petroleum containing a significant proportion of naphthenic compounds.

naphthine *See* hatchettite.

naphthionic acid [ORG CHEM] $C_{10}H_6(NH_2)SO_3H$ White powder or crystals that decompose when heated; used to manufacture dyes. Also known as 1-aminonaphthalene-4-sulfonic acid; 4-amino-1-naphthalene sulfonic acid; 1-naphthylamine-4-sulfonic acid.

1-naphthol *See* α-naphthol.

2-naphthol *See* β-naphthol.

α-naphthol [ORG CHEM] $C_{10}H_7OH$ Colorless to yellow powder, melting at 96°C; used to make dyes and perfumes, and in synthesis of organic molecules. Also known as 1-hydroxynaphthalene; 1-naphthol.

β-naphthol [ORG CHEM] $C_{10}H_7OH$ White crystals that melt at 121.6°C; insoluble in water; used to make pigments, dyes, and antioxidants. Also known as 2-hydroxynaphthalene; 2-naphthol.

3-naphthol-2-carboxylic acid *See* β-hydroxynaphthoic acid.

1,4-naphthoquinone [ORG CHEM] $C_{10}H_6O_2$ Greenish-yellow powder soluble in organic solvents, slightly soluble in water; melts at 123–126°C; used as an antimycotic agent, in synthesis, and as a rubber polymerization regulator. Also known as α-naphthoquinone.

α-naphthoquinone *See* 1,4-naphthoquinone.

1-naphthylacetic acid *See* α-naphthaleneacetic acid.

naphthylamine [ORG CHEM] $C_{10}H_7NH_2$ White, toxic crystals, soluble in alcohol and ether; used in dyes; the two forms are α-naphthylamine, boiling at 301°C, and β-naphthylamine, boiling at 306°C.

1-naphthylamine-4-sulfonic acid *See* naphthionic acid.

2-naphthylamine-6-sulfonic acid *See* Brönner's acid.

2,5-naphthylamine sulfonic acid *See* gamma acid.

napier *See* neper.

Napier diagram [NAV] A diagram on which compass deviation is plotted for various headings, and the points are connected by a smooth curve, permitting deviation problems to be solved quickly without interpolation; it consists of a vertical line, usually in two parts, each part being graduated for 180° of heading, and two additional sets of lines at an angle of 60° to each other and to the vertical lines.

Napier's analogies [MATH] Formulas which enable one to study the relationships between the sides and the angles of a spherical triangle.

Napier's rules [MATH] Two rules which give the formulas necessary in the solution of right spherical triangles.

napiform [BOT] Turnip-shaped, referring to roots.

Naples yellow *See* lead antimonite.

nappe [GEOL] A sheetlike, allochthonous rock unit formed by thrust faulting or recumbent folding or both. [MATH] One of the two parts of a conical surface defined by the vertex.

napped leather *See* suede.

narbonnais [METEOROL] A wind coming from Narbonne; a north wind in the Roussillon region of southern France resembling the tramontana; if associated with an influx of arctic air, it may be very stormy with heavy falls of rain or snow.

narceine [ORG CHEM] $C_{23}H_{27}O_8N \cdot 3H_2O$ White, odorless crystals with bitter taste; soluble in alcohol and water, insoluble in ether; melts at 170°C; used in medicine.

narcissism [PSYCH] Excessive self-love.

narcissus oil *See* jonquil oil.

narco- [MED] Combining form meaning numbness, narcosis, or stupor.

narcoanalysis [PSYCH] Induction of a reversible sleep by intravenous injections of drugs such as amobarbital or thiopental sodium in order to elicit memories and feelings not expressed by the person in a wakeful state because of resistance.

narcolepsy [MED] A disorder of sleep mechanism characterized by two or more of four distinct symptoms: uncontrollable periods of daytime drowsiness, cataleptic attacks of muscular weakness, sleep paralysis, and vivid nocturnal or hypnogogic hallucinations.

narcomania [MED] Morbid physiologic or psychologic craving for narcotics to avoid painful stimuli.

Narcomedusae [INV ZOO] A suborder of hydrozoan coelenterates in the order Trachylina; the hydroid generation is represented by an actinula larva.

narcosis [MED] Drug-produced state of profound stupor, unconsciousness, or arrested activity.

narcosis therapy [MED] Prolonged, drug-induced sleep as treatment for certain mental disorders. Also known as sleep therapy.

narcospasm [MED] Spasm accompanied by stupor.

narcosynthesis [MED] Psychotherapeutic treatment under partial anesthesia, in which abreaction is a significant factor in obtaining positive results.

narcotic [PHARM] A drug which in therapeutic doses diminishes awareness of sensory impulses, especially pain, by the brain; in large doses, it causes stupor, coma, or convulsions.

narcotine *See* noscapine.

nari *See* caliche.

naris *See* nostril.

narrow-angle glaucoma [MED] Increased intraocular tension due to a block of the angle of the anterior chamber from contact of the iris by the trabecula. Also known as obstructive glaucoma.

narrow-band amplifier [ELECTR] An amplifier which increases the magnitude of signals over a band of frequencies whose bandwidth is small compared to the average frequency of the band.

narrow-band frequency modulation [COMMUN] Frequency-modulation broadcasting system used primarily for two-way voice communication, having a maximum permissible deviation of 15 kilohertz or less.

narrow-band-pass filter [ELECTR] A band-pass filter in which the band of frequencies transmitted by the filter has a bandwidth which is small compared to the average frequency of the band.

narrow-band path [COMMUN] A communications path having a bandwith of less than 20 kilohertz.

narrow-band pyrometer [ENG] A pyrometer in which light from a source passes through a color filter, which passes only a limited band of wavelengths, before falling on a photoelectric detector. Also known as spectral pyrometer.

narrow-beam antenna [ELECTROMAG] An antenna which radiates most of its power in a cone having a radius of only a few degrees.

narrow gage [CIV ENG] A railway gage narrower than the standard gage of 4 feet 8½ inches (143.51 centimeters).

narrow-gap spark chamber [NUCLEO] A type of spark chamber in which the plates are only 6 to 10 millimeters apart, so that sparks usually follow the electric field perpendicular to the plates, and coincide with the particle track at one point; the great majority of spark chambers are of this type.

narrows [GEOGR] A navigable narrow part of a bay, strait, or river.

narrow-sector recorder [ELECTR] A radio direction finder with which atmospherics are received from a limited sector related to the position of the antenna; this antenna is usually rotated continuously and the bearings of the atmospherics recorded automatically.

narrow-spectrum antibiotic [MICROBIO] An antibiotic effective against a limited number of microorganisms.

narsarsukite [MINERAL] Na$_2$(Ti,Fe)Si$_4$(O,F) Mineral composed of sodium titanium iron fluoride and silicate.

narwhal [VERT ZOO] *Monodon monoceros.* An arctic whale characterized by lack of a dorsal fin, and by possession in the male of a long, twisted, pointed tusk (or rarely, two tusks) which is a source of ivory.

N-ary code [COMMUN] Code employing *N* distinguishable types of code elements.

n-ary composition [MATH] A function that associates an element of a set with every sequence of *n* elements of the set.

N-ary pulse-code modulation [COMMUN] Pulse-code modulation in which the code for each element consists of any one of *N* distinguishable types of elements.

nasal [ANAT] Of or pertaining to the nose.

nasal base breadth [ANTHRO] The distance measured across the alae (wings) of the nose, when they are in rest position.

nasal bone [ANAT] Either of two rectangular bone plates forming the bridge of the nose; they articulate with the frontal, ethmoid, and maxilla bones.

nasal bridge breadth [ANTHRO] The width measured between the junctures of the cheekbone and nasal bone, just inside the internal canthi.

nasal bridge salient [ANTHRO] The distance measured from the tip of the bony bridge in the midline of the nose to the juncture of the bony sidewall with the cheek.

nasal cavity [ANAT] Either of a pair of cavities separated by a septum and located between the nasopharynx and anterior nares.

nasal crest [ANAT] **1.** The linear prominence on the medial border of the palatal process of the maxilla. **2.** The linear prominence on the medial border of the palatine bone. **3.** The linear prominence on the internal border of the nasal bone and forming part of the nasal septum.

nasal height [ANTHRO] The height of the nose measured from the nasion to the middle of the lower margin of the anterior nares.

nasal index [ANTHRO] The ratio, × 100, of the greatest width of the anterior nasal openings of the skull to the height of the nasal skeleton.

nasal pit *See* olfactory pit.

nasal process of the frontal bone [ANAT] The downward projection of the nasal part of the frontal bone which terminates as the nasal spine.

nasal process of the maxilla [ANAT] Frontal process of the maxilla.

nasal root breadth [ANTHRO] The distance measured between junctures of the cheekbone and the nasal bone, just inside the internal canthi.

nasal root salient [ANTHRO] The distance measured between the internal canthus and the nasion.

nasal septum [ANAT] The partition separating the two nasal cavities.

nasal tip height [ANTHRO] The distance measured between the subnasale and the pronasale.

nasal tip salient [ANTHRO] The distance measured from the nasal wing to the pronasale.

nascent [CHEM] Pertaining to an atom or simple compound at the moment of its liberation from chemical combination, when it may have greater activity than in its usual state.

n'aschi [METEOROL] A northeast wind which occurs in winter on the Iranian coast of the Persian Gulf, especially near the entrance to the Gulf, and also on the Makran (West Pakistan) coast; it is probably associated with an outflow from the central Asiatic anticyclone which extends over the high land of Iran.

Nasellina [INV ZOO] The equivalent name for Monopylina.

nasion [ANTHRO] Midpoint of the nasofrontal suture.

nasion-menton [ANTHRO] The distance measured between the nasion and the midpoint of the lower edge of the chin.

nasolacrimal canal [ANAT] The bony canal that lodges the nasolacrimal duct. Also known as lacrimal canal.

nasolacrimal duct [ANAT] The membranous duct lodged within the nasolacrimal canal; it gives passage to the tears from the lacrimal sac to the inferior meatus of the nose.

nasolacrimal groove [EMBRYO] The furrow, the maxillary, and the lateral nasal processes of the embryo.

nasonite [MINERAL] Ca$_4$Pb$_6$Si$_6$O$_{21}$Cl$_2$ A white mineral composed of silicate and chloride of calcium and lead and occurring in granular masses.

nasopalatine cyst *See* median maxillary cyst.

nasopalatine duct [EMBRYO] A canal between the oral and nasal cavities of the embryo at the point of fusion of the maxillary and palatine processes.

nasopharynx [ANAT] The space behind the posterior nasal orifices, above a horizontal plane through the lower margin of the palate.

nastic movement [BOT] Movement of a flat plant part, oriented relative to the plant body and produced by diffuse stimuli causing disproportionate growth or increased turgor pressure in the tissues of one surface.

nasturan *See* pitchblende.

Nasutitermitinae [INV ZOO] A subfamily of termites in the family Termitidae, characterized by having the cephalic glands open at the tip of an elongated tube which projects anteriorly.

Natalidae [VERT ZOO] The funnel-eared bats, a monogeneric family of small, tropical American insectivorous bats (Chiroptera) with large, funnellike ears.

Natantia [INV ZOO] A suborder of decapod crustaceans comprising shrimp and related forms characterized by a long rostrum and a ventrally flexed abdomen.

Nathansohn's theory [OCEANOGR] The theory that nutrient salts in the lighted surface layers of the ocean are consumed by plants, accumulate in the deep ocean through sinking of dead plant and animal bodies, and eventually return to the euphotic layer through diffusion and vertical circulation of the water.

Naticacea [INV ZOO] A superfamily of gastropod mollusks in the order Prosobranchia.

Naticidae [INV ZOO] A family of gastropod mollusks in the order Pectinibranchia comprising the moon-shell snails.

native [BIOL] Grown, produced or originating in a specific region or country. [GEOCHEM] Pertaining to an element found in nature in a nongaseous state.

native asphalt [GEOL] Exudations or seepages of asphalt occurring in nature in a liquid or semiliquid state. Also known as natural asphalt.

native metal [GEOCHEM] A metallic native element; includes silver, gold, copper, iron, mercury, iridium, lead, palladium, and platinum.

native paraffin *See* ozocerite.

native uranium [GEOCHEM] Uranium as found in nature; a mixture of the fertile uranium-238 isotope (99.3%), the fissionable uranium-235 isotope (0.7%), and a minute percentage of other uranium isotopes. Also known as natural uranium; normal uranium.

natremia [MED] Excessive amounts of sodium in the blood.

natrium [CHEM] Latin name for sodium; source of the symbol Na.

natriuretic [PHARM] A medicinal agent which inhibits reabsorption of cations, particularly sodium, from urine.

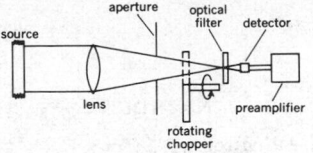

NARROW-BAND PYROMETER

Components of a narrow-band pyrometer.

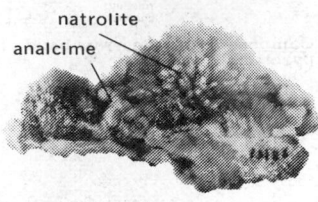

NATROLITE

natrolite
analcime

|— 5 cm —|

Natrolite crystals on analcime, Cape d'Or, Nova Scotia. *(American Museum of Natural History specimen)*

NATURAL-DRAFT COOLING TOWER

Photograph of natural-draft cooling tower. *(Haman, Inc.)*

natroalunite [MINERAL] $NaAl_3(SO_4)_2(OH)_6$ Mineral composed of basic sodium aluminum sulfate. Also known as almerite.

natrochalcite [MINERAL] $NaCu_2(SO_4)(OH)\cdot H_2O$ An emerald-green mineral composed of hydrous basic sulfate of sodium and copper.

natrolite [MINERAL] $Na_2Al_2Si_3O_{10}\cdot 2H_2O$ A zeolite mineral composed of hydrous silicate of sodium and aluminum; usually occurs in slender acicular or prismatic crystals.

natromontebrasite [MINERAL] $(Na,Li)Al(PO_4)(OH,F)$ Mineral composed of hydrous basic phosphate of sodium, lithium, and aluminum; it is isomorphous with montebrasite and amblygonite. Also known as fremontite.

natron [MINERAL] $Na_2CO_3\cdot 10H_{20}$ A white, yellow, or gray mineral that crystallizes in the monoclinic system, is soluble in water, and generally occurs in solution or in saline residues.

natronborocalcite *See* ulexite.

natron lake *See* soda lake.

natrophilite [MINERAL] $NaMn(PO_4)$ A mineral composed of sodium manganese phosphate.

natural aging [MET] Spontaneous aging at room temperature of a supersaturated metallic solid solution.

natural antenna frequency [ELECTROMAG] Lowest resonant frequency of an antenna without added inductance or capacitance.

natural arch *See* natural bridge.

natural asphalt *See* native asphalt.

natural bitumen [GEOL] Native mineral pitch, tar, or asphalt.

natural boundary [MATH] Those points of the boundary of a region where an analytic function is defined through which the function cannot be continued analytically.

natural bridge [GEOL] An archlike rock formation spanning a ravine or valley and formed by erosion. Also known as natural arch.

natural cement [MATER] Hydraulic cement made from pulverized and heated limestone containing clay, magnesia, and iron.

natural circulation reactor [NUCLEO] A reactor in which the coolant (usually water) circulates without pumping, owing to the different densities of its cold and reactor-heated portions.

natural convection [THERMO] Convection in which fluid motion results entirely from the presence of a hot body in the fluid, causing temperature and hence density gradients to develop, so that the fluid moves under the influence of gravity. Also known as free convection.

natural coordinates [FL MECH] An orthogonal, or mutually perpendicular, system of curvilinear coordinates for the description of fluid motion, consisting of an axis t tangent to the instantaneous velocity vector and an axis n normal to this velocity vector to the left in the horizontal plane, to which a vertically directed axis z may be added for the description of three-dimensional flow; such a coordinate system often permits a concise formulation of atmospheric dynamical problems, especially in the Lagrangian system of hydrodynamics.

natural draft [FL MECH] Unforced gas flow through a chimney or vertical duct, directly related to chimney height and the temperature difference between the ascending gases and the atmosphere, and not dependent upon the use of fans or other mechanical devices.

natural-draft cooling tower [MECH ENG] A cooling tower that depends upon natural convection of air flowing upward and in contact with the water to be cooled.

natural fiber [TEXT] A textile fiber of mineral, plant, or animal origin.

natural frequency [ELECTR] The lowest resonant frequency of an antenna, circuit, or component. [PHYS] The frequency with which a system oscillates in the absence of external forces; or, for a system with more than one degree of freedom, the frequency of one of the normal modes of vibration.

natural fuel reactor *See* natural uranium reactor.

natural function [MATH] A trigonometric function, as opposed to its logarithm.

natural gas [MATER] A combustible, gaseous mixture of low-molecular-weight paraffin hydrocarbons, generated below the surface of the earth; contains mostly methane and ethane with small amounts of propane, butane, and higher hydrocarbons, and sometimes nitrogen, carbon dioxide, hydrogen sulfide, and helium.

natural gasoline [MATER] The liquid paraffin hydrocarbon contained in natural gas and recovered by compression, distillation, and absorption.

natural-gasoline plant [CHEM ENG] Compression, distillation, and absorption process facility used to remove natural gasoline (mostly butanes and heavier components) from natural gas.

natural glass [GEOL] An amorphous, vitreous inorganic material that has solidified from magma too quickly to crystallize.

natural harbor [GEOGR] A harbor where the configuration of the coast provides the necessary protection.

natural immunity [IMMUNOL] Native immunity possessed by the individuals of a race, strain, or species.

natural interference [COMMUN] Electromagnetic interference arising from natural terrestrial phenomena (called atmospheric interference), or electromagnetic interference caused by natural disturbances originating outside the atmosphere of the earth (called galactic and solar noise).

natural language [ADP] A computer language whose rules reflect and describe current rather than prescribed usage; it is often loose and ambiguous in interpretation, meaning different things to different hearers.

natural levee [GEOL] An elongate embankment compounded of sand and silt and deposited along both banks of a river channel during times of flood.

natural logarithm *See* logarithm.

natural numbers [MATH] The integers 1,2,3, ...

natural period [PHYS] Period of the free oscillation of a body or system; when the period varies with amplitude, the natural period is the period when the amplitude approaches zero.

natural pressure cycle [MET] A cycle in which pressure buildup conforms proportionately to the buildup of stresses due to forming.

natural radiation *See* background radiation.

natural radioactivity [NUCLEO] Radioactivity exhibited by naturally occurring radionuclides.

natural radio-frequency interference [GEOPHYS] Natural terrestrial phenomena of an electromagnetic nature, or natural electromagnetic disturbances originating outside the atmosphere, which interfere with radio communications.

natural red *See* purpurin.

natural remanent magnetization [GEOPHYS] The magnetization of rock which exists in the absence of a magnetic field and has been acquired from the influence of the earth's magnetic field at the time of their formation or, in certain cases, at later times. Abbreviated NRM.

natural resonance [PHYS] Resonance in which the period or frequency of the applied agency maintaining oscillation is the same as the natural period of oscillation of a system.

natural resource [MATER] A deposit of minerals, water, or other materials furnished by nature.

natural scale [GRAPHICS] The ratio between the linear dimensions of a chart or drawing and the actual dimensions represented, expressed as a proportion; for example, 1 inch on a chart of natural scale 1:2,000,000 represents 2,000,000 inches on the earth. Also known as representative fraction.

natural science [SCI TECH] Collectively, the branches of science dealing with objectively measurable phenomena pertaining to the transformations and relationships of energy and matter; includes biology, physics, and chemistry.

natural selection [EVOL] Darwin's theory of evolution, according to which organisms tend to produce progeny far above the means of subsistence; in the struggle for existence that ensues, only those progeny with favorable variations survive; the favorable variations accumulate through subsequent generations, and descendants diverge from their ancestors.

natural splitting [MIN ENG] In mine ventilation, a flow of air dividing among the branches, of its own accord and without regulation, in inverse relation to the resistance of each airway.

natural steel [MET] **1.** Steel made directly from cast iron. **2.** Steel, such as wootz, made directly from the ore.

natural transmission [COMMUN] The signal transmitted over a line when no traffic is flowing.

natural uranium   *See* native uranium.

natural uranium reactor   [NUCLEO] A nuclear reactor in which natural unenriched uranium is the principal fissionable material. Also known as natural fuel reactor.

natural ventilation   [MIN ENG] The weak and varying ventilation in a mine caused by the difference in air density between shafts.

natural ventilation pressure   [MIN ENG] A pressure difference across the shaft-bottom doors caused by a lack of balance in the two vertical air columns.

natural wavelength   [ELECTROMAG] Wavelength corresponding to the natural frequency of an antenna or circuit.

natural width of energy level   [PHYS] A measure of the spread in energy of an excited state of a quantized system due to spontaneous transitions to other states; quantitatively, it is the difference between the energies for which the intensity of emission from or absorption by the state, or of the scattering cross section associated with it, is one-half its maximum value, in the absence of any external influence on the system.

Naucoridae   [INV ZOO] A family of hemipteran insects in the superfamily Naucoroidea.

Naucoroidea   [INV ZOO] The creeping water bugs, a superfamily of hemipteran insects in the subdivision Hydrocorisae; they are suboval in form, with chelate front legs.

naumannite   [MINERAL] Ag_2Se An iron-black mineral that crystallizes in the isometric system; consists of silver selenide, and occurs massive or in crystals; specific gravity is 8.

nauplius   [INV ZOO] A larval stage characteristic of many groups of Crustacea; the oval, unsegmented body has three pairs of appendages: uniramous antennules, biramous antennae, and mandibles.

nausea   [MED] Feeling of discomfort in the stomach region, accompanied by aversion to food and a tendency to vomit.

nautical almanac   [NAV] A book published annually by the governments of the principal maritime nations which contains the astronomical data required for navigation by observations of celestial objects; an abridged version is known as the abridged nautical almanac.

nautical astronomy   [NAV] The science of determining position and direction of a ship by observation of celestial objects.

nautical chain   [MECH] A unit of length equal to 15 feet or 4.572 meters.

nautical chart   [NAV] A graphic representation on a plane surface of a section of the earth's sea surface constructed to include known dangers and aids to navigation. Also known as marine chart.

nautical mile   [NAV] A unit of distance used principally in navigation; for practical consideration it is usually considered the length of 1 minute of any great circle of the earth, the meridian being the great circle most commonly used; the International Hydrographic Bureau in 1929 proposed a standard length of 1852 meters, which is known as the international nautical mile.

nautical twilight   [ASTRON] The interval of incomplete darkness between sunrise or sunset and the time at which the center of the sun's disk is 12° below the celestial horizon.

Nautilidae   [INV ZOO] A monogeneric family of cephalopod mollusks in the order Nautiloidea; *Nautilus pompilius* is the only well-known living species.

Nautiloidea   [INV ZOO] A primitive order of tetrabranchiate cephalopods; shells are external and smooth, being straight or coiled and chambered with curved transverse septa.

navaglobe   [NAV] The portion of the navarho navigation system that provides bearing; it utilizes three antennas located at the apexes of a triangle.

Navaho   [ORD] A U.S. Air Force surface-to-surface long-range supersonic strategic missile, powered by two ramjet engines.

Navajo sandstone   [GEOL] A fossil dune formation of Jurassic age found in the Colorado Plateau of the United States.

naval architect   [NAV ARCH] An individual who designs ships and makes alterations in them.

naval architecture   [ENG] The study of the physical characteristics and the design and construction of buoyant structures, such as ships, boats, barges, submarines, and floats, which operate in water; includes the construction and operation of the power plant and other mechanical equipment of these structures.

naval armament   [ORD] The combat equipment used in naval ships and by naval aircraft.

naval brass   [MET] Brass composed of 60–62% copper, 37–39% zinc, and 0.75–1% tin; relatively resistant to corrosion by seawater. Also known as naval bronze.

naval bronze   *See* naval brass.

naval meteorology   [METEOROL] The branch of meteorology which studies the interaction between the ocean and the overlying air mass, and which is concerned with atmospheric phenomena over the oceans, the effect of the ocean surface on these phenomena, and the influence of such phenomena on shallow and deep seawater.

naval mine   [ORD] An item designed to be located under water and exploded by means of propeller vibration, magnetic attraction, contact, or remote control.

naval ship   [NAV ARCH] A ship designed primarily for use in warfare, either directly in combat operations or to provide services and support such operations.

navar   [NAV] A coordinated series of radar air navigation and traffic-control aids utilizing transmissions at wavelengths of 10 and 60 centimeters to provide in an aircraft the distance and bearing from a given point, along with a display of other aircraft in the vicinity and commands from the ground; the system also provides on the ground a display of all aircraft in the vicinity, with their altitudes, identities, and means for transmitting certain commands. Derived from navigation and ranging.

navarho   [NAV] A long-distance, low-frequency continuous-wave navigation system providing simultaneous bearing and distance information; the portion providing bearing is termed navaglobe.

Navascreen   [NAV] System for displaying and computing air-traffic control data, using information obtained from radar and other sources.

NAVEAM   [NAV] A British-originated radio navigational warning of dangers in the eastern Atlantic, Mediterranean Sea, and Red Sea.

navel   [ANAT] The umbilicus.

navel height   [ANTHRO] The vertical distance, of a standing subject, measured from the center of the navel to the floor.

navicular   [ANAT] A boat-shaped bone, especially the lateral bone on the radial side of the proximal row of the carpus. [BIOL] Resembling or having the shape of a boat.

navicular cells   [PATH] Boat-shaped squamous epithelial cells filled with glycogen and prominent in the exfoliated cells of the uterine cervix of pregnant women.

Navier-Stokes equations   [FL MECH] The equations of motion for a viscous fluid which may be written $d\mathbf{V}/dt = -(1/\rho)\nabla p + \mathbf{F} + \nu\nabla^2\mathbf{V} + (1/3)\nu\nabla(\nabla\cdot\mathbf{V})$, where p is the pressure, ρ the density, F the total external force per unit mass, $\mathbf{V}$ the fluid velocity, and ν the kinematic viscosity; for an incompressible fluid, the term in $\nabla\cdot\mathbf{V}$ (divergence) vanishes, and the effects of viscosity then play a role analogous to that of temperature in thermal conduction and to that of density in simple diffusion.

navigable airspace   [NAV] Airspace at and above the minimum safe flight level, including airspace needed for safe takeoff and landing.

navigable semicircle   [METEOROL] That half of a cyclonic storm area in which the rotary and progressive motions of the storm tend to counteract each other, and the winds are in such a direction as to blow a vessel away from the storm track.

navigating bridge   *See* flying bridge.

navigating officer   [NAV] An officer serving as a navigator.

navigating sextant   [NAV] A sextant designed and used for observing the altitudes of celestial bodies, as contrasted with a hydrographic sextant.

navigation   [ENG] The process of directing the movement of a craft so that it will reach its intended destination; subprocesses are position fixing, dead reckoning, pilotage, and homing.

navigational aid   [NAV] An instrument, device, chart, method, or such, intended to assist in the navigation of a craft; this expression should not be confused with "aid to navigation," which refers only to devices external to a craft.

navigational planets   [NAV] The four planets commonly ob-

NAUPLIUS

Nauplius of the shrimp *Penaeus*; in addition to their sensory and feeding functions, the three pairs of appendages are organs of locomotion for this free-swimming larva. (*From T. I. Storer and R. L. Usinger, General Zoology, 4th ed., McGraw-Hill, 1965*)

NAUTILIDAE

Shell of *Nautilus pompilius*, which may be up to 10 inches (25 centimeters) in diameter.

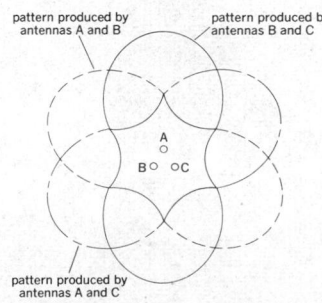

NAVAGLOBE

pattern produced by antennas A and B

pattern produced by antennas B and C

pattern produced by antennas A and C

Navaglobe antenna location and field pattern.

served for obtaining data for use in celestial navigation: Venus, Mars, Jupiter, and Saturn.

navigational plot [NAV] A plot of the movements of a craft.

navigational satellite [AERO ENG] An artificial earth-orbiting satellite designed for use in at least four widely different navigational systems.

navigational triangle [NAV] In celestial navigation, the spherical triangle solved in computing altitude and azimuth and great-circle sailing problems.

navigation computer [NAV] A computer that uses electronic or electric circuits to compute two or more navigation factors such as altitude, direction, and velocity, or to receive such data and compute course information.

navigation dome *See* astrodome.

navigation head [NAV] A transshipment point on a waterway where loads are transferred between water carriers and land carriers.

navigation lights [NAV] Statutory lights shown by aircraft and vessels during the hours between sunset and sunrise, in accordance with international agreements.

navigation receiver [ELECTR] An electronic device that determines a ship's position by receiving and comparing radio signals from transmitters at known locations.

navigator [NAV] A person who navigates or is directly responsible for the navigation of a craft.

navite [MINERAL] A porphyritic basalt containing phenocrysts of altered olivine, augite, and basic plagioclase in a groundmass of labradorite and augite.

Navy Electronics Laboratory International Algol Compilers *See* NELIAC.

Navy Heavy *See* bunker C fuel oil.

Nb *See* niobium.

NBR *See* nitrile rubber.

NC *See* numerical control.

N/C *See* numerical control.

N curve [ELECTR] A plot of voltage against current for a negative-resistance device; its slope is negative for some values of current or voltage.

Nd *See* neodymium.

n-dimensional space [MATH] A vector space whose basis has n vectors.

N display [ELECTR] Radar display in which the target appears as a pair of vertical deflections from a horizontal time base; direction is indicated by relative amplitude of the blips; target distance is determined by moving an adjustable pedestal signal along the base line until it coincides with the horizontal position of the blips; the pedestal control is calibrated in distance.

NDRO *See* nondestructive readout.

Ne *See* neon.

neallotype [SYST] A type specimen that, compared with the holotype, is of the opposite sex, and was collected and described later.

Neanderthal man [PALEON] A type of fossil human that is a subspecies of *Homo sapiens* and is distinguished by a low broad braincase, continuous arched browridges, projecting occipital region, short limbs, and large joints.

neap high water *See* mean high-water neaps.

neap low water *See* mean low-water neaps.

neap range [OCEANOGR] The mean semidiurnal range of tide when neap tides are occurring; the mean difference in height between neap high water and neap low water. Also known as mean neap range.

neap rise [OCEANOGR] The height of neap high water above the chart datum.

neaps *See* neap tide.

neap tidal currents [OCEANOGR] Tidal currents of decreased speed occurring at the time of neap tides.

neap tide [OCEANOGR] Tide of decreased range occurring about every 2 weeks when the moon is in quadrature, that is, during its first and last quarter. Also known as neaps.

Nearctic fauna [ECOL] The indigenous animal communities of the Nearctic zoogeographic region.

Nearctic zoogeographic region [ECOL] The zoogeographic region that includes all of North America to the edge of the Mexican Plateau.

near-end crosstalk [COMMUN] A type of interference that may occur at carrier telephone repeater stations when output signals of one repeater leak into the same end of the other repeater.

nearest approach [NAV] The least distance between two objects having relative motion with respect to each other.

nearest neighbors [CRYSTAL] Any pair of atoms in a crystal lattice which are as close to each other, or closer to each other, than any other pair.

near field [ACOUS] The acoustic radiation field that is close to an acoustic source such as a loudspeaker. [ELECTROMAG] The electromagnetic field that exists within one wavelength of a source of electromagnetic radiation, such as a transmitting antenna.

near-infrared radiation [ELECTROMAG] Infrared radiation having a relatively short wavelength, between 0.75 and about 2.5 micrometers (some scientists place the upper limit from 1.5 to 3 micrometers), at which radiation can be detected by photoelectric cells, and which corresponds in frequency range to the lower electronic energy levels of molecules and semiconductors. Also known as photoelectric infrared radiation.

near-infrared spectrophotometry [ANALY CHEM] Spectrophotometry at wavelengths in the near-infrared region, generally using instruments with quartz prisms in the monochromators and lead sulfide photoconductor cells as detectors to observe absorption bands which are harmonics of bands at longer wavelengths.

near miss [ORD] The strike of an explosive missile, especially of an aerial bomb, near but not on the object of attack, and usually close enough to cause effective damage.

near point [PHYSIO] The smallest distance from the eye at which a small object can be seen without blurring.

nearshore circulation [OCEANOGR] Ocean circulation consisting of both the nearshore currents and the coastal currents.

nearsightedness *See* myopia.

near stars [ASTRON] Those stars in the celestial neighborhood of the sun, sometimes taken as those 22 stars within 13 light-years of the sun.

nearthrosis [MED] A type of nonunion of broken ends of bones in which a cystic space resembling a joint cavity develops between poorly joined ends.

near-ultraviolet radiation [ELECTROMAG] Ultraviolet radiation having relatively long wavelength, in the approximate range from 300 to 400 nanometers.

near wilt [PL PATH] A fungus disease of peas caused by *Fusarium oxysporum pisi;* affects scattered plants and develops more slowly than true wilt.

neat cement grout [MATER] Grout made from a mixture of cement and water.

neat line [CIV ENG] The line to which a masonry wall should generally conform. [MAP] That border line which indicates the limits of an area shown on a map or chart.

neat plaster [MATER] A base-coat plaster, having sand added at the job location.

neatsfoot oil [MATER] Pale-yellow oil with unusual odor; soluble in organic solvents and kerosine; obtained by boiling shinbones and hoofless feet of cattle; used to treat leather, as a lubricant, and to oil wool. Also known as bubulum oil; hoof oil.

neat soap [MATER] Soap in the molten state formed during manufacture, especially after fitting and settling out of nigre and lye.

Nebaliacea [INV ZOO] A small, marine order of Crustacea in the subclass Leptostraca distinguished by a large bivalve shell, without a definite hinge line, an anterior articulated rostrum, eight thoracic and seven abdominal somites, a pair of articulated furcal rami, and the telson.

Nebraskan drift [GEOL] Rock material transported during the Nebraskan glaciation; it is buried below the Kansan drift in Iowa.

Nebraskan glaciation [GEOL] The first glacial stage of the Pleistocene epoch in North America, beginning about 1,000,-000 years ago, and preceding the Aftonian interglacial stage.

nebula [ASTRON] Interstellar clouds of gas or small particles; an example is the Horsehead Nebula in Orion.

nebular hypothesis [ASTROPHYS] A theory, proposed in 1796 by Laplace, supposing that the planets originated from the solar nebula surrounding the proto-sun; as the sun cooled, it contracted, rotated faster, and thus caused a ringlike bulging at the equator; this bulge eventually broke off and formed the planets; Laplace further theorized that the sun and other stars formed from clouds of nebulous matter; the theory in this form is not accepted.

nebular lines [ASTROPHYS] The spectral lines formed in the glow of bright nebulae; they arise from forbidden atomic transition which can take place because of the very low pressure in the nebula itself.

nebular red shift [ASTROPHYS] A systematic shift observed in the spectra of all distant galaxies; the wavelength shift toward the red increases with the distance of the galaxies from the earth.

nebular transitions [ASTROPHYS] Those electronic transitions for doubly ionized argon and chlorine that yield the nebular lines seen in the spectra of gaseous nebulae.

nebulite [PETR] A chorismite in which one of the textural elements occurs in nebulitic lenticular masses.

nebulitic [PETR] **1.** Having indistinct boundaries between textural elements. **2.** Of or pertaining to a nebulite.

nebulosus [METEOROL] A cloud species with the appearance of a nebulous veil, showing no distinct details; found principally in the genera cirrostratus and stratus.

necessary bandwidth [COMMUN] For a given class of emission, the minimum value of the occupied bandwidth sufficient to ensure the transmission of information at the rate and with the quality required for the system employed, under specified conditions.

neck [ANAT] The usually constricted communicating column between the head and trunk of the vertebrate body. [ENG] The part of a furnace where the flame is contracted before reaching the stack. [GEOGR] A narrow strip of land, especially one connecting two larger areas. [GEOL] **1.** A vertical, pipelike intrusion representing the formed vent of a volcano. **2.** *See* pipe. [MET] In a tensile test, that portion of the metal at which fracture is imminent during the later stages of plastic deformation in a tensile test. [OCEANOGR] The narrow band of water forming the part of a rip current where feeder currents converge and flow swiftly through the incoming breakers and out to the head.

neck breadth [ANTHRO] The diameter of the neck measured halfway between the otobasion inferior and the shoulder.

neck depth [ANTHRO] The diameter of the neck between the tip of the thyroid cartilage and the back of the neck, measured perpendicular to the axis of the neck, with contact only.

Neckeraceae [BOT] A family of mosses in the order Isobryales distinguished by undulate leaves.

necking [MET] Reducing the diameter or cross-sectional area of a tube or other piece of metal by stretching.

necking down [MET] Localized reduction in cross-sectional area of a specimen during tensile deformation.

neck rot [PL PATH] A fungus disease of onions caused by species of *Botrytis* and characterized by rotting of the leaves just above the bulb.

necr-, necro- [MED] Combining form denoting death.

necrobiosis [MED] Death of a cell or group of cells under either normal or pathologic conditions.

Necrolestidae [PALEON] An extinct family of insectivorous marsupials.

necrophagous [ZOO] Feeding on dead bodies.

necrophile [PSYCH] A person affected with necrophilia.

necrophilia [PSYCH] **1.** Longing for death. **2.** *See* necrophilism.

necrophilism [PSYCH] Also known as necrophilia. **1.** Unnatural obsession with and usually erotic attraction for dead bodies. **2.** Sexual violation of a corpse.

necrophobia [PSYCH] Abnormal dread of death and of dead bodies.

necropsy [MED] To perform an autopsy.

necrosis [MED] Death of a cell or group of cells as a result of injury, disease, or other pathologic state.

necrotic [MED] Pertaining to, causing, or undergoing necrosis.

necrotic enteritis [VET MED] A bacterial infection of young swine caused by *Salmonella suipestifer* or *S. choleraesuis* and

characterized by fever and necrotic and ulcerative inflammation of the intestine.

necrotic ring spot [PL PATH] A virus leaf spot of cherries marked by small, dark water-soaked rings which may drop out, giving the leaf a tattered appearance.

necrotize [MED] To undergo necrosis; to become necrotic.

necrozoospermia [MED] A condition in which spermatozoa are immobile.

nectar [BOT] A sugar-containing liquid secretion of the nectaries of many flowers.

nectarine [BOT] A smooth-skinned, fuzzless fruit originating as a spontaneous somatic mutation of the peach, *Prunus persica* and *P. persica* var. *nectarina*.

nectary [BOT] A secretory organ or surface modification of a floral organ in many flowers, occurring on the receptacle, in and around ovaries, on stamens, or on the perianth; secretes nectar.

nectocalyx [INV ZOO] A swimming bell of a siphonophore. Also known as nectophore.

Nectonematoidea [INV ZOO] A monogeneric order of worms belonging to the class Nematomorpha, characterized by dorsal and ventral epidermal chords, a pseudocoele, and dorsal and ventral rows of bristles; adults are parasites of true crabs and hermit crabs.

nectophore *See* nectocalyx.

nectosome [INV ZOO] The part of a complex siphonophore that bears swimming bells.

Nectridea [PALEON] An order of extinct lepospondylous amphibians characterized by vertebrae in which large fan-shaped hemal arches grow directly downward from the middle of each caudal centrum.

Nectrioidaceae [MYCOL] The equivalent name for Zythiaceae.

Necturus [VERT ZOO] A genus of mud puppies in the family Proteidae.

need [PSYCH] An acquired or physiological lack or deficit within the individual.

need complementarity [PSYCH] The concept that people having different needs like each other because they provide each other with mutual satisfaction of opposed needs.

needle [ADP] A slender rod or probe used to sort decks of edge-punched cards by inserting it through holes along the margin of the deck and vibrating the deck so that cards having that particular hole are retained, but those having a notch cut at that hole position drop out. [BOT] A slender-pointed leaf, as of the firs and other evergreens. [DES ENG] **1.** A device made of steel pointed at one end with a hole at the other; used for sewing. **2.** A device made of steel with a hook at one end; used for knitting. [ENG] **1.** A piece of copper or brass about ½ inch (13 millimeters) in diameter and 3 or 4 feet (90 or 120 centimeters) long, pointed at one end, thrust into a charge of blasting powder in a borehole and then withdrawn, leaving a hole for the priming, fuse, or squib. Also known as pricker. **2.** A thin pointed indicator on an instrument dial. [ENG ACOUS] *See* stylus.

needlebar [TEXT] A bar for mounting needles on a sewing or knitting machine.

needle beam [CIV ENG] A temporary member thrust under a building or a foundation for use in underpinning.

needle board [TEXT] A board that holds needles in a loom.

needle dam [CIV ENG] A barrier made of horizontal bars across a pass through a dam or of planks that can be removed in case of flooding.

needle file [DES ENG] A small file with an extended tang that serves as a needle.

needle gap [ELECTR] Spark gap in which the electrodes are needle points.

needle ice *See* frazil ice.

needle nozzle [MECH ENG] A streamlined hydraulic turbine nozzle with a movable element for converting the pressure and kinetic energy in the pipe leading from the reservoir to the turbine into a smooth jet of variable diameter and discharge but practically constant velocity.

needle ore *See* aikinite.

needle scratch *See* surface noise.

needle test point [ELEC] A sharp steel probe connected to a test cord for making contact with a conductor.

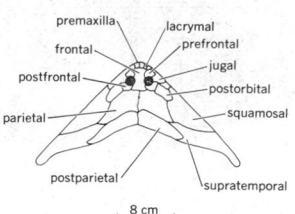

NECTRIDEA

premaxilla
frontal
postfrontal
parietal
lacrymal
prefrontal
jugal
postorbital
squamosal
postparietal
supratemporal

8 cm

Skull of a lower Permian nectridian, *Diplocaulus*. (*From E. H. Colbert, Evolution of the Vertebrates, Wiley, 1955*)

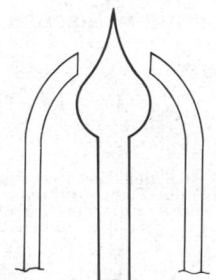

NEEDLE NOZZLE

Needle nozzle, the movement of the needle (in the center) backward and forward changes the size of the jet.

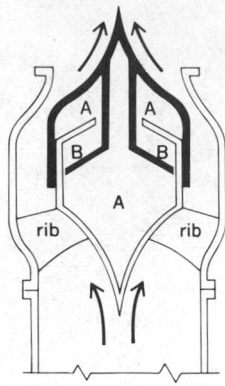

Large needle valve is actuated hydraulically by pressure in chambers A to close and in annular chambers B to open.

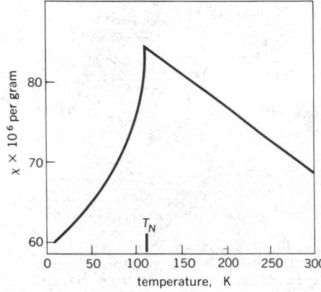

Magnetic susceptibility of powdered manganese oxide. T_N is the Néel temperature. *(After H. Bizette, C. F. Squire, and B. Tsai, 1938)*

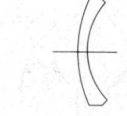

Shape of a negative meniscus lens. *(From F. A. Jenkins and H. E. White, Fundamentals of Optics, 3d ed., McGraw-Hill, 1957)*

needle tubing [ENG] Stainless steel tubing with outside diameters from 0.014 to 0.203 inch (0.36 to 5.16 millimeters); used for surgical instruments and radon implanters.

needle valve [DES ENG] A slender, pointed rod fitting in a hole or circular or conoidal seat; used in hydraulic turbines and hydroelectric systems.

needle weir [CIV ENG] A type of frame weir in which the wooden barrier is constructed of vertical square-section timbers placed side by side against the iron frames.

needling [CIV ENG] Underpinning the upper part of a building with horizontally placed timber or steel beams.

need-to-know [ORD] A criterion used in security procedures in the United States that requires a person requesting classified information to establish the need to know such information in terms of the pertinent mission.

Néel ferromagnetism *See* ferrimagnetism.

Néel point *See* Néel temperature.

Néel's theory [SOLID STATE] A theory of the behavior of antiferromagnetic and other ferrimagnetic materials in which the crystal lattice is divided into two or more sublattices; each atom in one sublattice responds to the magnetic field generated by nearest neighbors in other sublattices, with the result that magnetic moments of all the atoms in any sublattice are parallel, but magnetic moments of two different sublattices can be different.

Néel temperature [SOLID STATE] A temperature, characteristic of certain metals, alloys, and salts, below which spontaneous nonparalleled magnetic ordering takes place so that they become antiferromagnetic, and above which they are paramagnetic. Also known as Néel point.

Néel wall [SOLID STATE] The boundary between two magnetic domains in a thin film in which the magnetization vector remains parallel to the faces of the film in passing through the wall.

neem oil [MATER] An aromatic oil from the seeds and fruit of the neem tree (*Melia azadirachta*); contains sulfur compounds; used as an anthelmintic and as an alcohol denaturant. Also known as margosa oil; nim oil.

neencephalon [ANAT] The neopallium and the phylogenetically new acquisitions of the cerebellum and thalamus collectively. Also spelled neoencephalon.

negation [MATH] The negation of a proposition P is a proposition which is true if and only if P is false; this is often written ~ P.

negative [ELEC] Having a negative charge. [GRAPHICS] The image on film in which the dark tones of the original appear transparent, and the light tones appear black and opaque. Also known as reversed image.

negative acceleration [MECH] Acceleration in a direction opposite to the velocity, or in the direction of the negative axis of a coordinate system.

negative afterimage [PHYSIO] An afterimage that is seen on a bright background and is complementary in color to the initial stimulus.

negative angle [MATH] The angle subtended by moving a ray in the clockwise direction.

negative area *See* negative element.

negative booster [ELEC] Booster used in connection with a ground-return system to reduce the difference of potential between two points on the grounded return.

negative catalysis [CHEM] A catalytic reaction such that the reaction is slowed down by the presence of the catalyst.

negative charge [ELEC] The type of charge which is possessed by electrons in ordinary matter, and which may be produced in a resin object by rubbing with wool. Also known as negative electricity.

negative crystal [OPTICS] A uniaxial crystal in which the extraordinary wave travels faster than the ordinary wave, such as calcite.

negative easement [CIV ENG] An easement that can be exercised to prevent the owner of a piece of land from using his property in certain ways that he would otherwise be entitled to.

negative effective mass amplifiers and generators [ELECTR] Class of solid-state devices for broad-band amplification and generation of electrical waves in the microwave region; these devices use the property of the effective masses of charge carriers in semiconductors becoming negative with sufficiently high kinetic energies.

negative electricity *See* negative charge.

negative electrode [ELEC] *See* negative plate. [ELECTR] The cathode of a tube.

negative electron *See* electron.

negative element [GEOL] A large structural feature or part of the earth's crust, characterized through a long geologic time period by frequent and conspicuous downward movement (subsidence) or by extensive erosion, or by an uplift that is considerably less rapid or less frequent than that of adjacent positive elements. Also known as negative area.

negative empathy [PSYCH] Empathy which takes place against a certain resistance or unwillingness.

negative feedback [CONT SYS] Feedback in which a portion of the output of a circuit, device, or machine is fed back 180° out of phase with the input signal, resulting in a decrease of amplification so as to stabilize the amplification with respect to time or frequency, and a reduction in distortion and noise. Also known as inverse feedback; reverse feedback; stabilized feedback. [SCI TECH] Feedback which tends to reduce the output in a system.

negative g [MECH] In designating the direction of acceleration on a body, the opposite of positive *g*; for example, the effect of flying an outside loop in the upright seated position.

negative glow [ELECTR] The luminous flow in a glow-discharge cold-cathode tube occurring between the cathode dark space and the Faraday dark space.

negative-grid generator [ELECTR] Conventional oscillator circuit in which oscillation is produced by feedback from the plate circuit to a grid which is normally negative with respect to the cathode, and which is designed to operate without drawing grid current at any time.

negative-grid thyratron [ELECTR] A thyratron with only one grid, which serves to prevent the flow of current until its potential relative to the cathode is made less negative than a certain critical value.

negative impedance [ELECTR] An impedance such that when the current through it increases, the voltage drop across the impedance decreases.

negative-impedance repeater [ELECTR] A telephone repeater that provides an effective gain for voice-frequency signals by insertion into the line of a negative impedance that cancels out line impedances responsible for transmission losses.

negative integer [MATH] The additive inverse of a positive integer relative to the additive group structure of the integers.

negative interference [GEN] A crossover exchange between homologous chromosomes which increases the likelihood of another in the same vicinity.

negative ion [CHEM] An atom or group of atoms which by gain of one or more electrons has acquired a negative electric charge. [PHYS] An electron or negatively charged subatomic particle.

negative-ion vacancy [CRYSTAL] A point defect in an ionic crystal in which a negative ion is missing from its lattice site.

negative lens *See* diverging lens.

negative meniscus lens [OPTICS] A lens having one convex and one concave surface, with the radius of curvature of the convex surface greater than that of the concave surface. Also known as diverging meniscus lens.

negative modulation [ELECTR] 1. Modulation in which an increase in brightness corresponds to a decrease in amplitude-modulated transmitter power; used in United States television transmitters and in some facsimile systems. 2. Modulation in which an increase in brightness corresponds to a decrease in the frequency of a frequency-modulated facsimile transmitter. Also known as negative transmission.

negative phase [IMMUNOL] The temporary quantitative reduction of serum antibodies immediately following a second inoculation of antigen.

negative phase-sequence relay [ELEC] Relay which functions in conformance with the negative phase-sequence component of the current, voltage, or power of the circuit.

negative picture phase [ELECTR] The video signal phase in which the signal voltage swings in a negative direction for an increase in brilliance.

negative pion [PARTIC PHYS] A pion having a negative electric charge.

negative plate [ELEC] The internal plate structure that is connected to the negative terminal of a storage battery. Also known as negative electrode.

negative potential [ELEC] An electrostatic potential which is lower than that of the ground, or of some conductor or point in space that is arbitrarily assigned to have zero potential.

negative pressure [PHYS] A way of expressing vacuum; a pressure less than atmospheric or the standard 760 mm Hg.

negative rain [METEOROL] Rain which exhibits a net negative electric charge.

negative rake [MECH ENG] The orientation of a cutting tool whose cutting edge lags the surface of the tooth face.

negative resistance [ELECTR] The resistance of a negative-resistance device.

negative-resistance device [ELECTR] A device having a range of applied voltages within which an increase in this voltage produces a decrease in the current.

negative-resistance oscillator [ELECTR] An oscillator in which a parallel-tuned resonant circuit is connected to a vacuum tube so that the combination acts as the negative resistance needed for continuous oscillation.

negative resistance repeater [ELECTR] Repeater in which gain is provided by a series negative resistance or a shunt negative resistance, or both.

negative shoreline *See* shoreline of emergence.

negative skewness [MATH] Skewness in which the mean is smaller than the mode.

negative staining [BIOL] A method in microscopy for demonstrating the form of cells, bacteria, and other small objects by staining the ground rather than the objects.

negative temperature coefficient [PHYS] Condition wherein the resistance, length, or some other characteristic of a material decreases when temperature increases.

negative terminal [ELEC] The terminal of a battery or other voltage source that has more electrons than normal; electrons flow from the negative terminal through the external circuit to the positive terminal.

negative thermion *See* thermoelectron.

negative-transconductance oscillator [ELECTR] Electron-tube oscillator in which the output of the tube is coupled back to the input without phase shift, the phase condition for oscillation being satisfied by the negative transconductance of the tube.

negative transfer [PSYCH] The harmful effect that occurs on learning in one situation because previous learning, in another situation, required different responses, incompatible with the new learning situation.

negative transmission *See* negative modulation.

negativism [PSYCH] Indifference, opposition, or resistance to suggestions, or persistent refusal to do as asked, without apparent or objective reasons.

negatron *See* dynatron; electron.

negentropy *See* information content.

negotiated contract [IND ENG] A purchase or sales agreement made by a United States government agency without normally employing techniques required by formal advertising.

Negri bodies [PATH] Acidophil cytoplasmic inclusion bodies in neurons, considered diagnostic of rabies.

Neididae [INV ZOO] A small family of thread-legged hemipteran insects in the superfamily Lygaeoidea.

neighbor [CRYSTAL] One of a pair of atoms or ions in a crystal which are close enough to each other for their interaction to be of significance in the physical problem being studied.

neighborhood of a point [MATH] A set in a topological space which contains an open set which contains the point; in euclidean space, an example of a neighborhood of a point is an open (without boundary) ball centered at that point.

Neil's parabola [MATH] The graph of the equation $y = ax^{3/2}$, where a is a constant.

Neisseria [MICROBIO] A genus of gram-negative, aerobic or facultatively anaerobic, parasitic cocci in the family Neisseriaceae.

Neisseria catarrhalis [MICROBIO] A nonpathogenic species

of nonchromogenic bacteria found in the human respiratory tract.

Neisseriaceae [MICROBIO] A family of parasitic, gram-negative cocci in the order Eubacteriales; several species are human pathogens.

Neisseria flavescens [MICROBIO] A bacterial species marked by greenish-yellow chromogenesis and found within the respiratory tract and in the spinal fluid in certain cases of meningitis.

Neisseria gonorrhoeae [MICROBIO] The bacterial organism that causes gonorrhea; the type species of the genus. Also known as *Diplococcus gonorrhoeae*; gonococcus.

Neisseria intracellularis *See* Neisseria meningitidis.

Neisseria meningitidis [MICROBIO] The bacterial species that causes epidemic cerebrospinal meningitis. Also known as *Diplococcus intracellularis meningitidis*; *Micrococcus meningitidis*; *Neisseria intracellularis*.

nekton [INV ZOO] Free-swimming aquatic animals, essentially independent of water movements.

N electron [ATOM PHYS] An electron in the fourth (*N*) shell of electrons surrounding the atomic nucleus, having the principal quantum number 4.

NELIAC [ADP] An early dialect of ALGOL, which was developed for a specific data-processing application but, unlike ALGOL, is not primarily concerned with being used for complex scientific and engineering calculations. Derived from Navy Electronics Laboratory International Algol Compilers.

Nelson diaphragm cell [CHEM ENG] Obsolete carbon-electrode type of electrolytic diaphragm cell once widely used to produce chlorine and caustic soda from brine.

nelsonite [PETR] A group of hypabyssal rocks composed mainly of ilmenite and apatite.

Nelumbonaceae [BOT] A family of flowering aquatic herbs in the order Nymphaeales characterized by having roots, perfect flowers, alternate leaves, and triaperturate pollen.

nemalite [MINERAL] A fibrous brucite that contains ferrous oxide.

Nemata [INV ZOO] A proposed equivalent name for Nematoda.

Nemataceae [BOT] A family of mosses in the order Hookeriales distinguished by having perichaetial leaves only.

Nemathelminthes [INV ZOO] A subdivision of the Amera which comprised the classes Rotatoria, Gastrotrichia, Kinorhyncha, Nematoda, Nematomorpha, and Acanthocephala.

nematicide [MATER] A chemical used to kill plant-parasitic nematodes.

nematic phase [PHYS CHEM] A phase of a liquid crystal in the mesomorphic state, in which the liquid has a single optical axis in the direction of the applied magnetic field, appears to be turbid and to have mobile threadlike structures, can flow readily, has low viscosity, and lacks a diffraction pattern.

nematoblastic [PETR] Pertaining to a metamorphic rock with a homeoblastic texture due to development during recrystallization of slender prismatic crystals.

Nematocera [INV ZOO] A series of dipteran insects in the suborder Orthorrhapha; adults have antennae that are usually longer than the head, and the flagellum consists of 10–65 similar segments.

nematocyst [INV ZOO] An intracellular effector organelle in the form of a coiled tube which may be rapidly everted in food gathering or defense by coelenterates.

Nematoda [INV ZOO] A group of unsegmented worms which have been variously recognized as an order, class, and phylum.

nematode [INV ZOO] 1. Any member of the Nematoda. 2. Of or pertaining to the Nematoda.

Nematodonteae [BOT] A group of mosses included in the subclass Eubrya in which there may be faint transverse bars on the peristome teeth.

nematogen [INV ZOO] A reproductive phase of the Dicyemida during which vermiform larvae are formed asexually from the germ cells in the axial cells.

Nematognathi [VERT ZOO] The equivalent name for Siluriformes.

Nematoidea [INV ZOO] An equivalent name for Nematoda.

nematology [INV ZOO] The study of nematodes.

NEMATOGEN

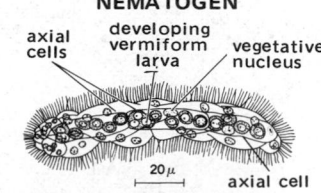

Young stem nematogen, with three axial cells, of *Dicyema schulzianum*. (After H. Nouvel)

NEODYMIUM

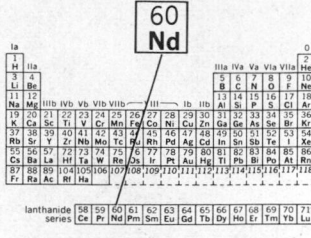

60
Nd

Periodic table of the chemical elements showing the position of neodymium.

NEON

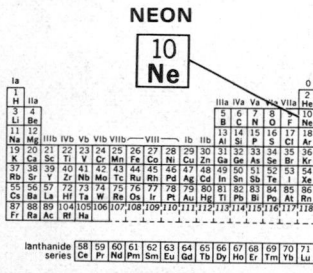

10
Ne

Periodic table of the chemical elements showing the position of neon.

NEON GLOW LAMP

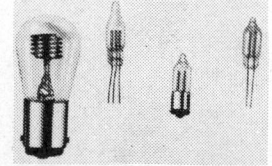

Examples of neon glow lamps.

Nematomorpha [INV ZOO] A group of the Aschelminthes or a separate phylum that includes the horsehair worms.

Nematophytales [PALEOBOT] A group of fossil plants from the Silurian and Devonian periods that bear some resemblance to the brown seaweeds (Phaeophyta).

Nematosporoideae [BOT] A subfamily of the Saccharomycetaceae containing parasitic yeasts; two genera have been studied in culture: *Nematospora* with asci that contain eight spindle-shaped ascospores, and *Metschnikowia* whose asci contain one or two needle-shaped ascospores.

nematozooid [INV ZOO] A zooid bearing organs of defense, in hydroids and siphonophores.

Nemedri [NAV] A British publication containing danger areas, routes, and instructions for Northwest European waters, the Baltic Sea, the Mediterranean Sea, and the Black Sea; a United States reprint is available to American shipping. Derived from North European and Mediterranean Routing Instructions.

nemere [METEOROL] In Hungary, a stormy, cold fall wind.

Nemertea [INV ZOO] An equivalent name for Rhynchocoela.

Nemertina [INV ZOO] An equivalent name for Rhynchocoela.

Nemertinea [INV ZOO] An equivalent name for Rhynchocoela.

Nemestrinidae [INV ZOO] The hairy flies, a family of dipteran insects in the series Brachycera of the suborder Orthorrhapha.

Nemichthyidae [VERT ZOO] A family of bathypelagic, eellike amphibians in the order Apoda.

Nemognathinae [INV ZOO] A subfamily of the coleopteran family Meloidae; members have greatly elongate maxillae that form a poorly fitted tube.

nemoral [ECOL] Pertaining to or inhabiting a grove or wooded area.

neo-, ne- [ORG CHEM] Prefix indicating hydrocarbons where a carbon is bonded directly to at least four other carbon atoms, such as neopentane. [SCI TECH] Prefix meaning new, or different in form; indicating a compound related to an older one, or a precursor.

Neoanthropinae [PALEON] A subfamily of the Hominidae in some systems of classification, set up to include *Homo sapiens* and direct ancestors of *H. sapiens.*

neoblast [INV ZOO] Any of various undifferentiated cells in annelids which migrate to and proliferate at sites of repair and regeneration.

Neocathartidae [PALEON] An extinct family of vulturelike diurnal birds of prey (Falconiformes) from the Upper Eocene.

neocerebellum [ANAT] Phylogenetically, the most recent part of the cerebellum; receives cerebral cortex impulses via the corticopontocerebellar tract.

Neocomian [GEOL] A European stage of Lower Cretaceous geologic time; includes Berriasian, Valanginian, Hauterivian, and Barremian.

neocortex [ANAT] Phylogenetically the most recent part of the cerebral cortex; includes all but the olfactory, hippocampal, and piriform regions of the cortex.

neodymium [CHEM] A metallic element, symbol Nd, with atomic weight 144.24, atomic number 60; a member of the rare-earth group of elements.

neodymium chloride [INORG CHEM] $NdCl_3 \cdot xH_2O$ Water- and acid-soluble, pink lumps; used to prepare metallic neodymium.

neodymium glass [MATER] A glass containing small amounts of neodymium oxide; used for color television filter plates since it transmits 90% of the blue, green, and red light rays and no more than 10% of the yellow.

neodymium oxide [INORG CHEM] Nd_2O_3 A hygroscopic, blue-gray powder; insoluble in water, soluble in acids; used to color glass and in ceramic capacitors.

neoencephalon *See* neencephalon.

Neogastropoda [INV ZOO] An order of gastropods which contains the most highly developed snails; respiration is by means of ctenidia, the nervous system is concentrated, an operculum is present, and the sexes are separate.

Neogene [GEOL] An interval of geologic time incorporating the Miocene and Pliocene of the Tertiary period; the Upper Tertiary.

neoglaciation [GEOL] The removal of glacier ice growth in certain mountain areas during the Little Ice Age, following its shrinkage or disappearance during the Altithermial interval.

Neognathae [VERT ZOO] A superorder of the avian order Neornithes, characterized as flying birds with fully developed wings and sternum with a keel, fused caudal vertebrae, and absence of teeth.

Neogregarinida [INV ZOO] An order of sporozoan protozoans in the subclass Gregarinia which are insect parasites.

neohexane [ORG CHEM] C_6H_{14} Volatile, flammable, colorless liquid boiling at 50°C; used as high-octane component of motor and aviation gasolines.

neohexane alkylation [CHEM ENG] A noncatalytic petroleum-refinery alkylation process that forms neohexane from a feed of ethylene and isobutane.

neomycin [MICROBIO] The collective name for several colorless antibiotics produced by a strain of *Streptomyces fradiae;* the commercial fraction ($C_{23}H_{46}N_6O_{13}$) has a broad spectrum of activity.

neon [CHEM] A gaseous element, symbol Ne, atomic number 10, atomic weight 20.183; a member of the family of noble gases in the zero group of the periodic table.

neonatal [MED] Pertaining to a newborn infant.

neonatal impetigo [MED] A type of impetigo occurring in the newborn, characterized by bullae and caused by staphylococci or sometimes streptococci.

neonatal line [ANAT] A prominent incremental line formed in the neonatal period in the enamel and dentin of a deciduous tooth or of a first permanent molar.

neonatal mortality rate [MED] The number of deaths reported among infants under 1 month of age in a calendar year per 1000 live births reported in the same year and place.

neonatal myasthenia [MED] Muscle weakness and ineffective motor activities in infants born of myasthenic mothers.

neonate [MED] A newborn infant.

neonatology [MED] The study of the newborn up to 2 months of age.

neon glow lamp [ELECTR] A glow lamp containing neon gas, usually rated between 1/25 and 3 watts, and producing a characteristic red glow; used as an indicator light and electronic circuit component.

neon-helium laser [OPTICS] A continuous-wave gas laser using a combination of neon and helium gases to obtain a 6328-angstrom visible red beam.

neon oscillator [ELECTR] Relaxation oscillator in which a neon tube or lamp serves as the switching element.

neon tube [ELECTR] An electron tube in which neon gas is ionized by the flow of electric current through long lengths of gas tubing, to produce a luminous red glow discharge; used chiefly in outdoor advertising signs.

neopallium [ANAT] Phylogenetically, the new part of the cerebral cortex; formed from the region between the pyriform lobe and the hippocampus, it comprises the nonolfactory region.

neopalynology [BOT] A field of palynology concerned with extant microorganisms and disassociated microscopic parts of megaorganisms.

neopentane [ORG CHEM] C_5H_{12} Colorless liquid boiling at 10°C; soluble in alcohol, insoluble in water; a hydrocarbon found as a minor component of natural gasoline. Also known as 2,2-dimethylpropane.

neophane glass [MATER] A glass containing neodymium oxide to reduce glare; used for yellow sunglasses or for windshield glass.

neoplasia [MED] **1.** Formation of a neoplasm or tumor. **2.** Formation of new tissue.

neoplasm [MED] An aberrant new growth of abnormal cells or tissues; a tumor.

neoprene [MATER] A synthetic rubber with outstanding resistance to ozone, weathering, various chemicals, oil, and flame, made by polymerization of chloroprene (2-chlorobutadiene-1,3); varies from amber to silver to cream in color; used in paints, putties, adhesives, shoe soles, tank linings, and rubber products.

Neopseustidae [INV ZOO] A family of Lepidoptera in the superfamily Eriocranioidea.

Neoptera [INV ZOO] A section of the insect subclass Pterygota; members have a muscular and articular mechanism allowing the wings to be flexed over the abdomen when at rest.

Neopterygii [VERT ZOO] An equivalent name for Actinopterygii.

Neorhabdocoela [INV ZOO] A group of the Rhabdocoela comprising fresh-water, marine, or terrestrial forms, with a bulbous pharynx, paired protonephridia, sexual reproduction, and ventral gonopores.

Neornithes [VERT ZOO] A subclass of the class Aves containing all known birds except the fossil *Archaeopteryx*.

neosilicate [MINERAL] A structural type of silicate mineral characterized by linkage of isolated SiO_4 tetrahedra by ionic bonding only; an example is olivine.

neossoptile [VERT ZOO] A downy feather on most newly hatched birds.

neostigmine [PHARM] A quaternary ammonium cation used as the bromide ($C_{12}H_{19}BrN_{20}O_2$) and methylsulfate ($C_{13}H_{22}N_2O_6S$) salts; has anticholinesterase activity.

neotenin [BIOCHEM] A hormone secreted by cells of the corpus allatum in arthropod larvae and nymphs; inhibits the development of adult characters. Also known as juvenile hormone.

neoteny [VERT ZOO] A phenomenon peculiar to some salamanders, in which large larvae become sexually mature while still retaining gills and other larval features.

Neotropical zoogeographic region [ECOL] A zoogeographic region that includes Mexico south of the Mexican Plateau, the West Indies, Central America, and South America.

neotype [SYST] A specimen selected as type subsequent to the original description when the primary types are known to be destroyed; a nomenclatural type.

neounitarian theory of hematopoiesis [HISTOL] A theory that under certain conditions, such as in tissue culture or in pathologic states, lymphocytes or cells resembling lymphocytes can become multipotent.

nep [TEXT] Any small entanglement of textile fibers that cannot be unraveled; formed during carding or ginning.

Nepenthaceae [BOT] A family of dicotyledonous plants in the order Sarraceniales; includes many of the pitcher plants.

neper [PHYS] Abbreviated Np. Also known as napier. **1.** A unit used for expressing the ratio of two currents, voltages, or analogous quantities; the number of nepers is the natural logarithm of this ratio. **2.** A unit used for expressing the ratio of two powers (even when this ratio is not the square of the corresponding current or voltage ratio); the number of nepers is the natural logarithm of the square root of this ratio; to avoid confusion, this usage should be accompanied by a specific statement.

nephanalysis [METEOROL] The analysis of a synoptic chart in terms of the types and amount of clouds and precipitation; cloud systems are identified both as entities and in relation to the pressure pattern, fronts, and other aspects.

nephcurve [METEOROL] In nephanalysis, a line bounding a significant portion of a cloud system, for example, a clear-sky line, precipitation line, cloud-type line, or ceiling-height line.

nepheline [MINERAL] (Na,K)AlSiO$_4$ A mineral of the feldspathoid group crystallizing in the hexagonal system and occurring as glassy or coarse crystals or colorless grains or green to brown masses of greasy luster in alkalic igneous rocks; hardness is 5.5–6 on Mohs scale. Also known as eleolite; nephelite.

nepheline basalt *See* olivine nephelinite.

nepheline monzonite [PETR] A nepheline syenite in which sodic plagioclase exceeds the quantity of alkali feldspar.

nepheline phonolite [PETR] The fine-grained equivalent of nepheline syenite.

nepheline syenite [PETR] A phaneritic plutonic rock with granular texture, composed largely of alkali feldspar, nepheline, and dark-colored materials.

nephelinite [PETR] A dark-colored, aphanitic rock of volcanic origin, composed essentially of nepheline and pyroxene; texture is usually porphyritic with large crystals of augite and nepheline in a very-fine-grained matrix.

nephelite *See* nepheline.

nephelometer [OPTICS] A type of instrument that measures, at more than one angle, the scattering function of particles suspended in a medium; information obtained may be used to determine the size of the suspended particles and the visual range through the medium.

nephelometry [OPTICS] **1.** The study of suspensoids using the techniques of light scattering. **2.** The study of the scattering properties of small samples of air and its suspensoids.

nepheloscope [ENG] An instrument for the production of clouds in the laboratory by condensation or expansion of moist air.

nephology [METEOROL] The study of clouds.

nephometer [ENG] A general term for instruments designed to measure the amount of cloudiness; an early type consists of a convex hemispherical mirror mapped into six parts; the amount of cloud coverage on the mirror is noted by the observer.

nephoscope [ENG] An instrument for determining the direction of cloud motion.

nephr-, nephro- [ANAT] Combining form denoting kidney.

nephrectomy [MED] Surgical removal of a kidney.

nephric tubule *See* uriniferous tubule.

nephridioblast [INV ZOO] An ecodermal precursor cell of a nephridium in certain animals.

nephridioduct [INV ZOO] The duct of a nephridium, sometimes serving as a common excretory and genital outlet.

nephridiopore [INV ZOO] The external opening of a nephridium.

nephridium [INV ZOO] Any of various paired excretory structures present in the Platyhelminthes, Rotifera, Rhynchocoela, Acanthocephala, Priapuloidea, Entoprocta, Gastrotricha, Kinorhyncha, Cephalochorda, and some Archiannelida and Polychaeta.

nephrite [MINERAL] An exceptionally tough, compact, fine-grained, greenish or bluish amphibole constituting the less valuable type of jade; formerly worn as a remedy for kidney diseases. Also known as greenstone; kidney stone.

nephritic [MED] **1.** Pertaining to or affected with nephritis. **2.** Pertaining to or affecting the kidney.

nephritis [MED] Inflammation of the kidney.

nephroabdominal [ANAT] Of or pertaining to the kidneys and abdomen.

nephroblastoma *See* Wilms' tumor.

nephrocalcinosis [PATH] Deposition of calcium salts in the kidney tubules.

nephrocoel [ANAT] The cavity of a nephrotome.

nephrodystrophy *See* nephrosis.

nephrogenic [EMBRYO] **1.** Having the potential to develop into kidney tissue. **2.** Of renal origin.

nephrogenic cord [EMBRYO] The longitudinal cordlike mass of mesenchyme derived from the mesomere or nephrostomal plate of the mesoderm, from which develop the functional parts of the pronephros, mesonephros, and metanephros.

nephrogenic tissue [EMBRYO] The tissue of the nephrogenic cord derived from the nephrotome plate that forms the blastema or primordium from which the embryonic and definitive kidneys develop.

nephrolithiasis [PATH] Formation of renal calculi.

nephrolithotomy [MED] Excision of renal calculi from the kidney.

nephrology [MED] The study of the kidney, including diseases.

nephrolysin [BIOCHEM] A toxic substance capable of disintegrating kidney cells.

nephrolysis [MED] **1.** Dissolution of kidney tissue by the action of a nephrolysin. **2.** Surgical detachment of a kidney from surrounding adhesions.

nephroma [MED] A tumor of the kidney.

nephromegaly [MED] Enlargement of the kidney.

nephromixium [INV ZOO] A compound nephridium composed of flame cells and the coelomic funnel; functions as both an excretory organ and a genital duct.

nephron [ANAT] The functional unit of a kidney, consisting of the glomerulus with its capsule and attached uriniferous tubule.

nephropathy [MED] **1.** Any disease of the kidney. **2.** *See* nephrosis.

nephropexy [MED] Fixation of a floating kidney by means of surgery.

Nephropidae [INV ZOO] The true lobsters, a family of decapod crustaceans in the superfamily Nephropidea.

Nephropidea [INV ZOO] A superfamily of the decapod section Macrura including the true lobsters and crayfishes, characterized by a rostrum and by chelae on the first three pairs of pereiopods, with the first pair being noticeably larger.

nephroptosis [MED] Prolapse of the kidney.

nephrorrhaphy [MED] **1.** The stitching of a floating kidney to the posterior wall of the abdomen or to the loin. **2.** Suturing a wound in the kidney.

nephros [ANAT] The kidney.

nephrosclerosis [MED] Sclerosis of the renal arteries and arterioles.

nephrosis [PATH] Degenerative or retrogressive renal lesions, distinct from inflammation (nephritis) or vascular involvement (nephrosclerosis), especially as applied to tubular lesions (tubular nephritis). Also known as nephrodystrophy; nephropathy.

nephrostome [INV ZOO] The funnel-shaped opening of a nephridium into the coelom.

nephrotic [MED] Pertaining to or affected by nephroses.

nephrotic edema [MED] A type of edema occurring in persons with chronic lepoid nephrosis or the nephrotic stage of glomerular nephritis.

nephrotic syndrome [MED] A complex of symptoms, including proteinuria, hyperalbuminemia, and hyperlipemia, resulting from damage to the basement membrane of glomeruli.

nephrotome [EMBRYO] The narrow mass of embryonic mesoderm connecting somites and lateral mesoderm, from which the pronephros, mesonephros, metanephros, and their ducts develop.

nephrotomy [MED] Incision of the kidney.

nephsystem *See* cloud system.

Nephtyidae [INV ZOO] A family of errantian annelids of highly opalescent colors, distinguished by an eversible pharynx.

Nepidae [INV ZOO] The water scorpions, a family of hemipteran insects in the superfamily Nepoidea, characterized by a long breathing tube at the tip of the abdomen, chelate front legs, and a short stout beak.

nepit *See* nit.

Nepoidea [INV ZOO] A superfamily of hemipteran insects in the subdivision Hydrocorisae.

Nepticulidae [INV ZOO] The single family of the lepidopteran superfamily Nepticuloidea.

Nepticuloidea [INV ZOO] A monofamilial superfamily of heteroneuran Lepidoptera; members are tiny moths with wing spines, and the females have a single genital opening.

Neptune [ASTRON] The outermost of the four giant planets, and the next to last planet, from the sun; it is 30 astronomical units from the sun, and the sidereal revolution period is 164.8 years.

neptune powder [MATER] An explosive consisting of nitroglycerin with a more or less explosive dope.

neptunianism *See* neptunism.

neptunian theory *See* neptunism.

neptunism [GEOL] The obsolete theory that all rocks of the earth's crust were deposited from or crystallized out of water. Also known as neptunianism; neptunian theory.

neptunite [MINERAL] $(Na,K)_2(Fe,Mn)TiSi_4O_{12}$ Black mineral composed of silicate of sodium, potassium, iron, manganese, and titanium.

neptunium [CHEM] A chemical element, symbol Np, atomic number 93, atomic weight 237.0482; a member of the actinide series of elements.

neptunium decay series [CHEM] Little-known radioactive elements with short lives; produced as successive series of decreasing atomic weight when uranium-237 and plutonium-241 decay radioactively through neptunium-237 to bismuth-209.

nepuite *See* garnierite.

Nereid [ASTRON] One of the two satellites of the planet

Neptune; it is the smaller, with a diameter of about 322 kilometers.

Nereidae [INV ZOO] A large family of mostly marine errantian annelids that have a well-defined head, elongated body with many segments, and large complex parapodia on most segments.

Nerillidae [INV ZOO] A family of archiannelids characterized by well-developed parapodia and setae.

Neritacea [INV ZOO] A superfamily of gastropod mollusks in the order Aspidobranchia.

neritic [OCEANOGR] Of or pertaining to the region of shallow water adjoining the seacoast and extending from low-tide mark to a depth of about 200 meters.

Neritidae [INV ZOO] A family of primitive marine, freshwater, and terrestrial snails in the order Archaeogastropoda.

Nernst approximation formula [THERMO] An equation for the equilibrium constant of a gas reaction based on the Nernst heat theorem and certain simplifying assumptions.

Nernst bridge [ELEC] A four-arm bridge containing capacitors instead of resistors, used for measuring capacitance values at high frequencies.

Nernst effect *See* Ettingshausen-Nernst effect.

Nernst equation [PHYS CHEM] The relationship showing that the electromotive force developed by a dry cell is determined by the activities of the reacting species, the temperature of the reaction, and the standard free-energy change of the overall reaction.

Nernst glower *See* Nernst lamp.

Nernst heat theorem [THERMO] The theorem expressing that the rate of change of free energy of a homogeneous system with temperature, and also the rate of change of enthalpy with temperature, approaches zero as the temperature approaches absolute zero.

Nernst lamp [ELEC] An electric lamp consisting of a short, slender rod of zirconium oxide in open air, heated to brilliant white incandescence by current. Also known as Nernst glower.

Nernst-Lindemann calorimeter [ENG] A calorimeter for measuring specific heats at low temperatures, in which the heat reservoir consists of a metal of high thermal conductivity such as copper, to promote rapid temperature equalization; none of the material under study is more than a few millimeters from a metal surface, and the whole apparatus is placed in an evacuated vessel and heated by current through a platinum heating coil.

Nernst-Simon statement of the third law of thermodynamics [THERMO] The statement that the change in entropy which occurs when a homogeneous system undergoes an isothermal reversible process approaches zero as the temperature approaches absolute zero.

Nernst-Thomson rule [PHYS CHEM] The rule that in a solvent having a high dielectric constant the attraction between anions and cations is small so that dissociation is favored, while the reverse is true in solvents with a low dielectric constant.

Nernst zero of potential [PHYS CHEM] An electrode potential corresponding to the reversible equilibrium between hydrogen gas at a pressure of 1 standard atmosphere and hydrogen ions at unit activity.

nerol [ORG CHEM] $C_{10}H_{17}OH$ Colorless liquid with roseneroli odor; derived from geraniol (a trans isomer); used in perfumery.

nerolidol [ORG CHEM] $C_{15}H_{26}O$ A straw-colored sesquiterpene alcohol; liquid with rose and apple aroma derived from cabreuva oil, oils of orange flower, and ylang ylang; soluble in alcohol; used in perfumery.

neroli oil *See* oil of orange blossoms.

nerve [ANAT] A bundle of nerve fibers or processes held together by connective tissue.

nerve block [PHYSIO] Interruption of impulse transmission through a nerve.

nerve cell *See* neuron.

nerve cord [INV ZOO] Paired, ventral cords of nervous tissue in certain invertebrates, such as insects or the earthworm. [ZOO] Dorsal, hollow tubular cord of nervous tissue in chordates.

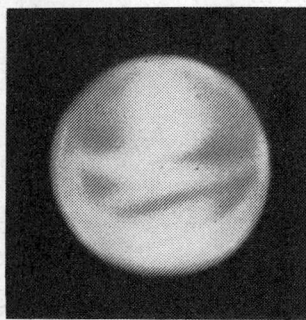

NEPTUNE

Telescopic appearance of Neptune.

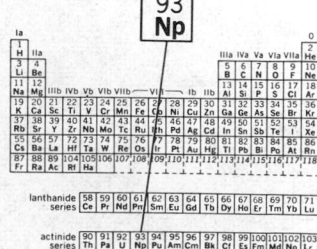

NEPTUNIUM

Periodic table of the chemical elements showing the position of neptunium.

nerve deafness [MED] Deafness due to an abnormality of the sense organs or of the nerves involved in hearing.

nerve ending [ANAT] **1.** The structure on the distal end of an axon. **2.** The termination of a nerve.

nerve fiber [CYTOL] The long process of a neuron, usually the axon.

nerve gas [MATER] Chemical agent which is absorbed into the body by breathing, by ingestion, or through the skin, and affects the nervous and respiratory systems and various body functions; an example is isopropylphosphonofluoridate.

nerve impulse [PHYSIO] The transient physicochemical change in the membrane of a nerve fiber which sweeps rapidly along the fiber to its termination, where it causes excitation of other nerves, muscle, or gland cells, depending on the connections and functions of the nerve.

nerve net [INV ZOO] A network of continuous nerve cells characterized by diffuse spread of excitation, local and equipotential autonomy, spatial attenuation of conduction, and facilitation; occurs in coelenterates and certain other invertebrates.

nerve tracing [MED] A method used by chiropractors by which nerves are located and their pathologies are studied.

nerve tract [ANAT] A bundle of nerve fibers having the same general origin and destination.

nervous [BIOL] **1.** Of or pertaining to nerves. **2.** Originating in or affected by nerves. **3.** Affecting or involving nerves. [PSYCH] A state or condition of nervousness.

nervousness [PSYCH] Hyperexcitability of the nervous system; characterized generally by restless or impulsive behavior, shaken mental poise, and an uncomfortable awareness of self.

nervous system [ANAT] A coordinating and integrating system which functions in the adaptation of an organism to its environment; in vertebrates, the system consists of the brain, brainstem, spinal cord, cranial and peripheral nerves, and ganglia.

nervous tissue [HISTOL] The nerve cells and neuroglia of the nervous system.

Nesiotinidae [INV ZOO] A family of bird-infesting biting lice (Mallophaga) that are restricted to penguins.

nesistor [ELECTR] A negative-resistance semiconductor device that is basically a bipolar field-effect transistor.

Nesophontidae [PALEON] An extinct family of large, shrewlike lipotyphlans from the Cenozoic found in the West Indies.

nesosilicate [MINERAL] A mineral (such as olivine) composed of independent silicon-oxygen tetrahedra bonded by ionic bonds, without sharing of oxygens.

Nessler's reagent [ANALY CHEM] Mercuric iodide–potassium iodide solution, used to analyze for small amounts of ammonia.

Nessler tubes [ANALY CHEM] Standardized glass tubes for filling with standard solution colors for visual color comparison with similar tubes filled with solution samples.

nest [ADP] To include data or subroutines in other items of a similar nature with a higher hierarchical level so that it is possible to access or execute various levels of data or routines recursively. [VERT ZOO] A bed, receptacle, or location in which the eggs of animals are laid and hatched.

nested sets [MATH] A family of sets where, given any two of its sets, one is contained in the other.

nesting [ADP] **1.** Inclusion of a routine wholly within another routine. **2.** Inclusion of a DO statement within a DO statement in FORTRAN.

Nestor [ASTRON] One of a group of asteroids whose period of revolution is approximately equal to that of Jupiter, or about 12 years (it is one of the Trojan planets).

net [COMMUN] A number of communication stations equipped for communicating with each other, often on a definite time schedule and in a definite sequence. [ENG] Threads or cords tied together at regular intervals to form a mesh. [GEOL] In structural petrology, coordinate network of meridians and parallels, projected from a sphere at intervals of 2°; used to plot points whose spherical coordinates are known and to study the distribution and orientation of planes and points. Also known as projection net; stereographic net. [MATH] A set whose members are indexed by elements from a directed set; this is a generalization of a sequence.

Net *See* Reticulum.

net blotch [PL PATH] A fungus disease of barley caused by *Helminthosporium teres* and marked by spotting of the foliage.

net call sign [COMMUN] A call sign that represents all stations within a net.

net control station [COMMUN] Communications station having the responsibility of clearing traffic and exercising circuit discipline within a net.

net floor area [BUILD] Gross floor area of a building, excluding the area occupied by walls and partitions, the circulation area (where people walk), and the mechanical area (where there is mechanical equipment).

net head [FL MECH] The difference in elevation between the last free water surface in a power conduit above the waterwheel and the first free water surface in the conduit below the waterwheel, less the friction losses in the conduit.

net heating value *See* low heat value.

net loss [COMMUN] The ratio of the power at the input of a transmission system to the power at the output; expressed in nepers, it is one-half the natural logarithm of this ratio, and in decibels it is 10 times the common logarithm of the ratio.

net positive suction head [MECH ENG] The minimum suction head required for a pump to operate; depends on liquid characteristics, total liquid head, pump speed and capacity, and impeller design. Abbreviated NPSH.

net power flow [ELECTROMAG] The difference between the power carried by electromagnetic waves traveling in a given direction along a waveguide and the power carried by waves traveling in the opposite direction.

net slip [GEOL] On a fault, the distance between two formerly adjacent points on either side of the fault; defines direction and relative amount of displacement. Also known as total slip.

net thrust [AERO ENG] The gross thrust of a jet engine minus the drag due to the momentum of the incoming air.

nettle [BOT] A prickly or stinging plant of the family Urticaceae, especially in the genus *Urtica*.

nettle cell *See* cnidoblast.

net ton *See* ton.

net tonnage [NAV ARCH] The volume of the interior of a ship measured in tons (100 cubic feet or approximately 2.8317 cubic meters per ton), excluding the space occupied by fuel, the engine room, navigation machinery, and the crew's quarters.

net-veined [BIOL] Having a network of veins, as a leaf or an insect wing.

network [COMMUN] A number of radio or television broadcast stations connected by coaxial cable, radio, or wire lines, so all stations can broadcast the same program simultaneously. [ELEC] A collection of electric elements, such as resistors, coils, capacitors, and sources of energy, connected together to form several interrelated circuits. Also known as electric network. [MATH] The name given to a graph in applications in management and the engineering sciences; to each segment linking points in the graph, there is usually associated a direction and a capacity on the flow of some quantity.

network admittance [ELEC] The admittance between two terminals of a network under specified conditions.

network analysis [ELEC] Derivation of the electrical properties of a network, from its configuration, element values, and driving forces.

network analyzer [ADP] An analog computer in which networks are used to simulate power line systems or physical systems and obtain solutions to various problems before the systems are actually built.

network constant [ELEC] One of the resistance, inductance, mutual inductance, or capacitance values involved in a circuit or network; if these values are constant, the network is said to be linear.

network filter [ELEC] A combination of electrical elements (for example, interconnected resistors, coils, and capacitors) that represents relatively small attenuation to signals of a certain frequency, and great attenuation to all other frequencies.

network flow [ELEC] Flow of current in a network.

network input impedance [ELEC] The impedance between

the input terminals of a network under specified conditions.

network master relay [ELEC] Relay that performs the chief functions of closing and tripping an alternating-current low-voltage network protector.

network phasing relay [ELEC] Relay which functions in conjunction with a master relay to limit closure of the network protector to a predetermined relationship between the voltage and the network voltage.

network relay [ELEC] Form of voltage, power, or other type of relay used in the protection and control of alternating-current low-voltage networks.

network structure [MET] A crystal structure in a metal in which one constituent occurs primarily at grain boundaries enveloping the grains made up of other constituents.

network synthesis [ELEC] Derivation of the configuration and element values of a network with given electrical properties.

network theory [ELEC] The systematizing and generalizing of the relations between the currents, voltages, and impedances associated with the elements of an electrical network.

network transfer admittance [ELEC] The current that would flow through a short circuit between one pair of terminals in a network if a unit voltage were applied across the other pair.

Neuberg blue [MATER] Pigment made up of a mixture of copper blue and iron blue.

Neumann bands *See* Neumann lines.

Neumann boundary condition [MATH] The boundary condition imposed on the Neumann problem in potential theory.

Neumann function [MATH] **1.** One of a class of Bessel functions arising in the study of the solutions to Bessel's differential equation. **2.** A harmonic potential function in potential theory occurring in the study of Neumann's problem.

Neumann-Kopp rule [THERMO] The rule that the heat capacity of 1 mole of a solid substance is approximately equal to the sum over the elements forming the substance of the heat capacity of a gram atom of the element times the number of atoms of the element in a molecule of the substance.

Neumann line [MATH] The generalization of the concept of a line occurring in Neumann's study of continuous geometry.

Neumann lines [MET] Mechanical deformation twins seen as straight, serrated narrow bands parallel to preferred planes in the crystals of an etched metal which has been strained, usually by sudden impact; most often observed along the 112 planes of body-centered-cubic ferrite. Also known as Neumann bands.

Neumann problem [MATH] The determination of a harmonic function within a finite region of three-dimensional space enclosed by a closed surface when the normal derivatives of the function on the surface are specified.

Neumann series *See* Liouville-Neumann series.

Neumann's principle [CRYSTAL] The principle that the symmetry elements of the point group of a crystal are included among the symmetry elements of any property of the crystal.

neural arc [PHYSIO] A nerve circuit consisting of effector and receptor with intercalated neurons between them.

neural arch *See* vertebral arch.

neural canal [EMBRYO] The embryonic vertebral canal.

neural crest [EMBRYO] Ectoderm composing the primordium of the cranial, spinal, and autonomic ganglia and adrenal medulla, located on either side of the neural tube.

neural ectoderm [EMBRYO] Embryonic ectoderm which will form the neural tube and neural crest.

neural fold [EMBRYO] Either of a pair of dorsal longitudinal folds of the neural plate which unite along the midline, forming the neural tube.

neuralgia [MED] Severe, stabbing, paroxysmal pain along the pathway of a nerve. Also known as neurodynia.

neural groove [EMBRYO] A longitudinal groove between the neural folds of the vertebrate embryo before the neural tube is completed.

neural lymphomatosis [VET MED] A form of the avian leukosis complex affecting primarily the sciatic nerve.

neural plate [EMBRYO] The thickened dorsal plate of ectoderm that differentiates into the neural tube.

neural spine [ANAT] The spinous process of a vertebra.

neural tube [EMBRYO] The embryonic tube that differentiates into brain and spinal cord.

neuraminic acid [BIOCHEM] $C_9H_{17}NO_8$ An amino acid, the aldol condensation product of pyruvic acid and *N*-acetyl-D-mannosamine, regarded as the parent acid of a family of widely distributed acyl derivatives known as sialic acids.

neuraminidase [MICROBIO] A bacterial enzyme that acts to split salic acid from neuraminic acid glycosides.

neurapophysis [EMBRYO] Either of two projections on each embryonic vertebra which unite to form the neural arch.

neurapraxia [MED] Injury to a nerve in which there is localized degeneration of the myelin sheath with transient nerve block.

neurasthenia [MED] A group of symptoms, now generally subsumed in the neurasthenic neurosis, formerly ascribed to debility or exhaustion of the nerve centers.

neurasthenic neurosis [PSYCH] A neurotic disorder characterized by chronic complaints of easy fatigability, lack of energy, weakness, various aches and pains, and sometimes exhaustion. Also known as psychophysiologic nervous system reaction.

neurectomy [MED] Surgical removal of a portion of a nerve.

neurenteric canal [EMBRYO] A temporary duct connecting the neural tube and primitive gut in certain vertebrate and tunicate embryos.

neurilemma [HISTOL] A thin tissue covering the axon directly, or covering the myelin sheath when present, of peripheral nerve fibers.

neurilemmoma [MED] A solitary, encapsulated benign tumor originating in the neurilemma of peripheral, cranial, and sympathetic nerves. Also known as schwannoma.

neurine [BIOCHEM] $CH_2=CHN(CH_3)_3OH$ A very poisonous, syrupy liquid with fishy aroma; soluble in water and alcohol; a product of putrefaction of choline in brain tissue and bile, and in cadavers. Also known as trimethylvinyl-ammonium hydroxide.

neurinomatosis *See* neurofibromatosis.

neuristor [ELECTR] A device that behaves like a nerve fiber in having attenuationless propagation of signals; one goal of research is development of a complete artificial nerve cell, containing many neuristors, that could duplicate the function of the human eye and brain in recognizing characters and other visual images.

neurite *See* axon.

neuritis [MED] Degenerative or inflammatory nerve lesions associated with pain, hypersensitivity, anesthesia or paresthesia, paralysis, muscular atrophy, and loss of reflexes in the innervated part of the body.

neuroanatomy [ANAT] The study of the anatomy of the nervous system and nerve tissue.

neuroarthropathy [MED] Joint disease associated with disease of the nervous system.

neuroastrocytoma [MED] Ganglioneuroma, especially when on the floor of the third brain ventricle and in the temporal lobes and exhibiting neuronal elements within predominant astrocytic elements.

neurobiotaxis [EVOL] Hypothetical migration of nerve cells and ganglia toward regions of maximum stimulation during phylogenetic development.

neuroblast [EMBRYO] Embryonic, undifferentiated neuron, derived from neural plate ectoderm.

neuroblastoma [MED] A malignant neoplasm composed of anaplastic sympathicoblasts; occurs usually in the adrenal medulla of children.

neuroblastomatosis *See* neurofibromatosis.

neurobrucellosis [MED] Brucellosis with neurologic involvement, manifested by signs and symptoms of meningitis, encephalitis, radiculitis, or neuritis.

neurochemistry [BIOCHEM] Chemistry of the nervous system.

neurochorioretinitis [MED] Chorioretinitis combined with optic neuritis.

neurochoroiditis [MED] Choroiditis combined with optic neuritis.

neurocirculatory [ANAT] Pertaining to both the nervous and the vascular systems.

neurocirculatory asthenia [MED] A syndrome characterized by dyspnea, palpitation, chest pain, fatigue, and faintness.

neurocoele [ANAT] The system of cavities and ventricles in the brain and spinal cord.

neurocranium [ANAT] The portion of the cranium which forms the braincase.

neurocutaneous [ANAT] **1.** Concerned with both the nerves and skin. **2.** Pertaining to innervation of the skin.

neurocyte [CYTOL] The body of a nerve cell.

neurodermatitis [MED] A skin disorder characterized by localized, often symmetrical, patches of pruritic dermatitis with lichenification, occurring in persons of nervous temperament.

neurodermatosis [MED] A skin disease which is presumed to have a psychogenic component or basis.

Neurodontiformes [PALEON] A suborder of Conodontophoridia having a lamellar internal structure.

neurodynia See neuralgia.

neuroelectricity [PHYSIO] A current or voltage generated in the nervous system.

neuroendocrine [BIOL] Pertaining to both the nervous and endocrine systems, structurally and functionally.

neuroendocrinology [BIOL] The study of the structural and functional interrelationships between the nervous and endocrine systems.

neuroepidermal [BIOL] Pertaining to both the nerves and epidermis, structurally and functionally.

neuroepithelioma [MED] A tumor resembling primitive medullary epithelium, containing cells of small cuboidal or columnar form with a tendency to form true rosettes, occurring in the retina, central nervous system, and occasionally in peripheral nerves. Also known as diktoma; esthesioneuroblastoma; esthesioneuroepithelioma.

neurofibril [CYTOL] A fibril of a neuron, usually extending from the processes and traversing the cell body.

neurofibroma [MED] A tumor characterized by the diffuse proliferation of peripheral nerve elements. Also known as endoneural fibroma; myxofibroma of nerve sheath; neurofibromyxoma; perineural fibroblastoma; perineural fibroma.

neurofibromatosis [MED] A hereditary disease characterized by the presence of neurofibromas in the skin or along the pathway of peripheral nerves. Also known as fibroma molluscum; multiple neurofibroma; multiple neurofibromatosis; neurinomatosis; neuroblastomatosis; Smith-Recklinghausen's disease.

neurofibromyxoma See neurofibroma.

neurofibrosarcoma [MED] A malignant tumor composed of interlacing bundles of anaplastic spindle-shaped cells which resemble those of nerve sheaths.

neurogenesis [EMBRYO] The formation of nerves.

neurogenic [BIOL] **1.** Originating in nervous tissue. **2.** Innervated by nerves. [MED] Caused or affected by a trauma, dysfunction, or disease of the nervous system.

neurogenic bladder [MED] A urinary bladder disorder due to lesions of the central or peripheral nervous system.

neurogenic shock [MED] Shock caused by vasodilation leading to low blood pressure and serious reduction in venous return and in cardiac output; due to such causes as injury to the central nervous system, spinal anesthesia, or reflex.

neuroglia [HISTOL] The nonnervous, supporting elements of the nervous system.

neurohemal organ [ZOO] Any of various structures in vertebrates and some invertebrates that consist of clusters of bulbous, secretion-filled axon terminals of neurosecretory cells which function as storage-and-release centers for neurohormones.

neurohormone [BIOCHEM] A hormone produced by nervous tissue.

neurohumor [BIOCHEM] A hormonal transmitter substance, such as acetylcholine, released by nerve endings in the transmission of impulses.

neurohypophysis [ANAT] The neural portion or posterior lobe of the hypophysis.

neurolathyrism See lathyrism.

neuroleptic [PHARM] **1.** A drug that is useful in the treatment of mental disorders, especially psychoses. **2.** Pertaining to the actions of such a drug.

neuroleptoanalgesia [MED] A state of analgesic consciousness produced by the administration of neuroleptic drugs, allowing painless surgery to be performed on a wakeful subject.

neurologist [MED] A person versed in neurology, usually a physician who specializes in the diagnosis and treatment of disorders of the nervous system and the study of its functioning.

neurology [MED] The study of the anatomy, physiology, and disorders of the nervous system.

neuroma [MED] A tumor of the nervous system.

neuromast [VERT ZOO] A lateral-line sensory organ in fishes and other lower vertebrates consisting of a cluster of receptor cells connected with nerve fibers.

neuromere [EMBRYO] An embryonic segment of the central nervous system in vertebrates.

neuromuscular [BIOL] Pertaining to both nerves and muscles, functionally and structurally.

neuromuscular junction See myoneural junction.

neuromyasthenia [MED] Fatigue, headache, intense muscle pain, slight or transient muscle weakness, mental disturbances, objective signs in neurologic examination but usually normal cerebrospinal fluid findings, occurring in epidemics and thought to be viral in origin. Also known as benign myalgic encephalomyelitis.

neuromyelitis [MED] Inflammation of the spinal cord and of nerves.

neuron [HISTOL] A nerve cell, including the cell body, axon, and dendrites.

neuron doctrine [BIOL] A doctrine that the neuron is the basic structural and functional unit of the nervous system, and that it acts upon another neuron through the synapse.

neuronitis [MED] Inflammation of a neuron; particularly, neuritis involving the cells and roots of spinal nerves.

neuropathy [MED] Any disease affecting neurons.

neuropharmacology [MED] The science dealing with the action of drugs on the nervous system.

neurophysiology [PHYSIO] The study of the functions of the nervous system.

neuropil [HISTOL] Nervous tissue consisting of a fibrous network of nonmyelinated nerve fibers; gray matter with few nerve cell bodies; usually a region of synapses between axons and dendrites.

neuroplasm [CYTOL] Protoplasm of nerve cells.

neuropodium [CYTOL] A terminal branch of an axon.

neuropore [EMBRYO] A terminal aperture of the neural tube before complete closure at the 20-25 somite stage.

neuropsychology [PSYCH] A system of psychology based on neurology.

neuropsychopathy [MED] A mental disease based upon or manifesting itself in disorders or symptoms of the nervous system.

Neuroptera [INV ZOO] An order of delicate insects having endopterygote development, chewing mouthparts, and soft bodies.

neuroradiology [MED] The roentgenology of neurologic disease.

neurorrhexis [MED] Surgical tearing away of a nerve from its origin, as in the treatment of neuralgia.

neurosarcoma [MED] A sarcoma composed of elements resembling those of the nervous system, or thought to be neurogenic.

neurosclerosis [MED] Hardening of nervous tissue.

neurosecretion [PHYSIO] The synthesis and release of hormones by nerve cells.

neurosis [PSYCH] A category of emotional maladjustments characterized by some impairment of thinking and judgment, with anxiety as the chief symptom.

neurosurgery [MED] Surgery of the nervous system.

neurosyphilis [MED] Syphilitic infection of the nervous system.

neuroticism [PSYCH] A neurotic condition, character, or trait.

neurotic personality [PSYCH] An individual who exhibits symptoms or manifestations intermediate between normal character traits and true neurotic features.

neurotoxin [BIOCHEM] A poisonous substance in snake venom that acts as a nervous system depressant.

neurotrophic ulcer [MED] A decubitus ulcer due to trophic disturbances following interruption or disease of afferent nerve fibers plus the factor of external trauma.

neurotropic [BIOL] Having an affinity for nerve tissue.

neurovaricosis [MED] The formation or the presence of a varicosity on a nerve fiber.

neurovascular [BIOL] Pertaining structurally and functionally to both the nervous and vascular structures.

neurulation [EMBRYO] Differentiation of nerve tissue and formation of the neural tube.

neuston [BIOL] Minute organisms that float or swim on surface water or on a surface film of water.

neutral atom [ATOM PHYS] An atom in which the number of electrons that surround the nucleus is equal to the number of protons in the nucleus, so that there is no net electric charge.

neutral axis [MECH] In a beam bent downward, the line of zero stress below which all fibers are in tension and above which they are in compression.

neutral beam [PHYS] A stream of uncharged particles.

neutral conductor [ELEC] A conductor of a polyphase circuit or of a single-phase, three-wire circuit which is intended to have a potential such that the potential differences between it and each of the other conductors are approximately equal in magnitude and are also equally spaced in phase.

neutral-density filter [OPTICS] An optical filter that reduces the intensity of light without appreciably changing its color; used on a camera when the lens cannot be stopped down sufficiently for use with a given film. Also known as gray filter; neutral filter.

neutral equilibrium [FL MECH] A property of the steady state of a system which exhibits neither instability nor stability according to the particular criterion under consideration; a disturbance introduced into such an equilibrium will thus be neither amplified nor damped. Also known as indifferent equilibrium.

neutral fiber [MECH] A line of zero stress in cross section of a bent beam, separating the region of compressive stress from that of tensile stress.

neutral filter *See* neutral-density filter.

neutral flame [CHEM] Gas flame produced by a mixture of fuel and oxygen so as to be neither oxidizing nor reducing.

neutral granulation [CHEM] Propellant granulation in which the surface area of a grain remains constant during burning.

neutral ground [ELEC] Ground connected to the neutral point or points of an electric circuit, transformer, rotating machine, or system.

neutralism [ECOL] A neutral interaction between two species, that is, one having no evident effect on either species.

neutralization equivalent [CHEM] For an acid or base, the same as equivalent weight; multiplication of the neutralization equivalent by the number of acidic or basic groups in the molecule gives the molecular weight.

neutralization fire [ORD] Fire which is delivered to cause casualties, to hamper and interrupt the firing of weapons, movement, or action, and to reduce the combat efficiency of enemy personnel.

neutralization number [ANALY CHEM] Petroleum product test; it is the milligrams of potassium hydroxide required to neutralize the acid in 1 gram of oil; used as an indication of oil acidity.

neutralize [CHEM] To make a solution neutral (neither acidic nor basic, pH of 7) by adding a base to an acidic solution, or an acid to a basic solution. [ELECTR] To nullify oscillation-producing voltage feedback from the output to the input of an amplifier through tube interelectrode capacitances; an external feedback path is used to produce at the input a voltage that is equal in magnitude but opposite in phase to that fed back through the interelectrode capacitance. [ORD] **1.** To destroy or reduce the effectiveness of enemy personnel and materiel by gunfire, bombing, or any other means. **2.** To make a toxic chemical agent harmless by chemical action. **3.** To disarm or otherwise render safe a mine, bomb, missile, or booby trap.

neutralized radio-frequency stage [ELECTR] Stage having an additional circuit connected to feed back, in the opposite phase, an amount of energy equivalent to what is causing the oscillation, thus neutralizing any tendency to oscillate and making the circuit function strictly as an amplifier.

neutralizing antibody [IMMUNOL] An antibody that reduces or abolishes some biological activity of a soluble antigen or of a living microorganism.

neutralizing capacitor [ELECTR] Capacitor, usually variable, employed in a radio receiving or transmitting circuit to feed a portion of the signal voltage from the plate circuit of a stage back to the grid circuit.

neutralizing circuit [ELECTR] Portion of an amplifier circuit which provides an intentional feedback path from plate to grid to prevent regeneration.

neutralizing power of a lens [OPTICS] The power of a lens, measured by neutralizing it with trial lenses of equal and opposite power; for a spectacle lens, it may differ significantly from back vertex power.

neutralizing transformer [ELECTROMAG] A transformer installed on a communication line which produces counter electromotive forces which largely cancel out induced longitudinal voltage and allow normal operation of the line during inductive disturbances.

neutralizing voltage [ELECTR] Voltage developed in the plate circuit (Hazeltine neutralization) or in the grid circuit (Rice neutralization), used to nullify or cancel the feedback through the tube.

neutral line *See* Busch lemniscate.

neutrally buoyant float *See* swallow float.

neutral molecule [PHYS CHEM] A molecule in which the number of electrons surrounding the nuclei is the same as the total number of protons in the nuclei, so that there is no net electric charge.

neutral oil [MATER] Light oil from dewaxed petroleum; medium or low viscosity, flash point 143–160°C.

neutral operation [COMMUN] System whereby marking signals are formed by current impulses of one polarity, either positive or negative, and spacing signals are formed by reducing the current to zero or nearly zero.

neutral particle [PARTIC PHYS] A particle that carries no electric charge.

neutral point [ELEC] Point which has the same potential as the point of junction of a group of equal nonreactive resistances connected at their free ends to the appropriate main terminals or lines of the system. [FL MECH] *See* hyperbolic point. [MET] In rolling mills, the point at which the speed of the work is equal to the peripheral speed of the rolls. [METEOROL] *See* col. [OPTICS] In atmospheric optics, one of several points in the sky for which the degree of polarization of diffuse sky radiation is zero.

neutral potassium phosphate *See* potassium phosphate.

neutral pressure *See* neutral stress.

neutral red [ORG CHEM] $(CH_3)_2NC_6H_3N_2C_6H_2CH_3NH_2 \cdot ClH$ Water- and alcohol-soluble green powder; used as pH 6.8–8.0 acid-base indicator, and as a dye to test stomach function. Also known as dimethyl diaminophenazine chloride; toluylene red.

neutral relay [ELEC] Relay in which the movement of the armature does not depend upon the direction of the current in the circuit controlling the armature. Also known as nonpolarized relay.

neutral return path [ELEC] A route from the load back to the power source, completing a circuit in an electric power distribution system, which is grounded, usually by connections to water pipes.

neutral stability [CONT SYS] Condition in which the natural motion of a system neither grows nor decays, but remains at its initial amplitude. [METEOROL] The state of an unsaturated or saturated column of air in the atmosphere when its environmental lapse rate of temperature is equal to the dry-adiabatic lapse rate or the saturation-adiabatic lapse rate respectively; under such conditions a parcel of air displaced vertically will experience no buoyant acceleration. Also known as indifferent equilibrium; indifferent stability.

neutral stress [HYD] The stress transmitted through the interstitial fluid of a soil or rock mass. Also known as neutral pressure; pore pressure; pore-water pressure.

neutral surface [MECH] A surface in a bent beam along which material is neither compressed nor extended.

neutral transmission [COMMUN] The signal transmitted over a line when no traffic is flowing.

neutral wave [PHYS] Any wave whose amplitude does not change with time; in most contexts the wave is referred to as a stable wave, the term "neutral wave" being used when it is important to emphasize that the wave is neither damped nor amplified.

neutral zone *See* dead band.

neutrino [PHYS] A neutral particle having zero rest mass and spin $1/2$ $(h/2\pi)$, where h is Planck's constant; experimentally, there are two such particles known as the e-neutrino (ν_e) and the μ-neutrino (ν_μ).

neutron [PHYS] An elementary particle which has approximately the same mass as the proton but lacks electric charge, and is a constituent of all nuclei having mass number greater than 1.

neutron absorber [NUCLEO] A material in which a significant number of neutrons passing through combine with nuclei and are not reemitted.

neutron absorption *See* neutron capture.

neutron age *See* Fermi age.

neutron albedo [NUCLEO] The probability, under specified conditions, that a neutron entering into a region through a surface will return through that surface.

neutron binding energy [NUC PHYS] The energy required to remove a single neutron from a nucleus.

neutron capture [NUC PHYS] A process in which the collision of a neutron with a nucleus results in the absorption of the neutron into the nucleus with the emission of one or more prompt gamma rays; in certain cases, beta decay or fission of the nucleus results. Also known as neutron absorption; neutron radiative capture.

neutron-capture cross section [NUC PHYS] The cross section for neutron capture by nuclei in a material; it is a measure of the probability that this reaction will occur.

neutron chopper [NUCLEO] A device that interrupts the output beam of neutrons from a nuclear reactor mechanically, to provide bursts or pulses of neutrons for research purposes.

neutron counter [NUCLEO] A neutron detector which counts the number of neutrons passing through the detecting medium.

neutron cross section [NUC PHYS] A measure of the probability that an interaction of a given kind will take place between a nucleus and an incident neutron; it is an area such that the number of interactions which occur in a sample exposed to a beam of neutrons is equal to the product of the number of nuclei in the sample and the number of neutrons in the beam that would pass through this area if their velocities were perpendicular to it.

neutron cycle [NUCLEO] The life history of the neutrons in a nuclear reactor, extending from the initial fission process until all the neutrons have been absorbed or have leaked out.

neutron detector [NUCLEO] Any device which detects passing neutrons, for example, by observing the charged particles or gamma rays released in nuclear reactions induced by the neutrons, or observing the recoil of charged particles caused by collisions with neutrons.

neutron diffraction [PHYS] The phenomenon associated with the interference processes which occur when neutrons are scattered by the atoms within solids, liquids, and gases.

neutron diffractometer [PHYS] A diffractometer in which a beam of neutrons is used for diffraction analysis, and the intensities of the diffracted beams at different angles are measured with an ionization chamber or radiation counter.

neutron economy [NUCLEO] The balance sheet describing the ways neutrons are used in a nuclear reactor, as in fission, leakage, and absorption.

neutron excess [NUC PHYS] The number of neutrons in a nucleus in excess of the number of protons. Also known as difference number; isotopic number.

neutron flux [NUCLEO] The intensity of neutron radiation, expressed as the number of neutrons passing through a unit area per unit time. Also known as neutron flux density.

neutron flux density *See* neutron flux.

neutron-gamma well logging *See* neutron well logging.

neutron hardening [NUCLEO] The increase which occurs in the average energy of thermal neutrons diffusing in a medium whose absorption cross section decreases as energy increases.

neutron howitzer [NUCLEO] A device which produces a neutron beam; consists of a neutron source contained in a block of moderating material with a small hole from the source to the surface of the block, from which the beam emerges.

neutron inelastic scattering reactions [NUCLEO] Emissions of low-energy neutrons when materials have been bombarded by fast neutrons during neutron activation analysis. Also known as n-n' reactions.

neutron irradiation [NUCLEO] The exposure of a material or object to neutrons.

neutron logging *See* neutron well logging.

neutron magnetic moment [NUC PHYS] A vector whose scalar product with the magnetic flux density gives the negative of the energy of interaction of a neutron with a magnetic field.

neutron multiplication factor *See* multiplication factor.

neutron number [NUC PHYS] The number of neutrons in the nucleus of an atom.

neutron optics [PHYS] The study of certain phenomena, for example, crystal diffraction, in which the wave character of neutrons dominates and leads to behavior similar to that of light.

neutron radiative capture *See* neutron capture.

neutron radiography [NUCLEO] Radiography that uses a neutron beam generated by a nuclear reactor; the neutrons are detected by placing a conventional x-ray film next to a converter screen composed of potentially radioactive materials or prompt emission materials which convert the neutron radiation to other types of radiation more easily detected by the film.

neutron reflection [PHYS] Specular reflection of neutrons, either from lattice planes of crystalline substances according to the Bragg law for their de Broglie wavelength, or from highly polished surfaces of certain substances at an angle smaller than their critical angle.

neutron scattering [NUCLEO] The change in direction of neutrons caused by collision with nuclei in a material.

neutron spectrometer [NUCLEO] An instrument used to determine the energies of neutrons and the relative intensities of neutrons of different energies in a neutron beam.

neutron spectrometry [NUC PHYS] A method of observing excited states of nuclei in which neutrons are used to bombard a target, causing nuclei to be transmuted into excited states by various nuclear reactions; the resultant excited states are determined by observing resonances in the reaction cross sections or by observing spectra of emitted particles or gamma rays. Also known as neutron spectroscopy.

neutron spectroscopy *See* neutron spectrometry.

neutron spectrum [NUC PHYS] A plot or display of the number of neutrons at various energies, such as the neutrons emitted in a nuclear reaction, or the neutrons in a nuclear reactor.

neutron star [ASTRON] A star that is supposed to occur in the final stage of stellar evolution; it consists of a superdense mass mainly of neutrons, and has a strong gravitational attraction from which only neutrinos and high-energy photons could escape so that the star is invisible.

neutron therapy [MED] Medical therapy involving irradiation with neutrons.

neutron transport theory [NUCLEO] Theory of diffusion of neutrons in a material based on the Boltzmann transport equation.

neutron velocity selector [NUCLEO] An instrument that isolates and detects neutrons having a particular range of velocities.

neutron well logging [ENG] Study of formation-fluid-content properties down a wellhole by neutron bombardment and detection of resultant gamma radiation. Also known as neutron-gamma well logging; neutron logging.

neutropenia [MED] Abnormally low number of neutrophils in the peripheral circulation.

neutrophil [HISTOL] A large granular leukocyte with a highly variable nucleus, consisting of three to five lobes, and cytoplasmic granules which stain with neutral dyes and eosin.

neutrophilia [BIOL] Affinity for neutral dyes. [MED] An abnormal increase in leukocytes in the tissues or peripheral circulation.

neutrophilic leukemia [MED] Granulocytic leukemia in

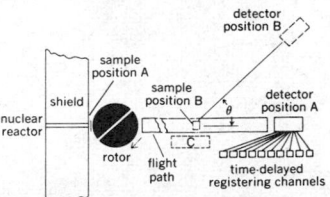

NEUTRON CHOPPER

Schematic diagram of a neutron chopper and time-of-flight apparatus. Registering channels record at successive times following passage of pulse of neutrons through rotor slit to measure neutron time-of-flight. Sample and detector position A used to measure total cross section. Sample and position B used to measure differential cross section for scattering through angle θ. Detector position C used to measure absorption cross section.

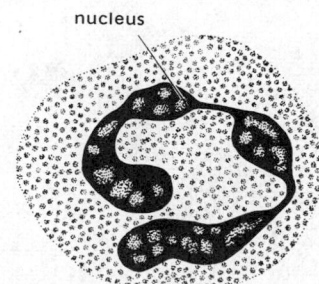

NEUTROPHIL

Diagram of neutrophil showing multilobed nucleus.

which the leukocytes resemble cells of the neutrophilic series.

neutrophilous [BIOL] Preferring an environment free of excess acid or base.

neutrosphere [METEOROL] The atmospheric shell from the earth's surface upward, in which the atmospheric constituents are for the most part un-ionized, that is, electrically neutral; the region of transition between the neutrosphere and the ionosphere is somewhere between 70 and 90 kilometers, depending on latitude and season.

nevada [METEOROL] A cold wind descending from a mountain glacier or snowfield, for example, in the higher valleys of Ecuador.

Nevadan orogeny [GEOL] Orogenic episode during Jurassic and Early Cretaceous geologic time in the western part of the North American Cordillera. Also known as Nevadian orogeny; Nevadic orogeny.

Nevadian orogeny *See* Nevadan orogeny.

Nevadic orogeny *See* Nevadan orogeny.

névé [GEOGR] A geographic area of perennial snow. [HYD] An accumulation of compacted, granular snow in transition from soft snow to ice; it contains much air; the upper portions of most glaciers and ice shelves are usually composed of névé.

nevocarcinoma [MED] A malignant melanoma.

nevus [MED] A lesion containing melanocytes.

nevus sebaceus [MED] A nevus formed by an aggregate of sebaceous glands. Also known as Jadassohn's nevus.

newberyite [MINERAL] $MgH(PO_4) \cdot 3H_2O$ A white, orthorhombic member of the brushite mineral group; it is isostructural with gypsum.

new blue [MATER] Any of several iron blue types of pigments; varieties are called mendola blue and prussian blue.

newborn [MED] Born recently; said of human infants less than a month old, especially of those only a few days old.

new candle *See* candela.

Newcastle disease [VET MED] An acute viral disease of fowls, with respiratory, gastrointestinal, and central nervous system involvement; may be transmitted to human beings as a mild conjunctivitis. Also known as avian pneumoencephalitis; avian pseudoplague; Philippine fowl disease.

Newcastle virus [VIROL] A ribonucleic acid hemagglutinating myxovirus responsible for Newcastle disease.

newel post [CIV ENG] **1.** A pillar at the end of an oblique retaining wall of a bridge. **2.** The post about which a circular staircase winds. **3.** A large post at the foot of a straight stairway or on a landing.

new global tectonics [GEOL] Comprehensive theory relating the formation of mountain belts, island arcs, and ocean trenches to the relative movement of regionally extensive lithospheric plates which are delineated by the major seismic belts of the earth.

New Jersey retort process *See* vertical retort process.

Newland's law of octaves [CHEM] An arrangement of the elements that predated Mendeleeff's periodic table; Newland's arrangement was a grouping of the elements in increasing atomic weights (starting with lithium) in horizontal rows of eight elements, with each new row directly beneath the previous one.

newly formed ice [HYD] Ice in the first stage of formation and development. Also known as fresh ice.

new moon [ASTRON] The moon at conjunction, when little or none of it is visible to an observer on the earth because the illuminated side is turned away.

New Red Sandstone [GEOL] The red sandstone facies of the Permian and Triassic systems exposed in the British Isles.

Newson's boring method [MIN ENG] A method of boring small shafts, up to 5½ feet (1.7 meters) in diameter using the principle of chilled-shot drilling on a large scale.

newsprint [MATER] The paper used in the publication of newspapers; an impermanent material made from mechanical wood pulp, with some chemical wood pulp.

newt [VERT ZOO] Any of the small, semiaquatic salamanders of the genus *Triturus* in the family Salamandridae; all have an aquatic larval stage.

newton [MECH] The unit of force in the meter-kilogram-second system, equal to the force which will impart an acceleration of 1 meter/second² to the International Proto-

type Kilogram mass. Abbreviated N. Formerly known as large dyne.

Newton-Cotes formulas [MATH] Approximation formulas for the integral of a function along a small interval in terms of the values of the function and its derivatives.

Newtonian attraction [MECH] The mutual attraction of any two particles in the universe, as given by Newton's law of gravitation.

Newtonian flow [FL MECH] Flow system in which the fluid performs as a Newtonian fluid, that is, shear stress is proportional to shear rate.

Newtonian fluid [FL MECH] A simple fluid in which the state of stress at any point is proportional to the time rate of strain at that point; the proportionality factor is the viscosity coefficient.

Newtonian friction law [FL MECH] The law that shear stress in a fluid is proportional to the shear rate; it holds only for some fluids, which are then called Newtonian. Also known as Newton formula for the stress.

Newtonian mechanics [MECH] The system of mechanics based upon Newton's laws of motion in which mass and energy are considered as separate, conservative, mechanical properties, in contrast to their treatment in relativistic mechanics.

Newtonian potential [PHYS] A potential which is associated with an inverse square law of force (such as an electrostatic force), and therefore varies with distance in the same manner as a gravitational potential.

Newtonian reference frame [MECH] One of a set of reference frames with constant relative velocity and within which Newton's laws hold; the frames have a common time, and coordinates are related by the Galilean transformation rule.

Newtonian speed of sound [ACOUS] An approximation to the speed of sound in a perfect gas given by the relation $c^2 = p/\rho$, where p is pressure and ρ the density.

Newtonian telescope [OPTICS] A reflecting telescope in which the light reflected from a concave mirror is reflected again by a plane mirror making an angle of 45° with the telescope axis, so that it passes through a hole in the side of the telescope containing the eyepiece.

Newtonian velocity [MECH] The velocity of an object in a Newtonian reference frame, S, which can be determined from the velocity of the object in any other such frame, S', by taking the vector sum of the velocity of the object in S' and the velocity of the frame S' relative to S.

Newtonian viscosity [FL MECH] The viscosity of a Newtonian fluid.

newton-meter of energy *See* joule.

newton-meter of torque [MECH] The unit of torque in the meter-kilogram-second system, equal to the torque produced by 1 newton of force acting at a perpendicular distance of 1 meter from an axis of rotation. Abbreviated N-m.

Newton-Raphson formula [MATH] If c is an approximate value of a root of the equation $f(x) = 0$, then a better approximation is the number $c - [f(c)/f'(c)]$.

Newton's alloy [MET] A fusible alloy made up of 50% bismuth, 31% lead, and 19% tin; melts at 95°C; used in applications where it is required to fall away at predetermined temperatures, as for automatic sprinkler links. Also known as Newton's metal.

Newton's equations of motion [MECH] Newton's laws of motion expressed in the form of mathematical equations.

Newton's first law [MECH] The law that a particle not subjected to external forces remains at rest or moves with constant speed in a straight line. Also known as first law of motion; Galileo's law of inertia.

Newton's identities [MATH] Relations between the coefficients of a polynomial and quantities s_k defined by $s_k = r_1^k + r_2^k + \ldots + r_n^k$, where $r_1, \ldots, r_n$ are the roots of the polynomial; for the polynomial $a_0 x^n + a_1 x^{n-1} + \ldots + a_n$, the relations are $a_0 s_k + a_1 s_{k-1} + \ldots + a_{k-1} s_1 + k a_k = 0$, for $k < n$, and $a_0 s_k + a_1 s_{k-1} + \ldots + a_n s_{k-n} = 0$, for $k > n$.

Newton's law of cooling [THERMO] The law that the rate of heat flow out of an object by both natural convection and radiation is proportional to the temperature difference between the object and its environment, and to the surface area of the object.

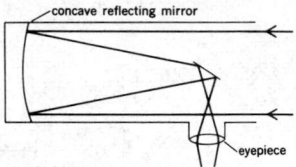

NEWTONIAN TELESCOPE

Simplified diagram of a Newtonian telescope.

Newton's law of gravitation [MECH] The law that every two particles of matter in the universe attract each other with a force that acts along the line joining them, and has a magnitude proportional to the product of their masses and inversely proportional to the square of the distance between them. Also known as law of gravitation.

Newton's law of resistance [FL MECH] The law that the force opposing the motion of an object through a fluid at moderate velocities is proportional to the square of the velocity.

Newton's laws of motion [MECH] Three fundamental principles (called Newton's first, second, and third laws) which form the basis of classical, or Newtonian, mechanics, and have proved valid for all mechanical problems not involving speeds comparable with the speed of light and not involving atomic or subatomic particles.

Newton's metal See Newton's alloy.

Newton's method [MATH] A technique to approximate the roots of an equation by the methods of the calculus.

Newton's rings [OPTICS] A series of circular bright and dark bands which appear about the point of contact between a glass plate and a convex lens which is pressed against it and illuminated with monochromatic light.

Newton's second law [MECH] The law that the acceleration of a particle is directly proportional to the resultant external force acting on the particle and is inversely proportional to the mass of the particle. Also known as second law of motion.

Newton's square-root method [MATH] A technique for the estimation of the roots of an equation exhibiting faster convergence than Newton's method; this involves calculus methods and the square-root function.

Newton's theory of lift [FL MECH] A theory of the forces acting on an airfoil in a fluid current in which these forces are assumed to result from the impact of particles of the fluid on the body.

Newton's theory of light See corpuscular theory of light.

Newton's third law [MECH] The law that, if two particles interact, the force exerted by the first particle on the second particle (called the action force) is equal in magnitude and opposite in direction to the force exerted by the second particle on the first particle (called the reaction force). Also known as law of action and reaction; third law of motion.

Neyman-Pearson theory [STAT] A theory that determines what is the best test to use to examine a statistical hypothesis.

Ney's chart See modified Lambert conformal chart.

Ney's projection [MAP] A modified Lambert conformal projection.

ng See nanogram.

NGU See nongonococcal urethritis.

NH propellant [MATER] A propellant which, by reason of its formulation or method of manufacture, does not absorb moisture from the air.

Ni See nickel.

niacin See nicotinic acid.

niacinamide See nicotinamide.

Niagaran [GEOL] A North American provincial geologic series, in the Middle Silurian.

nib [ENG] A small projecting point.

nibbling [MECH ENG] Contour cutting of material by the action of a reciprocating punch that takes repeated small bites as the work is passed beneath it.

nicad battery See nickel-cadmium battery.

nicarbing See carbonitriding.

niccolite [MINERAL] NiAs A pale-copper-red, hexagonal mineral with metallic luster; an important ore of nickel; hardness is 5–5.5 on Mohs scale. Also known as arsenical nickel; copper nickel; nickeline.

Nichol's chart [CONT SYS] A plot of curves along which the magnitude M or argument α of the frequency control ratio is constant on a graph whose ordinate is the logarithm of the magnitude of the open-loop transfer function, and whose abscissa is the open-loop phase angle.

Nichols radiometer [ENG] An instrument, used to measure the pressure exerted by a beam of light, in which there are two small, silvered glass mirrors at the ends of a light rod that is suspended at the center from a fine quartz fiber within an evacuated enclosure.

nick See knickpoint.

nickel [CHEM] A chemical element, symbol Ni, atomic number 28, atomic weight 58.71. [MET] A silver-gray, ductile, malleable, tough metal; used in alloys, plating, coins (to replace silver), ceramics, and electronic circuits.

nickel-63 [NUC PHYS] Radioactive nickel with beta radiation and 92-year half-life; derived by pile-irradiation of nickel; used in radioactive composition studies and tracer studies.

nickel acetate [ORG CHEM] $Ni(OOCCH_3)_2 \cdot 4H_2O$ Efflorescent green crystals that decompose upon heating; soluble in alcohol and water; used as textile dyeing mordant.

nickel-aluminum bronze [MET] An alloy composed of an 8–10% aluminum bronze with nickel added to increase strength, corrosion resistance, and heat resistance; used for dies, molds, cast propellers, and valve seats.

nickel arsenate [INORG CHEM] $Ni_3(AsO_4)_2 \cdot H_2O$ Poisonous yellow-green powder; soluble in acids, insoluble in water; used as a fat-hardening catalyst in soapmaking. Also known as nickelous arsenate.

nickel brass See nickel silver.

nickel bronze [MET] Bronze containing nickel; a common type contains 88% copper, 5% tin, 5% nickel, and 2% zinc.

nickel-cadmium battery [ELEC] A sealed storage battery having a nickel anode, a cadmium cathode, and an alkaline electrolyte; widely used in cordless appliances; without recharging, it can serve as a primary battery. Also known as cadmium-nickel storage cell; nicad battery.

nickel carbonate [INORG CHEM] $NiCO_3$ Light-green crystals that decompose upon heating; soluble in acid, insoluble in water; used in electroplating.

nickel carbonyl [INORG CHEM] $Ni(CO)_4$ Colorless, flammable, poisonous liquid boiling at 43°C; soluble in alcohol and concentrated nitric acid, insoluble in water; used in gas plating (vapor decomposes at 60°C) and to produce metallic nickel. Also known as nickel tetracarbonyl.

nickel cast iron [MET] An improved-strength alloy cast iron containing a small percentage of nickel (2–5%); in larger amounts (15–36%) nickel primarily imparts corrosion resistance.

nickel-chromium steel [MET] Steel containing nickel (0.2–3.75%) and chromium (0.3–1.5%) as alloying elements.

nickel cyanide [INORG CHEM] $Ni(CN)_2 \cdot 4H_2O$ Poisonous, water-insoluble apple-green powder; melts and loses water at 200°C, decomposes at higher temperatures; used for electroplating and metallurgy.

nickel formate [ORG CHEM] $Ni(HCOO)_2 \cdot 2H_2O$ Water-soluble green crystals; used in hydrogenation catalysts.

nickel glance See gersdorffite.

nickeline See niccolite.

nickel iodide [INORG CHEM] NiI_2 or $NiI_2 \cdot 6H_2O$ Hygroscopic black or blue-green solid; soluble in water and alcohol; sublimes when heated. Also known as nickelous iodide.

nickel-iron battery See Edison battery.

nickel-molybdenum iron [MET] An alloy containing 20–40% molybdenum and up to 60% nickel with some carbon added; has high acid resistance.

nickel-molybdenum steel [MET] Steel containing 0.2–0.3% molybdenum and 1.65–3.75% nickel as alloying elements.

nickel nitrate [INORG CHEM] $Ni(NO_3)_2 \cdot 6H_2O$ Fire-hazardous oxidant; deliquescent, green, water- and alcohol-soluble crystals; used for nickel plating and brown ceramic colors, and in nickel catalysts.

nickelous arsenate See nickel arsenate.

nickelous iodide See nickel iodide.

nickelous oxide See nickel oxide.

nickelous phosphate See nickel phosphate.

nickel oxide [INORG CHEM] NiO Green powder; soluble in acids and ammonium hydroxide; insoluble in water; used to make nickel salts and for porcelain paints. Also known as green nickel oxide; nickelous oxide.

nickel phosphate [INORG CHEM] $Ni_3(PO_4)_2 \cdot 7H_2O$ A light-green powder; soluble in acids and ammonium hydroxide, insoluble in water; used for electroplating and production of yellow nickel. Also known as nickelous phosphate; trinickelous orthophosphate.

nickel plating [MET] Electrolytic deposition of a metallic nickel coating.

nickel pyrites See millerite.

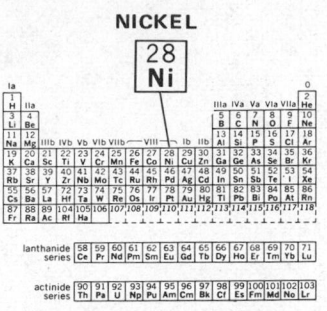

NICKEL

Periodic table of the chemical elements showing the position of nickel.

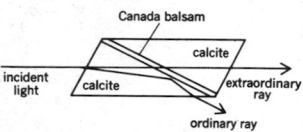

NICOL PRISM

Drawing of path of incident light through a Nicol prism. The extraordinary ray is plane-polarized.

NICOTINE

Structural formula of nicotine.

NICOTINIC ACID

Structural formula of nicotinic acid.

nickel rhodium [MET] Nickel alloy with 25–80% rhodium; can also contain other metals, such as platinum or molybdenum; used for pen points, reflectors, electrodes, and chemical equipment.

nickel silver [MET] A silver-white alloy composed of 52–80% copper, 10–35% zinc, and 5–35% nickel; sometimes also contains a few percent of lead and tin. Also known as German silver; nickel brass.

nickel steel [MET] Carbon steel containing up to 9% nickel as a major alloying element.

nickel tetracarbonyl See nickel carbonyl.

nickel-vanadium steel [MET] A nickel steel containing about 1.5% nickel, 1% manganese, 0.28% carbon, and 0.10% vanadium; used for high-strength cast parts.

nickel vitriol See morenosite.

nickpoint See knickpoint.

Nicoletiidae [INV ZOO] A family of the insect order Thysanura proper.

Nicol prism [OPTICS] A device for producing plane-polarized light, consisting of two pieces of transparent calcite (a birefringent crystal) which together form a parallelogram and are cemented together with Canada balsam.

Nicomachinae [INV ZOO] A subfamily of the limnivorous sedentary annelids in the family Maldanidae.

nicotinamide [BIOCHEM] $C_6H_6ON_2$ Crystalline basic amide of the vitamin B complex that is interconvertible with nicotinic acid in the living organism; the amide of nicotinic acid. Also known as niacinamide.

nicotinamide adenine dinucleotide See diphosphopyridine nucleotide.

nicotine [ORG CHEM] $C_{10}H_{14}N_2$ A liquid alkaloid that is soluble in water and forms water soluble salts; it is found in the tobacco leaf and is toxic. Also known as β-pyridyl-α-N-methylpyrrolidine.

nicotinic acid [BIOCHEM] $C_6H_5NO_2$ A component of the vitamin B complex; a white, water-soluble powder stable to heat, acid, and alkali; used for the treatment of pellagra. Also known as niacin.

nictitating membrane [VERT ZOO] A membrane of the inner angle of the eye or below the eyelid in many vertebrates, and capable of extending over the eyeball.

nidamental gland [ZOO] Any of various structures that secrete covering material for eggs or egg masses.

nidicolous [ZOO] **1.** Spending a short time in the nest after hatching. **2.** Sharing the nest of another species.

nidus [MED] A focus of infection. [ZOO] A nest or breeding place.

niello [MATER] Mixture of sulfides of copper, silver, and lead, with black metallic appearance; used in ornamental inlays engraved on metals such as silver.

Niemann-Pick disease [MED] A hereditary sphingolipidosis due to an enzyme deficiency resulting in abnormal accumulation of sphingomyelin; symptoms include anemia, enlargement of the liver, spleen, and lymph nodes, gastrointestinal disturbances, and various neurologic deficits. Also known as lipid hysticytosis; Pick's disease.

nieve penitente [GEOL] A jagged pinnacle or spike of snow or firn, up to several meters in height. Also known as penitent.

NIF See noise improvement factor.

night [ASTRON] The period of darkness between sunset and sunrise.

night ape See bushbaby.

night blindness [MED] Reduced dark adaptation resulting from vitamin A deficiency or from retinitis pigmentosa or other peripheral retinal disease. Also known as nyctalopia.

night effect See polarization error.

nightglow [GEOPHYS] A subdivision of airglow in which energy comes from reactions of atomic oxygen between 70 and 100 kilometers, and from ionic recombination around 300 kilometers.

night order book [NAV] A notebook in which the commanding officer of a ship writes orders with respect to courses and speeds, any special precautions concerning the speed and navigation of the ship, and all other orders for the night for the officer of the deck.

night-sky light See airglow.

night-sky luminescence See airglow.

night sweat [MED] Drenching perspiration occurring at night or during sleep in the course of certain febrile diseases.

nighttime visual range See night visual range.

night visual range [OPTICS] The greatest distance at which a point source of light of a given candlepower can be perceived at night by an observer under given atmospheric conditions. Also known as nighttime visual range; penetration range; transmission range.

night wind [METEOROL] Dry squalls which occur at night in southwest Africa and the Congo; the term is loosely applied to other diurnal local winds such as mountain wind, land breeze, and midnight wind.

nigre [CHEM ENG] Dark-colored layer formed between neat soap and lye during soap manufacture; contains more soap than lye, and a high concentration of salts and colored impurities.

nigrescent [BIOL] Blackish.

nigrite [MATER] A mixture of rubber with ozocerite distillation residue; used as a substitute for gutta-percha.

nigrosine [MATER] Any of a group of blue or black azine dyes used for coloring inks, shoe polish, leather, and wood; can be water-, alcohol-, or oil-soluble.

nihilism [MED] Pessimism in regard to the efficacy of treatment, particularly the use of drugs. [PSYCH] The content of delusions encountered in depressed or melancholic states; the patient insists that his inner organs no longer exist, and that his relatives have passed away.

Nike [ORD] A system of U.S. Army surface-to-air guided missiles designed to seek out, intercept, and destroy enemy aircraft; several weapons of the system are Nike-Ajax, Nike-Hercules, and Nike-Zeus.

Nike-Ajax [ORD] A U.S. Army surface-to-air guided missile that has a range of approximately 25 miles (40 kilometers) and a speed of approximately Mach 2; it is liquid-fueled and rocket-boosted, and directed by command guidance.

Nike-Hercules [ORD] A U.S. Army surface-to-air guided missile that has a range of approximately 70 miles (110 kilometers), a speed of over Mach 3, and nuclear warhead capability; it employs solid fuel, and is directed by command radar guidance.

Nike-Zeus [ORD] A U.S. Army surface-to-air guided missile that is an antimissile weapon with nuclear warhead capability, for attacking intercontinental ballistic missiles; uses solid fuel and radar guidance.

nikkel oil [MATER] A bright-yellow liquid with a lemon and cinnamon odor obtained from the leaves and twigs of a laurel tree, *Cinnamomum zeylanicum;* contains citral and cineol; used in perfumery.

Nilionidae [INV ZOO] The false ladybird beetles, a family of coleopteran insects in the superfamily Tenebrionoidea.

nilmanifold [MATH] The factor space of a connected nilpotent Lie group by a closed subgroup.

nilpotent [MATH] An element of some algebraic system which vanishes when raised to a certain power.

nimbostratus [METEOROL] A principal cloud type, or cloud genus, gray-colored and often dark, rendered diffuse by more or less continuously falling rain, snow, or sleet of the ordinary varieties, and not accompanied by lightning, thunder, or hail; in most cases the precipitation reaches the ground.

nimbus [METEOROL] A characteristic rain cloud; the term is not used in the international cloud classification except as a combining term, as cumulonimbus.

nim oil See neem oil.

N indicator See N scope.

nine-j-symbol [QUANT MECH] A coefficient used in the general recoupling of four angular momenta, as in a transformation from L-S to j-j coupling in a two-electron system. Also known as X coefficient.

nine-light indicator [ENG] A remote indicator for wind speed and direction used in conjunction with a contact anemometer and a wind vane; the indicator consists of a center light, connected to the contact anemometer, surrounded by eight equally spaced lights which are individually connected to a set of similarly spaced electrical contacts on the wind vane; wind speed is determined by counting the number of flashes of the center light during an interval of time; direction, indicated by the position of illuminated outer bulbs, is given to points of the compass.

nine's complement [ADP] The radix-minus-1 complement of a numeral whose radix is 10.

ninety-column card [ADP] A card, punched or to be punched, divided in half horizontally, each half containing 45 columns, and each column containing six punch positions.

niningerite [MINERAL] $(Mg,Fe,Mn)S$ A mineral found only in meteorites.

niobe oil *See* methyl benzoate.

niobic acid [INORG CHEM] $Nb_2O_5 \cdot nH_2O$ Family of hydrates; white precipitate, soluble in inorganic acids and bases, insoluble in water; its formation is part of the analytical determination of niobium.

niobium [CHEM] A chemical element, symbol Nb, atomic number 41, atomic weight 92.906. [MET] A platinum-gray, ductile metal with brilliant luster; used in alloys, especially stainless steels. Also known as columbium.

nip [GEOL] **1.** A small, low cliff or break in slope produced by wavelets at the high-water mark. **2.** The point on the bank of a meander lake where erosion takes place due to crowding of the stream current toward the lake. **3.** Thinning of a coal seam, particularly if caused by tectonic movements. Also known as want. [MET] *See* angle of nip. [MIN ENG] *See* squeeze.

Nipher shield [ENG] A conically shaped, copper, rain-gage shield; used to prevent the formation of vertical wind eddies in the vicinity of the mouth of the gage, thereby making the rainfall catch a representative one.

Nipkow disk [ADP] In optical character recognition, a disk having one or more spirals of holes around the outer edge, with successive openings positioned so that rotation of the disk provides mechanical scanning, as of a document.

nippers [DES ENG] Small pincers or pliers for cutting or gripping.

nipping [NAV] The forcible closing of ice around a vessel so that it is held fast by ice under pressure.

nipple [ANAT] The conical projection in the center of the mamma, containing the outlets of the milk ducts. [DES ENG] A short piece of tubing, usually with an internal or external thread at each end, used to couple pipes.

Nippotaeniidea [INV ZOO] An order of tapeworms of the subclass Cestoda including some internal parasites of certain fresh-water fishes; the head bears a single terminal sucker.

Nissen stamp [MIN ENG] **1.** Machine used in crushing rock to sand sizes. **2.** An individual stamp worked in its own circular mortar box.

Nissl bodies [CYTOL] Chromophil granules of nerve cells which ultrastructurally are composed of large ribosomes.

nit [COMMUN] A unit of information content such that the information content of a symbol or message in nits is the negative of the natural logarithm of the probability of selecting that symbol or message from all the symbols or messages which could have been chosen. Also known as nepit. [OPTICS] A unit of luminance, equal to 1 candela per square meter. Abbreviated nt.

Nitelleae [BOT] A tribe of stoneworts, order Charales, characterized by 10 cells in two tiers of five each composing the apical crown.

niter *See* potassium nitrate.

niter balls [MATER] A pellet form of potassium nitrate, used as a fertilizer. Also known as sal prunella; throat balls.

niter cake *See* sodium bisulfate.

Nitidulidae [INV ZOO] The sap-feeding beetles, a large family of coleopteran insects in the superfamily Cucujoidea; individuals have five-jointed tarsi and antennae with a terminal three-jointed clavate expansion.

meta-**nitraniline** *See meta*-nitraniline.

ortho-**nitraniline** *See ortho*-nitraniline.

para-**nitraniline** *See para*-nitraniline.

nitrate [CHEM] **1.** A salt or ester of nitric acid. **2.** Any compound containing the NO_3^- radical.

nitrate mineral [MINERAL] Any of several generally rare minerals characterized by a fundamental ionic structure of NO_3^-; examples are soder niter, niter, and nitrocalcite.

nitratine *See* soda niter.

nitrating acid [INORG CHEM] Sulfuric-nitric acid mix used to

nitrate cellulosics and aromatic chemicals. Also known as mixed acid.

nitration [ORG CHEM] Introduction of an NO_2^- group into an organic compound.

nitric acid [INORG CHEM] HNO_3 Strong oxidant that is fire-hazardous; colorless or yellowish liquid, miscible with water; boils at 86°C; used for chemical synthesis, explosives, and fertilizer manufacture, and in metallurgy, etching, engraving, and ore flotation. Also known as aqua fortis.

nitric oxide [INORG CHEM] NO A colorless gas that, at room temperature, reacts with oxygen to form nitrogen dioxide (NO_2, a reddish-brown gas); may be used to form other compounds.

nitride [INORG CHEM] Compound of nitrogen and a metal, such as Mg_3N_2.

nitride nuclear fuel [NUCLEO] Fissionable nuclear fuel in nitride form, such as UN and UN_2.

nitriding [MET] Surface hardening of steel by formation of nitrides; nitrogen is introduced into the steel usually by heating in gaseous ammonia.

nitrification [MICROBIO] Formation of nitrous and nitric acids or salts by oxidation of the nitrogen in ammonia; specifically, oxidation of ammonium salts to nitrites and oxidation of nitrites to nitrates by certain bacteria.

nitrifying bacteria [MICROBIO] Members of the family Nitrobacteraceae.

nitrile [ORG CHEM] $RC \equiv N$ Cyanide derived by removal of water from an acid amide.

nitrile-butadiene rubber *See* nitrile rubber.

nitrile rubber [MATER] A synthetic rubber formed by polymerization of acrylonitrile with butadiene; the structure of the polymer is $-CH_2CH=CHCH_2CH_2CH(CN)-$. Also known as acrylonitrile-butadiene rubber; acrylonitrile rubber; NBR; nitrile-butadiene rubber; NR.

nitrilotriacetic acid [ORG CHEM] $N(CH_2COOH)_3$ A white powder, melting point 240°C, with some decomposition; soluble in water; it is toxic, and birth abnormalities may result from ingestion; may be used as a chelating agent in the laboratory. Also known as NTA; TGA; triglycine; triglycolamic acid.

nitrite [CHEM] A compound containing the radical NO_2^-; can be organic or inorganic.

nitro- [CHEM] Chemical prefix showing the presence of the NO_2^- radical.

meta-**nitroaniline** [ORG CHEM] $NO_2C_6H_4NH_2$ Yellow crystals that melt at 112.5°C; a toxic material; used as a dye intermediate. Also known as *meta*-nitraniline.

ortho-**nitroaniline** [ORG CHEM] $NO_2C_6H_4NH_2$ Orange-red crystals that melt at 69.7°C, soluble in ethanol; a toxic material; used to manufacture dyes. Also known as *ortho*-nitraniline.

para-**nitroaniline** [ORG CHEM] $NO_2C_6H_4NH_2$ Yellow crystals that melt at 148°C; insoluble in water, soluble in ethanol; a toxic material; used to make dyes, and as a corrosion inhibitor. Also known as *para*-nitraniline.

nitroaromatic [ORG CHEM] A nitrated benzene or benzene derivative, such as nitrobenzene, $C_6H_5NO_2$, or nitrobenzoic acid, $NO_2 \cdot C_6H_4 \cdot COOH$.

Nitrobacter [MICROBIO] A genus of rod-shaped bacteria in the family Nitrobacteraceae; organisms oxidize nitrites to nitrates.

Nitrobacteraceae [MICROBIO] The nitrifying bacteria, a family of the order Pseudomonadales which live autotrophically, derive energy from nitrification, and obtain carbon for growth from carbon dioxide.

nitrobarite *See* barium nitrate.

nitrobenzene [ORG CHEM] $C_6H_5NO_2$ Greenish crystals or a yellowish liquid, melting point 5.70°C; a toxic material; used in aniline manufacture. Also known as oil of mirbane.

nitrocalcite *See* calcium nitrate.

nitrocellulose *See* cellulose nitrate.

nitrocellulose propellant [MATER] A single-base propellant whose main constituent is nitrocellulose, with only minor percentages of additives for stabilizing and other purposes.

nitrocotton [MATER] A chemical combination of ordinary cotton fiber with nitric acid; explosive and highly flammable.

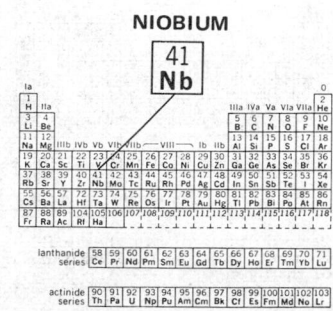

NIOBIUM

Periodic table of the chemical elements showing the position of niobium.

nitro dye [ORG CHEM] A dye with the NO_2 chromophore group in the molecules.

nitroethane [ORG CHEM] $CH_3CH_2NO_2$ A colorless liquid, slightly soluble in water; boils at 114°C; used as a solvent for cellulosics, resins, waxes, fats, and dyestuffs, and as a chemical intermediate.

nitro explosive [ORG CHEM] Explosive compound containing one or more NO_2^- groups, such as nitroglycerine, $C_3H_5(ONO_2)_3$, or trinitrotoluene, $C_6H_2(CH_3)(NO_2)_3$.

nitrogelatin See gelatin dynamite.

nitrogen [CHEM] A chemical element, symbol N, atomic number 7, atomic weight 14.0067; it is a gas, diatomic (N_2) under normal conditions; about 78% of the atmosphere is N_2; in the combined form the element is a constituent of all proteins.

nitrogen acid anhydride See nitrogen pentoxide.

nitrogenase [BIOCHEM] An enzyme that catalyzes a six-electron reduction of N_2 in the process of nitrogen fixation.

nitrogenated oil [MATER] A class of essential oils containing carbon, hydrogen, oxygen, and nitrogen; an example is oil of bitter almonds.

nitrogen balance [PHYSIO] The difference between nitrogen intake (as protein) and total nitrogen excretion for an individual.

nitrogen cycle See carbon-nitrogen cycle.

nitrogen dioxide [INORG CHEM] NO_2 A reddish-brown gas; it exists in varying degrees of concentration in equilibrium with other nitrogen oxides; used to produce nitric acid. Also known as dinitrogen tetroxide; liquid dioxide; nitrogen peroxide; nitrogen tetroxide.

nitrogen fixation [CHEM ENG] Conversion of atmospheric nitrogen into compounds such as ammonia, calcium cyanamide, or nitrogen oxides by chemical or electric-arc processes. [MICROBIO] Assimilation of atmospheric nitrogen by heterotrophic bacteria. Also known as dinitrogen fixation.

nitrogen monoxide See nitrous oxide.

nitrogen mustard [ORG CHEM] Any of the substituted mustard gases in which the sulfur is replaced by an amino nitrogen, such as for methyl bis(2-chloroethyl)amine, $(CH_2ClCH_2)_2NCH_3$; useful in cancer research.

nitrogen narcosis [MED] Narcosis caused by gaseous nitrogen at high pressure in the blood; produced in divers breathing air at depths of 100 feet (30 meters) or more. Also known as rapture of the deep.

nitrogenous fertilizer [MATER] Fertilizer materials, natural or synthesized, containing nitrogen available for fixation by vegetation, such as potassium nitrate, KNO_3, or ammonium nitrate, NH_4NO_3.

nitrogen pentoxide [INORG CHEM] N_2O_5 Colorless crystals, soluble in water (forms HNO_3); decomposes at 46°C. Also known as nitrogen acid anhydride.

nitrogen peroxide See nitrogen dioxide.

nitrogen solution [INORG CHEM] Mixture used to neutralize super-phosphate in fertilizer manufacture; consists of 60% ammonium nitrate, and the balance a 50% aqua ammonia solution.

nitrogen tetroxide See nitrogen dioxide.

nitrogen trioxide [INORG CHEM] N_2O_3 Green, water-soluble liquid; boils at 3.5°C.

nitroglycerin [ORG CHEM] $CH_2NO_3CHNO_3CH_2NO_3$ Highly unstable, explosive, flammable pale-yellow liquid; soluble in alcohol; freezes at 13°C and explodes at 260°C; used as an explosive, to make dynamite, and in medicine.

nitroglycerin powder [MATER] Any explosive characterized by a low nitroglycerin content, up to 10%, and a high ammonium nitrate content of 80–85%, with carbonaceous material forming the remainder.

nitroguanidine [ORG CHEM] $H_2NC(NH)NHNO_2$ Explosive yellow solid, soluble in alcohol; melts at 246°C; used in explosives and smokeless powder.

nitromagnesite [MINERAL] $Mg(NO_3)_2 \cdot 6H_2O$ Mineral consisting of magnesium nitrate, occurring as an efflorescence in limestone caverns.

nitromannite See mannitol hexanitrate.

nitromannitol See mannitol hexanitrate.

nitrometer [ANALY CHEM] Glass apparatus used to collect and measure nitrogen and other gases evolved by a chemical reaction. Also known as azotometer.

nitromethane [ORG CHEM] CH_3NO_2 A liquid nitroparaffin compound; oily and colorless; boils at 101°C; used as a monopropellant for rockets, in chemical synthesis, and as an industrial solvent for cellulosics, resins, waxes, fats, and dyestuffs.

nitron [ORG CHEM] $CN_4(C_6H_5)_3CH$ Yellow crystals, soluble in chloroform and acetone; used as reagent to detect NO_3 ion in dilute solutions. Also known as diphenylenedianilino-hydrotriazole.

nitronium [CHEM] Positively charged NO_2 ion, believed to be formed from HNO_3.

nitroparaffin [ORG CHEM] Any organic compound in which one or more hydrogens of a paraffinic hydrocarbon are replaced by a nitro, or NO_2^-, group, such as nitromethane, CH_3NO_2, or nitroethane, $C_2H_5NO_2$; used as a solvent and a chemical intermediate.

para-nitrophenylazosalicylate sodium See alizarin yellow.

nitroso [CHEM] The radical NO^- with trivalent nitrogen. Also known as hydroximino; oximido.

Nitrosomonas [MICROBIO] A genus of ellipsoidal bacteria occurring singly, in pairs, short chains, or irregular masses; included in the family Nitrobacteraceae.

nitrosophenylhydroxylamine See cupferron.

nitrostarch [ORG CHEM] $C_{12}H_{12}(NO_2)_8O_{10}$ Orange powder, soluble in ethyl alcohol; used in explosives. Also known as starch nitrate.

meta-nitrotoluene [ORG CHEM] $NO_2C_6H_4CH_3$ Yellow powder that melts at 15°C; insoluble in water; used in organic synthesis. Also known as meta-methylnitrobenzene.

ortho-nitrotoluene [ORG CHEM] $NO_2C_6H_4CH_3$ A yellow liquid boiling at 220.4°C; insoluble in water; used to produce toluidine and dyes. Also known as ortho-methylnitrobenzene.

para-nitrotoluene [ORG CHEM] $NO_2C_6H_4CH_3$ Yellow crystals that melt at 51.7°C; insoluble in water, soluble in ethanol; used to produce toluidine and to manufacture dyes. Also known as para-methylnitrobenzene.

nitrourea [ORG CHEM] $NH_2CONHNO_2$ Highly explosive white crystals, melting at 159°C; soluble in ether and alcohol, slightly soluble in water; used as a chemical intermediate.

nitrous acid [INORG CHEM] HNO_2 Aqueous solution of nitrogen trioxide, N_2O_3.

nitrous oxide [INORG CHEM] N_2O. Colorless, sweet-tasting gas, boiling at -90°C; slightly soluble in water, soluble in alcohol; used as a food aerosol, and as an anesthetic in dentistry and surgery. Also known as laughing gas; nitrogen monoxide.

nitroxanthic acid See picric acid.

nitryl halide [INORG CHEM] NO_2X Compound containing a halide (X) and a nitro group (NO_2).

nival [ECOL] **1.** Characterized by or living in or under the snow. **2.** Of or pertaining to a snowy environment.

nival surface [GEOL] The hypothetical planar surface containing all of the different snowlines of the same geologic time period.

nivation [GEOL] Rock or soil erosion beneath a snowbank or snow patch, due mainly to frost action but also involving chemical weathering, solifluction, and meltwater transport of weathering products. Also known as snow patch erosion.

nivation cirque See nivation hollow.

nivation glacier [HYD] A small, newly formed glacier; represents the initial stage of glaciation. Also known as snowbank glacier.

nivation hollow [GEOL] A small, shallow depression formed, and occupied during part of the year, by a snow patch or snowbank that, through nivation, is thought to initiate glaciation. Also known as nivation cirque; snow niche.

nivation ridge See winter-talus ridge.

nivenite [MINERAL] UO_2 A velvet-black member of the uranite group; contains rare-earth metals cerium and yttrium; a source of uranium.

niveous See niveus.

niveus [BIOL] Snow-white in color. Also spelled niveous.

NIXIE display indicator See NIXIE Tube.

NIXIE Tube [ELECTR] Trademark of Burroughs Corporation

NITROGEN

Periodic table of the chemical elements showing the position of nitrogen.

for a cold-cathode gas readout tube having a common anode and 10 different metallic cathodes, each formed in the shape of a different numeral, alphabetic character, or special symbol; the desired character is surrounded with a brilliant glow when the corresponding cathode is energized. Also known as NIXIE display indicator.

N-level address [ADP] A multilevel address specifying N levels of addressing.

N-level logic [ELECTR] An arrangement of gates in a digital computer in which not more than N gates are connected in series.

NLGI number [ENG] One of a series of numbers developed by the National Lubricating Grease Institute and used to classify the consistency range of lubricating greases; NLGI numbers are based on the ASTM cone penetration number.

N line [SPECT] One of the characteristic lines in an atom's x-ray spectrum, produced by excitation of an N electron.

N-m *See* newton-meter of torque.

NMR *See* nuclear magnetic resonance.

nn junction [ELECTR] In a semiconductor, a region of transition between two regions having different properties in *n*-type semiconducting material.

n-n′ reactions *See* neutron inelastic scattering reactions.

No *See* nobelium.

no-address instruction [ADP] An instruction which a computer can carry out without using an operand from storage.

no-atmospheric control [AERO ENG] Any device or system designed or set up to control a guided rocket missile, rocket craft, or the like outside the atmosphere or in regions where the atmosphere is of such tenuity that it will not affect aerodynamic controls.

nobelium [CHEM] A chemical element, symbol No, atomic number 102, atomic weight 254 when the element is produced in the laboratory; a synthetic element, in the actinium series.

noble gas [CHEM] A gas in group 0 of the periodic table of the elements; it is monatomic and, with limited exceptions, chemically inert. Also known as inert gas.

noble metal [MET] A metal, or alloy, such as gold, silver, or platinum having high resistance to corrosion and oxidation; used in the construction of thin-film circuits, metal-film resistors, and other metal-film devices.

noble potential [PHYS CHEM] A potential equaling or approaching that of the noble elements, such as gold, silver, or copper, of the electromotive series.

no-bottom sounding [ENG] A sounding in the ocean in which the bottom is not reached.

no-break power [ELEC] Power system designed to fulfill load requirements during the interval between the failure of the primary power and the time the auxiliary power can be made available.

Nocardia [MICROBIO] A genus of slender, filamentous or rod-shaped, aerobic, gram-positive bacteria in the family Actinomycetaceae.

nocardiosis [MED] Infection by species of the fungus *Nocardia* characterized by spreading granulomatous lesions.

nocioceptive reflex *See* flexion reflex.

noct-, nocti-, nocto-, noctu- [SCI TECH] Combining form meaning night.

noctalbuminuria [MED] Excretion of protein in night urine only.

Noctilionidae [VERT ZOO] The fish-eating bats, a tropical American monogeneric family of the Chiroptera having small eyes and long, narrow wings.

noctilucent cloud [METEOROL] A cloud of unknown composition which occurs at great heights and high altitudes; photometric measurements have located such clouds between 75 and 90 kilometers; they resemble thin cirrus, but usually with a bluish or silverish color, although sometimes orange to red, standing out against a dark night sky.

noctiphobia [PSYCH] Abnormal fear of night or darkness.

Noctuidae [INV ZOO] A large family of dull-colored, medium-sized moths in the superfamily Noctuoidea; larva are mostly exposed foliage feeders, representing an important group of agricultural pests.

Noctuoidea [INV ZOO] A large superfamily of lepidopteran insects in the suborder Heteroneura; most are moderately large moths with reduced maxillary palpi.

nocturnal emission [PHYSIO] Normal, involuntary seminal discharge occurring during sleep in males after puberty.

nocturnal enuresis [MED] Involuntary nocturnal urination during sleep.

nocturnal radiation *See* effective terrestrial radiation.

no-cut rounds [MIN ENG] Set of holes drilled straight into the face for blasting underground.

nodal line [PHYS] 1. A line or curve in a two-dimensional standing-wave system, such as a vibrating diaphragm, where some specified characteristic of the wave, such as velocity of pressure, does not oscillate. 2. A line which remains fixed during some deformation or rotation of a body or coordinate system.

nodal points [ELEC] Junction points in a transmission system; the automatic switches and switching centers are the nodal points in automated systems. [OPTICS] A pair of points on the axis of an optical system such that an incident ray passing through one of them results in a parallel emergent ray passing through the other.

nodal rhythm [PHYSIO] A cardiac rhythm characterized by pacemaker function originating in the atrioventricular node, with a heart rate of 40-70 per minute.

nodal tachycardia [MED] A cardiac arrhythmia characterized by a heart rate of 140-220 per minute.

nodal tissue [HISTOL] 1. Tissue from the sinoatrial node, and the atrioventricular node and bundle and its branches, composed of a dense network of Purkinje fibers. 2. Tissue from a lymph node.

node [ANAT] 1. A knob or protuberance. 2. A small, rounded mass of tissue, such as a lymph node. 3. A point of constriction along a nerve. [ELEC] *See* branch point. [ELECTR] A junction point in a network. [MATH] *See* crunode. [PHYS] A point, line, or surface in a standing-wave system where some characteristic of the wave has essentially zero amplitude.

node of Ranvier [ANAT] The region of a local constriction in a myelinated nerve; formed at the junction of two Schwann cells.

nodical month [ASTRON] The average period of revolution of the moon about the earth with respect to the moon's ascending node, a period of 27 days 5 hours 5 minutes 35.8 seconds, or approximately $27\frac{1}{4}$ days. Also known as draconitic month.

Nodosariacea [INV ZOO] A superfamily of Foraminiferida in the suborder Rotaliina characterized by a radial calcite test wall with monolamellar septa, and a test that is coiled, uncoiled, or spiral about the long axis.

nodose [BIOL] Having many or noticeable protuberances; knobby.

nodular [SCI TECH] Occurring in the form of small, rounded lumps.

nodular cast iron [MET] Cast iron treated in the molten state with a master alloy containing an element such as magnesium which favors formation of spheroidal graphite. Also known as ductile iron; spheroidal graphite cast iron.

nodular chert [GEOL] Chert occurring as nodular or concretionary segregations (chert nodules).

nodular goiter *See* adenomatous goiter.

nodular powder [MET] Irregularly shaped metal powder particles.

nodule [ANAT] 1. A small node. 2. A small aggregation of cells. [GEOL] A small, hard mass or lump of a mineral or mineral aggregate characterized by a contrasting composition from and a greater hardness than the surrounding sediment or rock matrix in which it is embedded. [MED] A primary skin lesion, seen as a circumscribed solid elevation.

nodules of the semilunar valves [ANAT] Small nodes in the midregion of the pulmonary and aortic semilunar valves.

nodulizing [ENG] Creation of spherical lumps from powders by working them together, coalescing them with binders, drying fluid-solid mixtures, heating, or chemical reaction.

nodulose [BIOL] Having minute nodules or fine knobs.

Noeggerathiales [PALEOBOT] A poorly defined group of fossil plants whose geologic range extends from Upper Carboniferous to Triassic.

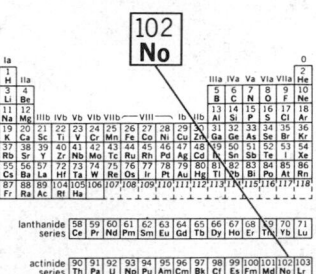

NOBELIUM

Periodic table of the chemical elements showing the position of nobelium.

Noetherian ring [MATH] A ring in which every nested family of ideals possesses only a finite number of members.

no-fines concrete [MATER] Concrete made without sand and therefore containing a high proportion of communicating pores which provide thermal insulation and drainage.

No. 6 fuel *See* bunker C fuel oil.

nog [MIN ENG] **1.** Roof support for stopes, formed of rectangular piles of logs squared at the ends and filled with waste rock. **2.** A wood block wedged tightly into the cut in a coal seam after the coal cutter has passed; it forms a temporary support.

nogalamycin [MICROBIO] An antineoplastic antibiotic produced by *Streptomyces nogalaster.*

no-go gage [ENG] A limit gage designed not to fit a part being tested; usually employed with a go gage to set the acceptable maximum and minimum dimension limits of the part.

noil [TEXT] Short-staple, wool fiber combings used in the manufacture of worsted fabrics.

noise [ACOUS] Sound which is unwanted, either because of its effect on humans, its effect on fatigue or malfunction of physical equipment, or its interference with the perception or detection of other sounds. [ELEC] Interfering and unwanted currents or voltages in an electrical device or system.

noise analysis [PHYS] Determination of the frequency components that make up a particular noise being studied.

noise analyzer [ELECTR] A device used for noise analysis.

noise-canceling microphone *See* close-talking microphone.

noise control [ACOUS] The process of obtaining an acceptable noise environment for a particular observation point or receiver, involving control of the noise source, transmission path, or receiver, or all three.

noise factor [ELECTR] The ratio of the total noise power per unit bandwidth at the output of a system to the portion of the noise power that is due to the input termination, at the standard noise temperature of 290K. Also known as noise figure.

noise figure *See* noise factor.

noise filter [ELECTR] **1.** A filter that is inserted in an alternating-current power line to block noise interference that would otherwise travel through the line in either direction and affect the operation of receivers. **2.** A filter used in a radio receiver to reduce noise, usually an auxiliary low-pass filter which can be switched in or out of the audio system.

noise generator [ELECTR] A device which produces (usually random) electrical noise, for use in tests of the response of electrical systems to noise, and in measurements of noise intensity. Also known as noise source.

noise grade [COMMUN] Number which defines the relative noise at a particular location with respect to other locations throughout the world.

noise improvement factor [COMMUN] In pulse modulation, the receiver output signal-to-noise ratio divided by the receiver input signal-to-noise ratio. Abbreviated NIF. Also known as improvement factor; signal-to-noise improvement factor.

noise jammer [ELECTR] **1.** An electronic jammer that emits a carrier modulated with recordings or synthetic reproductions of natural atmospheric noise; the radio-frequency carrier may be suppressed; used to discourage the enemy by simulating naturally adverse communications conditions. **2.** During World War II, a powerful transmitter modulated with white noise tuned to the approximate frequency of an enemy transmitter and used to obscure intelligible output at the receiver.

noise killer [ELECTR] **1.** Device installed in a circuit to reduce its interference to other circuits. **2.** *See* noise suicide circuit.

noiseless channel [COMMUN] In information theory, a communications channel in which the effects of random influences are negligible, and there is essentially no random error.

noise level [PHYS] The intensity of unwanted sound, or the magnitude of unwanted currents or voltages, averaged over a specified frequency range and time interval, and weighted with frequency in a specified manner; usually expressed in decibels relative to a specified reference.

noise limiter [ELECTR] A limiter circuit that cuts off all noise peaks that are stronger than the highest peak in the desired signal being received, thereby reducing the effects of atmospheric or man-made interference. Also known as noise silencer; noise suppressor.

noise measurement [ACOUS] The process of quantitatively determining one or more properties of acoustic noise. [ELECTR] Any of a wide range of measurements of random and nonrandom electrical noise, but usually noise-power measurement.

noise-metallic [ELECTR] In telephone communications, weighted noise current in a metallic circuit at a given point when the circuit is terminated at that point in the nominal characteristic impedance of the circuit.

noise-modulated jamming [ELECTR] Random electronic noise that appears at the radar receiver as background noise and tends to mask the desired radar echo or radio signal.

noise pollution [ACOUS] Excessive noise in the human environment.

noise-power measurement [ELECTR] Measurement of the power carried by electrical noise averaged over some brief interval of time, usually by amplifying noise from the source in a linear amplifier and then using a quadratic detector followed by a low-pass filter and an indicating device.

noise rating number [ACOUS] The perceived noise level of the noise that can be tolerated under specified conditions; for example, the noise rating number of a bedroom is 25, that of a workshop is 65.

noise-reducing antenna system [ELECTROMAG] Receiving antenna system so designed that only the antenna proper can pick up signals; it is placed high enough to be out of the noise-interference zone, and is connected to the receiver with a shielded cable or twisted transmission line that is incapable of picking up signals.

noise reduction [ENG ACOUS] A process whereby the average transmission of the sound track of a motion picture print, averaged across the track, is decreased for signals of low level; since background noise introduced by the sound track is less at low transmission, this process reduces noise during soft passages.

noise reduction coefficient [ACOUS] The average over the logarithm of frequency, in the frequency range from 256 to 2048 hertz inclusive, of the sound absorption coefficient of a material.

noise silencer *See* noise limiter.

noise source *See* noise generator.

noise suicide circuit [ELECTR] A circuit which reduces the gain of an amplifier for a short period whenever a sufficiently large noise pulse is received. Also known as noise killer.

noise suppression [ELECTR] Any method of reducing or eliminating the effects of undesirable electrical disturbances, as in frequency modulation whenever the signal carrier level is greater than the noise level.

noise suppressor [ELECTR] **1.** A circuit that blocks the audio-frequency amplifier of a radio receiver automatically when no carrier is being received, to eliminate background noise. Also known as squelch circuit. **2.** A circuit that reduces record surface noise when playing phonograph records, generally by means of a filter that blocks out the higher frequencies where such noise predominates. **3.** *See* noise limiter.

noise temperature [ELEC] The temperature at which the thermal noise power of a passive system per unit bandwidth would be equal to the actual noise at the actual terminals; the standard reference temperature for noise measurements is 290K.

noise testing [ELECTR] The measurement of the power dissipated in a resistance termination of given value joined to one end of a telephone or telegraph circuit when no test power is applied to the circuit.

noise tube [ELECTR] A gas tube used as a source of white noise.

noise weighting [ELECTR] Use of an electrical network to obtain a weighted average over frequency of the noise power, which is representative of the relative disturbing effects of noise in a communications system at various frequencies.

noisy channel [COMMUN] In information theory, a communications channel in which the effects of random influences cannot be dismissed.

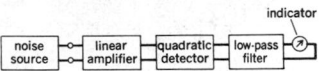

noisy mode [ADP] A floating-point arithmetic procedure associated with normalization in which "1" bits, rather than "0" bits, are introduced in the low-order bit position during the left shift.

no-load current [ELEC] The current which flows in a network when the output is open-circuited.

no-load loss [ELEC] The power loss of a device that is operated at rated voltage and frequency but is not supplying power to a load.

noma [MED] Spreading gangrene beginning in a mucous membrane; considered to be a malignant form of infection by fusospirochetal organisms. Also known as gangrenous stomatitis.

Nomarski microscope [OPTICS] A type of interference microscope that is used to study reflecting specimens, such as metallic surfaces or metallized replicas of surfaces, and that gives a true relief image uncomplicated by variations in refractive index.

nomenclature [SCI TECH] A systematic arrangement of the distinctive names employed in any science.

nomenclature plate [ORD] Plate, usually made of metal, which is conspicuously mounted on military equipment, giving model letters, symbols, and numbers, together with other pertinent information.

nomen dubium [SYST] A proposed taxonomic name invalid because it is not accompanied by a definition or description of the taxon to which it applies.

nomen nudum [SYST] A proposed taxonomic name invalid because the accompanying definition or description of the taxon cannot be interpreted satisfactorily.

nominal band [COMMUN] Frequency band of a facsimile-signal wave equal in width to that between zero frequency and maximum modulating frequency; the frequency band occupied in the transmitting medium will, in general, be greater than the nominal band.

nominal bandwidth [COMMUN] The interval between the assigned frequency limits of a channel.

nominal decline rate [PETRO ENG] The negative slope of the curve representing the hydrocarbon production rate versus time for an oil gas reservoir.

nominal diameter [GEOL] The diameter computed for a hypothetical sphere which would have the same volume as the calculated volume for a specific sedimentary particle. Also known as equivalent diameter.

nominal impedance [ELEC] Impedance of a circuit under conditions at which it was designed to operate; normally specified at center of operating frequency range.

nominal line width [COMMUN] 1. In television, the reciprocal of the number of lines per unit length in the direction of line progression. 2. In facsimile transmission, the average separation between centers of adjacent scanning or recording lines.

nominal size [DES ENG] Size used for purposes of general identification; the actual size of a part will be approximately the same as the nominal size but need not be exactly the same; for example, a rod may be referred to as ¼ inch, although the actual dimension on the drawing is 0.2495 inch, and in this case ¼ inch is the nominal size.

nominal stress [MET] The stress calculated by simple elasticity theory, ignoring stress raisers and plastic flow; in tensile testing of a notched specimen, the load applied at the notch divided by the initial cross-sectional area at the notch.

nominal value [ELEC] The value of some property (such as resistance, capacitance, or impedance) of a device at which it is supposed to operate, under normal conditions, as opposed to actual value.

nomogram See nomograph.

nomograph [MATH] A chart which represents an equation containing three variables by means of three scales so that a straight line cuts the three scales in values of the three variables satisfying the equation. Also known as abac; alignment chart; nomogram.

nonacosane [ORG CHEM] $C_{29}H_{60}$ Colorless hydrocarbon, melting at 63°C; found in beeswax and the fat of cabbage leaves.

nonadecane [ORG CHEM] $CH_3(CH_2)_{17}CH_3$ Flammable crystals, soluble in ether and alcohol, insoluble in water; melts at 32°C; used as a chemical intermediate.

nonagon [MATH] A nine-sided polygon.

nonambiguity [COMMUN] The property of a code in which any character can be recognized uniquely without reference to preceding characters or the spatial position of a character.

nonane [ORG CHEM] $CH_3(CH_2)_7CH_3$ Flammable, colorless liquid, boiling at 151°C; soluble in alcohol, insoluble in water; used as a chemical intermediate. Also known as nonyl hydride.

nonanedioic acid See azelaic acid.

n-nonanoic acid See pelargonic acid.

nonanol See n-nonyl alcohol.

nonanticipatory system See causal system.

nonaqueous [CHEM] Pertaining to a liquid or solution containing no water.

nonarithmetic shift See cyclic shift.

nonasphaltic road oil [MATER] Any of the nonhardening petroleum distillates or residual oils used to lay road dust; they have viscosities low enough to be applied without prior heating.

nonassociated-gas reservoir [PETRO ENG] Formation in which gaseous hydrocarbons exist as a free phase in a reservoir that is not commercially productive of crude oil.

nonatomic Boolean algebra [MATH] A Boolean algebra in which there is no element x with the property that if $y \cdot x = y$ for some y, then $y = 0$.

nonatomic measure space [MATH] A measure space in which no point has positive measure.

nonbanded coal [GEOL] Coal without lustrous bands, composed mainly of clarain or durain without nitrain.

nonbearing wall [CIV ENG] A wall that bears no vertical weight other than its own.

nonblackbody [THERMO] A body that reflects some fraction of the radiation incident upon it; all real bodies are of this nature.

nonblocking access [COMMUN] Connection of the incoming line or trunk made within the switching center at all times, provided that the required outgoing line or trunk is not busy.

non-bore-safe [ORD] Pertaining to a fuse or booster that does not include a safety device to prevent the explosion of the main charge of a projectile prematurely, while it is still in the bore of the gun.

noncaking coal [GEOL] Hard or dull coal that does not cake when heated. Also known as free-burning coal.

Noncalcic Brown soil [GEOL] A great soil group having a slightly acidic, light-pink or reddish-brown A horizon and a light-brown or dull-red B horizon, and developed under a mixture of grass and forest vegetation in a subhumid climate. Also known as Shantung soil.

noncentral distribution [STAT] A distribution of random variables which is not normal.

noncentral force [PHYS] A force between two particles, other than an attraction or repulsion, directed along the line connecting them; for example, a tensor force between two nucleons.

noncentral quadric [MATH] A quadric surface that does not have a point about which the surface is symmetrical; namely, an elliptic or hyperbolic paraboloid, or a quadric cylinder.

nonclastic [GEOL] Of the texture of a sediment or sedimentary rock, formed chemically or organically and showing no evidence of a derivation from preexisting rock or mechanical deposition. Also known as nonmechanical.

noncohesive See cohesionless.

noncombat vehicle [ORD] An unofficial term used to distinguish a vehicle not classed as a combat vehicle; that is, the vehicle lacks means for engaging in combat.

noncommunicating hydrocephaly See obstructive hydrocephaly.

noncompulsory reporting points [NAV] In air operations, those reports established by air-traffic control without the requirement for rule-making action.

noncondensable gas [MATER] A gas from chemical or petroleum processing units (such as distillation columns or steam ejectors) that is not easily condensed by cooling; consists mostly of nitrogen, light hydrocarbons, carbon dioxide, or other gaseous materials.

nonconformity [GEOL] A type of unconformity in which rocks below the surface of unconformity are either igneous or metamorphic.

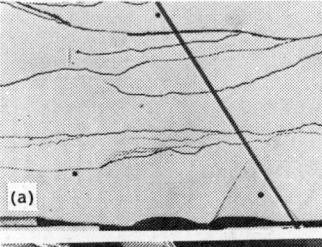
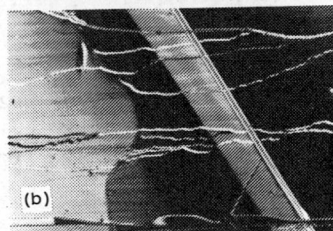
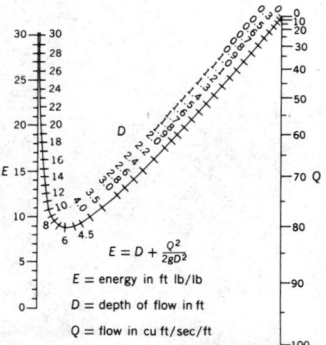
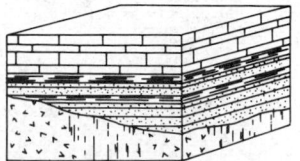

nonconsumable electrode [MET] An electrode, such as of carbon or tungsten, that is not consumed during a welding or melting operation.

noncontacting piston *See* choke piston.

noncontacting plunger *See* choke piston.

noncontributing area [HYD] An area with closed drainage.

noncoring bit [ENG] A general type of bit made in many shapes which does not produce a core and with which all the rock cut in a borehole is ejected as sludge; used mostly for blasthole drilling and in the unmineralized zones in a borehole where a core sample is not wanted. Also known as borehole bit; plug bit.

noncorrosive flux [MET] A soldering flux composed of rosin or of rosin in a volatile solvent; the residue is nonhygroscopic, noncorrosive, and nonconducting; suitable for soldering electronic components. Also known as activated rosin flux.

noncrossing rule [PHYS CHEM] The rule that when the potential energies of two electronic states of a diatomic molecule are plotted as a function of distance between the nuclei, the resulting curves do not cross, unless the states have different symmetry.

nondeforming steel [MET] A group of alloy steels which do not easily deform when heat-treated. Also known as nonshrinking steel.

nondegenerate amplifier [ELECTR] Parametric amplifier that is characterized by a pumping frequency considerably higher than twice the signal frequency; the output is taken at the signal input frequency; the amplifier exhibits negative impedance characteristics, indicative of infinite gain, and is therefore capable of oscillation.

nondegenerative basic feasible solution [ADP] In linear programming, a basic feasible solution with exactly m positive variables x_i, where m is the number of constraint equations.

nondelay fuse [ORD] Fuse that functions as a result of inertia of the firing pin (or primer) as the missile is retarded during penetration of the target; the inertia causes the firing pin to strike the primer (or the primer, the firing pin), initiating fuse action.

nondepositional unconformity *See* paraconformity.

nondestructive read [ADP] A reading process that does not erase the data in memory; the term sometimes includes a destructive read immediately followed by a restorative writeback. Also known as nondestructive readout (NDRO).

nondestructive readout *See* nondestructive read.

nondestructive testing [ENG] Any testing method which does not involve damaging or destroying the test sample; includes use of x-rays, ultrasonics, radiography, magnetic flux, and so on.

nondeterministic [SCI TECH] Unpredictable in terms of observable antecedents and known laws; this is a relative term pertaining to a given state of knowledge but not necessarily implying ultimate unpredictability.

nondeviated absorption [PHYS] Absorption that occurs without any appreciable slowing up of waves.

nondimensional parameter *See* dimensionless number.

nondirectional *See* omnidirectional.

nondirectional antenna *See* omnidirectional antenna.

nondirectional beacon [NAV] A beacon that provides navigational guidance over a 360° azimuth.

nondirective therapy [PSYCH] Type of psychotherapy in which the patient is in the dominant position and is given complete freedom to express himself.

nondisjunction mosaic [GEN] A population of cells with different chromosome numbers produced when one chromosome is lost during mitosis or when both members of a pair of chromosomes are included in the same daughter nucleus; can occur during embryogenesis or adulthood.

nondissipative muffler *See* reactive muffler.

nondissipative stub [ELECTROMAG] Nondissipative length of waveguide or transmission line.

nondivergent flow [OCEANOGR] Fluid flow in which the divergence of the ocean current field is zero.

nonene *See* 1-nonylene.

nonequilibrium thermodynamics [THERMO] A quantitative treatment of irreversible processes and of rates at which they occur. Also known as irreversible thermodynamics.

nonerasable storage [ADP] A device that permits a nonde-structive read, such as punched cards, electrically conductive sheets, or paper tape.

noneuclidean geometry [MATH] A geometry in which one or more of the axioms of euclidean geometry are modified or discarded.

nonexpendable [ENG] Pertaining to a supply item or piece of equipment that is not consumed, and does not lose its identity, in use, as a weapon, vehicle, machine, tool, piece of furniture, or instrument.

nonferrous metal [MET] Any metal other than iron and its alloys.

nonferrous metallurgy [MET] A branch of metallurgy that deals with metals other than iron and iron-base alloys.

nonfragmenting projectile [ORD] Projectile for antiaircraft gun practice, containing a smoke-producing substance, available in various colors, which makes it possible to observe the burst without a close burst destroying the target.

nonfreezing explosive [MATER] An explosive to which 15–20% of nitroethylene glycol has been added.

nonfrontal squall line *See* prefrontal squall line.

nonfunctional packages software [ADP] General-purpose software which permits the user to handle his particular applications requirements with little or no additional program or systems design work, or to perform certain specialized computational functions.

nongonococcal urethritis [MED] Human urethral inflammation not associated with common bacterial pathogens; thought to be caused by bacteria of the Bedsonia group. Abbreviated NGU.

nongranular leukocyte [HISTOL] A white blood cell, such as a lymphocyte or monocyte, with clear homogeneous cytoplasm.

nonholonomic constraints [MATH] A nonintegrable set of differential equations which describe the restrictions on the motion of a system or in optimization.

nonholonomic system [MECH] A system of particles which is subjected to restraints of such a nature that the system cannot be described by independent coordinates; examples are a rolling hoop, or an ice skate which must point along its path.

nonhoming [CONT SYS] Not returning to the starting or home position, as when the wipers of a stepping relay remain at the last-used set of contacts instead of returning to their home position.

nonhoming tuning system [ELECTR] Motor-driven automatic tuning system in which the motor starts up in the direction of previous rotation; if this direction is incorrect for the new station, the motor reverses, after turning to the end of the dial, then proceeds to the desired station.

nonhypergolic [CHEM] Not capable of igniting spontaneously upon contact; used especially with reference to rocket fuels.

***n*-nonic acid** *See* pelargonic acid.

nonideal gas [STAT MECH] A gas whose molecules have significant interaction, more than that needed to bring about the equilibrium.

nonideal solution [PHYS CHEM] A solution whose behavior does not conform to that of an ideal solution; that is, the behavior is not predictable over a wide range of concentrations and temperatures by the use of Raoult's law.

nonimpinging injector [AERO ENG] An injector used in rocket engines which employs parallel streams of propellant usually emerging normal to the face of the injector.

noninductive [ELEC] Having negligible or zero inductance.

noninductive capacitor [ELEC] A capacitor constructed so it has practically no inductance; foil layers are staggered during winding, so an entire layer of foil projects at either end for contact-making purposes; all currents then flow laterally rather than spirally around the capacitor.

noninductive resistor [ELEC] A wire-wound resistor constructed to have practically no inductance, either by using a hairpin winding or be reversing connections to adjacent sections of the winding.

noninductive winding [ELEC] A winding constructed so that the magnetic field of one turn or section cancels the field of the next adjacent turn or section.

nonine *See* nonyne.

nonintelligible crosstalk [COMMUN] Crosstalk which cannot be understood regardless of its received volume, but which because of its syllabic nature is more annoying subjectively than thermal-type noise.

noninteracting control [CONT SYS] A feedback control in a system with more than one input and more than one output, in which feedback transfer functions are selected so that each input influences only one output.

noninverting parametric device [ELECTR] Parametric device whose operation depends essentially upon three frequencies, a harmonic of the pump frequency and two signal frequencies, of which one is the sum of the other plus the pump harmonic.

Nonionacea [INV ZOO] A superfamily of Foraminiferida in the suborder Orbitoidacea, characterized by a granular calcite test wall with monolamellar septa, and a planispiral to trochospiral test.

nonionic detergent [MATER] A detergent with molecules that do not ionize in aqueous solution, for example, detergents derived from condensation products of long-chain glycols and octyl or nonyl phenols.

nonlinear [PHYS] Pertaining to a response which is other than directly or inversely proportional to a given variable.

nonlinear amplifier [ELECTR] An amplifier in which a change in input does not produce a proportional change in output.

nonlinear capacitor [ELEC] Capacitor having a mean charge characteristic or a peak charge characteristic that is not linear, or a reversible capacitance that varies with bias voltage.

nonlinear circuit [ELEC] A circuit in which the current and voltage in any element that results from two sources of energy acting together is not equal to the sum of the currents or voltages that result from each of the sources acting alone.

nonlinear circuit component [ELECTR] An electrical device for which a change in applied voltage does not produce a proportional charge in current. Also known as nonlinear device; nonlinear element.

nonlinear coil [ELECTROMAG] Coil having an easily saturable core, possessing high impedance at low or zero current and low impedance when current flows and saturates the core.

nonlinear control [CONT SYS] An on-line process control which can stabilize operations by means of nonlinear compensatory functions.

nonlinear coupler [ELECTR] A type of frequency multiplier which uses the nonlinear capacitance of a junction diode to couple energy from the input circuit, which is tuned to the fundamental, to the output circuit, which is tuned to the desired harmonic.

nonlinear crystal [SOLID STATE] A crystal in which some influence (such as stress, electric field, or magnetic field) produces a response (such as strain, electric polarization, or magnetization) which is not proportional to the influence.

nonlinear detection [ELECTR] Detection based on the curvature of a tube characteristic, such as square-law detection.

nonlinear device See nonlinear circuit component.

nonlinear dielectric [ELEC] A dielectric whose polarization is not proportional to the applied electric field.

nonlinear distortion [ELECTR] Distortion in which the output of a system or component does not have the desired linear relation to the input. [ENG ACOUS] The ratio of the total root-mean-square (rms) harmonic distortion output of a microphone to the rms value of the fundamental component of the output.

nonlinear element See nonlinear circuit component.

nonlinear equation [MATH] An equation in variables $x_1, \ldots x_n, y$ which cannot be put into the form $a_1x_1 + \ldots + a_nx_n = y$.

nonlinear feedback control system [CONT SYS] Feedback control system in which the relationships between the pertinent measures of the system input and output signals cannot be adequately described by linear means.

nonlinear inductance [ELEC] The behavior of an inductor for which the voltage drop across the inductor is not proportional to the rate of change of current, such as when the inductor has a core of magnetic material in which magnetic induction is not proportional to magnetic field strength.

nonlinearity [SCI TECH] The deviation of any functional relationship from direct proportionality.

nonlinear material [PHYS] A material in which some specified influence (such as stress, electric field, or magnetic field) produces a response (such as strain, electric polarization, or magnetization) which is not proportional to the influence.

nonlinear molecule [ORG CHEM] A branched-chain molecule, that is, one whose atoms do not all lie along a straight line. Also known as isomolecule.

nonlinear network [ELEC] A network in which the current or voltage in any element that results from two sources of energy acting together is not equal to the sum of the currents or voltages that result from each of the sources acting alone.

nonlinear optics [OPTICS] The study of the interaction of radiation with matter in which certain variables describing the response of the matter (such as electric polarization or power absorption) are not proportional to variables describing the radiation (such as electric field strength or energy flux).

nonlinear oscillator [ELECTR] A radio-frequency oscillator that changes frequency in response to an audio signal; it is the basic circuit used in eavesdropping devices.

nonlinear programming [MATH] A branch of applied mathematics concerned with finding the maximum or minimum of a function of several variables, when the variables are constrained to yield values of other functions lying in a certain range, and either the function to be maximized or minimized, or at least one of the functions whose value is constrained, is nonlinear.

nonlinear reactance [ELECTR] The behavior of a coil or capacitor whose voltage drop is not proportional to the rate of change of current through the coil, or the charge on the capacitor.

nonlinear regression See curvilinear regression.

nonlinear resistance [ELECTR] The behavior of a substance (usually a semiconductor) which does not obey Ohm's law but has a voltage drop across it that is proportional to some power of the current.

nonlinear system [MATH] A system in which the interrelationships among the quantities involved are expressed by equations, some of which are not linear. [SCI TECH] A system in which outputs are not linear functions of vectors whose components represent the inputs.

nonlinear taper [ELEC] Nonuniform distribution of resistance throughout the element of a potentiometer or rheostat.

nonlinear vibration [MECH] A vibration whose amplitude is large enough so that the elastic restoring force on the vibrating object is not proportional to its displacement.

nonlinear viscoelasticity [FL MECH] The behavior of a fluid which does not obey a first-order differential equation in stress and strain.

non-load-bearing tile [MATER] Tile unable to carry superimposed loads.

nonloaded Q [ELEC] Of an electric impedance, the Q value of the impedance without external coupling or connection. Also known as basic Q.

nonlocalized electron [PHYS] An electron whose wave function is not confined to the vicinity of one or two nuclei, but is spread out over a molecule or a crystal lattice.

nonmagnetic [ELECTROMAG] Not magnetizable, and therefore not affected by magnetic fields.

nonmagnetic steel [MET] A steel alloy containing about 12% manganese and sometimes a small quantity of nickel; it is practically nonmagnetic at ordinary temperatures.

nonmaintenance time [ADP] The elapsed time during scheduled working hours between the determination of a machine failure and placement of the equipment back into operation.

nonmechanical See nonclastic.

nonmetallic sheathed cable [ELEC] Assembly of two or more rubber-covered conductors in an outer sheath of nonconducting fibrous material that has been treated to make it flame-resistant and moisture-repellent.

non-minimum-phase system [CONT SYS] A linear system whose transfer function has one or more poles or zeros with positive, nonzero real parts.

nonmultiple switchboard [ELEC] Manual telephone switchboard in which each subscriber line is attached to only one jack.

non-Newtonian fluid [FL MECH] A fluid whose flow behavior departs from that of a Newtonian fluid, so that the rate of

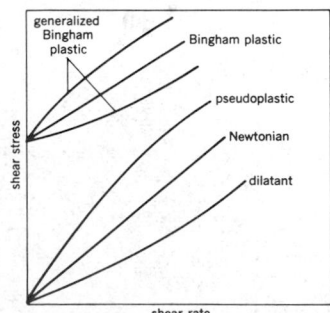

NON-NEWTONIAN FLUID

Typical flow curves, giving rate of shear as a function of shear stress, for a Newtonian fluid and for the three types of time-independent non-Newtonian fluid: Bingham plastic, pseudoplastic fluid, and dilatant fluid.

shear is not proportional to the corresponding stress. Also known as non-Newtonian system.

non-Newtonian fluid flow [FL MECH] The flow behavior of non-Newtonian fluids, whose study has applications in many important problems of practical significance such as flow in tubes, extrusion, flow through dies, coating operations, rolling operations, and mixing of fluids.

non-Newtonian system *See* non-Newtonian fluid.

non-Newtonian viscosity [FL MECH] The behavior of a fluid which, when subjected to a constant rate of shear, develops a stress which is not proportional to the shear.

nonnitroglycerin explosive [MATER] An explosive which contains TNT instead of nitroglycerin to sensitize ammonium nitrate, and a little aluminum powder may also be added to increase the power and sensitivity.

nonnumeric character [ADP] Any character except a digit.

nonodontogenic cyst [MED] Any oral cyst that develops from epithelium which has been sequestered in bony or soft-tissue suture lines during embryonic development.

nonoic acid [ORG CHEM] $C_8H_{17}COOH$ Any of a family of acids which are mixed isomers produced in the Fischer-Tropsch process; pelargonic acid is the straight-chain member; used as a chemical intermediate.

nonosteogenic fibroma [MED] A tumor of bone, usually in the shaft of long bones; characterized by whorls of spindle-shaped connective tissue cells.

nonparalytic poliomyelitis [MED] Infection by poliomyelitis virus accompanied by upper respiratory or gastrointestinal symptoms, muscular pain and stiffness, and mild fever.

nonparametric statistics [STAT] A class of statistical methods applicable to a large set of probability distributions used to test for correlation, location, independence, and so on.

nonpareil [GRAPHICS] A designation to indicate a specific amount of space surrounding the type, for example, interlinear space; the space is 6 points in the American point system.

nonpenetrative [GEOL] Of a type of deformation, affecting only part of a rock, such as kink bands.

nonpersistent war gas [ORD] A chemical agent, that is, a war gas, normally effective in the open for 10 minutes or less at the point of dispersion.

nonplunging fold [GEOL] A fold with a horizontal axial surface. Also known as horizontal fold; level fold.

nonpolar [CHEM] Pertaining to an element or compound which has no permanent electric dipole moment.

nonpolarized relay *See* neutral relay.

nonprecision [NAV] In air operations, pertaining to a navigational facility without a glide slope; does not imply an unacceptable quality of course guidance.

nonprint code [ADP] A bit combination which is interpreted as no printing, no spacing.

nonpriority interrupt [ADP] Any one of a group of interrupts which may be disregarded by the central processing unit.

nonprotein nitrogen [BIOCHEM] The nitrogen fraction in the body tissues, excretions, and secretions, not precipitated by protein precipitants.

nonpsychotic organic brain syndrome [MED] Organic brain syndrome in which there is no apparent psychosis.

nonquantum mechanics [MECH] The classical mechanics of Newton and Einstein as opposed to the quantum mechanics of Heisenberg, Schrödinger, and Dirac; particles have definite position and velocity, and they move according to Newton's laws.

nonreactive [ELEC] Pertaining to a circuit, component, or load that has no capacitance or impedance, so that an alternating current is in phase with the corresponding voltage.

nonrecording rain gage [ENG] A rain gage which indicates but does not record the amount of precipitation.

nonrecurring issue [ORD] An ordnance issue made on a one-time basis with no foreseeable subsequent demand from the consignee to whom issue was made.

nonrelativistic approximation [PHYS] The approximation in which it is assumed that speeds of objects are small compared to the speed of light.

nonrelativistic kinematics [MECH] The study of motions of systems of objects at speeds which are small compared to the speed of light, without reference to the forces which act on the system.

nonrelativistic mechanics [MECH] The study of the dynamics of systems in which all speeds are small compared to the speed of light.

nonrelativistic particle [RELAT] A particle whose velocity is small with respect to that of light.

nonrelativistic quantum mechanics [QUANT MECH] The modern theory of matter and its interaction with radiation, applicable to systems of material particles which move slowly compared to the speed of light, which are neither created or destroyed, and whose internal structure (except for spin) either does not change or is irrelevant to the description of the system.

nonremovable discontinuity [MATH] A point at which a function is not continuous or is undefined, and cannot be made continuous by being given a new value at the point.

nonrenewable fuse unit [ELEC] Fuse unit that cannot be readily restored for service after operation.

nonreproducible drawing [GRAPHICS] An opaque drawing which cannot be used to make copies by any contact (print-through) process.

nonreproducing code [ADP] A code which normally does not appear as such in a generated output but will result in a function such as paging or spacing.

nonresonant antenna [ELECTROMAG] A long-wire or traveling-wave antenna which does not have natural frequencies of oscillation, and responds equally well to radiation over a broad range of frequencies.

nonresonant line [ELECTROMAG] Transmission line having no reflected waves, and neither current nor voltage standing waves.

non-return-to-zero [ADP] A mode of recording and readout in which it is not necessary for the signal to return to zero after each item of recorded data.

nonrigid plastic [MATER] A plastic with modulus of elasticity not greater than 50,000 psi (3.45×10^8 newtons per square meter) at 25°C, according to standard ASTM test procedures.

nonrotational strain *See* irrotational strain.

nonsaccharine sorghum *See* grain sorghum.

nonsegregated reservoir [PETRO ENG] Solution-gas-drive oil reservoir in which the gas does not separate from the oil as a function of height or upward movement.

nonselective mining [MIN ENG] Mining methods permitting low cost, generally by using a cheap stoping method combined with large-scale operations; can be used in deposits where the individual stringers, bands, or lenses of high-grade ore are numerous and so irregular in occurrence and separated by such thin lenses of waste that a selective method cannot be employed.

nonselective radiator *See* graybody.

nonsense mutation [MOL BIO] A mutation that changes a codon that codes for one amino acid into a codon that does not specify any amino acid (a nonsense codon).

nonshared control unit [ADP] A control unit which controls only one device. Also known as unipath.

nonshattering glass [MATER] Two sheets of plate glass with a sheet of transparent resinoid between, the whole molded together under heat and pressure; it will crack without shattering. Also known as laminated glass; shatterproof glass.

nonshorting contact switch [ELEC] Selector switch in which the width of the movable contact is less than the distance between contact clips, so that the old circuit is broken before the new circuit is completed.

nonshrinking steel *See* nondeforming steel.

nonsingular matrix [MATH] A matrix which has an inverse; equivalently, its determinant is not zero.

nonsingular transformation [MATH] A linear transformation which has an inverse; equivalently, it has null space kernel consisting only of the zero vector.

nonsinusoidal waveform [ELEC] The representation of a wave which does not vary in a sinusoidal manner, and which therefore contains harmonics.

nonskid [CIV ENG] Pertaining to a surface that is roughened to reduce slipping, as a concrete floor treated with iron filings or carborundum powder, or indented while wet.

nonslip concrete [MATER] Rough-surface concrete made by applying oxide grains to the mixture before it hardens; used for steps.

nonsoap grease [MATER] Mineral oil thickened with solid lubricants such as graphite, mica, talc, molybdenum sulfide, asbestos fiber, uncombined fats, or rosin oils; used as a lubricant.

nonsorted polygon [GEOL] A form of patterned ground which has a dominantly polygonal mesh and an unsorted appearance due to the absence of border stones, and whose borders are generally marked by wedge-shaped fissures narrowing downward.

nonspecific [MED] Not attributable to any one definite cause, as a disease not caused by one particular microorganism, or an immunity not conferred by a specific antibody. [PHARM] Of medicines or therapy, not counteracting any one causative agent.

nonspecific hepatitis See interstitial hepatitis.

nonspecific immunity [IMMUNOL] Resistance attributable to factors other than specific antibodies, including genetic, age, or hormonal factors.

nonspherical nucleus [NUC PHYS] A nucleus which appears to have a permanent ellipsoidal shape in its ground state, as suggested by a large electric quadrupole moment.

nonstandard [ORD] Differing from specifications, conditions, or procedures that have been prescribed or established; for instance, weather conditions different from those assumed in firing tables are nonstandard.

nonstorage camera tube [ELECTR] Television camera tube in which the picture signal is, at each instant, proportional to the intensity of the illumination on the corresponding area of the scene.

nonstranded rope [DES ENG] A wire rope with the wires in concentric sheaths instead of in strands, and in opposite directions in the different sheaths, giving the rope nonspinning properties. Also known as nonspinning rope.

nonstriated muscle fiber See smooth muscle fiber.

nonsymmetrical aquifer [PETRO ENG] In an oil reservoir formation, a water-containing part that is irregular, being neither radial nor linear.

nonsynchronous [ELEC] Not related in phase, frequency, or speed to other quantities in a device or circuit.

nonsynchronous timer [ELECTR] A circuit at the receiving end of a communications link which restores the time relationship between pulses when no timing pulses are transmitted.

nonsynchronous vibrator [ELECTR] Vibrator that interrupts a direct-current circuit at a frequency unrelated to the other circuit constants and does not rectify the resulting stepped-up alternating voltage.

nonthermal decimetric emission [ELECTROMAG] A radio-wave emission above the 4-centimeter wavelength from the planet Jupiter that has a nearly constant flux between 5-centimeter and 1-meter wavelength. Also known as DIM.

nonthermal radiation [PHYS] Electromagnetic radiation emitted by accelerated charged particles that are not in thermal equilibrium; aurora light and fluorescent-lamp light are examples.

nontidal current [OCEANOGR] Any current due to causes other than tidal, as a permanent ocean current.

nontrivial solution [MATH] A solution to a homogeneous equation which is not the zero solution.

nontronite [MINERAL] $Na(Al,Fe,Si)O_{10}(OH)_2$ An iron-rich clay mineral of the montmorillonite group that represents the end member in which the replacement of aluminum by ferric ion is essentially complete. Also known as chloropal; gramenite; morencite; pinguite.

nonuniform flow [FL MECH] Fluid flow which does not have the same velocity at all points in a medium, at a given instant.

nonvenereal syphilis [MED] Syphilis not acquired during sexual intercourse.

nonviscous flow See inviscid flow.

nonviscous fluid See inviscid fluid.

nonviscous neutral [MATER] A neutral petroleum-derived oil with viscosity less than 135 SSU (Saybolt Seconds Universal) at 100°F (38°C).

nonvolatile memory See nonvolatile storage.

nonvolatile storage [ADP] A computer storage medium that retains information in the absence of power, such as a magnetic tape, drum, or core. Also known as nonvolatile memory.

nonwetting [MET] Of a metal or alloy, when it is molten, not adhering to or wetting the surface of a diamond which is to be set.

nonwetting phase [PETRO ENG] The oil phase contained in a reservoir pore structure when the reservoir fluid is two-phase (oil and water) or three-phase (oil, water, and gas).

nonwoven fabric [TEXT] Cloth produced from a random arrangement or matting of natural or synthetic fibers held together by adhesives, heat and pressure, or needling; felt is an example.

nonyl acetate [ORG CHEM] $C_9H_{19}OOCCH_3$ Any of a family of isomers, such as n-nonyl acetate and diisobutyl carbinyl acetate, which are products of Fischer-Tropsch and oxo syntheses.

n-nonyl acetate [ORG CHEM] $CH_3COO(CH_2)_8CH_3$ Alcohol-soluble, colorless liquid with pungent odor; boiling point 208–212°C; used in perfumery.

n-nonyl alcohol [ORG CHEM] $CH_3(CH_2)_7CH_2OH$ One of a family of $C_9H_{19}OH$ isomers; a colorless liquid with rose aroma; boils at 215°C; insoluble in water, soluble in alcohol; used in perfumery and flavorings. Also known as alcohol C-9; octyl carbinol; nonanol; pelargonic alcohol.

nonyl benzene [ORG CHEM] $C_9H_{19}C_6H_5$ Liquid boiling at 245–252°C; straw-colored with aromatic aroma; used to make surface-active agents.

nonyl carbinol See decyl alcohol.

1-nonylene [ORG CHEM] C_9H_{18} Colorless liquid boiling at 150°C; soluble in alcohol, insoluble in water; used as a chemical intermediate. Also known as nonene.

nonyl hydride See nonane.

n-nonylic acid See pelargonic acid.

nonyl phenol [ORG CHEM] $C_9H_{19}C_6H_4OH$ Pale-yellow liquid boiling at 283–302°C; soluble in organic solvents, insoluble in water; a mixture of monoalkyl phenol isomers, mostly para-substituted; used to make surface-active agents, resins, and plasticizers.

nonyne [ORG CHEM] $CH_3(CH_2)_6{\equiv}CCH$ Water-insoluble, colorless liquid boiling at 160°C. Also known as n-heptyl-acetylene; nonine.

noon [ASTRON] The instant at which a time reference is over the upper branch of the reference meridian.

noon constant [NAV] A predetermined value added to a meridian or ex-meridian sextant altitude to determine the latitude.

noon interval [ASTRON] The predicted time interval between a given instant, usually the time of a morning observation, and local apparent noon; it is used to predict the time for observing the sun on the celestial meridian.

Noon pollen unit [BIOL] The quantity of pollen extract which contains 0.00001 milligram of total nitrogen.

noon sight [NAV] **1.** Measurement of the altitude of the sun at local apparent noon. **2.** The altitude so measured.

NO OP [ADP] An instruction telling the computer to do nothing, except to proceed to the next instruction in sequence. Also known as no-operation instruction.

no-operation instruction See NO OP.

noosphere See anthroposphere.

nopinene See pinene.

nor- [CHEM] Chemical formula prefix for normal; indicates a parent for another compound to be formed by removal of one or more carbons and associated hydrogens.

NOR [ADP] A logic operator having the property that if P, Q, R, . . . are statements, then the NOR of P, Q, R, . . . is true if all statements are false, false if at least one statement is true. Derived from NOT-OR.

noradrenaline See norepinephrine.

norbergite [MINERAL] $Mg_3SiO_4(F,OH)_2$ A yellow or pink orthorhombic mineral composed of magnesium silicate with fluoride and hydroxyl; it is a member of the humite group.

NOR circuit [ELECTR] A circuit in which output voltage appears only when signal is absent from all of its input terminals.

Norden bombsight [ORD] A gyroscopically stabilized syn-

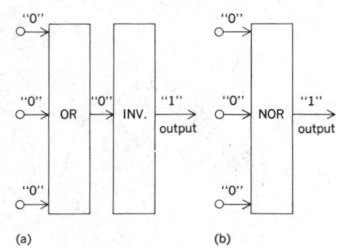

NOR CIRCUIT

NOR circuit. (a) OR gate followed by an inverter; if all inputs are "0," the output is "1." (b) OR-inverter (NOR) combination shown as a single symbol.

chronizing bombsight used mainly for synchronous bombing but useful for fixed-angle bombing.

Nordenskjöld line [CLIMATOL] The line connecting all places at which the mean temperature of the warmest month is equal (in degrees Celsius) to $9 - 0.1k$, where k is the mean temperature of the coldest month (in degrees Fahrenheit it becomes $51.4 - 0.1k$).

Nordheim's rule [SOLID STATE] The rule that the residual resistivity of a binary alloy that contains mole fraction x of one element and $1 - x$ of the other is proportional to $x(1 - x)$.

nor'easter *See* northeaster.

norepinephrine [BIOCHEM] $C_8H_{11}O_3N$ A hormone produced by chromaffin cells of the adrenal medulla; acts as a vasoconstrictor and mediates transmission of sympathetic nerve impulses. Also known as noradrenaline.

norite [PETR] A coarse-grained plutonic rock composed principally of basic plagioclase with orthopyroxene (hypersthene) as the dominant mafic material. Also known as hypersthenfels.

norm [MATH] A scalar valued function on a vector space with properties analogous to those of the modulus of a complex number; namely: the norm of the zero vector is zero, all other vectors have positive norm, the norm of a scalar times a vector equals the absolute value of the scalar times the norm of the vector, and the norm of a sum is less than or equal to the sum of the norms. [PETR] The theoretical mineral composition of a rock expressed in terms of standard mineral molecules as determined by means of chemical analyses. [QUANT MECH] **1.** The square of the modulus of a Schrödinger-Pauli wave function, integrated over the space coordinates and summed over the spin coordinates of the particles it describes. **2.** The square root of this quantity.

Norma [ASTRON] A southern constellation; right ascension 16 hours, declination 50°S. Also known as Rule.

normal [METEOROL] The average value of a meteorological element over any fixed period of years that is recognized as standard for the country and element concerned.

normal acceleration [MECH] **1.** The component of the linear acceleration of an aircraft or missile along its normal, or Z, axis. **2.** The usual or typical acceleration.

normal aeration [GEOL] The complete renewal of soil air to a depth of 20 centimeters about once each hour.

normal aiming error [ORD] In bombing, an aiming error that falls within the boundary for gross errors.

normal axis [MECH] The vertical axis of an aircraft or missile.

normal barometer [ENG] A barometer of such accuracy that it can be used for the determination of pressure standards; an instrument such as a large-bore mercury barometer is usually used.

normal benzine [MATER] A mixture of hydrocarbons; clear, colorless, water-insoluble liquid distilled from petroleum; boils at 65-95°C; density 0.695-0.705. Also known as benzoline.

normal chart [METEOROL] Any chart that shows the distribution of the official normal values of a meteorological element. Also known as normal map.

normal contour *See* accurate contour.

normal coordinates [MECH] A set of coordinates for a coupled system such that the equations of motion each involve only one of these coordinates.

normal curvature [MATH] The normal curvature at a point on a surface is the curvature of the normal section to the point.

normal curve *See* Gaussian curve.

normal cycle [GEOL] A cycle of erosion whereby a region is reduced to base level by running water, especially by the action of rivers. Also known as fluvial cycle of erosion.

normal density function [STAT] A normally distributed frequency distribution of a random variable x with mean e and variance σ is given by $(1/\sigma\sqrt{2\pi}) \exp[-(x-e)^2/2\sigma^2]$.

normal derivative [MATH] The derivative at a point of a function taken in the direction of the normal at the point.

normal dip *See* regional dip.

normal direction flow [ADP] The direction from left to right or top to bottom in flow charting.

normal dispersion [OPTICS] Dispersion in which the refrac-

tive index decreases monotonically and continuously with increasing wavelength.

normal displacement *See* dip slip.

normal distribution [STAT] The most commonly occurring probability distributions have the form

$$(1/\sigma\sqrt{2\pi}) \int_{-\infty}^{u} \exp(-u^2/2)du, \quad u = (x-e)/\sigma$$

where e is the mean and σ is the variance. Also known as Gauss' error curve; Gaussian distribution.

normal divisor *See* normal subgroup.

normal effort [IND ENG] The effort expended by the average operator in performing manual work with average skill and application.

normal electrode [ELEC] Standard electrode used for measuring electrode potentials.

normal elemental time [IND ENG] The selected or average elemental time adjusted to obtain the elemental time used by an average qualified operator. Also known as base time; leveled elemental time.

normal erosion [GEOL] Erosion effected by prevailing agencies of the natural environment, including running water, rain, wind, waves, and organic weathering. Also known as geologic erosion.

normal family [MATH] A family of complex functions analytic in a common domain where every sequence of these functions has a subsequence converging uniformly on compact subsets of the domain to an analytic function on the domain or to $+\infty$.

normal fault [GEOL] A fault, usually of 45-90°, in which the hanging wall appears to have shifted downward in relation to the footwall. Also known as gravity fault; normal slip fault; slump fault.

normal fluid [CRYO] The component of liquid helium II, postulated in the two-fluid theory, that has viscosity and behaves like an ordinary fluid.

normal fold *See* symmetrical fold.

normal functions [MATH] Two functions whose inner product is zero; the inner product is usually given by an integral $\int f g d\mu$.

normal horizontal separation *See* offset.

normal hydrostatic pressure [HYD] In porous strata or in a well, the pressure at a given point that is approximately equal to the weight of a column of water extending from the surface to that point.

normal impact [MECH] **1.** Impact on a plane perpendicular to the trajectory. **2.** Striking of a projectile against a surface that is perpendicular to the line of flight of the projectile.

normal impedance *See* free impedance.

normal-incidence pyrheliometer [ENG] An instrument that measures the energy in the solar beam; it usually measures the radiation that strikes a target at the end of a tube equipped with a shutter and baffles to collimate the beam.

normal incidence reflectivity [ELECTROMAG] The ratio of the energy of electromagnetic radiation reflected from the interface between two media to the energy of the incident radiation when the incident radiation travels in a direction perpendicular to the surface.

normal induction [ELECTROMAG] Limiting induction, either positive or negative, in a magnetic material that is under the influence of a magnetizing force which varies between two specific limits.

normality [CHEM] Measure of the number of gram-equivalent weights of a compound per liter of solution. Abbreviated N.

normalize [MATH] To multiply a quantity by a suitable constant or scalar so that it then has norm one; that is, its norm is then equal to one. [MET] To heat a ferrous alloy to some temperature above the transformation range, followed by air cooling. [QUANT MECH] To multiply a wave function by a constant so that its norm is equal to unity. [STAT] To carry out a normal transformation on a variate.

normalized admittance [ELECTROMAG] The reciprocal of the normalized impedance.

normalized coupling coefficient [ELECTROMAG] Mutual inductance, expressed on a scale running from zero to one.

normalized current [ELECTROMAG] The current divided by

1015

the square root of the characteristic admittance of a waveguide or transmission line.

normalized function [MATH] A function with norm one; the norm is usually given by an integral ($\int |f|^p d\mu)^{1/p}$, $1 \leq p < \infty$.

normalized impedance [ELECTROMAG] An impedance divided by the characteristic impedance of a transmission line or waveguide.

normalized Q [ELEC] The ratio of the reactive component of the impedance of a filter section to the resistive component.

normalized susceptance [ELECTROMAG] The susceptance of an element of a waveguide or transmission line divided by the characteristic admittance.

normalized voltage [ELECTROMAG] The voltage divided by the square root of the characteristic impedance of a waveguide or transmission line.

normally distributed observations [STAT] Any set of observations whose histogram looks like the normal curve.

normal magnetization curve [ELECTROMAG] Curve traced on a graph of magnetic induction versus magnetic field strength in an originally unmagnetized specimen, as the magnetic field strength is increased from zero. Also known as magnetization curve.

normal map *See* normal chart.

normal matrix [MATH] A matrix is normal if multiplying it on the right by its adjoint is the same as multiplying it on the left.

normal mode [ADP] Operation of a computer in which it executes its own instructions rather than those of a different computer.

normal-mode helix [ELECTROMAG] A type of helical antenna whose diameter and electrical length are considerably less than a wavelength, and which has a radiation pattern with greatest intensity normal to the helix axis.

normal mode of vibration [MECH] Vibration of a coupled system in which the value of one of the normal coordinates oscillates and the values of all the other coordinates remain stationary.

normal operator [MATH] A linear operator where composing it with its adjoint operator in either order gives the same result. Also known as normal transformation.

normal orientation [ADP] In optical character recognition, that determinate position which indicates that the line elements of an inputted source document appear parallel with the document's leading edge.

normal pace [IND ENG] The manual pace achieved by normal effort.

normal pitch [MECH ENG] The distance between working faces of two adjacent gear teeth, measured between the intersections of the line of action with the faces.

normal-plate anemometer [ENG] A type of pressure-plate anemometer in which the plate, restrained by a stiff spring, is held perpendicular to the wind; the wind-activated motion of the plate is measured electrically; the natural frequency of this system can be made high enough so that resonance magnification does not occur.

normal polarity [GEOPHYS] Natural remanent magnetism nearly identical to the present ambient field.

normal potassium pyrophosphate *See* potassium pyrophosphate.

normal pressure *See* standard pressure.

normal range [ADP] An interval within which results are expected to fall during normal operations.

normal reaction [MECH] The force exerted by a surface on an object in contact with it which prevents the object from passing through the surface; the force is perpendicular to the surface, and is the only force that the surface exerts on the object in the absence of frictional forces.

normal saline [PHYSIO] U.S. Pharmacopoeia title for a sterile solution of sodium chloride in purified water, containing 0.9 gram of sodium chloride in 100 milliliters; isotonic with body fluids. Also known as isotonic sodium chloride solution; normal salt solution; physiological saline; physiological salt solution; physiological sodium chloride solution; sodium chloride solution.

normal salt [CHEM] A salt in which all of the acid hydrogen atoms have been replaced by a metal, or the hydroxide radicals of a base are replaced by an acid radical; for example, Na_2CO_3.

normal salt solution *See* normal saline.

normal section [MATH] Relative to a surface, this is a planar section produced by a plane containing the normal to a point.

normal segregation [MET] Segregation of the lower-melting-point constituents of an alloy primarily near the center of a casting (last portion to solidify).

normal silver sulfate *See* silver sulfate.

normal slip fault *See* normal fault.

normal soil [GEOL] A soil having a profile that is more or less in equilibrium with the environment.

normal solution [CHEM] An aqueous solution containing one equivalent of the active reagent in grams in 1 liter of the solution.

normal space [MATH] A topological space in which any two disjoint closed sets may be covered respectively by two disjoint open sets.

normal-stage punching [ADP] A system of card punching in which only even number rows of the card are used.

normal state [NUC PHYS] A term sometimes used for ground state.

normal subgroup [MATH] A subgroup N of a group G where every expression $g^{-1}ng$ is in N for every g in G and every n in N. Also known as invariant subgroup; normal divisor.

normal surface [OPTICS] The surface that is generated by taking, at each point of the ray surface, the intersection of the tangent plane to the ray surface at that point with the perpendicular from the origin to this plane.

normal thorium sulfate *See* thorium sulfate.

normal time [IND ENG] **1.** The time required by a trained worker to perform a task at a normal pace. **2.** The total of all the normal elemental times constituting a cycle or operation. Also known as base time; leveled time.

normal to a curve [MATH] The normal to a curve at a point is the line perpendicular to the tangent line at the point.

normal to a surface [MATH] The normal to a surface at a point is the line perpendicular to the tangent plane at that point.

normal transformation [MATH] *See* normal operator. [STAT] A transformation on a variate that converts it into a variate which has a normal distribution.

normal uranium *See* native uranium.

normal volume *See* standard volume.

normal water [OCEANOGR] Water whose chlorinity lies between 19.30 and 19.50 parts per thousand and has been determined to within ±0.001 per thousand. Also known as Copenhagen water; standard seawater.

normative mineral *See* standard mineral.

normed linear space [MATH] A vector space which has a norm defined on it. Also known as normed vector space.

normed vector space *See* normed linear space.

normoblast [HISTOL] The smallest of the nucleated precursors of the erythrocyte; slightly larger than a mature adult erythrocyte. Also known as acidophilic erythroblast; arthochromatic erythroblast; eosinophilic erythroblast; karyocyte; metakaryocyte; metanitricyte; metarubricyte.

normochromatic [CYTOL] Pertaining to cells of the erythrocytic series which have a normal staining color; attributed to the presence of a full complement of hemoglobin.

normochromic [CYTOL] Pertaining to erythrocytes which have a mean corpuscular hemoglobin (MCH), or color, index and a mean corpuscular hemoglobin concentration (MCHC), or saturation, index within two standard deviations above or below the mean normal.

normocyte [HISTOL] An erythrocyte having both a diameter and a mean corpuscular volume (MCV) within two standard deviations above or below the mean normal.

norm of a complex number *See* absolute value of a complex number.

norm of a matrix [MATH] The square root of the sum of the squares of the moduli of the matrix entries.

normothermia [PHYSIO] A state of normal body temperature.

Norris-Eyring reverberation formula [ACOUS] The reverberation time of a chamber, in seconds, is equal to 0.05 times its volume, in cubic feet, divided by the product of its surface area, in square feet, and the negative of the natural logarithm of 1 minus the absorption coefficient averaged over the surface.

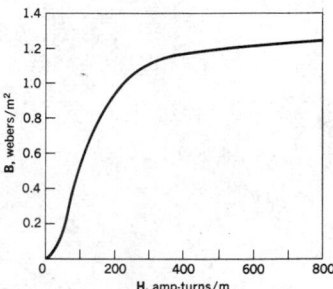

NORMAL MAGNETIZATION CURVE

Normal magnetization curve; the magnetic induction is the ordinate and the magnetic field strength is the abscissa.

norte [METEOROL] **1.** The winter north wind in Spain. **2.** A strong, cold northeasterly wind which blows in Mexico and on the shores of the Gulf of Mexico, and results from an outbreak of cold air from the north; actually, the Mexican extension of a norther.

north [GEOD] The direction of the north terrestrial pole; the primary reference direction on the earth; the direction indicated by 000° in any system other than relative.

North America [GEOGR] The northern of the two continents of the New World or Western Hemisphere, extending from narrow parts in the tropics to progressively broadened portions in middle latitudes and Arctic polar margins.

North American anticyclone *See* North American high.

North American blastomycosis [MED] A type of blastomycosis caused by the diphasic fungus *Blastomyces dermatitidis*; two recognized forms are cutaneous and systemic.

North American high [METEOROL] The relatively weak general area of high pressure which, as shown on mean charts of sea-level pressure, covers most of North America during winter. Also known as North American anticyclone.

North American Nebula [ASTRON] A cloud of dust and gas in the constellation Cygnus; the density of this gas and dust is possibly a thousand times greater than the average density of interstellar gas; a much denser cloud of dust between the nebula and earth obscures portions of the emission nebula to create the appearance of the "Gulf of Mexico" and the "Atlantic Ocean."

North Atlantic Current [OCEANOGR] A wide, slow-moving continuation of the Gulf Stream originating in the region east of the Grand Banks of Newfoundland.

northbound node *See* ascending node.

North Cape Current [OCEANOGR] A warm current flowing northeastward and eastward around northern Norway, and curving into the Barents Sea.

Northeast Drift Current [OCEANOGR] A North Atlantic Ocean current flowing northeastward toward the Norwegian Sea, gradually widening and, south of Iceland, branching and continuing as the Irminger Current and the Norwegian Current; it is the northern branch of the North Atlantic Current.

northeaster [METEOROL] A northeast wind, particularly a strong wind or gale. Also spelled nor'easter.

northeast storm [METEOROL] A cyclonic storm of the east coast of North America, so called because the winds over the coastal area are from the northeast; they may occur at any time of year but are most frequent and most violent between September and April.

northeast trades [METEOROL] The trade winds of the Northern Hemisphere.

North Equatorial Current [OCEANOGR] Westward ocean currents driven by the northeast trade winds blowing over tropical oceans of the Northern Hemisphere. Also known as Equatorial Current.

norther [METEOROL] A northerly wind.

northerly turning error [AERO ENG] An acceleration error in the magnetic compass of an aircraft in a banked attitude during a turn, so called because it was first noted and is most pronounced during turns made from initial north-south courses; during a turn the magnetic needle is tilted from the horizontal, due to acceleration and the banking of the aircraft; in this position the compass needle will be acted upon by the vertical as well as the horizontal component of the earth's magnetic field; in addition, the compass needle is mechanically restricted in movement, due to tilt. Also known as turning error.

northern anthracnose [PL PATH] A fungus disease of red and crimson clovers in North America, Asia, and Europe caused by *Kabatulla caulivora*; depressed, linear brown lesions form on the stems and petioles.

Northern Cross *See* Cygnus.

Northern Crown *See* Corona Borealis.

Northern Hemisphere [GEOGR] The half of the earth north of the Equator.

northern lights *See* aurora borealis.

north foehn [METEOROL] A foehn condition sustained by wind flow across the Alps from north to south.

north frigid zone [GEOGR] That part of the earth north of the Arctic Circle.

north geographic pole *See* North Pole.

north geomagnetic pole *See* north pole.

Northill anchor [NAV ARCH] A lightweight kedge anchor with large flukes and a stock added to the crown perpendicular to them; used in sizes up to 100 pounds.

northing [NAV] The difference in latitude to the north between the present reckoning and the previous reckoning.

north magnetic pole *See* north pole.

North Pacific Current [OCEANOGR] The warm branch of the Kuroshio Extension flowing eastward across the Pacific Ocean.

north pole [ASTRON] The north celestial pole that indicates the zenith of the heavens when viewed from the north geographic pole. [ELECTROMAG] The pole of a magnet at which magnetic lines of force are considered as leaving the magnet; the lines enter the south pole; if the magnet is freely suspended, its north pole points toward the north geomagnetic pole. [GEOPHYS] The geomagnetic pole in the Northern Hemisphere, at approximately latitude 78.5°N, longitude 69°W. Also known as north magnetic pole; north geomagnetic pole.

North Pole [GEOGR] The geographic pole located at latitude 90°N in the Northern Hemisphere of the earth; it is the northernmost point of the earth, and the northern extremity of the earth's axis of rotation. Also known as north geographic pole.

north-stabilized plan-position indicator [ENG] A heading-upward plan-position indicator; this term is deprecated because it may be confused with azimuth-stabilized plan-position indicator, a north-upward plan-position indicator.

North Star *See* Polaris.

north temperate zone [CLIMATOL] That part of the earth between the Tropic of Cancer and the Arctic Circle.

northupite [MINERAL] $Na_3MgCl(CO_3)_2$ A white, yellow, gray, or colorless isometric mineral composed of magnesium sodium carbonate; occurs in octahedral crystals.

north-upward plan-position indicator [ENG] A plan-position indicator on which north is maintained at the top of the indicator, regardless of the heading of the craft.

north-upward presentation [NAV] An indicator presentation, such as that of a plan position indicator or a radio direction finder, on which north is at the top of the indicator regardless of the heading of the craft.

northwester [METEOROL] A northwest wind. Also spelled nor'wester.

Norton's theorem [ELEC] The theorem that the voltage across an element that is connected to two terminals of a linear network is equal to the short-circuit current between these terminals in the absence of the element, divided by the sum of the admittances between the terminals associated with the element and the network respectively.

Norway Current [OCEANOGR] A continuation of the North Atlantic Current, which flows northward along the coast of Norway. Also known as Norwegian Current.

Norwegian Current *See* Norway Current.

nor'wester *See* northwester.

noscapine [PHARM] $C_{22}H_{23}NO_7$ A white, colorless, tasteless alkaloid obtained from opium; used as a nonaddicting antitussive. Formerly known as narcotine.

nose [ENG] The foremost point or section of a bomb, missile, or the like. [ANAT] The nasal cavities and the structures surrounding and associated with them in all vertebrates. [FL MECH] The dense, forward part of a turbidity current. [GEOL] A plunging anticline that is short and without closure.

nosean *See* noselite.

nose cone [AERO ENG] A protective cone-shaped case for the nose section of a missile or rocket; may include the warhead, fusing system, stabilization system, heat shield, and supporting structure and equipment.

nose fuse [ORD] A fuse for use in the forward end (nose) of a bomb or other missile.

nose-heavy [AERO ENG] Pertaining to an airframe in which the nose tends to sink when the longitudinal control is released in any attitude of normal flight.

nose height [ANTHRO] The distance measured between the nasion and the subnasale.

nose leaf [VERT ZOO] A leaflike expansion of skin on the nose of certain bats; believed to have a tactile function.

nose length [ANTHRO] The distance measured between the nasion and the pronasale.

noselite [MINERAL] $Na_4Al_3Si_3O_{12} \cdot SO_4$ A gray, blue or brown mineral of the sodalite group; similar to haüynite; hardness is 5.5 on Mohs scale. Also known as nosean.

nose spray [ORD] Fragments of a bursting projectile that are thrown forward in the line of flight, in contrast to base spray and side spray.

nosing [BUILD] Projection of a tread of a stair beyond the riser below it.

nosocomial [MED] 1. Pertaining to a hospital. 2. Of disease, caused or aggravated by hospital life.

Nosodendridae [INV ZOO] The wounded-tree beetles, a small family of coleopteran insects in the superfamily Dermestoidea.

nosophobia [PSYCH] An abnormal fear of disease.

nostopathy [PSYCH] Pathogenic homecoming, that is, stress-precipitated illness as observed in individuals who have spent a considerable length of time in institutions such as hospitals or prisons.

nostril [ANAT] One of the external orifices of the nose. Also known as naris.

nostrum [PHARM] A quack medicine.

Notacanthidae [VERT ZOO] A family of benthic, deep-sea teleosts in the order Notacanthiformes, including the spiny eel.

Notacanthiformes [VERT ZOO] An order of actinopterygian fishes whose body is elongated, tapers posteriorly, and has no caudal fin.

NOTAM [COMMUN] The aviation communications code word for "notice to airmen"; used to provide timely information or conditions which are essential to flight operations; these notices are controlled by the National Flight Data Center.

NOT-AND *See* NAND.

notation [ADP] *See* positional notation. [MATH] The use of symbols to denote quantities or operations.

notch [ELECTR] Rectangular depression extending below the sweep line of the radar indicator in some equipment. [ENG] A V-shaped indentation or cut in a surface or edge. [GEOL] A deep, narrow cut near the high-water mark at the base of a sea cliff. [GEOGR] A narrow passage between mountains or through a ridge, hill, or mountain.

notch acuity [MET] The severity of the stress concentration produced by a given notch in a structure; it is expressed as the ratio of the notch depth to the notch radius (depth is small compared to width, or diameter, of the narrowest cross section).

notch antenna [ELECTROMAG] Microwave antenna in which the radiation pattern is determined by the size and shape of a notch or slot in a radiating surface.

notch brittleness [MET] Susceptibility of a material to brittle fracture at areas of stress concentration; in notch tensile testing, a material has notch brittleness if the notch strength lies below the tensile strength.

notch depth [MET] The distance from the surface of a metal specimen to the bottom of the notch.

notch ductility [MET] Percentage reduction in area of the specimen after failure in a notched tensile test.

notched-bar test [MET] Test in which a notched metal specimen is bent with the notch in tension.

notch filter [ELECTR] A band-rejection filter that produces a sharp notch in the frequency response curve of a system; used in television transmitters to provide attenuation at the low-frequency end of the channel, to prevent possible interference with the sound carrier of the next lower channel.

notch graft [BOT] A plant graft in which the scion is inserted in a narrow slit in the stock.

notching [ELEC] Term indicating that a predetermined number of separate impulses are required to complete operation of a relay. [MECH ENG] Cutting out various shapes from the ends or edges of a workpiece.

notching press [MECH ENG] A mechanical press for notching straight or rounded edges.

notch sensitivity [MET] A measure of the reduction in strength of a metal caused by the presence of a notch.

notch strength [MET] The ratio of maximum tensional load required to fracture a notched specimen to the original minimum cross-sectional area.

notch test [MET] A tensile or creep test of a metal to determine the effect of a surface notch.

NOT circuit [ELECTR] A logic circuit with one input and one output that inverts the input signal at the output; that is, the output signal is a logical 1 if the input signal is a logical 0, and vice versa. Also known as inverter circuit.

note [ACOUS] 1. A conventional sign indicating the pitch of a musical sound by its position on a staff, and the duration of the sound by its shape. 2. The sound indicated by this sign.

Noteridae [INV ZOO] The burrowing water beetles, a small family of coleopteran insects in the suborder Adephaga.

NOT function [MATH] A logical operator having the property that if P is a statement, then the NOT of P is true if P is false, and false if P is true.

Nothosauria [PALEON] A suborder of chiefly marine Triassic reptiles in the order Sauropterygia.

Notice to Mariners [NAV] A weekly publication prepared jointly by the U.S. Coast Guard and the U.S. Naval Oceanographic Office, giving information on changes in aids to navigation (lights, buoyage, harbor construction), dangers to navigation (rocks, shoals, banks, bars), important new soundings, and in general, all such information as affects mariners' charts, manuals, and sailing directions.

Notidanoidea [VERT ZOO] A suborder of rare sharks in the order Selachii; all retain the primitive jaw suspension of the order.

Notiomastodontinae [PALEON] A subfamily of extinct elephantoid proboscidean mammals in the family Gomphotheriidae.

Notioprogonia [PALEON] A suborder of extinct mammals comprising a diversified archaic stock of Notoungulata.

notochord [VERT ZOO] An elongated dorsal cord of cells which is the primitive axial skeleton in all chordates; persists in adults in the lowest forms (*Branchiostoma* and lampreys) and as the nuclei pulposi of the intervertebral disks in adult vertebrates.

notochordal canal [EMBRYO] A canal formed by a continuation of the primary pit into the head process of mammalian embryos; provides a temporary connection between the yolk sac and amnion.

notochordal plate [EMBRYO] A plate of cells representing the root of the head process of the embryo after the embryo becomes vesiculated.

Notodelphyidiformes [INV ZOO] A tribe of the Gnathostoma in some systems of classification.

Notodelphyoida [INV ZOO] A small group of crustaceans bearing a superficial resemblance to many insect larvae as a result of uniform segmentation, comparatively small trunk appendages, and crowding of inconspicuous oral appendages into the anterior portion of the head.

Notodontidae [INV ZOO] The puss moths, a family of lepidopteran insects in the superfamily Noctuoidea, distinguished by the apparently three-branched cubitus.

Notogaean [ECOL] Pertaining to or being a biogeographic region including Australia, New Zealand, and the southwestern Pacific islands.

Notommatidae [INV ZOO] A family of rotifers in the order Monogonota including forms with a cylindrical body covered by a nonchitinous cuticle and with a slender posterior foot.

Notomyotina [INV ZOO] A suborder of echinoderms in the order Phanerozonida in which the upper marginals alternate in position with the lower marginals, and each tube foot has a terminal sucking disk.

Notonectidae [INV ZOO] The backswimmers, a family of aquatic, carnivorous hemipteran insects in the superfamily Notonectoidea; individuals swim ventral side up, aided in breathing by an air bubble.

Notonectoidea [INV ZOO] A superfamily of Hemiptera in the subdivision Hydrocorisae.

notopodium [INV ZOO] The dorsal branch of a parapodium in certain annelids.

Notopteridae [VERT ZOO] The featherbacks, a family of actinopterygian fishes in the order Osteoglossiformes; bodies are tapered and compressed, with long anal fins that are continuous with the caudal fin.

NOT-OR *See* NOR.

Notoryctidae [PALEON] An extinct family of Australian insectivorous mammals in the order Marsupialia.

NOTACANTHIFORMES

Spiny eel (*Notacanthus nasus*). (*After D. S. Jordan and B. W. Evermann, The Fishes of North and Middle America, U.S. Nat. Mus. Bull. no. 47, 1900*)

NOT CIRCUIT

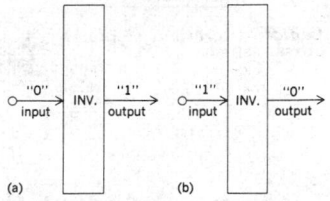

NOT circuit, also known as inverter circuit. (a) A "0" input produces a "1" output. (b) A "1" input produces a "0" output.

NOTHOSAURIA

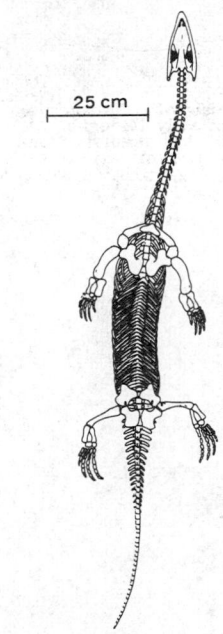

Ceresiosaurus calcagnii, ventral view; skeleton of a nothosaur with elongated neck; Triassic, Switzerland. (*From B. Peyer, 1944*)

NOTODELPHYOIDA

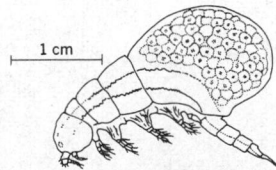

Doropygus psyllus Thorell, female; eggs are incubated in swollen part of thorax.

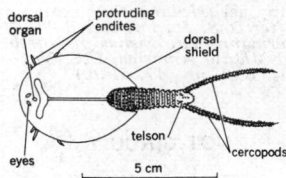

NOTOSTRACA

Lepidurus arcticus, female, dorsal aspect.

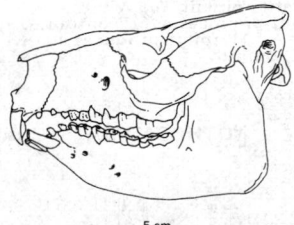

NOTOUNGULATA

Skull and jaw of *Adinotherium ovinum,* an early Miocene toxodontid notoungulate from the Santa Cruz formation of Patagonia, Argentina. *(After W. Scott)*

Notostigmata [INV ZOO] The single suborder of the Opilioacriformes, an order of mites.

Notostigmophora [INV ZOO] A subclass or suborder of the Chilopoda, including those centipedes embodying primitive as well as highly advanced characters, distinguished by dorsal respiratory openings.

Notostraca [INV ZOO] The tadpole shrimps, an order of crustaceans generally referred to the Branchiopoda, having a cylindrical trunk that consists of 25–44 segments, a dorsoventrally flattened dorsal shield, and two narrow, cylindrical cercopods on the telson.

Nototheniidae [VERT ZOO] A family of perciform fishes in the suborder Blennioidei, including most of the fishes of the permanently frigid waters surrounding Antarctica.

Notoungulata [PALEON] An extinct order of hoofed herbivorous mammals, characterized by a skull with an expanded temporal region, primitive dentition, and primitive feet with five toes, the weight borne mainly by the third digit.

noumeite *See* garnierite.

nova [ASTRON] A star that suddenly becomes explosively bright, the term is a misnomer because it does not denote a new star but the brightening of an existing faint star.

novaculite [GEOL] A siliceous sedimentary rock that is dense, hard, even-textured, light-colored, and characterized by dominance of microcrystalline quartz over chalcedony. Also known as razor stone.

novaculitic chert [GEOL] A gray chert that fragments into slightly rough, splintery pieces.

novar [ELECTR] Beam-power tube having a nine-pin base.

novobiocin [MICROBIO] $C_{30}H_{36}O_{11}N_2$ A moderately broad-spectrum antibiotic produced by strains of *Streptomyces niveus* and *S. spheroides*; it is a dibasic acid and is converted either to the monosodium salt or to the calcium acid salt for pharmaceutical use.

Novocain [PHARM] A trade name for procaine hydrochloride.

novolac resin [MATER] Any of the thermoplastic phenol-formaldehyde resins made with an excess of phenol in the reaction; used in varnishes.

nowhere dense set [MATH] A set in a topological space whose closure has empty interior.

no-wind heading [NAV] A heading determined without allowance for wind, and hence equal to the course.

no-wind position *See* air position.

nox [OPTICS] A unit of illumination, used in measuring low-level illumination, equal to 10^{-3} lux.

noy [ACOUS] A unit of perceived noisiness equal to the perceived noisiness of random noise occupying the frequency band 910–1090 hertz at a sound pressure level of 40 decibels above 0.0002 microbar; a sound that is *n* times as noisy as this sound has a perceived noisiness of *n* noys, under the assumption that the perceived noisiness of a sound increases with physical intensity at the same rate as the loudness.

nozzle [DES ENG] A tubelike device, usually streamlined, for accelerating and directing a fluid, whose pressure decreases as it leaves the nozzle.

nozzle blade [AERO ENG] Any one of the blades or vanes in a nozzle diaphragm. Also known as nozzle vane.

nozzle-contraction-area ratio [DES ENG] Ratio of the cross-sectional area for gas flow at the nozzle inlet to that at the throat.

nozzle-divergence loss factor [FL MECH] The ratio between the momentum of the gases in a nozzle and the momentum of an ideal nozzle.

nozzle efficiency [MECH ENG] The efficiency with which a nozzle converts potential energy into kinetic energy, commonly expressed as the ratio of the actual change in kinetic energy to the ideal change at the given pressure ratio.

nozzle exit area [DES ENG] The cross-sectional area of a nozzle available for gas flow measured at the nozzle exit.

nozzle-expansion ratio [DES ENG] Ratio of the cross-sectional area for gas flow at the exit of a nozzle to the cross-sectional area available for gas flow at the throat.

nozzle-mix gas burner [ENG] A burner in which injection nozzles mix air and fuel gas at the burner tile.

nozzle process [NUCLEO] A method of separating isotopes by allowing a gaseous compound to exhaust through a properly shaped nozzle.

nozzle throat [DES ENG] The portion of a nozzle with the smallest cross section.

nozzle throat area [DES ENG] The area of the minimum cross section of a nozzle.

nozzle thrust coefficient [AERO ENG] A measure of the amplification of thrust due to gas expansion in a particular nozzle as compared with the thrust that would be exerted if the chamber pressure acted only over the throat area. Also known as thrust coefficient.

nozzle vane *See* nozzle blade.

Np *See* neptunium.

npin transistor [ELECTR] An *npn* transistor which has a layer of high-purity germanium between the base and collector to extend the frequency range.

N-plus-one address instruction [ADP] An instruction with N + 1 address parts, one of which gives the location of the next instruction to be carried out.

npnp diode *See* pnpn diode.

npnp transistor [ELECTR] An *npn*-junction transistor having a transition or floating layer between *p* and *n* regions, to which no ohmic connection is made. Also known as *pnpn* transistor.

npn semiconductor [ELECTR] Double junction formed by sandwiching a thin layer of *p*-type material between two layers of *n*-type material of a semiconductor.

npn transistor [ELECTR] A junction transistor having a *p*-type base between an *n*-type emitter and an *n*-type collector; the emitter should then be negative with respect to the base, and the collector should be positive with respect to the base.

NPO-body [ELEC] Referring to a series of temperature-compensating capacitors that have an invariant dielectric constant over a specified temperature range.

np semiconductor [ELECTR] Region of transition between *n*- and *p*-type material.

NPSH *See* net positive suction head.

NR *See* nitrile rubber.

NRM *See* natural remanent magnetization.

NRM wind scale [METEOROL] A wind scale adapted by the United States Forest Service for use in the forested areas of the Northern Rocky Mountains (NRM); it is an adaptation of the Beaufort wind scale; the difference between these two scales lies in the specification of the visual effects of the wind; the force numbers and the corresponding wind speeds are the same in both.

N rod bit [DES ENG] A Canadian standard noncoring bit having a set diameter of 2.940 inches (74.676 millimeters).

N scan *See* N scope.

N scope [ELECTR] A cathode-ray scope combining the features of K and M scopes. Also known as N indicator; N scan.

N shell [ATOM PHYS] The fourth layer of electrons about the nucleus of an atom, having electrons characterized by the principal quantum number 4.

N star [ASTRON] An obsolete classification for a star in the carbon sequence; has about the same temperature as an M star in the Draper catalog.

nt *See* nit.

NTA *See* nitrilotriacetic acid.

n-type conduction [ELECTR] The electrical conduction associated with electrons, as opposed to holes, in a semiconductor.

N-type crystal rectifier [ELECTR] Crystal rectifier in which forward current flows when the semiconductor is negative with respect to the metal.

n-type germanium [ELECTR] Germanium to which more impurity atoms of donor type (with valence 5, such as antimony) than of acceptor type (with valence 3, such as indium) have been added, with the result that the conduction electron density exceeds the hole density.

n-type semiconductor [ELECTR] An extrinsic semiconductor in which the conduction electron density exceeds the hole density.

nub [TEXT] A planned knot or tangle in yarn to give a desired irregular texture to a fabric.

nucellus [BOT] The oval central mass of tissue in the ovule; contains the embryo sac.

nucha [ANAT] The nape of the neck.

nuchal ligament [ANAT] An elastic ligament extending from the external occipital protuberance and middle nuchal line to the spinous process of the seventh cervical vertebra. Also known as ligamentum nuchae.

nuchal organ [INV ZOO] Any of various sense organs on the prostomium of many annelids, which are sensitive to changes in the immediate environment of the individual.

nuchal rigidity [MED] Stiffness in the nape of the neck, often accompanied by pain and spasm on attempts to move the head; the most common sign of meningitis.

nuchal tentacle [INV ZOO] Any of various filiform or thick, fleshy tactoreceptors on anterior segments of many annelids.

nuclear [CHEM] A group of atoms joined directly to the central group of atoms or central ring of a molecule. [NUCLEO] Pertaining to nuclear energy. [NUC PHYS] Pertaining to the atomic nucleus.

nuclear absorption [NUC PHYS] Absorption of energy by the nucleus of an atom.

nuclear age determination See radiometric dating.

nuclear angular momentum See nuclear spin.

nuclear atom [CHEM] An atomic structure consisting of dense, positively charged nucleus (neutrons and protons) surrounded by a corresponding set of negatively charged electrons.

nuclear battery [NUCLEO] A primary battery in which the energy of radioactive material is converted into electric energy by solar cells or other energy converters. Also known as atomic battery; radioisotope battery; radioisotopic generator.

nuclear binding energy [NUC PHYS] The energy required to separate an atom into its constituent protons, neutrons, and electrons.

nuclear boiler [NUCLEO] A nuclear reactor in which water is the primary coolant and is converted to steam, such as a pressurized-water reactor or a boiling-water reactor.

nuclear bomb See atomic bomb.

nuclear breeder [NUCLEO] A nuclear reactor in which more fissionable material is formed in each generation than is used up in fission.

nuclear capture [NUC PHYS] Any process in which a particle, such as a neutron, proton, electron, muon, or alpha particle, combines with a nucleus.

nuclear chain reaction [NUCLEO] A succession of generation after generation of acts of nuclear division such that the neutrons set free in the nuclear disruptions of the nth generation split the fissile nuclei (U^{233}, U^{235}, Pu^{239}) of the $(n + 1)$st generation.

nuclear chemistry [ATOM PHYS] Study of the atomic nucleus, including fission and fusion reactions and their products.

nuclear cloud [NUCLEO] The cloud of dust and gas formed from the debris of a nuclear explosion.

nuclear collision [NUC PHYS] A collision between an atomic nucleus and another nucleus or particle.

nuclear converter See converter.

nuclear cross section [NUC PHYS] A measure of the probability for a reaction to occur between a nucleus and a particle; it is an area such that the number of reactions which occur in a sample exposed to a beam of particles equals the product of the number of nuclei in the sample and the number of incident particles which would pass through this area if their velocities were perpendicular to it.

nuclear decay mode [NUC PHYS] One of the ways in which a nucleus can undergo radioactive decay, distinguished from other decay modes by the resulting isotope and the particles emitted.

nuclear density [NUC PHYS] The mass per unit volume of a nucleus as a function of distance from the center of the nucleus, as determined by a number of different types of experiments which are in reasonably good agreement.

nuclear device [NUCLEO] A nuclear explosive used for peaceful purposes, tests, or experiments.

nuclear electric power generation [ELEC] Large-scale generation of electric power in which the source of energy is nuclear fission, generally in a nuclear reactor, or nuclear fusion.

nuclear-electric propulsion [AERO ENG] A system of propulsion utilizing a nuclear reactor to generate electricity which is then used in an electric propulsion system or as a heat source for the working fluid.

nuclear-electric rocket engine [AERO ENG] A rocket engine in which a nuclear reactor is used to generate electricity that is used in an electric propulsion system or as a heat source for the working fluid.

nuclear emulsion [NUCLEO] A photographic emulsion specially designed to register individual tracks of ionizing particles.

nuclear emulsion counter [NUCLEO] A device used to measure the intensity of ionizing radiation by counting the number of tracks in an emulsion which has been exposed to the radiation.

nuclear energy [NUCLEO] Energy released by nuclear fission or nuclear fusion. Also known as atomic energy.

nuclear engine [NUCLEO] A type of thermal engine utilizing nuclear fission or fusion reactions to heat a working fluid for propulsive purposes.

nuclear engineering [NUCLEO] The branch of technology that deals with the utilization of the nuclear fission process, and is concerned with the design and construction of nuclear reactors and auxiliary facilities, the development and fabrication of special materials, and the handling and processing of reactor products.

nuclear excavation [ENG] The use of nuclear explosions to remove earth for constructing harbors, canals, and other facilities.

nuclear explosion [NUCLEO] An explosion for which the energy is produced by a nuclear transformation, either fission or fusion.

nuclear fission See fission.

nuclear force [NUC PHYS] That part of the force between nucleons which is not electromagnetic; it is much stronger than electromagnetic forces, but drops off very rapidly at distances greater than about 10^{-13} centimeter; it is responsible for holding the nucleus together.

nuclear fuel [NUCLEO] A fissionable or fertile isotope with a reasonably long half-life, used as a source of energy in a nuclear reactor. Also known as fission fuel; reactor fuel.

nuclear fuel cycle See reactor fuel cycle.

nuclear fuel element [NUCLEO] A piece of nuclear fuel which has been formed and coated, and is ready to be placed in a reactor fuel assembly. Also known as reactor fuel element.

nuclear fuel form [NUCLEO] Any chemical form of fissionable nuclear fuel, such as UO_2.

nuclear fuel pebble See nuclear fuel pellet.

nuclear fuel pellet [NUCLEO] A piece of nuclear fuel usually in the shape of a sphere or cylinder, used in pebble-bed reactors, inserted in graphite blocks, or used in metallic tubular fuel elements. Also known as fuel ball; nuclear fuel pebble; reactor fuel pellet.

nuclear fuel plate [NUCLEO] A flat or curved sandwich of metallic cladding, with nuclear fuel inside.

nuclear fuel reprocessing [NUCLEO] The periodic chemical, physical, and metallurgical treatment of materials used as fuel elements in nuclear reactors, to recover and purify the residual fissionable and fertile materials.

nuclear fusion See fusion.

nuclear ground state [NUC PHYS] The stationary state of lowest energy of an isotope.

nuclear gyroscope [ENG] A gyroscope in which the conventional spinning mass is replaced by the spin of atomic nuclei and electrons; one version uses optically pumped mercury isotopes, and another uses nuclear magnetic resonance techniques.

nuclear heat [NUCLEO] The heat released in a nuclear reactor due to the fission process.

nuclear induction [PHYS] Magnetic induction originating in the magnetic moments of nuclei; the effect depends on the unequal population of energy states available when the material is placed in a magnetic field.

nuclear magnetic moment [NUC PHYS] The magnetic dipole moment of an atomic nucleus; a vector whose scalar product with the magnetic flux density gives the negative of the energy of interaction of a nucleus with a magnetic field.

nuclear magnetic resonance [PHYS] A phenomenon exhibited by a large number of atomic nuclei, in which nuclei in a static magnetic field absorb energy from a radio-frequency

NUCLEAR DENSITY

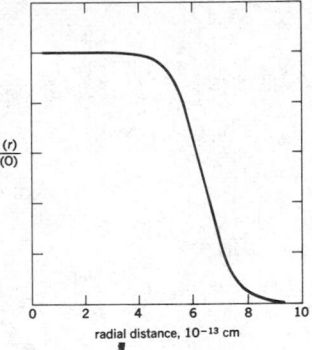

Distribution of charge in a nucleus of gold; this is believed to be approximately proportional to the nuclear mass density, on the assumption that the distributions of protons and neutrons are approximately the same; $\rho(r)$ = density at distance r from the center; $\rho(O)$ = density at the center.

NUCLEAR EXPLOSION

Medium-height nuclear detonation (air burst) evolving into characteristic mushroom shape.

NUCLEAR FORCE

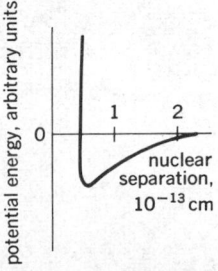

The potential operating between two nucleons due to the nuclear forces.

field at certain characteristic frequencies. Abbreviated NMR. Also known as magnetic nuclear resonance.

nuclear magnetism [PHYS] The phenomena associated with the magnetic dipole, octupole, and higher moments of a nucleus, including the magnetic field generated by the nucleus, the force on the nucleus in an inhomogeneous magnetic field, and the splitting of nuclear energy levels in a magnetic field.

nuclear magnetometer [ENG] Any magnetometer which is based on the interaction of a magnetic field with nuclear magnetic moments, such as the proton magnetometer.

nuclear magneton [NUC PHYS] A unit of magnetic dipole moment used to express magnetic moments of nuclei and baryons; equal to the electron charge times Planck's constant divided by the product of 4π, the proton mass, and the speed of light.

nuclear mass [NUC PHYS] The mass of an atomic nucleus, which is usually measured in atomic mass units; it is less than the sum of the masses of its constituent protons and neutrons by the binding energy of the nucleus divided by the square of the speed of light.

nuclear membrane [CYTOL] The envelope surrounding the cell nucleus, separating the nucleoplasm from the cytoplasm; composed of two membranes and contains numerous pores.

nuclear moment [NUC PHYS] One of the various static electric or magnetic multipole moments of a nucleus.

nuclear number See mass number.

nuclear paramagnetism [PHYS] Paramagnetism in which a substance develops a net magnetic moment because the magnetic moments of nuclei tend to point in the direction of the field.

nuclear physics [PHYS] The study of the characteristics, behavior, and internal structures of the atomic nucleus.

nuclear pile See nuclear reactor.

nuclear polarization [NUC PHYS] For a nucleus in a mixed state, with spin I and probability $p(I_z)$ that the I_z substate is populated, the polarization is the sum over allowed values of I_z of $I_z p(I_z)/I$.

nuclear potential [NUC PHYS] The potential energy of a nuclear particle as a function of its position in the field of a nucleus or of another nuclear particle.

nuclear potential energy [NUC PHYS] The average total potential energy of all the protons and neutrons in a nucleus due to the nuclear forces between them, excluding the electrostatic potential energy.

nuclear potential scattering [NUC PHYS] That part of elastic scattering of particles by a nucleus which may be treated by studying the scattering of a wave which obeys the Schrödinger equation with a potential determined by the properties of the nucleus.

nuclear power [NUCLEO] Power whose source is nuclear fission or fusion

nuclear power plant [NUCLEO] A power plant in which nuclear energy is converted into heat for use in producing steam for turbines, which in turn drive generators that produce electric power for commercial use.

nuclear propulsion [NAV ARCH] Propulsion of a ship or submarine by an engine driven by steam generated by nuclear energy in a reactor, rather than combustion of fuel in a boiler.

nuclear quadrupole moment [NUC PHYS] The electric quadrupole moment of an atomic nucleus.

nuclear quadrupole resonance [PHYS] The phenomenon in which certain nuclei in a static, inhomogeneous electric field absorb energy from a radio-frequency field.

nuclear radiation [NUC PHYS] A term used to denote alpha particles, neutrons, electrons, photons, and other particles which emanate from the atomic nucleus as a result of radioactive decay and nuclear reactions.

nuclear radiation spectroscopy [NUC PHYS] Study of the distribution of energies or momenta of particles emitted by nuclei.

nuclear radius [NUC PHYS] The radius of a sphere within which the nuclear density is large, and at the surface of which it falls off sharply.

nuclear reaction [NUC PHYS] A reaction involving a change in an atomic nucleus, such as fission, fusion, neutron capture, or radioactive decay, as distinct from a chemical reaction, which is limited to changes in the electron structure surrounding the nucleus. Also known as reaction.

nuclear reactor [NUCLEO] A device containing fissionable material in sufficient quantity and so arranged as to be capable of maintaining a controlled, self-sustaining nuclear fission chain reaction. Also known as atomic pile (deprecated usage); atomic reactor; fission reactor; nuclear pile (deprecated); pile (deprecated); reactor.

nuclear relaxation [PHYS] The approach of a system of nuclear spins to a steady-state or equilibrium condition over a period of time, following a change in the applied magnetic field.

nuclear resonance [NUC PHYS] **1.** An unstable excited state formed in the collision of a nucleus and a bombarding particle, and associated with a peak in a plot of cross section versus energy. **2.** The absorption of energy by nuclei from radio-frequency fields at certain frequencies when these nuclei are also subjected to certain types of static fields, as in magnetic resonance and nuclear quadrupole resonance.

nuclear rocket See atomic rocket.

nuclear scattering [NUC PHYS] The change in directions of particles as a result of collisions with nuclei.

nuclear sclerosis [MED] Hardening of the ocular lens nucleus.

nuclear ship [NAV ARCH] A ship in which nuclear energy in a reactor, rather than combustion of fuel in a boiler, generates steam which drives the engine to propel the ship.

nuclear species See nuclide.

nuclear spectrum [NUC PHYS] **1.** The relative number of particles emitted by atomic nuclei as a function of energy or momenta of these particles. **2.** The graphical display of data from devices used to measure these quantities.

nuclear spin [NUC PHYS] The total angular momentum of an atomic nucleus, resulting from the coupled spin and orbital angular momenta of its constituent nuclei. Also known as nuclear angular momentum. Symbolized I.

nuclear spontaneous reaction See radioactive decay.

nuclear stability [NUC PHYS] The ability of an isotope to resist decay or fission.

nuclear star See star.

nuclear submarine [NAV ARCH] A submarine in which nuclear energy in a reactor generates steam which drives the engine to propel the ship.

nuclear superheating [NUCLEO] Superheating the steam produced in a reactor by using additional heat from a reactor; two methods may be used: recirculating the steam through the same core in which it is first produced (integral superheating) or passing the steam through a second and separate reactor.

nuclear thermionic converter [NUCLEO] A thermionic converter whose heat source is a nuclear reactor or radioisotope.

nuclear transformation See transmutation.

nuclear triode detector [ELECTR] A type of junction detector that has two outputs which together determine the precise location on the detector where the ionizing radiation was incident, as well as the energy of the ionizing particle.

nuclear warhead [ORD] A warhead that contains fissionable or fissionable-fusionable material.

nuclear weapon See atomic weapon.

nuclear Zeeman effect [SPECT] A splitting of atomic spectral lines resulting from the interaction of the magnetic moment of the nucleus with an applied magnetic field.

nuclease [BIOCHEM] An enzyme that catalyzes the splitting of nucleic acids to nucleotides, nucleosides, or the components of the latter.

nucleate boiling [CHEM ENG] Boiling in which bubble formation is at the liquid-solid interface rather than from external or mechanical devices; occurs in kettle-type and natural-circulation heaters or reboilers.

nucleated glass [MATER] Glass treated with a nucleating agent to transform it into a crystalline material.

nucleation [CHEM] In crystallization processes, the formation of new crystal nuclei in supersaturated solutions.

nucleic acid [BIOCHEM] A large, acidic, chainlike molecule containing phosphoric acid, sugar, and purine and pyrimidine bases; two types are ribonucleic acid and deoxyribonucleic acid.

nuclein [BIOCHEM] Any of a poorly defined group of nucleic acid protein complexes occurring in cell nuclei.

nucleocytoplasmic ratio [CYTOL] The ratio between the measured cross-sectional area or estimated volume of the nucleus of a cell to the volume of its cytoplasm. Also known as karyoplasmic ratio.

nucleolar [CYTOL] Of or pertaining to the nucleolus.

nucleolus [CYTOL] A small, spherical body composed principally of protein and located in the metabolic nucleus. Also known as plasmosome.

nucleon [PHYS] A collective name for a proton or a neutron; these particles are the main constituents of atomic nuclei, have approximately the same mass, have a spin of $\frac{1}{2}$, and can transform into each other through the process of beta decay.

nucleonics [ENG] The technology based on phenomena of the atomic nucleus such as radioactivity, fission, and fusion; includes nuclear reactors, various applications of radioisotopes and radiation, particle accelerators, and radiation-detection devices.

nucleonium [ATOM PHYS] A bound state of a nucleus and an antinucleus.

nucleon number *See* mass number.

nucleoplasm [CYTOL] The protoplasm of a nucleus. Also known as karyoplasm.

nucleoprotein [BIOCHEM] Any member of a class of conjugated proteins in which molecules of nucleic acid are closely associated with molecules of protein.

nucleoreticulum [CYTOL] Any type of network found within a nucleus.

nucleosidase [BIOCHEM] An enzyme that catalyzes the hydrolysis of a nucleoside to its component pentose and its purine or pyrimidine base.

nucleoside [BIOCHEM] The glycoside resulting from removal of the phosphate group from a nucleotide; consists of a pentose sugar linked to a purine or pyrimidine base.

nucleosome [CYTOL] A morphologically repeating unit of DNA containing 190 base pairs of DNA folded together with eight histone molecules. Also known as v-body.

nucleospindle [CYTOL] A mitotic spindle derived from nuclear material.

nucleotidase [BIOCHEM] Any of a group of enzymes which split phosphoric acid from nucleotides, leaving nucleosides.

nucleotide [BIOCHEM] An ester of a nucleoside and phosphoric acid; the structural unit of a nucleic acid.

nucleus [ADP] 1. That portion of the control program that must always be present in main storage. 2. The main storage area used in the nucleus (first definition) and other transient control program routines. [ASTRON] The small permanent body of a comet, believed to have a diameter between one and a few tens of kilometers, and to be composed of water and volatile hydrocarbons. [CYTOL] A small mass of differentiated protoplasm rich in nucleoproteins and surrounded by a membrane; found in most animal and plant cells, contains chromosomes, and functions in metabolism, growth, and reproduction. [HISTOL] A mass of nerve cells in the central nervous system. [HYD] A particle of any nature upon which, or a locus at which, molecules of water or ice accumulate as a result of a phase change to a more condensed state. [NUC PHYS] The central, positively charged, dense portion of an atom. Also known as atomic nucleus. [SCI TECH] A central mass about which accretion takes place.

nucleus counter [ENG] An instrument which measures the number of condensation nuclei or ice nuclei per sample volume of air.

nuclide [NUC PHYS] A species of atom characterized by the number of protons, number of neutrons, and energy content in the nucleus, or alternatively by the atomic number, mass number, and atomic mass; to be regarded as a distinct nuclide, the atom must be capable of existing for a measurable lifetime, generally greater than 10^{-10} second. Also known as nuclear species; species.

Nuda [INV ZOO] A class of the phylum Ctenophora distinguished by the lack of tentacles.

Nudechiniscidae [INV ZOO] A family of heterotardigrades in the suborder Echiniscoidea characterized by a uniform cuticle.

Nudibranchia [INV ZOO] A suborder of the Opisthobranchia containing the sea slugs; these mollusks lack a shell and a mantle cavity, and the gills are variable in size and shape.

nudism [PSYCH] Intolerance for wearing clothing, with a morbid tendency for the individual to remove his clothing.

nuée ardente [GEOL] A turbulent, rapidly flowing, and sometimes incandescent gaseous cloud erupted from a volcano and containing ash and other pyroclastics in its lower part. Also known as glowing cloud; Pelean cloud.

nugget [GEOL] A small mass of metal found free in nature. [MET] A weld bead.

null [MATH] A term meaning that an object is nonexistent or a quantity is zero. [NAV] 1. The azimuth or elevation reading on a navigational device indicated by minimum signal output. 2. Any of the nodal points on the radiation patterns of some antennas.

nullary composition [MATH] The selection of a particular element of a set.

null-balance recorder [ENG] An instrument in which a motor-driven slide wire in a measuring circuit is continuously adjusted so that the voltage or current to be measured will be balanced against the voltage or current from this circuit; a pen linked to the slide wire makes a graphical record of its position as a function of time.

null-current circuit [ELECTR] A circuit used to measure current, in which the unknown current is opposed by a current resulting from applying a voltage controlled by a slide wire across a series resistor, and the slide wire is continuously adjusted so that the resulting current, as measured by a direct-current detector amplifier, is equal to zero.

null-current measurement [ELECTR] Measurement of current using a null-current circuit.

null detection [ELEC] Altering of adjustable bridge circuit components, to obtain zero current. [NAV] Method of determining the radio direction by altering the antenna position to obtain minimum signal strength.

null detector *See* null indicator.

null geodesic [MATH] In a Riemannian space, a minimal geodesic curve. [RELAT] A curve in space-time which has the property that the infinitesimal interval between any two neighboring points on the curve equals zero; it represents a possible path of a light ray. Also known as zero geodesic.

null hypothesis [STAT] The hypothesis that there is no validity to the specific claim that two variations (treatments) of the same thing can be distinguished by a specific procedure.

null indicator [ENG] A galvanometer or other device that indicates when voltage or current is zero; used chiefly to determine when a bridge circuit is in balance. Also known as null detector.

nullity [MATH] The dimension of the null space of a linear transformation.

null matrix [MATH] The matrix all of whose entries are zero.

null method [ENG] A method of measurement in which the measuring circuit is balanced to bring the pointer of the indicating instrument to zero, as in a Wheatstone bridge, and the settings of the balancing controls are then read. Also known as balance method; zero method.

null reference system [NAV] A type of instrument landing system in which the antenna array consists of two elements, one twice as high as the other; the lower element radiates with maximum intensity along the glide path, while the upper element produces a signal null at the glide path angle.

null sequence [MATH] A sequence of numbers or functions which converges to the number zero or the zero function.

null set [MATH] The empty set; the set which contains no elements.

null space [MATH] For a linear transformation, the vector subspace of all vectors which the transformation sends to the zero vector. Also known as kernel.

null vector [MATH] A vector whose invariant length, that is, the sum over the coordinates of the vector space of the product of its covariant component and contravariant component, is equal to zero. [RELAT] In special relativity, a four vector whose spatial part in any Lorentz frame has a magnitude equal to the speed of light multiplied by its time part in that frame; a special case of the mathematics definition.

number [MATH] 1. Any real or complex number. 2. The number of elements in a set is the cardinality of the set.

number class modulo N [MATH] The class of all numbers which differ from a given number by a multiple of N.

NUMERICAL CONTROL

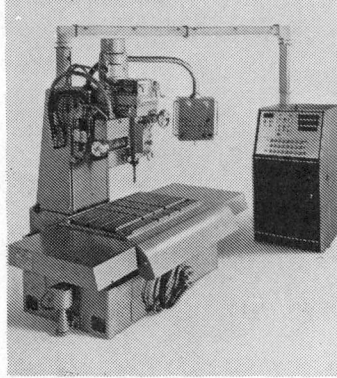

A typical numerically controlled machine tool, showing the numerical control system or "director" at the right. *(Cintimatic Div., Cincinnati Milacron Co.)*

NUMERICAL INDICATOR TUBE

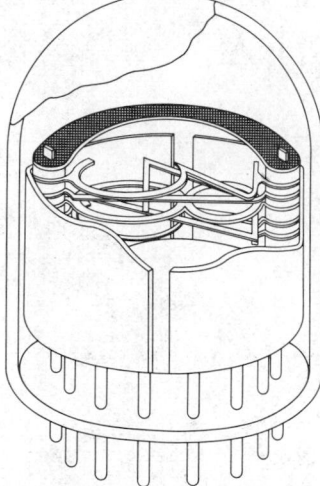

Cold-cathode gas-type numerical indicator tube.

NUTMEG

Nutmeg *(Myristica fragrans)*, mature fruits. *(USDA)*

number record printers [COMMUN] Printers in a relay station to provide a complete automatic written record of channel numbers and the fixed routing line associated with each message that is relayed through that particular station.

number scale [MATH] Representation of points on a line with numbers arranged in some order.

number theory [MATH] The study of integers and relations between them.

numeral [MATH] A symbol used to denote a number.

numeral system *See* numeration system.

numeration [MATH] The listing of numbers in their natural order.

numeration system [MATH] An orderly method of representing numbers by numerals in which each numeral is associated with a unique number. Also known as numeral system.

numerator [MATH] In a fraction a/b, the numerator is the quantity a.

numeric [ADP] In computers, composed wholly or partly of digits, as distinct from alphabetic.

numerical [MATH] Pertaining to numbers.

numerical analysis [MATH] The study of approximation techniques using arithmetic for solutions of mathematical problems.

numerical aperture [OPTICS] A measure of the resolving power of a microscope objective, equal to the product of the refractive index of the medium in front of the objective and the sine of the angle between the outermost ray entering the objective and the optical axis. Abbreviated N.A.

numerical control [CONT SYS] A control system for machine tools and some industrial processes, in which numerical values corresponding to desired positions of tools or controls are recorded on punched paper tapes, punched cards, or magnetic tapes so that they can be used to control the operation automatically. Abbreviated NC; N/C.

numerical decrement *See* decrement.

numerical display device [ELECTR] Any device for visually displaying numerical figures, such as a numerical indicator tube, a device utilizing electroluminescence, or a device in which any one of a stack of transparent plastic strips engraved with digits can be illuminated by a small light at the edge of the strip.

numerical equation [MATH] An equation all of whose constants and coefficients are numbers.

numerical forecasting [METEOROL] The forecasting of the behavior of atmospheric disturbances by the numerical solution of the governing fundamental equations of hydrodynamics, subject to observed initial conditions. Also known as dynamic forecasting; mathematical forecasting; numerical weather prediction; physical forecasting.

numerical indicator tube [ELECTR] An electron tube capable of visually displaying numerical figures; some varieties also display alphabetical characters and commonly used symbols.

numerical system [TEXT] A grade scale for calculating the quality of wool. Also known as count system.

numerical tape [ADP] The tape required by a computer operating a machine tool.

numerical taxonomy [SYST] The numerical evaluation of the affinity or similarity between taxonomic units and the ordering of these units into taxa on the basis of their affinities.

numerical value of a real number *See* absolute value of a real number.

numerical weather prediction *See* numerical forecasting.

numeric character *See* digit.

numeric coding [ADP] Code in which only digits are used, usually binary or octal.

numeric control [ADP] The action of programs written for specialized computers which operate machine tools.

numeric printer [ADP] Old type of printer which positioned its keys to print a field in one operation, rather than one digit at a time.

numeric punch [ADP] Punching of holes in a column of a computer card so that only one hole in rows zero through nine is punched in the column.

Numididae [VERT ZOO] A family of birds in the order Galliformes commonly known as guinea fowl; there are few if any feathers on the neck or head, but there may be a crest of feathers and various fleshy appendages.

nunatak [GEOL] An isolated hill, knob, ridge, or peak of bedrock projecting prominently above the surface of a glacier and completely surrounded by glacial ice.

nun buoy [NAV] A red buoy made of two conical or truncated cone-shaped sections joined at the base; marks the starboard side when entering a channel from the sea.

N unit [OPTICS] A unit of index of refraction; a mathematical simplification designed to replace rather awkward numbers involved in the values of the index of refraction n for the atmosphere; it is defined by the relation $N = (n-1)10^6$.

nuplex [NUCLEO] Proposed nuclear reactor which would both produce electrical power and heat water for desalination and other industrial benefits.

nurse cell [HISTOL] A cell type of the ovary of many animals which nourishes the developing egg cell.

nurse graft [BOT] A plant graft in which the scion remains united with the stock only until roots develop on the scion.

nursing [MED] The application of the principles of physical, biological, and social sciences in the physical and mental care of people.

Nusselt equation [THERMO] Dimensionless equation used to calculate convection heat transfer for heating or cooling of fluids outside a bank of 10 or more rows of tubes to which the fluid flow is normal.

Nusselt number [PHYS] A dimensionless number used in the study of mass transfer, equal to the mass-transfer coefficient times the thickness of a layer through which mass transfer is taking place divided by the molecular diffusivity. Symbolized Nu_m; N_{Nu_m}. Also known as Sherwood number (N_{Sh}). [THERMO] A dimensionless number used in the study of forced convection which gives a measure of the ratio of the total heat transfer to conductive heat transfer, and is equal to the heat-transfer coefficient times a characteristic length divided by the thermal conductivity. Symbolized N_{Nu}.

nut [BOT] **1.** A fruit which has at maturity a hard, dry shell enclosing a kernel consisting of an embryo and nutritive tissue. **2.** An indehiscent, one-celled, one-seeded, hard fruit derived from a single, simple, or compound ovary. [DES ENG] An internally threaded fastener for bolts and screws.

nutating antenna [ENG] An antenna system used in conical scan radar, in which a dipole or feed horn moves in a small circular orbit about the axis of a paraboloidal reflector without changing its polarization.

nutating-disk meter [ENG] An instrument for measuring flow of a liquid in which liquid passing through a chamber causes a disk to nutate, or roll back and forth, and the total number of rolls is mechanically counted.

nutation [BOT] Rhythmic change in the position of growing plant organs caused by variation in the growth rates on different sides of the growing apex. [MECH] A bobbing or nodding up-and-down motion of a spinning rigid body, such as a top, as it precesses about its vertical axis.

nutator [ENG] A mechanical or electrical device used to move a radar beam in a circular, conical, spiral, or other manner periodically to obtain greater air surveillance than could be obtained with a stationary beam.

nutgall [PL PATH] A nutlike gall.

nutmeg [BOT] *Myristica fragrans.* A dark-leafed evergreen tree of the family Myristicaceae cultivated for the golden-yellow fruits which resemble apricots; a delicately flavored spice is obtained from the kernels inside the seeds.

nutmeg liver [MED] Chronic passive hyperemia of the liver; the cut surface of the diseased organ resembles the cut surface of a nutmeg.

nutmeg oil [MATER] A pale-yellow or colorless essential oil with spicy taste and nutmeg aroma; obtained from nutmegs; soluble in alcohol, carbon bisulfide, and glacial acetic acid; chief constituents are myristin, pinene, and dipentene; used in flavors, perfumes, and medicines. Also known as myristica oil.

nutrient [BIOL] Providing nourishment.

nutrient foramen [ANAT] The opening into the canal which gives passage to the blood vessels of the medullary cavity of a bone.

nutrition [BIOL] The science of nourishment, including the study of nutrients that each organism must obtain from its environment in order to maintain life and reproduce.

nutritional anemia [MED] Anemia resulting from certain nutritional deficiencies.

nutritional dystrophy *See* kwashiorkor.

nutritional edema [MED] Edema resulting from starvation or malnutrition.

nutritional hypochromic anemia *See* iron-deficiency anemia.

Nuttalliellidae [INV ZOO] A family of ticks (Ixodides) containing one rare African species, *Nuttalliella namaqua*, morphologically intermediate between the families Argasidae and Ixodidae.

nu value [OPTICS] The reciprocal of the dispersive power of a medium. Also known as constringence.

nuvistor [ELECTR] Electron tube in which all electrodes are cylindrical, placed one inside the other with close spacing, in a ceramic envelope.

nux vomica [BOT] The seed of *Strychnos nux-vomica*, an Indian tree of the family Loganiaceae; contains the alkaloid strychnine, and was formerly used in medicine.

n value *See* field index.

Nyctaginaceae [BOT] A family of dicotyledonous plants in the order Caryophyllales characterized by an apocarpous, monocarpous, or syncarpous gynoecium, sepals joined to a tube, a single carpel, and a cymose inflorescence.

nyctalopia *See* night blindness.

Nycteridae [VERT ZOO] The slit-faced bats, a monogeneric family of insectivorous chiropterans having a simple, well-developed nose leaf, and large ears joined together across the forehead.

Nyctibiidae [VERT ZOO] A family of birds in the order Caprimulgiformes including the neotropical potoos.

nyctinasty [BOT] A nastic movement in higher plants associated with diurnal light and temperature changes.

nyctophobia [PSYCH] An abnormal fear of night and darkness.

Nyctribiidae [INV ZOO] The bat tick flies, a family of myodarian cyclorrhaphous dipteran insects in the subsection Acalypteratae.

Nylander reagent [CHEM] A solution of Rochelle salt (potassium sodium tartrate), potassium or sodium hydroxide, and bismuth subnitrate in water; used to test for sugar in urine.

nylon [MATER] Generic name for long-chain polymeric amide molecules in which recurring amide groups are part of the main polymer chain; used to make fibers, fabrics, sheeting, and extruded forms.

nylon 6 [MATER] Nylon made by polycondensation of caprolactam.

nylon 66 [MATER] Nylon made by condensation of hexamethylene diamine with adipic acid.

nylon 610 [MATER] Nylon made by the condensation of hexamethylene diamine with sebacic acid.

nymph [INV ZOO] Any immature larval stage of various hemimetabolic insects.

Nymphaeaceae [BOT] A family of dicotyledonous plants in the order Nymphaeales distinguished by the presence of roots, perfect flowers, alternate leaves, and uniaperturate pollen.

Nymphaeales [BOT] An order of flowering aquatic herbs in the subclass Magnoliidae; all lack cambium and vessels and have laminar placentation.

Nymphalidae [INV ZOO] The four-footed butterflies, a family of lepidopteran insects in the superfamily Papilionoidea; prothoracic legs are atrophied, and the well-developed patagia are heavily sclerotized.

Nymphalinae [INV ZOO] A subfamily of the lepidopteran family Nymphalidae.

nymphomania [PSYCH] Excessive sexual desire on the part of a woman. Also known as hysteromania.

Nymphonidae [INV ZOO] A family of marine arthropods in the subphylum Pycnogonida; members have chelifores, five-jointed palpi, and ten-jointed ovigers.

Nymphulinae [INV ZOO] A subfamily of the lepidopteran family Pyralididae which is notable because some species are aquatic.

Nyquist contour [CONT SYS] A directed closed path in the complex frequency plane used in constructing a Nyquist diagram, which runs upward, parallel to the whole length of the imaginary axis at an infinitesimal distance to the right of it, and returns from $+j\infty$ to $-j\infty$ along a semicircle of infinite radius in the right half-plane.

Nyquist diagram [CONT SYS] A plot in the complex plane of the open-loop transfer function as the complex frequency is varied along the Nyquist contour; used to determine stability of a control system.

Nyquist interval [COMMUN] Maximum separation in time which can be given to regularly spaced instantaneous samples of a wave of specified bandwidth for complete determination of the waveform of the signal.

Nyquist rate [COMMUN] The maximum rate at which code elements can be unambiguously resolved in a communications channel with a limited range of frequencies; equal to twice the frequency range.

Nyquist stability criterion *See* Nyquist stability theorem.

Nyquist stability theorem [CONT SYS] The theorem that the net number of counterclockwise rotations about the origin of the complex plane carried out by the value of an analytic function of a complex variable, as its argument is varied around the Nyquist contour, is equal to the number of poles of the variable in the right half-plane minus the number of zeros in the right half-plane. Also known as Nyquist stability criterion.

Nyquist's theorem [ELECTR] The mean square noise voltage across a resistance in thermal equilibrium is four times the product of the resistance, Boltzmann's constant, the absolute temperature, and the frequency range within which the voltage is measured.

Nysmyth's membrane [ANAT] The primary enamel cuticle which is the transitory remnants of the enamel organ and oral epithelium covering the enamel of a tooth after eruption.

Nyssaceae [BOT] A family of dicotyledonous plants in the order Cornales characterized by perfect or unisexual flowers with imbricate petals, a solitary ovule in each locule, a unilocular ovary, and more stamens than petals.

nystagmus [MED] Involuntary oscillatory movement of the eyeballs.

nystatin [MICROBIO] $C_{46}H_{77}NO_{19}$ An antifungal antibiotic produced by *Streptomyces noursei*; used for the treatment of infections caused by *Candida (Monilia) albicans*.

NYLON

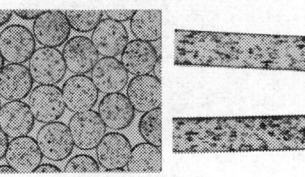

Cross-sectional (left) and longitudinal (right) views of nylon fiber.

NYMPHAEACEAE

A common eastern American species of water lily (*Nymphaea odorata*) in the family Nymphaeaceae of the order Nymphaeales. The large and indefinite number of tepals, stamens, and carpels is characteristic of the family and a large part of the order. *(Photograph by Hugh Spencer, National Audubon Society)*

O *See* oxygen.

oak [BOT] Any tree of the genus *Quercus* in the order Fagales, characterized by simple, usually lobed leaves, scaly winter buds, a star-shaped pith, and its fruit, the acorn, which is a nut; the wood is tough, hard, and durable, generally having a distinct pattern.

oakmoss resin [MATER] Concrete oleoresin from the oakmoss lichens *Evernia prunastri* and *E. furfuracea*; used as a perfume fixative. Also known as mousse de chêne.

oakum [MATER] Old hemp or jute fiber, loosely twisted and impregnated with tar or a tar derivative, used to caulk sides and decks of ships and to pack joints of pipes and caissons.

oak wilt [PL PATH] A fungus disease of oak trees caused by *Chalara quercina*, characterized by wilting and yellow and red discoloration of the leaves progressing from the top downward and inward.

O antigen [MICROBIO] A somatic antigen of certain flagellated microorganisms.

oasis [GEOGR] An isolated fertile area, usually limited in extent and surrounded by desert, and marked by vegetation and a water supply.

oat [BOT] Any plant of the genus *Avena* in the family Graminae, cultivated as an agricultural crop for its seed, a cereal grain, and for straw.

oatmeal paper [MATER] A paper in which fine sawdust is added to produce a sheet with a coarse texture; it can be used as an inexpensive sketching paper for work in pastels and charcoal, or as wallpaper.

O attenuator [ELECTR] A dissipative attenuator in which the circuit has the form of a ladder with two rungs, and the resistances across the rungs are unequal, so that the impedances across the two pairs of terminals are unequal.

OBA *See* octave-band analyzer.

obclavate [BIOL] Inversely clavate.

obcordate [BOT] Referring to a leaf, heart-shaped with the notch apical.

obdiplastemonous [BOT] Having the stamens arranged in two whorls, with members of the outer whorl positioned opposite the petals.

obduction [MED] The act or instance of performing a postmortem examination.

obelion [ANTHRO] The point where the line which joins the parietal foramens crosses the sagittal suture.

obelisk [ARCH] A four-sided pillar, tapering toward the top. [MATH] A frustrum of a regular, rectangular pyramid.

Obermayer's reagent [CHEM] A 0.4% solution of ferric chloride in concentrated hydrochloric acid; used to test for indican in urine, with a pale-blue or deep-violet color indicating positive.

Oberon [ASTRON] One of the five satellites of Uranus; diameter about 1400 kilometers.

oberwind [METEOROL] A night wind from mountains or the upper ends of lakes; a wind of Salzkammergut in Austria.

obese [ANAT] Extremely fat.

obfuscation [PSYCH] Mental confusion.

object [OPTICS] A collection of points which may be regarded as a source of light rays in an optical system, whether it actually has this function (as in a real object) or does not (as in a virtual object).

object computer [ADP] The computer processing an object program; the same computer compiling the source, program could, therefore, be called the source computer, such terminology is seldom used in practice.

object deck [ADP] The set of machine-readable computer instructions produced by a compiler, either in absolute format (that is, containing only fixed addresses) or, more frequently, in relocatable format.

objective [OPTICS] The first lens, lens system, or mirror through which light passes or from which it is reflected in an optical system; many scientists exclude mirrors from the definition.

objective basic research [SCI TECH] Basic research in fields recognized as having a potential technological importance.

objective function [MATH] In nonlinear programming, the function, expressing given conditions for a system, which one seeks to minimize subject to given constraints.

objective grating [OPTICS] A series of equally spaced parallel wires placed over the objective lens of a telescope; photographic magnitudes of stars are calculated from the relative brightnesses of images in the resulting diffraction pattern.

objective plane [ORD] Plane tangent to the ground or coinciding with the surface of the target, especially such a plane at the point of impact of a bomb or projectile.

objective prism [OPTICS] A large prism, usually having a small angle, which is placed in front of the objective of a photographic telescope to make spectroscopic observations.

objective sign [MED] A sign which can be detectable by someone other than the patient.

object language [ADP] The intended and desired output language in the translation or conversion of information from one language to another.

object library *See* object program library.

object module [ADP] The computer language program prepared by an assembler or a compiler after acting on a programmer-written source program.

object program [ADP] The computer language program prepared by an assembler or a compiler after acting on a programmer-written source program. Also known as object routine; target program; target routine.

object program library [ADP] A collection of computer programs in the form of relocatable instructions, which reside on, and may be read from, a mass storage device. Also known as object library.

object relationship [PSYCH] The attitudes and responses of one person toward another; the capacity of an individual to react appropriately to and to accept and love other people.

object routine *See* object program.

object space [OPTICS] The region of space where objects are located so that a given optical system can form images of them.

object tape [ADP] A paper or magnetic, containing the machine language instructions resulting from a compiler or assembler, often found in minicomputer environments.

object time [ADP] The time during which execution of an object program is carried out.

oblanceolate [SCI TECH] Inversely lanceolate.

oblate ellipsoid *See* oblate spheroid.

oblate spheroid [MATH] The surface or ellipsoid generated by rotating an ellipse about one of its axes so that the diameter of its equatorial circle exceeds the length of the axis of revolution. Also known as oblate ellipsoid.

obligate [BIOL] Restricted to a specified condition of life, as an obligate parasite.

oblique [ANAT] Referring to a muscle, positioned obliquely and having one end that is not attached to bone. [BOT] Referring to a leaf, having the two sides of a blade unequal. [SCI TECH] Having a slanted direction or position.

oblique angle [MATH] An angle that is neither a right angle nor a multiple of a right angle.

oblique ascension [ASTRON] The arc of the celestial equator, or the angle at the celestial pole, between the hour circle of the vernal equinox and the hour circle through the intersection of the celestial equator and the eastern horizon at the instant a point on the oblique sphere rises, measured eastward from the hour circle of the vernal equinox through 24 hours.

OAK

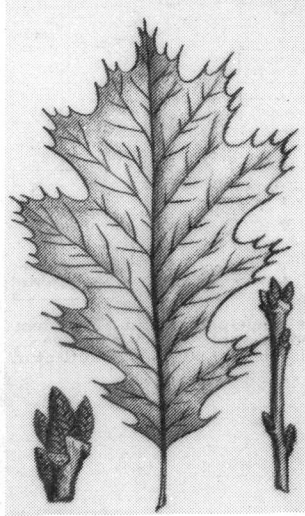

Terminal bud, leaf, and twig of white oak (*Quercus alba*).

OBLATE SPHEROID

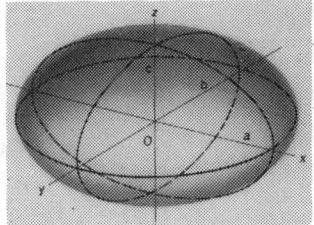

Drawing of an oblate spheroid generated by rotating an ellipse about its minor axis lying along z-axis of coordinate system, with center of ellipse at origin of coordinates, O. Diameters 2a and 2b along x and y axes are equal to each other, and greater than axis of revolution 2c.

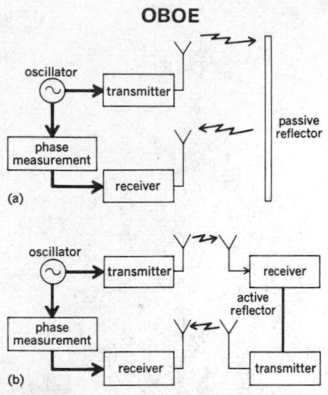

OBOE

The single-path round-trip system used in oboe. *(a)* With passive reflector. *(b)* With active reflector.

oblique chart [MAP] A chart on an oblique projection.

oblique coordinates [MATH] Magnitudes defining a point relative to two intersecting nonperpendicular lines, called axes; the magnitudes indicate the distance from each axis, measured along a parallel to the other axis; oblique coordinates are a form of cartesian coordinates.

oblique cylindrical orthomorphic projection *See* oblique Mercator projection.

oblique equator [MAP] A great circle, the plane of which is perpendicular to the axis of an oblique projection; an oblique equator serves as the origin for measurement of oblique latitude; on an oblique Mercator projection, the oblique equator is the tangent great circle.

oblique extinction *See* inclined extinction.

oblique fault *See* diagonal fault.

oblique fire [ORD] Fire placed on a target from a direction diagonal to the long dimension of the target, or on an enemy from a direction between his front and his flank.

oblique graticule [MAP] A fictitious graticule based upon an oblique projection.

oblique-incidence reflectivity [OPTICS] The reflectivity of an interface between two media when the direction of propagation of the incident electromagnetic radiation is not perpendicular to the interface; it differs for the component whose electric vector lies in the plane containing the perpendicular to the surface and the propagation direction, and the component for which this vector is perpendicular to this plane.

oblique-incidence transmission [COMMUN] Transmission of a radio wave obliquely up to the ionosphere and down again.

oblique joint *See* diagonal joint.

oblique lines [MATH] Lines that are neither perpendicular nor parallel.

oblique Mercator projection [MAP] A conformal cylindrical map projection in which points on the surface of a sphere or spheroid, such as the earth, are conceived as developed by Mercator principles on a cylinder tangent along an oblique great circle. Also known as oblique cylindrical orthomorphic projection.

oblique meridian [MAP] A great circle perpendicular to an oblique equator; the reference oblique meridian is called the prime oblique meridian.

oblique parallel [MAP] A circle or line parallel to an oblique equator, connecting all points of equal oblique latitude.

oblique pole [MAP] One of the two points 90° from an oblique equator.

oblique projection [MAP] A map projection with its axis at an oblique angle to the plane of the equator.

oblique rhumb line [MAP] **1.** A line making the same oblique angle with all fictitious meridians of an oblique Mercator projection; oblique parallels and meridians may be considered special cases of the oblique rhumb line. **2.** Any rhumb line, real or fictitious, making an oblique angle with its meridians; in this sense the expression is used to distinguish such rhumb lines from parallels and meridians, real or fictitious, which may be included in the expression "rhumb line."

oblique shock *See* oblique shock wave.

oblique shock wave [FL MECH] A shock wave inclined at an oblique angle to the direction of flow in a supersonic flow field. Also known as oblique shock.

oblique sphere [ASTRON] The celestial sphere as it appears to an observer between the equator and the pole, where celestial bodies appear to rise obliquely to the horizon.

oblique triangle [MATH] A triangle that does not contain a right angle.

oblique visibility *See* oblique visual range.

oblique visual range [OPTICS] The greatest distance at which a specified target can be perceived when viewed along a line of sight inclined to the horizontal. Also known as oblique visibility; slant visibility.

obliquity factor [OPTICS] A function which is proportional to the amplitudes of secondary waves propagating in various directions according to Huygens' principle; it is $1 + \cos\theta$, where θ is the angle between the normal to the original wavefront and the normal to the secondary wavefront.

obliquity of the ecliptic [ASTRON] The acute angle between the plane of the ecliptic and the plane of the celestial equator, about 23°27′.

obliterating endarteritis *See* endarteritis obliterans.

obliteration [MED] **1.** Complete removal of an organ or other body part by disease or surgical excision. **2.** Closure of a lumen. **3.** Loss of memory or consciousness of specific events.

obliterative appendicitis [MED] Obliteration of the lumen of the appendix by fibrofatty tissue.

oblong mesh *See* rectangular mesh.

oboe [NAV] An electronic navigation system utilizing a single-path round-trip system for determination of transmission times and distance; used for bombing in World War II.

Obolellida [PALEON] A small order of Early and Middle Cambrian inarticulate brachiopods, distinguished by a shell of calcium carbonate.

obovate [BIOL] Inversely ovate.

obscuration [METEOROL] In United States weather observing practice, the designation for the sky cover when the sky is completely hidden by surface-based obscuring phenomena, such as fog. Also known as obscured sky cover.

obscured sky cover *See* obscuration.

obscuring phenomenon [METEOROL] In United States weather observing practice, any atmospheric phenomenon (not including clouds) which restricts the vertical visibility or slant visibility, that is, which obscures a portion of the sky from the point of observation.

obsequent [GEOL] Of a stream, valley, or drainage system, being in a direction opposite to that of the original consequent drainage.

obsequent fault-line scarp [GEOL] A fault-line scarp which faces in the direction opposite to that of the original fault scarp or in which the structurally upthrown block is topographically lower than the downthrown block.

observability [CONT SYS] Property of a system for which observation of the output variables at all times is sufficient to determine the initial values of all the state variables.

observational day [GEOPHYS] Any 24-hour period selected as the basis for climatological or hydrological observations.

observation of fire [ORD] Act of watching artillery fire in order to locate the burst or impact of projectiles in respect to the target and to correct firing data; surveillance of fire; may be made from the ground or from the air.

observed fire [ORD] Fire for which the points of impact or burst can be seen by an observer on the ground, in aircraft, or on a naval vessel; it can be controlled and adjusted on the basis of the observations.

observed latitude [NAV] Latitude determined from observations which result in lines of position extending in a generally east-west direction.

observed longitude [NAV] Longitude determined from observations which result in lines of position extending in a generally north-south direction.

observed position [ORD] Position of a moving target at the instant of observation.

observer [METEOROL] Anyone who takes a weather observation.

observing angle [ORD] Angle at the target between a line to the observer and a line to the gun or battery; the angular distance of an observer from the gun or battery.

obsession [PSYCH] Persistence of or anxious preoccupation with an idea or emotion recognized as unreasonable by the individual.

obsessive-compulsive neurosis [PSYCH] A neurotic disorder in which anxiety relates to obsessions against which the individual fights but which he cannot control and by which he is dominated.

obsessive-compulsive personality [PSYCH] A behavioral disorder in which a person is generally characterized by chronic, excessive concern with conformity or adherence to standards, resulting in inhibited, inflexible behavior, inability to relax, and the performance of an inordinate amount of work.

obsidian [GEOL] A jet-black volcanic glass, usually of rhyolitic composition, formed by rapid cooling of viscous lava; generally forms the upper parts of lava flows. Also known as hyalopsite; Iceland agate; mountain mahogany.

obsidianite *See* tektite.

obsolescence [ENG] Decreasing value of functional and physical assets or value of a product or facility from technological changes rather than deterioration.

obsolete [BIOL] A part of an organism that is imperfect or indistinct, compared with a corresponding part of similar organisms. [ENG] No longer satisfactory for the purpose for which obtained, due to improvements or revised requirements.

obstacle [NAV] In air operations, a natural or artificial object which by its geographical location requires that its vertical dimension be considered in establishing the permissible flight paths in an area; this leads to the concept of the fictitious height for various obstacles; for example, a moving vehicle 17 feet high is assumed to be on an interstate highway 15 feet high, and thus the area is assumed to have an obstacle 32 feet high.

obstacle clearance [NAV] In air operations, the vertical distance between the lowest and the authorized height of operation.

obstacle clearance box [NAV] A rectangle on an aeronautical chart which encloses a number indicating the obstacle clearance required (in feet) on an approach segment where the box appears.

obstetric [MED] 1. Of or pertaining to obstetrics. 2. Of or pertaining to pregnancy and childbirth.

obstetrical analgesia [MED] Analgesia induced to diminish or obliterate the pain of childbirth.

obstetric forceps [MED] A perforated, double-bladed traction forceps which can be applied to the fetal head in cases of difficult labor.

obstetrician [MED] One who practices obstetrics.

obstetrics [MED] The branch of medicine that deals with pregnancy, labor, and the puerperium.

obstipation [MED] Constipation that is difficult to relieve.

obstruction [MED] Occlusion or stenosis of hollow viscera, ducts, and vessels. [NAV] Anything that hinders or prevents movement, particularly anything that endangers or prevents passage of a vessel or aircraft; usually refers to an isolated danger to navigation, such as a submerged rock or pinnacle in the case of marine navigation, and a tower, tall building, mountain peak, and so forth, in the case of air navigation.

obstruction beacon [NAV] A beacon marking an obstruction or hazard. Also known as hazard beacon.

obstruction buoy [NAV] A buoy marking an obstruction.

obstruction light [NAV] A light indicating the presence of an obstruction.

obstruction marker [NAV] A marker indicating the presence of an obstruction.

obstruction to vision [METEOROL] In United States weather observing practice, one of a class of atmospheric phenomena, other than the weather class of phenomena, which may reduce horizontal visibility at the earth's surface; examples are fog, smoke, and blowing snow.

obstructive atelectasis [MED] Collapse of all or a portion of the lung due to bronchial obstruction or occlusion. Also known as absorption atelectasis.

obstructive dysmenorrhea See mechanical dysmenorrhea.

obstructive emphysema [MED] Overdistension of the lung due to partial obstruction of the air passages, which permits air to enter the alveoli but which resists expiration of the air.

obstructive glaucoma See narrow-angle glaucoma.

obstructive hydrocephaly [MED] Accumulation of cerebrospinal fluid in the brain ventricles caused by obstruction of the passage of the fluid from the ventricles to the subarachnoid space. Also known as internal hydrocephaly; noncommunicating hydrocephaly.

obstructive jaundice [MED] Jaundice caused by mechanical obstruction of the biliary passages, preventing the outflow of bile.

obtund [MED] To make dull or reduce, as to obtund sensibility.

obturating cup [ORD] An inverted cup, sealing against passage of gases of explosion, which inverts due to pressure and actuates a firing pin or other mechanism.

obturation [MED] 1. The closing of an opening or passage. 2. A form of intestinal obstruction in which the lumen is occupied by its normal contents or by foreign bodies. [ORD] Sealing of the breech of a gun to prevent escape of propellant gases in firing.

obturator [ANAT] 1. Pertaining to that which closes or stops up, as an obturator membrane. 2. Either of two muscles, originating at the pubis and ischium, which rotate the femur laterally. [MED] A solid wire or rod contained within a hollow needle or cannula. [ORD] 1. Assembly of steel spindle, mushroom head, obturator rings, and a gas-check or obturator pad of tough plastic material used as a seal to prevent the escape of propellant gases around the breechblock of guns using separate-loading ammunition, and therefore not having the obturation provided by a cartridge case. 2. A device incorporated in a projectile to make the tube of a weapon gastight, preventing escape of gas until the projectile has left the muzzle of the weapon.

obturator artery [ANAT] A branch of the internal iliac; it branches into the pubic and acetabular arteries.

obturator foramen [ANAT] A large opening in the pelvis, between the ischium and the pubis, that gives passage to vessels and nerves; it is partly closed by a fibrous obturator membrane.

obturator membrane [ANAT] 1. A fibrous membrane closing the obturator foramen of the pelvis. 2. A thin membrane between the crura and foot plates of the stapes.

obturator nerve [ANAT] A mixed nerve arising in the lumbar plexus; innervates the adductor, gracilis, and obturator externus muscles, and the skin of the medial aspect of the thigh, hip, and knee joints.

obturator pad [ORD] Pad of tough plastic material, forming part of an obturator.

obturator spindle [ORD] Part of the breechblock assembly of a gun which fires separate-loading ammunition; it extends through the breechblock and holds in position the various parts of the obturator, while permitting the breechblock to rotate independently about these parts.

obtuse [BOT] Of a leaf, having a blunt or rounded free end.

obtuse angle [MATH] An angle of more than 90° and less than 180°.

obtuse triangle [MATH] A triangle having one of its angles obtuse.

obvallate [BIOL] Surrounded by or as if by a wall.

obvolute [BIOL] Overlapping.

occasional fog signal [NAV] A fog signal not sounded regularly in fog.

occasional light [NAV] A light not regularly exhibited for navigation purpose.

occipital arch [INV ZOO] A part of an insect cranium lying between the occipital suture and postoccipital suture.

occipital artery [ANAT] A branch of the external carotid which branches into the mastoid, auricular, sternocleidomastoid, and meningeal arteries.

occipital bone [ANAT] The bone which forms the posterior portion of the skull, surrounding the foramen magnum.

occipital condyle [ANAT] An articular surface on the occipital bone which articulates with the atlas. [INV ZOO] A projection on the posterior border of an insect head which articulates with the lateral neck plates.

occipital crest [ANAT] Either of two transverse ridges connecting the occipital protuberances with the foramen magnum.

occipital ganglion [INV ZOO] One of a pair of ganglia located just posterior to the brain in insects.

occipital lobe [ANAT] The posterior lobe of the cerebrum having the form of a three-sided pyramid.

occipital pole [ANAT] The tip of the occipital lobe of the brain.

occipital protuberance [ANAT] A prominence on the surface of the occipital bone to which the ligamentum nuchae is attached.

occipitofrontalis [ANAT] A muscle in two parts, the frontal (inserting in the skin of the forehead) and the occipital (inserting in the galea sponeurotica).

occiput [ZOO] The back of the head of an insert or vertebrate.

occluded cyclone [METEOROL] Any cyclone (or low) within which there has developed an occluded front.

occluded front [METEOROL] A composite of two fronts, formed as a cold front overtakes a warm front or quasi-stationary front. Also known as frontal occlusion; occlusion.

occluded gases [MIN ENG] Gases entering the mine atmosphere from feeders and blowers, and also from blasting operations.

Outboard view of *Star III*, an oceanographic submersible vessel, 24.5 feet (7.7 meters) long.

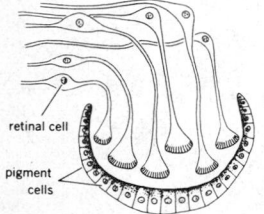

retinal cell

pigment cells

Simple pigment cup ocellus of flatworm. (*After L. Hyman; from R. D. Barnes, Invertebrate Zoology, 2d ed., W. B. Saunders Co., 1968*)

occlusal disharmony [MED] Increased or maldirected occlusal force on individual teeth or groups of teeth causing a malposition or functional aberration.

occlusion [ANAT] The relationship of the masticatory surfaces of the maxillary teeth to the masticatory surfaces of the mandibular teeth when the jaws are closed. [MED] A closing or shutting up. [METEOROL] *See* occluded front. [PHYS] Adhesion of gas or liquid on a solid mass, or the trapping of a gas or liquid within a mass. [PHYSIO] The deficit in muscular tension when two afferent nerves that share certain motoneurons in the central nervous system are stimulated simultaneously, as compared to the sum of tensions when the two nerves are stimulated separately.

occultation [ASTRON] The disappearance of the light of a celestial body by intervention of another body of larger apparent size; especially, a lunar eclipse of a star or planet.

occult blood [PATH] Blood in body products such as feces, not detectable on gross examination.

occult hydrocephaly [MED] A syndrome in which the brain ventricular system is enlarged while cerebrospinal fluid pressure remains normal, causing dementia, disturbances of equilibrium, and disorders of sphincter control.

occulting light [NAV] A navigational light totally interrupted at intervals by periods of darkness; the duration of the light period is equal to or greater than that of the dark period.

occulting quick-flashing light [NAV] A light showing flashes for several seconds, followed by a shorter period of darkness.

occult mineral [MINERAL] A mineral component of rock which cannot be seen through a microscope, but whose presence can be detected by chemical analyses.

occult virus [VIROL] A virus whose presence is assumed but which cannot be recovered.

occupational acne [MED] Acne acquired from regular exposure to acnegenic materials in certain industries; disappears when the cause is removed.

occupational disease [MED] A functional or organic disease caused by factors arising from the operations or materials of an individual's industry, trade, or occupation.

occupational medicine [MED] The branch of medicine which deals with the relationship of humans to their occupations, for the purpose of the prevention of disease and injury and the promotion of optimal health, productivity, and social adjustment.

occupational neurosis [PSYCH] Any neurotic disorder manifested by the individual's inability to use those parts of his body commonly employed in his occupation, such as a writer's inability to write due to a painful feeling of fatigue in the hand.

occupational therapy [MED] The teaching of skills or the use of selected occupations for therapeutic or rehabilitation purposes.

occupied bandwidth [COMMUN] Frequency bandwidth such that, below its lower and above its upper frequency limits, the mean powers radiated are each equal to 0.5% of the total mean power radiated by a given emission.

OC curve *See* operating characteristic curve.

ocean [GEOGR] A major primary subdivision of the intercommunicating body of salt water occupying the depressions of the earth's surface; bounded by continents and imaginary lines. Also known as sea.

ocean basin [GEOL] The great depression occupied by the ocean on the surface of the lithosphere.

ocean circulation [OCEANOGR] **1.** Water current flow in a closed circular pattern within an ocean. **2.** Large-scale horizontal water motion within an ocean.

ocean current [OCEANOGR] A net transport of ocean water along a definable path.

ocean floor [GEOL] The near-horizontal surface of the ocean basin.

ocean floor spreading *See* sea floor spreading.

Oceanian [ECOL] Of or pertaining to the zoogeographic region that includes the archipelagos and islands of the central and south Pacific.

oceanic anticyclone *See* subtropical high.

oceanic basalt [PETR] Rocks of the oceanic island volcanoes.

oceanic climate *See* marine climate.

oceanic crust [GEOL] A thick mass of igneous rock which lies under the ocean floor.

oceanic heat flow [GEOPHYS] The amount of thermal energy escaping from the earth through the ocean floor per unit area and unit time.

oceanic high *See* subtropical high.

oceanic island [GEOL] Any island which rises from the deep-sea floor rather than from shallow continental shelves.

oceanicity [CLIMATOL] The degree to which a point on the earth's surface is in all respects subject to the influence of the sea; it is the opposite of continentality; oceanicity usually refers to climate and its effects; one measure for this characteristic is the ratio of the frequencies of maritime to continental types of air mass. Also known as oceanity.

oceanic ridge *See* mid-oceanic ridge.

oceanic rise [GEOL] A long, broad elevation of the bottom of the ocean.

oceanic stratosphere *See* cold-water sphere.

oceanic zone [OCEANOGR] The biogeographic area of the open sea.

oceanite [PETR] A picritic basalt in which olivine is a great deal more abundant than plagioclase.

oceanity *See* oceanicity.

oceanization [GEOL] Process by which continental crust (sial) is converted into oceanic crust (sima).

oceanodromous [VERT ZOO] Of a fish, migratory in salt water.

oceanographic model [OCEANOGR] A theoretical representation of the marine environment which relates physical, chemical, geological, biological, and other oceanographic properties.

oceanographic platform [ENG] A man-made structure with a flat horizontal surface higher than the water, on which oceanographic equipment is suspended or installed.

oceanographic ship *See* research ship.

oceanographic station [OCEANOGR] A geographic location at which oceanographic observations are taken from a stationary ship.

oceanographic submersible [NAV ARCH] Any small research vessel designed for undersea operations.

oceanographic survey [OCEANOGR] A study of oceanographic conditions with reference to physical, chemical, biological, geological, and other properties of the ocean.

oceanographic vessel [NAV ARCH] A research ship or other manned vehicle used in oceanography.

oceanography [GEOPHYS] The scientific study and exploration of the oceans and seas in all their aspects. Also known as oceanology.

oceanology *See* oceanography.

ocean station vessel [NAV ARCH] A vessel assigned to patrol at a specified geographical position on the ocean; such vessels are specially equipped to make comprehensive meteorological and some oceanographic observations, and to provide meteorological observations, navigation assistance, and other services to transiting aircraft.

ocean weather station [METEOROL] As defined by the World Meteorological Organization, a specific maritime location occupied by a ship equipped and staffed to observe weather and sea conditions and report the observations by international exchange.

ocellar [PETR] Of the texture of an igneous rock, having crystalline aggregates of phenocrysts arranged radially or tangentially around larger euhedral crystals or which form rounded branching forms.

ocellus [INV ZOO] A small, simple invertebrate eye composed of photoreceptor cells and pigment cells.

ocelot [VERT ZOO] *Felis pardalis.* A small arboreal wild cat, of the family Felidae, characterized by a golden head and back, silvery flanks, and rows of somewhat metallic spots on the body.

ocher [MINERAL] A yellow, brown, or red earthy iron oxide, or any similar earthy, pulverulent metallic oxides used as pigments.

ochlophobia [PSYCH] An abnormal fear of crowds.

Ochnaceae [BOT] A family of dicotyledonous plants in the order Theales, characterized by simple, stipulate leaves, a mostly gynobasic style, and anthers that generally open by terminal pores.

Ochoan [GEOL] A North American provincial series: uppermost Permian (above Guadalupian, below Lower Triassic).

Ochotonidae [VERT ZOO] A family of the mammalian order Lagomorpha; members are relatively small, and all four legs are about equally long.

ochroleucous [BIOL] Pale ocher or buff colored.

ochronosis [MED] A blue or brownish-blue discoloration of cartilage and connective tissue, especially around joints, caused by melanotic pigment.

Ochteridae [INV ZOO] The velvety shorebugs, the single family of the hemipteran superfamily Ochteroidea.

Ochteroidea [INV ZOO] A monofamilial tropical and subtropical superfamily of hemipteran insects in the subdivision Hydrocorisae; individuals are black with a silky sheen, and the antennae are visible from above.

ocotea oil [MATER] A volatile essential oil obtained from the wood of a Brazilian tree, *Ocotea cymbarum*; a source of safrole; used to make heliotropin and other technical preparations.

OCR *See* optical character recognition.

ocrea [BOT] A tubular stipule or pair of coherent stipules.

9,12-octadecadienoic acid *See* linoleic acid.

n-octadecane [ORG CHEM] $C_{18}H_{38}$ Colorless liquid boiling at 318°C; soluble in alcohol, acetone, ether, and petroleum, insoluble in water; used as a solvent and chemical intermediate.

n-octadecanoic acid *See* stearic acid.

1-octadecanol *See* stearyl alcohol.

octadeca-9,11,13-trienoic acid *See* eleostearic acid.

9,12,15-octadecatrienoic acid *See* linolenic acid.

9-octadecen-1,12-diol *See* ricinoleyl alcohol.

1-octadecene [ORG CHEM] $C_{18}H_{36}$ Colorless liquid boiling at 180°C; soluble in alcohol, acetone, ether, and petroleum, insoluble in water; used as a chemical intermediate.

trans-9-octadecenoic acid *See* elaidic acid.

octadecyl alcohol *See* stearyl alcohol.

octagon [MATH] A polygon with eight sides.

octahedral borax *See* tincalconite.

octahedral cleavage [CRYSTAL] Crystal cleavage in the four planes parallel to the face of the octahedron.

octahedral coordination [MINERAL] An atomic structure where six cations surround every anion, and vice versa.

octahedral iron ore *See* magnetite.

octahedral plane [CRYSTAL] The plane in a cubic lattice having three numerically equal Miller indices.

octahedrite [GEOL] The most common iron meteorite, containing 6-18% nickel in the metal phase and having intimate intergrowths lying parallel to the octahedral planes.

octahedron [MATH] A polyhedron having eight faces, each of which is an equilateral triangle.

octal [MATH] Pertaining to the octal number system.

octal base [ELECTR] Tube base having a central aligning key and positioned for eight equally spaced pins.

octal debugger [ADP] A simple debugging program which permits only octal (instead of symbolic) address references.

octal digit [MATH] The symbol 0, 1, 2, 3, 4, 5, 6, or 7 used as a digit in the octal number system.

octal loading program [ADP] Computer utility program providing a method for making changes in programs and tables existing in core memory or drum storage, by reading in words coded in octal notation on punched cards or tape.

octal number system [MATH] A number system in which a number r is written as $n_k n_{k-1} \ldots n_1$ where $r = n_1 8^0 + n_2 8^1 + \ldots + n_k 8^{k-1}$.

octanal *See* octyl aldehyde.

n-octane [ORG CHEM] C_8H_{18} Colorless liquid boiling at 126°C; soluble in alcohol, acetone, and ether, insoluble in water; used as a solvent and chemical intermediate.

octanedioic acid *See* suberic acid.

octane number [ENG] A rating that indicates the tendency to knock when a fuel is used in a standard internal combustion engine under standard conditions; n-heptane is 0, isooctane is 100; different test methods give research octane, motor octane, and road octane. [MECH ENG] The fuel octane number needed for efficient operation (without knocking or spark retardation) of an internal combustion engine.

octane scale [ENG] Series of arbitrary numbers from 0 to 120.3 used to rate the octane number of a gasoline; n-heptane is 0 octane, isooctane is 100, and isooctane + 6 milliliters TEL (tetraethyl lead) is 120.3.

n-octanoic acid *See* caprylic acid.

octanone-3 *See* ethyl amyl ketone.

octant [MATH] **1.** One of the eight regions into which three-dimensional euclidean space is divided by the coordinate planes of a cartesian coordinate system. **2.** A unit of plane angle equal to 45° or $\pi/8$ radians. [NAV] A double-reflecting instrument used primarily for measuring altitudes of celestial bodies from a moving ship or airplane; has a range of 90°, and while a sextant has the capability up to 120°, the common practice applies the name sextant to both types of instruments.

octant error *See* sextant error.

octave [ACOUS] The interval in pitch between two tones such that one tone may be regarded as duplicating at the next higher pitch the basic musical import of the other tone; the sounds producing these tones then have a frequency ratio of 2 to 1. [PHYS] The interval between any two frequencies having a ratio of 2 to 1.

octave-band analyzer [ENG ACOUS] A portable sound analyzer which amplifies a microphone signal, feeds it into one of several band-pass filters selected by a switch, and indicates the magnitude of sound in the corresponding frequency band on a logarithmic scale; all the bands except the highest and lowest span an octave in frequency. Abbreviated OBA.

octave-band filter [ENG ACOUS] A band-pass filter in which the upper cutoff frequency is twice the lower cutoff frequency.

octave-band oscillator [ELECTR] An oscillator that can be tuned over a frequency range of 2 to 1, so that its highest frequency is twice its lowest frequency.

octave frequency band [PHYS] A band of frequencies whose highest frequency is twice its lowest frequency.

1-octene [ORG CHEM] $CH_3(CH_2)_5CHCH_2$ A colorless, flammable liquid; used as a plasticizer and in synthesis of organic compounds. Also known as 1-caprylene; 1-octylene.

2-octene [ORG CHEM] $CH_3(CH_2)_4CHCHCH_3$ A colorless, flammable liquid, with trans and cis forms; used to manufacture lubricants and to synthesize organic materials.

octet [ATOM PHYS] A collection of eight valence electrons in an atom or ion, which form the most stable configuration of the outermost, or valence, electron shell. [PARTIC PHYS] A multiplet of eight elementary particles, corresponding to a representation of the approximate unitary symmetry (SU₃) of the strong interactions.

octillion [MATH] The number 10^{27}.

octine *See* octyne.

Octocorallia [INV ZOO] The equivalent name for Alcyonaria.

octode [ELECTR] An eight-electrode electron tube containing an anode, a cathode, a control electrode, and five additional electrodes that are ordinarily grids.

octoid [DES ENG] Pertaining to a gear tooth form used to generate the teeth in bevel gears; the octoid form closely resembles the involute form.

octomethylene *See* cyclooctane.

octonary signaling [COMMUN] A communications mode in which information is passed by the presence and absence of plus and minus variation of eight discrete levels of one parameter of the signaling medium.

Octopoda [INV ZOO] An order of the dibranchiate cephalopods, characterized by having eight arms equipped with one to three rows of suckers.

Octopodidae [INV ZOO] The octopuses, in family of cephalopod mollusks in the order Octopoda.

octopus [INV ZOO] Any member of the genus *Octopus* in the family Octopodidae; the body is round with a large head and eight partially webbed arms, each bearing two rows of suckers, and there is no shell.

octupole [PHYS] **1.** Two electric or magnetic quadrupoles having charge distributions of opposite signs and separated from each other by a small distance. **2.** Any device for controlling beams of electrons or other charged particles, consisting of eight electrodes or magnetic poles arranged in a circular pattern, with alternating polarities; commonly used to correct aberrations of quadrupole systems.

OCTOPUS

Octopus bairdi with a body length of 3 inches (7.6 centimeters) and arms up to 40 inches (1 meter) long, webbed for one-third of their length.

octyl- [ORG CHEM] Prefix indicating the eight-carbon hydrocarbon radical ($C_8H_{17}-$).

octylacetic acid See capric acid.

octyl alcohol See 2-ethylhexyl alcohol.

octyl aldehyde [ORG CHEM] $C_8H_{16}O$ A liquid aldehyde boiling at 172°C; found in essential oils of many plants; used in perfume compositions. Also known as caprylaldehyde; octanal.

octyl carbinol See n-nonyl alcohol.

1-octylene See 1-octene.

octylic acid See caprylic acid.

n-octyl mercaptan [ORG CHEM] $C_8H_{17}SH$ Clear, colorless liquid boiling at 199°C; used as a chemical intermediate and polymerization conditioner.

octyl phenol [ORG CHEM] $C_8H_{17}C_6H_4OH$ White flakes, congealing at 73°C; soluble in organic solvents, insoluble in water; used to make surfactants, plasticizers, and antioxidants. Also known as diisobutyl phenol.

octyne [ORG CHEM] $CHC(CH_2)_5CH_3$ Colorless hydrocarbon liquid, boiling at 125°C. Also known as caprilydene; hexylacetylene; octine.

ocuba wax [MATER] A waxy fat obtained from the fruit of the myrtle Myristica ocuba; melting point, 40°C; used in candles.

ocular [BIOL] Of or pertaining to the eye. [OPTICS] See eyepiece.

ocular prism [OPTICS] The prism employed in a range finder to bend the line of sight through the instrument into the eyepiece.

ocular skeleton [VERT ZOO] A rigid structure in most submammalian vertebrates consisting of a cup of hyaline cartilage enclosing the posterior part of the eye, and a thin-walled ring of intramembranous bones in the edge of the sclera at its junction with the cornea.

oculist See ophthalmologist.

oculoagravic illusion See agravic illusion.

oculoglandular tularemia [MED] Infection by Pasteurella tularensis which, in addition to the usual symptoms of tularemia, causes swollen eyelids, conjunctivitis, swollen lymph nodes, and ulcers on the conjunctivae.

oculomotor [PHYSIO] Pertaining to eye movement. 2. Pertaining to the oculomotor nerve.

oculomotor nerve [ANAT] The third cranial nerve; a paired somatic motor nerve arising in the floor of the midbrain, which innervates all extrinsic eye muscles except the lateral rectus and superior oblique, and furnishes autonomic fibers to the ciliary and pupillary sphincter muscles within the eye.

oculomotor nucleus [ANAT] A nucleus in the floor of the midbrain that gives rise to motor fibers of the oculomotor nerve.

oculomotor paralysis [MED] Paralysis of the oculomotor nerve.

Oculosida [INV ZOO] An order of the protozoan subclass Radiolaria; pores are restricted to certain areas in the central capsule, and an olive-colored material is always present near the astropyle.

OD See optical density; outside diameter.

odd-even check [ADP] A means of detecting certain kinds of errors in which an extra bit, carried along with each word, is set to zero or one so that the total number of zeros or ones in each word is always made even or always made odd. Also known as parity check.

odd function [MATH] A function $f(x)$ is odd if, for every x, $f(-x) = -f(x)$.

odd-leg caliper [DES ENG] A caliper in which the legs bend in the same direction instead of opposite directions.

odd number [MATH] A natural number not divisible by 2.

odd-odd nucleus [NUC PHYS] A nucleus that has an odd number of protons and an odd number of neutrons.

odd parity [ADP] Property of an expression in binary code which has an odd number of ones. [QUANT MECH] Property of a system whose state vector is multiplied by −1 under the operation of space inversion, that is, the simultaneous reflection of all spatial coordinates through the origin.

odd-pinnate [BOT] Of a compound leaf, having a single leaflet at the tip of the petiole with leaflets on both sides of the petiole. Also known as imparipinnate.

odds ratio [STAT] The ratio of the probability of occurrence of an event to the probability of the event not occurring.

ODONATA

Adult dragonfly showing the four membranous wings.

odd term [ATOM PHYS] A term of an atom or molecule for which the sum of the angular-momentum quantum numbers of all the electrons is odd, so that the states have odd parity; designated by a superscript o or u.

Odessey protractor [NAV] A device used in plotting shoran, EPI (electronic position indicator), and Raydist positions; it consists of a transparent plate on which closely spaced concentric circles are drawn; these circles represent distances from the center in terms of miles, microseconds, or lanes, constructed to fit the electronic system in use and the scale at which the survey is to be plotted.

Odiniidae [INV ZOO] A family of cyclorrhaphous myodarian dipteran insects in the subsection Acalypteratae.

Odobenidae [VERT ZOO] A family of carnivorous mammals in the suborder Pinnipedia; contains a single species, the walrus (Odobenus rosmarus).

odograph [ENG] An instrument installed in a vehicle to automatically plot on a map the course and distance traveled by the vehicle.

odometer [ENG] 1. An instrument for measuring distance traversed, as of a vehicle. 2. The indicating gage of such an instrument. 3. A wheel pulled by surveyors to measure distance traveled.

O'Donahue's theory [MIN ENG] A mine subsidence theory, with subsidence regarded as taking place in two stages: first, a breaking of the rocks in which the lines of fracture tend to run at right angles to the stratification; then, an inward movement from the sides, resulting in a pull or draw beyond the edges of the workings.

Odonata [INV ZOO] The dragonflies, an order of the class Insecta, characterized by a head with large compound eyes, and wings with clear or transparent membranes traversed by networks of veins.

odont-, odonto- [VERT ZOO] A combining form meaning tooth.

odontalgia See toothache.

odontectomy [MED] Surgical excision of a tooth.

odontexesis [MED] Removal of deposits from the surface of teeth.

odontoblast [HISTOL] One of the elongated, dentin-forming cells covering the dental papilla.

odontoblastoma See ameloblastic odontoma.

Odontoceti [VERT ZOO] The toothed whales, a suborder of cetacean mammals distinguished by a single blowhole.

odontoclast [HISTOL] A multinuclear cell concerned with resorption of the roots of milk teeth.

odontogenesis [EMBRYO] Formation of teeth.

odontogenic [HISTOL] 1. Pertaining to the origin and development of teeth. 2. Originating in tissues associated with teeth.

odontogenic cyst [MED] A cyst originating in tissues associated with teeth.

odontogenic fibroma [MED] A benign tumor originating in the mesenchymal derivatives of the tooth germ.

Odontognathae [PALEON] An extinct superorder of the avian subclass Neornithes, including all large, flightless aquatic forms and other members of the single order Hesperornithiformes.

odontoid [BIOL] Toothlike.

odontoid process [ANAT] A toothlike projection on the anterior surface of the axis vertebra with which the atlas articulates.

odontology [VERT ZOO] A branch of science that deals with the formation, development, and abnormalities of teeth.

odontoma [MED] A benign tumor representing a developmental excess, composed of mesodermal or octodermal tooth-forming tissue, alone or in association with the calcified derivatives of these structures.

odontophobia [PSYCH] An abnormal fear of teeth.

Odontostomatida [INV ZOO] An order of the protozoan subclass Spirotrichia; individuals are compressed laterally and possess very little ciliature.

odorant [MATER] Material added to odorless fuel gases to give them a distinctive odor for safety purposes; usually a sulfur- or mercaptan-containing compound. Also known as malodorant; stench; warning agent.

odoriferous homing [ELECTR] Homing on the ionized air produced by the exhaust gases of a snorkeling submarine.

odorize [CHEM ENG] To add an unpleasant odor as a safety measure to an odorless material such as fuel gas.

Oecophoridae [INV ZOO] A family of small to moderately small moths in the lepidopteran superfamily Tineoidea, characterized by a comb of bristles, the pecten, on the scape of the antennae.

Oedemeridae [INV ZOO] The false blister beetles, a large family of coleopteran insects in the superfamily Tenebrionoidea.

oedipal [PSYCH] Pertaining to the Oedipus complex.

Oedipus complex [PSYCH] In psychoanalytic theory, the attraction and attachment of the child to the parent of the opposite sex, accompanied by feelings of envy and hostility toward the parent of the child's sex, whose displeasure and punishment the child so fears that the child represses his feelings toward the parent of opposite sex.

Oedogoniales [BOT] An order of fresh-water algae in the division Chlorophyta; characterized as branched or unbranched microscopic filaments with a basal holdfast cell.

Oegophiurida [INV ZOO] An order of echinoderms in the subclass Ophiuroidea, represented by a single living genus; members have few external skeletal plates and lack genital bursae, dorsal and ventral arm plates, and certain jaw plates.

Oegopsida [INV ZOO] A suborder of cephalopod mollusks in the order Decapoda of one classification system, and in the order Teuthoidea of another system.

Oehman's survey instrument [ENG] A drill-hole surveying apparatus that makes a photographic record of the compass and clinometer readings.

O electron [ATOM PHYS] An electron in the fifth (O) shell of electrons surrounding the atomic nucleus, having the principal quantum number 5.

Oepikellacea [PALEON] A dimorphic superfamily of extinct ostracods in the order Paleocopa, distinguished by convex valves and the absence of any trace of a major sulcus in the external configuration.

Oerlikon gun [ORD] Any of certain Swiss-developed, 20-millimeter automatic aircraft or antiaircraft cannons that shoot greased ammunition; used by various nations.

oersted [ELECTROMAG] The unit of magnetic field strength in the centimeter-gram-second electromagnetic system of units, equal to the field strength at the center of a plane circular coil of one turn and 1-centimeter radius, when there is a current of $1/2\pi$ abamp in the coil.

Oestridae [INV ZOO] A family of cyclorrhaphous myodarian dipteran insects in the subsection Calypteratae.

Oetling freezing method [MIN ENG] A method of shaft sinking by freezing the wet ground in sections as the sinking proceeds.

off-airways [AERO ENG] Pertaining to any aircraft course or track that does not lie within the bounds of prescribed airways.

off-carriage fire control [ORD] Process of controlling fire on a target with the aid of a sighting device which is not mounted directly on the weapon.

off-carriage fire control equipment [ORD] Fire control items, such as directors, aiming circles, and observation instruments, which are required for directing the fire of the weapon and which are separately transported.

off-center plan-position indicator [ELECTR] A plan-position indicator in which the center of the display that represents the location of the radar can be moved from the center of the screen to any position on the face of the PPI.

offense against the sine condition [OPTICS] A numerical measure of coma, equal to the sagittal coma divided by the perpendicular distance from the image point to the optical axis.

offensive grenade [ORD] A hand grenade having a nonmetallic container, designed to kill or injure by blast and concussion.

off-highway truck [MIN ENG] A truck of such size, weight, or dimensions that is cannot be used on public highways.

off-hook [COMMUN] The active state (closed loop) of a subscriber or PBX user loop.

off-hook service [COMMUN] Priority telephone service for key personnel that affords a connection from caller to receiver by the simple expedient of removing the phone from its cradle or hook.

office automation [ADP] Use of an electronic computer or computing system for routine clerical jobs.

official master drawing [GRAPHICS] That drawing which is established as the official drawing upon which any revision will originally be made and from which copies of the current drawing will originally be reproduced.

officials' inspection lamp [MIN ENG] A portable combined electric lamp and battery, fitted with a reflector to provide directional illumination.

offing [NAV] That part of the visible sea a considerable distance from the shore, or that part just beyond the limits of the area in which a pilot is needed.

offlap [GEOL] The successive lateral contraction extent of strata (in an upward sequence) due to their deposition in a shrinking sea or on the margin of a rising landmass. Also known as regressive overlap.

off-line [ADP] Describing equipment not connected to a computer, or temporarily disconnected from one. [ENG] **1.** A condition existing when the drive rod of the drill swivel head is not centered and parallel with the borehole being drilled. **2.** A borehole that has deviated from its intended course. **3.** A condition existing wherein any linear excavation (shaft, drift, borehole) deviates from a previously determined or intended survey line or course. [IND ENG] State in which an equipment or subsystem is in standby, maintenance, or mode of operation other than on-line.

off-line cipher [COMMUN] Method of encrypting which is not associated with a particular transmission system and in which the resulting encrypted message can be transmitted by any means.

off-line computer See course-line computer.

off-line equipment [ADP] Peripheral equipment or devices not in direct communication with the central processing unit of a computer.

off-line mode [ADP] Any operation such as printing, punching, converting which does not involve the main computer.

off-line operation [ADP] Operation of peripheral equipment in conjunction with, but not under the control of, the central processing unit.

off-line processing [ADP] Any processing which takes place independently of the central processing unit; for instance, card-to-tape conversion using auxiliary equipment.

off-line storage [ADP] A storage device not under control of the central processing unit.

off-line unit [ADP] Any operation device which is not attached to the main computer.

off-peak [SCI TECH] Not at the maximum.

off-punch [ADP] A hole which is not punched in precise position on a computer card.

off-reef facies [GEOL] Facies of the inclined strata made up of reef detritus deposited along the seaward margin of a reef.

offset [CONT SYS] The steady-state difference between the desired control point and that actually obtained in a process control system. [ENG] A short perpendicular distance measured to a traverse course or a surveyed line or principal line of measurement in order to locate a point with respect to a point on the course or line. [GEOL] **1.** The movement of an upcurrent part of a shore to a more seaward position than a downcurrent part. **2.** A spur from a mountain range. **3.** A level terrace on the side of a hill. **4.** The horizontal displacement component in a fault, measured parallel to the stroke of the fault. Also known as normal horizontal separation. [MECH] The value of strain between the initial linear portion of the stress-strain curve and a parallel line that intersects the stress-strain curve of an arbitrary value of strain; used as an index of yield stress; a value of 0.2% is common. [MIN ENG] **1.** A short drift or crosscut driven from a main gangway or level. **2.** The horizontal distance between the outcrops of a dislocated bed. [NAV ARCH] One of a series of measurements of the perpendicular distance of various points on a ship's hull from the centerline and above the molded baseline; used in ship construction. [ORD] The horizontal distance of forward travel covered by the missile after it strikes the ground; this distance is measured from the center of the hole of entry to the most forward part of the missile.

offset bombing [ORD] Any bombing procedure which employs a reference or aiming point other than the actual target.

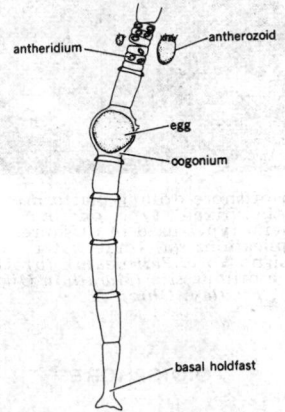

OEDOGONIALES

Oedogonium, an attached, unbranched, microscopic filament arising from a basal holdfast cell.

OFFSHORE DRILLING

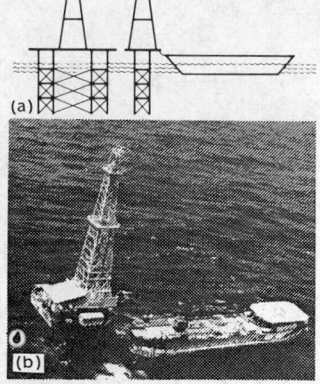

An offshore drilling platform of the "fixed" type, one of several types used in offshore applications. (a) Underwater design (World Petroleum). (b) Rig on a drilling site (Marathon Oil Co., Findlay, Ohio).

OIDIOPHORE

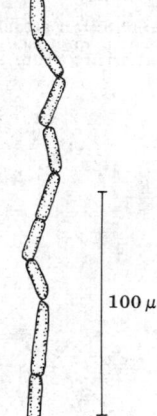

100μ

An oidiophore type of hypha showing the divisions that are the beginning of the oidia.

OIDIUM

10μ

Two oidia showing the flat ends of this thin-walled spore.

OIKOMONADIDAE

20μ

Oikomonas, a genus in the Oikomonadidae, showing the single flagellum characteristic of the family.

offset cab [ENG] Operator's cab positioned to one side of earthmoving equipment for greater visibility and safety.

offset-center plan-position indicator See off-center plan-position indicator.

offset-course computer See course-line computer.

offset cylinder [MECH ENG] A reciprocating part in which the crank rotates about a center off the centerline.

offset drilling [PETRO ENG] The drilling of a well on property under which oil is being drained away by a well on adjacent property to make up for the loss of oil from the first property.

offset duplicating [GRAPHICS] A method of duplicating in which an ink-receptive image is typed or drawn on a paper master or produced photographically on a sensitized metal plate; the master or plate is then placed on the master cylinder of an offset press; the inked image built up on the master is transferred onto a blanket cylinder; the paper picks up the inked image as it passes between the blanket cylinder and an impression cylinder.

offset ground zero [ORD] A case where the desired ground zero in an atomic explosion is displaced some distance and direction from the target center.

offset lithography [GRAPHICS] A system of printing that depends on the principle that the printing area accepts greasy ink while the nonprinting area is dampened with water and repels the ink; in practice, the image from a plate is offset onto the rubber blanket of an impression cylinder, and transferred to a sheet of paper.

offset paper [MATER] Paper with a certain degree of porosity as a result of coating with an alkali-swelling resin; used for offset printing.

offset plan-position indicator See off-center plan-position indicator.

offset plotting [ORD] Method of plotting firing data when different ranges and azimuths must be sent to each gun of a battery.

offset press [GRAPHICS] A printing press for offset lithography printing.

offset ridge [GEOL] A ridge consisting of resistant sedimentary rock that has been made discontinuous as a result of faulting.

offset screwdriver [DES ENG] A screwdriver with the blade set perpendicular to the shank for access to screws in otherwise awkward places.

offset stacker [ADP] A card stacker which is capable, under machine control, of stacking cards so that some of them stick out from the deck, permitting their identification.

offset yield strength [MECH] That stress at which the strain surpasses by a specific amount (called the offset) an extension of the initial proportional portion of the stress-strain curve; usually expressed in pounds per square inch.

offshore [GEOL] The comparatively flat zone of variable width extending from the outer margin of the shoreface to the edge of the continental shelf.

offshore bar See longshore bar.

offshore beach See barrier beach.

offshore current [OCEANOGR] 1. A prevailing nontidal current usually setting parallel to the shore outside the surf zone. 2. Any current flowing away from shore.

offshore drilling [PETRO ENG] The drilling of oil or gas wells into water-covered locations, usually on submerged continental shelves.

offshore gas [PETRO ENG] Nautral gas produced from reservoirs under the offshore continental shelves.

offshore navigation [NAV] Navigation at a distance from a coast, in contrast with coastwise navigation in the vicinity of a coast.

offshore oil [PETRO ENG] Oil produced from reservoirs under the offshore continental shelves.

offshore survey [PETRO ENG] Seismic geophysical survey procedures conducted over water-covered continental-shelf areas in the search for possible oil reservoirs.

offshore water [OCEANOGR] Water adjacent to land in which the physical properties are slightly influenced by continental conditions.

offshore wind [METEOROL] Wind blowing from the land toward the sea.

off-site facility [CHEM ENG] In a chemical process plant, any supporting facility that is not a direct part of the reaction train, such as utilities, steam, and waste-treatment facilities.

off soundings [NAV] Of a vessel, navigating beyond the 100-fathom (183-meter) curve; in earlier times, the term was applied to a vessel in water deeper than could be sounded with the sounding lead.

off-the-road equipment [MIN ENG] Tires and earthmoving equipment designed for off-highway duty in surface mines and quarries.

off-the-road hauling [MIN ENG] Hauling off the public highways, and generally on the mining site or excavation site.

off-the-shelf [IND ENG] Available for immediate shipment. [ORD] Referring to those items required by the military services which are generally used throughout the civilian economy and which are available through normal commercial distribution channels.

off time [MET] In resistance welding, usually in repetitive cycles, the time that the electrode is not in contact with the work.

ogee [ARCH] A reverse curve, shaped like an elongated letter S, as the outline of an ogee molding.

ogive [ARCH] 1. An arch or rib placed diagonally across a Gothic vault. 2. A pointed arch. [GEOL] One of a periodically repeated series of dark, curved structures occurring down a glacier that resemble a pointed arch. [ORD] The curved or tapered front of a projectile.

Ohio sampler [MIN ENG] A single tube or pipe with a thread on top, and the bottom beveled and hardened for driving into the ground to obtain a soil sample.

ohm [ELEC] The unit of electrical resistance in the rationalized meter-kilogram-second system of units, equal to the resistance through which a current of 1 ampere will flow when there is a potential difference of 1 volt across it.

ohmic contact [ELEC] A region where two materials are in contact, which has the property that the current flowing through it is proportional to the potential difference across it.

ohmic dissipation [ELECTR] Loss of electric energy when a current flows through a resistance due to conversion into heat.

ohmic resistance [ELEC] Property of a substance, circuit, or device for which the current flowing through it is proportional to the potential difference across it.

ohmmeter [ENG] An instrument for measuring electric resistance; scale may be graduated in ohms or megohms.

Ohm's law [ELEC] The law that the direct current flowing in an electric circuit is directly proportional to the voltage applied to the circuit; it is valid for metallic circuits and many circuits containing an electrolytic resistance.

ohms per volt [ENG] Sensitivity rating for measuring instruments, obtained by dividing the resistance of the instrument in ohms at a particular range by the full-scale voltage value at that range.

OHV engine See overhead-valve engine.

-oic [ORG CHEM] Suffix indicating the presence of a $-COOH$ group, as in ethyloic ($-CH_2-COOH$).

oidiophore [MYCOL] A hypha that produces oidia.

oidium [MYCOL] One of the small, thin-walled spores with flat ends produced by autofragmentation of the vegetative hyphae in certain Eumycetes.

Oikomonadidae [INV ZOO] A family of protozoans in the order Kinetoplastida containing organisms that have a single flagellum.

oikophobia [PSYCH] An abnormal fear of the home, or of a house.

oil [GEOL] See petroleum. [MATER] Any of various viscous, combustible, water-immiscible liquids that are soluble in certain organic solvents, as ether and naphtha; may be of animal, vegetable, mineral, or synthetic origin; examples are fixed oils, volatile or essential oils, and mineral oils.

oil accumulation See oil pool.

oil ammoniac [MATER] A yellow liquid distilled from gum ammoniac; boiling point is 275°C; soluble in alcohol and benzene.

oil asphalt [MATER] Water-insoluble, heavy black residue left after removing the tar tailings during the distillation of petroleum; used in roofing, paints, and coatings.

oil-base mud [PETRO ENG] Drilling mud made with oil as the solvent carrier for the solids content.

oil bath [ENG] **1.** Oil, in a container, within which a mechanism works or into which it dips. **2.** Oil in which a piece of apparatus is submerged. **3.** Oil poured on a cutting tool. [MET] Oil used in tempering.

oil blue [INORG CHEM] Violet-blue copper sulfide pigment used in varnishes.

oil buffer [ORD] A mechanism on certain types of automatic weapons, especially the caliber-.50 guns, for absorbing the shock of recoil and regulating the speed of firing.

oil burner [ENG] Liquid-fuel burner device using a mixture of air and vaporized or atomized oil for combustion.

oil cake [MATER] Solid residue after removal of vegetable oils from oil-bearing seeds (such as soya beans) by expression or solvent extraction; used as fertilizer and animal feed.

oil circuit breaker [ELECTR] A high-voltage circuit breaker in which the arc is drawn in oil to dissipate the heat and extinguish the arc; the intense heat of the arc decomposes the oil, generating a gas whose high pressure produces a flow of fresh fluid through the arc that furnishes the necessary insulation to prevent a restrike of the arc.

oilcloth [TEXT] **1.** A fabric coated with a mixture of oil and clay, used as a waterproof covering. **2.** A floor covering made of a heavy fabric treated with oil paint.

oil cup [ENG] A permanently mounted cup used to feed lubricant to a gear, usually with some means of regulating the flow.

oil derrick [PETRO ENG] Tower structure used during oil well drilling to aid in raising and lowering of drill and piping strings.

oil dilution valve [MECH ENG] A valve used to mix gasoline with engine oil to permit easier starting of the gasoline engine in cold weather.

oiled silk [TEXT] Soft woven silk made waterproof by treatment with boiled linseed oil, followed by drying.

oil emulsion [MATER] Suspension of oil droplets in another liquid in which the oil is insoluble.

oil-extended rubber [MATER] Synthetic rubber into which 25–50% of a petroleum oil emulsion has been incorporated to decrease cost and increase low-temperature flexibility and resilience.

oil field [PETRO ENG] The surface boundaries of an area from which petroleum is obtained; may correspond to an oil pool or may be circumscribed by political or legal limits.

oil-field brine [HYD] Connate waters, usually containing a high concentration of calcium and sodium salts and found during deep rock penetration by the drill.

oil field model [PETRO ENG] Laboratory simulation of steady-state fluid flow through porous reservoir media by electrical (Ohm's-law system), electronic (graphite-impregnated cloth), or electrolytic (gelatin, blotter, potentiometric-liquid) models.

oil-field separator See gas-oil separator.

oil-filled cable [ELEC] Cable having insulation impregnated with an oil which is fluid at all operating temperatures and provided with facilities such as longitudinal ducts or channels and with reservoirs; by this means positive oil pressure can be maintained within the cable at all times, incipient voids are promptly filled during periods of expansion, and all surplus oil is adequately taken care of during periods of contraction.

oil filter [ENG] Cartridge-type filter used in automotive oil-lubrication systems to remove metal particles and products of heat decomposition from the circulating oil.

oil furnace [MECH ENG] A combustion chamber in which oil is the heat-producing fuel.

oil gas [MATER] A heating gas made by interaction of petroleum oil vapors and steam in a process similar to the water-gas reaction.

oil-gas process [CHEM ENG] Process to manufacture high-caloric-value fuel gas by the destructive distillation of high-boiling petroleum oils.

oil-gas separator See gas-oil separator.

oil gland See uropygial gland.

oil groove [DES ENG] One of the grooves in a bearing which distribute and collect lubricating oil.

oil-harden [MET] To harden steel by immersing the hot steel in oil. Also known as oil-quench.

oil-hardened steel [MET] A carbon steel hardened by rapid cooling from elevated temperatures in an oil bath.

oil hole [ENG] A small hole for injecting oil for a bearing.

oil-hole drill [DES ENG] A twist drill containing holes through which oil can be fed to the cutting edges.

oiliness [ENG] The effect of a lubricant to reduce friction between two solid surfaces in contact; the effect is more than can be accounted for by viscosity alone.

oil isoperms [PETRO ENG] Reservoir-map plotting areas of equal oil permeability.

oil length [MATER] The ratio of oil to resin in varnish; expressed as gallons of oil per 100 pounds (45.3 kilograms) of resin.

oilless bearing [MECH ENG] A self-lubricating bearing containing solid or liquid lubricants in its material.

oil lift [MECH ENG] Hydrostatic lubrication of a journal bearing by using oil at high pressure in the area between the bottom of the journal and the bearing itself so that the shaft is raised and supported by an oil film whether it is rotating or not.

oil of amber [MATER] Brown essential oil distilled from amber; miscible with alcohol; has balsamic aroma.

oil of cloves [MATER] A thin, colorless to pale-yellow liquid distilled from cloves; thickens and darkens with time; boils at 250–260°C; has aromatic aroma and pungent taste; soluble in ether and chloroform; chief component is eugenol; used in medicine, perfumery, flavoring, and soaps. Also known as caryophyllus oil; clove oil.

oil of coriander [MATER] A colorless or pale-yellow essential oil extracted from the dried seed of the coriander plant (*Coriandrum sativum*); used in medicines, beverages, and flavoring extracts.

oil of cubeb [MATER] A camphorous tasting, colorless, or pale green or yellow liquid with a peppery odor; distilled from cubebs; soluble in ether, alcohol, and chloroform; boils at 175–180°C; chief components are sesquiterpenes, cadinene, and dipentane; used in medicine. Also known as cubeb oil.

oil of grapefruit See grapefruit oil.

oil of mirbane See nitrobenzene.

oil of orange blossoms [MATER] Bitter-tasting, fluorescent, pale-yellow essential oil with orange aroma; soluble in alcohol; main components are limonene, linalool, and geraniol; used for perfumes and flavors. Also known as neroli oil; orange flower oil.

oil of peppermint [MATER] Colorless or slightly yellow essential oil with minty aroma and taste; soluble in ether, alcohol, and chloroform; derived from leaves and flowering tops of the peppermint plant (*Mentha piperita*); has high menthol content; used in medicines, flavors, perfumes, and liqueurs. Also known as peppermint oil.

oil of sage [MATER] **1.** An alcohol-soluble yellow oil with sage aroma; obtained from leaves of the common sage (*Salvia officinalis*); used chiefly in flavors. Also known as dalmatian sage oil; sage oil; salvia oil. **2.** A pale-yellow oil with ambergris aroma obtained from the flowers of *Salvia sclarea*; used chiefly in perfumes.

oil of sassafras [MATER] A pungent, aromatic, yellowish or reddish-yellow liquid obtained from the bark of American sassafras (*Sassafras albidum*); soluble in organic solvents, glacial acetic acid, and carbon disulfide; used for flavors, perfumes, and medicines. Also known as clary sage oil; sassafras oil.

oil of shaddock See grapefruit oil.

oil of turpentine [MATER] Water-insoluble, colorless, volatile essential oil distilled from turpentine; contains pinene, sylvestrene, and dipentene; used as a carminative, solvent, paint vehicle, and disinfectant.

oil of vitriol See sulfuric acid.

oil paint [MATER] A paint made with a vegetable oil as the filmogen.

oil painting [GRAPHICS] A method of painting in which pigments are bound together, and to the canvas, by a drying oil; the oil, usually linseed, is thinned with solvents such as turpentine and mineral spirits; the pigments should dry in linseed oil to form an acceptably strong paint film.

oil pool [GEOL] An accumulation of petroleum locally confined by subsurface geologic features. Also known as oil accumulation; oil reservoir.

OIL CIRCUIT BREAKER

Bulk oil circuit breakers for 138-kilovolt application.

OKRA

Okra (*Hibiscus esculentus*),
branch with pods.

OLDHAMINIDINA

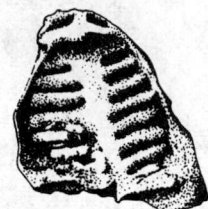

Dorsal view of the shell of
Leptodus.

oil pump [MECH ENG] A pump of the gear, vane, or plunger type, usually an integral part of the automotive engine; it lifts oil from the sump to the upper level in the splash and circulating systems, and in forced-feed lubrication it pumps the oil to the tubes leading to the bearings and other parts.

oil-quench *See* oil-harden.

oil reservoir *See* oil pool.

oil-reservoir water *See* formation water.

oil ring [MECH ENG] 1. A ring located at the lower part of a piston to prevent an excess amount of oil from being drawn up onto the piston during the suction stroke. 2. A ring on a journal, dipping into an oil bath for lubrication.

oil rock [GEOL] A rock stratum containing oil.

oil sand [GEOL] An unconsolidated, porous sand formation or sandstone containing or impregnated with petroleum or hydrocarbons.

oil saturation [PETRO ENG] Measurement of the degree of saturation of reservoir pore structure by reservoir oil.

oil seal [ENG] 1. A device for preventing the entry or return of oil from a chamber. 2. A device using oil as the sealing medium to prevent the passage of fluid from one chamber to another.

oil seed [MATER] Seeds of plants from which oil can be derived by expression or solvent extraction, such as soya beans.

oil separator *See* gas-oil separator.

oil shale [GEOL] A finely layered brown or black shale that contains kerogen and from which liquid or gaseous hydrocarbons can be distilled. Also known as kerogen shale.

oilskin [TEXT] A fabric made of cotton or linen and waterproofed with linseed oil.

oil-soluble resin [MATER] A resin that, at moderate temperatures, dissolves in, disperses in, or reacts with drying oils to produce a homogeneous film of modified characteristics.

oil stain [MATER] Thin oil paint with very little pigment, used to stain wood surfaces.

oilstone [MATER] A whetstone used with oil.

oil switch [ELEC] A switch whose contacts are immersed in oil in order to suppress the arc and prevent the contacts from being damaged.

oil tanker [NAV ARCH] A very large ship which carries crude oil or other petroleum products in big tanks.

oil trap [GEOL] An accumulation of petroleum which, by a combination of physical conditions, is prevented from escaping laterally or vertically. Also known as trap.

oiluvium *See* Pleistocene.

oil varnish [MATER] A varnish composed of resins dissolved in oil.

oil-water contact *See* oil-water surface.

oil-water interface *See* oil-water surface.

oil-water surface [GEOL] The datum of a two-dimensional oil-water interface. Also known as oil-water contact; oil-water interface.

oil well [PETRO ENG] A hole drilled (usually vertically) into an oil reservoir for the purpose of recovering the oil trapped in porous formations.

oil-well cement [MATER] A type of hydraulic cement which has a slow setting rate under the high temperatures obtained in oil wells; uses include support of tubing and bypassing of unwanted zones.

oil-well drive *See* reservoir drive mechanism.

oil-well pump [PETRO ENG] Device for artificial or secondary (non-gas-lift) oil production; about 85% of the production is by ground-level sucker-rod pumps, the remainder by downhole hydraulic lift pumps.

oil white [MATER] Mixture of lithopone and zinc white or white lead; used as a house-paint pigment.

O indicator *See* O scope.

ointment [PHARM] A semisolid preparation used for a protective and emollient effect or as a vehicle for the local or endermic administration of medicaments; ointment bases are composed of various mixtures of fats, waxes, animal and vegetable oils, and solid and liquid hydrocarbons.

oiticica oil [MATER] A light-yellowish oil obtained from seeds of the Brazilian oiticica tree (*Licania rigida*); raw oil becomes buttery unless heat-treated (semipolymerized); used

principally in paint and varnish as a drying oil as a substitute for tung oil or with tung oil.

okapi [VERT ZOO] *Okapia johnstoni*. An artiodactylous mammal in the family Giraffidae; has a hazel coat with striped hindquarters, and the head shape, lips, and tongue are the same as those of the giraffe, but the neck is not elongate.

okenite [MINERAL] $CaSi_2O_4(OH)_2 \cdot H_2O$ A whitish mineral consisting of calcium silicate and occurring in fibrous masses.

okonite [MATER] Insulating material made from the vulcanization of ozokerite and resin with rubber and sulfur.

okra [BOT] *Hibiscus esculentus*. A tall annual plant grown for its edible immature pods. Also known as gumbo.

-ol [ORG CHEM] Chemical suffix for an $-OH$ group in organic compounds, such as phenol (C_6H_5OH).

OL *See* only loadable.

Olacaceae [BOT] A family of dicotyledonous plants in the order Santalales characterized by dry or fleshy indehiscent fruit, the presence of petals, stamens, and chlorophyll, and a 2-5-celled ovary.

Olber's paradox [ASTRON] If the universe were static, of infinite age, and the galaxies distributed isotropically, the distance attenuation of their light would be exactly balanced by the increase in number in successive spherical shells centered at the earth; hence the night sky would be of daylight brightness instead of dark.

old age [GEOL] The last stage of the erosion cycle in the development of the topography of a region in which erosion has reduced the surface almost to base level and the land forms are marked by simplicity of form and subdued relief. Also known as topographic old age.

Oldham coupling *See* slider coupling.

Oldhaminidina [PALEON] A suborder of extinct articulate brachiopods in the order Strophomenida distinguished by a highly lobate brachial valve seated within an irregular convex pedicle valve.

oldhamite [MINERAL] CaS A pale-brown mineral known only from meteorites; unstable under earth conditions; member of the galena group with face-centered isometric structure.

Oldham-Wheat lamp [MIN ENG] A cap lamp designed for full self-service.

Old Red Sandstone [GEOL] A Devonian formation in Great Britain and northwestern Europe, of nonmarine, predominantly red sedimentary rocks, consisting principally of sandstone, conglomerates, and shales.

old snow [HYD] Deposited snow in which the original crystalline forms are no longer recognizable, such as firn or spring snow. Also known as firn snow.

old wives' summer [METEOROL] A period of calm, clear weather, with cold nights and misty mornings but fine warm days, which sets in over central Europe toward the end of September; comparable to Indian summer.

old workings [MIN ENG] Mines which have been abandoned, allowed to collapse, and sometimes sealed off.

Oleaceae [BOT] A family of dicotyledonous plants in the order Scrophulariales characterized generally by perfect flowers, two stamens, axile to parietal or apical placentation, a four-lobed corolla, and two ovules in each locule.

oleandomycin [MICROBIO] $C_{35}H_{61}O_{12}N$ A macrolide antibotic produced by *Streptomyces antibioticus*; active mainly against gram-positive microorganisms. Also known as matromycin.

oleate [ORG CHEM] Salt made up of a metal or alkaloid with oleic acid; used for external medicines and in soaps and paints.

olecranon [ANAT] The large process at the distal end of the ulna that forms the bony prominence of the elbow and receives the insertion of the triceps muscle.

olefiant gas *See* ethylene.

olefin [ORG CHEM] C_nH_{2n} A family of unsaturated, chemically active hydrocarbons with one carbon-carbon double bond; includes ethylene and propylene.

olefin copolymer [ORG CHEM] Polymer made by the interreaction of two or more kinds of olefin monomers, such as butylene and propylene.

olefin resin [ORG CHEM] Long-chain polymeric material produced by the chain reaction of olefinic monomers, such as polyethylene from ethylene, or polypropylene from propylene.

oleic acid [ORG CHEM] $C_{17}H_{33}COOH$ Yellowish, unsaturated fatty acid with lardlike aroma; soluble in organic solvents, slightly soluble in water; boils at 286°C (100 mm Hg); the main component of olive and cooking oils; used in soaps, ointments, cosmetics, and ore beneficiation. Also known as red oil.

olein [ORG CHEM] $(C_{17}H_{33}COO)_3C_3H_5$ Oleic acid triglyceride; yellow liquid melting at -5°C; slightly soluble in alcohol, soluble in chloroform, ether, and carbon tetrachloride; found in most fats and oils; used in textile lubrication.

olemeter [ENG] 1. A device for measuring specific gravity of oils. 2. An instrument for determining the proportion of oil in a substance.

Olenellidae [PALEON] A family of extinct arthropods in the class Trilobita.

oleo oil [MATER] Yellow liquid fat used to make oleomargarine; consists of liquid olein and palmitin from cold-pressed tallow.

oleoresin [MATER] A resin-essential oil mixture with pungent taste; extracted from various plants; used in pharmaceutical preparations; examples are Peru, tulu, and styrax balsams.

oleoresinous varnish [MATER] A varnish made by compounding the resin with oxidizable oil, such as linseed oil.

oleostearin [MATER] Edible solid fat from tissues of cattle (genus *Bos*); the solid remaining after oleo oil or tallow oil is removed from tallow. Also known as beef stearin.

oleo strut [MECH ENG] A shock absorber consisting of a telescoping cylinder that forces oil into an air chamber, thereby compressing the air; used on aircraft landing gear.

Olethreutidae [INV ZOO] A family of moths in the superfamily Tortricoidea whose hindwings usually have a fringe of long hairs along the basal part of the cubitus.

oleum [CHEM] Latin name for oil.

oleum gossypii seminis *See* cotton oil.

oleum morrhuae *See* cod-liver oil.

oleum theobromatis *See* cocoa butter.

oleyl alcohol [ORG CHEM] $C_{18}H_{35}OH$ Clear liquid, boiling at 282–349°C; fatty alcohol derived from oleic acid; commercial grade 80–90% pure; used to make resins and surface-active agents, and as a chemical intermediate.

olfaction [PHYSIO] 1. The function of smelling. 2. The sense of smell.

olfactoreceptor [PHYSIO] A structure which is a receptor for the sense of smell.

olfactory aura [MED] Prodromal disagreeable olfactory sensation preceding or characterizing an epileptic attack.

olfactory bulb [VERT ZOO] The bulbous distal end of the olfactory tract located beneath each anterior lobe of the cerebrum; well developed in lower vertebrates.

olfactory cell [PHYSIO] One of the sensory nerve cells in the olfactory epithelium.

olfactory foramen [ANAT] Any of the openings in the cribriform plate of the ethmoid bone through which pass the fila olfactoria of the olfactory nerves.

olfactory gland [PHYSIO] A type of serous gland in the nasal mucous membrane.

olfactory lobe [VERT ZOO] A lobe projecting forward from the inferior surface of the frontal lobe of each cerebral hemisphere, including the olfactory bulb, tracts, and trigone; well developed in most vertebrates, but reduced in man.

olfactory nerve [ANAT] The first cranial nerve; a paired sensory nerve with its origin in the olfactory lobe and formed by processes of the olfactory cells which lie in the nasal mucosa; greatly reduced in man.

olfactory organ [PHYSIO] Any of the small chemoreceptors in the mucous membrane lining the upper part of the nasal gravity which receive stimuli interpreted as odors.

olfactory pit [EMBRYO] A depression near the olfactory placode in the embryo that develops into part of the nasal cavity. Also known as nasal pit.

olfactory region [ANAT] The area on and above the superior conchae and on the adjoining nasal septum where the mucous membrane has olfactory epithelium and olfactory glands.

olfactory stalk [ANAT] The structure that connects the olfactory bulb to the cerebrum of the vertebrate brain.

olfactory tract [ANAT] A narrow tract of white nerve fibers originating in the olfactory bulb and extending posteriorly to the anterior perforated substance, where it enlarges to form a lateral root (olfactory trigone).

olibanum [MATER] A gum resin distilled from the dried exudation of African and Arabian trees of the genus *Boswellia*; used as a perfume, fixative, in incense, in fumigants, and in pharmacy. Also known as frankincense; gum thus.

olibanum oil [MATER] Colorless oil with balsamic aroma; distilled from olibanum; soluble in ether, chloroform, and carbon disulfide; main components are pinene, phellandrene, and dipentene; used in medicine. Also known as frankincense oil.

olig-, oligo- [SCI TECH] A combining form denoting few, scant, or deficiency.

oligemia [MED] A state in which the total blood volume is reduced.

Oligobrachiidae [INV ZOO] A monotypic family of the order Athecanephria.

Oligocene [GEOL] The third of the five major worldwide divisions (epochs) of the Tertiary period (Cenozoic era), extending from the end of the Eocene to the beginning of the Miocene.

Oligochaeta [INV ZOO] A class of the phylum Annelida including worms that exhibit both external and internal segmentation, and setae which are not borne on parapodia.

oligochromemia *See* anemia.

oligoclase [MINERAL] A plagioclase feldspar mineral with a composition ranging from $Ab_{90}An_{10}$ to $Ab_{70}An_{30}$, where $Ab = NaAlSi_3O_8$ and $An = CaAl_2O_8$.

oligocythemia [MED] A reduction in the total number of red blood cells in the body.

oligodendroglia [HISTOL] Small neuroglial cells with spheroidal or ovoid nuclei and fine cytoplasmic processes with secondary divisions.

oligodendroglioma [MED] A slowly growing, large, well-defined cerebral glioma, composed of small cells with richly chromatic nuclei and scanty, poorly staining cytoplasm.

oligomenorrhea [MED] Abnormally infrequent menstruation.

Oligomera [INV ZOO] A subphylum of the phylum Vermes comprising groups with two or three coelomic divisions.

oligomerous [BOT] Having one or more whorls with fewer members than other whorls of the flower.

oligomictic [PETR] Of a clastic sedimentary rock, composed of a single rock type.

oligomycin [MICROBIO] $C_{25}H_{40-42}O_7$ An antifungal antibiotic produced by an actinomycete resembling *Streptomyces diastachromogenes*; the colorless, hexagonal crystals are soluble in many organic solvents.

oligophagous [ZOO] Eating only a limited variety of foods.

Oligopygidae [PALEON] An extinct family of exocyclic Euechinoidia in the order Holectypoida which were small ovoid forms of the Early Tertiary.

oligosaccharide [BIOCHEM] A sugar composed of two or more monosaccharide units joined by glycosidic bonds. Also known as compound sugar.

oligospermia [MED] Scarcity of spermatozoa in the semen.

Oligotrichida [INV ZOO] A minor order of the Spirotrichia; the body is round in cross section, and the adoral zone of membranelles is often highly developed at the oral end of the organism.

oligotrophic [HYD] Of a lake, lacking plant nutrients and usually containing plentiful amounts of dissolved oxygen without marked stratification.

oliguria [MED] Diminished excretion of urine.

olistolith [GEOL] An exotic block or other rock mass that has been transported by submarine gravity sliding or slumping and is included in the binder of an olistostrome.

olistostrome [GEOL] A sedimentary deposit composed of a chaotic mass of heterogeneous material that is intimately mixed; accumulated in the form of a semifluid body by submarine gravity sliding or slumping of unconsolidated sediments.

olivaceous [BIOL] 1. Resembling an olive. 2. Olive colored.

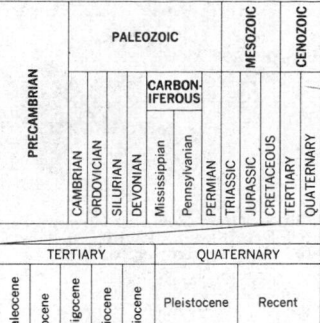

OLIGOCENE

											MESOZOIC	CENOZOIC
PRECAMBRIAN	PALEOZOIC											
						CARBONIFEROUS						
	CAMBRIAN	ORDOVICIAN	SILURIAN	DEVONIAN	Mississippian	Pennsylvanian	PERMIAN	TRIASSIC	JURASSIC	CRETACEOUS	TERTIARY	QUATERNARY

TERTIARY					QUATERNARY	
Paleocene	Eocene	Oligocene	Miocene	Pliocene	Pleistocene	Recent

Chart showing position of Oligocene epoch in relation to the eras and periods of geologic time.

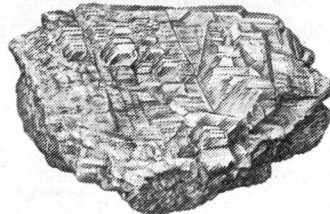

OLIGOCLASE

| 2.5 cm |

Oligoclase from North Carolina. (*Specimen from Department of Geology, Bryn Mawr College*)

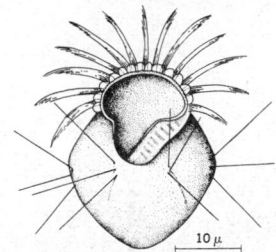

OLIGOTRICHIDA

10 μ

Halteria, an example of an oligotrichid. The long bristles are used in a kind of jumping motion.

OLIVE

Olive (*Olea europeae*) branches.
(a) Bearing small white flowers.
(b) Bearing drupes, or fruit.

OMMATIDIUM

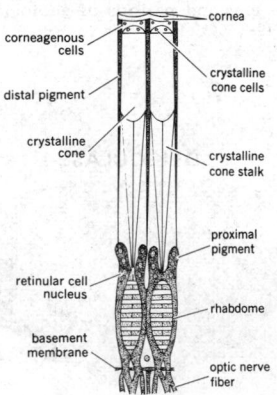

Two ommatidia from the compound eye of the crayfish *Astacus*. (*After H. Bernhards, from R. D. Barnes, Invertebrate Zoology, 2d ed., W. B. Saunders Co., 1968*)

olivary nucleus [ANAT] A prominent, convoluted gray band that opens medially and occupies the upper two-thirds of the medulla oblongata.

olive [BOT] Any plant of the genus *Olea* in the order Schrophulariales, especially the evergreen olive tree (*O. europeae*) cultivated for its drupaceous fruit, which is eaten ripe (black olives) and unripe (green), and is of high oil content.

olive hole [HOROL] In a watch, a jeweled bearing hole in which the sharp corners have been smoothed by grinding to reduce friction between the sides of the hole and the pivot that turns in it.

olive infused oil [MATER] A synthetic olive oil made by infusing corn oil with a paste of finely ground, partly dehydrated ripe olives; contains carotene.

olive knot [PL PATH] A bacterial disease of the olive caused by *Pseudomonas sevastonoi* and characterized by excrescences on the foliage and branches, and sometimes on the trunk. Also known as olive tubercle.

olivenite [MINERAL] $Cu_2(AsO_4)(OH)$ An olive-green, dull-brown, gray, or yellow mineral crystallizing in the orthorhombic system and consisting of a basic arsenate of copper. Also known as leucochalcite; wood copper.

olive oil [MATER] Pale- or greenish-yellow edible oil; main components are olein and palmitin; soluble in ether, chloroform, and carbon disulfide; derived from the pulp of olive tree fruit; used in foods, ointments, linaments, and soaps, as a lubricant, and for tanning. Also known as Florence oil; lucca oil; sweet oil.

Oliver filter [MIN ENG] A continuous-type filter made in the form of a cylindrical drum with filter cloth stretched over the convex surface of the drum.

olivette [ELEC] Standing floodlight used in the wings for lighting stage entrances and acting areas at fairly close range; bulb wattage ranges from 500 to 1500 watts.

olive tubercle *See* olive knot.

Olividae [INV ZOO] A family of snails in the gastropod order Neogastropoda.

olivine [MINERAL] $(Mg,Fe_2)SiO_4$ A neosilicate group of olive-green magnesium-iron silicate minerals crystallizing in the orthorhombic system and having a vitreous luster; hardness is $6\frac{1}{2}$–7 on Mohs scale; specific gravity is 3.27–3.37. Also known as chrysolite.

olivine basalt [PETR] Any of a group of olivine-bearing basalts.

olivine diabase [PETR] An igneous rock composed principally of olivine and formed from tholeiitic magmas by differentiation in thick sills.

olivine nephelinite [PETR] An extrusive igneous rock differing in composition from nephelinite only by the presence of olivine. Also known as ankaratrite; nepheline basalt.

Ollier's disease *See* enchondromatosis.

Olsen ductility test [MET] A cupping test in which a piece of sheet metal is deformed at the center by a steel ball until fracture occurs; ductility is measured by the height of the cup at the time of failure.

omasum [VERT ZOO] The third chamber of the ruminant stomach where the contents are mixed to a more or less homogeneous state. Also known as manyplies; psalterium.

ombrometer *See* rain gage.

ombrophilous [ECOL] Able to thrive in areas of abundant rainfall.

ombrophobia [PSYCH] An abnormal fear of rain.

ombrophobous [ECOL] Unable to live in the presence of long, continuous rain.

ombroscope [ENG] An instrument consisting of a heated, water-sensitive surface which indicates by mechanical or electrical techniques the occurrence of precipitation; the output of the instrument may be arranged to trip an alarm or to record on a time chart.

Omega [NAV] A worldwide radio navigation system providing navigational parameters by phase comparison of very-low-frequency (10 to 14 kilohertz), continuous-wave radio signals.

omega hyperon [PARTIC PHYS] A semistable baryon with a mass of approximately 1672 MeV (million electron volts), negative charge, spin of 3/2, and positive parity; constitutes an isoptic spin singlet. Also known as omega particle. Symbolized Ω^-.

omega meson [PARTIC PHYS] An unstable, neutral vector meson having a mass of about 783 MeV (million electron volts), a width of about 12 MeV, and negative charge parity and G parity. Symbolized $\omega(783)$.

omega particle *See* omega hyperon.

omegatron [ELECTR] A miniature mass spectrograph, about the size of a receiving tube, that can be sealed to another tube and used to identify the residual gases left after evacuation.

omentum [ANAT] A fold of the peritoneum connecting or supporting abdominal viscera.

omission factor [ADP] In information retrieval, the ratio obtained in dividing the number of nonretrieved relevant documents by the total number of relevant documents in the file.

omission solid solution [CRYSTAL] A crystal with certain atomic sites incompletely filled.

ommatidium [INV ZOO] The structural unit of a compound eye, composed of a cornea, a crystalline cone, and a receptor element connected to the optic nerve.

ommatophore [INV ZOO] A movable peduncle that bears an eye, as in snails.

omnibearing [NAV] The magnetic bearing of an omnidirectional radio range.

omnibearing beacon [NAV] A beacon transmitter capable of providing bearing information in all horizontal directions, particularly a very-high-frequency omnidirectional radio range.

omnibearing converter [ENG] An electromechanical device which combines an omnirange signal with heading information to furnish electrical signals for the operation of the pointer of a radio magnetic indicator.

omnibearing-distance navigation [NAV] Navigation based upon polar coordinates relative to a reference point. Also known as r-theta navigation; rho-theta navigation.

omnibearing-distance station [NAV] A radio station having an omnidirectional radio range and a distance-measuring-equipment (DME) transponder in combination.

omnibearing indicator [ENG] An instrument providing automatic and continuous indication of omnibearing.

omnibearing line [NAV] One of an infinite number of straight lines radiating from the geographical location of a VHF omnirange.

omnibearing selector [ENG] A device capable of being set manually to any desired omnibearing, or its reciprocal, to control a course-line deviation indicator. Also known as radial selector.

omnidirectional [ELECTR] Radiating or receiving equally well in all directions. Also known as nondirectional.

omnidirectional antenna [ELECTROMAG] An antenna that has an essentially circular radiation pattern in azimuth and a directional pattern in elevation. Also known as nondirectional antenna.

omnidirectional range *See* omnirange.

omnidistance [NAV] The distance indicated on the meter of an aircraft's distance-measuring equipment when this equipment is interrogating and receiving replies from a transponder beacon located at an omnirange station; thus it is the slant distance.

omnigraph [COMMUN] An instrument that converts Morse code signals on punched tape into corresponding buzzer-produced audio signals for training purposes.

omnimeter [ENG] A theodolite with a microscope that can be used to observe vertical angular movement of the telescope.

omnirange [NAV] A radio aid to navigation providing direct indication of the magnetic bearing (omnibearing) of that station from any direction. Also known as omnidirectional range.

omnivore [ZOO] An organism that eats both animal and vegetable matter.

omohyoid [ANAT] **1.** Pertaining conjointly to the scapula and the hyoid bone. **2.** A muscle attached to the scapula and the hyoid bone.

Omophronidae [INV ZOO] The savage beetles, a small family of coleopteran insects in the suborder Adephaga.

omphacite [MINERAL] A grassy- to pale-green, granular or foliated, high-temperature aluminous clinopyroxene mineral with a vitreous luster that commonly occurs in the rock eclogite; a variety of augite.

omphalitis [MED] Inflammation of the umbilicus.

omphalomesenteric artery *See* vitelline artery.

omphalomesenteric duct *See* vitelline duct.

omphalomesenteric vein *See* vitelline vein.

omphaloproptosis [MED] Abnormal protrusion of the navel.

Omphralidae [INV ZOO] A family of orthorrhaphous dipteran insects in the series Nematocera.

OMS *See* ovonic memory switch.

Onagraceae [BOT] A family of dicotyledonous plants in the order Myrtales characterized generally by an inferior ovary, axile placentation, twice as many stamens as petals, a four-nucleate embryo sac, and many ovules.

on-call circuit [COMMUN] A permanently designated circuit that is activated only upon request of the user; this type of circuit is usually provided when a full-period circuit cannot be justified and the duration of use cannot be anticipated; during unactivated periods, the communications facilities required for the circuit are available for other requirements.

on-carriage equipment [ORD] Items of supply which, although not part of the cannon carriage proper, are mounted on the carriage and remain in their respective mounts, brackets, or containers on the carriage when it is being towed.

on-carriage fire control [ORD] Process of controlling fire on a target with the aid of a sighting device mounted directly on the weapon.

on-carriage fire control equipment [ORD] Fire control items, such as telescope mounts, which are built into the cannon carriage or mount, or which are carried on the carriage or mount in the traveling position.

once-through boiler [MECH ENG] A boiler in which water flows, without recirculation, sequentially through the economizer, furnace wall, and evaporating and superheating tubes.

Onchidiidae [INV ZOO] An intertidal family of sluglike pulmonate mollusks of the order Systellommatophora in which the body is oval or lengthened, with the convex dorsal integument lacking a mantle cavity or shell.

onchocerciasis [MED] Infection with the filaria *Onchocerca volvulus*; results in skin tumors, papular dermatitis, and ocular complications.

onchogryposis [MED] A thickened, ridged, and curved condition of a nail.

oncholysis [MED] A slow process of loosening of a nail from its bed, beginning at the free edge and progressing gradually toward the root.

onchomycosis [MED] Any fungus disease of the nail.

onchosphere [INV ZOO] The hexacanth embryo identified as the earliest differentiated stage of cyclophyllidean tapeworms.

oncocyte [HISTOL] A columnar-shaped cell with finely granular eosinophilic cytoplasm, found in salivary and certain endocrine glands, nasal mucosa, and other locations.

oncocytoma [MED] A benign tumor composed principally of oncocytes; usually occurs in salivary glands.

oncogenesis [MED] Processes of tumor formation.

oncology [MED] The study of the causes, development, characteristics, and treatment of tumors.

Oncopoda [INV ZOO] A phylum of the superphylum Articulata.

oncotic pressure [PHYSIO] Also known as colloidal osmotic pressure. **1.** The osmotic pressure exerted by colloids in a solution. **2.** The pressure exerted by plasma proteins.

oncotomy [MED] Surgical incision of a tumor, abscess, or other swelling.

ondograph [ELECTR] An instrument that draws the waveform of an alternating-current voltage step by step; a capacitor is charged momentarily to the amplitude of a point on the voltage wave, then discharged into a recording galvanometer, with the action being repeated a little further along on the waveform at intervals of about 0.01 second.

ondoscope [ELECTR] A glow-discharge tube used to detect high-frequency radiation, as in the vicinity of a radar transmitter; the radiation ionizes the gas in the tube and produces a visible glow.

-one [ORG CHEM] Chemical suffix indicating a ketone, a substance related to starches and sugars, or an alkone.

one-address code [ADP] In computers, a code using one-address instructions.

one-address instruction [ADP] A digital computer programming instruction that explicitly describes one operation and one storage location. Also known as single-address instruction.

one-dimensional flow [FL MECH] Fluid flow in which all flow is parallel to some straight line, and characteristics of flow do not change in moving perpendicular to this line.

one-dimensional lattice [CRYSTAL] A simplified model of a crystal lattice consisting of particles lying along a straight line at either equal or periodically repeating distances.

one-group model [NUCLEO] A neutron-behavior model in which neutrons of all energies are treated as having the same characteristics.

one-hundred-percent premium plan [IND ENG] A wage incentive plan wherein each unit produced by an employee in excess of standard is compensated at the same rate paid for each unit of standard production. Also known as straight piecework system; straight proportional system.

one-level address [ADP] In digital computers, an address that directly indicates the location of an instruction or some data.

one-level code [ADP] Any code using absolute addresses and absolute operation codes.

one-line adapter [ADP] A unit connecting central processes and permitting high-speed transfer of data under program control.

one-part code [COMMUN] Code in which the plain text elements are arranged in alphabetical or numerical order, accompanied by their code groups also arranged in alphabetical, numerical, or other systematic order.

one-particle exchange [PARTIC PHYS] A model for the interaction of two particles in which the interaction results entirely from a single virtual particle being emitted by one interacting particle and absorbed by the other.

one-pass operation [ADP] An operating method, now standard, which produces an object program from a source program in one pass.

one-piece set [MIN ENG] A single stick of timber used as a post, stull, or prop.

one-plus-one address instruction [ADP] A digital computer instruction whose format contains two address parts; one address designates the operand to be involved in the operation; the other indicates the location of the next instruction to be executed.

one-point perspective *See* parallel perspective.

one-quadrant multiplier [ELECTR] Of an analog computer, a multiplier in which operation is restricted to a single sign of both input variables.

one-sample problem [STAT] The problem of testing the hypothesis that the average of a sequence of observations or measurement of the same kind has a specified value.

one's complement [ADP] A numeral in binary notation, derived from another binary number by simply changing the sense of every digit.

ones-complement code [ADP] A number coding system used in some computers, where, for any number x, $x = (1 - 2^{n-1}) \cdot a_0 + 2^{n-2}a_1 + \ldots + a_{n-1}$, where $a_i = 1$ or 0.

one-shot molding [ENG] Production of urethane-plastic foam in which the isocynate, polyol, and catalyst and other additives are mixed directly together and a foam is produced immediately.

one-shot multivibrator *See* monostable multivibrator.

one-sided acceptance sampling test [IND ENG] A test against a single specification only, in which permissible values in one direction are not limited.

one-sided limit [MATH] Either a limit on the left or a limit on the right.

one-sided test [STAT] A test statistic T which rejects a hypothesis only for $T \geq d$ or $T \leq c$ but not for both (here d and c are critical values).

O network [ELEC] Network composed of four impedance branches connected in series to form a closed circuit, two adjacent junction points serving as input terminals, the remaining two junction points serving as output terminals.

ONION

Onion flower and an oblate type of bulb.

ONISCOIDEA

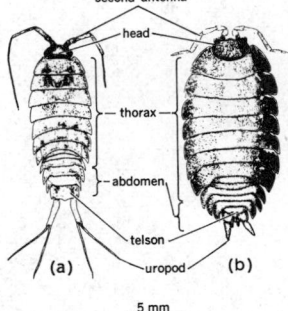

Genera of oniscoideans. (a) *Ligia*, the shore slater found above tidewater mark. (b) *Porcellio*, the sow bug or wood louse.

one-way slab [CIV ENG] A concrete slab in which the reinforcing steel runs perpendicular to the supporting beams, that is, one way.

one-way trunk [ELEC] Trunk between two central offices, used for calls that originate at one of those offices, but not for calls that originate at the other. Also known as outgoing trunk.

one-year ice [OCEANOGR] Sea ice formed the previous season, not yet 1 year old.

on grade [CIV ENG] **1.** At ground level. **2.** Supported directly on the ground.

on-hook [COMMUN] The idle state (open loop) of a subscriber or PBX user loop.

onion [BOT] **1.** *Allium cepa.* A biennial plant in the order Liliales cultivated for its edible bulb. **2.** Any plant of the genus *Allium*.

onion oil [MATER] A sharp-odored yellow liquid; derived from the bulb of the onion *Allium cepa*; soluble in ether, chloroform, and carbon disulfide; main component is allyl propyl disulfide; used in flavorings.

onion scab *See* onion smudge.

onionskin paper [MATER] A lightweight, durable bond paper; usually quite translucent, resembling the dry outer skin of an onion; used for duplicate typewriter copies and in interleaving order books.

onion smudge [PL PATH] A fungus disease of the onion caused by *Colletotrichum circinans* and characterized by black concentric integral rings or smutty spots on the bulb scales. Also known as onion scab.

onion smut [PL PATH] A fungus disease of onion, especially seedlings, caused by *Urocystis cepulae* and characterized by elongate black blisters on the scales and foliage.

Oniscoidea [INV ZOO] A terrestrial suborder of the Isopoda; the body is either dorsoventrally flattened or highly vaulted, and the head, thorax, and abdomen are broadly joined.

-onium [INORG CHEM] Chemical suffix indicating a complex cation, as for oxonium, $(H_3O)^+$.

onlap [GEOL] A type of overlap characterized by regular and progressive pinching out of the strata toward the margins of a depositional basin; each unit transgresses and extends beyond the point of reference of the underlying unit. Also known as transgressive overlap.

on-line [ADP] Pertaining to equipment capable of interacting with a computer. [ELECTR] The state in which a piece of equipment or a subsystem is connected and powered to deliver its proper output to the system.

on-line analyzer [MIN ENG] An instrument which monitors the content of materials at various stages in flotation or other mineral-processing flow sheets.

on-line cipher [COMMUN] A method of encryption directly associated with a particular transmission system, whereby messages may be encrypted and simultaneously transmitted from one station to one or more stations where reciprocal equipment is automatically operated.

on-line computer system [ADP] A computer system which is adapted to on-line operation.

on-line cryptographic operation *See* on-line operation.

on-line data reduction [ADP] The processing of information as rapidly as it is received by the computing system.

on-line disk file [ADP] A magnetic disk directly connected to the central processing unit, thereby increasing the memory capacity of the computer.

on-line equipment [ADP] The equipment or devices in a system whose operation is under control of the central processing unit, and in which information reflecting current activity is introduced into the data-processing system as soon as it occurs.

on-line mode [ADP] Mode of operation in which all devices are responsive to the central processor.

on-line operation [ADP] Computer operation in which input data are fed into the computer directly from observing instruments or other input equipment, and computer results are obtained during the progress of the event. [COMMUN] A method of operation whereby messages are encrypted and simultaneously transmitted from one station to one or more other stations where reciprocal equipment is automatically operated to permit reception and simultaneous decryptment

of the message. Also known as on-line cryptographic operation.

on-line secured communications system [COMMUN] Any combination of interconnected communications centers partially or wholly equipped for on-line cryptographic operation and capable of relaying or switching message traffic using on-line cryptographic procedures.

on-line storage [ADP] Storage controlled by the central processing unit of a computer.

only loadable [ADP] Attribute of a load module which can be brought into main memory only by a LOAD macroinstruction given from another module. Abbreviated OL.

on-off control [CONT SYS] A simple control system in which the device being controlled is either full on or full off, with no intermediate operating positions. Also known as on-off system.

on-off keying [COMMUN] Binary form of amplitude modulation in which one of the states of the modulated wave is the absence of energy in the keying interval.

on-off switch [ELEC] A switch used to turn a receiver or other equipment on or off; often combined with a volume control in radio and television receivers.

on-off system *See* on-off control.

on-off tests [ELEC] Tests conducted to determine the source of interference by switching various suspected sources on and off while observing the victim receiver.

Onsager equation [PHYS CHEM] An equation which relates the measured equivalent conductance of a solution at a certain concentration to that of the pure solvent.

Onsager reciprocal relations [THERMO] A set of conditions which state that the matrix, whose elements express various fluxes of a system (such as diffusion and heat conduction) as linear functions of the various conjugate affinities (such as mass and temperature gradients) for systems close to equilibrium, is symmetric when certain definitions are chosen for these fluxes and affinities.

Onsager theory of dielectrics [ELEC] A theory for calculating the dielectric constant of a material with polar molecules in which the local field at a molecule is calculated for an actual spherical cavity of molecular size in the dielectric using Laplace's equation, and the polarization catastrophe of the Lorentz field theory is thereby avoided.

onsetter [MIN ENG] The worker in charge of hoisting coal, ore, men, or materials in a mine shaft.

onshore [GEOGR] Pertaining to, in the direction toward, or located on the shore. Also known as shoreside.

onshore wind [METEOROL] Wind blowing from the sea toward the land.

on soundings [NAV] Of a vessel, navigating within the 100-fathom curve; in earlier times, the term was applied to a vessel in water sufficiently shallow for sounding by sounding lead.

on stream [CHEM ENG] Of plant or process-operations, unit, being in operation.

on-stream time [CHEM ENG] In plant or process operations, the actual time that a unit is operating and producing product.

on the beam [ELECTR] Centered on a beam of, or an equisignal zone of, radiant energy, as a radio range. [NAV] Bearing approximately 090° relative (on the starboard beam) or 270° relative (on the port beam); the expression is often used loosely for broad on the beam, or bearing exactly 090° or 270° relative. Also known as abeam.

on the bow [NAV] Bearing approximately 045° relative (on the starboard bow) or 315° relative (on the port bow); the expression is often used loosely for broad on the bow, or bearing exactly 045° or 315° relative.

on-the-fly printer [ADP] A high-speed line printer using continuously rotating print wheels and fast-acting hammers to print the letters contained on one line of text so rapidly that the characters appear to be printed simultaneously.

on the quarter [NAV] Bearing approximately 135° relative (on the starboard quarter) or 225° relative (on the port quarter); the expression is often used loosely for broad on the quarter, or bearing exactly 135° or 225° relative.

on the run [MIN ENG] Manner of working a seam of coal when there is sufficient inclination to cause the coal, as worked toward the rise, to fall by gravity to the gangways for loading into cars.

on the solid [MIN ENG] **1.** Pertaining to the practice of blasting heavy charges of explosives, in lieu of undercutting or channeling. **2.** *See* into the solid.

on the way [ORD] An expression sent over the communication system from the firing position to the observation position when the weapon is fired, thus warning the observers to be on the alert for spotting the impact.

ontogeny [EMBRYO] The origin and development of an organism from conception to adulthood.

on-top flight [AERO ENG] Flight above an overcast.

Onuphidae [INV ZOO] A family of tubicolous, herbivorous, scavenging errantian annelids in the superfamily Eunicea.

on-vehicle materiel [ORD] Items of supply which, although not part of the vehicle proper, are issued with, and carried on, ordnance or other technical service vehicles; the items are required for vehicular maintenance, armament, fire protection, or communication, to complete the major combination; examples are guns, gun mounts, radios, flashlights, and fire extinguishers.

onych-, onycho- [ZOO] A combining form denoting claw or nail.

onychia [MED] Inflammation of the nail matrix.

Onychodontidae [PALEON] A family of Lower Devonian lobefin fishes in the order Osteolepiformes.

onychomycosis [MED] A fungus disease of the nails.

Onychopalpida [INV ZOO] A suborder of mites in the order Acarina.

Onychophora [INV ZOO] A phylum of wormlike animals that combine features of both the annelids and the arthropods.

onychorrhexis [MED] Longitudinal striation of the nail plate, with or without the formation of fissures.

Onygenaceae [MYCOL] A family of ascomycetous fungi in the order Eurotiales comprising forms that inhabit various animal substrata, such as horns and hoofs.

onyx [MINERAL] **1.** Banded chalcedonic quartz, in which the bands are straight and parallel; natural colors are usually red or brown with white, although black is occasionally encountered. **2.** *See* onyx marble.

onyx marble [MINERAL] A hard, compact, dense, generally translucent variety of calcite resembling true onyx and usually banded. Also known as alabaster; Algerian onyx; Gibraltar stone; Mexican onyx; onyx; oriental alabaster.

oocyst [INV ZOO] The encysted zygote of some Sporozoa.

oocyte [HISTOL] An egg before the completion of maturation.

oogamete [BIOL] A large, nonmotile female gamete containing reserve materials.

oogamous [BIOL] Of sexual reproduction, characterized by fusion of a motile sperm with an oogamete.

oogenesis [PHYSIO] Processes involved in the growth and maturation of the ovum in preparation for fertilization.

oogonium [BOT] The unisexual female sex organ in oogamous algae and fungi. [HISTOL] A descendant of a primary germ cell which develops into an oocyte.

ookinete [INV ZOO] The elongated, mobile zygote of certain Sporozoa, as that of the malaria parasite.

oolemma *See* zona pellucida.

oolite [PETR] A sedimentary rock, usually a limestone, composed principally of cemented ooliths. Also known as eggstone; roestone.

oolith [PETR] A small (0.25–2.0 millimeters), rounded accretionary body in a sedimentary rock; generally formed of calcium carbonate by inorganic precipitation or by replacement; ooliths generally exhibit concentric or radial internal structure.

oolitic chert [PETR] Chert composed chiefly of ooliths.

oolitic limestone [PETR] An even-textured limestone made up almost entirely of calcareous ooliths with essentially no matrix.

oology [VERT ZOO] A branch of zoology concerned with the study of eggs, especially bird eggs.

Oomycetes [MYCOL] A class of the Phycomycetes comprising the biflagellate water molds and downy mildews.

oophagous [ZOO] Feeding or living on eggs.

oophoritis [MED] Inflammation of the ovaries.

ooplasm [CYTOL] Cytoplasm of an egg.

Oort's Cloud [ASTRON] A cloud of comets at distances from 75,000 to 150,000 astronomical units from the sun, which has been proposed as a source of comets that pass near the sun.

oospore [BOT] A spore which is produced by heterogamous fertilization and from which the sporophytic generation develops.

oostegite [INV ZOO] In many crustaceans, a platelike expansion of the basal segment of a thoracic appendage that aids in forming an egg receptacle.

ooze [GEOL] **1.** A soft, muddy piece of ground, such as a bog, usually resulting from the flow of a spring or brook. **2.** A marine pelagic sediment composed of at least 30% skeletal remains of pelagic organisms, the rest being clay minerals. **3.** Soft mud or slime, typically covering the bottom of a lake or river.

opacifier [MATER] A substance, used to treat a solid rocket propellant, that absorbs light and heat and protects the propellant from deterioration until ready for use.

opacity [OPTICS] The light flux incident upon a medium divided by the light flux transmitted by it.

opacus [METEOROL] A variety of cloud (sheet, layer, or patch), the greater part of which is sufficiently dense to obscure the sun; found in the genera altocumulus, altostratus, stratocumulus, and stratus; cumulus and cumulonimbus clouds are inherently opaque.

opal [MINERAL] A natural hydrated form of silica; it is amorphous, usually occurs in botryoidal or stalactic masses, has a hardness of 5–6 on Mohs scale, and specific gravity is 1.9–2.2.

opal-CT [PETR] A poorly ordered crystalline form of silica thought to be the intermediate phase in quartz chert formation.

opalescence [OPTICS] The milky, iridescent appearance of a dense, transparent medium or colloidal system when it is illuminated by polychromatic radiation in the visible range, such as sunlight.

opal glass [MATER] Translucent or opaque glass, often milky white, made by adding impurities such as fluorine compounds to the melt; it appears white by reflected light but shows color images through thin sections; used for ornamental glass and as an efficient light diffuser.

Opalinata [INV ZOO] A superclass of the subphylum Sarcomastigophora containing highly specialized forms which resemble ciliates.

opalized wood *See* silicified wood.

opaque attritus [GEOL] Attritus that does not contain large quantities of transparent humic degradation matter.

opaque medium [OPTICS] A medium which is impervious to rays of light, that is, not transparent to the human eye. [PHYS] **1.** A medium which does not transmit electromagnetic radiation of a specified type, such as that in the infrared, x-ray, ultraviolet, and microwave regions. **2.** A medium which prevents the passage of particles of a specified type.

opaque projector [OPTICS] A projector designed to project the image of an opaque object, or of graphic material on an opaque support, by reflected light.

opaque sky cover [METEOROL] In United States weather observing practice, the amount (in tenths) of sky cover that completely hides all that might be above it; opposed to transparent sky cover.

OPDAR [ENG] A laser system for measuring elevation angle, azimuth angle, and slant range of a missile during its firing period. Derived from optical direction and ranging. Also known as optical radar.

Opegraphaceae [BOT] A family of the Hysteriales characterized by elongated ascocarps; members are crustose on bark and rocks.

open [ELEC] **1.** Condition in which conductors are separated so that current cannot pass. **2.** Break or discontinuity in a circuit which can normally pass a current.

open ammunition space [ORD] Ground area prepared or improvised for storage of ammunition in open areas to supplement magazine space.

open-angle glaucoma [MED] Bilateral, increased intraocular tension due to reduced aqueous outflow but with the angle open and the aqueous in free contact with the trabecula.

open-arc furnace [MET] An electrosmelting furnace in which the arc is generated above the level of the furnace feed.

open ball [MATH] In a metric space, an open set about a point x which consists of all points within a fixed distance from x.

open-belt drive [MECH ENG] Pulley-belt arrangement where

3 mm

Oolitic limestone, from Volksen, Deister Mountains, Germany, showing the fragments of shell encased by fragments of microcrystalline calcite (dark stippling). It is often an important ore reservoir, and is quarried for building purposes.

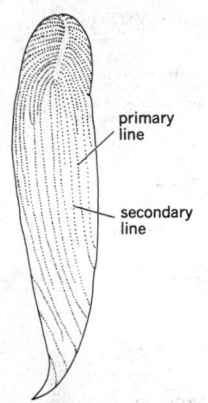

primary line

secondary line

An opalinid (*Protoopalina axonucleata*) showing primary and secondary lines of blepharoplasts.

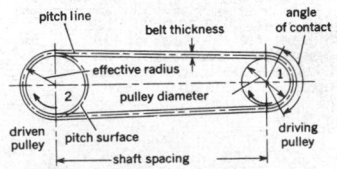

Diagram of open-belt drive; 1 and 2 indicate two pulleys whose shafts are parallel.

OPEN CAISSON

Underside of open caisson for Greater New Orleans bridge over Mississippi River. Note prefabricated steel cutting edge. (Dravo Corp.)

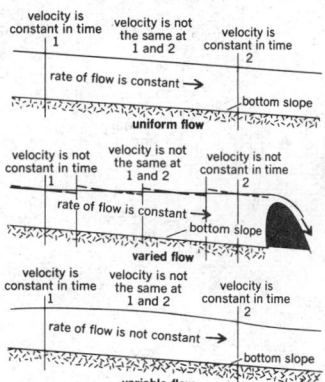

OPEN CHANNEL

Three types of flow in open channels (rivers, streams, canals, flumes).

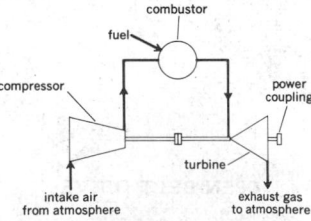

OPEN-CYCLE GAS TURBINE

Schematic diagram of the open-cycle gas turbine showing that the inlet and exhaust are open to the atmosphere.

---

the belt passes over two pulleys on parallel shafts which turn in the same direction.

open berth [CIV ENG] An anchorage berth in an open roadstead.

open caisson [CIV ENG] A caisson in the form of a cylinder or shaft that is open at both ends; it is set in place, pumped dry, and filled with concrete.

open-cast mining *See* open-pit mining.

open-cell foam [MATER] Foamed material, natural or synthetic, rigid or flexible, organic or metallic, in which there is interconnection between the cells.

open-center plan-position indicator [ENG] A plan-position indicator on which no signal is displayed within a set distance from the center.

open channel [SCI TECH] Any natural or artificial, covered or uncovered conduit in which liquid (usually water) flows with its top surface bounded by the atmosphere.

open circuit [ELEC] An electric circuit that has been broken, so that there is no complete path for current flow.

open-circuited line [ELECTROMAG] A microwave discontinuity which reflects an infinite impedance.

open-circuit grinding [MECH ENG] Grinding system in which material passes through the grinder without classification of product and without recycle of oversize lumps; in contrast to closed-circuit grinding.

open-circuit impedance [ELEC] Of a line or four-terminal network, the driving-point impedance when the far end is open.

open-circuit jack [ELEC] Jack that normally leaves its circuit open; the circuit can be closed only through a circuit connected to the plug that is inserted in the jack.

open-circuit potential [PHYS CHEM] The steady-state or equilibrium potential of an electrode in absence of external current flow to or from the electrode.

open-circuit signaling [COMMUN] Type of signaling in which no current flows while the circuit is in the idle condition.

open cluster [ASTRON] One of the groupings of stars that are concentrated along the central plane of the Milky Way; most have an asymmetrical appearance and are loosely assembled, and the stars are concentrated in their central region; they may contain from a dozen to many hundreds of stars. Also known as galactic cluster.

open coast [GEOGR] A coast that is not sheltered from the sea.

opencut [CIV ENG] An open trench, such as across a hill. [MIN ENG] **1.** To drive headings out, or to commence working in the coal after sinking the shafts. **2.** To commence longwall working. **3.** To increase the size of a shaft when it intersects a drift.

opencut mining *See* open-pit mining.

open cycle [THERMO] A thermodynamic cycle in which new mass enters the boundaries of the system and spent exhaust leaves it; the automotive engine and the gas turbine illustrate this process.

open-cycle engine [MECH ENG] An engine in which the working fluid is discharged after one pass through boiler and engine.

open-cycle gas turbine [MECH ENG] A gas turbine prime mover in which air is compressed in the compressor element, fuel is injected and burned in the combustor, and the hot products are expanded in the turbine element and exhausted to the atmosphere.

open-cycle reactor system [NUCLEO] A reactor system in which the coolant passes through the reactor core only once and is then discarded.

open-delta connection [ELEC] An unsymmetrical transformer connection which is employed when one transformer of a bank of three single-phase delta-connected units must be cut out, because of failure. Also known as V connection.

open die [MET] A forming or forging die in which there is little or no restriction to the lateral flow of metal within the die set.

open-die forging [MET] Forging performed with open dies.

open-ended [ADP] Of techniques, designed to facilitate or permit expansion, extension, or increase in capability; the opposite of closed-in and artificially constrained.

open-ended system [ADP] In character recognition, a system in which the input data to be read are derived from sources

---

other than the computer with which the character reader is associated.

open-end method [MIN ENG] A technique of mining pillars in which no stump is left.

open-end wrench [DES ENG] A wrench consisting of fixed jaws at one or both ends of a handle.

open fire [MIN ENG] A fire at a roadway or at the coal face in a mine.

open-flow capacity [PETRO ENG] Fluid flow rate from a gas well flowing open to the atmosphere.

open-flow potential [PETRO ENG] Gas flow rate in thousands of cubic feet of gas per 24 hours that would be produced by a well if the only pressure against the face of the producing formation wellbore were atmospheric pressure.

open-flow test [PETRO ENG] The flowing of wells wide open to the atmosphere with simultaneous measurement of gas-flow rate and pressure drop; used to analyze the potential of a reservoir to deliver gas to the wellbore.

open-flow well [PETRO ENG] Gas well flowing open to the atmosphere.

open fold [GEOL] A fold having only moderately compressed limbs.

open form [CRYSTAL] A crystal form in which the crystal faces do not entirely enclose a space.

open-fuse cutout [ELEC] Enclosed fuse cutout in which the fuse support and fuse holder are exposed.

open harbor [GEOGR] An unsheltered harbor exposed to the sea.

open-hearth furnace [MET] A reverberatory melting furnace with a shallow hearth and a low roof, in which the charge is heated both by direct flame and by radiation from the roof and walls of the furnace.

open-hearth process [MET] A steel-making process carried out in an open-hearth furnace in which selected pig iron and malleable scrap iron are melted, with the addition of pure iron ore.

open ice [OCEANOGR] On navigable waters, ice that has broken apart sufficiently to permit passage of vessels.

opening [GEOGR] A break in a coastline or a passage between shoals, and so forth. [MIN ENG] **1.** A widening of a crevice, in consequence of a softening or decomposition of the adjacent rock, so as to leave a vacant space. **2.** A short heading driven between two or more parallel headings or levels for ventilation. **3.** An area in a coal mine between pillars, or between pillars and ribs. [OCEANOGR] Any break in sea ice which reveals the water.

opening die [MECH ENG] A die head for cutting screws that opens automatically to release the cut thread.

opening material [MATER] A material added to plastic clay in ceramic making to speed drying and reduce shrinking.

open interval [MATH] An open interval of real numbers, denoted by (a,b), consists of all numbers strictly greater than a and strictly less than b.

open lead [NAV] A navigable passage through ice or between rocks or shoals not covered with ice.

open-link fuse [ELEC] A simple type of fuse that consists of a strip of fuse material bolted to open terminal blocks.

open-loop control system [CONT SYS] A control system in which the system outputs are controlled by system inputs only, and no account is taken of actual system output.

open map [MATH] A function between two topological spaces which sends each open set of one to an open set of the other.

open mapping theorem [MATH] A continuous linear function between Banach spaces which has closed range must be an open map.

open pack ice [OCEANOGR] Floes of sea ice that are seldom in contact with each other, generally covering between four-tenths and six-tenths of the sea surface.

open-phase protection [ELEC] Effect of a device operating on the loss of current in one phase of a polyphase circuit to cause and maintain the interruption of power in the circuit.

open-phase relay [ELEC] Relay which functions by reason of the opening of one or more phases of a polyphase circuit, when sufficient current is flowing in the remaining phase or phases.

open-pit mining [MIN ENG] Extracting metal ores and minerals that lie near the surface by removing the overlying material and breaking and loading the ore. Also known as open-cast mining; opencut mining.

open-pit quarry [MIN ENG] A quarry in which the opening is the full size of the excavation.

open plan [BUILD] Arrangement of the interior of a building without distinct barriers such as partitions.

open plug [ELEC] Plug designed to hold jack springs in their open position.

open roadstead [NAV] An area near the shore where vessels anchor with relatively little protection from the sea.

open routine [ADP] A routine which can be inserted directly into a larger routine without a linkage or calling sequence.

open sea [GEOGR] **1.** That part of the ocean not enclosed by headlands, not within narrow straits, and so on. **2.** That part of the ocean outside the territorial jurisdiction of any country.

open set [MATH] A set included in a topology; equivalently, a set which is a neighborhood of each of its points; a topology on a space is determined by a collection of subsets which are called open.

open shop [IND ENG] A shop in which employment is not restricted to members of a labor union.

openside planer [DES ENG] A planer constructed with one upright or housing to support the crossrail and tools.

open-side tool block [DES ENG] A toolholder on a cutting machine consisting of a T-slot clamp, a C-shaped block, and two or more tool clamping screws. Also known as heavy-duty tool block.

open sight [ORD] A rear gunsight having a notch.

open stope [MIN ENG] Underground working place that is unsupported, or supported by timbers or pillars of rock.

open-stope method [MIN ENG] Stoping in which no regular artificial method of support is employed, although occasional props or cribs may be used to hold local patches of insecure ground.

open storage [ORD] The storage of certain ordnance materiel outdoors.

open subroutine [ADP] A set of computer instructions that collectively perform some particular function and are inserted directly into the program each and every time that particular function is required.

open system [THERMO] A system across whose boundaries both matter and energy may pass.

open-timbered roof [BUILD] A roof in which the supporting timbers are left uncovered, forming part of the ceiling.

open timbering [MIN ENG] A method of supporting the ground in a mine shaft or tunnel; supports are several feet apart, with the ground between them secured by struts.

open traverse [ENG] A surveying traverse in which the last leg, because of error, does not terminate at the origin of the first leg.

open tuberculosis [MED] Tuberculosis in which tubercle bacilli are being discharged from the body; tuberculosis capable of transmission to other persons.

open water [OCEANOGR] Water less than one-tenth covered with floating ice.

open-web girder *See* lattice girder.

open-window unit *See* sabin.

open wire [ELEC] A conductor supported above the ground, separate from other conductors.

open-wire carrier system [COMMUN] A system for carrier telephony, using an open-wire line.

open-wire feeder *See* open-wire transmission line.

open-wire loop [ELEC] Branch line on a main open-wire line.

open-wire transmission line [ELEC] A transmission line consisting of two spaced parallel wires supported by insulators, at the proper distance to give a desired value of surge impedance. Also known as open-wire feeder.

open workings [MIN ENG] Surface workings, for example, a quarry or open-cast mine.

opera glasses [OPTICS] Small binocular telescopes, usually of the Galilean type, adapted for use where magnification and field of view are secondary to compactness and cost.

operand [ADP] Any one of the quantities entering into or arising from an operation.

operand-precision register [ADP] A special register found in some minicomputers which can be programmed from 8- to 32-bit precision.

operant conditioning [PSYCH] A form of learning in which the subject, in a given situation, tends to respond in a way that produces rewarding effects, reinforcing previous pleasurable experiences. Also known as instrumental conditioning; reinforcement conditioning.

operate time [ADP] The phase of computer operation when an instruction is being carried out. [ELEC] Total elapsed time from application of energizing current to a relay coil to the time the contacts have opened or closed.

operating angle [ELECTR] Electrical angle of the input signal (for example, portion of a cycle) during which plate current flows in a vacuum tube amplifier.

operating characteristic curve [STAT] In hypothesis testing, a plot of the probability of accepting the hypothesis against the true state of nature. Abbreviated OC curve.

operating handle [ORD] Handle or bar with which the operating lever of a gun is operated to open and close the breech.

operating lever [ORD] Lever device on a gun with which the breech is opened and closed.

operating line [CHEM ENG] In the graphical solution of equilibrium processes (such as distillation absorption extraction), the actual liquid-vapor relationship of a key component, in contrast to a true equilibrium relationship.

operating point [ELECTR] Point on a family of characteristic curves of a vacuum tube or transistor where the coordinates of the point represent the instantaneous values of the electrode voltages and currents for the operating conditions under study or consideration.

operating position [COMMUN] Terminal of a communications channel which is attended by an operator; usually the term refers to a single operator, such as a radio operator's position or a telephone operator's position; however, certain terminals may require more than one operating position.

operating power [ELECTROMAG] Power that is actually supplied to a radio transmitter antenna.

operating pressure [ENG] The system pressure at which a process is operating.

operating range [NAV] The maximum distance at which reliable service is provided by an aid to navigation.

operating ratio [ADP] The time during which computer hardware operates and gives reliable results divided by the total time scheduled for computer operation.

operating slide [ORD] Mechanism in a Browning machine gun that permits opening the breech for loading, unloading, and clearing out stoppages, and closing the breech for firing.

operating stress [MECH] The stress to which a structural unit is subjected in service.

operating system *See* executive system.

operating system supervisor [ADP] The control program of a set of programs which guide a computer in the performance of its tasks and which assist the program with certain supporting functions.

operation [ADP] **1.** A process or procedure that obtains a unique result from any permissible combination of operands. **2.** The sequence of actions resulting from the execution of one digital computer instruction. [IND ENG] A job, usually performed in one location, and consisting of one or more work elements.

operational [ENG] Of equipment such as aircraft or vehicles, being in such a state of repair as to be immediately usable.

operational advantage [NAV] In air operations, an improvement which benefits the users of an instrument procedure, such as the ability to use lower minimums.

operational amplifier [ELECTR] An amplifier having high direct-current stability and high immunity to oscillation, generally achieved by using a large amount of negative feedback; used to perform analog-computer functions such as summing and integrating.

operational analysis *See* operational calculus.

operational calculus [MATH] A technique by which problems in analysis, in particular differential equations, are transformed into algebraic problems, usually the problem of solving a polynomial equation. Also known as operational analysis.

operational game *See* management game.

operational label [ADP] A combination of letters and digits at the beginning of the tape which uniquely identify the tape required by the system.

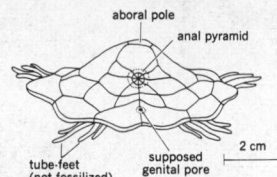

OPHIOCISTIOIDEA

Volchovia volborthi (Ordovician, Soviet Union).

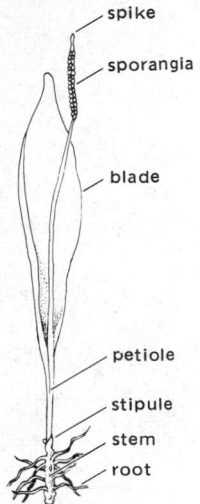

OPHIOGLOSSIDAE

Adder's-tongue fern (*Ophioglossum*) with a single leaf.

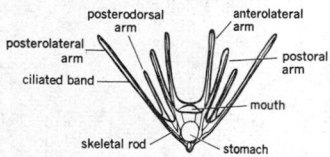

OPHIOPLUTEUS

Bilateral symmetrical larva of the brittle stars.

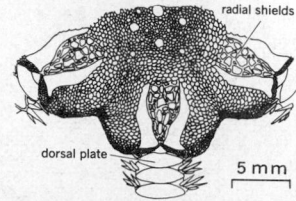

OPHIUROIDEA

Pectinura cylindrica, ophiuroid showing central disk and a part of one of the arms with the dorsal plate showing.

operational maintenance [ENG] The cleaning, servicing, preservation, lubrication, inspection, and adjustment of equipment; it includes that minor replacement of parts not requiring high technical skill, internal alignment, or special locative training.

operational standby program [ADP] The program operating in the standby computer when in the duplex mode of operation.

operational weather limits [METEOROL] The limiting values of ceiling, visibility, and wind, or runway visual range, established as safety minima for aircraft landings and takeoffs.

operation analysis [IND ENG] An analysis of all procedures concerned with the design or improvement of production, the purpose of the operation, inspection standards, materials used and the manner of handling them. the setup, tool equipment, and working conditions and methods.

operation analysis chart [IND ENG] A form that lists all the essential factors influencing the effectiveness of an operation.

operation breakdown *See* job breakdown.

operation code [ADP] A field or portion of a digital computer instruction that indicates which action is to be performed by the computer. Also known as command code.

operation cycle [ADP] The portion of a memory cycle required to perform an operation; division and multiplication usually require more than one memory cycle to be completed.

operation number [ADP] **1.** Number designating the position of an operation, or its equivalent subroutine, in the sequence of operations composing a routine. **2.** Number identifying each step in a program stated in symbolic code.

operation part [ADP] That portion of a digital computer instruction which is reserved for the operation code.

operation process chart [IND ENG] A graphic representation that gives an overall view of an entire process, including the points at which materials are introduced, the sequence of inspections, and all operations not involved in material handling.

operation register [ADP] A register used to store and decode the operation code for the next instruction to be carried out by a computer.

operations research [MATH] The mathematical study of systems with input and output from the viewpoint of optimization subject to given constraints. [SCI TECH] The application of objective and quantitative criteria to decision making previously undertaken by empirical methods.

operative ankylosis *See* arthrodesis.

operator [ADP] Anything that designates an action to be performed, especially the operation code of a computer instruction. [ENG] A person whose duties include operation, adjustment, and maintenance of a piece of equipment. [GEN] A sequence at one end of an operon on which a repressor acts, thus regulating the transcription of the operon. [MATH] A function between vector spaces.

operator interrupt [ADP] A step whereby control is passed to the monitor, and a message, usually requiring a typed answer, is printed on the console typewriter.

operator process chart [IND ENG] A chart of the time relationship of the movements made by the body members of a workman performing an operation.

operator productivity [IND ENG] The ratio of standard hours to actual hours for a given task.

operator's console [ADP] Equipment which provides for manual intervention and monitoring computer operation.

operator theory [MATH] The general qualitative study of operators in terms of such concepts as eigenvalues, range, domain, and continuity.

operator training [IND ENG] The process used to prepare the employee to make his expected contribution to his employer, usually involving the teaching of specialized skills.

operator utilization [IND ENG] The ratio of working time to total clock time; a ratio of 1.00 (or 100%) indicates full utilization of the operator's work time.

operculum [ANAT] **1.** The soft tissue partially covering the crown of an erupting tooth. **2.** That part of the cerebrum which borders the lateral fissure. [BIOL] **1.** A lid, flap, or valve. **2.** A lidlike body process.

operon [GEN] A functional unit composed of a number of adjacent cistrons on the chromosome; its transcription is regulated by a receptor sequence, the operator, and a repressor.

Opheliidae [INV ZOO] A family of limivorous worms belonging to the annelid group Sedentaria.

Ophiacodonta [VERT ZOO] A suborder of extinct reptiles in the order Pelycosauria, including primitive, partially aquatic carnivores.

Ophidiidae [VERT ZOO] A family of small actinopterygian fishes in the order Gadiformes, comprising the cusk eels and brotulas.

ophidiophobia [PSYCH] An abnormal fear of snakes.

Ophiocanopidae [INV ZOO] A family of asterozoan echinoderms in the subclass Ophiuroidea.

Ophiocistioidea [PALEON] A small class of extinct Echinozoa in which the domed aboral surface of the test was roofed by polygonal plates and carried an anal pyramid.

Ophioglossales [BOT] An order of ferns in the subclass Ophioglossidae.

Ophioglossidae [BOT] The adder's-tongue ferns, a small subclass of the class Polypodiopsida; the plants are homosporous and eusporangiate and are distinguished by the arrangement of the sporogenous tissue in the characteristic fertile spike of the sporophyte.

ophiolite [PETR] A group of mafic and ultramafic igneous rocks, including spilite, basalt, gabbro, peridotite, and their metamorphic alternation products such as serpentine.

ophiolitic eclogite [PETR] Any of the eclogites which are products of early orogenic volcanism and which by later metamorphism transformed into rocks of the high-pressure facies series.

Ophiomyxidae [INV ZOO] The single family of the echinoderm suborder Ophiomyxina distinguished by a soft, unprotected integument.

Ophiomyxina [INV ZOO] A monofamilial suborder of ophiuroid echinoderms in the order Phrynophiurida.

ophiopluteus [INV ZOO] The pluteus larva of brittle stars.

ophite [PETR] A diabase in which the ophitic structure is retained even though the pyroxene is altered to uralite.

ophitic [PETR] Of the holocrystalline, hypidiomorphic-granular texture of an igneous rock, exhibiting lath-shaped plagioclase crystals partly or wholly included within pyroxene crystals.

Ophiurida [INV ZOO] An order of echinoderms in the subclass Ophiuroidea in which the vertebrae articulate by means of ball-and-socket joints, and the arms, which do not branch, move mainly from side to side.

Ophiuroidea [INV ZOO] The brittle stars, a subclass of the Asterozoa in which the arms are usually clearly demarcated from the central disk and perform whiplike locomotor movements.

ophryon [ANTHRO] The point where the sagittal plane intersects an arc drawn horizontally from the frontotemporalis across the frontal bone.

ophthalmectomy [MED] Excision, or enucleation, of the eye.

ophthalmia [MED] Inflammation of the eye, especially involving the conjunctiva.

ophthalmia neonatorum [MED] Inflammation of the eyes in the newborn contracted during passage through the birth canal; may be gonorrheal or purulent.

ophthalmia nodosa [MED] Inflammation of the eye due to lodging of caterpillar hairs in the conjunctiva, cornea, or iris.

ophthalmic [ANAT] Of or pertaining to the eye.

ophthalmic nerve [ANAT] A sensory branch of the trigeminal nerve which supplies the lacrimal glands, upper eyelids, skin of the forehead, and anterior portion of the scalp, meninges, nasal mucosa, and frontal, ethmoid, and sphenoid air sinuses.

ophthalmodynamometer [MED] An instrument which measures the pressure necessary to collapse the retinal arteries.

ophthalmological [PHARM] Any drugs used in the treatment of eye disease.

ophthalmologist [MED] A physician who specializes in ophthalmology. Also known as oculist.

ophthalmology [MED] The study of the anatomy, physiology, and diseases of the eye.

ophthalmomalacia [MED] Abnormal softness or subnormal tension of the eye.

ophthalmometer [OPTICS] **1.** An instrument for measuring refractive errors, especially astigmatism. **2.** An instrument for measuring the capacity of the chamber of the eye. **3.** An instrument for measuring the eye as a whole.

ophthalmorrhexis [MED] Rupture of the eyeball.

ophthalmoscope [OPTICS] An instrument, consisting essentially of a concave mirror with a hole in it and fitted with lenses of different powers, for examining the interior of the eye through the pupil.

opianyl *See* meconin.

opiate [PHARM] **1.** A sleep-inducing drug. **2.** Any narcotic. **3.** An opium preparation. **4.** Any tranquilizing agent.

Opilioacaridae [INV ZOO] The single family of moderately large mites of the suborder Notostigmata which comprises the Opiliocariformes.

Opilioacariformes [INV ZOO] A small monofamilial order of the Acari comprising large mites characterized by long legs and by the possession of a pretarsus on the pedipalp, with prominent claws.

opisometer [ENG] An instrument for measuring the length of curved lines, such as those on a map; a wheel on the instrument is traced over the line.

opisthaptor [INV ZOO] A posterior adhesive organ in monogenetic trematodes.

Opisthobranchia [INV ZOO] A subclass of the class Gastropoda containing the sea hares, sea butterflies, and sea slugs; generally characterized by having gills, a small external or internal shell, and two pairs of tentacles.

Opisthocoela [VERT ZOO] A suborder of the order Anura; members have opisthocoelous trunk vertebrae, and the adults typically have free ribs.

opisthocoelous [ANAT] Of, related to, or being a vertebra with the centrum convex anteriorly and concave posteriorly.

Opisthocomidae [VERT ZOO] A family of birds in the order Galliformes, including the hoatzins.

opisthocranion [ANTHRO] The point, wherever it may lie in the sagittal plane on the occipital bone, which marks the posterior extremity of the longest diameter of the skull, measured from the glabella.

opisthognathous [INV ZOO] Having the mouthparts ventral and posterior to the cranium. [VERT ZOO] Having retreating jaws.

opisthonephros [VERT ZOO] The fundamental adult kidney in amphibians and fishes.

Opisthopora [INV ZOO] An order of the class Oligochaeta distinguished by meganephridiostomal, male pores opening posteriorly to the last testicular segment.

opisthotic [ANAT] Of, relating to, or being the posterior and inferior portions of the bony elements in the inner ear capsule.

opisthotonus [MED] A condition, caused by a tetanic spasm of the back muscles, in which the trunk is arched forward while the head and lower limbs are bent backward.

opium [PHARM] A narcotic obtained from the unripe capsules of the opium poppy (*Papaver somniferum*); crude extract contains alkaloids such as morphine (5-15%), narcotine (2-8%), and codeine (0.1-2.5%).

Opomyzidae [INV ZOO] A family of cyclorrhaphous myodarian dipteran insects in the subsection Acalypteratae.

opossum [VERT ZOO] Any member of the family Didelphidae in the order Marsupialia; these mammals are arboreal and mainly omnivorous, and have many incisors, with all teeth pointed and sharp.

Oppenauer oxidation [ORG CHEM] The oxidation of a primary or secondary hydroxyl compound to form the corresponding carbonyl compound; aluminum alkoxide and an excess amount of a carbonyl hydrogen acceptor, such as benzophenone or acetone, are required.

Oppenheimer-Phillips reaction [NUC PHYS] A type of stripping reaction which can occur when a deuteron passes near a nucleus, in which the proton in the deuteron experiences Coulomb repulsion from the nucleus while the neutron is attracted to the nucleus by nuclear forces, with the result that the neutron-proton bond in the deuteron is broken, the neutron is absorbed into the nucleus, and the proton is repelled.

Oppenheim's disease *See* amyotonia congenita.

opponent-colors theory [OPTICS] A theory of color vision

according to which various processes in the visual system are capable of responding in two opposite ways; the Hering theory is an example.

opposed engine [MECH ENG] A reciprocating engine having the pistons on opposite sides of the crankshaft, with the piston strokes on each side working in a direction opposite to the direction of the strokes on the other side.

opposite [BOT] **1.** Located side by side. **2.** Of leaves, being in pairs on an axis with each member separated from the other of the pair by half the circumference of the axis.

opposition [ASTRON] The situation of two celestial bodies having either celestial longitudes or sidereal hour angles differing by 180°; the term is usually used only in relation to the position of a superior planet or the moon with reference to the sun. [PHYS] The condition in which the phase difference between two periodic quantities having the same frequency is 180°, corresponding to one half-cycle.

opsonic action [IMMUNOL] The effect produced upon susceptible microorganisms and other cells by opsonins, which renders them vulnerable to phagocytes.

opsonic index [IMMUNOL] A numerical measure of the opsonic activity of sera, expressed as the ratio of the average number of bacteria engulfed per phagocytic cell in immune serum compared with the corresponding value for normal serum.

opsonin [IMMUNOL] A substance in blood serum that renders bacteria more susceptible to phagocytosis by leukocytes.

opsonize [IMMUNOL] To render microorganisms susceptible to phagocytosis.

-opsy [MED] A combining form denoting examination, or denoting a condition of vision.

optical aberration [OPTICS] Deviation from perfect image formation by an optical system; examples are spherical aberration, coma, astigmatism, curvature of field, distortion, and chromatic aberration.

optical activity [OPTICS] The behavior of substances which rotate the plane of polarization of plane-polarized light, as it passes through them. Also known as rotary polarization.

optical air mass [GEOPHYS] A measure of the length of the path through the atmosphere to sea level traversed by light rays from a celestial body, expressed as a multiple of the path length for a light source at the zenith.

optical analysis [OPTICS] Study of properties of a substance or medium, such as its chemical composition or the size of particles suspended in it, through observation of effects on transmitted light, such as scattering, absorption, refraction, and polarization.

optical anisotropy [OPTICS] The behavior of a medium, or of a single molecule, whose effect on electromagnetic radiation depends on the direction of propagation of the radiation.

optical anomaly [PHYS CHEM] The phenomenon in which an organic compound has a molar refraction which does not agree with the value calculated from the equivalents of atoms and other structural units composing it.

optical antipode *See* enantiomorph.

optical aspherical surface [OPTICS] An optical surface that does not form part of a sphere, such as a paraboloidal or ellipsoidal surface.

optical axis [ANAT] An imaginary straight line passing through the midpoint of the cornea (anterior pole) and the midpoint of the retina (posterior pole). [OPTICS] **1.** A line passing through a radially symmetrical optical system such that rotation of the system about this line does not alter it in any detectable way. **2.** *See* optic axis.

optical bench [ENG] A rigid horizontal bar or track for holding optical devices in experiments; it allows device positions to be changed and adjusted easily.

optical branch [SOLID STATE] The vibrations of an optical mode plotted on a graph of frequency versus wave number; it is separated from, and has higher frequencies than, the acoustic branch.

optical calcite [MINERAL] The type of calcite used to make Nicol prisms.

optical center [OPTICS] A point on the axis of a lens so that, for any ray passing through this point, the incident part and the emergent part are parallel.

optical character recognition [ADP] That branch of charac-

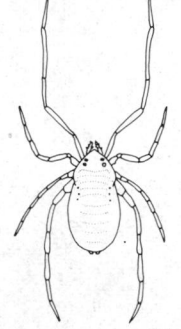

OPILIOACARIDAE

An opilioacarid mite, dorsal view. (*Acarology Laboratory, Ohio State University*)

OPOSSUM

Common opossum, with long prehensile tail used to climb and to gather twigs and leaves for nest building.

ter recognition concerned with the automatic identification of handwritten or printed characters by any of various photoelectric methods. Abbreviated OCR. Also known as electrooptical character recognition.

optical coating [OPTICS] Either a mirror coating, or a film of the proper thickness and refractive index applied to the airglass surface of a lens to reduce reflection.

optical communication [COMMUN] The use of electromagnetic waves in the region of the spectrum near visible light for the transmission of signals representing speech, pictures, data pulses, or other information, usually in the form of a laser beam modulated by the information signal.

optical comparator [ENG] Any comparator in which movement of a measuring plunger tilts a small mirror which reflects light in an optical system. Also known as visual comparator.

optical contact [OPTICS] Contact between two surfaces in which the surfaces are separated by a distance much less than a wavelength of light, so that interference fringes are not formed.

optical coupling [ELECTR] Coupling between two circuits by means of a light beam or light pipe having transducers at opposite ends, to isolate the circuits electrically.

optical crystal [CRYSTAL] Any natural or synthetic crystal, such as sodium chloride, calcium fluoride, silver chloride, potassium iodide, or stilbene, that is used in infrared and ultraviolet optics and for its piezoelectric effects.

optical density [OPTICS] The degree of opacity of a translucent medium expressed by $\log I_0/I$, where I_0 is the intensity of the incident ray, and I is the intensity of the transmitted ray. Abbreviated OD.

optical depth *See* optical thickness.

optical diffraction velocimeter *See* diffraction velocimeter.

optical direction and ranging *See* OPDAR.

optical dispersion [OPTICS] Separation of different colors of light such as occurs when it passes from one medium to another or is reflected from a diffraction grating.

optical Doppler effect [ELECTROMAG] A change in the observed frequency of light or other electromagnetic radiation caused by relative motion of the source and observer.

optical double star [ASTRON] Two stars not formerly a physical system but that appear to be a typical double star; a false binary star whose components happen to lie nearby in the same line of sight.

optical electronic reproducer *See* optical sound head.

optical element [OPTICS] A part of an optical instrument which acts upon the light passing through the instrument, such as a lens, prism, or mirror.

optical exaltation [PHYS CHEM] Optical anomaly in which the observed molar refraction exceeds the calculated one; most cases of optical anomaly are in this category.

optical fiber [OPTICS] A long, thin thread of fused silica, or other transparent substance, used to transmit light.

optical filter *See* filter.

optical flat [OPTICS] A disk of high-grade quartz glass approximately 2 centimeters thick, with a deviation in flatness usually not exceeding 0.05 micrometer all over, and a surface quality of 5 microfinish or less; used in determinations of surface contour and in comparison of lineal measurement.

optical fluid-flow measurement [ENG] Any method of measuring the varying densities of a fluid in motion, such as schlieren, interferometer, or shadowgraph, which depends on the fact that light passing through a flow field of varying density is retarded differently through the field, resulting in refraction of the rays, and in a relative phase shift among different rays.

optical frequency [PHYS] A frequency comparable to that of electromagnetic waves in the optical region, above about 3×10^{11} hertz.

optical gage [ENG] A gage that measures an image of an object, and does not touch the object itself.

optical galaxy [ASTRON] One of the galaxies that appear as nearly starlike, generally having compact nuclei.

optical glass [MATER] A type of glass which is free from imperfections, such as unmelted particles, bubbles, and chemical inhomogeneities, which would affect its transmission of light.

optical harmonic [SOLID STATE] Light, generated by passing a laser beam with a power density on the order of 10^{10} watts/cm² or more through certain transparent materials, which has a frequency which is an integral multiple of that of the incident laser light.

optical haze *See* terrestrial scintillation.

optical horizon [GEOD] Locus of points at which a straight line from the given point becomes tangential to the earth's surface.

optical indicator [ENG] An instrument which makes a plot of pressure in the cylinder of an engine as a function of piston (or volume) displacement, making use of magnification by optical systems and photographic recording; for example, the small motion of a pressure diaphragm may be transmitted to a mirror to deflect a beam of light.

optical indicatrix *See* index ellipsoid.

optical instrument [OPTICS] An optical system which acts on light in some desired way, such as to form a real or virtual image, to form an optical spectrum, or to produce light with a specified polarization or wavelength.

optical interference [OPTICS] Interference of light waves.

optical isomer *See* enantiomorph.

optical isomerism [PHYS CHEM] Existence of two forms of a molecule such that one is a mirror image of the other; the two molecules differ in that rotation of light is equal but in opposite directions.

optical landing system [NAV] A shipboard gyro-stabilized device or a short-based device which indicates to the pilot his displacement from a preselected glide path.

optical lantern [ENG] A device for projecting positive transparent pictures from glass or film onto a reflecting screen; it consists of a concentrated source of light, a condenser system, a holder (or changer) for the slide, a projection lens, and (usually) a blower for cooling the slide. Also known as slide projector.

optical length *See* optical path.

optically effective atmosphere [GEOPHYS] That portion of the atmosphere lying below the altitude (50–60 kilometers) from which scattered light at twilight still reaches the observer with sufficient intensity to be discerned. Also known as effective atmosphere.

optical maser *See* laser.

optical mask [ELECTR] A thin sheet of metal or other substance containing an open pattern, used to suitably expose to light a photoresistive substance overlaid on a semiconductor or other surface to form an integrated circuit.

optical material [MATER] A material which is transparent to light or to infrared, ultraviolet, or x-ray radiation, such as glass and certain single crystals, polycrystalline materials (chiefly for the infrared), and plastics.

optical measurement [OPTICS] Measurement of the intensity, spectral distribution, polarization, or other characteristics of light or of infrared or ultraviolet radiation, which is emitted by or reflected from an object or passes through some medium.

optical meteor [OPTICS] Any phenomenon of the atmosphere explained in terms of optical laws, such as a mirage or a halo.

optical microscope [OPTICS] An instrument used to obtain an enlarged image of a small object, utilizing visible light; in general it consists of a light source, a condenser, an objective lens, and an ocular or eyepiece, which can be replaced by a recording device. Also known as light microscope; photon microscope.

optical mode [SOLID STATE] A type of vibration of a crystal lattice whose frequency varies with wave number only over a limited range, and in which neighboring atoms or molecules in different sublattices move in opposition to each other.

optical model *See* cloudy crystal-ball model.

optical modulator [COMMUN] A device used for impressing information on a light beam, such as a blinker system or a device which electrically changes the properties of a material through which light is being transmitted.

optical moment [OPTICS] For a ray of light passing through an optical system, the triple product of a vector from an arbitrary origin on the optical axis to a point on the ray, a vector tangent to the ray at that point whose length equals the refractive index, and a unit vector along the optical axis; it does not depend on the point on the ray.

optical monochromator [SPECT] A monochromator used to observe the intensity of radiation at wavelengths in the visible, infrared, or ultraviolet regions.

optical null method [SPECT] In infrared spectrometry, the adjustment of a reference beam's energy transmission to match that of a beam that has been passed through a sample being analyzed.

optical parametric oscillator [OPTICS] A device, employing a nonlinear dielectric, which when pumped by a laser can generate coherent light whose wavelength can be varied continuously over a wide range.

optical path [OPTICS] For a ray of light traveling along a path between two points, the optical path is the integral, over elements of length along the path, of the refractive index. Also known as optical length.

optical-path difference *See* retardation.

optical phenomena [ELECTROMAG] Phenomena associated with the generation, transmission, and detection of electromagnetic radiation in the visible, infrared, or ultraviolet regions.

optical phonon [SOLID STATE] A quantum of an optical mode of vibration of a crystal lattice.

optical plastic [MATER] A plastic which is transparent to light, occasionally used in optical systems for reasons of economy, special index-dispersion relation, light weight, and nonbrittleness.

optical printing [GRAPHICS] Any printing technique in which a process camera is used in one step to form the end item; enlargements, reductions, or same-size negatives may be formed.

optical prism *See* prism.

optical projection system [OPTICS] An optical system which forms a real image of a suitably illuminated object so that it can be viewed, photographed, or otherwise observed. Also known as optical projector; projector.

optical projector *See* optical projection system.

optical properties [ELECTROMAG] The effects of a substance or medium on light or other electromagnetic radiation passing through it, such as absorption, scattering, refraction, and polarization.

optical pumping [OPTICS] The process of causing strong deviations from thermal equilibrium populations of selected quantized states of different energy in atomic or molecular systems by the use of electromagnetic radiation in or near the visible region.

optical pyrometer [ENG] An instrument which determines the temperature of a very hot surface from its incandescent brightness; the image of the surface is focused in the plane of an electrically heated wire, and current through the wire is adjusted until the wire blends into the image of the surface.

optical quenching [OPTICS] Reduction in the intensity of luminescent radiation by long-wavelength, visible or infrared radiation.

optical radar *See* OPDAR.

optical rangefinder [ENG] An optical instrument for measuring distance, usually from its position to a target point, by measuring the angle between rays of light from the target, which enter the rangefinder through the windows spaced apart, the distance between the windows being termed the baselength of the rangefinder; the two types are coincidence and stereoscopic.

optical recording [ENG] Production of a record by focusing on photographic paper a beam of light whose position on the paper depends on the quantity to be measured, as in a light-beam galvanometer.

optical reflectometer [ENG] An instrument which measures on surfaces the reflectivity of electromagnetic radiation at wavelengths in or near the visible region.

optical-righting reflex *See* visual-righting reflex.

optical rotary dispersion [OPTICS] Specific rotation, considered as a function of wavelength. Abbreviated ORD.

optical rotation [OPTICS] Rotation of the plane of polarization of plane-polarized light, or of the major axis of the polarization ellipse of elliptically polarized light by transmission through a substance or medium.

optical scanner *See* flying-spot scanner.

optical sight [OPTICS] A sight with lenses, prisms, or mirrors that is used in laying weapons, for aerial bombing, or for surveying.

optical sound head [ELECTR] The assembly in motion picture projection which reproduces photographically recorded sound; light from an incandescent lamp is focused on a slit, light from the slit is in turn focused on the optical sound track of a film, and the light passing through the film is detected by a photoelectric cell. Also known as optical electronic reproducer.

optical sound recorder *See* photographic sound recorder.

optical sound reproducer *See* photographic sound reproducer.

optical spectra [SPECT] Electromagnetic spectra for wavelengths in the ultraviolet, visible and infrared regions, ranging from about 10 nanometers to 1 millimeter, associated with excitations of valence electrons of atoms and molecules, and vibrations and rotations of molecules.

optical spectrograph [SPECT] An optical spectroscope provided with a photographic camera or other device for recording the spectrum made by the spectroscope.

optical spectrometer [SPECT] An optical spectroscope that is provided with a calibrated scale either for measurement of wavelength or for measurement of refractive indices of transparent prism materials.

optical spectroscope [SPECT] An optical instrument, consisting of a slit, collimator lens, prism or grating, and a telescope or objective lens, which produces an optical spectrum arising from emission or absorption of radiant energy by a substance, for visual observation.

optical spectroscopy [SPECT] The production, measurement, and interpretation of optical spectra arising from either emission or absorption of radiant energy by various substances.

optical spherical surface [OPTICS] An optical surface which forms part of a sphere.

optical square [ENG] A surveyor's hand instrument used for laying of right angles; employs two mirrors at a 45° angle.

optical staining *See* Rheinberg illumination.

optical storage [ADP] Storage of large amounts of data in permanent form on photographic film or its equivalent, for nondestructive readout by means of a light source and photodetector.

optical superposition principle [OPTICS] The principle that the optical rotation produced by a compound which is made up of two radicals of opposite optical activity is the algebraic sum of the rotations of each radical alone; not always valid.

optical surface [OPTICS] An interface between two media, such as between air and glass, which is used to reflect or refract light.

optical system [OPTICS] A collection of mirrors, lens, prisms, and other devices, placed in some specified configuration, which reflect, refract, disperse, absorb, polarize, or otherwise act on light.

optical thickness [METEOROL] 1. In calculations of the transfer of radiant energy, the mass of a given absorbing or emitting material lying in a vertical column of unit cross-sectional area and extending between two specified levels. Also known as optical depth. 2. Subjectively, the degree to which a cloud prevents light from passing through it; depends upon the physical constitution (crystals, drops, droplets), the form, the concentration of particles, and the vertical extent of the cloud. [OPTICS] The thickness of an optical material times its index of refraction.

optical-to-optical interface device [OPTICS] A device that converts a noncoherently illuminated image into a coherently illuminated object, which can then be used as input to certain types of data processor. Abbreviated OTTO.

optical tracking [ENG] The determination of spatial positions of distant airplanes, missiles, and artificial satellites as a function of time, or the recording of engineering events, by precise time-correlated observations with various types of telescopes or ballistic cameras.

optical transition [PHYS] A process in which an atom or molecule changes from one energy state to another and emits or absorbs electromagnetic radiation in the visible, infrared, or ultraviolet region.

optical twinning [CRYSTAL] Growing together of two crystals which are the same except that the structure of one is the

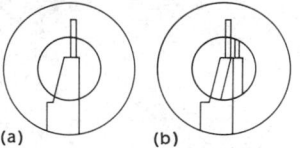
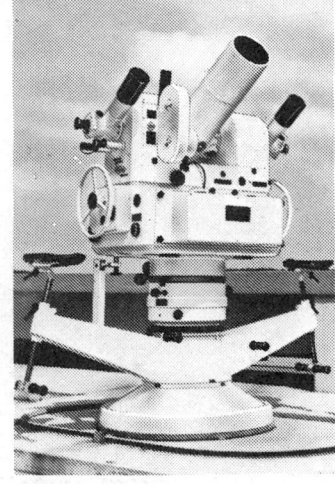

mirror image of the structure of the other. Also known as chiral twinning.

optic angle   *See* axial angle.

optic apraxia   [MED] A form of apraxia in which the individual fails to represent spatial relations correctly in drawing or construction by other means. Also known as constructional apraxia.

optic-axial angle   *See* axial angle.

optic axis   [OPTICS] The axis in a doubly refracting medium in which the ordinary and extraordinary waves propagate with the same velocity, and double refraction vanishes. Also known as optical axis; principal axis.

optic canal   [ANAT] The channel at the apex of the orbit, the anterior termination of the optic groove, just beneath the lesser wing of the sphenoid bone; it gives passage to the optic nerve and ophthalmic artery.

optic capsule   [VERT ZOO] A cartilaginous capsule that develops around the eye in elasmobranchs and higher vertebrate embryos.

optic chiasma   [ANAT] The partial decussation of the optic nerves on the undersurface of the hypothalamus.

optic cup   [EMBRYO] A two-layered depression formed by invagination of the optic vesicle from which the pigmented and sensory layers of the retina will develop.

optic disk   [ANAT] The circular area in the retina that is the site of the convergence of fibers from the ganglion cells of the retina to form the optic nerve.

optic gland   [INV ZOO] Either of a pair of endocrine glands in the octopus and squid which are found near the brain and produce a substance which causes gonadal maturation.

optician   [ENG] A maker of optical instruments or lenses.

optic lobe   [ANAT] One of the anterior pair of colliculi of the mammalian corpora quadrigemina. [INV ZOO] A lateral lobe of the forebrain in certain arthropods. [VERT ZOO] Either of the corpora bigemina of lower vertebrates.

optic nerve   [ANAT] The second cranial nerve; a paired sensory nerve technically consisting of three layers of special nerve cells in the retina of the eye; fibers converge to form the optic tracts.

optics   [PHYS] **1.** Narrowly, the science of light and vision. **2.** Broadly, the study of the phenomena associated with the generation, transmission, and detection of electromagnetic radiation in the spectral range extending from the long-wave edge of the x-ray region to the short-wave edge of the radio region, or in wavelength from about 1 nanometer to about 1 millimeter.

optic stalk   [EMBRYO] The constriction of the optic vesicle which connects the embryonic eye and forebrain in vertebrates.

optic tectum   [VERT ZOO] The roof of the mesencephalon constituting a major visual center and association area of the brain of premature vertebrates.

optic tract   [ANAT] The band of optic nerve fibers running from the optic chiasma to the lateral geniculate body and midbrain.

optic vesicle   [EMBRYO] An evagination of the lateral wall of the forebrain in vertebrate embryos which precedes formation of the optic cup.

optimal control theory   [MATH] A generalized calculus of variations dealing with the analysis of dynamical systems with the viewpoint of finding optimizing conditions.

optimal policy   [MATH] In optimization problems of systems, a sequence of decisions changing the states of a system in such a manner that a given criterion function is minimized.

optimal systems   [MATH] Systems where the variables representing the various states are so determined that a given criterion function is minimized subject to given constraints.

optimization   [MATH] The maximizing or minimizing of a given function possibly subject to some type of constraints. [SYS ENG] **1.** Broadly, the efforts and processes of making a decision, a design, or a system as perfect, effective, or functional as possible. **2.** Narrowly, the specific methodology, techniques, and procedures used to decide on the one specific solution in a defined set of possible alternatives that will best satisfy a selected criterion. Also known as system optimization.

optimization theory   [MATH] The specific methodology, techniques, and procedures used to decide on the one specific solution in a defined set of possible alternatives that will best satisfy a selected criterion; includes linear and nonlinear programming, stochastic programming, and control theory. Also known as mathematical programming.

optimize   [ADP] To rearrange the instructions or data in storage so that a minimum number of time-consuming jumps or transfers are required in the running of a program.

optimum array current   [ELECTROMAG] The current distribution in a broadside antenna array which is such that for a specified side-lobe level the beam width is as narrow as possible, and for a specified first null the side-lobe level is as small as possible.

optimum bunching   [ELECTR] Bunching condition required for maximum output in a velocity modulation tube.

optimum charge   [ORD] Propelling charge, with web and propellant weight combination, which produces maximum velocity at a specified pressure.

optimum code   [ADP] A computer code which is particularly efficient with regard to a particular aspect; for example, minimum time of execution, minimum or efficient use of storage space, and minimum coding time.

optimum coupling   *See* critical coupling.

optimum cure   [CHEM ENG] The degree of vulcanization at which maximum desired property is reached.

optimum filter   [ELECTR] An electric filter in which the mean square value of the error between a desired output and the actual output is at a minimum.

optimum flight   [AERO ENG] An aircraft flight so planned and navigated that it is completed under the optimum conditions of minimum time and minimum exposure to dangerous flying weather.

optimum reverberation time   [ACOUS] The reverberation time which is most desirable for a given room size and a given use, such as speech, chamber music, or symphony orchestra.

optimum-track ship routing   [NAV] The selection of an optimum track for a transoceanic crossing by the application of long-range predictions of winds, waves, and currents to the knowledge of how the routed ship reacts to these variables.

optimum traffic frequency   *See* optimum working frequency.

optimum working frequency   [COMMUN] The most effective frequency at a specified time for ionospheric propagation of radio waves between two specified points. Also known as frequency optimum traffic; optimum traffic frequency.

optoelectronic modulator   [ELECTR] A device in which variations in voltage applied to an electrooptic material modulate a laser beam passing through the material.

optoelectronics   *See* electrooptics.

optoelectronic scanner   [ELECTR] A scanner in which lenses, mirrors, or other optical devices are used between a light source or image and a photodiode or other photoelectric device.

optoelectronic shutter   [ENG] A shutter that uses a Kerr cell to modulate a beam of light.

optometrist   [MED] One who measures the degrees of visual powers, without the aid of cycloplegic or mydriatic agents.

optometry   [MED] Measurement of visual powers.

optophone   [ENG ACOUS] A device with a photoelectric cell to convert ordinary printed letters into a series of sounds; used by the blind.

or   [ADP] An instruction which performs the logical operation "or" on a bit-by-bit basis for its two or more operand words, usually storing the result in one of the operand locations. Also known as OR function. [MATH] A logical operation whose result is false (or zero) only if every one of its operands is false, and true (or one) otherwise. Also known as inclusive or.

ora   [METEOROL] A regular valley wind at Lake Garda in Italy.

oral   [ANAT] Of or pertaining to the mouth.

oral arm   [INV ZOO] In a jellyfish, any of the prolongations of the distal end of the manubrium.

oral cavity   [ANAT] The cavity of the mouth.

oral character   [PSYCH] A Freudian term applied to persons who have undergone an unusual degree of oral stimulation during the developmental period and are characterized by an attitude of carefree indifference and by dependence on a mother figure.

oral contraceptive   [PHARM] Any medication taken by mouth that renders a woman nonfertile as long as the medication is continued.

oral disc [INV ZOO] The flattened upper or free end of the body of a polyp that has the mouth in the center and tentacles around the margin.

oral erotic stage [PSYCH] In psychoanalysis, the first, or receptive, part of the oral phase of psychosexual development, dominated by sucking and lasting for the first 6 to 9 months of life.

oral erotism [PSYCH] The primordial pleasurable experience of nursing, reappearing in usually disguised and sublimated form in later life.

oral groove [INV ZOO] A depressed, groovelike peristome.

oral personality [PSYCH] An individual who is mouth-centered far beyond the age when the oral phase should have been passed, and who exhibits oral erotism and sadism in disguised and sublimated form.

oral phase [PSYCH] In psychoanalysis, the initial stage in the psychosexual development of a child, extending from birth to about 18 months, in which the mouth is the focus of sensations, interests, and activities and hence the source of gratification and security.

orange [BOT] Any of various evergreen trees of the genus *Citrus*, cultivated for the edible fruit, which is a berry with an aromatic, leathery rind containing numerous oil glands. [OPTICS] The hue evoked in an average observer by monochromatic radiation having a wavelength in the approximate range from 597 to 622 nanometers; however, the same sensation can be produced in a variety of other ways.

orange flower oil *See* oil of orange blossoms.

orange lake [MATER] Any of various transparent orange pigments from the precipitation of an orange dyestuff on aluminum hydrate or other base; used to produce transparent coatings for metal cans and bottle caps.

orange oil *See* bitter orange oil; sweet orange oil.

orangeophile [HISTOL] A type of acidophile cell of the anterior lobe of the adenohypophysis, presumed to elaborate growth hormone.

orange oxide *See* uranium trioxide.

orange peel [MATER] A pebbled film surface, resembling an orange skin, on lacquer or enamel as a result of too rapid drying after spraying, or failure to exhibit the desired leveling effects. [MET] A rough, pebble-grained metal surface resulting from either plastic deformation or electropolishing. Also known as alligator effect; pebbling.

orange-peel bucket [DES ENG] A type of grab bucket that is multileaved and generally round in configuration.

orange sapphire [MINERAL] An orange variety of gem corundum (sapphire). Also known as padparadsha.

orange spectrometer [SPECT] A type of beta-ray spectrometer that consists of a number of modified double-focusing spectrometers employing a common source and a common detector, and has exceptionally high transmission.

orange toner [ORG CHEM] A diazo dyestuff coupled to diacetoacetic acid anhydride; contains no sulfonic or carboxylic groups; used for printing inks.

orangutan [VERT ZOO] *Pongo pygmaeus*. The largest of the great apes, a long-armed primate distinguished by long sparse reddish-brown hair, naked face and hands and feet, and a large laryngeal cavity which appears as a pouch below the chin.

O ray [OPTICS] The ray which vibrates at right angles to the optic axis in uniaxial crystals. Also known as ordinary ray; ordinary wave.

orbicular [PETR] Of the structure of a rock, containing large quantities of orbicules. [SCI TECH] Having the form of a sphere or orb.

orbicule [GEOL] A nearly spherical body, up to 2 or more centimeters in diameter, in which the components are arranged in concentric layers.

Orbiniidae [INV ZOO] A family of polychaete annelids belonging to the Sedentaria; the prostomium is exposed, and the thorax and abdomen are weakly separated.

Orbiniinae [INV ZOO] A subfamily of sedentary polychaete annelids in the family Orbiniidae.

orbit [ANAT] The bony cavity in the lateral front of the skull beneath the frontal bone which contains the eyeball. Also known as eye socket. [PHYS] **1.** Any closed path followed by

a particle or body, such as the orbit of a celestial body under the influence of gravity, the elliptical path followed by electrons in the Bohr theory, or the paths followed by particles in a circular particle accelerator. **2.** More generally, any path followed by a particle, such as helical paths of particles in a magnetic field, or the parabolic path of a comet.

orbital [ATOM PHYS] The space-dependent part of the Schrödinger wave function of an electron in an atom or molecule in an approximation such that each electron has a definite wave function, independent of the other electrons.

orbital angular momentum [MECH] The angular momentum associated with the motion of a particle about an origin, equal to the cross product of the position vector with the linear momentum. Also known as orbital momentum. [QUANT MECH] The angular momentum operator associated with the motion of a particle about an origin, equal to the cross product of the position vector with the linear momentum, as opposed to the intrinsic spin angular momentum. Also known as orbital moment.

orbital curve [AERO ENG] One of the tracks on a primary body's surface traced by a satellite that orbits about it several times in a direction other than normal to the primary body's axis of rotation; each track is displaced in a direction opposite and by an amount equal to the degrees of rotation between each satellite orbit.

orbital decay [AERO ENG] The lessening of the eccentricity of the elliptical orbit of an artificial satellite. [ATOM PHYS] A change of an atom from one energy state to another of lower energy in which the orbital of one of the electrons changes.

orbital direction [AERO ENG] The direction that the path of an orbiting body takes; in the case of an earth satellite, this path may be defined by the angle of inclination of the path to the equator.

orbital electron [ATOM PHYS] An electron which has a high probability of being in the vicinity (at distances on the order of 10^{-10} meter or less) of a particular nucleus, but has only a very small probability of being within the nucleus itself.

orbital elements [PHYS] A set of seven parameters defining the orbit of a body attracted by a central, inverse-square force.

orbital fossa [INV ZOO] A depression from which the eyestalk arises on the front of the carapace of crustaceans.

orbital index [ANTHRO] The ratio of the orbital height, taken at right angles to the orbital width between the upper and lower orbital margins, times 100, to the orbital width, taken between the maxillofrontale and lateral margin of the orbit in such a manner that the line of the orbital width bisects the plane of the orbital entrance.

orbital magnetic moment [QUANT MECH] The magnetic dipole moment associated with the motion of a charged particle about an origin, rather than with its intrinsic spin.

orbital moment *See* orbital angular momentum.

orbital momentum *See* orbital angular momentum.

orbital motion [PHYS] Continuous motion of a body in a closed path, such as a circle or an ellipse, about some point.

orbital node [ASTRON] One of the two points at which the orbit of a planet or satellite crosses the plane of the ecliptic or equator.

orbital parity [QUANT MECH] The parity associated with the wave function of a particle, or system of particles, as a function of spatial coordinates; it is opposed to intrinsic parity; if the orbital angular momentum quantum number is l, the orbital parity is $(-1)^l$.

orbital period [ASTRON] The interval between successive passages of a satellite through the same specified point in its orbit.

orbital plane [MECH] The plane which contains the orbit of a body or particle in a central force field; it passes through the center of force.

orbital rendezvous [AERO ENG] **1.** The meeting of two or more orbiting objects with zero relative velocity at a preconceived time and place. **2.** The point in space at which such an event occurs.

orbital symmetry [PHYS CHEM] The property of certain molecular orbitals of being carried into themselves or into the negative of themselves by certain geometrical operations, such as a rotation of 180° about an axis in the plane of the molecule, or reflection through this plane.

orbital velocity [ASTRON] The instantaneous velocity at

ORANGE

Foliage, flowers, and fruit of *Citrus sinensis.*

ORANGUTAN

The orangutan (*Pongo pygmaeus*), a long-armed primate that frequently moves about on all fours.

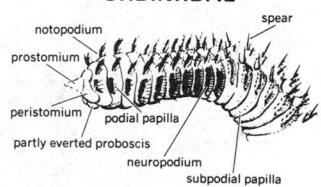

ORBINIIDAE

Anterior parts of the body of *Phylo*, a genus of Orbiniidae.

which an earth satellite or other orbiting body travels around the origin of its central force field.

Orbitoidacea [INV ZOO] A superfamily of foraminiferan protozoans in the suborder Rotaliina characterized by a low trochospire or a planispiral, uncoiled or branching test composed of radial calcite with bilamellar septa.

orbit point [AERO ENG] A geographically defined reference point over land or water, used in stationing airborne aircraft.

orch-, orchi-, orchid-, orchido-, orchio- [ZOO] A combining form denoting testis.

orchard [AGR] A group of fruit-bearing, nut-bearing, or sugar maple trees under cultivation.

orchid [BOT] Any member of the family Orchidaceae; plants have complex, specialized irregular flowers usually with only one or two stamens.

Orchidaceae [BOT] A family of monocotyledonous plants in the order Orchidales characterized by irregular flowers and a great number of microscopic seeds having no endosperm contained within a capsule.

Orchidales [BOT] An order of monocotyledonous plants in the subclass Liliidae; plants are mycotropic and sometimes nongreen with numerous tiny seeds that have an undifferentiated embryo and little or no endosperm.

orchiectomy [MED] Surgical removal of one or both testes.

orchil [MATER] Dark-brownish-red coloring matter derived from lichens as paste or aqueous extract; main components are orcin and orcein; used as carpet-yarn dye. Also known as orseille.

orchiopexy [MED] Surgical fixation of a testis.

orcin [ORG CHEM] $CH_3C_6H_3(OH)_2 \cdot H_2O$ White crystals with strong, sweet, unpleasant taste; soluble in water, alcohol, and ether; extracted from lichens; used in medicine and as an analytical reagent.

OR circuit See OR gate.

ORD See optical rotary dispersion.

order [CHEM] A classification of chemical reactions, in which the order is described as first, second, third, or higher, according to the number of molecules (one, two, three, or more) which appear to enter into the reaction; decomposition of H_2O_2 to form water and oxygen is a first-order reaction. [PHYS] A range of magnitudes of a quantity (and of all other quantities having the same physical dimensions) extending from some value of the quantity to some small multiple of the quantity (usually 10). Also known as order of magnitude. [SYST] A taxonomic category ranked below the class and above the family, made up either of families, subfamilies, or suborders.

order-disorder transition [SOLID STATE] The transition of an alloy or other solid solution between a state in which atoms of one element occupy certain regular positions in the lattice of another element, and a state in which this regularity is not present.

ordered field [MATH] A field with an ordering as a set analogous to the properties of less than or equal for real numbers relative to addition and multiplication.

ordered pair [MATH] A pair of elements x and y from a set, written (x,y), where x is distinguished as first and y as second.

ordered rings [MATH] Rings which have an ordering on them as sets in a manner analogous to the behavior of the usual ordering of the real numbers relative to addition and multiplication.

ordering [MATH] A binary relation $\leq$ among the elements of a set such that $a \leq b$ and $b \leq c$ implies $a \leq c$, and $a \leq b$, $b \leq a$ implies $a = b$; it need not be the case that either $a \leq b$ or $b \leq a$. Also known as order relation; partial ordering. [SOLID STATE] A solid-state transformation in certain solid solutions, in which a random arrangement in the lattice is transformed into a regular ordered arrangement of the atoms with respect to one another; a so-called superlattice is formed.

order of a differential equation [MATH] An equation has order n if the derivatives of a function appear up to the nth derivative.

order of a group [MATH] For a group G, the order is the least positive integer n for which g^n equals the identity element for every g in G; equivalently, the number of elements in the group.

order of a matrix [MATH] A square matrix with n rows and n columns has order n.

order of an elliptic function [MATH] The number of poles an elliptic function has in a parallelogram region where it repeats its values.

order of an infinitesimal [MATH] A characteristic of infinitesimals used in their comparison.

order of a polynomial [MATH] The largest exponent appearing in the polynomial.

order of degeneracy See degree of degeneracy.

order of interference [OPTICS] The difference in the number of wavelengths along the paths of two constructively interfering rays of light.

order of magnitude See order.

order of phase transition [THERMO] A phase transition in which there is a latent heat and an abrupt change in properties, such as in density, is a first-order transition; if there is not such a change, the order of the transition is one greater than the lowest derivative of such properties with respect to temperature which has a discontinuity.

order point [IND ENG] The inventory level at which a replenishment order must be placed.

order quantity [IND ENG] The number of pieces ordered to replenish the inventory.

order relation See ordering.

order tone [COMMUN] Tone sent over a trunk to indicate that the trunk is ready to receive an order or, to the receiving operator, that an order is about to arrive.

ordinal number [MATH] A generalized number which expresses the size of a set, in the sense of "how many" elements.

ordinary cut [MIN ENG] The arrangement of drill holes in a mine in which the drill holes are symmetrical with respect to the vertical center-line of the section, extend horizontally, and make a large angle with the working face.

ordinary differential equation [MATH] An equation involving functions of one variable and their derivatives.

ordinary gear train [MECH ENG] A gear train in which all axes remain stationary relative to the frame.

ordinary index [OPTICS] The index of refraction of the ordinary ray in a crystal.

ordinary point of a curve [MATH] A point where a curve does not cross itself and where there is a smoothly turning tangent.

ordinary ray [CRYSTAL] See O ray. [OPTICS] One of two rays into which a ray incident on an anisotropic uniaxial crystal is split; it obeys the ordinary laws of refraction, in contrast to the extraordinary ray.

ordinary sheathed explosive [MATER] A permitted explosive (one passing certain safety tests) whose safety has been further increased by a sheath of sodium bicarbonate; when the explosive is detonated, the sheath forms carbon dioxide, which tends to extinguish the flame around the detonator wave.

ordinary wave See O ray.

ordinary-wave component [GEOPHYS] One of the two components into which an electromagnetic wave entering the ionosphere is divided under the influence of the earth's magnetic field; it has characteristics more nearly like those expected in the absence of a magnetic field. Also known as O-wave component. [OPTICS] The component of electromagnetic radiation propagating in an anisotropic uniaxial crystal whose electric displacement vector is perpendicular to the optical axis and the direction normal to the wavefront; gives rise to the ordinary ray.

ordinate [MATH] The perpendicular distance of a point (x,y) of the plane from the x-axis.

ordnance [ENG] Military materiel, such as combat weapons of all kinds, with ammunition and equipment for their use, vehicles, and repair tools and machinery.

Ordnance Corps [ORD] A technical service of the U.S. Army, charged with the design, construction, testing, and supply of ordnance materiel; the corps provides guns, ammunition, missiles, armored and track-laying vehicles, and apparatus for sighting and firing guns; it maintains arsenals and depots for the design, manufacture, testing, storage, and issue of such material; it also maintains an extensive research program.

ordnance service [ORD] All activities necessary to maintain in usable condition the ordnance equipment of a command and such other equipment as directed by proper authority.

ORCHID

A typical orchid.

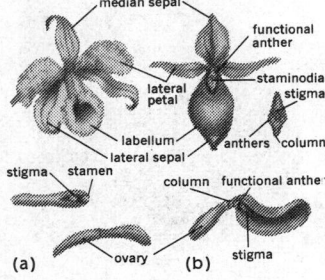

ORCHIDACEAE

median sepal

functional anther

lateral petal

staminodia stigma

labellum

lateral sepal

anthers column

stigma stamen

column functional anther

(a) ovary (b) stigma

The two common floral types found in orchids. (a) *Cattleya*. (b) *Cypripedium*.

ordnance stores [ORD] All commodities and materials used by the U.S. Army Ordnance Corps in the design, manufacture, testing, preservation, and overhaul of ordnance property or supplies.

ordnance supplies [ORD] All military supplies assigned to the U.S. Army Ordnance Corps for storage, issue, and maintenance; ordnance supplies consist of all raw materials, completely manufactured articles, and parts of such articles assigned to the corps.

ordnance troops [ORD] Technically trained troops assigned or attached to a tactical unit to provide ordnance maintenance, supply, or technical service; they also give instruction in the use, maintenance, and adjustment of ordnance materiel.

Ordovician [GEOL] The second period of the Paleozoic era, above the Cambrian and below the Silurian, from approximately 500 million to 440 million years ago.

ore [GEOL] **1.** The naturally occurring material from which economically valuable minerals can be extracted. **2.** Specifically, a natural mineral compound of the elements, of which one element at least is a metal. **3.** More loosely, all metalliferous rock, though it contains the metal in a free state. **4.** Occasionally, a compound of nonmetallic substances, as sulfur ore.

ore bed [GEOL] An economic aggregation of minerals occurring between or in rocks of sedimentary origin.

ore bin [MIN ENG] A receptacle for ore awaiting treatment or shipment.

ore block [MIN ENG] A vein of ore which is bound above, below, and at one or both ends; it is ready for excavation.

ore blocked out [MIN ENG] Ore exposed on three sides within a reasonable distance of each other.

orebody [GEOL] Generally, a solid and fairly continuous mass of ore, which may include low-grade ore and waste as well as pay ore, but is individualized by form or character from adjoining country rock.

ore bridge [MIN ENG] A gantry crane used to load and unload stockpiles of ore.

ore car [MIN ENG] A mine car for carrying ore or waste rock.

ore carrier [NAV ARCH] A vessel designed to carry ore in bulk, and similar in construction to a collier.

ore chimney *See* pipe.

ore chute [MIN ENG] An inclined passage for the transfer of ore to a lower level.

ore cluster [GEOL] A group of interconnected ore bodies.

ore control [GEOL] A geologic feature that has influenced the ore deposition.

ore crusher [MIN ENG] A machine for breaking up masses of ore, usually previous to passing through other size-reduction equipment.

Orectolobidae [VERT ZOO] An ancient isurid family of galeoid sharks, including the carpet and nurse sharks, which are primarily bottom feeders with small teeth and a blunt rostrum with barbels near the mouth.

ore deposit [GEOL] Rocks containing minerals of economic value in such amount that they can be profitably exploited.

ore developed [MIN ENG] Ore exposed on four sides in blocks variously prescribed.

ore district [GEOL] A combination of several ore deposits into one common whole or system.

ore dressing [MIN ENG] The cleaning of ore by the removal of certain valueless portions, as by jigging, cobbing, or vanning.

ore expectant [MIN ENG] The whole or any part of the ore below the lowest level or beyond the range of vision.

ore faces [MIN ENG] Those ore bodies that are exposed on one side, or show only one face.

ore flotation promoter [MATER] Material that gives a water-repellent surface to mineral particles so that air bubbles will adhere and cause selective flotation.

oregano [FOOD ENG] A spice prepared from leaves of various aromatic mints, especially wild marjoram.

ore grader [MIN ENG] In metal mining, a person who directs the storage of iron ores in bins at shipping docks so that the various grades in each bin will contain approximate percentages of iron.

oreide bronze [MET] A series of brass compositions containing 68–87% copper, 10–32% zinc, and sometimes small amounts of tin; used for hardware.

ore in sight [MIN ENG] **1.** Ore exposed on at least three sides within reasonable distance of each other. **2.** Ore which may be reasonably assumed to exist, though not actually blocked out. **3.** *See* developed reserves.

ore intersection [MIN ENG] **1.** The point at which a borehole, crosscut, or other underground opening encounters an ore vein or deposit. **2.** The thickness of the ore-bearing deposit so traversed.

ore-lead age [GEOL] Measurement of the age of the earth by comparing the relative progress of the two radioactive decay schemes U^{235}–Pb^{207} and U^{238}–Pb^{206}.

ore microscopy [MINERAL] The use of a reflecting microscope to study polished sections of ore minerals. Also known as mineragraphy; mineralography.

oreodont [PALEON] Any member of the family Merycoidodontidae.

ore of sedimentation *See* placer.

ore pass [MIN ENG] A vertical or inclined passage for the downward transfer of ore.

ore pipe *See* pipe.

ore pocket [MIN ENG] **1.** Excavation near the hoisting shaft into which ore from stopes is moved, preliminary to hoisting. **2.** An unusual concentration of ore in the lode.

ore reduction [MIN ENG] The size reduction of solids by crushing and grinding in mineral processing plants.

ore reserve [MIN ENG] The total tonnage and average value of proved ore, plus the total tonnage and value (assumed) of the probable ore.

ore sampling [MIN ENG] The process in which a portion of ore is selected so that its composition will represent the average composition of the entire bulk of ore.

ore shoot [GEOL] **1.** A large, generally vertical, pipelike ore body that is economically valuable. Also known as shoot. **2.** A large and usually rich aggregation of mineral in a vein.

OR function *See* or.

organ [ANAT] A differentiated structure of an organism composed of various cells or tissues and adapted for a specific function.

organelle [CYTOL] A specialized subcellular structure, such as a mitochondrion, having a special function; a condensed system showing a high degree of internal order and definite limits of size and shape.

organic [ORG CHEM] Of chemical compounds, based on carbon chains or rings and also containing hydrogen with or without oxygen, nitrogen, or other elements.

organic acid [ORG CHEM] A chemical compound with one or more carboxyl radicals (COOH) in its structure; examples are butyric acid, $CH_3(CH_2)_2COOH$, maleic acid, HOOCCH-CHCOOH, and benzoic acid, C_6H_5COOH.

organic bonded wheel [DES ENG] A grinding wheel in which organic bonds are used to hold the abrasive grains.

organic brain syndrome [MED] A mental condition of multiple etiologies, resulting in diffuse impairment of brain tissue function and manifested by a complex of symptoms including impaired judgment and intellectual function, and often somatic and motor dysfunctions.

organic chemistry [CHEM] The study of the composition, reactions, and properties of carbon-chain or carbon-ring compounds or mixtures thereof.

organic coating [MATER] Material used to protect metal surfaces from chemical or atmospheric attack; includes latex paints, plastics, asphaltic materials, rubbers, and elastomers.

organic-cooled reactor [NUCLEO] A reactor that uses organic chemicals, such as mixtures of polyphenyls (diphenyls and terphenyls), as coolant.

organic electrolyte cell [ELEC] A type of wet cell that is based on the use of particularly reactive metals such as lithium, calcium, or magnesium in conjunction with organic electrolytes; the best-known type is the lithium–cupric fluoride cell.

organic evolution [EVOL] The processes of change in organisms by which descendants come to differ from their ancestors, and a history of the sequence of such changes.

organic glass [MATER] An amorphous, solid, glasslike material made of transparent plastic.

ORDOVICIAN

PRECAMBRIAN	
CAMBRIAN	
ORDOVICIAN	
SILURIAN	PALEOZOIC
DEVONIAN	
Mississippian	CARBON-IFEROUS
Pennsylvanian	
PERMIAN	
TRIASSIC	
JURASSIC	MESOZOIC
CRETACEOUS	
TERTIARY	CENOZOIC
QUATERNARY	

Chart showing the position of the Ordovician period in relation to the other periods and to the eras of geologic time.

organicism   *See* holism.

organic-moderated reactor   [NUCLEO] A nuclear reactor in which organic compounds are used as moderator and coolant.

organic pigment   [ORG CHEM] Any of the materials with organic-chemical bases used to add color to dyes, plastics, linoleum, tones, and lakes.

organic quantitative analysis   [ANALY CHEM] Quantitative determination of elements, functional groups, or molecules in organic materials.

organic reaction mechanism   [ORG CHEM] A pathway of chemical states traversed by an organic chemical system in its passage from reactants to products.

organic reef   [GEOL] A sedimentary rock structure of significant dimensions erected by, and composed almost exclusively of the remains of, corals, algae, bryozoans, sponges, and other sedentary or colonial organisms.

organic rock   [PETR] A sedimentary rock composed principally of the remains of plants and animals.

organic salt   [ORG CHEM] The reaction product of an organic acid and an inorganic base, such as sodium acetate (CH_3COONa) from the reaction of acetic acid (CH_3COOH) and sodium hydroxide (NaOH).

organic semiconductor   [MATER] An organic material having unusually high conductivity, often enhanced by the presence of certain gases, and other properties commonly associated with semiconductors; an example is anthracene.

organic soil   [GEOL] Any soil or soil horizon consisting chiefly of, or containing at least 30% of, organic matter; examples are peat soils and muck soils.

organic solvent   [ORG CHEM] Liquid organic compound with the power to dissolve solids, gases, or liquids (miscibility); examples are methanol (methyl alcohol), CH_3OH, and benzene, C_6H_6.

organism   [BIOL] An individual constituted to carry out all life functions.

organismic psychology   [PSYCH] A movement in psychology based on the theory that the individual is made up of elements composing a single organized system, and an element in the system cannot be evaluated independently of its position within the system.

organisol   [MATER] A dispersion of very finely divided resin particles that are suspended in an organic-liquid mixture which cannot dissolve the resin at normal temperatures.

organization chart   [IND ENG] Graphic representation of the interrelationships within an organization, depicting lines of authority and responsibility and provisions for control.

organized ferment   *See* intracellular enzyme.

organizer   [EMBRYO] Any part of the embryo which exerts a morphogenetic stimulus on an adjacent part or parts, as in the induction of the medullary plate by the dorsal lip of the blastopore.

organizing pneumonia   [MED] Pneumonia in which the healing process is characterized by organization and cicatrization of the exudate rather than by resolution and resorption. Also known as unresolved pneumonia.

organ of Corti   [ANAT] A specialized structure located on the basilar membrane of the mammalian cochlea, which contains rods of Corti and hair cells connected to ganglia of the cochlear nerve. Also known as spiral organ.

organ of Leydig   [VERT ZOO] Two large accumulations of lymphoid tissues which run longitudinally the length of the esophagus in selachian fishes.

organogenesis   [EMBRYO] The formation of an organ.

organometallic compound   [ORG CHEM] Molecules containing carbon-metal linkage; a compound containing an alkyl or aryl radical bonded to a metal, such as tetraethyllead, $Pb(C_2H_5)_4$.

organophosphate   [ORG CHEM] A soluble fertilizer material made up of organic phosphate esters such as glucose, glycol, or sorbitol; useful for providing phosphorus to deep-root systems.

organosol   [MATER] Finely divided or colloidal suspension of insoluble material in a suspending organic liquid; known as plastisol when the solid is a synthetic resin suspended in an organic liquid; used for coatings, moldings, and casting of films.

organosulfur compound   [ORG CHEM] One of a group of substances which contain both carbon and sulfur.

organotropic   [MICROBIO] Of microorganisms, localizing in or entering the body by way of the viscera or, occasionally, somatic tissue.

organs of Zuckerkandl   *See* aortic paraganglion.

orgasm   [PHYSIO] The intense, diffuse, and subjectively pleasurable sensation experienced during sexual intercourse or genital manipulation, culminating in the male with seminal ejaculation and in the female with uterine contractions, warm suffusion, and pelvic throbbing sensations.

orgasmolepsy   [MED] Sudden loss of muscle tone during orgasm, accompanied by a transitory loss of consciousness.

OR gate   [ELECTR] A multiple-input gate circuit whose output is energized when any one or more of the inputs is in a prescribed state; performs the function of the logical inclusive-or; used in digital computers. Also known as OR circuit.

Oribatei   [INV ZOO] A heavily sclerotized group of free-living mites in the suborder Sarcoptiformes which serve as intermediate hosts of tapeworms.

Oribatulidae   [INV ZOO] A family of oribatid mites in the suborder Sarcoptiformes.

orient   [ADP] To change relative and symbolic addresses to absolute form.

orientability of a sound signal   [ACOUS] The property of a sound signal by virtue of which a listener can estimate the direction of the location of the apparatus producing the signal.

oriental alabaster   *See* onyx marble.

oriental amethyst   [MINERAL] A violet to purple variety of sapphire.

oriental topaz   [MINERAL] A yellow variety of corundum, used as a gem.

Oriental zoogeographic region   [ECOL] A zoogeographic region which encompasses tropical Asia from the Iranian Peninsula eastward through the East Indies to, and including, Borneo and the Philippines.

orientation   [CRYSTAL] The directions of the axes of a crystal lattice relative to the surfaces of the crystal, to applied fields, or to some other planes or directions of interest.   [ELECTROMAG] The physical positioning of a directional antenna or other device having directional characteristics.   [ENG] Establishment of the correct relationship in direction with reference to the points of the compass.   [MATH] A choice of sense or direction in a topological space.   [PHYS] **1.** The direction of some vector or set of vectors, such as the direction of the electric vector and the propagation direction of plane polarized light, or the direction of a preponderance of nuclear spins in a crystal near absolute zero, relative to some other directions of interest. **2.** Any process in which vectors associated with atoms or molecules in the substance are organized relative to some direction, rather than pointed at random; examples include dipole moments of polar molecules in an electric field, and nuclear spins in a crystal in a magnetic field at temperatures near absolute zero.   [PHYS CHEM] The arrangement of radicals in an organic compound in relation to each other and to the parent compound.   [PSYCH] Determination of one's relation to the environment.

orientation diagram   [GEOL] Any point or contour diagram used in structural petrology.

orientation effect   [ELEC] Those bulk properties of a material which result from orientation polarization.   [PHYS CHEM] A method of determining attractive forces among molecules, or components of these forces, from the interaction energy associated with the relative orientation of molecular dipoles.

orientation polarization   [ELEC] Polarization arising from the orientation of molecules which have permanent dipole moments arising from an asymmetric charge distribution. Also known as dipole polarization.

oriented   [GEOL] Pertaining to a specimen that is so marked as to show its exact, original position in space.

oriented-strategic research   [SCI TECH] That background research considered to be required in order to achieve a specified objective.

orifice   [ELECTROMAG] Opening or window in a side or end wall of a waveguide or cavity resonator through which energy is transmitted.   [SCI TECH] An aperture or hole.

orifice mixer [MECH ENG] Arrangement in which two or more liquids are pumped through an orifice constriction to cause turbulence and consequent mixing action.

orifice plate [DES ENG] A disk, with a hole, placed in a pipeline to measure flow.

orifice well tester [PETRO ENG] Velocity-type meter used to measure gas flow quantity from a gas well; static pressure differences before and after a sharp-edged orifice are converted to flow values.

origanum oil [MATER] A light-yellow essential oil obtained from herbs of the genus *Origanum*; contains carvacrol and cymene plus other components; used in flavors and pharmaceuticals.

origin [ADP] The absolute storage address in relative coding to which addresses in a region are referenced. [MATH] The center of a coordinate system, where all coordinate axes meet, usually denoted by $(O,O, \ldots ,O)$.

original dip *See* primary dip.

origin of rifling [ORD] The position in a rifled gun bore at which the rifling begins; more specifically, the plane, perpendicular to the axis of the gun bore, in which the rifling starts.

origin of the trajectory [ORD] Center of the muzzle of a gun at the instant when the projectile leaves it.

O ring [DES ENG] A flat ring made from synthetic rubber, used as an airtight seal or a seal against high pressures.

Orion [ASTRON] A northern constellation near the celestial equator, right ascension 5 hours, declination 5° north. Also known as Warrior.

Orionids [ASTRON] A meteor shower seen in October in the northern hemisphere; its radiant lies in the constellation Orion.

Orion Nebula [ASTRON] A luminous cloud surrounding Ori, the northern star in Orion's dagger; visible to the naked eye as a hazy object. Also known as Great Nebula of Orion.

Orion [TEXT] Trade name for an acrylic fiber having high bulk and a soft, warm hand.

orlop deck [NAV ARCH] The lowest continuous deck of a ship having four or more decks.

ormer *See* abalone.

Ormyridae [INV ZOO] A small family of hemipteran insects in the superfamily Chalcidoidea.

Orneodidae [INV ZOO] A small family of lepidopteran insects in the superfamily Tineoidea; adults have each wing divided into six featherlike plumes.

ornithine [BIOCHEM] $C_5H_{12}O_2N_2$ An amino acid occurring in the urine of some birds, but not found in native proteins.

Ornithischia [PALEON] An order of extinct terrestrial reptiles, popularly known as dinosaurs; distinguished by a four-pronged pelvis, and a median, toothless predentary bone at the front of the lower jaw.

ornithology [VERT ZOO] The study of birds.

Ornithopoda [PALEON] A suborder of extinct reptiles in the order Ornithischia including all bipedal forms in the order.

Ornithorhynchidae [VERT ZOO] A monospecific order of monotremes containing the semiaquatic platypus; characterized by a duck-billed snout, horny plates instead of teeth in the adult, and a flattened, well-developed tail.

ornithosis [MED] Any form of psittacosis originating in birds other than psittacines.

orogen *See* orogenic belt.

orogene *See* orogenic belt.

orogenesis *See* orogeny.

orogenic belt [GEOL] A linear region that has undergone folding or other deformation during the orogenic cycle. Also known as fold belt; orogen; orogene.

orogenic cycle [GEOL] A time interval during which a mobile belt evolved into an orogenic belt, passing through preorogenic, orogenic, and postorogenic stages. Also known as geotectonic cycle.

orogeny [GEOL] The process or processes of mountain formation, especially the intense deformation of rocks by folding and faulting which, in many mountainous regions, has been accompanied by metamorphism, invasion of molten rock, and volcanic eruption; in modern usage, orogeny produces the internal structure of mountains, and epeirogeny produces the

mountainous topography. Also known as orogenesis; tectogenesis.

orographic cloud [METEOROL] A cloud whose form and extent is determined by the disturbing effects of orography upon the passing flow of air; because these clouds are linked with the form of the terrestrial relief, they generally move very slowly, if at all, although the winds at the same level may be very strong.

orographic lifting [METEOROL] The lifting of an air current caused by its passage up and over surface elevations.

orographic occlusion [METEOROL] An occluded front in which the occlusion process has been hastened by the retardation of the warm front along the windward slopes of a mountain range.

orographic precipitation [METEOROL] Precipitation which results from the lifting of moist air over an orographic barrier such as a mountain range; strictly, the amount so designated should not include that part of the precipitation which would be expected from the dynamics of the associated weather disturbance, if the disturbance were over flat terrain.

orography [GEOGR] The branch of geography dealing with mountains.

Oromericidae [VERT ZOO] An extinct family of camellike tylopod ruminants in the superfamily Cameloidea.

orometer [ENG] A barometer with a scale that indicates elevation above sea level.

oropharynx [ANAT] The oral pharynx, located between the lower border of the soft palate and the larynx.

orophyte [ECOL] Any plant that grows in the subalpine region.

orotic acid [BIOCHEM] $C_4H_4O_4N_2$ A crystalline acid which is a growth factor for certain bacteria and is also a pyrimidine precursor.

Oroya fever [MED] The severe form of Carrion's disease, characterized by a sudden, severe, and rapid course, often fatal anemia, and remittent fever.

orpiment [MINERAL] As_2S_3 A lemon-yellow mineral, crystallizing in the monoclinic system, and generally occurring in foliated or columnar masses; luster is resinous and pearly on the cleavage surface, hardness is 1.5–2 on Mohs scale, and specific gravity is 3.49. Also known as yellow arsenic.

orris [MATER] The fragrant powder from the root of the plants *Iris florentinea, I. germanica,* and *I. pallida;* used in perfume, medicine, and tooth powder. Also known as orris root.

orris oil [MATER] A yellow, fatty, semisolid, fragrant essential oil obtained from roots of the Florentine iris; melts at 44–50°C; soluble in ether, chloroform, and alcohol; main components are myristic acid, oleic acid, and irone; used in flavoring and perfumes.

orris root *See* orris.

Orr's white *See* lithopone.

Orsat analyzer [ANALY CHEM] Gas analysis apparatus in which various gases are absorbed selectively (volumetric basis) by passing them through a series of preselected solvents.

orseille *See* orchil.

Orthacea [PALEON] An extinct group of articulate brachiopods in the suborder Orthidina in which the delthyrium is open.

orthamine *See* phenylenediamine.

Ortheziinae [INV ZOO] A subfamily of homopteran insects in the superfamily Coccoidea having abdominal spiracles present in all stages and a flat anal ring bearing pores and setae in immature forms and adult females.

orthicon [ELECTR] A camera tube in which a beam of low-velocity electrons scans a photoemissive mosaic that is capable of storing a pattern of electric charges; has higher sensitivity than the iconoscope.

Orthida [PALEON] An order of extinct articulate brachiopods which includes the oldest known representatives of the class.

Orthidina [PALEON] The principal suborder of the extinct Orthida, including those articulate brachiopods characterized by biconvex, finely ribbed shells with a straight hinge line and well-developed interareas on both valves.

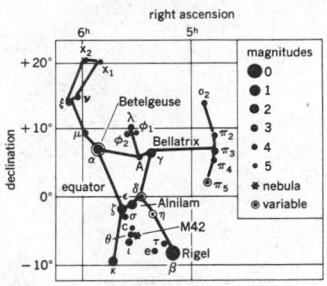

ORION

Line pattern of the constellation Orion. The grid lines represent the coordinates of the sky. The apparent brightness, or magnitude, of the stars is shown by the size of the dots, graded by appropriate numbers.

ORPIMENT

Foliated masses of orpiment found at Mercer, Utah. (*Specimen from Department of Geology, Bryn Mawr College*)

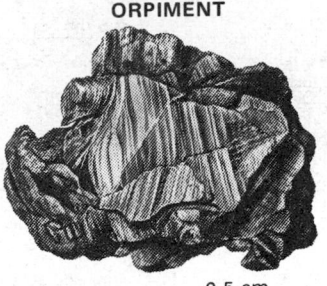

ORTHACEA

Shell of *Hesperothis.* (*a*) Pedicle valve posterior. (*b*) Brachial valve interior.

ORTHOCLASE

|← 2.5 cm →|

Twinned orthoclase crystals from Gunnison County, Colorado. *(American Museum of Natural History Specimens)*

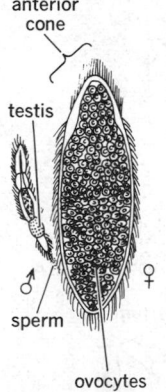

ORTHONECTIDA

anterior cone

testis

sperm

♂   ♀

ovocytes

The orthonectid *Rhopalura ophiocomae*, male discharging sperm near genital pore of female.

orthite [MINERAL] Allanite in the form of slender prismatic or acicular crystals.

ortho acid [ORG CHEM] **1.** Aromatic acid with a carboxyl group in the ortho position (1,2 position). **2.** Organic acid with one added molecule of water in chemical combination; for example, $HC(OH)_3$, orthoformic acid, in contrast to $HCOOH$, formic acid; $H_3PO_4(P_2O_5 \cdot 3H_2O)$, orthophosphoric acid, in contrast to the less hydrated form, metaphosphoric acid, $HPO_3(P_2O_5 \cdot H_2O)$.

orthoarsenic acid *See* arsenic acid.

orthoaxis [CRYSTAL] The diagonal or lateral axis perpendicular to the vertical axis in the monoclinic system.

orthobaric density [PHYS] The density of a liquid and of a saturated vapor with which it is at equilibrium at a given temperature.

orthoboric acid *See* boric acid.

orthocenter [MATH] The point at which the altitudes of a triangle intersect.

orthocephaly [ANTHRO] The condition of having the skull with a vertical index of 70.1–75.

orthochem [GEOCHEM] A precipitate formed within a depositional basin or within the sediment itself by direct chemical action.

orthochromatic [BIOL] Having normal staining characteristics. [GRAPHICS] Pertaining to sensitized materials that can be exposed by ultraviolet, blue, and green light, but not deep orange or red.

orthoclase [MINERAL] $KAlSi_3O_8$ A colorless, white, cream-yellow, flesh-reddish, or gray potassium feldspar that usually contains some sodium feldspar, either as albite or analbite or in some intermediate state; it is or appears to be monoclinic. Also known as common feldspar; orthose; pegmatolite.

orthoconglomerate [GEOL] A conglomerate with an intact gravel framework held together by mineral cement and deposited by ordinary water currents.

orthodontics [MED] A branch of dentistry that deals with the prevention and treatment of malocclusion.

orthoferrosilite [MINERAL] An orthopyroxene consisting of the orthorhombic silicate $FeSiO_3$.

orthogeosyncline [GEOL] A linear geosynclinal belt lying between continental and oceanic cratons, and having internal volcanic belts (eugeosynclinal) and external nonvolcanic belts (miogeosynclinal). Also known as geosynclinal couple; primary geosyncline.

orthognathic [ANTHRO] Pertaining to a condition of the upper jaw in which it is in an approximately vertical relationship to the profile of the facial skeleton, when the skull is oriented on the Frankfort horizontal plane; having a gnathic index of 97.9 or less.

orthogneiss [GEOL] Gneiss originating from igneous rock.

orthogonal [MATH] Perpendicular, or some concept analogous to it.

orthogonal antennas [ELECTROMAG] In radar, a pair of transmitting and receiving antennas, or a single transmitting-receiving antenna, designed for the detection of a difference in polarization between the transmitted energy and the energy returned from the target.

orthogonal basis [MATH] A basis for an inner product space consisting of mutually orthogonal vectors.

orthogonal complement [MATH] In an inner product space, the orthogonal complement of a vector $\mathbf{v}$ consists of all vectors orthogonal to $\mathbf{v}$; the orthogonal complement of a subset S consists of all vectors orthogonal to each vector in S.

orthogonal family *See* orthogonal system.

orthogonal functions [MATH] Two real-valued functions are orthogonal if their inner product vanishes.

orthogonal group [MATH] The group of matrices arising from the orthogonal transformations of a euclidean space.

orthogonality [MATH] Two geometric objects have this property if they are perpendicular.

orthogonalization [MATH] A procedure in which, given a set of linearly independent vectors in an inner product space, a set of orthogonal vectors is recursively obtained so that each set spans the same subspace.

orthogonal lines [MATH] Lines which are perpendicular.

orthogonal matrix [MATH] A matrix whose inverse and transpose are identical.

orthogonal parity check [ADP] A parity checking system involving both a lateral and a longitudinal parity check.

orthogonal polynomial [MATH] Orthogonal polynomials are various families of polynomials, which arise as solutions to differential equations related to the hypergeometric equation, and which are mutually orthogonal as functions.

orthogonal projection [MATH] A continuous linear map P of a Hilbert space H onto a subspace M such that if $\mathbf{h}$ is any vector in H, $\mathbf{h} = P\mathbf{h} + \mathbf{w}$, where $\mathbf{w}$ is in the orthogonal complement of M. Also known as orthographic projection.

orthogonal system [MATH] **1.** A system made up of n families of curves on an n-dimensional manifold in an $n + 1$ dimensional euclidean space, such that exactly one curve from each family passes through every point in the manifold, and, at each point, the tangents to the n curves that pass through that point are mutually perpendicular. **2.** A set of real-valued functions, the inner products of any two of which vanish. Also known as orthogonal family.

orthogonal trajectory [MATH] A curve that intersects all the curves of a given family at right angles.

orthogonal transformation [MATH] A linear transformation between inner product spaces which preserves the length of vectors.

orthogonal vectors [MATH] In an inner product space, two vectors are orthogonal if their inner product vanishes.

orthographic chart [MAP] A chart on the orthographic projection.

orthographic projection [MAP] A perspective azimuthal projection of one hemisphere produced by straight parallel lines from any point desired from an infinite distance; it is true to scale at the center only. [MATH] *See* orthogonal projection.

orthokinesis [BIOL] Random movement of a motile cell or organism in response to a stimulus.

orthomagmatic stage [GEOL] The principal stage in the crystallization of silicates from a typical magma; up to 90% of the magma may crystallize during this stage. Also known as orthotectic stage.

orthomorphic [MAP] Preserving the correct shape.

orthomorphic chart [MAP] A chart on which very small shapes are correctly represented.

orthomorphic map projection *See* conformal map projection.

Orthonectida [INV ZOO] An order of Mesozoa; orthonectids parasitize various marine invertebrates as multinucleate plasmodia, and sexually mature forms are ciliated organisms.

Orthonik [MET] A magnetic alloy composed of 45–50% nickel, with the remainder iron, having a grainy structure, high permeability, and a rectangular hysteresis loop; used in magnetic cores.

orthonormal coordinates [MATH] In an inner product space, the coordinates for a vector expressed relative to an orthonormal basis.

orthonormal functions [MATH] Orthogonal functions $f_1, f_2, \ldots$ with the additional property that the inner product of $f_n(x)$ with itself is 1.

orthonormal vectors [MATH] A collection of mutually orthogonal vectors, each having length 1.

orthopedics [MED] The branch of surgery concerned with corrective treatment of musculoskeletal deformities, diseases, and ailments by manual and instrumental measures.

Orthoperidae [INV ZOO] The minute fungus beetles, a family of coleopteran insects in the superfamily Cucujoidea.

orthophosphate [INORG CHEM] One of the possible salts of orthophosphoric acid; the general formula is M_3PO_4, where M may be potassium as in potassium orthophosphate, K_3PO_4.

orthophosphoric acid *See* phosphoric acid.

orthophyric [PETR] Of the texture of the matrix of certain igneous rocks, having feldspar crystals with quadratic or short and stumpy rectangular cross sections.

orthopinacoid *See* front pinacoid.

orthopnea [MED] A condition in which there is difficulty in breathing except when sitting or standing upright.

orthopositronium [PARTIC PHYS] The state of positronium in which the positron and electron have parallel spins.

Orthopsida [INV ZOO] An order of echinoderms in the subclass Euechinoidea.

Orthopsidae [PALEON] A family of extinct echinoderms in

the order Hemicidaroida distinguished by a camarodont lantern.

Orthoptera [INV ZOO] A heterogeneous order of generalized insects with gradual metamorphosis, chewing mouthparts, and four wings.

orthopyroxene [MINERAL] A series of pyroxene minerals crystallizing in the orthorhombic system; members include enstatite, bronzite, hypersthene, ferrohypersthene, eulite, and orthoferrosilite.

orthoquartzite [PETR] A clastic sedimentary rock composed almost entirely of detrital quartz grains; a quartzite of sedimentary origin. Also known as orthoquartzitic sandstone; sedimentary quartzite.

orthoquartzitic conglomerate [GEOL] A lithologically homogeneous, light-colored orthoconglomerate composed of quartzose residues that is commonly interbedded with pure quartz sandstone. Also known as quartz-pebble conglomerate.

orthoquartzitic sandstone *See* orthoquartzite.

orthorhombic lattice [CRYSTAL] A crystal lattice in which the three axes of a unit cell are mutually perpendicular, and no two have the same length. Also known as rhombic lattice.

orthorhombic pyroxene [MINERAL] A member of the mineral series enstatite ($Mg_2[Si_2O_6]$)-orthoferrosilite ($Fe_2[Si_2O_6]$), crystallizing in the orthorhombic system, space group *Pbca*.

orthorhombic system [CRYSTAL] A crystal system characterized by three axes of symmetry that are mutually perpendicular and of unequal length. Also known as rhombic system.

Orthorrhapha [INV ZOO] A suborder of the Diptera; in this group of flies, the adult escapes from the puparium through a T-shaped opening.

orthoschist [PETR] A schist derived from igneous rocks.

orthoscope [MED] **1.** An instrument for examination of the eye through a layer of water, whereby the curvature and hence the refraction of the cornea is neutralized. **2.** An instrument used in drawing the projections of skulls.

orthoscopic eyepiece [OPTICS] An eyepiece that consists of a single lens, made up of three cemented elements, to which a planoconvex lens is added; designed to minimize distortion and spherical aberration.

orthoscopic system [OPTICS] An optical system that has been corrected so that distortion and spherical aberration are eliminated. Also known as rectilinear system.

orthose *See* orthoclase.

orthostatic [MED] Pertaining to or caused by standing upright.

orthotectic stage *See* orthomagmatic stage.

orthotomic system [OPTICS] An optical system in which all the rays may be interesected at right angles by a suitably chosen surface.

orthotonus [MED] Tetanic muscle spasm in which the body assumes a posture of rigid straightness.

orthotronic error control [ADP] An error check carried out to ensure correct transmission, which uses lateral and longitudinal parity checks.

orthotropic [MECH] Having elastic properties such as those of timber, that is, with considerable variations of strength in two or more directions perpendicular to one another.

orthotropic deck [CIV ENG] A bridge deck constructed typically of flat steel plate and longitudinal and transverse ribs; functions in carrying traffic and acting as top flanges of floor beams.

orthotropism [BOT] The tendency of a plant to grow with the longer axis oriented vertically.

orthotropous [BOT] Having a straight ovule with the micropyle at the end opposite the stalk.

orthotungstic acid *See* tungstic acid.

Orussidae [INV ZOO] A small family of hymenopteran insects in the superfamily Siricoidea.

Os *See* osmium.

Osagean [GEOL] A provincial series of geologic time in North America; Lower Mississippian (above Kinderhookian, below Meramecian).

osage orange [BOT] *Maclura pomifera.* A tree in the mulberry family of the Urticales characterized by yellowish bark, milky sap, simple entire leaves, strong axillary thorns, and aggregate green fruit about the size and shape of an orange.

osar *See* esker.

osbornite [MINERAL] TiN A mineral found only in meteorites.

O scan *See* O scope.

osciducer [ELECTR] Transducer in which information pertaining to the stimulus is provided in the form of deviation from the center frequency of an oscillator.

oscillating conveyor [MECH ENG] A conveyor on which pulverized solids are moved by a pan or trough bed attached to a vibrator or oscillating mechanism. Also known as vibrating conveyor.

oscillating granulator [MECH ENG] Solids size-reducer in which particles are broken by a set of oscillating bars arranged in cylindrical form over a screen of suitable mesh.

oscillating magnetic field [ELECTROMAG] A magnetic field which varies periodically in time.

oscillating screen [MECH ENG] Solids separator in which the sifting screen oscillates at 300 to 400 revolutions per minute in a plane parallel to the screen.

oscillation [CONT SYS] *See* cycling. [PHYS] Any effect that varies periodically back and forth between two values.

oscillation of a function [MATH] The oscillation of a real-valued function on an interval is the difference between its least upper bound and greatest lower bound there. Also known as saltus.

oscillation ripple *See* oscillation ripple mark.

oscillation ripple mark [GEOL] A symmetric ripple mark having a sharp, narrow, and relatively straight crest between broadly rounded troughs, formed by the motion of water agitated by oscillatory waves on a sandy base at a depth shallower than wave base. Also known as oscillation ripple; oscillatory ripple mark; wave ripple mark.

oscillator [ELECTR] **1.** An electronic circuit that converts energy from a direct-current source to a periodically varying electric output. **2.** The stage of a superheterodyne receiver that generates a radio-frequency signal of the correct frequency to mix with the incoming signal and produce the intermediate-frequency value of the receiver. **3.** The stage of a transmitter that generates the carrier frequency of the station or some fraction of the carrier frequency. [PHYS] Any device (mechanical or electrical) which, in the absence of external forces, can have a periodic back-and-forth motion, the frequency determined by the properties of the oscillator.

oscillator harmonic interference [ELECTR] Interference occurring in a superheterodyne receiver due to the interaction of incoming signals with harmonics (usually the second harmonic) of the local oscillator.

Oscillatoriales [BOT] An order of blue-green algae (Cyanophyceae) which are filamentous and truly multicellular.

oscillator-mixer-first detector *See* converter.

oscillator strength [ATOM PHYS] A quantum-mechanical analog of the number of dispersion electrons having a given natural frequency in an atom, used in an equation for the absorption coefficient of a spectral line; it need not be a whole number. Also known as f value; Ladenburg f value.

oscillatory circuit [ELEC] Circuit containing inductance or capacitance, or both, and resistance, connected so that a voltage impulse will produce an output current which periodically reverses or oscillates.

oscillatory discharge [ELEC] Alternating current of gradually decreasing amplitude which, under certain conditions, flows through a circuit containing inductance, capacitance, and resistance when a voltage is applied.

oscillatory extinction *See* undulatory extinction.

oscillatory ripple mark *See* oscillation ripple mark.

oscillatory shear [FL MECH] Application of small-amplitude oscillations to produce shear in viscoelastic fluids for the study of dynamic viscosity.

oscillatory surge [ELEC] Surge which includes both positive and negative polarity values.

oscillatory twinning [CRYSTAL] Repeated, parallel twinning.

oscillistor [ELECTR] A bar of semiconductor material, such as germanium, that will oscillate much like a quartz crystal when it is placed in a magnetic field and is carrying direct current that flows parallel to the magnetic field.

oscillogram [ENG] The permanent record produced by an

ORTHORHOMBIC PYROXENE

2.5 cm

A black crystal of enstatite, which is an end member of the orthorhombic pyroxene series, found in Bomlo, Norway. *(Specimen from Department of Geology, Bryn Mawr College)*

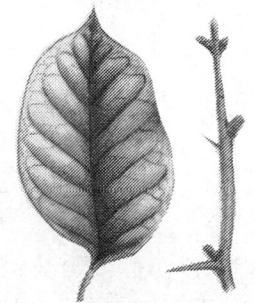

OSAGE ORANGE

Leaf and branch of *Maclura pomifera.*

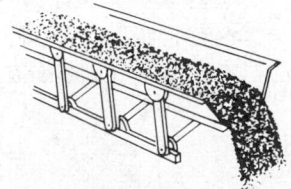

OSCILLATING CONVEYOR

Oscillating conveyor, with whole trough oscillating.

oscillograph, or a photograph of the trace produced by an oscilloscope.

oscillograph [ENG] A measurement device for determining waveform by recording the instantaneous values of a quantity such as voltage as a function of time.

oscillographic polarography [PHYS CHEM] A type of voltammetry using a dropping mercury electrode with oscillographic scanning of the applied potential; used to measure the concentration of electroactive species in solutions.

oscillograph tube [ELECTR] Cathode-ray tube used to produce a visible pattern, which is the graphical representation of electric signals, by variations of the position of the focused spot or spots according to these signals.

oscillometric titration [PHYS CHEM] Radio-frequency technique used for conductometric and dielectrometric titrations; the changes in conductance or dielectric properties changes the solution capacity and thus the frequency of the connected oscillator circuit.

oscillometry [PHYS CHEM] Electrode measurement of oscillation-frequency changes to detect the progress of a titration of electrolytic solutions.

oscilloscope See cathode-ray oscilloscope.

Oscillospiraceae [MICROBIO] A family of large, gram-negative, motile bacteria of the order Caryophanales; motility is lost on exposure to oxygen.

oscine See scopoline.

Oscines [VERT ZOO] The songbirds, a suborder of the order Passeriformes.

O scope [ELECTR] An A scope modified by the inclusion of an adjustable notch for measuring range. Also known as O indicator; O scan.

osculating orbit [ASTRON] The orbit which would be followed by a body such as an asteroid or comet if, at a given time, all the planets suddenly disappeared, and it then moved under the gravitational force of the sun alone.

osculating plane [MATH] For a curve C at some point p, this is the limiting plane obtained from taking planes through the tangent to C at p and containing some variable point p' and then letting p' approach p along C.

osculum [INV ZOO] An excurrent orifice in Porifera.

Osgood-Schlatter disease See osteochondrosis.

O shell [ATOM PHYS] The fifth layer of electrons about the nucleus of an atom, having electrons characterized by the principal quantum number 5.

Osler-Rendu-Weber disease See hereditary hemorrhagic telangiectasia.

osmate [INORG CHEM] A salt or ester of osmic acid, containing the osmate radical, $OsO_4{}^{--}$; for example, potassium osmate (K_2OsO_4).

osmic acid anhydride [INORG CHEM] OsO_4 Poisonous yellow crystals with disagreeable odor; melts at 40°C; soluble in water, alcohol, and ether; used in medicine, photography, and catalysis. Also known as osmium oxide; osmium tetroxide.

osmium [CHEM] A chemical element, symbol Os, atomic number 76, atomic weight 190.2. [MET] A hard white metal of rare natural occurrence.

osmium oxide See osmic acid anhydride.

osmium tetroxide See osmic acid anhydride.

osmolality [CHEM] The molality of an ideal solution of a nondissociating substance that exerts the same osmotic pressure as the solution being considered.

osmolarity [CHEM] The molarity of an ideal solution of a nondissociating substance that exerts the same osmotic pressure as the solution being considered.

osmole [CHEM] 1. The unit of osmolarity equal to the osmolarity of a solution that exerts an osmotic pressure equal to that of an ideal solution of a nondissociating substance that has a concentration of 1 mole of solute per liter of solution. 2. The unit of osmolality equal to the osmolality of a solution that exerts an osmotic pressure equal to that of an ideal solution of a nondissociating substance that has a concentration of 1 mole of solute per kilogram of solvent.

osmometer [ANALY CHEM] A device for measuring molecular weights by measuring the osmotic pressure exerted by solvent molecules diffusing through a semipermeable membrane.

osmophobia [PSYCH] An abnormal fear of odors.

osmoreceptor [PHYSIO] One of a group of structures in the hypothalamus which respond to changes in osmotic pressure of the blood by regulating the secretion of the neurohypophyseal antidiuretic hormone.

osmoregulatory mechanism [PHYSIO] Any physiological mechanism for the maintenance of an optimal and constant level of osmotic activity of the fluid in and around the cells.

osmosis [PHYS CHEM] The transport of a solvent through a semipermeable membrane separating two solutions of different solute concentration, from the solution that is dilute in solute to the solution that is concentrated.

osmotic fragility [PHYSIO] Susceptibility of red blood cells to lyses when placed in dilute (hypotonic) salt solutions.

osmotic gradient See osmotic pressure.

osmotic pressure [PHYS CHEM] 1. The applied pressure required to prevent the flow of a solvent across a membrane which offers no obstruction to passage of the solvent, but does not allow passage of the solute, and which separates a solution from the pure solvent. 2. The applied pressure required to prevent passage of a solvent across a membrane which separates solutions of different concentration, and which allows passage of the solute, but may also allow limited passage of the solvent. Also known as osmotic gradient.

osmotic shock [PHYSIO] The bursting of cells suspended in a dilute salt solution.

Osos wind [METEOROL] In California, a strong northwest wind blowing from the Loa Osos valley to the San Luis valley.

osphradium See asphradium.

osseine [MATER] The organic residue formed when bone is dissolved in hydrochloric acid; used to make glue and gelatin.

osseous [ANAT] Bony; composed of or resembling bone.

osseous system [ANAT] The skeletal system of the body.

osseous tissue [HISTOL] Bone tissue.

ossicle [ANAT] Any of certain small bones, as those of the middle ear. [INV ZOO] Any of various calcareous bodies.

ossify [PHYSIO] To form or turn into bone.

ossifying fibroma [MED] A benign bone tumor derived from ossiferous connective tissue. Also known as fibrous osteoma; osteogenic fibroma.

ost-, oste-, osteo- [ANAT] A combining form meaning bone.

Ostariophysi [VERT ZOO] A superorder of actinopterygian fishes distinguished by the structure of the anterior four or five vertebrae which are modified as an encasement for the bony ossicles connecting the inner ear and swim bladder.

Osteichthyes [VERT ZOO] The bony fishes, a class of fishlike vertebrates distinguished by having a bony skeleton, a swim bladder, a true gill cover, and mesodermal ganoid, cycloid, or ctenoid scales.

osteitis [MED] Inflammation of bone.

osteitis fibrosa cystica [MED] Generalized skeletal demineralization due to an increased rate of bone destruction resulting from hyperparathyroidism. Also known as Engel-Recklinghausen disease; osteitis fibrosa generalisata.

osteitis fibrosa generalisata See osteitis fibrosa cystica.

osteoarthritis See degenerative joint disease.

osteoarthropathy [MED] Any disease of bony articulations.

osteoblast [HISTOL] A bone-forming cell of mesenchymal origin.

osteochondritis [MED] Inflammation of both bone and cartilage.

osteochondroma [MED] A benign hamartomatous tumor originating in bone or cartilage.

osteochondrosis [MED] A disease characterized by avascular necrosis of ossification centers followed by regeneration. Also known as Calvé's disease; Kienböck's disease; Kohler's disease; Osgood-Schlatter disease; Scheuermann's disease.

osteoclasis [MED] Forcible fracture of a long bone without open operation, to correct a deformity. [PHYSIO] 1. Destruction of bony tissue. 2. Bone resorption.

osteoclast [HISTOL] A large multinuclear cell associated with bone resorption. [MED] A large surgical apparatus through which leverage can be exerted to effect osteoclasis.

osteocyte [HISTOL] A bone cell.

osteodermia [MED] A condition characterized by ossification within the skin.

osteodystrophy [MED] Any defective bone formation, as in rickets or dwarfism.

osteofibrosis [MED] Fibrosis of bone.

osteogenesis [PHYSIO] Formation or histogenesis of bone.

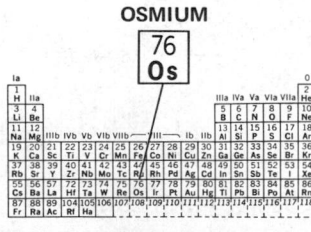

OSMIUM

Periodic table of the chemical elements showing the position of osmium.

osteogenesis imperfecta [MED] A disease inherited as an autosomal dominant and characterized by hypoplasia of osteoid tissue and collagen, resulting in bone fractures.

osteogenesis imperfecta congenita [MED] A form of osteogenesis imperfecta in which fractures occur at or before birth.

osteogenic fibroma *See* ossifying fibroma.

osteogenic sarcoma *See* osteosarcoma.

Osteoglossidae [VERT ZOO] The bony tongues, a family of actinopterygian fishes in the order Osteoglossiformes.

Osteoglossiformes [VERT ZOO] An order of soft-rayed, actinopterygian fishes distinguished by paired, usually bony rods at the base of the second gill arch, a single dorsal fin, no adipose fin, and a usually abdominal pelvic fin.

osteoid [HISTOL] The young hyaline matrix of true bone in which the calcium salts are deposited.

osteolathyrism [MED] Degeneration of bone collagen resulting from experimental administration of β-aminoproprionitrile.

Osteolepidae [PALEON] A family of extinct fishes in the order Osteolepiformes.

Osteolepiformes [PALEON] A primitive order of fusiform lobefin fishes, subclass Crossopterygii, generally characterized by rhombic bony scales, two dorsal fins placed well back on the body, and a well-ossified head covered with large dermal plating bones.

osteology [ANAT] The study of anatomy and structure of bone.

osteolysis [MED] Degeneration of bone tissue. [PHYSIO] Resorption of bone.

osteoma [MED] A benign bone tumor, especially in membrane bones of the skull.

osteomalacia [MED] Failure of bone to ossify due to a reduced amount of available calcium. Also known as adult rickets.

osteometry [ANAT] The study of the size and proportions of the osseous system.

osteomyelitis [MED] Inflammation of bone tissue and bone marrow.

osteon [HISTOL] A Haversian system with its lamellae and Haversian canal.

osteonephropathy [MED] Any syndrome involving bone changes accompanying kidney disease.

osteopath [MED] A physician who specializes in osteopathy.

osteopathy [MED] **1.** A school of healing which teaches that the body is a vital mechanical organism whose structural and functional integrity are coordinate and interdependent, the abnormality of either constituting disease. **2.** Any disease of bone.

osteopetrosis [MED] A rare developmental error of unknown cause but of familial tendency, characterized chiefly by excessive radiographic density of most or all of the bones. Also known as marble bone disease.

osteopoikilosis [MED] A bone affection of unknown cause and no symptoms, characterized by ellipsoidal, dense foci in all bones.

osteoporosis [MED] Deossification with absolute decrease in bone tissue, resulting in enlargement of marrow and Haversian spaces, decreased thickness of cortex and trabeculae, and structural weakness.

osteosarcoma [MED] A malignant tumor principally composed of anaplastic cells of mesenchymal derivation. Also known as osteogenic sarcoma.

osteosclerotic anemia *See* myelophthisic anemia.

Osteostraci [PALEON] An order of extinct jawless vertebrates; they were mostly small, with the head and part of the body encased in a solid armor of bone, and the posterior part of the body and the tail covered with thick scales.

osteotomy [MED] **1.** Surgical division of a bone. **2.** Making a section of a bone for the purpose of correcting a deformity.

ostiole [BIOL] A small orifice or pore.

ostium [BIOL] A mouth, entrance, or aperture.

Ostomidae [INV ZOO] The bark-gnawing beetles, a family of coleopteran insects in the superfamily Cleroidea.

Ostracoda [INV ZOO] A subclass of the class Crustacea containing small, bivalved aquatic forms; the body is unsegmented and there is no true abdominal region.

ostracoderm [PALEON] Any of various extinct jawless vertebrates covered with an external skeleton of bone which

together with the Cyclostomata make up the class Agnatha.

Ostreidae [INV ZOO] A family of bivalve mollusks in the order Anisomyaria containing the oysters.

ostria [METEOROL] A warm southerly wind on the Bulgarian coast; it is considered a precursor of bad weather. Also known as auster.

ostrich [VERT ZOO] *Struthio camelus*. A large running bird with soft plumage, naked head, neck and legs, small wings, and thick powerful legs with two toes on each leg; the only living species of the Struthioniformes.

Ostwald dilution law [PHYS CHEM] The law that for a sufficiently dilute solution of univalent electrolyte, the dissociation constant approximates $a^2c/(1-a)$, where c is the concentration of electrolyte and a is the degree of dissociation.

Ostwald process [CHEM ENG] An industrial preparation of nitric acid by the oxidation of ammonia; the oxidation takes place in successive stages to nitric oxide, nitrogen dioxide, and nitric acid; a catalyst of platinum gauze is used and high temperatures are needed.

Ostwald ripening [CHEM] Solution-crystallizer phenomenon in which small crystals, more soluble than large ones, dissolve and reprecipitate onto larger particles.

Ostwald's adsorption isotherm [THERMO] An equation stating that at a constant temperature the weight of material adsorbed on an adsorbent dispersed through a gas or solution, per unit weight of adsorbent, is proportional to the concentration of the adsorbent raised to some constant power.

Ostwald viscometer [ENG] A viscometer in which liquid is drawn into the higher of two glass bulbs joined by a length of capillary tubing, and the time for its meniscus to fall between calibration marks above and below the upper bulb is compared with that for a liquid of known viscosity.

osumilite [MINERAL] $(K,Na)(Mg,Fe^{2+})_2(Al,Fe^{3+})_3(Si,Al)_{12}$-$O_{30} \cdot H_2O$ A mineral that crystallizes in the hexagonal system and is commonly mistaken for cordierite.

Oswald diagram [ANALY CHEM] Diagram used in fuel Orsat analyses by plotting percent by volume CO_2 maximum in the fuel (ordinate) versus percent by volume O_2 in air (abscissa); O_2 and CO_2 Orsat readings should fall on a line connecting these maximum values if the analysis is proceeding properly.

ot-, oto- [ANAT] A combining form meaning ear.

otalgia [MED] Pain in the ear.

Otariidae [VERT ZOO] The sea lions, a family of carnivorous mammals in the superfamily Canoidea.

otavite [MINERAL] $CdCO_3$ A mineral that crystallizes in the hexagonal system and is isostructural with calcite.

Othmer process [CHEM ENG] Process to produce acetic acid from pyroligneous acid by azeotropic distillation.

Othniidae [INV ZOO] The false tiger beetles, a small family of coleopteran insects in the superfamily Tenebrionoidea.

otic [ANAT] Of or pertaining to the ear or a part thereof.

-otic [SCI TECH] A suffix meaning of, pertaining to, characterized by, or causing the process.

otic capsule [EMBRYO] A cartilaginous capsule surrounding the auditory vesicle during development, later fusing with the spheroid and occipital cartilages.

otic ganglion [ANAT] The nerve ganglion located immediately below the foramen ovale of the sphenoid bone.

Otitidae [INV ZOO] A family of cyclorrhaphous myodarian dipteran insects in the subsection Acalypteratae.

otitis [MED] Inflammation of the ear.

otitis externa [MED] Inflammation of the external ear.

otitis media [MED] Inflammation of the middle ear.

otocyst [EMBRYO] The auditory vesicle of vertebrate embryos. [INV ZOO] An auditory vesicle, otocell, or otidium in some invertebrates.

otolaryngology [MED] A branch of medicine that deals with the ear, nose, and throat. Also known as otorhinolaryngology.

otolith [ANAT] A calcareous concretion on the end of a sensory hair cell in the vertebrate ear and in some invertebrates.

otology [MED] A branch of medicine that deals with the ear and its diseases.

otomycosis [MED] Fungus infection of the external ear, usually caused by *Aspergillus niger* and *A. fumigatus*.

Otopheidomenidae [INV ZOO] A family of parasitic mites in the suborder Mesostigmata.

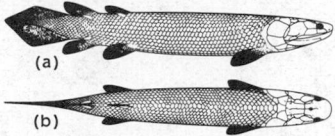
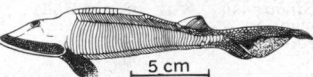

otorhinolaryngology   *See* otolaryngology.

otosclerosis   [MED] Sclerosis of the inner ear, causing a progressive increase in deafness.

otoscope   [MED] An apparatus designed for examination of the ear and for rendering the tympanic membrane visible.

OTS   *See* ovonic threshold switch.

otter   [ENG] *See* paravane.   [VERT ZOO] Any of various members of the family Mustelidae, having a long thin body, short legs, a somewhat flattened head, webbed toes, and a broad flattened tail; all are adapted to aquatic life.

otter trawl   [NAV ARCH] A large commercial fishing trawl which uses kitelike wooden boards at the corners of the mouth of the net.

Otto cycle   [THERMO] A thermodynamic cycle for the conversion of heat into work, consisting of two isentropic phases interspersed between two constant-volume phases. Also known as spark-ignition combustion cycle.

Otto engine   [MECH ENG] An internal combustion engine that operates on the Otto cycle, where the phases of suction, compression, combustion, expansion, and exhaust occur sequentially in a four-stroke-cycle or two-stroke-cycle reciprocating mechanism.

Otto-Lardillon method   [MECH] A method of computing trajectories of missiles with low velocities (so that drag is proportional to the velocity squared) and quadrant angles of departure that may be high, in which exact solutions of the equations of motion are arrived at by numerical integration and are then tabulated.

otto of rose oil   *See* rose oil.

ottrelite   [MINERAL] A gray to black variety of chloritoid containing manganese.

O-type backward-wave oscillator   [ELECTR] A backward-wave tube in which an electron gun produces an electron beam focused longitudinally throughout the length of the tube, a slow-wave circuit interacts with the beam, and at the end of the tube a collector terminates the beam. Also known as O-type carcinotron; type-O carcinotron.

O-type carcinotron   *See* O-type backward-wave oscillator.

O-type star   [ASTRON] A spectral-type classification in the Draper catalog of stars; a star having spectral type O; a very hot, blue star in which the spectral lines of ionized helium are prominent.

ouabain   [ORG CHEM] $C_{29}H_{44}O_{12}\cdot 8H_2O$ White crystals that melt with decomposition at 190°C, soluble in water and ethanol; used in medicine.

ouachitite   [PETR] A biotite monchiquite with no olivine and a glassy or analcime groundmass.

ouari   [METEOROL] A south wind of Somaliland, Africa; it is similar to the khamsin.

Oudeman law   [PHYS CHEM] The law that the molecular rotations of the various salts of an acid or base tend toward an identical limiting value as the concentration of the solution is reduced to zero.

ounce   [MECH] **1.** A unit of mass in avoirdupois measure equal to 1/16 pound or to approximately 0.0283495 kilogram. Abbreviated oz. **2.** A unit of mass in either troy or apothecaries' measure equal to 480 grains or exactly 0.0311034768 kilogram. Also known as apothecaries' ounce or troy ounce (abbreviations are oz ap and oz t in the United States, and oz apoth and oz tr in the United Kingdom).

ouncedal   [MECH] A unit of force equal to the force which will impart an acceleration of 1 foot per second per second to a mass of 1 ounce; equal to 0.0086409346485 newton.

ounce metal   [MET] An alloy composed of 1 ounce each of lead, tin, and zinc to 1 pound of copper. Also known as composition metal.

ouricury wax   [MATER] A hard brown wax obtained from leaves of the ouricury palm (*Cocos coronapa*); similar to carnauba wax in use and properties.

outage   [PETRO ENG] The difference between the full or rated capacity of a barrel, tank, or tank car as compared to actual content.

outage method   [PETRO ENG] Deduction of the liquid content of a tank by measurement of the distance from the top of the tank to the surface of the liquid; in contrast to the innage method.

outboard   [NAV ARCH] Toward the outside of a vessel; outside the hull.

OTTER

North American otter (*Lutra canadensis*), with supple body, webbed feet, and thick water-resistant fur.

outboard engine   [NAV ARCH] A unit assembly of engine, propeller, and vertical drive shaft used to propel a boat and usually clamped to the boat transom; power of various models ranges from 1 horsepower (approximately 750 watts) to well over 100 horsepower. Also known as outboard motor.

outboard motor   *See* outboard engine.

outboard profile   [NAV ARCH] A plan representing the longitudinal exterior of a vessel showing the starboard side of the shell, all deck erections, masts, yards, rigging, rails, and so on.

outbreeding   *See* exogamy.

outburst   [MIN ENG] The sudden issue of gases, chiefly methane (sometimes accompanied by coal dust), from the working face of a coal mine.

outby   [MIN ENG] Toward the mine entrance or shaft and therefore away from the working face.

outcrop   [GEOL] Exposed stratum or body of ore at the surface of the earth. Also known as cropout.

outer atmosphere   [METEOROL] Very generally, the atmosphere at a great distance from the earth's surface; possibly best usage of the term is as an approximate synonym for exosphere.

outer automorphism   [MATH] An automorphism of a group which is not an inner automorphism.

outer bottom   [NAV ARCH] A part of a double-bottom ship, specifically, the bottom shell plating.

outer bremsstrahlung   [PHYS] Bremsstrahlung involving the acceleration of a charged particle coming from outside the atom whose nucleus produces the acceleration, and in which the energy loss by radiation is much greater than that by ionization, usually seen in electrons with energies greater than about 50 MeV (million electron volts).

outer fix   [NAV] A fix in the destination terminal area, other than the approach fix, to which aircraft are normally cleared by an air route traffic-control center or a terminal area traffic-control facility, and from which aircraft are cleared to the approach fix or final approach course.

outer harbor   [GEOGR] The part of a harbor toward the sea, through which a vessel enters the inner harbor.

outer keel   [NAV ARCH] The outer plate of a double flat-plate keel.

outer mantle   *See* upper mantle.

outer marker   [NAV] That instrument landing system (ILS) marker farthest from the approach end of the instrument runway; the marker is located 4.5 miles ± 1000 feet (7240 ± 305 meters) from the end of the runway and not more than 250 feet (75 meters) from the extended center line of the runway.

outer measure   [MATH] A function with the same properties as a measure except that its value on a disjoint countable union of sets is less than or equal to the sum of its values on each set, and it is usually defined on the collection of subsets of a given set.

outer orbital complex   [PHYS CHEM] A metal coordination compound in which the *d* orbital used in forming the coordinate bond is at the same energy level as the *s* and *p* orbitals.

outer planets   [ASTRON] The planets with orbits larger than that of Mars; Jupiter, Saturn, Uranus, Neptune, and Pluto.

outer-shell electron   *See* conduction electron.

outer space   [ASTRON] A general term for any region that is beyond the earth's atmosphere.

outface   *See* dip slope.

outfall   [CIV ENG] The point at which a sewer or drainage channel discharges to the sea or to a river.   [HYD] The narrow part of a stream, lake, or other body of water where it drops away into a larger body.

outflow   [CHEM ENG] Flow of fluid product out of a process facility.

outgassing   [ENG] The release of adsorbed or occluded gases or water vapor, usually by heating, as from a vacuum tube or other vacuum system.

outgoing trunk   *See* one-way trunk.

outlet   [ELEC] A power line termination from which electric power can be obtained by inserting the plug of a line cord. Also known as convenience receptacle; electric outlet; receptacle.

outlet box   [ELEC] A box at which lines in an electric wiring system terminate, so that electric appliances or fixtures may be connected.

outlet glacier [HYD] A stream of ice from an ice cap to the sea.

outlier [GEOL] A group of rocks separated from the main mass and surrounded by outcrops of older rocks. [STAT] In a set of data, a value so far removed from other values in the distribution that its presence cannot be attributed to the random combination of chance causes.

out of phase [PHYS] Having waveforms that are of the same frequency but do not pass through corresponding values at the same instant.

out-of-service jack [ELEC] Jack associated with a test jack which removes the circuit from service when a shorted plug is inserted.

outpatient [MED] A patient who comes to the hospital or clinic for diagnosis and treatment but who does not occupy a bed in the institution.

output [ADP] **1.** The data produced by a data-processing operation, or the information that is the objective or goal in data processing. **2.** The data actively transmitted from within the computer to an external device, or onto a permanent recording medium (paper, microfilm). **3.** The activity of transmitting the generated information. **4.** The readable storage medium upon which generated data are written, as in hard-copy output. [ELECTR] **1.** The current, voltage, power, driving force, or information which a circuit or device delivers. **2.** Terminals or other places where a circuit or device can deliver current, voltage, power, driving force, or information. [SCI TECH] The product of a system.

output area [ADP] A part of storage that has been reserved for output data. Also known as output block.

output block [ADP] **1.** A portion of the internal storage of a computer that is reserved for receiving, processing, and transmitting data to be transferred out. **2.** See output area.

output capacitance [ELECTR] Of an *n*-terminal electron tube, the short-circuit transfer capacitance between the output terminal and all other terminals, except the input terminal, connected together.

output equipment [ADP] Equipment, such as punched cards, printers, and magnetic tape recorders, which transfers data from within a computer to its outside environment.

output gap [ELECTR] An interaction gap by means of which usable power can be abstracted from an electron stream in a microwave tube.

output impedance [ELECTR] The impedance presented by a source to a load.

output indicator [ENG] A meter or other device that is connected to a radio receiver to indicate variations in output signal strength for alignment and other purposes, without indicating the exact value of output.

output link [COMMUN] The last link in a communications chain.

output magazine [ADP] A mechanism that accumulates cards after they have passed through a machine.

output meter [ENG] An alternating-current voltmeter connected to the output of a receiver or amplifier to measure output signal strength in volume units or decibels.

output-meter adapter [ENG] Device that can be slipped over the plate prong of the output tube of a radio receiver to provide a conventional terminal to which an output meter can be connected during alignment.

output monitor interrupt [ADP] A data-processing step in which control is passed to the monitor to determine the precedence order for two requests having the same priority level.

output power [ELEC] Power delivered by a system or transducer to its load.

output rating See carrier power output rating.

output resistance [ELECTR] The resistance across the output terminals of a circuit or device.

output shaft [MECH ENG] The shaft that transfers motion from the prime mover to the driven machines.

output stage [ELECTR] The final stage in any electronic equipment.

output standard See standard time.

output transformer [ELECTR] The iron-core audio-frequency transformer used to match the output stage of a radio receiver or an amplifier to its loudspeaker or other load.

output tube [ELECTR] Power-amplifier tube designed for use in an output stage.

output unit [ADP] In computers, a unit which delivers information from the computer to an external device or from internal storage to external storage.

output winding [ELECTROMAG] Of a saturable reactor, a winding, other than a feedback winding, which is associated with the load, and through which power is delivered to the load.

output word [ADP] Any running word into which an input word is to be translated.

outrigger [ENG] A steel beam or lattice girder extending from a crane to provide stability by widening the base.

outside caliper [DES ENG] A caliper having two curved legs which point toward each other; used for measuring outside dimensions of a workpiece.

outside diameter [DES ENG] The outer diameter of a pipe, including the wall thickness; usually measured with calipers. Abbreviated OD.

outside extension [COMMUN] Telephone extension on premises separated from the main station.

outside fix [NAV] The fix position determined by the method of bisectors when the lines of position result from observations of objects or celestial bodies lying within a 180° arc of the horizon.

outside strake [NAV ARCH] The outer strake of an in-and-out system of shell plating; a strake which laps on the inner strake and which is the thickness of the plating outside the molded frame line.

outward bound [NAV] Heading for the open sea.

outwash [GEOL] Sand and gravel transported away from a glacier by streams of meltwater and either deposited as a floodplain along a preexisting valley bottom or broadcast over a preexisting plain in a form similar to an alluvial fan. Also known as glacial outwash; outwash drift; overwash.

outwash apron See outwash plain.

outwash cone [GEOL] A cone-shaped deposit consisting chiefly of sand and gravel found at the edge of shrinking glaciers and ice sheets.

outwash drift See outwash.

outwash plain [GEOL] A broad, outspread flat or gently sloping alluvial deposit of outwash in front of or beyond the terminal moraine of a glacier. Also known as apron; frontal apron; frontal plain; marginal plain; morainal apron; morainal plain; outwash apron; overwash plain; sandur; wash plain.

outwash train See valley train.

ouvarovite See uvarovite.

oval [MATH] A curve shaped like a section of an egg.

ovalbumin [BIOCHEM] The major, conjugated protein of egg-white.

oval of Cassini [MATH] An ovallike curve similar to a lemniscate obtained as the locus corresponding to a general type of quadratic equation in two variables.

oval window [ANAT] The membrane-covered opening into the inner ear of tetrapods, to which the ossicles of the middle ear are connected.

ovarian [ANAT] Of or pertaining to the ovaries.

ovarian agenesis [MED] Failure of the ovaries to develop.

ovarian dysmenorrhea [MED] Dysmenorrhea caused by ovarian disease.

ovarian follicle [HISTOL] An ovum and its surrounding follicular cells, found in the ovarian cortex.

ovarian insufficiency [MED] Deficient functioning of the ovaries, leading to amenorrhea, oligomenorrhea, or abnormal dysfunctional uterine bleeding.

ovariectomy [MED] Excision of an ovary.

ovariole [INV ZOO] The tubular structural unit of an insect ovary.

ovary [ANAT] A glandular organ that produces hormones and gives rise to ova in female vertebrates. [BOT] The enlarged basal portion of a pistil that bears the ovules in angiosperms.

oven [ENG] A heated enclosure for baking, heating, or drying.

over [ORD] A bomb or projectile hit beyond or past the target.

overaging [MET] Aging at a higher temperature or for a

OUTSIDE CALIPER

Outside caliper. *(From R. J. Sweeney, Measurement Techniques in Mechanical Engineering, Wiley, 1953)*

longer time than is required to produce maximum or optimum properties.

overall drilling time [MIN ENG] The total time for rock drilling including time for setting up, withdrawing, and moving drills, time for mechanical delays, and the time for the actual drilling.

overall efficiency [AERO ENG] The efficiency of a jet engine, rocket engine, or rocket motor in converting the total heat energy of its fuel first into available energy for the engine, then into effective driving energy.

overall plate efficiency [CHEM ENG] For a specified liquid-mixture separation in a fractionation (or distillation) tower, the ratio of actual to theoretical plates (or trays) required.

overall response [ELECTR] The ratio between system input and output.

overall stability constant [ANALY CHEM] Reaction equilibrium constant for the reaction forming soluble complexes during compleximetric titration.

overarching weight [MIN ENG] The pressure of the rocks over the active mine workings.

overbanking [HOROL] A malfunction in which the escape wheel unlocks prematurely, without the fork contacting the roller jewel.

overbreak [CIV ENG] Rock excavated in excess of the neat lines of a tunnel or cutting. Also known as backbreak.

overbunching [ELECTR] In velocity-modulated streams of electrons, the bunching condition produced by the continuation of the bunching process beyond the optimum condition.

overburden [GEOL] **1.** Rock material overlying a mineral deposit or coal seam. Also known as baring; top. **2.** Material of any nature, consolidated or unconsolidated, that overlies a deposit of useful materials, ores, or coal, especially those deposits that are mined from the surface by open cuts. **3.** Loose soil, sand, or gravel that lies above the bedrock. [MIN ENG] To charge in a furnace too much ore and flux in proportion to the amount of fuel.

overcast [METEOROL] **1.** Pertaining to a sky cover of 1.0 (95% or more) when at least a portion of this amount is attributable to clouds or obscuring phenomena aloft, that is, when the total sky cover is not due entirely to surface-based obscuring phenomena. **2.** Cloud layer that covers most or all of the sky; generally, a widespread layer of clouds such as that which is considered typical of a warm front. [MIN ENG] **1.** An enclosed airway to permit one air current to pass over another without interruption. **2.** To move overburden removed from coal mined from surface mines to an area from which the coal has been mined.

overcast bombing [ORD] The bombing of a target through an overcast above the target, using radar or other equipment to aid in sighting through the overcast.

overcompound [ELEC] To use sufficiently many series turns in a compound-wound generator so that the terminal voltage at rated load is greater than at no load, usually to compensate for increased line drop.

overconsolidation [GEOL] Consolidation of sedimentary material exceeding that which is normal for the existing overburden.

overcoupled circuits [ELECTR] Two resonant circuits which are tuned to the same freqency but coupled so closely that two response peaks are obtained; used to attain broad-band response with substantially uniform impedance.

overcritical electric field [ATOM PHYS] An electric field so strong that an electron-positron pair is created spontaneously; quantum electrodynamics predicts that this will happen near a nucleus having more than approximately 173 protons.

overcuring [CHEM ENG] A condition resulting from vulcanizing longer than necessary to achieve full development of physical strength; causes softness or brittleness and impaired age-resisting quality of the material.

overcurrent [ELECTR] An abnormally high current, usually resulting from a short circuit.

overcurrent protection *See* overload protection.

overcut [MIN ENG] A machine cut made along the top or near the top of a coal seam; sometimes used in a thick seam or a seam with sticky coal.

overcutting machine [MIN ENG] A coal-cutting machine designed to make the cut at a desired place in the coal seam some distance above the floor.

overdamping [PHYS] Damping greater than that required for critical damping.

overdeepening [GEOL] The erosive process by which a glacier deepens and widens an inherited preglacial valley to below the level of the subglacial surface.

overdevelop [GRAPHICS] To process a photographic film or plate for too long a time or in a solution that is too concentrated.

overdominance [GEN] Monohybrid heterosis, that is, the phenomenon of the phenotype being more pronounced in the heterozygote than in either homozygote with respect to a single specified pair of alleles.

overdose [MED] An excessive dose of medicine.

overdraft [MET] Upward curving of a piece of metal after leaving the rolls during forming, due to higher speed of the lower roll.

overdrilling [ENG] The act or process of drilling a run or length of borehole greater than the core-capacity length of the core barrel, resulting in loss of the core.

overdrive [MECH ENG] An automobile engine device that lowers the gear ratio, thereby reducing fuel consumption.

overdriven amplifier [ELECTR] Amplifier stage which is designed to distort the input-signal waveform by permitting the grid signal to drive the stage beyond cutoff or plate-current saturation.

overexpose [GRAPHICS] To expose a photographic film or plate with too much light or for too long a time.

overfall dam *See* overflow dam.

overfalls [OCEANOGR] Short, breaking waves occurring when a strong current passes over a shoal or other submarine obstruction or meets a contrary current or wind.

overfeed [MIN ENG] To attempt to make a diamond- or rock-drill bit penetrate rock at a rate in excess of that at which the optimum economical performance of the bit is attained, needlessly damaging the bit and shortening its life.

overflow [ADP] **1.** The condition that arises when the result of an arithmetic operation exeeds the storage capacity of the indicated result-holding storage. **2.** That part of the result which exceeds the storage capacity. [CIV ENG] Any device or structure that conducts excess water or sewage from a conduit or container. [SCI TECH] Excess liquid which overflows its given limits.

overflow capacity [ENG] Capacity of a container measured to its top, or to the point of overflow.

overflow channel [CIV ENG] An artificial waterway for conducting water away from an overflowing structure such as a reservoir or canal.

overflow check indicator *See* overflow indicator.

overflow dam [CIV ENG] A dam built with a crest to allow the overflow of water. Also known as overfall dam; spillway dam.

overflow error [ADP] The condition in which the numerical result of an operation exceeds the capacity of the register.

overflow groove [ENG] Small groove on a plastics mold that allows material to flow freely, to prevent weld lines and low density in the finished product and to dispose of excess material.

overflow indicator [ADP] A bistable device which changes state when an overflow occurs in the register associated with it, and which is designed so that its condition can be determined, and its original condition restored. Also known as overflow check indicator.

overflow pipe [ENG] Open pipe protruding above the surface of a liquid in a container, such as a distillation or absorption column or a toilet tank, to control the height of the liquid; excess liquid enters the pipe's open end and drains away.

overflow storage [COMMUN] Additional storage provided in a store-and-forward-switching center to prevent the loss of messages (or parts of messages) offered to a completely filled line store.

overfold [GEOL] A fold that is overturned.

overgear [MECH ENG] A gear train in which the angular

velocity ratio of the driven shaft to driving shaft is greater than unity, as when the propelling shaft of an automobile revolves faster than the engine shaft.

overglaze color [MATER] Any of the mixtures of ground pigment and low-melting glass melting at 704–816°C; used for decorative designs fired onto china and ceramics.

overgrinding [MIN ENG] Grinding an ore to a smaller particle size than that necessary to free the desired mineral from other materials.

overgrowth [CRYSTAL] A crystal growth in optical and crystallographic continuity around another crystal of different composition. [MINERAL] A mineral deposited on and growing in oriented, crystallographic directions on the surface of another mineral.

overhand cut and fill [MIN ENG] A method of mining ore in which material removed from the roof of a drive drops through chutes to a lower drive, from which the material is removed.

overhand stope [MIN ENG] A stope in which the ore above the point of entry is attacked, so that severed ore tends to gravitate toward discharge chutes, and the stope is self-draining.

overhand stoping [MIN ENG] A method of mining in which the ore is blasted from a series of ascending stepped benches; both horizontal and vertical holes may be employed.

overhang [GEOL] The part of a salt plug that projects from the top.

overhaul [ENG] A maintenance procedure for machinery involving disassembly, the inspecting, refinishing, adjusting, and replacing of parts, and reassembly and testing.

overhauling [MET] Removing scale and surface defects from metal castings or slabs by cutting away surface layers.

Overhauser effect [ATOM PHYS] The effect whereby, if a radio frequency field is applied to a substance in an external magnetic field, whose nuclei have spin ½ and which has unpaired electrons, at the electron spin resonance frequency, the resulting polarization of the nuclei is as great as if the nuclei had the much larger electron magnetic moment.

overhead [CHEM ENG] Pertaining to fluid (gas or liquid) effluent from the top of a process vessel, such as a distillation column.

overhead cableway [MIN ENG] A type of equipment for the removal of soil or rock, consisting of a strong overhead cable which is usually attached to towers at either end, and on which a car or traveler may run back and forth; from this car a pan or bucket may be lowered to the surface, then raised and locked to the car and transported to any position on the cable where it is desired to dump.

overhead compass *See* inverted compass.

overhead cost *See* fixed cost.

overhead fire [ORD] Fire that is delivered over the heads of friendly troops.

overhead shovel [MECH ENG] A tractor which digs with a shovel at its front end, swings the shovel rearward overhead, and dumps the shovel at its rear end.

overhead-traveling crane [MECH ENG] A hoisting machine with a bridgelike structure moved on wheels along overhead trackage which is usually fixed to the building structure.

overhead-valve engine [MECH ENG] A four-stroke-cycle internal combustion engine having its valves located in the cylinder head, operated by pushrods that actuate rocker arms. Abbreviated OHV engine. Also known as valve-in-head engine.

overheat [MET] To heat a metal or alloy to such high temperatures that its physical properties are impaired.

overhit [ORD] To hit a target with more destructive force than necessary to accomplish the desired amount of damage.

overite [MINERAL] $Ca_3Al_8(PO_4)_8(OH)_6 \cdot 15H_2O$ A mineral composed of hydrous basic calcium aluminum phosphate.

overland flow [HYD] Water flowing over the ground surface toward a channel; upon reaching the channel, it is called surface runoff. Also known as surface flow.

overlap [ADP] To perform some or all of an operation concurrently with one or more other operations. [COMMUN] **1.** In teletypewriter practice, the selecting of another code group while the printing of a previously selected code group is taking place. **2.** Amount by which the effective height of the scanning facsimile spot exceeds the nominal width of the scanning line. [GEOL] **1.** Movement of an upcurrent part of a shore to a position extending seaward beyond a downcurrent part. **2.** Extension of strata over or beyond older underlying rocks. **3.** The horizontal component of separation measured parallel to the strike of a fault. [MET] **1.** Projection of the weld metal beyond the bond at the toe of the weld. **2.** Extension of one sheet over another in spot, seam, or projection welding.

overlap integral [QUANT MECH] The integral over space of the product of the wave function of a particle and the complex conjugate of the wave function of another particle.

overlapped memories [ADP] An arrangement of computer memory banks in which, to cut down access time, successive words are taken from different memory banks, rewriting in one bank being overlapped by logic operations in another bank, with memory access in still another bank.

overlapping [ADP] An operation whereby, if the processor determines that the current instruction and the next instruction lie in different storage modules, the two words may be retrieved in parallel.

overlapping orbitals [ATOM PHYS] Two orbitals (usually of electrons associated with different atoms in a molecule) for which there is a region of space where both are of appreciable magnitude.

overlap radar [ENG] Radar located in one sector whose area of useful radar coverage includes a portion of another sector.

overlay [ADP] A technique for bringing routines into high-speed storage from some other form of storage during processing, so that several routines will occupy the same storage locations at different times; overlay is used when the total storage requirements for instructions exceed the available main storage. [CIV ENG] A repair topping of asphalt or concrete placed on a worn roadway. [ENG] **1.** Nonwoven fibrous mat (glass or other fiber) used as the top layer in a cloth or mat lay-up to give smooth finish to plastic products or to minimize the fibrous pattern on the surface. Also known as surfacing mat. **2.** An ornamental covering, as of wood or metal. [GRAPHICS] **1.** A sheet attached to copy or artwork and containing special instructions about reproduction or arrangement. **2.** A transparent or translucent film attached to artwork and carrying additional detail to be reproduced; in multicolor printing these overlays may represent the separation of the colors to be printed.

overlay transistor [ELECTR] Transistor containing a large number of emitters connected in parallel to provide maximum power amplification at extremely high frequencies.

overline [COMMUN] In teletypewriter practice, the printing of one group of characters over another.

overload [ELECTR] A load greater than that which a device is designed to handle; may cause overheating of power-handling components and distortion in signal circuits.

overload capacity [ELEC] Current, voltage, or power level beyond which permanent damage occurs to the device considered.

overload current [ELECTR] A current greater than that which a circuit is designed to carry; may melt wires or damage elements of the circuit.

overloader [MIN ENG] A loading machine which digs with a bucket, raises the bucket, and swings it in a wide horizontal arc to the dumping point.

overload level [ELEC] Level above which operation ceases to be satisfactory as a result of signal distortion, overheating, damage, and so forth.

overload protection [ELEC] Effect of a device operative on excessive current, but not necessarily on short circuit, to cause and maintain the interruption of current flow to the device governed. Also known as overcurrent protection.

overmatching plate [ORD] Armor plate whose thickness exceeds the diameter of the projectile.

overmatching projectile [ORD] A projectile whose diameter exceeds the thickness of the armor plate.

overmodulation [COMMUN] Amplitude modulation greater than 100%, causing distortion because the carrier voltage is reduced to zero during portions of each cycle.

overpass [CIV ENG] **1.** A grade separation in which traffic at the higher level is raised, and traffic at the lower level moves at approximately its original level. **2.** The upper part of a grade crossing.

overpotential *See* overvoltage.

overpressure [FL MECH] The transient pressure, usually expressed in pounds per square inch, exceeding existing atmospheric pressure and manifested in the blast wave from an explosion.

overprint [GRAPHICS] **1.** To imprint over something that has been printed. **2.** To apply a varnish or lacquer to printed matter from a type or litho process, by means of a brush, spray, or roller coating. **3.** To print in a primary color over an existing color print to obtain a compound shade.

overpunch [ADP] A hole punch in any one of the three rows above the zero row in an punch card, usually combined with a second punch on the same column to represent a character.

override [CONT SYS] To cancel the influence of an automatic control by means of a manual control.

overriding process control [CONT SYS] Process control in which any one of several controllers associated with one control valve can be made to override another in accordance with a priority requirement of the process.

overrun [CIV ENG] A cleared area extending beyond the end of a runway.

overrunning [METEOROL] A condition existing when an air mass is in motion aloft above another air mass of greater density at the surface; this term usually is applied in the case of warm air ascending the surface of a warm front or quasistationary front.

overrunning clutch [MECH ENG] A clutch that allows the driven shaft to turn freely only under certain conditions; for example, a clutch in an engine starter that allows the crank to turn freely when the engine attempts to run.

oversaturated *See* silicic.

overseeding [METEOROL] Cloud seeding in which an excess of nucleating material is released; as the term is normally used, the excess is relative to that amount of nucleating material which would, theoretically, maximize the precipitation received at the ground.

overshoot [ELECTROMAG] The reception of microwave signals where they were not intended, due to an unusual atmospheric condition that sets up variations in the index of refraction. [ENG] **1.** An initial transient response to a unidirectional change in input which exceeds the steady-state response. **2.** The maximum amount by which this transient response exceeds the steady-state response.

overshot [ENG] **1.** A fishing tool for recovering lost drill pipe or casing. **2.** *See* bullet.

overshot wheel [MECH ENG] A horizontal-shaft waterwheel with buckets around the circumference; the weight of water pouring into the buckets from the top rotates the wheel.

oversize control screen [MIN ENG] A screen used to prevent the entry into a machine of coarse particles which might interfere with its operation. Also known as check screen; guard screen.

oversize powder [MET] A metal powder having coarser particles than the maximum permitted.

oversize rod *See* guide rod.

overspin [MECH] In a spin-stabilized projectile, the overstability that results when the rate of spin is too great for the particular design of projectile, so that its nose does not turn downward as it passes the summit of the trajectory and follows the descending branch. Also known as overstabilization.

overstability [PL PHYS] Condition in which the restoring forces acting on an oscillation of a plasma or other conducting fluid drive the fluid back to its equilibrium state at a speed greater than its original outward speed, resulting in continually greater oscillation.

overstabilization *See* overspin.

oversteepening [GEOL] The process by which an eroding alpine glacier steepens the sides of an inherited preglacial valley.

overstressed area [MIN ENG] In strata control, an area where the force is concentrated on pillars.

overstressing [ENG] Cyclically stressing a material at a level higher than that used at the end of a fatigue test.

over-the-horizon propagation *See* scatter propagation.

over-the-horizon radar [ELECTROMAG] Long-range radar in which the transmitted and reflected beams are bounced off the ionosphere layers to achieve ranges far beyond line of sight.

overthrow distortion [COMMUN] Distortion caused when the maximum amplitude of the signal wavefront exceeds the steady state of amplitude of the signal wave.

overthrust [GEOL] **1.** A thrust fault that has a low dip or a net slip that is large. Also known as low-angle thrust; overthrust fault. **2.** A thrust fault with the active element being the hanging wall.

overthrust black *See* overthrust nappe.

overthrust fault *See* overthrust.

overthrust nappe [GEOL] The body of rock making up the hanging wall of a large-scale overthrust. Also known as overthrust block; overthrust sheet; overthrust slice.

overthrust sheet *See* overthrust nappe.

overthrust slice *See* overthrust nappe.

overtone [ACOUS] **1.** A component of a complex sound whose frequency is an integral multiple, greater than 1, of the fundamental frequency. **2.** A component of a complex tone having a pitch higher than that of the fundamental pitch. [MECH] One of the normal modes of vibration of a vibrating system whose frequency is greater than that of the fundamental mode. [PHYS] A harmonic other than the fundamental component.

overtone band [SPECT] The spectral band associated with transitions of a molecule in which the vibrational quantum number changes by 2 or more.

overtone crystal [ELECTR] Quartz crystal cut in such a manner that it will operate at a higher order than its fundamental frequency, or operate at two frequencies simultaneously as in a synthesizer.

overtravel [ORD] In machine guns, the distance the firing notch overrides the sear notch in cocking, to ensure positive engagement of the two notches.

overtub system [MIN ENG] An endless-rope system in which the rope runs over the tubs or cars in the center of the rails.

overturn [HYD] Renewal of bottom water in lakes and ponds in regions where winter temperatures are cold; in the fall, cooled surface waters become denser and sink, until the whole body of water is at 4°C; in the spring, the surface is warmed back to 4°C, and the lake is homothermous. Also known as convective overturn.

overturned [GEOL] Of a fold or the side of a fold, tilted beyond the perpendicular. Also known as inverted; reversed.

overvoltage [ELEC] A voltage greater than that at which a device or circuit is designed to operate. Also known as overpotential. [ELECTR] The amount by which the applied voltage exceeds the Geiger threshold in a radiation counter tube. [PHYS CHEM] The difference between electrode potential under electrolysis conditions and the thermodynamic value of the electrode potential in the absence of electrolysis for the same experimental conditions. Also known as overpotential.

overwash *See* outwash.

overwash plain *See* outwash plain.

overwind [ENG] To wind a spring, rope, or cable too tightly or too far.

ovicell [INV ZOO] A broad chamber in certain bryozoans.

ovicyst [INV ZOO] The pouch of a tunicate in which the eggs develop.

oviduct [ANAT] A tube that serves to conduct ova from the ovary to the exterior or to an intermediate organ such as the uterus. Also known in mammals as Fallopian tube; uterine tube.

oviger [INV ZOO] A modified leg used for carrying eggs in some pycnogonids.

ovine encephalomyelitis *See* louping ill.

oviparous [VERT ZOO] Producing eggs that develop and hatch externally.

ovipositor [INV ZOO] A specialized structure in many insects for depositing eggs. [VERT ZOO] A tubular extension of the genital orifice in most fishes.

ovography [GRAPHICS] A printing method based upon the use of a drum coated with ovonic memory material, in which a permanent but alterable image can be "written" onto the drum by a computer-controlled energy beam, such as a laser,

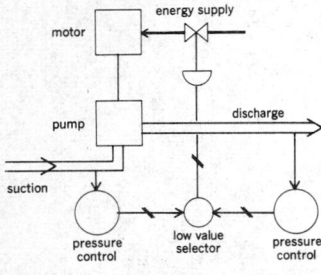

**OVERRIDING
PROCESS CONTROL**

Diagram of overriding process control in an oil pipeline. Motor driving the pipeline is controlled by suction pressure when this pressure approaches the atmospheric value, and by the discharge pressure when this is close to the maximum allowable value. The low-value selector passes the lower of its inputs, in order to prevent the noncritical controlled value from increasing the speed of the pump.

and printout can be achieved by electrostatic printing techniques.

ovonic device *See* glass switch.

ovonic memory switch [ELECTR] A glass switch which, after being brought from the highly resistive state to the conducting state, remains in the conducting state until a current pulse returns it to its highly resistive state. Abbreviated OMS.

ovonic threshold switch [ELECTR] A glass switch which, after being brought from the highly resistive state to the conducting state, returns to the highly resistive state when the current falls below a holding current value. Abbreviated OTS.

ovotesticular hermaphroditism [MED] A rare form of hermaphroditism in which an ovotestis is present on one or both sides.

ovoviviparous [VERT ZOO] Producing eggs that develop internally and hatch before or soon after extrusion.

Ovshinsky effect [ELECTR] The characteristic of a special thin-film solid-state switch that responds identically to both positive and negative polarities so that current can be made to flow in both directions equally.

ovulation [PHYSIO] Discharge of an ovum or ovule from the ovary.

ovule [BOT] A structure in the ovary of a seed plant that develops into a seed following fertilization.

ovum [CYTOL] A female gamete. Also known as egg.

O-wave component *See* ordinary-wave component.

Owen bridge [ELECTR] A four-arm alternating-current bridge used to measure self-inductance in terms of capacitance and resistance; bridge balance is independent of frequency.

Oweniidae [INV ZOO] A family of limivorous polychaete annelids of the Sedentaria.

owl [VERT ZOO] Any of a number of diurnal and nocturnal birds of prey composing the order Strigiformes; characterized by a large head, more or less forward-directed large eyes, a short hooked bill, and strong talons.

OW unit *See* sabin.

oxalate [ORG CHEM] Salt of oxalic acid; contains the $(COO)_2$ radical; examples are sodium oxalate, $Na_2C_2O_4$, ammonium oxalate, $(NH_4)_2C_2O_4 \cdot H_2O$, and ethyl oxalate, C_2H_5-$(C_2O_4)C_2H_5$.

oxaldehydic acid *See* glyoxalic acid.

oxalic acid [ORG CHEM] $HOOCCOOH \cdot 2H_2O$ Poisonous, transparent, colorless crystals melting at 187°C; soluble in water, alcohol, and ether; used as a chemical intermediate and a bleach, and in polishes and rust removers.

Oxalidaceae [BOT] A family of dicotyledonous plants in the order Geraniales, generally characterized by regular flowers, two or three times as many stamens as sepals or petals, a style which is not gynobasic, and the fruit which is a beakless, loculicidal capsule.

oxalite *See* humboldtine.

oxalosis [MED] A rare hereditary metabolic disorder, inherited as an autosomal recessive, in which glyoxylic acid metabolism is impaired, resulting in overproduction of oxalic acid and deposition of calcium oxalate in body tissues.

oxaluria [MED] The presence of oxalic acid or oxalates in the urine.

oxalyl chloride [INORG CHEM] $(COCl)_2$ Toxic, colorless liquid boiling at 64°C; soluble in ether, benzene, and chloroform; used as a chlorinating agent and for military poison gas. Also known as ethanedioyl chloride.

oxalylurea *See* parabanic acid.

oxamide [ORG CHEM] $NH_2COCONH_2$ Water-insoluble white powder, melting at 419°C; used as a stabilizer for nitrocellulose products.

Oxamycin [MICROBIO] A trade name for the antibiotic cycloserine.

oxatyl *See* carboxyl.

oxazole [ORG CHEM] C_3H_3ON A structure that consists of a five-membered ring containing oxygen and nitrogen in the 1 and 3 position; a colorless liquid (boiling point 69–70°C) that is miscible with organic solvents and water; used to prepare other organic compounds.

oxbow [HYD] 1. A closely looping, U-shaped stream meander whose curvature is so extreme that only a neck of land remains between the two parts of the stream. Also known as horseshoe bend. 2. *See* oxbow lake. [GEOL] The abandoned, horseshoe-shaped channel of a former stream meander after the stream formed a neck cutoff. Also known as abandoned channel.

oxbow lake [HYD] The crescent-shaped body of water located alongside a stream in an abandoned oxbow after a neck cutoff is formed and the ends of the original bends are silted up. Also known as crescentic lake; cutoff lake; horseshoe lake; loop lake; moat; mortlake; oxbow

Oxfordian [GEOL] A European stage of geologic time, in the Upper Jurassic (above Callovian, below Kimmeridgean). Also known as Divesian.

Oxford unit [MICROBIO] The minimum quantity of penicillin which, when dissolved in 50 milliliters of a meat broth, is sufficient to inhibit completely the growth of a test strain of *Micrococcus aureus*; equivalent to the specific activity of 0.6 microgram of the master standard penicillin.

oxidase [BIOCHEM] An enzyme that catalyzes oxidation reactions by the utilization of molecular oxygen as an electron acceptor.

oxidate [GEOL] A sediment made up of iron and manganese oxides and hydroxides crystallized from aqueous solution.

oxidation [CHEM] 1. A chemical reaction that increases the oxygen content of a compound. 2. A chemical reaction in which a compound or radical loses electrons, that is in which the positive valence is increased.

oxidation number [CHEM] 1. Numerical charge on the ions of an element. 2. *See* oxidation state.

oxidation potential [PHYS CHEM] The difference in potential between an atom or ion and the state in which an electron has been removed to an infinite distance from this atom or ion.

oxidation-reduction indicator [ANALY CHEM] A compound whose color in the oxidized-state differs from that in the reduced state.

oxidation-reduction potential *See* redox potential.

oxidation-reduction reaction [CHEM] An oxidizing chemical change, where an element's positive valence is increased (electron loss), accompanied by a simultaneous reduction of an associated element (electron gain).

oxidation state [CHEM] The number of electrons to be added (or subtracted) from an atom in a combined state to convert it to elemental form. Also known as oxidation number.

oxide [CHEM] Binary chemical compound in which oxygen is combined with a metal (such as Na_2O; basic) or nonmetal (such as NO_2; acidic).

oxide-coated cathode [ELECTR] A cathode that has been coated with oxides of alkaline-earth metals to improve electron emission at moderate temperatures.

oxide fuel reactor [NUCLEO] A nuclear fission reactor with fuel in the form UO_2 or PuO_2.

oxide mineral [MINERAL] A naturally occurring material in oxide form such as silicon dioxide, SiO_2, magnetite, Fe_3O_4, or lime, CaO.

oxide nuclear fuel [NUCLEO] The fissionable nuclear fuel UO_2 or PuO_2.

oxidized cellulose *See* oxycellulose.

oxidized microcrystalline wax [MATER] Refined, oxidized wax from bottoms of storage tanks for solvent-extracted petroleum; used in floor polishes.

oxidized shale *See* burnt shale.

oxidized zone [GEOL] A region of mineral deposits which has been altered by oxidizing surface waters.

oxidizing agent [CHEM] Compound that gives up oxygen easily, removes hydrogen from another compound, or attracts negative electrons.

oxidizing atmosphere [CHEM] Gaseous atmosphere in which an oxidation reaction occurs; usually refers to the oxidation of solids.

oxidizing flame [CHEM] A flame, or the portion of it, that contains an excess of oxygen.

oxidoreductase [BIOCHEM] An enzyme catalyzing a reaction in which two molecules of a compound interact so that one molecule is oxidized and the other reduced, with a molecule of water entering the reaction.

oxime [ORG CHEM] Compound containing the CH(:NOH)

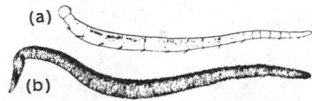

OWENIIDAE

Myriochele. (a) Entire ovigerous individual. *(b)* The tapering tube in which the mud-swallowing worm is securely contained.

OWL

Great horned owl (*Bubo virginianus*), a nocturnal bird of prey.

OXAZOLE

Structure and numbering system for a representative oxazole, 1,3-oxazole.

radical; condensation product of hydroxylamine with aldehydes or ketones.

oximeter [MED] A photoelectric photometer used to measure the oxygenated fraction of the hemoglobin in blood which is either circulating in a particular tissue of an intact animal or human being, or during, or shortly after, its withdrawal from the vascular system, by observation of the absorption of light transmitted through or reflected from the blood.

oximetry [PHYSIO] Measurement of the degree of oxygen saturation of the blood.

oximido *See* nitroso.

oxine [ORG CHEM] C_9H_6NOH White powder that darkens when exposed to light; slightly soluble in water, dissolves in ethanol, acetone, and benzene; used to prepare fungicides and to separate metals by precipitation. Also known as 8-hydroxyquinoline; oxyquinoline; 8-quinolinol.

oxo- [ORG CHEM] Chemical prefix designating the keto group, C:O.

S-oxodiallyl sulfide *See* allicin.

oxoethanoic acid *See* glyoxalic acid.

***para*-oxon** [ORG CHEM] $(C_2H_5O)_2P(O)OC_6H_4NO_2$ A toxic, yellow oil, the oxygen analog of parathion; slightly soluble in water, soluble in organic solvents; used as an insecticide. Also known as diethyl *para*-nitrophenyl phosphate.

oxonium ion *See* hydronium ion.

oxo process [CHEM ENG] Catalytic process for production of alcohols, aldehydes, and other oxygenated organic compounds by reaction of olefin vapors with carbon monoxide and hydrogen.

oxosilane *See* siloxane.

oxy- [CHEM] **1.** Prefix indicating the oxygen radical $(-O-)$ in a chemical compound. **2.** Prefix incorrectly used as a substitute for hydroxy-.

oxyacetylene cutting [ENG] The flame cutting of ferrous metals in which the preheating of the metal is accomplished with a flame produced by an oxyacetylene torch. Also known as acetylene cutting.

oxyacetylene torch [ENG] An acetylene gas-mixing and burning tool that produces a hot flame for the welding or cutting of metal. Also known as acetylene torch.

oxyacetylene welding [MET] A welding process in which the heat is supplied by an oxyacetylene flame. Also known as acetylene welding.

Oxyaenidae [PALEON] An extinct family of mammals in the order Deltatheridea; members were short-faced carnivores with powerful jaws.

oxybiotite [MINERAL] Phenocrystic biotite with increased amounts of Fe(III).

oxycellulose [MATER] Cellulose mixed with reaction products from oxidation of cellulose in the presence of steam or alkalies or by strong sunlight. Also known as oxidized cellulose.

oxycephaly [MED] A condition in which the head assumes a roughly conical shape due to premature closure of the coronal or lambdoid sutures, or to artificial pressure on the frontal and occipital regions of the infant's head. Also known as acrocephaly.

oxychloride cement [MATER] A strong, hard cement composed of magnesium chloride and calcined magnesia; used for floors and stucco. Also known as Sorel cement.

oxyethylsulfonic acid *See* isethionic acid.

oxygen [CHEM] A gaseous chemical element, symbol O, atomic number 8, and atomic weight 15.9994; an essential element in cellular respiration and in combustion processes; the most abundant element in the earth's crust, and about 20% of the air by volume.

oxygen-18 [NUC PHYS] Oxygen isotope with atomic weight 18; found 8 parts to 10,000 of oxygen-16 in water, air, and rocks; used in tracer experiments. Also known as heavy oxygen.

oxygen absorbent [CHEM] Any material that will absorb (dissolve) oxygen into its body without reacting with it.

oxygenase [BIOCHEM] An oxidoreductase that catalyzes the direct incorporation of oxygen into its substrate.

oxygenate [CHEM] To treat, infuse, or combine with oxygen.

oxygenated oil [MATER] A class of essential oils containing carbon, hydrogen, and oxygen; an example is oil of cassia.

oxygen bomb calorimeter [ENG] Device to measure heat of combustion; the sample is burned with oxygen in a closed vessel, and the temperature rise is noted.

oxygen cell *See* aeration cell.

oxygen corrosion [MET] The reaction of oxygen with metallic surfaces to form an oxide of the metal or alloy.

oxygen distribution [OCEANOGR] The concentration of dissolved oxygen in ocean water as a function of depth, ranging from as much as 5 milliliters of oxygen per liter at the surface to a fraction of that value at great depths.

oxygen-flask method [ANALY CHEM] Technique to determine the presence of combustible elements; the sample is burned with oxygen in a closed flask, and combustion products are absorbed in water of dilute alkali with subsequent analysis of the solution.

oxygen-free copper [MET] Pure copper having a conductivity greater than that of copper containing impurities such as cuprous oxide; used for the construction of high-power electron tubes because it does not release appreciable gas when hot.

oxygen furnace steel [MET] Steel made by a process in which oxygen under pressure is directed onto or into the molten metal.

oxygen gouging *See* flame gouging.

oxygen-kerosine burner [ENG] Liquid-fuel device using a mixture of oxygen and vaporized or atomized kerosine for combustion.

oxygen lance [MET] A pipe used to direct oxygen under pressure into a bath of molten steel.

oxygen mask [ENG] A mask that covers the nose and mouth and is used to administer oxygen.

oxygen minimum layer [HYD] A subsurface layer in which the content of dissolved oxygen is very low (or absent), lower than in the layers above and below.

oxygen point [THERMO] The temperature at which liquid oxygen and its vapor are in equilibrium, that is, the boiling point of oxygen, at standard atmospheric pressure; it is taken as a fixed point on the International Practical Temperature Scale of 1968, at -182.962°C.

oxygen propellant [MATER] A propellant having a minimum assay by volume of 99.9% oxygen when gasified.

oxygen ratio *See* acidity coefficient.

oxygen steelmaking [MET] The manufacture of steel from molten pig iron and steel scrap by methods which employ pure oxygen gas (99 + %) and suitable fluxes to remove carbon and phosphorus (and in part, sulfur) without introducing nitrogen or hydrogen.

oxyhemoglobin [BIOCHEM] The red crystalline pigment formed in blood by the combination of oxygen and hemoglobin, without the oxidation of iron.

oxyhornblende *See* basaltic hornblende.

oxyhydrogen flame [CHEM] A flame obtained from the combustion of a mixture of oxygen and hydrogen.

oxyhydrogen welding [MET] Welding with an oxyhydrogen flame.

oxylophyte [ECOL] A plant that thrives in or is restricted to acid soil.

oxyl process [CHEM ENG] Modified Fischer-Tropsch process used to make alcohols, other oxygenated compounds, paraffins, and olefin hydrocarbons from carbon monoxide and hydrogen.

Oxymonadida [INV ZOO] An order of xylophagous protozoans in the class Zoomastigophorea; colorless flagellate symbionts in the digestive tract of the roach *Cryptocercus* and of certain termites.

β-oxynaphthoic acid *See* β-hydroxynaphthoic acid.

oxyneurine *See* betaine.

oxyntic cell *See* parietal cell.

4-oxypentanoic acid *See* levulinic acid.

oxypetalous [BOT] Having sharp-pointed petals.

oxyphilia *See* eosinophilia.

oxyphytia [ECOL] Discordant habitat control due to an excessively acidic substratum.

oxyquinoline *See* oxine.

oxysphere *See* lithosphere.

Oxystomata [INV ZOO] A subsection of the Brachyura, including those true crabs in which the first pair of pereiopods

OXYGEN

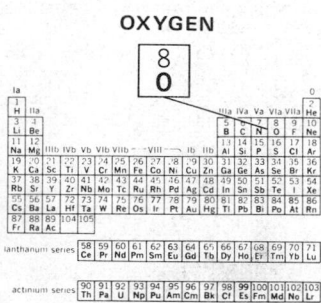

Periodic table of the chemical elements showing the position of oxygen.

OXYMONADIDA

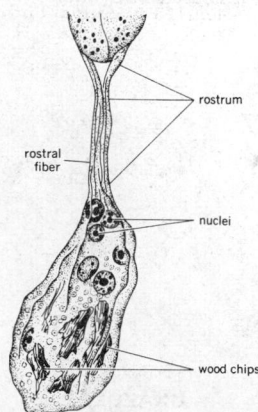

rostrum

rostral fiber

nuclei

wood chips

An oxymonad, *Microrhopalodina inflata*. The pliable necklike rostrum attaches the organism to the host's intestinal wall. The ingested wood chips taken in by the host are shown. (*After McCracken*)

is chelate, and the mouth frame is triangular and forward.

Oxystomatidae [INV ZOO] A family of free-living marine nematodes in the superfamily Enoploidea, distinguished by amphids that are elongated longitudinally.

oxytetracycline [MICROBIO] $C_{22}H_{24}O_9N_2$ A crystalline, amphoteric, broad-spectrum antibiotic produced by *Streptomyces rimosus*; produced commercially by fermentation.

oxytocic [MED] Hastening parturition. [PHARM] A drug that hastens parturition.

oxytocin [BIOCHEM] $C_{43}H_{66}O_{12}N_{12}S_2$ A polypeptide hormone secreted by the neurohypophysis that stimulates contraction of the uterine muscles.

Oxyurata [INV ZOO] The equivalent name for Oxyurina.

Oxyuridae [INV ZOO] A family of the nematode superfamily Oxyuroidea.

Oxyurina [INV ZOO] A suborder of nematodes in the order Ascaridida.

Oxyuroidea [INV ZOO] A superfamily of the class Nematoda.

Oyashio [OCEANOGR] A cold current flowing from the Bering Sea southwest along the coast of Kamchatka, past the Kuril Islands, continuing close to the northeast coast of Japan, and reaching nearly 35°N.

oyster [INV ZOO] Any of various bivalve mollusks of the family Ostreidae; the irregular shell is closed by a single adductor muscle, the foot is small or absent, and there is no siphon.

oz *See* ounce.

oz ap *See* ounce.

oz apoth *See* ounce.

Ozawainellidae [PALEON] A family of extinct protozoans in the superfamily Fusulinacea.

ozocerite [GEOL] A natural, brown to jet black paraffin wax occurring in irregular veins; consists principally of hydrocarbons, is soluble in water, and has a variable melting point. Also known as ader wax; earth wax; fossil wax; mineral wax; native paraffin; ozokerite.

ozokerite *See* ozocerite.

ozone [CHEM] O_3 Unstable blue gas with pungent odor; an allotropic form of oxygen; a powerful oxidant boiling at -112°C; used as an oxidant, bleach, and water purifier, and to treat industrial wastes.

ozone cloud [METEOROL] A limited region in which the total ozone content of the ozonosphere is greater than normal.

ozone generator [ENG] Apparatus that converts oxygen, O_2, into ozone, O_3, by subjecting the oxygen to an electric-brush discharge. Also known as ozonizer.

ozone layer *See* ozonosphere.

ozonide [ORG CHEM] Any of the oily, thick, unstable compounds formed by reaction of ozone with unsaturated compounds; an example is oleic ozonide from the reaction of oleic acid and ozone.

ozonization [CHEM] The process of treating, impregnating, or combining with ozone.

ozonizer *See* ozone generator.

ozonosphere [METEOROL] The general stratum of the upper atmosphere in which there is an appreciable ozone concentration and in which ozone plays an important part in the radiative balance of the atmosphere; lies roughly between 10 and 50 kilometers, with maximum ozone concentration at about 20 to 25 kilometers. Also known as ozone layer.

oz t *See* ounce.

oz tr *See* ounce.

OYSTER

Valve of an oyster.

OZAWAINELLIDAE

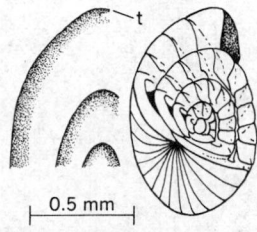

0.5 mm

A representative of the Ozawainellidae. The initial wall, the tectum (t), is shown on the left, and the calcareous shell on the right.

P

p- *See* para-.

P *See* phosphorus; poise.

P$_1$ [GEN] The parental generation; parents of the F$_1$ generation.

pA *See* picoampere.

Pa *See* pascal; protactinium.

Paal-Knorr synthesis [ORG CHEM] A method of converting a 1,4-dicarbonyl compound by cyclization with ammonia or a primary amine to a pyrrole.

Paar turbidimeter [ANALY CHEM] A visual-extinction device for measurement of solution turbidity; the length of the column of liquid suspension is adjusted until the light filament can no longer be seen.

PABA *See* para-aminobenzoic acid.

PABX *See* private automatic branch exchange.

paca [VERT ZOO] Any of several rodents of the genus *Cuniculus*, especially *C. paca*, with a white-spotted brown coat, found in South and Central America.

Pacchionian bodies *See* arachnoidal granulations.

pacemaker [MED] An electric device that functions to regulate the pace of the heartbeat. [PHYSIO] Any body structure, such as the sinoatrial node, that functions in the establishment and maintenance of a rhythmic activity. [SCI TECH] Any material or object that influences the rate of a process or a reaction.

pace rating *See* effort rating.

pachnolite [MINERAL] NaCaAlF$_6$·H$_2$O Colorless to white mineral composed of hydrous sodium calcium aluminum fluoride, occurring in monoclinic crystals.

pachoidal structure *See* flaser structure.

Pachuca tank [CHEM ENG] Air-agitated, solid-liquid mixing vessel in which the air is injected into the bottom of a center draft tube; air and solids rise through the tube, with solids exiting the top of the tube and falling through the bulk of the liquid.

pachycephalosaur [PALEON] A bone-headed dinosaur, composing the family Pachycephalosauridae.

Pachycephalosauridae [PALEON] A family of ornithischian dinosaurs characterized by a skull with a solid rounded mass of bone 10 centimeters thick above the minute brain cavity.

pachyderm [VERT ZOO] Any of various nonruminant hooved mammals characterized by thick skin, including the elephants, hippopotamuses, rhinoceroses, and others.

pachydermatous [MED] Abnormally thick-skinned.

pachydermia [MED] Abnormal thickening of the skin.

pachyglossal [VERT ZOO] Of lizards, having a thick tongue.

pachyglossia [MED] Abnormal thickness of the tongue.

pachygyria [MED] A malformation of the brain characterized by its being too broad in form.

pachymeningitis [MED] Inflammation of the dura mater.

pachymeter [ENG] An instrument used to measure the thickness of a material, for example, a sheet of paper.

pachynema *See* pachytene.

pachytene [CYTOL] The third stage of meiotic prophase during which paired chromosomes thicken, each chromosome splits into chromatids, and breakage and crossing over between nonsister chromatids occur. Also known as pachynema.

Pacific anticyclone *See* Pacific high.

Pacific Equatorial Countercurrent [OCEANOGR] The Equatorial Countercurrent flowing east across the Pacific Ocean between 3° and 10°N.

Pacific faunal region [ECOL] A marine littoral faunal region including offshore waters west of Central America, running from the coast of South America at about 5° south latitude to the southern tip of California.

Pacific high [METEOROL] The nearly permanent subtropical high of the North Pacific Ocean, centered, in the mean, at 30–40°N and 140–150°W. Also known as Pacific anticyclone.

Pacific North Equatorial Current [OCEANOGR] The North Equatorial Current which flows westward between 10° and 20°N in the Pacific Ocean.

Pacific Ocean [GEOGR] The largest division of the hydrosphere, having an area of 165,000,000 square kilometers and covering 46% of the surface of the total extent of the oceans and seas; it is bounded by Asia and Australia on the west and North and South America on the east.

Pacific South Equatorial Current [OCEANOGR] The South Equatorial Current flowing westward between 3°N and 10°S in the Pacific Ocean.

Pacific Standard Time *See* Pacific time.

Pacific suite [PETR] A large group of igneous rocks characterized by calcic and calc-alkalic rocks, especially in the region of the circum-Pacific orogenic belt. Also known as anapeirean; circum-Pacific province.

Pacific temperate faunal region [ECOL] A marine littoral faunal region including a narrow zone in the North Pacific Ocean, from Indochina to Alaska and along the west coast of the United States to about 40° north latitude.

Pacific time [ASTRON] The time for a given time zone that is based on the 120th meridian and is the eighth zone west of Greenwich. Also known as Pacific Standard Time.

Pacific-type continental margin [GEOL] A continental margin typified by that of the western Pacific where oceanic lithosphere descends beneath an adjacent continent and produces an intervening island arc system.

Pacinian corpuscle [ANAT] An encapsulated lamellar sensory nerve ending that functions as a kinesthetic receptor.

pack [ADP] To reduce the amount of storage required to hold information by changing the method of encoding the data. [IND ENG] To provide protection for an article or group of articles against physical damage during shipment; packing is accomplished by placing articles in a shipping container, and blocking, bracing, and cushioning them when necessary, or by strapping the articles or containers on a pallet or skid. [ORD] Part of a parachute assembly in which the canopy and shroud lines are folded and carried. Also known as pack assembly. [MIN ENG] **1.** A pillar built in the waste area or roadside within a mine to support the mine roof; constructed from loose stones and dirt. **2.** Waste rock or timber used to support the roof or underground workings or used to fill excavations. Also known as fill. [OCEANOGR] *See* pack ice.

packaged circuit *See* rescap.

packaged magnetron [ELECTR] Integral structure comprising a magnetron, its magnetic circuit, and its output matching device.

package freight [IND ENG] Freight shipped in lots insufficient to fill a complete car; billed by the unit instead of by the carload.

package power reactor [NUC PHYS] A small nuclear power plant designed to be crated in packages small enough for transportation to remote locations.

packaging [ELEC] The process of physically locating, connecting, and protecting devices or components.

packaging density [ELECTR] The number of components per unit volume in a working system or subsystem.

pack artillery [ORD] Artillery weapons designed for transport in sections by animals or delivery by parachute; the weapon and carriage are partially disassembled for transport and reassembled for firing from ground positions.

pack assembly *See* pack.

pack builder [MIN ENG] **1.** One who builds packs or pack walls. **2.** In anthracite and bituminous coal mining, one who fills worked-out rooms, from which coal has been mined, with

PACHYCEPHALOSAURIDAE

Head of *Stegocerus*, a member of the Pachycephalosauridae, Late Cretaceous, Canada. Dome above level of eyes is solid bone.

rock, slate, or other waste to prevent caving of walls and roofs, or who builds rough walls and columns of loose stone, heavy boards, timber, or coal along haulageways and passageways and in rooms where coal is being mined to prevent caving of roof or walls during mining operations. Also known as packer; pillar man; timber packer; waller.

pack carburizing [MET] A method of surface hardening of steel in which parts are packed in a steel box with the carburizing compound and heated to elevated temperatures.

packed bed [CHEM ENG] A fixed layer of small particles or objects arranged in a vessel to promote intimate contact between gases, vapors, liquids, solids, or various combinations thereof; used in catalysis, ion exchange, sand filtration, distillation, absorption, and mixing.

packed decimal [ADP] A means of representing two digits per character, to reduce space and increase transmission speed.

packed tower [CHEM ENG] A fractionating or absorber tower filled with small objects (packing) to bring about intimate contact between rising fluid (vapor or liquid) and falling liquid.

packed tube [CHEM ENG] A pipe or tube filled with high-heat-capacity granular material; used to heat gases when tubes are externally heated.

packer *See* pack builder; production packer.

packet [BIOL] A cluster of organisms in the form of a cube resulting from cell division in three planes.

packet gland [INV ZOO] A cluster of gland cells opening through the epidermis of nemertines.

packfong [MET] Chinese name for an alloy of nickel, zinc, and copper, which resembles nickel silver.

pack ice [OCEANOGR] Any area of sea ice, except fast ice, composed of a heterogeneous mixture of ice of varying ages and sizes, and formed by the packing together of pieces of floating ice. Also known as ice canopy; ice pack; pack.

packing [CRYSTAL] The arrangement of atoms or ions in a crystal lattice. [ENG] *See* stuffing. [ENG ACOUS] Excessive crowding of carbon particles in a carbon microphone, produced by excessive pressure or by fusion particles due to excessive current, and causing lowered resistance and sensitivity. [GEOL] The arrangement of solid particles in a sediment or in sedimentary rock.

packing density [ADP] The amount of information per unit of storage medium, as characters per inch on tape, bits per inch or drum, or bits per square inch in photographic storage.

packing fraction [NUC PHYS] The quantity $(M-A)/A$, where M is the mass of an atom in atomic mass units and A is its atomic number.

packing house [FOOD ENG] **1.** A food processing plant generally requiring the use of refrigeration. **2.** A building in which livestock are slaughtered and processed, and the meat products and by-products are packed.

packing house pitch [MATER] Dark-brown to black by-product residue from manufacturing soap and candle stock or from refining vegetable oils, refuse, or wool grease; soluble in naphtha and carbon disulfide; used to make paints, varnishes, and tar paper, and in marine caulking and waterproofing. Also known as fatty-acid pitch.

packing index [CRYSTAL] The volume of ion divided by the volume of the unit cell in a crystal.

packing material [MET] In powder metallurgy, a material in which compacts are embedded during presintering or sintering operations.

packing radius [CRYSTAL] One-half the smallest approach distance of atoms or ions.

packing ring *See* piston ring.

packing routine [ADP] A subprogram which compresses data so as to eliminate blanks and reduce the storage needed for a file.

pack rolling [MET] Hot rolling of two or more sheets of metal packed together; a thin surface oxide film prevents their welding.

pack unit [COMMUN] A compact, combination radio transmitter-receiver that can be carried or strapped on the back; some pack units are popularly known as walkie-talkies.

pack wall [MIN ENG] A wall of dry stone built along the side of a roadway, or in the waste area, of a coal or metal mine to

help support the roof and to retain the packing material and prevent its spreading into the roadway.

pactamycin [MICROBIO] An antitumor and antibacterial antibiotic produced by *Streptomyces pactum* var. *pactum*.

pad [AERO ENG] *See* launch pad. [ANAT] A small circumscribed mass of fatty tissue, as in terminal phalanges of the fingers or the underside of the toes of an animal, such as a dog. [ELECTR] **1.** An arrangement of fixed resistors used to reduce the strength of a radio-frequency or audio-frequency signal by a desired fixed amount without introducing appreciable distortion. Also known as fixed attenuator. **2.** *See* terminal area. [ENG] **1.** A layer of material used as a cushion or for protection. **2.** A projection of excess metal on a casting forging, or welded part. **3.** A takeoff or landing point for a helicopter or space vehicle. [MET] The brickwork that is beneath the molten iron at the base of a blast furnace.

padder [ELECTR] A trimmer capacitor inserted in series with the oscillator tuning circuit of a superheterodyne receiver to control calibration at the low-frequency end of a tuning range.

paddle [AERO ENG] A large, flat, paddle-shaped support for solar cells, used on some satellites. [DES ENG] Any of various implements consisting of a shaft with a broad, flat blade or bladelike part at one or both ends.

paddle wheel [MECH ENG] **1.** A device used to propel shallow-draft vessels, consisting of a wheel with paddles or floats on its circumference, the wheel rotating in a plane parallel to the ship's length. **2.** A wheel with paddles used to move leather in a processing vat.

paddle-wheel steamer [NAV ARCH] A steamer propelled by a wheel or wheels having long paddles, some of which are curved and feathering; the wheel revolves in a vertical plane parallel to the length of the ship; one type of craft has a wide stern wheel, and the other type has two narrow side wheels.

pad dyeing [TEXT] A process in which fabrics are dyed by passing them between rollers.

Padé table [MATH] A table associated to a power series having in its pth row and qth column the ratio of a polynomial of degree q by one of degree p so that this fraction expanded into a power series agrees with the original up to the $p + q$ term.

padlock [DES ENG] An unmounted lock with a shackle that can be opened and closed; the shackle is usually passed through an eye, then closed to secure a hasp.

padparadsha *See* orange sapphire.

paedogamy [INV ZOO] A type of autogamy in certain protozoans whereby there is mutual fertilization of gametes derived from a single cell.

paedomorphosis [EVOL] Phylogenetic change in which adults retain juvenile characters, accompanied by an increased capacity for further change; indicates potential for further evolution.

Paenungulata [VERT ZOO] A superorder of mammals, including proboscideans, xenungulates, and others.

Paeoniaceae [BOT] A monogeneric family of dicotyledonous plants in the order Dilleniales; members are mesophyllic shrubs characterized by cleft leaves, flowers with an intrastaminal disk, and seeds having copious endosperm.

paesa [METEOROL] A violent north-northeast wind of Lake Garda in Italy.

paesano [METEOROL] A northerly night breeze, blowing down from the mountains, of Lake Garda in Italy.

page [ADP] **1.** A standard quantity of main-memory capacity, usually 512 to 4096 bytes or words, used for memory allocation and for partitioning programs into control sections. **2.** A standard quantity of source program coding, usually 8 to 64 lines, used for displaying the coding on a cathode-ray tube.

page boundary [ADP] The address of the first (lowest) word or byte within a page of memory.

page printer [COMMUN] A high-speed printer used to transpose messages received on paper tape to full-page format, by printing characters one at a time.

page proof [GRAPHICS] A proof received from a compositor after the galley, and having the form of the final page, usually including any illustrations.

page reader [ADP] In character recognition, a character

PAEONIACEAE

The garden peony *(Paeonia lactiflora)*, member of the family Paeoniaceae in the order Dilleniales. *(Courtesy of F. E. Westlake, from National Audubon Society)*

reader capable of processing cut-form documents of varying sizes; sometimes capable of reading information in reel forms.

page table [ADP] A key element in the virtual memory technique; a table of addresses where entries are adjusted for easy relocation of pages.

Paget's cells [PATH] Large, epithelial cells with clear cytoplasm found in certain breast and skin cancers.

Paget's disease [MED] **1.** A type of carcinoma of the breast that involves the nipple or areola and the larger ducts, characterized by the presence of Paget's cells. **2.** Osseous hyperplasia simultaneous with accelerated deossification. **3.** An apocrine gland skin cancer, composed principally of Paget's cells.

page turning [ADP] **1.** The process of moving entire pages of information between main memory and auxiliary storage, usually to allow several concurrently executing programs to share a main memory of inadequate capacity. **2.** In conversational time-sharing systems, the moving of programs in and out of memory on a round-robin, cyclic schedule so that each program may use its allotted share of computer time.

pagination [GRAPHICS] The art of planning page format to allow sequence page numbering.

paging [ADP] The scheme used to locate pages, to move them between main storage and auxiliary storage, or to exchange them with pages of the same or other computer programs; used in computers with virtual memories.

paging system [COMMUN] A system which gives an indication to a particular individual that he is wanted at the telephone, such as by sounding a number on musical gongs, calling by name over a loudspeaker, or producing an audible signal in a radio receiver carried in the individual's pocket.

pagodite *See* agalmatolite.

Paguridae [INV ZOO] The hermit crabs, a family of decapod crustaceans belonging to the Paguridea.

Paguridea [INV ZOO] A group of anomuran decapod crustaceans in which the abdomen is nearly always asymmetrical, being either soft and twisted or bent under the thorax.

pahoehoe [GEOL] A type of lava flow whose surface is glassy, smooth, and undulating; the lava is basaltic, glassy, and porous. Also known as ropy lava.

paigeite [MINERAL] $(Fe,Mg)FeBO_5$ A black mineral composed of iron magnesium borate, occurring as fibrous aggregates.

pail [DES ENG] A cylindrical or slightly tapered container.

pain [PHYSIO] Patterns of somesthetic sensation, generally unpleasant, or causing suffering or distress.

pain spot [PHYSIO] Any of the small areas of skin overlying the endings of either very small myelinated (delta) or unmyelinated (C) nerve fibers whose stimulation, depending on the intensity and duration, results in the sensation of either pain or itching.

paint [ELECTR] Vernacular for a target image on a radarscope. [MATER] A mixture of a pigment and a vehicle, such as oil or water, that together form a liquid or paste that can be applied to a surface to provide an adherent coating that imparts color to and often protects the surface.

paint clay [MATER] A light-yellow to dark-reddish-brown iron- or manganese-bearing clay that mixes easily with linseed oil.

painter [METEOROL] A fog frequently experienced on the coast of Peru; the brownish deposit which it often leaves upon exposed surfaces is sometimes called Peruvian paint. Also known as Callao painter.

pain threshold [PHYSIO] The lowest limit for the perception of pain sensations.

paint pot [GEOL] A mud pot containing multicolored mud.

paint remover [MATER] Liquid or paste formulation used to remove dried paint, varnish, enamel, or lacquer; contains solvents such as methanol, ethyl alcohol, acetone, toluene, benzene, and ethyl acetate.

paint vehicle [MATER] The liquid constituent of paint; consists of volatile solvent or thinner and a film-forming component.

pair [ELEC] Two like conductors employed to form an electric circuit. [MECH ENG] Two parts in a kinematic mechanism that mutually constrain relative motion; for example, a sliding pair composed of a piston and cylinder. [SCI TECH] A

set of two things that are identical or nearly so, or are designed to function as a unit.

paired cable [ELEC] Cable in which the single conductors are twisted together in groups of two, none of which is arranged with others to form quads.

pairing [ELECTR] In television, imperfect interlace of lines composing the two fields of one frame of the picture; instead of having the proper equal spacing, the lines appear in groups of two.

pairing element [MECH ENG] Either of two machine parts connected to permit motion.

pair production [PHYS] The conversion of a photon into an electron and a positron when the photon traverses a strong electric field, such as that surrounding a nucleus or an electron.

Palaeacanthaspidoidei [PALEON] A suborder of extinct, placoderm fishes in the order Rhenanida; members were primitive, arthrodire-like species.

Palaeacanthocephala [INV ZOO] An order of the Acanthocephala including parasitic worms characterized by fragmented nuclei in the hypodermis, lateral placement of the chief lacunar vessels, and proboscis hooks arranged in long rows.

Palaechinoida [PALEON] An extinct order of echinoderms in the subclass Perischoechinoidea with a rigid test in which the ambulacra bevel over the adjoining interambulacra.

Palaemonidae [INV ZOO] A family of decapod crustaceans in the group Caridea.

Palaeocaridacea [INV ZOO] An order of crustaceans in the superorder Syncarida.

Palaeocaridae [INV ZOO] A family of the crustacean order Palaeocaridacea.

Palaeoconcha [PALEON] An extinct order of simple, smooth-hinged bivalve mollusks.

Palaeocopida [PALEON] An extinct order of crustaceans in the subclass Ostracoda characterized by a straight hinge and by the anterior location for greatest height of the valve.

Palaeognathae [VERT ZOO] The ratites, making up a superorder of birds in the subclass Neornithes; merged with the Neognathae in some systems of classification.

Palaeoisopus [PALEON] A singular, monospecific, extinct arthropod genus related to the pycnogonida, but distinguished by flattened anterior appendages.

Palaeomastodontinae [PALEON] An extinct subfamily of elaphantoid proboscidean mammals in the family Mastodontidae.

Palaeomerycidae [PALEON] An extinct family of pecoran ruminants in the superfamily Cervoidea.

Palaeonemertini [INV ZOO] A family of the class Anopla distinguished by the two- or three-layered nature of the body-wall musculature.

Palaeonisciformes [PALEON] A large extinct order of chondrostean fishes including the earliest known and most primitive ray-finned forms.

Palaeoniscoidei [PALEON] A suborder of extinct fusiform fishes in the order Palaeonisciformes with a heavily ossified exoskeleton and thick rhombic scales on the body surface.

Palaeopantopoda [PALEON] A monogeneric order of extinct marine arthropods in the subphylum Pycnogonida.

Palaeopneustidae [INV ZOO] A family of deep-sea echinoderms in the order Spatangoida characterized by an oval test, long spines, and weakly developed fascioles and petals.

Palaeopterygii [VERT ZOO] An equivalent name for the Actinopterygii.

Palaeoryctidae [PALEON] A family of extinct insectivorous mammals in the order Deltatheridia.

Palaeospondyloidea [PALEON] An ordinal name assigned to the single, tiny fish *Palaeospondylus,* known only from Middle Devonian shales in Carthness, Scotland.

Palaeotheriidae [PALEON] An extinct family of perissodactylous mammals in the superfamily Equoidea.

palaeotheriodont [VERT ZOO] Being or having lophodont teeth with longitudinal external tubercles that are connected with inner tubercles by transverse oblique crests.

palaeotropical *See* paleotropical.

palagonite [GEOL] A brown to yellow altered basaltic glass found as interstitial material or amygdules in pillow lavas.

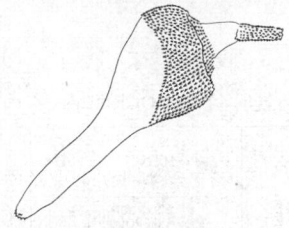

PALAEACANTHOCEPHALA

Corynosoma reductum, a palaeacanthocephalan. The anterior end of the body is provided with spines of various forms which aid in attachment to the host animal. (From H. J. Van Cleave, Acanthocephala of North American Mammals, University of Illinois Press, 1953)

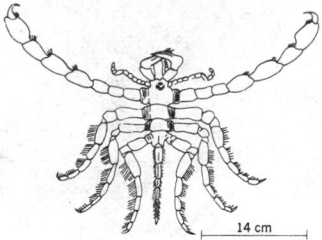

PALAEOISOPUS

Palaeoisopus problematicus. (After Broili)

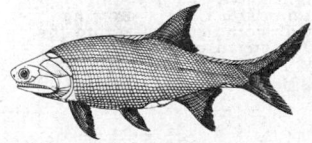

PALAEONISCOIDEI

Elonichthys robisoni Hibbert (Palaeoniscoidei), a Lower Carboniferous palaeoniscoid from Scotland, which attained a length of 12 inches (30 centimeters). (After R. H. Traquair, The ganoid fishes of the British Carboniferous formations, Palaeontographical Society, vol. 55, 1901)

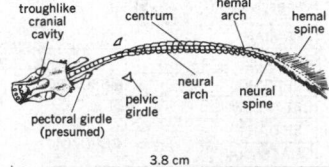

PALAEOSPONDYLOIDEA

Skeleton of Middle Devonian *Palaeospondylus,* shown dorsally. (After J. A. Moy-Thomas)

PALEOCENE

PRECAMBRIAN	PALEOZOIC						CARBON-IFEROUS				MESOZOIC			CENOZOIC	
	CAMBRIAN	ORDOVICIAN	SILURIAN	DEVONIAN	Mississippian	Pennsylvanian		PERMIAN	TRIASSIC	JURASSIC	CRETACEOUS			TERTIARY	QUATERNARY

TERTIARY					QUATERNARY	
Paleocene	Eocene	Oligocene	Miocene	Pliocene	Pleistocene	Recent

Chart showing the relationship of the Paleocene to the eras and periods of geologic time.

PALEOCOPA

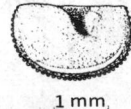

1 mm

Exterior and interior views of right valve of *Eurychilina subradiata* Ulrich, showing wide complete frill; from the Ordovician of Minnesota.

PALEOZOIC

PRECAMBRIAN		
CAMBRIAN	PALEOZOIC	
ORDOVICIAN		
SILURIAN		
DEVONIAN		
Mississippian	CARBON-IFEROUS	
Pennsylvanian		
PERMIAN		
TRIASSIC	MESOZOIC	
JURASSIC		
CRETACEOUS		
TERTIARY	CENOZOIC	
QUATERNARY		

Chart showing the relationship of the Paleozoic to the other eras and to the periods of geologic time.

palagonite tuff [PETR] A pyroclastic rock composed of angular fragments of palagonite.

palama [VERT ZOO] The membranous web on the feet of aquatic birds.

palasite [GEOL] The most abundant of the intermediate types of meteorites, consisting of olivine enclosed in a nickel-iron matrix.

palatal index [ANTHRO] The ratio, multiplied by 100, of the length to the breadth of the hard palate.

palate [ANAT] The roof of the mouth.

palatine bone [ANAT] Either of a pair of irregularly L-shaped bones forming portions of the hard palate, orbits, and nasal cavities.

palatine canal [ANAT] One of the canals in the palatine bone, giving passage to branches of the descending palatine nerve and artery.

palatine gland [ANAT] Any of numerous small oral glands on the palate of mammals.

palatine process [ANAT] A thick process that projects horizontally mediad from the medial aspect of the maxilla. [EMBRYO] An outgrowth on the ventromedial aspect of the maxillary process that develops into the definite palate.

palatine suture [ANAT] The median suture joining the bones of the palate.

palatine tonsil [ANAT] Either of a pair of almond-shaped aggregations of lymphoid tissue embedded between folds of tissue connecting the pharynx and posterior part of the tongue with the soft palate. Also known as faucial tonsil; tonsil.

palatomaxillary arch [ANAT] An arch formed by the palatine, maxillary, and premaxillary bones. Also known as maxillary arch.

palatomaxillary index [ANTHRO] An index denoting the form of the dental arch and palate, expressed by the formula: palatomaxillary width multiplied by 100, divided by the palatomaxillary length.

palatoquadrate [VERT ZOO] A series of bones or a cartilaginous rod constituting part of the roof of the mouth or upper jaw of most nonmammalian vertebrates.

palau [MET] A palladium-gold alloy; used as a platinum substitute in analytical chemistry.

palba wax [MATER] A grayish-yellow wax from older green leaves of the palm tree *Copernicia cerifera*.

palea [BOT] **1.** The upper, enclosing bract of a grass flower. **2.** A chaffy scale found on the receptacle of the disk flowers of some composite plants. [INV ZOO] One of the enlarged flattened setae forming the operculum of the tube of certain polychaete worms.

Paleaeodonta [VERT ZOO] A suborder of artiodactylous mammals including piglike forms such as the extinct "giant pigs" and the hippopotami.

Palearctic [ECOL] Pertaining to a biogeographic region including Europe, northern Asia and Arabia, and Africa north of the Sahara.

paleate [BOT] Having a covering of chaffy scales, as some rhizomes.

pale catechu See gambir.

paleoanthropology [ANTHRO] A branch of anthropology concerned with the study of fossil man.

paleobiochemistry [PALEON] The study of chemical processes used by organisms that lived in the geologic past.

paleobioclimatology [PALEON] The study of climatological events affecting living organisms for millennia or longer.

paleobotany [PALEON] The branch of paleontology concerned with the study of ancient and fossil plants and vegetation of the geologic past.

Paleocene [GEOL] A major worldwide division (epoch) of geologic time of the Tertiary period; extends from the end of the Cretaceous period to the Eocene epoch.

Paleocharaceae [PALEOBOT] An extinct group of fossil plants belonging to the Charophyta distinguished by sinistrally spiraled gyrogonites.

paleoclimate [GEOL] The climate of a given period of geologic time. Also known as geologic climate.

paleoclimatic sequence [GEOL] The sequence of climatic changes in geologic time; it shows a succession of oscillations between warm periods and ice ages, but superimposed on this are numerous shorter oscillations.

paleoclimatology [GEOL] The study of climates in the geologic past, involving the interpretation of glacial deposits, fossils, and paleogeographic, isotopic, and sedimentologic data.

Paleocopa [PALEON] An order of extinct ostracods distinguished by a long, straight hinge.

paleocrystic ice [HYD] Sea ice generally considered to be at least 10 years old, especially well-weathered polar ice.

paleocurrent [GEOL] Ancient fluid current flow whose orientation can be inferred by primary sedimentary structures and textures.

paleoecology [PALEON] The ecology of the geologic past.

Paleogene [GEOL] A geologic time interval comprising the Oligocene, Eocene, and Paleocene of the lower Tertiary period. Also known as Eogene.

paleogeography [GEOL] The geography of the geologic past; concerns all physical aspects of an area that can be determined from the study of the rocks.

paleogeologic map [GEOL] An areal map of the geology of an ancient surface immediately below a buried unconformity, showing the geology as it appeared at some time in the geologic past at the time the surface of unconformity was completed and before the overlapping strata were deposited.

paleogeology [GEOL] The geology of the past, applied particularly to the interpretation of the rocks at a surface of unconformity.

paleoherpetology [PALEON] The study of fossil reptiles.

paleoichnology [PALEON] The study of trace fossils in the fossil state. Also spelled palichnology.

pale oil [MATER] A petroleum lubricating or process oil refined until its color (measured by transmitted light) is straw to pale yellow.

paleolithologic map [GEOL] A paleogeologic map indicating lithologic variations at a buried horizon or within a restricted zone at a specific time in the geologic past.

paleomagnetics [GEOPHYS] The study of the direction and intensity of the earth's magnetic field throughout geologic time.

paleomagnetic stratigraphy [GEOPHYS] The use of natural remanent magnetization in the identification of stratigraphic units. Also known as magnetic stratigraphy.

Paleonthropinae [PALEON] A former subfamily of fossil man in the family Hominidae; set up to include the Neanderthalers together with Rhodesian man.

paleontology [BIOL] The study of life of the past as recorded by fossil remains.

paleopalynology [PALEON] A field of palynology concerned with fossils of microorganisms and of dissociated microscopic parts of megaorganisms.

Paleoparadoxidae [PALEON] A family of extinct hippopotamuslike animals in the order Desmostylia.

Paleoptera [INV ZOO] A section of the insect subclass Pterygota including primitive forms that are unable to flex their wings over the abdomen when at rest.

paleoslope [GEOL] The direction of initial dip of a former land surface, such as an ancient continental slope.

paleosol [GEOL] A soil horizon that formed on the surface during the geologic past, that is, an ancient soil. Also known as buried soil; fossil soil.

paleotectonic map [GEOL] Regional map that shows the structural patterns that existed during a particular period of geologic time, for example, the Lower Cretaceous in western Canada.

paleotemperature [GEOL] **1.** The temperature at which a geologic process took place in ancient past. **2.** The mean climatic temperature at a given time or place in the geologic past.

paleotopography [GEOL] The topography of a given area in the geologic past.

paleotropical [ECOL] Of or pertaining to a biogeographic region that includes the Oriental and Ethiopian regions. Also spelled palaeotropical.

Paleozoic [GEOL] The era of geologic time from the end of the Precambrian (600 million years before present) until the

1069

beginning of the Mesozoic era (225 million years before present).

paleozoology [PALEON] The branch of paleontology concerned with the study of ancient animals as recorded by fossil remains.

palette [GEOL] A broad sheet of calcite representing a solutional remnant in a cave. Also known as shield.

palichnology *See* paleoichnology.

palilalia [MED] Pathologic repetition of words or phrases.

palimpsest [PETR] Of a metamorphic rock, having remnants of the original structure or texture preserved.

palingenesis [EMBRYO] Unaltered recapitulation of ancestral features by the developing stages of an organism. [PETR] In-place formation of new magma by the melting of preexisting rock material.

palinspastic map [GEOL] A paleogeographic or paleotectonic map showing restoration of the features to their original geographic positions, before thrusting or folding of the crustal rocks.

Palinuridae [INV ZOO] The spiny lobsters or langoustes, a family of macruran decapod crustaceans belonging to the Scyllaridea.

palisade cell [BOT] One of the columnar cells of the palisade mesophyll which contain numerous chloroplasts.

Palisade disturbance [GEOL] Appalachian orogenic episode occurring during Triassic time which produced a series of faultlike basins.

palisade mesophyll [BOT] A tissue system of the chlorenchyma in well-differentiated broad leaves composed of closely spaced palisade cells oriented parallel to one another, but with their long axes perpendicular to the surface of the blade.

palisades [GEOL] A series of sharp cliffs.

Palladian window [ARCH] An arched central window with a narrower, square-headed window on each side.

palladium [CHEM] A chemical element, symbol Pd, atomic number 46, atomic weight 106.4. [MET] A white, ductile malleable metal that resembles platinum and follows it in abundance and importance of applications; does not tarnish at normal temperatures.

palladium amalgam *See* potarite.

palladium bichloride *See* palladium chloride.

palladium chloride [INORG CHEM] $PdCl_2$ or $PdCl_2 \cdot 2H_2O$ Dark-brown, deliquescent powder that decomposes at 501°C; soluble in water, alcohol, acetone, and hydrochloric acid; used in medicine, analytical chemistry, photographic chemicals, and indelible inks. Also known as palladium bichloride; palladous chloride.

palladium iodide [INORG CHEM] PdI_2 Black powder that decomposes above 100°C; soluble in potassium iodide solution, insoluble in water and alcohol. Also known as palladous iodide.

palladium monoxide *See* palladium oxide.

palladium nitrate [INORG CHEM] $Pd(NO_3)_2$ Brown, water-soluble, deliquescent salt; used as an analytical reagent. Also known as palladous nitrate.

palladium oxide [INORG CHEM] PdO Amber or black-green powder that decomposes at 750°C; soluble in dilute acids; used in chemical synthesis as a reduction catalyst. Also known as palladium monoxide.

palladous chloride *See* palladium chloride.

palladous iodide *See* palladium iodide.

palladous nitrate *See* palladium nitrate.

pallanesthesia [MED] Absence of pallesthesia, or vibration sense.

Pallas [ASTRON] An asteroid located between Mars and Jupiter, with a diameter about 480 kilometers (300 miles).

pallasite shell *See* lower mantle.

pallet [BUILD] A flat piece of wood laid in a wall to which woodwork may be securely fastened. [ENG] **1.** A lever that regulates or drives a ratchet wheel. **2.** A hinged valve on a pipe organ. **3.** A tray or platform used in conjunction with a fork lift for lifting and moving materials. [INV ZOO] One of a pair of plates on the siphon tubes of certain Bivalvia. [GRAPHICS] An instrument consisting of a flat blade with a handle, used in clay work. [MECH ENG] One of the disks or pistons in a chain pump.

palletize [IND ENG] To package material for convenient handling on a pallet or lift truck.

palletized ship [NAV ARCH] A ship designed to carry palletized cargo.

pallet stone [HOROL] A stone or jewel face on a pallet; reduces friction and wear.

pallial artery [INV ZOO] The artery that supplies blood to the mantle of a mollusk.

pallial chamber [INV ZOO] The mantle cavity in mollusks.

pallial line [INV ZOO] A mark on the inner surface of a bivalve shell caused by attachment of the mantle.

pallial nerve [INV ZOO] One of the pair of dorsal nerves that innervate the mantle in mollusks.

pallial sinus [INV ZOO] An inward bend in the posterior portion of the pallial line in bivalve mollusks.

palliative [PHARM] **1.** Having a soothing or relieving quality. **2.** A drug that soothes or relieves symptoms of a disease.

pallium [ANAT] The cerebral cortex. [INV ZOO] The mantle of a mollusk or brachiopod.

Pallopteridae [INV ZOO] A family of myodarian cyclorrhaphous dipteran insects in the subsection Acalypteratae.

pallor [MED] Paleness, especially of the skin and mucous membranes.

pall ring [CHEM ENG] A specially shaped steel ring used as packing for distillation columns.

palm [ANAT] The flexor or volar surface of the hand. [BOT] Any member of the monocotyledonous family Arecaceae; most are trees with a slender, unbranched trunk and a terminal crown of large leaves that are folded between the veins.

Palmales [BOT] An equivalent name for Arecales.

palmar [ANAT] Of or pertaining to the palm of the hand.

palmar aponeurosis [ANAT] Bundles of fibrous connective tissue which radiate from the tendons of the deep fascia of the forearm toward the proximal ends of the fingers.

palmar arch *See* deep palmar arch; superficial palmar arch.

palmarosa oil [MATER] Colorless to light-yellow, volatile essential oil with roselike aroma obtained from a rosha grass (*Cymbopogon martinii* var. *motia*); soluble in alcohol and mineral oils; main component is geraniol; used in soaps and perfumes. Also known as East Indian geranium oil; Indian grass oil; Rusa oil; Turkish geranium oil.

palmar reflex [PHYSIO] Flexion of the fingers when the palm of the hand is irritated.

palmate [BOT] Having lobes, such as on leaves, that radiate from a common point. [VERT ZOO] Having webbed toes. [ZOO] Having the distal portion broad and lobed, resembling a hand with the fingers spread.

palm butter [MATER] A reddish-yellow edible fatty oil expressed from putrid or fermented fruit pulp of the African oil palm (*Elaesis quineensis*); soluble in alcohol, ether, carbon disulfide, and chloroform; main components are palmitic acid, stearic acid, and glycerides of palmitic and oleic acids; melts at 27–42°C; used to make soaps and candles, as a lubricant, a color for butter substitutes, and an emollient. Also known as palm grease; palm oil.

Palmer scan [ELECTR] Combination of circular or raster and conical radar scans; the beam is swung around the horizon, and at the same time a conical scan is performed.

palmetto fiber [BOT] Brush or broom fiber obtained from young leafstalks of the cabbage palm tree (*Sabal palmetto*).

palm grease *See* palm butter.

palmierite [MINERAL] $(K,Na)_2Pb(SO_4)_2$ A white hexagonal mineral composed of potassium sodium lead sulfate.

palmitate [ORG CHEM] A derivative ester or salt of palmitic acid.

palmitic acid [ORG CHEM] $C_{15}H_{31}COOH$ A fatty acid; white crystals, soluble in alcohol and ether, insoluble in water; melts at 63.4°C, boils at 271.5°C (100 mm Hg); derived from spermaceti; used to make metallic palmitates and in soaps, waterproofing, and lube oils. Also known as cetylic acid; palmitinic acid.

palmitin *See* tripalmitin.

palmitinic acid *See* palmitic acid.

palmitoleic acid [ORG CHEM] $C_{16}H_{30}O_2$ An unsaturated fatty acid, found in marine animal oils; it is a clear liquid used

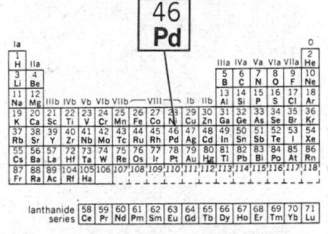

PALLADIUM

Periodic table of the chemical elements showing the position of palladium.

PALM

The common coconut palm (*Cocos nucifera*). The coconut fruit can be seen on the tree hanging below the branches.

as a standard in chromatography. Also known as *cis*-9-hexadecenoic acid.

palm kernel oil *See* palm nut oil.

palm nut [BOT] The edible seed of the African oil palm (*Elaeis guineensis*).

palm nut oil [MATER] Yellowish fatty oil; soluble in alcohol, either, carbon disulfide, and chloroform; main components are triolein and triglycerides of stearic, palmitic, myristic, lauric, and other fatty acids; used to make soap, chocolate products, margarine, cosmetics, and candles, and as an illuminant and a color for butter substitutes. Also known as palm kernel oil.

palm oil *See* palm butter.

palm wax [MATER] A yellow wax from the Ecuadoran palm (*Ceroxylon andicola*); used as a beeswax substitute.

Palmyridae [INV ZOO] A mongeneric family of errantian polychaete annelids.

palouser [METEOROL] A dust storm of northwestern Labrador.

palp [INV ZOO] Any of various sensory, usually fleshy appendages near the oral aperture of certain invertebrates.

palpable [MED] 1. Capable of being felt or touched. 2. Evident.

palpal organ [INV ZOO] An organ on the terminal joint of each pedipalp of a male spider which functions to convey sperm to the female genital orifice.

palpation [MED] Diagnostic examination by touch.

Palpatores [INV ZOO] A suborder of long-legged arachnids in the order Phalangida.

palpebra [ANAT] The eyelid.

palpebral disk [VERT ZOO] A scale, often transparent, covering the eyelid of certain lizards.

palpebral fissure [ANAT] The opening between the eyelids.

palpebral fold [ANAT] A fold formed by the reflection of the conjunctiva from the eyelid onto the eye.

Palpicornia [INV ZOO] The equivalent name for Hydrophiloidea.

palpiger [INV ZOO] The palpi-bearing portion of an insect labium.

Palpigradida [INV ZOO] An order of rare tropical and warm-temperate arachnids; all are minute, whitish, eyeless animals with an elongate body that terminates in a slender, multisegmented flagellum set with setae.

palpitate [MED] To flutter, or beat abnormally fast; applied especially to the rate of the heartbeat.

palpocil [INV ZOO] A fine, filamentous tactile hair.

palpus [INV ZOO] 1. A process on a mouthpart of an arthropod that has a tactile or gustatory function. 2. Any similar process on other invertebrates.

palsy [MED] Any of various special types of paralysis, such as cerebral palsy.

palustrine [ECOL] Being, living, or thriving in a marsh.

palygorskite [MINERAL] 1. A chain-structure type of clay mineral. 2. A group of lightweight, tough, fibrous clay minerals showing extensive substitution of aluminum for magnesium.

palynology [PALEON] The study of spores, pollen, microorganisms, and microscopic fragments of megaorganisms that occur in sediments.

PAM *See* pulse-amplitude modulation.

pamabrom [PHARM] $C_{11}H_{18}BrN_5O_3$ A water-soluble, fine white powder, decomposing at 300°C; used in medicine as a diuretic. Also known as 2-amino-2-methyl-1-propanol-8-bromotheophyllinate.

pamaquine [PHARM] $C_{19}H_{29}N_3O$ An antimalarial drug, used as the salt.

pampa [ECOL] An extensive plain in South America, usually covered with grass.

pampero [METEOROL] A wind of gale force blowing from the southwest across the pampas of Argentina and Uruguay, often accompanied by squalls, thundershowers, and a sudden drop of temperature; it is comparable to the norther of the plains of the United States.

Pamphiliidae [INV ZOO] The web-spinning sawflies, a family of hymenopteran insects in the superfamily Megalodontoidea.

pampiniform [ANAT] Of the network of veins in the spermatic cord and in the broad ligament, having the form of a tendril.

pampiniform plexus [ANAT] A venous network in the spermatic cord in the male, and in the broad ligament in the female.

pamprodactylous [VERT ZOO] Having the toes turned forward, as of certain birds.

pan [COMMUN] To tilt or otherwise move a television or movie camera vertically and horizontally to keep it trained on a moving object or to secure a panoramic effect. [GEOL] A shallow, natural depression or basin containing a body of standing water. 2. A hard, cementlike layer, crust, or horizon of soil within or just beneath the surface; may be compacted, indurated, or very high in clay content. 3. *See* pancake ice. [MIN ENG] 1. A shallow, circular, concave steel or porcelain dish in which drillers or samplers wash the drill sludge to gravity-separate the particles of heavy, dense minerals from the lighter rock powder as a quick visual means of ascertaining if the rocks traversed by the borehole contain minerals of value. 2. The act or process of performing the above operation.

panabase *See* tetrahedrite.

panacinar emphysema [MED] Emphysema characterized by diffuse destruction of one lung.

panadapter *See* panoramic adapter.

panagglutinin [IMMUNOL] An agglutinin lacking specificity, which agglutinates erythrocytes of various types.

Panama disease [PL PATH] A fungus disease of banana caused by invasion of the vascular system by *Fusarium oxysporum cubense*, resulting in yellowing and wilting of the foliage and ultimate death of the shoots.

pan-amalgamation process [MIN ENG] A process for extracting gold or silver from their ores; the ore is crushed and mixed with salt, copper sulfate, and mercury, and the gold or silver amalgamize with the mercury.

Pan-American jig [MIN ENG] Mineral jig developed to treat alluvial sands; the jig cell is pulsated vertically on a flexible diaphragm seated above the stationary hutch.

panarteritis [MED] 1. Arteritis involving all the coats of an artery. 2. *See* polyarteritis.

panarthritis [MED] Inflammation of several joints.

panas oetara [METEOROL] A strong, warm, dry north wind in February in Indonesia.

Panavision [GRAPHICS] A process for making wide-screen motion picture films which uses 65-millimeter film in the camera, and an anamorphic lens which gives a compression of 1.33:1 in the width.

pan bolt [DES ENG] A bolt with a head resembling an upside-down pan.

pancake [MIN ENG] A concrete disk employed in stope support. [OCEANOGR] *See* pancake ice.

pancake auger [DES ENG] An auger having one spiral web, 12 to 15 inches (30 to 38 centimeters) in diameter, attached to the bottom end of a slender central shaft; used as removable deadman to which a drill rig or guy line is anchored.

pancake coil [ELEC] A coil having the shape of a pancake, usually with the turns arranged in the form of a flat spiral.

pancake engine [MECH ENG] A compact engine with cylinders arranged radially.

pancake forging [MET] A rough, flat, forged shape made quickly with a minimum of tooling.

pancake ice [OCEANOGR] One or more small, newly formed pieces of sea ice, generally circular with slightly raised edges and about 1–10 feet (0.3 to 3 meters) across. Also known as lily-pad ice; pan; pancake; pan ice; plate ice.

pancake landing [AERO ENG] Landing of an aircraft at a low forward speed and at a very high rate of descent.

pancarditis [MED] Carditis involving the endocardium, myocardium, and pericardium.

Pancarida [INV ZOO] A superorder of the subclass Malacostraca; the cylindrical, cruciform body lacks an external division between the thorax and pleon and has the cephalon united with the first thoracomere.

panchromatic [GRAPHICS] Of a photographic emulsion, film, or plate, sensitive to all wavelengths within the visible spectrum, though not uniformly so.

PANCARIDA

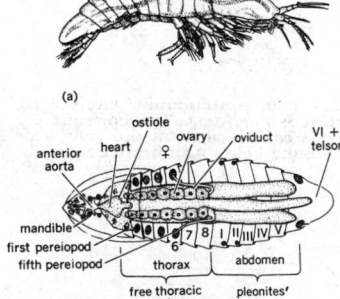

(a)

(b)

anterior aorta · heart · ♀ · ovary · oviduct · ostiole · VI + telson · mandible · first pereiopod · fifth pereiopod · thorax · abdomen · free thoracic somites · pleonites'

Female *Thermosbaena mirabilis* Monod. (a) Lateral view (after A. Bzuun). (b) Ventral section (after R. Siewing).

panchromatic film [GRAPHICS] Photographic film that is sensitive to light of all colors.

pan coefficient [METEOROL] The ratio of the amount of evaporation from a large body of water to that measured in an evaporation pan.

pan conveyor [MECH ENG] A conveyor consisting of a series of pans. [MIN ENG] *See* jigging conveyor.

pancreas [ANAT] A composite gland in most vertebrates that produces and secretes digestive enzymes, as well as at least two hormones, insulin and glucagon.

pancreatectomy [MED] Surgical removal of the pancreas.

pancreatic diarrhea [MED] Diarrhea due to deficiency of pancreatic digestive enzymes, characterized by the passage of large, greasy stools having a high fat and nitrogen content.

pancreatic diverticulum [EMBRYO] One of two diverticula (dorsal and ventral) from the embryonic duodenum or hepatic diverticulum that form the pancreas or its ducts.

pancreatic duct [ANAT] The main duct of the pancreas formed from the dorsal and ventral pancreatic ducts of the embryo.

pancreatic juice [PHYSIO] The thick, transparent, colorless secretion of the pancreas.

pancreatic lipase *See* steapsin.

pancreatin [BIOCHEM] A cream-colored, amorphous powder obtained from the fresh pancreas of a hog; contains amylopsin, trypsin, steapsin, and other enzymes.

pancreatitis [MED] Inflammation of the pancreas.

pancreozymin [BIOCHEM] A crude extract of the intestinal mucosa that stimulates secretion of pancreatic juice.

pan crusher [MECH ENG] Solids-reduction device in which one or more grinding wheels or mullers revolve in a pan containing the material to be pulverized.

pancytopenia [MED] Abnormally low numbers of all formed elements in the blood.

panda [VERT ZOO] Either of two Asian species of carnivores in the family Procyonidae; the red panda (*Ailurus fulgens*) has long, thick, red fur, with black legs; the giant panda (*Ailuropoda melanoleuca*) is white, with black legs and black patches around the eyes.

Pandanaceae [BOT] The single, pantropical family of the plant order Pandanales.

Pandanales [BOT] A monofamilial order of monocotyledonous plants; members are more or less arborescent and sparingly branched, with numerous long, firm, narrow, parallel-veined leaves that usually have spiny margins.

Pandaridae [INV ZOO] A family of dimorphic crustaceans in the suborder Caligoida; members are external parasites of sharks.

pandemic [MED] Epidemic occurring over a widespread geographic area.

pandermite *See* priceite.

Pandionidae [VERT ZOO] A monospecific family of birds in the order Falconiformes; includes the osprey (*Pandion haliaetus*), characterized by a reversible hindtoe, well-developed claws, and spicules on the scales of the feet.

pandurate [BOT] Of a leaf, having the outline of a fiddle.

pane [BUILD] A sheet of glass in a window or door. [DES ENG] One of the sides on a nut or on the head of a bolt.

panel [ADP] The face of the console, which is normally equipped with lights, switches, and buttons to control the machine, correct errors, determine the status of the various CPU (central processing unit) parts, and determine and revise the contents of various locations. Also known as control panel; patch panel. [CIV ENG] **1.** One of the divisions of a lattice girder. **2.** A sheet of material held in a frame. **3.** A distinct, usually rectangular, raised or sunken part of a construction surface or a material. [ENG] A metallic or nonmetallic sheet on which operating controls and dials of an electronic unit or other equipment are mounted. [MIN ENG] **1.** A system of coal extraction in which the ground is laid off in separate districts or panels, pillars of extra-large size being left between. **2.** A large rectangular block or pillar of coal.

panel board [ELECTR] *See* control board. [ENG] A drawing board with an adjustable outer frame that is forced over the drawing paper to hold and strain it. [MATER] A rigid paperboard used for paneling in buildings and automobile bodies.

panel code [COMMUN] Prearranged code designed for visual

communications between ground units and friendly aircraft.

panel coil *See* plate coil.

panel cooling [CIV ENG] A system in which the heat-absorbing units are in the ceiling, floor, or wall panels of the space which is to be cooled.

panel display [ELECTR] An unconventional method of displaying color television pictures in which luminescent conversion devices, such as light-emitting diodes or electroluminescent devices, are arranged in a matrix array, forming a flat-panel screen, and are controlled by signals sent over vertical and horizontal wires connected to both electrodes of the devices.

panel heating [CIV ENG] A system in which the heat-emitting units are in the ceiling, floor, or wall panels of the space which is to be heated.

panel length [CIV ENG] The distance between adjacent joints on a truss, measured along the upper or lower chord.

panel point [CIV ENG] The point in a framed structure where a vertical or diagonal member and a chord intersect.

panel system [BUILD] A wall composed of factory-assembled units connected to the building frame and to each other by means of anchors.

panel wall [BUILD] A nonbearing partition between columns or piers.

panendoscope [MED] A modification of the cystoscope, utilizing a Foroblique lens system, permitting adequate visualization of both the urinary bladder and the urethra.

Paneth cells *See* cells of Paneth.

panethite [MINERAL] A phosphate mineral known only in meteorites; contains sodium, potassium, magnesium, calcium, iron, and manganese.

Paneth's adsorption rule [PHYS CHEM] The rule that an element is strongly absorbed on a preciptiate which has a surface charge opposite in sign to that carried by the element, provided that the resulting adsorbed compound is very sparingly soluble in the solvent.

panfan *See* pediplain.

Pangea [GEOL] Postulated former supercontinent supposedly composed of all the continental crust of the earth, and later fragmented by drift into Laurasia and Gondwana.

pangene [CYTOL] A hypothetical heredity-controlling protoplasmic particle proposed by Darwin.

pangenesis [BIOL] Darwin's comprehensive theory of heredity and development, according to which all parts of the body give off gemmules which aggregate in the germ cells; during development, they are sorted out from one another and give rise to parts similar to those of their origin.

pangolin [VERT ZOO] Any of seven species composing the mammalian family Manidae; the entire dorsal surface of the body is covered with broad, horny scales, the small head is elongate, and the mouth is terminal in the snout.

pan head [DES ENG] The head of a screw or rivet in the shape of a truncated cone.

panhypopituitarism *See* hypopituitarism.

pan ice *See* pancake ice.

panicle [BOT] A branched or compound raceme in which the secondary branches are often racemose as well.

panmixis [BIOL] Random mating within a breeding population; in a closed population this results in a high degree of uniformity.

Panmycin [MICROBIO] A trade name for tetracycline.

panniculitis [MED] Inflammation of the layer of subcutaneous fat, especially in the abdomen.

panniculus [ANAT] A membrane or layer.

pannier *See* gabion.

pannose [BIOL] Having a felty or woolly texture.

pannus [MED] **1.** Vascularization accompanied by deposition of connective tissue beneath the cornal epithelium. **2.** Overgrowth of connective tissue on the articular surface of a diarthrodial joint. [METEOROL] Numerous cloud shreds below the main cloud; may constitute a layer separated from the main part of the cloud or attached to it.

panophthalmitis [MED] Inflammation of all the tissues of the eyeball.

panoramic adapter [ELECTR] A device designed to operate with a search receiver to provide a visual presentation on an

PANDA

The giant panda, noted for its black and white patches of fur.

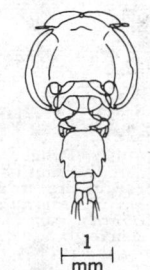

PANDARIDAE

1 mm

Pandarus satyrus Dana, male.

PANGOLIN

Giant pangolin (*Manis gigantea*), a native to tropical Africa, maximum length 5 feet (1.5 meters).

PANICLE

Drawing of a panicle showing the branching of this type of inflorescence.

oscilloscope screen of a band of frequencies extending above and below the center frequency to which the search receiver is tuned. Also known as panadapter.

panoramic display [ELECTR] A display that simultaneously shows the relative amplitudes of all signals received at different frequencies.

panoramic radar [ENG] Nonscanning radar which transmits signals over a wide beam in the direction of interest.

panoramic receiver [ELECTR] Radio receiver that permits continuous observation on a cathode-ray-tube screen of the presence and relative strength of all signals within a wide frequency range.

pan out [MIN ENG] To give a result, especially as compared with expectations; for example, in mining, the gravel may be said to pan out.

panplain [GEOL] A broad, level plain formed by coalescence of several adjacent flood plains. Also spelled panplane.

panplane See panplain.

pan-range [ELECTR] Intensity-modulated, A-type radar indication with a slow vertical sweep applied to video; stationary targets give solid vertical deflection, and moving targets give broken vertical deflection.

panspermia [BIOL] A 19th-century theory in opposition to the theory of spontaneous generation and proposing that all reproductive bodies are universal, developing wherever there is a favorable environment.

pansporoblast [INV ZOO] A sporont of cnidosporan protozoans that contains two sporoblasts.

pan tank See rundown tank.

pantellerite [PETR] A green to black extrusive rock characterized by acmite-augite or diopside, anorthoclase, and cossyrite phenocrysts in an acmite or feldspar matrix that is either pumiceous, partly glassy, fine-grained holocrystalline trachytic, or microlitic.

panting [NAV ARCH] A series of pulsations resulting from repeated minor explosions in the furnace of a ship's boiler or from vibration of a ship's plating due to sea loads.

panting beam [NAV ARCH] A beam fitted athwartship in the bow or stern of a vessel to prevent panting of the sides.

Pantodonta [PALEON] An extinct order of mammals which included the first large land animals of the Tertiary.

Pantodontidae [VERT ZOO] A family of fishes in the order Osteoglossiformes; the single, small species is known as African butterflyfish because of its expansive pectoral fins.

pantograph [ENG] A device that sits on the top of an electric locomotive or cars in an electric train and picks up electricity from overhead wires to run the train. [GRAPHICS] A drawing instrument used for copying and consisting of four rigid bars linked together in a parallelogram form; one arm, equipped with a pencil, is connected through the bars to a pointer that is used to trace the original drawing.

pantography [ENG] System for transmitting and automatically recording radar data from an indicator to a remote point.

Pantolambdidae [PALEON] A family of middle to late Paleocene mammals of North America in the superfamily Pantolambdoidea.

Pantolambdodontidae [PALEON] A family of late Eocene mammals of Asia in the superfamily Pantolambdoidea.

Pantolambdoidea [PALEON] A superfamily of extinct mammals in the order Pantodonta.

Pantolestidae [PALEON] An extinct family of large aquatic insectivores referred to the Proteutheria.

pantometer [ENG] An instrument that measures all the angles necessary for determining distances and elevations.

pantophagous [ZOO] Feeding on a variety of foods.

pantophobia [PSYCH] An abnormal fear of everything.

Pantophthalmidae [INV ZOO] The wood-boring flies, a family of orthorrhaphous dipteran insects in the series Brachycera.

Pantopoda [INV ZOO] The equivalent name for Pycnogonida.

pantothenate [BIOCHEM] A salt or ester of pantothenic acid.

pantothenic acid [BIOCHEM] $C_9H_{17}O_5N$ A member of the vitamin B complex that is essential for nutrition of some animal species. Also known as vitamin B_3.

Pantotheria [PALEON] An infraclass of carnivorous and insectivorous Jurassic mammals; early members retained many reptilian features of the jaws.

pan-type car [MIN ENG] A vehicle for removing material from quarries; it is doorless, is reversible in direction, and can be dumped from either side.

panuveitis [MED] Inflammation of the entire uveal tract.

Panzer-Forderer snaking conveyor [MIN ENG] An armored conveyor that is moved forward behind the coal plough by means of a traveling wedge pulled along by the plough or by means of jacks or compressed-air-operated rams attached at intervals to the conveyor structure.

panzootic [VET MED] Affecting many animals of different species.

papagayo [METEOROL] A violent, northeasterly fall wind on the Pacific coast of Nicaragua and Guatemala; it consists of the cold air mass of a norte which has overridden the mountains of Central America and, being a descending wind, it brings fine, clear weather.

papain [BIOCHEM] An enzyme preparation obtained from the juice of the fruit and leaves of the papaya (*Carica papaya*); contains proteolytic enzymes.

Papanicolaou's stains [CHEM] A group of stains used on exfoliated cells, particularly those from the vagina, for examination and diagnosis.

Papanicolaou test [PATH] A technique for the detection of precancerous and early noninvasive cancer by the staining and examination of exfoliated cells; used especially in the diagnosis of uterine cervical and endometrial cancer. Also known as Pap test.

Papaveraceae [BOT] A family of dicotyledonous plants in the order Papaverales, with regular flowers, numerous stamens, and a well-developed latex system.

Papaverales [BOT] An order of dicotyledonous plants in the subclass Magnoliidae, marked by a syncarpous gynoecium, parietal placentation, and only two sepals.

papaverine [ORG CHEM] $C_{20}H_{21}O_4N$ A white, crystalline alkaloid, melting at 147°C; soluble in acetone and chloroform, insoluble in water; used as a smooth muscle relaxant and weak analgesic, usually as the water-soluble hydrochloride salt. Also known as 6,7-dimethoxy-1-veratrylisoquinoline.

paper [MATER] Felted or matted sheets of cellulose fibers, formed on a fine-wire screen from a dilute water suspension, and bonded together as the water is removed and the sheet is dried.

paperboard [MATER] A composition board available in varying thicknesses and degrees of rigidity.

paper capacitor [ELEC] A capacitor whose dielectric material consists of oiled paper sandwiched between two layers of metallic foil.

paper chromatography [ANALY CHEM] Procedure for analysis of complex chemical mixtures by the progressive absorption of the components of the unknown sample (in a solvent) on a special grade of paper.

paper clay [MATER] A special-grade clay that is mixed with paper pulp to add body, weight, and finish to paper products.

paper coating [MATER] Surface coating for paper; made from suspension of clays, starches, casein, rosin, polymers, wax, or various combinations; used to give strength and special surface qualities.

paper cutter [DES ENG] A hand-operated paper cutter and trimmer, consisting of a cutting blade bolted at one end to a ruled board; when the blade is drawn flush with the board, which has a metal strip at the cutting edge, a shearing action takes place which cuts the paper cleanly and evenly.

paper electrochromatography [ANALY CHEM] Variation of paper electrophoresis in which the electrolyte-impregnated absorbent paper is suspended vertically and the electrodes are connected to the sides of the paper, producing a current at right angles to the downward movement of the unknown sample.

paper electrophoresis [ANALY CHEM] A variation of paper chromatography in which an electric current is applied to the ends of the electrolyte-impregnated absorbent paper, thus moving chargeable molecules of the unknown sample toward the appropriate electrode.

paper insulation [MATER] Electrical insulation made of paper, chiefly from coniferous woods but also from rags, rope, and other materials, which are chemically treated, beaten into

---

PAPAVERACEAE

Oriental poppy (*Papaver orientale*), of the family Papaveraceae in the order Papaverales. (*Photograph by John H. Gerard, from National Audubon Society*)

PAPER CHROMATOGRAPHY

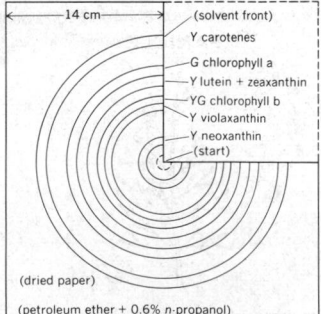

Paper chromatogram of chlorophyll showing ring separation. The circular type shown is one kind of paper chromatogram. *G* signifies green; *Y*, yellow.

a dispersed pulp, formed into a loose sheet by filtering on a moving wire screen, and compacted into paper by calendering with heated rolls.

paper machine [MECH ENG] A synchronized series of mechanical devices for transforming a dilute suspension of cellulose fibers into a dry sheet of paper.

paper mill [IND ENG] A building or complex of buildings housing paper machines.

paperoid [MATER] A heavy composition board, generally made of rope pulp and having a reddish color; used for large expandible filing envelopes.

paper tape [ADP] A paper ribbon in which data may be represented by means of partially or completely punched holes.

paper-tape chemical analyzer [ANALY CHEM] Chemically treated paper tape that is continuously unreeled, exposed to the sample, and viewed by a phototube to measure the color change that is empirically related to changes in the sample's chemical composition.

paper-tape code [ADP] The system by which data are represented by means of holes punched on a paper tape.

paper-tape punch [ADP] Device which places binary characters on a paper tape by punching holes in appropriate channels on the tape; a binary one is placed on the tape by punching a hole; a zero is indicated by the absence of a punched hole.

paper-tape reader [ADP] A reader used to sense information punched in paper tape as a series of holes. Also known as punched-tape reader; tape reader.

paper-tape unit [ADP] The mechanism which handles punched paper tape and usually consists of a paper-tape transport, sensing and recording or perforating heads, and associated electrical and electronic equipment.

papier maché [MATER] A lightweight molding material made from paper pulped with glue and other additives; dries to a hard finish that can be drilled, sanded, or painted.

Papilionidae [INV ZOO] A family of lepidopteran insects in the superfamily Papilionoidea; members are the only butterflies with fully developed forelegs bearing an epiphysis.

Papilionoidea [INV ZOO] A superfamily of diurnal butterflies (Lepidoptera) with clubbed antennae, which are rounded at the tip, and forewings that always have two or more veins.

Papilionoideae [BOT] A subfamily of the family Leguminosae with characteristic irregular flowers that have a banner, two wing petals, and two lower petals united to form a boat-shaped keel.

papilla [BIOL] A small, nipplelike eminence.

papilla of Vater See ampulla of Vater.

papillary adenoma of ovary See serous cystadenoma.

papillary carcinoma [MED] A carcinoma characterized by fingerlike outgrowths.

papillary muscle [ANAT] Any of the muscular eminences in the ventricles of the heart from which the chordae tendineae arise.

papillate [BIOL] **1.** Having or covered with papillae. **2.** Resembling a papilla.

papilledema [MED] Edema of the optic disk. Also known as choked disk.

papillocystoma See serous cystadenoma.

papilloma [MED] A growth pattern of epithelial tumors in which the proliferating epithelial cells grow outward from a surface, accompanied by vascularized cores of connective tissue, to form a branching structure.

papillomatosis [MED] Widespread formation of papillomas.

papillomatous [MED] Characterized by or pertaining to a papilloma.

papite [MATER] A poison gas composed of acrolein with stannic chloride.

papovavirus [VIROL] A deoxyribonucleic acid–containing group of animal viruses, including papilloma and vacuolating viruses.

pappataci fever See phlebotomus fever.

Pappotheriidae [PALEON] A family of primitive, tenreclike Cretaceous insectivores assigned to the Proteutheria.

pappus [BOT] An appendage or group of appendages consisting of a modified perianth on the ovary or fruit of various seed plants; adapted to dispersal by wind and other means.

Pappus' theorems [MATH] Theorems which give the area of a surface of revolution and volume of a solid of revolution.

paprika [BOT] *Capsicum annuum*. A type of pepper with nonpungent flesh, grown for its long red fruit from which a dried, ground condiment is prepared.

Pap test See Papanicolaou test.

papula [BIOL] A small papilla.

papule [MED] A solid circumscribed elevation of the skin varying from less than 0.1 to 1 centimeter in diameter.

papulonecrotic [MED] Papule formation with a tendency to central necrosis; applied especially to a variety of skin tuberculosis.

papyrus [MATER] A paperlike material made by pressing the pith of the papyrus plant in water.

PAR See precision approach radar.

para- [ORG CHEM] Chemical prefix designating the positions of substituting radicals on the opposite ends of a benzene nucleus, for example, paraxylene, $CH_3C_6H_4CH_3$. Abbreviated *p-*.

paraaortic body [ANAT] One of the small masses of chromaffin tissue lying along the abdominal aorta. Also known as glomus aorticum.

paraballoon [ELECTROMAG] Air-inflated radar antenna.

parabanic acid [ORG CHEM] $C_3H_2O_3N_2$ A water-soluble cyclic compound that decomposes when heated to about 227°C; used in organic synthesis. Also known as imidazoletrione; oxalylurea.

parabasal body See kinetoplast.

parabiosis [BIOL] Experimental joining of two individuals to study the effects of one partner upon the other.

parabola [MATH] The U-shaped curve in the plane given by an equation of the form $y = ax^2 + bx + c$.

parabolic antenna [ELECTROMAG] Antenna with a radiating element and a parabolic reflector that concentrates the radiated power into a beam.

parabolic dune [GEOL] A long, scoop-shaped sand dune having a ground plan approximating the form of a parabola, with the horns pointing windward (upwind).

parabolic equations [MATH] Partial differential equations which can be reduced to an equation of the form $\partial^2 u/\partial^2 x = \partial u/\partial t$.

parabolic flight [AERO ENG] A space flight occurring in a parabolic orbit.

parabolic microphone [ENG ACOUS] A microphone used at the focal point of a parabolic sound reflector to give improved sensitivity and directivity, as required for picking up a band marching down a football field.

parabolic orbit [ASTRON] An orbit whose overall shape is like a parabola; the orbit represents the least eccentricity for escape from an attracting body.

parabolic point of a surface [MATH] A point on a surface where the total curvature vanishes.

parabolic reflector [ELECTROMAG] An antenna having a concave surface which is generated either by translating a parabola perpendicular to the plane in which it lies (in a cylindrical parabolic reflector), or rotating it about its axis of symmetry (in a paraboloidal reflector). Also known as dish. [OPTICS] See paraboloidal reflector.

parabolic segment [MATH] The line segment given by a chord perpendicular to the axis of a parabola.

parabolic spiral [MATH] The curve whose equation in polar coordinates is $r^2 = a\theta$.

parabolic velocity [ASTRON] The velocity attained by a celestial body in a parabolic orbit.

paraboloid [ENG] A reflecting surface which is a paraboloid of revolution and is used as a reflector for sound waves and microwave radiation. [MATH] A surface where sections through one of its axes are ellipses, and sections through the other are parabolas.

paraboloidal antenna See paraboloidal reflector.

paraboloidal reflector [ELECTROMAG] An antenna having a concave surface which is a paraboloid of revolution; it concentrates radiation from a source at its focal point into a beam. Also known as paraboloidal antenna. [OPTICS] A concave mirror which is a paraboloid of revolution and produces parallel rays of light from a source located at the focus of the parabola. Also known as parabolic reflector.

PARABOLOIDAL REFLECTOR

Parallel rays of light produced from paraboloidal reflector; dark circle is the light source.

paraboloid of revolution [MATH] The surface obtained by spinning a parabola about its axis.

parabomb [ENG] An equipment container with a parachute which is capable of opening automatically after a delayed drop.

paracaisson [ORD] Small, two-wheeled, hand-drawn vehicle whose body forms an aerial delivery container for artillery ammunition and which, upon being assembled, becomes a utility cart.

Paracanthopterygii [VERT ZOO] A superorder of teleost fishes, including the codfishes and allied groups.

paracentesis [MED] Puncture of the wall of a fluid-filled cavity by means of a hollow needle to draw off the contents.

paracentric [DES ENG] Pertaining to a key and keyway with longitudinal ribs and grooves that project beyond the center, as used in pin-tumbler cylinder locks to deter lockpicking.

paracentric inversion [GEN] A type of chromosomal alteration that occurs within one arm of a chromosome and does not span the centromere.

PARACENTRIC INVERSION

a bc de f g → a bd ce f g

Schematic drawing of the change taking place in paracentric inversion.

paracetaldehyde See paraldehyde.

parachor [PHYS] The molecular weight of a liquid times the fourth root of its surface tension, divided by the difference between the density of the liquid and the density of the vapor in equilibrium with it; essentially constant over wide ranges of temperature.

parachute [AERO ENG] **1.** A contrivance that opens out somewhat like an umbrella and catches the air so as to retard the movement of a body attached to it. **2.** The canopy of this contrivance. [MIN ENG] A kind of safety catch for mine shaft cages.

parachute flare [ENG] Pyrotechnic device attached to a parachute and designed to provide intense illumination for a short period; it may be discharged from aircraft or from the surface.

parachute fragmentation bomb [ORD] A fragmentation bomb adapted for drop by parachute, used in low-level bombing to give the bombing plane time to escape.

parachute-opening shock [AERO ENG] The shock or jolt exerted on a suspended parachute load when the parachute fully catches the air.

parachute troops [ORD] Troops organized and trained to be carried into battle by transport aircraft and dropped by parachute, as distinguished especially from airborne infantry. Also known as paratroops.

parachute weather buoy [ENG] A general-purpose automatic weather station which can be air-dropped; it is 10 feet (3 meters) long and 22 inches (56 centimeters) in diameter, and is designed to operate for 2 months on a 6-hourly schedule, transmitting station identification, wind speed, wind direction, barometric pressure, air temperature, and sea-water temperature.

paracoccidioidomycosis See South American blastomycosis.

paracolon bacteria [MICROBIO] A group of bacteria intermediate between the *Escherichia-Aerobacter* genera and the *Salmonella-Shigella* group.

paracompact space [MATH] A Hausdorff space with the property that for every open covering *F* there is a locally finite open covering *G*, such that every element of *G* is a subset of an element of *F*.

paracondyloid [VERT ZOO] A process on the outer side of each condyle of the occipital bone in the skull of certain mammals.

paracone [VERT ZOO] **1.** The anterior cusp of a primitive tricuspid upper molar. **2.** The principal anterior, external cusp of an upper molar in higher forms.

paraconformity [GEOL] A type of unconformity in which strata are parallel; there is little apparent erosion and the unconformity surface resembles a simple bedding plane. Also known as nondepositional unconformity; pseudoconformity.

paraconid [VERT ZOO] **1.** The cusp of a primitive lower molar corresponding to the paracone. **2.** The anterior, internal cusp of a lower molar in higher forms.

paracoquimbite [MINERAL] $Fe_2(SO_4)_3 \cdot 9H_2O$ A pale-violet rhombohedral mineral composed of hydrous ferric iron sulfate; it is dimorphous with coquimbite.

paracrate [ENG] Rigid equipment container for dropping equipment from an airplane by parachute.

Paracrinoidea [PALEON] A class of extinct Crinozoa characterized by the numerous, irregularly arranged plates, uniserial armlike appendages, and no clear distinction between adoral and aboral surfaces.

Paracucumidae [INV ZOO] A family of holothurian echinoderms in the order Dendrochirotida; the body is invested with plates and has a simplified calcareous ring.

paracyanic acid See fulminic acid.

paracystitis [MED] Inflammation of the connective tissue surrounding the urinary bladder.

paradidymis [ANAT] Atrophic remains of the paragenital tubules of the mesonephros, occurring near the convolutions of the epididymal duct.

paradox [SCI TECH] An argument which gives a contradictory conclusion.

paradoxical embolus [MED] An embolus which is transported to the circulation in peripheral arteries through septal defect in the heart, usually a patent foramen ovale.

paraesophageal cyst [MED] A bronchogenic cyst intimately connected with the esophageal wall, containing cartilage, and usually filled with a mucoid material and desquamated epithelial cells.

paraffin [MATER] See paraffin wax. [ORG CHEM] Any of the saturated aliphatic hydrocarbons of the methane series C_nH_{2n+2}.

paraffin-base crude [MATER] Crude petroleum oil that contains predominately paraffin hydrocarbons, as contrasted with asphaltic- or naphthenic-base crudes; used as a source of fuels and high-grade lubricating oils.

paraffin distillate [MATER] In a petroleum refinery, the distillate oils ready for pressing to produce crystalline paraffin wax and paraffin oil.

paraffinicity [ORG CHEM] The paraffinic nature or composition of crude petroleum or its products.

paraffin jelly [MATER] A light, amber-colored petrolatum; used for medicinal purposes.

paraffin oil [MATER] A viscous, pale to yellow oil made from petroleum; used as a lubricant, medicine, and leather dressing.

paraffinum liquidum See white mineral oil.

paraffin wax [MATER] A solid, crystalline hydrocarbon mixture derived from the paraffin distillate portion of crude petroleum; used in paper coating, candles, creams, emollients, and lipsticks. Also known as ceresin wax; paraffin.

paraform See paraformaldehyde.

paraformaldehyde [ORG CHEM] $(HCHO)_n$ Polymer of formaldehyde where *n* is greater than 6; white, alkali-soluble solid, insoluble in alcohol, ether, and water; used as a disinfectant, fumigant, and fungicide, and to make resins. Also known as paraform.

paraganglion [ANAT] Any of various isolated chromaffin bodies associated with structures such as the abdominal aorta, heart, kidney, and gonads. Also known as chromaffin body.

paragastric [ANAT] Located near the stomach. [INV ZOO] A cavity in Porifera into which radial canals open, and which opens to the outside through the cloaca.

paragenesis [MINERAL] **1.** The association and order of crystallization of minerals in a rock or vein. **2.** The effect of one mineral on the development of another. Also known as mineral sequence; paragenetic sequence.

paragenetic mineralogy [MINERAL] The study of mineral paragenesis, usually accompanying the analysis of the general geologic structures within and around the ore body.

paragenetic sequence See paragenesis.

parageosyncline [GEOL] An epeirogenic geosynclinal basin located within a craton or stable area.

paragglutination [IMMUNOL] Agglutination of colon bacteria with the serum of a patient infected, or recovering from an infection, with dysentery bacilli.

paraglesia [MED] Sensation which is abnormal or disordered.

paraglomerate [GEOL] A conglomerate which contains more matrix than gravel-sized fragments and was deposited by subaqueous turbidity flows and glacier ice rather than normal aqueous flow. Also known as conglomeratic mudstone.

paragnath [INV ZOO] **1.** One of the paired leaflike lobes of the metastoma situated behind the mandibles in most crustaceans. **2.** One of the paired lobes of the hypopharynx in certain insects. **3.** One of the small, sharp and hard jaws of certain annelids.

paragneiss [GEOL] A gneiss showing a sedimentary parentage.

paragonimiasis [MED] Presence of the fluke *Paragonimus westermani* in the lungs or other tissues of humans.

paragonite [MINERAL] $NaAl_2(AlSi_3)O_{10}(OH)_2$ A yellowish or greenish mica species that contains sodium and usually occurs in metamorphic rock. Also known as soda mica.

paragranuloma [MED] The least aggressive form of Hodgkin's disease.

parahelium [ATOM PHYS] Those states of helium in which the spins of the two electrons are antiparallel, in contrast to orthohelium. Also spelled parhelium.

parahilgardite [MINERAL] $Ca_8(B_6O_{11})_3Cl \cdot 4H_2O$ A triclinic mineral composed of hydrous borate and chloride of calcium; it is dimorphous with hilgardite.

parahopeite [MINERAL] $Zn_3(PO_4)_2 \cdot 4H_2O$ A colorless mineral composed of hydrous phosphate of zinc, occurring in tabular triclinic crystals; it is dimorphous with hopeite.

parahydrogen [ATOM PHYS] Those states of hydrogen molecules in which the spins of the two nuclei are antiparallel.

parainfluenza [MED] A viral condition similar to or resulting from influenza. [MICROBIO] An organism exhibiting growth characteristics of *Hemophilus influenzae*.

parakeet [VERT ZOO] Any of various small, slender species of parrots with long tails in the family Psittacidae.

parakeratosis [PATH] Incomplete keratinization of epidermal cells characterized by retention of nuclei of cells attaining the level of the stratum corneum.

paralalia [MED] Disturbance of the faculty of speech, characterized by distortion of sounds, or the habitual substitution of one sound for another.

paralaurionite [MINERAL] $PbCl(OH)$ A white mineral composed of basic lead chloride; it is dimorphous with laurionite.

paraldehyde [ORG CHEM] $C_6H_{12}O_3$ Acetaldehyde polymer; colorless, flammable, toxic liquid, miscible with most organic solvents, soluble in water; melts at $12.6°C$, boils at $124.5°C$; used as a chemical intermediate, in medicine, and as a solvent. Also known as *para*-acetaldehyde; paracetaldehyde.

paraldol [ORG CHEM] $(CH_3CHOHCH_2CHO)_2$ Water-soluble, white crystals, boiling at $90-100°C$; used as a chemical intermediate, to make resins, and in cadmium plating baths.

paralexia [MED] Transposition or substitution of words or syllables in reading.

paralgesia [MED] **1.** Paresthesia characterized by pain. **2.** Any perverted and disagreeable cutaneous sensation, as of formication, cold, or burning.

paraliageosyncline [GEOL] A geosyncline developing along a present-day continental margin, such as the Gulf Coast geosyncline.

paralic [GEOL] Pertaining to deposits laid down on the landward side of a coast.

paralic coal basin [GEOL] Coal deposits formed along the margin of the sea.

paralic swamp *See* marine swamp.

parallactic angle *See* position angle.

parallactic ellipse [ASTRON] An annual apparent elliptical course of a celestial body on the celestial sphere about its mean position; caused by the elliptical orbital motion of the earth.

parallactic equation [ASTRON] An inequality in the moon's motion caused by the sun's perturbing effect on the moon being greater in that half of the moon's apparent orbit around the earth when at new moon rather than at full moon.

parallactic motion [ASTRON] An apparent motion of stars away from the point in the celestial sphere toward which the sun is moving.

parallactic orbit [ASTRON] The apparent orbit of a star as it appears to move once around in the sky each year; the motion is caused by the earth's orbital motion around the sun.

parallax [OPTICS] The change in the apparent relative orientations of objects when viewed from different positions.

parallax age *See* age of parallax inequality.

parallax correction [NAV] In celestial navigation, that sex-

tant altitude correction made necessary because of the difference between the apparent direction from a point on the surface of the earth to a celestial body and the apparent direction from the center of the earth to the same body.

parallax error [OPTICS] Error in reading an instrument employing a scale and pointer because the observer's eye and pointer are not in a line perpendicular to the plane of the scale.

parallax in altitude [NAV] Geocentric parallax of a celestial body at any altitude; the term is used to distinguish the parallax at the given altitude from the horizontal parallax, when the body is in the horizon.

parallax inequality [OCEANOGR] The variation in the range of tide or in the speed of tidal currents due to the continual change in the distance of the moon from the earth.

parallel [ADP] Simultaneous transmission of, storage of, or logical operations on the parts of a word, character, or other subdivision of a word in a computer, using separate facilities for the various parts. [ELEC] Connected to the same pair of terminals. Also known as multiple; shunt. [GEOD] A circle on the surface of the earth, parallel to the plane of the equator and connecting all points of equal latitude, or a circle parallel to the primary great circle of a sphere or spheroid. [MATH] **1.** Lines are parallel in a euclidean space if they lie in a common plane and do not intersect. **2.** Planes are parallel in a euclidean three-dimensional space if they do not intersect. [PHYS] Of two or more displacements or other vectors, having the same direction.

parallel access [ADP] Transferral of information to or from a storage device in which all elements in a unit of information are transferred simultaneously. Also known as simultaneous access.

parallel baffle muffler [DES ENG] A muffler constructed of a series of ducts placed side by side in which the duct cross section is a narrow but long rectangle.

parallel buffer [ELECTR] Electronic device (magnetic core or flip-flop) used to temporarily store digital data in parallel, as opposed to series storage.

parallel by character [ADP] The handling of all the characters of a machine word simultaneously in separate lines, channels, or storage cells.

parallel circuit [ELEC] An electric circuit in which the elements, branches (having elements in series), or components are connected between two points, with one of the two ends of each component connected to each point.

parallel compensation *See* feedback compensation.

parallel course computer *See* course-line computer.

parallel cut [ENG] A group of parallel holes, not all charged with explosive, to create the initial cavity to which the loaded holes break in blasting a development round. Also known as burn cut.

parallel digital computer [ADP] Computer in which the digits are handled in parallel; mixed serial and parallel machines are frequently called serial or parallel, according to the way arithmetic processes are performed; an example of a parallel digital computer is one which handles decimal digits in parallel, although it might handle the bits constituting a digit either serially or in parallel.

parallel drainage pattern [HYD] A drainage pattern characterized by regularly spaced streams flowing parallel to one another over a large area.

parallel drum [DES ENG] A cylindrical form of drum on which the haulage or winding rope is coiled.

parallel entry [MIN ENG] An intake airway parallel to the haulageway.

parallelepiped [MATH] A polyhedron all of whose faces are parallelograms.

parallel evolution [EVOL] Evolution of similar characteristics in different groups of organisms.

parallel extinction [OPTICS] Nearly total absorption of light that is propagating in an anisotropic crystal in a direction parallel to crystal outlines or traces of cleavage planes.

parallel feed [ELECTR] Application of a direct-current voltage to the plate or grid of a tube in parallel with an alternating-current circuit, so that the direct-current and the alternating-current components flow in separate paths. Also known as shunt feed.

PARALLEL CIRCUIT

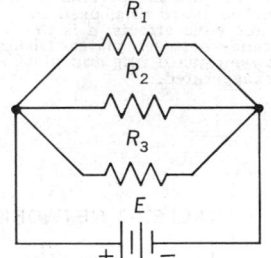

Schematic diagram of a simple parallel circuit in which the resistors, R_1, R_2, and R_3, are connected in parallel between terminals of battery which supplies voltage E.

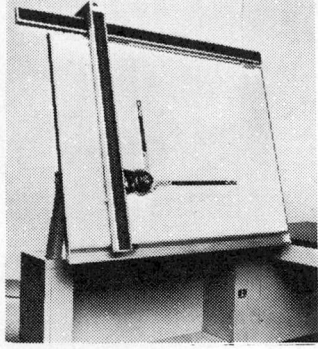

PARALLEL MOTION PROTRACTOR

Parallel motion protractor using vertical and horizontal bars. (Keuffel and Esser Co.)

PARALLEL-PLATE CAPACITOR

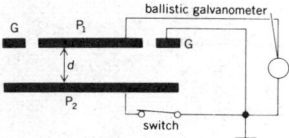

Cross section of parallel-plate capacitor with all parts of capacitor at ground potential. P_1 and P_2 are the parallel plates; G is the guard ring, used to reduce edge effects; d is the distance between plates. Distance between guard ring and plate P_1 is exaggerated.

PARALLEL-T NETWORK

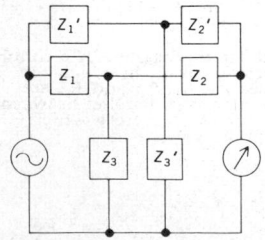

Schematic drawing of the network showing the two sets of three impedances: Z_1, Z_2, Z_3, and Z_1', Z_2', Z_3'.

parallel firing [ENG] A method of connecting together a number of detonators which are to be fired electrically in one blast.

parallel fold *See* concentric fold.

parallel growth *See* parallel intergrowth.

parallel impedance [ELEC] One of two or more impedances that are connected to the same pair of terminals.

paralleling reactor [ELECTROMAG] Reactor for correcting the division of load between parallel-connected transformers which have unequal impedance voltages.

parallel intergrowth [CRYSTAL] Intergrowth of two or more crystals in such a way that one or more axes in each crystal are approximately parallel. Also known as parallel growth.

parallel laminate [MATER] A laminate in which all the layers of material are set approximately parallel with respect to a particular characteristic, such as the grain or the direction of tension.

parallel linkage [MECH ENG] A linkage system in which reciprocating motion is amplified.

parallel middle body *See* dead flat.

parallel motion protractor [GRAPHICS] An instrument consisting essentially of a protractor and one or more arms attached to a parallel motion device, so that the movement of the arms is everywhere parallel; the protractor can be rotated and set at any position so that it can be oriented to a chart. Also known as drafting machine.

parallel muscle [ANAT] Any muscle having the long fibers arranged parallel to each other.

parallel of altitude [ASTRON] A circle on the celestial sphere parallel to the horizon connecting all points of equal altitude. Also known as almucantar; altitude circle.

parallel of declination [ASTRON] A small circle of the celestial sphere parallel to the celestial equator. Also known as celestial parallel; circle of equal declination.

parallel of latitude *See* circle of latitude.

parallel of longitude *See* circle of longitude.

parallelogram [MATH] A four-sided polygon with each pair of opposite sides parallel.

parallel operation [ADP] Performance of several actions, usually of a similar nature, by a computer system simultaneously through provision of individual similar or identical devices. [ELECTR] The connecting together of the outputs of two or more batteries or other power supplies so that the sum of their output currents flows to a common load.

parallelotope [MATH] A parallelepiped with sides in proportion of 1, ½, and ¼.

parallel padding [ELEC] Method of parallel operation for two or more power supplies in which their current limiting or automatic crossover output characteristic is employed so that each supply regulates a portion of the total current, each parallel supply adding to the total and padding the output only when the load current demand exceeds the capability, or limit setting, of the first supply.

parallel perspective [GRAPHICS] A perspective such that the faces and edges of represented objects are either perpendicular or parallel to the plane of the picture. Also known as one-point perspective.

parallel-plate capacitor [ELEC] A capacitor consisting of two parallel metal plates, with a dielectric filling the space between them.

parallel-plate laser [OPTICS] A laser which has two small parallel plates facing each other at a distance which is large compared with their diameters; one of them reflects light and the other is partially reflecting, so that light can bounce back and forth between the plates enough to build up a strong pulse.

parallel-plate waveguide [ELECTROMAG] Pair of parallel conducting planes used for propagating uniform circularly cylindrical waves having their axes normal to the plane.

parallel processor *See* multiprocessor.

parallel programming [ADP] A method for performing simultaneously the normally sequential steps of a computer program, using two or more processors. [ELECTR] Method of parallel operation for two or more power supplies in which their feedback terminals (voltage control terminals) are also paralleled; these terminals are often connected to a separate programming source.

parallel radio tap [COMMUN] A telephone tapping procedure in which a battery-powered miniature radio transmitter is bridged across the target pair.

parallel rectifier [ELECTR] One of two or more rectifiers that are connected to the same pair of terminals, generally in series with small resistors or inductors, when greater current is desired than can be obtained with a single rectifier.

parallel representation [ADP] The simultaneous appearance of the different bits of a digital variable on parallel bus lines.

parallel resonance Also known as antiresonance. [ELEC] **1.** The frequency at which the inductive and capacitive reactances of a parallel resonant circuit are equal. **2.** The frequency at which the parallel impedance of a parallel resonant circuit is a maximum. **3.** The frequency at which the parallel impedance of a parallel resonant circuit has a power factor of unity.

parallel resonant circuit [ELEC] A circuit in which an alternating-current voltage is applied across a capacitor and a coil in parallel. Also known as antiresonant circuit.

parallel resonant interstage [ELECTR] A coupling between two amplifier stages achieved by means of a parallel-tuned LC circuit.

parallel ripple mark [GEOL] A ripple mark characterized by a relatively straight crest and an asymmetric profile.

parallel-rod oscillator [ELECTR] Ultra-high-frequency oscillator circuit in which parallel rods or wires of required length and dimensions form the tank circuits.

parallel rulers [GRAPHICS] An instrument for transferring a line parallel to itself; in its most common form it consists essentially of two parallel bars or rulers connected in such manner that when one is held in place, the other may be moved, remaining parallel to its original position.

parallel running [ADP] **1.** The running of a newly developed system in a data-processing area in conjunction with the continued operation of the current system. **2.** The final step in the debugging of a system; this step follows a system test.

parallel sailing [NAV] A method of converting departure into difference of longitude, or vice versa, when the true course is 090° or 270°.

parallel search storage [ADP] A device for very rapid search of a volume of stored data to permit finding a specific item.

parallel series [ELEC] Circuit in which two or more parts are connected together in parallel to form parallel circuits, and in which these circuits are then connected together in series so that both methods of connection appear. [ENG] *See* multiple series.

parallel sphere [ASTRON] The celestial sphere as it appears to an observer at the pole, where celestial bodies appear to move parallel to the horizon.

parallel storage [ADP] A storage device in which words (or characters or digits) can be read in or out simultaneously.

parallel stripline *See* strip transmission line.

parallel-T network [ELEC] A network used in capacitance measurements at radio frequencies, having two sets of three impedances, each in the form of the letter T, with the arms of the two Ts joined to common terminals, and the source and detector each connected between two of these terminals. Also known as twin-T network.

parallel transfer [ADP] Simultaneous transfer of all bits in a storage location constituting a character or word.

parallel transmission [ADP] The transmission of characters of a word over different lines, usually simultaneously; opposed to serial transmission.

parallel-tuned circuit [ELEC] A circuit with two parallel branches, one having an inductance and a resistance in series, the other a capacitance and a resistance in series.

parallel-veined [BOT] Of a leaf, having the veins parallel, or nearly parallel, to each other.

parallel wire method [MIN ENG] An electrical prospecting method employing equipotential lines or curves in searching for ore bodies.

parallel wires [ELEC] Two conductors which are parallel to each other; often used in transmission lines.

paralutein cells [HISTOL] Epithelioid cells of the corpus luteum.

paralysis [MED] Complete or partial loss of motor or sensory function.

paralysis agitans   *See* parkinsonism.

paralytic secretion   [PHYSIO] Glandular secretion occurring in a denervated gland.

paralytic spinal poliomyelitis   [MED] An acute inflammatory virus disease chiefly involving the anterior horns of the gray matter of the spinal cord.

paramagnetic   [ELECTROMAG] Exhibiting paramagnetism.

paramagnetic alloy   [MET] An alloy whose permeability is slightly greater than that of vacuum and is independent of the magnetic field strength, such as intermetallic compounds of nickel and titanium.

paramagnetic amplifier   *See* maser.

paramagnetic analytical methods   [ANALY CHEM] Analysis of fluid mixtures by measurement of the paramagnetic (versus diamagnetic) susceptibilities of materials when exposed to a magnetic field.

paramagnetic cooling   *See* adiabatic demagnetization.

paramagnetic crystal   [ELECTROMAG] A crystal whose permeability is slightly greater than that of vacuum and is independent of the magnetic field strength.

paramagnetic Faraday effect   [OPTICS] The Faraday effect observed in paramagnetic salts at frequencies near an absorption line of the salt which is split due to splitting of the lower energy level.

paramagnetic iron   [MET] Iron which has been transformed from a ferromagnetic to a paramagnetic substance by application of a pressure somewhat greater than 10^5 bars.

paramagnetic material   [ELECTROMAG] A material within which an applied magnetic field is increased by the alignment of electron orbits.

paramagnetic relaxation   [ELECTROMAG] The approach of a system, which displays paramagnetism because of electronic magnetic moments of atoms or ions, to an equilibrium or steady-state condition over a period of time, following a change in the magnetic field.

paramagnetic resonance   *See* electron paramagnetic resonance.

paramagnetic salt   [ELECTROMAG] A salt whose permeability is slightly greater than that of vacuum and is independent of magnetic field strength; used in adiabatic demagnetization.

paramagnetic spectra   [SPECT] Spectra associated with the coupling of the electronic magnetic moments of atoms or ions in paramagnetic substances, or in paramagnetic centers of diamagnetic substances, to the surrounding liquid or crystal environment, generally at microwave frequencies.

paramagnetic susceptibility   [ELECTROMAG] The susceptibility of a paramagnetic substance, which is a positive number and is, in general, much smaller than unity.

paramagnetism   [ELECTROMAG] A property exhibited by substances which, when placed in a magnetic field, are magnetized parallel to the field to an extent proportional to the field (except at very low temperatures or in extremely large magnetic fields).

Parameciidae   [INV ZOO] A family of ciliated protozoans in the order Holotrichia; the body has differentiated anterior and posterior ends and is bounded by a hard but elastic pellicle.

paramecium   [INV ZOO] A single-celled protozoan belonging to the family Parameciidae.

Paramecium   [INV ZOO] The genus of protozoans composing the family Parameciidae.

paramedical   [MED] Having a supplementary or secondary relation to medicine.

paramelaconite   [MINERAL] A black tetragonal mineral composed of cupric and cuprous oxides, occurring in pyramidal crystals.

Paramelina   [VERT ZOO] An order of marsupials that includes the bandicoots in some systems of classification.

paramere   [BIOL] One half of a bilaterally symmetrical animal or somite.   [INV ZOO] Any of several paired structures of an insect, especially those on the ninth abdominal segment.

paramesonephric duct   [EMBRYO] An embryonic genital duct; in the female, it is the anlage of the oviducts, uterus, and vagina; in the male, it degenerates, leaving the appendix testes. Also known as Müllerian duct.

parameter   [CRYSTAL] Any of the axial lengths or interaxial angles that define a unit cell.   [ELEC] **1.** The resistance, capacitance, inductance, or impedance of a circuit element. **2.** The value of a transistor or tube characteristic.   [MATH] An arbitrary constant or variable so appearing in a mathematical expression that changing it gives various cases of the phenomenon represented.   [PHYS] A quantity which is constant under a given set of conditions, but may be different under other conditions.

parameterization   [SCI TECH] The representation, in a dynamic model, of physical effects in terms of admittedly oversimplified parameters, rather than realistically requiring such effects to be consequences of the dynamics of the system.

parameter tags   [ADP] Constants that are used by several computer programs.

paramethadione   [PHARM] $C_7H_{11}NO_3$ An anticonvulsant primarily useful in the treatment of petit mal epilepsy.

parametric amplifier   [ELECTR] A highly sensitive ultra-high-frequency or microwave amplifier having as its basic element an electron tube or solid-state device whose reactance can be varied periodically by an alternating-current voltage at a pumping frequency. Also known as mavar; paramp; reactance amplifier.   [OPTICS] A device consisting of an optically nonlinear crystal in which an optical or infrared beam draws power from a laser beam at a higher frequency and is amplified.

parametric converter   [ELECTR] Inverting or noninverting parametric device used to convert an input signal at one frequency into an output signal at a different frequency.

parametric curves   [MATH] On a surface determined by equations $x = f(u,v)$, $y = g(u,v)$, and $z = h(u,v)$, these are families of curves obtained by setting the parameters u and v equal to various constants.

parametric device   [ELECTR] Electronic device whose operation depends essentially upon the time variation of a characteristic parameter usually understood to be a reactance.

parametric down-converter   [ELECTR] Parametric converter in which the output signal is at a lower frequency than the input signal.

parametric equation   [MATH] An equation where coordinates of points appear dependent on parameters such as the parametric equation of a curve or a surface.

parametric excitation   [ENG] The method of exciting and maintaining oscillation in either an electrical or mechanical dynamic system, in which excitation results from a periodic variation in an energy storage element in a system such as a capacitor, inductor, or spring constant.

parametric mixing   [OPTICS] In a medium possessing optical nonlinearities, the mixing of electromagnetic waves to form waves with frequencies linearly related to the frequency of incident radiation.

parametric oscillator   [ELECTR] An oscillator in which the reactance parameter of an energy-storage device is varied to obtain oscillation.   [OPTICS] A device consisting of an optically nonlinear crystal surrounded by a pair of mirrors to which is applied a relatively high-frequency laser beam and a relatively low-frequency signal, resulting in a low-frequency output whose frequency can be varied, usually by varying the indices of refraction.

parametric phase-locked oscillator   *See* parametron.

parametric up-converter   [ELECTR] Parametric converter in which the output signal is at a higher frequency than the input signal.

parametron   [ELECTR] A resonant circuit in which either the inductance or capacitance is made to vary periodically at one-half the driving frequency; used as a digital computer element, in which the oscillation represents a binary digit. Also known as parametric phase-locked oscillator; phase-locked oscillator; phase-locked subharmonic oscillator.

paramo   [ECOL] A biological community, essentially a grassland, covering extensive high areas in equatorial mountains of the Western Hemisphere.

paramorph   [MINERAL] A mineral exhibiting paramorphism.

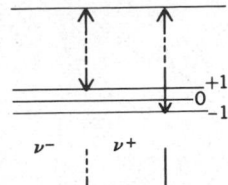

**PARAMAGNETIC
FARADAY EFFECT**

Schematic energy level diagram of ions of a paramagnetic salt. The lengths of the arrows are proportional to the frequencies of absorption lines for right-handed (ν^+) and left-handed (ν^-) circularly polarized light.

PARAMECIUM

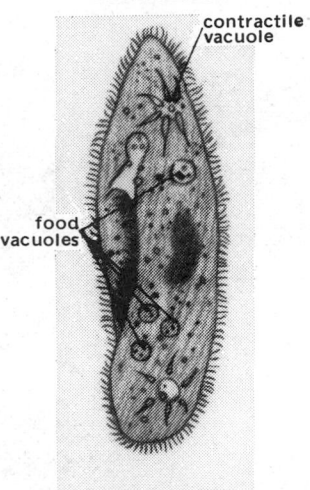

A paramecium. Two contractile vacuoles and numerous food vacuoles are shown.

PARAMETRIC AMPLIFIER

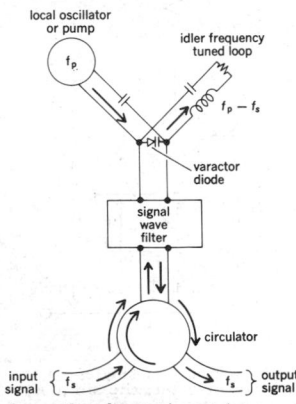

Schematic of negative-resistance-type parametric amplifier that used a varactor diode as the variable reactor. f_p = pumping frequency. Input signal, with frequency f_s, is reflected back down transmission line with increased power. Wave at idler frequency $f_p - f_s$ is not utilized outside the amplifier.

paramorphism [MINERAL] The property of a mineral whose internal structure has changed without change in composition or external form. Formerly known as allomorphism.

paramp *See* parametric amplifier.

paramutation [GEN] A mutation in which one member of a heterozygous pair of alleles permanently changes its partner allele.

paramylum [BIOCHEM] A reserve, starchlike carbohydrate of various protozoans and algae.

paramyotonia congenita [MED] A heredofamilial condition characterized by recurrent muscular stiffness and weakness (myotonia) on exposure to cold, as well as on mechanical irritation; transmitted as an autosomal dominant and considered to be a variety of the hyperkalemic form of periodic paralysis. Also known as Eulenburg's disease; myotonia congenita intermittens.

paramyxovirus [VIROL] A subgroup of myxoviruses, including the viruses of mumps, measles, parainfluenza, and Newcastle disease; all are ribonucleic acid–containing viruses and possess an ether-sensitive lipoprotein envelope.

paranasal sinus [ANAT] Any of the paired sinus cavities of the human face; includes the frontal, ethmoid, and sphenoid sinuses.

paranephritis [MED] **1.** Inflammation of the adrenal gland. **2.** Inflammation of the connective tissue adjacent to the kidney.

paranesthesia [MED] Anesthesia of the body below the waist.

paranitraniline red *See* para red.

paranoia [PSYCH] A rare form of paranoid psychosis characterized by the slow development of a complex, internally logical system of persecutory or grandiose delusions.

paranoid personality [PSYCH] An individual characterized by the tendency to be hypersensitive, rigid, extremely self-important, and jealous, to project hostile feelings so that he easily becomes suspicious of others and is quick to blame them or attribute evil motives to them.

paranoid schizophrenia [PSYCH] A form of schizophrenia in which delusions of persecution or grandeur (or both), hallucinations, and ideas of reference predominate and sometimes are systematized.

paranoid state [PSYCH] Any of various psychotic disorders in which the patient exhibits persistent delusions, usually of persecution or grandeur.

paranosmia [MED] A deviation in odor sensitivity involving change in odor quality.

paranthelion [ASTRON] A refraction phenomenon similar to a parhelion, but occurring generally at a distance of 120° (occasionally 90° and 140°) from the sun, on the parhelic circle. Also known as mock sun.

paranthropophytia [ECOL] Discrepant control of regions or areas due to immediate and continuous or periodic interference, as by certain cultivation practices.

Paranyrocidae [PALEON] An extinct family of birds in the order Anseriformes, restricted to the Miocene of South Dakota.

Paraonidae [INV ZOO] A family of small, slender polychaete annelids belonging to the Sedenteria.

parapack [ENG] A package or bundle with a parachute attached for dropping from an aircraft.

Paraparchitacea [PALEON] A superfamily of extinct ostracods in the suborder Kloedenellocopina including nonsulcate, nondimorphic forms.

paraparesis [MED] Partial paralysis of the lower extremities.

parapertussis [MED] An acute bacterial respiratory infection similar to mild pertussis and caused by *Bordetella pertussis*.

parapet [ARCH] A low retaining wall at the edge of a roof, bridge, porch, or other structure. [ORD] An elevation of earth or material which is thrown up in front of a trench or emplacement to protect the occupants from fire and observation, and over which fire may be delivered.

paraphase amplifier [ELECTR] An amplifier that provides two equal output signals 180° out of phase.

paraphimosis [MED] **1.** Retraction and constriction, especially of the prepuce behind the glans penis. Also known as Spanish collar. **2.** Retraction of the eyelid behind the eyeball.

paraphysis [BOT] A sterile filament borne among the sporogenous or gametogenous organs in many cryptogams. [VERT ZOO] A median evagination of the roof of the telencephalon of some lower vertebrates.

paraplasm *See* hyaloplasm.

paraplegia [MED] Paralysis of the lower limbs.

parapodium [INV ZOO] **1.** One of the short, paired processes on the sides of the body segments in certain annelids. **2.** A lateral expansion of the foot in gastropod mollusks.

parapolar cell [INV ZOO] Either of the first two trunk cells in the development of certain Mesozoa.

parapositronium [PARTIC PHYS] The state of positronium in which the positron and electron have antiparallel spins.

parapsychology [PSYCH] The study of the phenomena of extrasensory perception and psychokinesis.

para red [ORG CHEM] $C_{10}H_6(OH)NNC_6H_4NO_2$ Red pigment derived from the coupling of β-naphthol with diazotized paranitroaniline. Also known as paranitraniline red.

pararipple [GEOL] A large, symmetric ripple whose surface slopes gently and which shows no assortment of grains.

pararosaniline [ORG CHEM] $HOC(C_6H_4NH_2)_3$ Red to colorless crystals, melting at 205°C; soluble in ethanol, in the hydrochloride salt; used as a dye.

pararosolic acid *See* aurin.

Parasaleniidae [INV ZOO] A family of echinacean echinoderms in the order Echinoida composed of oblong forms with trigeminate ambulacral plates.

paraschist [PETR] A schist derived from sedimentary rocks.

paraselene [ASTRON] A weakly colored lunar halo identical in form and optical origin to the solar parhelion; paraselenae are observed less frequently than are parhelia, because of the moon's comparatively weak luminosity. Also known as mock moon.

paraselenic circle [ASTRON] A halo phenomenon consisting of a horizontal circle passing through the moon, corresponding to the parhelic circle through the sun, and produced by reflection of moonlight from ice crystals.

Paraselloidea [INV ZOO] A group of the Asellota that contains forms in which the first pleopods of the male are coupled along the midline, and are lacking in the female.

Paraseminotidae [PALEON] A family of Lower Triassic fishes in the order Palaeonisciformes.

parasexual cycle [GEN] A series of events leading to genetic recombination in vegetative or somatic cells; it was first described in filamentous fungi; there are three essential steps: heterokaryosis; fusion of unlike haploid nuclei in the heterokaryon to yield heterozygous diploid nuclei; and recombination and segregation at mitosis by two independent processes, mitotic crossing-over and loss of chromosomes.

parasheet [AERO ENG] A simple form of parachute in which the canopy is a single piece of material or two or more pieces sewed together; it may have any geometrical form, such as square or hexagonal, and the hem may be gathered to assist in the development of a crown when the parasheet is opened.

parasite [BIOL] An organism that lives in or on another organism of different species from which it derives nutrients and shelter. [ELEC] Current in a circuit, due to some unintentional cause, such as inequalities of temperature or of composition; particularly troublesome in electrical measurements.

parasite drag [FL MECH] The portion of the total drag of an aircraft exclusive of the induced drag of the wings.

parasitemia [MED] The presence of parasites in the blood.

Parasitica [INV ZOO] A group of hymenopteran insects that includes four superfamilies of the Apocrita: Ichneumonoidea, Chalcidoidea, Cynipoidea, and Proctotrupoidea; some are pytophagous, while others are parasites of other insects.

parasitic absorption *See* parasitic capture.

parasitic antenna *See* parasitic element.

parasitic capture [NUCLEO] Any absorption of a neutron that does not result in a fission or the production of a desired element. Also known as parasitic absorption.

parasitic castration [BIOL] Destruction of the reproductive organs by parasites.

parasitic cone *See* adventive cone.

parasitic current [ELEC] An eddy current in a piece of electrical machinery; gives rise to energy losses.

PARAPHASE AMPLIFIER

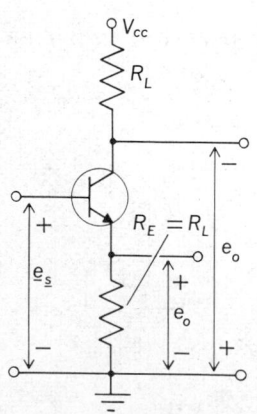

Circuit diagram of single-transistor-inverter, the simplest paraphase amplifier. $V_{CC} =$ collector supply voltage, $R_E =$ emitter resistance, $R_L =$ collector load, $e_S =$ input signal voltage, $e_O =$ output voltages, equal in magnitude and 180° out of phase.

parasitic element [ELECTROMAG] An antenna element that serves as part of a directional antenna array but has no direct connection to the receiver or transmitter and reflects or reradiates the energy that reaches it, in a phase relationship such as to give the desired radiation pattern. Also known as parasitic antenna; parasitic reflector; passive element.

parasitic oscillation [ELECTR] An undesired self-sustaining oscillation or a self-generated transient impulse in an oscillator or amplifier circuit, generally at a frequency above or below the correct operating frequency.

parasitic reflector *See* parasitic element.

parasitic stomatitis *See* thrush.

parasitic suppressor [ELECTR] A suppressor, usually in the form of a coil and resistor in parallel, inserted in a circuit to suppress parasitic high-frequency oscillations.

parasitoidism [BIOL] Systematic feeding by an insect larva on living host tissues so that the host will live until completion of larval development.

parasitology [BIOL] A branch of biology which deals with those organisms, plant or animal, which have become dependent on other living creatures.

parasitophobia [PSYCH] An abnormal fear of parasites.

parasphenoid [VERT ZOO] A bone in the base of the skull of many vertebrates.

parastacidae [INV ZOO] A family of crayfishes assigned to the Nephropoidea.

parastate [ATOM PHYS] A state of a diatomic molecule in which the spins of the nuclei are antiparallel.

parastyle [VERT ZOO] A small cusp anterior to the paracone of an upper molar.

Parasuchia [PALEON] The equivalent name for Phytosauria.

parasympathetic [ANAT] Of or pertaining to the craniosacral division of the autonomic nervous system.

parasympathetic nervous system [ANAT] The craniosacral portion of the autonomic nervous system, consisting of preganglionic nerve fibers in certain sacral and cranial nerves, outlying ganglia, and postganglionic fibers.

parasympatholytic [PHARM] Blocking the action of parasympathetic nerve fibers.

parasympathomimetic [PHARM] Of drugs, having an effect similar to that produced when the parasympathetic nerves are stimulated.

paratacamite [MINERAL] $Cu_2(OH)_3Cl$ Rhombohedral mineral composed of basic copper chloride; it is dimorphous with tacamite.

parataxic distortion [PSYCH] A perceptual or judgmental distortion of interpersonal relations resulting from the observer's need to pattern his responses on previous experiences and thus defend himself against anxiety.

parathelioma *See* interstitial endometrosis.

parathion [ORG CHEM] $(C_2H_5O)_2PSOC_6H_4NO_2$ Poisonous, deep-brown to yellow liquid boiling at 157–162°C; soluble in most organic solvents; used as an insecticide and acaricide.

parathormone [BIOCHEM] A polypeptide hormone that functions in regulating calcium and phosphate metabolism. Also known as parathyroid hormone.

Parathuramminacea [PALEON] An extinct superfamily of foraminiferans in the suborder Fusulinina, with a test having a globular or tubular chamber and a simple, undifferentiated wall.

parathyroidectomy [MED] Excision of a parathyroid gland.

parathyroid gland [ANAT] A paired endocrine organ located within, on, or near the thyroid gland in the neck region of all vertebrates except fishes.

parathyroid hormone *See* parathormone.

para toner [MATER] Water-insoluble red pigment made from β-naphthol and paranitroaniline; used in paints, in printing, and to make para lakes.

paratrachoma *See* inclusion conjunctivitis.

paratrichosis [MED] A condition in which the hair is either imperfect in growth or develops in abnormal places.

paratroch [INV ZOO] The ciliated band encircling the anus in certain trichophore larvae.

paratroops *See* parachute troops.

paratuberculosis *See* Johne's disease.

paratype [SYST] A specimen other than the holotype which is before the author at the time of original description and which is designated as such or is clearly indicated as being one of the specimens upon which the original description was based.

paratyphoid fever [MED] A bacterial disease of man resembling typhoid fever and caused by *Salmonella paratyphi*.

parauterine organ [INV ZOO] In certain tapeworms, a pouchlike sac which receives and retains the embryos.

paravane [ENG] A torpedo-shaped device with sawlike teeth along its forward end, towed with a wire rope underwater from either side of the bow of a ship to cut the cables of anchored mines. Also known as otter.

paravauxite [MINERAL] $FeAl_2(PO_4)_2(OH)_2\cdot8H_2O$ A colorless mineral composed of hydrous basic iron aluminum phosphate; contains more water than vauxite.

parawollastonite [MINERAL] $CaSiO_3$ A monoclinic mineral composed of silicate of calcium; it is dimorphous with wollastonite.

paraxial [SCI TECH] Lying near the axis.

paraxial rays [OPTICS] Rays which are close enough to the opical axis of a system, and thus whose directions are sufficiently close to being parallel to it, so that sines of angles between the rays and the optical axis may be replaced by the angles themselves in calculations.

paraxonic [VERT ZOO] Pertaining to a state or condition wherein the axis of the foot lies between the third and fourth digits.

Parazoa [INV ZOO] A name proposed for a subkingdom of animals which includes the sponges (Porifera).

parcel method [METEOROL] A method of testing for instability in which a displacement is made from a steady state under the assumption that only the parcel or parcels displaced are affected, the environment remaining unchanged.

parchment [MATER] **1.** The skin of a goat or sheep that has been treated so that it can be used to write upon. **2.** A drawing or written text on this material.

parchment paper [MATER] Paper that has been manufactured so that its appearance resembles parchment.

paregoric [PHARM] Camphorated opium tincture, a preparation of opium, camphor, benzoic acid, anise oil, glycerin, and diluted alcohol; used mainly as an antiperistaltic.

Pareiasauridae [PALEON] A family of large, heavy-boned terrestrial reptiles of the late Permian, assigned to the order Cotylosauria.

parenchyma [BOT] A tissue of higher plants consisting of living cells with thin walls that are agents of photosynthesis and storage; abundant in leaves, roots, and the pulp of fruit, and found also in leaves and stems. [HISTOL] The specialized epithelial portion of an organ, as contrasted with the supporting connective tissue and nutritive framework.

parenchymal jaundice [MED] Any of various forms of jaundice in which the disease is due in part to damaged liver cells.

parenchymella *See* diploblastula.

parenchymula [INV ZOO] The flagellate larva of calcinean sponges in which there is a cavity filled with gelatinous connective tissue.

parent [NUC PHYS] A radionuclide that upon disintegration yields a specified nuclide, the daughter, either directly or indirectly as a later member of a radioactive series.

parental magma [GEOL] The naturally occurring mobile rock material from which a particular igneous rock solidified or from which another magma was derived.

parent compound [CHEM] A chemical compound that is the basis for one or more derivatives; for example, ethane is the parent compound for ethyl alcohol and ethyl acetate.

parenteral [MED] Outside the intestine; not via the alimentary tract.

parent figure [PSYCH] A person who represents essential but not necessarily ideal attributes of a father or mother and who is the object of the attitudes and responses of an individual in a parent-child relationship.

parent material [GEOL] The unconsolidated mineral or organic material from which the true soil develops.

parent name [CHEM] That part of a chemical compound's name from which the name of a derivative comes; for example, ethane is the parent name for ethanol.

parent of a state [QUANT MECH] If an n-electron state is written as a sum of products of 1-electron states and $(n-1)$-

electron states, the $(n - 1)$-electron states are called the parents of the n-electron states.

parent rock [GEOL] **1.** The rock mass from which parent material is derived. **2.** *See* source rock.

paresis [MED] **1.** A slight paralysis. **2.** Incomplete loss of muscular power. **3.** Weakness of a limb.

paresthesia [MED] Tingling, crawling, or burning sensation of the skin.

Pareulepidae [INV ZOO] A monogeneriç family of errantian polychaete annelids.

parfocal eyepieces [OPTICS] Eyepieces whose lower focal points lie in the same plane, so that they can be interchanged without changing the focus of the instrument with which they are used.

parging [CIV ENG] A thin coating of mortar or plaster on a brick or stone surface.

parhelic circle [ASTRON] A halo consisting of a faint white circle passing through the sun and running parallel to the horizon for as much as 360° of azimuth. Also known as mock sun ring.

parhelion [ASTRON] Either of two colored luminous spots that appear at points 22° (or somewhat more) on both sides of the sun and at the same elevation as the sun; the solar counterpart of the lunar paraselene. Also known as mock sun; sun dog.

parhelium *See* parahelium.

Parian cement [MATER] Gypsum plaster containing borax, which dries to a hard finish.

parietal [ANAT] Of or situated on the wall of an organ or other body structure. [BOT] Of a plant part, having a peripheral location or orientation; in particular, attached to the main wall of an ovary.

parietal block *See* intraventricular heart block.

parietal bone [ANAT] The bone that forms the side and roof of the cranium.

parietal cell [HISTOL] One of the peripheral, hydrochloric acid-secreting cells in the gastric fundic glands. Also known as acid cell; delomorphous cell; oxyntic cell.

Parietales [BOT] An order of plants in the Englerian system; families are placed in the order Violales in other systems.

parietal lobe [ANAT] The cerebral lobe of the brain above the lateral cerebral sulcus and behind the central sulcus.

parietal peritoneum [ANAT] The portion of the peritoneum lining the interior of the body wall.

parietal pleura [ANAT] The pleura lining the inner surface of the thoracic cavity.

parisite [MINERAL] $(Ce,La)_2Ca(CO_3)_3F_2$ A brownish-yellow secondary mineral composed of a carbonate and a fluoride of calcium, cerium, and lanthanum.

parity [ADP] The use of a self-checking code in a computer employing binary digits in which the total number of 1's or 0's in each permissible code expression is always even or always odd. [MATH] Two integers have the same parity if they are both even or both odd. [QUANT MECH] A physical property of a wave function which specifies its behavior under an inversion, that is, under simultaneous reflection of all three spatial coordinates through the origin; if the wave function is unchanged by inversion, its parity is 1 (or even); if the function is changed only in sign, its parity is -1 (or odd).

parity bit [COMMUN] An additional nondata bit that is attached to a set of data bits to check their validity; it is set so that the sum of one-bits in the augmented set is always odd or always even.

parity check *See* odd-even check.

parity conservation *See* conservation of parity.

parity error [ADP] A machine error in which an odd number of bits are accidentally changed, so that the error can be detected by a parity check.

parity selection rules [QUANT MECH] Rules which specify whether or not a change in parity occurs during a given type of transition of an atom, molecule, or nucleus; for example, the Laporte selection rule, or the rule that there is no parity change in an allowed β-decay transition of a nucleus.

parity transformation [COMMUN] A change in value of a transmitted character denoting the number of one-bits. [PHYS] *See* inversion.

parker *See* rep.

PARROT

A parrot, with short hooked beak.

parkerite [MINERAL] $Ni_3(Bi,Pb)_2S_2$ A bright-bronze mineral composed of nickel bismuth lead sulfide.

parkerizing [MET] Trade name for a process for the production of phosphate coating on steel articles by immersion in an aqueous solution of manganese or zinc acid with phosphate.

Parker-Washburn boundary [SOLID STATE] A surface which separates two regions in a solid in which the crystal axes point in different directions, and which is made up of a single array of dislocations.

Parkes process [MET] A process for recovering precious metals from lead by stirring about 2% zinc into the melt to form zinc compounds with gold and silver which can then be skimmed off the surface.

parking apron [CIV ENG] A hard-surfaced area used for parking aircraft.

parking lot [CIV ENG] An outdoor lot for parking automobiles.

parking orbit [AERO ENG] A temporary earth orbit during which the space vehicle is checked out and its trajectory carefully measured to determine the amount and time of increase in velocity required to send it into a final orbit or into space in the desired direction.

Parkinje effect *See* Purkinje effect.

parkinsonism [MED] A clinical state characterized by tremor at a rate of three to eight tremors per second, with "pill-rolling" movements of the thumb common, muscular rigidity, dyskinesia, hypokinesia, and reduction in number of spontaneous and autonomic movements; produces a masked facies, disturbances of posture, gait, balance, speech, swallowing, and muscular strength. Also known as paralysis agitans; Parkinson's disease.

Parkinson's disease *See* parkinsonism.

parkland *See* temperate woodland; tropical woodland.

parkway [CIV ENG] A broad landscaped expressway which is not open to commercial vehicles.

Parmeliaceae [BOT] The foliose shield lichens, a family of the order Lecanorales.

Parnidae [INV ZOO] The equivalent name for Dryopidae.

paromomycin [MICROBIO] A broad-spectrum antibiotic produced by *Streptomyces rimosus forma paromomycinus*; it is effective in the treatment of intestinal amebiasis in humans.

paronychia [MED] A suppurative inflammation about the margin of a nail.

paronychium *See* perionychium.

parotid duct [ANAT] The duct of the parotid gland. Also known as Stensen's duct.

parotidectomy [MED] Excision of a parotid gland.

parotid gland [ANAT] The salivary gland in front of and below the external ear; the largest salivary gland in humans; a compound racemose serous gland that communicates with the mouth by Steno's duct.

parotitis [MED] Inflammation of the parotid glands.

parovarian cyst [MED] A cyst of mesonephric origin arising between the layers of the mesosalpinx, adjacent to the ovary.

parovarium *See* epoophoron.

paroxysm [MED] **1.** A sudden attack, or the periodic crisis in the progress of a disease. **2.** A spasm, convulsion, or seizure. [PSYCH] A sudden, uncontrollable emotional outburst.

paroxysmal eruption *See* Vulcanian eruption.

parrot [VERT ZOO] Any member of the avian family Psittacidae, distinguished by the short, stout, strongly hooked beak.

Parry arcs [OPTICS] A class of halos appearing as faintly colored arcs above and below the sun; these refraction phenomena are produced by ice crystals which exhibit a preferred orientation, and are correspondingly more unusual than those associated with randomly oriented crystals.

pars anterior [ANAT] The major secretory portion of the anterior lobe of the adenohypophysis. Also known as pars distalis.

pars distalis *See* pars anterior.

parsec [ASTRON] The distance at which a star would have a parallax equal to 1 second of arc; 1 parsec equals 3.258 light-years or 3.08572×10^{13} kilometers.

parsettensite [MINERAL] $Mn_5Si_6O_{13}(OH)_8$ Copper-red mineral composed of hydrous silicate of manganese.

Parseval equation [MATH] The general case when Bessel's inequality actually is an equality.

Parseval's identity [MATH] An identity which gives the

length of a vector in an inner product space in terms of its inner product with the members of an orthonormal basis.

Parseval's theorem [MATH] A theorem that gives the integral of a product of two functions, $f(x)$ and $f'(x)$, in terms of their respective Fourier coefficients; if the coefficients are defined by

$$a_n = (1/\pi) \int_0^{2\pi} f(x) \cos nx\, dx,$$

$$b_n = (1/\pi) \int_0^{2\pi} f(x) \sin nx\, dx,$$

and similarly for $f'(x)$, the relationship is

$$\int_0^{2\pi} f(x) f'(x) dx = \pi[\tfrac{1}{2} a_0 a_0' + \sum_{n=1}^{\infty} (a_n a_n' + b_n b_n')].$$

Parshall flume [ENG] A calibrated device for measuring the flow of liquids in open conduits by measuring the upper and lower beads at a specified distance from an obstructing sill.

pars intermedia [ANAT] The intermediate lobe of the adenohypophysis.

parsley [BOT] *Petroselinum crispum.* A biennial herb of European origin belonging to the order Umbellales; grown for its edible foliage.

parsley oil [MATER] Colorless or pale-greenish-yellow liquid with parsley aroma; soluble in alcohol, ether, and chloroform; distilled from parsley seeds; used in medicine.

pars nervosa [ANAT] The inferior subdivision of the neurohypophysis. Also known as pars neuralis.

pars neuralis *See* pars nervosa.

parsnip [BOT] *Pastinaca sativa.* A biennial herb of Mediterranean origin belonging to the order Umbellales; grown for its edible thickened taproot.

parsonsite [MINERAL] $Pb_2(UO_2)(PO_4)_2 \cdot 2H_2O$ A pale-yellow to brownish mineral composed of hydrous lead uranyl phosphate, occurring as a powder.

Parsons-stage steam turbine [MECH ENG] A reaction-type steam-turbine stage in which the pressure drop occurs partially across the stationary nozzles and partly across the rotating blades.

pars tuberalis [ANAT] A pair of processes that grow forward or upward along the stalk of the adenohypophysis.

part [ENG] An element of a subassembly, not normally useful by itself and not amenable to further disassembly for maintenance purposes.

parthenocarpy [BOT] Production of fruit without fertilization.

parthenogenesis [INV ZOO] A special type of sexual reproduction in which an egg develops without entrance of a sperm; common among rotifers, aphids, thrips, ants, bees, and wasps.

parthenomerogony [EMBRYO] Development of a nucleated fragment of an unfertilized egg following parthenogenetic stimulation.

partial [ACOUS] Also known as partial tone. **1.** A simple sinusoidal physical component of a complex tone. **2.** A sound sensation component that is distinguishable as a simple tone, cannot be further analyzed by the ear, and contributes to the character of the complex sound; the frequency of a partial may be higher or lower than the basic frequency and may be an integral multiple or submultiple of the basic frequency.

partial bulkhead [NAV ARCH] A partition wall that does not extend across a compartment; used to strengthen the structure.

partial carry [ADP] A word composed of the carries generated at each position when adding many digits in parallel.

partial cleavage [EMBRYO] Cleavage in which only part of the egg divides into blastomeres.

partial common battery [COMMUN] Type of telephone system in which the talking battery is supplied by each individual telephone, and the signaling and supervisory battery is supplied by the switchboard.

partial condensation [CHEM ENG] The cooling (or pressurization) of a saturated vapor until a part of it is condensed out as liquid.

partial derivative [MATH] A derivative of a function of several variables taken with respect to one variable while holding the others fixed.

partial dislocation [CRYSTAL] The line at the edge of an extended dislocation where a slip through a fraction of a lattice constant has occurred.

partial-duration series [GEOPHYS] A series composed of all events during the period of record which exceed some set criterion; for example, all floods above a selected base, or all daily rainfalls greater than a specified amount.

partial eclipse [ASTRON] An eclipse in which only part of the source of light is obscured.

partial fractions [MATH] A collection of fractions which when added are a given fraction whose numerator and denominator are usually polynomials; the partial fractions are usually constants or linear polynomials divided by factors of the denominator of the given fraction.

partially ordered set [MATH] A set on which a partial order is defined.

partial molal quantity [PHYS CHEM] The molal concentration of one component of a mixture of components as related to total molal concentration for all components in the mixture.

partial molar volume [PHYS CHEM] That portion of the volume of a solution or mixture related to the molar content of one of the components within the solution or mixture.

partial obscuration [METEOROL] In United States weather observing practice, the designation for sky cover when part (0.1 to 0.9) of the sky is completely hidden by surface-based obscuring phenomena.

partial ordering *See* ordering.

partial penetration [ORD] Penetration obtained when a projectile fails to pass through the target far enough for either the projectile itself or light from its penetration to be seen from the back of the target.

partial pluton [GEOL] That part of a composite intrusion representing a single intrusive episode.

partial potential temperature [METEOROL] The temperature that the dry-air component of an air parcel would attain if its actual partial pressure were changed to 1000 millibars.

partial pressure [PHYS] The pressure that would be exerted by one component of a mixture of gases if it were present alone in a container.

partial pressure maintenance [PETRO ENG] The partial replacement of produced gas in an oil reservoir by gas injection to maintain a portion of the initial reservoir pressure.

partial-read pulse [ELECTR] Current pulse that is applied to a magnetic memory to select a specific magnetic cell for reading.

partial-select output [ELECTR] The voltage response produced by applying partial-read or partial-write pulses to an unselected magnetic cell.

partial sum [MATH] A partial sum of an infinite series is the sum of its first n-terms for some n.

partial tide [OCEANOGR] One of the harmonic components composing the tide at any point. Also known as tidal component; tidal constituent.

partial tone *See* partial.

particle [MECH] *See* material particle. [PARTIC PHYS] *See* elementary particle. [PHYS] **1.** Any very small part of matter, such as a molecule, atom, or electron. **2.** Any relatively small subdivision of matter, ranging in diameter from a few angstroms (as with gas molecules) to a few millimeters (as with large raindrops).

particle accelerator [NUCLEO] A device which accelerates electrically charged atomic or subatomic particles, such as electrons, protons, or ions, to high energies. Also known as accelerator; atom smasher.

particle beam [PHYS] A concentrated, nearly unidirectional flow of particles.

particle board [MATER] Construction board made with wood particles impregnated with low-molecular-weight resin and then cured.

particle counting [ANALY CHEM] Microscopic or photomicrographic technique for the visual counting of the numbers of particles in a known quantity of a solid-liquid suspension.

particle detector [NUCLEO] A device used to indicate the presence of fast-moving charged atomic or nuclear particles by observation of the electrical disturbance created by a particle as it passes through the device. Also known as radiation detector.

particle diameter [GEOL] The diameter of a sedimentary particle considered as a sphere.

particle dynamics [MECH] The study of the dependence of the motion of a single material particle on the external forces acting upon it, particularly electromagnetic and gravitational forces.

particle emission [NUC PHYS] The ejection of a particle other than a photon from a nucleus, in contrast to gamma emission.

particle energy [MECH] For a particle in a potential, the sum of the particle's kinetic energy and potential energy. [RELAT] For a relativistic particle the sum of the particle's potential energy, kinetic energy, and rest energy; the last is equal to the product of the particle's rest mass and the square of the speed of light.

particle lens [PHYS] An electric or magnetic field, or a combination thereof, which acts upon an electron beam in a manner analogous to that in which an optical lens acts upon a light beam.

particle mechanics [MECH] The study of the motion of a single material particle.

particle multiplet *See* isospin multiplet.

particle physics [PHYS] The branch of physics concerned with understanding the properties and behavior of elementary particles, especially through study of collisions or decays involving energies of hundreds of Mev (million electron volts) or more. Also known as high-energy physics.

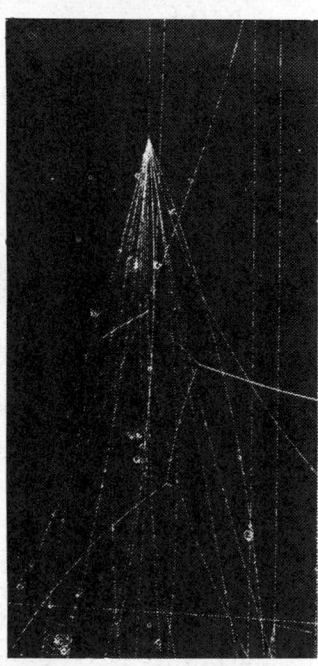

PARTICLE TRACK

Photograph of particle tracks consisting of trails of bubbles in a liquid-hydrogen bubble chamber, produced by negative pions with momentum 16 GeV/c, where c is the speed of light. (CERN)

particle properties [PARTIC PHYS] The various quantities which characterize the behavior of an elementary particle, such as mass, charge, baryon number, spin, parity, hypercharge, and isospin.

particle-scattering factor [ANALY CHEM] Factor in light-scattering equations used to compensate for the loss in scattered light intensity caused by destructive interference during the analysis of macromolecular compounds.

particle size [GEOL] The general dimensions of the particles or mineral grains in a rock or sediment based on the premise that the particles are spheres; commonly measured by sieving, by calculating setting velocities, or by determining areas of microscopic images. [MET] The average and controlling lineal dimension of an individual particle of metal powder as determined by suitable screens or other methods of analysis.

particle-size analysis [ENG] Determination of the proportion of particles of a specified size in a granular or powder sample. [GEOL] A determination of the distribution of particles in a series of size classes of a soil, sediment, or rock. Also known as size analysis; size-frequency analysis.

particle-size distribution [ENG] The percentages of each fraction into which a granular or powder sample is classified, with respect to particle size, by number or weight.

particle spectrum *See* mass spectrum.

particle-thickness technique [PHYS CHEM] Microscopic technique for visual measurement of the thickness of a fine particle (in the 3–100 micrometer range).

particle track [PHYS] Any visible phenomenon along the path of an ionizing particle, such as a trail of bubbles, water droplets, or sparks in a bubble chamber, cloud chamber, or spark chamber respectively, or of altered material in an emulsion or in glass.

particle velocity [ACOUS] The instantaneous velocity of a given infinitesimal part of a medium, with reference to the medium as a whole, due to the passage of a sound wave. [OCEANOGR] In ocean wave studies, the instantaneous velocity of a water particle undergoing orbital motion.

parti-colored buoy [NAV] A buoy whose visible surface is painted more than one color.

particulate mass analyzer [ENG] A unit which measures dust concentrations in emissions from furnaces, kilns, cupolas, and scrubbers.

particulate matter [PHYS] Matter in the form of small liquid or solid particles.

particulates [MATER] Fine solid particles which remain individually dispersed in gases and stack emissions.

parting [GEOL] 1. A bed or bank of waste material dividing mineral veins or beds. 2. A soft, thin sedimentary layer following a surface of separation between thicker strata of different lithology. 3. A surface along which a hard rock can be readily separated or is naturally divided into layers. [MET] 1. Recovery of gold (or occasionally another metal) from its alloys by a corrosion process. 2. Zone of separation between cope and drag portions of mold or flask in sand casting. 3. In

sand molding, a composition to facilitate removal of the pattern. 4. A shearing operation to produce two or more parts from a stamping. [MINERAL] Fracturing a mineral along planes weakened by deformation or twinning.

parting agent *See* release agent.

parting line [MET] 1. The line along which a mold is separated. 2. A line or seam on a casting corresponding to the joint of mold parts.

parting lineation [GEOL] A small-scale primary sedimentary structure made up of a series of parallel ridges and grooves formed parallel to the current. Also known as current lineation.

parting sand [MET] Fine, dry sand applied to the faces of a sand mold to allow disassembly.

parting stop [BUILD] A thin strip of wood that separates the sashes in a double-hung window.

partition [BUILD] An interior wall having a height of one story or less, which divides a structure into sections. [IND ENG] A slotted sheet of paperboard that can be assembled with similar sheets to form cells for holding goods during shipment.

partition chromatography [ANALY CHEM] Chromatographic procedure in which the stationary phase is a high-boiling liquid spread as a thin film on an inert support, and the mobile phase is a vaporous mixture of the components to be separated in an inert carrier gas.

partition coefficient [ANALY CHEM] In the equilibrium distribution of a solute between two liquid phases, the constant ratio of the solute's concentration in the upper phase to its concentration in the lower phase. Symbolized K.

partitioned data set [ADP] A single data set, divided internally into a directory and one or more sequentially organized subsections called members, residing on a direct access for each device, and commonly used for storage or program libraries.

partition function [STAT MECH] 1. The integral, over the phase space of a system, of the exponential of $(-E/kT)$, where E is the energy of the system, k is Boltzmann's constant, and T is the temperature; from this function all the thermodynamic properties of the system can be derived. 2. In quantum statistical mechanics, the sum over allowed states of the exponential of $(-E/kT)$. Also known as sum of states; sum over states.

partition noise [ELECTR] Noise that arises in an electron tube when the electron beam is divided between two or more electrodes, as between screen grid and anode in a pentode.

partition of a positive integer [MATH] Any sum of positive integers equaling the integer.

partition of a set [MATH] A finite collection of disjoint sets whose union is the given set.

partition of unity [MATH] On a topological space X, this is a covering by open sets U_α with continuous functions f_α from X to [0,1], where each f_α is zero on all but a finite number of the U_α, and the sum of all these f_α at any point equals 1.

partiversal [GEOL] Pertaining to formations that dip in different directions roughly as far as a semicircle.

partly cloudy [METEOROL] 1. The character of a day's weather when the average cloudiness, as determined from frequent observations, has been from 0.1 to 0.5 for the 24-hour period. 2. In popular usage, the state of the weather when clouds are conspicuously present, but do not completely dull the day or the sky at any moment.

parton [PARTIC PHYS] One of the very singular (or hard), small charged particles of which hadrons are proposed to be constructed, according to a theory developed to account for the scattering of very-high-energy electrons from protons at large angles and with large momentum transfers.

part operation [ADP] The part in an instruction that specifies the kind of arithmetical or logical operation to be performed, but not the address of the operands.

partridge [VERT ZOO] Any of the game birds comprising the genera *Alectoris* and *Perdix* in the family Phasianidae.

parts kit [ENG] A group of parts, not all having the same basic name, used for repair or replacement of the worn broken parts of an item; it may include instruction sheets and material, such as sandpaper, tape, cement, and gaskets.

parts list [ENG] One or more printed sheets showing a manufacturer's parts or assemblies of an end item by illustration or a numerical listing of part numbers and names; it does not outline any assembly, maintenance, or operating instructions, and it may or may not have a price list cover sheet.

parturient [MED] **1.** In labor; giving birth. **2.** Of or pertaining to parturition.

parturifacient [MED] **1.** Inducing labor. **2.** An agent that induces labor.

parturiometer [MED] An instrument to determine the progress of labor by measuring the expulsive force of the uterus.

parturition [MED] The process of giving birth.

party line [COMMUN] A subscriber line arranged to serve more than one station, with discriminatory ringing for each station.

party-line bus [ADP] Parallel input/output bus lines to which are wired all external devices, connected to a processor register by suitable logic.

party-line carrier system [COMMUN] A single-frequency carrier telephone system in which the carrier energy is transmitted directly to all other carrier terminals of the same channel.

party wall [BUILD] A wall providing joint service between two buildings.

parulis [MED] A subperiosteal abscess arising from dental structures.

parvafacies [GEOL] A body of rock constituting the part of any magnafacies that occurs between designated time-stratigraphic planes or key beds traced across the magnafacies.

parvovirus [VIROL] The equivalent name for picodnavirus.

parylene [ORG CHEM] Polyparaxylylene, used in ultrathin plastic films for capacitor dielectrics, and as a pore-free coating.

parylene capacitor [ELEC] A highly stable fixed capacitor using parylene film as the dielectric; it can be operated at temperatures up to 170°C, as well as at cryogenic temperatures.

PAS *See para*-aminosalicylic acid.

pascal [MECH] A unit of pressure equal to the pressure resulting from a force of 1 newton acting uniformly over an area of 1 square meter. Abbreviated Pa. Also known as tor.

Pascal rules [PHYS CHEM] Rules which give the diamagnetic susceptiblity of a complex molecule in terms of the sum of the susceptibilities of its constituent atoms, and a correction factor which depends on the type of bonds linking the atoms.

Pascal's law [FL MECH] The law that a confined fluid transmits externally applied pressure uniformly in all directions, without change in magnitude.

Pascal's limacon *See* limacon.

Pascal's theorem [MATH] The theorem that when one inscribes a simple hexagon in a conic, the three pairs of opposite sides meet in collinear points.

Pascal's triangle [MATH] A triangular array of the binomial coefficients, bordered by ones, where the sum of two adjacent entries from a row equals the entry in the next row directly below. Also known as binomial array.

Paschen-Back effect [SPECT] An effect on spectral lines obtained when the light source is placed in a very strong magnetic field; the anomalous Zeeman effect obtained with weaker fields changes over to what is, in a first approximation, the normal Zeeman effect.

Paschen bodies [PATH] Accumulations of the elementary bodies of smallpox.

Paschen-Runge mounting [SPECT] A diffraction grating mounting in which the slit and grating are fixed, and photographic plates are clamped to a fixed track running along the corresponding Rowland circle.

Paschen series [SPECT] A series of lines in the infrared spectrum of atomic hydrogen whose wave numbers are given by $R_H [(1/9) - (1/n^2)]$, where R_H is the Rydberg constant for hydrogen, and n is any integer greater than 3.

Paschen's law [ELECTR] The law that the sparking potential between two parallel plate electrodes in a gas is a function of the product of the gas density and the distance between the electrodes. Also known as Paschen's rule.

Paschen's rule *See* Paschen's law.

pascoite [MINERAL] $Ca_2V_6O_{17} \cdot 11H_2O$ A dark-red-orange to yellow-orange mineral composed of hydrous vanadate of calcium.

pasmo [PL PATH] A destructive fungus disease of flax caused by *Mycosphaerella linorum*.

pass [ADP] A complete cycle of reading, processing, and writing in a computer. [AERO ENG] **1.** A single circuit of the earth made by a satellite; it starts at the time the satellite crosses the equator from the Southern Hemisphere into the Northern Hemisphere. **2.** The period of time in which a satellite is within telemetry range of a data acquisition station. [GEOGR] **1.** A natural break, depression, or other low place providing a passage through high terrain, such as a mountain range. **2.** A navigable channel leading to a harbor or river. **3.** A narrow opening through a barrier reef, atoll, or sand bar. [MET] **1.** Passage of a metal bar between rolls. **2.** Open space between two grooved rolls through which metal is processed. **3.** Weld metal deposited in one trip along the axis of a weld. [MIN ENG] **1.** A mine opening through which coal or ore is delivered from a higher to a lower level. **2.** A passage left in old workings for workers to travel as they move from one level to another. **3.** A treatment of the whole ore sample in a sample divider. **4.** A passage of an excavation or grading machine. **5.** In surface mining, a complete excavator cycle in removing overburden.

passage [GEOGR] A navigable channel, especially one through reefs or islands. [NAV] A transit from one place to another; one leg of a voyage.

passage bed [GEOL] A stratum marking a transition from rocks of one geological system to those of another.

passageway [SCI TECH] A way that permits passage between two places or points.

Passalidae [INV ZOO] The peg beetles, a family of tropical coleopteran insects in the superfamily Scarabaeoidea.

passband [ELECTR] A frequency band in which the attenuation of a filter is essentially zero.

pass-by [ENG] The double-track part of any single-track system of rail transport. [MIN ENG] A passage around the working part of a shaft.

pass element [ELECTR] Controlled variable resistance device, either a vacuum tube or power transistor, in series with the source of direct-current power; the pass element is driven by the amplified error signal to increase its resistance when the output needs to be lowered or to decrease its resistance when the output must be raised.

passenger car [ENG] **1.** A railroad car in which passengers are carried. **2.** An automobile for carrying as many as nine passengers.

passenger ship [NAV ARCH] A ship used primarily to carry passengers.

Passeres [VERT ZOO] The equivalent name for Oscines.

Passeriformes [VERT ZOO] A large order of perching birds comprising two major divisions: Suboscines and Oscines.

Passifloraceae [BOT] A family of dicotyledonous, often climbing plants in the order Violales; flowers are polypetalous and hypognous with a corona, and seeds are arillate with an oily endosperm.

passing point [MIN ENG] The point at which two vehicles, such as coal cars or mine elevators, pass each other while going in opposite directions.

passing track [ENG] A sidetrack with switches at both ends.

passivation [ELECTR] Growth of an oxide layer on the surface of a semiconductor to provide electrical stability by isolating the transistor surface from electrical and chemical conditions in the environment; this reduces reverse-current leakage, increases breakdown voltage, and raises power dissipation rating. [MET] To render passive; to reduce the reactivity of a chemically active metal surface by electrochemical polarization or by immersion in a passivating solution.

passive-active cell [MET] An electrochemical corrosion cell established between passive and active areas on a metal surface.

passive-aggressive personality [PSYCH] A personality disorder characterized by the passive expression of hostility and aggressiveness, as by stubbornness, pouting, or inefficiency.

passive anaphylaxis [IMMUNOL] Anaphylaxis elicited by temporary sensitization with antibodies followed by injection of the corresponding sensitizing antigen.

passive AND gate [ELECTR] *See* AND gate. [ENG] A fluidic device which achieves an output signal, by stream interaction, only when both of two control signals appear simultaneously.

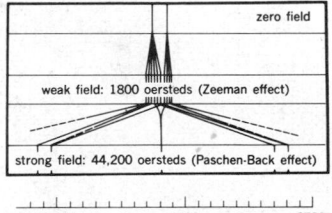

PASCHEN-BACK EFFECT

Zeeman and Paschen-Back effects of red lithium doublet, whose natural separation is 0.175 angstrom.

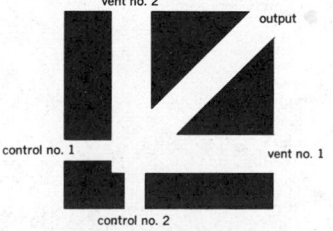

PASSIVE AND GATE

Passive AND gate, a type of fluidic element. When only one control signal appears it attaches to opposite wall and vents, producing no output. Output is achieved when both controls appear simultaneously.

passive antiroll system [NAV ARCH] A system of antiroll tanks connected by a channel that is sized so that the flow of the liquid is out of phase with the roll of the ship.

passive armor [ORD] A protective device against shaped charge ammunition; designed to absorb the energy of a shaped charge.

passive communications satellite [AERO ENG] A satellite that reflects communications signals between stations, without providing amplification; an example is the Echo satellite. Also known as passive satellite.

passive component See passive element.

passive congestion [MED] An increased content of blood in an organ or other body part due to impaired return of venous blood.

passive corner reflector [ELECTROMAG] A corner reflector that is energized by a distant transmitting antenna; used chiefly to improve the reflection of radar signals from objects that would not otherwise be good radar targets.

passive cutaneous anaphylaxis [IMMUNOL] The vascular reaction at the site of intradermally injected antibody when, 3 hours later, the specific antigen, usually mixed with Evans blue dye, is injected intravenously.

passive-dependent personality [PSYCH] A character disorder marked by a behavioral pattern characterized by a lack of self-confidence, indecisiveness, and a tendency to cling to and seek support from others.

passive detection [ORD] The detection of a target or other object by means that do not reveal the position of the detecting instrument.

passive electronic countermeasures [ELECTR] Electronic countermeasures that do not radiate energy, including reconnaissance or surveillance equipment that detects and analyzes electromagnetic radiation from radar and communications transmitters, and devices such as chaff which return spurious echoes to enemy radar.

passive element [ELEC] An element of an electric circuit that is not a source of energy, such as a resistor, inductor, or capacitor. Also known as passive component. [ELECTROMAG] See parasitic element.

passive filter [ELEC] An electric filter composed of passive elements, such as resistors, inductors, or capacitors, without any active elements, such as vacuum tubes or transistors.

passive front See inactive front.

passive homing [NAV] The homing of an aircraft or spacecraft wherein the craft directs itself toward its destination by receiving radio emission from a base station without the necessity of initiating the transmission by sending an interrogating signal from the craft.

passive immunity [IMMUNOL] **1.** Immunity acquired by injection of antibodies in another individual or in an animal. **2.** Immunity acquired by the fetus by the transfer of maternal antibodies through the placenta.

passive jamming [ELECTR] Use of confusion reflectors to return spurious and confusing signals to enemy radars. Also known as mechanical jamming.

passive junction [ELECTROMAG] A waveguide junction that does not have a source of energy.

passive metal [MET] A metal on which a surface film forms by natural process or by immersion in a passivating solution, making the metal resistant to corrosion.

passive network [ELEC] A network that has no source of energy.

passive permafrost [GEOL] Permafrost that will not refreeze under present climatic conditions after being disturbed or destroyed. Also known as fossil permafrost.

passive radar [ENG] A technique for detecting objects at a distance by picking up the microwave electromagnetic energy that is both radiated and reflected by all bodies.

passive reflector [ELECTROMAG] A flat reflector used to change the direction of a microwave or radar beam; often used on microwave relay towers to permit placement of the transmitter, repeater, and receiver equipment on the ground, rather than at the tops of towers. Also known as plane reflector.

passive satellite See passive communications satellite.

passive sonar [ENG] Sonar that uses only underwater listening equipment, with no transmission of location-revealing pulses.

passive system [ELECTR] Electronic system which emits no energy, and does not give away its position or existence.

passive transducer [ELECTR] A transducer containing no internal source of power.

passivity [CHEM] A state of chemical inactivity, especially of a metal that is relatively resistant to corrosion due to loss of chemical activity. [MET] The property of a metal that has been made passive.

paste [ELEC] In batteries, the medium in the form of a paste or jelly, containing an electrolyte; it is positioned adjacent to the negative electrode of a dry cell; in an electrolytic cell, the paste serves as one of the conducting plates. [MATER] An adhesive mixture with a characteristic plastic consistency, a high order of yield value, and a low bond strength; for example, a paste prepared by heating a starch and water mixture, then cooling the hydrolyzed product. [MET] Finely divided particles of ferromagnetic material in paste form used in the wet method of magnetic particle inspection.

pasteboard [MATER] A type of thin cardboard made from gluing together two or more sheets of paper.

pasted-plate storage battery See Faure storage battery.

paste ink [MATER] A pastelike mixture of pigment or dye with oil and other additives, such as resins, driers, tackifiers, and adhesives; used in paper and textile printing and ballpoint pens.

pastel [METER] A chalk or crayon made of a finely ground pigment and a minimum of nongreasy binder, such as gum tragacanth or methylcellulose; since pastels are blended on the painting itself, a larger assortment of shades and tints is required than with oil or other color mediums.

pastel fixative [MATER] A thin varnish material for the simple and even application to drawings; only enough fixative should be used to keep the pastel particles from falling off the paper.

paste mixer [ENG] Device for the blending together of solid particles and a liquid, with the final formation of a single paste phase.

paste resin [MATER] Solventless, fluid or semisolid mixture of powdered resin and plasticizer.

paste solder [MET] Finely powdered solder metal combined with a flux.

paste-up See mechanical.

Pasteur-Chamberland filter [MICROBIO] A porcelain filter with small pores used in filtration sterilization.

Pasteur effect [MICROBIO] Inhibition of fermentation by supplying an abundance of oxygen to replace anerobic conditions.

Pasteurella [MICROBIO] A genus of bacteria in the family Brucellaceae distinguished by the tendency for bipolar staining, and fermentation of carbohydrates.

pasteurellosis See hemorrhagic septicemia.

Pasteuriaceae [MICROBIO] A family of stalked bacteria in the order Hyphomicrobiales.

pasteurization [FOOD ENG] The application of heat for a specified time to a liquid food or beverage to enhance its keeping properties by destroying harmful microorganisms.

pasteurizer [ENG] An apparatus used for pasteurization of fluids.

Pasteur's salt solution [ANALY CHEM] Laboratory reagent consisting of potassium phosphate and calcium phosphate, magnesium sulfate, and ammonium tartrate in distilled water.

past of an event [RELAT] All events from which a signal could be emitted that could reach the event in question by traveling at speeds less than or equal to the speed of light.

patagium [VERT ZOO] **1.** A membrane or fold of skin extending between the forelimbs and hindlimbs of flying squirrels, flying lizards, and other arboreal animals. **2.** A membrane or fold of skin on a bird's wing anterior to the humeral and radioulnar bones.

patch [ADP] **1.** To modify a program or routine by inserting a machine language correction in an object deck, or by inserting it directly into the computer through the console. **2.** The section of coding inserted in this way. [ELEC] A temporary connection between jacks or other terminations on a patch board.

patch board [ELEC] A board or panel having a number of jacks at which circuits are terminated; patch cords are plugged into the jacks to connect various circuits temporarily as required in broadcast, communication, and computer work.

patch bolt [DES ENG] A bolt with a countersunk head having a square knob that twists off when the bolt is screwed in tightly; used to repair boilers and steel ship hulls.

patch budding [BOT] Budding in which a small rectangular patch of bark bearing the scion (a bud) is fitted into a corresponding opening in the bark of the stock.

patch cord [ELEC] A cord equipped with plugs at each end, used to connect two jacks on a patch board.

patchouli oil [MATER] A brownish essential oil with strong camphor aroma; soluble in ether, chloroform, alcohol, and oils; derived from dried leaves of the patchouli (*Pogostemon patchouly*); main components are patchouli alcohol, eugenol, and cinnamic aldehyde; used to perfume toiletries.

patch panel *See* control panel; panel.

patch reef [GEOL] **1.** A small, irregular organic reef with a flat top forming a part of a reef complex. **2.** A small, thick, isolated lens of limestone or dolomite surrounded by rocks of different facies. **3.** *See* reef patch.

patch test [IMMUNOL] A test in which material is applied and left in contact with intact skin surfaces for 48 hours in order to demonstrate tissue sensitivity.

patella [ANAT] A sesamoid bone in front of the knee, developed in the tendon of the quadriceps femoris muscle. Also known as kneecap.

Patellacea [PALEON] An extinct superfamily of gastropod mollusks in the order Aspidobranchia which developed a cap-shaped shell and were specialized for clinging to rock.

Patellidae [INV ZOO] The true limpets, a family of gastropod mollusks in the order Archeogastropoda.

patent [IND ENG] A certificate of grant by a government of an exclusive right with respect to an invention for a limited period of time. Also known as letters patent. [MED] Open; exposed.

patented claim [MIN ENG] A mining claim to which a patent has been secured from the government by compliance with the laws relating to such claims.

patent foramen ovale [MED] Persistence, usually functional, of the fetal foramen ovale after birth.

patenting [MET] A process used in the production of high-strength steel wire containing 0.35–0.85% carbon, in which the wire is heated to above the transformation temperature, then quenched in molten lead or molten salt, or cooled in air.

patent leveling *See* stretcher leveling.

patent log [NAV] Any mechanical log; for example, an instrument that is dragged from the stem of a vessel to measure the speed or the distance traveled.

patent medicine [PHARM] A medicine, generally trademarked, whose composition is incompletely disclosed.

Patera process [MET] A method used in the 19th century for extracting silver from ore: the ore was roasted with sodium chloride, a solution of sodium thiosulfate was then added to leach out the silver chloride, and sodium sulfide precipitated the silver as silver sulfide.

Paterinida [PALEON] A small extinct order of inarticulated brachiopods, characterized by a thin shell of calcium phosphate and convex valves.

paternity test [IMMUNOL] Identification of the blood groups of a mother, her child, and a putative father in order to establish the probability of paternity or nonpaternity; actually, only nonpaternity can be established.

Paterson-Kelly syndrome *See* Plummer-Vinson syndrome.

path [MATH] In a topological space, a path is a continuous curve joining two points. [NAV] A line connecting a series of points and constituting a proposed or traveled route.

path attenuation [COMMUN] Power loss between transmitter and receiver, due to any cause.

path coefficient [COMMUN] The ratio of the power transmitted over some designated path to that transmitted over the most direct path.

pathergy [IMMUNOL] Either a subnormal response to an allergen or an unusually intense one in which the individual becomes sensitive not only to the specific substance but to others.

pathogen [MED] A disease-producing agent; usually refers to living organisms.

pathogenesis [MED] The origin and course of development of disease.

pathogenic [MED] **1.** Producing or capable of producing disease. **2.** Pertaining to pathogenesis.

pathognomonic [MED] Characteristic of a given disease, enabling it to be distinguished from other diseases.

pathologic anatomy [PATH] The study of structural changes caused by disease. Also known as marked anatomy.

pathologist [MED] A person who specializes in the study and practice of pathology.

pathology [MED] The study of the causes, nature, and effects of diseases and other abnormalities.

pathophobia [PSYCH] Exaggerated dread of death.

pathopsychology [PSYCH] The branch of science dealing with mental processes, particularly as manifested by abnormal cognitive, perceptual, and intellectual functioning, during the course of mental disorders.

path plotting [ELECTROMAG] In laying out a microwave system, the plotting of the path followed by the microwave beam on a profile chart which indicates the earth's curvature.

patina [GEOL] A thin, colored film produced on a rock surface by weathering. [MET] The greenish product, usually basic copper sulfate, formed on copper and copper-rich alloys as a result of prolonged atmospheric corrosion.

patio process [MET] A crude chemical method of reducing silver from its ores, followed by amalgamation in low heaps with the aid of salt and copper sulfate.

patronite [MINERAL] A black vanadium sulfide mineral; mined as a vanadium ore in Minasragra, Peru.

pattern [AERO ENG] The flight path flown by an aircraft, or prescribed to be flown, as in making an approach to a landing. [ENG] A form designed and used as a model for making things. [GRAPHICS] A design or form. [ORD] The distribution of a series of shots fired from one gun or a battery of guns under conditions as nearly identical as possible, the points of impact of the projectiles being dispersed about a point called the center of impact.

pattern bombing [ORD] A method of bombing in which the bombs are made to strike the target in a certain pattern.

pattern card [TEXT] A perforated card used in jacquard weaving.

pattern chain [TEXT] A device on a loom or knitting machine that controls the pattern in the material.

patterned ground [GEOL] Any of several well-defined, generally symmetrical forms, such as circles, polygons, and steps, that are characteristic of surficial material subject to intensive frost action.

pattern flood [PETRO ENG] Waterflood of a petroleum reservoir in which there is an areal pattern of injection wells located so as to sweep oil from the area toward the bores of producing wells.

pattern generator [ELECTR] A signal generator used to generate a test signal that can be fed into a television receiver to produce on the screen a pattern of lines having usefulness for servicing purposes.

pattern harmonization [ORD] The adjusting of a fighter aircraft's fixed guns so that they will produce the largest uniform pattern of lethal density possible at a given range.

pattern recognition [ADP] The automatic identification of figures, characters, shapes, forms, and patterns without active human participation in the decision process.

pattern-sensitive fault [ADP] A fault that appears only in response to one pattern or sequence of data, or certain patterns or sequences.

pattern shooting [ENG] In seismic prospecting, firing of explosive charges arranged in geometric pattern.

Patterson function [SOLID STATE] A function of three spatial coordinates, constructed in the Patterson-Harker method, which has peaks at all vectors between two atoms in a crystal, the heights of the peaks being approximately proportional to the product of the atomic numbers of the corresponding atoms.

Patterson-Harker method [SOLID STATE] A method of analyzing the structure of a crystal from x-ray diffraction results; a Fourier series involving squares of the absolute values of the structure factors, which are directly observable, is used to construct a vectorial representation of interatomic distances in the crystal (Patterson map).

PAUROPODA

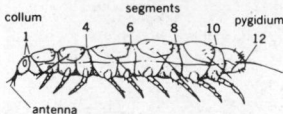

Pauropus silvaticus Tiegs, fully extended adult, enlarged, showing the 9 pairs of legs and the 12 trunk segments indicated by number.

PAWL

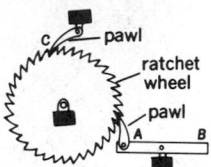

Pawls with a ratchet wheel. Driving pawl at *A*, forced upward by lever *B*, engages teeth of ratchet wheel and rotates it counterclockwise. Holding pawl *C* prevents clockwise rotation of the wheel.

PAXILLA

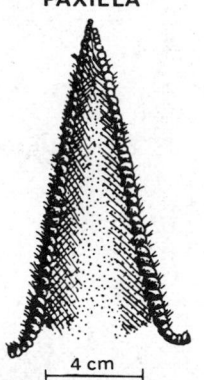

Arm of a starfish, *Persephonaster neozelanicus*, showing modified spines covering the surface.

PEA

Typical pea plant with pods hanging from the vines. *(Dumas Seed Co., Inc., Moscow, Idaho)*

Patterson map [SOLID STATE] A contour chart of the Patterson function.

Pattinson process [MET] A method for separating silver from its alloys rich in lead by slow cooling of the melt so that silver-poor lead crystals separate out and are removed.

paua *See* abalone.

Paucituberculata [VERT ZOO] An order of marsupial mammals in some systems of classification, including the opossum, rats, and polydolopids.

Paul-Bunnell test *See* heterophile agglutination test.

Pauli anomalous moment term [QUANT MECH] An additional term inserted in the Dirac equation to provide for a *g*-value of the particle different from 2.

Pauli electron correlation [QUANT MECH] Correlation in space of electrons as a result of the Pauli exclusion principle.

Pauli exclusion principle *See* exclusion principle.

Pauli-Fermi principle [QUANT MECH] The principle that each level of a quantized system can include one, two, or no electrons; if there are two electrons, they must have spins in opposite directions.

Pauli g-permanence rule [ATOM PHYS] For given *L*, *S*, and M_J in *LS* coupling, the sum, over *J*, of the weak-field *g*-factors is equal to the sum of the strong-field-factors.

Pauli g-sum rule [ATOM PHYS] For all the states arising from a given electron configuration, the sum of the *g*-factors for levels with the same *J* value is a constant, independent of the coupling scheme.

paulin [MATER] A fabricated textile item generally used as a weather protection cover for various items or materials during storage or transit.

paulingite [MINERAL] An isometric zeolite mineral consisting of an aluminosilicate of potassium, calcium, and sodium.

Pauling rule [SOLID STATE] A rule governing the number of ions of opposite charge in the neighborhood of a given ion in an ionic crystal, in accordance with the requirement of local electrical neutrality of the structure.

Pauli spin matrices [QUANT MECH] Three anticommuting matrices, each having two rows and two columns, which represent the components of the electron spin operator:

$$\sigma_x = \begin{pmatrix} 0 & 1 \\ 1 & 0 \end{pmatrix}, \quad \sigma_y = \begin{pmatrix} 0 & -i \\ i & 0 \end{pmatrix}, \quad \sigma_z = \begin{pmatrix} 1 & 0 \\ 0 & -1 \end{pmatrix}.$$

Pauli spin space [QUANT MECH] A two-dimensional vector space over the complex numbers, whose vectors describe orientations of the electron spin.

Pauli spin susceptibility [SOLID STATE] The susceptibility of free electrons in a metal due to the tendency of their spins to align with a magnetic field.

Pauli-Weisskopf equation [QUANT MECH] The equation resulting from second quantization of the Klein-Gordon equation.

paulopost *See* deuteric.

paunch [ANAT] In colloquial usage, the cavity of the abdomen and its contents. [VERT ZOO] *See* rumen.

paurometabolous metamorphosis [INV ZOO] A simple, gradual, direct metamorphosis in which immature forms resemble the adult except in size and are referred to as nymphs.

Pauropoda [INV ZOO] A class of the Myriapoda distinguished by bifurcate antennae, 12 trunk segments with 9 pairs of functional legs, and the lack of eyes, spiracles, tracheae, and a circulatory system.

Paussidae [INV ZOO] The flat-horned beetles, a family of coleopteran insects in the suborder Adephaga.

pavement [BUILD] A hard floor of concrete, brick, tiles, or other material. [CIV ENG] A paved surface.

pavement light [CIV ENG] A window built into the surface of a pavement to admit daylight to a space below ground level.

paver [MECH ENG] Any of several machines which, moving along the road, carry and lay paving material.

paving brick [MATER] A vitrified clay brick used in the construction of pavements.

paving-brick clay [MATER] Impure shale or fire clay with good tensile strength and plasticity; used to make paving bricks.

Pavlov's pouch [PHYSIO] A small portion of stomach, completely separated from the main stomach, but retaining its vagal nerve branches, which communicates with the exterior;

used in the long-term investigation of gastric secretion, and particularly in the study of conditioned reflexes.

Pavo [ASTRON] A southern constellation; right ascension 20 hours, declination 65°S. Also known as Peacock.

pavonite [MINERAL] $AgBi_3S_5$ A mineral composed of silver bismuth sulfide.

Pavy's solution [ANALY CHEM] Laboratory reagent used to determine the concentration of sugars in solution by color titration; contains copper sulfate, sodium potassium tartrate, sodium hydroxide, and ammonia in water solution.

paw [VERT ZOO] The foot of an animal, especially a quadruped having claws.

pawl [MECH ENG] The driving link or holding link of a ratchet mechanism, permits motion in one direction only.

PAX *See* private automatic exchange.

paxilla [INV ZOO] A pillarlike spine in certain starfishes that sometimes has a flattened summit covered with spinules.

Paxillosida [INV ZOO] An order of the Asteroidea in some systems of classification, equivalent to the Paxillosina.

Paxillosina [INV ZOO] A suborder of the Phanerozonida with pointed tube feet which lack suckers, and with paxillae covering the upper body surface.

pay dirt [MIN ENG] Profitable mineral-rich earth or ore.

paying [NAV ARCH] **1.** Filling the seams between planks with pitch, marine glue, or other material after the calking has been inserted. **2.** Slackening away on rope or chain.

pay load [AERO ENG] That which an aircraft, rocket, or the like carries over and above what is necessary for the operation of the vehicle in its flight. [MIN ENG] The weight of coal, ore, or mineral handled, as distinct from dirt, stone, or gangue.

pay load–mass ratio [AERO ENG] Of a rocket, the ratio of the effective propellant mass to the initial vehicle mass.

payoff matrix [MATH] A matrix arising from certain two-person games which gives the amount gained by a player.

pay ore [MIN ENG] Ore which can be mined, concentrated, or smelted at current cost of exploitation profitably at ruling market value of products.

payout time [IND ENG] A measurement of profitability or liquidity of an investment, being the time required to recover the original investment in depreciable facilities from profit and depreciation; usually, but not always, calculated after income taxes.

pay streak [MIN ENG] A layer of oil, ore, or other mineral that can be mined profitably.

pay television *See* subscription television.

Pb *See* lead.

P band [COMMUN] A band of radio frequencies extending from 225 to 390 megahertz, corresponding to wavelengths of 133.3 to 76.9 centimeters.

PBI *See* protein-bound iodine.

Pb-I-Pb junction *See* lead-I-lead junction.

PBI test *See* protein-bound iodine test.

P blood group [IMMUNOL] A system of immunologically distinct, genetically determined erythrocyte antigens first defined by their reaction with anti-P, and immune rabbit antiserum, and later broadened to include related antigens.

PBX *See* private branch exchange.

PC *See* phosphocreatine.

p chart [IND ENG] A chart of the fraction defective, either observed in the sample or in some production period.

PCM *See* pulse-code modulation.

PCP *See* primary control program.

Pd *See* palladium.

PD *See* potential difference.

PDA *See* postacceleration.

P display *See* plan position indicator.

pdl-ft *See* foot-poundal.

PDM *See* pulse-duration modulation.

PDR *See* precision depth recorder.

pea [BOT] **1.** *Pisum sativum.* The garden pea, an annual leafy leguminous vine cultivated for its smooth or wrinkled, round edible seeds which are borne in dehiscent pods. **2.** Any of several related or similar cultivated plants.

peach [BOT] *Prunus persica.* A low, spreading, freely branching tree of the order Rosales, cultivated in less rigorous parts of the temperate zone for its edible fruit, a juicy drupe with a single large seed, a pulpy yellow or white mesocarp, and a thin firm epicarp.

peachblossom ore   *See* erythrite.

peach kernel oil   *See* persic oil.

Peacock   *See* Pavo.

peacock blue   [ORG CHEM] $HSO_3C_6H_4COH$ $[C_6H_4N(C_2H_5)CH_2C_6H_4SO_3Na]_2$ Blue pigment used in inks for multicolor printing.

peacock copper   *See* peacock ore.

peacock ore   [MINERAL] A copper mineral, such as bornite, having an iridescent tarnished surface upon exposure to air. Also known as peacock copper.

peak   [METEOROL] The point of intersection of the cold and warm fronts of a mature extra-tropical cyclone.   [SCI TECH] The maximum instantaneous value of a quantity.

peak amplitude   [PHYS] The maximum amplitude of an alternating quantity, measured from its zero value.

peak attenuation   [COMMUN] The diminution of response to a modulated wave experienced on modulation crests.

peak cathode current   [ELECTR] 1. Maximum instantaneous value of a periodically recurring cathode current. 2. Highest instantaneous value of a randomly recurring pulse of cathode current. 3. Highest instantaneous value of a nonrecurrent pulse of cathode current occurring under fault conditions.

peak clipper   *See* limiter.

peak clipping   *See* limiting.

peak distortion   [COMMUN] Largest total distortion of telegraph signals noted during a period of observation.

peaked roof   [ARCH] A roof that rises to a point or ridge.

peak enthalpimetry   [ANALY CHEM] A thermochemical analytical procedure applicable to biochemical and chemical analyses; the salient feature is rapid mixing of a reagent stream with an isothermal solvent stream into which discrete samples are intermittently injected; peak enthalpograms result which exhibit the response characteristics of genuine differential detectors.

peak envelope power   [ELECTR] Of a radio transmitter, the average power supplied to the antenna transmission line by a transmitter during one radio-frequency cycle at the highest crest of the modulation envelope, taken under conditions of normal operation.

peaker   [ELECTR] A small fixed or adjustable inductance used to resonate with stray and distributed capacitances in a broad-band amplifier to increase the gain at the higher frequencies.

peak factor   *See* crest factor.

peak gust   [METEOROL] After United States weather observing practice, the highest instantaneous wind speed recorded at a station during a specified period, usually the 24-hour observational day; therefore, a peak gust need not be a true gust of wind.

peaking circuit   [ELECTR] A circuit used to improve the high-frequency response of a broad-band amplifier; in shunt peaking, a small coil is placed in series with the anode load; in series peaking, the coil is placed in series with the grid of the following stage.

peaking network   [ELECTR] Type of interstage coupling network in which an inductance is effectively in series (series-peaking network), or in shunt (shunt-peaking network), with the parasitic capacitance to increase the amplification at the upper end of the frequency range.

peaking transformer   [ELEC] A transformer in which the number of ampere-turns in the primary is high enough to produce many times the normal flux density values in the core; the flux changes rapidly from one direction of saturation to the other twice per cycle, inducing a highly peaked voltage pulse in a secondary winding.

peak inverse anode voltage   [ELECTR] Maximum instantaneous anode voltage in the direction opposite to that in which the tube or other device is designed to pass current.

peak inverse voltage   [ELECTR] Maximum instantaneous anode-to-cathode voltage in the reverse direction which is actually applied to the diode in an operating circuit.

peakless pumping   [MIN ENG] Spreading the pumping load over the entire day in a mine.

peak limiter   *See* limiter.

peak load   [ELEC] The maximum instantaneous load or the maximum average load over a designated interval of time. Also known as peak power.   [ENG] The maximum quantity of a specified material to be carried by a conveyor per minute in a specified period of time.

peak overpressure   [ORD] The highest overpressure resulting from a blast wave; peak overpressures near the fireball of an atomic explosion are very high but drop off rapidly as the blast wave travels along the ground outward from ground zero.

peak power   [ELEC] *See* peak load.   [ELECTROMAG] The maximum instantaneous power of a transmitted radar pulse.

peak pressure   [ORD] The maximum pressure reached in the bore of a weapon during the burning of the propellant.

peak signal level   [ELECTR] Expression of the maximum instantaneous signal power or voltage as measured at any point in a facsimile transmission system; this includes auxiliary signals.

peak tank   [NAV ARCH] A tank which is at the bow or stern end of a ship and is low in the ship, usually kept empty and dry but sometimes used to carry potable water.

peak-to-peak amplitude   [PHYS] Amplitude of an alternating quantity measured from positive peak to negative peak.

peak-to-valley ratio   [COMMUN] The ratio of the largest amplitude of a modulated wave to its smallest value.

peak value   [ELEC] The maximum instantaneous value of a varying current, voltage, or power during the time interval under consideration. Also known as crest value.

peak width   [ANALY CHEM] In a gas chromatogram (plot of eluent rise and fall versus time), the width of the base (time duration) of a symmetrical peak (rise and fall) of eluent.

Peano's postulates   [MATH] The five axioms by which the natural numbers may be formally defined.

peanut   [BOT] *Arachis hypogaea.* A low, branching, self-pollinated annual legume cultivated for its edible seed, which is a one-loculed legume formed beneath the soil in a pod.

peanut oil   [MATER] A yellow to greenish-yellow fatty oil obtained from peanuts; soluble in ether, chloroform, benzine, and carbon disulfide; main components are glycerides of oleic and linoleic acids; used as an edible substitute for olive oil, and in soaps and medicine. Also known as arachis oil; earthnut oil; katchung oil.

pear   [BOT] Any of several tree species of the genus *Pyrus* in the order Rosales, cultivated for their fruit, a pome that is wider at the apical end and has stone cells throughout the flesh.

pearceite   [MINERAL] $Ag_{16}As_2S_{11}$ A black mineral composed of sulfide of arsenic and silver.

pearl   [MATER] A dense, more or less round, white or light-colored concretion having various degrees of luster formed within or beneath the mantle of various mollusks by deposition of thin concentric layers of nacre about a foreign particle.   [PATH] 1. Rounded masses of concentrically arranged squamous epithelial cells, seen in some carcinomas. 2. Mucous casts of the bronchi or bronchioles found in the sputum of asthmatic persons.

pearl ash   [MATER] An impure substance derived from potash following partial purification from wood ash.

pearl essence   [MATER] A brilliant, translucent, lustrous material obtained from fish scales; used in pearl lacquers and to make artificial pearls. Also known as pearl white.

pearl hardening   [INORG CHEM] Commerical name for a crystallized grade of calcium sulfate; used as a paper filler.

pearlite   [GEOL] *See* perlite.   [MET] A lamellar aggregate of ferrite (almost pure iron) and cementite (Fe_3C) often occurring in carbon steels and in cast iron.

Pearl-Reed curve   *See* logistic curve.

pearl sinter   *See* siliceous sinter.

pearl spar   [MINERAL] A crystalline carbonate having a pearly luster; an example is ankerite.

pearlstone   *See* perlite.

pearl white   *See* pearl essence.

pearly   *See* nacreous.

pearly tumor   *See* cholesteatoma.

Pearson Type I distribution   *See* beta distribution.

pea-soup fog   [METEOROL] Any particularly dense fog.

peat   [GEOL] A dark-brown or black residuum produced by the partial decomposition and disintegration of mosses,

PEACH

Flowers and fruit of Chinese peach.

PEAR

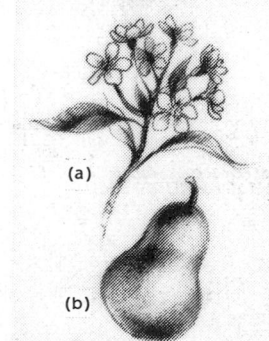

Pear *(Pyrus communis).*
(a) Flower cluster. *(b)* Fruit.

PECAN

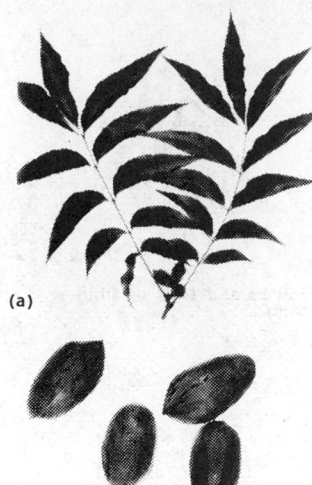

Pecan. (a) Leaves and fruit.
(b) Hulled nuts.

PECTINIBRANCHIA

2 mm

Shell of the snail *Orthonema*,
from the Pennsylvanian-Permian.
*(From R. R. Shrock and W. H.
Twenhofel, Principles of
Invertebrate Paleontology, 2d
ed., McGraw-Hill, 1953)*

PECTOLITE

11 cm

Globular masses of needlelike
pectolite crystals with calcite at
left, found in Paterson, New
Jersey. *(American Museum of
Natural History Specimen)*

sedges, trees, and other plants that grow in marshes and other wet places.

peat bed *See* peat bog.

peat bog [GEOL] A bog in which peat has formed under conditions of acidity. Also known as peat bed; peat moor.

peat formation [GEOCHEM] Decomposition of vegetation in stagnant water with small amounts of oxygen, under conditions intermediate between those of putrefaction and those of moldering.

peat moor *See* peat bog.

peat moss [ECOL] Moss, especially sphagnum moss, from which peat has been produced.

peat-sapropel [GEOL] A product of the degradation of organic matter that is transitional between peat and sapropel. Also known as sapropel-peat.

peat soil [GEOL] Soil containing a large amount of peat; it is rich in humus and gives an acid reaction.

peat wax [MATER] A hard, waxy material extracted from peat; it is similar to, and a substitute for, montan wax.

peau de soie [TEXT] Soft silk satin-textured fabric having a slight luster and faintly showing fine ribs in the filling.

pebble [GEOL] A clast, larger than a granule and smaller than a cobble having a diameter in the range of 4–64 millimeters. Also known as pebblestone. [MINERAL] *See* rock crystal.

pebble armor [GEOL] A desert armor made up of rounded pebbles.

pebble-bed reactor [NUCLEO] A nuclear reactor in which the fuel consists of small spheres or pellets stacked in the core; the reaction rate is controlled by coolant flow and by loading and unloading pellets.

pebble coal [GEOL] Coal that is transitional between peat and brown coal.

pebble heater [CHEM ENG] Gas-heating device (for air, hydrogen, methane, and steam) in which heat is transferred to the gas via a countercurrent movement of preheated pebbles.

pebble mill [MECH ENG] A solids size-reduction device with a cylindrical or conical shell rotating on a horizontal axis, and with a grinding medium such as balls of flint, steel, or porcelain.

pebbles [MATER] Grinding media for pebble mills, usually balls of hard flint or hard burned white porcelain.

pebblestone *See* pebble.

pebble-weave [TEX] Of a material, having an irregular texture produced by weaving with shrunken and twisted yarn.

pebbling *See* orange peel.

pebbly mudstone [GEOL] A delicately laminated till-like conglomeratic mudstone.

pebbly sand [GEOL] An unconsolidated sedimentary deposit containing at least 75% sand and up to a maximum of 25% pebbles.

pébrine [INV ZOO] A contagious protozoan disease of silkworms and other caterpillars caused by *Nosema bombycis*.

pecan [BOT] *Carya illinoensis*. A large deciduous hickory tree in the order Fagales which produces an edible, oblong, thin-shelled nut.

peccary [VERT ZOO] Either of two species of small piglike mammals in the genus *Tayassu*, composing the family Tayassuidae.

peck [MECH] Abbreviated pk. 1. A unit of volume used in the United States for measurement of solid substances, equal to 8 dry quarts, or ¼ bushel, or 537.605 cubic inches, or 0.00880976754172 cubic meter. 2. A unit of volume used in the United Kingdom for measurement of solid and liquid substances, although usually the former, equal to 2 gallons, or approximately 0.00909218 cubic meter.

pecking order [PSYCH] A social hierarchy of prestige, dominance, or authority. [VERT ZOO] A hierarchy of social dominance within a flock of poultry where each bird is allowed to peck another lower in the scale and must submit to pecking by one of higher rank.

Peclet number [CHEM ENG] Dimensionless group used to determine the chemical reaction similitude for the scale-up from pilot-plant data to commercial-sized units; incorporates heat capacity, density, fluid velocity, and other pertinent physical parameters.

Pecora [VERT ZOO] An infraorder of the Artiodactyla; includes those ruminants with a reduced ulna and usually with antlers, horns, or deciduous horns.

pecten [ZOO] Any of various comblike structures possessed by animals.

Pectenidae [INV ZOO] A family of bivalve mollusks in the order Anisomyaria; contains the scallops.

pectic acid [BIOCHEM] A complex acid, partially demethylated, obtained from the pectin of fruits.

pectin [BIOCHEM] A purified carbohydrate obtained from the inner portion of the rind of citrus fruits, or from apple pomace; consists chiefly of partially methoxylated polygalacturonic acids.

Pectinariidae [INV ZOO] The cone worms, a family of polychaete annelids belonging to the Sedentaria.

pectinase [BIOCHEM] An enzyme that catalyzes the transformation of pectin into sugars and galacturonic acid.

pectinesterase [BIOCHEM] An enzyme that catalyzes the hydrolytic breakdown of pectins to pectic acids.

pectineus [ANAT] A muscle arising from the pubis and inserted on the femur.

Pectinibranchia [INV ZOO] An order of gastropod mollusks which contains many families of snails; respiration is by means of ctenidia, the nervous system is not concentrated, and sexes are separate.

Pectobothridia [INV ZOO] A subclass of parasitic worms in the class Trematoda, characterized by caudal hooks or hard posterior suckers or both.

pectolite [MINERAL] $NaCa_2Si_3O_8(OH)$ A colorless, white, or gray inosilicate, crystallizing in the monoclinic system and having a vitreous to silky luster; hardness is 5 on Mohs scale, and specific gravity is 2.75.

pectoral fin [VERT ZOO] One of the pair of fins of fishes corresponding to forelimbs of a quadruped.

pectoral girdle [ANAT] The system of bones supporting the upper or anterior limbs in vertebrates. Also known as shoulder girdle.

pectoralis major [ANAT] The large muscle connecting the anterior aspect of the chest with the shoulder and upper arm.

pectoralis minor [ANAT] The small, deep muscle connecting the third to fifth ribs with the scapula.

peculiar part [ORD] A part of ordnance for which the design is controlled by a single manufacturer, and the use is restricted to items produced by a single manufacturer.

peculiar star [ASTRON] A star that does not fit into a standard spectral classification.

peculiar velocity [ASTRON] Superposed on the systematic rotation of the galaxy are individual motions of the stars; each star moves in a somewhat elliptical orbit and therefore shows a velocity of its own (peculiar velocity) to the local standard of rest, the standard moving in a circular orbit around the galactic center.

pedal [BIOL] Of or pertaining to the foot. [DES ENG] A lever operated by foot.

pedal disk [INV ZOO] The broad, flat base of many sea anemones, used for attachment to a substrate.

pedalfer [GEOL] A soil in which there is an accumulation of sesquioxides; it is characteristic of a humid region.

pedal ganglion [INV ZOO] One of the paired ganglia supplying nerves to the foot muscles in most mollusks.

pedal gland *See* foot gland.

pedate [BIOL] 1. Having toelike parts. 2. Having a foot. 3. Having tube feet.

pedestal [CIV ENG] 1. The support for a column. 2. A metal support carrying one end of a bridge truss or girder and transmitting any load to the top of a pier or abutment. [ELECTR] *See* blanking level. [ENG] A supporting part or the base of an upright structure, such as a radar antenna. [GEOL] A relatively slender column of rock supporting a wider rock mass and formed by undercutting as a result of wind abrasion or differential weathering. Also known as rock pedestal.

pedestal boulder [GEOL] A rock mass supported on a rock pedestal. Also known as pedestal rock.

pedestal level *See* blanking level.

pedestal pile [CIV ENG] A concrete pile with a bulbous enlargement at the bottom.

pedestal rock *See* pedestal boulder.

pedestal sight [ORD] In an aircraft gunnery system, a sight mounted on a pedestal for remote control of the guns.

pediatrician [MED] A physician who specializes in pediatrics.

pediatrics [MED] The branch of medicine that deals with the growth and development of the child through adolescence, and with the care, treatment, and prevention of diseases, injuries, and defects of children.

pedicel [BOT] **1.** The stem of a fruiting or sporebearing organ. **2.** The stem of a single flower. [ZOO] A short stalk in an animal body.

pedicellaria [INV ZOO] In echinoids and starfishes, any of various small grasping organs in the form of a beak carried on a stalk.

pedicellate [BIOL] Having a pedicel.

Pedicellinea [INV ZOO] The single order of the class Calyssozoa, including all entoproct bryozoans.

pedicle [ANAT] A slender process acting as a foot or stalk (as the base of a tumor), or the basal portion of an organ that is continuous with other structures.

pediculophobia [PSYCH] Abnormal fear of pediculosis.

pediculosis [MED] Infestation with lice, especially of the genus *Pediculus*.

pedigree [GEN] The ancestral line of an individual.

Pedilidae [INV ZOO] The false ant-loving flower beetles, a family of coleopteran insects in the superfamily Tenebrionoidea.

pediment [ARCH] A triangular face forming the gable of a two-pitched roof. [GEOL] A piedmont slope formed from a combination of processes which are mainly erosional; the surface is chiefly bare rock but may have a covering veneer of alluvium or gravel. Also known as conoplain; piedmont interstream flat.

pedimentation [GEOL] The actions or processes by which pediments are formed.

pediment pass [GEOL] A flat, narrow tongue that extends from a pediment on one side of a mountain to join a pediment on the other side.

Pedinidae [INV ZOO] The single family of the order Pedinoida.

Pedinoida [INV ZOO] An order of Diadematacea making up those forms of echinoderms which possess solid spines and a rigid test.

Pediococcus [MICROBIO] A genus of nonmotile, gram-positive cocci in the family Lactobacillaceae which occur singly or as tetrads, pairs, or short chains and which produce acidification and clouding of beer and wort.

pedion [CRYSTAL] A crystal form with only one face; member of the asymmetric class of the triclinic system.

Pedionomidae [VERT ZOO] A family of quaillike birds in the order Gruiformes.

Pedipalpida [INV ZOO] Former order of the Arachnida; these animals are now placed in the orders Uropygi and Amblypygi.

pedipalpus [INV ZOO] One of the second pair of appendages of an arachnid.

pediplain [GEOL] A rock-cut erosion surface formed in a desert by the coalescence of two or more pediments. Also known as desert peneplain; desert plain; panfan.

pediplanation [GEOL] The actions or processes by which pediplanes are formed.

pediplane [GEOL] Any planate erosion surface formed in the piedmont area of a desert, either bare or covered with a veneer of alluvium.

pedocal [GEOL] A soil containing a concentration of carbonates, usually calcium carbonate; it is characteristic of arid or semiarid regions.

pedodontics [MED] The branch of dentistry concerned with the care of children's teeth.

pedogenesis *See* soil genesis.

pedogeochemical survey [GEOCHEM] A geochemical prospecting survey in which the materials sampled are soil and till.

pedology [GEOL] *See* soil science. [MED] The science of the study of the physiological as well as the psychological aspects of childhood.

pedometer [ENG] **1.** An instrument for measuring and weighing a newborn child. **2.** An instrument that registers the number of footsteps in walking.

pedophilia [PSYCH] Love of children by adults for sexual purposes.

pedophobia [PSYCH] An abnormal fear or dislike of children.

peduncle [ANAT] A band of white fibers joining different portions of the brain. [BOT] **1.** A flower-bearing stalk. **2.** A stalk supporting the fruiting body of certain thallophytes. [INV ZOO] The stalk supporting the whole or a large part of the body of certain crinoids, brachiopods, and barnacles.

pedunculate [BIOL] **1.** Having or growing on a peduncle. **2.** Being attached to a peduncle.

peekaboo system *See* Batten system.

peel-back [ENG] The separation of two bonded materials, one or both of which are flexible, by stripping or pulling the flexible material from the mating surface at a 90 or 180° angle to the plane in which it is adhered.

peeling [MATER] Stripping or detaching a rubber coating from a metal, cloth, or other material.

peel test [ENG] A test to ascertain the adhesive strength of bonded strips of metals by peeling or pulling the metal strips back and recording the adherence values.

peen [DES ENG] The end of a hammer head with a hemispherical, wedge, or other shape; used to bend, indent, or cut.

peening [MET] Surface-hardening a piece of metal by hammering or by bombarding with hard shot.

peephole masks [ADP] In character recognition, a set of characters (each character residing in the character reader in the form of strategically placed points) which theoretically render all input characters as being unique regardless of their style.

peep sight [ORD] A rear gunsight having a small hole in which the front sight is centered in aiming; distinguished from an open sight.

peep slot [ORD] A vision slot in a combat vehicle which can readily be opened or closed from within.

peesweep storm [METEOROL] An early-spring storm in Scotland and England.

peg [ENG] **1.** A small pointed or tapered piece, often cylindrical, used to pin down or fasten parts. **2.** A projection used to hang or support objects. [MET] *See* plug.

Pegasidae [VERT ZOO] The single family of the order Pegasiformes.

Pegasiformes [VERT ZOO] The sea moths or sea dragons, a small order of actinopterygian fishes; the anterior of the body is encased in bone, and the nasal bones are enlarged to form a rostrum that projects well forward of the mouth.

Pegasus [ASTRON] A northern constellation; right ascension 22 hours, declination 20°N. Also known as Winged Horse.

peg count meter [ENG] A meter or register that counts the number of trunks tested, the number of circuits passed busy, the number of test failures, or the number of repeat tests completed.

pegging rammer [MET] A rod with an oblong piece of iron at its end; used to compact sand in a mold.

peg graft [BOT] A graft made by driving a scion of leafless dormant wood with wedge-shaped base into an opening in the stock and sealing with wax or other material.

pegmatite [PETR] Any extremely coarse-grained, igneous rock with interlocking crystals; pegmatites are relatively small, are relatively light colored, and range widely in composition, but most are of granitic composition; they are principal sources for feldspar, mica, gemstones, and rare elements. Also known as giant granite; granite pegmatite.

pegmatitization [GEOL] Formation of or replacement by a pegmatite.

pegmatolite *See* orthoclase.

peg model [GEOL] Three-dimensional model used to illustrate and study stratigraphic and structural conditions of subsurface geology; consists of a flat platform onto which vertical pegs of varying heights are mounted to represent the contours of various strata.

Peierls-Nabarro force [SOLID STATE] The force required to displace a dislocation along its slip plane.

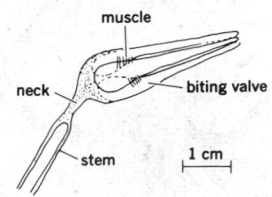

PEDICELLARIA

Tridentate type of echinoid pedicellaria showing the beak made of three movable jaws.

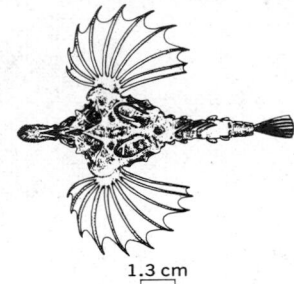

PEGASIFORMES

1.3 cm

Sea moth (*Pegasus draconis*), of the order Pegasiformes. (*After D. S. Jordan and J. O. Snyder, vol. 24, Leland Stanford University Contributions to Biology, 1901*)

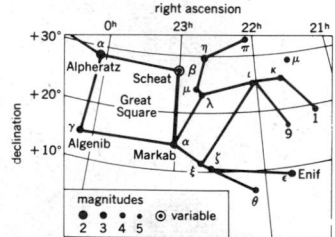

PEGASUS

Line pattern of the constellation Pegasus. The grid lines represent the coordinates of the sky. The apparent brightness, or magnitude, of the stars is shown by the sizes of the dots, graded by appropriate numbers.

PEKING MAN

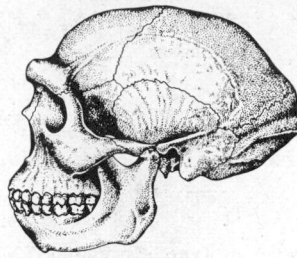

Reconstruction of female *Homo erectus pekinensis* skull, showing the heavy browridges. *(After F. Weidenreich, from M. F. Ashley Montagu, An Introduction to Physcial Anthropology, 2d ed., Charles C. Thomas, 1951)*

Peisidicidae [INV ZOO] A monogeneric family of polychaete annelids belonging to the Errantia.

Peking man [PALEON] *Sinanthropus pekinensis*. An extinct human type; the braincase was thick, with a massive basal and occipital torus structure and heavy browridges.

pekovskite [MINERAL] CaTiO₃ Lustrous, yellowish to gray-black, isometric or orthorhombic crystals with a hardness of 5.5; decomposes in hot sulfuric acid.

pelagic [GEOL] Pertaining to regions of a lake at depths of 10–20 meters or more, characterized by deposits of mud or ooze and by the absence of vegetation. Also known as eupelagic. [OCEANOGR] Pertaining to water of the open portion of an ocean, above the abyssal zone and beyond the outer limits of the littoral zone.

pelagic limestone [GEOL] A fine-textured limestone formed in relatively deep water by the concentration of calcareous tests of pelagic Foraminifera.

pelagism *See* seasickness.

pelargonic acid [ORG CHEM] $CH_3(CH_2)_7CO_2H$ A colorless or yellowish oil, boiling at 254°C; soluble in ether and alcohol, insoluble in water; used as a chemical intermediate and flotation agent, in lacquers, pharmaceuticals, synthetic flavors and aromas, and plastics. Also known as *n*-nonanoic acid; *n*-nonoic acid; *n*-nonylic acid.

pelargonic alcohol *See n*-nonyl alcohol.

pelargonidin [BIOCHEM] An anthocyanidin pigment obtained by hydrolysis of pelargonin in the form of its redbrown crystalline chloride, $C_{15}H_{11}ClO_5$.

pelargonin [BIOCHEM] An anthocyanin obtained from the dried petals of red pelargoniums or blue cornflowers in the form of its red crystalline chloride, $C_{27}H_{31}ClO_{15}$.

Pelean cloud *See* nuée ardente.

Pel-Ebstein fever [MED] A relapsing fever characteristic of Hodgkin's disease.

Pelecanidae [VERT ZOO] The pelicans, a family of aquatic birds in the order Pelecaniformes.

Pelecaniformes [VERT ZOO] An order of aquatic, fish-eating birds characterized by having all four toes joined by webs.

Pelecanoididae [VERT ZOO] The diving petrels, a family of oceanic birds in the order Procellariiformes.

Pelecinidae [INV ZOO] The pelecinid wasps, a monospecific family of hymenopteran insects in the superfamily Proctotrupoidea.

p electron [ATOM PHYS] In the approximation that each electron has a definite central-field wave function, an atomic electron that has an orbital angular momentum quantum number of unity.

Pelecypoda [INV ZOO] The equivalent name for Bivalvia.

pelelith [GEOL] Vesicular or pumiceous lava in the throat of a volcano.

Pele's hair [GEOL] A spun volcanic glass formed naturally by blowing out during quiet fountaining of fluid lava. Also known as capillary ejecta; filiform lapilli; lauoho o pele.

Pele's tears [GEOL] Volcanic glass in the form of small, solidified drops which precede pendants of Pele's hair.

Pelger anomaly [MED] A hereditary anomaly of granulocytes in the peripheral blood with no more than one or two nuclear lobes and unusually coarse nuclear chromatin.

pelican [VERT ZOO] Any of several species of birds composing the family Pelecanidae, distinguished by the extremely large bill which has a distensible pouch under the lower mandible.

pelican hook [NAV ARCH] A hinged hook shaped like a pelican's beak, held closed by a ring that can be instantaneously released; used to secure chain, cable, or cargo gear on a ship.

pelite [GEOL] A sediment or sedimentary rock, such as mudstone, composed of fine, clay- or mud-size particles. Also spelled pelyte.

pelitic [GEOL] Pertaining to, characteristic of, or derived from pelite.

pelitic hornfels [PETR] A fine-grained metamorphic rock derived from pelite.

pelitic schist [PETR] A foliated crystalline metamorphic rock derived from pelite.

pellagra [MED] A disease caused by nicotinic acid deficiency characterized by skin lesions, inflammation of the soft tissues of the mouth, diarrhea, and central nervous system disorders.

Pell equation [MATH] The diophantine equation $x^2 - Dy^2 = 1$, with D a positive integer that is not a perfect square.

pellet [AGR] A small, cylindrical, compressed mass of livestock feed. [GEOL] A fine-grained, sand-size, spherical to elliptical aggregate of clay-sized calcareous material, devoid of internal structure, and contained in the body of a well-sorted carbonate rock. [ORD] A small stone or metal ball used as a missile in firearms. [PHARM] A small pill. [SCI TECH] A small spherical or cylindrical body. [VERT ZOO] A mass of undigestible material regurgitated by a carnivorous bird.

pellet cooler [CHEM ENG] Gas-cooled, gravity-bed device for the cooling and drying of extruded pellets and briquets.

pelleting [ENG] Method of accelerating solidification of cast explosive charges by blending precast pellets of the explosives into the molten charge.

pelletization [MIN ENG] Forming aggregates of about 1/2 inch (13 millimeter) diameter from finely divided ore or coal.

pellet mill [MECH ENG] Device for injecting particulate, granular or pasty feed into holes of a roller, then compacting the feed into a continuous solid rod to be cut off by a knife at the periphery of the roller.

pellet technique *See* potassium bromide–disk technique.

pellicle [CYTOL] A plasma membrane. [INV ZOO] A thin protective membrane, as on certain protozoans.

pellicularia disease [PL PATH] A fungus disease of coffee and other tropical plants caused by *Pellicularia koleroga* and characterized by leaf spots.

pellicular water [HYD] Films of groundwater adhering to particles or cavities above the water table.

Pellon [TEXT] Trade name of nonwoven materials of the Pellon Corporation made by bonding the fibers together using chemicals and heat.

pellotine [ORG CHEM] $C_{13}H_{19}O_3N$ A colorless, crystalline alkaloid, derived from the dried cactus pellote, *Lophophora williamsi* (Mexico), slightly soluble in water; used as a hypnotic.

Pelmatozoa [INV ZOO] A division of the Echinodermata made up of those forms which are anchored to the substrate during at least part of their life history.

pelmicrite [GEOL] A limestone containing less than 25% each of intraclasts and ooliths, having a volume ratio of pellets to fossils greater than 3 to 1, and with the micrite matrix more abundant than the sparry-calcite cement.

Pelobatidae [VERT ZOO] A family of frogs in the suborder Anomocoela, including the spadefoot toads.

Pelodytidae [VERT ZOO] A family of frogs in the suborder Anomocoela.

pelogloea [GEOL] Marine detrital slime from settled plankton.

Pelogonidae [INV ZOO] The equivalent name for Ochteridae.

Pelomedusidae [VERT ZOO] The side-necked or hidden-necked turtles, a family of the order Chelonia.

Pelopidae [INV ZOO] A family of oribatid mites, order Sarcoptiformes.

Peloridiidae [INV ZOO] The single family of the homopteran series Coleorrhyncha.

pelorus [NAV] A compass card suitably mounted and provided with vanes to permit observation of the true or relative bearings; it may not include a magnetic compass, in which case reference must be made to a local compass to give true or magnetic bearings.

pelorus card [NAV] The part of a pelorus on which the direction graduations are placed, usually in the form of a thin disk or annulus graduated in degrees, clockwise from 0° at the reference direction to 360°.

pelotherapy [MED] Therapeutic treatment with earth or mud.

pelouse *See* meadow.

pelsparite [PETR] A limestone containing less than 25% each of intraclasts and ooliths, having a volume ratio of pellets to fossils greater than 3 to 1, and with the sparry-calcite cement more abundant than the micrite matrix.

peltate [BOT] Of leaves, having the petiole attached to the lower surface instead of the base.

Peltier coefficient [PHYS] The ratio of the rate at which heat

PELICAN

Great white pelican *(Pelicanus erythrorhynchos).*

is evolved or absorbed at a junction of two metals in the Peltier effect to the current passing through the junction.

Peltier effect [PHYS] Heat is evolved or absorbed at the junction of two dissimilar metals carrying a small current, depending upon the direction of the current.

Pelton turbine *See* Pelton wheel.

Pelton wheel [MECH ENG] An impulse hydraulic turbine in which pressure of the water supply is converted into velocity by a few stationary nozzles, and the water jets then impinge on the buckets mounted on the rim of a wheel; usually limited to high head installations, exceeding 500 feet (150 meters). Also known as Pelton turbine.

pelvic cavity *See* pelvis.

pelvic fin [VERT ZOO] One of the pair of fins of fishes corresponding to the hindlimbs of a quadruped.

pelvic girdle [ANAT] The system of bones supporting the lower limbs, or the hindlimbs, of vertebrates.

pelvic index [ANAT] The ratio of the anteroposterior diameter to the transverse diameter of the pelvis.

pelvis [ANAT] **1.** The main, basin-shaped cavity of the kidney into which urine is discharged by nephrons. **2.** The basin-shaped structure formed by the hipbones together with the sacrum and coccyx, or caudal vertebrae. **3.** The cavity of the bony pelvis. Also known as pelvic cavity.

pelviscope [MED] An endoscope for examination of the pelvic organs of the female.

Pelycosauria [PALEON] An extinct order of primitive, mammallike reptiles of the subclass Synapsida, characterized by a temporal fossa that lies low on the side of the skull.

pelyte *See* pelite.

pemphigus [MED] An acute or chronic disease of the skin characterized by the appearance of bullae, which develop in crops or in continuous succession.

pemphigus contageosus [MED] A vesicular dermatitis endemic in tropical areas, chiefly affecting the armpits and groin.

pen [ENG] **1.** A small place for confinement, storage, or protection. **2.** A device for writing with ink.

Penaeidea [INV ZOO] A primitive section of the Decapoda in the suborder Natantia; in these forms, the pleurae of the first abdominal somite overlap those of the second, the third legs are chelate, and the gills are dendrobranchiate.

penalty function [MATH] A function used in treating maxima and minima problems subject to constraints.

pencil [ENG] An implement for writing or making marks with a solid substance; the three basic kinds are graphite, carbon, and colored. [MATH] A family of geometric objects which share a common property.

pencil beam [ELECTROMAG] A beam of radiant energy concentrated in an approximately conical or cylindrical portion of space of relatively small diameter; this type of beam is used for many revolving navigational lights and radar beams.

pencilbeam antenna [ELECTROMAG] Unidirectional antenna designed so that cross sections of the major lobe formed by planes perpendicular to the direction of maximum radiation are approximately circular.

pencil gneiss [GEOL] A gneiss that splits into thin, rodlike quartz-feldspar crystal aggregates.

pencil ore [GEOL] Hard, fibrous masses of hematite that can be broken up into splinters.

pencil stone *See* pyrophyllite.

pencil tube [ELECTR] A small tube designed especially for operation in the ultra-high-frequency band; used as an oscillator or radio-frequency amplifier.

pendant *See* roof pendant.

pendant atomizer *See* hanging-drop atomizer.

pendant cloud *See* tuba.

pendant-drop method [PHYS] Method for the measurement of liquid surface tension by the elongation of a hanging drop of the liquid.

pendant post [BUILD] A post on a solid support and set against a wall to support a collar beam or other part of a roof.

pendular ring [PETRO ENG] Distribution of two nonmiscible liquids in a porous system; the pendular ring is a state of reservoir saturation in which the wetting phase is not continuous and the nonwetting phase is in contact with some of the solid surface.

pendulous gyroscope [MECH] A gyroscope whose axis of rotation is constrained by a suitable weight to remain horizontal; it is the basis of one type of gyrocompass.

pendulum [PHYS] A rigid body mounted on a fixed horizontal axis, about which it is free to rotate under the influence of gravity. Also known as gravity pendulum.

pendulum anemometer [ENG] A pressure-plate anemometer consisting of a plate which is free to swing about a horizontal axis in its own plane above its center of gravity; the angular deflection of the plate is a function of the wind speed; this instrument is not used for station measurements because of the false reading which results when the frequency of the wind gusts and the natural frequency of the swinging plate coincide.

pendulum clock [HOROL] A clock having a swinging pendulum as the means for producing a regularly recurring action.

pendulum day [PHYS] The time required for the plane of a freely suspended (Foucault) pendulum to complete an apparent rotation about the local vertical.

pendulum press [MECH ENG] A punch press actuated by a swinging treadle operated by the foot.

pendulum saw [MECH ENG] A circular saw that swings in a vertical arc for crosscuts.

pendulum scale [ENG] Weight-measurement device in which the load is balanced by the movement of one or more pendulums from vertical (zero weight) to horizontal (maximum weight).

pendulum seismograph [ENG] A seismograph that measures the relative motion between the ground and a loosely coupled inertial mass; in some instruments, optical magnification is used whereas others exploit electromagnetic transducers, photocells, galvanometers, and electronic amplifiers to achieve higher magnification.

pendulum sextant [NAV] A sextant provided with a pendulum to indicate the horizontal.

penecontemporaneous [GEOL] Of a geologic process or the structure or mineral that is formed by the process, occurring immediately following deposition but before consolidation of the enclosing rock.

peneplain *See* base-leveled plain.

peneplanation [GEOL] The actions or processes by which peneplains are formed.

penetrance [GEN] The proportion of individuals carrying a dominant gene in the heterozygous condition or a recessive gene in the homozygous condition in which the specific phenotypic effect is apparent. Also known as gene penetrance.

penetrant [INV ZOO] A large barbed nematocyst that pierces the body of the prey and injects a paralyzing agent. [MATER] A liquid with low surface tension, usually containing a dye or fluorescent chemical; when flowed over a metal surface, it is used to determine the existence and extent of cracks and other discontinuities.

penetrating oil [MATER] Low-viscosity oil that can penetrate between closely fitted parts, such as the leaves of springs and screw threads; used to loosen rusted parts.

penetration [AERO ENG] That phase of the letdown from high altitude to a specified approach altitude. [MET] **1.** The distance from the original surface of the base metal to that point at which weld fusion ends. **2.** A surface defect on a casting caused by molten metal filling voids in the sand mold. [ORD] Distance to which a projectile sinks into a target.

penetration ballistics [MECH] A branch of terminal ballistics concerned with the motion and behavior of a missile during and after penetrating a target.

penetration depth [CRYO] The depth beneath the surface of superconductor in a magnetic field at which the magnetic field strength has fallen to $1/e$ of its value at the surface. [ELEC] In induction heating, the thickness of a layer, extending inward from a conductor's surface, whose resistance to direct current equals the resistance of the whole conductor to alternating current of a given frequency.

penetration frequency *See* critical frequency.

penetration gland [INV ZOO] A gland at the anterior end of certain cercariae that secretes a histolytic substance.

penetration hardness *See* indentation hardness.

penetration log [ENG] A record of the speed with which a

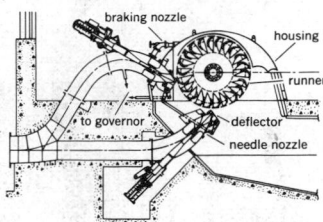

PELTON WHEEL

Cross section of a Pelton type of hydraulic-turbine installation.

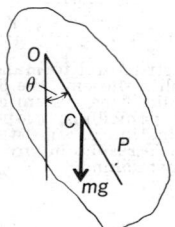

PENDULUM

Schematic diagram of a pendulum. O = axis about which pendulum rotates, C = center of mass, and P = center of oscillation. Line OC makes instantaneous angle θ with the vertical. Gravitational force on pendulum is equivalent to downward force at C with magnitude mg, where m is pendulum mass, g is acceleration of gravity.

PENGUIN

Emperor penguin with its young. The largest species of penguins, it may reach a height of 4 feet (1.2 meters) and weigh up to 100 pounds (45 kilograms).

PENICILLIN

$$H \qquad S$$
$$R-Y-N-CH-CH \quad C-(CH_3)_2$$
$$\qquad | \qquad | \quad |$$
$$\qquad C-N \quad C-COOH$$
$$\qquad || \qquad | $$
$$\qquad O \qquad H$$

Basic structural formula for penicillin showing the β-lactam-thiazolide ring system common to all penicillins; proper combination of substituents at R and Y results in any one of nine penicillins.

PENNSYLVANIAN

PRECAMBRIAN		
CAMBRIAN		
ORDOVICIAN		
SILURIAN		PALEOZOIC
DEVONIAN		
Mississippian	CARBON-IFEROUS	
Pennsylvanian		
PERMIAN		
TRIASSIC		
JURASSIC		MESOZOIC
CRETACEOUS		
TERTIARY		CENOZOIC
QUATERNARY		

Chart showing position of the Pennsylvanian in relation to the periods and eras of geologic time.

drill penetrates a bore, including such factors as hole size, bit size, mud pressure, rotation speed, and force on the bit; used to determine the thickness of coal and dirt bands in the bore.

penetration macadam [MATER] A paving material consisting of crushed stone in two sizes bound together by asphalt or tar.

penetration number [ENG] The consistency of greases, waxes, petrolatum, and asphalt or other bituminous materials expressed as the distance that a standard needle penetrates the sample under specified ASTM test conditions.

penetration of fractures [PETRO ENG] The depth to which artificially produced fractures penetrate the fractured reservoir formation.

penetration probability [QUANT MECH] The probability that a particle will pass through a potential barrier, that is, through a finite region in which the particle's potential energy is greater than its total energy. Also known as transmission coefficient.

penetration range *See* night visual range.

penetration rate [MECH ENG] The actual rate of penetration of drilling tools.

penetration speed [MECH ENG] The speed at which a drill can cut through rock or other material.

penetration test [ENG] A test to determine the relative values of density of noncohesive sand or silt at the bottom of boreholes.

penetration twin *See* interpenetration twin.

penetrometer [ENG] **1.** An instrument that measures the penetrating power of a beam of x-rays or other penetrating radiation. **2.** An instrument used to determine the consistency of a material by measurement of the depth to which a standard needle penetrates into it under standard conditions.

Penex process [CHEM ENG] A continuous, nonregenerative petroleum-refinery process for isomerization of C_5 or C_6 fractions in the presence of hydrogen and a platinum catalyst.

penfieldite [MINERAL] $Pb_2(OH)Cl_3$ A white hexagonal mineral composed of basic chloride of lead, occurring in hexagonal prisms.

penguin [VERT ZOO] Any member of the avian order Sphenisciformes; structurally modified wings do not fold and they function like flippers, the tail is short, feet are short and webbed, and the legs are set far back on the body.

penicillate [BIOL] Having a tuft of fine hairs.

penicillin [MICROBIO] **1.** The collective name for salts of a series of antibiotic organic acids produced by a number of *Penicillium* and *Aspergillus* species; active against most gram-positive bacteria and some gram-negative cocci. **2.** *See* benzyl penicillin sodium.

penicillinase [BIOCHEM] A bacterial enzyme that hydrolyzes and inactivates penicillin.

penicillin G₁ potassium *See* benzyl penicillin potassium.

Penicillium [MYCOL] A genus of fungi in the family Moniliaceae characterized by erect branching conidiophores having terminal tufts of club-shaped cells from which conidia are formed.

Penicillium chrysogenum [MYCOL] A penicillin-producing fungus.

Penicillium griseofulvum [MYCOL] The fungus that produces griseofulvin.

Penicillium notatum [MYCOL] A penicillin-producing fungus.

Peniculina [INV ZOO] A suborder of the Hymenostomatida.

peninsula [GEOGR] A body of land extending into water from the mainland, sometimes almost entirely separated from the mainland except for an isthmus.

penis [ANAT] The male organ of copulation in man and certain other vertebrates. Also known as phallus.

penis envy [PSYCH] The envy of the young female child for the penis which she does not possess, or which she thinks she has lost.

penitent *See* nieve penitente.

penitent ice [HYD] A jagged spike or pillar of compacted firn caused by differential melting and evaporation; necessary for this formation are air temperature near freezing, dew point much below freezing, and strong insolation.

penitent snow [HYD] A jagged spike or pillar of compacted snow caused by differential melting and evaporation.

penna *See* contour feather.

Pennales [BOT] An order of diatoms (Bacillariophyceae) in which the form is often circular, and the markings on the valves are radial.

pennant [METEOROL] A means of representing wind speed in the plotting of a synoptic chart; it is a triangular flag, drawn pointing toward lower pressure from a wind-direction shaft.

pennantite [MINERAL] $Mn_9Al_6Si_5O_{20}(OH)_{16}$ Orange mineral composed of basic manganese aluminum silicate; member of the chlorite group; it is isomorphous with thuringite.

pennate [BIOL] **1.** Wing-shaped. **2.** Having wings. **3.** Having feathers.

Pennatulacea [INV ZOO] The sea pens, an order of the subclass Alcyonaria; individuals lack stolons and live with their bases embedded in the soft substratum of the sea.

Pennellidae [INV ZOO] A family of copepod crustaceans in the suborder Caligoida; skin-penetrating external parasites of various marine fishes and whales.

penniculus [ANAT] A tuft of arterioles in the spleen. [BIOL] A brush-shaped structure.

penninite [MINERAL] $(Mg,Fe,Al)_6(Si,Al)_4O_{11}(OH)_8$ An emerald-green, olive-green, pale-green, or bluish mineral of the chlorite group crystallizing in the monoclinic system, with a hardness of 2-2.5 on Mohs scale, and specific gravity of 2.6-2.85.

Pennsylvania-base crude [MATER] A type of crude petroleum produced in Pennsylvania, New York, West Virginia, and parts of Ohio; contains a high percentage of paraffin-base lube-oil stock.

Pennsylvanian [GEOL] A division of late Paleozoic geologic time, extending from 320 to 280 million years ago, varyingly considered to rank as an independent period or as an epoch of the Carboniferous period; named for outcrops of coal-bearing rock formations in Pennsylvania.

Pennsylvania truss [CIV ENG] A truss characterized by subdivided panels, curved top chords on through trusses, and curved bottom chords on deck spans; used on long bridge spans.

pennyroyal oil [MATER] An essential oil distilled from the dried leaves and tops of the small pennyroyal plants *Hedeoma pulegioides* or *Mentha pulegium;* used as a counterirritant in liniments, in insect repellents, and for the production of methanol.

pennyweight [MECH] A unit of mass equal to ¹⁄₂₀ troy ounce or to 1.55517384 grams; the term is employed in the United States and in England for the valuation of silver, gold, and jewels. Abbreviated dwt; pwt.

Penokean *See* Animikean.

pen recorder [ENG] A device in which the varying inputs (electrical, pneumatic, mechanical) are marked by a signal-controlled pen onto a continuous recorder chart (circular or roll chart).

Pensky-Martens closed tester [CHEM ENG] Device to determine the ASTM flash point of fuel oils and cutback asphalts and other viscous materials and suspensions of solids.

penstock [CIV ENG] A valve or sluice gate for regulating water or sewage flow. [ENG] A closed water conduit controlled by valves and located between the intake and the turbine in a hydroelectric plant.

pentabasic [CHEM] A description of a molecule that has five hydrogen atoms that may be replaced by metals or bases.

pentaborane [INORG CHEM] B_5H_9 Flammable liquid boiling at 48°C; ignites spontaneously in air; proposed as high-energy fuel for aircraft and missiles.

pentachloride [CHEM] A molecule containing five atoms of chlorine in its structure.

pentachloroethane [ORG CHEM] $CHCl_2CCl_3$ Colorless, water-insoluble liquid, boiling at 159°C; used as a solvent to degrease metals. Also known as pentalin.

pentachlorophenol [ORG CHEM] C_6Cl_5OH A toxic white powder, decomposing at 310°C, melting at 190°C; soluble in alcohol, acetone, ether, and benzene; used as a fungicide, bactericide, algicide, herbicide, and chemical intermediate.

pentacite [MATER] Alkyd resin in which pentaerythritol is the polyhydric alcohol; used in coatings and printing inks.

pentacosane [ORG CHEM] $C_{25}H_{52}$ A water-insoluble hydrocarbon derived from beeswax.

pentacrinoid [INV ZOO] The larva of a feather star.

pentactinal [ZOO] Having five rays or branches.

pentacula [INV ZOO] The five-tentacled stage in the life history of echinoderms.

pentacyanium bis(methyl sulfate) [PHARM] $C_{29}H_{45}N_3O_9S_2$ A water-soluble, crystalline compound obtained from ethanol solution, and melting at 173–175°C; used in medicine in the dichloride form as an antihypertensive agent.

pentad [CLIMATOL] A period of 5 consecutive days, often preferred to the week for climatological purposes since it is an exact factor of the 365-day year.

pentadactyl [VERT ZOO] Having five digits on the hand or foot.

pentadecane [ORG CHEM] $C_{15}H_{32}$ A colorless, water-insoluble liquid, boiling at 270.5°C; soluble in alcohol; used as a chemical intermediate.

pentadelphous [BOT] Having the stamens in five sets with the filaments more or less united within each set.

pentadiene [ORG CHEM] C_5H_8 Any of several straight-chain liquid diolefins: $1,2\text{-}CH_2=C=CHCH_3$, a colorless liquid boiling at 45°C, also known as ethylallene; $1,3\text{-}CH_3=CHCH=CHCH_3$, a colorless liquid boiling at 43°C, also known as α-methyl bivinyl, piperylene; $1,4\text{-}(CH_2=CH)_2CH$, a colorless liquid boiling at 26°C.

pentaerythritol [ORG CHEM] $(CH_2OH)_4C$ A white crystalline solid, melting at 261–262°C; moderately soluble in cold water, freely soluble in hot water; used to make the explosive pentaerythritol tetranitrate (PETN) and in the manufacture of alkyol resins and other coating compounds. Also known as tetramethylolmethane.

pentaerythritol tetranitrate [ORG CHEM] $C(CH_2ONO_2)_4$ A white crystalline compound, melting at 139°C; explodes at 205–215°C; soluble in acetone, insoluble in water; used in medicines and explosives. Also known as penthrite; PETN.

Pentafining [CHEM ENG] Proprietary petroleum-refining process to isomerize pentane; uses regenerable platinum catalyst on a silica-alumina support.

pentafluoride [CHEM] A chemical compound onto which five fluoride atoms are bonded.

pentagon [MATH] A polygon with five sides.

pentagonal dodecahedron See pyritohedron.

pentagonal prism [MATH] A prism with two pentagonal sides, parallel and congruent.

pentahydrite [MINERAL] $MgSO_4\cdot5H_2O$ A triclinic mineral composed of hydrous magnesium sulfate; it is isostructural with chalcanthite.

pentahydroxyhexoic acid See galactonic acid.

pentalin See pentachloroethane.

pentalogy [MED] Five symptoms or defects which together characterize a disease or syndrome.

pentalogy of Fallot [MED] A syndrome consisting of the tetralogy of Fallot plus an interatrial septal defect.

Pentamerida [PALEON] An extinct order of articulate brachiopods.

Pentameridina [PALEON] A suborder of extinct brachiopods in the order Pentamerida; dental plates associated with the brachiophores were well developed, and their bases enclosed the dorsal adductor muscle field.

pentamerous [BOT] Having each whorl of the flower consisting of five members, or a multiple of five.

pentamethine See dicarbocyanine.

Δ¹-pentamethylene See cyclopentene.

pentamethylene glycol See 1,5-pentanediol.

pentanamide See valeramide.

pentandrous [BOT] Having five stamens.

n-pentane [ORG CHEM] $CH_3(CH_2)_3CH_3$ Colorless, flammable, water-insoluble hydrocarbon liquid, freezing at −130°C, boiling at 36°C; soluble in hydrocarbons and ethers; used as a chemical intermediate, solvent, and anesthetic.

pentane candle [OPTICS] A unit of luminous intensity equal to one-tenth of the luminous intensity of a standard pentane lamp, and approximately equal to 1 candela.

1,5-pentanediol [ORG CHEM] $HOCH_2(CH_2)_3CH_2OH$ Colorless, water-miscible liquid boiling at 242.5°C; used as a hydraulic fluid, lube-oil additive, and antifreeze, and in manufacture of polyester and polyurethane resins. Also known as pentamethylene glycol.

2,4-pentanedione See acetylacetone.

pentane insolubles [ANALY CHEM] Insoluble matter that can be separated from used lubricating oil in solution in n-pentane; may include resinous bitumens produced from the oxidation of oil and fuel; used in an ASTM test.

pentane lamp [ENG] A pentane-burning lamp formerly used as a standard for photometry.

n-pentanoic acid See valeric acid.

pentanol [ORG CHEM] $C_5H_{11}OH$ A toxic organic alcohol; 1-pentanol is n-amyl alcohol, primary; 2-pentanol is methylpropylcarbinol; 3-pentanol is diethylcarbinol; tert-pentanol is tert-amyl alcohol; pentanols are used in pharmaceuticals, as chemical intermediates, and as solvents.

pentanone [ORG CHEM] Either of two isomeric ketones derived from pentane: $CH_3COC_3H_7$ is a flammable, colorless, clear liquid, a mixture of methyl propyl and diethyl ketones; insoluble in water, soluble in ether and alcohol; used as a solvent; also known as ethyl acetone, methyl propyl ketone (MPK), 2-pentanone. $C_2H_5COC_2H_5$ is a colorless, flammable liquid with acetone aroma, boiling at 101°C; soluble in alcohol and ether; used in medicine and organic synthesis; also known as diethyl ketone, 3-pentanone, propione.

2-pentanone See pentanone.

3-pentanone See pentanone.

pentaquine [PHARM] $C_{18}H_{27}N_3O$ An antimalarial drug generally given with quinine; used as the phosphate salt.

Pentastomida [INV ZOO] A class of bloodsucking parasitic arthropods; the adult is vermiform, and there are two pairs of hooklike, retractile claws on the cephalothorax.

pentastyle [ARCH] Having five columns across the front.

Pentatomidae [INV ZOO] The true stink bugs, a family of hemipteran insects in the superfamily Pentatomoidea.

Pentatomoidea [INV ZOO] A subfamily of the hemipteran group Pentatomorpha distinguished by marginal trichobothria and by antennae which are usually five-segmented.

Pentatomorpha [INV ZOO] A large group of hemipteran insects in the subdivision Geocorisae in which the eggs are nonoperculate, a median spermatheca is present, accessory salivary glands are tubular, and the abdomen has trichobothria.

pentavalent [CHEM] An atom or radical that exhibits a valency of 5.

pentene [ORG CHEM] C_5H_{12} Colorless, flammable liquids derived from natural gasoline; isomeric forms are α-n-amylene and β-n-amylene.

penthouse [BUILD] 1. An enclosed space built on a flat roof to cover a stairway, elevator, or other equipment. 2. A dwelling built on top of the main roof. 3. A sloping shed or roof attached to a wall or building.

penthrite See pentaerythritol tetranitrate.

pentice [MIN ENG] 1. A rock pillar left, or a heavy timber bulkhead placed, in the bottom of a deep shaft of two or more compartments; the shaft is then further sunk through the pentice. 2. In shaft sinking, a solid rock pillar left in the bottom of the shaft for overhead protection of miners while the shaft is being extended by sinking.

2-pentine See pentyne.

pentlandite [MINERAL] $(Fe,Ni)_9S_8$ A yellowish-bronze mineral having a metallic luster and crystallizing in the isometric system; hardness is 3.5–4 on Mohs scale, and specific gravity is 4.6–5.0; the major ore of nickel.

pentobarbital sodium [PHARM] $C_{11}H_{17}N_2NaO_3$ A short- to intermediate-acting barbiturate; used as a hypnotic and sedative drug. Also known as sodium pentobarbitone.

pentode [ELECTR] A five-electrode electron tube containing an anode, a cathode, a control electrode, and two additional electrodes that are ordinarily grids.

pentode transistor [ELECTR] Point-contact transistor with four-point-contact electrodes; the body serves as a base with three emitters and one collector.

pentolite [MATER] A high explosive composed of pentaerythritol and trinitrotoluene.

pentose [BIOCHEM] Any one of a class of carbohydrates containing five atoms of carbon.

pentose phosphate pathway [BIOCHEM] A pathway by which glucose is metabolized or transformed in plants and microorganisms; glucose-6-phosphate is oxidized to 6-phosphogluconic acid, which then undergoes oxidative decar-

PENTODE

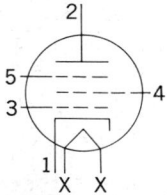

Circuit symbol for a pentode; the five electrodes are indicated by number. X's indicate leads to heater for indirectly heated cathode.

boxylation to form ribulose-5-phosphate, which is ultimately transformed to fructose-6-phosphate.

pentosuria [MED] The presence of pentose in the urine.

Pentothal [PHARM] Trademark for the ultra-short-acting barbiturate thiopental sodium. Also known as Pentothal sodium; sodium Pentothal.

Pentothal sodium See Pentothal.

pentoxide [INORG CHEM] A compound that is binary and has five atoms of oxygen; for example, phosphorus pentoxide, P_2O_5.

***para*-pentoxyphenol** See *para*-pentyloxyphenol.

pentyl See amyl.

pentylenetetrazol [PHARM] $C_6H_{10}N_4$ A central nervous system stimulant used as an analeptic.

pentylformic acid See caproic acid.

***para*-pentyloxyphenol** [ORG CHEM] $C_{11}H_{16}O_2$ Compound melting at 49-50°C; used as a bactericide. Also known as *para*-amoxyphenol; hydroquinone monopentyl ether; *para*-pentoxyphenol.

pentyne [ORG CHEM] C_5H_8 Either of two normal isometric acetylene hydrocarbons: $HC \equiv C(CH_2)_2CH_3$, colorless liquid boiling at 40°C, also known as pentine, propylacetylene; $CH_3C \equiv CC_2H_5$, liquid boiling at 56°C, also known as ethylmethylacetylene, 2-pentine.

penumbra [OPTICS] That portion of a shadow illuminated by only part of a radiating source.

penumbral eclipse See lunar appulse.

Penutian [GEOL] A North American stage of geologic time: lower Eocene (above Bulitian, below Ulatasian).

peperite [GEOL] A brecciatike material in marine sedimentary rock, considered to be either a mixture of lava with sediment, or shallow intrusions of magma into wet sediment.

pepo [BOT] A fleshy indehiscent berry with many seeds and a hard rind; characteristic of the Cucurbitaceae.

pepper [BOT] Any of several warm-season perennials of the genus *Capsicum* in the order Polemoniales, especially *C. annum* which is cultivated for its fruit, a many-seeded berry with a thickened integument. [FOOD ENG] Any of various spices and condiments obtained from the fruits of plants of the genus *Piper*.

peppermint [BOT] Any of various aromatic herbs of the genus *Mentha* in the family Labiatae, especially *M. piperita*.

peppermint camphor See menthol.

peppermint oil See oil of peppermint.

pepper sludge [MATER] Fine particles of sludge produced during the acid treatment of lubricating oils and other petroleum products.

pepsin [BIOCHEM] A proteolytic enzyme found in the gastric juice of mammals, birds, reptiles, and fishes.

pepsinogen [BIOCHEM] The precursor of pepsin, found in the stomach mucosa.

peptic [PHYSIO] 1. Of or pertaining to peptin. 2. Of or pertaining to digestion.

peptic ulcer [MED] An ulcer involving the mucosa, submucosa, and muscular layer on the lower esophagus, stomach, or duodenum, due in part at least to the action of acid-pepsin gastric juice.

peptidase [BIOCHEM] An enzyme that catalyzes the hydrolysis of peptides to amino acids.

peptide [BIOCHEM] A compound of two or more amino acids joined by peptide bonds.

peptide bond [ORG CHEM] A bond in which the carboxyl group of one amino acid is condensed with the amino group of another to form a $-CO \cdot NH-$ linkage. Also known as peptide linkage.

peptide linkage See peptide bond.

peptized fuel [MATER] Thickened flamethrower fuel to which water or other chemical is added before mixing, to reduce mixing time and to increase storage stability.

peptone [BIOCHEM] A water-soluble mixture of proteoses and amino acids derived from albumin, meat, or milk; used as a nutrient and to prepare nutrient media for bacteriology.

per- [CHEM] Chemical prefix meaning: 1. Complete, as in hydrogen peroxide. 2. Extreme, or the presence of the peroxy $(-O-O-)$ group. 3. Exhaustive (complete) substitution, as in perchloroethylene.

Peracarida [INV ZOO] A superorder of the Eumalacostraca;

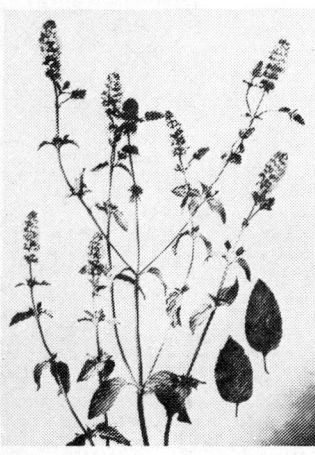

PEPPERMINT

Flowering peppermint (*Mentha piperita*) branch and two separate leaves showing the characteristic shape.

these crustaceans have the first thoracic segment united with the head, the cephalothorax usually larger than the abdomen, and some thoracic segments free from the carapace.

peracetic acid [ORG CHEM] CH_3COOOH A toxic, colorless liquid with strong aroma; boils at 105°C; explodes at 110°C; miscible with water, alcohol, glycerin, and ether; used as an oxidizer, bleach, catalyst, bactericide, fungicide, epoxyresin precursor, and chemical intermediate. Also known as peroxyacetic acid.

peracid [CHEM] Acid containing the peroxy $(-O-O-)$ group, such as peracetic acid or perchloric acid.

peralcohol [ORG CHEM] Chemical compound containing the peroxy group $(-O-O-)$, such as peracetic acid and perchromic acid.

peralkaline [PETR] Of igneous rock, having a molecular proportion of aluminum lower than that of sodium oxide and potassium oxide combined.

peraluminous [PETR] Of igneous rock, having a molecular proportion of aluminum oxide greater than that of sodium oxide and potassium oxide combined.

Peramelidae [VERT ZOO] The bandicoots, a family of insectivorous mammals in the order Marsupialia.

P-E ratio See precipitation-evaporation ratio.

perazine [PHARM] $C_{20}H_{25}N_3S$ A crystalline compound, melting at 51-53°C; used in medicine as a tranquilizer. Also known as *N*-methylpiperazinyl-*N'*-propylphenothiazine.

perbenzoic acid [ORG CHEM] $C_6H_5CO_2OH$ A crystalline compound forming leaflets from benzene solution, melting at 41-45°C, freely soluble in organic solvents; used in analysis of unsaturated compounds and to change ethylinic compounds into oxides. Also known as benzylhydroperoxide; peroxybenzoic acid.

percale [TEXT] A plain-weave, medium-weight fabric made of cotton or cotton and polyester.

percaline [TEXT] Fine cotton with a glossy or moiré surface.

perceived noise decibel [ACOUS] A unit of perceived noise level. Abbreviated PNdB.

perceived noise level [ACOUS] In perceived noise decibels, the noise level numerically equal to the sound pressure level, in decibels, of a band of random noise of width one-third to one octave centered on a frequency of 1000 hertz which is judged by listeners to be equally noisy.

percent [MATH] A quantitative term whereby *n*-percent of a number is *n* one-hundredths of the number. Symbolized %.

percentage [MATH] The result obtained by taking a given percent of a given quantity.

percentage depletion [PETRO ENG] Oil- or gas-reservoir depletion allowance calculated on the basis of unit sales and initial depletable leasehold cost.

percentage differential relay [ELECTR] Differential relay which functions when the difference between two quantities of the same nature exceeds a fixed percentage of the smaller quantity. Also known as biased relay; ratio-balance relay; ratio-differential relay.

percentage extraction [MIN ENG] The proportion of a coal seam or other ores which is removed from a mine.

percentage map [PETRO ENG] Contoured map in which the percentage of one component (such as sand) is compared with the total unit (such as sand plus shale).

percentage modulation See percent modulation.

percentage ripple [ELECTR] Ratio of the effective value of the ripple voltage to the average value of the total voltage, expressed as a percentage.

percentage subsidence [MIN ENG] The measured amount of subsidence expressed as a percentage of the thickness of coal extracted.

percentage support [MIN ENG] The percentage of the total wall area which will actually be covered by supports.

percent compaction [ENG] The ratio, expressed as a percentage, of dry unit weight of a soil to maximum unit weight obtained in a laboratory compaction test.

percent defective [IND ENG] The ratio of defective pieces per lot or sample, expressed as a percentage.

percent distortion [COMMUN] The ratio of the amplitude of a harmonic component to the fundamental component multiplied by 100.

percentile [STAT] A value in the range of a set of data which separates the range into two groups so that a given percentage of the measures lies below this value.

percent make [ELECTR] **1.** In pulse testing, the length of time a circuit stands closed compared to the length of the test signal. **2.** Percentage of time during a pulse period that telephone dial pulse springs are making contact.

percent modulation [COMMUN] The modulation factor expressed as a percentage. Also known as percentage modulation.

percept [PSYCH] A mental image of something perceived.

perception [PHYSIO] Recognition in response to sensory stimuli; the act or process by which the memory of certain qualities of an object is associated with other qualities impressing the senses, thereby making possible recognition of the object.

perceptron [ADP] A pattern recognition machine, based on an analogy to the human nervous system, capable of learning by means of a feedback system which reinforces correct answers and discourages wrong ones.

perch [MECH] Also known as pole; rod. **1.** A unit of length, equal to 5.5 yards, or 16.5 feet, or 5.0292 meters. **2.** A unit of area, equal to 30.25 square yards, or 272.25 square feet, or 25.29285264 square meters. [NAV] A staff placed on top of a buoy, rock, or shoal as a mark for navigators; a ball or cage is sometimes placed at the top of the perch, as an identifying mark. [VERT ZOO] **1.** Any member of the family Percidae. **2.** The common name for a number of unrelated species of fish belonging to the Centrarchidae, Anabantoidei, and Percopsiformes.

perched block [GEOL] A large, detached rock fragment presumed to have been transported and deposited by a glacier, and perched in a conspicuous and precarious position on the side of a hill. Also known as balanced rock; perched boulder; perched rock.

perched boulder *See* perched block.

perched groundwater *See* perched water.

perched rock *See* perched block.

perched spring [HYD] A spring that arises from a body of perched water.

perched stream [HYD] A stream whose surface level is above that of the water table and that is separated from underlying groundwater by an impermeable bed in the zone of aeration.

perched water [HYD] Groundwater that is unconfined and separated from an underlying main body of groundwater by an unsaturated zone. Also known as perched groundwater.

perched water table [HYD] The water table or upper surface of a body of perched water. Also known as apparent water table.

perchlorate [INORG CHEM] A salt of perchloric acid containing the ClO_4^- radical; for example, potassium perchlorate, $KClO_4$.

perchloric acid [INORG CHEM] $HClO_4$ Strongly oxidizing, corrosive, colorless, hygroscopic liquid, boiling at 16°C (8 mm Hg); soluble in water; unstable in pure form, but stable when diluted in water; used in medicine, electrolytic baths, electropolishing, explosives, and analytical chemistry, and as a chemical intermediate. Also known as Fraude's reagent.

perchloroethane *See* hexachloroethane.

perchloroethylene [ORG CHEM] CCl_2CCl_2 Stable, colorless liquid, boiling at 121°C; nonflammable and nonexplosive, with low toxicity; used as a dry-cleaning and industrial solvent, in pharmaceuticals and medicine, and for metal cleaning. Also known as tetrachloroethylene.

perchloromethyl mercaptan [ORG CHEM] $ClSCCl_3$ Poisonous, yellow oil with disagreeable aroma; decomposes at 148°C; used as a chemical intermediate, granary fumigant, and military poison gas.

Percidae [VERT ZOO] A family of fresh-water actinopterygian fishes in the suborder Percoidei; comprises the true perches.

Perciformes [VERT ZOO] The typical spiny-rayed fishes, comprising the largest order of vertebrates; characterized by fin spines, a swim bladder without a duct, usually ctenoid scales, and 17 or fewer caudal fin rays.

Perco copper sweetening [CHEM ENG] Proprietary petroleum-refinery process to desulfurize gasoline by mixing it with air and filtering it through an absorbent bed impregnated with copper reagent (copper sulfate and sodium chloride); mostly replaced by catalytic hydrogenation processes.

Perco HF alkylation [CHEM ENG] Proprietary petroleum-refinery process using hydrofluoric acid as a catalyst to unite olefins with isobutane to produce high-octane motor gasoline.

Percoidei [VERT ZOO] A large, typical suborder of the order Perciformes; includes over 50% of the species in this order.

percolation [HYD] Gravity flow of groundwater through the pore spaces in rock or soil. [MIN ENG] Gentle movement of a solvent through an ore bed in order to extract a mineral. [SCI TECH] Slow movement of a liquid through a porous material.

percolation filtration [CHEM ENG] A continuous petroleum-refinery process in which lubricating oils and waxes are percolated through a clay bed to improve color, odor, and stability.

percolation leaching [MIN ENG] The selective removal of a mineral by causing a suitable solvent to seep into and through a mass or pile of material containing the desired soluble mineral.

Percomorphi [VERT ZOO] An equivalent, ordinal name for the Perciformes.

Percopsidae [VERT ZOO] A family of fishes in the order Percopsiformes.

Percopsiformes [VERT ZOO] A small order of actinopterygian fishes characterized by single, ray-supported dorsal and anal fins and a subabdominal pelvic fin with three to eight soft rays.

percussion [MED] The act of striking or firmly tapping the surface of the body with a finger or a small hammer to elicit sounds, or vibratory sensations, of diagnostic value. [ORD] Setting off an explosive charge by a sharp blow, as of a gun hammer.

percussion bit [MECH ENG] A rock-drilling tool with chisel-like cutting edges, which when driven by impacts against a rock surface drills a hole by a chipping action.

percussion drill [MECH ENG] A drilling machine usually using compressed air to drive a piston that delivers a series of impacts to the shank end of a drill rod or steel and attached bit.

percussion drilling [MECH ENG] A drilling method in which hammer blows are transmitted by the drill rods to the drill bit.

percussion figure [CRYSTAL] Radiating lines on a crystal section produced by a sharp blow.

percussion fire [ORD] Fire with fuses set to burst on impact.

percussion firing mechanism [ORD] Any firing mechanism which fires the primer by percussion.

percussion hammer firing mechanism [ORD] Firing mechanism in which a hammer, actuated by a pull of a lanyard, strikes the firing pin and fires the weapon.

percussion mark [GEOL] A small, crescent-shaped scar produced on a hard, dense pebble by a blow.

percussion powder [MATER] Powder so composed as to ignite by a slight percussion.

percussion side-wall sampling [PETRO ENG] A method of side-wall core sampling from wellbores in softer reservoir formations.

percussion table *See* concussion table.

percussion welding [MET] Resistance welding with arc heat and simultaneously applied pressure from a hammerlike blow.

percylite [MINERAL] $PbCuCl_2(OH)_2$ Mineral made up of a basic chloride of copper and lead and occurring as cubic blue crystals, with a hardness of 2.5.

pereletok [GEOL] A frozen layer of ground, at the base of the active layer, which may persist for one or several years. Also known as intergelisol.

perennial [BOT] A plant that lives for an indefinite period, dying back seasonally and producing new growth from a perennating part.

perennial stream [HYD] A stream which contains water at all times except during extreme drought.

perezone [GEOL] A zone in which sediments accumulate along coastal lowlands; includes lagoons and brackish-water bays.

perfect cube [MATH] A number or polynomial which is the exact cube of another number or polynomial.

PERCH

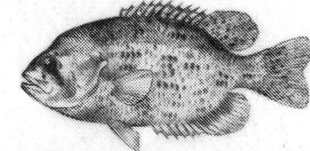

Rock bass (*Ambloplites rupestris*), a perch of the family Centrarchidae.

PERCOPSIFORMES

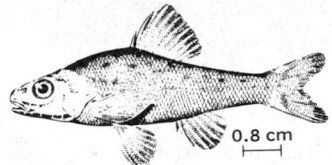

0.8 cm

Sand roller (*Percopsis transmontana*). (*After D. S. Jordan and B. W. Evermann, The Fishes of North and Middle America, U.S. Nat. Mus. Bull. no. 47, 1900*)

perfect dielectric   See ideal dielectric.

perfect flower   [BOT] A flower having both stamens and pistils.

perfect fluid   See inviscid fluid.

perfect gas   See ideal gas.

perfectly diffuse radiator   [OPTICS] A body that emits radiant energy in accordance with Lambert's law.

perfectly diffuse reflector   [OPTICS] A body that reflects radiant energy in such a manner that the reflected energy may be treated as if it were being emitted (radiated) in accordance with Lambert's law.

perfectly inelastic collision   [PHYS] A collision in which as much translational kinetic energy is converted into internal energy of the colliding systems as is consistent with the conservation of momentum. Also known as completely inelastic collision.

perfect number   [MATH] An integer which equals the sum of all its factors other than itself.

perfect prognostic   [METEOROL] The observed pressure pattern at the verifying time of a forecast of some element other than pressure; used in objective forecast studies in which a forecast of the element is based on a simultaneous relation between this element and the pressure pattern plus a forecast of the pressure pattern at some future time.

perfect set   [MATH] A set in a topological space which equals its set of accumulation points.

perfect square   [MATH] A number or polynomial which is the exact square of another number or polynomial.

perfect vacuum   See absolute vacuum.

perfluoroethylene   See tetrafluoroethylene.

perfoliate   [BOT] Pertaining to the form of a leaf having its base united around the stem.   [INV ZOO] Pertaining to the form of certain insect antennae having the terminal joints expanded and flattened to form plates which encircle the stalk.

perforate   [SCI TECH] To pierce or puncture; particularly, to make a line or series of holes for such purposes as identification, decoration, or easy separation.

perforated film   [GRAPHICS] Roll film containing perforations on one or both sides.

perforated metal   [MATER] Sheet metal with round, square, diamond, or rectangular perforations; used for screens and for construction.

perforated-pipe distributor   [CHEM ENG] Liquid distribution device consisting of a length of piping or tubing with holes at spaced intervals along the length; used in spray columns, liquid-vapor contactors, and spray driers. Also known as a sparger.

perforated plate   [CHEM ENG] Flat plate with series of holes used to control fluid distribution, as in a perforated-plate (distillation) column.

perforated-plate column   [CHEM ENG] Distillation column in which vapor-liquid contact is provided by perforated plates instead of bubble-cap trays.

perforated-plate distributor   [CHEM ENG] **1.** A perforated plate or screen used to even out liquid-flow fluctuations through flow channels. **2.** A perforated plate as used in a distillation column or liquid-liquid extraction column.

perforated-plate extractor   [CHEM ENG] A liquid-liquid extraction vessel in which perforated plates are used to bring about contact between the two or more liquid phases.

perforating   [PETRO ENG] Special oil-well downhole procedure to make holes in tubing walls and surrounding cement; used to allow formation oil or gas to enter the wellbore tubing, or to allow water to be forced out into the formation to cause fracturing (hydraulic fracturing).

perforation   [ORD] Passage of a missile completely through an object.   [SCI TECH] Any hole made by boring, punching, or piercing.

perforation rate   [ADP] The rate at which characters, rows, or words are punched in a paper tape.

perforations   [GRAPHICS] Small rectangular sprocket holes along the edge of film.   [PETRO ENG] Downwell holes made in well tubing, usually by shot-and-explosive or shaped-charge techniques; used for oil or gas production from desired horizons, or for injection of acidizing or fracturing fluids into the formation at predetermined depths.

perforator   [COMMUN] In telegraph practice, a device for punching code signals in paper tape for application to a tape transmitter.

perforatorium   See acrosome.

performance bond   [ENG] A bond that guarantees performance of a contract.

performance characteristic   [ENG] A characteristic of a piece of equipment, determined during its test or during its operation.

performance chart   [ENG] A graph used in evaluating the performance of any device, for example, the performance of an electrical or electronic device, such as a graph of anode voltage versus anode current for a magnetron.

performance curves   [ENG] Graphical representations showing the abilities of rotating equipment at various operating conditions; for example, the performance curve for a compressor would include rotor speed for various intake and outlet pressures versus gas flow rate adjusted for temperature, density, viscosity, head, and other factors.

performance data   [ENG] Data on the manner in which a given substance or piece of equipment performs during actual use.

performance evaluation   [IND ENG] The analysis in terms of initial objectives and estimates, and usually made on site, of accomplishments using an automatic data-processing system, to provide information on operating experience and to identify corrective actions required, if any.

performance index   [IND ENG] The ratio of standard hours to the hours of work actually used; a ratio exceeding 1.00 (or 100%) indicates standard output is being exceeded.

performance number   [ENG] One of a series of numbers (constituting the PN, or performance-number, scale) used to convert fuel antiknock values in terms of a reference fuel into an index which is an indication of relative engine performance; used mostly to rate aviation gasolines with octane values greater than 100.

performance rating   See effort rating.

performance sampling   [IND ENG] A technique in work measurement used to determine the leveling factor to be applied to an operator or a group of operators by short, randomly spaced observations of the performance index.

perfume base   [MATER] Any natural or synthetic material used by perfumers as a starting point for perfume manufacture.

perfume oil   [MATER] Any volatile oil distilled or extracted from the leaves, flowers, gums, or woods of plant life (but occasionally of animal origin) and used in making perfume; examples are linalyl acetate (from citral), crab apple, wisteria, lavendar, and attar of rose.

perfusion   [PHYSIO] The pumping of a fluid through a tissue or organ by way of an artery.

pergelation   [HYD] The act or process of forming permafrost.

pergelisol table   See permafrost table.

Pergidae   [INV ZOO] A small family of hymenopteran insects in the superfamily Tenthredinoidea.

perhumid climate   [CLIMATOL] As defined by C. W. Thornthwaite in his climatic classification, a type of climate which has humidity index values of +100 and above; this is his wettest type of climate (designated A), and compares closely to the "wet climate" which heads his 1931 grouping of humidity provinces.

perhydro-   [ORG CHEM] Prefix designating a completely saturated aromatic compound, as for decalin ($C_{10}H_{18}$), also known as perhydronaphthalene.

perianal   [ANAT] Situated or occurring around the anus.

perianth   [BOT] The calyx and corolla considered together.

periapical cyst   See radicular cyst.

periappendicitis   [MED] Inflammation of the tissue around the vermiform process, or of the serosal region of the vermiform appendix.

periapsis   [ASTRON] The orbital point nearest the center of attraction of an orbiting body.

periarteritis   [MED] Inflammation of the outer coat of an artery and of the periarterial tissues.

periarteritis nodosa   See polyarteritis nodosa.

periastron   [ASTRON] The coordinates and time when the two

PERFOLIATE

Shape of a perfoliate leaf.

stars of a binary star system are nearest to each other in their orbits.

periblast [EMBRYO] The nucleated layer of cytoplasm that surrounds the blastodisk of an egg undergoing discoidal cleavage.

periblastula [EMBRYO] The blastula of a centrolecithal egg, formed by superficial segmentation.

periblem [BOT] A layer of primary meristem which produces the cortical cells.

pericardial [ANAT] 1. Of or pertaining to the pericardium. 2. Located around the heart.

pericardial cavity [ANAT] A potential space between the inner layer of the pericardium and the epicardium of the heart.

pericardial fluid [PHYSIO] The fluid in the pericardial cavity.

pericardial organ [INV ZOO] One of the neurohemal organs associated with the pericardial cavity in crustaceans.

pericarditis [MED] Inflammation of the pericardium.

pericardium [ANAT] The membranous sac that envelops the heart; it contains 5-20 grams of clear serous fluid.

pericarp [BOT] The wall of a fruit, developed by ripening and modification of the ovarian wall.

pericenter [PHYS] That point on any orbit nearest to the center of attraction.

pericentric inversion [GEN] A type of chromosome aberration in which chromosome material involving both arms of the chromosome is inverted, thus spanning the centromere.

pericholangitis [MED] Inflammation of the tissues around the bile ducts or the interlobular bile capillaries.

perichondrium [ANAT] The fibrous connective tissue covering cartilage, except at joints.

periclase [MINERAL] MgO Native magnesia; a mineral occurring in granular forms or isometric crystals, with hardness of 6 on Mohs scale, and specific gravity of 3.67–3.90. Also known as periclasite.

periclasite See periclase.

periclinal chimera [GEN] A plant carrying a mixture of cells of two distinct species.

pericline [GEOL] A fold characterized by central orientation of the dip of the beds. [MINERAL] A variety of albite elongated, and often twinned, along the *b*-axis.

pericline twin law [CRYSTAL] A parallel twin law in triclinic feldspars, in which the *b*-axis is the twinning axis and the composition surface is a rhombic section.

pericolitis [MED] Inflammation of the peritoneum or tissues around the colon.

pericoronitis [MED] Inflammation of the tissue surrounding the coronal portion of the tooth, usually a partially erupted third molar.

pericranium [ANAT] The periosteum on the outer surface of the cranial bones.

pericycle [BOT] The outer boundary of the stele of plants; may not be present as a distinct layer of cells.

pericystium [ANAT] The tissues surrounding a bladder. [MED] The vascular wall of a cyst.

pericyte [HISTOL] A mesenchymal cell found around a capillary; it may or may not be contractile.

periderm [BOT] A group of secondary tissues forming a protective layer which replaces the epidermis of many plant stems, roots, and other parts; composed of cork cambium, phelloderm, and cork. [EMBRYO] The superficial transient layer of epithelial cells of the embryonic epidermis.

peridium [BOT] The outer investment of the sporophore of many fungi.

peridot [MINERAL] 1. A gem variety of olivine that is transparent to translucent and pale-, clear-, or yellowish-green in color. 2. A variety of tourmaline approaching olivine in color.

peridotite [PETR] A dark-colored, ultrabasic phaneritic igneous rock composed largely of olivine, with smaller amounts of pyroxene or hornblende.

peridotite shell See upper mantle.

peridynamic loudspeaker [ENG ACOUS] Box-type loudspeaker baffle designed to give good bass response by minimizing acoustic standing.

perifollicular [HISTOL] Surrounding a follicle.

perigean range [OCEANOGR] The average range of tide at the

time of perigean tides, when the moon is near perigee; the perigean range is greater than the mean range.

perigean tidal currents [OCEANOGR] Tidal currents of increased speed occurring at the time of perigean tides.

perigean tide [OCEANOGR] Tide of increased range occurring when the moon is near perigee.

perigee [ASTRON] The point in the orbit of the moon or other satellite when it is nearest the earth.

perigee-to-perigee period See anomalistic period.

periglacial [GEOL] Of or pertaining to the outer perimeter of a glacier, particularly to the fringe areas immediately surrounding the great continental glaciers of the geologic ice ages, with respect to environment, topography, areas, processes, and conditions influenced by the low temperature of the ice.

periglacial climate [CLIMATOL] The climate which is characteristic of the regions immediately bordering the outer perimeter of an ice cap or continental glacier; the principal climatic feature is the high frequency of very cold and dry winds off the ice area; it is also thought that these regions offer ideal conditions for the maintenance of a belt of intense cyclonic activity.

perigonium [BOT] The perianth of a liverwort. [INV ZOO] The sac containing the generative bodies in the gonophore of a hydroid.

perigynium [BOT] A fleshy cup- or tubelike structure surrounding the archegonium of various bryophytes.

perigynous [BOT] Bearing the floral organs on the rim of an expanded saucer- or cup-shaped receptacle or hypanthium.

perihelion [ASTRON] That orbital point nearest the sun when the sun is the center of attraction.

perihepatitis [MED] Inflammation of the peritoneum and tissues surrounding the liver.

peri-infarction block See intraventricular heart block.

perikaryon [CYTOL] 1. The body of a nerve cell, containing the nucleus. 2. A cytoplasmic mass surrounding a nucleus.

Perilampidae [INV ZOO] A family of hymenopteran insects in the superfamily Chalcidoidea.

perilla oil [MATER] A light-yellow drying oil derived from seeds of mints of the genus *Perilla*; soluble in alcohol, benzene, carbon disulfide, ether, and chloroform; used as a substitute for linseed oil, as an edible oil, and in manufacture of varnishes and artificial leather.

perilymph [PHYSIO] The fluid separating the membranous from the osseous labyrinth of the internal ear.

perimeter [MATH] The total length of a closed curve; for example, the perimeter of a polygon is the total length of its sides.

perimeter blasting [MIN ENG] A method of blasting in tunnels, drifts, and raises, designed to minimize overbreak and leave clean-cut solid walls; the outside holes are loaded with very light continuous explosive charges and fired simultaneously, so that they shear from one hole to the other.

perimeter of airway [MIN ENG] In mine ventilation, the linear distance in feet of the airway perimeter rubbing surface at right angles to the direction of the airstream.

perimetrium [ANAT] The serous covering of the uterus.

perimysium [ANAT] The connective tissue sheath enveloping a muscle or a bundle of muscle fibers.

perineum [ANAT] 1. The portion of the body included in the outlet of the pelvis, bounded in front by the pubic arch, behind by the coccyx and sacrotuberous ligaments, and at the sides by the tuberosities of the ischium. 2. The region between the anus and the scrotum in the male, between the anus and the posterior commissure of the vulva in the female.

perineural [ANAT] Situated around nervous tissue or a nerve.

perineural fibroblastoma See neurofibroma.

perineural fibroma See neurofibroma.

periocular [ANAT] Surrounding the eye.

period [CHEM] A family of elements with consecutive atomic numbers in the periodic table and with closely related properties; for example, chromium through copper. [GEOL] A unit of geologic time constituting a subdivision of an era; the fundamental unit of the standard geologic time scale. [NUCLEO] The time required for exponentially rising or falling neutron flux in a nuclear reactor to change by a factor of

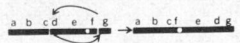
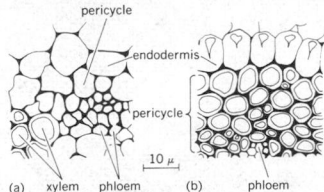
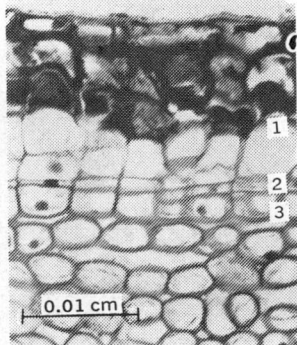
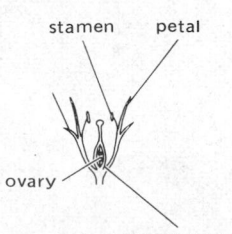

e (2.71828). [PHYS] The duration of a single repetition of a cyclic phenomenon.

periodate [INORG CHEM] A salt of periodic acid, HIO_4, for example, potassium periodate, KIO_4.

periodic [SCI TECH] Repeating itself identically at regular intervals.

periodic acid [INORG CHEM] $HIO_4 \cdot 2H_2O$ Water- and alcohol-soluble white crystals; loses water at 100°C; used as an oxidant.

periodic antenna [ELECTROMAG] An antenna in which the input impedance varies as the frequency is altered.

periodic current [OCEANOGR] Current produced by the tidal influence of moon and sun or by any other oscillatory forcing function.

periodic damping [PHYS] Damping which is less than critical damping.

periodic disease *See* familial Mediterranean fever.

periodic duty [ELEC] Intermittent duty in which the load conditions are regularly recurrent.

periodic field focusing [ELECTR] Focusing of an electron beam where the electrons follow a trochoidal path and the focusing field interacts with them at selected points.

periodic function [MATH] A function $f(x)$ of a real or complex variable is periodic with period T if $f(x + T) = f(x)$ for every value of x.

periodicity [MATH] The property of periodic functions.

periodic kiln [ENG] A kiln in which the cycle of setting ware in the kiln, heating up, "pouking" or holding at peak temperature for some time, "cooling," and removing or "drawing" the ware is repeated for each batch.

periodic lattice *See* lattice.

periodic law [CHEM] The law that the properties of the chemical elements and their compounds are a periodic function of their atomic weights.

periodic line [ELEC] Line consisting of successive and identical sections, similarly oriented, the electrical properties of each section not being uniform throughout; the periodicity is in space and not in time; an example of a periodic line is the loaded line with loading coils uniformly spaced.

periodic motion [MECH] Any motion that repeats itself identically at regular intervals.

periodic peritonitis *See* familial Mediterranean fever.

periodic perturbation [ASTRON] Small deviations from the computed orbit of a planet or satellite; the deviations extend through cycles that generally do not exceed a century. [MATH] A perturbation which is periodic as a function.

periodic quantity [PHYS] Oscillating quantity, the values of which recur for equal increments of the independent variable.

periodic table [CHEM] A table of the elements, written in sequence in the order of atomic number or atomic weight and arranged in horizontal rows (periods) and vertical columns (groups) to illustrate the occurrence of similarities in the properties of the elements as a periodic function of the sequence.

periodic wave [PHYS] A wave whose displacement has a periodic variation with time or distance, or both.

period-luminosity relation [ASTRON] Relation between the periods of Cepheid variable stars and their absolute magnitude; the absolutely brighter the star, the longer the period.

period of a variable star [ASTRON] The average time interval for a variable star to complete a cycle of its variations.

period of vibration [PHYS] The time for one complete cycle of a vibration.

periodontal [ANAT] **1.** Surrounding a tooth. **2.** Of or pertaining to the periodontium.

periodontitis [MED] Inflammation of the periodontium.

periodontium [ANAT] The tissues surrounding a tooth.

periodontoclasia [MED] Any periodontal disease that results in the destruction of the periodontium.

periodontosis [MED] A degenerative disturbance of the periodontium, characterized by degeneration of connective-tissue elements of the periodontal ligament and by bone resorption.

perionychium [ANAT] The border of epidermis surrounding an entire nail. Also known as paronychium.

periosteum [ANAT] The fibrous membrane enveloping

bones, except at joints and the points of tendonous and ligamentous attachment.

periostitis [MED] Inflammation of the periosteum.

periostracum [INV ZOO] A protective layer of chitin covering the outer portion of the shell in many mollusks, especially fresh-water forms.

periotic [ANAT] **1.** Situated about the ear. **2.** Of or pertaining to the parts immediately about the internal ear.

peripheral [ANAT] Pertaining to or located at or near the surface of a body or an organ. [SCI TECH] Remote from the center; marginal; on the periphery.

peripheral buffer [ADP] A device acting as a temporary storage when transmission occurs between two devices operating at different transmission speeds.

peripheral equipment [ADP] Equipment that works in conjunction with a computer but is not part of the computer itself, such as a card or paper-tape reader or punch, magnetic-tape handler, or line printer.

peripheral initiation [ORD] Simultaneous initiation of detonation around the entire periphery of a cylindrical explosive charge; it may be accomplished from point initiation by inserting a disk of inert material, of proper dimensions, in the explosive column.

peripheral milling [MET] Removing metal from a surface parallel to the axis of a milling cutter.

peripheral nervous system [ANAT] The autonomic nervous system, the cranial nerves, and the spinal nerves including their associated sensory receptors.

peripheral processor [ADP] Auxilary computer performing specific operations under control of the master computer.

peripheral speed *See* cutting speed.

peripheral support computer [ADP] A computer (on-line or off-line) used primarily for card-to-tape, tape-to-card, and tape-to-printer conversions, control of on-line devices, sorting, merging, and other similar auxiliary operations in support of a large automatic data-processing complex, and compatible with the larger computer to the extent that no required data interchange involves any necessity for conversion.

peripheral vision [PHYSIO] The act of seeing images that fall upon parts of the retina outside the macula lutea. Also known as indirect vision.

periphery [MATH] The bounding curve of a surface or the surface of a solid.

periphlebitis [MED] Inflammation of the tissues around a vein or of the adventitia of a vein.

periphyton [ECOL] Sessile biotal components of a fresh-water ecosystem.

periplast [CYTOL] **1.** A cell membrane. **2.** A pellicle covering ectoplasm. [HISTOL] The stroma of an animal organ. [INV ZOO] The ectoplasm of a flagellate.

periproct [INV ZOO] The area surrounding the anus of echinoids.

Periptychidae [PALEON] A family of extinct herbivorous mammals in the order Condylarthra distinguished by specialized, fluted teeth.

Peripylina [INV ZOO] An equivalent name for Porulosida.

perisarc [INV ZOO] The outer integument of a hydroid.

Periscelidae [INV ZOO] A family of myodarian cyclorrhaphous dipteran insects in the subsection Acalypteratae.

Perischoechinoidea [INV ZOO] A subclass of principally extinct echinoderms belonging to the Echinoidea and lacking stability in the number of columns of plates that make up the ambulacra and interambulacra.

periscope [OPTICS] An optical instrument used to provide a raised line of vision where it may not be practical or possible, as in entrenchments, tanks, or submarines; the raised line of vision is obtained by the use of mirrors or prisms within the structure of the item; it may have single or dual optical systems.

periscopic sextant [NAV] **1.** A sextant designed to be mounted inside an aircraft, with a tube extending vertically upward through the skin of the aircraft. **2.** A sextant designed to be used in conjunction with the periscope of a submarine.

periscopic sight [ORD] Gunsight made in the form of a periscope, permitting a gunner to see over an **obstacle**.

perisperm [BOT] In a seed, the nutritive tissue that is **derived**

from the nucellus and deposited on the outside of the embryo sac.

Perissodactyla [VERT ZOO] An order of exclusively herbivorous mammals distinguished by an odd number of toes and mesaxonic feet, that is, with the axis going through the third toe.

peristalsis [PHYSIO] The rhythmic progressive wave of muscular contraction in tubes, such as the intestine, provided with both longitudinal and transverse muscular fibers.

peristerite [MINERAL] A gem variety of albite (An₂–An₂₄) that resembles moonstone and has a blue or bluish-white luster characterized by sharp internal reflections of blue, green, and yellow.

peristome [BOT] The fringe around the opening of a moss capsule. [INV ZOO] The area surrounding the mouth of various invertebrates.

peristyle [ARCH] A surrounding series of columns; the space enclosed by such columns.

peritectic [PHYS CHEM] An isothermal reversible reaction in which a liquid phase reacts with a solid phase during cooling to produce a second solid phase.

peritectic point [PHYS CHEM] In a binary two-phase heteroazeotropic system at constant pressure, that point up to which the boiling point has remained constant until one of the phases has boiled away.

peritectoid [PHYS CHEM] An isothermal reversible reaction in which a solid phase on cooling reacts with another solid phase to form a third solid phase.

perithecium [MYCOL] A spherical, cylindrical, or oval ascocarp which usually opens by a terminal slit or pore.

peritoneal cavity [ANAT] The potential space between the visceral and parietal layers of the peritoneum.

peritoneoscope [MED] A long, slender endoscope equipped with sheath, obturator, biopsy forceps, a sphygmomanometer bulb and tubing, scissors, and a syringe; introduced into the peritoneal cavity through a small incision in the abdominal wall permitting visualization of the gas-inflated peritoneal cavity.

peritoneoscopy See laparoscopy.

peritoneum [ANAT] The serous membrane enveloping the abdominal viscera and lining the abdominal cavity.

peritonitis [MED] Inflammation of the peritoneum.

peritonsillar abscess [MED] An abscess forming in acute tonsillitis around one or both tonsils.

Peritrichia [INV ZOO] A specialized subclass of the class Ciliatea comprising both sessile and mobile forms.

Peritrichida [INV ZOO] The single order of the protozoan subclass Peritrichia.

peritrichous [INV ZOO] Of certain protozoans, having spirally arranged cilia around the oral disk. [MICROBIO] Of bacteria, having a uniform distribution of flagella on the body surface.

perityphlitis [MED] Inflammation of the peritoneum surrounding the cecum and vermiform appendix.

perivitelline space [CYTOL] In mammalian ova, the space formed between the ovum and the zona pellucida at the time of maturation, into which the polar bodies are given off.

Perkins' reaction [ORG CHEM] The formation of unsaturated cinnamic-type acids by the condensation of aromatic aldehydes (Ar) with fatty acids in the presence of acetic anhydride; for example, ArCHO + CH₃COONa→ ArCH = CHCOONa + H₂O.

perlèche [MED] An inflammatory condition occurring at the angles of the mouth with resultant fissuring.

perlite [GEOL] A rhyolitic glass with abundant spherical or convolute cracks that cause it to break into small pearllike masses or pebbles, usually less than a centimeter across; it is commonly gray or green with a pearly luster and has the composition of rhyolite. Also known as pearlite; pearlstone.

perlitic [PETR] **1.** Of the texture of a glassy igneous rock, exhibiting small spheruloids formed from cracks due to contraction during cooling. **2.** Pertaining to or characteristic of perlite.

perlucidus [METEOROL] A cloud variety, usually of the species stratiformis, in which distinct spaces between its elements permit the sun, moon, blue sky, or higher clouds to be seen.

perm [PETRO ENG] A unit indicating the degree of permeability of a porous reservoir structure; the unit is expressed as bbl day⁻¹ ft⁻² psi⁻¹ ft cp or ft³ day⁻¹ ft⁻² psi⁻¹ ft cp.

permafil [MATER] Polymerizable mixture that cures without any evaporation.

permafrost [GEOL] Perennially frozen ground, occurring wherever the temperature remains below 0°C for several years, whether the ground is actually consolidated by ice or not and regardless of the nature of the rock and soil particles of which the earth is composed.

permafrost drilling [ENG] Boreholes drilled in subsoil and rocks in which the contained water is permanently frozen.

permafrost table [GEOL] The upper limit of permafrost. Also known as pergelisol table.

permalloy [MET] A trade name for any of several highly magnetically permeable iron-base alloys containing about 45–80% nickel. Also known as permeability alloy.

permanent anode [MET] A very-corrosion-resistant anode, made of a material such as a carbon, aluminum, or lead alloy, or 14.5% silicon iron; used in cathodic protection against corrosion.

permanent aurora See airglow.

permanent-completion packer [PETRO ENG] A packer able to withstand large pressure differentials to allow for its permanent installation in a producing well.

permanent current [OCEANOGR] A current which continues with relatively little periodic or seasonal change.

permanent echo [ELECTROMAG] A signal reflected from an object that is fixed with respect to a radar site.

permanent emplacement See fixed emplacement.

permanent gas [THERMO] A gas at a pressure and temperature far from its liquid state.

permanent hardness [CHEM] The hardness of water persisting after boiling.

permanent ice foot [HYD] An ice foot that does not melt completely in summer.

permanent ink [MATER] Ink that contains up to 1% dissolved iron to prevent fading or washing away when dried.

permanent magnet [ELECTROMAG] A piece of hardened steel or other magnetic material that has been strongly magnetized and retains its magnetism indefinitely. Abbreviated PM.

permanent-magnet dynamic loudspeaker See permanent-magnet loudspeaker.

permanent-magnet focusing [ELECTR] Focusing of the electron beam in a television picture tube by means of the magnetic field produced by one or more permanent magnets mounted around the neck of the tube.

permanent-magnet loudspeaker [ENG ACOUS] A moving-conductor loudspeaker in which the steady magnetic field is produced by a permanent magnet. Also known as permanent-magnet dynamic loudspeaker.

permanent-magnet moving-coil instrument [ENG] An ammeter or other electrical instrument in which a small coil of wire, supported on jeweled bearings between the poles of a permanent magnet, rotates when current is carried to it through spiral springs which also exert a restoring torque on the coil; the position of the coil is indicated by an attached pointer.

permanent-magnet moving-iron instrument [ENG] A meter that depends for its operation on a movable iron vane that aligns itself in the resultant magnetic field of a permanent magnet and adjacent current-carrying coil.

permanent-magnet stepper motor [ELEC] A stepper motor in which the rotor is a powerful permanent magnet and each stator coil is energized independently in sequence; the rotor aligns itself with the stator coil that is energized.

permanent mold [MET] A reusable metal mold for the production of many castings of the same kind.

permanent monument [MIN ENG] A monument of a lasting character for marking a mining claim; it may be a mountain, hill, or ridge.

permanent pump [MIN ENG] A pump on which the mine depends for the final disposal of its drainage.

permanent set [MECH] Permanent plastic deformation of a structure or a test piece after removal of the applied load. Also known as set.

permanent starch [MATER] An emulsion of polyvinyl acetate

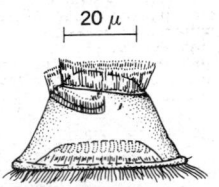

PERITRICHIA

20 μ

Trichodina, a mobile peritrich.

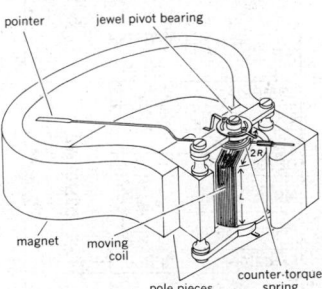

**PERMANENT-MAGNET
MOVING-COIL INSTRUMENT**

pointer · jewel pivot bearing · magnet · moving coil · pole pieces · counter-torque spring

Mechanism of a permanent-magnet moving-coil instrument. L = length of active conductors, R = radius of action of these conductors. (*Weston Instruments, Inc.*)

used for starching clothing and textiles; it is not removed by washing.

permanent storage [ADP] A means of storing data for rapid retrieval by a computer; does not permit changing the stored data.

permanent teeth [ANAT] The second set of teeth of a mammal, following the milk teeth; in humans, the set of 32 teeth consists of 8 incisors, 4 canines, 8 premolars, and 12 molars.

permanent thermocline *See* main thermocline.

permanent wave [FL MECH] A wave (in a fluid) which moves with no change in streamline pattern, and which, therefore, is a stationary wave relative to a coordinate system moving with the wave.

permanganate [INORG CHEM] A purple salt of permanganic acid containing the MnO_4^- radical; used as an oxidizing agent and a disinfectant.

permanganic acid [INORG CHEM] $HMnO_4$ An unstable acid that exists only in dilute solutions; decomposes to manganese dioxide and oxygen.

Permasyn motor [ELEC] A synchronous motor which has permanent magnets embedded in the squirrel-cage rotor to provide an equivalent direct-current field.

permatron [ELECTR] Thermionic gas-discharge diode in which the start of conduction is controlled by an external magnetic field.

permeability [ELECTROMAG] A factor, characteristic of a material, that is proportional to the magnetic induction produced in a material divided by the magnetic field strength; it is a tensor when these quantities are not parallel. [FL MECH] **1.** The ability of a membrane or other material to permit a substance to pass through it. **2.** Quantitatively, the amount of substance which passes through the material under given conditions. [GEOL] The capacity of a porous rock, soil, or sediment for transmitting a fluid without damage to the structure of the medium. Also known as perviousness. [NAV ARCH] The percentage of a given space in a ship that can be occupied by water.

permeability alloy *See* permalloy.

permeability-block method [PETRO ENG] Calculation method for oil recovery from water-drive oil fields in which there are variable-permeability distributions.

permeability coefficient [FL MECH] The rate of water flow in gallons per day through a cross section of 1 square foot under a unit hydraulic gradient, at the prevailing temperature or at 60°F (16°C). Also known as coefficient of permeability; hydraulic conductivity; Meinzer unit.

permeability number [ENG] A numbered value assigned to molding materials indicating the relative ease of passage of gases through them.

permeability profile [PETRO ENG] A graphical plot of porous reservoir permeability versus distance down the wellbore.

permeability tuning [ELEC] Process of tuning a resonant circuit by varying the permeability of an inductor; it is usually accomplished by varying the amount of magnetic core material of the inductor by slug movement.

permeable bed [GEOL] A porous reservoir formation through which hydrocarbon fluids (oil or gas) or water (waterflood or interstitial) can flow.

permeable membrane [CHEM] A thin sheet or membrane of material through which selected liquid or gas molecules or ions will pass, either through capillary pores in the membrane or by ion exchange; used in dialysis, electrodialysis, and reverse osmosis.

permeameter [ENG] **1.** Device for measurement of the average size or surface area of small particles; consists of a powder bed of known dimension and degree of packing through which the particles are forced; pressure drop and rate of flow are related to particle size, and pressure drop is related to surface area. **2.** A device for measuring the coefficient of permeability by measuring the flow of fluid through a sample across which there is a pressure drop produced by gravity. **3.** An instrument for measuring the magnetic flux or flux density produced in a test specimen of ferromagnetic material by a given magnetic intensity, to permit computation of the magnetic permeability of the material.

permeametry [ANALY CHEM] Determination of the average size of fine particles in a fluid (gas or liquid) by passing the

mixture through a powder bed of known dimensions and recording the pressure drop and flow rate through the bed.

permeance [ELECTROMAG] A characteristic of a portion of a magnetic circuit, equal to magnetic flux divided by magnetomotive force; the reciprocal of reluctance. Symbolized P.

permease [BIOCHEM] Any of a group of enzymes which mediate the phenomenon of active transport.

permeation [CHEM] The movement of atoms, molecules, or ions into or through a porous or permeable substance (such as zeolite or a membrane).

Permendur [MET] A magnetic alloy which is composed of equal parts of iron and cobalt and has an extremely high permeability when saturated.

permenorm alloy [MET] An alloy containing 50% nickel and 50% iron; used as magnet core material and in magnetic amplifiers.

Permian [GEOL] The last period of geologic time in the Paleozoic era, from 280 to 225 million years ago.

per mille [SCI TECH] Per thousand or 10^{-3}. Symbolized $^\circ/_{\circ\circ}$.

permineralization [GEOL] A fossilization process whereby additional minerals are deposited in the pore spaces of originally hard animal parts.

permissible [MIN ENG] Said of equipment completely assembled and conforming in every respect with the design formally approved by the United States Bureau of Mines for use in gassy and dusty mines.

permissible dose [NUCLEO] The amount of radiation that may be safely received by an individual within a specified period. Formerly known as tolerance dose.

permissible explosive [MATER] An explosive approved by the U.S. Bureau of Mines as safe for blasting in gassy and dusty mines.

permissible lamp [MIN ENG] A lamp that meets the standards of the U.S. Bureau of Mines.

permissible length [NAV ARCH] The floodable length of a ship times the factor of subdivision.

permissible machine [MIN ENG] A machine, such as a drill, mining machine, loading machine, conveyor, or locomotive, that meets the standards of the U.S. Bureau of Mines.

permissible velocity [CIV ENG] The highest velocity at which water is permitted to pass through a structure or conduit without excessive damage.

permissive stop [CIV ENG] A railway signal indicating the train must stop but can proceed slowly and cautiously after a specified interval, usually 1 minute.

permittivity [ELEC] The dielectric constant multiplied by the permittivity of empty space, where the permittivity of empty space (ε_0) is a constant appearing in Coulomb's law, having the value of 1 in centimeter-gram-second electrostatic units, and of 8.854×10^{-12} farad/meter in rationalized meter-kilogram-second units. Symbolized ε.

Permo-Carboniferous [GEOL] **1.** The Permian and Carboniferous periods considered as one unit. **2.** The Permian and Pennsylvanian periods considered as a single unit. **3.** The rock unit, or the period of geologic time, transitional between the Upper Pennsylvanian and the Lower Permian periods.

perm-plug method [PETRO ENG] Laboratory method of measuring the permeability of reservoir core samples (or plugs) by the measurement of airflow through the sample at several flow rates.

permutation [MATH] A function which rearranges a finite number of symbols; more precisely, a one-to-one function of a finite set onto itself.

permutation group [MATH] The group whose elements are permutations of some set of symbols where the product of two permutations is the permutation arising from successive application of the two.

permutation modulation [COMMUN] Proposed method of transmitting digital information by means of band-limited signals in the presence of additive white gaussian noise; pulse-code modulation and pulse-position modulation are considered simple special cases of permutation modulation.

permutation table [COMMUN] In computers, a table designed for the systematic construction of code groups; it may also be used to correct garbles in groups of code text.

permutation tensor *See* determinant tensor.

pernetti [ENG] **1.** Small iron pins or tripods that support ware

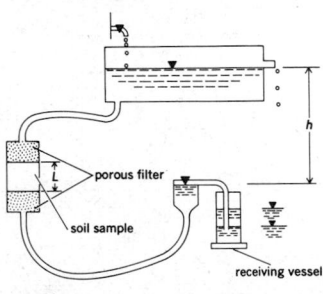

PERMEAMETER

Constant-head permeameter (def. 2). Head water level, h, is kept constant. Water percolates through soil sample of thickness L. Porous filters hold soil in place. The tail water level is kept constant by overflow. Volume of discharge is measured in receiving vessel. (*From G. P. Tschebotarioff, Soil Mechanics, Foundations and Earth Structures, McGraw-Hill, 1951*)

PERMIAN

PRECAMBRIAN		
CAMBRIAN		
ORDOVICIAN		
SILURIAN		
DEVONIAN		PALEOZOIC
Mississippian	CARBONIFEROUS	
Pennsylvanian		
PERMIAN		
TRIASSIC		
JURASSIC		MESOZOIC
CRETACEOUS		
TERTIARY		CENOZOIC
QUATERNARY		

Chart showing position of Permian period in relation to other periods and to the eras of geologic time.

while it is being fired in a kiln. **2.** The marks left on baked pottery by these supporting pins.

pernicious anemia [MED] A megaloblastic macrocytic anemia resulting from lack of vitamin B$_{12}$, secondary to gastric atrophy and loss of intrinsic factor necessary for vitamin B$_{12}$ absorption, and accompanied by degeneration of the posterior and lateral columns of the spinal cord.

pernicious malaria *See* falciparum malaria.

perniosis [MED] Any dermatitis resulting from chilblain.

Perognathinae [VERT ZOO] A subfamily of the rodent family Heteromyidae, including the pocket and kangaroo mice.

Peronosporales [MYCOL] An order of aquatic and terrestrial phycomycetous fungi with a hyphal thallus and zoospores with two flagella.

Perothopidae [INV ZOO] A small family of coleopteran insects in the superfamily Elateroidea found only in the United States.

perovskite [MINERAL] Ca[TiO$_3$] A natural, yellow, brownish-yellow, reddish, brown, or black mineral and a structure type which includes no less than 150 synthetic compounds; the crystal structure is ideally cubic, it occurs as rounded cubes modified by the octahedral and dodecahedral forms, luster is subadamantine to submetallic, hardness is 5.5 on Mohs scale, and specific gravity is 4.0.

peroxidase [BIOCHEM] An enzyme that catlyzes reactions in which hydrogen peroxide is an electron acceptor.

peroxide [CHEM] **1.** A compound containing the peroxy ($-O-O-$) group, as in hydrogen peroxide. **2.** *See* hydrogen peroxide.

peroxide number [ANALY CHEM] Measure of millimoles of peroxide (or milliequivalents of oxygen) taken up by 1000 grams of fat or oil; used to measure rancidity. Also known as peroxide value.

peroxide value *See* peroxide number.

peroxyacetic acid *See* peracetic acid.

peroxybenzoic acid *See* perbenzoic acid.

peroxydol *See* sodium perborate.

peroxymonosulfuric acid *See* Caro's acid.

perpendicular [MATH] Geometric objects are perpendicular if they intersect in an angle of 90°.

perpendiculars [NAV ARCH] Vertical lines through the intersections of the fore and aft ends of a ship with the design waterline.

perpendicular slip [GEOL] The component of a fault slip measured at right angles to the trace of the fault on any intersecting surface.

perpendicular throw [GEOL] The distance between two points which were formerly adjacent in a faulted bed, vein, or other surface, measured at right angles to the surface.

perpetual calendar [ASTRON] A table or mechanical device used to determine the day of the week corresponding to any given date over a period of many years.

perpetual frost climate [CLIMATOL] The climate of the ice cap regions of the world; thus, it requires temperatures sufficiently cold so that the annual accumulation of snow and ice is never exceeded by ablation. Also known as ice-cap climate.

perpetual motion machine of the first kind [PHYS] A mechanism which, once set in motion, continues to do useful work without an input of energy, or which produces more energy than is absorbed in its operation; it violates the principle of conservation of energy.

perpetual motion machine of the second kind [PHYS] A device that extracts heat from a source and then converts this heat completely into other forms of energy; it violates the second law of thermodynamics.

perpetual motion machine of the third kind [PHYS] A device which has a component that can continue moving forever; an example is a superconductor.

Perron-Frobenius theory [MATH] A study of differential equations with singularities.

perry [METEOROL] In England, a sudden, heavy fall of rain; a squall, sometimes referred to as "half a gale."

perryite [MINERAL] (Ni,Fe)$_5$(Si,P)$_2$ A mineral found only in meteorites.

Perseids [ASTRON] A meteor shower whose radiant lies in the constellation Perseus; it reaches a maximum about August 12.

Perseus [ASTRON] A northern constellation; right ascension 3 hours; declination 45°N.

Pershing [ORD] Name given to a U.S. Army tactical ballistic missile, using a solid propellant and with a range capability of 600 miles (970 kilometers); successor to the Redstone.

Persian ammoniac *See* ammoniac.

Persian melon [BOT] A variety of muskmelon (*Cucumis melo*) in the order Violales; the fruit is globular and without sutures, and has dark-green skin, thin abundant netting, and firm, thick, orange flesh.

Persian red [INORG CHEM] Red pigment made from basic lead chromate or ferric oxide.

persic oil [MATER] A pale-yellow to red fatty oil; soluble in ether, chloroform, and carbon disulfide; taste and aroma are similar to almond oil; the oil is expressed from blanched seeds of peaches or apricots; used as a flavoring, in medicine, and as a nutrient similar to olive and almond oils. Also known as apricot kernel oil; peach kernel oil.

persilicic *See* silicic.

persistence [ELECTR] A measure of the length of time that the screen of a cathode-ray tube remains luminescent after excitation is removed; ranges from 1 for short persistence to 7 for long persistence. [METEOROL] With respect to the long-term nature of the wind at a given location, the ratio of the magnitude of the mean wind vector to the average speed of the wind without regard to direction. Also known as constancy; steadiness.

persistence forecast [METEOROL] A forecast that the future weather condition will be the same as the present condition; often used as a standard of comparison in measuring the degree of skill of forecasts prepared by other methods.

persistence of vision [PHYSIO] The ability of the eye to retain the impression of an image for a short time after the image has disappeared.

persistent [BOT] Of a leaf, withering but remaining attached to the plant during the winter.

persistent current [CRYO] A magnetically induced current that flows undiminished in a superconducting material or circuit.

persistent war gas [MATER] War gas that is normally effective in the open at the point of dispersion for more than 10 minutes.

persistron [ELECTR] A device in which electroluminescence and photoconductivity are used in a single panel capable of producing a steady or persistent display with pulsed signal input.

personal equation [METEOROL] A systematic observational error due to the characteristics of the observer; the uncertainty in a reading made by an observer may be ascertained by a statistical analysis of his readings.

personality disorder [PSYCH] Any of various disorders characterized by abnormal behavior rather than by neurotic, psychotic, or mental disturbances.

personal locator beacon [NAV] A locator beacon capable of providing homing signals to help search-and-rescue operations.

personal probability [STAT] A number between 0 and 1 assigned to an event based upon personal views concerning whether the event will occur or not; it is obtained by deciding whether one would accept a bet on the event at odds given by this number. Also known as subjective probability.

personnel carrier [ORD] A motor vehicle, sometimes armored, used for transportation of troops and their equipment.

personnel monitoring [NUCLEO] Determination of the degree of radioactive contamination on individuals, using standard survey meters, and determination of the dose received by means of dosimeters.

Persoz's reagent [MATER] Chemical reagent used to detect the presence of silk with wool (only silk dissolves); consists of zinc chloride and zinc oxide in water.

perspective [GRAPHICS] The technique representing a figure or the space relationships of natural objects, on either a plane or curved surface, by means of projecting lines emanating from a single point, which may be infinity.

perspective axis *See* axis of homology.

perspective chart [MAP] A chart on a perspective projection.

PEROVSKITE

1.3 cm

Perovskite crystals, Magnet Cove, Arkansas. *(Specimens from Department of Geology, Bryn Mawr College)*

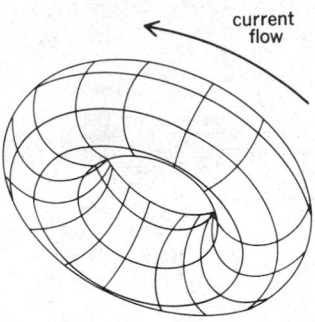

PERPETUAL MOTION MACHINE OF THE THIRD KIND

current flow

If a direct current is caused to flow in a superconducting ring, this current will continue to flow undiminished in time without application of any external force.

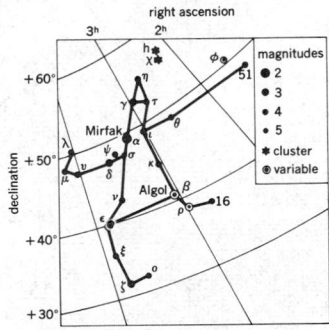

PERSEUS

Line pattern of the constellation Perseus. The grid lines in the chart represent the coordinates of the sky. The apparent brightness, or magnitude, of the stars is shown by the size of the dots, which are graded by appropriate numbers as indicated.

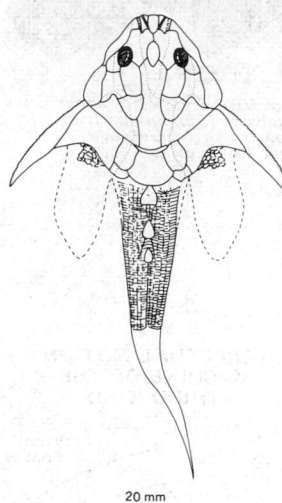

PETALICHTHYIDA

20 mm

Lunaspis broilii from the Lower Devonian of Germany; restoration in dorsal aspect. (*After Walter Gross*)

perspiration [PHYSIO] 1. The secretion of sweat. 2. *See* sweat.

Pers sunshine recorder [ENG] A type of sunshine recorder in which the time scale is supplied by the motion of the sun.

persuader [ELECTR] Element of storage tube which directs secondary emission to electron multiplier dynodes.

persulfate [INORG CHEM] Salt derived from persulfuric acid and containing the radical $S_2O_8^{--}$; made by electrolysis of sulfate solutions.

persulfuric acid [INORG CHEM] $H_2S_2O_8$ Acid formed in lead-cell batteries by electrolyzing sulfuric acid; strong oxidizing agent.

PERT [SYS ENG] A management control tool for defining, integrating, and interrelating what must be done to accomplish a desired objective on time; a computer is used to compare current progress against planned objectives and give management the information needed for planning and decision making. Derived from program evaluation and review technique.

perthite [GEOL] A parallel to subparallel intergrowth of potassium and sodium feldspar; the potassium-rich phase is usually the host from which the sodium-rich phase evolves.

perthitic [GEOL] Of a texture produced by perthite, exhibiting sodium feldspar as small strings, blebs, films, or irregular veinlets in a host of potassium feldspar.

pertinency factor [ADP] In information retrieval, the ratio obtained in dividing the total number of relevant documents retrieved by the total number of documents retrieved.

perturbation [ASTRON] A deviation of an astronomical body from its computed orbit because of the attraction of another body or bodies. [MATH] A function which produces a small change in the values of some given function. [PHYS] Any effect which makes a small modification in a physical system, especially in case the equations of motion could be solved exactly in the absence of this effect.

perturbation equation [PHYS] Any equation governing the behavior of a perturbation; often this will be a linear differential equation.

perturbation motion [PHYS] The motion of a disturbance (usually but not necessarily assumed infinitesimal), as opposed to the motion of the system on which the perturbation is superimposed.

perturbation quantity [PHYS] Any characteristic of a system which may be assumed to be a perturbation from an established value.

perturbation theory [MATH] The study of the solutions of differential and partial differential equations from the viewpoint of perturbation of solutions. [PHYS] The theory of obtaining approximate solutions to the equations of motion of a physical system when these equations differ by a small amount from equations which can be solved exactly.

pertussis [MED] An infectious inflammatory bacterial disease of the air passages, caused by *Hemophilus pertussis* and characterized by explosive coughing ending in a whooping inspiration. Also known as whooping cough.

Peru balsam [MATER] A dark, viscous liquid with bitter taste and pleasant aroma; soluble in alcohol, ether, acetone, chloroform, benzene, and glacial acetic acid; derived from the tropical American tree *Myroxylon pereirae;* main components are esters of cinnamic and benzoic acids; used in perfumery, medicine, and chocolate manufacture. Also known as balsam of Peru; black balsam; China oil; Chinese oil; Indian balsam; Peruvian balsam.

Peru Current [OCEANOGR] The cold ocean current flowing north along the coasts of Chile and Peru. Also known as Humboldt Current.

Peru saltpeter *See* soda niter.

Peruvian balsam *See* Peru balsam.

perveance [ELECTR] The space-charge-limited cathode current of a diode divided by the 3/2 power of the anode voltage.

perviousness *See* permeability.

pessary [MED] An appliance of varied form placed in the vagina for uterine support or contraception. [PHARM] Any suppository or other form of medication placed in the vagina for therapeutic purposes.

pessulus [VERT ZOO] A bar composed of cartilage or bone that crosses the windpipe of a bird at its division into bronchi.

pesticide [MATER] A chemical agent that destroys pests. Also known as biocide.

pestilence [MED] 1. Any epidemic contagious disease. 2. Infection with the plague organism *Pasteurella pestis*.

petal [BOT] One of the sterile, leaf-shaped flower parts that make up the corolla.

Petalichthyida [PALEON] A small order of extinct dorsoventrally flattened fishes belonging to the class Placodermi; the external armor is in two shields of large plates.

petaling [ORD] Condition of armor plate where the metal around the penetration is forced into a leaflike or petal form.

petalite [MINERAL] $LiAlSi_4O_{10}$ A white, gray, or colorless monoclinic mineral composed of silicate of lithium and aluminum, occurring in foliated masses or as crystals.

Petalodontidae [PALEON] A family of extinct cartilaginous fishes in the order Bradyodonti distinguished by teeth with deep roots and flattened diamond-shaped crowns.

Petaluridae [INV ZOO] A family of dragonflies in the suborder Anisoptera.

petasma [INV ZOO] A modified endopodite of the first abdominal appendage in a male decapod crustacean.

petechiae [MED] Hemorrhages the size of the head of a pin.

peter out [ENG] To fail gradually in size, quantity, or quality; for example, a mine may be said to have petered out.

Petersen grab [ENG] A bottom sampler consisting of two hinged semicylindrical buckets held apart by a cocking device which is released when the grab hits the ocean floor.

Peters' formula [STAT] An approximate formula for the probable error in the value of a quantity determined from several equally careful, independent measurements of the value of the quantity.

petiole [BOT] The stem which supports the blade of a leaf.

petit mal [MED] A generalized epileptic seizure of the absence type, that is, characterized by different degrees of impaired consciousness.

PETN *See* pentaerythritol tetranitrate.

petrel [VERT ZOO] A sea bird of the families Procellariidae and Hydrobatidae, generally small to medium-sized with long wings and dark plumage with white areas near the rump.

petri dish [MICROBIO] A shallow glass or plastic dish with a loosely fitting overlapping cover used for bacterial plate cultures and plant and animal tissue cultures.

petrifaction [GEOL] A fossilization process whereby inorganic matter dissolved in water replaces the original organic materials, converting them to a stony substance.

petrified wood *See* silicified wood.

Petriidae [INV ZOO] A small family of coleopteran insects in the superfamily Tenebrionoidea.

petrochemicals [ORG CHEM] Chemicals made from feedstocks derived from petroleum or natural gas; examples are ethylene, butadiene, most large-scale plastics and resins, and petrochemical sulfur. Also known as petroleum chemicals.

petrochemistry [GEOCHEM] An aspect of geochemistry that deals with the study of the chemical composition of rocks. [ORG CHEM] The chemistry and reactions of materials derived from petroleum, natural gas, or asphalt deposits.

petrofabric *See* fabric.

petrofabric analysis *See* structural petrology.

petrofabric diagram *See* fabric diagram.

petrofabrics *See* structural petrology.

petrofacies *See* petrographic facies.

petrogenesis [PETR] That branch of petrology dealing with the origin of rocks, particularly igneous rocks. Also known as petrogeny.

petrogenic grid [PETR] A diagram whose coordinates are parameters of the rock-forming environment on which equilibrium curves are plotted indicating the limits of the stability fields of specific minerals and mineral assemblages.

petrogeny *See* petrogenesis.

petrogeometry *See* structural petrology.

petrographer [GEOL] An individual who does petrography.

petrographic facies [GEOL] Facies distinguished principally by composition and appearance. Also known as petrofacies.

petrographic microscope [OPTICS] A polarizing microscope used for analysis of petrographic thin sections.

petrographic province [GEOL] A broad area in which similar igneous rocks are formed during the same period of igneous activity. Also known as comagmatic region; igneous province; magma province.

petrography [GEOL] The branch of geology that deals with the description and systematic classification of rocks, especially by means of microscopic examination.

petrol See gasoline.

petrolatum [MATER] A smooth, semisolid blend of mineral oil with waxes crystallized from the residual type of petroleum lubricating oil; the wax molecules contain 30-70 carbon atoms and are straight chains with a few branches or naphthene rings; used as a lubricant, as a carrier in polishes and cosmetics, and as a rust preventive.

petroleum [GEOL] A naturally occurring complex liquid hydrocarbon which after distillation yields combustible fuels, petrochemicals, and lubricants; can be gaseous (natural gas), liquid (crude oil, crude petroleum), solid (asphalt, tar, bitumen), or a combination of states.

petroleum asphalt [MATER] Asphalt recovered or made from petroleum.

petroleum benzin [MATER] Colorless, volatile, flammable, toxic petroleum distillate with boiling range of 35-80°C; a special grade is ligroin (petroleum ether; boiling range 20-135°C), used as an extractive solvent, especially for drugs. Also known as benzin; benzine (archaic usage).

petroleum chemicals See petrochemicals.

petroleum coke [MATER] A carbonaceous solid material made by the destructive heating of high-molecular-weight petroleum-refining residues.

petroleum engineer [PETRO ENG] An engineer whose primary objective is to find and produce oil or gas from petroleum reserves.

petroleum engineering [ENG] The application of almost all types of engineering to the drilling for and production of oil, gas, and liquefiable hydrocarbons.

petroleum ether [MATER] A volatile fraction of petroleum consisting chiefly of pentanes and hexanes. Also known as ligroin.

petroleum geology [GEOL] The branch of economic geology dealing with the origin, occurrence, movement, accumulation, and exploration of hydrocarbon fuels.

petroleum isomerization process [CHEM ENG] A fixed-bed, vapor-phase petroleum-refinery process using a precious-metal catalyst and external hydrogen; feedstocks include natural gas, pentane, and hexane cuts; the product is high-octane blending stock.

petroleum microbiology [MICROBIO] Those aspects of microbiological science and engineering of interest to the petroleum industry, including the role of microbes in petroleum formation, and the exploration, production, manufacturing, storage, and food synthesis from petroleum.

petroleum processing [CHEM ENG] The recovery and processing of various usable fractions from the complex crude oils; usable fractions include gasoline, kerosine, diesel oil, fuel oil, and asphalt. Also known as petroleum refining.

petroleum products [MATER] Materials derived from petroleum, natural gas, or asphalt deposits; includes gasolines, diesel and heating fuels, liquefied petroleum gases (LPG and bugas), lubricants, waxes, greases, petroleum coke, petrochemicals, and (from sour crudes and natural gases) sulfur.

petroleum refining See petroleum processing.

petroleum secondary engineering [PETRO ENG] The process of removing oil from its native reservoirs by the use of supplemental energies after the natural energies causing oil production have been depleted.

petroleum sulfonates [MATER] Sulfonated petroleum products made as by-products of SO_3 treatment of white oil or lube-stock; used as lube-oil additives, textile-processing emulsifiers, and rust preventives.

petroleum tailings See wax tailings.

petroleum tar [MATER] A viscous, black or dark-brown product of petroleum refining; yields substantial quantity of solid residue when partly evaporated or fractionally distilled.

petroleum trap [GEOL] Stable underground formation (geological or physical) of such nature as to trap and hold liquid or gaseous hydrocarbons; usually consists of sand or porous rock surrounded by impervious rock or clay formations.

petroleum wax [MATER] A wax occurring naturally in various fractions of crude petroleum; there are two groups: paraffin wax and microcrystalline wax.

petrologen See kerogen.

petrologist [GEOL] An individual who studies petrology.

petrology [GEOL] The branch of geology concerned with the origin, occurrence, structure, and history of rocks, principally igneous and metamorphic rock.

petromict [GEOL] Of a sediment, composed of metastable rock fragments.

petromorphology See structural petrology.

Petromyzonida [VERT ZOO] The lampreys, an order of eel-like, jawless vertebrates (Agnatha) distinguished by a single, dorsal nasal opening, and the mouth surrounded by an oral disk and provided with a rasping tongue.

Petromyzontiformes [VERT ZOO] The equivalent name for Petromyzonida.

petrosal nerve [ANAT] Any of several small nerves passing through the petrous part of the temporal bone and usually attached to the geniculate ganglion.

petrosal process [ANAT] A sharp process of the sphenoid bone located below the notch for the passage of the abducens nerve, which articulates with the apex of the petrous portion of the temporal bone and forms the medial boundary of the foramen lacerum.

Petrosaviaceae [BOT] A small family of monocotyledonous plants in the order Triuridales characterized by perfect flowers, three carpels, and numerous seeds.

Petterson-Nansen water bottle See Nansen bottle.

petticoat insulator [ELEC] Insulator having an outward-flaring lower part that is hollow inside to increase the length of the surface leakage path and keep part of the path dry at all times.

Pettit truss [CIV ENG] A bridge truss in which the panel is subdivided by a short diagonal and a short vertical member, both intersecting the main diagonal at its midpoint.

petzite [MINERAL] Ag_3AuTe_2 A steel-gray to iron-black mineral consisting of a silver gold telluride; hardness is 2.5-3 on Moh's scale, and specific gravity is 8.7-9.0.

Petzval condition [OPTICS] The condition whereby an optical system will eliminate the aberration of curvature of field only if the Petzval curvature vanishes.

Petzval curvature [OPTICS] The axial curvature of the image of a plane object produced by an optical system, equal to the sum over all the optical surfaces in the system of $R(1/n' - 1/n)$, where R is the curvature of the surface, and n and n' are the refractive indices before and after the surface. Also known as Petzval sum.

Petzval lens [OPTICS] A photographic objective which consists of four lenses ordered in two pairs widely separated from each other, with the first pair cemented together and the second usually having a small air space.

Petzval sum See Petzval curvature.

Petzval surface [OPTICS] A paraboloidal surface on which point images of point objects are formed by a doublet lens whose separation is such that astigmatism is eliminated.

peuroseite [MINERAL] $(Ni,Cu,Pb)Se_2$ A gray mineral composed of nickel copper lead selenide, occurring in columnar masses.

pewter [MET] An alloy that typically contained tin as the principal component and some antimony and copper; older produced pewter typically contains lead along with the other components.

Peyer's patches [HISTOL] Aggregates of lymph nodules beneath the epithelium of the ileum.

Peyronie's disease [MED] Scarring of the shaft of the penis; may interfere with normal erections.

pF See picofarad.

pf See power factor.

Pfaffian differential equation [MATH] The first-order linear total differential equation $P(x,y,z)dx + Q(x,y,z)dy + R(x,y,z)dz = 0$, where the functions P, Q, and R are continuously differentiable.

Pfeiffer's bacillus See Hemophilus influenzae.

Pfeiffer's disease See infectious mononucleosis.

PFM See pulse-frequency modulation.

Pfund series [SPECT] A series of lines in the infrared spec-

PHAEODICTYOSPORAE

Characteristic dark muriform spores of *Stemphylium*, a member of the Phaeodictyosporae.

PHAEODIDYMAE

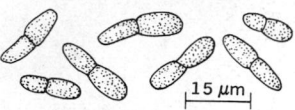

Characteristic spores of *Diplodia* in the Phaeodidymae.

PHALANGERIDAE

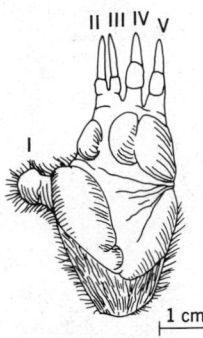

Hindfoot of *Trichosurus*, representing the five-toed, syndactylous condition (digits II and III are bound in one web of skin).

PHANTASTRON

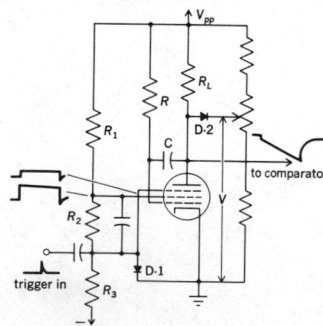

Circuit diagram of screen-coupled phantastron delay circuit. Resistors R_1, R_2, and R_3 form divider which prevents plate current from flowing before trigger pulse is applied. D-1, D-2 are diodes. V is voltage from which plate starts, determined by divider to which diode D-2 is connected. R_L = load resistor, V_{pp} = plate supply voltage.

trum of atomic hydrogen whose wave numbers are given by $R_H[(1/25) - (1/n^2)]$, where R_H is the Rydberg constant for hydrogen, and n is any integer greater than 5.

PGA *See* folic acid.

PGR *See* precision depth recorder.

pH [CHEM] A term used to describe the hydrogen-ion activity of a system; it is equal to $-\log a_{H}{}^+$; here $a_H{}^+$ is the activity of the hydrogen ion; in dilute solution, activity is essentially equal to concentration and pH is defined as $-\log_{10}[H^+]$, where $[H^+]$ is hydrogen-ion concentration in moles per liter; a solution of pH 0 to 7 is acid, pH of 7 is neutral, pH over 7 to 14 is alkaline.

phacella *See* gastric filament.

phacellite *See* kaliophilite.

phacolith [GEOL] A minor, concordant, lens-shaped, and usually granitic intrusion into folded sedimentary strata.

Phaenocephalidae [INV ZOO] A monospecific family of coleopteran insects in the superfamily Cucujoidea, found only in Japan.

Phaenothontidae [VERT ZOO] The tropic birds, a family of fish-eating aquatic forms in the order Pelecaniformes.

Phaeocoleosporae [MYCOL] A spore group of the Fungi Imperfecti with dark filiform spores.

Phaeodictyosporae [MYCOL] A spore group of the Fungi Imperfecti with dark muriform spores.

Phaeodidymae [MYCOL] A spore group of the Fungi Imperfecti with dark two-celled spores.

Phaeodorina [INV ZOO] The equivalent name for Tripylina.

Phaeohelicosporae [MYCOL] A spore group of the Fungi Imperfecti with dark, spirally coiled, septate spores.

Phaeophragmiae [MYCOL] A spore group of the Fungi Imperfecti with dark three- to many-celled spores.

Phaeophyta [BOT] The brown algae, constituting a division of plants; the plant body is multicellular, varying from a simple filamentous form to a complex, sometimes branched body having a basal attachment.

Phaeosporae [MYCOL] A spore group of Fungi Imperfecti characterized by dark one-celled, nonfiliform spores.

Phaeostaurosporae [MYCOL] A spore group of the Fungi Imperfecti with dark star-shaped or forked spores.

phage *See* bacteriophage.

phage cross [VIROL] Multiple infection of a single bacterium by phages that differ at one or more genetic sites, leading to the production of recombinant progeny phage.

phagocyte [CYTOL] An ameboid cell that engulfs foreign material.

phagocytic vacuole *See* food vacuole.

phagocytin [BIOCHEM] A type of bactericidal agent present within phagocytic cells.

phagocytosis [CYTOL] A mechanism by which certain animal cells, such as protozoans and leukocytes, engulf and carry particles into the cytoplasm.

phagosome [CYTOL] A closed intracellular vesicle containing material captured by phagocytosis.

phagotroph [INV ZOO] An organism that ingests nutrients by phagocytosis.

Phalacridae [INV ZOO] The shining flower beetles, a family of coleopteran insects in the superfamily Cucujoidea.

Phalacrocoracidae [VERT ZOO] The cormorants, a family of aquatic birds in the order Pelecaniformes.

Phalaenidae [INV ZOO] The equivalent name for Noctuidae.

Phalangeridae [VERT ZOO] A family of marsupial mammals in which the marsupium is well developed and opens anteriorly, the hindfeet are syndactylous, and the hallux is opposable and lacks a claw.

Phalangida [INV ZOO] An order of the class Arachnida characterized by an unsegmented cephalothorax broadly joined to a segmented abdomen, paired chelate chelicerae, and paired palpi.

phalanx [ANAT] One of the bones of the fingers or toes.

Phalaropodidae [VERT ZOO] The phalaropes, a family of migratory shore birds characterized by lobate toes and by reversal of the sex roles with respect to dimorphism and care of the young.

Phallostethidae [VERT ZOO] A family of actinopterygian fishes in the order Atheriniformes.

Phallostethiformes [VERT ZOO] An equivalent name for Atheriniformes.

phallus [ANAT] *See* penis. [EMBRYO] An undifferentiated embryonic structure derived from the genital tubercle that differentiates into the penis in males and the clitoris in females.

phanerite [PETR] An igneous rock having phaneritic texture.

phaneritic [PETR] Of the texture of an igneous rock, being visibly crystalline. Also known as coarse-grained; phanerocrystalline; phenocrystalline.

phanerocryst *See* phenocryst.

phanerocrystalline *See* phaneritic.

phanerophyte [ECOL] A perennial tree or shrub with dormant buds borne on aerial shoots.

Phanerorhynchidae [PALEON] A family of extinct chondrostean fishes in the order Palaeonisciformes having vertical jaw suspension.

Phanerozoic [GEOL] The part of geologic time for which there is abundant evidence of life, especially higher forms, in the corresponding rock, essentially post-Precambrian.

Phanerozonida [INV ZOO] An order of the Asteroidea in which the body margins are defined by two conspicuous series of plates and in which pentamerous symmetry is generally constant.

Phanodermatidae [INV ZOO] A family of free-living nematodes in the superfamily Enoploidea.

phanotron [ELECTR] A hot-filament diode rectifier tube utilizing an arc discharge in mercury vapor or an inert gas, usually xenon.

phantastron [ELECTR] A monostable pentode circuit used to generate sharp pulses at an adjustable and accurately timed interval after receipt of a triggering signal.

phantom [GEOL] A bed or member that is absent from a specific stratigraphic section but is usually present in a characteristic position in a sequence of similar geologic age. [NUCLEO] A volume of material approximating as closely as possible the density and effective atomic number of living tissue, used in biological experiments involving radiation. [PETR] *See* ghost.

phantom bottom [OCEANOGR] A false bottom indicated by an echo sounder, some distance above the actual bottom; such an indication, quite common in the deeper parts of the ocean, is due to large quantities of small organisms.

phantom circuit [COMMUN] A communication circuit derived from two other communication circuits or from one other circuit and ground, with no additional wire lines.

phantom-circuit loading coil [ELEC] Loading coil for introducing a desired amount of inductance into a phantom circuit, and a minimum amount of inductance into its constituent circuits.

phantom-circuit repeating coil [ELEC] Repeating coil used at a terminal of a phantom circuit, in the terminal circuit extending from the midpoints of the associated side-circuit repeating coils.

phantom crystal [CRYSTAL] A crystal containing an earlier stage of crystallization outlined by dust, minute inclusions, or bubbles. Also known as ghost crystal.

phantom group [ELEC] 1. Group of four open-wire conductors suitable for the derivation of a phantom circuit. 2. Three circuits which are derived from simplexing two physical circuits to form a phantom circuit.

phantom horizon [GEOL] In seismic reflection prospecting, a line constructed so that it is parallel to the nearest actual dip segment at all points along a profile.

phantom repeating coil [ELEC] A side-circuit repeating coil or a phantom-circuit repeating coil when discrimination between these two types is not necessary.

phantom signals [ELECTR] Signals appearing on the screen of a cathode-ray-tube indicator, the cause of which cannot readily be determined and which may be caused by circuit fault, interference, propagation anomalies, jamming, and so on.

phantom target *See* echo box.

Pharetronida [INV ZOO] An order of calcareous sponges in the subclass Calcinea characterized by a leuconoid structure.

pharmaceutical chemistry [CHEM ENG] The chemistry of drugs and of medicinal and pharmaceutical products.

pharmacodynamics [PHARM] The science that deals with the actions of drugs.

pharmacogenetics [GEN] The science of genetically determined variations in drug responses.

pharmacognosy [PHARM] The science of crude drugs.

pharmacolite [MINERAL] $CaH(AsO_4) \cdot 2H_2O$ A white to grayish monoclinic mineral composed of hydrous acid arsenate of calcium, occurring in fibrous form. Also known as arsenic bloom.

pharmacologic pyrogen [PHARM] A naturally occurring pharmacologic agent, such as serotonin or a catecholamine that controls body temperature; it can cause fever when injected under experimental conditions.

pharmacology [CHEM] The science dealing with the nature and properties of drugs, particularly their actions.

pharmacophobia [PSYCH] Abnormal fear of medicine.

pharmacopoeia [PHARM] A book containing a selected list of medicinal substances and their dosage forms, providing also a description and the standards for purity and strength for each.

pharmacosiderite [MINERAL] $Fe_3(AsO_4)_2(OH)_3 \cdot 5H_2O$ Green or yellowish-green mineral composed of a hydrous basic iron arsenate and commonly found in cubic crystals. Also known as cube ore.

pharmacotherapy [MED] The treatment of disease by means of drugs.

pharmacy [MED] **1.** The art and science of the preparation and dispensation of drugs. **2.** A place where drugs are dispensed.

pharyngeal aponeurosis [ANAT] The fibrous submucous layer of the pharynx.

pharyngeal bursa [EMBRYO] A small pit caudal to the pharyngeal tonsil, resulting from the ingrowth of epithelium along the course of the degenerating tip of the notochord of the vertebrate embryo.

pharyngeal cleft [EMBRYO] One of the paired open clefts on the sides of the embryonic pharynx between successive visceral arches in vertebrates.

pharyngeal plexus [ANAT] **1.** A nerve plexus innervating the pharynx. **2.** A plexus of veins situated at the side of the pharynx.

pharyngeal pouch [EMBRYO] One of the five paired sacculations in the lateral aspect of the pharynx in vertebrate embryos. Also known as visceral pouch.

pharyngeal tonsil See adenoid.

pharyngeal tooth [VERT ZOO] A tooth developed on the pharyngeal bone in many fishes.

pharyngitis [MED] Inflammation of the pharynx.

Pharyngobdellae [INV ZOO] A family of leeches in the order Arhynchobdellae distinguished by the lack of jaws.

pharyngology [MED] The science of the pharyngeal mechanism, functions, and diseases.

pharyngoscope [MED] An instrument for examining the pharynx.

pharynx [ANAT] A chamber at the oral end of the vertebrate alimentary canal, leading to the esophagus.

phase [ASTRON] One of the cyclically repeating appearances of the moon or other orbiting body as seen from earth. [CHEM] Portion of a physical system (liquid, gas, solid) that is homogeneous throughout, has definable boundaries, and can be separated physically from other phases. [MATH] The argument of a trigonometric function. [MET] A constituent of an alloy that is physically distinct and is homogeneous in chemical composition. [PHYS] **1.** The fractional part of a period through which the time variable of a periodic quantity (alternating electric current, vibration) has moved, as measured at any point in time from an arbitrary time origin; usually expressed in terms of angular measure, with one period being equal to 360° or 2π radians. **2.** For a sinusoidally varying quantity, the phase (first definition) with the time origin located at the last point at which the quantity passed through a zero position from a negative to a positive direction. **3.** The argument of the trigonometric function describing the space and time variation of a sinusoidal disturbance, $y = A \cos[(2\pi/\lambda)(x - vt)]$, where x and t are the space and time coordinates, v is the velocity of propagation,

and λ is the wavelength. [THERMO] The type of state of a system, such as solid, liquid, or gas.

phase advancer [ELEC] Phase modifier which supplies leading reactive volt-amperes to the system to which it is connected; may be either synchronous or asynchronous.

phase age See age of phase inequality.

phase-alternation line system [COMMUN] A color television system used in Europe, in which the phase of the color subcarrier is changed from scanning line to scanning line, requiring transmission of a line switching signal as well as a color burst. Abbreviated PAL system.

phase angle [PHYS] The difference between the phase of a sinusoidally varying quantity and the phase of a second quantity which varies sinusoidally at the same frequency. Also known as phase difference.

phase-angle meter See phase meter.

phase-balance relay [ELEC] Relay which functions by reason of a difference between two quantities associated with different phases of a polyphase circuit.

phase behavior [PETRO ENG] The equilibrium relationships between water, liquid hydrocarbons, and dissolved or free gas, either in reservoirs or as liquids and gases are separated above ground in gas-oil separator systems.

phase boundary [PHYS] The interface between two or more separate phases, such as liquid-gas, liquid-solid, gas-solid, or, for immiscible materials, liquid-liquid or solid-solid.

phase change [PHYS] **1.** The metamorphosis of a material or mixture from one phase to another, such as gas to liquid, solid to gas. **2.** See phase shift.

phase coherence [PHYS] The existence of a statistical or time coherence between the phases of two or more waves.

phase-comparison relaying [ELEC] A method of detecting faults in an electric power system in which signals are transmitted from each of two terminals every half cycle so that a continuous signal is received at an intermediate point if there is no fault between the terminals, while a periodic signal is received if there is a fault.

phase conductor [ELEC] In a polyphase circuit, any conductor other than the neutral conductor.

phase constant [ELECTROMAG] A rating for a line or medium through which a plane wave of a given frequency is being transmitted; it is the imaginary part of the propagation constant, and is the space rate of decrease of phase of a field component (or of the voltage or current) in the direction of propagation, in radians per unit length.

phase control [ELECTR] **1.** A control that changes the phase angle at which the alternating-current line voltage fires a thyratron, ignitron, or other controllable gas tube. Also known as phase-shift control. **2.** See hue control.

phase converter [ELEC] A converter that changes the number of phases in an alternating-current power source without changing the frequency.

phase-correcting network See phase equalizer.

phase correction [COMMUN] Process of keeping synchronous telegraph mechanisms in substantially correct phase relationship.

phase crossover [CONT SYS] A point on the plot of the loop ratio at which it has a phase angle of 180°.

phased array [ELECTROMAG] An array of dipoles on a radar antenna in which the signal feeding each dipole is varied so that antenna beams can be formed in space and scanned very rapidly in azimuth and elevation.

phase delay [COMMUN] Ratio of the total phase shift (radians) of a sinusoidal signal in transmission through a system or transducer, to the frequency (radians/second) of the signal.

phase detector [ELECTR] A circuit that provides a direct-current output voltage which is related to the phase difference between an oscillator signal and a reference signal, for use in controlling the oscillator to keep it in synchronism with the reference signal. Also known as phase discriminator.

phase deviation [COMMUN] The peak difference between the instantaneous angle of a modulated wave and the angle of the sine-wave carrier.

phase diagram [MET] See constitution diagram. [PHYS CHEM] A graphical representation of the equilibrium relationships between phases (such as vapor-liquid, liquid-solid) of a chemical compound, mixture of compounds, or solution.

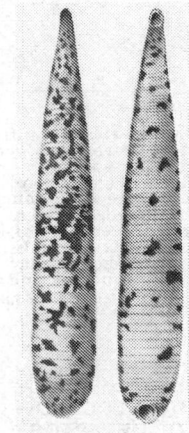

PHARYNGOBDELLAE

Dorsal and ventral view of *Erpobdella punctato*, a jawless leech common in lakes and streams in the Northern Hemisphere.

PHASE ANGLE

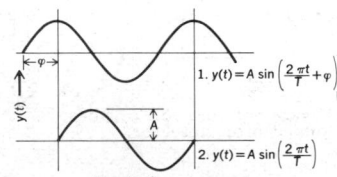

$$1. \ y(t) = A \sin\left(\frac{2\pi t}{T} + \varphi\right)$$

$$2. \ y(t) = A \sin\left(\frac{2\pi t}{T}\right)$$

An illustration of the meaning of phase for a sinusoidal wave, $y(t)$. The difference in phase between waves 1 and 2 is φ and is called the phase angle. For each wave, A is the amplitude and T is the period.

PHASE INVERTER

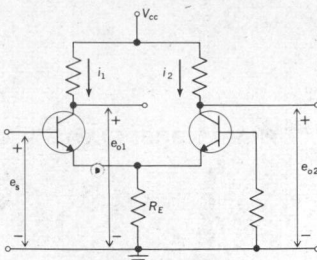

Circuit diagram for an emitter-coupled phase inverter. V_{cc} = collector supply voltage, e_s = input signal voltage. When emitter resistance R_E is much larger than impedance seen looking into the emitter of each transistor, current i_1 equals i_2, and output voltages e_{O1} and e_{O2} are equal and opposite, and therefore 180° out of phase.

PHASE-LOCKED LOOP DETECTOR

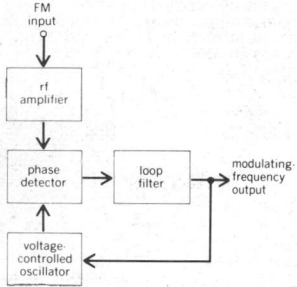

Basic configuration of phase-locked loop detector.

PHASE-MODULATION DETECTOR

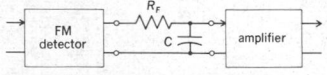

A resistance-capacitance filter used to convert a frequency-modulation detector into a phase-modulation detector.

PHASE MODULATOR

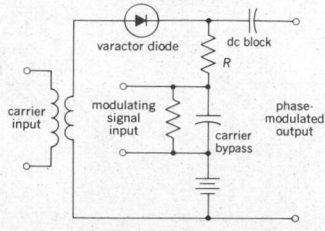

Circuit diagram for a simple phase modulator.

[THERMO] **1.** A graph showing the pressures at which phase transitions between different states of a pure compound occur, as a function of temperature. **2.** A graph showing the temperatures at which transitions between different phases of a binary system occur, as a function of the relative concentrations of its components.

phase difference   *See* phase angle.

phase discriminator   *See* phase detector.

phase distortion   [COMMUN] 1. The distortion which occurs in an instrument when the relative phases of the input signal differ from those of the output signal. 2. *See* phase-frequency distortion.

phase equalizer   [ELECTR] A network designed to compensate for phase-frequency distortion within a specified frequency band. Also known as phase-correcting network.

phase equilibria   [PHYS CHEM] The equilibrium relationships between phases (such as vapor, liquid, solid) of a chemical compound or mixture under various conditions of temperature, pressure, and composition.

phase excursion   [COMMUN] In angle modulation, the difference between the instantaneous angle of the modulated wave and the angle of the carrier.

phase factor   [ELEC] *See* power factor.   [SOLID STATE] The argument (phase) of a structure factor; it cannot be directly observed.

phase-frequency distortion   [COMMUN] Distortion occurring because phase shift is not proportional to frequency over the frequency range required for transmission. Also known as phase distortion.

phase front   [PHYS] A surface of constant phase (or phase angle) of a propagating wave disturbance.

phase inequality   [OCEANOGR] Variations in the tide or tidal currents associated with changes in the phase of the moon.

phase integral   *See* action.

phase inversion   [ELECTR] Production of a phase difference of 180° between two similar wave shapes of the same frequency.

phase inverter   [ELECTR] A circuit or device that changes the phase of a signal by 180°, as required for feeding a push-pull amplifier stage without using a coupling transformer, or for changing the polarity of a pulse; a triode is commonly used as a phase inverter. Also known as inverter.

phase lag   [OCEANOGR] Angular retardation of the maximum of a constituent of the observed tide behind the corresponding maximum of the same constituent of the hypothetical equilibrium tide. Also known as tidal epoch.   [PHYS] *See* lag angle.

phase lead   *See* lead angle.

phase localizer   [NAV] A localizer used by aircraft for landing guidance in which the on-course line is defined by the phase reversal of the emission from the side-band antenna system; phase is determined by comparison with a carrier signal which is transmitted for this purpose.

phase lock   [ELECTR] Technique of making the phase of an oscillator signal follow exactly the phase of a reference signal by comparing the phases between the two signals and using the resultant difference signal to adjust the frequency of the reference oscillator.

phase-locked communication   [COMMUN] A system in which oscillators at the receiver and transmitter are locked in phase, and intelligence is conveyed by phase shifts of the modulating signal.

phase-locked loop detector   [ELECTR] A type of locked-oscillator detector in which a local voltage-controlled oscillator and the frequency-modulated input signal feed a phase detector followed by a low-pass filter whose output is proportional to phase difference of the oscillator and input; the output is fed back to the voltage-controlled oscillator to lock it to the input signal.

phase-locked oscillator   *See* parametron.

phase-locked subharmonic oscillator   *See* parametron.

phase-locked system   [ENG] A radar system, having a stable local oscillator, in which information regarding the target is gained by measuring the phase shift of the echo.

phase-lock loop   [ELECTR] Electronic servo system incorporating phase lock, and used either as a tracking filter or as a frequency discriminator.

phase magnet   [COMMUN] Magnetically operated latch used

to phase a facsimile transmitter or recorder. Also known as trip magnet.

phase margin   [CONT SYS] The difference between 180° and the phase of the loop ratio of a stable system at the gain-crossover frequency.

phase meter   [ENG] An instrument for the measurement of electrical phase angles. Also known as phase-angle meter.

phase modifier   [ELEC] Machine whose chief purpose is to supply leading or lagging reactive volt-amperes to the system to which it is connected; may be either synchronous or asynchronous.

phase modulation   [COMMUN] A special kind of modulation in which the linearly increasing angle of a sine wave has added to it a phase angle that is proportional to the instantaneous value of the modulating wave (message to be communicated). Abbreviated PM.

phase-modulation detector   [ELECTR] A device which recovers or detects the modulating signal from a phase-modulated carrier.

phase-modulation transmitter   [ELECTR] A radio transmitter used to broadcast a phase-modulated signal.

phase modulator   [ELECTR] An electronic circuit that causes the phase angle of a modulated wave to vary (with respect to an unmodulated carrier) in accordance with a modulating signal.

phase plane analysis   [CONT SYS] A method of analyzing systems in which one plots the time derivative of the system's position (or some other quantity characterizing the system) as a function of position for various values of initial conditions.

phase portrait   [CONT SYS] A graph showing the time derivative of a system's position (or some other quantity characterizing the system) as a function of position for various values of initial conditions.

phase quadrature   *See* quadrature.

phaser   [COMMUN] Facsimile device for adjusting equipment so the recorded elemental area bears the same relation to the record sheet as the corresponding transmitted elemental area bears to the subject copy in the direction of the scanning line. [ELECTROMAG] Microwave ferrite phase shifter employing a longitudinal magnetic field along one or more rods of ferrite in a waveguide.

phase resonance   [PHYS] The frequency at which the angular phase difference between the fundamental components of an oscillation and of the applied agency is 90° ($\pi/2$ radians). Also known as velocity resonance.

phase response   [ELECTR] A graph of the phase shift of a network as a function of frequency.

phase reversal   [PHYS] A change of 180°, or one half-cycle, in phase.

phase reversal modulation   [COMMUN] Form of pulse modulation in which reversal of signal phase serves to distinguish between the two binary states used in data transmission.

phase-rotation relay   *See* phase-sequence relay.

phase-sequence relay   [ELEC] Relay which functions according to the order in which the phase voltages successively reach their maximum positive values. Also known as phase-rotation relay.

phase shift   [ELECTR] The phase angle between the input and output signals of a network or system.   [PHYS] 1. A change in the phase of a periodic quantity. Also known as phase change. 2. A change in the phase angle between two periodic quantities.   [QUANT MECH] For a partial wave of a particle scattered by a spherically symmetric potential, the phase shift is the difference between the phase of the wave function far from the scatterer and the corresponding phase of a free particle.

phase-shift circuit   [ELECTR] A network that provides a voltage component which is shifted in phase with respect to a reference voltage.

phase-shift control   *See* phase control.

phase-shift discriminator   [ELECTR] A discriminator that uses two similarly connected diodes, fed by a transformer that is tuned to the center frequency; when the frequency-modulated or phase-modulated input signal swings away from this center frequency, one diode receives a stronger signal than the other; the net output of the diodes is then proportional to the frequency displacement. Also known as Foster-Seeley discriminator.

phase shifter [ELEC] A device used to change the phase relation between two alternating-current values.

phase-shifting transformer [ELEC] A transformer which produces a difference in phase angle between two circuits.

phase-shift keying [COMMUN] A form of phase modulation in which the modulating function shifts the instantaneous phase of the modulated wave between predetermined discrete values. Abbreviated PSK.

phase-shift oscillator [ELECTR] An oscillator in which a network having a phase shift of 180° per stage is connected between the output and the input of an amplifier. Also known as phase shifter.

phase solubility [PHYS CHEM] The different solubilities of a sample's solid constituents (phases) in a selected solvent.

phase-solubility analysis [ANALY CHEM] Solvent technique used to determine the amount and number of components in a solid substance; the weight of sample added to the solvent is plotted against the weight of sample dissolved, with breakpoints in the curve occurring with each progressive saturation of the solvent with respect to each of the components; can be combined with extraction and recrystallization procedures.

phase space [MATH] In a dynamical system or transformation group, this is the topological space whose points are being moved about by the given transformations. [STAT MECH] For a system with n degrees of freedom, a Euclidean space with $2n$ dimensions, one dimension for each of the generalized coordinates and one for each of the corresponding momenta.

phase speed *See* phase velocity.

phase splitter [ELECTR] A circuit that takes a single input signal voltage and produces two output signal voltages 180° apart in phase.

phase stability [NUCLEO] A principle governing the stability of motion of particles in a synchrotron; the charged particle must be accelerated in each cycle at a time slightly earlier than the peak value of the accelerating potential.

phase titration [ANALY CHEM] Analysis of a binary mixture of miscible liquids by titrating with a third liquid that is miscible with only one of the components, using the ternary phase diagram to determine the end point.

phase transformation [ELEC] A change of polyphase power from three-phase to six-phase, from three-phase to twelve-phase, and so forth, by means of transformers. [PHYS] *See* phase transition.

phase transition [PHYS] A change of a substance from one phase to another. Also known as phase transformation.

phase undervoltage relay [ELEC] Relay which functions by reason of the reduction of one phase voltage in a polyphase circuit.

phase velocity [PHYS] The velocity of a point that moves with a wave at constant phase. Also known as celerity; phase speed; wave celerity; wave speed; wave velocity.

Phasianidae [VERT ZOO] A family of game birds in the order Galliformes; typically, members are ground feeders, have bare tarsi and copious plumage, and lack feathers around the nostrils.

phasing *See* framing.

phasing line [ELECTR] That portion of the length of scanning line set aside for the phasing signal in a television or facsimile system.

phasing signal [ELECTR] A signal used to adjust the picture position along the scanning line in a facsimile system.

phasitron [ELECTR] An electron tube used to frequency-modulate a radio-frequency carrier; internal electrodes are designed to produce a rotating disk-shaped corrugated sheet of electrons; audio input is applied to a coil surrounding the glass envelope of the tube, to produce a varying axial magnetic field that gives the desired phase or frequency modulation of the r-f carrier input to the tube.

phasmajector *See* monoscope.

phasmid [INV ZOO] One of a pair of lateral caudal pores which function as chemoreceptors in certain nematodes.

Phasmidae [INV ZOO] A family of the insect order Orthoptera including the walking sticks and leaf insects.

Phasmidea [INV ZOO] An equivalent name for Secernentea.

Phasmidia [INV ZOO] An equivalent name for Secernentea.

phasor [PHYS] **1.** A rotating line used to represent a sinusoidally varying quantity; the length of the line represents the magnitude of the quantity, and its angle with the x-axis at any instant represents the phase. **2.** Any quantity (such as impedance or admittance) which is a complex number.

phasotron *See* cyclotron.

pheasant [VERT ZOO] Any of various large sedentary game birds with long tails in the family Phasianidae; sexual dimorphism is typical of the group.

pH electrode [ANALY CHEM] Membrane-type glass electrode used as the hydrogen-ion sensor of most pH meters; the pH-response electrode surface is a thin membrane made of a special glass.

phellem [BOT] Cork; the outer tissue layer of the periderm.

phelloderm [BOT] Layers of parenchymatous cells formed as inward derivatives of the phellogen.

phellogen [BOT] The meristematic portion of the periderm, consisting of one layer of cells that initiate formation of the cork and secondary cortex tissue.

phenacite *See* phenakite.

Phenacodontidae [PALEON] An extinct family of large herbivorous mammals in the order Condylarthra.

phenacyl chloride *See* chloroacetophenone.

phenakite [MINERAL] Be_2SiO_4 A colorless, white, wine-yellow, pink, blue, or brown glassy mineral that crystallizes in the rhombohedral system; used as a minor gemstone. Also spelled phenacite.

phenanthrene [ORG CHEM] $C_{14}H_{10}$ A colorless, crystalline hydrocarbon; melts at about 100°C; the nucleus is produced by the degradation of certain alkaloids; used in the synthesis of dyes and drugs.

phenanthroline [ORG CHEM] $C_{12}H_8N_2$ Any of three nitrogen bases related to phenanthrene; the ortho form is an oxidation-reduction indicator, turning blue when oxidized.

phenanthroline indicator [ANALY CHEM] A sensitive, red-colored specific reagent for iron. Also known as bathophenanthroline; 4, 7-diphenyl-1,10-phenanthroline.

phenarsazine chloride [ORG CHEM] $C_{12}H_9AsClN$ A yellow, crystalline compound obtained as a precipitate from carbon tetrachloride solutions; it sublimes readily, and is slightly soluble in xylene, benzene, and carbon tetrachloride; used as a war gas. Also known as adamsite; 10-chloro-5,10-dihydrophenarsazine.

phenazine [ORG CHEM] $C_6H_4N_2C_6H_4$ Yellow crystals, melting at 170°C; slightly soluble in water, soluble in alcohol and ether; used as chemical intermediate and to make dyes. Also known as azophenylene.

phenethyl alcohol [ORG CHEM] $C_8H_{10}O$ A liquid with a floral odor found in many natural essential oils; soluble in 50% alcohol; used in perfumes and flavors, and in medicine as an antibacterial agent in diseases of the eye. Also known as benzyl carbinol; 2-phenylethanol; β-phenylethyl alcohol.

phenetidine [ORG CHEM] $NH_2C_6H_4OC_2H_5$ Either of two toxic, oily liquids that darken when exposed to light and air; soluble in alcohol, insoluble in water; the ortho form boils at 228–230°C, is used to make dyes, and is also known at 2-aminophenetole; the para form boils at 253–255°C, is used to make dyes and in pharmaceuticals, and is also known as 4-aminophenetole.

Phengodidae [INV ZOO] The fire beetles, a New World family of coleopteran insects in the superfamily Cantharoidea.

phengophobia [PSYCH] An abnormal fear of daylight.

phenicochroite *See* phoenicochroite.

phenobarbital [PHARM] $C_{12}H_{12}N_2O_3$ A crystalline compound, 5-ethyl-5-phenylbarbituric acid, with a slightly bitter taste, melting at 174–178°C; soluble in water, alcohol, chloroform, and ether; used in medicine as a long-acting sedative, anticonvulsant, and hypnotic. Also known as phenobarbitone; phenylethylmalonylurea.

phenobarbitone *See* phenobarbital.

phenoclasts [PETR] The larger, conspicuous fragments in a sediment or sedimentary rock, such as cobbles in a conglomerate.

phenocopy [GEN] The nonhereditary alteration of a pheno-

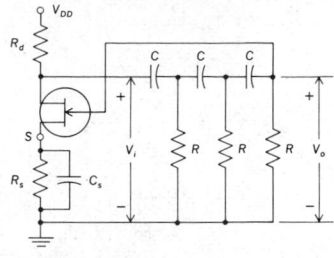

Circuit diagram of an FET phase-shift oscillator; V_i = input voltage, V_O = output voltage, V_{DD} = drain supply voltage, R_d = drain resistor, S = source, R_s = source resistor, C_s = source capacitor.

PHASITRON

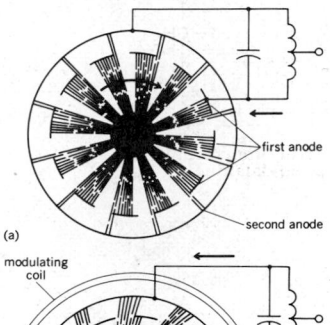

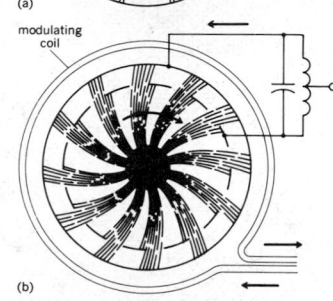

Principle of operation of phasitron tube. (a) Rotating electron beams originate from a central cathode (spokelike configuration). Arrows indicate currents and direction of rotation of electron beams. (b) Bending of the electron beams by an axial magnetic field, supplied by the modulating coil, advances or retards arrival of electrons to the second anode. (From R. Adler, *A new system of frequency modulation, Proc. IRE, 35(1), 1947*)

PHASMIDAE

Walking stick (*Diapheromera femorata*). (From Illinois Natural History Survey)

PHENACODONTIDAE

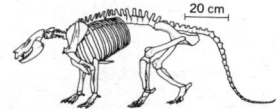

Skeleton of *Ectoconus*, an early Paleocene phenacodont condylarth. (After G. G. Simpson)

type to a form imitating a mutant trait; caused by external conditions during development.

phenocryst [PETR] A large, conspicuous crystal in a porphyritic rock. Also known as phanerocryst.

phenocrystalline *See* phaneritic.

phenogenetics [GEN] The study of the phenotypic effects of the genetic material. Also known as physiological genetics.

phenol [ORG CHEM] 1. C_6H_5OH White, poisonous, corrosive crystals with sharp, burning taste; melts at 43°C, boils at 182°C; soluble in alcohol, water, ether, carbon disulfide, and other solvents; used to make resins and weed killers, and as a solvent and chemical intermediate. Also known as carbolic acid; phenylic acid. 2. A chemical compound based on the substitution product of phenol, for example, ethylphenol ($C_2H_4C_4H_5OH$), the ethyl substitute of phenol.

phenol coefficient [ANALY CHEM] Number scale for comparison of antiseptics, using the efficacy of phenol as unity.

phenol-coefficient method [CHEM] A method for evaluating water-miscible disinfectants in which a test organism is added to a series of dilutions of the disinfectant; the phenol coefficient is the number obtained by dividing the greatest dilution of the disinfectant killing the test organism by the greatest dilution of phenol showing the same result.

β-phenoldiazine *See* phthalazine.

phenol extraction [CHEM ENG] Petroleum-refinery solvent-extraction process using phenol as the solvent to remove aromatic, unsaturated and naphthenic constituents from lubricating-oil stocks.

phenol-formaldehyde resin [ORG CHEM] Thermosetting resin made by the reaction of phenol and formaldehyde; has good strength and chemical resistance and low cost; used as a molding material for mechanical and electrical parts. Originally known as Bakelite.

phenolic laminate [MATER] Canvas, linen, kraft paper, glass fiber, or other substrate impregnated with 30% or more of thermosetting phenolic resin and cured; used for structural, mechanical, and electrical purposes.

phenolic plastic [MATER] A thermosetting plastic material available in many combinations of phenol and formaldehyde, often with added fillers to provide a broad range of physical, electrical, chemical, and molding properties.

phenology [CLIMATOL] The science which treats of periodic biological phenomena with relation to climate, especially seasonal changes; from a climatologic viewpoint, these phenomena serve as bases for the interpretation of local seasons and the climatic zones, and are considered to integrate the effects of a number of bioclimatic factors.

phenolphthalein [ORG CHEM] $(C_6H_4OH)_2COC_6H_4CO$ Pale-yellow crystals; soluble in alcohol, ether, and alkalies, insoluble in water; used as an acid-base indicator (carmine-colored to alkalies, colorless to acids) for titrations, as a laxative and dye, and in medicine.

phenol red *See* phenolsulfonphthalein.

phenolsulfonic acid [ORG CHEM] $C_6H_5SO_3H$ Water- and alcohol-soluble mixture of *ortho*- and *para*-phenolsulfonic acids; yellowish liquid that turns brown when exposed to air; used as a chemical intermediate and in water analysis. Also known as sulfocarbolic acid.

phenolsulfonphthalein [ORG CHEM] $C_{19}H_{14}O_5S$ A bright-red, crystalline compound, soluble in water, alcohol, and acetone; used as a pH indicator, and to test for kidney function in dogs. Also known as phenol red.

phenothiazine [ORG CHEM] $C_{12}H_9N$ A yellow, crystalline compound, forming rhomboid leaflets or diamond-shaped plates, obtained from toluene or butanol solution; soluble in hot acetic acid, benzene, and ether; used as an insecticide and in pharmaceutical manufacture. Also known as dibenzothiazine; thiophenylamine.

phenotole [ORG CHEM] $C_6H_5OC_2H_5$ Combustible, colorless liquid, boiling at 172°C; soluble in alcohol and ether, insoluble in water.

phenotype [GEN] The observable characters of an organism.

phenotypic lag [GEN] Delay in the expression of a newly acquired character.

phenoxybenzamine hydrochloride [ORG CHEM] $C_{18}H_{22}$-ONCl·HCl White crystals, slightly soluble in water, melting at 139°C; used in medicine.

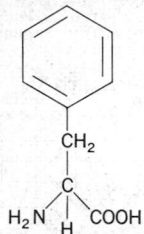

PHENYLALANINE

Structural formula of phenylalanine.

phenoxy resin [ORG CHEM] A high-molecular-weight thermoplastic polyether resin based on bisphenol-A and epichlorohydrin with bisphenol-A terminal groups; used for injection molding, extrusion, coatings, and adhesives.

phentolamine hydrochloride [ORG CHEM] $C_{17}H_{19}ON_2$·HCl White, water-soluble crystals, melting at 240°C; a sympatholytic; used in medicine.

α-phenylacetamide [ORG CHEM] C_8H_9NO An alcohol-soluble crystalline compound, forming bimorphous plates or leaflets, melting at 155°C; used in the pharmaceutical manufacture of penicillin G. Also known as α-toluamide.

phenylacetonitrile *See* benzyl cyanide.

phenylacrolein *See* cinnamaldehyde.

β-phenylacrylic acid *See* cinnamic acid.

phenylacrylyl chloride *See* cinnamoyl chloride.

phenylalanine [BIOCHEM] $C_9H_{11}O_2N$ An essential amino acid, obtained in the levo form by hydrolysis of proteins (as lactalbumin); converted to tyrosine in the normal body. Also known as α-aminohydrocinnamic acid; α-amino-β-phenyl-propionic acid; β-phenylalanine.

β-phenylalanine *See* phenylalanine.

phenylallylic alcohol *See* cinnamic alcohol.

phenylaniline *See* diphenylamine.

phenylazobenzene *See* azobenzene.

phenylbenzene *See* biphenyl.

N-phenyl benzoic acid amide *See* benzanilide.

phenyl benzoyl carbinol *See* benzoin.

phenylbutazone [ORG CHEM] $C_{19}H_{20}O_2N_2$ White or light-yellow powder with aromatic aroma and bitter taste; melts at 107°C; slightly soluble in water, soluble in acetone; used in medicine as an analgesic and antipyretic. Also known as butazolidine; 4-butyl-1,2-diphenyl-3,5-pyrazolindinedione.

phenylchloroform *See* benzotrichloride.

phenyl cyanide *See* benzonitrile.

phenylene blue *See* indamine.

phenylenediamine [ORG CHEM] $C_6H_4(NH_2)_2$ Also known as diaminobenzene. Any of three toxic isomeric crystalline compounds that are diamino derivatives of benzene; the ortho form, toxic colorless crystals melting at 102–104°C and soluble in alcohol, ether, water, and chloroform, is used to manufacture dyes, in photographic developers, and as a chemical intermediate, and is also known as *ortho*-diamino-benzene, orthamine; the meta form, colorless crystals unstable in air, melting at 63°C, and soluble in alcohol, ether, and water, is used to manufacture dyes, in textile dyeing, and as a nitrous acid detector; the para form, white to purple crystals melting at 147°C, soluble in alcohol and ether, and irritating to the skin, is used to manufacture dyes, in chemical analysis, and in photographic developers, and is also known as *para*-diaminobenzene.

phenylephrine [PHARM] $C_9H_{13}NO_2$ A sympathomimetic amine, used in its hydrochloride salt form as a vasoconstrictor.

2-phenylethanol *See* phenethyl alcohol.

β-phenylethyl alcohol *See* phenethyl alcohol.

phenylethylene *See* styrene.

phenylethylmalonylurea *See* phenobarbital.

N-phenylglycine [ORG CHEM] $C_6H_5NHCH_2COOH$ A crystalline compound, moderately soluble in water, melting at 127–128°C; used in dye manufacture (indigo). Also known as anilinoacetic acid.

phenylglycolic acid *See* mandelic acid.

phenylhydrazine [ORG CHEM] $C_6H_5NHNH_2$ Poisonous, oily liquid, boiling at 244°C; soluble in alcohol, ether, chloroform, and benzene, slightly soluble in water; used in analytical chemistry to detect sugars and aldehydes, and as a chemical intermediate. Also known as hydrazinobenzene.

α-phenylhydroxyacetic acid *See* mandelic acid.

phenyl ketone *See* benzophenone.

phenylketonuria [MED] A hereditary disorder of metabolism, transmitted as an autosomal recessive, in which there is a lack of the enzyme phenylalanine hydroxylase, resulting in excess amounts of phenylalanine in the blood and of excess phenylpyruvic and other acids in the urine. Abbreviated PKU. Also known as phenylpyruvic oligophrenia.

phenyl mercaptan *See* thiophenol.

phenylmethyl acetate *See* benzyl acetate.

phenyl methyl ketone   *See* acetophenone.

α-phenyl phenacyl   *See* desyl.

phenylphenol   [ORG CHEM] $C_6H_5C_6H_4OH$ Almost white crystals, soluble in alcohol, insoluble in water; the ortho form, melting at 56–58°C, is used to manufacture dyes, as germicide and fungicide, and in the rubber industry; the para form, melting at 164–165°C, is used to manufacture dyes, resins, and rubber chemicals, and as a fungicide, and is also known as *para*-hydroxydiphenyl, *para*-xenol.

phenylpropane   *See* propyl benzene.

3-phenylpropenal   *See* cinnamaldehyde.

3-phenyl-2-propen-1-ol   *See* cinnamic alcohol.

3-phenylpropionic acid   *See* hydrocinnamic acid.

phenylpyruvic acid   [BIOCHEM] $C_6H_5CH_2 \cdot CO \cdot COOH$ A keto acid, occurring as a metabolic product of phenylalanine.

phenylpyruvic oligophrenia   *See* phenylketonuria.

phenyl salicylate   *See* salol.

phenylthiocarbamide   *See* phenylthiourea.

phenylthiourea   [ORG CHEM] $C_6H_5NHCSNH_2$ A crystalline compound that has either a bitter taste or is tasteless, depending on the heredity of the taster; used in human genetics studies. Also known as phenylthiocarbamide.

pheochromoblast   [HISTOL] A precursor of a pheochromocyte.

pheochromocytoma   [MED] A tumor of the sympathetic nervous system composed principally of chromaffin cells; found most often in the adrenal medulla.

pheromone   [PHYSIO] Any substance secreted by an animal which influences the behavior of other individuals of the same species.

phialospore   [MYCOL] One of a chain of spores produced successively on phialides.

phi grade scale   [GEOL] A logarithmic transformation of the Wentworth grade scale in which the diameter value of the particle is replaced by the negative logarithm to the base 2 of the particle diameter (in millimeters).

Philadelphia chromosome   [PATH] An abnormally small G-group chromosome found in the hematopoietic cells of most patients with chronic granulocytic leukemia.

Philippine fowl disease   *See* Newcastle disease.

Philips Hot-air engine   [MECH ENG] A compact hot-air engine that is a Philips Research Lab (Holland) design; it uses only one cylinder and piston, and operates at 3000 revolutions per minute, with hot-chamber temperature of 1200°F (650°C) maximum pressure of 50 atmospheres, and mean effective pressure of 14 atmospheres.

Philips ionization gage   [ELECTR] An ionization gage in which a high voltage is applied between two electrodes, and a strong magnetic field deflects the resulting electron stream, increasing the length of the electron path and thus increasing the chance for ionizing collisions of electrons with gas molecules. Abbreviated pig. Also known as cold-cathode ionization gage.

philipstadite   [MINERAL] $Ca_2(Fe,Mg)_5(Si,Al)_8O_{22}(OH)_2$ Monoclinic mineral composed of basic silicate of calcium, iron, magnesium, and aluminum; member of the amphibole group.

phillipsite   [MINERAL] $(K_2,Na_2CA)Al_2Si_4O_{12} \cdot H_2O$ A white or reddish zeolite mineral crystallizing in the orthorhombic system; occurs in complex fibrous crystals, which make up a large part of the red-clay sediments in the Pacific Ocean.

Phillips screw   [DES ENG] A screw having in its head a recess in the shape of a cross; it is inserted or removed with a Phillips screwdriver that automatically centers itself in the screw.

Philomycidae   [INV ZOO] A family of pulmonate gastropods composed of slugs.

Philopteridae   [INV ZOO] A family of biting lice (Mallophaga) that are parasitic on most land birds and water birds.

philosopher's wool   *See* zinc oxide.

-philous   [SCI TECH] Suffix meaning having an affinity for.

philtrum-otobasion inferior   [ANTHRO] A measure of the distance from the midpoint of the philtrum to the otobasion inferior.

phi meson   [PARTIC PHYS] A neutral vector meson resonance, having a mass of about 1019 MeV (million electron volts), a width of about 4.2 MeV, and negative charge parity and G parity.

phimosis   [MED] Elongation of the prepuce and constriction of the orifice, so that the foreskin cannot be retracted to uncover the glans penis.

phlebite   [PETR] Roughly banded or veined metamorphite or migmatite.

phlebitis   [MED] Inflammation of a vein.

phlebography   [MED] **1.** X-ray photography of a vein or veins following intravenous injection of a radiopaque substance. **2.** Recording of venous pulsations.

phlebolith   [MED] A calculus in a vein.

phlebosclerosis   [MED] **1.** Sclerosis of a vein. **2.** Chronic phlebitis.

phlebotaxis   [BIOL] Movement of a simple motile organism in response to the presence of blood.

phlebothrombosis   [MED] A venous thrombus not associated with inflammation of the vein.

phlebotomus fever   [MED] An acute viral infection, transmitted by the fly *Phlebotomus papatosii* and characterized by fever, pains in the head and eyes, inflammation of the conjunctiva, leukopenia, and general malaise. Also known as Chitral fever; pappataci fever; sandfly fever; three-day fever.

phlegm   [PHYSIO] A viscid, stringy mucus, secreted by the mucosa of the air passages.

phlegmasia alba dolens   [MED] A painful swelling of the leg usually seen postpartum, due to femoral vein thrombophlebitis or lymphatic obstruction. Also known as milk leg.

phlegmon   [MED] Pyogenic inflammation with infiltration and spread in the tissues, seen with invasive organisms which produce hyaluronidase and fibrinolysins.

phleomycin   [MICROBIO] An antibacterial antibiotic produced by *Streptomyces verticillatus;* antitumor activity has also been demonstrated.

Phloeidae   [INV ZOO] The bark bugs, a small neotropical family of hemipteran insects in the superfamily Pentatomoidea.

phloem   [BOT] A complex, food-conducting vascular tissue in higher plants; principal conducting cells are sieve elements. Also known as bast; sieve tissue.

phloem necrosis   [PL PATH] A pathological state in a plant in which the phloem undergoes brown discoloration and disintegration.

phlogopite   [MINERAL] $K_2[Mg,Fe(II)]_6(Si_6,Al_2)O_{20}(OH)_4$ A yellow-brown to copper mineral of the mica group occurring in disseminated flakes, foliated masses, or large crystals; hardness is 2.5–3.0 on Mohs scale, and specific gravity is 2.8–3.0. Also known as bronze mica; brown mica.

pH measurement   [ANALY CHEM] Determination of the hydrogen-ion concentration in an ionized solution by means of an indicator solution (such as phenolphthalein) or a pH meter.

pH meter   [ENG] An electronic voltmeter using a pH-responsive electrode that gives a direct conversion of voltage differences to differences of pH at the temperature of the measurement.

phobia   [PSYCH] A disproportionate, obsessive, persistent, and unrealistic fear of an external situation or object, symbolically taking the place of an internal unconscious conflict.

phobic neurosis   [PSYCH] A neurotic disorder characterized by a persistent phobia which frequently interferes with the individual's activities and creates tension and anxiety, sometimes accompanied by physical manifestation.

phobophobia   [PSYCH] An abnormal fear of developing a phobia.

Phobos   [ASTRON] A satellite of Mars; it is the larger of the two satellites, with a diameter of about 24 kilometers.

Phocaenidae   [VERT ZOO] The porpoises, a family of marine mammals in the order Cetacea.

Phocidae   [VERT ZOO] The seals, a pinniped family of carnivoran mammals in the superfamily Canoidea.

Phodilidae   [VERT ZOO] A family of birds in the order Strigiformes; the bay owl (*Pholidus bodius*) is the single species.

Phoebe   [ASTRON] A satellite of the planet Saturn; its diameter is judged to be about 320 kilometers; it has an eccentric orbit and retrograde revolution.

phoenicite   *See* phoenicochroite.

phoenicochroite   [MINERAL] Pb_2CrO_5 A red mineral composed of basic chromate of lead, occurring in crystals and

PHLOGOPITE

Sheets of phlogopite from Otty Lake, Canada. *(Specimen from Department of Geology, Bryn Mawr College)*

2.5 cm

PHOLIDOPHORIDAE

Pholidophorus bechei Agassiz, Lower Jurassic of England: length to 8 inches (20 centimeters). *(From D. Rayner, The structure of certain Jurassic holostean fishes with special reference to their neurocrania, Phil. Trans. Roy. Soc. London, Ser. B, no. 601, Cambridge University Press, 1948)*

PHORONIDA

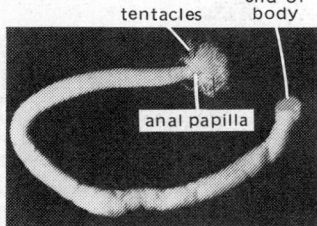

Phoronopsis harmeri removed from its tube, just below the surface in intertidal or subtidal mud flats. Length is about 20 centimeters.

masses. Also known as beresovite; phenicochroite; phoenicite.

Phoenicopteridae [VERT ZOO] The flamingos, a family of long-legged, long-necked birds in the order Ciconiiformes.

Phoenicopteriformes [VERT ZOO] An order comprising the flamingos in some systems of classification.

Phoeniculidae [VERT ZOO] The African wood hoopoes, a family of birds in the order Coraciiformes.

Phoenix [ASTRON] A southern constellation; right ascension 1 hour, declination 50°S.

Pholadidae [INV ZOO] A family of bivalve mollusks in the subclass Eulamellibranchia; individuals may have one or more dorsal accessory plates, and the visceral mass is attached to the valves in the dorsal portion of the body.

Pholidophoridae [PALEON] A generalized family of extinct fishes belonging to the Pholidophoriformes.

Pholidophoriformes [PALEON] An extinct actinopterygian group composed of mostly small fusiform marine and freshwater fishes of an advanced holostean level.

Pholidota [VERT ZOO] An order of mammals comprising the living pangolins and their fossil predecessors; characterized by an elongate tubular skull with no teeth, a long protrusive tongue, strong legs, and five-toed feet with large claws.

Phomaceae [MYCOL] The equivalent name for Sphaerioidaceae.

Phomales [MYCOL] The equivalent name for Sphaeropsidales.

phon [ACOUS] A unit of loudness level; the loudness level, in phons, of a sound is numerically equal to the sound pressure level, in decibels, of a 1000-hertz reference tone which is judged by listeners to be equally loud to the sound under evaluation.

phonation [PHYSIO] Production of speech sounds.

phone *See* headphone.

phone patch [ELECTR] A device connecting an amateur or citizens'-band transceiver temporarily to a telephone system.

phone plug [ELEC] A standard plug having a ¾-inch-diameter (19-millimeter-diameter) shank, used with headphones, microphones, and other audio equipment; usually designed for use with either two or three conductors. Also known as telephone plug.

phonetic alphabet [COMMUN] A list of standard words used for positive identification of letters in a voice message transmitted by radio or telephone.

phonocardiography [MED] A graphic recording of heart sounds.

phonograph [ENG ACOUS] An instrument for recording or reproducing acoustical signals, such as voice or music, by transmission of vibrations from or to a stylus that is in contact with a groove in a rotating disk.

phonograph cartridge *See* phonograph pickup.

phonograph cutter *See* cutter.

phonograph needle *See* stylus.

phonograph pickup [ENG ACOUS] A pickup that converts variations in the grooves of a phonograph record into corresponding electric signals. Also known as cartridge; phonograph cartridge.

phonograph record [ENG ACOUS] A shellac-composition or vinyl-plastic disk, usually 7 or 12 inches (18 or 30 centimeters) in diameter, on which sounds have been recorded as modulations in grooves. Also called disk; disk recording; platter.

phono jack [ELECTR] A jack designed to accept a phono plug and provide a ground connection for the shield of the conductor connected to the plug.

phonolite [PETR] A light-colored, aphanitic rock of volcanic origin, composed largely of alkali feldspar, feldspathoids, and smaller amounts of mafic minerals.

phonon [SOLID STATE] A quantum of an acoustic mode of thermal vibration in a crystal lattice.

phonon-electron interaction [SOLID STATE] An interaction between an electron and a vibration of a lattice, resulting in a change in both the momentum of the particle and the wave vector of the vibration.

phonon emission [SOLID STATE] The production of a phonon in a crystal lattice, which may result from the interaction of other phonons via anharmonic lattice forces, from scattering

of electrons in the lattice, or from scattering of x-rays or particles which bombard the crystal.

phonophobia [PSYCH] An abnormal fear of sound.

phono plug [ELECTR] A plug designed for attaching to the end of a shielded conductor, for feeding audio-frequency signals from a phonograph or other a-f source to a mating phono jack on a preamplifier or amplifier.

phonoreception [PHYSIO] The perception of sound through specialized sense organs.

phonotelemeter [ENG] A device consisting essentially of a stopwatch, for estimating the distance of guns in action by measuring the interval between the flash and the arrival of the sound waves from the discharge.

phony disease [PL PATH] A virus disease of the peach characterized by dwarfing, abnormal darkening of leaves, and a light crop of small, highly colored fruit; tree stops bearing fruit after a few years.

phony mine [ORD] Harmless object used to simulate a mine or to give false signals in detectors; used in phony mine fields.

phoresy [ECOL] A type of commensalism which involves the transporation of one organism (the guest) by a larger organism (the host) of a different species.

Phoridae [INV ZOO] The hump-backed flies, a family of cyclorrhaphous dipteran insects in the series Aschiza.

Phoronida [INV ZOO] A small, homogeneous group, or phylum, of animals having an elongate body, a crown of tentacles surrounding the mouth, and the anus occurring at the level of the mouth.

phosgene [ORG CHEM] $COCl_2$ A highly toxic, colorless gas that condenses at 0°C to a fuming liquid; used as a war gas and in manufacture of organic compounds. Also known as carbonyl chloride; chloroformyl chloride.

phosgenite [MINERAL] $Pb_2Cl_2(CO_3)$ A white, yellow, or grayish mineral that crystallizes in the tetragonal system, has adamantine luster, hardness of 3 on Mohs scale, and specific gravity of 6-6.3. Also known as cromfordite; horn lead.

phosphatase [BIOCHEM] An enzyme that catalyzes the hydrolysis and synthesis of phosphoric acid esters and the transfer of phosphate groups from phosphoric acid to other compounds.

phosphate [CHEM] **1.** Generic term for any compound containing a phosphate group (PO_4^{3-}), such as potassium phosphate, K_3PO_4. **2.** Generic term for a phosphate-containing fertilizer material. [MINERAL] A mineral compound characterized by a tetrahedral ionic group of phosphate and oxygen, PO_4^{3-}.

phosphate anion [INORG CHEM] PO_4^{3-} The negative ion of phosphoric acid.

phosphate buffer [ANALY CHEM] Laboratory pH reference solution made of KH_2PO_4 and Na_2HPO_4; when 0.025 molal (equimolal of the K and Na salts), the pH is 6.865 at 25°C.

phosphate coating [MET] A conversion coating on metal, usually steel, produced by dipping in an aqueous solution of zinc or manganese acid phosphate; used to furnish a black finish to small arms, artillery, or automotive components to provide resistance to corrosion.

phosphate desulfurization [CHEM ENG] A continuous, regenerative petroleum-refinery process using a tripotassium phosphate solution to remove hydrogen sulfide from natural gas, refinery gas, or liquid hydrocarbons.

phosphate fertilizer [MATER] Fertilizer compound or mixture containing available (soluble) phosphate; examples are phosphate rock (phosphorite), superphosphates or triple superphosphates, nitrophosphate, potassium phosphates, or N-P-K mixtures.

phosphate glass [MATER] Glass in which phosphorus pentoxide is a major component; resistant to hydrofluoric acid.

phosphate recovery process [MIN ENG] A process developed by the U.S. Bureau of Mines for recovering phosphate from low-grade phosphorus-bearing shales.

phosphate rock *See* phosphorite.

phosphatic nodule [GEOL] A dark, usually black, earthy mass or pebble of variable size and shape, having a hard shiny surface and occurring in marine strata.

phosphatide *See* phospholipid.

phosphating [MET] Forming a phosphate coating on a metal. Also known as phosphatizing.

phosphatizing   *See* phosphating.

phosphaturia   *See* hyperphosphaturia.

phosphide   [INORG CHEM] Binary compound of trivalent phosphorus, as in Na_3P.

phosphine   [INORG CHEM] PH_3 Poisonous, colorless, spontaneously flammable gas with garlic aroma; soluble in alcohol, slightly soluble in cold water; boils at $-85°C$; used in organic reactions. Also known as hydrogen phosphide; phosphoretted hydrogen.

phosphinic acid   [ORG CHEM] Organic derivative of hypophosphorous acid; contains the radical $-H_2PO_2$ or $=HPO_2$; examples are methylphosphinic acid, CH_3HPOOH, and dimethyl phosphinic acid, $(CH_3)_2POOH$.

phosphite   [INORG CHEM] Salt of phosphorous acid; contains the radical $PO_3{}^{3-}$; an example is normal sodium phosphite, Na_3PO_3

phosphocreatine   [BIOCHEM] $C_4H_{10}N_3O_5P$ Creatine phosphate, a phosphoric acid derivative of creatine which contains an energy-rich phosphate bond; it is present in muscle and other tissues, and during the anaerobic phase of muscular contraction it hydrolyzes to creatine and phosphate and makes energy available. Abbreviated PC.

phosphoenolpyruvic acid   [BIOCHEM] $CH_2=O(OPO_3H_2)-COOH$ A high-energy phosphate formed by dehydration of 2-phosphoglyceric acid; it reacts with adenosinediphosphate to form adenosinetriphosphate and enolpyruvic acid.

phosphoferrite   [MINERAL] $(Fe,Mn)_3(PO_4)_2 \cdot 3H_2O$ A white or greenish orthorhombic mineral composed of hydrous phosphate of ferrous iron manganese phosphate; exhibits micalike cleavage.

phosphoglucoisomerase   [BIOCHEM] An enzyme that catalyzes the conversion of galactose-1-phosphate to glucose-1-phosphate.

phosphoglucomutase   [BIOCHEM] An enzyme that catalyzes the conversion of glucose-1-phosphate to glucose-6-phosphate.

phospholipase   [BIOCHEM] An enzyme that catalyzes a hydrolysis of a phospholipid, especially a lecithinase that acts in this manner on a lecithin.

phospholipid   [BIOCHEM] Any of a class of esters of phosphoric acid containing one or two molecules of fatty acid, an alcohol, and a nitrogenous base. Also known as phosphatide.

phosphomolybdic acid   [INORG CHEM] $H_3PO_4 \cdot 12MoO_3 \cdot xH_2O$ Yellow crystals; soluble in alcohol, ether, and water; used as an alkaloid reagent and a pigment. Abbreviated PMA.

phosphomonoesterase   [BIOCHEM] An enzyme catalyzing hydrolysis of phosphoric acid esters containing one ester linkage.

phosphonic acid   [ORG CHEM] $ROP(OH)_2$, where R is an organic radical such as C_6H_5-, as in phenylphosphonic acid.

phosphophyllite   [MINERAL] $Zn_2(FeMn)(PO_4)_2 \cdot 4H_2O$ Colorless to pale-blue mineral composed of hydrous zinc ferrous iron manganese phosphate; exhibits micalike cleavage.

phosphor   *See* luminophor.

phosphor bronze   [MET] A hard copper-base alloy containing several percent tin, and sometimes smaller percentages of lead, deoxidized with phosphorus.

phosphor dot   [ELECTR] One of the tiny dots of phosphor material that are used in groups of three, one group for each primary color, on the screen of a color television picture tube.

phosphorescence   [ATOM PHYS] 1. Luminescence that persists after removal of the exciting source. Also known as afterglow. 2. Luminescence whose decay, upon removal of the exciting source, is temperature-dependent.

phosphorescent paint   [MATER] A luminous paint containing phosphors or phosphorogens which requires activation from an outside source of light, depending upon the ability of the chemical to absorb light energy, and to emit it in the form of photons of light.

phosphoretted hydrogen   *See* phosphine.

phosphoric acid   [INORG CHEM] H_3PO_4 Water-soluble, transparent crystals, melting at $42°C$; used as a fertilizer, in soft drinks and flavor syrups, pharmaceuticals, water treatment, and animal feeds and to pickle and rust-proof metals. Also known as orthophosphoric acid.

phosphoric acid polymerization   [CHEM ENG] A petroleum-refinery process using phosphoric acid catalyst to convert propylene, butylene, or both, into high-octane gasoline or petrochemical polymers.

phosphoric anhydride   [INORG CHEM] P_2O_5 A flammable, dangerous, soft-white deliquescent powder; used as a dehydrating agent, in medicine and sugar refining, and as a chemical intermediate and analytical reagent. Also known as anhydrous phosphoric acid; phosphoric oxide; phosphorus pentoxide.

phosphoric chloride   *See* phosphorus pentachloride.

phosphoric oxide   *See* phosphoric anhydride.

phosphoric perbromide   *See* phosphorus pentabromide.

phosphoric perchloride   *See* phosphorus pentachloride.

phosphoric sulfide   *See* phosphorus pentasulfide.

phosphorimetry   [ANALY CHEM] Low-temperature, analytical procedure related to fluorometry; based on the nature and intensity of the phosphorescent light emitted by an appropriately excited molecule.

phosphorite   [PETR] A sedimentary rock composed chiefly of phosphate minerals. Also known as phosphate rock; rock phosphate.

phosphorized copper   [MET] A phosphorus deoxidized copper.

phosphorogen   [PHYS] A substance that promotes phosphorescence in another substance, as manganese does in zinc sulfide.

phosphorolysis   [BIOCHEM] A reaction by which elements of phosphoric acid are incorporated into the molecule of a compound.

phosphorous acid   [INORG CHEM] H_3PO_3 Alcohol- and water-soluble deliquescent white or yellowish crystals; decomposes at $200°C$; used as an analytical reagent and reducing agent.

phosphor tin   [MET] A master alloy of tin and phosphorus, usually containing up to 5% phosphorus; used to make phosphor bronze.

phosphorus   [CHEM] A nonmetallic element, symbol P, atomic number 15, atomic weight 30.98; used to manufacture phosphoric acid, in phosphor bronzes, incendiaries, pyrotechnics, matches, and rat poisons; the white (or yellow) allotrope is a soft waxy solid melting at $44.5°C$, is soluble in carbon disulfide, insoluble in water and alcohol, and is poisonous and self-igniting in air; the red allotrope is an amorphous powder subliming at $416°C$, igniting at $260°C$, is insoluble in all solvents, and is nonpoisonous; the black allotrope comprises lustrous crystals similar to graphite, and is insoluble in most solvents.

phosphorus bromide   *See* phosphorus pentabromide.

phosphorus nitride   [INORG CHEM] P_3N_5 Amorphous white solid that decomposes in hot water; insoluble in cold water, soluble in organic solvents; used to dope semiconductors.

phosphorus-nitrogen ratio   [OCEANOGR] The proportion, by weight, of phosphorus to nitrogen in seawater or in plankton; the ratio is approximately 7:1.

phosphorus oxide   [INORG CHEM] An oxygen compound of phosphorus; examples are phosphorus monoxide (P_2O), phosphorus trioxide (P_2O_3), phosphorus suboxide (P_4O).

phosphorus oxychloride   [INORG CHEM] $POCl_3$ Toxic, colorless, fuming liquid with pungent aroma; boils at $107°C$; decomposes in water or alcohol; causes skin burns; used as a catalyst, chlorinating agent, and in manufacture of various anhydrides. Also known as phosphoryl chloride.

phosphorus pentabromide   [INORG CHEM] PBr_5 Yellow crystals, decomposing at $106°C$ and in water; used in organic synthesis. Also known as phosphoric perbromide; phosphoric bromide; phosphorus bromide.

phosphorus pentachloride   [INORG CHEM] PCl_5 Toxic, yellowish crystals with irritating aroma; an eye irritant; sublimes on heating, but will melt at $148°C$ under pressure; soluble in carbon disulfide; decomposes in water; used as a catalyst and chlorinating agent. Also known as phosphoric chloride; phosphoric perchloride.

phosphorus pentasulfide   [INORG CHEM] P_2S_5 Flammable, hygroscopic, yellow crystals, melting at $281°C$; decomposes in moist air; soluble in alkali hydroxides; used to make lube-oil additives, rubber additives, and flotation agents. Also

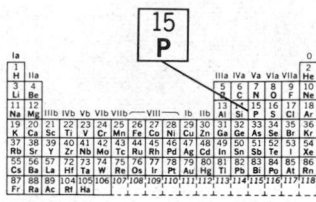

PHOSPHORUS

Periodic table of the chemical elements showing the position of phosphorus.

known as phosphoric sulfide; phosphorus persulfide; thiophosphoric anhydride.

phosphorus pentoxide *See* phosphoric anhydride.

phosphorus persulfide *See* phosphorus pentasulfide.

phosphorus sesquisulfide [INORG CHEM] P_4S_3 Flammable, yellow crystals, melting at 172°C; decomposed by hot water, insoluble in water, soluble in carbon disulfide; used as chemical intermediate and to make matches. Also known as tetraphosphorus trisulfide.

phosphorus sulfide *See* phosphorus trisulfide.

phosphorus thiochloride [INORG CHEM] $PSCl_3$ Yellow liquid, boiling at 125°C; used to make insecticides and oil additives.

phosphorus tribromide [INORG CHEM] PBr_3 A corrosive, fuming, colorless liquid with penetrating aroma; soluble in acetone, alcohol, carbon disulfide, and hydrogen sulfide; decomposes in water; used as an analytical reagent to test for sugar and oxygen.

phosphorus trichloride [INORG CHEM] PCl_3 A colorless, fuming liquid that decomposes rapidly in moist air and water; soluble in ether, benzene, carbon disulfide, and carbon tetrachloride; boils at 76°C; used as a chlorinating agent, phosphorus solvent, and in saccharin manufacture.

phosphorus triiodide [INORG CHEM] PI_3 Hygroscopic, red crystals, melting at 61°C; soluble in alcohol and carbon disulfide; decomposes in water; used in organic syntheses.

phosphorus trisulfide [INORG CHEM] P_2S_3 or P_4S_6 Grayish-yellow, tasteless, odorless solid that burns in air; soluble in alcohol, carbon disulfide, and ether; melts at 290°C; used as an analytical reagent. Also known as phosphorus sulfide.

phosphorylase [BIOCHEM] An enzyme that catalyzes the formation of glucose-1-phosphate (Cori ester) from glycogen and inorganic phosphate; it is widely distributed in animals, plants, and microorganisms.

phosphorylation [ORG CHEM] The esterification of compounds with phosphoric acid.

phosphoryl chloride *See* phosphorus oxychloride.

phosphotransacetylase [BIOCHEM] An enzyme that catalyzes the reversible transfer of an acetyl group from acetyl coenzyme A to a phosphate, with formation of acetyl phosphate.

phosphotungstic acid [INORG CHEM] $H_3PO_4 \cdot 12WO_3 \cdot xH_2O$ Heavy-greenish, water- and alcohol-soluble crystals; used as an analytical reagent and a pigment. Also known as heavy acid; phosphowolframic acid; PTA.

phosphowolframic acid *See* phosphotungstic acid.

phosphuranylite [MINERAL] $(UO_2)(PO_4)_2 \cdot 6H_2O$ A yellow secondary mineral composed of hydrous uranyl phosphate, occurring in powder form; it is phosphorescent when exposed to radium emanations.

phot [OPTICS] A unit of illumination equal to the illumination of a surface, 1 square centimeter in area, on which there is a luminous flux of 1 lumen, or the illumination on a surface all points of which are at a distance of 1 centimeter from a uniform point source of 1 candela. Also known as centimeter-candle (deprecated usage).

photic region [ECOL] The uppermost layer of a body of water that receives enough sunlight to permit the occurrence of photosynthesis.

Photidae [INV ZOO] A family of amphipod crustaceans in the suborder Gammaridea.

photoautotrophic [BIOL] Pertaining to organisms which derive energy from light and manufacture their own food.

Photobacterium [MICROBIO] A genus of luminescent coccobacilli and occasional rods in the family Pseudomonadaceae.

photocathode [ELECTR] A photosensitive surface that emits electrons when exposed to light or other suitable radiation; used in phototubes, television camera tubes, and other light-sensitive devices.

photocell [ELECTR] A solid-state photosensitive electron device whose current-voltage characteristic is a function of incident radiation. Also known as electric eye; photoelectric cell.

photocell relay [ELECTR] A relay actuated by a signal received when light falls on, or is prevented from falling on, a photocell.

photochemical oxidant [CHEM] Any of the chemicals which enter into oxidation reactions in the presence of light or other radiant energy.

photochemical reaction [PHYS CHEM] A chemical reaction influenced or initiated by light, particularly ultraviolet light, as in the chlorination of benzene to produce benzene hexachloride.

photochemistry [PHYS CHEM] The study of the effects of light on chemical reactions.

photochromic compound [CHEM] A chemical compound that changes in color when exposed to visible or near-visible radiant energy; the effect is reversible; used to produce very-high-density microimages.

photochromic glass [MATER] A glass that darkens when exposed to light but regains its original transparency a few minutes after light is removed; the rate of clearing increases with temperature.

photocoagulator [MED] An instrument that uses a xenon flash lamp and an associated train of optics to focus an intense beam of light on a detached retina for the purpose of inducing coagulation and a lesion that welds the retina back into position.

photoconduction [SOLID STATE] An increase in conduction of electricity resulting from absorption of electromagnetic radiation.

photoconductive cell [ELECTR] A device for detecting or measuring electromagnetic radiation by variation of the conductivity of a substance (called a photoconductor) upon absorption of the radiation by this substance. Also known as photoresistive cell; photoresistor.

photoconductive device [ELECTR] Any device that makes use of photoconductivity, such as a photoconductive cell.

photoconductive film [ELECTR] A film of material whose current-carrying ability is enhanced when illuminated.

photoconductive gain factor [ELECTR] The ratio of the number of electrons per second flowing through a circuit containing a cube of semiconducting material, whose sides are of unit length, to the number of photons per second absorbed in this volume.

photoconductive meter [ELECTR] An exposure meter in which a battery supplies power through a photoconductive cell to a milliammeter.

photoconductivity [SOLID STATE] The increase in electrical conductivity displayed by many nonmetallic solids when they absorb electromagnetic radiation.

photoconductor [SOLID STATE] A nonmetallic solid whose conductivity increases when it is exposed to electromagnetic radiation.

photoconductor diode *See* photodiode.

photocopying process [GRAPHICS] Any of the means by which a copy is created on a sensitized surface (generally paper, film, or metal plate) by the action of radiant energy.

photodetector [ELECTR] A detector that responds to radiant energy; examples include photoconductive cells, photodiodes, photoresistors, photoswitches, phototransistors, phototubes, and photovoltaic cells. Also known as light-sensitive cell; light-sensitive detector; light sensor photodevice; photodetector; photoelectric detector; photosensor.

photodevice *See* photodetector.

photodiffusion effect *See* Dember effect.

photodiode [ELECTR] A semiconductor diode in which the reverse current varies with illumination; examples include the alloy-junction photocell and the grown-junction photocell. Also known as photoconductor diode.

photodisintegration [NUC PHYS] The breakup of an atomic nucleus into two or more fragments as a result of bombardment by gamma radiation. Also known as Chadwick-Goldhaber effect.

photodissociation [PHYS CHEM] The removal of one or more atoms from a molecule by the absorption of a quantum of electromagnetic energy.

photodosimetry [NUCLEO] Determination of the cumulative dose of ionizing radiation by use of photographic film.

photodraft [DES ENG] A photographic reproduction of a master layout or design on a specially prepared emulsion-coated piece of sheet metal; used as a master in a tool-construction department.

photoecology [ENG] The application of air photography to ecology, integrated land resource studies, and forestry.

photoelastic effect [OPTICS] Changes in optical properties of a transparent dielectric when it is subjected to mechanical stress, such as mechanical birefringence. Also known as photoelasticity.

photoelasticity [OPTICS] **1.** An experimental technique for the measurement of stresses and strains in material objects by means of the phenomenon of mechanical birefringence. **2.** See photoelastic effect.

photoelectret [SOLID STATE] An electret produced by the removal of light from an illuminated photoconductor in an electric field.

photoelectric [ELECTR] Pertaining to the electrical effects of light, such as the emission of electrons, generation of voltage, or a change in resistance when exposed to light.

photoelectric absorption [ELECTR] Absorption of photons in one of the several photoelectric effects.

photoelectric absorption analysis [ANALY CHEM] Type of activation analysis in which the γ-photon gives all of its energy to an electron in the crystal under analysis, generating a maximum-sized pulse for that particular γ-energy.

photoelectric cell See photocell.

photoelectric color comparator See color comparator.

photoelectric colorimeter [ENG] A colorimeter that uses a phototube or photocell, a set of color filters, an amplifier, and an indicating meter for quantitative determination of color.

photoelectric colorimetry [ANALY CHEM] Measurement of the colorant concentration in a solution by means of the tristimulus values of three primary light filter-photocell combinations.

photoelectric constant [ELECTR] The ratio of the frequency of radiation causing emission of photoelectrons to the voltage corresponding to the energy absorbed by a photoelectron; equal to Planck's constant divided by the electron charge.

photoelectric control [ELECTR] Control of a circuit or piece of equipment by changes in incident light.

photoelectric counter [ELECTR] A photoelectrically actuated device used to record the number of times a given light path is intercepted by an object.

Photoelectric cutoff register control [ELECTR] Use of a photoelectric control system as a longitudinal position regulator to maintain the position of the point of cutoff with respect to a repetitive pattern of moving material.

photoelectric densitometer [ENG] An electronic instrument used to measure the density or opacity of a film or other material; a beam of light is directed through the material, and the amount of light transmitted is measured with a photocell and meter.

photoelectric detector See photodetector.

photoelectric door opener [CONT SYS] A control system that employs a photocell or other photo device, used to open and close a power-operated door.

photoelectric effect See photoelectricity.

photoelectric electron-multiplier tube See multiplier phototube.

photoelectric flame-failure detector [CONT SYS] A photoelectric control that cuts off duel flow when the fuel-consuming flame is extinguished.

photoelectric fluorometer [ENG] Device using a photoelectric cell to measure fluorescence in a chemical sample that has been excited (one or more electrons have been raised to higher energy level) by ultraviolet or visible light; used for anlaysis of chemical mixtures.

photoelectric infrared radiation See near-infrared radiation.

photoelectric intrusion detector [ELECTR] A burglar-alarm system in which interruption of a light beam by an intruder reduces the illumination on a phototube and thereby closes an alarm circuit.

photoelectricity [ELECTR] The liberation of electrons by electromagnetic radiation incident on a substance; includes photoemission, photoionization, photoconduction, the photovoltaic effect, and the Auger effect (an internal photoelectric process). Also known as photoelectric effect; photoelectric process.

photoelectric lighting control [ELECTR] Use of a photoelectric relay actuated by a change in illumination in a given area or at a given point.

photoelectric liquid-level indicator [ENG] A level indicator in which rising liquid interrupts the light beam of a photoelectric control system; used in a tank or process vessel.

photoelectric loop control [CONT SYS] A photoelectric control system used as a position regulator for a loop of material passing from one strip-processing line to another that may travel at a different speed. Also known as loop control.

photoelectric photometer [ENG] A photometer that uses a photocell, phototransistor, or phototube to measure the intensity of light. Also known as electronic photometer.

photoelectric photometry [OPTICS] In contrast to the methods of visual photometry, an objective approach to the problems of photometry, wherein any of several types of photoelectric devices are used to replace the human eye as the sensing element.

photoelectric plethysmograph [MED] A medical instrument for measuring and recording ear opacity by means of a tiny phototube and lamp clipped to the ear, as a measure of the state of fullness of blood vessels; also worn by aircraft pilots during high-altitude flights, as an alarm indicating the need for more oxygen.

photoelectric process See photoelectricity.

photoelectric pyrometer [ENG] An instrument that measures high temperatures by using a photoelectric arrangement to measure the radiant energy given off by the heated object.

photoelectric reader [ADP] A device for reading information stored on paper tape or cards; data are read by sensing the presence or absence of holes.

photoelectric reflectometer [ENG] A reflectometer that uses a photocell or phototube to measure the diffuse reflection of surfaces, powders, pastes, and opaque liquids.

photoelectric register control [CONT SYS] A register control using a light source, one or more phototubes, a suitable optical system, an amplifier, and a relay to actuate control equipment when a change occurs in the amount of light reflected from a moving surface due to register marks, dark areas of a design, or surface defects. Also known as photoelectric scanner.

photoelectric relay [ELECTR] A relay combined with a phototube and amplifier, arranged so changes in incident light on the phototube make the relay contacts open or close. Also known as light relay.

photoelectric scanner [ADP] A device that scans punched cards by photoelectric means, as opposed to the standard brushes or "feelers," or mechanical plungers. [CONT SYS] See photoelectric register control.

photoelectric smoke-density control [CONT SYS] A photoelectric control system used to measure, indicate, and control the density of smoke in a flue or stack.

photoelectric sorter [CONT SYS] A photoelectric control system used to sort objects according to color, size, shape, or other light-changing characteristics.

photoelectric transmissometer [ENG] A device to measure the runway visibility at an airport by measuring the degree to which a light beam falling on a photocell is obscured by clouds or fog.

photoelectric tube See phototube.

photoelectric turbidimeter [ENG] Device for measurement of solution turbidity by use of photocells to detect the loss of intensity of light beamed through the solution.

photoelectromagnetic effect [ELECTR] The effect whereby, when light falls on a flat surface of an intermetallic semiconductor located in a magnetic field that is parallel to the surface, excess hole-electron pairs are created, and these carriers diffuse in the direction of the light but are deflected by the magnetic field to give a current flow through the semiconductor that is at right angles to both the light rays and the magnetic field.

photoelectromotive force [ELECTR] Electromotive force caused by photovoltaic action.

photoelectron [ELECTR] An electron emitted by the photoelectric effect.

photoelectron spectroscopy [SPECT] The branch of electron spectroscopy concerned with the energy analysis of photoelectrons ejected from a substance as the direct result of bombardment by ultraviolet radiation or x-radiation.

photoemission [ELECTR] The ejection of electrons from a

PHOTOEMISSION

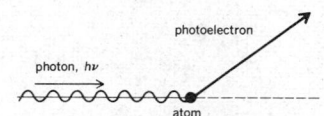

Albert Einstein's approach to photoemission. Light beam behaves like a stream of photons, each of energy $h\nu$, where h is Planck's constant, and ν the frequency of the photon. When a photon interacts with an electron, the electron absorbs entire photon energy and is ejected from emitter if this energy exceeds a well-defined minimum value.

solid (or less commonly, a liquid) by incident electromagnetic radiation. Also known as external photoelectric effect.

photoemissive cell [ELECTR] A device which detects or measures radiant energy by measurement of the resulting emission of electrons from the surface of a photocathode.

photoemissive tube photometer [ENG] A photometer which uses a tube made of a photoemissive material; it is highly accurate, but requires electronic amplification, and is used mainly in laboratories.

photoemissivity [ELECTR] The property of a substance that emits electrons when struck by light.

photoengraving [GRAPHICS] The technique of producing relief plates such as halftones or zinc etchings by photography; a metal plate is coated with a photosensitive emulsion and exposed to light under a reversed positive; the picture is developed by dissolving away the portion of the emulsion not acted upon by the light, and the plate is etched.

photoengraving zinc [MET] Pure zinc mixed with a small amount of iron to reduce grain size, and alloyed with a maximum of 0.2% each of cadmium, manganese, and magnesium; used for printing plates.

photoenlarger See enlarger.

photo finishing [GRAPHICS] The commercial processing of photographs.

photoflash bomb [ENG] A missile dropped from aircraft; it contains a photoflash mixture and a means for ignition at a distance above the ground, to produce a brilliant light of short duration for photographic purposes.

photoflash composition [MATER] A pyrotechnic material which, when loaded in a suitable casing and ignited, produces a flash of sufficient intensity and duration for photographic purposes; used as the filler in photoflash bombs and cartridges.

photoflash lamp [ELEC] A lamp consisting of a glass bulb filled with finely shredded aluminum foil in an atmosphere of oxygen; when the foil is ignited by a low-voltage dry cell, it burns with a burst of high-intensity light of short time duration and with definitely regulated time characteristics.

photoflash unit [ELECTR] A portable electronic light source for photographic use, consisting of a capacitor-discharge power source, a flash tube, a battery for charging the capacitor, and sometimes also a high-voltage pulse generator to trigger the flash.

photoflood lamp [ELEC] An incandescent lamp used in photography which has a high-temperature filament, so that it gives high illumination and high color temperature for a short lifetime.

photofluorography See fluorography.

photogelatin printing [GRAPHICS] A method of printing that produces extremely fine tones on uncoated papers by means of a photogelatin printing plate.

photogelatin printing plate [GRAPHICS] An aluminum or heavy-glass base coated with a layer of light-sensitive gelatin, exposed to light through an unscreened photographic negative of the copy to be reproduced; after treatment, the printing image varies in thickness, moisture content, and ability to hold or repel printing ink. Also known as albertype; artotype; collotype; heliotype; phototype.

photogelatin process [GRAPHICS] Photomechanical process for producing prints directly from a hardened colloid film; the sensitized film is exposed under a reversed negative, desensitized, and then soaked in glycerin and salt water so that the parts that have not been exposed to light undergo swelling; the swelled parts are ink-repellent and the unswelled parts ink-receptive, thereby forming a printing surface of the lithographic sort. Also known as collotype.

photogeology [GEOL] The geologic interpretation of landforms by means of aerial photographs.

photoglow tube [ELECTR] Gas-filled phototube used as a relay by making the operating voltage sufficiently high so that ionization and a flow discharge occur, with considerable current flow, when a certain illumination is reached.

photoglycine See glycin.

photogoniometer [ENG] A goniometer that uses a phototube or photocell as a sensing device for studying x-ray spectra and x-ray diffraction effects in crystals.

photogram [GRAPHICS] A design or pattern produced on regular photographic paper without the use of a negative or a lens system; opaque or transparent objects are assembled on the paper, and the paper is exposed to light and processed in the usual way.

photographic barograph [ENG] A mercury barometer arranged so that the position of the upper or lower meniscus may be measured photographically.

photographic emulsion [GRAPHICS] Microscopic grains of light-sensitive silver halide suspended in a gelatin surface on paper, plastic, metal, or glass; used to coat photographic film.

photographic field [OPTICS] The area covered or "seen" by the lens of a camera.

photographic film [GRAPHICS] Sensitized material (emulsion) coated on a flexible support, usually a transparent plastic material.

photographic fixing [GRAPHICS] The process by which the unexposed and unreduced silver halide in a negative is removed in an exposed film; sodium thiosulfate (hypo) is the chemical usually used.

photographic meteor [ASTRON] A meteor which has been photographed for the purpose of determining its origin, velocity, and other characteristics.

photographic photometry [SPECT] The use of a comparator-densitometer to analyze a photographed spectrograph spectrum by emulsion density measurements.

photographic plane [GRAPHICS] **1.** The flat area on which a document is photographed. **2.** The surface at which an image is formed by a lens for recording on a photosensitive material.

photographic recording [COMMUN] Facsimile recording in which a photosensitive surface is exposed to a signal-controlled light beam or spot.

photographic sound recorder [ELECTR] A sound recorder having means for producing a modulated light beam and means for moving a light-sensitive medium relative to the beam to give a photographic recording of sound signals. Also known as optical sound recorder.

photographic sound reproducer [ELECTR] A sound reproducer in which an optical sound record on film is moved through a light beam directed at a light-sensitive device, to convert the recorded optical variations back into audio signals. Also known as optical sound reproducer.

photographic surveying [ENG] Photographing of plumb bobs, clinometers, or magnetic needles in borehole surveying to provide an accurate permanent record.

photography [GRAPHICS] The process of forming visible images directly or indirectly by the action of light or other forms of radiation on sensitive surfaces.

photogravure [GRAPHICS] A method of making intaglio engravings in copper from photographs in which a gelatin image is used as an acid resist; it is considered high quality for halftones, but not very good for type.

photoionization [PHYS CHEM] The removal of one or more electrons from an atom or molecule by absorption of a photon of visible or ultraviolet light. Also known as atomic photoelectric effect.

photoisland grid [ELECTR] Photosensitive surface in the storage-type, Farnsworth dissector tube for television cameras.

photojunction battery [NUCLEO] A nuclear-type battery in which a radioactive material such as promethium-147 irradiates a phosphor which converts nuclear energy into light; the light is then converted to electrical energy by a small silicon junction.

photology [OPTICS] The scientific study of light.

photoluminescence [ATOM PHYS] Luminescence stimulated by visible, infrared, or ultraviolet radiation.

photomacrography [GRAPHICS] Making large pictures of small subjects by using a short-focal-length lens on a long-bellows camera.

photomagnetic effect [PHYS] The direct effect of light on the magnetic susceptibility of certain substances.

photometer [ENG] An instrument used for making measurements of light or electromagnetic radiation, in the visible range.

photometric titration [ANALY CHEM] A titration in which the titrant and solution cause the formation of a metal complex accompanied by an observable change in light absorbance by the titrated solution.

photometry [OPTICS] The calculation and measurement of quantities describing light, such as luminous intensity, luminous flux, luminous flux density, light distribution, color, absorption factor, spectral distribution, and the reflectance and transmittance of light; sometimes taken to include measurement of near-infrared and near-ultraviolet radiation as well as visible light. Also known as light measurement.

photomicrograph [GRAPHICS] A micrograph produced by photography.

photomicrography [GRAPHICS] The photography of the image formed by the microscope.

photomorphogenesis [BOT] The control exerted by light over growth, development, and differentiation of plants that is independent of photosynthesis.

photomosaic [GRAPHICS] Composite photograph (usually aerial) made up of individual, small-area photographs placed side by side.

photomultiplier *See* multiplier phototube.

photomultiplier cell [ELECTR] A transistor whose *pn*-junction is exposed so that it conducts more readily when illuminated.

photomultiplier counter [ELECTR] A scintillation counter that has a built-in multiplier phototube.

photomultiplier tube *See* multiplier phototube.

photon [OPTICS] *See* troland. [QUANT MECH] A massless particle, the quantum of the electromagnetic field, carrying energy, momentum, and angular momentum. Also known as light quantum.

photon coupled isolator [ELECTR] Circuit coupling device, consisting of an infrared emitter diode coupled to a photon detector over a short shielded light path, which provides extremely high circuit isolation.

photon coupling [ELECTR] Coupling of two circuits by means of photons passing through a light pipe.

photon curve [PETRO ENG] A graphical plot of depth versus gamma radiation (photon) scatter during the radioactive logging of a well bore; used to detect differences in density at various reservoir depths.

photonegative [ELECTR] Having negative photoconductivity, hence decreasing in conductivity (increasing in resistance) under the action of light; selenium sometimes exhibits photonegativity.

photon emission spectrum [PHYS] The relative numbers of optical photons emitted by a scintillator material per unit wavelength as a function of wavelength; the emission spectrum may also be given in alternative units such as wave number, photon energy, or frequency.

photonephelometer [ENG] A nephelometer that uses a photocell or phototube to measure the amount of light transmitted by a suspension of particles.

photoneutron [NUC PHYS] A neutron released from a nucleus in a photonuclear reaction.

photon flux [OPTICS] The number of photons in a light beam reaching a surface, such as the surface of the photocathode of a photomultiplier tube, in a unit of time.

photon gas [STAT MECH] An electromagnetic field treated as a collection of photons; it behaves as any other collection of bosons, except that the particles are emitted or absorbed without restriction on their number.

photon sail *See* solar sail.

photon theory [QUANT MECH] A theory of photoemission developed by Einstein, according to which a light beam behaves like a stream of particles (called photons) when it delivers energy to a substance displaying photoemission, the particles each having an energy equal to Planck's constant times the frequency of the light.

photonuclear reaction [NUC PHYS] A nuclear reaction resulting from the collision of a photon with a nucleus.

photoperiodism [PHYSIO] The physiological responses of an organism to the length of night or day or both.

photophilic [BIOL] Thriving in full light.

photophobia [PSYCH] An abnormal fear of light.

photophobic [BIOL] **1.** Avoiding light. **2.** Exhibiting negative phototropism.

photophore gland [VERT ZOO] A highly modified integumentary gland which develops into a luminous organ composed of a lens and a light-emitting gland; occurs in deep-sea teleosts and elasmobranchs.

photophoresis [PHYS] Production of unidirectional motion in a collection of very fine particles, suspended in a gas or falling in a vacuum, by a powerful beam of light.

photophosphorylase [BIOCHEM] An enzyme that is associated with the surface of a thylakoid membrane and is involved in the final stages of adenosinetriphosphate production by photosynthetic phosphorylation.

photophosphorylation [BIOCHEM] Phosphorylation induced by light energy in photosynthesis.

photophygous [BIOL] Thriving in shade.

photopic vision *See* foveal vision.

photopigment [BIOCHEM] A pigment that is unstable in the presence of light of appropriate wavelengths, such as the chromophore pigment which combines with opsins to form rhodopsin in the rods and cones of the vertebrate eye.

photopositive [ELECTR] Having positive photoconductivity, hence increasing in conductivity (decreasing in resistance) under the action of light; selenium ordinarily has photopositivity.

photoproton [NUC PHYS] A proton released from a nucleus in a photonuclear reaction.

photoreception [PHYSIO] The process of absorption of light energy by plants and animals and its utilization for biological functions, such as photosynthesis and vision.

photoreceptor [PHYSIO] A highly specialized, light-sensitive cell or group of cells containing photopigments.

photoresist [GRAPHICS] A light-sensitive coating that is applied to a substrate or board, exposed, and developed prior to chemical etching; the exposed areas serve as a mask for selective etching.

photoresistive cell *See* photoconductive cell.

photoresistor *See* photoconductive cell.

photorespiration [BIOCHEM] Respiratory activity taking place in plants during the light period; CO_2 is released and O_2 is taken up, but no useful form of energy, such as adenosinetriphosphate, is derived.

photosensitive *See* light-sensitive.

photosensitive glass [GRAPHICS] Glass containing submicroscopic metallic particles; when ultraviolet light passes through a negative on the glass, it precipitates the particles, with shadowed areas of the negative permitting deeper penetration into the glass than highlight areas, giving the picture three dimensions and color; photograph is developed by heating the glass to 1000°F (538°C).

photosensor *See* photodetector.

photosphere [ASTRON] The intensely bright portion of the sun visible to the unaided eye; it is a shell a few hundred miles in thickness marking the boundary between the dense interior gases of the sun and the more diffuse cooler gases in the outer portions of the sun.

photospheric granulation *See* granulation.

Photostat [GRAPHICS] A registered trademark of the Photostat Corporation for a photographic copying machine.

photosynthesis [BIOCHEM] Synthesis of chemical compounds in light, especially the manufacture of organic compounds (primarily carbohydrates) from carbon dioxide and a hydrogen source (such as water), with simultaneous liberation of oxygen, by chlorophyll-containing plant cells.

phototaxis [BIOL] Movement of a motile organism or free plant part in response to light stimulation.

phototelegraphy *See* facsimile.

phototheodolite [ENG] A device for measuring and recording the horizontal and vertical angles to a missile while photographing its flight.

photothermoelasticity [OPTICS] Changes in optical properties of a transparent dielectric when it is subjected to mechanical stress, which is, in turn, induced by temperature gradients.

phototransistor [ELECTR] A junction transistor that may have only collector and emitter leads or also a base lead, with the base exposed to light through a tiny lens in the housing; collector current increases with light intensity, as a result of amplification of base current by the transistor structure.

PHOTOVOLTAIC CELL

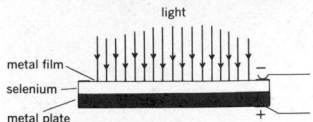

Schematic of a photovoltaic cell; metal plate usually is iron, and thin metal film is either gold or platinum.

PHOTOVOLTAIC METER

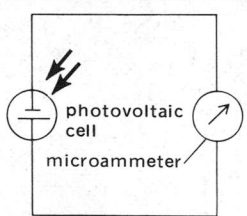

Circuit diagram of photovoltaic type of exposure meter. Heavy arrows represent light falling on cell.

PHREATOICOIDEA

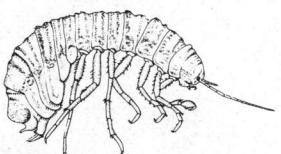

Onchotelson brevicaudatus (Smith), adult male.

phototroph [BIOL] An organism that utilizes light as a source of metabolic energy.

phototropism [BOT] A growth-mediated response of a plant to stimulation by visible light. [SOLID STATE] A reversible change in the structure of a solid exposed to light or other radiant energy, accompanied by a change in color. Also known as phototropy.

phototropy *See* phototropism.

phototube [ELECTR] An electron tube containing a photocathode from which electrons are emitted when it is exposed to light or other electromagnetic radiation. Also known as electric eye; light-sensitive tube; photoelectric tube.

phototube cathode [ELECTR] The photoemissive surface which is the most negative element of a phototube.

phototube relay [ELECTR] A photoelectric relay in which a phototube serves as the light-sensitive device.

phototype *See* photogelatin printing plate.

phototypesetter [GRAPHICS] Machine for placing individual characters on photographic film.

phototype setting [GRAPHICS] A method of composing text matter by successively projecting the images of characters on light-sensitive film or on photographic paper.

photovaristor [ELECTR] Varistor in which the current-voltage relation may be modified by illumination, for example, one in which the semiconductor is cadmium sulfide or lead telluride.

photoviscoelasticity [OPTICS] Changes in optical properties of a transparent, viscoelastic substance when it is subjected to stress.

photovoltaic [ELECTR] Capable of generating a voltage as a result of exposure to visible or other radiation.

photovoltaic cell [ELECTR] A device that detects or measures electromagnetic radiation by generating a potential at a junction (barrier layer) between two types of material, upon absorption of radiant energy. Also known as barrier-layer cell; barrier-layer photocell; boundary-layer photocell; photronic photocell.

photovoltaic effect [ELECTR] The production of a voltage in a nonhomogeneous semiconductor, such as silicon, or at a junction between two types of material, by the absorption of light or other electromagnetic radiation.

photovoltaic meter [ELECTR] An exposure cell in which a photovoltaic cell produces a current proportional to the light falling on the cell, and this current is measured by a sensitive microammeter.

photox cell [ELECTR] Type of photovoltaic cell in which a voltage is generated between a copper base and a film of cuprous oxide during exposure to visible or other radiation.

photronic cell [ELECTR] Type of photovoltaic cell in which a voltage is generated in a layer of selenium during exposure to visible or other radiation.

photronic photocell *See* photovoltaic cell.

Phoxichilidiidae [INV ZOO] A family of marine arthropods in the subphylum Pycnogonida; typically, chelifores are present, palpi are lacking, and ovigers have five to nine joints in males only.

Phoxocephalidae [INV ZOO] A family of amphipod crustaceans in the suborder Gammaridea.

Phractolaemidae [VERT ZOO] A family of tropical African fresh-water fishes in the order Gonorynchiformes.

Phragmobasidiomycetes [MYCOL] An equivalent name for Heterobasidiomycetidae.

phragmoid [BOT] Having septae perpendicular to the long axis, as the conidia of certain fungi.

phragmoplast [CYTOL] A thin barrier which is formed across the spindle equator in late cytokinesis in plant cells and within which the cell wall is laid down.

phragmosome [CYTOL] A differentiated cytoplasmic partition in which the phragmoplast and cell plate develop during cell division in plant cells.

Phragmosporae [MYCOL] A spore group of the Fungi Imperfecti with three- to many-celled spores.

phreatic [GEOL] Of a volcanic explosion of material such as steam or mud, not being incandescent.

phreatic surface *See* water table.

phreatic water [HYD] Groundwater in the zone of saturation.

phreatic-water discharge *See* groundwater discharge.

phreatic zone *See* zone of saturation.

Phreatoicidae [INV ZOO] A family of isopod crustaceans in the suborder Phreatoicoidea in which only the left mandible retains a lacinia mobilis.

Phreatoicoidea [INV ZOO] A suborder of the Isopoda having a subcylindrical body that appears laterally compressed, antennules shorter than the antennae, and the first thoracic segment fused with the head.

phreatophyte [ECOL] A plant with a deep root system which obtains water from the groundwater or the capillary fringe above the water table.

phrenectomy [MED] Resection of a section of a phrenic nerve or removal of an entire phrenic nerve.

phrenic nerve [ANAT] A nerve, arising from the third, fourth, and fifth cervical (cervical plexus) segments of the spinal cord; innervates the diaphragm.

phrynoderma [MED] Dryness of the skin with follicular hyperkeratosis, caused by vitamin A deficiency.

Phrynophiurida [INV ZOO] An order of the Ophiuroidea in which the vertebrae usually articulate by means of hourglass-shaped surfaces, and the arms are able to coil upward or downward in the vertical plane.

pH(S) *See* pH standard.

pH standard [ANALY CHEM] Five standard laboratory solutions available from the U. S. National Bureau of Standards, each solution having a known pH value; the standards cover pH ranges from 3.557 to 8.833. Abbreviated pH(S).

phthalate [ORG CHEM] A salt of phthalic acid; contains the radical $C_6H_4(COO)_2^{--}$; an example is dibutylphthalate, $C_{16}H_{22}O_4$; used as a plasticizer in plastics, and as a buffer in standard laboratory solutions.

phthalate buffer [ANALY CHEM] Laboratory pH reference solution made of potassium hydrogen phthalate, $KHC_8H_4O_4$; at 0.05 molal, the pH is 4.008 at 25°C.

phthalate ester [ORG CHEM] Any of a group of plastics plasticizers made by the direct action of alcohol on phthalic anhydride; generally characterized by moderate cost, good stability, and good general properties.

phthalazine [ORG CHEM] $C_6H_4CHN_2CH$ Colorless crystals, melting at 91°C; soluble in alcohol. Also known as 2, 3-benzodiazine; β-benzo-*o*-diazine; β-phenodiazine.

phthalic acid [ORG CHEM] $C_6H_4(CO_2H)_2$ Any of three isomeric benzene dicarboxylic acids; the ortho form is usually called phthalic acid, comprises alcohol-soluble, colorless crystals decomposing at 191°C, slightly soluble in water and ether, is used to make dyes, medicine, and synthetic perfumes, and as a chemical intermediate, and is also known as benzene orthodicarboxylic acid; the para form, known as terephthalic acid, is used to make polyester resins (Dacron) and as poultry feed additives; the meta form is isophthalic acid.

***meta*-phthalic acid** *See* isophthalic acid.

***ortho*-phthalic acid** *See* phthalic acid.

***para*-phthalic acid** *See* terephthalic acid.

phthalic anhydride [ORG CHEM] $C_6H_4(CO)_2O$ White crystals, melting at 131°C; sublimes when heated; slightly soluble in ether and hot water, soluble in alcohol; used to make dyes, resins, plasticizers, and insect repellents. Also known as acid phthalic anhydride.

phthalocyanine pigments [ORG CHEM] A group of light-fast organic pigments with four isoindole groups, $(C_6H_4)_2N$, linked by four nitrogen atoms to form a conjugated chain; included are phthalocyanine (blue-green), copper phthalocyanine (blue), chlorinated copper phthalocyanine (green), and sulfonated copper phthalocyanine (green); used in enamels, plastics, linoleum, inks, wallpaper, and rubber goods.

phthalocyanine Q switching [OPTICS] Laser Q switching in which a solution of metal-organic compounds known as phthalocyanines is placed in a cell between an uncoated ruby laser crystal and a high-reflectivity mirror; when the incident ruby light reaches a certain level, the solution suddenly becomes almost perfectly transparent to this light, permitting the release of all the energy stored in the ruby as a giant pulse.

phthisis [MED] **1.** Any disease characterized by emaciation and loss of strength. **2.** *See* tuberculosis.

phugoid [AERO ENG] Pertaining to variations in the longitudinal motion or course of the center of mass of an aircraft.

phycite *See* erythritol.

Phycitinae [INV ZOO] A large subfamily of moths in the family Pyralididae in which the frenulum of the female is a simple spine rather than a bundle of bristles.

phycobilin [BIOCHEM] Any of various protein-bound pigments which are open-chain tetrapyrroles and occur in some groups of algae.

phycocyanin [BIOCHEM] A blue phycobilin.

phycocyanobilin [BIOCHEM] $C_{31}H_{38}O_2N_4$ Phycobilin with an ethylidene side chain ($=CH-CH_3$) and only one asymmetric carbon atom (C_1).

phycoerythrin [BIOCHEM] A red phycobilin.

phycoerythrobilin [BIOCHEM] $C_{31}H_{38}O_2N_4$ Phycobilin with seven conjugated double bonds, an ethylidine side chain ($=CH-CH_3$), and two asymmetric carbon atoms (C_1 and C_7).

Phycomycetes [MYCOL] A primitive class of true fungi belonging to the Eumycetes; they lack regularly spaced septa in the actively growing portions of the plant body, and have the sporangiospore, produced in the sporangium by cleavage, as the fundamental, asexual reproductive unit.

Phycosecidae [INV ZOO] A small family of coleopteran insects of the superfamily Cucujoidea, including five species found in New Zealand, Australia, and Egypt.

Phylactolaemata [INV ZOO] A class of fresh-water ectoproct bryozoans; individuals have lophophores which are U-shaped in basal outline, and relatively short, wide zooecia.

phyletic evolution [EVOL] The gradual evolution of population without separation into isolated parts.

phyllary [BOT] A bract of the involucre of a composite plant.

phyllite [PETR] A metamorphic rock intermediate in grade between slate and schist, and derived from argillaceous sediments; has a silky sheen on the cleavage surface.

Phyllobothrioidea [INV ZOO] The equivalent name for Tetraphyllidea.

phyllobranchiate gill [INV ZOO] A type of decapod crustacean gill with flattened branches, or lamellae usually arranged in two opposite series.

phylloclade [BOT] A flattened stem that fulfills the same functions as a leaf.

phyllode [BOT] A broad, flat petiole that replaces the blade of a foliage leaf. [INV ZOO] A petal-shaped group of ambulacra near the mouth of certain echinoderms.

Phyllodocidae [INV ZOO] A leaf-bearing family of errantian annelids in which the species are often brilliantly iridescent and are highly motile.

Phyllogoniaceae [BOT] A family of mosses in the order Isobryales in which the leaves are equitant.

Phyllolepida [PALEON] A monogeneric order of placoderms from the late Upper Devonian in which the armor is broad and low with a characteristic ornament of concentric and transverse ridges on the component plates.

phyllonite [PETR] A metamorphic rock occupying an intermediate position between phyllite and mylonite.

Phyllophoridae [INV ZOO] A family of dendrochirotacean holothurians in the order Dendrochirotida having a rather naked skin and a complex calcareous ring.

phyllosilicate [MINERAL] A structural type of silicate mineral in which flat sheets are formed by the sharing of three of the four oxygen atoms in each tetrahedron with neighboring tetrahedrons. Also known as layer silicate; sheet mineral; sheet silicate.

phyllosoma [INV ZOO] A flat, transparent, long-legged larval stage of various spiny lobsters.

phyllospondylous [VERT ZOO] Of vertebrae, having a hypocentrum but no pleurocentra; the neural arch extends ventrad to enclose the notochord and form transverse processes which articulate with the ribs.

Phyllosticales [MYCOL] An equivalent name for Sphaeropsidales.

Phyllostomatidae [VERT ZOO] The New World leaf-nosed bats (Chiroptera), a large tropical and subtropical family of insect- and fruit-eating forms with narrow, pointed ears.

phyllotaxy [BOT] The arrangement of leaves on a stem.

Phylloxerinae [INV ZOO] A subfamily of homopteran insects in the family Chermidae in which the sexual forms lack mouthparts, and the parthenogenetic females have a beak but the digestive system is closed, and no honeydew is produced.

phylogeny [EVOL] The evolutionary or ancestral history of organisms.

phylum [SYST] A major taxonomic category in classifying animals (and plants in some systems), composed of groups of related classes.

phyma [MED] **1.** A tumor of new growth of varying size, composed of any of the structures of the skin or subcutaneous tissue. **2.** A localized plastic exudate larger than a tubercle; a circumscribed swelling of the skin.

Phymatidae [INV ZOO] A family of carnivorous hemipteran insects characterized by strong, thick forelegs.

phymatosis [MED] Any disease characterized by the formation of phymas or nodules.

Phymosomatidae [INV ZOO] A family of echinacean echinoderms in the order Phymosomatoida with imperforate crenulate tubercles; one surviving genus is known.

Phymosomatoida [INV ZOO] An order of Echinacea with a stirodont lantern and diademoid ambulacral plates.

physa [INV ZOO] The rounded basal portion of the body of certain sea anemones.

Physalopteridae [INV ZOO] A family of parasitic nematodes in the superfamily Spiruroidea.

Physaraceae [MYCOL] A family of slime molds in the order Physarales.

Physarales [MYCOL] An order of Myxomycetes in the subclass Myxogastromycetidae.

physiatrics [MED] The science and practice of physical therapy.

physical anthropology [ANTHRO] The science that deals with the biological aspects of man and their relation to his historical or cultural aspects.

physical climate [CLIMATOL] The actual climate of a place, as distinguished from a hypothetical climate, such as the solar climate or mathematical climate.

physical climatology [CLIMATOL] The major branch of climatology, which deals with the explanation of climate, rather than with presentation of it (climatography).

physical compatibility [ENG] The ability of two or more materials, substances, or chemicals to be used together without ill effect.

physical constant [PHYS] A physical quantity which has a fixed and unchanging numerical value.

physical electronics [ELECTR] The study of physical phenomena basic to electronics, such as discharges, thermionic and field emission, and conduction in semiconductors and metals.

physical forecasting *See* numerical forecasting.

physical geography [GEOGR] The branch of geography which deals with the description, analysis, classification, and genetic interpretation of the natural features and phenomena of the earth's surface.

physical geology [GEOL] That branch of geology concerned with understanding the composition of the earth and the physical changes occurring in it, based on the study of rocks, minerals, and sediments, their structures and formations, and their processes of origin and alteration.

physical input/output control system *See* PIOCS.

physical law [PHYS] A property of a physical phenomenon, or a relationship between the various quantities or qualities which may be used to describe the phenomenon, that applies to all members of a broad class of such phenomena, without exception.

physical libration of the moon *See* lunar libration.

physical measurement [PHYS] Quantitative information on a physical condition, property, or relation, generally in the form of the ratio of the measured quantity to a standard quantity, or to some fixed multiple or fraction thereof.

physical medicine [MED] A consultative, diagnostic, and therapeutic medical specialty, coordinating and integrating the use of physical and occupational therapy and physical reconditioning in the professional management of the diseased and injured.

physical metallurgy [MET] The branch of metallurgy concerned with physical and mechanical properties of metals as affected by composition, mechanical working, and heat treatment.

physical meteorology [METEOROL] That branch of meteorol-

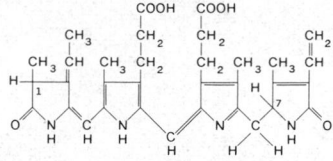

PHYCOCYANOBILIN

Structural formula of phycocyanobilin.

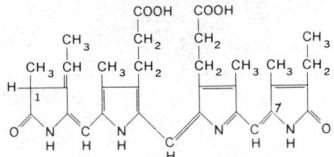

PHYCOERYTHROBILIN

Structural formula of phycoerythrobilin.

PHYLLOBRANCHIATE GILL

The phyllobranchiate gill of the mud shrimp (*Thalassina*). (*Smithsonian Institution*)

PHYLLOLEPIDA

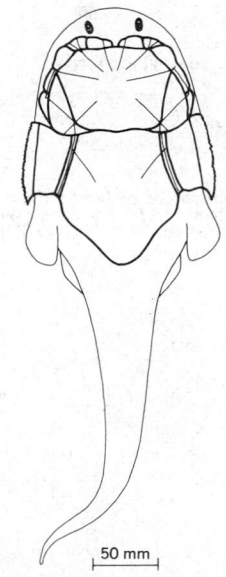

50 mm

Phyllolepis orvini from the Upper Devonian of Greenland. The restoration of the dermal armor is shown in dorsal aspect. The outline of the front of the head and the body is hypothetical, but shows how the fish might have appeared in life. (*After Erik Stensiö*)

ogy which deals with optical, electrical, acoustical, and thermodynamic phenomena of the atmosphere, its chemical composition, the laws of radiation, and the explanation of clouds and precipitation.

physical oceanography [OCEANOGR] The study of the physical aspects of the ocean, the movements of the sea, and the variability of these factors in relationship to the atmosphere and the ocean bottom.

physical optics [OPTICS] The study of the interaction of electromagnetic waves in the optical frequency range with material systems.

physical property [CHEM] Property of a compound that can change without involving a change in chemical composition; examples are the melting point and boiling point.

physical realizability [CONT SYS] For a transfer function, the possibility of constructing a network with this transfer function.

physical record [ADP] A set of adjacent data characters recorded on some storage medium, physically separated from other physical records that may be on the same medium by means of some indication that can be recognized by a simple hardware test. Also known as record block.

physical system *See* causal system.

physical testing [ENG] Determination of physical properties of materials based on observation and measurement.

physical theory [PHYS] An attempt to explain a certain class of physical phenomena by deducing them as necessary consequences of some primitive assumptions.

physical therapy [MED] The treatment of disease and injury by physical means.

physician [MED] An individual authorized to practice medicine.

physicist [PHYS] A person who does research in physics.

physics [SCI TECH] The study of those aspects of nature which can be understood in a fundamental way in terms of elementary principles and laws.

physiognomy [PSYCH] The prediction of personality functioning from facial appearances and expression.

physiographic province [GEOL] A region having a pattern of relief features or landforms that differs significantly from that of adjacent regions.

physiological biophysics [BIOPHYS] An area of biophysics concerned with the use of physical mechanisms to explain the behavior and the functioning of living organisms or parts thereof, and with the response of living organisms to physical forces.

physiological dead space *See* dead space.

physiological ecology [ECOL] The study of biological processes and growth under natural or simulated environments.

physiological genetics *See* phenogenetics.

physiological psychology [PSYCH] The study of the physiological mechanisms or correlates of behavior.

physiological saline *See* normal saline.

physiological salt solution *See* normal saline.

physiological sodium chloride solution *See* normal saline.

physiologic diplopia [PHYSIO] A normal phenomenon in which there is formation of images in noncorresponding retinal points, giving a perception of depth. Also known as introspective diplopia.

physiologic tremor [PHYSIO] A tremor in normal individuals, caused by fatigue, apprehension, or overexposure to cold.

physiology [BIOL] The study of the basic activities that occur in cells and tissues of living organisms by using physical and chemical methods.

Physopoda [INV ZOO] The equivalent name for Thysanoptera.

Physosomata [INV ZOO] A superfamily of amphipod crustaceans in the suborder Hyperiidea; the eyes are small or rarely absent, and the inner plates of the maxillipeds are free at the apex.

physostigmine [ORG CHEM] $C_{15}H_{21}O_2N_3$ An alkaloid; poisonous, colorless-to-pinkish crystals; soluble in alcohol and dilute acids; melts at 86°C; used as a source of salicylate and sulfate forms. Also known as calabarine; eserine.

physostigmine salicylate [ORG CHEM] $C_{15}H_{21}O_2N_3 \cdot C_7H_6O_3$ Poisonous, colorless-to-yellow crystals; soluble in water,

PHYTOSAURIA

Nicrosaurus, a quadrupedal thecodont. (After E. H. Colbert, Evolution of Vertebrates, Wiley, 1955)

alcohol, and chloroform; melts at 182°C; used for medicines.

physostigmine sulfate [ORG CHEM] $(C_{15}H_{21}O_2N_3)_2 \cdot H_2SO_4$ Poisonous, white crystals; soluble in water, alcohol, and chloroform; melts at 150°C; used for medicines.

Phytalmiidae [INV ZOO] A family of myodarian cyclorrhaphous dipteran insects in the subsection Acalypteratae.

Phytamastigophorea [INV ZOO] A class of the subphylum Sarcomastigophora, including green and colorless phytoflagellates.

phytase [BIOCHEM] An enzyme occurring in plants, especially cereals, which catalyzes hydrolysis of phytic acid to inositol and phosphoric acid.

phyteral [GEOL] Morphologically recognizable forms of vegetal matter in coal.

phytic acid [ORG CHEM] $C_6H_6[(OPO(OH)_2]_6$ An acid found in seeds of plants as the insoluble calcium magnesium salt (phytin); derived from corn steep liquor; inhibits calcium absorption in intestine; used to treat hard water, to remove iron and copper from wines, and to inactivate trace-metal contaminants in animal and vegetable oils.

phytochorology *See* plant geography.

phytochrome [BIOCHEM] A protein plant pigment which serves to direct the course of plant growth and development in response variously to the presence or absence of light, to photoperiod, and to light quality.

phytoclimatology [CLIMATOL] The study of the microclimate in the air space occupied by plant communities, on the surfaces of the plants themselves and, in some cases, in the air spaces within the plants.

phytogeography *See* plant geography.

phytohemagglutinin [BIOCHEM] A mucoprotein derivative of the stringbean (*Phaseolus vulgaris*); stimulates mammalian lymphocytes to mitosis.

phytohormone *See* plant hormone.

phytolith [PALEON] A fossilized part of a living plant that secreted mineral matter.

Phytomastigina [INV ZOO] The equivalent name for Phytamastigophorea.

phytometer [ENG] A device for measuring transpiration, consisting of a vessel containing soil in which one or more plants are rooted and sealed so that water can escape only by transpiration from the plant.

Phytomonadida [INV ZOO] The equivalent name for Volvocida.

phytopathogen [ECOL] An organism that causes a disease in a plant.

phytophagous [ZOO] Feeding on plants.

phytoplankton [ECOL] Planktonic plant life.

Phytosauria [PALEON] A suborder of Late Triassic long-snouted aquatic thecodonts resembling crocodiles but with posteriorly located external nostrils, absence of a secondary palate, and a different structure of the pelvic and pectoral girdles.

Phytoseiidae [INV ZOO] A family of the suborder Mesostigmata.

phytosterol [BIOCHEM] Any of various sterols obtained from plants, including ergosterol and stigmasterol.

phytotoxin [BIOCHEM] **1.** A substance toxic to plants. **2.** A toxin produced by plants.

pi [MATH] The irrational number which is the ratio of the circumference of any circle to its radius; an approximation is 3.14159. Symbolized π.

pia arachnoid [VERT ZOO] The outer meninx of certain submammalian forms having two membranes covering the brain and spinal cord.

Piacention *See* Plaisancian.

pia mater [ANAT] The vascular membrane covering the surface of the brain and spinal cord.

piano wire [MET] High-tensile-strength, 0.75 to 0.85% carbon steel wire cold-drawn to uniform thickness.

piastrenemia *See* thrombocytosis.

pi attenuator [ELEC] An attenuator consisting of a pi network whose impedances are all resistances.

pibal *See* pilot-balloon observation.

pi bonding [PHYS CHEM] Covalent bonding in which the greatest overlap between atomic orbitals is along a plane perpendicular to the line joining the nuclei of the two atoms.

Picard method [MATH] A method of successive substitution for solving differential equations.

Picard's theorem [MATH] A nonconstant entire function of the complex plane assumes every value save at most one.

Picatinny test [ENG] An impact test used in the United States for evaluating the sensitivity of high explosives; a small sample of the explosive is placed in a depression in a steel die cup and capped by a thin brass cover, a cylindrical steel plug is placed in the center of the cover, and a 2-kilogram weight is dropped from varying heights on the plug; the reported sensitivity figure is the minimum height, in inches, at which at least 1 firing results from 10 trials.

Piche evaporimeter [ENG] A porous-paper-wick atmometer.

Picidae [VERT ZOO] The woodpeckers, a large family of birds in the order Piciformes; adaptive modifications include a long tongue and hyoid mechanism, and stiffened tail feathers.

Piciformes [VERT ZOO] An order of birds characterized by the peculiar arrangement of the tendons of the toes.

Picinae [VERT ZOO] The true woodpeckers, a subfamily of the Picidae.

pick [ADP] To select the next card from an input stack for feeding into a card machine. [DES ENG] **1.** The steel cutting points used on a coal-cutter chain. **2.** A miner's steel or iron digging tool with sharp points at each end. [ENG] **1.** To dress the sides of a shaft or other excavation. **2.** To remove shale, dirt, and such from coal. [TEXT] *See* filling.

pick-a-back conveyor [MIN ENG] A short conveyor that advances with a loader or continuous miner at the face of a mine and loads coal on the main haulage system.

Pickard core barrel [MIN ENG] A type of double-tube core barrel; the distinguishing feature of the barrel is that when blocked the inner barrel slides upward into the head, closing the water ports and stopping the flow of the circulating liquid, without irreparably damaging the bit until the barrel is pulled and the blocked inner tube cleared.

pickax [DES ENG] A pointed steel or iron tool mounted on a wooden handle and used for breaking earth and stone.

picker [MIN ENG] **1.** An employee who picks or discards slate and other foreign matter from the coal in an anthracite breaker or at a picking table. **2.** A mechanical arrangement for removing slate from coal. [TEXT] A machine used to pull apart and separate cotton fibers.

pickeringite [MINERAL] $MgAl_2(SO_4)_4 \cdot 22H_2O$ A white or faintly colored mineral composed of hydrous sulfate of magnesium and aluminum, occurring in fibrous masses.

picker knives [ADP] The narrow edges of a moving slide which will pick the bottom card from a stack and feed it to a card reader.

picket ship [NAV ARCH] A radar-equipped ship, generally anchored, that is used to extend radar early-warning coverage seaward.

pick glass [TEXT] Calibrated magnifying glass used to count warp and filling threads in a square inch of fabric.

pick hammer [DES ENG] A hammer with a point at one end of the head and a blunt surface at the other end.

picking [MIN ENG] **1.** Removal of waste material from an ore. **2.** Extraction of the lightest-grade ore from a mine. **3.** Emission of particles from the roof of a mine on the verge of collapse.

picking conveyor [MIN ENG] A continuous belt or apron conveyor used to carry a relatively thin bed of material past pickers who hand-sort or pick the material being conveyed.

picking table [MIN ENG] A flat or slightly-inclined platform on which the coal or ore is run to be picked free from slate or gangue.

pick lacing [DES ENG] The pattern to which the picks are set in a cutter chain.

pickle liquor [MET] A spent pickling solution.

pickle patch [MET] A coating of oxide or scale that remains adherent after pickling.

pickle stain [MET] Discoloration of a metal surface due to chemical cleaning but without adequate washing and drying.

pickling [MET] Preferential removal of oxide or mill scale from the surface of a metal by immersion usually in an acidic or alkaline solution.

pickling acid [CHEM] Any of the acids used in pickling solutions, such as hydrochloric, sulfuric, nitric, phosphoric, or hydrofluoric acid.

pick miner [MIN ENG] **1.** In anthracite and bituminous coal mining, one who uses hand tools to extract coal in underground working places. **2.** One who cuts out a channel under the bottom of the working face of coal with a pick.

pickoff [ELECTR] A device used to convert mechanical motion into a proportional electric signal. [MECH ENG] A mechanical device for automatic removal of the finished part from a press die.

Pick's disease [MED] **1.** A form of presenile dementia characterized by severe atrophy of the frontal and temporal lobes of the cerebrum. **2.** A recurrent or progressive form of ascites with little or no edema. **3.** *See* constrictive pericarditis. **4.** *See* Niemann-Pick disease. **5.** *See* polyserositis.

pickup [AERO ENG] A potentiometer used in an automatic pilot to detect the motion of the airplane around the gyro and initiate corrective adjustments. [ELEC] **1.** A device that converts a sound, scene, measurable quantity, or other form of intelligence into corresponding electric signals, as in a microphone, phonograph pickup, or television camera. **2.** The minimum current, voltage, power, or other value at which a relay will complete its intended function. **3.** Interference from a nearby circuit or system. [MET] Transfer of metal from the work to the tool, or from the tool to the work, during a forming operation. [NUC PHYS] A type of nuclear reaction in which the incident particle takes a nucleon from the target nucleus and proceeds with this nucleon bound to itself.

pickup tube *See* camera tube.

pickup voltage [ELEC] Of a magnetically operated device, the voltage at which the device starts to operate.

pico- [MATH] A prefix meaning 10^{-12}; used with metric units. Also known as micromicro-.

picoammeter [ENG] An ammeter whose scale is calibrated to indicate current values in picoamperes.

picoampere [ELEC] A unit of current equal to 10^{-12} ampere, or one-millionth of a microampere. Abbreviated pA.

picodnavirus [VIROL] A group of deoxyribonucleic acid-containing animal viruses including the adeno-satellite viruses.

picofarad [ELEC] A unit of capacitance equal to 10^{-12} farad, or one-millionth of a microfarad. Also known as micromicrofarad (deprecated usage); puff (British usage). Abbreviated pF.

picoline [ORG CHEM] $C_5H_4N(CH_3)$ Family of colorless liquid isomers, soluble in water and alcohol; the alpha form, boiling at 129°C, is used as a solvent and chemical intermediate, and is also known as 2-methyl pyridine; the beta form, boiling at 143.5°C, is used as a solvent for chemical synthesis reactions, to make nicotinic acid, and in fabric waterproofing, and is also known as 3-methyl pyridine; the gamma form, boiling at 143.1°C, is used as a solvent for chemical synthesis reactions and in fabric waterproofing.

picolinic acid [ORG CHEM] $C_{10}H_8N_4O_5$ An alcohol-soluble crystalline compound, forming yellow leaflets that melt at 116-117°C; used as a reagent in phenylalanine, tryptophan, and alkaloids production, and for the quantitative detection of calcium. Also known as 3-methyl-4-nitro-1-(p-nitrophenyl)-2-pyrazoline-5-one.

picornavirus [VIROL] A viral group made up of small (18-30 nanometers), ether-sensitive viruses that lack an envelope and have a ribonucleic acid genome; among subgroups included are enteroviruses and rhinoviruses, both of human origin.

picosecond [MECH] A unit of time equal to 10^{-12} second, or one-millionth of a microsecond. Abbreviated ps; psec. Formerly known as micromicrosecond.

picotite [MINERAL] A dark-brown variety of hercynite that contains chromium and is commonly found in dunites. Also known as chrome spinel.

picowatt [MECH] A unit of power equal to 10^{-12} watt, or one-millionth of a microwatt. Abbreviated pW. Formerly known as micromicrowatt.

picramic acid [ORG CHEM] $C_6H_5N_3O_5$ A crystalline acid, forming dark red needles from alcohol solutions, melting at 169-170°C; used in dye manufacture and as a reagent in tests for albumin. Also known as 2-amino-4,6-dinitrophenol; 4,6-dinitro-2-aminophenol.

picratol [MATER] A binary explosive composed of 52% ammonium picrate and 48% TNT (trinitrotoluene); it can be melt-loaded; less sensitive than TNT, it was developed for use in armor-piercing bombs.

picric acid [ORG CHEM] $C_6H_2(NO_2)_3OH$ Poisonous, explosive, highly oxidative yellow crystals with bitter taste; soluble in water, alcohol, chloroform, benzene, and ether; melts at 122°C; used in explosives, in external medicines; to make dyes, matches, and batteries, and to etch copper. Also known as carbazotic acid; nitroxanthic acid; picronitric acid; trinitrophenol.

picrite [PETR] A medium- to fine-grained igneous rock composed chiefly of olivine, with smaller amounts of pyroxene, hornblende, and plagioclase felspar.

Picrodendraceae [BOT] A small family of dicotyledonous plants in the order Juglandales characterized by unisexual flowers borne in catkins, four apical ovules in a superior ovary, and trifoliate leaves.

picrolite *See* antigorite.

picromerite [MINERAL] $K_2Mg(SO_4)_2 \cdot 6H_2O$ A white mineral composed of hydrous sulfate of magnesium and potassium, occurring as crystalline incrustations.

picronitric acid *See* picric acid.

picropharmacolite [MINERAL] $(Ca,Mg)_3(AsO_4)_2 \cdot 6H_2O$ Mineral composed of hydrous calcium magnesium arsenate.

picrotoxin [BIOCHEM] $C_{30}H_{34}O_{13}$ A poisonous, crystalline plant alkaloid found primarily in *Cocculus indicus*; used as a stimulant and convulsant drug.

Pictet's liquid [MATER] Liquid mixture of carbon dioxide and sulfur dioxide; used to produce low temperatures.

Pictor [ASTRON] A southern constellation; right ascension 6 hours, declination 55°S. Also known as Easel.

picture [ADP] In COBOL, a symbolic description of each data element or item according to specified rules concerning numerals, alphanumerics, location of decimal points, and length. [COMMUN] The image on the screen of a television receiver. [GRAPHICS] The image produced by a photographic process.

picture black *See* black signal.

picture carrier [COMMUN] A carrier frequency located 1.25 megahertz above the lower frequency limit of a standard National Television Systems Committee television signal; in color television, it is used for transmitting color information. Also known as luminance carrier.

picture element [ELECTR] 1. That portion, in facsimile, of the subject copy which is seen by the scanner at any instant; it can be considered a square area having dimensions equal to the width of the scanning line. 2. In television, any segment of a scanning line, the dimension of which along the line is exactly equal to the nominal line width; the area which is being explored at any instant in the scanning process. Also known as critical area; elemental area; recording spot; scanning spot.

picture frequency [COMMUN] A frequency that results solely from scanning of subject copy in a facsimile system. [ELECTR] *See* frame frequency.

picture signal [COMMUN] The signal resulting from the scanning process in television or facsimile.

picture synchronizing pulse *See* vertical synchronizing pulse.

picture transmission [COMMUN] Electric transmission of a picture having a gradation of shade values.

picture transmitter *See* visual transmitter.

picture tube [ELECTR] A cathode-ray tube used in television receivers to produce an image by varying the electron-beam intensity as the beam is deflected from side to side and up and down to scan a raster on the fluorescent screen at the large end of the tube. Also known as kinescope; television picture tube.

picture-tube brightener [ELECTR] A small step-up transformer that can be inserted between the socket and base of a picture tube to increase the heater voltage and thereby increase picture brightness to compensate for normal aging of the tubes.

picture white *See* white signal.

picture window [BUILD] A large window framing an exterior view.

Picumninae [VERT ZOO] The piculets, a subfamily of the avian family Picidae.

Pidgeon process [MET] A method for producing magnesium from calcined dolomite by reduction with ferrosilicon. Also known as ferrosilicon process; silicothermic process.

piece [ORD] An artillery weapon, a machine gun, a rifle, or any firearm.

piece mark [ENG] Identification number for an individual part, subassembly, or assembly; shown on the drawing, but not necessarily on the part.

piece rate [IND ENG] Wages paid per unit of production.

piece-root grafting [BOT] Grafting in which each piece of a cut seedling root is used as a stock.

piecewise-linear [MATH] A continuous curve or function obtained by joining a finite number of linear pieces.

piecewise-linear system [CONT SYS] A system for which one can divide the range of values of input quantities into a finite number of intervals such that the output quantity is a linear function of the input quantity within each of these intervals.

piecewise linear topology *See* combinatorial topology.

piecework [IND ENG] Work paid for in accordance with the amount done rather than the hours taken.

piedmont [GEOL] Lying or formed at the base of a mountain or mountain range, as a piedmont terrace or a piedmont pediment.

piedmont alluvial plain *See* bajada.

piedmont bench *See* piedmont step.

piedmont benchland [GEOL] One of several successions or systems of piedmont steps. Also known as piedmont stairway; piedmont treppe.

piedmont bulb [HYD] The lobe or fan of ice formed when a glacier spreads out on a plain at the lower end of a valley.

piedmont flat *See* piedmont step.

piedmont glacier [HYD] A thick, continuous ice sheet formed at the base of a mountain range by the spreading out and coalescing of valley glaciers from higher mountain elevations.

piedmont ice [HYD] An ice sheet formed by the joining of two or more glaciers on a comparatively level plain at the base of the mountains down which the glaciers descended; it may be partly afloat.

piedmont interstream flat *See* pediment.

piedmontite *See* piemontite.

piedmont plain *See* bajada.

piedmont scarp [GEOL] A small, low cliff formed in alluvium on a piedmont slope at the foot of a steep mountain range; due to dislocation of the surface, especially by faulting. Also known as scarplet.

piedmont slope *See* bajada.

piedmont stairway *See* piedmont benchland.

piedmont step [GEOL] A terracelike or benchlike piedmont feature that slopes outward or downvalley. Also known as piedmont bench; piedmont flat.

piedmont treppe *See* piedmont benchland.

pi electron [PHYS CHEM] An electron which participates in pi bonding.

piemontite [MINERAL] $Ca_2(Al,Mn^{3+},Fe)_3Si_3O_{12}(OH)$ Reddish-brown epidote mineral that contains manganese. Also known as manganese epidote; piedmontite.

pier [CIV ENG] 1. A vertical, rectangular or circular support for concentrated loads from an arch or bridge superstructure. 2. A structure with a platform projecting from the shore into navigable waters for mooring vessels.

piercement *See* diapir.

piercement dome *See* diapir.

Pierce oscillator [ELECTR] Oscillator in which a piezoelectric crystal unit is connected between the grid and the plate of an electron tube, in what is basically a Colpitts oscillator, with voltage division provided by the grid-cathode and plate-cathode capacitances of the circuit.

Pierce's disease [PL PATH] A virus disease of grapes in which there is mottling between the veins of leaves, early

PICTURE TUBE

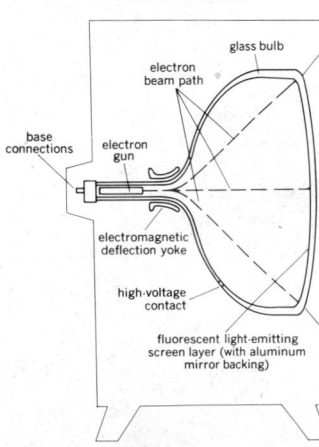

glass bulb
electron beam path
base connections
electron gun
electromagnetic deflection yoke
high-voltage contact
fluorescent light-emitting screen layer (with aluminum mirror backing)

Schematic drawing of a black-and-white picture tube.

defoliation, and early ripening and withering of the fruit.

piercing *See* fusion piercing.

piercing fold *See* diapir.

pier foundation *See* caisson foundation.

pierhead line [CIV ENG] The line in navigable waters beyond which construction is prohibited; open-pier construction may extend outward from the bulkhead line to the pierhead line.

Pieridae [INV ZOO] A family of lepidopteran insects in the superfamily Papilionoidea including white, sulfur, and orange-tip butterflies; characterized by the lack of a prespiracular bar at the base of the abdomen.

Piesmatidae [INV ZOO] The ash-gray leaf bugs, a family of hemipteran insects belonging to the Pentatomorpha.

Piesmidae [INV ZOO] A small family of hemipteran insects in the superfamily Lygaeoidea.

pièze [MECH] A unit of pressure equal to 1 sthène per square meter, or to 1000 pascals. Abbreviated pz.

piezochemistry [CHEM] The field of chemical reactions under high pressures.

piezoelectric [SOLID STATE] Having the ability to generate a voltage when mechanical force is applied, or to produce a mechanical force when a voltage is applied, as in a piezoelectric crystal.

piezoelectric crystal [SOLID STATE] A crystal which exhibits the piezoelectric effect; used in crystal loudspeakers, crystal microphones, and crystal cartridges for phono pickups.

piezoelectric effect [SOLID STATE] **1.** The generation of electric polarization in certain dielectric crystals as a result of the application of mechanical stress. **2.** The reverse effect, in which application of a voltage between certain faces of the crystal produces a mechanical distortion of the material.

piezoelectric element [ELECTR] A piezoelectric crystal used in an electric circuit, for example, as a transducer to convert mechanical or acoustical signals to electric signals, or to control the frequency of a crystal oscillator.

piezoelectric gage [ENG] A pressure-measuring gage that uses a piezoelectric material to develop a voltage when subjected to pressure; used for measuring blast pressures resulting from explosions and pressures developed in guns.

piezoelectric hysteresis [SOLID STATE] Behavior of a piezoelectric crystal whose electric polarization depends not only on the mechanical stress to which the crystal is subjected, but also on the previous history of this stress.

piezoelectricity [SOLID STATE] Electricity or electric polarization resulting from the piezoelectric effect.

piezoelectric loudspeaker *See* crystal loudspeaker.

piezoelectric microphone *See* crystal microphone.

piezoelectric oscillator *See* crystal oscillator.

piezoelectric pickup *See* crystal pickup.

piezoelectric resonator *See* crystal resonator.

piezoelectric semiconductor [SOLID STATE] A semiconductor exhibiting the piezoelectric effect, such as quartz, Rochelle salt, and barium titanate.

piezoelectric transducer [ELECTR] A piezoelectric crystal used as a transducer, either to convert mechanical or acoustical signals to electric signals, as in a microphone, or vice versa, as in ultrasonic metal inspection.

piezoelectric vibrator [SOLID STATE] An element cut from piezoelectric material, usually in the form of a plate, bar, or ring, with electrodes attached to or supported near the element to excite one of its resonant frequencies.

piezometer opening *See* pressure tap.

piezometric surface *See* potentiometric surface.

piezotropic [FL MECH] Characterized by piezotropy.

piezotropy [FL MECH] The property of a fluid in which processes are characterized by a functional dependence of the thermodynamic functions of state: $d\rho/dt = b(dp/dt)$, where ρ is the density, p the pressure, and b a function of the thermodynamic variables, called the coefficient of piezotropy.

pig [ELECTR] **1.** An ion source based on the same principle as the Philips ionization gage. **2.** *See* Philips ionization gage. [ENG] In-line scraper (brush, blade cutter, or swab) forced through pipelines by fluid pressure; used to remove scale, sand, water, and other foreign matter from the interior surfaces of the pipe. [MET] A crude metal casting prepared for storage, transportation, or remelting. [NUCLEO] A heavily shielded container, usually lead, used to ship or store

radioisotopes and other radioactive materials. [VERT ZOO] Any wild or domestic mammal of the superfamily Suoidea in the order Artiodactyla; toes terminate in nails which are modified into hooves, the tail is short, and the body is covered sparsely with hair which is frequently bristlelike.

pigeon [VERT ZOO] Any of various stout-bodied birds in the family Columbidae having short legs, a bill with a horny tip, and a soft cere.

pigeonite [MINERAL] $(Mg,Fe^{2+},Ca)(MgFe^{2+})Si_2O_6$ Clinopyroxene mineral species intermediate in composition between clinoenstatite and diopside, found in basic igneous rocks.

pigeon milk [PHYSIO] A milky glandular secretion of the crop of pigeons that is regurgitated to feed newly hatched young.

piggyback twistor [ELECTR] Electrically alterable nondestructive-readout storage device that uses a thin narrow tape of magnetic material wound spirally around a fine copper conductor to store information; another similar tape is wrapped on top of the first, piggyback fashion, to sense the stored information; a binary digit or bit is stored at the intersection of a copper strap and a pair of these twistor wires.

pig iron [MET] **1.** Crude, high-carbon iron produced by reduction of iron ore in a blast furnace. **2.** Cast iron in the form of pigs.

pigment [BIOCHEM] Any coloring matter in plant or animal cells. [MATER] A solid that reflects light of certain wavelengths while absorbing light of other wavelengths, without producing appreciable luminescence; used to impart color to other materials.

pigmentation [PHYSIO] The normal color of the body and its organs, resulting from a summation of the natural color of the tissue, the pigments deposited therein, and the pigments carried through the blood bathing the tissue.

pigment cell [CYTOL] Any cell containing deposits of pigment.

pigtail [ELEC] A short, flexible wire, usually stranded or braided, used between a stationary terminal and a terminal having a limited range of motion, as in relay armatures.

pigtail splice [ELEC] A splice made by twisting together the bared ends of parallel conductors.

pika [VERT ZOO] Any member of the family Ochotonidae, which includes 14 species of lagomorphs resembling rabbits but having a vestigial tail and short, rounded ears.

pike [VERT ZOO] Any of about five species of predatory fish which compose the family Esocidae in the order Clupeiformes; the body is cylindrical and compressed, with cycloid scales that have deeply scalloped edges.

Pilacraceae [MYCOL] A family of Basidiomycetes.

Pilargidae [INV ZOO] A family of small, short, depressed errantian polychaete annelids.

pilaster [CIV ENG] A vertical rectangular architectural member that is structurally a pier and architecturally a column.

pilchard oil [MATER] Pale-yellow oil expressed from pickled pilchards (of the herring family); used to make paints and potash soft soap.

pile [ENG] A long, heavy timber, steel, or reinforced concrete post that has been driven, jacked, jetted, or cast vertically into the ground to support a load. [NUCLEO] *See* nuclear reactor. [TEXT] Loops on a fabric surface.

pile beacon [NAV] In marine operation, a beacon formed of one or more piles.

pile bent [CIV ENG] A row of timber or concrete bearing piles with a pile cap forming that part of a trestle which carries the adjacent ends of timber stringers or concrete slabs.

pile cap [CIV ENG] A mass of reinforced concrete cast around the head of a group of piles to ensure that they act as a unit to support the imposed load.

pile dike [CIV ENG] A dike consisting of a group of piles braced and lashed together along a riverbank.

pile dolphin [NAV] A minor light structure consisting of a number of piles driven into the offshore bottom in a circular pattern and drawn together with a light mounted at the top.

pile driver [MECH ENG] A hoist and movable steel frame equipped to handle piles and drive them into the ground.

pile extractor [MECH ENG] **1.** A pile hammer which strikes the pile upward so as to loosen its grip and remove it from the

PIG

The Yorkshire pig, a domestic breed with a high-quality carcass, raised for its excellent bacon.

PIKA

A pika, a rock-dwelling diurnal mammal which grows to the size of a guinea pig.

PIKE

Esox lucius. One of two species of fish community called a pike.

ground. **2.** A vibratory hammer which loosens the pile by high-frequency jarring.

pile formula [MECH] An equation for the forces acting on a pile at equilibrium: $P = pA + tS + Sn \sin \phi$, where P is the load, A is the area of the pile point, p is the force per unit area on the point, S is the embedded surface of the pile, t is the force per unit area parallel to S, n is the force per unit area normal to S, and ϕ is the taper angle of the pile.

pile foundation [CIV ENG] A substructure supported on piles.

pile hammer [MECH ENG] The heavy weight of a pile driver that depends on gravity for its striking power and is used to drive piles into the ground. Also known as drop hammer.

pile lighthouse [NAV] A lighthouse built on piles.

pile shoe [CIV ENG] A cast-iron point on the foot of a timber or concrete driven pile to facilitate penetration of the ground.

pileum [VERT ZOO] The top of a bird's head, from the nape to the bill.

pileup [ELECTR] A set of moving and fixed contacts, insulated from each other, formed as a unit for incorporation in a relay or switch. Also known as stack.

pileus [BIOL] The umbrella-shaped upper cap of mushrooms and other basidiomycetous fungi. [METEOROL] An accessory cloud of small horizontal extent, often cirriform, in the form of a cap, hood, or scarf, which occurs above or attached to the top of a cumuliform cloud that often pierces it; several pileus clouds fairly often are observed above each other. Also known as scarf cloud.

pile weave [TEXT] Type of weave made by using two warp yarns and one filling yarn, or one warp yarn and two filling yarns.

Pilger tube-reducing process [MET] A tube-reducing process in which pierced billets are forced over a mandrel between two rolls with inclined axes and given a rotary forging treatment; the tube is advanced during the gap in each revolution.

Pilidae [INV ZOO] A family of fresh-water snails in the order Pectinibranchia.

Pilifera [VERT ZOO] Collective designation for animals with hair, that is, mammals.

pill [ELECTROMAG] A microwave stripline termination. [PHARM] A small, solid dosage form of a globular, ovoid, or lenticular shape, containing one or more medicinal substances.

pillar [MIN ENG] An area of coal or ore left to support the overlying strata or hanging wall in a mine.

pillar-and-breast system [MIN ENG] A system of coal mining in which the working places are rectangular rooms usually five or ten times as long as they are broad, opened on the upper side of the gangway.

pillar-and-room system [MIN ENG] A system of mining whereby solid blocks of coal are left on either side of working places to support the roof until first-mining has been completed, when the pillar coal is then recovered.

pillar-and-stall system [MIN ENG] A system of working coal and other minerals where the first stage of excavation is accomplished with the roof sustained by coal or ore.

pillar bolt [DES ENG] A bolt projecting from a part so as to support it.

pillar buoy [NAV] A buoy composed of a tall central structure mounted on a broad flat base; not used in United States waters.

pillar burst [MIN ENG] A failure of a pillar, by crushing.

pillar crane [MECH ENG] A crane whose mechanism can be rotated about a fixed pillar.

pillar drive [MIN ENG] A wide irregular drift or entry, in firm dry ground, in which the roof is supported by pillars of the natural earth, or by artificial pillars of stone, no timber being used.

pillar extraction [MIN ENG] Removal of the pillars of coal left over from mining by the pillar-and-stall method. Also known as pillar mining.

pillaring [MIN ENG] The process of extracting pillars. Also called pillar robbing; pulling pillars; robbing pillars. [ORD] The rapid vertical movement of smoke which sometimes results, for instance, from the explosion of a white phosphorus bomb or projectile; the effect is undesirable because it does not produce obscuration over a desirably large area.

pillar ladder [NAV ARCH] A ladder formed by fitting rungs extending out from a pillar or stanchion; commonly used as a means of providing access to cargo holds.

pillar line [MIN ENG] Air currents which have definitely coursed through an inaccessible abandoned panel or area or which have ventilated a pillar line or a pillar area, regardless of the methane content, or absence of methane, in such air.

pillar man *See* pack builder.

pillar mining *See* pillar extraction.

pillar press [MECH ENG] A punch press framed by two upright columns; the driving shaft passes through the columns, and the slide operates between them.

pillar robbing *See* pillaring.

pillar split [MIN ENG] An opening or crosscut driven through a pillar in the course of extraction of ore.

pillbox [ORD] Small, low fortification that houses machine guns or antitank weapons; usually made of concrete, steel, or sandbags.

pillbox antenna [ELECTROMAG] Cylindrical parabolic reflector enclosed by two plates perpendicular to the cylinder, spaced to permit the propagation of only one mode in the desired direction of polarization.

pillow lava [GEOL] Any lava characterized by pillow structure and presumed to have formed in a subaqueous environment. Also known as ellipsoidal lava.

pillow structure [GEOL] A primary sedimentary structure that resembles a pillow in size and shape. Also known as mammillary structure. [PETR] A pillow-shaped structure visible in some extrusive lavas attributed to the congealment of lava under water.

pilmer [METEOROL] In England, a heavy shower of rain.

pilocarpine [ORG CHEM] $C_{11}H_{16}N_2O_2$ An alkaloid, in either oil or crystal form, melting at 34°C; soluble in chloroform, water, and alcohol; used in medicine.

pilomotor nerve [ANAT] A nerve causing contraction of one of the arrectoris pilorum muscles.

pilomotor reflex [PHYSIO] Erection of the hairs of the skin (gooseflesh) in response to chilling or irritation of the skin or to an emotional stimulus.

pilosebaceous [ANAT] Pertaining to the hair follicles and sebaceous glands, as the pilosebaceous apparatus, comprising the hair follicle and its attached gland.

pilosis [MED] The abnormal or excessive development of hair.

pilot [AERO ENG] **1.** A person who handles the controls of an aircraft or spacecraft from within the craft, and guides or controls the craft in flight. **2.** A mechanical system designed to exercise control functions in an aircraft or spacecraft. [COMMUN] **1.** In a transmission system, a signal wave, usually single frequency, transmitted over the system to indicate or control its characteristics. **2.** Instructions, in tape relay, appearing in routing line, relative to the transmission or handling of that message. [MECH ENG] A cylindrical steel bar extending through, and about 8 inches beyond the face of, a reaming bit; it acts as a guide that follows the original unreamed part of the borehole and hence forces the reaming bit to follow, and be concentric with, the smaller-diameter, unreamed portion of the original borehole. [NAV] **1.** A person who directs the movements of a vessel through pilot waters, usually a person who has demonstrated extensive knowledge of channels, aids to navigation, dangers to navigation, and so on, in a particular area and is licensed for that area. **2.** A book of sailing directions; for waters of the United States and its possessions, the books are prepared by the U.S. Coast and Geodetic Survey, and are called coast pilots. **3.** The person who flies aircraft.

pilotage [NAV] **1.** A marine pilot's fee. **2.** The process of directing a vehicle by reference to recognizable landmarks or soundings; these observations may be made by optical, aural, mechanical, or electronic means.

pilotage waters *See* pilot waters.

pilot balloon [ENG] A small balloon whose ascent is followed by a theodolite in order to obtain data for the computation of the speed and direction of winds in the upper air.

pilot-balloon observation [METEOROL] A method of winds-aloft observation, that is, the determination of wind speeds and directions in the atmosphere above a station; involves reading the elevation and azimuth angles of a theodolite while visually tracking a pilot balloon. Also known as pibal.

pilot bit [DES ENG] A noncoring bit with a cylindrical diamond-set plug of somewhat smaller diameter than the bit proper, set in the center and projecting beyond the main face of the bit.

pilot boat [NAV] A small vessel used by a pilot to go to or from a vessel employing his services. Also known as pilot vessel.

pilot briefing [METEOROL] Oral comment on the observed and forecast weather conditions along a route, given by a forecaster to the pilot, navigator, or other air crew member prior to takeoff. Also known as briefing; flight briefing; flight-weather briefing.

pilot cell [ELEC] Selected cell of a storage battery whose temperature, voltage, and specific gravity are assumed to indicate the condition of the entire battery.

pilot chart [NAV] A chart of a major ocean area published, for the benefit of mariners, by the U.S. Naval Oceanographic Office in cooperation with the U.S. Weather Bureau; these charts contain information required in planning safe routes, including ocean currents, ice at sea, wind roses, storm tracks, isotherms, magnetic variation, and recommended routes or steamer tracks.

pilot chute [AERO ENG] A small parachute canopy attached to a larger canopy to actuate and accelerate the opening of the load-bearing canopy.

pilot drill [MECH ENG] A small drill to start a hole to ensure that a larger drill will run true to center.

pilot flood [PETRO ENG] A test waterflood operation (water-injection well) designed to evaluate procedures and to give advance information prior to instituting an extensive, multi-point waterflood.

pilot hole [ENG] A small hole drilled ahead of a larger borehole.

pilothouse [NAV ARCH] A compartment on or near the bridge of a ship that contains the steering wheel and other controls, compass, charts, navigating equipment, and means of communicating with the engine room and other parts of the ship. Also known as wheelhouse.

pilot lamp [ELEC] A small lamp used to indicate that a circuit is energized. Also known as pilot light.

pilotless aircraft [AERO ENG] An aircraft adapted to control by or through a preset self-reacting unit or a radio-controlled unit, without the benefit of a human pilot.

pilot light [ELEC] See pilot lamp. [ENG] A small, constantly burning flame used to ignite a gas burner.

pilot lightship [NAV] A lightship which also serves as a pilot station.

pilot line operation [IND ENG] Minimum production of an item in order to preserve or develop the art of its production.

pilot materials [IND ENG] A minimum quantity of special materials, partially finished components, forgings, and castings, identified with specific production equipment and processes and required for the purpose of proofing, tooling, and testing manufacturing processes to facilitate later reactivation.

pilot model [IND ENG] An early production model of a product used to debug the manufacturing process.

pilot motor [ELEC] A small motor used in the automatic control of an electric current.

pilot plant [IND ENG] A small version of a planned industrial plant, built to gain experience in operating the final plant.

pilot production [PETRO ENG] Limited or test production of oil or gas from a field to determine reservoir and product characteristics before commencing full-scale recovery operations.

pilot reaming bit See reaming bit.

pilot relaying [ELEC] A system for protecting transmission consisting of protective relays at line terminals and a communication channel between relays which is used by the relays to determine if a fault is within the protected line section, in which case all terminals are tripped simultaneously at high speed, or outside it, in which case tripping is blocked.

pilot report [METEOROL] A report of in-flight weather by an aircraft pilot or crew member; a complete pilot report includes the following information in this order: location or extent of reported weather phenomena, time of observation, description of phenomena, altitude of phenomena, type of aircraft (only with reports of turbulence or icing). Also known as aircraft report; pirep.

pilot rules [NAV] Regulations supplementing the Inland Rules of the Road.

pilot-scale chemical reaction [CHEM ENG] Small-scale chemical reaction used to test operating conditions and product yields; used as a pilot for design of large-scale reaction systems.

pilot station [NAV] At seaports, the office or headquarters of marine pilots; the place where the services of a pilot may be obtained. [NAV ARCH] Position on the bridge of a ship where the pilot stands to steer or to give directions for steering a ship into and out of a harbor.

pilot streamer [GEOPHYS] A relatively slow-moving, nonluminous lightning streamer, the existence of which has been postulated to help account for the observed mode of advance of a stepped leader as it initiates a lightning discharge.

pilot tone [COMMUN] Single frequency transmitted over a narrow channel to operate an alarm or automatic control.

pilot tunnel [ENG] A small tunnel or shaft excavated in advance of the main drivage in mining and tunnel building to gain information about the ground, create a free face, and thus simplify the blasting operations.

pilot vessel See pilot boat.

pilot waters [NAV] 1. Areas in which the services of a marine pilot are essential. 2. Waters in which navigation is by piloting. Also known as pilotage waters.

pilot wire regulator [CONT SYS] Automatic device for controlling adjustable gains or losses associated with transmission circuits to compensate for transmission changes caused by temperature variations, the control usually depending upon the resistance of a conductor or pilot wire having substantially the same temperature conditions as the conductors of the circuits being regulated.

Piltdown man [PALEON] An alleged fossil man based on fragments of a skull and mandible that were eventually discovered to constitute a skillful hoax.

pilus [ANAT] A hair. [BIOL] A fine, slender, hairlike body. [MICROBIO] Any filamentous appendage other than flagella on certain gram-negative bacteria. Also known as fimbria.

pimaricin [ORG CHEM] $C_{33}H_{47}NO_{13}$ A compound crystallizing from a methanol-water solution, decomposing at about 200°C; soluble in water and organic solvents; used in medicine as an antifungal agent for *Candida albicans* vaginitis. Also known as tennecetin.

pimelic acid [ORG CHEM] $HOOC(CH_2)_5COOH$ Crystals melting at 105°C; slightly soluble in water, soluble in alcohol and ether; used in biochemical research. Also known as heptanedioic acid.

pimelinketone See cyclohexanone.

pimenta oil [MATER] A yellow to brownish essential oil with spicy aroma and pungent taste; derived from allspice (*Pimenta officinalis*); main components are eugenol, cineol, and phellandrene; used in medicine and flavors. Also known as allspice oil; pimento oil.

pimento [BOT] *Capsicum annuum*. A type of pepper in the order Polemoniales grown for its thick, sweet-fleshed red fruit.

pimento oil See pimenta oil.

pi meson [PARTIC PHYS] 1. Collective name for three semistable mesons which have charges of $+1$, 0, and -1 times the proton charge, and form a charge multiplet, with an approximate mass of 138 MeV (million electron volts), spin 0, negative parity, negative G parity, and positive charge parity (for the neutral meson). Also known as pion. Symbolized π. 2. Any meson belonging to an isospin triplet with hypercharge 0, negative G parity, and positive charge parity (for the neutral meson).

pi mode [ELECTR] Of a magnetron, the mode of operation for which the phases of the fields of successive anode openings facing the interaction space differ by pi radians.

pimple [MATER] A small, conical elevation on the surface of a plastic. [MED] A small pustule or papule.

pimple mound [GEOL] A low, flattened, roughly circular or elliptical dome consisting of sandy loam that is entirely

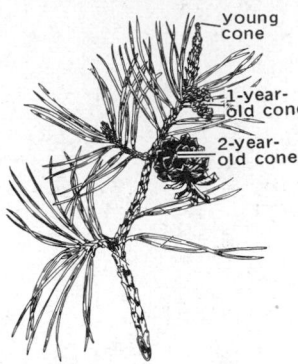

PINALES

young cone

1-year-old cone

2-year-old cone

Branch of pine with leaves and ovulate cones. *(From J. B. Hill et al., Botany, 3d ed., McGraw-Hill, 1960)*

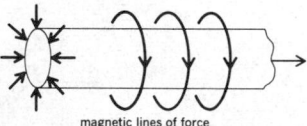

PINCH EFFECT

magnetic lines of force

Example of pinch effect. Current *I*, uniformly distributed in cylindrical wire, gives rise to compression force indicated by arrows at left.

PINEAPPLE

Pineapple *(Ananas sativus)* fruit with tuft of leaves. *(USDA)*

distinct from the surrounding soil; peculiar to the Gulf coast of eastern Texas and southwestern Louisiana.

pimple plain [GEOL] A plain distinguished by the presence of numerous, conspicuous pimple mounds.

pin [DES ENG] **1.** A cylindrical fastener made of wood, metal, or other material used to join two members or parts with freedom of angular movement at the joint. **2.** A short, pointed wire with a head used for fastening fabrics, paper, or similar materials. [ELECTR] A terminal on an electron tube, semiconductor, integrated circuit, plug, or connector. Also known as base pin; prong.

piña [BOT] A fiber obtained from the large leaves of the pineapple plant. Also known as pineapple fiber.

pinacocyte [INV ZOO] A flattened polygonal cell occurring in the dermal epithelium of sponges, and lining the exhalant canals.

pinacoid [CRYSTAL] An open crystal form that comprises two parallel faces.

pinakiolite [MINERAL] $Mg_3Mn_3B_2O_{10}$ A black mineral composed of borate of magnesium and manganese; it is polymorphous with orthopinakiolite.

Pinales [BOT] An order of gymnospermous woody trees and shrubs in the class Pinopsida, including pine, spruce, fir, cypress, yew, and redwood; the largest plants are the conifers.

Pinatae *See* Pinopsida.

pin bar [MET] A small-diameter, case-hardened steel rod used for making dowel pins.

pinboard [ADP] A board or panel containing an array of uniform holes into which pins may be inserted to control the operation of equipment.

pincer [INV ZOO] A grasping apparatus, as on the anterior legs of a lobster, consisting of two hinged jaws.

pinch [ENG] The closing-in of borehole walls before casing is emplaced, resulting from rock failure when drilling in formations having a low compressional strength. [GEOL] Thinning of a rock layer, as where a vein narrows. [MIN ENG] *See* horseback; squeeze.

pinch-and-swell structure [GEOL] A structural condition common in pegmatites and veins of quartz in metamorphosed rocks; the vein is pinched at frequent intervals, leaving expanded parts between.

pinch bar [DES ENG] A pointed lever, used somewhat like a crowbar, to roll heavy wheels.

pinch effect [ELEC] Manifestation of the magnetic self-attraction of parallel electric currents, such as constriction of ionized gas in a discharge tube, or constriction of molten metal through which a large current is flowing. Also known as cylindrical pinch; magnetic pinch; rheostriction.

pinch graft [MED] A small, full-thickness graft lifted from the donor area by a needle, and cut free with a razor.

pinch-off voltage [ELECTR] Of a field-effect transistor, the voltage at which the current flow between source and drain is blocked because the channel between these electrodes is completely depleted.

pinch pass [MET] A cold rolling of sheet metal to effect a very small reduction in thickness and to produce a piece of accurate dimensions.

pinch trimming [MET] Trimming a tubular or hollow part by pinching the flange or lip over the cutting edge of a punch.

pinch-tube process [ENG] A plastics blow-molding process in which the extruder drops a tube between mold halves, and the tube is pinched off when the mold closes.

pincushion distortion [ELECTR] Distortion in which all four sides of a received television picture are concave (curving inward). [OPTICS] Aberration in which the magnification produced by an optical system increases with the distance of the object point from the optical axis, so that the image of a square has concave sides.

pin-diode [ELECTR] A diode consisting of a silicon wafer containing nearly equal *p*-type and *n*-type impurities, with additional *p*-type impurities diffused from one side and additional *n*-type impurities from the other side; this leaves a lightly doped intrinsic layer in the middle, to act as a dielectric barrier between the *n*-type and *p*-type regions. Also known as power diode.

pine [BOT] Any of the cone-bearing trees composing the

genus *Pinus*; characterized by evergreen leaves (needles), usually in tight clusters of two to five.

pineal body [ANAT] An unpaired, elongated, club-shaped, knoblike or threadlike organ attached by a stalk to the roof of the vertebrate forebrain. Also known as conarium; epiphysis.

pineapple [BOT] *Ananas sativus.* A perennial plant of the order Bromeliales with long, swordlike, usually rough-edged leaves and a dense head of small abortive flowers; the fruit is a sorosis that develops from the fleshy inflorescence and ripens into a solid mass, covered by the persistent bracts and crowned by a tuft of leaves.

pineapple fiber *See* piña.

pinene [ORG CHEM] $C_{10}H_{16}$ Either of two colorless isomeric unsaturated bicyclic terpene hydrocarbon liquids derived from sulfate wood turpentine; 95% of the alpha form boils in the range 156–160°C, and of the beta form boils in the range 164–169°C; used as solvents for coatings and wax formulations, as chemical intermediates for resins, and as lube-oil additives. Also known as nopinene.

pine needle oil [MATER] An essential oil derived from various pines; colorless to yellowish oil with balsamic aroma; soluble in alcohol; used in perfumes and medicines. Also known as Douglas fir oil; fir wood oil.

pinene hydrochloride *See* terpene hydrochloride.

pine nut [BOT] The edible seed borne in the cone of various species of pine (*Pinus*), such as stone pine (*P. pinea*) and piñon pine (*P. cembroides* var. *edulis*).

pine oil [MATER] Any of a group of volatile essential oils with pinaceous aromas distilled from cones, needles, or stumps of various pine or other conifer species; used as solvents, emulsifying agents, wetting agents, deodorants, germicides, and sources of chemicals.

pine oleoresin [MATER] Fused solid blend of turpentine and rosin.

piner [METEOROL] In England, a rather strong breeze from the north or northeast.

pine tar [MATER] A viscous black mass obtained as a by-product in the distillation of pine wood; used for roofing.

pine tar pitch [MATER] Viscous residue resulting from distillation of volatile oils from pine tar.

pine-tree array [ELECTROMAG] Array of dipole antennas aligned in a vertical plane known as the radiating curtain, behind which is a parallel array of dipole antennas forming a reflecting curtain.

pi network [ELEC] An electrical network which has three impedance branches connected in series to form a closed circuit, with the three junction points forming an output terminal, an input terminal, and a common output and input terminal.

pin expansion test [MET] A test for determining tube expandability or for revealing longitudinal weaknesses by forcing a tapered pin into the open end of the tube.

pinfeather [VERT ZOO] A young, underdeveloped feather, especially one still enclosed in a cylindrical horny sheath which is afterward cast off.

ping [ELECTR] A sonic or ultrasonic pulse sent out by an echo-ranging sonar.

pingo [HYD] A frost mound resembling a volcano, being a relatively large and conical mound of soil-covered ice, elevated by hydrostatic pressure of water within or below the permafrost of arctic regions.

pingo ice [HYD] Clear or relatively clear ice that occurs in permafrost; originates from groundwater under pressure.

ping-pong [ADP] The programming technique of using two magnetic tape units for multiple reel files and switching automatically between the two units until the complete file is processed.

pinguecula [ANAT] A small patch of yellowish-white connective tissue located on the conjunctiva, between the cornea and the canthus of the eye.

pinguite *See* nontronite.

pinhole [MET] A material fault resulting from small blisters that have burst in a casting or that have formed during electroplating.

pinhole camera [OPTICS] A camera which has no lenses, but consists essentially of a darkened box with a small hole in one side, so that an inverted image of outside objects is projected on the opposite side where it is recorded on photographic film.

pinhole detector [ENG] A photoelectric device that detects extremely small holes and other defects in moving sheets of material.

pinic acid [ORG CHEM] $C_9H_{14}O_4$ A crystalline dicarboxylic acid derived from α-pinene; used to make diesters for plasticizers and lubricants.

Pinicae [BOT] A large subdivision of the Pinophyta, comprising woody plants with a simple trunk and excurrent branches, simple, usually alternative, needlelike or scalelike leaves, and wood that lacks vessels and usually has resin canals.

pinion [MECH ENG] The smaller of a pair of gear wheels or the smallest wheel of a gear train. [VERT ZOO] The distal portion of a bird's wing.

pinite [MINERAL] A compact gray, green, or brown mica, chiefly muscovite derived from other minerals such as cordierite.

pin jack [ELEC] Single conductor jack having an opening for the insertion of a plug of very small diameter.

pin joint [DES ENG] A joint made with a pin hinge which has a removable pin.

pin-junction [ELECTR] A semiconductor device having three regions: p-type impurity, intrinsic (electrically pure), and n-type impurity.

pink disease [PL PATH] A fungus disease of the bark of rubber, cacao, citrus, coffee, and other trees caused by *Corticium salmonicolor* and characterized by a pink covering of hyphae on the stems and branches.

pinkeye [MED] 1. A contagious, mucopurulent conjunctivitis. 2. *See* catarrhal conjunctivitis.

pink root [PL PATH] A fungus disease of onion and garlic caused by various organisms, especially species of *Phoma* and *Fusarium*; marked by red discoloration of the roots.

pink rot [PL PATH] 1. A fungus disease of potato tubers caused by *Phytophtora erythroseptica* and characterized by wet rot and pink color of the cut surfaces of the tuber upon exposure to air. 2. A rot disease of apples caused by the fungus *Tricothecium roseum*. 3. A watery soft rot of celery caused by the fungus *Sclerotinia sclerotiorum*.

pin metal [MET] Brass with a composition of 63% copper and 37% zinc, used as cold-drawn wire for making ordinary dressmaking pins.

pinna [ANAT] The cartilaginous, projecting flap of the external ear of vertebrates. Also known as auricle.

pinnacle [GEOL] 1. A sharp-pointed rock rising from the bottom, which may extend above the surface of the water, and may be a hazard to surface navigation; due to the sheer rise from the sea floor, no warning is given by sounding. 2. Any high tower or spire-shaped pillar of rock, alone or cresting a summit.

pinnacled iceberg [OCEANOGR] An iceberg weathered in such manner as to produce spires or pinnacles. Also known as irregular iceberg; pyramidal iceberg.

pinnate [BOT] Having parts arranged like a feather, branching from a central axis.

pinnate joint *See* feather joint.

pinnate muscle [ANAT] A muscle having a central tendon onto which many short, diagonal muscle fibers attach at rather acute angles.

Pinnipedia [VERT ZOO] A suborder of aquatic mammals in the order Carnivora, including walruses and seals.

pinnoite [MINERAL] $Mg(BO_2)_2 \cdot 3H_2O$ A yellow mineral composed of hydrous borate of magnesium, occurring in nodular masses.

Pinnotheridae [INV ZOO] The pea crabs, a family of decapod crustaceans belonging to the Brachygnatha.

pinnulate [BIOL] Having pinnules.

pinnule [BIOL] The secondary branch of a plumelike or pinnate organ.

pinocytosis [CYTOL] A form of active transport of water and dissolved substances across a cell membrane involving the internalization of a fluid-filled vacuole.

Pinophyta [BOT] The gymnosperms, a division of seed plants characterized as vascular plants with roots, stems, and leaves, and with seeds that are not enclosed in an ovary but are borne on cone scales or exposed at the end of a stalk.

Pinopsida [BOT] A class of gymnospermous plants in the subdivision Pinicae characterized by entire-margined or slightly toothed, narrow leaves.

pinosome [CYTOL] A closed intracellular vesicle containing material captured by pinocytosis.

pinpoint [NAV] 1. A precisely identified point, especially on the ground, that locates a very small target, a reference point for rendezvous or for other purposes; the coordinates that define this point. 2. The ground position of aircraft determined by direct observation of the ground. 3. To establish (position) with great accuracy.

pin rod [DES ENG] A rod designed to connect two parts so they act as one.

pin sensing [ADP] Device using a punched card, sensing the opening and closing of switches to generate digital data.

pint [MECH] Abbreviated pt. 1. A unit of volume, used in the United States for measurement of liquid substances, equal to $\frac{1}{8}$ U.S. gallon, or $29\frac{7}{8}$ cubic inches, or $4.73176473 \times 10^{-4}$ cubic meters. Also known as liquid pint (liq pt). 2. A unit of volume used in the United States for measurement of solid substances, equal to $\frac{1}{64}$ U.S. bushel, or 107,521/3200 cubic inches, or approximately 5.50610×10^{-4} cubic meters. Also known as dry pint (dry pt). 3. A unit of volume, used in the United Kingdom for measurement of liquid and solid substances, although usually the former, equal to $\frac{1}{8}$ imperial gallon, or approximately 5.68261×10^{-4} cubic meters. Also known as imperial pint.

pinta [MED] A disease of the skin seen most frequently in tropical America, characterized by dyschromic changes and hyperkeratosis in patches of the skin; caused by the spirochete *Treponema carateum*. Also known as carate; mal de pinto; piquite; purupuru; quitiqua.

pin timbering [MIN ENG] A method of mine roof support in which bolts are driven up into strong material, thus supporting lower weak layers.

pintle [DES ENG] A vertical pivot pin, as on a rudder or a gun carriage.

pintle center [ORD] An assumed center of a weapon on which all firing data computations are based.

pintle chain [DES ENG] A chain with links held together by pivot pins; used with sprocket wheels.

pintle hitch [ORD] A frame, secured to the rear of a tank or combat vehicle, that carries a quick-release pintle assembly with a cable running over a pulley into the tank; used for towing trailers.

pin-type mill [MECH ENG] Solids pulverizer in which protruding pins on high-speed rotating disk provide the breaking energy.

pinulus [INV ZOO] A sponge spicule, usually with five rays, one of which develops numerous small spines.

Pinus [BOT] The type genus of the family Pinaceae; the true pines, coniferous trees of north temperate regions having early deciduous primary leaves, needlelike secondary leaves, and cones with woody scales.

pinworm [INV ZOO] *Enterobius vermicularis.* A phasmid nematode of the superfamily Oxyuroidea; causes enterobiasis. Also known as human threadworm; seatworm.

Piobert lines *See* Lüders lines.

PIOCS [ADP] An extension of the hardware, constituting an interface between programs and data channels; opposed to LIOCS, logical input/output control system. Derived from physical input/output control system.

pion *See* pi meson.

pioneer [ECOL] An organism that is able to establish itself in a barren area and begin an ecological cycle.

pioneer tunnel [MIN ENG] A small tunnel parallel to but ahead of a main tunnel and used to make crosscuts to the path that the main tunnel will follow.

Piophilidae [INV ZOO] The skipper flies, a family of myodarian cyclorrhaphous dipteran insects in the subsection Acalypteratae.

piotine *See* saponite.

pip *See* blip.

pipe [DES ENG] A tube made of metal, clay, plastic, wood, or

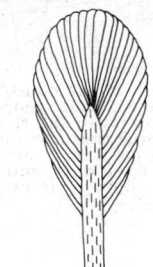

PINNATE MUSCLE

Diagram of pinnate muscle showing central tendon.

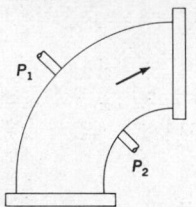

PIPE ELBOW METER

The meter measures flow rate by using the difference in pressure ($P_1 - P_2$) between pressure connections on the inside and outside of the bend in a pipe elbow.

PIPTOBLAST

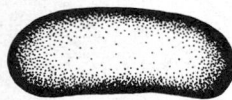

Piptoblast of *Fredericella sultana*.

PISCES

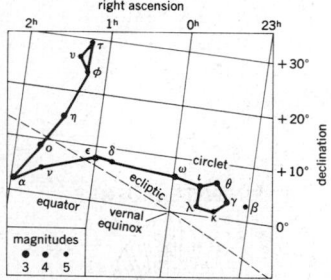

Line pattern in the constellation Pisces. The grid lines represent the coordinates of the sky. The apparent brightness, or magnitude, of the various stars is shown by the sizes of the dots, which are graded by appropriate numbers as indicated.

PI SECTION FILTER

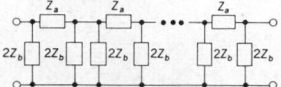

Schematic diagram of pi section filter. Z_a and Z_b represent impedances. These are chosen so that $|Z_a| \gg |Z_b|$ in the frequency range which is to be a stop band and $|Z_a| \ll |Z_b|$ in the desired pass band.

concrete and used to conduct a fluid, gas, or finely divided solid. [GEOL] **1.** A vertical, cylindrical ore body. Also known as chimney; neck; ore chimney; ore pipe; stock. **2.** A tubular cavity of varying depth in calcareous rocks, often filled with sand and gravel. **3.** A vertical conduit through the crust of the earth below a volcano, through which magmatic materials have passed. Also known as breccia pipe. [MET] **1.** The central cavity in an ingot or casting formed by contraction of the metal during solidification. **2.** An extrusion defect caused by the oxidized surface of the billet flowing toward the center of the rod at the back end.

pipe amygdule [GEOL] An elongate amygdule occurring toward the base of a lava flow, probably formed by the generation of gases or vapor from the underlying material.

pipe bit [DES ENG] A bit designed for attachment to standard coupled pipe for use in socketing the pipe in bedrock.

pipe clamp [DES ENG] A device similar to a casing clamp, but used on a pipe to grasp it and facilitate hoisting or suspension.

pipe culvert [CIV ENG] A buried pipe for carrying a watercourse below ground level.

pipe cutter [DES ENG] A hand tool consisting of a clamplike device with three cutting wheels which are forced inward by screw pressure to cut into a pipe as the tool is rotated around the pipe circumference.

pipe elbow meter [ENG] A variable-head meter for measuring flow around the bend in a pipe.

pipe fitting [ENG] A piece, such as couplings, unions, nipples, tees, and elbows for connecting lengths of pipes.

pipe flow [ENG] Conveyance of fluids in closed conduits.

pipe laying [ENG] The placing of pipe into position in a trench, as with buried pipelines for oil, water, or chemicals.

pipeline [ENG] A line of pipe connected to valves and other control devices, for conducting fluids, gases, or finely divided solids.

pipe pile [CIV ENG] A steel pipe 6–30 inches (15–76 centimeters) in diameter, usually filled with concrete and used for underpinning.

Piperaceae [BOT] A family of dicotyledonous plants in the order Piperales characterized by alternate leaves, a solitary ovule, copious perisperm, and scanty endosperm.

Piperales [BOT] An order of dicotyledonous herbaceous plants marked by ethereal oil cells, uniaperturate pollen, and reduced crowded flowers with orthotropous ovules.

piperazidine *See* piperazine.

piperazine [ORG CHEM] $C_4H_{10}N_2$ A cyclic compound; colorless, deliquescent crystals, melting at 104–107°C; soluble in water, alcohol, glycerol, and glycols; absorbs carbon dioxide from air; used in medicine. Also known as diethylenediamine; ethyleneamine; piperazidine.

2,5-piperazine-dione *See* diketopiperazine.

piperidine [ORG CHEM] $C_5H_{11}N$ A cyclic compound, and strong base; colorless liquid with pepper aroma; boils at 106°C; soluble in water, alcohol, and ether; used as a chemical intermediate and rubber accelerator, and in medicine.

piperoxan [PHARM] $C_{14}H_{19}NO_2$ An adrenergic blocking agent that has been used, as the hydrochloride salt, for diagnosis of pheochromocytoma.

piperylene *See* pentadiene.

pipe scale [ENG] Rust and corrosion products adhering to the inner surfaces of pipes; serve to decrease ability to transfer heat and to increase the pressure drop for flowing fluids.

pipe still [CHEM ENG] A petroleum-refinery still in which heat is applied to the oil while it is being pumped through a coil or pipe arranged in a firebox, the oil then running to a fractionator with continuous removal of overhead vapor and liquid bottoms.

pipestone [PETR] A pink or mottled argillaceous stone; carved by the Indians into tobacco pipes.

pipet [CHEM] Graduated or calibrated tube which may have a center reservoir (bulb); used to transfer known volumes of liquids from one vessel to another; types are volumetric or transfer, graduated, and micro.

pipe tap [ENG] A small threaded hole or entry made into the wall of a pipe; used for sampling of pipe contents, or connection of control devices or pressure-drop-measurement devices.

pipe-thread protector *See* thread protector.

pipe-to-soil potential [ELEC] The voltage potential (emf) generated between a buried pipe and its surrounding soil, the result of electrolytic action and a cause of electrolytic corrosion of the pipe.

pipe train [ENG] In the extrusion of plastic pipe, the entire equipment assembly used to fabricate the pipe (such as the extruder, die, cooling bath, haul-off, and cutter).

pipe wrench [DES ENG] A tool designed to grip and turn a pipe or rod about its axis in one direction only.

Pipidea [VERT ZOO] A family of frogs sometimes included in the suborder Opisthocoela, but more commonly placed in its own suborder, Aglossa; a definitive tongue is lacking, and free ribs are present in the tadpole but they fuse to the vertebrae in the adult.

piping [ENG] A system of pipes provided to carry a fluid. [HYD] Erosive action of water passing through or under a dam, which may result in leakage or failure.

pi point [ELEC] Frequency at which the insertion phase shift of an electric structure is 180° or an integral multiple of 180°.

pipper [OPTICS] A small hole in the reticle of an optical sight or computing sight.

pipper image [OPTICS] A spot of light projected through the pipper in an optical or computing sight, used in aiming.

Pipridae [VERT ZOO] The manakins, a family of colorful, neotropical suboscine birds in the order Passeriformes.

piptoblast [INV ZOO] A statoblast that is free but has no float.

piqué [TEXT] A cotton, polyester and cotton, rayon, or silk fabric with ribs running lengthwise, sometimes forming a waffle weave.

piquite *See* pinta.

piracy *See* capture.

Pirani gage [PHYS] A thermal conductivity gage (where the thermal conductivity of a gas heated by a hot wire varies with pressure) connected to a Wheatstone bridge to measure the resistance of the hot wire, thus the gas pressure; used to measure pressure from 1 to 10^{-3} mm Hg.

pirep *See* pilot report.

piriformis [ANAT] A muscle arising from the front of the sacrum and inserted into the greater trochanter of the femur.

Piroplasmea [INV ZOO] A class of parasitic protozoans in the superclass Sarcodina; includes the single genus *Babesia*.

pirssonite [MINERAL] $Na_2Ca(CO_3)_2 \cdot 2H_2O$ A colorless or white orthorhombic mineral composed of hydrous carbonate of sodium and calcium.

pisanite [MINERAL] $(Fe,Cu)SO_4 \cdot 7H_2O$ A blue mineral composed of hydrous sulfate of copper and iron; it is isomorphous with kirovite and melanterite.

Pisces [ASTRON] A northern constellation; right ascension 1 hour, declination 15°N. Also know as Fishes. [VERT ZOO] The fish and fishlike vertebrates, including the classes Agnatha, Placodermi, Chondrichthyes, and Osteichthyes.

Pisces Volan *See* Volan.

Piscis Australis [ASTRON] A southern constellation; right ascension 22 hours, declination 30°S. Also known as Southern Fish.

α Piscis Australis [ASTRON] The brightest star in the southern constellation Piscis Australis. Also known as Formalhaut.

piscivorous [ZOO] Feeding on fishes.

D-piscose *See* allulose.

pi section filter [ELEC] An electric filter made of several pi networks connected in series.

Pisionidae [INV ZOO] A small family of errantian polychaete annelids; allies of the scale bearers.

pisolite [PETR] A sedimentary rock composed principally of pisoliths.

pisolith [GEOL] Small, more or less spherical particles found in limestones and dolomites, having a diameter of 2–10 millimeters and often formed of calcium carbonate.

pisolitic [PETR] Pertaining to pisolite or to the characteristic texture of such a rock.

pisolitic tuff [GEOL] Of a tuff, composed of accretionary lapilli or pisolites.

pistachio [BOT] *Pistacia vera.* A small, spreading dioecious evergreen tree with leaves that have three to five broad leaflets, and with large drupaceous fruit; the edible seed consists of a single green kernel covered by a brown coat and enclosed in a tough shell.

pistil [BOT] The ovule-bearing organ of angiosperms; consists of an ovary, a style, and a stigma.

pistillate [BOT] **1.** Having a pistil. **2.** Having pistils but no stamens.

pistol [ORD] A short automatic or semiautomatic firearm aimed and fired from one hand, using the force of recoil to eject the empty shell and to insert a new round into the firing chamber.

pistol lanyard [ORD] An assembly of a cord, slides, and a fastening device, generally used by military police; it is worn looped over the shoulder with the end attached to the pistol.

piston [ELECTROMAG] A sliding metal cylinder used in waveguides and cavities for tuning purposes or for reflecting essentially all of the incident energy. Also known as plunger; waveguide plunger. [MECH ENG] A sliding metal cylinder that reciprocates in a tubular housing, either moving against or moved by fluid pressure.

piston attenuator [ELECTROMAG] A microwave attenuator inserted in a waveguide to introduce an amount of attenuation that can be varied by moving an output coupling device along its longitudinal axis.

piston blower [MECH ENG] A piston-operated, positive-displacement air compressor used for stationary, automobile, and marine duty.

piston corer [MECH ENG] A steel tube which is driven into the sediment by a free fall and by lead attached to the upper end, and which is capable of recovering undistorted vertical sections of sediment.

piston displacement [MECH ENG] The volume which a piston in a cylinder displaces in a single stroke, equal to the distance the piston travels times the internal cross section of the cylinder.

piston drill [MECH ENG] A heavy percussion-type rock drill mounted either on a horizontal bar or on a short horizontal arm fastened to a vertical column; drills holes to 6 inches (15 cm) in diameter. Also known as reciprocating drill.

piston engine [MECH ENG] A type of engine characterized by reciprocating motion of pistons in a cylinder. Also known as displacement engine; reciprocating engine.

piston flow [FL MECH] Two-phase (vapor-liquid) flow in which the gas flows as large plugs; occurs for gas superficial velocities from about 2 to 30 feet per second (60 to 900 centimeters per second). Also known as plug flow; slug flow.

piston meter [ENG] A variable-area, constant-head fluid-flow meter in which the position of the piston, moved by the buoyant force of the liquid, indicates the flow rate. Also known as piston-type area meter.

piston pump [MECH ENG] A pump in which motion and pressure are applied to the fluid by a reciprocating piston in a cylinder. Also known as reciprocating pump.

piston ring [DES ENG] A sealing ring fitted around a piston and extending to the cylinder wall to prevent leakage. Also known as packing ring.

piston rod [MECH ENG] The rod which is connected to the piston, and moves or is moved by the piston.

piston-type area meter See piston meter.

piston valve [MECH ENG] A cylindrical type of steam engine slide valve for admission and exhaust of steam.

piston viscometer [ENG] A device for the measurement of viscosity by the timed fall of a piston through the liquid being tested.

pit [BOT] **1.** A cavity in the secondary wall of a plant cell, formed where secondary deposition has failed to occur, and the primary wall remains uncovered; two main types are simple pits and bordered pits. **2.** The stone of a drupaceous fruit. [MET] A small hole in the surface of a metal; usually caused by corrosion or formed during electroplating operations. [MIN ENG] **1.** A coal mine; the term is not commonly used by the coal industry, except in reference to surface mining where the workings may be known as a strip pit. **2.** Any quarry, mine, or excavation area worked by the open-cut method to obtain material of value.

pitch [ACOUS] That psychological property of sound characterized by highness or lowness, depending primarily upon frequency of the sound stimulus, but also upon its sound

pressure and waveform. [ADP] The distance between the centerlines of adjacent rows of hole positions in punched paper tape. [ARCH] The ratio of the rise of a roof to its span. [DES ENG] The distance between similar elements arranged in a pattern or between two points of a mechanical part, as the distance between the peaks of two successive grooves on a disk recording or on a screw. [GEOL] See plunge. [MATER] A dark heavy liquid or solid substance obtained as a residue after distillation of tar, oil, and such materials; occurs naturally as asphalt. [MECH] **1.** Of an aerospace vehicle, an angular displacement about an axis parallel to the lateral axis of the vehicle. **2.** The rising and falling motion of the bow of a ship or the tail of an airplane as the craft oscillates about a transverse axis. [SCI TECH] The inclination or degree of slope of an object or structure.

pitch acceleration [MECH] The angular acceleration of an aircraft or missile about its lateral, or Y, axis.

pitch attitude [MECH] The attitude of an aircraft, rocket, or other flying vehicle, referred to the relationship between the longitudinal body axis and a chosen reference line or plane as seen from the side.

pitch axis [MECH] A lateral axis through an aircraft, missile, or similar body, about which the body pitches. Also known as pitching axis.

pitchblende [MINERAL] A massive, brown to black, and fine-grained, amorphous, or microcrystalline variety of uraninite which has a pitchy to dull luster and contains small quantities of uranium. Also known as nasturan; pitch ore.

pitch circle [DES ENG] In toothed gears, an imaginary circle concentric with the gear axis which is defined at the thickest point on the teeth and along which the tooth pitch is measured.

pitch coal See bituminous lignite.

pitch coke [MATER] Coke made from coal tar pitch, characterized by high carbon and low ash content, and used mainly for production of electrode carbon.

pitch cone [DES ENG] A cone representing the pitch surface of a bevel gear.

pitch cylinder [DES ENG] A cylinder representing the pitch surface of a spur gear.

pitch diameter [DES ENG] The diameter of the pitch circle of a gear.

pitcher plant [BOT] Any of various insectivorous plants of the families Sarraceniaceae and Nepenthaceae; the leaves form deep pitchers in which water collects and insects are drowned and digested.

pitch indicator [AERO ENG] An instrument for indicating the existence and approximate magnitude of the angular velocity about the lateral axis of an airframe.

pitching axis See pitch axis.

pitching moment [MECH] A moment about a lateral axis of an aircraft, rocket, or airfoil.

pitch line See cam profile.

pitch mining [MIN ENG] Mining coal beds with steep slopes.

pitch ore See pitchblende.

pitchover [AERO ENG] The programmed turn from the vertical that a rocket under power takes as it describes an arc and points in a direction other than vertical.

pitch-row [ADP] The distance between two adjacent holes in a paper tape.

pitchstone [GEOL] A type of volcanic glass distinguished by a waxy, dull, resinous, pitchy luster. Also known as fluolite.

pit furnace [MET] A low-temperature furnace in which steel is tempered.

pith [BOT] A central zone of parenchymatous tissue that occurs in most vascular plants and is surrounded by vascular tissue.

pi theorem See Buckingham's π theorem.

pit limits [MIN ENG] The vertical and lateral extent to which the mining of a mineral deposit by open pitting may be carried economically.

pitman [ENG] A worker in or near a pit, as in a quarry, mine, garage, or foundry.

pitometer [ENG] Reversed pitot-tube-type flow-measurement device with one pressure opening facing upstream and the other facing downstream.

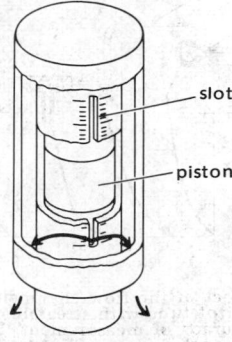

PISTON METER

Cutaway diagram of a piston meter; arrows indicate flow of liquid. Buoyant force of liquid carries piston upward until sufficient area has been uncovered in slot to allow liquid to flow through slot.

PIT

Diagrams of pit pairs in the secondary walls of plants. (a) Simple. (b) Bordered.

PITCH

Pitch level of sounds of different frequency, in mels. A tone of 1000 hertz at an intensity level of 40 decibels above absolute threshold has a pitch level of 1000 mels. (After S. S. Stevens and J. Volkman, The relation of pitch to frequency, Amer. J. Psychol., 53(3):329–353, 1940)

PITCH CIRCLE

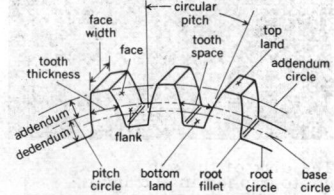

Drawing of principal features of gear teeth showing the pitch circle.

PITOT TUBE

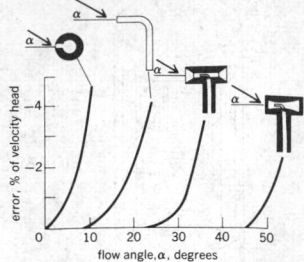

Effect of the flow alignment of a pitot tube with streamline on accuracy of measurement. Here α = angle between direction of flow and direction of tube opening.

PIVOT-BUCKET CONVEYOR-ELEVATOR

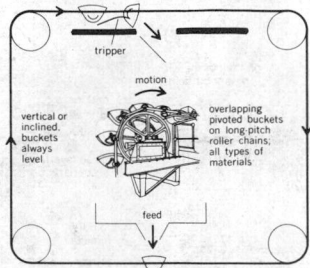

Components of pivot-bucket conveyor-elevator and typical path of travel.

PLACODONTIA

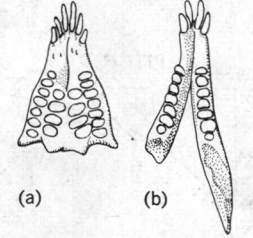

(a)   (b)

Dentition of *Paraplacodus broilii*, Triassic, Switzerland. (a) Upper jaw. (b) Lower jaw. (*After B. Peyer, 1935*)

PLAGE

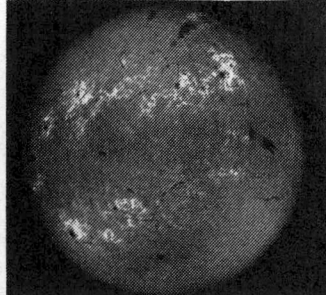

The disk of the sun photographed in H-α light, showing bright plages in centers of activity and dark filaments (prominences). (*Sacramento Peak Observatory*)

pitometer log [ENG] A log consisting essentially of a pitot tube projecting into the water, and suitable registering devices.

pitot pressure [FL MECH] Pressure at the open end of a pitot tube.

pitot tube [ENG] An instrument that measures the stagnation pressure of a flowing fluid, consisting of an open tube pointing into the fluid and connected to a pressure-indicating device. Also known as impact tube.

pitot-tube anemometer [ENG] A pressure-tube anemometer consisting of a pitot tube mounted on the windward end of a wind vane and a suitable manometer to measure the developed pressure, and calibrated in units of wind speed.

pitot-venturi flow element [ENG] Liquid-flow measurement device in which a pair of concentric venturi elements replaces the pitot-tube probe.

pit-run gravel [GEOL] A natural deposit of a mixture of gravel, sand, and foreign materials.

pit sampling [MIN ENG] Using small untimbered pits to gain access to shallow alluvial deposits or ore dumps for purpose of testing or valuation.

pit slope [MIN ENG] The angle at which the wall of an open pit or cut stands as measured along an imaginary plane extended along the crests of the berms or from the slope crest to its toe.

pitted outwash plain [GEOL] An outwash plain characterized by numerous depressions such as kettles, shallow pits, and potholes.

Pittidae [VERT ZOO] The pittas, a homogeneous family of brightly colored suboscine birds with an erectile crown of feathers, in the suborder Tyranni.

pitting [MED] **1.** The formation of pits; in the fingernails, a consequence and sign of psoriasis. **2.** The preservation for a short time of indentations on the skin made by pressing with the finger; seen in pitting edema. [MET] Selective localized formation of rounded cavities in a metal surface due to corrosion or to nonuniform electroplating. [MIN ENG] The act of digging or sinking a pit.

pitting edema [MED] Edema of such degree that the skin can be temporarily indented by pressure with the fingers.

pitting potential [MET] The electrochemical potential in a given environment above which, but not below, a corrosion pit initiates in a metal surface.

pi-T transformation *See* Y-delta transformation.

pituicyte [HISTOL] The characteristic cell of the neurohypophysis; these cells are pigmented and fusiform and are probably derived from neuroglial cells.

pituitary [ANAT] Of or pertaining to the hypophysis. [PHYSIO] Secreting phlegm or mucus (archaic usage).

pituitary dwarfism [MED] Stunted growth due to a deficiency of the primary growth hormone; characterized clinically by growth failure in early life, and in older persons by deficient subcutaneous fat with loose, wrinkled skin and precocious senility.

pituitary gland *See* hypophysis.

Pityaceae [PALEOBOT] A family of fossil plants in the order Cordaitales known only as petrifactions of branches and wood.

pityriasis [MED] A fine, branny desquamation of the skin.

Pitzer equation [PHYS CHEM] Equation for the approximation of data for heats of vaporization for organic and simple inorganic compounds; derived from temperature and reduced temperature relationships.

pivot [MECH] A short, pointed shaft forming the center and fulcrum on which something turns, balances, or oscillates.

pivotal fault *See* rotary fault.

pivot bridge [CIV ENG] A bridge in which a span can open by pivoting about a vertical axis.

pivot-bucket conveyor-elevator [MECH ENG] A bucket conveyor having overlapping pivoted buckets on long-pitch roller chains; buckets are always level except when tripped to discharge materials.

pivoted window [BUILD] A window having a section which is pivoted near the center so that the top of the section swings in and the bottom swings out.

pivoting point [NAV ARCH] The point about which a ship pivots when turning, usually somewhat forward of amidships.

pivot joint [ANAT] A diarthrosis that permits a rotation of one bone around another; an example is the articulation of the atlas with the axis. Also known as trochoid.

pk *See* peck.

PK *See* psychokinesis.

PKU *See* phenylketonuria.

Pl *See* poiseuille.

PL/1 [ADP] A multipurpose programming language, developed by IBM for the Model 360 systems, which can be used for both commercial and scientific applications.

place [MATH] A position corresponding to a given power of the base in positional notation. Also known as column.

placebo [MED] A preparation, devoid of pharmacologic effect, given to patients for psychologic effect, or as a control in evaluating a medicinal believed to have a pharmacologic action.

placenta [BOT] A plant surface bearing a sporangium. [EMBRYO] A vascular organ that unites the fetus to the wall of the uterus in all mammals except marsupials and monotremes.

placenta accreta [MED] A placenta that has partially grown into the myometrium of the uterus.

placental barrier [EMBRYO] The tissues intervening between the maternal and the fetal blood of the placenta, which prevent or hinder certain substances or organisms from passing from mother to fetus.

placentation [BOT] The arrangement of ovules and placenta in the ovary of a plant. [EMBRYO] The formation and fusion of the placenta to the uterine wall.

placer [GEOL] A mineral deposit at or near the surface of the earth, formed by mechanical concentration of mineral particles from weathered debris. Also known as ore of sedimentation.

placer claim [MIN ENG] A mining claim located upon gravel or ground whose mineral contents are extracted by the use of water, as by sluicing, or hydraulicking.

placer dredge [MIN ENG] A dredge for mining metals from placer deposits; it consists of a chain of closely connected buckets passing over an idler tumbler and an upper or driving tumbler, mounted on a structural-steel ladder which carries a series of rollers.

placer location [MIN ENG] Location of a tract of land for the sake of loose mineral-bearing or other valuable deposits on or near its surface, rather than within lodes or veins in rock in place.

placer mining [MIN ENG] **1.** The extraction and concentration of heavy metals from placers. **2.** Mining of gold by washing the sand, gravel, or talus.

placode [EMBRYO] A platelike epithelial thickening, frequently marking, in the embryo, the anlage of an organ or part.

Placodermi [PALEON] A large and varied class of Paleozoic fishes characterized by a complex bony armor covering the head and the front portion of the trunk.

Placodontia [PALEON] A small order of Triassic marine reptiles of the subclass Euryapsida characterized by flat-crowned teeth in both the upper and lower jaws and on the palate.

Placothuriidae [INV ZOO] A family of holothurian echinoderms in the order Dendrochirotida; individuals are invested in plates and have a complex calcareous ring mechanism.

plage [ASTRON] One of the luminous areas that appear in the vicinity of sunspots or disturbed areas on the sun; they may be seen distinctively in spectroheliograms taken in the calcium K line.

Plagiaulacida [PALEON] A primitive, monofamilial suborder of multituberculate mammals distinguished by their dentition (dental formula I 3/0 C 0/0 Pm 5/4 M 2/2), having cutting premolars and two rows of cusps on the upper molars.

Plagiaulacidae [PALEON] The single family of the extinct mammalian suborder Plagiaulacida.

plagiocephaly [MED] A type of strongly asymmetric cranial deformation, in which the anterior portion of one side and the posterior portion of the opposite side of the skull are developed more than their counterparts so that the maximum length of the skull is not in the midline but on a diagonal.

plagioclase [MINERAL] **1.** A type of triclinic feldspars having the general formula $(Na,Ca)Al(Si,Al)Si_2O_8$; they are common rock-forming minerals. **2.** A series in the plagioclase group which can be divided into a number of varieties based on the relative proportion of the solid solution end members, albite and anorthite (An): albite (An 0–10) oligoclase (An 10–30), andesine (An 30–50), labradorite (An 50–70), bytownite (An 70–90), and anorthite (An 90–100). Also known as sodium-calcium feldspar.

plagiodont [VERT ZOO] Of a snake, having obliquely set, or two converging series of, palatal teeth.

plagiohedral [CRYSTAL] Pertaining to obliquely arranged spiral faces; in particular, to a member of a group in the isometric system with 13 axes but no center or planes.

plagionite [MINERAL] $Pb_5Sb_8S_{17}$ A lead-gray mineral with metallic appearance, composed of sulfide of lead and antimony.

Plagiosauria [PALEON] An aberrant Triassic group of labyrinthodont amphibians.

plague [MED] 1. An infectious bacterial disease of rodents and humans caused by *Pasteurella pestis*, transmitted to man by the bite of an infected flea (*Xenopsylla cheopis*) or by inhalation. Also known as black death; bubonic plague. 2. Any contagious, malignant, epidemic disease.

plain [GEOGR] An extensive, broad tract of level or rolling, almost treeless land with a shrubby vegetation, usually at a low elevation. [GEOL] A flat, gently sloping region of the sea floor. Also known as submarine plain.

plain concrete [CIV ENG] Concrete without reinforcement but often with light steel to reduce shrinkage and temperature cracking.

plain-laid [DES ENG] Pertaining to a rope whose strands are twisted together in a direction opposite to that of the twist in the strands.

plain milling cutter [DES ENG] A cylindrical milling cutter with teeth on the periphery only; used for milling plain or flat surfaces. Also known as slab cutter.

plain-sawing *See* backsawing.

plain weave [TEXT] A weave made by passing the filling yarns over one warp yarn and under the next, continuing alternately across each row.

Plaisancian [GEOL] A European stage of geologic time: lower Pliocene (above Pontian of Miocene, below Astian). Also known as Piacention; Plaisanzian.

Plaisanzian *See* Plaisancian.

plait point [CHEM] Composition conditions in which the three coexisting phases of partially soluble components of a three-phase liquid system approach each other in composition.

plan [GRAPHICS] 1. An orthographic drawing on a horizontal plane, as of an instrument, a horizontal section, or a layout. 2. A large-scale map or chart of a small area.

planar [MATH] Lying in or pertaining to a euclidean plane.

planar array [ELECTR] An array of ultrasonic transducers that can be mounted in a single plane or sheet, to permit closer conformation with the hull design of a sonar-carrying ship.

planar ceramic tube [ELECTR] Electron tube having parallel planar electrodes and a ceramic envelope.

planar cross-bedding [GEOL] Cross-bedding characterized by planar surfaces of erosion in the lower bounding surface.

planar device [ELECTR] A semiconductor device having planar electrodes in parallel planes, made by alternate diffusion of *p*- and *n*-type impurities into a substrate.

planar diode [ELECTR] A diode having planar electrodes in parallel planes.

planar flow structure *See* platy flow structure.

planaria [INV ZOO] Any flatworm of the turbellarian order Tricladida; the body is broad and dorsoventrally flattened, with anterior lateral projections, the auricles, and a pair of eyespots on the anterior dorsal surface.

planar photodiode [ELECTR] A vacuum photodiode consisting simply of a photocathode and an anode; light enters through a window sealed into the base, behind the photocathode.

planar process [ENG] A silicon-transistor manufacturing process in which a fractional-micrometer-thick oxide layer is grown on a silicon substrate; a series of etching and diffusion steps is then used to produce the transistor inside the silicon substrate.

planar transistor [ELECTR] A transistor constructed by an etching and diffusion technique in which the junction is never exposed during processing, and the junctions reach the surface in one plane; characterized by very low leakage current and relatively high gain.

planation [GEOL] Erosion resulting in flat surfaces, caused by meandering streams, waves, ocean currents, wind, or glaciers.

planchet [MATER] A milled metal disk ready for coining.

planck [PHYS] A unit of action equal to the product of an energy of 1 joule and a time of 1 second.

Planck distribution law *See* Planck radiation formula.

Planck oscillator [QUANT MECH] An oscillator which can absorb or emit energy only in amounts which are integral multiples of Planck's constant times the frequency of the oscillator. Also known as radiation oscillator.

Planck radiation formula [STAT MECH] A formula for the intensity of radiation emitted by a blackbody within a narrow band of frequencies (or wavelengths), as a function of frequency, and of the body's temperature. Also known as Planck distribution law; Planck's law.

Planck's constant [QUANT MECH] A fundamental physical constant, the elementary quantum of action; the ratio of the energy of a photon to its frequency, it is equal to $6.62620 \pm 0.00005 \times 10^{-34}$ joule-second. Symbolized h.

Planck's law [QUANT MECH] A fundamental law of quantum theory stating that energy associated with electromagnetic radiation is emitted or absorbed in discrete amounts which are proportional to the frequency of radiation. [STAT MECH] *See* Planck radiation formula.

plane [ELECTR] Screen of magnetic cores; planes are combined to form stacks. [DES ENG] A tool consisting of a smooth-soled stock from the face of which extends a wide-edged cutting blade for smoothing and shaping wood. [MATH] A surface containing any straight line through any two of its points.

plane angle [MATH] An angle between lines in the euclidean plane.

plane atmospheric wave [METEOROL] An atmospheric wave represented in two-dimensional rectangular cartesian coordinates, in contrast to a wave considered on the spherical earth.

plane curve [MATH] Any curve lying entirely within a plane.

plane cyclic curve *See* cyclic curve.

plane dendrite *See* plane-dendritic crystal.

plane-dendritic crystal [CRYSTAL] An ice crystal exhibiting an elaborately branched (dendritic) structure of hexagonal symmetry, with its much larger dimension lying perpendicular to the principal (*c*-axis) of the crystal. Also known as plane dendrite; stellar crystal.

plane earth [ELECTROMAG] Earth that is considered to be a plane surface as used in ground-wave calculations.

plane-earth attenuation [ELECTROMAG] Attenuation of an electromagnetic wave over an imperfectly conducting plane earth in excess of that over a perfectly conducting plane.

plane field *See* field of planes on a manifold.

plane geometry [MATH] The geometric study of the figures in the euclidean plane such as lines, triangles, and polygons.

plane mirror [OPTICS] A mirror whose surface lies in a plane; it forms an image of an object such that the mirror surface is perpendicular to and bisects the line joining all corresponding object-image points.

plane of departure [MECH] Vertical plane containing the path of a projectile as it leaves the muzzle of the gun.

plane of fire [MECH] Vertical plane containing the gun and the target, or containing a line of site.

plane of incidence [PHYS] A plane containing the direction of propagation of a wave striking a surface and a line perpendicular to the surface. Also known as incidence plane.

plane of mirror symmetry Also known as mirror plane of symmetry; plane of symmetry; reflection plane; symmetry plane. [CRYSTAL] In certain crystals, a symmetry element whereby reflection of the crystal through a certain plane leaves the crystal unchanged. [MATH] An imaginary plane which divides an object into two halves, each of which is the mirror image of the other in this plane.

plane of polarization [ELECTROMAG] Plane containing the electric vector and the direction of propagation of electromagnetic wave.

Planaria, a bilaterally symmetrical, dorsoventrally flattened, triploblastic organism.

**PLANCK
RADIATION FORMULA**

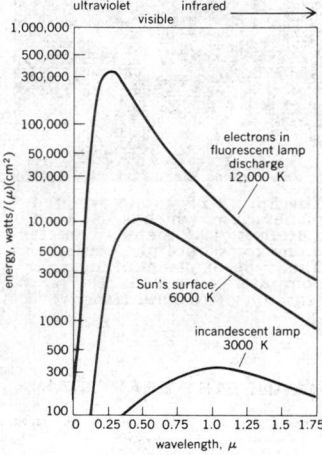

Graphs of the Planck radiation formula for various temperatures.

PLANETARIUM

The Spitz STP public oriented planetarium, which is also used in a 40-foot (12.2-meter) diameter dome for school planetariums. This type of planetarium is computer-controlled. *(Denver Museum of Natural History)*

PLANETARY GEAR TRAIN

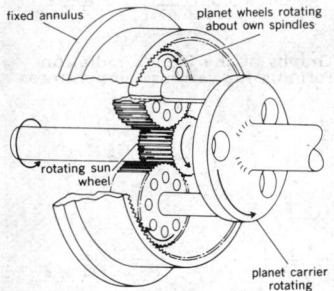

fixed annulus

planet wheels rotating about own spindles

rotating sun wheel

planet carrier rotating

Cutaway isometric of mode of operation of a simple planetary gear train with a fixed or locked annulus.

PLANOCONCAVE LENS

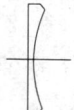

Shape of a planoconcave lens. *(From F. A. Jenkins and H. E. White, Fundamentals of Optics, 3d ed., McGraw-Hill, 1957)*

PLANOCONVEX LENS

Shape of a planoconvex lens. *(From F. A. Jenkins and H. E. White, Fundamentals of Optics, 3d ed., McGraw-Hill, 1957)*

plane of reflection [CRYSTAL] *See* plane of mirror symmetry. [MATH] *See* plane of mirror symmetry. [OPTICS] A plane containing the direction of propagation of radiation reflected from a surface, and the normal to the surface. Also known as reflection plane.

plane of saturation *See* water table.

plane of symmetry *See* plane of mirror symmetry.

plane of yaw [MECH] The plane determined by the tangent to the trajectory of a projectile in flight and the axis of the projectile.

plane Poiseuille flow [FL MECH] Rheological (viscosity) measurement in which the fluid of interest is propelled through a narrow slot, and the volumetric flow rate and the pressure gradient are measured simultaneously to determine viscosity.

plane polarization *See* linear polarization.

plane-polarized wave [ELECTROMAG] An electromagnetic wave whose electric field vector at all times lies in a fixed plane that contains the direction of propagation through a homogeneous isotropic medium.

plane polygon [MATH] A polygon lying in the euclidean plane.

plane quadrilateral [MATH] A four-sided polygon lying in the euclidean plane.

plan equation [MECH ENG] The mathematical statement that horsepower = $plan/33,000$, where p = mean effective pressure (psi), l = length of piston stroke (ft), a = net area of piston (in.^{2}), and n = number of cycles completed per minute.

planer [MECH ENG] A machine for the shaping of long, flat, or flat contoured surfaces by reciprocating the workpiece under a stationary single-point tool or tools. [MIN ENG] A fixed-blade device for continuous longwall mining of narrow seams of friable coal; the machine is operated along the coal face, planing a narrow cut from the solid coal as it travels.

plane reflector *See* passive reflector.

plane sailing [NAV] A method of solving the various problems involving course, distance, difference of latitude, and departure, in which the earth or a smaller area is considered a plane.

plane surveying [ENG] Measurement of areas on the assumption that the earth is flat.

planet [ASTRON] A relatively small, solid celestial body circulating around a star, in particular the star known as the sun (which has nine planets).

plane table [ENG] A surveying instrument consisting of a drawing board mounted on a tripod and fitted with a compass and a straight-edge ruler; used to graphically plot survey lines directly from field observations.

plane-table method [MIN ENG] A method of measuring areas of mine roadways; a drawing board is set up on a tripod in the plane of the mine section to be measured; the distance from a central point on the board to the perimeter of the roadway is measured with a tape along various offsets; the distance measured is scaled on the drawing board along the proper offset line.

planetarium [ASTRON] **1.** A projection device which accurately portrays the position of the stars and planets at any time in the past, present, or future from any point on the earth or the near region of space; the modern planetarium instrument is a mechanical-electrical analog of space. **2.** The name given to the building and gear associated with this device.

planetary atmosphere [ASTRON] The outer shell of gas around some planets.

planetary boundary layer [METEOROL] That layer of the atmosphere from the earth's surface to the geostrophic wind level, including, therefore, the surface boundary layer and the Ekman layer; above this layer lies the free atmosphere.

planetary circulation *See* general circulation.

planetary gear train [MECH ENG] An assembly of meshed gears consisting of a central gear, a coaxial internal or ring gear, and one or more intermediate pinions supported on a revolving carrier.

planetary nebula [ASTRON] An oval or round nebula of expanding concentric rings of gas associated with a hot central star.

planetary orbit [ASTRON] The path that a planet has as it revolves about the sun.

planetary perturbation [ASTRON] A deviation of a planet from its computed orbit because of the attraction of another celestial body or bodies.

planetary physics [ASTROPHYS] The study of the structure, composition, and physical and chemical properties of the planets of the solar system, including their atmospheres and immediate cosmic environment.

planetary precession [ASTRON] A comparatively small eastward motion of the equinoxes caused by the action of other planets in altering the plane of the earth's orbit.

planetary vorticity effect [GEOPHYS] The effect of the variation of the earth's vorticity with latitude in altering the relative vorticity of a flow with a meridional component; a fluid with a free surface in a rotating cylinder exhibits a corresponding effect, owing to the shrinking or stretching of radially displaced columns.

planetary wave *See* long wave; Rossby wave.

planet gear [MECH ENG] A pinion in a planetary gear train.

planetoid *See* asteroid.

planetology [ASTRON] Scientific study of the planets, in particular their surface markings.

plane trigonometry [MATH] The study of triangles in the euclidean plane.

plane wave [PHYS] Wave in which the wavefront is a plane surface; a wave whose equiphase surfaces form a family of parallel planes.

plane-wave initiation [ORD] Simultaneous initiation at all points of the rear surface of the main explosive charge by a flat detonation wave, usually accomplished by a composite explosive charge of proper dimensions.

planform [AERO ENG] The shape or form of an object, such as an airfoil, as seen from above, as in a plan view.

planidium [INV ZOO] A first-stage legless larva of various insects in the orders Diptera and Hymenoptera.

planigraphy *See* sectional radiography.

planimeter [ENG] A device used for measuring the area of any plane surface by tracing the boundary of the area.

planimetric map [MAP] A map indicating only the horizontal positions of features, without regard to elevation, in contrast with a topographic map, which indicates both horizontal and vertical positions. Also known as line map.

planimetric method [MET] A method of measuring grain size by counting the number of grains in a given area.

planing [ENG] Smoothing or shaping the surface of wood, metal, or plastic workpieces.

Planipennia [INV ZOO] A suborder of insects in the order Neuroptera in which the larval mandibles are modified for piercing and for sucking.

planishing [MECH ENG] Smoothing the surface of a metal by a rapid series of overlapping, light hammerlike blows or by rolling in a planishing mill.

planisphere [MAP] A representation, on a plane, of the celestial sphere, especially one on a polar projection, with means provided for making certain measurements such as altitude and azimuth.

plank [MATER] A heavy board with thickness of 2–4 inches (5–10 centimeters) and a width of at least 8 inches (20 centimeters).

planking [NAV ARCH] The wood decks and outside planks in wood or composite ships.

plankton [ECOL] Passively floating or weakly motile aquatic plants and animals.

plankton net [ENG] A net for collecting plankton.

planning chart [NAV] A chart designed for use in planning voyages or flight operations, or investigating areas of marine or aviation activities.

planoblast [INV ZOO] The medusa form of a hydrozoan.

planocaine base *See* procaine base.

planoconcave lens [OPTICS] A lens for which one surface is plane and the other is concave.

planoconvex lens [OPTICS] A lens for which one surface is plane and the other is convex.

planoconvex spotlight [ELEC] A light that can be used as a sharply defined spotlight or for soft-edged lighting; ranges in power from 100 to 2000 watts.

planographic process [GRAPHICS] Nonrelief printing process (such as collotype and lithography), in which the areas of the plate to receive ink are on the same level or plane as those that remain uninked.

Planosol [GEOL] An intrazonal, hydromorphic soil having a clay pan or hardpan covered with a leached surface layer; developed in a humid to subhumid climate.

planospiral [INV ZOO] Having the shell coiled in one plane, used particularly of foraminiferans and mollusks.

plan position indicator [ELECTR] A radarscope display in which echoes from various targets appear as bright spots at the same locations as they would on a circular map of the area being scanned, the radar antenna being at the center of the map. Abbreviated PPI. Also known as P display.

plan position indicator repeater [ELECTR] Unit which repeats a plan position indicator (PPI) at a location remote from the radar console. Also known as remote plan position indicator.

plant [BOT] Any organism belonging to the kingdom Plantae, generally distinguished by the presence of chlorophyll, a rigid cell wall, and abundant, persistent, active embryonic tissue, and by the absence of the power of locomotion. [IND ENG] The land, buildings, and equipment used in an industry.

Plantae [BOT] The plant kingdom.

Plantaginaceae [BOT] The single family of the plant order Plantaginales.

Plantaginales [BOT] An order of dicotyledonous herbaceous plants in the subclass Asteridae, marked by small hypogynous flowers with a persistent regular corolla and four petals.

plantar [ANAT] Of or relating to the sole of the foot.

plantaris [ANAT] A small muscle of the calf of the leg; origin is the lateral condyle of the femur, and insertion is the calcaneus; flexes the knee joint.

plantar reflex [PHYSIO] Flexion of the toes in response to stroking of the outer surface of the sole, from heel to little toe.

Plante cell [ELEC] A type of lead-acid cell in which the active material is formed on the plates by electrochemical means during repeated charging and discharging, instead of being applied as a prepared paste.

plant factor [ELEC] The ratio of the average power load of an electric power plant to its rated capacity. Also known as capacity factor.

plant fermentation [BIOCHEM] A form of plant metabolism in which carbohydrates are partially degraded without the consumption of molecular oxygen.

plant geography [BOT] A major division of botany, concerned with all aspects of the spatial distribution of plants. Also known as geographical botany; phytochorology; phytogeography.

plant hormone [BIOCHEM] An organic compound that is synthesized in minute quantities by one part of a plant and translocated to another part, where it influences physiological processes. Also known as phytohormone.

plantigrade [VERT ZOO] Pertaining to walking with the whole sole of the foot touching the ground.

plant key [BOT] An analytical guide to the identification of plants, based on the use of contrasting characters to subdivide a group under study into branches.

plant kingdom [BOT] The worldwide array of plant life constituting a major division of living organisms.

plant layout [IND ENG] The location of equipment and facilities in a manufacturing plant.

plant pathology [BOT] The branch of botany concerned with diseases of plants.

plant physiology [BOT] The branch of botany concerned with the processes which occur in plants.

plant protection [IND ENG] That portion of industrial security which concerns the safeguarding of industrial installations, resources, utilities, and materials by physical measures such as guards, fences, and lighting designation of restricted areas.

plant societies [ECOL] Assemblages of plants which constitute structural parts of plant communities.

plantula [INV ZOO] A small, cushionlike structure on the ventral surface of the segments of insect tarsi.

plant virus [VIROL] A virus that replicates only within plant cells.

planula [INV ZOO] The ciliated, free-swimming larva of coelenterates.

Planuloidea [INV ZOO] The equivalent name for Moruloidea.

plaque [MED] **1.** A patch, or an abnormal flat area on any internal or external body surface. **2.** A localized area of atherosclerosis. [VIROL] A clear area representing a colony of viruses on a plate culture formed by lysis of the host cell.

plasma [HISTOL] The fluid portion of blood or lymph. [PL PHYS] **1.** A highly ionized gas which contains equal numbers of ions and electrons in sufficient density so that the Debye shielding length is much smaller than the dimensions of the gas. **2.** A completely ionized gas, composed entirely of a nearly equal number of positive and negative free charges (positive ions and electrons).

plasma accelerator [PL PHYS] An accelerator that forms a high-velocity jet of plasma by using a magnetic field, an electric arc, a traveling wave, or other similar means.

plasma-arc welding [MET] Welding metal in a gas stream heated by a tungsten arc to temperatures approaching 60,000°F (33,315°C).

plasma cathode [ELECTR] A cathode in which the source of electrons is a gas plasma rather than a solid.

plasma cell *See* plasmacyte.

plasmacyte [HISTOL] A fairly large, generally ovoid cell with a small, eccentrically placed nucleus; the chromatin material is adherent to the nuclear membrane and the cytoplasm is agranular and deeply basophilic everywhere except for a clear area adjacent to the nucleus in the area of the cytocentrum. Also known as plasma cell.

plasma diode [ELECTR] A diode used for converting heat directly into electricity; it consists of two closely spaced electrodes serving as cathode and anode, mounted in an envelope in which a low-pressure cesium vapor fills the interelectrode space; heat is applied to the cathode, causing emission of electrons.

plasma engine [AERO ENG] An engine for space travel in which neutral plasma is accelerated and directed by external magnetic fields that interact with the magnetic field produced by current flow through the plasma. Also known as plasma jet.

plasma frequency *See* Langmuir plasma frequency.

plasmagel [CYTOL] The outer, gelated zone of protoplasm in a pseudopodium.

plasma generator [ELECTR] Any device that produces a high-velocity plasma jet, such as a plasma accelerator, engine, oscillator, or torch.

plasma gun [ELECTR] A machine, such as an electric-arc chamber, that will generate very high heat fluxes to convert neutral gases into plasma. [ELECTROMAG] An electromagnetic device which creates and accelerates bursts of plasma.

plasma instability [PL PHYS] A sudden change in the quasistatic distribution of positions or velocities of particles constituting a plasma, and a sudden change in the accompanying electromagnetic field.

plasma jet *See* plasma engine.

plasma-jet excitation [SPECT] The use of a high-temperature plasma jet to excite an element to provide measurable spectra with many ion lines similar to those from spark-excited spectra.

plasmalemma *See* cell membrane.

plasma membrane *See* cell membrane.

plasmapause [GEOPHYS] The sharp outer boundary of the plasmasphere, at which the plasma density decreases by a factor of 100 or more.

plasmapheresis [MED] The withdrawal of blood from a donor to obtain plasma, its components, or the nonerythrocytic formed elements of blood, followed by the return of the erythrocytes to the donor.

plasma physics [PHYS] The study of highly ionized gases.

plasma pinch [PL PHYS] Application of the pinch effect to plasma in attempts to produce controlled nuclear fusion.

plasma propulsion [AERO ENG] Propulsion of spacecraft and other vehicles by using electric or magnetic fields to accelerate both positively and negatively charged particles (plasma) to a very high velocity.

plasma radiation [PL PHYS] Electromagnetic radiation emitted from a plasma, primarily by free electrons undergoing transitions to other free states or to bound states of atoms and

PLAN POSITION INDICATOR

azimuth

Plan position indicator type of radar display; range is measured radially from center.

ions, but also by bound electrons as they undergo transitions to other bound states.

plasma rocket [AERO ENG] A rocket that is accelerated by means of a plasma engine.

plasma sheath [ELECTR] An envelope of ionized gas that surrounds a spacecraft or other body moving through an atmosphere at hypersonic velocities; affects transmission, reception, and diffraction of radio waves.

plasma sheet [GEOPHYS] A region of relatively hot plasma outside the plasmasphere, which reaches, during quiet times, from an altitude of about 50,000 kilometers to at least past the moon's orbit in a long tail extending away from the sun; composed of particles with typical thermal energies of 2 to 4 kilo electron volts.

plasmasol [CYTOL] The inner, solated zone of protoplasm in a pseudopodium.

plasmasphere [GEOPHYS] A region of relatively dense, cold plasma surrounding the earth and extending out to altitudes of approximately 2 to 6 earth radii, composed predominantly of electrons and protons, with thermal energies not exceeding several electron volts.

plasma thromboplastin antecedent *See* factor XI.

plasma thromboplastin component *See* Christmas factor.

plasma torch [ENG] A torch in which temperatures as high as 50,000°C are achieved by injecting a plasma gas tangentially into an electric arc formed between electrodes in a chamber; the resulting vortex of hot gases emerges at very high speed through a hole in the negative electrode, to form a jet for welding, spraying of molten metal, and cutting of hard rock or hard metals.

plasma wave [PL PHYS] A disturbance of a plasma involving oscillation of its constituent particles and of an electromagnetic field, which propagates from one point in the plasma to another without net motion of the plasma.

plasmid [GEN] An extrachromosomal genetic element found among various strains of *Escherichia coli* and other bacteria.

plasmin [BIOCHEM] A proteolytic enzyme in plasma which can digest many proteins through the process of hydrolysis. Also known as fibrinolysin.

plasminogen [BIOCHEM] The inert precursor, or zymogen, of plasmin. Also known as profibrinolysin.

plasmodesma [CYTOL] An intercellular bridge, thought to be strands of cytoplasm connecting two cells.

Plasmodiidae [INV ZOO] A family of parasitic protozoans in the suborder Haemosporina inhabiting the erythrocytes of the vertebrate host.

Plasmodiophorida [INV ZOO] An order of the protozoan subclass Mycetozoia occurring as endoparasites of plants.

Plasmodiophoromycetes [MYCOL] A class of the Fungi.

plasmoditrophoblast *See* syncytiotrophoblast.

plasmodium [MICROBIO] The noncellular, multinucleate, jellylike, ameboid, assimilative stage of the Myxomycetes.

Plasmodium [INV ZOO] A genus of protozoans in the family Plasmodiidae in which all the true malarial parasites are placed.

Plasmodroma [INV ZOO] A subphylum of the Protozoa, including Mastigophora, Sarcodina, and Sporozoa, in some taxonomic systems.

plasmogamy [INV ZOO] Fusion of protoplasts, without nuclear fusion, to form a multinucleate mass; occurs in certain protozoans.

plasmoid [PHYS] An isolated collection of electrons, ions, and neutral particles which holds together for a duration many times as long as the collision times between particles.

plasmolysis [PHYSIO] Shrinking of the cytoplasm away from the cell wall due to exosmosis by immersion of a plant cell in a solution of higher osmotic activity.

plasmon [SOLID STATE] A quantum of a collective longitudinal wave in the electron gas of a solid.

plasmosome *See* nucleolus.

plasmotomy [INV ZOO] Subdivision of a plasmodium into two or more parts.

plaster [MATER] A plastic mixture of various materials, such as lime or gypsum, and water which sets to a hard, coherent solid.

plaster bat [GRAPHICS] Basic working surface on which clay is turned or modeled.

plasterboard [MATER] A large, thin sheet of pulpboard, paper, or felt bonded to a hardened gypsum plaster core and used as a wall backing or as a substitute for plaster.

plaster coat [BUILD] A thin layer of plaster lining walls in buildings.

plaster of paris [INORG CHEM] White powder consisting essentially of the hemihydrate of calcium sulfate ($CaSO_4 \cdot \frac{1}{2}H_2O$ or $2CaSO_4 \cdot H_2O$), produced by calcining gypsum until it is partially dehydrated; forms with water a paste that quickly sets; used for casts and molds, building materials, and surgical bandages. Also known as calcined gypsum.

plaster shooting [ENG] A surface blasting method used when no rock drill is necessary or one is not available; consists of placing a charge of gelignite, primed with safety fuse and detonator, in close contact with the rock or boulder and covering it completely with stiff damp clay.

plastic [MATER] A polymeric material (usually organic) of large molecular weight which can be shaped by flow; usually refers to the final product with fillers, plasticizers, pigments, and stabilizers included (versus the resin, the homogeneous polymeric starting material); examples are polyvinyl chloride, polyethylene, and urea-formaldehyde. [MECH] Displaying, or associated with, plasticity.

plasticate [ENG] To soften a material by heating or kneading. Also known as plastify.

plastic bonding [ENG] The joining of plastics by heat, solvents, adhesives, pressure, or radio frequency.

plastic bronze [MET] A copper alloy containing lead, usually on the order of 30%, of sufficient plasticity to make a good bearing.

plastic cement [MATER] A plastic material used to seal narrow openings in buildings.

plastic clay [MATER] Fireclay which forms a moldable mass when mixed with water.

plastic deformation [MECH] Permanent change in shape or size of a solid body without fracture resulting from the application of sustained stress beyond the elastic limit.

plastic design *See* ultimate-load design.

plastic dielectric [MATER] A plastic used in an application in which its high resistance, dielectric strength, or other electrical properties are important, such as for electrical insulation or in a capacitor.

plastic explosive *See* high-explosive plastic.

plastic film [MATER] Film with thickness from 0.0015 to 0.006 inches (3.8×10^{-3} to 15×10^{-3} cm); made from polyvinyl chloride, polyethylene, polypropylene, polystyrene, Mylar, and other resins; used for wrapping, sealing, garment waterproofing, and coating wood, paper, or fabric.

plastic film capacitor [ELEC] A capacitor constructed by stacking, or forming into a roll, alternate layers of foil and a dielectric which consists of a plastic, such as polystyrene or Mylar, either alone or as a laminate with paper.

plastic flow [PHYS] Rheological phenomenon in which flowing behavior of the material occurs after the applied stress reaches a critical (yield) value, such as with putty.

plastic foam *See* expanded plastic.

plasticity [MECH] The property of a solid body whereby it undergoes a permanent change in shape or size when subjected to a stress exceeding a particular value, called the yield value.

plasticity index [GEOL] The percent difference between moisture content of soil at the liquid and plastic limits.

plasticize [ENG] To soften a material to make it plastic or moldable by adding a plasticizer or by using heat.

plasticizer *See* flexibilizer.

plasticizing oil [MATER] Coal tar distillate or solvent naphthas distilling in a wide range above 300°C; used with plastics as a plasticizer.

plastic limit [GEOL] The water-content limit of a sediment.

plasticorder [ENG] Laboratory device used to predict the performance of a plastic material by measurement of temperature, viscosity, and shear-rate relationships. Also known as plastigraph.

plasticoviscosity [MECH] Plasticity in which the rate of deformation of a body subjected to stresses greater than the yield stress is a linear function of the stress.

plastic paint [MATER] Paint composed of a plastic (such as vinyl or nitrocellulose) in a solvent.

plastic plate [ELECTR] A plate of plastic dielectric material used as a base for a semiconductor device. [GRAPHICS] A direct printing plate formed on a plastic base.

plastic semiconductor [MATER] An organic plastic resin with a conjugated double-bond structure, such as polyacetylene; the material is a semiconductor due to resistance of electrons to transfer from one molecule to another.

plastic surgery [MED] Surgical repair, replacement, or alteration of lost, injured, or deformed parts of the body by transfer of tissue.

plastic wood [MATER] Wood flour or wood cellulose compounded with a synthetic resin of high molecular weight; it is adhesive but does not penetrate wood, and is used to fill cavities or seams in wood products.

plastid [CYTOL] One of the specialized cell organelles containing pigments or protein materials, often serving as centers of special metabolic activities; examples are chloroplasts and leukoplasts.

plastify *See* plasticate.

plastigel [MATER] A plastisol with gellike flow properties achieved by adding a thixotropic agent (such as bentonite) to the plastisol.

plastigraph *See* plasticorder.

plastisol [MATER] A vinyl resin dissolved in a plasticizer to make a pourable liquid.

plastometer [ENG] Instrument used to determine the flow properties of a thermoplastic resin by forcing molten resin through a specified die opening or orifice at a given pressure and temperature.

plastron [INV ZOO] The ventral plate of the cephalothorax of spiders. [VERT ZOO] The ventral portion of the shell of tortoises and turtles.

plat [MAP] A plan that shows land ownership, boundaries, and subdivisions together with data for description and identification of various parts.

Platanaceae [BOT] A small family of monoecious dicotyledonous plants in which flowers have several carpels which are separate, three or four stamens, and more or less orthotropous ovules, and leaves are stipulate.

Plataspidae [INV ZOO] A family of shining, oval hemipteran insects in the superfamily Pentatomoidea.

plate [DES ENG] A rolled, flat piece of metal of some arbitrary minimum thickness and width depending on the type of metal. [ELEC] **1.** One of the conducting surfaces in a capacitor. **2.** One of the electrodes in a storage battery. [ELECTR] *See* anode. [GEOL] **1.** A smooth, thin, flat fragment of rock, such as a flagstone. **2.** A large rigid, but mobile, block involved in plate tectonics; thickness ranges from 50 to 250 kilometers and includes both crust and a portion of the upper mantle. [GRAPHICS] **1.** In etching, the piece of copper, zinc, or other metal that constitutes the base from which prints are made. **2.** In photography, a sheet of glass coated with a sensitized emulsion. **3.** In printing, the reproduction of type or cuts in metal or other material; a plate may bear a relief, intaglio, or planographic printing surface. [MET] A thick flat particle of metal powder.

plate amalgamation [MET] Use of copper-alloy plates or copper coated with mercury in order to trap gold from crushed ore pulp as it flows over the plates.

plate anemometer *See* pressure-plate anemometer.

plateau [ELECTR] The portion of the plateau characteristic of a counter tube in which the counting rate is substantially independent of the applied voltage. [GEOGR] An extensive, flat-surfaced upland region, usually more than 150–300 meters in elevation and considerably elevated above the adjacent country and limited by an abrupt descent on at least one side. [GEOL] A broad, comparatively flat and poorly defined elevation of the sea floor, commonly over 200 meters in elevation.

plateau basalt [GEOL] One or a succession of high-temperature basaltic lava flows from fissure eruptions which accumulate to form a plateau. Also known as flood basalt.

plateau characteristic [ELECTR] The relation between counting rate and voltage for a counter tube when radiation is constant, showing a plateau after the rise from the starting voltage to the Geiger threshold. Also known as counting rate–voltage characteristic.

plateau glacier [HYD] A highland glacier that overlies a generally flat mountain tract; usually overflows its edges in hanging glaciers.

plateau problem [MATH] The problem of finding a minimal surface having as boundary a given curve.

plate bearing test [ENG] Former method to estimate the bearing capacity of a soil; a rigid steel plate about 1 foot (30 centimeters) square was placed on the foundation level and then loaded until the foundation failed, as evidenced by rapid sinking of the plate.

plate-belt feeder *See* apron feeder.

plate budding [BOT] Plant budding by inserting a rectangular scion with a bud under a flap of bark on the stock in such a manner that the exposed wood on the stock is covered.

plate burning [GRAPHICS] After a negative has been prepared for an engraving, it is placed on the plate in a vacuum frame; the rays from the light source "burn" the plate, passing through the transparent areas of the negative and hardening the plate surface coating; this coating is later developed and the unexposed areas washed away; plates are also burned for offset printing.

plate cam [MECH ENG] A flat, open cam that imparts a sliding motion.

plate circuit *See* anode circuit.

plate-circuit detector *See* anode-circuit detector.

plate coil [MECH ENG] Heat-transfer device made from two metal sheets held together, one or both plates embossed to form passages between them for a heating or cooling medium to flow through. Also known as panel coil.

plate conveyor [MECH ENG] A conveyor with a series of steel plates as the carrying medium; each plate is a short trough, all slightly overlapped to form an articulated band, and attached to one center chain or to two side chains; the chains join rollers running on an angle-iron framework and transmit the drive from the driveheads, installed at intermediate points and sometimes also at the head or tail ends.

plate crystal [HYD] An ice crystal exhibiting typical hexagonal (rarely triangular) symmetry and having comparatively little thickness parallel to its principal axis (*c*-axis); as such crystals fall through the clouds in which they form, they may encounter conditions causing them to develop dendritic extensions, that is, to become plane-dendritic crystals.

plate current *See* anode current.

plated circuit [ELECTR] A printed circuit produced by electrodeposition of a conductive pattern on an insulating base. Also known as plated printed circuit.

plate detector *See* anode detector.

plate dissipation *See* anode dissipation.

plated printed circuit *See* plated circuit.

plated wire memory [ADP] A nonvolatile magnetic memory utilizing small zones of thin films plated on wires; such memories are characterized by very fast access and nondestructive readout.

plate efficiency *See* anode efficiency.

plate feeder *See* apron feeder.

plate-fin exchanger [MECH ENG] Heat-transfer device made up of a stack or layers, with each layer consisting of a corrugated fin between flat metal sheets sealed off on two sides by channels or bars to form passages for the flow of fluids.

plate girder [CIV ENG] A riveted or welded steel girder having a deep vertical web plate with a pair of angles along each edge to act as compression and tension flanges.

plate girder bridge [CIV ENG] A bridge whose structural elements have an I-shaped cross section and are riveted together from plates or angles, or welded from plates alone.

plate glass [MATER] Flat, high-quality glass with plane, parallel surfaces.

plate ice *See* pancake ice.

plate impedance *See* anode impedance.

plate input power *See* anode input power.

platelet *See* thrombocyte.

plate-load impedance *See* anode impedance.

plate-making [GRAPHICS] The forming of printing plates; the plate may be an offset plate, gravure plate, or a photogelatin plate.

plate modulation   *See* anode modulation.

platen   [ENG] **1.** A flat plate against which something rests or is pressed. **2.** The rubber-covered roller of a typewriter against which paper is pressed when struck by the typebars.

plate neutralization   *See* anode neutralization.

platen press   [GRAPHICS] A type of printing press with a flat surface bearing the inked type; another flat surface, bearing the paper, is pressed against the type; small hand presses are ordinarily of this sort.

plate proof   [GRAPHICS] A proof obtained from a printing plate.

plate pulse modulation   *See* anode pulse modulation.

plate resistance   *See* anode resistance.

plate saturation   *See* anode saturation.

plate-shear test   [ENG] A method used to get true shear data on a honeycomb core by bonding the core between two thick steel plates and subjecting the core to shear by displacing the plates relative to each other by loading in either tension or compression.

plate tectonics   [GEOL] Global tectonics based on a model of the earth characterized by a small number (10–25) of semirigid plates which float on some viscous underlayer in the mantle; each plate moves more or less independently and grinds against the others, concentrating most deformation, volcanism, and seismic activity along the periphery. Also known as raft tectonics.

plate theory   [ANALY CHEM] In gas chromatography, the theory that the column operates similarly to a distillation column; for example, chromatographic columns are considered as consisting of a number of theoretical plates, each performing a partial separation of components.

plate tower   [CHEM ENG] A distillation tower along the internal height of which is a series of transverse plates (bubble-cap or sieve) to force intimate contact between downward flowing liquid and upward flowing vapor.

plate-type exchanger   [MECH ENG] Heat-exchange device similar to a plate-and-frame filter press; fluids flow between the frame-held plates, transferring heat between them.

plate wave   [ACOUS] A type of ultrasonic vibration generated in a thin solid, such as a sheet of metal having a thickness of less than one wavelength, and usually consisting of a variety of simultaneous modes having different velocities; it is used in metal inspection. Also known as Lamb wave.

platform   [GEOL] **1.** Any level or almost level surface; a small plateau. **2.** A continental area covered by relatively flat or gently tilted, mainly sedimentary strata which overlay a basement of rocks consolidated during earlier deformations; platforms and shields together constitute cratons.   [MIN ENG] A wooden floor on the side of a gangway at the bottom of an inclined seam, to which the coal runs by gravity, and from which it is shoveled into mine cars.   [ORD] **1.** Temporary or permanent solid bed on which artillery pieces are supported to give greater stability. **2.** Metal stand at the base of some types of guns upon which the gun crew stands while serving the gun.

platform balance   [ENG] A weighing device with a flat plate mounted above a balanced beam.

platform blowing   [ENG] Special technique for blow-molding large parts made of plastic without sagging of the part being formed.

platform conveyor   [MECH ENG] A single- or double-strand conveyor with plates of steel or hardwood forming a continuous platform on which the loads are placed.

platform deck   [NAV ARCH] A partial deck fitted in the hold of a ship.

platform erection   [NAV] In the alignment of inertial navigation equipment, the alignment of the stable platform vertical axis with the local vertical.

platform facies   *See* shelf facies.

Platforming   [CHEM ENG] A proprietary petroleum-refining process for naphtha reforming; used a platinum-containing catalyst to produce high-octane gasoline or BTX aromatics.

platina   [MET] A white brittle brass containing 75% zinc and 25% copper; used for jewelry.

plating   [MET] Forming a thin, adherent layer of metal on an object. Also known as metal plating.

plating rack   [MET] A fixture that holds, and conducts current to, a piece of work during electrodeposition.

platinic chloride   *See* chloroplatinic acid.

platinic sulfate   *See* platinum sulfate.

platiniridium   [MINERAL] A silver-white cubic mineral composed of platinum, iridium, and related metals, occurring in grains.

platinite   *See* platynite.

platinochloride   *See* chloroplatinate.

platinocyanide   [INORG CHEM] A double salt of platinous cyanide and another cyanide, such as $K_2Pt(CN)_4$; used in photography and fluorescent x-ray screens. Also known as cyanoplatinate.

platinoid   [MET] **1.** Resembling or related to platinum. **2.** A copper-nickel-zinc alloy used for electrical resistance wire.

platinotron   [ELECTR] A microwave tube that may be used as a high-power saturated amplifier or oscillator in pulsed radar applications; requires permanent magnet just as does a magnetron.

platinous chloride   *See* platinum dichloride.

platinous iodide   *See* platinum iodide.

platinum   [CHEM] A chemical element, symbol Pt, atomic number 78, atomic weight 195.09.   [MET] A soft, ductile, malleable, grayish white noble metal with relatively high electric resistance; used in alloys, in electrical and electronic devices, and in jewelry.

platinum bichloride   *See* platinum dichloride.

platinum black   [MET] Black-colored, finely divided metallic platinum; soluble in aqua regia; used as a catalyst, as an absorbent for gases (hydrogen, oxygen), and for gas ignition. Also known as platinum Mohr.

platinum chloride   [INORG CHEM] $PtCl_4$ or $PtCl_4 \cdot 5H_2O$ A brown solid or red crystals; soluble in alcohol and water; decomposes when heated (loses $4H_2O$ at 100°C); used as an analytical reagent.

platinum dichloride   [INORG CHEM] $PtCl_2$ Water-insoluble, green-gray powder; decomposes to platinum at red heat; used to make platinum salts. Also known as platinous chloride; platinum bichloride.

platinum diiodide   *See* platinum iodide.

platinum electrode   [PHYS CHEM] A solid platinum wire electrode used during voltammetric analyses of electrolytes.

platinum iodide   [INORG CHEM] PtI_2 Water- and alkali-insoluble black powder; slightly soluble in hydrochloric acid; decomposes at 300–350°C. Also known as platinous iodide; platinum diiodide.

platinum-iridium alloy   [MET] An alloy with 1–30% iridium; as concentration of iridium increases, so do hardness, chemical resistance, and melting point; used in jewelry, electrical contacts, and hypodermic needles.

platinum Mohr   *See* platinum black.

platinum oxide   [INORG CHEM] An oxide of platinum; examples are platinum monoxide (or platinous oxide), PtO, and platinum dioxide (or platinic oxide), PtO_2.

platinum resistance thermometer   [ENG] The basis of the International Practical Temperature Scale of 1968 from 259.35° to 630.74°C; used in industrial thermometers in the range 0 to 650°C; capable of high accuracy because platinum is noncorrosive, ductile, and nonvolatile, and can be obtained in a very pure state. Also known as Callendar's thermometer.

platinum-rhodium alloy   [MET] An alloy with up to 40% rhodium; as concentration of rhodium increases, so do chemical resistance and hardness (although less hard than for platinum-iridium alloys); used as a catalyst to make nitric acid, in thermocouples, and in rayon spinnerets.

platinum sponge   [MET] Porous, grayish-black mass of finely divided platinum; soluble in aqua regia; used as a catalyst, and for ignition of combustible gases.

platinum sulfate   [INORG CHEM] $Pt(SO_4)_2$ A hygroscopic, dark mass; soluble in alcohol, ether, water, and dilute acids; used in microanalysis for halogens. Also known as platinic sulfate.

platmeric   [ANTHRO] Of a thighbone, being flattened laterally, with a platymeric index of 75 to 85.

platonic year   *See* great year.

PLATINUM

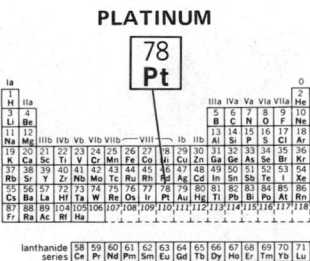

Periodic table of the chemical elements showing the position of platinum.

Platreating [CHEM ENG] Proprietary petroleum-refinery process to hydrogenate and refine concentrates of aromatics to recover pure benzene and toluene.

platter *See* phonograph record.

plattnerite [MINERAL] PbO_2 An iron-black mineral consisting of lead dioxide, occurring in masses with submetallic luster.

Plattner's process [MET] A process for extracting gold in which a charge of gold-bearing pulp is placed in a revolving iron drum lined with lead, and a stream of chlorine gas is conducted through the pulp, producing chloride of gold, which is soluble in water.

Platyasterida [INV ZOO] An order of Asteroidea in which traces of metapinnules persist, the ossicles of the arm skeleton being arranged in two growth gradient systems.

Platybelondoninae [PALEON] A subfamily of extinct elephantoid mammals in the family Gomphotheriidae consisting of species with digging specializations of the lower tusks.

platycelous [VERT ZOO] Of a vertebra, having a flat or concave ventral surface and a convex dorsal surface.

Platycephalidae [VERT ZOO] The flatheads, a family of perciform fishes in the suborder Cottoidei.

Platyceratacea [PALEON] A specialized superfamily of extinct gastropod mollusks which adapted to a coprophagous life on crinoid calices.

platycnemic index [ANTHRO] The ratio, multiplied by 100, of the anteroposterior diameter to the lateral diameter of the shinbone.

Platycopa [INV ZOO] A suborder of ostracod crustaceans in the order Podocopida including marine forms with two pairs of thoracic legs.

Platycopina [INV ZOO] The equivalent name for Platycopa.

Platyctenea [INV ZOO] An order of the ctenophores whose members are sedentary or parasitic; adults often lack ribs and are flattened due to shortening of the main axis.

platy flow structure [PETR] Structure of an igneous rock characterized by tabular sheets which suggest stratification, and formation by contraction during cooling. Also known as linear flow structure; planar flow structure.

Platygasteridae [INV ZOO] A family of hymenopteran insects in the superfamily Proctotrupoidea.

Platyhelminthes [INV ZOO] A phylum of invertebrates composed of bilaterally symmetrical, nonsegmented, dorsoventrally flattened worms characterized by lack of coelom, anus, circulatory and respiratory systems, and skeleton.

platymeric index [ANTHRO] The index, multiplied by 100, of the anteroposterior diameter to the lateral diameter of the femur.

platymyarian [INV ZOO] In nematodes, pertaining to flat muscle cells with the fibrillar region limited to a basal zone.

platynite [MINERAL] $PbBi_2(Se,S)_3$ An iron-black mineral composed of selenide and sulfide of lead and bismuth; occurs in thin metallic plates resembling graphite. Also spelled platinite.

platypellic [ANTHRO] Referring to or descriptive of a broad pelvis having a pelvic index of less than 90.

Platypodidae [INV ZOO] The ambrosia beetles, a family of coleopteran insects in the superfamily Curculionoidea.

Platypsyllidae [INV ZOO] The equivalent name for Leptinidae.

platypus [VERT ZOO] *Ornithorhynchus anatinus.* A monotreme, making up the family Ornithorhynchidae, which lays and incubates eggs in a manner similar to birds, and retains some reptilian characters; the female lacks a marsupium. Also known as duckbill platypus.

platysma [ANAT] A subcutaneous muscle of the neck, extending from the face to the clavicle; muscle of facial expression.

Platysomidae [PALEON] A family of extinct palaeonisciform fishes in the suborder Platysomoidei; typically, the body is laterally compressed and rhombic-shaped, with long dorsal and anal fins.

Platysomoidei [PALEON] A suborder of extinct deep-bodied marine and fresh-water fishes in the order Palaeonisciformes.

platyspondylia [MED] A rare congenital skeletal defect marked by abnormally shaped vertebrae.

Platysternidae [VERT ZOO] The big-headed turtles, a family of Asiatic fresh-water Chelonia with a single species (*Platysternon megacephalum*), characterized by a large head, hooked mandibles, and a long tail.

playa [GEOL] A low, essentially flat part of a basin or other undrained area in an arid region.

playa lake [HYD] A shallow temporary sheet of water covering a playa in the wet season.

playback [ENG ACOUS] Reproduction of a sound recording.

playback head [ELECTR] A head that converts a changing magnetic field on a moving magnetic tape into corresponding electric signals. Also known as reproduce head.

Playfair's law [GEOL] The law that each stream cuts its own valley, the valley being proportional in size to its stream, and the stream junctions in the valley are accordant in level.

play of color [OPTICS] An optical phenomenon consisting of a rapid succession of flashes of a variety of prismatic colors as certain minerals or cabochon-cut gems are moved about; caused by diffraction of light from spherical particles of amorphous silica stacked in an orderly three-dimensional pattern. Also known as schiller.

play therapy [PSYCH] A form of treatment, used particularly with children, in which a child's play, as with dolls in the presence of a therapist, is used as a medium for expression and communication.

pleasure principle [PSYCH] The instinctive attempt to avoid pain, discomfort, or unpleasant situations; the desire to obtain maximum gratification with minimum effort.

pleated cartridge [DES ENG] A filter cartridge made into a convoluted form that resembles the folds of an accordian.

Plecoptera [INV ZOO] The stoneflies, an order of primitive insects in which adults differ only slightly from immature stages, except for wings and tracheal gills.

plectane [INV ZOO] A cuticular plate supporting papillae in some nematodes.

Plectascales [MYCOL] An equivalent name for Eurotiales.

Plectognathi [VERT ZOO] The equivalent name for Tetraodontiformes.

Plectoidea [INV ZOO] A superfamily of small, free-living nematodes characterized by simple spiral amphids or variants thereof, elongate cylindroconoid stoma, and reflexed ovaries.

plectostele [BOT] A protostele that has the xylem divided into plates.

Pleiades [ASTRON] An open cluster of a few hundred stars in the constellation Taurus; six of the stars are easily visible to the naked eye.

Pleidae [INV ZOO] A family of hemipteran insects in the superfamily Pleoidea.

pleiotropy [GEN] The quality of a gene having more than one phenotypic effect.

Pleistocene [GEOL] An epoch of geologic time of the Quaternary period, following the Tertiary and before the Holocene. Also known as Ice Age; Oiluvium.

plenum [ENG] A condition in which air pressure within an enclosed space is greater than that in the outside atmosphere.

plenum chamber [ENG] An enclosed space in which a plenum condition exists; air is forced into it for slow distribution through ducts.

pleochroic halos [OPTICS] Halos of color or color differences that are sometimes observed around inclusions in minerals, resulting from irradiation by alpha particles.

pleochroism [OPTICS] Phenomenon exhibited by certain transparent crystals in which light viewed through the crystal has different colors when it passes through the crystal in different directions. Also known as polychroism.

pleocytosis [MED] Increase of cells in the cerebrospinal fluid.

pleodont [VERT ZOO] Having solid teeth.

Pleoidea [INV ZOO] A superfamily of suboval hemipteran insects belonging to the subdivision Hydrocoriseae.

pleomorphism [BIOL] The occurrence of more than one distinct form of an organism in a single life cycle. [CRYSTAL] *See* polymorphism.

pleonaste *See* ceylonite.

pleopod [INV ZOO] An abdominal appendage in certain crustaceans that is modified for swimming.

Pleosporales [BOT] The equivalent name for the lichenized Pseudosphaeriales.

plerocercoid [INV ZOO] The infective metacestode of certain

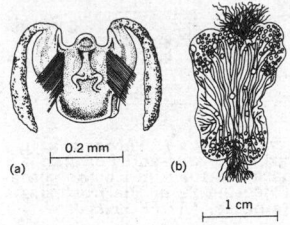

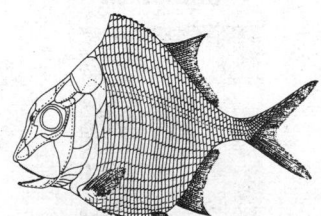

PLEURACANTHODII

30 cm

Xenacanthus (Pleuracanthus), Carboniferous and Permian sharklike form; note the spine projecting from the posterior braincase. *(After Fritsch)*

PLEURONECTIFORMES

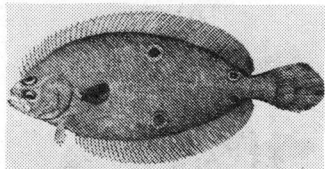

Fourspot flounder *(Paralichthys oblongus)*, of the Pleuronectiformes. *(After G. B. Goode, Fishery Industries of the United States, sect. 1, 1884)*

PLEUROPNEUMONIALIKE ORGANISM

Electron micrograph of a goat strain PPLO. *(From E. Klieneberger-Nobel and F. W. Cuckow, J. Gen. Microbiol., 12:95–99, 1955)*

PLIOCENE

	PALEOZOIC							MESOZOIC			CENOZOIC	
PRECAMBRIAN	CAMBRIAN	ORDOVICIAN	SILURIAN	DEVONIAN	Mississippian	Pennsylvanian	PERMIAN	TRIASSIC	JURASSIC	CRETACEOUS	TERTIARY	QUATERNARY
					CARBON-IFEROUS							

TERTIARY					QUATERNARY	
Paleocene	Eocene	Oligocene	Miocene	Pliocene	Pleistocene	Recent

Chart showing the relationship of the Pliocene to the eras and periods of geologic time.

cyclophyllidean tapeworms; distinguished by a solid body.

plerome [BOT] Central core of primary meristem which gives rise to all cells of the stele from the pericycle inward.

plesioaster [INV ZOO] A type of poriferan microscleric monaxonic spicule.

Plesiocidaroida [PALEON] An extinct order of echinoderms assigned to the Euechinoidea.

plesiomorph [EVOL] The original character of a branching phyletic lineage, found in the ancestral forms.

Plesiosauria [PALEON] A group of extinct reptiles in the order Sauropterygia constituting a highly specialized offshoot of the nothosaurs.

plesiotype [SYST] A specimen or specimens on which subsequent descriptions are based.

Plessy's green [INORG CHEM] $CrPO_4 \cdot xH_2O$ Deep-green pigment made of chromium phosphate mixed with chromium oxide and calcium phosphate.

Plethodontidae [VERT ZOO] A large family of salamanders in the suborder Salamandroidea characterized by the absence of lungs and the presence of a fine groove from nostril to upper lip.

plethysmograph [MED] An instrument for measuring changes in the size of a part of the body by measuring changes in the amount of blood in that part.

pleura [ANAT] The serous membrane covering the lung and lining the thoracic cavity.

Pleuracanthodii [PALEON] An order of Paleozoic sharklike fishes distinguished by two-pronged teeth, a long spine projecting from the posterior braincase, and direct backward extension of the tail.

pleural cavity [ANAT] The potential space included between the parietal and visceral layers of the pleura.

pleural rib *See* ventral rib.

pleurapophysis [ANAT] One of the lateral processes of a vertebra, corresponding morphologically to a rib.

pleurisy [MED] Inflammation of the pleura. Also known as pleuritis.

pleuritis *See* pleurisy.

pleurobranchia [INV ZOO] A gill that arises from the lateral wall of the thorax in certain arthropods.

pleurocarpous [BOT] Having the sporophyte in leaf axils along the side of the stem or on lateral branches; refers specifically to mosses.

Pleuroceridae [INV ZOO] A family of fresh-water snails in the order Pectinibranchia.

Pleurocoelea [INV ZOO] An extinct superfamily of gastropod mollusks of the order Opisthobranchia in which the shell, mantle cavity, and gills were present.

Pleurodira [VERT ZOO] A suborder of turtles (Chelonia) distinguished by spines on the posterior cervical vertebrae so that the head is retractile laterally.

pleurodontia [VERT ZOO] Attachment of the teeth to the inner surface of the jawbone.

pleurodynia [MED] Severe paroxysmal pain and tenderness of the intercostal muscles.

pleurolophocercous cercaria [INV ZOO] A larval digenetic trematode distinguished by a long, powerful tail with a pair of fin folds, a protrusible oral sucker, and pigmented dorsal eyespots.

Pleuromeiaceae [PALEOBOT] A family of plants in the order Pleuromiales, but often included in the Isoetales due to a phylogenetic link.

Pleuromeiales [PALEOBOT] An order of Early Triassic lycopods consisting of the genus *Pleuromeia*; the upright branched stem had grasslike leaves and a single terminal strobilus.

pleuron [INV ZOO] The lateral portion of a single thoracic segment in arthropods.

Pleuronectiformes [VERT ZOO] The flatfishes, an order of actinopterygian fishes distinguished by the loss of bilateral symmetry.

Pleuronematina [INV ZOO] A suborder of the Hymenostomatida.

pleuroperitoneal cavity [VERT ZOO] The body cavity containing both the lungs and the abdominal viscera in all pulmonate vertebrates except mammals.

pleuropneumonia [MED] Combined pleurisy and pneumonia. [VET MED] An infectious disease of cattle producing

pleural and lung inflammation, caused by *Mycoplasma* species.

pleuropneumonialike organism [MICROBIO] Any of a poorly defined group of microorganisms classified in the order Mycoplasmatales, including the smallest organisms capable of independent life, and comparable in size to the large filterable viruses. Abbreviated PPLO.

Pleurostigmophora [INV ZOO] A subclass of the centipedes, in some taxonomic systems, distinguished by lateral spiracles.

Pleurotomariacea [PALEON] An extinct superfamily of gastropod mollusks in the order Aspidobranchia.

Plexiglas [MATER] Trade name for transparent plastic material made from methyl methacrylate, produced in sheets, rods, and tubes.

plexus [ANAT] A network of interlacing nerves or anastomosing vessels.

pli [MECH] A unit of line density (mass per unit length) equal to 1 pound per inch, or approximately 17.8580 kilograms per meter.

plica [BIOL] A fold, as of skin or a leaf.

plied yarn [TEXT] A yarn composed of two or more strands.

Pliensbachian [GEOL] A European stage of geologic time: Lower Jurassic (above Sinemurian, below Toarcian).

pliers [DES ENG] A small instrument with two handles and two grasping jaws, usually long and roughened, working on a pivot; used for holding small objects and cutting, bending, and shaping wire.

Plimsoll mark [NAV ARCH] 1. A circular disk with a horizontal line placed on the side of a ship to indicate the summer load line. 2. A series of marks on the side of a ship indicating the load line for various seasons and geographical areas.

Plinian eruption *See* Vulcanian eruption.

plinth [ARCH] 1. The block forming the lowest member or base of a column or pedestal. 2. A slight widening at the base of a column or wall.

Pliocene [GEOL] A worldwide epoch of geologic time of the Tertiary period, extending from the end of the Miocene to the beginning of the Pleistocene.

Pliohyracinae [PALEON] An extinct subfamily of ungulate mammals in the family Procaviidae.

pliotron [ELECTR] General term for any hot-cathode vacuum tube having one or more grids.

ploidy [GEN] Number of complete chromosome sets in a nucleus.

Plokiophilidae [INV ZOO] A small family of predacious hemipteran insects in the superfamily Cimicoidea; individuals live in the webs of spiders and embiids.

plot [CIV ENG] A measured piece of land.

Plotosidae [VERT ZOO] A family of Indo-Pacific salt-water catfishes (Siluriformes).

plotter [ENG] A visual display or board on which a dependent variable is graphed by an automatically controlled pen or pencil as a function of one or more variables.

plotting board [ENG] The surface portion of a plotter, on which graphs are recorded. Also known as plotting table.

plotting chart [NAV] A chart designed primarily for plotting lines of position obtained from dead reckoning, celestial observations, or radio aids, and other sources of navigational data.

plotting head *See* reflection plotter.

plotting interval [NAV] The elapsed time between the first and last readings of radar range and bearing of an object being tracked.

plotting paper [GRAPHICS] Drawing paper laminated to both sides of an aluminum sheet or foil to ensure better dimensional stability; it is often used in drawing the color separations used in mapmaking.

plotting sheet [NAV] A blank chart, usually on the Mercator projection, showing only the graticule and a compass rose, so that the plotting sheet can be used for any longitude. [ORD] An item having lines forming rectangular grids printed on one side; the grid lines are approximately parallel with and extending to the edges of the sheet, and are approximately centered thereon; used in graphical determination of artillery ranges and deflection angles to targets from plotted gun position.

plotting table *See* plotting board.

plough [MIN ENG] **1.** A continuous mining machine in which cutting blades, moved over the face being worked, bite into the coal as they are pulled along and discharge it on an accompanying conveyor. **2.** A V-shaped scraper that presses against the return belt of a conveyor, removing coal and debris from it.

Plough [ASTRON] The name used in Britain for the constellation Big Dipper (Ursa Major).

plough deflector [MIN ENG] A steel plate at the end of a cutter-loader to deflect cut coal onto the face conveyor.

ploughed-and-tongued joint *See* feather joint.

plough wind *See* plow wind.

plow [AGR] An implement consisting of a share, moldboard, and landside attached to a frame; used to cut, lift, turn, and pulverize soil in preparation of a seedbed.

plowshare [DES ENG] The pointed part of a plow moldboard, which penetrates and cuts the soil first.

Plowshare [NUCLEO] The U.S. Atomic Energy Commission program of research and development on peaceful uses of nuclear explosives; possible uses include large-scale excavation, such as for canals and harbors, crushing ore bodies, and producing heavy transuranic isotopes.

plow steel [MET] High-quality, high-strength steel with 0.5 to 0.95% carbon content, used for wire rope.

plow wind [METEOROL] A term used in the midwestern United States to describe strong, straight-line winds associated with squall lines and thunderstorms; resulting damage is usually confined to narrow zones like that caused by tornadoes; however, the winds are all in one direction. Also spelled plough wind.

plucking [GEOL] A process of glacial erosion which involves the penetration of ice or rock wedges into subglacial niches, crevices, and joints in the bedrock; as the glacier moves, it plucks off pieces of jointed rock and incorporates them. Also known as glacial plucking; quarrying.

plug [ELEC] The half of a connector that is normally movable and is generally attached to a cable or removable subassembly; inserted in a jack, outlet, receptacle, or socket. [GEOL] **1.** A vertical pipelike magmatic body representing the conduit to a former volcanic vent. **2.** A crater filling of lava, the surrounding material of which has been removed by erosion. **3.** A mass of clay, sand, or other sediment filling the part of a stream channel abandoned by the formation of a cutoff. [MET] **1.** A rod or mandrel over which a pierced tube is forced, or that fills a tube as it is drawn through a die. **2.** A punch or mandrel over which a cup is drawn. **3.** A protruding portion of a die impression for forming a corresponding recess in the forging. **4.** A false bottom in a die. Also known as peg. [MIN ENG] A watertight seal in a shaft formed by removing the lining and inserting a concrete dam, or by placing a plug of clay over ordinary debris used to fill the shaft up to the location of the plug. [SCI TECH] **1.** A piece of material used to fill a hole. **2.** A small segment of material removed from a larger object.

plug adapter lamp holder [ELEC] A device that can be inserted in a lamp holder to act as a lamp holder and one or more receptacles. Also known as current tap.

plug-and-feather hole [ENG] A hole drilled in quarries for the purpose of splitting a block of stone by the plug-and-feather method.

plug-and-feather method [MIN ENG] A method of breaking large quarry stones into smaller blocks; a row of holes is drilled in the stone along a line where the break is desired; a pair of feathers (semicircular cross-section rods) is inserted in each hole; a plug (steel wedge) is inserted between each feather pair; the plugs are hammered in succession until the stone fractures.

plug bit *See* noncoring bit.

plugboard *See* control panel.

plug cock *See* plug valve.

plug die *See* floating plug.

plug drawing [MET] Drawing tubing over a plug or mandrel and through a die simultaneously to reduce diameter and thickness and to produce a smooth symmetrical bore surface.

plug flow *See* piston flow.

plug forming [ENG] Thermoforming process for plastics molding in which a plug or male mold is used to partially preform the part before forming is completed, using vacuum or pressure.

plug fuse [ELEC] A fuse designed for use in a standard screw-base lamp socket.

plug gage [DES ENG] A steel gage that is used to test the dimension of a hole; may be straight or tapered, plain or threaded, and of any cross-sectional shape.

plugging [ELEC] Braking an electric motor by reversing its connections, so it tends to turn in the opposite direction; the circuit is opened automatically when the motor stops, so the motor does not actually reverse. [ENG] The formation of a barrier (plug) of solid material in a process flow system, such as a pipe or reactor. [MIN ENG] *See* blinding.

plugging agent [PETRO ENG] A chemical used to plug or block off selected permeable zones of a reservoir formation; used during formation acidizing to direct the acid to the tighter (less permeable) zones; examples are viscous gels, suspensions of graded solids, and finely ground vegetable material.

plughole [MIN ENG] **1.** A passageway left open while an old portion of a mine is sealed off, to help maintain normal ventilation; it is sealed when the work is finished. **2.** A hole for an explosive charge or for a bolt.

plug-in unit [ELEC] A component or subassembly having plug-in terminals so all connections can be made simultaneously by pushing the unit into a suitable socket.

plug program patching [ADP] A relatively small auxiliary plugboard patched with a specific variation of a portion of a program and designed to be plugged into a relatively larger plugboard patched with the main program.

plug valve [DES ENG] A valve fitted with a plug that has a hole through which fluid flows and that is rotatable through 90° for operation in the open or closed position. Also known as plug cock.

plug weld [MET] A circular fusion weld made in the hole of a slotted lap or tee joint.

plum [BOT] Any of various shrubs or small trees of the genus *Prunus* that bear smooth-skinned, globular to oval, drupaceous stone fruit.

plumage [VERT ZOO] The entire covering of feathers of a bird.

Plumatellina [INV ZOO] The single order of the ectoproct bryozoan class Phylactolaemata.

Plumbaginaceae [BOT] The leadworts, the single family of the order Plumbaginales.

Plumbaginales [BOT] An order of dicotyledonous plants in the subclass Caryophyllidae; flowers are pentamerous with fused petals, trinucleate pollen, and a compound ovary containing a single basal ovule.

plumbago *See* graphite.

plumb bob [ENG] A weight suspended on a string to indicate the direction of the vertical.

plumb bond [CIV ENG] A masonry bond in which corresponding joints (for example, on alternate courses) are aligned.

plumbing [CIV ENG] The system of pipes and fixtures concerned with the introduction, distribution, and disposal of water in a building. [ELECTROMAG] Slang term for the pipelike waveguide circuit elements used in microwave radio and radar equipment.

plumbism [MED] Lead poisoning.

plumb line [ENG] The string on which a plumb bob hangs.

plum blotch [PL PATH] A fungus disease of plums caused by *Phyllosticta congesta* and characterized by minute brown or gray angular leaf spots and brown or gray blotches on the fruit.

plumboferrite [MINERAL] $PbFe_4O_7$ A dark hexagonal mineral composed of lead iron oxide.

plumbogummite [MINERAL] **1.** $PbAl_3(PO_4)_2(OH)_5 \cdot H_2O$ A mineral composed of hydrous basic lead aluminum phosphate. **2.** A group of isostructural minerals, that includes gorceixite, goyazite, crandallite, deltaite, florencite, and dussertite, as well as plumbogummite.

plumbojarosite [MINERAL] $PbFe_6(SO_4)_4(OH)_{12}$ A mineral composed of basic lead iron sulfate; it is isostructural with jarosite.

PLUG VALVE

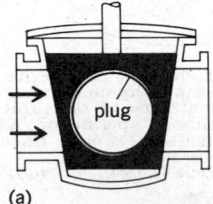

(a)

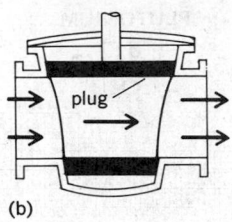

(b)

Plug valve. *(a)* Closed. *(b)* Open. Arrows show direction of fluid flow.

PLUM

Fruit on branch of Damson plum.

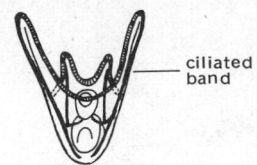

PLUTEUS

ciliated band

Drawing of a pluteus larva.

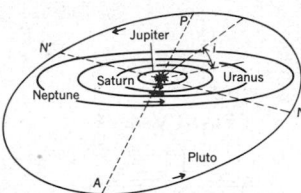

PLUTO

Orbit of Pluto. A perspective view to show the inclination *i* and eccentricity of the orbit. *A*, aphelion; *P*, perihelion; *NN'*, line of nodes. (From L. Rudaux and G. de Vaucouleurs, Astronomie, Larousse, 1948)

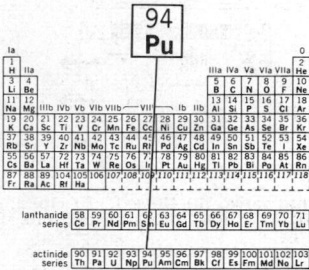

PLUTONIUM

Periodic table of the chemical elements showing the position of plutonium.

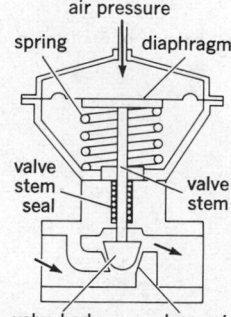

PNEUMATIC CONTROL VALVE

air pressure

spring   diaphragm

valve stem seal

valve stem

valve body   valve seat

Simplified cross section of typical pneumatic control valve. Arrows indicate direction of fluid stream.

plumbous oxide   *See* lead monoxide.

plumbous sulfide   *See* lead sulfide.

plumbum   [CHEM] Latin name for lead; source of the element symbol Pb.

plume   *See* column.

plumicome   [BIOL] A spicule with plumelike tufts.

Plummer-Vinson syndrome   [MED] Dysphagia koilonychia, gastric achlorhydria, glossitis, and hypochromic microcytic anemia caused by iron deficiency. Also known as Paterson-Kelly syndrome; sideropenic dysphagia.

plummet   [ENG] A loose-fitting metal plug in a tapered rotameter tube which moves upward (or downward) with an increase (or decrease) in fluid flow rate upward through the tube. Also known as float.

plumose   [VERT ZOO] Having feathers or plumes.

plum pocket   [PL PATH] A mild fungus disease of plums, caused either by *Taphrina pruni* or *T. communia,* in which the stone of the fruit is aborted.

plumule   [BOT] The primary bud of a plant embryo.   [VERT ZOO] A down feather.

plunge   [GEOL] The inclination of a geologic structure, especially a fold axis, measured by its departure from the horizontal. Also known as pitch; rake.

plunge grinding   [MECH ENG] Grinding in which the wheel moves radially toward the work.

plunge point   [OCEANOGR] The point at which a plunging wave curls over and falls as it moves toward the shore.

plunger   [DES ENG] A wooden shaft with a large rubber suction cup at the end, used to clear plumbing traps and waste outlets.   [ELECTROMAG] *See* piston.   [MECH ENG] The long rod or piston of a reciprocating pump.

plunger jig washer   [MIN ENG] A machine for washing ore, coal, or stones in which water is forced alternately up or down by a plunger.

plunger overtravel   [PETRO ENG] Excessive upward or downward movement in a reciprocating-plunger-type sucker-rod oil-well pump.

plunger pump   [MECH ENG] A reciprocating pump where the packing is on the stationary casing instead of the moving piston.

plunger-type instrument   [ENG] Moving-iron instrument in which the pointer is attached to a long and specially shaped piece of iron that is drawn into or moved out of a coil carrying the current to be measured.

plunging breaker   [OCEANOGR] A breaking wave whose crest curls over and collapses suddenly. Also known as spilling breaker; surging breaker.

plunging fire   [ORD] Gunfire that strikes the earth's surface at a high angle of fall, that is, greater than 45°.

plus   [MATH] A mathematical symbol; *A* plus *B*, where *A* and *B* are mathematical quantities, denotes the quantity obtained by taking their sum in an appropriate context.

plus-90 orientation   [ADP] In optical character recognition, that determinate position which indicates that the line elements of an inputted source document appear perpendicular with the leading edge of the optical reader.

plush copper ore   *See* chalcotrichite.

plus sieve   [MET] That portion of a powder sample that is retained by a standard sieve of a specified number.

plus zone   [ADP] The bit positions in a computer code which represent the algebraic plus sign.

pluteus   [INV ZOO] The free-swimming, bilaterally symmetrical, easel-shaped larva of ophiuroids and echinoids.

Pluto   [ASTRON] The most distant planet in the solar system; mean distance to the sun is about 5.6×10^9 kilometers; it has no known satellite, and its sidereal revolution period is 248.4 years.

pluton   [GEOL] 1. An igneous intrusion. 2. A body of rock formed by metasomatic replacement.

plutonian   *See* plutonic.

plutonic   [GEOL] Pertaining to rocks formed at a great depth. Also known as abyssal; deep-seated; plutonian.

plutonic breccia   [GEOL] Breccia consisting of older annular rock fragments enclosed in younger plutonic rock.

plutonic rock   [GEOL] A rock formed at considerable depth by crystallization of magma or by chemical alteration.

plutonism   [GEOL] 1. Pertaining to the processes associated

with pluton formation. 2. The theory that the earth formed by solidification of a molten mass.   [MED] A disease caused by exposure to plutonium, manifested in experimental animals by graying of the hair, liver degeneration, and tumor formation.

plutonium   [CHEM] A reactive metallic element, symbol Pu, atomic number 94, in the transuranium series of elements; the first isotope to be identified was plutonium-239; used as a nuclear fuel, to produce radioactive isotopes for research, and as the fissile agent in nuclear weapons.

plutonium-238   [NUC PHYS] The first synthetic isomer made of plutonium; similar chemically to uranium and neptunium; atomic number 94; formed by bombardment of uranium with deuterons.

plutonium-239   [NUC PHYS] A synthetic element chemically similar to uranium and neptunium; atomic number 94; made by bombardment of uranium-238 with slow electrons in a nuclear reactor; used as nuclear reactor fuel and an ingredient for nuclear weapons.

plutonium bomb   [ORD] An atomic bomb using plutonium.

plutonium oxide   [INORG CHEM] PuO_2 A radioactively poisonous pyrophoric oxide of plutonium; particles may be easily airborne.

plutonium reactor   [NUCLEO] A nuclear reactor in which plutonium is the principal fissionable material.

pluvial   [GEOL] Of a geologic process or feature, effected by rain action.   [METEOROL] Pertaining to rain, or more broadly, to precipitation, particularly to an abundant amount thereof.

pluviilignosa   [ECOL] A tropical rain forest.

pluviograph   *See* recording rain gage.

pluviometer   *See* rain gage.

pluviometric coefficient   [METEOROL] For any month at a given station, the ratio of the monthly normal precipitation to one-twelfth of the annual normal precipitation. Also known as hyetal coefficient.

ply   [MATER] A thin sheet of wood or other material bonded to one or more additional thin sheets, as in plywood.   [TEXT] A strand of yarn made by twisting together two or more strands.

plymetal   [MATER] 1. A material consisting of layers of dissimilar metals bonded together. 2. Plywood faced with aluminum on both sides.

plywood   [MATER] A material composed of thin sheets of wood glued together, with the grains of adjacent sheets oriented at right angles to each other.

Pm   *See* promethium.

PM   *See* permanent magnet; phase modulation.

PMA   *See* phosphomolybdic acid.

PMR   *See* projection microradiography.

pNa   [CHEM] Logarithm of the sodium-ion concentration in a solution; that is, $pNa = -\log a_{Na^+}$, where a_{Na^+} is the sodium-ion concentration.

PNdB   *See* perceived noise decibel.

pneumatic   [ENG] Pertaining to or operated by air or other gas.

pneumatic atomizer   [MECH ENG] An atomizer that uses compressed air to produce drops in the diameter range of 5–100 micrometers.

pneumatic caisson   [CIV ENG] A caisson having a chamber filled with compressed air at a pressure equal to the pressure of the water outside.

pneumatic controller   [MECH ENG] A device for the mechanical movement of another device (such as a valve stem) whose action is controlled by variations in pneumatic pressure connected to the controller.

pneumatic control valve   [MECH ENG] A valve in which the force of compressed air against a diaphragm is opposed by the force of a spring to control the area of the opening for a fluid stream.

pneumatic conveyor   [MECH ENG] A conveyor which transports dry, free-flowing, granular material in suspension, or a cylindrical carrier, within a pipe or duct by means of a high-velocity airstream or by pressure of vacuum generated by an air compressor. Also known as air conveyor.

pneumatic deception device   [ORD] A dummy tank, vehicle, or weapon made of inflatable material; used for deceiving the enemy intelligence as to location of friendly installations.

pneumatic drill [MECH ENG] Compressed-air drill worked by reciprocating piston, hammer action, or turbo drive.

pneumatic drilling [MECH ENG] Drilling a hole when using air or gas in lieu of conventional drilling fluid as the circulating medium; an adaptation of rotary drilling.

pneumatic filling [MIN ENG] A filling method using compressed air to blow filling material into the mined-out stope.

pneumatic gun charger [ORD] A gun charger that operates by compressed air or gas.

pneumatic hammer [MECH ENG] A hammer in which compressed air is utilized for producing the impacting blow.

pneumatic hoist *See* air hoist.

pneumatic injection [MIN ENG] A method for fighting underground coal fires, developed by the U.S. Bureau of Mines; this air-blowing technique involves the injection of incombustible mineral, like rock wool or dry sand, through 6-inch boreholes drilled from the surface to intersect underground passageways in the mines.

pneumatic lighting [MIN ENG] Lighting of underground chambers by a compressed-air turbomotor driving a small dynamo.

pneumatic loudspeaker [ENG ACOUS] A loudspeaker in which the acoustic ouput results from controlled variation of an airstream.

pneumatic method [MIN ENG] A method of flotation in which gas is introduced near the bottom of the flotation vessel.

pneumatic riveter [MECH ENG] A riveting machine having a rapidly reciprocating piston driven by compressed air.

pneumatics [FL MECH] Fluid statics and behavior in closed systems when the fluid is a gas.

pneumatic servo *See* valve positioner.

pneumatic servomechanism [CONT SYS] A servomechanism in which power is supplied and transmission of signals is carried out through the medium of compressed air.

pneumatic steelmaking [MET] Any steelmaking process which employs air or oxygen and for which all heat is derived from the initial heat content of the charge materials and from the thermal energy of the refining reactions.

pneumatic stowing [MIN ENG] A method of filling used mine cavities with crushed rock; which is forced by compressed air into the cavity.

pneumatic tank switcher [PETRO ENG] Pneumatic actuated valving system for oil-field tanks to shut off crude-oil flow to a filled tank and then to direct incoming crude flow to the next available empty tank.

pneumatic telemetering [ENG] The transmission of a pressure impulse by means of pneumatic pressure through a length of small-bore tubing; used for remote transmission of signals from primary process-unit sensing elements for pressure, temperature, flow rate, and so on.

pneumatic test [ENG] Pressure testing of a process vessel by the use of air pressure.

pneumatic transmission lag [ELEC] The time delay in a pneumatic transmission line between the generation of an impulse at one end and the resultant reaction at the other end.

pneumatic weighing system [ENG] A system for weight measurement in which the load is detected by a nozzle and balanced by modulating the air pressure in an opposing capsule.

pneumatocele [MED] 1. Herniation of the lung. 2. A sac or tumor containing gas; especially the scrotum filled with gas.

pneumatocodon [INV ZOO] Exumbrellar surface of the float or pneumatophore of siphonophorans.

pneumatolysis [GEOL] Rock alteration or mineral crystallization effected by gaseous emanations from solidifying magma.

pneumatolytic [GEOL] Formed by gaseous agents.

pneumatolytic metamorphism [PETR] Contact metamorphism by the chemical action of magmatic gases.

pneumatophore [BOT] 1. An air bladder in marsh plants. 2. A submerged or exposed erect root that functions in the respiration of certain marsh plants. [INV ZOO] The air sac of a siphonophore.

pneumatosaccus [INV ZOO] Subumbrellar surface of the float or pneumatophore of siphonophorans.

pneumatosis [MED] The presence of air or gas in abnormal situations in the body.

pneumaturia [MED] The voiding of urine containing free gas.

pneumoangiography [MED] The outline of the vessels of the lung by means of radiopaque material, for roentgenographic visualization.

pneumobacillus *See* Klebsiella pneumoniae.

pneumococci *See* Diplococcus pneumoniae.

pneumoconiosis [MED] Any lung disease caused by dust inhalation.

pneumocystis carinii pneumonia [MED] A lung infection in man caused by the protozoans *Pneumocystis carinii.* Also known as interstitial plasma-cell pneumonia.

pneumoencephalography [MED] A method of visualizing the ventricular system and subarachnoid pathways of the brain by roentgenography after removal of spinal fluid followed by the injection of air or gas into the subarachnoid space.

pneumoenteritis [MED] Inflammation of the lungs and of the intestine.

pneumography [MED] 1. Roentgenography of the lung. 2. The recording of the respiratory excursions.

pneumohemothorax [MED] The presence of air or gas and blood in the thoracic cavity.

pneumolithiasis [MED] The occurrence of calculi or concretions in a lung.

pneumomycosis [MED] Any disease of the lungs caused by a fungus.

pneumonectomy [MED] Surgical removal of an entire lung.

pneumonia [MED] An acute or chronic inflammation of the lungs caused by numerous microbial, immunological, physical, or chemical agents, and associated with exudate in the alveolar lumens.

pneumonic plague [MED] A virulent type of plague in man, with lung involvement.

pneumonitis [MED] Inflammation of the lung.

pneumonolysis [MED] The loosening of any portion of lung adherent to the chest wall; a form of collapse therapy used in the treatment of pulmonary tuberculosis.

pneumopericardium [MED] The presence of air in the pericardial cavity.

pneumoperitoneum [MED] 1. The presence of air or gas in the peritoneal cavity. 2. Injection of a gas into the peritoneal cavity as a diagnostic or therapeutic measure.

pneumostome [INV ZOO] The respiratory aperture of gastropod mollusks.

pneumothorax [MED] The presence of air or gas in the pleural cavity.

pn hook transistor *See* hook collector transistor.

pnigophobia [PSYCH] An abnormal fear of choking.

pnip-transistor [ELECTR] An intrinsic junction transistor in which the intrinsic region is sandwiched between the *n*-type base and the *p*-type collector.

pn junction [ELECTR] The interface between two regions in a semiconductor crystal which have been treated so that one is a *p*-type semiconductor and the other is an *n*-type semiconductor; it contains a permanent dipole charge layer.

pnpn-diode [ELECTR] A semiconductor device consisting of four alternate layers of *p*-type and *n*-type semiconductor material, with terminal connections to the two outer layers. Also known as *npnp*-diode.

pnp-transistor [ELECTR] A junction transistor having an *n*-type base between a *p*-type emitter and a *p*-type collector.

Po *See* polonium.

Poaceae [BOT] The equivalent name for Gramineae.

Poales [BOT] The equivalent name for Cyperales.

pock [MED] A pustule of an eruptive fever, especially of smallpox.

Pockels effect [OPTICS] Changes in the refractive properties of certain crystals in an applied electric field, which are proportional to the first power of the electric field strength.

pocket [GEOL] A localized enrichment. [MIN ENG] A receptacle from which coal, ore, or waste is loaded into wagons or cars.

pocket gopher *See* gopher.

pod [AERO ENG] An enclosure, housing, or detachable container of some kind on an airplane or space vehicle, as an engine pod. [BOT] A dry dehiscent fruit; a legume. [DES

PNEUMATIC WEIGHING SYSTEM

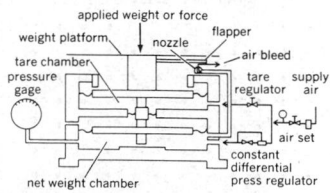

Elements of a pneumatic weighing system. *(From D. M. Considine, ed., Process Instruments and Controls Handbook, McGraw-Hill, 1957)*

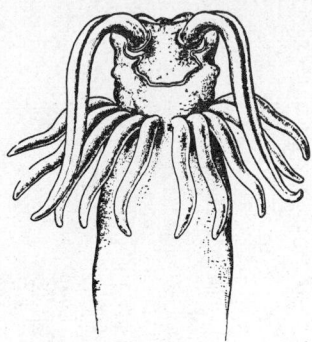

POEOBIIDAE

Anterior end of *Poeobius*
(Poeobiidae) in dorsal view.

POINSOT'S METHOD

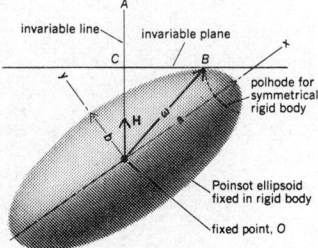

Inertia ellipsoid has axes with
lengths *a*, *b*, and *c* (not shown)
along principal axes of body,
x, *y*, and *z* (not shown).
OA = invariable line drawn
in direction of fixed vector
angular momentum **H**. Inertia
ellipsoid rolls on invariable
plane which is perpendicular
to *OA* and intersects it at *C*.
Ellipsoid is tangent to plane at
B, where *OB* = vector angular
momentum **ω**.

POINT DEFECT

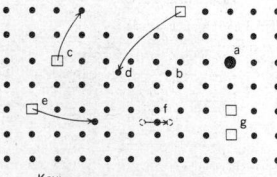

Key:
a = substitutional impurity
b = interstitial impurity
c = vacancy
d = interstitial
e = Frenkel pair
f = interstitial impurity moving from
    one site through a saddle point to
    a neighboring site
g = divacancy

Point defects in a lattice. The
squares indicate positions that
were formerly occupied by
atoms displaced as shown by
the arrows.

ENG] **1.** The socket for a bit in a brace. **2.** A straight groove in the barrel of a pod auger.

POD analysis [ANALY CHEM] A precision laboratory distillation procedure used to separate low-boiling hydrocarbon fractions quantitatively for analytical purposes. Also known as Podbielniak analysis.

Podargidae [VERT ZOO] The heavy-billed frogmouths, a family of Asian and Australian birds in the order Caprimulgiformes.

Podbielniak analysis *See* POD analysis.

Podbielniak extractor [CHEM ENG] A solvent-extraction device in which centrifugal action enhances liquid-liquid contact and increases resultant separation efficiency.

pod blight [PL PATH] A fungus disease of legumes caused by *Diaporthe* species.

podiatrist *See* chiropodist.

Podicipedide [VERT ZOO] The single family of the avian order Podicipediformes.

Podicipediformes [VERT ZOO] The grebes, an order of swimming and diving birds distinguished by dense, silky plumage, a rudimentary tail, and toes that are individually broadened and lobed.

Podicipitiformes [VERT ZOO] The equivalent name for Podicipediformes.

podite [INV ZOO] A segment of a limb of an arthropod.

podium [INV ZOO] The terminal portion of a body wall appendage in certain echinoderms.

podobranch [INV ZOO] A gill of a crustacean attached to the basal segment of a thoracic limb.

Podocopa [INV ZOO] A suborder of fresh-water ostracod crustaceans in the order Podocopida in which the inner lamella has a calcified rim joining the outer lamella along a chitinous zone of concrescence, and the two valves fit together firmly.

Podocopida [INV ZOO] An order of the Ostracoda; contains all fresh-water ostracods and is divided into the suborders Podocopa, Metacopina, and Platycopina.

Podocopina [INV ZOO] The equivalent name for Podocopa.

podocyst [INV ZOO] A sinus in the foot of certain gastropod mollusks.

Podogona [INV ZOO] The equivalent name for Ricinuleida.

Podostemaceae [BOT] The single family of the order Podostemales.

Podostemales [BOT] An order of dicotyledonous plants in the subclass Rosidae; plants are submerged aquatics with modified, branching shoots and small, perfect flowers having a superior ovary and united carpels.

pod rot [PL PATH] A fungus disease of cacao caused by *Monilia roreri* and characterized by lesions on the pods.

Podzol [GEOL] A soil group characterized by mats of organic matter in the surface layer and thin horizons of organic minerals overlying gray, leached horizons and dark-brown illuvial horizons; found in coal forests to temperate coniferous or mixed forests.

Poeciliidae [VERT ZOO] A family of fishes in the order Atheriniformes including the live-bearers, such as guppies, swordtails, and mollies.

Poecilosclerida [INV ZOO] An order of sponges of the class Demospongiae in which the skeleton includes two or more types of megascleres.

Poeobiidae [INV ZOO] A monotypic family of spioniform worms (*Poeobius meseres*) belonging to the Sedentaria and found in the North Pacific Ocean.

Poetsch process [MIN ENG] Shaft sinking in which brine at subzero temperature is circulated through boreholes to freeze running water through which a shaft or tunnel is to be driven, during development of a waterlogged mine.

Poggendorff's first method *See* constant-current dc potentiometer.

Poggendorff's second method *See* constant-resistance dc potentiometer.

pogonip *See* ice fog.

pogonochore [BOT] A type of plant that produces plumed disseminules.

Pogonophora [INV ZOO] The single class of the phylum Brachiata; the elongate body consists of three segments, each with a separate coelom; there is no mouth, anus, or digestive canal, and sexes are separate.

Pogson scale [ASTRON] An index of brightness used in star catalogs; it is the ratio of 2.512 to 1 between the brightness of successive magnitudes.

Pohlé air lift pump [MECH ENG] A pistonless pump in which compressed air fills the annular space surrounding the uptake pipe and is free to enter the rising column at all points of its periphery.

poidometer [ENG] An automatic weighing device for use on belt conveyors.

poikilitic [PETR] Of the texture of an igneous rock, having small crystals of one mineral randomly scattered without common orientation in larger crystals of another mineral.

poikiloblastic [PETR] Of a metamorphic texture, simulating the poikilitic texture of igneous rocks in having small idioblasts of one constituent lying within larger xenoblasts. Also known as sieve texture.

poikilocytosis [MED] A condition in which erythrocytes are distorted in shape.

poikilotherm [ZOO] An animal, such as reptiles, fishes, and invertebrates, whose body temperature varies with and is usually higher than the temperature of the environment; a cold-blooded animal.

Poincaré conjecture [MATH] The question as to whether a compact, simply connected three-dimensional manifold without boundary must be homeomorphic to the three-dimensional sphere.

Poincaré electron [ELECTROMAG] A classical model of the electron in which nonelectromagnetic forces hold the electron together so that it has zero self-stress; it is unstable and has infinite self-energy in the case of a point electron.

Poincaré recurrence theorem [MATH] **1.** A volume preserving homeomorphism T of a finite dimensional euclidean space will have, for almost all points x, infinitely many points of the form $T^i(x)$, $i = 1,2, \ldots$ within any open set containing x. **2.** A measure preserving transformation on a space with finite measure is recurrent.

Poinsot ellipsoid *See* inertia ellipsoid.

Poinsot motion [MECH] The motion of a rigid body with a point fixed in space and with zero torque or moment acting on the body about the fixed point.

Poinsot's method [MECH] A method of describing Poinsot motion, by means of a geometrical construction in which the inertia ellipsoid rolls on the invariable plane without slipping.

point [GEOGR] A tapering piece of land projecting into a body of water; it is generally less prominent than a cape. [LAP] A unit of mass, used in measuring precious stones, equal to 0.01 metric carat, or to 2 milligrams. [MATH] **1.** An element in a topological space. **2.** One of the basic undefined elements of geometry possessing position but no nonzero dimension. **3.** In positional notation, the character or the location of an implied symbol that separates the integral part of a numerical expression from its fractional part; for example, it is called the binary point in binary notation and the decimal point in decimal notation. [NAV] In marine operation, one thirty-second of a circle, or 11¼ degrees.

point angle [DES ENG] The angle at the point or edge of a cutting tool.

point bar [GEOL] One of a series of low, arcuate sand and gravel ridges formed on the inside of a growing meander by the gradual addition of accretions. Also known as meander bar.

point-bearing pile *See* end-bearing pile.

point-blank range [MECH] Distance to a target that is so short that the trajectory of a bullet or projectile is practically a straight, rather than a curved, line.

point contact [ELECTR] A contact between a specially prepared semiconductor surface and a metal point, usually maintained by mechanical pressure but sometimes welded or bonded.

point-contact transistor [ELECTR] A transistor having a base electrode and two or more point contacts located near each other on the surface of an *n*-type semiconductor.

point defect [CRYSTAL] A departure from crystal symmetry which effects only one, or, in some cases, two lattice sites.

point diagram [PETR] A fabric diagram in which a point represents the preferred orientation of each individual fabric element. Also known as scatter diagram.

pointed bracket *See* angle bracket.

pointer [ENG] The needle-shaped rod that moves over the scale of a meter.

point fire [ORD] Concentrated fire from a number of guns, directed at a single point or small area.

point function [MATH] A function whose values are points.

point group [CRYSTAL] A group consisting of the symmetry elements of an object having a single fixed point; 32 such groups are possible.

point harmonization [ORD] The adjusting of a fighter aircraft's fixed guns so that they converge upon a point at a given range.

pointing [CIV ENG] 1. Finishing a mortar joint. 2. Pressing mortar into a raked joint. [MATER] The material (mortar) used in pointing. [MET] Reducing the diameter and tapering a short length at the end of a rod, wire, or tube. Also known as metal pointing. [ORD] The operation of giving the piece a designated elevation and direction.

pointing trowel [ENG] A tool used to apply pointing to the joints between bricks.

point initiation [ENG] Application of the initial impulse from the detonator to a single point on the main charge surface; for a cylindrical charge this point is usually the center of one face.

point jammer [ELECTR] Any electronic jammer directed against a specific enemy installation operating on a specific frequency.

point-junction transistor [ELECTR] Transistor having a base electrode and both point-contact and junction electrodes.

point mutation [GEN] Mutation of a single gene due to addition, loss, replacement, or change of sequence in one or more base pairs of the deoxyribonucleic acid of that gene.

point of arrival [NAV] The position which a craft will reach after following specified courses for specified distances from a point of departure.

point of contraflexure [MECH] A point at which the direction of bending changes. Also known as point of inflection.

point of control [IND ENG] Fraction defective in those lots that have a probability of .50 of acceptance according to a specific sampling acceptance plan.

point of departure *See* departure.

point of destination [NAV] The point at which the final course from the point of departure ends, exclusive of the courses needed to reach a berth or runway.

point of fall [MECH] The point in the curved path of a falling projectile that is level with the muzzle of the gun. Also known as level point.

point of frog [CIV ENG] The place of intersection of the gage lines of the main track and a turnout.

point of impact [ORD] The point at which a bullet, bomb, or projectile strikes.

point of inflection [MATH] A point where a plane curve changes from the concave to the convex relative to some fixed line; equivalently, if the function determining the curve has a second derivative, this derivative changes sign at this point. Also known as inflection point. [MECH] *See* point of contraflexure.

point of intersection [CIV ENG] The point at which two straight sections or tangents to a road curve or rail curve meet when extended.

point of no return [NAV] A point along an aircraft track beyond which the aircraft's fuel supply will not permit it to return to its own or some other associated base but must continue in attempt to reach its destination.

point of switch [CIV ENG] That place in a track where a car passes from the main track to a turnout.

point of tangency [CIV ENG] The point at which a road curve or railway curve becomes straight or changes its curvature. Also known as tangent point.

point rainfall [METEOROL] The rainfall during a given time interval (or often one storm) measured in a rain gage, or an estimate of the amount which might have been measured at a given point.

point set topology [MATH] The general study of topological spaces.

points of the compass *See* compass points.

point source [PHYS] A source of radiation having definite position but no extension in space; this is an ideal which is a good approximation for distances from the source sufficiently large compared to the dimensions of the source.

point-source light [ELEC] A special lamp in which the radiating element is concentrated in a small physical area.

point spectrum [MATH] Those eigenvalues in the spectrum of a linear operator between Banach spaces whose corresponding eigenvectors are nonzero and of finite norm.

point system [IND ENG] 1. A system of job evaluation wherein job requirements are rated according to a scale of point values. 2. A wage incentive plan based on points instead of man minutes.

point target [ELECTROMAG] In radar, an object which returns a target signal by reflection from a relatively simple discrete surface; such targets are ships, aircraft, projectiles, missiles, and buildings. [ORD] A precise target of small dimensions.

point-to-point communication [COMMUN] Radio communication between two fixed stations.

point transposition [ELEC] Transposition, usually in an open-wire line, which is executed within a distance comparable to the wire separation, without material distortion of the normal wire configuration outside this distance.

pointwise convergence [MATH] A sequence of functions f_1, $f_2, \ldots$ defined on a set S converges pointwise to a function f if the sequence $f_1(x), f_2(x), \ldots$ converges to $f(x)$ for each x in S.

poise [FL MECH] A unit of dynamic viscosity equal to the dynamic viscosity of a fluid in which there is a tangential force 1 dyne per square centimeter resisting the flow of two parallel fluid layers past each other when their differential velocity is 1 centimeter per second per centimeter of separation. Abbreviated P.

poiseuille [FL MECH] A unit of dynamic viscosity of a fluid in which there is a tangential force of 1 newton per square meter resisting the flow of two parallel layers past each other when their differential velocity is 1 meter per second per meter of separation; equal to 10 poise; used chiefly in France. Abbreviated Pl.

Poiseuille flow [FL MECH] The steady flow of an incompressible fluid parallel to the axis of a circular pipe of infinite length, produced by a pressure gradient along the pipe.

Poiseuille's law [FL MECH] The law that the volume flow of an incompressible fluid through a circular tube is equal to $\pi/8$ times the pressure differences between the ends of the tube, times the fourth power of the tube's radius divided by the product of the tube's length and the dynamic viscosity of the fluid.

poison [ATOM PHYS] A substance which reduces the phosphorescence of a luminescent material. [CHEM] A substance that exerts inhibitive effects on catalysts, even when present only in small amounts; for example, traces of sulfur or lead will poison platinum-based catalysts. [ELECTR] A material which reduces the emission of electrons from the surface of a cathode. [MATER] A substance that in relatively small doses has an action that either destroys life or impairs seriously the functions of organs or tissues. [NUCLEO] A substance that absorbs neutrons without any fission resulting, and thereby lowers the reactivity of a nuclear reactor.

poison gas [MATER] A substance employed in chemical warfare to disable enemy troops; may be a gas, or a liquid or solid that gives off a gas. An example is mustard gas.

poison gland [VERT ZOO] Any of various specialized glands in certain fishes and amphibians which secrete poisonous mucuslike substances.

poison hemlock [BOT] *Conium maculatum.* A branching biennial poisonous herb that contains a volatile alkaloid, coniine, in its fruits and leaves.

poison ivy [BOT] Any of several climbing, shrubby, or arborescent plants of the genus *Rhus* in the sumac family (Anacardiaceae); characterized by ternate leaves, greenish flowers, and white berries that produce an irritating oil.

poison oak [BOT] Any of several bushy poison ivy plants or shrubby poison sumacs.

poisonous plant [BOT] Any of about 400 species of vascular

POISON HEMLOCK

Flowering branch of the poison hemlock. *(Adapted from Webster's New International Dictionary, 2d ed., Merriam, 1959)*

POISON IVY

Three-foliolate leaf of poison ivy.

POISON SUMAC

Poison sumac fruits and compound leaf (*Rhus vernix*).

plants containing principles which initiate pathological conditions in man and animals.

poison sumac [BOT] *Rhus vernix.* A tall bush of the sumac family (Anacardiaceae) bearing pinnately compound leaves with 7–13 entire leaflets, and drooping, axillary clusters of white fruits that produce an irritating oil.

Poisson binomial trials model *See* generalized binomial trials model.

Poisson bracket [MECH] For any two dynamical variables, X and Y, the sum, over all degrees of freedom of the system, of $(\partial X/\partial q)(\partial Y/\partial p) - (\partial X/\partial p)(\partial Y/\partial q)$, where q is a generalized coordinate and p is the corresponding generalized momentum.

Poisson constant [PHYS] The ratio k of the gas constant R to the specific heat at constant pressure C_p.

Poisson density functions [STAT] Density functions corresponding to Poisson distributions.

Poisson distribution [STAT] A probability distribution whose mean and variance have a common value k, and whose frequency is $f(x) = k^x e^{-k}/x!$, for $x = 0,1,2, \ldots$.

Poisson formula [MATH] If the infinite series of functions $f(2\pi k + t)$, k ranging from $-\infty$ to ∞, converges uniformly to a function of bounded variation, then the infinite series with term $f(2\pi k)$, k ranging from $-\infty$ to ∞, is identical to the series with term the integral of $f(x)e^{-ikx}dx$, k ranging from $-\infty$ to ∞.

Poisson integral formula [MATH] This formula gives a solution function for the Dirichlet problem in terms of integrals; an integral representation for the Bessel functions.

Poisson process [STAT] A process given by a discrete random variable which has a Poisson distribution.

Poisson ratio [MECH] The ratio of the transverse contracting strain to the elongation strain when a rod is stretched by forces which are applied at its ends and which are parallel to the rod's axis.

Poisson's equation [MATH] The partial differential equation which states that the Laplacian of an unknown function is equal to a given function.

Poisson transform [MATH] An integral transform which transforms the function $f(t)$ to the function

$$F(x) = (2/\pi) \int_0^\infty [t/(x^2 + t^2)]f(t)\,dt$$

Also known as potential transform.

poke welding *See* push welding.

polacke [METEOROL] A cold, dry, northeasterly katabatic wind in Bohemia descending from the Sudeten Mountains (from the direction of Poland).

polar [MATH] **1.** For a conic section, the polar of a point is the line that passes through the points of contact of the two tangents drawn to the conic from the point. **2.** For a quadric surface, the polar of a point is the plane that passes through the curve which is the locus of the points of contact of the tangents drawn to the surface from the point. **3.** For a quardric surface, the polar of a line is the line of intersection of the planes which are tangent to the surface at its points of intersection with the original line.

polar air [METEOROL] A type of air whose characteristics are developed over high latitudes; there are two types: continental polar air and maritime polar air.

polar anticyclone *See* arctic high; subpolar high.

polar automatic weather station [METEOROL] An automatic weather station which measures meteorological elements and transmits them by radio; the station is designed to function primarily in frigid or polar climates in order to fill the need for weather reports from inaccessible regions where manned stations are not practicable; since the equipment is designed to operate on ice or slush, the main structure is in the form of a sled with external pontoons for added stability.

polar axis [MATH] The directed straight line relative to which the angle is measured for a representation of a point in the plane by polar coordinates.

polar bear [VERT ZOO] *Thalarctos maritimus.* A large aquatic carnivore found in the polar regions of the Northern Hemisphere.

polar body [CYTOL] One of the small bodies cast off by the oocyte during maturation.

polar cap [HYD] An ice sheet centered at one of the poles of the earth.

polar-cap ice *See* polar ice.

polar chart [MAP] **1.** A chart of polar areas. **2.** A chart on a polar projection; the projections most used for polar charts are the gnomonic, stereographic, azimuthal equidistant, transverse Mercator, and modified Lambert conformal.

polar circle [GEOD] A parallel of latitude whose distance from the pole is equal to the obliquity of the ecliptic (approximately 23° 27′).

polar climate [CLIMATOL] The climate of a geographical polar region, most commonly taken to be a climate which is too cold to support the growth of trees. Also known as arctic climate; snow climate.

polar compound [CHEM] Molecules which contain polar covalent bonds; they can ionize when dissolved or fused; polar compounds include inorganic acids, bases, and salts.

polar continental air [METEOROL] Air of an air mass that originates over land or frozen ocean areas in the polar regions; characterized by low temperature, stability, low specific humidity, and shallow vertical extent.

polar-coordinate navigation system [NAV] A system in which one or more signals are emitted from a facility (or co-located facilities) to produce simultaneous indication of bearing and distance.

polar coordinates [MATH] A point in the plane may be represented by coordinates (r,θ), where θ is the angle between the positive x-axis and the ray from the origin to the point, and r the length of that ray.

polar covalent bond [PHYS CHEM] A bond in which a pair of electrons is shared in common between two atoms, but the pair is held more closely by one of the atoms.

polar crystal *See* ferroelectric crystal.

polar cyclone *See* polar vortex.

polar diagram [PHYS] A diagram employing polar coordinates to show the magnitude of a quantity in some or all directions from a point; examples include directivity patterns and radiation patterns.

polar distance [ASTRON] Angular distance from a celestial pole; the arc of an hour circle between a celestial pole, usually the elevated pole, and a point on the celestial sphere, measured from the celestial pole through 180°.

polar easterlies [METEOROL] The rather shallow and diffuse body of easterly winds located poleward of the subpolar low-pressure belt; in the mean in the Northern Hemisphere, these easterlies exist to an appreciable extent only north of the Aleutian low and Icelandic low.

polar-easterlies index [METEOROL] A measure of the strength of the easterly wind between the latitudes of 55° and 70°N; the index is computed from the average sea-level pressure difference between these latitudes and is expressed as the east to west component of geostrophic wind in meters and tenths of meters per second.

polar electrojet [GEOPHYS] An intense current that flows in a relatively narrow band of the auroral zone ionosphere during disturbances of the magnetosphere.

polar form [MATH] A complex number $z + iy$ has as polar form $re^{i\theta}$, where (r,θ) are the polar coordinates corresponding to the point of the plane with rectangular coordinates (x,y), that is, $r = \sqrt{x^2 + y^2}$ and $\theta = \arctan y/x$.

polar front [METEOROL] The semipermanent, semicontinuous front separating air masses of tropical and polar origin; this is the major front in terms of air mass contrast and susceptibility to cyclonic disturbance.

polar-front theory [METEOROL] A theory whereby a polar front, separating air masses of polar and tropical origin, gives rise to cyclonic disturbances which intensify and travel along the front, passing through various phases of a characteristic life history.

polar glacier [HYD] A glacier whose temperature is below freezing throughout its mass, and on which there is no melting during any season.

polar high *See* arctic high; subpolar high.

polar ice [OCEANOGR] Sea ice that is more than 1 year old; the thickest form of sea ice. Also known as polar-cap ice.

polarimeter [OPTICS] An instrument used to determine the rotation of the plane of polarization of plane polarized light when it passes through a substance; the light is linearly polarized by a polarizer (such as a Nicol prism), passes through the material being analyzed, and then passes through an analyzer (such as another Nicol prism).

polarimetric analysis [ANALY CHEM] A method of chemical analysis based on the optical activity of the substance being determined; the measurement of the extent of the optical rotation of the substance is used to identify the substance or determine its quantity.

polarimetry [OPTICS] Determination of the rotation of the plane of polarization of plane polarized light when it passes through a substance, using a polarimeter.

Polaris [ASTRON] A creamy supergiant star of stellar magnitude 2.0, spectral classification F8, in the constellation Ursa Minor; marks the north celestial pole, being about 1° from this point; the star Ursae Minoris. Also known as North Star; Pole Star.

polariscope [OPTICS] Any of several instruments used to determine the effects of substances on polarized light, in which linearly or elliptically polarized light passes through the substance being studied, and then through an analyzer.

Polaris correction [NAV] A correction to be applied to the corrected sextant altitude of Polaris to obtain latitude; this correction for the offset of Polaris from the north celestial pole varies with the local hour angle of Aries, latitude, and date.

Polaris missile [ORD] A U.S. Navy surface-to-surface intermediate-range ballistic missile designed to be launched from submarines and surface ships for accurate bombardment of small target areas with conventional or nuclear warheads at ranges up to 1,500 miles (2,400 kilometers).

polarity [COMMUN] 1. The direction in which a direct current flows, in a teletypewriter system. 2. The sense of the potential of a portion of a television picture signal representing a dark area of a scene relative to the potential of a portion of the signal representing a light area. [MATH] Property of a line segment whose two ends are distinguishable. [PHYS] Property of a physical system which has two points with different (usually opposite) characteristics, such as one which has opposite charges or electric potentials, or opposite magnetic poles.

polarity epoch [GEOPHYS] A period of time during which the earth's magnetic field was predominantly of a single polarity.

polarizability [ELEC] The electric dipole moment induced in a system, such as an atom or molecule, by an electric field of unit strength.

polarizability catastrophe [ELEC] According to a theory using the Lorentz field concept, the phenomenon where, at a certain temperature, the dielectric constant of a material becomes infinite.

polarizability ellipsoid *See* index ellipsoid.

polarization [ELEC] 1. The process of producing a relative displacement of positive and negative bound charges in a body by applying an electric field. 2. A vector quantity equal to the electric dipole moment per unit volume of a material. Also known as dielectric polarization; electric polarization. 3. A chemical change occurring in dry cells during use, increasing the internal resistance of the cell and shortening its useful life. [PHYS] 1. Phenomenon exhibited by certain electromagnetic waves and other transverse waves in which the direction of the electric field or the displacement direction of the vibrations is constant or varies in some definite way. 2. The direction of the electric field or the displacement vector of a wave exhibiting polarization (first definition). 3. The process of bringing about polarization (first definition) in a transverse wave. 4. Property of a collection of particles with spin, in which the majority have spin components pointing in one direction, rather than at random.

polarization charge *See* bound charge.

polarization diversity [COMMUN] A method of transmission and reception used to minimize the effects of selective fading of the horizontal and vertical components of a radio signal; it is usually accomplished through the use of separate vertically and horizontally polarized receiving antennas.

polarization ellipse [PHYS] The ellipse traced out by the tip of the electric field vector or the displacement vector of a polarized wave at a fixed point in space in the course of time.

polarization error [NAV] An error in a radio direction-finder

bearing or the course indicated by a radio beacon because of a change in the polarization of the radio waves between the transmitter and receiver on being reflected from the ionosphere. Formerly referred to as night effect because it occurs principally during the night.

polarization fading [COMMUN] Fading as the result of changes in the direction of polarization in one or more of the propagation paths of waves arriving at a receiving point.

polarization isocline [METEOROL] A locus of all points at which the inclination to the vertical of the plane of polarization of the diffuse sky radiation has the same value.

polarization potential [ELECTROMAG] One of two vectors from which can be derived, by differentiation, an electric scalar potential and magnetic vector potential satisfying the Lorentz condition. Also known as Hertz vector. [PHYS CHEM] The reverse potential of an electrolytic cell which opposes the direct electrolytic potential of the cell.

polarized electrolytic capacitor [ELEC] An electrolytic capacitor in which the dielectric film is formed adjacent to only one metal electrode; the impedance to the flow of current is then greater in one direction than in the other.

polarized electromagnetic radiation [ELECTROMAG] Electromagnetic radiation in which the direction of the electric field vector is not random.

polarized light [OPTICS] Polarized electromagnetic radiation whose frequency is in the optical region.

polarized meter [ENG] A meter having a zero-center scale, with the direction of deflection of the pointer depending on the polarity of the voltage or the direction of the current being measured.

polarized neutrons [PHYS] A collection of neutrons in which the majority have spin pointing in one direction rather than at random.

polarized plug [ELEC] A plug that can be inserted in its receptacle only when in a predetermined position.

polarized receptacle [ELEC] A receptacle designed for use with a polarized plug, to ensure that the grounded side of an alternating-current line or the positive side of a direct-current line is always connected to the same terminal on a piece of equipment.

polarized relay [ELEC] Relay in which the movement of the armature depends upon the direction of the current in the circuit controlling the armature. Also known as polar relay.

polarized-vane ammeter [ENG] An ammeter of only moderate accuracy in which the current to be measured passes through a small coil, distorting the field of a circular permanent magnet, and an iron vane aligns itself with the axis of the distorted field, the deflection being roughly proportional to the current.

polarizer [OPTICS] A device which produces polarized light, such as a Nicol prism or Polaroid sheet.

polarizing angle *See* Brewster's angle.

polarizing filter [OPTICS] A device which selectively absorbs components of electromagnetic radiation passing through it, so that light emerging from it is plane-polarized.

polarizing microscope [OPTICS] A microscope in which an object is viewed in polarized light.

polar keying [COMMUN] Telegraph signal in which circuit current flows in one direction for spacing.

polar low *See* polar vortex.

polar maritime air [METEOROL] Air of an air mass that originates in the polar regions and is then modified by passing over a relatively warm ocean surface; characterized by moderately low temperature, moderately high surface specific humidity, and a considerable degree of vertical instability.

polar meteorology [METEOROL] The application of meteorological principles to a study of atmospheric conditions in the earth's high latitudes or polar-cap regions, northern and southern.

polar migration *See* polar wandering.

polar navigation [NAV] Navigation in polar regions, where unique considerations and techniques are applied; no definite limit for these regions is recognized, but polar navigation techniques are usually used from about latitude 70° to the nearest pole, north or south.

POLARISCOPE

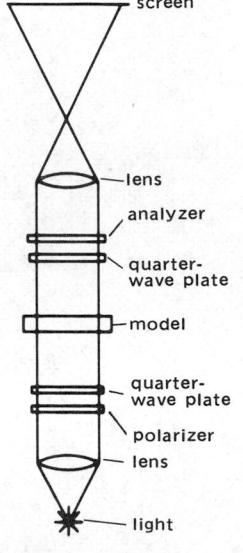

A polariscope used in measuring photoelastic stress.

(labels: screen, lens, analyzer, quarter-wave plate, model, quarter-wave plate, polarizer, lens, light)

POLARIS MISSILE

Polaris missile being fired from a submerged submarine. (*Official U.S. Navy photograph*)

polar night [ASTRON] The period of winter darkness in the polar regions, both northern and southern.

polar nucleus [BOT] One of the two nuclei in the center of the embryo sac of a seed plant which fuse to form the endosperm nucleus.

polarogram [ANALY CHEM] Plotted output (current versus electrode voltage) for polarographic analysis of an electrolyte.

polarographic analysis [ANALY CHEM] An electroanalytical technique in which the current through an electrolysis cell is measured as a function of the applied potential; the apparatus consists of a potentiometer for adjusting the potential, a galvanometer for measuring current, and a cell which contains two electrodes, a reference electrode whose potential is constant and an indicator electrode which is commonly the dropping mercury electrode. Also known as polarography.

polarographic cell [ANALY CHEM] Device for polarographic (voltammetric) analysis of an electrolyte solution; a known voltage is applied to the solution, and the ensuing current that passes through the cell (to an electrode) is measured.

polarographic maximum [ANALY CHEM] A deceptively high voltage buildup on an electrode during polarographic analysis of an electrolyte; caused by a reduction or oxidation process at the electrode.

polarography *See* polarographic analysis.

Polaroid [OPTICS] Trademark of Polaroid Corporation for a sheet material that produces plane-polarized light.

polar orbit [AERO ENG] A satellite orbit running north and south, so the satellite vehicle orbits over both the North Pole and the South Pole.

polar outbreak [METEOROL] The movement of a cold air mass from its source region; almost invariably applied to a vigorous equatorward thrust of cold polar air, a rapid equatorward movement of the polar front. Also known as cold-air outbreak.

polar projection [MAP] A map projection centered on a pole.

polar radiation pattern [ELECTROMAG] Diagram showing the relative strength of the radiation from an antenna in all directions in a given plane. [ENG ACOUS] Diagram showing the strength of sound waves radiated from a loudspeaker in various directions in a given plane, or a similar response pattern for a microphone.

polar regions [GEOGR] The regions near the geographic poles; no definite limit for these regions is recognized.

polar relay *See* polarized relay.

polar sequence [ASTRON] A compilation of 96 brightness-standard stars within 2° of the North Pole.

polar solvent [MATER] A solvent in whose molecules there is either a permanent separation of positive and negative charges, or the centers of positive and negative charges do not coincide; these solvents have high dielectric constants, are chemically active, and form coordinate covalent bonds; examples are alcohols and ketones.

polar transmission [COMMUN] **1.** A method of signaling in teletypewriter transmission in which direct currents flowing in opposite directions represent a mark and a space respectively, and absence of current indicates a no-signal condition. **2.** By extension, any system of signaling that uses three conditions, representing a mark, a space, or a no-signal condition.

polar triangle [MATH] A triangle associated to a given spherical triangle obtained from three directed lines perpendicular to the planes associated with the sides of the original triangle.

polar trough [METEOROL] In tropical meteorology, a wave trough in the circumpolar westerlies having sufficient amplitude to reach the tropics in the upper air; at the surface it is reflected as a trough in the tropical easterlies, but at moderate elevations it is characterized by westerly winds.

polar vortex [METEOROL] The large-scale cyclonic circulation in the middle and upper troposphere centered generally in the polar regions; specifically, the vortex has two centers in the mean, one near Baffin Island and another over northeastern Siberia; the associated cyclonic wind system comprises the westerlies of middle latitudes. Also known as circumpolar whirl; polar cyclone; polar low.

polar wandering [GEOL] Migration during geologic time of the earth's poles of rotation and magnetic poles. Also known as Chandler motion; polar migration.

polar westerlies *See* westerlies.

polder [CIV ENG] Land reclaimed from the sea or other body of water by the construction of an embankment to restrain the water.

pole [ELEC] **1.** One of the electrodes in an electric cell. **2.** An output terminal on a switch; a double-pole switch has two output terminals. [ELECTROMAG] *See* magnetic pole. [MATH] **1.** An isolated singular point z_0 of a complex function whose Laurent series expansion about z_0 will include finitely many terms of the form $a_n(z-z_0)^{-n}$. **2.** For a great circle on a sphere, the pole of the circle is a point of intersection of the sphere and a line that passes through the center of the sphere and is perpendicular to the plane of the circle. **3.** For a conic section, the pole of a line is the intersection of the tangents to the conic at the points of intersection of the conic with the line. **4.** For a quadric surface, the pole of a plane is the vertex of the cone which is tangent to the surface along the curve where the plane intersects the surface. [MECH] **1.** A point at which an axis of rotation or of symmetry passes through the surface of a body. **2.** *See* perch.

pole blight [PL PATH] A destructive disease of white pines characterized by shortening of the needle-bearing stems, yellowing and shortening of needles, and copious flow of resin.

pole dominance [PARTIC PHYS] Property of a scattering amplitude, analytically continued to complex values of energy and scattering angle, whose behavior is dominated by the term or terms of negative power in the Laurent series of a nearby pole.

pole-face winding [ELECTROMAG] Winding in the pole face of a motor or generator used to neutralize the cross-magnetizing armature reaction under the pole faces, which would otherwise cause a nonuniform distribution of voltage between commutator segments. Also known as compensated winding.

pole lathe [MECH ENG] A simple lathe in which the work is rotated by a cord attached to a treadle.

Polemoniaceae [BOT] A family of autotrophic dicotyledonous plants in the order Polemoniales distinguished by lack of internal phloem, corolla lobes that are convolute in the bud, three carpels, and axile placentation.

Polemoniales [BOT] An order of dicotyledonous plants in the subclass Asteridae, characterized by sympetalous flowers, a regular, usually five-lobed corolla, and stamens equal in number and alternate with the petals.

pole of inaccessibility *See* ice pole.

pole shoe [ELECTROMAG] Portion of a field pole facing the armature of the machine; it may be separable from the body of the pole.

Pole Star *See* Polaris.

polestar recorder [ENG] An instrument used to determine approximately the amount of cloudiness during the dark hours; consists of a fixed long-focus camera positioned so that Polaris is permanently within its field of view; the apparent motion of the star appears as a circular arc on the photograph and is interrupted as clouds come between the star and the camera.

pole strength *See* magnetic pole strength.

pole tide [OCEANOGR] An ocean tide, theoretically 6 millimeters in amplitude, caused by the Chandler wobble of the earth; has a period of 428 days.

pole-zero configuration [CONT SYS] A plot of the poles and zeros of a transfer function in the complex plane; used to study the stability of a system, its natural motion, its frequency response, and its transient response.

polhode [MECH] For a rotating rigid body not subject to external torque, the closed curve traced out on the inertia ellipsoid by the intersection with this ellipsoid of an axis parallel to the angular velocity vector and through the center.

polian vesicle [INV ZOO] Interradial reservoirs connecting with the ring vessel in most asteroids and holothuroids.

poling [ELEC] Adjustment of polarity; specifically, in wire-line practice, the use of transpositions between transposition sections of open wire or between lengths of cable, to cause the residual cross-talk couplings in individual sections or lengths to oppose one another. [MET] A technique used in the refining of copper that consists of the thrusting of green-wood poles into the molten metal in order to generate the reducing gases that react with the oxides in the metal. [MIN ENG] The act or process of temporarily protecting the face of a level, drift, or cut by driving poles or planks along the sides of the yet unbroken ground.

poling back [MIN ENG] Carrying out excavation behind timbering already in place.

poling board [CIV ENG] A timber plank driven into soft soil to support the sides of an excavation.

poliomyelitis [MED] An acute infectious viral disease which in its most serious form involves the central nervous system and, by destruction of motor neurons in the spinal cord, produces flaccid paralysis. Also known as Heine-Medin disease; infantile paralysis.

poliovirus vaccine [IMMUNOL] A vaccine prepared from one or all three types of polioviruses in a live or attenuated state.

polish [MATER] A powder, liquid, or semiliquid used to give smoothness, surface protection, or decoration to finishes; for example, finely ground red oxide (rouge) is used to polish plate glass, mirror backs, and optical glass; solvent-wax liquids and pastes are used to protect and enhance leather and wood surfaces; nitrocellulose lacquers are used to paint finger- and toenails.

polished cotton [TEXT] Cotton fabric, usually of plain weave, with a glazed finish produced with resin.

polished-joint hanger [PETRO ENG] A type of tubing hanger that is slipped over or assembled around the top tubing joint in an oil well tubing string.

polishing [CHEM ENG] In petroleum refining, removal of final traces of impurities, as for a lubricant, by clay adsorption or mild hydrogen treating. [MECH ENG] Smoothing and brightening a surface such as a metal or a rock through the use of abrasive materials.

polishing roll [MECH ENG] A roll or series of rolls on a plastics mold; has highly polished chrome-plated surfaces; used to produce a smooth surface on a plastic sheet as it is extruded.

polishing wheel [DES ENG] An abrasive wheel used for polishing.

Polish notation [ADP] A notation system for digital-computer logic, developed by J. Lukasiewicz, in which each operator is a binary or unary operator in the sense that it operates on not more than two operands.

pollen [BOT] The small male reproductive bodies produced in pollen sacs of the seed plants.

pollen count [BOT] The number of grains of pollen that collect on a specified area (often taken as 1 square centimeter) in a specified time.

pollen tube [BOT] The tube produced by the wall of a pollen grain which enters the embryo sac and provides a passage through which the male nuclei reach the female nuclei.

pollination [BOT] The transfer of pollen from a stamen to a pistil; fertilization in flowering plants.

polling [COMMUN] A process that involves interrogating in succession every terminal on a shared communications line to determine which of the terminals are in need of servicing.

polling list [COMMUN] A roster of transmitting devices sequentially scanned in a time-sharing system.

pollucite [MINERAL] $(Cs,Na)_2Al_2Si_4O_{12} \cdot H_2O$ A colorless, transparent zoolite mineral composed of hydrous silicate of cesium, sodium, and aluminum, occurring massive or in cubes; used as a gemstone. Also known as pollux.

pollution [ECOL] Destruction or impairment of the purity of the environment. [PHYSIO] Emission of semen at times other than during coitus.

pollux See pollucite.

Pollux [ASTRON] A giant orange-yellow star with visual brightness of 1.16, a little less than 35 light-years from the sun, spectral classification K0-III, in the constellation Gemini; the star β Geminorum.

polonium [CHEM] A chemical element, symbol Po, atomic number 84; all polonium isotopes are radioactive; polonium-210 is the naturally occurring isotope found in pitchblende.

polonium-210 [NUC PHYS] Radioactive isotope of polonium; mass 210, half-life 140 days, α-radiation; used to calibrate radiation counters, and in oil well logging and atomic batteries. Also known as radium F.

poly- [ORG CHEM] A chemical prefix meaning many; for example, a polymer is made of a number of single molecules known as monomers, as polyethylene is made from ethylene.

polyacetals See acetal resins.

polyacrylamide [ORG CHEM] $(CH_2CHCONH_2)_x$ A white, water-soluble high polymer based on acrylamide; used as a thickening or suspending agent in water-base formulations.

polyacrylate [ORG CHEM] A polymer of an ester or salt of acrylic acid.

polyacrylic acid [ORG CHEM] $(CH_2CHCOOH)_x$ An acrylic or acrylate resin formed by the polymerization of acrylic acid; water-soluble; used as a suspending and textile-sizing agent, and in adhesives, paints, and hydraulic fluids.

polyacrylic fiber [ORG CHEM] Continuous-strand fiber extruded from an acrylate resin.

polyacrylonitrile [ORG CHEM] Polymer of acrylonitrile; semiconductive; used like an inorganic oxide catalyst to dehydrogenate *tert*-butyl alcohol to produce isobutylene and water.

polyalcohol See polyhydric alcohol.

polyallomer [ORG .CHEM] A copolymer of propylene with other olefins.

polyamide See polyamide resin.

polyamide resin [ORG CHEM] Product of polymerization of amino acid or the condensation of a polyamine with a polycarboxylic acid; an example is the nylons. Also known as polyamide.

Polyangiaceae [MICROBIO] A family of bacteria in the order Myxobacterales in which microcysts are not spherical or surrounded by a wall but are contained in round or oval cysts; usually, complex fruiting bodies are formed.

polyargyrite [MINERAL] $Ag_{24}Sb_2S_{15}$ A gray to black mineral composed of antimony silver sulfide.

polyarteritis [MED] Inflammation of several arteries simultaneously. Also known as panarteritis.

polyarteritis nodosa [MED] A systemic disease characterized by widespread inflammation of small- and medium-sized arteries in which some of the foci are nodular. Also known as disseminated necrotizing periarteritis; periarteritis nodosa.

polyarthritis [MED] Inflammation of several joints simultaneously.

polyatomic molecule [CHEM] A chemical molecule with three or more atoms.

polyaxon [INV ZOO] A spicule that is laid down along several axes.

polybasic [CHEM] A chemical compound in solution that yields two or more H^+ions per molecule, such as sulfuric acid, H_2SO_4.

polybasite ˙[MINERAL] $(Ag,Cu)_{16}Sb_2S_{11}$ An iron-black to steel-gray metallic-looking mineral; an ore of silver.

polyblend [MATER] A mechanical mixture of two or more polymers, such as polystyrene and rubber.

Polybrachiidae [INV ZOO] A family of sedentary marine animals in the order Thecanephria.

polybutadiene [ORG CHEM] Oil-extendable synthetic elastomer polymer made from butadiene; resilience is similar to natural rubber; it is blended with natural rubber for use in tire and other rubber products. Also known as butadiene rubber.

polybutene [ORG CHEM] A polymer of isobutene, $(CH_3)_2CCH_2$; made in varying chain lengths to give a wide range of properties from oily to solid; used as a lube-oil additive, in adhesives, and in rubber products.

polybutylene [ORG CHEM] A polymer of one or more butylenes whose consistency ranges from a viscous liquid to a rubbery solid.

polycarbonate [ORG CHEM] $[OC_6H_4C(CH_3)_2C_6H_4OCO]_x$ A linear polymer of carbonic acid which is a thermoplastic synthetic resin made from bisphenol and phosgene; used in emulsion coatings with glass fiber reinforcement.

polycarboxylic [ORG CHEM] Prefix for a compound containing two or more carboxyl ($-COOH$) groups.

Polychaeta [INV ZOO] The largest class of the phylum Annelida, distinguished by paired, lateral, fleshy appendages (parapodia) provided with setae, on most segments.

polychroism See pleochroism.

Polycirrinae [INV ZOO] A subfamily of polychaete annelids in the family Terebellidae.

Polycladida [INV ZOO] A class of marine Turbellaria whose

POLLUCITE

Pollucite crystals with spodumene, Paris, Maine. *(Specimen from Department of Geology, Bryn Mawr College)*

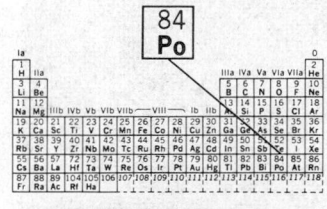

POLONIUM

Periodic table of the chemical elements showing the position of polonium.

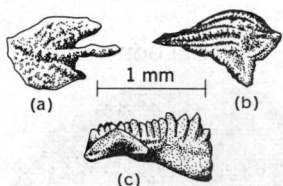

leaflike bodies have a central intestine with radiating branches, many eyes, and tentacles in most species.

polyclimax [ECOL] A climax community under the controlling influence of many environmental factors, including soils, topography, fire, and animal interactions.

polyclonal mixed cryoglobulin [BIOCHEM] A cryoglobulin made of heterogeneous immunoglobin molecules belonging to two or more different classes, and sometimes additional serum proteins.

polycondensation [ORG CHEM] A chemical condensation leading to the formation of a polymer by the linking together of molecules of a monomer and the releasing of water or a similar simple substance.

polyconic chart [MAP] A chart on the polyconic projection.

polyconic projection [MAP] A conic map projection in which the surface of a sphere or spheroid, such as the earth, is conceived as developed on a series of tangent cones, which are then spread out to form a plane; a separate cone is used for each small zone.

Polycopa [INV ZOO] The single family of the suborder Cladocopa.

polycrase [MINERAL] $(Y,Ca,Ce,U,Th)(Ti,Cb,Ta)_2O_6$ Black mineral composed of titanate, columbate, and tantalate of yttrium-group metals; it is isomorphous with euxenite and occurs in granite pegmatites.

polycrystal [MATER] A polycrystalline solid.

polycrystalline [MATER] 1. Pertaining to a material composed of aggregates of individual crystals. 2. Characterized by variously oriented crystals.

Polyctenidae [INV ZOO] A family of hemipteran insects in the superfamily Cimicoidea; the individuals are bat ectoparasites which resemble bedbugs but lack eyes and have ctenidia and strong claws.

polycyclic [ORG CHEM] A molecule that contains two or more closed atomic rings; can be aromatic (such as DDT), aliphatic (bianthryl), or mixed (dicarbazyl).

polycyclic hydrocarbon See polynuclear hydrocarbon.

polycyesis [MED] Multiple pregnancy.

polycystic kidney [MED] A usually hereditary, congenital, and bilateral disease in which a large number of cysts are present on the kidney.

polycythemia [MED] A condition characterized by an increased number of erythrocytes in the circulation.

polycythemia vera [MED] An absolute increase in all blood cells derived from bone marrow, especially erythrocytes.

polydactyly [MED] The condition of having supernumerary fingers or toes.

polydipsia [MED] Excessive thirst.

polydispersity [CHEM] Molecular-weight nonhomogeneity in a polymer system; that is, there is some molecular-weight distribution throughout the body of the polymer.

Polydolopidae [PALEON] A Cenozoic family of rodentlike marsupial mammals.

polydymite [MINERAL] Ni_3S_4 A mineral of the linnaeite group consisting of nickel sulfide.

polyelectrolyte [ORG CHEM] A natural or synthetic electrolyte with high molecular weight, such as proteins, polysaccharides, and alkyl addition products of polyvinyl pyridine; can be a weak or strong electrolyte; when dissociated in solution, it does not give uniform distribution of positive and negative ions (the ions of one sign are bound to the polymer chain while the ions of the other sign diffuse through the solution).

polyembryony [ZOO] A form of sexual reproduction in which two or more offspring are derived from a single egg.

polyene [ORG CHEM] Compound containing many double bonds, such as the carotenoids.

polyester fiber [TEXT] A fiber filament made from a material that is 85% or more thermoplastic polyester resin.

polyester film [MATER] Thin film made of polyester resin; used for packaging food and other products.

polyester laminate [MATER] Glass fabric or fiber mat impregnated with a polyester resin slurry, and cured; used to make sheets, bars, and structural shapes.

polyester-reinforced urethane [MATER] A poromeric material which may have a urethane impregnation or a silicone coating; used for shoe uppers and as a substitute for industrial leathers.

polyester resin [ORG CHEM] A thermosetting or thermoplastic synthetic resin made by esterification of polybasic organic acids with polyhydric acids; examples are Dacron and Mylar; the resin has high strength and excellent resistance to moisture and chemicals when cured.

polyester rubber See polyurethane rubber.

polyestrous [PHYSIO] Having several periods of estrus in a year.

polyether resin [ORG CHEM] A polymer that contains $-(CH_2CHRO-)_x$ in the main-chain or side-chain linkage.

polyethylene See ethylene resin.

polyethylene glycol [ORG CHEM] Any of a family of colorless, water-soluble liquids with molecular weights from 200 to 6000; soluble also in aromatic hydrocarbons (not aliphatics) and many organic solvents; used to make emulsifying agents and detergents, and as plasticizers, humectants, and water-soluble textile lubricants.

polyethylene resin See ethylene resin.

polyethylene terephthalate [ORG CHEM] A thermoplastic polyester resin made from ethylene glycol and terephthalic acid; melts at 265°C; used to make films or fibers.

polyforming [CHEM ENG] A noncatalytic, petroleum-refinery process charging C_3 and C_4 gases with naphtha or gas oil at high temperature to produce high-quality gasoline and fuel oil; mostly replaced by catalytic reforming; the product is known as polyformdistillate.

Polygalaceae [BOT] A family of dicotyledonous plants in the order Polygalales distinguished by having a bicarpellate pistil and monadelphous stamens.

polygalacturonase [BIOCHEM] An enzyme that catalyzes the hydrolysis of glycosidic linkage of polymerized galacturonic acids.

Polygalales [BOT] An order of dicotyledonous plants in the subclass Rosidae characterized by its simple leaves and usually irregular, hypogynous flowers.

polygamous [BOT] Having both perfect and imperfect flowers on the same plant. [VERT ZOO] Having more than one mate at one time.

polygen See polyvalent.

polygene [GEN] One of a group of nonallelic genes that collectively control a quantitative character.

polyglycol [ORG CHEM] A dihydroxy ether derived from the dehydration (removal of a water molecule) of two or more glycol molecules; an example is diethylene glycol, $CH_2OH\text{-}CH_2OCH_2CH_2OH$.

Polygnathidae [PALEON] A family of Middle Silurian to Cretaceous conodonts in the suborder Conodontiformes, having platforms with small pitlike attachment scars.

polygon [MATH] A figure in the plane given by points $p_1, p_2, \ldots, p_n$ and line segments $p_1 p_2, p_2 p_3, \ldots, p_{n-1} p_n, p_n p_1$.

Polygonaceae [BOT] The single family of the order Polygonales.

Polygonales [BOT] An order of dicotyledonous plants in the subclass Caryophyllidae characterized by well-developed endosperm, a unilocular ovary, and often trimerous flowers.

polygonal ground [GEOL] A ground surface consisting of polygonal arrangements of rock, soil, and vegetation formed on a level or gently sloping surface by frost action. Also known as cellular soil.

polygonal method [MIN ENG] A method of estimating ore reserves in which it is assumed that each drill hole has an area of influence extending halfway to the neighboring drill holes.

polygonization [SOLID STATE] A phenomenon observed during the annealing of plastically bent crystals in which the edge dislocations created by cold working organize themselves vertically above each other so that polygonal domains are formed.

polygraph See lie detector.

polyhalite [MINERAL] $K_2MgCa_2(SO_4)_4 \cdot 2H_2O$ A sulfate mineral usually found in fibrous brick-red masses due to iron.

polyhedral angle [MATH] The shape formed by the lateral faces of a polyhedron which have a common vertex.

polyhedral disease See polyhedrosis.

polyhedron [MATH] A solid bounded by planar polygons.

polyhedrosis [INV ZOO] Any of several virus diseases of insect larvae characterized by the breakdown of tissues and presence of polyhedral granules. Also known as polyhedral disease.

polyhidrosis   *See* hyperhidrosis.

polyhydramnios   [MED] An excessive volume of amniotic fluid. Also known as hydramnios.

polyhydric alcohol   [ORG CHEM] An alcohol with many hydroxyl (−OH) radicals, such as glycerol, $C_3H_5(OH_3)$. Also known as polyalcohol; polyol.

polyhydric phenol   [ORG CHEM] A phenolic compound containing two or more hydroxyl groups, such as diphenol, $C_6H_4(OH)_2$.

polyimide resin   [ORG CHEM] An aromatic polyimide made by reacting pyromellitic dianhydride with an aromatic diamine; has high resistance to thermal stresses; used to make components of internal combustion engines.

polyisoprene   [ORG CHEM] $(C_5H_8)_x$ The basis of natural rubber, balata, gutta-percha, and other rubberlike materials; can also be made synthetically; the stereospecific forms are *cis*-1,4- and *trans*-1,4-polyisoprene; the polymer is thermoplastic.

polylactic resin   [ORG CHEM] A soft, elastic resin made by the heat reaction of lactic acid with castor oil or other fatty oils; used to produce tough, water-resistant coatings.

polymenorrhea   *See* metrorrhagia.

polymer   [ORG CHEM] **1.** Substance made of giant molecules formed by the union of simple molecules (monomers); for example polymerization of ethylene forms a polyethylene chain, or condensation of phenol and formaldehyde (with production of water) forms phenol-formaldehyde resins.

Polymera   [INV ZOO] Formerly a subphylum of the Vermes; equivalent to the phylum Annelida.

polymerase   [BIOCHEM] An enzyme that links nucleotides together to form polynucleotide chains.

polymer gasoline   [MATER] A product of polymerization of normally gaseous hydrocarbons to form high-octane liquid hydrocarbons boiling in the gasoline range.

polymerization   [CHEM] **1.** The bonding of two or more monomers to produce a polymer. **2.** Any chemical reaction that produces such a bonding.

polymer paint   [MATER] A paint made of acrylic resin or vinyl resin, or a combination of both resins, in a liquid form with water as the base; it spreads out in a layer, and the water evaporates to leave a continuous, flexible, and waterproof film of plastic.

polymer plastic   [MATER] The product of a high polymer with or without additives, such as plasticizers, autoxidants, colorants, or fillers; can be sprayed, shaped, molded, extruded, cast or foamed, depending on whether it is thermoplastic or thermosetting.

polymethyl methacrylate   [ORG CHEM] A thermoplastic polymer derived from methyl methacrylate, $CH_2=C(CH_3)COOCH_3$; transparent solid with excellent optical qualities and water resistance; used for aircraft domes, lighting fixtures, optical instruments, and surgical appliances.

polymictic   [HYD] Pertaining to or characteristic of a lake having no stabile thermal stratification.   [PETR] Of a clastic sedimentary rock, being made up of many rock types or of more than one mineral species.

polymignite   *See* polymignyte.

polymignyte   [MINERAL] $(Ca,Fe,Y,Zr,Th)(Nb,Ti,Ta)O_4$ A black mineral composed of niobate, titanate, and tantalate of cerium-group metals, with calcium and iron. Also spelled polymignite.

polymorph   [BIOL] An organism that exhibits polymorphism. [CRYSTAL] A crystal form of a polymorphic material. Also known as polymorphic modification. [HISTOL] *See* granulocyte.

polymorphic modification   *See* polymorph.

polymorphism   [BIOL] **1.** Occurrence of different forms of individual in a single species. **2.** Occurrence of different structural forms in a single individual at different periods in the life cycle. [CRYSTAL] The property of a chemical substance crystallizing into two or more forms having different structures, such as diamond and graphite. Also known as pleomorphism. [GEN] The coexistence of genetically determined distinct forms in the same population, even the rarest of them being too common to be maintained solely by mutation; human blood groups are an example.

polymorphonuclear leukocyte   *See* granulocyte.

polymyarian   [INV ZOO] Referring to the cross-sectional appearance of muscle cells in a nematode, having many cells in each quadrant.

polymyositis   [MED] Inflammation of many muscles simultaneously.

polymyxin   [MICROBIO] Any of the basic polypeptide antibiotics produced by certain strains of *Bacillus polymyxa*.

Polynemidae   [VERT ZOO] A family of perciform shore fishes in the suborder Mugiloidei.

polyneuritis   [MED] Degenerative or inflammatory lesions of several nerves simultaneously, usually symmetrical. Also known as multiple neuritis.

Polynoidae   [INV ZOO] The largest family of polychaetes, included in the Errantia and having a body of varying size and shape that is covered with elytra.

polynomial   [MATH] A polynomial in the quantities $x_1, x_2, \ldots, x_n$ is an expression involving a finite sum of terms of the form $bx_1^{p_1}x_2^{p_2} \ldots x_n^{p_n}$, where b is some number, and $p_1, \ldots, p_n$ are integers.

Polynosic   [TEXT] Trade name of a low-modulus rayon fiber manufactured by American Enka Corporation.

polynuclear hydrocarbon   [ORG CHEM] Hydrocarbon molecule with two or more nuclei; examples are naphthalene, $C_{10}H_8$, with two benzene rings side by side, or diphenyl, $(C_6H_5)_2$, with two bond-connected benzene rings. Also known as polycyclic hydrocarbon.

polynucleotide   [BIOCHEM] A linear sequence of nucleotides.

polyn'ya   [OCEANOGR] A Russian term for a water area, other than a lead, lane, or crack, which is surrounded by sea ice; the term "window" is sometimes used for a similar open area in river ice. Also known as ice clearing.

Polyodontidae   [INV ZOO] A family of tubicolous, often large-bodied errantian polychaetes with characteristic cephalic and parapodial structures.

polyol   *See* polyhydric alcohol.

polyolefin   [ORG CHEM] A resinous material made by the polymerization of olefins, such as polyethylene from ethylene, polypropylene from propylene, or polybutene from butylene.

polyolefin fiber   [MATER] Continuous-strand fiber made from a polyolefin.

polyoma virus   [VIROL] A small deoxyribonucleic acid virus normally causing inapparent infection in mice, but experimentally capable of producing parotid tumors and a wide variety of other tumors.

polyopia   [MED] A condition in which more than one image of an object is formed upon the retina.

Polyopisthocotylea   [INV ZOO] An order of the trematode subclass Monogenea having a solid posterior holdfast bearing suckers or clamps.

polyorganosiloxane   *See* polysiloxane.

polyorrhymenitis   *See* polyserositis.

polyoxyalkylene resin   [ORG CHEM] Condensation polymer produced from an oxyalkene, such as polyethylene glycol from oxyethylene or ethylene glycol.

polyp   [INV ZOO] A sessile coelenterate individual having a hollow, somewhat cylindrical body, attached at one end, with a mouth surrounded by tentacles at the free end; may be solitary (hydra) or colonial (coral). [MED] A smooth, rounded or oval mass projecting from a membrane-covered surface.

polypectomy   [MED] Surgical excision of a polyp.

polypeptide   [BIOCHEM] A chain of amino acids linked together by peptide bonds but with a lower molecular weight than a protein; obtained by synthesis, or by partial hydrolysis of protein.

polypetalous   [BOT] Having distinct petals, in reference to a flower or a corolla. Also known as choripetalous.

Polyphaga   [INV ZOO] A suborder of the order Coleoptera; members are distinguished by not having the hind coxae fused to the metasternum and by lacking notopleural sutures.

polyphagous   [ZOO] Feeding on many different kinds of plants or animals.

polyphase   [ELEC] Having or utilizing two or more phases of an alternating-current power line.

polyphase circuit   [ELEC] Group of alternating-current circuits (usually interconnected) which enter (or leave) a de-

POLYNOIDAE

Harmothoe of the Polynoidae, dorsal view. (After McIntosh)

POLYPHASE RECTIFIER

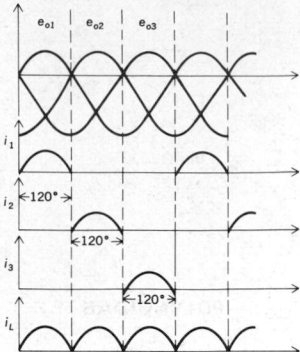

Graphs of transformer voltages e_{o1}, e_{o2}, e_{o3}, diode currents i_1, i_2, i_3, and load current i_{L1} versus time, in a three-phase half-wave rectifier.

POLYSOME

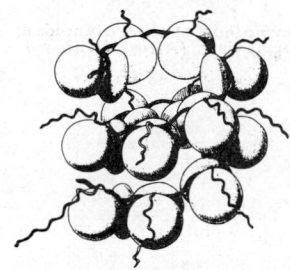

A helical free polysome is bound together by a single messenger-RNA molecule, which follows the helix. Successive ribosomes along the messenger have proceeded farther in production of polypeptide chains.

POLYSTOMATOIDEA

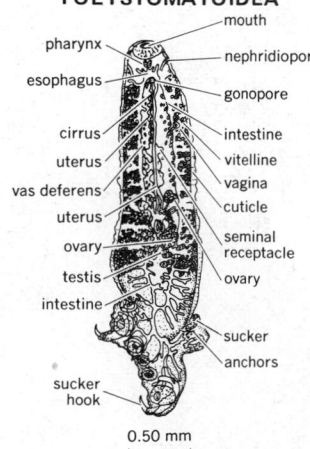

mouth
pharynx
nephridiopore
esophagus
gonopore
cirrus
intestine
uterus
vitelline
vas deferens
vagina
uterus
cuticle
ovary
seminal receptacle
testis
ovary
intestine
sucker
anchors
sucker hook

0.50 mm

Ventral view of *Heteronchocotyle leucas* Hargis from the bull shark (*Carcharhinus leucas*).

limited region at more than two points of entry; they are intended to be so energized that, in the steady state, the alternating currents through the points of entry, and the alternating potential differences between them, all have exactly equal periods, but have differences in phase, and may have differences in waveform.

polyphase meter [ENG] An instrument which measures some electrical quantity, such as power factor or power, in a polyphase circuit.

polyphase rectifier [ELECTR] A rectifier which utilizes two or more diodes (usually three), each of which operates during an equal fraction of an alternating-current cycle to achieve an output current which varies less than that in an ordinary half-wave or full-wave rectifier.

polyphase synchronous generator [ELEC] Generator whose alternating-current circuits are so arranged that two or more symmetrical alternating electromotive forces with definite phase relationships are produced at its terminals.

polyphase wattmeter [ENG] An instrument that measures electric power in a polyphase circuit.

polyphenol oxidase [BIOCHEM] A copper-containing enzyme that catalyzes the oxidation of phenol derivatives to quinones.

polyphenyl [ORG CHEM] Any of a group of direct colors used to dye cotton and wool.

polyphenylene oxide [ORG CHEM] A polyether resin of 2, 6-dimethylphenol, $(CH_3)_2C_6H_3OH$; useful temperature range is -275 to $375°F$ (-168 to $191°C$), with intermittent use possible up to $400°F$ ($204°C$).

polyphobia [PSYCH] An abnormal fear of many different things.

polyphosphoric acid [INORG CHEM] $H_6P_4O_{13}$ Viscous, water-soluble, hygroscopic, water-white liquid; used wherever concentrated phosphoric acid is needed.

polyphyodont [VERT ZOO] Having teeth which may be constantly replaced.

polypide [INV ZOO] The internal contents of an ectoproct bryozoan zooid.

Polyplacophora [INV ZOO] The chitons, an order of mollusks in the class Amphineura distinguished by an elliptical body with a dorsal shell that comprises eight calcareous plates overlapping posteriorly.

polyploidy [GEN] The occurrence of related forms possessing chromosome numbers which are three or more times the haploid number.

Polypodiales [BOT] The true ferns; the largest order of modern ferns, distinguished by being leptosporangiate and by having small sporangia with a definite number of spores.

Polypodiatae See Polypodiopsida.

Polypodiophyta [BOT] The ferns, a division of the plant kingdom having well-developed roots, stems, and leaves that contain xylem and phloem and show well-developed alternation of generations.

Polypodiopsida [BOT] A class of the division Polypodiophyta; stems of these ferns bear several large, spirally arranged, compound leaves with sporangia grouped in sori on their undermargins.

polypropylene [ORG CHEM] $(C_3H_5)_x$ A crystalline, thermoplastic resin made by the polymerization of propylene, C_3H_6; the product is hard and tough, resists moisture, oils, and solvents, and withstands temperatures up to $170°C$; used to make molded articles, fibers, film, rope, printing plates, and toys.

polypropylene glycol [ORG CHEM] $CH_3CHOH(CH_2OCH-CH_3)_xCH_2OH$ Polymeric material similar to polyethylene glycol, but with greater oil solubility and less water solubility; used as a solvent for vegetable oils, waxes, and resins, in hydraulic fluids and as a chemical intermediate.

Polypteridae [VERT ZOO] The single family of the order Polypteriformes.

Polypteriformes [VERT ZOO] An ancient order of actinopterygian fishes distinguished by thick, rhombic, ganoid scales with an enamellike covering, a slitlike spiracle behind the eye, a symmetrical caudal fin, and a dorsal series of free, spinelike finlets.

polyribosome See polysome.

polyrod antenna [ELECTROMAG] End-fire directional dielec-

tric antenna consisting of a polystyrene rod energized by a section of waveguide.

polysaccharide [BIOCHEM] A carbohydrate composed of many monosaccharides.

polysepalous [BOT] Having separate sepals. Also known as chorisepalous.

polyserositis [MED] Widespread, chronic fibrosing inflammation of serous membranes, especially in the upper abdomen. Also known as chronic hyperplastic perihepatitis; Concato's disease; multiple serositis; Pick's disease; polyorrhymenitis.

polysiloxane [ORG CHEM] $(R_2SiO)_n$ A polymer in which the chain contains alternate silicon and oxygen atoms; in the formula, R can be H or an alkyl or aryl group; commercially, the R is usually CH_3 (the methylsiloxanes); properties vary with molecular weight, from oils to greases to rubbers to plastics. Also known as polyorganosiloxane.

polysome [CYTOL] A complex of ribosomes bound together by a single messenger ribonucleic acid molecule. Also known as polyribosome.

polysomy [GEN] The occurrence in a nucleus of one or more individual chromosomes in a number higher than that of the remainder.

polyspermy [PHYSIO] Penetration of the egg by more than one sperm.

polystele [BOT] A stele consisting of vascular units in the parenchyma.

Polystomatoidea [INV ZOO] A superfamily of monogeneid trematodes characterized by strong suckers and hooks on the posterior end.

Polystylifera [INV ZOO] A suborder of the Hoplonemertini distinguished by many stylets.

polystyrene [ORG CHEM] $(C_6H_5CHCH_2)_x$ A water-white, tough synthetic resin made by polymerization of styrene; soluble in aromatic and chlorinated hydrocarbon solvents; used for injection molding, extrusion or casting for electrical insulation, fabric lamination, and molding of plastic objects.

polystyrene capacitor [ELEC] A capacitor that uses film polystyrene as a dielectric between rolled strips of metal foil.

polystyrene dielectric [ELEC] Polystyrene used in applications where its very high resistivity, good dielectric strength, and other electrical properties are important, such as for electrical insulation or in dielectrics.

polysulfide rubber [ORG CHEM] A synthetic polymer made by the reaction of sodium polysulfide with an organic dichloride; resistant to light, oxygen, oils, and solvents; impermeable to gases; poor tensile strength and abrasion resistance.

polysulfide treating [CHEM ENG] A petroleum-refinery process used to remove elemental sulfur from refinery liquids by contacting them with a nonregenerable solution of sodium polysulfide.

polysulfone resin [MATER] A thermoplastic polymer containing the sulfone linkage ($O=S=O$); have exceptional high-temperature, low-creep, and arc-resistance properties, and are self-extinguishing.

polysynthetic twinning [CRYSTAL] Repeated twinning that involves three or more individual crystals according to the same twin law and on parallel twin planes.

polytene chromosome [GEN] A giant, multistranded, cable-like chromosome composed of many identical chromosomes having their chromomeres in register and produced by polyteny.

polyterpene resin [ORG CHEM] A thermoplastic resin or viscous liquid from polymerization of turpentine; used in paints, polishes, and rubber plasticizers, and to cure concrete and impregnate paper.

polythene [ORG CHEM] Common name for polyethlylene in Great Britain.

Polytrichales [BOT] An order of ascocarpous perennial mosses; rigid, simple stems are highly developed and arise from a prostrate subterranean rhizome.

polytrifluorochloroethylene resin See chlorotrifluoroethylene polymer.

polytropic atmosphere [METEOROL] A model atmosphere in hydrostatic equilibrium with a constant nonzero lapse rate.

polytropic compression curve [PHYS] Graphical relationship between pressure p and volume V for various values of specific-heat ratios n in the compression formula $pV^n = K$.

polytropic process [THERMO] An expansion or compression of a gas in which the quantity pV^n is held constant, where p and V are the pressure and volume of the gas, and n is some constant.

polyunsaturated acid [ORG CHEM] A fatty acid with two or more double bonds per molecule, such as linoleic or linolenic acid.

polyurethane foam [MATER] A solid or spongy cellular material produced by the reaction of a polyester (such as glycerin) with a diisocyanate (such as toluene diisocyanate) while carbon dioxide is liberated by the reaction of a carboxyl with the isocyanate; used for thermal insulation, soundproofing, and padding.

polyurethane resin [ORG CHEM] Any resins resulting from the reaction of diisocyanates (such as toluene diisocyanate) with a phenol, amine, or hydroxylic or carboxylic compound to produce a polymer with free isocyanate groups; used as protective coatings, potting or casting resins, adhesives, rubbers, and foams, and in paints, varnishes, and adhesives.

polyurethane rubber [ORG CHEM] A synthetic polyurethane-resin elastomer made by the reaction of a diisocyanate to a polyester (such as the glycol-adipic acid ester); has high resistance to abrasion, oil, ozone, and high temperatures. Also known as polyester rubber.

polyuria [MED] The passage of copious amounts of urine.

polyvalent [CHEM] An ion or radical with more than one valency, such as the sulfate ion, SO_4^{--}. Also known as multivalent; polygen. [IMMUNOL] **1.** Of antigens, having many combining sites or determinants. **2.** Pertaining to vaccines composed of mixtures of different organisms, and to the resulting mixed antiserum.

polyvalent number [ADP] A number, consisting of several figures, used for description, wherein each figure represents one of the characteristics being described.

polyvinyl acetal resin *See* vinyl acetal resin.

polyvinyl acetate [ORG CHEM] $(H_2CCHOOCCH_3)_x$ A thermoplastic polymer; insoluble in water, gasoline, oils, and fats, soluble in ketones, alcohols, benzene, esters, and chlorinated hydrocarbons; used in adhesives, films, lacquers, inks, latex paints, and paper sizes. Abbreviated PVA; pVAc.

polyvinyl alcohol [ORG CHEM] Water-soluble polymer made by hydrolysis of a polyvinyl ester (such as polyvinyl acetate); used in adhesives, as textile and paper sizes, and for emulsifying, suspending, and thickening of solutions. Abbreviated PVA.

polyvinyl carbazole [ORG CHEM] Thermoplastic resin made by reaction of acetylene with carbazole; softens at 150°C; has good electrical properties and heat and chemical stabilities; used as a paper-capacitor impregnant and as a substitute for electrical mica.

polyvinyl chloride [ORG CHEM] $(H_2CCHCl)_x$ Polymer of vinyl chloride; tasteless, odorless; insoluble in most organic solvents; a member of the family of vinyl resins; used to make fibers and film. Abbreviated PVC.

polyvinyl chloride acetate [ORG CHEM] Thermoplastic copolymer of vinyl chloride, CH_2CHCl, and vinyl acetate, $CH_3COOCH=CH_2$; colorless solid with good resistance to water, concentrated acids, and alkalies; compounded with plasticizers, it yields a flexible material superior to rubber in aging properties; used for cable and wire coverings and protective garments.

polyvinyl fluoride [ORG CHEM] $(-H_2CCHF-)_x$ Vinyl fluoride polymer; has superior resistance to weather, chemicals, oils, and stains, and has high strength; used for packaging (but not of food) and electrical equipment.

polyvinyl formate resin [ORG CHEM] $(CH_2=CHOOCH)_x$ Clear-colored resin that is hard and solvent-resistant; used to make clear, hard plastics.

polyvinylidene chloride [ORG CHEM] Thermoplastic polymer of vinylidene chloride, $H_2C=CCl_2$; white powder softening at 185–200°C; used to make soft-flexible to rigid products.

polyvinylidene fluoride [ORG CHEM] $H_2C=CF_2$ Fluorocarbon polymer made from vinylidene fluoride; has good tensile and compressive strength and high impact strength; used in chemical equipment for gaskets, impellers, and other pump parts, and for drum linings and protective coatings.

polyvinylidene resin *See* vinylidene resin.

polyvinyl isobutyl ether [ORG CHEM] $[-CH_2CHO-CH_2CH(CH_3)_2-]_x$ An odorless synthetic resin; elastomer to viscous liquid depending on molecular weight; soluble in hydrocarbons, esters, ethers, and ketones, insoluble in water; used in adhesives, waxes, plasticizers, lubricating oils, and surface coatings. Abbreviated PVI.

polyvinyl pyrrolidone [ORG CHEM] $(C_6H_9NO)_x$ A water-soluble, white, resinous solid; used in pharmaceuticals, cosmetics, detergents, and foods, and as a synthetic blood plasma. Abbreviated PVP.

Polyzoa [INV ZOO] The equivalent name for Bryozoa.

Pomacentridae [VERT ZOO] The damselfishes, a family of perciform fishes in the suborder Percoidei.

Pomadasyidae [VERT ZOO] The grunts and sweetlips, a family of perciform fishes in the suborder Percoidei.

Pomatiasidae [INV ZOO] A family of land snails in the order Pectinibranchia.

Pomatomidae [VERT ZOO] A monotypic family of the Perciformes containing the bluefish (*Pomatomus saltatrix*).

pomegranate [BOT] *Punica granatum.* A small, deciduous ornamental tree of the order Myrtales cultivated for its fruit, which is a reddish, pomelike berry containing numerous seeds embedded in crimson pulp.

Pomeranchuk pole *See* Pomeron.

Pomeranchuk theorem [PARTIC PHYS] The theorem that if the total cross section both for scattering of a particle by a given target particle and for scattering of its antiparticle by the same target particle, approach a limit at high energies, and do so sufficiently rapidly, then these limits must be the same.

Pomeron [PARTIC PHYS] A Regge pole which is located at $+1$ in the angular momentum plane when the momentum transfer in the crossed channel equals zero, corresponding to the fact that total cross sections of reactions are observed to approach constants at high energies. Also known as Pomeranchuk pole.

Pompilidae [INV ZOO] The spider wasps, the single family of the superfamily Pompiloidea.

Pompiloidea [INV ZOO] A monofamilial superfamily of hymenopteran insects in the suborder Apocrita with oval abdomen and strong spinose legs.

pom-pom [ORD] **1.** A rack of antiaircraft cannon, usually mounted in fours, as on the deck of a ship. **2.** An automatic cannon.

PONA analysis [ENG] ASTM analysis of paraffins (P), olefins (O), naphthenes (N) and aromatics (A) in gasolines.

poncelet [PHYS] A unit of power equal to the power delivered by a force of 100 kilograms-force when the point at which the force is applied is moved at a rate of 1 meter per second in the direction of the force; equal to 980.665 watts.

Ponchon-Savarit method [CHEM ENG] Graphical solution on an enthalpy-concentration diagram of liquid-vapor equilibrium values between trays of a distillation column.

pond [GEOGR] A small natural body of standing fresh water filling a surface depression, usually smaller than a lake. [MECH] *See* gram-force.

pondage [HYD] Water held in a reservoir for short periods to regulate natural flow, usually for hydroelectric power.

ponded stream [HYD] A stream in which a pond forms due to an interruption of the normal streamflow.

ponderosa pine [BOT] *Pinus ponderosa.* A hard pine tree of western North America; attains a height of 150–225 feet (46–69 meters) and has long, dark-green leaves in bundles of two to five and tawny, yellowish bark.

ponding [BUILD] An accumulation of water on a flat roof because of clogged or inadequate drains.

ponente [METEOROL] A west wind on the French Mediterranean coast, the northern Roussillon region, and Corsica.

Ponerinae [INV ZOO] A subfamily of tropical carnivorous ants (Formicidae) in which pupae characteristically form in cocoons.

pongee [TEXT] A rough-textured, plain-woven fabric of raw silk with irregular filling yarns and generally an ecru color.

Pongidae [VERT ZOO] A family of anthropoid primates in the

PONGIDAE

Typical anthropoid ape of the family Pongidae.

superfamily Hominoidea; includes the chimpanzee, gorilla, and orangutan.

poniente [METEOROL] The west wind in the Straits of Gibraltar.

pons [ANAT] **1.** A process or bridge of tissue connecting two parts of an organ. **2.** A convex white eminence located at the base of the brain; consists of fibers receiving impulses from the cerebral cortex and sending fibers to the contralateral side of the cerebellum.

pontianak gum [MATER] A grayish-white copal from various jelutongs of the genus *Dyera*, indigenous to Malacca and Borneo; used in rubber manufacture, chewing gum, adhesives, paints, and varnishes. Also known as jelutong.

pontine flexure [EMBRYO] A flexure in the embryonic brain concave dorsally, occurring in the region of the myelencephalon.

pontocaine *See* tetracaine hydrochloride.

Pontodoridae [INV ZOO] A monotypic family of pelagic polychaetes assigned to the Errantia.

pontoon [AERO ENG] A float on an airplane. [NAV ARCH] **1.** A low, flat-bottomed ship, similar to a barge, carrying cranes, capstans, and other machinery used to lift weights, to lean ships on their sides for repairs, and to perform other such operation. **2.** A wooden flat-bottomed boat or other float, used particularly to make temporary bridges.

pontoon bridge [CIV ENG] A bridge which floats on pontoons moored to the bed of a river or lake.

pontoon-tank roof [ENG] A type of floating tank roof, supported by buoyant floats on the liquid surface of a tank; the roof rises and falls with the liquid level in the tank; used to minimize vapor space above the liquid, thus reducing vapor losses during tank filling and emptying.

Pontryagin's maximum principle [MATH] A theorem giving a necessary condition for the solution of optimal control problems: let $\theta(\tau)$, $\tau_0 \leq \tau \leq T$ be a piecewise continuous vector function satisfying certain constraints; in order that the scalar function $S = \Sigma c_i x_i(T)$ be minimum for a process described by the equation $\partial x_i / \partial \tau = (\partial H / \partial z_i)[z(\tau), x(\tau), \theta(\tau)]$ with given initial conditions $x(\tau_0) = x^0$ it is necessary that there exist a nonzero continuous vector function $z(\tau)$ satisfying $dz_i / d\tau = -(\partial H / \partial x_i)[z(\tau), x(\tau), \theta(\tau)]$, $z_i(T) = -c_i$, and that the vector $\theta(\tau)$ be so chosen that $H[z(\tau), x(\tau), \theta(\tau)]$ is maximum for all τ, $\tau_0 \leq \tau \leq T$.

pony set [MIN ENG] A small timber set or frame incorporated in the main sets of a haulage level to accommodate an ore chute or other equipment from above or below.

pony truss [CIV ENG] A truss too low to permit overhead braces.

pool [CIV ENG] A body of water contained in a reservoir, by a dam, or by the gates of a lock. [HYD] A small deep body of water, often fed by a spring. [MIN ENG] **1.** To wedge for splitting in quarrying or mining. **2.** To undermine or undercut.

pool cathode [ELECTR] A cathode at which the principal source of electron emission is a cathode spot on a liquid-metal electrode, usually mercury.

pool-cathode mercury-arc rectifier [ELECTR] A pool tube connected in an electric circuit; its rectifying properties result from the fact that only the mercury-pool cathode, and not the anode, can emit electrons. Also known as mercury-pool rectifier.

pool-cathode tube *See* pool tube.

pool reactor [NUCLEO] A research nuclear reactor in which the core is suspended in a large pool of water that serves as moderator, reflector, coolant, and radiation shield. Also known as swimming-pool reactor.

pool stage [HYD] As used along the Ohio and upper Mississippi Rivers of the United States, a low-water condition with the navigation dams up so that the river is a series of shallow pools; when this condition exists, the river is said to be "in pool"; river depth is regulated by the dams so as to be adequate for navigation.

pool tube [ELECTR] A gas-discharge tube having a mercury-pool cathode. Also known as mercury tube; pool-cathode tube.

poop bulkhead [NAV ARCH] A partition wall that extends

between the upper and lower poopdecks at the poop's forward end.

poop deck [NAV ARCH] A partial deck above the main deck in a ship's afterpart.

pop [MIN ENG] A drill hole blasted to reduce larger pieces of rock or to trim a working face. Also known as pop hole; pop shot.

pop hole *See* pop.

poplar [BOT] Any tree of the genus *Populus*, family Salicaceae, marked by simple, alternate leaves, scaly buds, bitter bark, and flowers and fruits in catkins.

poplin [TEXT] A corded fabric made of cotton, polyester, wool, nylon, or blend of fibers.

popliteal artery [ANAT] A continuation of the femoral artery in the posterior portion of the thigh above the popliteal space and below the buttock.

popliteal nerve [ANAT] Either of two branches of the sciatic nerve in the lower part of the thigh; larger branch continues as the tibial nerve, and the smaller branch continues as the peroneal nerve.

popliteal space [ANAT] A diamond-shaped area behind the knee joint.

popliteal vein [ANAT] A vein passing through the popliteal space, formed by merging of the tibial veins and continuing to become the femoral vein.

popliteus [ANAT] **1.** The ham or hinder part of the knee joint. **2.** A muscle on the back of the knee joint.

poppet [CIV ENG] One of the timber and steel structures supporting the fore and aft ends of a ship for launching from sliding ways. [DES ENG] A spring-loaded ball engaging a notch; a ball latch. [MIN ENG] A pulley frame or the headgear over a mine shaft.

poppet valve [MECH ENG] A cam-operated or spring-loaded reciprocating-engine mushroom-type valve used for control of admission and exhaust of working fluid; the direction of movement is at right angles to the plane of its seat.

popping [MIN ENG] Exploding a stick of dynamite on a boulder so as to break it for easy removal from a quarry or opencast mine.

poppy [BOT] Any of various ornamental herbs of the genus *Papaver*, family Papaveraceae, with large, showy flowers; opium is obtained from the fruits of the opium poppy (*P. somniferum*).

poppy oil [MATER] Golden-yellow drying oil with pleasant taste and aroma; soluble in carbon disulfide, ether, and chloroform; expressed from poppy seeds, especially of the opium poppy; used as a food oil; in artist's colors, soaps, varnishes, and lubricants. Also known as poppy-seed oil.

poppy-seed oil *See* poppy oil.

pop shot *See* pop.

population [BIOL] A group of organisms occupying a specific geographic area or biome.

population covariance [STAT] The number $(1/N)[(v_1 - \bar{v}) \cdot (w_1 - \bar{w}) + \ldots + (v_N - \bar{v})(w_N - \bar{w})]$, where v_i and w_i, $i = 1, 2, \ldots, N$, are the values obtained from two populations, and $\bar{v}$ and $\bar{w}$ are the respective means.

population dispersal [BIOL] The process by which groups of living organisms expand the space or range within which they live.

population dispersion [BIOL] The spatial distribution at any particular moment of the individuals of a species of plant or animal.

population dynamics [BIOL] The aggregate of processes that determine the size and composition of any population.

population genetics [GEN] The study of both experimental and theoretical consequences of Mendelian heredity on the population level; includes studies of gene frequencies, genotypes, phenotypes, and mating systems.

population mean [STAT] The average of the numbers obtained for all members in a population by measuring some quantity associated with each member.

population of levels [STAT MECH] The number of members of an ensemble which are in each of the allowed energy states of a system.

population variance [STAT] The arithmetic average of the numbers $(v_1 - \bar{v})^2, \ldots, (v_N - \bar{v})^2$, where v_i are numbers obtained

---

POPPET VALVE

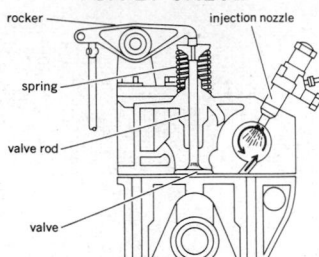

Poppet valve for internal combustion engine. *(From T. Baumeister, ed., Standard Handbook for Mechanical Engineers, 7th ed., McGraw-Hill, 1967)*

POPPY

Flower and leaves of the opium poppy *(Papaver somniferum)*.

from a population with N members, one for each member, and $\bar{v}$ is the population mean.

p orbital [ATOM PHYS] The orbital of an atomic electron with an orbital angular momentum quantum number of unity.

porcelain [MATER] A high-grade ceramic ware characterized by high strength, a white color, very low absorption, good translucency, and a hard glaze. Also known as European porcelain; hard paste porcelain; true porcelain.

porcelain capacitor [ELEC] A fixed capacitor in which the dielectric is a high grade of porcelain, molecularly fused to alternate layers of fine silver electrodes to form a monolithic unit that requires no case or hermetic seal.

porcelain cement [MATER] A cement for bonding porcelain to porcelain, such as a mixture of gutta-percha and shellac.

porcelain enamel *See* vitreous enamel.

porcelain insulator [MATER] An electrical insulator made from porcelain; the porcelain is often made in a one-fire process, the glaze being applied to the green or unfired ware, in contrast to the two-fire process used in making ordinary porcelain.

porcelainite *See* mullite.

porcelaneous [GEOL] Resembling unglazed porcelain.

Porcellanasteridae [INV ZOO] A family of essentially deep-water forms in the suborder Paxillosina.

Porcellanidae [INV ZOO] The rock sliders, a family of decapod crustaceans of the group Anomura which resemble true crabs but are distinguished by the reduced, chelate fifth pereiopods and the well-developed tail fan.

porcellanite [PETR] A hard, dense siliceous rock, such as impure chert or indurated clay or shale.

porcupine [VERT ZOO] Any of about 26 species of rodents in two families (Hystricidae and Erethizontidae) which have spines or quills in addition to regular hair.

pore [BIOL] Any minute opening by which matter passes through a wall or membrane. [GEOL] An opening or channelway in rock or soil. [MET] A minute cavity in a powder compact, metal casting, or electroplated coating.

pore compressibility [GEOL] The fractional change in reservoir-rock pore volume with a unit change in pressure upon that rock.

pore diameter [DES ENG] The average or effective diameter of the openings in a membrane, screen, or other porous material.

pore diffusion [FL MECH] The movement of fluids (gas or liquid) into the interstices of porous solids or membranes; occurs in membrane separation, zeolite adsorption, dialysis, and reverse osmosis.

pore fungus [MYCOL] The common name for members of the families Boletaceae and Polyporaceae in the group Hymenomycetes; sporebearing surfaces are characteristically within tubes or pores.

pore ice [HYD] Ice which fills or partially fills pore spaces in permafrost; forms by freezing soil water in place, with no addition of water.

porencephaly [MED] A condition in which the cavity of a lateral ventricle extends to the surface of the cerebral hemisphere; may result from brain tissue destruction or maldevelopment.

pore pressure *See* neutral stress.

pore-size distribution [GEOL] Variations in pore sizes in reservoir formations; each type of rock has its own typical pore size and related permeability.

pore space [GEOL] The pores in a rock or soil considered collectively. Also known as pore volume.

pore volume *See* pore space.

pore-water pressure *See* neutral stress.

poriaz [METEOROL] Violent northeast winds on the Black Sea near the Bosporus in the Soviet Union.

Porifera [INV ZOO] The sponges, a phylum of the animal kingdom characterized by the presence of canal systems and chambers through which water is drawn in and released; tissues and organs are absent.

Porlezzina [METEOROL] An east wind on Lake Lugano (Italy and Switzerland), blowing from the Gulf of Porlezza.

porocyte [INV ZOO] One of the perforated, tubular cells which constitute the wall of the incurrent canals in certain Porifera.

porogamy [BOT] Passage of the pollen tube through the micropyle of an ovule in a seed plant.

poromeric material [TEXT] A fabric made of polyurethane strengthened by polyester.

porosimeter [ENG] Laboratory compressed-gas device used for measurement of the porosity of reservoir rocks.

porosis [MED] Condition characterized by increased porosity, as of bone.

porosity [PHYS] **1.** Property of a solid which contains many minute channels or open spaces. **2.** The fraction as a percent of the total volume occupied by these channels or spaces; for example, in petroleum engineering the ratio (expressed in percent) of the void space in a rock to the bulk volume of that rock.

porosity feet [PETRO ENG] Reservoir porosity fraction multiplied by net pay in feet, where porosity fraction is the portion of the reservoir that is porous, and net pay is the depth and areal extent of the hydrocarbons-containing reservoir.

porosity trap *See* stratigraphic trap.

porous [MATER] **1.** Filled with pores. **2.** Capable of absorbing liquids.

porous bearing [DES ENG] A bearing made from sintered metal powder impregnated with oil by a vacuum treatment.

porous carbon [MATER] Plates, tubes, or disks of uniform carbon particles pressed together without a binder; used for the filtration of corrosive liquids and gases.

porous graphite [MATER] Plates, tubes, or disks of uniform graphite particles pressed together without a binder; more resistant to oxidation but lower in strength than porous carbon.

porous metals [MET] Metals, made by powder metallurgy, having uniformly distributed controlled pore sizes, in the form of sheets, tubes, and shapes; used for filtering liquids and gases at elevated temperatures.

porous mold [ENG] A plastic-forming mold made from bonded or fused aggregates (such as powdered metal or coarse pellets) so that the resulting mass contains numerous open interstices through which air or liquids can pass.

porous reservoir model [PETRO ENG] Scaled laboratory model of porous reservoir used for the study of reservoir areal waterflood efficiencies.

porous wheel [DES ENG] A grinding wheel having a porous structure and a vitrified or resinoid bond.

Poroxylaceae [PALEOBOT] A monogeneric family of extinct plants included in the Cordaitales.

porphin [BIOCHEM] A heterocyclic ring consisting of four pyrrole rings linked by methine ($-CH=$) bridges; the basic structure of chlorophyll, hemoglobin, the cytochromes, and certain other related substances.

porphobilinogen [BIOCHEM] $C_{10}H_{14}O_4N_2$ Dicarboxylic acid derived from pyrrole; a product of hemoglobin breakdown that gives the urine a Burgundy-red color.

porphrite *See* porphyry.

porphyria [MED] A usually hereditary, pathologic disorder of porphyrin metabolism characterized by porphyrinuria and photosensitivity.

porphyrin [BIOCHEM] A class of red-pigmented compounds with a cyclic tetrapyrrolic structure in which the four pyrrole rings are joined through their α-carbon atoms by four methene bridges ($=C-$); the porphyrins form the active nucleus of chlorophylls and hemoglobin.

porphyrinuria [MED] The excretion of large quantities of porphyrin in the urine.

porphyritic [PETR] Pertaining to or resembling porphyry.

porphyroblast [PETR] A relatively large crystal formed in a metamorphic rock.

porphyroclastic structure *See* mortar structure.

porphyry [PETR] An igneous rock in which large phenocrysts are enclosed in a very-fine-grained to aphanitic matrix. Formerly known as porphrite.

porpoise [VERT ZOO] Any of several species of marine mammals of the family Phocaenidae which have small flippers, a highly developed sonar system, and smooth, thick, hairless skin.

porpoise oil [MATER] A pale-yellow fatty oil obtained from blubber of the brown porpoise; soluble in ether, benzene,

PORCUPINE

The Canadian porcupine (*Erethizon dorsatum*), about 3½ feet (1 meter) long, with long white hairs scattered through the brownish-black fur.

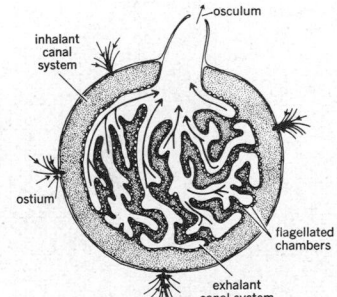

PORIFERA

osculum
inhalant canal system
ostium
flagellated chambers
exhalant canal system

Diagram of the canal system of a young fresh-water sponge. (*After Ankel, 1950*)

PORPOISE

The common porpoise, a cetacean found in North American and European coastal waters.

PORTUGUESE MAN-OF-WAR

Float and tentacles of a Portuguese man-of-war. *(From T. I. Storer and R. L. Usinger, General Zoology, 3d ed., McGraw-Hill, 1957)*

carbon disulfide, and chloroform; used as a lubricant, leather dressing, and illuminating oil, and in soap stock.

Porro prism [OPTICS] One of two identical prisms used in the Porro prism erecting system; it is a right-angle prism with the corners rounded to minimize breakage and simplify assembly.

Porro prism erecting system [OPTICS] A prism erecting system, designed by M. Porro, in which there are four reflections to completely erect the image; two porro prisms are employed; the line of sight is bent through 360°, is displaced, but is not deviated; used in prism binoculars and some telescope systems.

port [ELEC] An entrance or exit for a network. [ELECTROMAG] An opening in a waveguide component, through which energy may be fed or withdrawn, or measurements made. [ENG] The side of a ship or airplane on the left of a person facing forward. [ENG ACOUS] An opening in a bass-reflex enclosure for a loudspeaker, designed and positioned to improve bass response. [GEOGR] *See* harbor. [NAV ARCH] An opening in a vessel to provide access for passengers, cargo handling, discharging water, and so forth. [NUCLEO] An opening in a research reactor through which objects are inserted for irradiation or from which beams of radiation emerge for experimental use.

portable [ENG] Capable of being easily and conveniently transported.

porta hepatis [ANAT] The transverse fissure of the liver through which the portal vein and hepatic artery enter the liver and the hepatic ducts leave.

portal [ANAT] **1.** Of or pertaining to the porta hepatis. **2.** Pertaining to the portal vein or system. [ENG] A redundant frame consisting of two uprights connected by a third member at the top. [MIN ENG] **1.** An entrance to a mine. **2.** The rock face at which a tunnel is started.

portal circulation [PHYSIO] The passage of venous blood through a portal system.

portal cirrhosis [MED] Replacement of normal liver structure by abnormal lobules of liver cells, often hyperplastic, delimited by bands of fibrous tissue, giving the gross appearance of a finely nodular surface. Also known as Laennec's cirrhosis.

portal crane [MECH ENG] A jib crane carried on a four-legged portal built to run on rails.

portal hypertension [MED] Portal venous pressure in excess of 20 mm Hg, resulting from intrahepatic or extrahepatic portal venous compression or occlusion.

portal system [ANAT] A system of veins that break into a capillary network before returning the blood to the heart.

portal vein [ANAT] Any vein that terminates in a network of capillaries.

porthole [DES ENG] The opening or passageway connecting the inside of a bit or core barrel to the outside and through which the circulating medium is discharged. [ENG] A circular opening in the side of a ship or airplane, usually serving as a window and containing one or more panes of glass.

porthole die [MET] An extrusion die having two or more sections in which metal is extruded separately in each section and welded before leaving the die to form intricate hollow shapes.

portico [ARCH] A colonnade or other sheltered place to walk in.

portland cement [MATER] A hydraulic cement resembling portland stone when hardened; made of pulverized, calcined argillaceous and calcareous materials; the proper name for ordinary cement.

Portlandian [GEOL] A European geologic stage of the Upper Jurassic, above Kimmeridgian, below Berriasian of Cretaceous.

portland-pozzolana cement [MATER] Portland cement to which pozzolana has been added, in the amount of about 20%, to reduce the liability of leaching.

port of entry [CIV ENG] A location for clearance of foreign goods and citizens through a customhouse.

port operations service [COMMUN] Maritime mobile communications service in or near a port, between coast stations and ship stations, or between ship stations, in which messages are restricted to those relating to the movement and safety of ships and, in an emergency, to the safety of persons.

Portuguese man-of-war [INV ZOO] Any of several brilliantly colored tropical siphonophores in the genus *Physalia* which possess a large float and extremely long tentacles.

Portulacaceae [BOT] A family of dicotyledonous plants in the order Caryophyllales distinguished by a syncarpous gynoecium, few, cyclic tepals and stamens, two sepals, and two to many ovules.

Portunidae [INV ZOO] The swimming crabs, a family of the Brachyura having the last pereiopods modified as swimming paddles.

port-wine stain [MED] A congenital hemangioma characterized by one or more red to purplish patches, usually on the face.

Porulosida [INV ZOO] An order of the protozoan subclass Radiolaria in which the central capsule shows many pores.

posistor [ELECTR] A thermistor having a large positive resistance-temperature characteristic.

position [NAV] A point defined by stated or implied coordinates, usually on the surface of the earth.

positional-error constant [CONT SYS] For a stable unity feedback system, the limit of the transfer function as its argument approaches zero.

positional notation [MATH] Any of several numeration systems in which the significance of each digit of a numeral depends on its position in the sequence as well as its numeric value, and successive digits are interpreted as coefficients of successive powers of an integer called the base. Also known as notation.

positional punch [ADP] The row position of a punched hole in a specific column of a punch card; in an 80-column punch card the rows are designated 0 to 9, or *x* and *y*; in a 90-column card the rows are designated 0, 1, 3, 5, 7, and 9.

positional servomechanism [CONT SYS] A feedback control system in which the mechanical position (as opposed to velocity) of some object is automatically maintained.

position angle [ASTRON] **1.** The angle formed by the great circle running through two celestial objects and the hour circle running through one of the objects. **2.** In measuring double stars, the angle formed between the great circle running through both components and the hour circle going through the primary measured from the north through the east from 0 to 360°. [NAV] That angle of the navigational triangle at the celestial body having the hour circle and the vertical circle at its sides. Also known as parallactic angle.

position approximate [NAV] Of inexact position; used principally on charts to indicate that the position of a wreck, shoal, and so on, has not been accurately determined or does not remain fixed.

position blocks [MIN ENG] Unproved mining claims that are in a position to contain a lode if the lode continues in the direction in which it has been proved in other claims.

position circle [NAV] **1.** A circle covering all possible points within which a search objective can be located. **2.** The line of position produced by a distance measurement such as those produced by a DME (distance-measuring equipment).

position correction [ORD] Correction applied to firing data to compensate for difference in location of individual pieces in a battery.

position correction grid [ORD] Transparent template superimposed on a large-scale plot of a battery position used to determine individual range and deflection corrections within the battery in order to obtain the type of sheaf (planned planes of fire) desired.

position dialing [COMMUN] Dialing over the regular position cord circuits by means of relay circuit under control of a dial of the regular cord circuits.

position doubtful [NAV] Of uncertain position; used principally on charts to indicate that a wreck, shoal, and so on, has been reported in various positions and not definitely determined in any.

positioned weld [MET] A weld made in a joint in which members have been positioned to facilitate welding.

position effect [GEN] **1.** Change in expressivity of a gene associated with chromosome aberrations. **2.** Inherent gene expression as influenced by neighboring genes.

position error [NAV] **1.** The error of a position. **2.** That error of an instrument reading due to location or orientation.

position finder [ORD] Optical or electrical instrument used in finding the range and position of a target.

position finding [ORD] The process of determining the position of a target with relation to the battery, and the determination of a future position upon which to direct the fire.

position firing [ORD] A method of defensive gunnery used by bombers, especially during World War II, in which definite amounts of deflection are prescribed for firing at attacking fighter planes.

position indicator [ENG] An electromechanical dead-reckoning computer, either an air-position indicator or a ground-position indicator.

positioning [MECH ENG] A tooling function concerned with manipulating the workpiece in relationship to the working tools.

positioning action [CONT SYS] Automatic control action in which there is a predetermined relation between the value of a controlled variable and the position of a final control element.

positioning band [ORD] A metal band on some recoilless ammunition, placed to ensure the proper positioning of the round inside the chamber and tube.

position line _See_ line of position.

position operator [QUANT MECH] The quantum-mechanical operator corresponding to the classical position variable of a particle.

position report [NAV] A radio message containing specified information regarding the position and progress of a craft.

position representation [QUANT MECH] A representation in which the state functions are eigenfunctions of the position operator. Also known as Schrödinger representation.

position sensor [ENG] A device for measuring a position and converting this measurement into a form convenient for transmission. Also known as position transducer.

position telemetering [ENG] A variation of voltage telemetering in which the system transmits the measurand by positioning a variable resistor or other component in a bridge circuit so as to produce relative magnitudes of electrical quantities or phase relationships.

position transducer _See_ position sensor.

position vector [MATH] The position vector of a point in euclidean space is a vector whose length is the distance from the origin to the point and whose direction is the direction from the origin to the point. Also known as radius vector.

positive [ELEC] Having fewer electrons than normal, and hence having the ability to attract electrons. [GRAPHICS] Having the same rendition of light and shade as in the original scene. [MATH] Having value greater than zero.

positive acceleration [MECH] 1. Accelerating force in an upward sense or direction, such as from bottom to top, or from seat to head; 2. The acceleration in the direction that this force is applied.

positive afterimage [PHYSIO] An afterimage persisting after the eyes are closed or turned toward a dark background, and of the same color as the stimulating light.

positive angle [MATH] The angle swept out by a ray moving in a counterclockwise direction.

positive area _See_ positive element.

positive axis [MATH] The segment of an axis arising from a cartesian coordinate system which is realized by positive values of the coordinate variables. [METEOROL] In tropical synoptic analysis, a locus of maximum streamline curvature in an easterly wave; used primarily in the analysis of waves that span the equatorial trough (equatorial waves); a positive axis corresponds to a trough line in the Northern Hemisphere and a ridge line in the Southern Hemisphere.

positive bias [ELECTR] A bias such that the control grid of an electron tube is positive with respect to the cathode.

positive charge [ELEC] The type of charge which is possessed by protons in ordinary matter, and which may be produced in a glass object by rubbing with silk.

positive clutch [MECH ENG] A clutch designed to transmit torque without slip.

positive column [ELECTR] The luminous glow, often striated, that occurs between the Faraday dark space and the anode in a glow-discharge tube.

positive crystal [OPTICS] 1. Uniaxial anisotropic crystal having the ordinary index of refraction greater than the extraordinary index. 2. Biaxial anisotropic crystal having the intermediate index of refraction beta closer in value to alpha, and with Z the acute bisectrix.

positive definite [MATH] 1. A matrix is positive definite if it is Hermitian and all of its eigenvalues are positive. 2. A linear transformation T on an inner product space is positive definite if $<Tu,u> \geq 0$ for all u in the space.

positive derail [MIN ENG] A device installed in or on a mine track to derail runaway cars or trips.

positive-displacement compressor [MECH ENG] A compressor that confines successive volumes of fluid within a closed space in which the pressure of the fluid is increased as the volume of the closed space is decreased.

positive-displacement meter [ENG] A fluid quantity meter that separates and captures definite volumes of the flowing stream one after another and passes them downstream, while counting the number of operations.

positive-displacement pump [MECH ENG] A pump in which a measured quantity of liquid is entrapped in a space, its pressure is raised, and then it is delivered; for example, a reciprocating piston-cylinder or rotary-vane, gear, or lobe mechanism.

positive draft [MECH ENG] Pressure in the furnace or gas passages of a steam-generating unit which is greater than atmospheric pressure.

positive drive belt _See_ timing belt.

positive electrode _See_ anode.

positive element [GEOGR] A large structural feature of the earth's crust characterized by long-term upward movement (uplift, emergence) or subsidence less rapid than that of adjacent negative elements. Also known as archibole; positive area.

positive feedback [CONT SYS] Feedback in which a portion of the output of a circuit or device is fed back in phase with the input so as to increase the total amplification. Also known as reaction (British usage); regeneration; regenerative feedback; retroaction (British usage).

positive-grid oscillator _See_ retarding-field oscillator.

positive image [GRAPHICS] A picture as normally seen on a television picture tube or in a photograph, having the same rendition of light and shade as in the original scene.

positive interference [GEN] The reduction, by one crossover exchange, of the likelihood of another crossover in its vicinity.

positive ion [CHEM] An atom or group of atoms which by loss of one or more electrons has acquired a positive electric charge; occurs on ionization of chemical compounds as H^+ from ionization of hydrochloric acid, HCl.

positive-ion sheath [ELECTR] Collection of positive ions on the control grid of a gas-filled triode tube.

positive logic [ELECTR] Pertaining to a logic circuit in which the more positive voltage (or current level) represents the 1 state; the less positive level represents the 0 state.

positive meniscus lens [OPTICS] A lens having one convex (bulging) and one concave (depressed) surface, with the radius of curvature of the convex surface smaller than that of the concave surface.

positive modulation [ELECTR] In an amplitude-modulated television system, that form of television modulation in which an increase in brightness corresponds to an increase in transmitted power.

positive mold [ENG] A plastics mold designed to trap all of the molding resin when the mold closes.

positive motion [MECH ENG] Motion transferred from one machine part to another without slippage.

positive movement [GEOL] 1. Uplift or emergence of the earth's crust relative to an adjacent area of the crust. 2. A relative rise in sea level with respect to land level.

positive ore [MIN ENG] Ore exposed on four sides in blocks of a size variously prescribed.

positive phase-sequence relay [ELEC] Relay which functions in conformance with the positive phase-sequence component of the current, voltage, or power of the circuit.

positive ray [ELECTR] A stream of positively charged atoms or molecules, produced by a suitable combination of ionizing agents, accelerating fields, and limiting apertures.

positive real function [MATH] An analytic function whose value is real when the independent variable is real, and whose real part is positive or zero when the real part of the independent variable is positive or zero.

positive shoreline *See* shoreline of submergence.

positive temperature coefficient [THERMO] The condition wherein the resistance, length, or some other characteristic of a substance increases when temperature increases.

positive terminal [ELEC] The terminal of a battery or other voltage source toward which electrons flow through the external circuit.

positive transfer [PSYCH] Rapid learning in a new situation because the stimuli or responses required are similar to those learned in an earlier situation.

positive transmission [COMMUN] Transmission of television signals in such a way that an increase in initial light intensity causes an increase in the transmitted power.

positive zero [ADP] The zero value reached by counting down from a positive number in the binary system.

positron [PARTIC PHYS] An elementary particle having mass equal to that of the electron, and having the same spin and statistics as the electron, but a positive charge equal in magnitude to the electron's negative charge.

positron emission [NUC PHYS] A β-decay process in which a nucleus ejects a positron and a neutrino.

positronium [PARTIC PHYS] The bound state of an electron and a positron.

possible ore [MIN ENG] A class of ore whose existence is a reasonable possibility, based upon geologic-mineralogic relationships and the extent of ore bodies already developed. Also known as extension ore.

possible reserves [PETRO ENG] Primary petroleum reserves that may exist, but available data do not confirm their presence.

post [CIV ENG] 1. A vertical support such as a pillar, upright, or fence stake. 2. A pole used as a boundary marker. [MIN ENG] 1. A mine timber, or any upright timber, more commonly the uprights which support the roof crosspieces. 2. The support fastened between the roof and floor of a coal seam, used with certain types of mining machines or augers. 3. A pillar of coal or ore. [NAV] A small beacon used for marking channels; it is usually more substantial than a perch.

postaccelerating electrode *See* intensifier electrode.

postacceleration [ELECTR] Acceleration of beam electrons after deflection in an electron-beam tube. Also known as postdeflection acceleration (PDA).

post-and-lintel [ARCH] Pertaining to construction employing vertical supports and horizontal beams instead of arches or vaults.

post brake [MECH ENG] A brake occasionally fitted on a steam winder or haulage, and consisting of two upright posts mounted on either side of the drum that operate on brake paths bolted to the drum cheeks.

postcard paper [MATER] Lightweight Bristol board, made from soda and sulfite pulps, that has a smooth, firm surface for writing with pen or pencil, or for ordinary printing.

postcentral gyrus [ANAT] The cerebral convolution that lies immediately posterior to the central sulcus and extends from the longitudinal fissure above the posterior ramus of the lateral sulcus.

postcentral sulcus [ANAT] The first sulcus of the parietal lobe of the cerebrum, lying behind and roughly parallel to the central sulcus.

postcure bonding [ENG] A method of postcuring at elevated temperatures of parts previously subjected to autoclave or press in order to obtain higher heat-resistant properties of the adhesive bond.

postcure finish [TEXT] A durable-press finish applied to a fabric by the mill but heat-set after the garment is completed.

postdecrementing *See* autodecrement addressing.

post-deflection accelerating electrode *See* intensifier electrode.

postdeflection acceleration *See* postacceleration.

post drill [ENG] An auger or drill supported by a post.

postedit [ADP] To edit the output data of a computer.

postemphasis *See* deemphasis.

postencephalitic parkinsonism [MED] The parkinsonian syndrome occurring as a sequel to lethargic encephalitis within a variable period, from days to many years, after the acute process.

postequalization *See* deemphasis.

poster board [MATER] A stiff cardboard used for show cards, posters, display advertising, and signs; it may be white or colored on one side.

posterior [ZOO] 1. The hind end of an organism. 2. Toward the back, or hinder end, of the body.

posterior chamber [ANAT] The space in the eye between the posterior surface of the iris and the ciliary body, and the lens.

posterior probabilities [STAT] Probabilities of the outcomes of an experiment after it has been performed and a certain event has occurred.

poster paint [MATER] A water paint with a gum binder, which is brilliant, opaque, and fast-drying, and usually sold in jars. Also known as show-card color.

poster paper [MATER] A strong, waterproofed paper used for billboard poster work; it is white or colored with nonfading pigments, and does not curl when paste is applied to it.

postforming [ENG] Forming, bonding, or shaping of heated, flexible thermoset laminates before the final thermoset reaction has occurred; upon cooling, the formed shape is held.

postglacial [GEOL] Referring to the interval of geologic time since the total disappearance of continental glaciers in middle latitudes or from a particular area.

posthitis [MED] Inflammation of the prepuce.

posthole [CIV ENG] A hole bored in the ground to hold a fence post.

postignition [CHEM] Surface ignition after the passage of the normal spark.

postincrementing *See* autoincrement addressing.

postindexing [ADP] Operation in which the contents of a register indicated by the index bits of an indirect address are added to the indirect address to form the effective address.

postmagmatic [GEOL] Pertaining to geologic reactions or events occurring after the bulk of the magma has crystallized.

postmeridian [SCI TECH] After noon, or the period of time between noon (1200) and midnight (2400).

postmortem [ADP] Any action taken after an operation is completed to help analyze that operation. [MED] Occurring after death.

postmortem dump [ADP] 1. The printout showing the state of all registers and the contents of main memory, taken after a computer run terminates normally or terminates owing to fault. 2. The program which generates this printout.

postnatal [MED] Subsequent to birth.

postnecrotic cirrhosis [MED] Cirrhosis, usually due to toxic agents or viral hepatitis, characterized by necrosis of liver cells, regenerating nodules of hepatic tissue, and the presence of large bands of connective tissue.

postoperative hernia *See* incisional hernia.

postorogenic [GEOL] Of a geologic process or event, occurring after a period of orogeny.

postpartum [MED] Following childbirth.

postprandial [MED] After a meal.

postprecipitation [CHEM] Precipitation of an impurity from a supersaturated solution onto the surface of an already present precipitate; used for analytical laboratory separations.

posttensioning [ENG] Compressing of cast concrete beams or other structural members to impart the characteristics of prestressed concrete.

posttraumatic hernia *See* incisional hernia.

postulate *See* axiom.

pot *See* potentiometer; pothole.

potable [SCI TECH] Suitable for drinking.

Potamogalinae [VERT ZOO] An aberrant subfamily of West African tenrecs (Tenrecidae).

Potamogetonaceae [BOT] A large family of monocotyledonous plants in the order Najadales characterized by a solitary, apical or lateral ovule, usually two or more carpels, flowers in spikes or racemes, and four each of tepals and stamens.

Potamogetonales [BOT] The equivalent name for Najadales.

Potamonidae [INV ZOO] A family of fresh-water crabs included in the Brachyura.

potamoplankton [BIOL] Plankton found in rivers.

potarite [MINERAL] PdHg A silver-white isometric mineral composed of palladium and mercury alloy. Also known as palladium amalgam.

potash *See* potassium carbonate.

potash alum *See* kalinite.

potash bentonite *See* potassium bentonite.

potash blue [INORG CHEM] A pigment made by oxidizing ferrous ferrocyanide; used in making carbon paper.

potash feldspar *See* potassium feldspar.

potash mica *See* muscovite.

potash regulations [MIN ENG] Rules governing the prospecting and exploitation of land containing potash.

potassium [CHEM] A chemical element, symbol K, atomic number 19, atomic weight 39.102; an alkali metal. Also known as kalium.

potassium-42 [NUC PHYS] Radioactive isotope with mass number of 42; half-life is 12.4 hours, with β- and γ-radiation; radiotoxic; used as radiotracer in medicine.

potassium acetate [ORG CHEM] $KC_2H_3O_2$ White, deliquescent solid; soluble in water and alcohol, insoluble in ether; melts at 292°C; used as analytical reagent, dehydrating agent, in medicine, and in crystal glass manufacture.

potassium acid carbonate *See* potassium bicarbonate.

potassium acid fluoride *See* potassium bifluoride.

potassium acid oxalate *See* potassium binoxalate.

potassium acid phosphate *See* potassium phosphate.

potassium acid sulfate *See* potassium bisulfate.

potassium acid sulfite *See* potassium bisulfite.

potassium alum *See* potassium aluminum sulfate.

potassium aluminate [INORG CHEM] $K_2Al_2O_4 \cdot 3H_2O$ Water-soluble, alcohol-insoluble, lustrous crystals; used as a dyeing and printing mordant, and as a paper sizing.

potassium aluminum sulfate [INORG CHEM] $KAl(SO_4)_2 \cdot 12H_2O$ White, odorless crystals that are soluble in water; used in medicines and baking powder, in dyeing, papermaking, and tanning. Also known as alum; aluminum potassium sulfate; potassium alum.

potassium antimonate [INORG CHEM] $KSbO_3$ White, water-soluble crystals. Also known as potassium stibnate.

potassium-argon dating [GEOL] Dating of archeological, geological, or organic specimens by measuring the amount of argon accumulated in the matrix rock through decay of radioactive potassium.

potassium arsenate [INORG CHEM] K_3AsO_4 Poisonous, colorless crystals; soluble in water, insoluble in alcohol; used as an insecticide, analytical reagent, and in hide preservation and textile printing. Also known as Macquer's salt.

potassium arsenite [INORG CHEM] $KH(AsO_2)_2$ Poisonous, hygroscopic, white powder; soluble in alcohol; decomposes slowly in air; used in medicine, on mirrors, and as an analytical reagent. Also known as potassium metarsenite.

potassium bentonite [GEOL] A clay of the illite group that contains potassium and is formed by alteration of volcanic ash. Also known as K bentonite; potash bentonite.

potassium benzyl penicillinate *See* benzyl penicillin potassium.

potassium bicarbonate [INORG CHEM] $KHCO_3$ A white powder or granules, or transparent colorless crystals; used in baking powder and in medicine as an antacid. Also known as potassium acid carbonate.

potassium bichromate *See* potassium dichromate.

potassium bifluoride [INORG CHEM] KHF_2 Colorless, corrosive, poisonous crystals; soluble in water and dilute alcohol; used to etch glass and as a metallurgy flux. Also known as Fremy's salt; potassium acid fluoride.

potassium binoxalate [ORG CHEM] $KHC_2O_4 \cdot H_2O$ A poisonous, white, odorless, crystalline compound; used to clean wood and remove ink stains, as a mordant in dyeing, and in photography. Also known as potassium acid oxalate; sal acetosella; salt of sorrel.

potassium bisulfate [INORG CHEM] $KHSO_4$ Water-soluble, colorless crystals, melting at 214°C; used in winemaking, fertilizer manufacture, and as a flux and food preservative. Also known as acid potassium sulfate; potassium acid sulfate.

potassium bisulfite [INORG CHEM] $KHSO_3$ White, water-soluble powder with sulfur dioxide aroma; insoluble in alcohol; decomposes when heated; used as an antiseptic and reducing chemical, and in analytical chemistry, tanning, and bleaching. Also known as potassium acid sulfite.

potassium bromate [INORG CHEM] $KBrO_3$ Water-soluble, white crystals, melting at 434°C; insoluble in alcohol; strong oxidizer and a fire hazard; used in analytical chemistry and as an additive for permanent-wave compounds.

potassium bromide [INORG CHEM] KBr White, hygroscopic crystals with bitter taste; soluble in water and glycerin, slightly soluble in alcohol and ether; melts at 730°C; used in medicine, soaps, photography, and lithography.

potassium bromide-disk technique [ANALY CHEM] Method of preparing an infrared spectrometry sample by grinding it and mixing it with a dry powdered alkali halide (such as KBr), then compressing the mixture into a tablet or pellet. Also known as pellet technique; pressed-disk technique.

potassium cadmium iodide *See* potassium tetraiodocadmate.

potassium carbonate [INORG CHEM] K_2CO_3 White, water-soluble, deliquescent powder, melting at 891°C; insoluble in alcohol; used in brewing, ceramics, explosives, fertilizers, and as a chemical intermediate. Also known as potash; salt of tartar.

potassium chloride [INORG CHEM] KCl Colorless crystals with saline taste; soluble in water, insoluble in alcohol; melts at 776°C; used as a fertilizer and in photography and pharmaceutical preparations. Also known as potassium muriate.

potassium chromate [INORG CHEM] K_2CrO_4 Yellow crystals, melting at 971°C; soluble in water, insoluble in alcohol; used as an analytical reagent and textile mordant, in enamels, inks, and medicines, and as a chemical intermediate.

potassium chromium sulfate *See* chrome alum.

potassium citrate [ORG CHEM] $K_3C_6H_5O_7 \cdot H_2O$ Odorless crystals with saline taste; soluble in water and glycerol, deliquesent and insoluble in alcohol; decomposes about 230°C; used in medicine.

potassium cyanide [INORG CHEM] KCN Poisonous, white, deliquescent crystals with bitter almond taste; soluble in water, alcohol, and glycerol; used for metal extraction, for electroplating, for heat-treating steel, and as an analytical reagent and insecticide.

potassium dichromate [INORG CHEM] $K_2Cr_2O_7$ Poisonous, yellowish-red crystals with metallic taste; soluble in water, insoluble in alcohol; melts at 396°C, decomposes at 500°C; used as an oxidizing agent and analytical reagent, and in explosives, matches, and electroplating. Also known as potassium bichromate; red potassium chromate.

potassium dihydrogen phosphate *See* potassium phosphate.

potassium diphosphate *See* potassium phosphate.

potassium ethyldithiocarbonate *See* potassium xanthate.

potassium ethylxanthogenate *See* potassium xanthate.

potassium feldspar [MINERAL] Any alkali feldspar (orthoclase, microcline, sonidine, adularia) containing the molecule $KAlSi_3O_8$. Incorrectly known as K feldspar; potash feldspar.

potassium ferricyanide [INORG CHEM] $K_3Fe(CN)_6$ Poisonous, water-soluble, bright-red crystals; decomposes when heated; used in calico printing and wool dyeing. Also known as red potassium prussiate; red prussiate of potash.

potassium ferrocyanide [INORG CHEM] $K_4Fe(CN)_6 \cdot 3H_2O$ Yellow crystals with saline taste; soluble in water, insoluble in alcohol; loses water at 60°C; used in medicine, dry colors, explosives, and as an analytical reagent. Also known as yellow prussiate of potash.

potassium fluoborate [INORG CHEM] KBF_4 White powder or gelatinous crystals that decompose at high temperatures; slightly soluble in water and hot alcohol; used as a sand agent to cast magnesium and aluminum, and in electrochemical processes.

potassium fluoride [INORG CHEM] KF or $KF \cdot 2H_2O$ Poisonous, white, deliquescent crystals with saline taste; soluble in water and hydrofluoric acid, insoluble in alcohol; melts at 846°C; used to etch glass and as a preservative and insecticide.

potassium fluosilicate [INORG CHEM] K_2SiF_6 An odorless, white crystalline compound; slightly soluble in water; used in vitreous frits, synthetic mica, metallurgy, and ceramics. Also known as potassium silicofluoride.

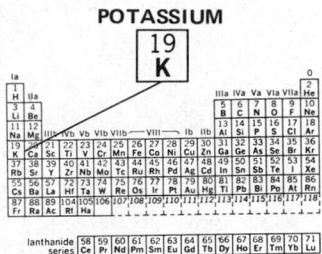

POTASSIUM

Periodic table of the chemical elements showing the position of potassium.

potassium gluconate [ORG CHEM] $KC_6H_{11}O_7$ An odorless, white crystalline compound with salty taste; soluble in water, insoluble in alcohol and benzene; used in medicine.

potassium hydrate *See* potassium hydroxide.

potassium hydrogen phosphate *See* potassium phosphate.

potassium hydroxide [INORG CHEM] KOH Toxic, corrosive, water-soluble, white solid, melting at 360°C; used to make soap and matches, and as an analytical reagent and chemical intermediate. Also known as caustic potash; potassium hydrate.

potassium hyperchlorate *See* potassium perchlorate.

potassium iodate [INORG CHEM] KIO_3 Odorless, white crystals; soluble in water, insoluble in alcohol; melts at 560°C; used as an analytical reagent and in medicine.

potassium iodide [INORG CHEM] KI Water- and alcohol-soluble, white crystals with saline taste; melts at 686°C; used in medicine and photography, and as an analytical reagent.

potassium linoleate [ORG CHEM] $C_{17}H_{31}COOK$ Light-tan, water-soluble paste; used as an emulsifying agent.

potassium manganate [INORG CHEM] K_2MnO_4 Water-soluble dark-green crystals, decomposing at 190°C; used as an analytical reagent, bleach, oxidizing agent, disinfectant, mordant for dyeing wool and in photography, printing, and water purification.

potassium metarsenite *See* potassium arsenite.

potassium monophosphate *See* potassium phosphate.

potassium muriate *See* potassium chloride.

potassium nitrate [INORG CHEM] KNO_3 Flammable, water-soluble, white crystals with saline taste; melts at 337°C; used in pyrotechnics, explosives, and matches, as a fertilizer, and as an analytical reagent. Also known as niter.

potassium nitrite [INORG CHEM] KNO_2 White, deliquescent prisms, melting at 297–450°C; soluble in water, insoluble in alcohol; strong oxidizer, exploding at over 550°C; used as an analytical reagent, in medicine, organic synthesis, pyrotechnics, and explosives.

potassium oxalate [ORG CHEM] $K_2C_2O_4 \cdot H_2O$ Odorless, efflorescent, water-soluble, colorless crystals; decomposes when heated; used in analytical chemistry and photography and as a bleach and oxalic acid source.

potassium oxide [INORG CHEM] K_2O Gray, water-soluble crystals; melts at red heat; forms potassium hydroxide in water.

potassium penicillin G₁ *See* benzyl penicillin potassium.

potassium perchlorate [INORG CHEM] $KClO_4$ Explosive, oxidative, colorless crystals; soluble in water, insoluble in alcohol; decomposes at 400°C; used in explosives, medicine, pyrotechnics, analysis, and as a reagent and oxidizing agent. Also known as potassium hyperchlorate.

potassium permanganate [INORG CHEM] $KMnO_4$ Highly oxidative, water-soluble, purple crystals with sweet taste; decomposes at 240°C; and explodes in contact with oxidizable materials; used as a disinfectant and analytical reagent, in dyes, bleaches, and medicines, and as a chemical intermediate. Also known as purple salt.

potassium phosphate [INORG CHEM] Any one of three orthophosphates of potassium. The monobasic form, KH_2PO_4, consists of colorless, water-soluble crystals melting at 253°C; used in sonar transducers, optical modulation, medicine, baking powders, and nutrient solutions; also known as potassium acid phosphate, potassium dihydrogen phosphate (KDP), potassium diphosphate, potassium orthophosphate. The dibasic form, K_2HOP_4, consists of white, water-soluble crystals; used in medicine, fermentation, and nutrient solutions; also known as potassium hydrogen phosphate, potassium monophosphate. The tribasic form, K_3PO_4, is a water-soluble, hygroscopic white powder, melting at 1340°C; used to purify gasoline, to soften water, and to make liquid soaps and fertilizers; also known as neutral potassium phosphate, tripotassium orthophosphate.

potassium pyrophosphate [INORG CHEM] $K_4P_2O_7 \cdot 3H_2O$ Water-soluble crystals; dehydrates below 300°C, melts at 1090°C; used in tin plating, china-clay purification, dyeing, oil-drilling muds, and synthetic rubber production. Also known as normal potassium pyrophosphate; tetrapotassium pyrophosphate.

potassium rhodanide *See* potassium thiocyanate.

potassium silicate [INORG CHEM] $SiO_2 = K_2O$ A compound existing in two forms, solution and solid (glass); as a solution, it is colorless to turgid in water, and is used in paints and coatings, as an arc-electrode binder and catalyst and in detergents; as a solid, it is colorless and water-soluble solid, and is used in glass manufacture and for dyeing and bleaching.

potassium silicofluoride *See* potassium fluosilicate.

potassium sorbate [ORG CHEM] $C_6H_7KO_2$ A crystalline compound, more soluble in water than in alcohol; decomposes above 270°C; used to inhibit mold and yeast growth in food. Also known as sorbic acid potassium salt.

potassium stannate [INORG CHEM] $K_2SnO_3 \cdot 3H_2O$ White crystals; soluble in water, insoluble in alcohol; used in textile printing and dyeing, and in tin-plating baths.

potassium stibnate *See* potassium antimonate.

potassium sulfate [INORG CHEM] K_2SO_4 Colorless crystals with bitter taste; soluble in water, insoluble in alcohol; melts at 1072°C; used as an analytical reagent, medicine, and fertilizer, and in aluminum and glass manufacture. Also known as salt of Lemery.

potassium sulfide [INORG CHEM] K_2S Moderately flammable, water-soluble, deliquescent red crystals; melts at 840°C; used in analytical chemistry, medicine, and depilatories. Also known as fused potassium sulfide; hepar sulfuris; potassium sulfuret.

potassium sulfite [INORG CHEM] $K_2SO_3 \cdot 2H_2O$ Water-soluble, white crystals; used in medicine and photography.

potassium sulfocyanate *See* potassium thiocyanate.

potassium sulfocyanide *See* potassium thiocyanate.

potassium sulfuret *See* potassium sulfide.

potassium tetraiodocadmate [INORG CHEM] $K_2(CdI_4) \cdot 2H_2O$ A crystalline compound; used in analytical chemistry for alkaloids, amines, and other compounds. Also known as cadmium potassium iodide; potassium cadmium iodide.

potassium thiocyanate [INORG CHEM] KCNS Water- and alcohol-soluble, colorless, odorless hygroscopic crystals with saline taste; decomposes at 500°C; used as an analytical reagent and in freezing mixtures, chemicals manufacture, textile printing and dyeing, and photographic chemicals. Also known as potassium rhodanide; potassium sulfocyanate; potassium sulfocyanide.

potassium xanthate [ORG CHEM] KC_2H_5OCSS Water- and alcohol-soluble, yellow crystals; used as an analytical reagent and soil-treatment fungicide. Also known as potassium ethyldithiocarbonate; potassium ethylxanthogenate; potassium xanthogenate.

potassium xanthogenate *See* potassium xanthate.

potato [BOT] *Solanum tuberosum.* An erect herbaceous annual that has a round or angular aerial stem, underground lateral stems, pinnately compound leaves, and white, pink, yellow, or purple flowers occurring in cymose inflorescences; produces an edible tuber which is a shortened, thickened underground stem having nodes (eyes) and internodes. Also known as Irish potato; white potato.

pot clay [MATER] Refractory clay used to make the pots in which glass is produced.

pot die forming [MECH ENG] Forming sheet or plate metal through a hollow die by the application of pressure which causes the workpiece to assume the contour of the die.

potential [ELEC] *See* electric potential. [PHYS] A function or set of functions of position in space, from whose first derivatives a vector can be formed, such as that of a static field intensity.

potential barrier [PHYS] The potential in a region in a field of force where the force exerted on a particle is such as to oppose the passage of the particle through the region. Also known as barrier; potential hill.

potential density [PHYS] The density that would be reached by a compressible fluid if it were adiabatically compressed or expanded to a standard pressure of the bar.

potential difference [ELEC] Between any two points, the work which must be done against electric forces to move a unit charge from one point to the other. Abbreviated PD.

potential divider *See* voltage divider.

potential drop [ELEC] The potential difference between two points in an electric circuit. [FL MECH] The difference in

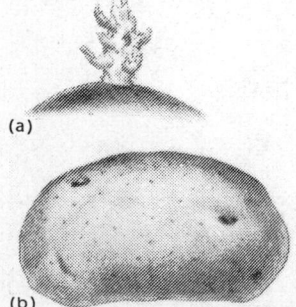

POTATO

Irish potato. *(a)* Flowering stem. *(b)* Tuber.

pressure head between one equipotential line and another.

potential energy [MECH] The capacity to do work that a body or system has by virtue of its position or configuration.

potential evaporation *See* evaporative power.

potential evaportranspiration [HYD] Generally, the amount of moisture which, if available, would be removed from a given land area by evapotranspiration; expressed in units of water depth.

potential flow [FL MECH] Flow in which the velocity of flow is the gradient of a scalar function, known as the velocity potential.

potential flow analyzer *See* electrolytic tank.

potential gradient [ELEC] Difference in the values of the voltage per unit length along a conductor or through a dielectric.

potential index of refraction [METEOROL] An atmospheric index of refraction so formulated that it would have no height variation in an adiabatic atmosphere. Also known as potential refractive index.

potential instability *See* convective instability.

potential refractive index *See* potential index of refraction.

potential scattering [QUANT MECH] Scattering of a particle which can be treated as the effect of a potential, representing the particle's potential energy, on the particle's Schrödinger wave function.

potential temperature [THERMO] The temperature that would be reached by a compressible fluid if it were adiabatically compressed or expanded to a standard pressure, usually 1 bar.

potential theory [MATH] The study of the functions arising from Laplace's equation, especially harmonic functions.

potential transform *See* Poisson transform.

potential transformer *See* voltage transformer.

potential transformer phase angle [ELEC] Angle between the primary voltage vector and the secondary voltage vector reversed; this angle is conveniently considered as positive when the reversed, secondary voltage vector leads the primary voltage vector.

potential vorticity [FL MECH] The product of the absolute vorticity and the static stability, conservative in adiabatic flow, given by the expression $(\eta/\theta)(\partial\theta/\partial p)$, where η is the absolute vorticity of a fluid parcel, θ the potential temperature, and p the pressure. Also known as absolute potential vorticity.

potential well [PHYS] For an object in a conservative field of force, a region in which the object has a lower potential energy than in all the surrounding regions.

potentiometer [ELEC] A resistor having a continuously adjustable sliding contact that is generally mounted on a rotating shaft; used chiefly as a voltage divider. Also known as pot (slang). [ENG] A device for the measurement of an electromotive force by comparison with a known potential difference.

potentiometric cell [ANALY CHEM] Container for the two electrodes and the electrolytic solution being titrated potentiometrically.

potentiometric model study [PETRO ENG] Analogic electrical-resistance model (electrolyte in a contoured container) of an underground reservoir based on Darcy's law that is, the steady-state flow of liquids through porous media is analogous to the flow of current through an electrical conductor; used to predict conditions in gas-condensate oil reservoirs.

potentiometric surface [HYD] An imaginary surface that represents the static head of groundwater and is defined by the level to which water will rise. Also known as isopotential level; piezometric surface; pressure surface.

potentiometric titration [ANALY CHEM] Solution titration in which the end point is read from the electrode-potential variations with the concentrations of potential-determining ions, following the Nernst concept. Also known as constant-current titration.

potentiometry [ELEC] Use of a potentiometer to measure electromotive forces, and the applications of such measurements.

potentiostat [ENG] An automatic laboratory instrument that controls the potential of a working electrode to within certain limits during coulometric (electrochemical reaction) titrations.

pot furnace [ENG] **1.** A furnace containing several pots in which glass is melted. **2.** A furnace in which the charge is contained in a pot or crucible.

pothole [CIV ENG] A pot-shaped hole in a pavement surface. [GEOL] **1.** A shaftlike cave opening upward to the surface. **2.** Any bowl-shaped, cylindrical, or circular hole formed by the grinding action of a stone in the rocky bed of a river or stream. Also known as churn hole; colk; eddy mill; evorsion hollow; kettle; pot. **3.** A vertical, or nearly vertical shaft in limestone. Also known as aven; cenote. **4.** A small depression with steep sides in a coastal marsh; contains water at or below low-tide level. Also known as rotten spot. [HYD] *See* moulin.

Potier diagram [ELEC] Vector diagram showing the voltage and current relations in an alternating-current generator.

potometer [ENG] A device for measuring transpiration, consisting of a small vessel containing water and sealed so that the only escape of moisture is by transpiration from a leaf, twig, or small plant with its cut end inserted in the water.

pot plunger [ENG] A plunger used to force softened plastic molding material into the closed cavity of a transfer mold.

potted circuit [ELEC] A pulse-forming network immersed in oil and enclosed in a metal container.

potted line [ELEC] Pulse-forming network immersed in oil and enclosed in a metal container.

Potter-Bucky grid [MED] An assembly of lead strips resembling an open venetian blind, placed between a patient being x-rayed and the screen or film, to reduce the effects of scattered radiation. Also known as Bucky diaphragm; grid.

potter's wheel [ENG] A revolving horizontal disk that turns when a treadle is operated; used to shape clay by hand.

pottery [MATER] Objects made of clay which may be nonvitreous, porous, opaque, and glazed or unglazed; also included is earthenware such as stoneware.

Pottiales [BOT] An order of mosses distinguished by erect stems, lanceolate to broadly ovate or obovate leaves, a strong, mostly percurrent or excurrent costa, and a cucullate calyptra.

potting [ELECTR] Process of filling a complete electronic assembly with a thermosetting compound for resistance to shock and vibration, and for exclusion of moisture and corrosive agents.

Pott's disease [MED] Abnormal backward curvature of the spine caused by tuberculous osteitis.

poultry [AGR] Domesticated fowl grown for their meat and eggs.

poumar [TEXT] A unit of line density (mass per unit length) used chiefly in the textile industry to measure yarn fiber, equal to 1 pound per 1,000,000 yards, or approximately 4.96055×10^{-7} kilogram per meter.

pounce [MATER] Pumice in the form of a very fine powder, used for preparing parchment and tracing cloth.

pounce wheel [GRAPHICS] An instrument having a toothed wheel at the tip, used to follow the lines in a master drawing in pouncing.

pouncing [GRAPHICS] A method of making copies of patterns and signs; a master drawing is made on paper of the design to be copied, and the lines are traced with a pounce wheel; the teeth perforate the paper with 12 to 24 holes per inch; the master drawing is then placed on the permanent support, and powder is forced through the holes, effectively transferring the drawing.

pouncing paper [MATER] Paper coated with pumice for polishing felt hats.

pound [MECH] **1.** A unit of mass in the English absolute system of units, equal to 0.45359237 kilogram. Abbreviated lb. Also known as avoirdupois pound; pound mass. **2.** A unit of force in the English gravitational system of units, equal to the gravitational force experienced by a pound mass when the acceleration of gravity has its standard value of 9.80665 meters per second per second (approximately 32.1740 ft/sec²) equal to 4.4482216152605 newtons. Abbreviated lb. Also spelled Pound (Lb). Also known as pound force (lbf). **3.** A unit of mass in the troy and apothecaries' systems, equal to 12 troy or apothecaries' ounces, or 5760 grains, or 5760/7000 avoirdupois pound, or 0.3732417216 kilogram. Also known

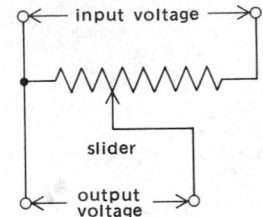

POTENTIOMETER

Schematic of potentiometer.

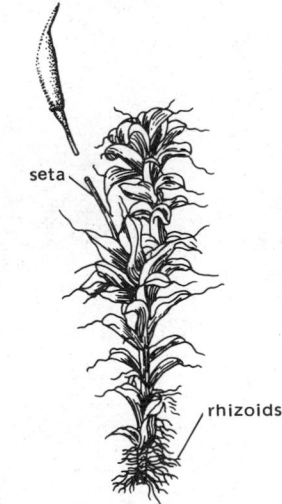

POTTIALES

The entire plant of the moss *Tortula ruralis.*

as apothecaries' pound (abbreviated lb ap in the US or lb apoth in the UK), troy pound (abbreviated lb UK).

poundal [MECH] A unit of force in the British absolute system of units equal to the force which will impart an acceleration of 1 ft/sec² to a pound mass, or to 0.138254954376 newton.

poundal-foot *See* foot-poundal.

pound-foot *See* foot-pound.

pound force *See* pound.

pound mass *See* pound.

pound per square foot [MECH] A unit of pressure equal to the pressure resulting from a force of 1 pound applied uniformly over an area of 1 square foot. Abbreviated psf.

pound per square inch [MECH] A unit of pressure equal to the pressure resulting from a force of 1 pound applied uniformly over an area of 1 square inch. Abbreviated psi.

pounds per square inch absolute [MECH] The absolute, thermodynamic pressure, measured by the number of pounds-force exerted on an area of 1 square inch. Abbreviated lbf in.⁻² abs; psia.

pounds per square inch differential [ENG] The difference in pressure between two points in a fluid-flow system, measured in pounds per square inch. Abbreviated psid.

pounds per square inch gage [MECH] The gage pressure, measured by the number of pounds-force exerted on an area of 1 square inch. Abbreviated psig.

Poupart's ligament *See* inguinal ligament.

pour depressant *See* pour-point depressant.

pouring [MET] Transferring molten metal to a mold or ladle.

pour-plate culture [MICROBIO] A technique for pure-culture isolation of bacteria; liquid, cooled agar in a test tube is inoculated with one loopful of bacterial suspension and mixed by rolling the tube between the hands; subsequent transfers are made from this to a second test tube, and from the second to a third; contents of each tube are poured into separate petri dishes; pure cultures can be isolated from isolated colonies appearing on the plates after incubation.

pour point [FL MECH] Lowest test temperature at which a liquid will flow. [MET] Temperature at which a molten alloy is cast.

pour-point depressant [MATER] An additive that lowers the pour point of a wax-containing petroleum-base lubricating oil by reducing the tendency of the wax to solidify. Also known as pour depressant; pour-point inhibitor.

pour-point inhibitor *See* pour-point depressant.

pour reversion [MATER] The difference between the original ASTM pour point of a lubricating oil and the relatively high solidification temperature observed in the field.

pour stability [MATER] Ability of a pour-depressant-treated petroleum lubricating oil to maintain its original ASTM pour point at low temperatures approximating winter conditions.

Pourtalesiidae [INV ZOO] A family of exocyclic Euechinoidea in the order Holasteroida, including those forms with a bottle-shaped test.

pour test [ENG] The chilling of a liquid under specified test conditions to determine the ASTM pour point.

powder [MATER] 1. A general term for explosives. 2. A loose grouping or aggregation of solid particles, usually smaller than 1000 micrometers.

powder box [MIN ENG] A wooden box used by miners to store explosive powder and blasting caps.

powder clutch [MECH ENG] A type of electromagnetic disk clutch in which the space between the clutch members is filled with dry, finely divided magnetic particles; application of a magnetic field coalesces the particles, creating friction forces between clutch members.

powder diffraction camera [CRYSTAL] A metal cylinder having a window through which an x-ray beam of known wavelength is sent by an x-ray tube to strike a finely ground powder sample mounted in the center of the cylinder; crystal planes in this powder sample diffract the x-ray beam at different angles to expose a photographic film that lines the inside of the cylinder; used to study crystal structure. Also known as x-ray powder diffractometer.

powdered-iron core *See* ferrite core.

powder flowmeter [ENG] A device used to measure the flow rate of a metal powder.

powder house [CIV ENG] A magazine for the temporary storage of explosives.

powder insulation [MATER] Thermal insulation material made up of a finely divided solid held between two surfaces (one hot, one cold); the powder reduces both convection and radiation heat flow between the surfaces.

powder keg [ENG] A small metal keg for black blasting powder.

powder lubricant [MET] An agent mixed with a powder metal to facilitate formation and ejection of a compact.

powderman [MIN ENG] A man in charge of explosives in an operation of any nature requiring their use.

powder metal [MET] Finely divided particles of a metal.

powder metallurgy [MET] The production of massive materials and shaped objects by pressing, binding, and sintering powdered metal.

powder method [SOLID STATE] A method of x-ray diffraction analysis in which a collimated, monochromatic beam of x-rays is directed at a sample consisting of an enormous number of tiny crystals having random orientation, producing a diffraction pattern that is recorded on film or with a counter tube. Also known as x-ray powder method.

powder mine [MIN ENG] An excavation filled with powder for the purpose of blasting rocks.

powder-moisture test [ENG] Determination of moisture in a propellant by drying under prescribed conditions; expressed as percentage by weight.

powder molding [ENG] Generic term for plastics-molding techniques to produce objects of varying sizes and shapes by melting polyethylene powder, usually against the heated inside of a mold.

powder pattern [ELECTROMAG] The pattern created by very fine powders or colloidal particles, spread over the surface of a magnetic material; reveals the magnetic domains in a single crystal of such material.

powder silk [TEXT] Special silk fabric formerly used in making propellant bags; it left no burning residue when the propellant burned; silk has now been largely replaced by other materials, and the fabric is called cartridge cloth. Also known as cartridge silk.

powder snow [HYD] A cover of dry snow that has not been compacted in any way.

powder train [ENG] 1. Train, usually of compressed black powder, used to obtain time action in older fuse types. 2. Train of explosives laid out for destruction by burning.

powdery mildew [MYCOL] A fungus characterized by production of abundant powdery conidia on the host; a member of the family Erysiphaceae or the genus *Oidium*. [PL PATH] A plant disease caused by a powdery mildew fungus.

powdery scab [PL PATH] A fungus disease of potato tubers caused by *Spongospora subterranea* and characterized by nodular discolored lesions, which burst and expose masses of powdery fungus spores.

powellite [MINERAL] Ca(WMo)O₄ A commercially important tungsten mineral, crystallizing in the tetragonal system; isomorphous with scheelite (CaWO₄).

power [MATH] 1. The value that is assigned to a mathematical expression and its exponent. 2. The power of a set is its cardinality. [PHYS] The time rate of doing work.

power amplification *See* power gain.

power amplifier [ELECTR] The final stage in multistage amplifiers, such as audio amplifiers and radio transmitters, designed to deliver maximum power to the load, rather than maximum voltage gain, for a given percent of distortion.

power amplifier tube *See* power tube.

power attenuation *See* power loss.

power barker *See* barker.

power brake [MECH ENG] An automotive brake with engine-intake-manifold vacuum used to amplify the atmospheric pressure on a piston operated by movement of the brake pedal.

power breeder [NUCLEO] A nuclear reactor designed to produce both useful power and fuel.

power car [AERO ENG] A suspended structure on an airship that houses an engine. [MECH ENG] 1. A railroad car with equipment for furnishing heat and electric power to a train.

2. A railroad car with controls, which can be operated by itself or as part of a train.

power control rod [NUCLEO] A control rod that produces only a small change in reactivity of a nuclear reactor, as required for controlling power level.

power cord *See* line cord.

power curve [STAT] The graph of the power of a test for various alternatives.

power density [ELECTROMAG] The amount of power per unit area in a radiated microwave or other electromagnetic field, usually expressed in watts per square centimeter. [NUCLEO] The power generation per unit volume of a nuclear-reactor core.

power-density spectrum *See* frequency spectrum.

power detection [ELECTR] Form of detection in which the power output of the detecting device is used to supply a substantial amount of power directly to a device such as a loudspeaker or recorder.

power detector [ELECTR] Detector capable of handling strong input signals without appreciable distortion.

power diode *See* pin-diode.

power divider [ELECTROMAG] A device used to produce a desired distribution of power at a branch point in a waveguide system.

power drill [MECH ENG] A motor-driven drilling machine.

power-driven [MECH ENG] Of a component or piece of equipment, moved, rotated, or operated by electrical or mechanical energy, as in a power-driven fan or power-driven turret.

power equation [MIN ENG] The relationship indicating that the natural ventilating power plus the power required to force air through a mine is equal to the power used in lifting water out of the mine plus the power lost in the kinetic energy of air leaving the mine plus the power converted to heat in overcoming friction.

power excursion [NUCLEO] A sudden increase in the power level of a nuclear reactor, caused by a sudden increase in reactivity.

power factor [ELEC] The ratio of the average (or active) power to the apparent power (root-mean-square voltage times rms current) of an alternating-current circuit. Abbreviated pf. Also known as phase factor.

power-factor meter [ENG] A direct-reading instrument for measuring power factor.

power factor regulator [ELEC] Regulator which functions to maintain the power factor of a line or an apparatus at a predetermined value, or to vary it according to a predetermined plan.

power flow [ELECTROMAG] The rate at which energy is transported across a surface by an electromagnetic field.

Powerforming [CHEM ENG] A fixed-bed petroleum-refinery process to catalytically reform petroleum naphtha to produce high-octane motor gasoline.

power frequency [ELEC] The frequency at which electric power is generated and distributed; in most of the United States it is 60 hertz.

power gain [ELECTR] The ratio of the power delivered by a transducer to the power absorbed by the input circuit of the transducer. Also known as power amplification. [ELECTROMAG] An antenna ratio equal to 4π (12.57) times the ratio of the radiation intensity in a given direction to the total power delivered to the antenna.

power generator [ELEC] A device for producing electric energy, such as an ordinary electric generator or a magnetohydrodynamic, thermionic, or thermoelectric power generator.

power grizzly [MIN ENG] Power-operated machine for removing dirt and fine particles from ore before it is crushed.

power-law fluid [FL MECH] A fluid in which the shear stress at any point is proportional to the rate of shear at that point raised to some power.

power-law profile [METEOROL] A formula for the variation of wind with height in the surface boundary layer.

power level [ELEC] The ratio of the amount of power being transmitted past any point in an electric system to a reference power value; usually expressed in decibels. [NUCLEO] The power production of a nuclear reactor in watts.

power line [ELEC] Two or more wires conducting electric power from one location to another. Also known as electric power line; line.

power-line carrier [ELEC] The use of transmission lines to transmit speech, metering indications, control impulses, and other signals from one station to another, without interfering with the lines' normal function of transmitting power.

power-line filter *See* line filter.

power loader [MIN ENG] Power-operated machine for loading ore, coal, or other material into a car, conveyor, or other collector.

power loss [ELECTR] The ratio of the power absorbed by the input circuit of a transducer to the power delivered to a specified load; usually expressed in decibels. Also known as power attenuation.

power meter *See* electric power meter.

power of a test [STAT] One minus the probability that a given test causes the acceptance of the null hypothesis when it is false due to the validity of an alternative hypothesis; this is the same as the probability of rejecting the null hypothesis by the test when the alternative is true.

power of the continuum [MATH] The cardinality of the set of real numbers.

power oil [MATER] Fluid used to actuate (power) hydraulic pumps, motors, and other power-type equipment; can be based on petroleum oils or synthetic materials.

power output [ELECTR] The alternating-current power in watts delivered by an amplifier to a load.

power output tube *See* power tube.

power pack [ELECTR] Unit for converting power from an alternating- or direct-current supply into an alternating- or direct-current power at voltages suitable for supplying an electronic device.

power package [MECH ENG] A complete engine and its accessories, designed as a single unit for quick installation or removal.

power plant [MECH ENG] Any unit that converts some form of energy into electrical energy, such as a hydroelectric or steam-generating station, a diesel-electric engine in a locomotive, or a nuclear power plant. Also known as electric power plant.

power rating [ELEC] The power available at the output terminals of a component or piece of equipment that is operated according to the manufacturer's specifications.

power ratio [ELECTROMAG] The ratio of the maximum power to the minimum power in a waveguide that is improperly terminated.

power reactor [NUCLEO] A nuclear reactor designed to provide useful power, as for submarines, aircraft, ships, vehicles, and power plants.

power rectifier [ELEC] A device which converts alternating current to direct current and operates at high power loads.

power relay [ELEC] Relay that functions at a predetermined value of power; may be an overpower relay, an underpower relay, or a combination of both.

power resistor [ELEC] A resistor used in electric power systems, ranging in size from 5 watts to many kilowatts, and cooled by air convection, air blast, or water.

power saw [MECH ENG] A power-operated woodworking saw, such as a bench or circular saw.

power series [MATH] An infinite series composed of functions having nth term of the form $a_n(x - x_0)^n$, where x_0 is some point and a_n some constant.

power set [MATH] The set consisting of all subsets of a given set.

power shovel [MECH ENG] A power-operated shovel that carries a short boom on which rides a movable dipper stick carrying an open-topped bucket; used to excavate and remove debris.

power-shovel mining [MIN ENG] A technique utilizing power shovels to mine ores by mining or stripping and taking away overburden.

power spectrum *See* frequency spectrum.

power steering [MECH ENG] A steering control system for a propelled vehicle in which an auxiliary power source assists the driver by providing the major force required to direct the road wheels.

power stroke [MECH ENG] The stroke in an engine during

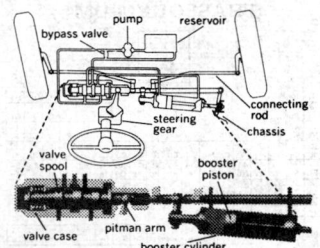

POWER STEERING

The linkage power steering reacts between the chassis and steering linkages. *(Ford Motor Co.)*

POWER SUPPLY

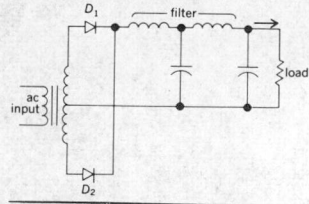

Schematic for a typical direct-current power supply. Alternating-current power line energizes primary of transformer whose secondary is connected to anodes of rectifier diodes D_1 and D_2.

POWER TRANSISTOR

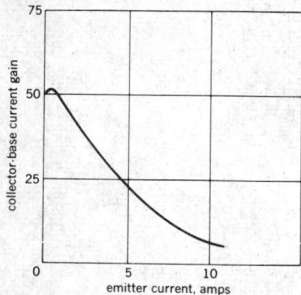

Variation of grounded-emitter current gain with emitter current for a typical *pnp* junction power transistor with a circular emitter and collector, showing that high current rating is obtained at sacrifice in gain.

PRAIRIE DOG

A prairie dog (*Cynomys ludovicianus*) has the habit of sitting erect at the entrance to its burrow.

PRASEODYMIUM

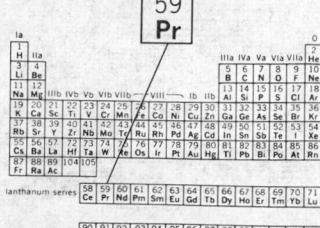

Periodic table of the chemical elements showing the position of praseodymium.

which pressure is applied to the piston by expanding steam or gases.

power supply [ELECTR] A source of electrical energy, such as a battery or power line, employed to furnish the tubes and semiconductor devices of an electronic circuit with the proper electric voltages and currents for their operation. Also known as electronic power supply.

power supply circuit [ELEC] An electrical network used to convert alternating current to direct current.

power switch [ELEC] An electric switch which energizes or deenergizes an electric load; ranges from ordinary wall switches to load-break switches and disconnecting switches in power systems operating at voltages of hundreds of thousands of volts.

power switchboard [ELEC] Part of a switch gear which consists of one or more panels upon which are mounted the switching control, measuring, protective, and regulatory equipment; the panel or panel supports may also carry the main switching and interrupting devices together with their connection.

power switching [ELEC] Switching between supplies of electrical energy at high levels of current and voltage.

power train [MECH ENG] The part of a vehicle connecting the engine to propeller or driven axle; may include drive shaft, clutch, transmission, and differential gear.

power transfer theorem [ELEC] The theorem that, in an electrical network which carries direct or sinusoidal alternating current, the greatest possible power is transferred from one section to another when the impedance of the section that acts as a load is the complex conjugate of the impedance of the section that acts as a source, where both impedances are measured across the pair of terminals at which the power is transferred, with the other part of the network disconnected.

power transformer [ELEC] An iron-core transformer having a primary winding that is connected to an alternating-current power line and one or more secondary windings that provide different alternating voltage values.

power transistor [ELECTR] A junction transistor designed to handle high current and power; used chiefly in audio and switching circuits.

power transmission line [ELEC] The facility in an electric power system used to transfer large amounts of power from one location to a distant location; distinguished from a subtransmission or distribution line by higher voltage, greater power capability, and greater length. Also known as electric main; main (both British usages).

power transmission tower [ELEC] A rigid steel tower supporting a high-voltage electric power transmission line, having a large enough spacing between conductors, and between conductors and ground, to prevent corona discharge.

power traverse [ORD] Turning of a gun to change the direction of fire by means of a power-driven mechanism, as in a tank, aircraft, or ship turret.

power tube [ELECTR] An electron tube capable of handling more current and power than an ordinary voltage-amplifier tube; used in the last stage of an audio-frequency amplifier or in high-power stages of a radio-frequency amplifier. Also known as power amplifier tube; power output tube.

power winding [ELEC] In a saturable reactor, a winding to which is supplied the power to be controlled; commonly the functions of the output and power windings are accomplished by the same winding, which is then termed the output winding.

pox [MED] A vesicular or pustular exanthematic disease that may leave pit scars.

poxvirus [VIROL] A deoxyribonucleic acid–containing animal virus group including the viruses of smallpox, molluscum contagiosum, and various animal pox and fibromas.

Poynting-Robertson effect [ASTRON] The gradual decrease in orbital velocity of a small particle such as a micrometeorite in orbit about the sun due to the absorption and reemission of radiant energy by the particle.

Poynting theorem [ELECTROMAG] A theorem, derived from Maxwell's equations, according to which the rate of loss of energy stored in electric and magnetic fields within a region of space is equal to the sum of the rate of dissipation of electrical energy as heat and the rate of flow of electromag-

netic energy outward through the surface of the region.

Poynting vector [ELECTROMAG] A vector, equal to the cross product of the electric-field strength and the magnetic-field strength (mks units) whose outward normal component, when integrated over a closed surface, gives the outward flow of electromagnetic energy through that surface.

pozzolan [GEOL] A finely ground burnt clay or shale resembling volcanic dust, found near Pozzuoli, Italy; used in cement because it hardens underwater. [MATER] Cement made by mixing and grinding together slaked lime and pozzolan without burning; sometimes used for concrete not exposed to the air.

PPI See plan position indicator.

pp-junction [ELECTR] A region of transition between two regions having different properties in *p*-type semiconducting material.

PPLO See pleuropneumonialike organism.

PPM See pulse-position modulation.

Pr See praseodymium.

practical astronomy [ASTRON] That part of astronomy concerned with the use of information acquired by an observer in the solution of problems determining latitude and longitude on sea or land and directions on the earth's surface by the help of celestial objects.

practical entropy See virtual entropy.

practical system See meter-kilogram–second-ampere system.

practical units [ELECTROMAG] The units of the meter-kilogram-second-ampere system.

practice ammunition [ORD] Ammunition used for target practice or similar types of training; for gun and rocket types of weapons, practice ammunition contains a propelling charge, and either an inert filler or a spotting charge in the projectile; other types of practice ammunition, such as bombs or mines, usually contain a spotting charge or some form of charge to indicate functioning.

practice mine [ORD] Imitation land mine used in training; it may contain a smoke-producing agent for maneuvers or for practice in observing the effects of mines against vehicles, or it may be simply a block of wood, metal, or concrete for practice in laying mine fields.

praecipitatio [METEOROL] Precipitation falling from a cloud and apparently reaching the earth's surface; this supplementary cloud feature is mostly encountered in altostratus, nimbostratus, stratocumulus, stratus, cumulus, and cumulonimbus.

Praesepe [ASTRON] A cluster of faint stars in the center of the constellation Cancer. Also known as Beehive; Manger.

prairie [GEOGR] An extensive level-to-rolling treeless tract of land in the temperate latitudes of central North America, characterized by deep, fertile soil and a cover of coarse grass and herbaceous plants.

prairie climate See subhumid climate.

prairie dog [VERT ZOO] The common name for three species of stout, fossorial rodents belonging to the genus *Cynomys* in the family Sciuridae; all have a short, flat tail, small ears, and short limbs terminating in long claws.

prairie wolf See coyote.

Prandtl number [FL MECH] A dimensionless number used in the study of diffusion in flowing systems, equal to the kinematic viscosity divided by the molecular diffusivity. Symbolized Pr_m. Also known as Schmidt number 1 (N_{Sc}). [THERMO] A dimensionless number used in the study of forced and free convection, equal to the dynamic viscosity times the specific heat at constant pressure divided by the thermal conductivity. Symbolized N_{Pr}.

praseodymium [CHEM] A chemical element, symbol Pr, atomic number 59, atomic weight 140.91; a metallic element of the rare-earth group.

Prasinovolvocales [BOT] An order of green algae in which there are lateral appendages in the flagellum.

pratincolous [ECOL] 1. Living in meadows. 2. Living in low grass.

Pratt truss [CIV ENG] A truss having both vertical and diagonal members between the upper and lower chords, with the diagonals sloped toward the center.

preadaptation [EVOL] Possession by an organism or group of organisms, specialized to one mode of life, of characters which favor easy adaptation to a new environment.

preamble [COMMUN] The portion of a commercial radio telegraph message that is sent first, containing the message number, office of origin, date, and other numerical data not part of the original message.

preamplifier [ELECTR] An amplifier whose primary function is boosting the output of a low-level audio-frequency, radio-frequency, or microwave source to an intermediate level so that the signal may be further processed without appreciable degradation of the signal-to-noise ratio of the system. Also known as preliminary amplifier.

preassembled [ENG] Assembled beforehand.

prebreaker [MECH ENG] Device used to break down large masses of solids prior to feeding them to a crushing or grinding device.

Precambrian [GEOL] All geologic time prior to the beginning of the Paleozoic era (before 600 million years ago); equivalent to about 90% of all geologic time.

precancerous [PATH] Pertaining to any pathological condition of a tissue which is likely to develop into cancer.

precast concrete [MATER] Concrete components which are cast and partly matured in a factory or on the site before being lifted into their final position on a structure.

precentral gyrus [ANAT] The cerebral convolution that lies between the precentral sulcus and the central sulcus and extends from the superomedial border of the hemisphere to the posterior ramus of the lateral sulcus.

precession [MECH] The angular velocity of the axis of spin of a spinning rigid body, which arises as a result of external torques acting on the body.

precession in declination [ASTRON] The component of general precession of the earth along a celestial meridian, amounting to about $20''$ per year.

precession in right ascension [ASTRON] The component of general precession of the earth along the celestial equator, amounting to about $46''.1$ per year.

precession of the equinoxes [ASTRON] A slow conical motion of the earth's axis about the vertical to the plane of the ecliptic, having a period of 26,000 years, caused by the attractive force of the sun, moon, and other planets on the equatorial protuberance of the earth; it results in a gradual westward motion of the equinoxes.

precharge [MET] The pressure introduced into the cavity of a mold before forming a part.

prechlorination [CIV ENG] Chlorination of water before filtration.

precious metal [MET] A relatively scarce, valuable metal, such as gold, silver, and members of the platinum group.

precious stone [MINERAL] 1. Any genuine gemstone. 2. A gemstone of high commercial value because of its beauty, rarity, durability, and hardness; examples are diamond, ruby, sapphire, and emerald.

precipitable water [METEOROL] The total atmospheric water vapor contained in a vertical column of unit cross-sectional area extending between any two specified levels, commonly expressed in terms of the height to which that water substance would stand if completely condensed and collected in a vessel of the same unit cross section. Also known as precipitable water vapor.

precipitable water vapor *See* precipitable water.

precipitant [CHEM] A chemical or chemicals that cause a precipitate to form when added to a solution.

precipitate [CHEM] 1. A substance separating, in solid particles, from a liquid as the result of a chemical or physical change; 2. To form a precipitate.

precipitation [CHEM] The process of producing a separable solid phase within a liquid medium; represents the formation of a new condensed phase, such as a vapor or gas condensing to liquid droplets; a new solid phase gradually precipitates within a solid alloy as a result of slow, inner chemical reaction; in analytical chemistry, precipitation is used to separate a solid phase in an aqueous solution. [METEOROL] 1. Any or all of the forms of water particles, whether liquid or solid, that fall from the atmosphere and reach the ground. 2. The amount, usually expressed in inches of liquid water

depth, of the water substance that has fallen at a given point over a specified period of time.

precipitation area [METEOROL] 1. On a synoptic surface chart, an area over which precipitation is falling. 2. In radar meteorology, the region from which a precipitation echo is received.

precipitation attenuation [ELECTROMAG] Loss of radio energy due to the passage through a volume of the atmosphere containing precipitation; part of the energy is lost by scattering, and part by absorption.

precipitation ceiling [METEOROL] After United States weather observing practice, a ceiling classification applied when the ceiling value is the vertical visibility upward into precipitation; this is necessary when precipitation obscures the cloud base and prevents a determination of its height.

precipitation cell [METEOROL] In radar meteorology, an element of a precipitation area over which the precipitation is more or less continuous.

precipitation clutter suppression [ELECTR] Technique of reducing, by one of the various devices integral to the radar system, clutter caused by rain in the radar range.

precipitation current [METEOROL] The downward transport of charge, from cloud region to earth, that occurs in a fall of electrically charged rain or other hydrometeors.

precipitation echo [METEOROL] A type of radar echo returned by precipitation.

precipitation effectiveness *See* precipitation-evaporation ratio.

precipitation electricity [GEOPHYS] 1. That branch of the study of atmospheric electricity concerned with the electric charges carried by precipitation particles and with the manner in which these charges are acquired. 2. The electric charge borne by precipitation particles.

precipitation-evaporation ratio [CLIMATOL] For a given locality and month, an empirical expression devised for the purpose of classifying climates numerically on the basis of precipitation and evaporation. Abbreviated P-E ratio. Also known as precipitation effectiveness.

precipitation gage [ENG] Any device that measures the amount of precipitation; principally, a rain gage or snow gage.

precipitation-generating element [METEOROL] In radar meteorology, a relatively small volume of supercooled cloud droplets in which ice crystals form and grow much more rapidly than in a lower, larger cloud mass.

precipitation hardening *See* age-hardening.

precipitation intensity [METEOROL] The rate of precipitation, usually expressed in inches per hour.

precipitation inversion [METEOROL] As found in some mountain areas, a decrease of precipitation with increasing elevation of ground above sea level. Also known as rainfall inversion.

precipitation noise [ELECTR] Noise generated in an antenna circuit, generally in the form of a relaxation oscillation, caused by the periodic discharge of the antenna or conductors in the vicinity of the antenna into the atmosphere.

precipitation number [ANALY CHEM] The number of milliliters of asphaltic precipitate formed when 10 milliliters of petroleum-lubricating oil is mixed with 90 milliliters of a special-quality petroleum naphtha, then centrifuged according to ASTM test conditions; used to determine the quantity of asphalt in petroleum-lubricating oil.

precipitation physics [METEOROL] The study of the formation and precipitation of liquid and solid hydrometeors from clouds; a branch of cloud physics and of physical meteorology.

precipitation static [COMMUN] Static interference due to the discharge of large charges built up on an aircraft or other object by rain, sleet, snow, or electrically charged clouds.

precipitation station [METEOROL] A station at which only precipitation observations are made.

precipitation titration [ANALY CHEM] Amperometric titration in which the potential of a suitable indicator electrode is measured during the titration.

precipitation trails *See* virga.

precipitation trajectory [METEOROL] In radar meteorology, a characteristic echo observed on range-height indicator scopes

PRECAMBRIAN

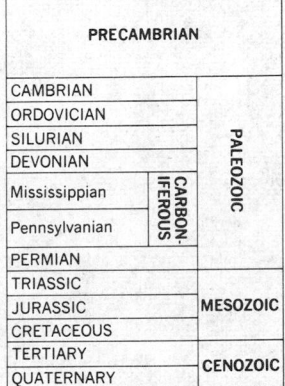

PRECAMBRIAN		
CAMBRIAN		
ORDOVICIAN		
SILURIAN		PALEOZOIC
DEVONIAN		
Mississippian	CARBONIFEROUS	
Pennsylvanian		
PERMIAN		
TRIASSIC		
JURASSIC		MESOZOIC
CRETACEOUS		
TERTIARY		CENOZOIC
QUATERNARY		

Chart showing position of the Precambrian in relation to the eras and periods of geologic time.

PRECESSION

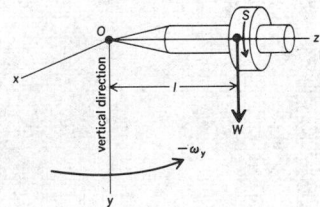

A fast-spinning top supported at point O and released in a horizontal plane precesses about the vertical y-axis. Here z = top's axis of symmetry; x = axis perpendicular to y and z; S = spin velocity of top about z-axis; W = weight of top; l = distance from O to top's center of mass; ω_y = angular velocity of precession of top about y-axis, given by $\omega_y = -Wl/I_zS$, where I_z = top's moment of inertia about z-axis.

and time-height sections which represents the height-range pattern of snow falling from isolated precipitation-generating elements of a few miles in diameter. Also known as mare's tail.

precipitator *See* electrostatic precipitator.

precipitin test [IMMUNOL] An immunologic test in which a specific reaction between antigen and antibody results in a visible precipitate.

precipitous terrain [NAV] In air operations, terrain characterized by steep or abrupt slopes.

precision [NAV] In air operations, pertaining to a navigational facility which provides a combined azimuth and glide slope guidance to a runway. [SCI TECH] The quality of being exactly or sharply defined or stated.

precision adjustment [ORD] A deliberate adjustment of the fire of one weapon for the purpose of placing the center of impact accurately on the target.

precision approach radar [NAV] A radar system located on an airfield for observation of the position of an aircraft with respect to an approach path, and specifically intended to provide guidance to the aircraft during its approach to the field; the system consists of a ground radar equipment which is alternately connected to two antenna systems; one antenna system sweeps a narrow beam over a 20° sector in the horizontal plane; the second sweeps a narrow beam over a 7° sector in the vertical plane; course correction is transmitted to the aircraft from the ground. Abbreviated PAR.

precision-balanced hybrid circuit [ELEC] Circuit used to interconnect a four-wire telephone circuit to a particular two-wire circuit, in which the impedance of the balancing network is adjusted to give a relatively high degree of balance.

precision block *See* gage block.

precision bombing [ORD] Horizontal bombing done with the appropriate precision instruments and equipment so as to strike a target of comparatively small bulk or area.

precision casting [MET] A metal casting of accurately reproducible dimensions.

precision depth recorder [ENG] A machine that plots sonar depth soundings on electrosensitive paper; can plot variations in depth over a range of 400 fathoms (730 meters) on a paper 18.85 inches (47.9 centimeters) wide. Abbreviated PDR. Also known as precision graphic recorder (PGR).

precision fire [ORD] Fire in which the center of impact is accurately placed on a limited target; fire which is based on precision adjustment.

precision graphic recorder *See* precision depth recorder.

precision grinding [MECH ENG] Machine grinding to specified dimensions and low tolerances.

precision net [ELEC] In a four-wire terminating set or similar device employing a hybrid coil, an artificial line designed and adjusted to provide an accurate balance for the loop and subscribers set or line impedance.

precision sweep [ELECTR] Delayed expanded radar sweep for high resolution and range accuracy.

precoat [MET] In casting, thin coating of refractory slurry applied to an expendable wax or plastic pattern as a base for the application of the main slurry.

precoat filter [ENG] A device designed to filter solid particles from a liquid-solid slurry after a precoat of buildup solid material (filter aid or filtered solid) has been applied to the inner surface of the filter medium.

precoating [ENG] The depositing of an inert material, such as filter aid, onto the filter medium prior to the filtration of suspended solids from a solid-liquid slurry.

precognition [PSYCH] A form of extrasensory perception involving foreknowledge of a future event.

pre-cold-frontal squall line *See* prefrontal squall line.

precollagenous fiber *See* reticular fiber.

precombustion chamber [MECH ENG] A small chamber before the main combustion space of a turbine or reciprocating engine in which combustion is initiated.

precompact set [MATH] A set in a metric space which can always be covered by open balls of any diameter about some finite number of its points. Also known as totally bounded set.

precomputation [NAV] The process of making navigational solutions in advance, applied particularly to the determination of computed altitude and azimuth before making a

celestial observation for a line of position; when this is done, the observation must be made at the time used for the computation, or a correction applied.

precomputed altitude [ASTRON] The altitude of a celestial body computed before observation with the sextant. Altitude corrections are included in the calculations but are applied with reversed sign.

precomputed curve [NAV] A graphical representation of the azimuth or altitude of a celestial body which is plotted against time for a given assumed position or positions, and which is computed for subsequent use with celestial observations.

preconduction current [ELECTR] Low value of plate current flowing in a thyratron or other grid-controlled gas tube prior to the start of conduction.

preconsolidation pressure [GEOL] The greatest effective stress exerted on a soil; result of this pressure from overlying materials is compaction. Also known as prestress.

precooler [MECH ENG] A device for reducing the temperature of a working fluid before it is used by a machine.

precure finish [TEXT] A durable-press finish applied to a fabric and heat-set by the mill.

predation [BIOL] The killing and eating of an individual of one species by an individual of another species.

predator [ZOO] An animal that preys on other animals as a source of food.

predazzite [PETR] A recrystallized limestone that resembles pencatite, but contains less brucite than calcite.

predetection combining [ELECTR] Method used to produce an optimum signal from multiple receivers involved in diversity reception of signals.

predicate [MATH] To affirm or deny, in mathematical logic, one or more subjects.

predictability [NAV] The measure of the accuracy with which the system can define the location of a point in terms of geographic coordinates rather than the lattice peculiar to that system.

predicted barrage *See* antiaircraft barrage.

predicted firing [ORD] Firing at the point at which a moving target is expected to be when the projectile reaches it, according to predictions based on observation.

predicted ground speed [NAV] The speed computed by applying estimated wind direction and speed to course and estimated airspeed.

predicted-wave signaling [COMMUN] Communications system in which detection is optimized in the presence of severe noise by using mechanical resonator filters and other circuits in the detector to take advantage of known information on the arrival and completion times of each pulse, as well as on pulse shape, pulse frequency and spectrum, and possible data content.

predicting dead time [ORD] The time allowed for calculation and applying firing data, from the time of observation to the instant of firing.

predicting interval [ORD] The interval between successive predictions of future positions of the target.

prediction [METEOROL] **1.** The act of making a weather forecast. **2.** The forecast itself.

predissociation [PHYS CHEM] The dissociation of a molecule that has absorbed energy before it can lose energy by radiation.

prednisolone [BIOCHEM] $C_{21}H_{28}O_5$ A glucocorticoid that is a dehydrogenated analog of hydrocortisone.

prednisone [PHARM] $C_{21}H_{26}O_5$ An adrenocortical steroid drug, obtained in crystalline form, that is an analog of cortisone.

predozzite [PETR] Limestone rich in periclase and brucite.

preece [ELEC] A unit of electrical resistivity equal to 10^{13} times the product of 1 ohm and 1 meter.

preeclampsia [MED] A toxemia occurring in the latter half of pregnancy, characterized by an acute elevation of blood pressure and usually by edema and proteinuria, but without convulsions or coma. Also known as toxemia of pregnancy.

preedit [ADP] To edit data before feeding it to a computer.

preemphasis [ELECTR] A process which increases the magnitude of some frequency components with respect to the magnitude of others to reduce the effects of noise introduced in subsequent parts of the system.

preemphasis network [ELECTR] An RC (resistance-capacitance) filter inserted in a system to emphasize one range of frequencies with respect to another. Also known as emphasizer.

preen gland *See* uropygial gland.

preferential shop [IND ENG] An establishment in which preference is given to union members in hiring, layoffs, and dismissals, with the understanding that nonunion workers may be employed without being required to join the union when the union cannot supply men.

preferred numbers [ELECTR] A series of numbers adopted by EIA and the military services for use as nominal values of resistors and capacitors, to reduce the number of different sizes that must be kept in stock for replacements. Also known as preferred values.

preferred orientation [PETR] The nonrandom orientation of planar or linear fabric elements in structural petrology.

preferred values *See* preferred numbers.

prefilter [ENG] Filter used to remove gross solid contaminants before the liquid stream enters a separator-filter.

prefix notation [ADP] That form of Polish notation in which operators precede the operands with which they are associated.

prefocus lamp [ELEC] A light bulb whose filaments are precisely positioned with respect to the lamp socket.

prefoliation *See* vernation.

preform [ENG ACOUS] The small slab of record stock material that is loaded into a press to be formed into a disk recording. Also known as biscuit (deprecated usage).

preforming [MET] **1.** Initial pressing of a powder metal to form a compact. **2.** Preliminary shaping of a refractory metal compact after presintering.

prefrontal [ANAT] Situated in the anterior part of the frontal lobe of the brain. [VERT ZOO] **1.** Of or pertaining to a bone of some vertebrate skulls, located anterior and lateral to the frontal bone. **2.** Of, pertaining to, or being a scale or plate in front of the frontal scale on the head of some reptiles and fishes.

prefrontal lobotomy *See* lobotomy.

prefrontal squall line [METEOROL] A squall line or instability line located in the warm sector of a wave cyclone, about 50 to 300 miles (80 to 480 kilometers) in advance of the cold front, usually oriented roughly parallel to the cold front, and moving in about the same manner as the cold front. Also called nonfrontal squall line; pre-cold-frontal squall line.

preglacial [GEOL] **1.** Pertaining to the geologic time immediately preceding the Pleistocene epoch. **2.** Of material, underlying glacial deposits.

Pregl procedure [ANALY CHEM] Microanalysis technique in which the sample is decomposed thermally, with subsequent oxidation of decomposition products.

pregnancy [MED] The state of being pregnant, from conception to childbirth.

pregnancy test [PATH] Any biologic or chemical procedure used to diagnose pregnancy.

pregnanediol [BIOCHEM] $C_{21}H_{36}O_2$ A metabolite of progesterone, present in urine during the progestational phase of the menstrual cycle and also during pregnancy.

pregnenolone [BIOCHEM] $C_{21}H_{32}O_2$ A steroid ketone that is formed as an oxidation product of cholesterol, stigmasterol, and certain other steroids.

preheater [MECH ENG] A device for preliminary heating of a material, substance, or fluid that will undergo further use or treatment by heating.

preheat roll [ENG] In plastic-extrusion coating, the heated roll between the pressure roll and the unwind roll; used to heat the substrate before it is coated.

prehensile [VERT ZOO] Adapted for seizing, grasping, or plucking, especially by wrapping around some object.

pre-Hilbert space [MATH] A linear space which has an inner product defined on it.

prehnite [MINERAL] $Ca_2Al_2Si_3O_{10}(OH)_2$ A light-green to white mineral sorosilicate crystallizing in the orthorhombic system and generally found in reniform and stalactitic aggregates with crystalline surface; it has a vitreous luster, hardness is 6–6.5 on Mohs scale, and specific gravity is 2.8–2.9.

preignition [MECH ENG] Ignition of the charge in the cylinder of an internal combustion engine before ignition by the spark.

preimpregnation [ENG] The mixing of a plastic resin with reinforcing material or substrate before molding takes place.

preindexing [ADP] Operation in which the address bits of a word are added to the contents of a specified register to determine the pointer address.

preliminary amplifier *See* preamplifier.

preloading [ENG] For back-pressure-control gas valves, a weight or spring device to control the gas pressure at which the valve will open or close.

prelogging [FOR] Cutting down and removing small trees before large trees are logged.

premaxilla [ANAT] Either of two bones of the upper jaw of vertebrates located in front of and between the maxillae.

premium motor oil [MATER] A lubricating oil with improved oxidation stability and having corrosion-preventive and detergent properties; used in internal combustion engines operating under severe conditions.

premix [ENG] In plastics molding, materials in which the resin, reinforcement, extenders, fillers, and so on have been premixed before molding.

premix gas burner [ENG] Fuel (gas or oil) burner in which fuel and air are premixed prior to ignition in the combustion chamber.

premolar [ANAT] In each quadrant of the permanent human dentition, one of the two teeth between the canine and the first molar; a bicuspid.

prenatal [MED] Existing or occurring before birth.

preorogenic [GEOL] The initial phase of an orogenic cycle during which geosynclines form.

preparation fire [ORD] Fire delivered on a target or predetermined point preparatory to an assault on the target; may be naval, ground, or air.

preparatory fire [ORD] Antiaircraft fire to determine or check corrections of firing data prior to conducting fire for effect.

preparatory work [MIN ENG] Various excavations within a deposit so that actual mining can begin; includes inclines, drives between levels, crosscuts, and chutes.

preparing salt *See* sodium stannate.

preplastication [ENG] Premelting of injection-molding powders in a chamber separate from the injection cylinder.

preplumbed system [ELECTROMAG] Fixed nontunable waveguides or coaxial transmission lines.

prepolymer [ORG CHEM] A plastic or resin intermediate whose molecular weight is between that of the original monomer or monomers and that of the final, cured polymer or resin.

prepolymer molding [ENG] A urethane-foam-producing system in which a portion of the polyol is prereacted with the isocyanate to form a liquid prepolymer with a pumpable viscosity; when combined with a second blend containing more polyol, catalyst, or blowing agent, the two components react and a foamed plastic results.

prepositioned [ORD] Actual placement of a weapon at the desired point of detonation and preparation for firing by some form of remote control or a timer mechanism.

prepreg [ENG] A reinforced-plastics term for the reinforcing material that contains or is combined with the full complement of resin before the molding operation.

preprogramming [ADP] The prerecording of instructions or commands for a machine, such as an automated tool in a factory.

prepsychotic [PSYCH] Of or pertaining to the mental state that precedes or is potentially capable of precipitating a psychotic disorder.

prepuce [ANAT] **1.** The foreskin of the penis, a fold of skin covering the glans penis. **2.** A similar fold over the glans clitoridis.

preputial gland *See* Tyson's gland.

prerinse [GRAPHICS] A water bath used prior to development of photographic film.

presbycusis [MED] A condition of diminished auditory acuity associated with old age.

presbyophrenia [PSYCH] A variety of senile dementia in which apparent mental alertness is combined with disorientation of place and loss of memory.

presbyopia [MED] Diminished ability to focus the eye on

PREHNITE

2.5 cm

Prehnite in reniform aggregates, Paterson, New Jersey. *(Specimen from Department of Geology, Bryn Mawr College)*

near objects due to gradual loss of elasticity of the crystalline lens with age.

prescutum [INV ZOO] The anterior part of the tergum of a segment of the thorax in insects.

preselection [ADP] A technique for saving time available in buffered computers by which a block of data is read into computer storage from the next input tape to be called upon before the data are required in the computer; the selection of the next input tape is determined by instructions to the computer.

preselector [ELEC] Device in automatic switching which performs its selecting operation before seizing an idle trunk. [ELECTR] A tuned radio-frequency amplifier stage used ahead of the frequency converter in a superheterodyne receiver to increase the selectivity and sensitivity of the receiver.

presence [ACOUS] The impression, as created by a recording or radio receiver, that the original program source is in the room.

present angular height [ORD] The element of data pertaining to the present position of the target; that is, the position at the instant the gun is fired.

presentation [ELECTR] Form which radar echo signals take on a cathode-ray-tube screen, which is dependent on the nature of the sweep circuit used.

present elevation [ORD] Elevation corresponding to the present position of the target.

present position [ORD] Position of a moving target at the instant of firing.

present range [ORD] The range between a gun and the present position of its target.

present value [MIN ENG] The sum of money which, if expended on a mine for purchase, development, and equipment, would produce over the life of the mine a return of the original investment plus a commensurate profit.

present-worth factor *See* discount factor.

preservative [MATER] A chemical added to foodstuffs to prevent oxidation, fermentation, or other deterioration, usually by inhibiting the growth of bacteria.

preset [ADP] **1.** Of a variable, having a value established before the first time it is used. **2.** To initialize a value of a variable before the value of the variable is used or tested.

preset guidance [ENG] Guidance in which a predetermined path is set into the guidance mechanism of a craft, drone, or missile and is not altered after launching.

preset parameter [ADP] In computers, a parameter which is fixed for each problem at a value set by the programmer.

preshrunk [TEXT] Pertaining to a fabric that has been shrunk during manufacture in order to minimize later shrinkage.

presintering [MET] Heating a compact to a temperature lower than the final sintering temperature to facilitate handling or to remove a binder or lubricant.

presoma [INV ZOO] The anterior portion of an invertebrate that lacks a definitive head structure.

press [MECH ENG] Any of various machines by which pressure is applied to a workpiece, by which a material is cut or shaped under pressure, by which a substance is compressed, or by which liquid is expressed.

press bonding [ENG] A method of bonding structures or materials through the application of pressure by a platen press or other tool.

press camera [OPTICS] A folding camera, usually of 4- by 5-inch (10.2- by 12.7-centimeter) format, once widely used in newspaper photography.

press drip [MATER] Oil that drips from the wax press after pressed petroleum distillate has been removed.

pressed brick [MATER] Brick subjected to pressure before burning to eliminate imperfections of shape and texture.

pressed density [MET] The density of a metal powder compact before sintering.

pressed-disk technique *See* potassium bromide–disk technique.

pressed distillate [MATER] Oil recovered when refinery paraffin distillate is pressed to separate the liquid from the solid wax.

pressed glass [MATER] Glass shaped by being poured into a mold under pressure or pressed into a mold in a plastic state.

pressed loading [ENG] A loading operation in which bulk

PRESSURE ALTIMETER

Photograph of a pressure altimeter. Windows at right indicate altimeter is adjusted to sea level pressure of 1013 millibars or 29.92 inches of mercury. Window at left indicates altitude between 41,500 and 41,600 feet; dial provides more precise indication of altitude.

material, such as an explosive in granular form, is reduced in volume by the application of pressure.

Pressfax Graphics [COMMUN] Facsimile system that transmits a newspaper page, including halftone pictures, in about 4 minutes.

press fit [ENG] An interference or force fit assembled through the use of a press. Also known as force fit.

press forging [MET] Forging hot metal between dies in a press.

pressing [ENG ACOUS] A phonograph record produced in a record-molding press from a master or stamper. [MET] **1.** Shallow-drawing metal sheet or plate. **2.** Using compressive force to form a metal powder compact.

pressolution *See* pressure solution.

press proof [GRAPHICS] Proof removed from the printing press to inspect line and color values and overall quality; this is the last proof before the complete printing is done.

press slide [MECH ENG] The reciprocating member of a power press on which the punch and upper die are fastened.

press teletype network [COMMUN] A large teletypewriter network employed by a press association or other news distributing organization, usually employing modern carrier telegraph circuits operating over both wire and radio facilities, and transmitting to as many as 2000 stations simultaneously.

press-to-talk switch [ELECTR] A switch mounted directly on a microphone to provide a convenient means for switching two-way radiotelephone equipment or electronic dictating equipment to the talk position.

pressure [MECH] A type of stress which is exerted uniformly in all directions; its measure is the force exerted per unit area.

pressure altimeter [ENG] A highly refined aneroid barometer that precisely measures the pressure of the air at the altitude an aircraft is flying, and converts the pressure measurement to an indication of height above sea level according to a standard pressure-altitude relationship. Also known as barometric altimeter.

pressure altitude [METEOROL] The height above sea level at which the existing atmospheric pressure would be duplicated in the standard atmosphere; atmospheric pressure expressed as height according to a standard scale.

pressure-altitude variation [METEOROL] The pressure difference, in feet or meters, between mean sea level and the standard datum plane.

pressure angle [MECH ENG] The angle that the line of force makes with a line at right angles to the center line of two gears at the pitch points.

pressure bag [ENG] A bag made of rubber, plastic, or other impermeable material that provides a flexible barrier between the pressure medium and the part being bonded.

pressure bar [MECH ENG] A bar that holds the edge of a metal sheet during press operations, such as punching, stamping, or forming, and prevents the sheet from buckling or becoming crimped.

pressure-base factor [CHEM ENG] Factor used in orifice pressure-drop calculations to allow for conditions where the pressure base used for calculating the orifice factor is not 14.73 psia; calculated as $F_{pb} = 14.73/\text{pressure base (absolute)}$.

pressure block [MIN ENG] The pressure on pillars, walls, and other supports in a mine caused by removal of surrounding formations from masses of rocks or by natural geological formations.

pressure bomb [PETRO ENG] Pipe-and-valve device used to capture downhole pressurized gas samples from oil wells; used to measure downhole pressure.

pressure breccia *See* tectonic breccia.

pressure broadening [SPECT] A spreading of spectral lines when pressure is increased, due to an increase in collision broadening.

pressure bump [MIN ENG] Sudden failure of a coal pillar overloaded by the weight of the rock above it.

pressure carburetor *See* injection carburetor.

pressure casting [MET] Making castings of molten or plastic metal in metal molds under applied pressures.

pressure center [METEOROL] **1.** On a synoptic chart (or on a mean chart of atmospheric pressure), a point of local minimum or maximum pressure; the center of a low or high. **2.** A center of cyclonic or anticyclonic circulation.

pressure chamber [ENG] A chamber in which an artificial environment is established at low or high pressures to test equipment under simulated conditions of operation. [MIN ENG] An enclosed space that seals off a part of a mine and in which the air pressure can be raised or lowered.

pressure-change chart [METEOROL] A chart indicating the change in atmospheric pressure of a constant-height surface over some specified interval of time. Also known as pressure-tendency chart.

pressure coefficient [THERMO] The ratio of the fractional change in pressure to the change in temperature under specified conditions, usually constant volume.

pressure contour [METEOROL] A line connecting points of equal height of a given barometric pressure; the intersection of a constant pressure surface by a plane parallel to mean sea level.

pressure control [ENG] Any device or system able to maintain, raise, or lower the pressure in a vessel or processing system as desired.

pressure decline [PETRO ENG] The loss or decline in reservoir pressure resulting from pressure drawdown during the production of gas or oil. Also known as pressure depletion.

pressure deflection [ENG] In a bourdon or bellows-type pressure gage, the deflection or movement of the primary sensing element when pressured by the fluid being measured.

pressure depletion See pressure decline.

pressure depth [OCEANOGR] The depth at which an ocean sample was taken, as inferred from the difference in readings on protected and unprotected thermometers on the sampler; the higher reading is on the unprotected thermometer due to the effect of pressure on the mercury column at the sampling depth.

pressure distillate [MATER] Light, gasoline-bearing distillate product from petroleum-refinery pressure stills; the product is cracked, as contrasted to virgin or straight-run stock; includes pressure naphtha.

pressure distribution [PETRO ENG] The relative pressures (pressure gradients) between various portions of a producing reservoir zone; lowest pressures are nearest the producing wellbores.

pressure drawdown [PETRO ENG] The drop in reservoir pressure related to the withdrawal of gas from a producing well; for low-permeability formations, pressures near the wellbores can be much lower than in the main part of the reservoir; leads to pressure decline in the reservoir and ultimate pressure depletion.

pressure drop [FL MECH] The difference in pressure between two points in a flow system, usually caused by frictional resistance to a fluid flowing through a conduit, filter media, or other flow-conducting system.

pressure-drop manometer [ENG] Manometer device (liquid-filled U tube) open at both ends, each end connected by tubing to a different location in a flow system (such as fluid- or gas-carrying pipe) to measure the drop in system pressure between the two points.

pressure effect [SPECT] The effect of changes in pressure on spectral lines in the radiation emitted or absorbed by a substance; namely, pressure broadening and pressure shift.

pressure elements [ENG] Those portions of a pressure-measurement gage which are moved or temporarily deformed by the gas or liquid of the system to which the gage is connected; the amount of movement or deformation is proportional to the pressure and is indicated by the position of a pointer or movable needle.

pressure-enthalpy chart [PHYS] A graph of the pressure versus the enthalpy of a substance at various values of temperature, specific volume, and entropy; especially useful in refrigeration calculations. Also known as enthalpy-pressure chart.

pressure-fall center [METEOROL] A point of maximum decrease in atmospheric pressure over a specified interval of time; on synoptic charts, a point of greatest negative pressure tendency. Also known as center of falls; isallobaric low; isallobaric minimum; katallobaric center.

pressure fan [MIN ENG] A fan that forces fresh air into a mine as distinguished from one that exhausts air from the mine.

pressure field [OCEANOGR] A representation of a pressure gradient as isobar contours, parallel to which ocean currents flow.

pressure force [FL MECH] The force due to differences of pressure within a fluid mass; the (vector) force per unit volume is equal to the pressure gradient $-\nabla p$, and the force per unit mass (specific force) is equal to the product of the volume force and the specific volume $-\alpha\nabla p$. [METEOROL] See pressure gradient-force.

pressure forming [ENG] A plastics thermoforming process using pressure to push the plastic sheet to be formed against the mold surface, as opposed to using vacuum to suck the sheet flat against the mold.

pressure front See shock front.

pressure gage [ENG] A device having a metallic sensing element such as a flexible curved tube (bourdon tube) or a flexible diaphragm which deforms under fluid pressure.

pressure gradient [FL MECH] The rate of decrease (that is, the gradient) of pressure in space at a fixed time; sometimes loosely used to denote simply the magnitude of the gradient of the pressure field. Also known as barometric gradient. [METEOROL] The change in atmospheric pressure per unit horizontal distance, usually measured along a line perpendicular to the isobars.

pressure gradient-force [METEOROL] The force due to differences of pressure within the atmosphere; it usually refers only to the horizontal component of the force. Also known as pressure force.

pressure head [FL MECH] Also known as head. **1.** The height of a column of fluid necessary to develop a specific pressure. **2.** The pressure of water at a given point in a pipe arising from the pressure in it.

pressure hydrophone [ENG ACOUS] A pressure microphone that responds to waterborne sound waves.

pressure ice [OCEANOGR] Ice, especially sea ice, which has been deformed or altered by the lateral stresses of any combination of wind, water currents, tides, waves, and surf; may include ice pressed against the shore, or one piece of ice upon another.

pressure ice foot [HYD] An ice foot formed along a shore by the freezing together of stranded pressure ice.

pressure interface [PETRO ENG] The interrelation of several individual reservoir pressures whose productions are supported by water influx from a common aquifer; pressure depletion from withdrawal of oil from one reservoir will affect the position of the common aquifer and thus affect the pressures and gas-oil or water-oil contacts in the other reservoirs.

pressure jump [METEOROL] A steady-state propagation of a sudden finite change of inversion height, in analogy to the shock wave in a compressible fluid or to a hydraulic jump; the prefrontal squall line has been interpreted as a pressure jump, with the cold front providing the initial pistonlike impetus.

pressure jump line [METEOROL] A fast-moving line of sudden rise in atmospheric pressure, followed by a higher pressure level than that which preceded the jump; under suitable moisture conditions, sudden instability of the atmosphere conducive to the formation of thunderstorms can result.

pressure line of position [NAV] A line of position parallel to the effective air path and at a distance perpendicular to it equal to the lateral displacement, determined by pressure pattern formulas.

pressure maintenance [PETRO ENG] The maintenance of gas pressure in a reservoir by an active water drive, water injection, gas injection, or a combination of the foregoing.

pressure mark [GRAPHICS] A defect found in processed film which may appear as reduced or increased density.

pressure measurement [ENG] Measurement of the internal forces of a process vessel, tank, or piping caused by pressurized gas or liquid; can be for a static or dynamic pressure, in English or metric units, either absolute (total) or gage (absolute minus atmospheric) pressure.

pressure melting [PHYS] The melting of ice due to applied pressure.

pressure microphone [ENG ACOUS] A microphone whose

output varies with the instantaneous pressure produced by a sound wave acting on a diaphragm; examples are capacitor, carbon, crystal, and dynamic microphones.

pressure naphtha [MATER] Petroleum naphtha made by cracking, as contrasted to virgin or straight-run naphtha; a special grade of pressure distillate.

pressure pad [ENG ACOUS] A felt pad mounted on a spring arm, used to hold magnetic tape in close contact with the head on some tape recorders.

pressure pattern [METEOROL] The general geometric characteristics of atmospheric pressure distribution as revealed by isobars on a constant-height chart; usually applied to cyclonic-scale features of a surface chart.

pressure pattern displacement *See* lateral displacement.

pressure pattern flying *See* aerologation.

pressure pickup [ELECTR] A device that converts changes in the pressure of a gas or liquid into corresponding changes in some more readily measurable quantity such as inductance or resistance.

pressure pillow [ENG] A mechanical-hydraulic snow gage consisting of a circular rubber or metal pillow filled with a solution of antifreeze and water, and containing either a pressure transducer or a riser pipe to record increase in pressure of the snow.

pressure plate [MECH ENG] The part of an automobile disk clutch that presses against the flywheel.

pressure-plate anemometer [ENG] An anemometer which measures wind speed in terms of the drag which the wind exerts on a solid body; may be classified according to the means by which the wind drag is measured. Also known as plate anemometer.

pressure point [PHYSIO] A point of marked sensibility to pressure or weight, arranged like the temperature spots, and showing a specific end apparatus arranged in a punctate manner and connected with the pressure sense.

pressure process [CHEM ENG] Treatment of timber to prevent decay by forcing a preservative such as creosote and zinc chloride into the cells of the wood.

pressure radius [PETRO ENG] The effective radius of increased reservoir pressure surrounding a water-injection well.

pressure rating [ENG] The operating (allowable) internal pressure of a vessel, tank, or piping used to hold or transport liquids or gases.

pressure-regulating valve [ENG] A valve that releases or holds process-system pressure (that is, opens or closes) either by preset spring tension or by actuation by a valve controller to assume any desired position between full open and full closed.

pressure regulator [ENG] Open-close device used on the vent of a closed, gas-pressured system to maintain the system pressure within a specified range.

pressure relief [ENG] A valve or other mechanical device (such as a rupture disk) that eliminates system overpressure by allowing the controlled or emergency escape of liquid or gas from a pressured system.

pressure ridge [GEOL] **1.** A seismic feature resulting from transverse pressure and shortening of the land surface. **2.** An elongate upward movement of the congealing crust of a lava flow. **3.** A ridge of glacier ice. [OCEANOGR] A ridge or wall of hummocks where one ice floe has been pressed against another.

pressure ring [MIN ENG] A ring about a large excavated area, evidenced by distortion of the openings near the main excavation.

pressure-rise center [METEOROL] A point of maximum increase in atmospheric pressure over a specified interval of time; on synoptic charts, a point of maximum positive pressure tendency. Also known as anallobaric center; center of rises; isallobaric high; isallobaric maximum.

pressure roll [ENG] In plastics-extrusion coating, the roll that with the chill roll applies pressure to the substrate and the molten extruded web.

pressure seal [ENG] A seal used to make pressure-proof the interface (contacting surfaces) between two parts that have frequent or continual relative rotational or translational motion.

pressure-sensitive adhesive [MATER] An adhesive that develops maximum bonding power when applied by a light pressure only.

pressure shift [SPECT] An increase in the wavelength at which a spectral line has maximum intensity, which takes place when pressure is increased.

pressure solution [PETR] In a sedimentary rock, solution occurring preferentially at the grain boundary surfaces. Also known as pressolution.

pressure-stabilized [AERO ENG] Referring to membrane-type structures that require internal pressure for maintenance of a stable structure.

pressure still [CHEM ENG] A continuous-flow, petroleum-refinery still in which heated oil (liquid and vapor) is kept under pressure so that it will crack (decompose into smaller molecules) to produce lower-boiling products (pressure distillate or pressure naphtha).

pressure storage [ENG] The storage of a volatile liquid or liquefied gas under pressure to prevent evaporation.

pressure suppression *See* vapor suppression.

pressure surface *See* potentiometric surface.

pressure survey [MIN ENG] A study to determine the pressure distribution or pressure losses along consecutive lengths or sections of a ventilation circuit. [PETRO ENG] The measurement of static bottomhole pressures in an oil field with producing wells shut in for a time interval sufficient for reservoir pressure buildup to stabilize.

pressure switch [ELEC] A switch that is actuated by a change in pressure of a gas or liquid.

pressure system [ENG] Any system of pipes, vessels, tanks, reactors, and other equipment, or interconnections thereof, operating with an internal pressure greater than atmospheric. [METEOROL] An individual cyclonic-scale feature of atmospheric circulation, commonly used to denote either a high or a low, less frequently a ridge or a trough.

pressure tank [CHEM ENG] A pressurized tank into which timber is inserted for impregnation with preservative. [CIV ENG] An airtight water tank in which air is compressed to exert pressure on the water and which is used in connection with a water distribution system.

pressure tap [ENG] A small perpendicular hole in the wall of a pressurized, fluid-containing pipe or vessel; used for connection of pressure-sensitive elements for the measurement of static pressures. Also known as piezometer opening; static pressure tap.

pressure tendency [METEOROL] The character and amount of atmospheric pressure change for a 3-hour or other specified period ending at the time of observation. Also known as barometric tendency.

pressure-tendency chart *See* pressure-change chart.

pressure tensor [PL PHYS] A tensor which plays a role in magnetohydrodynamics analogous to that of the pressure in ordinary fluid mechanics.

pressure thrust [AERO ENG] In rocketry, the product of the cross-sectional area of the exhaust jet leaving the nozzle exit and the difference between the exhaust pressure and the ambient pressure.

pressure topography *See* height pattern.

pressure transducer [ENG] An instrument component that detects a fluid pressure and produces an electrical signal related to the pressure. Also known as electrical pressure transducer.

pressure-travel curve [MECH] Curve showing pressure plotted against the travel of the projectile within the bore of the weapon.

pressure traverse [PETRO ENG] Measurement of reservoir pressures at progressive depths.

pressure treater [CHEM ENG] Any chemical treating device operated at higher-than-atmospheric pressure, as in the chemical and petroleum industries.

pressure-tube anemometer [ENG] An anemometer which derives wind speed from measurements of the dynamic wind pressures; wind blowing into a tube develops a pressure greater than the static pressure, while wind blowing across a tube develops a pressure less than the static; this pressure difference, which is proportional to the square of the wind speed, is measured by a suitable manometer.

pressure-tube reactor [NUCLEO] A nuclear reactor in which the fuel elements are located inside numerous tubes containing coolant circulating at high pressure; the tube assembly is surrounded by a tank containing the moderator at low pressure.

pressure tunnel [CIV ENG] A waterway tunnel under pressure because the hydraulic gradient lies above the tunnel crown.

pressure ulcer See decubitus ulcer.

pressure vessel [ENG] A metal container generally cylindrical or spheroid, capable of withstanding bursting pressures.

pressure viscosity [FL MECH] Property of petroleum lubricating oils to increase in viscosity when subjected to pressure.

pressure welding [MET] Welding of metal surfaces by the application of pressure; examples are percussion welding, resistance welding, seam welding, and spot welding.

pressurization [ENG] **1.** Use of an inert gas or dry air, at several pounds above atmospheric pressure, inside the components of a radar system or in a sealed coaxial line, to prevent corrosion by keeping out moisture, and to minimize high-voltage breakdown at high altitudes. **2.** The act of maintaining normal atmospheric pressure in a chamber subjected to high or low external pressure.

pressurize [ENG] To maintain normal atmospheric pressure in a chamber subjected to high or low external pressures.

pressurized blast furnace [ENG] A blast furnace operated under pressure above the ambient; pressure is obtained by throttling the off-gas line, which permits a greater volume of air to be passed through the furnace at a lower velocity, and results in increase in smelting rate.

pressurized stoppings [MIN ENG] Stoppings which are erected in the intake and return roadways of a district to isolate an open fire or spontaneous heating and in which the pressures on both sides of each stopping are made equal by the use of auxiliary fans.

pressurized water reactor [NUCLEO] A nuclear reactor in which water is circulated under enough pressure to prevent it from boiling, while serving as moderator and coolant for the uranium fuel; the heated water is then used to produce steam for a power plant. Abbreviated PWR.

presswork [GRAPHICS] In printing, the actual operation of putting ink on paper; this activity is preceded by composition and perhaps platemaking, and is followed by binding.

prester [METEOROL] A whirlwind or waterspout accompanied by lightning in the Mediterranean Sea and Greece.

prestore [ADP] To store a quantity in an available computer location before it is required in a routine.

prestress [ENG] To apply a force to a structure to condition it to withstand its working load more effectively or with less deflection. [GEOL] See preconsolidation pressure.

prestressed concrete [MATER] Concrete compressed with heavily loaded wires or bars to reduce or eliminate cracking and tensile forces.

presumptive address See address constant.

pretensioning [ENG] Process of precasting concrete beams with tensioned wires embedded in them. Also known as Hoyer method of prestressing.

pre-transmit-receive tube See pre-TR tube.

pretravel [CONT SYST] The distance or angle through which the actuator of a switch moves from the free position to the operating position.

pretrigger [ELECTR] Trigger used to initiate sweep ahead of transmitted pulse.

pre-TR tube [ELECTR] Gas-filled radio-frequency switching tube used in some radar systems to protect the TR tube from excessively high power and the receiver from frequencies other than the fundamental. Derived from pre-transmit-receive tube.

prevailing visibility [METEOROL] In United States weather observing practice, the greatest horizontal visibility equaled or surpassed throughout half of the horizon circle; in the case of rapidly varying conditions, it is the average of the prevailing visibility while the observation is being taken.

prevailing westerlies [METEOROL] The prevailing westerly winds on the poleward sides of the subtropical high-pressure belts.

prevailing wind See prevailing wind direction.

prevailing wind direction [METEOROL] The wind direction most frequently observed during a given period; the periods most often used are the observational day, month, season, and year. Also known as prevailing wind.

preventive maintenance [ENG] A procedure of inspecting, testing, and reconditioning a system at regular intervals according to specific instructions, intended to prevent failures in service or to retard deterioration.

previewing [ADP] In character recognition, a process of attempting to gain prior information about the characters that appear on an incoming source document; this information, which may include the range of ink density, relative positions, and so forth, is used as an aid in the normalization phase of character recognition.

previous element coding [COMMUN] System of signal coding, used for digital television transmission, whereby each transmitted picture element is dependent upon the similarity of the preceding picture element.

Prevost's theory [THERMO] A theory according to which a body is constantly exchanging heat with its surroundings, radiating an amount of energy which is independent of its surroundings, and increasing or decreasing its temperature depending on whether it absorbs more radiation than it emits, or vice versa.

prewhitening filter See whitening filter.

PRF See pulse repetition rate.

pri See primary winding.

priapism [MED] Persistent erection of the penis, usually unaccompanied by sexual desire, as seen in certain pathologic conditions.

Priapulida [INV ZOO] A minor phylum of wormlike marine animals; the body is made up of three distinct portions (proboscis, trunk, and caudal appendage) and is often covered with spines and tubercles, and the mouth is surrounded by concentric rows of teeth.

Priapuloidea [INV ZOO] An equivalent name for Priapulida.

priceite [MINERAL] $Ca_4B_{10}O_{19} \cdot 7H_2O$ A snow-white earthy mineral composed of hydrous calcium borate, occurring as a massive. Also known as pandermite.

Price meter [ENG] The ocean current meter in use in the United States: six conical cups, mounted around a vertical axis, rotate and cause a signal in a set of headphones with each rotation; tail vanes and a heavy weight stabilize the instrument.

pricker See needle.

prickly heat See miliaria.

prick punch [DES ENG] A tool that has a sharp point ground to an angle of 30–60°; used to punch holes in a workpiece after laying out lines.

priest [OPTICS] The Z tristimulus value.

prill [CHEM ENG] To form pellet-sized crystals or agglomerates of material by the action of upward-blowing air on falling hot solution; used in the manufacture of ammonium nitrate and urea fertilizers. [MATER] Spherical particles about the size of buckshot. [MIN ENG] **1.** The best ore after cobbing. **2.** A circular particle about the size of buckshot. **3.** Compressed and sized explosives such as ammonium nitrate.

Primacord [MATER] A fuse consisting of an explosive core within a textile or plastic covering.

Primacord-Bickford fuse [MATER] A detonating fuse using an explosive of pentaerythritetetranitrate (PETN); used in large-scale blasting work, especially in quarries.

primaquine [PHARM] $C_{15}H_{21}N_3O$ An ether-soluble viscous liquid, used as the diphosphate salt in medicine to cure malaria.

primary [ASTRON] **1.** A planet with reference to its satellites, or the sun with reference to its planets. **2.** The brighter star of a double star system. [CHEM] A term used to distinguish basic compounds from similar or isomeric forms; in organic compounds, for example, RCH_2OH is a primary alcohol, R_1R_2CHOH is a secondary alcohol, and $R_1R_2R_3COH$ is a tertiary alcohol; in inorganic compounds, for example, NaH_2PO_4 is primary sodium phosphate, Na_2HPO_4 is the secondary form, and Na_3PO_4 is the tertiary form. [ELEC] **1.** See primary winding. **2.** One of the high-voltage conductors of a power

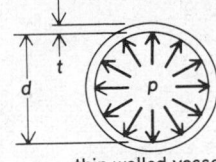

PRESSURE VESSEL

thin-walled vessel
$t < d/10$

thick-walled vessel
$t > d/10$

Pressure vessels for moderate and for high pressures; t = wall thickness; d = vessel diameter; p = pressure; d_i, d_o = inside and outside diameters.

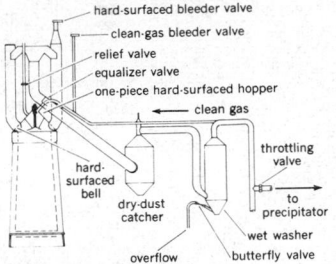

PRESSURIZED BLAST FURNACE

hard-surfaced bleeder valve
clean-gas bleeder valve
relief valve
equalizer valve
one-piece hard-surfaced hopper
clean gas
hard-surfaced bell
dry-dust catcher
throttling valve
to precipitator
wet washer
butterfly valve
overflow

Flow diagram of a pressurized blast furnace.

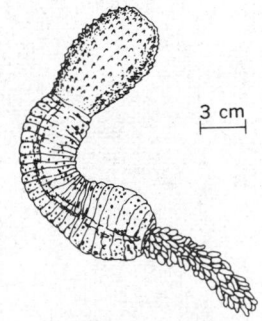

PRIAPULIDA

3 cm

Adult form of *Priapulus*.

distribution system. [GEOL] **1.** A young shoreline whose features are produced chiefly by nonmarine agencies. **2.** Of a mineral deposit, unaffected by supergene enrichment. [MET] Of a metal, obtained directly from ore. [VERT ZOO] Of or pertaining to quills on the distal joint of a bird wing.

primary alcohol [ORG CHEM] An alcohol whose molecular structure may be written as RCH_2OH, rather than as R_1R_2CHOH (secondary) or $R_1R_2R_3COH$ (tertiary).

primary amine [ORG CHEM] An amine whose molecular structure may be written as RNH_2, instead of R_1R_2NH (secondary) or $R_1R_2R_3N$ (tertiary).

primary area [NAV] In terminal operations, an area within a segment in which full obstacle clearance is applied.

primary atypical pneumonia See Eaton agent pneumonia.

primary battery [ELEC] A battery consisting of one or more primary cells.

primary body [ASTRON] The celestial body or central force field about which a satellite or other body orbits, or from which it is escaping, or toward which it is falling.

primary breaker [MECH ENG] A machine which takes over the work of size reduction from blasting operations, crushing rock to maximum size of about 2-inch diameter; may be a gyratory crusher or jaw breaker. Also known as primary crusher.

primary cell [ELEC] A cell that delivers electric current as a result of an electrochemical reaction that is not efficiently reversible, so that the cell cannot be recharged efficiently.

primary center [COMMUN] A telephone office having lower rank than a sectional center and higher rank than a toll center; connects toll centers and may also serve as a toll center for nearby end offices.

primary circle See primary great circle.

primary circulation [METEOROL] The prevailing fundamental atmospheric circulation on a planetary scale which must exist in response to radiation differences with latitude, to the rotation of the earth, and to the particular distribution of land and oceans, and which is required from the viewpoint of conservation of energy.

primary clay See residual clay.

primary coil [ELEC] The input coil in an induction coil or transformer.

primary colors [OPTICS] Three colors, red, yellow, and blue, which can be combined in various proportions to produce any other color.

primary constriction See centromere.

primary control program [ADP] The program which provides the sequential scheduling of jobs and basic operating systems functions. Abbreviated PCP.

primary cosmic rays See cosmic rays.

primary creep [MECH] The initial high strain-rate region in a material subjected to sustained stress.

primary crusher See primary breaker.

primary cyclone [METEOROL] Any cyclone (or low), especially a frontal cyclone, within whose circulation one or more secondary cyclones have developed. Also known as primary low.

primary detector See sensor.

primary dip [GEOL] The slight dip assumed by a bedded deposit at its moment of deposition. Also known as depositional dip; initial dip; original dip.

primary drilling [ENG] The process of drilling holes in a solid rock ledge in preparation for a blast by means of which the rock is thrown down.

primary electron [ELECTR] An electron which bombards a solid surface, causing secondary emission.

primary emission [ELECTR] Emission of electrons due to primary causes, such as heating of a cathode, and not to secondary effects, such as electron bombardment.

primary excavation [ENG] Digging performed in undisturbed soil.

primary explosive [MATER] Explosive or explosive mixture sensitive to shock and friction; used in primers and detonators to initiate explosion. Also known as initiating explosive.

primary extinction [SOLID STATE] A weakening of the stronger beams produced in x-ray diffraction by a very perfect crystal, as compared with the weaker.

primary fabric See apposition fabric.

primary fault [ELEC] In an electric circuit, the initial breakdown of the insulation of a conductor, usually followed by a flow of power current.

primary fission products See fission fragments.

primary flow [ELECTR] The current flow that is responsible for the major properties of a semiconductor device.

primary focus [OPTICS] In an astigmatic system, a line at which some of the bundle of rays from an off-axis point meet; this line is perpendicular to a plane which contains the point and the optical axis, and has a smaller image distance than the secondary focus. Also known as meridional focus; tangential focus.

primary frequency [COMMUN] Frequency assigned for normal use on a particular circuit or communications channel.

primary frequency standard [COMMUN] The national standard of frequency as maintained by the National Bureau of Standards, Washington, D.C.; the operating frequency of a radio station is determined by comparison with multiples or submultiples of this standard frequency as broadcast by station WWV.

primary front [METEOROL] The principal, and usually original, front in any frontal system in which secondary fronts are found.

primary fuel cell [ELEC] A fuel cell in which the fuel and oxidant are continuously consumed.

primary geosyncline See orthogeosyncline.

primary great circle [GEOD] A great circle used as the origin of measurement of a coordinate; particularly, such a circle 90° from the poles of a system of spherical coordinates, as the equator. Also known as fundamental circle; primary circle.

primary gun [ORD] Principal or main gun, especially of a tank or other armored vehicle.

primary haulage [MIN ENG] A short haul in which there is no secondary- or main-line haulage. Also known as face haulage.

primary high explosive [MATER] An explosive which is extremely sensitive to heat and shock and is normally used to initiate a secondary high explosive; examples are mercury fulminate, lead azide, lead styphnate, and tetracene.

primary hypertension See essential hypertension.

primary isoamyl alcohol See isobutyl carbinol.

primary lateral sclerosis [MED] A sclerotic disease of the crossed pyramidal tracts of the spinal cord, characterized by paralysis of the limbs, with rigidity, increased tendon reflexes, and absence of sensory and nutritive disorders. Also known as lateral sclerosis.

primary lead [MET] Lead recovered from ore, as contrasted with recycled scrap (secondary) lead.

primary lesion [MED] **1.** In syphilis, tuberculosis, and cowpox, a chancre. **2.** In dermatology, the earliest clinically recognizable manifestation of cutaneous disease, such as a macule, papule, vesicle, pustule, or wheal.

primary lights [OPTICS] Any three lights used in a system of tristimulus colorimetric analysis of solutions.

primary low See primary cyclone.

primary magma [GEOL] A magma that originates below the earth's crust.

primary measuring element [ENG] The portion of a measuring or sensing device that is in direct contact with the variables being measured (such as temperature, pressure, pH, or velocity).

primary meristem [BOT] Meristem which is derived directly from embryonic tissue and which gives rise to epidermis, vascular tissue, and the cortex.

primary mineral [MINERAL] A mineral that is formed at the same time as the rock in which it is contained, and that retains its original form and composition.

primary phloem [BOT] Phloem derived from apical meristem.

primary plasticizer [MATER] A plasticizer material for plastics formulations that has sufficient affinity to a polymer or resin so that it is considered compatible and therefore may be used as the sole plasticizer.

primary porosity [GEOL] Natural porosity in petroleum reservoir sands or rocks.

primary power cable [ELEC] Power service cables connecting the outside power source to the main-office switch and metering equipment.

primary radar [ENG] Radar in which the incident beam is reflected from the target to form the return signal.

primary radiation [PHYS] Radiation arriving directly from its source without interaction with matter.

primary rainbow [OPTICS] The most common of the principal rainbow phenomena, which appears as an arc of angular radius of about 42° about the observer's antisolar point; it is the inner of two rainbows, whose light undergoes only one internal reflection, and which is narrower and brighter than the outer, or secondary, rainbow.

primary reference fuel [MATER] 1. Gasoline; isooctane, *n*-heptane, or mixtures thereof used in the ASTM-CFR gasoline test engine to determine the octane rating of commercial gasoline. 2. Diesel fuel; cetane, α-methylnaphthalene, or mixtures thereof used in ASTM-CFR diesel test engines to rate the cetane number of commercial diesel fuels.

primary relay [ELEC] Relay that produces the initial action in a sequence of operations.

primary reserve [PETRO ENG] Petroleum reserve recoverable commercially at current prices and costs by conventional methods and equipment as a result of the natural energy inherent in the reservoir.

primary root [BOT] The first plant root to develop; derived from the radicle.

primary scattering [PHYS] Any scattering process in which radiation is received at a detector, such as the eye, after having been scattered just once; distinguished from multiple scattering.

primary sedimentary structure [GEOL] A sedimentary structure produced during deposition, such as ripple marks and graded bedding.

primary service area [COMMUN] The area in which the ground wave of a broadcast station is not subject to objectionable interference or fading.

primary sewage sludge [CIV ENG] A semiliquid waste resulting from sedimentation with no additional treatment.

primary skip zone [ELECTROMAG] Area around a transmitter beyond the ground wave but within the skip distance.

primary standard [SCI TECH] A unit directly defined and established by some authority, against which all secondary standards are calibrated; for example, in analytical chemistry, reference substances or solutions of known chemical purity and concentration are used to standardize laboratory solutions prior to volumetric analysis or titration.

primary storage [ADP] Main internal storage of a computer.

primary stratigraphic trap [GEOL] A stratigraphic trap formed by the deposition of clastic materials (such as shoe-string sands, lenses, sand patches, bars, or cocinas) or through chemical deposition (such as organic reefs or biostromes).

primary stress field *See* ambient stress field.

primary structure [AERO ENG] The main framework, of an aircraft including fittings and attachments; any structural member whose failure would seriously impair the safety of the missile is a part of the primary structure. [GEOL] A structure, in an igneous rock, that formed at the same time as the rock, but before its final consolidation.

primary succession *See* prisere.

primary syphilis [MED] The first stage of the venereal disease, characterized clinically by a painless ulcer, or chancre, at the point of infection and painless, discrete regional adenopathy.

primary tectonite [PETR] A tectonite with depositional fabric.

primary tissue [BOT] Plant tissue formed during primary growth. [HISTOL] Any of the four fundamental tissues composing the vertebrate body.

primary treatment [CIV ENG] Removal of floating solids and suspended solids, both fine and coarse, from raw sewage.

primary voltage [ELEC] The voltage applied to the terminals of the primary winding of a transformer.

primary wave [COMMUN] A radio wave traveling by a direct path, as contrasted with skips.

primary weapon [ORD] The principal arm of a combat unit; the rifle is the primary or basic weapon for an infantry rifle company, as compared with grenades or chemical projectiles, which are secondary or auxiliary weapons in such an organization.

primary winding [ELEC] The transformer winding that re-

ceives signal energy or alternating-current power from a source. Also known as primary. Abbreviated pri. Symbolized P.

primary xylem [BOT] Xylem derived from apical meristem.

Primates [VERT ZOO] The order of mammals to which man belongs; characterized in terms of evolutionary trends by retention of a generalized limb structure and dentition, increasing digital mobility, replacement of claws by flat nails, development of stereoscopic vision, and progressive development of the cerebral cortex.

prime [ENG] 1. Main or primary, as in prime contractor. 2. In blasting, to place a detonator in a cartridge or charge of explosive. 3. To treat wood with a primer or penetrant primer.

prime coat *See* primer.

prime contractor [ENG] A contractor having a direct contract for an entire project; he may in turn assign portions of the work to subcontractors.

prime fictitious meridian [NAV] The reference meridian (real or fictitious) used as the origin for measurement of fictitious longitude.

prime grid meridian [NAV] The reference meridian of a grid; in polar regions it is usually the 180°-0° geographic meridian, used as the origin for measuring grid longitude.

prime ideal [MATH] A principal ideal of a ring given by a single element that has properties analogous to those of the prime numbers.

prime inverse meridian *See* prime transverse meridian.

prime meridian [GEOD] The meridian of longitude 0°, used as the origin for measurement of longitude; the meridian of Greenwich, England, is almost universally used for this purpose.

prime mover [MECH ENG] The component of a power plant that transforms energy from the thermal or the pressure form to the mechanical form.

prime navaids facilities [NAV] Navigational aid facilities that, under certain operational environments (for example, weather or mission) are considered to be essential to the safe completion of departure and approach-to-land maneuvering by aircraft; these facilities may include RAPCON, GCA, ILS, TACAN, VOR, and associated critical air-traffic control communications equipments.

prime number [MATH] A positive integer having no divisors except itself and the integer 1.

prime oblique meridian [NAV] The reference fictitious meridian of an oblique graticule.

prime polynomial [MATH] A polynomial whose only factors are itself and constants.

primer [ENG] In general, a small, sensitive initial explosive train component which on being actuated initiates functioning of the explosive train, and will not reliably initiate high explosive charge; classified according to the method of initiation, for example, percussion primer, electric primer, or friction primer. [MATER] A prefinishing coat applied to a wood surface that is to be painted or otherwise finished. Also known as prime coat.

primer cup [ENG] A small metal cup, into which the primer mixture is loaded.

primer-detonator [ENG] A unit, in a metal housing, in which are assembled a primer, a detonator, and when indicated, an intervening delay charge.

primer leak [ENG] Defect in a cartridge which allows partial escape of the hot propelling gases in a primer, caused by faulty construction or an excessive charge.

primer mixture [MATER] An explosive mixture containing a sensitive explosive and other ingredients, used in a primer.

primer pouch [ORD] Container that holds the primer used in firing with separate-loading ammunition.

primer seat [ORD] The chamber, in the breech mechanism of a gun that uses separate-loading ammunition, into which the primer is set.

primer setback [ORD] The backward movement of a primer cup in a cartridge case which occurs when the base of the cup is not properly supported by the bolt face or breechblock.

primes [MET] High-quality metal products, particularly sheet and plate, that are free from visible defects.

prime transverse meridian [NAV] The reference meridian of

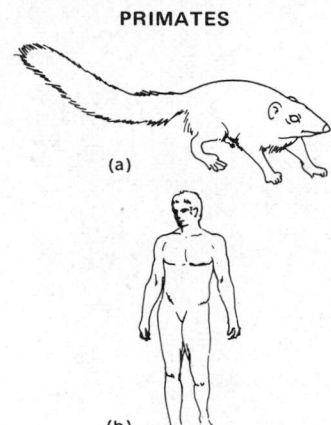

PRIMATES

(a)

(b)

Representative Primates.
(a) Member of the family Tupaiidae. *(b)* Man, of the family Hominidae.

a transverse graticule. Also known as prime inverse meridian.

prime vertical circle [GEOD] The vertical circle through the east and west points of the horizon.

prime-white kerosine [MATER] A kerosine with an off-white color, between water-white and standard-white kerosine.

primibrach [INV ZOO] In crinoids, the brachials of the unbranched arm.

priming composition [MATER] A physical mixture of materials that is very sensitive to impact or percussion and, when so exploded, undergoes very rapid autocombustion, producing hot gases and incandescent solid particles; priming compositions are used for the ignition of primary high explosives, black powder igniter charges, and propellants in small arms ammunition.

priming of the tides [OCEANOGR] The acceleration in the times of occurrence of high and low tides when the sun's tidal effect comes before that of the moon.

priming pump [MECH ENG] A device on motor vehicles and tanks, providing a means of injecting a spray of fuel into the engine to facilitate starting.

Primitiopsacea [PALEON] A small dimorphic superfamily of extinct ostracods in the suborder Beyrichicopina; the velum of the male was narrow and uniform, but that of the female was greatly expanded posteriorly.

primitive cell [CRYSTAL] A parallelepiped whose edges are defined by the primitive translations of a crystal lattice; it is a unit cell of minimum volume.

primitive equations [FL MECH] The Eulerian equations of motion of a fluid in which the primary dependent variables are the fluid's velocity components; these equations govern a wide variety of fluid motions and form the basis of most hydrodynamical analysis; in meteorology, these equations are frequently specialized to apply directly to the cyclonic-scale motions by the introduction of filtering approximations.

primitive gut [EMBRYO] The tubular structure in embryos which differentiates into the alimentary canal.

primitive polynomial [MATH] A polynomial with integer coefficients which have 1 as their greatest common divisor.

primitive root [MATH] An nth root of unity that is not an mth root of unity for any m less than n.

primitive streak [EMBRYO] A dense, opaque band of ectoderm in the bilaminar blastoderm associated with the morphogenetic movements and proliferation of the mesoderm and notochord; indicates the first trace of the vertebrate embryo.

primitive translation [CRYSTAL] For a space lattice, one of three translations which can be repeatedly applied to generate any translation which leaves the lattice unchanged.

primordial gut See archenteron.

primordium See anlage.

Primulaceae [BOT] A family of dicotyledonous plants in the order Primulales characterized by a herbaceous habit and capsular fruit with two to many seeds.

Primulales [BOT] An order of dicotyledonous plants in the subclass Dilleniidae distinguished by sympetalous flowers, stamens located opposite the corolla lobes, and a compound ovary with a single style.

principal axis [CRYSTAL] The longest axis in a crystal. [ENG ACOUS] A reference direction for angular coordinates used in describing the directional characteristics of a transducer; it is usually an axis of structural symmetry or the direction of maximum response. [MECH] One of three perpendicular axes in a rigid body such that the products of inertia about any two of them vanish. [OPTICS] See optic axis.

principal axis of strain [MECH] One of the three axes of a body that were mutually perpendicular before deformation. Also known as strain axis.

principal axis of stress [MECH] One of the three mutually perpendicular axes of a body that are perpendicular to the principal planes of stress. Also known as stress axis.

principal branch [MATH] For complex valued functions such as the logarithm which are multiple-valued, a selection of values so as to obtain a genuine single-valued function.

principal E plane [ELECTROMAG] Plane containing the direction of radiation of electromagnetic waves and arranged so that the electric vector everywhere lies in the plane.

principal focus See focal point.

principal H plane [ELECTROMAG] Plane that contains the direction of radiation and the magnetic vector, and is everywhere perpendicular to the E plane.

principal ideal [MATH] An ideal in a ring consisting of all powers and scalar multiples of a single element.

principal item [ENG] Item which, because of its major importance, requires detailed analysis and examination of all factors affecting its supply and demand, as well as an unusual degree of supervision; its selection is based upon such criteria as strategic importance, high monetary value, unusual complexity of issue, and procurement difficulties.

principal meridian [CIV ENG] One of the meridians established by the United States government as a reference for subdividing public land.

principal mode See fundamental mode.

principal moments [PHYS CHEM] The three moments of inertia of a rigid molecule calculated with respect to the principal axes.

principal normal [MATH] The line perpendicular to a space curve at some point which also lies in the osculating plane at that point.

principal plane [OPTICS] 1. Two planes perpendicular to the optical axis such that objects in one plane form images in the other with a lateral magnification of unity. 2. See principal section.

principal point [OPTICS] The intersection of a principal plane with the optical axis.

principal quantum number [ATOM PHYS] A quantum number for orbital electrons, which, together with the orbital angular momentum and spin quantum numbers, labels the electron wave function; the energy level and the average distance of an electron from the nucleus depend mainly upon this quantum number.

principal root [MATH] The positive real root of a positive number, or the negative real root in the case of odd roots of negative numbers.

principal section [OPTICS] A plane in a crystal that contains the crystal's optic axis and the ray of light under consideration. Also known as principal plane.

principal series [SPECT] A series occurring in the line spectra of many atoms and ions with one, two, or three electrons in the outer shell, in which the total orbital angular momentum quantum number changes from 1 to 0.

principal stress [MECH] A stress occurring at right angles to a principal plane of stress.

principal value [MATH] A limiting value that an integral takes at a point where it is not actually defined.

principal vertical circle [GEOD] The vertical circle through the north and south points of the horizon, coinciding with the celestial meridian.

principle of covariance [RELAT] 1. In classical physics and in special relativity, the principle that the laws of physics take the same mathematical form in all inertial reference frames. 2. In general relativity, the principle that the laws of physics take the same mathematical form in all conceivable curvilinear coordinate systems.

principle of duality See duality principle.

principle of equivalence See equivalence principle.

principle of insufficient reason [STAT] The principle that cases are equally likely to occur unless reasons to the contrary are known.

principle of least action [MECH] The principle that, for a system whose total mechanical energy is conserved, the trajectory of the system in configuration space is that path which makes the value of the action stationary relative to nearby paths between the same configurations and for which the energy has the same constant value. Also known as least-action principle.

principle of reciprocity See reciprocity theorem.

principle of superposition [ELEC] 1. The principle that the total electric field at a point due to the combined influence of a distribution of point charges is the vector sum of the electric field intensities which the individual point charges would produce at that point if each acted alone. 2. The principle that, in a linear electrical network, the voltage or current in any element resulting from several sources acting together is

1171

the sum of the voltages or currents resulting from each source acting alone. Also known as superposition theorem. [MECH] The principle that when two or more forces act on a particle at the same time, the resultant force is the vector sum of the two. [PHYS] Also known as superposition principle. **1.** A general principle applying to many physical systems which states that if a number of independent influences act on the system, the resultant influence is the sum of the individual influences acting separately. **2.** In all theories characterized by linear homogeneous differential equations, such as optics, acoustics, and quantum theory, the principle that the sum of any number of solutions to the equations is another solution.

principle of the maximum [MATH] The principle that for a nonconstant complex analytic function defined in a domain, the absolute value of the function cannot attain its maximum at any interior point of the domain.

principle of the minimum [MATH] The principle that for a nonvanishing nonconstant complex analytic function defined in a domain, the absolute value of the function cannot attain its minimum at any interior point of the domain.

principle of uniformity *See* uniformitarianism.

principle of virtual work [MECH] The principle that the total work done by all forces acting on a system in static equilibrium is zero for any infinitesimal displacement from equilibrium which is consistent with the constraints of the system. Also known as virtual work principle.

printed circuit [ELECTR] A conductive pattern that may or may not include printed components, formed in a predetermined design on the surface of an insulating base in an accurately repeatable manner.

printed-wiring armature [ELEC] An armature in which the conductors consist of printed-wiring strips on both sides of a thin insulating disk, to give a low-inertia armature for servomotors and other variable high-speed applications.

printer [ADP] A computer output mechanism that prints characters one at a time or one line at a time.

print film [GRAPHICS] A fine-grain, high-resolving-power, yellow-dyed positive photographic film used for continuous contact printing.

printing ink [MATER] Ink generally made from carbon black, lampblack, or other pigment suspended in an oil vehicle, with a resin, solvent, adhesive, and drier; available in many variations.

printing press [GRAPHICS] A machine for the production of printed impressions on paper and other materials.

printing-telegraph code [COMMUN] A five- or seven-unit code used for operation of a teleprinter, teletypewriter, and similar telegraph printing devices.

printing telegraphy [COMMUN] Method of telegraph operation in which the received signals are automatically recorded in printed characters.

printout [ADP] A printed output of a data-processing machine or system.

printout paper [GRAPHICS] A paper used to make prints by light alone, without any processing involved; it has a reddish color, is often used as a proof, and can be made permanent by toning.

print processes [GRAPHICS] Three general print processes: relief process, in which the printing areas stand above the surface of the plate; intaglio process, in which the printing areas fall below the surface of the plate; and planographic process, in which there is no appreciable difference between the level of the printing areas and the general surface.

printthrough [ADP] Transfer of signals from one recorded layer of magnetic tape to the next on a reel.

Prioniodidae [PALEON] A family of conodonts in the suborder Conodontiformes having denticulated bars with a large denticle at one end.

Prioniodinidae [PALEON] A family of conodonts in the suborder Conodontiformes characterized by denticulated bars or blades with a large denticle in the middle third of the specimen.

prionodont [VERT ZOO] Having many simple, similar teeth set in a row like sawteeth.

prior-art search [ENG] **1.** A search for prior art which may possibly anticipate an invention which is being considered for patentability. **2.** A similar search but for the purpose of

determining what the status of existing technology is before going ahead with new research; it is done to avoid unwittingly retracing new steps taken by other workers in the field.

priorite [MINERAL] $(Y,Ca,Th)(Ti,Nb)_2O_6$ A mineral composed of titanoniobate of rare-earth metals; it is isomorphous with eschynite. Also known as blomstrandine.

priority-arbitration circuit [ADP] A logic circuit which combines all interrupts but allows only the highest-priority request to enable its active flipflop.

priority interrupt [ADP] An interrupt procedure in which control is passed to the monitor, the required operation is initiated, and then control returns to the running program, which never knows that it has been interrupted.

priority phase [ADP] Phase consisting of execution of operations in response to instruments or process interrupts other than clock interrupts.

prior probabilities [STAT] Probabilities of the outcomes of an experiment before the experiment has been performed.

priscol *See* tolazoline hydrochloride.

prisere [ECOL] The ecological succession of vegetation that occurs in passing from barren earth or water to a climax community. Also known as primary succession.

prism [CRYSTAL] A crystal which has three, four, six, eight, or twelve faces, with the face intersection edges parallel, and which is open only at the two ends of the axis parallel to the intersection edges. [GEOL] A long, narrow, wedge-shaped sedimentary body with a width-thickness ratio greater than 5 to 1 but less than 50 to 1. [MATH] A polyhedron with two parallel, congruent faces and all other faces parallelograms. [OPTICS] An optical system consisting of two or more usually plane surfaces of a transparent solid or embedded liquid at an angle with each other. Also known as optical prism.

prismatic astrolabe [ENG] A surveying instrument used to make the celestial observations needed in establishing an astronomical position.

prismatic coefficient *See* longitudinal coefficient.

prismatic compass [ENG] A hand compass used by surveyors which is equipped with a prism that allows the compass to be read while the site is being taken.

prismatic error [OPTICS] That error due to lack of parallelism of the two faces of an optical element, such as a mirror or a shade glass.

prismatic jointing *See* columnar jointing.

prismatic plane [MET] Any plane parallel to the principal c axis in noncubic metals.

prismatic structure *See* columnar jointing.

prismatic surface [MATH] A surface generated by moving a straight line which always meets a broken line lying in a given plane and which is always parallel to some given line not in that plane.

prismatoid [MATH] A polyhedron whose vertices all are in one or the other of two parallel planes.

prism diopter [OPTICS] A unit used in measuring the deviating power of a prism; this power in prism diopters is 100 times the tangent of the angle of deviation of a ray of light.

prism level [ENG] A surveyor's level with prisms that allow the levelman to view the level bubble without moving his eye from the telescope.

prismoid [MATH] A prismatoid whose two parallel faces are polygons having the same number of sides while the other faces are trapezoids or parallelograms.

prism spectrograph [SPECT] Analysis device in which a prism is used to give two different but simultaneous light wavelengths derived from a common light source; used for the analysis of materials by flame photometry.

Pristidae [VERT ZOO] The sawfishes, a family of modern sharks belonging to the batoid group.

Pristiophoridae [VERT ZOO] The saw sharks, a family of modern sharks often grouped with the squaloids which have a greatly extended rostrum with enlarged denticles along the margins.

privacy system [COMMUN] A device or method for scrambling overseas telephone conversations handled by radio links in order to make them unintelligible to outside listeners. Also known as privacy transformation; secrecy system.

privacy transformation *See* privacy system.

private aid to navigation [NAV] In marine operation, an aid

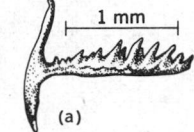

PRIONIODIDAE

Three examples of Prioniodidae:
*(a) Ligonodina. (b) Hibbardella.
(c) Hindeodella.*

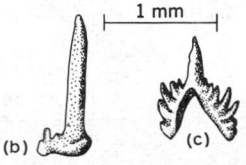

PRIONIODINIDAE

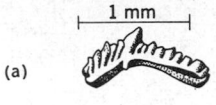

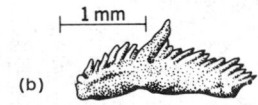

Two examples of the
Prioniodinidae: *(a) Bryantodus;
(b) Lewistownella.*

PRISMATIC ASTROLABE

A prismatic astrolabe, used to
make celestial observations. *(U.S.
Naval Oceanographic Office)*

to navigation authorized by the U.S. Coast Guard but established, operated, and maintained by private interests in the navigable waters of the United States.

private automatic branch exchange [COMMUN] A private branch exchange in which connections are made by remote-controlled switches. Abbreviated PABX.

private automatic exchange [COMMUN] A private telephone exchange in which connections are made by remote-controlled switches. Abbreviated PAX.

private branch exchange [COMMUN] A telephone exchange serving a single organization, having a switchboard and associated equipment, usually located on the customer's premises; provides for switching calls between any two extensions served by the exchange or between any extension and the national telephone system via a trunk to a central office. Abbreviated PBX. Also known as telephone private branch exchange.

private branch exchange access line [ELEC] Circuit that connects a main private branch exchange (PBX) to a switching center.

private exchange [COMMUN] Telephone exchange serving a single organization and having no means for connecting to a public telephone system.

private line [COMMUN] A line, channel, or service reserved solely for one user.

private line service [COMMUN] Service provided by United States common carrier engaged in domestic or international wire, radio, and cable communications for the intercity communications purposes of a customer; this service is provided over integrated communications pathways, including facilities or local channels, which are integrated components of intercity private line services, and station equipment between specified locations for a continuous period or for regularly recurring periods at stated hours.

privileged direction [OPTICS] One of two mutually perpendicular directions for the plane of polarization of a beam of plane-polarized light falling on a plate of anisotropic material such that the light which emerges from the plate is also plane-polarized.

proaccelerin [BIOCHEM] A labile procoagulant in normal plasma but deficient in the blood of patients with parahemophilia; essential for rapid conversion of prothrombin to thrombin. Also known as factor V; labile factor.

proamnion [EMBRYO] The part of the embryonic area at the sides and in front of the head of the developing amniote embryo, which remains without mesoderm for a considerable period.

Proanura [PALEON] Triassic forerunners of the Anura.

probability [STAT] The probability of an event is the ratio of the number of times it occurs to the large number of trials that take place; the mathematical model of probability is a positive measure which gives the measure of the space the value 1.

probability amplitude See Schrödinger wave function.

probability current density [QUANT MECH] A vector whose component normal to a surface gives the probability that a particle will cross a unit area of the surface during a unit time.

probability density [QUANT MECH] The square of the absolute value of the Schrödinger wave function for a particle at a given point; gives the probability per unit volume of finding the particle at that point.

probability density function [STAT] A real-valued function whose integral over any set gives the probability that a random variable has values in this set. Also known as density function; frequency function.

probability deviation See probable error.

probability distribution See distribution of random variable.

probability forecast [METEOROL] A forecast of the probability of occurrence of one or more of a mutually exclusive set of weather contingencies, as distinguished from a series of categorical statements.

probability ratio test [STAT] Testing a simple hypothesis against a simple alternative by using the ratio of the probability of each simple event under the alternative to the probability of the event under the hypothesis.

probability theory [MATH] The study of the mathematical structures and constructions used to analyze the probability of a given set of events from a family of outcomes.

probable [ORD] **1.** An instance in which a hostile aircraft is probably destroyed. **2.** The hostile aircraft so designated.

probable error [STAT] The error that is exceeded by a variable with a probability of ½. Also known as probability deviation.

probable maximum precipitation [METEOROL] The theoretically greatest depth of precipitation for a given duration that is physically possible over a particular drainage area at a certain time of year; in practice, this is derived over flat terrain by storm transposition and moisture adjustment to observed storm patterns.

probable reserves [PETRO ENG] Primary petroleum reserves based on limited evidence, but not proved by a commercial oil-production rate.

probe [AERO ENG] An instrumented vehicle moving through the upper atmosphere or space or landing upon another celestial body in order to obtain information about the specific environment. [COMMUN] To determine a radio interference by obtaining the relative interference level in the immediate area of a source by the use of a small, insensitive antenna in conjunction with a receiving device. [ELECTRO-MAG] A metal rod that projects into but is insulated from a waveguide or resonant cavity; used to provide coupling to an external circuit for injection or extraction of energy or to measure the standing-wave ratio. Also known as waveguide probe. [ENG] A small tube containing the sensing element of electronic equipment, which can be lowered into a borehole to obtain measurements and data. [PHYS] A small device which can be brought into contact with or inserted into a system in order to make measurements on the system; ordinarily it is designed so that it does not significantly disturb the system.

probe coil [ELECTROMAG] In eddy-current nondestructive tests, a type of test coil which is placed on the surface of an object.

proband [GEN] The clinically affected individual who is being studied with respect to a pedigree of interest to human genetics.

probertite [MINERAL] $NaCaB_5O_9 \cdot 5H_2O$ A colorless mineral crystallizing in the monoclinic system, consisting of hydrous sodium calcium borate.

probe-type liquid-level meter [ENG] Device to sense or measure the level of liquids in storage or process vessels by means of an immersed electrode or probe.

problem check [ADP] One or more tests used to assist in obtaining the correct machine solution to a problem.

problem definition [ADP] The art of compiling logic in the form of general flow charts and logic diagrams which clearly explain and present the problem to the programmer in such a way that all requirements involved in the run are presented.

problem file See run book.

problem folder See run book.

problem mode [ADP] A condition of computer operation in which, in contrast to supervisor mode, the privileged instructions cannot be executed, preventing the program from upsetting the supervisor program or any other program.

problem-oriented language [ADP] A language designed to facilitate the accurate expression of problems belonging to specific sets of problem types.

Proboscidea [VERT ZOO] An order of herbivorous placental mammals characterized by having a proboscis, incisors enlarged to become tusks, and pillarlike legs with five toes bound together on a broad pad.

proboscis [INV ZOO] A tubular organ of varying form and function on a large number of invertebrates, such as insects, annelids, and tapeworms. [VERT ZOO] The flexible, elongated snout of certain mammals.

Proca equations [QUANT MECH] A set of equations, analogous to Maxwell's equations, relating a four-vector potential and a second-rank tensor field describing a particle of spin 1 and nonzero mass.

procaine See procaine base.

procaine base [ORG CHEM] $C_6H_4NH_2COOCH_2CH_2N(C_2H_5)_2$ Water-insoluble, light-sensitive, odorless, white powder, melting at 60°C; soluble in alcohol, ether, chloroform, and benzene; used in medicine as a local anesthetic. Also known as para-aminobenzoyldiethylaminoethanol base; planocaine base; procaine.

Procampodeidae [INV ZOO] A family of the insect order Diplura.

Procaviidae [VERT ZOO] A family of mammals in the order Hyracoidea including the hyraxes.

Procaviinae [VERT ZOO] A subfamily of ungulate mammals in the family Procaviidae.

procedural language [ADP] A programming language made up of macroinstructions, each macroinstruction usually written in assembly language.

procedure [ADP] A sequence of actions (or computer instructions) which collectively accomplish some desired task.

procedure division [ADP] The section of a program (written in the COBOL language) in which a programmer specifies the operations to be performed with the data names appearing in the program.

procedure-oriented language [ADP] A language designed to facilitate the accurate description of procedures, algorithms, or routines belonging to a certain set of procedures.

procedure track [NAV] The path specified for making an instrument approach to an airport and for pull-up in case of a missed approach; this information is indicated on instrument approach charts by lines both in plan and profile.

procedure turn [NAV] A constant-rate turn of an aircraft in flight; used for computing the radius of turn and time required for its execution when very accurate navigation is required in controlling time or maintaining accurate, predetermined tracks.

procedure work log system [NAV] In air operations, a system utilized for logging the flight procedures in airspace data and establishing the source of documentation for automatic data processing. Abbreviated PROWL.

proceed-to-select signal [COMMUN] Signal returned from distant automatic equipment over the backward signaling path, in response to a calling signal, to indicate that selecting information can be transmitted; in certain signaling systems, both signals can be the same.

proceed to transmit signal [COMMUN] Signal returned from a distant manual switchboard over the backward signaling path, in response to a calling signal, to indicate that the teleprinter of the distant operator is connected to the circuit.

Procellariidae [VERT ZOO] A family of birds in the order Procellariiformes comprising the petrels, fulmars, and shearwaters.

Procellariiformes [VERT ZOO] An order of oceanic birds characterized by tubelike nostril openings, webbed feet, dense plumage, compound horny sheath of the bill, and, often, a peculiar musky odor.

procephalon [INV ZOO] The part of an insect's head that lies anteriorly to the segment in which the mandibles are located.

procercoid [INV ZOO] The solid parasitic larva of certain eucestodes, such as pseudophyllideans, that develops in the body of the intermediate host.

process [ADP] To assemble, compile, generate, interpret, compute, and otherwise act on information in a computer. [ENG] A system or series of continuous or regularly occurring actions taking place in a predetermined or planned manner; for example, as oil refining process or chemicals manufacturing process.

process annealing [MET] Softening a ferrous alloy by heating to a temperature close to but below the lower limit of the transformation range and then cooling.

process camera [OPTICS] Large camera used to produce materials for reproduction in printing; permits a large range of enlargement and reduction.

process chart [IND ENG] A graphic representation of events occurring during a series of actions or operations.

process color [GRAPHICS] Method of reproducing full-color originals such as paintings and color photographs; four-color process plates print in yellow, magenta, cyan, and black.

process control [ENG] Manipulation of the conditions of a process to bring about a desired change in the output characteristics of the process.

process-control chart [IND ENG] A tabulated graphical arrangement of test results and other pertinent data for each production assembly unit, arranged in chronological sequence for the entire assembly.

process control engineering [ENG] A field of engineering dealing with ways and means by which conditions of continuous processes are automatically kept as close as possible to desired values or within a required range.

process control system [CONT SYS] The automatic control of a continuous operation.

process dynamics [ENG] The dynamic response interrelationships between components (units) of a complex system, such as in a chemical process plant.

process engineering [ENG] A service function of production engineering that involves selection of the processes to be used, determination of the sequence of all operations, and requisition of special tools to make a product.

process furnace [CHEM ENG] Furnace used to heat process-stream materials (liquids, gases, or solids) in a chemical-plant operation; types are direct-fired, indirect-fired, and pebble heaters.

process heater [CHEM ENG] Equipment for the heating of chemical process streams (gases, liquids, or solids); usually refers to furnaces, in contrast to heat exchangers.

process heat reactor [NUCLEO] A nuclear reactor that produces heat for use in manufacturing processes.

processing [COMMUN] Further handling, manipulation, consolidation, compositing, and so on, of information to convert it from one format to another or to reduce it to manageable or intelligible information. [ENG] The act of converting material from one form into another desired form.

processing interrupt [ADP] The interruption of the batch processing mode in a real-time system when live data are entered in the system.

processing section [ADP] The computer unit that does the actual changing of input into output; includes the arithmetic unit and intermediate storage.

process lapse rate [METEOROL] The rate of decrease of the temperature of an air parcel as it is lifted, expressed as $-dT/dz$, where z is the altitude, or occasionally dT/dp, where p is pressure; the concept may be applied to other atmospheric variables, such as the process lapse rate of density.

process lens [OPTICS] A highly corrected, apochromatic lens used for precise color-separation work.

process metallurgy [MET] The branch of metallurgy concerned with the extraction of metals from ore, and with the refining of metals; usually synonymous with extractive metallurgy.

process monitoring [CHEM ENG] The observation of chemical process variables by means of pressure, temperature, flow, and other types of indicators; usually occurs in a central control room.

processor [ADP] **1.** A device that performs one or many functions, usually a central processing unit. **2.** A program that transforms some input into some output, such as an assembler, compiler, or linkage editor.

processor stack pointer [ADP] A programmable register used to access all temporary-storage words related to an interrupt-service routine which was halted when a new service routine was called in.

processor status word [ADP] A word comprising a set of flag bits and the interrupt-mask status.

process planning [IND ENG] Determining the conditions necessary to convert material from one state to another.

process research [SCI TECH] Applied research with a new or improved process in view.

process schizophrenia [PSYCH] Schizophrenia having a slow, insidious onset; the person exhibits poor adjustment before hospitalization.

process time [IND ENG] **1.** Time needed for completion of the machine-controlled portion of a work cycle. **2.** Time required for completion of an entire process.

process variable [CHEM ENG] Any of those varying operational and physical conditions associated with a chemical processing operation, such as temperature, pressure, flowrate, density, pH, viscosity, or chemical composition.

procoagulant [BIOCHEM] Any of blood clotting factors V to VIII; accelerates the conversion of prothrombin to thrombin in the presence of thromboplastin and calcium.

Procoela [VERT ZOO] A suborder of the Anura characterized by a procoelous vertebral column and a free coccyx articulating with a double condyle.

procoelous [VERT ZOO] The form of a vertebra that is concave anteriorly and convex posteriorly.

Procolophonia [PALEON] A subclass of extinct cotylosaurian reptiles.

proctiger [INV ZOO] The cone-shaped, reduced terminal segment of the abdomen of an insect which contains the anus.

proctitis [MED] Inflammation of the anus or rectum.

proctology [MED] A branch of medicine concerned with the structure and disease of the anus, rectum, and sigmoid colon.

proctoscope [MED] An instrument for inspecting the anal canal and rectum.

proctosigmoidectomy [MED] The abdominoperineal excision of the anus and rectosigmoid, usually with the formation of an abdominal colostomy.

Proctotrupidae [INV ZOO] A family of hymenopteran insects in the superfamily Proctotrupoidea.

Proctotrupoidea [INV ZOO] A superfamily of parasitic Hymenoptera in the suborder Apocrita.

procumbent [BOT] Having stems that lie flat on the ground but do not root at the nodes. [SCI TECH] **1.** Lying stretched out. **2.** Slanting forward. **3.** Lying face down.

procurement [ORD] **1.** The complete action or process of acquiring or obtaining personnel, materiel, services, or property from outside a military service by means authorized in pertinent directives. **2.** More specifically, the action or process of acquiring or obtaining materiel, property, or services at the operational level, for example, purchasing, contracting, and negotiating directly with the source of supply.

procurement lead time [ORD] The time elapsing between the initiation of procurement action and the receipt into the supply system of the materiel procured.

Procyon [ASTRON] A star of magnitude 0.3, of spectral type F5, and 11 light-years from earth; one of a binary. Also known as α Canis Minoris.

Procyonidae [VERT ZOO] A family of carnivoran mammals in the superfamily Canoidea, including raccoons and their allies.

prod *See* test prod.

prod cast [GEOL] The cast of a prod mark. Also known as impact cast.

prodelta [GEOL] The part of a delta lying beyond the delta front, and sloping gently down to the basin floor of the delta; it is entirely below the water level.

prodelta clay [GEOL] Fine sand, silt, and clay transported by the river and deposited on the floor of a sea or lake beyond the main body of a delta.

Prodinoceratinae [PALEON] A subfamily of extinct herbivorous mammals in the family Untatheriidae; animals possessed a carnivorelike body of moderate size.

prod mark [GEOL] A short tool mark oriented parallel to the current and gradually deepening downcurrent. Also known as impact mark.

prodrome [MED] **1.** An early or premonitory manifestation of impending disease before the specific symptoms begin. **2.** An aura.

producer [ECOL] An autotrophic organism of the ecosystem; any of the green plants.

producer gas [MATER] Fuel gas high in carbon monoxide and hydrogen, produced by burning a solid fuel with a deficiency of air or by passing a mixture of air and steam through a bed of incandescent fuel; used as a cheap, low-Btu industrial fuel.

producer's risk [IND ENG] The probability that in an acceptance sampling plan, material of an acceptable quality level will be rejected.

producing gas-oil ratio [PETRO ENG] The ratio of gas to oil (GOR, or gas-oil ratio) from a producing well; an increase in GOR is a danger signal in the efficient control of reservoir performance.

producing reserves [PETRO ENG] Developed (proved) petroleum reserves to be produced by existing wells in that portion of a reservoir subjected to full-scale secondary-recovery operations.

product [CHEM ENG] *See* discharge liquor. [IND ENG] **1.** An item or goods made by an industrial firm. **2.** The total of such items or goods. [MATH] **1.** The product of two algebraic quantities is the result of their multiplication relative to an operation analogous to multiplication of real numbers. **2.** The product of a collection of sets $A_1, A_2, \ldots, A_n$ is the set of all elements of the form $(a_1, a_2, \ldots, a_n)$ where each a_i is an element of A_i for each $i = 1, 2, \ldots, n$.

product demodulator [ELECTR] A receiver demodulator whose output is the product of the input signal voltage and a local oscillator signal voltage at the input frequency. Also known as product detector.

product design [DES ENG] The determination and specification of the parts of a product and their interrelationship so that they become a unified whole.

product detector *See* product demodulator.

Productinida [PALEON] A suborder of extinct articulate brachiopods in the order Strophomenida characterized by the development of spines.

production [ENG] Output, such as units made in a factory, oil from a well, or chemicals from a processing plant.

production control [IND ENG] The procedure for planning, routing, scheduling, dispatching, and expediting the flow of materials, parts, subassemblies, and assemblies within a plant, from the raw state to the finished product, in an orderly and efficient manner.

production-decline curve [PETRO ENG] A graphical means to estimate the ultimate recovery (oil or gas) from a reservoir; cumulative production is plotted against time, the curve being extrapolated to an end point (that is, ultimate recovery).

production engineering [IND ENG] The planning and control of the mechanical means of changing the shape, condition, and relationship of materials within industry toward greater effectiveness and value.

production model [IND ENG] A model in its final mechanical and electrical form of final production design made by production tools, jigs, fixtures, and methods.

production packer [PETRO ENG] A downhole tool used to assist in the efficient production of oil and gas from a well having one or more productive horizons; the function is to provide a seal between the outside of the tubing and the inside of the casing to prevent movement of fluids past that point. Also known as packer.

production reactor [NUCLEO] A nuclear reactor designed primarily for large-scale production of transmutation products, such as plutonium.

production requirements [IND ENG] The sum of authorized stock levels and pipeline needs less stocks expected to become available, stock on hand, stocks due in, returned stocks, and stocks from salvage, reclamation, rebuild, and other sources.

production standard *See* standard time.

production time [ADP] Good computing time, including occasional duplication of one case for a check or rerunning of the test run; also including duplication requested by the sponsor, any reruns caused by misinformation or bad data supplied by sponsor, and error studies using different intervals, convergence criteria, and so on.

productive time [IND ENG] Time during which useful work is performed in an operation or process.

productivity [AGR] The yield of a given crop per unit of land. [IND ENG] **1.** The effectiveness with which labor and equipment are utilized. **2.** The production output per unit of effort. [PETRO ENG] Measure of an oil well's ability to produce liquid or gaseous hydrocarbons; categories include relative, specific, ultimate, and fractured-well productivity.

productivity index [PETRO ENG] The number of barrels of oil produced per day per decline in well bottom-hole pressure in pounds per square inch.

productivity ratio [PETRO ENG] **1.** The amount of damage or improvement to reservoir formation permeability adjacent to the borehole (due to invasion or reduction of drilling mud present, drilling-fluid filtrate water, swollen clay particles, or salt or wax deposition). **2.** The ratio of permeability calculated from the productivity index to the permeability calculated from reservoir buildup pressure.

productivity test [PETRO ENG] Graphical relation of bottom-hole static pressure (calculated or measured) versus producing pressure for various gas flow rates; used to predict future oil well behavior.

product line [IND ENG] **1.** The range of products offered by a

PROCUMBENT

The procumbent stem of purslane.

firm. **2.** A group of basically similar products, differentiated only by such characteristics as color, style, or size.

product measure [MATH] A measure on a product of measure spaces constructed from the measures on each of the individual spaces by taking the measure of the product of a finite number of measurable sets, one from each of the measure spaces in the product, to be the product of the measures of these sets.

product model [STAT] A model for independent repetition of an experiment, or independent performance of several experiments, obtained by taking the cartesian product of the probability spaces representing the experiments.

product modulator [ELECTR] Modulator whose modulated output is substantially equal to the carrier and the modulating wave; the term implies a device in which intermodulation between components of the modulating wave does not occur.

product of inertia [MECH] Relative to two rectangular axes, the sum of the products formed by multiplying the mass (or, sometimes, the area) of each element of a figure by the product of the coordinates corresponding to those axes.

product research [SCI TECH] Applied research with a new or improved product in view.

product topology [MATH] A topology on a product of topological spaces whose open sets are constructed from cartesian products of open sets from the individual spaces.

proenzyme *See* zymogen.

proerythroblast of Ferrata *See* pronormoblast.

proestrus [PHYSIO] The beginning of the follicular phase of estrus.

profibrinolysin *See* plasminogen.

profile [GEOL] **1.** The outline formed by the intersection of the plane of a vertical section and the ground surface. Also known as topographic profile. **2.** Data recorded by a single line of receivers from one shot point in seismic prospecting. [GEOPHYS] A graphic representation of the variation of one property, such as gravity, usually as ordinate, with respect to another property, usually linear, such as distance. [HYD] A vertical section of a potentiometric surface, such as a water table. [PETR] In structural petrology, a cross section of a homoaxial structure.

profile chart [ELECTROMAG] A vertical cross-section drawing of a microwave path between two stations, indicating terrain, obstructions, and antenna height requirements.

profiled keyway [DES ENG] A keyway for a straight key formed by an end-milling cutter. Also known as end-milled keyway.

profile drag [FL MECH] That part of the airfoil drag that results from the skin friction and the shape of the airfoil as indicated by the airfoil profile.

profile of equilibrium [GEOL] **1.** The slope of the floor of a sea, ocean, or lake, taken in a vertical plane, when deposition of sediment is balanced by erosion. **2.** The longitudinal profile of a graded stream. Also known as equilibrium profile; graded profile.

profile thickness [AERO ENG] The maximum distance between the upper and lower contours of an airfoil, measured perpendicular to the mean line of the profile.

profiling machine [MECH ENG] A machine used for milling irregular profiles; the cutting tool is guided by the contour of a model.

profilograph [ENG] An instrument for measuring and recording roughness of the surface over which it travels.

profit in sight [MIN ENG] Probable gross profit from a mine's ore reserves, as distinct from the ground that is still to be blocked out.

profit sharing [IND ENG] Sharing of company profits with the employees.

profunda [ANAT] Deep-seated; applied to certain arteries.

profundal zone [ECOL] The region occurring below the limnetic zone and extending to the bottom in lakes deep enough to develop temperature stratification.

Proganosauria [PALEON] The equivalent name for Mesosauria.

progeny [BIOL] Offspring; descendants.

progestational hormone [BIOCHEM] **1.** The natural hormone progesterone, which induces progestational changes of the uterine mucosa. **2.** Any derivative or modification or progesterone having similar actions.

progesterone [BIOCHEM] $C_{21}H_{30}O_2$ A steroid hormone produced in the corpus luteum, placenta, testes, and adrenals; plays an important physiological role in the luteal phase of the menstrual cycle and in the maintenance of pregnancy; it is an intermediate in the biosynthesis of androgens, estrogens, and the corticoids.

proglacial [GEOL] Of streams, deposits, and other features, being immediately in front of or just beyond the outer limits of a glacier or ice sheet, and formed by or derived from glacier ice.

proglottid [INV ZOO] One of the segments of a tapeworm.

prognathic [ANTHRO] A condition of the upper jaw in which it projects anteriorly with respect to the profile of the facial skeleton, when the skull is oriented on the Frankfort horizontal plane; having a gnathic index of 103.0 or more.

prognosis [MED] A prediction as to the course and outcome of a disease, injury, or developmental abnormality.

prognostic chart [METEOROL] A chart showing, principally, the expected pressure pattern (or height pattern) of a given synoptic chart at a specified future time; usually, positions of fronts are also included, and the forecast values of other meteorological elements may be superimposed.

prognostic equation [METEOROL] Any equation governing a system which contains a time derivative of a quantity and therefore can be used to determine the value of that quantity at a later time when the other terms in the equation are known (for example, the vorticity equation).

progradation [GEOL] Seaward buildup of a beach, delta, or fan by nearshore deposition of sediments transported by a river, by accumulation of material thrown up by waves, or by material moved by longshore drifting.

prograding shoreline [GEOL] A shoreline that is being built seaward by accumulation or deposition.

program [ADP] A detailed and explicit set of directions for accomplishing some purpose, the set being expressed in some language suitable for input to a computer, or in machine language. [AERO ENG] In missile guidance, the planned flight path events to be followed by a missile in flight, including all the critical functions, preset in a program device, which control the behavior of the missile. [COMMUN] A sequence of audio signals alone, or audio and video signals, transmitted for entertainment or information.

program check [ADP] A built-in check system in a program to determine that the program is running correctly.

program counter *See* instruction counter.

program device [CONT SYS] In missile guidance, tha automatic device used to control time and sequence of events of a program.

program element [ADP] Part of a central computer system that carries out the instruction sequence scheduled by the programmer.

program evaluation and review technique *See* PERT.

program failure alarm [COMMUN] Signal-operated radio or television vacuum-tube relay that gives a visual and aural alarm when the program fails on the line being monitored; a time delay is provided to prevent the relay from operating and giving a false alarm during station identification periods or other short periods of silence in program continuity.

program generator [ADP] A program that permits a computer to write other programs automatically.

program language [ADP] A language which is used by programmers to write computer routines.

program level [ENG ACOUS] The level of the program signal in an audio system, expressed in volume units.

program library [ADP] An organized set of computer routines and programs.

programmed check [ADP] **1.** An error-detecting operation programmed by instructions rather than built into the hardware. **2.** A computer check in which a sample problem with known answer, selected for having a program similar to that of the next problem to be run, is put through the computer.

programmed marginal check [ADP] Computer program that varies its own voltage to check some piece of electronic computer equipment during a preventive maintenance check.

programmed operators [ADP] Computer instructions which

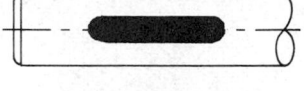

PROFILED KEYWAY

End-milled or profiled type of keyway. *(From P. H. Black and O. E. Adams, Jr., Machine Design, 3d ed., 1968)*

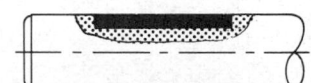

PROGESTERONE

Structural formula of progesterone.

enable subroutines to be accessed with a single programmed instruction.

programmed turn [AERO ENG] The automatically controlled turn of a ballistic missile into the curved path that will lead to the correct velocity and vector for the final portion of the trajectory.

programmer [ADP] A person who prepares sequences of instructions for a computer, without necessarily converting them into the detailed codes. [CONT SYS] A device used to control the motion of a missile in accordance with a predetermined plan.

programming [ADP] Preparing a detailed sequence of operating instructions for a particular problem to be run on a digital computer. Also known as computer programming.

program monitor [COMMUN] A monitor used to observe the quality of a radio or television broadcast.

program parameter [ADP] In computers, an adjustable parameter in a subroutine which can be given a different value each time the subroutine is used.

program register [ADP] The register in the control unit of a digital computer that stores the current instruction of the program and controls the operation of the computer during the execution of that instruction. Also known as computer control register.

program-sensitive fault [ADP] A hardware malfunction that appears only in response to a particular sequence (or kind of sequence) of program instructions.

program status word [ADP] An internal register to the central processing unit denoting the state of the computer at a moment in time.

program step [ADP] In computers, some part of a program, usually one instruction.

program stop [ADP] A stop instruction built into the program that will automatically stop the machine under certain conditions, or upon reaching the end of the processing or completing the solution of a problem.

program storage [ADP] Portion of the internal storage reserved for the storage of programs, routines, and subroutines; in many systems, protection devices are used to prevent inadvertent alteration of the contents of the program storage; contrasted with temporary storage.

program tape [ADP] Tape containing the sequence of computer instructions for a given problem.

program test [ADP] A system of checking before running any problem in which a sample problem of the same type with a known answer is run.

program testing time [ADP] The machine time expended for program testing, debugging, and volume and compatibility testing.

program time [ADP] The phase of computer operation when an instruction is being interpreted so that it can be carried out.

progress chart [IND ENG] A graphical representation of the degree of completion of work in progress.

progression [MATH] A sequence or series of mathematical objects or quantities, each entry determined from its predecessors by some algorithm.

progressive aging [MET] Aging of metals achieved by increasing the temperature in stages or by continuous elevation of the temperature.

progressive bonding [ENG] A method of curing a resin adhesive wherein heat and pressure are applied in successive steps.

progressive die [MET] A die in which two or more operations are performed sequentially at different positions.

progressive forming [MET] Sequential forming at consecutive stations either with a single die or with separate dies.

progressive granulation [MATER] Propellant granulation in which the surface area of a grain increases during burning.

progressive lateral sclerosis [MED] Amyotrophic lateral sclerosis with primary involvement of the pyramidal tracts.

progressive metamorphism [GEOL] Systematic change in metamorphic grade from lower to higher in any metamorphic terrain.

progressive muscular dystrophy [MED] Chronic progressive dystrophy of the skeletal muscles.

progressive powder [MATER] A slow-burning explosive.

progressive proofs [GRAPHICS] The proofs from color plates showing each color alone, and also in combination with each succeeding color in printing rotation; the order is yellow, magenta, cyan, and black.

progressive sand wave [GEOL] A sand wave characterized by downcurrent migration.

progressive scanning [COMMUN] Scanning all lines in sequence, without interlace, so all picture elements are included during one vertical sweep of the scanning beam. Also known as sequential scanning.

progressive wave [METEOROL] A wave or wavelike disturbance which moves relative to the earth's surface. [PHYS] A wave which transfers energy from one part of a medium to another, in contrast to a standing wave. Also known as free-traveling wave.

progressive-wave antenna *See* traveling-wave antenna.

Progymnospermopsida [PALEON] A class of plants intermediate between ferns and gymnosperms; comprises the Denovian genus *Archaeopteris*.

prohaptor [INV ZOO] The anterior attachment organ of a typical monogenetic trematode.

prohibited area [NAV] 1. An airspace of defined dimensions identified by an area on the surface of the earth within which flight is prohibited, except as authorized by governmental authority. 2. An area within which no vessels may be navigated, except as authorized by governmental authority.

Projapygidae [INV ZOO] A family of wingless insects in the order Diplura.

project [ENG] A specifically defined task within a research and development field, which is established to meet a single requirement, either stated or anticipated, for research data, an end item of material, a major component, or a technique.

projected planform [AERO ENG] The contour of the planform as viewed from above.

project engineering [ENG] 1. The engineering design and supervision (coordination) aspects of building a manufacturing facility. 2. The engineering aspects of a specific project, such as development of a product or solution to a problem.

projectile [ORD] A special device for use at proving grounds, usually consisting of a solid, blunt-nosed projectile of low cost; it duplicates the standard projectile in weight, and location and type of rotating band so that it may be used in developing propelling charges as well as for proof firing of guns.

projectile ogive [ORD] A hollow, conical, metallic item, designed to enclose the forward portion of a projectile to reduce air resistance during flight.

projection [MAP] A system for presenting on a plane surface the spherical surface of the earth or the celestial sphere; some of these systems are conic, cylindrical, gnomonic, Mercator, orthographic, and stereographic. [MATH] 1. The continuous map for a fiber bundle. 2. Geometrically, the image of a geometric object or vector superimposed on some other. 3. A linear map P from a linear space to itself such that $P \circ P$ is equal to P. [PSYCH] Ascribing one's motives to someone else to disguise a source of conflict in oneself.

projection area [ANAT] An area of the cortex connected with lower centers of the brain by projection fibers.

projection cathode-ray tube [ELECTR] A television cathode-ray tube designed to produce an intensely bright but relatively small image that can be projected onto a large viewing screen by an optical system.

projection chamber *See* projection spark chamber.

projection fibers [ANAT] Fibers joining the cerebral cortex to lower centers of the brain, and vice versa.

projection microradiography [PHYS] Microradiography in which an electron beam, focused into an extremely fine pencil, generates a point source of x-rays, and enlargement is achieved by placing the sample very near this source, and several centimeters from the recording material. Abbreviated PMR. Also known as shadow microscopy; x-ray projection microscopy.

projection microscope [PHYS] An x-ray microscope which magnifies by image projection, either in contact microradiography or in projection microradiography.

projection net *See* net.

projection optics *See* Schmidt system.

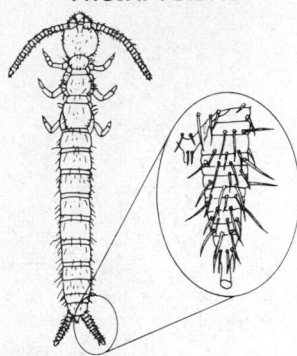

PROJAPYGIDAE

Anajapyx vesiculosus Silvestri, showing the cerci which contain a silk gland.

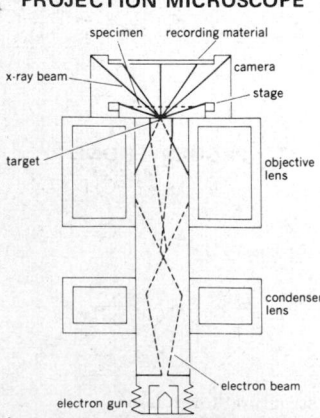

PROJECTION MICROSCOPE

specimen — recording material — camera — x-ray beam — stage — target — objective lens — condenser lens — electron gun — electron beam

Projection microscope used in projection microradiography, showing components. (*After V. E. Cosslett, A. Engström and H. H. Pattee, eds., X-Ray Microscopy and Microradiography, Academic Press, 1957*)

projection plan position indicator [ELECTR] Unit in which the image of a 4-inch dark-trace cathode-ray tube is projected on a 24-inch horizontal plotting surface; the echoes appear as magenta-colored arcs on white background.

projection printer [OPTICS] An optical, image-enlarging device, used in enlarging photographs.

projection printing [GRAPHICS] The production of photographic prints with the use of an enlarger.

projection spark chamber [NUCLEO] A spark chamber in which the track of the particle is perpendicular, or nearly so, to the electric field, so that each electron of the track produces a streamer across the gap; the resulting curtain contains information only as to the projection of the track perpendicular to the electric field. Also known as projection chamber.

projection thermography [ENG] A method of measuring surface temperature in which thermal radiation from a surface is imaged by an optical system on a thin screen of luminescent material, and the pattern formed corresponds to the heat radiation of the surface.

projection welding [MET] Resistance welding in which the welds are localized at projections, intersections, and overlaps on the parts.

projective geometry [MATH] The study of those properties of geometric objects which are invariant under projection.

projective group [MATH] A group of transformations arising in the general theory of projective geometry.

projective line [MATH] The line obtained from the stereographic projection of the circle.

projective plane [MATH] The topological space obtained from the two-dimensional sphere by identifying antipodal points; the space of all lines through the origin in euclidean space.

projective point [MATH] The point from which a projection by rays is performed, as in stereographic projection.

projective space [MATH] The topological space obtained from the n-dimensional sphere under identification of antipodal points.

projective technique [PSYCH] A procedure used to identify and evaluate an individual's characteristic modes of thought and behavior, his personality traits, attitudes, and motivation, by means of an objective test.

projective test [PSYCH] Observation of a subject's responses to various test materials presented in a relatively unstructured, yet standard situation.

projector [ENG ACOUS] 1. A horn designed to project sound chiefly in one direction from a loudspeaker. 2. *See* underwater sound projector. [OPTICS] *See* optical projection system. [ORD] 1. Any apparatus for launching a projectile, such as a gun or rocket launcher. 2. Smooth-bore-type barrel or other unrifled weapon from which pyrotechnic signals, grenades, and certain mortar projectiles are fired. 3. A rack for launching target rockets. 4. Special type of gun for projecting antisubmarine projectiles.

projector compass [NAV] A magnetic compass in which the lubber's line and compass card, or a portion thereof, are viewed as an image projected through a system of lenses upon a screen adjacent to the helmsman's position.

prokaryote [CYTOL] 1. A primitive nucleus, where the deoxyribonucleic acid–containing region lacks a limiting membrane. 2. Any cell containing such a nucleus, such as the bacteria and the blue-green algae.

Prolacertiformes [PALEON] A suborder of extinct terrestrial reptiles in the order Eosuchia distinguished by reduction of the lower temporal arcade.

prolactin [BIOCHEM] A protein hormone produced by the adenohypophysis; stimulates lactation and promotes functional activity of the corpus luteum. Also known as lactogenic hormone; luteotropic hormone; mammary-stimulating hormone; mammogen; mammogenic hormone; mammotropin.

prolamin [BIOCHEM] Any of the simple proteins, such as zein, found in plants; soluble in strong alcohol, insoluble in absolute alcohol and water.

prolapse [MED] The falling or sinking down of a part or organ.

prolate ellipsoid *See* prolate spheroid.

prolate spheroid [MATH] The ellipsoid or surface obtained by revolving an ellipse about one of its axes so that the equatorial circle has a diameter less than the length of the axis of revolution. Also known as prolate ellipsoid.

proliferative arthritis *See* rheumatoid arthritis.

proline [BIOCHEM] $C_5H_9O_2$ A heterocyclic amino acid occurring in essentially all proteins, and as a major constituent in collagen proteins.

prolinemia [MED] A rare hereditary disease caused by absence of the degradative enzymes that convert proline to glutamic acid, and characterized by a high content of proline in blood and urine with consequent mental retardation and renal malfunction.

prolonge [ORD] Rope, with a hook or loop at one end, with which soldiers can move a vehicle or gun carriage into position.

promenade deck [NAV ARCH] An upper deck, or part of a deck, of a passenger ship where passengers walk about. Also known as hurricane deck.

prometaphase [CYTOL] A stage between prophase and metaphase in mitosis in which the nuclear membrane disappears and the spindle forms.

promethium [CHEM] A chemical element, symbol Pm, atomic number 61; atomic weight of the most abundant isotope is 147; a member of the rare-earth group of metals.

promethium-147 [NUC PHYS] Artificially produced rare-earth element with atomic number 61 and mass 147; produced during fission of U^{235}. Also known as florentium; illinium.

prominence [ASTROPHYS] A volume of luminous, predominantly hydrogen gas that appears on the sun above the chromosphere; occurs only in the region of horizontal magnetic fields because these fields support the prominences against solar gravity.

promoter [CHEM] A chemical which itself is a feeble catalyst, but greatly increases the activity of a given catalyst.

prompt critical [NUCLEO] Capable of sustaining a chain reaction without the aid of delayed neutrons.

prompt neutron [NUC PHYS] A neutron released coincident with the fission process, as opposed to neutrons subsequently released.

prompt radiation [NUC PHYS] Radiation emitted within a time too short for measurement, including γ-rays, characteristic x-rays, conversion and Auger electrons, prompt neutrons, and annihilation radiation.

promyelocyte [HISTOL] The earliest myelocyte stage derived from the myeloblast.

pronate [ANAT] 1. To turn the forearm so that the palm of the hand is down or toward the back. 2. To turn the sole of the foot outward with the lateral margin of the foot elevated; to evert.

pronator [PHYSIO] A muscle which pronates, as the muscles of the forearm attached to the ulna and radius.

pronephros [EMBRYO] One of the anterior pair of renal organs in higher vertebrate embryos; the pair initiates formation of the archinephric duct.

prong *See* pin.

pronghorn [VERT ZOO] *Antilocapra americana.* An antelope-like artiodactyl composing the family Antilocapridae; the only hollow-horned ungulate with branched horns present in both sexes.

pronormoblast [HISTOL] A nucleated erythrocyte precursor with scanty basophilic cytoplasm without hemoglobin. Also known as lymphoid hemoblast of Pappenheim; macroblast of Naegeli; megaloblast of Sabin; proerythroblast of Ferrata; prorubricyte; rubriblast; rubricyte.

pronucleus [CYTOL] One of the two nuclear bodies of a newly fertilized ovum, the male pronucleus and the female pronucleus, the fusion of which results in the formation of the germinal (cleavage) nucleus.

prony brake [MECH ENG] An absorption dynamometer that applies a friction load to the output shaft by means of wood blocks, a flexible band, or other friction surface.

proof [FOOD ENG] The strength of the ethyl alcohol in distilled spirits; in the United States, each degree of proof is equal to ½% of alcohol by volume. [GRAPH] The inked impression of composed type or a plate; used for inspection purposes or for pasting up with other artwork. [MATH] A deductive demonstration of a mathematical statement.

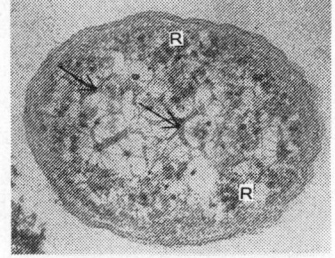

PROKARYOTE

Electron micrograph of a bacterial cell, showing a typical prokaryote nucleus. No membranes separate the DNA fibers (indicated by arrows) from the rest of the cell. Clusters of ribosomes (R) are visible at the edge of the cell. *(Photograph by D. Bastia)*

PROLACERTIFORMES

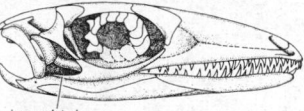

incomplete lower temporal arcade

Lateral view of *Prolacerta* skull, Lower Triassic. *(From A. S. Romer, Vertebrate Paleontology, 3d ed., University of Chicago Press, 1966)*

PROMETHIUM

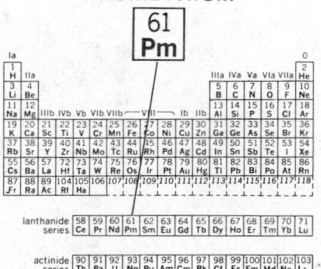

Periodic table of the chemical elements showing the position of promethium.

PRONY BRAKE

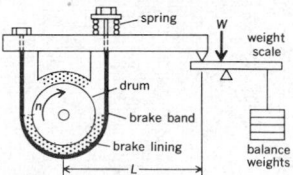

Diagram of a prony brake. Brake horsepower = $2\pi n L(W-W_0)/33,000$, where n = shaft speed in revolutions per minute, L = distance shown on drawing, W = scale weight with brake operating, and W_0 = scale weight with brake free.

proof firing [ORD] The firing of certain rounds for the purpose of testing the serviceability of a weapon or its mount.

proof load [ENG] A predetermined test load, greater than the service load, to which a specimen is subjected before acceptance for use.

proof mark [ORD] Distinguishing mark on a weapon to indicate inspection and proof firing.

proof resilience [MECH] The tensile strength necessary to stretch an elastomer from zero elongation to the breaking point, expressed in foot-pounds per cubic inch of original dimension.

proof stress [MECH] **1.** The stress that causes a specified amount of permanent deformation in a material. **2.** A specified stress to be applied to a member or structure in order to assess its ability to support service loads.

prop [MIN ENG] Underground supporting post set across the lode, seam, bed, or other opening.

propadiene *See* allene.

propagated blast [ENG] A blast of a number of unprimed charges of explosives plus one hole primed, generally for the purpose of ditching, where each charge is detonated by the explosion of the adjacent one, the shock being transmitted through the wet soil.

propagated error [ADP] An error which takes place in one operation and spreads through succeeding operations.

propagation *See* wave motion.

propagation anomaly [PHYS] Change in propagation characteristics due to a resonance in the medium of propagation.

propagation constant [ELECTROMAG] A rating for a line or medium along or through which a wave of a given frequency is being transmitted; it is a complex quantity; the real part is the attenuation constant in nepers per unit length, and the imaginary part is the phase constant in radians per unit length.

propagation delay [ELECTR] The time required for a signal to pass through a given complete operating circuit; it is generally of the order of nanoseconds, and is of extreme importance in computer circuits.

propagation forecasting [METEOROL] Forecasting in which the known or predicted vertical distribution of the index of refraction over an area is used to forecast the propagation performance of radars or any microwave radio equipment operating in that area.

propagation loss [COMMUN] The attenuation of signals passing between two points on a transmission path.

propagation notice [COMMUN] A forecast of propagation conditions for long-distance radio communications, broadcast at regular intervals over radio stations operated by the National Bureau of Standards.

propagation path [COMMUN] A path between receiver and transmitter including direct tropospheric scatter, ionospheric scatter, E-layer skip, and F_1-layer and F_2-layer skip and echo.

propagation rate [CHEM] The speed at which a flame front progresses through the body of a flammable fuel-oxidizer mixture, such as gas and air.

propagation time delay [COMMUN] The time required for a wave to travel between two points on a transmission path.

propagation velocity [ELECTROMAG] Velocity of electromagnetic wave propagation in the medium under consideration.

propagule [BOT] **1.** A reproductive structure of brown algae. **2.** A propagable shoot.

Propalticidae [INV ZOO] A family of coleopteran insects of the superfamily Cucujoidea found in Old World tropics and Pacific islands.

propane [ORG CHEM] $CH_3CH_2CH_3$. A heavy, colorless, gaseous petroleum hydrocarbon gas of the paraffin series; boils at $-44.5°C$; used as a solvent, refrigerant, and chemical intermediate. Also known as dimethylmethane.

propane bubble chamber [NUCLEO] A bubble chamber whose liquid is propane; used when it is desirable to have a dense, heavy target in experiments such as detection of neutrinos or measurements of the decay products of neutral pions, not usually visible in a hydrogen bubble chamber.

propane deasphalting [CHEM ENG] Petroleum-refinery solvent process using propane to remove and precipitate asphalt from petroleum stocks, such as for lubricating oils.

propane decarbonizing [CHEM ENG] Petroleum-refinery solvent process using propane to recover catalytic-cracking feedstock from heavy-fuel residues; when butane or butane-propane solvent is used, the process is called solvent decarbonizing.

propane dewaxing [CHEM ENG] Petroleum-refinery solvent process using propane to remove waxes from lubricating oils to lower the lube-oil pour point.

1, 2-propanediol *See* propylene glycol.

propane fractionation [CHEM ENG] Continuous, petroleum-refinery solvent process using liquid propane to segregate long-vacuum residue into two or more grades of lube-oil stock (such as heavy neutral stock or bright stock) and asphalt.

propanoic acid *See* propionic acid.

propanol *See* propyl alcohol.

2-propanone *See* acetone.

propargyl alcohol [ORG CHEM] $HCCCH_2OH$ Colorless, water- and alcohol-soluble liquid, boiling at $114°C$; used as a chemical intermediate, stabilizer, and corrosion inhibitor. Also known as 2-propyn-1-ol.

prop-crib timbering [MIN ENG] Shaft timbering with cribs kept apart at the proper distance by means of props.

propellant [MATER] A combustible substance that produces heat and supplies ejection particles as in a rocket engine.

propellant 23 *See* fluoroform.

propellant-actuated device [ENG] A device that employs the energy supplied by the gases produced by burning propellants to accomplish or initiate a mechanical action other than propelling a projectile.

propellant-actuated exactor [ORD] A propellant-actuated device intended to release the safety mechanism of another propellant-actuated device.

propellant additive [MATER] Any material added to the basic formulation of a solid propellant to accomplish some special purpose such as to increase or decrease the rate of burning.

propellant binder [MATER] An elastomeric fuel used in a composite propellant so that the propellant may be cast directly into the combustion chamber, where the binder cures to a rubber and the propellant grain is then supported by adhesion to the walls; used in missiles.

propellant fouling [ORD] Bits of unburned or partially burned propellant left in the bore after firing.

propellant grain [MATER] An elongated molding or extrusion, often of intricate shape, of solid propellant for a rocket, regardless of size.

propellant injector [AERO ENG] A device for injecting propellants, which include fuel and oxidizer, into the combustion chamber of a rocket engine.

propellant mass fraction *See* propellant mass ratio.

propellant mass ratio [AERO ENG] Of a rocket, the ratio of the effective propellant mass to the initial vehicle mass. Also known as propellant mass fraction.

propellant powder [MATER] A low explosive of fine granulation which, through burning, produces gases at a controlled rate to provide the energy for propelling a projectile.

propellant weight fraction [AERO ENG] The weight of the solid propellant charge divided by weight of the complete solid propellant propulsion unit.

propeller [MECH ENG] A bladed device that rotates on a shaft to produce a useful thrust in the direction of the shaft axis.

propeller anemometer [ENG] A rotation anemometer which is encased in a strong glass outer shell that protects it against hydrostatic pressure.

propeller blade [DES ENG] One of two or more plates radiating out from the hub of a propeller and normally twisted to form part of a helical surface.

propeller boss [DES ENG] The central portion of the screw propeller which carries the blades, and forms the medium of attachment to the propeller shaft. Also known as propeller hub.

propeller cavitation [FL MECH] Formation of vapor-filled and air-filled bubbles or cavities in water at or on the surface of a rotating propeller, occurring when the pressure falls below the vapor pressure of water.

propeller efficiency [MECH ENG] The ratio of the thrust horsepower delivered by the propeller to the shaft horsepower as delivered by the engine to the propeller.

propeller fan [MECH ENG] An axial-flow blower, with or without a casing, using a propeller-type rotor to accelerate the fluid.

propeller horsepower [NAV ARCH] The horsepower delivered to the propeller of a ship.

propeller hub *See* propeller boss.

propeller meter [ENG] A quantity meter in which the flowing stream rotates a propellerlike device and revolutions are counted.

propeller post [NAV ARCH] The forward post of a stern frame on vessels having a center-line propeller.

propeller pump *See* axial-flow pump.

propeller racing [NAV ARCH] The sudden increase in the number of revolutions made by the engine when the propeller blades are lifted clear of the water, or nearly so, due to the roll or pitch of the ship.

propeller shaft [MECH ENG] A shaft, carrying a screw propeller at its end, that transmits power from an engine to the propeller.

propeller slip angle [MECH ENG] The angle between the plane of the blade face and its direction of motion.

propeller strut *See* strut.

propeller thrust [NAV ARCH] The effort delivered by a propeller in pushing a vessel ahead.

propeller tip clearance [NAV ARCH] Generally, the shortest distance between the skin of a vessel and the circle swept by the propeller tips.

propeller tip speed [MECH ENG] The speed in feet per minute swept by the propeller tips.

propeller turbine [MECH ENG] A form of reactive-type hydraulic turbine using an axial-flow propeller rotor.

propeller windmill [MECH ENG] A windmill that extracts wind power from horizontal air movements to rotate the blades of a propeller.

propenal *See* acrolein.

propene *See* propylene.

propenol *See* allylalcohol.

***para*-propenylanisole** *See* anethole.

α-propenyldichlorohydrin *See* α-dichlorohydrin.

properdin [IMMUNOL] A macroglobin of normal plasma capable of killing various bacteria and viruses in the presence of complement and magnesium ions.

proper fraction [MATH] A fraction a/b where the absolute value of a is less than the absolute value of b.

proper Lorentz transformation [REL] A Lorentz transformation which can be represented by a matrix whose determinant is $+1$.

proper motion [ASTRON] That component of the space motion of a celestial body perpendicular to the line of sight, resulting in the change of a star's apparent position relative to that of other stars; expressed in angular units.

proper subset [MATH] A set X is a proper subset of a set Y if there is an element of Y which is not in X while X is a subset of Y.

property detector [ADP] In character recognition, that electronic component of a character reader which processes the normalized signal for the purpose of extracting from it a set of characteristic properties on the basis of which the character can be subsequently identified.

proper value [MATH] A proper value of a linear transformation of a matrix is any eigenvalue.

prop-free [MIN ENG] A face with no posts between the coal and the conveyor used to remove it in longwall mining of a coal seam.

prop-free front [MIN ENG] In coal mining, longwall working in which support to the roof is given by roof beams cantilevered from behind the working face.

prophage [VIROL] Integrated unit formed by union of the provirus into the bacterial genome.

prophase [CYTOL] The initial stage of mitotic or meiotic cell division in which chromosomes are condensed from the nuclear material and split logitudinally to form pairs.

prophylaxis [MED] An artificial substitute or replacement for a missing part of the body.

propine *See* allylene.

β-propiolactone [ORG CHEM] $C_3H_4O_2$ Water-soluble liquid that decomposes rapidly at boiling point (155°C); miscible

with ethanol, acetone, chloroform, and ether; reacts with alcohol; used as a chemical intermediate.

propionaldehyde [ORG CHEM] C_2H_5CHO Flammable, water-soluble, water-white liquid, with suffocating aroma; boils at 48.8°C; used to manufacture acetals, plastics, and rubber chemicals, and as a disinfectant and preservative.

propionate [ORG CHEM] A salt of propionic acid, CH_3CH_2COOH; an example is sodium propionate, CH_3CH_2COONa.

propione *See* pentanone.

Propionibacteriaceae [MICROBIO] A family of nonmotile, nonsporulating, gram-positive anaerobic rod-shaped bacteria in the order Eubacteriales that ferment glucose and other substrates; propionic acid is a characteristic end product of the fermentation.

Propionibacterium [MICROBIO] A genus of mostly nonpathogenic bacteria in the family Propionibacteriaceae that form propionic acid as a major product of metabolism.

propionic acid [ORG CHEM] CH_3CH_2COOH Water- and alcohol-soluble, clear, colorless liquid with pungent aroma; boils at 140.7°C; used to manufacture various propionates, in nickel-electroplating solutions, for perfume esters and artificial flavors, for pharmaceuticals, and as a cellulosics solvent. Also known as methylacetic acid; propanoic acid.

propionic ether *See* ethyl propionate.

propionitrile *See* ethyl cyanide.

proplastid [BOT] Precursor body of a cell plastid.

proplatinum [MET] A nickel-silver-bismuth alloy used as a substitute for platinum.

propleuron [INV ZOO] A pleuron of the prothorax in insects.

propodite [INV ZOO] The sixth leg joint of certain crustaceans. Also known as propodus.

propodus *See* propodite.

proportion [MATH] The proportion of two quantities is their ratio.

proportional band [CONT SYS] The range of values of the controlled variable that will cause a controller to operate over its full range.

proportional control [CONT SYS] Control in which the amount of corrective action is proportional to the amount of error; used, for example, in chemical engineering to control pressure, flow rate, or temperature in a process system.

proportional controller [CONT SYS] A controller whose output is proportional to the error signal.

proportional counter [NUCLEO] A radiation counter consisting of a proportional counter tube and its associated circuits; resembles a Geiger-Müller counter, but with a different counting gas (argon methane) and a lower voltage on the tube; used to measure α, β, and x-rays; has low sensitivity for γ-radiation.

proportional counter tube [NUCLEO] A radiation-counter tube operated at voltages high enough to produce ionization by collision and adjusted so the total ionization per count is proportional to the ionization produced by the initial ionizing event.

proportional dividers [DES ENG] Dividers with two legs, pointed at both ends, and an adjustable pivot; distances measured by the points at one end can be marked off in proportion by the points at the other end.

proportional elastic limit [MECH] The greatest stress intensity for which stress is still proportional to strain.

proportional ionization chamber [ELECTR] An ionization chamber in which the initial ionization current is amplified by electron multiplication in a region of high electric-field strength, as in a proportional counter; used for measuring ionization currents or charges over a period of time, rather than for counting.

proportional limit [MECH] The greatest stress a material can sustain without departure from linear proportionality of stress and strain.

proportional navigation [NAV] Homing guidance of a guided missile toward a moving target in which the missile turning rate is proportional to the rate of change of line of sight between missile and target.

proportional parts [MATH] Numbers in the same proportion as a set of given numbers; such numbers are used in an auxiliary interpolation table based on the assumption that the

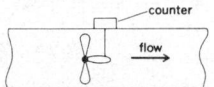
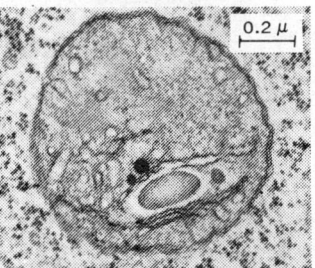

tabulated quantity and entering arguments differ in the same proportion.

proportional-plus-derivative control [CONT SYS] Control in which the control signal is a linear combination of the error signal and its derivative.

proportional-plus-integral control [CONT SYS] Control in which the control signal is a linear combination of the error signal and its integral.

proportional-plus-integral-plus-derivative control [CONT SYS] Control in which the control signal is a linear combination of the error signal, its integral, and its derivative.

proportional reducer [GRAPHICS] A solution that is used to lower density in a negative or positive image in proportion to the silver present, with a certain amount of loss in shadow detail.

proportional spacing [GRAPHICS] Spacing in which each character occupies space according to its width.

proportional-speed control *See* floating control.

proportioning pump *See* metering pump.

proportioning reactor [ELECTROMAG] A saturable-core reactor used for regulation and control; increasing the input control current from zero to rated value makes output current increase in proportion from cutoff up to full load value.

proposition [MATH] Any problem or theorem.

propositional calculus [MATH] The mathematical study of logical connectives between propositions and deductive inference.

propped cantilever [CIV ENG] A beam having one built-in support and one simple support.

proprioception [PHYSIO] The reception of internal stimuli.

proprioceptor [PHYSIO] A sense receptor that signals spatial position and movements of the body and its members, as well as muscular tension.

prop root [BOT] A root that serves to support or brace the plant. Also known as brace root.

propterygium [VERT ZOO] The anterior of the three principal basal cartilages forming a support of one of the paired fins of sharks, rays, and certain other fishes.

proptosis [MED] A falling downward or forward, especially of an eyeball.

propulsion [MECH] The process of causing a body to move by exerting a force against it.

propulsion system [MECH ENG] For a vehicle moving in a fluid medium, such as an airplane or ship, a system that produces a required change in momentum in the vehicle by changing the velocity of the air or water passing through the propulsive device or engine; in the case of a rocket-propelled vehicle operating without a fluid medium, the required momentum change is produced by using up some of the propulsive device's own mass, called the propellant.

propulsive coefficient [NAV ARCH] The ratio between the effective horsepower (ehp) and the shaft horsepower (shp) at any given speed.

propyl- [ORG CHEM] The $CH_3CH_2CH_2-$ radical, derived from propane; found, for example, in 1-propanol.

n-propyl acetate [ORG CHEM] $C_3H_7OOCCH_3$ Colorless liquid with pleasant aroma; miscible with alcohols, ketones, esters, and hydrocarbons; boils at 96-102°C; used for flavors and perfumes, in organic synthesis, and as a solvent.

propylacetone *See* methyl butyl ketone.

propyl alcohol [ORG CHEM] $CH_3CH_2CH_2OH$ A colorless liquid made by oxidation of aliphatic hydrocarbons; boils at 97°C; used as a solvent and chemical intermediate. Also known as ethyl carbinol; propanol.

n-propylamine [ORG CHEM] $C_3H_7NH_2$ Colorless, flammable liquid, boiling at 46-51°C; used as a sedative.

propyl benzene [ORG CHEM] $C_6H_5C_3H_7$ Water-insoluble, colorless liquid, boiling at 158°C. Also known as phenylpropane.

propylene [ORG CHEM] $CH_3CH=CH_2$ Colorless unsaturated hydrocarbon gas, with boiling point of −47°C; used to manufacture plastics and as a chemical intermediate. Also known as methyl ethylene; propene.

propylene aldehyde *See* crotonaldehyde.

propylene carbonate [ORG CHEM] $C_3H_6CO_3$ Odorless, colorless liquid, boiling at 242°C; miscible with acetone, benzene, and ether; used as a solvent, extractant, plasticizer, and chemical intermediate.

propylene chloride *See* propylene dichloride.

propylene dichloride [ORG CHEM] $CH_3CHClCH_2Cl$ Water-insoluble, colorless, moderately flammable liquid, with chloroform aroma; boils at 96.3°C; miscible with most common solvents; used as a solvent, dry-cleaning fluid, metal degreaser, and fumigant. Also known as 1,2-dichloropropane; propylene chloride.

propylene glycol [ORG CHEM] $CH_3CHOHCH_2OH$ A viscous, colorless liquid, miscible with water, alcohol, and many solvents; boils at 188°C; used as a chemical intermediate, antifreeze, solvent, lubricant, plasticizer, and bactericide. Also known as 1,2-dihydroxypropane; methyl ethylene glycol; methyl glycol; 1,2-propanediol.

propylene oxide [ORG CHEM] C_3H_6O . Colorless, flammable liquid, with etherlike aroma; soluble in water, alcohol, and ether; boils at 33.9°C; used as a solvent and fumigant, in lacquers, coatings, and plastics, and as a chemical intermediate.

propylene tetramer *See* dodecane.

propylformic acid *See* butyric acid.

propylite [PETR] A modified andesite, altered by hydrothermal processes, resembling a greenstone and consisting of calcite, epidote, serpentine, quartz, pyrite, and iron ore.

propylization [PETR] A hydrothermal process by which propylite is formed from andesite by the introduction of or replacement by an assemblage of minerals.

2-propyn-1-ol *See* propargyl alcohol.

Prorastominae [PALEON] A subfamily of extinct dugongs (Dugongidae) which occur in the Eocene of Jamaica.

prorennin *See* renninogen.

Prorhynchidae [INV ZOO] A family of turbellarians in the order Alloeocoela.

prorubricyte *See* pronormoblast.

Prosauropoda [PALEON] A division of the extinct reptilian suborder Sauropodomorpha; they possessed blunt teeth, long forelimbs, and extremely large claws on the first finger of the forefoot.

Prosobranchia [INV ZOO] The largest subclass of the Gastropoda; generally, respiration is by means of ctenidia, an operculum is present, there is one pair of tentacles, and the sexes are separate.

prosodus [INV ZOO] A canal leading from an incurrent canal to a flagellated chamber in Porifera.

prosoma [INV ZOO] The anterior part of the body of mollusks and other invertebrates; primitive segmentation is not apparent.

prosopite [MINERAL] $CaAl_2(F,OH)_8$ A colorless mineral composed of basic calcium aluminum fluoride.

Prosopora [INV ZOO] An order of the class Oligochaeta comprising mesonephridiostomal forms in which there are male pores in the segment of the posterior testes.

prosopyle [INV ZOO] The opening into a flagellated chamber from an inhalant canal in sponges.

prospect [MIN ENG] **1.** To search for minerals or oil by looking for surface indications, by drilling boreholes, or both. **2.** A plot of ground believed to be mineralized enough to be of economic importance.

prospector [MIN ENG] A person engaged in exploring for valuable minerals, or in testing supposed discoveries of the same.

Prospector [AERO ENG] A specific unmanned spacecraft designed to make a soft landing on the moon to take measurements, photographs, and soil samples, and then return to earth.

prostaglandin [BIOCHEM] Any of various physiologically active compounds containing 20 carbon atoms and formed from essential fatty acids; found in highest concentrations in normal human semen; activities affect the nervous system, circulation, female reproductive organs, and metabolism.

prostate [ANAT] A glandular organ that surrounds the urethra at the neck of the urinary bladder in the male.

prostatectomy [MED] Surgical removal of all or part of the prostate.

prostatitis [MED] Inflammation of the prostate.

prosthecae [MICROBIO] Appendages that are part of the wall in bacteria in the genus *Caulobacter*.

prosthecate bacteria [MICROBIO] Single-celled microorganisms that differ from typical unicellular bacteria in having one or more appendages which extend from the cell surface; the best-known genus is *Caulobacter*.

prosthesis [MED] An artificial substitute for a missing part of the body, such as a substitute hand, leg, eye, or denture.

prosthodontics [MED] The science and practice of replacement of missing dental and oral structures.

Prostigmata [INV ZOO] The equivalent name for Trombidiformes.

prostomium [INV ZOO] The portion of the head anterior to the mouth in annelids and mollusks.

protactinium [CHEM] A chemical element, symbol Pa, atomic number 91; the third member of the actinide group of elements; all the isotopes are radioactive; the longest-lived isotope is protactinium-231.

protamine [BIOCHEM] Any of the simple proteins that are combined with nucleic acid in the sperm of certain fish, and that upon hydrolysis yield basic amino acids; used in medicine to control hemorrhage, and in the preparation of an insulin form to control diabetes.

protandry [PHYSIO] That condition in which an animal is first a male and then becomes a female.

protanomaly *See* protanopia.

protanopia [MED] Partial color blindness in which there is defective red vision; green sightedness. Also known as protanomaly.

Proteaceae [BOT] A large family of dicotyledonous plants in the order Proteales, notable for often having a large cluster of small or reduced flowers.

Proteales [BOT] An order of dicotyledonous plants in the subclass Rosidae marked by its strongly perigynous flowers, a four-lobed, often corollalike calyx, and reduced or absent true petals.

protease [BIOCHEM] An enzyme that digests proteins.

protected location [ADP] A storage cell arranged so that access to its contents is denied under certain circumstances, in order to prevent programming accidents from destroying essential programs and data.

protected reversing thermometer [ENG] A reversing thermometer encased in a strong glass shell that protects it against hydrostatic pressure.

protected thermometer [ENG] A reversing thermometer which is encased in a strong glass outer shell that protects it against hydrostatic pressure.

protection [NUCLEO] Any provision to reduce exposure of persons to radiation; for example, protective barriers to reduce external radiation or measures to prevent inhalation of radioactive materials.

protection key [ADP] An indicator, usually 1 to 6 bits in length, associated with a program and intended to grant the program access to those sections of memory which the program can use but to deny the program access to all other parts of memory.

protective action guide [NUCLEO] The absorbed dose of ionizing radiation in individuals in the general population which would warrant protective action following a contaminating event, such as a nuclear explosion.

protective clothing [NUCLEO] Special clothing worn by a radiation worker to prevent contamination of the body or personal clothing.

protective coloration [ZOO] A color pattern that blends with the environment and increases the animal's probability of survival.

protective device *See* electric protective device.

protective finish [ENG] A coating applied to equipment to protect it from corrosion and wear; many substances, including metals, glass, and ceramics, are used.

protective fire [ORD] Fire delivered by supporting guns and directed against the enemy to hinder his fire or movement against friendly forces.

protective grounding [ELEC] Grounding of the neutral conductor of a secondary power-distribution system, and of all metal enclosures for conductors, to protect persons from dangerous currents.

protective relay [ELEC] A relay whose principal function is to protect service from interruption or to prevent or limit damage to apparatus.

protective resistance [ELECTR] Resistance used in series with a gas tube or other device to limit current flow to a safe value.

protective survey [NUCLEO] An evaluation of the radiation hazards incidental to the production, use, or existence of radioactive materials or other sources of radiation under a specific set of conditions.

protector [ELEC] Device to protect equipment or personnel from high voltages or currents.

protector block [ELEC] Rectangular piece of carbon with an insulated metal insert, or porcelain with a carbon insert, constituting an element of a protector; it forms a gap which will break down and provide a path to ground for voltages over 350 volts.

protector gap [ELEC] A device designed to limit or equalize voltage in order to protect telephone and telegraph equipment; consists of two carbon blocks with an air gap between them, which are brought into contact when there is a steady-state discharge across the gap. Also known as gap.

protector tube [ELECTR] A glow-discharge cold-cathode tube that becomes conductive at a predetermined voltage, to protect a circuit against overvoltage.

protectoscope [OPTICS] Device in a tank or armored car, similar to the periscope of a submarine; it enables a soldier to see around a shield without exposing himself to enemy gunfire directed at the ports of the vehicle.

Proteeae [MICROBIO] A tribe of the Enterobactereaceae comprising the genus *Proteus*; characteristically, representatives are motile, ferment dextrose with gas production, and produce urease.

Proteida [VERT ZOO] A suborder coextensive with Proteidae in some classification systems.

Proteidae [VERT ZOO] A family of the amphibian suborder Salamandroidea; includes the neotenic, aquatic *Necturus* and *Proteus* species.

protein [BIOCHEM] Any of a class of high-molecular-weight polymer compounds composed of a variety of α-amino acids joined by peptide linkages.

proteinaceous [MATER] 1. Pertaining to any material having a protein base. 2. Pertaining to adhesive materials having a protein base such as animal glue, casein, and soya.

proteinase [BIOCHEM] A type of protease which acts directly on native proteins in the first step of their conversion to simpler substances.

protein-bound iodine [BIOCHEM] Iodine bound to blood protein. Abbreviated PBI.

protein-bound iodine test [PATH] A test of thyroid function that reflects the level of circulating thyroid hormone by determination of the level of protein-bound iodine in the blood. Abbreviated PBI test.

proteinometer *See* hand sugar refractometer.

proteinuria [MED] The presence of protein in the urine.

Proteocephalidae [INV ZOO] A family of tapeworms in the order Proteocephaloidea in which the reproductive organs are within the central mesenchyme of the segment.

Proteocephaloidea [INV ZOO] An order of tapeworms of the subclass Cestoda in which the holdfast organ bears four suckers and, frequently, a suckerlike apical organ.

proteolysin [BIOCHEM] A lysin that produces proteolysis.

proteolysis [BIOCHEM] Fragmentation of a protein molecule by addition of water to the peptide bonds.

proteolytic enzyme [BIOCHEM] Any enzyme that catalyzes the breakdown of protein.

Proteomyxida [INV ZOO] The single order of the Proteomyxidia.

Proteomyxidia [INV ZOO] A subclass of Actinopodea including protozoan organisms which lack protective coverings or skeletal elements and have reticulopodia, or filopodia.

proteoplast [CYTOL] A type of cell plastid containing crystalline, fibrillar, or amorphous masses of protein.

proteose [BIOCHEM] One of a group of derived proteins intermediate between native proteins and peptones; soluble in water, not coagulable by heat, but precipitated by saturation with ammonium or zinc sulfate.

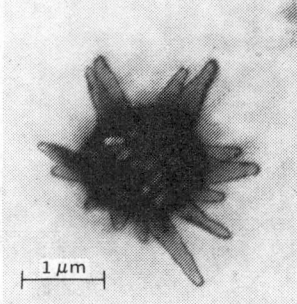

PROSTHECATE BACTERIA

Photograph of a bacterial cell taken with the electron microscope. Note the 14 prosthecae extending from the cell and the transparent gas vesicles inside the cell.

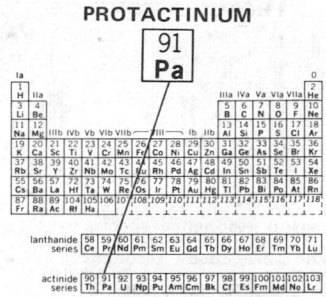

PROTACTINIUM

Periodic table of the chemical elements showing the position of protactinium.

PROTON MAGNETOMETER

The water container and biasing coil assembly of the proton vector magnetometer. Instrument measures frequency of voltage induced in coil by the protons in water. (*U.S. Coast and Geodetic Survey*)

PROTON RESONANCE

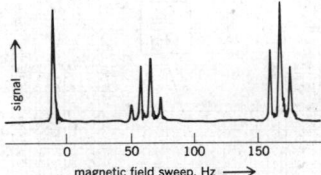

Proton resonance spectra of ethyl alcohol at 40 megahertz. The three main resonance frequencies are due to protons in the OH, CH_2, and CH_3 groups respectively. (*From J. D. Roberts, Nuclear Magnetic Resonance, McGraw-Hill, 1959*)

PROTON SYNCHROTRON

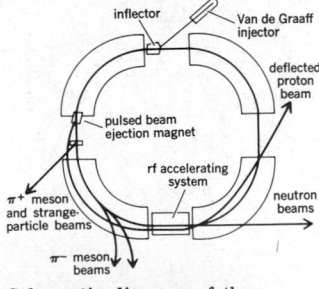

Schematic diagram of the principal components of a proton synchrotron. Note the various possibilities of external neutral beams, charged beams, and extracted primary beams.

Proterostomia [ZOO] That part of the animal kingdom in which cleavage of the egg is of the determinate type; includes all bilateral phyla except Echinodermata, Chaetognatha, Pogonophora, Hemichordata, and Chordata.

Proterosuchia [PALEON] A suborder of moderate-sized thecodont reptiles with lightly built triangular skulls, downturned snouts, and palatal teeth.

Proterotheriidae [PALEON] A group of extinct herbivorous mammals in the order Litopterna which displayed an evolutionary convergence with the horses in their dentition and in reduction of the lateral digits of their feet.

Proterozoic *See* Algonkian.

Proteutheria [VERT ZOO] A group of primatelike insectivores that contains the living tree shrews.

prothallium [BOT] The gametophyte of a pteridophyte in the form of a flat green thallus with thizoids.

prothoracic gland [INV ZOO] One of the paired glands in the prothorax of insects which produce ecdysone.

prothorax [INV ZOO] The first thoracic segment of an insect; bears the first pair of legs.

prothrombin [BIOCHEM] An inactive plasma protein precursor of thrombin. Also known as factor II; thrombinogen.

prothrombin factor *See* vitamin K.

prothrombin time [PATH] A one-stage clotting test based on the time required for clotting to occur after the addition of tissue thromboplastin and calcium to decalcified plasma.

Protista [BIOL] A proposed kingdom to include all unicellular organisms lacking a definite cellular arrangement, such as bacteria, algae, diatoms, and fungi.

protium [NUC PHYS] The lightest hydrogen isotope, having a mass number of 1 and consisting of a single proton and electron. Also known as light hydrogen.

Protoariciinae [INV ZOO] A subfamily of polychaete annelids in the family Orbiniidae.

Protobranchia [INV ZOO] A small and primitive order in the class Bivalvia; the hinge is taxodont in all but one family, there is a central ligament pit, and the anterior and posterior adductor muscles are nearly equal in size.

Protoceratidae [PALEON] An extinct family of pecoran ruminants in the superfamily Traguloidea.

Protochordata [INV ZOO] The equivalent name for Hemichordata.

protoclastic [PETR] Of igneous rocks, characterized by granulation and deformation of the earlier-formed minerals due to differential flow of the magma before solidification.

Protococcaceae [BOT] A monogeneric family of green algae in the suborder Ulotrichineae in which reproduction is entirely vegetative.

Protococcida [INV ZOO] A small order of the protozoan subclass Coccidia; all are invertebrate parasites, and only sexual reproduction is known.

Protocucujidae [INV ZOO] A small family of coleopteran insects in the superfamily Cucujoidea found in Chile and Australia.

protoderm *See* dermatogen.

Protodonata [PALEON] An extinct order of huge dragonflylike insects found in Permian rocks.

Protodrilidae [INV ZOO] A family of annelids belonging to the Archiannelida.

protoenstatite [MINERAL] An artificial, unstable, altered form of $MgSiO_3$ produced by thermal decomposition of talc; convertible to enstatite by grinding or heating to a high temperature.

Protoeumalacostraca [PALEON] The stem group of the crustacean series Eumalacostraca.

protogalaxy [ASTRON] The theoretical precursor of the Galaxy; suggested by James Jeans to be an initial structureless gas cloud, held together by its own gravitation, that broke up into a number of fragments.

protogyny [PHYSIO] A condition in hermaphroditic or dioecious organisms in which the female reproductive structures mature before the male structures.

Protomastigida [INV ZOO] The equivalent name for Kinetoplastida.

Protomonadina [INV ZOO] An order of flagellates, subclass Mastigophora, with one or two flagella, including many species showing protoplasmic collars ringing the base of the flagellum.

Protomonida [INV ZOO] The equivalent name for Protomonadina.

protomylonite [PETR] A mylonitic rock that develops from contact-metamorphosed rock; granulation and flowage are caused by overthrusts following the contact surfaces between the intrusion and the country rock.

Protomyzostomidae [INV ZOO] A family of parasitic polychaetes belonging to the Myzostomaria and known for three species from Japan and the Murman Sea.

proton [PHYS] An elementary particle that is the positively charged constituent of ordinary matter and, together with the neutron, is a building stone of all atomic nuclei; its mass is approximately 938 MeV (million electron volts) and spin ½.

proton accelerator [NUCLEO] A particle accelerator which accelerates protons to high energies, as opposed to one which accelerates heavier ions or electrons.

proton capture [NUC PHYS] A nuclear reaction in which a proton combines with a nucleus.

proton-electron-proton reaction [NUC PHYS] A nuclear reaction in which two protons and an electron react to form a deuteron and a neutrino; it is an important source of detectable neutrinos from the sun. Abbreviated PeP reaction.

protonema [BOT] A green, filamentous structure that originates from an asexual spore of mosses and some liverworts and that gives rise by budding to a mature plant.

protonephridium [INV ZOO] 1. A primitive excretory tube in many invertebrates. 2. The duct of a flame cell.

protonium [ATOM PHYS] A bound state of a proton and an antiproton.

proton magnetometer [ELECTROMAG] A highly sensitive magnetometer which measures the frequency of the proton resonance in ordinary water.

proton microscope [ELECTR] A microscope that is similar to the electron microscope but uses protons instead of electrons as the charged particles.

proton moment [NUC PHYS] The magnetic dipole moment of the proton, a physical constant equal to $(1.41062 \pm 0.00001) \times 10^{-23}$ erg per gauss.

proton-proton chain [NUC PHYS] An energy-releasing nuclear reaction chain which is believed to be of major importance in energy production in hydrogen-rich stars. Also known as deuterium cycle.

proton-recoil counter [NUCLEO] A counter for measuring fast neutrons.

proton resonance [SPECT] A phenomenon in which protons absorb energy from an alternating magnetic field at certain characteristic frequencies when they are also subjected to a static magnetic field; this phenomenon is used in nuclear magnetic resonance quantitative analysis technique.

proton-stability constant [PHYS CHEM] The reciprocal of the dissociation constant of a weak base in solution.

proton synchrotron [NUCLEO] A device for accelerating protons in circular orbits in a time-varying magnetic field, in which the orbit radius is kept constant.

Protophyta [BOT] A division of the plant kingdom, according to one system of classification, set up to include the bacteria, the blue-green algae, and the viruses.

protoplasm [CYTOL] The colloidal complex of protein that composes the living material of a cell.

protoplast [CYTOL] The living portion of a cell considered as a unit; includes the cytoplasm, the nucleus, and the plasma membrane.

protopodite [INV ZOO] The basal segment of a crustacean limb bearing an endopodite or exopodite, or both, at its distal extremity.

Protopteridales [PALEOBOT] An extinct order of ferns, class Polypodiatae.

protoquartzite [PETR] A well-sorted sandstone that is intermediate in composition between subgraywacke and orthoquartzite, consisting of 75–95% quartz and chert, with less than 15% detrital clay matrix and 5–25% unstable materials in which there is a greater abundance of rock fragments than feldspar grains. Also known as quartzose subgraywacke.

Protosireninae [PALEON] An extinct superfamily of sirenian mammals in the family Dugongidae found in the middle Eocene of Egypt.

Protospondyli [VERT ZOO] An equivalent name for Semionotiformes.

protostar [ASTRON] A flattened mass of gas in space that is hypothesized to form into a star.

protostele [BOT] A stele consisting of a solid rod of xylem surrounded by phloem.

Protostomia [INV ZOO] A major division of bilateral animals; includes most worms, arthropods, and mollusks.

Protosuchia [PALEON] A suborder of extinct crocodilians from the Late Triassic and Early Jurassic.

Prototheria [VERT ZOO] A small subclass of Mammalia represented by a single order, the Monotremata.

prototroch [INV ZOO] The band of cilia characteristic of a trochophore larva.

prototrophic [MICROBIO] Pertaining to bacteria with the nutritional properties of the wild type, or the strains found in nature.

Prototrupoidea [INV ZOO] A superfamily of the Hymenoptera.

prototype [ENG] A model suitable for use in complete evaluation of form, design, and performance.

protoxylem [BOT] The part of the primary xylem that differentiates from the procambium and is formed during elongation of an embryonic plant organ.

Protozoa [INV ZOO] A diverse phylum of eukaryotic microorganisms; the structure varies from a simple uninucleate protoplast to colonial forms, the body is either naked or covered by a test, locomotion is by means of pseudopodia or cilia or flagella, there is a tendency toward universal symmetry in floating species and radial symmetry in sessile types, and nutrition may be phagotrophic or autotrophic or saprozoic.

protozoology [INV ZOO] That branch of biology which deals with the Protozoa.

Protrachaeta [INV ZOO] The equivalent name for Onychophora.

protractor [ENG] An instrument used to construct and measure angles formed by lines of a plane; the midpoint of the diameter of the semicircle is marked and serves as the vertex of angles contructed or measured.

protrypsin *See* trypsinogen.

Protura [INV ZOO] An order of primitive wingless insects belonging to the subclass Apterygota; individuals are elongate and eyeless, lack antennae, and are from pale amber to white in color; anamorphosis is characteristic of the group.

proustite [MINERAL] Ag$_3$AsS$_3$ A cochineal-red mineral that crystallizes in the rhombohedral system, consists of silver arsenic sulfide, is isomorphous with pyrargyrite, and occurs massively and in crystals. Also known as light-red silver ore; light-ruby silver.

Prout's hypothesis [PHYS CHEM] The hypothesis that all atoms are built up from hydrogen atoms.

proved ore [MIN ENG] Ore in which there is practically no risk of failure of continuity.

proved reserves [PETRO ENG] Reserves (primary or secondary) that have been proved by production at commercial flow rates.

provenance [GEOL] The location, topography, and composition of the source area for any sedimentary rock. Also known as source area; sourceland.

proventriculus [INV ZOO] **1.** A sac anterior to the gizzard in earthworms. **2.** A dilation of the foregut anterior to the midgut of Mandibulata. [VERT ZOO] The true stomach of a bird, usually separated from the gizzard by a constriction.

proving ring [DES ENG] A ring used for calibrating test machines; the diameter of the ring changes when a force is applied along a diameter.

provirus [VIROL] The phage genome.

provitamin [BIOCHEM] A vitamin precursor; assumes vitamin activity upon activation or chemical change.

provitrain [GEOL] Vitrain in which some plant structure can be discerned by microscope. Also known as telain.

prow [NAV ARCH] The bow of a ship, especially the part above the waterline.

PROWL *See* procedure work log system.

Proxima Centauri [ASTRON] The star that is the sun's nearest neighbor; stellar magnitude is ll, and it is 2° from the bright star α Centauri.

proximal [ANAT] Near the body or the median line of the body. [GEOL] Of a sedimentary deposit, composed of coarse clastics and formed near the source.

proximal convoluted tubule [ANAT] The convoluted portion of the vertebrate nephron lying between Bowman's capsule and the loop of Henle; functions in the resorption of sugar, sodium and chloride ions, and water.

proximate analysis [CHEM ENG] A technique that separates and identifies categories of compounds in a mixture; reported are moisture and ash content, the extracts of the mixture made with alcohol, petroleum ether, water, hydrochloric acid and resins, starches, reducing sugars, proteins, fats, esters, free acids, and so on; this type of analysis of solid fuels allows a prediction to be made as to how the fuel will behave in a furnace.

proximity detector [ENG] A sensing device that produces an electrical signal when approached by an object or when approaching an object.

proximity effect [ELEC] Redistribution of current in a conductor brought about by the presence of another conductor.

proximity fuse [ORD] A fuse that detonates a warhead when the target is within some specified region near the fuse; radio, radar, photoelectric, or other devices may be used as activating elements. Also known as influence fuse; variable-time fuse; vt fuse.

proximoceptor [PHYSIO] An exteroceptor involved in taste or cutaneous sensations.

PRR *See* pulse repetition rate.

prudent limit of endurance [AERO ENG] The time during which an aircraft can remain airborne and still retain a given safety margin of fuel.

pruritus [MED] Localized or generalized itch due to irritation of sensory nerve endings.

prussic acid *See* hydrocyanic acid.

ps *See* picosecond.

psalterium *See* omasum.

psammite *See* arenite.

psammitic [GEOL] Pertaining to, characteristic of, or derived from a grenite (psammite).

Psammodontidae [PALEON] A family of extinct cartilaginous fishes in the order Bradyodonti in which the upper and lower dentitions consisted of a few large quadrilateral plates arranged in two rows meeting in the midline.

Psammodrilidae [INV ZOO] A small family of spioniform worms belonging to the Sedentaria.

psammoma [MED] A tumor, usually a meningioma, which contains psammoma bodies.

psammoma bodies [PATH] Laminated calcific spherical structures found in certain benign and malignant tumors.

psammomatus papilloma *See* serous cystadenoma.

psammophyte [ECOL] Thriving (as a plant) on sandy soil.

psec *See* picosecond.

Pselaphidae [INV ZOO] The ant-loving beetles, a large family of coleopteran insects in the superfamily Staphylinoidea.

Psephenidae [INV ZOO] The water penny beetles, a small family of coleopteran insects in the superfamily Dryopoidea.

psephite [GEOL] A sediment or sedimentary rock composed of fragments that are coarser than sand and which are set in a qualitatively and quantitatively varying matrix; equivalent to a rudite or, generally, a conglomerate.

Pseudaliidae [INV ZOO] A family of roundworms belonging to the Strongyloidea which occur as parasites of whales and porpoises.

pseudoadiabat [METEOROL] On a thermodynamic diagram, a line representing a pseudoadiabatic expansion of an air parcel; in practice, approximate computations are employed, and the resulting lines represent, ambiguously, pseudoadiabats and saturation adiabats.

pseudoadiabatic chart *See* Stuve chart.

pseudoadiabatic expansion [GEOPHYS] A saturation-adiabatic process in which the condensed water substance is removed from the system, and which therefore is best treated by the thermodynamics of open systems; meteorologically, this process corresponds to rising air from which the moisture is precipitating.

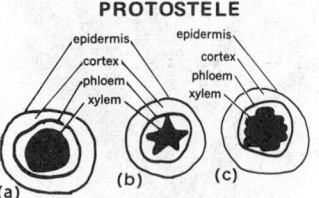

PROTOSTELE

epidermis epidermis
cortex   cortex
phloem   phloem
xylem   xylem

(a) (b) (c)

Cross sections of three types of protostele. *(a)* Haplostele. *(b)* Actinostele. *(c)* Plectostele.

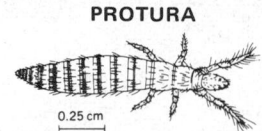

PROTURA

0.25 cm

Acerentulus barberi, a proturan insect. *(From H. S. Ewing, Ann. Entomol. Soc. Amer., 33(3):497, 1940)*

PROUSTITE

2.5 cm

Proustite crystals, Marienberg, Germany. *(Specimen from Department of Geology, Bryn Mawr College)*

PSEUDOBORNIALES

Pseudobornia node with a whorl of three leaves; has rhizomes and stems up to 6 centimeters wide and over a meter long. (*Modified from A. G. Nathorst*)

PSEUDOSCORPIONIDA

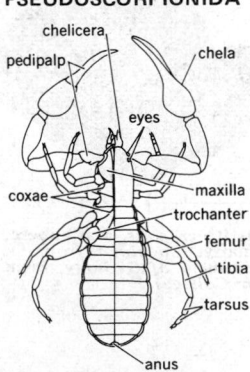

chelicera
chela
pedipalp
eyes
coxae
maxilla
trochanter
femur
tibia
tarsus
anus

Adult Pseudoscorpionida: ventral view, left side; dorsal view, right side. (*From H. S. Pratt, A Manual of the Common Invertebrate Animals, rev. ed., McGraw-Hill, 1951*)

pseudoalleles [GEN] Closely linked genes that behave as alleles and can be separated by crossing over.

Pseudoborniales [PALEOBOT] An order of fossil plants found in Middle and Upper Devonian rocks.

pseudobreccia [PETR] Limestone that is partially and irregularly dolomitized and is characterized by a mottled, breccialike appearance. Also known as recrystallization breccia.

pseudobrookite [MINERAL] Fe_2TiO_5 A brown or black mineral consisting of iron titanium oxide and occurring in orthorhombic crystals; specific gravity is 4.4–4.98.

pseudocarburizing *See* blank carburizing.

pseudocentrum [VERT ZOO] A centrum formed by fusion of the dorsal or dorsal and ventral arcualia, as in tailed amphibians.

pseudochrysolite *See* moldavite.

pseudocode *See* interpretive language.

pseudocoele [INV ZOO] A space between the body wall and internal organs that is not formed by gastrulation and lacks a cellular lining.

pseudocoelocyte [INV ZOO] In nematodes, a mesenchymal cell in the pseudocoelom.

Pseudocoelomata [INV ZOO] A group comprising the animal phyla Entoprocta, Aschelminthes, and Acanthocephala; characterized by a pseudocoelom.

pseudo cold front *See* pseudo front.

pseudocolumella [INV ZOO] In anthozoans, a type of axial structure.

pseudoconformity *See* paraconformity.

pseudocritical properties [CHEM] Effective (empirical) values for the critical properties (such as temperature, pressure, and volume) of a multicomponent chemical system.

pseudocumene [ORG CHEM] C_9H_{12} Water-insoluble, hydrocarbon liquid, boiling at 168°C; soluble in alcohol, benzene, and ether; used to manufacture perfumes and dyes, and as a catgut sterilant. Also known as pseudocumol; unstrimethylbenzene.

pseudocumol *See* pseudocumene.

Pseudocycnidae [INV ZOO] A family of the Caligoida which comprises external parasites on the gills of various fishes.

pseudocyesis [MED] A condition characterized by amenorrhea, enlargement of the abdomen, and other symptoms simulating gestation, due to an emotional disorder.

Pseudodiadematidae [INV ZOO] A family of Jurassic and Cretaceous echinoderms in the order Phymosomatoida which had perforate crenulate tubercles.

pseudoequivalent temperature *See* equivalent temperature.

pseudo front [METEOROL] A small-scale front, formed in association with organized severe convective activity, between a mass of rain-cooled air from the thunderstorm clouds and the warm surrounding air. Also known as pseudo cold front.

pseudogalena *See* sphalerite.

pseudoglanders [VET MED] An infectious bacterial lymphangitis of horses and other equines caused by *Corynebacterium pseudotuberculosis* and characterized by ulcerating nodules of the lymph nodes in the legs.

pseudohermaphroditism [PHYSIO] A condition in humans which simulates hermaphroditism, with gynandry in females and androgyny in males.

pseudoinstruction [ADP] **1.** A symbolic representation in a compiler or interpreter. **2.** *See* quasi-instruction.

pseudoleucite [MINERAL] A pseudomorph after leucite consisting of a mixture of nepheline, orthoclase, and analcime.

pseudoliquid density [PETRO ENG] For reservoir studies, the calculated value of a pseudodensity for reservoir liquid at atmospheric conditions, followed by application of suitable correction factors to obtain an approximate value for actual liquid density in the reservoir.

pseudometric *See* semimetric.

Pseudomonadaceae [MICROBIO] A family of rod-shaped to ovoid, gram-negative bacteria in the suborder Pseudomonadineae in which pigments, other than photosynthetic pigments, are common.

Pseudomonadales [MICROBIO] An order of ovoid, rod-shaped, comma-shaped, or spiral bacteria in the class Schizo-

mycetes; cells are characteristically rigid and motile by means of polar flagella.

Pseudomonadineae [MICROBIO] A suborder of bacteria of the order Pseudomonadales including those families whose cells do not contain photosynthetic pigments.

Pseudomonas [MICROBIO] The type genus of the Pseudomonadaceae; most species are aerobic and include cellulose decomposers and human, animal, and plant pathogens.

pseudomorph [MINERAL] An altered mineral whose crystal form has the outward appearance of another mineral species. Also known as false form.

pseudomucinous cystadenoma [MED] A benign ovarian tumor composed of columnar mucin-producing cells lining multilocular cysts filled with mucinous material.

pseudo-operation [ADP] An operation which is not part of the computer's operation repertoire as realized by hardware; hence, an entension of the set of machine operations.

pseudoparalysis [MED] An apparent motor paralysis that is caused by voluntary inhibition of motor impulses because of pain or other organic or psychic causes.

pseudoparasite [MED] Something in the blood that is mistaken for a parasite.

Pseudophoracea [INV ZOO] An extinct superfamily of gastropod mollusks in the order Aspidobranchia.

Pseudophyllidea [INV ZOO] An order of tapeworms of the subclass Cestoda, parasitic principally in the intestine of cold-blooded vertebrates.

pseudoplasmodium [INV ZOO] An aggregate of amebas resembling a plasmodium.

pseudoplastic fluid [FL MECH] A fluid whose apparent viscosity or consistency decreases instantaneously with an increase in shear rate.

pseudopodium [BOT] A slender, leafless branch of the gametophyte in certain Bryatae. [CYTOL] Temporary projection of the protoplast of ameboid cells in which cytoplasm streams actively during extension and withdrawal. [INV ZOO] Foot of a rotifer.

pseudoporphyritic [PETR] Pertaining to a rock that is not a true porphyry, but resembles one because of rapid growth of some of the crystals.

pseudorandom numbers [ADP] Numbers produced by a definite arithmetic process, but satisfying one or more of the standard tests for randomness.

pseudoreduced compressibility [CHEM] The compressibility factor for a multicomponent gaseous system, calculated at reduced conditions using the pseudoreduced properties of the mixture.

pseudoreduced properties [CHEM] Reduced-state relationships (such as reduced pressure, reduced temperature, and reduced volume) calculated for multicomponent chemical systems by using pseudocritical properties.

pseudo ripple mark [GEOL] A bedding-plane feature that resembles a ripple mark but is formed by lateral pressure caused by slumping or by local, small-scale tectonic deformation.

pseudoscalar [PHYS] A quantity which has magnitude only, and which acts, under Lorentz transformation, like a scalar but with a sign change under space reflection or time reflection, or both.

pseudoscalar coupling [PARTIC PHYS] A type of interaction postulated between a nucleon and a pion in which the interaction energy is a product of the pion's pseudoscalar field and a bilinear pseudoscalar function of the nucleon fields.

pseudoscalar meson [PARTIC PHYS] A meson, such as the pion, which has spin 0 and negative parity, and may be described by a field which is a pseudoscalar quantity. Also known as pseudoscalar particle.

pseudoscalar particle *See* pseudoscalar meson.

Pseudoscorpionida [INV ZOO] An order of terrestrial Arachnida having the general appearance of miniature scorpions without the postabdomen and sting.

Pseudosphaeriales [BOT] An order of the class Ascolichenes, shared by the class Ascomycetes; the ascocarp is flask-shaped and lined with a layer of interwoven, branched pseudoparaphyses.

Pseudosporidae [INV ZOO] A family of the protozoan subclass Proteomyxidia; flagellated stages invade Volvocidae and filamentous algae and become amebas.

pseudostatic spontaneous potential [PETRO ENG] Theoretical maximum spontaneous potential current that can be measured in a downhole, mud-column log in shaly sand. Abbreviated pseudostatic SP; PSP.

pseudo steady-state pressure distribution [PETRO ENG] The condition when the declining pressure distribution within a reservoir system closed at one boundary is declining at a uniform rate everywhere (or, there is constant pressure gradient within the reservoir).

pseudostratification *See* sheeting structure.

pseudostratified epithelium [HISTOL] A type of epithelium in which all cells reach to the basement membrane but some extend toward the surface only part way, while others reach the surface.

Pseudosuchia [PALEON] A suborder of extinct reptiles of the order Thecodontia comprising bipedal, unarmored or feebly armored forms which resemble dinosaurs in many skull features but retain a primitive pelvis.

pseudotachylite [PETR] A black rock that resembles tachylite; carries fragmental enclosures and shows evidence of having been at high temperature.

pseudotensor [PHYS] **1.** A quantity which transforms as a tensor under space rotations, but which transforms as a tensor, together with a change in sign, under space inversion. **2.** A quantity which transforms as a tensor under Lorentz transformations, but with an additional sign change under space reflection or time reflection or both.

Pseudothelphusidae [INV ZOO] A family of fresh-water crabs belonging to the Brachyura.

Pseudotriakidae [VERT ZOO] The false catsharks, a family of galeoids in the carcharinid line.

pseudotuberculosis [MED] A bacterial infection in man and many animals caused by *Pasteurella pseudotuberculosis;* may be severe in man with septicemia and symptoms resembling typhoid fever.

pseudovector [PHYS] **1.** A quantity which transforms as a vector under space rotations but which transforms as a vector, together with a change in sign, under a space inversion. Also known as axial vector. **2.** A quantity which transforms as a four-vector under Lorentz transformations, but with an additional sign change under space reflection or time reflection or both.

pseudovector coupling [PARTIC PHYS] A type of interaction postulated between a nucleon and another particle in which the expression for the interaction energy contains a bilinear pseudovector function of nucleon fields.

pseudovector meson [PARTIC PHYS] A meson which has spin quantum number 1 and positive parity, and may be described by a field which is a pseudovector quantity.

psf *See* pound per square foot.

P shell [ATOM PHYS] The sixth layer of electrons about the nucleus of an atom, having electrons whose principal quantum number is 6.

psi *See* pound per square inch.

psia *See* pounds per square inch absolute.

psid *See* pounds per square inch differential.

psi function [MATH] The special function of a complex variable which is obtained from differentiating the logarithm of the gamma function. [QUANT MECH] *See* Schrödinger wave function.

psig *See* pounds per square inch gage.

psilate [BOT] Lacking ornamentation; generally applied to pollen.

Psilidae [INV ZOO] The rust flies, a family of myodarian cyclorrhaphous dipteran insects in the subsection Acalypteratae.

Psilophytales [PALEOBOT] A group formerly recognized as an order of fossil plants.

Psilophytineae [PALEON] The equivalent name for Rhyniopsida.

Psilopsida [BOT] A subdivision of the Tracheophyta.

Psilorhynchidae [VERT ZOO] A small family of actinopterygian fishes belonging to the Cyprinoidei.

psilosis [MED] Falling out of the hair.

Psilotales [BOT] The equivalent name for Psilotophyta.

Psilotatae [BOT] A class of the Psilotophyta.

Psilotophyta [BOT] A division of the plant kingdom represented by three living species; the life cycle is typical of the vascular cryptogams.

Psittacidae [VERT ZOO] The single family of the Psittaciformes.

Psittaciformes [VERT] The parrots, a monofamilial order of birds that exhibit zygodactylism and have a strong hooked bill.

psittacinite *See* mottramite.

psittacosis [MED] Pneumonia and generalized infection of man and of birds caused by agents of the PLT-Bedsonia group; transmitted to man by psittacine birds.

PSK *See* phase-shift keying.

psoas [ANAT] Either of two muscles: psoas major which arises from the bodies and transverse processes of the lumbar vertebrae and is inserted into the lesser trochanter of the femur, and psoas minor which arises from the bodies and transverse processes of the lumbar vertebrae and is inserted on the pubis.

Psocoptera [INV ZOO] An order of small insects in which wings may be present or absent, tarsi are two- or three-segmented, cerci are absent, and metamorphosis is gradual.

Psolidae [INV ZOO] A family of echinoderms in the order Dendrochirotida characterized by a ventral adhesive sucker and a U-shaped gut, with the mouth and anus opening upward on the adoral surface.

Psophiidae [VERT ZOO] The trumpeters, a family of birds in the order Gruiformes.

psophometer [ENG] An instrument for measuring noise in electric circuits; when connected across a 600-ohm resistance in the circuit under study, the instrument gives a reading that by definition is equal to half of the psophometric electromotive force actually existing in the circuit.

psophometric electromotive force [ELECTR] The true noise voltage that exists in a circuit.

psophometric voltage [ELECTR] The noise voltage as actually measured in a circuit under specified conditions.

psoriasis [MED] A usually chronic, often acute inflammatory skin disease of unknown cause; characterized by dull red, well-defined lesions covered by silvery scales which when removed disclose tiny capillary bleeding points.

psorosis [PL PATH] A virus disease of tangerine, grapefruit, and sweet orange trees characterized by scaly bark, a gummy exudate, retarded growth, small yellow leaves, and dieback of twigs. Also known as scaly bark.

PSP *See* pseudostatic spontaneous potential.

psyche [PSYCH] The mind or self as a functional entity.

psychedelic [PSYCH] Of, pertaining to, or producing a psychic state.

psychiatric genetics [GEN] The study of the genetic causes and modes of inheritance that underlie the generally recognized mental illnesses.

psychiatric social worker [PSYCH] A specialist who utilizes the techniques of both social work and psychiatry to serve the community.

psychiatrist [MED] A person who specializes in psychiatry; a licensed physician trained in psychiatry.

psychiatry [MED] The medical science that deals with the origins, diagnosis, and treatment of mental and emotional disorders.

Psychidae [INV ZOO] The bagworms, a family of lepidopteran insects in the superfamily Tineoidea; males are large, hairy moths, but females are degenerate, wingless, and legless and live in bag-shaped cases.

psychoanalysis [PSYCH] A technique used in the treatment of neuroses and other emotional disorders which relies upon free associations of the patient to bring ideas and experiences from the unconscious to the conscious divisions of the psyche.

psychobiology [PSYCH] The school of psychiatry and psychology in which the individual is considered as the sum of his environment as well as being considered a physical organism.

Psychodidae [INV ZOO] The moth flies, a family of orthorrhaphous dipteran insects in the series Nematocera.

psychodynamics [PSYCH] The study of human behavior from the point of view of motivation and drives, depending

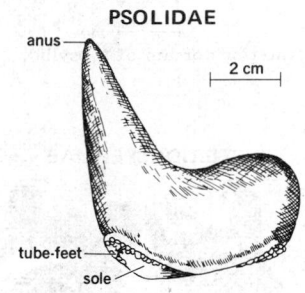

PSEUDOSTRATIFIED EPITHELIUM

Arrangement of cells in pseudostratified epithelium.

PSOLIDAE

anus

2 cm

tube-feet

sole

Psolus phantapus. (After T. Mortensen, 1927)

PSYCHIDAE

(a)

(b)

Thyridopteryx ephemeraeformis Haworth. *(a)* Male. *(b)* Male case.

PSYLLIDAE

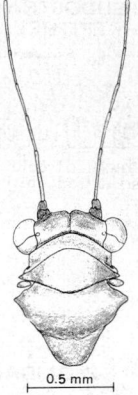

0.5 mm

Anterior dorsum of a psyllid.

PTERIDOSPERMAE

1 m

Reconstruction of a seed fern. Note appearance of adventitious roots on the lower portion of the stem.

PTEROBRANCHIA

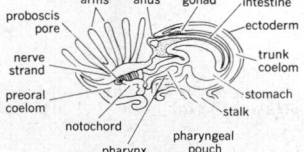

Cephalodiscus. (After W. Patten, from H. V. Neal and H. W. Rand, Comparative Anatomy, Blakiston-McGraw-Hill, 1936)

largely on the functional significance of emotion, and based on the assumption that an individual's total personality and reactions at any given time are the product of the interaction between his genetic constitution and his environment.

psychogalvanic reflex [PHYSIO] A variation in the electric conductivity of the skin in response to emotional stimuli, which cause changes in blood circulation, secretion of sweat, and skin temperature.

psychogalvanometer [ENG] An instrument for testing mental reaction by determining how skin resistance changes when a voltage is applied to electrodes in contact with the skin.

psychogenic [MED] Of psychic origin.

psychointegroammeter *See* lie detector.

psychokinesis [PSYCH] The alleged ability of an individual to exert a mental influence on physical events in advance of their occurrence. Abbreviated PK.

psychologist [PSYCH] **1.** An individual who has made a professional study of, and usually thereafter professionally engages in, psychology. **2.** Specifically, an individual with the minimum professional qualifications set forth by an intraprofessionally recognized psychological association.

psychology [BIOL] **1.** The science that deals with the functions of the mind and the behavior of an organism in relation to its environment. **2.** The mental activity characteristic of a person or a situation.

psychometrics [PSYCH] The mathematical and statistical treatment of psychological data.

psychomotor [PSYCH] Pertaining to both mental and motor activity.

psychopathic personality [PSYCH] An emotionally immature individual characterized by pronounced defects in judgment and prone to impulsive, generally amoral or antisocial behavior.

psychopathology [PSYCH] The systematic study of mental diseases.

psychopharmacology [PSYCH] The science that deals with the action of drugs on mental function.

psychophysics [PSYCH] **1.** The study of mental processes by physical methods. **2.** The study of the relations of stimuli to the sensations they produce.

psychophysiologic nervous system reaction *See* neurasthenic neurosis.

psychosis [PSYCH] An impairment of mental functioning to the extent that it interferes grossly with an individual's ability to meet the ordinary demands of life, characterized generally by severe affective disturbance, profound introspection, and withdrawal from reality, formation of delusions or hallucinations, and regression presenting the appearance of personality disintegration.

psychosomatic [MED] Of or pertaining to the interrelationship between mental processes and somatic functions.

psychosomatograph [ENG] An instrument for recording muscular action currents or physical movements during tests of mental-physical coordination.

psychosurgery [MED] The branch of medicine that deals with the treatment of various psychoses, severe neuroses, and chronic painful conditions by means of operative procedures on the brain.

psychotaxis [PSYCH] Involuntary adjustment of a person's thoughts and behavior in order to retain the agreeable and avoid the disagreeable; an ego defense mechanism.

psychotherapy [PSYCH] The use of psychological means in the treatment of emotional and mental disorders.

psychotomimetic [PSYCH] **1.** Mimicking a psychotic disorder. **2.** Pertaining to any drug or compound, such as lysergic acid diethylamide or mescaline, which can induce a psychoticlike state.

psychotropic [PSYCH] Pertaining to any drug or agent having a particular affinity for or effect on the psyche.

psychromatic ratio [THERMO] Ratio of the heat-transfer coefficient to the product of the mass-transfer coefficient and humid heat for a gas-vapor system; used in calculation of humidity or saturation relationships.

psychrometer [ENG] A device comprising two thermometers, one a dry bulb, the other a wet or wick-covered bulb, used in determining the moisture content or relative humidity

of air or other gases. Also known as wet and dry bulb thermometer.

psychrometric calculator [ENG] A device for quickly computing certain psychrometric data, usually the dew point and the relative humidity, from known values of the dry- and wet-bulb temperatures and the atmospheric pressure.

psychrometric chart [THERMO] A graph each point of which represents a specific condition of a gas-vapor system (such as air and water vapor) with regard to temperature (horizontal scale) and absolute humidity (vertical scale); other characteristics of the system, such as relative humidity, wet bulb temperature, and latent heat of vaporization, are indicated by lines on the chart.

psychrometric formula [THERMO] The semiempirical relation giving the vapor pressure in terms of the barometer and psychrometer readings.

psychrometric tables [THERMO] Tables prepared from the psychrometric formula and used to obtain vapor pressure, relative humidity, and dew point from values of wet-bulb and dry-bulb temperature.

psychrometry [ENG] The science and techniques associated with measurements of the water vapor content of the air or other gases.

psychrophile [BIOL] An organism that thrives at low temperatures.

psychrophobia [PHYSIO] Abnormal sensitivity to cold. [PSYCH] An abnormal fear of the cold.

psychrophyte [ECOL] A plant adapted to the climatic conditions of the arctic or alpine regions.

Psyllidae [INV ZOO] The jumping plant lice, a family of the Homoptera in the series Sternorrhyncha in which adults have a transverse head with protuberant eyes and three ocelli, 6- to 10-segmented antennae, and wings with reduced but conspicuous venation.

pt *See* pint.

Pt *See* platinum.

PTA *See* factor XI; phosphotungstic acid.

ptarmigan [VERT ZOO] Any of various birds of the genus *Lagopus* in the family Tetraonidae; during the winter, plumage is white and hairlike feathers cover the feet.

PT boat *See* motor torpedo boat.

PTC *See* Christmas factor.

Pteraspidomorphi [VERT ZOO] The equivalent name for Diplorhina.

Pterasteridae [INV ZOO] A family of deep-water echinoderms in the order Spinulosida distinguished by having webbed spine fins.

Pteridophyta [BOT] The equivalent name for Polypodiophyta.

Pteridospermae [PALEOBOT] Seed ferns, a class of the Cycadicae comprising extinct plants characterized by naked seeds borne on large fernlike fronds.

Pteridospermophyta [PALEOBOT] The equivalent name for Pteridospermae.

Pteriidae [INV ZOO] Pearl oysters, a family of bivalve mollusks which have nacreous shells.

pterinophore [CYTOL] A yellow to orange chromatophore that contains pterine pigment.

pterion [ANTHRO] The region surrounding the sphenoparietal suture where the frontal bone, parietal bone, squama temporalis, and greater wing of the sphenoid bone come together most closely.

Pterobranchia [INV ZOO] A group of small or microscopic marine animals regarded as a class of the Hemichordata; all are sessile, tubicolous organisms with a U-shaped gut and three body segments.

pterochore [BOT] A type of plant that produces winged disseminules.

Pteroclidae [VERT ZOO] The sandgrouse, a family of graminivorous birds in the order Columbiformes; mainly an Afro-Asian group resembling pigeons and characterized by cryptic coloration, usually corresponding with the soil color of the habitat.

pterodactyl [PALEON] The common name for members of the extinct reptilian order Pterosauria.

Pterodactyloidea [PALEON] A suborder of Late Jurassic and Cretaceous reptiles in the order Pterosauria distinguished by lacking tails and having increased functional wing length due to elongation of the metacarpels.

pteroic acid [BIOCHEM] $C_{14}H_{12}N_6O_3$ A crystalline amino acid formed by hydrolysis of folic acid or other pteroylglutamic acids.

Pteromalidae [INV ZOO] A family of hymenopteran insects in the superfamily Chalcidoidea.

Pteromedusae [INV ZOO] A suborder of hydrozoan coelenterates in the order Trachylina characterized by a modified, bipyramidal medusae.

Pterophoridae [INV ZOO] The plume moths, a family of the lepidopteran superfamily Pyraloidea in which the wings are divided into featherlike plumes, maxillary palpi are lacking, and the legs are long.

Pteropidae [INV ZOO] The fruit bats, a large family of the Chiroptera found in Asia, Australia, and Africa.

Pteropoda [INV ZOO] The sea butterflies, an order of pelagic gastropod mollusks in the subclass Opisthobranchia in which the foot is modified into a pair of large fins and the shell, when present, is thin and glasslike.

Pteropodidae [VERT ZOO] A family of fruit-eating bats in the suborder Megachiroptera, characterized by primitive ears and by shoulder joints.

pteropod ooze [GEOL] A pelagic sediment containing at least 45% calcium carbonate in the form of tests of marine animals, particularly pteropods.

Pteropsida [BOT] A large group of vascular plants characterized by having parenchymatous leaf gaps in the stele and by having leaves which are thought to have originated in the distant past as branched stem systems.

Pterosauria [PALEON] An extinct order of flying reptiles of the Mesozoic era belonging to the subclass Archosauria; the wing resembled that of a bat, and a large heeled sternum supported strong wing muscles.

pterostigma [INV ZOO] Opaque thickened spot occurring on the costal margin of an insect wing.

pteroylglutamic acid *See* folic acid.

pterygium [MED] **1.** A triangular mass of mucous membrane growing on the conjunctiva, usually near the inner canthus. **2.** Overgrowth of the cuticle forward on the nail. [VERT ZOO] A generalized vertebrate limb.

pterygoid bone [VERT ZOO] A rodlike bone or group of bones forming a portion of the palatoquadrate arch in lower vertebrates.

pterygopalatine fossa [ANAT] The gap between the pterygoid process of the sphenoid bone and the maxilla and palatine bone.

pterygoquadrate [EMBRYO] Of, pertaining to, or being the first branchial arch in lower vertebrate embryos; gives rise to most of the upper jaw.

Ptiliidae [INV ZOO] The feather-winged beetles, a family of coleopteran insects in the superfamily Staphylinoidea.

Ptilodactylidae [INV ZOO] The toed-winged beetles, a family of the Coleoptera in the superfamily Dryopoidea.

Ptilodontoidea [PALEON] A suborder of extinct mammals in the order Multituberculata.

ptilolite *See* mordenite.

Ptinidae [INV ZOO] The spider beetles, a family of coleopteran insects in the superfamily Bostrichoidea.

PTM *See* pulse-time modulation.

Ptolemaic system [ASTRON] The movements of the solar system according to Ptolemy; supposedly, the earth was a fixed center, with the sun and moon revolving about it in circular orbits; planets revolved in small circles (epicycles) whose centers revolved about the earth in larger circles (deferents).

ptosis [MED] Prolapse, abnormal depression, or falling down of an organ or part; applied especially to drooping of the upper eyelid, from paralysis of the third cranial nerve.

ptyalase *See* ptyalin.

ptyalin [BIOCHEM] A diastatic enzyme found in saliva which catalyzes the hydrolysis of starch to dextrin, maltose, and glucose, and the hydrolysis of sucrose to glucose and fructose. Also known as ptyalase; salivary amylase; salivary diastase.

Ptychodactiaria [INV ZOO] An order of the zoantharian anthozoans of the phylum Coelenterata known only from two genera, *Ptychodactis* and *Dactylanthus.*

Ptychomniaceae [BOT] A family of mosses in the order Isobryales distinguished by an eight-ribbed capsule.

ptychotis oil *See* ajowan oil.

Ptyctodontida [PALEON] An order of Middle and Upper Devonian fishes of the class Placodermi in which both the head and trunk shields are present, and the joint between them is a well-differentiated and variable structure.

ptygma [GEOL] Pegmatitic material with migmatite or gneiss, resembling disharmonic folds. Also known as ptygmatic fold.

ptygmatic fold *See* ptygma.

p-type conductivity [ELECTR] The conductivity associated with holes in a semiconductor, which are equivalent to positive charges.

p-type crystal rectifier [ELECTR] Crystal rectifier in which forward current flows when the semiconductor is positive with respect to the metal.

p-type semiconductor [ELECTR] An extrinsic semiconductor in which the hole density exeeds the conduction electron density.

p⁺-type semiconductor [ELECTR] A *p*-type semiconductor in which the excess mobile hole concentration is very large.

p-type silicon [ELECTR] Silicon to which more impurity atoms of acceptor type (with valence of 3, such as boron) than of donor type (with valence of 5, such as phosphorus) have been added, with the result that the hole density exceeds the conduction electron density.

Pu *See* plutonium.

puberty [PHYSIO] The period at which the generative organs become capable of exercising the function of reproduction; signalized in the boy by a change of voice and discharge of semen, in the girl by the appearance of the menses.

puberulic acid [BIOCHEM] $(HO)_3(C_7H_2O)COOH$ A keto acid formed as a metabolic product of certain species of *Penicillium;* has some germicidal activity against gram-positive bacteria.

pubic arch [ANAT] The arch formed by the conjoined rami of the pubis and ischium.

pubic crest [ANAT] The crest extending from the pubic tubercle to the medial extremity of the pubis.

pubic height [ANTHRO] A measure of the vertical distance from the pubis to the floor taken when the subject is in a standing position.

pubic symphysis [ANAT] The fibrocartilaginous union of the pubic bones. Also known as symphysis pubis.

pubis [ANAT] The pubic bone, the portion of the hipbone forming the front of the pelvis.

public address system *See* sound reinforcement system.

public area [BUILD] The total nonrentable area of a building, such as public conveniences and rest rooms.

public communications service [COMMUN] Telephone or telegraph service provided for the transmission of unofficial communications for the public.

public correspondence [COMMUN] Any telecommunications which offices and stations at the disposal of the public must accept for transmission.

public health [MED] **1.** The state of health of a community or of a population. **2.** The art and science dealing with the protection and improvement of community health.

public radio communications services [COMMUN] Land, mobile, and fixed services, the stations of which are open to public correspondence.

public safety frequency bands [COMMUN] Radio-frequency bands allocated in the United States for communication on land between base stations and mobile stations or between mobile stations by police, fire, highway, forestry, and emergency services.

public-safety radio service [COMMUN] Any service of radio communication essential to either the discharge of non-Federal governmental functions relating to public safety responsibilities or the alleviation of an emergency endangering life or property, the radio transmitting facilities of which are defined as fixed, land, or mobile stations.

public switched network [COMMUN] A national telephone network as contrasted with private lines and private exchanges.

public utility [IND ENG] A business organization considered

PTEROPHORIDAE

Platypilia carduidactyla Riley, showing plumelike wings and long legs.

PTYCTODONTIDA

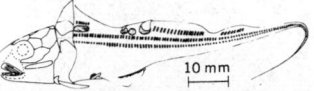

Rhamphodopsis threiplandi from the Middle Devonian of Scotland. Restoration of skeleton with outline of body in lateral aspect. *(After Roger S. Miles)*

by law to be vested with public interest and subject to public regulation.

public works [IND ENG] Government-owned and financed works and improvements for public enjoyment or use.

pucherite [MINERAL] $BiVO_4$ A reddish-brown orthorhombic mineral composed of bismuth vanadate, occurring as small crystals.

puckering [MET] Corrugations in metal parts resulting from pressing or drawing.

pudding ball *See* armored mud ball.

puddingstone [GEOL] In Great Britain, a conglomerate consisting of rounded pebbles whose colors are in marked contrast with the matrix, giving a section of the rock the appearance of a raisin pudding.

puddling [MET] A process for the production of wrought iron by agitation of a bath of molten pig iron with iron oxide in order to reduce the carbon, silicon, phosphorus, and manganese content.

puddling furnace [MET] A coal-fired reverberatory furnace for puddling pig iron.

puelche [METEOROL] An east wind which has crossed the Andes; the Andean foehn of the South American west coast.

puerperal sepsis [MED] A toxic condition caused by infection in the birth canal, occurring as a complication or sequel of pregnancy.

puerperium [MED] **1.** The state of a woman during labor or immediately after delivery. **2.** The period from delivery to the time when the uterus returns to normal size, about 6 weeks.

puff *See* picofarad.

puff cone *See* mud cone.

puffer [MIN ENG] A small stationary engine used in coal mines for hoisting material.

puff of wind [METEOROL] A slight local breeze which causes a patch of ripples on the surface of the sea.

pug mill [MECH ENG] A machine for mixing and tempering a plastic material by the action of blades revolving in a drum or trough.

puking [CHEM ENG] In a distillation column, the foaming and rising of liquid so that part of it is driven out of the vessel through the vapor line.

pull-apart [GEOL] A precompaction sedimentary structure having the appearance of boudinage and consisting of beds that have been stretched and pulled apart into relatively short slabs.

pull crack [MET] A crack in a casting caused by contraction strains during cooling and resulting from the shape of the casting.

puller [MECH ENG] A lever-operated chain or wire-rope hoist for lifting or pulling at any angle, which has a reversible ratchet mechanism in the lever permitting short-stroke operation for both tensioning and relaxing, and which holds the loads with a Weston-type friction brake or a releasable ratchet. Also known as come-along.

pulley [DES ENG] A wheel with a flat, round, or grooved rim that rotates on a shaft and carries a flat belt, V-belt, rope, or chain to transmit motion and energy.

pulley lathe [MECH ENG] A lathe for turning pulleys.

pulley stile [BUILD] The upright part of a window frame which holds the pulley and guides the sash.

pulling [ELECTR] An effect that forces the frequency of an oscillator to change from a desired value; causes include undesired coupling to another frequency source or the influence of changes in the oscillator load impedance.

pulling figure [ELECTR] The total frequency change of an oscillator when the phase angle of the reflection coefficient of the load impedance varies through 360°, the absolute value of this reflection coefficient being constant at 0.20.

pulling pillars *See* pillaring.

pull-in torque [MECH ENG] The largest steady torque with which a motor will attain normal speed after accelerating from a standstill.

pullorum disease [VET MED] A highly fatal disease of chickens and other birds caused by *Salmonella pullorum*, characterized by weakness, lassitude, lack of appetite, and whitish or yellowish diarrhea. Also known as bacillary white diarrhea; white diarrhea.

PULLEY

(a) (b) (c)

Some typical examples of pulleys. *(a)* Multigroove V-belt sheave. *(b)* Solid-hub crown face pulley; *(c)* brass pulley for round belt *(Boston Gear Div., Rockwell International).*

pull-out torque [MECH ENG] Th largest torque under which a motor can operate without sharply losing speed.

pull rope [MIN ENG] **1.** The rope that pulls a journey of loaded cars on a haulage plane. **2.** The rope that pulls the loaded scoop or bucket in a scraper loader layout. Also known as main rope.

pullshovel *See* backhoe.

pull strength [MECH] A unit in tensile testing; the bond strength in pounds per square inch.

pull tube [PETRO ENG] Tube used in rod-type, traveling-barrel oil well pumps to connect the pump plunger with the seating anchor.

pulmonary anthrax [MED] A form of anthrax in man caused by the inhalation of dust containing *Bacillus anthracis* spores.

pulmonary artery [ANAT] A large artery that conducts venous blood from the heart to the lungs of tetrapods.

pulmonary aspergillosis [MED] A form of aspergillosis caused by *Aspergillus fumigatus*, with symptoms resembling chronic tuberculosis.

pulmonary circulation [PHYSIO] The circulation of blood through the lungs for the purpose of oxygenation and the release of carbon dioxide. Also known as lesser circulation.

pulmonary edema [MED] An effusion of fluid into the alveoli and interstitial spaces of the lungs.

pulmonary plexus [ANAT] A nerve plexus composed chiefly of vagal fibers situated on the anterior and posterior aspects of the bronchi and accompanying them into the substance of the lung.

pulmonary stenosis [MED] Narrowing of the orifice of the pulmonary artery.

pulmonary valve [ANAT] A valve consisting of three semilunar cusps situated between the right ventricle and the pulmonary trunk.

pulmonary vein [ANAT] A large vein that conducts oxygenated blood from the lungs to the heart in tetrapods.

Pulmonata [INV ZOO] A subclass of the gastropod mollusks containing the "lung"-bearing snails; gills have been lost and in their place the mantle cavity has become a pulmonary sac.

pulp [ANAT] A mass of soft spongy tissue in the interior of an organ. [BOT] The soft succulent portion of a fruit. [ENG] *See* slime. [MATER] The cellulosic material produced by reducing wood mechanically or chemically and used in making paper and cellulose products. Also known as wood pulp.

pulpboard [MATER] Chipboard to which is added a percentage of mechanical wood pulp.

pulp cavity [ANAT] The space within the central part of a tooth containing the dermal pulp and made up of the pulp chamber and a root canal.

pulp chamber [ANAT] The coronal portion of the central cavity of a tooth.

pulper [MECH ENG] A machine that converts materials to pulp, for example, one that reduces paper waste to pulp.

pulping [ENG] Reducing wood to pulp.

pulp molding [ENG] A plastics-industry process in which a resin-impregnated pulp material is preformed by application of a vacuum, after which it is oven-cured and molded.

pulpotomy [MED] Surgical removal of the pulp of a tooth.

pulpstone [MATER] A block of sandstone cut into wheels for grinding, especially wood pulp in paper manufacture.

pulpwood [MATER] Any wood that can be reduced to pulp.

pulsar [ASTROPHYS] A celestial radio source, emitting intense short bursts of radio emission; the periods of known pulsars range between 33 milliseconds and 3.75 seconds, and pulse durations range from 2 to about 150 milliseconds with longer-period pulsars generally having a longer pulse duration.

pulsatance [PHYS] Angular velocity in radians, equal to 2π times frequency in hertz.

pulsating current [ELEC] Periodic direct current.

pulsating electromotive force [ELEC] Sum of a direct electromotive force and an alternating electromotive force. Also known as pulsating voltage.

pulsating flow [ENG] Irregular fluid flow in a piping system often resulting from the pressure variations of reciprocating compressors or pumps within the system.

pulsating star [ASTRON] Variable star whose luminosity fluctuates as the star expands and contracts; the variation in brightness is thought to come from the periodic change of radiant energy to gravitational energy and back.

pulsating voltage *See* pulsating electromotive force.

pulsation [PHYSIO] A beating or throbbing, usually rhythmic, as of the heart or an artery.

pulsation dampening [ENG] Device installed in a fluid piping system (gas or liquid) to eliminate or even out the fluid-flow pulsations caused by reciprocating compressors, pumps, and such.

pulsation welding *See* multiple-impulse welding.

pulse [PHYS] A variation in a quantity which is normally constant; has a finite duration and is usually brief compared to the time scale of interest. [PHYSIO] **1.** The regular, recurrent, palpable wave of arterial distention due to the pressure of the blood ejected with each contraction of the heart. **2.** A single wave.

pulse altimeter [ENG] A device which is used to measure the distance of an aircraft above the ground by sending out radar signals in short pulses and measuring the time delay between the leading edge of the transmitted pulse and that of the pulse returned from the ground.

pulse amplitude [PHYS] The peak, average, effective, instantaneous, or other magnitude of a pulse, usually with respect to the normal constant value; the exact meaning should be specified when giving a numerical value.

pulse-amplitude discriminator [ENG] Electronic instrument used to investigate the amplitude distribution of the pulses produced in a nuclear detector.

pulse-amplitude modulation [COMMUN] Amplitude modulation of a pulse carrier. Abbreviated PAM.

pulse-amplitude modulation-frequency modulation [COMMUN] System in which pulse-amplitude-modulated subcarriers are used to frequency-modulate a second carrier; binary digits are formed by the absence or presence of a pulse in an assigned position.

pulse analyzer [ELECTR] An instrument used to measure pulse widths and repetition rates, and to display on a cathode-ray screen the waveform of a pulse.

pulse bandwidth [COMMUN] The bandwidth outside of which the amplitude of a pulse-frequency spectrum is below a prescribed fraction of the peak amplitude.

pulse beating [COMMUN] Intermodulation of pulse-modulated waves.

pulse cable [COMMUN] A communications cable, capable of transmitting pulses without unacceptable distortion.

pulse carrier [COMMUN] A pulse train used as a carrier.

pulse circuit [ELECTR] An active electrical network designed to respond to discrete pulses of current or voltage.

pulse code [COMMUN] A code consisting of various combinations of pulses, such as the Morse code, Baudot code, and the binary code used in computers.

pulse-code modulation [COMMUN] Modulation in which the peak-to-peak amplitude range of the signal to be transmitted is divided into a number of standard values, each having its own three-place code; each sample of the signal is then transmitted as the code for the nearest standard amplitude. Abbreviated PCM.

pulse-code modulation television [COMMUN] A television system in which digital signals using a pulse code are transmitted, rather than analog signals.

pulse coder *See* coder.

pulse coding and correlation [COMMUN] A general technique concerning a variety of methods used to change the transmitted waveform and then decode upon its reception; pulse compression is a special form of pulse coding and correlation.

pulse column [CHEM ENG] Continuous-phase process column (such as liquid only or gas only) in which the flow-through is pulsating; used to increase mass-transfer rates, as in a liquid-liquid extraction operation.

pulse communication [COMMUN] Radio communication using pulse modulation.

pulse compression [ELECTR] A matched filter technique used to discriminate against signals which do not correspond to the transmitted signal.

pulse counter [ELECTR] A device that indicates or records

the total number of pulses received during a time interval.

pulsed-bed sorption [CHEM ENG] Solid-liquid countercurrent adsorption process (such as an ion-exchange process) in which the granulated solids bed and the solution flow alternately, in opposite directions.

pulse decay time [COMMUN] The interval of time required for the trailing edge of a pulse to decay from 90% to 10% of the peak pulse amplitude.

pulse-delay network [ELECTR] A network consisting of two or more components such as resistors, coils, and capacitors, used to delay the passage of a pulse.

pulse demoder [COMMUN] Circuit that responds only to pulse signals having the specified spacing between pulses for which the device is adjusted. Also known as constant-delay discriminator.

pulse discriminator [ELECTR] A discriminator circuit that responds only to a pulse having a particular duration or amplitude.

pulsed laser [OPTICS] A laser which generates light in a pulse lasting a few hundred microseconds or less, with a peak power of tens of kilowatts or more.

pulsed light [OPTICS] A beam of light whose intensity is modulated in some prescribed manner; analogous to a radar pulse.

pulsed-light ceilometer *See* pulsed-light cloud-height indicator.

pulsed-light cloud-height indicator [ENG] An instrument used for the determination of cloud heights; it operates on the principle of pulse radar, employing visible light rather than radio waves. Also known as pulsed-light ceilometer.

pulse-Doppler radar [ENG] Pulse radar that uses the Doppler effect to obtain information about the velocity of a target.

pulsed oscillator [ELECTR] An oscillator that generates a carrier-frequency pulse or a train of carrier-frequency pulses as the result of self-generated or externally applied pulses.

pulsed reactor [NUCLEO] A research nuclear reactor in which continual short, intense surges of power and radiation can be produced; the neutron flux during the surge is much higher than could be tolerated during steady-state operation.

pulsed ruby laser [OPTICS] A laser in which ruby is used as the active material; the extremely high pumping power required is obtained by discharging a bank of capacitors through a special high-intensity flash tube, giving a coherent beam that lasts for about 0.5 millisecond.

pulsed transfer function [CONT SYS] The ratio of the z transform of the output of a system to the z transform of the input, when both input and output are trains of pulses. Also known as discrete transfer function; z transfer function.

pulse duration [COMMUN] The time interval between the first and last instants at which the instantaneous amplitude reaches a stated fraction of the peak pulse amplitude. Also known as pulse length; pulse width (both deprecated usages).

pulse-duration coder *See* coder.

pulse-duration discriminator [ELECTR] A circuit in which the sense and magnitude of the output are a function of the deviation of the pulse duration from a reference.

pulse-duration modulation [COMMUN] Modulation of a pulse carrier wherein the value of each instantaneous sample of a modulating wave produces a pulse of proportional duration by varying the leading, trailing, or both edges of a pulse. Abbreviated PDM. Also known as pulse-length modulation; pulse-width modulation.

pulse-duration modulation-frequency modulation [COMMUN] System in which pulse-duration-modulated subcarriers are used to frequency-modulate a second carrier. Also known as pulse-width modulation-frequency modulation.

pulse-echo method [MET] A nondestructive test in which pulses of energy are directed into a part, and the time for the echo to return from one or more surfaces is measured.

pulse form [PHYS] The amplitude of a pulse plotted as a function of time.

pulse-forming network [ELECTR] A network used to shape the leading or trailing edge of a pulse.

pulse-frequency modulation [COMMUN] A form of pulse-time modulation in which the pulse repetition rate is the characteristic varied. Abbreviated PFM.

pulse-frequency spectrum *See* pulse spectrum.

PULSE-DURATION MODULATION

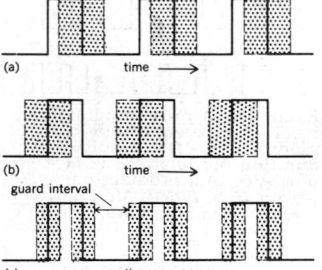

Types of pulse-duration modulation. *(a)* Trailing edge modulated, leading edge fixed. *(b)* Leading edge modulated, trailing edge fixed. *(c)* Both edges modulated. Solid lines indicate duration of unmodulated pulses, shaded areas limits of maximum modulation. *(From H. S. Black, Modulation Theory, Van Nostrand, 1953)*

PULSE GENERATOR

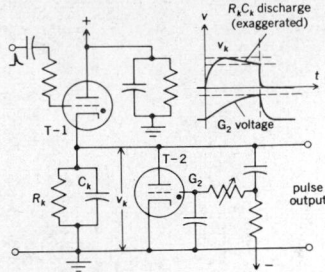

Thyratron pulse generator, an example of a pulse generator having a controllable pulse width. Tube T-1 fires when positive triggering pulse is applied, charging capacitor C_k to some positive potential v_k which then slowly decreases as C_k discharges through resistor R_k. Grid G_2 of tube T-2 rises until it becomes sufficiently positive to fire T-2, which then acts as a low impedance across C_k discharging it to near ground potential.

PULSE MODULATION

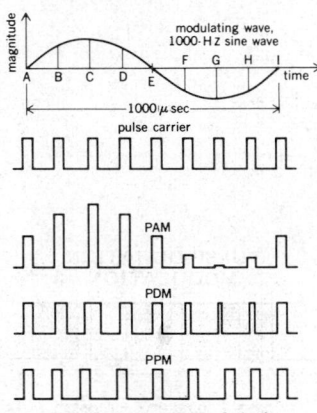

Examples of pulse modulation. PAM = pulse-amplitude modulation, PDM = pulse-duration modulation, PPM = pulse-position modulation. *(From H. S. Black, Modulation Theory, Van Nostrand, 1953)*

PULSE-POSITION MODULATION

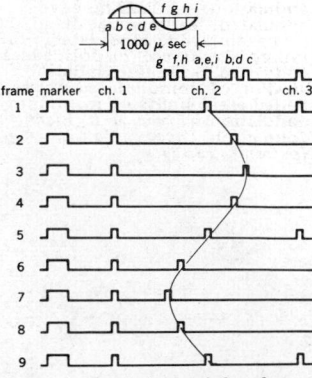

Modulation of channel 2 pulse by a sine wave. Successive diagrams show relative position in time of channel 2 pulse for nine successive samples, at phases of sine wave indicated by letters *a–i*. *(From H. S. Black, Modulation Theory, Van Nostrand, 1953)*

pulse generator [ELEC] *See* impulse generator. [ELECTR] A generator that produces repetitive pulses or signal-initiated pulses.

pulse group *See* pulse train.

pulse height [ELECTR] The strength or amplitude of a pulse, measured in volts.

pulse-height analyzer [NUCLEO] An instrument capable of indicating the number of occurrences of pulses falling within each of one or more specified amplitude ranges; used to obtain the energy spectrum of nuclear radiations. Also known as kick-sorter (British usage); multichannel analyzer.

pulse-height discriminator [ELECTR] A circuit that produces a specified output pulse when and only when it receives an input pulse whose amplitude exceeds an assigned value. Also known as amplitude discriminator.

pulse-height selector [ELECTR] A circuit that produces a specified output pulse only when it receives an input pulse whose amplitude lies between two assigned values. Also known as amplitude selector; diffractional pulse-height discriminator.

pulse-height spectrum [PHYS] Distribution of various pulse wavelengths and strengths (heights) developed during activation analysis.

pulse improvement threshold [COMMUN] In a constant-amplitude pulse-modulation system, the condition in which the peak pulse voltage is greater than twice the peak noise voltage, after selection and before nonlinear processes such as amplitude clipping and limiting.

pulse integrator [ELECTR] An RC (resistance-capacitance) circuit which stretches in time duration a pulse applied to it.

pulse interference eliminator [ELECTR] Device which removes pulsed signals which are not precisely on the radar operating frequency.

pulse interference separator and blanker [ELECTR] Automatic interference blanker that will blank all video signals not synchronous with the radar pulse-repetition frequency.

pulse interleaving [COMMUN] A process in which pulses from two or more sources are combined in time-division multiplex for transmission over a common path.

pulse interval *See* pulse spacing.

pulse-interval modulation *See* pulse-spacing modulation.

pulsejet engine [AERO ENG] A type of compressorless jet engine in which combustion occurs intermittently so that the engine is characterized by periodic surges of thrust; the inlet end of the engine is provided with a grid to which are attached flap valves; these can be sucked inward by a negative differential pressure to allow a regulated amount of air to flow inward to mix with the fuel. Also known as aeropulse engine.

pulse jitter [COMMUN] A relatively small variation of the pulse spacing in a pulse train; the jitter may be random or systematic, depending on its origin, and is generally not coherent with any pulse modulation imposed.

pulse length *See* pulse duration.

pulse-length modulation *See* pulse-duration modulation.

pulse-link repeater [ELECTR] Arrangement of apparatus used in telephone signaling systems for receiving pulses from one E and M signaling circuit, and retransmitting corresponding pulses into another E and M signaling circuit.

pulse-modulated jamming [COMMUN] Use of jamming pulses of various widths and repetition rates.

pulse-modulated radar [ENG] Form of radar in which the radiation consists of a series of discrete pulses.

pulse modulation [COMMUN] A system of modulation in which the amplitude, duration, position, or mere presence of discrete pulses may be so controlled as to represent the message to be communicated.

pulse noise [ELECTR] Noise due to a succession of separated pulses. Also known as impulse noise.

pulse-numbers modulation [COMMUN] Modulation in which a pulse carrier's pulse density per unit time varies in accordance with a modulating wave, by making systematic omissions without changing the phase or amplitude of the transmitted pulses; as an example, the omission of every other pulse could correspond to zero modulation; the reinsertion of some or all pulses then corresponds to positive modulation, and the omission of more than every other pulse corresponds to negative modulation.

pulse operation [ELECTR] For microwave tubes, a method of operation in which the energy is delivered in pulses.

pulse period [COMMUN] In telephony, time required for one opening and closing of the loop of a calling telephone; for example, the time required to open and close the dial pulse springs once. Also known as impulse period.

pulse-phase modulation *See* pulse-position modulation.

pulse-position modulation [COMMUN] Modulation of a pulse carrier wherein the value of each instantaneous sample of a modulating wave varies the position in time of a pulse relative to its unmodulated time of occurrence. Abbreviated PPM. Also known as pulse-phase modulation.

pulse pressure [PHYSIO] The difference between the systolic and diastolic blood pressure.

pulser [CHEM ENG] Device used to create a pulsating fluid flow through a process vessel, such as a liquid-liquid or vapor-liquid extraction tower; used to increase contact and mass transfer rates. [ELECTR] A generator used to produce high-voltage, short-duration pulses, as required by a pulsed microwave oscillator or a radar transmitter.

pulse radar [ENG] Radar in which the transmitter sends out high-power pulses that are spaced far apart in comparison with the duration of each pulse; the receiver is active for reception of echoes in the interval following each pulse.

pulse rate [PHYSIO] The number of pulsations of an artery per minute.

pulse recurrence rate *See* pulse repetition rate.

pulse recurrence time [COMMUN] Time elapsing between the start of one transmitted pulse and the next pulse; the reciprocal of the pulse repetition rate.

pulse regeneration [ELECTR] The process of restoring pulses to their original relative timings, forms, and magnitudes.

pulse repeater [ELECTR] Device used for receiving pulses from one circuit and transmitting corresponding pulses into another circuit; it may also change the frequencies and waveforms of the pulses and perform other functions.

pulse repetition frequency *See* pulse repetition rate.

pulse repetition rate [ELECTR] The number of times per second that a pulse is transmitted. Abbreviated PRR. Also known as pulse recurrence rate; pulse repetition frequency (PRF).

pulse rise time [COMMUN] The interval of time required for the leading edge of a pulse to rise from 10% to 90% of the peak pulse amplitude.

pulse scaler [ELECTR] A scaler that produces an output signal when a prescribed number of input pulses has been received.

pulse selector [ELECTR] A circuit or device for selecting the proper pulse from a sequence of telemetering pulses.

pulse shaper [ELECTR] A transducer used for changing one or more characteristics of a pulse, such as a pulse regenerator or pulse stretcher.

pulse spacing [PHYS] Time between corresponding points of successive pulses. Also known as pulse interval.

pulse-spacing modulation [COMMUN] A form of pulse-time modulation in which the pulse spacing is varied. Also known as pulse-interval modulation.

pulse spectrum [PHYS] The frequency distribution of the sinusoidal components of a pulse in relative amplitude and in relative phase. Also known as pulse-frequency spectrum.

pulse stretcher [ELECTR] A pulse shaper that produces an output pulse whose duration is greater than that of the input pulse and whose amplitude is proportional to the peak amplitude of the input pulse.

pulse subcarrier [COMMUN] One of a number of frequency-modulation carriers modulating a radio-frequency carrier, each of which is in turn pulse-modulated.

pulse synthesizer [ELECTR] A circuit used to supply pulses that are missing from a sequence due to interference or other causes.

pulse-time-modulated radiosonde [ENG] A radiosonde which transmits the indications of the meteorological sensing elements in the form of pulses spaced in time; the meteorological data are evaluated from the intervals between the pulses. Also known as time-interval radiosonde.

pulse-time modulation [COMMUN] Modulation in which the time of occurrence of some characteristic of a pulse carrier is varied from the unmodulated value; examples include pulse-duration, pulse-interval, and pulse-position modulation. Abbreviated PTM.

pulse tracking system [ENG] Tracking system which uses a high-energy, short-duration pulse radiated toward the target from which the velocity, direction, and range are determined by the characteristics of the reflected pulse.

pulse train [PHYS] A series of regularly recurrent pulses having similar characteristics. Also known as pulse group.

pulse-train analysis [COMMUN] A Fourier analysis of a pulse train.

pulse transformer [ELECTR] A transformer capable of operating over a wide range of frequencies, used to transfer nonsinusoidal pulses without materially changing their waveforms.

pulse transmitter [ELECTR] A pulse-modulated transmitter whose peak-power-output capabilities are usually large with respect to the average-power-output rating.

pulse-type altimeter *See* radar altimeter.

pulse-type telemetering [COMMUN] Signal transmission system with pulses as a function of time, but independent of electrical magnitude; in a pulse-counting system the number of pulses per unit time corresponds to the measured variable; in pulse-width or pulse-duration types, the length of the pulse is controlled by the measured variable.

pulse width *See* pulse duration.

pulse-width discriminator [ELECTR] Device that measures the pulse length of video signals and passes only those whose time duration falls into some predetermined design tolerance.

pulse-width modulation *See* pulse-duration modulation.

pulse-width modulation-frequency modulation *See* pulse-duration modulation-frequency modulation.

pulsing key [COMMUN] 1. Method of passing voice frequency pulses over the line under control of a key at the original office; used with E and M supervision on intertoll dialing. 2. System of signaling where numbered keys are depressed instead of using a dial.

pulsing transformer [ELEC] Transformer designed to supply pulses of voltage or current.

pulverization *See* comminution.

pulverizer [MECH ENG] Device for breaking down of solid lumps into a fine material by cleavage along crystal faces.

pulvillus [INV ZOO] A small cushion or cushionlike pad, often covered with short hairs, on an insect's foot between the claws of the last segment.

pulvinus [BOT] A cushionlike enlargement of the base of a petiole which functions in turgor movements of leaves.

puma [VERT ZOO] *Felis concolor.* A large, tawny brown wild cat (family Felidae) once widespread over most of the Americas. Also known as American lion; catamount; cougar; mountain lion.

pumice [GEOL] A rock froth, formed by the extreme puffing up of liquid lava by expanding gases liberated from solution in the lava prior to and during solidification. Also known as foam; pumice stone; pumicite; volcanic foam.

pumice fall [GEOL] Pumice falling from a volcano eruption cloud.

pumice stone *See* pumice.

pumicite *See* pumice.

pump [ELECTR] Of a parametric device, the source of alternating-current power which causes the nonlinear reactor to behave as a time-varying reactance. [MECH ENG] A machine that draws a fluid into itself through an entrance port and forces the fluid out through an exhaust port.

pumpability [MATER] 1. The property of a lubricating grease that causes it to flow under pressure through lines, nozzles, and fittings. 2. The ability of any liquid, slurry, or suspension to be moved through a flow conduit by pressure from a pump.

pumpability test [ENG] Standard test to ascertain the lowest temperature at which a petroleum fuel oil may be pumped.

pumparound [CHEM ENG] A system or process vessel that moves liquid out of and back into the vessel at a new location; for example, in a bubble tower, the withdrawing of liquid from a plate or tray, followed by cooling, and returning to another plate to induce condensation of vapors.

pump bob [MECH ENG] A device such as a crank that converts rotary motion into reciprocating motion.

pumped hydroelectric storage [ELEC] A method of energy storage in which excess electrical energy produced at times of low demand is used to pump water into a reservoir, and this water is released at times of high demand to operate hydroelectric generators.

pumped tube [ELECTR] An electron tube that is continuously connected to evacuating equipment during operation; large pool-cathode tubes are often operated in this manner.

pumpellyite [MINERAL] $Ca_2Al_3Si_3O_{12}(OH)$ A greenish epidotelike mineral that is probably related to clinozoisite. Also known as lotrite; zonochlorite.

pumpellyite-prehnite-quartz facies [PETR] A variety of low-temperature, moderate-pressure metamorphism.

pumphouse [CIV ENG] A building in which are housed pumps that supply an irrigation system, a power plant, a factory, a reservoir, a farm, a home, and so on.

pumping [FL MECH] Unsteadiness of the mercury in the barometer, caused by fluctuations of the air pressure produced by a gusty wind or due to the motion of a vessel. [PHYS] 1. The application of optical, infrared, or microwave radiation of appropriate frequency to a laser or maser medium so that absorption of the radiation increases the population of atoms or molecules in higher energy states. Also known as electronic pumping. 2. The removal of gases and vapors from a vacuum system.

pumping frequency [ELECTR] Frequency at which pumping is provided in a maser, quadrupole amplifier, or other amplifier requiring high-frequency excitation.

pumping pressure [PETRO ENG] Pressure required to inject (pump) water, gas, or acid into a pressurized petroleum reservoir.

pumping radiation [PHYS] Electromagnetic radiation applied to a laser or maser in the process of pumping.

pumping station [CIV ENG] A building in which two or more pumps operate to supply fluid flowing at adequate pressure to a distribution system.

pumping well [PETRO ENG] A producing oil well in which liquid products are recovered from the reservoir by means of a pump, rather than by gas lift.

pumpkin [BOT] Any of several prickly vines with large lobed leaves and yellow flowers in the genus Cucurbita of the order Violales; the fruit is orange-colored and large, with a firm rind.

pump oscillator [ELECTR] Alternating-current generator that supplies pumping energy for maser and parametric amplifiers; operates at twice or some higher multiple of the signal frequency.

puna [ECOL] An alpine biological community in the central portion of the Andes Mountains of South America characterized by low-growing, widely spaced plants that lack much green color most of the year.

punch [MECH ENG] A tool that forces metal into a die for extrusion or similar operations.

punch card [ADP] A medium by means of which data are fed into a computer in the form of rectangular holes punched in the card. Also known as punched card.

punched card *See* punch card.

punched-card accounting machines [ADP] Machines and equipment, primarily electromechanical in operation, using punched cards as input/output media to record, sort, list, tabulate, select, merge, interpret, or total data.

punched-card equipment [ADP] Data-processing equipment which is essentially card-oriented and is predominately electromechanical, such as encoders, key punchers, key tapes, mechanical sorters, collators, and tabulators.

punched-card field [ADP] A set of columns fixed as to number and position into which the same item or items of data are regularly entered.

punched-card interpreter *See* interpreter.

punched-card reader *See* card reader.

punched-card reproducer *See* reproducer.

punched-card sorter *See* card sorter.

punched-plate screen [ENG] Flat, perforated plate with round, square, hexagonal, or elongated openings; used for screening (size classification) of crushed or pulverized solids.

punched tape *See* punch tape.

punched-tape reader *See* paper-tape reader.

PUNA

Landscape of the puna in the Lake Titicaca area, Bolivia.

punching [ENG] **1.** A piece removed from a sheet of metal or other material by a punch press. **2.** A method of extrusion, cold heading, hot forging, or stamping in a machine for which mating die sections determine the shape or contour of the work.

punching rate [ADP] The number of cards, characters, blocks, fields, or words of information placed in the form of holes distributed on cards, or paper tape per unit of time.

punch position [ADP] The location of the row in a columnated card; for example, in an 80-column card the rows or punch position may be 0 to 9 or X and Y corresponding to positions 11 and 12.

punch press [MECH ENG] **1.** A press consisting of a frame in which slides or rams move up and down, of a bed to which the die shoe or bolster plate is attached, and of a source of power to move the slide. Also known as drop press. **2.** Any mechanical press.

punch radius [DES ENG] The radius on the bottom end of the punch over which the metal sheet is bent in drawing.

punch tape [ADP] A paper or plastic ribbon in which data may be represented by means of partially or completely punched holes; it generally has one row of small sprocket-feed holes and five, seven, or eight rows of larger data-representing holes. Also known as punched tape.

punch-tape code [ADP] A code used to represent data on punch tape.

punch-through [ELECTR] An emitter-to-collector breakdown which can occur in a junction transistor with very narrow base region at sufficiently high collector voltage when the space-charge layer extends completely across the base region.

punctate [BIOL] Dotted; full of minute points.

puncture [ELEC] Disruptive discharge through insulation involving a sudden and large increase in current through the insulation due to complete failure under electrostatic stress. [SCI TECH] To pierce or indent.

puncture voltage [ELEC] The voltage at which a test specimen is electrically punctured.

punt [NAV ARCH] **1.** A heavily built boat of rectangular shape used by workmen employed in painting, cleaning, or repairing a ship's topsides when in sheltered waters. **2.** A square-ended boat used on shallow rivers and lakes, often propelled by poles.

pupa [INV ZOO] The quiescent, intermediate form assumed by an insect that undergoes complete metamorphosis; it follows the larva and precedes the adult stages and is enclosed in a hardened cuticle or a cocoon.

pupate [INV ZOO] **1.** To develop into a pupa. **2.** To pass through a pupal stage.

pupil [ANAT] The contractile opening in the iris of the vertebrate eye.

pupillary reflex [PHYSIO] **1.** Contraction of the pupil in response to stimulation of the retina by light. Also known as Whytt's reflex. **2.** Contraction of the pupil on accommodation for close vision, and dilation of the pupil on accommodation for distant vision. **3.** Contraction of the pupil on attempted closure of the eye. Also known as Westphal-Pilcz reflex; Westphal's pupillary reflex.

Pupin coil *See* loading coil.

Pupipara [INV ZOO] A section of cyclorrhaphous dipteran insects in the Schizophora series in which the young are born as mature maggots ready to become pupae.

pup jack *See* tip jack.

Puppis [ASTRON] A southern constellation; right ascension 8 hours, declination 40° south. Also known as Stern.

Purbeckian [GEOL] A stage of geologic time in Great Britain: uppermost Jurassic (above Bononian, below Cretaceous).

pure coal *See* vitrain.

pure culture [MICROBIO] A culture that contains cells of one kind, all progeny of a single cell.

pure forest [FOR] A forest in which one species makes up 80% or more of the total number of trees.

pure geometry [MATH] Geometry studied from the standpoint of its axioms and postulates rather than its objects.

pure imaginary number [MATH] A complex number $z = x + iy$, where $x = 0$.

pure mathematics [MATH] The intrinsic study of mathematical structures, with no consideration given as to the utility of the results for practical purposes.

PURINE

Structural formula of purine.

pure projective geometry [MATH] The axiomatic study of geometric systems which exhibit invariance relative to a notion of projection.

pure research *See* basic research.

pure tone *See* simple tone.

purga [METEOROL] A severe storm similar to the blizzard and buran, which rages in the tundra regions of northern Siberia in winter.

purging [MED] The condition in which there is rapid and continuous evacuation of the bowels. [SCI TECH] The act or process of cleaning and purifying.

purified ozokerite *See* ceresin wax.

purify [ENG] To remove unwanted constituents from a substance.

purine [BIOCHEM] A heterocyclic compound containing fused pyrimidine and imidazole rings; adenine and guanine are the purine components of nucleic acids and coenzymes.

purity [ANALY CHEM] *See* excitation purity. [CHEM] The state of a chemical compound when no impurity can be detected by any experimental method; absolute purity is never reached in practice. [OPTICS] The degree to which a primary color is pure and not mixed with the other two primary colors.

purity coil [ELECTR] A coil mounted on the neck of a color picture tube, used to produce the magnetic field needed for adjusting color purity; the direct current through the coil is adjusted to a value that makes the magnetic field orient the three individual electron beams so each strikes only its assigned color of phosphor dots.

purity control [ELECTR] A potentiometer or rheostat used to adjust the direct current through the purity coil.

purity magnet [ELECTR] An adjustable arrangement of one or more permanent magnets used in place of a purity coil in a color television receiver.

purity of state [STAT MECH] Property of a system which is definitely in a certain quantum state, rather than having a certain probability of being in any of several quantum states.

Purkinje cell [HISTOL] Any of the cells of the cerebral cortex with large, flask-shaped bodies forming a single cell layer between the molecular and granular layers.

Purkinje effect [PHYSIO] When illumination is reduced to a low level, slowly enough to allow adaptation by the eye, the sensation produced by the longer-wave stimuli (red, orange) decreases more rapidly than that produced by shorter-wave stimuli (violet, blue). Also spelled Parkinje effect.

Purkinje fibers [HISTOL] Modified cardiac muscle fibers composing the terminal portion of the conducting system of the heart.

purlin [BUILD] A horizontal roof beam, perpendicular to the trusses or rafters; supports the roofing material or the common rafters.

puromycin [MICROBIO] $C_{22}H_{29}O_5N_7$ A colorless, crystalline broad-spectrum antibiotic produced by a strain of *Streptomyces*.

purple bacteria [MICROBIO] Any of various photosynthetic bacteria that contain bacteriochlorophyll, distinguished by purplish or reddish-brown pigments.

purple blende *See* kermesite.

purple blotch [PL PATH] A fungus disease of onions, garlic, and shallots caused by *Alternaria porri* and characterized by small white spots which become large purplish blotches.

purple boundary [OPTICS] A straight line connecting the ends of the spectrum locus on the chromaticity diagram.

purple lakes [MATER] A class of lake (pigment) used in printing inks; derived from a combination of such compounds as β-hydroxynaphthoic acid and 2-diazonaphthalene-1-sulfonic acid.

purple light [GEOPHYS] The faint purple glow observed on clear days over a large region of the western sky after sunset and over the eastern sky before sunrise.

purple nonsulfur bacteria [MICROBIO] Any of various purple photosynthetic bacteria, especially members of the family Athiorhodaceae, that utilize organic hydrogen donor compounds.

purple plague [ELECTR] A compound formed by intimate contact of gold and aluminum, which appears on silicon planar devices and integrated circuits using gold leads bonded to aluminum thin-film contacts and interconnections, and which seriously degrades the reliability of semiconductor devices.

purple salt *See* potassium permanganate.

purple sulfur bacteria [MICROBIO] Any of various anaerobic photosynthetic purple bacteria, especially in the family Thiorhodaceae, that utilize H_2S and other inorganic sulfur compounds as a source of hydrogen, while the carbon source can be carbon monoxide.

purple-top [PL PATH] A virus disease of potato plants characterized by purplish or chlorotic discoloration of the top shoots, swelling of axillary branches, and severe wilting.

purpura [MED] Spontaneous hemorrhages into tissues such as joints, skin, and mucosal surfaces.

purpurin [ORG CHEM] $C_{14}H_8O_5$ A compound crystallizing as long orange needles from dilute alcohol solutions; used in the manufacture of dyes, and as a reagent for the detection of boron. Also known as natural red; 1,2,4-trihydroxyanthraquinone.

purpurite [MINERAL] $(Mn,Fe)PO_4$ A dark-red or purple mineral composed of ferric-manganic phosphate; it is isomorphous with heterosite.

purse seine [ENG] A net that can be dropped by two boats to encircle a school of fish, then pulled together at the bottom and raised, thereby catching the fish.

purse seiners [NAV ARCH] Fishing boats equipped to fish with a purse seine.

purulent [MED] Consisting of, containing, or forming pus.

purupuru *See* pinta.

pus [MED] A viscous, creamy, pale-yellow or yellow-green fluid produced by liquefaction necrosis in a neutrophil-rich exudate.

push-bar conveyor [MECH ENG] A type of chain conveyor in which two endless chains are cross-connected at intervals by push bars which propel the load along a stationary bed or trough of the conveyor.

push bench [MECH ENG] A machine used for drawing tubes of moderately heavy gage by cupping metal sheet and applying pressure to the inside bottom of the cup to force it through a die.

push-button dialing [ELECTR] Dialing a number by pushing buttons on the telephone rather than turning a circular wheel; each depressed button causes a transistor oscillator to oscillate simultaneously at two different frequencies, generating a pair of audio tones which are recognized by central-office (or PBX) switching equipment as digits of a telephone number. Also known as tone dialing; touch call; Touch-Tone.

push-button switch [ELEC] A master switch that is operated by finger pressure on the end of an operating button.

push-button tuner [ELECTR] A device that automatically tunes a radio receiver or other piece of equipment to a desired frequency when the button assigned to that frequency is pressed.

push-down list [ADP] An ordered set of data items so constructed that the next item to be retrieved is the item most recently stored; in other words, last-in, first-out (LIFO).

pusher grade *See* helper grade.

pusher tractor [MIN ENG] A bulldozer exerting pressure on the rear of a scraper-loader while the loader is digging and loading unconsolidated ground during opencast mining.

push fit [DES ENG] A hand-tight sliding fit between a shaft and a hole.

pushing [ELEC] A change in the resonant frequency of a circuit due to changes in the applied voltages.

push-pull amplifier [ELECTR] A balanced amplifier employing two similar electron tubes or equivalent amplifying devices working in phase opposition.

push-pull currents *See* balanced currents.

push-pull electret transducer [ELECTR] A type of transducer in which a foil electret is sandwiched between two electrodes and is specially treated or arranged so that the electrodes exert forces in opposite directions on the diaphragm, and the net force is a linear function of the applied voltage.

push-pull magnetic amplifier [ELECTR] A realization of a push-pull amplifier using magnetic amplifiers.

push-pull oscillator [ELECTR] A balanced oscillator employing two similar electron tubes or equivalent amplifying devices in phase opposition.

push-pull sound track [ENG ACOUS] A sound track having two recordings so arranged that the modulation in one is 180° out of phase with that in the other.

push-pull transformer [ELECTR] An audio-frequency transformer having a center-tapped winding and designed for use in a push-pull amplifier.

push-pull transistor [ELECTR] **1.** A realization of a push-pull amplifier using transistors. **2.** A Darlington circuit in which the two transistors required for a push-pull amplifier exist in a single substrate.

push-pull voltages *See* balanced voltages.

push-push amplifier [ELECTR] An amplifier employing two similar electron tubes with grids connected in phase opposition and with anodes connected in parallel to a common load; usually used as a frequency multiplier to emphasize even-order harmonics; transistors may be used in place of tubes.

push rod [MECH ENG] A rod, as in an internal combustion engine, which is actuated by the cam to open and close the valves.

push-to-talk circuit [ELEC] Simplex circuit in which change-over from the receive to transmit state is accomplished by depressing a single spring-return switch, and releasing the switch returns the circuit to the receive state; the push-to-talk switch is located on microphones and telephone handsets; it is most often applied to radio circuits.

push-up list [ADP] An ordered set of data items so constructed that the next item to be retrieved will be the item that was inserted earliest in the list, resulting in a first-in, first-out (FIFO) structure.

push welding [MET] Spot or projection welding in which the force is applied manually by one electrode; the work takes the place of the other electrode. Also known as poke welding.

pustule [MED] A small, circumscribed, pus-filled elevation on the skin. [PL PATH] A blisterlike mark on a leaf due to rupture of surface tissues overlying spore masses of a parasitic fungus.

Pustulosa [PALEON] An extinct suborder of echinoderms in the order Phanerozonida found in the Paleozoic.

pusule [INV ZOO] A noncontractile fluid-filled vacuole emptied by means of a duct; found in dinoflagellates.

putlog [CIV ENG] A crosspiece in a scaffold or formwork; supports the soffits and is supported by the ledgers.

putrefaction [BIOCHEM] Decomposition of organic matter, particularly the anaerobic breakdown of proteins by bacteria, with the production of foul-smelling compounds.

putty [MATER] A cement of dough consistency made of whiting and boiled linseed oil and used in fastening glass in sashes and sealing crevices in woodwork.

putty knife [DES ENG] A knife with a broad flexible blade, used to apply and smooth putty.

putty oil [MATER] Petroleum oil that is added to putty; serves to lubricate the putty and keep it soft after the linseed oil dries.

PVA *See* polyvinyl acetate; polyvinyl alcohol.

PVAc *See* polyvinyl acetate.

PVC *See* polyvinyl chloride.

PVI *See* polyvinyl isobutyl ether.

PVP *See* polyvinyl pyrrolidone.

pW *See* picowatt.

P wave *See* compressional wave.

PWR *See* pressurized water reactor.

pwt *See* pennyweight.

pyarthrosis [MED] Suppuration involving a joint.

pycnidiospore [MYCOL] A conidium produced by a pycnidium.

pycnidium [MYCOL] A cavity that bears pycnidiospores in certain fungi.

pycniospore [MYCOL] A haploid spore of a rust fungus that fuses with a haploid hypha of opposite sex to produce dikaryotic aeciospores.

pycnium [MYCOL] A flask-shaped fruit body of a rust fungus formed in clusters just beneath the surface of a host tissue.

Pycnodontiformes [PALEON] An extinct order of specialized fishes characterized by a laterally compressed, disk-shaped body, long dorsal and anal fins, and an externally symmetrical tail.

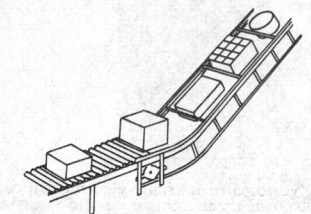

PUSH-BAR CONVEYOR

Push-bar conveyor, a basic type of chain conveyor.

PUSH-BUTTON DIALING

Desk-type telephone equipped for push-button dialing.

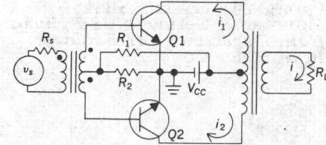

PUSH-PULL AMPLIFIER

Circuit diagram of two transistors, $Q1$ and $Q2$, in a push-pull arrangement. Here, v_S = source voltage, R_S = source resistance, R_L = load resistance, V_{CC} = collector supply voltage.

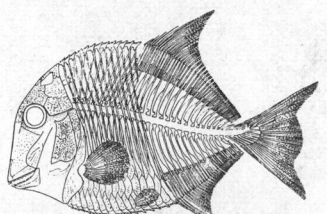

PYCNODONTIFORMES

Coelodus costae Heckel, a pycnodont from Lower Cretaceous of Italy; length to 4 inches (10 centimeters). *(After A. S. Woodward)*

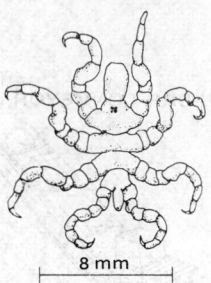

PYCNOGONIDA

Pycnogonum stearnsi, a common littoral pycnogonid of the Pacific Coast.

PYRARGYRITE

Pyrargyrite crystals, Harz Mountains, Germany. *(Specimen from Department of Geology, Bryn Mawr College)*

PYRIMIDINE

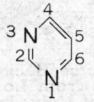

Structural formula of pyrimidine.

Pycnogonida [INV ZOO] The sea spiders, a subphylum of marine arthropods in which the body is reduced to a series of cylindrical trunk somites supporting the appendage.

Pycnogonidae [INV ZOO] A family of the Pycnogonida lacking both chelifores and palpi and having six to nine jointed ovigers in the male only.

pycnometer [ENG] A container whose volume is precisely known, used to determine the density of a liquid by filling the container with the liquid and then weighing it.

pycnometry [PHYS] The determination of liquid density by weighing the liquid in a container (pycnometer) of known volume.

pyelitis [MED] Inflammation of the renal pelvis.

pyelolithotomy [MED] Excision of a renal calculus through an incision in the renal pelvis.

pyelonephritis [MED] The disease process resulting from the effects of infections of the parenchyma and the pelvis of the kidney. Also known as interstitial nephritis.

pyemia [MED] A disease state due to the presence of pyogenic microorganisms in the blood and the formation, wherever these organisms lodge, of embolic or metastatic abscesses.

Pygasteridae [PALEON] The single family of the extinct order Pygasteroida.

Pygasteroida [PALEON] An order of extinct echinoderms in the superorder Diadematacea having four genital pores, noncrenulate tubercles, and simple ambulacral plates.

pygidium [INV ZOO] **1.** A caudal shield on the abdomen of some Arthropoda. **2.** The terminal body segment of many invertebrates.

Pygopodidae [VERT ZOO] The flap-footed lizards, a family of the suborder Sauria.

pyknosis [PATH] The polymerization and contraction of the nuclear chromosomal components.

pylephlebitis [MED] Inflammation of the portal vein.

pylome [INV ZOO] An aperture for emission of pseudopodia and intake of food in some Sarcodina.

pylon [AERO ENG] A suspension device externally installed under the wing or fuselage of an aircraft; it is aerodynamically designed to fit the configuration of specific aircraft, thereby creating an insignificant amount of drag; it includes means of attaching to accommodate fuel tanks, bombs, rockets, torpedoes, rocket motors, or the like. [CIV ENG] **1.** A massive structure, such as a truncated pyramid, on either side of an entrance. **2.** A tower supporting a wire over a long span. **3.** A tower or other structure marking a route for an airplane.

pyloric caecum [INV ZOO] **1.** One of the tubular pouches that open into the ventriculus of an insect. **2.** One of the paired tubes having lateral glandular diverticula in each ray of a starfish. [VERT ZOO] One of the tubular pouches that open from the pyloric end of the stomach into the alimentary canal of most fishes.

pyloric sphincter [ANAT] The thickened ring of circular smooth muscle at the lower end of the pyloric canal of the stomach.

pyloric stenosis [MED] Obstruction of the pyloric opening of the stomach due to hypertrophy of the pyloric sphincter.

pylorospasm [MED] Spasm of the pylorus.

pylorus [ANAT] The orifice of the stomach communicating with the small intestine.

pyobacillosis [VET MED] A bacterial infection of sheep, swine, or rarely cattle caused by *Corynebacterium pyogenes;* usually marked by abscess formation, but in sheep takes the form of chronic purulent pneumonia.

pyocyanin [MICROBIO] $C_{13}H_{10}N_{20}$ An antibiotic substance forming blue crystals, produced by *Pseudomonas aeruginosa;* active against many bacteria and fungi.

pyoderma [MED] Any pus-producing skin lesion or lesions, used in reference to groups of furuncles, pustules, or even carbuncles.

pyonephritis [MED] Suppurative inflammation of a kidney.

pyonephrosis [MED] Replacement of renal tissue by abscesses.

pyorrhea [MED] A purulent discharge.

Pyralidae [INV ZOO] The equivalent name for Pyralididae.

Pyralididae [INV ZOO] A large family of moths in the lepidopteran superfamily Pyralidoidea; the labial palpi are well developed, and the legs are usually long and slender.

Pyralidinae [INV ZOO] A subfamily of the Pyralididae.

Pyralidoidea [INV ZOO] A superfamily of the Lepidoptera belonging to the Heteroneura and including long-legged, slender-bodied moths with well-developed maxillary palpi.

pyramid [CRYSTAL] An open crystal having three, four, six, eight, or twelve nonparallel faces that meet at a point. [MATH] A polyhedron with one face a polygon and all other faces triangles with a common vertex.

pyramidal area *See* motor area.

pyramidal horn [ENG] Horn whose sides form a pyramid.

pyramidal iceberg *See* pinnacled iceberg.

pyramidal surface [MATH] A surface generated by a line passing through a fixed point and moving along a broken line in a plane not containing that point.

pyramidal system [ANAT] The corticospinal and corticobulbar tracts.

Pyramidellidae [INV ZOO] A family of gastropod mollusks in the order Tectibranchia; the operculum is present in this group.

pyramid of numbers [ECOL] The concept that an organism making up the base of a food chain is numerically abundant while each succeeding member of the chain is represented by successively fewer individuals; uses feeding relationship as a basis for the quantitative analysis of an ecological system.

pyranometer [ENG] An instrument used to measure the combined intensity of incoming direct solar radiation and diffuse sky radiation; compares heating produced by the radiation on blackened metal strips with that produced by an electric current. Also known as solarimeter.

pyrargyrite [MINERAL] Ag_3SbS_3 A deep ruby-red to black mineral, crystallizing in the hexagonal system, occurring in massive form and in disseminated grains, and having an adamantine luster; hardness is 2.5 on Mohs scale, and specific gravity is 5.85; an important silver ore. Also known as dark-red silver ore; dark ruby silver.

Pyraustinae [INV ZOO] A large subfamily of the Pyralididae containing relatively large, economically important moths.

pyrazolone dye [ORG CHEM] An acid dye containing both $-N=N-$ and $=C=C=$ chromophore groups, such as tartrazine; used for silk and wool.

pyrenoid [BOT] A colorless body found within the chromatophore of certain algae; a center for starch formation and storage.

Pyrenolichenes [BOT] The equivalent name for Pyrenulales.

Pyrenulaceae [BOT] A family of the Pyrenulales; all species are crustose and most common on tree bark in the tropics.

Pyrenulales [BOT] An order of the class Ascolichenes including only those lichens with perithecia that contain true paraphyses and unitunicate asci.

pyrethrum [MATER] A toxicant obtained in the form of dried powdered flowers of the plant of the same name; mixed with petroleum distillates, it is used as an insecticide.

pyrexia [MED] Elevation of temperature above the normal; fever.

pyrgeometer [ENG] An instrument for measuring radiation from the surface of the earth into space.

Pyrgotidae [INV ZOO] A family of myodarian cyclorrhaphous dipteran insects in the subsection Acalypteratae.

pyrheliometer [ENG] An instrument for measuring the total intensity of direct solar radiation received at the earth.

pyridine [ORG CHEM] C_5H_5N Organic base; flammable, toxic yellowish liquid, with penetrating aroma and burning taste; soluble in water, alcohol, ether, benzene, and fatty oils; boils at 116°C; used as an alcohol denaturant, solvent, in paints, medicine, and textile dyeing.

pyridine-4-carboxylic acid *See* isonicotinic acid.

pyridoxal hydrochloride *See* pyridoxine hydrochloride.

pyridoxal phosphate *See* codecarboxylase.

pyridoxine hydrochloride [BIOCHEM] $C_8H_{11}NO_3 \cdot HCl$ A crystalline compound, decomposing at about 208°C; used in medicine in vitamin therapy. Also known as pyridoxal hydrochloride; vitamin B_6 hydrochloride.

β-pyridyl-α-N-methylpyrrolidine *See* nicotine.

pyrimidine [BIOCHEM] $C_4H_4N_2$ A heterocyclic organic compound containing nitrogen atoms at positions 1 and 3;

naturally occurring derivatives are components of nucleic acids and coenzymes.

pyrite [MINERAL] FeS_2 A hard, brittle, brass-yellow mineral with metallic luster, crystallizing in the isometric system; hardness is 6–6.5 on Mohs scale, and specific gravity is 5.02. Also known as common pyrite; fool's gold; iron pyrite; mundic.

pyrite roasting [MIN ENG] Thermal processing of iron pyrite (FeS_2, iron disulfide) in the presence of air to produce iron oxide sinter (used in steel mills) and elemental sulfur.

pyritization [GEOL] A common process of hydrothermal alteration involving introduction of or replacement by pyrite.

pyritobitumen [GEOL] Any of various dark-colored, relatively hard, nonvolatile hydrocarbon substances often associated with mineral matter, which decompose upon heating to yield bitumens. Also known as pyrobitumen.

pyritohedron [CRYSTAL] A dodecahedral crystal with 12 irregular pentagonal faces; it is characteristic of pyrite. Also known as pentagonal dodecahedron; pyritoid; regular dodecahedron.

pyritoid *See* pyritohedron.

pyro- [CHEM] A chemical prefix for compounds formed by heat, such as pyrophosphoric acid, an inorganic acid formed by the loss of one water molecule from two molecules of an ortho acid.

pyroaurite [MINERAL] $Mg_6Fe_2(OH)_{16} \cdot CO_3 4H_2O$ A gold-like or brownish rhombohedral mineral composed of hydrous basic magnesium iron carbonate.

pyrobelonite [MINERAL] $PbMn(VO_4)(OH)$ A fire-red to deep brilliant-red mineral composed of basic vanadate of manganese and lead, occurring as crystal needles.

pyrobitumen *See* pyritobitumen.

pyroborate *See* borax.

pyrocatechol *See* catechol.

pyrocatechuic acid *See* catechol.

pyrocellulose [ORG CHEM] Highly nitrated cellulose; used to make explosives; originally called guncotton in the United States, cordite in England.

pyroceram [MATER] Hard, strong, opaque-white nucleated glass with nonporous crystalline structure; softens at 2460°F (1350°C); has high flexural strength and shock resistance; used for molded mechanical and electrical parts, heat-exchanger tubes, and coatings.

pyrochlore [MINERAL] $(Na,Ca)_2(Nb,Ta)_2O_6(OH,F)$ Pale-yellow, reddish, brown, or black mineral, crystallizing in the isometric system, and occurring in pegmatites derived from alkalic igneous rocks. Also known as pyrrhite.

pyrochroite [MINERAL] $Mn(OH)_2$ A hexagonal mineral composed of naturally occurring manganese hydroxide; it is white when fresh, but darkens upon exposure.

pyroclast [GEOL] An individual pyroclastic fragment or clast.

pyroclastic flow [GEOL] Ash flow not involving high-temperature conditions.

pyroclastic ground surge [GEOL] The relatively thin mantle of rock found around a volcanic vent; the thickness is not uniform, the internal stratification is not parallel to the top and bottom of the layer, and the extent is a few kilometers from the source.

pyroclastic rock [PETR] A rock that is composed of fragmented volcanic products ejected from volcanoes in explosive events.

pyroconductivity [SOLID STATE] Electrical conductivity that develops in a material only at high temperature, chiefly at fusion, in solids that are practically nonconductive at atmospheric temperatures.

pyroelectric crystal [SOLID STATE] A crystal exhibiting pyroelectricity, such as tourmaline, lithium sulfate monohydrate, cane sugar, and ferroelectric barium titanate.

pyroelectricity [SOLID STATE] The property of certain crystals to produce a state of electrical polarity by a change of temperature.

pyrogallic acid [ORG CHEM] $C_6H_3(OH)_3$ Lustrous, light-sensitive white crystals, melting at 133°C; soluble in alcohol, ether, and water; used for photography, dyes, drugs, medicines, and process engravings, and as an analytical reagent

and protective colloid. Also known as pyrogallol; 1,2,3,-trihydroxybenzene.

pyrogallol *See* pyrogallic acid.

pyrogallolphthalein *See* gallein.

pyrogen [BIOCHEM] A group of substances thought to be polysaccharides of microbial origin that produce an increase in body temperature when injected in man and some animals.

pyrolithic acid [ORG CHEM] $HOC(NCOH)_2N \cdot 2H_2O$ Colorless monoclinic crystals, slightly soluble in water. Also known as cyanuric acid.

pyrolusite [MINERAL] MnO_2 An iron-black mineral that crystallizes in the tetragonal system and is the most important ore of manganese; hardness is 1–2 on Mohs scale, and specific gravity is 4.75.

pyrolysis [CHEM] The breaking apart of complex molecules into simpler units by the use of heat, as in the pyrolysis of heavy oil to make gasoline.

pyromania [PSYCH] A monomania for setting or watching fires.

pyrometallurgy [MET] High-temperature process metallurgy.

pyrometamorphism [PETR] Contact metamorphism at temperatures near the melting points of the component minerals.

pyrometasomatism [PETR] Forming of contact-metamorphic mineral deposits at high temperatures by emanations from the intrusive rock, involving replacement of the enclosing rock with the addition of materials.

pyrometer [ENG] Any of a broad class of temperature-measuring devices; they were originally designed to measure high temperatures, but some are now used in any temperature range; includes radiation pyrometers, thermocouples, resistance pyrometers, and thermistors.

pyrometric cone *See* Seger cone.

pyrometry [THERMO] The science and technology of measuring high temperatures.

pyromorphite [MINERAL] $Pb_5(PO_4)_3Cl$ A green, yellow, brown, gray, or white mineral of the apatite group, crystallizing in the hexagonal system; a minor ore of lead. Also known as green lead ore.

pyromucic acid *See* furoic acid.

pyron [PHYS] A unit of area-density of power, equal to the area-density of power resulting from a power of one international table calorie per minute acting uniformly over an area of 1 square centimeter; equal to 697.8 watts per square meter.

pyrone detector [ELECTR] Crystal detector in which rectification occurs between iron pyrites and copper or other metallic points.

pyrope [MINERAL] $Mg_3Al_2(SiO_4)_3$ A mineral species of the garnet group characterized by a deep fiery-red color and occurring in basic and ultrabasic igneous rocks.

pyrophane *See* fire opal.

pyrophanite [MINERAL] $MnTiO_3$ A blood-red rhombohedral mineral consisting of manganese titanate; it is isomorphous with ilmenite.

pyrophobia [PSYCH] An abnormal fear of fires.

pyrophoric alloy [MET] **1.** An alloy such as ferrocerium that produces a spark when struck with metal (steel) at an angle; used for automatic cigarette lighters. **2.** An alloy in powder form that spontaneously oxidizes in air, reaching high temperatures.

pyrophoric propellant [MATER] A propellant combination of a liquid fuel and a fluid oxidizer (usually air) that will quickly react when brought into intimate contact and achieve ignition temperature.

pyrophosphatase [BIOCHEM] An enzyme catalyzing hydrolysis of esters containing two or more molecules of phosphoric acid to form a simpler phosphate ester.

pyrophosphoric acid [INORG CHEM] $H_4P_2O_7$ Water-soluble, syrupy liquid melting at 61°C; used as a catalyst and to make organic phosphate esters.

pyrophyllite [MINERAL] $AlSi_2O_5(OH)$ A white, greenish, gray, or brown phyllosilicate mineral that resembles talc and occurs in a foliated form or in compact masses in quartz veins, granites, and metamorphic rocks. Also known as pencil stone.

pyrosin *See* tetraiodofluorescein.

pyrosmalite [MINERAL] $(Mn,Fe)_4Si_3O_7(OH,Cl)_6$ A color-

PYRITE

A pyrite crystal from Rio Marina, Elba, Italy. *(American Museum of Natural History Specimens)*

├── 2.6 cm ──┤

PYROMORPHITE

├── 2.5 cm ──┤

Pyromorphite with hexagonal crystals, from Nassau. *(Specimen from Department of Geology, Bryn Mawr College)*

PYROSOMIDA

Colony of *Pyrosoma atlanticum.*
*(From Metcalf and Hopkins,
after Ritter)*

PYROTHERIA

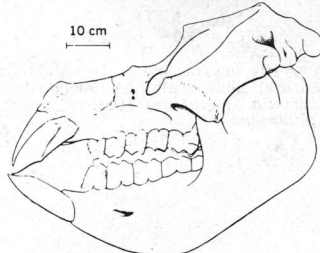

10 cm

Skull and jaw of *Pyrotherium
sorondi,* an early Oligocene
pyrothere from South America.
(After F. Loomis, 1914)

PYRRHOTITE

2.5 cm

Pyrrhotite crystals from
Chihuahua, Mexico. *(Specimen
from Department of Geology,
Bryn Mawr College)*

PYTHON

A python, which kills its prey
by constriction.

less, pale-brown, gray, or gray-green mineral composed mainly of basic iron manganese silicate with chlorine.

Pyrosomida [INV ZOO] An order of pelagic tunicates in the class Thaliacea in which species form tubular swimming colonies and are often highly luminescent.

pyrostat [ENG] **1.** A sensing device that automatically actuates a warning or extinguishing mechanism in case of fire. **2.** A high-temperature thermostat.

pyrostibite *See* kermesite.

pyrostilpmite [MINERAL] Ag_3SbS_3 A hyacinth-red mineral composed of silver antimony sulfide, occurring in monoclinic crystal tufts; it is polymorphous with pyrargerite.

pyrotechnic code [COMMUN] Significant arrangement of the various colors and patterns of fireworks, signal lights, or signal smokes used for communication between units or between ground and air.

pyrotechnic pistol [ENG] A single-shot device designed specifically for projecting pyrotechnic signals.

pyrotechnics [ENG] Art and science of preparing and using fireworks. [MATER] Items which are used for both military and nonmilitary purposes to produce a bright light for illumination, or colored lights or smoke for signaling, and which are consumed in the process.

pyrotechnic signal [COMMUN] Signal designed for military use to produce a colored light or smoke, for the purpose of transmitting information.

Pyrotheria [PALEON] An extinct monofamilial order of primitive, mastodonlike, herbivorous, hoofed mammals restricted to the Eocene and Oligocene deposits of South America.

Pyrotheriidae [PALEON] The single family of the Pyrotheria.

pyrouric acid *See* cyanuric acid.

pyroxene [MINERAL] A family of diverse and important rock-forming minerals having infinite (Si_2O_6) single inosilicate chains as their principal motif; colors range from white through yellow and green to brown and greenish black; hardness is 5.5-6 on Mohs scale, and specific gravity is 3.2-4.0.

pyroxenite [PETR] A heavy, dark-colored, phaneritic igneous rock composed largely of pyroxene with smaller amounts of olivine and hornblende, and formed by crystallization of gabbraic magma.

pyroxenoids [MINERAL] A mineral group (including wollastonite and rhodonite) compositionally similar to pyroxene, but SiO_4 tetrahedrons are connected in rings rather than chains.

pyroxylin [ORG CHEM] $[C_{12}H_{16}O_6(NO_3)_4]_x$ Any member of the group of commercially available nitrocelluloses that are used for properties other than their combustibility; the term is commonly used to identify products that are principally made from nitrocellulose, such as pyroxylin plastic or pyroxylin lacquer. Also known as collodion cotton; soluble gun-cotton; soluble nitrocellulose.

pyroxylin cement [MATER] A solution of nitrocellulose in a chemical solvent, compounded with a resin, or plasticized with a gum or synthetic; dries by evaporation of the solvent.

pyrrhite *See* pyrochlore.

Pyrrhocoridae [INV ZOO] A family of hemipteran insects belonging to the superfamily Pyrrhocoroidea.

Pyrrhocoroidea [INV ZOO] A superfamily of the Pentatomorpha.

Pyrrhophyta [BOT] A small division of motile, generally unicellular flagellate algae characterized by the presence of yellowish-green to golden-brown plastids and by the general absence of cell walls.

pyrrhotite [MINERAL] $Fe_{1-x}S$ A common reddish-brown to brownish-bronze mineral that occurs as rounded grains to large masses, more rarely as tabular pseudohexagonal crystals and rosettes; hardness is 4 on Mohs scale, and specific gravity is 4.6 (for the composition Fe_7S_8).

pyrrole [ORG CHEM] C_4H_5N Water-insoluble, yellowish oil, with pungent taste; soluble in alcohol, ether, and dilute acids; boils at 130°C; polymerizes in light; used to make drugs.

2-pyrrolidone [ORG CHEM] C_4H_7ON Combustible, light-yellow liquid, boiling at 245°C; soluble in ethyl alcohol, water, chloroform, and carbon disulfide; used as a plasticizer and polymer solvent, in insecticides and specialty inks, and as a nylon-4 precursor.

pyrrone [ORG CHEM] A polyimidazopyrrolone synthesized from dianhydrides and tetramines; soluble in sulfuric acid; resists temperatures to 600°C.

pyruric acid *See* cyanuric acid.

pyruvate [BIOCHEM] Salt of pyruvic acid, such as sodium pyruvate, $NaOOCCOCH_3$.

pyruvic acid [BIOCHEM] Important intermediate in protein and carbohydrate metabolism; liquid with acetic-acid aroma; melts at 11.8°C; miscible with alcohol, ether, and water; used in biochemical research. Also known as acetyl formic acid; α-ketopropionic acid; pyroracemic acid.

Pythagorean numbers [MATH] Positive integers x, y, and z which satisfy the equation $x^2 + y^2 = z^2$.

Pythagorean scale [ACOUS] A musical scale such that the frequency intervals are represented by the ratios of integral powers of the numbers 2 and 3.

Pythagorean theorem [MATH] In a right triangle the square of the length of the hypotenuse equals the sum of the squares of the lengths of the other two sides.

Pythidae [INV ZOO] An equivalent name for Salpingidae.

python [VERT ZOO] The common name for members of the reptilian subfamily Pythoninae.

Pythoninae [VERT ZOO] A subfamily of the reptilian family Boidae distinguished anatomically by the skull structure and the presence of a pair of vestigial hindlegs in the form of stout, movable spurs.

pyuria [MED] The presence of pus in the urine.

pyxidium [BOT] A capsular fruiting body dehiscing around its circumference, thus causing the upper part to fall off.

Pyxis [ASTRON] A southern constellation; right ascension 9 hours, declination 30° south. Also known as Malus.

pz *See* pièze.

Q [NUC PHYS] *See* disintegration energy. [PHYS] A measure of the ability of a system with periodic behavior to store energy equal to 2π times the average energy stored in the system divided by the energy dissipated per cycle. Also known as Q factor; quality factor; storage factor.

Q band [ELECTROMAG] A radio-frequency band of 36 to 46 gigahertz.

Q factor [ORD] A correction factor applied to a bombsight setting to help account for the differing winds between flight altitude and the ground; this factor corrects only for that component of the differential ballistic wind that is parallel to the actual wind at flight level. [PHYS] *See* Q.

Q fever [MED] An acute, febrile infectious disease of man, characterized by sudden onset and patchy pneumonitis, and caused by a bacterialike organism, *Coxiella burneti*.

Q-machine plasma [PLASMA PHYS] A plasma column in a magnetic field created by surface ionization of a cesium beam on a hot tungsten plate.

Q meter [ENG] A direct-reading instrument which measures the Q of an electric circuit at radio frequencies by determining the ratio of inductance to resistance, and which has also been developed to measure many other quantities. Also known as quality-factor meter.

Q multiplier [ELECTR] A filter that gives a sharp response peak or a deep rejection notch at a particular frequency, equivalent to boosting the Q of a tuned circuit at that frequency.

Q point [ENG] Data point describing the position and movement of a radar target, based on two or more radar observations.

qr *See* quarter.

QRS complex [MED] The electrocardiographic deflection representing ventricular depolarization; the initial downward deflection is termed a Q wave; the initial upward deflection, an R wave; and the downward deflection following the R wave, an S wave. Also known as ventricular depolarization complex.

qr tr *See* quarter.

Q signal [COMMUN] A three-letter abbreviation starting with Q, used in the International List of Abbreviations for radiotelegraphy to represent complete sentences. [ELECTR] The quadrature component of the chrominance signal in color television, having a bandwidth of 0 to 0.5 megahertz; it consists of $+0.48(R-Y)$ and $+0.41(B-Y)$, where Y is the luminance signal, R is the red camera signal, and B is the blue camera signal.

QSO *See* quasar.

Q-switched laser [OPTICS] A laser whose Q factor is kept at a low value while an ion population inversion is built up, and then is suddenly switched to a high value just before instability occurs, resulting in a very high rate of stimulated emission. Also known as giant pulse laser.

qt *See* quart.

quad [ELEC] A series of four separately insulated conductors, generally twisted together in pairs. [ELECTR] A series-parallel combination of transistors; used to obtain increased reliability through double redundancy, because failure of one transistor will not disable the entire circuit. [GRAPHICS] One of the small pieces of metal used in typesetting to space or to fill out a line of characters; used mostly to fill space when indenting the first line and to fill out the last line.

quadded cable [ELEC] Cable in which at least some of the conductors are arranged in the form of quads.

quadrangle [CIV ENG] 1. A four-cornered, four-sided courtyard, usually surrounded by buildings. 2. The buildings surrounding such a courtyard. 3. A four-cornered, four-sided building. [MATH] A geometric figure bounded by four straight-line segments called sides, each of which intersects each of two adjacent sides in points called vertices, but fails

to intersect the opposite sides. Also known as quadrilateral.

quadrant [ANAT] One of the four regions into which the abdomen may be divided for purposes of physical diagnosis. [ELECTROMAG] *See* international henry. [ENG] **1.** An instrument for measuring altitudes, used, for example, in astronomy, surveying, and gunnery; employs a sight that can be moved through a graduated 90° arc. **2.** A lever that can move through a 90° arc. [MATH] **1.** A quarter of a circle; either an arc of 90° or the area bounded by such an arc and the two radii. **2.** Any of the four regions into which the plane is divided by a pair of coordinate axes. [MECH ENG] A device for converting horizontal reciprocating motion to vertical reciprocating motion. [NAV] One of the four areas between consecutive equisignal zones of a four course radio range station. [NAV ARCH] A casting, forging, or built-up frame in the shape of a sector of a circle attached to the rudder stock and through which the steering gear leads turn the rudder. [OPTICS] A double-reflecting instrument for measuring angles, used primarily for measuring altitudes of celestial bodies; the instrument was replaced by the sextant. [PHYSIO] A sector of one-fourth of the field of vision of one or both eyes.

quadrantal correctors [NAV] Masses of soft iron placed near a magnetic compass to correct for quadrantal deviation.

quadrantal deviation [NAV] Deviation which changes its sign (E or W) approximately each 90° change of heading; caused by induced magnetism in horizontal soft iron present near the compass.

quadrantal error [NAV] The angular error in a measured bearing that is due to the presence of metal in the vicinity of the direction-finding antenna, such as the metal structure and engines of an airplane or the hull of a ship.

quadrantal point *See* intercardinal point.

quadrantal spheres [NAV] Two hollow spheres of soft iron placed near a magnetic compass to correct for quadrantal deviation.

quadrant angle of fall [MECH] The vertical acute angle at the level point, between the horizontal and the line of fall of a projectile.

quadrant electrometer [ENG] An instrument for measuring electric charge by the movement of a vane suspended on a wire between metal quadrants; the charge is introduced on the vane and quadrants in such a way that there is a proportional twist to the wire.

quadrant elevation [ORD] Vertical angle between the base and the axis of the bore of a gun which exists just prior to firing.

Quadrantids [ASTRON] A meteor shower whose radiant-right ascension of 15 hours and declination of +48° is in the constellation Boötes; velocity is 43 kilometers per second, and the strength is medium.

quadrant mount [ORD] A device on a gun that holds the gunner's quadrant while the gun is being laid in elevation.

quadrant sight [ORD] Sighting instrument on a gun that is used in laying the gun in elevation; used in conjunction with a telescope, which lays the gun for direction.

quadraphonic sound [ENG ACOUS] A system for reproducing sound by means of four loudspeakers properly situated in the listening room, usually at the four corners of a square, with each loudspeaker being fed its own identifiable segment of the program signal. Also known as four-channel stereophonic sound.

quadrate bone [VERT ZOO] A small element forming part of the upper jaw joint on each side of the head in vertebrates below mammals.

quadratic [MATH] Any second-degree expression.

quadratic equation [MATH] Any second-degree polynomial equation.

quadratic form [MATH] Any second-degree, homogeneous polynomial.

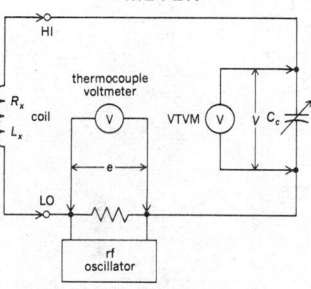

Q METER

Simplified measurement circuit of a Q meter. Coil being measured, with inductance L_x and resistance R_x, is connected into the circuit by external terminals HI and LO. Calibrated capacitor C_c is tuned to bring the coil into resonance. An input voltage e is supplied by a radio-frequency oscillator and measured by a thermocouple voltmeter. A vacuum-tube voltmeter (VTVM) measures voltage V across calibrated capacitor. Q of coil is determined from the equation $V/e = (1 + Q^2)^{1/2}$.

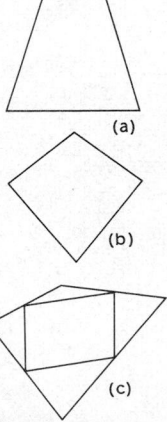

QUADRANGLE

Quadrangles. (a) Isosceles trapezoid. (b) Kite. (c) Parallelogram in quadrangle.

quadratic formula [MATH] A formula giving the roots of a quadratic equation in terms of the coefficients; for the equation $ax^2 + bx + c = 0$, the roots are $x = (-b \pm \sqrt{b^2 - 4ac})/2a$.

quadratic polynomial [MATH] A polynomial where the highest degree of any of its terms is 2.

quadratic programming [MATH] A body of techniques developed to find extremal points for systems of quadratic inequalities.

quadratic Stark effect [ATOM PHYS] A splitting of spectral lines of atoms in an electric field in which the energy levels shift by an amount proportional to the square of the electric field, and all levels shift to lower energies; observed in lines resulting from the lower energy states of many-electron atoms.

quadratic Zeeman effect [ATOM PHYS] A splitting of spectral lines of atoms in a magnetic field in which the energy levels shift by an amount proportional to the square of the magnetic field.

quadratojugal [VERT ZOO] A small bone connecting the quadrate and jugal bones on each side of the skull in many lower vertebrates.

quadrature [ASTRON] **1.** Two celestial bodies are in quadrature when there is a difference in celestial longitude of 90°; for example, the position of Mars, Jupiter, Saturn, Uranus, Neptune, or Pluto when the planet's elongation is 90°, east or west, depending on its direction from the sun. **2.** When the moon is in the first or last quarter, it is in quadrature with the sun. Also known as phase quadrature. [MATH] The construction of a square whose area is equal to that of a given surface. [PHYS] State of being separated in phase by 90°, or one quarter-cycle. Also known as phase quadrature.

quadrature amplifier [ELECTR] An amplifier that shifts the phase of a signal 90°; used in a color television receiver to amplify the 3.58-megahertz chrominance subcarrier and shift its phase 90° for use in the Q demodulator.

quadrature component [ELEC] **1.** A vector representing an alternating quantity which is in quadrature (at 90°) with some reference vector. **2.** *See* reactive current.

quadrature current *See* reactive current.

quadrature modulation [COMMUN] Modulation of two carrier components 90° apart in phase by separate modulating functions.

quadric cone [MATH] A conical surface whose directrices are conic curves.

quadriceps [ANAT] Four-headed, as a muscle.

quadriceps femoris [ANAT] The large extensor muscle of the thigh, combining the rectus femoris and vastus muscles.

quadrics [MATH] Homogeneous, second-degree expressions.

quadric surface [MATH] A surface whose equation is a second-degree algebraic equation.

quadridentate ligand [CHEM] A group which forms a chelate and has four points of attachment.

quadrigeminal body *See* corpora quadrigemina.

Quadrijugatoridae [PALEON] A monomorphic family of extinct ostracods in the superfamily Hollinacea.

quadrilateral *See* quadrangle.

quadrille paper [MATER] A good-quality white ledger paper with light-blue lines ruled on it.

quadrinomial distribution [STAT] A multinomial distribution with four possible outcomes.

quadriplegia [MED] Paralysis affecting the four extremities of the body; may be spastic or flaccid.

quadripuntal [ADP] Having four punches; specifically, having four random punches on an IBM or Hollerith-type punched card.

quadruped [VERT ZOO] An animal that has four legs.

quadruple-diversity system [COMMUN] Receiving system that simultaneously uses space diversity and frequency diversity techniques.

quadruple point [PHYS CHEM] Temperature at which four phases are in equilibrium, such as a saturated solution containing an excess of solute.

quadruple vector product [MATH] **1.** For any four vectors, the dot product of two derived vectors, one of which is the cross product of two of the original vectors, and the other of which is the cross product of the other two. **2.** For any four vectors, the cross product of two derived vectors, one of which is the cross product of two of the original vectors, and the other of which is the cross product of the other two.

quadruplex circuit [ELEC] Telegraph circuit designed to carry two messages in each direction at the same time.

quadrupole [ELECTROMAG] A distribution of charge or magnetization which produces an electric or magnetic field equivalent to that produced by two electric or magnetic dipoles whose dipole moments have the same magnitude but point in opposite directions, and which are separated from each other by a small distance.

quadrupole amplifier [ELECTR] A low-noise parametric amplifier consisting of an electron-beam tube in which quadrupole fields act on the fast cyclotron wave of the electron beam to produce high amplification at frequencies in the range of 400–800 megahertz.

quadrupole field [ELECTROMAG] **1.** An electric or magnetic field equivalent to that produced by two electric or magnetic dipoles whose dipole moments have the same magnitude but point in opposite directions, and which are separated from each other by a small distance. **2.** The field produced by a quadrupole lens.

quadrupole lens [ELECTROMAG] A device for focusing beams of charged particles which has four electrodes or magnetic poles of alternating sign arranged in a circle about the beam; used in instruments such as electron microscopes and particle accelerators.

quadrupole moment [ELECTROMAG] A quantity characterizing a distribution of charge or magnetization; it is given by integrating the product of the charge density or divergence of magnetization density, the second power of the distance from the origin, and a spherical harmonic Y^*_{2m} over the charge or magnetization distribution.

quadrupole spectrometer [ANALY CHEM] A type of mass spectroscope in which ions pass along a line of symmetry between four parallel cylindrical rods; an alternating potential superimposed on a steady potential between pairs of rods filters out all ions except those of a predetermined mass. Also known as Massenfilter.

quad word [ADP] A word 16 bytes long.

quagmire *See* bog.

quail [VERT ZOO] Any of several migratory game birds in the family Phasianidae.

quaking bog [GEOL] A peat bog floating or growing over water-saturated land which shakes or trembles when walked on.

qualification test [ENG] A formally defined series of tests by which the functional, environmental, and reliability performance of a component or system may be evaluated in order to satisfy the engineer, contractor, or owner as to its satisfactory design and construction prior to final approval and acceptance.

qualified name [ADP] A name that is further identified by associating it with additional names, usually the names of things that contain the thing being named.

qualifier [ADP] A name that is associated with another name to give additional information about the latter and distinguish it from other things having the same name.

qualitative analysis [ANALY CHEM] The analysis of a gas, liquid, or solid sample or mixture to identify the elements, radicals, or compounds composing the sample.

quality analysis [IND ENG] Examination of the quality goals of a product or service.

quality assurance [IND ENG] Testing and inspecting all of or a portion of the final product to ensure that the desired quality level of product reaches the consumer.

quality control [IND ENG] Inspection, analysis, and action applied to a portion of the product in a manufacturing operation to estimate overall quality of the product and determine what, if any, changes must be made to achieve or maintain the required level of quality.

quality control chart [IND ENG] A control chart used to indicate and control the quality of a product.

quality factor [NUCLEO] The factor by which absorbed dose is to be multiplied to obtain a quantity that expresses on a common scale, for all ionizing radiations, the irradiation incurred by exposed persons. [PHYS] *See* Q.

quality-factor meter *See* Q meter.

quality of snow [METEOROL] The amount of ice in a snow

sample expressed as a percent of the weight of the sample. Also known as thermal quality of snow.

quantal response [STAT] Response to treatment which has only two outcomes, all or none.

quantasome [CYTOL] One of the highly ordered array of units that has a "cobblestone" appearance in electron micrographs of the lamella of chloroplasts, and thought to be the most probable site of the light reaction in photosynthesis.

quantic [MATH] A homogeneous algebraic polynomial with more than one variable.

quantification [SCI TECH] The act of quantifying, that is, of giving a numerical value to a measurement of something, as in computer applications, psychology, or market research.

quantifier [MATH] Either of the phrases "for all" and "there exists"; these are symbolized respectively by an inverted A and a backward E.

quantitative analysis [ANALY CHEM] The analysis of a gas, liquid, or solid sample or mixture to determine the precise percentage composition of the sample in terms of elements, radicals, or compounds.

quantitative genetics [GEN] The study of continuously varying traits, such as those of the intellect and personality, which cannot be categorized as dichotomies.

quantitative inheritance [GEN] The acquisition of characteristics, such as height, weight, and intelligence, which show a quantitative and continuous type of variation.

quantity [ADP] In computers, a positive or negative real number in the mathematical sense; the term quantity is preferred to the term number in referring to numerical data; the term number is used in the sense of natural number and reserved for "the number of digits," the "number of operations," and so forth. [MATH] Any expression which is concerned with value rather than relations.

quantity-distance tables [ORD] The regulations pertaining to the amounts and kinds of explosives that can be stored and the proximity of such storage to buildings, highways, railways, magazines, or other installations.

quantity meter [ENG] A type of fluid meter used to measure volume of flow.

quantity of electricity *See* charge.

quantization [COMMUN] Division of the range of values of a wave into a finite number of subranges, each of which is represented by an assigned or quantized value within the subrange. [QUANT MECH] **1.** The restriction of an observable quantity, such as energy or angular momentum, to a discrete set of values. **2.** The transition from a description of a system of particles or fields in the classical approximation where canonically conjugate variables commute, to a description where these variables are treated as noncommuting operators; quantization (first definition) is a result of this procedure.

quantization distortion [COMMUN] Inherent distortion introduced in the process of quantization of a waveform. Also known as quantization noise; quantumization distortion; quantumization noise.

quantization level [COMMUN] Discrete value of the output designating a particular subrange of the input.

quantization noise *See* quantization distortion.

quantized frequency modulation [COMMUN] Frequency modulation that involves quantization; it uses time and frequency redundancy within a voice frequency channel during each transmitted symbol; used to combat distortion due to multipath, selection fading, and noise spikes.

quantized pulse modulation [COMMUN] Pulse modulation that involves quantization, such as pulse-numbers modulation and pulse-code modulation.

quantized spin wave *See* magnon.

quantized vortex [CRYO] A circular flow pattern observed in superfluid helium and type II superconductors, in which a superfluid flows about a normal (nonsuperfluid) cylindrical region or core which has the form of a thin line, and either the circulation or the magnetic flux is quantized.

quantizer [ELECTR] A device that measures the magnitude of a time-varying quantity in multiples of some fixed unit, at a specified instant or specified repetition rate, and delivers a proportional response that is usually in pulse code or digital form.

quantum [QUANT MECH] **1.** For certain physical quantities, a

unit such that the values of the quantity are restricted to integral multiples of this unit; for example, the quantum of angular momentum is Planck's constant divided by 2π. **2.** An entity resulting from quantization of a field or wave, having particlelike properties such as energy, mass, momentum and angular momentum; for example, the photon is the quantum of an electromagnetic field, and the phonon is the quantum of a lattice vibration.

quantum chemistry [PHYS CHEM] A branch of physical chemistry concerned with the explanation of chemical phenomena by means of the laws of quantum mechanics.

quantum detector [PHYS] A detector of electromagnetic radiation which converts a quantum of the radiation into a proportionate signal by some process which is insensitive to quanta of less than a certain energy; examples include photographic emulsions, photoelectric cells, and Geiger counters.

quantum efficiency [ELECTR] The average number of electrons photoelectrically emitted from a photocathode per incident photon of a given wavelength in a phototube.

quantum electrodynamics [QUANT MECH] The quantum theory of electromagnetic radiation, synthesizing the wave and corpuscular pictures, and of the interaction of radiation with electrically charged matter, in particular with atoms and their constituent electrons. Also known as quantum theory of light; quantum theory of radiation.

quantum electronics [ELECTR] The branch of electronics associated with the various energy states of matter, motions within atoms or groups of atoms, and various phenomena in crystals; examples of practical applications include the atomic hydrogen maser and the cesium atomic-beam resonator.

quantum evolution [EVOL] A special but extreme case of phyletic evolution; the rapid evolution that takes place when relatively sudden and drastic change occurs in the environment or when organisms spread into new habitats where conditions differ from those to which they are adapted; the organisms must then adapt quickly to the new conditions if they are to survive.

quantum field theory [QUANT MECH] Quantum theory of physical systems possessing an infinite number of degrees of freedom, such as the electromagnetic field, gravitation field, or wave fields in a medium.

quantum hydrodynamics [CRYO] The mechanics of a superfluid, such as helium II, investigating phenomena such as the fountain effect and second sound.

quantum hypothesis [QUANT MECH] A hypothesis that some physical quantity can assume only a certain discrete set of values; examples are Planck's law, and the condition in the Bohr-Sommerfield theory that the action integral of a system must be an integral multiple of Planck's constant.

quantumization distortion *See* quantization distortion.

quantumization noise *See* quantization distortion.

quantum-mechanical operator [QUANT MECH] A linear, Hermitian operator associated with some physical quantity; for a physical system in any state, the expectation value of the physical quantity equals the integral over configuration space of $\psi^*(A\psi)$, where $A\psi$ is the result of the operator acting on the wave function of the system, and ψ^* is the complex conjugate of the wave function.

quantum mechanics [PHYS] The modern theory of matter, of electromagnetic radiation, and of the interaction between matter and radiation; it differs from classical physics, which it generalizes and supersedes, mainly in the realm of atomic and subatomic phenomena. Also known as quantum theory.

quantum number [QUANT MECH] One of the quantities, usually discrete with integer or half-integer values, needed to characterize a quantum state of a physical system; they are usually eigenvalues of quantum-mechanical operators or integers sequentially assigned to these eigenvalues.

quantum state [QUANT MECH] **1.** The condition of a physical system as described by a wave function; the function may be simultaneously an eigenfunction of one or more quantum-mechanical operators; the eigenvalues are then the quantum numbers that label the state. **2.** *See* energy state.

quantum statistics [STAT MECH] The statistical description of particles or systems of particles whose behavior must be

QUANTASOME

1000A

Membranes containing chlorophyll taken from a spinach chloroplast. This chromium-shadowed preparation shows that the membrane is composed of a highly ordered array of units, or quantasomes. *(After R. B. Park, courtesy of Science, 144(3621), 1964)*

described by quantum mechanics rather than classical mechanics.

quantum theory *See* quantum mechanics.

quantum theory of heat capacity [STAT MECH] Application of quantum statistics to calculate heat capacities of various substances; an important result of the theory is the decrease of specific heats at low temperatures to values smaller than their classical values as a result of energy quantization.

quantum theory of light *See* quantum electrodynamics.

quantum theory of radiation [QUANT MECH] **1.** The theory of heat radiation based on Planck's law; its principal result is the Planck radiation formula. **2.** *See* quantum electrodynamics.

quantum theory of spectra [QUANT MECH] The contemporary theory of spectra, based on the idea that an atom, molecule, or nucleus can exist only in certain allowed energy states, that it emits or absorbs energy as it changes from one state to another, and that the frequency of the associated electromagnetic radiation equals the difference in energies of two states divided by Planck's constant.

quantum theory of valence [PHYS CHEM] The theory of valence based on quantum mechanics; it accounts for many experimental facts, explains the stability of a chemical bond, and allows the correlation and prediction of many different properties of molecules not possible in earlier theories.

quantum-wave equation [QUANT MECH] A partial differential equation which relates the spatial and time dependences of the wave function of a system of one or more atomic or subatomic particles; examples are the Schrödinger equation in nonrelativistic quantum mechanics, and the Klein-Gordon, Dirac, Rarita-Schwinger and Proca equations in relativistic quantum mechanics.

quantum yield [PHYS CHEM] For a photochemical reaction, the number of moles of a stated reactant disappearing, or the number of moles of a stated product produced, per einstein of light of the stated wavelength absorbed.

quaquaversal [GEOL] Of strata and geologic structures, dipping outward in all directions away from a central point.

quarantine [MED] Limitation of freedom of movement of susceptible individuals who have been exposed to communicable disease, for a period of time equal to the incubation period of the disease.

quarantine anchorage [CIV ENG] An area where a vessel anchors when satisfying quarantine regulations.

quarantine buoy [NAV] A buoy marking the location of a quarantine anchorage.

quark [PARTIC PHYS] One of the hypothetical basic particles, having charges whose magnitudes are ⅓ or ⅔ of the electron charge, from which many of the elementary particles may, in theory, be built up; for example, nucleons may be formed from three quarks and mesons from quark-antiquark combinations; no experimental evidence for the actual existence of free quarks has been found.

quarry [ENG] An open or surface working or excavation for the extraction of building stone, ore, coal, gravel, or minerals.

quarry bar [ENG] A horizontal bar with legs at each end, used to carry machine drills.

quarry face [MIN ENG] The freshly split face of ashlar, squared off for the joints only and used for massive work.

quarrying [ENG] The surface exploitation and removal of stone or mineral deposits from the earth's crust. [GEOL] *See* plucking.

quarrying machine [MECH ENG] Any machine used to drill holes or cut tunnels in native rock, such as a gang drill or tunneling machine; most commonly, a small locomotive bearing rock-drilling equipment operating on a track.

quarry powder [MATER] Ammonium nitrate dynamites used in quarrying where blasts of several tons of explosives are needed.

quart [MECH] Abbreviated qt. **1.** A unit of volume used for measurement of liquid substances in the United States, equal to 2 pints, or ¼ gallon, or 57¾ cubic inches, or $9.46352946 \times 10^{-4}$ cubic meter. **2.** A unit of volume used for measurement of solid substances in the United States, equal to 2 dry pints, or ¹⁄₃₂ bushel, or 107,521/1,600 cubic inches, or approximately 1.10122×10^{-3} cubic meter. **3.** A unit of volume used for measurement of both liquid and solid substances, although mainly the former, in the United Kingdom, equal to

2 U.K. pints, or ¼ U.K. gallon, or approximately 1.13652×10^{-3} cubic meter.

quartation *See* inquartation.

quarter [MECH] **1.** A unit of mass in use in the United States, equal to ¼ short ton, or 500 pounds, or 226.796185 kilograms. **2.** A unit of mass in troy measure, equal to ¼ troy hundredweight, or 25 troy pounds, or 9.33104304 kilograms. Abbreviated qr tr. **3.** A unit of mass used in the United Kingdom, equal to ¼ hundredweight, or 28 pounds, or 12.70058636 kilograms. Abbreviated qr. **4.** A unit of volume used in the United Kingdom for measurement of liquid and solid substances, equal to 8 bushels, or 64 gallons, or approximately 0.290950 cubic meter. [NAV ARCH] Portions of a vessel's sides about midway between the stem and the middle and between the middle and the stern.

quarter deck [NAV ARCH] The after portion of a weather deck.

quartering machine [MECH ENG] A machine that bores parallel holes simultaneously in such a way that the center lines of adjacent holes are 90° apart.

quartering sea [NAV] Waves moving in a direction approximately 45° from a vessel's heading, striking the vessel on the quarter.

quarter-phase *See* two-phase.

quarter-square multiplier [ADP] A device used to carry out function multiplication in an analog computer by implementing the algebraic identity $xy = \frac{1}{4}[(x+y)^2 - (x-y)^2]$.

quarter-turn drive [MECH ENG] A belt drive connecting pulleys whose axes are at right angles.

quarter-wave [ELECTROMAG] Having an electrical length of one quarter-wavelength.

quarter-wave antenna [ELECTROMAG] An antenna whose electrical length is equal to one quarter-wavelength of the signal to be transmitted or received.

quarter-wave attenuator [ELECTROMAG] Arrangement of two wire gratings, spaced an odd number of quarter-wavelengths apart in a waveguide, used to attenuate waves traveling through in one direction.

quarter-wave line *See* quarter-wave stub.

quarter-wave matching section *See* quarter-wave transformer.

quarter-wave plate [OPTICS] A thin sheet of mica or other doubly refracting crystal material of such thickness as to introduce a phase difference of one quarter-cycle between the ordinary and the extraordinary components of light passing through; such a plate converts circularly polarized light into plane-polarized light.

quarter-wave stub [ELECTROMAG] A section of transmission line that is one quarter-wavelength long at the fundamental frequency being transmitted; when shorted at the far end, it has a high impedance at the fundamental frequency and all odd harmonics, and a low impedance for all even harmonics. Also known as quarter-wave line; quarter-wave transmission line.

quarter-wave termination [ELECTROMAG] Metal plate and a wire grating spaced about ¼ of a wavelength apart in a waveguide, with the plate serving as the termination of the guide; waves reflected from the metal plate are canceled by waves reflected from the grating so that all energy is absorbed (none is reflected) by the quarter-wave termination.

quarter-wave transformer [ELECTROMAG] A section of transmission line approximately one quarter-wavelength long, used for matching a transmission line to an antenna or load. Also known as quarter-wave matching section.

quarter-wave transmission line *See* quarter-wave stub.

quartic *See* biquadratic.

quartic equation [MATH] Any fourth-degree polynomial equation.

quartz [MINERAL] SiO_2 A colorless, transparent rock-forming mineral with vitreous luster, crystallizing in the trigonal trapezohedral class of the rhombohedral subsystem; hardness is 7 on Mohs scale, and specific gravity is 2.65; the most abundant and widespread of all minerals.

quartzarenite [PETR] A quartz-rich sandstone with framework grains separated predominantly by cement rather than matrix; essentially an orthoquartzite.

quartz basalt [PETR] An igneous rock with more than 5% quartz.

QUARK

hypercharge, Y

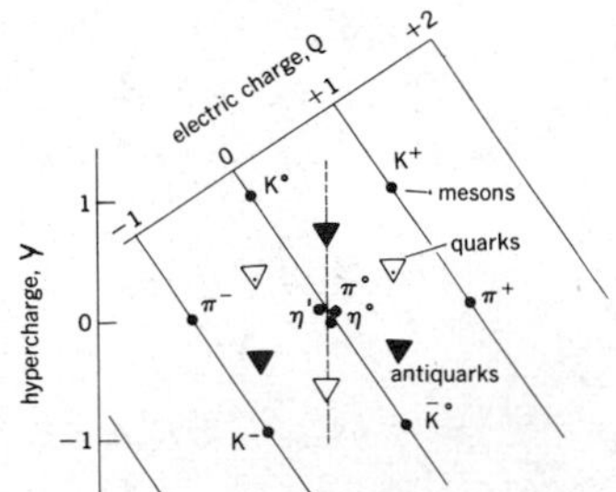

Charge-hypercharge plot showing the Gell-Mann-Nishijima loci of the elementary particles together with the positions of the pseudoscalar meson nonet and the quarks and the antiquarks.

QUARTZ

5 cm

Smoky quartz crystal with microcline, found in Florissant, Colorado. (*American Museum of Natural History Specimen*)

quartz-bearing diorite   *See* quartz diorite.

quartz claim   [MIN ENG] In the United States, a mining claim containing ore in veins or lodes, as contrasted with placer claims carrying mineral, usually gold, in alluvium.

quartz clock   [HOROL] A clock using the piezoelectric property of a quartz crystal, in which the crystal is introduced into an oscillating electric circuit having a frequency nearly equal to the natural frequency of vibration of the crystal.

quartz crystal   [ELECTR] A natural or artificially grown piezoelectric crystal composed of silicon dioxide, from which thin slabs or plates are carefully cut and ground to serve as a crystal plate.   [MINERAL] *See* rock crystal.

quartz-crystal filter   [ELECTR] A filter which utilizes a quartz crystal; it has a small bandwidth, a high rate of cutoff, and a higher unloaded Q than can be obtained in an ordinary resonator.

quartz-crystal resonator   [ELECTR] A quartz plate whose natural frequency of vibration is used to control the frequency of an oscillator. Also known as quartz resonator.

quartz delay line   [ELECTR] An acoustic delay line in which quartz is used as the medium of sound transmission.

quartz diorite   [PETR] A group of plutonic rocks having the composition of diorite but with large amounts of quartz (greater than 20%). Also known as quartz-bearing diorite; tonalite.

quartz-fiber electroscope   [ELECTR] Electroscope in which a gold-plated quartz fiber serves the same function as the gold leaf of a conventional electroscope.

quartz horizontal magnetometer   [ENG] A type of relative magnetometer used as a geomagnetic field instrument and as an observatory instrument for routine calibration of recording equipment.

quartzite   [PETR] A granoblastic metamorphic rock consisting largely or entirely of quartz; most quartzites are formed by metamorphism of sandstone.

quartzitic sandstone   [PETR] Sandstone consisting of 100% quartz grains cemented with silica.

quartz lamp   [ELECTR] A mercury-vapor lamp having a transparent envelope made from quartz instead of glass; quartz resists heat, permitting higher currents, and passes ultraviolet rays that are absorbed by ordinary glass.

quartz lattice   *See* rhyodacite.

quartz monzonite   [PETR] Granitic rock in which 10–50% of the felsic constituents are quartz, and in which the ratio of alkali feldspar to total feldspar is between 35% and 65%. Also known as adamellite.

quartz oscillator   [ELECTR] An oscillator in which the frequency of the output is determined by the natural frequency of vibration of a quartz crystal.

quartzose sandstone   [PETR] Sandstone consisting of more than 95% clear quartz grains and less than 5% matrix. Also known as quartz sandstone.

quartzose subgraywacke   *See* protoquartzite.

quartz-pebble conglomerate   *See* orthoquartzitic conglomerate.

quartz plate   *See* crystal plate.

quartz porphyry   [PETR] A porphyritic extrusive or hypabyssal rock containing quartz and alkali feldspar phenocrysts embedded in a microcrystalline or cryptocrystalline matrix. Also known as granite porphyry.

quartz resonator   *See* quartz-crystal resonator.

quartz sandstone   *See* quartzose sandstone.

quartz strain gage   [ELECTR] A device used to measure small deformations of a substance by determining the resulting voltage that develops in a quartz attached to it.

quartz syenite   [PETR] A group of plutonic rocks having the characteristics of syenite but with a greater amount of quartz (5–20%).

quasar   [ASTRON] Quasi-stellar astronomical object, often a radio source; all quasars have large red shifts; they have small optical diameter, but may have large radio diameter. Also known as quasi-stellar object (QSO).

quasi-atom   [ATOM PHYS] A system formed by two colliding atoms whose nuclei approach each other so closely that, for a very short time, the atomic electrons arrange themselves as if they belonged to a single atom whose atomic number equals the sum of the atomic numbers of the colliding atoms.

quasi-fission   [NUC PHYS] A nuclear reaction induced by heavy ions in which the two product nuclei have kinetic energies typical of fission products, but have masses close to those of the target and projectile, individually. Also known as deep inelastic transfer; incomplete fusion; relaxed peak process; strongly damped collision.

quasi–F martingale   [MATH] A stochastic process which is the sum of an F martingale and an F process having bounded variation on every finite time interval.

quasi-free-electron theory   [SOLID STATE] A modification of the free-electron theory of metals to take into account the periodic variation of the potential acting on a conduction electron, in which these electrons are assigned an effective scalar mass which differs from their real mass.

quasi-hydrostatic approximation   [METEOROL] The use of the hydrostatic equation as the vertical equation of motion, thus implying that the vertical accelerations are small without constraining them to be zero. Also known as quasi-hydrostatic assumption.

quasi-hydrostatic assumption   *See* quasi-hydrostatic approximation.

quasi-instruction   [ADP] An expression in a source program which resembles an instruction in form, but which does not have a corresponding machine instruction in the object program, and is directed to the assembler or compiler. Also known as pseudoinstruction.

quasi-linear feedback control system   [CONT SYS] Feedback control system in which the relationships between the pertinent measures of the system input and output signals are substantially linear despite the existence of nonlinear elements.

quasi-linear system   [CONT SYS] A control system in which the relationships between the input and output signals are substantially linear despite the existence of nonlinear elements.

quasi-molecule   [ATOM PHYS] The structure formed by two colliding atoms when their nuclei are close enough for the atoms to interact, but not so close as to form a quasi-atom.

quasi-particle   [PHYS] An entity used in the description of a system of many interacting particles which has particlelike properties such as mass, energy, and momentum, but which does not exist as a free particle; examples are phonons and other elementary excitations in solids, and "dressed" helium-3 atoms in Landau's theory of liquid helium-3.

quasi-random code generator   [COMMUN] High-speed pulse-code-modulation information source used in the design and evaluation of wide-band communications links by providing a means of closed loop testing.

quasi-reflection   [OPTICS] A term applied to the very strong return of light produced by dust particles and other suspensoids whose diameters are large compared to the wavelength of the incident radiation.

quasi-stable elementary particle   [PARTIC PHYS] An elementary particle which cannot decay into other particles through strong interactions; has a mean life longer than 10^{-10} second, in contrast to unstable particles which have mean lives on the order of 10^{-23} second. Also known as semistable elementary particle.

quasi-static process   *See* reversible process.

quasi-stationary front   [METEOROL] A front which is stationary or nearly so; conventionally, a front which is moving at a speed less than about 5 knots is generally considered to be quasi-stationary. Commonly known as stationary front.

quasi-stellar object   *See* quasar.

Quaternary   [GEOL] The second period of the Cenozoic geologic era, following the Tertiary, and including the last 2–3 million years.

quaternary alloy   [MET] An alloy containing four principal elements apart from accidental impurities.

quaternary ammonium base   [ORG CHEM] Ammonium hydroxide (NH_4OH) with the ammonium hydrogens replaced by organic radicals, such as $(CH_3)_4NOH$.

quaternary ammonium salt   [ORG CHEM] A nitrogen compound in which a central nitrogen atom is joined to four organic radicals and one acid radical, for example, hexamethonium chloride; used as an emulsifying agent, corrosion inhibitor, and antiseptic.

quaternary phase equilibria   [PHYS CHEM] The solubility re-

**QUARTZ
HORIZONTAL
MAGNETOMETER**

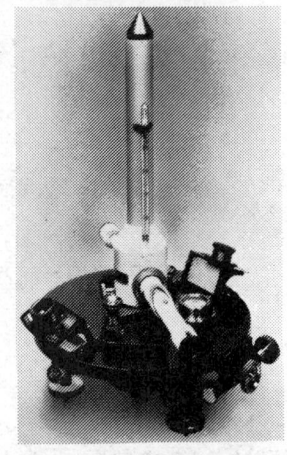

Photograph of quartz horizontal magnetometer. *(U.S. Coast and Geodetic Survey)*

QUATERNARY

PRECAMBRIAN		
CAMBRIAN		
ORDOVICIAN		PALEOZOIC
SILURIAN		
DEVONIAN		
Mississippian	CARBONIFEROUS	
Pennsylvanian		
PERMIAN		
TRIASSIC		
JURASSIC		MESOZOIC
CRETACEOUS		
TERTIARY		CENOZOIC
QUATERNARY		

Chart showing the position of the Quaternary in relation to other periods and to the eras of geologic time.

lationships in any liquid system with four nonreactive components with varying degrees of mutual solubility.

quaternary signaling [COMMUN] An electrical communications mode in which information is passed by the presence and absence, or plus and minus variations of four discrete levels of one parameter of the signaling medium.

quaternary system [PHYS CHEM] An equilibrium relationship between a mixture of four (four phases, four components, and so on).

quaternion [MATH] The division algebra over the real numbers generated by elements i, j, k subject to the relations $i^2 = j^2 = k^2 = -1$ and $ij = -ji = k$, $jk = -kj = i$, and $ki = -ik = j$. Also known as hypercomplex number.

quay [CIV ENG] A solid embankment or structure parallel to a waterway; used for loading and unloading ships.

quebracho [BOT] Any of a number of South American trees belonging to different genera in the order Sapindales, but all being a valuable source of wood, bark, and tannin.

quebracho bark [BOT] Bark of the white quebracho tree *Aspidosperma quebracho* of Chile and Argentina; main components are aspidospermine, tannin, and quebrachine; used in medicine and tanning.

queen [INV ZOO] A mature, fertile female in a colony of ants, bees, or termites, whose function is to lay eggs.

queen post [CIV ENG] Either of two vertical members, one on each side of the apex of a triangular truss.

Queensland tick typhus [MED] A benign infectious disease of humans found in rural northeastern Australia, caused by the bacterialike microorganism *Rickettsia australis* and presumed to be carried by the tick *Ixodes holocyclus*.

queen's metal [MET] An alloy consisting principally of tin to which antimony, zinc, and lead or copper are added.

Queenston shale [GEOL] A red bed series from the Ordovician found in Niagara Gorge; it is composed of deltaic red shale.

queenstownite *See* Darwin glass.

quellung reaction [MICROBIO] Swelling of the capsule of a bacterial cell, caused by contact with serum containing antibodies capable of reacting with polysaccharide material in the capsule; applicable to *Pneumococcus*, *Klebsiella*, and *Hemophilus*.

quench aging [MET] Aging of metal induced by rapid cooling from solution heat-treatment temperatures.

quench annealing [MET] Annealing an austenitic ferrous alloy by heating followed by quenching from solution temperatures.

quench bath [ENG] A liquid medium, such as oil, fused salt, or water, into which a material is plunged for heat-treatment purposes.

quenched spark gap [ELEC] A spark gap having provisions for rapid deionization; one form consists of many small gaps between electrodes that have relatively large mass and are good radiators of heat; the electrodes serve to cool the gaps rapidly and thereby stop conduction.

quench frequency [ELECTR] Number of times per second that a circuit is caused to go in and out of oscillation.

quench hardening [MET] The hardening of a ferrous alloy by quenching from a temperature above the transformation range.

quenching [ATOM PHYS] Phenomenon in which a very strong electric field, such as a crystal field, causes the orbit of an electron in an atom to precess rapidly so that the average magnetic moment associated with its orbital angular momentum is reduced to zero. [ELECTR] **1.** The process of terminating a discharge in a gas-filled radiation-counter tube by inhibiting reignition. **2.** Reduction of the intensity of resonance radiation resulting from deexcitation of atoms, which would otherwise have emitted this radiation, in collisions with electrons or other atoms in a gas. [ENG] Shock cooling by immersing liquid or molten material into a cooling medium (liquid or gas); used in metallurgy, plastics forming, and petroleum refining. [SOLID STATE] Reduction in the intensity of sensitized luminescence radiation when energy migrating through a crystal by resonant transfer is dissipated in crystal defects or impurities rather than being reemitted as radiation.

quenching oil [MET] Animal, vegetable, or mineral oil, such as fish oil, cottonseed oil, or lard, used in quenching baths for carbon and alloy steels; removes heat from the steel more slowly and uniformly than water.

quenching stress [MET] Internal stresses set up in a metal as a result of quenching.

quench oscillator [ELECTR] Circuit in a superregenerative receiver which produces the frequency signal.

quench-tank extrusion [ENG] Plastic-film or metal extrusion that is cooled in a quenching medium.

quench temperature [ENG] The temperature of the medium used for quenching.

quercetin [BIOCHEM] $C_{15}H_5O_2(OH)_5$ A yellow, crystalline flavonol obtained from oak bark and Douglas fir bark; used as an antioxidant and absorber of ultraviolet rays, and in rubber, plastics, and vegetable oils.

quercimelin *See* quercitrin.

quercitrin [ORG CHEM] $C_{21}H_{20}O_{11}$ The 3-rhamnoside of quercitin, forming yellow crystals from dilute ethanol or methanol solution, melting at 176–179°C, soluble in alcohol; used as a textile dye. Also known as quercimelin; quercitroside.

quercitroside *See* quercitrin.

quetsch [TEXT] **1.** A vat containing rollers, in which chemical solutions are applied to yarns or fabrics. **2.** One of the rollers in such a vat.

queue *See* waiting line.

queuing [ENG] The movement of discrete units through channels, such as programs or data arriving at a computer, or movement on a highway of heavy traffic.

queuing theory [MATH] The area of stochastic processes emphasizing those processes modeled on the situation of individuals lining up for service.

Quevenne scale [CHEM] Arbitrary scale used with hydrometers or lactometers in the determination of the specific gravity of milk; degrees Quevenne = 1000 (specific gravity − 1).

quibinary [ADP] A numeration system, used in data processing, in which each decimal digit is represented by seven binary digits, a group of five which are coefficients of 8, 6, 4, 2, and 0, and a group of two which are coefficients of 1 and 0.

quick-break fuse [ELEC] A fuse designed to draw out the arc and break the circuit rapidly when the fuse wire melts, generally by separating the broken ends with a spring.

quick-break switch [ELEC] A switch that breaks a circuit rapidly, independently of the rate at which the switch handle is moved, to minimize arcing.

quick clay [GEOL] Clay that loses its shear strength after being disturbed.

quick flashing light [NAV] In marine operations, a light showing short flashes at the rate of not less than 60 per minute; it has a cautionary significance.

quicklime *See* calcium oxide.

quick-make switch [ELEC] Switch or circuit breaker which has a high contact-closing speed, independent of the operator.

quick malleable iron [MET] Malleable iron containing 2.2% carbon, 1.5% silicon, 0.30–0.60% manganese, and 0.75–1% copper.

quickmatch [ENG] Fast-burning fuse made from a cord impregnated with black powder.

quickness [ORD] General term, expressing the mass rate of gas evolution of a propellant in a quantitative sense; basically a function of the propellant geometry.

quick return [MECH ENG] A device used in a reciprocating machine to make the return stroke faster than the power stroke.

quicksand [GEOL] A highly mobile mass of fine sand consisting of smooth, rounded grains with little tendency to mutual adherence, usually thoroughly saturated with upward-flowing water; tends to yield under pressure and to readily swallow heavy objects on the surface. Also known as running sand. [MATER] A loose sand mixture with a high proportion of water, thus having a low bearing pressure.

quicksilver *See* mercury.

quicksilver vermilion *See* mercuric sulfide.

QUICKTRAN [ADP] A time-sharing language developed by IBM for use on an IBM 1050 terminal connected to an IBM

QUEBRACHO

Quebracho, branch with fruit.

QUERCETIN

Structural formula for quercetin.

7044 computer; as a programming system, it provides concurrent computer access to a maximum of 50 remotely located terminals from a centralized data-processing system having tape, drum, and disk storage.

quiescent [ELECTR] Condition of a circuit element which has no input signal, so that it does not perform its active function. [ENG] State of a body at rest, or inactive, such as an undisturbed liquid in a storage or process vessel. [MED] Inactive, latent, or dormant, referring to a disease or pathological process.

quiescent-carrier telephony [COMMUN] A radiotelephony system in which the carrier is suppressed whenever there are no voice signals to be transmitted.

quiescent period [COMMUN] Resting period, or the period between pulse transmissions.

quiescent point [ELECTR] The point on the characteristic curve of an amplifier representing the conditions that exist when the input signal equals zero.

quiescent push-pull [ELECTR] Push-pull output stage so arranged in a radio receiver that practically no current flows when no signal is being received.

quiescent value [ELECTR] The voltage or current value for an electron-tube electrode when no signals are present.

quiet automatic volume control *See* delayed automatic gain control.

quiet battery [ELECTR] Source of energy of special design or with added filters which is sufficiently quiet and free from interference that it may be used for speech transmission. Also known as talking battery.

quieting sensitivity [ELECTR] Minimum signal input to a frequency-modulated receiver which is required to give a specified output signal-to-noise ratio under specified conditions.

quiet sun [ASTROPHYS] The sun when it is free from unusual radio wave or thermal radiation such as that associated with sunspots.

quiet sun noise [ASTROPHYS] Electromagnetic noise originating in the sun at a time when there is little or no sunspot activity.

quiet tuning [ELECTR] Circuit arrangement for silencing the output of a radio receiver, except when it is accurately tuned to an incoming carrier wave.

quill [DES ENG] A hollow shaft into which another shaft is inserted in mechanical devices. [TEXT] A shaft or spool on which filling yarn is wound before insertion into a shuttle. [VERT ZOO] The hollow, horny shaft of a large stiff wing or tail feather.

quill drive [MECH ENG] A drive in which the motor is mounted on a nonrotating hollow shaft surrounding the driving-wheel axle; pins on the armature mesh with spokes on the driving wheels, thereby transmitting motion to the wheels; used on electric locomotives.

quill gear [MECH ENG] A gear mounted on a hollow shaft.

quillwort [BOT] The common name for plants of the genus *Isoetes*.

quinacrine [PHARM] $C_{23}H_{30}ClN_3O$ Formerly an important antimalarial drug but now used in the treatment of giardiasis, tapeworm infections, amebiasis, and a variety of other conditions.

quinalbarbitone *See* secobarbital.

quince [BOT] *Cydonia oblonga*. A deciduous tree of the order Rosales characterized by crooked branching, leaves that are densely hairy on the underside and solitary white or pale-pink flowers; fruit is an edible pear- or apple-shaped tomentose pome.

quinhydrone [ORG CHEM] $C_6H_4O_2 \cdot C_6H_4(OH)_2$ Green, water-soluble powder, subliming at 171°C; a compound of quinone and hydroquinone dissociating in solution.

quinhydrone electrode [ANALY CHEM] A platinum wire in a saturated solution of quinhydrone; used as a reversible electrode standard in pH determinations.

quinic acid [ORG CHEM] $C_6H_7(OH)_4COOH \cdot H_2O$ Ether-insoluble, white crystals with acid taste; melts at 162°C; soluble in alcohol, water, and glacial acetic acid; used in medicine. Also known as chinic acid; kinic acid.

quinidine [ORG CHEM] $C_{20}H_{24}N_2O_2$ A crystalline alkaloid that melts at 171.5°C and that may be derived from the bark of cinchona; used as the salt in medicine. Also known as chinidine; β-quinine.

quinine [PHARM] $C_{20}H_{24}N_2O_2$ An alkaloid of cinchona, used principally as an antimalarial drug.

β-quinine *See* quinidine.

quinoa [BOT] *Chenopodium quinoa*. An annual herb of the family Chenopodiaceae grown at high altitudes in South America for the highly nutritious seeds.

quinol *See* hydroquinone.

quinoline [ORG CHEM] C_9H_7N Water-soluble, aromatic nitrogen compound; colorless, hygroscopic liquid; also soluble in alcohol, ether, and carbon disulfide; boils at 238°C; used in medicine and as a chemical intermediate. Also known as chinoline; leucoline; leukol.

quinoline blue *See* cyanine dye.

8-quinolinol *See* 8-hydroxyquinoline; oxine.

quinone [ORG CHEM] $CO(CHCH)_2CO$ Yellow crystalline compound with irritating aroma; melts at 116°C; soluble in alcohol, alkalies, and ether; used to make dyes and hydroquinone. Also known as benzoquinone; chinone.

quinoxaline [ORG CHEM] $C_8H_6N_2$ Bicyclic organic base; colorless powder, soluble in water and organic solvents; melts at 30°C; used in organic synthesis. Also known as 1,4-benzodiazine; benzo-*para*-diazine.

quinquefoliate [BOT] Of a leaf, having five leaflets.

quintal *See* hundredweight; metric centner.

quintic [MATH] A fifth-degree expression.

quintic equation [MATH] A fifth-degree polynomial equation.

quintuplet [BIOL] One of five children who have been born at one birth.

quire [MATER] Twenty-five sheets of paper, or one-twentieth of a 500-sheet ream.

quitclaim [MIN ENG] Legal release of a claim, right, title, or interest by one person or estate to another.

quitiqua *See* pinta.

Q unit [THERMO] A unit of energy, used in measuring the heat energy of fuel reserves, equal to 10^{18} British thermal units, or approximately 1.055×10^{21} joules.

quoin [BUILD] One of the members forming an outside corner or exterior angle of a building, and differentiated from the wall by color, texture, size, or projection. [GRAPHICS] One of the wedge-shaped devices made of steel, generally triangular, and less than type-high; used to lock type and plates in chases for the press.

quoin post [CIV ENG] The vertical member at the jointed end of a gate in a navigation lock.

quotient [MATH] The result of dividing one quantity by another.

quotient field [MATH] The smallest field containing a given integral domain; obtained by formally introducing all quotients of elements of the integral domain.

quotient group [MATH] A group G/H whose elements are the cosets gH of a given normal subgroup H of a given group G, and the group operation is defined as $g_1H \cdot g_2H \equiv (g_1 \cdot g_2)H$. Also known as factor group.

quotient ring [MATH] A ring R/I whose elements are the cosets rI of a given ideal I in a given ring R, where the additive and multiplicative operations have the form: $r_1I + r_2I \equiv (r_1 + r_2)I$ and $r_1I \cdot r_2I \equiv (r_1 \cdot r_2)I$.

quotient set [MATH] The set of all the equivalence classes relative to a given equivalence relation on a given set.

quotient space [MATH] The topological space Y which is the set of equivalence classes relative to some given equivalence relation on a given topological space X; the topology of Y is canonically constructed from that of X.

quotient topology [MATH] If X is a topological space, X/R the quotient space by some equivalence relation on X, the quotient topology on X/R is the smallest topology which makes the function which assigns to each element of X its equivalence class in X/R a continuous function.

Q value *See* disintegration energy.

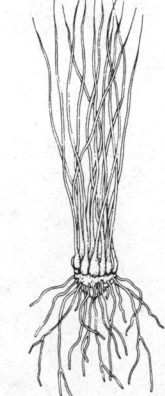

QUILLWORT

A quillwort (*Isoetes nuttallii*). (From A. W. Haupt, *Plant Morphology*, McGraw-Hill, 1953)

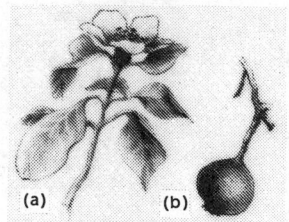

QUINCE

Quince. (a) Flower is borne on wood of the same season, not from an autumn fruit bud. (b) Short shoots start from the winter buds, and the flowers and fruits are produced on the ends of these shoots.

R

r *See* roentgen.

R *See* roentgen.

Raabe's convergence test [MATH] An infinite series with positive terms a_n where, for each n, $a_{n+1}/a_n = 1/(1 + b_n)$ will converge if, after a certain term, nb_n always exceeds a fixed number greater than 1 and will diverge if nb_n always is less than a fixed number less than or equal to 1.

rabal [METEOROL] A method of winds-aloft observation, that is, the determination of wind speeds and directions in the atmosphere above a station; it is accomplished by recording the elevation and azimuth angles of the balloon at specified time intervals while visually tracking a radiosonde balloon with the theodolite.

rabbet [ENG] **1.** A groove cut into a part. **2.** A strip applied to a part as, for example, a stop or seal. **3.** A joint formed by fitting one member into a groove, channel, or recess in the face or edge of a second member.

rabbet plane [DES ENG] A plane with the blade extending to the outer edge of one side that is open.

rabbit [NUCLEO] A small container that is propelled, usually pneumatically or hydraulically, through a tube into a nuclear reactor; used to expose samples to the radiation, especially neutron flux, then remove them rapidly for measurements of radioactive atoms having short half-lives. Also known as shuttle. [VERT ZOO] Any of a large number of burrowing mammals in the family Leporidae.

rabbit ear [MET] A recess in the corner of a die allowing wrinkling or folding of the blank.

rabbling [ENG] Stirring a molten charge, as of metal or ore.

rabies [VET MED] An acute, encephalitic viral infection transmitted to humans by the bite of a rabid animal. Also known as hydrophobia.

Racah coefficient [QUANT MECH] A coefficient that appears in the transformation between the modes of coupling eigenfunctions of three angular momenta; they differ only by, at most, a sign from the six-j coefficients. Also known as W coefficient.

raccoon [VERT ZOO] Any of 16 species of carnivorous nocturnal mammals belonging to the family Procyonidae; all are arboreal or semiarboreal and have a bushy, long ringed tail.

race [ANTHRO] **1.** A distinctive human type possessing characteristic traits that are transmissible by descent. **2.** Descendants of a common ancestor. [BIOL] **1.** An infraspecific taxonomic group of organisms, such as subspecies or microspecies. **2.** A fixed variety or breed. [DES ENG] Either of the concentric pair of steel rings of a ball bearing or roller bearing. [ENG] A channel transporting water to or away from hydraulic machinery, as in a power house. [OCEANOGR] A rapid current, or a constricted channel in which such a current flows; the term is usually used only in connection with a tidal current, which may be called a tide race.

race condition [ELEC] An ambiguous condition occurring in control counters when one flip-flop changes to its next state before a second one has had sufficient time to latch.

racemase [BIOCHEM] Any of a group of enzymes that catalyze racemization reactions.

raceme [BOT] An inflorescence on which flowers are borne on stalks of equal length on an unbranched main stalk that continues to grow during flowering.

racemic acid [ORG CHEM] $C_2H_4O_2(COOH)_2 \cdot H_2O$ Colorless crystals, melting at 205°C; soluble in water, slightly soluble in alcohol; used as a chemical intermediate. Also known as inactive tartaric acid.

racemic mixture [CHEM] A compound which is a mixture of equal quantities of dextrorotatory and levorotatory isomers of the same compound, and therefore is optically inactive.

racemic 1-phenyl-2-aminopropane *See* amphetamine.

racemose [ANAT] Of a gland, compound and shaped like a bunch of grapes, with freely branching ducts that terminate in acini. [BOT] Bearing, or occurring in the form of, a raceme.

race track [NUCLEO] An assembly of several calutron isotope separators in the shape of a race track, having a common magnetic field. Also known as track.

raceway [ELEC] A channel used to hold and protect wires, cables, or busbars. Also known as electric raceway.

rachiglossate radula [INV ZOO] A radula of certain gastropod mollusks which has one or three longitudinal series of teeth, each of which may bear many cusps.

rachilla [BOT] The axis of a grass spikelet.

rachis [ANAT] The vertebral column. [BIOL] An axial structure such as the axis of an inflorescence, the central petiole of a compound leaf, or the central cord of an ovary in Nematoda.

rack [AERO ENG] A suspension device permanently fixed to an aircraft; it is designed for attaching, arming, and releasing one or more bombs; it may also be utilized to accommodate other items such as mines, rockets, torpedoes, fuel tanks, rescue equipment, sonobuoys, and flares. [CIV ENG] A fixed screen composed of parallel bars placed in a waterway to catch debris. [DES ENG] *See* relay rack. [ENG] A frame for holding or displaying articles. [MECH ENG] A bar containing teeth on one face for meshing with a gear. [MIN ENG] An inclined trough or table for washing or separating ore.

rack and snail [HOROL] A mechanism in a striking timepiece that permits the hands to be advanced without waiting for the striking to be completed.

racking [CIV ENG] Setting back the end of each course of brick or stone from the end of the preceding course. [MET] Suspending work from a frame that holds and conducts current to one or more electrodes for electroplating and other electrochemical operations. [TEXT] Lateral movements of the needles or needle bed of a knitting machine in order to compact a fabric and enable it to hold its shape.

rack panel [ELECTR] A panel designed for mounting on a relay rack; its width is 19 inches (48.26 centimeters), height is a multiple of 1¾ inches (4.445 centimeters), and the mounting notches are standardized as to size and position.

rack railway [CIV ENG] A railway with a rack between the rails which engages a gear on the locomotive; used on steep grades.

racon *See* radar beacon.

rad [NUCLEO] The standard unit of absorbed dose, equal to energy absorption of 100 ergs per gram (0.01 joule per kilogram); supersedes the roentgen as the unit of dosage.

radappertization [FOOD ENG] The use of radiation for sterilizing foods.

radar [ENG] **1.** A system using beamed and reflected radio-frequency energy for detecting and locating objects, measuring distance or altitude, navigating, homing, bombing, and other purposes; in detecting and ranging, the time interval between transmission of the energy and reception of the reflected energy establishes the range of an object in the beam's path. Derived from radio detection and ranging. **2.** *See* radar set.

radar advisory [COMMUN] The term used to indicate that the provision of advice and information is based on radar observation.

radar altimeter [NAV] A radio altimeter, useful at altitudes much greater than the 5000-foot (1500 meter) limit of frequency-modulated radio altimeters, in which simple pulse-type radar equipment is used to send a pulse straight down from an aircraft and to measure its total time of travel to the surface and back to the aircraft. Also known as high-altitude radio altimeter; pulse-type altimeter.

radar altimetry area [NAV] A large and comparatively level terrain area with a defined elevation which can be used in

RABBIT

The snowshoe hare (*Lepus americanus*) undergoes seasonal color changes, being white in the winter and brown in the summer. Ears are always black-tipped.

RACCOON

A semiarboreal raccoon which uses tree hollows for breeding and nesting.

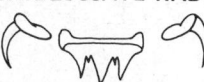

RACHIGLOSSATE RADULA

Rachiglossate radula, from *Murex* (marine gastropod). (*From R. R. Shrock and W. H. Twenhofel, Principles of Invertebrate Paleontology, 2d ed., McGraw-Hill, 1953*)

determining the altitude of airborne equipment by the use of radar.

radar altitude [NAV] The altitude of an aircraft or spacecraft as determined by a radio altimeter; thus, the actual vertical distance from the terrain. Also known as radio altitude.

radar and television aid to navigation [NAV] Device which converts a circular scan radar presentation to a horizontally scanned television presentation; it provides a continuous bright display with target trails for course and speed indications of moving targets.

radar antenna [ELECTROMAG] A device which radiates radio-frequency energy in a radar system, concentrating the transmitted power in the direction of the target, and which provides a large area to collect the echo power of the returning wave.

radar antijamming [ELECTR] Measures taken to counteract radar jamming.

radar approach control [NAV] A facility providing radar approach control service by use of airport surveillance radar and precision approach radar equipment.

radar astronomy [ASTRON] The study of astronomical bodies and the earth's atmosphere by means of radar pulse techniques, including tracking of meteors and the reflection of radar pulses from the moon and the planets.

radar attenuation [ELECTROMAG] Ratio of the power delivered by the transmitter to the transmission line connecting it with the transmitting antenna, to the power reflected from the target which is delivered to the receiver by the transmission line connecting it with the receiving antenna.

radar beacon [NAV] A radar receiver-transmitter that transmits a strong coded radar signal whenever its radar receiver is triggered by an interrogating radar on an aircraft or ship; the coded beacon reply can be used by the navigator to determine his own position in terms of bearing and range from the beacon. Also known as racon; radar transponder.

radar beam [ELECTROMAG] The movable beam of radio-frequency energy produced by a radar transmitting antenna; its shape is commonly defined as the loci of all points at which the power has decreased to one-half of that at the center of the beam.

radar bearing [NAV] A bearing obtained by radar.

radar bombing [ORD] Bombing in which radar is used to locate the target or aiming point, to aid in positioning the bombing aircraft at the proper release point for bombing, or to release bombs automatically, especially under conditions of poor visibility.

radar bombsight [ENG] An airborne radar set used to sight the target, solve the bombing problem, and drop bombs.

radar buoy [NAV] A buoy having corner reflectors designed into the superstructure, but maintaining the characteristic shape of the buoy in order to differentiate a radar buoy from a buoy on which a corner reflector is mounted.

radar camera [OPTICS] A special manual or automatic camera used to photograph images on a radarscope. Also known as radarscope camera.

radar camouflage [ORD] The use of special coverings or surfaces that reduce the reflection of radio-frequency energy back to a radar set, to minimize chances for detection of the object by an enemy radar set.

radar cell [ELECTROMAG] Volume whose dimensions are one radar pulse length by one radar beam width.

radar chart [NAV] A special map used in radar navigation that emphasizes objects which give prominent radar echoes.

radar climatology [CLIMATOL] The statistics in time and space of radar weather echoes.

radar clutter See clutter.

radar command guidance [ENG] A missile guidance system in which radar equipment at the launching site determines the positions of both target and missile continuously, computes the missile course corrections required, and transmits these by radio to the missile as commands.

radar confusion reflector See confusion reflector.

radar conspicuous object [ELECTROMAG] An object which returns a strong radar echo.

radar constant [ELECTR] One of those terms of the radar equation or radar storm-detection equation which are functions of the particular radar to which the equations are applied; these include peak power, antenna gain or aperture, beam width, pulse length, pulse repetition frequency, wavelength, polarization, and noise level of the receiver.

radar contact [ENG] Recognition and identification of an echo on a radar screen; an aircraft is said to be on radar contact when its radar echo can be seen and identified on a PPI (plan-position indicator) display.

radar control [ELECTR] Guidance, direction, or employment exercised over an aircraft, guided missile, gun battery, or the like, by means of, or with the aid of, radar.

radar controller [NAV] An air-traffic controller or other responsible person proficient in the use and interpretation of radar and capable of performing one or more of the following functions: surveillance controller, traffic director, or final controller.

radar countermeasure [ELECTR] An electronic countermeasure used against enemy radar, such as jamming and confusion reflectors. Abbreviated RCM.

radar coverage [ENG] The limits within which objects can be detected by one or more radar stations.

radar coverage indicator [ENG] Device that shows how far a given aircraft should be tracked by a radar station, and also provides a reference (detection) range for quality control; takes into account aircraft size, altitude, screening angle, site elevation, type radar, antenna radiation pattern, and antenna tilt.

radar cross section See echo area.

radar data [NAV] Data originating from a radar facility as distinct from air-traffic control data facility or other source. [ORD] Either long-range or gap-filler radar information modified to a form which a computer can use; the two types of data are as follows: correlated, associated with a track; and uncorrelated, not associated with a track.

radar data filtering [ELECTR] Quality analysis process that causes the computer to reject certain radar data and to alert personnel of mapping and surveillance consoles to the rejection.

radar deception [ORD] Radiation, reradiation, alteration, absorption, or reflection of electromagnetic energy in a manner intended to mislead an enemy in interpretation or use of information received by his radar systems.

radar decoy [ORD] A reflecting object used in radar deception, having the same reflective characteristics as the target.

radar display [ELECTR] The pattern representing the output data of a radar set, generally produced on the screen of a cathode-ray tube. Also known as radar presentation.

radar distribution switchboard [ELECTR] Switching panel for connecting video, trigger, and bearing from any one of five systems, to any or all of 20 repeaters; also contains order lights, bearing cutouts, alarms, test equipment, and so forth.

radar dome [ENG] Weatherproof cover for a primary radiating element of a radar or radio device which is transparent to radio-frequency energy, and which permits active operation of the radiating element, including mechanical rotation or other movement as applicable.

radar echo See echo.

radar equation [ELECTROMAG] An equation that relates the transmitted and received powers and antenna gains of a primary radar system to the echo area and distance of the radar target.

radar fire [ORD] Gunfire aimed at a target which is tracked by radar.

radar fire control [ORD] Fire control by means of radar.

radar fix [NAV] A determination of position by means of radar.

radar frequency band [ELECTROMAG] A frequency band of microwave radiation in which radar operates.

radar gun-layer [ENG] A radar device which tracks a target and aims a gun or guns automatically.

radar homing [ENG] Homing in which a missile-borne radar locks onto a target and guides the missile to that target. [NAV] Homing on the source of a radar beam.

radar horizon [NAV] The distance to which a radar's operation is limited by the quasi-optical characteristics of the radio waves employed.

radar image [ELECTR] The image of an object which is produced on a radar screen.

radar indicator [ELECTR] A cathode-ray tube and associated equipment used to provide a visual indication of the echo signals picked up by a radar set.

radar intelligence [COMMUN] **1.** Intelligence concerning radar or intelligence derived from the use of radar equipment. **2.** Organization or activity that deals with such intelligence.

radar intelligence item [ELECTR] A feature which is radar significant but which cannot be identified exactly at the moment of its appearance as homogeneous.

radar jamming [ELECTR] Radiation, reradiation, or reflection of electromagnetic waves so as to impair the usefulness of radar used by the enemy.

radar line of position [NAV] A line of position determined by radar.

radar marker [ENG] A fixed facility which continuously emits a radar signal so that a bearing indication appears on a radar display.

radar meteorological observation [METEOROL] An evaluation of the echoes which appear on the indicator of a weather radar, in terms of orientation, coverage, intensity, tendency of intensity, height, movement, and unique characteristics of echoes, that may be indicative of certain types of severe storms (such as hurricanes, tornadoes, or thunderstorms) and of anomalous propagation. Also known as radar weather observation.

radar meteorology [METEOROL] The study of the scattering of radar waves by all types of atmospheric phenomena and the use of radar for making weather observations and forecasts.

radar mile [ELECTROMAG] The time for a radar pulse to travel from the radar to a target 1 mile distant and return, equal to 10.75 microseconds.

radar nautical mile [ELECTROMAG] The time interval of approximately 12.355 microseconds that is required for the radio-frequency energy of a radar pulse to travel 1 nautical mile (1852 meters) and return.

radar netting [ENG] The linking of several radars to a single center to provide integrated target information.

radar netting station [ENG] A center which can receive data from radar tracking stations and exchange these data among other radar tracking stations, thus forming a radar netting system.

radar netting unit [ELECTR] Optional electronic equipment that converts the operations central of certain air defense fire distribution systems to a radar netting station.

radar paint [MATER] A coating that absorbs radar waves.

radar photography [GRAPHICS] Making a photograph of a radar display.

radar picket [ENG] A ship or aircraft equipped with early-warning radar and operating at a distance from the area being protected, to extend the range of radar detection.

radar picket escort ship [NAV ARCH] Any of the escort ships modified to give an increased combat information center, electronic countermeasures, and electronic search facilities.

radar prediction [ENG] A graphic portrayal of the estimated radar intensity, persistence, and shape of the cultural and natural features of a specific area.

radar presentation *See* radar display.

radar pulse [ELECTROMAG] Radio-frequency radiation emitted with high power by a pulse radar installation for a period of time which is brief compared to the interval between such pulses.

radar range [ELECTROMAG] The maximum distance at which a radar set is ordinarily effective in detecting objects.

radar range equation [ELECTROMAG] The equation stating that the radar range equals $(P\sigma^2 A^2 F^2/4\pi S\lambda^2)^{1/4}$, where P is the transmitted power, A is the area of the antenna aperture, F is a dimensionless constant (equal to 0.6 for a paraboloidal reflector), S is the minimum detectable signal power, σ is the radar cross section of the target, and λ is the signal wavelength.

radar range marker *See* distance marker.

radar receiver [ELECTR] A high-sensitivity radio receiver that is designed to amplify and demodulate radar echo signals and feed them to a radarscope or other indicator.

radar receiver-transmitter [ELECTR] A single component having the dual functions of generating electromagnetic energy for transmission, and of receiving, demodulating, and sometimes presenting intelligence from the reflected electromagnetic energy.

radar reconnaissance [ORD] Reconnaissance by means of radar to determine the location, disposition, and strength of enemy forces, and to determine the nature of the terrain.

radar reflection [ELECTROMAG] The return of electromagnetic waves, generated by a radar installation, from an object on which the waves are incident.

radar reflection interval [ELECTROMAG] The time required for a radar pulse to travel from the source to the target and return to the source, taking the velocity of radio propagation to be equal to the velocity of light.

radar reflectivity [ELECTROMAG] The fraction of electromagnetic energy generated by a radar installation which is reflected by an object.

radar reflector [ELECTROMAG] A device that reflects or deflects radar waves.

radar relay [ENG] **1.** Equipment for relaying the radar video and appropriate synchronizing signal to a remote location. **2.** Process or system by which radar echoes and synchronization data are transmitted from a search radar installation to a receiver at a remote point.

radar repeater [ELECTR] A cathode-ray indicator used to reproduce the visible intelligence of a radar display at a remote position; when used with a selector switch, the visible intelligence of any one of several radar systems can be reproduced.

radar report [METEOROL] The encoded and transmitted report of a radar meteorological observation; these reports usually give the azimuth, distance, altitude, intensity, shape and movement, and other characteristics of precipitation echoes observed by the radar. Also known as rain area report. Abbreviated RAREP.

radar return [NAV] The signal indication of an object which has reflected energy that was transmitted by a primary radar. Also known as radio echo.

radar scanning [ENG] The process or action of directing a radar beam through a space search pattern for the purpose of locating a target.

radarscope [ELECTR] Cathode-ray tube, serving as an oscilloscope, the face of which is the radar viewing screen. Also known as scope.

radarscope camera *See* radar camera.

radarscope overlay [ENG] A transparent overlay placed on a radarscope for comparison and identification of radar returns.

radar selector switch [ELECTR] Manual or motor-driven switch which transfers a plan-position indicator repeater from one system to another, switching video, trigger, and bearing data.

radar set [ENG] A complete assembly of radar equipment for detecting and ranging, consisting essentially of a transmitter, antenna, receiver, and indicator. Also known as radar.

radar shadow [ELECTROMAG] A region shielded from radar illumination by an intervening reflecting or absorbing medium such as a hill.

radar signal film [GRAPHICS] The film on which is recorded all the reflected signals acquired by a coherent radar, and which must be viewed or processed through an optical correlator to permit interpretation.

radar signal spectrograph [ELECTR] An electronic device in the form of a scanning filter which provides a frequency analysis of the amplitude-modulated back-scattered signal.

radar silence [ORD] A period of time during which radar transmission is stopped, generally for security reasons.

radarsonde [ENG] **1.** An electronic system for automatically measuring and transmitting high-altitude meteorological data from a balloon, kite, or rocket by pulse-modulated radio waves when triggered by a radar signal. **2.** A system in which radar techniques are used to determine the range, elevation, and azimuth of a radar target carried aloft by a radiosonde.

radar station [ENG] The place, position, or location from which, or at which, a radar set transmits or receives signals.

radar station pointer [NAV] A transparent plotting chart inscribed with radial lines from 0° to 360°; used to plot radar echoes to the scale of the chart in use; it is also used to assist in identifying radar responses with charted features.

radar storm detection [METEOROL] The detection of certain storms or stormy conditions by means of radar; liquid or frozen water drops within the storm reflect radar echoes.

radar storm-detection equation [METEOROL] The equation which relates the variables involved in the radar detection of precipitation.

radar surveying [ENG] Surveying in which airborne radar is used to measure accurately the distance between two ground radio beacons positioned along a baseline; this eliminates the need for measuring distance along the baseline in inaccessible or extremely rough terrain.

radar target [ELECTROMAG] An object belonging to a desired class which reflects back a signal sufficient to produce a fluorescent mark on the radar screen.

radar telescope [ENG] A large radar antenna and associated equipment used for radar astronomy.

radar threshold limit [ENG] For a given radar and specified target, the point in space relative to the focal point of the antenna at which initial detection criteria can be satisfied.

radar tracking [ENG] Tracking a moving object by means of radar.

radar tracking information [NAV] Information issued to alert an aircraft to any radar targets observed on the controller's radar display which may be in a dangerous situation.

radar tracking station [ENG] A radar facility which has the capability of tracking moving targets.

radar transmitter [ELECTR] The transmitter portion of a radar set.

radar transponder *See* radar beacon.

radar triangulation [ENG] A radar system of locating targets, usually aircraft, in which two or more separate radars are employed to measure range only; the target is located by automatic trigonometric solution of the triangle composed of a pair of radars and the target in which all three sides are known.

radar upper band *See* upper bright band.

radar volume [ELECTROMAG] The volume in space that is irradiated by a given radar; for a continuous-wave radar it is equivalent to the antenna radiation pattern; for a pulse radar it is a function of the cross-section area of the beam of the antenna and the pulse length of the transmitted pulse.

radar weather observation *See* radar meteorological observation.

radar wind [METEOROL] Wind of which the movement, speed, and direction is observed or determined by a radar tracking of a balloon carrying a radiosonde, a radio transmitter, or a radar reflector.

radar wind system [ENG] Apparatus in which radar techniques are used to determine the range, elevation, and azimuth of a balloon-borne target, and hence to compute upper-air wind data.

Radechon [ELECTR] A storage tube having a single electron gun and a dielectric storage medium consisting of a sheet of mica sandwiched between a continuous metal backing plate and a fine-mesh screen; used in simple delay schemes, signal-to-noise improvement, signal comparison, and conversion of signal-time bases. Also known as barrier-grid storage tube.

radiac [NUCLEO] Detection, identification, and measurement of the intensity of nuclear radiation in an area. Derived from radioactivity detection, identification, and computation.

radiac instrument *See* radiac set.

radiacmeter *See* radiac set.

radiac set [NUCLEO] A complete system for detecting, identifying, and measuring radioactivity. Also known as radiac instrument; radiacmeter.

radial [NAV] One of a number of radial lines of position defined by an azimuthal radio navigation facility and identified in terms of the bearing (usually magnetic) of all points on that line from the facility. [SCI TECH] 1. Directed or diverging from a center. 2. Raylike.

radial acceleration *See* centripetal acceleration.

radial artery [ANAT] A branch of the brachial artery in the forearm; principal branches are the radial recurrent and the main artery of the thumb.

radial band pressure [MECH] The pressure which is exerted on the rotating band by the walls of the gun tube, and hence

against the projectile wall at the band seat, as a result of the engraving of the band by the gun rifling.

radial-beam tube [ELECTR] A vacuum tube in which a radial beam of electrons is rotated past circumferentially arranged anodes by an external rotating magnetic field; used chiefly as a high-speed switching tube or commutator.

radial bearing [MECH ENG] A bearing with rolling contact in which the direction of action of the load transmitted is radial to the axis of the shaft.

radial canal [INV ZOO] 1. One of the numerous canals that radiate from the spongocoel in certain Porifera. 2. Any of the canals extending from the coelenteron to the circular canal in the margin of the umbrella in jellyfishes. 3. A canal radiating from the circumoral canal along each ambulacral area in many echinoderms.

radial chromatography [ANALY CHEM] A circular disk of absorbent paper which has a strip (wick) cut from edge to center to dip into a solvent; the solvent climbs the wick, touches the sample, and resolves it into concentric rings (the chromatogram). Also known as circular chromatography; radial-paper chromatography.

radial cleavage [EMBRYO] A cleavage pattern characterized by formation of a mass of cells that show radial symmetry.

radial displacement [GRAPHICS] On vertical photographs, the apparent "leaning out," or the apparent displacement of the top of any object having height in relation to its base.

radial distribution function [MATH] A function $F(r)$ equal to the average of a given function of the three coordinates over a sphere of radius r centered at the origin of the coordinate system. [PHYS CHEM] A function $\rho(r)$ equal to the average over all directions of the number density of molecules at distance r from a given molecule in a liquid.

radial Doppler effect [ELECTROMAG] The part of the optical Doppler effect which depends on the direction of the relative velocity of source and observer, and is analogous to the acoustical Doppler effect, in contrast to the transverse Doppler effect.

radial drainage pattern [GEOL] A drainage pattern characterized by radiating streams diverging from a high central area. Also known as centrifugal drainage pattern.

radial draw forming [MECH ENG] A metal-forming method in which tangential stretch and radial compression are applied gradually and simultaneously.

radial drill [MECH ENG] A drilling machine in which the drill spindle can be moved along a horizontal arm which itself can be rotated about a vertical pillar.

radial drilling [ENG] The drilling of several holes in one plane, all radiating from a common point.

radial engine [MECH ENG] An engine characterized by radially arranged cylinders at equiangular intervals around the crankshaft.

radial error [ORD] An error associated with delivery of munitions on a target; it is the distance between the desired point of impact and actual point of impact, both points projected and measured on an imaginary plane drawn perpendicular to the flight path of the munition.

radial faults [GEOL] Faults arranged like the spokes of a wheel, radiating from a central point.

radial-flow [ENG] Having the fluid working substance flowing along the radii of a rotating tank. [PETRO ENG] Pertaining to spokelike flow of reservoir fluids radially inward toward a wellbore focal area.

radial-flow turbine [MECH ENG] A turbine in which the gases flow primarily in a radial direction.

radial gate *See* tainter gate.

radial grating [ELECTROMAG] Conformal wire grating consisting of wires arranged radially in a circular frame, like the spokes of a wagon wheel, and placed inside a circular waveguide to obstruct E waves of zero order while passing the corresponding H waves.

radial lead [ELEC] A wire lead coming from the side of a component rather than axially from the end.

radial locating [MECH ENG] One of the three locating problems in tooling to maintain the desired relationship between the workpiece, the cutter, and the body of the machine tool; the other two locating problems are concentric and plane locating.

RADECHON

Circuit diagram of the Radechon, a signal-converter storage tube. T is mica sheet; P is metal backing plate; S is fine-mesh screen; i_{pr} is electron beam formed by electron gun with cathode K and anode A; G is collector that collects reading signal i_c; R_o is output resistor; R_i is input resistor.

RADIAL CLEAVAGE

Radial cleavage pattern found in invertebrates.

RADIAL DRILL

Large radial drilling machine set up for deep boring. (*Cincinnati Bickford*)

radial motion [MECH] Motion in which a body moves along a line connecting it with an observer or reference point; for example, the motion of stars which move toward or away from the earth without a change in apparent position.

radial nerve [ANAT] A large nerve that arises in the brachial plexus and branches to enervate the extensor muscles and skin of the posterior aspect of the arm, forearm, and hand.

radial-paper chromatography *See* radial chromatography.

radial-parallel search [NAV] Search by ship or aircraft on a series of radial course lines to a given line, beyond which parallel course lines are followed.

radial percussive coal cutter [MIN ENG] A heavy coal cutter having a percussive drill, with extension rods; used in headings and rooms in pillar methods of working and for drilling shot-firing holes.

radial-ply [DES ENG] Pertaining to the construction of a tire in which the cords run straight across the tire, and an additional layered belt of fabric is placed around the circumference between the plies and the tread.

radial rake [MECH ENG] The angle between the cutter tooth face and a radial line passing through the cutting edge in a plane perpendicular to the cutter axis.

radial saw [MECH ENG] A power saw that has a circular blade suspended from a transverse head mounted on a rotatable overarm.

radial search [NAV] Search by ship or aircraft on a series of radial course lines.

radial selector *See* omnibearing selector.

radial stress [MECH] Tangential stress at the periphery of an opening.

radial symmetry [SCI TECH] An arrangement of usually similar parts in a regular pattern around a central axis.

radial velocity [ASTRON] The component of velocity a celestial body has when the body is moving toward or away from the observer; the radial velocities of stars are valuable in determining the structure and dynamics of the Galaxy. Also known as line-of-sight velocity. [MECH] The component of the velocity of a body that is parallel to a line from an observer or reference point to the body; the radial velocities of stars are valuable in determining the structure and dynamics of the Galaxy. Also known as line-of-sight velocity.

radial wave equation [MECH] Solutions to wave equations with spherical symmetry can be found by separation of variables; the ordinary differential equation for the radial part of the wave function is called the radial wave equation.

radian [MATH] The central angle of a circle determined by two radii and an arc joining them, all of the same length.

radiance [OPTICS] The radiant flux per unit solid angle per unit of projected area of the source; the usual unit is the watt per steradian per square meter. Also known as steradiancy.

radiancy *See* radiant emittance.

radian frequency *See* angular frequency.

radian length [PHYS] Distance, in a sinusoidal wave, between phases differing by an angle of 1 radian; it is equal to the wavelength divided by 2π.

radiant [ASTRON] A point on the celestial sphere through which pass the backward extensions of the trail of a meteor as observed at various locations, or the backward extensions of trails of a number of meteors traveling parallel to each other. [PHYS] **1.** Pertaining to motion of particles or radiation along radii from a common point or a small region. **2.** A point, region, substance, or entity from which particles or radiations are emitted.

radiant density [PHYS] The instantaneous amount of radiant energy contained in a unit volume of propagation medium.

radiant emittance [ELECTROMAG] The radiant flux per unit area that emerges from a surface. Also known as radiancy; radiant exitance.

radiant energy *See* radiation.

radiant-energy thermometer *See* radiation pyrometer.

radiant exitance *See* radiant emittance.

radiant flux [OPTICS] The time rate of flow of radiant energy.

radiant flux density [ELECTROMAG] The amount of radiant power per unit area that flows across or onto a surface. Also known as irradiance.

radiant heating [ENG] Any system of space heating in which the heat-producing means is a surface that emits heat to the surroundings by radiation rather than by conduction or convection.

radiant intensity [ELECTROMAG] The energy emitted per unit time per unit solid angle about the direction considered; usually expressed in watts per steradian.

radiant power [ELECTROMAG] The energy carried across or onto a surface by electromagnetic radiation per unit time, or the total radiant energy emitted by a source of electromagnetic radiation per unit time.

radiant reflectance [ELECTROMAG] Ratio of reflected radiant power to incident radiant power.

radiant superheater [MECH ENG] A superheater designed to transfer heat from the products of combustion to the steam primarily by radiation.

radiant transmittance [ELECTROMAG] Ratio of transmitted radiant power to incident radiant power.

radiant-type boiler [MECH ENG] A water-tube boiler in which boiler tubes form the boundary of the furnace.

Radiata [INV ZOO] Members of the Eumetazoa which have a primary radial symmetry; includes the Coelenterata and Ctenophora.

radiated interference [COMMUN] Interference which is transmitted through the atmosphere according to the laws of electromagnetic wave propagation; the term is generally considered to include the transfer of interfering energy in inductive or capacitive coupling.

radiated power [ELECTROMAG] The total power emitted by a transmitting antenna.

radiating curtain [ELECTROMAG] Array of dipoles in a vertical plane, positioned to reinforce each other; it is usually placed $1/4$ wavelength ahead of a reflecting curtain of corresponding half-wave reflecting antennas.

radiating element [ELECTROMAG] Basic subdivision of an antenna which in itself is capable of radiating or receiving radio-frequency energy.

radiating guide [ELECTROMAG] Waveguide designed to radiate energy into free space; the waves may emerge through slots or gaps in the guide, or through horns inserted in the wall of the guide.

radiating power *See* emittance.

radiating scattering [PHYS] The diversion of radiation (thermal, electromagnetic, or nuclear) from its orginal path as a result of interactions or collisions with atoms, molecules, or larger particles in the atmosphere or other media between the source of radiation (for example, a nuclear explosion) and a point at some distance away.

radiation [PHYS] **1.** The emission and propagation of waves transmitting energy through space or through some medium; for example, the emission and propagation of electromagnetic, sound, or elastic waves. **2.** The energy transmitted by waves through space or some medium; when unqualified, usually refers to electromagnetic radiation. Also known as radiant energy. **3.** A stream of particles, such as electrons, neutrons, protons, α-particles, or high-energy photons, or a mixture of these.

radiation accident [NUCLEO] Any accident resulting in the spread of radioactive materials or in the exposure of individuals to radiation.

radiational cooling [METEOROL] The cooling of the earth's surface and adjacent air, accomplished (mainly at night) whenever the earth's surface suffers a net loss of heat due to terrestrial radiation.

radiation angle [ELECTROMAG] The vertical angle between the line of radiation emitted by a directional antenna and the horizon.

radiation area [NUCLEO] Any accessible area in which the level of radiation is such that a major portion of an individual's body could receive in any 1 hour a dose in excess of 5 millirem or in any 5 consecutive days a dose in excess of 150 millirem.

radiation biochemistry [BIOCHEM] The study of the response of the constituents of living matter to radiation.

radiation biology *See* radiobiology.

radiation biophysics [BIOPHYS] The study of the response of organisms to ionizing radiations and to ultraviolet light.

radiation budget [GEOPHYS] A quantitative statement of the

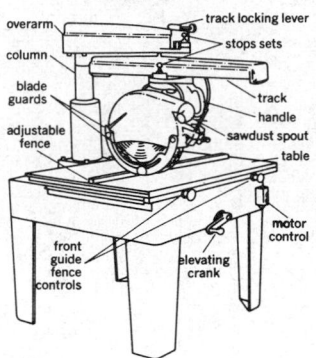

RADIAL-PLY

Construction of a radial-ply tire. *(Goodyear Tire and Rubber Co.)*

RADIAL SAW

overarm
column
blade guards
adjustable fence
track locking lever
stops sets
track
handle
sawdust spout
table
front guide fence controls
elevating crank
motor control

Radial saw allows wood to be clamped in position while it is being cut. *(Delta)*

amounts of radiation entering and leaving a given region of the earth.

radiation burn [MED] A burn caused by overexposure to radiant energy.

radiation catalysis [CHEM] The use of radiation (such as gamma, neutron, proton, electron, or x-ray) to activate or speed up a chemical or physical change; for example, radiation alone can initiate polymerization without heat, pressure, or chemical catalysts.

radiation cataract *See* irradiation cataract.

radiation characteristic [COMMUN] One of the identifying features of a radiating signal, such as frequency and pulse width.

radiation chart [GEOPHYS] Any chart or diagram which permits graphical solution of the (generally unintegrable) flux integrals arising in problems of atmospheric infrared radiation transfer.

radiation chemistry [NUCLEO] The branch of chemistry that is concerned with the chemical effects, including decomposition, of energetic radiation or particles on matter.

radiation cooling [ELECTR] Cooling of an electrode resulting from its emission of heat radiation.

radiation corrosion [MET] Accelerated corrosion of a metal caused by radiation.

radiation counter [NUCLEO] An instrument used for detecting or measuring nuclear radiation by counting the resultant ionizing events; examples include Geiger counters and scintillation counters. Also known as counter.

radiation counter tube *See* counter tube.

radiation cytology [CYTOL] An aspect of biology that deals with the effects of radiations on living cells.

radiation damage [NUCLEO] Harmful changes in the properties of liquids, gases, and solids caused by any type of radiation.

radiation damping [ELECTROMAG] Damping of a system which loses energy through electromagnetic radiation. [QUANT MECH] Damping which arises in quantum electrodynamics from the virtual interaction of a particle with its zero point field.

radiation decontamination [NUCLEO] The removal of unwanted radioactive material.

radiation dermatitis *See* radiodermatitis.

radiation detection instrument [NUCLEO] Any device that detects and records the characteristics of ionizing radiation.

radiation detector *See* particle detector.

radiation dose [NUCLEO] The total amount of ionizing radiation absorbed by material or tissues, in the sense of absorbed dose (expressed in rads), exposure dose (expressed in roentgens), or dose equivalent (expressed in rems).

radiation dose rate [NUCLEO] The radiation dosage absorbed per unit of time; a radiation dose rate can be set at some particular unit of time (for example, H-hour plus 1 hour) and would be called H-hour plus 1 radiation dose rate.

radiation dosimetry *See* dosimetry.

radiation effects [BIOL] The harmful effects of ionizing radiation on humans and other animals, such as production of cancers, cataracts, and radiation ulcers, loss of hair, reddening of skin, sterilization, nausea, vomiting, mucous or bloody diarrhea, purpura, epilation, and agranulocytic infections.

radiation efficiency [ELECTROMAG] Of an antenna, the ratio of the power radiated to the total power supplied to the antenna at a given frequency.

radiation field [ELECTROMAG] The electromagnetic field that breaks away from a transmitting antenna and radiates outward into space as electromagnetic waves; the other type of electromagnetic field associated with an energized antenna is the induction field.

radiation filter [ELECTROMAG] Selectively transparent body, which transmits only certain wavelength ranges.

radiation fog [METEOROL] A major type of fog, produced over a land area when radiational cooling reduces the air temperature to or below its dew point; thus, strictly, a nighttime occurrence, although the fog may begin to form by evening twilight and often does not dissipate until after sunrise.

radiation gage [NUCLEO] An instrument for measuring radiation quantity and intensity.

radiation genetics [GEN] The study of the genetic effects of radiation. Also known as radiogenetics.

radiation hazard [MED] Health hazard arising from exposure to ionizing radiation.

radiation intensity [ELECTROMAG] The power radiated from an antenna per unit solid angle in a given direction. [NUCLEO] The quantity of radiant energy passing perpendicularly through a specified location of unit area in unit time; reported as a number of particles or photons per square centimeter per second, or in energy units such as ergs per square centimeter per second.

radiation ionization [PHYS] Ionization of the atoms or molecules of a gas or vapor by the action of electromagnetic radiation.

radiation laws [PHYS] **1.** The four physical laws which, together, fundamentally describe the behavior of blackbody radiation, Kirchhoff's law, Planck's law, Stefan-Boltzmann law, and Wien's law. **2.** All of the more inclusive assemblage of empirical and theoretical laws describing all manifestations of radiative phenomena.

radiation length [NUCLEO] The mean path length required to reduce the energy of relativistic charged particles by the factor $1/e$, or 0.368, as they pass through matter. Also known as cascade unit; radiation unit.

radiationless transition [PHYS] A transition of a system between two energy states in which energy is given to or taken up from another system or particle, rather than being emitted or absorbed in electromagnetic radiation; examples include internal conversion, the Auger effect, and excitation or deexcitation of atoms or molecules in collisions with other atoms or molecules.

radiation lobe *See* lobe.

radiation microbiology [MICROBIO] A field of basic and applied radiobiology concerned chiefly with the damaging effects of radiation on microorganisms.

radiation monitoring [NUCLEO] Continuous or periodic determination of the amount of radiation present in a given area.

radiation noise *See* electromagnetic noise.

radiation oscillator *See* Planck oscillator.

radiation oven [ENG] Heating chamber relying on tungsten-filament infrared lamps with reflectors to create temperatures up to 600°F (315°C); used to dry sheet and granular material and to bake surface coatings.

radiation pattern [ELECTROMAG] Directional dependence of the radiation of an antenna. Also known as antenna pattern; directional pattern; field pattern.

radiation preservation [FOOD ENG] Exposure of food products to ionizing radiation, such as electrons, x-rays, and γ-rays, in order to destroy microorganisms and thereby aid preservation.

radiation pressure [ACOUS] The average pressure exerted on a surface or interface between two media by a sound wave. [ELECTROMAG] The pressure exerted by electromagnetic radiation on objects on which it impinges.

radiation protection [NUCLEO] **1.** Legislation and regulations to protect the public and laboratory or industrial workers against radiation. **2.** Measures to reduce exposure to radiation.

radiation protection guide [NUCLEO] The officially determined radiation doses which should not be exceeded without careful consideration of the reasons for doing so; these standards, established by the Federal Radiation Council, are equivalent to what was formerly called the maximum permissible dose or maximum permissible exposure.

radiation pyrometer [ENG] An instrument which measures the temperature of a hot object by focusing the thermal radiation emitted by the object and making some observation on it; examples include the total-radiation, optical, and ratio pyrometers. Also known as radiant-energy thermometer; radiation thermometer.

radiation quality [PHYS] The spectrum of radiant energy produced by a given radiation source with respect to its penetration or its suitability for a specific application.

radiation resistance [ACOUS] For a medium, the acoustic impedance of a plane wave in that medium. [ELECTROMAG] The total radiated power of an antenna divided by the square

RADIATION PYROMETER

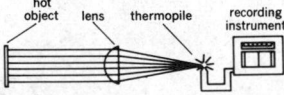

hot object — lens — thermopile — recording instrument

Diagram of an elementary radiation pyrometer. Lens focuses part of radiation emitted by hot object onto thermopile. Resulting heating of thermopile causes it to generate electrical signal which is displayed on recording instrument. *(From D. M. Considine, ed., Process Instruments and Controls Handbook, McGraw-Hill, 1957)*

of the effective antenna current measured at the point where power is supplied to the antenna.

radiation safety [NUCLEO] Protection of personnel against harmful effects of ionizing radiation by taking steps to ensure that people will not receive excessive doses of radiation and by monitoring all sources of radiation to which they may be exposed.

radiation scattering [PHYS] The diversion of radiation (thermal, electromagnetic, or nuclear) from its original path as a result of interactions or collisions with atoms, molecules, or larger particles in the atmosphere or other media.

radiation shelter *See* fallout shelter.

radiation shield [ENG] A shield or wall of material interposed between a source of radiation and a radiation-sensitive body, such as a person, radiation-detection instrument, or photographic film, to protect the latter.

radiation sickness [MED] **1.** Illness, usually manifested by nausea and vomiting, resulting from the effects of therapeutic doses of radiation. **2.** Radiation injury following exposure to excessive doses of radiation, such as the explosion of an atomic bomb.

radiation source [NUCLEO] Usually a man-made, sealed source of radioactivity used in teletherapy, radiography, as a power source for batteries, or in various types of industrial gages; machines such as accelerators, and radioisotopic generators and natural radionuclides may also be considered as sources.

radiation standards [NUCLEO] Exposure standards, permissible concentrations, rules for safe handling, regulations for transportation, regulations for industrial control of radiation, and control of radiation exposure by legislative means.

radiation sterilization [NUCLEO] Exposure of a material, object, or body to ionizing radiation in order to destroy microorganisms.

radiation survey meter [NUCLEO] Portable device to measure the intensity of nuclear radiations in a given region, in such applications as health physics (atomic radiation safety) or supervision of radioactively hot areas.

radiation therapy [MED] The use of ionizing radiation or radioactive substances to treat disease. Also known as actinotherapy; radiotherapy.

radiation thermocouple [ELEC] An infrared detector consisting of several thermocouples connected in series, arranged so that the radiation falls on half of the junctions, causing their temperature to increase so that a voltage is generated.

radiation thermometer *See* radiation pyrometer.

radiation unit *See* radiation length.

radiation vacuum gage [ENG] Vacuum (reduced-pressure) measurement device in which gas ionization from an α-source of radiation varies measurably with changes in the density (molecular concentration) of the gas being measured.

radiation warning symbol [NUCLEO] A standard symbol used on posters displayed in locations where radiation hazards exist; consists of a magenta trefoil printed on a yellow background.

radiation well logging *See* radioactive well logging.

radiation zone *See* Fraunhofer region.

radiative capture [NUC PHYS] A nuclear capture process whose prompt result is the emission of electromagnetic radiation only.

radiative diffusivity [METEOROL] A characteristic property of a given layer of the atmosphere which governs the rate at which that layer will warm or cool as a result of the transfer, within it, of infrared radiation; the radiative diffusivity is dependent upon the temperature and water-vapor content of the layer of air and upon the pressure within the layer.

radiative equilibrium [ASTROPHYS] Energy transfer through a star by radiation, absorption, and reradiation at a rate such that each section of the star is maintained at the appropriate temperature.

radiative recombination [PHYS] Recombination of parts of an atom during which electromagnetic radiation is emitted.

radiative transfer [THERMO] The transmission of heat by electromagnetic radiation.

radiative transition [QUANT MECH] A change of a quantum-mechanical system from one energy state to another in which electromagnetic radiation is emitted.

radiator [ACOUS] A vibrating element of a transducer which radiates sound waves. [ELECTROMAG] **1.** The part of an antenna or transmission line that radiates electromagnetic waves either directly into space or against a reflector for focusing or directing. **2.** A body that emits radiant energy. [ENG] Any of numerous devices, units, or surfaces that emit heat, mainly by radiation, to objects in the space in which they are installed. [PHYS] **1.** In general, a body which emits particles or radiation in any form. **2.** A body placed in a beam of ionizing radiation which, as a result, emits radiation of another kind.

radiator temperature drop [MECH ENG] In internal combustion engines, the difference in temperature of the coolant liquid entering and leaving the radiator.

radiatus [METEOROL] A cloud variety whose elements are arranged in straight parallel bands; owing to the effect of perspective, these bands seem to converge toward a point on the horizon, or, when the bands cross the entire sky, toward two opposite points.

radical [BOT] **1.** Of, pertaining to, or proceeding from the root. **2.** Arising from the base of a stem or from an underground stem. [CHEM] A stable group of atoms found as part of the molecules of a number of compounds, organic or inorganic; examples are CH_3^- (methyl), $C_2H_5^-$ (ethyl), SO_4^{--} (sulfate), and ClO_3^- (chlorate). [MATH] **1.** In a ring, the intersection of all maximal ideals of a quantity. **2.** An indicated root of a quantity. Symbolized $\sqrt{\ }$.

radical axis [MATH] The line passing through the two points of intersection of a pair of circles.

radical equation *See* irrational equation.

radicidation [FOOD ENG] Destruction by radiation of microorganisms in food that are significant to public health; an example is *Salmonella* species.

radicle [BOT] The embryonic root of a flowering plant.

radicotomy *See* rhizotomy.

radicular cyst [MED] A cyst arising from a granuloma of the root of a tooth. Also known as periapical cyst.

radiculitis [MED] Inflammation of a nerve root.

radiculoneuropathy [MED] Disease of the peripheral spinal nerves and their roots.

radio [COMMUN] The transmission of signals through space by means of electromagnetic waves; usually applied to the transmission of sound and code signals, although television and radar also depend on electromagnetic waves. [ELECTR] *See* radio receiver.

radio- [ELECTROMAG] A prefix denoting the use of radiant energy, particularly radio waves. [NUCLEO] Chemical prefix designating radiation or radioactivity; used to designate radioactive elements (such as radiocarbon) and substances containing them (such as radiochemicals, radiocolloids, or radio compounds).

radioacoustic position finding *See* radioacoustic ranging.

radioacoustic ranging [ENG] A method for finding the position of a vessel at sea; a bomb is exploded in the water, and the sound of the explosion transmitted through water is picked up by the vessel and by shore stations, other vessels, or buoys whose positions are known; the received sounds are transmitted instantaneously by radio to the surveying vessel, and the elapsed times are proportional to the distances to the known positions. Abbreviated RAR. Also known as radioacoustic position finding; radioacoustic sound ranging.

radioacoustics [COMMUN] Study of the production, transmission, and reproduction of sounds carried from one place to another by radiotelephony.

radioacoustic sound ranging *See* radioacoustic ranging.

radioactinium [NUC PHYS] Conventional name for the isotope of thorium which has mass number 227 and is in the actinium series. Symbolized RdAc.

radioactive [NUC PHYS] Exhibiting radioactivity or pertaining to radioactivity.

radioactive age determination *See* radiometric dating.

radioactive carbon dating *See* radiocarbon dating.

radioactive chain *See* radioactive series.

RADIO

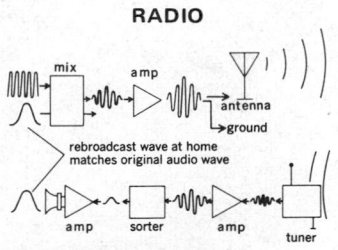

Transmission of audio information by radio.

radioactive clock [NUC PHYS] A radioactive isotope such as potassium-40 which spontaneously decays to a stable end product at a constant rate, allowing absolute geologic age to be determined.

radioactive cloud [NUCLEO] A mass of air and vapor in the atmosphere carrying radioactive debris from a nuclear explosion.

radioactive cobalt [NUC PHYS] Radioactive form of cobalt, such as cobalt-60 with a half-life of 5.3 years.

radioactive contaminant [NUCLEO] A radioactive material which has spread to places where it may harm persons, spoil experiments, or make products or equipment unsuitable or unsafe for some specific purpose.

radioactive dating *See* radiometric dating.

radioactive debris [NUCLEO] Radioactive material which is carried through the air from the site of a nuclear explosion.

radioactive decay [NUC PHYS] The spontaneous transformation of a nuclide into one or more different nuclides, accompanied by either the emission of particles from the nucleus, nuclear capture or ejection of orbital electrons, or fission. Also known as decay; nuclear spontaneous reaction; radioactive disintegration; radioactive transformation; radioactivity.

radioactive decay constant *See* decay constant.

radioactive decay product *See* daughter.

radioactive decay series *See* radioactive series.

radioactive disintegration *See* radioactive decay.

radioactive displacement law [NUC PHYS] The statement of the changes in mass number A and atomic number Z that take place during various nuclear transformations. Also known as displacement law.

radioactive element [NUC PHYS] An element all of whose isotopes spontaneously transform into one or more different nuclides, giving off various types of radiation; examples include promethium, radium, thorium, and uranium.

radioactive emanation [NUC PHYS] A radioactive gas given off by certain radioactive elements; all of these gases are isotopes of the element radon. Also known as emanation.

radioactive fallout *See* fallout.

radioactive half-life *See* half-life.

radioactive heat [THERMO] Heat produced within a medium as a result of absorption of radiation from decay of radioisotopes in the medium, such as thorium-232, potassium-40, uranium-238, and uranium-235.

radioactive isotope *See* radioisotope.

radioactive material [NUCLEO] A material having one or more constituents that exhibit significant radioactivity.

radioactive metal [NUC PHYS] A luminous metallic element, such as actinium, radium, or uranium, that spontaneously and continuously emits radiation capable in some degree of penetrating matter impervious to ordinary light.

radioactive mineral [MINERAL] Any mineral species that contains uranium or thorium as an essential part of the chemical composition; examples are uraninite, pitchblende, carnotite, coffinite, and autunite.

radioactive paint [MATER] A luminous paint that gives off light without being activated.

radioactive salt [NUCLEO] A salt whose molecules have radioactive atoms, such as radium bromide or mesothorium bromide; used in trace amounts to energize self-luminous paints.

radioactive series [NUC PHYS] A succession of nuclides, each of which transforms by radioactive disintegration into the next until a stable nuclide results. Also known as decay chain; decay family; decay series; disintegration chain; disintegration family; disintegration series; radioactive chain; radioactive decay series; series decay; transformation series.

radioactive snow gage [ENG] A device which automatically and continuously records the water equivalent of snow on a given surface as a function of time; a small sample of a radioactive salt is placed in the ground in a lead-shielded collimator which directs a beam of radioactive particles vertically upward; a Geiger-Müller counting system (located above the snow level) measures the amount of depletion of radiation caused by the presence of the snow.

radioactive source [NUCLEO] Any quantity of radioactive material intended for use as a source of ionizing radiation.

radioactive standard [NUCLEO] A sample of radioactive material which contains a known number and type of radioactive atoms at some definite time; used to calibrate radiation measuring instruments.

radioactive thickness gage [NUCLEO] An instrument for measuring the thickness of the metal wall of a pipe or tank from one side, by directing a beam of γ-rays through the wall at an angle and measuring the amount of backscattering with a radiation detector.

radioactive tracer [NUCLEO] A radioactive isotope which, when attached to a chemically similar substance or injected into a biological or physical system, can be traced by radiation detection devices, permitting determination of the distribution or location of the substance to which it is attached. Also known as radiotracer.

radioactive transformation *See* radioactive decay.

radioactive waste [NUCLEO] Liquid, solid, or gaseous waste resulting from mining of radioactive ore, production of reactor fuel materials, reactor operation, processing of irradiated reactor fuels, and related operations, and from use of radioactive materials in research, industry, and medicine.

radioactive-waste disposal [NUCLEO] The disposal of waste radioactive materials and of equipment contaminated by radiation; the two basic disposal methods are concentration for burial underground or in the sea, and dilution for controlled dispersion; reprocessing of reactor fuel is a major source of radioactive waste. Also known as waste disposal.

radioactive well logging [ENG] The recording of the differences in radioactive content (natural or neutron-induced) of the various rock layers found down an oil well borehole; types include γ-ray, neutron, and photon logging. Also known as radiation well logging; radioactivity prospecting.

radioactivity [NUC PHYS] **1.** A particular type of radiation emitted by a radioactive substance, such as α-radioactivity. **2.** *See* radioactive decay. **3.** *See* activity.

radioactivity analysis *See* activation analysis.

radioactivity concentration guide [NUCLEO] The concentration of radioactive material in an environment which would result in doses equal, over a period of time, to those in the radiation protection guide; this Federal Radiation Council term replaces the former maximum permissible concentration.

radioactivity equilibrium [NUC PHYS] A condition which may arise in the decay of a radioactive parent with short-lived descendants, in which the ratio of the activity of a parent to that of a descendant remains constant.

radioactivity log [ENG] Record of radioactive well logging.

radioactivity prospecting *See* radioactive well logging.

radio aid to navigation [ELECTR] An aid to navigation which utilizes the propagation characteristics of radio waves to furnish navigation information.

radio altimeter [ENG] An absolute altimeter that depends on the reflection of radio waves from the earth for the determination of altitude, as in a frequency-modulated radio altimeter and a radar altimeter. Also known as electronic altimeter; reflection altimeter.

radio altitude *See* radar altitude.

radio and wire integration [COMMUN] The combining of wire circuits with radio facilities.

radio antenna *See* antenna.

radio approach aids [NAV] Equipment that uses radio or radar to furnish guidance to an aircraft with required accuracy from the time it is in the vicinity of an airfield until it reaches a position from which a landing can be made.

radio astronomy [ASTRON] The study of celestial objects by measurement and analysis of their emitted electromagnetic radiation in the wavelength range from roughly 1 millimeter to 30 millimeters.

radio atmometer [ENG] An instrument designed to measure the effect of sunlight upon evaporation from plant foliage; consists of a porous-clay atmometer whose surface has been blackened so that it absorbs radiant energy.

radio attenuation [ELECTROMAG] For one-way propagation,

radio attenuation is the ratio of the power delivered by the transmitter to the transmission line connecting it with the transmitting antenna to the power delivered to the receiver by the transmission line connecting it with the receiving antenna.

radioautography See autoradiography.

radio autopilot coupler [ENG] Equipment providing means by which an electrical navigational signal operates an automatic pilot.

radio B battery [ELEC] A B-type battery used in a radio set, usually consisting of 15 to 30 permanently connected cells.

radio beacon [NAV] A nondirectional radio transmitting station in a fixed geographic location, emitting a characteristic signal from which bearing information can be obtained by a radio direction finder on a ship or aircraft. Also known as aerophare; radiophare.

radio-beacon buoy [NAV] A buoy equipped with a marker radio beacon; such a buoy is usually used to mark an important entrance to a channel; the beacon is of low power and provides a signal for a short range.

radio-beacon monitor station [COMMUN] A station which monitors the signal from one or more remotely located marine radio beacons.

radio beam [ELECTROMAG] A concentrated stream of radio-frequency energy as used in radio ranges, microwave relays, and radar.

radio bearing [NAV] The bearing of a radio transmitter from a receiver as determined by a radio direction finder.

radiobiology [BIOL] Study of the scientific principles, mechanisms, and effects of the interaction of ionizing radiation with living matter. Also known as radiation biology.

radio blackout [COMMUN] A fadeout that may last several hours or more at a particular frequency. Also known as blackout.

radio broadcasting [COMMUN] Radio transmission intended for general reception.

radiocarbon See carbon-14.

radiocarbon dating [NUCLEO] A method of estimating the age of carbon-bearing materials which have formed within the last 70,000 years and which have utilized carbon dioxide from the atmosphere or some other portion of the earth's dynamic reservoir, based on the decay of carbon-14. Also known as carbon-14 dating; radioactive carbon dating.

radiocesium See cesium-137.

radiochemical laboratory [CHEM] A specially equipped and shielded chemical laboratory designed for conducting radiochemical studies without danger to the laboratory personnel.

radiochemistry [CHEM] That area of chemistry concerned with the study of radioactive substances.

radio climatology [CLIMATOL] The study of regional and seasonal variations in the manner of propagation of radio energy through the atmosphere.

radio command [ELECTR] A radio control signal to which a guided missile or other remote-controlled vehicle or device responds.

radio communication [COMMUN] Communication by means of radio waves, such as by radio facsimile, radiotelegraph, radiotelephone, and radioteletypewriter.

radio communications guard See radio guard.

radio compass See automatic direction finder.

radio control [ELECTR] The control of stationary or moving objects by means of signals transmitted through space by radio.

radio countermeasures [ELECTR] Electrical or other techniques depriving the enemy of the benefits which would ordinarily accrue to him through the use of any technique employing the radiation of radio waves; it includes benefits derived from radar and intercept services.

radio deception [COMMUN] The use of radio to deceive the enemy, as by sending false dispatches or using enemy call signs.

radiodermatitis [MED] Degenerative changes in the skin following excessive exposure to ionizing radiation. Also known as radiation dermatitis.

radio detection [ENG] The detection of the presence of an object by radiolocation without precise determination of its position.

radio detection and location [ENG] Use of an electronic system to detect, locate, and predict future positions of earth satellites.

radio detection and ranging See radar.

radio direction finder [NAV] A radio aid to navigation that uses a rotatable loop or other highly directional antenna arrangement to determine the direction of arrival of a radio signal. Abbreviated RDF. Also known as direction finder.

radio direction-finder station [COMMUN] A land-based radio station equipped with special apparatus for determining the direction of radio signals transmitted by ships and other stations.

radio direction finding [NAV] A procedure for determining the bearing at a receiving point of the source of a radio signal by observing the direction of arrival of the wave front.

radio Doppler [ENG] Direct determination of the radial component of the relative velocity of an object by an observed frequency change due to such velocity.

radio duct [GEOPHYS] An atmospheric layer, typically shallow and almost horizontal, in which radio waves propagate in an anomalous fashion; ducts occur when, due to sharp inversions of temperature or humidity, the vertical gradient of the radio index of refraction exceeds a critical value.

radio echo See radar return.

radioecology [ECOL] The interdisciplinary study of organisms, radionuclides, ionizing radiation, and the environment.

radioelectric meteorology See radio meteorology.

radio element [NUC PHYS] A radioactive isotope of an element, or a sample consisting of one or more radioactive isotopes of an element.

radio emission [ELECTROMAG] The emission of radio-frequency electromagnetic radiation by oscillating charges or currents.

radio energy [ELECTROMAG] The energy carried by radio-frequency electromagnetic radiation.

radio engineering [ENG] The field of engineering that deals with the generation, transmission, and reception of radio waves and with the design, manufacture, and testing of associated equipment.

radio facility chart See enroute chart.

radio-facsimile system [COMMUN] A facsimile system in which signals are transmitted by radio rather than by wire.

radio fadeout [COMMUN] Increased absorption of radio waves passing through the lower layers of the ionosphere due to a sudden and abnormal increase in ionization in these regions; signals at receivers then fade out or disappear. Also known as fadeout.

radio fan-marker beacon See fan-marker beacon.

radio field intensity [ELECTROMAG] Electric or magnetic field intensity at a given location associated with the passage of radio waves.

radio field-to-noise ratio [ELECTROMAG] Ratio, at a given location, of the radio field intensity of the desired wave to the noise field intensity.

radio fix [COMMUN] Determination of the position of the source of radio signals by obtaining cross bearings on the transmitter with two or more radio direction finders in different locations, then computing the position by triangulation. [NAV] **1.** Determination of the position of a vessel or aircraft equipped with direction-finding equipment by ascertaining the direction of radio signals received from two or more transmitting stations of known location and then computing the position by triangulation. **2.** Determination of position of an aircraft in flight by identification of a radio beacon or by locating the intersection of two radio beams.

radio fixing aid [NAV] Equipment making use of radio to assist in the determination of a geographical position.

radio frequency [ELECTROMAG] A frequency at which coherent electromagnetic radiation of energy is useful for communication purposes; roughly the range from 10 kilohertz to 100 gigahertz. Abbreviated rf.

radio-frequency alternator [ELEC] A rotating-type alternator designed to produce high power at frequencies above power-line values but generally lower than 100,000 hertz; used chiefly for high-frequency heating.

radio-frequency amplifier [ELECTR] An amplifier that ampli-

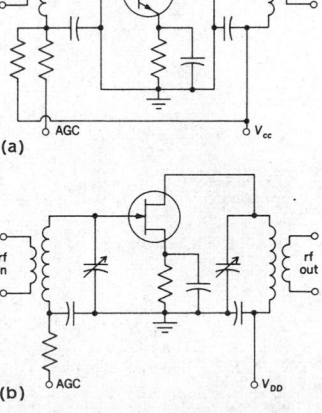

RADIO-FREQUENCY AMPLIFIER

(a)

(b)

Typical radio-frequency amplifier circuits with (a) bipolar transistor and (b) field-effect transistor. AGC = automatic gain control; V_{CC} = collector supply voltage; V_{DD} = drain supply voltage.

fies the high-frequency signals commonly used in radio communications.

radio-frequency bandwidth [COMMUN] Band of frequencies comprising 99% of the total radiated power extended to include any discrete frequency on which the power is at least 0.25% of the total radiated power.

radio-frequency cable [ELECTROMAG] A cable having electric conductors separated from each other by a continuous homogeneous dielectric or by touching or interlocking spacer beads; designed primarily to conduct radio-frequency energy with low losses. Also known as RG line.

radio-frequency cavity preselector [ELECTROMAG] A turnable cavity resonator in an ultra-high-frequency circuit, which is similar in function to a tuned resonant circuit.

radio-frequency choke [ELEC] A coil designed and used specifically to block the flow of radio-frequency current while passing lower frequencies or direct current.

radio-frequency component [COMMUN] Portion of a signal or wave which consists only of the radio-frequency alternations, and not including its audio rate of change in amplitude frequency.

radio-frequency current [ELEC] Alternating current having a frequency higher than 10,000 hertz.

radio-frequency filter [ELECTR] An electric filter which enhances signals at certain radio frequencies or attenuates signals at undesired radio frequencies.

radio-frequency generator [ELECTR] A generator capable of supplying sufficient radio-frequency energy at the required frequency for induction or dielectric heating.

radio-frequency head [ENG] Unit consisting of a radar transmitter and part of a radar receiver, the two contained in a package for ready removal and installation.

radio-frequency heating *See* electronic heating.

radio-frequency interference [COMMUN] Interference from sources of energy outside a system or systems, as contrasted to electromagnetic interference generated inside systems. Abbreviated RFI.

radio-frequency line *See* radio-frequency transmission line.

radio-frequency measurement [ELECTR] The precise measurement of frequencies above the audible range by any of various techniques, such as a calibrated oscillator with some means of comparison with the unknown frequency, a digital counting or scaling device which measures the total number of events occurring during a given time interval, or an electronic circuit for producing a direct current proportional to the frequency of its input signal.

radio-frequency oscillator [ELECTR] An oscillator that generates alternating current at radio frequencies.

radio-frequency power supply [ELECTR] A high-voltage power supply in which the output of a radio-frequency oscillator is stepped up by an air-core transformer to the high voltage required for the second anode of a cathode-ray tube, then rectified to provide the required high direct-current voltage; used in some television receivers.

radio-frequency preheating [ENG] Preheating of plastics-molding materials by radio frequencies of 10–100 megahertz per second to facilitate the molding operation or to reduce the molding-cycle time. Abbreviated rf preheating.

radio-frequency pulse [COMMUN] A radio-frequency carrier that is amplitude-modulated by a pulse; the amplitude of the modulated carrier is zero before and after the pulse. Also known as radio pulse.

radio-frequency reactor [ELECTR] A reactor used in electronic circuits to pass direct current and offer high impedance at high frequencies.

radio-frequency resistance *See* high-frequency resistance.

radio-frequency shift *See* frequency shift.

radio-frequency signal generator [ELECTR] A test instrument that generates the various radio frequencies required for alignment and servicing of radio, television, and electronic equipment. Also known as service oscillator.

radio-frequency spectrometer [SPECT] An instrument which measures the intensity of radiation emitted or absorbed by atoms or molecules as a function of frequency at frequencies from 10^5 to 10^9 hertz; examples include the atomic-beam apparatus, and instruments for detecting magnetic resonance.

radio-frequency spectroscopy [SPECT] The branch of spectroscopy concerned with the measurement of the intervals between atomic or molecular energy levels that are separated by frequencies from about 10^5 to 10^9 hertz, as compared to the frequencies that separate optical energy levels of about 6×10^{14} hertz.

radio-frequency spectrum *See* radio spectrum.

radio-frequency transformer [ELECTROMAG] A transformer having a tapped winding or two or more windings designed to furnish inductive reactance or to transfer radio-frequency energy from one circuit to another by means of a magnetic field; may have an air core or some form of ferrite core. Also known as radio transformer.

radio-frequency transmission line [ELECTROMAG] A transmission line designed primarily to conduct radio-frequency energy, consisting of two or more conductors supported in a fixed spatial relationship along their own length. Also known as radio-frequency line.

radio-frequency welding *See* high-frequency welding.

radio galaxy [ASTROPHYS] A galaxy that is emitting much energy in radio frequencies often from regions devoid of visible matter.

radiogenetics *See* radiation genetics.

radiogenic [NUC PHYS] Pertaining to a material produced by radioactive decay, as the production of lead from uranium decay.

radiogenic age determination *See* radiometric dating.

radiogenic dating *See* radiometric dating.

radiogoniometer [ELECTR] A goniometer used as part of a radio direction finder.

radiogoniometry [ENG] Science of locating a radio transmitter by means of taking bearings on the radio waves emitted by such a transmitter.

radiogram [COMMUN] A message transmitted by radio.

radiograph [GRAPHICS] The photographic image produced in radiography. Also known as shadowgraph.

radiographic equivalence factor [NUCLEO] The reciprocal of the thickness of a specified material having the same radiographic absorption as a unit thickness of a standard material.

radiographic film [GRAPHICS] The photographic film used in radiography, which must be properly selected for contrast, latitude, and sensitivity.

radiographic sensitivity [NUCLEO] A measure of radiographic quality whereby the minimum discontinuity that may be detected on the film is expressed as a percentage of the base thickness.

radiography [GRAPHICS] The technique of producing a photographic image of an opaque specimen by transmitting a beam of x-rays or γ-rays through it onto an adjacent photographic film; the image results from variations in thickness, density, and chemical composition of the specimen; used in medicine and industry.

radio guard [COMMUN] Ship, aircraft, or radio station designed to listen for and record transmission, and to handle traffic on a designated frequency for a certain unit or units. Also known as radio communications guard.

radio guidance [ELECTR] Guidance of a flight-borne missile or other vehicle from a ground station by means of radio signals.

radio guided bomb [ORD] A bomb, such as the azon, guided by radio control from outside the missile.

radio hole [GEOPHYS] Strong fading of the radio signal at some position in space along an air-to-air or air-to-ground path; the effect is caused by the abnormal refraction of radio waves.

radio homing aid [NAV] Radio equipment which permits an aircraft to reach a way point or its destination by maintaining constant some parameter which is derived through the use of the equipment.

radio homing beacon *See* homing beacon.

radio horizon [COMMUN] The locus of points at which direct rays from a transmitter become tangential to the surface of the earth; the distance to the radio horizon is affected by atmospheric refraction.

radiohumeral index [ANTHRO] The ratio, multiplied by 100, of the length of the radius to the length of the humerus of the human arm.

radio-inertial guidance system [ENG] A command type of

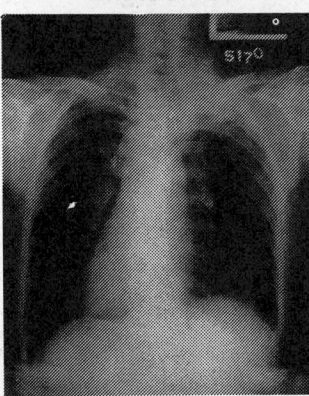

RADIOGRAPH

Chest radiograph of foundry worker made with intense beam from rotating-target x-ray tube, showing nodules in lungs which are caused by silicosis, and shadows of skeleton, heart, and stomach.

missile guidance system consisting essentially of a radar tracking unit; a computer that accepts missile position and velocity information from the tracking system and furnishes to the command link appropriate signals to steer the missile; the command link, which consists of a transmitter on the ground and an antenna and receiver on the missile; and an inertial system for partial guidance in case of radio guidance failure.

radio intelligence [COMMUN] Information regarding the enemy obtained by interception and interpretation of enemy radio transmissions.

radio interception [COMMUN] Tuning in on a radio message not intended for the listener.

radio interference *See* interference.

radio interferometer [ENG] Radiotelescope or radiometer employing a separated receiving antenna to measure angular distances as small as 1 second of arc; records the result of interference between separate radio waves from celestial radio sources.

radioiodine [NUC PHYS] Any radioactive isotope of iodine, especially iodine-131; used as a tracer to determine the activity and size of the thyroid gland, and experimentally, to destroy the thyroid glands of animals.

radioisotope [NUC PHYS] An isotope which exhibits radioactivity. Also known as radioactive isotope; unstable isotope.

radioisotope assay [ANALY CHEM] An analytical technique including procedures for separating and reproducibly measuring a radioactive tracer.

radioisotope battery *See* nuclear battery.

radioisotope thermoelectric generator [NUCLEO] A device for converting nuclear energy to electrical energy in which the heat produced by radioactivity of a radioisotope is used to produce a voltage in a thermocouple circuit; chief use has been in space vehicles and in instruments left on the lunar surface. Abbreviated RTG.

radioisotopic generator *See* nuclear battery.

radio knife [MED] A surgical knife that uses a high-frequency electric arc at its tip to cut tissue and simultaneously sterilize the edges of the wound.

radio landing aid [NAV] Equipment using radio waves to assist aircraft in carrying out their actual landings.

radio landing beam [NAV] A radio beam used for vertical guidance of aircraft during descent to a landing surface.

Radiolaria [INV ZOO] A subclass of the protozoan class Actinopodea whose members are noted for their siliceous skeletons and characterized by a membranous capsule which separates the outer from the inner cytoplasm.

radiolarian chert [GEOL] A homogeneous cryptocrystalline radiolarite with a well-developed matrix.

radiolarian earth [GEOL] A porous, unconsolidated siliceous sediment formed from the opaline silica skeletal remains of Radiolaria; formed from radiolarian ooze.

radiolarian ooze [GEOL] A siliceous ooze containing the skeletal remains of the Radiolaria.

radiolarite [GEOL] **1.** A whitish, hard, consolidated equivalent of radiolarian earth. **2.** Radiolarian ooze that has been indurated.

radiole [INV ZOO] A spine on a sea urchin.

radio line of position [NAV] A line of position determined by any of a series of radio navigation aids.

radio link [COMMUN] A radio system used to provide a communication or control channel between two specific points.

radiolocation [ENG] Determination of relative position of an object by means of equipment operating on the principle that propagation of radio waves is at a constant velocity and rectilinear.

radio log [COMMUN] A log of radio messages sent and received, together with other pertinent information, maintained by radio operators.

radiological [NUCLEO] Pertaining to nuclear radiation, radioactivity, and atomic weapons.

radiological agent [NUCLEO] Any of a family of substances that produce casualties by emitting radiation.

radiological defense [ORD] The methods, plans, and procedures involved in establishing and exercising defensive measures against the radiation effects of an attack by atomic weapons or radiological warfare agents.

radiological dose [NUCLEO] The total amount of ionizing radiation absorbed by an individual exposed to any radiating source.

radiological survey [NUCLEO] Determination of the distribution and dose rates of radiation in an area.

radiological warfare [ORD] The employment of agents or weapons to produce residual radioactive contamination, as distinguished from the initial effects of a nuclear explosion (blast, thermal, and initial nuclear radiation).

radiologist [MED] A physician who specializes in the use of radiant energy in the diagnosis and treatment of disease.

radiology [MED] The medical science concerned with radioactive substances, x-rays, and other ionizing radiations, and the application of the principles of this science to diagnosis and treatment of disease.

radiolucent [ELECTROMAG] Transparent to x-rays and radio waves.

radioluminescence [PHYS] Luminescence produced by x-rays or γ-rays, or by particles emitted in radioactive decay.

radiolysis [PHYS CHEM] The dissociation of molecules by radiation; for example, a small amount of water in a reactor core dissociates into hydrogen and oxygen during operation.

radio magnetic indicator [NAV] An aircraft instrument which presents a combined display of omnibearing vehicle heading and relative bearings of a radio station.

radio marker beacon station [NAV] Station marking a definite location on the ground as an aid to aircraft at altitude; the ICAO beacons operate at a single frequency of 75 megahertz.

radio mast [ENG] A tower, pole, or other structure for elevating an antenna.

radio metal locator *See* metal detector.

radio meteor [ASTRON] A meteor which has been detected by the reflection of a radio signal from the meteor trail of relatively high ion density (ion column); such an ion column is left behind a meteoroid when it reaches the region of the upper atmosphere between about 80 and 120 kilometers, although occasionally radio meteors are detected at higher altitudes.

radio-meteorograph [COMMUN] A device for the automatic radio transmission of the indications of a set of meteorological instruments.

radio meteorology [METEOROL] That branch of the science of meteorology which embraces the propagation of radio energy through the atmosphere, and the use of radio and radar equipment in meteorology; this is the most general term and includes radar meteorology. Also known as radioelectric meteorology.

radiometer [ELECTR] A receiver for detecting microwave thermal radiation and similar weak wide-band signals that resemble noise and are obscured by receiver noise; examples include the Dicke radiometer, subtraction-type radiometer, and two-receiver radiometer. Also known as microwave radiometer; radiometer-type receiver. [ENG] An instrument for measuring radiant energy; examples include the bolometer, microradiometer, and thermopile.

radiometer-type receiver *See* radiometer.

radiometric age [GEOL] Geologic age expressed in years determined by quantitatively measuring radioactive elements and their decay products.

radiometric analysis [ANALY CHEM] Quantitative chemical analysis that is based on measurement of the absolute disintegration rate of a radioactive component having a known specific activity.

radiometric dating [NUCLEO] A technique for measuring the age of an object or sample of material by determining the ratio of the concentration of a radioisotope to that of a stable isotope in it; for example, the ratio of carbon-14 to carbon-12 reveals the approximate age of bones, pieces of wood, and other archeological specimens. Also known as isotopic age determination; nuclear age determination; radioactive age determination; radioactive dating; radiogenic age determination; radiogenic dating.

radiometric magnitude [ASTRON] A celestial body's magnitude as calculated from the total amount of radiant energy of all the wavelengths that reach the earth's surface.

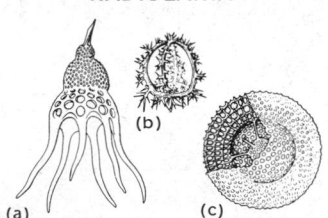

RADIOLARIA

Radiolaria. Skeletons representing (a, b) certain Monopylina (or Nasellina) and (c) certain Periphylina (or Spumellina). *(After Haeckel)*

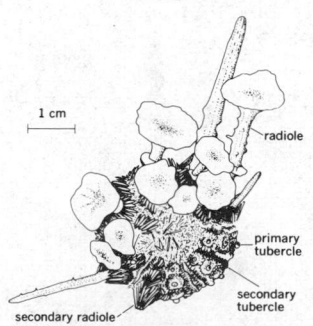

RADIOLE

Morphological features of *Goniocidaris parasol* showing radioles.

radiometric titration [ANALY CHEM] Use of radioactive indicator to track the transfer of material between two liquid phases in equilibrium, such as titration of $Ag^{110}NO_3$ against potassium chloride.

radiometry [PHYS] The detection and measurement of radiant electromagnetic energy, especially that associated with infrared radiation.

radiomicrometer See microradiometer.

radiomimetic activity [BIOL] The radiationlike effects of certain chemicals, such as nitrogen mustard, urethane, and fluorinated pyrimidines.

radiomimetic substances [CHEM] Chemical substances which cause biological effects similar to those caused by ionizing radiation.

radio mirage [ELECTROMAG] The detection of radar targets at phenomenally long range due to radio ducting.

radiomutation [GEN] A mutation which is the result of exposure of living-tissue chromosomes to ionizing radiation.

radio navigation [NAV] The use of apparatus operating at radio frequencies to determine parameters useful for navigation; it includes radio direction finding, radio ranges, radio compasses, radio homing beacons, and loran.

radio nebula [ASTROPHYS] A nebula that emits nonthermal radio-frequency radiation; derives its luminosity from collisions with the surrounding interstellar medium, or from processes associated with the magnetic fields presumably involved within the nebula; examples are the network nebulae in Cygnus and NGC 443.

radionecrosis [PATH] Destruction of living tissue by radiation.

radio net [COMMUN] System of radio stations operating with each other; a military net usually consists of a radio station of a superior unit and stations of all subordinate or supporting units.

radio noise [ELECTROMAG] Electromagnetic noise having radio frequencies.

radio-noise field strength [ELECTROMAG] A quantity which is proportional, or related in a known manner, to the field strength of electromagnetic waves of an interfering character at a point, such as a radio receiver. Also known as noise field strength.

radionuclide [NUC PHYS] A nuclide that exhibits radioactivity.

radiopaque [ELECTROMAG] Not appreciably penetrable by x-rays or other forms of radiation.

radiopaque agent [MED] A substance, such as barium sulfate, which is opaque to x-rays, administered orally to a patient to provide contrast in x-ray photographs of his gastrointestinal system.

radiopasteurization [ENG] Pasteurization by surface treatment with low-energy irradiation.

radiophare See radio beacon.

radiophone See radiotelephone.

radiophoto See facsimile.

radiophotoluminescence [PHYS] Luminescence exhibited by minerals such as fluorite and kunzite as a result of irradiation with β- and γ-rays followed by exposure to light.

radio pill [ELECTR] A device used in biotelemetry for monitoring the physiologic activity of an animal, such as pH values of stomach acid; an example is the Heidelberg capsule.

radio position finding [ENG] Process of locating a radio transmitter by plotting the intersection of its azimuth as determined by two or more radio direction finders.

radio-positioning land station [COMMUN] Station in the radiolocation service, other than a radio-navigation station, not intended for operation while in motion.

radio-positioning mobile station [COMMUN] Station in the radiolocation service, other than a radio-navigation station, intended to be used while in motion or during halts at unspecified points.

radio prospecting [ENG] Use of radio and electric equipment to locate mineral or oil deposits.

radio proximity fuse [ORD] A proximity fuse that contains a miniature radio transmitter and uses radio echoes from a target to trigger the fuse within predetermined limits of distance from the target.

radio pulse [COMMUN] See radio-frequency pulse. [ELEC-TROMAG] An intense burst of radio-frequency energy lasting for a fraction of a second.

radio range [COMMUN] A radio facility, usually land-based, that emits signals which when received by appropriate equipment provide navigational information. Also known as range.

radio range finding [NAV] Radiolocation in which the distance of an object is determined by measuring the time of arrival of its radio emissions, whether independent, reflected, or retransmitted on the same or other wavelength.

radio range station [NAV] Radio-navigation land station in the aeronautical, radio-navigation service providing radio equisignal zones.

radio receiver [ELECTR] A device that converts radio waves into intelligible sounds or other perceptible signals. Also known as radio; radio set; receiving set.

radio recognition [COMMUN] Determination by radio means of the friendly or enemy character, or the individuality, of another radio station.

radio relay satellite See communications satellite.

radio relay system [COMMUN] A radio transmission system in which intermediate radio stations or radio repeaters receive and retransmit radio signals. Also known as relay system.

radio repeater [COMMUN] A repeater that acts as an intermediate station in transmitting radio communications signals or radio programs from one fixed station to another; serves to extend the reliable range of the originating station; a microwave repeater is an example.

radioresistance [BIOL] The resistance of organisms or tissues to the harmful effects of various radiations.

radio scattering See scattering.

radiosensitive [MATER] Sensitive to damage by radiant energy.

radio set See radio transmitter.

radio sextant [ELECTROMAG] An antenna with a high-resolution beam pattern that measures the angle between local direction references and an astronomical radio signal source such as an artificial satellite, the sun, the moon, or a radio star.

radio shielding [ELEC] Metallic covering over all electric wiring and ignition apparatus, which is grounded at frequent intervals for the purpose of eliminating electric interference with radio communications.

radio signal [COMMUN] A signal transmitted by radio.

radio-signal reporting code [COMMUN] A code for reporting the quality of radiotelephone or radiotelegraph transmission, consisting of a code word followed by a group of numbers rating various characteristics. Also known as signal reporting code.

radio silence [COMMUN] Period during which all or certain radio equipment capable of radiation is kept inoperative.

radiosonde [ENG] A balloon-borne instrument for the simultaneous measurement and transmission of meteorological data; the instrument consists of transducers for the measurement of pressure, temperature, and humidity, a modulator for the conversion of the output of the transducers to a quantity which controls a property of the radio-frequency signal, a selector switch which determines the sequence in which the parameters are to be transmitted, and a transmitter which generates the radio-frequency carrier.

radiosonde balloon [AERO ENG] A balloon used to carry a radiosonde aloft; it is considerably larger than a pilot balloon or a ceiling balloon.

radiosonde commutator [ELECTR] A component of a radiosonde consisting of a series of alternate electrically conducting and insulating strips; as these are scanned by a contact, the radiosonde transmits temperature and humidity signals alternately.

radiosonde observation [METEOROL] An evaluation in terms of temperature, relative humidity, and pressure aloft, of radio signals received from a balloon-borne radiosonde; the height of each mandatory and significant pressure level of the observation is computed from these data. Also known as raob.

radiosonde-radio-wind system [ENG] An apparatus consisting of a standard radiosonde and radiosonde ground equipment to obtain upper-air data on pressure, temperature, and humidity, and a self-tracking radio direction finder to provide the elevation and azimuth angles of the radiosonde so that the wind vectors may be obtained.

radiosonde set [ENG] A complete set for automatically measuring and transmitting high-altitude meteorological data by radio from such carriers as a balloon or rocket.

radio sonobuoy *See* sonobuoy.

radio source [ASTROPHYS] A source of extragalactic or interstellar electromagnetic emission in radio wavelengths.

radio spectrum [COMMUN] The entire range of frequencies in which useful radio waves can be produced, extending from the audio range to about 300,000 megahertz. Also known as radio-frequency spectrum.

radio spectrum allocation [COMMUN] The specification of the frequencies of the radio spectrum which are available for use by the various radio services.

radio star [ASTROPHYS] A discrete celestial radio source.

radio station [COMMUN] A station equipped to engage in radio communication or radio broadcasting.

radio storm [ASTROPHYS] A prolonged period of disturbed emission or reception that lasts for periods of hours up to days.

radio sun [ASTROPHYS] The sun as defined by its electromagnetic radiation in the radio portion of the spectrum.

radiotelegraphy [COMMUN] Telegraphy involving the use of radio waves in place of wire lines.

radio telemetry [COMMUN] The presentation of data at a location remote from the source of the data, using radio-frequency electromagnetic radiation as the means of transmission.

radiotelephone [COMMUN] **1.** Pertaining to telephony over radio channels. **2.** A radio transmitter and radio receiver used together for two-way telephone communication by radio. Also known as radiophone.

radiotelephony [COMMUN] Two-way transmission of sounds by means of modulated radio waves, without interconnecting wires.

radio telescope [ENG] An astronomical instrument used to measure the amount of radio energy coming from various directions in the sky, consisting of a highly directional antenna and associated electronic equipment.

radioteletype [COMMUN] A teletypewriter and the associated equipment needed for operation over a radio channel rather than over wires. Abbreviated RTTY.

radioteletypewriter [COMMUN] A teletypewriter and the associated equipment needed for operation over a radio channel rather than over wires.

radiotherapy *See* radiation therapy.

radiothorium [NUC PHYS] Conventional name of the isotope of thorium which has mass number 228. Symbolized RdTh.

radio time signal [COMMUN] A time signal sent by radio broadcast.

radio tower [COMMUN] A tower, usually several hundred meters tall, either guyed or freestanding, on which a transmitting antenna is mounted to increase the range of radio transmission; in some cases, the tower itself may be the antenna.

radiotracer *See* radioactive tracer.

radio tracking [ENG] The process of keeping a radio or radar beam set on a target and determining the range of the target continuously.

radio transmission [COMMUN] The transmission of signals through space at radio frequencies by means of radiated electromagnetic waves.

radio transmitter [ELECTR] The equipment used for generating and amplifying a radio-frequency carrier signal, modulating the carrier signal with intelligence, and feeding the modulated carrier to an antenna for radiation into space as electromagnetic waves. Also known as radio set; transmitter.

radio transponder [ELECTR] A transponder which receives and transmits radio waves, in contrast to a sonar transponder, which receives and transmits acoustic waves.

radio tube *See* electron tube.

radio watch *See* watch.

radio wave [ELECTROMAG] An electromagnetic wave produced by reversal of current in a conductor at a frequency in the range from about 10 kilohertz to about 300,000 megahertz.

radio wavefront distortion [ELECTROMAG] Change in the direction of advance of a radio wave.

radio-wave propagation [ELECTROMAG] The transfer of energy through space by electromagnetic radiation at radio frequencies.

radio window [GEOPHYS] A band of frequencies extending from about 6 to 30,000 megahertz, in which radiation from the outer universe can enter and travel through the atmosphere of the earth.

radish [BOT] *Raphanus sativus*. **1.** An annual or biennial crucifer belonging to the order Capparales. **2.** The edible, thickened hypocotyl of the plant.

radist [NAV] Radio-navigation system in which the comparison of arrival times of transmitted pulses, at three or more ground stations, indicates the position of the vehicle.

radium [CHEM] A radioactive member of group IIA, symbol Ra, atomic number 88; the most abundant naturally occurring isotope has mass number 226 and a half-life of 1620 years. A highly toxic solid that forms water-soluble compounds; decays by emission of α, β, and γ-radiation; melts at 700°C, boils at 1140°C; turns black in air; used in medicine, in industrial radiography, and as a source of neutrons and radon.

radium age [NUCLEO] The age of a mineral as calculated from the numbers of radium atoms present originally, now, and when equilibrium is established with ionium.

radium bromide [INORG CHEM] $RaBr_2$ Water-soluble, poisonous, radioactive white powder, corrosive to skin or flesh; melts at 728°C; used in medicine, physical research, and luminous paint.

radium carbonate [INORG CHEM] $RaCO_3$ Water-insoluble, poisonous, radioactive, white powder; used in medicine.

radium cell [NUCLEO] A sealed thin-wall tube containing radium.

radium chloride [INORG CHEM] $RaCl_2$ Water- and alcohol-soluble, poisonous, radioactive, yellow-white crystals; corrosive effect on skin and flesh; melts at 1000°C; used in medicine, physical research, and luminous paint.

radium F *See* polonium-210.

radium needle [NUCLEO] A radium cell in the form of a needle, usually of platinum-iridium or gold alloy, designed primarily for insertion in tissue.

radium plaque [NUCLEO] A radium container in which the radium is distributed over a surface; the shielding is usually small in one direction so as to permit transmission of β-rays as well as γ-rays.

radium sulfate [INORG CHEM] $RaSO_4$ Water-insoluble, radioactive, poisonous, white crystals; used in medicine.

radium therapy [MED] Radiotherapy using the radiations from radium.

radius [ANAT] The outer of the two bones of the human forearm or of the corresponding part in vertebrates other than fish. [MATH] **1.** A line segment joining the center and a point of a circle or sphere. **2.** The length of such a line segment.

radius of action [ENG] The maximum distance a ship, aircraft, or other vehicle can travel away from its base along a given course with normal load and return without refueling, but including the fuel required to perform those maneuvers made necessary by all safety and operating factors.

radius of convergence [MATH] The positive real number corresponding to a power series expansion about some number a with the property that if $x - a$ has absolute value less than this number the power series converges at x, and if $x - a$ has absolute value greater than this number the power series diverges at x.

radius of curvature [MATH] The radius of the circle of curvature at a point of a curve.

radius of damage [ORD] The distance from ground zero of a nuclear blast at which there is a 0.50 probability of achieving the desired damage.

radius of gyration [MATH] The square root of the ratio of the moment of inertia of a plane figure about a given axis to its area. [MECH] The square root of the ratio of the moment of inertia of a body about a given axis to its mass.

radius of protection [ENG] The radius of the circle within which a lightning discharge will not strike, due to the presence of an elevated lightning rod at the center.

RADIO TELESCOPE

Installation of electronics at the focus of an 85-foot (26-meter) diameter radio telescope.

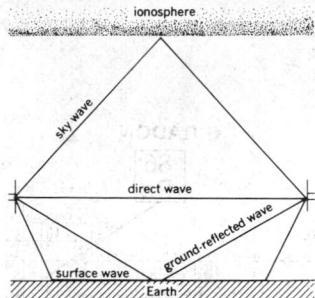

RADIO-WAVE PROPAGATION

Possible transmission paths of electromagnetic radiation at radio frequencies.

RADISH

Radish (*Raphanus sativus*), cultivar Red Boy. (*Joseph Harris Co., Rochester, N.Y.*)

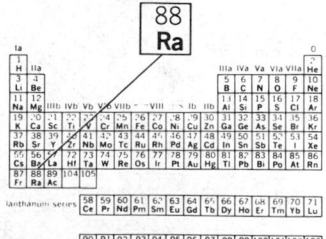

RADIUM

Periodic table of the chemical elements showing the position of radium.

radius of rupture [ORD] Greatest distance from the center of an underground explosive charge at which the explosion will be destructive.

radius of safety [ORD] The horizontal distance from ground area beyond which the weapon effects on friendly troops are acceptable.

radius of visibility [NAV] The radius of a circle limiting the area in which an objective can be seen under specified conditions.

radius ratio [PHYS CHEM] The ratio of the radius of a cation to the radius of an ion; relative ionic radii are pertinent to crystal lattice structure, particularly the determination of coordination number.

radius rod [ENG] A rod which restricts movement of a part to a given arc.

radius vector *See* position vector.

radix *See* base of a number system; root.

radix complement [MATH] A numeral in positional notation that can be derived from another by subtracting the original numeral from the numeral of highest value with the same number of digits, and adding 1 to the difference. Also known as complement; true complement.

radix-minus-one complement [MATH] A numeral in positional notation of base (or radix) B derived from a given numeral by subtracting the latter from the highest numeral with the same number of digits, that is, from $B - 1$; it is 1 less than the radix complement.

radix notation [MATH] A positional notation in which the successive digits are interpreted as coefficients of successive integral powers of a number called the radix or base; the represented number is equal to the sum of this power series. Also known as base notation.

radix point [MATH] A dot written either on or slightly above the line, used to mark the point at which place values change from positive to negative powers of the radix in a number system; a decimal point is a radix point for radix 10.

radome [ELECTROMAG] A strong, thin shell, made from a dielectric material that is transparent to radio-frequency radiation, and used to house a radar antenna, or a space communications antenna of similar structure.

radon [CHEM] A chemical element, symbol Rn, atomic number 86; all isotopes are radioactive, the longest half-life being 3.82 days for mass number 222; it is the heaviest element of the noble-gas group, produced as a gaseous emanation from the radioactive decay of radium. [NUC PHYS] The conventional name for the isotope of the element radon having mass number 222, which is a member of the uranium series.

radon breath analysis [MED] Examination of exhaled air for the presence of radon to determine the presence and quantity of radium in the human body.

radula [INV ZOO] A filelike ribbon studded with horny or chitinous toothlike structures, found in the mouth of all classes of mollusks except Bivalvia.

raffiche [METEOROL] In the Mediterranean region, gusts from the mountains; violent gusts of the bora.

raffinase [BIOCHEM] An enzyme that hydrolyzes raffinose, yielding fructose in the reaction.

raffinate [CHEM ENG] In solvent refining, that portion of the treated liquid mixture that remains undissolved and is not removed by the selective solvent. Also known as good oil to petroleum-refinery operators.

raffinose [BIOCHEM] $C_{18}H_{32}O_{16} \cdot 5H_2O$ A white, crystalline trisaccharide found in sugarbeets, cottonseed meal, and molasses; yields glucose, fructose, and galactose on complete hydrolysis. Also known as gossypose; melitose; melitriose.

Rafflesiales [BOT] A small order of dicotyledonous plants; members are highly specialized, nongreen, rootless parasites which grow from the roots of the host.

raft [ENG] A quantity of timber or lumber secured together by means of ropes, chains, or rods and used for transportation by floating.

rafted ice [OCEANOGR] A form of pressure ice composed of overlying pieces of ice floe.

rafter [BUILD] A roof-supporting member immediately beneath the roofing material.

rafter dam [CIV ENG] A dam made of horizontal timbers that meet in the center of the stream like rafters in a roof.

raft foundation [CIV ENG] A continuous footing that supports an entire structure, such as a floor. Also known as foundation mat.

rafting [GEOL] Transporting of rock by floating ice or floating organic materials (such as logs) to places not reached by water currents. [OCEANOGR] The process of forming rafted ice.

raft tectonics *See* plate tectonics.

ragged ceiling *See* indefinite ceiling.

raggiatura [METEOROL] Land squalls descending with great force from ravines and valleys in high land in Italy; they extend only a short distance off the west coast.

rag papers [MATER] The most expensive papers, made wholly or partly from cotton or linen rags; rag content is expressed as 25, 50, 75, or 100%; pure rag papers are the strongest and most resistant to the discoloration and deterioration due to age.

rail [ENG] 1. A bar extending between posts or other supports as a barrier or guard. 2. A steel bar resting on the crossties to provide track for railroad cars and other vehicles with flanged wheels. [MECH ENG] A high-pressure manifold in some fuel injection systems.

rail anchor [CIV ENG] A device that prevents tracks from moving longitudinally and maintains the proper gap between sections of rail.

rail bender [ENG] A portable appliance for bending rails for track or for straightening bent or curved rails.

rail capacity [CIV ENG] The maximum number of trains which can be planned to move in both directions over a specified section of track in a 24-hour period.

rail clip [CIV ENG] 1. A plate that holds a rail at its base. 2. A device used to fasten a derrick or crane to the rails of a track to prevent tipping. 3. A support on a track rail, used for holding a detector bar.

rail crane *See* locomotive crane.

rail-fence jammer *See* continuous-wave jammer.

railhead [CIV ENG] 1. The topmost part of a rail, supporting the wheels of railway vehicles. 2. A point at which railroad traffic originates and terminates. 3. The temporary ends of a railroad line under construction.

railing [CIV ENG] A barrier consisting of a rail and supports. [ELECTR] Radar pulse jamming at high recurrence rates (50 to 150 kilohertz); it results in an image on a radar indicator resembling fence railing.

rail joint [CIV ENG] A rigid connection of the ends of two sections of railway track.

Raillietiellidae [INV ZOO] A small family of parasitic arthropods in the order Cephalobaenida.

railroad [CIV ENG] A permanent line of rails forming a route for freight cars and passenger cars drawn by locomotives.

railroad engineering [CIV ENG] That part of transportation engineering involved in the planning, design, development, operation, construction, maintenance, use, or economics of facilities for transportation of goods and people in wheeled units of rolling stock running on, and guided by, rails normally supported on crossties and held to fixed alignment. Also known as railway engineering.

railroad ferry [NAV ARCH] A ship having a deck with railroad tracks, for carrying railroad cars between two ports.

railroad jack [MECH ENG] 1. A hoist used for lifting locomotives. 2. A portable jack for lifting heavy objects. 3. A hydraulic jack, either powered or lever-operated.

railroad thermit [MET] Red thermit to which is added up to 16% nickel, manganese, and steel.

rail steel [MET] Steel used to make rail track.

railway artillery [ORD] Heavy artillery, given mobility by mounting on railway mounts or carriages; no longer in use by U.S. forces.

railway carriage *See* railway mount.

railway dry dock [CIV ENG] A railway dock consisting of tracks built on an incline on a strong foundation, and extending from a sufficient distance in shore to allow a vessel to be hauled out of the water.

railway end-loading ramp [CIV ENG] A sloping platform situated at the end of a track and rising to the level of the floor of the railcars (wagons).

railway engineering *See* railroad engineering.

railway mount [ORD] Mount for railway artillery; no longer

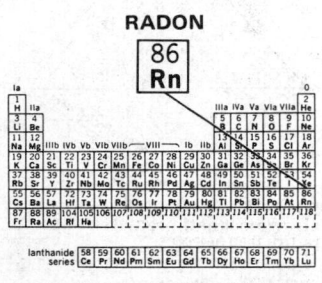

RADON

86
Rn

Periodic table of the chemical elements showing the position of radon.

in use by U.S. forces, but still of interest for possible future application. Also known as railway carriage.

rain [METEOROL] Precipitation in the form of liquid water drops with diameters greater than 0.5 millimeter, or if widely scattered the drops may be smaller; the only other form of liquid precipitation is drizzle.

rain and snow mixed [METEOROL] Precipitation consisting of a mixture of rain and wet snow; usually occurs when the temperature of the air layer near the ground is slightly above freezing.

rain area report See radar report.

rain attenuation [COMMUN] Attenuation of radio waves when passing through moisture-bearing cloud formations or areas in which rain is falling; increases with the density of the moisture in the transmission path.

rainbow [ELECTR] Technique which applies pulse-to-pulse frequency changing to identifying and discriminating against decoys and chaff. [OPTICS] Colored arc seen in the sky when the sun or moon is illuminating large numbers of falling raindrops.

rainbow granite [PETR] A type of granite having either a black or dark-green background with pink, yellowish, or reddish mottling, or a pink background with dark mottling.

rainbow roof [ARCH] A pitched roof with slightly arched slopes.

rain cloud [METEOROL] Any cloud from which rain falls; a popular term having no technical denotation.

rain crust [HYD] A type of snow crust, formed by refreezing after surface snow crystals have been melted and wetted by liquid precipitation; composed of individual ice particles such as firn.

raindrop [METEOROL] A drop of water of diameter greater than 0.5 millimeter falling through the atmosphere.

raindrop impressions See rain prints.

raindrop imprints See rain prints.

Rainey's corpuscle [INV ZOO] The sickle-shaped spore of an encysted sarcosporidian.

rain factor [HYD] A coefficient designed to measure the combined effect of temperature and moisture on the formation of soil humus; it is obtained by dividing the annual rainfall (in millimeters) by the mean annual temperature (in degrees centigrade).

rainfall [METEOROL] The amount of precipitation of any type; usually taken as that amount which is measured by means of a rain gage (thus a small, varying amount of direct condensation is included).

rainfall frequency [CLIMATOL] The number of times, during a specified period of years, that precipitation of a certain magnitude or greater occurs or will occur at a station; numerically, the reciprocal of the frequency is usually given.

rainfall inversion See precipitation inversion.

rainfall regime [CLIMATOL] The character of the seasonal distribution of rainfall at any place; the chief rainfall regimes, as defined by W. G. Kendrew, are equatorial, tropical, monsoonal, oceanic and continental westerlies, and Mediterranean.

rainforest [ECOL] A forest of broad-leaved, mainly evergreen, trees found in continually moist climates in the tropics, subtropics, and some parts of the temperate zones.

rainforest climate See wet climate.

rain gage [ENG] An instrument designed to collect and measure the amount of rain that has fallen. Also known as ombrometer; pluviometer.

rain-gage shield [ENG] A device which surrounds a rain gage and acts to maintain horizontal flow in the vicinity of the funnel so that the catch will not be influenced by eddies generated near the gage. Also known as wind shield.

rain gush See cloudburst.

rain gust See cloudburst.

raininess [METEOROL] Generally, the quantitative character of rainfall for a given place.

rain-intensity gage [ENG] An instrument which measures the instantaneous rate at which rain is falling on a given surface. Also known as rate-of-rainfall gage.

rainmaking [METEOROL] Popular term applied to all activities designed to increase, through any artificial means, the amount of precipitation released from a cloud.

rain prints [GEOL] Small, shallow depressions formed in soft sediment or mud by the impact of falling raindrops. Also known as raindrop impressions; raindrop imprints.

rain rot [VET MED] Weeping dermatitis accompanied by swelling of the skin and loss of hair in sheep exposed to prolonged periods of rain.

rain shadow [METEOROL] An area of diminished precipitation on the lee side of mountains or other topographic obstacles.

rainsquall [METEOROL] A squall associated with heavy convective clouds, frequently the cumulonimbus type; usually sets in shortly before the thunderstorm rain, blowing outward from the storm and generally lasting only a short time. Also known as thundersquall.

rain stage [METEOROL] The thermodynamic process of condensation of water from moist air in an idealized saturation-adiabatic or pseudoadiabatic lifting, at temperatures above the freezing point; begins at the condensation level.

rainy climate [CLIMATOL] In W. Koppen's climatic classification, any climate type other than the dry climates; however, it is generally understood that this refers principally to the tree climates and not the polar climates.

rainy season [CLIMATOL] In certain types of climate, an annually recurring period of one or more months during which precipitation is a maximum for that region. Also known as wet season.

raise [MIN ENG] A shaftlike mine opening, driven upward from a level to connect with a level above, or to the surface.

raise boring machine [MIN ENG] A machine that is used to drill pilot holes between levels in a mine, and to ream the pilot hole to the finished dimension of the raise.

raised beach [GEOL] An ancient beach raised to a level above the present shoreline by uplift or by lowering of the sea level; often bounded by inland cliffs.

raise drill [MIN ENG] A circular raise driving machine which bores a pilot hole and reams it to finished raise diameter.

raise driller [MIN ENG] A person who works in a raise. Also known as raiseman.

raiseman See raise driller.

Rajakaruna engine [MECH ENG] A rotary engine that uses a combustion chamber whose sides are pin-jointed together at their ends.

Rajidae [VERT ZOO] The skates, a family of elasmobranchs included in the batoid group.

Rajiformes [VERT ZOO] The equivalent name for Batoidea.

rake [DES ENG] A hand tool consisting of a long handle with a row of projecting prongs at one end; for example, the tool used for gathering leaves or grass on the ground. [ENG] The angle between an inclined plane and the vertical. [GEOL] See plunge. [MECH ENG] The angle between the tooth face or a tangent to the tooth face of a cutting tool at a given point and a reference plane or line. [NAV ARCH] The angle between the vertical direction and a part of a ship, such as a mast, funnel, bow, stern, rudder, or sternpost. [ORD] To sweep a target, especially a ship or a column of troops, with gun or cannon fire.

raked joint [CIV ENG] A mortar, or masonry, joint from which the mortar has been scraped out to about ¾ inch (20 millimeters).

râle [MED] An abnormal sound accompanying the normal sounds of respiration within the air passages and heard on auscultation of the chest.

Rallidae [VERT ZOO] A large family of birds in the order Gruiformes comprising rails, gallinules, and coots.

ralstonite [MINERAL] $NaMgAl_5F_{12}(OH)_6 \cdot 3H_2O$ A colorless, white, or yellowish mineral composed of hydrous basic sodium magnesium aluminum fluoride, occurring in octahedral crystals.

ram [AERO ENG] The forward motion of an air scoop or air inlet through the air. [MECH ENG] A plunger, weight, or other guided structure for exerting pressure or drawing something by impact. [MIN ENG] See barney. [VERT ZOO] A male sheep or goat.

Ram See Aries.

Raman effect [OPTICS] A phenomenon observed in the scattering of light as it passes through a transparent medium; the light undergoes a change in frequency and a random alteration in phase due to a change in rotational or vibrational

Diagram of the steps in the formation of rain. *(Based on photodisplay in Willetts Memorial Weather Exhibit, Hayden Planetarium, New York City)*

RAISE BORING MACHINE

Raise boring machine to drill pilot holes between levels in Idaho silver-lead mine. *(Hecla Mining Co.)*

RAJAKARUNA ENGINE

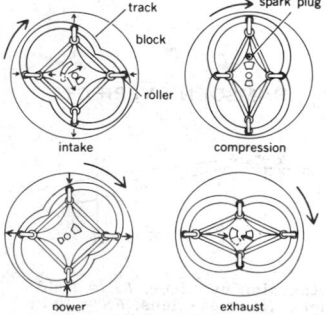

Cross sections of the Rajakaruna engine, a type of revolving block rotary engine, showing the four stages of one combustion cycle.

RAMAN EFFECT

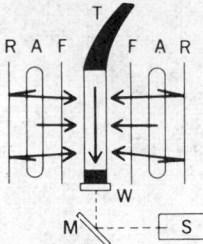

Arrangement for excitation of the Raman effect with a noncoherent source. Mercury vapor arcs A are surrounded by appropriate reflectors R to increase intensity of available radiation. Light from arcs passes through filters F which transmit one of several monochromatic frequencies of mercury radiation. Monochromatic radiation enters circular cylindrical tube T in which liquid or vapor is to be studied. Light scattered in direction of vertical arrow passes through window W and to the spectrograph S through optical system M.

RAMIE

Branch of the ramie shrub (*Boehmeria nivea*); grows to a height of 4–7 feet (1.2–2 meters).

RAMSDEN EYEPIECE

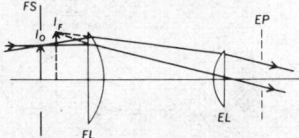

Ramsden eyepiece. *FL* = field lens; *EL* = eye lens; *FS* = field stop; *EP* = exit pupil or eye point; *I$_O$* = image (real or virtual) formed by the preceding system; *I$_F$* = image (virtual or real) formed by the preceding system and the field lens.

energy of the scattering molecules. Also known as Raman scattering.

Raman-Rayleigh ratio [OPTICS] The ratio of Raman scattering (of a light beam passing through a transparent medium) to Rayleigh scattering (of a light beam horizontal to the medium).

Raman scattering *See* Raman effect.

Raman spectrophotometry [SPECT] The study of spectral-line patterns on a photograph taken at right angles through a substance illuminated with a quartz mercury lamp.

Raman spectroscopy [SPECT] Analysis of the intensity of Raman scattering of monochromatic light as a function of frequency of the scattered light.

Raman spectrum [SPECT] A display, record, or graph of the intensity of Raman scattering of monochromatic light as a function of frequency of the scattered light.

Ramapithecinae [PALEON] A subfamily of Hominidae including the protohominids of the Miocene and Pliocene.

Ramapithecus [PALEON] The genus name given to a fossilized upper jaw fragment found in the Siwalik hills, India; closely related to the family of man.

ramark [NAV] A radio marker beacon which continuously transmits pulsed signals appearing as radial lines on a ship's PPI (plan position indicator); the line indicates the direction to the beacon.

ramate [BIOL] Having branches.

ramdohrite [MINERAL] $Pb_3Ag_2Sb_6S_{13}$ A dark-gray mineral composed of a lead silver antimony sulfur compound.

ram effect [MECH ENG] The increased air pressure in a jet engine or in the manifold of a piston engine, due to ram.

ramentum [BOT] A thin brownish scale consisting of a single layer of cells and occurring on the leaves and young shoots of many ferns.

ramicolous [BOT] Living on twigs.

ramie [BOT] *Boehmeria nivea.* A shrub or half-shrub of the nettle family (Urticaceae) cultivated as a source of a tough, strong, durable, lustrous natural woody fiber resembling flax, obtained from the phloem of the plant; used for high-quality papers and fabrics. Also known as China grass; rhea.

ramjet engine [AERO ENG] A type of jet engine with no mechanical compressor, consisting of a specially shaped tube or duct open at both ends, the air necessary for combustion being shoved into the duct and compressed by the forward motion of the engine; the air passes through a diffuser and is mixed with fuel and burned, the exhaust gases issuing in a jet from the rear opening.

ramjet exhaust nozzle [AERO ENG] The discharge nozzle in a ramjet engine; hot gas is ejected rearward through this nozzle.

rammelsbergite [MINERAL] $NiAs_2$ A gray mineral composed of nickel diarsenide; it is dimorphous with pararammelsbergite. Also known as white nickel.

rammer [ORD] **1.** Device for driving a projectile into position in a gun; may be hand- or power-operated. **2.** Tool used to remove a live projectile from the bore of a gun.

ramming [ENG] Packing a powder metal or sand into a compact mass.

ramoff [MET] A defect in a casting due to improper ramming of the sand.

Ramon flocculation test [IMMUNOL] A method of standardizing antitoxins; a toxin-antitoxin flocculation that is a precipitin reaction in which the end point is the zone of optimal proportion; that is, the zone in which there is no uncombined antigen or antibody.

ramose [BIOL] Having lateral divisions or branches.

ramp [ENG] **1.** A uniformly sloping platform, walkway, or driveway. **2.** A stairway which gives access to the main door of an airplane. [HYD] An accumulation of snow forming an inclined plane between land or land ice and sea ice or shelf ice. Also known as drift ice foot. [MIN ENG] A slope between levels in open-pit mining.

ramp generator [ELECTR] A circuit that generates a sweep voltage which increases linearly in value during one cycle of sweep, then returns to zero suddenly to start the next cycle.

Ramphastidae [VERT ZOO] The toucans, a family of birds with large, often colorful bills in the order Piciformes.

ramp mining [MIN ENG] The development of moderately inclined accessways from the surface to mining levels for haulage of ore, materials, waste, men, and equipment.

ramp sight [ORD] Type of metallic gun sight in which the aperture is raised or lowered by moving it forward or backward on an inclined ramp.

ramp valley [GEOL] A trough between faults, forced downward by lateral pressure.

ramp weight [AERO ENG] The static weight of a mission aircraft determined by adding operating weight, payload, flight plan fuel load, and fuel required for ground turbine power unit, taxi, runup, and takeoff.

ram recovery *See* recovery.

ram rocket [AERO ENG] **1.** A rocket motor mounted coaxially in the open front end of a ramjet, used to provide thrust at low speeds and to ignite the ramjet fuel. **2.** The entire unit or power plant consisting of the ramjet and such a rocket.

Ramsauer effect [ATOM PHYS] The vanishing of the scattering cross section of electrons from atoms of a noble gas at some value of the electron energy, always below 25 electron-volts.

Ramsay-Shields-Eötvös equation [THERMO] An elaboration of the Eötvös rule which states that at temperatures not too near the critical temperature the molar surface energy of a liquid is proportional to $t_c - t - 6K$, where t is the temperature and t_c is the critical temperature.

Ramsay-Young rule [THERMO] An empirical relationship which states that the ratio of the absolute temperatures at which two chemically similar liquids have the same vapor pressure is independent of this vapor pressure.

Ramsbottom coke test [ANALY CHEM] A laboratory test for carbon residue in petroleum products.

Ramsbottom safety valve [ENG] A steam safety valve with provision for manually checking the pressure adjustment.

ramsdellite [MINERAL] MnO_2 An orthorhombic mineral composed of manganese dioxide; it is dimorphous with pyrolusite.

Ramsden circle [OPTICS] A sharp, bright circle of light which appears on a sheet of white paper held near the eyepiece of a telescope focused for infinity and pointed at a bright sky.

Ramsden eyepiece [OPTICS] An eyepiece consisting of two planoconvex lenses with their plane sides facing outward, having the same power and focal length, and separated by a distance equal to their common focal length.

ram-type turret lathe [MECH ENG] A horizontal turret lathe in which the turret is mounted on a ram or slide which rides on a saddle.

ramus [ANAT] A slender bone process branching from a large bone. [VERT ZOO] The barb of a feather. [ZOO] The branch of a structure such as a blood vessel, nerve, arthropod appendage, and so on.

random access [ADP] **1.** A data-storage device having the property that the time required to access (read or write) a randomly selected datum does not depend on the time of the last access or the location of the most recently accessed datum. Also known as random-access storage; random storage. **2.** A process in which data are accessed in nonsequential order and possibly at irregular intervals of time. Also known as single reference.

random-access discrete address [COMMUN] Communications technique in which radio users share one wide band instead of each user getting an individual narrow band.

random-access disk file [ADP] A file which is contained on a disk having one head per track and in which consecutive records are not necessarily in consecutive locations.

random-access input/output [ADP] A technique which minimizes seek time and overlaps with processing.

random-access programming [ADP] Programming without regard for the time required for access to the storage positions called for in the program, in contrast to minimum-access programming.

random-access storage *See* random access.

random copolymer [ORG CHEM] Resin copolymer in which the molecules of each monomer are randomly arranged in the polymer backbone.

random diffusion chamber *See* reverberation chamber.

random error [STAT] An error that can be predicted only on a statistical basis.

random experiments [STAT] Experiments which do not always yield the same result when repeated under the same conditions.

random forecast [METEOROL] A forecast in which one of a set of meteorological contingencies is selected on the basis of chance; it is often used as a standard of comparison in determining the degree of skill of another forecast method.

random function [MATH] A function whose domain is an interval of the extended real numbers and has range in the set of random variables on some probability space; more precisely, a mapping of the cartesian product of an interval in the extended reals with a probability space to the extended reals so that each section is a random variable.

randomization [STAT] Assigning subjects to treatment groups by use of tables of random numbers.

randomized blocks [STAT] An experimental design in which the various treatments are reproduced in each of the blocks and are randomly assigned to the units within the blocks, permitting unbiased estimates of error to be made.

randomized jitter [ELECTR] Jitter by means of noise modulation.

randomized test [STAT] Acceptance or rejection of the null hypothesis by use of a random variable to decide whether an observation causes rejection or acceptance.

randomizing scheme [ADP] A technique of distributing records among storage modules to ensure even distribution and seek time.

random noise [MATH] A form of random stochastic process arising in control theory. [PHYS] Noise characterized by a large number of overlapping transient disturbances occurring at random, such as thermal noise and shot noise. Also known as fluctuation noise.

random number generator [ADP] **1.** A mathematical program which generates a set of numbers which pass a randomness test. **2.** An analog device that generates a randomly fluctuating variable, and usually operates from an electrical noise source.

random numbers [MATH] A listing of numbers which is nonrepetitive and satisfies no algorithm.

random ordered sample [STAT] An ordered sample of size s drawn from a population of size N such that the probability of any particular ordered sample is the reciprocal of the number of permutations of N things taken s at a time.

random process *See* stochastic process.

random pulsing [COMMUN] Continuous, varying, pulse-repetition rate, accomplished by noise modulation or continuous frequency change.

random sampling [STAT] A sampling from some population where each entry has an equal chance of being drawn.

random-sampling voltmeter [ENG] A sampling voltmeter which takes samples of an input signal at random times instead of at a constant rate; the synchronizing portions of the instrument can then be simplified or eliminated.

random sequence [MET] A longitudinal sequence of weld beads deposited in random increments.

random storage *See* random access.

random superimposed coding [ADP] A system of coding in which a set of random numbers is assigned to each concept to be encoded; with punched cards, each number corresponds to some one hole to be punched in a given field.

random variable [MATH] A measurable function on a probability space; usually real valued, but possibly with values in a general measurable space.

random vector *See* diverse vector.

random vibration [MECH] A varying force acting on a mechanical system which may be considered to be the sum of a large number of irregularly timed small shocks; induced typically by aerodynamic turbulence, airborne noise from rocket jets, and transportation over road surfaces.

random walk [MATH] A succession of movements along line segments where the direction and possibly the length of each move is randomly determined.

random winding [ELEC] A coil winding in which the turns are positioned haphazardly rather than in layers.

Raney nickel [MET] A nickel powder prepared from an alloy of nickel and aluminum in equal parts by preferentially dissolving the aluminum in a warm solution of sodium hydroxide.

rang [PETR] A unit of subdivision in the C.I.P.W. (Cross-Iddings-Pirsson-Washington) classification of igneous rocks.

range [CIV ENG] Any series of contiguous townships of the U.S. Public Land Survey system. [COMMUN] **1.** In printing telegraphy, that fraction of a perfect signal element through which the time of selection may be varied to occur earlier or later than the normal time of selection without causing errors while signals are being received; the range of a printing telegraph receiving device is commonly measured in percent of a perfect signal element by adjusting the indicator. **2.** Upper and lower limits through which the index arm of the range-finder mechanism of a teletypewriter may be moved and still receive correct copy. [ENG] **1.** The distance capability of an aircraft, missile, gun, radar, or radio transmitter. **2.** A line defined by two fixed landmarks, used for missile or vehicle testing and other test purposes. [MATH] The range of a function f from a set X to a set Y consists of those elements y in Y for which there is an x in X with $f(x) = y$. [MECH] The horizontal component of a projectile displacement at the instant it strikes the ground. [NAV] **1.** A line of bearing defined by a radio range. **2.** *See* radio range. [NUCLEO] The distance that a given ionizing particle can penetrate a given medium before its energy drops to the point that the particle no longer produces ionization. [PHYS] The greatest distance between two particles at which a given force between them is appreciable. [STAT] The difference between the maximums and minimums of a variable quantity.

range adjustment [ORD] Correction of firing data so that the impact or burst will be on the target with respect to range.

range-amplitude display [ELECTR] Radar display in which a time base provides the range scale from which echoes appear as deflections normal to the base.

range angle [ORD] Angle between the aircraft-target line and the vertical line from the aircraft to the ground at the instant a bomb is released. Also known as dropping angle.

range attenuation [ELECTROMAG] In radar terminology, the decrease in power density (flux density) caused by the divergence of the flux lines with distance, this decrease being in accordance with the inverse-square law.

range-bearing display *See* B display.

range calibration [ENG] Adjustment of a radar set so that when on target the set will indicate the correct range.

range calibrator [ELECTR] **1.** A device with which the operator of a transmitter calculates the distance over which the signal will extend intelligibly. **2.** A device for adjusting radar range indications by use of known range targets or delayed signals, so when on target the set will indicate the correct range.

range coding [ENG] Method of coding a radar transponder beacon response so that it appears as a series of illuminated bars on a radarscope; the coding provides identification.

range control [AERO ENG] The operation of an aircraft to obtain the optimum flying time.

range control chart [AERO ENG] A graph kept in flight on which actual fuel consumption is plotted against distance flown for comparison with planned fuel consumption.

range correction [ORD] Changes of firing data necessary to allow for deviations of range due to weather, material, or ammunition.

range correction board [ORD] Device with which the correction to be applied to a gun is computed mechanically; the correction obtained allows for all nonstandard conditions, such as variations in weather and ammunition.

range corrector setting [ENG] Degree to which the range scale of a position-finding apparatus must be adjusted before use.

range delay [ELECTROMAG] A control used in radars which permits the operator to present on the radarscope only those echoes from targets which lie beyond a certain distance from the radar; by using range delay, undesired echoes from nearby targets may be eliminated while the indicator range is increased.

range determination [ORD] Process of finding the distance between a gun and a target, usually by firing the gun, by estimating with the eye, by using a range-finding instrument, or by plotting.

range deviation [MECH] Distance by which a projectile strikes beyond, or short of, the target; the distance as mea-

RANGER PROGRAM

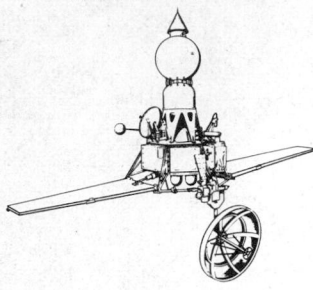

Drawing of space probe used in Ranger program.

RANKINE CYCLE

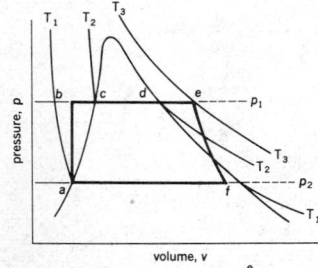

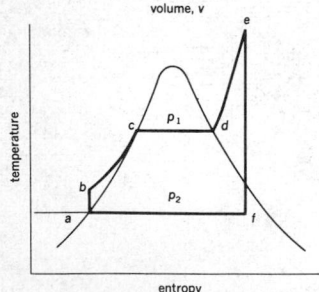

Rankine-cycle diagrams (pressure-volume and temperature-entropy) for a steam power plant using superheated steam. Typically, cycle has four phases: (1) heat addition b, c, d, e in a boiler at constant pressure p_1, changing water at b to superheated steam at e; (2) isentropic expansion e, f in a prime mover from initial pressure p_1 to back pressure p_2; (3) heat rejection f, a in a condenser at constant pressure p_3 with steam at f converted to a saturated liquid at a; (4) isentropic compression of a, b at water in feed pump from pressure p_2 to pressure p_1. T_1–T_3 are lines of constant temperature.

sured along the gun-target line or along a line parallel to the gun-target line.

range discrimination *See* distance resolution.

range disk [ORD] Graduated disk, used for range setting, connected mechanically with the elevating mechanism of a gun.

range dispersion diagram [ORD] Chart indicating the expected percentage of shots fired with the same data which will fall into each of eight areas within the dispersion pattern for ranges.

range drum [ORD] Graduated indicator of cylinder type, connected mechanically with the elevating mechanism of a gun, used for range setting.

range error [ORD] Difference between the range to the point of impact of a particular projectile and the range to the center of impact of a group of shots fired with the same data.

range estimation [ORD] A rough estimate of the distance to a target or other object.

range finder [COMMUN] A movable, calibrated unit of the receiving mechanism of a teletypewriter by means of which the selecting interval may be moved with respect to the start signal. [ELECTR] A device which determines the distance to an object by measuring the time it takes for a radio wave to travel to the object and return. [ENG] *See* optical range finder.

range flag [ORD] Red flag displayed on or near a target range during firing practice as a warning that firing is being conducted.

range gate capture [ELECTR] Electronic countermeasure technique using a spoofer radar transmitter to produce a false target echo that can make a fire-control tracking radar move off the real target and follow the false one.

range gating [ELECTR] The process of selecting those radar echoes that lie within a small range interval.

range-height indicator [ENG] A scope which simultaneously indicates range and height of a radar target; this presentation is commonly used by height finders.

range-height indicator display [ELECTR] A radar display that presents visually the scalar distance between a reference point and a target, along with the vertical distance between a reference plane and the target. Abbreviated RHI display.

range ladder [ORD] A naval term used to describe a method of adjusting gunfire by firing successive volleys; starting with a range which is assuredly beyond or short and applying small uniform range corrections to the successive volleys until the target is crossed.

range lights [NAV] Two or more lights in the same horizontal direction, and installed as aids to nautical navigation, particularly those lights placed to mark any line of importance to vessels, as a channel.

range marker *See* distance marker.

range mark offset [ELECTR] Displacement of range mark on a type B-indicator.

range of a loop [ADP] The set of instructions contained between the opening and closing statements of a DO loop.

range of tide [OCEANOGR] The difference in height between consecutive high and low tides at a place.

range of visibility [NAV] The extreme distance at which an object or light can be seen.

range oil [MATER] Kerosine similar in properties to a No. 1 distillate heating oil; used for space heating.

range probable error [ORD] Error in range that a gun or other weapon may be expected to exceed as often as not; given in the firing tables for the gun, and may be taken as an index of accuracy of the piece.

range rake [ORD] A T-shaped device with pegs set in the cross; the distance between pegs subtends a definite angle at the base of the T; by sighting with a range rake, a flank observer can get a quick angular measurement of range deviation.

range rate [ELECTR] The rate at which the distance from the measuring equipment to the target or signal source being tracked is changing with respect to time.

range recorder [ENG] An item which makes a permanent representation of distance, expressed as range, versus time. [ENG ACOUS] A display used in sonar in which a stylus sweeps across a paper moving at a constant rate and chemically

treated so that it is darkened by an electrical signal from the stylus; the stylus starts each sweep as a sound pulse is emitted so that the distance along the trace at which the echo signal appears is a measure of the range to the target.

range resolution *See* distance resolution.

range ring [ELECTR] Accurate, adjustable, ranging mark on a plan position indicator corresponding to a range step on a type M indicator.

Ranger program [AERO ENG] A series of nine spacecraft, launched in 1961–1965, designed to transmit photographs back to earth while on a collision course with the moon; the first six Rangers failed, but *Rangers 7, 8,* and *9* successfully transmitted high-resolution television pictures of the lunar surface up to the instant of impact.

range safety [ORD] On a rocket test range, the planning and execution of each test so as to ensure the maximum safety of all personnel and property within the range boundaries.

range scale [ORD] 1. Scale on the arm of a plotting board where the observed range of a moving target is recorded in finding firing data. 2. Graduated scale on the sight or mount of a gun, used to show the elevation of the gun. 3. Table of firing data giving elevation settings corresponding to various ranges for the standard charges.

range selection [ELECTR] Control on a radar indicator for selection of range scale.

range sensing [ORD] Observing the location of the striking or bursting point of a projectile with respect to range, and reporting it as a hit, over, short, lost, or doubtful; range sensing does not include accurate determination of distances.

range spotting [ORD] Watching the burst or impact of shots to note their deviation beyond, or short of, the target.

range step [ELECTR] Vertical displacement on M-indicator sweep to measure range.

range strobe [ELECTROMAG] An index mark which may be displayed on various types of radar indicators to assist in the determination of the exact range of a target.

range surveillance [ENG] Surveillance of a missile range by means of electronic and other equipment.

range sweep [ELECTR] A sweep intended primarily for measurement of range.

range table [ORD] Prepared table that gives elevations corresponding to ranges for a gun or other weapon, under various conditions; part of a firing table.

range-tracking element [ELECTR] An element in a radar set which measures range and its time derivative; by means of the latter, a range gate is actuated slightly before the predicted instant of signal reception.

range unit [ELECTR] Radar system component used for control and indication (usually counters) of range measurements.

range wind [ORD] Horizontal component of true wind in the vertical plane through the line of fire of an artillery piece.

range zero [ELECTR] Alignment of start sweep trace with zero range.

range zone [GEOL] Formal biostratigraphic zone made up of a body of strata comprising the total horizontal (geographic) and vertical (stratigraphic) range of occurrence of a specified taxon of a group of taxa.

ranging [ORD] 1. Wide-scale scouting, especially by aircraft, designed to search an area systematically. 2. Locating an enemy gun by watching its flash, listening to its report, or other similar means.

ranging oscillator [ELECTR] Oscillator circuit containing an LC (inductor-capacitor) resonant combination in the cathode circuit, usually used in radar equipment to provide range marks.

Ranidae [VERT ZOO] A family of frogs in the suborder Diplasiocoela including the large, widespread genus *Rana*.

rank [GEOL] 1. A coal classification based on degree of metamorphism. 2. *See* stack. [MATH] 1. The rank of a matrix is its maximum number of linearly independent rows. 2. The rank of a system of homogeneous linear equations equals the rank of the matrix of its coefficients. 3. A tensor in an n-dimensional space is of rank r if it has n^r components.

rank correlation [STAT] A nonparametric test of statistical dependence for a random sample of paired observations.

Rankine cycle [THERMO] An ideal thermodynamic cycle consisting of heat addition at constant pressure, isentropic expansion, heat rejection at constant pressure, and isentropic compression; used as an ideal standard for the performance of heat-engine and heat-pump installations operating with a condensable vapor as the working fluid, such as a steam power plant. Also known as steam cycle.

Rankine efficiency [MECH ENG] The efficiency of an ideal engine operating on the Rankine cycle under specified conditions of steam temperature and pressure.

Rankine-Hugoniot equations [THERMO] Equations, derived from the laws of conservation of mass, momentum, and energy, which relate the velocity of a shock wave and the pressure, density, and enthalpy of the transmitting fluid before and after the shock wave passes.

Rankine temperature scale [THERMO] A scale of absolute temperature; the temperature in degrees Rankine (°R) is equal to $\frac{9}{5}$ of the temperature in kelvins and to the temperature in degrees Fahrenheit plus 459.67.

Rankine vortex [FL MECH] A vortex with a vertical axis and circular motion, in which the motion is that of a rotating solid cylinder inside some fixed radius, and the circulation is constant outside this radius.

ranking method [IND ENG] A system of job evaluation wherein each job as a whole is given a rank with respect to all the other jobs, and no attempt is made to establish a measure of value.

rankinite [MINERAL] $Ca_3Si_2O_7$ A monoclinic mineral composed of calcium silicate.

rank of an observation [STAT] The number assigned to an observation if a collection of observations is ordered from smallest to largest and each observation is given the number corresponding to its place in the order.

rank tests [STAT] Tests which use the ranks of observations with respect to one another rather than the observations themselves.

Ranney oil-mining system [PETRO ENG] A method used to get oil from oil sands that involves driving mine galleries from shafts communicating to the surface in impermeable strata above and below the oil strata; holes drilled at short intervals along the galleries into the oil sands drain the oil or gas through pipes sealed in the drill holes into tanks from which the gas or oil is pumped to the surface.

Ranney well [CIV ENG] A well that has a center caisson with horizontal perforated pipes extending radially into an aquifer; particularly applicable to the development of thin aquifers at shallow depths.

ransomite [MINERAL] $Cu(Fe,Al)_2(SO_4)_4 \cdot 7H_2O$ A sky-blue mineral composed of hydrous copper iron aluminum sulfate.

ranula [MED] A retention cyst of a salivary gland.

Ranunculaceae [BOT] A family of dicotyledonous herbs in the order Ranunculales distinguished by alternate leaves with net venation, two or more distinct carpels, and numerous stamens.

Ranunculales [BOT] An order of dicotyledons in the subclass Magnoliidae characterized by its mostly separate carpels, triaperturate pollen, herbaceous or only secondarily woody habit, and frequently numerous stamens.

raob *See* radiosonde observation.

Raoult's law [PHYS CHEM] The law that the vapor pressure of a solution equals the product of the vapor pressure of the pure solvent and the mole fraction of solvent.

rapakivi [PETR] Granite or quartz monzonite characterized by orthoclase phenocrysts mantled with plagioclase. Also known as wiborgite.

rapakivi texture [PETR] An igneous and metamorphic rock texture in which spherical potassium feldspar crystals are surrounded by a rim of sodium feldspar, both within a finer-grained matrix.

rape [BOT] *Brassica napus.* A plant of the cabbage family in the order Capparales; the plant does not form a compact head, the leaves are bluish-green, deeply lobed, and curled, and the small flowers produce black seeds; grown for forage.

rape oil [MATER] A fatty, nondrying or semidrying, viscous, dark-brown to yellow oil with unpleasant taste and aroma, obtained from the seed of rape and turnip; soluble in ether, carbon disulfide, and chloroform; solidifies at -2 to $-10°C$; used to make lubricants and rubber substitutes, as an illumi-

nant, and in steel heat treatment. Also known as colza oil; rape-seed oil.

rape-seed oil *See* rape oil.

raphania [MED] A disease thought to be due to chronic ingestion of the poison in seeds of the wild radish.

raphe [ANAT] A broad seamlike junction between two lateral halves of an organ or other body part. [BOT] **1.** The part of the funiculus attached along its full length to the integument of an anatropous ovule, between the chalaza and the attachment to the placenta. **2.** The longitudinal median line or slit on a diatom valve.

Raphidae [VERT ZOO] A family of birds in the order Columbiformes that included the dodo (*Raphus calcullatus*); completely extirpated during the 17th and early 18th centuries.

raphide [BOT] One of the long, needle-shaped crystals, usually consisting of calcium oxalate, occurring as a metabolic by-product in certain plant cells.

rapid [HYD] A portion of a stream in swift, disturbed motion, but without cascade or waterfall; usually used in the plural.

rapid access loop [ADP] A small section of storage, particularly in drum, tape, or disk storage units, which has much faster access than the remainder of the storage.

rapid-curing asphalt [MATER] A liquid asphalt composed of asphalt cement and a gasoline- or naphtha-type diluent. Abbreviated RC asphalt.

rapid-eye-movement sleep [PSYCH] That part of the sleep cycle during which the eyes move rapidly, accompanied by a loss of muscle tone and a low-amplitude encephalogram recording; most dreaming occurs during this stage of sleep. Abbreviated REM sleep.

rapid fire [ORD] Rate of firing small arms or automatic weapons, faster than slow fire, but slower than quick fire.

rapid memory *See* rapid storage.

rapid sand filter [CIV ENG] A system for purifying water, which is forced through layers of sand and gravel under pressure.

rapid selector [ADP] A device which scans codes recorded on microfilm; microimages of the documents associated with the codes may also be recorded on the film.

rapid sequence camera [OPTICS] A conventional camera in most respects except that it is designed to permit a number of photographs to be obtained in rapid succession with one winding of the shutter.

rapid storage [ADP] In computers, storage with a very short access time; rapid access is generally gained by limiting storage capacity. Also known as high-speed storage; rapid memory.

raptorial [ZOO] **1.** Living on prey. **2.** Adapted for snatching or seizing prey, as birds of prey.

rapture of the deep *See* nitrogen narcosis.

rare-earth alloy [MET] An alloy containing rare-earth materials.

rare-earth element [CHEM] The name given to any of the group of chemical elements with atomic numbers 58 to 71; the name is a misnomer since they are neither rare nor earths; examples are cerium, erbium, and gadolinium.

rare-earth garnet [MATER] A synthetic garnet having the general structure of grossularite, but with calcium replaced by a rare-earth metal, and aluminum and silicon replaced by iron; used for electronic applications.

rare-earth magnet [ELECTROMAG] Any of several types of magnets made with rare-earth elements, such as rare-earth-cobalt magnets, which have coercive forces up to ten times that of ordinary magnets; used for computers and signaling devices.

rare-earth mineral [MINERAL] A mineral having a high concentration of rare-earth elements; examples are monazite, xenotime, and bastnaesite.

rare-earth salts [INORG CHEM] Salts derived from monazite, and with rare earths in similar proportions as in monazite; contains La, Ce, Pr, Nd, Sm, Gd, and Y as acetates, carbonates, chlorides, fluorides, nitrates, sulfates, and so on.

rarefaction [ACOUS] The instantaneous, local reduction in density of a gas resulting from passage of a sound wave, or the region in which the density is reduced at some instant.

rarefaction wave [FL MECH] A pressure wave or rush of air or water induced by rarefaction; it travels in the opposite

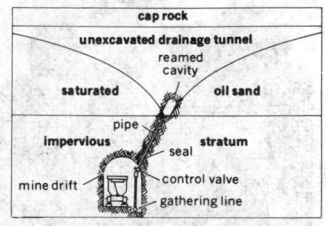

RASPBERRY

A red raspberry branch, Loudon variety. *(USDA)*

RAT

The brown Norway or common rat *(Rattus norvegicus)*. One of the worst pests in the United States, it destroys billions of dollars worth of farm products annually.

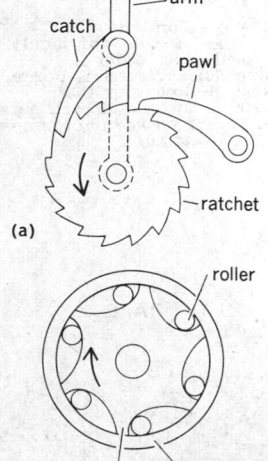

RATCHET

(a) arm, catch, pawl, ratchet

(b) roller, driver, follower

Ratchet mechanisms. *(a)* Toothed ratchet is driven in direction of arrow by catch when arm moves to left; pawl holds ratchet during return stroke of catch. *(b)* In roller ratchet, rollers become wedged between driver and follower when driver turns faster than follower in direction of arrow.

direction to that of a shock wave directly following an explosion. Also known as suction wave.

rarefied gas [FL MECH] A gas whose pressure is much less than atmospheric pressure.

rare gas *See* noble gas.

rare metal [MET] Any metal that is difficult to extract from ore and is rare and expensive commercially; includes masurium, alabamine, and virginium.

RAREP *See* radar report.

Rarita-Schwinger equation [QUANT MECH] A partial differential equation, similar in form to the Dirac equation, relating the spatial and time dependence of a 16-component wave function describing a free relativistic particle with intrinsic spin $3/2$, and its antiparticle.

Raschig process [CHEM ENG] A method for production of phenol that begins with a first-stage chlorination of benzene, using an air-hydrochloric acid mixture.

Raschig ring [CHEM ENG] A type of packing in the shape of a short pipe; used in columns for absorption operations, and to a limited extent for distillation operations.

rash [MED] A lay term for nearly any skin eruption, but more commonly for acute inflammatory dermatoses.

RA size [ENG] One of a series of sizes to which untrimmed paper is manufactured; for reels of paper, the standard sizes in millimeters are 430, 610, 860, and 1220; for sheets of paper, the sizes are RA0, 860 × 1220; RA1, 610 × 860; RA2, 430 × 610; RA sizes correspond to A sizes when trimmed.

rasorial [ZOO] Adapted for scratching for food; applied to birds.

rasorite *See* kernite.

rasp [DES ENG] A metallic tool with a rough surface of small points used for shaping and finishing metal, plaster, stone, and wood; designed in a number of useful curved shapes.

raspberry [BOT] Any of several species of upright shrubs of the genus *Rubus*, with perennial roots and prickly biennial stems, in the order Rosales; the edible black or red juicy berries are aggregate fruits, and when ripe they are easily separated from the fleshy receptacle.

raspite [MINERAL] $PbWO_4$ A yellow or brownish-yellow mineral composed of lead tungstate, occurring as monoclinic crystals.

raster [ELECTR] A predetermined pattern of scanning lines that provides substantially uniform coverage of an area; in television the raster is seen as closely spaced parallel lines, most evident when there is no picture.

raster scanning [ELECTR] Radar scan very similar to electron-beam scanning in an ordinary television set; horizontal sector scan that changes in elevation.

Rast method [ANALY CHEM] The melting-point depression method often used for the determination of the molecular weight of organic compounds.

rat [VERT ZOO] The name applied to over 650 species of mammals in several families of the order Rodentia; they differ from mice in being larger and in having teeth modified for gnawing.

RAT *See* rocket-assisted torpedo.

rataria larva [INV ZOO] The second, hourglass-shaped, free-swimming larva of the siphonophore *Velella*.

rat-bite fever [MED] Either of two diseases transmitted by the bite of a rat: spirillary rat-bite fiver and streptobacillary fever.

ratchet [DES ENG] A wheel, usually toothed, operating with a catch or a pawl so as to rotate in only a single direction.

ratchet coupling [MECH ENG] A coupling between two shafts that uses a ratchet to allow the driven shaft to be turned in one direction only, and also to permit the driven shaft to overrun the driving shaft.

ratchet jack [DES ENG] A jack operated by a ratchet mechanism.

ratchet tool [DES ENG] A tool in which torque or force is applied in one direction only by means of a ratchet.

rate [SCI TECH] The amount of change of some quantity during a time interval divided by the length of the time interval.

rate action *See* derivative action.

Rateau formula [FL MECH] A formula, $m = A_2p_1(16.367 - 0.96 \log p_1)/1000$, for determining the discharge m of saturated steam in pounds per second through a well-rounded conver-

gent orifice; A_2 is the area of the orifice in square inches, and p_1 is the reservoir pressure in pounds per square inch.

rate climb [AERO ENG] The climb of an aircraft to higher altitudes at a constant rate.

rate constant [PHYS CHEM] Numerical constant in a rate-of-reaction equation; for example, $r_A = kC_A{}^aC_B{}^bC_C{}^c$, where C_A, C_B and C_C are reactant concentrations, k is the rate constant (specific reaction rate constant), and $a, b,$ and c are empirical constants.

rated engine speed [MECH ENG] The rotative speed of an engine specified as the allowable maximum for continuous reliable performance.

rate descent [AERO ENG] An aircraft descent from higher altitudes at a constant rate.

rated flow [ENG] **1.** Normal operating flow rate at which a fluid product is passed through a vessel or piping system. **2.** Flow rate for which a vessel or process system is designed.

rated horsepower [MECH ENG] The normal maximum, allowable, continuous power output of an engine, turbine motor, or other prime mover.

rated load [MECH ENG] The maximum load a machine is designed to carry.

rate effect [ELECTR] The phenomenon of a *pnpn* device switching to a high-conduction mode when anode voltage is applied suddenly or when high-frequency transients exist.

rate feedback [ELECTR] The return of a signal, proportional to the rate of change of the output of a device, from the output to the input.

rate-grown transistor [ELECTR] A junction transistor in which both impurities (such as gallium and antimony) are placed in the melt at the same time and the temperature is suddenly raised and lowered to produce the alternate *p*-type and *n*-type layers of rate-grown junctions. Also known as graded-junction transistor.

rate gyroscope [MECH ENG] A gyroscope that is suspended in just one gimbal whose bearings form its output axis and which is restrained by a spring; rotation of the gyroscope frame about an axis perpendicular to both spin and output axes produces precession of the gimbal within the bearings proportional to the rate of rotation.

rate integrating gyroscope [MECH ENG] A single-degree-of-freedom gyro having primarily viscous restraint of its spin axis about the output axis; an output signal is produced by gimbal angular displacement, relative to the base, which is proportional to the integral of the angular rate of the base about the input axis.

rate meter *See* counting rate meter.

rate of approach [AERO ENG] The relative speed of two aircraft when the distance between them is decreasing.

rate of change [MATH] The limit of the ratio of the increment of a function to the increment of its argument taken as this increment approaches zero.

rate of change of acceleration [MECH] Time rate of change of acceleration; this rate is a factor in the design of some items of ammunition that undergo large accelerations.

rate of climb [AERO ENG] Ascent of aircraft per unit time, usually expressed as feet per minute.

rate-of-climb indicator [AERO ENG] A device used to indicate changes in the vertical position of an aircraft by comparing the actual outside air pressure to a reference volume that lags the outside pressure because a calibrated restrictor imposes a lag-time constant to the reference pressure volume. Also known as rate-of-descent indicator; vertical speed indicator.

rate of departure [AERO ENG] The relative speed of two aircraft when the distance between them is increasing.

rate-of-descent indicator *See* rate-of-climb indicator.

rate of detonation [ORD] Rate at which detonation of an explosive progresses; usually expressed in meters or yards per second.

rate of flow *See* flow rate.

rate-of-flow control valve *See* flow control valve.

rate-of-rainfall gage *See* rain-intensity gage.

rate of return [AERO ENG] Aircraft relative to its base, either fixed or moving.

rate of strain hardening [MET] Rate of change of true stress with respect to true strain in the plastic range. Also known as modulus of strain hardening.

rate receiver [ELECTROMAG] A guidance antenna that receives a signal from a launched missile as to its rate of speed.

rate response [ENG] Quantitative expression of the output rate of a control system as a function of its input signal.

rate servomechanism *See* velocity servomechanism.

rate test [ADP] A test that verifies that the time constants of the integrators are correct; used in analog computers.

rate transmitter [ELECTR] A transmitter in a missile being launched, used with a ground receiver to indicate the rate of speed increase.

ratfish [VERT ZOO] The common name for members of the chondrichthyan order Chimaeriformes.

rat guard [NAV ARCH] A circular or conical metal shield placed on a mooring line to keep rats from boarding or leaving a vessel.

rathite [MINERAL] $Pb_{13}As_{18}S_{40}$ A dark-gray mineral with metallic luster composed of sulfide of lead and arsenic; occurs as orthorhombic crystals.

Rathke's pouch *See* craniobuccal pouch.

rathole [MIN ENG] A shallow, small-diameter, auxiliary hole alongside the main borehole, drilled at an angle to the main hole; after core drilling is completed, the rathole is reamed out and the larger-size hole is advanced, usually by some noncoring method.

rating [ENG] A designation of an operating limit for a machine, apparatus, or device used under specified conditions.

rating curve [HYD] For a given point on a stream, a graph of discharge versus stage.

rating nut [HOROL] A nut used to vary the effective length of a pendulum in adjusting the rate of a clock.

ratio [MATH] A ratio of two quantities or mathematical objects A and B is their quotient or fraction A/B.

ratio arm circuit [ELEC] Two adjacent arms of a Wheatstone bridge, designed so they can be set to provide a variety of indicated resistance ratios.

ratio-balance relay *See* percentage differential relay.

ratio control system [CONT SYS] Control system in which two process variables are kept at a fixed ratio, regardless of the variation of either of the variables, as when flow rates in two separate fluid conduits are held at a fixed ratio.

ratio delay study *See* work sampling.

ratio detector [ELECTR] A frequency-modulation detector circuit that uses two diodes and requires no limiter at its input; the audio output is determined by the ratio of two developed intermediate-frequency voltages whose relative amplitudes are a function of frequency.

ratio deviation *See* modulation index.

ratio-differential relay *See* percentage differential relay.

ratio meter [ENG] A meter that measures the quotient of two electrical quantities; the deflection of the meter pointer is proportional to the ratio of the currents flowing through two coils.

rational function [MATH] A function which is a quotient of polynomials.

rational horizon *See* celestial horizon.

rationalization [PSYCH] A defense mechanism against difficult and unpleasant situations in which the individual attempts to use plausible means to justify or defend the unacceptable situations.

rational number [MATH] A number which is the quotient of two integers.

ratio of expansion [MECH ENG] The ratio of the volume of steam in the cylinder of an engine when the piston is at the end of a stroke to that when the piston is in the cutoff position.

ratio of reduction [ENG] The ratio of the maximum size of the stone which will enter a crusher, to the size of its product.

ratio of rise [OCEANOGR] The ratio of the height of tide at two places.

ratio of specific heats [PHYS CHEM] The ratio of specific heat at constant pressure to specific heat at constant volume, $\gamma = C_p/C_v$.

ratio of transformation [ELEC] Ratio of the secondary voltage of a transformer to the primary voltage under no-load conditions, or the corresponding ratio of currents in a current transformer.

ratio of transformer [ELEC] Ratio of the number of turns in one winding of a transformer to the number of turns in the other, unless otherwise specified.

ratio print [GRAPHICS] A print the scale of which has been changed from that of the negative by photographic enlargement or reduction.

ratio resistor [ELEC] One of the resistors in a Wheatstone or Kelvin bridge whose resistances appear in a pair of ratios which are equal in a balanced bridge.

ratio test *See* Cauchy ratio test.

ratites [VERT ZOO] A group of flightless, mostly large, running birds comprising several orders and including the emus, cassowaries, kiwis, and ostriches.

rato [AERO ENG] A rocket system providing additional thrust for takeoff of an aircraft. Derived from rocket-assisted takeoff.

rat race [ELECTR] A particular type of radar waveguide configuration which allows the handling of greater power.

rattail [MET] A small irregular line marking a minor buckle on the surface of a casting.

rattail file [DES ENG] A round tapering file used for smoothing or enlarging holes.

rattan [BOT] Any of several long-stemmed, climbing palms, especially of the genera *Calamus* and *Daemonothops*; stem material is used to make walking sticks, wickerwork, and cordage.

rattlesnake [VERT ZOO] Any of a number of species of the genera *Sistrurus* or *Crotalus* distinguished by the characteristic rattle on the end of the tail.

rat typhus *See* murine typhus.

Raunkiaer system [BOT] A classification system for plant life-forms based on the position of perennating buds in relation to the soil surface.

Rauschelback rotor [ENG] A free-turning S-shaped propeller used to measure ocean currents; the number of rotations per unit time is proportional to the flow.

Rauwolfia [BOT] A genus of mostly poisonous, tropical trees and shrubs of the dogbane family (Apocynaceae); certain species yield substances used as emetics and cathartics, while *R. serpentina* is a source of alkaloids used as tranquilizers.

ravelly ground [GEOL] Rock that breaks into small pieces when drilled and tends to cave or slough into the hole when the drill string is pulled, or binds the drill string by becoming wedged or locked between the drill rod and the borehole wall.

ravine [GEOGR] A small and narrow valley with steeply sloping sides.

raw [METEOROL] Colloquially descriptive of uncomfortably cold weather, usually meaning cold and damp, but sometimes cold and windy.

raw data [SCI TECH] Data that have not been processed; may be in machine-readable form.

rawhide [MATER] Untanned animal hide that has been dried or treated with a preservative.

raw humus *See* ectohumus.

rawin [METEOROL] A method of winds-aloft observation, that is, the determination of wind speeds and directions in the atmosphere above a station; accomplished by tracking a balloon-borne radar target, responder, or radiosonde transmitter with either radar or a radio direction finder.

rawinsonde [METEOROL] A method of upper-air observation consisting of an evaluation of the wind speed and direction, temperature, pressure, and relative humidity aloft by means of a balloon-borne radiosonde tracked by a radar or radio direction finder.

rawin target [ELECTROMAG] A special type of radar target tied beneath a free balloon, and designed to be an efficient reflector of radio energy; such targets usually consist of a corner reflector and are made of some reflecting material stretched over light wooden or metal struts.

raw material [IND ENG] A crude, unprocessed or partially processed material used as feedstock for a processing operation; for example, crude petroleum is the raw material from which naphtha is obtained; naphtha is the raw material from which (by catalytic reforming) benzene-toluene-xylene aromatics are obtained.

raw sewage [CIV ENG] Untreated waste materials.

raw silk [TEXT] A stage in the production of silk from the

RATTLESNAKE

Rattlesnake *(Crotalus horridus)*, one of the crotalid vipers.

RAUWOLFIA

Branch and root of *Rauwolfia serpentina*.

RAY

Devil ray, which is named for the anterior cephalic fins, extensions of the pectoral fins.

RAYLEIGH INTERFEROMETER

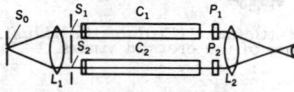

Schematic drawing of the Rayleigh interferometer. S_1, S_2 are slits illuminated with parallel light from lens L_1 whose focus is a single narrow slit S_0. Second lens L_2 brings the two beams to a focus, where interference fringes become visible. Cells (C_1, C_2) are placed in front of each slit; material of interest is placed in one cell and reference material in other. One of the plates of glass (P_1, P_2) is tilted until effect of cells on interference fringes is cancelled.

RAYLEIGH PRISM

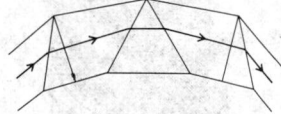

Rayleigh prism system; the arrow shows the path of light through the prism.

RAYON

American Viscose rayon continuous filament is used in the construction of many different fabrics.

cocoons of cultivated silkworms when the silk filament consists of 80% fibroin and 20% sericin.

raw sludge [CIV ENG] Sewage sludge preliminary to primary and secondary treatment processes.

raw water [CIV ENG] Water that has not been purified.

ray [ASTRON] One of the broad streaks that radiate from some craters on the moon, especially Copernicus and Tycho; they consist of material of high reflectivity and are seen from earth best at full moon. [MATH] A straight-line segment emanating from a point. Also known as half line. [OPTICS] A curve whose tangent at any point lies in the direction of propagation of a light wave. [PHYS] A moving particle or photon of ionizing radiation. [VERT ZOO] Any of about 350 species of the elasmobranch order Batoidea having flattened bodies with large pectoral fins attached to the side of the head, ventral gill slits, and long, spikelike tails.

ray acoustics [ACOUS] The study of the behavior of sound under the assumption that sound traversing a homogeneous medium travels along straight lines or rays. Also known as geometrical acoustics.

Raydist [NAV] A radio navigation system used in hydrographic and geophysical surveying and ships' trials which employs phase comparison techniques.

ray ellipsoid *See* Fresnel ellipsoid.

rayfin fish [VERT ZOO] The common name for members of the Actinopterygii.

ray flower [BOT] One of the small flowers with a strap-shaped corolla radiating from the margin of the head of a capitulum.

ray initial [BOT] One of the cells of the cambium which divide to produce new phloem and xylem ray cells.

Raykin fender [CIV ENG] Sandwich-type fender buffer to protect docks from the impact of mooring ships; made of a connected series of steel plates cemented to layers of rubber.

rayl [ACOUS] A unit of specific acoustical impedance, equal to a sound pressure of 1 dyne per square centimeter divided by a sound particle velocity of 1 centimeter per second. Also known as specific acoustical ohm (Ω_s); unit-area acoustical ohm.

rayleigh [OPTICS] A unit of brightness, used to measure the brightness of the night sky and aurorae, equal to $10^{10}/4\pi$ quanta per square meter per second per steradian.

Rayleigh atmosphere [METEOROL] An idealized atmosphere consisting of only those particles, such as molecules, that are smaller than about one-tenth of the wavelength of all radiation incident upon that atmosphere; in such an atmosphere, simple Rayleigh scattering would prevail.

Rayleigh balance [ELECTROMAG] An apparatus for assigning the value of the ampere in which the force exerted on a movable circular coil by larger circular coils above and below, but coaxial with, the movable coil is compared with the gravitational force on a known mass.

Rayleigh cycle [ELECTROMAG] A cycle of magnetization that does not extend beyond the initial portion of the magnetization curve, between zero and the upward bend.

Rayleigh disk [ACOUS] An acoustic radiometer, used to measure particle velocity, consisting of a thin disk set at an angle of 45° to a sound beam; the particle velocity is calculated from the resulting torque on the disk.

Rayleigh distribution [STAT] A normal distribution of two uncorrelated variates with the same variance.

Rayleigh flow [FL MECH] An idealized type of gas flow in which heat transfer may occur, satisfying the assumptions that the flow takes place in constant-area cross section and is frictionless and steady, that the gas is perfect and has constant specific heat, that the composition of the gas does not change, and that there are no devices in the system which deliver or receive mechanical work.

Rayleigh interferometer [OPTICS] An optical interferometer in which two rays of light, emanating from a single slit, are collimated by a lens, pass through separate slits and cells, and are brought to focus by a second lens so that interference fringes become visible. Also known as Rayleigh refractometer.

Rayleigh-Jeans law [STAT MECH] A law giving the intensity of radiation emitted by a blackbody within a narrow band of wavelengths; it states that this intensity is proportional to the temperature divided by the fourth power of the wavelength;

it is a good approximation to the experimentally verified Planck radiation formula only at long wavelengths.

Rayleigh line [SPECT] Spectrum line in scattered radiation which has the same frequency as the corresponding incident radiation.

Rayleigh loop [ELECTROMAG] A parabolic approximation to a magnetic hysteresis loop.

Rayleigh number 1 [FL MECH] A dimensionless number used in studying the breakup of liquid jets, equal to Weber number 2. Symbolized N_{Ra_1}.

Rayleigh number 2 [THERMO] A dimensionless number used in studying free convection, equal to the product of the Grashof number and the Prandtl number. Symbolized R'_2.

Rayleigh number 3 [THERMO] A dimensionless number used in the study of combined free and forced convection in vertical tubes, equal to Rayleigh number 2 times the Nusselt number times the tube diameter divided by its entry length. Symbolized Ra_3.

Rayleigh prism [OPTICS] A system of prisms used to produce greater dispersion of light than would be produced by a single prism.

Rayleigh ratio [OPTICS] Light-scattering relationship defined by the ratio of intensities of incident and scattered light at a specified distance; used in photometric and refractometric analyses.

Rayleigh reciprocity theorem [ELECTROMAG] Reciprocal relationship for an antenna when it is transmitting or receiving; the effective heights, radiation resistance, and the radiation pattern are alike, whether the antenna is transmitting or receiving.

Rayleigh refractometer *See* Rayleigh interferometer.

Rayleigh-Ritz method [MATH] An approximation method for finding solutions of functional equations in terms of finite systems of equations.

Rayleigh scattering [ELECTROMAG] Scattering of electromagnetic radiation by independent particles which are much smaller than the wavelength of the radiation.

Rayleigh's dissipation function [MECH] A function which enters into the equations of motion of a system undergoing small oscillations and represents frictional forces which are proportional to velocities; given by a positive definite quadratic form in the time derivatives of the coordinates. Also known as dissipation function.

Rayleigh-Taylor instability [FL MECH] The instability of the interface separating two fluids having different densities when the lighter fluid is accelerated toward the heavier fluid.

Rayleigh wave [GEOPHYS] In seismology, a surface wave with a retrograde, elliptical motion at the free surface. Also known as R wave. [MECH] A wave which propagates on the surface of a solid; particle trajectories are ellipses in planes normal to the surface and parallel to the direction of propagation.

Raymond concrete pile [CIV ENG] A pile made by driving a thin steel shell into the ground with a tapered mandrel and filling it with concrete.

Raynaud's disease [MED] A usually bilateral disease of blood vessels, especially of the extremities; excited by cold or emotion, characterized by intermittent pallor, cyanosis, and redness, and generally accompanied by pain.

rayon [TEXT] A fiber made from regenerated cellulose by the viscose or cuprammonium process.

rayon coning oil [MATER] Lubricant oil used to reduce static in yarns being wound on cones; composed of low-viscosity mineral oils emulsifiable in water.

rayon staple [TEXT] Rayon fibers of spinnable lengths; continuous filaments from a number of spinnerettes are collected into a loose, ropelike form and cut into suitable lengths.

rayon tow [TEXT] Large number of continuous rayon filaments that give the appearance of untwisted rope; when cut into suitable lengths, this material becomes rayon staple.

ray path [COMMUN] Straight, geometric path between the transmitting and receiving location.

Raysistor [ELECTR] A device which contains a photosensitive semiconductor and a light source; light source can be used to control the conductivity of the semiconductor.

ray surface [OPTICS] The locus of points reached in a unit time in an anisotropic medium by an electromagnetic disturbance that starts from the origin.

ray tracing [OPTICS] Calculation of the paths followed by rays of light through an optical system, using Snell's law and trigonometrical formulas.

razon [ORD] A kind of glide bomb having movable control surfaces in the tail adjusted by radio signals to control the bomb in range and in azimuth.

Rb *See* rubidium.

RBE *See* relative biological effectiveness.

R-C amplifier *See* resistance-capacitance coupled amplifer.

RC asphalt *See* rapid-curing asphalt.

R-C circuit *See* resistance-capacitance circuit.

R-C constant *See* resistance-capacitance constant.

R-C coupled amplifier *See* resistance-capacitance coupled amplifier.

R-C coupling *See* resistance coupling.

RCM *See* radar countermeasure.

R-C network *See* resistance-capacitance network.

R-C oscillator *See* resistance-capacitance oscillator.

rd *See* rutherford.

RdAc *See* radioactinium.

RDC extractor *See* rotary-disk contactor.

RDF *See* radio direction finder.

R display [ELECTR] Radar display, essentially an expanded A display, in which an echo can be expanded for more detailed examination.

RDS *See* respiratory distress syndrome of newborn.

RDX *See* cyclonite.

Re *See* rhenium.

reach [CIV ENG] A portion of a waterway between two locks or gages. [ENG] The length of a channel, uniform with respect to discharge, depth, area, and slope. [GEOGR] **1.** A continuous, unbroken surface of land or water. Also known as stretch. **2.** A bay, estuary, or other arm of the sea extending up into the land. [HYD] A straight, continuous, or extended part of a river, stream, or restricted waterway.

reach rod [MECH ENG] A rod motion in a link used to transmit motion from the reversing rod to the lifting shaft.

reactance [ELEC] The imaginary part of the impedance of an alternating-current circuit.

reactance amplifier *See* parametric amplifier.

reactance frequency multiplier [ELECTR] Frequency multiplier whose essential element is a nonlinear reactor.

reactance grounded [ELEC] Grounded through a reactance.

reactance relay [ELEC] Form of impedance relay, the operation of which is a function of the reactance of a circuit.

reactance tube [ELECTR] Vacuum tube operated in a way that it presents almost a pure reactance to the circuit.

reactance-tube modulator [ELECTR] An electron-tube circuit, used to produce phase or frequency modulation, in which the reactance is varied in accordance with the instantaneous amplitude of the modulating voltage.

reactant ratio [AERO ENG] The ratio of the weight flow of oxidizer to fuel in a rocket engine.

reactants [CHEM] The molecules that act upon one another to produce a new set of molecules (products); for example, in the reaction $HCl + NaOH \rightarrow NaCl + H_2O$, the HCl and NaOH are the reactants.

reaction [CONT SYS] *See* positive feedback. [MECH] The equal and opposite force which results when a force is exerted on a body, according to Newton's third law of motion. [NUC PHYS] *See* nuclear reaction.

reaction border *See* reaction rim.

reaction energy *See* disintegration energy.

reaction engine [AERO ENG] An engine that develops thrust by its reaction to a substance ejected from it; specifically, such an engine that ejects a jet or stream of gases created by the addition of energy to the gases in the engine. Also known as reaction motor.

reaction enthalpy number [PHYS CHEM] A dimensionless number used in the study of interphase transfer in chemical reactions, equal to the enthalpy of reaction per unit mass of a specified compound produced in a reaction, times the mass fraction of that compound, divided by the product of the specific heat at constant pressure and the temperature change during the reaction.

reaction formation [PSYCH] A defense mechanism, characterized by the development of conscious, socially acceptable activity which is the antithesis of repressed or rejected unconscious desires.

reaction kinetics *See* chemical kinetics.

reaction motor [AERO ENG] *See* reaction engine. [ELEC] A synchronous motor whose rotor contains salient poles but which has no windings and no permanent magnets.

reaction pair [MINERAL] Any two minerals, one of which is formed at the expense of the other by reaction with liquid.

reaction principle [MINERAL] The concept of a reaction series for the principal rock-forming minerals.

reaction propulsion [AERO ENG] Propulsion by means of reaction to a jet of gas or fluid projected rearward, as by a jet engine, rocket engine, or rocket motor.

reaction rim [PETR] A surficial rim around one mineral produced by the reaction of the core mineral with the surrounding magma. Also known as reaction border.

reaction series [MINERAL] Any series of minerals in which early formed varieties react with the melt to yield new minerals; two different types of reaction series exist, continuous and discontinuous.

reaction time [PHYSIO] The interval between application of a stimulus and the beginning of the response.

reaction turbine [MECH ENG] A power-generation prime mover utilizing the steady-flow principle of fluid acceleration, where nozzles are mounted on the moving element.

reaction wheel [MECH ENG] A device capable of storing angular momentum which may be used in a space ship to provide torque to effect or maintain a given orientation.

reaction wood [BOT] An abnormal development of a tree and therefore its wood as the result of unusual forces acting on it, such as an atypical gravitational pull.

reaction zone [CHEM ENG] In a catalytic reactor vessel, the location or zone within the vessel where the bulk of the chemical reaction takes place.

reactive [ELEC] Pertaining to either inductive or capacitance reactance; a reactive circuit has a high value of reactance in comparison with resistance.

reactive bond [CHEM] A bond between atoms that is easily invaded (reacted to) by another atom or radical; for example, the double bond in $CH_2 = CH_2$ (ethylene) is highly reactive to other ethylene molecules in the reaction known as polymerization to form polyethylene.

reactive current [ELEC] In the phasor representation of alternating current, the component of the current perpendicular to the voltage, which contributes no power but increases the power losses of the system. Also known as quadrature component; quadrature current; wattless current.

reactive dye [MATER] Dye that reacts with the textile fiber to produce both a hydroxyl and an oxygen linkage, the chlorine combining with the hydroxyl to form a strong ether linkage; gives fast, brilliant colors.

reactive factor [ELEC] The ratio of reactive power to apparent power.

reactive fluid [PETRO ENG] Any fluid that alters the internal geometry of a reservoir's porosity; for example, water is a reactive fluid when it causes swelling of clays and consequent changes in porosity.

reactive load [ELEC] A load having inductive or capacitive reactance.

reactive muffler [ENG] A muffler that attenuates by reflecting sound back to the source. Also known as nondissipative muffler.

reactive power [ELEC] The power value obtained by multiplying together the effective value of current in amperes, the effective value of voltage in volts, and the sine of the angular phase difference between current and voltage. Also known as wattless power.

reactive volt-ampere *See* volt-ampere reactive.

reactive volt-ampere hour *See* var hour.

reactive volt-ampere meter *See* varmeter.

reactivity [CHEM] The relative capacity of an atom, molecule, or radical to combine chemically with another atom, molecule, or radical. [NUCLEO] A measure of the deviation of a nuclear reactor from the critical state at any instant of time such that positive and negative values correspond to reactors above and below critical, respectively; measured in percent k, millikays, dollars, or in-hours.

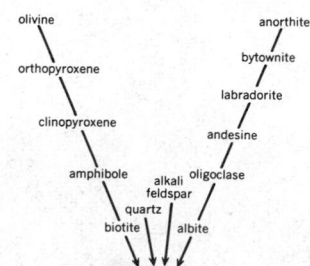

REACTION SERIES

Continuous reaction series showing the course of crystallization of the common magmas; early crystals are at top of diagram.

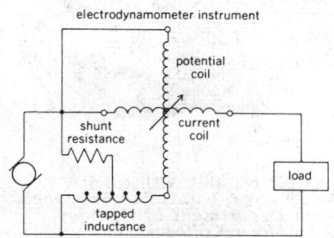

REACTIVE POWER

Diagram of a circuit for single-phase reactive power measurement which uses a tapped inductance. This inductance and shunt resistance are adjusted until currents in potential coil and current coil are in quadrature.

reactor [CHEM ENG] Device or process vessel in which chemical reactions (catalyzed or noncatalyzed) take place during a chemical conversion type of process. [ELEC] A device that introduces either inductive or capacitive reactance into a circuit, such as a coil or capacitor. Also known as electric reactor. [NUC PHYS] *See* nuclear reactor.

reactor fuel *See* nuclear fuel.

reactor fuel cycle [NUCLEO] The processes of preparing fuel elements and assemblies for use in a reactor, using these elements in reactor operation, recovering radioactive byproducts from spent fuel, and reprocessing remaining fissionable material into new fuel elements. Also known as fuel cycle; nuclear fuel cycle.

reactor fuel element *See* nuclear fuel element.

reactor fuel pellet *See* nuclear fuel pellet.

reactor physics [NUCLEO] The science of the interaction of the elementary particles and radiations characteristic of nuclear reactors with matter in bulk.

reactor vessel [NUCLEO] A large tanklike structure built to prevent radioactive materials from escaping from the reactor and associated equipment.

read [ADP] **1.** To acquire information, usually from some form of storage in a computer. **2.** To convert magnetic spots, characters, or punched holes into electrical impulses. [COMMUN] To understand clearly, as in radio communication. [ELECTR] To generate an output corresponding to the pattern stored in a charge storage tube.

read-around number *See* read-around ratio.

read-around ratio [ADP] The number of times that a particular bit in electrostatic storage may be read without seriously affecting nearby bits. Also known as read-around number.

read-back check *See* echo check.

Read diode [ELECTR] A high-frequency semiconductor diode consisting of an avalanching *pn*-junction, biased to fields of several hundred thousand volts per centimeter, at one end of a high-resistance carrier serving as a drift space for the charge carriers.

reader [ADP] A device that converts information from one form to another, as from punched paper tape to magnetic tape. [GRAPHICS] A projection device for viewing an enlarged microimage with the unaided eye.

reader-interpreter [ADP] A service routine that reads an input string, stores programs and data on random-access storage for later processing, identifies the control information contained in the input string, and stores this control information separately in the appropriate control lists.

reader-punch equipment [ADP] An input/output unit which can punch computer results on cards and read card data into the computer.

read head [ADP] A device that converts digital information stored on a magnetic tape, drum, or disk into electrical signals usable by the computer arithmetic unit.

read-in [ADP] To sense information contained in some source and transmit this information to an internal storage.

readiness review [ADP] An on-site examination of the adequacy of preparations for effective utilization upon installation of a computer, and to identify any necessary corrective actions.

readiness time [ENG] The length of time required to obtain a stabilized system ready to perform its intended function (readiness time includes warm-up time); the time is measured from the point when the system is unassembled or uninstalled to such time as it can be expected to perform as accurately as at any later time; maintenance time is excluded from readiness time.

reading [ENG] **1.** The indication shown by an instrument. **2.** Observation of the readings of one or more instruments.

reading rate [ADP] Number of characters, words, fields, blocks, or cards sensed by an input sensing device per unit of time.

read-in program [ADP] Computer program that can be put into a computer in a simple binary form and allows other programs to be read into the computer in more complex forms.

read-only memory *See* read-only storage.

read-only storage [ADP] A medium for storing data in permanent, or nonerasable, form; usually a high-speed static

storage mechanism is indicated, for punched paper tape and other such memories are not generally referred to as read-only storage. Also known as read-only memory.

readout [ADP] **1.** The presentation of output information by means of lights, printed or punched tape or cards, or other methods. **2.** To sense information contained in some computer internal storage and transmit this information to a storage external to the computer.

readout station [COMMUN] A recording or receiving radio station at which data are received, as the transmitter in a missile, probe, satellite, or other spacecraft reads the data out.

read-punch unit [ADP] An input-output unit of a computing system which punches computed results into cards, reads input information into the system, and segregates output cards; the read-punch unit generally consists of a card feed, a read station, a punch station, another read station, and two output card stackers.

read screen [ADP] In optical character recognition (OCR), the transparent component part of most character readers through which appears the input document to be recognized.

read time *See* access time.

read-while-writing [ADP] The reading of a record or group of records into storage from tape at the same time another record or group of records is written from storage to tape.

read/write check indicator [ADP] A device incorporated in certain computers to indicate upon interrogation whether or not an error was made in reading or writing; the machine can be made to stop, retry the operation, or follow a special subroutine, depending upon the result of the interrogation.

read/write head [ADP] A magnetic head that both senses and records data. Also known as combined head.

ready [ORD] The term used to indicate that a weapon is aimed, loaded, and prepared to fire.

ready-mixed concrete [MATER] Concrete mixed away from the construction site and delivered in readiness for placing.

ready-to-receive signal [COMMUN] Signal sent back to a facsimile transmitter to indicate that a facsimile receiver is ready to accept the transmission.

reagent [ANALY CHEM] A substance, chemical, or solution used in the laboratory to detect, measure, or otherwise examine other substances, chemicals, or solutions; grades include ACS (American Chemical Society standards), reagent (for analytical reagents), CP (chemically pure), USP (U.S. Pharmacopeia standards), NF (National Formulary standards), and purified, technical (for industrial use).

reagent chemicals [ANALY CHEM] High-purity chemicals used for analytical reactions, for testing of new reactions where the effect of impurities are unknown, and, in general, for chemical work where impurities must either be absent or at a known concentration.

reagin [IMMUNOL] **1.** An antibody which occurs in human atopy, such as hay fever and asthma, and which readily sensitizes the skin. **2.** An antibody which reacts in various serologic tests for syphilis.

real axis [MATH] The horizontal axis of the cartesian coordinate system for the euclidean or complex plane.

real fluid flow [FL MECH] The flow in which effects of tangential or shearing forces are taken into account; these forces give rise to fluid friction, because they oppose the sliding of one particle past another.

realgar [MINERAL] AsS A red to orange mineral crystallizing in the monoclinic system, having a resinous luster and found in short, vertical striated crystals; specific gravity is 3.48, and hardness is 1.5–2 on Mohs scale. Also known as red arsenic; red orpiment; sandarac.

real gas [THERMO] A gas, as considered from the viewpoint in which deviations from the ideal gas law, resulting from interactions of gas molecules, are taken into account. Also known as imperfect gas.

real image [OPTICS] An optical image such that all the light from a point on an object that passes through an optical system actually passes close to or through a point on the image.

reality principle [PSYCH] The concept that the pleasure principle is normally modified by the demands of the external environment and that the individual adjusts to these inescap-

REALGAR

realgar   calcite

|— 2.5 cm —|

Realgar crystals with calcite, Felsöbanya, Rumania. *(Specimen from Department of Geology, Bryn Mawr College)*

REAL IMAGE

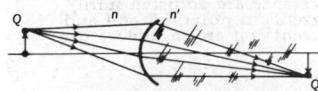

Formation of real image. Rays leaving object point *Q* and passing through the refracting surface separating media *n* and *n'* are brought to a focus at the image point *Q'*. *(Modified from F. A. Jenkins and H. E. White, Fundamentals of Optics, 3d ed., McGraw-Hill, 1957)*

able requirements so that he ultimately secures satisfaction of his instinctual wishes.

realizability [CONT SYS] Property of a transfer function that can be realized by a network that has only resistances, capacitances, inductances, and ideal transformers.

realization of a stochastic process [STAT] A probability space whose points are sample paths of the stochastic process and whose probability is obtained from the joint probability distributions of the random variables in the process.

real number [MATH] Any member of the unique (to within isomorphism) complete ordered field.

real orthogonal group [MATH] The group composed of orthogonal matrices having real number entries.

real part [MATH] The real part of a complex number $z = x + iy$ is the real number x.

real power [ELEC] The component of apparent power that represents true work; expressed in watts, it is equal to volt-amperes multiplied by the power factor.

real precession [NAV] In marine gyroscopes, that component of the total precession caused by bearing friction, torque unbalance, and other manufacturing or design defects. Also known as induced precession.

real-time [ADP] Pertaining to a data-processing system that controls an ongoing process and delivers its outputs (or controls its inputs) not later than the time when these are needed for effective control; for instance, airline reservations booking and chemical processes control.

real-time clock [ADP] A pulse generator which operates at precise time intervals to determine time intervals between events and initiate specific elements of processing.

real-time control system [ADP] A computer system which controls an operation in real time, such as a rocket flight.

real-time operation [ADP] **1.** Of a computer or system, an operation or other response in which programmed responses to an event are essentially simultaneous with the event itself. **2.** An operation in which information obtained from a physical process is processed to influence or control the physical process.

real-time processing [ADP] The handling of input data at a rate sufficient to ensure that the instructions generated by the computer will influence the operation under control at the required time.

real-time programming [ADP] Programming for a situation in which results of computations will be used immediately to influence the course of ongoing physical events.

real unimodular group [MATH] The group of all square $n \times n$ matrices with real number entries and of determinant 1.

real variable [MATH] A variable that assumes real numbers for its values.

ream [ENG] To enlarge or clean out a hole. [MATER] **1.** A layer of nonhomogeneous material in flat glass. **2.** Five hundred sheets of paper; a printer's ream consists of 516 sheets.

reamed extrusion ingot [MET] A hollow extrusion ingot whose original inside surface has been removed by reaming.

reamer [DES ENG] A tool used to enlarge, shape, smooth, or otherwise finish a hole.

reaming bit [DES ENG] A bit used to enlarge a borehole. Also known as broaching bit; pilot reaming bit.

rear area [ORD] The area in the rear of the combat and forward areas.

rearming [ORD] **1.** An operation that replenishes the prescribed stores of ammunition, bombs, and other armament items for an aircraft, naval ship, tank, or armored vehicle, including replacement of defective ordnance equipment, in order to make it ready for combat service. **2.** Resetting the fuse on a bomb, or on an artillery, mortar, or rocket projectile, so that it will detonate at the desired time.

rear-projection [ELECTR] Pertaining to television system in which the picture is projected on a ground-glass screen for viewing from the opposite side of the screen.

rearrangement reaction [NUC PHYS] A nuclear reaction in which nucleons are exchanged between nuclei.

rear sight [ORD] An item attached to the breech end and integral to a carbine, machine gun, pistol, rifle, or the like; it may be a fixed or adjustable cross blade with a U- or V-shaped notch or aperture, or it may have elevation and windage adjustment knobs, slides, and graduated scales and be provided with aperture disks.

Reaumur temperature scale [THERMO] Temperature scale where water freezes at 0°R and boils at 80°R.

rebat [METEOROL] The lake breeze of Lake Geneva, Switzerland; it blows from about 10 a.m. to 4 p.m.

rebecca [NAV] An electronic navigation system that has at least one radio transmitter and one radio receiver; the transmitter emits pulses which travel over a single path to a transponder and return to the interrogation receiver.

rebecca-eureka system [NAV] An aircraft radar homing system in which an airborne interrogator-responsor (rebecca) homes on a ground radar beacon (eureka) that has been dropped or set up in advance; the system can also give the distance from the rebecca radar to the eureka beacon.

reboiler [CHEM ENG] An auxiliary heating unit for a fractionating tower designed to supply additional heat to the lower portion of the tower; liquid withdrawn from the side or bottom of the tower is reheated by heat exchange, then reintroduced into the tower.

rebound [GEOL] The isostatic readjustment upward of a landmass depressed by glacial loading.

rebound clip [DES ENG] A clip surrounding the back and one or two other leaves of a leaf spring, to distribute the load during rebounds.

rebound leaf [DES ENG] In a leaf spring, a leaf placed over the master leaf to limit the rebound and help carry the load imposed by it.

reboyo [METEOROL] A persistent (day-long) storm from the southwest during the rainy season on the Brazilian coast.

rebreather [ENG] A closed-loop oxygen supply system consisting of gas supply and face mask.

rebroadcast [COMMUN] Repetition of a radio or television program at a later time.

rebuild [ENG] To restore to a condition comparable to new by disassembling the item to determine the condition of each of its component parts, and reassembling it, using serviceable, rebuilt, or new assemblies, subassemblies, and parts.

recalescence [MET] Brightening (reglowing) of iron on cooling through the γ- to α-phase transformation temperature caused by liberation of the latent heat of transformation.

recall [COMMUN] A flashing signal to the attendant's switchboard; the operator may be recalled by the subscriber operating the switch hook of the subscriber's set.

recall factor [ADP] A measure of the efficiency of an information retrieval system, equal to the number of retrieved relevant documents divided by the total number of relevant documents in the file.

recapitulation theory [BIOL] The biological theory that an organism passes through developmental stages resembling various stages in the phylogeny of its group; ontogeny recapitulates phylogeny. Also known as biogenetic law; Haeckel's law.

recarburize [MET] **1.** To increase the carbon content of molten steel or cast iron. **2.** To carburize a metal part, making up for any loss of carbon during processing.

receding log [ORD] That portion of the target's course line in which the slant range increases for successive target positions.

received power [ELECTROMAG] **1.** The total power received at an antenna from a signal, such as a radar target signal. **2.** In a mobile communications system, the root-mean-square value of power delivered to a load which properly terminates an isotropic reference antenna.

receiver [CHEM ENG] Vessel, container, or tank used to receive and collect liquid material from a process unit, such as the distillate receiver from the overhead condenser of a distillation column. [ELECTR] The complete equipment required for receiving modulated radio waves and converting them into the original intelligence, such as into sounds or pictures, or converting to desired useful information as in a radar receiver. [MECH ENG] An apparatus placed near the compressor to equalize the pulsations of the air as it comes from the compressor to cause a more uniform flow of air through the pipeline and to collect moisture and oil carried in the air.

receiver bandwidth [ELECTR] Spread, in frequency, between the halfpower points on the receiver response curve.

receiver gating [ELECTR] Application of operating voltages to one or more stages of a receiver only during that part of a cycle of operation when reception is desired.

receiver incremental tuning [ELECTR] Control feature to permit receiver tuning (of a transceiver) up to 3 kilohertz to either side of the transmitter frequency.

receiver lockout system *See* lockout.

receiver noise threshold [ELECTR] External noise appearing at the front end of a receiver, plus the noise added by the receiver itself, determines a noise threshold that has to be exceeded by the minimum discernible signal.

receiver primary *See* display primary.

receiver radiation [ELECTROMAG] Radiation of interfering electromagnetic fields by the oscillator of a receiver.

receiver synchro *See* synchro receiver.

receiving antenna [ELECTROMAG] An antenna used to convert electromagnetic waves to modulated radio-frequency currents.

receiving area [ELECTROMAG] The factor by which the power density must be multiplied to obtain the received power of an antenna, equal to the gain of the antenna times the square of the wavelength divided by 4π.

receiving gage [ENG] A fixed gage designed to inspect a number of dimensions and also their reaction to each other.

receiving house [CHEM ENG] A building where liquid streams from petroleum-refining-process condensers are observed through a look box, and samples are taken for testing, and also where products are diverted to storage tanks or to other processing units.

receiving loop loss [COMMUN] In telephones, that part of the repetition equivalent assignable to the station set, subscriber line, and battery supply circuit that are on the receiving end.

receiving set *See* radio receiver.

receiving station [MECH ENG] The location or device on conveyor systems where bulk material is loaded or otherwise received onto the conveyor.

receiving tank *See* rundown tank.

receiving tube [ELECTR] A low-voltage and low-power vacuum tube used in radio receivers, computers, and sensitive control and measuring equipment.

Recent [GEOL] An epoch of geologic time (late Quaternary) following the Pleistocene; referred to as Holocene in several European countries.

receptacle *See* outlet.

reception [COMMUN] The conversion of modulated electromagnetic waves or electric signals, transmitted through the air or over wires or cables, into the original intelligence, or into desired useful information (as in radar), by means of antennas and electronic equipment.

receptive aphasia *See* sensory aphasia.

receptor [BIOCHEM] A site or structure in a cell which combines with a drug or other biological to produce a specific alteration of cell function. [PHYSIO] A sense organ.

recess [ENG] A surface groove or depression. [GEOL] **1.** An indentation occurring in a surface, bounded by a straight line. **2.** An area having the axial traces of folds concave toward the outer edge of the folded belt.

recessional moraine [GEOL] **1.** An end moraine formed during a temporary halt in the final retreat of a glacier. **2.** A moraine formed during a minor readvance of the ice front during a period of glacial recession. Also known as stadial moraine.

recession of galaxies [ASTROPHYS] The increase in the velocity of recession (red shift) of galaxies with distance from an observer on earth.

recessive [GEN] **1.** An allele that is not expressed phenotypically when present in the heterozygous condition. **2.** An organism homozygous for a recessive gene.

rechamber [ORD] To rebore or otherwise alter the chamber of a small arm, normally for the purpose of adapting it to cartridges for which not originally designed.

recharge [HYD] **1.** The processes involved in the replenishment of water to the zone of saturation. **2.** The amount of water added or absorbed. Also known as groundwater increment; groundwater recharge; groundwater replenishment; increment; intake.

rechargeable battery *See* storage battery.

recharge well [HYD] A well used as a source of water in the process of artificial recharge. Also known as injection well.

reciprocal [MATH] The reciprocal of a number A is the number $1/A$.

reciprocal bearing *See* back bearing.

reciprocal differences [MATH] An interpolation technique using successive quotients of a function with its values so as to obtain a continued fraction expansion approximating the given function by a rational function.

reciprocal ellipsoid *See* index ellipsoid.

reciprocal ferrite switch [ELECTROMAG] A ferrite switch that can be inserted in a waveguide to switch an input signal to either of two output waveguides; switching is done by a Faraday rotator when acted on by an external magnetic field.

reciprocal impedance [ELEC] Two impedances Z_1 and Z_2 are said to be reciprocal impedances with respect to an impedance Z (invariably a resistance) if they are so related as to satisfy the equation $Z_1 Z_2 = Z^2$.

reciprocal inhibition [PHYSIO] In muscular movement, the simultaneous relaxation of one muscle and the contraction of its antagonist. [PSYCH] The modification of a behavioral pattern by the conditioning of responses that are incompatible with the response to be eliminated.

reciprocal junction [ELECTROMAG] A waveguide junction in which the transmission coefficient from the i-th port to the j-th port is the same as that from the j-th port to the i-th port; that is, the S matrix is symmetrical.

reciprocal lattice [CRYSTAL] A lattice array of points formed by drawing perpendiculars to each plane *(hkl)* in a crystal lattice through a common point as origin; the distance from each point to the origin is inversely proportional to spacing of the specific lattice planes; the axes of the reciprocal lattice are perpendicular to those of the crystal lattice.

reciprocal laying [ORD] Method of making the planes of fire of two guns parallel by pointing the guns in parallel direction; in reciprocal laying, the two guns sight on each other, then swing out through supplementary angles to produce equal deflections from the base line connecting the two pieces.

reciprocal leveling [CIV ENG] A variant of straight differential leveling applied to long distances in which levels are taken on two points, and the average of the two elevation differences is the true difference.

reciprocal ohm *See* mho.

reciprocal ohm centimeter *See* roc.

reciprocal ohm meter *See* rom.

reciprocal space *See* wave-vector space.

reciprocal spiral *See* hyperbolic spiral.

reciprocal transducer [ELECTR] Transducer which satisfies the principle of reciprocity.

reciprocal translocation [CYTOL] The special case of translocation in which two segments exchange positions.

reciprocal velocity region [NUC PHYS] The energy region in which the capture cross section for neutrons by a given element is inversely proportional to neutron velocity.

reciprocating compressor [MECH ENG] A positive-displacement compressor having one or more cylinders, each fitted with a piston driven by a crankshaft through a connecting rod.

reciprocating drill *See* piston drill.

reciprocating engine *See* piston engine.

reciprocating flight conveyor [MECH ENG] A reciprocating beam or beams with hinged flights that advance materials along a conveyor trough.

reciprocating-plate column *See* reciprocating-plate extractor.

reciprocating-plate extractor [CHEM ENG] A liquid-liquid contactor in which equally spaced perforated plates (as in a distillation column) move up and down rapidly over a short distance to cause liquid agitation and mixing. Also known as reciprocating-plate column.

reciprocating-plate feeder [MECH ENG] A back-and-forth shaking tray used to feed abrasive materials, such as pulverized coal, into process units.

reciprocating pump *See* piston pump.

RECENT

PRECAMBRIAN	PALEOZOIC						MESOZOIC	CENOZOIC
	CARBON-IFEROUS							
	CAMBRIAN	ORDOVICIAN	SILURIAN	DEVONIAN	Mississippian	Pennsylvanian	PERMIAN	TRIASSIC	JURASSIC	CRETACEOUS	TERTIARY	QUATERNARY

TERTIARY					QUATERNARY	
Paleocene	Eocene	Oligocene	Miocene	Pliocene	Pleistocene	Recent

Chart showing the position of the Recent epoch in relation to the various periods of geologic time.

reciprocating screen [MECH ENG] Horizontal solids-separation screen (sieve) oscillated back and forth by an eccentric gear; used for solids classification.

reciprocation [ELECTR] In electronics, a process of deriving a reciprocal impedance from a given impedance, or finding a reciprocal network for a given network.

reciprocity law [GRAPHICS] The law that, with standard development, the optical density of an exposed emulsion is proportional to the exposure time and to the illumination.

reciprocity theorem Also known as principle of reciprocity. [ACOUS] The theorem that, in an acoustic system consisting of a fluid medium with boundary surfaces and subject to no impressed body forces, if p_1 and p_2 are the pressure fields produced respectively by the components of the fluid velocities V_1 and V_2 normal to the boundary surfaces, then the integral over the boundary surfaces of $p_1 V_2 - p_2 V_1$ vanishes. [ELEC] 1. The electric potentials V_1 and V_2 produced at some arbitrary point, due to charge distributions having total charges of q_1 and q_2 respectively, are such that $q_1 V_2 = q_2 V_1$. 2. In an electric network consisting of linear passive impedances, the ratio of the electromotive force introduced in any branch to the current in any other branch is equal in magnitude and phase to the ratio that results if the positions of electromotive force and current are exchanged. [ELECTROMAG] Given two loop antennas, a and b, then $I_{ab}/V_a = I_{ba}/V_b$, where I_{ab} denotes the current received in b when a is used as transmitter, and V_a denotes the voltage applied in a; I_{ba} and V_b are the corresponding quantities when b is the transmitter, a the receiver; it is assumed that the frequency and impedances remain unchanged. [ENG ACOUS] The sensitivity of a reversible electroacoustic transducer when used as a microphone divided by the sensitivity when used as a source of sound is independent of the type and construction of the transducer. [PHYS] In general, any theorem that expresses various reciprocal relations for the behavior of some physical systems, in which input and output can be interchanged without altering the response of the system to a given excitation.

recirculator [ENG] A self-contained underwater breathing apparatus that recirculates an oxygen supply (mix-gas or pure) to the diver until the oxygen is depleted.

reclaimed oil [MATER] Used lubricating oil that is collected, reprocessed, and sold for reuse. Also known as recovered oil.

reclaimed rubber [MATER] Scrap rubber (natural or synthetic) prepared for reuse; fragmented scrap is digested in hot caustic solution to which reclaiming agents have been added; reclaimed rubber is used to blend with virgin rubber, or for low-grade rubber products.

reclaim rinse [MET] A nonflowing rinse used to recover dragout in electroplating operations.

reclamation [CIV ENG] 1. The recovery of land or other natural resource that has been abandoned because of fire, water, or other cause. 2. Reclaiming dry land by irrigation.

reclassify [ORD] To change the security classification of a document, piece of equipment, or the like.

reclined fold See recumbent fold.

reclosing relay [ELEC] Form of voltage, current, power, or other type of relay which functions to reclose a circuit.

recognition [ADP] The act or process of identifying (or associating) an input with one of a set of possible known alternatives, as in character recognition and pattern recognition.

recognition gate [ADP] A logic circuit used to select devices identified by a binary address code. Also known as decoding gate.

recoil See gun reaction.

recoil adapter [ORD] A device fastened between a gun, especially an aircraft machine gun, and its mount to adapt the gun for mounting and to absorb the recoil.

recoil booster [ORD] Component of a machine gun which traps some of the gas from the barrel and acts to ensure positive recoil action when the gun is fired at angles other than the usual horizontal.

recoil brake [ORD] That part of the recoil mechanism that actually absorbs the energy of recoil and stops the rearward movement of the recoiling parts.

recoil click [HOROL] A device in a timepiece that prevents a

mainspring from being wound too tightly; uses a pawl that recoils after winding.

recoil electron [PHYS] An electron that has been set into motion by a collision.

recoil escapement See anchor escapement.

recoiling mass [ORD] The mass of the recoiling parts of a weapon.

recoiling parts [ORD] Those parts of a weapon which move in recoil, usually including the tube, breech housing, breech-block assembly, and parts of the recoil mechanism.

recoilless [ORD] Built so as to eliminate or cancel out recoil; most recoilless guns are designed to let part of the propellant gases escape to the rear.

recoilless ammunition [ORD] Ammunition intended for use in recoilless rifles; provision is made in the ammunition for release of propellant gases in the manner and quantity necessary to produce the recoilless action.

recoilless gun [ORD] A smooth bore, open breech, launcher-type artillery weapon constructed of lightweight metals and employing a muzzle-inserted propellant; it is designed with a firing mechanism activated electrically or mechanically by remote control.

recoilless rifle [ORD] A weapon consisting of a light artillery tube of the recoilless type and a very light mount.

recoil mechanism [ORD] A hydraulic, pneumatic, or spring type shock absorber that decreases the energy of the recoil gradually and so avoids violent movement of the gun.

recoil oil [MATER] A neutral, constant-viscosity oil used in hydropneumatic and hydrospring recoil systems.

recoil particle [PHYS] A particle that has been set into motion by a collision or by a process involving the ejection of another particle.

recoil pit [ORD] Pit dug near the breech of a gun to provide space for the breech when it moves backward during recoil.

recoil velocity [ORD] Velocity in recoil of the recoiling parts of a gun.

recombination [GEN] 1. The occurrence of gene combinations in the progeny that differ from those of the parents as a result of independent assortment, linkage, and crossing-over. 2. The production of genetic information in which there are elements of one line of descent replaced by those of another line, or additional elements. [PHYS] The combination and resultant neutralization of particles or objects having unlike charges, such as a hole and an electron or a positive ion and a negative ion.

recombination coefficient [ELECTR] The rate of recombination of positive ions with electrons or negative ions in a gas, per unit volume, divided by the product of the number of positive ions per unit volume and the number of electrons or negative ions per unit volume.

recombination electroluminescence See injection electroluminescence.

recombination energy [PHYS] The energy released when two oppositely charged portions of an atom or molecule rejoin to form a neutral atom or molecule.

recombination mosaic [GEN] A mosaic produced as the result of somatic crossing-over.

recombination radiation [SOLID STATE] The radiation emitted in semiconductors when electrons in the conduction band recombine with holes in the valence band.

recombination velocity [ELECTR] On a semiconductor surface, the ratio of the normal component of the electron (or hole) current density at the surface to the excess electron (or hole) charge density at the surface.

recompletion [PETRO ENG] Redrilling an oil well to a new producing zone (new depth) when the current zone is depleted.

recomposed granite [PETR] An arkose composed of consolidated feldspathic residue that has been reworked and decomposed so slightly that upon cementation the rock resembles granite except that its grain is less even and it contains a greater percentage of quartz. Also known as reconstructed granite.

recomputed point of turn [NAV] An altered dead-reckoning position of an aircraft at a turning point, determined after wind has been established by drift observations made before and after the turn.

recon [GEN] The smallest deoxyribonucleic acid unit capable of recombination.

reconditioned carrier reception [ELECTR] Method of reception in which the carrier is separated from the sidebands to eliminate amplitude variations and noise, and is then added at an increased level to the sideband, to obtain a relatively undistorted output.

reconditioning [ENG] Restoration of an object to a good condition.

reconnaissance [ENG] A mission to secure data concerning the meteorological, hydrographic, or geographic characteristics of a particular area. [ORD] A mission undertaken to obtain, by visual observation or other detection methods, information about the activities and resources of an enemy or potential enemy.

reconnaissance drone [AERO ENG] An unmanned aircraft guided by remote control, with photographic or electronic equipment for providing information about an enemy or potential enemy.

reconnaissance spacecraft [AERO ENG] A satellite put into orbit about the earth and containing electronic equipment designed to pick up and transmit back to earth information pertaining to activities such as military.

reconstituted mica [MATER] Mica sheets or shaped objects made by breaking up scrap natural mica, combining with a binder, and pressing into forms suitable for use as electrical insulating material.

reconstructed coal [MATER] Coal formed from crushed or powdered, briquetted lignite or coal, waterproofed with a coating of pitch.

reconstructed granite See recomposed granite.

recontrol time See deionization time.

record [ADP] A group of adjacent data items in a computer system, manipulated as a unit. [SCI TECH] 1. To preserve for later reproduction or reference. 2. See recording.

record block See physical record.

record changer [ENG ACOUS] A record player that plays a number of records automatically in succession.

record density See bit density; character density.

recorder See recording instrument.

record gap [ADP] An area in a storage medium, such as magnetic tape or disk, which is devoid of information; it delimits records, and, on tape, allows the tape to stop and start between records without loss of data. Also known as interrecord gap (IRG).

recording [SCI TECH] 1. Any process for preserving signals, sounds, data, or other information for future reference or reproduction, such as disk recording, facsimile recording, ink-vapor recording, magnetic tape or wire recording, and photographic recording. 2. The end product of a recording process, such as the recorded magnetic tape, disk, or record sheet. Also known as record.

recording balance [ANALY CHEM] An analytical balance equipped to record weight results by electromagnetic or servomotor-driven accessories.

recording-completing trunk [ELEC] Trunk for extending a connection from a local line to a toll operator, used for recording the call and for completing the toll connection.

recording head [ELECTR] A magnetic head used only for recording. Also known as record head. [ENG ACOUS] See cutter.

recording instrument [ENG] An instrument that makes a graphic record of one or more quantities as a function of another variable, usually time. Also known as recorder.

recording lamp [ELECTR] A lamp whose intensity can be varied at an audio-frequency rate, for exposing variable-density sound tracks on motion picture film and for exposing paper or film in photographic facsimile recording.

recording level [ELECTR] Amplifier output level required to secure a satisfactory recording.

recording noise [ELECTR] Noise that is introduced during a recording process.

recording optical tracking instrument [ENG] Optical system used for recording data in connection with missile flights.

recording rain gage [ENG] A rain gage which automatically records the amount of precipitation collected, as a function of time. Also known as pluviograph.

recording spot See picture element.

recording storage tube [ELECTR] Type of cathode-ray tube in which the electric equivalent of an image can be stored as an electrostatic charge pattern on a storage surface; there is no visual display, but the stored information can be read out at any later time as an electric output signal.

recording thermometer See thermograph.

recording trunk [ELEC] Trunk extending from a local central office or private branch exchange to a toll office, which is used only for communications with toll operators and not for completing toll connections.

record length [ADP] The number of characters required for all the information in a record.

record mark [ADP] A symbol that signals a record's beginning or end.

record observation [METEOROL] A type of aviation weather observation; the most complete of all such observations and usually taken at regularly specified and equal intervals (hourly, usually on the hour). Also known as hourly observation.

record player [ENG ACOUS] A motor-driven turntable used with a phonograph pickup to obtain audio-frequency signals from a phonograph record.

record storage mark [ADP] A special character which appears only in the record storage unit of the card reader to limit the length of the record read into storage.

recoupling [QUANT MECH] A transformation between eigenfunctions of total angular momentum resulting from coupling eigenfunctions of three or more angular momenta in some order, and eigenfunctions of total angular momentum resulting from coupling of the same eigenfunctions in a different order.

recoverable shear [FL MECH] Measure of the elastic content of a fluid, related to elastic recovery (mechanicallike property of elastic recoil); found in unvulcanized, unfilled natural rubber, and certain polymer solutions, soap gels, and biological fluids.

recovered oil See reclaimed oil.

recovery [AERO ENG] 1. The procedure or action that obtains when the whole of a satellite, or a section, instrumentation package, or other part of a rocket vehicle, is retrieved after a launch. 2. The conversion of kinetic energy to potential energy, such as in the deceleration of air in the duct of a ramjet engine. Also known as ram recovery. 3. In flying, the action of a lifting vehicle returning to an equilibrium attitude after a nonequilibrium maneuver. [MET] 1. The percentage of valuable material obtained from a processed ore. 2. Reduction or elimination of work-hardening effects, usually by heat treatment. [MIN ENG] The proportion or percentage of coal or ore mined from the original seam or deposit. [PETRO ENG] Term designating the removal (recovery) of oil or gas from reservoir formations.

recovery area [AERO ENG] An area in which a satellite, satellite package, or spacecraft is recovered after reentry.

recovery capsule [AERO ENG] A space capsule designed to be recovered after reentry.

recovery interrupt [ADP] A type of interruption of program execution which provides the computer with access to subroutines to handle an error and, if successful, to continue with the program execution.

recovery package [AERO ENG] A package attached to a reentry or other body designed for recovery, containing devices intended to locate the body after impact.

recovery party [ORD] A form of contact party whose purpose is the recovery of disabled ordnance materiel from predesignated collecting points, and the transportation of this materiel to the ordnance shops for repairs.

recovery room [MED] A hospital room in which surgical patients are kept during the period immediately following an operation for care and recovery from anesthesia.

recovery ship [NAV ARCH] A naval vessel designed to participate in the retrieval of a satellite, instrument package, or spaceship after it has reentered the atmosphere and landed in the ocean.

recovery temperature See adiabatic recovery temperature.

recovery time [ELECTR] 1. The time required for the control electrode of a gas tube to regain control after anode-current interruption. **2.** The time required for a fired TR (transmit-receive) or pre-TR tube to deionize to such a level that the attenuation of a low-level radio-frequency signal transmitted through the tube is decreased to a specified value. **3.** The time required for a fired ATR (anti-transmit-receive) tube to deionize to such a level that the normalized conductance and susceptance of the tube in its mount are within specified ranges. **4.** The interval required, after a sudden decrease in input signal amplitude to a system or component, to attain a specified percentage (usually 63%) of the ultimate change in amplification or attenuation due to this decrease. **5.** The time required for a radar receiver to recover to half sensitivity after the end of the transmitted pulse, so it can receive a return echo. [NUCLEO] The minimum time from the start of a counted pulse to the instant a succeeding pulse can attain a specific percentage of the maximum value of the counted pulse in a Geiger counter.

recovery vehicle [MECH ENG] A special-purpose vehicle equipped with winch, hoist, or boom for recovery of vehicles.

recruitment [PHYSIO] A serial discharge from neurons innervating groups of muscle fibers.

recrystallization [CHEM] Repeated crystallization of a material from fresh solvent to obtain an increasingly pure product. [MET] A process which takes place in metals and alloys following distortion and fragmentation of constituent crystals by severe mechanical deformation, in which some fragments grow at the expense of others, so that larger, strain-free grains are formed; it progresses slowly at room temperature, but is greatly speeded by annealing. [PETR] The formation of new mineral grains in crystalline form in a rock under the influence of metamorphic processes.

recrystallization annealing [MET] Producing a new grain structure without phase change by annealing cold-worked metal.

recrystallization breccia *See* pseudobreccia.

recrystallization temperature [MET] The minimum temperature at which complete recrystallization occurs in an annealed cold-worked metal within a specified time.

rectangle [MATH] A plane quadrilateral having four interior right angles and opposite sides of equal length.

rectangular cartesian coordinate system *See* cartesian coordinate system.

rectangular cavity [ELECTROMAG] A resonant cavity having the shape of a rectangular parallelepiped.

rectangular chart [MAP] **1.** A chart in a rectangular shape. **2.** A chart on the rectangular map projection.

rectangular coordinates *See* cartesian coordinates.

rectangular cross ripple mark [GEOL] An oscillation cross ripple mark consisting of two sets of ripples which intersect at right angles, enclosing a rectangular pit.

rectangular distribution *See* uniform distribution.

rectangular drainage pattern [GEOL] A drainage pattern characterized by many right-angle bends in both the main streams and their tributaries. Also known as lattice drainage pattern.

rectangular hyperbola [MATH] A hyperbola whose major and minor axes are equal.

rectangular map projection [MAP] A cylindrical map projection with uniform spacing of the parallels.

rectangular mesh [TEXT] Cloth with a different mesh count in the fill than in the warp. Also known as oblong mesh.

rectangular parallelepiped [MATH] A parallelepiped with bases as rectangles all perpendicular to its lateral faces. Also known as rectangular solid.

rectangular projection [MAP] A cylindrical map projection with uniform spacing of the parallels; used for the star chart in the Air Almanac.

rectangular pulse [ELECTR] A pulse in which the wave amplitude suddenly changes from zero to another value at which it remains constant for a short period of time, and then suddenly changes back to zero.

rectangular scanning [ELECTR] Two-dimensional sector scanning in which a slow sector scanning in one direction is superimposed on a rapid sector scanning in a perpendicular direction.

rectangular search [NAV] A search of three legs from a moving point, the first and third legs being perpendicular to

the base course of the moving point, and the second leg being parallel to it.

rectangular solid *See* rectangular parallelepiped.

rectangular wave [ELECTR] A periodic wave that alternately and suddenly changes from one to the other of two fixed values. Also known as rectangular wave train.

rectangular waveguide [ELECTROMAG] A waveguide having a rectangular cross section.

rectangular wave train *See* rectangular wave.

rectangular weir [CIV ENG] A weir with a rectangular notch at top for measurement of water flow in open channels; it is simple, easy to make, accurate, and popular.

Rectenna [ELECTR] A device that converts microwave energy in direct-current power; consists of a number of small dipoles, each having its own diode rectifier network, which are connected to direct-current buses.

Recticornia [INV ZOO] A family of amphipod crustaceans in the superfamily Genuina containing forms in which the first antennae are straight, arise from the anterior margin of the head, and have few flagellar segments.

rectifiable curve [MATH] A curve whose length can be computed and is finite.

rectification [ELEC] The process of converting an alternating current to a unidirectional current.

rectification distillation [CHEM ENG] A distillation technique in which a rectifying column is used.

rectification factor [ELECTR] Quotient of the change in average current of an electrode by the change in amplitude of the alternating sinusoidal voltage applied to the same electrode, the direct voltages of this and other electrodes being maintained constant.

rectified value of an alternating quantity [ELEC] Average of all the positive (or negative) values of the quantity during an integral number of periods.

rectifier [ELEC] A nonlinear circuit component that allows more current to flow in one direction than the other; ideally, it allows current to flow in one direction unimpeded but allows no current to flow in the other direction.

rectifier filter [ELECTR] An electric filter used in smoothing out the voltage fluctuation of an electron tube rectifier, and generally placed between the rectifier's output and the load resistance.

rectifier instrument [ENG] Combination of an instrument sensitive to direct current and a rectifying device whereby alternating current (or voltages) may be rectified for measurement.

rectifier rating [ELECTR] A performance rating for a semiconductor rectifier, usually on the basis of the root-mean-square value of sinusoidal voltage that it can withstand in the reverse direction and the average current density that it will pass in the forward direction.

rectifier stack [ELECTR] A dry-disk rectifier made up of layers or stacks of disks of individual rectifiers, as in a selenium rectifier or copper-oxide rectifier.

rectifier transformer [ELECTR] Transformer, the secondary of which supplies energy to the main anodes of a rectifier.

rectifying column [CHEM ENG] Portion of a distillation column above the feed tray in which rising vapor is enriched by interaction with a countercurrent falling stream of condensed vapor; contrasted to the stripping column section below the column feed tray.

rectilinear [MATH] Consisting of or bounded by lines.

rectilinear generators [MATH] Straight lines which generate ruled surfaces.

rectilinear motion [MECH] A continuous change of position of a body so that every particle of the body follows a straight-line path. Also known as linear motion.

rectilinear scanning [ELECTR] Process of scanning an area in a predetermined sequence of narrow parallel strips.

rectilinear system *See* orthoscopic system.

recto [GRAPHICS] A right-hand book page, which always bears an odd number.

rectocele [MED] Bulging or herniation of the rectum into the vagina. Also known as vaginal protocele.

rectrix [VERT ZOO] One of the stiff tail feathers used by birds to control direction of flight.

RECTANGLE

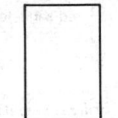

Shape of a rectangle.

RECTILINEAR MOTION

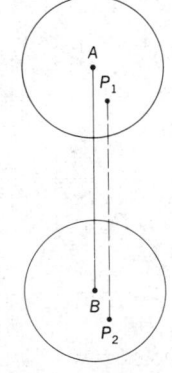

Diagram of rectilinear motion. Motion of center of mass of a body from *A* to *B* is along the straight line connecting these points. Path of any particle in the body is a straight line parallel to or coinciding with *AB*, such as the straight line connecting P_1 and P_2.

rectum [ANAT] The portion of the large intestine between the sigmoid flexure and the anus.

rectus [ANAT] Having a straight course, as certain muscles.

rectus abdominis [ANAT] The long flat muscle of the anterior abdominal wall which, as vertical fibers, arises from the pubic crest and symphysis, and is inserted into the cartilages of the fifth, sixth, and seventh ribs.

rectus femoris [ANAT] A division of the quadriceps femoris inserting in the patella and ultimately into the tubercle of the tibia.

rectus oculi [ANAT] Any of four muscles (superior, inferior, lateral, and medial) of the eyeball, running forward from the optic foramen and inserted into the sclerotic coat.

recumbent [BOT] Of or pertaining to a plant or plant part that tends to rest on the surface of the soil.

recumbent fold [GEOL] An overturned fold with a nearly horizontal axial surface. Also known as reclined fold.

recuperability [COMMUN] Ability to continue to operate after a partial or complete loss of the primary communications facility resulting from sabotage, enemy attack, or other disaster.

recuperative air heater [ENG] An air heater in which the heat-transferring metal parts are stationary and form a separating boundary between the heating and cooling fluids.

recuperator [ENG] An apparatus in which heat is conducted from the combustion products to incoming cooler air through a system of thin-walled ducts.

recurrence rate *See* repetition rate.

recurring decimal *See* repeating decimal.

recurring demand [IND ENG] A request made periodically or anticipated to be repetitive by an authorized requisitioner for material for consumption or use, or for stock replenishment.

recurring issue [ORD] An issue made on a cyclic basis to replenish materiel consumed or worn out through fair wear and tear in operations, with each issue being made to a consignee eligible to request further replenishment when required, in the forseeable future.

recursion [ADP] A technique in which an apparently circular process is used to perform an iterative process.

recursion formula [MATH] An algorithm allowing computation of a succession of quantities. Also known as recursion relation.

recursion relation *See* recursion formula.

recursive functions [MATH] Functions that can be obtained by a finite number of operations, computations, or algorithms.

recursive macro call [ADP] A call to a macroinstruction already called when used in conjunction with conditional assembly.

recursive procedure [ADP] A method of calculating a function by deriving values of it which become more elementary at each step; recursive procedures are explicitly outlawed in most systems with the exception of a few which use languages such as ALGOL and LISP.

recursive subroutine [ADP] A reentrant subroutine whose partial results are stacked, with a processor stack pointer advancing and retracting as the subroutine is called and completed.

recurvature [METEOROL] With respect to the motion of severe tropical cyclones (hurricanes and typhoons), the change in direction from westward and poleward to eastward and poleward; such recurvature of the path frequently occurs as the storm moves into middle latitudes.

recurved [SCI TECH] Curving inward to backward.

recurved spit [GEOGR] A spit having the outer end curved toward the land. Also known as hook; hooked spit.

recycle base [ORD] A base used by returning mission aircraft for servicing and maintenance, before the next mission launch.

recycle mixing [CHEM ENG] The mixing of a portion of a product stream (fluid or solid) from a processing unit with incoming raw feed.

recycle ratio [CHEM ENG] In a continuous chemical process, the ratio of recycle stock to fresh feed.

recycle stock [CHEM ENG] That portion of a feedstock that has passed through a processing unit and is recirculated (recycled) back through the process.

recycling [ELECTR] Returning to an original condition, as to

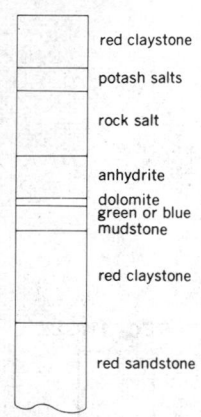

REDBED

red claystone
potash salts
rock salt
anhydrite
dolomite
green or blue mudstone
red claystone
red sandstone

Sequence of redbeds with evaporites, characteristic of Permian formations in southwestern United States.

0 or 1 in a counting circuit. [ENG] The extraction and recovery of valuable materials from scrap or other discarded materials. [NUCLEO] Reuse of fissionable nuclear reactor fuel by chemical processing, reenriching, and refabricating into new fuel elements.

red [OPTICS] The hue evoked in an average observer by monochromatic radiation having a wavelength in the approximate range from 622 to 770 nanometers; however, the same sensation can be produced in a variety of other ways.

red acetate *See* mordant rouge.

red algae [BOT] The common name for members of the phylum Rhodophyta.

red antimony *See* kermesite.

red arsenic *See* realgar.

red azimuth tables [NAV] Popular title for Hydrographic Office Publication No. 260, "Azimuths of the Sun," used in navigation.

redbed [GEOL] Continentally deposited sediment composed principally of sandstone, siltsone, and shale; red in color due to the presence of ferric oxide (hematite). Also known as red rock.

red blood cell *See* erythrocyte.

red brass [MET] A brass containing 85% copper and 15% zinc.

red charcoal [MATER] An impure charcoal made by heating wood to 300°C; much of the oxygen and hydrogen is retained.

red clay [GEOL] A fine-grained, reddish-brown pelagic deposit consisting of relatively large proportions of windblown particles, meteoric and volcanic dust, pumice, shark teeth, manganese nodules, and debris transported by ice. Also known as brown clay.

red cobalt *See* erythrite.

red dwarf star [ASTRON] A red star of low luminosity, so designated by E. Hertzsprung; dwarf stars are commonly those main-sequence stars fainter than an absolute magnitude of +1, and red dwarfs are the faintest and coolest of the dwarfs.

Redeye [ORD] A man-transportable guided missile, fired from the shoulder, designed to provide combat troops with the capability of destroying low-flying aircraft.

red giant star [ASTRON] A star whose evolution has progressed to the point where hydrogen core burning has been completed, the helium core has become denser and hotter than originally, and the envelope has expanded to perhaps 100 times its initial size.

red glass [MATER] Soda-zinc glass to which small amounts of cadmium and selenium are added.

red gum *See* eucalyptus gum.

red hematite *See* hematite.

redia [INV ZOO] A larva produced within the miracidial sporocyst of certain digenetic trematodes which may give rise to daughter rediae or to cercariae.

redifferentiation [PHYSIO] The return to a position of greater specialization in actual and potential functions, or the developing of new characteristics.

redingtonite [MINERAL] $(Fe,Mg,Ni)(Cr,Al)_2(SO_4)_4 \cdot 22H_2O$ A pale-purple mineral composed of a hydrous sulfate of iron, magnesium, nickel, chromium, and aluminum.

red iron ore *See* hematite.

redistribution [ELECTR] The alteration of charges on an area of a storage surface by secondary electrons from any other area of the surface in a charge storage tube or television camera tube.

red lake C pigment [MATER] An organic azo pigment; made by coupling the diazonium salt (barium or sodium) of ortho-chloro-meta-toluidine-para-sulfonic acid with β-naphthol; used to color inks, plastics, and rubbers.

red lead *See* lead tetroxide.

red leaf [PL PATH] Any of various nonparasitic plant diseases marked by red discoloration of the foliage.

Redler conveyor [MECH ENG] A conveyor in which material is dragged through a duct by skeletonized or U-shaped impellers which move the material in which they are submerged because the resistance to slip through the element is greater than the drag against the walls of the duct.

red liquor *See* mordant rouge.

red magnetism [GEOPHYS] The magnetism of the north-seeking end of a freely suspended magnet; this is the magnetism of the earth's south magnetic pole.

red mercury sulfide *See* mercuric sulfide.

red metal [MET] A copper matte having a copper content of about 48%.

red mud [GEOL] A reddish terrigenous mud composed of up to 25% calcium carbonate and deriving its color from the presence of ferric oxide; found on the sea floor near deserts and near the mouths of large rivers. [MET] An iron oxide-rich residue obtained in purifying bauxite in the Bayer process.

red nitric acid [MATER] A type of liquid bipropellant with boiling point 104°F (40°C), freezing point −80°F (−62°C), and density 1.58 grams per milliliter; used to supply power for jet propulsion.

red nucleus [HISTOL] A mass of reticular fibers in the gray matter of the tegmentum of the mesencephalon of higher vertebrates; it receives fibers from the cerebellum of the opposite side and gives rise to rubrospinal tract fibers of the opposite side.

red ocher *See* ferric oxide; hematite.

red oil [MATER] **1.** Any intermediate-grade petroleum lubricating oil that is red in color by transmitted light; includes so-called red engine oils, bearing oils, and machinery oils. **2.** *See* oleic acid.

red orpiment *See* realgar.

redox cell [ELEC] Cell designed to convert the energy of reactants to electrical energy; an intermediate reductant, in the form of liquid electrolyte, reacts at the anode in a conventional manner; it is then regenerated by reaction with a primary fuel.

red oxide of zinc *See* zincite.

redox potential [PHYS CHEM] Voltage difference at an inert electrode immersed in a reversible oxidation-reduction system; measurement of the state of oxidation of the system. Also known as oxidation-reduction potential.

redox potentiometry [ANALY CHEM] Use of neutral electrode probes to measure the solution potential developed as the result of an oxidation or reduction reaction.

redox system [CHEM] A chemical system in which reduction and oxidation (redox) reactions occur.

redox titration [ANALY CHEM] A titration characterized by the transfer of electrons from one substance to another (from the reductant to the oxidant) with the end point determined colorimetrically or potentiometrically.

red phosphorus [CHEM] An allotropic form of the element phosphorus; violet-red, amorphous powder subliming at 416°C, igniting at 260°; insoluble in all solvents,; nonpoisonous.

red potassium chromate *See* potassium dichromate.

red potassium prussiate *See* potassium ferricyanide.

red precipitate *See* mercuric oxide.

red prussiate of potash *See* potassium ferricyanide.

red prussiate of soda *See* sodium ferricyanide.

red ring [PL PATH] A nematode disease of the coconut palm caused by *Aphelenchoides cocophilus* and marked by the appearance of red rings in the stem cross section.

red rock *See* redbed.

red rot [PL PATH] Any of several fungus diseases of plants characterized by red patches on stems or leaves; common in sugarcane, sisal, and various evergreen and deciduous trees.

red rust [PL PATH] An algal disease of certain subtropical plants, such as tea and citrus, caused by the green alga *Cephaleuros virescens* and characterized by a rusty appearance of the leaves or twigs.

Red Sea [GEOGR] A body of water that lies between Arabia and northeastern Africa, about 2000 kilometers long, 300 kilometers wide, and a maximum depth of about 2300 meters.

red sector [NAV] A sector of the circle of visibility of a navigational light in which a red light is exhibited; such sectors are designated by their limiting bearing, as observed at some point other than the light; red sectors are often located so that they warn of danger to vessels.

red shift [ASTROPHYS] A systematic displacement toward longer wavelengths of lines in the spectra of distant galaxies and also of the continuous portion of the spectrum; increases with distance from the observer. Also known as Hubble effect.

red snow [HYD] A snow surface of reddish color caused by the presence within it of certain microscopic algae or particles of red dust.

Red Spot [ASTRON] A semipermanent marking of the planet Jupiter; some fluctuations in visibility exist; it does not rotate uniformly with the planet, indicating that it is a disturbance of Jupiter's atmosphere.

red stele [PL PATH] A fatal fungus disease of strawberries caused by *Phytophthora fragariae* invading the roots, producing redness and rotting with consequent dwarfing and wilting of the plant.

Redstone [ORD] A U.S. Army surface-to-surface guided missile having inertial guidance and a range of about 350 miles (560 kilometers); can carry either conventional or nuclear warheads; used to extend and supplement the range and firepower of artillery.

red stripe [PL PATH] **1.** A fungus decay of timber caused by *Polyporus vaporarius* and characterized by reddish streaks. **2.** A bacterial disease of sugarcane caused by *Xanthomonas rubrilineans* and characterized by red streaks in the young leaves, followed by invasion of the vascular system.

red-tape operation *See* bookkeeping operation.

red thermit [MET] Thermit made with red iron oxide.

red thread [PL PATH] A fungus disease of turf grasses caused by *Cortecium fuciforme* and characterized by the appearance of red stromata on the pinkish hyphal threads.

red tide [BIOL] A reddish discoloration of coastal surface waters due to concentrations of certain toxin-producing dinoflagellates. Also known as red water.

redtop grass [BOT] One of the bent grasses, *Agrostis alba* and its relatives, which grow on a wide variety of soils; it is a perennial, spreads slowly by rootstocks, and has top growth 2–3 feet (60–90 centimeters) tall.

reduce [ORD] To clear a stoppage in a weapon.

reduced crude [MATER] In petroleum refining, a residual product remaining after removal by distillation or other means of an appreciable quantity of the more volatile components of crude oil.

reduced equation of state [PHYS] An equation relating the reduced pressure, reduced volume, and reduced temperature of a substance.

reduced frequency *See* Strouhal number.

reduced nickel [MET] Nickel obtained by the precipitation of nickel hydroxide or nickel carbonate onto kieselguhr, then reducing the precipitate by heating it with hydrogen.

reduced pressure [METEOROL] The calculated value of atmospheric pressure at mean sea level or some other specified level, as derived (reduced) from station pressure or actual pressure; thus, sea level pressure is nearly always a reduced pressure. [THERMO] The ratio of the pressure of a substance to its critical pressure.

reduced-pressure distillation *See* vacuum distillation.

reduced property *See* reduced value.

reduced telemetry [COMMUN] Raw telemetry data transformed into a usable form.

reduced temperature [THERMO] The ratio of the temperature of a substance to its critical temperature.

reduced value [THERMO] The actual value of a quantity divided by the value of that quantity at the critical point. Also known as reduced property.

reduced viscosity [ENG] In plastics processing, the ratio of the specific viscosity to concentration.

reduced volume [THERMO] The ratio of the specific volume of a substance to its critical volume.

reducer [BIOL] *See* decomposer. [CHEM] *See* reducing agent. [DES ENG] A fitting having a larger size at one end than at the other and threaded inside, unless specifically flanged or for some special joint. [GRAPHICS] A solution capable of dissolving silver; used to cut down the contrast or density of a negative or positive image.

reducible polynomial [MATH] A polynomial relative to some field which can be written as the product of two polynomials of degree at least 1.

reducing agent [CHEM] Also known as reducer. **1.** A material that adds hydrogen to an element or compound. **2.** A material that adds an electron to an element or compound, that is, decreases the positiveness of its valence.

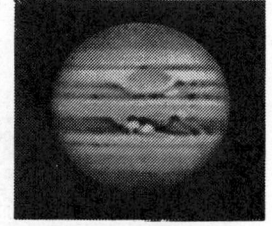

reducing atmosphere [CHEM] An atmosphere of hydrogen (or other substance that readily provides electrons) surrounding a chemical reaction or physical device; the effect is the opposite to that of an oxidizing atmosphere.

reducing coupling [ENG] A coupling used to connect a smaller pipe to a larger one.

reducing flame [CHEM] A flame having excess fuel and being capable of chemical reduction, such as extracting oxygen from a metallic oxide.

reducing glass [OPTICS] A double-concave lens that reduces the apparent size of objects viewed through it; used by illustrators and painters to create an artificial sense of distance from their work.

reducing sugar [ORG CHEM] Any of the sugars that because of their free or potentially free aldehyde or ketone groups, possess the property of readily reducing alkaline solutions of many metallic salts such as copper, silver, or bismuth; examples are the monosaccharides and most of the disaccharides, including maltose and lactose.

reductant [MET] Coal or other reducing materials introduced in a smelting process to remove oxygen from ores or concentrates.

reduction [ADP] Any process by which data are condensed, such as changing the encoding to eliminate redundancy, extracting significant details from the data and eliminating the rest, or choosing every second or third out of the totality of available points. [CHEM] **1.** Reaction of hydrogen with another substance. **2.** Chemical reaction in which an element gains an electron (has a decrease in positive valence). [NAV] The process of substituting for an observed value one derived therefrom, as the calculation of the corresponding meridian altitude from an observation of a celestial body near the meridian, or the derivation from a celestial observation of the information needed for establishing a line of position.

reduction cell [CHEM] A vessel in which aqueous solutions of salts or fused salts are reduced electrolytically.

reduction gear [MECH ENG] A gear train which lowers the output speed.

reduction of area [MET] In tensile testing, the percentage of decrease in cross-sectional area of a specimen at the point of rupture.

reduction of tidal current [OCEANOGR] The processing of observed tidal current data to obtain mean values of tidal current constants.

reduction of tides [OCEANOGR] The processing of observed tidal data to obtain mean values of tidal constants.

reduction potential [PHYS CHEM] The potential drop involved in the reduction of a positively charged ion to a neutral form or to a less highly charged ion, or of a neutral atom to a negatively charged ion.

reduction ratio [ENG] Ratio of feed size to product size for a mill (crushing or grinding) operation; measured by lump and sieve sizes.

reduction roll [MET] A roller used to reduce the thickness of a piece of metal.

reduction to the meridian [NAV] The process of applying a correction to an altitude observed when a body is near the celestial meridian of the observer, for the purpose of finding the altitude at meridian transit.

reduction to the sun [ASTRON] In the spectroscopic determination of a star's radial motion referred to the sun, the correction that is needed to be applied to the observed radial velocity of the star to compensate for the motion of the earth with respect to the sun.

redundancy [ADP] Any deliberate duplication or partial duplication of circuitry or information to decrease the probability of a system or communication failure. [COMMUN] In the transmission of information, the fraction of the gross information content of a message which can be eliminated without loss of essential information. [GEN] **1.** Repetition of a specified deoxyribonucleic acid sequence in a nucleus. **2.** Multiplicity of codons for individual amino acids. [MATH] A repetitive statement. [MECH] A statically indeterminate structure.

redundancy check [ADP] A forbidden-combination check that uses redundant digits called check digits to detect errors made by a computer.

redundant bombing [ORD] A term used in reference to the bombing of a target or targets which have already been made inoperative or useless by attacks on other targets upon which the former are dependent for their operation.

redundant character [ADP] A character specifically added to a group of characters to ensure conformity with certain rules which can be used to detect computer malfunction.

redundant code [COMMUN] A code which uses more signal elements than are needed to represent the information it transmits.

redundant digit [ADP] Digit that is not necessary for an actual computation but serves to reveal a malfunction in a digital computer.

Reduviidae [INV ZOO] The single family of the hemipteran group Reduvioidea; nearly all have a stridulatory furrow on the prosternum, ocelli are generally present, and the beak is three-segmented.

Reduvioidea [INV ZOO] The assassin bugs or conenose bugs, a monofamilial group of hemipteran insects in the subdivision Geocorisae.

reduzate [GEOL] A sediment accumulated under reducing conditions and consequently rich in organic carbon and in iron sulfide minerals; examples are coal and black shale.

red water [BIOL] See red tide. [VET MED] **1.** A babesiasis of cattle characterized by hematuria following release of hemoglobin by destruction of erythrocytes. **2.** A chronic disease of cattle attributed to oxalic acid in the forage; hematuria results from escape of blood from lesions in the bladder. **3.** An acute febrile septicemia of cattle, and sometimes horses, sheep, and swine, caused by the bacterium *Clostridium hemolyticum* and characterized by hemoglobinuria and sometimes intestinal hemorrhages.

redwater fever [VET MED] Any of several diseases of cattle, such as babesiasis or blackwater, caused by species of *Babesia* and characterized by hematuria.

red white dwarf star [ASTRON] A star type that is considered an anomaly; these are objects 10,000 times fainter than the sun, with surface temperature below 4000K so that surface radiation has cooled the star at an unexpectedly rapid rate.

redwood [BOT] *Sequoia sempervirens.* An evergreen tree of the pine family; it is the tallest tree in the Americas, attaining 350 feet (107 meters); its soft heartwood is a valuable building material.

Redwood viscometer [ENG] A standard British-type viscometer in which the viscosity is determined by the time, in seconds, required for a certain quantity of liquid to pass out through the orifice under given conditions; used for determining viscosities of petroleum oils.

red zinc ore See zincite.

reed [BOT] Any tall grass with a slender jointed stem. [TEXT] A comblike loom attachment that keeps the warp yarns apart and pushes the filling thread against the woven fabric.

reed frequency meter See vibrating-reed frequency meter.

Reed-Frost theory [MED] An epidemiological theory concerned with acute communicable diseases, based on an expression of the probable number of cases at time $T + 1$ in terms of known facts at time T.

reed horn [ENG ACOUS] A horn that produces sound by means of a steel reed vibrated by air under pressure.

reeding [ENG] Corrugating or serrating, as in coining or embossing.

reedmergnerite [MINERAL] $NaBSi_3O_8$ A colorless, triclinic borate mineral that represents the boron analog of albite.

reed pen [GRAPHICS] A short section of bamboo sharpened to a point at both ends and used for ink drawing.

reed relay [ELECTROMAG] A relay having contacts mounted on magnetic reeds sealed into a length of small glass tubing; an actuating coil is wound around the tubing or wound on an auxiliary ferrite-core structure, to provide the magnetic field required for relay operation.

Reed-Sternberg cell [PATH] An anaplastic reticuloendothelial cell constituting the characteristic microscopic feature of Hodgkin's disease.

reef [GEOL] **1.** A ridge- or moundlike layered sedimentary rock structure built almost exclusively by organisms. **2.** An offshore chain or range of rock or sand at or near the surface

REDWOOD

Leaves and cone of redwood
(*Sequoia sempervirens*).

of the water. [MIN ENG] A major ore trend or ore body.

reef conglomerate   *See* reef talus.

reef debris   *See* reef detritus.

reef detritus   [GEOL] Fragmental material derived from the erosion of an organic reef. Also known as reef debris.

reef flat   [GEOL] A flat expanse of dead reef rock which is partly or entirely dry at low tide; shallow pools, potholes, gullies, and patches of coral debris and sand are features of the reef flat.

reef limestone   [PETR] Limestone composed of the remains of sedentary organisms such as sponges, and of sediment-binding organic constituents such as calcareous algae. Also known as coral rock.

reef patch   [GEOL] A single large colony of coral formed independently on a shelf at depths less than 70 meters in the lagoon of a barrier reef or of an atoll. Also known as patch reef.

reef pinnacle   [GEOL] A small, isolated spire of rock or coral, especially a small reef patch.

reef rock   [PETR] A hard, unstratified rock composed of sand, shale, and the calcareous remains of sedentary organisms, cemented by calcium carbonate.

reef talus   [GEOL] Massive inclined strata composed of reef detritus deposited along the seaward margin of an organic reef. Also known as reef conglomerate.

reel   [DES ENG] A revolving spool-shaped device used for storage of hose, rope, cable, wire, magnetic tape, and so on.

reeled silk   [TEXT] The raw silk strand made by combining several filaments from separate cocoons.

reel locomotive   [MIN ENG] A trolley locomotive with a wire-rope reel for drawing mining cars out of rooms.

reenterable   [ADP] The attribute that describes a program or routine which can be shared by several tasks concurrently.

reentrant   [ENG] Having one or more sections directed inward, as in certain types of cavity resonators.

reentrant winding   [ELEC] Armature winding that returns to its starting point, thus forming a closed circuit.

reentry   [AERO ENG] The event when a spacecraft or other object comes back into the sensible atmosphere after being in space.

reentry angle   [AERO ENG] That angle of the reentry body trajectory and the sensible atmosphere at which the body reenters the atmosphere.

reentry body   [AERO ENG] That part of a space vehicle that reenters the atmosphere after flight above the sensible atmosphere.

reentry nose cone   [AERO ENG] A nose cone designed especially for reentry, consists of one or more chambers protected by a heat sink.

reentry system   [ADP] In character recognition, a system in which the input data to be read are printed by the computer with which the reader is associated.   [ORD] That portion of a ballistic missile or spacecraft designed to place one or more reentry vehicles on terminal trajectories so as to arrive at selected targets; penetration aids, spacers, deployment modules, and associated programming, and control and sensing devices are included in the reentry system.

reentry trajectory   [AERO ENG] That part of a rocket's trajectory that begins at reentry and ends at the target or at the surface.

reentry vehicle   [AERO ENG] Any payload-carrying vehicle designed to leave the sensible atmosphere and then return through it to earth.

reentry window   [AERO ENG] The area, at the limits of the earth's atmosphere, through which a spacecraft in a given trajectory can pass to accomplish a successful reentry for a landing in a desired region.

reepithelialization   [MED] **1.** Regrowth of epithelial tissue over a denuded surface. **2.** Surgical placement of a graft of epithelium over a denuded surface.

reeve   [NAV ARCH] **1.** To pass a rope through a hole or opening in a block, ringbolt, or similar device. **2.** To fasten something by passing a rope or line through or around it.

reevesite   [MINERAL] $Na_6Fe_2(OH)_{16}(CO_3)\cdot 4H_2O$ Hydrous oxide mineral known only in meteorites.

refer   [ORD] To bring the gunsights on a chosen aiming point without moving an artillery piece which has been laid for direction.

reference acoustic pressure   [ACOUS] Magnitude of any complex sound that will produce a sound-level meter reading equal to that produced by a sound pressure of 0.0002 dyne per square centimeter at 1000 hertz. Also known as reference sound level.

reference address   *See* address constant.

reference angle   [ELECTROMAG] Angle formed between the center line of a radar beam as it strikes a reflecting surface and the perpendicular drawn to that reflecting surface.

reference block   [ADP] A block within a computer program governing a numerically controlled machine which has enough data to allow resumption of the program following an interruption.

reference burst   *See* color burst.

reference craft   [NAV] A craft to which relative movement of other craft is referred; on a maneuvering board this first craft is placed at the center; it may be a reference ship, reference aircraft, or reference vehicle, depending upon the type of circumstances.

reference dimension   [DES ENG] In dimensioning, a dimension without tolerance used for informational purposes only, and does not govern machining operations in any way; it is indicated on a drawing by writing the abbreviation REF directly following or under the dimension.

reference dipole   [ELECTROMAG] Straight half-wave dipole tuned and matched for a given frequency, and used as a unit of comparison in antenna measurement work.

reference direction   [NAV] A direction used as a basis for comparison of other directions.

reference electrode   [PHYS CHEM] A nonpolarizable electrode that generates highly reproducible potentials; used for pH measurements and polarographic analyses; examples are the calomel electrode, silver-silver chloride electrode, and mercury pool.

reference ellipsoid   [MAP] A reference surface used to represent the size and shape of the earth for cartography.

reference frame   *See* frame of reference.

reference frequency   [COMMUN] Frequency having a fixed and specified position with respect to the assigned frequency.

reference fuel   [MATER] One of the standardized laboratory engine fuels, blends of which are used to determine the octane numbers of motor gasoline and the cetane numbers of diesel fuels.

reference level   [ENG ACOUS] The level used as a basis of comparison when designating the level of an audio-frequency signal in decibels or volume units. Also known as reference signal level.   [OCEANOGR] **1.** Level of no motion. **2.** A level for which current is known; allows determination of absolute current from relative current.

reference line   [NAV] A line from which angular or linear measurements are reckoned. Also known as datum line.

reference lot   [IND ENG] A lot of select components, used as a standard.

reference mark   [ELECTR] One of the marks used in a design of a printed circuit, giving scale dimensions and indicating the edges of the circuit board.

reference noise   [ELECTR] The power level used as a basis of comparison when designating noise power expressed in decibels above reference noise (dBrn); the reference usually used is 10^{-12} watt (-90 decibels above 1 milliwatt; dBm) at 1000 hertz.

reference plane   [ENG] *See* datum plane.   [MECH ENG] The plane containing the axis and the cutting point of a cutter.

reference point   [NAV] A point to which other points, lines, and so forth are referred, usually in terms of distance or direction, or both.

reference range   [ENG] Range obtained from the radar coverage indicator for a given penetrating aircraft.

reference record   [ADP] Output of a compiler that lists the operations and their positions in the final specific routine and contains information describing the segmentation and storage allocation of the routine.

reference rounds   [ORD] Ammunition rounds of known performance which are fired during ballistic tests of ammunition for comparative purposes.

reference signal level   *See* reference level.

reference sound level   *See* reference acoustic pressure.

reference station   [OCEANOGR] **1.** A place for which independent daily predictions are given in the tide or current tables, from which corresponding predictions are obtained for other stations by means of differences or factors. **2.** A place for which tidal or tidal current constants have been determined and which is used as a standard for the comparison of simultaneous observations at a second station. Also known as standard station.

reference time   [ADP] The instant near the beginning of switching that is chosen as a reference for time measurements in a digital computer.

reference tone   [ENG] Stable tone of known frequency continuously recorded on one track of multitrack signal recordings and intermittently recorded on signal track recordings by the collection equipment operators for subsequent use by the data analysts as a frequency reference.

reference vehicle   [NAV] A vehicle serving as a reference craft.

reference voltage   [ELEC] An alternating-current voltage used for comparison, usually to identify an in-phase or out-of-phase condition in an ac circuit.

reference volume   [ACOUS] The audio volume level that gives a reading of 0 vu (volume units) on a standard volume indicator; the sensitivity of the volume indicator is adjusted so reference volume or 0 is read when the instrument is connected across a 600-ohm resistance to which is delivered a power of 1 milliwatt at 1000 hertz.

reference white level   [ELECTR] In television, the level at the point of observation corresponding to the specified maximum excursion of the picture signal in the white direction.

referred pain   [MED] Pain felt in one area but originating in another area.

refine   [ENG] To free from impurities, as the separation of petroleum, ores, or chemical mixtures into their component parts.

refined kerosine   *See* deodorized kerosine.

refined lard   [FOOD ENG] A purified lard, produced by removing with alkali the free fatty acids, coloring matter, and mucilaginous gums from rendered lard.

refined lecithin   *See* lecithin.

refined oil   [MATER] Class term for a petroleum oil used for home lighting and cooking purposes. Also known as burning oil.

refined paraffin wax   [MATER] A grade of paraffin wax; a hard, crystalline hydrocarbon wax derived from mixed-base or paraffin-base crude oils.

refinery   [CHEM ENG] System of process units used to convert crude petroleum into fuels, lubricants, and other petroleum-derived products.   [MET] System of process units used to convert nonferrous-metal ores into pure metals, such as copper or zinc.

refinery gas   [MATER] Gas produced in petroleum refineries by cracking, reforming, and other processes; principally methane, ethane, ethylene, butanes, and butylenes.

refining temperature   [MET] The temperature just above the transformation range employed in the heat treatment of steel in order to refine grain size.

reflectance   [ADP] In optical character recognition, the relative brightness of the inked area that forms the printed or handwritten character; distinguished from background reflectance and brightness.   [ELEC] *See* reflection factor.   [PHYS] *See* reflectivity.

reflectance spectrophotometry   [SPECT] Measurement of the ratio of spectral radiant flux reflected from a light-diffusing specimen to that reflected from a light-diffusing standard substituted for the specimen.

reflected binary   [ADP] A particular form of gray code which is constructed according to the following rule: Let the first 2^N code patterns be given, for any N greater than 1; the next 2^N code patterns are derived by changing the $(N+1)$-th bit from the right from 0 to 1 and repeating the original 2^N patterns in reverse order in the N rightmost positions. Also known as reflected code.

reflected code   *See* reflected binary.

reflected impedance   [ELEC] **1.** Impedance value that ap-

pears to exist across the primary of a transformer due to current flowing in the secondary. **2.** Impedance which appears at the input terminals as a result of the characteristics of the impedance at the output terminals.

reflected pressure   [PHYS] The pressure from an explosion (especially an airburst bomb), which is reflected back from a solid object or surface, rather than dissipated in the air.

reflected ray   [PHYS] A ray extending outward from a point of reflection.

reflected resistance   [ELEC] Resistance value that appears to exist across the primary of a transformer when a resistive load is across the secondary.

reflected signal indicator   [ENG] Pen recorder which presents the radar signals within frequency gates; these recordings enable the operator to determine that an airborne object has penetrated the Doppler link and its direction of penetration.

reflected ultraviolet method   [GRAPHICS] A method of ultraviolet photography in which an ultraviolet source is used and the camera is provided with a filter which permits only reflected ultraviolet light to reach the film.

reflected wave   [PHYS] A wave reflected from a surface, discontinuity, or junction of two different media, such as the sky wave in radio, the echo wave from a target in radar, or the wave that travels back to the source end of a mismatched transmission line.

reflecting antenna   [ELECTROMAG] An antenna used to achieve greater directivity or desired radiation patterns, in which a dipole, slot, or horn radiates toward a larger reflector which shapes the radiated wave to produce the desired pattern; the reflector may consist of one or two plane sheets, a parabolic or paraboloidal sheet, or a paraboloidal horn.

reflecting curtain   [ELECTROMAG] A vertical array of half-wave reflecting antennas, generally used one quarter-wavelength behind a radiating curtain of dipoles to form a high-gain antenna.

reflecting electrode   [ELECTR] Tabular outer electrode or the repeller plate in a microwave oscillator tube, corresponding in construction but not in function to the plate of an ordinary triode; used for generating extremely high frequencies.

reflecting galvanometer   *See* mirror galvanometer.

reflecting grating   [ELECTROMAG] Arrangement of wires placed in a waveguide to reflect one desired wave while allowing one or more other waves to pass freely.

reflecting microscope   [OPTICS] A microscope whose objective is composed of two mirrors, one convex and the other concave; its imaging properties are independent of the wavelength of light, allowing it to be used even for infrared and ultraviolet radiation.

reflecting nephoscope   *See* mirror nephoscope.

reflecting prism   [OPTICS] A prism used in place of a mirror for deviating light, usually designed so that there is no dispersion of light; the light undergoes at least one internal reflection.

reflecting sign   [CIV ENG] A road sign painted with reflective paint so as to be easily visible in the light of a headlamp.

reflecting spectrograph   [OPTICS] A solar spectrograph in which the collimator and camera element are long-focus concave mirrors.

reflecting telescope   [OPTICS] A telescope in which a concave parabolic mirror gathers light and forms a real image of an object. Also known as reflector telescope.

reflection   [PHYS] The return of waves or particles from surfaces on which they are incident.

reflection altimeter   *See* radio altimeter.

reflection angle   *See* angle of reflection.

reflection coefficient   [PHYS] The ratio of the amplitude of a wave reflected from a surface to the amplitude of the incident wave. Also known as coefficient of reflection.

reflection density   [OPTICS] The common logarithm of the ratio of the luminance of a nonabsorbing perfect diffuser to that of the surface under consideration, when both are illuminated at an angle of 45° to the normal and the direction of measurement is perpendicular to the surface.

reflection diffraction   [PHYS] Type of electron diffraction analysis in which the electron beam grazes the sample surface.

REFLECTING PRISM

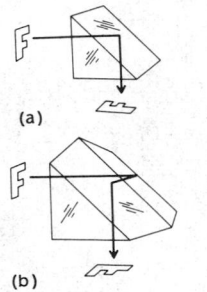

(a)

(b)

Examples of reflecting prism.
(a) Right-angle. *(b)* Amici roof prism.

reflection factor [ELEC] Ratio of the load current that is delivered to a particular load when the impedances are mismatched to that delivered under conditions of matched impedances. Also known as mismatch factor; reflectance; transition factor.

reflection law [PHYS] When a wave, such as electromagnetic radiation or sound, is reflected from a surface in a sharply defined direction, the reflected and incident waves travel in directions that make the same angle with a perpendicular to the surface and lie in a common plane with it. Also known as law of reflection.

reflection lobes [ELECTROMAG] Three-dimensional sections of the radiation pattern of a directional antenna, such as a radar antenna, which results from reflection of radiation from the earth's surface.

reflection loss [ELEC] **1.** Reciprocal of the ratio, expressed in decibels, of the scalar values of the volt-amperes delivered to the load to the volt-amperes that would be delivered to a load of the same impedance as the source. **2.** Apparent transmission loss of a line which results from a portion of the energy being reflected back toward the source due to a discontinuity in the transmission line.

reflection nebula [ASTRON] A type of bright diffuse nebula composed mainly of cosmic dust; it is visible because of starlight from nearby stars or nebula stars that is scattered by the dust particles.

reflection plane *See* plane of mirror symmetry; plane of reflection.

reflection plotter [NAV] An attachment fitted to a radar plan position indicator which provides a plotting surface permitting plotting without parallax errors; any mark made on the plotting surface will be reflected on the radarscope directly below. Also known as plotting head.

reflection rainbow [OPTICS] A rainbow formed by light rays which have been reflected from an extended water surface; not to be confused with a reflected rainbow whose image may be seen in a still body of water.

reflection seismology *See* reflection shooting.

reflection shooting [ENG] A procedure in seismic prospecting based on the measurement of the travel times of waves which, originating from an artificially produced disturbance, have been reflected back to detectors from subsurface boundaries separating media of different elastic-wave velocities; used primarily for oil and gas exploration. Also known as reflection seismology.

reflection survey [ENG] Study of the presence, depth, and configuration of underground formations; a ground-level explosive charge (shot) generates vibratory energy (seismic rays) that strike formation interfaces and are reflected back to ground-level sensors. Also known as seismic survey.

reflection x-ray microscopy [ENG] A technique for producing enlarged images in which a beam of x-rays is successively reflected at grazing incidence, from two crossed cylindrical surfaces; resolution is about 0.5–1 micrometer.

reflective coating *See* mirror coating.

reflective code *See* gray code.

reflective insulation [MATER] An insulating material used to retard the flow of heat by reflecting heat radiation; usually made of aluminum foil or sheets, although coated steel sheets, aluminized paper, gold and silver surfaces, and refractory metals at higher temperatures are also used.

reflectivity [PHYS] The ratio of the energy carried by a wave which is reflected from a surface to the energy carried by the wave which is incident on the surface. Also known as reflectance.

reflectometer [ELECTROMAG] *See* microwave reflectometer. [ENG] A photoelectric instrument for measuring the optical reflectance of a reflecting surface.

reflector [ELECTR] *See* repeller. [ELECTROMAG] **1.** A single rod, system of rods, metal screen, or metal sheet used behind an antenna to increase its directivity. **2.** A metal sheet or screen used as a mirror to change the direction of a microwave radio beam. [NUCLEO] A layer of water, graphite, beryllium, or other scattering material placed around the core of a nuclear reactor to reduce the loss of neutrons. Also known as tamper.

reflector characteristic [ELECTR] A chart of power output

and frequency deviation of a reflex klystron as a function of reflector voltage.

reflector compass [NAV] A magnetic compass in which the image of the compass card is viewed by direct reflection in a mirror adjacent to the helmsman's position.

reflector microphone [ENG ACOUS] A highly directional microphone which has a surface that reflects the rays of impinging sound from a given direction to a common point at which a microphone is located, and the sound waves in the speech-frequency range are in phase at the microphone.

reflector plate [OPTICS] A transparent mirror in a computing gunsight, or in some types of optical gunsights and bombsights, that reflects the reticle image or images to the eye.

reflector satellite [AERO ENG] Satellite so designed that radio or other waves bounce off its surface.

reflector telescope *See* reflecting telescope.

reflector voltage [ELECTR] Voltage between the reflector electrode and the cathode in a reflex klystron.

reflex [PHYSIO] An automatic response mediated by the nervous system.

reflex angle [MATH] An angle greater than 180° and less then 360°.

reflex arc [ANAT] A chain of neurons composing the anatomical substrate or pathway of the unconditioned reflex.

reflex baffle [ENG ACOUS] A loudspeaker baffle in which a portion of the radiation from the rear of the diaphragm is propagated forward after controlled shift of phase or other modification, to increase the overall radiation in some portion of the audio-frequency spectrum. Also known as vented baffle.

reflex bladder [MED] A urinary bladder controlled only by the simple reflex arc through the sacral part of the spinal cord.

reflex bunching [ELECTR] The bunching that occurs in an electron stream which has been made to reverse its direction in the drift space.

reflex camera [OPTICS] A camera in which a mirror is used to reflect a full-size image of a scene on a ground glass so that the composition and focus may be judged.

reflex circuit [ELECTR] A circuit in which the signal is amplified twice by the same amplifier tube or tubes, once as an intermediate-frequency signal before detection and once as an audio-frequency signal after detection.

reflexed [BOT] Turned abruptly backward.

reflex epilepsy [MED] Seizure brought on by specific sensory stimuli.

reflexive relation [MATH] A relation among the elements of a set such that every element stands in that relation to itself.

reflex klystron [ELECTR] A single-cavity klystron in which the electron beam is reflected back through the cavity resonator by a repelling electrode having a negative voltage; used as a microwave oscillator. Also known as reflex oscillator.

reflex oscillator *See* reflex klystron.

reflex printing [GRAPHICS] A method of copying in which a sensitized film or paper is placed face down on the material to be copied; light passes through the base of the sensitized material and is reflected from the light and dark portions of the original back to the emulsion.

reflex sight [OPTICS] An optical or computing sight that reflects a reticle image or images onto a reflector plate for superimposition on the target by the eye.

reflowing [ENG] Melting and resolidifying an electrodeposited or other type coating.

reflux [CHEM ENG] In a chemical process, that part of the product stream that may be returned to the process to assist in giving increased conversion or recovery, as in distillation or liquid-liquid extraction.

reflux condenser [CHEM ENG] An auxiliary vessel for a distillation column that constantly condenses vapors and returns liquid to the column.

reflux ratio [CHEM ENG] The quantity of liquid reflux per unit quantity of product removed from the process unit, such as a distillation tower or extraction column.

refolding [GEOL] A process by which folds of one generation are subjected to and stressed by a force of different orientation.

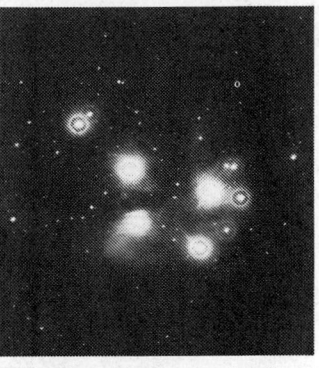
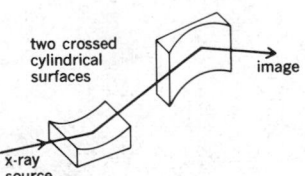
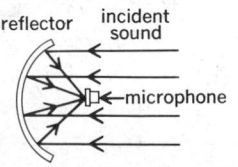

refoliation [GEOL] A foliation that is subsequent to and oriented differently from an earlier foliation.

reforestation [FOR] Establishment of a new forest by seeding or planting seedlings on forest land that fails to restock naturally.

reformate [MATER] Product from a petroleum-refinery reforming process; types are thermal reformate (from thermal reforming), and cat or catalytic reformate (from catalytic reforming).

Reformatsky reaction [ORG CHEM] A condensation-type reaction between ketones and α-bromoaliphatic acids in the presence of zinc or magnesium, such as $R_2CO + BrCH_2 \cdot COOR + Zn \rightarrow (ZnO \cdot HBr) + R_2C(OH)CH_2COOR$.

reformed gas [MATER] A lower-thermal-value fuel gas made by pyrolysis and steam decomposition of high-thermal-value natural and refinery gases.

reformed gasoline [MATER] Gasoline made by a catalytic or thermal reforming process.

reforming [CHEM ENG] The thermal or catalytic conversion of petroleum naphtha into more volatile products of higher octane number; represents the total effect of numerous simultaneous reactions, such as cracking, polymerization, dehydrogenation and isomerization.

refracted ray [PHYS] A ray extending onward from the point of refraction.

refracted wave [PHYS] That portion of an incident wave which travels from one medium into a second medium. Also known as transmitted wave.

refracting sphere [OPTICS] A sphere made of a transparent material whose index of refraction differs from the medium surrounding it, so that it refracts light passing through it.

refracting telescope [OPTICS] A telescope in which a lens gathers light and forms a real image of an object. Also known as refractor telescope.

refraction [PHYS] The change of direction of propagation of any wave, such as an electromagnetic or sound wave, when it passes from one medium to another in which the wave velocity is different, or when there is a spatial variation in a medium's wave velocity.

refraction coefficient [OCEANOGR] The square root of the ratio of the spacing between orthogonals in deep water and in shallow water; it is a measure of the effect of refraction in diminishing wave height by increasing the length of the wave crest.

refraction diagram [OCEANOGR] A chart showing the position of the wave crests at a particular time, or the successive positions of a particular wave crest as it moves shoreward.

refraction error [NAV] An error due to refraction, particularly such an error in a sextant altitude, due to atmospheric refraction.

refraction loss [ELECTROMAG] Portion of the transmission loss that is due to refraction resulting from nonuniformity of the medium.

refraction process [ENG] Seismic (reflection) survey in which the distance between the explosive shot and the receivers (sensors) is large with respect to the depths to be mapped.

refractive index *See* index of refraction.

refractive modulus *See* modified index of refraction.

refractivity [ELECTROMAG] **1.** Some quantitative measure of refraction, usually a measure of the index of refraction. **2.** The index of refraction minus 1.

refractometer [ENG] An instrument used to measure the index of refraction of a substance in any one of several ways, such as measurement of the refraction produced by a prism, measurement of the critical angle, observation of an interference pattern produced by passing light through the substance, and measurement of the substance's dielectric constant.

refractometry [OPTICS] The measurement of the index of refraction of a substance; it is an important tool in analytical chemistry.

refractor telescope *See* refracting telescope.

refractory [MATER] **1.** A material of high melting point. **2.** The property of resisting heat.

refractory cement [MATER] Any of a variety of mixtures, such as fireclay-silica-ganister mixture, or fireclay mixed with crushed brick, or fireclay and silica sand, with a refractory range of 2600–2800°F (1412–1523°C); used for furnace and oven linings and for fillers.

refractory clay [MATER] Clay with a melting point above 1600°C; used to make firebrick and linings for furnaces and reactors.

refractory coating [MATER] A coating composed of a refractory material.

refractory concrete [MATER] Heat-resistant concrete made with high-alumina or calcium-aluminate cement and a refractory aggregate.

refractory enamel [MATER] An enamel for coating and protecting metals against attack by hot gases.

refractory hard metals [CHEM] True chemical compounds composed of two or more metals in the crystalline form, and having a very high melting point and high hardness.

refractory metal [MET] A metal or alloy that is heat-resistant, having a high melting point.

refracture index *See* index of refraction.

refrangible [PHYS] Capable of being refracted.

refrigerant [MATER] A substance that by undergoing a change in phase (liquid to gas, gas to liquid) releases or absorbs a large latent heat in relation to its volume, and thus effects a considerable cooling effect; examples are ammonia, sulfur dioxide, ethyl or methyl chloride (these are no longer widely used), and the fluorocarbons, such as Freon, Ucon, and Genetron.

refrigerant 23 *See* fluoroform.

refrigerated truck [MECH ENG] An insulated truck equipped and used as a refrigerator to transport fresh perishable or frozen products.

refrigeration [MECH ENG] The cooling of a space or substance below the environmental temperature.

refrigeration condenser [MECH ENG] A vapor condenser in a refrigeration system, where the refrigerant is liquefied and discharges its heat to the environment.

refrigeration cycle [THERMO] A sequence of thermodynamic processes whereby heat is withdrawn from a cold body and expelled to a hot body.

refrigeration oil [MATER] A mineral oil with all moisture and wax removed; used for lubricating refrigerating machinery.

refrigerator [MECH ENG] An insulated, cooled compartment.

refrigerator car [MECH ENG] An insulated freight car constructed and used as a refrigerator.

Refsum's disease [MED] A familial disorder characterized by visual disturbances, ataxia, neuritic changes, and cardiac damage, associated with high blood level of phytinic acid.

regelation [HYD] Phenomenon in which ice melts at the bottom of droplets of highly concentrated saline solution that are trapped in ice which has frozen over polar waters, and freezes at the top of these droplets, so that the droplets move downward through the ice, leaving it hard and clear. [THERMO] Phenomenon in which ice (or any substance which expands upon freezing) melts under intense pressure and freezes again when this pressure is removed; accounts for phenomena such as the slippery nature of ice and the motion of glaciers.

regenerate [CHEM ENG] To clean of impurities and make reusable as in regeneration of a catalytic cracking catalyst by burning off carbon residue, regeneration of clay adsorbent by washing free of adherents, or regeneration of a filtration system by cleaning off the filter media. [ELECTR] **1.** To restore pulses to their original shape. **2.** To restore stored information to its original form in a storage tube in order to counteract fading and disturbances.

regenerated cellulose [MATER] **1.** Rayon in which the raw cellulose is changed physically but not chemically, such as viscose, cuprammonium, and nitrocellulose rayons. **2.** A transparent cellulose plastic material made by mixing cellulose xanthate with a dilute sodium hydroxide solution to form a viscose.

regenerated fiber [TEXT] Textile fiber produced by dissolving a natural material (such as cellulose), then regenerating it by extrusion and precipitation, as with viscose.

regenerated glacier [HYD] A glacier that becomes active after a period of stagnation.

regeneration [BIOL] The replacement by an organism of tissues or organs which have been lost or severely injured.

[CONT SYS] *See* positive feedback. [ELECTR] Replacement or restoration of charges in a charge storage tube to overcome decay effects, including loss of charge by reading. [NUCLEO] Restoration of contaminated nuclear fuel to a usable condition.

regenerative air heater [MECH ENG] An air heater in which the heat-transferring members are alternately exposed to heat-surrendering gases and to air.

regenerative amplifier [ELECTR] An amplifier that uses positive feedback to give increased gain and selectivity.

regenerative braking [ELEC] A system of dynamic braking in which the electric drive motors are used as generators and return the kinetic energy of the motor armature and load to the electric supply system.

regenerative chaff [ORD] An electronic countermeasure technique in which a missile or powered decoy sequentially ejects clouds of chaff.

regenerative clipper [ELECTR] A type of monostable multivibrator which is a modification of a Schmitt trigger; used for pulse generation.

regenerative cycle [MECH ENG] *See* bleeding cycle. [THERMO] An engine cycle in which low-grade heat that would ordinarily be lost is used to improve the cyclic efficiency.

regenerative detector [ELECTR] A vacuum-tube detector circuit in which radio-frequency energy is fed back from the anode circuit to the grid circuit to give positive feedback at the carrier frequency, thereby increasing the amplification and sensitivity of the circuit.

regenerative divider [ELECTR] Frequency divider which employs modulation, amplification, and selective feedback to produce the output wave.

regenerative engine [AERO ENG] 1. A jet or rocket engine that utilizes the heat of combustion to preheat air or fuel entering the combustion chamber. 2. Specifically applied to a type of rocket engine in which one of the propellants is used to cool the engine by passing through a jacket prior to combustion.

regenerative feedback *See* positive feedback.

regenerative fuel cell [ELEC] A fuel cell in which the reaction product is processed to regenerate the reactants.

regenerative pump [MECH ENG] Rotating-vane device that uses a combination of mechanical impulse and centrifugal force to produce high liquid heads at low volumes. Also known as turbine pump.

regenerative reactor [NUCLEO] A nuclear reactor that produces fissionable material in addition to energy; when loaded with fissionable uranium-235, and nonfissionable uranium-238 or thorium, it converts the uranium-238 or thorium into fissionable materials which are then used as fuel in the core of the reactor.

regenerative receiver [ELECTR] A radio receiver that uses a regenerative detector.

regenerative repeater [COMMUN] A repeater that performs pulse regeneration to restore the original shape of a pulse signal used in teletypewriter and other code circuits.

regenerative track [ADP] Track on a magnetic drum with interconnected reading and writing heads; information stored on these tracks is continuously read from the drum, transmitted round a closed circuit, and recorded back on the drum; consequently, access times to these data are short; operation is analogous to that of the acoustic delay line.

regenerator [CHEM ENG] Device or system used to return a system or a component of it to full strength in a chemical process; examples are a furnace to burn carbon from a catalyst, a tower to wash impurities from clay, and a flush system to clean off the surface of filter media. [MECH ENG] A device used with hot-air engines and gas-burning furnaces which transfers heat from effluent gases to incoming air or gas.

Reggeism [PARTIC PHYS] An attempt to account for and correlate hadron resonances and the asymptotic behavior of scattering amplitudes of hadrons at high energies in terms of Regge poles.

Regge pole [PARTIC PHYS] A pole singularity of a scattering amplitude in the complex angular momentum plane; the scattering amplitude is formed by continuing partial wave amplitudes from positive integer values of the angular momentum to the complex plane.

Regge recurrence [PARTIC PHYS] One of a sequence of hadrons, with successive hadrons increasing by one in spin and also increasing in mass, but with the same values of other quantum numbers, except for parity, charge parity and G parity, which alternate in sign; it is believed that they are rotationally excited states of a particle, and that they alternate between two Regge trajectories.

Regge trajectory [PARTIC PHYS] 1. The path followed by a Regge pole in the complex angular momentum plane as the center-of-mass energy is varied. 2. The relationship between the spin and mass of a sequence of hadrons, with successive hadrons increasing by 2 in spin and also increasing in mass, but with the same values of other quantum numbers; the hadrons are thought to correspond to energies at which a Regge pole passes near positive integers (or half integers).

regime [GEOL] The existence in a stream channel of a balance between erosion and deposition over a period of years.

regimen [HYD] 1. The behavior characteristic of the total amount of water involved in a drainage basin. 2. Analysis of the total volume of water involved with a lake, including water losses and gains, over a period of a year. 3. The flow characteristics of a stream with respect to velocity, volume, form of and alterations in the channel, capacity to transport sediment, and the amount of material supplied for transportation.

region [ADP] A group of machine addresses which refer to a base address. [MATH] *See* domain.

regional address [ADP] An address of a machine instruction within a series of consecutive addresses; for example, R18 and R19 are specific addresses in an R region of N consecutive addresses, where all addresses must be named.

regional anatomy [ANAT] The detailed study of the anatomy of a part or region of the body of an animal.

regional anesthesia *See* regional block anesthesia.

regional block anesthesia [MED] Anesthesia limited to a part or region of the body by injecting an anesthetic around the nerves that supply the area to block conduction from the area. Also known as regional anesthesia.

regional center [COMMUN] A long-distance telephone office which has the highest rank in routing of telephone calls; there are 12 such offices in the United States and Canada.

regional dip [GEOL] The nearly uniform and generally low-angle inclination of strata over a wide area. Also known as normal dip.

regional enteritis [MED] Inflammation of isolated segments of the small intestine, with involved parts becoming thick-walled and rigid. Also known as regional enterocolitis; regional ileitis.

regional enterocolitis *See* regional enteritis.

regional forecast *See* area forecast.

regional geology [GEOL] The geology of a large region, treated from the viewpoint of the spatial distribution and position of stratigraphic units, structural features, and surface forms.

regional ileitis *See* regional enteritis.

regional metamorphism [GEOL] Geological metamorphism affecting an extensive area.

regional migration [PETRO ENG] Horizontal movement of gas or oil through a reservoir formation as a result of artificial pressure differences created by withdrawal of gas or oil at well sites.

regional slope [GEOL] The generally uniform dip of rock strata or land surface over a wide area.

regional slope deposit [GEOL] A sedimentary deposit widely distributed as a thin sheet over a regional slope.

regional unconformity [GEOL] A continuous unconformity extending throughout a wide region that may be nearly continentwide, and usually represents a long period of time.

region of escape *See* exosphere.

register [ADP] The computer hardware for storing one machine word. [COMMUN] Part of an automatic switching telephone system which receives and stores the dialing pulses which control the further operations necessary in establishing a telephone connection. [ENG] Also known as registration. 1. The accurate matching or superimposition of two or more images, such as the three color images on the screen of a color television receiver, or the patterns on opposite sides of a printed circuit board, or the colors of a design on a printed sheet. **2.** The alignment of positions relative to a specified reference or coordinate, such as hole alignments in punched cards, or positioning of images in an optical character recognition device. [GRAPHICS] **1.** Exact agreement in the position of printed material on both sides of a sheet or on all pages of a book or pamphlet. **2.** Exact overprinting of colorplates, or other subsequent plates, so that all printed detail is correctly combined; proper color overprinting is checked by the exact superimposition of the register marks that are printed with each color run. **3.** In flat preparation, the exact agreement between color or complementary flats. [ORD] **1.** To adjust fire on a visible point, called a check point, and compute accurate adjusted data so that firing data for later targets may be computed with reference to that check point. **2.** To adjust fire on several selected points in order that they may serve later as auxiliary targets.

register breadth [NAV ARCH] The breadth of beam dimension of a vessel; specifically, the breadth of the broadest part on the outside of the vessel.

register capacity [ADP] The upper and lower limits of the numbers which may be processed in a register.

register circuit [ELECTR] A switching circuit with memory elements that can store from a few to millions of bits of coded information; when needed, the information can be taken from the circuit in the same code as the input, or in a different code.

register control [CONT SYS] Automatic control of the position of a printed design with respect to reference marks or some other part of the design, as in photoelectric register control.

register glass [GRAPHICS] In photography, a glass plate at the focal plane, against which the film is pressed during exposure.

register length [ADP] The number of digits, characters, or bits, which a register can store.

register mark [ENG] A mark or line printed or otherwise impressed on a web of material for use as a reference to maintain register. [GRAPHICS] One of a set of marks, usually cross-shaped, placed on the margin of colored originals to be photographed for colorplates; the marks are used to register the images when the plates are printed.

register-sender [COMMUN] A unit that generates and recognizes the supervisory signals to make connection to a circuit switching unit.

register ton *See* ton.

register tonnage [NAV ARCH] The available gross tonnage of a ship for transport of freight or passengers.

registration *See* register.

registration fire [ORD] That artillery fire delivered to obtain corrections for increasing the accuracy of subsequent artillery.

registration mark [ADP] In character recognition, a preprinted indication of the relative position and direction of various elements of the source document to be recognized.

regmagenesis [GEOL] Diastrophic production of regional strike-slip displacements.

regolith [GEOL] The layer rock or blanket of unconsolidated rocky debris of any thickness that overlies bedrock and forms the surface of the land. Also known as mantle rock.

Regosol [GEOL] In early United States soil classification systems, one of an azonal group of soils that form from deep, unconsolidated deposits and have no definite genetic horizons.

regression [OCEANOGR] Retreat of the sea from land areas, and the consequent evidence of such withdrawal. [PSYCH] A mental state and a mode of adjustment to difficult and unpleasant situations, characterized by behavior of a type that had been satisfying and appropriate at an earlier stage of development but which no longer befits the age and social status of the individual. [STAT] Given two stochastically dependent random variables, regression functions measure the mean expectation of one relative to the other.

regression of nodes [ASTRON] The westward movement of the nodes of the moon's orbit; one cycle is completed in about 18.6 years.

REGOLITH

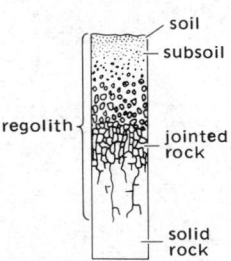

Cross section through the regolith showing the components.

REGULAR LAY

Drawing of wire rope wound in regular lay showing position of the wires and strands.

regret criterion *See* Savage principle.

regula falsi [MATH] A method of calculating an unknown quantity by first making an estimate and then using this and the properties of the unknown to obtain it.

regular [BOT] Having radial symmetry, referring to a flower. [ELECTROMAG] In a definite direction; not diffused or scattered, when applied to reflection, refraction, or transmission.

regular connective tissue [HISTOL] Connective tissue in which the fibers are arranged in definite patterns.

regular dodecahedron [CRYSTAL] *See* pyritohedron. [MATH] A regular polyhedron of twelve faces.

regular element [IND ENG] An element that occurs with a fixed frequency in each work cycle. Also known as repetitive element.

regular function [MATH] An analytic function of one or more complex variables.

Regularia [INV ZOO] An assemblage of echinoids in which the anus and periproct lie within the apical system; not considered a valid taxon.

regular icosohedron [MATH] A 20-sided regular polyhedron, having five equilateral triangles meeting at each face.

regularization [QUANT MECH] A formal procedure used to eliminate ambiguities which arise in evaluating certain integrals in a quantized field theory; corresponds to adding extra fields whose masses are allowed to approach infinity.

regular lay [DES ENG] The lay of a wire rope in which the wires in the strand are twisted in directions opposite to the direction of the strands.

regular-lay left twist *See* left-laid.

regular motor oil [MATER] A petroleum lubricating oil suitable for use in internal combustion engines under normal operating conditions.

regular octahedron [MATH] A regular polyhedron of eight faces.

regular polygon [MATH] A polygon with congruent sides and congruent interior angles.

regular polyhedron [MATH] A polyhedron all of whose faces are regular polygons, and whose polyhedral angles are congruent.

regular reflection *See* specular reflection.

regular representation [MATH] A regular representation of a finite group is an isomorphism of it with a group of permutations.

regular sampling [MIN ENG] The continuous or intermittent sampling of the same coal or coke received regularly at a given point.

regular tetrahedron [MATH] A regular polyhedron of four faces.

regular topological space [MATH] A topological space where any point and a closed set not containing it can be enclosed in disjoint open sets.

regular ventilating circuit [MIN ENG] All places in the mine through which there is a positive natural flow of air.

regulated item [ORD] An item requiring special control in handling distribution and use, because of its highly technical, costly, or hazardous nature, or because of some other specified concern.

regulated power supply [ELEC] A power supply containing means for maintaining essentially constant output voltage or output current under changing load conditions.

regulated split [MIN ENG] In mine ventilation, a split where it is necessary to control the volumes in certain low-resistance splits to cause air to flow into the splits of high resistance.

regulating reservoir [CIV ENG] A reservoir that regulates the flow in a water-distributing system.

regulating rod [NUCLEO] A control rod intended to accomplish rapid, fine, and sometimes continuous adjustment of the reactivity of a nuclear reactor; it usually can move much more rapidly than a shim rod but makes a smaller change in reactivity.

regulating station [ORD] A command agency established to control all movements of personnel and supplies into or out of a given area.

regulating system *See* automatic control system.

regulating transformer [ELEC] Transformer having one or more windings excited from the system circuit or a separate source and one or more windings connected in series with the system circuit for adjusting the voltage or the phase relation or both in steps, usually without interrupting the load.

regulating winding [ELEC] Of a transformer, a supplementary winding connected in series with one of the main windings to change the ratio of transformation or the phase relation, or both, between circuits.

regulation [CONT SYS] The process of holding constant a quantity such as speed, temperature, voltage, or position by means of an electronic or other system that automatically corrects errors by feeding back into the system the condition being regulated; regulation thus is based on feedback, whereas control is not. [ELEC] The change in output voltage that occurs between no load and full load in a transformer, generator, or other source. [ELECTR] The difference between the maximum and minimum tube voltage drops within a specified range of anode current in a gas tube.

regulation of constant-current transformer [ELEC] Maximum departure of the secondary current from its rated value expressed in percent of the rated secondary current, with rated primary voltage and frequency applied.

regulative egg [EMBRYO] An egg in which unfertilized fragments develop as complete, normal individuals.

regulator [CONT SYS] A device that maintains a desired quantity at a predetermined value or varies it according to a predetermined plan. [MIN ENG] An opening in a wall or door in the return airway of a district to increase its resistance and reduce the volume of air flowing.

regulator gene [GEN] A gene that controls the rate of transcription of one or more other genes.

regulator pin [HOROL] A pin in a watch to adjust the effective length of the hairspring and thereby adjust the speed of the watch.

regulus [MET] Impure metal formed beneath the slag during smelting or reduction of ores.

Regulus [ASTRON] A star of stellar magnitude 1.34, about 67 light-years from the sun, spectral classification B8, in the constellation Leo; the star α Leonis. Also known as Little Ruler. [ORD] A United States surface-to-surface, jet-powered guided missile; it is equipped with a nuclear warhead, and launched from a surfaced submarine or a cruiser.

regurgitation [MED] Reverse circulation of blood in the heart due to defective functioning of the valves. [PHYSIO] Bringing back into the mouth undigested food from the stomach.

rehabilitation [MED] The restoration to a disabled individual of maximum independence commensurate with his limitations by developing his residual capacity.

Rehbock weir formula [FL MECH] Probably the most accurate formula for the rate of flow of water over a rectangular suppressed weir; it includes a correction for the velocity of approach for normal, or fairly uniform, velocity distribution in the upstream channel; the formula is $Q = [3.234 + 5.347/(320h - 3) + 0.428h/d_0]lh^{3/2}$, where Q is the flow rate in cubic feet per second, l is the width of the weir in feet, h is the head of water above the crest of the weir in feet, and d_0 is the height of weir or depth of water at zero head in feet.

reheating [THERMO] A process in which the gas or steam is reheated after a partial isentropic expansion to reduce moisture content. Also known as resuperheating.

Rehfuss tube [MED] A stomach tube designed for the removal of specimens of gastric contents for analysis after administration of a test meal.

Reichert-Meissl number [ANALY CHEM] An indicator of the measure of volatile soluble fatty acids.

Reichert's cartilage [EMBRYO] The cartilage of the hyoid arch in a human embryo.

Reich process [CHEM ENG] Process to purify carbon dioxide produced during fermentation; organic impurities in the gas are oxidized and absorbed, then the gas is dehydrated.

Reid equation [PETRO ENG] Relation of gas-well flow rate to pitot-tube readings for various impact pressures.

Reid vapor pressure [ENG] A measure in a test bomb of the vapor pressure in pounds pressure of a sample of gasoline at 100°F (37.8°C).

Reighardiidae [INV ZOO] A monotypic family of arthropods in the order Cephalobaenida; the posterior end of the organism is rounded, without lobes, and the cuticula is covered with minute spines.

reimbursed time [ADP] The machine time which is loaned or rented to another office, agency, or organization, either on a reimbursable or reciprocal basis.

Reimer-Tiemann reaction [ORG CHEM] Formation of phenolic aldehydes by reaction of phenol with chloroform in the presence of an alkali.

Reinartz crystal oscillator [ELECTR] Crystal-controlled vacuum-tube oscillator in which the crystal current is kept low by placing a resonant circuit in the cathode lead tuned to half the crystal frequency; the resulting regeneration at the crystal frequency improves efficiency without the danger of uncontrollable oscillation at other frequencies.

reindeer [VERT ZOO] *Rangifer tarandus*. A migratory ruminant of the deer family (Cervidae) which inhabits the Arctic region and has a circumpolar distribution; characteristically, both sexes have antlers and are brown with yellow-white areas on the neck and chest.

Reinecke's salt [ANALY CHEM] $[(NH_3)_2Cr(SCN)_4]NH_4 \cdot H_2O$ A reagent to detect mercury (gives a red color or a precipitate), and to isolate organic bases (such as proline or histidine).

reinfection [MED] A second infection after recovery from an earlier infection with the same kind of organism.

reinforced beam [CIV ENG] A concrete beam provided with steel bars for longitudinal tension reinforcement and sometimes compression reinforcement and reinforcement against diagonal tension.

reinforced brickwork [CIV ENG] Brickwork strengthened by expanded metal, steel-wire mesh, hoop iron, or thin rods embedded in the bed joints.

reinforced column [CIV ENG] **1.** A long concrete column reinforced with longitudinal bars with ties or circular spirals. **2.** A composite column. **3.** A combination column.

reinforced concrete [CIV ENG] Concrete containing reinforcing steel rods or wire mesh.

reinforced molding compound [MATER] A compound containing polymer or resin and a reinforcing filler, supplied in the form of ready-to-use material as distinguished from premix.

reinforced plastic [MATER] High-strength filled plastic product used for mechanical, construction, and electrical products, automotive components, and ablative coatings; filling can be whiskers of glass, metal, boron, or other materials.

reinforcement [CIV ENG] Strengthening concrete, plaster, or mortar by embedding steel rods or wire mesh in it.

reinforcement conditioning See operant conditioning.

reinforcement of weld [MET] Weld metal that extends beyond the surface or plane of the weld joint.

reinforcing bars [CIV ENG] Steel rods that are embedded in building materials such as concrete for reinforcement.

Reinsch test [ANALY CHEM] A test for detecting small amounts of arsenic, silver, bismuth, and mercury.

reinserter See direct-current restorer.

reinsertion of carrier [ELECTR] Combining a locally generated carrier signal in a receiver with an incoming signal of the suppressed carrier type.

Reissner's membrane [ANAT] The anterior wall of the cochlear duct, which separates the cochlear duct from the scala vestibuli. Also known as vestibular membrane of Reissner.

Reiter's syndrome [MED] The triad of idiopathic nongonococcal urethritis, conjunctivitis, and subacute or chronic polyarthritis. Also known as arthritis urethritica; idiopathic arthritis; infectious uroarthritis.

rejection band Also known as stop band. [ELECTROMAG] The band of frequencies below the cutoff frequency in a uniconductor waveguide. [PHYS] A frequency band within which electrical or electromagnetic signals are reduced or eliminated.

rejector See trap.

rejector circuit See band-stop filter.

rejector impedance See dynamic impedance.

rejuvenate [GEOL] The act of stimulating a stream to renewed erosive activity either by tectonic uplift or a drop in sea level.

rejuvenated fault scarp [GEOL] A fault scarp revived by renewed movement along an old fault line after partial

REINDEER

Reindeer do not display sexual dimorphism; males are larger than females, and reach a length of about 6 feet (1.8 meters) and a shoulder height of about 4 feet (1.2 meters).

dissection or erosion of the initial scarp. Also known as revived fault scarp.

rejuvenated stream [HYD] A mature stream that has reverted to the behavior and forms of a more youthful stage due to rejuvenation, usually as a result of uplift. Also known as revived stream.

rejuvenation [GEOL] **1.** The development or restoration of youthful features of a landform or landscape in an area previously worn down to base level. **2.** Renewal of a geologic process. [HYD] **1.** The stimulation of a stream to renew erosive activity. **2.** The renewal of youthful vigor in a mature stream.

rejuvenation head See knickpoint.

rel [ELECTROMAG] Unit of reluctance equal to 1 ampere-turn per magnetic line of force.

relapsing fever [MED] An acute infectious disease caused by various species of the spirochete *Borrelia,* characterized by episodes of fever which subside spontaneously and recur over a period of weeks.

relation [MATH] A set of ordered pairs.

relative [NAV] **1.** Related to a moving point; apparent, as relative wind, relative movement. **2.** Related to or measured from the heading, as relative bearing.

relative address [ADP] The numerical difference between a desired address and a known reference address.

relative attenuation [ELECTR] The ratio of the peak output voltage of an electric filter to the voltage at the frequency being considered.

relative azimuth [NAV] Azimuth relative to heading of a craft.

relative bandwidth [ELECTR] For an electric filter, the ratio of the bandwidth being considered to a specified reference bandwidth, such as the bandwidth between frequencies at which there is an attenuation of 3 decibels.

relative bearing [NAV] Bearing relative to heading or to the craft; the horizontal direction of a terrestrial point from a craft expressed as the angular difference between the heading and the direction.

relative biological effectiveness [NUCLEO] A factor used to compare the biological effectiveness of different types of ionizing radiation; it is the inverse ratio of the amount of absorbed radiation required to produce a given effect to a standard (or reference) radiation required to produce the same effect. Abbreviated RBE.

relative coding [ADP] A form of computer programming in which the address part of an instruction indicates not the desired address but the difference between the location of the instruction and the desired address.

relative compaction [ENG] The percentage ratio of the field density of soil to the maximum density as determined by standard compaction.

relative contour See thickness line.

relative coordinate system [PHYS] Any coordinate system which is moving with respect to an inertial coordinate system.

relative current [OCEANOGR] The current which is a function of the dynamic slope of an isobaric surface and which is determined from an assumed layer of no motion.

relative damping ratio See damping ratio.

relative deflection See astrogeodetic deflection.

relative dielectric constant See dielectric constant.

relative direction [NAV] Horizontal direction expressed as angular distance from heading.

relative distance [NAV] Distance relative to a specified reference point, usually one in motion.

relative divergence See development index.

relative erythrocytosis [MED] A form of erythrocytosis that occurs when the concentration of erythrocytes in the circulating blood increases through loss of plasma.

relative force [ENG] Ratio of the force of a test propellant to the force of a standard propellant, measured at the same initial temperature and loading density in the same closed chamber.

relative frequency [STAT] The ratio of the number of occurrences of a given type of event or the number of members of a population in a given class to the total number of events or the total number of members of the population.

relative gain of an antenna [ELECTROMAG] Gain of an an-

tenna in a given direction when the reference antenna is a half-wave, loss-free dipole isolated in space whose equatorial plane contains the given direction.

relative geologic time [GEOL] Nonabsolute geological time in which events may be placed relatively to one another.

relative humidity [METEOROL] The (dimensionless) ratio of the actual vapor pressure of the air to the saturation vapor pressure. Abbreviated RH.

relative hypsography See thickness pattern.

relative interference effect [ENG ACOUS] Of a single-frequency electric wave in an electroacoustic system, the ratio, usually expressed in decibels, of the amplitude of a wave of specified reference frequency to that of the wave in question when the two waves are equal in interference effects.

relative ionospheric opacity meter See riometer.

relative isohypse See thickness line.

relative luminosity factor See luminosity function.

relatively closed set [MATH] A subset of a topological space is relatively closed if it is a closed set in some relative topology of a subset.

relatively compact set See conditionally compact set.

relatively open set [MATH] A subset of a topological space is relatively open if it is an open set in some relative topology of a subset.

relatively prime [MATH] Integers m and n are relatively prime if there are integers p and q so that $pm + qn = 1$; equivalently, if they have no common factors other than 1.

relative mach number See Mach number.

relative magnetometer [ENG] Any magnetometer which must be calibrated by measuring the intensity of a field whose strength is accurately determined by other means; opposed to absolute magnetometer.

relative momentum [MECH] The momentum of a body in a reference frame in which another specified body is fixed.

relative motion [MECH] The continuous change of position of a body with respect to a second body, that is, in a reference frame where the second body is fixed.

relative movement line [NAV] A line connecting successive positions of a maneuvering craft relative to a reference craft.

relative permeability [GEOL] Specific permeability of a porous rock formation to a particular phase (oil, water, gas) at a particular saturation and a particular saturation distribution; for example, ratio of effective permeability to a specified phase to the rock's absolute permeability.

relative permittivity See dielectric constant.

relative plot [NAV] A plot of the successive positions of a craft relative to a reference point, which is usually in motion.

relative position [NAV] A point defined with reference to another position, either fixed or moving; the coordinates of such a point are usually bearing, true or relative, and distance from an identified reference point.

relative power gain [ELECTROMAG] Of one transmitting or receiving antenna over another, the measured ratio of the signal power one produces at the receiver input terminals to that produced by the other, the transmitting power level remaining fixed.

relative quickness [ORD] Ratio of the quickness of a test propellant to the quickness of a standard propellant, measured at the same initial temperature and loading density in the same closed chamber.

relative resistance [ELEC] The ratio of the resistance of a piece of a material to the resistance of a piece of specified material, such as annealed copper, having the same dimensions and temperature.

relative roughness factor [FL MECH] Roughness of pipe-wall interior (distance from peaks to valleys) divided by pipe internal diameter; used to modify Reynolds number calculations for fluid flow through pipes.

relative scatter intensity [OPTICS] For scattering of radiation under any given set of physical conditions, the ratio of the radiant intensity scattered in any given direction to the radiant intensity scattered in the direction of the incident beam.

relative search [NAV] A search in which the area to be searched is defined relative to a point which is moving over the surface of the earth.

relative sector search [NAV] A search of three legs, the first and third being on courses such as to produce relative movement along radial lines of equal length from a moving point, and the middle leg connecting the other two.

relative square search [NAV] Search by a series of course lines such as to produce relative movement lines of increasing length, each change of course being such as to produce a change of 90° in the direction of relative movement, all changes being in the same direction (right or left), so that the pattern of the search is an expanding square relative to a moving point.

relative stability test [ANALY CHEM] A color test using methylene blue that indicates when the oxygen present in a sewage plant's effluent or polluted water is exhausted.

relative sunspot number [ASTRON] A measure of sunspot activity, computed from the formula $R = k (10g + f)$, where R is the relative sunspot number, f the number of individual spots, g the number of groups of spots, and k a factor that varies with the observer (his personal equation), the seeing, and the observatory (location and instrumentation). Also known as sunspot number; sunspot relative number; Wolf number; Wolf-Wolfer number; Zurich number.

relative topography See thickness pattern.

relative topology [MATH] In a topological space X any subset A has a topology on it relative to the given one by intersecting the open sets of X with A to obtain open sets in A.

relative triple precision [ADP] The retention of three times as many digits of a quantity as the computer normally handles; for example, a computer whose basic word consists of 10 decimal digits is called upon to handle 30 decimal digit quantities.

relative volatility [CHEM] The volatility of a standard material whose relative volatility is by definition equal to unity.

relative wind [NAV] The speed and relative direction from which the wind appears to blow with reference to a moving point. Also known as apparent wind.

relativistic beam [RELAT] A beam of particles traveling at a speed comparable with the speed of light.

relativistic electrodynamics [ELECTROMAG] The study of the interaction between charged particles and electric and magnetic fields when the velocities of the particles are comparable with that of light.

relativistic kinematics [RELAT] A description of the motion of particles compatible with the special theory of relativity, without reference to the causes of motion.

relativistic mass [RELAT] The mass of a particle moving at a velocity exceeding about one-tenth the velocity of light; it is significantly larger than the rest mass.

relativistic mechanics [RELAT] **1.** Any form of mechanics compatible with either the special or the general theory of relativity. **2.** The nonquantum mechanics of a system of particles or of a fluid interacting with an electromagnetic field, in the case when some of the velocities are comparable with the speed of light.

relativistic particle [RELAT] A particle moving at a speed comparable with the speed of light.

relativistic quantum theory [QUANT MECH] The quantum theory of particles which is consistent with the special theory of relativity, and thus can describe particles moving arbitrarily close to the speed of light.

relativistic theory [PHYS] Any theory which is consistent with the special or general theory of relativity.

relativity [PHYS] Theory of physics which recognizes the universal character of the propagation speed of light and the consequent dependence of space, time, and other mechanical measurements on the motion of the observer performing the measurements; it has two main divisions, the special theory and the general theory.

relaxation [GEOL] In experimental structural geology, the diminution of applied stress with time, as the result of any of various creep processes. [MATH] See relaxation method. [MECH] **1.** Relief of stress in a strained material due to creep. **2.** The lessening of elastic resistance in an elastic medium under an applied stress resulting in permanent deformation. [PHYS] A process in which a physical system approaches a steady state after conditions affecting it have been suddenly changed, and in which the presence of dissipative agents

prevents the system from overshooting and then oscillating about this state.

relaxation circuit [ELECTR] Circuit arrangement, usually of vacuum tubes, reactances, and resistances, which has two states or conditions, one, both, or neither of which may be stable; the transient voltage produced by passing from one to the other, or the voltage in a state of rest, can be used in other circuits.

relaxation inverter [ELECTR] An inverter that uses a relaxation oscillator circuit to convert direct-current power to alternating-current.

relaxation method [MATH] A successive approximation method for solving systems of equations where the errors from an initial approximation are viewed as constraints to be minimized or relaxed within a toleration limit. Also known as relaxation.

relaxation oscillations [PHYS] Oscillations having a sawtooth waveform in which the displacement increases to a certain value and then drops back to zero, after which the cycle is repeated.

relaxation oscillator [ELECTR] An oscillator whose fundamental frequency is determined by the time of charging or discharging a capacitor or coil through a resistor, producing waveforms that may be rectangular or sawtooth.

relaxation test [ENG] A creep test in which the decrease of stress with time is measured while the total strain (elastic and plastic) is maintained constant.

relaxation time [PHYS] For many physical systems undergoing relaxation, a time τ such that the displacement of a quantity from its equilibrium value at any instant of time t is the exponential of $-t/\tau$. [SOLID STATE] The travel time of an electron in a metal before it is scattered and loses its momentum.

relaxed peak process See quasi-fission.

relaxin [BIOCHEM] A hormone found in the serum of humans and certain other animals during pregnancy; probably acting with progesterone and estrogen, it causes relaxation of pelvic ligaments in the guinea pig.

relay [COMMUN] A microwave or other radio system used for passing on a signal from one radio communication link to another. [ELEC] A device that is operated by a variation in the conditions in one electric circuit and serves to make or break one or more connections in the same or another electric circuit. Also known as electric relay.

relay contact [ELEC] One of the pair of contacts that are closed or opened by the movement of the armature of a relay.

relay control system [CONT SYS] A control system in which the error signal must reach a certain value before the controller reacts to it, so that the control action is discontinuous in amplitude.

relay haulage [MIN ENG] Single-track, high-speed mine haulage from one relay station to another. Also known as intermediate haulage.

relay rack [DES ENG] A standardized steel rack designed to hold 19-inch (48.26-centimeter) panels of various heights, on which are mounted radio receivers, amplifiers, and other units of electronic equipment. Also known as rack.

relay satellite See communications satellite.

relay selector [ELEC] Relay circuit associated with a selector, consisting of a magnetic impulse counter, for registering digits and holding a circuit.

relay station See repeater station.

relay system [COMMUN] See radio relay system. [ELEC] Dial switching equipment that does not use mechanical switches, but is made up principally of relays.

release [MECH ENG] A mechanical arrangment of parts for holding or freeing a device or mechanism as required.

release agent [MATER] A lubricant, such as wax or silicone oil, used to coat a mold cavity to prevent the molded piece from sticking when removed. Also known as mold release; parting agent.

release altitude [AERO ENG] Altitude of an aircraft above the ground at the time of release of bombs, rockets, missiles, tow targets, and so forth.

released mineral [MINERAL] A mineral formed during the crystallization of a magma due to failure of an earlier phase to react with the liquid portion of the magma.

RELAY

A multicontrol telphone-type auxiliary relay. When applied current or voltage in electric coil exceeds a threshold value, the coil activates an armature which opens or closes contacts. (C. P. Clare and Co.)

release fracture [GEOL] A fracture formed as a result of a decrease in the maximum principal stress.

release joint *See* sheeting structure.

reliability [ENG] The probability that a component part, equipment, or system will satisfactorily perform its intended function under given circumstances, such as environmental conditions, limitations as to operating time, and frequency and thoroughness of maintenance for a specified period of time. [STAT] **1.** The amount of credence placed in a result. **2.** The precision of a measurement, as measured by the variance of repeated measurements of the same object.

relic [GEOL] **1.** A landform that remains intact after decay or disintegration or that remains after the disappearance of the major portion of its substance. **2.** A vestige of a particle in a sedimentary rock, such as a trace of a fossil fragment.

relict [BIOL] A persistent, isolated remnant of a once-abundant species.

relict dike [GEOL] In a granitized mass, a tabular, crystalloblastic body that represents a dike which was emplaced prior to, and which was relatively resistant to, the granitization process.

relict mineral [MINERAL] A mineral of a rock that persists from an earlier rock.

relict permafrost [GEOL] Permafrost formed in the past which persists in areas where it would not form today.

relict sediment [GEOL] A sediment which was in equilibrium with its environment when first deposited but which is unrelated to its present environment even though it is not buried by later sediments, such as a shallow-marine sediment on the deep ocean floor.

relict soil [GEOL] A soil formed on a preexisting landscape but not subsequently buried under younger sediments.

relict texture [GEOL] In mineral deposits, an original texture that persists after partial replacement.

relief [CRYSTAL] The apparent topography exhibited by minerals in thin section as a consequence of refractive index. [GEOD] The configuration of a part of the earth's surface, with reference to altitude and slope variations and to irregularities of the land surface. [MECH ENG] **1.** A passage made by cutting away one side of a tailstock center so that the facing or parting tool may be advanced to or almost to the center of the work. **2.** Clearance provided around the cutting edge by removal of tool material.

relief angle [MECH ENG] The angle between a relieved surface and a tangential plane at a cutting edge.

relief frame [MECH ENG] A frame placed between the slide valve of a steam engine and the steam chest cover; reduces pressure on the valve and thereby reduces friction.

relief hole [ENG] Any of the holes fired after the cut holes and before the lifter holes in breaking ground for tunneling or shaft sinking.

relief map [MAP] A map of an area showing the topographic relief.

relief model [MAP] A three-dimensional relief map.

relief press *See* letterpress.

relief printing [GRAPHICS] A method of printing in which the type or other images stand above the printing surface; even though the lower areas may have ink in them, they do not print because the paper does not contact them.

relief valve [ENG] A valve that opens automatically for the release of steam or fluid under excess pressure.

relief well [CIV ENG] A well that drains a pervious stratum, to relieve waterlogging at the surface.

relieving [MECH ENG] Treating an embossed metal surface with an abrasive to reveal the base-metal color on the elevations or highlights of the surface.

relieving anode [ELECTR] Of a pool-cathode tube, an auxiliary anode which provides an alternative conducting path for reducing the current to another electrode.

relieving arch *See* discharging arch.

relieving platform [CIV ENG] A deck on the land side of a retaining wall to transfer loads vertically down to the wall.

relighter flame safety lamp [MIN ENG] A locked spirit-burning lamp fitted with an internal relighting device.

reline [ORD] To replace a worn liner of a gun to give it the ballistics of a new weapon.

Relizean stage [GEOL] A subdivision of the Miocene in the California-Oregon-Washington area.

reloadable control storage [ADP] The control storage made up of unit control words necessary for channel multiplexing.

relocatable code [ADP] A code generated by an assembler or compiler, and in which all memory references needing relocation are either specially marked or relative to the current program-counter reading.

relocatable emulator [ADP] An emulator which does not require a stand-alone machine but executes in a multiprogramming environment.

relocatable program [ADP] A program coded in such a way that it may be located and executed in any part of memory.

relocate [ADP] To establish or change the location of a program routine while adjusting or modifying the address references within the instructions to correctly indicate the new locations.

relogging [FOR] An operation in which small trees are salvaged, often for pulpwood, after the large trees are logged.

reluctance [ELECTROMAG] A measure of the opposition presented to magnetic flux in a magnetic circuit, analogous to resistance in an electric circuit; it is equal to magnetomotive force divided by magnetic flux. Also known as magnetic reluctance.

reluctance microphone *See* magnetic microphone.

reluctance motor [ELEC] A synchronous motor, similar in construction to an induction motor, in which the member carrying the secondary circuit has salient poles but no direct-current excitation; it starts as an induction motor but operates normally at synchronous speed.

reluctance pickup *See* variable-reluctance pickup.

reluctance pressure transducer [ENG] Pressure-measurement transducer in which pressure changes activate equivalent magnetic-property changes.

reluctivity [PHYS] The reciprocal of magnetic permeability; the reluctivity of empty space is unity. Also known as magnetic reluctivity; specific reluctance.

rem [NUCLEO] A unit of ionizing radiation, equal to the amount that produces the same damage to humans as 1 roentgen of high-voltage x-rays. Derived from roentgen equivalent man.

remainder [MATH] **1.** The remaining integer when a division of an integer by another is performed; if $l = m \cdot p + r$, where l, m, p, and r are integers and r is less than p, then r is the remainder when l is divided by p. **2.** The remaining polynomial when division of a polynomial is performed; if $l = m \cdot p + r$, where l, m, p, and r are polynomials, and the degree of r is less than that of p, then r is the remainder when l is divided by p. **3.** The remaining part of a convergent infinite series after a computation, for some n, of the first n terms.

remainder formula [MATH] A formula by which the remainder resulting from an approximation of a function can be computed or analyzed.

remainder theorem [MATH] Dividing a polynomial $p(x)$ by $(x - a)$ gives a remainder equaling the number $p(a)$.

remaining velocity [MECH] Speed of a projectile at any point along its path of fire.

Remak's ganglion [ANAT] A ganglion near the junction of the coronary sinus and the right atrium.

remanence [ELECTROMAG] The magnetic flux density that remains in a magnetic circuit after the removal of an applied magnetomotive force; if the magnetic circuit has an air gap, the remanence will be less than the residual flux density.

remanent magnetization [GEOPHYS] That component of a rock's magnetization whose direction is fixed relative to the rock and which is independent of moderate, applied magnetic fields.

remedial maintenance [ADP] The maintenance performed by the contractor following equipment failure; therefore, it is performed as required, on an unscheduled basis.

remedial operation [CHEM ENG] In a chemical process operation, the revision of operating conditions so as to correct the overall operation and bring the product into desired rote or specification limits. Also known as corrective operation.

remex *See* flight feather.

remodulator [ELECTR] A circuit that converts amplitude modulation to audio frequency-shift modulation for transmission of facsimile signals over a voice-frequency radio channel. Also known as converter.

remoistening adhesive [MATER] Any adhesive material, such as dextrin, animal glue, or gum arabic, which is reactivated with the application of water upon the adhesive surface.

remote batch computing [ADP] The running of programs, usually during nonprime hours, or whenever the demands of real-time or time-sharing computing slacken sufficiently to allow less pressing programs to be run.

remote batch processing [ADP] Batch processing in which an input device is located at a distance from the main installation and has access to a computer through a communication link.

remote control [CONT SYS] Control of a quantity which is separated by an appreciable distance from the controlling quantity; examples include master-slave manipulators, telemetering, telephone, and television.

remote-cutoff tube See variable-mu tube.

remote gun control [ORD] Pointing a gun in azimuth and elevation by means of a remote control system, which automatically keeps the gun pointed according to the firing data.

remote-indicating compass [NAV] A compass equipped with one or more indicators to repeat at a distance the readings of the master compass.

remote indicator [ELECTR] 1. An indicator located at a distance from the data-gathering sensing element, with data being transmitted to the indicator mechanically, electrically over wires, or by means of light, radio, or sound waves. 2. See repeater.

remote inquiry [ADP] Interrogation of the content of an automatic data-processing equipment storage unit from a device remotely displaced from the storage unit site.

remote manipulation [ENG] Use of mechanical equipment controlled from a distance to handle materials, such as radioactive materials.

remote metering See telemetering.

remote pickup [COMMUN] Picking up a radio or television program at a remote location and transmitting it to the studio or transmitter over wire lines or a shortwave or microwave radio link.

remote plan position indicator See plan position indicator repeater.

remote sensing [ELEC] Sensing, by a power supply, of voltage directly at the load, so that variations in the load lead drop do not affect load regulation.

remote subscriber [COMMUN] Subscriber to a network that does not have direct access to the switching center, but has access to the circuit through a facility such as a base message center.

removable discontinuity [MATH] A point where a function is discontinuous, but it is possible to redefine the function at this point so that it will be continuous there.

REM sleep See rapid-eye-movement sleep.

renal artery [ANAT] A branch of the abdominal or ventral aorta supplying the kidneys in vertebrates.

renal calculus [MED] A concretion in the kidney.

renal-cell carcinoma [MED] A malignant renal tumor composed principally of large, often hyalin, polygonal cells. Also known as clear-cell carcinoma; Grawitz's tumor; hypernephroma.

renal corpuscle [ANAT] The glomerulus together with its Bowman's capsule in the renal cortex. Also known as Malpighian corpuscle.

renal dwarfism [MED] Dwarfism due to any of various chronic diseases of the kidney in children.

renal failure [MED] Severe malfunction of the kidneys, producing uremia and the resulting constitutional symptoms.

renal papilla [ANAT] A fingerlike projection into the renal pelvis through which the collecting tubules discharge.

renal pyramid [ANAT] Any of the conical masses composing the medullary substance of the kidney. Also known as Malpighian pyramid.

renal rickets [MED] The metabolic bone disease due to increased bone resorption resulting from the acidosis and secondary hyperparathyroidism of renal insufficiency.

renal tubular acidosis [MED] Defective hydrogen-ion excretion in the renal tubules, resulting in hyperchloremic acidosis and inadequate acidification of the urine.

renal tubule [ANAT] One of the glandular tubules which elaborate urine in the kidneys.

renal vein [ANAT] A vein which returns blood from the kidney to the vena cava.

renardite [MINERAL] $Pb(UO_2)_4(PO_4)_2(OH)_4 \cdot 7H_2O$ A yellow mineral composed of hydrous basic lead uranyl phosphate.

rendezvous [AERO ENG] 1. The event of two or more objects meeting with zero relative velocity at a preconceived time and place. 2. The point in space at which such an event takes place, or is to take place.

Rendzina [GEOL] One of an intrazonal, calcimorphic group of soils characterized by a brown to black, friable surface horizon and a light-gray or yellow, soft underlying horizon; found under grasses or forests in humid to semiarid climates.

renette [INV ZOO] An excretory cell found in certain nematodes.

reniform [SCI TECH] Bean- or kidney-shaped, as describing the structure of a crystal in which rounded masses occur at the ends of radiating crystals, or certain structures in animals and plants.

renin [BIOCHEM] A proteolytic enzyme produced in the afferent glomerular arteriole which reacts with the plasma component hypertensinogen to produce angiotensin II.

rennet [VERT ZOO] The lining of the stomach of certain animals, especially the fourth stomach in ruminants.

rennin [BIOCHEM] An enzyme found in the gastric juice of the fourth stomach of calfs; used for coagulating milk casein in cheesemaking. Also known as chymosin.

renninogen [BIOCHEM] The zymogen of rennin. Also known as prorennin.

Renn-Walz process [MET] A method of reclaiming iron and other metals from the waste materials produced in the smelting of zinc and lead ores; this material is brought up to 1000°C in the preheating zone of the kiln by the countercurrent gases, and the oxidized metal vapors are carried off in the flue gases, from which they are subsequently filtered.

renovation [ORD] 1. Process of restoring materiel to, or nearly to, its original condition by cleaning, painting, or similar methods. 2. Restoration of ammunition to serviceable condition by operations more extensive or hazardous than routine ammunition maintenance; normally involves replacement of components.

reorder cycle [IND ENG] The interval between successive reorder (procurement) actions.

reorder point [IND ENG] An arbitrary level of stock on hand plus stock due in, at or below which routine requisitions for replenishment purposes are submitted in accordance with established requisitioning schedules.

reovirus [VIROL] A group of ribonucleic acid–containing animal viruses, including agents of encephalitis and phlebotomus fever.

rep [NUCLEO] A unit of ionizing radiation, equal to the amount that causes absorption of 93 ergs per gram of soft tissue. Derived from roentgen equivalent physical. Also known as parker; tissue roentgen.

repair [ENG] To restore that which is unserviceable to a serviceable condition by replacement of parts, components, or assemblies.

repair cycle [ENG] The period that elapses from the time the item is removed in a reparable condition to the time it is returned to stock in a serviceable condition.

repair dock [CIV ENG] A graving dock or floating dry dock built primarily for ship repair.

repair forecast [ENG] The quantity of items estimated to be repaired or rebuilt for issue during a stated future period.

repair kit [ENG] A group of parts and tools, not all having the same basic name, used for repair or replacement of the worn or broken parts of an item; it may include instruction sheets and material, such as sandpaper, tape, cement, gaskets, and the like.

repair parts list [ENG] List approved by designated authorities, indicating the total quantities of repair parts, tools, and

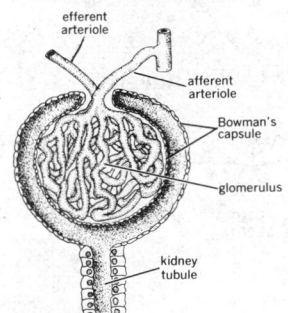

RENAL CORPUSCLE

efferent arteriole

afferent arteriole

Bowman's capsule

glomerulus

kidney tubule

Drawing of renal corpuscle. *(From C. K. Weichert, Anatomy of the Chordates, 4th ed., McGraw-Hill, 1970)*

equipment necessary for the maintenance of a specified number of end items for a definite period of time.

repair synthesis [MOL BIO] Enzymatic excision and replacement of regions of damaged deoxyribonucleic acid, as in repair of thymine dimers by ultraviolet irradiation.

repand [BOT] Having a margin that undulates slightly, referring to a leaf.

repeatability [NAV] In a navigation system, the measure of the accuracy with which the system permits the operator to return to a specific point as defined only in terms of the lattice peculiar to that system.

repeated load [MECH] A force applied repeatedly, causing variation in the magnitude and sometimes in the sense, of the internal forces.

repeated measurements model [STAT] A product model in which each factor is the same.

repeated reflection See multiple reflection.

repeated root See multiple root.

repeater [ELEC] See repeating coil. [ELECTR] 1. An amplifier or other device that receives weak signals and delivers corresponding stronger signals with or without reshaping of waveforms; may be either a one-way or two-way repeater. 2. An indicator that shows the same information as is shown on a master indicator. Also known as remote indicator.

repeater compass See compass repeater.

repeater jammer [ELECTR] A jammer that intercepts an enemy radar signal and reradiates the signal after modifying it to incorporate erroneous data on azimuth, range, or number of targets.

repeater station [COMMUN] A station containing one or more repeaters. Also known as relay station.

repeat glass [OPTICS] Used by textile and wallpaper designers, a device consisting of four lenses formed in one piece of glass; when a single drawing or pattern is viewed, the subject matter is repeated four times.

repeating coil [ELEC] A transformer used to provide inductive coupling between two sections of a telephone line when a direct connection is undesirable. Also known as repeater.

repeating-coil bridge cord [ELEC] In telephony, a method of connecting the common office battery to the cord circuits by connecting the battery to the midpoints of a repeating coil, bridged across the cord circuit.

repeating decimal [MATH] A decimal that is either finite or infinite with a finite block of digits repeating indefinitely. Also known as recurring decimal.

repeat operator [ADP] A pseudo instruction using two arguments, a count p and an increment 9: the word immediately following the instruction is repeated p times, with the values $0,n,2n, \ldots ,(p-1)n$ added to the successive words.

repeat point See double-spot tuning.

repellency [CHEM] Ability to repel water, or being hydrophobic; opposite to water wettability.

repeller [ELECTR] An electrode whose primary function is to reverse the direction of an electron stream in an electron tube. Also known as reflector.

repent [BOT] Of a stem, creeping along the ground and rooting at the nodes.

reperforation [PETRO ENG] Creation of new perforations (holes) in oil well tubing opposite to oil-bearing reservoir zones; creates more opportunity for fluid to drain from the formation into the wellbore.

reperforator See tape reperforator.

reperforator switching center [COMMUN] Message-relaying center at which incoming messages are received on a reperforator which perforates a storage tape from which the message is retransmitted into the proper outgoing circuit.

reperforator/transmitter [COMMUN] A tape reperforator and a tape transmitter operating independently in a teletypewriter system; used in relaying, to establish compatibility between different rates of reception and transmission, and in temporary queuing.

repetition equivalent [COMMUN] In a complete telephone connection, a measure of the grade of transmission experienced by the subscribers using the connection; it includes the combined effects of volume, distortion, noise, and all other subscriber reactions and usages.

repetition frequency See repetition rate.

repetition instruction [ADP] An instruction that causes one or more other instructions to be repeated a specified number of times, usually with systematic address modification occurring between repetitions.

repetition rate [COMMUN] The rate at which recurrent signals are produced or transmitted by radar. Also known as recurrence rate; repetition frequency.

repetitive analog computer [ADP] An analog computer which repeatedly carries out the solution of a problem at a rapid rate (10 to 60 times a second) while an operator may vary parameters in the problem.

repetitive element See regular element.

repetitive time method [IND ENG] A technique where the stopwatch is read and simultaneously returned to zero at each break point. Also known as snapback method.

repetitive unit [ADP] A type of circuit which appears more than once in a computer.

replacement [GEOL] Growth of a new or chemically different mineral in the body of an old mineral by simultaneous capillary solution and deposition.

replacement bit See reset bit.

replacement demand [ENG] A demand representing replacement of items consumed or worn out.

replacement deposit [MINERAL] A mineral deposit formed by the in-position replacement of one mineral for another.

replacement dike [GEOL] A dike which is made by gradual transformation of wall rock by solutions along fractures or permeable zones.

replacement factor [ENG] The estimated percentage of equipment or repair parts in use that will require replacement during a given period.

replacement study [IND ENG] An economic analysis involving the comparison of an existing facility and a proposed replacement facility.

replacement texture [GEOL] The texture exhibited where one mineral has replaced another.

replacement transfusion See exchange transfusion.

replacement vein [GEOL] A mineral vein formed by the gradual transformation of an original vein by secondary fluids.

replenisher [ORD] A cylinder containing recoil oil and a spring-actuated piston, which allows for expansion of oil in the recoil system which has become heated during firing of the weapon; it also returns oil to the recoil system after firing, when the temperature (and volume) of the oil is down.

replica [ENG] A thin plastic or inorganic film which is formed on a surface and then removed from it for study in an electron microscope.

replica grating [OPTICS] A diffraction grating made by flowing a plastic solution over an original grating, evaporating the solvent, and removing the resulting film, which has the lines of the original grating impressed on it.

replica plating [MICROBIO] A method for the isolation of nutritional mutants in microorganisms; colonies are grown from a microorganism suspension previously exposed to a mutagenic agent, on a complete medium in a petri dish; a velour surface is used to transfer the impression of all these colonies to a petri dish containing a minimal medium; colonies that do not grow on the minimal medium are the mutants.

replicating fork [MOL BIO] The Y-shaped region of a chromosome that is a growing point in the replication of deoxyribonucleic acid.

replication [ANALY CHEM] The formation of a faithful mold or replica of a solid that is thin enough for penetration by an electron microscope beam; can use plastic (such as collodion) or vacuum deposition (such as of carbon or metals) to make the mold. [MOL BIO] Duplication, as of a nucleic acid, by copying from a molecular template. [VIROL] Multiplication of phage in a bacterial cell.

replicon [GEN] A genetic element characterized by possession of the structural gene that controls a specific initiator and receptor locus (replicator) for its action.

replum [BOT] A thin wall separating the two valves or chambers of certain fruits.

reply [COMMUN] A radio-frequency signal or combination of signals transmitted by a transponder in response to an interrogation. Also known as response.

REPEATER STATION

A typical TD-2 radio repeater station. *(American Telephone and Telegraph Co.)*

REPENT

Ground ivy showing repent type of stem.

report [ACOUS] Sharp explosive sound, as of a shot, bursting bomb, or projectile. [ADP] An output document prepared by a data-processing system.

report generator [ADP] A routine which produces a complete data-processing report, given only a description of the desired content and format, plus certain information concerning the input file.

reporting point [NAV] **1.** A specified point in relation to which a craft reports its position. **2.** In air operations, a geographical point established for use by air-traffic control in the movement and separation of aircraft.

reporting time interval [COMMUN] The time for transmission of data or a report from the originating terminal to the end receiver.

report of survey [ORD] An official report prepared on a standardized form, which records the circumstance concerning the loss, damage, or destruction of property, and which serves as, or supports, a voucher for dropping the property from property records.

repoussage [GRAPHICS] In etching, when a plate is repaired, scraped, and rubbed, there is a hollow left that will print a gray smudge; repoussage involves hammering out the hollow by putting the plate face down on a flat steel surface and tapping it gently from the back with a repoussage hammer.

repoussé [GRAPHICS] A method of producing a relief design on a thin sheet of metal or leather by hammering or otherwise working the reverse side.

Reppe process [CHEM ENG] A family of high-pressure, catalytic acetylene-reaction processes yielding (depending upon what the acetylene reacts with) butadiene, allyl alcohol, acrylonitrile, vinyl ethers and derivatives, acrylic acid esters, cyclooctatetraene, and resins.

representation of groups [MATH] A representation of a group is given by a homomorphism of it onto some group either of matrices or unitary operators of a Hilbert space.

representation theory [QUANT MECH] Quantum-mechanical device in which one selects the common eigenfunctions of a complete set of quantum-mechanical operators as a basis of vectors in a Hilbert space, and expresses wave functions and operators in terms of column matrices and square matrices respectively which correspond to this basis.

representative calculating time [ADP] The time required to perform a specified operation or series of operations.

representative fraction *See* natural scale.

representative sample [MIN ENG] In testing or valuation of a mineral deposit, a sample so large and average in composition as to be considered representative of a specified volume of the surrounding ore body.

repressing [MET] Applying pressure to a pressed and sintered compact to improve some physical property.

repression [PSYCH] A defense mechanism whereby ideas, feelings, or desires, in conflict with the individual's conscious self-image or motives, are unconsciously dismissed from consciousness.

repressor [BIOCHEM] An end product of metabolism which represses the synthesis of enzymes in the metabolic pathway. [GEN] The product of a regulator gene that acts to repress the transcription of another gene.

repro *See* reproduction proof.

reprocessing [NUCLEO] Processing of fuel after use in a nuclear reactor, to permit reuse.

reproduce head *See* playback head.

reproducer [ADP] A punched-card machine that reads a punched card and duplicates part or all of its contents by punching another card. Also known as punched-card reproducer.

reproducing stylus *See* stylus.

reproducing system *See* sound-reproducing system.

reproducing unit [ADP] An electromechanical device which will duplicate a deck of cards.

reproduction [BIOL] The mechanisms by which organisms give rise to other organisms of the same kind.

reproduction proof [GRAPHICS] A page proof made with great care on smooth paper and used for photographic reproduction; it smudges easily and must be handled with care. Also known as repro; slick paper proof.

reproduction speed [COMMUN] Area of copy recorded per unit time in facsimile transmission.

reproductive behavior [ZOO] The behavior patterns in different types of animals by means of which the sperm is brought to the egg and the parental care of the resulting young insured.

reproductive distribution [ECOL] The range of areas where conditions are favorable to maturation, spawning, and early development of marine animals.

reproductive system [ANAT] The structures concerned with the production of sex cells and perpetuation of the species.

Reptantia [INV ZOO] A suborder of the crustacean order Decapoda including all decapods other than shrimp.

reptile [VERT ZOO] Any member of the class Reptilia.

Reptilia [VERT ZOO] A class of terrestrial vertebrates composed of turtles, tuatara, lizards, snakes, and crocodileans; characteristically they lack hair, feathers, and mammary glands, the skin is covered with scales, they have a three-chambered heart, and the pleural and peritoneal cavities are continuous.

repulsion [MECH] A force which tends to increase the distance between two bodies having like electric charges, or the force between atoms or molecules at very short distances which keeps them apart. Also known as repulsive force.

repulsion-induction motor [ELEC] A repulsion motor that has a squirrel-cage winding in the rotor in addition to the repulsion-motor winding.

repulsion motor [ELEC] An alternating-current motor having stator windings connected directly to the source of a-c power and rotor windings connected to a commutator; brushes on the commutator are short-circuited and are positioned to produce the rotating magnetic field required for starting and running.

repulsion-start induction motor [ELEC] An alternating-current motor that starts as a repulsion motor; at a predetermined speed the commutator bars are short-circuited to give the equivalent of a squirrel-cage winding for operation as an induction motor with constant-speed characteristics.

repulsive force *See* repulsion.

request/grant logic [ADP] Logic circuitry which, in effect, selects the interrupt line with highest priority.

request repeat system [COMMUN] System using an error-detecting code, and so arranged that a signal detected as being in error automatically initiates a request for retransmission.

reradiation [COMMUN] Undesirable radiation of signals generated locally in a radio receiver, causing interference or revealing the location of the receiver.

rerailer [ENG] A small, lightweight Y-shaped device, used to retrack railroad cars and locomotives; as the car is pulled across the device, the derailed wheels are channeled back onto the tracks. Also known as retracker.

rering locked in [COMMUN] Universal cord circuit feature by which, on magnetic lines, the called or calling party may rering the operator, causing the supervisory lamps of the cord circuit to remain lighted until the operator answers.

rerun [ADP] To run a program or a portion of it again on a computer. Also known as rollback. [COMMUN] To repeat a complete transmission.

rerunning [CHEM ENG] The distillation of a liquid material that has already been distilled; usually implies taking a large proportion of the charge stock overhead.

rerun point [ADP] A location in a program from which the program may be started anew after an interruption of the computer run.

rerun routine [ADP] A routine designed to be used in the wake of a computer malfunction or a coding or operating mistake to reconstitute a routine from the last previous rerun point.

RES *See* reticuloendothelial system.

resaw [ENG] To cut lumber to boards of final thickness.

resbenzophenone *See* benzoresorcinol.

rescap [ELEC] A capacitor and resistor assembly manufactured as a packaged encapsulated circuit. Also known as capacitor-resistor unit; capristor; packaged circuit; resistor-capacitor unit.

research [SCI TECH] Scientific investigation aimed at discovering and applying new facts, techniques, and natural laws.

research method [ENG] A standard test to determine the research octane number (or rating) of fuels for use in spark-ignition engines.

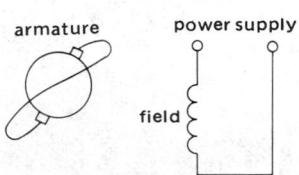

REPULSION MOTOR

armature   power supply

field

Schematic of a repulsion motor showing stator windings connected to the ac power and the rotor or armature windings short-circuited through the commutator and brushes.

research octane number [ENG] An expression for the antiknock rating of a motor gasoline as a guide to how vehicles will operate under mild conditions associated with low engine speeds.

research reactor [NUCLEO] A reactor primarily designed to supply neutrons or other ionizing radiation for experimental purposes; it may also be used for training, materials testing, and production of radioisotopes. Also known as teaching reactor.

research rocket [AERO ENG] A rocket-propelled vehicle used to collect scientific data.

research ship [NAV ARCH] A ship used to carry out oceanographic research; conventional research ships carry on multidiscipline research, either on a single cruise or on successive cruises, while other research ships carry out special tasks, such as fisheries research ships, ships to tend and carry deepsubmergence vehicles, and research drilling ships. Also known as oceanographic ship.

réseau [METEOROL] The term adopted by the World Meteorological Organization for the worldwide network of meteorological stations which have been chosen to represent the meteorology of the globe (*réseau mondial*).

Resedaceae [BOT] A family of dicotyledonous herbs in the order Capparales having irregular, hypogynous flowers.

resequent fault-line scarp [GEOL] A fault-line scarp which faces in the same direction as the original fault scarp or in which the downthrown block is topographically lower than the upthrown block.

resequent stream [HYD] A stream whose direction follows an original consequent stream but is generally lower; resequent streams are generally tributary to a subsequent stream.

reserpine [PHARM] $C_{33}H_{40}N_2O_9$ An alkaloid extracted from certain species of *Rauwolfia* and used as a sedative and antihypertensive.

reserve aircraft [AERO ENG] Those aircraft which have been accumulated in excess of immediate needs for active aircraft and are retained in the inventory against possible future needs.

reserve battery [ELEC] A battery which is inert until an operation is performed which brings all the cell components into the proper state and location to become active.

reserve buoyancy [NAV ARCH] That part of the volume of a ship which is above the water surface and is watertight, so that it will increase buoyancy if the ships sinks deeper into the water.

reserve cell [HISTOL] **1.** One of the small, undifferentiated epithelial cells at the base of the stratified columnar lining of the bronchial tree. **2.** A chromophobe cell.

reserved minerals [MIN ENG] Economic minerals that belong to the state, which confers the right to prospect for and to mine them on any applicant.

reserves [MIN ENG] The quantity of workable mineral or of gas or oil which is calculated to lie within given boundaries.

reserves-decline relationship [PETRO ENG] Relationship between production-rate decline over a period of time to the total remaining hydrocarbon reserves in a reservoir.

reservoir [CIV ENG] A pond or lake built for storage of water, usually by the construction of a dam across a river. [GEOL] A subsurface accumulation of crude oil or natural gas under adequate trap conditions.

reservoir cycling [PETRO ENG] Repressuring of an oil reservoir by reinjection of dry gas (gas with liquids stripped out) into the formation.

reservoir drive mechanism [PETRO ENG] The physical action by which hydrocarbons (gas or liquid) are moved through the porous reservoir structure; for example, gas drive or water drive. Also known as oil-well drive.

reservoir dynamics [PETRO ENG] Fluid-flow performance within an oil or gas reservoir.

reservoir fluid [GEOL] The subterranean fluid trapped by a reservoir formation; can include natural gas, liquid and vapor petroleum hydrocarbons, and interstitial water.

reservoir rock [GEOL] Friable, porous sandstone containing deposits of oil or gas.

reset *See* clear.

reset action [CONT SYS] Floating action in which the final control element is moved at a speed proportional to the extent of proportional-position action.

reset bit [DES ENG] A diamond bit made by reusing diamonds salvaged from a used bit and setting them in the crown attached to a new bit blank. Also known as replacement bit.

reset input [ADP] The act of resetting the original conditions of a problem when running on an analog computer.

reset pulse [ELECTR] **1.** A drive pulse that tends to reset a magnetic cell in the storage section of a digital computer. **2.** A pulse used to reset an electronic counter to zero or to some predetermined position.

reset rate [ENG] The number of times per minute that the effect of the proportional-position action upon the final control element is repeated by the proportional-speed floating action.

resettability [ELECTR] The ability of the tuning element of an oscillator to retune the oscillator to the same operating frequency for the same set of input conditions.

reshabar [METEOROL] A strong, very turbulent, dry northeast wind of bora type which blows down mountain ranges in southern Kurdistan in Persia; it is dry and hot in summer and cold in winter.

resid *See* residual oil.

residence time [CHEM ENG] The average length of time a particle of reactant spends within a process vessel or in contact with a catalyst. [NUCLEO] The time during which radioactive material remains in the atmosphere following the detonation of a nuclear explosive; it is usually expressed as a half-time, since the time for all material to leave the atmosphere is not well known.

resident executive [ADP] The portion of the executive program (sometimes called monitor system) which is permanently stored in core. Also known as resident monitor.

resident monitor *See* resident executive.

residual [GEOL] **1.** Of a mineral deposit, formed by either mechanical or chemical concentration. **2.** Pertaining to a residue left in place after weathering of rock. **3.** Of a topographic feature, representing the remains of a formerly great mass or area and rising above the surrounding surface.

residual air *See* residual volume.

residual anticline [GEOL] In salt tectonics, a relative structural high resulting from the depression of two adjacent rim synclines. Also known as residual dome.

residual charge [ELEC] The charge remaining on the plates of a capacitor after initial discharge.

residual clay [GEOL] Very finely divided clay material formed in place by weathering of rock. Also known as primary clay.

residual current [ELECTR] Current flowing through a thermionic diode when there is no anode voltage, due to the velocity of the electrons emitted by the heated cathode.

residual deviation [NAV] Deviation of a magnetic compass after adjustment or compensation.

residual dome *See* residual anticline.

residual elements [MET] Elements present in small amounts in a metal or alloy, not added intentionally.

residual error rate *See* undetected error rate.

residual error ratio [PHYS] The difference between an optimum result derived from experience or experiment and a supposedly exact result derived from theory.

residual field [ELECTROMAG] The magnetic field left in an iron core after excitation has been removed.

residual flux density [ELECTROMAG] The magnetic flux density at which the magnetizing force is zero when the material is in a symmetrically and cyclically magnetized condition. Also known as residual induction; residual magnetic induction; residual magnetism.

residual free gas [PETRO ENG] Free gas-cap gas in equilibrium with residual liquid hydrocarbons in a depleted reservoir, such as a reservoir at the end of its primary or economic producing life.

residual fuel oil [MATER] Topped crude petroleum or viscous residuums from refinery operations; commercial grades of burner oils Nos. 5 and 6 are residual oils, and include the bunker fuels.

residual induction *See* residual flux density.

residual ionization [PHYS] Ionization of air or other gas in a closed chamber, not accounted for by recognizable neighboring agencies; now attributed to cosmic rays.

residual magnetic induction   *See* residual flux density.

residual magnetism   *See* residual flux density.

residual method   [MET] Magnetic particle inspection in which particles are supplied to a specimen after the magnetizing force has been removed.

residual modulation   *See* carrier noise.

residual nuclear radiation   [NUCLEO] Lingering radiation, or radiation emitted by radioactive material remaining after a nuclear explosion; it is arbitrarily designated as that emitted more than 1 minute after the explosion.

residual ochre   [GEOL] An earthy, red, yellow, or brownish iron oxide powder of iron oxide (usually the mineral limonite) produced during chemical weathering.

residual oil   [MATER] Petroleum-refinery term for combustible, viscous or semiliquid bottoms product from crude oil distillation; used in adhesives, roofing compounds, asphalt manufacture, low-grade fuel oils, and sealants. Also known as liquid asphalt; resid; residuum; tailings.

residual penetration   [ORD] As pertains to shaped charge ammunition, penetration of the jet into a backing target of some standard material after passage through a thickness of target material under test; this penetration is a measure of the effectiveness of the tested material against the jet.

residual radiation   [NUCLEO] Nuclear radiation emitted by radioactive material deposited after an atomic burst, including fission products, unfissioned nuclear material, and material in which radioactivity may have been induced by neutron bombardment. Also known as reststrahlen. [OPTICS] The nearly monochromatic radiation resulting from several reflections of light or other radiation from polished surfaces of certain substances such as quartz and rock salt, due to high reflectivity of these substances in certain bands of wavelengths.

residual resistance   [SOLID STATE] The value to which the electrical resistance of a metal drops as the temperature is lowered to near absolute zero, caused by imperfections and impurities in the metal rather than by lattice vibrations.

residual sediment   *See* resistate.

residual set   [MATH] In a topological space, the complement of a set which is a countable union of nowhere dense sets.

residual spectrum   [MATH] Those members λ of the spectrum of a linear operator A on a Banach space X for which $(A - \lambda I)^{-1}$, I the identity operator, is unbounded with domain not dense in X.

residual stress   *See* internal stress.

residual stress field   *See* ambient stress field.

residual vibration   *See* zero-point vibration.

residual voltage   [ELEC] Vector sum of the voltages to ground of the several phase wires of an electric supply circuit.

residual volume   [PHYSIO] Air remaining in the lungs after the most complete expiration possible; it is elevated in diffuse obstructive emphysema and during an attack of asthma. Also known as residual air.

residuary resistance   [NAV ARCH] The sum of wavemaking resistance and eddy resistance opposing the motion of a ship through the water; the resistance which remains when frictional resistance is subtracted from total fluid resistance or drag.

residue   [CHEM ENG] 1. The substance left after distilling off all but the heaviest components from crude oil in petroleum refinery operations. Also known as bottoms; residuum. 2. Solids deposited onto the filter medium during filtration. Also known as cake; discharged solids. [GEOL] The in-place accumulation of rock debris which remains after weathering has removed all but the least soluble constituent. [MATH] The residue of a complex function $f(z)$ at an isolated singularity z_0 is given by $(1/2\pi i) \int f(z)\, dz$ along a simple closed curve interior to an annulus about z_0; equivalently, the coefficient of the term $(z - z_0)^{-1}$ in the Laurent series expansion of $f(z)$ about z_0.

residue check   *See* modulo N check.

residue class   [MATH] A set of numbers satisfying a congruency relation.

residue theorem   [MATH] The value of the integral of a complex function, taken along a simple closed curve enclosing at most a finite number of isolated singularities, is given by $2\pi i$ times the sum of the residues of the function at each of the singularities.

residuum   *See* residual oil; residue.

resilience   [MECH] 1. Ability of a strained body, by virtue of high yield strength and low elastic modulus, to recover its size and form following deformation. 2. The work done in deforming a body to some predetermined limit, such as its elastic limit or breaking point, divided by the body's volume.

resin   [ORG CHEM] Any of a class of solid or semisolid organic products of natural or synthetic origin with no definite melting point, generally of high molecular weight; most resins are polymers.

resin-anchored bolt   [ENG] A bolt is anchored in the resin placed at the back of the hole in a glass cartridge, which ruptures when the bolt is inserted.

resin duct   [BOT] A canal (intercellular space) lined with secretory cells that release resins into the canal; common in gymnosperms.

resin emulsion   [MATER] Stable emulsion of a resin in a solvent carrier, such as the latex emulsions used in water-based latex paints.

resin finish   [TEXT] A synthetic-resin finish produced by impregnating the fiber with resin and then baking it.

resin-in-pulp ion exchange   [CHEM ENG] Combination of coarse anion-exchange resin with a slurry of finely ground uranium ore in an acid-leach liquor.

resinite   [GEOL] A variety of exinite composed of resinous compounds, often in elliptical or spindle-shaped bodies.

resin of copper   *See* cuprous chloride.

resinography   [CHEM] Science of resins, polymers, plastics, and their products; includes study of morphology, structure, and other characteristics relatable to composition or treatment.

resinoid   [ORG CHEM] A thermosetting synthetic resin either in its initial (temporarily fusible) or in its final (infusible) state.

resinoid wheel   [DES ENG] A grinding wheel bonded with a synthetic resin.

resinol   [MATER] Heat- and oxidation-sensitive, benzene-soluble coal tar fraction containing phenols; insoluble in light petroleum.

resinous cement   [MATER] An acid-proof cement with a base of synthetic resin.

resinous coal   [GEOL] Coal in which large proportions of resinous material are contained in the attritus.

resin roof bolting   [MIN ENG] The fixation of metal roof bolts in rock holes with a bonding resin.

resin tin   *See* rosin tin.

resist   [GRAPHICS] A protective layer applied to the image, or other parts of a plate, to protect that portion of the metal from the action of an etching bath or a sandblasting operation. [MATER] An acid-resistant nonconducting coating used to protect desired portions of a wiring pattern from the action of the etchant during manufacture of printed wiring boards. [MET] An insulating material, for example lacquer, applied to the surface of work to prevent electroplating or electrolytic action at the coated area. Also known as stopoff.

resistance   [ACOUS] *See* acoustic resistance.   [FL MECH] *See* fluid resistance.   [ELEC] 1. The opposition that a device or material offers to the flow of direct current, equal to the voltage drop across the element divided by the current through the element. Also known as electrical resistance. 2. In an alternating-current circuit, the real part of the complex impedance. [MECH] In damped harmonic motion, the ratio of the frictional resistive force to the speed. Also known as damping coefficient; damping constant; mechanical resistance.

resistance box   [ELEC] A box containing a number of precision resistors connected to panel terminals or contacts so that a desired resistance value can be obtained by withdrawing plugs (as in a post-office bridge) or by setting multicontact switches.

resistance brazing   [MET] Brazing employing the heat developed by an electric current, the joint being part of the electric circuit.

resistance bridge   *See* Wheatstone bridge.

RESISTANCE-CAPACITANCE COUPLED AMPLIFIER

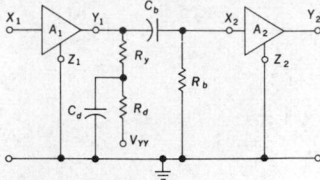

Circuit diagram of resistance-capacitance coupled amplifier. Amplifier stages A_1 and A_2 have inputs X_1 and X_2 and outputs Y_1 and Y_2. Y_1 is coupled to X_2 by blocking capacitor C_b. For vacuum-tube amplifier, Z_1 and Z_2 = cathode of A_1 and A_2; R_b = grid-leak resistor; R_y = plate resistor; V_{YY} = plate supply voltage. Decoupling filter R_d and C_d is used for compensation.

RESISTANCE THERMOMETER

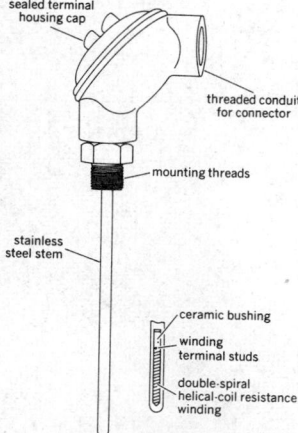

sealed terminal housing cap

threaded conduit for connector

mounting threads

stainless steel stem

ceramic bushing

winding terminal studs

double-spiral helical-coil resistance winding

Industrial-type resistance thermometer. *(From D. M. Considine, ed., Process Instruments and Controls Handbook, McGraw-Hill, 1957)*

resistance-capacitance circuit [ELEC] A circuit which has a resistance and a capacitance in series, and in which inductance is negligible. Abbreviated *R-C* circuit.

resistance-capacitance constant [ELEC] Time constant of a resistive-capacitive circuit; equal in seconds to the resistance value in ohms multiplied by the capacitance value in farads. Abbreviated *R-C* constant.

resistance-capacitance coupled amplifier [ELECTR] An amplifier in which a capacitor provides a path for signal currents from one stage to the next, with resistors connected from each side of the capacitor to the power supply or to ground; it can amplify alternating-current signals but cannot handle small changes in direct currents. Also known as *R-C* amplifier; *R-C* coupled amplifier; resistance-coupled amplifier.

resistance-capacitance network [ELEC] Circuit containing resistances and capacitances arranged in a particular manner to perform a specific function. Abbreviated *R-C* network.

resistance-capacitance oscillator [ELECTR] Oscillator in which the frequency is determined by resistance and capacitance elements. Abbreviated *R-C* oscillator.

resistance coefficient 1 [FL MECH] A dimensionless number used in the study of flow resistance, equal to the resistance force in flow divided by ½ the product of fluid density, the square of fluid velocity, and the square of a characteristic length. Symbolized c_f.

resistance coefficient 2 *See* Darcy number 1.

resistance commutation [ELEC] Commutation of an electric rotating machine in which brushes with relatively high resistance span at least one commutator segment, in order to achieve a linear variation of current with time, and thereby minimize self-inductive voltage in the coils.

resistance-coupled amplifier *See* resistance-capacitance coupled amplifier.

resistance coupling [ELECTR] Coupling in which resistors are used as the input and output impedances of the circuits being coupled; a coupling capacitor is generally used between the resistors to transfer the signal from one stage to the next. Also known as *R-C* coupling; resistance-capacitance coupling; resistive coupling.

resistance drop [ELEC] The voltage drop occurring between two points on a conductor due to the flow of current through the resistance of the conductor; multiplying the resistance in ohms by the current in amperes gives the voltage drop in volts. Also known as *IR* drop.

resistance element [ELEC] An element of resistive material in the form of a grid, ribbon, or wire, used singly or built into groups to form a resistor for heating purposes, as in an electric soldering iron.

resistance factor *See* R factor.

resistance furnace [ENG] An electric furnace in which the heat is developed by the passage of current through a suitable internal resistance that may be the charge itself, a resistor embedded in the charge, or a resistor surrounding the charge. Also known as electric resistance furnace.

resistance grounding [ELEC] Electrical grounding in which lines are connected to ground by a resistive (totally dissipative) impedance.

resistance heating [ELEC] The generation of heat by electric conductors carrying current; degree of heating is proportional to the electrical resistance of the conductor; used in electrical home appliances, home or space heating, and heating ovens and furnaces.

resistance lamp [ELEC] Electric lamp used to prevent the current in a circuit from exceeding a desired limit.

resistance loss [ELEC] Power loss due to current flowing through resistance; its value in watts is equal to the resistance in ohms multiplied by the square of the current in amperes.

resistance magnetometer [ENG] A magnetometer that depends for its operation on variations in the electrical resistance of a material immersed in the magnetic field to be measured.

resistance material [ELEC] Material having sufficiently high resistance per unit length or volume to permit its use in the construction of resistors.

resistance measurement [ELEC] The quantitative determination of that property of an electrically conductive material, component, or circuit called electrical resistance.

resistance meter [ENG] Any instrument which measures electrical resistance. Also known as electrical resistance meter.

resistance methanometer [ENG] A catalytic methanometer, with platinum used as the filament, which both heats the detecting element and acts as a resistance-type thermometer.

resistance noise *See* thermal noise.

resistance pyrometer *See* resistance thermometer.

resistance-rate flowmeter *See* resistive flowmeter.

resistance seam welding [MET] Resistance welding process which produces a series of individual spot welds, overlapping spot welds, or a continuous nugget weld made by circular or wheel-type electrodes.

resistance spot welding [MET] Resistance welding process in which the parts are lapped and held in place under pressure; the size and shape of the electrodes (usually circular) control the size and shape of the welds.

resistance-start motor [ELEC] A split-phase motor having a resistance connected in series with the auxiliary winding; the auxiliary circuit is opened when the motor attains a predetermined speed.

resistance strain gage [ELECTR] A strain gage consisting of a strip of material that is cemented to the part under test and that changes in resistance with elongation or compression.

resistance thermometer [ENG] A thermometer in which the sensing element is a resistor whose resistance is an accurately known function of temperature. Also known as electrical resistance thermometer; resistance pyrometer.

resistance transfer factor [GEN] A carrier of genetic information in bacteria which is considered to control the ability of self-replication and conjugal transfer of R factors. Abbreviated RTF.

resistance welding [MET] Joining metals together under pressure by making use of heat developed by an electric current, the work being part of the electrical circuit.

resistance wire [MET] Wire made from a metal or alloy having high resistance per unit length, such as Nichrome; used in wire-wound resistors and heating elements.

resistate [GEOL] A sediment consisting of minerals that are chemically resistant and are enriched in the residues of weathering processes. Also known as residual sediment.

resist-dyeing [TEXT] A cross-dyeing method in which a chemical is applied to certain yarns before weaving so that, when the material is dyed, only the untreated yarns take the dye, producing a colorful pattern; the chemical is later removed from the fabric.

resisting moment [MECH] A moment produced by internal tensile and compressive forces that balances the external bending moment on a beam.

resistive coupling *See* resistance coupling.

resistive flowmeter [ENG] Liquid flow-rate measurement device in which flowrates are read electrically as the result of the rise or fall of a conductive differential-pressure manometer fluid in contact with a resistance-rod assembly. Also known as resistance-rate flowmeter.

resistive unbalance [ELEC] Unequal resistance in the two wires of a transmission line.

resistivity *See* electrical resistivity.

resistivity index [PETRO ENG] Ratio of the true electrical resistivity of a rock system at a specified water saturation, to the resistivity of the rock itself; used for calculation of electrical well-logging data.

resistivity well logging [PETRO ENG] The measurement of subsurface electrical resistivities (normal and lateral to the borehole) during electrical logging of oil wells.

resistor [ELEC] A device designed to have a definite amount of resistance; used in circuits to limit current flow or to provide a voltage drop. Also known as electrical resistor.

resistor bulb [ENG] A temperature-measurement device inside of which is a resistance winding; changes in temperature cause corresponding changes in resistance, varying the current in the winding.

resistor-capacitor-transistor logic [ELECTR] A resistor-transistor logic with the addition of capacitors that are used to enhance switching speed.

resistor-capacitor unit *See* rescap.

resistor color code [ELEC] Code adopted by the Electronic

Industries Association to mark the values of resistance on resistors in a readily recognizable manner; the first color represents the first significant figure of the resistor value, the second color the second significant figure, and the third color represents the number of zeros following the first two figures; a fourth color is sometimes added to indicate the tolerance of the resistor.

resistor core [ELEC] Insulating support on which a resistor element is wound or otherwise placed.

resistor element [ELEC] That portion of a resistor which possesses the property of electric resistance.

resistor furnace [ENG] An electric furnace in which heat is developed by the passage of current through distributed resistors (heating units) mounted apart from the charge.

resistor network [ELEC] An electrical network consisting entirely of resistances.

resistor oven [ENG] Heating chamber relying on an electrical-resistance element to create temperatures of up to 800°F (430°C); used for drying and baking.

resistor-transistor logic [ELECTR] One of the simplest logic circuits, having several resistors, a transistor, and a diode. Abbreviated RTL.

resite *See* C-stage resin.

resitol *See* B-stage resin.

resnatron [ELECTR] A microwave-beam tetrode containing cavity resonators, used chiefly for generating large amounts of continuous power at high frequencies.

resolution [ELECTR] In television, the maximum number of lines that can be discerned on the screen at a distance equal to tube height; this ranges from 350 to 400 for most receivers. [ELECTROMAG] In radar, the minimum separation between two targets, in angle or range, at which they can be distinguished on a radar screen. Also known as resolving power. [PHYS] **1.** For a measurement of energy or momentum of a collection of particles, the difference between the highest and lowest energies at which the response of an instrument to a beam of monoenergetic particles is at least half its maximum value, divided by the energy of the particles. **2.** The procedure of breaking up a vectorial quantity into its components. [OPTICS] *See* resolving power. [SPECT] *See* resolving power.

resolution chart [COMMUN] *See* test pattern. [OPTICS] A device to test resolving power; usually alternate black and white lines of equal width arranged in groups of decreasing line width, identified as the number of line pairs per millimeter.

resolution factor [ADP] In information retrieval, the ratio obtained in dividing the total number of documents retrieved (whether relevant or not to the user's needs) by the total number of documents available in the file.

resolution in azimuth [ENG] The angle by which two targets must be separated in azimuth in order to be distinguished by a radar set when the targets are at the same range.

resolution in range [ENG] Distance by which two targets must be separated in range in order to be distinguished by a radar set when the targets are on the same azimuth line.

resolution of a vector [MATH] The determination of vectors parallel to specified (usually perpendicular) axes such that their sum equals a given vector.

resolution of the identity [MATH] A family of linear projection operators on a Banach space used in studying the spectra of linear operators.

resolution reading [OPTICS] A number indicating how many lines per millimeter are contained in the finest group which can be distinguished on a resolution chart.

resolution wedge [COMMUN] On a television test pattern, a group of gradually converging lines used to measure resolution.

resolvent kernel [MATH] A function appearing as an integrand in an integral representation for a solution of a linear integral equation which often completely determines the solutions.

resolvent of an operator [MATH] The function, defined on the complement of the spectrum of a linear operator T of Banach spaces, given by $(T - \lambda I)^{-1}$ for each λ in this complement, where I is the identity operator; this enables a study of T relative to its eigenvalues.

resolvent set [MATH] Those scalars λ for which the operator $T - \lambda I$ has a bounded inverse, where T is some linear

operator on a Banach space, and I is the identity operator.

resolver [ELEC] A synchro or other device whose rotor is mechanically driven to translate rotor angle into electrical information corresponding to the sine and cosine of rotor angle; used for interchanging rectangular and polar coordinates. Also known as sine-cosine generator; synchro resolver. [ELECTR] **1.** A synchro or other device whose input is the angular position of an object, such as the rotor of an electric machine, and whose output is electric signals, usually proportional to the sine and cosine of an angle, and often in digital form; used to interchange rectangular and polar coordinates, and in servomechanisms to report the orientation of controlled objects. Also known as angular resolver. **2.** A device that accepts a single vector-valued analog input and produces for output either analog or digital signals proportional to two or three orthogonal components of the vector. Also known as vector resolver.

resolving cell [ELECTROMAG] In radar, volume in space whose diameter is the product of slant range and beam width, and whose length is the pulse length.

resolving power [ELECTROMAG] **1.** The reciprocal of the beam width of a unidirectional antenna, measured in degrees. **2.** *See* resolution. [OPTICS] A quantitative measure of the ability of an optical instrument to produce separable images of different points on an object; usually, the smallest angular or linear separation of two object points for which they may be resolved according to the Rayleigh criterion. Also known as resolution. [PHYS] A measure of the ability of a mass spectroscope to separate particles of different masses, equal to the ratio of the average mass of two particles whose mass spectrum lines can just be completely separated, to the difference in their masses. [SPECT] A measure of the ability of a spectroscope or interferometer to separate spectral lines of nearly equal wavelength, equal to the average wavelength of two equally strong spectral lines whose images can barely be separated, divided by the difference in wavelengths; for spectroscopes, the lines must be resolved according to the Rayleigh criterion; for interferometers, the wavelengths at which the lines have half of maximum intensity must be equal. Also known as resolution.

resolving time [ADP] In computers, the shortest permissible period between trigger pulses for reliable operation of a binary cell. [ENG] Minimum time interval, between events, that can be detected; resolving time may refer to an electronic circuit, to a mechanical recording device, or to a counter tube.

resonance [ELEC] A phenomenon exhibited by an alternating-current circuit in which there are relatively large currents near certain frequencies, and a relatively unimpeded oscillation of energy from a potential to a kinetic form; a special case of the physics definition. [PHYS] A phenomenon exhibited by a physical system acted upon by an external periodic driving force, in which the resulting amplitude of oscillation of the system becomes large when the frequency of the driving force approaches a natural free oscillation frequency of the system. [QUANT MECH] *See* resonance absorption; resonance level.

resonance absorption [NUCLEO] The absorption of neutrons having a narrow range of energies corresponding to a nuclear resonance level of the absorber in a nuclear reactor. [QUANT MECH] The absorption of electromagnetic radiation by a quantum-mechanical system at a characteristic frequency satisfying the Bohr frequency condition. Also known as resonance.

resonance bridge [ELEC] A four-arm alternating-current bridge used to measure inductance, capacitance, or frequency; the inductor and the capacitor, which may be either in series or in parallel, are tuned to resonance at the frequency of the source before the bridge is balanced.

resonance capture [NUC PHYS] The combination of an incident particle and a nucleus in a resonance level of the resulting compound nucleus, characterized by having a large cross section at and very near the corresponding resonance energy.

resonance curve [ELEC] Graphical representation illustrating the manner in which a tuned circuit responds to the various frequencies in and near the resonant frequency.

resonance fluorescence [ATOM PHYS] *See* resonance radi-

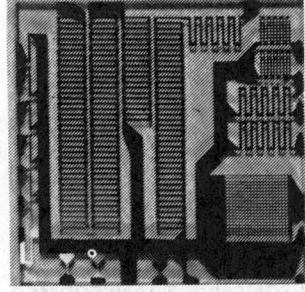

RESISTOR NETWORK

A thin-film resistor network. *(TRW Inc.)*

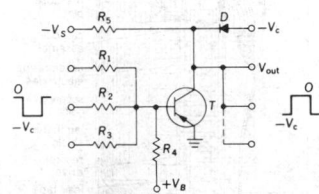

RESISTOR-TRANSISTOR LOGIC

Resistor-transistor logic circuit: R_1, R_2, R_3 are resistors coupling the logic circuit to preceding logic circuits. With resistor R_4 and positive supply voltage $+V_B$ they form the gate circuit. D is the diode and T the transistor. V_s = input signal voltage; V_c = collector voltage; V_{out} = output voltage.

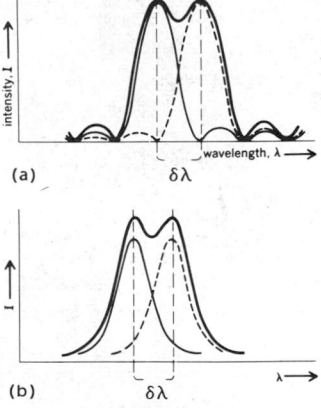

RESOLVING POWER

Two equally strong spectrum lines which are said to be barely separated (spectrocopy definition), *(a)* when the shape is determined by diffraction (Rayleigh criterion), and *(b)* when the shape follows the Airy formula, applicable to multiple-beam interferometers. $\delta\lambda$ = difference in wavelength of lines. Light solid curve and broken curve represent intensities of lines taken separately; heavy curve represents combined intensity.

ation. [NUC PHYS] Resonant scattering from an atomic nucleus.

resonance frequency [PHYS] A frequency at which some measure of the response of a physical system to an external periodic driving force is a maximum; three types are defined, namely, phase resonance, amplitude resonance, and natural resonance, but they are nearly equal when dissipative effects are small. Also known as resonant frequency. [QUANT MECH] A characteristic frequency, satisfying the Bohr frequency condition, at which a quantum-mechanical system absorbs radiation.

resonance hybrid [CHEM] A molecule that may be considered an intermediate between two or more valence bond structures.

resonance lamp [ATOM PHYS] An evacuated quartz bulb containing mercury, which acts as a source of radiation at the wavelength of the pure resonance line of mercury when irradiated by a mercury-arc lamp.

resonance level [QUANT MECH] An unstable state of a compound system capable of being formed in a collision between two particles, and associated with a peak in a graph of cross section versus energy for the scattering of the particles. Also known as resonance.

resonance luminescence *See* resonance radiation.

resonance method [ELEC] A method of determining the impedance of a circuit element, in which resonance frequency of a resonant circuit containing the element is measured.

resonance radiation [ATOM PHYS] The emission of radiation by a gas or vapor as a result of excitation of atoms to higher energy levels by incident photons at the resonance frequency of the gas or vapor; the radiation is characteristic of the particular gas or vapor atom but is not necessarily the same frequency as the absorbed radiation. Also known as resonance fluorescence; resonance luminescence.

resonance spectrum [SPECT] An emission spectrum resulting from illumination of a substance (usually a molecular gas) by radiation of a definite frequency or definite frequencies.

resonance transformer [ELEC] A high-voltage transformer in which the secondary circuit is tuned to the frequency of the power supply. [ELECTR] An electrostatic particle accelerator, used principally for acceleration of electrons, in which the high-voltage terminal oscillates between the two voltages which are equal in magnitude and opposite in sign.

resonance trough [METEOROL] A large-scale pressure trough which forms at an appropriate wavelength away from a dominant trough; for example, the mean trough over the Mediterranean in winter is often considered a resonance trough between the two dynamically active troughs along the east coasts of North America and Asia.

resonance vibration [MECH] Forced vibration in which the frequency of the disturbing force is very close to the natural frequency of the system, so that the amplitude of vibration is very large.

resonant antenna [ELECTROMAG] An antenna for which there is a sharp peak in the power radiated or intercepted by the antenna at a certain frequency, at which electric currents in the antenna form a standing-wave pattern.

resonant capacitor [ELEC] A tubular capacitor that is wound to have inductance in series with its capacitance.

resonant cavity *See* cavity resonator.

resonant-cavity maser [PHYS] A maser in which the paramagnetic active material is placed in a cavity resonator.

resonant chamber *See* cavity resonator.

resonant-chamber switch [ELECTROMAG] Waveguide switch in which a tuned cavity in each waveguide branch serves the functions of switch contacts; detuning of a cavity blocks the flow of energy in the associated waveguide.

resonant circuit [ELEC] A circuit that contains inductance, capacitance, and resistance of such values as to give resonance at an operating frequency.

resonant coupling [ELEC] Coupling between two circuits that reaches a sharp peak at a certain frequency.

resonant detector [PHYS] A detector of electromagnetic radiation which is responsive to radiation only at certain frequencies at which resonance is created in the detector.

resonant diaphragm [ELECTROMAG] Diaphragm, in wave-

guide technique, so proportioned as to introduce no reactive impedance at the design frequency.

resonant element *See* cavity resonator.

resonant frequency *See* resonance frequency.

resonant gate transistor [ELECTR] Surface field-effect transistor incorporating a cantilevered beam which resonates at a specific frequency to provide high-Q-frequency discrimination.

resonant helix [ELECTROMAG] An inner helical conductor in certain types of transmission lines and resonant cavities, which carries currents with the same frequency as the rest of the line or cavity.

resonant iris [ELECTROMAG] A resonant window in a circular waveguide; it resembles an optical iris.

resonant jet [AERO ENG] A pulsejet engine, exhibiting intensification of power under the rhythm of explosions and compression waves within the engine.

resonant line [ELECTROMAG] A transmission line having values of distributed inductance and distributed capacitance so as to make the line resonant at the frequency it is handling.

resonant line oscillator [ELECTR] Oscillator in which one or more sections of transmission lines are employed as resonant elements.

resonant-line tuner [ELECTR] A television tuner in which resonant lines are used to tune the antenna, radio-frequency amplifier, and radio-frequency oscillator circuits; tuning is achieved by moving shorting contacts that change the electrical lengths of the lines.

resonant-reed relay [ELEC] A reed relay in which the reed switch closes only when the required frequency is applied to the operating coil, to make one of the reeds vibrate until its amplitude is sufficient to make contact with the other reed; used in selective paging systems.

resonant resistance [ELEC] Resistance value to which a resonant circuit is equivalent.

resonant scattering [QUANT MECH] Scattering of a photon by a quantum-mechanical system (usually an atom or nucleus) in which the system first absorbs the photon by undergoing a transition from one of its energy states to one of higher energy, and subsequently reemits the photon by the exact inverse transition.

resonant voltage step-up [ELEC] Ability of an inductor and a capacitor in a series resonant circuit to deliver a voltage several times greater than the input voltage of the circuit.

resonant wavelength [ELECTROMAG] The wavelength in free space of electromagnetic radiation having a frequency equal to a natural resonance frequency of a cavity resonator.

resonant window [ELECTROMAG] A parallel combination of inductive and capacitive diaphragms, used in a waveguide structure to provide transmission at the resonant frequency and reflection at other frequencies.

resonate [ELEC] To bring to resonance, as by tuning.

resonating cavity [ELECTROMAG] Short piece of waveguide of adjustable length, terminated at either or both ends by a metal piston, an iris diaphragm, or some other wave-reflecting device; it is used as a filter, as a means of coupling between guides of different diameters, and as impedance networks corresponding to those used in radio circuits.

resonator [PHYS] A device that exhibits resonance at a particular frequency, such as an acoustic resonator or cavity resonator.

resonator grid [ELECTR] Grid that is attached to a cavity resonator in velocity-modulated tubes to provide coupling between the resonator and the electron beam.

resonator wavemeter [ELECTROMAG] Any resonant circuit used to determine wavelength, such as a cavity-resonator frequency meter.

resorcin *See* resorcinol.

resorcinol [ORG CHEM] $C_6H_4(OH)_2$ Sweet-tasting, white, toxic crystals; soluble in water, alcohol, ether, benzene, and glycerol; melts at 111°C; used for resins, dyes, pharmaceuticals, and adhesives, and as a chemical intermediate. Also known as *meta*-dihydroxybenzene; resorcin.

resorcinol acetate [ORG CHEM] $HOC_6H_4OCOCH_3$ A viscous, combustible, yellow to amber liquid with burning taste; soluble in alcohol and solvent; boils at 283°C; used in

RESONANCE TRANSFORMER

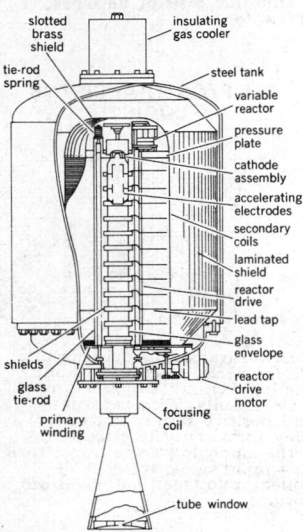

Schematic diagram of 1-MeV resonance transformer that is used as an electrostatic particle accelerator. *(General Electric Co.)*

cosmetics and medicine. Also known as resorcinol monoacetate.

resorcinol monoacetate   *See* resorcinol acetate.

β-resorcylic acid   [ORG CHEM] $(OH)_2C_6H_3COOH$ Combustible, white needles; soluble in alcohol and ether, very slightly soluble in water; decomposes at 220°C; used as a dyestuff and a pharmaceutical intermediate, and in the manufacture of fine chemicals. Also known as BRA; 4-carboxy-resorcinol; 2,4-dihydroxybenzene carboxylic acid; 2,4-dihydroxybenzoic acid; 4-hydroxysalicylic acid.

resorption   [PETR]   The process by which a magma redissolves previously crystallized minerals.   [PHYS]   Absorption or, less commonly, adsorption of material by a body or system from which the material was previously released.

resource   [SCI TECH]   A reserve source of supply, such as a material or mineral.

respiration   [PHYSIO]   **1.** The processes by which tissues and organisms exchange gases with their environment. **2.** The act of breathing with the lungs, consisting of inspiration and expiration.

respirator   [ENG]   A device for maintaining artificial respiration to protect the respiratory tract against irritating and poisonous gases, fumes, smoke, and dusts, with or without equipment supplying oxygen or air; some types have a fitting which covers the nose and mouth.

respiratory arrest   [MED]   Sudden cessation of spontaneous respiration due to failure of the respiratory center.

respiratory center   [PHYSIO]   A large area of the brain involved in regulation of respiration.

respiratory distress syndrome of newborn   [MED]   A disease occurring during the first days of life, characterized by respiratory distress and cyanosis; a hyaline membrane lines the alveoli when the disease persists for more than several hours. Abbreviated RDS.

respiratory epithelium   [HISTOL]   The ciliated pseudostratified epithelium lining the respiratory tract.

respiratory pigment   [BIOCHEM]   Any of various conjugated proteins that function in living organisms to transfer oxygen in cellular respiration.

respiratory quotient   [PHYSIO]   The ratio of volumes of carbon dioxide evolved and oxygen consumed during a given period of respiration. Abbreviated RQ.

respiratory system   [ANAT]   The structures and passages involved with the intake, expulsion, and exchange of oxygen and carbon dioxide in the vertebrate body.

respiratory tree   [ANAT]   The trachea, bronchi, and bronchioles.   [INV ZOO]   Either of a pair of branched tubular appendages of the cloaca in certain holothurians that is thought to have a respiratory function.

respirometer   [ENG]   **1.** An instrument for studying respiration. **2.** A diver's helmet containing a compressed air supply for replenishing oxygen used by the diver.

responder   [ELECTR]   The transmitter section of a radar beacon.

responder beacon   [ELECTR]   The radar beacon that serves to emit the signals of the responder in a transponder.

response   [COMMUN]   *See* reply.   [CONT SYS]   A quantitative expression of the output of a device or system as a function of the input. Also known as system response.   [ENG]   *See* frequency response.   [STAT]   The value of some measurable quantity after a treatment has been applied.

response characteristic   [CONT SYS]   The response as a function of an independent variable, such as direction or frequency, often presented in graphical form.

response time   [ADP]   The delay experienced in time sharing between request and answer, a delay which increases when the number of users on the system increases.   [CONT SYS] The time required for the output of a control system or element to reach a specified fraction of its new value after application of a step input or disturbance.   [ELEC]   The time it takes for the pointer of an electrical or electronic instrument to come to rest at a new value, after the quantity it measures has been abruptly changed.

responsor   [ELECTR]   The receiving section of an interrogator-responsor.

restart   [ADP]   To go back to a specific planned point in a routine, usually in the case of machine malfunction, for the purpose of rerunning the portion of the routine in which the error occurred; the length of time between restart points in a given routine should be a function of the mean free error time of the machine itself.   [AERO ENG]   The act of firing a stage of a rocket after a previous powered flight.

rest density   [RELAT]   The density of a small portion of a fluid in a Lorentz frame in which that portion of the fluid is at rest.

rest frame   [RELAT]   The Lorentz frame in which the total momentum of a system equals zero; for an accelerated system, the rest frame varies from instant to instant.

restiform body   *See* inferior cerebellar peduncle.

resting cell   [CYTOL]   An interphase cell.

resting frequency   *See* carrier frequency.

resting potential   [PHYSIO]   The potential difference between the interior cytoplasm and the external aqueous medium of the living cell.

resting spore   [BIOL]   A spore that remains dormant for long periods before germination, withstanding adverse conditions; usually invested in a thickened cell wall.

Restionaceae   [BOT]   A large family of monocotyledonous plants in the order Restionales characterized by unisexual flowers, wholly cauline leaves, unilocular anthers, and a more or less open inflorescence.

Restionales   [BOT]   An order of monocotyledonous plants in the subclass Commelinidae having reduced flowers and a single, pendulous, orthotropous ovule in each of the one to three locules of the ovary.

restitution coefficient   *See* coefficient of restitution.

rest mass   [RELAT]   The mass of a particle in a Lorentz reference frame in which it is at rest.

restoration   [ECOL]   A conservation measure involving the correction of past abuses that have impaired the productivity of the resources base.

restore   [ADP]   In computers, to regenerate, to return a cycle index or variable address to its initial value, or to store again. [ELECTR]   Periodic charge regeneration of volatile computer storage systems.

restorer   *See* direct-current restorer.

restorer pulses   [ELECTR]   In computers, pairs of complement pulses, applied to restore the coupling-capacitor charge in an alternating-current flip-flop.

rest potential   [ELEC]   Residual potential difference remaining between an electrode and an electrolyte after the electrode has become polarized.

restrainer   [GRAPHICS]   The ingredient of a developer which prevents too rapid development and chemical fog.

restraint of loads   [ENG]   The process of binding, lashing, and wedging items into one unit onto or into its transporter in a manner that will ensure immobility during transit.

restricted adhesive   [MATER]   An adhesive which for any reason cannot satisfactorily pass its evaluation test; as a result, the maximum time required for curing, that is, its usable life, cannot be assigned; it cannot be used for structural bonding.

restricted air cargo   [IND ENG]   Cargo which is not highly dangerous under normal conditions, but which possesses certain qualities which require extra precautions in packing and handling.

restricted area   [NAV]   **1.** An airspace of defined dimensions identified by an area on the surface of the earth within which the flight of aircraft is subject to restrictions. **2.** An area within which vessels may be navigated subject to prohibitions with respect to anchoring, trawling, fishing, and so on.

restricted basin   [GEOL]   A depression in the floor of the ocean in which the water circulation is topographically restricted and therefore generally is oxygen-depleted. Also known as barred basin; silled basin.

restricted internal rotation   [PHYS CHEM]   Restrictions on the rotational motion of molecules or parts of molecules in some substances, such as solid methane, at certain temperatures.

restricted waters   [NAV]   Areas which for navigational reasons, such as the presence of shoals or other dangers, confine the movements of shipping within narrow limits.

restricted work   [IND ENG]   Manual or machine work where the work pace is only partially under the control of the worker.

restriction of ego   [PSYCH]   A defense mechanism for escaping anxiety by avoiding situations consciously perceived as dangerous or uncomfortable.

rest stick *See* mahlstick.

reststrahlen *See* residual radiation.

resultant of forces [MECH] A system of at most a single force and a single couple whose external effects on a rigid body are identical with the effects of the several actual forces that act on that body.

resultant rake [MECH ENG] The angle between the face of a cutting tooth and an axial plane through the tooth point measured in a plane at right angles to the cutting edge.

resultant wind [CLIMATOL] The vectorial average of all wind directions and speeds for a given level at a given place for a certain period, such as a month.

resuperheating *See* reheating.

resupinate [BOT] Inverted, usually through 180°, so as to appear upside down or reversed.

resupply [IND ENG] The act of replenishing stocks in order to maintain required levels of supply.

resuscitation [MED] Restoration of consciousness or life functions after apparent death.

resuscitator [ENG] A device for supplying oxygen to and inducing breathing in asphyxiation victims.

ret [CHEM] The reduction or digestion of fibers (usually linen) by enzymes.

retained water [HYD] The water remaining in rock or soil after gravity groundwater has been drained out.

retainer [ENG] A device that holds a mechanical component in place.

retainer wall [ENG] A wall, usually earthen, around a storage tank or an area of storage tanks (tank farm); used to hold (retain) liquid in place if one or more tanks begin to leak.

retaining ring [DES ENG] **1.** A shoulder inside a reaming shell that prevents the core lifter from entering the core barrel. **2.** A steel ring between the races of a ball bearing to maintain the correct distribution of the balls in the races.

retaining wall [CIV ENG] A wall designed to maintain differences in ground elevations by holding back a bank of material.

retardant *See* retarder.

retardation [MED] Slow mental or physical functioning. [NAV] The amount of delay in time or phase angle introduced by the resistivity of the surface over which the radio wave in radio navigation is passing. [OCEANOGR] The amount of time by which corresponding tidal phases grow later day by day, averaging approximately 50 minutes. [OPTICS] In interference microscopy, the difference in optical path between the light passing through the specimen and the light bypassing the specimen. Also known as optical-path difference.

retardation coil [ELECTROMAG] A high-inductance coil used in telephone circuits to permit passage of direct current or low-frequency ringing current while blocking the flow of audio-frequency currents.

retardation sheet *See* wave plate.

retardation theory [OPTICS] General methods of calculating the effect of one or more wave plates on light which is normally incident on the plates and which is initially polarized in some fashion.

retarded acid [PETRO ENG] Oil well acidizing solution whose reactivity is slowed by addition of artificial gums and thickening agents, so that the acid penetrates deeper into the formation before being spent.

retarded field [ELECTROMAG] An electric or magnetic field strength as found from the retarded potentials.

retarded potentials [ELECTROMAG] The electromagnetic potentials at an instant in time t and a point in space r as a function of the charges and currents that existed at earlier times at points on the past light cone of the event r,t.

retarder [MATER] A material that inhibits the action of another substance, such as flameproofing agents or substances added to cement to retard setting time. Also known as retardant. [MECH ENG] A braking device used to control the speed of railroad cars moving along the classification tracks in a hump yard.

retarding basin [CIV ENG] A basin designed and operated to provide temporary storage and thus reduce the peak flood flows of a stream.

retarding conveyor [MECH ENG] Any type of conveyor used to restrain the movement of bulk materials, packages, or objects where the incline is such that the conveyed material tends to propel the conveying medium.

retarding-field oscillator [ELECTR] An oscillator employing an electron tube in which the electrons oscillate back and forth through a grid that is maintained positive with respect to both the cathode and anode; the field in the region of the grid exerts a retarding effect through the grid in either direction. Also known as positive-grid oscillator.

retarding potential [PHYS] A potential which causes the speed of a moving particle to be reduced.

retard transmitter [ELECTRO] Transmitter in which a delay period is introduced between the time of actuation and the time of transmission.

rete cord [EMBRYO] One of the deep, anastomosing strands of cells of the medullary cords of the vertebrate embryo, that form the rete testis or the rete ovarii.

rete mirabile [VERT ZOO] A network of small blood vessels that are formed by the branching of a large vessel and that usually reunite into a single trunk; believed to have an oxygen-storing function in certain aquatic fauna.

retene [ORG CHEM] $C_{18}H_{18}$ A cyclic hydrocarbon, melting at 100.5–101°C, soluble in benzene and hot ethanol; used in organic syntheses. Also known as 7-isopropyl-1-methyl phenanthrene; methylisopropylphenanthrene.

retention cyst [MED] A cyst caused by obstructed outflow of secretion from a gland.

retention index [ANALY CHEM] In gas chromatography, the relationship of retention volume with arbitrarily assigned numbers to the compound being analyzed; used to indicate the volume retention behavior during analysis.

retention time [ANALY CHEM] In gas chromatography, the time at which the center, or maximum, of a symmetrical peak occurs on a gas chromatogram. [ELECTR] The maximum time between writing into a storage tube and obtaining an acceptable output by reading. Also known as storage time.

retention volume [ANALY CHEM] In gas chromatography, the product of retention time and flow rate.

retentivity [ELECTROMAG] The residual flux density corresponding to the saturation induction of a magnetic material.

rete ovarii [ANAT] Vestigial tubules or cords of cells near the hilus of the ovary, corresponding with the rete testis, but not connected with the mesonephric duct.

rete testis [ANAT] The network of anastomosing tubules in the mediastinum testis.

Retgers' law [SOLID STATE] The law that the properties of crystalline mixtures of isomorphous substances are continuous functions of the percentage composition.

reticle [OPTICS] A series of intersecting fine lines, wires, or the like which are placed in the focus of the objective of an optical instrument to aid in measurement of angles or distances.

reticle image [OPTICS] A light image of the reticle in a computing gunsight or in certain types of optical gunsights and bombsights, cast on a reflector plate and superimposed on the target.

reticular cell *See* reticulocyte.

reticular degeneration [PATH] Rupture of epidermal cells with formation of multilocular bullae due to intracellular edema.

reticular density [MATH] The number of points per unit area in a two-dimensional lattice, such as the plane of a crystal lattice.

reticular fiber [HISTOL] Any of the delicate, branching argentophile fibers conspicuous in the connective tissue of lymphatic tissue, myeloid tissue, the red pulp of the spleen, and most basement membranes. Also known as argentaffin fiber; argyrophil lattice fiber; precollagenous fiber.

reticular formation [ANAT] The portion of the central nervous system which consists of small islands of gray matter separated by fine bundles of nerve fibers running in every direction.

Reticulariaceae [MYCOL] A family of plasmodial slime molds in the order Liceales.

reticular system *See* reticuloendothelial system.

reticular tissue [HISTOL] Connective tissue having reticular fibers as the principal element.

reticulate [BIOL] Having or resembling a network of fibers, veins, or lines. [GEN] Of or relating to evolutionary change resulting from genetic recombination between strains in an interbreeding population.

reticulated glass [MATER] Ornamental glassware containing interlacing sets of lines.

reticulation [GRAPHICS] The contraction or wrinkling of a film emulsion because of sudden changes of temperature during processing.

reticulin [BIOCHEM] A protein isolated from reticular fibers.

reticulocyte [HISTOL] Also known as reticular cell. **1.** A large, immature red blood cell, having a reticular appearance when stained due to retention of portions of the nucleus. **2.** A cell of reticular tissue.

reticuloendothelial granulomatosis [MED] A group of rare diseases characterized by generalized reticuloendothelial hyperplasia with or without intracellular lipid deposition.

reticuloendothelial system [ANAT] The macrophage system, including all phagocytic cells such as histiocytes, macrophages, reticular cells, monocytes, and microglia, except the granular white blood cells. Abbreviated RES. Also known as reticular system; hematopoietic system.

reticulopodia [INV ZOO] Pseudopodia in the form of a branching network.

Reticulosa [PALEON] An order of Paleozoic hexactinellid sponges with a branching form in the subclass Hexasterophora.

reticulospinal tract [ANAT] Nerve fibers descending from large cells of the reticular formation of the pons and medulla into the spinal cord.

reticulum [BIOL] A fine network. [VERT ZOO] The second stomach in ruminants.

Reticulum [ASTRON] A southern constellation, right ascension 4 hours, declination 60° south. Also known as Net.

retina [ADP] In optical character recognition, a scanning device. [ANAT] The photoreceptive layer and terminal expansion of the optic nerve in the dorsal aspect of the vertebrate eye.

retina character reader [ADP] A character reader that operates in the manner of the human retina in recognizing identical letters in different type fonts.

retinaculum [INV ZOO] **1.** A clasp on the forewing of certain moths for retaining the frenulum of the hindwing. **2.** An appendage on the third abdominal somite of springtails that articulates with the furcula.

retinal astigmatism [MED] Astigmatism due to changes in the localization of the fixation point.

retinal illuminance [OPTICS] A psychophysiological quantity which is a measure of the brightness of a visual sensation; it is measured in trolands.

retinal pigment *See* rhodopsin.

retinal retinitis *See* vascular retinopathy.

retinene [BIOCHEM] A pigment extracted from the retina, which turns yellow by the action of light; the chief carotenoid of the retina.

retinite [MINERAL] A fossil resin, such as glessite, krantzite, muckite, and ambrite, composed of 6–15% oxygen, lacking succinic acid, and found in brown coals and peat.

retinitis [MED] Inflammation of the retina.

retinitis pigmentosa [MED] A hereditary affection inherited as a sex-linked recessive and characterized by slowly progressing atrophy of the retinal nerve layers, and clumping of retinal pigment, followed by attenuation of the retinal arterioles and waxy atrophy of the optic disks.

retinoblastoma [MED] A malignant tumor of the sensory layer of the retina.

retinochoroiditis [MED] Inflammation of the retina and choroid.

retinol *See* vitamin A.

retinopathy [MED] Any pathologic condition involving the retina.

retinoschisis [MED] **1.** Separation with hole formation of the layers composing the retina. **2.** A congenital anomaly characterized by cleavage of the retina.

retinula [INV ZOO] The receptor element at the inner end of the ommatidium in a compound eye.

retire [NAV] To move a line of position back, parallel to itself, along a course line to obtain a line of position at an earlier time.

retired line of position [NAV] A line of position which has been moved backward along the course line to correspond with a time previous to that at which the line was established.

retort [CHEM ENG] **1.** A closed refractory chamber in which coal is carbonized for manufacture of coal gas. **2.** A vessel for the distillation or decomposition of a substance.

Retortamonadida [INV ZOO] An order of parasitic flagellate protozoans belonging to the class Zoomastigophorea, having two or four flagella and a complex blepharoplast-centrosome-axostyle apparatus.

retouch colors [GRAPHICS] Colors used to correct defects in black-and-white and color photographs; they adhere without crawling and can be used with brushes or airbrush on matte and glossy prints.

retrace *See* flyback.

retrace blanking [ELECTR] Blanking a television picture tube during vertical retrace intervals to prevent retrace lines from showing on the screen.

retrace line [ELECTR] The line traced by the electron beam in a cathode-ray tube in going from the end of one line or field to the start of the next line or field. Also known as return line.

retracker *See* rerailer.

retract [MATH] A subset R of a topological space X is a retract of X if there is a continuous map f from X to R, with $f(r) = r$ for all points r of R.

retractor [ANAT] A muscle that draws a limb or other body part toward the body. [MED] A clawlike instrument for holding tissues away from the surgical field.

retransmission unit [ELECTR] Control unit used at an intermediate station for feeding one radio receiver-transmitter unit for two-way communication.

retreat [MIN ENG] Workings in the opposite direction of advance work which, when completed, will permit the area to be abandoned as finished.

retreater [ENG] A defective maximum thermometer of the liquid-in-glass type in which the mercury flows too freely through the constriction; such a thermometer will indicate a maximum temperature that is too low.

retreat gun *See* evening gun.

retrievable inner barrel [ENG] The inner barrel assembly of a wire-line core barrel, designed for removing core from a borehole without pulling the rods.

retrieve [ADP] To find and select specific information.

retroaction *See* positive feedback.

retroactive refit *See* retrofit.

retrocerebral gland [INV ZOO] Any of various endocrine glands located behind the brain in insects which function in postembryonic development and metamorphosis.

retrodirective mirror [OPTICS] **1.** An optical system consisting of two mutually perpendicular plane mirrors; it reflects any beam of light which lies in a plane perpendicular to the mirrors into a direction antiparallel to its original direction. **2.** An optical system consisting of three mutually perpendicular plane mirrors; it reflects any beam of light into a direction antiparallel to its original direction.

retrofire time [AERO ENG] The computed starting time and duration of firing of retrorockets to decrease the speed of a recovery capsule and make it reenter the earth's atmosphere at the correct point for a planned landing.

retrofit [ENG] A modification of equipment to incorporate changes made in later production of similar equipment; it may be done in the factory or field. Derived from retroactive refit.

retroflexion [ANAT] The state of being bent backward. [MED] A condition in which the uterus is bent backward on itself, producing a sharp angle in its longitudinal axis at the junction of the cervix and the fundus.

retrogradation [CHEM] **1.** Generally, a process of deterioration; a reversal or retrogression to a simpler physical form. **2.** A chemical reaction involving vegetable adhesives, which revert to a simpler molecular structure.

retrograde amnesia [MED] Loss of memory for events occur-

RETORTAMONADIDA

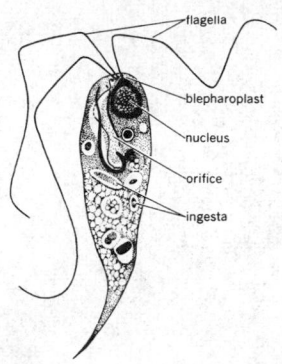

Chilomastix aulastomi.

Labels: flagella, blepharoplast, nucleus, orifice, ingesta

ring prior to, but not after, the onset of a current disease or trauma.

retrograde condensation [ORG CHEM] Phenomenon associated with the behavior of a hydrocarbon mixture in the critical region wherein, at constant temperature, the vapor phase in contact with the liquid may be condensed by a decrease in pressure; or at constant pressure, the vapor is condensed by an increase in temperature.

retrograde evaporation [ORG CHEM] Phenomenon associated with the behavior of a hydrocarbon mixture in the critical region wherein, at constant temperature, the liquid phase in contact with the vapor may be vaporized by an increase in pressure; or at constant pressure, the liquid is evaporated by a decrease in temperature.

retrograde gas-condensate reservoir See dew-point reservoir.

retrograde metamorphism [PETR] Formation of metamorphic minerals of a lower grade of metamorphism at the expense of minerals which are characteristic of a higher grade. Also known as diaphthoresis; retrogressive metamorphism.

retrograde motion [ASTRON] An apparent backward motion of a planet among the stars resulting from the observation of the planet from the planet earth which is also revolving about the sun at a different velocity.

retrograde orbit [ASTRON] Motion in an orbit opposite to the usual orbital direction of celestial bodies within a given system; specifically, of a satellite, motion in a direction opposite to the direction of rotation of the primary.

retrograde reservoir [GEOL] Hydrocarbon reservoir in which hydrocarbons are initially in the vapor phase; as pressure is reduced, the bubble-point line is passed and liquids are formed; upon further pressure reduction, a vapor phase is again formed.

retrograde wave [METEOROL] An atmospheric wave which moves in a direction opposite to that of the flow in which the wave is embedded; retrogression of a particular wave on daily charts is rarely seen, but is frequently observed on 4-day or monthly mean charts.

retrograding shoreline [GEOL] A shoreline that is being moved landward by wave erosion.

retrogression [MED] Going backward, as in degeneration or atrophy of tissues. [METEOROL] The motion of an atmospheric wave or pressure system in a direction opposite to that of the basic flow in which it is embedded. [PSYCH] Return to infantile behavior.

retrogressive metamorphism See retrograde metamorphism.

retrolental fibroplasia [MED] An oxygen-induced disease of the retina in premature infants characterized by formation of an opaque membrane behind the lens of the eye.

retroreflection [PHYS] Reflection wherein the reflected rays of radiation return along paths parallel to those of their corresponding incident rays.

retroreflector [PHYS] Any instrument used to cause reflected radiation to return along paths parallel to those of their corresponding incident rays; one type, the corner reflector, is an efficient radar target.

retrorocket [AERO ENG] A rocket fitted on or in a spacecraft, satellite, or the like to produce thrust opposed to forward motion. Also known as braking rocket.

retrorse [BIOL] Bent downward or backward.

retrostalsis [PHYSIO] Reverse peristalsis.

retroussage [GRAPHICS] The method of bringing ink up from incised lines in an intaglio plate; dragging a soft cloth across the ink-filled lines prior to printing makes the lines wider and renders certain passages darker and richer.

retroversion [ANAT] A turning back. [MED] A condition in which the uterus is tilted backward without any change in the angle of its longitudinal axis.

retry [ADP] When a central processing unit error is detected during execution of an instruction, the computer will execute this instruction unless a register was altered by the operation.

retting [CHEM ENG] Soaking vegetable stalks to decompose the gummy material and release the fibers.

return [ADP] 1. To return control from a subroutine to the calling program. 2. To go back to a planned point in a computer program and rerun a portion of the program,

usually when an error is detected; rerun points are usually not more than 5 minutes apart. [ELECTR] See echo.

return bend [DES ENG] A pipe fitting, equal to two ells, used to connect parallel pipes so that fluid flowing into one will return in the opposite direction through the other.

return busy tone [COMMUN] An appropriate signal returned to the register-sender that, in turn, returns a busy indication to the calling station.

return connecting rod [MECH ENG] A connecting rod whose crankpin end is located on the same side of the crosshead as the cylinder.

return difference [CONT SYS] The difference between 1 and the loop transmittance.

return idler [MECH ENG] The idler or roller beneath the cover plates on which the conveyor belt rides after the load which it was carrying has been dumped.

return interval [ELECTR] Interval corresponding to the direction of sweep not used for delineation.

return jump [ADP] A jump instruction in a subroutine which passes control to the first statement in the program which follows the instruction called the subroutine.

return line See retrace line.

return loss [COMMUN] 1. The difference between the power incident upon a discontinuity in a transmission system and the power reflected from the discontinuity. 2. The ratio in decibels of the power incident upon a discontinuity to the power reflected from the discontinuity.

return streamer [GEOPHYS] The intensely luminous streamer which propagates upward from earth to cloud base in the last phase of each lightning stroke of a cloud-to-ground discharge. Also known as main stroke; return stroke.

return stroke See return streamer.

return to zero mode [ADP] Computer readout mode in which the signal returns to zero between each bit indication.

return trace See flyback.

return wall [BUILD] An interior wall of about the same height as the outside wall of a building; distinct from a partition or a low wall.

return water [PETRO ENG] In a water-injection operation (waterflood) for an oil reservoir, the reinjection of salt water that is produced along with the oil.

return wire [ELEC] The ground wire, common wire, or negative wire of a direct-current power circuit.

retuse [BOT] Having a rounded apex with a slight, central notch.

reusable [ADP] Of a program, capable of being used by several tasks without having to be reloaded; it is a generic term, including reenterable and serially reusable.

reveille gun [ORD] The firing of a gun at the first note of reveille or at sunrise. Also known as morning gun.

revenue cutter [NAV ARCH] A ship used by the government mainly for enforcement of revenue laws and prevention of smuggling.

reverberation [ACOUS] The prolongation of sound at a given point after direct reception from the source has ceased, due to such causes as reflections from bounding surfaces, scattering from inhomogeneities in a medium, and vibrations excited by the original sound.

reverberation chamber [ACOUS] An enclosure with heavy surfaces which randomly reflect as great an amount of sound as possible; used in acoustic measurements. Also known as random diffusion chamber.

reverberation time [ACOUS] The time in seconds required for the average sound-energy density at a given frequency to reduce to one-millionth of its initial steady-state value after the sound source has been stopped; this corresponds to a decrease of 60 decibels.

reverberatory furnace [ENG] A furnace in which heat is supplied by burning of fuel in a space between the charge and the low roof.

reversal film [GRAPHICS] A type of film designed to yield a positive image directly when reversal processed.

reversal of dip [GEOL] Change in the dip direction of bedding near a fault such that the beds curve toward the fault surface in a direction exactly opposite that of the drag folds. Also known as dip reversal.

reversal process [GRAPHICS] A method in which a positive

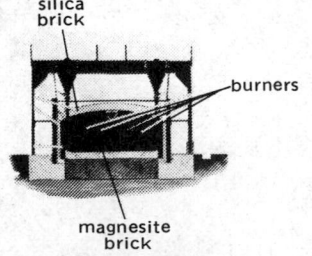

REVERBERATORY FURNACE

silica
brick

burners

magnesite
brick

Cross section of a reverberatory furnace for smelting copper.

is produced from the plate or film exposed in a camera; the negative and positive stages are often produced from the same emulsion layer, with no separate printing material required; only one copy can be made at a time.

reversal spectrum [SPECT] A spectrum which may be observed in intense white light which has traversed luminous gas, in which there are dark lines where there were bright lines in the emission spectrum of the gas.

reversal temperature [SPECT] The temperature of a blackbody source such that, when light from this source is passed through a luminous gas and analyzed in a spectroscope, a given spectral line of the gas disappears, whereas it appears as a bright line at lower blackbody temperatures, and a dark line at higher temperatures.

reverse bias [ELECTR] A bias voltage applied to a diode or a semiconductor junction with polarity such that little or no current flows; the opposite of forward bias.

reverse-blocking tetrode thyristor See silicon-controlled switch.

reverse-blocking triode thyristor See silicon-controlled rectifier.

reverse Brayton cycle [THERMO] A refrigeration cycle using air as the refrigerant but with all system pressures above the ambient. Also known as dense-air refrigeration cycle.

reverse Carnot cycle [THERMO] An ideal thermodynamic cycle consisting of the processes of the Carnot cycle reversed and in reverse order, namely, isentropic expansion, isothermal expansion, isentropic compression, and isothermal compression.

reverse cell [METEOROL] A circulating fluid system in which the circulation in a vertical plane is thermally indirect; that is, cooler air rises relative to warmer air.

reverse circulation drilling [MIN ENG] 1. A variation of the rotary drilling method in which the cuttings are pumped up and out of the drill pipe, an advantage in certain large diameter holes. 2. In diamond core drilling, when the water is injected through a stuffing box into the annular space around the drill rods and thus forced up special large drill rods.

reverse code dictionary [ADP] Alphabetic or alphanumeric arrangement of codes associated with their corresponding English words or terms.

reverse current [ELECTR] Small value of direct current that flows when a semiconductor diode has reverse bias.

reverse-current cleaning See anodic cleaning.

reverse-current protection [ELEC] A device which senses when there is a reversal in the normal direction of current in an electric power system, indicating an abnormal condition of the system, and which initiates appropriate action to prevent damage to the system.

reverse-current relay [ELEC] Relay that operates whenever current flows in the reverse direction.

reverse curve [MATH] An S-shaped curve, that is, one having two arcs with their centers on opposite sides of the curve. Also known as S curve.

reversed See overturned.

reversed air-blast process [CHEM ENG] A gasmaking process in which, after a short period of the ordinary blow, the air blast is reversed so as to enter the top of the superheater, and passes back to the top of the generator and down.

reversed arc [GEOL] A curved belt of islands which is concave toward the open ocean, the opposite of most island arcs.

reversed image [GRAPHICS] 1. A mirror image in which the right and left sides of the picture are interchanged. 2. See negative. [OPTICS] See inverted image.

reverse-direction flow [ADP] A logical path that runs upward or to the left on a flowchart.

reversed-phase partition chromatography [ANALY CHEM] Paper chromatography in which the low-polarity phase (such as paraffin, paraffin jelly, or grease) is put onto the support (paper) and the high-polarity phase (such as water, acids, or organic solvents) is allowed to flow over it.

reversed polarity [GEOPHYS] Natural remanent magnetism opposite that of the present geomagnetic field.

reverse drawing [MET] Drawing for a second time, in a direction opposite to the original drawing.

reverse fault See thrust fault.

reverse feedback See negative feedback.

reverse flange [ENG] A flange made by shrinking.

reverse graft [BOT] A plant graft made by inserting the scion in an inverted position.

reverse key [ELEC] Key used in a circuit to reverse the polarity of that circuit.

reverse lay [DES ENG] The lay of a wire rope with strands alternating in a right and left lay.

reverse mutation [GEN] A mutation in a mutant allele which makes it capable of producing the nonmutant phenotype; may actually restore the original deoxyribonucleic-acid sequence of the gene or produce a new one which has a similar effect. Also known as back mutation.

reverse osmosis [CHEM ENG] A technique used in desalination and waste-water treatment; pressure is applied to the surface of a saline (or waste) solution, forcing pure water to pass from the solution through a membrane (hollow fibers of cellulose acetate or nylon) that will not pass sodium or chloride ions.

reverse passive anaphylaxis [IMMUNOL] Hypersensitivity produced when the antigen is injected first, then followed in several hours by the specific antibody, causing shock.

reverse pinocytosis See emiocytosis.

reverse pitch [MECH ENG] A pitch on a propeller blade producing thrust in the direction opposite to the normal one.

reverse polarity [MET] An arc-welding circuit in which the electrode is connected to the positive terminal.

reverse process [GRAPHICS] The conversion of a negative to a positive image by chemical means.

reverse-roll coating [ENG] Substrate coating that is premetered between rolls and then wiped off on the web; amount of coating is controlled by the metering gap and the rotational speed of the roll.

reverse slip fault See thrust fault.

reverse slope [GEOL] A hill descending away from a ridge.

reverse voltage [ELEC] In the case of two opposing voltages, voltage of that polarity which produces the smaller current.

reversibility principle [OPTICS] The principle that if a beam of light is reflected back on itself, it will traverse the same path or paths as it did before reversal. [STAT MECH] See microscopic reversibility.

reversible booster [ELEC] Booster capable of adding to and subtracting from the voltage of a circuit.

reversible capacitance [ELECTR] Limit, as the amplitude of an applied sinusoidal capacitor voltage approaches zero, of the ratio of the amplitude of the resulting in-phase fundamental-frequency component of transferred charge to the amplitude of the applied voltage, for a given constant bias voltage superimposed on the sinusoidal voltage.

reversible chemical reaction [CHEM] A chemical reaction that can be made to proceed in either direction by suitable variations in the temperature, volume, pressure, or quantities of reactants or products.

reversible electrode [PHYS CHEM] An electrode that owes its potential to unit charges of a reversible nature, in contrast to electrodes used in electroplating and destroyed during their use.

reversible engine [THERMO] An ideal engine which carries out a cycle of reversible processes.

reversible motor [ELEC] A motor in which the direction of rotation can be reversed by means of a switch that changes motor connections when the motor is stopped.

reversible path [THERMO] A path followed by a thermodynamic system such that its direction of motion can be reversed at any point by an infinitesimal change in external conditions; thus the system can be considered to be at equilibrium at all points along the path.

reversible-pitch propeller [MECH ENG] A type of controllable-pitch propeller; of either controllable or constant speed, it has provisions for reducing the pitch to and beyond the zero value, to the negative pitch range.

reversible process [THERMO] An ideal thermodynamic process which can be exactly reversed by making an indefinitely small change in the external conditions. Also known as quasistatic process.

reversible steering gear [MECH ENG] A steering gear for a vehicle which permits road shock and wheel deflections to

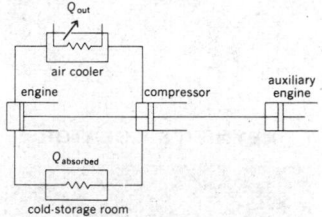
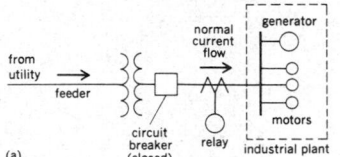
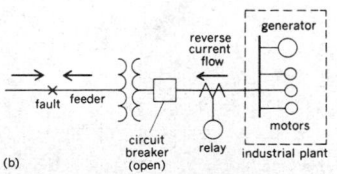

come through the system and be felt in the steering control.

reversible tramway   *See* jig back.

reversible transducer   [ELECTR] Transducer whose loss is independent of transmission direction.

reversing current   [OCEANOGR] Any current that changes direction, with a period of slack water at each reversal of direction.

reversing layer   [ASTROPHYS] Cooler layer of thinner gases in the sun's atmosphere, just above the photosphere; it produces the dark lines in the sun's spectrum.

reversing mill   [MET] A rolling mill in which the workpiece is passed forward and backward through a given pair of rolls.

reversing motor   [ELEC] A motor for which the direction of rotation can be reversed by changing electric connections or by other means while the motor is running at full speed; the motor will then come to a stop, reverse, and attain full speed in the opposite direction.

reversing switch   [ELEC] A switch intended to reverse the connections of one part of a circuit.

reversing thermometer   [ENG] A mercury-in-glass thermometer which records temperature upon being inverted and thereafter retains its reading until returned to the first position.

reversing water bottle   *See* Nansen bottle.

revetment   [CIV ENG] A facing made on a soil or rock embankment to prevent scour by weather or water.   [ORD] A retaining wall with a facing such as concrete or stone, commonly used for fortifications or to protect against explosions.

revived fault scarp   *See* rejuvenated fault scarp.

revived stream   *See* rejuvenated stream.

revolute   [BOT] Rolled backward and downward.

revolution   [GEOL] A little-used term to describe a time of profound crustal movements, on a continentwide or worldwide scale, which led to abrupt geographic, climatic, and environmental changes that were related to changes in forms of life.

revolution counter   [ENG] An instrument for registering the number of revolutions of a rotating machine. Also known as revolution indicator.

revolution indicator   *See* revolution counter.

revolution per minute   [MECH] A unit of angular velocity equal to the uniform angular velocity of a body which rotates through an angle of 360° (2π radians), so that every point in the body returns to its original position, in 1 minute. Abbreviated rpm.

revolution per second   [MECH] A unit of angular velocity equal to the uniform angular velocity of a body which rotates through an angle of 360° (2π radians), so that every point in the body returns to its original position, in 1 second. Abbreviated rps.

revolution table   [NAV ARCH] A table giving the number of shaft revolutions corresponding to various speeds of a vessel.

revolver   [NAV] The pair of horizontal angles between three points, as observed at any place on the circle defined by the three points; this is the one situation in which such angles do not establish a fix. Also known as swinger.   [ORD] A firearm with a cylinder of several chambers so arranged as to revolve on an axis and be discharged in succession by the same lock.

revolving-block engine   [MECH ENG] Any of various engines which combine reciprocating piston motion with rotational motion of the entire engine block.

revolving door   [BUILD] A door consisting of four leaves that revolve together on a central vertical axis within a circular vestibule.

revolving shovel   [MECH ENG] A digging machine, mounted on crawlers or on rubber tires, that has the machinery deck and attachment on a vertical pivot so that it can swing freely.

revolving storm   [METEOROL] A cyclonic storm, or one in which the wind revolves about a central low-pressure area.

rewind   [ELECTR] **1.** The components on a magnetic tape recorder that serve to return the tape to the supply reel at high speed. **2.** To return a magnetic tape to its starting position.

rework   [GEOL] Any geologic material that has been removed or displaced by natural agents from its origin and incorporated in a younger formation.

rewrite   [ADP] The process of restoring a storage device to its

state prior to reading; used when the information-storing state may be destroyed by reading.

Rexforming   [CHEM ENG] A proprietary petroleum-refinery process combining Platforming with aromatics extraction, wherein low-octane raffinate is recycled to the Platformer.

reyn   [FL MECH] A unit of dynamic viscosity equal to the dynamic viscosity of a fluid in which there is a tangential force of 1 poundal per square foot resisting the flow of two parallel fluid layers past each other when their differential velocity is 1 foot per second per foot of separation; equal to approximately 14.8816 poise.

Reynier's isolator   [ENG] A mechanical barrier made of steel that surrounds the area in which germ-free vertebrates and accessory equipment are housed; has electricity for light and power, an exit-entry opening with a steam barrier, a means for sterile air exchange, glass viewing port, and neoprene gloves which allow handling of the animals.

Reynolds analogy   [CHEM ENG] Relationship showing the similarity between the transfer of mass, heat, and momentum.

Reynolds criterion   [FL MECH] The principle that the type of fluid motion, that is, laminar flow or turbulent flow, in geometrically similar flow systems depends only on the Reynolds number; for example, in a pipe, laminar flow exists at Reynolds numbers less than 2000, turbulent flow at numbers above about 3000.

Reynolds effect   [METEOROL] A process of drop growth in clouds which involves net evaporation from cloud drops warmer than others and net condensation on the cooler drops.

Reynolds equation   [FL MECH] A form of the Navier-Stokes equation which is $\rho\partial u/\partial t = (\partial/\partial x)(p_{xx} - \rho u^2) + (\partial/\partial y)(p_{xy} - \rho uv) + (\partial/\partial z)(p_{xz} - \rho uw)$, where ρ is the fluid density, u, v, and w are the components of the fluid velocity, and p_{xx}, p_{xy}, and p_{xz} are normal and shearing stresses.

Reynolds model   [OCEANOGR] A laboratory model of ocean currents in which inertial forces and frictional forces predominate, and in which the Reynolds number is used extensively in calculations.

Reynolds number   [FL MECH] A dimensionless number which is significant in the design of a model of any system in which the effect of viscosity is important in controlling the velocities or the flow pattern of a fluid; equal to the density of a fluid, times its velocity, times a characteristic length, divided by the fluid viscosity. Symbolized N_{Re}. Also known as Damköhler number V (DaV).

Reynolds stress   [FL MECH] The net transfer of momentum across a surface in a turbulent fluid because of fluctuations in fluid velocity. Also known as eddy stress.

Reynolds stress tensor   [FL MECH] A tensor whose components are the components of the Reynolds stress across three mutually perpendicular surfaces.

rezbanyite   [MINERAL] $Pb_3Cu_2Bi_{10}S_{19}$ A metallic-gray mineral composed of sulfide of lead, copper, and bismuth.

rf   *See* radio frequency.

R factor   [GEN] A self-replicating, infectionlike agent that carries genetic information and transmits drug resistance from bacterium to bacterium by conjugation of cell. Also known as resistance factor.

RFI   *See* radio-frequency interference.

rf preheating   *See* radio-frequency preheating.

RG line   *See* radio-frequency cable.

RH   *See* relative humidity.

Rhabdiasoidea   [INV ZOO] An order or superfamily of parasitic nematodes.

rhabdion   [INV ZOO] One of the sclerotized segments lining the buccal cavity of nematodes.

rhabdite   [INV ZOO] A small rodlike or fusiform body secreted by epidermal or parenchymal cells of certain turbellarians and trematodes.   [MINERAL] *See* schreibersite.

Rhabditia   [INV ZOO] A subclass of nematodes in the class Secernentea.

Rhabditidia   [INV ZOO] An order of nematodes in the subclass Rhabditia including parasites of man and domestic animals.

Rhabditoidea   [INV ZOO] A superfamily of small to moderate-sized nematodes in the order Rhabditidia with small, pore-like, anteriorly located amphids, and esophagus with corpus, isthmus, and valvulated basal bulb.

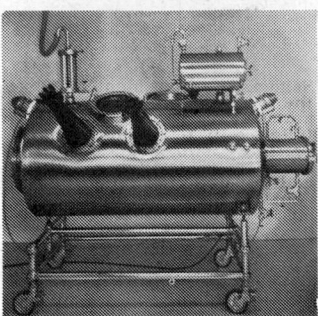

Rhabdocoela [INV ZOO] Formerly an order of the Turbellaria, and now divided into three orders, Catenulida, Macrostomida, and Neorhabdocoela.

rhabdolith [BOT] A minute coccolith having a shield surmounted by a long stem and found at all depths in the ocean, from the surface to the bottom.

rhabdome [INV ZOO] The central translucent cylinder in the retinula of a compound eye.

rhabdomyoblastoma *See* rhabdomyosarcoma.

rhabdomyoma [MED] A benign tumor of skeletal muscle.

rhabdomyosarcoma [MED] A malignant tumor of skeletal muscle in the extremities composed of anaplastic muscle cells. Also known as malignant rhabdomyoma; rhabdomyoblastoma.

Rhabdophorina [INV ZOO] A suborder of ciliates in the order Gymnostomatida.

rhabdosome [INV ZOO] A colonial graptolite that develops from a single individual.

rhabdovirus [VIROL] A group of ribonucleic acid–containing animal viruses, including rabies virus and certain infective agents of fish and insects.

rhabdus [INV ZOO] A uniaxial sponge spicule.

Rhachitomi [PALEON] A group of extinct amphibians in the order Temnospondyli in which pleurocentra were retained.

rhachitomous [VERT ZOO] Being, having, or pertaining to vertebrae with centra whose parts do not fuse.

Rhacophoridae [VERT ZOO] A family of arboreal frogs in the suborder Diplasiocoela.

Rhacopilaceae [BOT] A family of mosses in the order Isobryales generally having dimorphous leaves with smaller dorsal leaves and a capsule that is plicate when dry.

Rhaetian [GEOL] A European stage of geologic time; the uppermost Triassic (above Norian, below Hettangian of Jurassic). Also known as Rhaetic.

Rhaetic *See* Rhaetian.

Rhagionidae [INV ZOO] The snipe flies, a family of predatory orthorrhaphous dipteran insects in the series Brachycera that are brownish or gray with spotted wings.

rhagon [INV ZOO] A pyramid-shaped, colonial sponge having an osculum at the apex and flagellated chambers in the upper wall only.

Rhamnaceae [BOT] A family of dicotyledonous plants in the order Rhamnales characterized by a solitary ovule in each locule, free stamens, simple leaves, and flowers that are hypogynous to perigynous or epigynous.

Rhamnales [BOT] An order of dicotyledonous plants in the subclass Rosidae having a single set of stamens, opposite the petals, usually a well-developed intrastamenal disk, and two or more locules in the ovary.

rhamnose [BIOCHEM] $C_6H_{12}O_5$ A deoxysugar occurring free in poison sumac, and in glycoside combination in many plants. Also known as isodulcitol.

rhamphoid [BIOL] Beak-shaped.

Rhamphorhynchoidea [PALEON] A Jurassic suborder of the Pterosauria characterized by long, slender tails with an expanded tip.

rhampotheca [VERT ZOO] The horny sheath covering a bird's beak.

Rh antigen *See* Rh factor.

Rh blocking serum [IMMUNOL] A serum that reacts with Rh-positive blood without causing agglutination, but which blocks the action of anti-Rh serums that are subsequently introduced.

Rh blocking test [IMMUNOL] A test for the detection of Rh antibody in plasma wherein erythrocytes having the Rh antigen are incubated in the patient's serum so that the antibodies may be adsorbed on these cells, which are then employed in the antiglobulin test. Also known as indirect Coombs test; indirect developing test.

Rh blood group [IMMUNOL] The extensive, genetically determined system of red blood cell antigens defined by the immune serum of rabbits injected with rhesus monkey erythrocytes, or by human antiserums. Also known as rhesus blood group.

rhe [FL MECH] **1.** A unit of dynamic fluidity, equal to the dynamic fluidity of a fluid whose dynamic viscosity is 1 centipoise. **2.** A unit of kinematic fluidity, equal to the kinematic fluidity of a fluid whose kinematic viscosity is 1 centistoke.

rhea [BOT] *See* ramie. [VERT ZOO] The common name for members of the avian order Rheiformes.

Rhea [ASTRON] A satellite of Saturn, estimated diameter is 1200 kilometers.

rhegmagenesis [GEOL] Orogeny characterized by the development of large-scale strike-slip faults.

rheid [GEOL] A substance (below its melting point) which deforms by viscous flow during applied stress at an order of magnitude at least three times that of elastic deformation under similar circumstances.

Rheidae [VERT ZOO] The single family of the avian order Rheiformes.

rheid fold [GEOL] A fold whose strata deform by viscous flow as if they were fluid.

rheidity [GEOL] Relaxation time of a substance, divided by 1000.

Rheiformes [VERT ZOO] The rheas, an order of South American running birds; called American ostriches, they differ from the true ostrich in their smaller size, feathered head and neck, three-toed feet, and other features.

Rheinberg illumination [OPTICS] An illumination technique used in optical microscopes that is a modification of the dark-field method; the central disk is transparent and colored; an annulus of a complementary color fills the remaining condenser aperture; the specimen is seen in the color of the annulus against the background of the central disk. Also known as optical staining.

Rhenanida [PALEON] An order of extinct marine fishes in the class Placodermi distinguished by mosaics of small bones between the large plates in the head shield.

rhenium [CHEM] A metallic element, symbol Re, atomic number 75, atomic weight 186.2; a transition element.

rhenium halide [INORG CHEM] Halogen compound of rhenium; examples are $ReCl_3$, $ReCl_4$, ReF_4, and ReF_6.

rheobase [PHYSIO] The intensity of the steady current just sufficient to excite a tissue when suddenly applied.

rheoelectroencephalograph [MED] Electroencephalograph for measuring differential blood flow in both sides of the brain and in any other part of the body.

rheogoniometry [MECH] Rheological tests to determine the various stress and shear actions on Newtonian and non-Newtonian fluids.

rheology [MECH] The study of the deformation and flow of matter, especially non-Newtonian flow of liquids and plastic flow of solids.

rheomorphism [PETR] Mobilization of a rock by at least partial fusion accompanied by, and sometimes promoted by, addition of new material by diffusion.

rheopectic fluid [FL MECH] A fluid for which the structure builds up on shearing; this phenomenon is regarded as the reverse of thixotropy.

rheopexy [PHYS CHEM] A property of certain sols, having particles shaped like rods or plates, which set to gel form more quickly when mechanical means are used to hasten the orientation of the particles.

rheophile [ECOL] Living or thriving in running water.

rheophilous bog [ECOL] A bog which draws its source of water from drainage.

rheoplankton [ECOL] Plankton found in flowing water.

rheostat [ELEC] A resistor constructed so that its resistance value may be changed without interrupting the circuit to which it is connected. Also known as variable resistor.

rheostatic braking [ENG] A system of dynamic braking in which direct-current drive motors are used as generators and convert the kinetic energy of the motor rotor and connected load to electrical energy, which in turn is dissipated as heat in a braking rheostat connected to the armature.

rheostriction *See* pinch effect.

rheotaxial growth [ENG] A chemical vapor deposition technique for producing silicon diodes and transistors on a fluid layer having high surface mobility.

rheotaxis [BIOL] Movement of a motile cell or organism in response to the direction of water currents.

rheotron *See* betatron.

rheotropic brittleness [MET] A low-temperature or high-

RHAMPHORHYNCHOIDEA

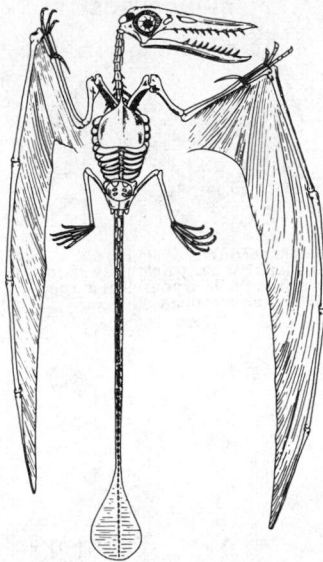

Restoration of Jurassic pterosaur skeleton of *Rhamphorhynchus* by S. W. Williston; length about 30 centimeters. (*After O. C. Marsh*)

RHENANIDA

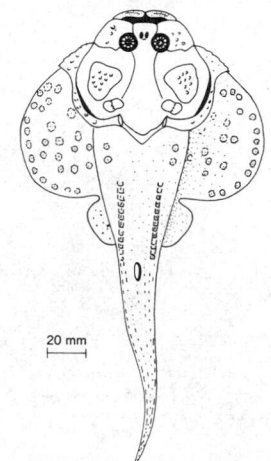

Gemuendina stuertzi from the Lower Devonian of Germany; restoration in dorsal aspect. (*After Walter Gross*)

RHENIUM

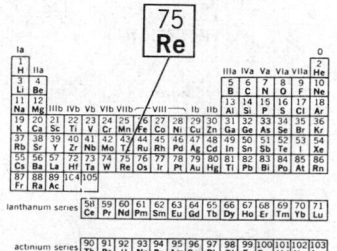

Periodic table of the chemical elements showing the position of rhenium.

RHINOCEROS

Black rhinoceros (*Diceros bicornis*) with thick gray skin and a mobile upper lip shaped like a parrot beak.

RHIZOMASTIGIDA

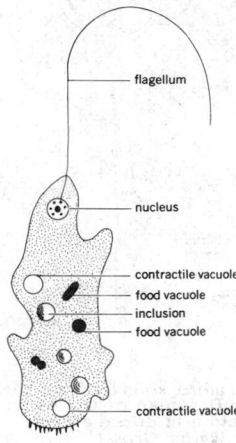

Mastigamoeba reptans, with lobose pseudopodia.

RHIZOME

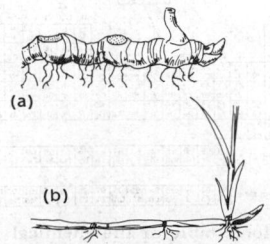

The underground stems of (*a*) Solomon's seal and (*b*) grass.

strain-rate brittleness that may be eliminated by prestraining under milder conditions.

rheotropism [BIOL] Orientation response of an organism to the stimulus of a flowing fluid, as water.

rhesus blood group *See* Rh blood group.

rhesus factor *See* Rh factor.

rhesus macaque *See* rhesus monkey.

rhesus monkey [VERT ZOO] *Macaque mulatta.* An agile, gregarious primate found in southern Asia and having a short tail, short limbs of almost equal length, and a stocky build. Also known as rhesus macaque.

rheumatic arteritis [MED] A type of allergic arteritis associated with acute rheumatic fever.

rheumatic carditis [MED] Inflammation of the heart resulting from rheumatic fever.

rheumatic encephalopathy [MED] An inflammatory reaction of the brain and the smaller arteries of the cerebral cortex associated with rheumatic fever.

rheumatic endocarditis [MED] Inflammation of the endocardium in acute rheumatic fever, usually involving heart valves.

rheumatic fever [MED] A febrile disease occurring in childhood as a delayed sequel of infection by *Streptococcus hemolyticus*, group A; characterized by arthritis, carditis, nosebleeds, and chorea.

rheumatic pneumonia [MED] Pneumonia associated with acute rheumatic fever.

rheumatism [MED] Any combination of muscle or joint pain, stiffness, or discomfort arising from nonspecific disorders.

rheumatoid arthritis [MED] A chronic systemic inflammatory disease of connective tissue in which symptoms and changes predominate in articular and related structures. Also known as atrophic arthritis; chronic infectious arthritis; proliferative arthritis.

rheumatoid nodules [MED] Subcutaneous lateral foci of fibrinoid degeneration or necrosis surrounded by mononuclear cells in a regular palisade arrangement, occurring usually in association with rheumatoid arthritis or rheumatic fever.

rheumatoid spondylitis [MED] A chronic progressive arthritis of young men, affecting mainly the spine and sacroiliac joints, leading to fusion and deformity. Also known as Marie-Strümpell disease.

rheum emodin *See* emodin.

Rh factor [IMMUNOL] Any of several red blood cell antigens originally identified in the blood of rhesus monkeys. Also known as Rh antigen; rhesus factor.

RHI display *See* range-height indicator display.

Rhigonematidae [INV ZOO] A family of nematodes in the superfamily Oxyuroidea.

Rhincodontidae [VERT ZOO] The whale sharks, a family of essentially tropical galeoid elasmobranchs in the isurid line.

rhinencephalon [ANAT] The anterior olfactory portion of the vertebrate brain.

rhinestone [MATER] A clear, colorless imitation of diamond, made of glass, paste, or gem quartz, backed with metallic foil.

rhinitis [MED] Inflammation of the mucous membranes in the nose.

Rhinobatidae [VERT ZOO] The guitarfishes, a family of elasmobranchs in the batoid group.

Rhinoceratidae [VERT ZOO] A family of perissodactyl mammals in the superfamily Rhinoceratoidea, comprising the living rhinoceroses.

Rhinoceratoidea [VERT ZOO] A superfamily of perissodactyl mammals in the suborder Ceratomorpha including living and extinct rhinoceroses.

rhinoceros [VERT ZOO] The common name for the odd-toed ungulates composing the family Rhinoceratidae, characterized by massive, thick-skinned limbs and bodies, and one or two horns which are composed of a solid mass of hairs attached to the bony prominence of the skull.

Rhinochimaeridae [VERT ZOO] A family of ratfishes, order Chimaeriformes, distinguished by an extremely elongate rostrum.

Rhinocryptidae [VERT ZOO] The tapaculos, a family of ground-inhabiting suboscine birds in the suborder Tyranni characterized by a large, movable flap which covers the nostrils.

rhinogenous [MED] Originating in the nose.

rhinolaryngology [MED] The science of the anatomy, physiology, and pathology of the nose and larynx.

rhinolaryngoscope [MED] A scope containing a mirror and a light, used to examine the nose and larynx.

rhinology [MED] The science of the anatomy, functions, and diseases of the nose.

Rhinolophidae [VERT ZOO] The horseshoe bats, a family of insect-eating chiropterans widely distributed in the Eastern Hemisphere and distinguished by extremely complex, horseshoe-shaped nose leaves.

rhinopharyngitis [MED] Inflammation of the nose and pharynx or of the nasopharynx.

rhinophore [INV ZOO] An olfactoreceptor of certain land mollusks, usually borne on a tentacle.

rhinoplasty [MED] A plastic operation on the nose.

Rhinopomatidae [VERT ZOO] The mouse-tailed bats, a small family of insectivorous chiropterans found chiefly in arid regions of northern Africa and southern Asia and characterized by long, wirelike tails and rudimentary nose leaves.

Rhinopteridae [VERT ZOO] The cow-nosed rays, a family of batoid sharks having a fleshy pad at the front end of the head and a well-developed poison spine.

rhinorrhea [MED] **1.** A mucous discharge from the nose. **2.** Escape of cerebrospinal fluid through the nose.

rhinoscleroma [MED] A chronic infectious bacterial disease caused by *Klebsiella rhinoscleromatis* and characterized by hard nodules and plaques of inflamed tissue in the nose and adjacent areas.

rhinoscope [MED] An instrument for examining the nasal cavities.

Rhinotermitidae [INV ZOO] A family of lower termites of the order Isoptera.

rhinotheca [VERT ZOO] The horny sheath on the upper part of a bird's bill.

rhinovirus [VIROL] A subgroup of the picornavirus group including small, ribonucleic acid–containing forms which are not inactivated by ether.

Rhipiceridae [INV ZOO] The cedar beetles, a family of coleopteran insects in the superfamily Elateroidea.

Rhipidistia [VERT ZOO] The equivalent name for Osteolepiformes.

rhipidium [BOT] A fan-shaped inflorescence with cymose branching in which branches lie in the same plane and are suppressed alternately on each side.

Rhipiphoridae [INV ZOO] The wedge-shaped beetles, a family of coleopteran insects in the superfamily Meloidea.

rhizanthous [BOT] Producing flowers directly from the root.

rhizautoicous [BOT] Of mosses, having the antheridial branch and the archegonial branch connected by rhizoids.

rhizic water *See* soil water.

rhizine [BOT] The rhizoid of a lichen.

Rhizobiaceae [MICRO] A family of rod-shaped, gram-negative, soil-inhabiting bacteria consisting of two genera, *Rhizobium* and *Agrobacterium*, in the order Eubacteriales.

rhizocarpous [BOT] Pertaining to perennial herbs having perennating underground parts from which stems and foliage arise annually.

Rhizocephala [INV ZOO] An order of crustaceans which parasitize other crustaceans; adults have a thin-walled sac enclosing the visceral mass and show no trace of segmentation, appendages, or sense organs.

Rhizochloridina [INV ZOO] A suborder of flagellate protozoans in the order Heterochlorida.

Rhizodontidae [PALEON] An extinct family of lobefin fishes in the order Osteolepiformes.

rhizoid [BOT] A rootlike structure which helps to hold the plant to a substrate; found on fungi, liverworts, lichens, mosses, and ferns.

Rhizomastigida [INV ZOO] An order of the protozoan class Zoomastigophorea; all species are microscopic and ameboid, and have one or two flagella.

Rhizomastigina [INV ZOO] The equivalent name for Rhizomastigida.

rhizome [BOT] An underground horizontal stem, often thickened and tuber-shaped, and possessing buds, nodes, and scalelike leaves.

rhizomorph [BOT] A rootlike structure, characteristic of many basidiomycetes, consisting of a mass of densely packed and intertwined hyphae.

Rhizophagidae [INV ZOO] The root-eating beetles, a family of minute coleopteran insects in the superfamily Cucujoidea.

Rhizophoraceae [BOT] A family of dicolyledonous plants in the order Cornales distinguished by opposite, stipulate leaves, two ovules per locule, folded or convolute bud petals, and a berry fruit.

rhizophore [BOT] A leafless, downward-growing dichotomous *Selaginella* shoot that has tufts of adventitious roots at the apex.

rhizoplast [CYTOL] A delicate fiber or thread running between the nucleus and the blepharoplast in cells bearing flagella.

rhizopod [INV ZOO] An anastomosing rootlike pseudopodium.

Rhizopodea [INV ZOO] A class of the protozoan superclass Sarcodina in which pseudopodia may be filopodia, lobopodia, or reticulopodia, or may be absent.

rhizosphere [GEOL] The soil region subject to the influence of plant roots and characterized by a zone of increased microbiological activity.

Rhizostomeae [INV ZOO] An order of the class Scyphozoa having the umbrella generally higher than it is wide with the margin divided into many lappets but not provided with tentacles.

rhizotomy [MED] Surgical division of any root, as of a nerve. Also known as radicotomy.

RHM See roentgen-per-hour-at-one-meter.

rhodamine B [ORG CHEM] $C_{28}H_{31}ClN_2O_3$ Red, green, or reddish-violet powder, soluble in alcohol and water; forms bluish red, fluorescent solution in water; used as red dye for paper, wool, and silk, and as an analytical reagent and biological stain.

rhodamine toner [MATER] Rhodamine dye and phosphotungstic or phosphomolybdic acid; red to maroon; used in printing inks.

rhodanate See thiocyanate.

rhodanic acid See thiocyanic acid.

rhodanide See thiocyanate.

rhodanine [ORG CHEM] $C_3H_3NOS_2$ A pale-yellow crystalline compound that may decompose violently when heated, giving off toxic by-products; used in organic synthesis. Also known as 2-thioxo-4-thiazolidinone.

L-rhodeose See L-fucose.

Rhodesian man [PALEON] A type of fossil man inhabiting southern and central Africa during the late Pleistocene; the skull was large and low, marked by massive browridges, with a cranial capacity of 1300 cubic centimeters or less.

Rhodesian trypanosomiasis [MED] A fulminating form of African sleeping sickness caused by *Trypanosoma rhodesiense*, transmitted by the tsetse fly, and characterized by parasitemia, edema, lymphadenitis, and myocarditis. Also known as East African sleeping sickness.

Rhodininae [INV ZOO] A subfamily of limivorous worms in the family Maldanidae.

rhodinol [MATER] Colorless, combustible liquid mixture of terpene alcohols with rose scent; soluble in mineral oil and alcohol; derived from geranium oil; used in perfumes and flavors.

rhodinyl acetate [MATER] Terpene-alcohol-acetates mixture; colorless-to-yellow, combustible liquid with rose scent; soluble in mineral oil, alcohol, and glycerin; used in perfumes and flavors.

rhodium [CHEM] A chemical element, symbol Rh, atomic number 45, atomic weight 102.905. [MET] A silver-white metal in the platinum family; sometimes alloyed with platinum for thermocouples or used as a tarnish-resistant electrode posit.

rhodium chloride [INORG CHEM] $RhCl_3$ Water-insoluble, brown-red powder, soluble in cyanides and alkalies; decomposes at 450–500°C. Also known as rhodium trichloride.

rhodium trichloride See rhodium chloride.

rhodizite [MINERAL] $CsAl_4Be_4B_{11}O_{25}(OH)_4$ A white mineral composed of a basic borate of cesium, aluminum, and beryllium, occurring as isometric crystals.

Rhodobacteriineae [MICROBIO] A suborder of the order Pseudomonadales comprising all of the photosynthetic, or phototrophic, bacteria except those of the genus *Rhodomicrobium*.

rhodochrosite [MINERAL] $MnCO_3$ A rose-red to pink or gray mineral form of manganese carbonate with hexagonal symmetry but occurring in massive or columnar form; isomorphous with calcite and siderite, has a hardness of 3.5–4 on Mohs scale, and a specific gravity of 3.7; a minor ore of manganese.

rhodolite [MINERAL] A violet-red garnet species composed of a mixture of almandite and pyrope in about a 3:1 ratio.

Rhodomicrobium [MICROBIO] A genus of pear-shaped phototrophic bacteria whose cells multiply by bud formation; assigned to the family Athiorhodaceae in some systems of classification.

rhodonite [MINERAL] $MnSiO_3$ A pink or brown mineral inosilicate crystallizing in the triclinic system and commonly found in cleavable to compact masses or in embedded grains; luster is vitreous, hardness is 5.5–6 on Mohs scale, and specific gravity is 3.4–3.7.

Rhodophyceae [BOT] A class of algae belonging to the division or subphylum Rhodophyta.

Rhodophyta [BOT] The red algae, a large diverse phylum or division of plants distinguished by having an abundance of the pigment phycoerythrin.

rhodoplast [BOT] A reddish chromatophore occurring in red algae.

Rhodopseudomonas [MICROBIO] A genus of ovoid to rod-shaped phototrophic bacteria in the family Athiorhodaceae which do not produce buds at the end of long, thin stalks.

rhodopsin [BIOCHEM] A deep-red photosensitive pigment contained in the rods of the retina of marine fishes and most higher vertebrates. Also known as retinal pigment; visual purple.

Rhodospirillum [MICROBIO] A genus of bacteria in the family Athiorhodaceae which includes the spiral-shaped members of the group.

rhodotorulic acid [ORG CHEM] A diketopiperazine composed of two molecules of N^δ-acetyl-N^δ-hydroxy-ornithine; a potent growth factor for microorganisms that require hemin or iron transport compounds, for example, *Arthrobacter* species.

rhodoxanthin [BIOCHEM] $C_{40}H_{50}O_2$ A xanthophyll carotenoid pigment.

Rhoipteleaceae [BOT] A monotypic family of dicotyledonous plants in the order Juglandales having pinnately compound leaves, and flowers in triplets with four sepals and six stamens, and the lateral flowers female but sterile.

rhomb See rhombohedron.

rhombencephalon [EMBRYO] The most caudal of the primary brain vesicles in the vertebrate embryo. Also known as hindbrain.

rhombic antenna [ELECTROMAG] A horizontal antenna having four conductors forming a diamond or rhombus; usually fed at one apex and terminated with a resistance or impedance at the opposite apex. Also known as diamond antenna.

rhombic lattice See orthorhombic lattice.

rhombic sulfur [CHEM] Crystalline sulfur with three unequal axes, all at right angles.

Rhombifera [PALEON] An extinct order of Cystoidea in which the thecal canals crossed the sutures at the edges of the plates, so that one-half of any canal lay in one plate and the other half on an adjoining plate.

rhomboclase [MINERAL] $HFe^{3+}(SO_4)_2 \cdot 4H_2O$ A colorless mineral composed of hydrous acid ferric sulfate, occurring in rhombic plates.

rhombogen [INV ZOO] A form of reproductive individual of the mesozoan order Dicyemida found in the sexually mature host which arises from nematogens and gives rise to free-swimming infusorigens.

rhombohedral [CRYSTAL] 1. Of or pertaining to the rhombohedral system. 2. Of or pertaining to crystal cleavage in or a centered lattice of the hexagonal system.

rhombohedral close packing [GEOL] A tight arrangement of uniform solid spheres in a cluster sediment or crystal lattice in which the unit cell has six planes passing through the centers of light spheres located at the corners of a regular

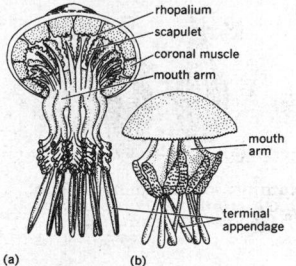
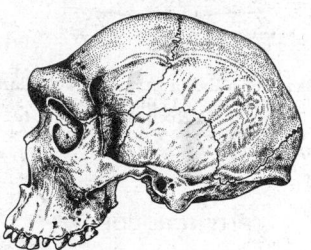
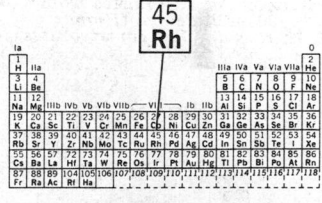

RHYNCHOBDELLAE

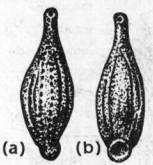

Examples of Rhynchobdellae.
(a) Glossiphonia complanata.
(b) Placobdella parasitica.

RHYNCHOCEPHALIA

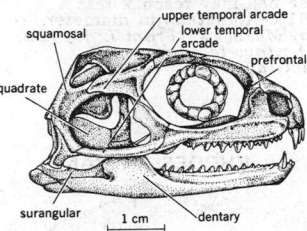

Skull of a living rhynchocephalian, *Sphenodon*, a typical diapsid. *(From A. S. Romer, Vertebrate Paleontology, 3d ed., University of Chicago Press, 1966)*

RHYNCHOCOELA

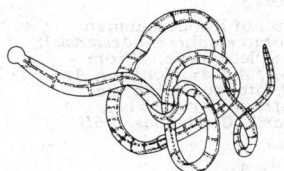

Tubulanus capistratus from the Pacific coast; size ranges from 5 to 30 millimeters. *(From L. H. Hyman, The Invertebrates, vol. 2, McGraw-Hill, 1951)*

RHYNCHONELLIDA

Hypothyridina; posterior view of shell showing posterior curved margin of shell. *(From R. C. Moore, ed., Treatise on Invertebrate Paleontology, pt. H, Geological Society of America, Inc., and University of Kansas Press, 1965)*

RHYNCHOTHERIINAE

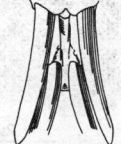

Arrangement of tusks in the gomphotheriid *Rhynchotherium.* *(After Osborn)*

rhombohedron; an aggregate so packed has minimum porosity.

rhombohedral iron ore   *See* hematite; siderite.

rhombohedral lattice   [CRYSTAL] A crystal lattice in which the three axes of a unit cell are of equal length, and the three angles between axes are the same, and are not right angles. Also known as trigonal lattice.

rhombohedral system   [CRYSTAL] A division of the trigonal crystal system in which the rhombohedron is the basic unit cell.

rhombohedron   [CRYSTAL] A trigonal crystal form that is a parallelepiped, the six identical faces being rhombs. Also known as rhomb.   [MATH] A prism with six parallelogram faces.

rhomboid   [MATH] A parallelogram whose adjacent sides are not equal.

rhomboidal prism   [OPTICS] A prism with four parallel sides and two slanting, or oblique, parallel ends; it will divert the path of light entering its ends without changing the form of the light.

rhomboid ripple mark   [GEOL] An aqueous current ripple mark characterized by a reticular arrangement of diamond-shaped tongues of sand, with each tongue having two acute angles, one pointing upcurrent and the other pointing downcurrent.

rhomboporoid cryptostome   [PALEON] Any of a group of extinct bryozoans in the order Cryptostomata that built twiglike colonies with zooecia opening out in all directions from the central axis of each branch.

rhomb-porphyry   [PETR] A porphyritic alkaline syenite composed of an alkali feldspar groundmass with augites having rhombohedral cross sections as the principal phenocryst minerals.

rhombus   [MATH] A parallelogram with all sides equal.

rho meson   [PARTIC PHYS] Collective name for vector meson resonances belonging to a charge multiplet with total isospin 1, hypercharge 0, negative charge conjugation parity, positive g-parity, mass of about 770 MeV (million electron volts), and width of about 146 MeV. Designated $\rho(770)$.

Rhopalidae   [INV ZOO] A family of pentatomorphan hemipteran insects in the superfamily Coreoidea.

rhopalium   [INV ZOO] A sense organ found on the margin of a discomedusan.

Rhopalocera   [INV ZOO] Formerly a suborder of Lepidoptera comprising those forms with clubbed antennae.

rhopalocercous cercaria   [INV ZOO] A free-swimming digenetic trematode larva distinguished by a very wide tail.

Rhopalodinidae   [INV ZOO] A family of holothurian echinoderms in the order Dactylochirotida in which the body is flask-shaped, the mouth and anus lying together.

Rhopalosomatidae   [INV ZOO] A family of hymenopteran insects in the superfamily Scolioidea.

rho-theta navigation   *See* omnibearing-distance navigation.

rhubarb   [BOT] *Rheum rhaponticum.* A herbaceous perennial of the order Polygoniales grown for its thick, edible petioles.

rhumbatron   *See* cavity resonator.

rhumb bearing   [NAV] The direction of a rhumb line through two terrestrial points, expressed as angular distance from a reference direction; usually measured from 000° at the reference direction clockwise through 360°. Also known as Mercator bearing.

rhumb direction   *See* mercator direction.

rhumb line   [MAP] A line on the surface of the earth making the same oblique angle with all meridians. Also known as loxodrome.

rhumb-line course   [NAV] The direction of the rhumb line from the point of departure to the destination, expressed as the angular distance from a reference direction, usually north. Also known as mercator course.

rhumb-line distance   [NAV] Distance along a rhumb line, usually expressed in nautical miles.

rhumb-line error   [NAV] An acceleration error, such as in a bubble sextant reading, due to an aircraft being in curved flight in space when proceeding along a rhumb line on the earth's surface.

rhumb-line sailing   [NAV] Any method of solving the various problems involving course, distance, difference of latitude,

difference of longitude, and departure, as they are related to a rhumb line.

rhyacolite   *See* sanidine.

Rhynchobdellae   [INV ZOO] An order of the class Hirudinea comprising leeches that possess an eversible proboscis and lack hemoglobin in the blood.

Rhynchocephalia   [VERT ZOO] An order of lepidosaurian reptiles represented by a single living species, *Sphenodon punctatus*, and characterized by a diapsid skull, teeth fused to the edges of the jaws, and an overhanging beak formed by the upper jaw.

rhynchocoel   [INV ZOO] A cavity that holds the inverted proboscis in nemertinean worms.

Rhynchocoela   [INV ZOO] A phylum of bilaterally symmetrical, unsegmented, ribbonlike worms having an eversible proboscis and a complete digestive tract with an anus.

rhynchodaeum   [INV ZOO] The part of the proboscis lying anterior to the brain in nemertinean worms.

Rhynchodina   [INV ZOO] A suborder of ciliate protozoans in the order Thigmotrichida.

Rhynchonellida   [INV ZOO] An order of articulate brachiopods; typical forms are dorsibiconvex, the posterior margin is curved, the dorsal interarea is absent, and the ventral one greatly reduced.

rhynchophorous   [ZOO] Having a beak.

Rhynchosauridae   [PALEON] An extinct family of generally large, stout, herbivorous lepidosaurian reptiles in the order Rhynchocephalea.

Rhynchotheriinae   [PALEON] A subfamily of extinct elaphantoid mammals in the family Gomphotheriidae comprising the beak-jawed mastodons.

Rhyniatae   *See* Rhyniopsida

Rhyniophyta   [PALEOBOT] A subkingdom of the Embryobionta including the relatively simple, uppermost Silurian-Devonian vascular plants.

Rhyniopsida   [PALEOBOT] A class of extinct plants in the subkingdom Rhyniophyta characterized by leafless, usually dichotomously branched stems that bore terminal sporangia.

Rhynochetidae   [VERT ZOO] A monotypic family of gruiform birds containing only the kagu of New Caledonia.

rhyodacite   [PETR] A group of extrusive porphyritic igneous rocks containing quartz, plagioclase, and biotite phenocrysts in a fine-grained to glassy groundmass composed of alkali feldspar and silica minerals. Also known as dellenite; quartz latite.

rhyolite   [PETR] A light-colored, aphanitic volcanic rock composed largely of alkali feldspar and free silica with minor amounts of mafic minerals; the extrusive equivalent of granite.

rhyolitic glass   [GEOL] Volcanic glass that is chemically equivalent to rhyolite.

rhyolitic lava   [GEOL] A highly viscous, silica-rich lava.

rhyolitic magma   [PETR] A type of magma formed by differentiation from basaltic magma in combination with assimilation of siliceous material, or by melting of portions of the earth's sialic layer.

rhyolitic tuff   [GEOL] A tuff composed of fragments of rhyolitic lava.

Rhysodidae   [INV ZOO] The wrinkled bark beetles, a family of coleopteran insects in the suborder Adephaga.

rhythmic accumulations   [GEOL] Regular patterns of ripples and cusps in sediment on the beach or the sea floor, formed by currents and waves.

rhythmic driving   [MIN ENG] Driving carried out between two shifts; that is, the drilling, loading, and blasting are carried out in one shift and the mucking and transportation in the following one.

rhythmic sedimentation   [GEOL] A repetitious, regular sequence of rock units formed by sedimentary succession and indicating a frequent, predictable recurrence of the same sequence of conditions.

rhythmic stratification   [GEOL] The occurrence of sediment layers in repetitive patterns, such as a regular alternation of layers of lime and clay.

rhythmite   [GEOL] An independent unit of a rhythmic succession or of beds that were developed by rhythmic sedimentation.

ria   [GEOGR] **1.** Any broad, estuarine river mouth. **2.** A long, narrow coastal inlet, except a fjord, whose depth and width gradually and uniformly diminish inland.

RIAA curve [ENG ACOUS] **1.** Recording Industry Association of America curve representing standard recording characteristics for long-play records. **2.** The corresponding equalization curve for playback of long-play records.

ria coast [GEOGR] A coast with several parallel rias extending far inland and alternating with ridgelike promontories.

ria shoreline [GEOGR] A type of coastline developed along a drowning landmass in which numerous long and narrow arms of the sea extend inland parallel with one another and perpendicular to the coastline.

rib [AERO ENG] A transverse structural member that gives cross-sectional shape and strength to a portion of an airfoil. [ANAT] One of the long curved bones forming the wall of the thorax in vertebrates. [BOT] A primary vein in a leaf. [MIN ENG] **1.** A solid pillar of coal or ore left for support. **2.** A thin stratum in a seam of coal.

rib-and-furrow [GEOL] The bedding-plane expression for micro-cross-bedding, consisting of sets of small, transverse arcuate markings confined to long, narrow, parallel grooves oriented parallel to the current flow and separated by narrow ridges.

rib arch [CIV ENG] An arch consisting of ribs placed side by side and extending from the springings on one end to those on the other end.

ribbed-clamp coupling [DES ENG] A rigid coupling which is split longitudinally and bored to shaft diameter, with a shim separating the two halves.

ribbon [MATH] The plane figure generated by a straight line which moves so that.it is always perpendicular to the path traced by its middle point. [PETR] One of a set of parallel bands in a rock or mineral.

ribbon conveyor [MECH ENG] A type of screw conveyor which has an open space between the shaft and a ribbon-shaped flight, used for wet or sticky materials which would otherwise build up on the spindle.

ribbon diagram [GEOL] A continuous geologic cross section that is drawn in perspective along a curved or sinuous line.

ribbon lightning [GEOPHYS] Ordinary streak lightning that appears to be spread horizontally into a ribbon of parallel luminous streaks when a very strong wind is blowing at right angles to the observer's line of sight; successive strokes of the lightning flash are then displaced by small angular amounts and may appear to the eye or camera as distinct paths. Also known as band lightning; fillet lightning.

ribbon microphone [ENG ACOUS] A microphone whose electric output results from the motion of a thin metal ribbon mounted between the poles of a permanent magnet and driven directly by sound waves; it is velocity-actuated if open to sound waves on both sides, and pressure-actuated if open to sound waves on only one side.

ribbon mixer [MECH ENG] Device for the mixing of particles, slurries, or pastes of solids by the revolution of an elongated helicoid (spiral) ribbon of metal.

ribbon parachute [AERO ENG] A type of parachute having a canopy consisting of an arrangement of closely spaced tapes; this parachute has high porosity with attendant stability and slight opening shock.

ribbon rock [PETR] A rock showing a succession of thin layers of differing composition or appearance.

ribbon structure [GEOL] A succession of thin layers of different mineralogy and texture often contorted and deformed.

ribbon vein *See* banded vein.

ribbon windows [ARCH] Windows in a continuous horizontal band.

rib hole [MIN ENG] One of the final holes fired in blasting ground at the sides of a shaft or tunnel. Also known as trimmer.

Riblet coupler *See* three-decibel coupler.

riboflavin [BIOCHEM] $C_{17}H_{20}N_4O_6$ A water-soluble, yellow-orange fluorescent pigment that is essential to human nutrition as a component of the coenzymes flavin mononucleotide and flavin adenine dinucleotide. Also known as lactoflavin; vitamin B_2; vitamin G.

riboflavin 5′-phosphate [BIOCHEM] $C_{17}H_{21}N_4O_9P$ The phosphoric acid ester of riboflavin. Also known as flavin phosphate; flavin mononucleotide; FMN; isoalloxazine mononucleotide; vitamin B_2 phosphate.

D-ribo-2-ketohexose *See* allulose.

D-riboketose *See* ribulose.

ribonuclease [BIOCHEM] $C_{587}H_{909}N_{171}O_{197}S_{12}$ An enzyme that catalyzes the depolymerization of ribonucleic acid.

ribonucleic acid [BIOCHEM] A long-chain, usually single-stranded nucleic acid consisting of repeating nucleotide units containing four kinds of heterocyclic, organic bases: adenine, cytosine, quanine, and uracil; they are conjugated to the pentose sugar ribose and held in sequence by phosphodiester bonds; involved intracellularly in protein synthesis. Abbreviated RNA.

ribonucleoprotein [BIOCHEM] Any of a large group of conjugated proteins in which molecules of ribonucleic acid are closely associated with molecules of protein.

ribose · [BIOCHEM] $C_5H_{10}O_5$ A pentose sugar occurring as a component of various nucleotides, including ribonucleic acid.

riboside [BIOCHEM] Any glycoside containing ribose as the sugar component.

ribosomal ribonucleic acid [BIOCHEM] Any of three large types of ribonucleic acid found in ribosomes: 5S RNA, with molecular weight 40,000; 14-16S RNA, with molecular weight 600,000; and 18-22S RNA with molecular weight 1,200,000. Abbreviated r-RNA.

ribosome [CYTOL] One of the small, complex particles composed of various proteins and three molecules of ribonucleic acid which synthesize proteins within the living cell.

rib pillar [MIN ENG] A pillar whose length is large compared with its width.

rib rifling [ORD] Rifling of the bore of a gun in which the lands and grooves are of equal width.

ribulose [BIOCHEM] $C_5H_{10}O_5$ A pentose sugar that exists only as a syrup; synthesized from arabinose by isomerization with pyridine; important in carbohydrate metabolism. Also known as D-erythropentose; D-riboketose.

ribulose diphosphate [BIOCHEM] $C_5H_{12}O_{11}P_2$ The phosphate ester of ribulose.

ribut [METEOROL] Sharp, short squalls during comparatively calm winds from May to November in Malaya.

Riccati-Bessel functions [MATH] Solutions of a second-order differential equation in a complex variable which have the form $zf(z)$, where $f(z)$ is a function in terms of polynomials and $\cos(z)$, $\sin(z)$.

Riccati's equation [MATH] A first-order differential equation having the form $y' = A_0(x) + A_1(x)y + A_2(x)y^2$; every second-order linear differential equation can be transformed into an equation of this form.

Ricci equations [MATH] Equations relating the components of the Ricci tensor, the curvature tensor, and an arbitrary tensor of a Riemann space. Also known as Ricci identities.

Ricci identities *See* Ricci equations.

Ricci tensor *See* contracted curvature tensor.

Ricci theorem [MATH] The covariant derivative vanishes for either of the fundamental tensors of a Riemann space.

rice [BOT] *Oryza sativa.* An annual cereal grass plant of the order Cyperales, cultivated as a source of human food for its carbohydrate-rich grain.

rice bran oil [MATER] Clear, combustible liquid, derived by solvent-extraction of oil from fresh rice bran; used to make soaps and animal feeds, salad and cooking oils, and hydrogenated shortening.

rice glue [MATER] A paste made from ground rice boiled in soft water; used in molded objects such as statuary.

rice grains [ASTRON] Bright patches that stand out against the darker background of the surface of the sun; they are short-lived, and the pattern changes in a matter of minutes.

rice neutralization [ELECTR] Development of voltage in the grid circuit of a vacuum tube in order to nullify or cancel feedback through the tube.

rice neutralizing circuit [ELECTR] Radio-frequency amplifier circuit that neutralizes the grid-to-plate capacitance of an amplifier tube.

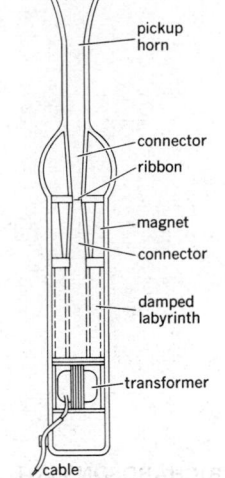

RIBBON MICROPHONE

pickup horn
connector
ribbon
magnet
connector
damped labyrinth
transformer
cable

Sectional view of ribbon-type pressure microphone.

RIBULOSE

$$CH_2OH$$
$$|$$
$$C{=}O$$
$$|$$
$$HCOH$$
$$|$$
$$HCOH$$
$$|$$
$$CH_2OH$$

Structural formula of ribulose.

rice paper [MATER] **1.** A product, not a true paper and not made from rice, but manufactured from the pith of a tree grown in Taiwan; tissue-thin sheets of the pith are peeled away as a cylindrical section of the wood rotates against a knife. **2.** Any of various oriental papers used in block printing.

Rice's bromine solution [ANALY CHEM] Analytical reagent for the quantitative analysis of urea; has 12.5% bromine and sodium bromide in aqueous solution.

Richardson automatic scale [ENG] An automatic weighing and recording machine for flowable materials carried on a conveyor; weighs batches from 200 to 1000 pounds (90 to 450 kilograms).

Richardson-Dushman equation [ELECTR] An equation for the current density of electrons that leave a heated conductor in thermionic emission.

Richardson effect *See* thermionic emission.

Richardson number [FL MECH] A dimensionless number used in studying the stratified flow of multilayer systems; equal to the acceleration of gravity times the density gradient of a fluid, divided by the product of the fluid's density and the square of its velocity gradient at a wall. Symbolized N_{Ri}.

Richardson plot [ELECTR] A graph of log (J/T^2) against $1/T$, where J is the current density of electrons leaving a heated conductor in thermionic emission, and T is the temperature of the conductor; according to the Richardson-Dushman equation, this is a straight line.

Richard's solder [MET] A yellow brass containing 3% aluminum and 3% phosphor tin.

rich concrete [MATER] Concrete with a high cement content.

richellite [MINERAL] $Ca_3Fe_{10}(PO_4)_8(OH,F)_{12} \cdot nH_2O$ A yellow mineral composed of hydrous basic iron calcium fluophosphate; occurs in masses.

rich mixture [CHEM] An air-fuel mixture that is high in its concentration of combustible component.

Richmondian [GEOL] A North American stage of geologic time: Upper Ordovician (above Maysvillian, below Lower Silurian).

rich oil [MATER] Natural-gasoline-plant absorption oil containing dissolved natural-gasoline fractions.

rich ore [MIN ENG] Relatively high grade ore.

richterite [MINERAL] $(Na,K)_2(Mg,Mn,Ca)_6 Si_8O_{22}(OH)_2$ A brown, yellow, or rose-red monoclinic mineral composed of basic silicate of sodium, potassium, magnesium, manganese, and calcium; a member of the amphibole group.

Richter scale [GEOPHYS] A scale of numerical values of earthquake magnitude ranging from 1 to 9.

ricin [MATER] White, poisonous powder derived from pressed castor oil bran.

Ricinidae [INV ZOO] A family of bird lice, order Mallophaga, which occur on numerous land and water birds.

ricinoleic acid [ORG CHEM] $C_{18}H_{34}O_3$ Unsaturated fatty acid; a combustible, water-insoluble, viscous liquid; soluble in most organic solvents; boils at 226°C (10 mm Hg); used as a chemical intermediate, in soaps and Turkey red oils, and for textile finishing. Also known as castor oil acid; *cis*-12-hydroxyoctadec-9-enoic acid; 12-hydroxyoleic acid.

ricinoleyl alcohol [ORG CHEM] $C_{18}H_{36}O_2$ Fatty alcohol of ricinoleic acid; a combustible, colorless, nondrying liquid, boiling at 170–328°C; used as a chemical intermediate, in protective coatings, surface-active agents, pharmaceuticals, and plasticizers. Also known as 9-octadecen-1,12-diol.

Ricinuleida [INV ZOO] An order of rare, ticklike arachnids in which the two anterior pairs of appendages are chelate, and the terminal segments of the third legs of the male are modified as copulatory structures.

rickardite [MINERAL] Cu_4Te_3 A deep-purple mineral composed of copper telluride, occurring in masses.

rickets [MED] A disorder of calcium and phosphorus metabolism affecting bony structures, due to vitamin D deficiency.

Rickettsiaceae [MICROBIO] A family of typhuslike and related agents in the order Rickettsiales.

Rickettsiales [MICROBIO] An order of very small bacteria which are mostly obligate intracellular parasites of animals; placed in the class Schizomycetes.

rickettsialpox [MED] An acute febrile disease caused by the rickettsial organism *Rickettsia akari* and transmitted from the mouse to humans by the mite *Allodermanyssus sanguineus;* characterized by rash, a primary ulcer, and often swelling of glands.

rickettsiosis [MED] Any disease caused by rickettsiae.

ricochet [MECH] **1.** Of a projectile, bomb, or the like, to skip, bounce, or fly off at an angle after striking an object or surface. **2.** An act or instance of ricocheting.

ricochet burst [ORD] Burst of a projectile in the air after it has hit and bounced; it is used effectively against enemy personnel.

ricochet fire [ORD] Fire in which the projectiles ricochet; sometimes used in artillery to obtain air bursts after initial impact.

rictus [VERT ZOO] The mouth aperture in birds.

riddle [DES ENG] A sieve used for sizing or for removing foreign material from foundry sand or other granular materials.

rider [MIN ENG] A steel or iron crossbeam which slides between the guides in a sinking shaft; it is carried by the hoppit and serves to guide and steady the hoppit during its movement up and down the shaft.

ridge [ARCH] The line on which the sides of a sloping roof meet. [GEOL] An elongate, narrow, steep-sided elevation of the earth's surface or the ocean floor. [METEOROL] An elongated area of relatively high atmospheric pressure, almost always associated with, and most clearly identified as, an area of maximum anticyclonic curvature of wind flow. Also known as wedge.

ridge aloft *See* upper-level ridge.

ridge board [BUILD] A horizontal board placed on edge at the apex of the roof.

ridge pole [BUILD] The horizontal supporting member placed along the ridge of a roof.

ridge waveguide [ELECTROMAG] A circular or rectangular waveguide having one or more longitudinal internal ridges that serve primarily to increase transmission bandwidth by lowering the cutoff frequency.

riding boom *See* boat boom.

riebeckite [MINERAL] $Na_2(Fe,Mg)_5Si_8O_{22}(OH)_2$ A blue or black monoclinic amphibole occurring as a primary constituent in some acid- or sodium-rich igneous rocks.

riebungsbreccia [GEOL] A breccia developed during folding.

Riecke's principle [MINERAL] The principle that solution of a mineral occurs most readily at points of greatest external pressure, and crystallization occurs most readily at points of least external pressure; applied to recrystallization in metamorphic rock.

Riedel's disease [MED] A form of chronic thyroiditis with irregular localized areas of stony, hard fibrosis.

Riefler clock [HOROL] A clock employing a pin escapement and a mercurial pendulum; used as a standard.

riegel [GEOL] A low, traverse ridge of bedrock on the floor of a glacial valley. Also known as rock bar; threshold; verrou.

Riegler's test [ANALY CHEM] Analytical technique for nitrous acid; uses sodium naphthionate and β-naphthol.

Rieke diagram [ELECTR] A chart showing contours of constant power output and constant frequency for a microwave oscillator, drawn on a Smith chart or other polar diagram whose coordinates represent the components of the complex reflection coefficient at the oscillator load.

Riemann-Christoffel tensor [MATH] The basic tensor used for the study of curvature of a Riemann space; it is a fourth-rank tensor, formed from Christoffel symbols and their derivatives, and its vanishing is a necessary condition for the space to be flat. Also known as curvature tensor.

Riemann function [MATH] A type of Green's function used in solving the Cauchy problem for a real hyperbolic partial differential equation.

Riemann hypothesis [MATH] The conjecture that the only zeros of the Riemann zeta function with positive real part must have their real part equal to ½.

Riemannian curvature [MATH] A general notion of space curvature at a point of a Riemann space which is directly obtained from orthonormal tangent vectors there.

Riemannian geometry *See* elliptic geometry.

Riemannian manifold [MATH] A differentiable manifold

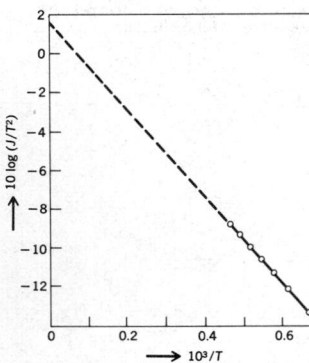

RICHARDSON PLOT

Richardson plot for tungsten, an important thermionic emitter. *(After G. Herrmann and S. Wagener, The Oxide-Coated Cathode, vol. 2, Chapman and Hall, 1951)*

where the tangent vectors about each point have an inner product so defined as to allow a generalized study of distance and orthogonality.

Riemann integral [MATH] The Riemann integral of a real function $f(x)$ on an interval (a,b) is the unique limit (when it exists) of the sum of $f(a_i)(x_i - x_{i-1})$, $i = 1, \ldots, n$, taken over all partitions of (a,b), $a = x_0 < a_1 < x_1 < \ldots < a_n < x_n = b$, as the maximum distance between x_i and x_{i-1} tends to zero.

Riemann-Lebesgue lemma [MATH] If the absolute value of a function is integrable over the interval where it has a Fourier expansion, then its Fourier coefficients a_n tend to zero as n goes to infinity.

Riemann mapping theorem [MATH] Any simply connected domain in the plane with boundary containing more than one point can be conformally mapped onto the interior of the unit disk.

Riemann method [MATH] A method of solving the Cauchy problem for hyperbolic partial differential equations.

Riemann space [MATH] A Riemannian manifold or subset of a euclidean space where tensors can be defined to allow a general study of distance, angle, and curvature.

Riemann sphere [MATH] The surface of a ball; the two-dimensional sphere in euclidean three-dimensional space obtained from all points whose distance from the origin is 1; this name is most often used in referring to a model for the compactified complex numbers.

Riemann surfaces [MATH] Sheets or surfaces obtained by analyzing multiple-valued complex functions and the various choices of principal branches.

Riemann tensors [MATH] Various types of tensors used in the study of curvature for a Riemann space.

Riemann zeta function [MATH] The complex function $\zeta(z)$ defined by an infinite series with nth term $e^{-z \log n}$. Also known as zeta function.

Riesz-Fischer theorem [MATH] The vector space of real functions whose absolute value function squared is integrable constitutes a complete inner product space.

rifampicin [MICROBIO] An antibacterial and antiviral antibiotic; action depends upon its preferential inhibition of bacterial ribonucleic acid polymerase over animal-cell RNA polymerase.

riffle [HYD] **1.** A shallows across a stream bed over which water flows swiftly and is broken into waves by submerged obstructions. **2.** Shallow water flowing over a riffle.

rifle [DES ENG] A drill core that has spiral grooves on its outside surface. [ENG] A borehole that is following a spiral course. [ORD] A firearm having spiral grooves upon the surface of its bore to impart rotary motion to a projectile, thereby stabilizing the projectile and ensuring greater accuracy of impact and longer range; it may fire projectiles automatically or semiautomatically, or successive rounds may be manually loaded.

rifle bracket [ORD] Metal clamp for holding the rifle in easy accessibility, usually on a motor vehicle.

rifle grenade [ORD] A grenade especially designed or adapted to be fired or launched from the muzzle of a rifle or carbine.

rifle grenade ogive [ORD] A hollow metallic item designed for attachment to the forward end of a practice rifle grenade; it cushions the impact and permits reuse of the grenade.

rifle range [ORD] Place for practice in shooting with a rifle.

rifling [MECH ENG] The technique of cutting helical grooves inside a rifle barrel to impart a spinning motion to a projectile around its long axis.

rift [GEOL] **1.** A narrow opening in a rock caused by cracking or splitting. **2.** A high, narrow passage in a cave.

rift-block mountain [GEOL] A mountain range which is a horst block bounded by normal faults.

rift-block valley [GEOL] A valley which occupies a graben.

rift saw [DES ENG] **1.** A saw for cutting wood radially from the log. **2.** A circular saw divided into toothed arms for sawing flooring strips from cants.

rift valley [GEOL] A deep, central cleft with a mountainous floor in the crest of a midoceanic ridge. Also known as central valley; midocean rift.

Rift Valley fever [MED] A toxic generalized febrile virus disease of man and animals in South and East Africa,

transmitted by a mosquito, and characterized by headache, photophobia, myalgia, and anorexia.

rig [MECH ENG] A tripod, derrick, or drill machine complete with auxiliary and accessory equipment needed to drill. [NAV ARCH] The method according to which spars and sails are designed and fitted.

Rigel [ASTRON] A multiple star of stellar magnitude 0.08, 650 light-years from the sun, spectral classification B8-Ia, in the constellation Orion; the star β Orionis.

rigging [AERO ENG] The shroud lines attached to a parachute. [NAV ARCH] Collectively, all the ropes and chains employed to support and work the masts, yards, booms, and sails of a vessel.

Righi experiment [OPTICS] An experiment in which a rotating Nicol prism, a Fresnel mirror, a quarter-wave plate, and a fixed Nicol prism are used to produce effects in light beams similar to beats between sounds with slightly different frequencies.

Righi-Leduc effect [PHYS] The phenomenon wherein, if a magnetic field is applied at right angles to the direction of a temperature gradient in a conductor, a new temperature gradient is produced perpendicular to both the direction of the original temperature gradient and to the magnetic field. Also known as Leduc effect.

right-and-left-hand chart [IND ENG] A graphic symbolic representation of the motions made by one hand in relation to those made by the other hand.

right angle [MATH] An angle of 90°.

right-angle prism [OPTICS] A type of prism used to turn a beam of light through a right angle (90°); it will invert (turn upside-down) or will revert (turn right for left), according to the position of the prism, any light reflected by it.

right ascension [ASTRON] A celestial coordinate; the angular distance taken along the celestial equator from the vernal equinox eastward to the hour circle of a given celestial body.

right astern *See* dead astern.

right bank [NAV] That bank of a stream or river on the right of the observer when he is facing in the direction of flow, or downstream.

right circular cylinder [MATH] A solid bounded by two parallel planes and by a cylindrical surface consisting of the straight lines perpendicular to the planes and passing through a circle in one of them.

right-hand cutting tool [DES ENG] A cutter whose flutes twist in a clockwise direction.

right-handed [CRYSTAL] Having a crystal structure with a mirror-image relationship to a left-handed structure. [DES ENG] **1.** Pertaining to screw threads that allow coupling only by turning in a clockwise direction. **2.** *See* right-laid.

right-handed coordinate system [MATH] **1.** A three-dimensional rectangular coordinate system such that when the thumb of the right hand extends in the positive direction of the first (or x) axis the fingers fold in the direction in which the second (or y) axis could be rotated about the first axis to coincide with the third (or z) axis. **2.** A Riemann space which has negative scalar density function.

right-hand helicity [QUANT MECH] Property of a particle whose spin is parallel to its momentum.

right-hand polarization [ELECTROMAG] In elementary particle discussions, circular or elliptical polarization of an electromagnetic wave in which the electric field vector at a fixed point in space rotates in the right-hand sense about the direction of propagation; in optics, the opposite convention is used; in facing the source of the beam, the electric vector is observed to rotate clockwise.

right-hand rule [ELECTROMAG] **1.** For a current-carrying wire, the rule that if the fingers of the right hand are placed around the wire so that the thumb points in the direction of current flow, the fingers will be pointing in the direction of the magnetic field produced by the wire. **2.** For a moving wire in a magnetic field, such as the wire on the armature of a generator, if the thumb, first, and second fingers of the right hand are extended at right angles to one another, with the first finger representing the direction of magnetic lines of force and the second finger representing the direction of current flow induced by the wire's motion, the thumb will be pointing in the direction of motion of the wire. Also known as Fleming's rule.

right-hand taper [ELEC] Taper in which there is greater resistance in the clockwise half of the operating range of a rheostat or potentiometer (looking from the shaft end) than in the counterclockwise half.

righting arm [NAV ARCH] The horizontal distance between the center of gravity and a vertical line through the center of buoyancy of a ship that is displaced from the upright position; knowledge of this quantity is necessary to determine the righting moment.

righting moment [NAV ARCH] The torque which tends to restore a vessel heeled over to its upright position; it is the product of the righting arm and the weight of the vessel.

right-laid [DES ENG] Rope or cable construction in which strands are twisted counterclockwise. Also known as right-handed.

right lang lay [DES ENG] Rope or cable in which the individual wires or fibers and the strands are twisted to the right.

right-lateral fault See dextral fault.

right-lateral slip fault See dextral fault.

right-of-way [CIV ENG] 1. Areas of land used for a road and along the side of the roadway. 2. A thoroughfare or path established for public use. 3. Land occupied and used by a railroad or a public utility.

right rudder [NAV] The operation of moving the rudder to starboard, and consequently turning the bow of the ship to the right.

right section [MATH] A right section of a surface is the plane section produced by a perpendicular plane.

right-slip fault See dextral fault.

right sphere [ASTRON] The appearance of the celestial sphere as seen by an observer at the earth's equator.

right triangle [MATH] A triangle one of whose angles is a right angle.

rigid body [MECH] An idealized extended solid whose size and shape are definitely fixed and remain unaltered when forces are applied.

rigid-body dynamics [MECH] The study of the motions of a rigid body under the influence of forces and torques.

rigid copper coaxial line [ELECTROMAG] A coaxial cable in which the central conductor and outer conductor are formed by joining rigid pieces of copper.

rigid coupling [MECH ENG] A mechanical fastening of shafts connected with the axes directly in line.

rigid frame [BUILD] A steel skeleton frame in which the end connections of all members are rigid so that the angles they make with each other do not change.

rigid insulation [ELEC] Electrical insulation that is part of a rigid structure, and must provide mechanical strength and stability of form as well as a dielectric barrier; mica, glass, porcelain, and thermosetting resins are the principal materials used.

rigidity [MECH] The quality or state of resisting change in form.

rigidity modulus See modulus of elasticity in shear.

rigid pavement [CIV ENG] A thick portland cement pavement on a gravel base and subbase, with steel reinforcement and often with transverse joints.

rigid PVC [MATER] Polyvinyl chloride or a polyvinyl chloride-acetate copolymer with a relatively high hardness; may be formulated with or without a small percentage of plasticizer; a rigid resin.

rigid resin [MATER] A resin with a modulus of 10,000 psi $(6.895 \times 10^7$ newtons per square meter) or greater.

rigor [MED] 1. Stiffness. 2. A chill associated with muscular contraction and tremor.

rigor mortis [PATH] Stiffening and rigidity of the musculature occurring after death, beginning within 5–10 hours, and disappearing after 3–4 days.

rill [ASTRON] A crooked, narrow crack on the moon's surface; may be a kilometer or more in width and a few to several hundred kilometers in length. [GEOL] A small, transient runnel. [HYD] A small brook or stream.

rill erosion [GEOL] The formation of numerous, closely spaced rills due to the uneven removal of surface soil by streamlets of running water. Also known as rilling; rill wash; rillwork.

rilling See rill erosion.

rill mark [GEOL] A small, dendritic channel formed on beach

mud or sand by a rill, especially if on the lee side of a partially buried obstruction.

rillstone See ventifact.

rill wash See rill erosion.

rillwork See rill erosion.

rim [DES ENG] 1. The outer part of a wheel, usually connected to the hub by spokes. 2. An outer edge or border, sometimes raised or projecting.

rim blight [PL PATH] A fungus disease of tea caused by members of the genus *Cladosporium* and characterized by yellow discoloration of the leaf margins followed by browning.

rim cement [GEOL] A thin layer of calcium carbonate, hematite, or silica developed on the surface of detrital grains during diagenesis.

rim clutch [MECH ENG] A frictional contact clutch having surface elements that apply pressure to the rim either externally or internally.

rim drive [ENG ACOUS] A phonograph or sound recorder drive in which a rubber-covered drive wheel is in contact with the inside of the rim of the turntable.

rime [HYD] A white or milky and opaque granular deposit of ice formed by the rapid freezing of supercooled water drops as they impinge upon an exposed object; composed essentially of discrete ice granules, and has densities as low as 0.2–0.3 grams per cubic centimeter.

rime fog See ice fog.

rimfire [ORD] Of a cartridge, having the primer mixture in the rim of the cartridge case base.

rimless [ORD] Pertaining to a cartridge case in which the extracting groove is machined into the body of the case; that is, no part of the case extends beyond the body.

rimmed [ORD] Pertaining to a cartridge case in which an extractor rim projects beyond the body of the case.

rimmed solution pool [GEOL] A pool in rock with a hardened rim resulting from deposition of lime during evaporation at low tide.

rimmed steel [MET] Low-carbon steel, partially deoxidized, which on cooling continuously, evolves sufficient carbon monoxide to form a case or rim of metal virtually free of voids.

rimrock [GEOL] A top layer of resistant rock on a plateau outcropping with vertical or near vertical walls.

rimstone [GEOL] A calcium-containing deposit ringing an overflowing basin such as a hot spring.

rind [BOT] 1. The bark of a tree. 2. The thick outer covering of certain fruits.

rinderpest [VET MED] An acute, contagious, and often fatal virus disease of cattle, sheep, and goats which is characterized by fever and the appearance of ulcers on the mucous membranes of the intestinal tract.

R indicator See R scope.

ring [DES ENG] A tie member or chain link; tension or compression applied through the center of the ring produces bending moment, shear, and normal force on radial sections. [MATH] 1. An algebraic system with two operations called multiplication and addition; the system is a commutative group relative to addition, and multiplication is associative, and is distributive with respect to addition. 2. A ring of sets is a collection of sets where the union and symmetric difference of any two members is also a member.

ring-and-bead sight [ORD] Type of gunsight in which the front sight is a bead or post and the rear sight a ring.

ring and circle shear [DES ENG] A rotary shear designed for cutting circles and rings where the edge of the metal sheet cannot be used as a start.

ring-around [COMMUN] Improper routing of a call back through a switching center already trying to complete the same call, thus tying up the trunks by repeating the cycle.

ringbolt [DES ENG] An eyebolt with a ring passing through the eye.

ring canal [INV ZOO] In echinoderms, the circular tube of the water-vascular system that surrounds the esophagus.

ring circuit [ELECTROMAG] In waveguide practice, a hybrid T junction having the physical configuration of a ring with radial branches.

ring counter [ELECTR] A loop of binary scalers or other bistable units so connected that only one scaler is in a

RIGID PAVEMENT

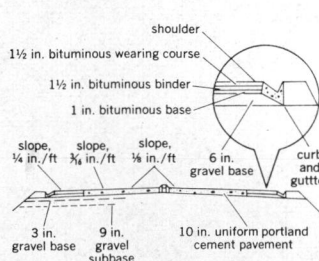

Rigid main roadway with flexible shoulders. The design is for maximum loads of 9 tons (8 metric tons) per axle. 1 inch = 2.5 centimeters; 1/8 inch/foot = 1.0 centimeter/meter.

specified state at any given time; as input signals are counted, the position of the one specified state moves in an ordered sequence around the loop.

ring crusher [MECH ENG] Solids-reduction device with a rotor having loose crushing rings held outwardly by centrifugal force, which crush the feed by impact with the surrounding shell.

ring deoxyribonucleic acid *See* circular deoxyribonucleic acid.

ring dike [GEOL] A roughly circular dike that is vertical or inclined away from the center of the arc. Also known as ring fracture intrusion.

ringdown [COMMUN] In telephone switching, that method of signaling an operator in which telephone ringing current is sent over the line to operate a drop or a self-locking relay and lamp.

ring-dyed fiber [TEXT] A synthetic fiber that is dipped in plastic to coat the surface fiber and to provide a surface on which the dye will adhere.

Ringelmann chart [ENG] A chart used in making subjective estimates of the amount of solid matter emitted by smoke stacks; the observer compares the grayness of the smoke with a series of shade diagrams formed by horizontal and vertical black lines on a white background.

ringent [BOT] Having widely separated, gaping lips. [ZOO] Gaping irregularly.

Ringer's solution [CHEM] A solution of 0.86 gram sodium chloride, 0.03 gram potassium chloride, and 0.033 gram calcium chloride in boiled, purified water, used topically as a physiological salt solution.

ring-fracture intrusion *See* ring dike.

ring gage [DES ENG] A cylindrical ring of steel whose inside diameter is finished to gage tolerance and is used for checking the external diameter of a cylindrical object.

ring gate [CIV ENG] A type of gate used to regulate and control the discharge of a morning-glory spillway; like a drum gate, it offers a minimum of interference to the passage of ice or drift over the gate and requires no external power for operation.

ring gear [MECH ENG] The ring-shaped gear in an automobile differential that is driven by the propeller shaft pinion and transmits power through the differential to the line axle.

ring holes [MIN ENG] The group of boreholes radially drilled from a common-center setup.

ringing [COMMUN] The production of an audible or visible signal at a station or switchboard by means of an alternating or pulsating current. [CONT SYS] An oscillatory transient occurring in the output of a system as a result of a sudden change in input.

ringing circuit [ELECTR] A circuit which has a capacitance in parallel with a resistance and inductance, with the whole in parallel with a second resistance; it is highly underdamped and is supplied with a step or pulse input.

ring isomerism [ORG CHEM] A type of geometrical isomerism in which bond lengths and bond angles prevent the existence of the trans structure if substituents are attached to alkenic carbons which are part of a cyclic system, the ring of which contains fewer than eight members; for example, 1,2-dichlorocyclohexene.

ring isomorphism [MATH] An isomorphism between rings.

ring jewel [DES ENG] A type of jewel used as a pivot bearing in a time-keeping device, gyro, or instrument.

ring lifter *See* split-ring core lifter.

ring-lifter case *See* lifter case.

ringlock nail [DES ENG] A nail ringed with grooves to provide greater holding power.

ring micrometer [OPTICS] A flat, thin ring in the focal plane of a telescope; used to measure differences in right ascension and declination.

ring modulator [ELECTR] A modulator in which four diode elements are connected in series to form a ring around which current flows readily in one direction; input and output connections are made to the four nodal points of the ring; used as a balanced modulator, demodulator, or phase detector.

Ring Nebula [ASTRON] A nebula in the summer constellation Lyra; it is an example of the planetary type of gaseous nebulae.

ring-oil [MECH ENG] To oil (a bearing) by conveying the oil to the point to be lubricated by means of a ring, which rests upon and turns with the journal, and dips into a reservoir containing the lubricant.

ring-roller mill [MECH ENG] A grinding mill in which material is fed past spring-loaded rollers that apply force against the sides of a revolving bowl. Also known as roller mill.

ring-rolling [MET] Producing a thin, large-diameter ring from a thicker, smaller-diameter ring by placing the ring between two rotating rolls.

ring rot [PL PATH] **1.** A fungus disease of the sweet potato root caused by *Rhizopus stolonifer* and marked by rings of dry rot. **2.** A bacterial disease of potatoes caused by *Corynebacterium sepedonicum* and characterized by brown discoloration of the annular vascular tissue.

ring shift *See* cyclic shift.

ring sight [ORD] A sight, especially a gunsight, in the shape of a ring or concentric rings, through which aim is taken and range is estimated.

ring silicate *See* cyclosilicate.

rings of Saturn [ASTRON] Circular rings that encircle the planet Saturn at its equator; there are four main regions to the ring system; theory and observations indicate that the rings are composed of separate particles which move independently in the four series of circular coplanar orbits.

ring spinner [TEXT] A spinning machine in which a loop slides on a bobbin ring to control the twist of the yarn.

ring spot [PL PATH] Any of various virus and fungus diseases of plants characterized by the appearance of a discolored, annular lesion.

ring stress [MIN ENG] The zone of stress in rock surrounding all development excavations.

ring structure [GEOL] An epigenetic structure having a ring-shaped trace in plan.

ring system [ORG CHEM] Arbitrary designation of certain compounds as closed, circular structures, as in the six-carbon benzene ring; common rings have four, five, and six members, either carbon or some combination of carbon, nitrogen, oxygen, sulfur, or other elements.

ringtail *See* cacomistle.

ring test [IMMUNOL] The simplest of the precipitin tests for antigen-antibody reaction; the solution containing antigen is layered on a solution containing antibody; a white disk or precipitate forms at the point where the two solutions diffuse until optimum concentration for precipitation is reached.

ring theory [MATH] The study of the structure of rings in algebra.

ring time [ELECTR] The length of time in microseconds required for a pulse of energy transmitted into an echo box to die out; a measurement of the performance of the radar.

ring vessel [INV ZOO] A part of the water-vascular system in echinoderms; it is the circular canal around the mouth into which the stone canal empties, and from which a radial water vessel traverses to each of five radii.

ring vortex *See* vortex ring.

ringworm [MED] A fungus infection of skin, hair, or nails producing annular lesions with elevated margins. Also known as tinea.

rinkite *See* mosandrite.

rinkolite *See* mosandrite.

rinneite [MINERAL] NaK_3FeCl_6 A colorless, pink, violet, or yellow mineral composed of sodium potassium iron chloride, occurring in granular masses.

rinse [GRAPHICS] To wash off; a liquid bath to remove foreign matter.

Riodininae [INV ZOO] A subfamily of the lepidopteran family Lycaenidae in which prothoracic legs are nonfunctional in the male.

riometer [ENG] An instrument that measures changes in ionospheric absorption of electromagnetic waves by determining and recording the level of extraterrestrial cosmic radio noise. Derived from relative ionospheric opacity meter.

riot-control agent [MATER] A chemical that produces temporary irritating or disabling effects when in contact with the eyes or when inhaled.

riot grenade [ORD] Grenade of plastic or other nonfrag-

RINGING CIRCUIT

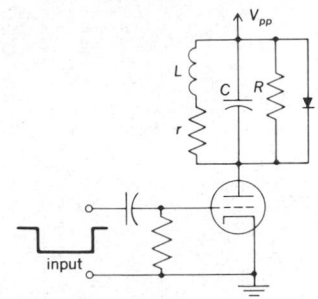

Circuit diagram of a ringing circuit used as a trigger source. V_{pp} = plate supply voltage, L = inductance, C = capacitance, R, r = resistances.

RING NEBULA

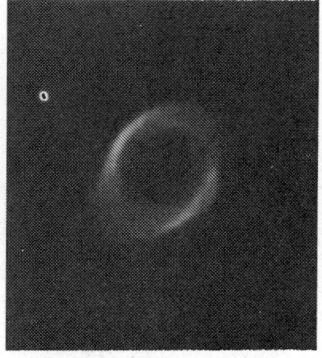

Ring Nebula NGC 6720. *(U.S. Navy Electrograph made with 61-inch astrometric reflector at Flagstaff, Ariz.)*

RINGS OF SATURN

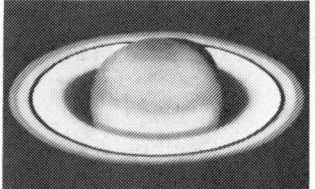

Telescopic appearance of Saturn at maximum inclination of ring system.

RING VESSEL

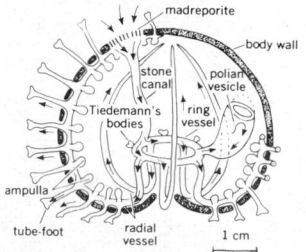

Schematic diagram of water-vascular system of an echinoid showing ring vessel in the lower center. Arrows show direction of water flow.

menting material, containing a charge of tear gas and a detonating fuse with short delay; the gas is released by a bursting action.

riot gun [ORD] Any shotgun with a short barrel, especially a short-barreled shotgun used in guard duty or to scatter rioters; usually has a 20-inch (50.8-centimeter) cylinder barrel.

Rio Tinto process [MIN ENG] Heap leaching of curiferous sulfides that have been oxidized to sulfates by prolonged atmospheric weathering.

rip [ENG] To saw wood with the grain. [MIN ENG] To break down the roof in mine roadways to increase the headroom for haulage, traffic, and ventilation. [OCEANOGR] A turbulent agitation of water generally caused by the interaction of currents and wind.

riparian [BIOL] Living or located on a riverbank.

ripbit *See* detachable bit; jackbit.

rip current [OCEANOGR] The return flow of water piled up on shore by incoming waves and wind.

ripe [BOT] Of fruit, fully developed, having mature seed and so usable as food. [FOR] Of timber or a forest, ready to be cut. [HYD] Descriptive of snow that is in a condition to discharge meltwater; ripe snow usually has a coarse crystalline structure, a snow density near 0.5, and a temperature near 32°F (0°C).

ripidolite [MINERAL] $(Mg,Fe^{2+})_9Al_6Si_5O_{20}(OH)_{16}$ A mineral of the chlorite group; consists of basic magnesium iron aluminum silicate. Also known as aphrosiderite.

rip panel [AERO ENG] A part of a manned free balloon; it is the panel to which the ripcord is attached and extends about ¼ to ⅕ of the circumference of the balloon along one of its meridians; it is torn open when the ripcord is pulled so that all the gas in the balloon escapes.

ripping bar [DES ENG] A steel bar with a chisel at one end and a curved claw for pulling nails at the other.

ripping face support [MIN ENG] A timber or steel support at the ripping lip.

ripping lip [MIN ENG] The end of the enlarged roadway section where work is proceeding.

ripping punch [DES ENG] A tool with a rectangular cutting edge, used in a punch press to crosscut metal plates.

ripple [ELEC] The alternating-current component in the output of a direct-current power supply, arising within the power supply from incomplete filtering or from commutator action in a d-c generator. [FL MECH] *See* capillary wave. [OCEANOGR] A small curling or undulating wave controlled to a significant degree by both surface tension and gravity.

ripple drift [GEOL] A pattern of cross-lamination formed by sedimentary deposits on both sides of a migrating ripple.

ripple filter [ELECTR] A low-pass filter designed to reduce ripple while freely passing the direct current obtained from a rectifier or direct-current generator. Also known as smoothing circuit; smoothing filter.

ripple index [GEOL] On a rippled surface, the ratio of the crest-to-crest distance to the crest-to-trough distance.

ripple load cast [GEOL] A load cast of a ripple mark showing evidence of penecontemporaneous deformation in the accumulation of its trough and crest and in the oversteepening of the component laminae.

ripple mark [GEOL] **1.** A surface pattern on incoherent sedimentary material, especially loose sand, consisting of alternating ridges and hollows formed by wind or water action. **2.** One of the ridges on a ripple-marked surface.

ripple quantity [PHYS] Alternating component of a pulsating quantity when this component is small relative to the constant component.

ripple tank [PHYS] A shallow tray containing a liquid and equipped with means for generating surface waves; used to illustrate several types of wave phenomena, such as interference and diffraction.

ripple voltage [ELEC] The alternating component of the unidirectional voltage from a rectifier or generator used as a source of direct-current power.

riprap [CIV ENG] A foundation or revetment in water or on soft ground made of irregularly placed stones or pieces of boulders; used chiefly for river and harbor work, for roadway filling, and on embankments.

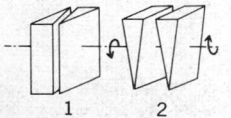

RISLEY PRISM SYSTEM

Diagram of Risley prism system. In orientation 1, combined deviation of prisms is zero; when both have been rotated 90° as in orientation 2, combined deviation is at a maximum.

rips [OCEANOGR] A turbulent agitation of water, generally caused by the interaction of currents and wind; in nearshore regions they may be currents flowing swiftly over an irregular bottom; sometimes referred to erroneously as tide rips.

ripsaw [MECH ENG] A heavy-tooth power saw used for cutting wood with the grain.

rise [ASTRON] Of a celestial body, to cross the visible horizon while ascending. [GEOL] A long, broad elevation which rises gently from its surroundings, such as the sea floor. [SCI TECH] The increase in the height or the value of something, such as a rise of tide or a rise of temperature.

rise of bottom *See* deadrise.

rise of floor *See* deadrise.

rise of tide [OCEANOGR] Vertical distance from the chart datum to a higher water datum.

riser [CHEM ENG] That portion of a bubble-cap assembly in a distillation tower that channels the rising vapor and causes it to flow downward to pass through the liquid held on the bubble plate. [CIV ENG] **1.** A board placed vertically beneath the tread of a step in a staircase. **2.** A vertical steam, water, or gas pipe. [GEOL] A steplike topographic feature, such as a steep slope between terraces. [MET] *See* feedhead. [TEXT] A raised spot in a weaving pattern where the warp traverses the weft or the filling.

riser plate [CIV ENG] A plate used to support a tapering switch rail above the base of the rail; used with a railroad gage or tie plate to maintain minimum gage.

rise time [CONT SYS] The time it takes for the output of a system to change from a specified small percentage (usually 5 or 10%) of its steady-state increment to a specified large percentage (usually 90 or 95%). [ELEC] The time for the pointer of an electrical instrument to make 90% of the change to its final value when electric power suddenly is applied from a source whose impedance is high enough that it does not affect damping.

rising hinge [BUILD] A hinge that raises a door slightly as it is opened.

rising limb [HYD] The rising portion of the hydrograph resulting from runoff of rainfall or snowmelt.

rising-sun magnetron [ELECTR] A multicavity magnetron in which resonators having two different resonant frequencies are arranged alternately for the purpose of mode separation; the cavities appear as alternating long and short radial slots around the perimeter of the anode structure, resembling the rays of the sun.

rising tide [OCEANOGR] The portion of the tide cycle between low water and the following high water.

Risley prism system [OPTICS] A type of dispersing prism used to test ocular convergence in ophthalmology; consists of two thin prisms mounted so that they can be rotated simultaneously in opposite directions.

Riss [GEOL] **1.** A European stage of geologic time: Pleistocene (above Mindel, below Würm). **2.** The third stage of glaciation of the Pleistocene in the Alps.

Rissoacea [PALEON] An extinct superfamily of gastropod mollusks.

Riss-Würm [GEOL] The third interglacial stage of the Pleistocene in the Alps, following the Riss glaciation and preceding the Würm glaciation.

ristocetin [MICROBIO] The generic name for an antibiotic produced by *Nocardia lurida* that is effective in test tube experiments against such microorganisms as *Diplococcus pneumoniae*, *Streptococcus hemolyticus*, and *Clostridium tetani*.

Ritchey-Chrétien optics [OPTICS] A modification of the Cassegrain optical system used in large optical telescopes; it has a hyperbolic image-forming primary mirror, no spherical aberration, and no coma; it has a larger usable field than either Newtonian or Cassegrain optical systems.

Ritchie wedge [OPTICS] A photometer in which a test source and a standard source of light illuminate two perpendicular white, diffusing surfaces which intersect in a movable wedge, and these surfaces are viewed from a direction perpendicular to a line connecting the sources.

Ritter reaction [ORG CHEM] A procedure for the preparation of amides by reacting alkenes or tertiary alcohols with nitriles in an acidic medium.

Rittinger's law [MECH ENG] The law that energy needed to reduce the size of a solid particle is directly proportional to the resultant increase in surface area.

Ritz formula [ATOM PHYS] A particular expansion of an equation used in studying the spectra of atoms.

Ritz method [MATH] A method of solving boundary value problems based upon reformulating the given problem as a minimization problem.

Ritz's combination principle [SPECT] The empirical rule that sums and differences of the frequencies of spectral lines often equal other observed frequencies. Also known as combination principle.

river [HYD] A large, natural freshwater surface stream having a permanent or seasonal flow and moving toward a sea, lake, or another river in a definite channel.

river basin [GEOL] The area drained by a river and all of its tributaries.

riverbed [GEOL] The channel which contains, or formerly contained, a river.

riverboat [NAV ARCH] A vessel used on a river, such as a shallow-draft passenger boat, barge, towboat, or rowboat.

river buoy [NAV] A lightweight nun or can buoy especially designed to withstand strong currents.

river capture See capture.

river engineering [CIV ENG] A branch of transportation engineering consisting of the physical measures which are taken to improve a river and its banks.

river forecast [HYD] A forecast of the expected stage or discharge at a specified time, or of the total volume of flow within a specified time interval, at one or more points along a stream.

river gage [ENG] A device for measuring the river stage; types in common use include the staff gage, the water-stage recorder, and wire-weight gage. Also known as stream gage.

river ice [HYD] Any ice formed in or carried by a river.

river mining [MIN ENG] Mining or excavating beds of existing rivers after deflecting their course, or by dredging without changing the flow of water.

river piracy See capture.

River Po See Eridanus.

river radar [NAV] A marine set especially designed for river pilotage, generally characterized by a high degree of resolution and a wide selection of range scales.

river system [HYD] The aggregate of stream channels draining a river basin.

river terrace See stream terrace.

river tide [HYD] A tide that occurs in rivers emptying directly into the sea, showing three characteristic modifications of ocean tides: the speed at which the tide travels upstream depends on the depth of the channel, the further upstream the longer the duration of the falling tide and shorter the duration of the rising tide, and the range of the tide decreases with distance upstream.

river water [HYD] Water having carbonate, sulfate, and calcium as its main dissolved constituents; distinguished from seawater by its chloride and sodium content.

rivet [DES ENG] A short rod with a head formed on one end; it is inserted through aligned holes in parts to be joined, and the protruding end is pressed or hammered to form a second head.

riveting [ENG] The permanent joining of two or more machine parts or structural members, usually plates, by means of rivets.

riveting hammer [MECH ENG] A hammer used for driving rivets.

riveting stake [HOROL] A block of steel with holes, used in watchmaking; arbors are inserted in the holes, and pinions or collets are placed on the edges of the holes for riveting.

rivet pitch [ENG] The center-to-center distance of adjacent rivets.

rivet weld [MET] A weld shaped like a countersunk rivet.

rivulet [HYD] A small stream; a brook.

rivulose [BOT] Marked by irregular, narrow lines.

R meson [PARTIC PHYS] A meson resonance observed in missing mass experiments involving scattering of negative pions by protons; it has been resolved into three peaks labeled R_1, R_2, and R_3, having masses of 1640, 1700, and 1750 MeV

(million electron volts) respectively, and is believed to have a spin of 3.

R meter [NUCLEO] An ionization instrument calibrated to indicate the intensity of γ-rays, x-rays, and other ionizing radiation in roentgens.

R Monocerotis [ASTRON] An irregular variable star at the tip of the small fan-shaped emission nebula NGC 2261.

rms value See root-mean-square value.

RNA See ribonucleic acid.

roach [INV ZOO] An insect of the family Blattidae; the body is wide and flat, the anterior part of the thorax projects over the head, and antennae are long and filiform, with many segments. Also known as cockroach.

road [CIV ENG] An open way for travel and transportation. [MIN ENG] Any mine passage or tunnel.

roadbed [CIV ENG] The earth foundation of a highway or a railroad.

road capacity [CIV ENG] The maximum traffic flow obtainable on a given roadway, using all available lanes, usually expressed in vehicles per hour or vehicles per day

road grade [CIV ENG] The level and gradient of a road, measured along its center way.

road map [MAP] A map showing the roadways of a specified area.

road net [CIV ENG] The system of roads available within a particular locality or area.

road octane [ENG] A numerical value for automotive antiknock properties of a gasoline; determined by operating a car over a stretch of level road or on a chassis dynamometer under conditions simulating those encountered on the highway.

road oil [MATER] A heavy residual petroleum oil, usually one of the slow-curing grades of liquid asphalt.

roadstead [GEOGR] An area near the shore, where vessels can anchor in safety; usually a shallow indentation in the coast.

road test [ENG] A motor-vehicle test conducted on the highway or on a chassis dynamometer to determine the performance of fuels or lubricants or the performance of the vehicle.

roadway [CIV ENG] The portion of the thoroughfare over which vehicular traffic passes.

roaring forties [METEOROL] A popular nautical term for the stormy ocean regions between 40° and 50° latitude; it usually refers to the Southern Hemisphere, where there is an almost completely uninterrupted belt of ocean with strong prevailing westerly winds.

roast [MET] To heat ore to effect some chemical change that will facilitate smelting.

roaster [ENG] Equipment for the heating of materials, such as in pyrite roasting; a furnace.

roasting regeneration [CHEM ENG] Regeneration of a processing (treating) clay by heating or burning it in contact with air to remove combustible impurities adsorbed onto the surface.

roast sintering See blast roasting.

rob [MIN ENG] To take out ore or coal from a mine with a view to immediate product, and not to subsequent working.

robber [MET] An extra cathode that reduces current density at local areas of the work being electroplated for the purpose of producing a more uniform thickness coating.

robbery See capture.

robbing pillars See pillaring.

Robertinacea [INV ZOO] A superfamily of marine, benthic foraminiferans in the suborder Rotaliina characterized by a trochospiral or modified test with a wall of radial aragonite, and having bilamellar septa.

Roberts evaporator See short-tube vertical evaporator.

Roberts' linkage [MECH ENG] A type of approximate straight-line mechanism which provided, early in the 19th century, a practical means of making straight metal guides for the slides in a metal planner.

Robin Hood's wind [METEOROL] In Great Britain, saturated air with temperatures near freezing; it is raw and penetrating.

Robin law [PHYS] The law that an increase in pressure on a system in chemical or physical equilibrium favors the system formed with a decrease in volume, and conversely, a change

RIVET

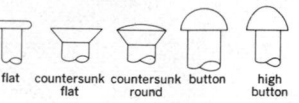

flat   countersunk   countersunk   button   high
       flat   round   button

Some types of standard rivet heads.

R MESON

(Mass)^2 plot for the prominent mesonic (MM) peaks observed in the reaction $\pi^- p \rightarrow p$ (MM)$^-$ with the Missing Mass spectrometer at CERN (Geneva), against their order in this sequence. The straight-line fit has generally been interpreted as a Regge trajectory, the ordinate being identified with the spin J of the resonance state. The R-meson peak has been resolved into several peaks, and spin 3 is associated at least with the peak R_2 at 1700 MeV; a Regge trajectory associated with the ρ-meson (spin 1) would then be expected to pass through the point R_2, as shown by the dashed line.

ROACH

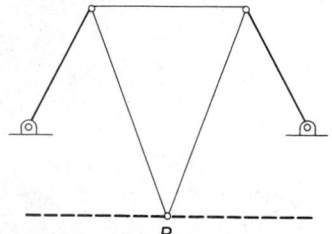

American roach (*Periplantea americana*) has long well-developed wings, and often flies.

ROBERTS' LINKAGE

Roberts' linkage, an example of straight-line mechanisms. Point P moves along a straight line.

in pressure does not affect a system formed with no change in volume.

Robins-Messiter system [MECH ENG] A stacking conveyor system in which material arrives on a conveyor belt and is fed to one or two wing conveyors.

robinsonite [MINERAL] $Pb_7Sb_{12}S_{25}$ A mineral composed of lead antimony sulfide.

Robitzsch actinograph [ENG] A pyranometer whose design utilizes three bimetallic strips which are exposed horizontally at the center of a hemispherical glass bowl; the outer strips are white reflectors, and the center strip is a blackened absorber; the bimetals are joined in such a manner that the pen of the instrument deflects in proportion to the difference in temperature between the black and white strips, and is thus proportional to the intensity of the received radiation; this instrument must be calibrated periodically.

robot [CONT SYS] A completely self-controlled electronic, electric, or mechanical device.

roc [ELEC] A unit of electrical conductivity equal to the conductivity of a material in which an electric field of 1 volt per centimeter gives rise to a current density of 1 ampere per square centimeter. Derived from reciprocal ohm centimeter.

Roccilaceae [BOT] A family of fruticose species of Hysterales that grow profusely on trees and rocks along the coastlines of Portugal, California, and western South America.

Rochelle salt crystal [SOLID STATE] A crystal of sodium potassium tartrate, having a pronounced piezoelectric effect; used in crystal microphones and crystal pickups.

roche moutonnée [GEOL] A small, elongate hillock of bedrock sculptured by a large glacier so that its long axis is oriented in the direction of ice movement; the upstream side is gently inclined, smoothly rounded, but striated, and the downstream side is steep, rough, and hackly.

Roche's limit [ASTROPHYS] The limiting distance below which a satellite orbiting a celestial body would be disrupted by the tidal forces generated by the gravitational attraction of the primary; the distance depends on the relative densities of the bodies and on the orbit of the satellite; it is computed by $R = 2.45(Lr)$, where L is a factor that depends on the relative densities of the satellite and the body, R is the radius of the satellite's orbit measured from the center of the primary body, and r is the radius of the primary body; if the satellite and the body have the same density, the relationship is $R = 2.45r$.

rock [PETR] **1.** A consolidated or unconsolidated aggregate of mineral grains consisting of one or more mineral species and having some degree of chemical and mineralogic constancy. **2.** In the popular sense, a hard, compact material with some coherence, derived from the earth.

rockair [AERO ENG] A high-altitude sounding system consisting of a small solid-propellant research rocket carried aloft by an aircraft; the rocket is fired while the aircraft is in vertical ascent.

rock asphalt See asphalt rock.

rock awash [OCEANOGR] In U.S. Coast and Geodetic Survey terminology, a rock exposed at any stage of the tide between the datum of mean high water and the sounding datum, or one just bare at these data.

rock bar See riegel.

rock bench See structural bench.

rock bit [ENG] Any one of many different types of roller bits used on rotary-type drills for drilling large-size holes in soft to medium-hard rocks.

rockbolt [ENG] A bar, usually constructed of steel, which is inserted into predrilled holes in rock and secured for the purpose of ground control.

rockbridgeite [MINERAL] $Fe^{+2}Fe_6^{+3}(PO_4)_4(OH)_8$ A basic phosphate mineral containing iron; isomorphous with frondelite.

rock-bulk compressibility [GEOL] One of three types of rock compressibility (matrix, bulk, and pore); the fractional change in volume of the bulk volume of the rock with a unit change in pressure.

rock bump [MIN ENG] The sudden release of the weight of the rocks over a coal seam or of enormous lateral stresses.

rockburst [MIN ENG] A sudden and violent rock failure around a mining excavation on a sufficiently large scale to be considered a hazard.

rock candy [FOOD ENG] Large, transparent, hydrated crystals of cane sugar.

rock channeler [MECH ENG] A machine used in quarrying for cutting an artificial seam in a mass of stone.

rock cleavage [PETR] The capacity of a rock to split along certain parallel surfaces more easily than along others.

rock control [GEOL] The influences of differences in earth materials on development of landforms.

rock cork See mountain cork.

rock creep [GEOL] A form of slow flowage in rock materials evident in the downhill bending of layers of bedded or foliated rock and in the slow downslope migration of large blocks of rock away from their parent outcrop.

rock crystal [MINERAL] A transparent, colorless form of quartz with low brilliance; used for lenses, wedges, and prisms in optical instruments. Also known as berg crystal; crystal; mountain crystal; pebble; quartz crystal.

rock cycle [GEOL] The interrelated sequence of events by which rocks are initially formed, altered, destroyed, and reformed as a result of magmatism, erosion, sedimentation, and metamorphism.

rock-defended terrace [GEOL] **1.** A river terrace having a ledge or outcrop of resistant rock at its base which serves as protection against undermining. **2.** A marine terrace having a mass of resistant rock at the base of the cliff which protects against wave erosion.

rock desert [GEOL] An upland desert in which bedrock is either exposed or is covered with a thin veneer of coarse rock fragments.

rock drill [MECH ENG] A machine for boring relatively short holes in rock for blasting purposes; motive power may be compressed air, steam, or electricity.

rock dust distributor See rock duster.

rock duster [MIN ENG] A machine that distributes rock dust over the interior surfaces of a coal mine by means of air to prevent coal dust explosions. Also known as rock dust distributor.

rock element [PETR] The coherent, intact piece of rock that is the basic constituent of the rock system and which has physical, mechanical, and petrographic properties that can be described or measured by laboratory tests.

rocker [CIV ENG] A support at the end of a truss or girder which permits rotation and horizontal movement to allow for expansion and contraction. [GRAPHICS] A type of chisel, with a sharp beveled edge, used in mezzotint engraving; it is set on the surface of a copper plate and rocked back and forth to produce a rough grain, which, when printed, produces a velvety black. [MIN ENG] A small digging bucket mounted on two rocker arms in which auriferous alluvial sands are agitated by oscillation, in water, to collect gold. [ORD] Movable, built-in support in a field gun carriage, between the trail and the cradle, that allows changes in elevation to be made without disturbing the angle of position setting.

rocker arm [MECH ENG] In an internal combustion engine, a lever that is pivoted near its center and operated by a pushrod at one end to raise and depress the valve stem at the other end.

rocker bearing [CIV ENG] A bridge support that is free to rotate but cannot move horizontally.

rocker bent [CIV ENG] A bent used on a bridge span; hinged at one or both ends to provide for the span's expansion and contraction.

rocker cam [MECH ENG] A cam that moves with a rocking motion.

rocker dump car [MIN ENG] A small-capacity mining car; the most popular and most widely used are the gravity dump types, designed so that the weight of the load tips the body when a locking latch is released by hand.

rocker panel [ENG] The part of the paneling on a passenger vehicle located below the passenger compartment doorsill.

rocker shovel [MIN ENG] A digging and loading machine consisting of a bucket attached to a pair of semicircular runners; lifts and dumps the bucket load into a car or another materials-transport unit behind the machine.

rocket [AERO ENG] **1.** Any kind of jet propulsion capable of operating independently of the atmosphere. **2.** A complete vehicle driven by such a propulsive system.

rocket airplane [AERO ENG] An airplane using a rocket or rockets for its chief or only propulsion.

rocket ammunition [ORD] Any type of ammunition incorporating rockets.

rocket antenna [ELECTROMAG] An antenna carried on a rocket, to receive signals controlling the rocket or to transmit measurements made by instruments aboard the rocket.

rocket assist [AERO ENG] An assist in thrust given an airplane or missile by use of a rocket motor or rocket engine during flight or during takeoff.

rocket-assisted takeoff *See* rato.

rocket-assisted torpedo [ORD] A torpedo designed to be fired into the air by rocket and to drop into the water by parachute; the torpedo then seeks its underwater target by a special homing device. Abbreviated RAT.

rocket astronomy [ASTRON] The discipline comprising measurements of the electromagnetic radiation from the sun, planets, stars, and other bodies, of wavelengths that are almost completely absorbed below the 250-kilometer level, by using a rocket to carry instruments above 250 kilometers to measure these phenomena.

rocket chamber [AERO ENG] A chamber for the combustion of fuel in a rocket; in particular, that section of the rocket engine in which combustion of propellants takes place.

rocket engine [AERO ENG] A reaction engine that contains within itself, or carries along with itself, all the substances necessary for its operation or for the consumption or combustion of its fuel, not requiring intake of any outside substance and hence capable of operation in outer space. Also known as rocket motor.

rocket fuel [MATER] Any of the substances or mixtures of substances that can burn rapidly with controlled combustion to produce large volumes of gas at high pressures and temperatures; includes monopropellants (hydrogen peroxide and hydrazine), liquid bipropellant fuels (organic fuel and oxidizer), and solid propellants (mixed oxidizer-fuel in a propellant grain).

rocket igniter [AERO ENG] An igniter for the propellant in a rocket.

rocket launcher [AERO ENG] A device for launching a rocket, wheel-mounted, motorized, or fixed for use on the ground; rocket launchers are mounted on aircraft, as under the wings, or are installed below or on the decks of ships.

rocket lightning [GEOPHYS] A rare form of lightning whose luminous channel seems to advance through the air with only the speed of a skyrocket.

rocket missile [ORD] A missile using rocket propulsion.

rocket motor *See* rocket engine.

rocket nose section [AERO ENG] The extreme forward portion of a rocket, designed to contain instrumentation, spotting charges, fusing or arming devices, and the like, but does not contain the payload.

rocket ogive [ORD] A hollow, conical, metallic shell covering for a rocket warhead, designed to reduce air resistance during flight.

rocket propellant [MATER] **1.** Any agent which is used for consumption or combustion in a rocket, and from which the rocket derives its thrust, such as a fuel, oxidizer, and additive. **2.** The ejected fluid in a nuclear rocket.

rocket propulsion [AERO ENG] Reaction propulsion by a rocket engine.

rocket ramjet [AERO ENG] A ramjet engine having a rocket mounted within the ramjet duct, the rocket being used to bring the ramjet up to the necessary operating speed. Also known as ducted rocket.

rocketry [AERO ENG] **1.** The science or study of rockets, embracing theory, research, development, and experimentation. **2.** The art and science of using rockets, especially rocket ammunition.

rocket sled [AERO ENG] A sled that runs on a rail or rails and is accelerated to high velocities by a rocket engine; the sled is used in determining g tolerances and for developing crash survival techniques.

rocket-sled testing [AERO ENG] A method of subjecting structures and devices to high accelerations or decelerations and aerodynamic flow phenomena under controlled conditions; the test object is mounted on a sled chassis running on precision steel rails and accelerated by rockets or decelerated by water scoops.

rocketsonde *See* meteorological rocket.

rocket staging [AERO ENG] The use of successive rocket sections or stages, each having its own engine or engines; each stage is a complete rocket vehicle in itself.

rocket station [ENG] A life-saving station equipped with line-carrying rocket apparatus.

rocket thrust [AERO ENG] The thrust of a rocket engine.

rocket tube [AERO ENG] **1.** A launching tube for rockets. **2.** A tube or nozzle through which rocket gases are ejected.

rockeye [ORD] A free-fall cluster bomb designed for use against tanks and armored vehicles with antipersonnel secondary effects.

rock fabric *See* fabric.

rock failure [GEOL] Fracture of a rock that has been stressed beyond its ultimate strength.

rockfall [GEOL] **1.** The fastest-moving landslide; free fall of newly detached bedrock segments from a cliff or other steep slope; usually occurs during spring thaw. **2.** The rock material moving in or moved by a rockfall.

rock fan [GEOL] A fan-shaped bedrock surface whose apex is where a mountain stream debouches upon a piedmont slope, and which occupies an area where a pediment meets the mountain slope.

rock-fill [CIV ENG] Composed of large, loosely placed rocks.

rock-fill dam [CIV ENG] A dam constructed of loosely placed rock or stone.

rock flour [GEOL] A fine, chemically unweathered powder of rock-forming minerals produced by pulverization of rock fragments during natural transport or crushing. Also known as glacial flour.

rockforming [GEOL] Referring to any minerals which commonly occur in important proportions in common rocks.

rock glacier [GEOL] Boulders and fine material cemented by ice about a meter below the surface. Also known as talus glacier.

rock gypsum [MINERAL] Massive, coarsely crystalline to earthy, finely granular type of gypsum found in gyp rock.

rocking bar [HOROL] **1.** The plate that holds the winding wheels in a watch; rocks to alternately engage the winding mechanism and the hand-setting mechanism. **2.** The part in an alarm timepiece that engages the time and alarm mainsprings.

rocking furnace [MECH ENG] A horizonal, cylindrical melting furnace that is rolled back and forth on a geared cradle.

rocking pier [CIV ENG] A pier that is hinged to allow for longitudinal expansion or contraction of the bridge.

rocking the sextant *See* swinging the arc.

rocking valve [MECH ENG] An engine valve in which a disk or cylinder turns in its seat to permit fluid flow.

rock island *See* meander core.

rock loader [MIN ENG] Any device or machine used for loading slate or rock inside a mine.

rock magnetism [GEOPHYS] The natural remanent magnetization of igneous, metamorphic, and sedimentary rocks resulting from the presence of iron oxide minerals.

rock matrix compressibility [GEOL] One of three types of rock compressibility (matrix, bulk, and pore); the fractional change in volume of the solid rock material (grains) with a unit change in pressure.

rock meal *See* rock milk.

rock mechanics [GEOPHYS] Application of the principles of mechanics and geology to quantify the response of rock when it is acted upon by environmental forces, particularly when man-induced factors alter the original ambient forces.

rock milk [MINERAL] A soft, white, earthy or powdery variety of calcite. Also known as agaric mineral; bergmehl; forril farina; rock meal.

rockoon [AERO ENG] A high-altitude sounding system consisting of a small solid-propellant research rocket carried aloft by a large plastic balloon.

rock pedestal *See* pedestal.

rock pediment [GEOL] A pediment formed on the surface of bedrock.

rock permeability [GEOL] The ability of a rock to receive, hold, or pass fluid materials (oil, water, and gas) by nature of the interconnections of its internal porosity.

rock pressure [GEOPHYS] **1.** Stress in underground geologic material due to weight of overlying material, residual stresses, and pressures resulting from swelling clays. **2.** *See* ground pressure.

rock salt *See* sodium chloride.

rock saw [MIN ENG] A type of mechanical miner that is used to remove large blocks of material; cuts narrow slots or channels by the action of a moving steel band or blade and a slurry of abrasive particles (sometimes diamonds) rather than teeth; small flame jets are also used.

rockshaft [MIN ENG] A shaft through which rock can be brought into a mine for filling, stopes, or other excavations.

rock shell [INV ZOO] The common name for a large number of gastropod mollusks composing the family Muridae and characterized by having conical shells with various sculpturing.

Rocksite program [MIN ENG] A U.S. Navy program concerned with undersea mining or consolidated mineral deposits; studied direct sea-floor access at remote sites through shafts drilled in the sea floor.

rockslide [GEOL] The sudden, rapid downward movement of newly detached bedrock segments over a surface of weakness, such as of bedding, jointing, or faulting. Also known as rock slip.

rock slip *See* rockslide.

rock step *See* knickpoint.

rock-stratigraphic unit [GEOL] A lithologically homogeneous body of strata characterized by certain observable physical features, or by the dominance of a certain rock type or combination of rock types; rock-stratigraphic units include groups, formations, members, and beds. Also known as geolith; lithologic unit; lithostratic unit; lithostratigraphic unit; rock unit.

rock stream [GEOL] Rocks moving (or already moved) in a mass down a slope under the influence of their own weight.

rock system [GEOPHYS] In rock mechanics, all natural environmental factors that can influence the behavior of that portion of the earth's crust that will become part of an engineering structure.

rock terrace [GEOL] A stream terrace on the side of a valley composed of resistant bedrock which remains during erosion of weaker overlying and underlying beds.

rock type [PETR] **1.** One of the three major rock groups: igneous, sedimentary, metamorphic. **2.** A rock having a unique, identifiable set of characters, such as basalt.

rock unit *See* rock-stratigraphic unit.

Rockwell hardness [ENG] A measure of hardness of a material as determined by the Rockwell hardness test.

Rockwell hardness test [ENG] One of the arbitrarily defined measures of resistance of a material to indentation under static or dynamic load; depth of indentation of either a steel ball or a 120° conical diamond with rounded point, $\frac{1}{16}, \frac{1}{8}, \frac{1}{4}$, or $\frac{1}{2}$ inch (1.5875, 3.175, 6.35, 12.7 millimeters) in diameter, called a brale, under prescribed load is the basis for Rockwell hardness; 60, 100, 150 kilogram load is applied with a special machine, and depth of impression under initial minor load is indicated on a dial whose graduations represent hardness number.

rock wool [MATER] A fibrous mineral material made by blowing a jet of steam through molten rock or slag; used as insulation.

Rocky Mountain spotted fever [MED] An acute, infectious, typhuslike disease of man caused by the rickettsial organism *Rickettsia rickettsi* and transmitted by species of hard-shelled ticks; characterized by sudden onset of chills, headache, fever, and an exanthem on the extremities. Also known as American spotted fever; tick fever; tick typhus.

rocky point effect [ELECTR] Transient but violent discharges between electrodes in high-voltage transmitting tubes.

rod [DES ENG] **1.** A bar whose end is slotted, tapered, or screwed for the attachment of a drill bit. **2.** A thin, round bar of metal or wood. [GEOL] A rodlike sedimentary particle characterized by a width-length ratio less than 2/3 and a thickness-width ratio more than 2/3. Also known as roller. [HISTOL] One of the rod-shaped sensory bodies in the retina which are sensitive to dim light. [MECH] *See* perch. [NUCLEO] A relatively long and slender body of material used in, or in conjunction with, a nuclear reactor; may contain fuel, absorber, or fertile material or other material in which activation or transmutation is desired.

rod bit [DES ENG] A bit designed to fit a reaming shell that is threaded to couple directly to a drill rod.

rod coupling [DES ENG] A double-pin-thread coupling used to connect two drill rods together.

rod dope [MATER] Grease or other material used to protect or lubricate drill rods. Also called gunk; rod grease.

rodent [VERT ZOO] The common name for members of the order Rodentia.

Rodentia [VERT ZOO] An order of mammals characterized by a single pair of ever-growing upper and lower incisors, a maximum of five upper and four lower cheek teeth on each side, and free movement of the lower jaw in an anteroposterior direction.

rodenticide [MATER] A chemical agent used to kill rodents.

rod gap [ELEC] **1.** A device that is usually formed of two $\frac{1}{2}$ square-inch (3-square-centimeter) rods, one grounded and the other connected to the line conductor, but may also have the shape of rings or horns, used to limit the magnitude of transient overvoltages on an electrical system as a result of lightning strikes. **2.** Spark gap in which the electrodes are two coaxial rods, with ends between which the discharge takes place, cut perpendicularly to the axis.

rod grease *See* rod dope.

rod mill [MECH ENG] A pulverizer operated by the impact of heavy metal rods. [MET] A mill for making metal rods.

rod pump [PETRO ENG] Type of oil well sucker-rod pump that can be inserted into or removed from oil well tubing without moving or disturbing the tubing itself. Also known as insert pump.

Rodriguez formula [MATH] The equation giving the nth function in a class of special functions in terms of the nth derivatives of some polynomial.

rod string [MECH ENG] Drill rods coupled to form the connecting link between the core barrel and bit in the borehole and the drill machine at the collar of the borehole.

rod stuffing box [ENG] An annular packing gland fitting between the drill rod and the casing at the borehole collar; allows the rod to rotate freely but prevents the escape of gas or liquid under pressure.

rod thermistor [ELECTR] A type of thermistor that has high resistance, long time constant, and moderate power dissipation; it is extruded as a long vertical rod 0.250–2.0 inches (0.63–5.1 centimeters) long and 0.050–0.110 inch (0.13–0.28 centimeter) in diameter, of oxide-binder mix and sintered; ends are coated with conducting paste and leads are wrapped on the coated area.

rod weeder [AGR] A type of equipment used to prepare the soil during harrowing; it is a power-driven rod, usually square in cross section, which also operates below the surface of loose soil, killing weeds and maintaining the soil in loose mulched condition; adapted to large operations and used in dry areas in the northwestern United States.

roedderite [MINERAL] $(Na,K)_2(Mg,Fe)_5Si_{12}O_{30}$ A silicate meteorite mineral.

roemerite [MINERAL] $FeFe_2(SO_4)_4 \cdot 14H_2O$ A rust-brown to yellow mineral composed of hydrous ferric and ferrous iron sulfate.

roentgen [NUCLEO] An exposure dose of x- or γ-radiation such that the electrons and positrons liberated by this radiation produce, in air, when stopped completely, ions carrying positive and negative charges of 2.58×10^{-4} coulomb per kilogram of air. Abbreviated R (formerly r). Also spelled röntgen.

roentgen diffractometry *See* x-ray crystallography.

roentgen equivalent man *See* rem.

roentgen equivalent physical *See* rep.

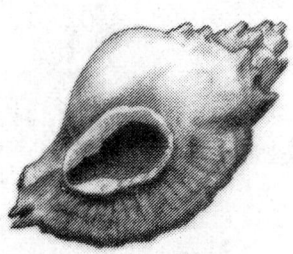

ROCK SHELL

The festive rock shell (*Murex festivus*). It has an elaborately ornamented shell and grows over 3 inches (8 centimeters) long.

ROCKSITE PROGRAM

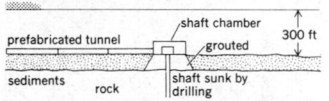

Proposed method of sea-floor access in the Rocksite program. 300 feet = 90 meters, 10 miles = 16 kilometers.

Roentgen meter [NUCLEO] A meter for measuring the cumulative quantity of x-rays or γ-rays, without reference to time.

roentgenography [PHYS] Radiography by means of x-rays.

roentgenoluminescence [PHYS] Luminescence which can be produced by x-rays.

roentgen optics See x-ray optics.

roentgenotherapy See x-ray therapy.

roentgen-per-hour-at-one-meter [NUCLEO] A unit of γ-ray source strength, corresponding to a dose rate of 1 roentgen per hour at a distance of 1 meter in air. Abbreviated RHM.

roentgen rays See x-rays.

roentgen spectrometry See x-ray spectrometry.

Roese-Gottlieb method [ANALY CHEM] A solvent extraction method used to obtain an accurate determination of the fat content of milk.

roesslerite [MINERAL] $MgH(AsO_4) \cdot 7H_2O$ A monoclinic mineral composed of hydrous acid magnesium arsenate; it is isomorphous with phosphorroesslerite.

roestone See oolite.

ROFOR [METEOROL] An international code word used to indicate a route forecast (along an air route).

ROFOT [METEOROL] An international code word used to indicate a route forecast, with units in the English system.

Rogallo wing [AERO ENG] A glider folded inside a spacecraft; to be deployed during the spacecraft's reentry like a parachute, gliding the spacecraft to a landing.

Rohrback solution [MATER] Toxic, clear, yellow liquid used for specific-gravity separation of minerals and microchemical detection of alkaloids.

roily oil [MATER] Crude petroleum oil that is more or less emulsified with water.

rolamite mechanism [MECH ENG] An elemental mechanism consisting of two rollers contained by two parallel planes and bounded by a fixed S-shaped band under tension.

roll [MECH] Rotational or oscillatory movement of an aircraft or similar body about a longitudinal axis through the body; it is called roll for any degree of such rotation. [MECH ENG] A cylinder mounted in bearings; used for such functions as shaping, crushing, moving, or printing work passing by it. [MIN ENG] See horseback. [TEXT] A continuous strand made by rolling, rubbing, or twisting fibers.

roll acceleration [MECH] The angular acceleration of an aircraft or missile about its longitudinal or X axis.

roll axis [MECH] A longitudinal axis through an aircraft, rocket, or similar body, about which the body rolls.

rollback See rerun.

roll cloud See rotor cloud.

roll compacting [MET] Compacting a metal powder by using a rolling mill.

roll control [ENG] The exercise of control over a missile so as to make it roll to a programmed degree, usually just before pitchover.

roll crusher [MECH ENG] A crusher having one or two toothed rollers to reduce the material.

rolled glass [MATER] Thick flat glass made by passing a roller over the molten glass.

rolled gold [MET] Same as gold-filled except that the proportion of gold alloy to total weight of the article may be less than 1/20; fineness of the gold alloy may not be less than 10 karat.

roller [DES ENG] A cylindrical device for transmitting motion and force by rotation. [GEOL] See rod. [OCEANOGR] A long, massive wave which usually retains its form without breaking until it reaches the beach or a shoal.

roller analyzer [ENG] Device for quantitative separation of fine particles (down to 5 micrometers) by use of the graduated lift of a variable-rate pneumatic stream.

roller bearing [MECH ENG] A shaft bearing characterized by parallel or tapered steel rollers confined between outer and inner rings.

roller bit See cone rock bit.

roller cam follower [MECH ENG] A follower consisting of a rotatable wheel at the end of the shaft.

roller chain [MECH ENG] A chain drive assembled from roller links and pin links.

roller coating [ENG] The application of paints, lacquers, or other coatings onto raised designs or letters by means of a roller.

roller conveyor [MECH ENG] A gravity conveyor having a track of parallel tubular rollers set at a definite grade, usually on antifriction bearings, at fixed locations, over which package goods which are sufficiently rigid to prevent sagging between rollers are moved by gravity or propulsion.

roller drying [CHEM ENG] A method used to dry milk for purposes other than human consumption; concentrated milk is fed between two heated and narrowly spaced stainless steel rollers, the adhering thin film of milk dries as the rollers turn and is scraped off the roller by a doctor blade.

roller gate [CIV ENG] A cylindrical, usually hollow crest gate that is raised and lowered by large toothed wheels running on sloping racks.

roller-hearth kiln [ENG] A type of tunnel kiln through which the ware is conveyed on ceramic rollers.

roller leveling [MECH ENG] Leveling flat stock by passing it through a machine having a series of rolls whose axes are staggered about a mean parallel path by a decreasing amount.

roller mill See ring-roller mill.

roller printing [TEXT] Printing patterns on fabrics by running them through copper rollers to which inks are applied.

roller pulverizer [MECH ENG] A pulverizer operated by the crushing action of rotating rollers.

rollers [OCEANOGR] Swells coming from a great distance and forming large breakers on exposed coasts.

roller stamping die [MECH ENG] An engraved roller used for stamping designs and other markings on sheet metal.

Rolle's theorem See first law of the mean.

roll film [GRAPH] A long strip of plastic material with a sensitive emulsion on one side and a layer of gelatin on the other; the gelatin layer reduces the tendency of the base material to curl; the film is attached to opaque backing paper that protects the film from light before and after it is exposed in the camera.

roll flattening See flattening.

roll forging [MET] Forging metal by using grooved rotating dies.

roll forming [MET] Metal forming by using contoured rolls.

roll in [ADP] To restore to main memory a section of program or data that had previously been rolled out.

rolling [MECH] Motion of a body across a surface combined with rotational motion of the body so that the point on the body in contact with the surface is instantaneously at rest. [MET] Reducing or changing the cross-sectional area of a workpiece by the compressive forces exerted by rotating rolls. Also known as metal rolling. [NAV ARCH] The oscillating motion of a vessel from side to side due to ground swell, heavy sea, or other causes.

rolling barrage [ORD] An artillery barrage that precedes infantry troops at a predetermined rate in their advance during the attack to protect them and facilitate their advance.

rolling contact [MECH] Contact between bodies such that the relative velocity of the two contacting surfaces at the point of contact is zero.

rolling-contact bearing [MECH ENG] A bearing composed of rolling elements interposed between an outer and inner ring.

rolling door [ENG] A door that moves up and down or from side to side by means of wheels moving along a track.

rolling friction [MECH] A force which opposes the motion of any body which is rolling over the surface of another.

rolling lift bridge [CIV ENG] A bridge having on the shore end of the lifting portion a segmental bearing that rolls on a flat surface.

rolling recoil [ORD] System formerly used for absorbing the recoil energy of some railway guns; with the brakes set, the entire car is allowed to roll back when the gun is fired.

rolling transposition [ELEC] Transposition in which the conductors of an open wire circuit are physically rotated in a substantially helical manner; with two wires, a complete transposition is usually executed in two consecutive spans.

roll mill [MECH ENG] A series of rolls operating at different speeds for grinding and crushing.

roll-off [ELECTR] Gradually increasing loss or attenuation with increase or decrease of frequency beyond the substantially flat portion of the amplitude-frequency response characteristic of a system or transducer.

roll-out [ADP] **1.** To make available additional main memory

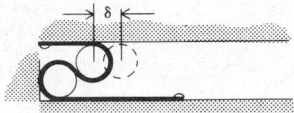

ROLAMITE MECHANISM

Drawing of rolamite mechanism showing components. δ represents deflection of upper roller during motion of mechanism.

ROLLER CAM FOLLOWER

Drawing showing components of roller cam follower.

ROLLER CHAIN

Section of a single-strand roller chain.

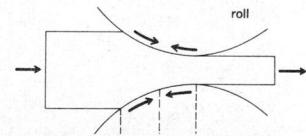

ROLLING

Direction of friction forces in the roll gap during the rolling process. At the neutral point (arrows meet), roll and metal strip have the same velocity; to the left of this point (entry side) the strip moves slower than the roll; and to the right of this point (exit) it moves faster.

for one task by copying another task onto auxiliary storage. **2.** To read a computer register or counter by adding a one to each digit column simultaneously until all have returned to zero, with a signal being generated at the instant a column returns to zero.

roll roofing [MATER] Composition sheet roofing supplied in rolls from which it is laid in overlapping strips.

roll-tube technique [MICROBIO] A pure-culture technique, employed chiefly in tissue culture, in which, during incubation, the test tubes are held in a wheellike instrument at an angle of about 15° from the horizontal and the wheel is rotated vertically about once every 2 minutes.

roll welding [MET] Forge welding by heating in a furnace and applying pressure with rolls.

rom [ELEC] A unit of electrical conductivity, equal to the conductivity of a material in which an electric field of 1 volt per meter gives rise to a current density of 1 ampere per square meter. Derived from reciprocal ohm meter.

Romanche trench [GEOL] A 7370-meter-deep trench in the Mid-Atlantic ridge near the equator.

Romberg's sign [MED] **1.** A sign for obturator hernia in which there is pain radiating to the knee. **2.** A sign for loss of position sense in which the patient cannot maintain equilibrium when standing with feet together and eyes closed.

romeite [MINERAL] $(Ca,Fe,Mn,Na)_2(Sb,Ti)_2O_6(O,OH,F)$ A honey-yellow to yellowish-brown mineral composed of oxide of calcium, iron, manganese, sodium, antimony, and titanium, occurring in minute octahedrons.

Römer method [OPTICS] A method of measuring the speed of light, in which apparent changes in the periods of satellites of another planet, such as Jupiter, whose distance from the earth is known, are observed throughout the year.

ROMET [METEOROL] An international code word denoting route forecast, with units in the metric system.

Ronchi test [OPTICS] An improvement on the Foucault knife-edge test for testing curved mirrors, in which the knife edge is replaced with a transmission grating with 15–80 lines per centimeter, and the pinhole source is replaced with a slit or a section of the same grating.

rondada [METEOROL] In Spain, a wind that shifts diurnally from northwest through north, east, south, and west.

röntgen See roentgen.

rood [MECH] A unit of area, equal to ¼ acre, or 10,890 square feet, or 1011.7141056 square meters.

roof [BUILD] The cover of a building or similar structure. [MIN ENG] The rock immediately above a coal seam; corresponds to a hanging wall in metal mining.

roof beam [BUILD] A load-bearing member in the roof structure.

roof bolt [MIN ENG] One of the long steel bolts driven into walls or roofs of underground excavations to strengthen the pinning of rock strata.

roof control [MIN ENG] The study of rock behavior when undermined by mining operations, and the most effective measures to control movements.

roof cut [MIN ENG] A machine cut made with a turret coal cutter in the roof immediately above the seam.

roof filter [ELECTR] Low-pass filter used in carrier telephone systems to limit the frequency response of the equipment to frequencies needed for normal transmission, thereby blocking unwanted higher frequencies induced in the circuit by external sources; improves runaround cross-talk suppression and minimizes high-frequency singing.

roof foundering [GEOL] Collapse of overlying rock into a magma chamber following excavation of a large quantity of magma.

roofing [MATER] Material used in roof construction, such as tar, tar paper, shingles, slate, and tin.

roofing copper [MET] Copper that has been hot-rolled to sheets in 14–32-ounce (400–900-gram) weights.

roofing felt [MATER] Thick asphalt-impregnated paper used for roofing.

roofing granules [MATER] Graded particles of crushed rock, slate, slag, porcelain, or tile, used as surfacing on asphalt roofing and shingles.

roofing nail [DES ENG] A nail used for attaching paper or shingle to roof boards; usually short with a barbed shank and a large flat head.

roofing putty [MATER] Heavy consistency asphalt solution with asbestos fibers; used for caulking metal roofs.

roofing slate [MATER] Hard varieties of slate varying in size from 12×6 inches (30×15 centimeters) to 24×14 inches (60×35 centimeters), and from ⅛ to ¾ inches (3 to 19 millimeters) in thickness.

roof jack [MIN ENG] A screw- or pump-type extension post used as a temporary roof support.

roof pendant [GEOL] Downward projection or sag into an igneous intrusion of the country rock of the roof. Also known as pendant.

roof stringer [MIN ENG] A lagging bar running parallel with the working place above the header in a weak or scaly top in narrow rooms or entries which have short life.

roof truss [BUILD] A truss used in roof construction; it carries the weight of roof deck and framing and of wind loads on the upper chord; an example is a Fink truss.

room [BUILD] A partitioned-off area inside a building or dwelling. [MIN ENG] **1.** Space driven off an entry in which coal is produced. **2.** Working place in a flat mine.

room acoustics [ACOUS] The study of the behavior of sound waves in an enclosed room.

room-and-pillar [MIN ENG] A system of mining in which the coal or ore is mined in rooms separated by narrow ribs or pillars; pillars are subsequently worked.

room conveyor [MIN ENG] Any conveyor which carries coal from the face of a room toward the mouth.

room crosscut See breakthrough.

room noise [COMMUN] Ambient noise in a telephone station.

rooster [AGR] An adult male domestic fowl. [VERT ZOO] An adult male of certain birds and fowl, such as pheasants and ptarmigans.

root [ADP] The origin or most fundamental point of a tree diagram. Also known as base. [BOT] The absorbing and anchoring organ of a vascular plant; it bears neither leaves nor flowers and is usually subterranean. [CIV ENG] The portion of a dam which penetrates into the ground where the dam joins the hillside. [DES ENG] The bottom of a screw thread. [GEOL] **1.** The lower limit of an ore body. Also known as bottom. **2.** The part of a fold nappe that was originally linked to its root zone. [MATH] **1.** A root of a given real or complex number is a number which when raised to some exponent equals that number. Also known as radix. **2.** A root of a polynomial $p(x)$ is a number a such that $p(a) = 0$.

root canal [ANAT] The cavity within the root of a tooth, occupied by pulp, nerves, and vessels.

root cap [BOT] A thick, protective mass of parenchymal cells covering the meristematic tip of the root.

root circle [DES ENG] A hypothetical circle defined at the bottom of the tooth spaces of a gear.

root clay See underclay.

root crack [MET] A crack in the weld or in the heat-affected zone at the root of the weld.

rooter [ENG] A heavy plowing device equipped with teeth and used for breaking up the ground surface; a towed scarifier.

root field See Galois field.

root fillet [DES ENG] The rounded corner at the angle of a gear tooth flank and the bottom land.

root hair [BOT] One of the hairlike outgrowths of the root epidermis that function in absorption.

root knot [PL PATH] Any of various plant diseases caused by root-knot nematodes which produce gall-like enlargements on the roots.

root locus plot [CONT SYS] A plot in the complex plane of values at which the loop transfer function of a feedback control system is a negative number.

root-mean-square current See effective current.

root-mean-square error [STAT] The square root of the second moment corresponding to the frequency function of a random variable.

root-mean-square sound pressure See effective sound pressure.

root-mean-square value Abbreviated rms value. [PHYS] The square root of the time average of the square of a quantity; for a periodic quantity the average is taken over one complete cycle. Also known as effective value. [STAT] The square root of the average of the squares of a series of related values.

root nodule [PL PATH] An abnormal nodular growth on a plant root system caused by the establishment of symbiotic nitrogen-fixing bacteria in the host tissue.

root of an equation [MATH] A number or quantity which satisfies a given equation.

root of joint [MET] The area of closest proximity between members of a joint to be welded.

root of weld [MET] The points at which the bottom of the weld and the base metal surfaces intersect.

root opening [MET] In welding, the distance between members at the root of the joint.

root pass [MET] The first weld bead deposited in a multiple pass weld. Also known as root sealer bead.

root penetration [MET] The depth of penetration of the weld metal into the root of a joint.

root rot [PL PATH] Any of various plant diseases characterized by decay of the roots.

Roots blower [MECH ENG] A compressor in which a pair of hourglass-shaped members rotate within a casing to deliver large volumes of gas at relatively low pressure increments.

root sealer bead *See* root pass.

root segment [ADP] The master or controlling segment of an overlay structure which always resides in the main memory of a computer.

rootstock [BOT] A root or part of a root used as the stock for grafting.

root-sum-square value [PHYS] The square root of the sum of the squares of a series of related values; commonly used to express total harmonic distortion.

root test [MATH] An infinite series of nonnegative terms a_n converges if, after some term, the ith root of a_i is less than a fixed number smaller than 1.

rootworm [INV ZOO] **1.** An insect larva that feeds on plant roots. **2.** A nematode that infests the roots of plants.

root zone [GEOL] **1.** The area where a low-angle thrust fault steepens and descends into the crust. **2.** The source of the root of a fold nappe.

ropak [OCEANOGR] An ice cake standing on edge as a result of excessive pressure. Also known as turret ice.

rope [MATER] A long, flexible object consisting of many strands of wire, plastic, or vegetable fiber such as manila. [ORD] A form of confusion reflector consisting of long strips of metal foil, sometimes used with tiny parachutes to reduce the rate of fall.

rope-and-button conveyor [MECH ENG] A conveyor consisting of an endless wire rope or cable with disks or buttons attached at intervals.

rope boring [ENG] A method similar to rod drilling except that rigid rods are replaced by a steel rope to which the boring tools are attached and allowed to fall by their own weight.

rope chaff [ORD] Chaff that contains one or more rope elements.

rope cutter *See* hook tender.

rope drive [MECH ENG] A system of ropes running in grooved pulleys or sheaves to transmit power over distances too great for belt drives.

rope-lay conductor [ELEC] Cable composed of a central core surrounded by one or more layers of helically laid groups of wires.

rope rider *See* trip rider.

rope sheave [DES ENG] A grooved wheel, usually made of cast steel or heat-treated alloy steel, used for rope drives.

rope socket [DES ENG] A drop-forged steel device, with a tapered hole, which can be fastened to the end of a wire cable or rope and to which a load may be attached.

ropewalk [ENG] A long walkway down which a worker carries and lays rope in a manufacturing plant.

ropeway [ENG] One or a pair of steel cables between several supporting towers which serve as tracks for transporting materials in mountainous areas or at sea.

Roproniidae [INV ZOO] A small family of hymenopteran insects in the superfamily Proctotrupoidea.

ropy lava *See* pahoehoe.

Rorschach test [PSYCH] A projective psychologic test in which the subject describes what he imagines seeing in a series of 10 standard inkblots of varying designs and colors.

Rosa and Dorsey method [ELECTROMAG] A method of measuring the speed of light by comparing the capacitance of a capacitor in electromagnetic units, as measured experimentally, with values of currents determined from a current balance, to the capacitance of the same capacitor in electrostatic units, as determined from its geometrical dimensions.

rosacea [MED] A chronic skin disorder of middle age characterized by redness, papules, and oiliness.

Rosaceae [BOT] A family of dicotyledonous plants in the order Rosales typically having stipulate leaves and perigynous flowers, numerous stamens, and several or many separate carpels.

Rosales [BOT] A morphologically diffuse order of dicotyledonous plants in the subclass Rosidae.

roscherite [MINERAL] $(Ca,Mn,Fe)_2Al(PO_4)(OH) \cdot 2H_2O$ A dark-brown mineral composed of hydrous basic phosphate of aluminum, calcium, manganese, and iron, occurring as monoclinic crystals.

roscoelite [MINERAL] $K(V,Al,Mg)_3Si_3O_{10}(OH)_2$ Tan, grayish-brown, or greenish-brown vanadium-bearing mica mineral occurring in minute scales or flakes.

rose [BOT] A member of the genus *Rosa* in the rose family (Rosaceae); plants are erect, climbing, or trailing shrubs, generally prickly stemmed, and bear alternate, odd-pinnate single leaves. [MATH] A graph consisting of loops shaped like rose petals arising from the equations in polar coordinates $r = a \sin n\theta$ or $r = a \cos n\theta$.

rose absolute [MATER] First filtrate from the cooled alcohol solvent solution of rose concrete (after removal of waxes) during perfume manufacture; pure oil of rose.

rose bit [DES ENG] A hardened steel or alloy noncore bit with a serrated face to cut or mill out bits, casing, or other metal objects lost in the hole.

rosebloom *See* false blossom.

rose chucking reamer [DES ENG] A machine reamer with a straight or tapered shank and a straight or spiral flute; cutting is done at the ends of the teeth only; produces a rough hole since there are few teeth.

rose concrete [MATER] Semisolid mixture of essential oils and waxes solvent-extracted from rose flower petals, leaves, and bark.

rose diagram [GEOL] A circular graph indicating values in several classes of vector properties of rocks such as crossbedding direction.

rose flower oil *See* rose oil.

rose hip [BOT] The ripened false fruit of a rose plant.

rosemary [BOT] *Rosmarinus officinalis.* A fragrant evergreen of the mint family from France, Spain, and Portugal; leaves have a pungent bitter taste and are used as an herb and in perfumes.

rosemary oil [MATER] Pungent, combustible, colorless-to-yellow essential oil with camphorlike aroma; soluble in ether, glacial acetic acid, and alcohol; derived from flowers of rosemary (*Rosmarinus officinalis*); used in flavors, perfumes, and medicines.

Rosenberg crossed-field generator [ELEC] A type of dynamoelectric amplifier which is self-regulating and can operate while the rotor varies in speed, the current never rising above a certain value.

Rosenmueller's organ *See* epoophoron.

Rosenmund reaction [ORG CHEM] Catalytic hydrogenation of an acid chloride to form an aldehyde; the reaction is in the presence of sulfur to prevent the subsequent hydrogenation of the aldehyde.

rose oil [MATER] Transparent, combustible, yellow-to-green or red essential oil with fragrant scent and sweet taste; solidifies at $18–37°C$; steam-distilled from rose flowers; used in flavors, perfumes, and medicines. Also known as attar of roses; otto of rose oil; rose flower oil.

roseola infantum *See* exanthem subitum.

roseola typhosa [MED] The rose-colored eruption characteristic of typhus or typhoid fever.

rose quartz [MINERAL] A pink variety of crystalline quartz;

ROSACEAE

A common eastern American species of wild rose (*Rosa carolina*), in the family Rosaceae of the order Rosales. (*Photograph by A. W. Ambler, from National Audubon Society*)

ROSE

Typical flowers of a rose plant.

commonly massive and used as a gemstone. Also known as Bohemian ruby.

Rose's metal [MET] An alloy of bismuth tin and lead; melts at 94°C.

rosette [BIOL] Any structure or marking resembling a rose. [MET] **1.** Rounded constituents in a microstructure arranged in whorls. **2.** Strain gages arranged to indicate at a single position the strains in three different directions. [MINERAL] Rose-shaped, crystalline aggregates of barite, marcasite, or pyrite formed in sedimentary rock. [PL PATH] Any of various plant diseases in which the leaves become clustered in the form of a rosette.

Rosette Nebula [ASTRON] A nebula classified as NGC 2237; this nebula contains numerous small dense clouds that have been photographed with large telescopes.

rose water [MATER] Aqueous solution with rose scent (from steam disillation of fresh flowers of rose plants); used in lotions, flavors, and perfumes.

Rosidae [BOT] A large subclass of the class Magnoliatae; most have a well-developed corolla with petals separate from each other, binucleate pollen, and ovules usually with two integuments.

rosieresite [MINERAL] A yellow to brown mineral composed of hydrous aluminum phosphate containing lead and copper, occurring in stalactitic masses.

rosin [MATER] A translucent yellow, umber, or reddish resinous residue from the distillation of crude turpentine from the sap of pine trees (gum rosin) or from an extract of the stumps and other parts of the tree (wood rosin); used in varnishes, lacquers, printing inks, adhesives, and soldering fluxes, in medical ointments, and as a preservative.

rosin-core solder [MATER] Solder made up in tubular or other hollow form, with the inner space filled with noncorrosive rosin flux.

rosin essence [MATER] That part of rosin that can be distilled off at a temperature below 360°C.

rosin ester See ester gum.

rosin-extended rubber [MATER] Cold rubber with up to 50% rosin.

rosin joint [ELEC] A soldered joint in which one of the wires is surrounded by an almost invisible film of insulating rosin, making the joint intermittently or continuously open even though it looks good.

rosin oil [MATER] Viscous, water-insoluble, white-to-brown liquid; soluble in ether, chloroform, carbon disulfide, and fatty oils; distilled from rosin; used as a lubricant and in adhesives, inks, and linoleum.

rosin size [MATER] An alkali-treated rosin used as a dry powder or emulsion to surface-size paper products.

rosin tin [MINERAL] A red or yellow variety of cassiterite. Also known as resin tin.

rosolic acid See aurin.

Ross Barrier [OCEANOGR] A wall of shelf ice bordering on the Ross Sea.

Rossby diagram [THERMO] A thermodynamic diagram, named after its designer, with mixing ratio as abscissa and potential temperature as ordinate; lines of constant equivalent potential temperature are added.

Rossby number [FL MECH] The nondimensional ratio of the inertial force to the Coriolis force for a given flow of a rotating fluid, given as $R_0 = U/fL$, where U is a characteristic velocity, f the Coriolis parameter (or, if the system is cylindrical rather than spherical, twice the system's rotation rate), and L a characteristic length.

Rossby parameter [FL MECH] The northward variation of the Coriolis parameter, arising from the sphericity of the earth. Also known as Rossby term.

Rossby regime [FL MECH] A type of flow pattern in a rotating fluid with differential radial heating in which the major radial transport of shear and momentum is effected by horizontal eddies of low wave-number; this regime occurs for low values of the Rossby number (of the order of 0.1).

Rossby term See Rossby parameter.

Rossby wave [METEOROL] A wave on a uniform current in a two-dimensional nondivergent fluid system, rotating with varying angular speed about the local vertical (beta plane); this is a special case of a barotropic disturbance, conserving

absolute vorticity; applied to atmospheric flow, it takes into account the variability of the Coriolis parameter while assuming the motion to be two-dimensional. Also known as planetary wave.

Rossel Current [OCEANOGR] A seasonal Pacific Ocean current flowing westward and north-westward along both the southern and northeastern coasts of New Guinea, the southern part flowing through Torres Strait and losing its identity in the Arafura Sea, and the northern part curving northeastward to join the equatorial countercurrent of the Pacific Ocean.

Ross feeder [MECH ENG] A chute for conveying bulk materials by means of a screen of heavy endless chains hung on a sprocket shaft; rotation of the shaft causes materials to slide.

Rossman drive [ENG] A method used to provide speed control of alternating-current motors; an induction motor stator is mounted on trunnion bearings and driven with an auxiliary motor, to provide the desired change in slip between the stator and rotor.

Ross objective [OPTICS] A type of wide-field lens objective in cameras used for astrometric work.

Ross Sea [GEOGR] Arm of the South Pacific Ocean off Antarctica.

rostellum [BIOL] The anterior, flattened region of the scolex of armed tapeworms.

rosterite See vorobyevite.

Rostratulidae [VERT ZOO] A small family of birds in the order Charadriiformes containing the painted snipe; females are more brightly colored than males.

rostrum [BIOL] A beak or beaklike process.

rot [PL PATH] Any plant disease characterized by breakdown and decay of plant tissue.

Rotaliacea [INV ZOO] A superfamily of foraminiferans in the suborder Rotaliina characterized by a planispiral or trochospiral test having apertural pores and composed of radial calcite, with secondarily bilamellar septa.

rotameter [ENG] A variable-area, constant-head, rate-of-flow volume meter in which the fluid flows upward through a tapered tube, lifting a shaped weight to a position where upward fluid force just balances its weight.

rotary [MECH ENG] **1.** A rotary machine, such as a rotary printing press or a rotary well-drilling machine. **2.** The turntable and its supporting and rotating assembly in a well-drilling machine.

rotary actuator [MECH ENG] A device that converts electric energy into controlled rotary force; usually consists of an electric motor, gear box, and limit switches.

rotary air heater [MECH ENG] A regenerative air heater in which heat-transferring members are moved alternately through the gas and air streams.

rotary amplifier See rotating magnetic amplifier.

rotary annular extractor [MECH ENG] Vertical, cylindrical shell with an inner, rotating cylinder; liquids to be contacted flow countercurrently through the annular space between the rotor and shell; used for liquid-liquid extraction processes.

rotary atomizer [MECH ENG] A hydraulic atomizer having the pump and nozzle combined.

rotary beam [ELECTROMAG] Short-wave antenna system highly directional in azimuth and altitude, mounted in such a manner that it can be rotated to any desired position, either manually or by an electric motor drive.

rotary belt cleaner [MECH ENG] A series of blades symmetrically spaced about the axis of rotation and caused to scrape or beat against the conveyor belt for the purpose of cleaning.

rotary blasthole drilling [MIN ENG] A term applied to two types of drilling: in quarrying and open pit mining it implies rotary drilling with roller-type bits, using compressed air for cuttings removal, either conventional rotary table drive or hydraulic motor to produce rotation, with hydraulic or wireline mechanisms to add part of the weight of the drill to the weight of the tools to increase bit pressure; and in underground mining and sometimes aboveground, it implies the drilling of small-diameter blastholes with a diamond drill, using either coring or noncoring diamond bits.

rotary blower [MECH ENG] Positive-displacement, rotating-impeller, air-movement device; can be straight-lobe, screw, sliding-vane, or liquid-piston type.

ROSS OBJECTIVE

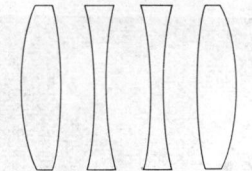

The four component lenses of the Ross objective.

ROTALIACEA

(a)　　　　　　(b)

|— 0.04 cm —|

Scanning electron micrograph of *Elphidium*, suborder Rotaliina, superfamily Rotaliacea, from Recent of Adriatic Sea. *(a)* Side view showing large sutural pores that open into canal system. *(b)* Edge view with series of apertural pores at lower margin of final chamber face and imperforate peripheral band. *(R. B. MacAdam, Chevron Oil Field Research Co.)*

ROTAMETER

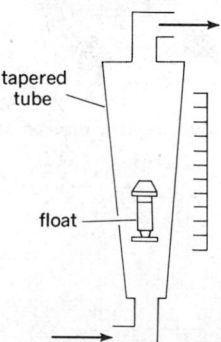

tapered tube

float

Diagram of a rotameter. Arrows indicate direction of fluid flow.

rotary boring [MECH ENG] A system of boring in which rock penetration is achieved by the rotation of the hollow cutting tool.

rotary breaker [MIN ENG] A breaking machine for coal or ore; consists of a trommel screen with a heavy steel shell fitted with lifts which raise and convey the coal and stone forward and break it; as the material is broken, the undersize passes through the apertures.

rotary bucket [MECH ENG] A 12- to 96-inch-diameter (30 to 244-centimeter-diameter) posthole augerlike device, the bottom end of which is equipped with cutting teeth used to rotary-drill large-diameter shallow holes to obtain samples of soil lying above the groundwater level.

rotary calculator [ADP] A type of mechanical desk calculator that is distinguished from key-driven calculators by virtue of a latching selection keyboard in which the multiplicand or divisor could be indexed and then repeatedly transferred to the accumulator, positively or negatively by turning a crank; the accumulator and a cycle-counting register were mounted in a carriage that could be shifted right or left, relative to the selection keyboard, to accommodate the successive digits of the multiplier or quotient.

rotary compressor [MECH ENG] A positive-displacement machine in which compression of the fluid is effected directly by a rotor and without the usual piston, connecting rod, and crank mechanism of the reciprocating compressor.

rotary converter *See* dynamotor.

rotary coupler *See* rotating joint.

rotary crane [MECH ENG] A crane consisting of a boom pivoted to a fixed or movable structure.

rotary crusher [MECH ENG] Solids-reduction device in which a high-speed rotating cone on a vertical shaft forces solids against a surrounding shell.

rotary-cup oil burner [ENG] Oil burner that uses centrifugal force to spray fuel oil from a rotary fuel atomizing cup into the combustion chamber.

rotary current [OCEANOGR] A current with the direction of flow rotating through all points of the compass.

rotary cutter [MECH ENG] Device used to cut tough or fibrous materials by the shear action between two sets of blades, one set on a rotating holder, the other stationary on the surrounding casing.

rotary-disk contactor [CHEM ENG] Liquid-liquid contactor, having a vertical cylindrical shell with vertical rotating shaft upon which are mounted a spaced series of flat disks; spinning of the disks forces liquid into shell-mounted baffles, causing mixing; used for liquid-liquid extraction processes. Also known as RDC extractor.

rotary dispersion [OPTICS] The change in the angle through which an optically active substance rotates the plane of polarization of plane polarized light as the wavelength of the light is varied. Also known as rotatory dispersion.

rotary drill [MECH ENG] Any of various drill machines that rotate a rigid, tubular string of rods to which is attached a rock cutting bit, such as an oil well drilling apparatus.

rotary drilling [MECH ENG] The act or process of drilling a borehole by means of a rotary-drill machine, such as in drilling an oil well.

rotary dryer [MECH ENG] A cylindrical furnace slightly inclined to the horizontal and rotated on suitable bearings; moisture is removed by rising hot gases.

rotary dump car [MIN ENG] A small mine car in which the car body is mounted on a rotary dumper.

rotary dumper [MIN ENG] A steel structure on which a mine car revolves and discharges the contents, usually sideways.

rotary engine [MECH ENG] A positive displacement engine (such as a steam or internal combustion type) in which the thermodynamic cycle is carried out in a mechanism that is entirely rotary and without the more customary structural elements of a reciprocating piston, connecting rods, and crankshaft.

rotary excavator *See* bucket-wheel excavator.

rotary fault [GEOL] A fault in which displacement is downward at one point and upward at another point. Also known as pivotal fault; rotational fault.

rotary feeder [MECH ENG] Device in which a rotating element or vane discharges powder or granules at a predetermined rate.

rotary filter *See* drum filter.

rotary gap *See* rotary spark gap.

rotary joint *See* rotating joint.

rotary kiln [ENG] A long cylindrical kiln lined with refractory, inclined at a slight angle, and rotated at a slow speed.

rotary-percussive drill [MECH ENG] Drilling machine which operates as a rotary machine by the action of repeated blows to the bit.

rotary phase converter [ELEC] Machine which converts power from an alternating-current system of one or more phases to an alternating-current system of a different number of phases, but of the same frequency.

rotary polarization *See* optical activity.

rotary press [GRAPHICS] A press utilizing two cylinders, one of which supports the paper while the other one prints on it; large rotary presses are web-fed and accept continuous strips of paper from large rolls; this web of paper is not cut or trimmed until after it is printed.

rotary pump [MECH ENG] A displacement pump that delivers a steady flow by the action of two members in rotational contact.

rotary rig [PETRO ENG] The collective equipment used with a rotary drill; includes prime movers or engines, derrick or mast, hoisting and rotating equipment, drill pipe, drill collars and bit, and the mud system used to circulate drilling fluid.

rotary shear [MECH ENG] A sheet-metal cutting machine having two rotary-disk cutters mounted on parallel shafts and driven in unison.

rotary shot drill [MECH ENG] A rotary drill used to drill blastholes.

rotary spark gap [ELEC] A spark gap in which sparks occur between one or more fixed electrodes and a number of electrodes projecting outward from the circumference of a motor-driven metal disk. Also known as rotary gap.

rotary stepping relay *See* stepping relay.

rotary stepping switch *See* stepping relay.

rotary swager [MECH ENG] A machine for reducing diameter or wall thickness of a bar or tube by delivering hammerlike blows to the surface of the work supported on a mandrel.

rotary switch [ELEC] A switch that is operated by rotating its shaft.

rotary transformer [ELEC] Term sometimes applied to a rotating machine used to transform direct-current power from one voltage to another.

rotary vacuum filter *See* drum filter.

rotary valve [MECH ENG] A valve for the admission or release of working fluid to or from an engine cylinder where the valve member is a ported piston that turns on its axis.

rotary-vane attenuator [ELECTROMAG] Device designed to introduce attenuation into a waveguide circuit by varying the angular position of a resistive material in the guide.

rotary-vane meter [ENG] A type of positive-displacement rate-of-flow meter having spring-loaded vanes mounted on an eccentric drum in a circular cavity; each time the drum rotates, a fixed volume of fluid passes through the meter.

rotary vibrating tippler [MIN ENG] A tippler designed to overcome the tendency for coal or dirt to stick to the bottom of the tubs so that when the tippler is inverted, the car rests upon a vibrating frame which frees any material tending to stick to the bottom.

rotary voltmeter [ENG] Type of electrostatic voltmeter used for measuring high voltages.

rotate [BOT] Of a sympetalous corolla, having a short tube and petals radiating like the spokes of a wheel.

rotating amplifier *See* rotating magnetic amplifier.

rotating-anode tube [ELECTR] An x-ray tube in which the anode rotates continuously to bring a fresh area of its surface into the beam of electrons, allowing greater output without melting the target.

rotating beacon [NAV] A navigational beacon with one or more beams that rotate.

rotating-beam ceilometer [ENG] An electronic, automatic-recording meteorological device which determines cloud height by means of triangulation.

rotating coordinate system [MECH] A coordinate system

ROTARY-VANE METER

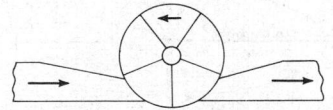

Schematic of rotary-vane meter. Arrows indicate direction of fluid flow.

whose axes as seen in an inertial coordinate system are rotating.

rotating crystal method [SOLID STATE] Any method of studying crystalline structures by x-ray or neutron diffraction in which a monochromatic, collimated beam of x-rays or neutrons falls on a single crystal that is rotated about an axis perpendicular to the beam.

rotating-cylinder method [FL MECH] A method of measuring the viscosity of a fluid in which the fluid fills the space between two concentric cylinders, and the torque on the stationary inner cylinder is measured when the outer cylinder is rotated at constant speed.

rotating-drum heat transfer [CHEM ENG] Procedure for solidifying layers of solids onto the outside surface of an inside-cooled drum that is partly immersed in a melt of the solids material.

rotating joint [ELECTROMAG] A joint that permits one section of a transmission line or waveguide to rotate continuously with respect to another while passing radio-frequency energy. Also known as rotary coupler; rotary joint.

rotating magnetic amplifier [ELEC] A prime-mover-driven direct-current generator whose power output can be controlled by small field input powers, to give power gain as high as 10,000. Also known as rotary amplifier; rotating amplifier.

rotating meter See velocity-type flowmeter.

rotating models [OCEANOGR] Laboratory models for studying ocean currents, the models being rotated to simulate in part the earth's rotation.

rotating platinum electrode [ANALY CHEM] Platinum wire sealed in a soft-glass tubing and rotated by a constant-speed motor; used as the electrode in amperometric titrations. Abbreviated RPE.

rotating Reynolds number [FL MECH] A nondimensional number arising in problems of a rotating viscous fluid. Also known as rotation Reynolds number.

rotating spreader [ENG] Plastics-molding injection device consisting of a finned torpedo that is rotated by a shaft extending through a tubular cross-section injection ram behind it.

rotating viscometer vacuum gage [ENG] Vacuum (reduced-pressure) measurement device in which the torque on a spinning armature is proportional to the viscosity (and the pressure) of the rarefied gas being measured; sensitive for absolute pressures of 1 millimeter of mercury, down to a few tens of micrometers.

rotating wedge [OPTICS] A circular optical wedge mounted to be rotated in the path of light to divert the line of sight to a limited degree.

rotation [MATH] See curl. [MECH] Also known as rotational motion. **1.** Motion of a rigid body in which either one point is fixed, or all the points on a straight line are fixed. **2.** Angular displacement of a rigid body. **3.** The motion of a particle about a fixed point.

rotational casting [ENG] Method to make hollow plastic articles from plastisols and lattices using a hollow mold rotated in one or two planes; the hot mold fuses the plastisol into a gel, which is then chilled and the product stripped out. Also known as rotational molding.

rotational constant [PHYS CHEM] That constant inversely proportioned to the moment of inertia of a linear molecule; used in calculations of microwave spectroscopy quantums.

rotational energy [MECH] The kinetic energy of a rigid body due to rotation. [PHYS CHEM] For a diatomic molecule, the difference between the energy of the actual molecule and that of an idealized molecule which is obtained by the hypothetical process of gradually stopping the relative rotation of the nuclei without placing any new constraint on their vibration, or on motions of electrons.

rotational fault See rotary fault.

rotational flow [FL MECH] Flow of a fluid in which the curl of the fluid velocity is not zero, so that each minute particle of fluid rotates about its own axis.

rotational impedance [MECH] A complex quantity, equal to the phasor representing the alternating torque acting on a system divided by the phasor representing the resulting angular velocity in the direction of the torque at its point of

application. Also known as mechanical rotational impedance.

rotational inertia See moment of inertia.

rotational molding See rotational casting.

rotational motion See rotation.

rotational position sensing [ADP] A fast disk search method whereby the control unit looks for a specified sector, and then receives the sector number required to access the record.

rotational reactance [MECH] The imaginary part of the rotational impedance. Also known as mechanical rotational reactance.

rotational resistance [MECH] The real part of rotational impedance; it is responsible for dissipation of energy. Also known as mechanical rotational resistance.

rotational spectrum [SPECT] The molecular spectrum resulting from transitions between rotational levels of a molecule which behaves as the quantum-mechanical analog of a rotating rigid body.

rotational stability [MECH] Property of a body for which a small angular displacement sets up a restoring torque that tends to return the body to its original position.

rotational strain [MECH] Strain in which the orientation of the axes of strain is changed.

rotational sum rule [SPECT] The rule that, for a molecule which behaves as a symmetric top, the sum of the line strengths corresponding to transitions to or from a given rotational level is proportional to the statistical weight of that level, that is, to $2J + 1$, where J is the total angular momentum quantum number of the level.

rotational transform [PL PHYS] Property possessed by a magnetic field, in a system used to confine a plasma, in which magnetic lines of force do not close in on themselves after making a circuit around the system, but are rotationally displaced.

rotational viscometer See Couette viscometer.

rotational wave See shear wave; S wave.

rotation anemometer [ENG] A type of anemometer in which the rotation of an element serves to measure the wind speed; rotation anemometers are divided into two classes: those in which the axis of rotation is horizontal, as exemplified by the windmill anemometer; and those in which the axis is vertical, such as the cup anemometer.

rotation axis [CRYSTAL] A symmetry element of certain crystals in which the crystal can be brought into a position physically indistinguishable from its original position by a rotation through an angle of $360°/n$ about the axis, where n is the multiplicity of the axis, equal to 2, 3, 4, or 6. Also known as symmetry axis.

rotation camera [SOLID STATE] An instrument for studying crystalline structure by x-ray or neutron diffraction, in which a monochromatic, collimated beam of x-rays or neutrons falls on a single crystal which is rotated about an axis perpendicular to the beam and parallel to one of the crystal axes, and the various diffracted beams are registered on a cylindrical film concentric with the axis of rotation.

rotation coefficients [MECH] Factors employed in computing the effects on range and deflection which are caused by the rotation of the earth; they are published only in firing tables involving comparatively long ranges.

rotation firing [ENG] Setting off explosions so that each hole throws its burden toward the space made by the preceding explosions.

rotation group [MATH] The group consisting of all orthogonal matrices or linear transformations having determinant 1.

rotation-inversion axis [CRYSTAL] A symmetry element of certain crystals in which a crystal can be brought into a position physically indistinguishable from its original position by a rotation through an angle of $360°/n$ about the axis followed by an inversion, where n is the multiplicity of the axis, equal to 1, 2, 3, 4, or 6. Also known as inversion axis.

rotation moment See torque.

rotation-reflection axis [CRYSTAL] A symmetry element of certain crystals in which a crystal can be brought into a position physically indistinguishable from its original position by a rotation through an angle of $360°/n$ about the axis followed by a reflection in the plane perpendicular to the axis, where n is the multiplicity of the axis, equal to 1, 2, 3, 4, or 6.

ROTATION CAMERA

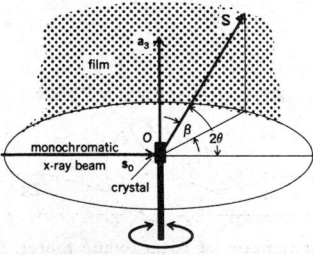

Schematic of rotation camera. Single crystal is located at O. 2θ represents angle between primary beam s_0 and diffracted beam S. β represents angle that S makes with plane perpendicular to crystal axis a_3 about which crystal is rotated.

rotation Reynolds number *See* rotating Reynolds number.

rotation spectrum [PHYS CHEM] Absorption-spectrum (absorbed electromagnetic energy) wavelengths produced if only the rotational energy of a molecule is affected during excitation.

rotation therapy [MED] Radiation therapy in which either the patient or the source of radiation is rotated, to permit a larger dose at the center of rotation within the patient's body than on any area of the skin.

rotation twin [CRYSTAL] A twin crystal in which the parts will coincide if one part is rotated 180° (sometimes 30, 60, or 120°).

rotation-vibration spectrum [PHYS CHEM] Absorption-spectrum (absorbed electromagnetic energy) wavelengths produced when both the energy of vibration and energy of rotation of a molecule are affected by excitation.

rotator [ELECTROMAG] A device that rotates the plane of polarization of a plane-polarized electromagnetic wave, such as a twist in a waveguide. [MECH] A rotating rigid body. [QUANT MECH] A molecule or other quantum-mechanical system which behaves as the quantum-mechanical analog of a rotating rigid body. Also known as top.

Rotatoria [INV ZOO] The equivalent name for Rotifera.

rotatory [OPTICS] Having the capability to rotate the plane of polarization of polarized electromagnetic radiation.

rotatory dispersion *See* rotary dispersion.

rotatory power [OPTICS] A substance's capability to rotate the plane of polarization of polarized electromagnetic radiation. [PHYS CHEM] The product of the specific rotation of an element or compound and its atomic or molecular weight.

röteln *See* rubella.

rotenturm wind [METEOROL] A warm south wind blowing through Rotenturm Pass in the Transylvanian Alps.

Rotifera [INV ZOO] A class of the phylum Aschelminthes distinguished by the corona, a retractile trochal disk provided with several groups of cilia and located on the head.

Rotliegende [GEOL] A European series of geologic time: Lower and Middle Permian.

rotoflector [ELECTROMAG] In radar, elliptically shaped, rotating reflector used to reflect a vertically directed radar beam at right angles so that it radiates in a horizontal direction.

rotogravure [GRAPHICS] A print made by the intaglio process, using a rotary press.

roton [PHYS] A quantum of rotational motion in a liquid, such as superfluid helium.

rotor [COMMUN] **1.** Disk with a set of input contacts and a set of output contacts, connected by any prearranged scheme designed to rotate within an electrical cipher machine. **2.** Disk whose rotation produces a variation of some cryptographic element in a cipher machine usually by means of lugs (or pins) in or on its periphery. [ELEC] The rotating member of an electrical machine or device, such as the rotating armature of a motor or generator, or the rotating plates of a variable capacitor. [MECH ENG] *See* impeller.

rotor cloud [METEOROL] Turbulent, altocumulus-type cloud formation found in the lee of some large mountain barriers, particularly in the Sierra Nevadas near Bishop, California; the air in the cloud rotates around an axis parallel to the range. Also known as roll cloud.

rotor plate [ELEC] One of the rotating plates of a variable capacitor, usually directly connected to the metal frame.

rotten ice [HYD] Any piece, body, or area of ice which is in the process of melting or disintegrating; it is characterized by honeycomb structure, weak bonding between crystals, or the presence of meltwater or sea water between grains. Also known as spring sludge.

rotten spot *See* pothole.

rottenstone [MATER] A soft, decomposed limestone, light gray to olive in color; used in powder form as a polishing material for metal and wood.

Rouche's theorem [MATH] If analytic functions $f(z)$ and $g(z)$ in a simply connected domain satisfy on the boundary $|g(z)| < |f(z)|$, then $f(z)$ and $f(z) + g(z)$ have the same number of zeros in the domain.

rouge [MATER] Finely divided, hydrated iron oxide, used in polishing glass, metal, or gems, and as a pigment.

rough air [AERO ENG] An aviation term for turbulence encountered in flight.

rough bark [PL PATH] **1.** Any of various virus diseases of woody plants characterized by roughening and often splitting of the bark. **2.** A fungus disease of apples caused by *Phomopsis mali* and characterized by rough cankers on the bark.

rough burning [AERO ENG] Pressure fluctuations frequently observed at the onset of burning and at the combustion limits of a ramjet or rocket.

roughcast [CIV ENG] A rough finish on a surface; in particular, a plaster made of lime and shells or pebbles, applied by throwing it against a wall with a trowel.

rough colony [MICROBIO] A flattened, irregular, and wrinkled colony of bacteria indicative of decreased capsule formation and virulence.

rougher cell [MIN ENG] Flotation cells in which the bulk of the gangue is removed from the ore.

rough grinding [MECH ENG] Preliminary grinding without regard to finish.

rough hardware [ENG] Utility items such as nails, sash balances, and studs, without attractive finished appearance.

roughing stand [MET] The first stand of rolls, or the last stand before the finishing rolls, through which a preheated billet is passed.

rough lumber [MATER] Sawed, undressed lumber.

rough machining [MECH ENG] Preliminary machining without regard to finish.

roughness [FL MECH] Distance from peaks to valleys in pipe-wall irregularities; used to modify Reynolds number calculations for fluid flow through pipes.

roughness elements [OCEANOGR] Structures attached to laboratory models to simulate the roughness of the ocean floor.

roughness factor [FL MECH] A correction factor used in fluid-flow calculations to allow for flow resistance caused by the roughness of the surface over which the fluid must flow.

roughness length *See* dynamic roughness.

roughness-width cutoff [MECH ENG] The maximum width of surface irregularities included in roughness height measurements.

rough turning [MECH ENG] The removal of excess stock from a workpiece as rapidly and efficiently as possible.

rouleau [PATH] A roll of erythrocytes resembling a stack of coins.

round [ENG] A series of shots fired either simultaneously or with delay periods between them. [NAV] To pass and alter direction of travel, as a vessel rounds a cape. [ORD] A single munition, missile, or device to be loaded on or in a delivery platform, vehicle, or device for purposes of expenditure; the configuration of the round may vary.

round-face bit [DES ENG] Any bit with a rounded cutting face.

round file [DES ENG] A file having a circular cross section.

round-head bolt [DES ENG] A bolt having a rounded head at one end.

round-head buttress dam [CIV ENG] A mass concrete dam built of parallel buttresses thickened at the upstream end until they meet.

rounding [MATH] Dropping or neglecting decimals after some significant place. Also known as truncation.

rounding error [MATH] The computational error due to always rounding numbers in a calculation. Also known as roundoff.

round ligament [ANAT] **1.** A flattened band extending from the fovea on the head of the femur to attach on either side of the acetabular notch between which it blends with the transverse ligament. **2.** A fibrous cord running from the umbilicus to the notch in the anterior border of the liver; represents the remains of the obliterated umbilical vein.

roundness [GEOL] The degree of abrasion of sedimentary particles; expressed as the radius of the average radius of curvature of the edges or corners to the radius of curvature of the maximum inscribed sphere.

round of beam [NAV ARCH] The arch or slope from side to side of a vessel's weather deck for water drainage. Also known as camber.

round of bearings [NAV] A group of bearings observed si-

ROUND-HEAD BOLT

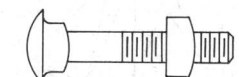

Drawing of round-head carriage bolt with a nut.

multaneously, or over a short period of time with no appreciable delay between the completion of one observation and the start of the next.

round off [MATH] To truncate the least significant digit or digits of a numeral, and adjust the remaining numeral to be as close as possible to the original number.

roundoff *See* rounding error.

round of sights [NAV] A group of celestial observations made with a sextant or other similar instrument over a short period of time, usually with no appreciable delay between the completion of one observation and the start of the next.

rounds complete [ORD] The term used to report that the number of rounds specified in fire for effect have been fired.

round strand rope [DES ENG] A rope composed generally of six strands twisted together or laid to form the rope around a core of hemp, sisal, or manila, or, in a wire-cored rope, around a central strand composed of individual wires.

round-the-world echo [COMMUN] A signal occurring every $1/7$ second when a radio wave repeatedly encircles the earth at its speed of 186,000 miles (300,000 kilometers) per second.

round trip echoes [ELECTROMAG] Multiple reflection echoes produced when a radar pulse is reflected from a target strongly enough so that the echo is reflected back to the target where it produces a second echo.

round wind [METEOROL] A wind that gradually changes direction through approximately 180° during the daylight hours.

round window [ANAT] A membrane-covered opening between the middle and inner ears in amphibians and mammals through which energy is dissipated after traveling in the membranous labyrinth.

roundworm [INV ZOO] The name applied to nematodes.

Rous sarcoma [VET MED] A fibrosarcoma that can be produced in chickens, pheasants, and ducklings inoculated with the filterable, ribonucleic acid Rous virus.

route [NAV] The prescribed course to be traveled from a specific point of origin to a specific destination.

route chart [MAP] A chart showing routes between various places, usually with distances indicated.

route component [METEOROL] The average forecast wind component parallel to the flight path at flight level for an entire route; it is positive if helping (tailwind), and negative if retarding (headwind).

route forecast [METEOROL] An aviation weather forecast for one or more specified air routes.

route locking [CIV ENG] Electrically locking in position switches, movable point frogs, or derails on the route of a train, after the train has passed a proceed signal.

router plane [DES ENG] A plane for cutting grooves and smoothing the bottom of grooves.

route survey [CIV ENG] A survey for the design and construction of linear works, such as roads and pipelines.

Routh's rule [MATH] A rule that the number of roots with positive real parts of an algebraic equation is equal to the number of changes of algebraic sign of a sequence whose terms are formed from coefficients of the equation in a specified manner. Also known as Routh test.

Routh's rule of inertia [MECH] The rule that the moment of inertia of a body about an axis of symmetry equals $M(a^2 + b^2 + c^2)/n$, where M is the body's mass, a, b, and c are the lengths of the body's perpendicular semiaxes, and n equals 3, 4, or 5 depending on whether the body is a rectangular parallelepiped, an elliptic cylinder, or an ellipsoid.

Routh table [MATH] An array of numbers each of which is formed from coefficients of an algebraic equation in a specified manner; the first row of this array constitutes the sequence used in Routh's rule.

Routh test *See* Routh's rule.

routine [ADP] A set of digital computer instructions designed and constructed so as to accomplish a specified function.

routine ammunition maintenance [ORD] Maintenance operations not involving disassembly of ammunition or replacement of components, and comprising chiefly cleaning and protecting exterior surfaces of individual items, packages of ammunition, ammunition components, and explosives.

routine library [ADP] Ordered set of standard and proven

computer routines by which problems or parts of problems may be solved.

routing [COMMUN] The assigning and establishing of a path for sending information between stations of a communications network.

routing indicator [COMMUN] **1.** A group of letters, engineered and assigned, to identify a station within a digital communications network. **2.** A group of letters assigned to indicate the geographic location of a station; a fixed headquarters of a command, activity, or unit at a geographic location; or the general location of a tape relay or tributary station to facilitate the routing of traffic over tape relay networks.

routing message [COMMUN] The function performed at a central message processor of selecting the route, or alternate route required, by which a message will proceed to the next point in reaching its destination.

Rovamycin [MICROBIO] A trade name for spiramycin.

Rover program [ORD] The entire experimental nuclear rocket program of the United States, including the Kiwi, Phoebus, and Nerva series.

roving [TEXT] Drawing out and slightly twisting cotton or wool fibers.

roving artillery [ORD] Artillery withdrawn from its regular position and assigned to special missions; usually moved about and fired from different positions to deceive the enemy as to position and strength.

roving gun [ORD] Gun that is moved about and fired from different positions to mislead or harass the enemy; generally used for registration when the location of the battery position must remain secret.

row [ADP] **1.** The characters, or corresponding bits of binary-coded characters, in a computer word. **2.** Equipment which simultaneously processes the bits of a character, the characters of a word, or corresponding bits of binary-coded characters in a word. **3.** Corresponding positions in a group of columns.

row address [ADP] An index array entry field which contains the main storage address of a data block.

row binary [ADP] A method of encoding binary information onto punched cards in which successive bits are punched row-wise onto the card.

row binary card [ADP] A card in which the binary data are punched along the rows.

roweite [MINERAL] $(Mn,Mg,Zn)Ca(BO_2)_2(OH)_2$ A light-brown mineral composed of basic borate of calcium, manganese, magnesium, and zinc.

rowland [SPECT] A unit of length, formerly used in spectroscopy, equal to 999.81/999.94 angstrom, or approximately 0.99987×10^{-10} meter.

Rowland circle [SPECT] A circle drawn tangent to the face of a concave diffraction grating at its midpoint, having a diameter equal to the radius of curvature of a grating surface; the slit and camera for the grating should lie on this circle.

Rowland ghost [SPECT] A false spectral line produced by a diffraction grating, arising from periodic errors in groove position.

Rowland mounting [SPECT] A mounting for a concave grating spectrograph in which camera and grating are connected by a bar forming a diameter of the Rowland circle, and the two run on perpendicular tracks with the slit placed at their junction.

Rowland ring [ELECTROMAG] A ring-shaped sample of magnetic material, generally surrounded by a coil of wire carrying a current.

rowlock [NAV ARCH] A U-shaped fitting with shank or socket attachment to the gunwale of a boat, through which an oar is swung.

rowlock arch [ARCH] An arch in which wedge-shaped blocks are arranged in concentric rings.

row order [ADP] The storage of a matrix $a(m,n)$ as $a(1,1),a(1,2), \ldots ,a(1,n),a(2,1),a(2,2), \ldots$ Also known as lexicographic order.

row pitch [ADP] The distance between the centerlines of adjacent rows of holes in punched cards and paper tapes.

row shooting [MIN ENG] Setting off a row of holes nearest the face first, and then other rows in succession behind it.

royal jelly [MATER] A protein complex high in vitamin B

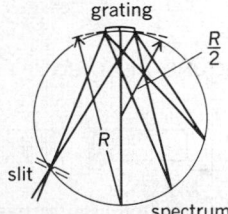

ROWLAND CIRCLE

grating

$\dfrac{R}{2}$

slit

R

spectrum

R represents a radius of the circle of which the curved surface of the grating forms a part. $R/2$ represents a radius of the Rowland circle. The camera may be located anywhere on this circle.

secreted by bees to nourish the egg of the queen bee; used in face creams.

RPE   *See* rotating platinum electrode.

rpm   *See* revolution per minute.

rps   *See* revolution per second.

RQ   *See* respiratory quotient.

RR Lyrae stars   [ASTRON] Pulsating variable stars with a period of 0.05-1.2 days in the halo population of the Milky Way Galaxy; color is white, and they are mostly stars of spectral class A and have been called cluster-type cepheids.

r-RNA   *See* ribosomal ribonucleic acid.

R scan   *See* R scope.

R scope   [ELECTR] An A scope presentation with a segment of the horizontal trace expanded near the target spot (pip) for greater accuracy in range measurement. Also known as R indicator; R scan.

R tectonite   [PETR] A tectonite in which the fabric is believed to have resulted from rotation.

RTF   *See* resistance transfer factor.

RTG   *See* radioisotope thermoelectric generator.

r-theta navigation   *See* omnibearing-distance navigation.

RTL   *See* resistor-transistor logic.

RTTY   *See* radioteletype.

Ru   *See* ruthenium.

rubber   [ORG CHEM] A natural, synthetic, or modified high polymer with elastic properties and, after vulcanization, elastic recovery; the generic term is elastomer.

rubber accelerator   [ORG CHEM] A substance that increases the speed of curing of rubber, such as thiocarbanilide.

rubber adhesive   [MATER] An adhesive made with a rubber base by using natural or synthetic rubber in an evaporative solvent; a tacky mixture of rubber and filler material, as used on pressure-sensitive tapes; or rubber-solvent-catalyst mixtures (usually two-part) that cure in place.

rubber-base paint   [MATER] A paint in which chlorinated rubber or synthetic latex is the nonvolatile vehicle.

rubber belt   [DES ENG] A conveyor belt that consists essentially of a rubber-covered fabric; fabric is cotton, or nylon or other synthetic fiber, with steel-wire reinforcement.

rubber blanket   [ENG] A rubber sheet used as a functional die in rubber forming.

rubber cement   [MATER] An adhesive composed of unvulcanized rubber in an organic solvent.

rubber-covered steel conveyor   [DES ENG] A steel conveyor band with a cover of rubber bonded to the steel.

rubber fiber   [MATER] A fiber composed of natural or synthetic rubber; used to make elastic yarn for clothing.

rubber foam   *See* rubber sponge.

rubber hydrochloride   [ORG CHEM] White, thermoplastic hydrochloric acid derivative of rubber; water-insoluble powder or clear film, soluble in aromatic hydrocarbons; softens at 110-120°C; used for protective coverings, food packaging, shower curtains, and rainwear.

rubber ice   [OCEANOGR] Newly formed sea ice which is weak and elastic.

rubber plating   [ENG] The laying down of a rubber coating onto metals by electrodeposition or by ionic coagulation.

rubber solvent   [MATER] Fast-evaporating petroleum distillate used as a solvent for tackifying rubber during plying (laminating) operations, and in compounding rubber cements.

rubber sponge   [MATER] Foamed, flexible rubber; produced by beating air into unvulcanized latex, or by incorporating a gas-producing ingredient (such as sodium bicarbonate) into a strongly masticated rubber stock; used for comfort cushioning, packaging, and shock insulation. Also known as cellular rubber; foam rubber; rubber foam; sponge rubber.

rubber tree   [BOT] *Hevea brasiliensis.* A tall tree of the spurge family (Euphorbiaceae) from which latex is collected and coagulated to produce rubber.

rubber wheel   [DES ENG] A grinding wheel made with rubber as the bonding agent.

rubbing   [GRAPHICS] An impression made by moistening a thin, tough paper and patting it into the incised parts of a carved or modeled surface, then rubbing it with pencil, chalk, inked pad, or watercolor; an actual-size image of the original is produced on the paper.

rubbing oil   [MATER] **1.** A low-viscosity petroleum oil used either with or without an abrasive to polish dried surfaces, such as paint. **2.** A nonviscous oil used for polishing wood furniture.

rubble   [CIV ENG] **1.** Rough, broken stones and other debris resulting from the deterioration and destruction of a building. **2.** Rough stone or brick used in coarse masonry or to fill the space in a wall between the facing courses. [GEOL] **1.** A loose mass of rough, angular rock fragments, coarser than sand. **2.** *See* talus.

rubella   [MED] A benign, infectious virus disease of humans characterized by coldlike symptoms, fever, and transient, generalized pale-pink rash. Also known as epidemic roseola; French measles; German measles; röteln.

rubellite   [MINERAL] The red to red-violet variety of the gem mineral tourmaline; hardness is 7-7.5 on Mohs scale, and specific gravity is near 3.04.

rubeola   *See* measles.

Rubiaceae   [BOT] The single family of the plant order Rubiales.

Rubiales   [BOT] An order of dicotyledonous plants marked by their inferior ovary, regular or nearly regular corolla, and opposite leaves with interpetiolar stipules or whorled leaves without stipules.

rubidium   [CHEM] A chemical element, symbol Rb, atomic number 37, atomic weight 85.47; a reactive alkali metal; salts of the metal may be used in glass and ceramic manufacture.

rubidium bromide   [INORG CHEM] RbBr Colorless, regular crystals, melting at 683°C; soluble in water; used as a nerve sedative.

rubidium chloride   [INORG CHEM] RbCl A water-soluble, white, lustrous powder melting at 715°C; used as a source for rubidium metal, and as a laboratory reagent.

rubidium halide   [INORG CHEM] Any of the halogen compounds of rubidium; examples are RbBr, RbCl, RbF, RbIBrCl, RbBr$_2$Cl, and RbIBr$_2$.

rubidium halometallate   [INORG CHEM] Halogen-metal-containing compounds of rubidium; examples are Rb$_2$GeF$_6$ (rubidium hexafluorogermanate), Rb$_2$PtCl$_6$ (rubidium chloroplatinate), and Rb$_2$PdCl$_5$ (rubidium palladium chloride).

rubidium magnetometer   *See* rubidium-vapor magnetometer.

rubidium-strontium dating   [GEOL] A method for determining the age of a mineral or rock based on the decay rate of rubidium-87 to strontium-87.

rubidium sulfate   [INORG CHEM] Rb$_2$SO$_4$ Colorless, water-soluble rhomboid crystals, melting at 1060°C; used as a cathartic.

rubidium-vapor frequency standard   [PHYS] An atomic frequency standard in which the frequency is established by a gas cell containing rubidium vapor and a neutral buffer gas.

rubidium-vapor magnetometer   [ENG] A highly sensitive magnetometer in which the spin precession principle is combined with optical pumping and monitoring for detecting and recording variations as small as 0.01 gamma (0.1 micro-oersted) in the total magnetic field intensity of the earth. Also known as rubidium magnetometer.

rubriblast   *See* pronormoblast.

rubricyte   *See* pronormoblast.

ruby   [MINERAL] The red variety of the mineral corundum; in its finest quality, the most valuable of gemstones.

ruby glass   [MATER] Glass of a rich red color produced by adding selenium or cadmium sulfide, or copper oxide to the glass.

ruby laser   [OPTICS] An optically pumped solid-state laser that uses a ruby crystal to produce an intense and extremely narrow beam of coherent red light.

ruby maser   [PHYS] A maser that uses a ruby crystal in the cavity resonator.

ruby mica   [MINERAL] The finest grade of Indian mica; used for electrical capacitors.

ruby silver   [MINERAL] Either of two red silver sulfide minerals: pyrorgyrite (dark-ruby silver) and proustite (light-ruby silver).

ruby zinc   *See* zincite.

rudaceous   [PETR] Of or pertaining to a sedimentary rock

RUBBER TREE

Branch of the rubber tree.

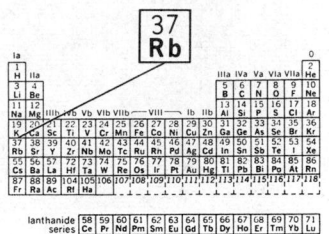

RUBIDIUM

Periodic table of the chemical elements showing the position of rubidium.

composed of a large quantity of fragments that are larger than sand grains (diameter greater than 2 millimeters).

rudder [ENG] **1.** A flat, usually foil-shaped movable control surface attached upright to the stern of a boat, ship, or aircraft, and used to steer the craft. **2.** *See* rudder angle.

rudder angle [ENG] The acute angle between a ship or plane's rudder and its fore-and-aft line. Also known as rudder.

rudder gudgeon [NAV ARCH] One of the lugs cast or forged on the sternpost for the purpose of hanging and hinging the rudder.

rudder lug [NAV ARCH] A projection that is cast or fitted to the forward edge of the rudder frame for the purpose of taking the pintles.

rudderpost [NAV ARCH] A second sternpost to which the rudder of a single-screw ship is joined.

rudderstock [NAV ARCH] A vertical shaft of a rudder having a yoke, quadrant, or tiller fitted to its upper portion by which it may be turned.

rudder stop [NAV ARCH] One of the fittings attached to the structure of the ship or to the shoulders of the sternpost, which limit the swing of the rudder to an angle of about 35°.

ruddy turnstone [VERT ZOO] *Arenaria interpes.* A member of the avian order Charadriiformes that perform transpacific flights during their migration.

ruderal [ECOL] **1.** Growing on rubbish, or waste or disturbed places. **2.** A plant that thrives in such a habitat.

rudistids [PALEON] Fossil sessile bivalves that formed reefs during the Cretaceous in the southern Mediterranean or the Tethyan belt.

rudite [GEOL] A sedimentary rock composed of fragments coarser than sand grains.

rue oil [MATER] Yellow, combustible essential oil, soluble in most fixed and mineral oils; derived from blooming plants of genus *Ruta;* used in perfumery and veterinary medicine, and as a chemical intermediate.

Ruffini cylinder [ANAT] A cutaneous nerve ending suspected as the mediator of warmth.

rufous [BOT] Having a reddish-brown color.

ruggedization [ELECTR] Making electronic equipment and components resistant to severe shock, temperature changes, high humidity, or other detrimental environmental influences.

ruggedized computer [ADP] A computer, especially a minicomputer, built so as to reduce vibrations, resist moisture, and remain unaffected by electromagnetic interferences such as are found in factory, military, or mobile environments.

ruggedness number [GEOL] A dimensionless number that expresses the geometric characteristics of a drainage system; derived from the product of maximum basin relief and drainage density within the drainage basin.

Rugosa [PALEON] An order of extinct corals having either simple or compound skeletons with internal skeletal structures consisting mainly of three elements, the septa, tabulae, and dissepiments.

rugose [BIOL] Having a wrinkled surface.

rugose mosaic [PL PATH] A virus disease of potatoes marked by dwarfed, wrinkled, and mottled leaves and resulting in premature death.

Rule *See* Norma.

ruled surface [MATH] A surface that can be generated by the motion of a straight line.

rule of approximation [MIN ENG] A rule applicable to placer mining locations and entries upon surveyed lands, to be applied on the basis of 10-acre (40,469-square-meter) legal subdivisions.

rule of sixty [NAV] An air navigation rule of thumb which states that for each mile off course in a 60-mile leg, the correction to parallel the original course is 1°.

ruler [ENG] A graduated strip of wood, metal, or other material, used to measure lines or as a guide in drawing lines.

ruling engine [SPECT] A machine operated by a long micrometer screw which rules equally spaced lines on an optical diffraction grating.

ruling pen [GRAPHICS] A draftsman's pen that has two adjustable pointed blades between which ink is contained.

rumble *See* turntable rumble.

rumen [VERT ZOO] The first chamber of the ruminant stomach. Also known as paunch.

ruminant [PHYSIO] Characterized by the act of regurgitation and rechewing of food. [VERT ZOO] A mammal belonging to the Ruminantia.

Ruminantia [VERT ZOO] A suborder of the Artiodactyla including sheep, goats, camels, and other forms which have a complex stomach and ruminate their food.

rumination [MED] Voluntary regurgitation of food from the stomach, followed by remastication and swallowing in emotionally or mentally disturbed persons. Also known as merycism. [PHYSIO] Regurgitation and remastication of food in preparation for true digestion in ruminants. [PSYCH] An obsessional preoccupation with a single idea or system of ideas.

run [ADP] A single, complete execution of a computer program, or one continuous segment of computer processing, used to complete one or more tasks for a single customer or application. Also known as machine run. [CHEM ENG] **1.** The amount of feedstock processed by a petroleum refinery unit during a given time; often used colloquially in relation to the type of stock being processed, as in crude run or naphtha run. **2.** A processing-cycle or batch-treatment operation. [MIN ENG] *See* slant. [NAV] The distance traveled by a craft during any given time interval, or since leaving a designated place. [NAV ARCH] The underwater portion of that part of the aft end of a ship where it curves inward and upward to the stern. [ORD] **1.** Steady, level flight of an aircraft across a target to enable bombs to be dropped accurately in horizontal bombing. **2.** Passing of a moving target once across the range.

run a line of soundings [ENG] To obtain a series of soundings along a course line.

runaround [MIN ENG] A bypass driven in the shaft pillar to permit safe passage from one side of the shaft to the other.

runaround crosstalk [COMMUN] Crosstalk resulting from coupling between the high-level end of one repeater and the low-level end of another repeater, as at a carrier telephone repeater station.

runaway effect [ELECTR] The phenomenon whereby an increase in temperature causes an increase in a collector-terminal current in a transistor, which in turn results in a higher temperature and, ultimately, failure of the transistor; the effect limits the power output of the transistor.

runaway electron [ELECTR] An electron, in an ionized gas to which an electric field is applied, that gains energy from the field faster than it loses energy by colliding with other particles in the gas.

runaway gun [ORD] Automatic weapon that continues firing after the trigger is released; caused by a defect in some part of its mechanism.

runback [CHEM ENG] A pipe through which all or part of a distillation column's overhead condensate can be run back into the column, instead of being drawn off as product. [ENG] **1.** To retract the drill feed mechanism to its starting position. **2.** To drill slowly downward toward the bottom of the hole when the drill string has been lifted off-bottom for rechucking.

run before the wind [NAV] To steer a course downwind.

run book [ADP] The collection of materials necessary to document a program run on a computer. Also known as problem file; problem folder.

runcinate [BOT] Pinnately cut with downward-pointing lobes.

run down a coast [NAV] To sail approximately parallel with the coastline.

rundown line [CHEM ENG] A line from a process unit that connects the look box in the receiving house with the tank in which the product is temporarily stored.

rundown tank [CHEM ENG] A tank in which the product from a still, agitator, or other processing equipment is received, and from which the product is pumped to larger storage tanks. Also known as pan tank; receiving tank.

Runge-Kutta method [MATH] A numerical approximation technique for solving differential equations.

Runge vector [MECH] A vector which describes certain unchanging features of a nonrelativistic two-body interaction obeying an inverse-square law, either in classical or quantum

RUFFINI CYLINDER

Longitudinal section through a Ruffini cylinder. *(From F. A. Geldard, Human Senses, Wiley, 1953)*

mechanics; its constancy is a reflection of the symmetry inherent in the inverse-square interaction.

run in [ENG] To lower the assembled drill rods and auxiliary equipment into a borehole.

R unit *See* German R unit; Solomon R unit.

runite *See* graphic granite.

run motor [ELEC] In facsimile equipment, a motor which supplies the power to drive the scanning or recording mechanisms; a synchronous motor is used to limit the speed.

runnel [GEOL] A troughlike hollow on a tidal sand beach which carries water drainage off the beach as the tide retreats.

runner [BOT] A horizontally growing, sympodial stem system; adventitious roots form near the apex, and a new runner emerges from the axil of a reduced leaf. Also known as stolon. [ENG] In a plastics injection or transfer mold, the channel (usually circular) that connects the sprue with the gate to the mold cavity. [MET] **1.** The part of a casting between itself and the gate assembly of the mold. **2.** A channel through which molten metal flows from one receptacle to another. [MIN ENG] A vertical timber sheet pile used to prevent collapse of an excavation.

runner box [MET] A box that divides the molten metal into several streams before it enters the cavity of the mold.

running bird [VERT ZOO] Any of the large, flightless, heavy birds usually categorized as ratites.

running block *See* traveling block.

running bond [CIV ENG] A masonry bond involving the placing of each brick as a stretcher and overlapping the bricks in adjoining courses.

running fit [DES ENG] The intentional difference in dimensions of mating mechanical parts that permits them to move relative to each other.

running fix [NAV] A position determined by crossing lines of position obtained at different times and advanced or retarded to a common time.

running gate [MET] A gate through which molten metal enters a mold.

running gear [MECH ENG] The means employed to support a truck and its load and to provide rolling-friction contact with the running surface.

running ground [MIN ENG] Insecure, incoherent ground that may be semiplastic or plastic and deforms readily under pressure.

running sand *See* quicksand.

running survey [MAP] A rough survey made by a vessel while coasting.

runoff [HYD] **1.** Surface streams that appear after precipitation. **2.** The flow of water in a stream, usually expressed in cubic feet per second; the net effect of storms, accumulation, transpiration, meltage, seepage, evaporation, and percolation. [MIN ENG] Collapse of a coal pillar in a mine.

runoff cycle [HYD] The part of the hydrologic cycle involving water between the time it reaches the land as precipitation and its subsequent evapotranspiration or runoff.

runoff pit [MIN ENG] Catchment area to which spillage from classifiers, thickeners, and slurry pumps can gravitate if it becomes necessary to dump their contents. Also known as spill pit.

run-of-mill [MIN ENG] Ore accepted for treatment, after waste and dense media rejection. Also known as mill-head ore.

run-of-mine *See* mine run.

run of the coast [GEOGR] The trend of the coast.

runout [MET] **1.** Escape of molten metal from a casting mold, crucible, or furnace. **2.** Defect in a casting caused by escape of metal from a mold.

runout table [MET] A roll table used to receive a rolled or extruded member.

run-out time [IND ENG] Time required by machine tools after cutting time is finished before tool and material are completely free of interference and before the start of the next sequence of operation.

run-up *See* swash.

runway [CIV ENG] A straight path, often hard-surfaced, within a landing strip, normally used for landing and takeoff of aircraft.

runway environment [NAV] This term is used to indicate the presence of a runway threshold or lighting aids which identify the runway.

runway lights [NAV] Aeronautical ground lights arranged along a runway indicating its direction or boundaries.

runway observation [METEOROL] An evaluation of certain meteorological elements observed at a specified point on or near an airport runway; temperature, wind speed and direction, ceiling, and visibility are among the elements frequently observed at such locations, because of the importance of these data to aircraft landing and takeoff operations.

runway temperature [METEOROL] The temperature of the air just above the runway at an airport (usually at about 4 feet but ideally at engine or wing height), used in the determination of density altitude; therefore, runway temperature observations are made and reported at airports when critical values of density altitude prevail.

runway visibility [METEOROL] The visibility along an identified runway, determined from a specified point on the runway with the observer facing in the same direction as a pilot using the runway.

runway visual range [METEOROL] The maximum distance along the runway at which the runway lights are visible to a pilot after touchdown.

rupicolous [ECOL] Living among or growing on rocks.

Rüping process [ENG] A system for preservative treatment of wood by using positive initial pressure, followed by introduction of the preservative and release of air, creating a vacuum.

rupture *See* fracture; hernia.

rupture disk *See* burst disk.

rural radio service [COMMUN] A radio service used to provide public message communication service between a central office and subscribers located in rural areas to which it is impracticable to run wire lines.

Rusa oil *See* palmarosa oil.

Rushton-Oldshue column [CHEM ENG] A mixing unit used for continuous pipeline blending in which two-phase contacting is desired; it is a column containing separation plates, baffles, and mixing impellers.

Russell bodies [PATH] Hyaline eosinophilic globules 4–5 micrometers in diameter, thought to be particles of antibody globulin, occurring in the cytoplasm of plasma cells in chronic inflammatory exudates.

Russell diagram *See* Hertzsprung-Russell diagram.

Russell effect [GRAPHICS] The formation of latent developable images on a photographic film or paper by an agent other than electromagnetic radiation, such as a resin, metal, volatile liquid, or printing ink. Also known as Vogel-Colson-Russell effect.

Russell flask [PETRO ENG] Device for volumetric determination of the true volume of sand grains within a unit bulk volume of grains plus voids.

Russell movable-wall oven [CHEM ENG] An oven for coal carbonization which cokes a 400-pound (180-kilogram) charge in a horizontal, 12-inch-wide (30-centimeter-wide) chamber, heated from both sides, but with one side floating and balanced against scales.

Russell-Saunders coupling [PHYS] A process for building many-electron single-particle eigenfunctions of orbital angular momentum and spin; the orbital functions are combined to make an eigenfunction of the total orbital angular momentum, the spin functions are combined to make an eigenfunction of the total spin angular momentum, and then the results are combined into eigenfunctions of the total angular momentum of the system. Also known as LS coupling.

Russell's paradox [MATH] The paradox concerning the concept of all sets which are not members of themselves which forces distinctions in set theory between sets and classes.

Russell's viper *See* tic polonga.

rust [MET] The iron oxides formed on corroded ferrous metals and alloys. [PL PATH] Any plant disease caused by rust fungi (Uredinales) and characterized by reddish-brown lesions on the plant parts.

rusting [MET] The formation of rust on ferrous metals and alloys.

rust prevention [ENG] Surface protection of ferrous structures or equipment to prevent formation of iron oxide; can be by coatings, surface treatment, plating, chemicals, cathodic arrangements, or other means.

rust preventive [MATER] One of a group of products, often

RUNNING BIRD

A kiwi with typical furlike plumage, vestigial wings, and a long, slender slightly curved bill.

RUTHENIUM

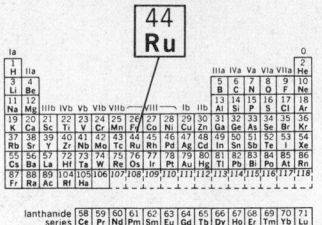

Periodic table of the chemical elements showing the position of ruthenium.

RYE

Spikes of Tetra Petkus, a variety of rye. *(Wisconsin Agricultural Experiment Station)*

RZEPPA JOINT

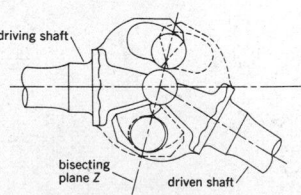

Drawing of Rzeppa joint showing the middle ball and two of the four transmitting elements.

with petroleum thinners, used to prevent corrosion to metal surfaces.

rusty blotch [PL PATH] A fungus disease of barley caused by *Helminthosporium californicum* and characterized by brown blotches on the foliage.

rusty mottle [PL PATH] A virus disease of cherry characterized by retarded development of blossoms and leaves in the spring, followed by necrotic spotting and shot-holing of the foliage with considerable defoliation.

rut [PHYSIO] The period during which the male animal has a heightened mating drive.

rutabaga [BOT] *Brassica napobrassica*. A biennial crucifer of the order Capparales probably resulting from the natural crossing of cabbage and turnip and characterized by a large, edible, yellowish fleshy root.

Rutaceae [BOT] A family of dicotyledonous plants in the order Sapindales distinguished by mostly free stamens and glandular-punctate leaves.

ruthenic chloride *See* ruthenium chloride.

ruthenium [CHEM] A chemical element, symbol Ru, atomic number 44, atomic weight 101.07. [MET] A hard, brittle, grayish-white metal used as a catalyst; workable only at high temperatures.

ruthenium chloride [INORG CHEM] $RuCl_3$ Black, deliquescent, water-insoluble solid that decomposes in hot water and above 500°C; used as a laboratory reagent. Also known as ruthenic chloride; ruthenium sesquichloride.

ruthenium halide [INORG CHEM] Halogen compound of ruthenium; examples are $RuCl_2$, $RuCl_3$, $RuCl_4$, $RuBr_3$, and RuF_5.

ruthenium red [INORG CHEM] $Ru_2(OH)_2Cl_4 \cdot 7NH_3 \cdot 3H_2O$ A water-soluble, brownish-red powder; used as an analytical reagent and stain. Also known as ammoniated ruthenium oxychloride.

ruthenium sesquichloride *See* ruthenium chloride.

ruthenium tetroxide [INORG CHEM] RuO_4 A yellow, toxic solid, melting at 25°C; used as an oxidizing agent.

rutherford [NUCLEO] Abbreviated rd. **1.** A unit used to express the decay rate of radioactive material, equal to 10^6 disintegrating atoms per second. **2.** That amount of a substance which is undergoing 10^6 disintegrations per second.

rutherfordine [MINERAL] $(UO_2)(CO_3)$ A yellow mineral composed of uranyl carbonate, occurring as masses of fibers.

Rutherford nuclear atom [ATOM PHYS] A theory of atomic structure in which nearly all the mass is concentrated in a small nucleus, electrons surrounding the nucleus fill nearly all the atom's volume, the number of these electrons equals the atomic number, and the positive charge on the nucleus is equal in magnitude to the negative charge of the electrons.

Rutherford scattering [ATOM PHYS] Scattering of heavy charged particles by the Coulomb field of an atomic nucleus.

rutile [MINERAL] TiO_2 A reddish-brown tetragonal mineral common in acid igneous rocks, in metamorphic rocks, and as residual grain in beach sand.

rutin [ORG CHEM] $C_{27}H_{32}O_{16}$ A hydroxyflavone glucorhamnoside derived from cowslip and other plants; yellow needles melting at 190°C; used to treat capillary disorders.

RV Tauri stars [ASTRON] A class of stars; they are long period pulsating variable types with periods from about 50 to 150 days; otherwise they are like the shorter period W Virginis stars; they are found in both the Milky Way Galaxy and the globular clusters.

RW Aurigal stars [ASTRON] A class of stars that are variable, and whose light variations are rapid and irregular.

R wave *See* Rayleigh wave.

ry *See* rydberg.

rydberg [ATOM PHYS] A unit of energy used in atomic physics, equal to the square of the charge of the electron divided by twice the Bohr radius; equal to 13.60583 ± 0.00004 electron volts. Abbreviated ry. [SPECT] *See* kayser.

Rydberg constant [ATOM PHYS] **1.** An atomic constant which enters into the formulas for wave numbers of atomic spectra, equal to $2\pi^2 me^4/ch^3$, where m and e are the rest mass and charge of the electron, c is the speed of light, and h is Planck's constant; equal to $109,737.31 \pm 0.01$ inverse centimeters. Symbolized R_∞. **2.** For any atom, the Rydberg constant (first definition) divided by $1 + m/M$, where m and M are the masses of an electron and of the nucleus.

Rydberg correction [ATOM PHYS] A term inserted into a formula for the energy of a single electron in the outermost shell of an atom to take into account the failure of the inner electron shells to screen the nuclear charge completely.

Rydberg series formula [SPECT] An empirical formula for the wave numbers of various lines of certain spectral series such as neutral hydrogen and alkali metals; it states that the wavenumber of the n-th member of the series is $\lambda_\infty - R/(n+a)^2$, where λ_∞ is the series limit, R is the Rydberg constant of the atom, and a is an empirical constant.

rye [BOT] *Secale cereale*. A cereal plant of the order Cyperales cultivated for its grain, which contains the most desirable gluten, next to wheat.

rye buckwheat *See* tartary buckwheat.

rye whiskey [FOOD ENG] One of the potable alcoholic beverages, obtained by distilling the alcohol-containing liquid resulting from the fermentation of a rye mash by yeast; the distilled product is aged for a number of years in new-charred oak barrels.

Rynchopidae [VERT ZOO] The skimmers, a family of birds in the order Charadriiformes distinguished by a knifelike lower beak that is longer and narrower than the upper one.

Rytiodinae [VERT ZOO] A subfamily of trichechiform sirenians in the family Dugongidae.

Rzeppa joint [MECH ENG] A special application of the Bendix-Weiss universal joint in which four large balls are transmitting elements, while a center ball acts as a spacer; it transmits constant angular velocity through a single universal joint.

s See second.

S See mho; secondary winding; stoke; sulfur.

S-2 See Tracker.

saba [BOT] A plant (*Musa sapientum* var. *compressa*) common in the Philippines; the fruit is a cooking banana. [TEXT] A textile made from fibers of the saba plant.

sabach See caliche.

sabadilla [MATER] The ripe seeds of the sabadilla plant (*Schoenocaulon officionale*) that have been dried; used as an insecticide on cattle. Also known as caustic barley; cevedilla.

Sabathé's cycle [MECH ENG] An internal combustion engine cycle in which part of the combustion is explosive and part at constant pressure.

Sabellariidae [INV ZOO] The sand-cementing worms, a family of polychaete annelids belonging to the Sedentaria and characterized by a compact operculum formed of setae of the first several segments.

Sabellidae [INV ZOO] A family of sedentary polychaete annelids often occurring in intertidal depths but descending to great abyssal depths; one of two families that make up the feather-duster worms.

Sabellinae [INV ZOO] A subfamily of the Sabellidae including the most numerous and largest members.

saber saw [MECH ENG] A portable saw consisting of an electric motor, a straight saw blade with reciprocating mechanism, a handle, baseplate, and other essential parts.

sabin [ACOUS] A unit of sound absorption for a surface, equivalent to 1 square foot (0.09290304 square meter) of perfectly absorbing surface. Also known as absorption unit; open window unit (OW unit); square-foot unit of absorption.

Sabine formula [ACOUS] An empirical equation for the reverberation time of sound in a room; its form is identical to that of the Franklin equation.

Sabin vaccine [IMMUNOL] A live-poliovirus vaccine that is administered orally.

sabkha See sebkha.

sable [VERT ZOO] *Martes zibellina.* A carnivore of the family Mustelidae; a valuable fur-bearing animal, quite similar to the American marten.

sable brush [GRAPHICS] A fine artist's brush, usually round and fairly long with tapered point, made from the hair of the Siberian mink.

sabot [ORD] Lightweight carrier in which a subcaliber projectile is centered to permit firing the projectile in the larger-caliber weapon; the sabot diameter fills the bore of the weapon from which the projectile is fired.

sabotage [ORD] Action by enemy agents or sympathizers with intent to stop or otherwise hinder a nation's war effort or to interfere with or obstruct the defense of a nation.

Sabouraud's agar [MICROBIO] A peptone-maltose agar used as a culture medium for pathogenic fungi, especially the dermatophytes.

sabrejet [AERO ENG] An airplane widely used by the U.S. Air National Guard and free world countries; it has a range of beyond 1000 miles (1600 kilometers) and a speed of over 650 miles (1050 kilometers) per hour, carries two 1000-pound (450-kilogram) bombs or sixteen 5-inch (12.7-centimeter) rockets, or a combination plus two additional 1000-pound bombs in lieu of fuel tanks; has a crew of one. Designated F-86.

sac [BIOL] A soft-walled cavity within a plant or animal, often containing a special fluid and usually having a narrow opening or none at all. [MAP] Indentation in the contour lines of equal depth showing submarine relief.

saccate [BOT] Having a saclike or pouchlike form.

saccharase [BIOCHEM] An enzyme that catalyzes the hydrolysis of disaccharide to monosaccharides, specifically of sucrose to dextrose and levulose. Also known as invertase; invertin; sucrase.

saccharimeter [ENG] An instrument for measuring the amount of sugar in a solution, often by determining the change in polarization produced by the solution.

saccharin [ORG CHEM] $C_6H_4COSO_2NH$ A sweet-tasting, white powder, soluble in acetates, benzene, and alcohol; slightly soluble in water and ether; melts at 228°C; used as a sugar substitute for syrups, and in medicines, foods, and beverages. Also known as benzosulfimide; gluside.

saccharoidal [PETR] The texture of a rock that is crystalline or granular. Also known as sucrosic; sugary.

saccharometer [ENG] An instrument for measuring the amount of sugar in a solution, by determining either the specific gravity or the gases produced by fermentation.

Saccharomyces [MYCOL] A genus of unicellular yeasts in the subfamily Saccharomycetoideae which have simple, budding cells, a pseudomycelium, or both, and one to four round to oval spores per ascus.

Saccharomycetaceae [MYCOL] The single family of the order Saccharomycetales.

Saccharomycetales [MYCOL] An order of the subclass Hemiascomycetidae comprising typical yeasts, characterized by the presence of naked asci in which spores are formed by free cells.

Saccharomycetoideae [MYCOL] A subfamily of Saccharomycetacae in which spores may be hat-, sickle-, or kidney-shaped, or round or oval.

saccharopinuria [MED] An inborn error of amino acid metabolism characterized by abnormally high levels of saccharopine in the urine.

saccharose See sucrose.

Saccoglossa [INV ZOO] An order of gastropod mollusks belonging to the Opisthobranchia.

Saccopharyngiformes [VERT ZOO] Formerly an order of actinopterygian fishes, the gulpers, now included in the Anguilliformes.

Saccopharyngoidei [VERT ZOO] The gulpers, a suborder of actinopterygian fishes in the order Anguilliformes having degenerative adaptations, including loss of swim bladder, opercle, branchiostegal ray, caudal fin, scales, and ribs.

saccular aneurysm [MED] A saclike arterial dilation communicating with the artery by a relatively small opening.

sacculus [ANAT] The smaller, lower saclike chamber of the membranous labyrinth of the vertebrate ear.

saccus See vesicle.

sac fungus [MYCOL] The common name for members of the class Ascomycetes.

sack cloth [TEXT] A coarse cloth made of animal hairs, cotton, flax, or hemp.

Sackur-Tetrode equation [STAT MECH] An equation for the translational entropy of an ideal gas made up of free fermions.

sacral block [MED] Anesthesia induced by injection of an anesthetic through the caudal hiatus.

sacral nerve [ANAT] Any of five pairs of spinal nerves in the sacral region which innervate muscles and skin of the lower back, lower extremities, and perineum, and branches to the hypogastric and pelvic plexuses.

sacral vertebrae [ANAT] Three to five fused vertebrae that form the sacrum in most mammals; amphibians have one sacral vertebra, reptiles usually have two, and birds have 10–23 fused in the synsacrum.

sacrificial metal [PHYS CHEM] A metal that can be used for a sacrificial anode.

sacrococcygeus [ANAT] One of two inconstant thin muscles extending from the lower sacral vertebrae to the coccyx.

sacroiliac [ANAT] Pertaining to the sacrum and the ilium.

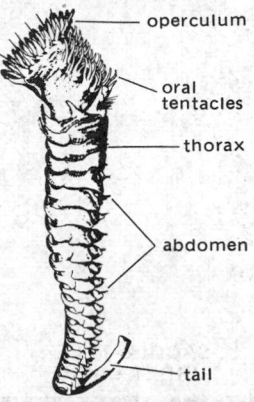

SABELLARIIDAE

operculum — oral tentacles — thorax — abdomen — tail

Sabellaria in right lateral view.

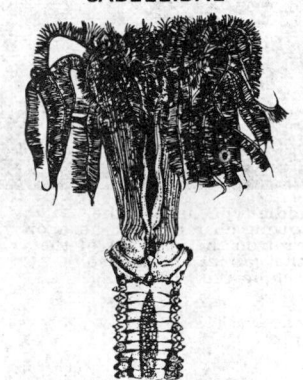

SABELLIDAE

Anterior end of *Sabella* in dorsal view, showing crown used as gill and a collecting organ.

SABLE

The sable (*Martes zibellina*), which has an elongated body reaching 25 inches (60 centimeters) in length, excluding the tail.

sacrospinous [ANAT] Pertaining to the sacrum and the spine of the ischium.

sacrum [ANAT] A triangular bone, consisting in man of five fused vertebrae, located below the last lumbar vertebra, above the coccyx, and between the hipbones.

saddle [DES ENG] A support shaped to fit the object being held. [GEOL] **1.** A gap that is broad and gently sloping on both sides. **2.** A relatively flat ridge that connects the peaks of two higher elevations. **3.** That part along the surface axis or axial trend of an anticline that is a low point or depression.

saddleback [GEOL] A hill or ridge with a concave outline along its crest. [METEOROL] The cloudless air between the "towers" of two cumulus congestus or cumulonimbus clouds and above a lower cloud mass.

saddle fold [GEOL] A flexural fold perpendicular to the parent fold and having an additional flexure at its crest.

saddle leather [MATER] **1.** Tanned cattlehide used in furnishings for saddle horses. **2.** Leather resembling saddle leather used in handbags and other leather goods.

saddle point [GEOL] *See* col. [MATH] A point where all the first partial derivatives of a function vanish but which is not a local maximum or minimum.

saddle-point azeotrope [PHYS CHEM] A rarely occurring azeotrope which is formed in ternary systems and for which the boiling point is intermediate between the highest and lowest boiling mixture in the system.

saddle point method *See* steepest descent method.

saddle point theory [MATH] The study of differentiable functions and their derivatives from the viewpoint of saddle points, especially applicable to the calculus of variations.

saddle-type turret lathe [MECH ENG] A turret lathe designed without a ram and with the turret mounted directly on a support (saddle) which slides on the bedways of the lathe.

saddling [MET] Forming a seamless ring by forging a pierced disk over a mandrel (or saddle).

sadism [PSYCH] Sexual perversion in which pleasure is derived from inflicting physical or mental cruelty upon another.

SAE EP lubricant tester [ENG] A machine that tests the extreme pressure (EP) properties of a lubricant under a combined rolling and sliding action, in which the revolving members are two bearing cups that rotate at different speeds.

Saefftigen's pouch [INV ZOO] An elongated pouch inside the genital sheath in many acanthocephalans.

SAE number [ENG] A classification of motor, transmission, and differential lubricants to indicate viscosities, standardized by the Society of Automotive Engineers; SAE numbers do not connote quality of the lubricant.

safe [ORD] Pertaining to ordnance constituted and set so as not to detonate or function accidentally.

safe burst height [ORD] The height of burst at or above which the level of fallout or damage to ground installations is at a predetermined level acceptable to the military commander.

safelight [GRAPHICS] A darkroom light of a particular color that will not affect the sensitive material being handled; for example, blue-sensitive materials require a red, orange, or yellow safelight.

safe load [MECH] The stress usually expressed in tons per square foot, which a soil or foundation can safely support.

safety [ENG] Methods and techniques of avoiding accident or disease. [ORD] A locking or cut-off device that prevents a weapon or missile from being fired accidentally.

safety belt [ENG] A strong strap or harness used to fasten a person to an object, such as the seat of an airplane or automobile.

safety block [ORD] A block which, in the safe position, prevents functioning of the fuse by limiting the motion of the firing pin.

safety board [PETRO ENG] A board placed in a derrick for a man to stand on when handling drill rods at single, double, triple, or quadruple levels; the boards are placed at suitable heights to handle a stand of drill rods for that number of joints.

safety bolt [CIV ENG] A bolt that can be opened from only one side of the door or gate it fastens.

safety button [NUCLEO] A device worn by workers exposed to nuclear radiation to warn of excessive exposure.

safety cable [MIN ENG] A mining machine cable designed to cut off power when the positive conductor insulation is damaged.

safety cage [MIN ENG] A cage, box, or platform used for lowering and hoisting miners, tools, and equipment into and out of mines.

safety can [ENG] A cylindrical metal container used for temporary storage or handling of flammable liquids, such as gasoline, naphtha, and benzine, in buildings not provided with properly constructed storage rooms; these cans are also used to transport such liquids for filling and supply purposes within local areas.

safety car [MIN ENG] Any mine car or hoisting cage provided with safety stops, catches, or other precautionary devices.

safety chain [MIN ENG] A chain connecting the first and last cars of a trip to prevent separation, if a coupling breaks.

safety chuck [DES ENG] Any drill chuck on which the heads of the set screws do not protrude beyond the outer periphery of the chuck.

safety door [MIN ENG] An extra door ready for use in the event of damage to the existing ventilation door or in any emergency, for example, explosion or fire.

safety engineer [IND ENG] A person who inspects all possible danger spots in a factory, mine, or other industrial building or plant.

safety engineering [IND ENG] The testing and evaluating of equipment and procedures to prevent accidents.

safety explosive [MATER] An explosive which may be handled safely under ordinary conditions; it requires a powerful detonating force.

safety factor [ELEC] The amount of load, above the normal operating rating, that a device can handle without failure. [MECH] *See* factor of safety. [ORD] **1.** Increase in range or elevation that must be set on a gun so that friendly troops, over whose heads fire is to be delivered, will not be endangered. **2.** Overload factor in design to ensure safe operation.

safety film [GRAPHICS] Films made from cellulose acetate, polyester, and other plastics that are not readily flammable.

safety flange [DES ENG] A type of flange with tapered sides designed to keep a wheel intact in the event of accidental breakage.

safety fork [ORD] Metal clip that fits over the collar of the fuse in a land mine and prevents the mine from being set off accidentally; its function is the same as that of a safety pin.

safety fuse [ENG] A train of black powder which is enclosed in cotton, jute yarn, and waterproofing compounds, and which burns at the rate of 2 feet (60 centimeters) per minute; it is used mainly for small-scale blasting.

safety gate [MIN ENG] An automatically operated gate at the top of a mine shaft or at landings both to guard the entrance and to prevent falling into the shaft.

safety glass [MATER] **1.** A glass that resists shattering (such as a glass containing a net of wire or constructed of sheets separated by plastic film). **2.** A glass that has been tempered so that when it shatters, it breaks up into grains instead of jagged fragments.

safety groove [ORD] A groove incorporated in an item or component of ammunition to ensure that any possible failure will occur at a selected location and thereby will be of a less hazardous nature.

safety hoist [MECH ENG] A hoisting gear that does not continue running when tension is released.

safety hook [DES ENG] A hoisting hook with a spring-loaded latch that prevents the load from accidentally slipping off the hook.

safety lamp [MIN ENG] In coal mining, a lamp that is relatively safe to use in atmospheres which may contain flammable gas.

safety lanes [NAV] Specified sea lanes designated for use in transit by submarines and surface ships to prevent attack by friendly forces.

safety level of supply [IND ENG] The quantity of material, in addition to the operating level of supply, required to be on hand to permit continuous operations in the event of minor interruption of normal replenishment or unpredictable fluctuations in demand.

safety lever [ORD] **1.** A metal piece forming part of a grenade

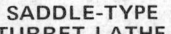

fuse that is restrained either by hand or by the projection adapter after the safety pin is removed; following the release of the grenade, the lever is discarded and the fuse train is initiated by the action of the released fuse firing pin. **2.** A lever that sets the safety mechanism on certain types of automatic weapons.

safety lock [ORD] Locking device that prevents a weapon from being fired accidentally.

safety match [ENG] A match that can be ignited only when struck against a specially made friction surface.

safety pin [ORD] **1.** A device designed to fit the mechanism of a fuse in order to prevent accidental arming or functioning of the fuse and to ensure transport safety; the device is removed just before employment of the fuse. **2.** *See* safety wire.

safety pinion [HOROL] A center pinion gear in a watch that protects delicate parts by unscrewing when the mainspring breaks, thereby preventing pressure from being applied to the train wheel teeth and arbor pivots.

safety plug [ENG] A protective device used on a heated pressure vessel (for example, a steam boiler), and containing a fusible element that melts at a predetermined safe temperature to prevent the buildup of excessive pressure. Also known as fusible plug.

safety post [MIN ENG] A timber placed near the face of workings to protect the workmen. Also known as safety prop.

safety prop *See* safety post.

safety rail *See* guard rail.

safety relief valve *See* safety valve.

safety rod [NUCLEO] A control rod capable of shutting down a reactor quickly in case of failure of the ordinary control system using regulating rods and shim rods; a safety rod may be suspended above the core by a magnetic coupling and allowed to fall in when power reaches a predetermined level. Also known as scram rod.

safety service [COMMUN] Radio communications service used permanently or temporarily for safeguarding human life and property.

safety shoe [ENG] A special shoe without spark-producing nails or plates, worn by personnel working around explosives.

safety stake [ORD] One of the stakes set in the ground to mark the right or left limit of safe fire of a weapon.

safety stop [MECH ENG] **1.** On a hoisting apparatus, a device by which the load may be prevented from falling. **2.** An automatic device on a hoisting engine designed to prevent overwinding.

safety valve [ENG] A spring-loaded, pressure-actuated valve that allows steam to escape from a boiler at a pressure slightly above the safe working level of the boiler; fitted by law to all boilers. Also known as safety relief valve.

safety wire [ORD] Wire set into the body of a fuse to lock all movable parts into safe positions so that the fuse will not be set off accidentally; it is pulled out just before firing. Also known as safety pin.

safe yield [CIV ENG] The maximum dependable draft that can be made continuously upon a source of water supply over a given period of time during which the probable driest period, and therefore period of greatest deficiency in water supply, is likely to occur.

safflorite [MINERAL] CoAs$_2$ A cobalt arsenide mineral that occurs in tin-white masses, and is dimorphous with smaltite; found in Canada, Morocco, and the United States.

safflower [BOT] *Carthamnus tinctorius.* An annual thistlelike herb belonging to the composite family (Compositae); the leaves are edible, flowers yield dye, and seeds yield a cooking oil.

safflower oil [MATER] Nonyellowing oil derived from safflower seed and similar to linseed oil; contains high proportion of linoleic acid; used as a drying oil and in food and medicine.

saffron [BOT] *Crocus sativus.* A crocus of the iris family (Iridaceae); the source of a yellow dye used for coloring food and medicine.

safing [ORD] As applied to weapons and ammunition, the changing from a state of readiness for initiation to a safe condition.

safranine [ORG CHEM] Any of a group of phenazine-based dyes; some are used as biological stains.

safrole [ORG CHEM] C$_3$H$_5$C$_6$H$_3$O$_2$CH$_2$ A toxic, water-insoluble, colorless oil that boils at 233°C; found in sassafras and camphorwood oils; used in medicine, perfumes, insecticides, and soaps, and as a chemical intermediate. Also known as 3,4-methylenedioxy-1-allyl benzene.

sag [ELEC] Slack introduced in an aerial cable or open-wire line to compensate for contraction during cold weather. [GEOL] **1.** A pass or gap in a ridge or mountain range shaped like a saddle. **2.** A shallow depression in a relatively flat land surface. **3.** A regional basin with gently sloping sides. [MET] Decrease in the section thickness of a casting caused by weakness of the sand mold.

Sagartiidae [INV ZOO] A family of zoantharians in the order Actiniaria.

sagathy [TEXT] A fine twilled worsted fabric, used at one time for curtains and clothes.

sag bolt [MIN ENG] A device to measure roof sag; a 12-foot (3.7-meter) unit installed without a bearing plate and securely anchored with the aid of a heavy nut, extending about 2 inches (5 centimeters) from the hole; three ½-inch (1.3 centimeter) strips of colored tape wrapped around the extending section of the bolt, green at the roof line followed by yellow and then red, help detect roof sag at a glance.

sage [BOT] *Salvia officinalis.* A half-shrub of the mint family (Labiatae); the leaves are used as a spice.

SAGE [ORD] An air defense system in which air surveillance data are processed for transmission to computers at direction centers, where the data is further processed, evaluated, and analyzed automatically to produce weapon assignment and guidance orders. Derived from semiautomatic ground environment.

sagebrush [BOT] Any of various hoary undershrubs of the genus *Artemisia* found on the alkaline plains of the western United States.

sagenite [MINERAL] A variety of rutile that is acicular and occurs in reticulated twin groups of crystals crossing at 60°.

sagenitic [GEOL] Containing acicular minerals.

Sagenocrinida [PALEON] A large order of extinct, flexible crinoids that occurred from the Silurian to the Permian.

sage oil *See* oil of sage.

saggar clay [MATER] A fire clay of which the case is made that is used for the firing of porcelain and pottery. Also spelled sagger clay.

sagger clay *See* saggar clay.

sagging [NAV ARCH] Deflection of the hull of a ship in which the middle of the keel is bowed downward.

Saghathiinae [PALEON] An extinct subfamily of hyracoids in the family Procaviidae.

sagitta [MATH] The distance between the midpoint of an arc and the midpoint of its chord. [VERT ZOO] The larger of two otoliths in the ear of most fishes.

Sagitta [ASTRON] A small constellation; right ascension 20 hours, declination 10° north. Also known as Arrow.

sagittal [ZOO] In the median longitudinal plane of the body, or parallel to it.

sagittal focus *See* secondary focus.

Sagittariidae [VERT ZOO] A family of birds in the order Falconiformes comprising a single species, the secretary bird, noted for its nuchal plumes resembling quill pens stuck behind an ear.

Sagittarius [ASTRON] A constellation whose major portion lies in the Milky Way; right ascension 19 hours, declination 25° south. Also known as Archer.

Sagittarius star cloud [ASTRON] A large star cloud within the Milky Way; its extension is about 1500 to 6000 light-years from the sun.

sagittate [BOT] Shaped like an arrowhead, especially referring to leaves.

sagittocyst [INV ZOO] A cyst in the epidermis of certain turbellarians containing a single spindle-shaped needle.

sago [MATER] A starch obtained from the trunks of certain tropical palms, such as the sago; used as a thickening agent in food and as textile stiffening.

Saha ionization [STAT MECH] The ionization of a gas which exists when the gas is in thermal equilibrium at a given

SAFFLOWER

Safflower (*Carthamnus tinctorius*). (USDA)

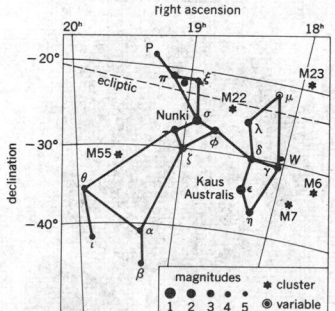

SAGE

Sage (*Salvia officinalis*). (USDA)

SAGITTARIUS

Line pattern of the constellation Sagittarius. The grid lines in the chart represent the coordinates of the sky. The apparent brightness, or magnitude, of the stars is shown by the sizes of the the dots, which are graded by appropriate numbers as indicated.

temperature, in the absence of external influences; it increases with increasing temperature. Also known as thermal ionization.

Saha's equation [STAT MECH] An equation for Saha ionization of a monatomic gas in terms of the temperature and pressure of the gas, the ionization potential, and statistical weights of ion, electron, and atom.

sahel [METEOROL] A strong dust-bearing desert wind in Morocco.

sahlite See salite.

sail [NAV ARCH] An article made of canvas and rope designed to be spread on spars in such a manner as to utilize the power of the wind in driving a vessel.

Sail See Vela.

sailboat [NAV ARCH] A boat fitted with one or more sails, and propelled by wind striking the sails.

sailfish [VERT ZOO] Any of several large fishes of the genus *Istiophorus* characterized by a very large dorsal fin that is highest behind its middle.

sailing [NAV] A method of solving the various problems involving course, distance, difference of latitude, difference of longitude, and departure; the various methods are collectively termed the sailings.

sailing chart [NAV] A small-scale nautical chart for offshore navigation.

sailing directions See coast pilot.

sail plan [NAV ARCH] A plan drawn to show the number, arrangement, and dimensions of the sails for a sailing vessel.

Saint Elmo's fire [ELEC] A visible electric discharge, sometimes seen on the mast of a ship, on metal towers, and on projecting parts of aircraft, due to concentration of the atmospheric electric field at such projecting parts.

Saint Hilaire method [NAV] The establishing of a line of position by assuming an altitude for a celestial body and then comparing this assumed altitude with that which is measured; the difference between the observed and computed altitudes and the azimuth gives the line of position. Also known as the intercept method.

Saint Joseph retort process [MET] An electrothermic retort process for processing zinc ore and zinc from secondary sources into zinc; heat of reaction between the sintered zinc concentrate and the coke mixture is supplied by passage of heavy electric current through the resistance of the charge.

Saint Louis encephalitis [MED] A mosquito-borne arbovirus infection of the central nervous system, occurring in the central and western United States and in Florida.

Saint Peter sandstone [GEOL] An artesian aquifer of early Lower Paleozoic age which underlies part of Minnesota, Wisconsin, Iowa, Illinois, and Indiana.

Saint Vitus dance [MED] Chorea associated with rheumatic fever. Also known as Sydenham's chorea.

Sakata-Taketani equation [QUANT MECH] A relativistic wave equation for a particle with spin 1 whose form resembles that of the nonrelativistic Schrödinger equation.

Sakmarian [GEOL] A European stage of geologic time; the lowermost Permian, above Stephanian of Carboniferous and below Artinskian.

sal See sial.

salable coal [MIN ENG] Total output of a coal mine, less the tonnage rejected or consumed during preparation for market.

sal acetosella See potassium binoxalate.

Salado formation [GEOL] A red-bed formation from the Permian found in southeast New Mexico; contains rock salt and potash salts.

salamander [VERT ZOO] The common name for members of the order Urodela.

salamander stove [ENG] A small portable stove used for temporary or emergency heat; for example, on construction sites or in greenhouses.

Salamandridae [VERT ZOO] A family of urodele amphibians in the suborder Salamandroidea characterized by a long row of prevomerine teeth.

Salamandroidea [VERT ZOO] The largest suborder of the Urodela characterized by teeth on the roof of the mouth posterior to the openings of the nostrils.

sal ammoniac [MINERAL] NH_4Cl A white, isometric, crystalline mineral composed of native ammonium chloride.

sal ammoniac cell [ELEC] Cell in which the electrolyte consists primarily of a solution of ammonium chloride.

Salangidae [VERT ZOO] A family of soft-rayed fishes, in the suborder Galaxioidei, which live in estuaries of eastern Asia.

salazosulfadimidine [ORG CHEM] $C_{19}H_{17}N_5O_5S$ A brown crystalline compound that melts at 207°C; used in medicine in cases of ulcerative colitis. Also known as salicylazosulfadimidine; salicylazosulfamethazine.

salband See selvage.

Saldidae [INV ZOO] The shore bugs, a family of predacious hemipteran insects in the superfamily Saldoidea.

Saldoidea [INV ZOO] A superfamily of the hemipteran group Leptopodoidea.

saléeite [MINERAL] $Mg(UO_2)_2(PO_4)_2 \cdot 10H_2O$ A lemon-yellow mineral composed of hydrous phosphate of magnesium and uranium.

Saleniidae [INV ZOO] A family of echinoderms in the order Salenioida distinguished by imperforate tubercles.

Salenioida [INV ZOO] An order of the Echinacea in which the apical system includes one or several large angular plates covering the periproct.

salesite [MINERAL] $Cu(IO_3)(OH)$ A bluish-green mineral composed of basic iodate of copper.

salic [GEOL] A soil horizon enriched with secondary salts, at least 2 percent, and measuring at least 15 centimeters in thickness. [MINERAL] Pertaining to certain light-colored minerals, such as quartz and feldspars, that are rich in silica or magnesium and commonly occur in igneous rock.

Salicaceae [BOT] The single family of the order Salicales.

Salicales [BOT] A monofamilial order of dicotyledonous plants in the subclass Dilleniidae; members are dioecious, woody plants, with alternate, simple, stipulate leaves and plumose-hairy mature seeds.

salicin [ORG CHEM] $C_{13}H_{18}O_7$ A glucoside; colorless crystals, soluble in water, alcohol, alkalies, and glacial acetic acid; melts at 199°C; used in medicine and as an analytical reagent.

salicylal See salicylaldehyde.

salicyl alcohol [ORG CHEM] $C_7H_8O_2$ A crystalline alcohol that forms plates or powder, melting at 86-87°C; used in medicine as a local anesthetic. Also known as *ortho*-hydroxybenzyl alcohol; saligenol.

salicylaldehyde [ORG CHEM] C_6H_4OHCHO Clear to dark-red oily liquid, with burning taste and almond aroma; soluble in alcohol, benzene, and ether, very slightly soluble in water; boils at 196°C; used in analytical chemistry, in perfumery, and for synthesis of chemicals. Also known as *ortho*-hydroxybenzaldehyde; salicylal; salicylic aldehyde.

salicylamide [ORG CHEM] $C_6H_4(OH)CONH_2$ Pinkish or white crystals; soluble in alcohol, ether, chloroform, and hot water; melts at 193°C; used in medicine as an analgesic, antipyretic, and antirheumatic drug. Also known as *ortho*-hydroxybenzamide.

salicylaniline [ORG CHEM] $HOC_6H_4CONHC_6H_5$ White or pinkish crystals; soluble in alcohol, ether, and chloroform, slightly soluble in water; melts at 134°C; used as fungicide and antimildew agent and in medicine.

salicylate [ORG CHEM] A salt of salicylic acid with the formula $C_6H_4(OH)COOM$, where M is a monovalent metal; for example, $NaC_7H_5O_3$, sodium salicylate.

salicylated mercury See mercuric salicylate.

salicylazosulfadimidine See salazosulfadimidine.

salicylazosulfamethazine See salazosulfadimidine.

salicylic acid [ORG CHEM] $C_6H_4(OH)(COOH)$ White crystals with sweetish taste; soluble in alcohol, acetone, ether, benzene, and turpentine, slightly soluble in water; discolored by light; melts at 158°C; used as a chemical intermediate and in medicine, dyes, perfumes, and preservatives. Also known as *ortho*-hydroxybenzoic acid.

salicylic acid phenyl ester See salol.

salicylic aldehyde See salicylaldehyde.

salicylism [MED] A syndrome produced by excessive doses of salicylates; characterized by dizziness, headache, and nausea.

Salientia [VERT ZOO] The equivalent name for Anura.

salient pole [ELECTROMAG] A structure of magnetic material on which is mounted a field coil of a generator, motor, or similar device.

salient-pole field winding [ELEC] A type of field winding in

SALICALES

Eastern American cottonwood (*Populus deltoides*), showing female catkins with ripening capsules and cotton seeds. Note characteristic three carpels in opened capsule at lower left; more common number in other species of the order is two. (*Photograph by John H. Gerard, from National Audubon Society*)

SALIENT-POLE FIELD WINDING

Salient-pole field winding on rotor of synchronous motor showing the pole core, lower right. (*Allis-Chalmers*)

electric machinery where the winding turns are concentrated around the pole core.

saligenol *See* salicyl alcohol.

salimeter [ENG] A hydrometer graduated to read directly the percentage of salt in a solution such as brine.

salina [GEOL] An area, such as a salt flat, in which deposits of crystalline salts are formed or found. [HYD] A body of water containing high concentrations of salt.

saline-water reclamation [CHEM ENG] Purification and removal of salts from brine or brackish water by ion exchange, crystallization, distillation, evaporation, and reverse osmosis.

salinity logging [PETRO ENG] Technique for measurement and recording of saltwater-bearing zones in an oil or gas reservoir; uses a combination of neutron logging with a chlorine curve.

salinity-temperature-depth recorder [ENG] An instrument consisting of sensing elements usually lowered from a stationary ship, and a recorder on board which simultaneously records measurements of temperature, salinity, and depth. Also known as CTD recorder; STD recorder.

Salisbury dark box [ELECTR] Isolating chamber used for test work in connection with radar equipment; the walls of the chamber are specially constructed to absorb all impinging microwave energy at a certain frequency.

salite [MINERAL] $(Mg,Fe)_2Si_2O_6$ A grayish-green to black mineral variety of diopside containing more magnesium than iron; member of the clinopyroxene group. Also spelled sahlite.

saliva [PHYSIO] The opalescent, tasteless secretions of the oral glands.

salivary amylase *See* ptyalin.

salivary diastase *See* ptyalin.

salivary gland [PHYSIO] A gland that secretes saliva, such as the sublingual or parotid.

salivary gland chromosomes [CYTOL] Polytene chromosomes found in the interphase nuclei of salivary glands in the larvae of Diptera; chromosomes in the larva undergo complete somatic pairing to form two homologous polytene chromosomes fused side by side.

salivation [MED] Mild mercury poisoning suffered by workmen in amalgamation plants. [PHYSIO] Excessive secretion of saliva.

Salk vaccine [IMMUNOL] A killed-virus vaccine administered for active immunization against poliomyelitis.

sallying ship [NAV ARCH] Producing rolling motion of a vessel by running a group of people in unison from side to side.

sally port [ORD] Large gate or passage in a fortified place.

salmon [VERT ZOO] The common name for a number of fish in the family Salmonidae which live in coastal waters of the North Atlantic and North Pacific and breed in rivers tributary to the oceans.

Salmonella [MICROBIO] A genus of gram-negative, rod-shaped pathogenic bacteria of the family Enterobacteriaceae that are usually motile by peritrichous flagella and lack proteolytic enzymes. Also known as *Eberthella*.

Salmonella aertrycke *See* Salmonella typhimurium.

Salmonella choleraesuis [MICROBIO] The organism that is a secondary invader of pigs in hog cholera; may cause localized lesions, enteric fever, or gastroenteritis in humans. Also known as *Salmonella suipestifer*.

Salmonella enteritidis [MICROBIO] An organism causing gastroenteritis in humans. Also known as *Bacterium enteritidis*.

Salmonella hirschfeldii [MICROBIO] An organism causing enteric fever in humans. Also known as *Bacterium paratyphosum* C; *Salmonella paratyphi* C.

Salmonella oranienburg [MICROBIO] An organism isolated from feces of normal carriers and of persons with food poisoning.

Salmonella paratyphi A [MICROBIO] An organism that causes gastroenteritis and enteric fever in humans. Also known as *Salmonella paratyphosa*.

Salmonella paratyphi B *See* Salmonella schottmülleri.

Salmonella paratyphi C *See* Salmonella hirschfeldii.

Salmonella paratyphosa *See* Salmonella paratyphi A.

Salmonella pullorum [MICROBIO] An organism causing pullorum disease or bacillary white diarrhea of chicks.

Salmonella schottmülleri [MICROBIO] An organism causing enteric fever in humans. Also known as *Bacterium paratyphosum* B; *Salmonella paratyphi* B.

Salmonella suipestifer *See* Salmonella choleraesuis.

Salmonella typhi *See* Salmonella typhosa.

Salmonella typhimurium [MICROBIO] An organism causing diarrhea in mice, rats, and birds, and gastroenteritis in humans. Also known as *Salmonella aertrycke*.

Salmonella typhosa [MICROBIO] The organism that causes typhoid fever in humans. Also known as *Bacterium typhosum; Eberthella typhosa; Salmonella typhi*.

Salmonelleae [MICROBIO] A tribe of the Enterobacteriaceae comprising the pathogenic genera *Salmonella* and *Shigella*.

salmonellosis [MED] Infection with any species of *Salmonella*.

Salmonidae [VERT ZOO] A family of soft-rayed fishes in the suborder Salmonoidei including the trouts, salmons, whitefishes, and graylings.

Salmoniformes [VERT ZOO] An order of soft-rayed fishes comprising salmon and their allies; the stem group from which most higher teleostean fishes evolved.

Salmonoidei [VERT ZOO] A suborder of the Salmoniformes comprising forms having an adipose fin.

salmon oil [MATER] Combustible, pale golden-yellow liquid with sweet taste; soluble in alcohol, ether, chloroform, and carbon disulfide; obtained from the waste in the canning of salmon; used in pet foods, soaps, and leather dressing.

salmonsite [MINERAL] A buff-colored mineral composed of hydrous phosphate of manganese and iron occurring in cleavable masses.

Salmopercae [VERT ZOO] An equivalent name for Percopsiformes.

salol [ORG CHEM] $C_6H_4OHCOOC_6H_5$ White powder with aromatic taste and aroma; soluble in alcohol, ether, chloroform, and benzene; slightly soluble in water; melts at 42°C; used in medicinals and as a preservative. Also known as phenyl salicylate; salicylic acid phenyl ester.

Salpida [INV ZOO] An order of tunicates in the class Thaliacea including transparent forms ringed by muscular bands.

Salpingidae [INV ZOO] The narrow-waisted bark beetles, a family of coleopteran insects in the superfamily Tenebrionoidea.

salpingitis [MED] 1. Inflammation of the fallopian tube. 2. Inflammation of the eustachian tube.

salpingo-oophoritis [MED] Inflammation of the fallopian tubes and ovaries.

sal prunella *See* niter balls.

sal soda [INORG CHEM] $Na_2CO_3 \cdot 10H_2O$ White, water-soluble crystals, insoluble in alcohol; melts and loses water at about 33°C; mild irritant to mucous membrane; used in cleansers and for washing textiles and bleaching linen and cotton. Also known as sodium carbonate decahydrate; washing soda.

salsoline [PHARM] $C_{11}H_{15}NO_2$ A compound that crystallizes from alcohol solution, melts at 221°C, soluble in hot alcohol and chloroform; used in medicine as an antihypertensive agent.

salt [CHEM] The reaction product when a metal displaces the hydrogen of an acid; for example, $H_2SO_4 + 2NaOH \rightarrow Na_2SO_4$ (a salt) $+ 2H_2O$. [ENG] To add an accelerator or retardant to cement. [MIN ENG] 1. To introduce extra amounts of a valuable or waste mineral into a sample to be assayed. 2. To artificially enrich, as a mine, usually with fraudulent intent.

salt-and-pepper sand [GEOL] A sand composed of a mixture of light- and dark-colored grains.

salt anticline [GEOL] A structure like a salt dome but with a linear salt core. Also known as salt wall.

saltation [GEOL] Transport of a sediment in which the particles are moved forward in a series of short intermittent bounces from a bottom surface.

saltatorial [ZOO] Adapted for leaping.

salt bath [MET] Molten salts in which steel is heated for hardening and tempering.

saltbox [ARCH] A type of house built in colonial New England, having a frame structure with two stories in the front and one in the rear and a double-sloping roof.

salt bridge [PHYS CHEM] A bridge of a salt solution, usually potassium chloride, placed between the two half-cells of a

SALMON

Coho salmon (*Oncorhynchus kisutch*).

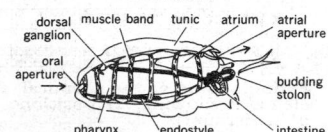

SALPIDA

Salp, *Thalia democratica* (Salpida), solitary asexual form. (*After Metcalf*)

galvanic cell, either to reduce to a minimum the potential of the liquid junction between the solutions of the two half-cells or to isolate a solution under study from a reference half-cell and prevent chemical precipitations.

salt cake [INORG CHEM] Impure sodium sulfate; used in soaps, paper pulping, detergents, glass, ceramic glaze, and dyes.

salt dome [GEOL] A diapiric or piercement structure in which there is a central, equidimensional salt plug.

salted weapon [ORD] A nuclear weapon which has, in addition to its normal components, certain elements or isotopes which capture neutrons at the time of the explosion and produce radioactive weapon debris.

salt flowers *See* ice flowers.

salt-fog test [MET] An accelerated corrosion test in which a piece of metal is subjected to a fine spray of sodium chloride solution. Also known as salt-spray test.

salt gland [VERT ZOO] A compound tubular gland, located around the eyes and nasal passages in certain marine turtles, snakes, and birds, which copiously secretes a watery fluid containing a high percentage of salt.

salt glaze [ENG] Glaze formed on the surface of stoneware by putting salt into the kiln during firing.

salt grainer [CHEM ENG] Type of evaporative crystallizer in which the solution is kept hot, and supersaturation is developed by evaporation rather than by cooling.

salt haze [METEOROL] A haze created by the presence of finely divided particles of sea salt in the air, usually derived from the evaporation of sea spray.

Salticidea [INV ZOO] The jumping spiders, a family of predacious arachnids in the suborder Dipneumonomorphae having keen vision and rapid movements.

salting-out effect [CHEM ENG] The growth of crystals of a substance on heated, liquid-holding surfaces of a crystallizing evaporator as a result of the decrease in solubility of the substance with increase in temperature.

salt lake [HYD] A confined inland body of water having a high concentration of salts, principally sodium chloride.

saltmarsh [ECOL] A maritime habitat found in temperate regions, but typically associated with tropical and subtropical mangrove swamps, in which excess sodium chloride is the predominant environmental feature.

salt mine [MIN ENG] A mine containing deposits of rock salt.

salt-mud combination log [PETRO ENG] Record of electrical logging of the mud in oil well boreholes in the presence of sodium chloride incursions from adjacent formations.

salt of Lemery *See* potassium sulfate.

salt of sorrel *See* potassium binoxalate.

salt of tartar *See* potassium carbonate.

salt pan [CHEM] A pool used for obtaining salt by the natural evaporation of sea water.

saltpeter *See* potassium nitrate.

saltpeter cave [GEOL] A cave in which there are deposits of saltpeter earth.

saltpeter earth [GEOL] A deposit containing calcium nitrate and found in caves.

salt-spray climax [ECOL] A climax community along exposed Atlantic and Gulf seacoasts composed of plants able to tolerate the harmful effects of salt picked up and carried by onshore winds from seawater.

salt-spray test *See* salt-fog test.

saltus *See* oscillation of a function.

salt velocity meter [ENG] A rate-of-flow volume meter used to find the transit time of passage between two fixed points of a small quantity of salt or radioactive isotope in a flowing stream by measuring electrical conductivity or radiation level at those points.

salt wall *See* salt anticline.

salt water *See* seawater.

saltwater intrusion [HYD] Displacement of fresh surface water or groundwater by salt water due to its greater density.

saltwater wedge [OCEANOGR] A wedge-shaped intrusion of salty ocean water into a freshwater estuary or tidal river; it slopes downward in the upstream direction, and salinity increases with depth.

saluting gun [ORD] Cannon used for firing salutes.

salvage procedure [ENG] The recovery, evacuation, and rec-

lamation of damaged, discarded, condemned, or abandoned allied or enemy materiel, ships, craft, and floating equipment for reuse, repair, refabrication, or scrapping.

salvage value [ENG] The net worth of diamonds recovered from a used diamond-inset tool.

salvage vessel [NAV ARCH] A ship equipped to rescue or save ships or material that has been sunk, wrecked, damaged, or discarded.

Salvarsan [PHARM] A trademark for arsephenamine, an early antisyphilitic drug.

salvia [MATER] The dried leaves of the sage, *Salvia officinalis;* contains volatile oil, resin, and tannin; used in food engineering as a flavoring agent and condiment, and in medicine as an antisecretory agent.

salvia oil *See* oil of sage.

Salviniales [BOT] A small order of heterosporous, leptosporangiate ferns (division Polypodiophyta) which float on the surface of the water.

salvo [ORD] **1.** A simultaneous, or nearly simultaneous, discharge of shots from two or more closely placed guns or launchers against the same target. **2.** The release of several bombs or rocket missiles simultaneously, or in close train, from one or more aircraft, at a single target. **3.** The aggregate of shots, bombs, or rockets so discharged or released.

SAM *See* surface-to-air missile.

samara [BOT] A dry, indehiscent, winged fruit usually containing a single seed, such as sugar maple (*Acer saccharum*).

samarium [CHEM] Group III rare-earth metal, atomic number 62, symbol Sm; melts at 1350°C, tarnishes in air, ignites at 200–400°C.

samarskite [MINERAL] $(Y,Ce,U,Ca,Fe,Pb,Th)(Nb,Ta,Ti,Sn)_2O_6$ A velvet-black to brown metamict orthorhombic mineral with splendent vitreous to resinous luster occurring in granite pegmatites. Also known as ampangabeite; uranotantalite.

Sambonidae [INV ZOO] A family of pentastomid arthropods in the suborder Porocephaloidea of the order Porocephalida.

SAMOS program [AERO ENG] A series of military reconnaissance satellites carrying cameras and other surveillance equipment; each satellite has a jettisonable camera package that is ejected from orbit and recovered in midair.

sample [SCI TECH] Representative fraction of material tested or analyzed in order to determine the nature, composition, and percentage of specified constituents, and possibly their reactivity. [STAT] A selection of a certain collection from a larger collection.

sample-and-hold circuit [ELECTR] A circuit that measures an input signal at a series of definite points in time, and whose output remains constant at a value corresponding to the most recent measurement until the next measurement is made.

sampled-data control system [CONT SYS] A form of control system in which the signal appears at one or more points in the system as a sequence of pulses or numbers usually equally spaced in time.

sampled grade [MIN ENG] The amount of valuable metal in the ore in place as determined by underground, surface, or drill-hole sampling.

sample function [STAT] A function or procedure which, when applied repeatedly to a given population, produces a collection of samples.

sampleite [MINERAL] $NaCaCu_5(PO_4)_4Cl \cdot 5H_2O$ A blue mineral composed of hydrous phosphate and chloride of sodium, calcium, and copper.

sample log [ENG] Record of core samples or drill cuttings; gives geological, visual, and hydrocarbon-content record versus depth of drilling.

sample path [MATH] If $\{X_t: t \text{ in } T\}$ is a stochastic process, a sample path for the process is the function on T to the range of the process which assigns to each t the value $X_t(w)$, where w is a previously given fixed point in the domain of the process.

sampler [CONT SYS] A device, used in sampled-data control systems, whose output is a series of impulses at regular intervals in time; the height of each impulse equals the value of the continuous input signal at the instant of the impulse. [ENG] A mechanical or other device designed to obtain small samples of materials for analysis; used in biology, chemistry, and geology.

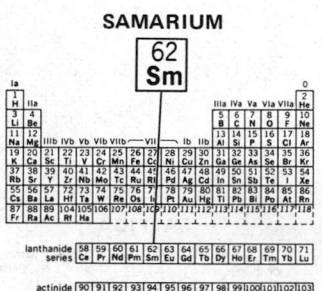

SAMARIUM

**62
Sm**

Periodic table of the chemical elements showing the position of samarium.

sample size [STAT] The number of objects in the sample.

sample splitter [ENG] An instrument, generally constructed of acrylic resin, designed to subdivide a total sample of marine plankton while maintaining a quantitatively correct relationship between the various phyla in the sample.

sampling [ENG] Process of obtaining a sequence of instantaneous values of a wave. [SCI TECH] The obtaining of small representative quantities of materials (gas, liquid, solid) for the purpose of analysis. [STAT] A drawing of a collection from a given population.

sampling area ratio [MIN ENG] The volume of the soil displaced in proportion to the volume of the sample; a well-designed tool has an area ratio of about 20 percent.

sampling bottle [ENG] A cylindrical container, usually closed at a chosen depth, to trap a water sample and transport it to the surface without introducing contamination.

sampling fraction [STAT] The ratio of the sample size to the population size.

sampling gate [ELECTR] A gate circuit that extracts information from the input waveform only when activated by a selector pulse.

sampling interval [CONT SYS] The time between successive sampling pulses in a sampled-data control system.

sampling pipe [MIN ENG] A small pipe built into a seal to take air samples in a sealed area.

sampling plan [IND ENG] A plan stating sample sizes and the criteria for accepting or rejecting items or taking another sample during inspection of a group of items.

sampling process [ENG] The process of obtaining a sequence of instantaneous values of some quantity that varies continuously with time.

sampling rate [ENG] The rate at which measurements of physical quantities are made; for example, if it is desired to calculate the velocity of a missile and its position is measured each millisecond, then the sampling rate is 1000 measurements per second.

sampling risk [IND ENG] In inspection procedure, the probability, under the sampling plan used, that acceptable material will be rejected or that unsatisfactory material will be accepted.

sampling spark chamber [NUCLEO] A spark chamber that yields as output data the coordinates of a single point on the track (or tracks) in each gap; all narrow-gap chambers are of this type.

sampling spoon [MIN ENG] A cylinder with a spoonlike cutting edge for taking soil samples.

sampling switch *See* commutator switch.

sampling techniques [STAT] The methods used in drawing samples from a population usually in such a manner that the sample will facilitate determination of some hypothesis concerning the population.

sampling theorem [COMMUN] The theorem that a signal that varies continuously with time is completely determined by its values at an infinite sequence of equally spaced times if the frequency of these sampling times is greater than twice the highest frequency component of the signal. Also known as Shannon's sampling theorem.

sampling theory [STAT] The mathematical study of sampling techniques.

sampling voltmeter [ENG] A special type of voltmeter that detects the instantaneous value of an input signal at prescribed times by means of an electronic switch connecting the signal to a memory capacitor; it is particularly effective in detecting high-frequency signals (up to 12 gigahertz) or signals mixed with noise.

samsonite [MINERAL] $Ag_4MnSb_2S_6$ A black mineral composed of sulfide of silver, manganese, and antimony occurring in monoclinic prismatic crystals.

samson post *See* kingpost.

Samythinae [INV ZOO] A subfamily of sedentary polychaete annelids in the family Ampharetidae having a conspicuous dorsal membrane.

SAN *See* styrene-acrylonitrile resin.

sanatron circuit [ELECTR] A variable time-delay circuit having two pentodes and two diodes, used to produce very short gate waveforms having time durations that vary linearly with a reference voltage.

sanbornite [MINERAL] $BaSi_2O_5$ A white triclinic mineral composed of barium silicate.

sand [GEOL] A loose material consisting of small mineral particles, or rock and mineral particles, distinguishable by the naked eye; grains vary from almost spherical to angular, with a diameter range from $\frac{1}{16}$ to 2 millimeters.

Sandalidae [INV ZOO] The equivalent name for Rhipiceridae.

sandal oil *See* sandalwood oil.

sandalwood [BOT] 1. Any species of the genus *Santalum* of the sandalwood family (Santalaceae) characterized by a fragrant wood. 2. *S. album*. A parasitic tree with hard, close-grained, aromatic heartwood used in ornamental carving and cabinetwork.

sandalwood oil [MATER] Pale-yellow essential oil with harsh taste and faint aromatic scent; soluble in fixed oils; insoluble in glycerin; derived from the sandalwood *Santalum album*; used in medicine, perfumes, and flavors. Also known as East Indian sandalwood oil; sandal oil; santal oil; santalwood oil.

sandarac *See* realgar.

sandarac gum [MATER] Yellow, brittle, water-insoluble, natural resin obtained from the African sandarac tree of Morocco; used in varnishes and lacquers.

sand auger *See* dust whirl.

sandbag [ENG] A bag filled with sand; used to build temporary protective walls.

sandbank [GEOL] A deposit of sand forming a mound, hillside, bar, or shoal.

sandbar [GEOL] A bar or low ridge of sand bordering the shore and built up, or near, to the surface of the water by currents or wave action. Also known as sand reef.

sandblasting [ENG] *See* grit blasting. [GEOL] Abrasion affected by the action of hard, windblown mineral grains.

sand boil *See* blowout.

sand-cast [MET] Made by pouring molten metal into a mold made of sand.

sand cay *See* sandkey.

sand control [MET] Control of the properties of foundry sand mixtures.

sand count [PETRO ENG] Determination of the total thickness of an oil or gas reservoir's permeable section (excluding shale streaks and other impermeable zones); can be derived from electrical logs.

sand crystal [GEOL] A large crystal loaded up to 60% with detrital sand inclusions formed in a sandstone during or as a result of cementation.

sand devil *See* dust whirl.

sand dollar [INV ZOO] The common name for the flat, disk-shaped echinoderms belonging to the order Clypeasteroida.

sand drain [CIV ENG] A vertical boring through a clay or silty soil filled with sand or gravel to facilitate drainage.

sand drift [GEOL] 1. Movement of windblown sand along the surface of a desert or shore. 2. An accumulation of sand against the leeward side of a fixed obstruction.

sand dune [GEOL] A mound of loose windblown sand commonly found along low-lying seashores above high-tide level.

sander [MECH ENG] 1. An electric machine used to sand the surface of wood, metal, or other material. 2. A device attached to a locomotive or electric rail car which sands the rails to increase friction on the driving wheels.

sandfall *See* slip face.

sand filter [CIV ENG] A filter consisting of graded layers of sand and aggregate for purifying domestic water.

sand finish [ENG] A smooth finish on a plaster surface made by rubbing the sand or mortar coat.

sandfly [INV ZOO] Any of various small biting Diptera, especially of the genus *Phlebotomus*, which are vectors for phlebotomus (sandfly) fever.

sandfly fever *See* phlebotomus fever.

sand-grain volume [PETRO ENG] In oil reservoir porosity calculations, the actual volume filled by sand grains, without allowance for spaces (voids) between the grains.

sandhog [ENG] A worker in compressed-air environments, as in driving tunnels by means of pneumatic caissons.

sand hopper [INV ZOO] The common name for gammaridean crustaceans found on beaches.

sanding [ENG] 1. Covering or mixing with sand. 2. Smooth-

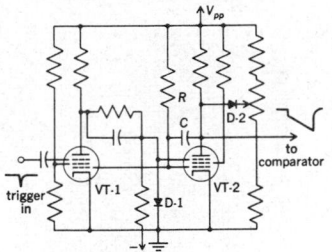

SANATRON CIRCUIT

Diagram of a simple sanatron circuit. VT-1 and VT-2 = pentodes; D-1 and D-2 = diodes; V_{pp} = plate supply voltage with respect to cathode.

SANDALWOOD

Sandalwood branch with foliage and fruit.

SAND DOLLAR

Sand dollar. The nearly circular body may reach 3 inches (7.6 centimeters) in diameter. The characteristic pattern of a five-part flower is seen on the aboral surface.

ing a surface with sandpaper or other abrasive paper or cloth.

sandkey [GEOL] A small sandy island parallel with the shore. Also known as sand cay.

sand levee *See* whaleback dune.

sand-lime brick [MATER] A clay building material made by molding a mixture of sand with 6% hydrated lime and water.

sand load [ELECTROMAG] An attenuator used as a power-dissipating terminating section for a coaxial line or waveguide; the dielectric space in the line is filled with a mixture of sand and graphite that acts as a matched-impedance load, preventing standing waves.

Sandmeyer's reaction [CHEM] Conversion of diazo compounds (in the presence of cuprous halogen salts) into halogen compounds.

sand mill [MECH ENG] Variation of a ball-type size-reduction mill in which grains of sand serve as grinding balls.

sandpaper [MATER] Paper with abrasive glued to the surface.

sand pile [CIV ENG] A compacted filling of sand in a deep round hole formed by ramming the sand with a pile; used for foundations in soft soil.

sandpiper [VERT ZOO] Any of various small birds that are related to plovers and that frequent sandy and muddy shores in temperate latitudes; bill is moderately long with a soft, sensitive tip, legs and neck are moderately long, and plumage is streaked brown, gray, or black above and is white below.

sandpit [CIV ENG] An excavation dug in sand, especially as a source of sand for construction materials.

sand plain [GEOL] A small outwash plain formed by deposition of sand transported by meltwater streams flowing from a glacier.

sand pump [MECH ENG] A pump, usually a centrifugal type, capable of handling sand- and gravel-laden liquids without clogging or wearing unduly; used to extract mud and cuttings from a borehole. Also known as sludge pump.

sand reef *See* sandbar.

sand return [PETRO ENG] The return of injected sand to the wellbore following formation fracturing; it constitutes a problem.

sand ridge *See* sand wave.

sands [MIN ENG] The coarser and heavier particles of crushed ore, of such size that they settle readily in water and may be leached by allowing the solution to percolate.

sand-shale ratio [GEOL] The ratio between the thickness or percentage of sandstone and that of shale in a geologic section.

sand shark [VERT ZOO] Any of various shallow-water predatory elasmobranchs of the family Carchariidae. Also known as tiger shark.

sand sheet [GEOL] A thin accumulation of coarse sand or fine gravel having a flat surface.

sand slinger [MECH ENG] A machine which delivers sand to and fills molds at high speed by centrifugal force.

sand snow [HYD] Snow that has fallen at very cold temperatures (of the order of $-25°C$); as a surface cover, it has the consistency of dust or light dry sand.

sandspit [GEOL] A spit consisting principally of sand.

sandstone [PETR] A detrital sedimentary rock consisting of individual grains of sand-size particles 0.06 to 2 millimeters in diameter either set in a fine-grained matrix (silt or clay) or bonded by chemical cement.

sandstone dike [GEOL] A dike made of sandstone or lithified sand.

sandstorm [METEOROL] A strong wind carrying sand through the air, the diameter of most particles ranging from 0.08 to 1 millimeter; in contrast to a duststorm, the sand particles are mostly confined to the lowest 2 meters above ground, rarely rising more than 11 meters.

sand trap [ENG] A device in a conduit for trapping sand or soil particles carried by the water.

sandur *See* outwash plain.

sand wave [GEOL] A large, ridgelike primary structure resembling a water wave on the upper surface of a sedimentary bed that is formed by high-velocity air or water currents. Also known as sand ridge.

sand wedge [GEOL] A wedge-shaped accumulation of sand with the apex downward formed by the filling in of winter contraction cracks.

sand wheel [MECH ENG] A wheel fitted with steel buckets around the circumference for lifting sand or sludge out of a sump to stack it at a higher level.

sandwich construction [DES ENG] Composite construction of alloys, plastics, wood, or other materials consisting of a foam or honeycomb layer laminated and glued between two hard outer sheets. Also known as sandwich laminate.

sandwich heating [ENG] Method for heating both sides of a thermoplastic sheet simultaneously prior to forming or shaping.

sandwich laminate *See* sandwich construction.

sandwich rolling [MET] Rolling strips of metal together to form a metallurgically bonded composite sheet.

sandy bentonite *See* arkosic bentonite.

sandy chert [PETR] Chert formed in sandy beds by replacement of cement, or the filling of pore spaces, with silica.

sane [PSYCH] Of sound mind.

Sanfilippo's syndrome [MED] A hereditary metabolic disorder, transmitted as an autosomal recessive, characterized by excessive amounts of heparitin sulfate in the urine, and manifested by minor skeletal changes and slight hepatomegaly.

Sangamon [GEOL] The third interglacial stage of the Pleistocene epoch in North America, following the Illinoian glacial and preceding the Wisconsin.

sanguivorous [ZOO] Feeding on blood.

sanidaster [INV ZOO] A rod-shaped spicule having spines at intervals along its length.

sanidine [MINERAL] $KAlSi_3O_8$ An alkali feldspar mineral occurring in clear, glassy crystals embedded in unaltered acid volcanic rocks; a high-temperature, disordered form. Also known as glassy feldspar; ice spar; rhyacolite.

sanidinite [PETR] A type of igneous rock composed chiefly of sanidine.

sanitary engineering [CIV ENG] A field of civil engineering concerned with works and projects for the protection and promotion of public health.

sanitary landfill [CIV ENG] The disposal of garbage by spreading it in layers covered with soil or ashes to a depth sufficient to control rats, flies, and odors.

sanitary nipper *See* latrine cleaner.

sanitary sewer [CIV ENG] A sewer which is restricted to carrying sewage and to which storm and surface waters are not admitted.

sanitation [CIV ENG] The act or process of making healthy environmental conditions.

sanitizer [MATER] Disinfectant formulated to clean food-processing equipment and dairy and eating utensils.

San Joaquin Valley fever *See* coccidioidomycosis.

sanmartinite [MINERAL] $ZnWO_4$ A mineral composed of zinc tungstate.

SA node *See* sinoauricular node.

sansar [METEOROL] A northwest wind of Persia.

Santa Ana [METEOROL] A hot, dry, foehnlike desert wind, generally from the northeast or east, especially in the pass and river valley of Santa Ana, California, where it is further modified as a mountain-gap wind.

Santalaceae [BOT] A family of parasitic dicotyledonous plants in the order Santalales characterized by dry or fleshy indehiscent fruit, plants with chlorophyll, petals absent, and ovules without integument.

Santalales [BOT] An order of dicotyledonous plants in the subclass Rosidae characterized by progressive adaptation to parasitism, accompanied by progressive simplification of the ovules.

santal oil *See* sandalwood oil.

santalwood oil *See* sandalwood oil.

Santa Rosa storm [METEOROL] In Argentina, an annual storm near the end of August.

sap [BOT] The fluid part of a plant which circulates through the vascular system and is composed of water, gases, salts, and organic products of metabolism.

saphenous nerve [ANAT] A somatic sensory nerve arising from the femoral nerve and innervating the skin of the medial aspect of the leg, foot, and knee joint.

Sapindaceae [BOT] A family of dicotyledonous plants in the order Sapindales distinguished by mostly alternate leaves, usually one and less often two ovules per locule, and seeds lacking endosperm.

Sapindales [BOT] An order of mostly woody dicotyledonous plants in the subclass Rosidae with compound or lobed leaves and polypetalous, hypogynous to perigynous flowers with one or two sets of stamens.

sapling [BOT] A young tree with a trunk less than 4 inches (10 centimeters) in diameter at a point approximately 4 feet (1.2 meters) above the ground.

saponification [CHEM] The process of converting chemicals into soap; involves the alkaline hydrolysis of a fat or oil, or the neutralization of a fatty acid.

saponification equivalent [CHEM] The quantity of fat in grams that can be saponified by 1 liter of normal alkalies.

saponification number [ANALY CHEM] Milligrams of potassium hydroxide required to saponify the fat, oil, or wax in a 1-gram sample of a given material, using a specific ASTM test method.

saponin [ORG CHEM] Any of numerous plant glycosides characterized by foaming in water and by producing hemolysis when water solutions are injected into the bloodstream; used as beverage foam producer, textile detergent and sizing, soap substitute, and emulsifier.

saponite [MINERAL] A soft, soapy, white or light-buff to bluish or reddish trioctahedral montmorillonitic clay mineral consisting of hydrous magnesium aluminosilicate and occurring in masses in serpentine and basaltic rocks. Also known as bowlingite; mountain soap; piotine; soapstone.

Sapotaceae [BOT] A family of dicotyledonous plants in the order Ebenales characterized by a well-developed latex system.

sappare See kyanite.

sapphire [MINERAL] Any of the gem varieties of the mineral corundum, especially the blue variety, except those that have medium to dark tones of red that characterize ruby; hardness is 9 on Mohs scale, and specific gravity is near 4.00.

sapphire quartz [MINERAL] An indigo-blue opaque variety of quartz.

sapphirine [MINERAL] $(MgFe)_{15}(Al,Fe)_{34}Si_7O_{80}$ A green or pale-blue mineral composed of silicate and oxide of magnesium, iron, and aluminum; usually occurs in granular form.

saprobic [BOT] Living on decaying organic matter; applied to plants and microorganisms.

saprogen [BIOL] An organism that lives on nonliving organic matter.

saprogenous ooze [GEOL] Ooze formed of putrefying organic matter.

Saprolegniales [MYCOL] An order of aquatic fungi belonging to the class Phycomycetes, having a mostly hyphal thallus and zoospores with two flagella.

saprolite [GEOL] A soft, earthy red or brown, decomposed igneous or metamorphic rock that is rich in clay and formed in place by chemical weathering. Also known as saprolith; sathrolith.

saprolith See saprolite.

sapropel [GEOL] A mud, slime, or ooze deposited in more or less open water.

sapropel-clay [GEOL] A sedimentary deposit in which the amount of clay is greater than that of sapropel.

sapropelic coal [GEOL] Coal formed by putrefaction of organic matter under anaerobic conditions in stagnant or standing bodies of water. Also known as sapropelite.

sapropelite See sapropelic coal.

sapropel-peat See peat-sapropel.

saprophage [BIOL] An organism that lives on decaying organic matter.

saprophyte [BOT] A plant that lives on decaying organic matter.

saprovore [ZOO] A detritus-eating animal.

saprozoic [ZOO] Feeding on decaying organic matter; applied to animals.

sapwood [BOT] The younger, softer, outer layers of a woody stem, between the cambium and heartwood. Also known as alburnum.

Sapygidae [INV ZOO] A family of hymenopteran insects in the superfamily Scolioidea.

sarah [NAV] Radio homing device originally designed for personnel rescue and now used in spacecraft recovery operations at sea. Derived from search and rescue and homing.

Saran [MATER] Trade name for a plastic based on a copolymer of vinyl chloride–vinylidene chloride; used as a chemical-resistant lining and tubing and in rot-proof fabrics.

sàrca [METEOROL] A violent north wind of Lake Garda in Italy.

sarcochore [BOT] A plant dispersing minute, light disseminules.

Sarcodina [INV ZOO] A superclass of Protozoa in the subphylum Sarcomastigophora in which movement involves protoplasmic flow, often with recognizable pseudopodia.

sarcoglia [CYTOL] The protoplasm occurring at a myoneural junction.

sarcoidosis [MED] A disease of unknown etiology characterized by granulomatous lesions, somewhat resembling true tubercles, but showing little or no necrosis, affecting the lymph nodes, skin, liver, spleen, heart, skeletal muscles, lungs, bones in distal parts of the extremities (osteitis cystica of Jüngling), and other structures, and sometimes by hyperglobulinemia, cutaneous anergy, and hypercalcinuria.

sarcolemma [HISTOL] The thin connective tissue sheath enveloping a muscle fiber.

sarcoleukemia See leukosarcoma.

sarcoma [MED] A malignant tumor arising in connective tissue and composed principally of anaplastic cells that resemble those of supportive tissues.

sarcoma botryoides [MED] A malignant mesenchymoma that forms grapelike structures; most common in the vagina of infants.

Sarcomastigophora [INV ZOO] A subphylum of Protozoa comprising forms that possess flagella or pseudopodia or both.

sarcomatosis [MED] Multiple growths resembling a sarcoma and found in various parts of the body.

sarcomere [HISTOL] One of the segments defined by Z disks in a skeletal muscle fibril.

Sarcophagidae [INV ZOO] A family of the myodarian orthorrhaphous dipteran insects in the subsection Calypteratae comprising flesh flies, blowflies, and scavenger flies.

sarcoplasm [HISTOL] Hyaline, semifluid interfibrillar substance of striated muscle tissue.

sarcoplasmic reticulum [CYTOL] Collectively, the cysternae of a single muscle fiber.

sarcopside [MINERAL] $(Fe,Mn,Mg)_3(PO_4)_2$ A mineral composed of a phosphate of manganese, magnesium, and iron.

Sarcopterygii [VERT ZOO] A subclass of Osteichthyes, including Crossopterygii and Dipnoi in some systems of classification.

Sarcoptiformes [INV ZOO] A suborder of the Acarina including minute globular mites without stigmata.

sarcosine [ORG CHEM] CH_3NHCH_2COOH Sweet-tasting, deliquescent crystals; soluble in water, slightly soluble in alcohol; decomposes at 210–215°C; used in toothpaste manufacture. Also known as methyl aminoacetic acid; methyl glycocol.

sarcosoma [INV ZOO] The fleshy portion of an anthozoan.

Sarcosporida [INV ZOO] An order of Protozoa of the class Haplosporea which comprises parasites in skeletal and cardiac muscle of vertebrates.

sarcosporidiosis [VET MED] A disease of mammals other than man caused by muscle infestation by sporozoans of the order Sarcosporida.

sarcostyle [INV ZOO] A fibril or column of muscular tissue.

sarcotubule [CYTOL] A tubular invagination of a muscle fiber.

sard [MINERAL] A translucent brown, reddish-brown, or deep orange-red variety of chalcedony. Also known as sardine; sardius.

sardine [MINERAL] See sard. [VERT ZOO] **1.** *Sardina pilchardus.* The young of the pilchard, a herringlike fish in the family Clupeidae found in the Atlantic along the European coasts. **2.** The young of any of various similar and related forms which are processed and eaten as sardines.

sardine oil [MATER] Combustible, yellow liquid obtained from sardines; soluble in alcohol, ether, and chloroform;

SARRACENIACEAE

An eastern American species of the pitcher plant *(Sarracenia purpurea)*, in the family Sarraceniaceae. *(Photograph of Henry M. Mayer, from National Audubon Society)*

SARSAPARILLA

Smilax aristolochiaefolia, which yields a flavoring material known as Mexican sarsaparilla.

SASSAFRAS

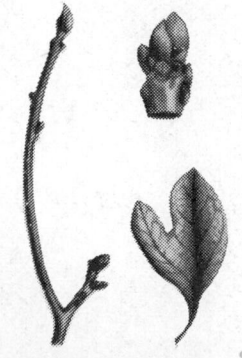

Sassafras albidum, twig, terminal bud, and leaf.

solidifies at about 30°C; used in soaps and pet foods, and as a lubricant.

sardius *See* sard.

sardonyx [MINERAL] An onyx characterized by parallel layers of sard, a deep orange-red variety of chalcedony, and a mineral of different color.

Sargasso Sea [GEOGR] A region of the North Atlantic Ocean; boundaries are defined in the west and north by the Gulf Stream, in the east by longitude 40°W, and in the south by latitude 20°N.

Sargassum [BOT] A genus of brown algae characterized by branching thalli with lateral outgrowths; they develop along tropical shores, then break away to drift in the ocean.

Sargent curve [NUC PHYS] A graph of logarithms of decay constants of radioisotopes subject to beta-decay against logarithms of the corresponding maximum beta-particle energies; most of the points fall on two straight lines.

Sargent cycle [THERMO] An ideal thermodynamic cycle consisting of four reversible processes: adiabatic compression, heating at constant volume, adiabatic expansion, and isobaric cooling.

sarking [BUILD] A layer of boards or bituminous felt placed beneath tiles or other roofing to provide thermal insulation or to prevent ingress of water.

sarkinite [MINERAL] $Mn_2(AsO_4)(OH)$ A flesh-red monoclinic mineral composed of hydrous manganese arsenate, occurring in crystals.

sarkomycin [MICROBIO] $C_7H_8O_3$ An antibiotic produced by an actinomycete which acts as a carcinolytic agent.

Sarmatian [GEOL] A European stage of geologic time: the upper Miocene, above Tortonian, below Pontian.

sarmentocymarin [BIOCHEM] A cardioactive, steroid glycoside from the seeds of *Strophanthus sarmentosus;* on hydrolysis it yields sarmentogenin and sarmentose.

sarmentogenin [BIOCHEM] $C_{23}H_{34}O_5$ The steroid aglycon of sarmentocymarin; isometric with digitoxigenin, and characterized by a hydroxyl group at carbon number 11.

sarmentose [BOT] Producing slender, prostrate stems or runners.

sarmientite [MINERAL] $Fe_2(AsO_4)(SO_4)(OH)\cdot5H_2O$ A yellow mineral composed of basic hydrous arsenate and sulfate of iron; it is isomorphous with diadochite.

saros [ASTRON] A cycle of time after which the centers of the sun and moon, and the nodes of the moon's orbit return to the same relative position; this period is 18 years 11⅓ days, or 18 years 10⅓ days if 5 rather than 4 leap years are included.

Sarothriidae [INV ZOO] The equivalent name for Jacobsoniidae.

Sarraceniaceae [BOT] A small family of dicotyledonous plants in the order Sarraceniales in which leaves are modified to form pitchers, placentation is axile, and flowers are perfect with distinct filaments.

Sarraceniales [BOT] An order of dicotyledonous herbs or shrubs in the subclass Dilleniidae; plants characteristically have alternate, simple leaves that are modified for catching insects, and grow in waterlogged soils.

sarsaparilla [BOT] Any of various tropical American vines of the genus *Smilax* (family Liliaceae) found in dense, moist jungles; a flavoring material used in medicine and soft drinks is obtained from the dried roots of at least four species.

sartorite [MINERAL] $PbAs_2S_4$ A dark-gray monoclinic mineral, occurring in crystalline form.

sartorius [ANAT] A large muscle originating in the anterior superior iliac spine and inserting in the tibia; flexes the hip and knee joints, and rotates the femur laterally.

SAS *See* stability augmentation system.

sash [BUILD] A frame for window glass.

sash bars [BUILD] Strips of wood which separate the panes in a window composed of several panes. Also known as muntins.

sash cord [BUILD] A cord or chain used to attach a counterweight to the window sash.

Sa spiral [ASTRON] A class of spiral galaxy, including those galaxies that have the largest center sections and closely wound galactic arms.

sassafras [BOT] *Sassafras albidum.* A medium-sized tree of the order Magnoliales recognized by the bright-green color and aromatic odor of the leaves and twigs.

sassafras oil *See* oil of sassafras.

sassoline *See* sassolite.

sassolite [MINERAL] H_3BO_3 A white or gray mineral consisting of native boric acid usually occurring in small pearly scales as an incrustation or as tabular triclinic crystals. Also known as sassoline.

sastruga [HYD] A ridge of snow up to 2 inches (5 centimeters) high formed by wind erosion and aligned parallel to the wind. Also known as skavl; zastruga.

satchel charge [ORD] A number of blocks of explosive taped to a board fitted with a rope or wire loop for carrying and attaching.

satellite [AERO ENG] *See* artificial satellite. [ASTRON] A small, solid body moving in an orbit around a planet; the moon is a satellite of earth. [CYTOL] A chromosome segment distant from but attached to the rest of the chromosome by an achromatic filament.

satellite and missile surveillance [ENG] The systematic observation of aerospace for the purpose of detecting, tracking, and characterizing objects, events, and phenomena associated with satellites and inflight missiles, friendly and enemy.

satellite antenna [ELECTROMAG] Antenna on an artificial satellite to receive command signals from earth, act as a beacon for tracking, or transmit scientific or other data to earth.

satellite band [MOL BIO] A fraction of the deoxyribonucleic acid (DNA) of an organism which has a different density from the rest and is therefore separable as a band in density gradient centrifugation; these bands are usually made up of highly repetitive sequences of DNA.

satellite cell [HISTOL] One of the neurilemmal cells surrounding nerve cells in the peripheral nervous system.

satellite communication [COMMUN] Use of communication satellites, passive reflecting belts of dipoles or needles, or reflecting orbiting balloons to extend the range of radio communication by returning signals to earth from the orbiting object, with or without amplification.

satellite computer [ADP] A computer which, under control of the main computer, handles the input and output routines, thereby allowing the main computer to be fully dedicated to computations.

satellite infrared spectrometer [SPECT] A spectrometer carried aboard satellites in the Nimbus series which measures the radiation from carbon dioxide in the atmosphere at several different wavelengths in the infrared region, giving the vertical temperature structure of the atmosphere over a large part of the earth. Abbreviated SIRS.

satellite reconnaissance [ORD] Strategic reconnaissance that is conducted by means of data obtained from a satellite.

satellite tracking [AERO ENG] Determination of the positions and velocities of satellites through radio and optical means.

satellitosis [MED] A condition, associated with inflammatory and degenerative diseases of the central nervous system, in which satellite cells increase around the nerve cells.

satelloid [AERO ENG] A vehicle that revolves about the earth or other celestial body, but at such altitudes as to require sustaining thrust to balance drag.

sathrolith *See* saprolite.

satin [TEXT] A closely woven fabric with a glossy face and a dull-finish back made by carrying the warp (or filling) uninterruptedly on the surface over many filling (or warp) yarns; made of silk, polyester, or other fibers.

satin finish [MET] A finish involving soft scratch-brushing of polished metal surfaces to produce a soft sheen. Also known as Butler finish; scratch-brush finish.

satin ice *See* acicular ice.

satin spar [MINERAL] A white, translucent, fine fibrous variety of gypsum having a silky luster. Also known as satin stone.

satin stone *See* satin spar.

saturable-core magnetometer [ENG] A magnetometer that depends for its operation on the changes in permeability of a ferromagnetic core as a function of the magnetic field to be measured.

saturable-core reactor *See* saturable reactor.

saturable reactor [ELECTROMAG] An iron-core reactor having an additional control winding that carries direct current whose value is adjusted to change the degree of saturation of the core, thereby changing the reactance that the alternating-current winding offers to the flow of alternating current; with appropriate external circuits, a saturable reactor can serve as a magnetic amplifier. Also known as saturable-core reactor; transductor.

saturable transformer [ELECTROMAG] A saturable reactor having additional windings to provide voltage transformation or isolation from the alternating-current supply.

saturated activity [NUCLEO] The maximum activity obtainable by activation in a definite flux in a nuclear reactor.

saturated air [METEOROL] Moist air in a state of equilibrium with a plane surface of pure water or ice at the same temperature and pressure; that is, air whose vapor pressure is the saturation vapor pressure and whose relative humidity is 100%.

saturated ammonia [CHEM] **1.** Liquid ammonia in a state in which adding heat at constant pressure causes the liquid to vaporize at constant temperature, and in which removing heat at constant pressure causes the temperature of the liquid to drop immediately. **2.** Ammonia vapor in a state in which adding heat at constant pressure causes an immediate temperature rise (superheating) and in which removing heat at constant pressure starts immediate condensation at constant temperature.

saturated color [OPTICS] A pure color not contaminated by white.

saturated compound [ORG CHEM] An organic compound with all carbon bonds satisfied; it does not contain double or triple bonds and thus cannot add elements or compounds.

saturated diode [ELECTR] A diode that is passing the maximum possible current, so further increases in applied voltage have no effect on current.

saturated hydrocarbon [ORG CHEM] A saturated carbon-hydrogen compound with all carbon bonds filled; that is, there are no double or triple bonds as in olefins and acetylenics.

saturated liquid [CHEM] A solution that contains enough of a dissolved solid, liquid, or gas so that no more will dissolve into the solution at a given temperature and pressure.

saturated mineral [MINERAL] A mineral that forms in the presence of free silica.

saturated rock [PETR] An igneous rock composed principally of saturated minerals.

saturated surface *See* water table.

saturated vapor [THERMO] A vapor whose temperature equals the temperature of boiling at the pressure existing on it.

saturated zone *See* zone of saturation.

saturating felt [MATER] A blanket made of felt, intended for impregnation with a waterproofing material such as asphalt.

saturating signal [ELECTR] In radar, a signal of an amplitude greater than the dynamic range of the receiving system.

saturation [ELECTR] **1.** The condition that occurs when a transistor is driven so that it becomes biased in the forward direction (the collector becomes positive with respect to the base, for example, in a *pnp* type of transistor). **2.** *See* anode saturation. **3.** *See* temperature saturation. [ELECTROMAG] *See* magnetic saturation. [NUCLEO] **1.** The condition in which the decay rate of a given radionuclide is equal to its rate of production in an induced nuclear reaction. **2.** The condition in which the voltage applied to an ionization chamber is high enough to collect all the ions formed by radiation but not high enough to produce ionization by collision. [OPTICS] *See* color saturation. [ORD] The striking of a target area with such numbers of missiles that no place in it remains untouched by destruction. [PHYS] The condition in which a further increase in some cause produces no further increase in the resultant effect.

saturation adiabat [METEOROL] On a thermodynamic diagram, a line of constant wet-bulb potential temperatures; in practice, approximate computations are usually employed, and the resulting lines represent, ambiguously, saturation adiabats and pseudoadiabats. Also known as moist adiabat; wet adiabat.

saturation-adiabatic lapse rate [METEOROL] A special case of process lapse rate, defined as the rate of decrease of temperature with height of an air parcel lifted in a saturation-adiabatic process through an atmosphere in hydrostatic equilibrium. Also known as moist-adiabatic lapse rate.

saturation-adiabatic process [METEOROL] An adiabatic process in which the air is maintained at saturation by the evaporation or condensation of water substance, the latent heat being supplied by or to the air respectively; the ascent of cloudy air, for example, is often assumed to be such a process.

saturation bombing [ORD] Intense area bombing intended to leave no place in a given area free from destructive effects; it may be achieved by dropping many small bombs, or by dropping a medium number of large bombs, or by dropping a single massive bomb.

saturation current [ELECTR] **1.** In general, the maximum current which can be obtained under certain conditions. **2.** In a vacuum tube, the space-charge-limited current, such that further increase in filament temperature produces no specific increase in anode current. **3.** In a vacuum tube, the temperature-limited current, such that a further increase in anode-cathode potential difference produces only a relatively small increase in current. **4.** In a gaseous-discharge device, the maximum current which can be obtained for a given mode of discharge. **5.** In a semiconductor, the maximum current which just precedes a change in conduction mode. [NUCLEO] The ionization current in a gas tube when the applied potential is large enough to collect all ions produced by ionizing radiation.

saturation deficit [METEOROL] **1.** The difference between the actual vapor pressure and the saturation vapor pressure at the existing temperature. **2.** The additional amount of water vapor needed to produce saturation at the current temperature and pressure, expressed in grams per cubic meter. Also known as vapor-pressure deficit.

saturation diving [PHYSIO] Diving in which the tissues exposed to high pressure at great ocean depths for 24 hours become saturated with gases, especially inert gases, thereby reaching a new equilibrium state.

saturation flux density *See* saturation induction.

saturation induction [ELECTROMAG] The maximum intrinsic induction possible in a material. Also known as saturation flux density.

saturation limiting [ELECTR] Limiting the minimum output voltage of a vacuum-tube circuit by operating the tube in the region of plate-current saturation (not to be confused with emission saturation).

saturation mixing ratio [METEOROL] A thermodynamic function of state; the value of the mixing ratio of saturated air at the given temperature and pressure; this value may be read directly from a thermodynamic diagram.

saturation of forces [PHYS] Property exhibited by forces between particles wherein each particle can interact strongly with only a limited number of other particles, as in the forces between atoms in a molecule, and between nucleons in a nucleus.

saturation ratio [METEOROL] The ratio of the actual specific humidity to the specific humidity of saturated air at the same temperature.

saturation scale [OPTICS] A series of colors which appear to have equal differences in color saturation.

saturation signal [ELECTROMAG] A radio signal (or radar echo) which exceeds a certain power level fixed by the design of the receiver equipment; when a receiver or indicator is "saturated," the limit of its power output has been reached.

saturation specific humidity [THERMO] A thermodynamic function of state; the value of the specific humidity of saturated air at the given temperature and pressure.

saturation spectroscopy [SPECT] A branch of spectroscopy in which the intense, monochromatic beam produced by a laser is used to alter the energy-level populations of a resonant medium over a narrow range of particle velocities, giving rise to extremely narrow spectral lines that are free from Doppler broadening; used to study atomic, molecular, and nuclear structure, and to establish accurate values for fundamental physical constants.

saturation vapor pressure [THERMO] The vapor pressure of a thermodynamic system, at a given temperature, wherein the vapor of a substance is in equilibrium with a plane surface of that substance's pure liquid or solid phase.

saturator [ENG] A device, equipment, or person that saturates one material with another; examples are a tank in which vapors become saturated with ammonia from coal (in carbon-

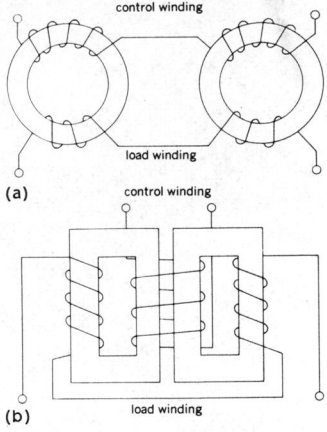

SATURABLE REACTOR

(a) control winding / load winding

(b) control winding / load winding

Typical construction of saturable reactors. *(a)* Two separate two-legged cores. *(b)* Three-legged core formed by placing two two-legged cores together.

ization of coal), a humidifier, and the operator of a machine for impregnating roofing felt with asphalt.

Saturn [AERO ENG] One of the very large launch vehicles built primarily for the Apollo program; begun by Army Ordnance but turned over to the National Aeronautics and Space Administration for the manned space flight program to the moon. [ASTRON] The second largest planet in the solar system (mass is 95.3 compared to earth's 1) and the sixth in the order of distance to the sun; it is visible to the naked eye as a yellowish first-magnitude star except during short periods near its conjunction with the sun; it is surrounded by a series of rings.

Saturniidae [INV ZOO] A family of medium- to large-sized moths in the superfamily Saturnioidea including the giant silkworms, characterized by reduced, often vestigial, mouthparts and strongly bipectinate antennae.

Saturnioidea [INV ZOO] A superfamily of medium- to very-large-sized moths in the suborder Heteroneura having the frenulum reduced or absent, reduced mouthparts, no tympanum, and pectinate antennae.

Satyrinae [INV ZOO] A large, cosmopolitan subfamily of lepidopterans in the family Nymphalidae, containing the wood nymphs, meadow browns, graylings, and arctics, characterized by bladderlike swellings at the bases of the forewing veins.

sauconite [MINERAL] The zinc-bearing end member of the montmorillonite group; a trioctahedral clay mineral.

Saunders air-lift pump [MECH ENG] A device for raising water from a well by the introduction of compressed air below the water level in the well.

Sauria [VERT ZOO] The lizards, a suborder of the Squamata, characterized generally by two or four limbs but sometimes none, movable eyelids, external ear openings, and a pectoral girdle.

Saurichthyidae [PALEON] A family of extinct chondrostean fishes bearing a superficial resemblance to the Aspidorhynchiformes.

Saurischia [PALEON] The lizard-hipped dinosaurs, an order of extinct reptiles in the subclass Archosauria characterized by an unspecialized, three-pronged pelvis.

Sauropoda [PALEON] A group of fully quadrupedal, seemingly herbivorous dinosaurs from the Jurassic and Cretaceous periods in the suborder Sauropodomorpha; members had small heads, spoon-shaped teeth, long necks and tails, and columnar legs.

Sauropodomorpha [PALEON] A suborder of extinct reptiles in the order Saurischia, including large, solid-limbed forms.

Sauropterygia [PALEON] An order of Mesozoic marine reptiles in the subclass Euryapsida.

Saururaceae [BOT] A family of dicotyledonous plants in the order Piperales distinguished by mostly alternate leaves, two to ten ovules per carpel, and carpels distinct or united into a compound ovary.

sausage instability *See* kink instability.

saussurite [MINERAL] A white or grayish, tough, compact mineral aggregate composed chiefly of a mixture of albite or oligoclase and zoisite or epidote.

saussuritization [GEOL] A metamorphic process involving replacement of plagioclase in basalts and gabbros by a fine-grained aggregate of zoisite, epidote, albite, calcite, sericite, and zeolites.

sauterelle [ENG] A device used by masons for tracing and forming angles.

Savage principle [MATH] A technique used in decision theory; a criterion is used to construct a regret matrix in which each outcome entry represents a regret defined as the difference between best possible outcome and the given outcome; the matrix is then used as in decision making under risk with expected regret as the decision-determining quality. Also known as regret criterion.

savane armée *See* thornbush.

savane épineuse *See* thornbush.

savanna [ECOL] Any of a variety of physiognomically or environmentally similar vegetation types in tropical and extratropical regions; all contain grasses and one or more species of trees of the families Leguminosae, Bombacaceae, Bignoniaceae, or Dilleniaceae.

savanna climate *See* tropical savanna climate.

SAWTOOTH WAVEFORM

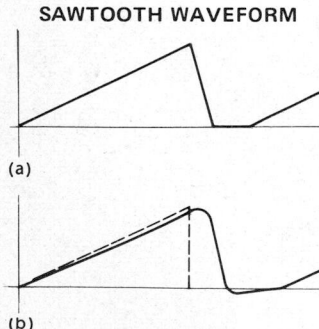

(a)

(b)

Sawtooth waveforms.
(a) An ideal linear sawtooth.
(b) Approximate sawtooth generated by actual circuits.

savanna-woodland *See* tropical woodland.

savart [ACOUS] A unit of pitch interval, such that the interval between two frequencies measured in savarts is equal to 1000 times the common logarithm of the ratio of the frequencies; one octave equals approximately 301.030 savarts.

Savart plate [OPTICS] A device consisting of a pair of calcite plates having the same thickness, cut along the natural cleavage faces, and mounted with corresponding faces perpendicular to each other; used to detect polarization of light by means of interference fringes.

Savart polariscope [OPTICS] A polariscope consisting of a specially constructed double-plate polarizer and a tourmaline plate analyzer; polarized light passing through the instrument is indicated by the presence of parallel colored fringes, while unpolarized light results in a uniform field.

savin oil [MATER] Pale-yellow, volatile oil, soluble in alcohol; derived from the fresh tops of the savin (*Juniperus sabina*); used in medicine.

Savonius rotor [MECH ENG] A rotor composed of two offset semicylindrical elements rotating about a vertical axis.

Savonius windmill [MECH ENG] A windmill composed of two semicylindrical offset cups rotating about a vertical axis.

saw [DES ENG] Any of various tools consisting of a thin, usually steel, blade with continuous cutting teeth on the edge.

saw bit [DES ENG] A bit having a cutting edge formed by teeth shaped like those in a handsaw.

sawdust [MATER] Wood fragments made by a saw in cutting.

sawdust concrete [MATER] Concrete containing sawdust as the principal aggregate.

sawfish [VERT ZOO] Any of several elongate viviparous fishes of the family Pristidae distinguished by a dorsoventrally flattened elongated snout with stout toothlike projections along each edge.

saw gumming [MECH ENG] Grinding away the punch marks in the spaces between the teeth in saw manufacture.

sawhorse [ENG] A wooden rack used to support wood that is being sawed.

sawing [ENG] Cutting with a saw.

sawmill [IND ENG] A plant housing sawing machines. [MECH ENG] A machine for cutting logs with a saw or a series of saws.

sawtooth barrel *See* basket.

sawtooth blasting [MIN ENG] The cutting of a series of slabs which, in plan, resemble sawteeth by blasting oblique horizontal holes along a face.

sawtooth crusher [MECH ENG] Solids crusher in which feed is broken down between two sawtoothed shafts rotating at different speeds.

sawtooth floor channeling [MIN ENG] A method of channeling inclined beds of marble by removing right-angle blocks in succession from the various beds, thus giving the floor a zigzag or sawtooth appearance.

sawtooth generator [ELECTR] A generator whose output voltage has a sawtooth waveform; used to produce sweep voltages for cathode-ray tubes.

sawtooth modulated jamming [ELECTR] Electronic countermeasure technique when a high level jamming signal is transmitted, thus causing large automatic gain control voltages to be developed at the radar receiver that, in turn, cause target pip and receiver noise to completely disappear.

sawtooth pulse [ELECTR] An electric pulse having a linear rise and a virtually instantaneous fall, or conversely, a virtually instantaneous rise and a linear fall.

sawtooth roof [ARCH] A roof form having a succession of monitors in sawtooth shape.

sawtooth stoping [MIN ENG] In the United States, overhand stoping in which the line of advance is up the dip, and benches are advanced in a line parallel with the drift.

sawtooth waveform [ELECTR] A waveform characterized by a slow rise time and a sharp fall, resembling a tooth of a saw.

sax [DES ENG] A tool for chopping away the edges of roof slates; it has a pick at one end for making nail holes.

saxicolous [ECOL] Living or growing among rocks.

Saxifragaceae [BOT] A family of dicotyledonous plants in the order Rosales which are scarcely or not at all succulent and have two to five carpels usually more or less united, and leaves not modified into pitchers.

saxophone [ELECTROMAG] Vertex-fed linear array antenna giving a cosecant-squared radiation pattern.

Saybolt chromometer [OPTICS] Device used to measure the color of undyed gasolines, jet fuels, naphthas, kerosines, petroleum waxes, and pharmaceutical white oils.

Saybolt color [ENG] A color standard for petroleum products determined with a Saybolt chromometer.

Saybolt Furol viscosimeter [ENG] An instrument for measuring viscosity of very thick fluids, for example, heavy oils; similar to the Saybolt Universal viscosimeter, but with a larger-diameter tube so that the efflux time is about one-tenth that of the Universal instrument.

Saybolt Furol viscosity [FL MECH] The time in seconds for 60 milliliters of fluid to flow through a capillary tube in a Saybolt Furol viscosimeter at specified temperatures between 70 and 210°F (21 and 99°C); used for high-viscosity petroleum oils, such as transmission and gear oils, and heavy fuel oils.

Saybolt Seconds Universal [FL MECH] A unit of measurement for Saybolt Universal viscosity. Abbreviated SSU.

Saybolt Universal viscosimeter [ENG] An instrument for measuring viscosity by the time it takes a fluid to flow through a calibrated tube; used for the lighter petroleum products and lubricating oils.

Saybolt Universal viscosity [FL MECH] The time in seconds for 60 milliliters of fluid to flow through a capillary tube in a Saybolt Universal viscosimeter at a given temperature.

Sayelle [TEXT] Trade name for acrylic fiber made by DuPont.

sb *See* stilb.

Sb *See* antimony.

S band [COMMUN] A band of radio frequencies extending from 1550 to 5200 megahertz, corresponding to wavelengths of 19.37 to 5.77 centimeters.

S-band hiran *See* shiran.

SBK catalytic reforming [CHEM ENG] A proprietary petroleum refinery process using SBK (Sinclair-Baker-Kellogg) regenerable platinum catalyst; used for catalytic reforming of gasoline-range hydrocarbons to produce high octanes.

SBR *See* styrene-butadiene rubber.

Sb spiral galaxy [ASTRON] A class of spiral galaxy characterized by smaller central bodies and more open, larger arms.

Sc *See* scandium.

SC *See* sectional center.

scab [MED] Crusty exudate covering a wound or ulcer during the healing process. [MET] A defect consisting of a flat, partially detached piece of metal joined to the surface of a casting or piece of rolled metal.

scabbard [ORD] A sheath with an open top; it is usually made of leather or canvas and is designed to protect edged weapons, rifles, carbines, and submachine guns from the elements and rough usage.

scabbing [ORD] Breaking off of fragments in the inside of a wall of hard material due to the impact or explosion of a projectile on the outside.

scabies [MED] A contagious skin disorder caused by the mite *Sarcoptes scabiei* burrowing beneath the skin, causing the formation of multiform lesions with intense itching.

scabland [GEOL] Elevated land that is essentially flat-lying and covered with basalt and has only a thin soil cover, sparse vegetation, and usually deep, dry channels.

scabrous [BIOL] Having a rough surface covered with stiff hairs or scales.

scacchite [MINERAL] $MnCl_2$ A mineral composed of native manganese chloride, found in volcanic regions.

scaffold [CIV ENG] A temporary or movable platform supported on the ground or suspended; used for working at considerable heights above the ground.

scaglia [GEOL] A dark, very-fine-grained, somewhat calcareous shale usually developed in the Upper Cretaceous and Lower Tertiary periods of the northern Apennines.

scala media [ANAT] The middle channel of the cochlea, filled with endolymph and bounded above by Reissner's membrane and below by the basilar membrane. Also known as cochlear duct.

scalar [MATH] One of the algebraic quantities which form a field, usually the real or complex numbers, by which the vectors of a vector space are multiplied. [PHYS] 1. A quantity which has magnitude only and no direction, in contrast to a vector. 2. A quantity which has magnitude only, and has the same value in every coordinate system.

scalar field [MATH] 1. The field consisting of the scalars of a vector space. 2. A function on a vector space into the scalars of the vector space. [PHYS] A field which is characterized by a function of position and time whose value at each point is a scalar.

scalar function [MATH] A function from a vector space to its scalar field. [PHYS] A function of position and time whose value at each point is a scalar.

scalar gradient [MATH] The gradient of a function.

scalariform [BIOL] Resembling a ladder; having transverse markings or bars.

scalar meson [PARTIC PHYS] A meson which has spin 0 and positive parity, and may be described by a scalar field.

scalar multiplication [MATH] The multiplication of a vector from a vector space by a scalar from the associated field; this usually contracts or expands the length of a vector.

scalar potential [PHYS] A scalar function whose negative gradient is equal to some vector field, at least when this field is time-independent; for example, the potential energy of a particle in a conservative force field, and the electrostatic potential.

scalar product *See* inner product.

scalar triple product [MATH] The scalar triple product of vectors v_1, v_2, and v_3 from euclidean three-dimensional space determines the volume of the parallelepiped with these vectors as edges; it is given by the determinant of the 3×3 matrix whose rows are the components of v_1, v_2, and v_3.

scala tympani [ANAT] The lowest channel in the cochlea of the ear; filled with perilymph.

scala vestibuli [ANAT] The uppermost channel of the cochlea; filled with perilymph.

scale [ACOUS] A series of musical notes arranged from low to high by a specified scheme of intervals suitable for musical purposes. [BOT] The bract of a catkin. [ENG] 1. A series of markings used for reading the value of a quantity or setting. 2. To change the magnitude of a variable in a uniform way, as by multiplying or dividing by a constant factor, or the ratio of the real thing's magnitude to the magnitude of the model or analog of the model. 3. A weighing device. 4. A ruler or other measuring stick. An indication of represented to actual distances on a map, chart or drawing. [MET] A thick metallic oxide coating formed usually by heating metals in air. [PHYS] 1. A one-to-one correspondence between numbers and the value of some physical quantity, such as the centigrade or Kelvin temperature scales on the API or Baumé scales of specific gravity. 2. To determine a quantity at some order of magnitude by using data or relationships which are known to be valid at other (usually lower) orders of magnitude. [VERT ZOO] A flat calcified or cornified platelike structure on the skin of most fishes and of some tetrapods.

scaleboard [MATER] Thin sheets of wood used as veneer.

scale drawing [GRAPHICS] A drawing of an object or structure showing all parts in the same proportion of their true size.

scale effect [AERO ENG] The necessary corrections applied to measurements of a model in a wind tunnel to ascertain corresponding values for a full-sized object. [FL MECH] An effect in fluid flow that results from changing the scale, but not the shape, of a body around which the flow passes; this effect is relevant to wind tunnel experiments.

scale factor [ENG] The factor by which the reading of an instrument or the solution of a problem should be multiplied to give the true final value when a corresponding scale factor is used initially to bring the magnitude within the range of the instrument or computer.

scale height [GEOPHYS] A measure of the thickness of an ionized layer in the upper atmosphere, using the equation $H = kT/mg = R^*T/Mg$, where k is the Boltzmann constant equal to 1.3804×10^{-16} erg/degree, T the absolute temperature, m and M the mean molecular mass and mean grammolecular weight respectively of the layer, g the acceleration of gravity, and R^* the universal gas constant.

scale insect [INV ZOO] Any of various small, structurally degenerate homopteran insects in the superfamily Coccoidea which resemble scales on the surface of a host plant.

scale model [GRAPHICS] A three-dimensional representation of an object or structure having all parts in the same proportion of their true size.

scalene triangle [MATH] A triangle where no two angles are equal.

scalenus [ANAT] One of three muscles in the neck, arising from the transverse processes of the cervical vertebrae, and inserted on the first two ribs.

scale-of-ten circuit *See* decade scaler.

scale-of-two circuit *See* binary scaler.

scaler [ELECTR] A circuit that produces an output pulse when a prescribed number of input pulses is received. Also known as counter; scaling circuit.

scale scar [BOT] A mark left on a stem after bud scales have fallen off.

scales of motion [OCEANOGR] A series of increasing characteristic magnitudes of motion, ranging from tiny eddies of turbulence to oceanwide currents, each member of the series interacting with the adjacent members.

scale-up [DES ENG] Design process in which the data of an experimental-scale operation (model or pilot plant) is used for the design of a large (scaled-up) unit, usually of commercial size.

scale wax [MATER] The paraffin wax derived by sweating the greater part of the oil from slack wax; contains up to 6% oil. Also known as crude scale; paraffin scale.

Scalibregmidae [INV ZOO] A family of mud-swallowing worms belonging to the Sedentaria and found chiefly in sublittoral and great depths.

scaling [BIOL] Removing scales from fishes. [ELECTR] Counting pulses with a scaler when the pulses occur too fast for direct counting by conventional means. [ENG] Removing scale (rust or salt) from a metal or other surface. [GRAPHICS] Using a scale to measure dimensions in a scale drawing. [MECH] Expressing the terms in an equation of motion in powers of nondimensional quantities (such as a Reynolds number), so that terms of significant magnitude under conditions specified in the problem can be identified, and terms of insignificant magnitude can be dropped. [MET] **1.** Forming of a thick layer of metallic oxide on metals at high temperatures. **2.** Depositing of solid inorganic solutes from a water on a metal surface, such as a cooling tube or boiler. [MIN ENG] Removing loose rocks and coal from the roof, walls, or face after blasting.

scaling circuit *See* scaler.

scaling factor [ELECTR] The number of input pulses per output pulse of a scaling circuit. Also known as scaling ratio. [ENG] Factor used in heat-exchange calculations to allow for the loss in heat conductivity of a material because of the development of surface scale, as inside pipelines and heat-exchanger tubes. [PHYS] A constant of proportionality which appears in a scaling law.

scaling law [PHYS] A law, stating that two quantities are proportional, which is known to be valid at certain orders of magnitude and is used to calculate the value of one of the quantities at another order of magnitude.

scaling ratio [ELECTR] *See* scaling factor. [ENG] The ratio of a certain property of a laboratory model to the same property in the natural prototype.

scallion *See* shallot.

scallop [INV ZOO] Any of various bivalve mollusks in the family Pectinidae distinguished by radially ribbed valves with undulated margins.

scalp [MET] To remove surface layers, and thereby defects, from ingots, billets, or slabs by machining. [MIN ENG] To remove undesirable fine material from broken ore, stone, or gravel.

scalped anticline *See* breached anticline.

scalped extrusion ingot [METEOROL] A cast, solid, or hollow extrusion ingot which has been machined to remove outside surface layers.

scalpel [DES ENG] A small, straight, very sharp knife (or detachable blade for a knife), used for dissecting.

Scalpellidae [INV ZOO] A primitive family of barnacles in the suborder Lepadomorpha having more than five plates.

scaly bark *See* psorosis.

scaly leg [VET MED] A highly contagious disease of poultry caused by the sarcoptid mite *Knemidokoptes mutans*.

scan [ENG] **1.** To examine an area, a region in space, or a

portion of the radio spectrum point by point in an ordered sequence; for example, conversion of a scene or image to an electric signal or use of radar to monitor an airspace for detection, navigation, or traffic control purposes. **2.** One complete circular, up-and-down, or left-to-right sweep of the radar, light, or other beam or device used in making a scan.

scan converter [ELECTR] Equipment that converts radar data images at a 3 kilohertz to 10 kilohertz sampling rate that can be sent over telephone line or narrow bandwidth radio circuits and converted into a slow-scan image, through a similar converter, at the receiving end.

scandent [BOT] Climbing by stem-roots or tendrils.

scandia *See* scandium oxide.

scandium [CHEM] A metallic group III element, symbol Sc, atomic number 21; melts at 1200°C; found associated with rare-earth elements.

scandium halide [INORG CHEM] A compound of scandium and a halogen; for example, scandium chloride, $ScCl_3$.

scandium oxide [INORG CHEM] Sc_2O_3 White powder, soluble in hot acids; used to prepare scandium. Also known as scandia.

scandium sulfate [INORG CHEM] $Sc_2(SO_4)_3$ Water-soluble, colorless crystals.

scandium sulfide [INORG CHEM] Sc_3S_3 Yellowish powder; decomposes in dilute acids and boiling water to give off hydrogen sulfide.

scanistor [ELECTR] Integrated semiconductor optical-scanning device that converts images into electrical signals; the output analog signal represents both amount and position of light shining on its surface.

scanner [ADP] In character recognition, a magnetic or photoelectric device which converts the input character into corresponding electric signals for processing by electronic apparatus. [COMMUN] That part of a facsimile transmitter which systematically translates the densities of the elemental areas of the subject copy into corresponding electric signals. [ELECTR] The motion, usually periodic, given to the major lobe of an antenna; the process of directing the radio-frequency beam successively over all points in a given region of space. [ENG] **1.** Any device that examines an area or region point by point in a continuous systematic manner, repeatedly sweeping across until the entire area or region is covered; for example, a flying-spot scanner. **2.** A device that automatically samples, measures, or checks a number of quantities or conditions in sequence, as in process control.

scanner selector [ADP] An electronic device interfacing computer and multiplexers when more than one multiplexer is used.

scanning circuit *See* sweep circuit.

scanning electron microscope [ELECTR] A type of electron microscope in which a beam of electrons, a few hundred angstroms in diameter, systematically sweeps over the specimen; the intensity of secondary electrons generated at the point of impact of the beam on the specimen is measured, and the resulting signal is fed into a cathode-ray-tube display which is scanned in synchronism with the scanning of the specimen.

scanning frequency *See* stroke speed.

scanning head [ELECTR] Light source and phototube combined as a single unit for scanning a moving strip of paper, cloth, or metal in photoelectric side-register control systems.

scanning line [COMMUN] **1.** In television, a single, continuous, narrow strip which is determined by the process of scanning. **2.** Path traced by the scanning or recording spot in one sweep across the subject copy or record sheet.

scanning linearity [ELECTR] In television, the uniformity of scanning speed during the trace interval.

scanning line frequency *See* stroke speed.

scanning loss [ELECTROMAG] In a radar system employing a scanning antenna, the reduction in sensitivity (usually expressed in decibels) due to scanning across the target, compared with that obtained when the beam is directed constantly at the target.

scanning sequence [ENG] The order in which the points in a region are scanned; for example, in television the picture is scanned horizontally from left to right and vertically from top to bottom.

scanning sonar [ENG] Sonar in which all targets of interest

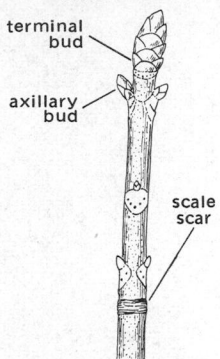

SCALE SCAR

terminal bud

axillary bud

scale scar

Position of scale scar on the twig of a buckeye tree.

SCALLOP

Typical ribbed valve of a scallop.

SCANDIUM

21
Sc

Periodic table of the chemical elements showing the position of scandium.

are shown simultaneously, as on a radar PPI (plan position indicator) display or sector display; the sound pulse may be transmitted in all directions simultaneously and picked up by a rotating receiving transducer, or transmitted and received in only one direction at a time by a scanning transducer.

scanning speed *See* spot speed.

scanning spot *See* picture element.

scanning switch *See* commutator switch.

scanning yoke *See* deflection yoke.

scansorial [BOT] Adapted for climbing.

scantlings [NAV ARCH] The dimensions and material thicknesses of frames, shell plating, deck plating, and other structures of a ship, together with the suitability of the means for protecting openings and making them sufficiently watertight or weathertight.

Scapanorhychidae [VERT ZOO] The goblin sharks, a family of deep-sea galeoids in the isurid line having long, sharp teeth and a long, pointed rostrum.

scapha [ANAT] The furrow of the auricle between the helix and the antihelix.

Scaphidiidae [INV ZOO] The shining fungus beetles, a family of coleopteran insects in the superfamily Staphylinoidea.

scaphocephaly [MED] A condition of the skull characterized by elongation and narrowing, and a projecting, keellike sagittal suture, caused by its premature closure.

scaphoid [ANAT] A boat-shaped bone of the carpus.

Scaphopoda [INV ZOO] A class of the phylum Mollusca in which the soft body fits the external, curved and tapering, nonchambered, aragonitic shell which is open at both ends.

scapolite [MINERAL] A white, gray, or pale-green complex aluminosilicate of sodium and calcium belonging to the tectosilicate group of silicate minerals; crystallizes in the tetragonal system and is vitreous; hardness is 5–6 on Mohs scale, and specific gravity is 2.65–2.74. Also known as wernerite.

scapolitization [GEOL] Introduction of or replacement by scapolite.

scapula [ANAT] The large, flat, triangular bone forming the back of the shoulder. Also known as shoulder blade.

scapulet [INV ZOO] In some medusae, fringed outgrowths on the outer surfaces of the arms near the bell.

scapulus [INV ZOO] A modified submarginal region in some sea anemones.

scapus [BIOL] The stem, shaft, or column of a structure.

scar [GEOL] **1.** A steep, rocky eminence, such as a cliff or precipice, where bare rock is well exposed. Also known as scaur; scaw. **2.** *See* shore platform. [MED] A permanent mark on the skin or other tissue, formed from connective-tissue replacement of tissue destroyed by a wound or disease process.

Scarabaeidae [INV ZOO] The lamellicorn beetles, a large cosmopolitan family of coleopteran insects in the superfamily Scarabaeoidea including the Japanese beetle and other agricultural pests.

Scarabaeoidea [INV ZOO] A superfamily of Coleoptera belonging to the suborder Polyphaga.

scarabiasis [MED] Invasion of the intestine by the dung beetle, characterized by anorexia, emaciation, and disturbance of the gastrointestinal tract.

scarfing [MET] Cutting away of surface defects on metals by use of a gas torch.

scarf joint [DES ENG] A joint made by the cutting of overlapping mating parts so that the joint is not enlarged and the patterns are complementary, and securing them by glue, fasteners, welding, or other joining method.

Scaridae [VERT ZOO] The parrotfishes, a family of perciform fishes in the suborder Percoidei which have the teeth of the jaw generally coalescent.

scarification [MED] The operation of making numerous small, superficial incisions in skin or other tissue.

scarifier [ENG] An implement or machine with downward projecting tines for breaking down a road surface 2 feet (60 centimeters) or less.

scarious [BOT] Having a thin, membranous texture.

scarlet *See* scarlet red.

scarlet fever [MED] An acute, contagious bacterial disease caused by *Streptococcus hemolyticus*; characterized by a papular, or rough, bright-red rash over the body, with fever, sore throat, headache, and vomiting occurring 2–3 days after contact with a carrier.

scarlet fever streptococcus antitoxin [IMMUNOL] A sterile aqueous solution of antitoxins obtained from the blood of animals immunized against group A beta hemolytic streptococci toxin; formerly used in the treatment of, and to produce immunity against, scarlet fever.

scarlet fever streptococcus toxin [IMMUNOL] Toxic filtrate of cultures of *Streptococcus pyogenes* responsible for the characteristic rash of scarlet fever; the toxin is used in the Dick test.

scarlet red [ORG CHEM] $CH_3C_6H_4H:NC_6H_3CH_3N:NC_{10}H_{15}OH$ A brown, water-insoluble powder, used as a dye in ointments. Also known as Biebrich red; scarlet.

scarp *See* escarpment.

Scarpa's fascia [ANAT] The deep, membranous layer of the superficial fascia of the lower abdomen.

scarped plain [GEOL] A terrain characterized by a succession of faintly inclined or gently folded strata.

scarp face *See* scarp slope.

scarplet *See* piedmont scarp.

scarpline [GEOL] A relatively straight line of cliffs of considerable extent, produced by faulting or erosion along a fault.

scarp slope [GEOL] The steep face of a cuesta, or asymmetric ridge, facing in an opposite direction to the dip of the strata. Also known as front slope; inface; scarp face.

scar tissue [MED] Contracted, dense connective tissue that is formed by the healing process of a wound or diseased tissue.

SC asphalt *See* slow-curing liquid asphaltic material.

Scatopsidae [INV ZOO] The minute black scavenger flies, a family of orthorrhaphous dipteran insects in the series Nematocera.

scatter angle *See* scattering angle.

scatter band [COMMUN] In pulse interrogation systems, the total bandwidth occupied by the frequency spread by numerous interrogations operating on the same nominal radio frequency.

scatter diagram [PETR] *See* point diagram. [STAT] A plot of the pairs of values of two variates in rectangular coordinates.

scattered [METEOROL] Descriptive of a sky cover of 0.1 to 0.5 (5 to 54%), applied only when clouds or obscuring phenomena aloft are present, not applied for surface-based obscuring phenomena.

scattering [ELECTROMAG] Diffusion of electromagnetic waves in a random manner by air masses in the upper atmosphere, permitting long-range reception, as in scatter propagation. Also known as radio scattering. [PHYS] **1.** The change in direction of a particle or photon because of a collision with another particle or a system. **2.** Diffusion of acoustic or electromagnetic waves caused by inhomogeneity or anisotropy of the transmitting medium. **3.** In general, causing a collection of entities to assume a less orderly arrangement.

scattering amplitude [QUANT MECH] A quantity, depending in general on the energy and scattering angle, which specifies the wave function of particles scattered in a collision, and whose squared modulus is proportional to the number of particles scattered in a given direction.

scattering angle [PHYS] The angle between the initial and final directions of motion of a scattered particle. Also known as scatter angle.

scattering coefficient [ELECTROMAG] One of the elements of the scattering matrix of a waveguide junction; that is, a transmission or reflection coefficient of the junction. [PHYS] The fractional decrease in intensity of a beam of electromagnetic radiation or particles per unit distance traversed, which results from scattering rather than absorption. Also known as dissipation coefficient.

scattering cross section [ELECTROMAG] The power of electromagnetic radiation scattered by an antenna divided by the incident power. [PHYS] The sum of the cross sections for elastic and inelastic scattering.

scattering function [ELECTROMAG] The intensity of scattered radiation in a given direction per lumen of flux incident upon the scattering material.

SCAPOLITE

Crystal of scapolite, a member of the tectosilicate minerals, taken from Pierrepont, New York. *(Specimen from Department of Geology, Bryn Mawr College)*

2.5 cm

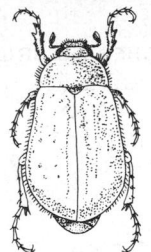

SCARABAEIDAE

A drawing of a lamellicorn beetle. *(From T. I. Storer and R. L. Usinger, General Zoology, 3d ed., McGraw-Hill, 1957)*

SCHEELITE

2.5 cm

Scheelite crystals with chalcopyrite ore from Piedmont, Italy. *(Specimen from Department of Geology, Bryn Mawr College)*

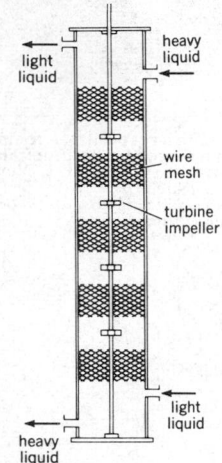

SCHEIBEL EXTRACTOR

Schematic of Scheibel extractor. *(From R. E. Treybal, Mass Transfer Operations, 2d ed., McGraw-Hill, 1968)*

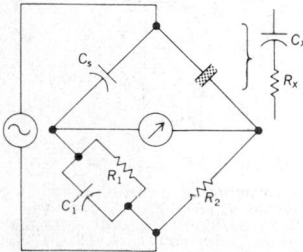

SCHERING BRIDGE

Circuit diagram of Schering bridge used to measure capacitance C_x and resistance R_x of equivalent series-circuit representation of capacitor. Standard capacitor C_s is assumed free from loss. Bridge is balanced when $C_x R_1 = C_s R_1$; $R_x C_s = R_2 C_1$.

scattering layer [OCEANOGR] A layer of organisms in the sea which causes sound to scatter and to return echoes.

scattering length [NUC PHYS] A parameter used in analyzing nuclear scattering at low energies; as the energy of the bombarding particle becomes very small, the scattering cross section approaches that of an impenetrable sphere whose radius equals this length. Also known as scattering power.

scattering loss [ELECTROMAG] The portion of the transmission loss that is due to scattering within the medium or roughness of the reflecting surface.

scattering matrix [ELECTROMAG] A square array of complex numbers consisting of the transmission and reflection coefficients of a waveguide junction. [QUANT MECH] A matrix which expresses the initial state in a scattering experiment in terms of the possible final states. Also known as collision matrix; S matrix.

scattering-matrix theory *See* S-matrix theory.

scattering power *See* scattering length.

scatter loading [ADP] The process of loading a program into main memory such that each section or segment of the program occupies a single, connected memory area but the several sections of the program need not be adjacent to each other.

scatterometer [ENG] Wide-sweep terrain-mapping radar.

scatter propagation [ELECTROMAG] Transmission of radio waves far beyond line-of-sight distances by using high power and a large transmitting antenna to beam the signal upward into the atmosphere and by using a similar large receiving antenna to pick up the small portion of the signal that is scattered by the atmosphere. Also known as beyond-the-horizon communication; forward-scatter propagation; over-the-horizon propagation.

scatter read [ADP] An input operation that places various segments of an input record into noncontiguous areas in central memory.

scatter reflections [ELECTROMAG] Reflections from portions of the ionosphere having different virtual heights, which mutually interfere and cause rapid fading.

scaur *See* scar.

scavenger [ECOL] An organism that feeds on carrion, refuse, and similar matter. [MET] A reactive metal added to a molten metal to combine with and remove dissolved gases.

scavenger system [ORD] A device for clearing smoke and gases from the chamber and bore of a firearm after firing.

scavenging [MECH ENG] Removal of spent gases from an internal combustion engine cylinder and replacement by a fresh charge or air. [MET] Removal of dissolved gases from molten metal. [ORD] The sweeping out, by a blast of air, of the gaseous products resulting from the firing of a gun.

scaw *See* scar.

Scelionidae [INV ZOO] A family of small, shining wasps in the superfamily Proctotrupoidea, characterized by elbowed, 11- or 12-segmented antennae.

scend [ENG] 1. The upward motion of the bow and stern of a vessel associated with pitching. 2. The lifting of the entire vessel by waves or swell. Also known as send.

scene paint [MATER] A paint used in theatrical scene painting; it is a dry pigment mixed with a glue-water mixture called size water.

scent gland [VERT ZOO] A specialized skin gland of the tubuloalveolar or acinous variety which produces substances having peculiar odors; found in many mammals.

scfh [FL MECH] Cubic feet per hour of gas flow at specified standard conditions of temperature and pressure.

scfm [FL MECH] Cubic feet per minute of gas flow at specified standard conditions of temperature and pressure.

Schaeffer's salt [ORG CHEM] $HOC_{10}SO_3Na$ A light-yellow to pink, water-soluble powder; the sodium salt formed from 2-naphthol-6-sulfonic acid; used as an intermediate in synthesis of organic compounds.

schafarzikite [MINERAL] $Fe_5Sb_4O_{11}$ A red to brown mineral composed of iron antimony oxide.

schairerite [MINERAL] $Na_3(SO_4)(F,Cl)$ A colorless rhombohedral mineral composed of sodium sulfate with fluorine and chlorine, occurring in crystals.

schalstein [PETR] A slaty rock formed by shearing basaltic or andesitic tuff or lava.

schappe [TEXT] Yarn or fabric made of spun silk or synthetic yarn resembling silk.

Schardinger dextrin [BIOCHEM] Any of several water-soluble, nonreducing, dextrorotatory polysaccharides that have a low molecular weight and that are obtained by the action of *Bacillus macerans* on a starch solution.

scharnitzer [METEOROL] A cold, northerly wind of long duration in Tyrol, Austria.

Scheat [ASTRON] A red giant, irregular, variable star, in the constellation Pegasus.

scheduled target [ORD] A planned target on which a nuclear weapon is to be delivered at a specific time during the operation of the supported force.

schedule of fire [ORD] Groups of fires or series of fires fired in a definite sequence according to a definite program.

scheelite [MINERAL] $CaWO_4$ A yellowish-white mineral crystallizing in the tetragonal system and occurring in tabular or massive form in pneumatolytic veins associated with quartz; an ore of tungsten.

Scheffel engine [MECH ENG] A type of multirotor engine that uses nine approximately equal rotors turning in the same clockwise sense.

schefferlite [MINERAL] $(Ca,Mn)(Mg,Fe,Mn)-Si_2O_6$ Brown to black variety of pyroxene that crystallizes in the monoclinic system and contains manganese and frequently iron.

Scheibel column *See* Scheibel extractor.

Scheibel extractor [CHEM ENG] Liquid-liquid contact vessel used in liquid-liquid extraction processes: a vertical cylinder with interspersed open spaces and wire-mesh packing along its height, with liquid agitators in the open spaces, or a vertical cylinder fully filled with wire-mesh packing. Also known as Scheibel column; Scheibel-York extractor; York-Scheibel column.

Scheibel-York extractor *See* Scheibel extractor.

Scheie's syndrome [MED] A hereditary disease transmitted as an autosomal recessive and characterized by high levels of chondroitin sulfate B in the urine, mild distortion of the facies, hypertrichosis, clouding of the cornea, and aortic valve disease.

schematic circuit diagram *See* circuit diagram.

schematic diagram [GRAPHICS] A presentation of the element-by-element relationship of all parts of a system.

scheme of maneuver [ORD] The tactical plan to be executed by a force in order to seize assigned objectives.

schemochrome [ZOO] A feather color that originates within the feather structures, through refraction of light independent of pigments.

Schenck's disease *See* sporotrichosis.

Schering bridge [ELEC] A four-arm alternating-current bridge used to measure capacitance and dissipation factor; bridge balance is independent of frequency.

scheteligite [MINERAL] $(Ca,Y,Sb,Mn)_2(Ti,Ta,Nb,W)_2O_6(O,OH)$ A mineral composed of oxide of calcium, rare-earth metals, antimony, manganese, titanium, columbium, and tantalum.

Scheuermann's disease *See* osteochondrosis.

Schick test [IMMUNOL] A skin test for determining susceptibility to diphtheria performed by the intradermal injection of diluted diphtheria toxin; a positive reaction, showing edema and scaling after 5 to 7 days, indicates lack of immunity.

Schiff base [ORG CHEM] $RR'C:NR''$ Any of a class of derivatives of the condensation of aldehydes or ketones with primary amines; colorless crystals, weakly basic; hydrolyzed by water and strong acids to form carbonyl compounds and amines; used as chemical intermediates and perfume bases, in dyes and rubber accelerators, and in liquid crystals for electronics.

schiffli [TEXT] A machine for putting embroidery and lace patterns on textiles.

Schiff's reagent [ANALY CHEM] An aqueous solution of rosaniline and sulfurous acid; used in the Schiff test.

Schiff test [ANALY CHEM] A test for aldehydes by using an aqueous solution of rosaniline and sulfurous acid.

Schilder's disease [MED] 1. A retrogressive disease of the white matter in the central nervous system characterized by diffuse loss of myelin. 2. Any of the progressive degenerative diseases of the white matter in the central nervous system.

schiller *See* play of color.

schillerization [OPTICS] Development of schiller in crystals due to the pattern of inclusions.

Schindleriidae [VERT ZOO] The single family of the order Schindlerioidei.

Schindlerioidei [VERT ZOO] A suborder of fishes in the order Perciformes composed of one monogeneric family comprising two tiny oceanic species that are transparent and neotenic.

schindylesis [ANAT] A synarthrosis in which a plate of one bone is fixed in a fissure of another.

schirmerite [MINERAL] $PbAg_4Bi_4S_9$ A mineral composed of lead, silver, and bismuth sulfide.

schist [GEOL] A large group of coarse-grained metamorphic rocks which readily split into thin plates or slabs as a result of the alignment of lamellar or prismatic minerals.

schistose [GEOL] Pertaining to rocks exhibiting schistosity.

schistosity [GEOL] A type of cleavage characteristic of metamorphic rocks, notably schists and phyllites, in which the rocks tend to split along parallel planes defined by the distribution and parallel arrangement of platy mineral crystals.

Schistosoma [INV ZOO] A genus of blood flukes infecting man.

schistosome dermatitis [MED] A dermatitis caused by penetration of the skin by certain schistosome cercariae. Also known as swamp itch; swimmer's itch.

schistosomiasis [MED] A disease in which humans are parasitized by any of three species of blood flukes: *Schistosoma mansoni, S. haematobium,* and *S. japonicum;* adult worms inhabit the blood vessels. Also known as snail fever.

Schistostegiales [BOT] A monospecific order of mosses; the small, slender, glaucous plants are distinguished by the luminous protonema.

schizaxon [ANAT] An axon that divides, in its course, into equal or nearly equal branches.

schizocarp [BOT] A dry fruit that separates at maturity into single-seeded indehiscent carpels.

Schizocoela [INV ZOO] A group of animal phyla, including Bryozoa, Brachiopoda, Phoronida, Sipunculoidea, Echiuroidea, Priapuloidea, Mollusca, Annelida, and Arthropoda, all characterized by the appearance of the coelom as a space in the embryonic mesoderm.

schizodont [INV ZOO] A multinucleate trophozoite that segments into merozoites.

schizogamy [BIOL] A form of reproduction involving division of an organism into a sexual and an asexual individual.

schizogenesis [BIOL] Reproduction by fission.

schizognathous [VERT ZOO] Descriptive of birds having a palate in which the vomer is small and pointed, the maxillopalatines are not united with each other or with the vomer, and the palatines articulate posteriorly with the rostrum.

Schizogoniales [BOT] A small order of the Chlorophyta containing algae that are either submicroscopic filaments or macroscopic ribbons or sheets a few centimeters wide and attached by rhizoids to rocks.

schizogony [INV ZOO] Asexual reproduction by multiple fission of a trophozoite; a characteristic of certain Sporozoa.

Schizomeridaceae [BOT] A family of green algae in the order Ulvales.

Schizomycetes [MICROBIO] A class of the division Protophyta which includes the bacteria.

Schizomycophyta [BOT] The designation for bacteria in those taxonomic systems that consider bacteria as plants.

schizont [INV ZOO] A multinucleate cell in certain members of the Sporozoa that is produced from a trophozoite in a cell of the host, and that segments into merozoites.

Schizopathidae [INV ZOO] A family of dimorphic zoantharians in the order Antipatharia.

schizopelmous [VERT ZOO] Having the two flexor tendons of the toes separate, as in certain birds.

Schizophora [INV ZOO] A series of the dipteran suborder Cyclorrhapha in which adults possess a frontal suture through which a distensible sac, or ptilinum, is pushed to help the organism escape from its pupal case.

schizophrenia [PSYCH] A group of mental disorders charac-

terized by withdrawal from reality and by alterations in thinking, feeling, and concept formations. Also known as dementia praecox.

Schizophyceae [MICROBIO] The blue-green algae, a class of the division Protophyta.

Schizophyta [BOT] The prokaryotes, a division of the plant subkingdom Thallobionta; includes the bacteria and blue-green algae.

schizopod [INV ZOO] **1.** Having the limbs split so that each has an endopodite and an exopodite, as in certain crustaceans. **2.** A biramous appendage.

Schizopteridae [INV ZOO] A family of minute ground-inhabiting hemipterans in the group Dipsocoeoidea; individuals characteristically live in leaf mold.

schizorhinal [VERT ZOO] Having a deep cleft on the posterior margin of the osseous external nares, as in certain birds.

schizothecal [VERT ZOO] Having the horny envelope of the tarsus divided into scalelike plates; refers to most birds.

schizothoracic [INV ZOO] Having a prothorax that is large and loosely articulated with the thorax.

Schlemm's canal [ANAT] A space or series of spaces at the junction of the sclera and cornea in the eye; drains aqueous humor from the anterior chamber.

schlempe *See* vinasse.

Schlernwind [METEOROL] East wind blowing down from the Schlern near Bozen in Tyrol, Austria.

Schlick vibration formula [NAV ARCH] A formula for calculation of hull vibration in ships.

schlieren [OPTICS] In atmospheric optics, parcels or strata of air having densities sufficiently different from that of their surroundings so that they may be discerned by means of refraction anomalies in transmitted light. [PETR] Irregular streaks with shaded borders in some igneous rocks, representing the segregation of light and dark minerals or altered inclusions, elongated by flow.

schlieren method [OPTICS] An optical technique that detects density gradients occurring in a fluid flow; in its simplest form, light from a slit is collimated by a lens and focused onto a knife-edge by a second lens, the flow pattern is placed between these two lenses, and the diffraction pattern that results on a screen or photographic film placed behind the knife-edge is observed.

Schlumberger dipmeter [ENG] An instrument that measures both the amount and direction of dip by readings taken in the borehole; it consists of a long, cylindrical body with two telescoping parts and three long, springy metal strips, arranged symmetrically round the body, which press outward and make contact with the walls of the hole.

Schlumberger photoclinometer [ENG] An instrument that measures simultaneously the amount and direction of the deviation of a borehole; the sonde, designed to lie exactly parallel to the axis of the borehole, is fitted with a small camera on the axis of a graduated glass bowl, in which a steel ball rolls freely and a compass is mounted in gimbals; the camera is electrically operated from the surface and takes a photograph of the bowl, the steel ball marks the amount of deviation, the position in relation to the image of the compass needle gives the direction of deviation.

Schmidt camera *See* Schmidt system.

Schmidt correction plate [OPTICS] In the Schmidt system, a glass plate with one face a plane and the other aspherical and deviating from a plane in such a way that it bends light, traveling to the system's spherical mirror, so as to correct for spherical aberration and coma.

Schmidt lines [NUC PHYS] Two lines, on a graph of nuclear magnetic moment versus nuclear spin, on which points describing all nuclides should lie, according to the independent particle model; experimentally, however, points describing nuclides are scattered between the lines.

Schmidt net [GEOL] A coordinate or reference system used to plot a Schmidt projection.

Schmidt number 1 *See* Prandtl number.

Schmidt number 2 *See* Semenov number 1.

Schmidt number 3 [PHYS CHEM] A dimensionless number used in electrochemistry, equal to the product of the dielectric susceptibility and the dynamic viscosity of a fluid divided by

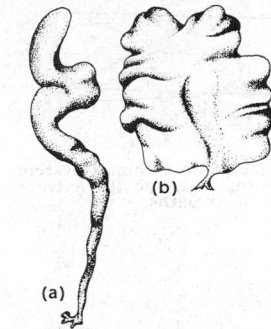

SCHIZOGONIALES

Thalli of *Prasiola* species: *(a)* ribbonlike; *(b)* sheetlike.

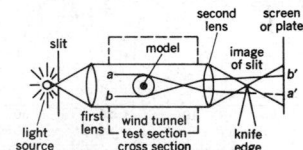

SCHLIEREN METHOD

Optical system used in the schlieren method. Light rays from *a* and *b* are bent by density gradients in test section. Ray from *a* is interrupted by knife-edge so that a dark spot occurs on the screen at *a'*; ray from *b* escapes knife-edge so that light spot occurs on screen at *b'*.

the product of the fluid density, electrical conductivity, and the square of a characteristic length. Symbolized Sc_3.

Schmidt objective   See Schmidt system.

Schmidt optics   See Schmidt system.

Schmidt projection   [GEOL] A Lambert azimuthal equal-area projection of the lower hemisphere of a sphere onto the plane of a meridian; used in structural geology.

Schmidt reflector   [OPTICS] A telescope employing the Schmidt system.

Schmidt system   [OPTICS] An optical system designed to eliminate spherical aberration and coma, which, in its original form, consists of a spherical mirror, a Schmidt correction plate near the focus of the mirror, and usually a curved reflecting plate at the focus of the mirror; used in astronomical telescopes with unusually wide fields of view and in spectroscopes, and to project a television image from a cathode-ray tube onto a screen. Also known as projection optics; Schmidt camera; Schmidt objective; Schmidt optics.

Schmitt circuit   [ELECTR] A bistable pulse generator in which an output pulse of constant amplitude exists only as long as the input voltage exceeds a certain value. Also known as Schmitt limiter; Schmitt trigger.

Schmitt limiter   See Schmitt circuit.

Schmitt trigger   See Schmitt circuit.

Schmitz bacillus   [MICROBIO] The common name for the dysentery bacillus *Shigella ambigua*.

Schneiderian membrane   [ANAT] The mucosa lining the nasal cavities and paranasal sinuses.

Schneider recoil system   [MECH ENG] A recoil system for artillery, employing the hydropneumatic principle without a floating piston.

Schneider's index   [MED] A test of general physical and circulatory efficiency, consisting of pulse and blood pressure observations under standard conditions of rest and exercise.

Schoelkopf's acid   [ORG CHEM] A dye of the following types: 1-naphthol-4,8-disulfonic acid, 1-naphthylamine-4,8-disulfonic acid, and 1-naphthylamine-8-sulfonic acid; may be toxic.

Schoenbiinae   [INV ZOO] A subfamily of moths in the family Pyralididae, including the genus *Acentropus*, the most completely aquatic Lepidoptera.

Schoenherr-Hessberger process   [CHEM ENG] A nitrogen-fixation process used in Norway; employs a very long (7 meters) alternating-current arc around which air moves in a helical path in a 746-kilowatt furnace.

schoepite   [MINERAL] $UO_3 \cdot 2H_2O$ A yellow secondary mineral composed of hydrous uranium oxide.

schönfelsite   [PETR] A form of basalt containing embedded crystals of olivine and augite in a complex, dense fine-grained groundmass.

Schönflies crystal symbols   [CRYSTAL] Symbols denoting the 32 crystal point groups or symmetry classes; capital letters indicate the general type of class, and subscripts the multiplicity of rotation axes and the existence of additional symmetries.

schooner   [NAV ARCH] A sailing vessel with two or more masts rigged fore and aft.

Schoop process   [ENG] A process for coating surfaces by spraying with high-velocity molten metal particles.

schorl   See schorlite.

schorlite   [MINERAL] The black, iron-rich, opaque variety of tourmaline. Also known as schorl.

schorlomite   [MINERAL] $Ca_3(Fe,Ti)_2(Si,Ti)_3O_{12}$ Black mineral of the garnet group that has a vitreous luster and usually occurs in masses; hardness is 7–7.5 on Mohs scale, and specific gravity is 3.81–3.88.

schott   [GEOGR] A shallow saline lake in southern Tunisia or on the plateaus of northern Algeria, which is usually dry during the summer.

Schotten-Baumann reaction   [ORG CHEM] An acylation reaction that uses an acid chloride in the presence of dilute alkali to acylate the hydroxyl and amino group of organic compounds.

Schottky barrier diode   [ELECTR] A *pn*-junction diode whose metal contacts a large area of the semiconductor. Also known as barrier diode.

Schottky defect   [SOLID STATE] **1.** A defect in an ionic crystal

in which a single ion is removed from its interior lattice site and relocated in a lattice site at the surface of the crystal. **2.** A defect in an ionic crystal consisting of the smallest number of positive-ion vacancies and negative-ion vacancies which leave the crystal electrically neutral.

Schottky effect   [SOLID STATE] The enhancement of the thermionic emission of a conductor resulting from an electric field at the conductor surface.

Schottky line   [SOLID STATE] A graph of the logarithm of the saturation current from a thermionic cathode as a function of the square root of anode voltage; it is a straight line according to the Schottky theory.

Schottky noise   See shot noise.

Schottky theory   [SOLID STATE] A theory describing the rectification properties of the junction between a semiconductor and a metal that result from formation of a depletion layer at the surface of contact.

schreibersite   [MINERAL] $(Fe,Ni)_3P$ A silver-white to tin-white magnetic meteorite mineral crystallizing in the tetragonal system and occurring in tables or plates as oriented inclusions in iron meteorites. Also known as rhabdite.

schreinerize   [TEXT] To give a lustrous finish to a cotton fabric by passing it under rollers engraved with fine lines.

Schrödinger equation   [QUANT MECH] A partial differential equation governing the Schrödinger wave function ψ of a system of one or more nonrelativistic particles; $\hbar (\partial \psi / \partial t) = H\psi$, where H is a linear operator, the Hamiltonian, which depends on the dynamics of the system, and $\hbar$ is Planck's constant divided by 2π.

Schrödinger-Klein-Gordon equation   See Klein-Gordon equation.

Schrödinger-Pauli equation   [QUANT MECH] A modification of the Schrödinger equation to describe a particle with spin of $\frac{1}{2}\hbar$, where $\hbar$ is Planck's constant divided by 2π; the wave function has two components, corresponding to the particle's spin pointing in either of two opposite directions.

Schrödinger picture   [QUANT MECH] A mode of description of a quantum-mechanical system in which dynamical states are represented by vectors which evolve in the course of time, and physical quantities are represented by stationary operators, in contrast to the Heisenberg picture.

Schrödinger representation   See position representation.

Schrödinger's wave mechanics   [QUANT MECH] The version of nonrelativistic quantum mechanics in which a system is characterized by a wave function which is a function of the coordinates of all the particles of the system and time, and obeys a differential equation, the Schrödinger equation; physical quantities are represented by differential operators which may act on the wave function, and expectation values of measurements are equal to integrals involving the corresponding operator and the wave function. Also known as wave mechanics.

Schrödinger wave function   [QUANT MECH] A function of the coordinates of the particles of a system and of time which is a solution of the Schrödinger equation and which determines the average result of every conceivable experiment on the system. Also known as probability amplitude; psi function.

schroeckingerite   [MINERAL] $NaCa_3(UO_2)(CO_3)(SO_4)F \cdot 10H_2O$ A yellowish secondary mineral composed of hydrous sodium calcium uranyl carbonate, sulfate, and fluoride.

Schroeder-Bernstein theorem   [MATH] If a set A has at least as many elements as another set B and B has at least as many elements as A, then A and B have the same number of elements.

schrund line   [GEOL] The base of the bergschrund, or deep crevasse, at a late stage in the excavation of a cirque; the schrund line separates the steep slope of the cirque wall from the gentler slope below.

Schubertellidae   [PALEON] An extinct family of marine protozoans in the superfamily Fusulinacea.

Schuler pendulum   [MECH] Any apparatus which swings, because of gravity, with a natural period of 84.4 minutes, that is, with the same period as a hypothetical simple pendulum whose length is the earth's radius; the pendulum arm remains vertical despite any motion of its pivot, and the apparatus is therefore useful in navigation.

Schuler tuning   [ENG] The designing of gyroscopic devices so

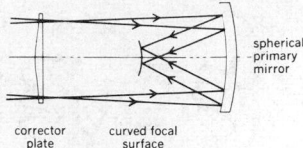

SCHMIDT SYSTEM

spherical primary mirror

corrector plate

curved focal surface

Optics of the Schmidt system showing axial and the extra-axial light paths.

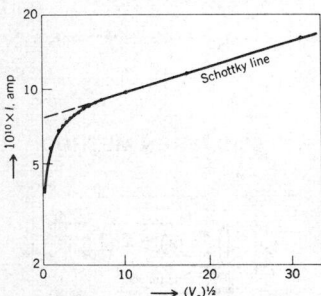

SCHOTTKY LINE

Schottky line

Graph of the logarithm of thermionic emission current I of tungsten as function of square root of anode voltage V_a. (*After W. B. Nottingham, Phys. Rev., 58:927–928, 1940*)

that their periods of oscillation will be about 84.4 minutes.

schultenite [MINERAL] $PbHAsO_4$ A colorless mineral composed of lead hydrogen arsenate occurring in tabular orthorhombic crystals.

Schultz-Charlton test [IMMUNOL] A skin test for the diagnosis of scarlet fever, performed by the intradermal injection of human scarlet fever immune serum; a positive reaction consists of blanching of the rash in the area surrounding the point of injection.

Schultz-Dale reaction [IMMUNOL] A method for demonstrating anaphylactic hypersensitivity outside the body by suspending an excised intestinal loop or uterine strip from a sensitized animal in an oxygenated, physiological salt solution; addition of the proper allergen causes contraction of the smooth muscle.

Schultze powder [MATER] A smokeless powder propellant, consisting of wood pellets impregnated with barium nitrate and potassium nitrate.

Schumann plate [GRAPHICS] A type of photographic plate in which the silver halide is held to the glass, using a special method, by a layer of gelatin so thin that it is transparent to ultraviolet light with wavelengths as short as 1200 angstroms, in contrast to the limit of about 2200 angstroms in a conventional photographic plate.

Schumann region [OPTICS] The most extreme ultraviolet portion of the electromagnetic spectrum that will affect a photographic plate.

schungite [GEOL] Amorphous carbon-rich material occurring in Precambrian schists.

schuppen structure *See* imbricate structure.

Schur's lemma [MATH] For certain types of modules M, the ring consisting of all homomorphisms of M to itself will be a division ring.

Schuster method [SPECT] A method for focusing a prism spectroscope without using a distant object or a Gauss eyepiece.

Schwagerinidae [PALEON] A family of fusulinacean protozoans that flourished during the Early and Middle Pennsylvanian and became extinct during the Late Permian.

Schwann cell [HISTOL] One of the cells that surround peripheral axons forming sheaths of the neurilemma.

schwannoma *See* neurilemmoma.

schwartzembergite [MINERAL] $Pb_5(IO_3)Cl_3O_3$ A mineral composed of lead iodate, chloride, and oxide.

Schwarzchild anastigmat [OPTICS] A Gregorian telescope whose surfaces are altered to reduce astigmatism.

Schwarzchild radius [RELAT] For a given body of matter, a distance equal to the mass of the body times the gravitational constant divided by the square of the speed of light. Also known as gravitational radius.

Schwarzchild solution [RELAT] An exact solution of the field equations of general relativity for the field generated by a point mass.

Schwarz-Christoffel transformations [MATH] Those complex transformations which conformally map the interior of a given polygon onto the portion of the complex plane above the real axis.

Schwarz' inequality *See* Cauchy-Schwarz inequality.

Schwarz-reflection principle [MATH] To obtain the analytic continuation of a given function $f(z)$ analytic in a region R, whose boundary contains a segment of the real axis, into a region reflected from R through this segment, one takes the complex conjugate function $f(\bar{z})$.

Schwarz's lemma [MATH] If an analytic function of the unit disk to itself sends the origin to the origin then it must be distance-decreasing.

Schwassman-Wachmann comet [ASTRON] A variable cometlike asteroid whose period is 16 years; its orbit is very nearly circular and lies between the orbits of Saturn and Jupiter.

Schweitzer's reagent [CHEM] An ammoniacal solution of cupric hydroxide; used to dissolve cellulose, silk, and linen, and to test for wool.

Sciaenidae [VERT ZOO] A family of perciform fishes in the suborder Percoidei, which includes the drums.

sciatica [MED] Neuralgic pain in the lower extremities, hips, and back caused by inflammation or injury to the sciatic nerve.

sciatic nerve [ANAT] Either of a pair of long nerves that originate in the lower spinal cord and send fibers to the upper thigh muscles and the joints, skin, and muscles of the leg.

science [SCI TECH] A branch of study in which facts are observed and classified, and, usually, quantitative laws are formulated and verified; involves the application of mathematical reasoning and data analysis to natural phenomena.

scientific computer [ADP] A computer which has a very large memory and is capable of handling extremely high-speed arithmetic and a very large variety of floating-point arithmetic commands.

scientific information [SCI TECH] That part of technical information concerned with the study of natural phenomena.

scientific method [SCI TECH] The systematic collection and classification of data and, usually, the formulation and testing of hypotheses based on the data.

scientific system [ADP] A system devoted principally to computations as opposed to business and data-processing systems, the main emphasis of which is on the updating of data records and files rather than the performance of calculations.

scientist [SCI TECH] A person having the training, ability, and desire to seek new knowledge, new principles, and new materials in some field of science.

scillarenin [PHARM] $C_{24}H_{32}O_4$ A crystalline compound that forms prisms from a methanol solution and melts at 232–238°C; used in medicine for cardiac disease.

Scincidae [VERT ZOO] The skinks, a family of the reptilian suborder Sauria which have reduced limbs and snakelike bodies.

Scinidae [INV ZOO] A family of bathypelagic, amphipod crustaceans in the suborder Hyperiidea.

scintillation [ELECTROMAG] **1.** A rapid apparent displacement of a target indication from its mean position on a radar display; one cause is shifting of the effective reflection point on the target. Also known as target glint; target scintillation; wander. **2.** Random fluctuation, in radio propagation, of the received field about its mean value, the deviations usually being relatively small. [NUCLEO] A flash of light produced in a phosphor by an ionizing particle or photon. [OPTICS] Rapid changes of brightness of stars or other distant, celestial objects caused by variations in the density of the air through which the light passes.

scintillation camera [NUCLEO] A camera that gives a complete image of radionuclide distribution in a particular area of the human body in one exposure, in contrast to line-scanning techniques.

scintillation counter [NUCLEO] A device in which the scintillations produced in a fluorescent material by an ionizing radiation are detected and counted by a multiplier phototube and associated circuits; used in medical and nuclear research and in prospecting for radioactive ores. Also known as scintillation detector; scintillometer.

scintillation counter crystal [NUCLEO] A substance (a fluor, such as thallium-activated calcium tungstate) that emits a flash of light (scintillation) when contacted by a high-energy particle, for example, an alpha, beta, or gamma ray.

scintillation detector *See* scintillation counter.

scintillation spectrometer [NUCLEO] A scintillation counter adapted to measuring the energy and intensity of gamma rays from radioactive elements.

scintillator [NUCLEO] A material that emits optical photons in response to ionizing radiation.

scintillometer *See* scintillation counter.

Sciomyzidae [INV ZOO] A family of myodarian cyclorrhaphous dipteran insects in the subsection Acalypteratae.

scion [BOT] A section of a plant, usually a stem or bud, which is attached to the stock in grafting.

sciophilous [ECOL] Capable of thriving in shade.

sciophyte [BOT] A plant that thrives at lowered light intensity.

scirrhous carcinoma [MED] A hard, poorly differentiated adenocarcinoma in which the anaplastic cells are surrounded by dense bundles of collagenous fibers.

scissor engine *See* cat-and-mouse engine.

scissor jack [MECH ENG] A lifting jack driven by a horizontal screw; the linkages of the jack are parallelograms whose

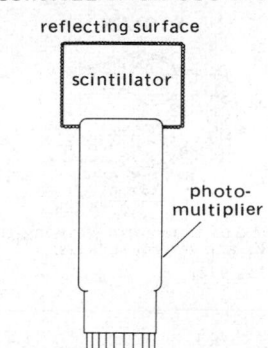

SCINTILLATION COUNTER

Diagram of a scintillation counter.

SCLERACTINIA

Solitary coral polyps, *Oulangia* species. *(After S. Hickson)*

SCOLECODONT

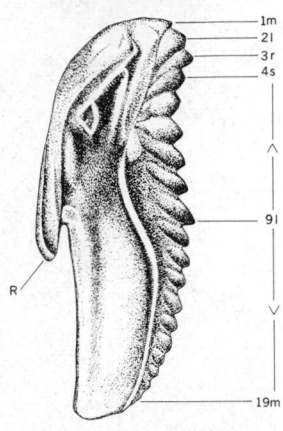

```
1m
2l
3r
4s
∧
9l
∨
R
19m
```

R19
1m – 2l – 3r – 4s < 9l > 19m

Dental formula for a large typical scolecodont. R = ramus, r = roder or largest denticle, l = large, s = small, m = minute. Teeth are numbered in sequence from anterior to posterior, and range in size from 1 to 2 millimeters.

SCOLEX

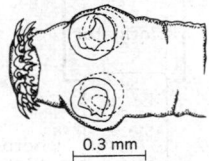

0.3 mm

Scolex of tapeworm showing the hooks and lateral suckers.

SCOLYTIDAE

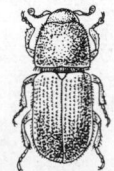

Drawing of a bark beetle. *(From T. I. Storer and R. L. Usinger, General Zoology, 3d ed., McGraw-Hill, 1957)*

horizontal diagonals are lengthened or shortened by the screw.

scissors bridge [CIV ENG] A light metal bridge that can be folded and carried by a military tank.

scissors crossover [CIV ENG] A scissor-shaped junction between two parallel railway tracks. Also called double crossover.

scissors fault [GEOL] A fault on which the offset or separation along the strike increases in one direction from an initial point and decreases in the other direction. Also known as differential fault.

scissors truss [BUILD] A roof truss in which the braces cross like scissors blades.

Scitaminales [BOT] An equivalent name for Zingiberales.

Scitamineae [BOT] An equivalent name for Zingiberales.

Sciuridae [VERT ZOO] A family of rodents including squirrels, chipmunks, marmots, and related forms.

Sciuromorpha [VERT ZOO] A suborder of Rodentia according to the classical arrangement of the order.

sclera [ANAT] The hard outer coat of the eye, continuous with the cornea in front and with the sheath of the optic nerve behind.

Scleractinia [INV ZOO] An order of the subclass Zoantharia which comprises the true, or stony, corals; these are solitary or colonial anthozoans which attach to a firm substrate.

Scleraxonia [INV ZOO] A suborder of coelenterates in the order Gorgonacea in which the axial skeleton has calcareous spicules.

sclereid [BOT] A thick-walled, lignified plant cell typically found in sclerenchyma.

sclerema neonatorum [MED] A disease of the newborn, particularly the premature or the undernourished, dehydrated, and debilitated, characterized by waxy-white hardening of subcutaneous tissue.

sclerenchyma [BOT] A supporting plant tissue composed principally of sclereids whose walls are often mineralized.

scleriasis *See* scleroderma.

sclerite [INV ZOO] One of the sclerotized plates of the integument of an arthropod.

scleroblast [INV ZOO] A spicule-secreting cell in Porifera.

scleroblastema [EMBRYO] Embryonic tissue from which bones are formed.

sclerocaulous [BOT] Having a hard, dry stem because of exceptional development of sclerenchyma.

sclerochore [BOT] A plant that disperses disseminules without apparent morphological adaptations.

Sclerodactylidae [INV ZOO] A family of echinoderms in the order Dendrochirotida having a complex calcareous ring.

scleroderma [MED] An abnormal increase in collagenous connective tissue in the skin. Also known as chorionitis; dermatosclerosis; scleriasis.

sclerodermatous [INV ZOO] Having a skeleton that is composed of scleroderm, as certain corals. [VERT ZOO] Having a hard outer covering, for example, hard plate or horny scale.

Sclerogibbidae [INV ZOO] A monospecific family of the hymenopteran superfamily Bethyloidea.

sclerometer [ENG] An instrument used to determine the hardness of a material by measuring the pressure needed to scratch or indent a surface with a diamond point.

sclerophyllous [BOT] Characterized by thick, hard foliage due to well-developed sclerenchymatous tissue.

scleroprotein [BIOCHEM] Any of a large class of fibrous proteins, such as collagen, keratin, and fibroin, which occur in tendons, bones, cartilages, ligaments, and other parts of the body. Also known as albuminoid.

scleroscope [ENG] An instrument used to determine the hardness of a material by measuring the height to which a standard ball rebounds from its surface when dropped from a standard height.

scleroscope hardness test *See* Shore scleroscope hardness test.

scleroseptum [INV ZOO] A calcareous radial septum in certain corals.

sclerosing adenomatosis [MED] A form of mammary dysplasia in which ductular structures are encased in fibrous tissue, simulating invading cancerous ductular structures. Also known as fibrosing adenomatosis.

sclerosing hemangioma [MED] A type of benign histiocytoma marked by prominence of the capillary channels.

sclerosis [PATH] Hardening of a tissue, especially by proliferation of fibrous connective tissue.

sclerotic [ANAT] Pertaining to the sclera. [MED] 1. Hard. 2. Of or pertaining to sclerosis.

sclerotinite [GEOL] A variety of inertinite composed of fungal sclerotia.

sclerotium [MICROBIO] The hardened, resting or encysted condition of the plasmodium of Myxomycetes. [MYCOL] A hardened, resting mass of hyphae, usually black on the outside, from which fructifications may develop.

sclerotome [EMBRYO] The part of a mesodermal somite which enters into the formation of the vertebrae. [MED] A knife used in sclerotomy. [VERT ZOO] The fibrous tissue separating successive myotomes in certain lower vertebrates.

sclerotomy [MED] Surgical incision of the sclera.

scobiform [BOT] Resembling sawdust.

scolecite [MINERAL] $CaAl_2Si_3O_{10}$ A zeolite mineral that occurs in delicate, radiating groups of white fibrous or acicular crystals; sometimes shows wormlike motion upon heating.

scolecodont [PALEON] Any of the paired, pincerlike jaws occurring as fossils of annelid worms.

Scolecosporae [MYCOL] A spore group of Fungi Imperfecti characterized by filiform spores.

scolex [INV ZOO] The head of certain tapeworms, typically having a muscular pad with hooks, and two pairs of lateral suckers.

Scoliidae [INV ZOO] A family of the Hymenoptera in the superfamily Scolioidea.

Scolioidea [INV ZOO] A superfamily of hymenopteran insects in the suborder Apocrita.

scoliosis [MED] Lateral curvature of the spine.

scolite [GEOL] Any of the small tubes in rock believed to be the fossilized burrows of worms.

scolop [INV ZOO] The thickened, distal tip of a vibration-sensitive organ in insects.

Scolopacidea [VERT ZOO] A large, cosmopolitan family of birds of the order Charadriiformes including snipes, sandpipers, curlews, and godwits.

Scolopendridae [INV ZOO] A family of centipedes in the order Scolopendromorpha which characteristically possess eyes.

Scolopendromorpha [INV ZOO] An order of the chilopod subclass Pleurostigmophora containing the dominant tropical forms, and also the largest of the centipedes.

scolophore *See* scolopophore.

scolopophore [INV ZOO] A spindle-shaped, bipolar nerve ending in the integument of insects, believed to be auditory in function. Also known as scolophore.

Scolytidae [INV ZOO] The bark beetles, a large family of coleopteran insects in the superfamily Curculionoidea characterized by a short beak and clubbed antennae.

Scombridae [VERT ZOO] A family of perciform fishes in the suborder Scombroidei including the mackerels and tunas.

Scombroidei [VERT ZOO] A suborder of fishes in the order Perciformes; all are moderate- to large-sized shore and oceanic fishes having fixed premaxillae.

scoop [DES ENG] 1. Any of various ladle-, shovel-, or bucket-like utensils or containers for moving liquid or loose materials. 2. A funnel-shaped opening for channeling a fluid into a desired path. [ELEC] *See* ellipsoidal floodlight. [MECH ENG] A large shovel with a scoop-shaped blade.

scoopfish *See* underway sampler.

scopa [INV ZOO] A brushlike arrangement of short stiff hairs on the body surface of certain insects.

scope *See* cathode-ray oscilloscope; radarscope.

Scopeumatidae [INV ZOO] The dung flies, a family of myodarian cyclorrhaphous dipteran insects in the subsection Calypteratae.

Scopidae [VERT ZOO] A family of birds in the order Ciconiiformes containing a single species, the hammerhead (*Scopus umbretta*) of tropical Africa.

scopolamine [PHARM] $C_{17}H_{21}O_4N$ An alkaloid derivative of several plants in the family Solanaceae, used as an anticholinergic drug; its hydrobromide salt is used as a sedative.

scopoline [ORG CHEM] $C_8H_{13}O_2H$ A white crystalline alkaloid that melts at 108–109°C, soluble in water and ethanol; used in medicine. Also known as oscine.

scopometer [OPTICS] An instrument used to measure the absorption or scattering of light in a solution containing solid particles by measuring the contrast between an illuminated line placed behind the solution and a field of constant brightness.

scopophilia [PSYCH] Sexual stimulation from looking at the unclad human figure; observed chiefly in normal adolescent and adult males where it takes the aim-inhibited form of "girl watching" or looking at nude or seminude females in magazines or as part of some stage performance, or where it may be sublimated as scientific curiosity; when present to a pathologic degree, it is deviant and called voyeurism.

scopula [ZOO] A tuft of hair, as on the feet and chelicerae of certain spiders.

scopulite [GEOL] A crystallite in the form of a rod with terminal brush or plume.

scorching [ENG] **1.** Burning an exposed surface so as to change color, texture, or flavor without consuming. **2.** Destroying by fire. [PL PATH] Browning of plant tissues caused by heat or parasites; may also be symptomatic of disease.

scoria [GEOL] Vesicular, cindery, dark lava formed by the escape and expansion of gases in basaltic or andesitic magma; generally denser and darker than pumice. [MATER] Refuse after melting metals or reducing ore.

scoria cone [GEOL] A volcanic cone composed of a vesicular, cindery crust on the surface of lava that is basaltic or andesitic in nature.

scoria mound [GEOL] A volcanic knoll composed of vesicular, cindery crust on the surface of lava that is basaltic or andesitic in nature.

scorification [MET] Concentration of precious metals, such as gold and silver, in molten lead by oxidation employing appropriate fluxes.

scoring [ENG] Scratching the surface of a material. [MATER] *See* attrition.

scoring test *See* L-2 test.

scorodite [MINERAL] $FeAsO_4 \cdot 2H_2O$ A pale leek-green or liver-brownish orthorhombic mineral consisting of ferric arsenate; isomorphous with mansfieldite and represents a minor ore of arsenic.

Scorpaenidae [VERT ZOO] The scorpion fishes, a family of Perciformes in the suborder Cottoidei, including many tropical shorefishes, some of which are venomous.

Scorpaeniformes [VERT ZOO] An order of fishes coextensive with the perciform suborder Cottoidei in some systems of classification.

scorpioid cyme [BOT] A cyme with a curved axis and flowers arising two-ranked on alternate sides of the axis.

scorpion [INV ZOO] The common name for arachnids constituting the order Scorpionida.

Scorpion [AERO ENG] An all-weather interceptor aircraft with twin turbojet engines; its armament consists of air-to-air rockets with nuclear or nonnuclear warheads. Designated F-89. [ASTRON] *See* Scorpius.

Scorpionida [VERT ZOO] The scorpions, an order of arachnids characterized by a shieldlike carapace covering the cephalothorax and by large pedipalps armed with chelae.

Scorpio X-1 [ASTROPHYS] The most intense celestial source of x-rays known, associated with a highly variable radio source and a variable optical source, ranging from the twelfth to fourteenth magnitude. Abbreviated Sco X-1.

Scorpius [ASTRON] A southern constellation, right ascension 16 hours, declination 40° south; the bright-red star Antares is located in it. Also known as Scorpion.

scorzalite [MINERAL] $FeAl_2(PO_4)_2(OH)_2$ A blue mineral composed of basic iron aluminum phosphate; it is isomorphous with lazulite.

scotch [ENG] A wooden stopblock or iron catch placed under a wheel or other curved object to prevent slipping or rolling.

Scotch bond *See* American bond.

Scotch derrick *See* stiffleg derrick.

scotch marine boiler [MECH ENG] A self-contained fire-tube steam boiler with internal firing and minimum brickwork; originally developed for use on ships.

Scotch mist [METEOROL] A combination of thick mist (or fog) and heavy drizzle occurring frequently in Scotland and in parts of England.

Scotch pine [BOT] *Pinus sylvestris.* A hard pine of North America having two short, bluish needles in a cluster.

Scotch whiskey [FOOD ENG] A distilled spirit made in Scotland primarily from malted barley.

Scotch yoke [MECH ENG] A type of four-bar linkage; it is employed to convert a steady rotation into a simple harmonic motion.

scotochromogen [MICROBIO] **1.** Any microorganism which produces pigment when grown without light as well as with light. **2.** A member of group II of the atypical mycobacteria.

scotodinia [MED] Dizziness and headache associated with the appearance of black spots before the eyes.

scotoma [MED] A blind spot or area of depressed vision in the visual field.

scotophor [MATER] A solid that exhibits reversible darkening and bleaching actions of tenebrescence under suitable irradiation.

scotopic vision [PHYSIO] Vision that is due to the activity of the rods of the retina only; it is the type of vision that occurs at very low levels of illumination, and it can detect differences of brightness but not of hue.

scotoscope [ELECTR] A telescope which employs an image intensifier to see in the dark.

Scott connection [ELECTR] A type of transformer which transmits power from two-phase to three-phase systems, or vice versa.

Scott-Darey process [CIV ENG] A chemical precipitation method used for fine solids removal in sewage plants; employs ferric chloride solution made by treating scrap iron with chlorine.

Scott top [ELEC] Transformers arranged in the Scott connection for converting electrical power from two-phase to three-phase, or vice versa.

scour *See* tidal scour.

scour and fill [GEOL] The process of first digging out and then refilling a channel instigated by the action of a stream or tide; refers particularly to the process that occurs during a period of flood.

scour depression [GEOL] A crescent-shaped hollow in the stream bed near the outside of the stream's bend, caused by water that scours below the grade of the stream.

scouring [ENG] Physical or chemical attack on process equipment surfaces, as in a furnace or fluid catalytic cracker. [GEOL] **1.** An erosion process resulting from the action of the flow of air, ice, or water. **2.** *See* glacial scour. [MATER] *See* attrition. [MECH ENG] Mechanical finishing or cleaning of a hard surface by using an abrasive and low pressure. [TEXT] **1.** Removal of grease and dirt from wool. **2.** The cleaning of fabric before the dyeing step.

scouring agent [TEXT] A cleaner used to remove oily substances from textile fibers; for example, natural oils and fats from raw wool, or lubricants from rayon yarns or fabrics during an operation such as winding or weaving.

scouring basin [CIV ENG] A basin containing impounded water which is released at about low water in order to maintain the desired depth in the entrance channel. Also known as sluicing pond.

scourway [GEOL] A channel created by a powerful water current, particularly the temporary channels formed by streams on the edge of a Pleistocene ice sheet.

scout [ENG] An engineer who makes a preliminary examination of promising oil and mining claims and prospects. [NAV] **1.** To search an area by following an orderly pattern of courses. **2.** A craft engaged in search.

Scout [AERO ENG] A four-stage all-solid-propellant rocket, used as a space probe and orbital test vehicle; first launched July 1, 1960, with a 150-pound (68-kilogram) payload.

scout boring [MIN ENG] A bore made to test a geologic formation being prospected.

scout car [ORD] Lightly armed and armored reconnaissance vehicle, either wheel or half-track, without turrets, adapted

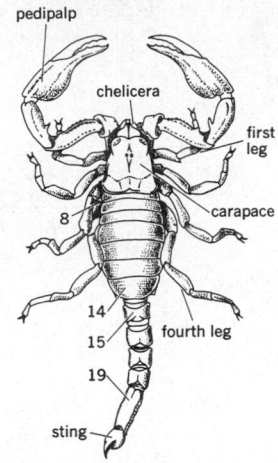

Dorsal aspect of a scorpion, *Chactas vanbenedein.* Some of the body segments have been numbered. (*From R. E. Snodgrass, A Textbook of Arthropod Anatomy, Cornell University Press, 1952*)

SCORPIUS

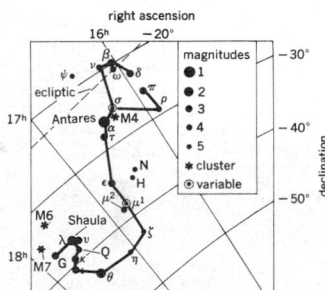

Line pattern of the constellation Scorpius. The grid lines represent the coordinates of the sky. The apparent brightness, or magnitude, of the stars is shown by the sizes of the dots, which are graded by appropriate numbers as indicated.

SCOTCH YOKE

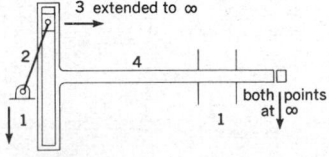

Scotch yoke, a type of four-bar linkage; the numbers indicate the linkage.

for high-speed operation on hard roads and for cross-country missions.

scout hole [MIN ENG] **1.** A borehole penetrating only the uppermost part of an ore body in order to delineate the surface configuration. **2.** A shallow borehole used to ascertain the presence of ore or to explore an area in a preliminary manner.

scouting distance [NAV] The distance between adjacent scouts on a scouting line.

scouting front [NAV] The distance along the scouting line between the outer limits of visibility of the two outboard scouts on the scouting line.

scouting interval [NAV] The distance between consecutive scouts following the same track.

scouting line [NAV] A line on which scouts are located while scouting in a formation suitable to cover a definite pattern.

Sco X-1 *See* Scorpio X-1.

SCR *See* silicon controlled rectifier.

scram [NUCLEO] **1.** A sudden shutting down of a nuclear reactor, usually by dropping safety rods, when a predetermined neutron flux or other dangerous condition occurs. **2.** To close down a reactor by bringing about a scram.

scramble [AERO ENG] To take off as quickly as possible (usually followed by course and altitude instructions). [COMMUN] To mix, in cryptography, in random or quasirandom fashion.

scrambler [ELECTR] A circuit that divides speech frequencies into several ranges by means of filters, then inverts and displaces the frequencies in each range so that the resulting reproduced sounds are unintelligible; the process is reversed at the receiving apparatus to restore intelligible speech. Also known as speech inverter; speech scrambler.

scram drive [MIN ENG] Underground drive above the tramming level, along which scrapers move ore to a discharge chute.

scramjet [AERO ENG] Essentially a ramjet engine, intended for flight at hypersonic speeds. Derived from supersonic combustion ramjet.

scram rod *See* safety rod.

scram system [NUCLEO] A system which causes a scram in a nuclear reactor when dangerous conditions arise.

scrap [ENG] Any solid material cutting or reject of a manufacturing operation, which may be suitable for recycling as feedstock to the primary operation; for example, scrap from plastic or glass molding or metalworking.

scraped-surface exchanger [CHEM ENG] A liquid-liquid heat-exchange device that has a rotating element with springloaded scraper blades to wipe the process-fluid exchange surfaces clean of crystals or other foulants; used in paraffin-wax processing.

scraper conveyor [MECH ENG] A type of flight conveyor in which the element (chain and flight) for moving materials rests on a trough.

scraper hoist [MECH ENG] A drum hoist that operates the scraper of a scraper loader.

scraper loader [MECH ENG] A machine used for loading coal or rock by pulling a scoop through the material to an apron or ramp, where the load is discharged onto a car or conveyor.

scraper ring [MECH ENG] A piston ring that scrapes oil from a cylinder wall to prevent it from being burnt.

scraper ripper [MIN ENG] A piece of strip-mine equipment with teeth on the lip to rip or break the coal and with a flight conveyor to remove the broken coal.

scraper trap [ENG] A device for the insertion or recovery of pigs, or scrapers, that are used to clean the inside surfaces of pipelines.

scrapie [VET MED] A transmissible, usually fatal, virus disease of sheep, characterized by degeneration of the central nervous system.

Scraptidae [INV ZOO] An equivalent name for Melandryidae.

scratchboard [GRAPHICS] A plain, white-coated drawing board, covered with india ink, which, when dry, is scratched with a scratch knife to produce a snow-white line; brilliant, contrasty black-and-white drawings can thus be made.

scratch-brush finish *See* satin finish.

scratch coat [ENG] The first layer of plaster applied to a surface; the surface is scratched to improve the bond with the next coat.

scratch filter [ENG ACOUS] A low-pass filter circuit inserted in the circuit of a phonograph pickup to suppress higher audio frequencies and thereby minimize needle-scratch noise.

scratch hardness [MATER] A measure of the resistance of minerals or metals to scratching; for minerals it is defined by comparison with 10 selected minerals which are numbered in order of increasing hardness according to the Mohs scale.

scratch hardness test [MET] A hardness test in which a cutting point under given pressure is drawn across the surface of a metal, and the width of the scratch is measured. [MINERAL] A determination of the resistance of a mineral to scratching by testing it with minerals on the Mohs scale.

scratch knife [GRAPHICS] A knife point that fits in a standard penholder and makes a fine to broad line on a scratchboard; used in the graphic arts for cutting, erasing, and retouching.

scratch-pad memory [ADP] A very fast intermediate storage (in the form of flip-flop register or semiconductor memory) which often supplements main core memory.

screaming [AERO ENG] A form of combustion instability, especially in a liquid-propellant rocket engine, of relatively high frequency, characterized by a high-pitched noise.

scree [GEOL] **1.** A mound of loose, angular material, less than 10 centimeters. **2.** *See* talus.

screeching [AERO ENG] A form of combustion instability, especially in an afterburner, of relatively high frequency, characterized by a harsh, shrill noise.

screed [CIV ENG] **1.** A straight-edged wood or metal template, fixed temporarily to a surface as a guide when plastering or concreting. **2.** An oscillating metal bar mounted on wheels and spanning a freshly placed road slab, used to strike off and smooth the surface.

screed wire *See* ground wire.

screen [ADP] To make a preliminary selection from a set of entities, selection criteria being based on a given set of rules or conditions. [ELECTR] **1.** The surface on which a television, radar, x-ray, or cathode-ray oscilloscope image is made visible for viewing; it may be a fluorescent screen with a phosphor layer that converts the energy of an electron beam to visible light, or a translucent or opaque screen on which the optical image is projected. Also known as viewing screen. **2.** *See* screen grid. [ELECTROMAG] Metal partition or shield which isolates a device from external magnetic or electric fields. [ENG] **1.** A large sieve of suitably mounted wire cloth, grate bars, or perforated sheet iron used to sort rock, ore, or aggregate according to size. **2.** A covering to give physical protection from light, noise, heat, or flying particles. **3.** A filter medium for liquid-solid separation.

screen analysis [ENG] A method for finding the particle-size distribution of any loose, flowing, conglomerate material by measuring the percentage of particles that pass through a series of standard screens with holes of various sizes.

screen angle [ELECTROMAG] Vertical angle bounded by a straight line from the radar antenna to the horizon and the horizontal at the antenna assuming a 4/3 earth's radius.

screen cloth [MATER] A woven material suitable for use in a screen deck.

screen deck [DES ENG] A surface provided with apertures of specified size, used for screening purposes.

screen dissipation [ELECTR] Power dissipated in the form of heat on the screen grid as the result of bombardment by the electron stream.

screen dryer *See* traveling-screen dryer.

screened pan [METEOROL] An evaporation pan the top of which is covered by wire-mesh screening (¼-inch or 6-millimeter mesh); the screening reduces air circulation and insolation, and results in a pan coefficient nearer to unity than that for unscreened pans.

screened trailing cable [ELEC] A flexible cable provided with a protective screen of conducting material, so applied as to enclose each power core separately or to enclose together all the cores of the cable.

screen grid [ELECTR] A grid placed between a control grid and an anode of an electron tube, and usually maintained at a fixed positive potential, to reduce the electrostatic influence of the anode in the space between the screen grid and the cathode. Also known as screen.

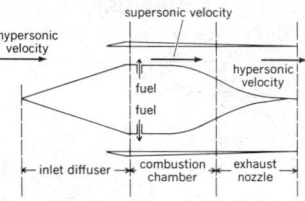

SCRAMJET

Diagram of scramjet engine.

screening [ATOM PHYS] The reduction of the electric field about a nucleus by the space charge of the surrounding electrons. [ELECTROMAG] *See* electric shielding. [ENG] **1.** The separation of a mixture of grains of various sizes into two or more size-range portions by means of a porous or woven-mesh screening media. **2.** The removal of solid particles from a liquid-solid mixture by means of a screen. **3.** The material that has passed through a screen. [IND ENG] The elimination of defective pieces from a lot by inspection for specified defects. Also known as detailing.

screening agent [ANALY CHEM] A nonchelating dye used to improve the colorimetric end point of a complexometric titration; a dye addition forms a complementary pair of colors with the metalized and unmetalized forms of the end-point indicator.

screening constant [ATOM PHYS] The difference between the atomic number of an element and the apparent atomic number for a given process; this difference results from screening.

screening smoke [ORD] A smoke cloud produced by chemical agents or smoke generators; used to conceal friendly troops or to deny observation to enemy troops.

screen memory [PSYCH] A consciously tolerable but usually unimportant memory recalled in place of an associated important one, which would be painful and disturbing.

screen mesh [ENG] A wire network or cloth mounted in a frame for separating and classifying materials.

screen printing [GRAPHICS] A method of printing in which ink is forced by a rubber squeegee through a silk, paint, or stencil screen (as through a sieve or strainer) onto a paper or a fabric. Also known as stencil printing.

screen process [GRAPHICS] A color photography process in which analysis and synthesis of color are accomplished by use of a screen containing a mosaic of very small color filters.

screen size [MIN ENG] A standard for determining the size of diamond particles; the size of the screened particle is determined by the size of the opening through which the diamond particle will not pass.

screw [DES ENG] **1.** A cylindrical body with a helical groove cut into its surface. **2.** A fastener with continuous ribs on a cylindrical or conical shank and a slotted, recessed, flat, or rounded head. Also known as screw fastener.

screw axis [CRYSTAL] A symmetry element of some crystal lattices, in which the lattice is unaltered by a rotation about the axis combined with a translation parallel to the axis and equal to a fraction of the unit lattice distance in this direction.

screw compressor [MECH ENG] A rotary-element gas compressor in which compression is accomplished between two intermeshing, counterrotating screws.

screw conveyor [MECH ENG] A conveyor consisting of a helical screw that rotates upon a single shaft within a stationary trough or casing, and which can move bulk material along a horizontal, inclined, or vertical plane. Also known as auger conveyor; spiral conveyor; worm conveyor.

screw dislocation [CRYSTAL] A dislocation in which atomic planes form a spiral ramp winding around the line of the dislocation.

screwdriver [DES ENG] A tool for turning and driving screws in place; a thin, wedge-shaped or fluted end enters the slot or recess in the head of the screw.

screw elevator [MECH ENG] A type of screw conveyor for vertical delivery of pulverized materials.

screw fastener *See* screw.

screwfeed [MECH ENG] A system or combination of gears, ratchets, and friction devices in the swivel head of a diamond drill, which controls the rate at which a bit penetrates a rock formation.

screw feeder [MECH ENG] A mechanism for handling bulk (pulverized or granulated solids) materials, in which a rotating helicoid screw moves the material forward, toward and into a process unit.

screw jack *See* jackscrew.

screw machine [MECH ENG] A lathe for making relatively small, turned metal parts in large quantities.

screw pile [CIV ENG] A pile having a wide helical blade at the foot which is twisted into position, for use in soft ground or other location requiring a large supporting surface.

screw plasticating injection molding [ENG] A plastic-mold-ing technique in which plastic is converted from pellets to a viscous (plasticated) melt by an extruder screw that is an integral part of the molding machine.

screw press [MECH ENG] A press having the slide operated by a screw mechanism.

screw propeller [MECH ENG] A marine and airplane propeller consisting of a streamlined hub attached outboard to a rotating engine shaft on which are mounted two to six blades; the blades form helicoidal surfaces in such a way as to advance along the axis about which they revolve.

screw pump [MECH ENG] A pump that raises water by means of helical impellers in the pump casing.

screw rivet [DES ENG] A short rod threaded along the length of the shaft that is set without access to the point.

screw spike [DES ENG] A large nail with a helical thread on the upper portion of the shank; used to fasten railroad rails to the ties.

screwstock [MECH ENG] Free-machining bar, rod, or wire.

screw thread [DES ENG] A helical ridge formed on a cylindrical core, as on fasteners and pipes.

screw-thread gage [DES ENG] Any of several devices for determining the pitch, major, and minor diameters, and the lead, straightness, and thread angles of a screw thread.

screw-thread micrometer [DES ENG] A micrometer used to measure pitch diameter of a screw thread.

scribe projection [COMMUN] Method of automatic information presentation; information is placed on a small metallic-coated glass slide by removing the coating with a movable fine-pointed scribe controlled by a servo system; light passing through the scribed area projects on the screen.

scriber [DES ENG] A sharp-pointed tool used for drawing lines on metal workpieces.

scribing [GRAPHICS] A method of preparing a map or chart by cutting the lines into a prepared coating.

Scribner log rule [FOR] A method of scaling logs to derive board-foot calculations; uses a table showing expected log output in board feet that originated from diagrams of 1-inch boards drawn to scale within cylinders of various sizes.

scrim [TEXT] A loose plain-woven fabric, generally cotton, with fine to coarse meshes; used in various finishes in clothing, curtains, and industry.

scrod [VERT ZOO] A young fish, especially cod.

scrofula [MED] Tuberculosis of cervical lymph nodes.

scroll [GEOL] One of a series of crescent-shaped sediments on the inner bank of a moving channel, deposited there by the stream.

scroll gear [DES ENG] A variable gear resembling a scroll with teeth on one face.

scroll meander [GEOL] A type of forced-cut meander, in which the scrolls built on the inner bank cause erosion of the outer bank.

scroll saw [ENG] A saw with a narrow blade, used for cutting curves or irregular designs.

Scrophulariaceae [BOT] A large family of dicotyledonous plants in the order Scrophulariales, characterized by a usually herbaceous habit, irregular flowers, axile placentation, and dry, dehiscent fruit.

Scrophulariales [BOT] An order of flowering plants in the subclass Asteridae distinguished by a usually superior ovary and, generally, either by an irregular corolla or by fewer stamens than corolla lobes, or commonly both.

scrotum [ANAT] The pouch containing the testes.

scrub [AERO ENG] To cancel a scheduled firing, either before or during countdown. [ECOL] A tract of land covered with a generally thick growth of dwarf or stunted trees and shrubs and a poor soil.

scrubber [ENG] A device for the removal, or washing out, of entrained liquid droplets or dust, or for the removal of an undesired gas component from process gas streams. Also known as wet collector. [MIN ENG] A device, such as a wash screen, wash trommel, log washer, and hydraulic jet or monitor, in which a coarse and sticky material, for example, ore or clay, is either washed free of adherents or mildly disintegrated.

scrubbing oil *See* absorption oil.

scrub typhus *See* tsutsugamushi disease.

scruff [MET] A mixture of tin oxide and iron-tin alloy formed as dross on a molten tin-coating bath.

SCREW CONVEYOR

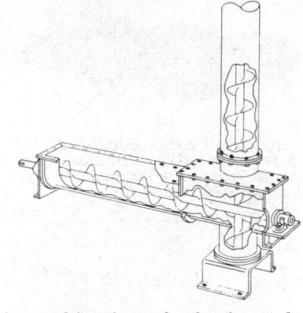

A combination of a horizontal and vertical screw conveyor.

SCREW PROPELLER

Stern view of the twin-screw SS *Florida*, showing the location of the three-bladed propeller and its shaft. The other propeller is hidden from view by the rudder. (*From D. Arnott, ed., Design and Construction of Steel Merchant Ships, Society of Naval Architects and Marine Engineers, 1955*)

SCREW THREAD

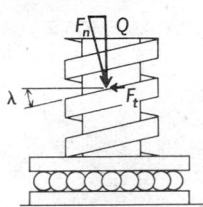

Diagram of a frictionless screw with square threads mounted on a ball thrust bearing, used to raise load Q. Here λ = screw lead angle, F_t = tangential force required to turn screw, F_n = normal force.

SCREW-THREAD GAGE

A plug gage, a type of screw-thread gage.

SCULPIN

The sculpin *Cottus cognatus*, with large mouth, winglike pectoral fins, and large anal and second dorsal fins.

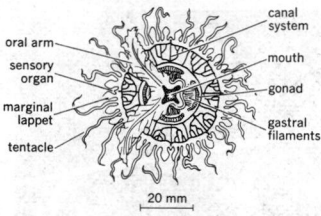

SCYPHOMEDUSA

Schema of scyphomedusa, oral view.

SEA ANEMONE

Sea anemone *(Sagartia)*, a translucent, sessile species that has retractile tentacles.

scruple [PHARM] A unit of mass in apothecaries' measure, equal to 20 grains or to 1.2959782 grams.

SCS *See* silicon controlled switch.

Sc spiral galaxy [ASTRON] A class of spiral galaxy characterized by spirals with the largest and most loosely coiled arms and the smallest central portion.

scuba diving [ENG] Any of various diving techniques using self-contained underwater breathing apparatus.

scud [METEOROL] Ragged low clouds, usually stratus fractus; most often applied when such clouds are moving rapidly beneath a layer of nimbostratus.

scuffing [ENG] The dull mark, sometimes the result of abrasion, on the surface of glazed ceramic or glassware.

scuffle hoe [DES ENG] A hoe having two sharp edges so that it can be pushed and pulled.

sculpin [VERT ZOO] Any of several species of small fishes in the family Cottidae characterized by a large head that sometimes has spines, spiny fins, broad mouth, and smooth, scaleless skin.

Sculptor [ASTRON] A southern constellation, right ascension 0 hours, declination 30° south. Also known as Sculptor's Apparatus; Workshop.

Sculptor's Apparatus *See* Sculptor.

sculpture [GRAPHICS] The art of carving or shaping stone, wood, metal, or other materials into figures, or of modeling figures in wax or clay to be cast in plaster, or bronze or other metals.

scum [MATER] **1.** A film of impurities that rises to or is formed on the surface of a liquid. **2.** A slimy film formed on the surface of a solid object.

scum chamber [CIV ENG] An enclosed compartment in an Imhoff tank, in which gas escapes from the scum which rises to the surface of sludge during sewage digestion.

scupper [NAV ARCH] A drain on or below the deck of a ship that guides water over or through the side.

scupper pipe [NAV ARCH] A pipe leading from the scuppers to the fitting in the ship's side, for draining water from the deckhouse roof.

scurf [MED] A branlike desquamation of the epidermis, especially from the scalp; dandruff.

S curve *See* reverse curve.

scurvy [MED] An acute or chronic nutritional disorder due to vitamin C deficiency; characterized by weakness, subcutaneous hemorrhages, and alterations of any tissue containing collagen, ground substance, dentine, intercellular cement, or osteoid.

scutching [TEXT] The breaking up and separating of the woody portions of flax stems from the retted fiber by means of rollers.

scute [INV ZOO] A cornified, epithelial, scalelike structure in lizards and snakes.

Scutechiniscidae [INV ZOO] A family of heterotardigrades in the suborder Echiniscoidea, with segmental and intersegmental thickenings of cuticle.

Scutelleridae [INV ZOO] The shield bugs, a family of Hemiptera in the superfamily Pentatomoidea.

scutellum [BOT] **1.** A rounded apothecium with an elevated rim found in certain lichens. **2.** The flattened cotyledon of a monocotyledonous plant embryo, such as a grass. [INV ZOO] The third of four pieces forming the upper part of the thoracic segment in certain insects. [VERT ZOO] One of the scales on the tarsi and toes of birds.

Scutigeromorpha [INV ZOO] The single order of notostigmophorous centipedes; members are distinguished by a dorsal respiratory opening, compound-type eyes, long flagellate multisegmental antennae, and long thin legs with multisegmental tarsi.

scuttlebutt [NAV ARCH] **1.** A cask on shipboard which holds a day's supply of drinking water. **2.** A drinking fountain on shipboard or at a marine installation.

scutum [INV ZOO] **1.** A bony, horny, or chitinous plate. **2.** The second of four pieces forming the upper part of the thoracic segment in certain insects. **3.** One or two lower opercular valves in certain barnacles.

Scutum [ASTRON] A southern constellation, right ascension 19 hours, declination 10° south. Also known as Shield.

Scydmaenidae [INV ZOO] The antlike stone beetles, a large cosmopolitan family of the Coleoptera in the superfamily Staphylinoidea.

Scyllaridae [INV ZOO] The Spanish, or shovel-nosed, lobsters, a family of the Scyllaridea.

Scyllaridea [INV ZOO] A superfamily of decapod crustaceans in the section Macrura including the heavily armored spiny lobsters and the Spanish lobsters, distinguished by the absence of a rostrum and chelae.

Scylliorhinidae [VERT ZOO] The catsharks, a family of the cacharinid group of galeoids; members exhibit the most exotic color patterns of all sharks.

scyphistoma [INV ZOO] A sessile, polyploid larva of many Scyphozoa which may produce either more scyphistomae or free-swimming medusae.

scyphomedusa [INV ZOO] A medusa of the scyphozoans.

Scyphomedusae [INV ZOO] A subclass of the class Scyphozoa characterized by reduced marginal tentacles, tetramerous medusae, and medusalike polyploids.

scyphopolyp [INV ZOO] A polyp of the scyphozoans.

Scyphozoa [INV ZOO] A class of the phylum Coelenterata; all members are marine and are characterized by large, well-developed medusae and by small, fairly well-organized polyps.

scythe [DES ENG] A tool with a long curved blade attached at a more or less right angle to a long handle with grips for both hands; used for cutting grass as well as grain and other crops.

Scythian stage [GEOL] A stage in the lesser Triassic series of the alpine facies. Also known as Werfenian stage.

SDA *See* automation source data.

Se *See* selenium.

sea [GEOGR] A usually salty lake lacking an outlet to the ocean. [OCEANOGR] **1.** A major subdivision of the ocean. **2.** A heavy swell or ocean wave still under the influence of the wind that produced it. **3.** *See* ocean.

sea-air temperature difference correction [NAV] A correction to sextant altitude readings made necessary by abnormal terrestrial refraction occurring when there is a nonstandard density lapse rate in the atmosphere due to a difference in the temperature of the water and of the air at the surface.

sea anchor [NAV ARCH] An object towed by a usually small vessel to keep the vessel end-on to a heavy sea or surf or to reduce drift; the usual form is a conical canvas bag whose large end is open, and, when towed with the large end in the forward position, the bag offers considerable resistance.

sea anemone [INV ZOO] Any of the 1000 marine coelenterates that constitute the order Actiniaria; the adult is a cylindrical polyp or hydroid stage with the free end bearing tentacles that surround the mouth.

sea bank *See* seawall.

Sea Bat [AERO ENG] An antisubmarine warfare helicopter equipped with active-passive sonar, acoustic homing torpedoes, and instrument–night flight capability. Designated SH-34G.

seabeach [GEOL] A beach along the margin of the sea.

seabed *See* sea floor.

sea bottom *See* sea floor.

sea breeze [METEOROL] A coastal, local wind that blows from sea to land, caused by the temperature difference when the sea surface is colder than the adjacent land; it usually blows on relatively calm, sunny summer days, and alternates with the oppositely directed, usually weaker, nighttime land breeze.

sea breeze of the second kind *See* cold-front-like sea breeze.

sea buoy [NAV] The outermost buoy marking the entrance to a channel or harbor. Also known as farewell buoy, since it is the last buoy passed by a vessel proceeding out to sea.

sea cave [GEOL] A split or hollow opening, usually at sea level, in the base of a sea cliff, formed by waves acting on weak parts of the weathered rock. Also known as marine cave; sea chasm.

sea channel [GEOL] A long, narrow, U-shaped or V-shaped shallow depression of the sea floor, usually occurring on a gently sloping plain or fan.

sea chasm *See* sea cave.

sea chest [NAV ARCH] **1.** A trunk in which a sailor stores his personal property. **2.** A pipe between a ship's side and a valve in the hull for draining water.

sea cliff [GEOL] An erosional landform, produced by wave action, which is either at the seaward edge of the coast or at the landward side of a wave-cut platform and which denotes the inner limit of the beach erosion.

sea clutter [ELECTROMAG] A clutter on an airborne radar due to reflection of signals from the sea. Also known as sea return; wave clutter.

seacoast [GEOGR] The land adjacent to the sea.

seacoast artillery [ORD] Class of artillery formerly used for seacoast defense; consisted of fixed guns, howitzers, and mortars defending the harbors.

sea-control operations [ORD] The employment of naval forces, supported by land and air forces, as appropriate, to achieve military objectives in vital sea areas; such operations include destruction of enemy naval forces, suppression of enemy sea commerce, protection of vital sea lanes, and establishment of local military superiority in areas of naval operations.

sea cucumber [INV ZOO] The common name for the echinoderms that make up the class Holothuroidea.

seadrome [CIV ENG] **1.** A designated area for landing and takeoff of seaplanes. **2.** A platform at sea for landing and takeoff of land planes.

sea echelon [ORD] A portion of the assault shipping which withdraws from or remains out of the transport area during an amphibious landing and operates in designated areas to seaward in an on-call or unscheduled status.

sea fan [GEOL] See submarine fan. [INV ZOO] A form of horny coral that branches like a fan.

sea floor [GEOL] The bottom of the ocean. Also known as seabed; sea bottom.

sea floor spreading [GEOL] The hypothesis that the ocean floor is spreading away from the midoceanic ridges and is being conveyed landward by convective cells in the earth's mantle, carrying the continental blocks as passive passengers; the ocean floor moves away from the midoceanic ridge at the rate of 1 to 10 centimeters per year and provides the source of power in the hypothesis of plate tectonics. Also known as ocean floor spreading; spreading concept; spreading floor hypothesis.

sea-foam See sepiolite.

sea fog [METEOROL] A type of advection fog formed over the ocean as a result of any of a variety of processes, as when air that has been lying over a warm water surface is transported over a colder water surface, resulting in a cooling of the lower layer of air below its dew point.

sea front [GEOGR] An area partly bounded by the sea.

sea frontier [ORD] The naval command of a coastal frontier, including the coastal zone in addition to the land area of the coastal frontier and the adjacent sea areas.

sea gate [CIV ENG] A gate which serves to protect a harbor or tidal basin from the sea, such as one of a pair of supplementary gates at the entrance to a tidal basin exposed to the sea. [GEOGR] A way giving access to the sea such as a gate, channel, or beach.

sea glow [OCEANOGR] The luminous, cobalt-blue appearance of very clear water in the open ocean, caused by upward-scattered light from which much of the red has been absorbed.

seagoing barge [NAV ARCH] A barge capable of sailing on the open sea.

seagoing tug [NAV ARCH] A tugboat capable of sailing on the open sea.

sea horse [INV ZOO] Any of about 50 species of tropical and subtropical marine fishes constituting the genus *Hippocampus* in the family Syngnathidae; the body is compressed, the head is bent ventrally and has a tubiform snout, and the tail is tapering and prehensile.

Sea Horse See Capricornus.

sea ice [OCEANOGR] **1.** Ice formed from seawater. **2.** Any ice floating in the sea.

sea-ice shelf [OCEANOGR] Sea ice floating in the vicinity of its formation and separated from fast ice, of which it may have been a part, by a tide crack or a family of such cracks.

seakeeping [NAV ARCH] The ability of a vessel to navigate safely at sea during heavy storms.

sea kindliness [NAV ARCH] Property of a vessel which is well adapted to being handled safely at sea in heavy weather.

sea knoll See knoll.

seal [ENG] **1.** Any device or system that creates a nonleaking

union between two mechanical or process-system elements; for example, gaskets for pipe connection seals, mechanical seals for rotating members such as pump shafts, and liquid seals to prevent gas entry to or loss from a gas-liquid processing sequence. **2.** A tight, perfect closure or joint. [VERT ZOO] Any of various carnivorous mammals of the suborder Pinnipedia, especially the families Phoridae, containing true seals, and Otariidae, containing the eared and fur seals.

sea-lane See seaway.

sea-launched ballistic missile [ORD] A missile launched from a submarine or surface ship.

seal coat [MATER] A layer of bituminous material applied to bituminous macadam or concrete to seal the surface.

sealed cabin [AERO ENG] The occupied space of an aircraft or spacecraft characterized by walls which do not allow gaseous exchange between the inner atmosphere and its surrounding atmosphere, and containing its own mechanisms for maintenance of the inside atmosphere.

sealed tube [ELECTR] Electron tube which is hermetically sealed.

sealer [MATER] A preliminary coating applied to seal the pores in a porous, uncoated surface, such as wood.

Seale rope [DES ENG] A wire rope with six or eight strands, each having a large wire core covered by nine small wires, which, in turn, are covered by nine large wires.

sea level [GEOL] The level of the surface of the ocean; especially, the mean level halfway between high and low tide, used as a standard in reckoning land elevation or sea depths.

sea-level chart See surface chart.

sea-level pressure [METEOROL] The atmospheric pressure at mean sea level, either directly measured or, most commonly, empirically determined from the observed station pressure.

sea-level pressure chart See surface chart.

sea lily [INV ZOO] The common name for those crinoids in which the body is flower-shaped and is carried at the tip of an anchored stem.

sealing [MET] **1.** Impregnation of porous castings with resins to overcome porosity. **2.** Reducing porosity of an anodic oxide film on aluminum and aluminum alloys by immersion in boiling water.

sealing compound [ELEC] A compound used in dry batteries, capacitor blocks, transformers, and other components to keep out air and moisture.

sealing tape [MATER] Gummed tape for sealing packages.

sealing wax [MATER] A colored, scented mixture of resins and shellac; used for sealing containers and documents.

sea lion [VERT ZOO] Any of several large, eared seals of the Pacific Ocean; related to fur seals but lack a valuable coat.

seal off [ENG] To close off, as a tube or borehole, by using a cement or other sealant to eliminate ingress or egress.

seal oil [MATER] A yellowish, liquid, fatty oil obtained from seal blubber; soluble in ether and chloroform; melts at 22–33°C; used in soapmaking, in dressing animal skins, and as a lubricant.

seam [ENG] **1.** A mechanical or welded joint. **2.** A mark on ceramic or glassware where matching mold parts join. [GEOL] **1.** A stratum or bed of coal or other mineral. **2.** A thin layer or stratum of rock. **3.** A very narrow coal vein. [MET] An unwelded fold or lap which appears as a crack on the surface of a casting or wrought product.

seamanite [MINERAL] $Mn_3(PO_4)(BO_3) \cdot 3H_2O$ A pale- to wine-yellow orthorhombic mineral that is a phosphate and borate of manganese; occurs in crystals.

sea manners [NAV] A consideration for the other ship or craft and the exercise of good judgment under certain conditions when ships or craft meet.

seamanship [NAV] The ability to handle, manage, and navigate a ship at sea.

seamark [NAV] An object designed as an aid to navigation and located with the express purpose of being visible from a distance to seaward; it is often erected in shoal water rather than on land.

sea marker [ENG] A patch of color on the ocean surface produced by releasing dye; used, for example, to attract the attention of the crew of a rescue airplane.

seam blast [MIN ENG] A blast made by placing powdered or

SEA HORSE

The sea horse *Hippocampus hudsonius*, which has gill tufts and 11 trunk rings and grows to a length of 6 inches (15 centimeters).

SEAL

Alaska fur seal (*Callorhinus ursinus*).

other explosive along and in a seam or crack between the solid wall and the stone or coal to be removed.

sea mile [NAV] An approximate mean value of the nautical mile equal to 6080 feet (1853.184 meters) or the length of a minute of arc along the meridian at latitude 48°.

seaming [MET] The joining of the edges of sheet-metal parts by interlocking folds.

sea mist *See* steam fog.

seamless ring rolling [MET] The hot-rolling of a circular blank, with a hole in the center, to form a seamless ring.

seamless tubing [MET] A tubing made by extrusion or by piercing and rolling a billet.

seamount [GEOL] An elevation of the sea floor that is either flat-topped or peaked, rising to about 3000–1000 feet (900–300 meters) or more, with the summit approximately 1000–6000 feet (300–1800 meters) below sea level.

seamount chain [GEOL] Several seamounts in a line with bases separated by a relatively flat sea floor.

seamount group [GEOL] Several closely spaced seamounts not in a line.

seamount range [GEOL] Three or more seamounts having connected bases and aligned along a ridge or rise.

seam weld [MET] **1.** A longitudinal weld joining of sheet-metal parts or in making tubing. **2.** Arc or resistance welding in which a series of overlapping spot welds is produced.

sea otter [VERT ZOO] *Enhydra lutris.* A large marine otter found close to the shoreline in the North Pacific; these animals are diurnally active and live in herds.

sea peak [GEOL] A peaked elevation of the sea floor, rising 1000 meters or more from the floor.

sea pen [INV ZOO] The common name for coelenterates constituting the order Pennatulacea.

seaplane [AERO ENG] An airplane that takes off from and alights on the water; it is supported on the water by pontoons, or floats, or by a hull which is a specially designed fuselage. Also known as airboat.

seaport [CIV ENG] A harbor or town that has facilities for seagoing ships and is active in marine activities.

seaquake [GEOPHYS] An earth tremor whose epicenter is beneath the ocean and can be felt only by ships in the vicinity of the epicenter. Also known as submarine earthquake.

sear [ORD] **1.** An item that retains the firing mechanism of a gun in the cocked position. **2.** A variety of lockwork in the firing mechanism of a propellant-actuated device which prevents motion of the firing pin until released.

sea rainbow *See* marine rainbow.

search [ADP] To seek a desired item or condition in a set of related or similar items or conditions, especially a sequentially organized or nonorganized set, rather than a multidimensional set. [ENG] To explore a region in space with radar. [NAV] An orderly arrangement of course lines used in searching an area.

search and rescue [ENG] The use of aircraft, surface craft, submarines, specialized rescue teams and equipment to search for and rescue personnel in distress on land or at sea.

search and rescue coordination center [COMMUN] A primary search and rescue facility suitably staffed by supervisory personnel and equipped for coordinating and controlling search and rescue operations.

search antenna [ELECTROMAG] A radar antenna or antenna system designed for search.

search attack unit [ORD] The designation given to one or more ships separately organized or detached from a formation as a tactical unit to search for and destroy submarines.

search coil *See* exploring coil.

search gate [ELECTR] A gate pulse used to search back and forth over a certain range.

searching control [ENG] A mechanism that changes the azimuth and elevation settings on a searchlight automatically and constantly, so that its beam is swept back and forth within certain limits.

searching fire [ORD] Fire distributed in depth by successive changes in the elevation of the gun.

searching lighting *See* horizontal scanning.

searchlight-control radar [ENG] A ground-based radar used to direct searchlights at aircraft.

searchlighting [NAV] The directing of a beam of radiant

energy in an oscillatory manner but continuously at a single target.

search radar [ENG] A radar intended primarily to cover a large region of space and to display targets as soon as possible after they enter the region; used for early warning, in connection with ground-controlled approach and interception, and in air-traffic control.

search time [ADP] Time required to locate a particular field of data in a computer storage device; requires a comparison of each field with a predetermined standard until an identity is obtained.

sea return *See* sea clutter.

searlesite [MINERAL] $NaB(SiO_3)_2 \cdot H_2O$ A white mineral composed of hydrous sodium borosilicate occurring as spherulites.

sea room [NAV] Space in which to maneuver without danger of grounding or colliding.

sea-run [VERT ZOO] Having the habit of ascending a river from the sea, especially to spawn, as salmon and brook trout.

sea salt [OCEANOGR] The salt remaining after the evaporation of seawater, containing sodium and magnesium chlorides and magnesium and calcium sulfates.

sea-salt nucleus [OCEANOGR] A condensation nucleus of a highly hygroscopic nature produced by partial or complete desiccation of particles of sea spray or of seawater droplets derived from breaking bubbles.

seascape [OCEANOGR] The surrounding sea as it appears to an observer.

seascarp [GEOL] A submarine cliff that is relatively long, high, and straight.

seashell [INV ZOO] The shell of a marine invertebrate, especially a mollusk.

seashore [GEOL] **1.** The strip of land that borders a sea or ocean. Also known as seaside; shore. **2.** The ground between the usual tide levels. Also known as seastrand.

seashore lake [GEOGR] A lake, containing either fresh or salt water, which lies along a seashore; it is separated from the sea by a river, a delta, or a wall of sediment.

seasickness [MED] Motion sickness occurring at sea. Also known as pelagism.

seaside *See* seashore.

sea slug [INV ZOO] The common name for the naked gastropods composing the suborder Nudibranchia.

sea smoke *See* steam fog.

season [CLIMATOL] A division of the year according to some regularly recurrent phenomena, usually astronomical or climatic.

seasonal balancing [CHEM ENG] A seasonal adjustment of the front-end boiling range (volatility) of a motor gasoline to control engine starting characteristics by compensating for seasonal temperature changes.

seasonal current [OCEANOGR] An ocean current which has large changes in speed or direction due to seasonal winds.

seasonal factors [COMMUN] Factors that are used to adjust skywave absorption data for seasonal variations; these variations are due primarily to seasonal fluctuations in the heights of the ionospheric layers.

seasonally frozen ground [GEOL] Ground that is frozen during low temperatures and remains so only during the winter season. Also known as frost zone.

seasonal stream [HYD] A stream whose flow is not constant because it has water in its course only during certain seasons.

seasonal thermocline [OCEANOGR] A thermocline which develops in the oceans in summer at relatively shallow depths due to surface heating and downward transport of heat caused by mixing of water generated by summer winds.

seasonal variation [GEOPHYS] The variation in the ionization of different parts of the ionosphere, and the resulting variation in transmission of radio signals over large distances, with the season.

season check [MATER] A longitudinal crack in wood, caused by uneven seasoning.

season crack [MET] A stress-corrosion crack produced in a copper-base alloy subject to a residual or applied tensile stress and exposed to a specific environment such as moist air containing traces of ammonia.

season cracking [ORD] An occurrence in brass cartridge cases and other brass parts in which cracking occurs because

SEASONAL THERMOCLINE

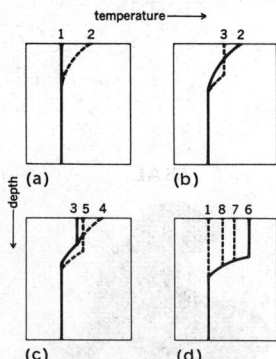

Diagrams showing temperature structure of seasonal thermocline during *(a)* first stage, *(b)* second stage, and *(c)* third stage of formation, and *(d)* breakup. Numbers 1–8 indicate sequence in this process.

of residual internal strains from the manufacturing operations.

seasoned lumber [MATER] Lumber which has been cured by drying to ensure a uniform moisture content.

seasoning [ELECTR] Overcoming a temporary unsteadiness of a component that may appear when it is first installed.

sea spider [INV ZOO] The common name for arthropods in the subphylum Pycnogonida.

sea squirt [INV ZOO] A sessile, marine tunicate of the class Ascidiacea; it squirts water from two openings in the unattached end when touched or disturbed.

sea state [OCEANOGR] The numerical or written description of ocean-surface roughness.

seastrand *See* seashore.

sea supremacy [ORD] That degree of sea superiority wherein the opposing force is incapable of effective interference.

sea-surface slope [OCEANOGR] A gradual change in the level of the sea surface with distance, caused by Coriolis and wind forces.

sea surveillance [ENG] The systematic observation of surface and subsurface sea areas by all available and practicable means primarily for the purpose of locating, identifying, and determining the movements of ships, submarines, and other vehicles, friendly and enemy, proceeding on or under the surface of seas and oceans.

seat [ORD] **1.** Support or holder for a mechanism, or for a part of one. **2.** To fit correctly in or on a holder, or prepared position, such as to seat a fuse in a bomb, a projectile in the bore of a gun, or a cartridge in a chamber.

seat clay *See* underclay.

seat earth *See* underclay.

sea terrace *See* marine terrace.

seating [ORD] The distance to which a projectile is rammed into the bore of a cannon, usually measured from the base of the projectile to the rear face of the breech.

seating-lock locking fastener [DES ENG] A locking fastener that locks only when firmly seated and is therefore free-running on the bolt.

sea turn [METEOROL] A wind coming from the sea, often bringing mist; the term is limited mainly to New England, United States.

sea turtle [VERT ZOO] Any of various marine turtles, principally of the families Cheloniidae and Dermochelidae, having paddle-shaped feet.

seatworm *See* pinworm.

sea urchin [INV ZOO] A marine echinoderm of the class Echinoidea; the soft internal organs are enclosed in and protected by a test or shell consisting of a number of close-fitting plates beneath the skin.

sea valley [GEOL] A relatively shallow, wide depression with gentle slopes in the sea floor, the bottom of which grades continuously downward.

sea van [IND ENG] Commercial or government-owned (or leased) shipping containers which are moved via ocean transportation; since wheels are not attached, they must be lifted on and off the ship.

seawall [CIV ENG] A concrete, stone, or metal wall or embankment constructed along a shore to reduce wave erosion and encroachment by the sea. Also known as sea bank. [GEOL] A steep-faced, long embankment situated by powerful storm waves along a seacoast at high-water mark.

seaward [NAV] The direction or side in which the sea is located; the direction or side away from the land.

seawater [OCEANOGR] Water of the seas, distinguished by high salinity. Also known as salt water.

seawater thermometer [ENG] A specially designed thermometer to measure the temperature of a sample of seawater; an instrument consisting of a mercury-in-glass thermometer protected by a perforated metal case.

seaway [NAV] **1.** The motion or rate of motion of a vessel. **2.** Headway of a vessel. **3.** The sea as a route of travel from one place to another; a shipping lane. Also known as sea-lane.

seaweed [BOT] A marine plant, especially algae.

seaworthiness [NAV ARCH] The fitness of a ship to sail on the sea and meet any usual condition, or to sail on a specific voyage.

sebaceous gland [PHYSIO] A gland, arising in association

with a hair follicle, which produces and liberates sebum.

sebacic acid [ORG CHEM] COOH(CH$_2$)$_8$COOH Combustible, white crystals; slightly soluble in water, soluble in alcohol and ether; melts at 133°C; used in perfumes, paints, and hydraulic fluids and to stabilize synthetic resins. Also known as decanedioic acid; sebacylic acid.

sebacylic acid *See* sebacic acid.

sebcha *See* sebkha.

Sebekidae [INV ZOO] A family of pentastomid arthropods in the suborder Porocephaloidea.

sebka *See* sebkha.

sebkha [GEOL] A geologic feature, in North Africa, which is a smooth, flat, plain usually high in salt; after a rain the plain may become a marsh or a shallow lake until the water evaporates. Also known as sabkha; sebcha; sebka; sibjet.

seborrheic dermatitis [MED] An acute inflammation of the skin, occurring usually on oily skin; characterized by scales, crusting yellowish patches, and itching.

sebum [PHYSIO] The secretion of sebaceous glands, composed of fat, cellular debris, and keratin.

sec *See* secant; second; secondary winding.

SEC *See* secondary electron conduction.

seca [METEOROL] A drought, or dry wind, in Brazil.

secalose [BIOCHEM] A polysaccharide consisting of fructose units; occurs in green rye and oats, and in rye flour.

sec-*n*-amyl alcohol *See* diethyl carbinol; methyl propyl carbinol.

secant [MATH] **1.** The function given by the reciprocal of the cosine function. Abreviated sec. **2.** The secant of an angle *A* is 1/cos *A*.

secant conic chart *See* conic chart with two standard parallels.

secant conic projection *See* conic projection with two standard parallels.

Secchi disk [ENG] An opaque white disk used to measure the transparency or clarity of seawater by lowering the disk into the water horizontally and noting the greatest depth at which it can be visually detected.

Secernentea [INV ZOO] A class of the phylum Nematoda in which the primary excretory system consists of intracellular tubular canals joined anteriorly and ventrally in an excretory sinus, into which two ventral excretory gland cells may also open.

sech *See* hyperbolic secant.

sechard [METEOROL] A dry, warm foehn wind over Lake Geneva in Switzerland.

seclusion [METEOROL] A special case of the process of occlusion, where the point at which the cold front first overtakes the warm front (or quasi-stationary front) is at some distance from the apex of the wave cyclone.

secobarbital [PHARM] C$_{12}$H$_{18}$N$_2$O$_3$ 5-Allyl-5 (1-methylbutyl) barbituric acid, a short-acting barbiturate; white powder with bitterish taste; very soluble in alcohol and ether, slightly soluble in water; melts at 82°C; used as a sedative and hypnotic, frequently as the sodium derivative. Also known as quinalbarbitone.

secodont [VERT ZOO] Having teeth adapted for cutting.

secohm *See* international henry.

Seconal [PHARM] A trade name for secobarbital.

second [MATH] A unit of plane angle, equal to 1/60 minute, or 1/3,600 degree, or π/648,000 radian. [MECH] The fundamental unit of time equal to 9,192,631,770 periods of the radiation corresponding to the transition between the two hyperfine levels of the ground state of an atom of cesium-133. Abbreviated s: sec.

secondary [ELEC] Low-voltage conductors of a power distributing system. [ELECTROMAG] *See* secondary winding. [GEOL] A term with meanings that changed from early to late in the 19th century, when the term was confined to the entire Mesozoic era; it was finally replaced by Mesozoic era.

secondary alcohol [ORG CHEM] An organic alcohol with molecular structure R$_1$R$_2$CHOH, where R$_1$ and R$_2$ designate either identical or different groups.

secondary amine [ORG CHEM] An organic compound that may be written R$_1$R$_2$NH, where R$_1$ and R$_2$ designate either identical or different groups.

secondary amyloidosis [MED] Amyloidosis that usually fol-

SEA SQUIRT

Sea squirts; these saclike tunicates serve as scavengers and as food for higher forms.

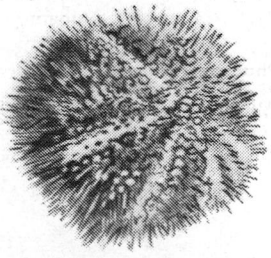

SEA URCHIN

The sea urchin shell, which protects the soft internal organs, is covered with spines arranged in five broad areas that are separated by narrow unprotected areas.

lows chronic suppurative, inflammatory diseases, such as tuberculosis, osteomyelitis, and bronchiectasis.

secondary area [NAV] In air operations, that area within a segment in which ROC (required obstacle clearance) is reduced as distance from the prescribed course is increased.

secondary armament [ORD] In ships with multiple-size guns installed, that battery consisting of guns next largest to those of the main battery.

secondary battery *See* storage battery.

secondary blast injuries [ORD] Those injuries sustained from the indirect effects, such as falling rubble from a collapsed building, or missiles (debris or objects) which have been picked up by the generated winds and hurled against an individual; also includes injuries resulting from individuals being hurled against stationary objects.

secondary bomb damage [ORD] Nonphysical bomb damage.

secondary bow *See* secondary rainbow.

secondary cambium [BOT] One of the tissue layers formed after the initial cambial layers in certain plant roots, and that produce a ring of tissue.

secondary cell *See* storage cell.

secondary circle *See* secondary great circle.

secondary coast [GEOL] A relatively stable seacoast or shoreline whose features are the result of present-day marine processes.

secondary cold front [METEOROL] A front which forms behind a frontal cyclone and within a cold air mass, characterized by an appreciable horizontal temperature gradient.

secondary cosmic rays [GEOPHYS] Radiation produced when primary cosmic rays enter the atmosphere and collide with atomic nuclei and electrons.

secondary creep [MECH] The change in shape of a substance under a minimum and almost constant differential stress, with the strain-time relationship a constant. Also known as steady-state creep.

secondary crusher [MECH ENG] Any of a group of crushing and pulverizing machines used after the primary treatment to further reduce the particle size of shale or other rock.

secondary cyclone [METEOROL] A cyclone which forms near or in association with a primary cyclone. Also known as secondary low.

secondary drilling [MIN ENG] The process of drilling the so-called popholes for the purpose of breaking the larger masses of rock thrown down by the primary blast.

secondary electron [ELECTR] **1.** An electron emitted as a result of bombardment of a material by an incident electron. **2.** An electron whose motion is due to a transfer of momentum from primary radiation.

secondary electron conduction [ELECTR] Transport of charge by secondary electrons moving through the interstices of a porous material under the influence of an externally applied electric field. Abbreviated SEC.

secondary emission [ELECTR] The emission of electrons from the surface of a solid or liquid into a vacuum as a result of bombardment by electrons or other charged particles.

secondary enlargement [MINERAL] Overgrowth by chemical deposition on a mineral grain of additional material of identical composition in optical and crystallographic continuity with the original grain; crystal faces characteristic of the original mineral often result. Also known as secondary growth.

secondary enrichment [GEOL] The addition to a vein or ore body of material that originated later in time from the oxidation of decomposed ore masses that overlie the vein.

secondary fermentation [FOOD ENG] A fermentation process produced by adding sugar to wine and used to create natural carbonation in, for example, champagne.

secondary flow [FL MECH] A field of fluid motion which can be considered as superposed on a primary field of motion through the action of friction usually in the vicinity of solid boundaries. Also known as frictional secondary flow.

secondary focus [OPTICS] In an astigmatic system, a line at which some of the bundle of rays from an off-axis point meet; this line lies in a plane which contains the point and the optical axis, and has a greater image distance than the primary focus. Also known as sagittal focus.

secondary fragment [ORD] Secondary fragmentation resulting from either breakup of controlled or uncontrolled fragments upon impact or from the creation of fragments from the target material when it is impacted by a projectile; an example is the spalling of armor plate when it is pierced by armor-piercing shot; the fragments or spalls are broken off the back of the plate and become significant kill mechanisms within the armored enclosure.

secondary front [METEOROL] A front which may form within a baroclinic cold air mass which itself is separated from a warm air mass by a primary frontal system; the most common type is the secondary cold front.

secondary great circle [GEOD] A great circle perpendicular to a primary great circle, as a meridian. Also known as secondary circle.

secondary grid emission [ELECTR] Electron emission from a grid resulting directly from bombardment of its surface by electrons or other charged particles.

secondary grinding [MECH ENG] A further grinding of material previously reduced to sand size.

secondary growth *See* secondary enlargement.

secondary hardening [MET] The hardening of certain alloy steels at moderate temperatures (250–650°C) by the precipitation of carbides; the resultant hardness is greater than that obtained by tempering the steel at some lower temperature for the same time.

secondary haulage [MIN ENG] That portion of the haulage system which collects coal from gathering-haulage delivery points and delivers it to the main portion of the system.

secondary high explosive [MATER] A high explosive which is relatively insensitive to heat and shock and is usually initiated by a primary high explosive; used for boosters and bursting charges.

secondary interstices [GEOL] Openings in a rock that formed after the enclosing rock was formed.

secondary ion mass analyzer [ENG] A type of secondary ion mass spectrometer that provides general surface analysis and depth-profiling capabilities.

secondary ion mass spectrometer [ENG] An instrument for microscopic chemical analysis, in which a beam of primary ions with an energy in the range 5–20 kilo-electron-volts bombards a small spot on the surface of a sample, and positive and negative secondary ions sputtered from the surface are analyzed in a mass spectrometer. Abbreviated SIMS. Also known as ion microprobe; ion probe.

secondary item [ORD] Any item, including end items, components, and spare parts, which has not been classified as a principal item.

secondary lobe *See* minor lobe.

secondary low *See* secondary cyclone.

secondary manganous phosphate *See* acid manganous phosphate.

secondary metal [MET] Metal recovered from scrap by remelting and refining.

secondary mineral [MINERAL] A mineral produced in an enclosing rock after the rock was formed as a result of weathering or metamorphic or solution activity, and usually at the expense of a primary material that came into existence earlier.

secondary oil recovery [PETRO ENG] Procedures used to increase the flow of oil from depleted or nearly depleted wells; includes fracturing, acidizing, waterflood, and gas injection.

secondary periderm [BOT] Any layer of the periderm except the first and outermost layer.

secondary phloem [BOT] Phloem produced by the cambium, consisting of two interpenetrating systems, the vertical or axial and the horizontal or ray.

secondary plasticizer [MATER] A plastics plasticizer that has insufficient affinity for a resin for it to be the sole plasticizer, and must be blended with a primary plasticizer. Also known as extender plasticizer.

secondary porosity [GEOL] The interstices that appear in a rock formation after it has formed, because of dissolution or stress distortion taking place naturally or artificially as a result of the effect of acid treatment or the injection of coarse sand.

secondary port [CIV ENG] A port with one or more berths,

SECONDARY PHLOEM

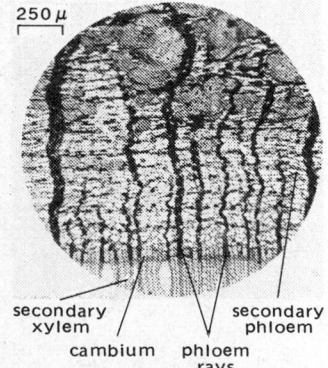

secondary xylem

secondary phloem

cambium

phloem rays

Photomicrograph of cross section of the secondary phloem of paper birch *(Betula papyrifera).* *(Forest Products Laboratory, USDA)*

normally at quays, which can accommodate oceangoing ships for discharge.

secondary radar [ELECTR] Radar which receives pulses transmitted by an interrogator and makes a return transmission (usually on a different frequency) by its transponder, as opposed to a primary radar which receives pulses returned from illuminated objects.

secondary radiation [PHYS] Particles or photons produced by the action of primary radiation on matter, such as Compton recoil electrons, delta rays, secondary cosmic rays, and secondary electrons.

secondary rainbow [OPTICS] A faint rainbow of angular radius about 50° which may appear outside the primary rainbow of 42° radius, and which has its colors in reverse order to those of the primary. Also known as secondary bow.

secondary reference fuel [MATER] A commercially produced internal combustion engine fuel which is acceptable for knock testing or cetane rating, and which has been calibrated against primary reference fuels by engine tests.

secondary reflection *See* multiple reflection; shoot.

secondary rescue facilities [ENG] Local airbase-ready aircraft, crash boats, and other air, surface, subsurface, and ground elements suitable for rescue missions, including government and privately operated units and facilities.

secondary reserves [PETRO ENG] Reserves recoverable commercially at current prices and costs as a result of artificial supplementation of the reserve's natural (gas-drive) energy.

secondary sewage sludge [CIV ENG] Sludge that includes activated sludge, mixed sludge, and chemically precipitated sludge.

secondary shaft [MIN ENG] The shaft which extends a mine downward from the bottom of, but not in line with, the primary shaft.

secondary splits [MIN ENG] Splits formed by separation of the main air splits.

secondary standard [PHYS] **1.** A unit, as of length, capacitance, or weight, used as a standard of comparison in individual countries or localities, but checked against the one primary standard in existence somewhere. **2.** A unit defined as a specified multiple or submultiple of a primary standard, such as the centimeter.

secondary station [COMMUN] Any station in a radio net other than the net control station. [ORD] Observation post at the end of a base line farthest from the gun or directing point.

secondary storage [ADP] Any means of storing and retrieving data external to the main computer itself but accessible to the program.

secondary stratification [GEOL] The layering that occurs when sediments that were at one time deposited are resuspended and redeposited. Also known as indirect stratification.

secondary stratigraphic trap *See* stratigraphic trap.

secondary structure [GEOL] A structure such as a fault, fold, or joint resulting from tectonic movement that started after the rock in which it is found was emplaced. [PALEON] A coarse structure usually between the thin sheets in the protective wall of a tintinnid.

secondary target [ORD] The target against which fire is directed when the main fire mission has been accomplished, or when it has become impossible or impracticable for the gun or battery to carry out the main fire mission.

secondary tide station [ENG] A place at which tide observations are made over a short period to obtain data for a specific purpose.

secondary twinning [CRYSTAL] Twinning of a crystal caused by an external influence, such as pressure in rock.

secondary tympanic membrane [ANAT] The membrane closing the fenestra cochleae.

secondary voltage [ELECTROMAG] The voltage across the secondary winding of a transformer.

secondary wall [BOT] The portion of a plant cell wall produced internal to and following deposition of the primary wall; usually consists of several anisotropic layers, and often has prominent internal rings, spirals, bars, or reticulations.

secondary wave [GEOPHYS] *See* S wave. [OPTICS] One of the waves that radiate from each point on a wavefront, according to Huygens' principle.

secondary weapon [ORD] A supporting or auxiliary weapon of a unit, vehicle, position, or aircraft; generally, it is a gun of smaller caliber than the primary weapon, and its purpose is to protect or supplement the fire of the primary weapon.

secondary winding [ELECTROMAG] A transformer winding that receives energy by electromagnetic induction from the primary winding; a transformer may have several secondary windings, and they may provide alternating-current voltages that are higher, lower, or the same as that applied to the primary winding. Abbreviated sec. Also known as secondary.

secondary xylem [BOT] Xylem produced by cambium, composed of two interpenetrating systems, the horizontal (ray) and vertical (axial).

second bottom [GEOL] The first terrace rising over a floodplain.

second breakdown [ELECTR] Destructive breakdown in a transistor, wherein structural imperfections cause localized current concentrations and uncontrollable generation and multiplication of current carriers; reaction occurs so suddenly that the thermal time constant of the collector regions is exceeded, and the transistor is irreversibly damaged.

second-channel interference *See* alternate-channel interference.

second crop [AGR] A crop succeeding one already harvested during a growing season; either a regrowth of the harvested crop, or a newly planted crop.

second-degree burn [MED] A burn that is more severe than a first-degree burn and is characterized by blistering as well as reddening of the skin, edema, and destruction of the superficial underlying tissues.

second detector [ELECTR] The detector that separates the intelligence signal from the intermediate-frequency signal in a superheterodyne receiver.

second-foot [HYD] A contraction of cubic foot per second (cfs), the unit of stream discharge commonly used in the United States.

second-foot day [HYD] The volume of water represented by a flow of 1 cubic foot per second for 24 hours; equal to 86,400 cubic feet (approximately 2446.58 cubic meters); used extensively as a unit of runoff volume or reservoir capacity, particularly in the eastern United States.

second generation *See* F_2.

second-generation computer [ADP] A computer characterized by the use of transistors rather than vacuum tubes, the execution of input/output operations simultaneously with calculations, and the use of operating systems.

second growth [FOR] New trees that naturally replace trees which have been removed from a forest by cutting or by fire.

second hand [HOROL] The hand that indicates seconds on a timepiece.

second law of motion *See* Newton's second law.

second law of thermodynamics [THERMO] A general statement of the idea that there is a preferred direction for any process; there are many equivalent statements of the law, the best known being those of Clausius and of Kelvin.

second-order climatological station [CLIMATOL] A station at which observations of atmospheric pressure, temperature, humidity, winds, clouds, and weather are made at least twice daily at fixed hours, and at which the daily maximum and minimum of temperature, the daily amount of precipitation, and the duration of bright sunshine are observed.

second-order equation [MATH] A differential equation where some term includes the second derivative of the unknown function and no derivative of higher order is present.

second-order reaction [PHYS CHEM] A reaction whose rate of reaction is determined by the concentration of two chemical species.

second-order station [METEOROL] After U.S. Weather Bureau practice, a station manned by personnel certified to make aviation weather observations or synoptic weather observations.

second-order transition [THERMO] A change of state through which the free energy of a substance and its first derivatives are continuous functions of temperature and pressure, or other corresponding variables.

second pilot [AERO ENG] A pilot, not necessarily qualified on type, who is responsible for assisting the first pilot to fly the aircraft and is authorized as second pilot.

second quantization [QUANT MECH] A procedure in which the dependent variables of a classical field or a quantum-mechanical wave function are regarded as operators on which commutation rules are imposed; this produces a formalism in which particles may be created and destroyed.

second radiation constant [STAT MECH] A constant appearing in the Planck radiation formula, equal to the speed of light times Planck's constant divided by Boltzmann's constant, or approximately 1.4388 degree-centimeters. Symbolized c_2; C_2.

second sound [CRYO] A type of wave propagated in the superfluid phase of liquid helium (helium II), in which temperature and entropy variations propagate with no appreciable variation in density or pressure.

seconds pendulum [HOROL] A pendulum measuring 99.353 centimeters between its center of suspension and center of oscillation, and which requires exactly 1 second to swing from one side to another at sea level and 45° latitude.

second strike [ORD] The first counterblow of a war (generally associated with nuclear operations).

second-strike capability [ORD] The ability to survive a first strike with sufficient resources to deliver an effective counterblow (generally associated with nuclear weapons).

second-time-around echo [ELECTR] A radar echo received after an interval exceeding the pulse interval. Also known as second-trip echo.

second-trip echo *See* second-time-around echo.

secrecy system *See* privacy system.

secretin [BIOCHEM] A basic polypeptide hormone produced by the duodenum in response to the presence of acid; acts to excite the pancreas to activity.

secretion [PHYSIO] **1.** The act or process of producing a substance which is specialized to perform a certain function within the organism or is excreted from the body. **2.** The material produced by such a process.

secretor gene [GEN] A dominant autosomal gene in man which controls secretion of A and B antigenic material in saliva, urine, plasma, and other body fluids; it is not linked to the ABO genes.

secretory granules [CYTOL] Accumulations of material produced within a cell for secretion outside the cell.

secretory structure [BOT] Plant cells or organizations of plant cells which produce a variety of secretions.

secret weapon [ORD] A weapon closely guarded or kept under concealment so as to be used with advantage before countermeasures can be taken against it.

section [CIV ENG] A piece of land usually 1 mile square (640 acres or approximately 2.58999 square kilometers) with boundaries conforming to meridians and parallels within established limits; 1 of 36 units of subdivision of a township in the U.S. Public Land survey system. [COMMUN] Each individual transmission span in a radio relay system; a system has one more section than it has repeaters. [GEOL] **1.** An inclined or vertical surface that is uncovered either naturally (as a sea cliff or stream bank) or artificially (as a strip mine or road cut) through a part of the earth's crust. **2.** A description or scale drawing of the successive rock units or geologic structures shown by the exposed surface, or their appearance if cut through by any intersecting plane. **3.** *See* columnar section. **4.** *See* geologic section. **5.** *See* type section. [PETR] *See* thin section.

sectional center [COMMUN] A long-distance telephone office which connects several primary centers and which is in class number 2; only a regional center has greater importance in routing telephone calls. Abbreviated SC.

sectional conveyor [MECH ENG] A belt conveyor that can be lengthened or shortened by the addition or the removal of interchangeable sections.

sectional core barrel [DES ENG] A core barrel whose length can be increased by coupling unit sections together.

sectionalized vertical antenna [ELECTROMAG] Vertical antenna that is insulated at one or more points along its length; the insertion of suitable reactances or applications of a driving voltage across the insulated points results in a modified current distribution giving a more desired radiation pattern in the vertical plane.

sectional radiography [ELECTR] The technique of making radiographs of plane sections of a body or an object; its purpose is to show detail in a predetermined plane of the body, while blurring the images of structures in other planes. Also known as laminography; planigraphy; tomography.

section house [CIV ENG] A building near a railroad section for housing railroad workers, or for storing maintenance equipment for the section.

section line [CIV ENG] A line representing the boundary of a section of land. [GRAPHICS] One of a series of parallel lines indicating a cut surface in a mechanical or architectural drawing.

section modulus [MECH] The ratio of the moment of inertia of the cross section of a beam undergoing flexure to the greatest distance of an element of the beam from the neutral axis.

sector [CIV ENG] A clearly defined area or airspace designated for a particular purpose. [ELECTROMAG] Coverage of a radar as measured in azimuth. [MATH] A portion of a circle bounded by two radii and an arc joining their end points. [METEOROL] Something resembling the sector of a circle, as a warm sector between the warm and cold fronts of a cyclone.

sectoral horn [ELECTROMAG] Horn with two opposite sides parallel and the two remaining sides which diverge.

sector disk [PHYS] A device used to reduce the intensity of a beam of light or other electromagnetic radiation by an accurately known amount; in its simplest form, it consists of a circular, opaque disk with one or more sectors cut out of it, rapidly rotating in the path of the beam.

sector display [ELECTR] A display in which only a sector of the total service area of a radar system is shown; usually the sector is selectable.

sectored light [NAV] A light having sectors of different colors or of the same color in specific sectors separated by dark sectors.

sector gate [CIV ENG] A horizontal gate with a pie-slice cross section used to regulate the level of water at the crest of a dam; it is raised and lowered by a rack and pinion mechanism.

sector gear [DES ENG] **1.** A toothed device resembling a portion of a gear wheel containing the center bearing and a part of the rim with its teeth. **2.** A gear having such a device as its chief essential feature. [MECH ENG] A gear system employing such a gear as a principal part.

sector of fire [ORD] An area which is required to be covered by fire by an individual, a weapon, or a unit.

sector scan [ELECTR] A radar scan through a limited angle, as distinguished from complete rotation.

sector search [NAV] A flight or sailing plan of three legs, the turning points being at equal distances along radial lines from a fixed or moving point.

sector wind [METEOROL] The average observed or computed wind (direction and speed) at flight level for a given sector of an air route; sectors for over-ocean flights usually consist of 10° of longitude.

secular [ENG] Of or pertaining to a long indefinite period of time.

secular equilibrium [NUCLEO] Radioactive equilibrium in which the parent has such a small decay constant that there has been no appreciable change in the quantity of parent present by the time the decay products have reached radioactive equilibrium.

secular parallax [ASTRON] An apparent angular displacement of a star, resulting from the sun's motion.

secular perturbations [ASTROPHYS] Changes in the orbit of a planet, or of a satellite, that operates in extremely long cycles.

secular variable [ASTRON] A star whose brightness appears to have slowly lessened or increased over a time period of centuries.

secular variation [GEOPHYS] The changes, measured in hundreds of years, in the magnetic field of the earth. Also known as geomagnetic secular variation.

secund [BOT] Having lateral members arranged on one side only.

secure [ORD] To gain possession of a position or terrain feature, with or without force, and to make such disposition

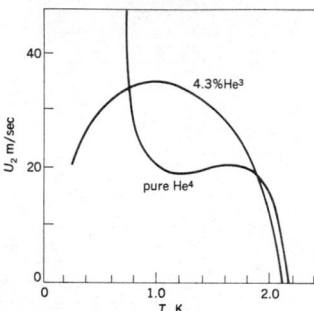

SECOND SOUND

Velocity of second sound U_2 as a function of temperature (degrees Kelvin) for pure He⁴ and a 4.3% He³ mixture. *(From J. C. King and H. A. Fairbank, Second sound in He³–He⁴ mixtures below 1°K, Phys. Rev., 93:21, 1954)*

as will prevent, as far as possible, its destruction or loss by enemy action.

secure visual communications [COMMUN] The transmission of an encrypted digital signal consisting of animated visual and audio information; the distance may vary from a few hundred feet to thousands of miles.

secure voice [COMMUN] Voice message that is scrambled or coded, therefore not transmitted in the clear.

securinine [PHARM] $C_{13}H_{15}NO_2$ A crystalline compound that forms yellow crystals from a methanol solution and melts at 142–143°C; used to make the nitrate compound for cardiac insufficiency.

securite explosive [MATER] A type of plastic explosive with a balanced oxygen content; it is built up on a nonexplosive, hydrophilic gel and contains oxygen-emitting salts, solid high explosive, and water.

security [ORD] **1.** Measures taken by a command to protect itself from espionage, observation, sabotage, annoyance, or surprise. **2.** A condition which results from the establishment and maintenance of protective measures which ensure a state of inviolability from hostile acts or influences. **3.** Protection of supplies or supply establishments against enemy attack, fire, theft, and sabotage.

security classification [ORD] A category or grade assigned to defense information or materiel to indicate the degree of danger to national security that would result from its unauthorized disclosure and the standard of protection required to guard against unauthorized disclosure.

security clearance [ORD] A clearance given to a person to permit access to classified material, equipment, or information up to and including a given classification, provided the person can establish a need-to-know.

security control officer [ORD] In the United States, an officer, warrant officer, or responsible civilian official appointed in each command or agency to exercise staff supervision over the safeguarding of defense information.

security reporting/alerting system [COMMUN] A rapid communications procedure that integrates all U.S. Air Force bases and commands, so that a significant happening at one location, or a pattern of seemingly unrelated happenings at several locations, can serve as a basis for swift security alerting or warning throughout the system.

SED *See* skin erythema dose.

sedation [MED] A state of lessened activity.

sedative [PHARM] An agent or drug that has a quieting effect on the central nervous system.

Sedentaria [INV ZOO] A group of families of polychaete annelids in which the anterior, or cephalic, region is more or less completely concealed by overhanging peristomial structures, or the body is divided into an anterior thoracic and a posterior abdominal region.

sedentary soil [GEOL] Soil that still lies on the rock from which it was formed.

sediment [GEOL] **1.** A mass of organic or inorganic solid fragmented material, or the solid fragment itself, that comes from weathering of rock and is carried by, suspended in, or dropped by air, water, or ice; or a mass that is accumulated by any other natural agent and that forms in layers on the earth's surface such as sand, gravel, silt, mud, fill, or loess. **2.** A solid material that is not in solution and either is distributed through the liquid or has settled out of the liquid.

sedimentary breccia [PETR] A rock composed of fragments that are larger than 2 millimeters in diameter and are the result of sedimentary processes; characterized by imperfect mechanical sorting of its materials and by a higher concentration of fragments from one local source or by a wide variety of materials mixed together in no particular pattern. Also known as sharpstone conglomerate.

sedimentary cycle *See* cycle of sedimentation.

sedimentary facies [GEOL] A stratigraphic facies differing from another part or parts of the same unit in both lithologic and paleontologic characters.

sedimentary injection *See* injection.

sedimentary intrusion *See* intrusion.

sedimentary petrology [PETR] The study of the composition, characteristics, and origin of sediments and sedimentary rocks.

sedimentary quartzite *See* orthoquartzite.

sedimentary rock [PETR] A rock formed by consolidated sediment deposited in layers. Also known as derivative rock; stratified rock.

sedimentary tuff [GEOL] A tuff containing a small amount of nonvolcanic detrital material.

sedimentary volcanism [GEOL] The expelling, extruding, or breaking through of overlying formations by a mixture of sediment, water, and gas, driven by the gas under pressure.

sedimentation [GEOL] **1.** The act or process of accumulating sediment in layers. **2.** The process of deposition of sediment. [MET] Classification of metal powders by the rate of settling in a fluid.

sedimentation balance [ANALY CHEM] A device to measure and record the weight of sediment (solid particles settled out of a liquid) versus time; used to determine particle sizes of fine solids.

sedimentation basin [GEOL] A depression in the ocean floor with a wide, flat bottom in which sediment accumulates.

sedimentation coefficient [PHYS CHEM] In the sedimentation of molecules in an accelerating field, such as that of a centrifuge, the velocity of the boundary between the solution containing the molecules and the solvent divided by the accelerating field. (In the case of a centrifuge, the accelerating field equals the distance of the boundary from the axis of rotation multiplied by the square of the angular velocity in radians per second.)

sedimentation constant [PHYS CHEM] A quantity used in studying the behavior of colloidal particles subject to forces, especially centrifugal forces; it is equal to $2r^2(\rho - \rho')/9\eta$, where r is the particle's radius, ρ and ρ' are reciprocals of partial specific volumes of particle and medium respectively, and η is the medium's viscosity.

sedimentation curve [GEOL] A curve showing cumulatively, and in successive units of time, the amount of sediment accumulated or removed from an originally uniform suspension.

sedimentation diameter [GEOL] The diameter of a sedimentary particle, determined from the measurement of a hypothetical sphere of the same gravity and settling velocity as those of a given sedimentary particle in the same fluid.

sedimentation equilibrium [ANALY CHEM] The equilibrium between the forward movement of a sample's liquid-sediment boundary and reverse diffusion during centrifugation; used in molecular-weight determinations.

sedimentation radius [GEOL] One-half of the sedimentation diameter.

sedimentation rate [PATH] The rate at which red blood cells settle out of anticoagulated blood.

sedimentation tank [ENG] A tank in which suspended matter is removed either by quiescent settlement or by continuous flow at high velocity and extended retention time to allow deposition.

sedimentation trough [GEOL] A depression in the ocean floor with a narrow U- or V-shaped bottom in which sediment accumulates.

sedimentation unit [GEOL] A sedimentary deposit formed during one distinct act of sedimentation.

sedimentation velocity [ANALY CHEM] The rate of movement of the liquid-sediment boundary in the sample holder during centrifugation; used in molecular-weight determinations.

sediment bulb [ENG] A bulb for holding sediment that settles from the liquid in a tank.

sediment corer [ENG] A heavy coring tube which punches out a cylindrical sediment section from the ocean bottom.

sediment discharge rating [HYD] A relationship between the discharge of sediment and the total discharge of the stream. Also known as silt discharge rating.

sedimentology [GEOL] The science concerned with the description, classification, origin, and interpretation of sediments and sedimentary rock.

sediment trap [ENG] A device for measuring the accumulation rate of sediment on the floor of a body of water.

sedoheptulose [BIOCHEM] A seven-carbon ketose sugar widely distributed in plants of the Crassulaceae family; a significant intermediary compound in the cyclic regeneration of D-ribulose.

Seebeck coefficient [ELECTR] The ratio of the open-circuit voltage to the temperature difference between the hot and cold junctions of a circuit exhibiting the Seebeck effect.

Seebeck effect [ELECTR] The development of a voltage due to differences in temperature between two junctions of dissimilar metals in the same circuit. [GRAPHICS] A photographic emulsion that is exposed until a faint visible image appears, and is then exposed to colored light and takes on the color of the light to which it is exposed.

seed [BOT] A fertilized ovule containing an embryo which forms a new plant upon germination. [CHEM] A small, single crystal of a desired substance added to a solution to induce crystallization. [SOLID STATE] A small, single crystal of semiconductor material used to start the growth of a large, single crystal for use in cutting semiconductor wafers.

seed charge [CHEM] A small amount of material added to a supersaturated solution to initiate precipitation.

seed coat [BOT] The envelope which encloses the seed except for a tiny pore, the micropyle.

seed core [NUCLEO] A reactor core that includes a relatively small volume of highly enriched uranium, surrounded by a much larger volume of natural uranium or thorium in a blanket; as a result of fission in the seed uranium, neutrons are supplied to the blanket, which is made to furnish a substantial fraction of the core power. Also known as spiked core.

seed down [AGR] To sow seeds for grass or forage legumes.

seed fern [PALEOBOT] The common name for the extinct plants classified as Pteridospermae, characterized by naked seeds borne on large, fernlike fronds.

seeding [AGR] The planting of seed. [CHEM] The adding of a seed charge to a supersaturated solution, or a single crystal of a desired substance to a solution of the substance to induce crystallization.

seedling [BOT] **1.** A plant grown from seed. **2.** A tree younger and smaller than a sapling. **3.** A tree grown from a seed.

seeing [ASTRON] The clarity and steadiness of an image of a star in a telescope. [ELECTR] The introduction of atoms with a low ionization potential into a hot gas to increase electrical conductivity.

seek [ADP] **1.** To position the access mechanism of a random-access storage device at a designated location or position. **2.** The command that directs the positioning to take place. [ORD] To go toward a target or other object in reaction to some influence such as heat, light, sound, or other radiation emitted by the target or object.

seeker [ORD] **1.** Any moving object, especially a missile, that finds its target by means of a device attracted to light, heat, radio waves, sound, or other radiation emitted by the target. **2.** The device used in such an object.

seen fire [ORD] Fire which is continuously aimed at the future position of an aircraft, the aim being derived from visual observation.

seepage [FL MECH] The slow movement of water or other fluid through a porous medium. [HYD] The slow movement of water through small openings and spaces in the surface of unsaturated soil into or out of a body of surface or subsurface water.

seersucker [TEXT] A washable fabric with crinkled stripes made by altering the tension of the warp threads during weaving.

Segas process [CHEM ENG] A process for the production of low-Btu gas by the catalytic method using a fixed bed catalyst, lime-bauxite mixture bonded with bentonite.

Seger cone [MATER] Any of a series of conical shaped thermometric devices made of materials that deform at specified temperatures; consists of mixtures of clay, salt, and other materials in such proportions that their softening temperatures vary progressively through the series; used to indicate temperatures of furnaces, particularly in ceramic industries. Also known as pyrometric cone.

segment [ADP] **1.** A single section of an overlay program structure, which can be loaded into the main memory when and as needed. **2.** In some direct-access storage devices, a hardware-defined portion of a track having fixed data capacity. [MATH] **1.** A segment of a line or curve is any connected piece. **2.** A segment of a circle is a portion of the circle bounded by a chord and an arc subtended by the chord. [NAV] In air operations, a basic functional division of an

instrument approach procedure; it bears a fixed orientation with respect to the course to be flown; it is assigned specific geometric coordinates which uniquely determine its position; the location of the segment is assigned with respect to the obstacles in the operations area.

segmental arch [ARCH] A round arch whose curved surface is less than a semicircle.

segmental gate *See* tainter gate.

segmental meter [ENG] A variable head meter whose orifice plate has an opening in the shape of a half circle.

segmental reflex [PHYSIO] A reflex arc having afferent inputs by way of the spinal dorsal roots, and efferent outputs over spinal ventral roots of the same or adjacent segments.

segmentation *See* metamerism.

segmentation cavity *See* blastocoele.

segment die [MET] A die made of parts that can be disassembled to facilitate removal of the workpiece. Also known as split die.

segment mark [ADP] A special character written on tape to separate one section of a tape file from another.

segment saw [MECH ENG] A saw consisting of steel segments attached around the edge of a flange and used for cutting veneer.

segregated ice [HYD] Ice films, seams, lenses, rods, or layers generally 1 to 150 millimeters thick that grow in permafrost by drawing in water as the ground freezes. Also known as Taber ice.

segregated vein [GEOL] A fissure filled with mineral matter derived from country rock by the action of percolating water. Also known as exudation vein.

segregation [ENG] **1.** The keeping apart of process streams. **2.** In plastics molding, a close succession of parallel, relatively narrow, and sharply defined wavy lines of color on the surface of a plastic that differ in shade from surrounding areas and create the impression that the components have separated. [GEN] The separation of alleles and homologous chromosomes during meiosis in the formation of gametes. [GEOL] The formation of a secondary feature within a sediment after deposition due to chemical rearrangement of minor constituents. [MET] The nonuniform distribution of alloying elements, impurities, or microphases, resulting in localized concentrations.

segregation banding [PETR] A compositional band in gneisses that is the result of segregation of material from an originally homogeneous rock.

segregation distorter [GEN] An abnormality of meiosis which produces a distortion of the 1:1 segregation ratio in a heterozygote.

seiche [FL MECH] An oscillation of a fluid body in response to the disturbing force having the same frequency as the natural frequency of the fluid system. [OCEANOGR] A standing-wave oscillation of an enclosed or semienclosed water body, continuing pendulum-fashion after cessation of the originating force, which is usually considered to be strong winds or barometric pressure changes.

Seidel aberrations [OPTICS] The five types of aberration of monochromatic light deduced from the Seidel theory, namely, spherical aberration, coma, astigmatism, curvature of field, and distortion.

Seidel method [MATH] A basic iterative procedure for solving a system of linear equations by reducing it to triangular form. Also known as Gauss-Seidel method.

Seidel theory [OPTICS] A theory of aberrations in which the sine of the angle which a ray makes with the optical axis is approximated by the first two terms in the sine's Taylor expansion (the first- and third-order terms), rather than the first term alone, as in a first-order theory.

seif dune [GEOL] A large, tapering, longitudinal dune or chain of sand dunes with a sharp crest that in profile consists of a succession of peaks and cols.

seignette-electric *See* ferroelectric.

seine net [ENG] A net used to catch fish by encirclement, usually by closure of the two ends and the bottom.

seismic activity *See* seismicity.

seismic area *See* earthquake zone.

seismic belt [GEOPHYS] An elongate seismic zone, such as that in the Circum-Pacific.

seismic detector [ENG] An instrument that receives seismic impulses.

seismic discontinuity [GEOPHYS] **1.** A surface at which velocities of seismic waves change abruptly. **2.** A boundary between seismic layers of the earth. Also known as interface; velocity discontinuity.

seismic exploration [ENG] The exploration for economic deposits by using seismic techniques, usually involving explosions, to map subsurface structures.

seismicity [GEOPHYS] The phenomena of earth movements. Also known as seismic activity.

seismic profiler [ENG] A continuous seismic reflection system used to study the structure beneath the sea floor to depths of 10,000 feet (3000 meters) or more, using a rotating drum to record reflections.

seismic prospecting [GEOPHYS] Geophysical prospecting based on the analysis of elastic waves generated in the earth by artificial means.

seismic sea wave *See* tsunami.

seismic survey *See* reflection survey.

seismic vertical [GEOL] **1.** The point on the earth's surface directly over the point within the earth from which an earthquake impulse originates. **2.** The vertical line between the surface point and the point of origin.

seismic wave [OCEANOGR] **1.** Any elastic sea wave, either body wave or surface wave, produced by earthquakes or by explosions. **2.** *See* tsunami.

seismochronograph [ENG] A chronograph for determining the time at which an earthquake shock appears.

seismograph [ENG] An instrument that records vibrations in the earth, especially earthquakes.

seismology [GEOPHYS] **1.** The study of earthquakes. **2.** The science of strain-wave propagation in the earth.

seismometer [ENG] An instrument that detects movements in the earth.

seismoscope [ENG] An instrument for recording only the occurrence or time of occurrence (not the magnitude) of an earthquake.

Seisonacea [INV ZOO] A monofamiliar order of the class Rotifera characterized by an elongated jointed body with a small head, a long slender neck region, a thick fusiform trunk, and an elongated foot terminating in a perforated disk.

Seisonidea [INV ZOO] The equivalent name for Seisonacea.

seistan [METEOROL] A strong wind of monsoon origin which blows from between the northwest and north-northwest and sets in about the end of May or early June in the historic Seistan district of eastern Iran and Afghanistan; it continues almost without cessation until about the end of September; because of its duration it is known as the wind of 120 days (bad-i-sad-o-bistroz).

Seitz filter [MICROBIO] A bacterial filter made of asbestos and used to sterilize solutions without the use of heat.

seizure [MED] **1.** The sudden onset or recurrence of a disease or an attack. **2.** Specifically, an epileptic attack, fit, or convulsion.

Sejournet process [MET] During hot extrusion, the lubrication and insulation of a metal billet with molten glass. Also known as Ugine Sejournet process.

Selachii [VERT ZOO] An order of elasmobranchs including all fossil sharks, except Cladoselachii and Pleuracanthodii.

Selaginellales [BOT] The plant order of small club mosses, containing one living genus, *Selaginella;* distinguished from other lycopods in being heterosporous and in having a ligule borne on the upper base of the leaf.

selatan [METEOROL] Strong, dry, southerly winds of the southeast monsoon in the Netherlands East Indies and the Celebes.

select [ADP] **1.** To choose a needed subroutine from a file of subroutines. **2.** To take one alternative if the report on a condition is of one state, and another alternative if the report on the condition is of another state. **3.** To pull from a mass of data certain items that require special attention; selection of individual cards is accomplished automatically by either the sorter or collator, according to the type of selection.

select bit [ADP] The bit (or bits) in an input/output instruction word which selects the function of a specified device. Also known as subdevice bit.

selected areas *See* Kapetyn selected areas.

selected mine [ORD] A controlled underwater mine which has been connected, through the selector assembly, to the control equipment at the shore station; it is exclusive of all other mines in its group and may be fired, tested, or disarmed independently of the remainder of the group.

selected time [IND ENG] An observed actual time value for an element, measured by time study, which is identified as being the most representative of the situation observed.

selecting circuit [ELEC] A simple switching circuit that receives the identity (the address) of a particular item and selects that item from among a number of similar ones.

selection [COMMUN] The process of addressing a call to a specific station in a selective calling system. [GEN] Any natural or artificial process which favors the survival and propagation of individuals of a given genotype in a population.

selection bias [STAT] A bias built into an experiment by the method used to select the subjects which are to undergo treatment.

selection check [ADP] Electronic computer check, usually automatic, to verify that the correct register, or other device, is selected in the performance of an instruction.

selection coefficient [GEN] A measure of the rate of transmission through successive generations of a given allele compared to the rate of transmission of another (usually the wild-type) allele.

selection pressure [EVOL] Those factors that influence the direction of natural selection.

selection rules [PHYS] Rules summarizing the changes that must take place in the quantum numbers of a quantum-mechanical system for a transition between two states to take place with appreciable probability; transitions that do not agree with the selection rules are called forbidden and have considerably lower probability.

selective absorption [ELECTROMAG] A greater absorption of electromagnetic radiation at some wavelengths (or frequencies) than at others.

selective acidizing [PETRO ENG] Oil-reservoir acid treatment (acidizing) in which the acid is injected into specific reservoir zones; contrasted with uncontrolled acidizing in which the acid solution is simply pumped down the casing and is forced into adjacent rock.

selective adsorbent [CHEM ENG] Material that will selectively adsorb (or reject) one or more specific components from a multicomponent mixture of gases or liquids; common adsorbents are silica gel, carbon and activated carbon, activated alumina, and synthetic or natural zeolites (molecular sieves).

selective breeding [BIOL] Breeding of animals or plants having desirable characters.

selective calling system [COMMUN] A radio communications system in which the central station transmits a coded call that activates only the receiver to which that code is assigned.

selective circuit [ELEC] A circuit that transmits certain types of signals and fails to transmit or attenuates others.

selective dump [ADP] An edited or nonedited listing of the contents of selected areas of memory or auxiliary storage.

selective fading [COMMUN] Fading that is different at different frequencies in a frequency band occupied by a modulated wave, causing distortion that varies in nature from instant to instant.

selective filling [MIN ENG] Filling by hand so that stone or dirt is rejected and only clean coal or ore is loaded.

selective flotation [MIN ENG] The surface or froth selecting of the valuable minerals rather than the gangue.

selective fracturing [PETRO ENG] Procedures for obtaining multiple formation fractures in a specific reservoir zone by plugging casing perforations or by isolating (with packers) the desired zone prior to fracturing operations.

selective fusion [GEOL] The fusion of only a portion of a mixture, such as a rock.

selective heating [MET] Heating only certain portions of a workpiece to impart desired properties.

selective identification feature [ELECTR] Airborne pulse-type transponder which provides automatic selective identification of aircraft in which it is installed to ground, ship-

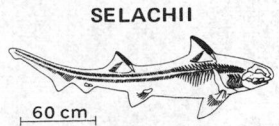

SELACHII

60 cm

Primitive selachian *Hybodus hauffianus,* a Mesozoic shark about 7½ feet (2.3 meters) long, showing claspers and the constricted base for paired fins that differentiates the Selachii from Cladoselachii. *(After Smith Woodward)*

SELAGINELLALES

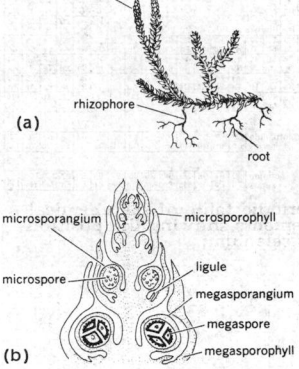

cone

rhizophore

(a)

root

microsporangium

microsporophyll

microspore

ligule

megasporangium

megaspore

(b)

megasporophyll

Selaginella, or small club moss. *(a)* Plant with creeping, dichotomous stems bearing numerous, tiny leaves and leafless branches, or rhizophores *(from H. J. Fuller and O. Tippo, College Botany, Holt, 1949). (b)* Diagram of a longitudinal section through a cone *(after Lyon from H. J. Fuller and O. Tippo, College Botany, rev. ed., Holt, 1954).*

SELECTING CIRCUIT

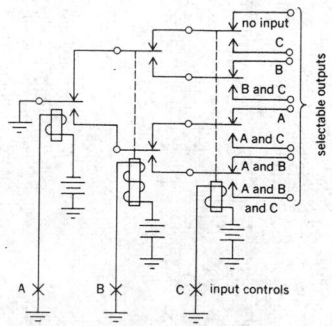

no input

C

B

B and C

A

A and C

A and B

A and B and C

selectable outputs

A   B   C input controls

Circuit diagram of a simple relay selecting circuit. Three relays (A, B, C) are used by the circuit to select one of eight outputs according to the combination in which the three relays are operated or not operated.

board, or airborne identification, friend or foe-selective identification feature recognition installations.

selective interference [COMMUN] Interference whose energy is concentrated in a narrow band of frequencies.

selective jamming [ELECTR] Jamming in which only a single radio channel is jammed.

selective loading [NAV ARCH] The arrangement and stowage of equipment and supplies aboard ship in a manner designed to facilitate issues to units.

selective medium [MICROBIO] A bacterial culture medium containing an individual organic compound as the sole source of carbon, nitrogen, or sulfur for growth of an organism.

selective mining [MIN ENG] A method of mining whereby ore of unwarranted high value is mined in such manner as to make the low-grade ore left in the mine incapable of future profitable extraction; in other words, the best ore is selected in order to make good mill returns, leaving the low-grade ore in the mine.

selective permeability [PHYS] The property of a membrane or other material that allows some substances to pass through it more easily than others.

selective polymerization [CHEM ENG] The polymerization of a single type of molecule in a mixture of monomers; for example, the production of diisobutylene from a mixture of butylenes.

selective quenching [MET] Quenching only certain portions of a piece of metal.

selective radiator [PHYS] An object that emits electromagnetic radiation whose spectral energy distribution differs from that of a blackbody with the same temperature.

selective reflection [ELECTROMAG] Reflection of electromagnetic radiation more strongly at some wavelengths (or frequencies) than at others.

selective replacement [GEOL] The replacement of one mineral by another, preferentially within an altered rock mass.

selective ringing [COMMUN] Telephone arrangement on party lines, in which only the bell of the called subscriber rings, with other bells on the party line remaining silent.

selective scattering [ELECTROMAG] Scattering of electromagnetic radiation more strongly at some wavelengths than at others.

selective solubility diffusion [CHEM ENG] The transmission of fluids through a nonporous, polymeric barrier (membrane) by an adsorption-solution-diffusion-desorption sequence.

selective solvent [CHEM ENG] A solvent that, at certain temperatures and ratios with other materials, preferentially dissolves more of one component of a liquid or solids mixture than of another, thereby permitting partial separation.

selective trace [ADP] A tracing routine wherein only instructions satisfying certain specified criteria are subject to tracing.

selective transmission [MECH ENG] A gear transmission with a single lever for changing from one gear ratio to another; used in automotive vehicles.

selectivity [ELECTR] The ability of a radio receiver to separate a desired signal frequency from other signal frequencies, some of which may differ only slightly from the desired value.

selectivity coefficient [ANALY CHEM] Ion equilibria relationship formula for ion-exchange-resin systems.

selectivity diagram [CHEM ENG] A triangular plot of solubilities in a ternary liquid system; used to calculate the ability of a solvent to extract a component from a mixture (its selectivity) at various concentration combinations.

selector [ADP] Computer device which interrogates a condition and initiates a particular operation dependent upon the report. [CIV ENG] A device that automatically connects the appropriate railroad signal to control the track selected. [ELEC] An automatic or other device for making connections to any one of a number of circuits, such as a selector relay or selector switch. [ENG] 1. A device for selecting objects or materials according to predetermined properties. 2. A device for starting or stopping at predetermined positions. [MECH ENG] 1. The part of the gearshift in an automotive transmission that selects the required gearshift bar. 2. The lever with which a driver operates an automatic gearshift. [MET] A converter that separates purified copper from residue in a single operation.

selector channel [ADP] A piece of hardware which allows more than one high-speed input/output device to be accessed by line computer.

selector switch [ELEC] A manually operated multiposition switch. Also called multiple-contact switch.

s-electron [ATOM PHYS] An atomic electron that is described by a wave function with orbital angular momentum quantum number of zero in the independent particle approximation.

selenic acid [INORG CHEM] H_2SeO_4 A highly toxic, water-soluble, white solid, melting point 58°C, decomposing at 260°C.

selenide [INORG CHEM] M_2Se A binary compound of divalent selenium, such as Ag_2Se, silver selenide. [ORG CHEM] An organic compound containing divalent selenium, such as $(C_2H_5)_2Se$, ethyl selenide.

selenious acid *See* selenous acid.

selenite [MINERAL] The clear, colorless variety of gypsum crystallizing in the monoclinic system and occurring in crystals or in crystal mass. Also known as spectacle stone.

selenium [CHEM] A highly toxic, nonmetallic element in group VI, symbol Se, atomic number 34; steel-gray color; soluble in carbon disulfide, insoluble in water and alcohol; melts at 217°C; and boils at 690°C; used in analytical chemistry, metallurgy, and photoelectric cells, and as a lube-oil stabilizer and chemicals intermediate.

selenium bromide [INORG CHEM] Any of three compounds of selenium and bromine: Se_2Br_2, a red liquid melting at −46°C, also known as selenium monobromide; $SeBr_2$, a brown liquid, also known as selenium dibromide; and $SeBr_4$, orange, carbon-disulfide-soluble crystals, also known as selenium tetrabromide.

selenium cell [ELECTR] A photoconductive cell in which a thin film of selenium is used between suitable electrodes; the resistance of the cell decreases when the illumination is increased.

selenium dibromide *See* selenium bromide.

selenium diode [ELECTR] A small area selenium rectifier which has characteristics similar to those of selenium rectifiers used in power systems.

selenium dioxide [INORG CHEM] SeO_2 Water- and alcohol-soluble, white to reddish, lustrous crystals; melts at 340°C; used in medicine, and as an oxidizing agent and catalyst. Also known as selenous acid anhydride; selenous anhydride; selenium oxide.

selenium halide [INORG CHEM] A compound of selenium and a halogen, for example, Se_2Br_2, $SeBr_2$, $SeBr_4$; Se_2Cl_2, $SeCl_2$, $SeCl_4$; Se_2I_2, SeI_4.

selenium monobromide *See* selenium bromide.

selenium nitride [INORG CHEM] Se_2N_2 A water-insoluble, yellow solid that explodes at 200°C.

selenium oxide *See* selenium dioxide.

selenium rectifier [ELECTR] A metallic rectifier in which a thin layer of selenium is deposited on one side of an aluminum plate and a conductive metal coating is deposited on the selenium.

selenium stainless steel [MET] Stainless steel to which about 0.1 percent or more selenium is added to improve machinability.

selenium tetrabromide *See* selenium bromide.

selenocentric [ASTRON] Pertaining to the moon's center.

selenodesy [ASTRON] The branch of applied mathematics that determines, by observation and measurement, the exact positions of points on the moon's surface, as well as the shape and size of the moon.

selenodetic [ASTRON] Of, or pertaining to, or determined by selenodesy.

selenodont [VERT ZOO] 1. Being or pertaining to molars having crescentic ridges on the crown. 2. A mammal with selenodont dentition.

selenography [ASTRON] Studies pertaining to the physical geography of the moon; specifically, referring to positions on the moon measured in latitude from the moon's equator and in longitude from a reference meridian.

selenology [ASTRON] A branch of astronomy that treats of the moon, including such attributes as magnitude, motion and constitution. Also known as lunar geology.

selenomorphology [ASTRON] The study of landforms on the moon, including their origin, evolution, and distribution.

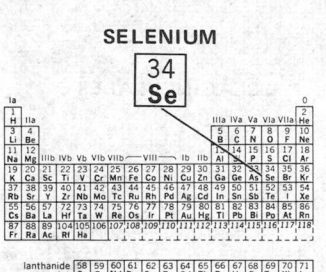

SELENIUM

Periodic table of the chemical elements showing the position of selenium.

selenone [ORG CHEM] A group of organic selenium compounds with the general formula R_2SeO_2.

selenonic acid [ORG CHEM] Any organic acid containing the radical $-SeO_3H$; analogous to a sulfonic acid.

selenosis [MED] Selenium poisoning.

selenotrope [ENG] A device used in geodetic surveying for reflecting the moon's rays to a distant point, to aid in long-distance observations.

selenous acid [INORG CHEM] H_2SeO_3 Colorless, transparent crystals; soluble in water and alcohol, insoluble in ammonia; decomposes when heated; used as an analytical reagent. Also spelled selenious acid.

selenous acid anhydride *See* selenium dioxide.

selenous anhydride *See* selenium dioxide.

selenoxide [ORG CHEM] A group of organic selenium compounds with the general formula R_2SeO.

self-absorption [NUCLEO] Absorption of ionizing radiation by the material that emits the radiation, thus reducing the radiation level against which further shielding must be provided. Also known as self-shielding. [SPECT] Reduction of the intensity of the center of an emission line caused by selective absorption by the cooler portions of the source of radiation. Also known as self-reduction.

self-acting door [MIN ENG] A ventilation door that is constructed of two halves which move on small pulleys and which are forced apart centrally as the trams come in contact with the converging beams that operate the door.

self-acting incline *See* gravity haulage.

self-adapting system [SYS ENG] A system which has the ability to modify itself in response to changes in its environment.

self-adjoint operator [MATH] A linear operator which is identical with its adjoint operator.

self-advancing supports [MIN ENG] An assembly of hydraulically operated steel hydraulic supports, on a long-wall face, which are moved forward as a unit. Also known as walking props.

self-analysis [PSYCH] The attempt to gain insight into one's own psychic state and behavior.

self-bias [ELECTR] A grid bias provided automatically by the resistor in the cathode or grid circuit of an electron tube; the resulting voltage drop across the resistor serves as the grid bias. Also known as automatic C bias; automatic grid bias.

self-bias transistor circuit [ELECTR] A transistor with a resistance in the emitter lead that gives rise to a voltage drop which is in the direction to reverse-bias the emitter junction; the circuit can be used even if there is zero direct-current resistance in series with the collector terminal.

self-centering chuck [MECH ENG] A drill chuck that, when closed, automatically positions the drill rod in the center of the drive rod of a diamond-drill swivel head.

self-chambering [ORD] Ability of a weapon to chamber cartridges without manual aid.

self-charge [QUANT MECH] A contribution to a particle's electric charge arising from the vacuum polarization in the neighborhood of the bare charge.

self-checking code [ADP] An encoding of data so designed and constructed that an invalid code can be rapidly detected; this permits the detection, but not the correction, of almost all errors. Also known as error-checking code; error-detecting code.

self-checking number [ADP] A number with a suffix figure related to the figure of the number, used to check the number after it has been transferred from one medium or device to another.

self-cleaning [ENG] Pertaining to any device that is designed to clean itself without disassembly, for example, a filter in which accumulated filter cake or sludge is removed by an internal scraper or by a blowdown or backwash action.

self-cleaning contact *See* wiping contact.

self-conjugate particle [PARTIC PHYS] An elementary particle which is identical to its antiparticle; it must have zero charge, lepton number, baryon number, and hypercharge.

self-consistent field method *See* Hartree method.

self-contained base-line system [ORD] System of target location whereby the target is located in azimuth and range by using a self-contained range finder.

self-contained breathing apparatus [ENG] A portable breathing unit which permits freedom of movement.

self-contained night attack [ORD] The capability of a single aircraft to accurately navigate to a target area, acquire and strike a designated target, and return to the operating base during the hours of darkness.

self-contained portable electric lamp [MIN ENG] An electric lamp which is operated by an electric battery and is specifically designed to be carried about by its user.

self-contained range finder [ENG] Instrument used for measuring range by direct observation, without using a base line; the two types are the coincidence range finder and the stereoscopic range finder.

self-destroying [ORD] In connection with a fuse or tracer, indicating that the projectile, rocket, or missile with which it is used will be destroyed (functioned) in flight prior to ground impact in case the target is missed.

self-destroying fuse [ORD] A fuse designed to burst a projectile before the end of its flight.

self-destruct charge [ORD] An explosive element which operates in conjunction with that part of the guided missile which, of itself or by command, senses a catastrophic flight malfunction and destroys the missile.

self-destruction [ORD] The event, due to fuse or tracer action without outside stimulus when provided for in the design, in which the fuse or tracer effects projectile or missile destruction, after flight to a range greater than that of the target.

self-destruction equipment [ORD] Any equipment that may be destroyed by a self-contained explosive.

self-differentiation [PHYSIO] The differentiation of a tissue, even when isolated, solely as a result of intrinsic factors after determination.

self-diffusion [SOLID STATE] The spontaneous movement of an atom to a new site in a crystal of its own species.

self-dumping cage [MIN ENG] A cage in which the deck is pivoted so that as the cage is lifted, toward the end of the lift, the deck tilts and the end door is lifted, discharging the coal.

self-dumping car [MIN ENG] A mine car which can be side-tipped while in motion on the rail track; it is fitted with a spherically contoured wheel which engages a ramp structure and gradually tilts the car.

self-energizing brake [MECH ENG] A brake designed to reinforce the power applied to it, such as a hand brake.

self-energy [PHYS] **1.** Classically, the contribution to the energy of a particle that arises from the interaction between different parts of the particle. **2.** In a quantized field theory, the contribution to the energy of a particle due to virtual emission and absorption of other particles, in particular, mesons and photons.

self-excited [ELEC] Operating without an external source of alternating-current power.

self-excited oscillator [ELECTR] An oscillator that depends on its own resonant circuits for initiation of oscillation and frequency determination.

self-excited vibration *See* self-induced vibration.

self-extinguishing [MATER] The ability of a material to cease burning once the source of the flame has been removed.

self-focusing fiber [OPTICS] A type of optical fiber in which the refractive index decreases continuously along the radius, but progressively more rapidly with distance from the radius, so that light rays which travel longer distances are speeded up, and nearly all light rays travel with the same net axial velocity.

self-hardening steel *See* air-hardening steel.

self-impedance *See* mesh impedance.

self-induced transparency [OPTICS] A phenomenon in which a pulse of coherent light, with a certain frequency, amplitude, and duration, is transmitted by a normally opaque medium; energy absorbed from the first half of the pulse, whose frequency is at or near the average resonance peak of a band of coherent atomic two-quantum-level optical oscillators, is returned to the last half of the pulse by the medium in the form of coherently emitted light. Abbreviated SIT.

self-induced vibration [MECH] The vibration of a mechanical system resulting from conversion, within the system, of nonoscillatory excitation to oscillatory excitation. Also known as self-excited vibration.

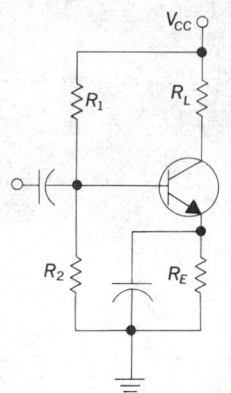

**SELF-BIAS
TRANSISTOR CIRCUIT**

Circuit diagram of a self-bias transistor circuit. R_L = load resistance; R_E = resistance in emitter lead; V_{CC} = collector supply voltage with respect to emitter.

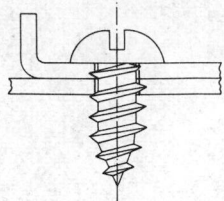

Diagram of a type of self-tapping screw.

self-inductance [ELECTROMAG] **1.** The property of an electric circuit whereby an electromotive force is produced in the circuit by a change of current in the circuit itself. **2.** Quantitatively, the ratio of the electromotive force produced to the rate of change of current in the circuit.

self-induction [ELECTROMAG] The production of a voltage in a circuit by a varying current in that same circuit.

self-loading [ORD] Pertaining to a firearm or gun that utilizes either the explosive gases or recoil to extract the empty case and to chamber the next round; self-loading firearms or guns include both semiautomatic and full-automatic types.

self-locking nut [DES ENG] A nut having an inherent locking action, so that it cannot readily be loosened by vibration.

self-organizing system [SYS ENG] A system that is able to effect or determine its own internal structure.

self-pollination [BOT] Transfer of pollen from the anther to the stigma of the same flower or of another flower on the same plant.

self-propelled [MECH ENG] Pertaining to a vehicle given motion by means of a self-contained motor. [ORD] **1.** Pertaining to a gun mounted on a vehicle that has its own motive power. **2.** Pertaining to a missile that is propelled by fuel carried by the missile itself, as in the case of a rocket. **3.** Pertaining to a military unit having self-propelled guns.

self-propelled artillery [ORD] Artillery weapons permanently installed on vehicles, which provide motive power for the piece; these weapons are fired from the vehicle.

self-pulsing [ELECTR] Special type of grid pulsing which automatically stops and starts the oscillations at the pulsing rate by a special circuit.

self-quenched detector [ELECTR] Superregenerative detector in which the time constant of the grid leak and grid capacitor is sufficiently large to cause intermittent oscillation above audio frequencies, serving to stop normal regeneration each time just before it spills over into a squealing condition.

self-quenching oscillator [ELECTR] Oscillator producing a series of short trains of radio-frequency oscillations separated by intervals of quietness.

self-reduction *See* self-absorption.

self-rescuer [MIN ENG] A small filtering device carried by a miner underground to provide immediate protection against carbon monoxide and smoke in case of a mine fire or explosion; used for escape purposes only, because it does not sustain life in atmospheres containing deficient oxygen.

self-reset [ELEC] Automatically returning to the original position when normal conditions are resumed; applied chiefly to relays and circuit breakers.

self-saturation [ELECTR] The connection of half-wave rectifiers in series with the output windings of the saturable reactors of a magnetic amplifier, to give higher gain and faster response.

self-scanned image sensor [ELECTR] A solid-state device, still in the early stages of development, which converts an optical image into a television signal without the use of an electron beam; it consists of an array of photoconductor diodes, each located at the intersection of mutually perpendicular address strips respectively connected to horizontal and vertical scan generators and video coupling circuits.

self-screening range [ELECTROMAG] Range at which a target can be detected by a radar in the midst of its jamming mask, with a certain specified probability.

self-sealing [ENG] A fluid container, such as a fuel tank or a tire, lined with a substance that allows it to close immediately over any small puncture or rupture.

self-selection bias [STAT] Bias introduced into an experiment by having the subjects decide themselves whether or not they will receive treatment.

self-shielding [NUCLEO] **1.** Shielding of the inner portion of the fuel in a nuclear reactor by the outer portion of the fuel. **2.** *See* self-absorption.

self-starter [MECH ENG] An attachment for automatically starting an internal combustion engine.

self-starting synchronous motor [ELEC] A synchronous motor provided with the equivalent of a squirrel-cage winding, to permit starting as an induction motor.

self-synchronous device *See* synchro.

self-synchronous repeater *See* synchro.

self-tapping screw [DES ENG] A screw with a specially hardened thread that makes it possible for the screw to form its own internal thread in sheet metal and soft materials when driven into a hole. Also known as sheet-metal screw; tapping screw.

self-timer [ENG] A device that delays the tripping of a camera shutter so that the photographer can be included in the photograph.

self-unloading ship [NAV ARCH] A ship equipped with endless belt or chain devices or with swinging booms, which enable it to be unloaded without the use of harbor facilities.

seligmannite [MINERAL] $PbCuAsS_3$ A metallic gray orthorhombic mineral, occurring in crystals.

Seliwanoff's test [ANALY CHEM] A color test helpful in the identification of ketoses, which develop a red color with resorcinol in hydrochloric acid.

sellaite [MINERAL] MgF_2 A colorless mineral composed of magnesium fluoride occurring in tetragonal prismatic crystals.

sella turcica [ANAT] A depression in the upper surface of the sphenoid bone in which the pituitary gland rests in vertebrates.

sellers hob [MECH ENG] A hob that turns on the centers of a lathe, the work being fed to it by the lathe carriage.

sellite [INORG CHEM] A solution of sodium sulfate that is used during the manufacture of trinitrotoluene.

Sellmeier's equation [ELECTROMAG] An equation for the index of refraction of electromagnetic radiation as a function of wavelength in a medium whose molecules have oscillators of different frequencies.

selsyn *See* synchro.

selsyn generator *See* synchro transmitter.

selsyn motor *See* synchro receiver.

selsyn receiver *See* synchro receiver.

selsyn system *See* synchro system.

selva *See* tropical rainforest.

selvage [PETR] The marginal zone of an igneous mass, generally characterized by a fine-grain, or sometimes glassy, texture. Also known as salband.

Selwood engine [MECH ENG] A revolving-block engine in which two curved pistons opposed 180° run in toroidal tracks, forcing the entire engine block to rotate.

selysn transmitter *See* synchro transmitter.

Semaeostomeae [INV ZOO] An order of the class Scyphozoa including most of the common medusae, characterized by a flat, domelike umbrella whose margin is divided into many lappets.

semanteme [COMMUN] A language element that expresses a definite idea or image, such as a word, base of a word, or data element. Also known as lexeme.

semeiography [MED] Description of the signs and symptoms of a disease.

semen [PHYSIO] The fluid that carries the male germ cells.

Semenov number 1 [PHYS CHEM] A dimensionless number used in reaction kinetics, equal to a mass transfer constant divided by a reaction rate constant. Symbolized S_m. Formerly known as Schmidt number 2.

Semenov number 2 [PHYS] The reciprocal of the Lewis number.

semiactive homing guidance [NAV] Guidance in which a craft or vehicle is directed toward a target by illuminating the target with radiation from a ground-based or shipborne radar; the missile or other vehicle is equipped with a direction-finding receiver which enables it to home on the target by use of the radar signal reflected from it.

semiactive tracking system [NAV] A tracking system used for semiactive homing guidance.

semianechoic room [ACOUS] A room having surfaces which reduce the reflection of sound to less than normal, although not to as great an extent as an anechoic room.

semianthracite [GEOL] Coal which is between bituminous coal and anthracite in metamorphic rank, and which has a fixed-carbon content of 86–92%.

semiarid climate *See* steppe climate.

semiautomatic [ORD] Pertaining to a firearm or gun that utilizes a part of the force of an exploding cartridge to extract the empty case and to chamber the next round, but requires a separate pull on the trigger to fire each round; examples are the semiautomatic rifle and the semiautomatic pistol.

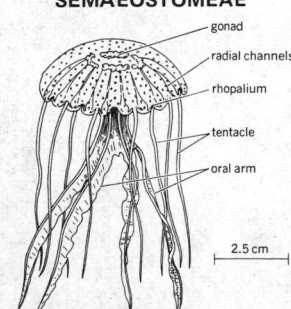

gonad
radial channels
rhopalium
tentacle
oral arm

2.5 cm

Pelagia, showing the tentacles arising between the lappets on the umbrella. *(From L. Hyman, The Invertebrates, vol. 1, McGraw-Hill, 1940)*

semiautomatic ground environment See SAGE.

semiautomatic keying circuits [COMMUN] Mechanization providing torn tape switching systems in teleprinter links; incoming and outgoing messages are converted to tapes and manually inserted into a teletypewriter distributor that mechanically keys the circuit automatically.

semiautomatic supply [ORD] System by which certain specified items of supplies needed by units, activities, or forces are shipped by the agencies responsible for supply on the basis of periodic reports of the status of stocks on hand and en route to the using agency; all other supplies are furnished on the basis of requisitions initiated by the using agency.

semiautomatic tape relay [COMMUN] Method of communication whereby messages are received and retransmitted in teletypewriter tape form involving manual intervention in transfer of the tape from receiving reperforator to automatic transmitter.

semiautomatic telephone system [COMMUN] Telephone system that limits automatic dialing to only those subscribers who are served by the same exchange as the calling subscriber.

semiautomatic transmission [MECH ENG] An automobile transmission that assists the driver to shift from one gear to another.

semiautomatic welding [MET] An arc-welding method in which the electrode, a long length of small-diameter bare wire, usually in coil form, is positioned and advanced by the operator from a hand-held welding gun which feeds the electrode through the nozzle.

semibatch chemical reactor [CHEM ENG] A reactor in which a constant liquid volume is maintained without any overflow, and with the continuous addition of one reactant, usually a gas.

semibituminous coal [GEOL] Coal that is harder and more brittle than bituminous coal, has a high fuel ratio, contains 10–20% volatile matter, and burns without smoke; ranks between bituminous and semianthracite coals.

semicarbazide hydrochloride [ORG CHEM] $CH_5ON_3 \cdot HCl$ Colorless prisms, soluble in water, decomposing at 175°C; used as an analytical reagent for aldehydes and ketones, and to recover constituents of essential oils. Also known as aminourea hydrochloride; carbamylhydrazine hydrochloride.

semicarbazone [ORG CHEM] $R_2C:N_2HCONH_2$ A condensation product of an aldehyde or ketone with semicarbazide.

semichemical pulp [MATER] Wood which has been pulped by the process of semichemical pulping.

semichemical pulping [CHEM ENG] A method of producing wood-fiber products in which the wood chips are merely softened by chemical treatment (neutral sodium sulfite solution), while the remainder of the pulping action is supplied by a disk attrition mill or by some similar mechanical device for separating the fibers.

semicircular canal [ANAT] Any of three loop-shaped tubular structures of the vertebrate labyrinth; they are arranged in three different spatial planes at right angles to each other, and function in the maintenance of body equilibrium.

semicircular deviation [NAV] A deviation of a magnetic compass which changes sign east or west approximately each 180° change of heading.

semiclosed-cycle gas turbine [MECH ENG] A heat engine in which a portion of the expanded gas is recirculated.

semicoma [MED] A mildly or partially comatose state in which the patient can be roused and responds to strong stimuli with some purposeful movements.

semicompreg [MATER] Resin-impregnated wood compressed to a density not exceeding 1.25.

semiconducting compound [SOLID STATE] A compound which is a semiconductor, such as copper oxide, mercury indium telluride, zinc sulfide, cadmium selenide, and magnesium iodide.

semiconducting crystal [SOLID STATE] A crystal of a semiconductor, such as silicon, germanium, or gray tin.

semiconductive loading tube [ENG] A loading tube for blast-hole explosives which dissipates static electric charges to prevent premature blasts.

semiconductor [SOLID STATE] A solid crystalline material whose electrical conductivity is intermediate between that of a metal and an insulator, ranging from about 10^5 mhos to 10^{-7} mho per meter, and is usually strongly temperature-dependent.

semiconductor device [ELECTR] Electronic device in which the characteristic distinguishing electronic conduction takes place within a semiconductor.

semiconductor diode [ELECTR] Also known as crystal diode; crystal rectifier; diode. **1.** A two-electrode semiconductor device that utilizes the rectifying properties of a *pn* junction or a point contact. **2.** More generally, any two-terminal electronic device that utilizes the properties of the semiconductor from which it is constructed.

semiconductor-diode parametric amplifier [ELECTR] Parametric amplifier using one or more varactors.

semiconductor doping See doping.

semiconductor intrinsic properties [SOLID STATE] Properties of a semiconductor that are characteristic of the ideal crystal.

semiconductor junction [ELECTR] Region of transition between semiconducting regions of different electrical properties, usually between *p*-type and *n*-type material.

semiconductor laser [OPTICS] A laser in which the wavelength of the coherent light beam is determined by a semiconductor compound.

semiconductor rectifier See metallic rectifier.

semiconductor thermocouple [ELECTR] A thermocouple made of a semiconductor, which offers the prospect of operation with high-temperature gradients, because semiconductors are good electrical conductors but poor heat conductors.

semiconductor trap See trap.

semiconservative replication [MOL BIO] Replication of deoxyribonucleic acid by longitudinal separation of the two complementary strands of the molecule, each being conserved and acting as a template for synthesis of a new complementary strand.

semicontrolled mosaic [MAP] A mosaic which is composed of photographs of approximately the same scale laid so that major ground features match their geographical coordinates.

semicrystalline See hypocrystalline.

semidiameter [ASTRON] Measured at the observer, half the angle subtended by the visible disk of a celestial body.

semidiameter correction [NAV] The sextant altitude correction that, when applied to the observation of the upper or lower limb of a celestial body, determines the altitude of the center of that body.

semidiesel engine [MECH ENG] **1.** An internal combustion engine of a type resembling the diesel engine in using heavy oil as fuel but employing a lower compression pressure and spraying it under pressure, against a hot (uncooled) surface or spot, or igniting it by the precombustion or supercompression of a portion of the charge in a separate member or uncooled portion of the combustion chamber. **2.** A true diesel engine that uses a means other than compressed air for fuel injection.

semidiurnal [ASTRON] Having a period of, occurring in, or related to approximately half a day.

semidiurnal current [OCEANOGR] A tidal current in which the tidal-day current cycle consists of two flood currents and two ebb currents, separated by slack water, or of two changes in direction of 360° of a rotary current; this is the most common type of tidal current throughout the world.

semidiurnal tide [OCEANOGR] A tide having two high waters and two low waters during a tidal day.

semidormancy [BOT] Decrease in plant growth rate; may be seasonal or associated with unfavorable environmental conditions.

semidouble [BOT] Pertaining to a flower that has more than the usual number of petals or disk florets while it retains some pollen-bearing stamens or some perfect disk florets.

semifinishing [MET] The preliminary finishing operations.

semifixed ammunition [ORD] Ammunition in which the cartridge case is not permanently fixed to the projectile, so that the zone charge within the cartridge case can be adjusted to obtain the desired range; it is loaded into the weapon as a unit.

semifloating axle [MECH ENG] A supporting member in motor vehicles which carries torque and wheel loads at its outer end.

semifusinite [GEOL] A coal maceral with a well-defined woody structure and optical properties intermediate between those of nitrinite and those of fusinite.

semigel [MATER] A cohesive powder used as an explosive.

semigloss [MATER] Pertaining to a surface finish intermediate between flat and glossy; used especially of paint and varnish.

semigroup [MATH] A set which is closed with respect to a given associative binary operation.

semigroup theory [MATH] The formal algebraic study of the structure of semigroups.

semi-invariants *See* cumulants.

semikilled steel [MET] Incompletely deoxidized steel containing enough dissolved oxygen to react with the carbon it contains to form carbon monoxide, the latter offsetting solidification shrinkage.

semilate [BOT] Pertaining to a plant whose growing season is intermediate between midseason forms and late forms.

semilethal gene [GEN] A mutant causing the death of some of the individuals of the relevant genotype, but never 100%. Also known as sublethal gene.

semilive skid [ENG] A platform having two fixed legs at one end and two wheels at the other; used for moving bulk materials.

semilogarithmic coordinate paper [MATH] Paper ruled with two sets of mutually perpendicular, parallel lines, one set being spaced according to the logarithms of consecutive numbers, and the other set uniformly spaced.

semilunar cartilage [ANAT] One of the two interarticular knee cartilages.

semilunar ganglion *See* Gasserian ganglion.

semilunar valve [ANAT] Either of two tricuspid valves in the heart, one at the orifice of the pulmonary artery and the other at the orifice of the aorta.

semimagnetic controller [ELEC] Electrical controller having only part of its basic functions performed by devices that are operated by electromagnets.

semimat [MATER] Intermediate between glossy and mat, as photographic paper.

semimember [CIV ENG] A part in a frame or truss that ceases to bear a load when the stress in it starts to reverse.

semimembranosus [ANAT] One of the hamstring muscles, arising from the ischial tuber, and inserted into the tibia.

semimetric [MATH] A real valued function $d(x,y)$ on pairs of points from a topological space which has all the same properties as a metric save that $d(x,y)$ may be zero even if x and y are distinct points. Also known as pseudometric.

semimicroanalysis [ANALY CHEM] A chemical analysis procedure in which the weight of the sample is between 10 and 100 milligrams.

semimobile artillery [ORD] Artillery weapons designed for movement but which require partial disassembly to be placed in a firing position; wheels or other suspension devices are removed from the mount so that it may rest on the ground.

semimonocoque [AERO ENG] A fuselage structure in which longitudinal members (stringers) as well as rings or frames which run circumferentially around the fuselage reinforce the skin and help carry the stress. Also known as stiffened-shell fuselage.

seminal bursa [INV ZOO] A sac which retains sperm for a period of time in turbellarians.

seminal groove [ZOO] A passage in many animals providing a pathway for sperm.

seminal receptacle *See* spermatheca.

seminal vesicle [ANAT] A saclike, glandular diverticulum on each ductus deferens in male vertebrates; it is united with the excretory duct and serves for temporary storage of semen.

seminiferous tubule [ANAT] Any of the tubercles of the testes which produce spermatozoa.

seminivorous [ZOO] Feeding on seeds.

seminoma [MED] A malignant tumor of the testes composed of large, uniform cells with clear cytoplasm.

Semionotiformes [VERT ZOO] An order of actinopterygian fishes represented by the single living genus *Lepisosteus*, the gars; the body is characteristically encased in a heavy armor of interlocking ganoid scales.

semipalmate [VERT ZOO] Having a web halfway down the toes.

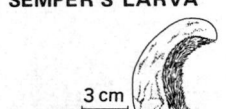

SEMPER'S LARVA

3 cm

(a)   (b)

Two types of pelagic Semper's larva. *(a) Zoanthina tentaculata. (b) Zoanthella galapagoensis.* *(After Senna)*

semiparasite *See* hemiparasite.

semipermanent mold [MET] A reusable metal mold with expendable sand cores.

semipermeable membrane [PHYS] A membrane which allows a solvent to pass through it, but not certain dissolved or colloidal substances.

semiplastic explosive [MATER] An explosive in which the quantities of liquid products are insufficient to render the mixture compressible.

semipositive mold [ENG] A plastics mold that allows a small amount of excess material to escape when it is closed.

semiprecious [LAP] Pertaining to gemstones whose value is lower than that of precious stones; in particular, stones whose hardness is less than 8.

semirefined wax [MATER] Commercial grades of petroleum wax which are inferior to fully refined grades but which meet specified requirements as to color and oil content.

semiregular variables [ASTRON] Red giant stars with absolute magnitude of about 0 or -1 that are variable and have quasi periods of from about 40 to 150 days.

semirigid plastic [MATER] A plastic that has a stiffness or apparent modulus of elasticity of between 10,000 and 100,000 psi (6.895×10^7 and 6.895×10^8 newtons per square meter) under prescribed test conditions.

semischist [PETR] A partly metamorphosed sedimentary rock, exhibiting some foliation.

semiselective ringing [COMMUN] In telephone service, party line ringing wherein the bells of two stations are rung simultaneously; the differentiation is made by the number of rings.

semisilica refractory [MATER] A silica refractory made from clay with a high silica (sand) content (over 70% total silica); characterized by dimensional stability when heated or fired.

semispinalis [ANAT] One of the deep longitudinal muscles of the back, attached to the vertebrae.

semisplint coal [GEOL] Banded coal that is intermediate between bright-banded and splint coal, and has 20–30% opaque attritus and more than 5% anthraxylon.

semistable elementary particle *See* quasi-stable elementary particle.

semisteel [MET] Low-carbon steel made by replacing about one-fourth of the pig iron in the cupola with steel scrap.

semitendinosus [ANAT] One of the hamstring muscles, arising from the ischium and inserted into the tibia.

semitone [ACOUS] The interval between two sounds whose frequencies have a ratio approximately equal to the twelfth root of 2. Also known as half step.

semitrailer [ENG] A cargo-carrying piece of equipment that has one or two axles at the rear; the load is carried on these axles and on the fifth wheel of the tractor that supplies motive power to the semitrailer.

semitransparent photocathode [ELECTR] Photocathode in which radiant flux incident on one side produces photoelectric emission from the opposite side.

semivitreous [MATER] Pertaining to ceramics whose glassy content is not sufficient to reduce porosity below 0.2%.

Semper's larva [INV ZOO] A cylindrical larva in the life history of certain zoanthid corals, characterized by a hole at each end and an annular or longitudinal band of long cilia.

Semple plunger [ORD] A centrifugal plunger which operates to maintain a fuse in a safe condition until centrifugal force unlocks and moves the firing pin into the armed position.

sems [DES ENG] A preassembled screw and washer combination.

semseyite [MINERAL] $Pb_9Sb_8S_{21}$ A gray to black mineral composed of lead antimony sulfide.

senarmontite [MINERAL] Sb_2O_3 A colorless or grayish mineral composed of native antimony trioxide occurring in masses or as octahedral crystals.

send *See* scend.

sender [COMMUN] Part of an automatic-switching telephone system that receives pulses from a dial or other source and, in accordance with them, controls the further operations necessary in establishing a telephone connection.

sending-end impedance [ELEC] Ratio of an applied potential difference to the resultant current at the point where the potential difference is applied; the sending-end impedance of a line is synonymous with the driving-point impedance of the line.

Sendust [MET] A magnetic alloy composed of 85% iron, 9.5% silicon, and 5.5% aluminum; used in high-frequency powders.

Sendzimir mill [MET] A mill having small-diameter working rolls, each backed by a pair of supporting rolls, and each pair of these supported by a cluster of three rolls; used for cold-rolling wide sheets of metal to close tolerance.

senescence [BIOL] The study of the biological changes related to aging, with special emphasis on plant, animal, and clinical observations which may apply to man. [GEOL] The part of the erosion cycle at which the stage of old age begins.

senescent arthritis *See* degenerative joint disease.

senglerite [MINERAL] $Cu(UO_2)_2(VO_4)_2 \cdot 8-10H_2O$ A yellowish-green mineral composed of hydrous copper uranyl vanadate.

senile [MED] Of, pertaining to, or caused by the aging process or by the infirmities of old age.

senile dementia [PSYCH] A chronic, organic brain syndrome associated with old age, characterized by intellectual deterioration, impairment of judgment, and gross emotional instability.

senile eczema [MED] A form of eczema associated with aging and caused by factors such as dryness of skin, soap sensitivity, or poor hygiene or diet.

senile emphysema [MED] Degenerative changes in the lungs and thoracic cage associated with aging.

senile gangrene [MED] A form of tissue death caused by deterioration of the blood supply to the extremities in the elderly.

senile psychosis [PSYCH] A severe form of senile dementia characterized by personality deterioration, progressive memory loss, eccentricity, irritability, and sometimes delusions and hallucinations.

senile vaginitis [MED] Inflammation of the vagina occurring in elderly women following chronic irritation of the thinned, atrophic mucosa.

senna [PHARM] The dried leaflets of plants of the *Cassia* genus; used in medicine as a cathartic.

Senn process [FOOD ENG] A butter-making process in which the cream is subjected to rapid agitation at 1500 revolutions per minute, decreasing to 20 revolutions per minute in the presence of 2-4 atmospheres ($2-4 \times 10^5$ newtons per square meter) of carbon dioxide.

sensation [PHYSIO] The subjective experience that results from the stimulation of a sense organ.

sensation level *See* level above threshold.

sensation unit [ACOUS] A unit of loudness, no longer in use; the loudness of a sound is $20 \log_{10}(p/p_0)$ sensation units above threshold, where p is the pressure level of the sound, and p_0 is the pressure of a sound which can just be detected by the ear.

sense [ADP] To read punched holes in tape or cards. [ENG] To determine the arrangement or position of a device or the value of a quantity. [NAV] The general direction from which a radio signal arrives; if a radio bearing is received by a simple loop antenna, there are two possible readings approximately 180° apart: the resolving of this ambiguity is called sensing of the bearing.

sense amplifier [ELECTR] Circuit used to determine either a phase or voltage change in communications-electronics equipment and to provide automatic control function.

sense antenna [ELECTROMAG] An auxiliary antenna used with a directional receiving antenna to resolve a 180° ambiguity in the directional indication. Also known as sensing antenna.

sense finder [NAV] Portion of a direction finder which permits resolution of the 180° ambiguity.

sense indicator *See* to-from indicator.

sense light [ADP] A light which can be turned on or off, its status being the determinant as to which path a program will select.

sense organ [PHYSIO] A structure which is a receptor for external or internal stimulation.

sense switch *See* alteration switch.

sensibility [PHYS] The ability of a magnetic compass card to align itself with the magnetic meridian after deflection.

sensible atmosphere [METEOROL] That part of the atmosphere that offers resistance to a body passing through it.

sensible heat *See* enthalpy.

sensible-heat factor [THERMO] The ratio of space sensible heat to space total heat; used for air-conditioning calculations. Abbreviated SHF.

sensible heat flow [METEOROL] In the atmosphere, the poleward transport of sensible heat (enthalpy) across a given latitude belt by fluid flow. [THERMO] The heat given up or absorbed by a body upon being cooled or heated, as the result of the body's ability to hold heat; excludes latent heats of fusion and vaporization.

sensible horizon [ASTRON] That circle of the celestial sphere formed by the intersection of the celestial sphere and a plane through any point, such as the eye of an observer, and perpendicular to the zenith-nadir line.

sensible temperature [METEOROL] The temperature at which air with some standard humidity, motion, and radiation would provide the same sensation of human comfort as existing atmospheric conditions.

sensillum [ZOO] A simple, epithelial sense organ composed of one cell or of a few cells.

sensing [NAV] The process of eliminating ambiguity in direction-finder bearings. [ORD] The direction of a point of burst or impact, or centers of burst or impact with respect to the target, such as over, short, air, or graze.

sensing antenna *See* sense antenna.

sensing element *See* sensor.

sensing zone technique [ANALY CHEM] Particle-size measurement in a dilute solution, with fine particles passed through a small zone (opening) so that individual particles may be observed and measured by electrolytic, photic, or sonic methods.

sensistor [ELECTR] Silicon resistor whose resistance varies with temperature, power, and time.

sensitive altimeter [ENG] An aneroid altimeter constructed to respond to pressure changes (altitude changes) with a high degree of sensitivity; it contains two or more pointers to refer to different scales, calibrated in hundreds of feet, thousands of feet, and so on.

sensitive switch *See* snap-action switch.

sensitive time [NUCLEO] The duration of supersaturation, sufficient for track formation, following expansion in a cloud chamber.

sensitive volume [NUCLEO] The portion of a radiation-counter tube that responds to a specific radiation.

sensitivity [ELECTR] **1.** The minimum input signal required to produce a specified output signal, for a radio receiver or similar device. **2.** Of a camera tube, the signal current developed per unit incident radiation, that is, per watt per unit area. [ENG] **1.** A measure of the ease with which a substance can be caused to explode. **2.** A measure of the effect of a change in severity of engine-operating conditions on the antiknock performance of a fuel; expressed as the difference between research and motor octane numbers. Also known as spread. [PHYSIO] The capacity for receiving sensory impressions from the environment. [SCI TECH] **1.** The ability of the output of a device, system, or organism to respond to an input stimulus. **2.** Mathematically, the ratio of the response or change induced in the output to a stimulus or change in the input.

sensitivity function [CONT SYS] The ratio of the fractional change in the system response of a feedback-compensated feedback control system to the fractional change in an open-loop parameter, for some specified parameter variation.

sensitivity time control [ELECTR] In a radar receiver a circuit which greatly reduces the gain at the time that the transmitter emits a pulse; following the pulse, the circuit increases the sensitivity; thus reflection from distant objects will be received and those from nearby objects will be prevented from saturating the receiver.

sensitization *See* activation.

sensitometer [ENG] An instrument for measuring the sensitivity of light-sensitive materials.

sensor [ENG] The generic name for a device that senses either the absolute value or a change in a physical quantity such as temperature, pressure, flow rate, or pH, or the intensity of light, sound, or radio waves and converts that change into a useful input signal for an information-gathering system; a television camera is therefore a sensor, and a transducer is a special type of sensor. Also known as primary detector; sensing element.

sensorium [PHYSIO] **1.** A center, especially in the brain, for receiving and integrating sensations. **2.** The entire sensory apparatus of an individual.

sensory aphasia [MED] A form of aphasia in which the perception of sounds as language is partially preserved, but the patient is unable to comprehend the meaning of words, and in speaking, words are evoked with difficulty, are used incorrectly, and do not convey ideas correctly, resulting frequently in other forms of language impairment, particularly in agrammatism. Also known as receptive aphasia.

sensory area [PHYSIO] Any area of the cerebral cortex associated with the perception of sensations.

sensory cell [PHYSIO] **1.** A neuron having its terminal processes connected with sensory nerve endings. **2.** A modified epithelial or connective tissue cell adapted for the reception and transmission of sensations.

sensory learning [PSYCH] Learning situations in which a person or animal is trained to respond to changes in or differences between some aspects of a physical stimulus presented to one of the sense organs.

sensory nerve [PHYSIO] A nerve that conducts afferent impulses from the periphery to the central nervous system.

sentinel [ADP] Symbol marking the beginning or end of an element of computer information such as an item or a tape.

sepal [BOT] One of the leaves composing the calyx.

separable extension [MATH] A field extension K of a field F is separable if every element of K is a root of a separable polynomial whose coefficients are elements of F.

separable polynomial [MATH] A polynomial with no multiple roots.

separable space [MATH] A topological space which has a countable subset that is dense.

separated aggregate [MATER] Aggregate for concrete that has been separated into fine and coarse constituents.

separated ammunition [ORD] Ammunition characterized by the arrangement of the propelling charge and the projectile for loading into the gun; the propelling charge, contained in a primed cartridge case that is sealed with a closing plug, and the projectile are loaded into the gun in one operation; separated ammunition is used when the ammunition is too large to handle as fixed ammunition.

separated sets [MATH] Sets A and B in a topological space are separated if both the closure of A intersected with B and the closure of B intersected with A are disjoint.

separate loading ammunition [ORD] Ammunition in which the projectile, propellant charge (bag-loaded), and primer are loaded separately into the gun; no cartridge case is utilized in this type of ammunition.

separate sewage system [CIV ENG] A drainage system in which sewage and groundwater are carried in separate sewers.

separating calorimeter [PHYS] A device for measuring the moisture content of steam.

separating power [CHEM ENG] The measure of the ability of a system (such as a rectifying system) to separate the components of a mixture, when the components have increasingly close boiling points.

separation [AERO ENG] The action of a fallaway section or companion body as it casts off from the remaining body of a vehicle, or the action of the remaining body as it leaves a fallaway section behind it. [CHEM ENG] The separation of liquids or gases in a mixture, as by distillation or extraction. [ENG] **1.** The action segregating phases, such as gas-liquid, gas-solid, liquid-solid. **2.** The segregation of solid particles by size range, as in screening. [GEOL] The apparent relative displacement on a fault, measured in any given direction. [MIN ENG] The removal of gangue from raw ores, as in frothing.

separation axioms [MATH] Properties of topological spaces such as Hansdorff, regular, and normal which reflect how points and closed sets may be enclosed in disjoint neighborhoods.

separation disk [BOT] A layer of gelatinous material between two adjacent negative cells in some blue-green algae; associated with hormogonium formation.

separation energy [NUC PHYS] The energy needed to remove a proton, neutron, or alpha particle from a nucleus.

separation factor [NUCLEO] The abundance ratio of two isotopes after processing, divided by their abundance ratio before processing.

separation filter [ELECTR] Combination of filters used to separate one band of frequencies from another.

separation layer [BOT] A structurally distinct layer of the abscission zone of a plant containing abundant starch and dense cytoplasm.

separation negatives [GRAPHICS] The negatives made from full-color originals and used in the preparation of colorplates; four negatives are made, for yellow, magenta, cyan, and black printing plates.

separation of variables [MATH] A technique where certain differential equations are rewritten in the form $f(x)dx = g(y)dy$ which is then solvable by integrating both sides of the equation.

separative work unit [NUC PHYS] A fundamental measure of work required to separate a quantity of isotopic mixture into two component parts, one having a higher percentage of concentration of the desired isotope and one having a lower percentage.

separator [ADP] A datum or character that denotes the beginning or ending of a unit of data. [ELEC] A porous insulating sheet used between the plates of a storage battery. [ELECTR] A circuit that separates one type of signal from another by clipping, differentiating, or integrating action. [ENG] **1.** A machine for separating materials of different specific gravity by means of water or air. **2.** Any machine for separating materials, as the magnetic separator. [MECH ENG] See cage. [PETRO ENG] See gas-oil separator.

separator-filter [ENG] A vessel that removes solids and entrained liquid from a liquid or gas stream, using a combination of a baffle or coalescer with a screening (filtering) element.

Separator-Nobel dewaxing See S-N dewaxing.

separatory funnel [CHEM] A funnel-shaped device used for the careful and accurate separation of two immiscible liquids; a stopcock on the funnel stem controls the rate and amount of outflow of the lower liquid.

sepia [MATER] A brown pigment prepared from the dried, inky exudation of a cuttlefish; used as a dye and in watercolors and ink.

sepia negative See vandyke.

Sepioidea [INV ZOO] An order of the molluscan subclass Coleoidea having a well-developed eye, an internal shell, fins separated posteriorly, and chromatophores in the dermis.

sepiolite [MINERAL] $Mg_4(Si_2O_5)_3(OH)_2 \cdot 6H_2O$ A soft, lightweight, absorbent, white to light-gray or light-yellow clay mineral, found principally in Asia Minor; used for tobacco pipe bowls and ornamental carvings. Also known as meerschaum; sea-foam.

Sepsidae [INV ZOO] The spiny-legged flies, a family of myodarian cyclorrhaphous dipteran insects in the subsection Acalypteratae; development takes place in decaying organic matter.

sepsis [MED] **1.** Poisoning by products of putrefaction. **2.** The severe toxic, febrile state resulting from infection with pyogenic microorganisms, with or without associated septicemia.

septal filament [INV ZOO] In anthozoans, the free edges of the septum containing gland cells and nematocysts.

septal ostium [INV ZOO] Any of the openings in septa of anthozoans.

septarian [GEOL] Pertaining to the irregular polygonal pattern of internal cracks developed in septaria.

septarian boulder See septarium.

septarian nodule See septarium.

septarium [GEOL] A large (80–90 centimeters in diameter), spheroidal concretion, usually composed of argillaceous carbonate, characterized by internal cracking into irregular polygonal blocks that become cemented together by crystalline minerals. Also known as beetle stone; septarian boulder; septarian nodule; turtle stone.

septate [BIOL] Having a septum.

septate coaxial cavity [ELECTROMAG] Coaxial cavity having a vane or septum, added between the inner and outer conductors, so that it acts as a cavity of a rectangular cross section bent transversely.

septate waveguide [ELECTROMAG] Waveguide with one or more septa placed across it to control microwave power transmission.

Septibranchia [INV ZOO] A small order of bivalve mollusks in which the anterior and posterior abductor muscles are about equal in size, the foot is long and slender, and the gills have been transformed into a muscular septum.

septic [MED] Of or pertaining to sepsis.

septic abortion [MED] An abortion complicated by acute infection of the endometrium.

septic embolus [MED] An embolus formed by bacteria.

septicemia [MED] A clinical syndrome in which infection is disseminated through the body in the bloodstream. Also known as blood poisoning.

septic tank [CIV ENG] A settling tank in which settled sludge is in immediate contact with sewage flowing through the tank while solids are decomposed by anaerobic bacterial action.

septinary number [MATH] A number in which the quantity represented by each figure is based on a radix of 7.

septulum [ANAT] A small septum.

septum [BIOL] A partition or dividing wall between two cavities. [ELECTROMAG] A metal plate placed across a waveguide and attached to the walls by highly conducting joints; the plate usually has one or more windows, or irises, designed to give inductive, capacitive, or resistive characteristics.

septum pellucidum [ANAT] A thin translucent septum forming the internal boundary of the lateral ventricles of the brain and enclosing between its two laminas the so-called fifth ventricle.

septum primum [EMBRYO] The first incomplete interatrial septum of the embryo.

septum secundum [EMBRYO] The second incomplete interatrial septum of the embryo, containing the foramen ovale; it develops to the right of the septum primum and fuses with it to form the adult interatrial septum.

Sequanian [GEOL] Upper Lower Jurassic (Upper Lusitanian) geologic time. Also known as Astartian.

sequence [ADP] To put a set of symbols into an arbitrarily defined order; that is, to select A if A is greater than or equal to B, or to select B if A is less than B. [ENG] An orderly progression of items of information or of operations in accordance with some rule. [GEOL] **1.** A sequence of geologic events, processes, or rocks, arranged in chronological order. **2.** A geographically discrete, major informal rock-stratigraphic unit of greater than group or supergroup rank. Also known as stratigraphic sequence. [MATH] A listing of mathematical entities $x_1, x_2 \ldots$ which is indexed by the positive integers; more precisely, a function whose domain is an infinite subset of the positive integers. [METEOROL] *See* collective.

sequence calling [ADP] The instructions used for linking a closed subroutine with a main routine; that is, standard linkage and a list of the parameters.

sequence check [ADP] To verify that correct precedence relationships are obeyed, usually by checking for ascending sequence numbers.

sequence-checking routine [ADP] In computers, a checking routine which records specified data regarding the operations resulting from each instruction.

sequence counter *See* instruction counter.

sequence monitor [ADP] The automatic step-by-step check by a computer of the manual actions required for the starting and shutdown of a computer.

sequence number [ADP] A number assigned to an item to indicate its relative position in a series of related items.

sequence of current [OCEANOGR] The order of occurrence of the tidal current strengths of a day, with special reference to whether the greater flood immediately precedes or follows the greater ebb.

sequence of tide [OCEANOGR] The order in which the tides of a day occur, with special reference to whether the higher high water immediately precedes or follows the lower low water.

sequencer [ADP] A machine which puts items of information into a particular order, for example, it will determine whether A is greater than, equal to, or less than B, and sort or

order accordingly. Also known as sorter. [ENG] A mechanical or electronic device that may be set to initiate a series of events and to make the events follow in a given sequence.

sequence register [ADP] A counter which contains the address of the next instruction to be carried out.

sequence timer [MET] A device used in resistance welding to control the sequence and duration of all elements of the weld cycle, except weld time or heat time.

sequence weld timer [MET] A sequence timer which also controls weld time or heat time.

sequencing [IND ENG] Designating the order of performance of tasks to assure optimal utilization of available production facilities.

sequencing equipment [COMMUN] Special selecting device that permits messages received from several teletypewriter circuits to be subsequently selected and retransmitted over a reduced number of trunks or circuits.

sequential access [ADP] A process that involves reading or writing data serially and, by extension, a data-recording medium that must be read serially, as a magnetic tape.

sequential analysis [STAT] The continuous analysis of data, obtained via sampling, performed as the amount of sampling increases.

sequential circuit [ELEC] A switching circuit whose output depends not only upon the present state of its input, but also on what its input conditions have been in the past.

sequential collation of range [ENG] Spherical, long-baseline, phase-comparison trajectory-measuring system using three or more ground stations, time-sharing a single transponder, to provide nonambiguous range measurements to determine the instantaneous position of a vehicle in flight.

sequential color television [COMMUN] A color television system in which the primary color components of a picture are transmitted one after the other; the three basic types are the line-sequential, dot-sequential, and field-sequential color television systems. Also known as sequential system.

sequential compactness [MATH] A topological space is sequentially compact if every sequence formed from its points has a convergent sequence contained in it.

sequential control [ADP] Manner of operating a computer by feeding orders into the computer in a given order during the solution of a problem.

sequential logic element [ELECTR] A circuit element having at least one input channel, at least one output channel, and at least one internal state variable, so designed and constructed that the output signals depend on the past and present states of the inputs.

sequential operation [ADP] The consecutive or serial execution of operations, without any simultaneity or overlap.

sequential organization [ADP] The write and read of records in a physical rather than a logical sequence.

sequential sampling [IND ENG] A sampling plan in which an undetermined number of samples are tested one by one, accumulating the results until a decision can be made.

sequential scanning *See* progressive scanning.

sequential scheduling system [ADP] A first-come, first-served method of selecting jobs to be run.

sequential selection [COMMUN] The selection of the elements of a message (such as letters) from a set of possible elements (such as the alphabet), one after another.

sequential system *See* sequential color television.

sequential trials [STAT] The outcome of each trial is known before the next trial is performed.

sequestering agent [CHEM] A substance that removes a metal ion from a solution system by forming a complex ion that does not have the chemical reactions of the ion that is removed; can be a chelating or a complexing agent.

sequestrum [MED] A piece of dead or detached bone within a cavity, abscess, or wound.

Sequoia [BOT] A genus of conifers having overlapping, scalelike evergreen leaves and vertical grooves in the trunk; the giant sequoia (*Sequoia gigantea*) is the largest and oldest of all living things.

serac [HYD] A sharp ridge or pinnacle of ice among the crevasses of a glacier.

serandite [MINERAL] $Na(Mn,Ca)_2Si_3O_8(OH)$ A rose-red

SEQUOIA

The giant sequoia tree, *Sequoia gigantea*, showing relative size of man at base.

mineral composed of a basic silicate of manganese, lime, potash, and soda occurring in monoclinic crystals.

Serber potential [NUC PHYS] A potential between nucleons, equal to $\frac{1}{2}(1 + M)V(r)$, where $V(r)$ is a function of the distance between the nucleons, and M is an operator which exchanges the spatial coordinates of the particles but not their spins (corresponding to the Majorana force).

sere [ECOL] A temporary community which occurs during a successional sequence on a given site.

serein [METEOROL] The doubtful phenomenon of fine rain falling from an apparently clear sky, the clouds, if any, being too thin to be visible; frequently, fine rain is observed with a clear sky overhead, but clouds to windward clearly indicate the source of the drops.

Sergestidae [INV ZOO] A family of decapod crustaceans including several species of prawns.

serial [ADP] Pertaining to the internal handling of data in sequential fashion. [IND ENG] An element or a group of elements within a series which is given a numerical or alphabetical designation for convenience in planning, scheduling, and control.

serial-access [ADP] **1.** Pertaining to memory devices having structures such that data storage sites become accessible for read/write in time-sequential order; circulating memories and magnetic tapes are examples of serial-access memories. **2.** Pertaining to a particular process or program that accesses data items sequentially, without regard to the capability of the memory hardware. **3.** Pertaining to character-by-character transmission from an on-line real-time keyboard.

serial bit [ADP] Digital computer storage in which the individual bits that make up a computer word appear in time sequence.

serial digital computer [ADP] A digital computer in which the digits are handled serially, although the bits that make up a digit may be handled either serially or in parallel.

serial homology [ZOO] The similarity between the members of a single series of structures, such as vertebrae, in an organism.

serial learning [PSYCH] The type of association in verbal learning involved in learning the alphabet; studied in the laboratory by giving the subject serial lists to learn, where each list would consist of a number of unrelated items.

serially reusable [ADP] An attribute possessed by a program that can be used for several tasks in sequence without having to be reloaded into main memory for each additional use.

serial observation [OCEANOGR] The procurement of water samples and temperature readings at a number of levels between the surface and the bottom of an ocean.

serial operation [ADP] The flow of information through a computer in time sequence, using only one digit, word, line, or channel at a time.

serial parallel [ADP] **1.** A combination of serial and parallel; for example, serial by character, parallel by bits comprising the character. **2.** Descriptive of a device which converts a serial input into a parallel output.

serial-parallel conversion [ADP] The transformation of a serial data representation as found on a disk or drum into the parallel data representation as exists in core.

serial printer [MECH ENG] A device that prints characters one at a time across a page, such as a typewriter.

serial processor [ADP] A computer in which data are handled sequentially by separate units of the system.

serial programming [ADP] In computers, programming in which only one operation is executed at one time.

serial station [OCEANOGR] An oceanographic station consisting of one or more Nansen casts.

serial storage [ADP] Computer storage in which time is one of the coordinates used to locate any given bit, character, or word; access time, therefore, includes a variable waiting time, ranging from zero to many word times.

serial transfer [ADP] Transfer of the characters of an element of information in sequence over a single path in a digital computer.

serial transmission [COMMUN] Transmission of groups of elements of a signal in time intervals that follow each other without overlapping.

seriate [GEOL] Having crystals that vary gradually in size.

sericeous [BOT] Of, pertaining to, or consisting of silk.

sericite [MINERAL] A white, fine-grained potassium mica, usually muscovite in composition, having a silky luster and found as small flakes in various metamorphic rocks.

sericitization [GEOL] A hydrothermal or metamorphic process involving the introduction of or replacement by sericite.

series [ELEC] An arrangement of circuit components end to end to form a single path for current. [GEOL] **1.** A number of rocks, minerals, or fossils that can be arranged in a natural sequence due to certain characteristics, such as succession, composition, or occurrence. **2.** A time-stratigraphic unit, below system and above stage, composed of rocks formed during an epoch of geologic time. [MATH] An expression of the form $x_1 + x_2 + x_3 + \ldots$, where x_i are real or complex numbers.

series circuit [ELEC] A circuit in which all parts are connected end to end to provide a single path for current.

series compensation *See* cascade compensation.

series connection [ELEC] A connection that forms a series circuit.

series decay *See* radioactive series.

series disintegration [NUC PHYS] The successive radioactive transformations in a radioactive series. Also known as chain decay; chain disintegration.

series excitation [ELEC] The obtaining of field excitation in a motor or generator by allowing the armature current to flow through the field winding.

series-fed vertical antenna [ELECTROMAG] Vertical antenna which is insulated from the ground and energized at the base.

series feed [ELECTR] Application of the direct-current voltage to the plate or grid of a vacuum tube through the same impedance in which the alternating-current flows.

series firing [ENG] The firing of detonators in a round of shots by passing the total supply current through each of the detonators.

series generator [ELEC] A generator whose armature winding and field winding are connected in series.

series loading [ELECTR] Loading in which reactances are inserted in series with the conductors of a transmission circuit.

series modulation [ELECTR] Modulation in which the plate circuits of a modulating tube and a modulated amplifier tube are in series with the same plate voltage supply.

series motor [ELEC] A commutator-type motor having armature and field windings in series; characteristics are high starting torque, variation of speed with load, and dangerously high speed on no-load. Also known as series-wound motor.

series multiple [ELEC] Type of switchboard jack arrangement in which a single line circuit appears before two or more operators, all appearances being connected in series.

series of lines [SPECT] A collection of spectral lines of an atom or ion for a set of transitions, with the same selection rules, to a single final state; often the frequencies have a general formula of the form $[R/(a + c_1)^2] - [R/(n + c_2)^2]$, where R is the Rydberg constant for the atom, a and c_1 and c_2 are constants, and n takes on the values of the integers greater than a for the various lines in the series.

series-parallel circuit [ELEC] A circuit in which some of the components or elements are connected in parallel, and one or more of these parallel combinations are in series with other components of the circuit.

series-parallel firing [ENG] The firing of detonators in a round of shots by dividing the total supply current into branches, each containing a certain number of detonators wired in series.

series-parallel switch [ELEC] A switch used to change the connections of lamps or other devices from series to parallel, or vice versa.

series peaking [ELECTR] Use of a peaking coil and resistor in series as the load for a video amplifier to produce peaking at some desired frequency in the passband, such as to compensate for previous loss of gain at the high-frequency end of the passband.

series production [IND ENG] The manufacture of a product or service by a group of operations sequenced so that all materials will be routed successively through each production state. Also known as batch production.

series radio tap [COMMUN] A telephone tapping procedure in which a miniature radio transmitter is inserted in series

---

SERIES GENERATOR

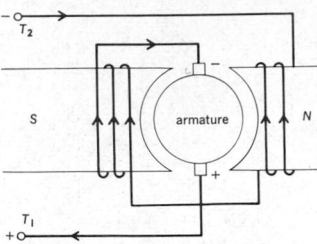

Armature and field winding in a series generator. N = north; S = south; T_1, T_2 = terminals connected to external load.

SERIES-PARALLEL CIRCUIT

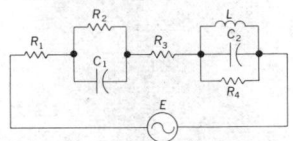

Circuit diagram of a typical series-parallel circuit. E = source of electromotive force; C_1, C_2 = capacitors; L = inductor; R_1, R_2, R_3 = resistors.

with one wire of the target pair so that the transmitter derives its power from the telephone central battery.

series reactor [ELEC] A reactor used in alternating-current power systems for protection against excessively large currents under short-circuit or transient conditions; it consists of coils of heavy insulated cable either cast in concrete columns or supported in rigid frames and mounted on insulators. Also known as current-limiting reactor.

series regulator [ELEC] A regulator that controls output voltage or current by automatically varying a resistance in series with the voltage source.

series repeater [ELEC] A type of negative impedance telephone repeater which is stable when terminated in an open circuit and oscillates when it is connected to a low impedance, in contrast to a shunt repeater.

series resonance [ELEC] Resonance in a series resonant circuit, wherein the inductive and capacitive reactances are equal at the frequency of the applied voltage; the reactances then cancel each other, reducing the impedance of the circuit to a minimum, purely resistive value.

series resonant circuit [ELEC] A resonant circuit in which the capacitor and coil are in series with the applied alternating-current voltage.

series shots [ENG] The connecting and firing of a number of loaded holes one after the other.

series-shunt network See ladder network.

series T junction See E-plane T junction.

series transistor regulator [ELECTR] A voltage regulator whose circuit has a transistor in series with the output voltage, a Zener diode, and a resistor chosen so that the Zener diode is approximately in the middle of its operating range.

series-tuned circuit [ELEC] A simple resonant circuit consisting of an inductance and a capacitance connected in series.

series ventilation [MIN ENG] A system of ventilating a number of faces consecutively by the same air current.

series welding [MET] Making two or more resistance welds simultaneously by using a single welding transformer with three or more electrodes forming a series circuit.

series winding [ELEC] A winding in which the armature circuit and the field circuit are connected in series with the external circuit.

series-wound motor See series motor.

serigraph [GRAPHICS] The silk-screen process when it is used as a fine-art reproduction method.

serine [BIOCHEM] $C_3H_7O_3N$ An amino acid obtained by hydrolysis of many proteins; a biosynthetic precursor of several metabolites, including cysteine, glycine, and choline.

serioscopy [NUCLEO] A radiographic technique enabling three-dimensional exploration by moving two of the three components of the system (tube, subject, film) in order to register the radiographic image of one plane in the object while images outside this slice have a continuous relative displacement and are blurred.

seritinous [ECOL] Of, pertaining to, or occurring during the latter, drier half of the summer.

seroche See mountain sickness.

serodiagnosis [MED] Diagnosis based upon the reaction of blood serum of a patient.

serofibrinous [PHYSIO] Composed of serum and fibrin.

Serolidae [INV ZOO] A family of isopod crustaceans which contains greatly flattened forms that live partially buried on sandy bottoms.

serology [BIOL] The branch of science dealing with the properties and reactions of blood sera.

Seromycin [MICROBIO] A trade name for cycloserine.

seronegative [PATH] **1.** Having a negative serologic test for some condition. **2.** Specifically, having a negative serologic test for syphilis.

seropositive [PATH] **1.** Having a positive serologic test for some condition. **2.** Specifically, having a positive serologic test for syphilis.

seropurulent [MED] Composed of serum and pus, as a seropurulent exudate.

seroresistance [PATH] Persistent positive serologic reaction for syphilis despite prolonged intensive treatment; the patient is said to be Wassermann-fast.

serosa [ANAT] The serous membrane lining the pleural,

peritoneal, and pericardial cavities. [EMBRYO] The chorion of reptile and bird embryos.

serotherapy [MED] The treatment of disease by means of human or animal serum containing antibodies. Also known as immunotherapy.

serotinous [BOT] Plants which flower or develop late in a season.

serotonin [BIOCHEM] $C_{10}H_{12}ON_2$ A compound derived from tryptophan which functions as a local vasoconstrictor, plays a role in neurotransmission, and has pharmacologic properties. Also known as 5-hydroxytryptamine.

serotype [MICROBIO] A serological type of intimately related microorganisms, distinguished on the basis of antigenic composition.

serous cystadenoma [MED] A benign cystic tumor of the ovary, made up of cylindrical cells resembling those of the uterine tube; psammoma bodies often appear in the wall of the cyst. Also known as endosalpingioma; papillary adenoma of ovary; papillary cystadenoma of ovary; papillocystoma; psammomatus papilloma; serous cystoma.

serous cystoma See serous cystadenoma.

serous gland [PHYSIO] A structure that secretes a watery, albuminous fluid.

serous membrane [HISTOL] A delicate membrane covered with flat, mesothelial cells lining closed cavities of the body.

serous plethora [MED] An increase in the watery part of the blood.

Serpens [ASTRON] A constellation, right ascension 17 hours, declination 0°. Also known as Serpent.

Serpent See Serpens.

Serpentes [VERT ZOO] The snakes, a suborder of the Squamata characterized by the lack of limbs and pectoral girdle and external ear openings, immovable eyelids, and a braincase that is completely bony anteriorly.

serpentine [MINERAL] $(Mg,Fe)_3Si_2O_5(OH)_4$ A group of green, greenish-yellow, or greenish-gray ferromagnesian hydrous silicate rock-forming minerals having greasy or silky luster and a slightly soapy feel; translucent varieties are used for gemstones as substitutes for jade.

serpentine cooler See cascade cooler.

serpentine curve [MATH] The curve given by the equation $x^2y + b^2y - a^2x = 0$, passing through and having symmetry about the origin while being asymptotic to the x-axis in both directions.

serpentine locomotion [VERT ZOO] The wavelike or undulating movements characteristic of snakes.

serpentine rock See serpentinite.

serpentinite [PETR] A rock composed almost entirely of serpentine minerals. Also known as serpentine rock.

serpentinization [GEOL] A hydrothermal process by which magnesium-rich silicate minerals are converted into or replaced by serpentine minerals.

serpent kame See esker.

serpierite [MINERAL] $(Cu,Zn,Ca)_5(SO_4)_2(OH)_6 \cdot 3H_2O$ A bluish-green mineral composed of hydrous basic sulfate of copper, zinc, and calcium; occurs in tabular crystals and tufts.

Serpulidae [INV ZOO] A family of polychaete annelids belonging to the Sedentaria including many of the featherduster worms which construct calcareous tubes in the earth, sometimes in such abundance as to clog drains and waterways.

Serranidae [VERT ZOO] A family of perciform fishes in the suborder Percoidei including the sea basses and groupers.

serrate [BIOL] Having a notched or toothed edge. [GEOL] Pertaining to topographic features having a notched or toothed edge, or a saw-edge profile.

serrated pulse [ELECTR] Vertical and horizontal synchronizing pulse divided into a number of small pulses, each of which acts for the duration of half a line in a television system.

serrate ridge See arête.

Serratieae [MICROBIO] A tribe of the Enterobacteriaceae containing the genus *Serratia*, with soil and water forms characterized by the production of a bright-orange to deep-red pigment, prodigiosin.

Serret-Frenet formulas See Frenet-Serret formulas.

Serridentinae [PALEON] An extinct subfamily of elephantoids in the family Gomphotheriidae.

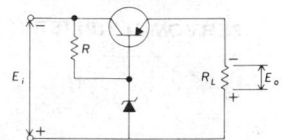

SERIES TRANSISTOR REGULATOR

Circuit diagram of a series transistor regulator. E_O = output voltage; E_i = input voltage; R = resistor; R_L = load resistance.

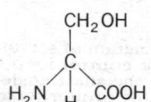

SERINE

Structural formula of serine.

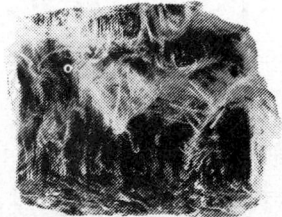

SERPENTINE

Serpentine, variety chrystile, taken from the Thetford Mines, Quebec, Canada. (*American Museum of Natural History Specimen*)

SERPULIDAE

Serpula in right lateral view.

SERVOMULTIPLIER

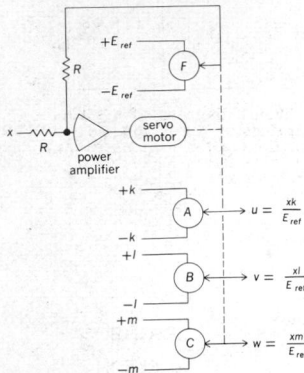

Circuit diagram of servomultiplier. Feedback control loop is used to make the shaft angle of servomotor proportional to drive signal x. Motor shaft is linked to potentiometers A, B, and C so that their output voltages u, v, and w are also proportional to x. In feedback loop servomotor moves slider shaft of an attenuator until signal from the feedback potentiometer F is equal (but opposite in sign) to x so that there is a zero error signal at the input of the power amplifier.

Serritermitidae [INV ZOO] A family of the Isoptera which contains the single monotypic genus *Serritermes*.

serrodyne [ELECTR] Phase modulator using transit time modulation of a traveling-wave tube or klystron.

Serropalpidae [INV ZOO] An equivalent name for Melandryidae.

serrulate [BIOL] Finely serrate.

Sertoli cell [HISTOL] One of the sustentacular cells of the seminiferous tubules.

serum [PHYSIO] The liquid portion that remains when blood clots spontaneously and the formed and clotting elements are removed by centrifugation; it differs from plasma by the absence of fibrinogen.

serum accident [IMMUNOL] A serious allergic reaction which immediately follows the introduction of a foreign serum into a hypersensitive individual; dyspnea and flushing occur, soon followed by shock and occasionally by fatal termination.

serum albumin [BIOCHEM] The principal protein fraction of blood serum and serous fluids.

serum globulin [BIOCHEM] The globulin fraction of blood serum.

serum hepatitis [MED] A form of viral hepatitis transmitted by parenteral injection of human blood or blood products contaminated with the virus.

serum shock [MED] An anaphylactic reaction following the injection of foreign serum into a sensitized individual.

serum sickness [MED] A syndrome manifested in 8-12 days after the administration of serum by an urticarial rash, edema, enlargement of lymph nodes, arthralgia, and fever.

service [ENG] To perform services of maintenance, supply, repair, installation, distribution, and so on, for or upon an instrument, installation, vehicle, or territory.

service agreement [ENG] A contract which agrees to provide mechanical maintenance of a machine for a fixed period of time at a stated charge.

service area [COMMUN] The area that is effectively served by a given radio or television transmitter, navigation aid, or other type of transmitter. Also known as coverage.

service band [COMMUN] Band of frequencies allocated to a given class of radio service.

service brake [MECH ENG] The brake used for ordinary driving in an automotive vehicle; usually foot-operated.

service ceiling [AERO ENG] The height at which, under standard atmospheric conditions, an aircraft is unable to climb faster than a specified rate (100 feet or 30 meters per minute in the United States, Great Britain, and Canada).

service compartment [MIN ENG] The section of a mine shaft that houses water pipes, compressed-air pipeline, cables and telephone wires, and signaling and similar arrangements.

service engineering [ENG] The function of determining the integrity of material and services in order to measure and maintain operational reliability, approve design changes, and assure their conformance with established specifications and standards.

service factor [ENG] For a chemical or a petroleum processing plant or its equipment, the measure of the continuity of an operation, computed by dividing the time on-stream (actual running time) by the total elapsed time.

service life [ENG] The length of time during which a machine, tool, or other apparatus or device can be operated or used economically or before breakdown.

service oscillator *See* radio-frequency signal generator.

service pipe [CIV ENG] A pipe linking a building to a main pipe.

service road [CIV ENG] A small road parallel to the main road for convenient access to shops and houses.

service routine [ADP] A section of a computer code that is used in so many different jobs that it cannot belong to any one job.

service shaft [MIN ENG] A shaft used only for hoisting men and materials to and from underground.

service test [ORD] A test of an item, system of materiel, or technique conducted under simulated or actual operational conditions to determine whether the specified military requirements or characteristics are satisfied.

service velocity [ORD] The muzzle velocity established as the velocity to be attained by a projectile of standard weight

and under standard conditions of temperature, when fired from a new gun of the designated type; the range tables are computed on the basis of this velocity.

service volume [NAV] That volume in airspace surrounding a VOR, Tacan, or Vortac facility within which a signal of usable strength exists and is free of interchannel and cochannel interference; the advertised service volume is defined as a simple cylinder of airspace, usually smaller than theoretical maximum, and is used in planning areas of air operations.

service wires [ELEC] The conductors that bring the electric power into a building.

servicing time [ADP] Machine down-time necessary for routine testing, for machine servicing due to breakdown, or for preventive servicing measures; includes all test time (good or bad) following breakdown and subsequent repair or preventive servicing.

serving [ELEC] A covering, such as thread or tape, that protects a winding from mechanical damage. Also known as coil serving.

servo *See* servomotor.

servo amplifier [ELECTR] An amplifier used in a servomechanism.

servo brake [MECH ENG] **1.** A brake in which the motion of the vehicle is used to increase the pressure on one of the shoes. **2.** A brake in which the force applied by the operator is augmented by a power-driven mechanism.

servolink [CONT SYS] A power amplifier, usually mechanical, by which signals at a low power level are made to operate control surfaces requiring relatively large power inputs, for example, a relay and motor-driven actuator.

servo loop *See* single-loop servomechanism.

servomechanism [CONT SYS] An automatic feedback control system for mechanical motion; it applies only to mose systems in which the controlled quantity or output is mechanical position or one of its derivatives (velocity, acceleration, and so on). Also known as servo system.

servomotor [CONT SYS] The electric, hydraulic, or other type of motor that serves as the final control element in a servomechanism; it receives power from the amplifier element and drives the load with a linear or rotary motion. Also known as servo.

servomultiplier [ELECTR] An electromechanical multiplier in which one variable is used to position one or more ganged potentiometers across which the other variable voltages are applied.

servonoise [ENG] Hunting action of the tracking servomechanism of a radar, which results from backlash and compliance in the gears, shafts, and structures of the mount.

servo system *See* servomechanism.

sesame oil [MATER] A combustible, yellow, optically active, semidrying fatty oil obtained from sesame seeds; soluble in ether, benzene, and carbon disulfide, slightly soluble in alcohol; melts at 20-25°C; used in edible food products, such as shortenings, salad oils, and margarine. Also known as benne oil; gingelly oil; teel oil.

sesamoid bone [MED] A small bone developed in a tendon subjected to much pressure.

sesquioxide [CHEM] A compound composed of a metal and oxygen in the ratio 2:3; for example, Al_2O_3.

sesquisideband transmission [COMMUN] Transmission of a carrier modulated by one full sideband and half of the other sideband.

sesquiterpene [ORG CHEM] Any terpene with the formula $C_{15}H_{24}$; that is, $1\frac{1}{2}$ times the terpene formula.

sessile [BOT] Attached directly to a branch or stem without an intervening stalk. [ZOO] Permanently attached to the substrate.

sessile dislocation [MET] A dislocation in a metal lattice that is relatively immobile, offering an obstacle to the movement of other dislocations.

sessile drop method [FL MECH] A method of measuring surface tension in which the depth and mass of a drop resting on a surface that it does not wet are measured; from this, the shape of the drop and, in turn, the surface tension are determined.

Sessilina [INV ZOO] A suborder of ciliates in the order Peritrichida.

sessoblast [INV ZOO] A statoblast that attaches to zooecial tubes or to the substratum.

seston [OCEANOGR] Minute living organisms and particles of nonliving matter which float in water and contribute to turbidity.

set [ASTRON] Of a celestial body, to cross the visible western horizon while descending. [CHEM] The hardening or solidifying of a plastic or liquid substance. [COMMUN] A radio or television receiver. [ELECTR] The placement of a storage device in a prescribed state, for example, a binary storage cell in the high or 1 state. [ENG] **1.** A combination of units, assemblies, and parts connected or otherwise used together to perform an operational function, such as a radar set. **2.** In plastics processing, the conversion of a liquid resin or adhesive into a solid state by curing or evaporation of solvent or suspending medium, or by gelling. [GRAPHICS] The fixing or drying of a printing ink on a printed sheet, so that, though not completely dry, the sheet can be handled without smudging. [MATER] The hardening or firmness displayed by some materials when left undisturbed. [MATH] A collection of objects which has the property that, given any thing, it can be determined whether or not the thing is in the collection. [MECH] *See* permanent set. [MIN ENG] *See* frame set. [NAV] The establishment of a course. [OCEANOGR] The direction toward which an oceanic current flows.

seta [BIOL] A slender, usually rigid bristle or hair. Also known as chaeta.

set analyzer *See* analyzer.

setback [MECH] The relative rearward movement of component parts in a projectile, missile, or fuse undergoing forward acceleration during its launching; these movements, and the setback force which causes them, are used to promote events which participate in the arming and eventual functioning of the fuse.

setback force [MECH] The rearward force of inertia which is created by the forward acceleration of a projectile or missile during its launching phase; the forces are directly proportional to the acceleration and mass of the parts being accelerated.

set bit [DES ENG] A bit insert with diamonds or other cutting media.

set casing [ENG] Introducing cement between the casing and the wall of the hole to seal off intermediate formations and prevent fluids from entering the hole.

set composite [ELEC] Signaling circuit in which two signaling or telegraph legs may be superimposed on a two-wire, interoffice trunk by means of one of a balanced pair of high-impedance coils connected to each side of the line with an associated capacitor network.

set copper [MET] An intermediate copper product obtained at the end of the oxidizing portion of the fire-refining cycle and containing about 3–4% cuprous oxide.

set forward [MECH] Relative forward movement of component parts which occurs in a projectile, missile, or bomb in flight when impact occurs; the effect is due to inertia and is opposite in direction to setback.

set forward force [MECH] The forward force of inertia which is created by the deceleration of a projectile, missile, or bomb when impact occurs; the forces are directly proportional to the deceleration and mass of the parts being decelerated. Also known as impact force.

set forward point [MECH] A point on the expected course of the target at which it is predicted the target will arrive at the end of the time of flight.

set hammer [DES ENG] **1.** A hammer used as a shaping tool by blacksmiths. **2.** A hollow-face tool used in setting rivets.

set mark [TEXT] A weaving defect in cloth; in particular, a transverse mark resulting from an improperly set loom or from an interruption in the operation of the loom.

set point [CONT SYS] The value selected to be maintained by an automatic controller.

set screw [DES ENG] A small headless machine screw, usually having a point at one end and a recessed hexagonal socket or a slot at the other end, used for such purposes as holding a knob or gear on a shaft.

set theory [MATH] The study of the structure and size of sets from the viewpoint of the axioms imposed.

setting angle [MECH ENG] The angle, usually 90°, between

the straight portion of the tool shank of the machined portion of the work.

setting circle [ENG] A coordinate scale on an optical pointing instrument, such as a telescope or surveyor's transit.

setting gage [ENG] A standard gage for testing a limit gage or setting an adjustable limit gage.

setting ring [ORD] Part of a mechanical fuse setter that takes hold of a fixed ring on the fuse of a projectile; it then rotates the entire projectile except a small ring, or setting element, in the fuse; this setting element is kept from turning by the adjusting ring in the fuse setter just long enough to make the desired change in the setting of the fuse.

setting temperature [ENG] The temperature at which a liquid resin or adhesive, or an assembly involving them, will set, that is, harden, gel, or cure.

setting time [ENG] The length of time that a resin or adhesive must be subjected to heat or pressure to cause them to set, that is, harden, gel, or cure.

settleable solids test [CIV ENG] A test used in examination of sewage to help determine the sludge-producing characteristics of sewage; a measurement of the part of the suspended solids heavy enough to settle is made in an Imhoff cone.

settled [METEOROL] Pertaining to weather, devoid of storms for a considerable period.

settled ground [MIN ENG] Ground which has ceased to subside over the waste area of a mine, having reached a state of full subsidence.

settler [ENG] A separator, such as a tub, pan, vat, or tank in which the partial separation of a mixture is made by density difference; used to separate solids from liquid or gas, immiscible liquid from liquid, or liquid from gas.

settling [ENG] The gravity separation of heavy from light materials; for example, the settling out of dense solids or heavy liquid droplets from a liquid carrier, or the settling out of heavy solid grains from a mixture of solid grains of different densities.

settling basin [CIV ENG] An artificial trap designed to collect suspended stream sediment before discharge of the stream into a reservoir. [IND ENG] A sedimentation area designed to remove pollutants from factory effluents.

settling chamber [ENG] A vessel in which solids or heavy liquid droplets settle out of a liquid carrier by gravity during processing or storage.

settling pond [MIN ENG] A natural or artificial pond for recovering the solids from an effluent.

settling reservoir [CIV ENG] A reservoir consisting of a series of basins connected in steps by long weirs; only the clear top layer of each basin is drawn off.

settling rounds [ORD] Rounds fired at varying angles of elevation to seat the spade and base plate of a gun mount firmly in the ground.

settling tank [ENG] A tank into which a two-phase mixture is fed and the entrained solids settle by gravity during storage.

settling time *See* correction time.

setup [ELECTR] The ratio between the reference black level and the reference white level in television, both measured from the blanking level; usually expressed as a percentage. [IND ENG] The preparation of a facility or a machine for a specific work method, activity, or process.

setup time [ADP] The time before, after, and between computer machine runs, in which manual tasks are carried out, such as changing tape reels or transporting tapes, cards, or supplies to and from the computer equipment, to prepare for a new run. [MATER] The time required for a cement or a gelatin to harden.

seven-eighths rule [NAV] A thumb rule which states that the approximate distance to an object broad on the beam equals 7/8 of the distance traveled by a craft while the relative bearing (right or left) changes from 30 to 60° or from 120 to 150°, neglecting current and wind.

seven-tenths rule [NAV] A thumb rule which states that the approximate distance to an object broad on the beam equals 7/10 of the distance traveled by a craft while the relative bearing (right or left) changes from 22°5 to 45° or from 135° to 157°5, neglecting current and wind.

seven-thirds rule [NAV] A thumb rule which states that the approximate distance to an object broad on the beam equals 7/3 of the distance traveled by a craft while the relative

SESSOBLAST

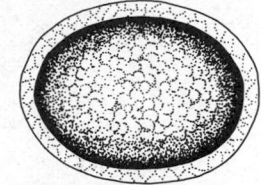

Sessoblast of *Stolella indica.*

SET SCREW

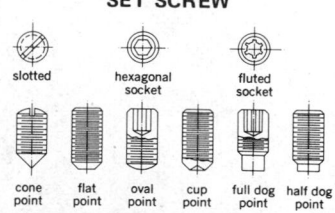

slotted	hexagonal socket	fluted socket

| cone point | flat point | oval point | cup point | full dog point | half dog point |

Types of set screws, with end views of three showing slot or socket. (*From W. J. Luzadder, Fundamentals of Engineering Drawing, 5th ed., Prentice-Hall, 1965*)

bearing (right or left) changes from 22°5 to 26°5, 67°5 to 90°, 90° to 112°5, or 153°5 to 157°5, neglecting current and wind.

seventy-five-degree line [ORD] Imaginary line between the final bomb release line and the vulnerable area upon which antiaircraft artillery guns are located; from such positions the guns at an elevation of 75° can still engage and deliver effective fire on the final bomb release line.

severe storm [METEOROL] In general, any destructive storm, but usually applied to a severe local storm, that is, an intense thunderstorm, hail storm, or tornado.

SEXTANT

A marine sextant.

severe-storm observation [METEOROL] An observation (and report) of the occurrence, location, time, and direction of movement of severe local storms.

severe weather [METEOROL] A more general term for severe storm.

severity factor [CHEM ENG] A measure of the severeness or intensity of overall reaction conditions in a chemical reaction; for example, the temperature, pressure, or conversion in a catalytic cracker or reformer.

sewage [CIV ENG] The fluid discharge from medical, domestic, and industrial sanitary appliances.

sewage disposal plant [CIV ENG] The land, building, and apparatus employed in the treatment of sewage by chemical precipitation or filtration, bacterial action, or some other method.

sewage farm [AGR] A farm in which sewage is used for irrigation and fertilizer.

sewage sludge [CIV ENG] A semiliquid waste with a solid concentration in excess of 2500 parts per million, obtained from the purification of municipal sewage. Also known as sludge.

sewage system [CIV ENG] Any of several drainage systems for carrying surface water and sewage for disposal.

sewage treatment [CIV ENG] A process for the purification of mixtures of human and other domestic wastes; the process can be aerobic or anaerobic.

sewer [CIV ENG] An underground pipe or open channel in a sewage system for carrying water or sewage to a disposal area.

sewerage [CIV ENG] The sewage system in a particular district.

sewer gas [MATER] The gas evolved from the decomposition of municipal sewage; it has a high content of methane and hydrogen sulfide, and can be used as a fuel gas.

sewing machine [MECH ENG] A mechanism that stitches cloth, leather, book pages, or other material by means of a double-pointed or eye-pointed needle.

sewing press [GRAPHICS] A wooden fixture used to stretch cords used in hand-sewn book bindings.

sex [BIOL] The state of condition of an organism which comes to expression in the production of germ cells.

sexadecimal See hexadecimal.

sexagesimal [MATH] Pertaining to a multiplicity of 60 distinct alternative states or conditions or, simply, a positional numeration system to radix (or base) 60.

sexagesimal counting table [MATH] A table for converting numbers using the 60 system into decimals, for example, minutes and seconds.

sexagesimal measure of angles [MATH] A system of angular units in which a complete revolution is divided into 360 degrees, a degree into 60 minutes, and a minute into 60 seconds.

sex chromosome [GEN] Either member of a pair of chromosomes responsible for sex determination of an organism.

sex cords [EMBRYO] Cordlike masses of epithelial tissue that invaginate from germinal epithelium of the gonad and give rise to seminiferous tubules and rete testes in the male, and primary ovarian follicles and rete ovarii in the female.

sex determination [GEN] The mechanisms by which sex is determined in a species.

sex factor See fertility factor.

sex hormone [BIOCHEM] Any hormone secreted by a gonad, but also found in other tissues.

sex-influenced inheritance [GEN] That part of the inheritance pattern on which sex differences operate to promote character differences.

sex-limited inheritance [GEN] Expression of a phenotype in only one sex; may be due to either a sex-linked or autosomal gene.

sex-linked inheritance [GEN] The transmission to successive generations of differences that are due to genes located in the sex chromosomes.

sex organs [ANAT] The organs pertaining entirely to the sex of an individual, both physiologically and anatomically.

sex ratio [BIOL] The relative proportion of males and females in a population.

Sextans [ASTRON] A constellation in the southern hemisphere, right ascension 10 hours, declination 0°. Also known as Sextant.

sextant [MATH] A unit of plane angle, equal to 60° or $\pi/3$ radians. [NAV] An optical instrument; a double reflecting instrument used in navigation for measuring angles, primarily altitudes of celestial bodies.

Sextant See Sextans.

sextant adjustment [NAV] In celestial navigation, the process of checking the accuracy of a sextant and removing or reducing its error.

sextant altitude [NAV] Altitude as indicated by a sextant or similar instrument before corrections are applied.

sextant altitude corrections [NAV] Corrections applied to sextant altitude readings which are made necessary by external physical phenomena, such as sea-air temperature difference, wave height, and sea tilt, personal errors such as tilt and height of eye, and sextant errors.

sextant error [NAV] The error in sextant readings caused by sources other than its operator or physical phenomena; some of these errors are prismatic, collimation, perpendicularity of horizon glass, and perpendicularity of index mirror. Also known as octant error.

sextic [MATH] Having the sixth degree or order.

sextile aspect [ASTRON] The position of two celestial bodies when they are 60° apart.

sexual cycle [PHYSIO] A cycle of physiological and structural changes associated with sex; examples are the estrous cycle and the menstrual cycle.

sexual dimorphism [BIOL] Diagnostic morphological differences between the sexes.

sexuality [BIOL] **1.** The sum of a person's sexual attributes, behavior, and tendencies. **2.** The psychological and physiological sexual impulses whose satisfaction affords pleasure. [PSYCH] The quality of being sexual, or the degree of a person's sexual attributes, attractiveness, and drives.

sexual reproduction [BIOL] Reproduction involving the paired union of special cells (gametes) from two individuals.

sexual spore [BIOL] A spore resulting from conjugation of gametes or nuclei of opposite sex.

seybertite See clintonite.

Seyfert galaxy [ASTRON] A galaxy that has a small, bright nucleus from which violent explosions may occur.

Seymouriamorpha [PALEON] An extinct group of labyrinthodont Amphibia of the Upper Carboniferous and Permian in which the intercentra were reduced.

Sezary syndrome [MED] Exfoliative erythroderma with a cutaneous infiltrate of atypical mononuclear cells; similar cells are also present in the peripheral blood.

S factor [ORD] The deflection change in mils required to keep the burst on the observer target line when the range is changed 100 yards (91.44 meters) along that line.

SFC See specific fuel consumption.

sferics See atmospheric interference.

sferics fix [METEOROL] The estimated location of a source of atmospherics, presumably a lightning discharge.

sferics observation [METEOROL] An evaluation, from one or more sferics receivers, of the location of weather conditions with which lightning is associated; such observations are more commonly obtained from networks of two or three widely spaced stations; simultaneous observations of the azimuth of the discharge are made at all stations, and the location of the storm is determined by triangulation.

sferics receiver [ELECTR] An instrument which measures, electronically, the direction of arrival, intensity, and rate of occurrence of atmospherics; in its simplest form, the instrument consists of two orthogonally crossed antennas, whose output signals are connected to an oscillograph so that one loop measures the north-south component while the other measures the east-west component; the signals are combined vertically to give the azimuth. Also known as lightning recorder.

shackle [DES ENG] An open or closed link of various shapes with extended legs; each leg has a transverse hole to accommodate a pin, bolt, or the like, which may or may not be furnished.

shackle bolt [DES ENG] A cylindrically shaped metal bar for connecting the ends of a shackle.

shade [OPTICS] The color of a mixture of pigments or dyes which has some black pigment or dye in it.

shaded-pole motor [ELEC] A single-phase induction motor having one or more auxiliary short-circuited windings acting on only a portion of the magnetic circuit; generally, the winding is a closed copper ring embedded in the face of a pole; the shaded pole provides the required rotating field for starting purposes.

shade error [NAV] That error of a sextant due to refraction in the shade glasses.

shade glass [OPTICS] A darkened transparency that can be moved into the line of sight of an optical instrument, such as a sextant, to reduce the intensity of light reaching the eye.

shading [ELECTR] Television process of compensating for the spurious signal generated in a camera tube during trace intervals.

shading coil *See* shading ring.

shading ring [ELECTROMAG] The copper ring used in a shaded-pole motor to produce a rotating magnetic field for starting purposes, or used around a part of the core of an alternating-current relay to prevent contact chatter. Also known as shading coil. [ENG ACOUS] A heavy copper ring sometimes placed around the central pole of an electrodynamic loudspeaker to serve as a shorted turn that suppresses the hum voltage produced by the field coil.

shading signal [ELECTR] Television camera signal that serves to increase the gain of the amplifier in the camera during those intervals of time when the electron beam is on an area corresponding to a dark portion of the scene being televised.

shadow [OPTICS] A region of darkness caused by the presence of an opaque object interposed between such a region and a source of light. [PHYS] A region which some type of radiation, such as sound or x-rays, does not reach because of the presence of an object, which the radiation cannot penetrate, interposed between the region and the source of radiation.

shadow attenuation [ELECTROMAG] Attenuation of radio waves over a sphere in excess of that over a plane when the distance over the surface and other factors are the same.

shadow bands [ASTRON] Rippling bands of shadow that appear on every white surface of flat terrestrial objects a few minutes before the total eclipse of the sun.

shadow effect [COMMUN] Reduction in the strength of an ultra-high-frequency signal caused by some object (such as a mountain or a tall building) between the points of transmission and reception.

shadow factor [ELECTROMAG] The ratio of the electric-field strength that would result from propagation of waves over a sphere to that which would result from propagation over a plane under comparable conditions. [OPTICS] A multiplication factor derived from the sun's declination, the latitude of the target and the time of photography, used in determining the heights of objects from shadow length. Also known as tan alt.

shadowgram [GRAPHICS] An x-ray photograph. [PHYS] A plot or display of a shadow.

shadowgraph [GRAPHICS] **1.** A photographic image in the form of a shadow. **2.** *See* radiograph. [OPTICS] A simple method of making visible the disturbances that occur in fluid flow at high velocity, in which light passing through a flowing fluid is refracted by density gradients in the fluid, resulting in bright and dark areas on a screen placed behind the fluid.

shadow line [GRAPHICS] A thick line in a line drawing of an object illuminated by light; indicates the edge farthest from the source of light.

shadow mask [ELECTR] A thin, perforated metal mask mounted just back of the phosphor-dot faceplate in a three-gun color picture tube; the holes in the mask are positioned to ensure that each of the three electron beams strikes only its intended color phosphor dot. Also known as aperture mask.

shadow microscopy *See* projection microradiography.

shadow region [ELECTROMAG] Region in which, under normal propagation conditions, the field strength from a given transmitter is reduced by some obstruction which renders effective radio reception of signals or radar detection of objects in this region improbable.

shadow zone [ACOUS] A region, usually under water or in the atmosphere, which sound waves will not reach, according to ray acoustics.

shadscale [BOT] *Atriplex confertifolia.* A small shiny shrub found in the Great Basin Desert.

shaft [MIN ENG] An excavation of limited area compared with its depth, made for finding or mining ore or coal, raising water, ore, rock, or coal, hoisting and lowering men and material, or ventilating underground workings; the term is often specifically applied to approximately vertical shafts as distinguished from an incline or an inclined shaft. [SCI TECH] A long, slender, usually cylindrical part.

shaft alley [NAV ARCH] The watertight trunk in a ship through which pass the propeller shafts between the propellers and the engine room.

shaft allowance [MIN ENG] The extra space between the excavation diameter and the finished diameter to accommodate the permanent shaft lining.

shaft balancing [MECH ENG] The process of compensating for the minor eccentricities and unbalance of a shaft.

shaft cable [MIN ENG] A specially armored cable of great mechanical strength running down a mine shaft.

shaft capacity [MIN ENG] The output of ore or coal that can be expected to be raised regularly and in normal circumstances.

shaft column [MIN ENG] A length of pipes installed in a mine shaft for pumping, for hydraulic stowing, or for compressed air.

shaft coupling *See* coupling.

shaft crusher [MIN ENG] A hard-rock crusher in a shaft, set to reduce large lumps of ore to a convenient size for delivery to the skip.

shaft deformation bar [MIN ENG] A length of 1½-inch (3.8-centimeter) pipe fitted at one end with a micrometer and at the other end with a hard-steel cone for measuring the deformation in the cross section of a shaft.

shaft drilling [MIN ENG] The drilling of small shafts up to about 5 feet (1.5 meters) in diameter with the shot drill.

shaft furnace [ENG] A vertical, refractory-lined cylinder in which a fixed bed (or descending column) of solids is maintained, and through which an ascending stream of hot gas is forced; for example, the pig-iron blast furnace and the phosphors-from-phosphate-rock furnace.

shaft hopper [MECH ENG] A hopper that feeds shafts or tubes to grinders, threaders, screw machines, and tube benders.

shaft horsepower [MECH ENG] The output power of an engine, motor, or other prime mover; or the input power to a compressor or pump.

shaft house [MIN ENG] A building at the mouth of a shaft, where ore or rock is received from the mine.

shafting [MECH ENG] The cylindrical machine element used to transmit rotary motion and power from a driver to a driven element; for example, a steam turbine driving a ship's propeller.

shaft kiln [ENG] A kiln in which raw material fed into the top, moves down through hot gases flowing up from burners on either side at the bottom, and emerges as a product from the bottom; used for calcining operations.

shaft lining [MIN ENG] The timber, steel, brick, or concrete structure fixed around a shaft to support the walls.

shaft pillar [MIN ENG] A large area of a coal or ore seam which is left unworked around the shaft bottom to protect the shaft from damage by subsidence.

shaft plumbing [MIN ENG] The operation of orienting two plumb bobs, both at surface and at depth in order to transfer the bearing underground.

shaft pocket [MIN ENG] Ore storage pocket, of one or more compartments, cut into the wall on one or both sides of a vertical shaft or in the hanging wall of an inclined shaft.

shaft-position encoder [ELECTR] An analog-to-digital con-

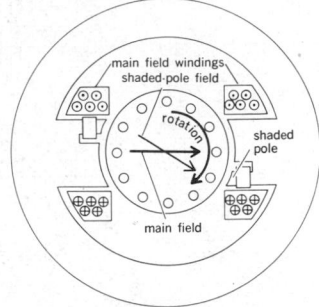

SHADED-POLE MOTOR

⊙ current toward reader
⊕ current away from reader

Cross-sectional view of shaded-pole motor.

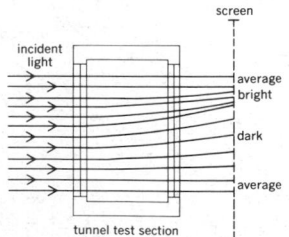

SHADOWGRAPH

Shadowgraph produced by light passing through fluid in test section indicates flow pattern.

verter in which the exact angular position of a shaft is sensed and converted to digital form.

shaft siding [MIN ENG] The station or landing place arranged for buckets or tubs at the bottom of the winding shaft.

shaft signaling [MIN ENG] The transmission of visible and audible signals between the onsetter or hitcher at the pit bottom and the banksman or hoistman at the pit top.

shaft sinking [MIN ENG] Excavating a shaft downward, usually from the surface, to the workable coal or ore.

shaft sinking drill [MIN ENG] A large-diameter drill with multiple rotary cones or cutting bits, used for shaft sinking.

shaft spillway [CIV ENG] A vertical shaft which has a funnel-shaped mouth and ends in an outlet tunnel, providing an overflow duct for a reservoir. Also known as morning glory spillway.

shaft station [MIN ENG] An enlargement of a level near a shaft from which ore, coal, or rock may be hoisted and supplies unloaded.

shaft strut [NAV ARCH] A term applied to a bracket supporting the after propeller shaft bearing external to the hull.

shagreen [MATER] 1. A leather made by pressing grains into the hide to create indentations. 2. The skin of certain sharks and rays containing small knobs.

shake [MATER] Separation between adjoining layers of wood, due to causes other than drying.

shake culture [MICROBIO] 1. A method for isolating anaerobic bacteria by shaking a deep liquid culture of an agar or gelatin to distribute the inoculum before solidification of the medium. 2. A liquid medium in a flask that has been inoculated with an aerobic microorganism and placed on a shaking machine; action of the machine continually aerates the culture.

shakedown test [ENG] An equipment test made during the installation work.

shakeout [MET] Removing a casting from a sand mold.

shaker [ELECTROMAG] An electromagnetic device capable of imparting known and usually controlled vibratory acceleration to a given object. Also known as electrodynamic shaker; shake table.

shaker conveyor [MIN ENG] A conveyor consisting of a length of metal troughs, with suitable supports, to which a reciprocating motion is imparted by drives.

shake table *See* shaker; vibration machine.

shake-table test [ENG] A laboratory test for vibration tolerance, in which the device to be tested is placed on a shake table.

shake wave *See* S wave.

shaking screen [MECH ENG] A screen used in separating material into desired sizes; has an eccentric drive or an unbalanced rotating weight to produce shaking.

shaking table *See* Wilfley table.

shale [PETR] A fine-grained laminated or fissile sedimentary rock made up of silt- or clay-size particles; generally consists of about one-third quartz, one-third clay materials, and one-third miscellaneous minerals, including carbonates, iron oxides, feldspars, and organic matter.

shale clay [MATER] A clay made from ground shale.

shale oil [MATER] Liquid obtained from the destructive distillation of kerogens in oil shale; further processing is needed to convert shale oil into products similar to petroleum oils.

shale reservoir [GEOL] Underground hydrocarbon reservoir in which the reservoir rock is a brittle, siliceous, fractured shale.

shallot [BOT] *Allium ascalonicum*. A bulbous onionlike herb. Also known as scallion.

shallow-focus earthquake [GEOPHYS] An earthquake whose focus is located within 70 kilometers of the earth's surface.

shallow fog [METEOROL] In weather-observing terminology, low-lying fog that does not obstruct horizontal visibility at a level 6 feet (1.8 meters) or more above the surface of the earth; this is, almost invariably, a form of radiation fog.

shallow fording [ORD] The ability of a vehicle or gun, equipped with built-in waterproofing with its suspension in contact with the ground, to negotiate a water obstacle without the use of a special waterproofing.

shallow inland seas [GEOL] Epeiric seas which periodically cover cratonic areas as a result of continental subsidence or eustatic rises in sea level.

shallow marginal seas [GEOL] Epeiric seas along the cratonic margins.

shallow water [HYD] Water of such a depth that bottom topography affects surface waves.

shallow-water wave [HYD] A progressive gravity wave in water whose depth is much less than the wavelength.

shaluk [METEOROL] Any hot desert wind other than simoom.

shaly [GEOL] Pertaining to, composed of, containing, or having the properties of shale, especially readily split along close-spaced bedding planes.

shaly bedding [GEOL] Laminated bedding varying between 2 and 10 millimeters in thickness.

shamal [METEOROL] The northwest wind in the lower valley of the Tigris and Euphrates and the Persian Gulf; it may set in suddenly at any time, and generally lasts from 1 to 5 days, dying down at night and freshening again by day; but in June and early July it continues almost without cessation (the great or 40-day shamal).

shandite [MINERAL] $Ni_3Pb_2S_2$ A rhombohedral mineral composed of nickel lead sulfide, occurring in crystals.

shank *See* bit blank.

shank-type cutter [DES ENG] A cutter having a shank to fit into the machine tool spindle or adapter.

Shannon formula [COMMUN] A theorem in information theory which states that the highest number of binary digits per second which can be transmitted with arbitrarily small frequency of error is equal to the product of the bandwidth and $\log_2 (1 + R)$, where R is the signal-to-noise ratio.

Shannon limit [COMMUN] Maximum signal-to-noise ratio improvement which can be achieved by the best modulation technique as implied by Shannon's theorem relating channel capacity to signal-to-noise ratio.

Shannon-McMillian-Breiman Theorem [MATH] Given an ergodic measure preserving transformation T on a probability space and a finite partition ζ of that space the limit as $n \to \infty$ of $1/n$ times the information function of the common refinement of ζ, $T^{-1}\zeta$, . . ., $T^{-n+1}\zeta$ converges almost everywhere and in the L_1 metric to the entropy of T given ζ.

Shannon's sampling theorem *See* sampling theorem.

Shannon's theorems [MATH] These results are foundational to the mathematical study of information; mathematically they link the concept of entropy with the amount of efficient transmittal and reception of information.

Shantung soil *See* Noncalcic Brown soil.

shape coding [DES ENG] The use of special shapes for control knobs, to permit recognition and sometimes also position monitoring by sense of touch.

shaped-beam antenna [ELECTROMAG] Antenna with a directional pattern which, over a certain angular range, is of special shape for some particular use.

shaped-chamber manometer [ENG] A flow measurement device that measures differential pressure with a uniform flow-rate scale with a specially shaped chamber.

shaped charge [ORD] An explosive charge with a shaped cavity that forces the impact of the explosion to the front so that there is an armor-piercing force. Also known as cavity charge.

shape factor [ELEC] *See* form factor. [FL MECH] The quotient of the area of a sphere equivalent to the volume of a solid particle divided by the actual surface of the particle; used in calculations of gas flow through beds of granular solids. [OPTICS] For a lens, the quantity $(R_2 + R_1)/(R_2 - R_1)$, where R_1 and R_2 are the radii of the first and second surface of the lens. Also known as Coddington shape factor.

shaper [MECH ENG] A machine tool for cutting flat-on-flat, contoured surfaces by reciprocating a single-point tool across the workpiece.

shaping [ELECTROMAG] The adjustment of a plan-position-indicator pattern set up by a rotating magnetic field. [MECH ENG] A machining process in which a reciprocating single-point tool moves over the work in straight, parallel lines to produce a flat surface.

shaping circuit *See* corrective network.

shaping dies [MECH ENG] A set of dies for bending, pressing, or otherwise shaping a material to a desired form.

shaping network *See* corrective network.

shapometer [ENG] A device used to measure the shape of sedimentary particles.

shared control unit [ADP] A control unit which controls several devices with similar characteristics, such as tape devices.

shark [VERT ZOO] Any of about 225 species of carnivorous elasmobranchs which occur principally in tropical and subtropical oceans; the body is fusiform with a heterocercal tail and a tough, usually gray, skin roughened by tubercles, and the snout extends beyond the mouth.

sharki *See* kaus.

shark liver oil [MATER] A yellow to brown oil with strong aroma, obtained from the livers of various sharks; insoluble in water, soluble in ether, benzene, and carbon disulfide; used as a vitamin A source, in biochemical research. Also known as dogfish oil; shark oil.

shark oil *See* shark liver oil.

sharkskin [TEXT] **1.** A smooth wool or worsted fabric with a basketweave or twill pattern in two tones. **2.** A smooth, dull-finished rayon fabric in a basketweave pattern used for dresses.

sharkskin pahoehoe [GEOL] A type of pahoehoe displaying numerous tiny spines or spicules on the surface.

shark-tooth projection [GEOL] Sharp pointed projections several centimeters in length, formed by the pulling apart of plastic lava.

sharp-crested weir [CIV ENG] A weir in which the water flows over a thin, sharp edge.

sharp-cutoff tube [ELECTR] An electron tube in which the control-grid openings are uniformly spaced; the anode current then decreases linearly as the grid voltage is made more negative, and cuts off sharply at a particular grid voltage.

sharp-edged gust [METEOROL] A gust that represents an instantaneous change in wind direction or speed.

sharpen [ENG] To give a thin keen edge or a sharp acute point to.

sharpening stone [ENG] A device such as a whetstone used for sharpening by hand.

sharp iron [ENG] A tool used to open seams for caulking.

sharpite [MINERAL] $(UO_2)(CO_3) \cdot H_2O$ A greenish-yellow mineral composed of hydrous basic uranyl carbonate.

sharpness of resonance [ELEC] The narrowness of the frequency band around the resonance at which the response of an electric circuit exceeds an arbitrary fraction of its maximum response, often 70.7%.

sharps [FOOD ENG] The bran from wheat kernels; used as cattle feed. Also known as middlings; wheat middlings.

sharp sand [GEOL] An angular-grain sand free of clay, loam, and other foreign particles.

sharp series [SPECT] A series occurring in the line spectra of many atoms and ions with one, two, or three electrons in the outer shell, in which the total orbital angular momentum quantum number changes from 0 to 1.

sharpstone [GEOL] Any rock fragment having angular edges and corners and being more than 2 millimeters in diameter.

sharpstone conglomerate *See* sedimentary breccia.

sharp tuning [ELEC] Having high selectivity; responding only to a desired narrow range of frequencies.

shatter cone [GEOL] A striated conical rock fragment along which fracturing has occurred.

shattercrack *See* flake.

shattering [FL MECH] One theory to explain homogenization or globule fractionation in milk; it is the effect that occurs when whole milk under high velocity strikes a flat surface, such as an impact ring.

shatterproof glass *See* nonshattering glass.

shattuckite [MINERAL] $Cu_5(SiO_3)_4H_2O$ A blue mineral composed of basic copper silicate, occurring in fibrous masses.

Shaula [ASTRON] A blue-white subgiant star of stellar magnitude 1.7, spectral classification B2-IV, in the constellation Scorpius; the star λ Scorpii.

shave hook [DES ENG] A plumber's or metalworker's tool composed of a sharp-edged steel plate on a shank; used for scraping metal.

shaving [ENG ACOUS] Removing material from the surface of a disk recording medium to obtain a new recording surface.

[MECH ENG] **1.** Cutting off a thin layer from the surface of a workpiece. **2.** Trimming uneven edges from stampings, forgings, and tubing.

Shaw process [MET] A foundry molding process which makes use of wood or metal patterns and a refractory mold bonded with an ethyl silicate base material.

sheaf [MATH] A fiber bundle with algebraic and topological structure usually associated to a differentiable manifold M which reflects the local behavior of differentiable functions on M. [ORD] Planned planes of fire which produce a desired pattern of bursts with rounds fired by two or more weapons.

sheaf of planes [MATH] All the planes passing through a given point.

shear [DES ENG] A cutting tool having two opposing blades between which a material is cut. [MECH] *See* shear strain. [MIN ENG] To make vertical cuts in a coal seam that has been undercut.

shear angle [MECH ENG] The angle made by the shear plane with the work surface.

shear burst [MIN ENG] The explosive breaking of wall rock in a deep mining field by the occurrence of a single shear crack parallel to the face in one of the walls, causing rock behind the shear plane to expand freely into the stope and then to disrupt and fill the place with debris.

shear center *See* center of twist.

shear diagram [MECH] A diagram in which the shear at every point along a beam is plotted as an ordinate.

shear fold [GEOL] A similar fold whose mechanism is shearing or slipping along closely spaced planes that are parallel to the fold's axial surface. Also known as glide fold; slip fold.

shear fracture [MECH] A fracture resulting from shear stress.

shear-gravity wave [GEOPHYS] A combination of gravity waves and a Helmholtz wave on a surface of discontinuity of density and velocity.

shearing [MECH ENG] Separation of material by the cutting action of shears. [MIN ENG] The vertical side cutting which, together with holing or horizontal undercutting, constitutes the attack upon a face of coal.

shearing die [MECH ENG] A die with a punch for shearing the work from the stock.

shearing field [PL PHYS] A special type of magnetic field, used to confine a plasma whose rotational transform angle changes with distance from the magnetic axis.

shearing instability *See* Helmholtz instability.

shearing machine [MECH ENG] A machine for cutting cloth or bars, sheets, or plates of metal or other material.

shearing punch [MECH ENG] A punch that cuts material by shearing it, with minimal crushing effect.

shearing stress [MECH] A stress in which the material on one side of a surface pushes on the material on the other side of the surface with a force which is parallel to the surface. Also known as shear stress; tangential stress.

shearing tool [DES ENG] A cutting tool (for a lathe, for example) with a considerable angle between its face and a line perpendicular to the surface being cut.

shear joint [GEOL] A joint that is a shear fracture; it is a potential plane of shear. Also known as slip joint.

shear line [METEOROL] A line or narrow zone across which there is an abrupt change in the horizontal wind component parallel to this line; a line of maximum horizontal wind shear.

shear lip [MET] An area or ridge at the edge of a shear fracture surface.

shear mark [ENG] A crease on a piece of pressed glass; results when the piece is sheared off for pressing.

shear modulus *See* modulus of elasticity in shear.

shear pin [DES ENG] **1.** A pin or wire provided in a fuse design to hold parts in a fixed relationship until forces are exerted on one or more of the parts which cause shearing of the pin or wire; the shearing is usually accomplished by setback or set forward (impact) forces; the shear member may be augmented during transportation by an additional safety device. **2.** In a propellant-actuated device, a locking member which is released by shearing. **3.** In a power train, such as a winch, any pin, as through a gear and shaft, which is designed to fail at a predetermined force in order to protect a mechanism.

shear plane [MECH] A confined zone along which fracture occurs in metal cutting.

SHARK

The mackerel shark *(Lamnia nasus)* has a horizontal heellike tail and gill slits on both sides; claspers and prolonged pelvic fins, occur on males.

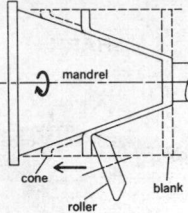

SHEAR SPINNING

Schematic of shear spinning showing the blank in three positions.

SHEEP

(a)

(b)

Examples of some prominent breeds of sheep. *(a)* Hampshire ram. *(b)* Southdown ram *(Southdown Association).*

shear rate [FL MECH] The relative velocities in laminar flow of parallel adjacent layers of a fluid body under shear force.

shear spinning [MECH ENG] A sheet-metal-forming process which forms parts with rotational symmetry over a mandrel with the use of a tool or roller in which deformation is carried out with a roller in such a manner that the diameter of the original blank does not change but the thickness of the part decreases by an amount dependent on the mandrel angle.

shear steel [MET] A cutlery steel made from short sheared lengths of blister steel; the lengths are heated, joined by rolling or hammering, and finished by hammering.

shear strain [MECH] Also known as shear. **1.** A deformation of a solid body in which a plane in the body is displaced parallel to itself relative to parallel planes in the body; quantitatively, it is the displacement of any plane relative to a second plane, divided by the perpendicular distance between planes. **2.** The force causing such deformation.

shear strength [MECH] **1.** The maximum shear stress which a material can withstand without rupture. **2.** The ability of a material to withstand shear stress.

shear stress *See* shearing stress.

shear structure [GEOL] A local structure in which earth stresses have been relieved by many small, closely spaced fractures.

shear test [ENG] Any of various tests to determine shear strength of soil samples.

shear thickening [FL MECH] Viscosity increase of non-Newtonian fluids (for example, complex polymers, proteins, protoplasm) that undergo viscosity increases under conditions of shear stress (that is, viscometric flow).

shear thinning [FL MECH] Viscosity reduction of non-Newtonian fluids (for example, polymers and their solutions, most slurries and suspensions, lube oils with viscosity-index improvers) that undergo viscosity reductions under conditions of shear stress (that is, viscometric flow).

shear-viscosity function [FL MECH] The expression of the viscometric flow of a purely viscous, non-Newtonian fluid in terms of velocity gradient and shear stress of the flowing fluid.

shearwater [VERT ZOO] Any of various species of oceanic birds of the genus *Puffinus* having tubular nostrils and long wings.

shear wave [GEOPHYS] *See* S wave. [MECH] A wave that causes an element of an elastic medium to change its shape without changing its volume. Also known as rotational wave.

shear zone [GEOL] A tabular area of rock that has been crushed and brecciated by many parallel fractures resulting from shear strain; often becomes a channel for underground solutions and the seat of ore deposition.

sheath [ELEC] A protective outside covering on a cable. [ELECTR] A space charge formed by ions near an electrode in a gas tube. [ELECTROMAG] The metal wall of a waveguide. [SCI TECH] A protective case or cover.

sheathed explosive [ENG] A permitted explosive enveloped by a sheath containing a noncombustible powder which reduces the temperature of the resultant gases of the explosion and, therefore, reduces the risk of these hot gases causing a firedamp ignition.

sheathing board [MATER] A composition board (for example, of fiber or gypsum cement) used instead of wood sheathing.

sheathing paper [MATER] A paper that is heavier and of better quality than the usual building paper.

sheath-reshaping converter [ELECTROMAG] In a waveguide, a mode converter in which the change of wave pattern is achieved by gradual reshaping of the sheath of the waveguide and of conducting metal sheets mounted longitudinally in the guide.

sheave [DES ENG] A grooved wheel or pulley.

shed [NUC PHYS] A unit of cross section, used in studying collisions of nuclei and particles, equal to 10^{-24} barn, or 10^{-48} square centimeter.

SHED *See* solar heat exchanger drive.

shedding [TEXT] A process performed by the heddle frame and heddles of a loom to raise and lower certain groups of alternate warp yarns so that the filling yarns alternate in passing under one group of warp yarns and over another.

shed dormer [ARCH] A dormer window which (unlike a gabled dormer) has a horizontal eave line.

Sheehan's syndrome [MED] Hypopituitarism resulting from postpartum adenohypophyseal necrosis.

sheep [VERT ZOO] Any of various mammals of the genus *Ovis* in the family Bovidae characterized by a stocky build and horns, when present, which tend to curl in a spiral.

sheepsfoot roller [DES ENG] A cylindrical steel drum to which knob-headed spikes are fastened; used for compacting earth.

sheep's-head clock [HOROL] A lantern clock having one hand, a crown escapement, and large dials overlapping the clock movement.

sheepskin wheel [DES ENG] A polishing wheel made of sheepskin disks or wedges either quilted or glued together.

sheer [NAV] To swerve off course to avoid collision. [NAV ARCH] The curvature of a ship's deck from bow to stern, with the lowest point of the deck usually amidships.

sheering batten [NAV ARCH] One of the long strips of wood attached to the frames of a ship to locate the strakes of the shell plating in relation to the sheer of the deck.

sheer line [NAV ARCH] The longitudinal curve of the rail or decks, which shows the variation in height above water or freeboard throughout the vessel's entire length.

sheer strake [NAV ARCH] The uppermost line of planking of a wooden ship, or the upper strake of plating at a steel ship's main deck.

sheer-strake plate [NAV ARCH] A plate forming part of a sheer strake.

sheet [GEOL] **1.** A thin flowstone coating of calcite in a cave. **2.** A tabular igneous intrusion, especially when concordant or only slightly discordant. [HYD] *See* sheetflood. [MATER] A wide, thin, and usually flexible piece of material, such as paper or metal, which is thinner than plate. [NAV ARCH] A rope or chain used to haul the clew of a sail out toward the yard arm or downward toward the deck and aft.

sheet asphalt [MATER] Asphalt which provides a smooth surface and is used for continuous road surfacing.

sheet cavitation [FL MECH] A type of cavitation in which cavities form on a solid boundary and remain attached as long as the conditions that led to their formation remain unaltered. Also known as steady-state cavitation.

sheet composting [AGR] Addition of large amounts of organic residue to a soil; extra nitrogen is usually added for faster decomposition.

sheet copper [MET] Copper rolled into sheets; for roofing sometimes used as it leaves the rolls, but for other purposes it is commonly employed after it has been cold-rolled to increase hardness and strength.

sheet deposit [GEOL] A stratiform mineral deposit that is more or less horizontal and extensive relative to its thickness.

sheeted fissure [GEOL] A closely spaced fissure.

sheeted vein [GEOL] A vein filling a shear zone.

sheeted zone [GEOL] An area of mineral deposits consisting of sheeted veins.

sheet erosion [GEOL] Erosion of thin layers of surface materials by continuous sheets of running water. Also known as sheetflood erosion; sheetwash; surface wash; unconcentrated wash.

sheet-fed press [GRAPHICS] A printing press that accepts paper in the form of sheets.

sheet film [GRAPHICS] Film that consists of a negative material with its emulsion coated on a base of plastic rather than glass.

sheetflood [HYD] A broad expanse of moving, storm-borne water that spreads as a thin, continuous, relatively uniform film over a large area for a short distance and duration. Also known as sheet; sheetwash.

sheetflood erosion *See* sheet erosion.

sheet forming [ENG] The process of producing thin, flat sections of solid materials; for example, sheet metal, sheet plastic, or sheet glass.

sheet frost [HYD] A thick coating of rime formed on windows and other surfaces.

sheet glass [MATER] Flat sections of glass made by drawing a continuous thin film of glass from a molten bath, then cooling and cutting it; used for common glazing.

sheet grating [ELECTROMAG] Three-dimensional grating consisting of thin, longitudinal, metal sheets extending along the inside of a waveguide for a distance of about a wavelength, and used to stop all waves except one predetermined wave that passes unimpeded.

sheet ice [HYD] A smooth, thin layer of ice formed by rapid freezing of the surface layer of a body of water.

sheeting [GEOL] The process by which thin sheets, slabs, scales, plates, or flakes of rock are successively broken loose or stripped from the outer surface of a large rock mass in response to release of load. Also known as exfoliation. [MATER] **1.** A continuous film of a material such as plastic. **2.** Steel or wood members used to face the walls of an excavation such as a basement or a trench.

sheeting caps [MIN ENG] A row of caps put on blocks about 14 inches high which are placed on top of the drift sets when constructing the permanent floor in the stope.

sheeting structure [GEOL] A fracture or joint formed by pressure-release jointing or exfoliation. Also known as exfoliation joint; expansion joint; pseudostratification; release joint; sheet joint; sheet structure.

sheet joint *See* sheeting structure.

sheet lightning [GEOPHYS] A diffuse, but sometimes fairly bright, illumination of those parts of a thundercloud that surround the path of a lightning flash, particularly a cloud discharge or cloud-to-cloud discharge. Also known as luminous cloud.

sheet line [MAP] The outermost border line of a map or chart.

sheet metal [MET] Thin sections of metal formed by rolling hot metal and usually less than 0.25 inch (6.35 millimeters) thick; when thicker than 0.25 inch, called plate.

sheet-metal gage [MET] A standard for expressing the thickness of metal sheets; some manufacturers, for example, Brown & Sharpe (B&S), Birmingham (BG), and Imperial, use code numbers with actual thickness in inches or millimeters.

sheet-metal screw *See* self-tapping screw.

sheet mineral *See* phyllosilicate.

sheet piling [CIV ENG] Closely spaced piles of wood, steel, or concrete driven vertically into the ground to obstruct lateral movement of earth or water, and often to form an integral part of the permanent structure.

sheet plastic [MATER] Flat sections of extruded, molded, or cast plastic, with a thickness greater than that for film, that is, greater than 0.05 inch (1.3 millimeters).

sheet polarizer [OPTICS] A mechanism for obtaining linear polarized light; there are several types, one of which is a microcrystalline polarizer in which small crystals of a dichroic material (quinine iodosulfate), oriented parallel to each other in a plastic medium, absorb one polarization and transmit the other.

Sheetrock [MATER] A plasterboard, usually made of two sheets of heavy paper separated by a layer of gypsum.

sheet rubber [MATER] Latex that has been rolled into sheets, either smooth or ribbed.

sheet sand *See* blanket sand.

sheet sandstone [GEOL] A thin, blanket-shaped deposit of sandstone of regional extent.

sheet separation [MET] The gap between faying surfaces surrounding the weld in spot, seam, or projection welding.

sheet silicate *See* phyllosilicate.

sheet steel [MET] Steel rolled in the form of sheet, usually used for deep-drawing applications.

sheet structure *See* sheeting structure.

sheet train [ENG] The entire assembly needed to produce plastic sheet; includes the extruder, die, polish rolls, conveyor, draw rolls, cutter, and stacker.

sheetwash [GEOL] **1.** The detritus deposited by a sheetflood. **2.** *See* sheet erosion. [HYD] **1.** A wide, moving expanse of water on an arid plain; the combined result of many streams issuing from the mountains. **2.** *See* sheetflood.

sheffer stroke *See* NAND.

Sheffield plate [MET] A cladding of silver rolled and fused on both sides of a copper sheet.

Shelby tube [ENG] A thin-shelled tube used to take deep-soil samples; the tube is pushed into the undisturbed soil at the bottom of the casting of the borehole driven into the ground.

shelf [GEOL] **1.** Solid rock beneath alluvial deposits. **2.** A flat, projecting ledge of rock. **3.** *See* continental shelf.

shelf angle [CIV ENG] A mild steel angle section, riveted or welded to the web of an I beam to support the formwork for hollow tiles or the floor or roof units, or to form a seat for precast concrete.

shelf channel [GEOL] A valley formed in a shelf by erosion.

shelf edge [GEOL] The demarcation, without dramatic change in gradient, between continental shelf and continental slope.

shelf facies [GEOL] A sedimentary facies characterized by carbonate rocks and fossil shells and produced in the neritic environments of marginal shelf seas. Also known as foreland facies; platform facies.

shelf ice [HYD] The ice of an ice shelf. Also known as barrier ice.

shelf life [ENG] The time that elapses before stored food, chemicals, batteries, and other materials or devices become inoperative or unusable due to age or deterioration.

shelf sea [OCEANOGR] A shallow marginal sea located on the continental shelf, usually less than 150 fathoms (275 meters) in depth; an example is the North Sea.

shell [ARCH] A reinforced concrete arched or domed roof used over unpartitioned areas. [ATOM PHYS] A set of orbital electron states that have the same principal quantum number and, therefore, have approximately the same energy level and average distance from the nucleus. [BUILD] A building without internal partitions or furnishings. [DES ENG] **1.** The case of a pulley block. **2.** A thin hollow cylinder. **3.** A hollow hemispherical structure. **4.** The outer wall of a vessel or tank. [GEOL] **1.** The crust of the earth. **2.** A thin hard layer of rock. [GRAPHICS] An engraved roller made of copper and used in calico printing. [MET] **1.** The outer wall of a metal mold. **2.** The hard layer of sand and thermosetting plastic formed over a pattern and used as a mold wall in shell molding. **3.** The metal sleeve remaining when a billet is extruded with a dummy block at smaller diameter. **4.** A tubular casting used in preparing seamless drawn tubes. **5.** A pierced forging. [ORD] **1.** A hollow metal projectile designed to be projected from a gun, containing or intended to contain, a high-explosive, chemical, atomic, or other charge. **2.** A shotgun cartridge or a cartridge for artillery of small arms. [ZOO] **1.** A hard, usually calcareous, outer covering on an animal body, as of bivalves and turtles. **2.** The hard covering of an egg. **3.** Chitinous exoskeleton of certain arthropods.

shellac [MATER] A natural, alcohol-soluble, water-insoluble, flammable resin; made from lac resin deposited on tree twigs in India by the lac insect (*Laccifer lecca*) used as an ingredient of wood coatings.

shellac varnish [MATER] A solution of shellac in denatured alcohol; used in wood finishing where a fast-drying, light-colored, hard finish is desired.

shellac wax [MATER] A hard wax with 3% shellac, from which it is extracted, and used in polishes and insulating materials.

shellac wheel [DES ENG] A grinding wheel having the abrasive bonded with shellac.

shell-and-tube exchanger [ENG] A device for the transfer of heat from a hot fluid to a cooler fluid; one fluid passes through a group (bundle) of tubes, the other passes around the tubes, through a surrounding shell.

shell capacity [ENG] The amount of liquid that a tank car or tank truck will hold when the liquid just touches the underside of the top of the tank shell.

shell clearance [DES ENG] The difference between the outside diameter of a bit or core barrel and the outside set or gage diameter of a reaming shell.

shell core [MET] A sand core formed by shell molding.

shell-destroying tracer [ORD] A tracer that includes an explosive element beyond the tracer element and that is designed to cause activation of the explosive by the tracer after the antiaircraft projectile has passed the target point, thus destroying the projectile to avoid impact in friendly territory.

shell filler [ORD] Explosive or other material used to make up the filler or charge in a projectile.

shellfish [INV ZOO] An aquatic invertebrate, such as a mollusk or crustacean, that has a shell or exoskeleton.

Shell fluid catalytic cracking [CHEM ENG] A two-stage, proprietary petroleum-refining process for catalytic cracking of gas oils; the catalyst is regenerated in one of the stages.

shell gland [INV ZOO] An organ that secretes the embryonic shell in many mollusks. [VERT ZOO] A specialized structure attached to the oviduct in certain animals that secretes the eggshell material.

shell ice [HYD] Ice, on a body of water, that remains as an unbroken surface when the water level drops so that a cavity is formed between the water surface and the ice.

shell innage [ENG] The depth of a liquid in a tank car or tank truck shell.

shell knocker [ENG] A device to strike the external surface of a horizontally rotating process vessel (for example, a kiln or a dryer) to loosen accumulations of solid materials from the inner walls or flights of the shell. Also known as knocker.

shell membrane [CYTOL] Either of a pair of membranes lining the inner surface of an egg shell; they allow free entry of oxygen but prevent rapid evaporation of moisture.

shell model [NUC PHYS] A model of the nucleus in which the shell structure is either postulated or is a consequence of other postulates; especially the model in which the nucleons act as independent particles filling a preassigned set of energy levels as permitted by the quantum numbers and Pauli principle.

shell molding [MET] Forming a rigid, porous, self-supporting refractory mold by sprinkling molding sand blended with thermosetting plastic or resin over a preheated metal pattern and then curing in an oven.

shell outage [ENG] The unfilled portion of a tank car or tank truck shell; the distance from the underside of the top of the shell to the level of the liquid in the shell.

shell plating [NAV ARCH] The plating forming the outer skin of a vessel; in addition to providing a watertight boundary, it contributes largely to the strength of the vessel.

shell pump [MECH ENG] A simple pump for removing wet sand or mud; consists of a hollow cylinder with a ball or clack valve at the bottom.

shell reamer [DES ENG] A machine reamer consisting of two parts, the arbor and the replaceable reamer, with straight or spiral flutes; designed as a sizing or finishing reamer.

shell roof [BUILD] A roof made of a thin, curved, platelike structure, usually of concrete but lumber and steel are also used.

shell room [ORD] A large room for the storage of projectiles.

shell still [CHEM ENG] A distillation device formerly used in petroleum refineries; oil was charged into a closed, cylindrical shell and heat was applied to the outside of the bottom by a firebox.

shell structure [NUC PHYS] Structure of the nucleus in which nucleons of each kind occupy quantum states which are in groups of approximately the same energy, called shells, the number of nucleons in each shell being limited by the Pauli exclusion principle.

shell-type transformer [ELECTROMAG] Transformer in which the magnetic circuit completely surrounds the windings.

shelly pahoehoe [GEOL] A type of pahoehoe characterized by open tubes and blisters on the surface.

shelterbelt [ECOL] A natural or planned barrier of trees or shrubs to reduce erosion and provide shelter from wind and storm activity.

shelter deck [NAV ARCH] A ship's deck which is above the principal deck and is continuous and of light construction.

shelterwood method [FOR] A method for ensuring tree reproduction; older trees are removed by successive cuttings so that the amount of light reaching the seedlings is gradually increased.

Shelton loader [MIN ENG] A modified coal-cutting machine in which the picks of the cutter chain are replaced by loading flights, which push the prepared coal up a ramp on to the face conveyor.

sherardizing [MET] Coating iron with zinc by tumbling the article in powdered zinc at about 250–375°C.

shergottite [GEOL] An achondritic stony meteorite that is composed chiefly of pigeonite and maskelynite.

sheridanite [MINERAL] $(Mg,Al)_6(Al,Si)_4O_{10}(OH)_8$ A pale-green to colorless talclike mineral composed of basic magnesium aluminum silicate.

Sheridan tank [ORD] The United States' light tank, M551, introduced into limited service in 1967; it can be transported by air and dropped by parachute, has a crew of four, weighs 15 tons (13,600 kilograms), and is armed with a 152-millimeter gun launcher and two machine guns.

Sherpa [TEXT] Trade name of a pile acrylic fabric resembling shearling and manufactured by the Collins and Aikman Corporation.

Sherritt Gordon process [MET] A method of Sherritt Gordon Mines, Ltd., for nickel extraction, in which the bulk flotation concentrate is separated into nickel and copper concentrate, and the nickel concentrate is treated by a pressure hydrometallurgical process.

sherry [FOOD ENG] A dry to sweet fortified wine with nutty flavor and ranging from pale to dark amber in color.

Sherwood number See Nusselt number.

Sherwood project [NUCLEO] The overall research program in the United States for producing useful power from nuclear fusion.

SHF See sensible-heat factor; superhigh frequency.

shield [ENG] An iron, steel, or wood framework used to support the ground ahead of the lining in tunneling and mining. [GEOL] See palette. [NUCLEO] The material placed around a nuclear reactor, or other source of radiation, to reduce escaping radiation or particles to a permissible level. Also known as shielding. [ORD] Armor plate mounted on a gun carriage to protect the operating mechanism and gun crew from enemy fire.

Shield See Scutum.

shield basalt [GEOL] A basaltic lava flow from a group of small, close-spaced shield-volcano vents that coalesced to form a single unit.

shield cone [GEOL] A cone or dome-shaped volcano built up by successive outpourings of lava.

shielded arc welding [MET] Arc welding in which the electric arc and the weld metal are protected by gas, decomposition products of the electrode covering, or a blanket of fusible flux.

shielded-conductor cable [ELEC] Cable in which the insulated conductor or conductors are enclosed in a conducting envelope or envelopes, constructed so that substantially every point on the surface of the insulation is at ground potential or at some predetermined potential with respect to ground.

shielded joint [ELEC] Cable joint having its insulation so enveloped by a conducting shield that substantially every point on the surface of the insulation is at ground potential, or at some predetermined potential with respect to ground.

shielded line [ELECTROMAG] Transmission line, the elements of which confine the propagated waves to an essentially finite space; the external conducting surface is called the sheath.

shielded metal-arc welding [MET] Arc welding in which heating with an electric arc between the electrode and the work produces fusion of the electrode covering which shields the work.

shielded wire [ELEC] Insulated wire covered with a metal shield, usually of tinned braided copper wire.

shield factor [COMMUN] Ratio of noise (or induced current or voltage) in a telephone circuit when a source of shielding is present to the corresponding quantity when the shielding is absent.

shield grid [ELECTR] A grid that shields the control grid of a gas tube from electrostatic fields, thermal radiation, and deposition of thermionic emissive material; it may also be used as an additional control electrode.

shield-grid thyratron [ELECTR] A thyratron having a shield grid, usually operated at cathode potential.

shielding [ELECTROMAG] See electric shielding. [MET] Placing a nonconducting object in an electrolytic bath during plating to alter the current distribution. [NUCLEO] **1.** Reducing the ionizing radiation reaching one region of space from another region by using a shield or other device. **2.** See shield.

shielding distance See Debye shielding length.

shielding factor [GEOPHYS] The ratio of the strength of the magnetic field at a directional compass to its strength if there were no disturbing material; usually expressed as a decimal.

shielding layer [METEOROL] The layer of air nearest the earth, with reference to the manner in which this layer shields the earth from activity in the free atmosphere above, or vice versa.

shielding ratio [ELECTROMAG] The ratio of a field in a specified region when electrical shielding is in place to the field in that region when the shielding is removed.

shield volcano [GEOL] A broad, low volcano shaped like a flattened dome and built of basaltic lava. Also known as basaltic dome; lava dome.

shift [ADP] A movement of data to the right or left, in a digital-computer location, usually with the loss of characters shifted beyond a boundary. [GEOL] The relative displacement of the units affected by a fault but outside the fault zone itself. [IND ENG] The number of hours or the part of any day worked. Also known as tour. [MECH ENG] To change the ratio of the driving to the driven gears to obtain the desired rotational speed or to avoid overloading and stalling an engine or a motor. [MET] A casting defect caused by malalignment of the mold parts.

shifting theorem [MATH] **1.** If the Fourier transform of $f(t)$ is $F(x)$, then the Fourier transform of $f(t-a)$ is $\exp(iax)F(x)$. **2.** If the Laplace transform of $f(x)$ is $F(y)$, then the Laplace transform of $f(x-a)$ is $\exp(-ay)F(y)$.

shift joint [BUILD] A vertical joint placed on a solid member of the course below.

shift of butts [NAV ARCH] The arrangement of butt joints in plating.

shift of spectral line [SPECT] A small change in the position of a spectral line that is due to a corresponding change in frequency which, in turn, results from one or more of several causes, such as the Doppler effect.

shift register [ADP] A computer hardware element constructed to perform shifting of its contained data.

shift work [IND ENG] Work paid for by day wage.

Shiga's bacillus [MICROBIO] Type A1 of *Shigella dysenteriae*.

Shigella [MICROBIO] The dysentery bacilli, a genus of the family Enterobacteriaceae.

Shigella dysenteriae [MICROBIO] The organism that causes dysentery in man and monkeys; found only in the feces of the sick.

shikimic acid [BIOCHEM] $C_7H_{10}O_5$ A crystalline acid that is a plant constituent, and an intermediate in the biochemical pathway from phosphoenolpyruvic acid to tyrosine.

Shillelagh [ORD] A United States weapon system including a gun launcher and a fire-control system mounted on the main battle tank and assault reconnaissance vehicle for employment against enemy armor, troops, and field fortifications.

shim [ENG] **1.** In the manufacture of plywood, a long, narrow patch glued into the panel or cemented into the lumber core itself. **2.** A thin piece of material placed between two surfaces to obtain a proper fit, adjustment, or alignment.

shimmer [METEOROL] To appear tremulous or wavering, due to varying atmospheric refraction in the line of sight.

shimmy [MECH] Excessive vibration of the front wheels of a wheeled vehicle causing a jerking motion of the steering wheel.

shim rod [NUCLEO] A control rod used for making occasional coarse adjustments in the reactivity of a nuclear reactor.

shingle [GEOL] Pebbles, cobble, and other beach material, coarser than ordinary gravel but roughly the same size and occurring typically on the higher parts of a beach. [MATER] A rectangular piece of wood, metal, or other material that is used like a tile and arranged in overlapping rows for covering roofs and walls.

shingle barchan [GEOL] A dunelike ridge formed of shingle perpendicular to the beach in shallow water.

shingle beach [GEOL] A narrow beach composed of shingle and commonly having a steep slope on both its landward and seaward sides. Also known as cobble beach.

shingle-block structure *See* imbricate structure.

shingle lap [DES ENG] A lap joint of tapered sections, the bottom of each section overlapping the top of the section below it.

shingle nail [DES ENG] A nail about a half to a full gage thicker than a common nail of the same length.

shingle rampart [GEOL] A rampart of shingle built along a reef on the seaward edge.

shingles *See* herpes zoster.

shingle structure *See* imbricate structure.

shingling *See* imbrication.

shin splints [MED] An injury and an inflammation of the lower leg muscles and bones of the lower and middle third of the tibia and fibula, seen in athletes such as runners or basketball and football players.

ship [NAV ARCH] Any large vessel which travels over the seas, rivers, or lakes.

Ship *See* Argo.

ship auger [DES ENG] An auger consisting of a spiral body having a single cutting edge, with or without a screw; there is no spur at the outer end of the cutting edge.

shipbuilding [CIV ENG] The construction of ships.

ship drift [OCEANOGR] A method of measuring ocean currents; the ship itself is used as a current tracer, its motions being measured by navigating equipment on board.

shipfitter [CIV ENG] A worker who builds the steel structure of a ship, including laying-off and fabricating the individual members, subassembly, and erection on the shipway.

ship heading marker [NAV] A mark on a direction compass which indicates the position of the ship's head, such as a lubber's line, or an electronic radial sweep line on a PPI (plan position indicator).

ship motion [ENG] Translational and rotational motions of a ship in a wave system which cause the center of gravity to deviate from simple straight-line motion; these motions are heave, surge, sway, roll, pitch, and yaw.

shipping [NAV ARCH] A term applied collectively to those ships which are used to transport personnel or cargo, or both; often modified to denote type, use, or force to which assigned.

shipping and storage container [IND ENG] A reusable noncollapsible container of any configuration designed to provide protection for a specific item against impact, vibration, climatic conditions, and the like, during handling, shipment, and storage.

shipping control [ORD] All matters pertaining to convoy organization, routing, reporting, and diversion of shipping of all nations under charter thereto; it does not include cognizance over the general employment and allocation of shipping, harbor movements, and loading and unloading, which are functions of other agencies.

shipping designator [COMMUN] A code word assigned to a particular overseas base, port, or area, for specific use as an address on shipments to the overseas location concerned; the code word is usually four letters and may be followed by a number to indicate a particular addressee.

shipping document [IND ENG] A document listing the items in a shipment, and showing other supply and transportation information that is required by agencies concerned in the movement of material.

shipping fever [VET MED] An acute, occasionally subacute, septicemic disease in cattle and sheep, probably caused by a combination of virus and *Pasteurella multocida* or *P. hemolytica*.

shipping lane [NAV] An established route traversed by ocean shipping.

shipping time [ENG] The time elapsing between the shipment of material by the supplying activity and receipt of material by the requiring activity.

shipping ton *See* ton.

ship report [METEOROL] The encoded and transmitted report of a marine weather observation.

ship's clock [HOROL] A clock that rings to indicate the time according to the system of bells used on shipboard.

ship's field error [NAV] That error in radio direction finder bearings due to the ship's directional antenna being located where the main component of the ship's field is not parallel to the ship's center line.

ship's head [NAV] The heading of a vessel in degrees.

ship's lines *See* lines.

ship synoptic code [METEOROL] A synoptic code for communicating marine weather observations; it is a modification of the international synoptic code.

ship-to-shore movement [ORD] That portion of the assault

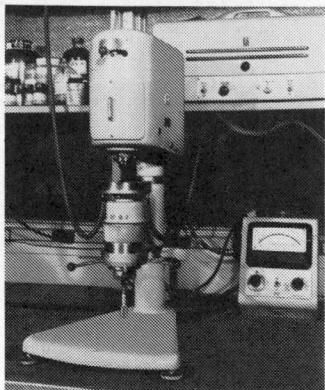

A Shirley-Ferranti viscometer.

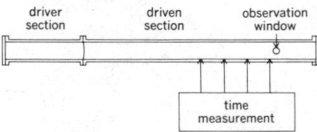

Schematic representation of a shock tube.

Schlieren photograph of supersonic flow over blunt object, showing approximately parabolic shock wave detached from object. *(Avco Everett Research Laboratory, Inc.)*

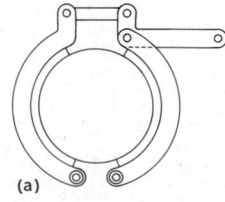

(a)

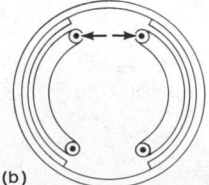

(b)

Two types of shoe brake. *(a)* External shoe brake; shoes are lined with frictional material. *(b)* Internal shoe brake with lining.

phase of an amphibious operation which includes the deployment of the landing force from the assault shipping to designated landing areas.

shipway [CIV ENG] **1.** The ways on which a ship is constructed. **2.** The supports placed underneath a ship in dry dock. [NAV] A channel through which ships pass.

shipworm [INV ZOO] Any of several bivalve mollusk species belonging to the family Teredinidae and which superficially resemble earthworms because the two valves are reduced to a pair of plates at the anterior of the animal or are used for boring into wood.

shipwright [CIV ENG] A worker whose responsibility is to ensure that the structure of a ship is straight and true and to the designed dimensions; the work starts with the laying down of the keel blocks and continues throughout the steelwork; applicable also to wood ship builders.

shipyard [CIV ENG] A facility adjacent to deep water where ships are constructed or repaired.

shipyard eye *See* keratoconjunctivitis.

shiran [ELECTR] Specially designed frequency-modulation continuous-wave distance-measuring equipment used for performing distance measurements of an accuracy comparable to first-order triangulation. Contraction of S-band hiran.

Shirley-Ferranti viscometer [ENG] An instrument used to determine an ink's resistance to flow.

shistosoma [INV ZOO] A genus of blood flukes infecting man.

shivering [MATER] Cracks and scales on a pottery glaze caused by unequal contraction during cooling.

SHM *See* harmonic motion.

shoal [GEOL] A submerged elevation rising from the bed of a shallow body of water and consisting of, or covered by, unconsolidated material, and may be exposed at low water.

shoaling [OCEANOGR] The bottom effect which influences the height of waves moving from deep to shallow water.

shoal patches [OCEANOGR] Individual and scattered elevations of the bottom, with depths of 10 fathoms (18 meters) or less, but composed of any material except rock or coral.

shoal reef [GEOL] A reef formed in irregular masses amid submerged shoals of calcareous reef detritus.

shoal water [OCEANOGR] Shallow water; over a shoal.

shock [MECH] A pulse or transient motion or force lasting thousandths to tenths of a second which is capable of exciting mechanical resonances; for example, a blast produced by explosives. [MED] Clinical manifestations of circulatory insufficiency, including hypotension, weak pulse, tachycardia, pallor, and diminished urinary output.

shock absorber [MECH ENG] A spring, a dashpot, or a combination of the two, arranged to minimize the acceleration of the mass of a mechanism or portion thereof with respect to its frame or support.

shock action [ORD] A method of attack by mobile units in which the suddenness, violence, and massed weight of the first impact produce the main effect; tank attacks usually rely on shock action.

shock bump [MIN ENG] A rock bump resulting from the sudden collapse of a strong deposit.

shock diamonds [PHYS] The shock waves that appear in the exhaust stream of a rocket; they are made visible by their luminosity and describe an approximate diamond configuration in side view.

shock excitation [ELEC] Excitation produced by a voltage or current variation of relatively short duration; used to initiate oscillation in the resonant circuit of an oscillator. Also known as impulse excitation.

shock front [PHYS] The outer side of a shock wave whose pressure rises from zero up to its peak value. Also known as pressure front.

shock heating [PHYS] The nonisentropic heating of a fluid which takes place when a shock wave passes through it.

shock isolation [MECH ENG] The application of isolators to alleviate the effects of shock on a mechanical device or system.

Shockley diode [ELECTR] A *pnpn* silicon controlled switch having characteristics that permit operation as a unidirectional diode switch.

Shockley partial dislocation [SOLID STATE] A partial dislocation in which the Burger's vector lies in the fault plane, so that

it is able to glide, in contrast to a Frank partial dislocation. Also known as glissile dislocation.

shock metamorphism [PETR] The complete permanent changes (physical, chemical, mineralogic, morphologic) in rocks caused by transient high-pressure shock waves that act over short-time intervals, ranging from a few microseconds to a fraction of a minute.

shock mount [MECH ENG] A mount used with sensitive equipment to reduce or prevent transmission of shock motion to the equipment.

shock organ [IMMUNOL] The organ or tissue that exhibits the most marked response to the antigen-antibody interaction in hypersensitivity, as the lungs in allergic asthma or the skin in allergic contact dermatitis.

shock resistance [ENG] The property which prevents cracking or general rupture when impacted.

shock strut [AERO ENG] The primary working part of any landing gear, which supplies the force as the airplane sinks toward the ground, turning the flight path from one intersecting the ground to one parallel to the ground.

shock test [ENG] The test to determine whether the armor sample will crack or spall under impact by kinetic energy or high-explosive projectiles.

shock therapy [PSYCH] The use of drugs, carbon dioxide, insulin, or electric current to induce coma in the treatment of psychiatric disorders.

shock tube [FL MECH] A long tube divided into two parts by a diaphragm; the volume on one side of the diaphragm constitutes the compression chamber, the other side is the expansion chamber; a high pressure is developed by suitable means in the compression chamber, and the diaphragm ruptured; the shock wave produced in the expansion chamber can be used for the calibration of air blast gages, or the chamber can be instrumented for the study of the characteristics of the shock wave.

shock tunnel [ENG] A hypervelocity wind tunnel in which a shock wave generated in a shock tube ruptures a second diaphragm in the throat of a nozzle at the end of the tube, and gases emerge from the nozzle into a vacuum tank with Mach numbers of 6 to 25.

shock wave [PHYS] A fully developed compression wave of large amplitude, across which density, pressure, and particle velocity change drastically.

shock wave lip [PHYS] The shock wave obtained from the lip of a free jet nozzle, because of failure to match the stream pressure and the ambient exhaust pressure.

shoe [ENG] In glassmaking, an open-ended crucible placed in a furnace for heating the blowing irons. [MECH ENG] **1.** A metal block used as a form or support in various bending operations. **2.** A replaceable piece used to break rock in certain crushing machines. **3.** *See* brake shoe. [MIN ENG] **1.** Pieces of steel fastened to a mine cage and formed to fit over the guides to guide it when it is in motion. **2.** The bottom wedge-shaped piece attached to tubbing when sinking through quicksand. **3.** A trough to convey ore to a crusher. **4.** A coupling of rolled, cast, or forged steel to protect the lower end of the casting or drivepipe in overburden, or the bottom end of a sampler when pressed into a formation being sampled.

shoe brake [MECH ENG] A type of brake in which friction is applied by a long shoe, extending over a large portion of the rotating drum; the shoe may be external or internal to the drum.

shoestring [GEOL] A long, relatively straight and narrow sedimentary body having a width/thickness ratio of less than 5:1, usually 1:1.

shoestring sand [GEOL] A shoestring composed of sand and usually buried in mud or shale, usually a sandbar or channel fill.

shonkinite [PETR] A dark-colored syenite composed principally of augite and orthoclase with some olivine, hornblende, biotite, and nepheline.

shoofly *See* slant.

shoot [BOT] **1.** The aerial portion of a plant, including stem, branches, and leaves. **2.** A new, immature growth on a plant. [ENG] To detonate an explosive, used to break coal loose from a seam or in blasting operation or in a borehole. [GEOL] *See*

ore shoot. [GEOPHYS] The energy that goes up through the strata from a seismic profiling shot and is reflected downward at the surface or at the base of the weathering; appears either as a single wave or unites with a wave train that is traveling downward. Also known as secondary reflection. [ORD] To project a missile with force; to fire a weapon, as a gun or cannon; to strike or hit something with a missile.

shooting board [ENG] **1.** A fixture used as a guide in planing boards; it is more accurate than a miter. **2.** A table and plane used for trimming printing plates.

shooting star [ASTRON] A small meteor that has the brief appearance of a darting, starlike object.

shop drawing [GRAPHICS] A scale drawing to be used as a design guide in the workshop of a manufacturer.

Shope papilloma [VET MED] A transmissible, virus-induced papilloma occurring naturally on the skin of rabbits.

shop fabrication [ENG] Making parts and materials in the shop rather than at the work site.

shop lumber [MATER] Softwood lumber graded and used in the factory for general cut-up purposes; similar to factory lumber but of a lower grade.

shop supplies [ENG] Expendable items consumed in operation and maintenance (for example, waste, oils, solvents, tape, packing, flux, or welding rod).

shop typhus *See* murine typhus.

shop weld [ENG] A weld made in the workshop prior to delivery to the construction site.

shoran *See* short-range navigation.

shore [ENG] Timber or other material used as a temporary prop for excavations or buildings; may be sloping, vertical, or horizontal. [GEOL] **1.** The narrow strip of land immediately bordering a body of water. **2.** *See* seashore.

shore bird [VERT ZOO] A general term applied to a large number of birds in 12 families of the suborder Charadrii which are always found near water, although the habitat and morphology is varied. Also known as wader.

shore current [HYD] A water current near a shoreline, often flowing parallel to the shore.

shore drift *See* littoral drift.

shore effect [ELECTROMAG] Bending of waves toward the shoreline when traveling over water near a shoreline, due to the slightly greater velocity of radio waves over water than over land; this effect causes errors in radio-direction-finder indications.

shoreface [GEOL] The narrow, steeply sloping zone between the seaward limit of the shore at low water and the nearly horizontal offshore zone.

shore fire-control party [ORD] A specially trained unit for control of naval gunfire in support of troops ashore, consisting of a spotting team to adjust fire and a naval gunfire liaison team to perform liaison functions for the supported battalion commander.

Shore hardness [ENG] A method of rating the hardness of a metal or of a plastic or rubber material.

shore ice [OCEANOGR] Sea ice that has been beached by wind, tides, currents, or ice pressure; it is a type of fast ice, and may sometimes be rafted ice.

shore lead [OCEANOGR] A lead between pack ice and fast ice or between floating ice and the shore; it may be closed by wind or currents so that only a tide crack remains.

shoreline [GEOL] The intersection of a specified plane of water, especially mean high water, with the shore; a limit which changes with the tide or water level. Also known as strandline; waterline.

shoreline cycle [GEOL] The cycle of changes through which sequential forms of coastal features pass during shoreline development, from the establishment of a water level to the time when the water can do no more work.

shoreline of emergence [GEOL] A straight or gently curving shoreline formed by the dominant relative emergence of the floor of an ocean or a lake. Also known as emerged shoreline; negative shoreline.

shoreline of submergence [GEOL] A shoreline, characterized by bays, promontories, and other minor features, formed by the dominant relative submergence of a landmass. Also known as positive shoreline; submerged shoreline.

shore party [ORD] A task organization of the landing force, formed for the purpose of facilitating the landing and movement off the beaches of troops, equipment, and supplies; for the evacuation from the beaches of casualties and prisoners of war; and for facilitating the beaching, retraction, and salvaging of landing ships and craft; it comprises elements of both the naval and landing forces.

shore platform [GEOL] The horizontal or gently sloping surface produced along a shore by wave erosion. Also known as scar.

shore protection [CIV ENG] Preventing erosion of the ground bordering a body of water.

Shore scleroscope [ENG] A device used in rebound hardness testing of rubber, metal, and plastic; consists of a small, conical hammer fitted with a diamond point and acting in a glass tube.

Shore scleroscope hardness test [MET] A rebound hardness test in which a metal body is dropped vertically down a glass tube onto the surface of the material being tested; the height of the rebound is a measure of the hardness. Also known as scleroscope hardness test.

shoreside *See* onshore.

shore tank [PETRO ENG] A shoreside storage tank for liquid petroleum products discharged by tankers.

shore terrace [GEOL] **1.** A terrace produced along the shore by wave and current action. **2.** *See* marine terrace.

shore-to-shore movement [ORD] The assault movement of personnel and materiel directly from a shore staging area to the objective, involving no further transfers between types of craft or ships incident to the assault movement.

shoring [ENG] Providing temporary support with shores to a building or an excavation.

short [ELEC] *See* short circuit. [ENG] In plastics injection molding, the failure to fill the mold completely. Also known as short shot. [ORD] A bomb or projectile hit short of the target.

short antenna [ELECTROMAG] An antenna shorter than about one-tenth of a wavelength, so that the current may be assumed to have constant magnitude along its length, and the antenna may be treated as an elementary dipole.

short-baseline system [AERO ENG] A trajectory measuring system using a baseline the length of which is very small compared with the distance of the object being tracked.

short circuit [ELEC] A low-resistance connection across a voltage source or between both sides of a circuit or line, usually accidental and usually resulting in excessive current flow that may cause damage. Also known as short.

short-circuit impedance [ELEC] Of a line or four-terminal network, the driving point impedance when the far-end is short-circuited.

short column [CIV ENG] A column in which both compression and bending are significant, generally having a slenderness ratio between 30 and 120–150.

shortcoming [DES ENG] An imperfection or malfunction occurring during the life cycle of equipment, which should be reported and which must be corrected to increase efficiency and to render the equipment completely serviceable.

short-contact switch [ELEC] Selector switch in which the width of the movable contact is greater than the distance between contact clips, so that the new circuit is contacted before the old one is broken; this avoids noise during switching.

short-crested wave [OCEANOGR] An ocean wave whose crest is of finite length; that is, the type actually found in nature.

short-delay blasting [ENG] A method of blasting by which explosive charges are detonated in a given sequence with short time intervals.

short-delay detonator *See* millisecond delay cap.

short-distance navigation [NAV] **1.** Navigation (air) served by aids located less than 200-miles (322-kilometers) distance (International Civil Aviation Organization). **2.** Navigation (marine) at distances less than 3 miles (4.8 kilometers) (International Meeting on Marine Navigation, 1958).

short flashing light [NAV] A flashing light having individual flashes of less than 2-seconds duration.

short fuse [ENG] **1.** Any fuse that is cut too short. **2.** The practice of firing a blast, the fuse on the primer of which is not sufficiently long to reach from the top of the charge to the collar of the borehole; the primer, with fuse attached, is dropped into the charge while burning.

short-gate gain [ELECTR] Video gain on short-range gate.

short-haul convoy [ORD] A convoy whose voyage lies in general in coastal waters and whose terminals of departure and arrival lie in different countries.

short hundredweight *See* hundredweight.

shortite [MINERAL] $Na_2Ca_2(CO_3)_3$ A mineral composed of sodium and calcium carbonate.

short leg [ENG] One of the wires on an electric blasting cap, which has been shortened so that when placed in the borehole, the two splices or connections will not come opposite each other and make a short circuit.

short-long flashing light [NAV] A light showing a short flash of about 0.4 second, and a long flash of four times that duration, this combination recurring about six to eight times per minute.

shortness [MET] A form of brittleness in metal, designated as hot, cold, or red to indicate the temperature range in which brittleness occurs.

short oil [MATER] Varnish containing a small percentage of oil.

short-path principle *See* Hittorf principle.

short-pulse laser [OPTICS] A laser designed to generate a pulse of light lasting on the order of nanoseconds or less, and having very high power, such as by Q switching or mode-locking.

short-range attack missile [ORD] An air-to-surface missile with a range capability up to about 600 nautical miles (1100 kilometers).

short-range ballistic missile [ORD] A ballistic missile with a range capability up to about 600 nautical miles (1100 kilometers).

short-range force [PHYS] A force between two particles which is negligible when the distance between the particles is greater than a certain amount; in particular, nuclear forces whose range is several times 10^{-15} meter.

short-range forecast [METEOROL] A weather forecast made for a time period generally not greater than 18 hours in advance.

short-range navigation [NAV] Navigation employing only aids usable at short ranges. Also known as shoran.

short-range order [PHYS] A regularity in the arrangement of atoms in a disordered solid or a liquid in which the probability of a given type of atom having neighbors of a given type is greater than one would expect on a purely random basis.

short-range radar [ENG] Radar whose maximum line-of-sight range, for a reflecting target having 1 square meter of area perpendicular to the beam, is between 50 and 150 miles (80 and 240 kilometers).

short residuum [CHEM ENG] A petroleum refinery term for residual oil from crude-oil distillation operations in which neutral oils are taken overhead with the distillate.

short round [ORD] **1.** Defective cartridge in which the bullet has been seated too deeply. **2.** A projectile which fails to travel the expected distance or range.

short run [MET] Pertaining to a mold or casting filled only partially with molten metal.

shorts [ENG] Oversize particles held on a screen after sieving the fines through the screen. [FOOD ENG] Wheat-husk parts that are finer than bran; used in flour milling.

short shipment [ENG] Freight listed or manifested but not received.

short shot *See* short.

short-slot coupler *See* three-decibel coupler.

short supply [IND ENG] An item is in short supply when the total of stock on hand and anticipated receipts during a given period is less than the total estimated demand during that period.

short takeoff and landing [AERO ENG] The ability of an aircraft to clear a 50-foot (15-meter) obstacle within 1500 feet (450 meters) of commencing takeoff, or in landing, to stop within 1500 feet after passing over a 50-foot obstacle. Abbreviated STOL.

Shortt clock [HOROL] An accurate clock manufactured by the Shortt-Synchronome Corporation; it is a pendulum clock in which the master pendulum is enclosed in an airtight, nearly evacuated chamber; this pendulum receives a small impulse to maintain its swing. Also known as free-pendulum clock.

short-time rating [ELEC] A rating defining the load that a machine, apparatus, or device can carry for a specified short time.

short ton *See* ton.

short-tube vertical evaporator [CHEM ENG] A liquid evaporation process unit with a vertical bundle of tubes 2–3 inches (5–8 centimeters) in diameter and 4–6 feet (1.2–1.8 meters) long; the heating fluid is inside the tubes, and the liquid to be evaporated is in the shell area outside the tubes; used mainly to evaporate cane-sugar juice. Also known as calandria evaporator; Roberts evaporator; standard evaporator.

shortwall [MIN ENG] **1.** A method of mining in which comparatively small areas are worked separately. **2.** A length of coal face between about 5 and 30 yards (4.6 and 27 meters), generally employed in pillar methods of working.

shortwall coal cutter [MIN ENG] A machine for undercutting coal which has a long, rigid chain jib fixed in relation to the main body of the machine and which cuts across a heading from right to left, being drawn across by means of a steel-wire rope.

shortwave broadcasting [COMMUN] Radio broadcasting at frequencies in the range from about 1600 to 30,000 kilohertz, above the standard broadcast band.

shortwave converter [ELECTR] Electronic unit designed to be connected between a receiver and its antenna system to permit reception of frequencies higher than those the receiver ordinarily handles.

shortwave propagation [COMMUN] Propagation of radio waves at frequencies in the range from about 1600 to 30,000 kilohertz.

shortwave radiation [ELECTROMAG] A term used loosely to distinguish radiation in the visible and near-visible portions of the electromagnetic spectrum (roughly 0.4 to 1.0 micrometer in wavelength) from long-wave radiation (infrared radiation).

short word [ADP] The fixed word of lesser length in computers capable of handling words of two different lengths; in many computers this is referred to as a half-word because the length is exactly the half-length of the full word.

shot [AERO ENG] An act or instance of firing a rocket, especially from the earth's surface. [ENG] **1.** A charge of some kind of explosive. **2.** Small spherical particles of steel. **3.** Small steel balls used as the cutting agent of a shot drill. **4.** The firing of a blast. **5.** In plastics molding, the yield from one complete molding cycle, including scrap. [MIN ENG] Coal broken by blasting or other methods. [ORD] **1.** A solid projectile for cannon, without a bursting charge; the term projectile is preferred for uniformity in nomenclature. **2.** A mass or load of numerous, relatively small, lead pellets used in a shotgun, as birdshot or buckshot.

shot bit [DES ENG] A short length of heavy-wall steel tubing with diagonal slots cut in the flat-faced bottom edge.

shot blasting [MET] Cleaning and descaling metal by shot peening or by means of a stream of abrasive powder blown through a nozzle under air pressure in the range 30–150 pounds per square inch (2×10^5 to 1.0×10^6 newtons per square meter).

shot boring [ENG] The act or process of producing a borehole with a shot drill.

shot capacity [ENG] The maximum weight of molten resin that an accumulator can push out with one forward stroke of the ram during plastics forming operations.

shot depth [ENG] The distance from the surface to the charge.

shot drill *See* calyx drill.

shot effect *See* shot noise.

shot elevation [ENG] Elevation of the dynamite charge in the shot hole.

shot feed [MECH ENG] A device to introduce chilled-steel shot, at a uniform rate and in the proper quantities, into the circulating fluid flowing downward through the rods or pipe connected to the core barrel and bit of a shot drill.

shot-firing cable [ELEC] A two-conductor cable which leads from the exploder to the detonator wires. Also known as firing cable.

shot-firing circuit [ELEC] The path taken by the electric current from the exploder along the shot-firing cable, the detonator wires, and finally the detonator when a shot is detonated.

shot-firing curtain [MIN ENG] A steel frame with chains about 6 inches (15 centimeters) apart suspended from the roof about 9 to 12 feet (2.7 to 3.7 meters) from the face of an advancing tunnel to intercept flying debris when shot-firing at the face.

shot group *See* shot pattern.

shotgun [ORD] A smooth-bore shoulder weapon; the usual classes are riot gun, skeet gun, and sporting gun.

shotgun cartridge [ORD] A container or capsule, usually of stiff paper with a brass base, containing primer, powder, wadding, and shot for use in a shotgun.

shothole [ENG] The borehole in which an explosive is placed for blasting.

shothole casing [ENG] A lightweight pipe, usually about 4 inches in diameter and 10 feet long, with threaded connections on both ends, used to prevent the shothole from caving and bridging.

shothole drill [MECH ENG] A rotary or churn drill for drilling shotholes.

shot noise [ELECTR] Noise voltage developed in a thermionic tube because of the random variations in the number and the velocity of electrons emitted by the heated cathode; the effect causes sputtering or popping sounds in radio receivers and snow effects in television pictures. Also known as Schottky noise; shot effect.

shot pattern [ORD] Design made on a surface by all the impacts of a series of shots fired under similar conditions. Also known as shot group.

shot peening [MET] Shot blasting with small steel balls driven by a blast of air.

shot point [ENG] The point at which an explosion (such as in seismic prospecting) originates, generating vibrations in the ground.

shot rock [ENG] Blasted rock.

shotting [MET] Making shot by pouring molten metal in finely divided streams; the particles solidify during descent and are cooled in a tank of water.

shot tongs [ORD] Device used to lift and convey heavy projectiles in a horizontal position.

shoulder [ANAT] **1.** The area of union between the upper limb and the trunk in humans. **2.** The corresponding region in other vertebrates.

shoulder blade *See* scapula.

shoulder-elbow height [ANTHRO] A measure of the distance taken from the top of the acromion to the tip of the elbow, as the subject sits erect, with the upper arm vertical and the forearm horizontal.

shoulder girdle *See* pectoral girdle.

shoulder guard [ORD] Any shield over the firing mechanism of a gun designed to protect the gunner from contact; particularly, such a shield on cannon mounted in tanks and in other cramped quarters.

shoulder-hand syndrome [MED] A syndrome characterized by severe constant intractable pain in the shoulder and arm, limited joint motion, diffuse swelling of the distal part of the upper extremity, fibrosis and atrophy of muscles, and decalcification of underlying bones; the cause is not well understood; it is similar to, or may be a form of, causalgia. Also known as hand-shoulder syndrome.

shoulder harness [ENG] A harness in a vehicle that fastens over the shoulders to prevent a person's being thrown forward in the seat.

shoulder stock [ORD] An item usually made of hardwood to which the barrel assembly and other parts of a shoulder-fired gun are attached; usually designed in one piece.

shoulder weapon [ORD] Any firearm designed to be braced on or against the shoulder when firing, as a rifle, carbine, or bazooka launcher.

shovel [DES ENG] A hand tool having a flattened scoop at the end of a long handle for moving soil, aggregate, cement, or other similar material. [MECH ENG] A mechanical excavator.

shovel dozer *See* tractor loader.

shovel loader [MECH ENG] A loading machine mounted on wheels, with a bucket hinged to the chassis which scoops up loose material, elevates it, and discharges it behind the machine.

Showalter stability index [METEOROL] A measure of the local static stability of the atmosphere, expressed as a numerical index.

show-card color *See* poster paint.

shower [ASTRON] *See* cosmic-ray shower. [METEOROL] Precipitation from a convective cloud; characterized by the suddenness with which it starts and stops, by the rapid changes of intensity, and usually by rapid changes in the appearance of the sky.

shower unit [NUCLEO] The mean path length required to reduce the energy of relativistic charged particles to half value as they pass through matter; one shower unit is equal to 0.693 radiation length.

shrapnel [ORD] Small lead or steel balls contained in a shrapnel case which is fired from an artillery piece; the balls are projected in a forward direction upon the functioning of the fuse.

shrew [VERT ZOO] Any of more than 250 species of insectivorous mammals of the family Soricidae; individuals are small with a moderately long tail, minute eyes, a sharp-pointed snout, and small ears.

Shrike [ORD] A U.S. Navy air-to-surface missile, used chiefly as an antiradar weapon.

shrimp [INV ZOO] The common name for a number of crustaceans, principally in the decapod suborder Natantia, characterized by having well-developed pleopods and by having the abdomen sharply bent in most species, producing a humped appearance.

shrinkage [ENG] **1.** Contraction of a molded material, such as metal or resin, upon cooling. **2.** Contraction of a plastics casting upon polymerizing.

shrinkage cavity [MET] A cavity resulting from shrinkage during casting.

shrinkage crack [MET] An irregular interdendritic crack in a casting caused by unequal contraction or inadequate feeding.

shrinkage rule *See* contraction rule.

shrinkage stoping [MIN ENG] A modification of overhead stoping, involving the use of a part of the ore for the purpose of support and as a working platform. Also known as back stoping.

shrink fit [DES ENG] A tight interference fit between mating parts made by shrinking-on, that is, by heating the outer member to expand the bore for easy assembly and then cooling so that the outer member contracts.

shrink forming [DES ENG] Forming metal wherein the piece undergoes shrinkage during cooling following the application of heat, cold upset, or pressure.

shrink-mixed concrete [MATER] Concrete that is partially mixed before being put in a truck mixer.

shrink ring [DES ENG] A heated ring placed on an assembly of parts, which on subsequent cooling fixes them in position by contraction.

shrink rule *See* contraction rule.

shrink wrapping [ENG] A technique of packaging with plastics in which the strains in plastics film are released by raising the temperature of the film, causing it to shrink-fit over the object being packaged.

shroud [ENG] A protective covering, usually of metal plate or sheet. [HOROL] The ends of lantern clock pinions that hold the pins. [NAV ARCH] A principal member of the standing rigging consisting of hemp or wire ropes which extend from or near a masthead to a vessel's side or to the rim of the top of the mast to afford lateral support for the mast.

shrouded propeller *See* ducted fan.

shrub [BOT] A low woody plant with several stems.

shrub-coppice dune [GEOL] A small dune formed on the leeward side of bush-and-clump vegetation.

Shubnikov–de Haas effect [SOLID STATE] Oscillations of the

SHREW

The Eurasian common shrew (*Sorex araneus*) has the soft, velvetlike fur characteristic of all shrews.

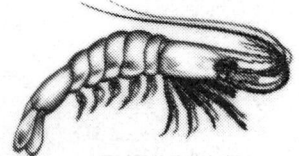

SHRIMP

Common shrimp (*Crangon vulgaris*), with a laterally compressed body and curved abdomen.

resistance or Hall coefficient of a metal or semiconductor as a function of a strong magnetic field, due to the quantization of the electron's energy.

Shubnikov groups [SOLID STATE] The point groups and space groups of crystals having magnetic moments. Also known as black-and-white groups; magnetic groups.

shuga [OCEANOGR] A spongy, rather opaque, whitish chunk of ice which forms instead of pancake ice if the freezing takes place in sea water which is considerably agitated.

shungite [GEOL] A hard, black, amorphous, coallike material composed of more than 98% carbon.

shunt [ELEC] **1.** A precision low-value resistor placed across the terminals of an ammeter to increase its range by allowing a known fraction of the circuit current to go around the meter. Also known as electric shunt. **2.** To place one part in parallel with another. **3.** *See* parallel. [ELECTROMAG] A piece of iron that provides a parallel path for magnetic flux around an air gap in a magnetic circuit. [MIN ENG] To shove or turn off to one side, as a car or train from one track to another.

shunt-excited [ELECTROMAG] Having field windings connected across the armature terminals, as in a direct-current generator.

shunt-excited antenna [ELECTROMAG] A tower antenna, not insulated from the ground at the base, whose feeder is connected at a point about one-fifth of the way up the antenna and usually slopes up to this point from a point some distance from the antenna's base.

shunt-fed vertical antenna [ELECTROMAG] Vertical antenna connected to the ground at the base and energized at a point suitably positioned above the grounding point.

shunt feed *See* parallel feed.

shunt generator [ELEC] A generator whose field winding and armature winding are connected in parallel, and in which the armature supplies both the load current and the field current.

shunting [ELEC] The act of connecting one device to the terminals of another so that the current is divided between the two devices in proportion to their respective admittances.

shunt loading [ELEC] Loading in which reactances are applied in shunt across the conductors.

shunt motor [ELEC] A direct-current motor whose field circuit and armature circuit are connected in parallel.

shunt neutralization *See* inductive neutralization.

shunt peaking [ELECTR] The use of a peaking coil in a parallel circuit branch connecting the output load of one stage to the input load of the following stage, to compensate for high-frequency loss due to the distributed capacitances of the two stages.

shunt reactor [ELEC] A reactor that has a relatively high inductance and is wound on a magnetic core containing an air gap; used to neutralize the charging current of the line to which it is connected.

shunt regulator [ELEC] A regulator that maintains a constant output voltage by controlling the current through a dropping resistance in series with the load.

shunt repeater [ELEC] A type of negative impedance telephone repeater which is stable when it is short-circuited, but oscillates when terminated by a high impedance, in contrast to a series repeater; it can be thought of as a negative admittance.

shunt T junction *See* H-plane T junction.

shunt valve [ENG] A valve that gives a fluid under pressure a more readily available escape route than the normal route.

shunt wound [ELEC] Having armature and field windings in parallel, as in a direct-current generator or motor.

shut-down circuit [ENG] An electronic, electric, or pneumatic system designed to shut off and close down process systems or equipment; can be used for routine or emergency situations.

shut height [MECH ENG] The distance in a press between the bottom of the slide and the top of the bed, indicating the maximum die height that can be accommodated.

shut-in pressure [PETRO ENG] The equilibrated reservoir pressure measured when all the gas or oil outflow has been shut off.

shut-in well [PETRO ENG] An oil or gas well that is closed off; the well is shut so that it does not produce a fluid product of any kind.

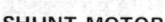
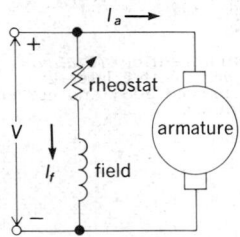

SHUNT MOTOR

Cicuit diagram showing connections of shunt motor. V = total impressed voltage from line; I_a = total armature current; I_f = field current.

shutoff [AERO ENG] In rocket propulsion, the intentional termination of burning by command from the ground or from a self-contained guidance system.

shutoff head [MECH ENG] The pressure developed in a centrifugal or axial flow pump when there is zero flow through the system.

shutter [NUCLEO] A movable plate of absorbing material used to cover a window or a beam hole in a reactor when radiation is not desired, or used to shut off a flow of slow neutrons. [OPTICS] A mechanical device that cuts off a beam of light by opening and closing at different rates of speed to expose film or plates; used in cameras and motion picture projectors. [ORD] A barrier in an explosive train used to stop a detonation wave; an interrupter which opens or closes as a shutter; often used to obtain fuse safety.

shutter dam [CIV ENG] A dam consisting of a series of pieces that can be lowered or raised by revolving them about their horizontal axis.

shuttered fuse [ORD] A fuse in which inadvertent initiation of the detonator will not initiate either the booster or the burst charge.

shuttering *See* formwork.

shuttle [MECH ENG] A back-and-forth motion of a machine which continues to face in one direction. [NUCLEO] *See* rabbit. [TEXT] A device on a loom that moves filling yarns between the warp yarns during weaving.

shuttle bombing [ORD] Bombing of objectives, utilizing two bases; a bomber formation bombs its target, flies on to its second base, reloads, and returns to its home base, again bombing a target if required.

shuttle box [TEXT] **1.** A case on a loom at either end of the lay; holds the shutter after it has been moved through the shed. **2.** A compartment for quick-access storage of shuttles containing threads of various colors.

shuttle car [MIN ENG] An electrically propelled vehicle on rubber tires or caterpillar treads used to transfer raw materials, such as coal and ore, from loading machines in trackless areas of a mine to the main transportation system.

shuttle conveyor [MECH ENG] Any conveyor in a self-contained structure movable in a defined path parallel to the flow of the material.

shuttling [ENG] A movement involving two or more trips or partial trips by the same motor vehicles between two points.

Shwartzman phenomenon [IMMUNOL] A type of local tissue reactivity in the skin in which a preparatory injection of the endotoxin is followed by an intravenous injection of the same or another endotoxin 24 hours later, producing immediate neutropenia and thrombopenia with the development of leukocyte-platelet thrombi with subsequent hemorrhage.

Si *See* silicon.

SI *See* International System of units.

Siacci method [MECH] An accurate and useful method for calculation of trajectories of high-velocity missiles with low quadrant angles of departure; basic assumptions are that the atmospheric density anywhere on the trajectory is approximately constant, and the angle of departure is less than about 15°.

sial [PETR] A petrologic term for the silica- and alumina-rich upper rock layers of the earth's crust; gives rise to granite magma; the bulk of the continental blocks is sialic. Also known as granitic layer; sal.

sialadenitis [MED] Inflammation of a salivary gland.

sialagogue [PHARM] A drug producing a flow of saliva.

sialic acid [BIOCHEM] Any of a family of amino sugars, containing nine or more carbon atoms, that are nitrogen- and oxygen-substituted acyl derivatives of neuraminic acid; as components of lipids, polysaccharides, and mucoproteins, they are widely distributed in bacteria and in animal tissues.

sialogram [MED] Roentgenogram of a salivary duct system after the injection of an opaque medium.

sialography [MED] Radiographic examination of a salivary gland following injection of an opaque substance into its duct.

sialolith [PATH] A salivary calculus.

sialolithiasis [MED] The presence of salivary calculi.

sialomucin [BIOCHEM] An acid mucopolysaccharide containing sialic acid as the acid component.

Siamese blow [ENG] In the plastics industry, the blow molding of two or more parts of a product in a single blow, then cutting them apart.

Siamese twins [MED] Viable conjoined twins.

Siberian anticyclone *See* Siberian high.

Siberian high [METEOROL] An area of high pressure which forms over Siberia in winter, and which is particularly apparent on mean charts of sea-level pressure; centered near lake Baikal. Also known as Siberian anticyclone.

Siberian tick typhus [MED] A relatively benign, rash- and eschar-producing spotted-fever-like disease in northern Asia, caused by *Rickettsia siberica;* transmitted by four species of *Dermacentor* and two of *Haemaphysalis.*

sibjet *See* sebkha.

sibling rivalry [PSYCH] Competition between siblings for parental love, or for some other recognition.

Siboglinidae [INV ZOO] A family of pogonophores in the order Athecanephria.

SIC *See* dielectric constant.

sickle [AGR] The cutting mechanism of a binder, reaper, or combine. [DES ENG] A hand tool consisting of a hooked metal blade with a short handle, used for cutting grain or other agricultural products. [TEXT] A hooked arm for guiding the thread in a spinning mule.

Sickle [ASTRON] A group of six stars in the constellation Leo that outline the head of the lion.

sickle-cell anemia [MED] A chronic, hereditary hemolytic and thrombotic disorder in which hypoxia causes the erythrocyte to assume a sickle shape; occurs in individuals homozygous for sickle-cell hemoglobin trait.

sickle-cell hemoglobin [PATH] The hemoglobin found in sickle-cell anemia, differing in electrophoretic mobility and other physiochemical properties from normal adult hemoglobin. Also known as hemoglobin S.

sicklerite [MINERAL] $(Li,Mn)(PO_4)$ A dark-brown mineral composed of hydrous lithium manganese phosphate occurring in cleavable masses.

sicula [INV ZOO] The cone-shaped chitinous skeleton of the first zooid of a graptolite colony.

SID *See* sudden ionospheric disturbance.

side arms [ORD] Weapons that are worn at the side or in the belt when not in use; examples are the bayonet, automatic pistol, and revolver.

sideband [ELECTROMAG] **1.** The frequency band located either above or below the carrier frequency, within which fall the frequency components of the wave produced by the process of modulation. **2.** The wave components lying within such bands.

sideband interference *See* adjacent-channel interference.

sideband splash *See* adjacent-channel interference.

side-boom dredge [NAV ARCH] A dredge that carries the discharge in a discharge pipe hung from a boom, a distance of from 200 to 500 feet (60 to 150 meters) directly to port or starboard of the vessel, and there discharges into the atmosphere, dropping vertically from a height of about 50 feet (15 meters) onto the surface of the sea.

side chain [ORG CHEM] A grouping of similar atoms (two or more, generally carbons, as in the ethyl radical, C_2H_5-) that branches off from a straight-chain or cyclic (for example, benzene) molecule.

side-channel spillway [CIV ENG] A dam spillway in which the initial and final flow are approximately perpendicular to each other. Also known as lateral flow spillway.

side-construction tile [MATER] A type of structural clay tile designed to receive its principal stress at right angles to the axis of the cells.

side-cut brick [MATER] Brick cut by taut wire along the long side, as opposed to the edge.

side direction [MECH] In stress analysis, the direction perpendicular to the plane of symmetry of an object.

side-discharge shovel [MIN ENG] A shovel loader having a 21-cubic-foot (0.59-cubic-meter) bucket, hinged to the chassis to dig, lift, and discharge the material sideways onto a scraper or a belt conveyor.

side drift *See* adit.

side dumper [MIN ENG] An ore, rock, or coal car that can be tilted sidewise and thus emptied.

side echo [ELECTROMAG] Echo due to a side lobe of an antenna.

side-end lines [MIN ENG] Limits of a mining claim looked on as boundary lines, especially in the case of veins originating within but extending outside the claim.

side error [NAV] That error in the reading of a marine sextant due to nonperpendicularity of the horizon glass to the frame.

side keelson [NAV ARCH] A term applied to the fore and aft girders running along the bottom of the ship parallel, or nearly so, to the keel.

side lobe *See* minor lobe.

side-lobe blanking [ELECTR] Radar technique which compares relative signal strengths between an omnidirectional antenna and the radar antenna.

side-lobe cancellation [ORD] Jamming countermeasure technique which is designed to exclude or greatly reduce the strength of jamming signals introduced through the side or back lobes of a receiving system.

side-looking radar [ENG] A high-resolution airborne radar having antennas aimed to the right and left of the flight path; used to provide high-resolution strip maps with photograph-like detail, to map unfriendly territory while flying along its perimeter, and to detect submarine snorkels against a background of sea clutter.

side milling [MECH ENG] Milling with a side-milling cutter to machine one vertical surface.

side-milling cutter [DES ENG] A milling cutter with teeth on one or both sides as well as around the periphery.

side oblique air photograph [GRAPHICS] An oblique photograph taken with the camera axis at right angles to the longitudinal axis of the aircraft.

side plates [MIN ENG] In timbering, where both a cap and a sill are used, and the posts act as spreaders, the cap and the sill are termed side plates.

sideraerolite *See* stony-iron meteorite.

side rake [MECH ENG] The angle between the tool face and a reference plane for a single-point turning tool.

side reaction [CHEM] A secondary or subsidiary reaction that takes place simultaneously with the reaction of primary interest.

sidereal [ASTRON] Referring to a quantity, such as time, to indicate that it is measured in relation to the apparent motion or position of the stars.

sidereal clock [HOROL] A clock set at 0 hours 0 minutes 0 seconds (midnight) as the vernal equinox crosses the meridian; it is the astronomical clock.

sidereal day [ASTRON] The time between two successive upper transits of the vernal equinox; this period measures one sidereal day.

sidereal hour angle [ASTRON] The angle along the celestial equator formed between the hour circle of a celestial body and the hour circle of the vernal equinox, measuring westward from the vernal equinox through 360°.

sidereal month [ASTRON] The time period of one revolution of the moon about the earth relative to the stars; this period varies because of perturbations, but it is a little less than 27 1/3 days.

sidereal noon [ASTRON] The instant in time that the vernal equinox is on the meridian.

sidereal period [ASTRON] The length of time required for one revolution of a celestial body about its primary, with respect to the stars.

sidereal table [ORD] A device for artillery; a servo-driven table which is used to cancel out earth's rotation; the axis of the table is aligned for the particular latitude of location.

sidereal time [ASTRON] Time based on diurnal motion of stars; it is used by astronomers but is not convenient for ordinary purposes.

sidereal year [ASTRON] The time period relative to the stars of one revolution of the earth around the sun; it is about 365.2564 mean solar days.

side relief angle [DES ENG] The angle that the portion of the flanks of a cutting tool below the cutting edge makes with a plane normal to the base.

siderite [MINERAL] $FeCO_3$ A brownish, gray, or greenish rhombohedral mineral composed of ferrous carbonate; hard-

SIDERITE

2.5 cm

Dark-brown rhombohedral siderite crystals with quartz from Cornwall, England. *(Specimen from Department of Geology, Bryn Mawr College)*

SIDETONE

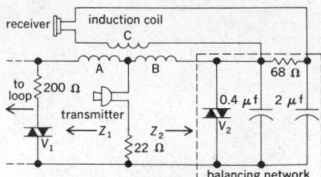

Circuit diagram of transmission circuit of a telephone set illustrating methods of eliminating sidetone. Voltage developed at transmitter is divided in windings A and B so that voltages induced in winding C are opposing. Voltage across network resistance arising from current in B opposes resultant voltages induced in C. Silicon carbide varistors V_1 and V_2 help to balance impedances Z_1 and Z_2.

SIDEWINDER

Sidewinder missile, shown with a jet pilot. (Official U.S. Navy photograph)

SIEMEN'S ELECTRODYNAMOMETER

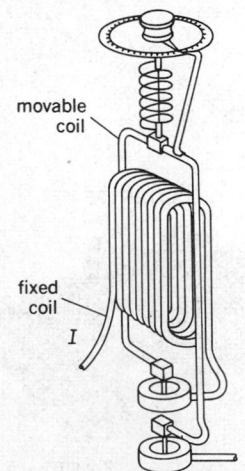

Drawing of Siemen's electrodynamometer, which can be used as ammeter. Current I flows through fixed and movable coils in series. (Weston Instruments, Inc.)

ness is 4 on Mohs scale, and specific gravity is 3.9. Also known as chalybite; iron spar; rhombohedral iron ore; siderose; sparry iron; spathic iron; white iron ore.

Siderocapsaceae [MICROBIO] A family of aquatic gram-negative bacteria of the order Pseudomonadales, found in iron-bearing waters; all members possess the ability to deposit iron or manganese compounds around the cells.

siderocyte [CYTOL] An erythrocyte which contains granules staining blue with the Prussian blue reaction.

side rod [MECH ENG] 1. A rod linking the crankpins of two adjoining driving wheels on the same side of a locomotive; distributes power from the main rod to the driving wheels. 2. One of the rods linking the piston-rod crossheads and the side levers of a side-lever engine.

siderofibrosis [MED] Fibrosis associated with deposits of iron-bearing pigment.

siderograph [ENG] An instrument that keeps the time of the Greenwich longitude; consists of a clock and a navigation instrument.

siderolite See stony-iron meteorite.

sideromelane [MINERAL] Any iron-rich mafic mineral.

sideronatrite [MINERAL] $Na_2Fe(SO_4)(OH) \cdot 3H_2O$ A yellow mineral composed of basic hydrous sodium iron sulfate occurring in fibrous masses.

sideropenic dysphagia See Plummer-Vinson syndrome.

siderophile element [CHEM] An element with a weak affinity for oxygen and sulfur and that is readily soluble in molten iron; includes iron, nickel, cobalt, platinum, gold, tin, and tantalum.

siderophyllite [MINERAL] An iron-rich variety of biotite.

siderophyre [GEOL] A stony-iron meteorite containing bronzite and tridymite crystals in a nickel-iron network. Also known as siderophyry.

siderophyry See siderophyre.

siderose See siderite.

siderosilicosis [MED] A pneumoconiosis resulting from prolonged inhalation of silica and iron dusts.

siderosis [MED] Pneumoconiosis due to prolonged inhalation of dust containing iron salt. Also known as arc-welder's disease. [PATH] The presence or concentration of stainable iron pigment in a tissue or organ.

siderostat [OPTICS] A more precise model of a heliostat; the siderostat uses a modified mirror mounting so that the image of a star is kept steady while the rest of the field is in rotation about the center.

side slope [ENG] A test course used to determine lateral stability of a vehicle as well as steering, carburetion, and other functions.

side spray [ORD] Fragments of a bursting projectile thrown sidewise from the line of flight, in contrast with base spray, thrown to the rear, and nose spray, thrown to the front.

side stream See tributary.

sidestream [CHEM ENG] A liquid stream taken from an intermediate point of a liquids-processing unit, for example, a distillation or extraction tower.

sidestream stripper [CHEM ENG] A device used to perform further distillation on a liquid stream (sidestream) from any one of the plates of a bubble tower, usually with the use of steam.

sidetone [COMMUN] The sound of the speaker's own voice as heard in his telephone receiver; the effect is undesirable and is usually suppressed by special circuits.

sidetone level [COMMUN] The ratio of the volume of the sidetone to the volume of the speaker's voice, usually expressed in decibels.

sidetone ranging [COMMUN] A method of introducing a time marker in a radio signal broadcast by a satellite for tracking or navigation, in which several audio tones of different frequencies are broadcast, and the phases of the tones transmitted from the satellite are compared with the tone phases received from the ground station or user craft.

sidetrack [CIV ENG] 1. To move railroad cars onto a siding. 2. See siding.

sidewalk [CIV ENG] 1. A walkway for pedestrians on the side of a street or road. 2. A foot pavement.

side-wall sampling [PETRO ENG] A technique for taking rock or sand samples from the sides of oil well boreholes.

Sidewinder [ORD] A Navy air-to-air missile having infrared homing guidance and a speed of over mach 2; when launched from an airplane, the missile seeks and hits an enemy bomber or fighter target by homing on the heat emitted by the target.

siding [CIV ENG] A short railroad track connected to the main track at one or more points and used to move railroad cars in order to free traffic on the main line or for temporary storage of cars. Also known as sidetrack. [MATER] Any wall cladding, except masonry or brick.

siduron [ORG CHEM] $C_{14}H_{20}N_2O$ A water-soluble solid that melts at 120–122°C; used as a herbicide.

siegbahn [SPECT] A unit of length, formerly used to express wavelengths of x-rays, equal to exactly 1/3029.45 of the spacing of the (200) planes of calcite at 18°C, or to $(1.00202 \pm 0.00003) \times 10^{-13}$ meter. Also known as X-unit; x-ray unit. Symbolized X; XU.

siege-howitzer [ORD] A short, heavy gun of large caliber, used for destruction of fortresses.

siegenite [MINERAL] $(Co,Ni)_3S_4$ A mineral composed of nickel cobalt sulfide.

siemens See mho.

Siemen's electrodynamometer [ELECTROMAG] An early type of electromagnetic instrument in which current flows through all the coils in series.

sienna [MATER] Any of various yellowish-brown earthy substances consisting of hydrated iron oxide occurring in limonite; becomes orange-brown when burnt and is generally darker and more transparent in oils than is ocher; used as pigment for oil paints and stains.

Sierolomorphidae [INV ZOO] A small family of hymenopteran insects in the superfamily Scolioidea.

sierozem [GEOL] A soil found in cool to temperate arid regions, characterized by a brownish-gray surface on a lighter layer based on a carbonate or hardpan layer.

sierra [GEOGR] A high range of hills or mountains with irregular peaks that give a sawtooth profile.

sieve [ENG] 1. A meshed or perforated device or sheet through which dry loose material is refined, liquid is strained, and soft solids are comminuted. 2. A meshed sheet with apertures of uniform size used for sizing granular materials.

sieve analysis [ENG] The size distribution of solid particles on a series of standard sieves of decreasing size, expressed as a weight percent. Also known as sieve classification; sieving.

sieve area [BOT] An area in the wall of a sieve-tube element, sieve cell, or parenchyma cell characterized by clusters of pores through which strands of cytoplasm pass to adjoining cells.

sieve cell [BOT] A long, tapering cell characteristic of phloem in gymnosperms and lower vascular plants, in which all the sieve areas are of equal specialization.

sieve classification See sieve analysis.

sieve fraction [ENG] That portion of solid particles which pass through a standard sieve of given number and is retained by a finer sieve of a different number.

sieve mesh [DES ENG] The standard opening in sieve or screen, defined by four boundary wires (warp and woof); the laboratory mesh is square and is defined by the shortest distance between two parallel wires as regards aperture (quoted in micrometers or millimeters), and by the number of parallel wires per linear inch as regards mesh; 60-mesh equals 60 wires per inch.

sieve of Eratosthenes [MATH] An iterative procedure which determines all the primes less than a given number.

sieve plate [CHEM ENG] A distillation-tower tray that is perforated so that the vapor emerges vertically through the tray, passing through the liquid holdup on top of the tray; used as a replacement for bubble-cap trays in distillation. Also known as sieve tray.

sievert [NUCLEO] A unit of radiation dose, equal to the dose delivered by a point source of 1 milligram of radium, enclosed in a platinum container with walls ½-millimeter thick, to a sample at a distance of 1 centimeter over a period of 1 hour; equal to approximately 8.38 roentgens. Also known as millicurie-of-intensity-hour.

sieve shaker [CHEM ENG] A device used to shake a stacked column of standard sieve-test trays to cause solids to sift progressively from the top (large openings) to the bottom (small openings and a final pan), according to particle size.

sieve texture   *See* poikiloblastic.

sieve tissue   *See* phloem.

sieve tray   *See* sieve plate.

sieve tube   [BOT] A phloem element consisting of a series of thin-walled cells arranged end to end, in which some sieve areas are more specialized than others.

sieving   *See* sieve analysis.

siffanto   [METEOROL] A southwest wind of the Adriatic Sea; it is often violent.

Sigalionidae   [INV ZOO] A family of scale-bearing polychaete annelids belonging to the Errantia.

Siganidae   [VERT ZOO] A small family of herbivorous perciform fishes in the suborder Acanthuroidei having minute concealed scales embedded in the skin and strong, sharp fin spines.

Sigatoka   [PL PATH] Leaf-spot disease of banana caused by the fungus *Mycosphaerella musicola.*

sight   [NAV] *See* celestial observation.   [ORD] **1.** Mechanical or optical device for aiming a firearm or for laying a gun or launcher in position. **2.** To aim at a target or aiming point. [PHYSIO] *See* vision.

sight aperture   [ORD] An irregularly shaped, adjustable, mechanical item usually integral to a rear sight; it functions as a peephole through which the sight at the opposite end of a gun is brought into view in aiming at a target or object.

sight base   [ORD] **1.** In gunnery, the distance between the eye and the rings of an optical ring sight. **2.** Mount for a gunsight.

sight bracket   [ORD] Clamp used to hold a detachable sight in position when mounted on a gun.

sight distance   [OPTICS] The distance from which an object at eye level remains visible to an observer.

sight extension   [ORD] Any device which raises the normal base or mount of a gunsight to provide improved or unobstructed sighting.

sight-feed   [ENG] Pertaining to piping in which the flowing liquid can be observed through a transparent tube or wall.

sight glass   [ENG] A glass tube or a glass-faced section on a process line or vessel; used for visual reading of liquid levels or of manometer pressures.

sighting   [OPTICS] **1.** The act or procedure of aiming with the aid of a sight. **2.** The action of bringing something into view; the action of seeing something.

sighting angle   [ORD] In bombing, the angle between the line of sight to the aiming point and the vertical.

sighting bar   [ORD] Wooden device with enlarged front and rear sight, eyepiece, and a movable target, used for training in the proper method of aiming a small-arms weapon; the eyepiece forces the student to hold the eye in proper position; because of the size of the sights, errors of aiming are very apparent.

sighting instrument   [ORD] A device or an instrument designed to aid in the pointing of a weapon.

sighting shot   [ORD] Trial shot, fired to find out whether the sights are properly adjusted.

sighting station   [ORD] A place from which remote-controlled weapons are sighted.

sighting system   [ORD] A mechanical or optical device for aiming a firearm or for laying a gun in position.

sighting tube   [ENG] A tube, usually ceramic, inserted into a hot chamber whose temperature is to be measured; an optical pyrometer is sighted into the tube to observe the interior end of the tube to give a temperature reading.

sight leaf   [ORD] The movable hinged part of a rear sight of a gun that can be raised and set to a desired range, or snapped down when not in use.

sight reduction   [NAV] The information needed for establishing a line of position, derived from the use of a sextant or an octant.

sight-reduction tables   [NAV] Tables for performing sight reduction obtained by the use of a sextant or octant, particularly those sights for determining computed altitude for comparison with the observed altitude of a celestial body to determine the altitude difference for establishing a line of position.

sight tracking line   [ORD] The line of sight from a computing gunsight reticle image to the target.

sight unit   [OPTICS] A compact sighting device composed of an elbow or panoramic telescope, mount, or adapter, usually used for pointing a weapon for direct or indirect fire; it may be attached to, or used in conjunction with, a weapon rocket launcher, or the like.

sigma   [INV ZOO] A C-shaped spicule.

sigma algebra   [MATH] A Boolean algebra of sets where any countable union of its members is also a member.

sigma finite   [MATH] A measure is sigma finite on a space X if X is a countable disjoint union of sets each of which is measurable and has finite measure.

sigma function   [THERMO] A property of a mixture of air and water vapor, equal to the difference between the enthalpy and the product of the specific humidity and the enthalpy of water (liquid) at the thermodynamic wet-bulb temperature; it is constant for constant barometric pressure and thermodynamic wet-bulb temperature.

sigma hyperon   [PARTIC PHYS] **1.** The collective name for three semistable baryons with charges of $+1$, 0, and -1 times the proton charge, designated Σ^+, Σ^0, Σ^-, all having masses of approximately 1193 million electron volts, spin of $\frac{1}{2}$, and positive parity; they form an isotopic spin multiplet with a total isotopic spin of 1 and a hypercharge of 0. **2.** Any baryon belonging to an isotopic spin multiplet having a total isotopic spin of 1 and a hypercharge of 0; designated $\Sigma_J P(m)$, where m is the mass of the baryon in million electron volts, and J and P are its spin and parity; the $\Sigma_{3/2^+}$ (1385) is sometimes designated Σ^*.

sigma phase   [MET] A brittle, nonmagnetic phase of tetragonal structure occurring in many transition-metal alloys; frequently encountered in high chromium stainless steels.

sigma pile   [NUCLEO] An assembly of moderating material containing a neutron source, used to study the absorption cross sections and other neutron properties of the material.

sigma ring   [MATH] A ring of sets where any countable union of its members is also a member.

sigmaspire   [INV ZOO] An S-shaped sponge spicule.

sigma-t   [OCEANOGR] An abbreviated value of the density of a sea-water sample of temperature T and salinity S: $\sigma T = [\rho(S, T) - 1] \times 10^3$, where $\rho(S, T)$ is the value of the sea-water density in centimeter-gram-second units at standard atmospheric pressure.

sigmatron   [NUCLEO] A cyclotron and betatron operating in tandem to produce billion-volt x-rays.

sigmoid   [BIOL] S-shaped.

sigmoidal fold   [GEOL] A recumbent fold having an axial surface which resembles the Greek letter sigma.

sigmoid colon   [ANAT] The S-shaped portion of the colon between the descending colon and the rectum.

sigmoiditis   [MED] Inflammation of the sigmoid flexure of the colon.

sigmoidoscope   [MED] An appliance for the inspection, by artificial light, of the sigmoid colon; it differs from the proctoscope in its greater length and diameter.

sign   [MATH] **1.** A symbol which indicates whether a quantity is greater than zero or less than zero; the signs are often the marks $+$ and $-$ respectively, but other arbitrarily selected symbols are used, especially in automatic data processing. **2.** A unit of plane angle, equal to 30° or $\pi/6$ radians.

signal   [COMMUN] **1.** A visual, aural, or other indication used to convey information. **2.** The intelligence, message, or effect to be conveyed over a communication system. **3.** *See* signal wave.

signal area   [NAV] That part of an airport used for the display of visual ground signals for the benefit of aircraft in flight.

signal bias   [COMMUN] Form of teletypewriter signal distortion brought about by the lengthening or shortening of pulses during transmission; when marking pulses are all lengthened, a marking signal bias results; when marking pulses are all shortened, a spacing signal bias results.

signal carrier   *See* carrier.

signal center   [COMMUN] A combination of signal communication facilities operated by the U.S. Army in the field and consisting of a communications center, telephone switching central, and appropriate means of signal communications.

signal channel   [COMMUN] A signal path for transmitting

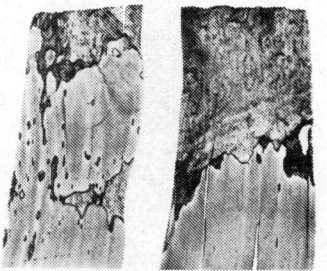

SIGATOKA

Sigatoka disease; banana with spotting, and necrosis of marginal areas on leaves of Gros Michel variety. *(Division of Tropical Research, United Brands Co.)*

electric signals; such paths may be separated by frequency division or time division.

signal conditioning [COMMUN] Processing the form or mode of a signal so as to make it intelligible to or compatible with a given device, such as a data transmission line, including such manipulation as pulse shaping, pulse clipping, digitizing, and linearizing.

Signal Corps [COMMUN] A technical service of the U.S. Army, charged with the development, maintenance, and operation of communications systems within the army.

signal detection theory [PSYCH] A theory which characterizes not only the acuity of an individual's discrimination but also the psychological factors that bias his judgment.

signal distance [ADP] The number of bits that are not the same in two binary words of equal length. Also known as hamming distance.

signal distortion generator [ELECTR] Instrument designed to apply known amounts of distortion on a signal for the purpose of testing and adjusting communications equipment such as teletypewriters.

signal flare [ENG] A pyrotechnic flare of distinct color and character used as a signal.

signal-flow graph [SYS ENG] An abbreviated block diagram in which small circles, called nodes, represent variables of the system, and the nodes are connected by lines, called branches, which represent one-way signal multipliers; an arrow on the line indicates direction of signal flow, and a letter near the arrow indicates the multiplication factor. Also known as flow graph.

signal generator [ENG] An electronic test instrument that delivers a sinusoidal output at an accurately calibrated frequency that may be anywhere from the audio to the microwave range; the frequency and amplitude are adjustable over a wide range, and the output usually may be amplitude- or frequency-modulated. Also known as test oscillator.

signaling key *See* key.

signaling rate [COMMUN] The rate at which signals are transmitted.

signal intensity [COMMUN] The electric-field strength of the electromagnetic wave transmitting a signal.

signal level [COMMUN] The difference between the level of a signal at a point in a transmission system and the level of an arbitrarily specified reference signal.

signal light [COMMUN] A light specifically designed for the transmission of code messages by means of visible light rays that are interrupted or deflected by electric or mechanical means. [ENG] A signal, illumination, or any pyrotechnic light used as a sign.

signal reporting code *See* radio-signal reporting code.

signal rocket [ORD] A rocket that gives off some characteristic color or display which has a meaning according to an established code.

signal-shaping network [ELECTR] Network inserted in a telegraph circuit, usually at the receiving end, to improve the waveform of the code signals.

signal speed [COMMUN] The rate at which code elements are transmitted by a communications system.

signal station [COMMUN] A place on shore at which signals are made to ships at sea.

signal strength [ELECTROMAG] The strength of the signal produced by a radio transmitter at a particular location, usually expressed as microvolts or millivolts per meter of effective receiving antenna height.

signal-strength meter [ELECTR] A meter that is connected to the automatic volume-control circuit of a communication receiver and calibrated in decibels or arbitrary S units to read the strength of a received signal. Also known as S meter; S-unit meter.

signal-to-noise improvement factor *See* noise improvement factor.

signal-to-noise ratio [ELECTR] The ratio of the amplitude of a desired signal at any point to the amplitude of noise signals at that same point; often expressed in decibels; the peak value is usually used for pulse noise, while the root-mean-square (rms) value is used for random noise. Abbreviated S/N; SNR.

signal tower [CIV ENG] A switch tower from which railroad signals are displayed or controlled.

signal tracer [ELECTR] An instrument used for tracing the progress of a signal through a radio receiver or an audio amplifier to locate a faulty stage.

signal voltage [ELEC] Effective (root-mean-square) voltage value of a signal.

signal wave [COMMUN] A wave whose characteristics permit some intelligence, message, or effect to be conveyed. Also known as signal.

signal-wave envelope [COMMUN] Contour of a signal wave which is composed of a series of wave cycles.

signal winding [ELEC] Control winding, of a saturable reactor, to which the independent variable (signal wave) is applied.

sign-and-magnitude code [ADP] The representation of an integer X by $(-1)^{a_0} (2^{n-2} a_1 + 2^{n-3} a_2 + \ldots + a_{n-1})$, where a_0 is 0 for X positive, and a_0 is 1 for X negative, and any a_j is either 0 or 1.

signature [ELECTR] The characteristic pattern of a target as displayed by detection and classification equipment. [GRAPHICS] A folded, printed sheet, usually consisting of 16 or 32 pages, that forms a section of a book or a pamphlet; the sheet may have fewer pages, but is always in multiples of four. [NAV ARCH] The graphic record of the magnetic properties of a vessel automatically traced as the vessel passes over the sensitive element of a recording instrument; more accurately called magnetic signature. [ORD] The identifying characteristics peculiar to each type of target which enable detecting apparatus, such as certain fuses, to sense and differentiate targets.

sign check indicator [ADP] An error checking device, indicating no sign or improper signing of a field used for arithmetic processes; the machine can, upon interrogation, be made to stop or enter into a correction routine.

sign-control flip-flop [ADP] In computers, a flip-flop in the arithmetic unit used for storing the sign of the result of an operation.

sign convention [OPTICS] A convention as to which quantities, such as angles, distances, and radii of curvature, are positive and which are negative in computations involving a lens or a mirror.

sign digit [ADP] A digit containing one to four binary bits, associated with a data item and used to denote an algebraic sign.

signet-ring cell [HISTOL] A cell with a large fat- or carbohydrate-filled vacuole that pushes the nucleus against the cell membrane.

significance [MATH] The arbitrary rank, priority, or order of relative magnitude assigned to a given position in a number.

significance level *See* level of significance of a test.

significance probability [STAT] The probability of observing a value of a test statistic as significant as, or even more significant than, the value actually observed.

significant digit *See* significant figure.

significant figure [MATH] A prescribed decimal place which determines the amount of rounding off to be done; this is usually based upon the degree of accuracy in measurement. Also known as significant digit.

significant wave [OCEANOGR] Statistically, a wave with the average height of the highest third of the waves of a given wave group.

sign of the zodiac [ASTRON] The zodiac is divided into 12 sections, called signs, in each of which the sun is situated for 1 month of the year; each sign, 30° in length, is named from a constellation with which the sign once coincided.

sign position [ADP] That position, always at or near the left or right end of a numeral, in which the algebraic sign of the number is represented.

sign test [STAT] A test which can be used whenever an experiment is conducted to compare a treatment with a control on a number of matched pairs, provided the two treatments are assigned to the members of each pair at random.

signum [MATH] The real function $sgn(x)$ defined for all x different from zero, where $sgn(x) = 1$ if $x > 0$ and $sgn(x) = -1$ if $x < 0$.

sigua [METEOROL] A straight-blowing monsoon gale of the Philippines.

siicon dioxide [INORG CHEM] SiO_2 Colorless, transparent crystals, soluble in molten alkalies and hydrofluoric acid; melts at 1710°C; used to make glass, ceramic products, abrasives, foundry molds, and concrete.

sikussak [OCEANOGR] Very old sea ice trapped in fjords; it resembles glacier ice because snowfall and snow drifts contribute to its formation.

SIL *See* speech interference level.

silage [AGR] Green or mature fodder that is fermented to retard spoilage and produce a succulent winter feed for livestock.

silane [INORG CHEM] Si_nH_{2n+2} A class of silicon-based compounds analogous to alkanes, that is, straight-chain, saturated paraffin hydrocarbons; they can be gaseous or liquid. Also known as silicon hydride.

Silastic [MATER] A trade name for any of several heat-resistant silicone rubbers.

silcrete [GEOL] A conglomerate of sand and gravel cemented by silica.

silent period [COMMUN] Period during each hour in which ship and shore radio stations must remain silent and listen for distress calls.

silent speed [ENG] The speed at which silent motion pictures are fed through a projector, equal to 16 frames per second (sound-film speed is 24 frames per second).

silent stock support [MECH ENG] A flexible metal guide tube in which the stock tube of an automatic screw machine rotates; it is covered with a casing which deadens sound and prevents transfer of noise and vibration.

silex [MATER] Heat- and shock-resistant glass containing about 98% quartz. [MINERAL] A pure or finely ground quartz.

silexite [GEOL] Chert occurring in calcareous beds. [PETR] Igneous rock composed mainly of primary quartz.

silhouette target [ORD] **1.** Target whose shape is outlined against a light background, although its body features cannot be clearly seen. **2.** Practice target consisting of the dark image of a person or object outlined against a light background.

silica [MINERAL] SiO_2 Naturally occurring silicon dioxide; occurs in five crystalline polymorphs (quartz, tridymite, cristobalite, coesite, and stishovite), in cryptocrystalline form (as chalcedony), in amorphous and hydrated forms (as opal), and combined in silicates.

silica aerogel [MATER] A colloidal silica powder whose grains have small pores; used as a low-temperature insulator.

silica brick [MATER] A type of refractory brick formed of at least 90% silica cemented with, for example, slurried lime; used to line furnace roofs.

silica cement [MATER] A mortar used with silica cement; it is a refractory material.

silica flour [MET] A sand additive for casting produced by pulverizing quartz sand.

silica gel [INORG CHEM] A colloidal, highly absorbent silica used as a dehumidifying and dehydrating agent, as a catalyst carrier, and sometimes as a catalyst.

silica glass [MATER] A translucent or transparent vitreous material consisting almost entirely of silica. Also known as fused silica; vitreous silica.

silica sand [GEOL] Sand having a very high percentage of silicon dioxide; a source of silicon.

silica stone [PETR] A sedimentary rock composed of siliceous minerals.

silicate [INORG CHEM] The generic term for a compound that contains silicon, oxygen, and one or more metals, and may contain hydrogen. [MINERAL] Any of a large group of minerals whose crystal lattice contains SiO_4 tetrahedra, either isolated or joined through one or more of the oxygen atoms.

silicate cement [MATER] The silicate of soda glue, used as an adhesive in cardboard and plywood boxes.

silicate grinding wheel [DES ENG] A mild-acting grinding wheel where the abrasive grain is bonded with sodium silicate and fillers.

silicate of soda *See* sodium silicate.

silicate paint [MATER] A paint in which the vehicle is water-soluble sodium silicate; used for painting mortar.

silication [GEOL] The conversion to or the replacement by silicates.

silicatization [MIN ENG] The sealing off of water by the injection of calcium silicate under pressure; sometimes used to reduce the leakage of water through defective lengths of tubing in a shaft.

siliceous [PETR] Describing a rock containing abundant silica, especially free silica.

siliceous dust [MIN ENG] The dust arising from the dry-working of sand, sandstone, trap, granite, and other igneous rocks; the dust is not soluble in the body fluids, and often results in a form of pneumoconiosis, known as silicosis.

siliceous earth [GEOL] A loose, friable, soft, porous, light-weight, fine-grained, and usually white siliceous sediment, usually derived from the remains of organisms.

siliceous ooze [GEOL] An ooze composed of siliceous skeletal remains of organisms, such as radiolarians.

siliceous sediment [GEOL] A sediment composed of fragmental, concretionary, or precipitated siliceous materials.

siliceous shale [PETR] A hard, fine-grained rock with the texture of shale and with as much as 85% silica.

siliceous sinter [MINERAL] A white, lightweight, porous, opaline variety of silica, deposited by a geyser or hot spring. Also known as fiorite; geyserite; pearl sinter; sinter.

silicic [PETR] Sard of magma or igneous rock rich in silica (usually at least 65%); granite is a silicic rock. Also known as oversaturated; persilicic.

silicic acid [INORG CHEM] $SiO_2 \cdot nH_2O$ A white, amorphous precipitate; used to bleach fats, waxes, and oils. Also known as hydrated silica.

silicification [GEOL] Introduction of or replacement by silica. Also known as silification.

silicified wood [GEOL] A material formed by the silicification of wood, generally in the form of opal or chalcedony, in such a manner as to preserve the original form and structure of the wood. Also known as agatized wood; opalized wood; petrified wood; woodstone.

silicle [BOT] A many-seeded capsule formed from two united carpels, usually of equal length and width, and divided on the inside by a replum.

silicoblast [INV ZOO] Poriferan amebocytes involved in formation of siliceous spicules.

Silicoflagellata [BOT] A class of unicellular flagellates of the plant division Chrysophyta represented by a single living genus, *Dictyocha*.

Silicoflagellida [INV ZOO] An order of marine flagellates in the class Phytamastigophorea which have an internal, siliceous, tubular skeleton, numerous yellow chromatophores, and a single flagellum.

silicomagnesiofluorite [MINERAL] $Ca_4Mg_3Si_2O_5(OH)_2F_{10}$ A mineral composed of basic calcium magnesium fluoride and silicate.

silicomanganese [MET] A crude alloy made up of 65–70% manganese, 16–25% silicon, and 1–2.5% carbon; used in the manufacture of low-carbon steel.

silicon [CHEM] A group IV nonmetallic element, symbol Si, with atomic number 14, atomic weight 28.086; dark-brown crystals that burn in air when ignited; soluble in hydrofluoric acid and alkalies; melts at 1410°C; used to make silicon-containing alloys, as an intermediate for silicon-containing compounds, and in rectifiers and transistors.

silicon bromide *See* silicon tetrabromide.

silicon bronze [MET] An alloy of copper with 1–5% silicon; it is corrosion-resistant and has good mechanical properties.

silicon capacitor [ELECTR] A capacitor in which a pure silicon-crystal slab serves as the dielectric; when the crystal is grown to have a p zone, a depletion zone, and an n zone, the capacitance varies with the externally applied bias voltage, as in a varactor.

silicon carbide [INORG CHEM] SiC Water-insoluble, bluish-black crystals, very hard and iridescent; soluble in fused alkalies; sublimes at 2210°C; used as an abrasive and a heat-refractory.

silicon chloride *See* silicon tetrachloride.

silicon-controlled rectifier [ELECTR] A semiconductor recti-

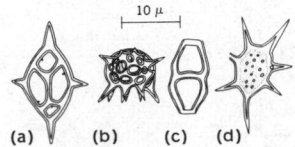

SILICOFLAGELLATA

Examples of fossil and modern Silicoflagellata. *(a) Dictyocha*, Cretaceous to Recent; *(b) Cannopilus*, Miocene; *(c) Naviculopsis*, Eocene to Miocene; and *(d) Vallacerta*, Upper Cretaceous.

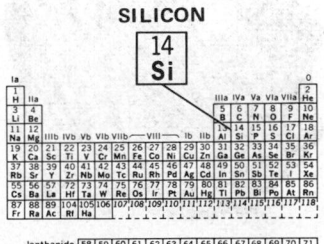

SILICON

Periodic table of the chemical elements showing the position of silicon.

SILICON-CONTROLLED RECTIFIER

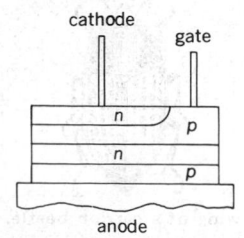

Diagrammatic view of typical silicon-controlled rectifier showing four alternate layers of p-type and n-type material.

fier that can be controlled; it is a *pnpn* four-layer semiconductor device that normally acts as an open circuit, but switches rapidly to a conducting state when an appropriate gate signal is applied to the gate terminal. Abbreviated SCR. Also known as reverse-blocking triode thyristor.

silicon-controlled switch [ELECTR] A four-terminal switching device having four semiconductor layers, all of which are accessible; it can be used as a silicon-controlled rectifier, gate-turnoff switch, complementary silicon-controlled rectifier, or conventional silicon transistor. Abbreviated SCS. Also known as reverse-blocking tetrode thyristor.

silicon copper [MET] An alloy containing 70-80% copper and 20-30% silicon, used as an addition to molten copper or brass.

silicon detector *See* silicon diode.

silicon diode [ELECTR] A crystal diode that uses silicon as a semiconductor; used as a detector in ultra-high- and super-high-frequency circuits. Also known as silicon detector.

silicon dioxide [INORG CHEM] SiO_2 Colorless, transparent crystals, soluble in molten alkalies and hydrofluoric acid; melts at 1710°C; used to make glass, ceramic products, abrasives, foundry molds, and concrete.

silicone [MATER] A fluid, resin, or elastomer; can be a grease, a rubber, or a foamable powder; the group name for heat-stable, water repellent, semiorganic polymers of organic radicals attached to the silicones, for example, dimethyl silicone; used in adhesives, cosmetics, and elastomers.

silicon fluoride *See* silicon tetrafluoride.

silicon halide [INORG CHEM] A compound of silicon and a halogen; for example, $SiBr_4$, Si_2Br_6, $SiCl_4$, Si_2Cl_6, Si_3Cl_8, SiF_4, Si_2F_6, SiI_4, and Si_2F_6.

silicon hydride *See* silane.

siliconizing [MET] Diffusing silicon into solid metal at an elevated temperature.

silicon nitride [INORG CHEM] Si_3N_4 A white, water-insoluble powder, resistant to thermal shock and to chemical reagents; used as a catalyst support and for stator blades of high-temperature gas turbines.

silicon rectifier [ELECTR] A metallic rectifier in which rectifying action is provided by an alloy junction formed in a high-purity silicon slab.

silicon reference diode *See* Zener diode.

silicon resistor [ELECTR] A resistor using silicon semiconductor material as a resistance element, to obtain a positive temperature coefficient of resistance that does not appreciably change with temperature; used as a temperature-sensing element.

silicon solar cell [ELECTR] A solar cell consisting of *p* and *n* silicon layers placed one above the other to form a *pn* junction at which radiant energy is converted into electricity.

silicon steel [MET] A steel that contains 0.5-4.5% silicon, used in electric transformer coils.

silicon-symmetrical switch [ELECTR] Thyristor modified by adding a semiconductor layer so that the device becomes a bidirectional switch; used as an alternating-current phase control, for synchronous switching and motor speed control.

silicon tetrabromide [INORG CHEM] $SiBr_4$ A fuming, colorless liquid that yellows in air; disagreeable aroma; boils at 153°C. Also known as silicon bromide.

silicon tetrachloride [INORG CHEM] $SiCl_4$ A clear, corrosive, fuming liquid with suffocating aroma; decomposes in water and alcohol; boils at 57.6°C; used in warfare smoke screens, to make ethyl silicate and silicones, and as a source of pure silicon and silica. Also known as silicon chloride.

silicon tetrafluoride [INORG CHEM] SiF_4 A colorless, suffocating gas absorbed readily by water, in which it decomposes; boiling point, −86°C; used in chemical analysis and to make fluosilicic acid. Also known as silicon fluoride.

silicon transistor [ELECTR] A transistor in which silicon is used as the semiconducting material.

silicosiderosis [MED] Pneumoconiosis caused by inhalation of silicate- and iron-containing dust.

silicosis [MED] Pneumoconiosis due to the inhalation of silica (SiO_2) particles.

silicospiegel [MET] A spiegeleisen pig iron containing 15-20% manganese and 8-15% silicon and up to 4% carbon with the balance iron; used in making steel.

silicothermic process *See* Pidgeon process.

silification *See* silicification.

silique [BOT] A silicle-like capsule, but usually at least four times as long as it is wide, which opens by sutures at either margin and has parietal placentation.

silk [INV ZOO] A continuous protein fiber consisting principally of fibroin and secreted by various insects and arachnids, especially the silkworm, for use in spinning cocoons, webs, egg cases, and other structures. [TEXT] A thread or fabric made from silk secretions of the silkworm.

silk cotton *See* kapok.

silk cotton tree *See* kapok tree.

silk gland [INV ZOO] A gland in certain insects which secretes a viscous fluid in the form of filaments known as silk; it is a salivary gland in insects and an abdominal gland in spiders.

silk paper [MATER] 1. A paper containing a small amount of silk fibers which give a mottled appearance. 2. A safety paper sometimes used for postage and revenue stamps.

silk-screen printing [GRAPHICS] The process of printing a flat color design through a piece of silk; the silk is tightly stretched on a wooden frame and a design is transferred to the silk; the parts of the design not to be printed are stopped out with a resist medium; the ink is pushed through the open mesh of the design area with a rubber-edged squeegee; only one color can be printed at a time.

silkworm [INV ZOO] The larva of various moths, especially *Bombyx mori,* that produces a large amount of silk for building its cocoon.

silky fracture [MET] A metal fracture in which the broken surface is fine in texture and dull in appearance; characteristic of tough, strong metals.

sill [BUILD] The lowest horizontal member of a framed partition or of a window or door frame. [CIV ENG] 1. A timber laid across the foot of a trench or a heading under the side truss. 2. The horizontal overflow line of a dam spillway or other weir structure. 3. A horizontal member on which a lift gate rests when closed. 4. A low concrete or masonry dam in a small stream to retard bottom erosion. [GEOL] 1. Submarine ridge in relatively shallow water that separates a partly closed basin from another basin or from an adjacent sea. 2. A tabular igneous intrusion that is oriented parallel to the planar structure of surrounding rock. [MIN ENG] 1. A piece of wood laid across a drift to constitute a frame to support uprights of timber sets and to carry the track of the tramway. 2. The floor of a gallery or passage in a mine.

sill depth [OCEANOGR] The maximum depth at which there is horizontal communication between an ocean basin and the open ocean. Also known as threshold depth.

silled basin *See* restricted basin.

sillenite [MINERAL] Bi_2O_3 A mineral composed of native bismuth oxide, is polymorphous with bismite, and occurs as earthy masses.

sillimanite [MINERAL] Al_2SiO_5 A brown, pale-green, or white neosilicate mineral with vitreous luster crystallizing in the orthorhombic system; commonly occurs in slender crystals, often in fibrous aggregates; hardness is 6-7 on Mohs scale, and specific gravity is 3.23. Also known as fibrolite.

silo [AERO ENG] A missile shelter that consists of a hardened vertical hole in the ground with facilities either for lifting the missile to a launch position, or for direct launch from the shelter. [CIV ENG] A large vertical, cylindrical structure, made of reinforced concrete, steel, or timber, and used for storing grain, cement, or other materials.

siloxane [ORG CHEM] R_2SiO Any of a family of silica-based polymers in which R is an alkyl group, usually methyl; these polymers exist as oily liquids, greases, rubbers, resins, or plastics. Also known as oxosilane.

Silphidae [INV ZOO] The carrion beetles, a family of coleopteran insects in the superfamily Staphylinoidea.

Silsbee effect [CRYO] The ability of an electric current to destroy superconductivity by means of the magnetic field that it generates, without raising the cryogenic temperature.

silt [GEOL] 1. A rock fragment or a mineral or detrital particle in the soil having a diameter of 0.002-0.05 millimeter that is, smaller than fine sand and larger than coarse clay. 2.

SILLIMANITE

2.5 cm

Sillimanite crystals from Chester, Connecticut. (*Specimen from Department of Geology, Bryn Mawr College*)

SILPHIDAE

Drawing of a carrion beetle.

Sediment carried or deposited by water. **3.** Soil containing at least 80% silt and less than 12% clay.

silt discharge rating *See* sediment discharge rating.

silting [CIV ENG] The filling up or raising of the bed of a body of water by depositing silt. [GEOL] The deposition or accumulation of stream-deposited silt that is suspended in a body of standing water.

silting index [ENG] The measurement of the tendency of a solids- or gel-carrying fluid to cause silting in close-tolerance devices, such as valves or other process-line flow constrictions.

siltite *See* siltstone.

siltstone [GEOL] Indurated silt having a shalelike texture and composition. Also known as siltite.

silttil [GEOL] A chemically decomposed and eluviated till consisting of a friable, brownish, open-textured silt that contains a few small siliceous pebbles.

Silurian [GEOL] **1.** A period of geologic time of the Paleozoic era, covering a time span of between 430–440 and 395 million years ago. **2.** The rock system of this period.

Siluridae [VERT ZOO] A family of European catfish in the suborder Siluroidei in which the adipose dorsal fin is rudimentary or lacking.

Siluriformes [VERT ZOO] The catfishes, a distinctive order of actinopterygian fishes in the superorder Ostariophysi, distinguished by a complex Weberian apparatus that involves the fifth vertebrae and one to four pair of barbels.

Siluroidei [VERT ZOO] A suborder of the Siluriformes.

Silvanidae [INV ZOO] An equivalent name for Cucujidae.

silver [CHEM] A white metallic element in group I, symbol Ag, with atomic number 47; soluble in acids and alkalies, insoluble in water; melts at 961°C, boils at 2212°C; used in photographic chemicals, alloys, conductors, and plating. [MET] A sonorous, ductile, malleable metal that is capable of a high degree of polish and that has high thermal and electric conductivity.

silver acetate [ORG CHEM] CH_3COOAg A white powder, moderately soluble in water and nitric acid; used in medicine.

silver alloy [MET] A metal consisting of silver and one or more additional metallic components.

silver arsenite [INORG CHEM] Ag_3AsO_3 A poisonous, light-sensitive, yellow powder; soluble in acids and alkalies, insoluble in water and alcohol; decomposes at 150°C; used in medicine.

silver brazing [MET] Brazing in which silver-base alloys are used as the filler metal.

silver-brazing alloy *See* silver solder.

silver bromate [INORG CHEM] $AgBrO_3$ A poisonous, light- and heat-sensitive, white powder; soluble in ammonium hydroxide, slightly soluble in hot water; decomposed by heat.

silver bromide [INORG CHEM] AgBr Yellowish, light-sensitive crystals; soluble in potassium bromide and potassium cyanide, slightly soluble in water; melts at 432°C; used in photographic films and plates.

silver-cadmium storage battery [ELEC] A storage battery that combines the excellent space and weight characteristics of silver-zinc batteries with long shelf life and other desirable properties of nickel-cadmium batteries.

silver carbonate [INORG CHEM] Ag_2CO_3 Yellowish, light-sensitive crystals; insoluble in water and alcohol, soluble in alkalies and acids; decomposes at 220°C; used as a reagent.

silver chloride [INORG CHEM] AgCl A white, poisonous, light-sensitive powder; slightly soluble in water, soluble in alkalies and acids; melts at 445°C; used in photography, photometry, silver plating, and medicine.

silver chromate [INORG CHEM] Ag_2CrO_4 Dark-colored crystals insoluble in water, soluble in acids and in solutions of alkali chromates; used as an analytical reagent.

silver coating *See* silver plating.

silver cyanide [INORG CHEM] AgCN A poisonous, white, light-sensitive powder; insoluble in water, soluble in alkalies and acids; decomposes at 320°C; used in medicine and in silver plating.

silver-disk pyrheliometer [ENG] An instrument used for the measurement of direct solar radiation; it consists of a silver disk located at the lower end of a diaphragmed tube which serves as the radiation receiver for a calorimeter; radiation

falling on the silver disk is periodically intercepted by means of a shutter located in the tube, causing temperature fluctuations of the calorimeter which are proportional to the intensity of the radiation.

silvered mica capacitor [ELECTR] A mica capacitor in which a coating of silver is deposited directly on the mica sheets to serve in place of conducting metal foil.

silverfish [INV ZOO] Any of over 350 species of insects of the order Thysanura; they are small, wingless forms with biting mouthparts.

silver fluoride [INORG CHEM] $AgF \cdot H_2O$ A light-sensitive, yellow or brownish solid, soluble in water; dehydrated form melts at 435°C; used in medicine. Also known as tachiol.

silver foil [MET] Silver or a silver-colored metal in very thin sheets.

silver frost [METEOROL] A deposit of glaze built up on trees, shrubs, and other exposed objects during a fall of freezing precipitation; the product of an ice storm. Also known as silver thaw.

silver halide [INORG CHEM] A compound of silver and a halogen; for example, silver bromide (AgBr), silver chloride (AgCl), silver fluoride (AgF), and silver iodide (AgI).

silver iodate [INORG CHEM] $AgIO_3$ A white powder, soluble in ammonium hydroxide and nitric acid, slightly soluble in water; melts above 200°C; used in medicine.

silver iodide [INORG CHEM] AgI A pale-yellow powder, insoluble in water, soluble in potassium iodide–sodium chloride solutions and ammonium hydroxide; melts at 556°C; used in medicine, photography, and artificial rainmaking.

silver lactate [ORG CHEM] $CH_3CHOHCOOAg \cdot H_2O$ Gray-to-white, light-sensitive crystals; slightly soluble in water and in alcohol; used in medicine. Also known as actol.

silverline system [INV ZOO] A series of superficial argentophilic lines in many protozoans, especially ciliates.

silver metallurgy [MET] The art and science of extracting silver metal economically from ores, and the reclamation of silver from industrial processes or from scrap metal.

silver migration [ELEC] A process, causing reduction in insulation resistance and dielectric failure; silver, in contact with an insulator, at high humidity, and subjected to an electrical potential, is transported ionically from one location to another.

silver nitrate [INORG CHEM] $AgNO_3$ Poisonous, corrosive, colorless crystals; soluble in glycerol, water, and hot alcohol; melts at 212°C; used in external medicine, photography, hair dyeing, silver plating, ink manufacture, and mirror silvering, and as a chemical reagent.

silver orthophosphate *See* silver phosphate.

silver oxide [INORG CHEM] Ag_2O An odorless, dark-brown powder with a metallic taste; soluble in nitric acid and ammonium hydroxide, insoluble in alcohol; decomposes above 300°C; used in medicine and in glass polishing and coloring, as a catalyst, and to purify drinking water.

silver oxide cell [ELEC] A primary cell in which depolarization is accomplished by an oxide of silver.

silver permanganate [INORG CHEM] $AgMnO_4$ Water-soluble, violet crystals that decompose in alcohol; used in medicine and in gas masks.

silver phosphate [INORG CHEM] Ag_3PO_4 A poisonous, yellow powder; darkens when heated or exposed to light; soluble in acids and in ammonium carbonate, very slightly soluble in water; melts at 849°C; used in photographic emulsions and in pharmaceuticals, and as a catalyst. Also known as silver orthophosphate.

silver picrate [ORG CHEM] $C_6H_2O(NO_2)_3Ag \cdot H_2O$ Yellow crystals, soluble in water, insoluble in ether and chloroform; used in medicine.

silver plating [MET] Electrolytically depositing a coating of metallic silver on a base metal. Also known as silver coating.

silverprint [GRAPHICS] A print made from photographic paper treated with silver halides.

silver protein [ORG CHEM] A brown, hygroscopic powder containing 7.5–8.5% silver; made by reaction of a silver compound with gelatin in the presence of an alkali; used as an antibacterial.

silver selenide [INORG CHEM] Ag_2Se A gray powder, insol-

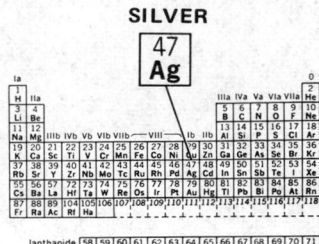

Chart showing the position of the Silurian in relation to the various periods of geologic time.

SILVER

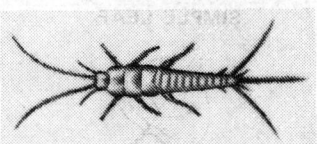

Periodic table of the chemical elements showing the position of silver.

SILVERFISH

A silverfish, a primitive insect pest.

uble in water, soluble in ammonium hydroxide; melts at 880°C.

silver solder [MET] A solder composed of silver, copper, and zinc, having a melting point lower than silver but higher than lead-tin solder. Also known as silver-brazing alloy.

silverstat regulator [ELEC] Multitapped resistor, the taps of which are connected to single-leaf silver contacts; variation of voltage causes a solenoid to open or close these contacts, shorting out more or less of the resistance in the exciter circuit as a means of regulating the output voltage to the desired value.

silver storm *See* ice storm.

silver suboxide [INORG CHEM] AgO A charcoal-gray powder that crystallizes in the cubic or orthorhombic system, and has diamagnetic properties; used in making silver oxide-zinc alkali batteries. Also known as argentic oxide.

silver sulfate [INORG CHEM] Ag_2SO_4 Light-sensitive, colorless, lustrous crystals; soluble in alkalies and acids, insoluble in alcohol; melts at 652°C; used as an analytical reagent. Also known as normal silver sulfate.

silver sulfide [INORG CHEM] Ag_2S A dark, heavy powder, insoluble in water, soluble in concentrated sulfuric and nitric acids; melts at 825°C; used in ceramics and in inlay metalwork.

silver thaw *See* silver frost.

silvery iron [MET] A variety of cast iron with a high silicon content, a light-gray color, and a fine grain.

silver-zinc storage battery [ELEC] A storage battery that gives higher current output and greater watt-hour capacity per unit of weight and volume than most other types, even at high discharge rates; used in missiles and torpedoes, where its high cost can be tolerated.

silviculture [FOR] The theory and practice of controlling the establishment, composition, and growth of stands of trees for any of the goods and benefits that they may be called upon to produce.

sima [PETR] A petrologic term for the lower layer of the earth's crust, composed of silica- and magnesia-rich rocks; source of basaltic magma; sima is equivalent to the lower part of the continental crust and the bulk of the oceanic crust. Also known as intermediate layer.

similar figures [MATH] Two figures or bodies that are identical except for size; similar figures can be placed in perspective, so that straight lines joining corresponding parts of the two figures will pass through a common point.

similar fold [GEOL] A fold in deformed beds in which the successive folds resemble each other.

similarity coefficient [SYST] In numerical taxonomy, a factor S used to calculate the similarity between organisms, according to the formula $S = n_s/(n_s + n_d)$, where n_s represents the number of positive features shared by two strains, and n_d represents the number of features positive for one strain and negative for the other.

similarity transformation [MATH] A transformation of a euclidean space obtained from such transformations as translations, rotations, and those which either shrink or expand the length of vectors.

similitude [ENG] A likeness or resemblance; for example, the scale-up of a chemical process from a laboratory or pilot-plant scale to a commercial scale. [PHYS] The use in scientific studies and engineering designs of the corresponding behavior between large and small objects or systems which are of similar nature and, more precisely, have geometrical, kinematic, and dynamical similarity.

Simmonds' disease [MED] Hypopituitarism with marked insufficiency of the target glands and profound cachexia. Also known as hypophyseal cachexia; hypopituitary cachexia.

simo chart [IND ENG] A basic motion-time chart used to show the simultaneous nature of motions; commonly a therblig chart for two-hand work with motion symbols plotted vertically with respect to time, showing the therblig abbreviation and a brief description for each activity, and individual times values and body-member detail. Also known as simultaneous motion-cycle chart.

Simon's theory [ENG] A theory of drilling which includes the effects of drilling by percussion and by vibration with a rotary (oil well) bit, cable tool, and pneumatic hammer; the rate of penetration of a chisel-shaped bit into brittle rock may be

SIMPLE LEAF

The one-blade type of simple leaf.

SIMPLE MICROSCOPE

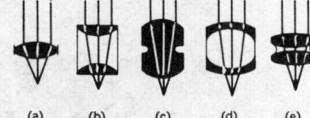

(a)　(b)　(c)　(d)　(e)

Types of lenses used in simple microscope. (a) Double convex. (b) Doublet. (c) Coddington. (d) Hastings triplet. (e) Achromat. *(From F. A. Jenkins and H. E. White, Fundamentals of Optics, 3d ed., McGraw-Hill, 1957)*

defined as follows: $R = NAf_v/\pi D$, where R equals the rate of advance of bit, N equals the number of wings of bit, f_v equals the number of impacts per unit time, D equals the diameter of the bit, and A equals the cross-sectional area of the crater at the periphery of the drill hole.

simoom [METEOROL] A strong, dry, dust-laden desert wind which blows in the Sahara, Israel, Syria, and the desert of Arabia; its temperature may exceed 130°F (54°C), and the humidity may fall below 10%.

simple [BIOL] **1.** Made up of one piece. **2.** Unbranched. **3.** Consisting of identical units, as a simple tissue.

simple alternative [STAT] An alternative to the null hypothesis which completely specifies the distribution of the observed random variables.

simple balance [ENG] An instrument for measuring weight in which a beam can rotate about a knife-edge or other point of support, the unknown weight is placed in one of two pans suspended from the ends of the beam and the known weights are placed in the other pan, and a small weight is slid along the beam until the beam is horizontal.

simple branched tubular gland [ANAT] A structure consisting of two or more unbranched, tubular secreting units joining a common outlet duct.

simple buffering [ADP] A technique for obtaining simultaneous performance of input/output operations and computing; it involves associating a buffer with only one input or output file (or data set) for the entire duration of the activity on that file (or data set).

simple closed curve [MATH] A closed curve which never crosses itself.

simple conic chart [MAP] A chart on a simple conic projection.

simple conic projection [MAP] A conic map projection in which the surface of a sphere or spheroid, such as the earth, is developed on a tangent cone which is then spread out to form a plane.

simple continuous distillation *See* equilibrium flash vaporization.

simple cubic lattice [CRYSTAL] A crystal lattice whose unit cell is a cube, and whose lattice points are located at the vertices of the cube.

simple dike [PETR] An igneous dike emplaced in a single episode.

simple engine [MECH ENG] An engine (such as a steam engine) in which expansion occurs in a single phase, after which the working fluid is exhausted.

simple function *See* step function.

simple gland [ANAT] A gland having a single duct.

simple goiter [MED] Diffuse enlargement of the thyroid gland, usually not associated with constitutional features.

simple harmonic current [ELEC] Alternating current, the instantaneous value of which is equal to the product of a constant, and the cosine of an angle varying linearly with time. Also known as sinusoidal current.

simple harmonic electromotive force [ELEC] An alternating electromotive force which is equal to the product of a constant and the cosine or sine of an angle which varies linearly with time.

simple harmonic motion *See* harmonic motion.

simple hypothesis [STAT] A hypothesis which completely specifies the distribution of the observed random variables.

simple leaf [BOT] A leaf having one blade, or a lobed leaf in which the separate parts do not reach down to the midrib.

simple machine [MECH ENG] Any of several elementary machines, one or more being incorporated in every mechanical machine; usually, only the lever, wheel and axle, pulley (or block and tackle), inclined plane, and screw are included, although the gear drive and hydraulic press may also be considered simple machines.

simple microscope [OPTICS] A diverging lens system, which can form an enlarged image of a small object. Also known as hand lens; magnifier; magnifying glass.

simple oscillator *See* harmonic oscillator.

simple pendulum [MECH] A device consisting of a small, massive body suspended by an inextensible object of negligible mass from a fixed horizontal axis about which the body and suspension are free to rotate.

simple protein [BIOCHEM] One of a group of proteins which, upon hydrolysis, yield exclusively amino acids; included are globulins, glutelins, histones, prolamines, and protamines.

simple results [STAT] Results of observations such that on each trial of an experiment one and only one of these results will occur.

simple salt [CHEM] One of four classes of salts in a classification system that depends on the character of completeness of the ionization; examples are NaCl, NaHCO$_3$, and Pb(OH)Cl.

simple shear [GEOPHYS] Strain caused by differential movements on one set of parallel planes which results in internal rotation of fabric elements.

simple stomach [ANAT] A stomach consisting of a single dilation of the alimentary canal, as found in man, the dog, and many higher and lower vertebrates.

simple tone [ACOUS] Also known as pure tone. 1. A sound wave whose instantaneous sound pressure is a simple sinusoidal function of time. 2. A sound sensation characterized by singleness of pitch.

simple tubular gland [ANAT] A gland consisting of a single, tubular secreting unit.

simplex [MATH] An n-dimensional simplex in a euclidean space consists of $n + 1$ linearly independent points $p_0, p_1, \ldots, p_n$ together with all line segments $a_0 p_0 + a_1 p_1 + \ldots + a_n p_n$ where the $a_i \geq 0$ and $a_0 + a_1 + \ldots + a_n = 1$; a triangle with its interior and a tetrahedron with its interior are examples.

simplex channel [COMMUN] A channel which permits transmission in one direction only.

simplex concrete pile [CIV ENG] A molded-in-place pile made by using a hollow cylindrical mandrel which is filled with concrete after having been driven to the desired depth and raised a few feet at a time, the concrete flowing out at the bottom and filling the hole in the earth.

simplex machine [TEXT] A warp knitting machine that is similar to a tricot machine, but employs two sets of needles and knits a double fabric.

simplex method [MATH] A finite iterative algorithm used in linear programming whereby successive solutions are obtained and tested for optimality.

simplex pump [MECH ENG] A pump with only one steam cylinder and one water cylinder.

simplex transmission [COMMUN] A mode of radio transmission in which communication takes place between two stations in only one direction at a time.

simplex uterus [ANAT] A uterus consisting of a single cavity, representing the greatest degree of fusion of the Müllerian ducts; found in man and apes.

simplicial complex [MATH] A set consisting of finitely many simplices where either two simplices are disjoint or intersect in a simplex which is a face common to each.

simplicial homology [MATH] A homology for a topological space where the nth group reflects how the space may be filled out by n-dimensional simplicial complexes and detects the presence of analogs of n-dimensional holes.

simplicial subdivision [MATH] A decomposition of the simplices composing a simplicial complex which results in a simplicial complex with a larger number of simplices.

simply connected region [MATH] A region having no holes; all closed curves can be shrunk to a point without passing through points in the complement of the region.

simply connected space [MATH] A topological space whose fundamental group consists of only one element; equivalently, all closed curves can be shrunk to a point.

simpsonite [MINERAL] AlTaO$_4$ A hexagonal mineral composed of aluminum tantalum oxide and occurring in short crystals.

Simpson's rule [MATH] A basic approximation formula for definite integrals. [PETRO ENG] A mathematical relationship for calculating the oil- or gas-bearing net-pay volume of a reservoir; uses the contour lines from a subsurface geological map of the reservoir, including gas-oil and gas-water contacts.

SIMS See secondary ion mass spectrometer.

SIMSCRIPT [ADP] A simulation programming language written for the U.S. Air Force by the Rand Corporation and readily available for many computers.

simulate [ENG] To mimic some or all of the behavior of one system with a different, dissimilar system, particularly with computers, models, or other equipment.

simulation language [ADP] A computer language used to write programs for the simulation of the behavior through time of such things as transportation and manufacturing systems; SIMSCRIPT is an example.

simulator [ADP] A routine which is executed by one computer but which imitates the operations of another computer. [ENG] A computer or other piece of equipment that simulates a desired system or condition and shows the effects of various applied changes, such as a flight simulator.

Simuliidae [INV ZOO] The black flies, a family of orthorrhaphous dipteran insects in the series Nematocera.

simultaneity [MECH] Two events have simultaneity, relative to an observer, if they take place at the same time according to a clock which is fixed relative to the observer.

simultaneous access See parallel access.

simultaneous altitudes [ASTRON] Altitudes of two or more celestial bodies observed at the same time.

simultaneous color television [ELECTR] A color television system in which the phosphors for the three primary colors are excited at the same time, not one after another; the shadow-mask color picture tube gives a simultaneous display.

simultaneous equations [MATH] A collection of equations considered to be a set of joint conditions imposed on the variables involved.

simultaneous lobing See monopulse radar.

simultaneous motion-cycle chart See simo chart.

sin A See sine.

Sinclair-Baker reforming [CHEM ENG] A proprietary petroleum-refinery catalytic-reforming process, using platinum-on-alumina as the catalyst.

Sinclair hydrogen treating [CHEM ENG] A proprietary petroleum-refinery hydrodesulfurization process, using cobalt-molybdenum-on-alumina as the catalyst.

sincosite [MINERAL] Ca(VO)$_2$(PO$_4$)$_2$·5H$_2$O A leek-green mineral composed of hydrous calcium vanadyl phosphate and occurring in tetragonal scales or plates.

sine [MATH] The sine of an angle A in a right triangle with hypotenuse of length c given by the ratio a/c, where a is the length of the side opposite A; more generally, the sine function assigns to any real number A the ordinate of the point on the unit circle obtained by moving from $(1,0)$ counterclockwise A units along the circle, or clockwise $|A|$ units if A is less than 0. Denoted sin A.

sine bar [DES ENG] A device consisting of a steel straight edge with two cylinders of equal diameter attached near the ends with their centers equidistant from the straightedge; used to measure angles accurately and to lay out work at a desired angle in relationship to a surface.

sine-cosine encoder [ELECTR] A shaft-position encoder having a special type of angle-reading code disk that gives an output which is a binary representation of the sine of the shaft angle.

sine-cosine generator See resolver.

sine galvonometer [ENG] A type of magnetometer in which a small magnet is suspended in the center of a pair of Helmholtz coils, and the rest position of the magnet is measured when various known currents are sent through the coils.

Sinemurian [GEOL] A European stage of geologic time; Lower Jurassic, above Hattangian and below Pliensbachian.

sine potentiometer [ELECTR] A potentiometer whose direct-current output voltage is proportional to the sine of the shaft angle; used as a resolver in computer and radar systems.

sine wave [PHYS] A wave whose amplitude varies as the sine of a linear function of time. Also known as sinusoidal wave.

sine-wave modulated jamming [ELECTR] Jamming signal produced by modulating a continuous wave signal with one or more sine waves.

sine-wave oscillator See sinusoidal oscillator.

sine-wave response See frequency response.

singeing [TEXT] Passing a fabric over heated plates or gas flames during finishing to remove lint or loose threads from the surface. Also known as gassing.

singing [CONT SYS] An undesired, self-sustained oscillation in a system or component, at a frequency in or above the

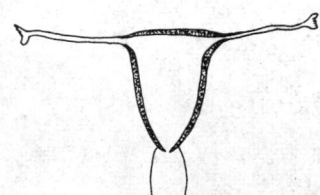

SIMPLEX UTERUS

The simplex uterus of the human. (From C. K. Weichert, Elements of Chordate Anatomy, 3d ed., McGraw-Hill, 1967)

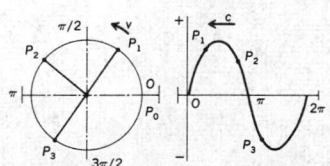

SINE WAVE

Relation of sine wave, simple harmonic motion, and uniform circular motion. Point P moves around the circle at constant tangential speed v; projection of P on vertical line moves up and down, executing simple harmonic motion. Plane on which the projection is plotted travels to the left at speed c, generating a sine wave.

passband of the system or component; generally due to excessive positive feedback.

singing margin [CONT SYS] The difference in level, usually expressed in decibels, between the singing point and the operating gain of a system or component.

singing point [CONT SYS] The minimum value of gain of a system or component that will result in singing.

singing sand *See* sounding sand.

singing-stovepipe effect [ELEC] Reception and reproduction of radio signals by ordinary pieces of metal in contact with each other, such as sections of stovepipe; it occurs when rusty bolts, faulty welds, or mechanically loose connections within strong radiated fields near transmitters produce intermodulation interference; the mechanically poor connections serve as nonlinear diodes.

single acting [MECH ENG] Acting in one direction only, as a single-acting plunger, or a single-acting engine (admitting the working fluid on one side of the piston only).

single action [ORD] A method of fire in some revolvers and shoulder arms in which the hammer must be cocked by hand, in contrast to double action in which a single pull of the trigger both cocks and fires the weapon.

single-action press [MECH ENG] A press having a single slide.

single-address instruction *See* one-address instruction.

single axis gyroscope [ENG] A gyroscope suspended in just one gimbal whose bearings form its output axis; an example is a rate gyroscope.

single-base powder [MATER] An explosive or propellant powder in which nitrocellulose is the only active ingredient.

single-bevel groove weld [MET] A groove weld in which one member has a joint edge beveled from one side.

single-blind technique [BIOL] A method in which the subjects of a drug experiment do not know whether or not they have received the drug.

single-block brake [MECH ENG] A friction brake consisting of a short block fitted to the contour of a wheel or drum and pressed up against the surface by means of a lever on a fulcrum; used on railroad cars.

single-button carbon microphone [ENG ACOUS] Microphone having a carbon-filled buttonlike container on only one side of its flexible diaphragm.

single-carrier theory [SOLID STATE] A theory of the behavior of a rectifying barrier which assumes that conduction is due to the motion of carriers of only one type; it can be applied to the contact between a metal and a semiconductor.

single-channel simplex [COMMUN] Simplex operation that provides nonsimultaneous radio communications between stations using the same frequency channel.

single-column punch [ADP] A system of coding whereby any one of the numerical values 0 to 11 can be represented by a single punch in a column.

single completion [PETRO ENG] An oil or gas well drilled to produce fluids from a single reservoir level or zone, and using a single tubing string.

single compound explosive [MATER] Explosive composed of a single chemical compound.

single crystal [CRYSTAL] A crystal, usually grown artificially, in which all parts have the same crystallographic orientation.

single-current transmission [COMMUN] Telegraph transmission in which a current flows, in only one direction, during marking intervals, and no current flows during spacing intervals.

single-degree-of-freedom gyro [MECH] A gyro the spin reference axis of which is free to rotate about only one of the orthogonal axes, such as the input or output axis.

single drift flight [NAV] A flight made by using a single drift correction angle based upon net drift between the point of departure and the point of destination.

single-edged push-pull amplifier circuit [ELECTR] Amplifier circuit having two transmission paths designed to operate in a complementary manner and connected to provide a single unbalanced output without the use of an output transformer.

single-end amplifier [ELECTR] Amplifier stage which normally employs only one tube or semiconductor or, if more than one tube or semiconductor is used, they are connected in

parallel so that operation is asymmetric with respect to ground. Also known as single-sided amplifier.

single-ended [ELEC] Unbalanced, as when one side of a transmission line or circuit is grounded.

single-frequency duplex [COMMUN] Duplex carrier communications that provide communications in opposite directions, but not simultaneously, over a single-frequency carrier channel, the transfer between transmitting and receiving conditions being automatically controlled by the voices of the communicating parties.

single-frequency simplex [COMMUN] Single-frequency carrier communications in which manual rather than automatic switching is used to change over from transmission to reception.

single-gun color tube [ELECTR] A color television picture tube having only one electron gun and one electron beam; the beam is sequentially deflected across phosphors for the three primary colors to form each color picture element, as in the chromatron.

single-hand drilling [ENG] A method of rock drilling in which the drill steel, which is held in the hand, is struck with a 4-pound (1.8-kilogram) hammer, the drill being turned between the blows.

single heading flight *See* aerologation.

single-hop transmission [COMMUN] Radio transmission in which radio waves are reflected from the ionosphere only once along their path from the transmitter to the receiver.

single-impulse welding [MET] Spot, projection, or upset welding by means of a single current impulse.

single-J groove weld [MET] A groove weld in which one member has a joint edge in the form of a J from one side.

single-layer bit *See* surface-set bit.

single-layer solenoid [ELECTROMAG] A solenoid which has only one layer of wire, wound in a cylindrical helix.

single-loop feedback [CONT SYS] A system in which feedback may occur through only one electrical path.

single-loop servomechanism [CONT SYS] A servomechanism which has only one feedback loop. Also known as servo loop.

single packing [MIN ENG] Strip packing on a longwall face in which the widest pack is along the roadside.

single-pass weld [MET] A weld made by depositing the filler metal with a single pass.

single-path system [NAV] An electronic navigation system in which a measurement is made on the delay incurred in transmission of electromagnetic energy from a radio transmitter to a radio receiver; the single-path system measures absolute transmission time and produces a circular line of position.

single-perforated grain [MATER] A cylindrical propellant grain with a single perforation located in its axis; this type of granulation is used in propelling charges for several calibers of guns, and in rockets.

single-phase [ELEC] Energized by a single alternating voltage.

single-phase circuit [ELEC] Either an alternating-current circuit which has only two points of entry, or one which, having more than two points of entry, is intended to be so energized that the potential differences between all pairs of points of entry are either in phase or differ in phase by 180°.

single-phase flow [CHEM ENG] The flow of a material, as a gas, single-phase liquid, or a solid, but not in any combination of the three.

single-phase meter [ENG] A type of power-factor meter that contains a fixed coil that carries the load current, and crossed coils that are connected to the load voltage; there is no spring to restrain the moving system, which takes a position to indicate the angle between the current and voltage.

single-phase motor [ELEC] A motor energized by a single alternating voltage.

single-phase rectifier [ELECTR] A rectifier whose input voltage is a single sinusoidal voltage, in contrast to a polyphase rectifier.

single-piece milling [MECH ENG] A milling method whereby one part is held and milled in one machine cycle.

single-point grounding [ELEC] Grounding system that attempts to confine all return currents to a network that serves as the circuit reference; to be effective, no appreciable current is allowed to flow in the circuit reference, that is, the sum of the return currents is zero.

single-polarity pulse [ELEC] Pulse in which the sense of the departure from normal is in one direction only.

single-polarity pulse amplitude modulation *See* unidirectional pulse amplitude modulation.

single-pole double-throw [ELEC] A three-terminal switch or relay contact arrangement that connects one terminal to either of two other terminals. Abbreviated SPDT.

single-pole single-throw [ELEC] A two-terminal switch or relay contact arrangement that opens or closes one circuit. Abbreviated SPST.

single reference *See* random access.

single sampling [IND ENG] A sampling inspection in which the lot is accepted or rejected on the basis of one sample.

single shot [ORD] Semiautomatic operation of an automatic gun, in which the trigger must be pulled for each shot fired.

single-shot blocking oscillator [ELECTR] Blocking oscillator modified to operate as a single-shot trigger circuit.

single-shot exploder [ENG] A magneto exploder operated by the twist action given by a half turn of the firing key.

single-shot multivibrator *See* monostable multivibrator.

single-shot probability [ORD] Probability that a single projectile fired against a target will hit that target under a given set of conditions.

single-shot survey [PETRO ENG] An oil well directional log or record with a single-reading device that is either run down into the drill pipe or positioned in a nonmagnetic drill collar.

single-shot trigger circuit [ELECTR] Trigger circuit in which one triggering pulse initiates one complete cycle of conditions ending with a stable condition. Also known as single-trip trigger circuit.

single-sideband [COMMUN] Pertaining to single-sideband communication. Abbreviated SSB.

single-sideband communication [COMMUN] A communication system in which one of the two sidebands used in amplitude-modulation is suppressed; the carrier wave may be either transmitted or suppressed.

single-sideband modulation [COMMUN] Modulation resulting from elimination of all components of one sideband from an amplitude-modulated wave.

single-sideband transmission [COMMUN] Transmission of a carrier and substantially only one sideband of modulation frequencies, as in television where only the upper sideband is transmitted completely for the picture signal; the carrier wave may be either transmitted or suppressed.

single-sided amplifier *See* single-end amplifier.

single-signal receiver [ELECTR] A highly selective superheterodyne receiver for code reception, having a crystal filter in the intermediate-frequency amplifier.

single-sized aggregate [MATER] Aggregate in which most of the particles lie between narrow size limits.

single-stage compressor [MECH ENG] A machine that effects overall compression of a gas or vapor from suction to discharge conditions without any sequential multiplicity of elements, such as cylinders or rotors.

single-stage pump [MECH ENG] A pump in which the head is developed by a single impeller.

single-stage rocket [AERO ENG] A rocket or rocket missile to which the total thrust is imparted in a single phase, by either a single or multiple thrust unit.

single-stand mill [MET] A rolling mill in which the product is in contact with only two rolls at a time.

single-station analysis [METEOROL] The analysis or reconstruction of the weather pattern from more or less continuous meteorological observations made at a single geographic location, or the body of techniques employed in such an analysis.

single-stub transformer [ELECTROMAG] Shorted section of a coaxial line that is connected to a main coaxial line near a discontinuity to provide impedance matching at the discontinuity.

single-stub tuner [ELECTROMAG] Section of transmission line terminated by a movable short-circuiting plunger or bar, attached to a main transmission line for impedance-matching purposes.

single-theodolite observation [METEOROL] The usual type of pilot-balloon observation, that is, using one theodolite.

single-throw switch [ELEC] A switch in which the same pair of contacts is always opened or closed.

single-tone keying [COMMUN] Form of keying in which the modulating function causes the carrier to be modulated with a single tone for one condition, which may be either marking or spacing, and the carrier is unmodulated for the other condition.

single-trip trigger circuit *See* single-shot trigger circuit.

single-tuned amplifier [ELECTR] An amplifier characterized by resonance at a single frequency.

single tuned circuit [ELEC] A circuit whose behavior is the same as that of a circuit with a single inductance and a single capacitance, together with associated resistances.

single tuned interstage [ELECTR] An interstage circuit which is resonant at a single frequency.

single-U groove weld [MET] A groove weld in which the joint edge of both members is prepared in the form of a J from one side, giving a final U form to the completed weld.

single-unit semiconductor device [ELECTR] Semiconductor device having one set of electrodes associated with a single carrier stream.

single-V groove weld [MET] A groove weld in which the joint edge of each member is beveled from the same side.

single welded joint [MET] A joint welded from one side only.

single-wire line [ELEC] Transmission line that uses the ground as one side of the circuit.

singular corresponding point [METEOROL] A center of elevation or depression on a constant-pressure chart (or a center of high or low pressure on a constant-height chart) considered as a reappearing characteristic of successive charts.

singular integral equation [MATH] An integral equation where the integral appearing either has infinite limits of integration or the kernel function has points where it is infinite.

singularity [MATH] A point where a function of real or complex variables is not differentiable or analytic. Also known as singular point of a function. [METEOROL] A characteristic meteorological condition which tends to occur on or near a specific calendar date more frequently than chance would indicate; an example is the January thaw.

singular matrix [MATH] A matrix which has no inverse; equivalently, its determinant is zero.

singular point of a differential equation [MATH] A point which is a singularity for at least one of the known functions appearing in the equation.

singular point of a function *See* singularity.

singular solution of a differential equation [MATH] A solution which is not generic, that is, not obtainable from the general solution.

singular transformation [MATH] A linear transformation which has no corresponding inverse transformation.

singular values of a matrix [MATH] For a matrix A these are the positive square roots of the eigenvalues of $A^{*}A$, where A^{*} denotes the adjoint matrix of A.

sinh *See* hyperbolic sine.

sinhalite [MINERAL] $MgAl(BO_4)$ A mineral composed of magnesium aluminum borate; sometimes used as a gem.

sinistral fault *See* left lateral fault.

sinistral fold [GEOL] An asymmetric fold whose long limb, when viewed along its dip, appears to have a leftward offset.

sinistrorse [BIOL] Twisting or coiling counterclockwise.

sink [ELECTROMAG] The region of a Rieke diagram where the rate of change of frequency with respect to phase of the reflection coefficient is maximum for an oscillator; operation in this region may lead to unsatisfactory performance by reason of cessation or instability of oscillations. [GEOL] A circular or ellipsoidal depression formed by collapse on the flank of or near to a volcano. [MIN ENG] **1.** To excavate strata downward in a vertical line for the purpose of winning and working minerals. **2.** To drill or put down a shaft or borehole. [PHYS] A device or system where some extensive entity is absorbed, such as a heat sink, a sink flow, a load in an electrical circuit, or a region in a nuclear reactor where neutrons are strongly absorbed.

sinker [MIN ENG] **1.** A person who sinks mine shafts and puts

SINK-FLOAT SEPARATION PROCESS

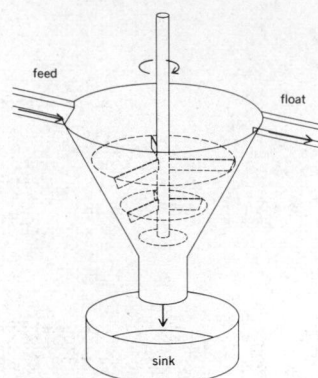

feed

float

sink

Feed particles are introduced into suspension, whose specific gravity is between that of mineral and gangue particles; particles of higher specific gravity sink while those of lower specific gravity float; stirrer prevents suspension from settling out on the walls.

SIPHONAPTERA

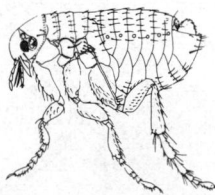

Human flea, *Pulex irritans.*
(From E. O. Essig, College Entomology, Macmillan, 1942)

SIPHONOCLADALES

Anadyomene, habit of thallus, showing expanded blades.

SIPHONOTRETACEA

Multispinula, external surface of pedicle valve. *(From R. C. Moore, ed., Treatis on Invertebrate Paleontology, pt. H, Geological Society of America, Inc. and University of Kansas Press, 1965)*

in framing. **2.** A special movable pump used in shaft sinking. **3.** *See* sinker drill.

sinker bar [MIN ENG] A short, heavy rod placed above the drill jars to increase the effect of the upward sliding jars in well-drilling with cable tools.

sinker drill [MIN ENG] A jackhammer type of rock drill used in shaft sinkings. Also known as sinker.

sink-float separation process [ENG] A simple gravity process used in ore dressing that separates particles of different sizes or composition on the basis of differences in specific gravity.

sink flow [FL MECH] **1.** In three-dimensional flow, a point into which fluid flows uniformly from all directions. **2.** In two-dimensional flow, a straight line into which fluid flows uniformly from all directions at right angles to the line.

sinkhead *See* feedhead.

sinkhole plain [GEOL] A regionally extensive plain or plateau characterized by well-developed karst features.

sinking [OCEANOGR] The downward movement of surface water generally caused by converging currents or when a water mass becomes denser than the surrounding water. Also known as downwelling. [OPTICS] In atmospheric optics, a refraction phenomenon, the opposite of looming, in which an object on, or slightly above, the geographic horizon apparently sinks below it.

sinking and walling scaffold [MIN ENG] A platform designed for use in shaft sinking to enable sinking and walling to be performed simultaneously. Also known as Galloway sinking and walling stage.

sinking bucket *See* hoppit.

sinking fund [IND ENG] A fund established by periodically depositing funds at compound interest in order to accumulate a given sum at a given future time for some specific purpose.

sinking pump [MIN ENG] A long, narrow, electrically driven centrifugal-type pump designed for keeping a shaft dry during sinking operations.

sinking tubing [MET] Drawing tubing through a die or passing it through rolls without the use of a tool in the bore to control the inside diameter.

sink mark [ENG] A shallow depression or dimple on the surface of an injection-molded plastic part due to collapsing of the surface following local internal shrinkage after the gate seals.

sinoatrial node [ANAT] A bundle of Purkinje fibers located near the junction of the superior vena cava with the right atrium which acts as a pacemaker for cardiac excitation. Abbreviated SA node. Also known as sinoauricular node.

sinoauricular node *See* sinoatrial node.

sinoite [MINERAL] Si_2N_2O A nitride mineral known only in meteorites.

sinter [MET] **1.** The product of a sintering operation. **2.** A shaped body composed of metal powders and produced by sintering with or without previous compacting. [MINERAL] *See* siliceous sinter. [PETR] A chemical sedimentary rock deposited by precipitation from mineral waters, especially siliceous sinter and calcareous sinter.

sintered copper [MET] Copper prepared by heating a compressed powder of the metal to form a solid mass.

sintered steel [MET] Steel prepared by heating compressed iron powder and graphite to form a solid.

sintering [MET] Forming a coherent bonded mass by heating metal powders without melting; used mostly in powder metallurgy.

sintering furnace [MET] A furnace in which presintering and sintering operations are carried out.

sinter setting *See* mechanical setting.

sinuate [BOT] Having a wavy margin with strong indentations.

sinus [BIOL] A cavity, recess, or depression in an organ, tissue, or other part of an animal body.

sinus gland [INV ZOO] An endocrine gland in higher crustaceans, lying in the eyestalk in most stalk-eyed species, which is the site of storage and release of a molt-inhibiting hormone.

sinus hairs *See* vibrissae.

sinusitis [MED] Inflammation of a paranasal sinus.

sinus of Morgagni [ANAT] The space between the upper border of the levator veli palatini muscle and the base of the skull.

sinusoid [ANAT] Any of the relatively large spaces comprising part of the venous circulation in certain organs, such as the liver.

sinusoidal angular modulation *See* angle modulation.

sinusoidal current *See* simple harmonic current.

sinusoidal function [MATH] The real or complex function $\sin(u)$ or any function with analogous continuous periodic behavior.

sinusoidal oscillator [ELECTR] An oscillator circuit whose output voltage is a sine-wave function of time. Also known as harmonic oscillator; sine-wave oscillator.

sinusoidal wave *See* sine wave.

sinus venosus [EMBRYO] The vessel in the transverse septum of the embryonic mammalian heart into which open the vitelline, allantoic, and common cardinal veins. [VERT ZOO] The chamber of the lower vertebrate heart to which the veins return blood from the body.

Siphinodentallidae [INV ZOO] A family of mollusks in the class Scaphopoda characterized by a subterminal epipodial ridge which is not slit dorsally and which terminates with a crenulated disk.

siphon [BOT] A tubular element in various algae. [ENG] A tube, pipe, or hose through which a liquid can be moved from a higher to a lower level by atmospheric pressure forcing it up the shorter leg while the weight of the liquid in the longer leg causes continuous downward flow. [INV ZOO] **1.** A tubular structure for intake or output of water in bivalves and other mollusks. **2.** The sucking-type of proboscis in many arthropods.

Siphonales [BOT] A large order of green algae (Chlorophyta) which are coenocytic, nonseptate, and mostly marine.

Siphonaptera [INV ZOO] The fleas, an order of insects characterized by a small, laterally compressed, oval body armed with spines and setae, three pairs of legs modified for jumping, and sucking mouthparts.

siphon barograph [ENG] A recording siphon barometer.

siphon barometer [ENG] A J-shaped mercury barometer in which the stem of the J is capped and the cusp is open to the atmosphere.

Siphonocladaceae [BOT] A family of green algae in the order Siphonocladales.

Siphonocladales [BOT] An order of green algae in the division Chlorophyta including marine, mostly tropical forms.

siphonogamous [BOT] In plants, especially seed plants, the accomplishment of fertilization by means of a pollen tube.

siphonoglyph [INV ZOO] A ciliated groove leading from the mouth to the gullet in certain anthozoans.

Siphonolaimidae [INV ZOO] A family of nematodes in the superfamily Monhysteroidea in which the stoma is modified into a narrow, elongate, hollow, spearlike apparatus.

Siphonophora [INV ZOO] An order of the coelenterate class Hydrozoa characterized by the complex organization of components which may be connected by a stemlike region or may be more closely united into a compact organism.

siphonosome [INV ZOO] The lower part of a siphonophore colony, bearing the nutritive and reproductive zooids.

siphonostele [BOT] A type of stele consisting of pith surrounded by xylem and phloem.

Siphonotretacea [PALEON] A superfamily of extinct, inarticulate brachiopods in the suborder Acrotretidina of the order Acrotretida having an enlarged, tear-shaped, apical pedicle valve.

siphonozooid [INV ZOO] A zooid of certain alcyonarians that lacks tentacles and gonads.

siphon recorder [ENG] A recorder in which a small siphon discharges ink to make the record; used in submarine telegraphy.

siphon spillway [CIV ENG] An enclosed spillway passing over the crest of a dam in which flow is maintained by atmospheric pressure.

siphuncle [INV ZOO] **1.** A honeydew-secreting tube (cornicle) in aphids. **2.** A tubular extension of the mantle extending through all the chambers to the apex of a shelled cephalopod.

Siphunculata [INV ZOO] The equivalent name for Anoplura.

siporex [MATER] A building material composed of sand, lime or cement, and aluminum powder which are mixed and cast into molds to be made into roof slabs, door lintels, and wall blocks which give excellent heat and sound insulation.

Sipunculida [INV ZOO] A phylum of marine worms which dwell in burrows, secreted tubes, or adopted shells; the mouth and anus occur close together at one end of the elongated body, and the jawless mouth, surrounded by tentacles, is situated in an eversible proboscis.

Sipunculoidea [INV ZOO] An equivalent name for Sipunculida.

siren [ENG ACOUS] An apparatus for generating sound by the mechanical interruption of the flow of fluid (usually air) by a perforated disk or cylinder.

Sirenia [VERT ZOO] An order of aquatic placental mammals which include the living manatees and dugongs; these are nearly hairless, thick-skinned mammals without hindlimbs and with paddlelike forelimbs.

Siricidae [INV ZOO] The horntails, a family of the Hymenoptera in the superfamily Siricoidea; females use a stout, hornlike ovipositor to deposit eggs in wood.

Siricoidea [INV ZOO] A superfamily of wasps of the suborder Symphala in the order Hymenoptera.

siriometer [ASTRON] A unit of length, formerly used in astronomical measurement, equal to 10^6 astronomical units, or 1.496×10^{17} meters.

Sirius [ASTRON] The brightest-appearing star in the sky; 8.7 light-years from the sun, spectral class A1V; it has a white dwarf companion. Also known as Dog Star.

sirocco [METEOROL] A warm south or southeast wind in advance of a depression moving eastward across the southern Mediterranean Sea or North Africa.

SIRS See satellite infrared spectrometer.

sisal [BOT] Agave sisalina. An agave of the family Amaryllidaceae indigenous to Mexico and Central America; a coarse, stiff yellow fiber produced from the leaves is used for making twine and brush bristles.

sisal-hemp wax [MATER] A hard wax derived from sisal waste; melts at 63°C, decomposes at 95°C. Also known as sisal wax.

sisal wax See sisal-hemp wax.

sister chromatids [CYTOL] The two daughter strands of a chromosome after it has duplicated.

sister hook [DES ENG] 1. Either of a pair of hooks which can be fitted together to form a closed ring. 2. A pair of such hooks.

SIT See self-induced transparency.

site [ADP] 1. A position available for the symbols of an inscription, for example, a digital place. 2. A location on a tally that can bear either a mark or a blank; for example, a location that can be punched or left unpunched on a card. [ENG] Position of anything; for example, the position of a gun emplacement.

sitting height [ANTHRO] A measure of the vertical distance (taken along the back) from the table surface to the crest of the head as the subject sits erectly on the table, knees pressed against the table edge and head in the eye-ear horizontal plane.

situation-display tube [ELECTR] Large cathode-ray tube used to display tabular and vector messages pertinent to the various functions of an air defense mission.

situation map [MAP] A map showing the tactical or the administrative situation at a particular time.

situation therapy See milieu therapy.

situs inversus [MED] Reversed location or position.

six-j-symbol [QUANT MECH] A coefficient that appears in the transformation between various modes of coupling eigenfunctions of three angular momenta; it is equal to the Racah coefficient, except perhaps in sign, and has greater symmetry than the Racah coefficient.

six-phase circuit [ELEC] Combination of circuits energized by alternating electromotive forces which differ in phase by $\frac{1}{6}$ of a cycle (60°).

six-phase rectifier [ELECTR] A rectifier in which transformers are used to produce six alternating electromotive forces which differ in phase by one-sixth of a cycle, and which feed six diodes.

Six's thermometer [ENG] A combination maximum thermometer and minimum thermometer; the tube is shaped in the form of a U with a bulb at either end; one bulb is filled with creosote which expands or contracts with temperature variation, forcing before it a short column of mercury having iron indexes at either end; the indexes remain at the extreme positions reached by the mercury column, thus indicating the maximum and minimum temperatures; the indexes can be reset with the aid of a magnet.

six-tenths factor [IND ENG] An empirical relationship between the cost and the size of a manufacturing facility; as size increases, cost increases by an exponent of six-tenths, that is $cost_1/cost_2 = (size_1/size_2)^{0.6}$.

sixty degrees Fahrenheit British thermal unit See British thermal unit.

six-vector [RELAT] An antisymmetrical, second-rank tensor in Minkowski space; that is, a tensor whose components, $T_{\mu\nu}$, with $\mu,\nu = 1,2,3,4$ satisfy $T_{\mu\nu} = -T_{\nu\mu}$; it has six independent components.

size [MATER] Materials used to surface-treat textiles, papers, and leathers; some of these materials are starch, gelatins, casein, water-soluble gums, and waxes.

size analysis See particle-size analysis.

size block See gage block.

size classification See sizing.

size control [ELECTR] A control provided on a television receiver for changing the size of a picture either horizontally or vertically.

size dimension [DES ENG] In dimensioning, a specified value of a diameter, width, length, or other geometrical characteristic directly related to the size of an object.

size effect [MET] The effect of the size of a piece of metal on its properties and manufacturing variables; in general, mechanical properties are lower for a larger size.

size enlargement [CHEM ENG] Making large particles out of small ones by crystallization, particle cementation, tableting, briquetting, agglomeration, flocculation, melting, casting, compaction and extrusion, and sintering or nodulizing.

size-frequency analysis See particle-size analysis.

size of weld [MET] 1. The joint penetration in a groove weld. 2. The lengths of the nominal legs of a fillet weld.

size reduction [MECH ENG] The breaking of large pieces of coal, ore, or stone by a primary breaker, or of small pieces by grinding equipment.

sizing [ENG] 1. Separating an aggregate of mixed particles into groups according to size, using a series of screens. Also known as size classification. 2. See sizing treatment. [MECH ENG] A finishing operation to correct surfaces and shapes to meet specified dimensions and tolerances. [MET] Final pressing of a metal powder compact after sintering.

sizing screen [DES ENG] A mesh sheet with standard-size apertures used to separate granular material into classes according to size; the Tyler standard screen is an example.

sizing treatment [ENG] Also known as sizing; surface sizing. 1. Application of material to a surface to fill pores and thus reduce the absorption of subsequently applied adhesive or coating; used for textiles, paper, and other porous materials. 2. Surface-treatment applied to glass fibers used in reinforced plastics.

sjogrenite [MINERAL] $Mg_6Fe_2(OH)_{16}(CO_3) \cdot 4H_2O$ A hexagonal mineral composed of hydrous basic magnesium iron carbonate.

Sjögren's syndrome [MED] A complex of symptoms including keratoconjunctivitis sicca, laryngopharyngitis sicca, rhinitis sicca, enlargement of the parotid gland, and polyarthritis. Also known as xerodermosteosis.

SK See Stefan number.

skarn [GEOL] A lime-bearing silicate derived from nearly pure limestone and dolomite with the introduction of large amounts of silcon, aluminum, iron, and magnesium.

skate [VERT ZOO] Any of various batoid elasmobranchs in the family Rajidae which have flat bodies with winglike pectoral fins and a slender tail with two small dorsal fins.

skatole [ORG CHEM] C_9H_9N A white, crystalline compound that melts at 93–95°C, dissolves in hot water, and has an unpleasant feceslike odor. Also known as 3-methylindole.

skavl See sastruga.

skeet gun [ORD] A classification of shotguns which includes

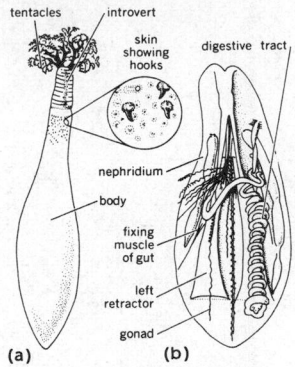

SIPUNCULIDA

Two examples of Sipunculida.
(a) Dendrostoma petraeum.
(b) Dendrostomum pyroides.
(After Chamberlin)

SIRICIDAE

Drawing of a horntail. (From T. I. Storer and R. L. Usinger. General Zoology, 3d ed., McGraw-Hill, 1957)

SISAL

Flower of the West Indian agave (Agave sisalina).

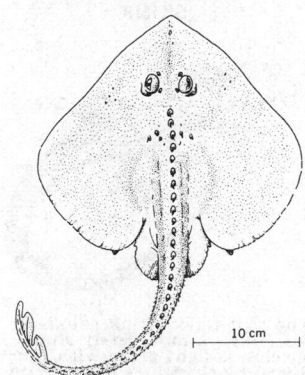

SKATE

Skate (Raja). (From H. B. Bigelow and W. C. Schroeder, Fishes of the Western North Atlantic, pt. 2, 1953)

those with 26-inch (66 centimeter) improved cylinder barrels.

skeg [NAV ARCH] An extension of the stern end of the keel, which in some cases passes under the propeller and supports the rudder post.

skeletal coding [ADP] A set of incomplete instructions in symbolic form, intended to be completed and specialized by a processing program written for that purpose.

skeletal muscle [ANAT] A striated, voluntary muscle attached to a bone and concerned with body movements.

skeletal system [ANAT] Structures composed of bone or cartilage or a combination of both which provide a framework for the vertebrate body and serve as attachment for muscles.

skeleton crystal [CRYSTAL] A crystal formed in microscopic outline with incomplete filling in of the faces.

skeleton framing [BUILD] Framing in which steel framework supports all the gravity loading of the structure; this system is used for skyscrapers.

skeleton texture [PETR] Descriptive of the texture of limestone that consists of an in-place accumulation of skeletal material, that is, the hard parts secreted by organisms.

skelic index [ANTHRO] The ratio, multiplied by 100, of the length of the leg to that of the trunk.

skelp [MET] A strip or sheet of steel which will be rolled and welded to form a tube.

skew [ADP] In character recognition, a condition arising at the read station whereby a character or a line of characters appears in a "twisted" manner in relation to a real or imaginary horizontal baseline. [ELECTR] 1. The deviation of a received facsimile frame from rectangularity due to lack of synchronism between scanner and recorder; expressed numerically as the tangent of the angle of this deviation. 2. The degree of nonsynchronism of supposedly parallel bits when bit-coded characters are read from magnetic tape. [MECH ENG] Gearing whose shafts are neither interesecting nor parallel. [SCI TECH] Deviating from rectangularity or a straight line.

skew back [CIV ENG] The beveled or inclined support at each end of a segmental arch.

skew bridge [CIV ENG] A bridge which spans a gap obliquely and is therefore longer than the width of the gap.

skewed density function [STAT] A density function which is not symmetrical, and which depends not only on the magnitude of the difference between the average value and the variate, but also on the sign of this difference.

skew failure [ADP] In character recognition, the condition that exists during document alignment whereby the document reference edge is not parallel to that of the read station.

skew field [MATH] A ring whose nonzero elements form a non-Abelian group with respect to the multiplicative operation.

skew Hermitian matrix [MATH] A square matrix which equals the negative of its adjoint.

skew level gear [DES ENG] A level gear whose axes are not in the same place.

skew lines [MATH] Lines which do not lie in the same plane in euclidean three-dimensional space.

skew product [MATH] A multiplicative operation or structure induced upon a cartesian product of sets, where each has some algebraic structure.

skew surface [MATH] A ruled surface whose total curvature vanishes everywhere.

skew-symmetric matrix *See* antisymmetric matrix.

skew-symmetric tensor [MATH] A tensor where interchanging two indices will only change the sign of the corresponding component.

skialith [PETR] A vague remnant of country rock assimilated in granite.

skiascope [OPTICS] An instrument used to study optical refraction within the eye.

skiatron *See* dark-trace tube.

skid [AERO ENG] The metal bar or runner used as part of the landing gear of helicopters and planes. [ENG] 1. A device attached to a chain and placed under a wheel to prevent its turning when descending a steep hill. 2. A timber, bar, rail, or log placed under a heavy object when it is being moved over bare ground. [MECH ENG] A brake for a power machine.

[MIN ENG] An arrangement upon which certain coal-cutting machines travel along the working faces.

skidding [FOR] The short distance movement of tree lengths or segments over unimproved terrain to loading points on transportation routes.

skid-mounted [ENG] Equipment or processing systems mounted on a portable platform.

skidway [FOR] A platform, usually inclined, for mounting logs for loading and sawing.

skill score [METEOROL] In synoptic meteorology, an index of the degree of skill of a set of forecasts, expressed with reference to some standard such as forecasts based upon chance, persistence, or climatology.

skim coat [BUILD] A finish coat of plaster composed of lime putty and fine white sand.

skim gate [MET] A gate used to prevent slag and other undesirable materials from passing into the casting.

skimmer [VERT ZOO] Any of various ternlike birds, members of the Rhynchopidae, that inhabit tropical waters around the world and are unique in having the knifelike lower mandible substantially longer than the wider upper mandible.

skimming plant [CHEM ENG] A petroleum refinery designed to remove and finish only the lighter constituents of crude oil, such as gasoline and kerosine; the heavy ends are sold as fuel oil or for further processing elsewhere.

skin [AERO ENG] The covering of a body, such as the covering of a fuselage, a wing, a hull, or an entire aircraft. [ANAT] The external covering of the vertebrate body, consisting of two layers, the outer epidermis and the inner dermis. [BUILD] The exterior wall of a building. [ENG] In flexible bag molding, a protective covering for the mold; it may consist of a thin piece of plywood or a thin hardwood. [MET] A thin outside layer of metal differing in composition, structure, or other characteristics from the main mass of metal but not formed by bonding or electroplating.

skin antenna [ELECTROMAG] Flush-mounted aircraft antenna made by using insulating material to isolate a portion of the metal skin of the aircraft.

skin depth [ELECTROMAG] The depth beneath the surface of a conductor, which is carrying current at a given frequency due to electromagnetic waves incident on its surface, at which the current density drops to one neper below the current density at the surface.

skin diving [ENG] Diving without breathing apparatus, using fins and faceplate only.

skin effect [ELEC] The tendency of alternating currents to flow near the surface of a conductor thus being restricted to a small part of the total sectional area and producing the effect of increasing the resistance. Also known as conductor skin effect; Kelvin skin effect. [PETRO ENG] The restriction to fluid flow through a reservoir adjacent to the borehole; calculated as a factor of reservoir pressure, product rate, formation volume and thickness, porosity, and other related parameters.

skin erythema dose [NUCLEO] A unit of radioactive dose resulting from exposure to electromagnetic radiation, equal to the dose which slightly reddens or browns the skin of 80% of all persons within 3 weeks after exposure; it is approximately 1000 roentgens for gamma rays, 600 roentgens for x-rays. Abbreviated SED.

skin friction [FL MECH] A type of friction force which exists at the surface of a solid body immersed in a much larger volume of fluid which is in motion relative to the body.

skin-friction coefficient [METEOROL] A dimensionless drag coefficient expressing the proportionality between the frictional force per unit area, or the shearing stress exerted by the wind at the earth's surface, and the square of the surface wind speed.

skink [VERT ZOO] Any of numerous small- to medium-sized lizards comprising the family Scincidae with a cylindrical body; short, sometimes vestigial, legs; cores of bone in the body scales; and pleurodont dentition.

skin lamination [MET] Surface rupture in flat-rolled metals due to exposure of a subsurface lamination.

skin resistance [ELEC] For alternating current of a given frequency, the direct-current resistance of a layer at the surface of a conductor whose thickness equals the skin depth.

SKIN FRICTION

The thin boundary layer, in which fluid flow is distorted by the passage of the body, fills the region between the body skin and the dashed line. $\delta(x)$ is the thickness of the boundary layer; u_1 is the velocity of fluid relative to the body outside the boundary layer; $u(y)$, the fluid velocity within the boundary at a distance y from the body skin, is graphed by the solid curved line. The friction force per unit surface area of the body, τ_w, equals the rate of deformation of an adjacent fluid element, $(\partial u/\partial y)_w$, times the dynamic viscosity μ.

SKINK

The blue-tailed skink (*Eumeces fasciatus*), a moderately stout species, is light green with four dorsal black stripes and one on each side between white ones.

[NAV ARCH] The frictional resistance existing between the shell or skin of a ship and the water flowing over it.

skin test [IMMUNOL] A procedure for evaluating immunity status involving the introduction of a reagent into or under the skin.

skintle [CIV ENG] To set bricks in an irregular fashion so that they are out of alignment with the face by $\frac{1}{4}$ inch (6 millimeters) or more.

skin tracking [ELECTROMAG] Tracking of an object by means of radar without using a beacon or other signal device on board the object being tracked.

skiograph [ELECTR] An instrument used to measure the intensity of x-rays.

skiou *See* morvan.

skip [ADP] In fixed-instruction-length digital computers, to bypass or ignore one or more instructions in an otherwise sequential process. [MECH ENG] *See* skip hoist.

skip bombing [ORD] Releasing one or more bombs from a plane flying at a low altitude, so that the bomb or bombs glance off the surface of the water or ground and strike the target.

skip cast [GEOL] The cast of a skip mark.

skip chain [ADP] A programming technique which matches a word against a set of test words; if there is a match, control is transferred (skipped) to a routine, otherwise the word is matched with the next test word in sequence.

skip distance [ELECTROMAG] The minimum distance that radio waves can be transmitted between two points on the earth by reflection from the ionosphere, at a specified time and frequency.

skip effect [COMMUN] The existence of a ring-shaped area around a radio transmitter within which no radio signals are received, because ground signals are received only inside the ring and sky-wave signals are received only outside the ring.

skip-fading [ELECTROMAG] Fading due to fluctuations of ionization density at the place in the ionosphere where the wave is reflected which causes the skip distance to increase or decrease.

skip flag [ADP] The 35th bit of a channel command word which suppresses the transfer of data to main storage.

skip hoist [MECH ENG] A basket, bucket, or open car mounted vertically or on an incline on wheels, rails, or shafts and hoisted by a cable; used to raise materials. Also known as skip.

skipjack *See* bluefish.

skip-keying [ELECTR] Reduction of radar pulse repetition frequency to submultiple of that normally used, to reduce mutual interference between radar or to increase the length of radar time base.

skip logging [ENG] A phenomenon during acoustical (sonic) logging in which the acoustical energy is attenuated by low-elasticity formations and lacks the energy to trip the second sonic receiver (skips a cycle). Also known as cycle skip.

skip mark [GEOL] A crescent-shaped mark that is one of a linear pattern of regularly spaced marks made by an object that skipped along the bottom of a stream.

skip shaft [MIN ENG] A mine shaft prepared for hauling a skip.

skip trajectory [MECH] A trajectory made up of ballistic phases alternating with skipping phases; one of the basic trajectories for the unpowered portion of the flight of a reentry vehicle or spacecraft reentering earth's atmosphere.

skip vehicle [AERO ENG] A reentry body which climbs after striking the sensible atmosphere in order to cool the body and to increase its range.

skip zone [ACOUS] A region in the air surrounding a source of sound in which no sound is heard, although the sound becomes audible at greater distances. Also known as zone of silence. [COMMUN] The area between the outer limit of reception of radio high-frequency ground waves and the inner limit of reception of sky waves, where no signal is received.

skiron [METEOROL] The Greek name for the northwest wind, which is cold in winter but hot and dry in summer.

skirting plate [ORD] A thin plate which is placed a considerable distance in front of the main armor plate and which acts as a passive form of resistance to the jet of shaped-charge ammunition.

skirt roof [BUILD] A false band of roofing projecting from between the stories of a building.

skiver [MATER] Thin, soft leather made from the grain side of a split sheepskin.

skiving [MECH ENG] **1.** Removal of material in thin layers or chips with a high degree of shear or slippage of the cutting tool. **2.** A machining operation in which the cut is made with a form tool with its face at an angle allowing the cutting edge to progress from one end of the work to the other as the tool feeds tangentially past ten rotating workpieces.

skot [OPTICS] A unit of luminance, used particularly to measure low-level luminance, equal to 10^{-3} apostilb, or $10^{-3}/\pi$ nit.

Skraup synthesis [ORG CHEM] A method for the preparation of commercial synthetic quinoline by heating aniline and glycerol in the presence of sulfuric acid and an oxidizing agent to form pyridine unsubstituted quinolines.

skull [ANAT] The bones and cartilages of the vertebrate head which form the cranium and the face. [MET] A layer of solidified metal or dross left in the pouring vessel after the molten metal has been poured.

skull cracker [ENG] A heavy iron or steel ball that can be swung freely or dropped by a derrick to raze buildings or to compress bulky scrap. Also known as wrecking ball.

skull crucible [MET] A consumable-electrode vacuum arc melting and casting furnace; used in the production of turbine buckets for aircraft jet engines using a nickel-base high-temperature alloy.

skunk [VERT ZOO] Any one of a group of carnivores in the family Mustelidae characterized by a glossy black and white coat and two musk glands at the base of the tail.

skutterudite [MINERAL] $(Co,Ni)As_3$ A tin-white mineral with metallic luster composed of cobalt and nickel arsenides; crystallizes in the isometric system but commonly is massive; hardness is 5.5–6 on Mohs scale, and specific gravity is 6.6; it is a minor ore of cobalt and nickel.

sky [ASTRON] In the daytime the apparent blue dome resting on the earth along the horizon circle; at night the blue becomes nearly black.

Skybolt [ORD] United States air-to-ground air-launched ballistic missile; designed for a range of 1000 miles (1600 kilometers); it is a two-stage, solid-propellant missile with hypersonic velocity.

sky compass [NAV] A type of astro compass, designed for use in the Arctic during long periods of twilight, which utilizes the polarization of sunlight in the sky; it operates whenever the zenith is clear whether or not the sun is visible; when the sun is more than $6°5$ below or $10°$ above the horizon, readings are uncertain.

sky cover [METEOROL] In surface weather observations, a term used to denote the amount of sky covered but not necessarily concealed by clouds or by obscuring phenomena aloft, the amount of sky concealed by obscuring phenomena that reach the ground, or the amount of sky covered or concealed by a combination of the two phenomena.

sky diagram [ASTRON] A diagram of the heavens, indicating the apparent positions of various celestial bodies with reference to the horizon system of coordinates.

Skyhawk [AERO ENG] A United States single-engine, turbojet attack aircraft designed to operate from aircraft carriers, and capable of delivering nuclear or nonnuclear weapons, providing troop support, or conducting reconnaissance missions; it can act as a tanker, and can itself be air refueled; it possesses a limited all-weather attack capability, and can operate from short, unprepared fields.

skyhook [MIN ENG] To drive bolts into the overhead rock of a mine in order to reinforce the ceiling.

skyhook balloon [AERO ENG] A large plastic constant-level balloon for duration flying at very high altitudes, used for determining wind fields and measuring upper-atmospheric parameters.

skylight [ASTROPHYS] *See* diffuse sky radiation. [ENG] An opening in a roof or ship deck that is covered with glass or plastic and designed to admit daylight.

skyline yarding [FOR] A technique that uses a skyline cable

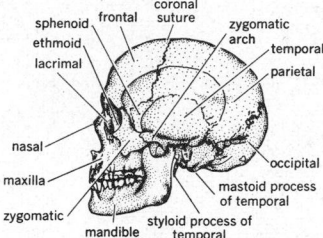

SKULL

Lateral view of human skull. *(From W. T. Foster, Anatomy, Foster Art Service)*

SKUNK

The spotted skunk (*Spilogale putorius*).

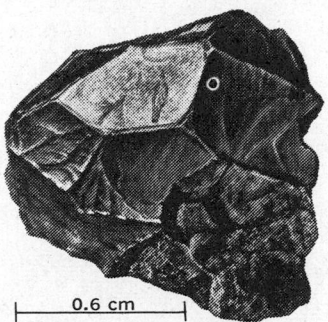

SKUTTERUDITE

Skutterudite crystal from Scutterude, Norway. *(Specimen from Department of Geology, Bryn Mawr College)*

system to transport logs from a forest to a clear space along a road for trucking to a sawmill; it requires fewer roads than older cable systems, suspends logs off the ground to reduce erosion, and reduces damage to young trees that are left.

sky map [ASTRON] A planar representation of areas of the sky showing positions of celestial bodies. [METEOROL] A pattern of variable brightness observable on the underside of a cloud layer, and caused by the different reflectivities of material on the earth's surface immediately beneath the clouds; this term is used mainly in polar regions.

sky noise [ELECTROMAG] Noise produced by radio energy from stars.

sky radiation See diffuse sky radiation.

Skyraider [AERO ENG] A United States single reciprocating-engine, general-purpose attack aircraft designed to operate from aircraft carriers; it is capable of relatively long-range, low-level nuclear and nonnuclear weapons delivery, minelaying, reconnaissance, torpedo delivery, and troop support. Designated A-1.

Skyray [AERO ENG] A United States single-engine, single-pilot, supersonic, limited all-weather jet fighter designed for operating from aircraft carriers for interception and destruction of enemy aircraft; armament includes the Sidewinder. Designated F-6.

skyscraper [BUILD] A very tall, multistory building.

Skywarrior [AERO ENG] A United States twin-engine, turbojet, tactical, all-weather attack aircraft designed to operate from aircraft carriers, and capable of delivering nuclear or nonnuclear weapons, conducting reconnaissance, or minelaying missions; its range can be extended by in-flight refueling, and it has a crew of four. Designated A-3.

sky wave [ELECTROMAG] A radio wave that travels upward into space and may or may not be returned to earth by reflection from the ionosphere. Also known as ionospheric wave.

sky-wave correction [ELECTR] The correction to be applied to the time difference readings of received sky waves to convert them to an equivalent ground-wave reading.

sky-wave-synchronized loran [NAV] A type of loran in which the transmitting stations are synchronized by signals reflected from the ionosphere; used to obtain greater range and more accurate nighttime navigation. Abbreviated ss loran.

sky-wave transmission delay [ELECTROMAG] Amount by which the time of transit from transmitter to receiver of a pulse carried by sky waves reflected once from the E layer exceeds the time of transit of the same pulse carried by ground waves.

slab [CIV ENG] That part of a reinforced concrete floor, roof, or platform which spans beams, columns, walls, or piers. [ELECTR] A relatively thick-cut crystal from which blanks are obtained by subsequent transverse cutting. [ENG] The outside piece cut from a log when sawing it into boards. [GEOL] A cleaved or finely parallel jointed rock, which splits into tabular plates from 1 to 4 inches (2.5 to 10 centimeters) thick. Also known as slabstone. [HYD] A layer in, or the whole-thickness of, a snowpack that is very hard and has the ability to sustain elastic deformation under stress. [MATER] A thin piece of concrete or stone. [MET] A piece of metal, intermediate between ingot and plate, with the width at least twice the thickness. [MIN ENG] A slice taken off the rib of an entry or room in a mine.

slabbing machine [MIN ENG] A coal-cutting machine designed to make cuts in the side of a room or entry pillar preparatory to slabbing.

slabbing method [MIN ENG] A method of mining pillars in which successive slabs are cut from one side or rib of the pillar after a room is finished, until as much of the pillar is removed as can safely be recovered.

slabbing mill [MET] A steel rolling mill for making slabs.

slab cutter See plain milling cutter.

slab entry [MIN ENG] An entry which is widened or slabbed to provide a working place for a second miner.

slab oil [MATER] White petroleum-based oil used by candymakers and bakers to oil the slab or surface on which the candy or pastry is worked.

slab pahoehoe [GEOL] A pahoehoe whose surface consists of a jumbled arrangement of slabs of flow crust.

slabstone See slab.

slack [ENG] Looseness or play in a mechanism, as the play in the trigger of a small-arms weapon.

slack barrel [PETRO ENG] A petroleum-industry container used for shipment of petroleum paraffin; generally contains 235 to 245 pounds (107 to 111 kilograms) net, and is of lighter construction than the ordinary oil barrel but of the same general shape.

slack ice [HYD] Ice fragments on still or slow-moving water.

slackline cableway [MECH ENG] A machine, widely used in sand-and-gravel plants, employing an open-ended dragline bucket suspended from a carrier that runs upon a track cable, which can dig, elevate, and convey materials in one continuous operation.

slack water [OCEANOGR] The interval when the speed of the tidal current is very weak or zero; usually refers to the period of reversal between ebb and flood currents.

slack wax [MATER] A soft, oily, crude wax obtained from the pressing of petroleum paraffin distillate or wax distillate.

slag [MET] A nonmetallic product resulting from the interaction of flux and impurities in the smelting and refining of metals.

slag cement [MATER] Cement produced by grinding blast-furnace slag and mixing it with lime, portland cement, or dehydrated gypsum.

slagging [MET] Freeing from or converting into slag.

slag inclusion [MET] Slag entrapped in solidified metal.

slag sand [MATER] Slag that has been finely crushed for use in mortar and concrete.

slaked lime See calcium hydroxide.

slaking [GEOL] 1. Crumbling and disintegration of earth materials when exposed to air or moisture. 2. The breaking up of dried clay when saturated with water.

slamming [NAV ARCH] The impact of the bottom of a ship's bow hitting the water during a severe downward pitch.

slant [MIN ENG] 1. Any short, inclined crosscut connecting the entry with its air course to facilitate the hauling of coal. Also known as shoofly. 2. A heading driven diagonally between the dip and the strike of a coal seam. Also known as run.

slant culture [MICROBIO] A method for maintaining bacteria in which the inoculum is streaked on the surface of agar that has solidified in inclined glass tubes.

slant depth [DES ENG] The distance between the crest and root of a screw thread measured along the angle forming the flank of the thread.

slant distance [NAV] The straight-line distance between two points not at the same elevation as contrasted with ground distance, ground range, or the great-circle distance between the two positions. Also known as slant range.

slant plane [ORD] In antiaircraft artillery, the plane containing the target course line and the pintle center of the gun.

slant range See slant distance.

slant visibility See oblique visual range.

slashing [TEXT] The stiffening of warp yarn before the spinning operation, using a solution of oil and starch.

slash-sawing See backsawing.

slat [AERO ENG] A movable auxiliary airfoil running along the leading edge of a wing, remaining against the leading edge in normal flight conditions, but lifting away from the wing to form a slot at certain angles of attack.

slat conveyor [MECH ENG] A conveyor consisting of horizontal slats on an endless chain.

slate [PETR] A group name for various very-fine-grained rocks derived from mudstone, siltstone, and other clayey sediment as a result of low-degree regional metamorphism; characterized by perfect fissility or slaty cleavage which is a regular or perfect planar schistosity.

Slater determinant [QUANT MECH] A quantum mechanical wave function for n fermions, which is an $n \times n$ determinant whose entries are n different one-particle wave functions depending on the coordinates of each of the particles in the system.

Slater's rule [ELECTR] The ratio of the cathode radius to the anode radius of a magnetron is approximately equal to

$(N - 4)/(N + 4)$, where N is the number of resonators.

slaty cleavage   See flow cleavage.

slave antenna   [ELECTROMAG] A directional antenna positioned in azimuth and elevation by a servo system; the information controlling the servo system is supplied by a tracking or positioning system.

slave clock   [HOROL] The part of a Shortt clock which consists of a complete clock with pendulum and indicating mechanism, and which receives trigger pulses from a master clock as synchronizing signals.

slaved gyro magnetic compass   [NAV] A directional gyro compass with an input from a flux valve to keep the gyro oriented to magnetic north.

slavery   [INV ZOO] An interspecific association among ants in which members of one species bring pupae of another species to their nest, which, when adult, become slave workers in the colony.

slave station   [NAV] In a radio navigation system such as Loran, the transmitting station controlled or triggered by the signal received from the master station.

slavikite   [MINERAL] $MgFe_3^{3+}(SO_4)_4(OH)_3 \cdot 18H_2O$ A greenish-yellow mineral composed of hydrous basic magnesium ferric sulfate and occurring as rhombohedral crystals.

sled   [ENG] An item equipped with runners and a suitable body designed to transport loads over ice and snow.

sledgehammer   [DES ENG] A large heavy hammer that is usually wielded with two hands; used for driving stakes or breaking stone.

sleep   [PHYSIO] A state of rest in which consciousness and activity are diminished.

sleeper   [CIV ENG] A timber, steel, or precast concrete beam placed under rails to hold them at the correct gage.

sleeping sickness   See African sleeping sickness; encephalitis lethargica.

sleep paralysis   [MED] Transient paralysis with spontaneous recovery occurring on falling asleep or on awakening.

sleep therapy   See narcosis therapy.

sleet   [METEOROL] Colloquially in some parts of the United States, precipitation in the form of a mixture of snow and rain.

sleeve   [ELEC] 1. The cylindrical contact that is farthest from the tip of a phone plug. 2. Insulating tubing used over wires or components. Also known as sleeving. [ENG] A cylindrical part designed to fit over another part.

sleeve antenna   [ELECTROMAG] A single vertical half-wave radiator, the lower half of which is a metallic sleeve through which the concentric feed line runs; the upper radiating portion, one quarter-wavelength long, connects to the center of the line.

sleeve bearing   [MECH ENG] A machine bearing in which the shaft turns and is lubricated by a sleeve.

sleeve brick   [MATER] Tube-shaped firebrick for lining slag vents.

sleeve burner   [ENG] A type of oil burner for domestic heating.

sleeve coupling   [DES ENG] A hollow cylinder which fits over the ends of two shafts or pipes, thereby joining them.

sleeve dipole antenna   [ELECTROMAG] Dipole antenna surrounded in its central portion by a coaxial cable.

sleeve valve   [MECH ENG] An admission and exhaust valve on an internal-combustion engine consisting of one or two hollow sleeves that fit around the inside of the cylinder and move with the piston so that their openings align with the inlet and exhaust ports in the cylinder at proper stages in the cycle.

sleeving   See sleeve.

sleigh   [ORD] Part of a gun carriage which supports the recoil mechanism and barrel of the gun and slides with the gun on recoil, guiding it in runways in the cradle.

slenderness ratio   [AERO ENG] A configuration factor expressing the ratio of a missile's length to its diameter. [CIV ENG] The ratio, L/r, of the length of a column L to the radius of gyration r about the principal axes of the section.

slewing   [ENG] Moving a radar antenna or a sonar transducer rapidly in a horizontal or vertical direction, or both. [NAV] In sea ice navigation, the act of forcing a ship through ice by pushing apart adjoining ice floes.

slewing mechanism   [ENG] Device which permits rapid traverse or change in elevation of a weapon or instrument.

slewing motor   [ELEC] A motor used to drive a radar antenna at high speed for slewing to pick up or track a target.

slice   [GEOL] An arbitrary section of some uniform standard, such as thickness of a stratigraphic unit that is otherwise indivisible for purposes of analytic study. [MIN ENG] 1. A thin broad piece cut off, as a portion of ore cut from a pillar or face. 2. To remove ore by successive slices.

slice bar   [ENG] A broad, flat steel blade used for chipping and scraping.

slice drift   [MIN ENG] In sublevel caving, the crosscuts driven between every other slice from 18 to 36 feet (5.5 to 11.0 meters) apart.

slice method   [METEOROL] A method of evaluating the static stability over a limited area at any reference level in the atmosphere; unlike the parcel method, the slice method takes into account continuity of mass by considering both upward and downward motion.

slicer   See amplitude gate; slitting mill.

slicer amplifier   See amplitude gate.

slicing   [ELECTR] Transmission of only those portions of a waveform lying between two amplitude values.

slicing method   [MIN ENG] Removal of a horizontal layer from a massive ore body.

slick   [OCEANOGR] Area in which capillary waves are absent or suppressed.

slickens   [GEOL] A layer of fine silt deposited by a flooding stream. [MIN ENG] The light earth removed by sluicing in hydraulic mining.

slickenside   [GEOL] A surface that is polished and smoothly striated and results from slippage along a fault plane.

slickolite   [GEOL] A vertically discontinuous slip-scratch surface made by slippage and shearing and developed on sharply dipping bedding planes of limestone that shapes the wall of a solution cavity.

slick paper proof   See reproduction proof.

slide   [ENG] 1. A sloping chute with a flat bed. 2. A sliding mechanism. [GEOL] 1. A vein of clay intersecting and dislocating a vein vertically, or the vertical dislocation itself. 2. A rotational or planar mass movement of earth, snow, or rock resulting from failure under shear stress along one or more surfaces. [MECH ENG] The main reciprocating member of a mechanical press, guided in a press frame, to which the punch or upper die is fastened. [MIN ENG] 1. An upright rail fixed in a shaft with corresponding grooves for steadying the cages. 2. A trough used to guide and to support rods in a tripod when drilling an angle hole. Also known as rod slide. [ORD] 1. Sliding part of the receiver of certain automatic weapons. 2. Sliding catch on the breech mechanism of certain weapons.

slide-back voltmeter   [ELECTR] An electronic voltmeter in which an unknown voltage is measured indirectly by adjusting a calibrated voltage source until its voltage equals the unknown voltage.

slide conveyor   [ENG] A slanted, gravity slide for the forward downward movement of flowable solids, slurries, liquids, or small objects.

slide gate   [CIV ENG] A crest gate which has high frictional resistance to opening because it slides on its bearings in opening and closing.

slide projector   See optical lantern.

slider   [ELEC] Sliding type of movable contact. [ORD] A fuse or exploder component that interrupts the explosive train when the device is in the unarmed condition, and that moves during arming in such a way as to render the explosive train operative.

slide rail   See guard rail.

slider coupling   [MECH ENG] A device for connecting shafts that are laterally misaligned. Also known as double-slider coupling; Oldham coupling.

slide rule   [MATH] A mechanical device, composed of a ruler with sliding insert, marked with various number scales, which facilitates such calculations as division, multiplication, finding roots, and finding logarithms.

slide-rule dial   [ENG] A dial in which a pointer moves in a straight line over long straight scales resembling the scales of a slide rule.

slide valve   [MECH ENG] A sliding mechanism to cover and

SLIDE RULE

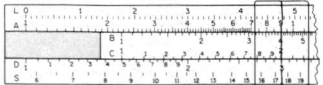

Left end of slide rule, showing position of sliding insert for multiplication of 1.5 by 2.

uncover ports for the admission of fluid, as in some steam engines.

slide-wire bridge [ELEC] A bridge circuit in which the resistance in one or more branches is controlled by the position of a sliding contact on a length of resistance wire stretched along a linear scale.

slide-wire potentiometer [ELEC] A potentiometer (variable resistor) which employs a movable sliding connection on a length of resistance wire.

sliding *See* gravitational sliding.

sliding-block linkage [MECH ENG] A mechanism in which a crank and sliding block serve to convert rotary motion into translation, or vice versa.

sliding-chain conveyor [MECH ENG] A conveying machine to handle cases, cans, pipes, or similar products on the plain or modified links of a set of parallel chains.

sliding contact *See* wiping contact.

sliding fit [DES ENG] A fit between two parts that slide together.

sliding friction [MECH] Rubbing of bodies in sliding contact.

sliding gear [DES ENG] A change gear in which speed changes are made by sliding gears along their axes, so as to place them in or out of mesh.

sliding-gear transmission [MECH ENG] A transmission system utilizing a pair of sliding gears.

sliding pair [MECH ENG] Two adjacent links, one of which is constrained to move in a particular path with respect to the other; the lower, or closed, pair is completely constrained by the design of the links of the pair.

sliding scale [METEOROL] A set of combinations of ceilings and visibilities which constitute the operational weather limits at an airport; as the observed value of one element increases, the limiting value of the other element decreases, and vice versa.

sliding-vane compressor [CHEM ENG] A rotary-element gas compressor in which spring-loaded sliding vanes (evenly spaced around a cylinder off-center in a surrounding chamber) pick up, compress, and discharge gas as the cylinder revolves.

sliding way [CIV ENG] One of the timbers which form the upper part of the cradle supporting a ship during its construction, and which slide over the ground ways with the ship when it is launched.

slime [ENG] Liquid slurry of very fine solids with slime- or mudlike appearance. Also known as mud; pulp; sludge.

slime bacteria [MICROBIO] The common name for bacteria in the order Myxobacterales, so named for the layer of slime deposited behind cells as they glide on a surface.

slime disease [PL PATH] Any of several diseases of plants characterized by slimy rot of the parts.

slime flux [PL PATH] The fluid or viscous outflow from the bark or wood of a deciduous tree that is indicative of injury or disease.

slime fungus *See* slime mold.

slime gland [ZOO] A glandular structure in many animals producing a mucous material.

slime mold [MYCOL] The common name for members of the Myxomycetes. Also known as slime fungus.

sliming [OCEANOGR] The formation of films of algae on submerged structures.

sling [ENG] A length of rope, wire rope, or chain used for attaching a load to a crane hook.

sling psychrometer [ENG] A psychrometer in which the wet- and dry-bulb thermometers are mounted upon a frame connected to a handle at one end by means of a bearing or a length of chain; the psychrometer may be whirled in the air for the simultaneous measurement of wet- and dry-bulb temperatures.

sling thermometer [ENG] A thermometer mounted upon a frame connected to a handle at one end by means of a bearing or length of chain, so that the thermometer may be whirled by hand.

slip [CIV ENG] A narrow body of water between two piers. [ELEC] **1.** The difference between synchronous and operating speeds of an induction machine. Also known as slip speed. **2.** Method of interconnecting multiple wiring between switching units by which trunk number 1 becomes the first choice for the first switch, trunk number 2 first choice for the second

switch, and trunk number 3 first choice for the third switch, and so on. [ELECTR] Distortion produced in the recorded facsimile image which is similar to that produced by skew but is caused by slippage in the mechanical drive system. [FL MECH] The difference between the velocity of a solid surface and the mean velocity of a fluid at a point just outside the surface. [GEOL] The actual relative displacement along a fault plane of two points which were formerly adjacent on either side of the fault. Also known as actual relative movement; total displacement. [MATER] A suspension of fine clay in water with a creamy consistency, the casting process and in decorating ceramic ware. Also known as slurry. [NAV ARCH] **1.** To part from an anchor by releasing the shackles from the anchor chain. **2.** The reduction in the distance a propeller advances, per unit time, due to yielding of the fluid.

slip additive [MATER] A plastics modifier that acts as an internal lubricant by exuding to the surface of the plastic during and immediately after processing to reduce friction and improve slip.

slipband [CRYSTAL] One of the microscopic parallel lines (Lüders' lines) on the surface of a crystalline material stretched beyond its elastic limit, located at the intersection of the surface with intracrystalline slip planes in the grains of the material. Also known as slip line.

slip bedding [GEOL] Convolute bedding formed as the result of subaqueous sliding.

slip casting [ENG] A process in the manufacture of shaped refractories, cermets, and other materials in which the slip is poured into porous plaster molds.

slip clay [MATER] Clay containing a high percentage of fluxing impurities; easily fusible and used in clayware to produce a natural glaze.

slip cleavage [GEOL] Cleavage that is superposed on slaty cleavage or schistosity, characterized by spaced cleavage with thin tabular bodies of rock between the cleavage planes. Also known as close-joints cleavage; crenulation cleavage; shear cleavage; strain-slip cleavage.

slip direction [CRYSTAL] The crystallographic direction in which the translation of slip occurs.

slip face [GEOL] The steeply sloping leeward surface of a sand dune. Also known as sandfall.

slip flow [FL MECH] A situation in which the mean free path of a gas is between 1 and 65% of the channel diameter; the gas layer next to the channel wall assumes a velocity of slip past the liquid, known as slip flow.

slip fold *See* shear fold.

slip form [CIV ENG] A narrow section of formwork that can be easily removed as concrete placing progresses.

slip forming [ENG] A plastics-sheet forming technique in which some of the sheet is allowed to slip through the mechanically operated clamping rings during stretch-forming operations.

slip friction clutch [MECH ENG] A friction clutch designed to slip when too much power is applied to it.

slip joint [CIV ENG] **1.** Contraction joint between two adjoining wall sections, or at the horizontal bearing of beams, slabs, or precast units, consisting of a vertical tongue fitted into a groove which allows independent movement of the two sections. **2.** A telescoping joint between two parts. [ENG] **1.** A method of laying-up plastic veneers in flexible-bag molding, wherein edges are beveled and allowed to overlap part or all of the scarfed area. **2.** A mechanical union that allows limited endwise movement of two solid items for example, pipe, rod, or duct with relation to each other. [GEOL] *See* shear joint.

slip line *See* slipband.

slip-off slope [GEOL] The long, low, gentle slope on the inside of the downstream face of a stream meander.

slippage [ENG] The leakage of fluid between the plunger and the bore of a pump piston. Also known as slippage loss. [PETRO ENG] The movement of gas past or through a liquid-phase reservoir front; this movement occurs instead of driving the liquid forward; it can exist in gas-drive reservoirs or in gas-lift oil-well bores. [TEXT] A fabric defect on which warp threads slip over filling threads or vice versa as a result of loose weaving or of unevenly matched yarns.

slippage loss [ENG] **1.** Unintentional movement between the faces of two solid objects. **2.** *See* slippage.

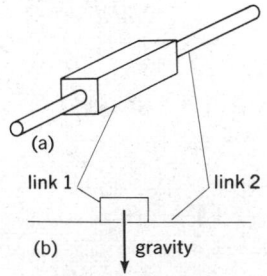

SLIDING PAIR

(a)

link 1   link 2

(b)   gravity

Two types of sliding pair. (a) Motion of link 1 is constrained by the design of the links. (b) Motion of link 1 is constrained by gravity.

slipped disk   *See* herniated disk.

slipper brake   [MECH ENG] **1.** A plate placed against a moving part to slow or stop it. **2.** A plate applied to the wheel of a vehicle or to the track roadway to slow or stop the vehicle.

slipping   [MIN ENG] Enlarging an excavation by breaking one or more walls.

slipping cut   [MIN ENG] A drill-hole pattern used in a wide tunnel face, in which each successive vertical line of shots breaks to the face made by the previous round, so that the relieving cut moves across the end being blasted. Also known as slabbing cut; swing cut.

slip plane   [CRYSTAL] *See* glide plane. [GEOL] A planar slip surface.

slip ratio   [MECH ENG] For a screw propeller, relates the actual advance to the theoretic advance determined by pitch and spin.

slip ring   [ELEC] A conductive rotating ring which, in combination with a stationary brush, provides a continuous electrical connection between rotating and stationary conductors; used in electric rotating machinery, synchros, gyroscopes, and scanning radar antennas.

slip sheet   [GEOL] A stratum or rock on the limb of an anticline that has slid down and away from the anticline; a gravity collapse structure.

slip speed   *See* slip.

slip surface   [GEOL] The displacement surface of a landslide.

slip tongue   [ENG] A pole on a horse-drawn wagon that is fastened by slipping it between two plates connected to the forecarriage.

slip velocity   [FL MECH] The difference in velocities between liquids and solids (or gases and liquids) in the vertical flow of two-phase mixtures through a pipe because of the slip between the two phases.

slipway   [CIV ENG] The space in a shipyard where a foundation for launching ways and keel blocks exists and which is occupied by a ship while under construction. [NAV ARCH] **1.** One of the timbers over which the cradle of a marine railway travels. **2.** A sloping passage in a whaling ship's stern, through which whales are hauled in.

slip-weld hanger   [PETRO ENG] A type of hanger used to suspend the lower strings of an oil-well casing pipe.

slit   [DES ENG] A long, narrow opening through which radiation or streams of particles enter or leave certain instruments.

slit scan   [ADP] In character recognition, a magnetic or photoelectric device that obtains the horizontal structure of an inputted character by vertically projecting its component elements at given intervals.

slitter   [MECH ENG] A synchronized feeder-knife variation of a rotary cutter; used for precision cutting of sheet material, such as metal, rubber, plastics, or paper, into strips.

slitting   [MECH ENG] The passing of sheet or strip material (metal, plastic, paper, or cloth) through rotary knives.

slitting mill   [LAP] A rotating disk used by gem cutters in slitting. Also known as slicer.

sliver   [MATER] A piece of propellant grain of triangular cross section which remains unburned when the web of multiperforated grains has been burned through. [MET] A thin, elongated fragment of metal that has been rolled onto the surface of the parent metal and is attached by only one end. [TEXT] A round, untwisted strand of fiber removed from the carding or combing machine and used to spin yarn.

slop   [CHEM ENG] A petroleum-refinery term for odds and ends of oil produced in the refinery; the slop must be rerun or further processed to make it suitable for use. Also known as slop oil.

slop culture   [BOT] A method of growing plants in which surplus nutrient fluid is allowed to run through the sand or other medium in which the plants are growing.

slope   [GEOL] The inclined surface of any part of the earth's surface. [MATH] **1.** The slope of a line through the points (x_1,y_1) and (x_2,y_2) is the number $(y_2 - y_1)/(x_2 - x_1)$. **2.** The slope of a curve at a point p is the slope of the tangent line to the curve at p. [NAV] The projection of a flight path in the vertical plane.

slope angle   [MATH] The angle of inclination of a line in the plane, where this angle is measured from the positive x-axis to the line in the counterclockwise direction.

slope control   [MET] Electronic production of a change in the welding current within set limits and a selected interval of time.

slope conveyor   [MECH ENG] A troughed belt conveyor used for transporting material on steep grades.

slope course   [ENG] A proving ground facility consisting of a large mound of earth with various sloping sides on which are roads having different grades; this slope course is used to measure the slope performance of military and other vehicles, including maximum speed on various grades, the most suitable gear for best performance, traction, and the holding ability of brakes.

slope deviation   [AERO ENG] The difference between planned and actual slopes of aircraft travel, expressed in either angular or linear measurement.

slope engineer   [MIN ENG] In anthracite and bituminous coal mining, one who operates a hoisting engine to haul loaded and empty mine cars along a haulage road in a mine.

slope failure   [GEOL] The downward and outward movement of a mass of soil beneath a natural slope or other inclined surface; four types of slope failure are rockfall, rock flow, plane shear, and rotational shear.

slope mine   [MIN ENG] A mine opened by a slope or an incline.

slope of fall   [MECH] Ratio between the drop of a projectile and its horizontal movement; tangent of the angle of fall.

slope stability   [GEOL] The resistance of an inclined surface to failure by sliding or collapsing.

slope stake   [MIN ENG] Stake set at the point where the finished side slope of an excavation or embankment meets the original grade.

slope wash   [GEOL] **1.** The mass-wasting process, assisted by nonchanneled running water, by which rock and soil is transported down a slope, specifically, sheet erosion. **2.** The material that is or has been transported.

slop oil   *See* slop.

slosh test   [ENG] A test to determine the ability of the control system of a liquid-propelled missile to withstand or overcome the dynamic movement of the liquid within its fuel tanks.

slot   [ADP] A punched-out area of a hand-sorted card to connect two or more guide holes; slots can be extended to the outside edge with notches. [AERO ENG] **1.** An air gap between a wing and the length of a slat or certain other auxiliary airfoils, the gap providing space for airflow or room for the auxiliary airfoil to be depressed in such a manner as to make for smooth air passage on the surface. **2.** Any of certain narrow apertures made through a wing to improve aerodynamic characteristics. [DES ENG] A narrow, vertical opening. [MIN ENG] To hole; to undercut or channel.

slot antenna   [ELECTROMAG] An antenna formed by cutting one or more narrow slots in a large metal surface fed by a coaxial line or waveguide.

slot coupling   [ELECTROMAG] Coupling between a coaxial cable and a waveguide by means of two coincident narrow slots, one in a waveguide wall and the other in the sheath of the coaxial cable.

slot distributor   [ENG] A long, narrow discharge opening (slot) in a pipe or conduit; used for the extrusion of sheet material, such as plastics.

slot dozing   [ENG] A method of moving large quantities of material with a bulldozer using the same path for each trip so that the spillage from the sides of the blade builds up along each side; afterward all material pushed into the slot is retained in front of the blade.

slot extrusion   [ENG] A method of extruding plastics-film sheet in which the molten thermoplastic compound is forced through a straight slot.

sloth   [VERT ZOO] Any of several edentate mammals in the family Bradypodidae found exclusively in Central and South America; all are slow-moving, arboreal species that cling to branches upside down by means of long, curved claws.

slot-mask picture tube   [ELECTR] An in-line gun-type picture tube in which the shadow mask is perforated by short, vertical slots, and the screen is painted with vertical phosphor stripes.

slot radiator   [ELECTROMAG] Primary radiating element in the form of a slot cut in the walls of a metal waveguide or cavity resonator or in a metal plate.

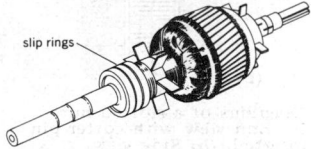

SLIP RING

slip rings

Slip rings on the rotor of electric rotating machinery.

SLOTH

The three-toed sloth *(Bradypus tridactylus)* in characteristic hanging position in which it spends most of its life.

SLOTTED NUT

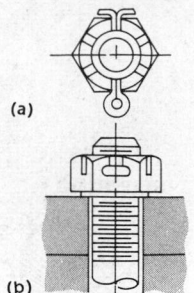

(a)

(b)

Diagrams of a slotted nut.
(a) End view with cotter pin
inserted. *(b)* Side view.

slotted-head screw [DES ENG] A screw fastener with a single groove across the diameter of the head.

slotted line *See* slotted section.

slotted nut [DES ENG] A regular hexagon nut with slots cut across the flats of the hexagon so that a cotter pin or safety wire can hold it in place.

slotted section [ELECTROMAG] A section of waveguide or shielded transmission line in which the shield is slotted to permit the use of a traveling probe for examination of standing waves. Also known as slotted line; slotted wave guide.

slotted wave guide *See* slotted section.

slotter [MECH ENG] A machine tool used for making a mortise or shaping the sides of an aperture.

slotting [MECH ENG] Cutting a mortise or a similar narrow aperture in a material using a machine with a vertically reciprocating tool.

slotting machine [MECH ENG] A vertically reciprocating planing machine, used for making mortises and for shaping the sides of openings.

slot washer [DES ENG] **1.** A lock washer with an indentation on its edge through which a nail or screw can be driven to hold it in place. **2.** A washer with a slot extending from its edge to the center hole to allow the washer to be removed without first removing the bolt.

slot weld [MET] Similar to plug weld, but the hole is elongated and may extend to the edge of a member without closing.

slough [ENG] The fragments of rocky material from the wall of a borehole. [HYD] A minor marshland or tidal waterway which usually connects other tidal areas; often more or less equivalent to a bayou. [MED] A necrotic mass of tissue in, or separating from, healthy tissue.

slow-acting relay [ELECTROMAG] A time-delay relay in which an interval of several seconds may exist between energizing of the coil and pulling up of the armature; the delay can be obtained electrically by placing a solid copper ring on the core of the relay. Also known as slow-operate relay.

slow-curing liquid asphaltic material [MATER] An asphalt cement blended with slow-volatilizing gas oil. Also known as SC asphalt.

slow death [ELECTR] The gradual change of transistor characteristics with time; this change is attributed to ions which collect on the surface of the transistor.

slowed-down video [ELECTR] Technique or method of transmitting radar data over narrow-bandwidth circuits; the procedure involves storing the radar video over the time required for the antenna to move through the beam width, and the subsequent sampling of this stored video at some periodic rate at which all of the range intervals of interest are sampled at least once each beam width or per azimuth quantum; the radar returns are quantized at the gap-filler radar site.

slow fire [ORD] Type of firing used in instructing beginners and in record practice, in which no time limit for completing a score is set.

slow igniter cord [ENG] An igniter cord made with a central copper wire around which is extruded a plastic incendiary material with an iron wire embedded to give greater strength; the whole is enclosed in a thin extruded plastic coating.

slowing-down [NUCLEO] A decrease in the energy of a particle as a result of collisions with nuclei.

slowing of clocks [RELAT] According to the special theory of relativity, a clock appears to tick less rapidly to an observer moving relative to the clock than to an observer who is at rest with respect to the clock. Also known as time dilation effect.

slow ion *See* large ion.

slow match [ENG] A match or fuse that burns at a known slow rate; used for igniting explosive charges.

slow memory *See* slow storage.

slow neutron [NUC PHYS] **1.** A neutron having low-kinetic energy, up to about 100 electron-volts. **2.** *See* thermal neutron.

slow-operate relay *See* slow-acting relay.

slow reactor [NUCLEO] A nuclear reactor in which fission is induced primarily by slow neutrons, as in a thermal reactor.

slow-release relay [ELECTROMAG] A time-delay relay in which there is an appreciable delay between deenergizing of the coil and release of the armature.

slow sand filter [CIV ENG] A bed of fine sand 20–48 inches (151–122 centimeters) deep through which water, being made suitable for human consumption and other purposes, is passed at a fairly low rate, 2,500,000 to 10,000,000 gallons per acre (23,000 to 94,000 cubic meters per hectare); an underdrain system of graded gravel and perforated pipes carries the water from the filters to the point of discharge.

slow-scan television [COMMUN] Television system using a slow rate of horizontal scanning suitable for transmitting printed matter, photographs, and illustrations.

slow-spiral drill *See* low-helix drill.

slow storage [ADP] In computers, storage with a relatively long access time. Also known as slow memory.

slow virus [VIROL] Any of a group of animal viruses characterized by prolonged periods of incubation and an extended clinical course lasting months or years.

slow wave [ELECTROMAG] A wave having a phase velocity less than the velocity of light, as in a ridge wave guide.

slow wave sleep *See* deep sleep.

slub [TEXT] A planned irregularity in yarn produced by alternately tightening and relaxing tensions during spinning.

sludge [CHEM ENG] **1.** Residue left after acid treatment of petroleum oils. **2.** Any semisolid waste from a chemical process. [CIV ENG] *See* sewage sludge. [ENG] **1.** Mud from a drill hole in boring. **2.** Sediment in a steam boiler. **3.** A precipitate from oils, such as the products from crankcase oils in engines. **4.** *See* slime. [GEOL] A soft or muddy bottom deposit as on tideland or in a stream bed. [OCEANOGR] A dense, soupy accumulation of new sea ice consisting of incoherent floating frazil crystals. Also known as cream ice; sludge ice; slush.

sludge acid [MATER] The residue from the sulfuric-acid treatment of petroleum lubricants. Also known as spent acid; waste acid.

sludge assay [MIN ENG] The chemical assaying of drill cuttings for a specific metal or group of metals.

sludge barrel *See* calyx.

sludge box [MIN ENG] A wooden box in which sludge settles from the mud flush.

sludge bucket *See* calyx.

sludge cake [OCEANOGR] An accumulation of sludge hardened into a cake strong enough to bear the weight of a man.

sludged blood [MED] The intracapillary aggregation of erythrocytes associated with decreased blood flow in the involved capillary bed.

sludge ice *See* sludge.

sludge lump [OCEANOGR] An irregular mass of sludge formed as a result of strong winds.

sludge pit *See* slushpit.

sludge pond *See* slushpit.

sludge pump [MECH ENG] *See* sand pump. [MIN ENG] A short iron pipe or tube fitted with a valve at the lower end, with which the sludge is extracted from a borehole.

sludging *See* solifluction.

sluff [ENG] The mud cake detached from the wall of a borehole. [MIN ENG] The falling of decomposed, soft rocks from the roof or walls of mine openings.

slug [ELECTROMAG] **1.** A heavy copper ring placed on the core of a relay to delay operation of the relay. **2.** A movable iron core for a coil. **3.** A movable piece of metal or dielectric material used in a wave guide for tuning or impedance-matching purposes. [GRAPHICS] A strip of metal used to space between lines of type. [INV ZOO] Any of a number of pulmonate gastropods which have a rudimentary shell and the body elevated toward the middle and front end where the mantle covers the lung region. [MECH] A unit of mass in the British gravitational system of units, equal to the mass which experiences an acceleration of 1 foot per second per second when a force of 1 pound acts on it; equal to approximately 32.1740 pound mass or 14.5939 kilograms. Also known as geepound. [MET] **1.** A small, roughly shaped piece of metal for subsequent processing, as by forging or extruding. **2.** The piece of material produced by piercing a hole in a sheet. [MIN ENG] To inject a borehole with cement, slurry, or various liquids containing shredded materials in an attempt

SLUG

Field gray slug (*Deroceras agreste*) has a thin, white-bordered shell; body length is about 1½ inches (4 centimeters). Eyes are borne at tips of the longer pair of feelers.

to restore lost circulation by sealing off the openings in the borehole-wall rocks. [NUCLEO] A short fuel rod inserted in a hole or channel in the active lattice of a nuclear reactor. [ORD] 1. As pertains to shaped charge ammunition, massive and relatively slow-moving remnant of the collapsed metal liner, as distinguished from the jet. 2. A solid cast iron projectile used in test firing.

slug bit See insert bit.

slug flow See piston flow.

slugging [MET] Adding a separate piece of material to a weld joint, resulting in a joint which does not meet specifications.

slug tuner [ELECTROMAG] Waveguide tuner containing one or more longitudinally adjustable pieces of metal or dielectric.

slug tuning [ELECTROMAG] Means of varying the frequency of a resonant circuit by introducing a slug of material into either the electric field or magnetic field, or both.

sluice [CIV ENG] 1. A passage fitted with a vertical sliding gate or valve to regulate the flow of water in a channel or lock. 2. A body of water retained by a floodgate. 3. A channel serving to drain surplus water.

sluice box [MIN ENG] A long, inclined trough or launder with riffles in the bottom that provide a lodging place for heavy minerals in ore concentration.

sluice gate [CIV ENG] The vertical slide gate of a sluice.

sluice tender [MIN ENG] In metal mining, a laborer who tends sluice boxes.

sluicing [MIN ENG] 1. Washing auriferous earth through sluices provided with riffles and other gold-saving appliances. 2. Separation of minerals in a flowing stream of water. 3. Moving earth, sand, gravel, or other rock or mineral materials by flowing water.

sluicing pond See scouring basin.

slump [GEOL] A type of landslide characterized by the downward slipping of a mass of rock or unconsolidated debris, moving as a unit or several subsidiary units, characteristically with backward rotation on a horizontal axis parallel to the slope; common on natural cliffs and banks and on the sides of artificial cuts and fills.

slump bedding [GEOL] Also known as slurry bedding. 1. Any disturbed bedding. 2. Convolute bedding produced by subaqueous slumping or lateral movement of newly deposited sediment.

slump fault See normal fault.

slump test [ENG] Determining the consistency of concrete by filling a conical mold with a sample of concrete, then inverting it over a flat plate and removing the mold; the amount by which the concrete drops below the mold height is measured and this represents the slump.

slurry [MATER] 1. A semiliquid refractory material, such as clay, used to repair furnace refractories. 2. A free-flowing, pumpable suspension of fine solid material in liquid. 3. An emulsion of a sulfonated soluble oil in water used to cool and lubricate metal during cutting operations. 4. See slip.

slurry bedding See slump bedding.

slurry blasting agent [MATER] A dense, insensitive, high-velocity explosive of great power and very high water resistance used principally for blasting hard rock or where blastholes are wet.

slurrying [ENG] The formation of a mud or a suspension from a liquid and nonsoluble solid particles.

slurry preforming [ENG] The preparation of reinforced plastics preforms by wet-processing techniques; similar to pulp molding.

slush [HYD] Snow or ice on the ground that has been reduced to a soft, watery mixture by rain, warm temperature, or chemical treatment. [OCEANOGR] See sludge.

slush casting [MET] Producing a hollow casting without a core in the mold by rotating a liquid alloy in a hollow metal mold until a solid layer chills onto the mold, and then pouring off the remaining liquid.

slusher [ENG] A method for the application of vitreous enamel slip to ware by dashing it on the ware to cover all its parts, excess then being removed by shaking the ware.

slush icing [METEOROL] The accumulation of ice and water on exposed surfaces of aircraft when the craft is flown through wet snow or snow and liquid drops at temperatures near 0°C.

slushing compound [MATER] A temporary, corrosion-protective coating for metals; made of nondrying oil, grease, or other similar material.

slushing grease [MATER] A special grade of grease used as a metal coating to prevent corrosion.

slushing oil [MATER] A nondrying oil which is strongly adhesive to metal and is applied to metal surfaces to minimize corrosion.

slush molding [ENG] A thermoplastic casting in which a liquid resin is poured into a hot, hollow mold where a viscous skin forms; excess slush is drained off, the mold is cooled, and the molded product is stripped out.

slushpit [ENG] An excavation or diked area to hold water, mud, sludge, and other discharged matter from an oil well. Also known as mud pit; sludge pit; sludge pond.

Sm See samarium.

Smalian's formula [FOR] A cubic volume formula used in log scaling, expressed cubic volume = $(B + b)/2L$, where B = the cross-sectional area at the large end of the log, b = the cross-sectional area at the small end of the log, and L = log length.

small arm [ORD] A gun of small character; includes such hand and shoulder weapons as pistols, carbines, rifles, and shotguns.

small-arms ammunition [ORD] Ammunition for use in small arms.

small-arms sling [ORD] An item made of leather, webbing, or other material, designed to be attached to a carbine, mortar, rifle, rocket launcher, shotgun, or submachine gun; used as a means of carrying a small-arms weapon or to steady the weapon for firing.

small-bore practice [ORD] Practice in firing with small arms using caliber-.22 ammunition instead of the standard service rounds.

small calorie See calorie.

small circle [GEOD] A circle on the surface of the earth, the plane of which does not pass through the earth's center. [NAV] A planned or accomplished route of travel which is generated by the intersection of a sphere and a plane which does not pass through the center of the sphere.

small-craft warning [METEOROL] A warning, for marine interests, of impending winds up to 28 knots (32 miles per hour or 52 kilometers per hour).

small-diameter blasthole [ENG] A blast hole 1½ to 3 inches (3.8 to 7.6 centimeters) in diameter, in low-face quarries.

small diurnal range [OCEANOGR] The difference in height between mean lower high water and mean higher low water.

small hail [METEOROL] Frozen precipitation consisting of small, semitransparent, roundish grains, each grain consisting of a snow pellet surrounded by a very thin ice covering, giving it a glazed appearance.

small ice floe [OCEANOGR] An ice floe of sea ice 30 to 600 feet (9 to 180 meters) across.

small intestine [ANAT] The anterior portion of the intestine in man and other mammals; it is divided into three parts, the duodenum, the jejunum, and the ileum.

small ion [METEOROL] An atmospheric ion of the type that has the greatest mobility; and hence, collectively, it is the principal agent of atmospheric conduction; evidence indicates that each ion is a singly-charged atmospheric molecule (or, rarely, an atom) about which a few other neutral molecules are held by the electrical attraction of the central ionized molecule; estimates of the number of satellite molecules range as high as 12. Also known as fast ion; light ion.

small-ion combination [METEOROL] Either of two processes by which small ions disappear: the union of a small ion and a neutral Aitken nucleus to form a new large ion, or the neutralization of a large ion by the small ion.

small-lot storage [IND ENG] Generally, a quantity of less than one pallet stack, stacked to maximum storage height; thus, the term refers to a lot consisting of from one container to two or more pallet loads, but is not of sufficient quantity to form a complete pallet column.

Small Magellanic Cloud [ASTRON] The smaller of the two star clouds near the south celestial pole; it is about 170,000 light-years away and contains a wide assortment of giant and variable stars, star clusters, and nebulae.

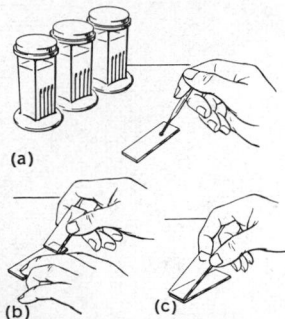

SMEAR

Making a smear preparation. *(a)* Place the drop about 1 inch (2.5 centimeters) from end of slide. *(b)* Apply a second slide just in front of the drop. *(c)* Push slide smoothly forward to spread the smear. *(From P. Gray, Handbook of Basic Microtechnique, 3d ed., McGraw-Hill, 1964)*

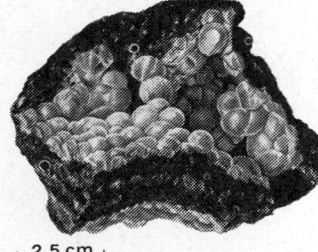

SMITHSONITE

⊢—— 2.5 cm ——⊣

Botryoidal aggregate of smithsonite in limestone, Larium, Greece. *(Specimen from Department of Geology, Bryn Mawr College)*

small of the stock [ORD] The part of the stock of a small-arms weapon ordinarily gripped by the right hand; the part of the stock immediately behind the receiver and trigger assembly.

small perturbation [PHYS] A disturbance imposed on a system in steady state, with amplitude assumed small of the first order; that is, the square of the amplitude is negligible in comparison with the amplitude, and the derivatives of the perturbation are assumed to be of the same order of magnitude as the perturbation.

smallpox [MED] An acute, infectious, viral disease characterized by severe systemic involvement and a single crop of skin lesions which proceeds through macular, papular, vesicular, and pustular stages. Also known as variola.

smallpox vaccine [IMMUNOL] A vaccine prepared from a glycerinated suspension of the exudate from cowpox vesicles obtained from healthy vaccinated calves or sheep. Also known as antismallpox vaccine; glycerinated vaccine virus; Jennerian vaccine; virus vaccinium.

small scale [GRAPHICS] A scale involving a large reduction in size, for example, 1:1,000,000; small-scale chart covers a large area.

small tropic range [OCEANOGR] The difference in height between tropic lower high water and tropic higher low water.

smalt [MATER] A blue glass made by fusing silica and potash with cobalt oxides; used as a pigment for glass, ceramics, paints, and dyes.

smaltite [MINERAL] $(Co,Ni)As_{3-x}$ A metallic-gray isometric mineral composed of nickel cobalt arsenide.

smaragd *See* emerald.

smaragdite [MINERAL] A green amphibole mineral that is pseudomorphous after pyroxene in rocks such as eclogite.

S matrix *See* scattering matrix.

S-matrix theory [PARTIC PHYS] A theory of elementary particles based on the scattering matrix, and on its properties such as unitarity and analyticity. Also known as scattering-matrix theory.

smear [BIOL] A preparation for microscopic examination made by spreading a drop of fluid, such as blood, across a slide and using the edge of another slide to leave a uniform film. [ELECTR] A television-picture defect in which objects appear to be extended horizontally beyond their normal boundaries in a blurred or smeared manner; one cause is excessive attenuation of high video frequencies in the television receiver.

smectic phase [PHYS CHEM] A form of the mesomorphic (liquid crystal) state in which liquid motion is more of a gliding than a flowing nature, drops are often formed which exhibit a series of fine lines, especially under polarized light, and x-ray diffraction patterns can be obtained in only one direction.

smectite [MINERAL] Dioctahedral (montmorillonite) and trioctahedral (saponite) clay minerals, and their chemical varieties characterized by swelling properties and high cation-exchange capacities.

smegma [PHYSIO] The sebaceous secretion that accumulates around the glans penis and the clitoris.

smell [PHYSIO] To perceive by olfaction.

smelling salts [PHARM] A preparation containing ammonium carbonate and ammonia water, usually scented with aromatic substances.

smell prism [PSYCH] A diagram that attempts to systematize an odor classification system that has six main odor qualities: fruity, flowery, resinous, spicy, foul, and burnt.

smell the bottom *See* feel the bottom.

smelter [MET] A furnace used for smelting.

smelting [MET] The heating of ore mixtures accompanied by a chemical change resulting in liquid metal.

S meter *See* signal-strength meter.

Smilacaceae [BOT] A family of monocotyledonous plants in the order Liliales; members are usually climbing, leafy-stemmed plants with tendrils, trimerous flowers, and a superior ovary.

Sminthuridae [INV ZOO] A family of insects in the order Collembola which have simple tracheal systems.

Smith-Baker microscope [OPTICS] A type of interference microscope in which a beam of polarized light is split by a birefringent calcite plate cemented to the front lens of the condenser, and reunited by another such plate cemented to the objective.

Smith chart [ELECTROMAG] A special polar diagram containing constant-resistance circles, constant-reactance circles, circles of constant standing-wave ratio, and radius lines representing constant line-angle loci; used in solving transmission-line and waveguide problems.

Smithell's burner [ENG] Two concentric tubes that can be added to a bunsen burner to separate the inner and outer flame cones.

Smitherm process [FOOD ENG] A direct heating process employing fat as the heating medium and used for deep-fat cooking of potato chips.

smith forging [MET] Manual forging of small, hot metal parts with flat or simple-shaped dies, as with a hammer and anvil.

Smith-Helmholtz law [OPTICS] For a single refracting surface of sufficiently small aperture, the product of the index of refraction, distance from the optical axis, and the angle which a light ray makes with the optical axis at the object point is equal to the corresponding product at the image point.

smithite [MINERAL] $AgAsS_2$ A red monoclinic mineral composed of silver arsenic sulfide and occurring as small crystals.

Smith-McIntyre sampler [MECH ENG] A device for taking samples of sediment from the ocean bottom; the digging and hoisting mechanisms are independent: the digging bucket is forced into the sediment before the hoisting action occurs.

Smith-Petersen nail [MED] A three-flanged nail used to fix fractures of the neck of the femur; it is inserted from just below the greater trochanter, through the neck, and into the head of the femur.

Smith-Recklinghausen's disease *See* neurofibromatosis.

smithsonite [MINERAL] $ZnCO_3$ White, yellow, gray, brown, or green secondary carbonate mineral associated with sphalerite and commonly reniform, botryoidal, stalactitic, or granular; hardness is 5 on Mohs scale, and specific gravity is 4.30–4.45; it is an ore of zinc. Also known as calamine; dry-bone ore; szaskaite; zinc spar.

smog [METEOROL] Air pollution consisting of smoke and fog.

smoke [ENG] Dispersions of finely divided (0.01–5.0 micrometers) solids or liquids in a gaseous medium.

smoke candle [ORD] Munition which produces smoke by vaporizing an oil.

smoke detector [ENG] A photoelectric system for an alarm when smoke in a chimney or other location exceeds a predetermined density.

smoke grenade [ORD] Hand grenade or rifle grenade containing a smoke-producing mixture, used for screening or signaling; sometimes charged with colored smoke such as red, green, yellow, or violet.

smoke horizon [METEOROL] The top of a smoke layer which is confined by a low-level temperature inversion in such a way as to give the appearance of the horizon when viewed from above against the sky; in such instances the true horizon is usually obscured by the smoke layer.

smokehouse [FOOD ENG] A building with cabinets generally built of sanitary stainless steel with excellent temperature and humidity control, in which volatilization and redeposition on food of certain components of hardwoods from burning of hardwood sawdust are accomplished.

smokeless [ORD] Indicating that ammunition is relatively smokeless when used in the weapon for which it is intended.

smokeless powder [MATER] Nitrocellulose containing 13.1 percent nitrogen with small amounts of stabilizers (amines) and plasticizers usually present, as well as various modifying agents (nitrotoluene and nitroglycerin salts); used in ammunition.

smoke point [ENG] The maximum flame height in millimeters at which kerosine will burn without smoking, tested under standard conditions; used as a measure of the burning cleanliness of jet fuel and kerosine.

smoke pot [ORD] A cylindrical metal munition designed to produce smoke for screening or signaling purposes, either by combustion of a smoke-producing mixture, or by combustion of a fuel mixture to vaporize a smoke-producing oil; it may be with or without an igniting device and a filling, and it is not intended for throwing or for firing from weapons.

smoke printing [GRAPHICS] A photocopy process in which electrically charged particles are deposited on paper or other material; the paper is held behind a sheet of glass which is backed with a thin metallic coating; the mist of particles is dispensed from behind the paper by an electrode which gives it a charge; a positive or negative print can be made.

smoke projectile [ORD] Any projectile containing a smoke-producing chemical agent with means for properly dispersing the agent.

smoke rocket [ORD] A rocket with a warhead which contains material to produce a smoke cloud on functioning.

smokes [METEOROL] Dense white haze and dust clouds common in the dry season on the Guinea coast of Africa, particularly at the approach of the harmattan.

smoke screen [ORD] A screen of smoke used to hide a maneuver, force, place, or activity.

smoke signal [ORD] A pyrotechnic item designed to produce a sign by means of smoke and used, for example, to provide identification, location, or warning.

smokestack [ENG] A chimney for the discharge of flue gases from a furnace operation such as in a steam boiler, powerhouse, heating plant, ship, locomotive, or foundry.

smokestone *See* smoky quartz.

smoke technique [FL MECH] A technique used to measure only very-low-speed air velocity; smoke enables the fluid motion to be observed with the eye, and the smoke is timed over a measured distance along an airway of constant cross section to determine the velocity of flow.

smoke test [ENG] A test used on kerosine to determine the highest point to which the flame can be turned before smoking occurs.

smoke washer [ENG] A device for removing particles from smoke by forcing it through a spray of water.

smoky quartz [MINERAL] A smoky-yellow, smoky-brown, or brownish-gray, often transparent variety of crystalline quartz containing inclusions of carbon dioxide; may be used as a semiprecious stone. Also known as cairngorm; smokestone.

S Monel [MET] An alloy similar to H Monel but containing 4% silicon.

smooth [OCEANOGR] Comparatively calm water between heavy seas. [STAT] To modify a sequential set of numerical data items in a manner designed to reduce the differences in value between adjacent items.

smooth blasting [ENG] Blasting to ensure even faces without cracks in the rock.

smooth bore [ORD] A bore that is smooth and without rifling, for example, in shotguns and mortars.

smooth drilling [ENG] Drilling in a rock formation in which a fast rotation of the drill stem, a fast rate of penetration, and a high recovery of core can be achieved with vibration-free rotation of the drill stem.

smoothing [ENG] Making a level, or continuously even, surface. [MATH] Approximating or perturbing a function by one which has a higher degree of differentiability.

smoothing choke [ELECTR] Iron-core choke coil employed as a filter to remove fluctuations in the output current of a vacuum-tube rectifier or direct-current generator.

smoothing circuit *See* ripple filter.

smoothing filter *See* ripple filter.

smoothing mill [MECH ENG] A revolving stone wheel used to cut and bevel glass or stone.

smoothing plane [DES ENG] A finely set hand tool, usually 5.5–10 inches (14.0–25.4 centimeters) long, for finishing small areas on wood.

smooth manifold [MATH] A differentiable manifold whose local coordinate systems depend upon those of euclidean space in an infinitely differentiable manner.

smooth map [MATH] An infinitely differentiable function.

smooth muscle [ANAT] The involuntary muscle tissue found in the walls of viscera and blood vessels, consisting of smooth muscle fibers.

smooth muscle fiber [HISTOL] Any of the elongated, nucleated, spindle-shaped cells comprising smooth muscles. Also known as involuntary fiber; nonstriated fiber; unstriated fiber.

smooth sea [OCEANOGR] Sea with waves no higher than ripples or small wavelets.

smother kiln [ENG] A kiln into which smoke can be introduced for blackening pottery.

smudge [PL PATH] Any of several fungus diseases of cereals and other plants characterized by dark, sooty discolorations.

smudge oil [MATER] A dark petroleum distillate or gas oil that is burned in citrus fruit-growing areas to prevent frost damage.

smudging [ENG] A frost-preventive measure used in orchards; properly, it means the production of heavy smoke, supposed to prevent radiational cooling, but it is generally applied to both heating and smoke production.

smut [MET] A reaction product left on the surface of a metal after pickling. [PL PATH] Any of various destructive fungus diseases of cereals and other plants characterized by large dusty masses of dark spores on the plant organs.

smut fungus [MYCOL] The common name for members of the Ustilaginales.

S/N *See* signal-to-noise ratio.

snagging [MECH ENG] Removing surplus metal or large surface defects by using a grinding wheel.

snail [INV ZOO] Any of a large number of gastropod mollusks distinguished by a spiral shell that encloses the body, a head, a foot, and a mantle.

snail fever *See* schistosomiasis.

snake [MET] 1. A twisted and bent hod rod formed before subsequent rolling operations. 2. A flexible mandrel used to prevent collapse of a shaped piece during bending operations. [VERT ZOO] Any of about 3000 species of reptiles which belong to the 13 living families composing the suborder Serpentes in the order Squamata.

snake hole [ENG] 1. A blasting hole bored directly under a boulder. 2. A drill hole used in quarrying or bench blasting.

snaking [ENG] Towing a load with a long cable.

snaking stream *See* meandering stream.

snap [METEOROL] A brief period of extreme (generally cold) weather setting in suddenly, as in a "cold snap."

SNAP [NUCLEO] A small nuclear power plant in which heat from radioisotope decay in a fuel such as strontium-90 is converted into electric energy, to provide power for spacecraft instrumentation, telemetry, and other applications. Derived from systems for nuclear auxiliary power.

snap-action switch [ELEC] A switch that responds to very small movements of its actuating button or lever and changes rapidly and positively from one contact position to the other; the trademark of one version is Micro Switch. Also known as sensitive switch.

snap-back forming [ENG] A plastic-sheet-forming technique in which an extended, heated, plastic sheet is allowed to contract over a form shaped to the desired final contour.

snapback method *See* repetitive time method.

snap fastener [DES ENG] A fastener consisting of a ball on one edge of an article that fits in a socket on an opposed edge, and used to hold edges together, such as those of a garment.

snap flask [MET] A foundry flask having its sides latched on one corner to allow removal of the flask from around the sand mold.

snap gage [DES ENG] A device with two flat, parallel surfaces spaced to control one limit of tolerance of an outside diameter or a length.

snap hook *See* spring hook.

snap-off diode [ELECTR] Planar epitaxial passivated silicon diode that is processed so a charge is stored close to the junction when the diode is conducting; when reverse voltage is applied, the stored charge then forces the diode to snap off or switch rapidly to its blocking state.

snap-on ammeter [ELEC] An ac ammeter having a magnetic core in the form of hinged jaws that can be snapped around the current-carrying wire. Also known as clamp-on ammeter.

snapper [ENG] A device for collecting samples from the ocean bottom, and which closes to prevent the sample from dropping out as it is raised to the surface.

snap ring [DES ENG] A form of spring used as a fastener; the ring is elastically deformed, put in place, and allowed to snap back toward its unstressed position into a groove or recess.

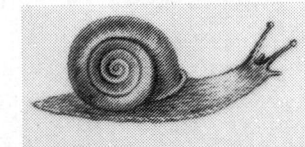

SNAIL

The white-lipped sand snail (*Triodopsis albalabris*); eyes are on the longer tentacles.

SNAKE

Garter snake (*Thamnophis ordinoides*).

snapshot dump [ADP] An edited printout of selected parts of the contents of main memory, performed at one or more times during the execution of a program without materially affecting the operation of the program.

snatch block [DES ENG] A pulley frame or sheave with an eye through which lashing can be passed to fasten it to a scaffold or pole.

snatch plate [ENG] A thick steel plate through which a hole about one-sixteenth of an inch larger than the outside diameter of the drill rod on which it is to be used is drilled; the plate is slipped over the drill rod and one edge is fastened to a securely anchored chain, and if rods must be pulled because high-pressure water is encountered, the eccentric pull of the chain causes the outside of the rods to be gripped and held against the pressure of water; the rod is moved a short distance out of the hole each time the plate is tapped.

S-N dewaxing [CHEM ENG] A proprietary, petroleum-refinery solvent process for removal of wax from petroleum oils. Also known as Separator-Nobel dewaxing.

S-N diagram [ENG] In fatigue testing, a graphic representation of the relationship of stress S and the number of cycles N before failure of the material.

sneak path [ADP] In computers, an undesired circuit through a series-parallel configuration.

sneeze [PHYSIO] A sudden, noisy, spasmodic expiration through the mouth and nose.

Snellen test [MED] A test for visual acuity presenting letters, numbers, or letter E's in various positions, with the symbols varying in size; if the smallest are read at a distance of 20 feet (6 meters), vision is recorded as 20/20, or normal.

Snell laws of refraction [OPTICS] When light travels from one medium into another the incident and refracted rays lie in one plane with the normal to the surface; are on opposite sides of the normal; and make angles with the normal whose sines have a constant ratio to one another. Also known as Descartes laws of refraction; laws of refraction.

snezhura *See* snow slush.

snifter valve [ENG] A valve on a pump that allows air to enter or escape, and accumulated water to be released.

sniperscope [ORD] A snooperscope which, when combined with a carbine or other firearm, enables the operator to see and shoot at targets in the dark.

snivet [ELECTR] Straight, jagged, or broken vertical black line appearing near the right-hand edge of a television receiver screen.

sno *See* elvegust.

SNOBOL [ADP] A computer programming language that has significant applications in program compilation and generation of symbolic equations. Derived from String-Oriented-Symbolic Language.

snooperscope [ELECTR] An infrared source, an infrared image converter, and a battery-operated high-voltage direct-current source constructed in portable form to permit a foot soldier or other user to see objects in total darkness; infrared radiation sent out by the infrared source is reflected back to the snooperscope and converted into a visible image on the fluorescent screen of the image tube.

snorkel [ENG] Any tube which supplies air for an underwater operation, whether it be for material or personnel. [NAV ARCH] A tube or pair of tubes for air intake and exhaust that can be extended above the surface of the water for operating submerged submarines.

snout [VERT ZOO] The elongated nose of various mammals.

snow [ELECTR] Small, random, white spots produced on a television or radar screen by inherent noise signals originating in the receiver. [METEOROL] The most common form of frozen precipitation, usually flakes of starlike crystals, matted ice needles, or combinations, and often rime-coated.

snow accumulation [METEOROL] The actual depth of snow on the ground at any instant during a storm, or after any single snowstorm or series of snowstorms.

snow avalanche [HYD] An avalanche of relatively pure snow; some rock and earth material may also be carried downward. Also known as snowslide.

snowbank glacier *See* nivation glacier.

snow banner [METEOROL] Snow being blown from a mountain crest. Also known as snow plume; snow smoke.

snow bin [ENG] A box for measuring the amount of snowfall; a type of snow gage.

snow blindness [MED] A transient visual impairment and actinic keratoconjunctivitis caused by exposure of the eyes to ultraviolet rays reflected from snow. Also known as solar photophthalmia.

snow blink [METEOROL] A bright, white glare on the underside of clouds, produced by the reflection of light from a snow-covered surface; this term is used in polar regions with reference to the sky map. Also known as snow sky.

snow blower [MECH ENG] A machine that removes snow from a road surface or pavement using a screw-type blade to push the snow into the machine and from which it is ejected at some distance.

snowbreak [CIV ENG] Any barrier designed to shelter an object or area from snow.

snowbridge [HYD] Snow bridging a crevasse in a glacier.

snow cap [HYD] 1. Snow covering a mountain peak when no snow exists at lower elevations. 2. Snow on the surface of a frozen lake.

snow climate *See* polar climate.

snow cloud [METEOROL] A popular term for any cloud from which snow falls.

snow course [HYD] An established line, usually from several hundred feet to as much as a mile long, traversing representative terrain in a mountainous region of appreciable snow accumulation; along this course, measurements of snow cover are made to determine its water equivalent.

snow cover [HYD] 1. All accumulated snow on the ground, including that derived from snowfall, snowslides, and drifting snow. Also known as snow mantle. 2. The extent, expressed as a percentage, of snow cover in a particular area.

snow-cover chart [METEOROL] A synoptic chart showing areas covered by snow and contour lines of snow depth.

snow crust [HYD] A crisp, firm, outer surface upon snow.

snow crystal [METEOROL] Any of several types of ice crystal found in snow; a snow crystal is a single crystal, in contrast to a snowflake which is usually an aggregate of many single snow crystals.

snow density [HYD] The ratio of the volume of meltwater that can be derived from a sample of snow to the original volume of the sample; strictly speaking, this is the specific gravity of the snow sample.

snowdrift [HYD] Snow deposited on the lee of obstacles, lodged in irregularities of a surface, or collected in heaps by eddies in the wind.

snowdrift glacier [HYD] A semipermanent mass of firn, formed by drifted snow in depressions in the ground or behind obstructions. Also known as catchment glacier; drift glacier.

snowdrift ice [HYD] Permanent or semipermanent masses of ice, formed by the accumulation of drifted snow in the lee of projections, or in depressions of the ground. Also known as glacieret.

snow eater [METEOROL] 1. Any warm wind blowing over a snow surface; usually applied to a foehn wind. 2. A fog over a snow surface; so called because of the frequently observed rapidity with which a snow cover disappears after a fog sets in.

snowfall [METEOROL] 1. The rate at which snow falls; in surface weather observations, this is usually expressed as inches of snow depth per 6-hour period. 2. A snow storm.

snow fence [CIV ENG] An open-slatted board fence usually 4 to 10 feet (1.2 to 3.0 meters) high, placed about 50 feet (15 meters) on the windward side of a railroad track or highway; the fence serves to disrupt the flow of the wind so that the snow is deposited close to the fence on the leeward side, leaving a comparatively clear, protected strip parallel to the fence and slightly farther downwind.

snowfield [HYD] 1. A broad, level, relatively smooth and uniform snow cover on ground or ice at high altitudes or in mountainous regions above the snow line. 2. The accumulation area of a glacier. 3. A small glacier or accumulation of perennial ice and snow too small to be designated a glacier.

snowflake [MET] *See* flake. [METEOROL] An ice crystal or, much more commonly, an aggregation of many crystals which falls from a cloud; simple snowflakes (single crystal) exhibit beautiful variety of form, but the symmetrical shapes reproduced so often in photomicrographs are not actually found frequently in snowfalls; broken single crystals, fragments, or clusters of such elements are much more typical of actual snows.

snow flurry [METEOROL] Popular term for snow shower, particularly of a very light and brief nature.

snow forest climate [CLIMATOL] A major category in W. Koppen's climatic classification, defined by a coldest-month mean temperature of less than 26.6°F (3°C) and a warmest-month mean temperature of greater than 50°F (10°C).

snow gage [HYD] An instrument for measuring the amount of water equivalent in a snowpack. Also known as snow sampler.

snow garland [HYD] A rare phenomenon in which snow is festooned from trees, fences, and so on, in the form of a rope of snow, several feet long and several inches in diameter; produced by surface tension acting in thin films of water bonding individual crystals; such garlands form only when the surface temperature is close to the melting point, for only then will the requisite films of slightly supercooled water exist.

snow geyser [METEOROL] Fine, powdery snow blown upward by a snow tremor.

snow grains [METEOROL] Precipitation in the form of very small, white opaque particles of ice; the solid equivalent of drizzle; the grains resemble snow pellets in external appearance, but are more flattened and elongated, and generally have diameters of less than 1 millimeter; they neither shatter nor bounce when they hit a hard surface. Also known as granular snow.

snow ice [HYD] Ice crust formed from snow, either by compaction or by the refreezing of partially thawed snow.

snow line [GEOGR] **1.** A transient line delineating a snow-covered area or altitude. **2.** An area with more than 50 percent snow cover. **3.** The altitude or geographic line separating areas in which snow melts in summer from areas having perennial ice and snow.

snow load [CIV ENG] The unit weight factor considered in the design of a flat or pitched roof for the probable amount of snow lying upon it.

snow mantle *See* snow cover.

snow mat [ENG] A device used to mark the surface between old and new snow, consisting of a piece of white duck 28 inches (71 centimeters) square, having in each corner triangular pockets in which are inserted slats placed diagonally to keep the mat taut and flat.

snowmelt [HYD] The water resulting from the melting of snow; it may evaporate, seep into the ground, or become a part of runoff.

snow-melting system [CIV ENG] A system of pipes containing a circulating nonfreezing liquid or electric-heating cables, embedded beneath the surface of a road, walkway, or other area to be protected from snow accumulation.

snow niche *See* nivation hollow.

snowpack [HYD] The amount of annual accumulation of snow at higher elevations in the western United States, usually expressed in terms of average water equivalent.

snow patch erosion *See* nivation.

snow pellets [METEOROL] Precipitation consisting of white, opaque, approximately round (sometimes conical) ice particles which have a snowlike structure and are about 2 to 5 millimeters in diameter; snow pellets are crisp and easily crushed, differing in this respect from snow grains, and they rebound when they fall on a hard surface and often break up. Also known as graupel; soft hail; tapioca snow.

snowplow [MECH ENG] A device for clearing away snow, as from a road or railway track.

snow plume *See* snow banner.

snow point [PHYS CHEM] Referring to a gas mixture, the temperature at which the vapor pressure of the sublimable component is equal to the actual partial pressure of that component in the gas mixture; analogous to dew point.

snowquake *See* snow tremor.

snow ripple *See* wind ripple.

snow roller [HYD] A mass of snow, shaped somewhat like a lady's muff, rather common in mountainous or hilly regions; it occurs when snow, moist enough to be cohesive, is picked up by wind blowing down a slope and rolled onward and downward until either it becomes too large or the ground levels off too much for the wind to propel it further; snow rollers vary in size from very small cylinders to some as large as 4 feet (1.2 meters) long and 7 feet (2.1 meters) in circumference.

snow sampler [ENG] A hollow tube for collecting a sample of snow in place. Also known as snow tube. [HYD] *See* snow gage.

snow scale *See* snow stake.

snowshed [CIV ENG] A structure to protect an exposed area as a road or rail line from snow.

snow sky *See* snow blink.

snowslide *See* snow avalanche.

snow sludge [OCEANOGR] Sludge formed mainly from snow.

snow slush [HYD] Slush formed from snow that has fallen into water that is at a temperature below that of the snow. Also known as snezhura.

snow smoke *See* snow banner.

snow stage [METEOROL] The thermodynamic process of sublimation of water vapor into snow in an idealized saturation-adiabatic or pseudoadiabatic expansion (lifting) of moist air; the snow stage begins at the condensation level when it is higher than the freezing level.

snow stake [ENG] A wood scale, calibrated in inches, used in regions of deep snow to measure its depth; it is bolted to a wood post or angle iron set in the ground. Also known as snow scale.

snow static [ELECTROMAG] Precipitation static caused by falling snow.

snowstorm [METEOROL] A storm in which snow falls.

snow survey [HYD] The process of determining depth and water content of snow at representative points, for example, along a snow course.

snow tremor [HYD] A disturbance in a snowfield, caused by the simultaneous settling of a large area of thick snow crust or surface layer. Also known as snowquake.

snow tube *See* snow sampler.

SNR *See* signal-to-noise ratio.

SNU *See* solar neutrino unit.

snub [MIN ENG] **1.** To increase the height of an undercut by means of explosives or otherwise. **2.** To check the descent of a car by the turn of a rope around a post.

snubber [MECH ENG] A mechanical device consisting essentially of a drum, spring, and friction band, connected between axle and frame, in order to slow the recoil of the spring and reduce jolting.

Snyder sampler [ENG] A mechanical device for obtaining small representative quantities from a moving stream of pulverized or granulated solids; it consists of a cast-iron plate revolving in a vertical plane on a horizontal axis with an inclined sample spout; the material to be sampled comes to the sampler by way of an inclined chute whenever the sample spout comes in line with the moving stream.

soak cleaning [MET] Cleaning the surface of a metal by immersion in a cleaning solution without electrolysis.

soaking [MET] Heating an alloy, usually an ingot, to a temperature not far below its melting temperature and holding it there for a long time to eliminate segregation that occurred on solidification.

soaking drum [CHEM ENG] A heated petroleum-refinery process vessel used in connection with petroleum thermal-cracking coils to furnish the residence time needed to complete the cracking reaction.

soaking pit [MET] A high-temperature, gas-fired, tightly covered, refractory-lined hole or pit into which a hot metal ingot (with liquid interior) is held at a fixed temperature until needed for rolling into sheet or other forms.

soap [MATER] **1.** A particular type of detergent, in which the water-solubilizing group is a carboxylate, COO−, and the positive ion is usually sodium, Na^+, or potassium, K^+. **2.** A

SODA NITER

Soda niter from La Noria, Chile. *(Specimen from Department of Geology, Bryn Mawr College)*

SODIUM

Periodic table of the chemical elements showing the position of sodium.

soap compound mixed with a fragrance and other ingredients and then cast into soap bars of different shapes.

soap builder [MATER] A substance mixed with soap to modify the alkali content, to add water-softening ability, or to improve otherwise the cleaning properties; examples are rosin and sodium phosphate.

soaprock *See* soapstone.

soapstone [MINERAL] 1. A mineral name applied to steatite or to massive talc. Also known as soaprock. 2. *See* saponite. [PETR] A metamorphic rock characterized by massive, schistose, or interlaced fibrous texture and a soft unctuous feel.

soar [AERO ENG] To fly without loss of altitude, using no motive power other than updrafts in the atmosphere.

sobole [BOT] An underground creeping stem.

social animal [ZOO] An animal that exhibits social behavior.

social anthropology [ANTHRO] The study of social organization among nonliterate peoples.

social behavior [ZOO] Any behavior on the part of an organism stimulated by, or acting upon, another member of the same species.

social hierarchy [VERT ZOO] The establishment of a dominance-subordination relationship among higher animal societies.

socialization [PSYCH] The process whereby a child learns to get along with and to behave similarly to other people in the group, largely through imitation as well as group pressure.

social parasitism [VERT ZOO] An aberrant type of parasitism occurring in some birds, in which the female of one species lays her eggs in the nests of other species and permits the foster parents to raise the young.

social psychiatry [PSYCH] Psychiatry especially concerned with the study of social influences on the cause and dynamics of emotional and mental illness, the use of the social environment in treatment, and preventive community programs, as well as the application of psychiatry to social issues, industry, law, education, and other such activities and organizations.

social psychology [PSYCH] The study of the manner in which the attitudes, personality, and motivations of the individual influence, and are influenced by, the structure, dynamics, and behavior of the social group with which the individual interacts.

society [ECOL] A secondary or minor plant community forming part of a community.

socked in [METEOROL] In the early days of aviation, pertaining to weather at an airport when ceiling or visibility were of such low values that the airport was effectively closed to aircraft operations.

socket [ELEC] A device designed to provide electric connections and mechanical support for an electronic or electric component requiring convenient replacement.

socket-head screw [DES ENG] A screw fastener with a geometric recess in the head into which an appropriate wrench is inserted for driving and turning, with consequent improved nontamperability.

socket wrench [DES ENG] A wrench with a socket to fit the head of a bolt or a nut.

soda *See* sodium carbonate.

soda-acid extinguisher [ENG] A fire-extinguisher from which water is expelled at a high rate by the generation of carbon dioxide, the result of mixing (when the extinguisher is tilted) of sulfuric acid and sodium bicarbonate.

soda ash [INORG CHEM] Na_2CO_3 The commercial grade of sodium carbonate; a powder soluble in water, insoluble in alcohol; used in glass manufacture and petroleum refining, and for soaps and detergents. Also known as anhydrous sodium carbonate; calcined soda.

soda blasting powder *See* B blasting powder.

sodaclase *See* albite.

soda crystals *See* metahydrate sodium carbonate.

soda-granite [PETR] 1. A granite in which soda is more abundant than potash. 2. A granite that contains soda-plagioclase instead of the orthoclase found in normal granite.

soda lake [HYD] An alkali lake rich in dissolved sodium salts, especially sodium carbonate, sodium chloride, and sodium sulfate. Also known as natron lake.

soda lime [MATER] A mixture of sodium or potassium hydroxide with calcium oxide; granules are used to absorb water vapor and carbon dioxide gas.

soda-lime glass [MATER] Glass made by fusion of sand with sodium carbonate, or sodium sulfate and lime, or limestone; used for window glass.

sodalite [MINERAL] $Na_2Al_3Si_3O_{12}Cl$ A blue or sometimes white, gray, or green mineral tectosilicate of the feldspathoid group, crystallizing in the isometric system, with vitreous luster, hardness of 5 on Mohs scale, and specific gravity of 2.2–2.4; used as an ornamental stone.

soda mica *See* paragonite.

sodamide *See* sodium amide.

soda niter [MINERAL] $NaNO_3$ A colorless to white mineral composed of sodium nitrate, crystallizing in the rhombohedral division of the hexagonal system; hardness is 1½ to 2 on Mohs scale and specific gravity is 2.266. Also known as nitratine; Peru saltpeter.

soda pulping process [CHEM ENG] The digestion of wood chips by caustic soda; used to manufacture pulp for paper products.

soddyite [MINERAL] $(UO_2)_{12}Si_5O_{22} \cdot 14H_2O$ A pale-yellow orthorhombic mineral composed of hydrous uranium silicate and occurring in fine-grained aggregates or crystals.

Soddy's displacement law [NUC PHYS] The atomic number of a nuclide decreases by 2 in alpha decay, increases by 1 in beta negatron decay, and decreases by 1 in beta positron decay and electron capture.

sodium [CHEM] A metallic element of group I, symbol Na, with atomic number 11, atomic weight 22.9898; silver-white, soft, and malleable; oxidizes in air; melts at 97.6°C; used as a chemical intermediate and in pharmaceuticals, petroleum refining, and metallurgy; the source of the symbol Na is natrium.

sodium-24 [NUC PHYS] A radioactive isotope of sodium, mass 24, half-life 15.5 hours; formed by deuteron bombardment of sodium; decomposes to magnesium with emission of beta rays.

sodium acetate [ORG CHEM] $NaC_2H_3O_2$ Colorless, efflorescent crystals, soluble in water and ether; melts at 324°C; used as a chemical intermediate and for pharmaceuticals, dyes, and dry colors.

sodium acid carbonate *See* sodium bicarbonate.

sodium acid chromate *See* sodium dichromate.

sodium acid fluoride *See* sodium bifluoride.

sodium acid sulfate *See* sodium bisulfate.

sodium acid sulfite *See* sodium bisulfite.

sodium alphanaphthylamine sulfonate *See* sodium naphthionate.

sodium aluminate [INORG CHEM] $Na_2Al_2O_4$ A white powder soluble in water, insoluble in alcohol; melts at 1800°C; used as a zeolite-type of material and a mordant, and in water purification, milkglass manufacture, and cleaning compounds.

sodium amalgam [INORG CHEM] Na_xHg_x A fire-hazardous, silver-white crystal mass that decomposes in water; used to make hydrogen and as an analytical reagent.

sodium amalgam–oxygen cell [ELEC] Fuel cell system in which materials functioning in the dual capacity of fuel and anode are consumed continuously; low operating temperatures and high power-to-weight ratios are significant characteristics of the system.

sodium amide [INORG CHEM] $NaNH_2$ White crystals that decompose in water; melts at 210°C; a fire hazard; used to make sodium cyanide. Also known as sodamide.

sodium aminophenylarsonate *See* sodium arsanilate.

sodium anilinearsonate *See* sodium arsanilate.

sodium antimony bis (pyrocatechol-2,4-disulfonate) *See* stibophen.

sodium arsanilate [ORG CHEM] $C_6H_4NH_2(AsO \cdot OH \cdot ONa)$ A white, water-soluble, poisonous powder with a faint saline taste, used in medicine and as a chemical intermediate. Also known as sodium aminophenylarsonate; sodium anilinearsonate.

sodium arsenate [INORG CHEM] $Na_3AsO_4 \cdot 12H_2O$ Water-soluble, poisonous, clear, colorless crystals with a mild alkaline taste; melts at 86°C; used in medicine, insecticides, dry colors, and textiles, and as a germicide and a chemical intermediate.

sodium arsenite [INORG CHEM] $NaAsO_2$ A poisonous, water-soluble, grayish powder; used in antiseptics, dyeing, insecticides, and soaps for taxidermy. Also known as sodium metaarsenite.

sodium ascorbate [ORG CHEM] $CH_2OH(CHOH)_2$ COHCOHCOONa White, odorless crystals; soluble in water, insoluble in alcohol; decomposes at 218°C; used in therapy for vitamin C deficiency.

sodium azide [INORG CHEM] NaN_3 Poisonous, colorless crystals; soluble in water and liquid ammonia; decomposes at 300°C; used in medicine and to make lead azide explosives.

sodium benzoate [ORG CHEM] $NaC_7H_5O_2$ Water- and alcohol-soluble, white, amorphous crystals with a sweetish taste; used as a food preservative and an antiseptic and in tobacco, pharmaceuticals, and medicine.

sodium benzoylacetone dihydrate [ORG CHEM] A metal chelate with low melting point (115°C) and slight solubility in acetone.

sodium benzyl penicillinate See benzyl penicillin sodium.

sodium bicarbonate [INORG CHEM] $NaHCO_3$ White, water-soluble crystals with an alkaline taste; loses carbon dioxide at 270°C; used as a medicine and a butter preservative, in food preparation, in effervescent salts and beverages, in ceramics, and to prevent timber mold. Also known as baking soda; sodium acid carbonate.

sodium bichromate See sodium dichromate.

sodium bifluoride [INORG CHEM] $NaHF_2$ Poisonous, water-soluble, white crystals; decomposes when heated; used as a laundry-rinse neutralizer, preservative, and antiseptic, and in glass etching and tin-plating. Also known as sodium acid fluoride.

sodium bismuthate [INORG CHEM] $NaBiO_3$ A yellow to brown amorphous powder; used as an analytical reagent and in pharmaceuticals.

sodium bisulfate [INORG CHEM] $NaHSO_4$ Colorless crystals, soluble in water; the aqueous solution is strongly acidic; decomposes at 315°C; used for flux to decompose minerals, as a disinfectant, and in dyeing and manufacture of magnesia, cements, perfumes, brick, and glue. Also known as niter cake; sodium acid sulfate.

sodium bisulfide See sodium hydrosulfide.

sodium bisulfite [INORG CHEM] $NaHSO_3$ A colorless, water-soluble solid; decomposes when heated. Also known as sodium acid sulfite.

sodium bisulfite test [ANALY CHEM] A test for aldehydes in which aldehydes form a crystalline salt upon addition of a 40% aqueous solution of sodium bisulfite.

sodium bitartrate [ORG CHEM] $NaHC_4H_5O_6 \cdot H_2O$ A white, combustible, water-soluble powder that loses water at 100°C, decomposes at 219°C; used in effervescing mixtures and as an analytical reagent. Also known as acid sodium tartrate.

sodium borate [INORG CHEM] $Na_2B_4O_7 \cdot 10H_2O$ A water-soluble, odorless, white powder; melts between 75 and 200°C; used in glass, ceramics, starch and adhesives, detergents, agricultural chemicals, pharmaceuticals, and photography; the impure form is known as borax. Also known as sodium pyroborate; sodium tetraborate.

sodium (1:2) borate See borax.

sodium boroformate [ORG CHEM] $NaH_2BO_3 \cdot 2HCOOH \cdot 2H_2O$ Water-soluble, white crystals, melting at 110°C; used in textile treating and in tanning, and as a buffering agent.

sodium borohydride [INORG CHEM] $NaBH_4$ A flammable, hygroscopic, white to gray powder; soluble in water, insoluble in ether and hydrocarbons; decomposes in damp air; used as a hydrogen source, a chemical reagent, and a rubber foaming agent.

sodium bromate [INORG CHEM] $NaBrO_3$ Odorless, white crystals; soluble in water, insoluble in alcohol; decomposes at 381°C; a fire hazard, used as an analytical reagent.

sodium bromide [INORG CHEM] NaBr White, water-soluble, crystals with a bitter, saline taste; absorbs moisture from air; melts at 758°C; used in photography and medicine, as a chemical intermediate, and to make bromides.

sodium cacodylate [ORG CHEM] $(CH_3)_2AsOONa \cdot 3H_2O$ A white crystalline compound that melts at approximately 60°C

and dissolves in ethanol and water; used as a herbicide. Also known as sodium dimethylarsenate.

sodium-calcium feldspar See plagioclase.

sodium carbolate See sodium phenate.

sodium carbonate [INORG CHEM] Na_2CO_3 A white, water-soluble powder that decomposes when heated to about 852°C; used as a reagent; forms a monohydrate compound, $Na_2CO_3 \cdot H_2O$, and a decahydrate compound, $Na_2CO_3 \cdot 10H_2O$. Also known as soda.

sodium carbonate decahydrate See sal soda.

sodium carboxymethylcellulose See carboxymethyl cellulose.

sodium caseinate [ORG CHEM] A tasteless, odorless, water-soluble, white powder; used in medicine, foods, emulsification, and stabilization; formed by dissolving casein in sodium hydroxide and then evaporating. Also known as casein sodium; nutrose.

sodium cellulose glycolate See carboxymethyl cellulose.

sodium chlorate [INORG CHEM] $NaClO_3$ Water- and alcohol-soluble, colorless crystals with a saline taste; melts at 255°C; used as a medicine, weed killer, defoliant, and oxidizing agent, and in matches, explosives, and bleaching.

sodium chloride [INORG CHEM] NaCl Colorless or white crystals; soluble in water and glycerol, slightly soluble in alcohol; melts at 804°C; used in foods and as a chemical intermediate and an analytical reagent. Also known as common salt; rock salt; table salt.

sodium chloride solution See normal saline.

sodium chlorite [INORG CHEM] $NaClO_2$ An explosive, white, mildly hygroscopic, water-soluble powder; decomposes at 175°C; used as an analytical reagent and oxidizing agent.

sodium chromate [INORG CHEM] $Na_2CrO_4 \cdot 10H_2O$ Water-soluble, translucent, yellow, efflorescent crystals that melt at 20°C; used as a rust preventive and in inks, dyeing, and leather tanning.

sodium citrate [ORG CHEM] $C_6H_5Na_3O_7 \cdot 2H_2O$ A white powder with the taste of salt; soluble in water, slightly soluble in alcohol; loses water at 150°C; decomposes at red heat; used in medicine as an anticoagulant, in soft drinks, cheesemaking, and electroplating. Also known as disodium citrate; disodium hydrogen citrate.

sodium cobaltinitrite [INORG CHEM] $Na_3Co(NO_2)_6 \cdot \frac{1}{2}H_2O$ Purple, water-soluble, hygroscopic crystals; used as a reagent for analysis of potassium.

sodium-cooled reactor [NUCLEO] A nuclear reactor in which sodium metal in liquid form is used as the coolant.

sodium cyanate [INORG CHEM] NaOCN A poisonous, white powder; soluble in water, insoluble in alcohol and ether; used as a chemical intermediate and for the manufacture of medicine and the heat-treating of steels.

sodium cyanide [INORG CHEM] NaCN A poisonous, water-soluble, white powder melting at 563°C; decomposes rapidly when standing; used to manufacture pigments, in heat treatment of metals, and as a silver- and gold-ore extractant.

sodium cyclamate [ORG CHEM] $C_6H_{11}NHSO_3Na$ White, water-soluble crystals; sweetness 30 times that of sucrose; formerly used as an artificial sweetener for foods, but now prohibited.

sodium diacetate [ORG CHEM] $CH_3COONa \cdot x(CH_3COOH)$ Combustible, white, water-soluble crystals with an acetic acid aroma; decomposes above 150°C; used to inhibit mold, and as a buffer, varnish hardener, sequestrant, and food preservative, and in mordants.

sodium dichromate [INORG CHEM] $Na_2Cr_2O_7 \cdot 2H_2O$ Poisonous, red to orange deliquescent crystals; soluble in water, insoluble in alcohol; melts at 320°C; loses water of hydration upon prolonged heating at 105°C; used as a chemical intermediate and corrosion inhibitor and in the manufacture of pigments, leather tanning, and electroplating. Also known as bichromate of soda; sodium acid chromate; sodium bichromate.

sodium diethyldithiocarbamate [ORG CHEM] $(C_2H_5)_2NCS_2Na$ A solid that is soluble in water and in alcohol; the trihydrate is used to determine small amounts of copper and to separate copper from other metals.

sodium dimethylarsenate See sodium cacodylate.

sodium *para*-diphenylaminoazobenzenesulfonate *See* tropeoline 00.

sodium dithionite *See* sodium hydrosulfite.

sodium ethoxide *See* sodium ethylate.

sodium ethylate [ORG CHEM] C_2H_5ONa A white powder formed from ethanol by replacement of the hydroxyl groups' hydrogen by monovalent sodium; used in organic synthesis. Also known as caustic alcohol; sodium ethoxide.

sodium feldspar *See* albite.

sodium ferricyanide [INORG CHEM] $Na_3Fe(CN)_6 \cdot H_2O$ A poisonous, deliquescent, red powder; soluble in water, insoluble in alcohol; used in printing and for the manufacture of pigments. Also known as red prussiate of soda.

sodium ferrocyanide [INORG CHEM] $Na_4Fe(CN)_6 \cdot 10H_2O$ Semitransparent crystals, soluble in water; insoluble in alcohol; used in photography, dyes, tanning, and blueprint paper. Also known as yellow prussiate of soda.

sodium fluoborate [INORG CHEM] $NaBF_4$ A white powder with a bitter taste; soluble in water, slightly soluble in alcohol; decomposes when heated, fuses below 500°C; used in electrochemical processes, as flux for nonferrous metals refining, and as an oxidation inhibitor.

sodium fluorescein *See* uranine.

sodium fluoride [INORG CHEM] NaF A poisonous, water-soluble, white powder, melting at 988°C; used as an insecticide and a wood and adhesive preservative, and in fungicides, vitreous enamels, and dentistry.

sodium fluoroacetate [ORG CHEM] FCH_2COONa A toxic, white, water-soluble powder; decomposes at 200°C; used as a rodenticide.

sodium fluosilicate [INORG CHEM] Na_2SiF_6 A poisonous, white, amorphous powder; slightly soluble in water; decomposes at red heat; used to fluoridate drinking water and to kill rodents and insects. Also known as sodium silicofluoride.

sodium folate [ORG CHEM] $C_{19}H_{18}N_7NaO_6$ A yellow to yellow-orange liquid; used in medicine for folic acid deficiency. Also known as folic acid sodium salt; sodium pteroylglutamate.

sodium formate [ORG CHEM] $HCOONa$ A mildly hygroscopic, white powder, soluble in water; has a formic acid aroma; melts at 245°C; used in medicine and as a chemical intermediate and reducing agent.

sodium gluconate [ORG CHEM] $C_6H_{11}NaO_7$ A water-soluble, yellow to white, crystalline powder, produced by fermentation; used in food and pharmaceutical industries, and as a metal cleaner. Also known as gluconic acid sodium salt.

sodium glucosulfone [PHARM] $C_{24}H_{34}N_2Na_2O_{18}S_3$ A leprostatic drug, and suppressant for dermatitis herpetiformis. Also known as glucosulfone sodium.

sodium glutamate [ORG CHEM] $COOH(CH_2)_2CH(NH_2)COONa$ A salt of an amino acid; a white powder, soluble in water and alcohol; used as a taste enhancer. Also known as monosodium glutamate.

sodium-graphite reactor [NUCLEO] A nuclear reactor that uses slightly enriched uranium as fuel, graphite as moderator, and liquid sodium as the coolant.

sodium halide [INORG CHEM] A compound of sodium with a halogen; for example, sodium bromide ($NaBr$), sodium chloride ($NaCl$), sodium iodide (NaI), and sodium fluoride (NaF).

sodium halometallate [INORG CHEM] A compound of sodium with halogen and a metal; for example, sodium platinichloride, $Na_2PtCl_6 \cdot 6H_2O$.

sodium hydrate *See* sodium hydroxide.

sodium hydride [INORG CHEM] NaH A white powder, decomposed by water, and igniting in moist air; used to make sodium borohydride and as a drying agent and a reagent.

sodium hydrogen phosphate [INORG CHEM] $NaH_2PO_4 \cdot H_2O$ Hygroscopic, transparent, water-soluble crystals; used as a purgative, reagent, and buffer.

sodium hydrogen sulfide *See* sodium hydrosulfide.

sodium hydrosulfide [INORG CHEM] $NaSH \cdot 2H_2O$ Toxic, colorless, water-soluble needles, melting at 55°C; used in pulping of paper, processing dyestuffs, hide dehairing, and bleaching. Also known as sodium bisulfide; sodium hydrogen sulfide; sodium sulfhydrate.

sodium hydrosulfite [INORG CHEM] $Na_2S_2O_4$ A fire-hazardous, lemon to whitish-gray powder; soluble in water, insoluble in alcohol; melts at 55°C; used as a chemical intermediate and catalyst and in ore flotation. Also known as sodium dithionite.

sodium hydroxide [INORG CHEM] $NaOH$ White, deliquescent crystals; absorbs carbon dioxide and water from air; soluble in water, alcohol, and glycerol; melts at 318°C; used as an analytical reagent and chemical intermediate, in rubber reclaiming and petroleum refining, and in detergents. Also known as sodium hydrate.

sodium hypochlorite [INORG CHEM] $NaOCl$ Air-unstable, pale-green crystals with sweet aroma; soluble in cold water, decomposes in hot water; used as a bleaching agent for paper pulp and textiles, as a chemical intermediate, and in medicine.

sodium hyposulfite *See* sodium thiosulfate.

sodium iodate [INORG CHEM] $NaIO_3$ A white, water- and acetone-soluble powder; used as a disinfectant and in medicine.

sodium iodide [INORG CHEM] NaI A white, air-sensitive powder, deliquescent, with bitter taste; soluble in water, alcohol, and glycerin; melts at 653°C; used in photography and in medicine and as an analytical reagent.

sodium iodide scintillator [NUCLEO] A sodium iodide crystal activated with thallium; used especially in the measurement of the energy of gamma rays, by measuring the amplitude of pulses of light generated by electrons in the crystal which are excited by the gamma rays.

sodium lactate [ORG CHEM] $CH_3CHOHCOONa$ A water-soluble, hygroscopic, yellow to colorless, syrupy liquid; solidifies at 17°C; used in medicine, as a corrosion inhibitor in antifreeze, and a hygroscopic agent.

sodium lauryl sulfate [ORG CHEM] $CH_3(CH_2)_{10}CH_2OSO_3Na$ A water-soluble salt, produced as a white or cream powder, crystals, or flakes; used in the textile industry as a wetting agent and detergent. Also known as dodecyl sodium sulfate.

sodium lead alloy [MATER] A highly toxic, explosion-prone alloy of lead and sodium; contains 10% sodium and 90% lead when used to make tetraethyllead, and 2% sodium and 98% lead when used as a deoxidizer and homogenizer in lead-containing nonferrous alloys; reacts with moisture, acids, and oxidizing agents.

sodium lead hyposulfate *See* lead sodium thiosulfate.

sodium lead thiosulfate *See* lead sodium thiosulfate.

sodium mercaptoacetate *See* sodium thioglycolate.

sodium metaarsenite *See* sodium arsenite.

sodium metaborate [INORG CHEM] $NaBO_2$ Water-soluble, white crystals, melting at 966°C; the aqueous solution is alkaline; made by fusing sodium carbonate with borax.

sodium metaphosphate [INORG CHEM] $(NaPO_3)_x$ Sodium phosphate groupings; cyclic forms range from $x = 3$ for the trimetaphosphate, to $x = 10$ for the decametaphosphate; sodium hexametaphosphate with $x = 10$ to 20 is probably a polymer; used for dental polishing, building detergents, and water softening, and as a sequestrant, emulsifier, and food additive.

sodium metasilicate *See* sodium silicate.

sodium methoxide [ORG CHEM] CH_3ONa A salt produced as a free-flowing powder, soluble in methanol and ethanol; used as an intermediate in organic synthesis. Also known as sodium methylate.

sodium methylate *See* sodium methoxide.

sodium molybdate [INORG CHEM] Na_2MoO_4 Water-soluble crystals, melting at 687°C; used as an analytical reagent, corrosion inhibitor, catalyst, and zinc-plating brightening agent, and in medicine.

sodium monoxide [INORG CHEM] Na_2O A strong basic white powder soluble in molten caustic soda; forms sodium hydroxide in water; used as a dehydrating and polymerization agent. Also known as sodium oxide.

sodium naphthionate [ORG CHEM] $NaC_{10}H_6(NH_2)SO_3 \cdot 4H_2O$ White, light-sensitive crystals, soluble in water and insoluble in ether; used in analysis (Riegler's reagent) for nitrous acid. Also known as sodium alphanaphthylamine sulfonate.

sodium nitrate [INORG CHEM] $NaNO_3$ Fire-hazardous, transparent, colorless crystals with bitter taste; soluble in glycerol and water; melts at 308°C; decomposes when heated;

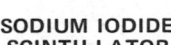

SODIUM IODIDE SCINTILLATOR

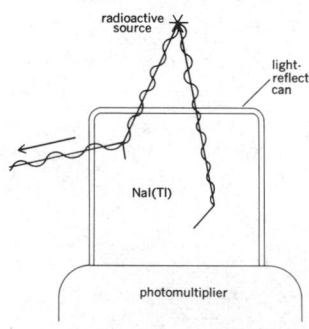

Illustration of γ-ray absorption in sodium iodide crystal used as scintillator. In event at left, photon coming from source undergoes Compton scattering in crystal and secondary quantum escapes unabsorbed. At right, photon is fully absorbed because of photoelectric effect.

used in manufacture of glass and pottery enamel and as a fertilizer and food preservative.

sodium nitrate gelignites [MATER] A group of explosives, modifications of blasting gelatin in which varying percentages of nitroglycerin are replaced by sodium nitrate and combustible material; characterized by plastic consistency, high densities, medium velocity of detonation, good resistance to the effects of water, and fume characteristics which are suitable for underground workings.

sodium nitrite [INORG CHEM] $NaNO_2$ A fire-hazardous, air-sensitive, yellowish powder, soluble in water; decomposes above 320°C; used as an intermediate for dyestuffs and for pickling meat, textiles dyeing, and rust-proofing, and in medicine.

sodium nitroferricyanide [INORG CHEM] $Na_2Fe(CN)_5NO \cdot 2H_2O$ Water-soluble, transparent, reddish crystals; slowly decomposes in water; used as an analytical reagent. Also known as sodium nitroprussiate; sodium nitroprusside.

sodium nitroprussiate *See* sodium nitroferricyanide.

sodium nitroprusside *See* sodium nitroferricyanide.

sodium oleate [ORG CHEM] $C_{17}H_{33}COONa$ A white powder with a tallow aroma; soluble in alcohol and water, with partial decomposition; used in medicine and textile waterproofing.

sodium oxalate [ORG CHEM] $Na_2C_2O_4$ A poisonous, white powder; soluble in water, insoluble in alcohol; used for leather tanning and as an analytical reagent.

sodium oxide *See* sodium monoxide.

sodium penicillin G₁ *See* benzyl penicillin sodium.

sodium pentaborate [INORG CHEM] $Na_2B_{10}O_{16} \cdot 10H_2O$ A white, water-soluble powder; used in glassmaking, weed killers, and fireproofing compositions.

sodium pentobarbitone *See* pentobarbital sodium.

sodium Pentothal *See* Pentothal.

sodium perborate [INORG CHEM] $NaBO_2 \cdot H_2O_2 \cdot 3H_2O$ A white powder with a saline taste; slightly soluble in water, decomposes in moist air; used in deodorants, in dental compositions and as a germicide. Also known as peroxydol.

sodium perchlorate [INORG CHEM] $NaClO_4$ Fire-hazardous, white, deliquescent crystals; soluble in water and alcohol; melts at 482°C; explosive when in contact with concentrated sulfuric acid; used in jet fuel, as an analytical reagent, and for explosives.

sodium permanganate [INORG CHEM] $NaMnO_4 \cdot 3H_2O$ A fire-hazardous, water-soluble, purple powder; decomposes when heated; used to make saccharin, as a disinfectant, and as an oxidizing agent.

sodium peroxide [INORG CHEM] Na_2O_2 A fire-hazardous, white powder that yellows with heating; decomposes when heated; causes ignition when in contact with water; used as an oxidizing agent and a bleach, and in medicinal soap.

sodium phenate [ORG CHEM] C_6H_5ONa White, deliquescent crystals, soluble in water and alcohol; decomposed by carbon dioxide in air; used as a chemical intermediate, antiseptic, and military gas absorbent. Also known as sodium carbolate; sodium phenolate.

sodium phenolate *See* sodium phenate.

sodium phosphate [INORG CHEM] A general term encompassing the following compounds: sodium hexametaphosphate, sodium metaphosphate, dibasic sodium phosphate, hemibasic sodium phosphate, monobasic sodium phosphate, tribasic sodium phosphate, sodium pyrophosphate, and acid sodium pyrophosphate.

sodium phosphite [INORG CHEM] $Na_2HPO_3 \cdot 5H_2O$ White, hygroscopic crystals, melting at 53°C; soluble in water, insoluble in alcohol; used in medicine.

sodium plumbite [INORG CHEM] $Na_2PbO_2 \cdot 3H_2O$ A toxic, corrosive solution of lead oxide (litharge) in sodium hydroxide; used (as doctor solution) to sweeten gasoline.

sodium polysulfide [INORG CHEM] Na_2S_x Yellow-brown granules, used to make dyes and colors, and insecticides, as a petroleum additive, and in electroplating.

sodium propionate [ORG CHEM] CH_3CH_2COONa Deliquescent, transparent crystals; soluble in water, slightly soluble in alcohol; used as a fungicide, and mold preventive.

sodium pteroylglutamate *See* sodium folate.

sodium pyroborate *See* sodium borate.

sodium pyrophosphate [INORG CHEM] $Na_4P_2O_7$ A white powder; soluble in water, insoluble in alcohol and ammonia; melts at 880°C; used as a water softener and a newsprint deinker, and to control drilling-mud viscosity. Also known as normal sodium pyrophosphate; tetrasodium pyrophosphate (TSPP).

sodium salicylate [ORG CHEM] HOC_6H_4COONa A shiny, white powder with sweetish taste and mild aromatic aroma; soluble in water, glycerol, and alcohol; used in medicine and as a preservative.

sodium selenate [INORG CHEM] $Na_2SeO_4 \cdot 10H_2O$ White, poisonous, water-soluble crystals; used as an insecticide.

sodium silicate [INORG CHEM] Na_2SiO_3 A gray-white powder; soluble in alkalies and water, insoluble in alcohol and acids; used to fireproof textiles, in petroleum refining and corrugated paperboard manufacture, and as an egg preservative. Also known as liquid glass; silicate of soda; sodium metasilicate; soluble glass; water glass.

sodium silicofluoride *See* sodium fluosilicate.

sodium stannate [INORG CHEM] $Na_2SnO_3 \cdot 3H_2O$ Water- and alcohol-insoluble, whitish crystals; used in ceramics, dyeing, and textile fireproofing, and as a mordant. Also known as preparing salt.

sodium stearate [ORG CHEM] $NaC_{18}H_{35}O_2$ A white powder with a fatty aroma; soluble in hot water and alcohol; used in medicine and toothpaste and as a waterproofing agent.

sodium succinate [ORG CHEM] $Na_2C_4H_4O_4 \cdot 6H_2O$ Water-soluble, white crystals; loses water at 120°C; used in medicine.

sodium sulfate [INORG CHEM] Na_2SO_4 Crystalline compound, melts at 888°C, soluble in water; used to make paperboard, kraft paper, glass, and freezing mixtures.

sodium sulfhydrate *See* sodium hydrosulfide.

sodium sulfide [INORG CHEM] Na_2S An irritating, water-soluble, yellow to red, deliquescent powder; melts at 1180°C; used as a chemical intermediate, solvent, photographic reagent, and analytical reagent. Also known as sodium sulfuret.

sodium sulfite [INORG CHEM] Na_2SO_3 White, water-soluble, crystals with a sulfurous, salty taste; decomposes when heated; used as a chemical intermediate and food preservative, in medicine and paper manufacturing, and for dyes and photographic developing.

sodium sulfite process [CHEM ENG] A process for the digestion of wood chips in a solution of magnesium, ammonium, or calcium disulfite containing free sulfur dioxide; used in papermaking.

sodium sulfocyanate *See* sodium thiocyanate.

sodium sulfuret *See* sodium sulfide.

sodium TCA *See* sodium trichloroacetate.

sodium tetraborate *See* sodium borate.

sodium tetrafluorescein *See* easin.

sodium tetrasulfide [INORG CHEM] Na_2S_4 Hygroscopic, yellow or dark-red crystals, melting at 275°C; used for insecticides and fungicides, ore flotation, and dye manufacture, and as a reducing agent.

sodium thiocyanate [INORG CHEM] $NaSCN$ A poisonous, water- and alcohol-soluble, deliquescent, white powder; melts at 287°C; used as an analytical reagent, solvent, and chemical intermediate, and for rubber treatment and textile dyeing and printing. Also known as sodium sulfocyanate.

sodium thioglycolate [ORG CHEM] $C_2H_3NaO_3S$ A water-soluble compound produced as hygroscopic crystals; used as an ingredient in bacteriology mediums, and in hair-waving solutions. Also known as sodium mercaptoacetate.

sodium thiosulfate [INORG CHEM] $Na_2S_2O_3 \cdot 5H_2O$ A deliquescent, water- and alcohol-soluble, white, translucent powder with a bitter taste; melts at 48°C; used as a dye mordant and an analytical reagent, in medicine and photographic chemicals, and for the extraction of silver from ore. Also known as hypo; sodium hyposulfite.

sodium trichloroacetate [ORG CHEM] CCl_3COONa A toxic material, used in herbicides and pesticides. Abbreviated sodium TCA.

sodium tungstate [INORG CHEM] $Na_2WO_4 \cdot 2H_2O$ Water-soluble, colorless crystals; lose water at 100°C, melts at 692°C; used as a chemical intermediate analytical reagent, and for fireproofing. Also known as sodium wolframate.

sodium-vapor lamp [ELECTR] A discharge lamp containing sodium vapor, used chiefly for outdoor illumination.

sodium wolframate *See* sodium tungstate.

sofar [NAV] A system of fixing a position at sea by exploding a charge under water, measuring the time for the shock waves to travel through water to three widely separated shore stations, and calculating the position of the explosive by triangulation; the explosive can be dropped from a lifeboat by survivors of air or sea disasters. Derived from sound fixing and ranging.

soffet [CIV ENG] The underside of a horizontal structural member, such as a beam or a slab.

soffione [GEOL] A jet of steam and other vapors issuing from the ground in a volcanic area.

soffosian knob *See* frost mound.

soft cataract [MED] A cataract, affecting the cortex of the lens of the eye, which is of soft consistency and has a milky appearance.

soft chancre *See* chancroid.

soft coal *See* bituminous coal.

soft coral [INV ZOO] The common name for coelenterates composing the order Alcyonacea; the colony is supple and leathery.

softening agent [MATER] 1. A substance that is added to another substance to increase softness; for example, stearic acid added to plastics, fat-liquoring agents to leather, and fatty alcohol to fabrics. 2. A chemical that softens hard water by removing or trapping calcium and magnesium ions.

softening point [PHYS] For a substance which does not have a definite melting point, the temperature at which viscous flow changes to plastic flow.

soft ground [MIN ENG] 1. A mineral deposit which can be mined without drilling and shooting hard rock. 2. The rock about underground openings that does not stand well and requires heavy timbering.

soft hail *See* snow pellets.

soft-iron ammeter [ENG] An ammeter in which current in a coil causes two pieces of magnetic material within the coil, one fixed and one attached to a pointer, to become similarly magnetized and to repel each other, moving the pointer; used for alternating-current measurement.

soft landing [AERO ENG] The act of landing on the surface of a planet or moon without damage to any portion of the vehicle or payload, except possibly the landing gear.

soft magnetic material [ELECTROMAG] A magnetic material which is relatively easily magnetized or demagnetized.

soft missile base [CIV ENG] A missile-launching base that is not protected against a nuclear explosion.

soft palate [ANAT] The posterior part of the palate which consists of an aggregation of muscles, the tensor veli palatini, levator veli palatini, azygos uvulae, palatoglossus, and palatopharyngeus, and their covering mucous membrane.

soft patch [ENG] A patch in a crack in a vessel such as a steam boiler consisting of a soft material inserted in the crack and covered by a metal plate bolted or riveted to the vessel.

soft phosphate [MATER] Powdery, impure tricalcium phosphate separated in fertilizer manufacture from rock and pebble phosphates.

soft point [ORD] A bullet with a soft point, intended to spread upon striking a target with some resistance, such as the flesh of game; not permitted in combat operations.

soft radiation [PHYS] Radiation whose particles or photons have a low energy, and, as a result, do not penetrate any type of material readily.

soft rime [HYD] A white, opaque coating of fine rime deposited chiefly on vertical surfaces, especially on points and edges of objects, generally in supercooled fog.

soft rock [MIN ENG] Rock that can be removed by air-operated hammers, but cannot be handled economically by a pick. [PETR] 1. A broad designation for sedimentary rock. 2. A rock that is relatively nonresistant to erosion.

soft rot [PL PATH] A mushy, watery, or slimy disintegration of plant parts caused by either fungi or bacteria.

soft-shell disease [INV ZOO] A disease of lobsters caused by a chitinous bacterium which extracts chitin from the exoskeleton.

SOFT-IRON AMMETER

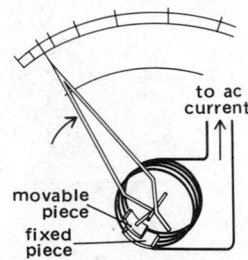

to ac current

movable piece

fixed piece

Schematic of the repulsion-type soft-iron ammeter showing the fixed and movable pieces of magnetic material.

soft solder [MET] Solder composed of an alloy of lead and tin. Also known as low melting solder.

soft soldering [MET] Soldering with a soft solder.

soft tube [ELECTR] 1. An x-ray tube having a vacuum of about 0.000002 atmosphere, the remaining gas being left in intentionally to give less-penetrating rays than those of a more completely evacuated tube. 2. *See* gassy tube.

software [ADP] The totality of programs usable on a particular kind of computer, together with the documentation associated with a computer or program, such as manuals, diagrams, and operating instructions.

software monitor [ADP] A system, used to evaluate the performance of computer software, that is similar to accounting packages, but can collect more data concerning usage of various components of a computer system and is usually part of the control program.

soft waste [TEXT] The waste from yarn manufacturing prior to spinning, including some spinning waste; usually reprocessed in the mill.

soft water [CHEM] Water that is free of magnesium or calcium salts.

soft wood [MATER] Wood from a coniferous tree.

soft x-ray [ELECTROMAG] An x-ray having a comparatively long wavelength and poor penetrating power.

soft x-ray absorption spectroscopy [SPECT] A spectroscopic technique which is used to get information about unoccupied states above the Fermi level in a metal or about empty conduction bands in an inoculator.

soft x-ray appearance potential spectroscopy [SPECT] A branch of electron spectroscopy in which a solid surface is bombarded with monochromatic electrons, and small but abrupt changes in the resulting total x-ray emission intensity are detected as the energy of the electrons is varied. Abbreviated SXAPS.

sogasoid [PHYS] A system of solid particles dispersed in a gas.

Sohm Abyssal Plain [GEOL] A basin in the North Atlantic, about 2400 fathoms deep, between Newfoundland and the Mid-Atlantic Ridge.

soil [GEOL] 1. Unconsolidated rock material over bedrock. 2. Freely divided rock-derived material containing an admixture of organic matter and capable of supporting vegetation.

soil air [GEOL] The air and other gases in spaces in the soil; specifically, that which is found within the zone of aeration. Also known as soil atmosphere.

soil atmosphere *See* soil air.

soil blister *See* frost mound.

soil-cement [MATER] A compacted mixture of soil, cement, and water used as a base course or surface for roads and airport paving.

soil chemistry [GEOCHEM] The study and analysis of the inorganic and organic components and the life cycles within soils.

soil colloid [GEOL] Colloidal complex of soils composed principally of clay and humus.

soil complex [GEOL] A mapping unit used in detailed soil surveys; consists of two or more recognized classifications.

soil conservation [ECOL] Management of soil to prevent or reduce soil erosion and depletion by wind and water.

soil creep [GEOL] The slow, steady downhill movement of soil and loose rock on a slope. Also known as surficial creep.

soil erosion [GEOL] The detachment and movement of topsoil by the action of wind and flowing water.

soil flow *See* solifluction.

soil fluction *See* solifluction.

soil formation *See* soil genesis.

soil genesis [GEOL] The mode by which soil originates, with particular reference to processes of soil-forming factors responsible for the development of true soil from unconsolidated parent material. Also known as pedogenesis; soil formation.

soil mechanics [ENG] The application of the laws of solid and fluid mechanics to soils and similar granular materials as a basis for design, construction, and maintenance of stable foundations and earth structures.

soil microbiology [MICROBIO] A study of the microorganisms in soil, their functions, and the effect of their activities on the character of the soil and the growth and health of plant life.

soil moisture *See* soil water.

soil physics [GEOPHYS] The study of the physical characteristics of soils; concerned also with the methods and instruments used to determine these characteristics.

soil pipe [CIV ENG] A vertical cast-iron or plastic pipe for carrying sewage from a building into the soil drain.

soil profile [GEOL] A vertical section of a soil, showing horizons and parent material.

soil rot [PL PATH] Plant rot caused by soil microorganisms.

soil science [GEOL] The study of the formation, properties, and classification of soil; includes mapping. Also known as pedology.

soil series [GEOL] A family of soils having similar profiles, and developing from similar original materials under the influence of similar climate and vegetation.

soil shear strength [GEOL] The maximum resistance of a soil to shearing stresses.

soil stabilizer [MATER] A chemical that alters the engineering property of a natural soil; used to stabilize soil slopes, to prepare for building foundations, and to prevent erosion.

soil stripes [GEOL] Alternating bands of fine and coarse material in a soil structure.

soil structure [GEOL] Arrangement of soil into various aggregates, each differing in the characteristics of its particles.

soil survey [GEOL] The systematic examination of soils, their description and classification, mapping of soil types, and the assessment of soils for various agricultural and engineering uses.

soil thermograph [ENG] A remote-recording thermograph whose sensing element may be buried at various depths in the earth.

soil thermometer [ENG] A thermometer used to measure the temperature of the soil, usually the mercury-in-glass thermometer. Also known as earth thermometer.

soil water [HYD] Water in the belt of soil water. Also known as rhizic water; soil moisture.

soil-water belt *See* belt of soil water.

soil-water zone *See* belt of soil water.

sol [CHEM] A colloidal solution consisting of a suitable dispersion medium, which may be gas, liquid, or solid, and the colloidal substance, the disperse phase, which is distributed throughout the dispersion medium.

Sol *See* sun.

solaire [METEOROL] A name generally applied to winds from an easterly direction (that is, from the rising sun) in central and southern France.

sol-air temperature [METEOROL] The temperature which, under conditions of no direct solar radiation and no air motion, would cause the same heat transfer into a house as that caused by the interplay of all existing atmospheric conditions.

Solanaceae [BOT] A family of dicotyledonous plants in the order Polemoniales having internal phloem, mostly numerous ovules and seeds on axile placentae, and mostly cellular endosperm.

solano [METEOROL] A southeasterly or easterly wind on the southeast coast of Spain in summer; usually an extension of the sirocco; it is hot and humid and sometimes brings rain; when dry, it is dusty.

solar absorption index [GEOPHYS] A relation of the sun's angle at various latitudes and local times with the ionospheric absorption.

solar activity [ASTRON] Disturbances on the surface of the sun; examples are sunspots, prominences, and solar flares.

solar air mass [METEOROL] The optical air mass penetrated by light from the sun for any given position of the sun.

solar apex [ASTRON] A point toward which the solar system is moving; it is about 10° southwest of the star Vega.

solar atmospheric tide [GEOPHYS] An atmospheric tide due to the thermal or gravitational action of the sun.

solar attachment [ENG] A device for determining the true meridian directly from the sun; used an an attachment on a surveyor's transit or compass.

solar battery [ELECTR] An array of solar cells, usually connected in parallel and series.

solar burst [ASTROPHYS] A sudden increase in the radio-frequency energy radiated by the sun, generally associated with visible solar flares.

solar calendar [ASTRON] A calendar based on the time period known as the tropical year, which has 365.24220 days.

solar cell [ELECTR] A *pn*-junction device which converts the radiant energy of sunlight directly and efficiently into electrical energy.

solar climate [CLIMATOL] The hypothetical climate which would prevail on a uniform solid earth with no atmosphere; thus, it is a climate of temperature alone and is determined only by the amount of solar radiation received.

solar constant [METEOROL] The rate at which energy from the sun is received just outside the earth's atmosphere on a surface normal to the incident radiation and at the earth's mean distance from the sun; it is 0.140 watt per square centimeter.

solar cooking [FOOD ENG] The preparation of food by concentrating solar radiation on a heater plate.

solar corona [ASTRON] The upper, rarefied solar atmosphere which becomes visible around the darkened sun during a total solar eclipse.

solar cosmic rays *See* energetic solar particles.

solar cycle [ASTRON] The periodic change in the number of sunspots; the cycle is taken as the interval between successive minima and is about 11.1 years.

solar day [ASTRON] A time measurement, the duration of one rotation of the earth on its axis with respect to the sun; this may be a mean solar day or an apparent solar day as the reference is to the mean sun or apparent sun.

solar dermatitis [MED] Any of various skin eruptions caused by exposure to the sun, excluding sunburn.

solar distillation [CHEM ENG] A procedure in which the sun's heat is used to evaporate seawater in order to produce sodium chloride and other salts or potable water.

solar eclipse [ASTRON] An eclipse that takes place when the new moon passes between the earth and the sun and the shadow formed reaches the earth; may be classified as total, partial, or annular.

solar energy [ASTROPHYS] The energy transmitted from the sun in the form of electromagnetic radiation.

solar engine [MECH ENG] An engine which converts thermal energy from the sun into electrical, mechanical, or refrigeration energy; may be used as a method of spacecraft propulsion, either directly by photon pressure on huge solar sails, or indirectly from solar cells or from a reflector-boiler combination used to heat a fluid.

solar evaporation [HYD] The evaporation of water due to the sun's heat.

solar-excited laser *See* sun-pumped laser.

solar faculae [ASTRON] Bright streaks or regions on the surface of the sun, especially near solar sunspots.

solar flare [ASTROPHYS] An abrupt increase in the intensity of the H-α and other emission near a sunspot region; the brightness may be many times that of the associated plage.

solar furnace [ENG] An image furnace in which high temperatures are produced by focusing solar radiation.

solar generator [ELEC] An electric generator powered by radiation from the sun and used in some satellites.

solar heat exchanger drive [AERO ENG] A proposed method of spacecraft propulsion in which solar radiation is focused on an area occupied by a boiler to heat a working fluid that is expelled to produce thrust directly. Abbreviated SHED.

solar heating [MECH ENG] The conversion of solar radiation into heat for technological, comfort-heating, and cooking purposes.

solar heat storage [ENG] The storage of solar energy for later use; usually accomplished by the heating of water or fusing a salt, although sand and gravel have been used as storage media.

solar house [BUILD] A house with large expanses of glass designed to catch solar radiation for heating.

solarimeter [ENG] 1. A type of pyranometer consisting of a Moll thermopile shielded from the wind by a bell glass. 2. *See* pyranometer.

solar magnetic field [ASTROPHYS] The magnetic field that pervades the ionized and highly conducting gas composing the sun.

solar magnetograph [ENG] An instrument that utilizes the

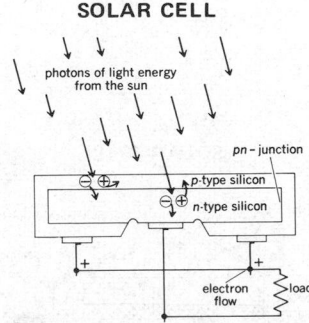

SOLAR CELL

Cross-sectional view of a silicon *pn*-junction solar cell, illustrating the creation of electron pairs by photons of light energy from the sun.

Zeeman effect to directly measure the strength and polarity of the complex patterns of magnetic fields at the sun's surface; comprises a telescope, a differential analyzer, a spectrograph, and a photoelectric or photographic means of differencing and recording.

solar month [ASTRON] A time interval equal to one-twelfth of the solar year.

solar motion [ASTRON] The two main motions of the sun: relative motion with respect to the neighboring stars, or motion due to the rotation of the Milky Way of which the sun is a part.

solar neutrino unit [ASTROPHYS] A unit for measuring the capture rate of neutrinos emanating from the sun, equal to 10^{-36} per second per atom. Abbreviated SNU.

solar noise See solar radio noise.

solar nutation [ASTRON] Nutation caused by the change in declination of the sun.

solar orbit [ASTRON] An orbit of a planet or other celestial body or satellite about the sun.

solar parallax [ASTRON] The sun's mean equatorial horizontal parallax p, which is the angle subtended by the equatorial radius r of the earth at mean distance a of the sun.

solar photophthalmia See snow blindness.

solar physics [ASTROPHYS] The scientific study of all physical phenomena connected with the sun; it overlaps with geophysics in the consideration of solar-terrestrial relationships, such as the connection between solar activity and auroras.

solar power [MECH ENG] The conversion of the energy of the sun's radiation to useful work.

solar probe [AERO ENG] A space probe whose trajectory passes near the sun so that instruments on board may detect and transmit back to earth data about the sun.

solar prominence [ASTRON] Sheets of luminous gas emanating from the sun's surface; they appear dark against the sun's disk but bright against the dark sky, and occur only in regions of horizontal magnetic fields.

solar propagation [BOT] A method of rooting plant cuttings involving the use of a modified hotbed; bottom heat is provided by radiation of stored solar heat from bricks or stones in the bottom of the hotbed frame.

solar propulsion [AERO ENG] Spacecraft propulsion with a system composed of a type of solar engine.

solar radiation [ASTROPHYS] The electromagnetic radiation and particles (electrons, protons, and rarer heavy atomic nuclei) emitted by the sun.

solar-radiation observation [GEOPHYS] An evaluation of the radiation from the sun that reaches an observation point; the observing instrument is usually a pyrheliometer or pyranometer.

solar radio emission [ASTROPHYS] Radio-frequency electromagnetic radiation emitted from the sun, and increasing greatly in intensity during sunspots and flares.

solar radio noise [ELECTROMAG] Radio noise originating at the sun, and increasing greatly in intensity during sunspots and flares; it is heard as a hissing noise on shortwave radio receivers. Also known as solar noise.

solar rocket [AERO ENG] A rocket designed to carry instruments to measure and transmit parameters of the sun.

solar sail [AERO ENG] A surface of a highly polished material upon which solar light radiation exerts a pressure. Also known as photon sail.

solar satellite [AERO ENG] A space vehicle designed to enter into orbit about the sun. Also known as sun satellite.

solar sensor [ELECTR] A light-sensitive diode that sends a signal to the attitude-control system of a spacecraft when it senses the sun. Also known as sun sensor.

solar spectrum [ASTROPHYS] The spectrum of the sun's electromagnetic radiation extending over the whole electromagnetic spectrum, from wavelengths of 10^{-9} centimeter to 30 kilometers.

solar spicule See spicule.

solar still [CHEM ENG] A device for evaporating seawater, in which water is confined in one or more shallow pools, over which is placed a roof-shaped transparent cover made of glass or plastic film; the sun's heat evaporates the water, leaving behind a residue of salt; the vapor from the evaporated water condenses on the surface of the cover and trickles down into gutters, which thus collect fresh water.

solar system [ASTRON] The sun and the celestial bodies moving about it; the bodies are planets, satellites of the planets, asteroids, comets, and meteor swarms.

solar telescope [OPTICS] An observational instrument of the solar astronomer; it is designed so that heating effects produced by the sun do not distort the images; the two classes consist of those designed for observations of the brilliant solar disk, and those designed for the study of the much fainter prominences and the still fainter corona through the relatively bright, scattered light of the sky.

solar-terrestrial phenomena [GEOPHYS] All observed physical effects that are caused by solar activity; the phenomena may be in the atmosphere or on the earth's surface; an example is the aurora borealis.

solar tide [OCEANOGR] The tide caused solely by the tide-producing forces of the sun.

solar time [ASTRON] Time based on the rotation of the earth relative to the sun.

solar-topographic theory [CLIMATOL] The theory that the changes of climate through geologic time (the paleoclimates) have been due to changes of land and sea distribution and orography, combined with fluctuations of solar radiation of the order of 10–20% on either side of the mean.

solar turboelectric drive [AERO ENG] A proposed method of spacecraft propulsion in which solar radiation is focused on an area occupied by a boiler to heat a working fluid that drives a turbine generator system, producing electrical energy. Abbreviated STED.

solar-type star [ASTRON] Any of the stars (yellow stars) of spectral type G, so called because the sun is in this class.

solar ultraviolet radiation [ASTROPHYS] That portion of the sun's electromagnetic radiation that has wavelengths from about 400 to about 4 nanometers; this radiation may sufficiently ionize the earth's atmosphere so that propagation of radio waves is affected.

solar wind [GEOPHYS] The supersonic flow of gas, composed of ionized hydrogen and helium, which continuously flows from the sun out through the solar system with velocities of 300 to 1000 kilometers per second; it carries magnetic fields from the sun.

Solasteridae [INV ZOO] The sun stars, a family of asteroid echinoderms in the order Spinulosida.

solation [PHYS CHEM] The change of a substance from a gel to a sol.

solder [MET] **1.** To join by means of solder. **2.** An alloy, such as of zinc and copper, or of tin and lead, used when melted to join metallic surfaces.

solder brazing [MET] Brazing by means of a relatively high-melting solder.

solder glass [MATER] A special glass having a relatively low softening point (below 500°C); used to join two pieces of higher-melting glass without softening and deforming them.

soldering embrittlement [MET] The reduction in mechanical properties of a metal due to the local penetration of solder along grain boundaries.

soldering flux [MET] A chemical substance which aids the flow of solder and serves to remove and prevent the formation of oxides on the pieces to be united.

soldering gun [ENG] A soldering iron shaped like a gun.

soldering iron [ENG] A rod of copper with a handle on one end and pointed or wedge-shaped at the other end, and used for applying heat in soldering.

soldering lug [ELEC] A stamped metal strip used as a terminal to which wires can be soldered.

solderless contact See crimp contact.

solderless wrapped connection See wire-wrap connection.

sole [ELECTR] Electrode used in magnetrons and backward-wave oscillators to carry a current that generates a magnetic field in the direction wanted. [GEOGR] The lowest part of a valley. [GEOL] **1.** The bottom of a sedimentary stratum. **2.** The middle and lower portion of the shear surface of a landslide. **3.** The underlying fault plane of a thrust nappe. Also known as sole plane. [HYD] The basal ice of a glacier, often dirty in appearance due to contained rock fragments.

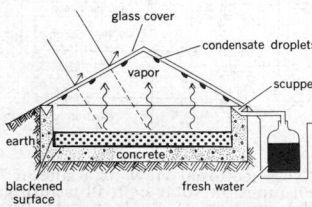

SOLAR STILL

Roof-type solar still. Dotted area above the concrete is sea water.

soleil compensator [OPTICS] A compensator which resembles the Babinet compensator but is constructed so that the phase change is constant over the entire field.

sole injection [GEOL] An igneous intrusion that was put in place along a thrust plane.

sole mark [GEOL] An irregularity or penetration on the undersurface of a sedimentary stratum.

Solemyidae [INV ZOO] A family of bivalve mollusks in the order Protobranchia.

Solenichthyes [VERT ZOO] An equivalent name for Gasterosteiformes.

solenium [INV ZOO] A diverticulum of the enteron in certain hydroids.

solenocyte [INV ZOO] Any of various hollow, flagellated cells in the nephridia of the larvae of certain annelids, mollusks, rotifers, and lancelets.

solenodon [VERT ZOO] Either of two species of insectivorous mammals comprising the family Solenodontidae; the almique (*Atopogale cubana*) is found only in Cuba, while the white agouta (*Solenodon paradoxus*) is confined to Haiti.

Solenodontidae [VERT ZOO] The solenodons, a family of insectivores belonging to the group Lipotyphla.

Solenogastres [INV ZOO] The equivalent name for Aplacophora.

solenoid [ELECTROMAG] Also known as electric solenoid. **1.** An electrically energized coil of insulated wire which produces a magnetic field within the coil. **2.** In particular, a coil that surrounds a movable iron core which is pulled to a central position with respect to the coil when the coil is energized by sending current through it. [METEOROL] *See* meteorological solenoid.

solenoidal [MATH] A vector field has this property in a region if its divergence vanishes at every point of the region.

solenoidal index [METEOROL] The difference between the mean virtual temperature from the surface to some specified upper level averaged around the earth at 55° latitude, and the mean virtual temperature for the corresponding layer averaged at 35° latitude.

solenoid brake [MECH ENG] A device that retards or arrests rotational motion by means of the magnetic resistance of a solenoid.

solenoid valve [MECH ENG] A valve actuated by a solenoid, for controlling the flow of gases or liquids in pipes.

Solenopora [PALEOBOT] A genus of extinct calcareous red algae in the family Solenoporaceae that appeared in the Late Cambrian and lasted until the Early Tertiary.

Solenoporaceae [PALEOBOT] A family of extinct red algae having compact tissue and the ability to deposit calcium carbonate within and between the cell walls.

solepiece [CIV ENG] One of two steel plates, port and starboard, whose forward parts are bolted to the ground ways supporting a ship about to be launched, while their aft parts are attached to the sliding ways; at the start of the launch, they are cut simultaneously with burning torches to release the ship. Also known as soleplate. [NAV ARCH] **1.** A member attached to the bottom of the rudder of a wooden ship, which brings it down to the false keel. **2.** A member which joins the sternpost and the rudderpost of a wooden ship. **3.** A projection from the keel of a ship.

sole plane *See* sole.

soleplate [BUILD] The plate on which stud bases butt in a stud partition. [CIV ENG] *See* solepiece. [ENG] **1.** The supporting base of a machine. **2.** A plate on which a bearing can be attached and, if necessary, adjusted slightly.

soleus [ANAT] A flat muscle of the calf; origin is the fibula, popliteal fascia, and tibia, and insertion is the calcaneus; plantar-flexes the foot.

solfatara [GEOL] A fumarole from which sulfurous gases are emitted.

solid [PHYS] **1.** A substance that has a definite volume and shape and resists forces that tend to alter its volume or shape. **2.** A crystalline material, that is, one in which the constituent atoms are arranged in a three-dimensional lattice, periodic in three independent directions.

solid angle [MATH] A surface formed by all rays joining a point to a closed curve.

solid asphalt [MATER] Asphalt with a penetration number of less than 10 under specified test conditions.

solid box [MECH ENG] A solid, unadjustable ring bearing lined with babbitt metal, used on light machinery.

solid car [MIN ENG] A mine car equipped with a swivel coupling and generally used with a rotary dump.

solid coupling [MECH ENG] A flanged-face or a compression-type coupling used to connect two shafts to make a permanent joint and usually designed to be capable of transmitting the full load capacity of the shaft; a solid coupling has no flexibility.

solid crib timbering [MIN ENG] Shaft timbering with cribs laid solidly upon one another.

solid cutter [DES ENG] A cutter made of a single piece of material.

solid die [DES ENG] A one-piece screw-cutting tool with internal threads.

solid-dielectric capacitor [ELEC] A capacitor whose dielectric is one of several solid materials such as ceramic, mica, glass, plastic film, or paper.

solid drilling [ENG] In diamond drilling, using a bit that grinds the whole face, without preserving a core for sampling.

solid-electrolyte battery [ELEC] A primary battery whose electrolyte is either a solid crystalline salt, such as silver iodide or lead chloride, or an ion-exchange membrane; in either case, conductivity is almost entirely ionic.

solid-electrolyte fuel cell [ELEC] Self-contained fuel cell in which oxygen is the oxidant and hydrogen is the fuel; the oxidant and fuel are kept separated by a solid electrolyte which has a crystalline structure and a low conductivity.

solid explosive [MATER] An explosive employed in the form of a powder, a light-running granulated mass, or as solid sticks.

solid helium [CRYO] A certain state which is not attained by helium under its own vapor pressure down to absolute zero, but which requires an external pressure of 25 atmospheres at absolute zero.

solidification [PHYS] The change of a fluid (liquid or gas) into the solid state.

solidification shrinkage [MET] Volume contraction of a metal during solidification.

solidified petroleum *See* jellied petroleum.

solid injection system [MECH ENG] A fuel injection system for a diesel engine in which a pump forces fuel through a fuel line and an atomizing nozzle into the combustion chamber.

solid insulator [ELEC] An electric insulator made of a solid substance, such as sulfur, polystyrene, rubber, or porcelain.

solid-liquid equilibrium [PHYS CHEM] **1.** The interrelation of a solid material and its melt at constant vapor pressure. **2.** The concentration relationship of a solid with a solvent liquid other than its melt. Also known as liquid-solid equilibrium.

solid logic technology [ELECTR] Miniaturized modules used in computers which result in faster circuitry because of reduced distance for current to travel.

solid moment of inertia [PHYS] The integral of the products of the mass of each of the infinitesimal elements of the solid with the square of their distance from a given axis.

solid-phase welding [MET] A welding method in which the weld is consummated by pressure or by heat and pressure without fusion.

solid-piled [MATER] Pertaining to plywood which is fresh from clamps or a hot press, and which is piled onto a solid flat base without stickers and weighted down until it reaches its normal temperature and moisture content. Also known as bulked-down; dead-piled.

solid propellant [MATER] A rocket propellant in solid form, usually containing both fuel and oxidizer combined or mixed, and formed into a monolithic (not powdered or granulated) grain. Also known as solid rocket fuel; solid rocket propellant.

solid propellant binder [MATER] The ingredient component of a propellant that is the agent for holding all the other ingredients together; contributes most to the physical or mechanical properties of the grain.

solid-propellant rocket engine [AERO ENG] A rocket engine fueled with a solid propellant; such motors consist essentially of a combustion chamber containing the propellant, and a nozzle for the exhaust jet.

SOLENODON

The solenodon uses its elongated snout to grub for food.

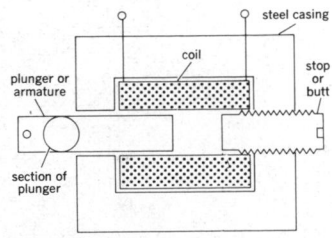

SOLENOID

A cross-sectional view of a steel-clad solenoid showing coil and movable core known as plunger or armature.

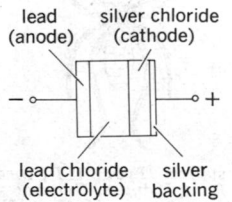

SOLID-ELECTROLYTE BATTERY

Schematic diagram of typical solid-electrolyte cell with solid crystalline salt electrolyte.

solid rocket [AERO ENG] A rocket that is propelled by a solid-propellant rocket engine.

solid rocket fuel *See* solid propellant.

solid rocket propellant *See* solid propellant.

solid shafting [MECH ENG] A solid round bar that supports a roller and wheel of a machine.

solid solution [PHYS] A homogeneous crystalline phase composed of several distinct chemical species, occupying the lattice points at random and existing in a range of concentrations.

solid state [PHYS] The condition of a substance in which it is a solid.

solid-state circuit [ELECTR] Complete circuit formed from a single block of semiconductor material.

solid-state circuit breaker [ELECTR] A circuit breaker in which a Zener diode, silicon controlled rectifier, or solid-state device is connected to sense when load terminal voltage exceeds a safe value.

solid-state component [ELECTR] A component whose operation depends on the control of electrical or magnetic phenomena in solids, such as a transistor, crystal diode, or ferrite device.

solid-state computer [ADP] A digital computer which uses diodes and transistors instead of vacuum tubes.

solid-state counter [NUCLEO] A radiation counter whose sensitive material is a crystalline solid; for example, a crystal counter or a scintillation counter.

solid-state device [ELECTR] A device, other than a conductor, which uses magnetic, electrical, and other properties of solid materials, as opposed to vacuum or gaseous devices.

solid-state laser [OPTICS] A laser in which a semiconductor material produces the coherent output beam.

solid-state maser [PHYSICS] A maser in which a semiconductor material produces the coherent output beam; two input waves are required: one wave, called the pumping source, induces upward energy transitions in the active material, and the second wave, of lower frequency, causes downward transitions and undergoes amplification as it absorbs photons from the active material.

solid-state physics [PHYS] The branch of physics centering about the physical properties of solid materials.

solid-state switch [ELECTR] A microwave switch in which a semiconductor material serves as the switching element; a zero or negative potential applied to the control electrode will reverse-bias the switch and turn it off, and a slight positive voltage will turn it on.

solid-state thyratron [ELECTR] A semiconductor device, such as a silicon controlled rectifier, that approximates the extremely fast switching speed and power-handling capability of a gaseous thyratron tube.

solid stowing [MIN ENG] The complete filling of the waste area behind a longwall face with stone and dirt.

solid tantalum capacitor [ELEC] A tantalum electrolytic capacitor that uses a solid tantalum wire as the electrode immersed in the electrolyte; for nonpolarized alternating-current capacitors, there will be two such electrodes.

solidus [PHYS CHEM] In a constitution or equilibrium diagram, the locus of points representing the temperature below which the various compositions finish freezing on cooling, or begin to melt on heating.

solidus curve [PHYS CHEM] A curve on the phase diagram of a system with two components which represents the equilibrium between the liquid phase and the solid phase.

solid-web girder [CIV ENG] A beam, such as a box girder, having a web consisting of a plate or other solid section but not a lattice.

solifluction [GEOL] A rapid soil creep, especially referring to downslope soil movement in periglacial areas. Also known as sludging; soil flow; soil fluction.

solifluction lobe [GEOL] An isolated, tongue-shaped feature of the land surface with a steep front and a smooth upper surface formed by more rapid solifluction on certain sections of the slope. Also known as solifluction tongue.

solifluction mantle [GEOL] The locally derived, unsorted material moved downslope by solifluction. Also known as flow earth.

solifluction sheet [GEOL] A broad deposit of a solifluction mantle.

solifluction stream [GEOL] A narrow, streamlike deposit of a solifluction mantle.

solifluction tongue *See* solifluction lobe.

solion [ELEC] An electrochemical device in which amplification is obtained by controlling and monitoring a reversible electrochemical reaction.

soliquid [PHYS CHEM] A system in which solid particles are dispersed in a liquid.

solitary wave [PHYS] A traveling wave in which a single disturbance is neither preceded by nor followed by other such disturbances, but which does not involve unusually large amplitudes or rapid changes in variables, in contrast to a shock wave.

solodize [GEOL] To improve a soil by removing alkalis from it.

Solod soil *See* Soloth soil.

Solo man [PALEON] A relative but primitive form of fossil man from Java; this form had a small brain, heavy horizontal browridges, and a massive cranial base.

Solomon R unit [NUCLEO] A unit of radiation dose rate due to x-rays, equal to 2100 roentgens per hour. Also known as R unit.

Solonchak soil [GEOL] One of an intrazonal, balamorphic group of light-colored soils rich in soluble salts.

Solonetz soil [GEOL] One of an intrazonal group of black alkali soils having a columnar structure.

solore [METEOROL] A cold, night wind of the mountains following the course of the Drome River in southeastern France.

Soloth soil [GEOL] One of an intrazonal halomorphic group of soils formed from saline material; the surface layer is soft and friable, and overlies a light-colored leached horizon which, in turn, overlies a dark horizon. Also known as Solod soil.

Solpugida [INV ZOO] The sun spiders, an order of nonvenomous, spiderlike, predatory arachnids having large chelicerae for holding and crushing prey.

solstice [ASTRON] The two days (actually, instants) during the year when the earth is so located in its orbit that the inclination (about 23½°) of the polar axis is toward the sun; the days are June 21 for the North Pole and December 22 for the South Pole; because of leap years, the dates vary a little.

solstitial colure [ASTRON] That great circle of the celestial sphere through the celestial poles and the solstices.

solstitial points [ASTRON] Those points of the ecliptic that are 90° from the equinoxes north or south at which the greatest declination of the sun is reached.

solstitial tidal currents [OCEANOGR] Tidal currents of especially large tropic diurnal inequality occurring at the time of solstitial tides.

solstitial tides [OCEANOGR] Tides occurring near the times of the solstices, when the tropic range is especially large.

solubility [PHYS CHEM] The ability of a substance to form a solution with another substance.

solubility coefficient [PHYS CHEM] The volume of a gas that can be dissolved by a unit volume of solvent at a specified pressure and temperature.

solubility curve [PHYS CHEM] A graph showing the concentration of a substance in its saturated solution in a solvent as a function of temperature.

solubility product constant [PHYS CHEM] A type of simplified equilibrium constant, K_{sp}, defined for and useful for equilibria between solids and their respective ions in solution; for example, the equilibrium

$$AgCl(s) \rightleftarrows Ag^+ + Cl^-, \ [Ag^+][Cl^-] \cong K_{sp},$$

where $[Ag^+]$ and $[Cl^-]$ are molar concentrations of silver ions and chloride ions.

solubility test [ANALY CHEM] **1.** A test for the degree of solubility of asphalts and other bituminous materials in solvents, such as carbon tetrachloride, carbon disulfide, or petroleum ether. **2.** Any test made to show the solubility of one material in another (such as liquid-liquid, solid-liquid, gas-liquid, or solid-solid).

soluble [CHEM] Capable of being dissolved.

soluble castor oil *See* Turkey red oil.

SOLPUGIDA

Sun spider. *(From T. I Storer and R. L. Usinger, General Zoology, 3d ed., McGraw-Hill, 1957)*

soluble cutting oil [MATER] A petroleum oil containing an emulsifying agent to make it mix easily with water; used as a coolant for metal-cutting tools.

soluble glass *See* sodium silicate.

soluble gun-cotton *See* pyroxylin.

soluble nitrocellulose *See* pyroxylin.

soluble oil [MATER] An oil that readily forms a stable emulsion or colloidal suspension in water. Also known as emulsifying oil.

soluble starch [MATER] A group of water-soluble polymers formed from starch, such as the starches derived from corn or potato, by acetylation, acid hydrolysis, chlorination, or by action of enzymes to form starch acetates, ethers, and esters; used as textile sizing agents, emulsifying agents, and paper coatings.

solum [GEOL] The upper part of a soil profile, composed of A and B horizons in mature soil. Also known as true soil.

solute [CHEM] The substance dissolved in a solvent.

solution [CHEM] A single, homogeneous liquid, solid, or gas phase that is a mixture in which the components (liquid, gas, solid, or combinations thereof) are uniformly distributed throughout the mixture.

solution ceramic [ELEC] A nonbrittle, inorganic ceramic insulating coating that can be applied to wires at a low temperature; examples include ceria, chromia, titania, and zirconia.

solution dyeing [TEXT] Adding dye to the chemical compound in the spinneret before extrusion. Also known as dope dyeing.

solution gas [PETRO ENG] Gaseous reservoir hydrocarbons dissolved in liquid reservoir hydrocarbons because of the prevailing pressures in the reservoir. Also known as dissolved gas.

solution-gas reservoir [PETRO ENG] Oil reservoir initially at or above the bubble-point pressure of the gas-oil mixture, and produced primarily by the expansion of the oil and its dissolved gas. Also known as dissolved-gas reservoir.

solution heat treatment [MET] Heating and holding an alloy at a temperature at which one (or more) constituent enters into solid solution, then cooling the alloy rapidly to prevent the constituent from precipitating.

solution mining [MIN ENG] The extraction of soluble minerals from subsurface strata by injection of fluids, and the controlled removal of mineral-laden solutions.

solution poison [NUCLEO] A soluble nuclear poison, such as boric acid, added to the coolant of a nuclear reactor for purposes of reactivity control; generally used only during shutdown periods, and chemically removed from the coolant prior to resuming operation.

solution pool [GEOL] A pool in a rock that is formed by the dissolution of the rock in ocean water.

solution porosity [PETRO ENG] A generic designation for reservoir-rock porosity created by solution action; some examples are crystalline limestone and dolomite, porous cap rock, and honeycombed anhydrite.

solution potholes [GEOL] Potholes produced in carbonate rocks by dissolution.

solution pressure [PHYS CHEM] **1.** A measure of the tendency of molecules or atoms to cross a bounding surface between phases and to enter into a solution. **2.** A measure of the tendency of hydrogen, metals, and certain nonmetals to pass into solution as ions.

solution transfer [GEOL] A process whereby pressure solution of detrital mineral grains at contact areas is followed by recrystallization on the less strained parts of the grain surfaces.

solutizer-air regenerative process [CHEM ENG] A petroleum refinery process that is identical to the solutizer-steam regeneration process, except for the regeneration step; the newer units use uncatalyzed air regeneration.

solutizer-steam regenerative process [CHEM ENG] A petroleum refinery process used to extract mercaptans from gasoline or naphtha; uses solutizers (potassium isobutyrate or potassium alkyl phenolate) in strong potassium hydroxide solution as the selective solvent.

solutizer-tannin process [CHEM ENG] A petroleum refinery process that is an early variation of the solutizer-air regenera-

tive process for extraction of mercaptans from gasoline; uses tannin-catalyzed oxidation for the regeneration step.

solutrope [CHEM] A ternary mixture with two liquid phases and a third component distributed between the phases, or selectively dissolved in one or the other of the phases; analogous to an azeotrope.

solvable group [MATH] A group G which has subgroups G_0, $G_1,\ldots,G_n$, where $G_0 = G$, $G_n =$ the identity element alone, and each G_i is a normal subgroup of G_{i-1} with the quotient group G_{i-1}/G_i Abelian.

solvation [CHEM] The process of swelling, gelling, or dissolving of a material by a solvent; for resins, the solvent can be a plasticizer.

Solvay process [CHEM ENG] The process to make sodium carbonate and calcium chloride by treating sodium chloride with ammonia and carbon dioxide.

solvent [CHEM] That part of a solution that is present in the largest amount, or the compound that is normally liquid in the pure state (as for solutions of solids or gases in liquids).

solvent deasphalting [CHEM ENG] A petroleum refinery process used to remove asphaltic and resinous materials from reduced crude oils, lubricating oil stocks, gas oils, or middle distillates through the extractive or precipitant action of solvents. Also known as solvent deresining.

solvent deresining *See* solvent deasphalting.

solvent dewaxing [CHEM ENG] A petroleum refinery process for solvent removal of wax from oils; the mixture of waxy oil and solvent is chilled, then filtered or centrifuged to remove the precipitated oil; the solvent is recovered for reuse.

solvent dyeing [TEXT] The dyeing of synthetic textiles by using chlorinated hydrocarbon solvents (such as trichloroethylene or perchloroethylene) instead of water; used for nylons, polyesters, and acrylics.

solvent extraction [CHEM ENG] The separation of materials of different chemical types and solubilities by selective solvent action; that is, some materials are more soluble in one solvent than in another, hence there is a preferential extractive action; used to refine petroleum products, chemicals, vegetable oils, and vitamins. [NUCLEO] A process for removing uranium fuel residue from used fuel elements of a reactor; it generally involves decay cooling under water for up to 6 months, removal of cladding, dissolution, separation of reusable fuel, decontamination, and disposal of radioactive wastes. Also known as liquid extraction.

solvent molding [ENG] A process to form thermoplastic articles by dipping a mold into a solution or dispersion of the resin and drawing off (evaporating) the solvent to leave a plastic film adhering to the mold.

solvent naphtha [MATER] Refined petroleum naphtha of restricted boiling range; used as a solvent and paint thinner and in dry cleaning; Stoddard Solvent is a special grade of solvent naphtha.

solvent recovery [CHEM ENG] For reuse purposes, the catching and recovery of solvent vapors from vent lines, process vessels, or other sources of evaporative loss, usually with a solid adsorbent material.

solvent-refined [CHEM ENG] Pertaining to any product material whose final quality and condition is in part the result of a solvent treatment during processing of the feedstock material.

solvent-refined oil [MATER] A lubricating oil that has been solvent-treated during refining, such as most motor, aircraft, diesel-engine, steam-turbine, and other high-quality oils.

solvent refining [CHEM ENG] The process of treating a mixed material with a solvent that preferentially dissolves and removes certain minor constituents (usually the undesired ones); common in the petroleum refining industry.

solvmanifold [MATH] A homogeneous space obtained by factoring a connected solvable Lie group by a closed subgroup.

solvolysis [CHEM] A reaction in which a solvent reacts with the solute to form a new substance.

solvus [PHYS CHEM] In a phase or equilibrium diagram, the locus of points representing the solid-solubility temperatures of various compositions of the solid phase.

soma [BIOL] The whole of the body of an individual, excluding the germ tract.

Somali Current *See* East Africa Coast Current.

Somasteroidea [INV ZOO] A subclass of Asterozoa comprising sea stars of generalized structure, the jaws often only partly developed, and the skeletal elements of the arm arranged in a double series of transverse rows termed metapinnules.

somatic aneuploidy [CYTOL] An irregular variation in number of one or more individual chromosomes in the cells of a tissue.

somatic cell [BIOL] Any cell of the body of an organism except the germ cells.

somatic copulation [MYCOL] A form of reproduction in ascomycetes and basidiomycetes involving sexual fusion of undifferentiated vegetative cells.

somatic crossing-over [CYTOL] Crossing-over during mitosis in somatic or vegetative cells.

somatic death [BIOL] The cessation of characteristic life functions.

somatic mesoderm [EMBRYO] The external layer of the lateral mesoderm associated with the ectoderm after formation of the coelom.

somatic nervous system [PHYSIO] The portion of the nervous system concerned with the control of voluntary muscle and relating the organism with its environment.

somatic pairing [CYTOL] The pairing of homologous chromosomes at mitosis in somatic cells; occurs in Diptera.

somatic reflex system [PHYSIO] An involuntary control system characterized by a control loop which includes skeletal muscles.

somatization [PSYCH] A type of neurosis manifested in neurasthenias, hypochondriacal symptoms, and conversion hysterias.

somatoblast [INV ZOO] **1.** An undifferentiated cleavage cell that gives rise to somatic cells in annelids. **2.** The outer cell layer of the nematogen in Dicyemida.

somatochrome [CYTOL] A nerve cell possessing a well-defined body completely surrounding the nucleus on all sides, the cytoplasm having a distinct contour, and readily taking a stain.

somatocyst [INV ZOO] A cavity filled with air in the float of certain Siphonophora.

somatometry [ANAT] Measurement of the human body with the soft parts intact.

somatophyte [BOT] A plant composed of distinct somatic cells that develop especially into mature or adult tissue.

somatopleure [EMBRYO] A complex layer of tissue consisting of the somatic layer of the mesoblast together with the epiblast, forming the body wall in craniate vertebrates and the amnion and chorion in amniotes.

somatopsychic [MED] Pertaining to both the body and mind.

somatotonia [PSYCH] The temperamental trait associated with a mesomorphic somatotype, characterized by an active, aggressive, and risk-taking approach to life.

somatotropin [BIOCHEM] The growth hormone of the pituitary gland.

somatotype [PSYCH] A basic body type; three primary components are ectomorph, mesomorph, and endomorph.

somesthesis [PHYSIO] The general name for all systems of sensitivity present in the skin, muscles and their attachments, visceral organs, and nonauditory labyrinth of the ear.

somite *See* metamere.

somma [GEOL] The rim of a volcano.

Sommelet process [ORG CHEM] The preparation of thiophene aldehydes by treatment of thiophene with hexamethylenetetramine.

Sommerfeld equation *See* Sommerfeld formula.

Sommerfeld fine-structure constant *See* fine-structure constant.

Sommerfeld formula [ELECTROMAG] An approximate formula for the field strength of electromagnetic radiation generated by an antenna at distances small enough so that the curvature of the earth may be neglected, in terms of radiated power, distance from the antenna, and various constants and parameters. Also known as Sommerfeld equation.

Sommerfeld law for doublets [ATOM PHYS] According to the Bohr-Sommerfeld theory, the splitting in frequency of regular or relativistic doublets is $\alpha^2 R(Z - \sigma)^4/n^3(l + 1)$, where α is the fine structure constant, R is the Rydberg constant of the atom, Z is the atomic number, σ is a screening constant, n is the principal quantum number, and l is the orbital angular momentum quantum number.

Sommerfeld model *See* free-electron theory of metals.

Sommerfeld theory *See* free-electron theory of metals.

Sommerfeld-Watson transformation *See* Watson-Sommerfeld transformation.

somnambulism [PHYSIO] **1.** Sleepwalking. **2.** The performance of any fairly complex act while in a sleeplike state or trance.

somnificant *See* hypnotic.

sonar [ENG] **1.** A system that uses underwater sound, at sonic or ultrasonic frequencies, to detect and locate objects in the sea, or for communication; the commonest type is echo-ranging sonar, other versions are passive sonar, scanning sonar, and searchlight sonar. Derived from sound navigation and ranging. **2.** *See* sonar set.

sonar array [ELECTR] An arrangement of several sonar transducers or sonar projectors, appropriately spaced and energized to give proper directional characteristics.

sonar attack plotter [ORD] A system that coordinates information from a sonar installation, a ship's gyrocompass, and related devices, and presents graphically the information needed to plan an antisubmarine attack.

sonar beacon [ENG ACOUS] An underwater beacon that transmits sonic or ultrasonic signals for the purpose of providing bearing information; it may have receiving facilities that permit triggering an external source.

sonar capsule [ENG ACOUS] A capsule that reflects high-frequency sound waves; the sonar capsule, if attached to a reentry body, may be used to locate the reentry body.

sonar countermeasures [ORD] Actions taken to prevent or reduce an enemy's effective use of sonar.

sonar detector *See* sonar receiver.

sonar dome [ENG] A streamlined, watertight enclosure that provides protection for a sonar transducer, sonar projector, or hydrophone and associated equipment, while offering minimum interference to sound transmission and reception.

sonar navigation *See* sonic navigation.

sonar projector [ENG ACOUS] An electromechanical device used under water to convert electrical energy to sound energy; a crystal or magnetostriction transducer is usually used for this purpose.

sonar receiver [ELECTR] A receiver designed to intercept and amplify the sound signals reflected by an underwater target and display the accompanying intelligence in useful form; it may also pick up other underwater sounds. Also known as sonar detector.

sonar resolver [ELECTR] A resolver used with echo-ranging and depth-determining sonar to calculate and record the horizontal range of a sonar target, as required for depth-bombing.

sonar self-noise [ELECTR] Unwanted sonar signals generated in the sonar equipment itself.

sonar set [ENG] A complete assembly of sonar equipment for detecting and ranging or for communication. Also known as sonar.

sonar target [ENG ACOUS] An object which reflects a sufficient amount of a sonar signal to produce a detectable echo signal at the sonar equipment.

sonar transducer [ENG ACOUS] A transducer used under water to convert electrical energy to sound energy and sound energy to electrical energy.

sonar transmission [ENG ACOUS] The process by which underwater sound signals generated by a sonar set travel through the water.

sonar transmitter [ELECTR] A transmitter that generates electrical signals of the proper frequency and form for application to a sonar transducer or sonar projector, to produce sound waves of the same frequency in water; the sound waves may carry intelligence.

sonar window [ENG ACOUS] The portion of a sonar dome or sonar transducer that passes sound waves at sonar frequencies with little attenuation while providing mechanical protection for the transducer.

sondage [ARCHEO] A trial excavation made prior to an archeological excavation.

sonde [ENG] An instrument used to obtain weather data during ascent and descent through the atmosphere, in a form

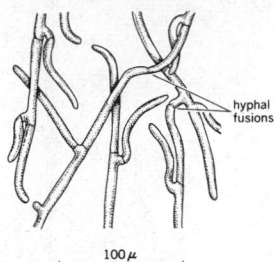

SOMATIC COPULATION

hyphal fusions

100 μ

Somatic copulation by sexual fusion of vegetative cells of hyphae.

suitable for telemetering to a ground station by radio, as in a radiosonde.

sone [ACOUS] A unit of loudness, equal to the loudness of a simple 1000-hertz tone with a sound pressure level 40 decibels above 0.0002 microbar; a sound that is judged by listeners to be n times as loud as this tone has a loudness of n sones.

sonic [ACOUS] **1.** Of or pertaining to the speed of sound. **2.** Pertaining to that which moves at acoustic velocity, as in sonic flow. **3.** Designed to operate or perform at the speed of sound, as in sonic leading edge.

sonic altimeter [ENG] An instrument for determining the height of an aircraft above the earth by measuring the time taken for sound waves to travel from the aircraft to the surface of the earth and back to the aircraft again.

sonic anemometer [ENG] An anemometer which measures wind speed by means of the properties of wind-borne sound waves; it operates on the principle that the propagation velocity of a sound wave in a moving medium is equal to the velocity of sound with respect to the medium plus the velocity of the medium.

sonic barrier [AERO ENG] A popular term for the large increase in drag that acts upon an aircraft approaching acoustic velocity; the point at which the speed of sound is attained and existing subsonic and supersonic flow theories are rather indefinite. Also known as sound barrier.

sonic bearing [NAV] A bearing determined by measuring the direction from which a sound wave is coming. Also known as acoustic bearing.

sonic boom [ACOUS] A noise caused by a shock wave that emanates from an aircraft or other object traveling at or above sonic velocity.

sonic chemical analyzer [ENG] A device to characterize the composition of a gas, liquid, or solid by the attenuation or change in the velocity of sound waves through a sample; the effect is related to molecular structure and intermolecular interactions.

sonic cleaning [ENG] Cleaning of contaminated materials by the action of intense sound in the liquid in which the material is immersed.

sonic delay line *See* acoustic delay line.

sonic depth finder [ENG] A sonar-type instrument used to measure ocean depth and to locate underwater objects; a sound pulse is transmitted vertically downward by a piezo-electric or magnetostriction transducer mounted on the hull of the ship; the time required for the pulse to return after reflection is measured electronically. Also known as echo sounder.

sonic drilling [MECH ENG] The process of cutting or shaping materials with an abrasive slurry driven by a reciprocating tool attached to an audio-frequency electromechanical transducer and vibrating at sonic frequency.

sonic fix [NAV] A fix established by means of sound waves. Also known as acoustic fix.

sonic flaw detection [ENG] The process of locating imperfections in solid materials by observing internal reflections or a variation in transmission through the materials as a function of sound-path location.

sonic frequency *See* audio frequency.

sonic line of position [NAV] A line of position determined by means of sound waves. Also known as acoustic line of position.

sonic liquid-level meter [ENG] A meter that detects a liquid level by sonic-reflection techniques.

sonic navigation [NAV] Navigation by means of sound waves whether or not they are within the audible range. Also known as acoustic navigation; sonar navigation.

sonic nucleation [CHEM ENG] In supersaturated solutions, the use of sonic or ultrasonic radiation to help bring about nucleation and corresponding crystallization of substances otherwise difficult to crystallize.

sonic pump [PETRO ENG] A type of lifting pump used in a shallow oil well to pump out the crude; consists of a string of tubing equipped with a check valve at each point, and mechanical means on the surface to vibrate the tubing string vertically; creates a harmonic condition that results in several hundred strokes per minute, with the strokes being a small fraction of an inch.

sonic radiation [ACOUS] Acoustic radiation with a frequency between about 16 hertz and about 20,000 hertz.

sonics [ACOUS] The technology of sound, or elastic wave motion, as applied to problems of measurement, control, and processing.

sonic sifter [MECH ENG] A high-speed vibrating apparatus used in particle size analysis.

sonic sounding [ENG] Determining the depth of the ocean bottom by measuring the time for an echo to return to a shipboard sound source.

sonic spark chamber [NUCLEO] A spark chamber in which the position of a spark is determined by measuring the time it takes for sound from the spark to arrive at each of two microphones.

sonic speed [MECH] Acoustic velocity; by extension, the speed of a body traveling at Mach 1.

sonic thermometer [ENG] A thermometer based upon the principle that the velocity of a sound wave is a function of the temperature of the medium through which it passes.

sonic velocity *See* speed of sound.

sonic well logging [ENG] A well logging technique that uses a pulse-echo system to measure the distance between the instrument and a sound-reflecting surface; used to measure the size of cavities around brine wells, and capacities of underground liquefied petroleum gas storage chambers.

Sonnar lens [OPTICS] A modified triplet lens used as a photographic objective.

Sonne [NAV] The German forerunner of the British Consol long-range, low-frequency navigation system.

Sonne dysentery [MED] An intestinal bacterial infection caused by *Shigella sonnei*.

Sonnenschein's reagent [ANALY CHEM] A solution of phosphomolybdic acid that forms a yellow precipitate with alkaloid sulfates.

Sonne's bacillus [MICROBIO] The bacterium *Shigella sonnei*.

sonobuoy [ENG] An acoustic receiver and radio transmitter mounted in a buoy that can be dropped from an aircraft by parachute to pick up underwater sounds of a submarine and transmit them to the aircraft; to track a submarine, several buoys are dropped in a pattern that includes the known or suspected location of the submarine, with each buoy transmitting an identifiable signal; an electronic computer then determines the location of the submarine by comparison of the received signals and triangulation of the resulting time-delay data. Also known as radio sonobuoy.

sonograph [ENG] **1.** An instrument for recording sound or seismic vibrations. **2.** An instrument for converting sounds into seismic vibrations.

Sonolog [PETRO ENG] An acoustical device used for sound-reflection logging of oil well boreholes to determine the fluid level in a pumping well.

sonoluminescence [PHYS] Luminescence produced by high-frequency sound waves or by phonons.

sonometer [ENG] An instrument for measuring rock stress; a piano wire is stretched between two bolts in the rock, and any change of pitch after destressing is observed and used to indicate stress.

sonora [METEOROL] A summer thunderstorm in the mountains and deserts of southern California and Baja California.

Sonotone battery [ELEC] A variety of nickel-cadmium battery in which the positive plates are nickel oxide when the battery is charged, and the negative plates are metallic cadmium; on discharge the positive plates are reduced to a state of lower oxidation, and the negative plates regain oxygen.

soot [MATER] Impure black carbon with oily compounds obtained from the incomplete combustion of resinous materials, oils, wood, or coal.

soot blower [ENG] A system of steam or air jets used to maintain cleanliness, efficiency, and capacity of heat-transfer surfaces by the periodic removal of ash and slag from the heat-absorbing surfaces.

soot luminosity [OPTICS] The portion of the luminosity of a flame attributable to soot particles in the flame.

sooty mold [MYCOL] Ascomycetous fungi of the family Capnodiaceae, with dark mycelium and conidia. [PL PATH] A plant disease, common on *Citrus* species, characterized by a

SONE

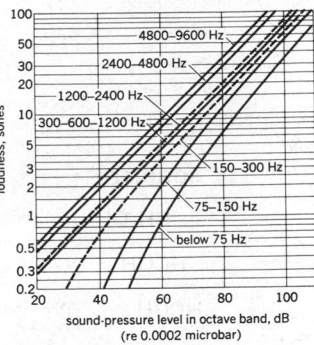

Functions relating loudness, in sones, to sound pressure, in decibels above 0.0002 microbar, for bands of noise one octave wide in a sound field.

SONNAR LENS

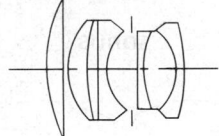

Components of a Sonnar camera lens.

dense velvety layer of a sooty mold on exposed parts of the plant.

sophisticated vocabulary [ADP] An advanced and elaborate set of instructions; a computer with a sophisticated vocabulary can go beyond the more common mathematical calculations such as addition, multiplication, and subtraction, and perform operations such as linearize, extract square root, and select highest number.

soporific *See* hypnotic.

sorbent [MATER] A material, compound, or system that can provide a sorption function, such as adsorption, absorption, or desorption.

sorbic acid [ORG CHEM] $CH_3CH=CHCH=CHCOOH$ A white, crystalline compound; soluble in most organic solvents, slightly soluble in water; melts at 135°C; used as a fungicide and food preservative, and in the manufacture of plasticizers and lubricants. Also known as 2,4-hexadienoic acid.

sorbic acid potassium salt *See* potassium sorbate.

sorbide [CHEM] The generic term for anhydrides derived from sorbitol.

sorbin *See* sorbose.

D-sorbite *See* sorbitol.

sorbitol [ORG CHEM] $C_6H_8(OH)_6$ Combustible, white, water-soluble, hygroscopic crystals with a sweet taste; melt at 93 to 97.5°C (depending on the form); used in cosmetic creams and lotions, toothpaste, and resins; as a food additive; and for ascorbic acid fermentation. Also known as D-sorbite; D-sorbitol.

D-sorbitol *See* sorbitol.

sorbose [BIOCHEM] $C_6H_{12}O_6$ A carbohydrate prepared by fermentation; produced as water-soluble crystals that melt at 165°C; used in the production of vitamin C. Also known as sorbin; L-sorbose.

L-sorbose *See* sorbose.

sordawalite *See* tachylite.

soredium [BOT] A structure comprising algal cells wrapped in the hyphal tissue of lichens, as in certain Lecanorales; when separated from the thallus, it grows into a new thallus.

Sorel cement *See* oxychloride cement.

Sörensen titration [ANALY CHEM] Titration with one of the Sörensen hydrogen-ion-concentration indicators.

sore shin [PL PATH] A fungus disease of cowpea, cotton, tobacco, and other plants, beyond the seedling stage, marked by annular growth of the pathogen on the stem at the groundline.

Soret coefficient [PHYS] A tabulated value used in binary thermal diffusion calculations in gaseous systems; expressed as $D'/D = \alpha\, X_1 X_2$, where D' is the coefficient of thermal diffusion, D is the coefficient of ordinary diffusion, α is the thermal diffusion constant, X_1 is the mole fraction of the lower-molecular-weight component, and X_2 is the mole fraction of the higher-molecular-weight component.

Soret effect [PHYS] Thermal diffusion in liquids.

sorghum [BOT] Any of a variety of widely cultivated grasses, especially *Sorghum bicolor* in the United States, grown for grain and herbage; growth habit and stem form are similar to Indian corn, but leaf margins are serrate and spikelets occur in pairs on a hairy rachis.

Soricidae [VERT ZOO] The shrews, a family of insectivorous mammals belonging to the Lipotyphla.

sorosilicate [MINERAL] A structural type of silicate whose crystal lattice has two SiO_4 tetrahedra sharing one oxygen atom.

sorotiite [GEOL] A type of meteorite similar to the pallasites, with troilite substituting for olivine.

sorption [PHYS CHEM] A general term used to encompass the processes of adsorption, absorption, desorption, ion exchange, ion exclusion, ion retardation, chemisorption, and dialysis.

sorption pumping [ENG] A technique used to reduce the pressure of gas in an atmosphere; the gas is adsorbed on a granular sorbent material such as a molecular sieve in a metal container; when this sorbent-filled container is immersed in liquid nitrogen, the gas is sorbed.

sorrel tree *See* sourwood.

sort [ADP] **1.** To rearrange a set of data items into a new

sequence, governed by specific rules of precedence. **2.** The program designed to perform this activity.

sorted polygon [GEOL] A patterned ground having a sorted appearance due to a border of stones and characterized by a polygonal mesh. Also known as stone polygon.

sorter *See* card sorter; sequencer.

sortie [AERO ENG] An operational flight by one aircraft. [ORD] A sudden attack made from a defensive position.

sortie number [ENG] A reference used to identify the images taken by all the sensors during one air reconnaissance sortie.

sortie plot [MAP] An overlay representing the area on a map covered by imagery taken during one sortie.

sorting [GEOL] The process by which similar in size, shape, or specific gravity sedimentary particles are selected and separated from associated but dissimilar particles by the agent of transportation.

sorting index [GEOL] A measure of the degree of sorting in a sediment based on the statistical spread of the frequency curve of particle sizes.

sorting table [ENG] Any horizontal conveyor where operators, along its side, sort bulk material, packages, or objects from the conveyor.

sort key [ADP] A key used as a basis for determining the sequence of items in a set.

sorus [BOT] **1.** A cluster of sporangia on the lower surface of a fertile fern leaf. **2.** A clump of reproductive bodies or spores in lower plants.

SOS [COMMUN] The distress signal in radiotelegraphy, consisting of the letters S, O, and S of the international Morse code.

sosoloid [PHYS CHEM] A system consisting of particles of a solid dispersed in another solid.

Sothic cycle [ASTRON] A time period of about 1460 years; this cycle is such that the New Year of the calendar used in ancient Egypt was in error by a whole year because the adopted year of 365 days is about a quarter of a day shorter than the mean solar year.

sou'easter *See* southeaster.

souma [VET MED] A disease caused by *Trypanosoma vivax* in domestic and wild animals; the insect vectors are the tsetse fly and the stable fly.

sound [ACOUS] **1.** An alteration of properties of an elastic medium, such as pressure, particle displacement, or density, that propagates through the medium, or a superposition of such alterations; sound waves having frequencies above the audible (sonic) range are termed ultrasonic waves; those with frequencies below the sonic range are called infrasonic waves. Also known as acoustic wave; sound wave. **2.** The auditory sensation which is produced by these alterations. Also known as sound sensation.

sound absorption [ACOUS] A process in which sound energy is reduced when sound waves pass through a medium or strike a surface. Also known as acoustic absorption.

sound absorption coefficient [ACOUS] The ratio of sound energy absorbed to that arriving at a surface or medium. Also known as acoustic absorption coefficient; acoustic absorptivity.

sound analyzer [ENG] An instrument which measures the amount of sound energy in various frequency bands; it generally consists of a set of fixed electrical filters or a tunable electrical filter, along with associated amplifiers and a meter which indicates the filter output.

sound and flash ranging [ORD] Two distinct and separate but supplementary systems of locating enemy weapons and, secondarily, adjusting friendly counterfire: by observation by sonic devices on the sound produced by the enemy weapon in firing or by the friendly projectile in exploding; or by visual observation of the flash produced or of the point of burst of the enemy weapon or friendly projectile.

sound attenuation [ACOUS] Diminution of the intensity of sound energy propagating in a medium; caused by absorption, spreading, and scattering.

sound band pressure level [ACOUS] The effective sound pressure for the sound energy in a given frequency band.

sound barrier *See* sonic barrier.

sound buoy [NAV] A buoy equipped with a characteristic sound signal, such as a bell or a whistle.

sound carrier [COMMUN] The television carrier that is fre-

SORBITOL

CH$_2$OH
|
HCOH
|
HOCH
|
HCOH
|
HCOH
|
CH$_2$OH

Structural formula for sorbitol.

SORUS

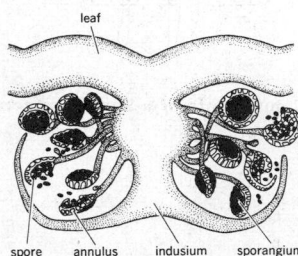

Diagram of section through a fern leaf, showing details of a sorus. (*From W. W. Robbins, T. E. Weier, and C. R. Stocking, Botany: An Introduction to Plant Science, 3d ed., Wiley, 1964*)

quency-modulated by the sound portion of a television program; the unmodulated center frequency of the sound carrier is 4.5 megahertz higher than the video carrier frequency for the same channel.

sound channel [ACOUS] A layer of seawater extending from about 700 meters down to about 1500 meters, in which sound travels at about 450 meters per second, the slowest it can travel in seawater; below 1500 meters the speed of sound increases as a result of pressure. [ELECTR] The series of stages that handles only the sound signal in a television receiver.

sound detection [ACOUS] The discrimination of a sound from background noise, either by the ear or by an electronic instrument such as a volume indicator.

sound effects [ENG ACOUS] Mechanical devices or recordings used to provide lifelike imitations of various sounds.

sound energy [ACOUS] The difference between the total energy and the energy which would exist if no sound waves were present. Also known as acoustic energy.

sound energy density [ACOUS] Sound energy per unit volume; the commonly used unit is the erg per cubic centimeter.

sound energy flux [ACOUS] Average over one period of the rate of flow of sound energy through any specified area; the unit is the erg per second.

sound exclusion [PETRO ENG] Several techniques used to prevent a borehole from sloughing in drilling a well through a reservoir rock that has an unconsolidated nature, similar to beach sand; the borehole can be lined with a screen, the sand consolidated with a binding material, or prepack gravel liner can be used.

sound film [ENG ACOUS] Motion picture film having a sound track along one side for reproduction of the sounds that are to accompany the film.

sound filmstrip [ENG ACOUS] A filmstrip that has accompanying sound on a separate disk or tape, which is manually or automatically synchronized with projection of the pictures in the strip.

sound fixing and ranging *See* sofar.

sound frequency *See* audio frequency.

sound gate [ENG ACOUS] The gate through which film passes in a sound-film projector for conversion of the sound track into audio-frequency signals that can be amplified and reproduced.

sound head [ENG ACOUS] The section of a sound motion picture projector that converts the photographic or magnetic sound track to audible sound signals.

sound image [ACOUS] The photographic image of a sound, as on a film sound track.

sounding [ENG] **1.** Determining the depth of a body of water by an echo sounder or sounding line. **2.** Measuring the depth of bedrock by driving a steel rod into the soil. **3.** Any penetration of the natural environment for scientific observation. [METEOROL] *See* upper-air observation. [MIN ENG] **1.** Knocking on a mine roof to see whether it is sound or safe to work under. **2.** Subsurface investigation by observing the penetration resistance of the subsurface material without drilling holes, by driving a rod into the ground or by using a penetrometer.

sounding balloon [ENG] A small free balloon used for carrying radiosonde equipment aloft.

sounding device [PETRO ENG] An acoustical device used to measure the liquid level in a wellbore; for example, a Sonolog or an Echometer.

sounding lead [ENG] A lead used for determining the depth of water.

sounding line [ENG] The line attached to a sounding lead. Also known as lead line.

sounding machine [ENG] An instrument for measuring the depth of water, consisting essentially of a reel of wire; to one end of this wire there is attached a weight which carries a device for measuring and recording the depth; a crank or motor reels in the wire.

sounding pipe [NAV ARCH] A pipe through which the depth of liquid in a water or oil tank on board a ship can be measured or sounded.

sounding pole [ENG] A pole or rod used for sounding in shallow water, and usually marked to indicate various depths.

sounding rocket [AERO ENG] A rocket that carries aloft equipment for making observations of or from the upper atmosphere.

sounding sand [GEOL] Sand that emits musical, humming, or crunching sounds when disturbed. Also known as singing sand.

sounding sextant *See* hydrographic sextant.

sounding velocity [ACOUS] The vertical velocity of sound in water, usually assumed to be constant at 800 to 820 fathoms per second for sounding measurements.

sounding wire [ENG] A wire used with a sounding machine in determining depth of water.

sound insulation [ACOUS] Any material or structure which is used to reduce the propagation of sound.

sound intensity [ACOUS] For a specified direction and point in space, the average rate at which sound energy is transmitted through a unit area perpendicular to the specified direction.

sound irradiator [ACOUS] A device for focusing sound waves so that sound of high intensity is produced at the focus.

sound lag [ACOUS] Time necessary for a sound wave to travel from its source to the point of reception.

sound level [ACOUS] The sound pressure level (in decibels) at a point in a sound field, averaged over the audible frequency range and over a time interval, with a frequency weighting and the time interval as specified by the American National Standards Association.

sound-level meter [ENG] An instrument used to measure noise and sound levels in a specified manner; the meter may be calibrated in decibels or volume units and includes a microphone, an amplifier, an output meter, and frequency-weighting networks.

sound locator [ENG ACOUS] A device formerly used to detect aircraft in flight by sound, consisting of four horns, or sound collectors (two for azimuth detection and two for elevation), together with their associated mechanisms and controls, which enabled the listening operator to determine the position and angular velocity of an aircraft.

sound masking [ACOUS] The ability of one sound to make the ear incapable of perceiving another sound.

sound navigation and ranging *See* sonar.

sound power [ACOUS] The total sound energy radiated by a source per unit time, generally expressed in ergs per second or watts. Also known as acoustic power.

sound-powered telephone [ENG ACOUS] A telephone operating entirely on current generated by the speaker's voice, with no external power supply; sound waves cause a diaphragm to move a coil back and forth between the poles of a powerful but small permanent magnet, generating the required audio-frequency voltage in the coil.

sound pressure *See* effective sound pressure.

sound pressure level [ACOUS] A value in decibels equal to 20 times the logarithm to the base 10 of the ratio of the pressure of the sound under consideration to a reference pressure; reference pressures in common use are 0.0002 microbar and 1 microbar. Abbreviated SPL.

sound production [ENG ACOUS] Conversion of energy from mechanical or electrical into acoustical form, as in a siren or loudspeaker.

soundproofing *See* damping.

sound ranging [ENG ACOUS] Determining the location of a gun or other sound source by measuring the travel time of the sound wave to microphones at three or more different known positions.

sound-ray diagram [ACOUS] A plot of the paths taken by sound rays in an acoustical system; analogous to a light-ray diagram in optics.

sound reception [ENG ACOUS] Conversion of acoustical energy into another form, usually electrical, as in a microphone.

sound recording [ENG ACOUS] The process of recording sound signals so they may be reproduced at any subsequent time, as on a phonograph disk, motion picture sound track, or magnetic tape.

sound reduction factor [ACOUS] A measure of the reduction in the intensity of sound when it crosses an interface, equal to 10 times the common logarithm of the reciprocal of the sound transmission coefficient of the surface.

sound reflection coefficient *See* acoustic reflectivity.

sound-reinforcement system [ENG ACOUS] An electronic

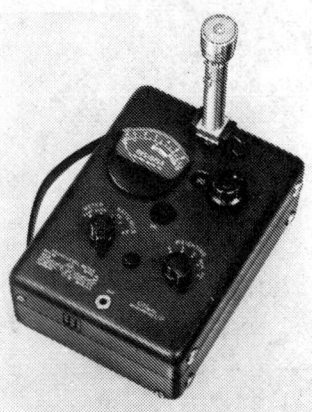

SOUND-LEVEL METER

General Radio Type 1551-B sound-level meter. *(General Radio Co.)*

**SOUND-REINFORCEMENT
SYSTEM**

Basic sound-reinforcement system.

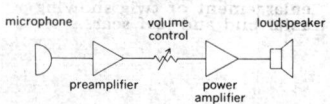

means for augmenting the sound output of a speaker, singer, or musical instrument in cases where it is either too weak to be heard above the general noise or too reverberant; basic elements of such a system are microphones, amplifiers, volume controls, and loudspeakers. Also known as public address system.

sound-reproducing system [ENG ACOUS] A combination of transducing devices and associated equipment for picking up sound at one location and time and reproducing it at the same or some other location and at the same or some later time. Also known as audio system; reproducing system; sound system.

sound reproduction [ENG ACOUS] The use of a combination of transducing devices and associated equipment to pick up sound at one point and reproduce it either at the same point or at some other point, at the same time or at some subsequent time.

sound sensation See sound.

sound signal [COMMUN] Any sound produced to convey intelligence, as a fog signal.

sound spectrum [ACOUS] A plot of the strength of a sound at various frequencies.

sound speed [ENG] The speed of sound motion picture film, standardized at 24 frames per second (silent film speed is 18 frames per second).

soundstripe [ENG ACOUS] A longitudinal stripe of magnetic material placed on some motion picture films for recording a magnetic sound track.

sound system See sound-reproducing system.

sound track [ENG ACOUS] A narrow band, usually along the margin of a sound film, that carries the sound record; it may be a variable-width or variable-density optical track or a magnetic track.

sound transducer See electroacoustic transducer.

sound transmission [ACOUS] Passage of a sound wave through a medium or series of media.

sound transmission coefficient [ACOUS] The ratio of transmitted to incident sound energy at an interface in a sound medium; the value depends on the angle of incidence of the sound. Also known as acoustic transmission coefficient; acoustic transmissivity.

sound velocity See speed of sound.

sound volume velocity [ACOUS] The rate at which a substance flows through a specified area as a result of a sound wave.

sound wave See sound.

sour [CHEM] The condition of containing large amounts of sulfur or sulfur compounds (such as mercaptans or hydrogen sulfide), as in crude oils, naphthas, or gasoline.

source [ELEC] The circuit or device that supplies signal power or electric energy or charge to a transducer or load circuit. [NUCLEO] A radioactive material packaged so as to produce radiation for experimental or industrial use. [PHYS] **1.** In general, a device that supplies some extensive entity, such as energy, matter, particles, or electric charge. **2.** A point, line, or area at which mass or energy is added to a system, either instantaneously or continuously. [SPECT] The arc or spark that supplies light for a spectroscope. [THERMO] A device that supplies heat.

source address [ADP] The first address of a two-address instruction (the sound address is known as the destination address).

source area See provenance.

source bed [GEOL] The original stratigraphic horizon from which secondary sulfide minerals were derived.

source data [SCI TECH] Data generated in the course of research.

source data automation See automation source data.

source data automation equipment [ADP] Equipment (except paper tape and magnetic tape cartridge typewriters acquired separately and not operated in support of a computer) which, as a by-product of its operation, produces a record in a medium which is acceptable by automatic data-processing equipment.

source flow [FL MECH] **1.** In three-dimensional flow, a point from which fluid issues at a uniform rate in all directions. **2.** In two-dimensional flow, a line normal to the planes of flow,

from which fluid flows uniformly in all directions at right angles to the line.

source-follower amplifier See common-drain amplifier.

source impedance [ELEC] Impedance presented by a source of energy to the input terminals of a device.

sourceland See provenance.

source language [ADP] The language in which a program (or other text) is originally expressed.

source level [ACOUS] The sound intensity, in decibels above a reference level, at a point which is a unit distance from a source and on an axis of the source.

source library [ADP] A collection of computer programs in compiler language or assembler language.

source material [NUCLEO] Material from which fissionable material can be extracted.

source module [ADP] An organized set of statements in any source language recorded in machine-readable form and suitable for input to an assembler or compiler.

source program [ADP] The form of a program just as the programmer has written it, often on coding forms or machine-readable media; a program expressed in a source-language form.

source program optimizer [ADP] A routine for examining the source code of a program under development and providing information about use of the various portions of the code, enabling the programmer to modify those sections of the target program that are most heavily used in order to improve performance of the final, operational program.

source region [METEOROL] An extensive area of the earth's surface characterized by essentially uniform surface conditions and so situated with respect to the general atmospheric circulation that an air mass may remain over it long enough to acquire its characteristic properties.

source rock [GEOL] **1.** Rock from which fragments have been derived which form a later, usually sedimentary rock. Also known as mother rock; parent rock. **2.** Sedimentary rock, usually shale and limestone, deposited together with organic matter which was subsequently transformed to liquid or gaseous hydrocarbons.

source time [ADP] The time involved in fetching the contents of the register specified by the first address of a two-address instruction.

sourcing [ELECTR] Redesign or the modification of existing equipment to eliminate a source of radio-frequency interference.

sour corrosion [PETRO ENG] Corrosion occurring in oil or gas wells where there is an iron sulfide corrosion product, and hydrogen sulfide is present in the produced reservoir fluid.

sour crude [MATER] Crude oil containing an abnormally large amount of sulfur compounds that, upon refining, liberate corrosive sulfur compounds; opposite to sweet crude.

sour gas [MATER] Natural gas that contains corrosive, sulfur-bearing compounds, such as hydrogen sulfide and mercaptans.

sourwood [BOT] *Oxydendrum arboreum.* A deciduous tree of the heath family (Ericaceae) indigenous along the Alleghenies and having long, simple, finely toothed, long-pointed leaves that have an acid taste, and white, urn-shaped flowers. Also known as sorrel tree.

south [GEOD] The direction 180° from north.

South African jade See Transvaal jade.

South African tick-bite fever [MED] An infectious tick-borne rickettsial disease of humans which is similar to fièvre boutonneuse.

South America [GEOGR] The southernmost of the Western Hemisphere continents, three-fourths of which lies within the tropics.

South American blastomycosis [MED] An infectious, yeast-like fungus disease of humans seen primarily in Brazil; caused by *Blastomyces brasiliensis* and characterized by massive enlargement of the cervical lymph nodes. Also known as paracoccidioidomycosis.

South American leishmaniasis See American mucocutaneous leishmaniasis.

South American trypanosomiasis See Chagas' disease.

South Atlantic Current [OCEANOGR] An eastward-flowing

SOURWOOD

Sourwood (Oxydendrum arboreum) leaf, twig, and enlargement of twig showing axial bud and leaf scar.

current of the South Atlantic Ocean that is continuous with the northern edge of the West Wind Drift.

South Australian faunal region [ECOL] A marine littoral region along the southwestern coast of Australia.

southbound node *See* descending node.

Southeast Drift Current [OCEANOGR] A North Atlantic Ocean current flowing southeastward and southward from a point west of the Bay of Biscay toward southwestern Europe and the Canary Islands, where it continues as the Canary Current.

southeaster [METEOROL] A southeasterly wind, particularly a strong wind or gale; for example, the winter southeast storms of the Bay of San Francisco. Also spelled sou'easter.

South Equatorial Current [OCEANOGR] Any of several ocean currents, flowing westward, driven by the southeast trade winds blowing over the tropical oceans of the Southern Hemisphere and extending slightly north of the equator. Also known as Equatorial Current.

souther [METEOROL] A south wind, especially a strong wind or gale.

southerly burster [METEOROL] A cold wind from the south in Australia.

Southern Cross *See* Crux.

Southern Crown *See* Corona Australis.

Southern Fish *See* Piscis Australis.

southern lights *See* aurora australis.

Southern Polar Front *See* Antarctic Convergence.

Southern Triangle *See* Triangulum Australe.

south foehn [METEOROL] A foehn condition sustained by a strong south-to-north airflow across a transverse mountain barrier; the south foehn of the Alps may well be the most striking foehn in the world.

south frigid zone [GEOGR] That part of the earth south of the Antarctic Circle.

south geographical pole [GEOGR] The geographical pole in the Southern Hemisphere, at latitude 90°S. Also known as South Pole.

south geomagnetic pole [GEOPHYS] The geomagnetic pole in the Southern Hemisphere at approximately 78.5°S, longitude 111°E, 180° from the north geomagnetic pole. Also known as south pole.

South Indian Current [OCEANOGR] An eastward-flowing current of the southern Indian Ocean that is continuous with the northern edge of the West Wind Drift.

southing [NAV] The distance a craft makes good to the south.

South Pacific Current [OCEANOGR] An eastward-flowing current of the South Pacific Ocean that is continuous with the northern edge of the West Wind Drift.

south point [ASTRON] That imaginary point on the celestial sphere at which the meridian intersects the horizon; it is due south of the observer.

south pole [ELECTROMAG] The pole of a magnet at which magnetic lines of force are assumed to enter. [GEOPHYS] *See* south geomagnetic pole.

South Pole *See* south geographical pole.

south temperate zone [GEOGR] That part of the earth between the Tropic of Capricorn and the Antarctic Circle.

south tropical disturbance [ASTRON] An elongated dark band seen on the surface of Jupiter at about the latitude of the Great Red Spot; it has at times exceeded 180° of longitude in length and, like the Red Spot, appears and disappears intermittently.

southwester [METEOROL] A southwest wind, particularly a strong wind or gale. Also spelled sou'wester.

sou'wester *See* southwester.

souzalite [MINERAL] $(Mg,Fe)_3(Al,Fe)_4(PO_4)_4(OH)_6 \cdot 2H_2O$ A green mineral composed of hydrous basic phosphate of magnesium, iron, and aluminum.

Sovafining [CHEM ENG] A proprietary petroleum refinery hydrogenation process for pretreatment of straight-run and thermal naphtha for feed to catalytic reforming units, and for desulfurization of fuel oils.

Sovaforming [CHEM ENG] A proprietary petroleum refinery process using a platinum catalyst in a multiple-reactor, fixed-bed catalytic system; used to catalytically reform gasoline-range hydrocarbons to produce high octanes.

sow [MET] 1. A mold of larger size than a pig. 2. A channel that conducts molten metal to molds in a pig bed. [VERT ZOO] An adult female swine.

sow block [MET] In forging, a removable block set into the hammer anvil to reduce wear on the anvil.

Soxhlet extractor [CHEM] A flask and condenser device for the continuous extraction of alcohol- or ether-soluble materials.

soybean [BOT] *Glycine max.* An erect annual legume native to China and Manchuria and widely cultivated for forage and for its seed.

soybean lecithin *See* lecithin.

soybean oil [MATER] An oil that is produced by solvent extraction from soybeans and that is used in shortening, margarine, and salad oil; it contains linoleic and linolenic acids.

soy lecithin *See* lecithin.

Soyuz Program [AERO ENG] A manned space-flight program begun in 1967 by the Soviet Union.

space [ASTRON] 1. Specifically, the part of the universe lying outside the limits of the earth's atmosphere. 2. More generally, the volume in which all celestial bodies, including the earth, move. [COMMUN] The open-circuit condition or the signal causing the open-circuit condition in telegraphic communication; the closed-circuit condition is called the mark. [MATH] In context, usually a set with a topology on it or some other type of structure.

space attenuation [ACOUS] Loss of energy, expressed in decibels, of a signal in free air; caused by such factors as absorption, reflection, scattering, and dispersion.

space biology [BIOL] A term for the various biological sciences and disciplines that are concerned with the study of living things in the space environment.

space capsule [AERO ENG] A container, manned or unmanned, used for carrying out an experiment or operation in space.

space character *See* blank character.

space charge [ELEC] The net electric charge within a given volume. [GEOPHYS] In atmospheric electricity, space charge refers to a preponderance of either negative or positive ions within any given portion of the atmosphere.

space-charge balanced flow [ELECTR] A method of focusing an electron beam in the interaction region of a traveling-wave tube; there is an axial magnetic field in the interaction region which is stronger than that in the gun region; at the transition between the two values of magnetic field strength, the beam is given a rotation in such a direction as to produce an inward force that counterbalances the outward forces from space charge and from the centrifugal forces set up by rotation.

space-charge debunching [ELECTR] A process in which the mutual interactions between electrons in a stream spread out the electrons of a bunch.

space-charge effect [ELECTR] Repulsion of electrons emitted from the cathode of a thermionic vacuum tube by electrons accumulated in the space charge near the cathode.

space-charge grid [ELECTR] Grid operated at a low positive potential and placed between the cathode and control grid of a vacuum tube to reduce the limiting effect of space charge on the current through the tube.

space-charge layer *See* depletion layer.

space-charge limitation [ELECTR] The current flowing through a vacuum between a cathode and an anode cannot exceed a certain maximum value, as a result of modification of the electric field near the cathode due to space charge in this region.

space-charge polarization [ELEC] Polarization of a dielectric which occurs when charge carriers are present which can migrate an appreciable distance through the dielectric but which become trapped or cannot discharge at an electrode. Also known as interfacial polarization.

space-charge region [ELECTR] Of a semiconductor device, a region in which the net charge density is significantly different from zero.

space cloth [CHEM ENG] Woven cloth or wire used for solids screening, and for which the openings between the fibers or strands are designated in terms of space or clear opening.

space communication [COMMUN] Communication between

a vehicle in outer space and the earth, using high-frequency electromagnetic radiation.

space coordinates [MATH] A three-dimensional system of cartesian coordinates by which a point is located by three magnitudes indicating distance from three planes which intersect at a point.

spacecraft [AERO ENG] Devices, manned and unmanned, which are designed to be placed into an orbit about the earth or into a trajectory to another celestial body. Also known as space ship; space vehicle.

spacecraft ground instrumentation [ENG] Instrumentation located on the earth for monitoring, tracking, and communicating with manned spacecraft, satellites, and space probes. Also known as ground instrumentation.

spacecraft launching [AERO ENG] The setting into motion of a space vehicle with sufficient force to cause it to leave the earth's atmosphere.

spacecraft propulsion [AERO ENG] The use of rocket engines to accelerate space vehicles.

spacecraft tracking [ENG] The determination of the positions and velocities of spacecraft through radio and optical means.

space current [ELECTR] Total current flowing between the cathode and all other electrodes in a tube; this includes the plate current, grid current, screen grid current, and any other electrode current which may be present.

spaced antenna [ELECTROMAG] Antenna system consisting of a number of separate antennas spaced a considerable distance apart, used to minimize local effects of fading at short-wave receiving stations.

spaced armor [ORD] An arrangement of armor plate, using two or morè thicknesses, each thickness spaced from the adjoining one; used as a protective device, particularly against shaped-charge ammunition.

space defense [ORD] All measures designed to reduce or nullify the effectiveness of hostile acts by vehicles (including missiles) while in space.

space detection and tracking system [ENG] System capable of detecting and tracking space vehicles from the earth, and reporting the orbital characteristics of these vehicles to a central control facility. Abbreviated SPADATS.

space diversity reception [ELECTROMAG] Radio reception involving the use of two or more antennas located several wavelengths apart, feeding individual receivers whose outputs are combined; the system gives an essentially constant output signal despite fading due to variable propagation characteristics, because fading affects the spaced-out antennas at different instants of time.

spaced loading [ENG] Loading shot holes so that cartridges are separated by open spacers which do not prevent the concussion from one charge from reaching the next.

space-dyed yarn [TEXT] Yarn dyed in one color for a specified length and in other colors for other lengths, the sequence being repeated.

space environment [ASTRON] The environment encountered by vehicles and living creatures in space, characterized by absence of atmosphere.

space factor [ELECTROMAG] **1.** The ratio of the space occupied by the conductors in a winding to the total cubic content or volume of the winding, or the similar ratio of cross sections. **2.** The ratio of the space occupied by iron to the total cubic content of an iron core.

space fixed reference [NAV] An oriented reference system in space independent of earth phenomena for positioning.

space flight [AERO ENG] Travel beyond the earth's sensible atmosphere; space flight may be an orbital flight about the earth or it may be a more extended flight beyond the earth into space.

space-flight trajectory [AERO ENG] The track or path taken by a spacecraft.

space frame [BUILD] A three-dimensional steel building frame which is stable against wind loads.

space group [CRYSTAL] A group of operations which leave the infinitely extended, regularly repeating pattern of a crystal unchanged; there are 230 such groups.

space group extinction [CRYSTAL] The absence of certain classes of reflections in the x-ray diffraction pattern of a crystal due to the existence of symmetry elements in the space group of the crystal which are not present in its point group.

space guidance [NAV] Guidance operations considered as consisting of the following three phases: ascent from earth to an orbit or space trajectory, operations in space requiring navigation or guidance, and descent to the surface of the earth or the moon.

space-hold [COMMUN] The transmission of a steady space signal over a transmission line which is carrying no traffic.

space inversion *See* inversion.

space lattice [BUILD] A space frame built of lattice girders. [CRYSTAL] *See* crystal lattice; lattice.

spacelike surface [RELAT] A three-dimensional surface in a four-dimensional space-time which has the property that no event on the surface lies in the past or the future of any other event on the surface.

spacelike vector [RELAT] A four vector in Minkowski space whose space component has a magnitude which is greater than the magnitude of its time component multiplied by the speed of light.

space medicine [MED] A branch of medicine that deals with the physiologic disturbances and diseases produced in man by high-velocity projection through and beyond the earth's atmosphere, flight through interplanetary space, and return to earth.

space mission [AERO ENG] A journey by a vehicle, manned or unmanned, beyond the earth's atmosphere, usually for the purpose of collecting scientific data.

space modulation [COMMUN] The combining of signals in space to form a signal of desired characteristics.

space motion [ASTRON] Motion of a celestial body through space.

space navigation [NAV] Determination of the three-dimensional position and velocity vector of a space vehicle relative to a selected frame of reference.

space perception [PHYSIO] The awareness of the spatial properties and relations of an object, or of one's own body, in space; especially, the sensory appreciation of position, size, form, distance, and direction of an object, or of the observer himself, in space.

space permeability [ELECTROMAG] Factor that expresses the ratio of magnetic induction to magnetizing force in a vacuum; in the centimeter-gram-second electromagnetic system of units, the permeability of a vacuum is arbitrarily taken as unity; in the meter-kilogram-second-ampere system, it is $4\pi \times 10^{-7}$.

space polar coordinates [MATH] A system of coordinates by which a point is located in space by its distance from a fixed point called the pole, the colatitude or angle between the polar axis (a reference line through the pole) and the radius vector (a straight line connecting the pole and the point), and the longitude or angle between a reference plane containing the polar axis and a plane through the radius vector and polar axis.

spaceport [AERO ENG] An installation used to test and launch spacecraft.

space power system [AERO ENG] An on-board assemblage of equipment to generate and distribute electrical energy on satellites and spacecraft.

space probe [AERO ENG] An instrumented vehicle, the payload of a rocket-launching system designed specifically for flight missions to other planets or the moon and into deep space, as distinguished from earth-orbiting satellites.

space quadrature [PHYS] A difference of a quarter-wavelength in the position of corresponding points of a wave in space.

spacer [ENG] **1.** A piece of metal wire twisted at one end to form a guard to keep the explosive in a shothole in place and twisted at the other end to form a guard to hold the tamping in its place. **2.** A piece of wood doweling interposed between charges to extend the column of explosive. **3.** A device for holding two members at a given distance from each other. Also known as spacer block. **4.** The tapered section of a pug joining the barrel to the die; clay is compressed in this section before it issues through the die.

spacer block *See* spacer.

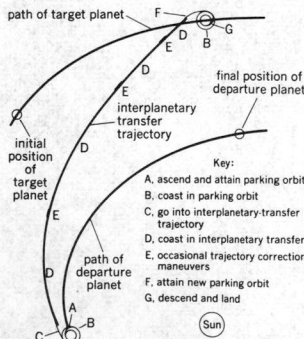

SPACE-FLIGHT TRAJECTORY

path of target planet

initial position of target planet

interplanetary transfer trajectory

final position of departure planet

path of departure planet

Sun

Key:
A, ascend and attain parking orbit
B, coast in parking orbit
C, go into interplanetary-transfer trajectory
D, coast in interplanetary transfer
E, occasional trajectory correction maneuvers
F, attain new parking orbit
G, descend and land

Simplified trajectory, a schematic diagram of an interplanetary mission.

space reconnaissance [AERO ENG] Reconnaissance of the surface of a planet from a space ship or satellite.

space research [AERO ENG] Research involving studies of all aspects of environmental conditions beyond the atmosphere of the earth.

spacer strip [MET] A strip or bar of metal placed in the root of a weld joint, prepared for a groove weld, to serve as backing and maintain root opening during welding.

space satellite [AERO ENG] A vehicle, manned or unmanned, for orbiting the earth.

space ship *See* spacecraft.

space shuttle [AERO ENG] A spacecraft designed to travel from the earth to a space station and to return to earth.

space simulator [AERO ENG] **1.** Any device which simulates one or more parameters of the space environment and which is used to test space systems or components. **2.** Specifically, a closed chamber capable of reproducing approximately the vacuum and normal environments of space.

space station [AERO ENG] An autonomous, permanent facility in space for the conduct of scientific and technological research, earth-oriented applications, and astronomical observations.

space suit [ENG] A pressure suit for wear in space or at very low ambient pressures within the atmosphere, designed to permit the wearer to leave the protection of a pressurized cabin.

space technology [AERO ENG] The systematic application of engineering and scientific disciplines to the exploration and utilization of outer space.

space-time [RELAT] A four-dimensional space used to represent the universe in the theory of relativity, with three dimensions corresponding to ordinary space and the fourth to time. Also known as space-time continuum.

space-time continuum *See* space-time.

space-to-mark transition [COMMUN] The transition from the space condition to the mark condition in telegraphic communication.

Space Tracking and Data Acquisition Network [ENG] A network of ground stations operated by the National Aeronautics and Space Administration, which tracks, commands, and receives telemetry for United States and foreign unmanned satellites. Abbreviated STADAN.

space vehicle *See* spacecraft.

space velocity [ASTRON] A star's true velocity with reference to the sun. [CHEM ENG] The relationship between feed rate and reactor volume in a flow process; defined as the volume or weight of feed (measured at standard conditions) per unit time per unit volume of reactor (or per unit weight of catalyst).

space walk [AERO ENG] The movement of an astronaut outside the protected environment of a spacecraft during a space flight; the astronaut wears a spacesuit.

space wave [ELECTROMAG] The component of a ground wave that travels more or less directly through space from the transmitting antenna to the receiving antenna; one part of the space wave goes directly from one antenna to the other; another part is reflected off the earth between the antennas.

space weapon [ORD] A weapon that travels through space and is directed against an enemy target whether on the ground, in the air, or in space.

spacing bias *See* bias telegraph distortion.

spacing pulse [COMMUN] In teletypewriter operation, the signal interval during which the selector unit is not operated.

spacing wave *See* back wave.

spacistor [ELECTR] A multiple-terminal solid-state device, similar to a transistor, that generates frequencies up to about 10,000 megahertz by injecting electrons or holes into a space-charge layer which rapidly forces these carriers to a collecting electrode.

spackling [ENG] The process of repairing a part of a plaster wall or mural by cleaning out the defective spot and then patching it with a plastering material.

SPADATS *See* space detection and tracking system.

spade [DES ENG] A shovellike implement with a flat oblong blade; used for turning soil by pushing against the blade with the foot.

spade bolt [DES ENG] A bolt having a spade-shaped flattened head with a transverse hole, used to fasten shielded coils, capacitors, and other components to a chassis.

spade drill [DES ENG] A drill consisting of three main parts: a cutting blade, a blade holder or shank, and a device, such as a screw, which fastens the blade to the holder; used for cutting holes over 1 inch in diameter.

spade grip [ORD] D-shaped handle for pointing a gun, fastened on the rear of the receiver of certain flexible automatic weapons.

spade lug [DES ENG] An open-ended flat termination for a wire lead, easily slipped under a terminal nut.

spadix [BOT] A fleshy spike that is enclosed in a leaflike spathe and is the characteristic inflorescence of palms and arums. [INV ZOO] A cone-shaped structure in male Nautiloidea formed of four modified tentacles, and believed to be homologous with the hectocotylus in male squids.

spaghetti [ELEC] Insulating tubing used over bare wires or as a sleeve for holding two or more insulated wires together; the tubing is usually made of a varnished cloth or a plastic.

spall [ENG] **1.** To reduce irregular stone blocks to an approximate size by chipping with a hammer. **2.** To break off thin chips from, and parallel to, the surface of a material, such as a metal or rock. [GEOL] **1.** A fragment removed from the surface of a rock by weathering. **2.** A relatively thin, sharp-edged fragment produced by exfoliation. **3.** A rock fragment produced by chipping with a hammer. [MIN ENG] To break ore. [ORD] A fragment torn from the surface of an armor plate.

spallation [NUC PHYS] A nuclear reaction in which the energy of each incident particle is so high that more than two or three particles are ejected from the target nucleus and both its mass number and atomic number are changed. Also known as nuclear spallation.

spallation reaction [NUC PHYS] A high-energy nuclear reaction which results in the release of large numbers of nucleons as reaction products.

spalling [GEOL] The chipping or fracturing with an upward heaving, of rock caused by a compressional wave at a free surface.

span [AERO ENG] **1.** The dimension of a craft measured between lateral extremities; the measure of this dimension. **2.** Specifically, the dimension of an airfoil from tip to tip measured in a straight line. [ENG] A structural dimension measured between certain extremities. [MATH] The span of a set of vectors is the set of all possible linear combinations of those vectors.

spandex [TEXT] An elastic synthetic fiber; usually provides a core around which other fibers are wound.

spandrel [BUILD] The part of a wall between the sill of a window and the head of the window below it.

spandrel wall [BUILD] A wall on the outer surface of a vault to fill the spandrels.

spangolite [MINERAL] $Cu_6Al(SO_4)(OH)_{12}Cl \cdot 3H_2O$ A dark-green hexagonal mineral composed of hydrous basic sulfate and chloride of aluminum and copper and occurring as crystals.

Spanish collar *See* paraphimosis.

Spanishing [GRAPHICS] The depositing of ink in the valleys of a grained or embossed substrate.

Spanish white *See* bismuth subnitrate.

spanned record [ADP] A logical record which covers more than one block, used when the size of a data buffer is fixed or limited.

spanner [DES ENG] A wrench with a semicircular head having a projection or hole at one end. [ENG] **1.** A horizontal brace. **2.** An artificial horizon attachment for a sextant.

spar [AERO ENG] A principal spanwise member of the structural framework of an airplane wing, aileron, stabilizer, and such; it may be of one-piece design or a fabricated section. [MIN ENG] A small clay vein in a coal seam. [MINERAL] Any transparent or translucent, nonmetallic, light-colored, readily cleavable, crystalline mineral; examples are calcspar and fluorspar. [NAV ARCH] A long, round stick of steel or wood, often tapered at one or both ends, and usually a part of a ship's masts or rigging.

sparagmite [GEOL] Late precambrian fragmental rocks of

Scandinavia, characterized by high proportions of microcline.

spar buoy [NAV] A long, thin, typically cylindrical buoy, ballasted at one end so that it floats in an approximately vertical position; used to mark the port side of a channel.

spar deck [NAV ARCH] A deck fitted from bow to stern on a superstructure having heavier scantlings than those under an awning deck.

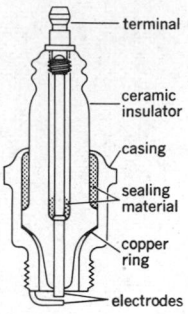

SPARK PLUG

- terminal
- ceramic insulator
- casing
- sealing material
- copper ring
- electrodes

Diagram of a typical spark plug.

spare part [ENG] In supply usage, any part, component, or subassembly kept in reserve for the maintenance and repair of major items of equipment.

spare parts list [ENG] List approved by designated authorities, indicating the total quantities of spare parts, tools, and equipment necessary for the maintenance of a specified number of major items for a definite period of time.

Sparganiaceae [BOT] A family of monocotyledonous plants in the order Typhales distinguished by the inflorescence of globose heads, a vestigial perianth, and achenes that are sessile or nearly sessile.

sparganosis [VET MED] An infection by the plerocercoid larva, or sparganum, of certain species of *Spirometra*; the adult form normally occurs in the intestine of dogs and cats.

sparganum [INV ZOO] The plerocercoid larva of a tapeworm.

sparger *See* perforated-pipe distributor.

Sparidae [VERT ZOO] A family of perciform fishes in the suborder Percoidei, including the porgies.

sparite *See* sparry calcite.

spark [ELEC] A short-duration electric discharge due to a sudden breakdown of air or some other dielectric material separating two terminals, accompanied by a momentary flash of light. Also known as electric spark; spark discharge; sparkover.

spark arrester [ENG] **1.** An apparatus that prevents sparks from escaping from a chimney. **2.** A device that reduces or eliminates electric sparks at a point where a circuit is opened and closed.

spark capacitor [ELEC] Capacitor connected across a pair of contact points, or across the inductance which causes the spark, for the purpose of diminishing sparking at these points.

spark chamber [NUCLEO] A particle-detecting device in which the trajectory of a charged particle is made visible by a series of sparks that are triggered by the particle as it passes through an array of spark gaps.

spark coil [ELECTROMAG] An induction coil for producing spark discharges, as to initiate combustion in an internal combustion engine.

spark counter [NUCLEO] A particle detector in which high-speed charged particles ionize a gas, consisting of argon mixed with an organic gas, triggering a spark between two plane parallel metal electrodes.

spark discharge *See* spark.

spark excitation [SPECT] The use of an electric spark (10,000 to 30,000 volts) to excite spectral line emissions from otherwise hard-to-excite samples; used in emission spectroscopy.

spark explosion method [ANALY CHEM] A technique for the analysis of hydrogen; the sample is mixed with an oxidant and exploded by a spark or hot wire, and the combustion products are then analyzed.

spark gap [ELEC] An arrangement of two electrodes between which a spark may occur; the insulation (usually air) between the electrodes is self-restoring after passage of the spark; used as a switching device, for example, to protect equipment against lightning or to switch a radar antenna from receiver to transmitter and vice versa.

spark-gap generator [ELEC] A high-frequency generator in which a capacitor is repeatedly charged to a high voltage and allowed to discharge through a spark gap into an oscillatory circuit, generating successive trains of damped high-frequency oscillations.

spark-ignition combustion cycle *See* Otto cycle.

spark-ignition engine [MECH ENG] An internal combustion engine in which an electrical discharge ignites the explosive mixture of fuel and air.

sparking potential *See* breakdown voltage.

sparking voltage *See* breakdown voltage.

spark killer *See* spark suppressor.

spark knock [MECH ENG] The knock produced in an internal

SPARROW

Sparrow, air-to-air missile launched from fighter aircraft. *(Official U.S. Navy photograph)*

combustion engine precedes the arrival of the piston at the top dead-center position.

spark lead [MECH ENG] The amount by which the spark precedes the arrival of the piston at its top (compression) dead-center position in the cylinder of an internal combustion engine.

sparkle metal [MET] A crude mixture of sulfides containing 74% copper produced by the smelting of copper ore.

spark machining [MET] Cutting metal by repetitive sparking between a tool (the cathode) and the workpiece (the anode).

sparkover *See* spark.

spark-over-initiated discharge machining [MECH ENG] An electromachining process in which a potential is impressed between the tool (cathode) and workpiece (anode) which are separated by a dielectric material; a heavy discharge current flows through the ionized path when the applied potential is sufficient to cause rupture of the dielectric.

spark plate [ELEC] A metal plate insulated from the chassis of an auto radio by a thin sheet of mica, and connected to the battery lead to bypass noise signals picked up by battery wiring in the engine compartment.

spark plug [ELEC] A device that screws into the cylinder of an internal combustion engine to provide a pair of electrodes between which an electrical discharge is passed to ignite the explosive mixture.

sparkproof [ENG] **1.** Treated with a material to prevent ignition or damage by sparks. **2.** Generating no sparks.

spark range [ORD] A firing range in which missiles in free flight can be photographed by the light from an electric spark which is triggered by passage of the projectile.

spark recorder [ENG] Recorder in which the recording paper passes through a spark gap formed by a metal plate underneath and a moving metal pointer above the paper; sparks from an induction coil pass through the paper periodically, burning small holes that form the record trace.

spark spectrum [SPECT] The spectrum produced by a spark discharging through a gas or vapor; with metal electrodes, a spectrum of the metallic vapor is obtained.

spark suppressor [ELEC] A device used to prevent sparking between a pair of contacts when the contacts open, such as a resistor and capacitor in series between the contacts, or, in the case of an inductive circuit, a rectifier in parallel with the inductor. Also known as spark killer.

spark transmitter [ELECTR] A radio transmitter that utilizes the oscillatory discharge of a capacitor through an inductor and a spark gap as the source of radio-frequency power.

spark voltage [ELEC] The voltage required to create an arc across the gap of a spark plug.

Sparnacean [GEOL] A European stage of geologic time; upper upper Paleocene, above Thanetian, below Ypresian of Eocene.

Sparrow [ORD] A Navy air-to-air guided missile having a speed of over 1500 miles (2400 kilometers) per hour, guided to its target by a radar beam transmitted by the launching airplane; several models exist, designated Sparrow I, Sparrow II, and Sparrow III.

sparry calcite [MINERAL] A clean, coarse-grained calcite crystal. Also known as calcsparite; sparite.

sparry cement [GEOL] Clear, relatively coarse-grained calcite in the interstices of any sedimentary rock.

sparry iron *See* siderite.

spartalite *See* zincite.

sparteine [ORG CHEM] $C_{15}H_{26}N_2$ A poisonous, colorless, oily alkaloid; soluble in alcohol and ether, slightly soluble in water; boils at 173°C; used in medicine. Also known as lupinidine.

spar varnish [MATER] A flammable varnish made of drying oils, resins, thinners, and driers to provide a durable, water-resistant coating for outside or other severe service.

spasmodic turbidity current [GEOPHYS] A single, rapidly developed turbidity current.

spasmophilia [MED] A morbid tendency to convulsions, and to tonic spasms, such as those observed in tetany, infantile spasms, or spasmus nutans.

spastic colon *See* irritable colon.

spastic diplegia [MED] **1.** Spastic paralysis of the arms and legs caused by diffuse lesions of the cerebral cortex. **2.** A form of cerebral palsy, possibly due to prenatal or perinatal hypoxia or other injuries resulting in atrophic lobar sclerosis, or to congenital or developmental abnormalities.

spastic ileus [MED] A form of ileus in which temporary obstruction is due to segmental intestinal spasm. Also known as dynamic ileus; hyperdynamic ileus.

spastic paralysis [MED] A condition in which a group of muscles manifest increased tone, exaggerated tendon reflexes, depressed or absent superficial reflexes, and sometimes clonus, due to an upper motor neuron lesion.

spastic paraplegia [MED] Paralysis of the lower limbs with increased muscular tone and hyperactive tendon reflexes; commonly seen in diseases and injuries involving pyramidal tracts of the spinal cord.

spastic strabismus [MED] A squint resulting from the contraction of an ocular muscle.

Spatangoida [INV ZOO] An order of exocyclic Euechinoidea in which the posterior ambulacral plates form a shield-shaped area behind the mouth.

spathe [BOT] A large, usually colored bract or pair of bracts enclosing an inflorescence, especially a spadix, on the same axis.

spathic iron *See* siderite.

spatial [PHYS] Of or pertaining to space; occupying space; occurring in, or conditioned by, space; considered with relation to space.

spatial dendrite [METEOROL] A complex ice crystal with fernlike arms that extend in many directions (spatially) from a central nucleus; its form is roughly spherical. Also known as spatial dendritic crystal.

spatial dendritic crystal *See* spatial dendrite.

spatiotemporal [PHYS] Of or pertaining to space time; having extent and duration.

spatter [MET] Particles of metal expelled during arc or gas welding.

spatter cone [GEOL] A low, steep-sided cone of small pyroclastic fragments built up on a fissure or vent. Also known as agglutinate cone; volcanello.

spatter dash [CIV ENG] 1. A finish put on stucco by dashing a mortar and sand mixture against it. 2. Paint spattered on a different-colored ground coat.

spatter rampart [GEOL] A low, circular ridge of pyroclastics built up around the margins of small volcanoes.

spatulate [BIOL] Shaped like a spoon.

spawn [ZOO] 1. The collection of eggs deposited by aquatic animals, such as fish. 2. To produce or deposit eggs or discharge sperm; applied to aquatic animals.

spay [VET MED] To remove the ovaries.

SPDT *See* single-pole double-throw.

speaker *See* loudspeaker.

spear [DES ENG] A rodlike fishing tool having a barbed-hook end, used to recover rope, wire line, and other materials from a borehole.

spearmint [BOT] *Mentha spicata.* An aromatic plant of the mint family, Labiatae; the leaves are used as a flavoring in foods.

spearmint oil [MATER] A colorless to yellowish essential oil obtained from spearmint with characteristic taste and scent; soluble in alcohol, ether, and chloroform; used as a flavor and a source of carvone.

special ammunition supply point [ORD] A mobile supply point where special ammunition is stored and issued to delivery units.

special atomic demolition munition [ORD] A very-low-yield, man-portable, atomic demolition munition which is detonated by a timer device.

special cargo [IND ENG] Cargo which requires special handling or protection, such as pyrotechnics, detonators, watches, and precision instruments.

special character [ADP] A computer-representable character that is not alphabetic, numeric, or blank.

special flight [AERO ENG] An air transport flight, other than a scheduled service, set up to move a specific load.

special functions [MATH] The various families of solution functions corresponding to cases of the hypergeometric equation or functions used in the equation's study, such as the gamma function.

special gelatin [MATER] The brand name of a series of ammonia-gelatin-type dynamites used in open-pit mining, underground metal mining, quarrying, and construction.

special job cover map [MAP] A small-scale map used to record progress on photographic reconnaissance tasks covering very large areas; as each portion of the task is completed, the area covered is outlined on the map.

special nuclear material [NUCLEO] Fissionable and related material controlled directly by the Atomic Energy Commission, such as uranium enriched in the isotopes U^{235} and U^{233}, and plutonium.

special observation [METEOROL] A category of aviation weather observation taken to report significant changes in one or more of the observed elements since the last previous record observation.

special-purpose buoy [NAV] A buoy for a special purpose which has no lateral significance, such as those buoys used to mark quarantine and anchorage areas, or dredging and survey operations.

special-purpose computer [ADP] A digital or analog computer designed to be especially efficient in a certain class of applications.

special-purpose item [ENG] In supply usage, any item designed to fill a special requirement, and having a limited application; for example, a wrench or other tool designed to be used for one particular model of a piece of machinery.

special-purpose vehicle [ENG] A vehicle having a special chassis, or a general-purpose chassis incorporating major modifications, designed to fill a specialized requirement; all tractors (except truck tractors) and tracklaying vehicles, regardless of design, size, or intended purpose, are classified as special-purpose vehicles.

special relativity [RELAT] The division of relativity theory which relates the observations of observers moving with constant relative velocities and postulates that natural laws are the same for all such observers.

special weapon [ORD] Any extraordinary modern weapon, such as an atomic, radiological, or biological weapon.

special weather report [METEOROL] The encoded and transmitted weather report of a special observation.

speciation [EVOL] The evolution of species.

species [NUC PHYS] *See* nuclide. [SYST] A taxonomic category ranking immediately below a genus and including closely related, morphologically similar individuals which actually or potentially interbreed.

species concept [EVOL] The idea that the diversity of nature is divisible into a finite number of definable species.

species number [OCEANOGR] The first number in the argument number in a Doodson tide schedule; indicates approximately the period of a component of tidal potential.

species population [ECOL] A group of similar organisms residing in a defined space at a certain time.

specific [PHYS] A term indicating the amount of a physical quantity per unit mass, weight, volume, or area, or the ratio of the quantity for the substance under consideration to the same quantity for a standard substance, such as water.

specific acoustical impedance [ACOUS] The ratio of the pressure phasor associated with a sound wave at any given point in a medium to the velocity phasor at that point.

specific acoustical ohm *See* rayl.

specific acoustical reactance [ACOUS] The magnitude of the imaginary part of the specific acoustical impedance.

specific acoustical resistance [ACOUS] The real part of the specific acoustical impedance.

specific activity [NUCLEO] 1. The activity of a radioisotope of an element per unit weight of element present in the sample. 2. The activity per unit mass of a pure radionuclide. 3. The activity per unit weight of a sample of radioactive material. In these three cases, specific activity can be expressed in such units as millicuries per gram, disintegrations per second per milligram, or counts per minute per milligram.

specification [ENG] An organized listing of basic requirements for materials of construction, product compositions, dimensions, or test conditions; a number of organizations publish standards (for example, ASME, API, ASTM), and many companies have their own specifications. Also known as specs. [IND ENG] A quantitative description of the required characteristics of a device, machine, structure, product, or process.

specific charge [ELEC] The ratio of a particle's charge to its mass.

specific conductance See conductivity.

specific cryptosystem [COMMUN] A general cryptosystem and a key or set of keys for controlling the cryptographic process.

specific energy [HYD] The energy at any cross section of an open channel, measured above the channel bottom as datum; numerically the specific energy is the sum of the water depth plus the velocity head, $v^2/2g$, where v is the velocity of flow and g the acceleration of gravity. [THERMO] The internal energy of a substance per unit mass.

specific fuel consumption [MECH ENG] The weight flow rate of fuel required to produce a unit of power or thrust, for example, pounds per horsepower-hour. Abbreviated SFC. Also known as specific propellant consumption.

specific gravity [MECH] The ratio of the density of a material to the density of some standard material, such as water at a specified temperature, for example, 4°C or 60°F, or (for gases) air at standard conditions of pressure and temperature. Abbreviated sp gr.

specific gravity hydrometer [ENG] A hydrometer which indicates the specific gravity of a liquid, with reference to water at a particular temperature.

specific heat [THERMO] 1. The ratio of the amount of heat required to raise a mass of material 1 degree in temperature to the amount of heat required to raise an equal mass of a reference substance, usually water, 1 degree in temperature; both measurements are made at a reference temperature, usually at constant pressure or constant volume. 2. The quantity of heat required to raise a unit mass of homogeneous material one degree in temperature in a specified way; it is assumed that during the process no phase or chemical change occurs.

specific humidity [METEOROL] In a system of moist air, the (dimensionless) ratio of the mass of water vapor to the total mass of the system.

specific impulse [AERO ENG] A performance parameter of a rocket propellant, expressed in seconds, equal to the thrust in pounds divided by the weight flow rate in pounds per second. Also known as specific thrust.

specific inductive capacity See dielectric constant.

specific insulation resistance See volume resistivity.

specific ionization [NUCLEO] The number of ion pairs formed per unit distance along the track of an ion passing through matter. Also known as total specific ionization.

specific locus test [GEN] A technique used to detect recessive induced mutations in diploid organisms; a strain which carries several known recessive mutants in a homozygous condition is crossed with a nonmutant strain treated to induce mutations in its germ cells; induced recessive mutations allelic with those of the test strain will be expressed in the progeny.

specific power [NUCLEO] The power produced per unit mass of fuel present in a nuclear reactor.

specific productivity index [PETRO ENG] Barrels per day of oil produced per pound decline in bottom-hole pressure per foot of effective reservoir thickness.

specific propellant consumption See specific fuel consumption.

specific reaction rate constant [PHYS CHEM] Empirical constant k in a chemical rate equation; for example, $r_A = kC_A{}^aC_B{}^b \ldots$, where r_A is the rate of reaction of component A, C_A, $C_B \ldots$ are concentrations of reacting components, and $a, b, \ldots$ are empirical constants.

specific reluctance See reluctivity.

specific repetition rate [ELECTR] The pulse repetition rate of a pair of transmitting stations of an electronic navigation system using various rates differing slightly from each other, as in loran.

specific resistance See electrical resistivity.

specific retention volume [ANALY CHEM] The relationship among retention volume, void volume, and adsorbent weight, used to standardize gas chromatography adsorbents by the elution of a standard solute by a standard eluent from the adsorbent under test.

specific rotation [OPTICS] The calculated rotation of light passing through a solution as related to the solution volume and depth, the amount of solute, and the observed optical rotation at a given wavelength and temperature.

specific routine [ADP] Computer routine to solve a particular data-handling problem in which each address refers to explicitly stated registers and locations.

specific speed [MECH ENG] A number, N_s, used to predict the performance of centrifugal and axial pumps or hydraulic turbines: for pumps, $N_s = N\sqrt{Q}/H^{3/4}$; for turbines, $N_s = N\sqrt{P}/H^{5/4}$, where N_s is specific speed, N is the rotational speed in revolutions per minute, Q is the rate of flow in gallons per minute, H is head in feet, and P is shaft horsepower.

specific surface [CHEM ENG] The surface area per unit weight or volume of a particulate solid; used in size-reduction (crushing and grinding) calculations.

specific susceptibility See mass susceptibility.

specific thrust See specific impulse.

specific viscosity [FL MECH] The specific viscosity of a polymer is the relative viscosity of a polymer solution of known concentration minus 1; usually determined at low concentration of the polymer; for example, 0.5 gram per 100 milliliters of solution, or less.

specific volume [MECH] The volume of a substance per unit mass; it is the reciprocal of the density. Abbreviated sp vol.

specific-volume anomaly [OCEANOGR] The excess of the actual specific volume of the sea water at any point in the ocean over the specific volume of sea water of salinity 35 parts per thousand (0/00) and temperature 0°C at the same pressure. Also known as steric anomaly.

specific weight [MECH] The weight per unit volume of a substance.

specific yield [HYD] The quantity of water which a unit volume of aquifer, after being saturated, will yield by gravity; it is expressed either as a ratio or as a percentage of the volume of the aquifer; specific yield is a measure of the water available to wells.

specimen [SCI TECH] 1. An item representative of others in the same class or group. 2. A sample selected for testing, examination, or display.

speck [PL PATH] A fungus or bacterial disease of rice characterized by speckled grains.

specs See specification.

spectacle [ZOO] A colored marking in the form of rings around the eyes, as in certain birds, reptiles, and mammals (as the raccoon).

spectacle frame [NAV ARCH] A frame at or close to the sternposts of a twin-screw ship, through which pass propeller shafts.

spectacle stone See selenite.

spectral bandwidth [SPECT] The minimum radiant-energy bandwidth to which a spectrophotometer is accurate; that is, 1–5 nanometers for better models.

spectral centroid [OPTICS] An average wavelength; specifically, for a light filter or other light-transmitting device, a weighted average of the spectral energy distribution of the incident light, the transmittance of the device, and the luminosity function.

spectral characteristic [OPTICS] The relation between wavelength and some other variable, such as between wavelength and emitted radiant power of a luminescent screen per unit wavelength interval.

spectral classification [ASTRON] A classification of stars by characteristics revealed by study of their spectra; the six classes B, A, F, G, K, and M include 99% of all known stars.

spectral color [OPTICS] 1. A color corresponding to light of a pure frequency; the basic spectral colors are violet, bluegreen, yellow, orange, and red. 2. A color that is represented by a point on the chromaticity diagram that lies on a straight line between some point on the spectral color (first definition) locus and the achromatic points; purple, for example, is not a spectral color.

spectral density [ELECTROMAG] See spectral energy distribution. [MATH] The density function for the spectral measure of a linear transformation on a Hilbert space. [SYS ENG] *See* frequency spectrum.

spectral directional reflectance factor [ANALY CHEM] In spectrophotometric colorimetry, the ratio of the energy diffused in any desired direction by the object under analysis to that energy diffused in the same direction by an ideal perfect (energy) diffuser.

spectral energy distribution [ELECTROMAG] The power carried by electromagnetic radiation within some small interval of wavelength (of frequency) of fixed amount as a function of wavelength (of frequency). Also known as spectral density.

spectral extinction [OPTICS] The selective absorption of different wavelengths of light as a function of depth in water.

spectral function [MATH] In the theory of stationary stochastic processes, the function

$$F(y) = (2/\pi) \int_0^\infty \rho(x)(\sin xy/x)(dx), 0 \leq y \leq \infty$$

where $\rho(x)$ is the autocorrelation function of a stationary time series.

spectral hygrometer [ENG] A hygrometer which determines the amount of precipitable moisture in a given region of the atmosphere by measuring the attenuation of radiant energy caused by the absorption bands of water vapor; the instrument consists of a collimated energy source, separated by the region under investigation and a detector which is sensitive to those frequencies that correspond to the absorption bands of water vapor.

spectral irradiance [OPTICS] The density of the radiant flux that is incident on a surface per unit of wavelength.

spectral line [SPECT] A discrete value of a quantity, such as frequency, wavelength, energy, or mass, whose spectrum is being investigated; one may observe a finite spread of values resulting from such factors as level width, Doppler broadening, and instrument imperfections. Also known as line; spectrum line.

spectral locus *See* spectrum locus.

spectral luminous efficiency *See* luminosity function.

spectral measure [MATH] A function on the spectrum of an operator on a Hilbert space whose values are projection operators there; spectral theorems concerning linear operators often give an integral representation of the operator in terms of these projection operators and measures.

spectral photography [OPTICS] A technique used in airborne surveys for mineral deposits; narrow-band-pass filters and special film are used to accentuate minor color effects caused by mineralization and alteration which would be undetectable by broad-band photography.

spectral pyrometer *See* narrow-band pyrometer.

spectral radiance factor [ANALY CHEM] A situation when the desired directions for analysis of energy diffused from (reflected from) an object under spectrophotometric colorimetric analysis are all substantially the same (a solid angle of nearly zero steradians).

spectral radius [MATH] For the spectrum of an operator, this is the least upper bound of the set of all $|\lambda|$, where λ is in the spectrum.

spectral reflectance [ANALY CHEM] Situation when the desired directions for analysis of energy from (reflected from) an object under spectrophotometric colorimetric analysis is diffused in all directions (not directed as a single beam).

spectral regions [SPECT] Arbitrary ranges of wavelength, some of them overlapping, into which the electromagnetic spectrum is divided, according to the types of sources that are required to produce and detect the various wavelengths, such as x-ray, ultraviolet, visible, infrared, or radio-frequency.

spectral response *See* spectral sensitivity.

spectral sensitivity [ELECTR] Radiant sensitivity, considered as a function of wavelength. [PHYSICS] The response of a device or material to monochromatic light as a function of wavelength. Also known as spectral response.

spectral series [SPECT] Spectral lines or groups of lines that occur in sequence.

spectral-shift reactor [NUCLEO] A reactor in which, for control or other purposes, the neutron spectrum may be adjusted by varying the properties or amount of moderator.

spectral theorems [MATH] Spectral theorems enable detailed study of various types of operators on Banach spaces by giving an integral or series representation of the operator in terms of its spectrum, eigenspaces, and simple projectionlike operators.

spectral transmission [OPTICS] The radiant flux which passes through a filter divided by the radiant flux incident upon it, for monochromatic light of a specified wavelength.

spectral type [ASTRON] A label used to indicate the physical and chemical characteristics of a star as indicated by study of the star's spectra; for example, the stars in the spectral type known as class B are blue-white, and are referred to as helium stars because the dominant lines in their spectra are the lines in helium spectra.

Spectra Pritchard photometer [OPTICS] A photoelectric instrument for measuring the luminance of surfaces; it has a telescopic viewing system for imaging the bright surface to be measured on the cathode of a photoemissive tube, and a separate unit that combines the power supply with the controls and readout meter.

spectrobolometer [SPECT] An instrument that measures radiation from stars; measurement can be made in a narrow band of wavelengths in the electromagnetic spectrum; the instrument itself is a combination spectrometer and bolometer.

spectrofluorometer [SPECT] A device used in fluorescence spectroscopy to increase the selectivity of fluorometry by passing emitted fluorescent light through a monochromator to record the fluorescence emission spectrum.

spectrogram [SPECT] The record of a spectrum produced by a spectrograph.

spectrograph [SPECT] A spectroscope provided with a photographic camera or other device for recording the spectrum.

spectrography [SPECT] The use of photography to record the electromagnetic spectrum displayed in a spectroscope.

spectroheliocinematograph [OPTICS] A camera used to make motion pictures of, for example, prominences of the sun; the camera utilizes monochromatic light; it is composed of a camera and a spectrohelioscope.

spectroheliogram [ASTRON] A photograph of the sun obtained by means of a spectroheliograph.

spectroheliograph [OPTICS] An instrument used to photograph the sun in one spectral band.

spectrohelioscope [OPTICS] An instrument based on the principle of the spectroheliograph but used for visual observation, and not for photography.

spectrometer [SPECT] **1.** A spectroscope that is provided with a calibrated scale either for measurement of wavelength or for measurement of refractive indices of transparent prism materials. **2.** A spectroscope equipped with a photoelectric photometer to measure radiant intensities at various wavelengths.

spectrometry [SPECT] The use of spectrographic techniques for deriving the physical constants of materials.

spectrophotometer [SPECT] An instrument that measures transmission or apparent reflectance of visible light as a function of wavelength, permitting accurate analysis of color or accurate comparison of luminous intensities of two sources or specific wavelengths.

spectrophotometric titration [ANALY CHEM] An analytical method in which the radiant-energy absorption of a solution is measured spectrophotometrically after each increment of titrant is added.

spectrophotometry [SPECT] A procedure to measure photometrically the wavelength range of radiant energy absorbed by a sample under analysis; can be by visible light, ultraviolet light, or x-rays.

spectropolarimeter [OPTICS] A device used to measure optical rotation in solutions for different light wavelengths.

spectropyrheliometer [SPECT] An astronomical instrument used to measure distribution of radiant energy from the sun in the ultraviolet and visible wavelengths.

spectroscope [SPECT] An optical instrument consisting of a slit, collimator lens, prism or grating, and a telescope or objective lens which produces a spectrum for visual observation.

spectroscopic binary star [ASTRON] A binary star that may be distinguished from a single star only by noting the Doppler

SPECTRA PRITCHARD PHOTOMETER

Spectra Pritchard photometer. *(a)* The viewing system. *(b)* The readout meter. *(Photo Research Corp.)*

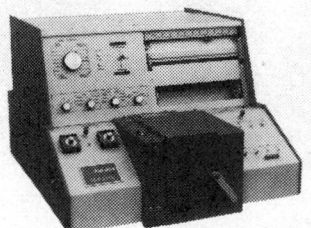

SPECTROFLUOROMETER

Turner 210 "Spectro," an advanced spectrofluorometer. *(G. K. Turner Associates)*

shift of the spectral lines of one or both stars as they revolve about their common center of mass.

spectroscopic displacement law [SPECT] The spectrum of an un-ionized atom resembles that of a singly ionized atom of the element one place higher in the periodic table, and that of a doubly ionized atom two places higher in the table, and so forth.

spectroscopic parallax [ASTRON] Parallax as determined from examination of a stellar spectrum; critical spectral lines indicate the star's absolute magnitude, from which the star's distance, or parallax, can be deduced.

spectroscopic splitting factor *See* Landé g factor.

spectroscopy [PHYS] The branch of physics concerned with the production, measurement, and interpretation of electromagnetic spectra arising from either emission or absorption of radiant energy by various substances.

spectrum [MATH] If T is a linear operator of a normed space X to itself and I is the identity transformation ($I(x) \equiv x$), the spectrum of T consists of all scalars λ for which either $T - \lambda I$ has no inverse or the range of $T - \lambda I$ is not dense in X. [PHYS] **1.** A display or plot of intensity of radiation (particles, photons, or acoustic radiation) as a function of mass, momentum, wavelength, frequency, or some related quantity. **2.** The set of frequencies, wavelengths, or related quantities, involved in some process; for example, each element has a characteristic discrete spectrum for emission and absorption of light. **3.** A range of frequencies within which radiation has some specified characteristic, such as audio-frequency spectrum, ultraviolet spectrum, or radio spectrum.

spectrum analysis [PHYS] The measurement of the amplitude of the components of a complex waveform throughout the frequency range of the waveform.

spectrum analyzer [ENG] Test instrument used to show the distribution of energy contained in the frequencies emitted by a pulse magnetron; also used to measure the Q of resonant cavities and lines, and to measure the cold impedance of a magnetron.

spectrum line *See* spectral line.

spectrum locus [OPTICS] The locus of points representing the chromaticities of spectrally pure stimuli in a chromaticity diagram. Also known as spectral locus.

spectrum of turbulence [ASTROPHYS] A relationship between the size of turbulent eddies in the sun's atmosphere and their average speed.

spectrum-selectivity characteristic [ELECTR] Measure of the increase in the minimum input signal power over the minimum detectable signal required to produce an indication on a radar indicator, if the received signal has a spectrum different from that of the normally received signal.

spectrum signature analysis [ELECTR] The evaluation of electromagnetic interference from transmitting and receiving equipment to determine operational and environment compatibility.

specular hematite [MINERAL] A variety of hematite with a blue-gray color and bright metallic luster.

specular iron *See* specularite.

specularite [MINERAL] A black or gray variety of hematite with brilliant metallic luster, occurring in micaceous or foliated masses, or in tabular or disklike crystals. Also known as gray hematite; iron glance; specular iron.

specular reflection [PHYS] Reflection of electromagnetic, acoustic, or water waves in which the reflected waves travel in a definite direction, and the directions of the incident and reflected waves make equal angles with a line perpendicular to the reflecting surface, and lie in the same plane with it. Also known as direct reflection; mirror reflection; regular reflection.

specular reflection factor [OPTICS] The ratio of the specularly reflected light to the incident light.

specular reflector [OPTICS] A reflecting surface (polished metal or silvered glass) that gives a direct image of the source, with the angle of reflection equal to the angle of incidence. Also known as regular reflector.

specular transmittance [ELECTROMAG] The ratio of the power carried by electromagnetic radiation which emerges from a body and is parallel to a beam entering the body, to the power carried by the beam entering the body.

speculum [MED] A tubular instrument for inserting into a passage or cavity of the body to facilitate visual inspection or medication. [OPTICS] An optical instrument reflector of polished metal or of glass with a film of metal.

speculum alloy [MET] A brilliant white, hard, brittle alloy composed of copper and tin in a 2:1 proportion and sometimes with further additions of other elements.

speech [PHYSIO] A set of audible sounds produced by disturbing the air through the integrated movements of certain groups of anatomical structures.

speech amplifier [ENG ACOUS] An audio-frequency amplifier designed specifically for amplification of speech frequencies, as for public-address equipment and radiotelephone systems.

speech bandwidth [COMMUN] The range of speech frequencies that can be transmitted by a carrier telephone system.

speech clipper [ENG ACOUS] A clipper used to limit the peaks of speech-frequency signals, as required for increasing the average modulation percentage of a radiotelephone or amateur radio transmitter.

speech clipping [ACOUS] In tests of the intelligibility of speech signals, the limiting of peak signals to a maximum value, or the reduction of signals of less than a certain value to zero.

speech coil *See* voice coil.

speech compression [COMMUN] Modulation technique that takes advantage of certain properties of the speech signal to permit adequate information quality, characteristics, and the sequential pattern of a speaker's voice to be transmitted over a narrower frequency band than would otherwise be necessary.

speech frequency *See* voice frequency.

speech intelligibility *See* intelligibility.

speech interference level [ACOUS] The average sound pressure, in decibels above 0.0002 microbar, in the frequency range from 600 to 4800 hertz. Abbreviated SIL.

speech interpolation [COMMUN] Method of obtaining more than one voice channel per voice circuit by giving each subscriber a speech path in the proper direction only at times when his speech requires it.

speech inverter *See* scrambler.

speech scrambler *See* scrambler.

speed [GRAPHICS] The sensitivity of a photographic film, expressed according to one of several scales. [MECH] The time rate of change of position of a body without regard to direction; in other words, the magnitude of the velocity vector. [OPTICS] **1.** The light-gathering power of a lens, expressed as the reciprocal of the f number. **2.** The time that a camera shutter is open. [PHYS] In general, the rapidity with which a process takes place.

speed circle [NAV] A circle having a radius equal to a given speed and drawn about a specified center; the term is used chiefly in connection with relative movement problems.

speed cone [MECH ENG] A cone-shaped pulley, or a pulley composed of a series of pulleys of increasing diameter forming a stepped cone.

speed control [ELEC] A control that changes the speed of a motor or other drive mechanism, as for a phonograph or magnetic tape recorder.

speed-course-latitude error [NAV] An error in both pendulous and nonpendulous types of gyro compasses resulting from movement of the gyro compass in other than an east-west direction. Also known as speed error.

speed density metering [AERO ENG] A type of aircraft carburetion in which the fuel feed is regulated by the parameters of engine feed and intake manifold pressure.

speed error [NAV] **1.** Acceleration error due to a change in the speed of a craft. **2.** *See* speed-course-latitude error.

speed-length ratio [NAV ARCH] The speed of a ship, in knots, divided by the square root of its length, in feet.

speed line [NAV] A line of position approximately perpendicular to the course, which can be used to determine the speed made good.

speed made good [NAV] The actual average speed in knots which was maintained in proceeding along the intended track to the ultimate destination or an intermediate point.

speed of advance [NAV] The average speed in knots which

SPECULAR REFLECTION

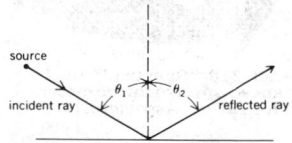

Specular reflection, such as from a polished surface. Angle θ_1 between incident ray and perpendicular to surface equals angle θ_2 between reflected ray and perpendicular.

must be maintained during a passage to arrive at a destination at an appointed time.

speed of light [ELECTROMAG] The speed of propagation of electromagnetic waves in a vacuum, which is a physical constant equal to $299,792.4580 \pm 0.0012$ kilometers per second. Also known as electromagnetic constant; velocity of light.

speed of response [PHYS] The time required for a system to react to some signal; for example, the delay time for a photon detector to react to a radiation pulse, or the time needed for a current or voltage in a circuit to reach a definite fraction of its final value as a result of an abrupt change in the electromotive force.

speed of sound [ACOUS] The phase velocity of a sound wave. Also known as sonic velocity; velocity of sound.

speed of travel [MET] The speed at which a weld is made along its longitudinal axis; measured in inches or spots per minute.

speedometer [ENG] An instrument that indicates the speed of travel of a vehicle in miles per hour, kilometers per hour, or knots.

speed over the ground [NAV] The speed of travel of a craft over the ground; usually called ground speed.

speed reducer [MECH ENG] A train of gears placed between a motor and the machinery which it will drive, to reduce the speed with which power is transmitted.

speed regulator [ELEC] A device that maintains the speed of a motor or other device at a predetermined value or varies it in accordance with a predetermined plan.

speed triangle [NAV] A vector diagram composed of vectors representing the actual courses and speeds of two craft and the relative course and speed; the third vector is the sum and represents them.

speiss [MET] A mixture of impure metal arsenides and antimonides resulting from the smelting of certain ores such as cobalt and lead.

Spelaeogriphacea [INV ZOO] A peracaridan order of the Malacostraca comprised of the single species *Spelaeogriphus lepidops*, a small, blind, transparent, shrimplike crustacean with a short carapace that coalesces dorsally with the first thoracic somite.

spelean [GEOL] Of or pertaining to a feature in a cave.

speleology [GEOL] The study and exploration of caves.

speleothem [GEOL] A secondary mineral deposited in a cave by the action of water. Also known as cave formation.

spelter [MET] A commercially pure grade of zinc used in galvanizing; contains lead or iron as impurities.

spelter shakes *See* metal fume fever.

spelter solder [MET] Brass composed of equal parts of copper and zinc; used in brazing as a filler metal. Also known as brazing brass.

spencerite [MINERAL] $Zn_4(PO_4)_2(OH)_2\cdot3H_2O$ A pearly white monoclinic mineral composed of hydrous basic zinc phosphate and occurring in scaly masses and small crystals.

spending beach [GEOL] In a wave basin, the beach on which the entering waves spend themselves, except for the small remainder entering the inner harbor.

spent acid *See* sludge acid.

spent fuel [NUCLEO] Nuclear reactor fuel that has been irradiated to the extent that it can no longer effectively sustain a chain reaction because its fissionable isotopes have been partially consumed and fission-product poisons have accumulated in it.

spent iron sponge [MATER] Iron sponge saturated with sulfur; prone to spontaneous heating. Also known as spent oxide.

spent liquor [MATER] The liquid effluent from the digestion of wood during pulping; contains wood chemicals (for example, lignin) and spent digestant (caustic, sulfite, or sulfate, depending on the process used).

spent oxide *See* spent iron sponge.

spergenite [GEOL] A biocalcarenite containing ooliths and fossil debris and having a maximum quartz content of 10%. Also known as Bedford limestone; Indiana limestone.

sperm *See* spermatozoon.

spermaceti [MATER] A white, crystalline, oily (waxy) solid that separates from sperm oil; soluble in ether, chloroform,

and carbon disulfide, insoluble in water; melts at 42 to 50°C; used in ointments, emulsions, candles, soaps, and cosmetics; and for linen finishing. Also known as spermaceti wax.

spermaceti wax *See* spermaceti.

spermatheca [ZOO] A sac in the female for receiving and storing sperm until fertilization; found in many invertebrates and certain vertebrates. Also known as seminal receptacle.

spermatic cord [ANAT] The cord consisting of the ductus deferens, epididymal and testicular nerves and blood vessels, and connective tissue that extends from the testis to the deep inguinal ring.

spermatid [HISTOL] A male germ cell immediately before assuming its final typical form.

spermatin [BIOCHEM] An albuminoid material occurring in semen.

spermatocele [MED] A cystic dilation of a duct in the head of the epididymis or in the rete testis.

spermatocyte [HISTOL] A cell of the last or next to the last generation of male germ cells which differentiates to form spermatozoa.

spermatogenesis [PHYSIO] The process by which spermatogonia undergo meiosis and transform into spermatozoa.

spermatogonium [HISTOL] A primitive male germ cell, the last generation of which gives rise to spermatocytes.

spermatophore [ZOO] A bundle or packet of sperm produced by certain animals, such as annelids, arthropods, and some vertebrates.

spermatorrhea [MED] Involuntary discharge of semen without orgasm.

spermatozoon [HISTOL] A mature male germ cell. Also known as sperm.

spermaturia [MED] The presence of sperm in the urine.

spermidine [BIOCHEM] $H_2N(CH_2)_3NH(CH_2)_4NH_2$ The triamine found in semen and other animal tissues.

spermine [BIOCHEM] $C_{10}H_{26}N_4$ A tetramine found in semen, blood serum, and other body tissues.

spermiogenesis [CYTOL] Nuclear and cytoplasmic transformation of spermatids into spermatozoa.

sperm nucleus [BOT] One of the two nuclei in a pollen grain that function in double fertilization in seed plants.

sperm oil [MATER] A combustible, yellowish oil found in the head cavities and blubber of the sperm whale; soluble in ether, chloroform, and benzene; used as a lubricant for precision machinery, for rustproofing metals, and in transmission fluids.

sperm whale [VERT ZOO] *Physeter catadon.* An aggressive toothed whale belonging to the group Odontoceti of the order Cetacea; it produces ambergris and contains a mixture of spermaceti and oil in a cavity of the nasal passage.

sperrylite [MINERAL] $PtAs_2$ A tin-white isometric mineral composed of platinum arsenide; the only platinum compound known to occur in nature; hardness is 6–7 on Mohs scale, and specific gravity is 10.60.

Sperry process [CHEM ENG] The electrolytic manufacture of basic lead carbonate (white lead) from desilverized lead that contains some bismuth; impure lead collects at the anode, and carbon dioxide is passed into the solution to convert the lead to carbonate.

spessartite [MINERAL] $Mn_3Al_2(SiO_4)_3$ A mineral composed of manganese aluminum silicate with small amounts of iron, magnesium, or other elements. [PETR] A lamprophyre composed of a sodic plagioclase groundmass in which green hornblende phenocrysts are embedded; also contains accessory olivine, biotite, apatite, and opaque oxides.

sp gr *See* specific gravity.

Sphaeractinoidea [PALEON] An extinct group of fossil marine hydrozoans distinguished in part by the relative prominence of either vertical or horizontal trabeculae and by the presence of long, tabulate tubes called autotubes.

Sphaeriales [MYCOL] An order of fungi in the subclass Euascomycetes characterized by hard, dark perithecia with definite ostioles.

Sphaeriidae [INV ZOO] The minute bog beetles, a small family of coleopteran insects in the suborder Myxophaga.

Sphaerioidaceae [MYCOL] A family of fungi of the order Sphaeropsidales in which the pycnidia are black or dark-

Speedometer dial. (*AC Spark Plug Div., General Motors*)

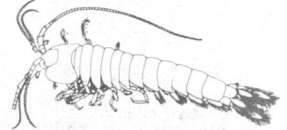

Spelaeogriphus lepidops Gordon, ovigerous female in dorsolateral aspect. (*British Museum of Natural History*)

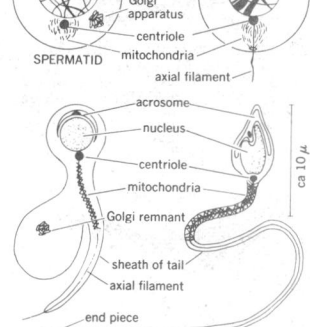

Diagram of spermiogenesis showing sequence of nuclear and cytoplasmic transformations of a spermatid into a spermatozoon; applies to the mammalian sperm.

The sperm whale (*Physeter catodon*) has a head that measures 20 feet (6 meters), or one-third of the total body length.

SPHAEROPLEINEAE

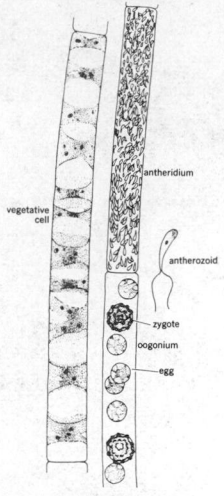

Representative cells of *Sphaeroplea* showing bandlike chloroplasts and heterogametes.

SPHALERITE

Sphalerite crystals in limestone from Joplin, Missouri. *(Specimen from Department of Geology, Bryn Mawr College)*

SPHECIDAE

Typical member of the family Sphecidae.

SPHENE

2.5 cm

Sphene crystal from Eganville, Ontario. *(American Museum of Natural History Specimen.*

colored and are flask-, cone-, or lens-shaped with thin walls and a round, relatively small pore.

sphaerite [MINERAL] Light-gray or bluish mineral composed of hydrous aluminum phosphate and occurring in global concretions.

Sphaeroceridae [INV ZOO] A family of myodarian cyclorrhaphous dipteran insects in the subsection Acalypteratae.

Sphaerodoridae [INV ZOO] A family of polychaete annelids belonging to the Errantia in which species are characterized by small bodies, and are usually papillated.

Sphaerolaimidae [INV ZOO] A family of free-living nematodes in the superfamily Monhysteroidea characterized by a spacious and deep stoma.

sphaerolitic *See* spherulitic.

Sphaeromatidae [INV ZOO] A family of isopod crustaceans in the suborder Flabellifera in which the body is broad and oval and the inner branch of the uropod is immovable.

Sphaerophoraceae [BOT] A family of the Ascolichenes in the order Caliciales which are fruticose with a solid thallus.

Sphaeropleineae [BOT] A suborder of green algae in the order Ulotrichales distinguished by long, coenocytic cells, numerous bandlike chloroplasts, and heterogametes produced in undifferentiated vegetative cells.

Sphaeropsidaceae [MYCOL] An equivalent name for Sphaerioidaceae.

Sphaeropsidales [MYCOL] An order of fungi of the class Fungi Imperfecti in which asexual spores are formed in pycnidia, which may be separate or joined to vegetative hyphae, conidiophores are short or absent, and conidia are always slime spores.

Sphagnaceae [BOT] The single monogeneric family of the order Sphagnales.

Sphagnales [BOT] The single order of mosses in the subclass Sphagnobrya containing the single family Sphagnaceae.

Sphagnobrya [BOT] A subclass of the Bryopsida; plants are grayish-green with numerous, spirally arranged branches and grow in deep tufts or mats, commonly in bogs and in other wet habitats.

sphagnum bog [ECOL] A bog composed principally of mosses of the genus *Sphagnum* (Sphagnales) but also of other plants, especially acid-tolerant species, which tend to form peat.

sphalerite [MINERAL] (Zn,Fe)S The low-temperature form and common polymorph of zinc sulfide; a usually brown or black mineral that crystallizes in the hextetrahedral class of the isometric system, occurs most commonly in coarse to fine, granular, cleanable masses, has resinous luster, hardness of 3.5 on Mohs scale, and specific gravity of 4.1. Also known as blende; false galena; jack; lead marcasite; mock lead; mock ore; pseudogalena; steel jack.

Sphecidae [INV ZOO] A large family of hymenopteran insects in the superfamily Sphecoidea.

Sphecoidea [INV ZOO] A superfamily of wasps belonging to the suborder Apocrita.

Sphenacodontia [PALEON] A suborder of extinct reptiles in the order Pelycosauria which were advanced, active carnivores.

sphene [MINERAL] CaTiSiO$_5$ A brown, green, yellow, gray, or black neosilicate mineral common as an accessory mineral in igneous rocks; it is monoclinic and has resinous luster; hardness is 5–5.5 on Mohs scale; specific gravity is 3.4–3.5. Also known as grothite; titanite.

sphenethmoid [VERT ZOO] A bone that surrounds the anterior portion of the brain in many amphibians.

Spheniscidae [VERT ZOO] The single family of the avian order Sphenisciformes.

Sphenisciformes [VERT ZOO] The penguins, an order of aquatic birds found only in the Southern Hemisphere and characterized by paddlelike wings, erect posture, and scalelike feathers.

Sphenodontidae [VERT ZOO] A family of lepidosaurian reptiles in the order Rhynchocephalia represented by a single living species, *Sphenodon punctatus*, a lizardlike form distinguished by lack of a penis.

sphenoid [CRYSTAL] An open crystal, occurring in monoclinic crystals of the sphenoidal class, and characterized by two nonparallel faces symmetrical with an axis of twofold symmetry. [SCI TECH] Wedge-shaped.

sphenoid bone [ANAT] The butterfly-shaped bone forming the anterior part of the base of the skull and portions of the cranial, orbital, and nasal cavities.

sphenoid sinus [ANAT] Either of a pair of paranasal sinuses located centrally between and behind the eyes, below the ethymoid sinus.

sphenolith [GEOL] A wedgelike igneous intrusion that is partly concordant and partly discordant.

sphenopalatine [ANAT] Of or pertaining to the region of or surrounding the sphenoid and palatine bones.

sphenopalatine foramen [ANAT] The space between the sphenoid and orbital processes of the palatine bone; it opens into the nasal cavity and gives passage to branches from the pterygopalatine ganglion and the sphenopalatine branch of the maxillary artery.

sphenoparietal index [ANTHRO] The ratio, multiplied by 100, of the breadth of the skull from stenion to stenion to its greatest breadth.

Sphenophyllatae *See* Sphenophyllopsida.

Sphenopsida [BOT] A group of vascular cryptogams characterized by whorled, often very small leaves and by the absence of true leaf gaps in the stele; essentially equivalent to the division Equisetophyta.

Sphenyllopsida [PALEOBOT] An extinct class of embryophytes in the division Equisetophyta.

spherator [PL PHYS] One of the class of low-β, low-density, quasi-steady-state closed devices (like Tokamak) used in studying production of electric power by fusion.

sphere [MATH] **1.** The set of all points in a euclidean space which are a fixed common distance from some given point; in euclidean three-dimensional space the Riemann sphere consists of all points (x,y,z) which satisfy the equation $x^2+y^2+z^2=1$. **2.** The set of points in a metric space whose distance from a fixed point is constant.

sphere gap [ELEC] A spark gap between two equal-diameter spherical electrodes.

sphere of attraction [PHYS CHEM] The distance within which the potential energy arising from mutual attraction of two molecules is not negligible with respect to the molecules' average thermal energy at room temperature.

sphere photometer *See* integrating-sphere photometer.

spheres of Eudoxus [ASTRON] A theory of Eudoxus from about 400 B.C.; the planets, sun, and moon were on a series of concentric spheres rotating inside one another on different axes.

spherical aberration [OPTICS] Aberration arising from the fact that rays which are initially at different distances from the optical axis come to a focus at different distances along the axis when they are reflected from a spherical mirror or refracted by a lens with spherical surfaces.

spherical antenna [ELECTROMAG] An antenna having the shape of a sphere, used chiefly in theoretical studies.

spherical capacitor [ELEC] A capacitor made of two concentric metal spheres with a dielectric filling the space between the spheres.

spherical coordinates [MATH] A system of curvilinear coordinates in which the position of a point in space is designated by its distance r from the origin or pole, called the radius vector, the angle ϕ between the radius vector and a vertically directed polar axis, called the cone angle or colatitude, and the angle θ between the plane of ϕ and a fixed meridian plane through the polar axis, called the polar angle or longitude.

spherical cyclic curve *See* cyclic curve.

spherical degree [MATH] A solid angle equal to 1/90 of a spherical right angle.

spherical-earth attenuation [ELECTROMAG] Attenuation over an imperfectly conducting spherical earth in excess of that over a perfectly conducting plane.

spherical-earth factor [ELECTROMAG] The ratio of the electric field strength that would result from propagation over an imperfectly conducting spherical earth to that which would result from propagation over a perfectly conducting plane.

spherical excess [MATH] The sum of the angles of a spherical triangle, minus 180 degrees.

spherical harmonics [MATH] Solutions of Laplace's equation in spherical coordinates.

spherical indicatrix of binormal to a curve [MATH] All the end points of those radii from the sphere of radius one which are parallel to the positive direction of the binormal to a space curve.

spherical indicatrix of tangent to a curve [MATH] Those points on the unit sphere traced out by a radius moving from point to point always parallel with the tangent to the curve.

spherical mirror [OPTICS] A mirror, either convex or concave, whose surface forms part of a sphere.

spherical pendulum [MECH] A simple pendulum mounted on a pivot so that its motion is not confined to a plane; the bob moves over a spherical surface.

spherical powder [MATER] A powder consisting of globular-shaped particles.

spherical sailing [NAV] Any of the sailing computation methods which are used to solve the problems of course, distance, difference of latitude, difference of longitude, and departure which take into account the spherical or spheroidal shape of the earth.

spherical sector [MATH] The cap and cone formed by the intersection of a plane with a sphere, the cone extending from the plane to the center of the sphere and the cap extending from the plane to the surface of the sphere.

spherical segment [MATH] A solid that is bounded by a sphere and two parallel planes which intersect the sphere or are tangent to it.

spherical separator [PETRO ENG] A gas-oil separator in the form of a spherical vessel.

spherical triangle [MATH] A three-sided surface on a sphere the sides of which are arcs of great circles.

spherical trigonometry [MATH] The study of spherical triangles from the viewpoint of angle, length, and area.

spherical wave [PHYS] A wave whose equiphase surfaces form a family of concentric spheres; the direction of travel is always perpendicular to the surfaces of the spheres.

spherical weathering *See* spheroidal weathering.

spherical wedge [MATH] The portion of a sphere bounded by two semicircles and a lune (the surface of the sphere between the semicircles).

sphericity [SCI TECH] The degree to which a shape approaches that of a sphere.

spherocyte [PATH] A spherical red blood cell.

spherocytosis [MED] Preponderance of spherocytes in the blood.

spheroid *See* ellipsoid of revolution.

spheroidal excess [MATH] The amount by which the sum of the three angles of a triangle on the surface of a spheroid exceeds 180°.

spheroidal galaxy *See* elliptical galaxy.

spheroidal graphite cast iron *See* nodular cast iron.

spheroidal group [CRYSTAL] A group in the tetragonal symmetry system; the sphenoid is the typical form.

spheroidal harmonics [MATH] Solutions to Laplace's equation when phrased in ellipsoidal coordinates.

spheroidal recovery [GEOPHYS] The hypothetical return of the earth to spheroid form after it has been distorted.

spheroidal weathering [GEOL] Chemical weathering in which concentric or spherical shells of decayed rock are successively separated from a block of rock; commonly results in the formation of a rounded boulder of decomposition. Also known as concentric weathering; spherical weathering.

spheroidized carbides [MET] Globular forms of carbide, as formed in spheroidized steel.

spheroidized steel [MET] Steel that has been heat-treated to produce a spheroidized carbide structure.

spheroidizing [MET] Heating steels just below Ae_1 until the shape of cementite particles becomes relatively spherical.

spherometer [ENG] A device used to measure the curvature of a spherical surface.

spherulite [GEOL] A spherical body or coarsely crystalline aggregate having a radial internal structure arranged about one or more centers.

spherulitic [PETR] Relating to the texture of a rock composed of numerous spherulites. Also known as globular; sphaerolitic.

sphincter [ANAT] A muscle that surrounds and functions to close an orifice.

sphincter of Oddi [ANAT] Sphincter of the hepatopancreatic ampulla.

Sphinctozoa [PALEON] A group of fossil sponges in the class Calcarea which have a skeleton of massive calcium carbonate organized in the form of hollow chambers.

Sphindidae [INV ZOO] The dry fungus beetles, a family of coleopteran insects in the superfamily Cucujoidea.

Sphingidae [INV ZOO] The single family of the lepidopteran superfamily Sphingoidea.

Sphingoidea [INV ZOO] A superfamily of Lepidoptera in the suborder Heteroneura consisting of the sphinx, hawk, or hummingbird moths; these are heavy-bodied forms with antennae that are thickened with a pointed apex, a well-developed proboscis, and narrow wings.

sphingolipid [BIOCHEM] Any lipid, such as a sphingomyelin, that yields sphingosine or one of its derivatives as a product of hydrolysis.

sphingolipidosis [MED] Any of a group of hereditary metabolic disorders characterized by excessive accumulations of certain glycolipids and phospholipids in various tissues of the body.

sphingomyelin [BIOCHEM] A phospholipid consisting of choline, sphingosine, phosphoric acid, and a fatty acid.

sphingosine [BIOCHEM] $C_{18}H_{37}O_2N$ A moiety of sphingomyelin, cerebrosides, and certain other phosphatides.

sphygmomanometer [MED] An instrument for measuring the arterial blood pressure.

sphygmophone [MED] A microphone attached to the wrist to pick up the sounds of the pulse.

Sphyraenidae [VERT ZOO] A family of shore fishes in the suborder Mugiloidei of the order Perciformes comprising the barracudas.

Sphyriidae [INV ZOO] A family of ectoparasitic Crustacea belonging to the group Lernaeopodoida; the parasite embeds its head and part of its thorax into the host.

Spica [ASTRON] A blue-white dwarf star of stellar magnitude 1.0, 160 light-years from the sun, spectral classification B1-V, in the constellation Virgo; the star α Virginis.

spice [FOOD ENG] An aromatic vegetable material used for food seasoning.

spicule [ASTRON] One of an irregular distribution of jets shooting up from the sun's chromosphere. Also known as solar spicule. [BOT] An empty diatom shell. [INV ZOO] A calcareous or siliceous, usually spikelike supporting structure in many invertebrates, particularly in sponges and alcyonarians.

spiculin [BIOCHEM] An organic material making up a portion of a spicule.

spiculite [PETR] A spindle-shaped belonite thought to have formed by the coalescence of globulites.

spiculum [INV ZOO] A bristlelike copulatory organ in certain nematodes. Also known as copulatory spicule.

spider [ENG ACOUS] A highly flexible perforated or corrugated disk used to center the voice coil of a dynamic loudspeaker with respect to the pole piece without appreciably hindering in-and-out motion of the voice coil and its attached diaphragm. [INV ZOO] The common name for arachnids comprising the order Araneida.

spider nevus [MED] A type of telangiectasis characterized by a central, elevated, tiny red dot, pinhead in size, from which blood vessels radiate like strands of a spider's web. Also known as stellar nevus.

spiderweb antenna [ELECTROMAG] All-wave receiving antenna having several different lengths of doublets connected somewhat like the web of a spider to give favorable pickup characteristics over a wide range of frequencies.

spiegeleisen [MET] An iron alloy containing 15–30% manganese and 5% carbon used in steelmaking.

Spiegler's test [PATH] A test for the presence of protein in urine performed by overlaying clear acidulated urine on Spiegler's reagent (mercuric chloride, tartaric acid, glycerin, distilled water); opalescence at the fluid junction indicates protein.

spigot mortar [ORD] A mortar which propels a warhead larger than the bore of the mortar by means of a closed tube (spigot) attached to the warhead and extending into the

SPICULE

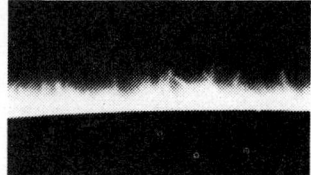

Large-scale photograph of the chromosphere in H-α light, with the disk of the sun artificially eclipsed and the hairy spicules projecting above the continuous chromosphere. The length of this section is about 70,000 kilometers. *(Photography by R. B. Dunn, through 15-inch telescope, Sacramento Peak Observatory)*

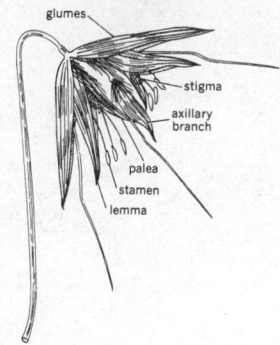

SPIKELET

glumes
stigma
axillary
branch
palea
stamen
lemma

Spikelet of wild oat (*Avena fatua*).

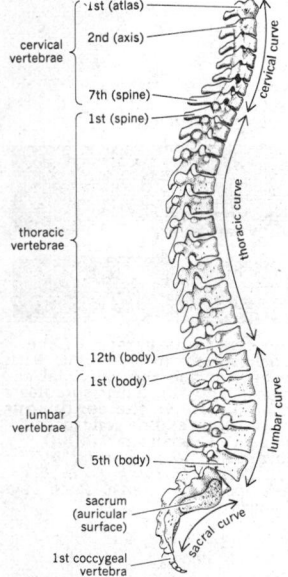

SPINE

1st (atlas)
cervical vertebrae
2nd (axis)
7th (spine)
cervical curve
1st (spine)
thoracic vertebrae
thoracic curve
12th (body)
1st (body)
lumbar vertebrae
lumbar curve
5th (body)
sacrum (auricular surface)
sacral curve
1st coccygeal vertebra

The human spine in lateral view. (*From W. J. Hamilton et al., Textbook of Human Anatomy, Macmillan, 1956*)

SPINEL

|⟵ 1.3 cm ⟶|

Spinel crystal from Franklin, New Jersey. (*Specimen from Department of Geology, Bryn Mawr College*)

mortar; the force of the propellant within the mortar acts upon the tube, thus propelling the warhead toward the target.

spike [BOT] An indeterminate inflorescence with sessile flowers. [DES ENG] A large nail, especially one longer than 3 inches (7.6 centimeters), and often of square section. [PHYS] A short-duration transient whose amplitude considerably exceeds the average amplitude of the associated pulse or signal.

spike antenna *See* monopole antenna.

spiked core *See* seed core.

spikelet [BOT] The compound inflorescence of a grass consisting of one or several bracteate spikes.

spike microphone [ENG ACOUS] A device for clandestine aural surveillance in which the sensor is a spike driven into the wall of the target area and mechanically coupled to the diaphragm of a microphone on the other side of the wall.

spike-tooth harrow [AGR] An implement with steel spikes extending downward from a frame and pulled by a tractor to pulverize and smooth plowed soil.

spile [MIN ENG] **1.** A temporary lagging driven ahead on levels in loose ground. **2.** A short piece of plank sharpened flatwise and used for driving into watery stratums as sheet piling to assist in checking the flow of water.

spiling *See* forepoling.

spilite [PETR] An altered basalt containing albitized feldspar accompanied by low-temperature, hydrous crystallization products such as chlorite, calcite, and epidote.

spill [NUCLEO] The accidental release of radioactive material.

spill box [CIV ENG] A device such as a flume that maintains a constant head on a measuring weir or orifice.

spilling [OCEANOGR] The process by which steep waves break on approaching the shore; white water appears on the crest and the wave top gradually rolls over, without a crash.

spilling breaker *See* plunging breaker.

spillover [COMMUN] The receiving of a radio signal of a different frequency from that to which the receiver is tuned, due to broad tuning characteristics. [METEOROL] That part of orographic precipitation which is carried along by the wind so that it reaches the ground in the nominal rain shadow on the lee side of the barrier.

spillover positions [COMMUN] When a send channel is unusually busy or inoperative, the resulting backlogged traffic can be switched to spillover (storage) positions where it is held for immediate transmission when a channel becomes available.

spill pit *See* runoff pit.

spillway [CIV ENG] A passage in or about a dam or other hydraulic structure for escape of surplus water.

spillway apron [CIV ENG] A concrete or timber floor at the bottom of a spillway to prevent soil erosion from heavy or turbulent flow.

spillway channel [CIV ENG] An outlet channel from a spillway.

spillway dam *See* overflow dam.

spillway gate [CIV ENG] A gate for regulating the flow from a reservoir.

spin [MECH] The rotation of a body about its axis. [QUANT MECH] The intrinsic angular momentum of an elementary particle or nucleus, which exists even when the particle is at rest, as distinguished from orbital angular momentum.

spina bifida [MED] A congenital anomaly characterized by defective closure of the vertebral canal with herniation of the spinal cord meninges.

spina bifida occulta [MED] An asymptomatic congenital anomaly consisting of incomplete fusion of the posterior arch of the vertebral canal without hernial protrusion of the meninges.

spinacene *See* squalene.

spinach [BOT] *Spinacia oleracea.* An annual potherb of Asiatic origin belonging to the order Caryophyllales and grown for its edible foliage.

spinal anesthesia [MED] **1.** Anesthesia due to a lesion of the spinal cord. **2.** Anesthesia produced by the injection of an anesthetic into the spinal subarachnoid space.

spinal column *See* spine.

spinal cord [ANAT] The cordlike posterior portion of the

central nervous system contained within the spinal canal of the vertebral column of all vertebrates.

spinal foramen [ANAT] Central canal of the spinal cord.

spinal ganglion [ANAT] Any one of the sensory ganglions, each associated with the dorsal root of a spinal nerve.

spinal nerve [ANAT] Any of the paired nerves arising from the spinal cord.

spinal reflex [PHYSIO] A reflex mediated through the spinal cord without the participation of the more cephalad structures of the brain or spinal cord.

spin axis [PHYS] The axis of rotation of a gyroscope.

spin compensation [MECH] Overcoming or reducing the effect of projectile rotation in decreasing the penetrating capacity of the jet in shaped-charge ammunition.

spin-decelerating moment [MECH] A couple about the axis of the projectile, which diminishes spin.

spin dependent force [PHYS] A force between two particles which depends in some way on the spin, possibly on the angle between their spin directions, or on the angles between their spin directions and a line joining the particles.

spindle [CYTOL] A structure formed of fiberlike elements just before metaphase that extends between the poles of the achromatic figure and is attached to the centromeric regions of the chromatid pairs. [DES ENG] A short, slender or tapered shaft. [NAV] A spar serving as a beacon.

spindle fiber [CYTOL] One of the fiberlike elements of the spindle; an aggregation of microtubules resulting from the polymerization of a series of small protein fibrils by primary $-S-S-$ linkages.

spindle tuber [PL PATH] A virus disease of the potato characterized by spindliness of the tops and tubers.

spine [ANAT] An articulated series of vertebrae forming the axial skeleton of the trunk and tail, and being a characteristic structure of vertebrates. Also known as backbone; spinal column; vertebral column. [BOT] A rigid sharp-pointed process in plants; many are modified leaves. [INV ZOO] One of the processes covering the surface of a sea urchin. [VERT ZOO] **1.** One of the spiny rays supporting the fins of most fishes. **2.** A sharp-pointed modified hair on certain mammals, such as the porcupine.

spin echo technique [NUC PHYS] A variation of the nuclear magnetic resonance technique in which the radio frequency field is applied in two pulses, separated by a time interval t, and a strong nuclear induction signal is observed at a time t after the second pulse.

spinel [MINERAL] **1.** $MgAl_2O_4$ A colorless, purplish-red, greenish, yellow, or black mineral, usually forming octahedral crystals, and characterized by great hardness; used as a gemstone. **2.** A group of minerals of general formula AB_2O_4, where A is magnesium, ferrous iron, zinc, or manganese, or a combination of them, and B is aluminum, ferric iron, or chromium.

spin-flip Raman laser [OPTICS] A type of tunable laser that uses a fixed-frequency laser to pump a semiconductor crystal at low temperature in a high magnetic field; the pump laser photons scatter from electrons in the semiconductor and are shifted in energy by an amount equal to the magnetic field splitting between the electron spin states; tuning is accomplished by varying the magnetic field strength.

spin-flip scattering [QUANT MECH] Scattering of a particle with spin ½ in which the direction of the particle's spin is reversed.

spin-lattice interaction [SOLID STATE] The state of a solid when the energy of electron spins is being shared with the thermal-vibration energy of the solid as a whole.

spin-lattice relaxation [SOLID STATE] Magnetic relaxation in which the excess potential energy associated with electron spins in a magnetic field is transferred to the lattice.

spin magnetism [SOLID STATE] Paramagnetism or ferromagnetism that arises from polarization of electron spins in a substance.

Spinnbarkeit relaxation [FL MECH] A rheological effect illustrated by the pulling away of liquid threads when an object that has been immersed in a viscoelastic fluid is pulled out.

spinner [ENG] **1.** Automatically rotatable radar antenna, together with directly associated equipment. **2.** Part of a mechanical scanner which rotates about an axis, generally restricted to cases where the speed of rotation is relatively high.

spinneret [ENG] An extrusion die with many holes through which plastic melt is forced to form filaments. [INV ZOO] An organ that spins fiber from the secretion of silk glands. [TEXT] A metal device with tiny holes through which a solution is forced at high speeds to make fine textile filaments.

spinning [ENG] The extrusion of a spinning solution (such as molten plastic) through a spinneret. [MECH ENG] Shaping and finishing sheet metal by rotating the workpiece over a mandrel and working it with a round-ended tool. Also known as metal spinning. [TEXT] Converting fibers or filaments into thread or yarn by drawing and twisting.

spinning band column [ANALY CHEM] An analytical distillation column inside of which is a series of driven, spinning bands; centrifugal action of the bands throws a layer of liquid onto the inner surface of the column; used as an aid in liquid-vapor contact.

spinning machine [MECH ENG] **1.** A machine that winds insulation on electric wire. **2.** A machine that shapes metal hollow ware. [TEXT] A machine that spins yarn from staple fiber or continuous filament.

spinoblast [INV ZOO] A statoblast having a float of air cells and barbs or hooks on the surface.

spinochrome [BIOCHEM] A type of echinochrome; an organic pigment that is known only from sea urchins and certain homopteran insects.

spinode *See* cusp.

spinor [MATH] **1.** A vector with two complex components, which undergoes a unitary unimodular transformation when the three-dimensional coordinate system is rotated; it can represent the spin state of a particle of spin ½. **2.** More generally, a spinor of order (or rank) n is an object with 2^n components which transform as products of components of n spinors of rank one. **3.** A quantity with four complex components which transforms linearly under a Lorentz transformation in such a way that if it is a solution of the Dirac equation in the original Lorentz frame it remains a solution of the Dirac equation in the transformed frame; it is formed from two spinors (definition 1). Also known as Dirac spinor.

spin-orbit coupling [QUANT MECH] The interaction between a particle's spin and its orbital angular momentum.

spin-orbit multiplet [PHYS] A collection of atomic or nuclear states which differ in energy only on account of spin-orbit coupling; the total spin angular momentum quantum number S and total orbital angular momentum quantum number L are the same for all states; the energy levels are labeled by the total angular momentum quantum number J.

spinous process [ANAT] Any slender, sharp-pointed projection on a bone.

spin paramagnetism [SOLID STATE] Paramagnetism that arises from the electron spins in a substance.

spin-parity [PARTIC PHYS] A combined symbol J^P for an elementary particle's spin J, and its intrinsic parity P.

spin quantum number [QUANT MECH] The ratio of the maximum observable component of a system's spin to Planck's constant divided by 2π; it is an integer or a half-integer.

spin resonance *See* magnetic resonance.

spin rocket [AERO ENG] A small rocket that imparts spin to a larger rocket vehicle or spacecraft.

spin safe [ORD] Referring to a fuse that is safe when experiencing a rotation equivalent to that attained during flight; that is, other arming forces are necessary to arm the fuse.

spin space [MATH] The two-dimensional vector space over the complex numbers, whose unitary unimodular transformations are a two-dimensional double-valued representation of the three-dimensional rotation group; its vectors can represent the various spin states of a particle with spin ½, and its unitary unimodular transformations can represent rotations of this particle.

spin-spin energy [PHYS] An interaction energy proportional to the dot product of the spin angular momenta of two systems.

spin-spin relaxation [SOLID STATE] Magnetic relaxation, observed after application of weak magnetic fields, in which the excess potential energy associated with electron spins in a magnetic field is redistributed among the spins, resulting in heating of the spin system.

spin stabilization [AERO ENG] Directional stability of a spacecraft obtained by the action of gyroscopic forces which result from spinning the body about its axis of symmetry.

spin state [QUANT MECH] Condition of a particle in which its total spin, and the component of its spin along some specified axis, have definite values; more precisely, the particle's wave function is an eigenfunction of the operators corresponding to these quantities.

spin temperature [SOLID STATE] For a system of electron spins in a lattice, a temperature such that the population of the energy levels of the spin system is given by the Boltzmann distribution with this temperature.

spinthariscope [ELECTR] An instrument for viewing the scintillations of alpha particles on a luminescent screen, usually with the aid of a microscope.

Spintheridae [INV ZOO] An amphinomorphan family of small polychaete annelids included in the Errantia.

Spinulosida [INV ZOO] An order of Asteroidea in which pedicellariae rarely occur, marginal plates bounding the arms and disk are small and inconspicuous, and spines occur in groups on the upper surface.

spin wave [SOLID STATE] A sinusoidal variation, propagating through a crystal lattice, of that angular momentum which is associated with magnetism (mostly spin angular momentum of the electrons).

spin welding [ENG] Fusion of two objects (for example, plastics) by forcing them together while one of the pair is spinning; frictional heat melts the interface, spinning is stopped, and the bodies are held together until they are frozen in place (welded).

spiny-rayed fish [VERT ZOO] The common designation for actinopterygian fishes, so named for the presence of stiff, unbranched, pointed fin rays, known as spiny rays.

Spionidae [INV ZOO] A family of spioniform annelid worms belonging to the Sedentaria.

spioniform worm [INV ZOO] A polychaete annelid characterized by the presence of a pair of short to long, grooved palpi near the mouth.

spiracle [INV ZOO] An external breathing orifice of the tracheal system in insects and certain arachnids. [VERT ZOO] **1.** The external respiratory orifice in cetaceous and amphibian larvae. **2.** The first visceral cleft in fishes.

spiral [MATH] A simple curve in the plane which continuously winds about itself either into some point or out from some point.

spiral arms [ASTRON] The shape of sections of certain galaxies called spirals; these sections are two so-called arms composed of stars, dust, and gas extending from the center of the galaxy and coiled about it.

spiral band [METEOROL] Spiral-shaped radar echoes received from precipitation areas within intense tropical cyclones (hurricanes or typhoons); they curve cyclonically in toward the center of the storm and appear to merge to form the wall around the eye of the storm. Also known as hurricane band; hurricane radar band.

spiral bevel gear [DES ENG] Bevel gear with curved, oblique teeth to provide gradual engagement and bring more teeth together at a given time than an equivalent straight bevel gear.

spiral binding [GRAPHICS] A book binding in which a spiral of wire or plastic strip is wound through holes at the edge of the book.

spiral chute [DES ENG] A gravity chute in the form of a continuous helical trough spiraled around a column for conveying materials to a lower level.

spiral cleavage [EMBRYO] A cleavage pattern characterized by formation of a cell mass showing spiral symmetry; occurs in mollusks.

spiral conveyor *See* screw conveyor.

spiral delay line [ELECTROMAG] A transmission line which has a helical inner conductor.

spiral flow tank [CIV ENG] An aeration tank of the activated sludge process into which air is diffused in a spiral helical movement guided by baffles and proper location of diffusers.

spiral flow test [ENG] The determination of the flow properties of a thermoplastic resin by measuring the length and weight of resin flowing along the path of a spiral cavity.

spiral four cable [ELEC] A quad cable in which the four

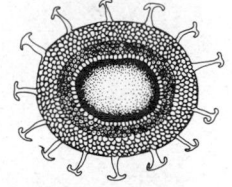

SPINOBLAST

Spinoblast of *Pectinatella magnifica*.

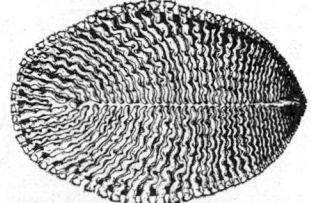

SPINTHERIDAE

Spinther, dorsal view.

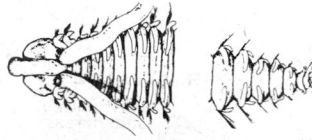

SPIONIDAE

Nerinides showing the anterior portion (left) and the tail section (right), in dorsal view.

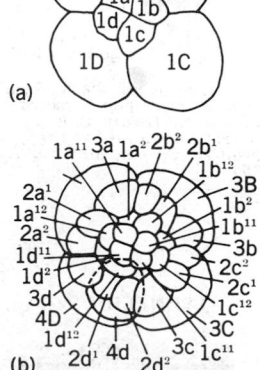

SPIRAL CLEAVAGE

Cell lineage for spiral cleavage pattern in egg of gastropod *Crepidula*, seen from the animal pole. (a) Eight-cell stage, showing the first quarter of micromeres (1a–1d) and the macromeres (1A–1D). (b) Twenty-five-cell stage. (*After Conklin, 1897*)

conductors are twisted about a common axis, the two sets of opposite conductors being used as pairs.

spiral gage *See* spiral pressure gage.

spiral galaxy [ASTRON] A type of galaxy classified on the basis of appearance of its photographic image; this type includes two main groups: normal spirals with circular symmetry of the nucleus and of the spiral arms, and barred spirals in which the dominant form is a luminous bar crossing the nucleus with spiral arms starting at the ends of the bar or tangent to a luminous rim on which the bar terminates.

spiral ganglion [ANAT] The ganglion of the cochlear part of the vestibulocochlear nerve embedded in the spiral canal of the modiolus.

spiral gear [MECH ENG] A helical gear that transmits power from one shaft to another, nonparallel shaft.

spiral-jaw clutch [MECH ENG] A modification of the square-jaw clutch permitting gradual meshing of the mating faces, which have a helical section.

spiral layer *See* Ekman layer.

spiral ligament [ANAT] The reticular connective tissue connecting the basilar membrane to the outer cochlear wall in the ear of mammals.

spiral of Archimedes [MATH] The curve spiraling into the origin which in polar coordinates is given by the equation $r = a\theta$. Also known as Archimedes' spiral.

spiral organ *See* organ of Corti.

spiral pipe [DES ENG] Strong, lightweight steel pipe with a single continuous welded helical seam from end to end.

spiral plate exchanger [CHEM ENG] A heat-transfer device made from a pair of plates rolled in a spiral to provide two relatively long, rectangular passages for heat-transfer between fluids in countercurrent flow.

spiral pressure gage [ENG] A device for measurement of pressures; a hollow tube spiral receives the system pressure which deforms (unwinds) the spiral in direct relation to the pressure in the tube. Also known as spiral gage.

spiral ramp system [MIN ENG] Development of a mine by driving moderately inclined haulageways from the surface to underground ore horizons.

spiral scanning [ENG] Scanning in which the direction of maximum radiation describes a portion of a spiral; the rotation is always in one direction; used with some types of radar antennas.

spiral spring [DES ENG] A spring bar or wire wound in an Archimedes spiral in a plane; each end is fastened to the force-applying link of the mechanism.

spiral thermometer [ENG] A bimetal (facing strips of two metals with different expansion coefficients) spiral that winds tighter or opens with changes in temperature.

spiral-tube heat exchanger [ENG] A countercurrent heat-exchange device made of a group of concentric spirally wound coils, generally connected by manifolds; used for cryogenic exchange in air-separation plants.

spiral valve [VERT ZOO] A spiral fold of mucous membrane in the small intestine of elasmobranchs and some primitive fishes which increases the surface area for absorption.

spiral welded pipe [DES ENG] A steel pipe made of long strips of steel plate fitted together to form helical seams, which are welded.

spiral wire column [ANALY CHEM] An analytical rectification (distillation) column with a wire spiral the length of the inside of the column to serve as a liquid-vapor contact surface.

spiramycin [MICROBIO] A complex of related antibiotics, which resemble erythromycin structurally and in antibacterial spectrum, produced by *Streptomyces ambofaciens*.

spiraster [INV ZOO] A spiral spicule bearing rays in Porifera.

spire [ARCH] As a landmark, a prominent, slender, pointed structure surmounting a building; a spire is seldom less than two-thirds of the entire height, and its lines are rarely broken by stages or other features. [BOT] A narrow, tapering blade or stalk.

spiricle [BOT] Any of the coiled threads in certain seed coats which uncoil when moistened.

Spiriferida [PALEON] An order of fossil articulate brachiopods distinguished by the spiralium, a pair of spirally coiled ribbons of calcite supported by the crura.

Spiriferidina [PALEON] A suborder of the extinct brachiopod order Spiriferida including mainly ribbed forms having laterally or ventrally directed spires, well-developed interareas, and a straight hinge line.

Spirillaceae [MICROBIO] A family of bacteria of the order Pseudomonadales comprising chemoheterotrophic, gram-negative, curved to spirally twisted rods, with rigid cell walls and polar flagella.

Spirillinacea [INV ZOO] A superfamily of foraminiferan protozoans in the suborder Rotaliina characterized by a planispiral or low conical test with a wall composed of radial calcite.

Spirillum [MICROBIO] A genus of aerobic bacteria in the family Spirillaceae having a tuft of polar flagella (5–20) and usually volutin granules in the cytoplasm.

spirit compass [NAV] A liquid compass using a mixture of alcohol and water.

spirit duplicating [GRAPHICS] A contact process used in reproducing maps in which a map with several colors in accurate position relative to each other can be produced in one run through the machine; the map is drawn on a paper master having a carbon backing; different-colored carbons can be substituted to prepare a multicolor image.

spirit level *See* level.

spirit stain [MATER] A dye dissolved in methylated spirits; used to stain wood surfaces.

spirit thermometer [ENG] A temperature-measurement device consisting of a closed capillary tube with a liquid (for example, alcohol) reservoir bulb at the bottom; as the bulb is heated, the liquid expands up into the capillary tubing, indicating the temperature of the bulb.

spirit varnish [MATER] An artificial varnish consisting of resin, asphalt, or a cellulose ester dissolved in a volatile solvent.

Spirobrachiidae [INV ZOO] A family of the Brachiata in the order Thecanephria.

Spirochaetaceae [MICROBIO] A family of the Spirochaetales including the large spirochetes, which are about 30 to 500 microns long.

Spirochaetales [MICROBIO] An order of bacteria characterized by elongate cells twisted three-dimensionally into a spiral shape.

spirochetal jaundice *See* Weil's disease.

spirochete [MICROBIO] A member of the Spirochaetales, distinguished by their spiral form and motility.

spirochetemia [MED] The presence of spirochetes in the blood.

spirocyst [INV ZOO] A thin-walled capsule that contains a long, unarmed, evertable, spirally coiled thread of uniform diameter; found in coelenterates.

spirograph [MED] An instrument for registering respiration.

Spiroid gear [DES ENG] A trade name of the Illinois Tool Works, it resembles a hypoid-type bevel gear but performs like a worm mesh; used to connect skew shafts.

spirometry [PHYSIO] The measurement, by a form of gas meter (spirometer), of volumes of air that can be moved in or out of the lungs.

spironolactone [PHARM] $C_{24}H_{32}O_4S$ A steroid having a lactone ring attached at carbon-17; used as a diuretic.

spiro ring system [ORG CHEM] A molecular structure with two ring structures having one atom in common; for example, spiropentane.

Spirotrichia [INV ZOO] A subclass of the protozoan class Ciliatea which contains those ciliates characterized by conspicuous, compound ciliary structures, known as cirri, and buccal organelles.

spirule [GRAPHICS] A plastic device consisting of a protractor and a pivoted arm with a logarithmic spiral, used to add angles between vectors and multiply magnitudes of vectors.

Spirulidae [INV ZOO] A family of cephalopod mollusks containing several species of squids.

Spiruria [INV ZOO] A subclass of nematodes in the class Secernentea.

Spirurida [INV ZOO] An order of phasmid nematodes in the subclass Spiruria.

Spiruroidea [INV ZOO] A superfamily of spirurid nematodes which are parasitic in the respiratory and digestive systems of vertebrates.

spit [ENG] To light a fuse. [GEOGR] A small point of land

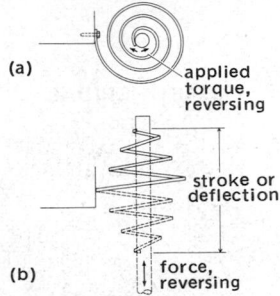

SPIRAL SPRING

(a)

applied torque, reversing

stroke or deflection

(b)

force, reversing

Spiral spring is unique in being able to respond to two types of forces. *(a)* In response to torsional force, spring remains flat spiral. *(b)* In response to translational force, perpendicular to spiral plane, spring is deformed into conical helix.

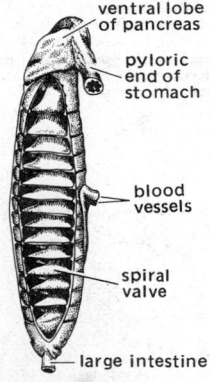

SPIRAL VALVE

ventral lobe of pancreas

pyloric end of stomach

blood vessels

spiral valve

large intestine

Small intestine of shark (*Squalus acanthias*) cut open to show spiral valve. (*From C. K. Weichert, Anatomy of the Chordates, 3d ed., McGraw-Hill, 1965*)

SPIRIFERIDINA

Dorsal view of *Spinocyrtia* valve.

commonly consisting of sand or gravel and which terminates in open water.

Spitsbergen Current [OCEANOGR] An ocean current flowing northward and westward from a point south of Spitsbergen, and gradually merging with the East Greenland Current in the Greenland Sea; the Spitsbergen Current is the continuation of the northwestern branch of the Norwegian Current.

spitted fuse [ENG] A slow-burning fuse which has been cut open at the lighting end for ease of ignition.

spitting rock [ENG] A rock mass under stress that breaks and ejects small fragments with considerable velocity.

SPL *See* sound pressure level.

splanchnic mesoderm [EMBRYO] The internal layer of the lateral mesoderm that is associated with the entoderm after the formation of the coelom.

splanchnic nerve [ANAT] A nerve carrying nerve fibers from the lower thoracic paravertebral ganglions to the collateral ganglions.

splanchnocranium [ANAT] Portions of the skull derived from the primitive skeleton of the gill apparatus.

splanchnopleure [EMBRYO] The inner layer of the mesoblast from which part of the wall of the alimentary canal and portions of the visceral organs are derived in coelomates.

splashdown [AERO ENG] **1.** The landing of a spacecraft or missile on water. **2.** The moment of impact of a spacecraft on water.

splash erosion [GEOL] Erosion resulting from the impact of falling raindrops.

splash lubrication [ENG] An engine-lubrication system in which the connecting-rod bearings dip into troughs of oil, splashing the oil onto the cylinder and piston rods.

splatter [COMMUN] Distortion due to overmodulation of a transmitter by peak signals of short duration, particularly sounds containing high-frequency harmonics; it is a form of adjacent-channel interference.

spleen [ANAT] A blood-forming lymphoid organ of the circulatory system, present in most vertebrates.

splenectomy [MED] Surgical removal of the spleen.

splenic fever *See* anthrax.

splenic flexure [ANAT] An abrupt turn of the colon beneath the lower end of the spleen, connecting the descending with the transverse colon.

splenium [ANAT] The rounded posterior extremity of the corpus callosum. [MED] A bandage.

splenomegaly [MED] Enlargement of the spleen.

splent coal *See* splint coal.

splice [ELEC] A joint used to connect two lengths of conductor with good mechanical strength and good conductivity. [ENG] To unite two parts, such as rope or wire, to form a continuous length. [GRAPHICS] To join two pieces of film together.

spliced [GEOL] Relating to veins that pinch out and are overlapped at that point by another parallel vein.

splice plate [CIV ENG] A plate for joining the web plates or the flanges of girders.

splicing tape [MATER] A pressure-sensitive nonmagnetic tape used for splicing magnetic tape and motion picture film; it has a hard adhesive that will not ooze and gum up the equipment or cause adjacent layers of tape or film on the reel to stick together.

spline [ENG] A strip of wood, metal, or plastic. [GRAPHICS] A flexible strip used in drawing curves.

spline broach [MECH ENG] A broach for cutting straight-sided splines, or multiple keyways in holes.

splined shaft [DES ENG] A shaft with longitudinal gearlike ridges along its interior or exterior surface.

splint *See* splint coal.

splint coal [GEOL] A hard, dull, blocky, grayish-black, banded bituminous coal characterized by an uneven fracture and a granular texture; burns with intense heat. Also known as splent coal; splint.

split [GEOL] A coal seam that cannot be mined as a single unit because it is separated by a parting of other sedimentary rock. Also known as coal split; split coal. [MIN ENG] **1.** To divide the air current into separate circuits to ventilate more than one section of the mine. **2.** Any division or branch of the ventilating current.

split-altitude profile [AERO ENG] Flight profile at two separate altitudes.

split-anode magnetron [ELECTR] A magnetron in which the cylindrical anode is divided longitudinally into halves, between which extremely high-frequency oscillations are produced.

split barrel [DES ENG] A core barrel that is split lengthwise so that it can be taken apart and the sample removed.

split-barrel sampler [DES ENG] A drive-type soil sampler with a split barrel.

split base concept [ORD] A principle which applies to a deployed tactical combat unit that divides its resources between two separate operating bases.

split bearing [DES ENG] A shaft bearing composed of two pieces bolted together.

split cameras [OPTICS] An assembly of two cameras disposed at a fixed overlapping angle relative to each other.

split cavity [ENG] A cavity, such as in a mold, made in sections.

split die [MET] **1.** A screw-thread die made in one piece with a longitudinal slit connecting the outside to the central hole which allows size adjustment. **2.** *See* segment die.

split fix [NAV] A fix by horizontal sextant angles obtained by measuring two angles between four objects, or suitable charted features, with no common center object being observed.

split flap [AERO ENG] A hinged plate forming the rear upper or lower portion of an airfoil; the lower portion may be deflected downward to give increased lift and drag; the upper portion may be raised over a portion of the wing for the purpose of lateral control.

split-lens interference [OPTICS] Interference produced by a Billet split lens.

split link [DES ENG] A metal link in the shape of a two-turn helix pressed together.

split personality [PSYCH] The type of human being in whom there is a separation of various components of the normal personality unit, and each component functions as an independent entity.

split-phase motor [ELEC] A single-phase induction motor having an auxiliary winding connected in parallel with the main winding, but displaced in magnetic position from the main winding so as to produce the required rotating magnetic field for starting; the auxiliary circuit is generally opened when the motor has attained a predetermined speed.

split pin [DES ENG] A pin with a split at one end so that it can spread to hold it in place.

split-ring core lifter [DES ENG] A hardened steel ring having an open slit, an outside taper, and an inside or outside serrated surface; in its expanded state it allows the core to pass through it freely, but when the drill string is lifted, the outside taper surface slides downward into the bevel of the bit or reaming shell, causing the ring to contract and grip tightly the core which it surrounds. Also known as core catcher; core gripper; core lifter; ring lifter; split-ring lifter; spring lifter.

split-ring lifter *See* split-ring core lifter.

split-ring mold [ENG] A plastics mold in which a split-cavity block is assembled in a chase to permit the forming of undercuts in a molded piece.

split-ring piston packing [MECH ENG] A metal ring mounted on a piston to prevent leakage along the cylinder wall.

split shovel [DES ENG] A shovel containing parallel troughs separated by slots; used for sampling ground ore.

split-stator variable capacitor [ELECTR] Variable capacitor having a rotor section that is common to two separate stator sections; used in grid and plate tank circuits of transmitters for balancing purposes.

splitter [CHEM ENG] A petroleum-refinery term for a fractionating tower that produces only an overhead and bottom stream.

splitter vanes [ENG] A group of curved, parallel vanes located in a sharp (for example, miter) bend of a gas conduit; the vane shape and its location help guide the moving gas around the bend.

splitting [ELECTR] In the scope presentation of the standard loran (2000 kilohertz), signals the slow diminution of the

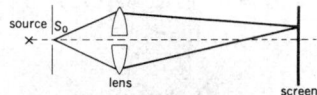

SPLIT-LENS INTERFERENCE

Billet split-lens interference. Light from source S_0 is transmitted through two separated parts of lens to screen, producing interference fringes on screen.

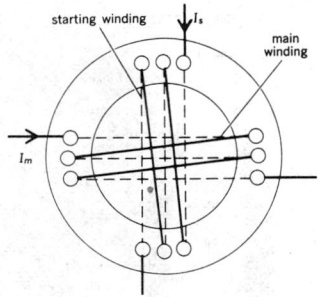

SPLIT-PHASE MOTOR

Windings of a split-phase motor; I_m is current of the main winding, and I_s is current of the starting winding.

SPODUMENE

2.5 cm

Spodumene crystal with pegmatite, Doshen, Massachusetts. *(Specimen from Department of Geology, Bryn Mawr College)*

SPONGILLIDAE

An encrusting spongillid sponge growing on a twig.

SPONGOCOEL

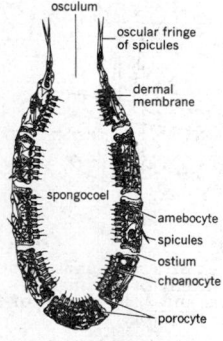

osculum

oscular fringe of spicules

dermal membrane

spongocoel

amebocyte

spicules

ostium

choanocyte

porocyte

Morphology of asconoid calcareous sponge—longitudinal section. *(After Hyman, 1940)*

SPONTANEOUS-POTENTIAL WELL LOGGING

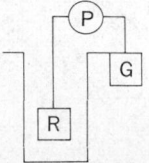

Basic pattern of spontaneous-potential electric well logging; P = potentiometer, G = grounded electrode, and R = recording electrode.

leading or lagging edge of the pulse so that it resembles two pulses and eventually a single pulse, which appears to be normal but which may be displaced in time by as much as 10,000 microseconds; this phenomenon is caused by shifting of the E_1 reflections from the ionosphere, and if the deformation is that of the leading edge and is not detected, it will cause serious errors in the reading of the navigational parameter. [MIN ENG] **1.** Lamina of mica with a maximum thickness of 0.0012 inch (30 micrometers), split from blocks and thins. **2.** One of a pair of horizontal level headings driven through a pillar, in pillar workings, in order to mine the pillar coal.

split vertical photography [GRAPHICS] Photographs taken simultaneously by two cameras mounted at an angle from the vertical, one tilted to the left and one to the right, to obtain a small side lap.

SP logging *See* spontaneous-potential well logging.

Spodosol [GEOL] A soil order characterized by a spodic or placic horizon overlying a fragipan.

spodumene [MINERAL] $LiAlSi_2O_6$ A white to yellowish-, purplish-, or emerald-green clinopyroxene mineral occurring in prismatic crystals; hardness is 6.5–7 on Mohs scale, and specific gravity 3.13–3.20; an ore of lithium. Also known as triphane.

spoil [MIN ENG] **1.** The overburden or nonore material from a coal mine. **2.** A stratum of coal and dirt mixed.

spoil bank [MIN ENG] **1.** In surface mining, the accumulation of overburden. **2.** The place where spoil is deposited. Also known as spoil heap.

spoil dam [MIN ENG] An earthen dike forming a depression, in which returns from a borehole can be collected and retained.

spoiler [AERO ENG] A plate, series of plates, comb, tube, bar, or other device that projects into the airstream about a body to break up or spoil the smoothness of the flow, especially such a device that projects from the upper surface of an airfoil, giving an increased drag and a decreased lift. [ELECTROMAG] Rod grating mounted on a parabolic reflector to change the pencil-beam pattern of the reflector to a cosecant-squared pattern; rotating the reflector and grating 90° with respect to the feed antenna changes one pattern to the other.

spoil heap *See* spoil bank.

spoke [DES ENG] A bar or rod radiating from the center of a wheel.

spokeshave [ENG] A small tool for planing convex or concave surfaces.

spondylitis [MED] Inflammation of the vertebrae.

spondylolisthesis [MED] Forward displacement of a vertebra upon the one below as a result of a bilateral defect in the vertebral arch, or erosion of the articular surface of the posterior facets due to degenerative joint disease.

sponge [INV ZOO] The common name for members of the phylum Porifera.

sponge gold *See* cake of gold.

sponge grease [MATER] Fibrous, spongy, soda-base grease.

sponge iron [MET] Iron in porous or powder form made without fusion by heating iron ore in a reducing gas or with charcoal.

sponge metal [MET] Any porous metal made by decomposition or reduction of a compound without melting.

sponge rubber *See* rubber sponge.

spongework [GEOL] A pattern of small irregular interconnecting cavities on walls of limestone caves.

Spongiidae [INV ZOO] A family of sponges of the order Dictyoceratida; members are encrusting, massive, or branching in form and have small spherical flagellated chambers which characteristically join the exhalant canals by way of narrow channels.

Spongillidae [INV ZOO] A family of fresh- and brackish water sponges in the order Haplosclerida which are chiefly gray, brown, or white in color, and encrusting, massive, or branching in form.

spongin [BIOCHEM] A scleroprotein, occurring as the principal component of skeletal fibers in many sponges.

spongioblast [EMBRYO] A primordial cell arising from the ectoderm of the embryonic neural tube which differentiates to form the neuroglia, the ependymal cells, the neurolemma sheath cells, the satellite cells of ganglions, and Müller's fibers of the retina.

spongiocyte [HISTOL] **1.** A neuroglia cell. **2.** A cell of the adrenal cortex which has a spongy appearance due to the solution of lipids during tissue preparation for microscopical examination.

Spongiomorphida [PALEON] A small, extinct Mesozoic order of fossil colonial Hydrozoa in which the skeleton is a reticulum composed of perforate lamellae parallel to the upper surface and of regularly spaced vertical elements in the form of pillars.

Spongiomorphidae [PALEON] The single family of extinct hydrozoans comprising the order Spongiomorphida.

spongocoel [INV ZOO] The branching, internal cavity of a sponge, connected to the outside by way of the osculum.

spongolite [GEOL] A rock or sediment composed chiefly of the remains of sponges. Also known as spongolith.

spongolith *See* spongolite.

spongy mesophyll [BOT] A system of loosely and irregularly arranged parenchymal cells with numerous intercellular spaces found near the lower surface in well-differentiated broad leaves. Also known as spongy parenchyma.

spongy parenchyma *See* spongy mesophyll.

sponson mount [ORD] A gun mount positioned on the sponson of a tank or combat vehicle. Practically abandoned on account of vulnerability and limited field of fire, although widely used in earlier tanks.

sponsor [COMMUN] The advertiser who pays part or all of the cost of a television or radio program.

spontaneous [PHYS] Occurring without application of an external agency, because of the inherent properties of an object.

spontaneous abortion [MED] An unexpected, premature expulsion of the fetus.

spontaneous amputation [MED] **1.** Congenital amputation. **2.** Amputation not caused by external trauma or injury, as in ainhum.

spontaneous fission [NUC PHYS] Nuclear fission in which no particles or photons enter the nucleus from the outside.

spontaneous generation *See* abiogenesis.

spontaneous heating [CHEM] The slow reaction of material with atmospheric oxygen at ambient temperatures; liberated heat, if undissipated, accumulates so that in the presence of combustible substances a fire will result.

spontaneous ignition [CHEM] Ignition which can occur when certain materials such as tung oil are stored in bulk, resulting from the generation of heat, which cannot be readily dissipated; often heat is generated by microbial action.

spontaneous magnetization [ELECTROMAG] Magnetization which a substance possesses in the absence of an applied magnetic field.

spontaneous mutation [GEN] A mutation that occurs naturally.

spontaneous nucleation [METEOROL] The nucleation of a phase change of a substance without the benefit of any seeding nuclei within or otherwise in contact with that substance; examples of such systems are a pure vapor condensing to its pure liquid state, a pure liquid freezing to its pure solid state, and a pure solution crystallizing to yield pure solute crystals.

spontaneous polarization [ELEC] Electric polarization that a substance possesses in the absence of an external electric field.

spontaneous-potential well logging [ENG] The recording of the natural electrochemical and electrokinetic potential between two electrodes, one above the other, lowered into a drill hole; used to detect permeable beds and their boundaries. Also known as SP logging.

spontaneous process [THERMO] A thermodynamic process which takes place without the application of an external agency, because of the inherent properties of a system.

spontaneous symmetry breaking [PHYS] A situation in which the solution of a set of physical equations fails to exhibit a symmetry possessed by the equations themselves; an example is a magnet, in which the underlying equations describing the metal do not distinguish any direction of space from any other, but the magnet certainly does, since it points in some definite direction.

spoofing [ELECTR] Deceiving or misleading the enemy in electronic operations, as by continuing transmission on a frequency after it has been effectively jammed by the enemy, using decoy radar transmitters to lead the enemy into a useless jamming effort, or transmitting radio messages containing false information for intentional interception by the enemy.

spool [MECH ENG] **1.** The drum of a hoist. **2.** The movable part of a slide-type hydraulic valve. [TEXT] A cylinder made of wood or other material on which yarn is held during spinning or weaving.

spool-type roller conveyor [MECH ENG] A type of roller conveyor in which the rolls are of conical or tapered shape with the diameter at the ends of the roll larger than that at the center.

spoon [DES ENG] A slender rod with a cup-shaped projection at right angles to the rod, used for scraping drillings out of a borehole. [MIN ENG] An instrument in which earth or pulp may be delicately tested by washing to detect gold or amalgam.

spoon nail *See* koilonychia.

sporadic E layer [GEOPHYS] A layer of intense ionization that occurs sporadically within the E layer; it is variable in time of occurrence, height, geographical distribution, penetration frequency, and ionization density.

sporadic meteor [ASTRON] A meteor which is not associated with one of the regularly recurring meteor showers or streams.

sporadic reflections [ELECTROMAG] Sharply defined reflections of substantial intensity from the sporadic E layer at frequencies greater than the critical frequency of the layer; they are variable with respect to time of occurrence, geographic location, and range of frequencies at which they are observed.

sporangiophore [BOT] A stalk or filament on which sporangia are borne.

sporangiospore [BOT] A spore that forms in a sporangium.

sporangium [BOT] A case in which asexual spores are formed and borne.

spore [BIOL] A uni- or multicellular, asexual, reproductive or resting body that is resistant to unfavorable environmental conditions and produces a new vegetative individual when the environment is favorable.

spore mother cell [BOT] One of the cells of the archespore of a sporebearing plant from which a spore, but usually a tetrad of spores, is produced.

Spörer's law [ASTRON] A relationship to indicate the frequency of occurrence of sunspots and their progressive movement to lower latitudes on the sun.

sporidium [MYCOL] A small spore, especially one formed on a promycelium.

sporinite [GEOL] A variety of exinite composed of spore exines which have been compressed parallel to the stratification.

sporoblast [INV ZOO] A sporozoan cell from which sporozoites arise.

Sporobolomycetaceae [MYCOL] The single family of the order Sporobolomycetales.

Sporobolomycetales [MYCOL] An order of yeastlike and moldlike fungi assigned to the class Basidiomycetes characterized by the formation of sterigmata, upon which the asexual ballistospores are formed.

sporocarp [BOT] Any multicellular structure in or on which spores are formed.

sporocyst [BOT] A unicellular resting body from which asexual spores arise. [INV ZOO] **1.** A resistant envelope containing an encysted sporozoan. **2.** An encysted sporozoan. **3.** The first reproductive form of a digenetic trematode in which rediae develop.

sporogenesis [BIOL] **1.** Reproduction by means of spores. **2.** Formation of spores.

sporogony [BIOL] Reproduction by means of spores. [INV ZOO] Propagative reproduction involving formation, by sexual processes, and subsequent division of a zygote.

sporont [INV ZOO] A stage in the life history of sporozoans which gives rise to spores.

sporophore [MYCOL] A structure on the thallus of fungi which produces spores.

sporophyll [BOT] A modified leaf that develops sporangia.

sporophyte [BOT] **1.** An individual of the spore-bearing generation in plants exhibiting alternation of generation. **2.** The spore-producing generation. **3.** The diplophase in a plant life cycle.

sporopollenin [BIOCHEM] A substance related to suberin and cutin but more resistant to decay that is found in the exine of pollen grains.

sporosac [INV ZOO] A degenerate gonophore in certain hydroid coelenterates.

sporotrichosis [MED] A granulomatous fungus disease caused by *Sporotrichum schenckii*, with cutaneous lesions along the lymph channels and occasionally involving the internal organs. Also known as de Beurmann-Gougerot disease; Schenk's disease.

Sporozoa [INV ZOO] A subphylum of parasitic Protozoa, typically producing spores during the asexual stages of the life cycle.

sporozoite [INV ZOO] A motile, infective stage of certain sporozoans, which is the result of sexual reproduction and which gives rise to an asexual cycle in the new host.

sporting gun [ORD] A classification of shotguns which includes those with 30-inch (762-millimeter) full choke barrels.

sporulation [BIOL] The act and process of spore formation.

spot [ELECTR] In a cathode-ray tube, the area instantaneously affected by the impact of an electron beam. [ORD] **1.** To determine, by observation, deviations of ordnance from the target for the purpose of supplying necessary information for adjustment of fire. **2.** To place ordnance in a proper location. **3.** To locate or espy something, as an aircraft or troop concentration.

spot blight *See* grease spot.

spot blotch [PL PATH] A fungus disease of barley caused by *Helminthosporium sativum* and characterized by the appearance of dark, elongated spots on the foliage.

spot check [IND ENG] A check or inspection of certain steps in an operation, process, or the like, of certain parts of a piece of equipment or of a representative lot of completed parts or articles; the steps or parts inspected would normally be only a small percentage of the total.

spot drilling [MECH ENG] Drilling a small hole or indentation in the surface of a material to serve as a centering guide in later machining operations.

spot elevation [MAP] Elevation of a point on a map or chart, usually indicated by a dot accompanied by a number indicating the vertical distance of the point from the reference datum; spot elevations are used principally to indicate points higher than their surroundings.

spot film [MED] A small, highly collimated radiograph of an anatomic part, usually obtained in conjunction with fluoroscopy.

spot gluing [ENG] Applying heat to a glued assembly by dielectric heating to make the glue set in spots that are more or less regularly distributed.

spot hover [AERO ENG] To remain stationary relative to a point on the ground while airborne.

spot jammer [ELECTR] A jammer that interferes with reception of a specific channel or frequency.

spotlight [ELEC] **1.** A strong beam of light that illuminates only a small area about an object. **2.** A lamp that has a strongly focused beam.

spot net [COMMUN] Radio communication net used by a spotter in calling fire.

spot noise figure [ELECTR] Of a transducer at a selected frequency, the ratio of the output noise power per unit bandwidth to a portion thereof attributable to the thermal noise in the input termination per unit bandwidth, the noise temperature of the input termination being standard (290K).

spot punch [ADP] A hand-operated device resembling a pair of pliers, for selectively punching holes in punch cards.

spot-size error [ELECTR] The distortion of the radar returns on the radarscope presentation caused by the diameter of the electron beam which displays the returns of the scope and the lateral radiation across the scope of part of the glow produced when the electron beam strikes the phosphorescent coating of the cathode-ray tube.

spot speed [COMMUN] **1.** In television, the product of the length (in units of elemental area, that is, in spots) of scanning line by the number of scanning lines per second. **2.** In facsimile transmission, the speed of the scanning or recording spot within the available line. Also known as scanning speed.

spotted wilt [PL PATH] A virus disease of various crop and wild plants, especially tomato, characterized by bronzing and downward curling of the leaves.

spotter [OPTICS] *See* dotter. [ORD] An observer stationed for the purpose of observing and reporting results of naval gunfire to the firing agency, or in designating targets.

spotter-tracer [ORD] In cartridge nomenclature, indicating that the bullet or projectile is equipped with a tracer and contains a filler suitable for spotting purposes.

spot test [ANALY CHEM] The addition of a drop of reagent to a drop or two of sample solution to obtain distinctive colors or precipitates; used in qualitative analysis.

spottiness [ELECTR] Bright spots scattered irregularly over the reproduced image in a television receiver, due to man-made or static interference entering the television system at some point.

spotting [ENG] Fitting one part of a die to another part by applying an oil color to the surface of the finished part and bringing this against the surface of the intended mating part, the high spots being marked by the transferred color. [MIN ENG] Bringing mine cars or surface wagons to the correct spot for loading, discharging, or any other purpose.

spotting board [ORD] Device for determining the direction and size of deviations from the target; it converts the readings of spotters into usable form for firing data.

spotting hoist [MIN ENG] A small haulage engine used for bringing mine cars into the correct position under a loading chute or feeder or some other point.

spotting line [ORD] Either the gun-target line, the observer-target line, or a reference line used by the observer or spotter in making spot corrections.

spotting pistol [ORD] A short automatic or semiautomatic firearm mounted coaxially on a larger-caliber gun, designed to conserve and to increase first-round hit probability of the ammunition used in the larger weapon; it employs a magazine and fires a spotter-tracer projectile which is ballistically matched with the trajectory of the projectile of the gun on which it is mounted.

spotting rifle [ORD] An auxiliary item mounted coaxially on a larger-caliber gun used to assist a gunner in determining range; it is usually a magazine-fed, semiautomatic gun with a rifled barrel that utilizes ammunition incorporating a tracer element providing a smoke puff on impact.

spotty ore [MIN ENG] Ore in which the valuable material is concentrated irregularly as small particles.

spot welding [MET] Resistance welding in which fusion is localized in small circular areas; sometimes also accomplished by various arc-welding processes.

spot wind [METEOROL] In air navigation, wind direction and speed, either observed or forecast if so specified, at a designated altitude over a fixed location.

spout hole [VERT ZOO] 1. A blowhole of a cetacean mammal. 2. A nostril of a walrus or seal.

spouting [ENG] A term used in the feeding or ejection of powdered or granulated solids by means of vertical or slanted discharge spouts.

sprag [ENG] A stake used as a brake for a vehicle by inserting it through the spokes of a wheel or digging it into the ground at an angle. [MIN ENG] A prop supporting the roof or ore in a mine.

spragger [MIN ENG] In coal mining, a laborer who rides trains of cars and controls their free movement down gently sloping inclines by throwing switches and by poking sprags between the wheel spokes to stop them.

sprag road [MIN ENG] A road so steep that sprags must be used on the wheels of ore cars during descent.

sprain [MED] A wrenching of a joint, producing a stretching or laceration of the ligaments.

sprain fracture [MED] An injury in which a tendon or ligament, together with a shell of bone, is torn from its attachment; occurs most commonly at the ankle.

spray [ENG] A mechanically produced dispersion of liquid into a gas stream; as drops are large, the spray is unstable and the liquid will fall free of the gas stream when velocity decreases.

spray chamber [MECH ENG] A compartment in an air conditioner where humidification is conducted.

spray dryer [MECH ENG] A machine for drying an atomized mist by direct contact with hot gases.

sprayed metal mold [ENG] A plastics mold made by spraying molten metal onto a master form until a shell of predetermined thickness is achieved; the shell is then removed and backed up with plaster, cement, or casting resin; used primarily in plastic sheet forming.

spray gun [MECH ENG] An apparatus shaped like a gun which delivers an atomized mist of liquid.

spray nozzle [MECH ENG] A device in which a liquid is subdivided to form a stream (mist) of small drops.

spray oil [MATER] A low-viscosity petroleum oil similar to lubricating oil; used to combat pests that attack trees and shrubbery.

spray painting [ENG] Applying a fine, even coat of paint by means of a spray nozzle.

spray point [ELEC] One of the sharp points arranged in a row and charged to a high direct-current potential, used to charge and discharge the conveyor belt in a Van de Graaff generator.

spray pond [ENG] An arrangement for cooling large quantities of water in open reservoirs or ponds; nozzles spray a portion of the water into the air for the evaporative cooling effect.

spray quenching [MET] Rapid cooling in a spray of water or oil.

spray region *See* fringe region.

spray tower [CHEM ENG] A vertical column, at the top of which is a liquid spray device; used to contact liquids with gas streams for absorption, humidification, or drying.

spray-up [ENG] A term for a number of techniques in which a spray gun is used as the processing tool; for example, in reinforced plastics manufacture, fibrous glass and resin can simultaneously be spray-deposited into a mold or onto a form.

spread [ENG] *See* sensitivity. [STAT] The range within which the values of a variable quantity occur.

spreader [ELEC] An insulating crossarm used to hold apart the wires of a transmission line or multiple-wire antenna. [MECH ENG] 1. A tool used in sharpening machine drill bits. 2. A machine which spreads dumped material with its blades. [MIN ENG] 1. A horizontal timber below the cap of a set, used to stiffen the legs, and to support the brattice when there are two air courses in the same gangway. 2. A piece of timber stretched across a shaft as a temporary support of the walls.

spreader beam [ENG] A rigid beam hanging from a crane hook and fitted with a number of ropes at different points along its length; employed for such purposes as lifting reinforced concrete piles or large sheets of glass.

spreader stoker [MECH ENG] A coal-burning system where mechanical feeders and distributing devices form a thin fuel bed on a traveling grate, intermittent-cleaning dump grate, or reciprocating continuous-cleaning grate.

spread footing [CIV ENG] A wide, shallow footing usually made of reinforced concrete.

spreading coefficient [THERMO] The work done in spreading one liquid over a unit area of another, equal to the surface tension of the stationary liquid, minus the surface tension of the spreading liquid, minus the interfacial tension between the liquids.

spreading concept *See* sea floor spreading.

spreading factor *See* hyaluronidase.

spreading fire [ORD] A notification by the spotter or the naval gunfire ship, depending on who is controlling the fire, to indicate that fire is about to be distributed over an area.

spreading floor hypothesis *See* sea floor spreading.

spread reflection [ELECTROMAG] Reflection of electromagnetism from a rough surface with large irregularities. Also known as mixed reflection.

spread spectrum transmission [ELECTR] Communications technique in which many different signal waveforms are transmitted in a wide band; power is spread thinly over the band so narrow-band radios can operate within the wideband without interference.

Sprengel pump [MECH ENG] An air pump that exhausts by trapping gases between drops of mercury in a tube.

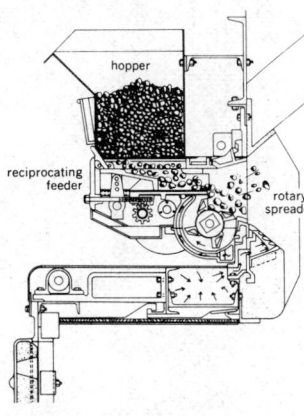

SPREADER STOKER

A type of spreader stoker utilizing a reciprocating coal feeder.

Sprigginidae [INV ZOO] An extinct family of annelid worms distinguished by a horseshoe-shaped head.

spring [ASTRON] The period extending from the vernal equinox to the summer solstice; comprises the transition period from winter to summer. [ENG] To enlarge the bottom of a drill hole by small charges of a high explosive in order to make room for the full charge; to chamber a drill hole. [HYD] A general name for any discharge of deep-seated, hot or cold, pure or mineralized water. [MECH ENG] An elastic, stressed, stored-energy machine element that, when released, will recover its basic form or position. Also known as mechanical spring.

springback [MET] **1.** Return of a metal part to its original shape after release of stress. **2.** The degree to which a metal returns to its original shape after forming operations. **3.** In flash, upset or pressure welding, the deflection in the welding machine caused by the upset pressure.

spring bearings [NAV ARCH] Bearings designed to take the weight of the propeller shaft inside a vessel.

spring bolt [DES ENG] A bolt which must be retracted by pressure and which is shot into place by a spring when the pressure is released.

spring brass [MET] Common brass containing 70-72% copper which has been cold-worked to make it stiffer.

spring clip [DES ENG] **1.** A U-shaped fastener used to attach a leaf spring to the axle of a vehicle. **2.** A clip that grips an inserted part under spring pressure; used for electrical connections.

spring clock [HOROL] A clock containing a spring that supplies the energy to operate the clock mechanism.

spring collet [DES ENG] A bushing that surrounds and holds the end of the work in a machine tool; the bushing is slotted and tapered, and when the collet is slipped over it, the slot tends to close and the bushing thereby grips the work.

spring contact [ELEC] A relay or switch contact mounted on a flat spring, usually of phosphor bronze.

spring cotter [DES ENG] A cotter made of an elastic metal that has been bent double to form a split pin.

spring coupling [MECH ENG] A flexible coupling with resilient parts.

spring crust [HYD] A type of snow crust, formed when loose firn is recemented by a decrease in temperature; it is most common in late winter and spring.

spring die [DES ENG] An adjustable die consisting of a hollow cylinder with internal cutting teeth, used for cutting screw threads.

spring equinox *See* vernal equinox.

spring faucet [ENG] A faucet that is kept closed by a spring; force must be exerted to open it, and it closes when the force is removed.

spring gravimeter [ENG] An instrument for making relative measurements of gravity; the elongation s of the spring may be considered proportional to gravity g, $s = (1/k)g$, and the basic formula for relative measurements is

$$g_2 - g_1 = k(s_2 - s_1).$$

spring hammer [MECH ENG] A machine-driven hammer actuated by a compressed spring or by compressed air.

spring high water *See* mean high-water springs.

spring hook [DES ENG] A hook closed at the end by a spring snap. Also known as snap hook.

spring-joint caliper [DES ENG] An outside or inside caliper having a heavy spring joining the legs together at the top; legs are opened and closed by a knurled nut.

spring lifter *See* split-ring core lifter.

spring-lifter case *See* lifter case.

spring line [NAV ARCH] A hawser run out from any part of a vessel to a point on shore, as a dock, to prevent the vessel from going ahead or astern.

spring-loaded regulator [MECH ENG] A pressure-regulator valve for pressure vessels or flow systems; the regulator is preloaded by a calibrated spring to open (or close) at the upper (or lower) limit of a preset pressure range.

spring low water *See* mean low-water springs.

spring modulus [MECH] The additional force necessary to deflect a spring an additional unit distance; if a certain spring has a modulus of 40 pounds per inch, a 40-pound weight will compress it 1 inch, an 80-pound weight 2 inches, and so on.

spring pin [MECH ENG] An iron rod which is mounted between spring and axle on a locomotive, and which maintains a regulated pressure on the axle.

spring range [OCEANOGR] The mean semidiurnal range of tide when spring tides are occurring; the mean difference in height between spring high water and spring low water. Also known as mean spring range.

spring rise [OCEANOGR] The height of mean high-water springs above the chart datum.

spring scale [ENG] A scale that utilizes the deflection of a spring to measure the load.

spring seepage [HYD] A spring of small discharge. Also known as weeping spring.

spring shackle [ENG] A shackle for supporting the end of a spring, permitting the spring to vary in length as it deflects.

spring sludge *See* rotten ice.

spring snow [HYD] A coarse, granular snow formed during spring by alternate freezing and thawing. Also known as corn snow.

spring steel [MET] Carbon or low-alloy steel which can be processed to give it the hardness and yield strength needed in springs.

spring stop-nut locking fastener [DES ENG] A locking fastener that functions by a spring action clamping down on the bolt.

spring switch [CIV ENG] A railroad switch that contains a spring to return it to the running position after it has been thrown over by trailing wheels moving on the diverging route.

spring temper [MET] **1.** A steel temper characterized by an increased upper limit of elasticity; obtained by hardening and tempering in the usual way, then reheating until the steel turns blue. **2.** A similar temper in brass obtained by cold rolling.

spring tidal currents [OCEANOGR] Tidal currents of increased speed occurring at the time of spring tides.

spring tide [OCEANOGR] Tide of increased range which occurs about every 2 weeks when the moon is new or full.

spring velocity [OCEANOGR] The average speed of the maximum flood and maximum ebb of a tidal current at the time of spring tides.

springwood [BOT] The portion of an annual ring that is formed principally during the growing season; it is softer, more porous, and lighter than summerwood because of its higher proportion of large, thin-walled cells.

sprinkle [METEOROL] A very light shower of rain.

sprinkler system [ENG] A fire-protection system of pipes and outlets in a building, mine, or other enclosure for delivering a fire extinguishing liquid or gas, usually automatically by the action of heat on the sprinkler head. Also known as fire sprinkling system.

Sprint [ORD] A United States guided, surface-to-air, high-acceleration, antimissile missile with nuclear warhead capability employed in the Nike X system.

s-process [NUC PHYS] The synthesis of elements, predominantly in the iron group, over long periods of time through the capture of slow neutrons which are produced mainly by the reactions of α-particles with carbon-13 and neon-21.

sprocket [DES ENG] A tooth on the periphery of a wheel or cylinder to engage in the links of a chain, the perforations of a motion picture film, or other similar device.

sprocket chain [MECH ENG] A continuous chain which meshes with the teeth of a sprocket and thus can transmit mechanical power from one sprocket to another.

sprocket pulse [ADP] **1.** A pulse generated by a magnetized spot which accompanies every character recorded on magnetic tape; this pulse is used during read operations to regulate the timing of the read circuits, and also to provide a count on the number of characters read from the tape. **2.** A pulse generated by the sprocket or driving hole in paper tape which serves as the timing pulse for reading or punching the paper tape.

sprocket wheel [DES ENG] A wheel with teeth or cogs, used for a chain drive or to engage the blocks on a cable.

spruce [BOT] An evergreen tree belonging to the genus *Picea* characterized by single, four-sided needles borne on small peglike projections, pendulous cones, and resinous wood.

SPRIGGINIDAE

Spriggina floundersi fossil showing characteristic horseshoe-shaped head.

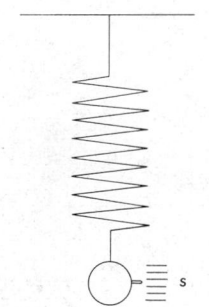

SPRING GRAVIMETER

Principle of a spring gravimeter. The variable length of the spring is *s*.

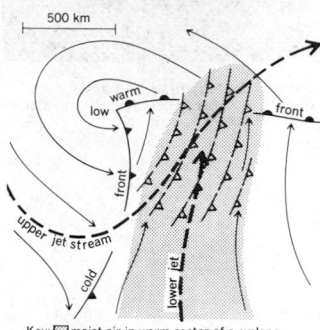

SQUALL LINE

500 km

Key: [shaded] moist air in warm sector of a cyclone

Map showing successive locations of squall line moving eastward through unstable northern portion of tongue of moist air in warm sector of a cyclone. Solid arrows show general flow in low levels; broken arrows, axes of strongest wind at about 1 kilometer (lower jet) aboveground and at 10–12 kilometers (upper jet stream).

SQUAMOUS EPITHELIUM

Cellular arrangement in squamous epithelial tissue.

spruce oil [MATER] A combustible, colorless to yellow essential oil, soluble in fixed and mineral oils; derived from spruce needles and branches; the main components are bornyl acetate, cadinene, and pinene; used in flavors and veterinary liniments, and as an odorant for soaps and cosmetics. Also known as hemlock oil.

spruce sulfite extract [MATER] A paper-manufacture byproduct from the sulfite-pulping process; used as a foundry core binder and road binder, and in tanning.

sprue [ENG] **1.** A feed opening or vertical channel through which molten material, such as metal or plastic, is poured in an injection or transfer mold. **2.** A slug of material that solidifies in the channel. [MED] A syndrome characterized by impaired absorption of food, water, and minerals by the small intestine; symptoms are the result of nutritional deficiencies.

sprung axle [MECH ENG] A supporting member for carrying the rear wheels of an automobile.

sprung weight [MECH ENG] The weight of a vehicle which is carried by the springs, including the frame, radiator, engine, clutch, transmission, body, load, and so forth.

SPST *See* single-pole single-throw.

spud [DES ENG] **1.** A diamond-point drill bit. **2.** An offset type of fishing tool used to clear a space around tools stuck in a borehole. **3.** Any of various spade- or chisel-shaped tools or mechanical devices. **4.** *See* grouser. [MIN ENG] A nail, resembling a horseshoe nail, with a hole in the head, driven into mine timbering or into a wooden plug inserted in the rock to mark a surveying station. [NAV ARCH] A foot piece to provide support for the legs of the A-frame of a floating dipper dredge.

spudded-in [MIN ENG] A borehole that has been started and has reached bedrock or in which the standpipe has been set.

Spumellina [INV ZOO] The equivalent name for Peripylina.

spun glass [MATER] Blown glass made of fine threads of glass.

spun rayon [TEXT] A yarn or fabric spun from rayon staple.

spun yarn [TEXT] A textile yarn spun and twisted from staple-length fiber, either natural or synthetic.

spur [BOT] **1.** A hollow process at the base of a petal or sepal. **2.** A short fruit-bearing tree branch. **3.** A short projecting root. [MATH] *See* trace. [ZOO] A stiff, sharp outgrowth, as on the legs of certain birds and insects.

spur blight [PL PATH] A fungus disease of raspberries and blackberries caused by *Didymella applanata* which kills the fruit spurs and causes dark spotting of the cane.

spur dike *See* groin.

spur gear [DES ENG] A toothed wheel with radial teeth parallel to the axis.

spurious disk [OPTICS] The nearly round image of perceptible diameter of a star as seen through a telescope, due to diffraction of light in the telescope.

spurious emission *See* spurious radiation.

spurious modulation [ELECTR] Undesired modulation occurring in an oscillator, such as frequency modulation caused by mechanical vibration.

spurious radiation [ELECTROMAG] Any emission from a radio transmitter at frequencies outside its frequency band. Also known as spurious emission.

spurious response [ELECTR] **1.** Response of a radio receiver to a frequency different from that to which the receiver is tuned. **2.** In electronic warfare, the undesirable signal images in the intercept receiver resulting from the mixing of the intercepted signal with harmonics of the local oscillators in the receiver.

spurrite [MINERAL] $Ca_5(SiO_4)_2(CO_3)$ A light-gray mineral occurring in granular masses.

spurt tone [COMMUN] Short audio-frequency tone used for signaling or dialing selection.

Sputnik program [AERO ENG] A series of Soviet earth-orbiting satellites; Sputnik I, launched on October 4, 1957, was the first artificial satellite.

sputtering [ELECTR] Also known as cathode sputtering. **1.** The ejection of atoms or groups of atoms from the surface of the cathode of a vacuum tube as the result of heavy-ion impact. **2.** The use of this process to deposit a thin layer of metal on a glass, plastic, metal, or other surface in vacuum.

sputter-ion pump *See* getter-ion pump.

sputum [PHYSIO] Material discharged from the surface of the respiratory passages, mouth, or throat; may contain saliva, mucus, pus, microorganisms, blood, or inhaled particulate matter in any combination.

sp vol *See* specific volume.

sq *See* square.

squadron [ORD] **1.** An organization consisting of two or more divisions of ships, or two or more divisions (U.S. Navy) or flights of aircraft; it is normally, but not necessarily, composed of ships or aircraft of the same type. **2.** The basic administrative aviation unit of the U.S. Army, Navy, Marine Corps, and Air Force.

squalene [BIOCHEM] $C_{30}H_{50}$ A liquid triterpene which is found in large quantities in shark liver oil, and which appears to play a role in the biosynthesis of sterols and polycyclic terpenes; used as a bactericide and as an intermediate in the synthesis of pharmaceuticals. Also known as spinacene.

Squalidae [VERT ZOO] The spiny dogfishes, a family of squaloid elasmobranchs recognized by their well-developed fin spines.

squall [METEOROL] A strong wind with sudden onset and more gradual decline, lasting for several minutes; in the United States observational practice, a squall is reported only if a wind speed of 16 knots or higher (8.23 meters per second) is sustained for at least 2 minutes.

squall cloud [METEOROL] A small eddy cloud sometimes formed below the leading edge of a thunderstorm cloud, between the upward and downward currents.

squall line [METEOROL] A line of thunderstorms near whose advancing edge squalls occur along an extensive front; the region of thunderstorms is typically 20 to 50 kilometers wide and a few hundred to 2000 kilometers long.

Squamata [VERT ZOO] An order of reptiles, composed of the lizards and snakes, distinguished by a highly modified skull that has only a single temporal opening, or none, by the lack of shells or secondary palates, and by possession of paired penes on the males.

squamodisk [INV ZOO] In monogenetic trematodes, a disk bearing concentric circles of spines, scales, or ridges, and located on the opisthaptor.

squamosal bone [ANAT] The part of the temporal bone in man corresponding with the squamosal bone in lower vertebrates. [VERT ZOO] A membrane bone lying external and dorsal to the auditory capsule of many vertebrate skulls.

squamous [BIOL] Covered with or composed of scales.

squamous-cell carcinoma [MED] A carcinoma composed principally of anaplastic, squamous epithelial cells. Also known as epidermoid carcinoma.

squamous epithelium [HISTOL] A single-layered epithelium consisting of thin, flat cells.

squamulose [BIOL] Covered with or composed of minute scales.

square [MATH] **1.** The square of a number r is the number r^2, that is, r times r. **2.** The plane figure with four equal sides and four interior right angles. [MECH] Denotes a unit of area; if x is a unit of length, a square x is the area of a square whose sides have a length of $1x$: for example, a square meter, or a meter squared, is the area of a square whose sides have a length of 1 meter. Abbreviated sq.

square degree [MATH] A unit of a solid angle equal to $(\pi/180)^2$ steradian, or approximately 3.04617×10^{-4} steradian.

square dome [ARCH] A spherical dome with a square base.

square-edged orifice [ENG] An orifice plate with straight-through edges for the hole through which fluid flows; used to measure fluid flow in fluid conduits by means of differential pressure drop across the orifice.

square engine [MECH ENG] An engine in which the stroke is equal to the cylinder bore.

square-foot unit of absorption *See* sabin.

square grade [MATH] A unit of solid angle equal to $(\pi/200)^2$ steradian, or approximately 2.46740×10^{-4} steradian.

square groove weld [MET] A groove weld in which the joint edges are square.

square-head bolt [DES ENG] A cylindrical threaded fastener with a square head.

square-jaw clutch [MECH ENG] A type of positive clutch consisting of two or more jaws of square section which mesh together when they are aligned.

square key [DES ENG] A machine key of square, usually uniform, but sometimes tapered, cross section.

square-law demodulator *See* square-law detector.

square-law detector [ELECTR] A demodulator whose output voltage is proportional to the square of the amplitude-modulated input voltage. Also known as square-law demodulator.

square-loop ferrite [ELECTROMAG] A ferrite that has an approximately rectangular hysteresis loop.

square matrix [MATH] A matrix with the same number of rows and columns.

square mesh [DES ENG] A wire-cloth textile mesh count that is the same in both directions.

squareness ratio [ELECTROMAG] **1.** The magnetic induction at zero magnetizing force divided by the maximum magnetic induction, in a symmetric cyclic magnetization of a material. **2.** The magnetic induction when the magnetizing force has changed half-way from zero toward its negative limiting value divided by the maximum magnetic induction in a symmetric cyclic magnetization of a material.

square-nose bit *See* flat-face bit.

square root [MATH] A square root of a real or complex number s is a number t for which $t^2 = s$.

square-root law [STAT] The standard deviation of the ratio of the number of successes to number of trials is inversely proportional to the square root of the number of trials.

square search [NAV] Search by a series of course lines such as to produce an expanding square pattern relative to either a fixed point or a moving point. Also known as expanding square search.

square-serif type [GRAPHICS] A contemporary type face; the serifs do not have any curvature and appear boxlike.

square set [MIN ENG] A set of timbers composed of a cap, girt, and post which meet so as to form a solid 90° angle; they are so framed at the intersection as to form a compression joint, and join with three other similar sets.

square-set block caving [MIN ENG] Block caving in which the ore is extracted through drifts supported by square sets.

square-set stopes [MIN ENG] The use of square-set timbering to support the ground as ore is extracted.

square wave [ELEC] An oscillation the amplitude of which shows periodic discontinuities between two values, remaining constant between jumps.

square-wave amplifier [ELECTR] Resistance-coupled amplifier, the circuit constants of which are to amplify a square wave with the minimum amount of distortion.

square-wave generator [ELECTR] A signal generator that generates a square-wave output voltage.

square wheel [DES ENG] A wheel with a flat spot on its rim.

squaring circuit [ELECTR] **1.** A circuit that reshapes a sine or other wave into a square wave. **2.** A circuit that contains nonlinear elements proportional to the square of the input voltage.

squaring shear [MECH ENG] A machine tool consisting of one fixed cutting blade and another mounted on a reciprocating crosshead; used for cutting sheet metal or plate.

squarrose [BOT] Having stiff divergent bracts, or other processes.

squarrulose [BOT] Mildly squarrose.

squash [BOT] Either of two plants of the genus *Cucurbita*, order Violales, cultivated for its fruit; some types are known as pumpkins.

Squatinidae [VERT ZOO] A group of squaloid elasmobranchs of uncertain affinity characterized by a greatly extended rostrum with enlarged denticles along the margins; maximum length is under 4 feet (1.2 meters).

squatting [NAV ARCH] The lowering of the stern of a ship when it is traveling at high speed.

squealing [ELECTR] A condition in which a radio receiver produces a high-pitched note or squeal along with the desired radio program, due to interference between stations or to oscillation in some receiver circuit.

squeegee [DES ENG] A device consisting of a handle with a blade of rubber or leather set transversely at one end and used for spreading, pushing, or wiping liquids off or across a surface.

squeezable waveguide [ELECTROMAG] A waveguide whose dimensions can be altered periodically; used in rapid scanning.

squeeze [ENG] **1.** To inject a grout into a borehole under high pressure. **2.** The plastic movement of a soft rock in the walls of a borehole or mine working that reduced the diameter of the opening. [MIN ENG] **1.** The settling, without breaking, of the roof over a considerable area of working. Also known as creep; nip; pinch. **2.** The gradual upheaval of the floor of a mine due to the weight of the overlying strata. **3.** The sections in coal seams that have become constricted by the squeezing in of the overlying or underlying rock. [PHYS] Increasing external pressure upon the ears and sinuses in diving.

squeeze roll [MECH ENG] A roller designed to exert pressure on material passing between it and a similar roller.

squeeze section [ELECTROMAG] Length of waveguide constructed so that alteration of the critical dimension is possible with a corresponding alteration in the electrical length.

squeeze time [MET] In resistance welding, the time between the initial application of the current and of the pressure.

squegger *See* blocking oscillator.

squegging [ELECTR] Condition of self-blocking in an electron-tube-oscillator circuit.

squegging oscillator *See* blocking oscillator.

squelch [ELECTR] To automatically quiet a receiver by reducing its gain in response to a specified characteristic of the input.

squelch circuit *See* noise suppressor.

squib [ENG] A small tube filled with fine-grained black powder; upon the lighting and burning of the ignition match, the squib assumes a rocket effect and darts back into the hole to ignite the powder charge. [ORD] A small explosive device, similar in appearance to a detonator, but loaded with low explosive, so that its output is primarily heat (flash); usually electrically initiated, and provided to initiate action of pyrotechnic devices and rocket propellants.

squid [INV ZOO] Any of a number of marine cephalopod mollusks characterized by a reduced internal shell, ten tentacles, an ink sac, and chromatophores.

Squillidae [INV ZOO] The single family of the eumalacostracan order Stomatopoda, the mantis shrimp.

squint [ELECTROMAG] **1.** The angle between the two major lobe axes in a radar lobe-switching antenna. **2.** The angular difference between the axis of radar antenna radiation and a selected geometric axis, such as the axis of the reflector. **3.** The angle between the full-right and full-left positions of the beam of a conical-scan radar antenna. [MED] *See* strabismus.

squirrel [VERT ZOO] Any of over 200 species of arboreal rodents of the families Sciuridae and Anomaluridae having a bushy tail and long, strong hind limbs.

squirrel-cage motor [ELEC] An induction motor in which the secondary circuit consists of a squirrel-cage winding arranged in slots in the iron core.

squirrel-cage rotor *See* squirrel-cage winding.

squirrel-cage winding [ELEC] A permanently short-circuited winding, usually uninsulated, around the periphery of the rotor and joined by continuous end rings. Also known as squirrel-cage rotor.

squirt can [ENG] An oil can with a flexible bottom and a tapered spout; pressure applied to the bottom forces oil out the spout.

squirt gun [ENG] A device with a bulb and nozzle; when the bulb is pressed, liquid squirts from the nozzle.

squitter [ELECTR] Random firing, intentional or otherwise, of the transponder transmitter in the absence of interrogation.

sr *See* steradian.

Sr *See* strontium.

SRA-size [ENG] One of a series of sizes to which untrimmed paper is manufactured; for reels of paper the standard sizes are 450, 640, 900, and 1280 millimeters; for sheets of paper the sizes are SRA0, 900 × 1280 mm; SRA1, 640 × 900 mm; and

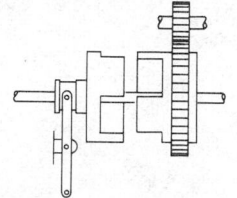

SQUARE-JAW CLUTCH

Square-jaw positive clutch.

SQUARE-SERIF TYPE

ABC

Example of square-serif type.

SQUASH

(a)

(b)

Two varieties of squash (*Cucurbita pepo*). (a) Summer Crookneck. (b) Summer Bergen.

SQUID

Dorsal view of a squid (*Loligo*), a cephalopod mollusk having worldwide distribution.

SRA2, 450 × 640 mm; SRA sizes correspond to A sizes when trimmed.

SR cylinder oil [MATER] A viscous, unfiltered lubrication oil, generally made from reduced petroleum crudes that have had the lighter lubricant fractions removed by direct steam heating; used to lubricate steam engine cylinders and valves. Also known as steam-refined cylinder oil.

SS *See* stainless steel.

SSB *See* single-sideband.

SS Cygni stars *See* U Geminorum stars.

SSD *See* steady-state distribution.

ss loran *See* sky-wave-synchronized loran.

SSM *See* surface-to-surface missile.

SSP *See* static spontaneous potential.

SST *See* supersonic transport.

S star [ASTRON] A spectral classification of stars, comprising red stars with surface temperature of about 2200K; prominent in the spectra is zirconium oxide.

s-state [QUANT MECH] A single-particle state whose orbital angular momentum quantum number is zero.

SSU *See* Saybolt Seconds Universal.

St *See* stoke.

stab culture [MICROBIO] A culture of anaerobic bacteria made by piercing a solid agar medium in a test tube with an inoculating needle covered with the bacterial inoculum.

stabilator [AERO ENG] A one-piece horizontal tail that is swept back and movable; movement is controlled by motion of the pilot's control stick; usually used in supersonic aircraft.

stability [CHEM] The property of a chemical compound which is not readily decomposed and does not react with other compounds. [CONT SYS] The property of a system for which any bounded input signal results in a bounded output signal. [ENG] The property of a body, as an aircraft, rocket, or ship, to maintain its attitude or to resist displacement, and, if displaced, to develop forces and moments tending to restore the original condition. [FL MECH] The resistance to overturning or mixing in the water column, resulting from the presence of a positive (increasing downward) density gradient. [GEOL] **1.** The resistance of a structure, spoil heap, or clay bank to sliding, overturning, or collapsing. **2.** Chemical durability, resistance to weathering. [MATER] Of a fuel, the capability to retain its characteristics in an adverse environment, for example, extreme temperature. [PHYS] **1.** The property of a system which does not undergo any change without the application of an external agency. **2.** The property of a system in which any departure from an equilibrium state gives rise to forces or influences which tend to return the system to equilibrium. Also known as static stability. **3.** *See* dynamic stability. [PL PHYS] The property of a plasma which maintains its shape against externally applied forces (usually pressure of magnetic fields) and whose constituents can pass through confining fields only by diffusion of individual particles.

stability augmentation system [AERO ENG] Automatic control devices which supplement a pilot's manipulation of the aircraft controls and are used to modify inherent aircraft handling qualities. Abbreviated SAS. Also known as stability augmentors.

stability augmentors *See* stability augmentation system.

stability chart [METEOROL] A synoptic chart that shows the distribution of a stability index.

stability constant [CHEM] Refers to the equilibrium reaction of a metal cation and a ligand to form a chelating mononuclear complex; the absolute-stability constant is expressed by the product of the concentration of products divided by the product of the concentrations of the reactants; the apparent-stability constant (also known as the conditional- or effective-stability constant) allows for the nonideality of the system because of the combination of the ligand with other complexing agents present in the solution.

stability criterion [CONT SYS] A condition which is necessary and sufficient for a system to be stable, such as the Nyquist criterion, or the condition that poles of the system's overall transmittance lie in the left half of the complex-frequency plane.

stability exchange principle [CONT SYS] In a linear system, which is either dynamically stable or unstable depending on the value of a parameter, the complex frequency varies with the parameter in such a way that its real and imaginary parts pass through zero simultaneously; the principle is often violated.

stability factor [ELECTR] A measure of a transistor amplifier's bias stability, equal to the rate of change of collector current with respect to reverse saturation current.

stability index [METEOROL] An indication of the local static stability of a layer of air.

stability of a system [MATH] Stability theory of systems of differential equations deals with those solution functions possessing some particular property which still maintain the property after a perturbation.

stability test [ENG] Accelerated test to determine the probable suitability of an explosive material for long-term storage.

stabilivolt [ELECTR] Gas tube that maintains a constant voltage drop across its terminals, essentially independent of current, over a relatively wide range.

stabilization [CHEM ENG] A petroleum-refinery process for separating light gases from petroleum or gasoline, thus leaving a stable (less volatile) liquid so that it can be handled or stored with less change in composition. [CONT SYS] *See* compensation. [ELECTR] Feedback introduced into vacuum tube or transistor amplifier stages to reduce distortion by making the amplification substantially independent of electrode voltages and tube constants. [ELECTROMAG] Treatment of a magnetic material to improve the stability of its magnetic properties. [ENG] Maintenance of a desired orientation independent of the roll and pitch of a ship or aircraft.

stabilized feedback *See* negative feedback.

stabilized flight [AERO ENG] Maintenance of desired orientation in flight.

stabilized gasoline [MATER] Gasoline from which the "wild" or low-boiling (high vapor pressure, volatile) hydrocarbons have been removed by stabilization.

stabilized platform *See* stable platform.

stabilized winding [ELEC] Auxiliary winding used particularly in star-connected transformers to stabilize the neutral point of the fundamental frequency voltages, to protect the transformer and the system from excessive third-harmonic voltages; and to prevent telephone interference caused by third-harmonic currents and voltages in the lines and earth. Also known as tertiary winding.

stabilizer [AERO ENG] Any airfoil or any combination of airfoils considered as a single unit, the primary function of which is to give stability to an aircraft or missile. [CHEM ENG] The fractionation column in a petroleum refinery used to stabilize (remove fractions from) hydrocarbon mixtures. [ENG] A hardened, splined bushing, sometimes freely rotating, slightly larger than the outer diameter of a core barrel and mounted directly above the core barrel back head. Also known as ferrule; fluted coupling. [MATER] **1.** Any powdered or liquid additive used as an agent in soil stabilization. **2.** An ingredient used in the formulation of some compounded plastics to maintain the physical and chemical properties at their initial values throughout the processing and service life of the material, for example, heat and RV stabilizers. [NAV ARCH] Any of the submerged fins used on ships to prevent rolling.

stabilizing magnetic field [PL PHYS] A magnetic field which is added to a device confining a plasma in order to increase the plasma's stability.

stabilizing treatment [MET] Any of various treatments intended to promote dimensional stability of a metal part or stabilize the structure of an alloy.

stabistor [ELECTR] A diode component having closely controlled conductance, controlled storage charge, and low leakage, as required for clippers, clamping circuits, bias regulators, and other logic circuits that require tight voltage-level tolerances.

stable [PHYS] Not subject to any change without the application of an external agency, such as radiation; said of a molecule, atom, nucleus, or elementary particle.

stable-base film [MATER] A particular type of film having high stability in regard to shrinkage and stretching.

stable element [ENG] Any instrument or device, such as a gyroscope, used to stabilize a radar antenna, turret, or other piece of equipment mounted on an aircraft or ship.

stable equilibrium [SCI TECH] Equilibrium in which any

departure from the equilibrium state gives rise to forces or influences which tend to return the system to equilibrium.

stable factor *See* factor VII.

stable homeomorphism conjecture [MATH] For dimension *n*, the assertion that each orientation-preserving homeomorphism of the real *n* space, R^n, into itself can be expressed as a composition of homeomorphisms, each of which is the identity on some nonempty open set in R^n.

stable isobar [NUC PHYS] One of two or more stable nuclides which have the same mass number but differ in atomic number.

stable isotope [NUC PHYS] An isotope which does not spontaneously undergo radioactive decay.

stable local oscillator *See* stalo.

stable nucleus [NUC PHYS] A nucleus which does not spontaneously undergo radioactive decay.

stable orbit *See* equilibrium orbit.

stable platform [AERO ENG] A gimbal-mounted platform, usually containing gyros and accelerometers, the purpose of which is to maintain a desired orientation in inertial space independent of craft motion. Also known as stabilized platform.

stable strobe [ELECTR] Series of strobes which behaves as if caused by a single jammer.

stable vertical [ENG] Vertical alignment of any device or instrument maintained during motion of the mount.

stack [ADP] Portion of a computer memory used to temporarily hold information. [BUILD] The portion of a chimney rising above the roof. [CHEM ENG] In gas works, a row of benches containing retorts. [ELECTR] *See* pileup. [ENG] **1.** To stand and rack drill rods in a drill tripod or derrick. **2.** Any structure or part thereof that contains a flue or flues for the discharge of gases. **3.** One or more filter cartridges mounted on a single column. **4.** Tall, vertical conduit (such as smokestack, flue) for venting of combustion or evaporation products or gaseous process wastes. **5.** The exhaust pipe of an internal combustion engine. [GEOL] An erosional, coastal landform that is a steep-sided, pillarlike rocky island or mass that has been detached by wave action from a shore made up of cliffs; applies particularly to a stack that is columnar in structure and has horizontal stratifications. Also known as marine stack; rank. [MET] The cone-shaped section of a blast furnace or cupola above the hearth and melting zone and extending to the throat. [NAV] To assign different altitudes by radio to aircraft awaiting their turns to land at an airport.

stack cutting [MET] Cutting a stack of metal plates with a single cut using a stream of oxygen.

stacked antennas [ELECTROMAG] Two or more identical antennas arranged above each other on a vertical supporting structure and connected in phase to increase the gain.

stacked array [ELECTROMAG] An array in which the antenna elements are stacked one above the other and connected in phase to increase the gain.

stacked-beam radar [ENG] Three-dimensional radar system that derives elevation by emitting narrow beams stacked vertically to cover a vertical segment, azimuth information from horizontal scanning of the beam, and range information from echo-return time.

stacked-dipole antenna [ELECTROMAG] Antenna in which directivity is increased by providing a number of identical dipole elements, excited either directly or parasitically; the resultant radiation pattern depends on the number of dipole elements used, the spacing and phase difference between the elements, and the relative magnitudes of the currents.

stacked-job processing [ADP] A technique of automatic job-to-job transition, with little or no operator intervention.

stacked loops [ELECTROMAG] Two or more loop antennas arranged above each other on a vertical supporting structure and connected in phase to increase the gain. Also known as vertically stacked loops.

stack effect [MECH ENG] The pressure difference between the confined hot gas in a chimney or stack and the cool outside air surrounding the outlet.

stacker [ADP] That part (or parts) of a punched-card handling device which arranges the processed cards into an orderly stack and holds them until they are removed by the operator. [MECH ENG] A machine for lifting merchandise on

a platform or fork and arranging it in tiers; operated by hand, or electric or hydraulic mechanisms.

stacker-reclaimer [MECH ENG] Equipment which transports and builds up material stockpiles, and recovers and transports material to processing plants.

stack gas [ENG] Gas passed through a chimney.

stacking [AERO ENG] The holding pattern of aircraft awaiting their turn to approach and land at an airport. [ELECTROMAG] The placing of antennas one above the other, connecting them in phase to increase the gain.

stacking fault [CRYSTAL] A defect in a face-centered cubic or hexagonal close-packed crystal in which there is a change from the regular sequence of positions of atomic planes. Also known as fault.

stack operation [ADP] A computer system in which flags, return address, and all temporary addresses are saved in the core in sequential order for any interrupted routine so that a new routine (including the interrupted routine) may be called in.

stack pointer [ADP] A register which contains the last address of a stack of addresses.

stack pollutants [ENG] Smokestack emissions subject to Environmental Protection Agency standards regulations, including sulfur oxides, particulates, nitrogen oxides, hydrocarbons, carbon monoxide, and photochemical oxidants.

stack welding [MET] Simultaneous spot welding of stacked plates.

STADAN *See* Space Tracking and Data Acquisition Network.

stade [GEOL] A substage of a glacial stage marked by a secondary advance of glaciers.

Stader's splint [MED] A metal bar with pins affixed at right angles; the pins are driven into the fragments of a fracture, and the bar maintains the alignment.

stadia [ENG] A surveying instrument consisting of a telescope with special horizontal parallel lines or wires, used in connection with a vertical graduated rod.

stadia hairs [ENG] Two horizontal lines in the reticule of a theodolite arranged symmetrically above and below the line of sight. Also known as stadia wires.

stadial moraine *See* recessional moraine.

stadia rod [ENG] A graduated rod used with a stadia to measure the distance from the observation point to the rod by observation of the length of rod subtended by the distance between the stadia hairs.

stadia tables [ENG] Mathematical tables from which may be found, without computation, the horizontal and vertical components of a reading made with a transit and stadia rod.

stadia wires *See* stadia hairs.

stadimeter [ENG] An instrument for determining the distance to an object, but its height must be known; the angle, subtended by the object's bottom and top as measured at the observer's position, is proportional to the object's height; the instrument is graduated directly in distance.

staff bead [BUILD] **1.** A bead between a wooden frame and adjacent masonry. **2.** A molded or beaded angle of wood or metal set into the corner of plaster walls.

Staffellidae [PALEON] An extinct family of marine protozoans (superfamily Fusulinacea) that persisted during the Pennsylvanian and Early Permian.

staff gage [ENG] A graduated scale placed in a position so that the stage of a stream may be read directly therefrom; a type of river gage.

stage [AERO ENG] A self-propelled separable element of a rocket vehicle or spacecraft. [ELECTR] A circuit containing a single section of an electron tube or equivalent device or two or more similar sections connected in parallel, push-pull, or push-push; it includes all parts connected between the control-grid input terminal of the device and the input terminal of the next adjacent stage. [GEOL] **1.** A developmental phase of an erosion cycle in which landscape features have distinctive characteristic forms. **2.** A phase in the historical development of a geologic feature. **3.** A major subdivision of a glacial epoch. **4.** A time-stratigraphic unit ranking below series and above substage. [HYD] The elevation of the water surface in a stream as measured by a river gage with reference to some arbitrarily selected zero datum. [MIN ENG] **1.** A certain length of underground roadway

STAFFELLIDAE

A representative of the fusulinacean family Staffellidae.

STALACTITE

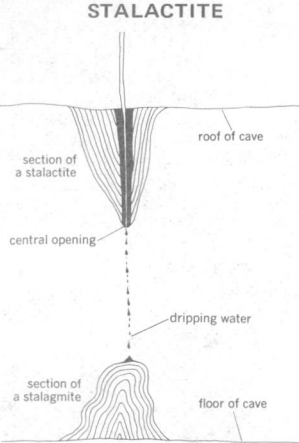

Diagram of stalactite and stalagmite formation. Deposits of calcium carbonate are made along the central opening of the stalactite by water dripping from the crack in the limestone of the roof of the cave. Water dripping to the floor from the stalactite deposits calcium carbonate to form a stalagmite. *(After E. B. Branson et al., Introduction to Geology, McGraw-Hill, 3d ed., 1952)*

worked by one horse. **2.** A narrow thin dike, especially one where the material of which the dike is composed is soft. **3.** A platform on which mine cars stand.

stage acidizing [PETRO ENG] An oil-reservoir acid treatment (acidizing) in which the formation is treated with two or more separate stages of acid, instead of a single large treatment.

staged crew [AERO ENG] An aircrew specifically positioned at an intermediate airfield to take over aircraft operating on an air route, thus relieving a complementary crew of flying fatigue and speeding up the traffic rate of the aircraft concerned.

stage gain [ELECTR] The ratio of the output power of an amplifier stage to the input power, usually expressed in decibels.

stage loader *See* feeder conveyor.

stage separation [PETRO ENG] A system for gas-oil separation of well fluids by a series of stages instead of a single operation.

stage theory [PHYSIO] A theory of color vision which proposes that there are three or more types of cone receptors whose responses are conducted to higher visual centers, and that interactions occur at some stage along the conducting paths so that strong activity in one type of response inhibits that of other response paths. Also known as zone theory.

stagger [COMMUN] Periodic error in the position of the recorded spot along a recorded facsimile line.

staggered blastholes [MIN ENG] Two rows of holes staggered to a triangular pattern to distribute the burden when shot-firing in thick coal seams.

staggered-intermittent fillet welding [MET] Making a line of intermittent fillet welds on each side of a joint in a manner such that the increments on one side are not opposite to those on the other side.

staggered-line drive [PETRO ENG] In the placement and drilling of water-injection wells for water flood recovery of oil, the staggered (versus in-line) areal arrangement of the injection wells.

staggered tuning [ELECTR] Alignment of successive tuned circuits to slightly different frequencies in order to widen the overall amplitude-frequency response curve.

staggering [COMMUN] Offsetting of two channels of different carrier systems from exact side-band frequency coincidence to avoid mutual interference.

staggering advantage [COMMUN] Effective reduction, in decibels, of interference between carrier channels, due to staggering.

staggers [VET MED] Any of various diseases of livestock, sheep, and horses manifested by lack of coordination in movement and a staggering gait.

stagger-tuned amplifier [ELECTR] An amplifier that uses staggered tuning to give a wide bandwidth.

stagger-tuned filter [ELECTR] A filter consisting of a cascade of amplifier stages with tuned coupling networks whose resonant frequencies and bandwidths may be easily adjusted to achieve an overall transmission function of desired shape (maximally flat or equal ripple).

staghead *See* witches'-broom disease.

staging [AERO ENG] The process or operation during the flight of a rocket vehicle whereby a full stage or half stage is disengaged from the remaining body and made free to decelerate or be propelled along its own flightpath.

staging area [ORD] A general locality, containing accommodations for troops, established for the concentration of troop units and transient personnel between movements over the lines of communication.

staging base [ORD] **1.** An advanced naval base for the anchoring, fueling, and refitting of transports and cargo ships, and for replenishing mobile service squadrons. **2.** A landing and takeoff area with minimum servicing, supply, and shelter provided for the temporary occupancy of military aircraft during the course of movement from one location to another.

stagnant glacier [HYD] A glacier which has ceased to move.

stagnant water [HYD] Motionless water, not flowing in a stream or current.

stagnation [HYD] **1.** The condition of a body of water unstirred by a current or wave. **2.** The condition of a glacier that has stopped flowing.

stagnation point [FL MECH] A point in a field of flow about a body where the fluid particles have zero velocity with respect to the body.

stagnation temperature *See* adiabatic recovery temperature.

stagnum [HYD] A pool of water with no outlet.

stain [MATER] **1.** A nonprotective coloring matter used on wood surfaces; imparts color without obscuring the wood grains. **2.** Any colored, organic compound used to stain tissues, cells, cell components, cell contents, or other biological substrates for microscopic examination.

stained glass [ENG] Glass colored by any of several means and assembled to produce a varicolored mosaic or representation.

stainierite *See* heterogenite.

stainless alloy [MET] Any of a large and complex group of corrosion-resistant iron-chromium alloys (containing 10% or more chromium), sometimes containing other elements, such as nickel, silicon, molybdenum, tungsten, and niobium. Commonly known as stainless steel (SS).

stainless-clad steel [MET] Steel clad on one or two sides with a stainless steel to provide a surface that is corrosion-resistant and attractive.

stainless iron *See* ferritic stainless steel.

stainless steel *See* stainless alloy.

stair [CIV ENG] A series of steps between levels or from floor to floor in a building.

stairway [CIV ENG] One or more flights of stairs connected by landings.

stake [ELEC] An iron peg used as a power electrode to transfer current into the ground in electrical prospecting. [ENG] **1.** To fasten back or prop open with a piece of chain or otherwise the valves or clacks of a water barrel in order that the water may run back into the sump when necessary. **2.** A pointed piece of wood driven into the ground to mark a boundary, survey station, or elevation.

staking [ENG] Joining two parts together by fitting a projection on one part against a mating feature in the other part and then causing plastic flow at the joint.

staking out [ENG] Driving stakes into the earth to indicate the foundation location of a structure to be built.

stalactite [GEOL] A conical or roughly cylindrical speleothem formed by dripping water and hanging from the roof of a cave; usually composed of calcium carbonate.

stalagmite [GEOL] A conical speleothem formed upward from the floor of a cave by the action of dripping water; usually composed of calcium carbonate.

stalked barnacle [INV ZOO] The common name for crustaceans composing the suborder Lepadomorpha.

stall [AERO ENG] **1.** The action or behavior of an airplane (or one of its airfoils) when by the separation of the airflow, as in the case of insufficient airspeed or of an excessive angle of attack, the airplane or airfoil tends to drop; the condition existing during this behavior. **2.** A flight performance in which an airplane is made to lose flying speed and to drop by pointing the nose steeply upward. **3.** An act or instance of stalling.

stall flutter [AERO ENG] A type of dynamic instability that takes place when the separation of flow around an airfoil occurs during the whole or part of each cycle of a flutter motion.

stalling angle *See* angle of stall.

stalling angle of attack *See* critical angle of attack.

stalling Mach number [AERO ENG] The Mach number of an aircraft when the coefficient of lift of the aerodynamic surfaces is the maximum obtainable for the pressure altitude, true airspeed, and angle of attack under which the craft is operated.

stallion [VERT ZOO] **1.** A mature male equine mammal. **2.** A male horse not castrated.

stall warning indicator [AERO ENG] A device that determines the critical angle of attack for a given aircraft; usually operates from vane sensors, airflow sensors, tabs on leading edges of wings, and computing devices such as accelerometers or airspeed indicators.

stalo [ELECTR] A highly stable local radio-frequency oscillator used for heterodyning signals to produce an intermediate frequency in radar moving-target indicators; only echoes that have changed slightly in frequency due to reflection from a moving target produce an output signal. Derived from stable local oscillator.

stalogometer [ENG] A device for measuring surface tension; a drop is suspended from a tube of known radius, and the radius of the drop is measured at the instant the drop detaches itself from the tube.

stamen [BOT] The male reproductive structure of a flower, consisting of an anther and a filament.

Stamey test [MED] A test of differential urinary excretion designed to detect unilateral renovascular disease.

staminode [BOT] A stamen with no functional anther.

stamp battery [MIN ENG] A machine for crushing very strong ores or rocks; consists essentially of a crushing member (gravity stamp) which is dropped on a die, the ore being crushed in water between the shoe and the die.

stamp copper [MIN ENG] A copper-bearing rock that is stamped and washed before it is smelted.

stamper [ENG ACOUS] A negative, generally made of metal by electroforming, used for molding phonograph records.

stamp head [MIN ENG] A heavy and nearly cylindrical cast-iron head fixed on the lower end of the stamp rod, shank, or lifter to give weight in stamping the ore.

stamping [ELECTR] A transformer lamination that has been cut out of a strip or sheet of metal by a punch press. [MECH ENG] Almost any press operation including blanking, shearing, hot or cold forming, drawing, bending, and coining. [MIN ENG] Reducing to the desired fineness in a stamp mill; the grain is usually not so fine as that produced by grinding in pans.

stamping mill [MIN ENG] A machine in which ore is finely crushed by descending pestles (stamps), usually operated by hydraulic power. Also known as crushing mill.

stamukha [OCEANOGR] An individual piece of stranded ice.

stanchion [NAV ARCH] An upright metal post on a ship.

stand [ECOL] A growth of plants, especially trees, on a given area. [OCEANOGR] The interval at high or low water when there is no appreciable change in the height of the tide. Also known as tidal stand.

stand-alone machine [ADP] A machine capable of functioning independently of a master computer, either part of the time or all of the time.

standard [PHYS] An accepted reference sample used for establishing a unit for the measurement of a physical quantity.

standard advanced base units [ORD] Personnel and materiel organized to function as advanced base units, including the functional components which are employed in the stops of naval advanced bases.

standard air munitions package [ORD] A package of air-to-ground conventional munitions required for the 30-day support of a specific type of aircraft; designed to be air transportable in three equal increments.

standard antenna [ELECTROMAG] An open single-wire antenna (including the lead-in wire) having an effective height of 4 meters.

standard arm [ORD] A passive homing air-to-surface antiradiation missile designed as a follow-on to the Shrike.

standard artillery atmosphere [METEOROL] A set of values describing atmospheric conditions on which ballistic computations are based, namely: no wind, a surface temperature of 15°C, a surface pressure of 1000 millibars, a surface relative humidity of 78%, and a lapse rate which yields a prescribed density-altitude relation.

standard artillery zone [METEOROL] A vertical subdivision of the standard artillery atmosphere; it may be considered a layer of air of prescribed thickness and altitude.

standard atmosphere [METEOROL] A hypothetical vertical distribution of atmospheric temperature, pressure, and density which is taken to be representative of the atmosphere for purposes of pressure altimeter calibrations, aircraft performance calculations, aircraft and missile design, and ballistic tables; the air is assumed to obey the perfect gas law and hydrostatic equation, which, taken together, relate temperature, pressure, and density variations in the vertical; it is further assumed that the air contains no water vapor, and that the acceleration of gravity does not change with height. [PHYS] See atmosphere.

standard ballistic conditions [MECH] A set of ballistic conditions arbitrarily assumed as standard for the computation of firing tables.

standard blocked F-format data set *See* FBS data set.

standard broadcast band *See* broadcast band.

standard broadcast channel [COMMUN] Band of frequencies occupied by the carrier and two side bands of a broadcast signal, with the carrier frequency at the center.

standard broadcasting [COMMUN] Broadcasting using amplitude modulation in the band of frequencies from 535 to 1605 kilohertz; carrier frequencies are placed 10 kilohertz apart.

standard calomel electrode [PHYS CHEM] A mercury-mercurous chloride electrode used as a reference (standard) measurement in polarographic determinations.

standard candle *See* international candle.

standard capacitor [ELEC] A capacitor constructed in such a manner that its capacitance value is not likely to vary with temperature and is known to a high degree of accuracy. Also known as capacitance standard.

standard cell [ELEC] A primary cell whose voltage is accurately known and remains sufficiently constant for instrument calibration purposes; the Weston standard cell has a voltage of 1.018636 volts at 20°C.

standard compass [NAV] A compass designated as the standard for a vessel; it is located in a favorable position, that is, a position with a minimum magnetic aberration, and is accurately calibrated.

standard conditions [PHYS] 1. A temperature of 0°C and a pressure of 1 atmosphere (760 torr). Also known as standard temperature and pressure (STP). 2. According to the American Gas Association, a temperature of 60°F (15 5/9°C) and a pressure of 762 millimeters (30 inches) of mercury. 3. According to the Compressed Gas Institute, a temperature of 20°C (68°F) and a pressure of 1 atmosphere. [SOLID STATE] The allotropic form in which a substance most commonly occurs.

standard depth-pressure recorder [PETRO ENG] A device for the measurement of pressures at the bottom of a well bore (that is, bottom-hole pressure); a spring-restrained piston moves a recording stylus on a pressure-sealed chart.

standard deviation [STAT] The positive square root of the expected value of the square of the difference between a random variable and its mean.

standard electrode potential [PHYS CHEM] The reversible or equilibrium potential of an electrode in an environment where reactants and products are at unit activity.

standard elemental time [IND ENG] A standard time for individual work elements.

standard error [STAT] A measure of the variability any statistical constant would be expected to show in taking repeated random samples of a given size from the same universe of observations.

standard evaporator *See* short-tube vertical evaporator.

standard fit [DES ENG] A fit whose allowance and tolerance are standardized.

standard free energy increase [THERMO] The increase in Gibbs free energy in a chemical reaction, when both the reactants and the products of the reaction are in their standard states.

standard-frequency signal [COMMUN] One of the highly accurate signals broadcast by government radio stations and used for testing and calibrating radio equipment all over the world; in the United States signals are broadcast by the National Bureau of Standards' radio stations WWV and WWVH.

standard gage [CIV ENG] A railroad gage measuring 4 feet 8½ inches (1.4351 meters). [DES ENG] A highly accurate gage used only as a standard for working gages.

standard gold [MET] A gold alloy containing 10% copper; at one time used for legal coinage in the United States.

standard gravity [MECH] A value of the acceleration of gravity equal to 9.80665 meters per second per second.

standard heat of formation [THERMO] The heat needed to produce one mole of a compound from its elements in their standard state.

standard hole [DES ENG] A hole with zero allowance plus a specified tolerance; fit allowance is provided for by the shaft in the hole.

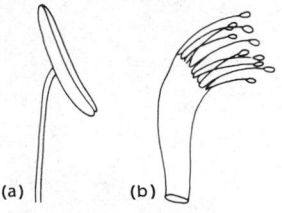

STAMEN

Two types of stamens:
(a) with versatile anther;
(b) monodelphous, with filaments united into one set.

standard hour [IND ENG] The quantity of output required of an operator to meet an hourly production quota. Also known as allowed hour.

standard hour plan [IND ENG] A wage incentive plan in which standard work times are expressed as standard hours and the worker is paid for standard hours instead of the actual work hours.

standard illuminants [OPTICS] Three standard sources of light, designated A, B, and C, used in specifying the light used when colors are matched; A is light from a filament at a color temperature of 2575°C, and B and C, representing noon sunlight and normal daylight respectively, are obtained by modifying A with rigorously specified filters.

standard inductor [ELECTROMAG] An inductor (coil) having high stability of inductance value, with little variation of inductance with current or frequency and with a low temperature coefficient; it may have an air core or an iron core; used as a primary standard in laboratories and as a precise working standard for impedance measurements.

standardization [DES ENG] The adoption of generally accepted uniform procedures, dimensions, materials, or parts that directly affect the design of a product or a facility. [ENG] The process of establishing by common agreement engineering criteria, terms, principles, practices, materials, items, processes, and equipment parts and components.

standardize [ADP] To replace any given floating point representation of a number with its representation in standard form; that is, to adjust the exponent and fixed-point part so that the new fixed-point part lies within a prescribed standard range.

standardized product [DES ENG] A product that conforms to specifications resulting from the same technical requirements.

standardized test statistic [STAT] A test statistic which has been reduced to standardized units.

standardized units [STAT] A random variable Z has been reduced to standardized units when it has zero expected value and standard deviation 1; this is accomplished by dividing the difference of Z and the expected value of Z by the standard deviation of Z.

standard lens [OPTICS] Usually the lens provided with a camera as standard equipment; in still cameras, the standard lens is one whose focal length is about equal to the length of the diagonal of the negative area normally provided by the camera; the normal field of view of a standard lens is about 53°.

standard load [DES ENG] A load which has been preplanned as to dimensions, weight, and balance, and designated by a number or some classification.

standard loran *See* loran A.

standard meridian [GEOD] The meridian used for reckoning standard time; throughout most of the world the standard meridians are those whose longitudes are exactly divisible by 15°. [MAP] A meridian of a map projection along which the scale is as stated.

standard mineral [MINERAL] A mineral that, on the basis of chemical analyses, is theoretically capable of being present in a rock. Also known as normative mineral.

standard muzzle velocity [MECH] The velocity at which a given projectile is supposed to leave the muzzle of a gun.

standard noise temperature [ELECTR] The standard reference temperature for noise measurements, equal to 290K.

standard noon [ASTRON] Twelve o'clock standard time, or the instant the mean sun is over the upper branch of the standard meridian.

standard output [IND ENG] The reciprocal of standard time.

standard parallel [MAP] A parallel on a map or chart along which the scale is as stated for that map or chart. [GEOD] The parallel or parallels of latitude used as control lines in the computation of a map projection.

standard performance [IND ENG] The performance of an individual or of a group on meeting standard output.

standard pitch [ACOUS] A musical pitch based on 440 hertz for tone A; with this standard, the frequency of middle C is 261 hertz.

standard plane [CRYSTAL] The crystal plane whose Miller indices are (111), that is, whose intercepts on the crystal axes are proportional to the corresponding sides of a unit cell.

standard potential [PHYS CHEM] The potential of an electrode composed of a substance in its standard state, in equilibrium with ions in their standard states compared to a hydrogen electrode.

standard preemphasis [COMMUN] Preemphasis in frequency-modulation and television aural broadcasting whose level lies between upper and lower limits specified by the Federal Communications Commission.

standard pressure [METEOROL] The arbitrarily selected atmospheric pressure of 1000 millibars to which adiabatic processes are referred for definitions of potential temperature, equivalent potential temperature, and so on. [PHYS] A pressure of 1 atmosphere, to which measurements of quantities dependent on pressure, such as the volume of a gas, are often referred. Also known as normal pressure.

standard project flood [HYD] The volume of streamflow expected to result from the most severe combination of meteorological and hydrologic conditions which are reasonably characteristic of the geographic region involved, excluding extremely rare combinations.

standard propagation [ELECTROMAG] Propagation of radio waves over a smooth spherical earth of specified dielectric constant and conductivity, under conditions of standard refraction in the atmosphere.

standard rate turn [AERO ENG] A turn in an aircraft in which heading changes at the rate of 3° per second.

standard refraction [ELECTROMAG] Refraction which would occur in an idealized atmosphere in which the index of refraction decreases uniformly with height at a rate of 39×10^{-6} per kilometer; standard refraction may be included in ground wave calculations by use of an effective earth radius of 8.5×10^6 meters, or $\frac{4}{3}$ the geometrical radius of the earth.

standard seawater *See* normal water.

standard shaft [DES ENG] A shaft with zero allowance minus a specified tolerance.

standard star [ASTRON] A star whose position or other data are precisely known so that it is used as a reference to calculate positions of other celestial bodies, or of objects on earth.

standard state [PHYS] The stable and pure form of a substance at standard pressure and ordinary temperature.

standard station *See* reference station.

standard subroutine [ADP] In computers, a subroutine which is applicable to a class of problems.

standard target [ELECTROMAG] A radar target which will produce an echo of known power under various conditions; smooth metal spheres or corner reflectors of known dimensions are such targets, and they may be used to calibrate a radar or check its performance.

standard temperature and pressure *See* standard conditions.

standard test-tone power [ELECTR] One milliwatt (0 decibels above one milliwatt) at 1000 hertz.

standard time [ASTRON] The mean solar time, based on the transit of the sun over a specified meridian, called the time meridian, and adopted for use over an area that is called a time zone. [IND ENG] A unit time value for completion of a work task as determined by the proper application of the appropriate work-measurement techniques. Also known as direct labor standard; output standard; production standard; time standard.

standard ton *See* ton.

standard trajectory [MECH] Path through the air that it is calculated a projectile will follow under given conditions of weather, position, and material, including the particular fuse, projectile, and propelling charge that are used; firing tables are based on standard trajectories.

standard visibility *See* meteorological range.

standard visual range *See* meteorological range.

standard volume [PHYS] The volume of 1 mole of a gas at a pressure of 1 atmosphere and a temperature of 0°C. Also known as normal volume.

standard waveguide [ELECTROMAG] Any one of several rectangular waveguides whose dimensions have been specified by various agencies and which are in general use.

standard wire rope [DES ENG] Wire rope made of six wire strands laid around a sisal core. Also known as hemp-core cable.

standby battery [ELEC] A storage battery held in reserve as an emergency power source in event of failure of regular power facilities at a radio station or other location.

standby computer [ADP] A computer in a duplex system that takes over when the need arises.

standby register [ADP] In computers, a register into which information can be copied to be available in case the original information is lost or mutilated in processing.

standby time [ADP] **1.** The time during which two or more computers are tied together and available to answer inquiries or process intermittent actions on stored data. **2.** The elapsed time between inquiries when the equipment is operating on an inquiry application.

stand fire [FOR] A forest fire igniting in the trunks of trees.

standing cloud [METEOROL] Any stationary cloud maintaining its position with respect to a mountain peak or ridge.

standing ground [MIN ENG] Ground that will stand firm without timbering.

standing knee height [ANTHRO] The vertical distance taken from the top of the kneecap to the floor as the subject stands.

standing-on-nines carry [ADP] In high-speed parallel addition of decimal numbers, an arrangement that causes carry digits to pass through one or more nine digits, while signaling that the skipped nines are to be reset to zero.

standing operating procedure [AERO ENG] A set of instructions covering those features of operations which lend themselves to a definite or standardized procedure without loss of effectiveness; the procedure is applicable unless prescribed otherwise in a particular case; thus, the flexibility necessary in special situations is retained.

standing rigging [NAV ARCH] Rigging that is permanently secured such as shrouds, stays, bob-stays, martingales, and mast pendants.

standing valve [PETRO ENG] A sucker-rod-pump (oil well) discharge valve that remains stationary during the pumping cycle, in contrast to a traveling valve.

standing wave [PHYS] A wave in which the ratio of an instantaneous value at one point to that at any other point does not vary with time. Also known as stationary wave.

standing-wave detector [ELECTROMAG] An electric indicating instrument used for detecting a standing electromagnetic wave along a transmission line or in a waveguide and measuring the resulting standing-wave ratio; it can also be used to measure the wavelength, and hence the frequency, of the wave. Also known as standing-wave indicator; standing-wave meter; standing-wave-ratio meter.

standing-wave indicator *See* standing-wave detector.

standing-wave loss factor [ELECTROMAG] The ratio of the transmission loss in an unmatched waveguide to that in the same waveguide when matched.

standing-wave meter *See* standing-wave detector.

standing-wave method [ELECTROMAG] Any method of measuring the wavelength of electromagnetic waves that involves measuring the distance between successive nodes or antinodes of standing waves.

standing-wave producer [ELECTROMAG] A movable probe inserted in a slotted waveguide to produce a desired standing-wave pattern, generally for test purposes.

standing-wave ratio [PHYS] **1.** The ratio of the maximum amplitude to the minimum amplitude of corresponding components of a wave in a transmission line or waveguide. **2.** The reciprocal of this ratio.

standing-wave-ratio meter *See* standing-wave detector.

standing ways *See* ground ways.

stand method [FOR] The practice of successively cutting trees of different ages so that ultimately the trees in the stand are new growth of a uniform age.

standoff insulator [ELEC] An insulator used to support a conductor at a distance from the surface on which the insulator is mounted.

standoff weapon [ORD] A weapon which may be launched at a distance sufficient to allow attacking personnel to evade defensive fire from the target area.

stand on [NAV] To proceed on the same course.

standpipe [ENG] A vertical tube filled with a material, for example, water, or in petroleum refinery catalytic cracking, a catalyst to serve as a seal between high- and low-pressure parts of the process equipment.

stanfieldite [MINERAL] $Ca_4(Mg,Fe,Mn)_5(PO_4)_6$ A phosphate mineral found only in meteorites.

Stanford achievement test [PSYCH] A group test employing a primary (grades 2 and 3) and an advanced (grades 4 through 9) examination to measure achievement at various grade levels.

stannane *See* tin hydride.

stannic acid *See* stannic oxide.

stannic anhydride *See* stannic oxide.

stannic bromide [INORG CHEM] $SnBr_4$ Water- and alcohol-soluble, white crystals that fume when exposed to air, and melt at 31°C; used in mineral separations. Also known as tin bromide; tin tetrabromide.

stannic chloride [INORG CHEM] $SnCl_4$ A colorless, fuming liquid; soluble in cold water, alcohol, carbon disulfide, and oil of turpentine; decomposed by hot water; boils at 114°C; used as a conductive coating and a sugar bleach, and in drugs, ceramics, soaps, and blueprinting. Also known as tin chloride; tin tetrachloride.

stannic chromate [INORG CHEM] $Sn(CrO_4)_2$ Toxic, brownish-yellow crystals, slightly soluble in water; used to decorate porcelain and china. Also known as tin chromate.

stannic iodide [INORG CHEM] SnI_4 Yellow-reddish crystals; insoluble in water, soluble in alcohol, ether, chloroform, carbon disulfide, and benzene; decomposed by water, melt at 144°C, sublime at 180°C. Also known as tin iodide; tin tetraiodide.

stannic oxide [INORG CHEM] SnO_2 A white powder; insoluble in water, soluble in concentrated sulfuric acid; melts at 1127°C; used in ceramic glazes and colors, special glasses, putty, and cosmetics, and as a catalyst. Also known as flowers of tin; stannic acid; stannic anhydride; tin dioxide; tin oxide; tin peroxide.

stannic sulfide [INORG CHEM] SnS_2 A yellow-brown powder; insoluble in water, soluble in alkaline sulfides; decomposes at red heat; used as a pigment and for imitation gilding. Also known as artificial gold; mosaic gold; tin bisulfide.

stannite [MINERAL] Cu_2FeSnS_4 A steel-gray or iron-black mineral crystallizing in the tetragonal system and occurring in granular masses; luster is metallic, hardness is 4 on Mohs scale, and specific gravity is 4.3–4.53. Also known as bell-metal ore; tin pyrites.

stannous bromide [INORG CHEM] $SnBr_2$ A yellow powder; soluble in water, alcohol, acetone, ether, and dilute hydrochloric acid; browns in air; melts at 215°C. Also known as tin bromide.

stannous chloride [INORG CHEM] $SnCl_2$ White crystals; soluble in water, alcohol, and alkalies; oxidized in air to the oxychloride; melt at 247°C; used as a chemical intermediate, reducing agent, and ink-stain remover, and for silvering mirrors. Also known as tin chloride; tin crystals; tin dichloride; tin salts.

stannous chromate [INORG CHEM] $SnCrO_4$ A brown powder; very slightly soluble in water; used to decorate porcelain. Also known as tin chromate.

stannous fluoride [INORG CHEM] SnF_2 A white, lustrous powder; slightly soluble in water; used to fluoridate toothpaste and as a medicine. Also known as tin difluoride; tin fluoride.

stannous oxide [INORG CHEM] SnO An air-unstable, brown to black powder; insoluble in water, soluble in acids and strong bases; decomposes when heated; used as a reducing agent and chemical intermediate, and for glass plating. Also known as tin oxide; tin protoxide.

stannous sulfate [INORG CHEM] $SnSO_4$ Heavy light-colored crystals; decomposes rapidly in water, loses SO_2 at 360°C; used for dyeing and tin-plating. Also known as tin sulfate.

stannous sulfide [INORG CHEM] SnS Dark crystals; insoluble in water, soluble (with decomposition) in concentrated hydrochloric acid; melts at 880°C; used as an analytical reagent and catalyst, and in bearing material. Also known as tin monosulfide; tin protosulfide; tin sulfide.

stannum [CHEM] The Latin name for tin, thus the symbol Sn for the element.

Stanton diagram [FL MECH] The plot of the airflow friction coefficient against the Reynolds number.

Stanton number [THERMO] A dimensionless number used in

the study of forced convection, equal to the heat-transfer coefficient of a fluid divided by the product of the specific heat at constant pressure, the fluid density, and the fluid velocity. Symbolized N_{St}. Also known as Margoulis number (M).

stapedius muscle [ANAT] The muscle which attaches to and controls the stapes in the inner ear.

stapes [ANAT] The stirrup-shaped middle-ear ossicle, articulating with the incus and the oval window. Also known as columella.

STAPHYLINIDAE

A representative member of the coeleopteran superfamily Staphylinidae.

Staphylinidae [INV ZOO] The rove beetles, a very large family of coleopteran insects in the superfamily Staphylinoidea.

Staphylinoidea [INV ZOO] A superfamily of Coleoptera in the suborder Polyphaga.

staphylococcal pneumonia [MED] A severe form of pneumonia caused by *Staphylococcus aureus*.

Staphylococcus [MICROBIO] A genus of bacteria in the family Micrococcaceae; all members are gram-positive, spherical pathogens, typically occurring in grapelike clusters.

Staphylococcus aureus [MICROBIO] A common human pathogen that causes furunculosis, pyemia, osteomyelitis, suppuration of wounds, and food poisoning. Also known as *Micrococcus pyogenes*.

staphylococcus toxoid [IMMUNOL] Formaldehyde-detoxified toxins of *Staphylococcus aureus* whose antigenicity is maintained; used prophylactically and therapeutically in the treatment of various staphylococcic pyodermas and localized pyogenic processes.

staphylomycin [MICROBIO] An antibiotic composed of three active components produced by a strain of *Actinomyces* that inhibits growth of gram-positive microorganisms and acid-fast bacilli.

staphylotoxin [BIOCHEM] Any of the various toxins elaborated by strains of *Staphylococcus aureus,* including hemolysins, enterotoxins, and leukocidin.

staple [DES ENG] A U-shaped loop of wire with points at both ends; used as a fastener. [TEXT] The average fiber length to be used in spinning a yarn.

stapler [ENG] **1.** A device for inserting wire staples into paper or wood. **2.** A hammer for inserting staples.

star [ASTRON] A celestial body consisting of a large, self-luminous mass of hot gas held together by its own gravity; the sun is a typical star. [NUCLEO] A star-shaped group of tracks made by ionizing particles originating at a common point in a nuclear emulsion, cloud chamber, or bubble chamber; some stars are produced by successive disintegrations of an atom in a radioactive series, and others by nuclear reactions of the spallation type, such as those due to cosmic-ray particles. Also known as nuclear star.

STAR *See* standard terminal arrival route.

star atlas [ASTRON] A series of star maps for different times, for example, for each month; the maps are generally drawn to a small scale and in book form.

starboard [NAV ARCH] The side of a ship or airplane which is on one's right when one faces forward.

star catalog [ASTRON] A comprehensive tabulation of data concerning the stars listed; the data may include, for example, apparent positions, brightness, motions, parallaxes, and other properties of stars.

starch [BIOCHEM] Any one of a group of carbohydrates or polysaccharides, of the general composition $(C_6H_{10}O_5)_n$, occurring as organized or structural granules of varying size and markings in many plant cells; it hydrolyzes to several forms of dextrin and glucose; its chemical structure is not completely known, but the granules consist of concentric shells containing at least two fractions: an inner portion called amylose, and an outer portion called amylopectin.

star chain [NAV] A group of synchronized, transmitting, navigation stations having the master station located at the center of a roughly circular area and surrounded by three or more slave stations located at some distance from the master.

star chart *See* star map.

starch nitrate *See* nitrostarch.

star cloud [ASTRON] An aggregation of thousands or of millions of stars spread over hundreds or thousands of light-years.

star cluster [ASTRON] A group of stars held together by gravitational attraction; the two chief types are open clusters

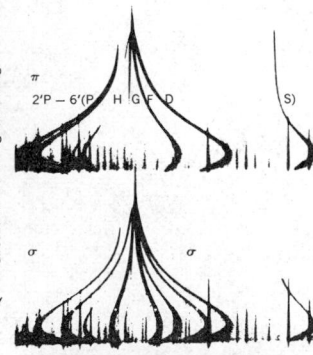

STARK EFFECT

An example of the linear Stark effect in the 4144- and 4169-angstrom lines of helium showing large symmetrical pattern. Electric field strengths range continuously from 0 to 85,000 volts per centimeter. Symbols π and σ refer respectively to light polarized parallel and perpendicular to the electric field.

(composed of from 12 to hundreds of stars) and globular clusters (composed of thousands to hundreds of thousands of stars).

star color [ASTRON] The color of a star as a function of its radiation and related to its temperature; colors range from blue-white to deep red.

star-connected circuit [ELEC] Polyphase circuit in which all the current paths within the region that delimits the circuit extend from each of the points of entry of the phase conductors to a common conductor (which may be the neutral conductor).

star connection *See* star network.

star count [ASTRON] A count of stars on a photographic plate.

star cut [LAP] A gem cut characterized by a hexagonal table surrounded by six facets in the form of an equilateral triangle.

star day [ASTRON] The time period between two successive passages of a star across the meridian.

star density [ASTRON] The average number of stars in a unit volume of space.

star drift [ASTRON] A description of two star groups in the Milky Way traveling through each other in opposite directions; individual stars have movements that are relative to each other. Also known as star stream.

star drill [DES ENG] A tool with a star-shaped point, used for drilling in stone or masonry.

Starfighter [AERO ENG] A United States supersonic, single-engine, turbojet-powered, tactical and air superiority fighter; the tactical version employs cannons or nuclear weapons for attack against surface targets, and is capable of providing close support for ground forces; the interceptor version employs Sidewinders or cannons.

star finder [ASTRON] A device such as a star map or celestial globe to facilitate the identification of stars. Also known as a star identifier.

starfish [INV ZOO] The common name for echinoderms belonging to the subclass Asteroidea.

star globe *See* celestial globe.

star grain [MATER] A rocket propellant grain with a cavity of star-shaped cross section.

star group [ASTRON] A number of stars that move in the same general direction at the same time.

star identifier *See* star finder.

Stark effect [SPECT] The effect on spectrum lines of an electric field which is either externally applied or is an internal field caused by the presence of neighboring ions or atoms in a gas, liquid, or solid. Also known as electric field effect.

Stark-Einstein law *See* Einstein photochemical equivalence law.

Stark-Lunelund effect [ELECTROMAG] The polarization of light emitted from a beam of moving atoms in a region where there are no electric or magnetic fields.

Stark number *See* Stefan number.

star lamp [ELEC] A high-pressure xenon arc, used in a planetarium, which produces a tiny, intense point of light focused through thousands of individual lenses and pinholes, and projected to the planetarium's dome.

Starlifter [AERO ENG] A United States large cargo transport powered by four turbofan engines, capable of intercontinental range with heavy payloads and airdrops.

Starling's law of the heart [PHYSIO] The energy associated with cardiac contraction is proportional to the length of the myocardial fibers in diastole.

star map [ASTRON] A map indicating the relative apparent positions of the stars. Also known as star chart.

star model *See* stellar model.

star motions [ASTRON] For the Milky Way, this includes rotation within the galaxy, motion which is described with respect to an external frame of reference; superposed on this systematic rotation are the individual motions of a star; each star moves in a somewhat elliptical orbit, with respect to the local standard of rest, the standard moving in a circular orbit around the galactic center.

star names [ASTRON] Nomenclature for the identification of stars; hundreds of stars have proper names that are traditional, for example, Betelgeuse; this star may be also identified as

α Orionis (Alpha Orionis), α for its being the brightest visual star in the constellation Orion.

star network [ELEC] A set of three or more branches with one terminal of each connected at a common node to give the form of a star. Also known as star connection; Y connection.

star ruby [MINERAL] An asteriated variety of ruby with normally six chatoyant rays.

star sapphire [MINERAL] A variety of sapphire exhibiting a six-pointed star resulting from the presence of microscopic crystals in various orientations within the gemstone.

star shell *See* illuminating projectile.

star stream *See* star drift.

start bit [ADP] The first bit transmitted in asynchronous data transmission to unequivocally indicate the start of the word.

start dialing signal [COMMUN] Signal transmitted from the incoming end of a circuit, following the receipt of a seizing signal, to indicate that the necessary circuit conditions have been established for receiving the numerical routine information.

start element [COMMUN] The first element of a character in certain serial transmissions, used to permit synchronization.

star telescope [OPTICS] An accessory of the marine navigational sextant designed primarily for star observations; it has a large objective to give a greater field of view and increased illumination; it is an erect telescope, that is, the object viewed is seen erect as opposed to the inverting telescope in which the object viewed is inverted.

starter [ELEC] **1.** A device used to start an electric motor and to accelerate the motor to normal speed. **2.** *See* engine starter. [ELECTR] An auxiliary control electrode used in a gas tube to establish sufficient ionization to reduce the anode breakdown voltage. Also known as trigger electrode. [ENG] A drill used for making the upper part of a hole, the remainder of the hole being made with a drill of smaller gage, known as a follower.

starting barrel [ENG] A short (12 to 24 inches or 30 to 60 centimeters) core barrel used to begin coring operations when the distance between the drill chuck and the bottom of the hole or to the rock surface in which a borehole is to be collared is too short to permit use of a full 5- or 10-foot-long (1.5- or 3.0-meter-long) core barrel.

starting box [ELEC] A device for providing extra resistance in the armature of a motor while it is being started.

starting friction *See* static friction.

starting mix [MATER] In pyrotechnic devices, an easily ignited mixture which transmits flame from an initiating device to a less readily ignitable composition.

starting motor *See* engine starter.

starting resistance [MECH ENG] The force needed to produce an oil film on the journal bearings of a train when it is at a standstill.

starting sheet [MET] A thin sheet of metal used as the initial cathode in electrowinning or electrorefining.

starting taper [DES ENG] A slight end taper on a reamer to aid in starting.

startle response [PHYSIO] The complex, involuntary, usually spasmodic psychophysiological response movement of an organism to a sudden unexpected stimulus.

startover [ADP] Program function that causes a computer that is not active to become active.

startover data transfer and processing program [ADP] Program which controls the transfer of startover data from the active to the standby machine and their subsequent processing by the standby machine.

star tracker *See* astrotracker.

start-stop multivibrator *See* monostable multivibrator.

start-stop printing telegraph [COMMUN] Form of printing telegraph in which the signal-receiving mechanisms, normally at rest, are started in operation at the beginning and stopped at the end of each character transmitted over the channel.

start-stop system [COMMUN] A telegraph system in which each group of code elements corresponding to a character is preceded by a start signal that prepares the receiving mechanism to receive and register a character, and is followed by a

stop signal that brings the receiving mechanism to rest in preparation for the reception of the next character.

start-up curve [IND ENG] A learning curve applied to a job for the purpose of adjusting work times that are longer than the standard because of the introduction of new jobs or new workers.

starved basin [GEOL] A sedimentary basin in which the rate of subsidence exceeds the rate of sedimentation.

starved joint [ENG] A glued joint containing insufficient or inadequate adhesive. Also known as hungry joint.

star wheel [ADP] The sensing device of a card punch, which is held in contact with the card under spring tension and which, detecting a hole, closes a contact point.

stasis [MED] A cessation of the normal flow of blood or other body fluids.

stasis dermatitis [MED] Chronic inflammation of the skin of the legs, resulting from poor circulation.

stat [NUCLEO] A unit of radioactive disintegration rate equal to the disintegration rate of that quantity of radon that gives rise to a charge of 1 statcoulomb in 1 second when situated in air.

statΩ *See* statohm.

stat$\mho$ *See* stat mho.

stat- [ELEC] A prefix indicating an electrical unit in the electrostatic centimeter-gram-second system of units; it is attached to the corresponding SI unit.

statA *See* statampere.

statampere [ELEC] The unit of electric current in the electrostatic centimeter-gram-second system of units, equal to a flow of charge of 1 statcoulomb per second; equal to approximately 3.3356×10^{-10} ampere. Abbreviated statA.

statC *See* statcoulomb.

statcoulomb [ELEC] The unit of charge in the electrostatic centimeter-gram-second system of units, equal to the charge which exerts a force of 1 dyne on an equal charge at a distance of 1 centimeter in a vacuum; equal to approximately 3.3356×10^{-10} coulomb. Abbreviated statC. Also known as franklin (Fr); unit charge.

state [CONT SYS] A minimum set of numbers which contain enough information about a system's history to enable its future behavior to be computed. [PHYS] The condition of a system which is specified as completely as possible by observations of a specified nature, for example, thermodynamic state, energy state. [QUANT MECH] The condition in which a system exists; the state may be pure and describable by a wave function or mixed and describable by a density matrix.

state equations [CONT SYS] Equations which express the state of a system and the output of a system at any time as a single valued function of the system's input at the same time and the state of the system at some fixed initial time.

statement [ADP] An elementary specification of a computer action or process, complete and not divisible into smaller meaningful units; it is analogous to the simple sentence of a natural language.

state of matter [PHYS] One of three fundamental conditions of matter: the solid, liquid, and gaseous states.

state of strain [MECH] A complete description, including the six components of strain, of the deformation within a homogeneously deformed volume.

state of stress [MECH] A complete description, including the six components of stress, of a homogeneously stressed volume.

state of the sea [OCEANOGR] A description of the properties of the wind-generated waves on the surface of the sea.

state of the sky [METEOROL] The aspect of the sky in reference to the cloud cover; the state of the sky is fully described when the amounts, kinds, directions of movement, and heights of all clouds are given.

state parameter *See* thermodynamic function of state.

state space [CONT SYS] The set of all possible values of the state vector of a system.

state transition matrix [CONT SYS] A matrix whose product with the state vector at an initial time gives the state vector at a later time.

state variable [CONT SYS] One of a minimum set of numbers which contain enough information about a system's history to enable computation of its future behavior. [THERMO] *See* thermodynamic function of state.

state vector [CONT SYS] A column vector whose components are the state variables of a system. [QUANT MECH] A vector in Hilbert space which corresponds to the state of a quantum-mechanical system.

statF *See* statfarad.

statfarad [ELEC] Unit of capacitance in the electrostatic centimeter-gram-second system of units, equal to the capacitance of a capacitor having a charge of 1 statcoulomb, across the plates of which the charge is 1 statvolt; equal to approximately 1.1126×10^{-12} farad. Abbreviated statF. Also known as centimeter.

statH *See* stathenry.

stathenry [ELECTROMAG] The unit of inductance in the electrostatic centimeter-gram-second system of units, equal to the self-inductance of a circuit or the mutual inductance between two circuits if there is an induced electromotive force of 1 statvolt when the current is changing at a rate of 1 statampere per second; equal to approximately 8.9876×10^{11} henry. Abbreviated statH.

static [COMMUN] A hissing, crackling, or other sudden sharp sound that tends to interfere with the reception, utilization, or enjoyment of desired signals or sounds. [PHYS] Without motion or change.

static bed [CHEM ENG] Refers to a layer of solids in a process vessel (absorber, catalytic reactor, packed distillation column, or granular filter bed) in which the particles rest upon one another at essentially the settled bulk density of the solids phase; contrasted to moving-solids or fluidized-solids beds.

static breeze *See* convective discharge.

static characteristic [ELECTR] A relation between a pair of variables, such as electrode voltage and electrode current, with all other operating voltages for an electron tube, transistor, or other amplifying device maintained constant.

static charge [ELEC] An electric charge accumulated on an object.

static check [ADP] Of a computer, one or more tests of computing elements, their interconnections, or both, performed under static conditions.

static debugging routine [ADP] A debugging routine which is used after the program being checked has been run and has stopped.

static discharger [ELEC] A rubber-covered cloth wick about 6 inches (15 centimeters) long, sometimes attached to the trailing edges of the surfaces of an aircraft to discharge static electricity in flight.

static dump [ADP] An edited printout of the contents of main memory or of the auxiliary storage, performed in a fixed way; it is usually taken at the end of a program run either automatically or by operator intervention.

static electricity [ELEC] 1. The study of the effects of macroscopic charges, including the transfer of a static charge from one object to another by actual contact or by means of a spark that bridges an air gap between the objects. 2. *See* electrostatics.

static eliminator [ELECTR] Device intended to reduce the effect of atmospheric static interference in a radio receiver.

static equilibrium *See* equilibrium.

static error [STAT] Error independent of the time-varying nature of a variable.

static firing [AERO ENG] The firing of a rocket engine in a hold-down position to measure thrust and to accomplish other tests.

static fluid column [FL MECH] An unchanging height of fluid in a vertical pipe, well bore, process vessel, or tank.

static friction [MECH] 1. The force that resists the initiation of sliding motion of one body over the other with which it is in contact. 2. The force required to move one of the bodies when they are at rest. Also known as starting friction.

static gearing ratio [AERO ENG] The ratio of the control-surface deflection in degrees to angular displacement of the missile which caused the deflection of the control surface.

static gel buildup [FL MECH] A method used to infer the degree of thixotropy of a fluid by viscometric measurement of its gel strength.

static granitization [PETR] The formation of a granitic rock by a metasomatic process in the absence of compressive forces or strains.

static grizzly [MIN ENG] A grizzly in the form of a stationary bar screen which either allows suitable pieces of rock or ore to pass over and unwanted small sizes to drop through, or rejects oversize pieces while allowing suitable material to drop through.

static head [FL MECH] Pressure of a fluid due to the head of fluid above some reference point.

staticize [ADP] 1. To capture transient data in stable form, thus converting fleeting events into examinable information. 2. To extract an instruction from the main computer memory and store the various component parts of it in the appropriate registers, preparatory to interpreting and executing it.

static level [FL MECH] Elevation of the water level or a pressure surface at rest. [HYD] The height to which water will rise in an artesian well; the static level of a flowing well is above the ground surface.

static line [AERO ENG] A line attached to a parachute pack and to a strop or anchor cable in an aircraft so that when the load is dropped the parachute is deployed automatically.

static load [MECH] A nonvarying load; the basal pressure exerted by the weight of a mass at rest, such as the load imposed on a drill bit by the weight of the drill-stem equipment or the pressure exerted on the rocks around an underground opening by the weight of the superimposed rocks. Also known as dead load.

static machine [ELEC] A machine for generating electric charges, usually by electric induction, sometimes used to build up high voltages for research purposes.

static magnetism *See* magnetostatics.

static metamorphism [GEOL] Regional metamorphism caused by heat and solvents at high lithostatic pressures. Also known as load metamorphism.

static moment [MECH] 1. A scalar quantity (such as area or mass) multiplied by the perpendicular distance from a point connected with the quantity (such as the centroid of the area or the center of mass) to a reference axis. 2. The magnitude of some vector (such as force, momentum, or a directed line segment) multiplied by the length of a perpendicular dropped from the line of action of the vector to a reference point.

static oceanography [OCEANOGR] Branch of oceanography that deals with the physical and chemical nature of water in the ocean and with the shape and composition of the ocean bottom.

static pressure [ACOUS] The pressure that would exist at a point in a medium if no sound waves were present. [FL MECH] 1. The normal component of stress, the force per unit area, exerted across a surface moving with a fluid, especially across a surface which lies in the direction of fluid flow. 2. The average of the normal components of stress exerted across three mutually perpendicular surfaces moving with a fluid.

static pressure tap *See* pressure tap.

static-pressure tube [ENG] A smooth tube with a rounded nose that has radial holes in the portion behind the nose and is used to measure the static pressure within the flow of a fluid.

static reaction [MECH] The force exerted on a body by other bodies which are keeping it in equilibrium.

static reflex [PHYSIO] Any one of a series of reflexes which are involved in the establishment of muscular tone for postural purpose.

static regulator [ELECTR] Transmission regulator in which the adjusting mechanism is in self-equilibrium at any setting and requires control power to change the setting.

statics [MECH] The branch of mechanics which treats of force and force systems abstracted from matter, and of forces which act on bodies in equilibrium.

static seal *See* gasket.

static sensitivity [ELECTR] In phototubes, quotient of the direct anode current divided by the incident radiant flux of constant value.

static SP *See* static spontaneous potential.

static split tracking *See* monopulse radar.

static spontaneous potential [PETRO ENG] Theoretical maximum spontaneous potential current that can be measured in a down-hole, mud log in clean sand. Abbreviated SSP; static SP.

static stability [METEOROL] The stability of an atmosphere in hydrostatic equilibrium with respect to vertical displacements, usually considered by the parcel method. Also known as convectional stability; hydrostatic stability; vertical stability. [PHYS] *See* stability.

static storage [ADP] Computer storage such that information is fixed in space and available at any time, as in flip-flop circuits, electrostatic memories, and coincident-current magnetic-core storage.

static subroutine [ADP] In computers, a subroutine which involves no parameters other than the addresses of the operands.

static switching [ELEC] Switching of circuits by means of magnetic amplifiers, semiconductors, and other devices that have no moving parts.

static test [AERO ENG] In particular, a test of a rocket or other device in a stationary or hold-down position, either to verify structural design criteria, structural integrity, and the effects of limit loads, or to measure the thrust of a rocket engine. [ENG] A measurement taken under conditions where neither the stimulus nor the environmental conditions fluctuate.

static trees [GRAPHICS] A pattern created on film by undesirable emulsion exposures to static electricity.

static tube [ENG] A device used to measure the static (not kinetic or total) pressure in a stream of fluid; consists of a perforated, tapered tube that is placed parallel to the flow, and has a branch tube that is connected to a manometer.

static universe [ASTRON] A postulated universe that has a finite static volume and is closed.

static weapon [ORD] A weapon which is used in place.

static work [IND ENG] Manual work performed with no significant motion.

station [ADP] One of a series of essentially similar positions or facilities in a data-processing system. [COMMUN] *See* broadcast station. [ELEC] An assembly line or assembly machine location at which a wiring board or chassis is stopped for insertion of one or more parts. [ELECTR] A location at which radio, television, radar, or other electric equipment is installed. [ENG] Any predetermined point or area on the seas or oceans which is patrolled by naval vessels. [MIN ENG] **1.** An enlargement in a mining shaft or gallery on any level used for a landing at any desired place and also for receiving loaded mine cars that are to be sent to the surface. **2.** An opening into a level which heads out of the side of an inclined plane; the point at which a surveying instrument is planted or observations are made. [SCI TECH] A geographic location at which scientific observations and measurements are made.

stationary cone classifier [MECH ENG] In a pulverizer directly feeding a coal furnace, a device which returns oversize coal to the pulverizing zone.

stationary distribution [PHYS] A time-independent distribution of a scalar quantity.

stationary engine [MECH ENG] A permanently placed engine, as in a power house, factory, or mine.

stationary ergodic noise [ELECTR] A stationary noise for which the probability that the noise voltage lies within any given interval at any time is nearly equal to the fraction of time that the noise voltage lies within this interval if a sufficiently long observation interval is recorded.

stationary field [PHYS] Field which does not change during the time interval under consideration. Also known as constant field.

stationary front *See* quasi-stationary front.

stationary noise [ELECTR] A random noise for which the probability that the noise voltage lies within any given interval does not change with time.

stationary orbit [AERO ENG] A circular, equatorial orbit in which the satellite revolves about the primary body at the angular rate at which the primary body rotates on its axis; from the primary body, the satellite thus appears to be stationary over a point on the primary body; a stationary orbit must be synchronous, but the reverse need not be true.

stationary phase [MICROBIO] The period following termination of exponential growth in a bacterial culture when the number of viable microorganisms remains relatively constant for a time.

stationary state *See* energy state.

stationary stochastic process [MATH] A stochastic process $x(t)$ is stationary if each of the probability distributions is unaffected by a change in the time parameter t.

stationary time series [STAT] A time series which as a stochastic process is unchanged by a uniform increment in the time parameter defining it.

stationary wave *See* standing wave.

station authentication [COMMUN] Security measure designed to establish the authenticity of a transmitting or receiving station.

station buoy [NAV] A buoy used to mark the approximate station (position) of an important buoy or lightship should it be carried away or temporarily removed. Also known as marker buoy; watch buoy.

station continuity chart [METEOROL] A chart or graph on which time is one coordinate, and one or more of the observed meteorological elements at that station is the other coordinate.

station elevation [METEOROL] The vertical distance above mean sea level that is adopted as the reference datum level for all current measurements of atmospheric pressure at the station.

station model [METEOROL] A specified pattern for entering, on a weather map, the meteorological symbols that represent the state of the weather at a particular observation station.

station pointer *See* three-arm protractor.

station pole [CIV ENG] One of various rods used in surveying to mark stations, to sight points and lines; or to measure elevation with respect to the transit.

station pressure [METEOROL] The atmospheric pressure computed for the level of the station elevation.

station stock level [ORD] Maximum quantity of supplies expressed in days of supply, permitted to be on hand or due in at any time at a military installation; this level is based on actual past issues and anticipated demands; it represents the requisitioning objective.

station time [AERO ENG] Time at which crews, passengers, and cargo are to be on board air transport and ready for the flight.

statistic [STAT] An estimate or piece of data, concerning some parameter, obtained from a sampling.

statistical analysis [STAT] The body of techniques used in statistical inference concerning a population.

statistical distribution *See* distribution of a random variable.

statistical forecast [METEOROL] A weather forecast based upon a systematic statistical examination of the past behavior of the atmosphere, as distinguished from a forecast based upon thermodynamic and hydrodynamic considerations.

statistical hypothesis [STAT] A statement about the way a random variable is distributed.

statistical independence [STAT] Two events are statistically independent if the probability of their occurring jointly equals the product of their respective probabilities.

statistical inference [STAT] The process of reaching conclusions concerning a population upon the basis of random samplings.

statistical map [MAP] A special type of map in which the variation in quantity of a factor such as rainfall, population, or crops in a geographic area is indicated; a dot map is one type.

statistical mechanics [PHYS] That branch of physics which endeavors to explain and predict the macroscopic properties and behavior of a system on the basis of the known characteristics and interactions of the microscopic constituents of the system, usually when the number of such constituents is very large. Also known as statistical thermodynamics.

statistical quality control [IND ENG] The use of statistical techniques as a means of controlling the quality of a product or process.

statistical thermodynamics *See* statistical mechanics.

statistics [MATH] A discipline dealing with methods of obtaining data, analyzing and summarizing it, and drawing

STATOR ARMATURE

Photograph of a stator armature of an alternating-current induction motor. *(Allis-Chalmers)*

STAUROLITE

|← 1.2 cm →|

Staurolite crystal from Taos County, New Mexico. *(American Museum of Natural History Specimen)*

STAUROMEDUSAE

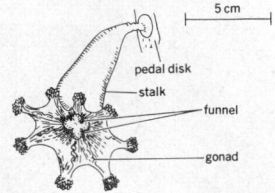

Medusa of *Lucernaria*; calyx is eight-sided and has eight groups of short, capped tentacles on its margin.

inferences from data samples by the use of probability theory.

statmho [ELEC] The unit of conductance, admittance, and susceptance in the electrostatic centimeter-gram-second system of units, equal to the conductance between two points of a conductor when a constant potential difference of 1 statvolt applied between the points produces in this conductor a current of 1 statampere, the conductor not being the source of any electromotive force; equal to approximately 1.1126×10^{-12} mho. Abbreviated stat℧. Also known as statsiemens (statS).

statoblast [INV ZOO] A chitin-encapsulated body which serves as a special means of asexual reproduction in the Phylactolaemata.

statocone [INV ZOO] One of the minute calcareous granules found in the statocyst of certain animals.

statocyst [BOT] A cell containing statoliths in a fluid medium. Also known as statocyte. [INV ZOO] A sensory vesicle containing statoliths and which functions in the perception of the position of the body in space.

statocyte *See* statocyst.

statohm [ELEC] The unit of resistance, reactance, and impedance in the electrostatic centimeter-gram-second system of units, equal to the resistance between two points of a conductor when a constant potential difference of 1 statvolt between these points produces a current of 1 statampere; equal to approximately 8.9876×10^{11} ohm. Abbreviated statΩ.

statokinetic [PHYSIO] Pertaining to the balance and posture of the body or its parts during movement, as in walking.

statolith [BOT] A sand grain or other solid inclusion which moves readily in the fluid contents of a statocyst, comes to rest on the lower surface of the cell, and is believed to function in gravity perception. [INV ZOO] A secreted calcareous body, a sand grain, or other solid inclusion contained in a statocyst.

stator [ELEC] The portion of a rotating machine that contains the stationary parts of the magnetic circuit and their associated windings. [MECH ENG] A stationary machine part in or about which a rotor turns.

stator armature [ELEC] A stator which includes the main current-carrying winding in which electromotive force produced by magnetic flux rotation is induced; it is found in most alternating-current machines.

statoreceptor [PHYSIO] A sense organ concerned primarily with equilibrium.

stator plate [ELEC] One of the fixed plates in a variable capacitor; stator plates are generally insulated from the frame of the capacitor.

statoscope [ENG] **1.** A barometer that records small variations in atmospheric pressure. **2.** An instrument that indicates small changes in an aircraft's altitude.

statS *See* statmho.

statsiemens *See* statmho.

statT *See* stattesla.

stattesla [ELECTROMAG] The unit of magnetic flux density in the electrostatic centimeter-gram-second system of units, equal to one statweber per square centimeter; equal to approximately 2.9979×10^6 tesla. Abbreviated statT.

statuary bronze [MET] Special bronze alloys used for casting statues and other ornamental objects; a typical bronze for statuary work contains 90% copper, 6% tin, 3% zinc, and 1% lead.

stature [ANTHRO] A measure of the distance from the floor to the vertex of the head, taken either front or back as the subject stands erectly with heels together.

status asthmaticus [MED] Intractable asthma lasting from a few days to a week or longer.

status epilepticus [MED] Occurrence of prolonged, generalized epileptic seizures in rapid succession with brief intervals of coma.

status word [ADP] A word indicating the state of the system or the diagnosis of a state into which the system has entered.

statute mile *See* mile.

statV *See* statvolt.

statvolt [ELEC] The unit of electric potential and electromotive force in the electrostatic centimeter-gram-second system

of units, equal to the potential difference between two points such that the work required to transport 1 statcoulomb of electric charge from one to the other is equal to 1 erg; equal to approximately 2.9979×10^2 volts. Abbreviated statV.

statWb *See* statweber.

statweber [ELECTROMAG] The unit of magnetic flux in the electrostatic centimeter-gram-second system of units, equal to the magnetic flux which, linking a circuit of one turn, produces in it an electromotive force of 1 statvolt as it is reduced to zero at a uniform rate in 1 second; equal to approximately 2.9979×10^2 weber. Abbreviated statWb.

stauractine [INV ZOO] A sponge spicule in which the four rays lie in one plane.

staurolite [MINERAL] $FeAl_4(SiO_4)_2(OH)_2$ A reddish-brown to black neosilicate mineral that crystallizes in the orthorhombic system, has resinous to vitreous luster, hardness is 7–7.5 on Mohs scale, and specific gravity is 3.7. Also known as cross-stone; fairy stone; grenatite; staurotide.

Stauromedusae [INV ZOO] An order of the class Scyphozoa in which the medusa is composed of a cuplike bell called a calyx and a stem that terminates in a pedal disk.

Staurosporae [MYCOL] A spore group of the Fungi Imperfecti characterized by star-shaped or forked spores.

staurotide *See* staurolite.

stave [DES ENG] **1.** A rung of a ladder. **2.** Any of the narrow wooden strips or metal plates placed edge to edge to form the sides, top, or lining of a vessel or structure, such as a barrel.

staybolt [DES ENG] A bolt with a thread along the entire length of the shaft; used to attach machine parts that are under pressure to separate.

stayed-cable bridge [CIV ENG] A modified cantilever bridge consisting of girders or trusses cantilevered both ways from a central tower and supported by inclined cables attached to the tower at the top or sometimes at several levels.

Staypak [MATER] Wood made more dense by pressure and heat, with the introduction of added resin.

stay time [AERO ENG] In rocket engine usage, the average value of the time spent by each gas molecule or atom within the chamber volume.

STD recorder *See* salinity-temperature-depth recorder.

steadiness *See* persistence.

steadite [MET] A hard structural constituent of cast iron consisting of the eutectic of ferrite and iron phosphide (Fe_3P); composition of the eutectic is 10.2% phosphorus and 89.8% iron; melts at 1049°C (1920°F).

steady bearing [NAV] An approaching or closing craft is said to be on a steady bearing if the compass bearing does not appreciably change.

steady pin [ENG] **1.** A retaining device such as a dowel, pin, or key that prevents a pulley from turning on its axis. **2.** A guide pin used to lift a cope or pattern.

steady state [PHYS] The condition of a body or system in which the conditions at each point do not change with time, that is after initial transients or fluctuations have disappeared.

steady-state cavitation *See* sheet cavitation.

steady-state conduction [THERMO] Heat conduction in which the temperature and heat flow at each point does not change with time.

steady-state creep *See* secondary creep.

steady-state current [ELEC] An electric current that does not change with time.

steady-state distribution [ANALY CHEM] The equilibrium condition between phases in each step of a multistage, countercurrent liquid-liquid extraction. Abbreviated SSD.

steady-state error [CONT SYS] The error that remains after transient conditions have disappeared in a control system.

steady-state flow [CHEM ENG] Fluid flow without any change in composition or phase equilibria relationships.

steady-state model [PETRO ENG] Electric or electrolytic analogs of a reservoir formation used to study the steady-state flow of fluids through porous media; includes gel, blotter, liquid, potentiometric, and similar models.

steady-state reactor [NUCLEO] A reactor in which conditions such as temperature, reaction rate, and neutron flux do not change appreciably with time.

steady-state wave motion [PHYS] Wave motion in which the wave quantities at each point in the region through which the wave is passing repeat themselves periodically.

steam [PHYS] Water vapor, or water in its gaseous state; the most widely used working fluid in external combustion engine cycles.

steam accumulator [MECH ENG] A pressure vessel in which water is heated by steam during off-peak demand periods and regenerated as steam when needed.

steam attemperation [MECH ENG] The control of the maximum temperature of superheated steam by water injection or submerged cooling.

steam bending [ENG] Forming wooden members to a desired shape by pressure after first softening by heat and moisture.

steam boiler [MECH ENG] A pressurized system in which water is vaporized to steam by heat transferred from a source of higher temperature, usually the products of combustion from burning fuels. Also known as steam generator.

steam bronze [MET] A leaded tin-bronze containing 88% copper, 6% tin, 4.5% zinc, 1.5% lead; used for steam valve bodies, gears, and bearings.

steam calorimeter See throttling calorimeter.

steam capstan [NAV ARCH] A capstan operated by a steam engine.

steam cock [ENG] A valve for the passage of steam.

steam condenser [MECH ENG] A device to maintain vacuum conditions on the exhaust of a steam prime mover by transfer of heat to circulating water or air at the lowest ambient temperature.

steam cracking [CHEM ENG] High-temperature cracking of petroleum hydrocarbons in the presence of steam.

steam cycle See Rankine cycle.

steam distillation [CHEM ENG] A distillation in which vaporization of the volatile constituents of a liquid mixture takes place at a lower temperature by the introduction of steam directly into the charge; steam used in this manner is known as open steam. Also known as steam stripping.

steam drive [MECH ENG] Any device which uses power generated by the pressure of expanding steam to move a machine or a machine part.

steam dryer [MECH ENG] A device for separating liquid from vapor in a steam supply system.

steam emulsion test [ENG] A test used for measuring the ability of oil and water to separate, especially for steam-turbine oil; after emulsification and separation, the time required for the emulsion to be reduced to 3 milliliters or less is recorded at 5-minute intervals.

steam engine [MECH ENG] A thermodynamic device for the conversion of heat in steam into work, generally in the form of a positive displacement, piston and cylinder mechanism.

steam engine indicator [ENG] An instrument that plots the steam pressure in an engine cylinder as a function of piston displacement.

steam fog [METEOROL] Fog formed when water vapor is added to air which is much colder than the vapor's source; most commonly, when very cold air drifts across relatively warm water. Also known as frost smoke; sea mist; sea smoke; steam mist; water smoke.

steam gage [ENG] A device for measuring steam pressure.

steam-generating furnace See boiler furnace.

steam generator See steam boiler.

steam hammer [MECH ENG] A forging hammer in which the ram is raised, lowered, and operated by a steam cylinder.

steam-hammer oil See tempering oil.

steam-heated evaporator [MECH ENG] A structure using condensing steam as a heat source on one side of a heat-exchange surface to evaporate liquid from the other side.

steam heating [MECH ENG] A system that used steam as the medium for a comfort or process heating operation.

steam jacket [MECH ENG] A casing applied to the cylinders and heads of a steam engine, or other space, to keep the surfaces hot and dry.

steam jet [ENG] A blast of steam issuing from a nozzle. [MIN ENG] A system of ventilating a mine by means of a number of jets of steam at high pressure kept constantly blowing off from a series of pipes in the bottom of the upcast shaft.

steam-jet cycle [MECH ENG] A refrigeration cycle in which

water is used as the refrigerant; high-velocity steam jets provide a high vacuum in the evaporator, causing the water to boil at low temperature and at the same time compressing the flashed vapor up to the condenser pressure level.

steam-jet ejector [MECH ENG] A fluid acceleration vacuum pump or compressor using the high velocity of a steam jet for entrainment.

steam locomotive [MECH ENG] A railway propulsion power plant using steam generally in a reciprocating, noncondensing engine.

steam loop [ENG] Two vertical pipes connected by a horizontal one, used to condense boiler steam so that it can be returned to the boiler without a pump or injector.

steam mist See steam fog.

steam nozzle [MECH ENG] A streamlined flow structure in which heat energy of steam is converted to the kinetic form.

steam point [THERMO] The boiling point of pure water whose isotopic composition is the same as that of sea water at standard atmospheric pressure; it is assigned a value of 100°C on the International Practical Temperature Scale of 1968.

steam pump [MECH ENG] A pump driven by steam acting on the coupled piston rod and plunger.

steam purifier See steam separator.

steam-refined asphalt [MATER] Petroleum-derived asphalt that has been refined in the presence of steam during the distillation of crude oil.

steam-refined cylinder oil See SR cylinder oil.

steam-refined stock [MATER] Unfiltered petroleum products distilled with heat applied in the form of steam, for example, for gear oils, lubricating oils, and cylinder oils.

steam refining [CHEM ENG] A petroleum refinery distillation process, in which the only heat used comes from steam in open and closed coils near the bottom of the still; used to produce gasoline and naphthas where odor and color are of prime importance; where open steam is used, it is known as steam distillation.

steam reheater [MECH ENG] A steam boiler component in which heat is added to intermediate-pressure steam, which has given up some of its energy in expansion through the high-pressure turbine.

steam roller [MECH ENG] A road roller driven by a steam engine.

steam separator [MECH ENG] A device for separating a mixture of the liquid and vapor phases of water. Also known as steam purifier.

steamship [NAV ARCH] A ship propelled by a steam engine.

steam shovel [MECH ENG] A power shovel operated by steam.

steam stripping See steam distillation.

steam superheater [MECH ENG] A boiler component in which sensible heat is added to the steam after it has been evaporated from the liquid phase.

steam tracing [ENG] A steam-carrying heater (such as tubing or piping) next to or twisted around a process-fluid or instrument-air line; used to keep liquids from solidifying or condensing.

steam trap [MECH ENG] A device which drains and removes condensate automatically from steam lines.

steam-tube dryer [MECH ENG] Rotary dryer with steam-heated tubes running the full length of the cylinder and rotating with the dryer shell.

steam turbine [MECH ENG] A prime mover for the conversion of heat energy of steam into work on a revolving shaft, utilizing fluid acceleration principles in jet and vane machinery.

steam valve [ENG] A valve used to regulate the flow of steam.

steam washer [ENG] A device for removing contaminants, such as silica, from the steam produced in a boiler.

steapsin [BIOCHEM] An enzyme in pancreatic juice that catalyzes the hydrolysis of fats. Also known as pancreatic lipase.

stearate [ORG CHEM] $C_{17}H_{35}COOM$ A salt or ester of stearic acid where M is a monovalent radical, for example, sodium stearate, $C_{17}H_{35}COONa$.

stearic acid [ORG CHEM] $CH_3(CH_2)_{16}COOH$ Nature's most common fatty acid, derived from natural animal and vegetable fats; colorless, waxlike solid, insoluble in water, soluble in alcohol, ether, and chloroform; melts at 70°C; used

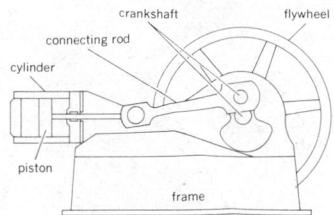

STEAM ENGINE

Principal parts of horizontal steam engine.

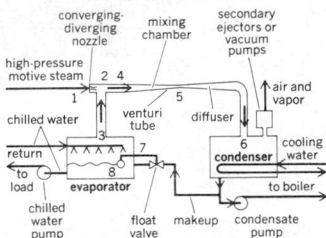

STEAM-JET CYCLE

Diagram of the basic steam-jet water-vapor cycle. High-pressure motive steam at 1 is expanded to low absolute pressure at 2. Water vapor in evaporator at 3 is entrained by motive steam at 4. Motive steam plus entrained moisture is forced through venturi tube at 5, in which velocity of the mixture is reduced and converted into pressure head at 6. Makeup water is added from 7 to 8.

STEAM LOCOMOTIVE

Steam locomotive uses high-pressure steam to move pistons which are connected to rods that force the driving wheels to rotate.

as a lubricant and in pharmaceuticals, cosmetics, and food packaging. Also known as *n*-octadecanoic acid.

stearin [ORG CHEM] $C_3H_5(C_{18}H_{35}O_2)_3$ A colorless combustible powder; insoluble in water, soluble in alcohol, chloroform, and carbon disulfide; melts at 72°C; used in metal polishes, pastes, candies, candles, and soap, and to waterproof paper. Also known as glyceryl tristearate; tristearin.

stearyl alcohol [ORG CHEM] $CH_3(CH_2)_{16}CH_2OH$ Oily white, combustible flakes; insoluble in water, soluble in alcohol, acetone, and ether; melt at 59°C; used in lubricants, resins, perfumes, and cosmetics, and as a surface-active agent. Also known as 1-octadecanol; octadecyl alcohol.

steatite [PETR] A compact, massive, fine-ground rock composed principally of talc, but with much other material.

steatization [GEOL] Introduction of or replacement by talc or steatite.

Steatornithidae [VERT ZOO] A family of birds in the order Caprimulgiformes which contains a single, South American species, the oilbird or guacharo (*Steatornis caripensis*).

steatorrhea [MED] **1.** Fatty stool. **2.** Increased flow of sebum.

Steckel rolling [MET] Cold metal rolling in which the strip is pulled through idler rolls by front tension only; the direction is reversed until the desired thickness is attained.

S tectonite [PETR] A tectonite whose fabric is dominated by planar surfaces of formation or deformation, such as slate.

STED *See* solar turboelectric drive.

steel [MET] An iron base alloy, malleable under proper conditions, containing up to about 2% carbon.

steel bronze [MET] A hardened bronze consisting of 92% copper and 8% tin; used as a substitute for steel in guns.

steel-cable conveyor belt [DES ENG] A rubber conveyor belt in which the carcass is composed of a single plane of steel cables.

steel-case [ORD] In cartridge nomenclature, indicating that the cartridge case is made of steel.

steel-clad rope [DES ENG] A wire rope made from flat strips of steel wound helically around each of the six strands composing the rope.

steel converter [MET] A retort in which cast iron is converted to steel; an example is the Bessemer converter.

steel emery [MET] An abrasive material composed of chilled iron produced by forcing iron through a steam jet; used in tumbling barrels and for grinding stones.

steel engraving [GRAPHICS] Engraving on a steel plate, as used in printing.

Steelflex coupling [MECH ENG] A flexible coupling made with two grooved steel hubs keyed to their respective shafts and connected by a specially tempered alloy-steel member called the grid.

steel foil [MET] A very thin sheet of steel, the thickness of which is measured in thousandths of an inch.

steel jack [MIN ENG] A screw jack suitable in mechanical mining; used for legs or upright timbers. [MINERAL] *See* sphalerite.

steel-jacket [ORD] In small-arms ammunition nomenclature, indicating that the bullet has a steel shell and core.

steelmaking [MET] Any of various processes for making steel from pig iron.

steel sets [MIN ENG] Steel beam used in main entries of coal mines and in shafts of metal mines; I beams for caps and H beams for posts or wall plates.

steel tunnel support [MIN ENG] One of the tunnel support systems made of steel; five types are continuous rib; rib and post; rib and wall plate; rib, wall plate, and post; and full circle rib.

steel wool [MET] Fine steel threads matted into a mass.

steelyard [ENG] A weighing device with a counterbalanced arm supporting the load to be weighed on the short end.

steen [CIV ENG] To line an excavation such as a cellar or well with stone, cement, or similar material.

Steenrod algebra [MATH] The cohomology groups of a topological space have additive operations on them, which can be added and multiplied so as to form the Steenrod algebra.

Steenrod squares [MATH] Operations which associate elements from different cohomology groups of a topological space and produce an element in another of the groups; these operations can be so added and multiplied as to produce the Steenrod algebra.

steepest descent method [MATH] Certain functions can be approximated for large values by an asymptotic formula derived from a Taylor series expansion about a saddle point. Also known as saddle point method.

steeple [ARCH] A tall, tapering structure atop a church tower.

steerable antenna [ELECTROMAG] A directional antenna whose major lobe can be readily shifted in direction.

steerage [NAV ARCH] The least desirable portions of a vessel used for accommodations for passengers who pay the lowest fare.

steerageway [NAV ARCH] The least speed (forward or reverse) to allow a vessel to be steered.

steering [METEOROL] Loosely used for any influence upon the direction of movement of an atmospheric disturbance exerted by another aspect of the state of the atmosphere; for example, a surface pressure system tends to be steered by isotherms, contour lines, or streamlines aloft, or by warm-sector isobars or the orientation of a warm front.

steering arm [MECH ENG] An arm that transmits turning motion from the steering wheel of an automotive vehicle to the drag link.

steering brake [MECH ENG] Means of turning, stopping, or holding a tracked vehicle by braking the tracks individually.

steering compass [NAV] A compass by which a craft is steered; it sometimes refers to a gyro repeater, which is used for the same purpose as the steering compass; the term steering repeater is preferable.

steering engine [NAV ARCH] A steam-, electric-, or hydraulic-power machine used for turning the rudder, and having its valves or operating gear actuated by leads from the pilot house.

steering gear [MECH ENG] The mechanism, including gear train and linkage, for the directional control of a vehicle or ship.

steering level [METEOROL] A hypothetical level, in the atmosphere, where the velocity of the basic flow bears a direct relationship to the velocity of movement of an atmospheric disturbance embedded in the flow.

steering repeater [NAV] A device which repeats at a distance the reading of the gyro compass; a craft is steered by reference to the device; sometimes called a steering compass.

steering wheel [MECH ENG] A hand-operated wheel for controlling the direction of the wheels of an automotive vehicle or of the rudder of a ship.

Stefan-Boltzmann constant [STAT MECH] The energy radiated by a blackbody per unit area per unit time divided by the fourth power of the body's temperature; equal to $(5.6696 \pm 0.0010) \times 10^{-8}$ (watts)(meter)$^{-2}$(degrees Kelvin)$^{-4}$.

Stefan-Boltzmann law [STAT MECH] The total energy radiated from a blackbody is proportional to the fourth power of the temperature of the body. Also known as fourth-power law; Stefan's law of radiation.

Stefan number [THERMO] A dimensionless number used in the study of radiant heat transfer, equal to the Stefan-Boltzmann constant times the cube of the temperature times the thickness of a layer divided by the layer's thermal conductivity. Symbolized $\overline{St}$. Also known as Stark number (Sk).

Stefan's formula [OCEANOGR] A formula for the growth of thickness h of an ice cover on the ocean at various freezing temperatures, expressed as

$$h \approx \sqrt{\left(\frac{2l}{\lambda_i \rho_i}\right)\psi}$$

where l is the coefficient of thermal conductivity, λ_i is the latent heat of fusion, ρ_i is the density of ice, and ψ is the cold sum (in degree days below 0°C).

Stefan's law of radiation *See* Stefan-Boltzmann law.

Steffen process [FOOD ENG] A process used in sugar factories to concentrate the dilute waste liquor from beet molasses and recover sugar for use in a fermentation process to make monosodium glutamate.

Steganopodes [VERT ZOO] Formerly, an order of birds that included the totipalmate swimming birds.

Stegodontinae [PALEON] An extinct subfamily of elephantoid proboscideans in the family Elephantidae.

Stegosauria [PALEON] A suborder of extinct reptiles of the order Ornithischia comprising the plated dinosaurs of the Jurassic which had tiny heads, great triangular plates arranged on the back in two alternating rows, and long spikes near the end of the tail.

steigerite [MINERAL] $4AlVO_4 \cdot 13H_2O$ A canary-yellow mineral composed of hydrous aluminum vanadate and occurring in masses.

Steinheil lens [OPTICS] A type of magnifier lens in which a biconvex crown lens is cemented between a pair of flint lenses.

Steinheim man [PALEON] A prehistoric man represented by a skull, without mandible, found near Stuttgart, Germany; the browridges are massive, the face is relatively small, and the braincase is similar in shape to that of *Homo sapiens*.

steinkern [GEOL] **1.** Rock material formed from consolidated mud or sediment that filled a hollow organic structure, such as a fossil shell. **2.** The fossil formed after dissolution of the mold. Also known as endocast; internal cast.

Stein-Leventhal syndrome [MED] A complex of symptoms characterized by amenorrhea or abnormal uterine bleeding or both, enlarged polycystic ovaries, frequently hirsutism, and occasionally retarded breast development.

Steinmann pin [MED] A surgical nail inserted in distal portions of such bones as the femur or tibia for skeletal tractions.

Steinmetz coefficient [ELECTROMAG] The constant of proportionality in Steinmetz's law.

Steinmetz's law [ELECTROMAG] The energy converted into heat per unit volume per cycle during a cyclic change of magnetization is proportional to the maximum magnetic induction raised to the 1.6 power, the constant of proportionality depending only on the material.

stele [BOT] The part of a plant stem including all tissues and regions of plants from the cortex inward, including the pericycle, phloem, cambium, xylem, and pith.

Stelenchopidae [INV ZOO] A family of polychaete annelids belonging to the Myzostomaria, represented by a single species from Crozet Island in the Antarctic Ocean.

stellar [ASTRON] Relating to or consisting of stars.

stellar association [ASTRON] A loose grouping of stars which may have had a common origin.

stellar atmosphere [ASTRON] The envelope of gas and plasma surrounding a star; consists of about 90% hydrogen atoms and 9% helium atoms, by number of atoms.

stellarator [PL PHYS] A device for confining a high-temperature plasma, consisting of a tube, which closes in on itself in a figure-eight or race-track configuration, and external coils which generate magnetic fields whose lines of force run parallel to the walls of the tube and prevent the plasma from touching the walls.

stellar crystal *See* plane-dendritic crystal.

stellar evolution [ASTROPHYS] The changes in spectrum and luminosity that take place in the life of a star.

stellar flare [ASTRON] Ejection of material from a star in an eruption that may last from a few minutes to an hour or more.

stellar guidance [NAV] Guidance by means of celestial bodies, particularly the stars.

stellar inertial guidance [NAV] The guidance of a flight-borne vehicle by a combination of celestial and inertial devices and techniques.

stellar interferometer [OPTICS] An optical interferometer for measuring angular diameters of stars; it is attached to a telescope and measures interference rings at the telescope's focus.

stellar lightning [GEOPHYS] Lightning consisting of several flashes seeming to radiate from a single point.

stellar luminosity [ASTRON] A star's brightness; it is measured either in ergs per second or in units of solar luminosity or in absolute magnitude.

stellar magnetic field [ASTROPHYS] A magnetic field, generally stronger than the earth's magnetic field, possessed by many stars.

stellar magnitude *See* magnitude.

stellar mass [ASTROPHYS] The mass of a star, usually expressed in terms of the sun's mass.

stellar model [ASTROPHYS] A mathematical characterization of the internal properties of a star. Also known as star model.

stellar nevus *See* spider nevus.

stellar parallax [ASTRON] The subtended angle at a star formed by the mean radius of the earth's orbit; it indicates distance to the star.

stellar photometry [ASTRON] The measurement of the brightness of stars.

stellar population [ASTRON] Either of two classes of stars, termed population I and population II; population I are relatively young stars, found in the arms of spiral galaxies, especially the blue stars of high luminosity; population II stars are the much older, more evolved stars of lower metallic content; many high luminosity red giants and many variable stars are members of population II.

stellar pulsation [ASTROPHYS] Expansion of a star followed by contraction so that its surface temperature and intrinsic brightness undergo periodic variation.

stellar rotation [ASTRON] Axial rotation of stars; surface rotational equatorial velocities of stars range from a few to 500 kilometers per second.

stellar scintillation *See* astronomical scintillation.

stellar spectroscopy [ASTRON] The techniques of obtaining spectra of stars and their study.

stellar spectrum [ASTRON] The spectrum of a star normally obtained with a slit spectrograph by black-and-white photography; the spectrum of a star in a large majority of cases shows absorption lines superposed on a continuous background.

stellar structure [ASTROPHYS] The mathematical study of a rotating, chemically homogeneous mass of gas held together by its own gravitation; a representative model of the observable properties of a star; thermonuclear reactions are postulated to be the main source of stellar energy.

stellar system [ASTRON] A gravitational system of stars.

stellar temperature [ASTROPHYS] Any temperature above several million degrees, such as occurs naturally in the interior of the sun and other stars.

stellate ganglion [ANAT] The ganglion formed by the fusion of the inferior cervical and the first thoracic sympathetic ganglions.

stellate reticulum [HISTOL] The part of the epithelial dental organ of a developing tooth which lies between the inner and the outer dental epithelium; composed of stellate cells with long, anastomosing processes in a mucoid fluid in the interstitial spaces.

Stelleroidea [INV ZOO] The single class of echinoderms in the subphylum Asterozoa; characters coincide with those of the subphylum.

Stellite [MET] A trademark used for the hard, wear-resistant and corrosion-resistant family of alloys containing 20-65% cobalt and 11-32% chromium, and 2-5% tungsten, with or without other metals.

stem [BOT] The organ of vascular plants that usually develops branches and bears leaves and flowers. [ENG] **1.** The heavy iron rod acting as the connecting link between the bit and the balance of the string of tools on a churn rod. **2.** To insert packing or tamping material in a shothole. [NAV] To make headway against an obstacle, as a current. [NAV ARCH] The foremost part of a ship's hull.

stem bag [MIN ENG] A fire-resisting paper bag filled with dry sand for stemming shotholes.

stem blight [PL PATH] Any of various fungus blights that affect the plant stem.

stem break *See* browning.

stem canker [PL PATH] Any canker disease affecting the stem.

stem cell [EMBRYO] A formative cell. [HISTOL] *See* hemocytoblast.

stem-cell leukemia [MED] A form of leukemia in which the predominant cell type is so poorly differentiated that its series cannot be identified.

stem correction [THERMO] A correction which must be made in reading a thermometer in which part of the stem, and the

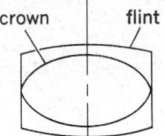

STEINHEIL LENS

crown   flint

Schematic of Steinheil aplanatic magnifier showing the crown lens within the flint lenses.

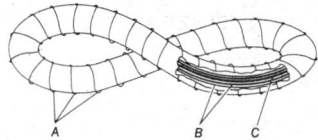

STELLARATOR

Diagram of a stellarator. External, closely spaced, current-carrying windings *A* produce a magnetic field whose lines of force *B* run parallel to the walls, confining plasma *C* in tube's center.

thermometric fluid within it, is at a temperature which differs from the temperature being measured.

stemming rod [ENG] A nonmetallic rod used to push explosive cartridges into position in a shothole and to ram tight the stemming.

Stemonitaceae [MYCOL] The single family of the order Stemonitales.

Stemonitales [MYCOL] An order of fungi in the subclass Myxogastromycetidae of the class Myxomycetes.

stem rust [PL PATH] Any of several fungus diseases, especially of grasses, affecting the stem and marked by black or reddish-brown lesions.

stem-winding [MECH ENG] Pertaining to a timepiece that is wound by an internal mechanism turned by an external knob and stem (the winding button of a watch).

stench *See* odorant.

stencil [GRAPHICS] A means of applying a pattern with ink or paint; the stencil itself is cut from thin metal or cardboard sheet, and the paint or ink is brushed or sprayed into the open areas.

stencil printing *See* screen printing.

Stenetrioidea [INV ZOO] A group of isopod crustaceans in the suborder Asellota consisting mostly of tropical marine forms in which the first pleopods are fused.

Stengel process [CHEM ENG] Manufacture of ammonium nitrate fertilizer from nitric acid and anhydrous ammonia.

Stenocephalidae [INV ZOO] A family of Old World, neotropical Hemiptera included in the Pentatomorpha.

stenocephaly [MED] Unusual narrowness of the head.

stenode circuit [ELECTR] Superheterodyne receiving circuit in which a piezoelectric unit is used in the intermediate-frequency amplifier to balance out all frequencies except signals at the crystal frequency, thereby giving very high selectivity.

Stenoglossa [INV ZOO] The equivalent name for Neogastropoda.

stenohaline [ECOL] In marine organisms, indicating the ability to tolerate only a narrow range of salinities.

Stenolaemata [INV ZOO] A class of marine ectoproct bryozoans having lophophores which are circular in basal outline and zooecia which are long, slender, tubular or prismatic, and gradually tapering to their proximal ends.

Stenomasteridae [PALEON] An extinct family of Euchinoidea in the order Holasteroida comprising oval and heart-shaped forms with fully developed pore pairs.

stenometer [ENG] An instrument for measuring distances; employs a telescope in which two target images a known distance apart are superimposed by turning a micrometer screw.

Stenopodidea [INV ZOO] A section of decapod crustaceans in the suborder Natantia which includes shrimps having the third pereiopods chelate and much longer and stouter than the first pair.

stenosis [MED] Constriction or narrowing, as of the heart or blood vessels.

Stenostomata [INV ZOO] The equivalent name for Cyclostomata.

Stenothoidae [INV ZOO] A family of amphipod crustaceans in the suborder Gammaridea containing semiparasitic and commensal species.

Stensen's duct *See* parotid duct.

Stensioellidae [PALEON] A family of Lower Devonian placoderms of the order Petalichthyida having large pectoral fins and a broad subterminal mouth.

Stenurida [PALEON] An order of Ophiuroidea, comprising the most primitive brittlestars, known only from Paleozoic sediments.

step [ADP] A single computer instruction or operation. [ENG] A small offset on a piece of core or in a drill hole resulting from a sudden sidewise deviation of the bit as it enters a hard, tilted stratum or rock underlying a softer rock. [GEOL] A hitch or dislocation of the strata. [MIN ENG] The portion of a longwall face at right angles to the line of the face formed when a place is worked in front of or behind an adjoining place.

step aeration [CIV ENG] An activated sludge process in which the settled sewage is introduced into the aeration tank at more than one point.

step-and-repeat camera [OPTICS] A type of camera providing a gridlike pattern of latent image frames in a given sequence.

step attenuator [ELECTR] An attenuator in which the attenuation can be varied in precisely known steps by means of switches.

step bearing [MECH ENG] A device which supports the bottom end of a vertical shaft. Also known as pivot bearing.

step-by-step switch [ELEC] A bank-and-wiper switch in which the wipers are moved by electromagnet ratchet mechanisms individual to each switch.

step-by-step system [COMMUN] *See* Strowger system. [CONT SYS] A control system in which the drive motor moves in discrete steps when the input element is moved continuously.

step-climb profile [AERO ENG] The aircraft climbs a specified number of feet whenever its weight reaches a predetermined amount, thus stepping to an optimum altitude as gross weight decreases.

step counter [ADP] In computers, a counter in the arithmetic unit used to count the steps in multiplication, division, and shift operations.

step cut [LAP] A cut for gems in which the facets decrease in length the farther they are from the girdle, giving the appearance of steps; an example is the emerald cut.

step-down transformer [ELEC] A transformer in which the alternating-current voltages of the secondary windings are lower than those applied to the primary winding.

step fault [GEOL] One of a set of closely spaced, parallel faults. Also known as distributive fault; multiple fault.

step function [MATH] A function f defined on an interval $[a,b]$ so that $[a,b]$ can be partitioned into a finite number of subintervals on each of which f is a constant. Also known as simple function.

step gage [DES ENG] 1. A plug gage containing several cylindrical gages of increasing diameter mounted on the same axis. 2. A gage consisting of a body in which a blade slides perpendicularly; used to measure steps and shoulders.

Stephanidae [INV ZOO] A small family of the Hymenoptera in the superfamily Ichneumonoidea characterized by many-segmented filamentous antennae.

stephanite [MINERAL] Ag_5SbS_4 An iron-black mineral crystallizing in the orthorhombic system and having a metallic luster; an ore of silver. Also known as black silver; brittle silver ore; goldschmidtine.

step-out well [PETRO ENG] A well drilled at a later time over remote, undeveloped portions of a partially developed continuous reservoir rock. Also known as delayed development well.

steppe [GEOGR] An extensive grassland in the semiarid climates of southeastern Europe and Asia; it is similar to but more arid than the prairie of the United States.

steppe climate [CLIMATOL] The type of climate in which precipitation though very slight, is sufficient for growth of short, sparse grass; typical of the steppe regions of south-central Eurasia. Also known as semiarid climate.

stepped footing [CIV ENG] A widening at the bottom of a wall consisting of a series of steps in the proportion of one horizontal to two vertical units.

stepped gear wheel [DES ENG] A gear wheel containing two or more sets of teeth on the same rim, with adjacent sets slightly displaced to form a series of steps.

stepped leader [GEOPHYS] The initial streamer of a lightning discharge; an intermittently advancing column of high ion density which established the channel for subsequent return streamers and dart leaders.

stepped screw [DES ENG] A screw from which sectors have been removed, the remaining screw surfaces forming steps.

stepper motor [ELEC] A motor that rotates in short and essentially uniform angular movements rather than continuously; typical steps are 30, 45, and 90°; the angular steps are obtained electromagnetically rather than by the ratchet and pawl mechanisms of stepping relays. Also known as magnetic stepping motor; stepping motor; step-servo motor.

stepping *See* zoning.

STEPPER MOTOR

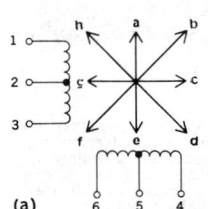

Rotor position	Windings to which voltage is applied
a	V_{1-2}
b	V_{1-2} and V_{4-5}
c	V_{4-5}
d	V_{4-5} and V_{3-2}
e	V_{3-2}
f	V_{3-2} and V_{6-5}
g	V_{6-5}
h	V_{6-5} and V_{1-2}

Permanent-magnet stepper motor. *(a)* Circuit diagram. *(b)* Excitation sequence for 45°-per-step clockwise rotation.

stepping motor *See* stepper motor.

stepping reflex [PHYSIO] A reflex response of the newborn and young infant, characterized by alternating stepping movements with the legs, as in walking, elicited when the infant is held upright so that both soles touch a flat surface while the infant is moved forward to accompany any step taken.

stepping relay [ELEC] A relay whose contact arm may rotate through 360° but not in one operation. Also known as rotary stepping relay; rotary stepping switch; stepping switch.

stepping switch *See* stepping relay.

step pulley [MECH ENG] A series of pulleys of various diameters combined in a single concentric unit and used to vary the velocity ratio of shafts. Also known as cone pulley.

step-recovery diode [ELECTR] A varactor in which forward voltage injects carriers across the junction, but before the carriers can combine, voltage reverses and carriers return to their origin in a group; the result is abrupt cessation of reverse current and a harmonic-rich waveform.

step response [CONT SYS] The behavior of a system when its input signal is zero before a certain time and is equal to a constant nonzero value after this time.

step rocket *See* multistage rocket.

step-servo motor *See* stepper motor.

step strobe marker [ELECTR] Form of strobe marker in which the discontinuity is in the form of a step in the time base.

step tablet *See* density step tablet.

step test [GRAPHICS] A series of exposures made to determine the optimum exposures of either film or paper prints.

step trench [ARCHEO] A trench cut in a mound in a series of steps.

step-up transformer [ELEC] Transformer in which the energy transfer is from a low-voltage winding to a high-voltage winding or windings.

step voltage regulator [ELEC] A type of voltage regulator used on distribution feeder lines; it provides increments or steps of voltage change.

step wedge [GRAPHICS] A series of tones in steps from white to black usually on film or a glass plate for testing purposes.

sterad *See* steradian.

steradian [MATH] The unit of measurement for solid angles; it is equal to the solid angle subtended at the center of a sphere by a portion of the surface of the sphere whose area equals the square of the sphere's radius. Abbreviated sr; sterad.

steradiancy *See* radiance.

sterba curtain [ELECTROMAG] Type of stacked dipole antenna array consisting of one or more phased half-wave sections with a quarter-wave section at each end; the array can be oriented for either vertical or horizontal radiation, and can be either center or end fed.

stercobilin [BIOCHEM] Urobilin as a component of the brown fecal pigment.

stercobilinogen [BIOCHEM] A colorless reduction product of stercobilin found in feces.

Stercorariidae [VERT ZOO] A family of predatory birds of the order Charadriiformes including the skuas and jaegers.

Sterculiaceae [BOT] A family of dicotyledonous trees and shrubs of the order Malvales distinguished by imbricate or contorted petals, bilocular anthers, and ten to numerous stamens arranged in two or more whorls.

stère [MECH] A unit of volume equal to 1 cubic meter; it is used mainly in France, and in measuring timber volumes.

steregon [MATH] The entire solid angle bounded by a sphere; equal to 4π steradians.

stereo *See* stereophonic; stereo sound system.

stereo- [PHYS] A prefix used to designate a three-dimensional characteristic.

stereo amplifier [ENG ACOUS] An audio-frequency amplifier having two or more channels, as required for use in a stereo sound system.

stereoblastula [EMBRYO] A blastula that lacks a cavity, making it unable to gastrulate.

stereo broadcasting [COMMUN] Broadcasting two sound channels for reproduction by a stereo sound system having a stereo tuner at its input.

stereo camera *See* stereoscopic camera.

stereochemistry [PHYS CHEM] The study of the spatial arrangement of atoms in molecules and the chemical and physical consequences of such arrangement.

stereocilia [CYTOL] **1.** Nonmotile tufts of secretory microvilli on the free surface of cells of the male reproductive tract. **2.** Homogeneous cilia within simple membrane coverings; found on the free-surface hair cells.

stereocomparagraph [OPTICS] A projection device in which two-dimensional aerial photographs taken at slightly different angles are combined so as to give the appearance of tridimensionality.

stereo comparator [OPTICS] An instrument that may be used to view two photographs taken of the stars in the same section of sky at different times; viewing the images stereoscopically may reveal stars that have moved between exposures or stars of varying brightness.

stereo effect [ACOUS] Reproduction of sound in such a manner that the listener receives the sensation that individual sounds are coming from different locations, just as did the original sounds reaching the stereo microphone system.

stereofluoroscopy [ELECTR] A fluoroscopic technique that gives three-dimensional images.

stereogastrula [EMBRYO] A gastrula that lacks a cavity.

stereognomogram [CRYSTAL] The projection resulting from the superposition of the projection planes of a stereogram and a gnomogram.

stereognosis [PSYCH] The recognition or identification of objects by the sense of touch.

stereogram [GRAPHICS] A stereoscopic set of photographs or drawings correctly oriented and mounted for stereoscopic viewing.

stereographic chart [MAP] A chart on the stereographic projection.

stereographic coverage [GRAPHICS] Photographic coverage with overlapping air photographs to provide a three-dimensional presentation of the picture; 60% overlap is considered normal, and 53% is generally regarded as the minimum.

stereographic net *See* net.

stereographic projection [MAP] A perspective conformal, azimuthal map projection in which points on the surface of a sphere or spheroid, such as the earth, are conceived as projected by radial lines from any point on the surface to a plane tangent to the antipode of the point of projection; circles project as circles through the point of tangency, except for great circles which project as straight lines; the principal navigational use of the projection is for charts of the polar regions. Also known as azimuthal orthomorphic projection. [MATH] The projection of the Riemann sphere onto the euclidean plane performed by emanating rays from the north pole of the sphere through a point on the sphere.

stereoisomers [ORG CHEM] Compounds whose molecules have the same number and kind of atoms and the same atomic arrangement, but differ in their spatial relationship.

stereomicrography [OPTICS] The taking of two microphotographs of the same field at different angles (a stereo pair), then viewing them simultaneously with a stereo viewer.

stereomicrometer [ENG] An instrument attached to an optical instrument (such as a telescope) to measure small angles.

stereo multiplex [COMMUN] Stereo broadcasting by a frequency-modulation station, in which the output of two microphones is transmitted on the same carrier by frequency-division multiplexing.

stereophonic [ENG ACOUS] Pertaining to three-dimensional pickup or reproduction of sound, as achieved by using two or more separate audio channels. Also known as stereo.

stereophonics [ENG ACOUS] The study of reproducing or reinforcing sound in such a way as to produce the sensation that the sound is coming from sources whose spatial distribution is similar to that of the original sound sources.

stereophonic sound system *See* stereo sound system.

stereo pickup [ENG ACOUS] A phonograph pickup designed for use with standard single-groove two-channel stereo records; the pickup cartridge has a single stylus that actuates two elements, one responding to stylus motion at 45° to the right of vertical and the other responding to stylus motion at 45° to the left of vertical.

stereoplanigraph [ENG] An instrument for drawing topo-

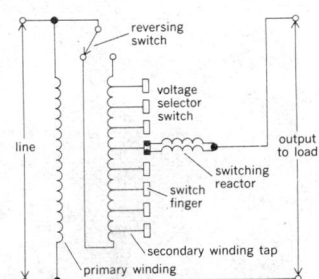
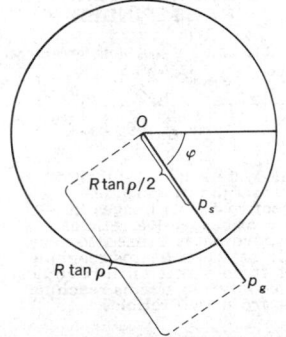

graphic maps from observations of stereoscopic aerial photographs with a stereocomparator.

stereo power [OPTICS] The magnifying power of a binoculars or other stereo system multiplied by the ratio of the distance between the objective axes to the distance between the eyepiece axes; it is a measure of the stereoscopic radius.

stereo preamplifier [ENG ACOUS] An audio-frequency preamplifier having two channels, used in a stereo sound system.

stereopsis *See* stereoscopy.

stereo rangefinder *See* stereoscopic rangefinder.

stereo record [ENG ACOUS] A single-groove disk record having V-shaped grooves at 45° to the vertical; each groove wall has one of the two recorded channels. Also known as stereo recording.

stereo recorded tape [ENG ACOUS] Recorded magnetic tape having two separate recordings, one for each channel of a stereo sound system. Also known as stereo recording.

stereo recording *See* stereo record; stereo recorded tape.

stereoregular polymer *See* stereospecific polymer.

stereorubber [ORG CHEM] Synthetic rubber, *cis*-polyisoprene, a polymer with stereospecificity.

stereoscope [OPTICS] An optical instrument in which each eye views one of two photographs taken with the camera or object of study displaced, or simultaneously with two cameras, or with a stereoscopic camera, so that a sensation of depth is produced.

stereoscopic camera [OPTICS] A camera which takes photographs simultaneously with two similar lenses a few inches apart, for use in a stereoscope or other optical system which gives a sensation of depth to the viewer. Also known as stereo camera.

stereoscopic heightfinder *See* stereoscopic rangefinder.

stereoscopic model [GRAPHICS] A mental impression of an area or object seen as being in three dimensions when viewed stereoscopically on photographs.

stereoscopic pair [GRAPHICS] Two photographs with sufficient overlap of detail to make possible stereoscopic examination of an object or an area common to both.

stereoscopic parallax *See* absolute stereoscopic parallax.

stereoscopic photography [OPTICS] A technique that simulates stereoscopic vision, in which two photographs are made with the camera or object of study displaced, or simultaneously with two cameras, or with a stereoscopic camera, and each of the photographs is viewed by one eye, using a stereoscope or other optical system.

stereoscopic radius [PHYSIO] The greatest distance at which there is a sensation of depth in vision due to the fact that the two eyes do not perceive exactly the same view.

stereoscopic rangefinder [OPTICS] An optical rangefinder which utilizes stereoscopic vision; it is essentially a large binoculars fitted with special reticles which allow a skilled user to superimpose the stereoscopic image formed by the pair of reticles over the images of the target seen in the eyepieces, so that the correct range is obtained when the reticle marks appear to be suspended over the target and at the same apparent distance. Also known as stereo rangefinder; stereoscopic heightfinder.

stereoscopic vision *See* stereoscopy.

stereoscopy [PHYSIO] The phenomenon of simultaneous vision with two eyes in which there is a vivid perception of the distances of objects from the viewer; it is present because the two eyes view objects in space from two points, so that the retinal image patterns of the same object are slightly different in the two eyes. Also known as stereopsis; stereoscopic vision.

stereo sound system [ENG ACOUS] A sound reproducing system in which a stereo pickup, stereo tape recorder, stereo tuner, or stereo microphone system feeds two independent audio channels, each of which terminates in one or more loudspeakers arranged to give listeners the same audio perspective that they would get at the original sound source. Also known as stereo; stereophonic sound system.

stereospecificity [ORG CHEM] The condition of a polymer whose molecular structure has a fixed spatial (geometric) arrangement of its constituent atoms, thus having crystalline properties; for example, synthetic natural rubber, *cis*-polyisoprene.

stereospecific polymer [ORG CHEM] A polymer with specific or definite order of arrangement of molecules in space, as in isotactic polypropylene; permits close packing of molecules and leads to a high degree of polymer crystallinity. Also known as stereoregular polymer.

stereospecific synthesis [ORG CHEM] Catalytic polymerization of monomer molecules to produce stereospecific polymers, as with Ziegler or Natta catalysts (derived from a transition metal halide and a metal alkyl).

Stereospondyli [PALEON] A group of labyrinthodont amphibians from the Triassic characterized by a flat body without pleurocentra and with highly developed intercentra.

stereo subcarrier [COMMUN] A subcarrier whose frequency is the second harmonic of the pilot subcarrier frequency used in frequency-modulation stereo broadcasting.

stereo tape recorder [ENG ACOUS] A magnetic-tape recorder having two stacked playback heads, used for reproduction of stereo recorded tape.

stereotaxis [BIOL] An orientation movement in response to stimulation by contact with a solid body. Also known as thigmotaxis.

stereotropism [BIOL] Growth or orientation of a sessile organism or part of an organism in response to the stimulus of a solid body. Also known as thigmotropism.

stereo tuner [ENG ACOUS] A tuner having provisions for receiving both channels of a stereo broadcast.

stereotype [GRAPHICS] A duplicate printing plate made from type and cuts; a paper matrix, or mat, is forced down over the type and cuts to form a mold, into which molten metal is poured, resulting in a new metal printing surface that exactly duplicates the original.

sterhydraulic [MECH ENG] Pertaining to a hydraulic press in which motion or pressure is produced by the introduction of a solid body into a cylinder filled with liquid.

steric anomaly *See* specific-volume anomaly.

steric effect [PHYS CHEM] The influence of the spatial configuration of reacting substances upon the rate, nature, and extent of reaction.

steric hindrance [CHEM] The prevention or retardation of chemical reaction because of neighboring groups on the same molecule; for example, ortho-substituted aromatic acids are more difficult to esterify than are the meta and para substitutions.

stericooling [FOOD ENG] A method of cooling or precooling fruit and vegetables prior to shipping; uses a cooling spray of water containing, in addition to salt in solution to lower freezing point, a fungicide or bactericide.

sterigma [BOT] A peg-shaped structure to which needles are attached in certain conifers. [MYCOL] A slender stalk arising from the basidium of some fungi, from the top of which basidiospores are formed by abstriction.

sterile distribution [ECOL] A range of areas in which marine animals may live and spawn, but in which eggs do not hatch and larvae do not survive.

sterility [PHYSIO] The inability to reproduce because of congenital or acquired reproductive system disorders involving lack of gamete production or production of abnormal gametes.

sterilization [MICROBIO] An act or process of destroying all forms of microbial life on and in an object.

sterilizer [ENG] An apparatus for sterilizing by dry heat, steam, or water.

sterling silver [MET] A silver alloy having a defined standard of purity of 92.5% silver and the remaining 7.5% usually of copper.

stern [NAV ARCH] The aftermost part of a ship.

Sternaspidae [INV ZOO] A monogeneric family of polychaete annelids belonging to the Sedentaria.

stern attack [AERO ENG] In air intercept, an attack by an interceptor aircraft which terminates with a heading crossing angle of 45° or less.

sternbergite [MINERAL] $AgFe_2S_3$ A dark-brown or black mineral composed of silver iron sulfide and occurring as tabular crystals or flexible laminae.

sternebra [VERT ZOO] A segment of the sternum in vertebrates.

stern frame [NAV ARCH] 1. The timbers making up the upper

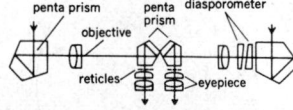

part of the stern of a wooden ship. **2.** A casting or forging, including the propeller post, sternpost, arch, and solepiece of a steel ship.

Stern-Gerlach effect [ATOM PHYS] The splitting of a beam of atoms passing through a strong, inhomogeneous magnetic field into several beams.

Sterninae [VERT ZOO] A subfamily of birds in the family Laridae, including the Arctic tern.

sternite [INV ZOO] **1.** The ventral part of an arthropod somite. **2.** The chitinous plate on the ventral surface of an abdominal segment of an insect.

stern layer [PHYS CHEM] One of two electrically charged layers of electrolyte ions, the layer of ions immediately adjacent to the surface, in the neighborhood of a negatively charged surface.

sternocleidomastoid [ANAT] A muscle of the neck that flexes the head; origin is the manubrium of the sternum and clavicle, and insertion is the mastoid process.

sternocostal [ANAT] Pertaining to the sternum and the ribs.

sternohyoid [ANAT] A muscle arising from the manubrium of the sternum and inserted into the hyoid bone.

Sternorrhyncha [INV ZOO] A series of the insect order Homoptera in which the beak appears to arise either between or behind the fore coxae, and the antennae are long and filamentous with no well-differentiated terminal setae.

sternothyroid [ANAT] Pertaining to the sternum and thyroid cartilage.

Sternoxia [INV ZOO] The equivalent name for Elateroidea.

stern post [NAV ARCH] The main member of the after end of a ship, usually upright, running from the keel up to the bottom of the hull.

stern tube [NAV ARCH] **1.** A long, circular bearing or bushing which supports the propeller shaft where it emerges from the stern of a ship. **2.** A torpedo tube at the stern of a ship.

sternum [ANAT] The bone, cartilage, or series of bony or cartilaginous segments in the median line of the anteroventral part of the body of vertebrates above fishes, connecting with the ribs or pectoral girdle.

sternum height [ANTHRO] Vertical distance taken from the lower tip of the sternum to the floor as the subject stands.

sternway [NAV] Making way through the water in a direction opposite to the heading.

stern wheel [NAV ARCH] A paddle wheel located at the vessel's stern and used to propel the vessel.

Stern-Zartman experiment [STAT MECH] An experiment in which the distribution in speed of atoms or molecules in a beam emitted from an opening in an oven is measured by having the beam impinge on a rotating cylindrical drum, with a slit cut parallel to the drum axis, and measuring the density of atoms or molecules deposited on the inner surface of the drum, roughly opposite the slit, as a function of distance from a point directly opposite the slit; it is used to test the Maxwell-Boltzmann distribution law.

steroid [BIOCHEM] A compound composed of a series of four carbon rings joined together to form a structural unit called cyclopentanoperhydrophenanthrene.

sterol [BIOCHEM] Any of the natural products derived from the steroid nucleus; all are waxy, colorless solids soluble in most organic solvents but not in water, and contain one alcohol functional group.

sterrometal [MET] Hard brass containing a small amount of iron and manganese; used for hydraulic cylinders and marine castings.

Stetefeldt furnace [MET] A furnace for desulfurizing and chloridizing silver ores; the ores are powdered, mixed with salt, and dropped through a hot atmosphere.

stethoscope [MED] An instrument for indirect auscultation for the detection and study of sounds arising within the body; sounds are conveyed to the ears of the examiner through rubber tubing connected to a funnel or disk-shaped endpiece.

stewartite [GEOL] A steel-gray, iron-containing variety of bort that has magnetic properties. [MINERAL] $Mn_3(DO)_2 \cdot 4H_2O$ A brownish-yellow mineral composed of hydrous manganese phosphate occurring in minute crystals or fibrous tufts in pegmatites.

sthène [MECH] The force which, when applied to a body whose mass is 1 metric ton, results in an acceleration of 1 meter per second per second; equal to 1000 newtons. Formerly known as funal.

Sthenurinae [PALEON] An extinct subfamily of marsupials of the family Diprotodontidae, including the giant kangaroos.

stibide *See* antimonide.

stibium [CHEM] The Latin name for antimony, thus the symbol Sb for the element.

stibnate *See* potassium antimonate.

stibnite *See* antimonite.

stibophen [PHARM] $C_{12}H_4Na_5O_{16}S_4Sb \cdot 7H_2O$ A crystalline compound that is soluble in water, insoluble in organic solvents; used in medicine for protozoan infections. Also known as sodium antimony bis(pyrocatechol-2,4-disulfonate).

Stichaeidae [VERT ZOO] The pricklebacks, a family of perciform fishes in the suborder Blennioidei.

Stichocotylidae [INV ZOO] A family of trematodes in the subclass Aspidogastrea in which adults are elongate and have a single row of alveoli.

Stichopodidae [INV ZOO] A family of the echinoderm order Aspidochirotida characterized by tentacle ampullae and by left and right gonads.

stichtite [MINERAL] $Mg_6Cr_2(CO_3)(OH)_{16} \cdot 4H_2O$ A lilac-colored rhombohedral mineral composed of hydrous basic carbonate of magnesium and chromium.

stick [ENG] **1.** A rigid bar hinged to the boom of a dipper or pull shovel and fastened to the bucket. **2.** A long slender tool bonded with an abrasive for honing or sharpening tools and for dressing of wheels. [ORD] A succession of missiles fired or released separately at predetermined intervals from a single aircraft.

stick gage [ENG] A suitably divided vertical rod, or stick, anchored in an open vessel so that the magnitude of rise and fall of the liquid level may be observed directly.

sticking [ADP] In computers, the tendency of a flip-flop to remain in, or to spontaneously switch to, one of its two stable states.

Stickland reaction [BIOCHEM] An amino acid fermentation involving the coupled decomposition of two or more substrates.

stickleback [VERT ZOO] Any fish which is a member of the family Gasterosteidae, so named for the variable number of free spines in front of the dorsal fin.

stick shellac [MATER] Shellac in the form of a solid stick and in a variety of colors; used for filling imperfections in wood.

sticky charge *See* sticky grenade.

sticky grenade [ORD] A small explosive charge covered with an adhesive, intended to be thrown or placed by hand where the adhesive will hold the charge in place until detonated by a time fuse. Also known as sticky charge.

stiction [MECH] Friction that tends to prevent relative motion between two movable parts at their null position.

Stieltjes integral [MATH] The Stieltjes integral of a real function $f(x)$ relative to a real function $g(x)$ of bounded variation on an interval $[a,b]$ is defined, analogously to the Riemann integral, as a limit of a sum of terms $f(a_i) [g(x_i) - g(x_{i-1})]$ taken as partitions of the interval shrink. Denoted

$$\int_a^b f(x)dg(x).$$

Stieltjes transform [MATH] A form of the Laplace transform of a function where the usual Riemann integral is replaced by a Stieltjes integral.

stiffened-shell fuselage *See* semimonocoque.

stiffener [CIV ENG] A steel angle or plate attached to a slender beam to prevent its buckling by increasing its stiffness.

stiffleg derrick [MECH ENG] A derrick consisting of a mast held in the vertical position by a fixed tripod of steel or timber legs. Also known as derrick crane; Scotch derrick.

stiffness [ACOUS] A characteristic of a medium equal to 2π times the acoustic reactance of a medium at a given frequency. Also known as acoustic stiffness. [MECH] The ratio of a steady force acting on a deformable elastic medium to the resulting displacement.

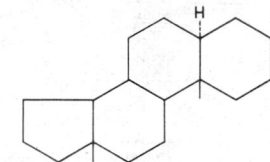

STEROID

Shorthand formulation for steroid skeleton; lines attached to the rings represent methyl groups.

STICKLEBACK

The three-spined stickleback. It is used for fish meal and oil in some Baltic countries.

STILBELLACEAE

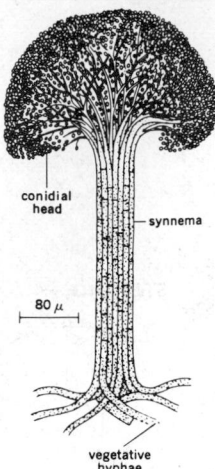

Graphium ulmi, cause of Dutch elm disease. (*After C. Ferdinandsen and C. A. Jörgensen, 1938–1939*)

STILBITE

Sheaflike aggregates of thin crystals of stilbite, found at Two Islands, Nova Scotia. (*American Museum of Natural History Specimens*)

STIRLING ENGINE

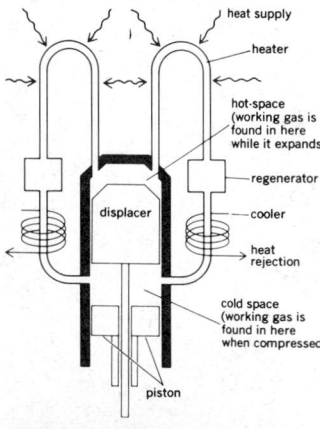

Diagram of Stirling engine, displacer type. Second piston, called displacer, transfers gas back and forth between hot space, at a fixed high temperature, and cold space, at a fixed low temperature.

stiffness coefficient [MECH] The ratio of the force acting on a linear mechanical system, such as a spring, to its displacement from equilibrium.

stiffness constant [MECH] Any one of the coefficients of the relations in the generalized Hooke's law used to express stress components as linear functions of the strain components. Also known as elastic constant.

stigma [BOT] The rough or sticky apical surface of the pistil for reception of the pollen. [INV ZOO] **1.** The eyespot of certain protozoans, such as *Euglena*. **2.** The spiracle of an insect or arthropod. **3.** A colored spot on many lepidopteran wings. [MECH] A unit of length used mainly in nuclear measurements, equal to 10^{-12} meter. Also known as bicron.

stigmatic [OPTICS] **1.** Property of an optical system whose focal power is the same in all meridians. **2.** *See* homocentric.

stigmatic concave grating [OPTICS] An optical element with many parallel grooves on a concave optical surface; combines the two functions of light dispersion and focusing in one dispersive element; used in space optics, in food and metal analysis, and as a dispersive element of spectrophotometers and spectrographs.

stigmatism [PHYSIO] A condition of the refractive media of the eye in which rays of light from a point are accurately brought to a focus on the retina.

stilb [OPTICS] A unit of luminance, equal to 1 candela per square centimeter. Abbreviated sb.

Stilbaceae [MYCOL] The equivalent name for Stilbellaceae.

Stilbellaceae [MYCOL] A family of fungi of the order Moniliales in which conidiophores are aggregated in long bundles or fascicles, forming synnemata or coremia, generally having the conidia in a head at the top.

stilbene [ORG CHEM] $C_6H_5CH{:}CHC_6H_5$ Colorless crystals soluble in ether and benzene, insoluble in water; melts at 124°C; used to make dyes and bleaches and as phosphors. Also known as diphenylethylene; toluylene.

stilbesterol *See* diethylstilbesterol.

stilbite [MINERAL] $Ca(Al_2Si_7O_{18}){\cdot}7H_2O$ A white, brown, or yellow mineral belonging to the zeolite family of silicates; crystallizes in the monoclinic system, occurs in sheaflike aggregates of tabular crystals, and has pearly luster; hardness is 3.5–4 on Mohs scale, and specific gravity is 2.1–2.2. Also known as desmine.

Stiles method [PETRO ENG] A technique for computing oil recovery by waterflood methods, taking into account the distribution of varying permeability stratums throughout the reservoir.

stiletto [ELECTR] An advanced electronic subsystem contained in United States strike aircraft type F-4D for detection, identification, and location of ground-based radars; the location of radar targets is determined by direction finding and passive ranging techniques; it is used for the delivery of guided and unguided weapons against the target radars under all weather conditions.

stillage [FOOD ENG] The residue grain from the manufacture of alcohol from grain; used as a feed supplement.

stillbirth [MED] Birth of a dead infant.

stilling basin [ENG] A depressed area in a channel or reservoir that is deep enough to reduce the velocity of the flow. Also known as stilling box.

stilling box *See* stilling basin.

stillingia oil [MATER] A combustible, toxic, pale-yellow drying oil with a linseed-oil scent and a mustard taste, derived from tallow tree seeds; used in lubricants, candles, textile dressing, and soap. Also known as tallow-seed oil.

Still's disease [MED] Juvenile rheumatoid arthritis in which involvement of the viscera is prominent. Also known as Chauffard-Still disease.

Stillson wrench [DES ENG] A trademark for an adjustable pipe wrench consisting of an L-shaped jaw in a sleeve, the sleeve being pivoted to a handle; pressure on the handle increases the grip of the jaw.

stillstand [GEOL] Referring to a land area, a continent, or an island, to remain stationary with respect to the interior of the earth or to sea level.

still water [HYD] A portion of a stream having a very slight gradient and no visible current.

still-water level [OCEANOGR] The level that the sea surface would assume in the absence of wind waves.

still wax *See* wax tailings.

still well [METEOROL] A device, used in evaporation pan measurements, which provides an undisturbed water surface and support for the hook gage; the U.S. Weather Bureau model consists of a brass cylinder, 8 inches (20.32 centimeters) high and 3.5 inches (8.89 centimeters) in diameter, mounted over a hole in a triangular galvanized iron base which is provided with leveling screws.

stilpnomelane [MINERAL] $K(Fe,Mg,Al)_3Si_4O_{10}(OH)_2{\cdot}H_2O$ A black or greenish-black mineral composed of basic hydrous potassium iron magnesium aluminum silicate; occurs as fibers, iron magnesium aluminum silicate; occurs as fibers, incrustations, and foliated plates.

stilt root [BOT] A prop root of a mangrove tree.

Stimson anchor [NAV ENG] A cast-iron anchor in the shape of an airfoil and bridled in such a way that the slightest horizontal force, combined with the weight of the anchor, tends to plow it into the soft sediments of the sea floor.

stimulated emission device [ELECTR] A device that uses the principle of amplification of electromagnetic waves by stimulated emission, namely, a maser or a laser.

stimulation deafness [MED] Deafness induced by noise; involves changes in the chemical interchange between the canals of the cochlea, as well as nerve destruction.

stimulation treatment [PETRO ENG] One of the techniques to increase (stimulate) oil- or gas-reservoir production, such as acidizing, fracturing, controlled underground explosions, or various cleaning techniques. Also known as well stimulation.

stimulator [MED] A neurosurgical device that supplies a controlled alternating-current voltage to two electrodes that are applied to a patient.

stimulus [CONT SYS] A signal that affects the controlled variable in a control system. [PHYSIO] An agent that produces a temporary change in physiological activity in an organism or in any of its parts.

stimulus filtering [PSYCH] The apparent awareness of only a few of the great number of stimuli bombarding an animal.

stinger [ZOO] A sharp piercing organ, as of a bee, stingray, or wasp, usually connected with a poison gland.

stinging cell *See* cnidoblast.

stingray [VERT ZOO] Any of various rays having a whiplike tail armed with a long serrated spine, at the base of which is a poison gland.

stink *See* gob stink.

stinkdamp [MIN ENG] The hydrogen sulfide that occurs in mines.

stinkstone [GEOL] A stone containing decomposing organic matter that gives off an offensive odor when rubbed or struck.

stipe [BOT] **1.** The petiole of a fern frond. **2.** The stemlike portion of the thallus in certain algae. [MYCOL] The short stalk or stem of the fruit body of a fungus, such as a mushroom.

stipoverite *See* stishovite.

stippling [GRAPHICS] Graduation of shading by numerous separate touches; shadow areas on charts, for instance, are sometimes indicated by numerous dots decreasing in density as the depth increases.

stipule [BOT] Either of a pair of appendages that are often present at the base of the petiole of a leaf.

stir-in resin [MATER] A vinyl resin that does not require grinding in order to disperse in a plastisol or organisol.

Stirling cycle [THERMO] A regenerative thermodynamic power cycle using two isothermal and two constant volume phases.

Stirling engine [MECH ENG] An engine in which work is performed by the expansion of a gas at high temperature; heat for the expansion is supplied through the wall of the piston cylinder.

stirling numbers [MATH] The coefficients which occur in the Stirling interpolation formula for a difference operator.

Stirling's formula [MATH] The expression $(n/e)^n\sqrt{2\pi n}$ is asymptotic to factorial n; that is, the limit as n goes to ∞ of their ratio is 1.

Stirling's interpolation formula [MATH] In interpolation theory, a formula to asymptotically realize a given difference operator by an infinite series.

Stirodonta [INV ZOO] Formerly, an order of Euechinoidea that included forms with stirodont dentition.

stirodont dentition [INV ZOO] In Echinoidea, the condition in which the teeth are keeled within and the foramen magnum is open.

stirring [PHYS] A turbulent process in which molecular diffusion and molecular heat conduction are speeded up.

stirring effect [ELECTROMAG] The circulation in a molten metal carrying electric current as a result of the combined forces of the pinch and motor effects.

stirrup [CIV ENG] In concrete construction, a U-shaped bar which is anchored perpendicular to the longitudinal steel as reinforcement to resist shear. [MIN ENG] **1.** A piece of steel hung from a gallows frame to engage the endgate hooks when a mine car is tilted over; used at dumps. **2.** A screw joint suspended from the brakestaff of a spring pole, by which the boring rods are adjusted to the depth of the borehole. Also known as temper screw.

stishovite [MINERAL] SiO_2 A polymorph of quartz, a dense, fine-grained mineral formed under very high pressure (about 1,000,000 pounds per square inch or 7×10^9 newtons per square meter); it is the only mineral in which the silicon atom has a coordination number of six; specific gravity is 4.28. Also known as stipoverite.

stitching [ENG] Progressive welding of thermoplastic materials (resins) by successive applications of two small, mechanically operated, radio-frequency-heated electrodes; the mechanism is similar to that of a normal sewing machine.

stitch rivet [ENG] One of a series of rivets joining the parallel elements of a structural member so that they act as a unit.

stitch welding [MET] A series of spaced spot resistance welds.

Stobbe reaction [ORG CHEM] A type of aldol condensation reaction represented by the reaction of benzophenone with dimethyl succinate and sodium methoxide to form monoesters of an α-alkylidene (or arylidene) succinic acid.

stochastic [MATH] Pertaining to random variables.

stochastic calculus [MATH] The mathematical theory of stochastic integrals and differentials, and its application to the study of stochastic processes.

stochastic chain rule [MATH] A generalization of the ordinary chain rule to stochastic processes; it states that the process $U_t = u(X_t^1, X_t^2, \ldots, X_t^n)$ satisfies

$$dU = \sum_i \partial_i u dX^i + \tfrac{1}{2} \sum_{i,j} \partial_i \partial_j u dX^i dX^j$$

with the conventions $(dt)^2 = 0$ and $dW^\alpha dW^\beta = \partial_{\alpha\beta} dt$, where the X^i are processes satisfying

$$dX^i = a_i^i dt + \sum_{\alpha=1}^{m} b_t^{i\alpha} dW_t^\alpha, i = 1, 2, \ldots, n;$$
$$\{W_t^\alpha, t \geq 0\}, \alpha = 1, 2, \ldots, m$$

are independent Wiener processes; the dW_t^α are the corresponding random disturbances occurring in the infinitesimal time interval dt; the a_t^i and $b_t^{i\alpha}$ are independent of future disturbances, and $u(x_1, x_2, \ldots, x_n)$ is a function whose derivatives $\partial_i u$ and $\partial_i \partial_j u$ are continuous. Also known as Itô's formula.

stochastic differential [MATH] An expression representing the random disturbances occurring in an infinitesimal time interval; it has the form dW_t, where $\{W_t, t \geq 0\}$ is a Wiener process.

stochastic integral [MATH] An integral used to construct the sample functions of a general diffusion process from those of a Wiener process; it has the form

$$\int_{W_0}^{W_s} a_t dW_t$$

where $\{W_t, t \geq 0\}$ is a Wiener process, dW_t represents the random disturbances occurring in an infinitesimal time interval dt, and a_t is independent of future disturbances. Also known as Itô's integral.

stochastic matrix [MATH] A square matrix with nonnegative real entries such that the sum of the entries of each row is equal to 1.

stochastic process [MATH] A family of random variables, dependent upon a parameter which usually denotes time. Also known as random process.

stock [GEOL] *See* pipe. [IND ENG] **1.** A product or material kept in storage until needed for use or transferred to some ultimate point for use, for example, crude oil tankage or paper-pulp feed. **2.** Designation of a particular material, such as bright stock or naphtha stock. [PETR] A usually discordant, batholithlike body of intrusive igneous rock not exceeding 40 square miles in surface exposure and usually discordant.

stock accounting [IND ENG] The establishment and maintenance of formal records of material in stock reflecting such information as quantities, values, or condition.

stockage objective [AERO ENG] The maximum quantities of material to be maintained on hand to sustain current operations; it will consist of the sum of stocks represented by the operating level and the safety level.

stock control [IND ENG] Process of maintaining inventory data on the quantity, location, and condition of supplies and equipment due in, on hand, and due out, to determine quantities of material and equipment available or required for issue and to facilitate distribution and management of material.

stock coordination [IND ENG] A supply management function exercised usually at department level which controls the assignment of material cognizance for items or categories of material to inventory managers.

stock-dye [TEXT] Dyeing fibers before they are spun into yarn.

stocking cutter [MECH ENG] **1.** A gear cutter having side rake or curved edges to rough out the gear-tooth spaces before they are formed by the regular gear cutter. **2.** A concave gear cutter ganged beside a regular gear cutter and used to finish the periphery of a gear blank by milling ahead of the regular cutter.

stock number [IND ENG] Number assigned to an item, principally to identify that item for storage and issue purposes.

stockpile [ENG] A reserve stock of material, equipment, raw material, or other supplies.

stock rail [CIV ENG] The fixed rail in a track, against which the switch rail operates.

stock record account [IND ENG] A basic record showing by item the receipt and issuance of property, the balances on hand, and such other identifying or stock control data as may be required by proper authority.

Stoddard solvent [MATER] A petroleum naphtha product with a comparatively narrow boiling range; used mostly for dry cleaning.

stoichiometry [PHYS CHEM] The numerical relationship of elements and compounds as reactants and products in chemical reactions.

stoke [FL MECH] A unit of kinematic viscosity, equal to the kinematic viscosity of a fluid with a dynamic viscosity of 1 poise and a density of 1 gram per cubic centimeter. Symbol St (formerly S). Also known as lentor (deprecated usage); stokes.

stoker [MECH ENG] A mechanical means, as used in a furnace, for feeding coal, removing refuse, controlling air supply, and mixing with combustibles for efficient burning.

stokes *See* stoke.

Stokes-Adams syndrome [MED] Syncopic or convulsive attacks occurring in patients with complete heart block.

Stokes drift [FL MECH] The drift of particles in a gravity wave, which arises from the fact that particle velocities are periodic with a mean which is not zero.

Stokes frequencies [OPTICS] Scattered (secondary) light in the Raman effect (when a high-intensity light beam passes through a transparent medium) that occurs at frequencies smaller than the frequency of the primary beam.

Stokes' integral theorem [MATH] The analog of Green's theorem in n-dimensional euclidean space; that is, a line integral of $F_1(x_1, x_2, \ldots, x_n)dx_1 + \ldots + F_n(x_1, x_2, \ldots, x_n)dx_n$ over a closed curve equals an integral of an expression containing various partial derivatives of $F_1, \ldots, F_n$ over a surface bounded by the curve.

stokesite [MINERAL] $CaSnSi_3O_9 \cdot 2H_2O$ A colorless orthorhombic mineral composed of hydrous calcium tin silicate occurring in crystals.

Stokes' law [FL MECH] At low velocities, the frictional force on a spherical body moving through a fluid at constant velocity is equal to 6π times the product of the velocity, the

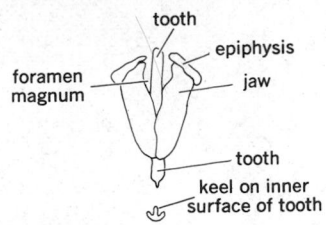

fluid viscosity, and the radius of the sphere. [SPECT] The wavelength of luminescence excited by radiation is always greater than that of the exciting radiation.

Stokes line [SPECT] A spectrum line in luminescent radiation whose wavelength is greater than that of the radiation which excited the luminescence, and thus obeys Stokes' law.

Stokes number 1 [FL MECH] A dimensionless number used in the study of the dynamics of a particle in a fluid, equal to the product of the dynamic viscosity of the fluid and the particle's vibration time, divided by the product of the fluid density and a characteristic length. Symbol St.

Stokes number 2 [ENG] A dimensionless number used in the calibration of rotameters, equal to $1.042\, m_f g \rho\, (1-\rho/\rho_f)R^3/\mu^2$, where ρ and μ are the density and dynamic viscosity of the fluid respectively, m_f and ρ_f are the mass and density of the float respectively, and R is the ratio of the radius of the tube to the radius of the float. Symbol St_2.

Stokes shift [SPECT] The displacement of spectral lines or bands of luminescent radiation toward longer wavelengths than those of the absorption lines or bands.

Stokes stream function [FL MECH] A one-component vector potential function used in analyzing and describing a steady, axially symmetric fluid flow; at any point it is equal to $1/2\pi$ times the mass rate of flow inside the surface generated by rotating the streamline on which the point is located about the axis of symmetry.

Stokes stretcher [MED] A basket-type stretcher constructed of tubular steel and strong wire mesh, and which acts as a splint for the entire body.

stoking [MET] Presintering or sintering a metal powder in such a way as to advance the compacts through the furnace at a fixed rate. Also known as continuous sintering.

STOL See short takeoff and landing.

STOL AIRCRAFT

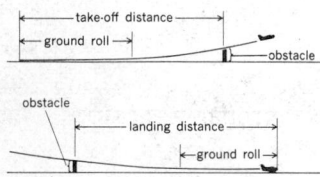

Takeoff and landing profile of STOL aircraft; obstacles are 15 meters high.

STOL aircraft [AERO ENG] Heavier-than-air craft that cannot take off and land vertically, but can operate within areas substantially more confined than those normally required by aircraft of the same size. Derived from short takeoff and landing aircraft.

stolon [BOT] See runner. [INV ZOO] An elongated projection of the body wall from which buds are formed giving rise to new zooids in Anthozoa, Hydrozoa, Bryozoa, and Ascidiacea. [MYCOL] A hypha produced above the surface and connecting a group of conidiophores.

Stolonifera [INV ZOO] An order of the Alcyonaria, lacking a coenenchyme; they form either simple (*Clavularia*) or rather complex colonies (*Tubipora*).

stolzite [MINERAL] $PbWO_4$ A tetragonal mineral composed of native lead tungstate; it is isomorphous with wulfenite and dimorphous with raspite.

stoma [BIOL] A small opening or pore in a surface. [BOT] One of the minute openings in the epidermis of higher plants which are regulated by guard cells and through which gases and water vapor are exchanged between internal spaces and the external atmosphere.

stomach [ANAT] The tubular or saccular organ of the vertebrate digestive system located between the esophagus and the intestine and adapted for temporary food storage and for the preliminary stages of food breakdown.

stomatitis [MED] Inflammation of the soft tissues in the mouth.

stomatoblastula [INV ZOO] A blastula stage in some sponges capable of engulfing maternal amebocytes for nutrition.

stomatology [MED] The branch of medical science that concerns the anatomy, physiology, pathology, therapeutics, and hygiene of the oral cavity, of the tongue, teeth, and adjacent structures and tissues, and of the relationship of that field to the entire body.

Stomatopoda [INV ZOO] The single order of the Eumalacostraca in the superorder Hoplocarida distinguished by raptorial arms, especially the second pair of maxillipeds.

Stomiatoidei [VERT ZOO] A suborder of fishes of the order Salmoniformes including the lightfishes and allies, which are of small size and often grotesque form and are equipped with photophores.

stomocnidae nematocyst [INV ZOO] A nematocyst which has an open-ended thread.

stomodaeum [EMBRYO] The anterior part of the embryonic alimentary tract formed as an invagination of the ectoderm.

STOLONIFERA

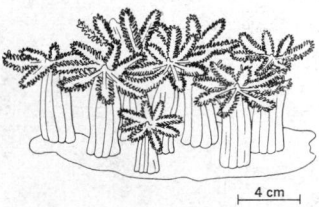

Clavularia garciae. (After Y. Delage)

stone [GEOL] 1. A small fragment of rock or mineral. 2. See stony meteorite. [LAP] A cut and polished natural gemstone. [MECH] A unit of mass in common use in the United Kingdom, equal to 14 pounds or 6.35029318 kilograms.

stoneboat [MIN ENG] A flat runnerless sled for transporting heavy material.

stone bubble See lithophysa.

stone canal [INV ZOO] A canal in many echinoderms that has a more or less calcified wall and that leads from the madreporite to the ring vessel.

stone cell See brachysclereid.

stone coal See anthracite.

stone dust [MIN ENG] Inert dust spread on roadways in coal mines as a defense against the danger of coal-dust explosions; effective because the stone dust absorbs heat.

stone-dust barrier [MIN ENG] A device erected in mine roadways to arrest explosions; consists of trays or Vee troughs loaded with stone dust, which are upset or overturned by the pressure wave in front of an explosion and the flame, producing a dense cloud of inert dust which blankets the flame and stops further propagation of the explosion.

stone fruit See drupe.

stone gobber [MIN ENG] In bituminous coal mining, one who removes stone and other refuse from coal mine floors and dumps the refuse into mine cars for disposal.

stone ice See ground ice.

stone polygon See sorted polygon.

stoner [FOOD ENG] A machine that removes stones and other undesirable material from coffee beans and other crops before the crops are processed further.

stone ring [GEOL] A ring of stones surrounding a central area of finer material; characteristic of sorted circle and sorted polygon.

Stone's representation theorem [MATH] This theorem determines the nature of all unitary representations of locally compact Abelian groups.

Stone's theorem [MATH] Every Boolean ring is isomorphic to a ring of subsets of some set.

stoneware [MATER] Vitrified ware with impermeable surface; used for corrosive materials in the laboratory and for some industrial operations.

Stone-Weierstrass theorem [MATH] Members from a collection of real-valued continuous functions on a compact space which form an algebra with certain properties can be used to approximate any continuous real-valued function on the space.

stonework [CIV ENG] A structure or the part of a structure built of stone.

stonewort [BOT] The common name for algae comprising the class Charophyceae, so named because most species are lime-encrusted.

stoney gate [CIV ENG] A crest gate which moves along a series of rollers traveling vertically in grooves in masonry piers, independently of the gate and piers.

stony coral [INV ZOO] Any coral characterized by a calcareous skeleton.

stony-iron meteorite [GEOL] Any of the rare meteorites containing at least 25% of both nickel-iron and heavy basic silicates. Also known as iron-stony meteorite; lithosiderite; sideraerolite; siderolite; syssiderite.

stony meteorite [GEOL] Any meteorite composed principally of silicate minerals, especially olivine, pyroxene, and plagioclase. Also known as aerolite; asiderite; meteoric stone; meteorolite; stone.

stooping [METEOROL] An atmospheric refraction phenomenon; a special case of sinking in which the curvature of light rays due to atmospheric refraction decreases with elevation so that the visual image of a distant object is foreshortened in the vertical.

stop [OPTICS] The aperture or useful opening of a lens, usually adjustable by means of a diaphragm.

stop and stay See absolute stop.

stop band See rejection band.

stop bath [GRAPHICS] When a negative or print is removed from the developer, it is usually placed in a stop bath to halt the action of the developer immediately; a common stop bath is a solution of 2 to 5% acetic or citric acid, or potassium metabisulfite.

stop bead [BUILD] A molding on the pulley stile of a window frame; forms one side of the groove for the inner sash.

stop bits [ADP] The last two bits transmitted in asynchronous data transmission to unequivocally indicate the end of a word.

stop cock [ENG] A small valve for stopping or regulating the flow of a fluid through a pipe.

stop down [OPTICS] To reduce the size of a lens aperture.

stope [MIN ENG] **1.** To excavate ore in a vein by driving horizontally upon it a series of workings, one immediately over the other, or vice versa; each horizontal working is called a stope because when a number of them are in progress, each working face under attack assumes the shape of a flight of stairs. **2.** Any subterranean extraction of ore except that which is incidentally performed in sinking shafts or driving levels for the purpose of opening the mine.

stope assay plan [MIN ENG] A plan that details assay value of ore exposures in a stope.

stope board [MIN ENG] A timber staging on the floor of a stope for setting a rock drill; the stage is tilted so that the bottom holes can be drilled in the same inclined direction.

stope fillings [MIN ENG] Broken waste material or low-grade matter from a lode or vein used to fill stopes on abandonment.

Stopehammer [MIN ENG] A trademark for an air-feed hammer drill.

stope hoist [MIN ENG] A small, portable, compressed-air hoist for operating a scraper-loader or for pulling heavy timbers into position, often used in narrow stopes.

stop element [COMMUN] The last element of a character in certain serial transmissions, used to ensure the recognition of the next start element.

stope pillar [MIN ENG] An ore column left in place to support the stope.

stoper *See* stoping drill.

stoping drill [MIN ENG] A small air or electric drill, usually mounted on an extensible column, for working stopes, raises, and narrow workings. Also known as stoper.

stoping ground [MIN ENG] Part of an ore body opened by drifts and raises, and ready for breaking down.

stoplog [CIV ENG] A log, plank, or steel or concrete beam that fits into a groove or rack between walls or piers to prevent the flow of water through an opening in a dam, conduit, or other channel.

stop nut [DES ENG] **1.** An adjustable nut that restricts the travel of an adjusting screw. **2.** A nut with a compressible insert that binds it so that a lock washer is not needed.

stopoff *See* resist.

stoppage [ORD] A jam in an automatic weapon; the condition of a weapon being jammed.

stopping [MIN ENG] A brattice, or more commonly, a masonry or brick wall built across old headings, chutes, or airways to confine the ventilating current to certain passages, or to lock up the gas in old workings, or to smother a mine fire. [NUCLEO] The decrease in kinetic energy of an ionizing particle as a result of energy losses along its path through matter.

stopping capacitor *See* coupling capacitor.

stopping off [MET] **1.** Local deposition of a protective coating, such as copper or fireclay, to prevent carburization, decarburization, or nitriding during heat treatment. **2.** Filling up a portion of the mold cavity to keep out molten metal. **3.** Applying a nonconducting layer to avoid electrodeposition in certain areas.

stopping potential [ELECTR] Voltage required to stop the outward movement of electrons emitted by photoelectric or thermionic action.

stopping power [NUCLEO] The energy lost by a charged particle passing through a substance per unit length of path; related concepts are mass, atomic, molecular, and relative stopping power. Also known as linear energy transfer (LET); linear stopping power.

stopping rule [STAT] A rule which specifies when observation is to be discontinued in sequential trials.

stop signal [COMMUN] Signal that initiates the transfer of facsimile equipment from active to standby conditions.

stop valve [ENG] A valve that can be opened or closed to regulate or stop the flow of fluid in a pipe.

stopwatch [HOROL] A watch with one or more sweep hands that can be started or stopped at will by pressing a small button; used to measure elapsed time.

stopwater [NAV ARCH] A canvas, backed with red lead or some other substance, fitted between metal parts of a ship to make it watertight.

storable propellant [MATER] A propellant capable of being placed and kept in a tank without benefit of special measures for temperature or pressure control.

storage [ADP] Any device that can accept, retain, and read back one or more times; the means of storing data may be chemical, electrical, magnetic, mechanical, or sonic.

storage allocation [ADP] The process of assigning storage locations to data or instructions in a digital computer.

storage and retrieval system [ADP] An organized method of putting items away in a manner which permits their recall or retrieval from storage. Also known as storetrieval system.

storage area [ADP] A specified set of locations in a storage unit. Also known as zone.

storage battery [ELEC] A connected group of two or more storage cells or a single storage cell. Also known as accumulator; accumulator battery; rechargeable battery; secondary battery.

storage battery locomotive [MIN ENG] An underground locomotive powered by storage batteries.

storage calorifier *See* cylinder.

storage camera *See* iconoscope.

storage capacity [ADP] The quantity of data that can be retained simultaneously in a storage device; usually measured in bits, digits, characters, bytes, or words. Also known as capacity; memory capacity.

storage cell [ADP] An elementary (logically indivisible) unit of storage; the storage cell can contain one bit, character, byte, digit (or sometimes word) of data. [ELEC] An electrolytic cell for generating electric energy, in which the cell after being discharged may be restored to a charged condition by sending a current through it in a direction opposite to that of the discharging current. Also known as secondary cell.

storage cycle [ADP] **1.** Periodic sequence of events occurring when information is transferred to or from the storage device of a computer. **2.** Storing, sensing, and regeneration from parts of the storage sequence.

storage density [ADP] The number of characters stored per unit-length of area of storage medium (for example, number of characters per inch of magnetic tape).

storage device [ADP] A mechanism for performing the function of data storage: accepting, retaining, and emitting (unchanged) data items. Also known as computer storage device.

storage dump [ADP] A printout of the contents of all or part of a computer storage. Also known as memory dump; memory print.

storage element [ADP] Smallest part of a digital computer storage used for storing a single bit.

storage equation [HYD] The equation of continuity applied to unsteady flow; it states that the fluid inflow to a given space during an interval of time minus the outflow during the same interval is equal to the change in storage; it is applied in hydrology to the routing of floods through a reservoir or a reach of a stream; the moisture continuity equation applied to the atmosphere is a modification of this.

storage factor *See* Q.

storage fill [ADP] Storing a pattern of characters in areas of a computer storage that are not intended for use in a particular machine run; these characters cause the machine to stop if one of these areas is erroneously referred to. Also known as memory fill.

storage hierachy [ADP] The sequence of storage devices, characterized by speed, type of access, and size for the various functions of a computer; for example, core storage from programs and data, disks or drums for temporary storage of massive amounts of data, tapes and cards for back up storage.

storage integrator [ADP] In an analog computer, an integrator used to store a voltage in the hold condition for future use while the rest of the computer assumes another computer control state.

storage key [ADP] A special set of bits associated with every word or character in some block of storage, which allows tasks having a matching set of protection key bits to use that block of storage.

storage location [ADP] A digital-computer storage position holding one machine word and usually having a specific address.

storage mark [ADP] The name given to a point location which defines the character space immediately to the left of the most significant character in accumulator storage.

storage medium [ADP] Any device or recording medium into which data can be copied and held until some later time, and from which the entire original data can be obtained.

storage print [ADP] In computers, a utility program that records the requested core image, core memory, or drum locations in absolute or symbolic form either on the line-printer or on the delayed-printer tape.

storage protection [ADP] Any restriction on access to storage blocks, with respect to reading, writing, or both. Also known as memory protection.

storage register [ADP] A register in the main internal memory of a digital computer storing one computer word. Also known as memory register.

storage reservoir *See* impounding reservoir.

storage rings [NUCLEO] Annular vacuum chambers in which charged particles can be stored, without acceleration, by a magnetic field of suitable focusing properties; they are used to stretch effectively the duty cycle of a particle accelerator or to produce colliding beams of particles, resulting in a greater possible center of mass energy.

storage ripple [ADP] A hardware function, used during maintenance periods, which reads or writes zeros or ones through available storage locations to detect a malfunctioning storage unit.

storage routing *See* flood routing.

storage surface [ADP] In computers, the surface (screen), in an electrostatic storage tube, on which information is stored.

storage tank *See* tank.

storage time *See* decay time.

storage tube [ELECTR] An electron tube employing cathode-ray beam scanning and charge storage for the introduction, storage, and removal of information. Also known as electrostatic storage tube; memory tube (deprecated usage).

storage-type camera tube *See* iconoscope.

store [ADP] **1.** To record data into a (static) data storage device. **2.** To preserve data in a storage device.

store and forward [COMMUN] A procedure in data communications in which data are stored at some point between the sender and the receiver and are later forwarded to the receiver.

stored-energy welding [MET] Welding by means of energy accumulated electrostatically, electrochemically, or electromagnetically at a low rate.

stored-program computer [ADP] A digital computer which executes instructions that are stored in main memory as patterns of data.

stored program logic [ADP] Program that is stored in a memory unit containing logical commands in order to perform the same processes on all problems.

stored routine [ADP] In computers, a series of instructions in storage to direct the step-by-step operation of the machine.

stored word [ADP] The actual linear combination of letters (or their machine equivalents) to be placed in the machine memory; this may be physically quite different from a dictionary word.

store transmission bridge [ELEC] Transmission bridge, which consists of four identical impedance coils (the two windings of the back-bridge relay and live relay of a connector, respectively) separated by two capacitors, which couples the calling and called telephones together electrostatically for the transmission of voice-frequency (alternating) currents, but separates the two lines for the transmission of direct current for talking purposes (talking current).

storetrieval system *See* storage and retrieval system.

stork [VERT ZOO] Any of several species of long-legged wading birds in the family Ciconiidae.

storm [METEOROL] An atmospheric disturbance involving perturbations of the prevailing pressure and wind fields on scales ranging from tornadoes (1 kilometer across) to extra-tropical cyclones (2–3000 km across); also the associated weather (rain storm or blizzard) and the like.

storm beach [GEOL] A ridge composed of gravel or shingle built up by storm waves at the inner margin of a beach.

storm cellar *See* cyclone cellar.

storm center [METEOROL] The area of lowest atmospheric pressure of a cyclone; this is a more general expression than eye of the storm, which refers only to the center of a well-developed tropical cyclone, in which there is a tendency of the skies to clear.

storm choke [PETRO ENG] A device installed in an oil-well tubing string below the surface to shut in the well when the flow reaches a predetermined rate; provides an automatic shutoff in case Christmas-tree or control valves are damaged. Also known as tubing safety valve.

storm delta *See* washover.

storm detection [METEOROL] Any of the methods and techniques used to detect the formation of severe storms, including procedures for locating, tracking, and forecasting; special tools adapted to this purpose are radar and satellites to supplement meteorological charts and visual observations.

storm drain [CIV ENG] A drain which conducts storm surface, or wash water, or drainage after a heavy rain from a building to a storm or a combined sewer. Also known as storm sewer.

storm ice foot [OCEANOGR] An ice foot produced by the breaking of a heavy sea or the freezing of wind-driven spray.

storm microseism [GEOPHYS] A microseism lasting 25 or more seconds, caused by ocean waves.

storm model [METEOROL] A physical, three-dimensional representation of the inflow, outflow, and vertical motion of air and water vapor in a storm.

storm sewage [CIV ENG] Refuse liquids and waste carried by sewers during or following a period of heavy rainfall.

storm sewer *See* storm drain.

storm surge [OCEANOGR] A rise above normal water level on the open coast due only to the action of wind stress on the water surface; includes the rise in level due to atmospheric pressure reduction as well as that due to wind stress. Also known as storm wave; surge.

storm tide [OCEANOGR] Height of a storm surge or hurricane wave above the astronomically predicted sea level.

storm track [METEOROL] The path followed by a center of low atmospheric pressure.

storm transposition [METEOROL] The transfer of precipitation patterns or DDA (depth-duration-area) values from the areas where they actually occurred to areas where they could occur; if necessary, the precipitation values are modified to account for differences in elevation or intervening barriers, and restrictions on change in shape or orientation of the storm may be imposed.

storm warning [METEOROL] A specially worded forecast of severe weather conditions, designed to alert the public to impending dangers; usually, this refers to a warning of potentially dangerous wind conditions for marine interests.

storm-warning signal [METEOROL] An arrangement of flags or pennants (by day) and lanterns (by night) displayed on a coastal storm-warning tower.

storm-warning tower [METEOROL] A tower, generally constructed of steel, for displaying coastal storm-warning signals.

storm wave *See* storm surge.

storm wind [METEOROL] In the Beaufort wind scale, a wind whose speed is from 56 to 63 knots (64 to 72 miles per hour or 104 to 117 kilometers per hour).

storm window [BUILD] A sash placed on the outside of an ordinary window to give added protection from the weather.

Storrow whirling hygrometer [ENG] A hygrometer in which the two thermometers are mounted side by side on a brass frame and fitted with a loose handle so that it can be whirled in the atmosphere to be tested; the instrument is whirled at some 200 revolutions per minute for about 1 minute and the readings on the wet- and dry-bulb thermometers are recorded; used in conjunction with Glaisher's or Marvin's hygrometrical tables.

story [BUILD] The space between two floors or between a floor and the roof.

storyboard [GRAPHICS] A series of small drawings intended to show the sequence and continuity of a proposed motion

---

STORM

Key:
→ hurricanes
→ extratropical cyclones (W-winter only)
----- summer position of intertropical convergence
■ subtropical high
L_s semipermanent summer-heat lows
H_w winter continental anticyclones

Principal tracks of extratropical cyclones and hurricanes with significantly associated features in the Northern Hemisphere.

STORM DETECTION

Hurricane Donna as observed at 0840 EST, September 10, 1960, on the WSR-57, 10-centimeter radar set at Key West, Florida. A spiral overlay of crossover angle $\alpha = 15°$ has been fitted to the precipitation bands in order to indicate the location of the storm center. *(ESSA photograph)*

picture, television production, or slide presentation; only key portions of the action or story are shown, which help to visualize the total idea.

stoss [GEOL] Of the side of a hill, knob, or prominent rock, facing the upstream side of a glacier.

stove [ENG] A chamber within which a fuel-air mixture is burned to provide heat, the heat itself being radiated outward from the chamber; used for space heating, process-fluid heating, and steel blast furnaces.

stove bolt [DES ENG] A coarsely-threaded bolt with a slotted head, which with a square nut is used to join metal parts.

stovepipe [ENG] Large-diameter pipe made of sheet steel.

stovewood [MATER] Firewood sawed into short lengths for use in a stove.

stowage diagram [NAV ARCH] A scaled drawing included in the loading plan of a ship for each deck or platform showing the exact location of all cargo, and specifying for each location such data as overall dimensions, location of obstructions, dimensions of the overhead hatch opening, dimensions of bow door or stern gate opening, minimum clearances to the overhead, bale cubic capacity, square feet of deck area, and the capacity of booms.

stowage factor [NAV ARCH] The number which expresses the space, in cubic feet, occupied by a long ton of any commodity as prepared for shipment, including all crating or packaging.

stowage plan [NAV ARCH] A completed stowage diagram showing each type of material that has been loaded and its exact stowage location; each port of discharge is indicated by a particular color or other symbol; deck and between-deck cargo is shown in perspective, cargo stowed in the lower hold is shown in profile, but vehicles are shown in perspective, regardless of stowage.

stowboard [MIN ENG] A mine heading used for storing waste.

STP See standard conditions.

strabismus [MED] Incoordinate action of the extrinsic ocular muscles resulting in failure of the visual axes to meet at the desired objective point. Also known as cast; heterotropia; squint.

straddle milling [MECH ENG] Face milling of two parallel vertical surfaces of a workpiece simultaneously by using two side-milling cutters.

strafe [ORD] To rake a body of troops or other persons with gunfire or rocket fire at close range and from a flying aircraft, or to attack a roadway, railyard, factory, or other installation with bullets, projectiles, or rockets fired from a low firing airplane.

straggling [PHYS] Random variations in some property associated with the passage of ions through matter.

straight bevel gear [DES ENG] A simple form of bevel gear having straight teeth which, if extended inward, would come together at the intersection of the shaft axes.

straight dynamite [MATER] Any of the powerful, quick-acting dynamites composed of nitroglycerin, a combustible such as wood meal, sodium nitrate, and an antacid such as calcium or magnesium carbonate; made in 15 to 60% strength, the percentage representing the proportion of nitroglycerin in the dynamite.

straightedge [DES ENG] A strip of wood, plastic, or metal with one or more long edges made straight with a desired degree of accuracy.

straightening vanes [ENG] Horizontal vanes mounted on the inside of fluid conduits to reduce the swirling or turbulent flow ahead of the orifice or the venturi meters.

straight filing [ENG] Filing by pushing a file in a straight line across the work.

straightforward circuit [COMMUN] Circuit in which signaling is automatic and in one direction.

straightforward trunking [COMMUN] In a manual telephone switchboard system, that method of operation in which the A operator gives the order to the B operator over the trunk on which talking later takes place.

straight-line coding [ADP] A digital computer program or routine (section of program) in which instructions are executed sequentially, without branching, looping, or testing.

straight-line mechanism [MECH ENG] A linkage so proportioned and constrained that some point on it describes over part of its motion a straight or nearly straight line.

straight piecework system See one-hundred-percent premium plan.

straight polarity [MET] Arrangement of an arc welding circuit in which the electrode is connected to the negative terminal.

straight proportional system See one-hundred-percent premium plan.

straight-run [CHEM ENG] Petroleum fractions derived from the straight distillation of crude oil without chemical reaction or molecular modification. Also known as virgin.

straight-run distillation [CHEM ENG] Continuous nonreactive distillation of petroleum oil to separate it into products in the order of their boiling points.

straight-run gasoline [MATER] Gasoline comprised of only natural ingredients from crude oil or natural-gas liquids; for example, no cracked, polymerized, alkylated, reformed, or visbroken stock.

straight-run stock See virgin stock.

straight sinus [ANAT] A sinus of the dura mater running from the inferior sagittal sinus along the junction of the falx cerebri and tentorium to the transverse sinus.

straight strap clamp [DES ENG] A clamp made of flat stock with an elongated slot for convenient positioning; held in place by a T bolt and nut.

straight-tube boiler [MECH ENG] A water-tube boiler in which all the tubes are devoid of curvature and therefore require suitable connecting devices to complete the circulatory system. Also known as header-type boiler.

straight vertical antenna [ELECTROMAG] An antenna consisting of a straight vertical wire.

straightway pump [MECH ENG] A pump with suction and discharge valves arranged to give a direct flow of fluid.

straight wheel [DES ENG] A grinding wheel whose sides or face are straight and not in any way changed from a cylindrical form.

strain [MECH] Change in length of an object in some direction per unit undistorted length in some direction, not necessarily the same; the nine possible strains form a second-rank tensor.

strain aging [MET] Change of mechanical properties of a metal by aging induced by plastic deformation.

strain axis See principal axis of strain.

strain bursts [MIN ENG] Rock bursts in which there is spitting, flaking, and sudden fracturing at the face, indicating increased pressure at the site.

strain ellipsoid [MECH] A mathematical representation of the strain of a homogeneous body by a strain that is the same at all points or of unequal stress at a particular point. Also known as deformation ellipsoid.

strain energy [MECH] The potential energy stored in a body by virtue of an elastic deformation, equal to the work that must be done to produce this deformation.

strainer [ENG] A porous or screen medium used ahead of equipment to filter out harmful solid objects and particles from a fluid stream; used for example, in river-water intakes for process plants or to remove decomposition products from the circulating fluid in a hydraulic system.

strain foil [ENG] A strain gage produced from thin foil by photoetching techniques; may be applied to curved surfaces, has low transverse sensitivity, exhibits negligible hysteresis under cycling loads, and creeps little under sustained loads.

strain gage [ENG] A device which uses the change of electrical resistance of a wire under strain to measure pressure.

strain gage accelerometer [ENG] Any accelerometer whose operation depends on the fact that the resistance in a wire changes when it is strained; these devices are classified as bonded or unbonded.

strain gage bridge [ENG] A bridge arrangement of four strain gages, cemented to a stressed part in such a way that two gages show increases in resistance and two show decreases when the part is stressed; the change in output voltage under stress is thus much higher than that for a single gage.

strain hardening [MET] Increasing the hardness and tensile strength of a metal by cold plastic deformation.

strain insulator [ELEC] An insulator used between sections of a stretched wire or antenna to break up the wire into insulated sections while withstanding the total pull of the wire.

STRAIGHT-TUBE BOILER

A straight-tube-type marine boiler, which has been fitted with a superheater and an air heater and arranged for oil firing. *(Babcock and Wilcox Co.)*

STRAIN ROSETTE

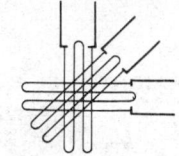

(a)

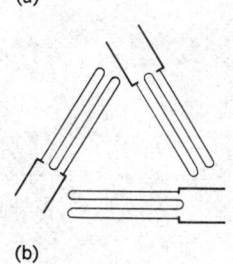

(b)

Strain rosettes. *(a)* 45° type; *(b)* 60° type.

STRAND BURNER

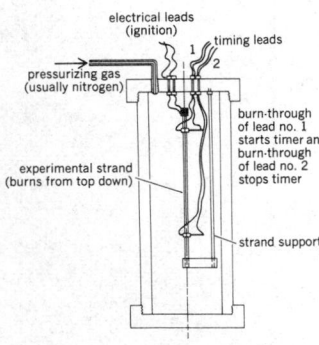

Strand burner apparatus.

strain rate [MECH] The time rate for the usual tensile test.

strain relief method [MIN ENG] A method for determining absolute strain and stress within rock in place by boring a smooth hole in the rock and inserting a gage capable of measuring diametral deformation, and overcoring the hole with a large coring bit; the change in the diameter of the hole when the rock cylinder is free to expand is a function of the original stress in the rock and its elastic modulus.

strain restoration method [MIN ENG] A method for determining absolute strain and stress within rock in place by the installation of strain gages on the rock surface, cutting of a slot in the rock between the strain gages so that the surface rock is free to expand, installation of a flat jack in the slot, and application of hydraulic pressure to the flat jack until the rock is restored to its original state of strain; the original stress in the rock is presumed to be equal to the final pressure in the flat jack.

strain rosette [MECH] A pattern of intersecting lines on a surface along which linear strains are measured to find stresses at a point.

strain seismograph [ENG] A seismograph that detects secular strains related to tectonic processes and tidal yielding of the solid earth; also detects strains associated with propagating seismic waves.

strain seismometer [ENG] A seismometer that measures relative displacement of two points in order to detect deformation of the ground.

strain shadow *See* undulatory extinction.

strain-slip [GEOL] A rock fracture resulting in a slight displacement.

strain-slip cleavage *See* slip cleavage.

strait [GEOGR] **1.** A neck of land. **2.** A narrow waterway connecting two larger bodies of water.

strake [MIN ENG] A relatively wide trough set at a slope and covered with a blanket or corduroy for catching comparatively coarse gold and any valuable mineral. [NAV ARCH] A continuous band of planking or plating running fore and aft along the hull of a ship.

strand [ENG] **1.** One of a number of steel wires twisted together to form a wire rope or cable or an electrical conductor. **2.** A thread, yarn, string, rope, wire, or cable of specified length. **3.** One of the fibers or filaments twisted or laid together into yarn, thread, rope, or cordage. [GEOL] A beach bordering a sea or an arm of an ocean. [NAV] To run aground; term strand usually refers to a serious grounding, while the term "ground" refers to any grounding, however slight. [TEXT] An element of a woven material.

strand burner [ENG] A device that determines the rate at which a propellant burns at various pressures by using a propellant strand.

stranded caisson *See* box caisson.

stranded conductor *See* stranded wire.

stranded floe ice foot *See* stranded ice foot.

stranded ice [OCEANOGR] Ice held in place by virtue of being grounded. Also known as grounded ice.

stranded ice foot [OCEANOGR] An ice foot formed by the stranding of floes or small icebergs along a shore; it may be built up by freezing spray or breaking seas. Also known as stranded floe ice foot.

stranded wire [ELEC] A conductor composed of a group of wires or a combination of groups of wires, usually twisted together. Also known as stranded conductor.

strandflat [GEOL] **1.** A low, flat, very wide wave-cut platform extending off the coast of western Norway. **2.** A discontinuous shelf of land inside a fjord.

strandline [GEOL] **1.** A beach raised above the present sea level. **2.** The level at which a body of standing water meets the land.

strandline *See* shoreline.

strangeness conservation [PARTIC PHYS] The principle that the sum of the strangeness numbers of the hadrons in an isolated system is constant; it is violated by the weak interactions.

strangeness number [PARTIC PHYS] A quantum number carried by hadrons, equal to the hypercharge minus the baryon number. Symbol S.

strange particle [PARTIC PHYS] A hadron whose strangeness number is not zero, for example, a K-meson or a Σ-hyperon.

stranger [AERO ENG] In air intercept, an unidentified aircraft, bearing, distance, and altitude as indicated relative to an aircraft.

strangulated hernia [MED] A hernia involving the intestine in which circulation of the blood and fecal current are blocked.

strangulation [MED] **1.** Asphyxiation due to obstruction of the air passages, as by external pressure on the neck. **2.** Constriction of a part producing arrest of the circulation, as strangulation of a hernia.

strap bolt [DES ENG] **1.** A bolt with a hook or flat extension instead of a head. **2.** A bolt with a flat center portion and which can be bent into a U shape.

strap hammer [MECH ENG] A heavy hammer controlled and operated by a belt drive in which the head is slung from a strap, usually of leather.

strap hinge [DES ENG] A hinge fastened to a door and the adjacent wall by a long hinge.

strapped-down inertial navigation equipment [NAV] Inertial navigation equipment in which a stable platform and gimbal system are not used: the inertial devices are attached or strapped directly to the carrier; a computer utilizing gyro information resolves accelerations sensed along the carrier axes and refers these accelerations to an inertial frame of reference. Also known as gimballess inertial navigation equipment.

strapped magnetron [ELECTR] A multicavity magnetron in which resonator segments having the same polarity are connected together by small conducting strips to suppress undesired modes of oscillation.

strapping [ELEC] Connecting two or more points in a circuit or device with a short piece of wire or metal. [ELECTR] Connecting together resonator segments having the same polarity in a multicavity magnetron to suppress undesired modes of oscillation. [PETRO ENG] A petroleum industry procedure in which storage tanks are strapped (measured) on their outside with steel measuring tapes to calculate the volumetric capacity of the tank for increments of height.

strapping table [PETRO ENG] A tabular record of tank volume versus height so that taped (strapped) measurements of liquid depth can be converted into liquid volumes.

Strasbourg turpentine [MATER] A balsam from the European white fir; a heavy, thick material, it is sometimes used in painting mediums and glazes, but an excessive amount causes smearing and slow drying of the paint.

Stratco alkylation [CHEM ENG] A proprietary petroleum refinery process for the manufacture of alkylate gasoline; uses sulfuric acid as catalyst and refrigeration of reactor effluent to control isobutane-olefin ratio in the reactor feedstream. Also known as effluent refrigeration alkylation.

strategic [ORD] Of or pertaining to military measures or actions taken against the enemy's war-making effort.

strategic airlift [AERO ENG] The continuous, sustained air movement of units, personnel, and materiel in support of all U.S. Department of Defense agencies between area commands.

strategic air warfare [ORD] Air combat and supporting operations designed to effect, through the systematic application of force to a selected series of vital targets, the progressive destruction and disintegration of the enemy's war-making capacity to a point where the ability or the will to wage war is no longer retained; vital targets may include key manufacturing systems, sources of raw material, critical material, stockpiles, power systems, transportation systems, communication facilities, concentrations of uncommitted elements of enemy armed forces, key agricultural areas, and other such target systems.

strategic attack [ORD] An attack by means of aerospace forces directed at selected vital targets of an enemy nation so as to destroy its war-making capacity or will to fight.

strategic concentration [ORD] The assembly of designated forces in areas from which it is intended that operations of the assembled forces shall begin so that they are best disposed to initiate the plan of campaign.

strategic map [MAP] A map of medium scale, or smaller, used for planning of operations, including the movement, concentration, and supply of troops.

strategic material [IND ENG] A material needed for the industrial support of a war effort.

strategic mobility [ORD] The capability of a unit, command, force, or thing that enables it to be readily moved in advance of engagement with hostile forces.

strategic nuclear weapon [ORD] A nuclear weapon which is programmed primarily for use against strategic targets in strategic nuclear war.

strategic research [SCI TECH] Research conducted to produce specific applied programs.

strategic reserve [ORD] That quantity of material which is placed in a particular geographic location due to strategic considerations or in anticipation of major interruptions in the supply distribution system; it is over and above the stockage objective.

strategic target [ORD] Any installation, network, group of buildings, or the like, considered vital to a country's war-making capacity and singled out for attack.

strategic transport aircraft [AERO ENG] Aircraft designed primarily for the carriage of personnel or cargo over long distances.

strategy [MATH] In game theory a strategy is a specified collection of moves, which cover all possible situations, for the complete play of a given game.

strath [GEOL] **1.** A broad, elongate depression with steep sides on the continental shelf. **2.** An extensive remnant of a broad, flat valley floor that has undergone degradation following uplift.

strath terrace [GEOL] An extensive remnant of a strath from a former erosion cycle.

stratification [GEOL] An arrangement or deposition of sedimentary material in layers, or of sedimentary rock in strata. [HYD] **1.** The arrangement of a body of water, as a lake, into two or more horizontal layers of differing characteristics, especially densities. **2.** The formation of layers in a mass of snow, ice, or firn.

stratification plane [GEOL] A demarcation between two layers of sedimentary rock, often signifying that the layers were deposited under different conditions.

stratified drift [GEOL] Fluvioglacial drift composed of material deposited by a meltwater stream or settled from suspension.

stratified flow [FL MECH] A two-phase flow in which a liquid flows along the bottom of a pipe and gas flows separately above it.

stratified fluid [FL MECH] A fluid having density variation along the axis of gravity, usually implying upward decrease of density, that is, a stratification characterized by static stability.

stratified ocean [OCEANOGR] An ocean where there is a vertical gradient of density.

stratified rock *See* sedimentary rock.

stratified sampling [STAT] A random sample of specified size is drawn from each stratum of a population.

stratified squamous epithelium [HISTOL] A multiple-layered epithelium composed of thin, flat superficial cells and cuboidal and columnar deeper cells.

stratiform [GEOL] **1.** Descriptive of a layered mineral deposit of either igneous or sedimentary origin. **2.** Consisting of parallel bands, layers, or sheets. [METEOROL] Description of clouds of extensive horizontal development, as contrasted to the vertically developed cumuliform types.

stratiformis [METEOROL] A cloud species consisting of a very extensive horizontal layer or layers which need not be continuous; this species is the most common form of the genera altocumulus and stratocumulus and is occasionally found in cirrocumulus.

stratigrapher [GEOL] A geologist who deals with stratified rocks, for example, the classification, nomenclature, correlation, and interpretation of rocks.

stratigraphic geology *See* stratigraphy.

stratigraphic map [GEOL] A map showing the areal distribution, configuration, or aspect of a stratigraphic unit or surface, such as an isopach map or a lithofacies map.

stratigraphic oil fields [GEOL] Hydrocarbon reserves in stratigraphic (sedimentary) traps formed by the positioning of clastic materials through chemical deposition.

stratigraphic separation *See* stratigraphic throw.

stratigraphic sequence *See* sequence.

stratigraphic throw [GEOL] The thickness of the strata which originally separated two beds brought into contact at a fault. Also known as stratigraphic separation.

stratigraphic trap [GEOL] Sealing of a reservoir bed due to lithologic changes rather than geologic structure. Also known as porosity trap; secondary stratigraphic trap.

stratigraphic unit [GEOL] A stratum of rock or a body of strata classified as a unit on the basis of character, property, or attribute.

stratigraphy [GEOL] A branch of geology concerned with the form, arrangement, geographic distribution, chronologic succession, classification, correlation, and mutual relationships of rock strata, especially sedimentary. Also known as stratigraphic geology.

Stratiomyidae [INV ZOO] The soldier flies, a family of orthorrhaphous dipteran insects in the series Brachycera.

stratocumulus [METEOROL] A principal cloud type predominantly stratiform, in the form of a gray or whitish layer of patch, which nearly always has dark parts.

Stratofortress [AERO ENG] A United States all-weather, intercontinental, strategic heavy bomber powered by eight turbojet engines; it is capable of delivering nuclear and nonnuclear bombs, air-to-surface missiles, and decoys; its range is extended by in-flight refueling. Designated B-52.

Stratofreighter [AERO ENG] A United States strategic, aerial tanker-freighter powered by four reciprocating engines; it is equipped for inflight refueling of bombers and fighters. Designated KC-97.

Stratojet [AERO ENG] A United States all-weather strategic medium bomber; it is powered by six turbojet engines, has intercontinental range through in-flight refueling, and is capable of delivering nuclear and nonnuclear bombs. Designated B-47.

stratopause [METEOROL] The boundary or zone of transition separating the stratosphere and the mesosphere; it marks a reversal of temperature change with altitude.

stratoscope [OPTICS] A balloon-borne astronomical telescope for taking solar or other celestial photographs at high altitudes; subsequently, the photos are transmitted to a ground receiving station.

stratosphere [METEOROL] The atmospheric shell above the troposphere and below the mesosphere; it extends, therefore, from the tropopause to about 55 kilometers, where the temperature begins again to increase with altitude.

stratosphere radiation [GEOPHYS] Any infrared radiation involved in the complex infrared exchange continually proceeding within the stratosphere.

stratospheric coupling [METEOROL] The interaction between disturbances in the stratosphere and those in the troposphere.

stratospheric steering [METEOROL] The steering of lower-level atmospheric disturbances along the contour lines of the tropopause, which lines are presumably roughly parallel to the direction of the wind at the tropopause level.

Stratotanker [AERO ENG] A United States multipurpose aerial tanker-transport powered by four turbojet engines; and equipped for high speed, high-altitude refueling of bombers and fighters. Designated KC-135.

stratovolcano [GEOL] A volcano constructed of lava and pyroclastics, deposited in alternating layers. Also known as composite volcano.

stratum *See* subpopulation.

stratum corneum [HISTOL] The outer layer of flattened keratinized cells of the epidermis.

stratum disjunctum [HISTOL] The outermost layer of desquamating keratinized cells of the stratum corneum.

stratum germinativum [HISTOL] The innermost germinative layer of the epidermis.

stratum granulosum [HISTOL] A layer of granular cells interposed between the stratum corneum and the stratum germinativum in the thick skin of the palms and soles.

stratum lucidum [HISTOL] A layer of irregular transparent epidermal cells with traces of nuclei interposed between the stratum corneum and stratum germinativum in the thick skin of the palms and soles.

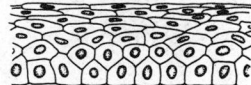

**STRATIFIED
SQUAMOUS EPITHELIUM**

Drawing of a section through stratified squamous epithelium showing arrangement of cells.

STRAUSS REACTION

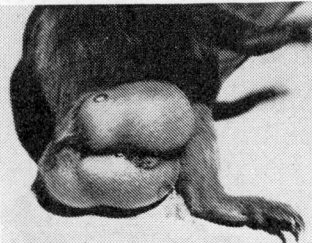

Strauss reaction in the guinea pig.

STRAWBERRY

Flower and fruit of a strawberry plant.

STREAMLINE FLOW

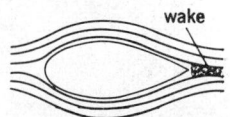

wake

Flow indicated about a streamlined body which is traveling at subsonic speed.

stratus [METEOROL] A principal cloud type in the form of a gray layer with a rather uniform base; a stratus does not usually produce precipitation, but when it does occur it is in the form of minute particles, such as drizzle, ice crystals, or snow grains.

Strauss reaction [IMMUNOL] The exudative swelling of the scrotum in male hamsters and guinea pigs upon subcutaneous or intraperitoneal inoculation of *Pseudomonas mallei,* the causative agent of glanders.

straw [AGR] Grain stalks after threshing and usually mixed with leaves and chaff. [BOT] A stem of grain, such as wheat or oats.

strawberry [BOT] A low-growing perennial of the genus *Fragaria,* order Rosales, that spreads by stolons; the juicy, usually red, edible fruit consists of a fleshy receptacle with numerous seeds in pits or nearly superficial on the receptacle.

strawberry hemangioma [MED] A vascular birthmark characterized by a soft, raised, bright-red, lobular appearance.

straw oil [MATER] A straw-colored petroleum paraffin oil; used for many process applications.

strawwalker [AGR] A set of reciprocating notched bars inside a thresher or combine that push the straw to the rear.

stray [GEOL] A lenticular rock formation encountered unexpectedly in drilling an oil or a gas well; it differs from an adjacent persistent formation in lithology and hardness.

stray current [ELEC] **1.** A portion of a current that flows over a path other than the intended path, and may cause electrochemical corrosion of metals in contact with electrolytes. **2.** An undesirable current generated by discharge of static electricity; it commonly arises in loading and unloading petroleum fuels and some chemicals, and can initiate explosions.

stray current corrosion [MET] Corrosion of metals caused by a stray current.

stray emission [PHYS] Emission of radiation that serves no useful purpose.

stray field [ELECTROMAG] Leakage of magnetic flux that spreads outward from a coil and does no useful work.

stray line [ENG] An ungraduated portion of the line connected to a current pole, used so that the pole will acquire the speed of the current before a measurement is begun.

strays *See* atmospheric interference.

stray sand [GEOL] A stray composed of sandstone.

streak [MINERAL] The color of a powdered mineral, obtained by rubbing the mineral on a streak plate.

streak lightning [GEOPHYS] Ordinary lightning, of a cloud-to-ground discharge, that appears to be entirely concentrated in a single, relatively straight lightning channel.

streak line [FL MECH] A line within a fluid which, at a given instant, is formed by those fluid particles which at some previous instant have passed through a specified fixed point in the fluid; an example is the line of color in a flow into which a dye is continuously introduced through a small tube, all dyed fluid particles having passed the tube's end.

streak photograph [GEOPHYS] A time-exposure photograph of tracer particles in a fluid; the photograph reveals the motion of each tracer particle in the form of a streak which may be interpreted as a velocity vector.

streak plate [MICROBIO] A method of culturing aerobic bacteria by streaking the surface of a solid medium in a petri dish with an inoculating wire or glass rod in a continuous movement so that most of the surface is covered; used to isolate majority members of a mixed population.

stream [HYD] A body of running water moving under the influence of gravity to lower levels in a narrow, clearly defined natural channel.

stream anchor [NAV ARCH] An anchor used in narrow channels to prevent the stern of the vessel moving with the tide.

stream capacity [GEOL] The ability of a stream to carry detritus, measured at a given point per unit of time.

stream capture *See* capture.

stream channel [GEOL] A long, narrow, sloping troughlike depression where a natural stream flows or may flow. Also known as streamway.

stream channel form ratio [GEOL] The mathematical relationship between a stream channel width, depth, and channel perimeter.

stream current [HYD] A steady current in a stream or river.

[OCEANOGR] A deep, narrow, well-defined fast-moving ocean current.

stream day [CHEM ENG] Denoting a 24-hour actual operation of a processing unit, in contrast to the hours actually operated during a calendar (24-hour) day.

streamer [GEOPHYS] A sinuous channel of very high ion-density which propagates itself through a gas by continual establishment of an electron avalanche just ahead of its advancing tip; in lightning discharges, the stepped leader, and return streamer all constitute special types of streamers.

stream erosion [GEOL] The progressive removal of exposed matter from the surface of a stream channel by a stream.

streamflow [HYD] A type of channel flow, applied to surface runoff moving in a stream.

streamflow routing *See* flood routing.

stream frequency [GEOL] A measure of topographic texture expressed as the ratio of the number of streams in a drainage basin to the area of the basin. Also known as channel frequency.

stream function *See* Lagrange stream function.

stream gage *See* river gage.

stream gradient [GEOL] The angle, measured in the direction of flow, between the water surface (for large streams) or the channel flow (for small streams) and the horizontal. Also known as stream slope.

stream gradient ratio [GEOL] Ratio of the stream gradient of a stream channel of one order to the stream gradient of the next higher order channel in the same drainage basin. Also known as channel gradient ratio.

streaming current [ELEC] The electric current which is produced when a liquid is forced to flow through a diaphragm, capillary, or porous solid.

streaming potential [ELEC] The difference in electric potential between a diaphragm, capillary, or porous solid and a liquid that is forced to flow through it.

stream-length ratio [HYD] Ratio of the mean length of a stream of a given order to the mean length of the next lower order stream in the same basin.

streamline [FL MECH] A line which is every where parallel to the direction of fluid flow at a given instant.

streamline flow [FL MECH] Flow of a fluid in which there is no turbulence: particles of the fluid follow well-defined continuous paths, and the flow velocity at a fixed point either remains constant or varies in a regular fashion with time.

streamlining [DES ENG] The contouring of a body to reduce its resistance to motion through a fluid.

stream order [HYD] The designation by a dimensionless integer series (1, 2, 3, . . .) of the relative position of stream segments in the network of a drainage basin. Also known as channel order.

stream piracy *See* capture.

stream profile [HYD] The longitudinal profile of a stream.

stream robbery *See* capture.

stream segment [HYD] The part of a stream extending between designated tributary junctions. Also known as channel segment.

stream slope *See* stream gradient.

stream takeoff [AERO ENG] Aircraft taking off in tail/column formation.

stream terrace [GEOL] One of a series of level surfaces on a stream valley flanking and parallel to a stream channel and above the stream level, representing the uneroded remnant of an abandoned floodplain or stream bed. Also known as river terrace.

stream tin [GEOL] The mineral cassiterite occurring as pebbles in alluvial deposits.

stream transport [GEOL] Movement of rock material in and by a stream.

stream tube [FL MECH] In fluid flow, an imaginary tube whose wall is generated by streamlines passing through a closed curve.

streamway *See* stream channel.

Streblidae [INV ZOO] The bat flies, a family of cyclorrhaphous dipteran insects in the section Pupipara; adults are ectoparasites on bats.

street [CIV ENG] A paved road for vehicular traffic in an urban area.

street elbow [DES ENG] A pipe elbow with an internal thread at one end and an external thread at the other.

Strelitziaceae [BOT] A family of monocotyledonous plants in the order Zingiberales distinguished by perfect flowers with five functional stamens and without an evident hypanthium, penniveined leaves, and symmetrical guard cells.

stremmatograph [ENG] An instrument for measuring longitudinal stress in rails as trains pass over.

strengite [MINERAL] $FePO_4 \cdot 2H_2O$ A pale-red mineral crystallizing in the orthorhombic system, isomorphous with variscite and dimorphous with phosphosiderite, and specific gravity 2.87.

strength [ACOUS] The maximum instantaneous rate of volume displacement produced by a sound source when emitting a wave with sinusoidal time variation. [MECH] The stress at which material ruptures or fails.

strength of current [OCEANOGR] **1.** The phase of a tidal current at which the speed is a maximum. **2.** The velocity of the current at this time.

strength of ebb [OCEANOGR] **1.** The ebb current at the time of maximum speed. **2.** The speed of the current at this time.

strength of ebb interval [OCEANOGR] The time interval between the transit (upper or lower) of the moon and the next maximum ebb current at a place.

strength of enemy forces [ORD] The description of an enemy unit of force in terms of men, weapons, and equipment.

strength of flood [OCEANOGR] **1.** The flood current at the time of maximum speed. **2.** The speed of the current at this time.

strength of flood interval [OCEANOGR] The time interval between the transit (upper or lower) of the moon and the next maximum flood current at a place.

strepaster [INV ZOO] A short, spiny microscleric, monaxonic spicule.

strepogenin [BIOCHEM] A factor, possibly a peptide derivative of glutamic acid, reported to exist in certain proteins, acting as a growth stimulant in bacteria and mice in the presence of completely hydrolyzed protein. Also known as streptogenin.

Strepsiptera [INV ZOO] An order of the Coleoptera that is coextensive with the family Stylopidae.

streptobacillary fever *See* Haverhill fever.

streptobiosamine [BIOCHEM] $C_{13}H_{23}NO_9$ A nitrogen-containing disaccharide, obtained when streptomycin undergoes acid hydrolysis; in the streptomycin molecule it is glycosidally linked to streptidine.

Streptococceae [MICROBIO] A tribe of the family Lactobacillaceae including cocci that occur in pairs, short chains, or tetrads and which generally obtain energy by fermentation of carbohydrates or related compounds.

Streptococcus [MICROBIO] A genus of the tribe Streptococceae including many pathogenic strains; the cells are round and gram-positive and occur in characteristic chains.

Streptococcus fecalis [MICROBIO] A species of enterococci normally found in the intestinal contents of man and animals; they are penicillin-resistant.

Streptococcus hemolyticus *See* Streptococcus pyogenes.

Streptococcus lactis [MICROBIO] A species of nonpathogenic bacteria which produce C substance of Lancefield group N; they coagulate milk and are used in the dairy industry.

Streptococcus pyogenes [MICROBIO] A species of beta-hemolytic streptococci constituting Lancefield group A; causes pus formation and septicemia in the human body. Also known as *Streptococcus hemolyticus*.

streptogenin *See* strepogenin.

streptokinase [BIOCHEM] An enzyme occurring as a component of fibrinolysin in cultures of certain hemolytic streptococci.

streptolysin [BIOCHEM] Any of a group of hemolysins elaborated by *Streptococcus pyogenes*.

Streptomyces [MICROBIO] A genus of the family Streptomycetaceae which are aerobic, nonacid-fast, saprophytic soil-inhabiting organisms with branching filaments and conidia produced in aerial hyphae in chains.

Streptomycetaceae [MICROBIO] A family of soil-inhabiting bacteria which form nonfragmenting vegetative mycelia with conidia borne on sporophores.

streptomycin [MICROBIO] $C_{21}H_{39}O_{12}N_7$ Water-soluble antibiotic obtained from *Streptomyces griseus* that is used principally in the treatment of tuberculosis.

streptothricin [MICROBIO] $C_{19}H_{34}O_7N_8$ An antibiotic produced by *Streptomyces lavendulae*; active against various gram-negative and gram-positive microorganisms.

stress [BIOL] A stimulus or succession of stimuli of such magnitude as to tend to disrupt the homeostasis of the organism. [MECH] The force acting across a unit area in a solid material in resisting the separation, compacting, or sliding that tends to be induced by external forces.

STRESS [ADP] A problem-oriented programming language used to solve structural engineering problems. Derived from structural engineering system solver.

stress amplitude [MECH ENG] One half the algebraic difference between the maximum and minimum stress in one fatigue test cycle.

stress analysis [PHYS] The determination of the stresses produced in a solid body when subjected to various external forces.

stress axis *See* principal axis of stress.

stress birefringence *See* mechanical birefringence.

stress concentration [MECH] A condition in which a stress distribution has high localized stresses; usually induced by an abrupt change in the shape of a member; in the vicinity of notches, holes, changes in diameter of a shaft, and so forth, maximum stress is several times greater than where there is no geometrical discontinuity.

stress concentration factor [MECH] A theoretical factor K_t expressing the ratio of the greatest stress in the region of stress concentration to the corresponding nominal stress.

stress corrosion [MET] Corrosion that is accelerated by stress, applied or residual, in a metal.

stress-corrosion cracking [MET] Failure by cracking under the conjoint action of a constant tensile stress, which is applied to residual, in certain chemical environments specific to the metal.

stress crack [MECH] An external or internal crack in a solid body (metal or plastic) caused by tensile, compressive, or shear forces.

stress difference [MECH] The difference between the greatest and the least of the three principal stresses.

stressed skin construction [CIV ENG] A type of construction in which the outer skin and the framework interact, thus contributing to the flexural strength of the unit.

stress ellipsoid [MECH] A mathematical representation of the state of stress at a point that is defined by the minimum, intermediate, and maximum stresses and their intensities.

stress mineral [MINERAL] Any mineral whose formation in metamorphosed rock is favored by shearing stress.

stress-optic law [OPTICS] In a transparent, isotropic plate subjected to a biaxial stress field, the relative retardation R_t between the two components produced by temporary double refraction is equal to $Ct(p - q)$, which in turn is equal to $n\lambda$; C is the stress-optic coefficient, t the plate thickness, p and q the principal stresses, n the number of fringes which have passed the point during application of the load, and λ the wavelength of the light.

stress raiser [MET] A notch, hole, or other discontinuity in contour or structure which causes localized stress concentration.

stress range [MECH] The algebraic difference between the maximum and minimum stress in one fatigue test cycle.

stress ratio [MECH] The ratio of minimum to maximum stress in fatigue testing, considering tensile stresses as positive and compressive stresses as negative.

stress relieving [MET] Low-temperature heating to reduce residual stress.

stress-strain curve *See* deformation curve.

stress tensor [MECH] A second-rank tensor whose components are stresses exerted across surfaces perpendicular to the coordinate directions.

stress trajectories *See* isostatics.

stretch [GEOGR] *See* reach. [PETRO ENG] The increase in length of oil-well casing or tubing when freely suspended in fluid mediums.

stretched pebbles [GEOL] Pebbles in a sedimentary rock

STRESS CONCENTRATION

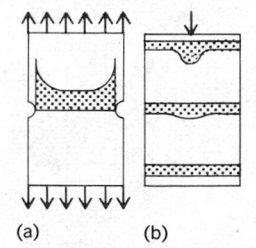

(a)   (b)

Stress concentrations. (a) Tensile stress distribution in a plate reduced by circular notches is shown qualitatively; stress at root of notch is about three times stress at end of plate. (b) Bar under concentrated end load; load concentrated at end of bar produces nonuniformly distributed normal stresses on adjacent sections with variation decreasing at more remote sections.

which have been elongated from their original shape by deformation.

stretcher [CIV ENG] A brick or block that is laid with its length paralleling the wall. [MED] A litter usually made of canvas stretched on a frame for carrying injured, disabled, or dead persons. [MIN ENG] A bar used for roof support on roadways and which is either wedged against or pocketed into the sides of the roadway without support of legs or struts.

stretcher bar [MIN ENG] A single screw column, capable of holding one machine drill; used in small drifts.

stretcher leveling [MET] Removing warp and distortion from a piece of metal by gripping it at each end and subjecting it to stress beyond the yield strength. Also known as patent leveling.

stretcher strains *See* Lüders' lines.

stretch fault *See* stretch thrust.

stretch former [MECH ENG] A machine used to stretch form materials, such as metals and plastics.

stretch forming [MECH ENG] Shaping metals and plastics by applying tension to stretch the heated sheet or part, wrapping it around a die, and then cooling it. Also known as wrap forming.

stretch out [IND ENG] A reduction in the delivery rate specified for a program without a reduction in the total quantity to be delivered.

stretch reflex [PHYSIO] Contraction of a muscle in response to a sudden, brisk, longitudinal stretching of the same muscle. Also known as myostatic reflex.

stretch thrust [GEOL] A reverse fault developed as a result of shear in the middle limb of an overturned fold. Also known as stretch fault.

stria [BIOL] A minute line, band, groove, or channel.

striated ground *See* striped ground.

striated muscle [HISTOL] Muscle tissue consisting of muscle fibers having cross striations.

striation [ELECTR] A succession of alternately luminous and dark regions sometimes observed in the positive column of a glow-discharge tube near the anode. [GEOL] One of a series of parallel or subparallel scratches, small furrows, or lines on the surface of a rock or rock fragment; usually inscribed by rock fragments embedded at the base of a moving glacier. [MINERAL] One of a series of parallel, shallow depressions or narrow bands on the cleavage face of a mineral caused either by growth twinning or oscillatory growth of different crystal faces.

striation technique [ACOUS] A technique for making sound waves visible by using their ability to refract light waves.

strich *See* millimeter.

striding compass [ENG] A compass mounted on a theodolite for orientation.

stridor [MED] A peculiar, harsh, vibrating sound produced during respiration.

stridulation [INV ZOO] Creaking and other audible sounds made by certain insects, produced by rubbing various parts of the body together.

Strigidae [VERT ZOO] A family of birds of the order Strigiformes containing the true owls.

Strigiformes [VERT ZOO] The order of birds containing the owls.

strigose [BIOL] Covered with stiff, pointed, hairlike scales or bristles.

strigovite [MINERAL] $Fe_3(Al,Fe)_3Si_3O_{11}(OH)_7$ A dark-green mineral of the chlorite group, composed of basic aluminum iron silicate; occurs as crystalline incrustations.

Strigulaceae [BOT] A family of Ascolichenes in the order Pyrenulales comprising crustose species confined to tropical evergreens, and which form extensive crusts on or under the cuticle of leaves.

strike [GEOL] The direction taken by a structural surface, such as a fault plane, as it intersects the horizontal. Also known as line of strike. [MET] **1.** A very thin, initially electroplated film or the plating solution with which to deposit such a film. **2.** A local crater in a metal surface due to accidental contact with the welding electrode. [ORD] Concerted air attack on a single objective.

strike board [MIN ENG] A board at the top of a shaft from which the bucket is tipped; used in shaft sinking; formerly, the beam or plank at the shaft top on which the baskets were landed. Also known as strike tree.

strike fault [GEOL] A fault whose strike is parallel with that of the strata involved.

strike force [ORD] A force composed of appropriate units necessary to conduct strikes, attack, or assault operations.

strike joint [GEOL] A joint that strikes parallel to the bedding or cleavage of the constituent rock.

strike-off board [ENG] A straight-edge board used to remove excess, freshly placed plaster, concrete, or mortar from a surface.

strike photography [GRAPHICS] Air photographs taken during an air strike.

strike plating [MET] Applying a thin electroplated film prior to depositing the principal electroplate.

striker [ORD] A firing pin or a projection on the hammer of a firearm, which strikes the primer to initiate a propelling charge explosive train or a fuse explosive train.

striker plate [ORD] A plate in the breech of a firearm or gun, which supports the base of the cartridge and which is pierced with a hole through which the striker or firing pin hits the primer.

strike separation [GEOL] The distance of separation on either side of a fault surface of two formerly adjacent beds.

strike-shift fault *See* strike-slip fault.

strike slip [GEOL] The component of the slip of a fault that is parallel to the strike of the fault. Also known as horizontal displacement; horizontal separation.

strike-slip fault [GEOL] A fault whose direction of movement is parallel to the strike of the fault. Also known as strike-shift fault.

strike stream *See* subsequent stream.

strike tree *See* strike board.

striking hammer [ENG] A hammer used to strike a rock drill.

striking potential [ELECTR] **1.** Voltage required to start an electric arc. **2.** Smallest grid-cathode potential value at which plate current begins flowing in a gas-filled triode.

striking velocity *See* impact velocity.

string [ADP] A set of consecutive, adjacent items of similar type; normally a bit string or a character string. [ENG] A piece of pipe, casing, or other down-hole drilling equipment coupled together and lowered into a borehole. [GEOL] A very small vein, either independent or occurring as a branch of a larger vein. Also known as stringer. [MECH] A solid body whose length is many times as large as any of its cross-sectional dimensions, and which has no stiffness.

string bead [MET] A continuous weld bead made without appreciable transverse oscillation.

string electrometer [ENG] An electrometer in which a conducting fiber is stretched midway between two oppositely charged metal plates; the electrostatic field between the plates displaces the fiber laterally in proportion to the voltage between the plates.

stringer [CIV ENG] **1.** A long horizontal member used to support a floor or to connect uprights in a frame. **2.** An inclined member supporting the treads and risers of a staircase. [GEOL] *See* string. [MET] An elongated mass of microconstituents or foreign material in wrought metal oriented in the direction of working.

stringer lode [GEOL] A lode that consists of many narrow veins in a mass of country rock.

stringer plate [NAV ARCH] One of the plates that make up the outer strake of the deck of a ship, and which are usually heavier than those making up the rest of the deck.

string galvanometer [ENG] A galvanometer consisting of a silver-plated quartz fiber under tension in a magnetic field, used to measure oscillating currents.

stringing [PETRO ENG] The connecting of lengths of pipe end to end (tubing or casing) to make a string long enough to reach to the desired depth in a well bore.

string milling [MECH ENG] A milling method in which parts are placed in a row and milled consecutively.

String-Oriented-Symbolic Language *See* SNOBOL.

string shot [PETRO ENG] An oil-well stimulation technique in which a string of explosive (for example, Prima Cord) is hung opposite to the producing zone down a wellbore and detonated; used to remove deposits (gypsum, mud, or paraffin) from the formation face.

strip [ENG] To remove insulation from a wire. [MIN ENG] To remove coal, stone, or other material from a quarry or from a working that is near the surface of the earth. [ORD] To dissassemble a piece of equipment, such as a gun, in order to clean, repair, or transport it.

strip-borer drill [MECH ENG] An electric or diesel skid- or caterpillar-mounted drill used at quarry or opencast sites to drill 3–6-inch-diameter (8–15-centimeter-diameter), horizontal blast holes up to 100 feet (30 meters) in length, without the use of flush water.

strip-chart recorder [ENG] A recorder in which one or more writing pens or other recording devices trace changes in a measured variable on the surface of a strip chart that is moved at constant speed by a time-clock motor.

strip-cropping [AGR] Growing separate crops in adjacent strips that follow the contour of the land as a method of reducing soil erosion.

striped ground [GEOL] A pattern of alternating stripes formed by frost action on a sloping surface. Also known as striated ground; striped soil.

striped soil *See* striped ground.

strip lights [NAV] Lights marking the edge of a landing strip.

strip line *See* strip transmission line.

strip-line circuit [ELECTROMAG] A circuit in which one or more strip transmission lines serve as filters or other circuit components.

strip method [FOR] A lumbering method in which timbers are cleared from a forest in strips; new growth in the strip results from seeds sown in the adjoining forest.

strip mine [MIN ENG] An opencut mine in which the overburden is removed from a coal bed before the coal is taken out.

strip mining [MIN ENG] The mining of coal by surface mining methods.

stripped atom [ATOM PHYS] An ionized atom which has appreciably fewer electrons than it has protons in the nucleus.

stripped illite *See* degraded illite.

stripped plane [GEOL] The upper, exposed surface of a resistant stratum that forms a stripped structural surface when extended over a considerable area.

stripped structural surface [GEOL] An erosion surface formed in an area underlain by horizontal or gently sloping strata of unequal resistance where the overlying softer beds have been removed by erosion. Also known as stripped surface.

stripped surface *See* stripped structural surface.

stripper [CHEM ENG] An evaporative device for the removal of vapors from liquids; can be in a bubble-tray distillation tower, a vacuum vessel, or an evaporator; if it is a part of a distillation column below the feed tray, it is called the stripping section. [ENG] A hand or motorized tool used to remove insulation from wires.

stripper punch [MET] In powder metallurgy, a device used as the bottom or top of the die cavity which can be pushed into the die to eject the formed compact.

stripper rubber [PETRO ENG] A pressure-actuated seal used to control gas pressure in the casing-tubing annulus of low-pressure wells while inserting (running) or withdrawing (pulling) tubing.

stripping [CHEM ENG] In petroleum refining, the removal (by flash evaporation or steam-induced vaporation) of the more volatile components from a cut or fraction; used to raise the flash point of kerosine, gas oil, or lubricating oil. [MET] Removing a coating from the surface of a metal.

stripping agent [TEXT] A substance used to remove dyes from fabrics so that they can be redyed, for example, sodium hydrosulfite, titanous sulfate, and sodium and zinc formaldehyde sulfoxylates.

stripping analysis [ANALY CHEM] An analytic process of solutions or concentrations containing ions, in which the ions are electrodeposited onto an electrode, stripped (dissolved) from the material from the electrode, and weighed.

stripping area [MIN ENG] In stripping operations, an area encompassing the pay material, its bottom depth, the thickness of the layer of waste, the slope of the natural ground surface, and the steepness of the safe slope of cuts.

stripping a shaft [MIN ENG] **1.** Removing the timber from an abandoned shaft. **2.** Trimming or squaring the sides of a shaft.

stripping film [GRAPHICS] **1.** The process of assembling photographic negatives or positives to make printing plates. **2.** A film with an emulsion that can be removed from its base and transferred to another one.

stripping ratio [MIN ENG] The unit amount of spoil or waste that must be removed to gain access to a similar unit amount of ore or mineral material.

stripping reaction [NUC PHYS] A nuclear reaction in which part of the incident nucleus combines with the target nucleus, and the other part proceeds with most of its original momentum in practically its original direction; especially the reaction in which the incident nucleus is a deuteron and only a proton emerges from the target.

stripping shovel [MIN ENG] A shovel with an especially long boom and stick, enabling it to reach further and pile higher.

strip pit [MIN ENG] **1.** A coal or other mine worked by stripping. **2.** An open-pit mine.

strip plot [MAP] A portion of a map or overlay on which a number of photographs taken along a flight line is delineated without defining the outlines of individual prints.

strip survey [FOR] A survey of the value of a strip of forest; used to estimate the value of a larger area of the forest.

strip transmission line [ELECTROMAG] A microwave transmission line consisting of a thin, narrow, rectangular metal strip that is supported above a ground-plane conductor or between two wide ground-plane conductors and is usually separated from them by a dielectric material. Also known as flat coaxial transmission line; microstrip; microwave stripline; parallel strip line; strip line.

strobe [ELECTR] **1.** Intensified spot in the sweep of a deflection-type indicator, used as a reference mark for ranging or expanding the presentation. **2.** Intensified sweep on a plan-position indicator or B-scope; such a strobe may result from certain types of interference, or it may be purposely applied as a bearing or heading marker. **3.** Line on a console oscilloscope representing the azimuth data generated by a jammed radar site.

strobe circuit [ELECTR] A circuit that produces an output pulse only at certain times or under certain conditions, such as a gating circuit or a coincidence circuit.

strobe marker [ELECTR] A small bright spot, or a short gap, or other discontinuity produced on the trace of a radar display to indicate that part of the time base which is receiving attention.

strobe photography *See* stroboscopic photography.

strobe pulse [ELECTR] Pulse of duration less than the time period of a recurrent phenomenon used for making a close investigation of that phenomenon; the frequency of the strobe pulse bears a simple relation to that of the phenomenon, and the relative timing is usually adjustable.

strobilation [INV ZOO] Asexual reproduction by segmentation of the body into zooids, proglottids, or separate individuals.

strobilocercus [INV ZOO] A larval tapeworm that has undergone strobilation.

strobilus [BOT] **1.** A conelike structure made up of sporophylls, or spore-bearing leaves, as in Equisetales. **2.** The cone of members of the Pinophyta.

stroboscope [ENG] An instrument for making moving bodies visible intermittently, either by illuminating the object with brilliant flashes of light or by imposing an intermittent shutter between the viewer and the object; a high-speed vibration can be made visible by adjusting the strobe frequency close to the vibration frequency.

stroboscopic direction finder [NAV] A radio direction finder employing a continuously rotating antenna or other directional element such as a goniometer; associated with the rotating element, either directly or remotely, is a mechanism which rotates a thin metal disk in synchronism with the antenna or directional element; the disk has a single thin radial slot which covers a neon tube so that its light can be seen only through the slot; the output of the receiver is connected to the neon tube so that illumination occurs at a reading corresponding to the bearing of the received signal.

stroboscopic disk [ENG] A printed disk having a number of concentric rings each containing a different number of dark and light segments; when the disk is placed on a phonograph turntable or rotating shaft and illuminated at a known

STRIP TRANSMISSION LINE

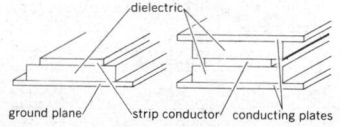

Components of strip transmission lines.

STROMATOLITE

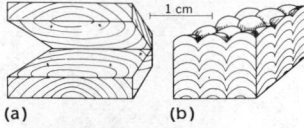

1 cm

(a)   (b)

Sectioned stromatolites.
(a) A large, strongly laminated
stromatolite. Laminae dip away
from the center of the
hemispheroidal mass if viewed
from above, and toward the
center if viewed from beneath.
(b) A small stromatolite
consisting of columns with
highly arched laminae. Lower
Ordovician of western Wisconsin.
*(From R. R. Shrock, Sequence
in Layered Rocks, McGraw-Hill,
1948)*

STRONTIANITE

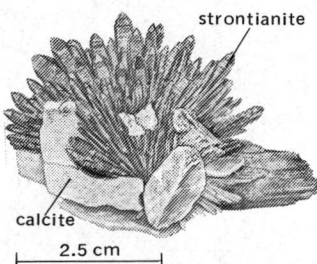

strontianite

calcite

2.5 cm

Strontianite crystals with calcite.
*(Specimen from Department of
Geology, Bryn Mawr College)*

STRONTIUM

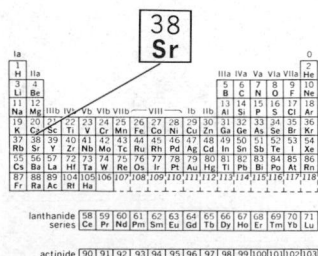

Periodic table of the chemical
elements showing the position
of strontium.

frequency by a flashing discharge tube, speed can be determined by noting which pattern appears to stand still or to
rotate slowly.

stroboscopic lamp   *See* flash lamp.

stroboscopic photography [GRAPHICS] The technique of
producing pictures of both single and multiple exposure
taken by flashes of light from electrical discharges. Also
known as flash photography; strobe photography.

stroboscopic tachometer [ENG] A stroboscope having a
scale that reads in flashes per minute or in revolutions per
minute; the speed of a rotating device is measured by
directing the stroboscopic lamp on the device, adjusting the
flashing rate until the device appears to be stationary, then
reading the speed directly on the scale of the instrument.

stroboscopic tube   *See* strobotron.

strobotron [ELECTR] A cold-cathode gas-filled arc-discharge
tube having one or more internal or external grids to initiate
current flow and produce intensely bright flashes of light for
a stroboscope. Also known as stroboscopic tube.

stroke [ADP] **1.** A key-depressing operation in keypunching.
2. In optical character recognition, straight or curved portion
of a letter, such as is commonly made with one smooth
motion of a pen. Also known as character stroke. **3.** That
segment of a printed or handwritten character which has been
temporarily isolated from other segments for the purpose of
analyzing it, particularly with regard to its dimensions and
relative reflectance. Also known as character stroke.
[ELECTR] The penlike motion of a focused electron beam in
cathode-ray-tube diplays. [MECH ENG] The linear movement, in either direction, of a reciprocating mechanical part.
[MED] A sudden cerebrovascular accident.

stroke analysis [ADP] In character recognition, a method
employed in character property detection in which an input
specimen is dissected into certain prescribed elements; the
sequence, relative positions, and number of detected elements
are then used to identify the characters.

stroke-bore ratio [MECH ENG] The ratio of the distance traveled by a piston in a cylinder to the diameter of the cylinder.

stroke center line [ADP] In character recognition, a line
midway between the two average-edge lines; the center line
describes the stroke's direction of travel. Also known as
center line.

stroke density [GEOPHYS] The areal density of lightning
discharges over a given region during some specified period of
time, as number per square mile per year.

stroke edge [ADP] In character recognition, a continuous
line, straight or otherwise, which traces the outermost part of
intersection of the stroke along the two sides of its greatest
dimension.

stroke speed [COMMUN] Number of times per minute that a
fixed line, perpendicular to the direction of scanning, is
crossed in one direction by a scanning or recording spot in a
facsimile system. Also known as scanning frequency; scanning line frequency.

stroke width [ADP] In character recognition, the distance
that obtains, at a given location, between the points of
intersection of the stroke edges and a line drawn perpendicular to the stroke center line.

stroma [ANAT] The supporting tissues of an organ, including
connective and nervous tissues and blood vessels.

stromal endometriosis   *See* interstitial endometrosis.

stromal myosis   *See* interstitial endometrosis.

Stromateidae [INV ZOO] A family of perciform fishes in the
suborder Stromateoidei containing the butterfishes.

Stromateoidei [VERT ZOO] A suborder of fishes of the order
Perciformes in which most species have teeth in pockets
behind the pharyngeal bone.

stromatite [GEOL] Chorismite having flat or folded parallel
layers of two or more textural elements. Also known as
stromatolith.

stromatolite [GEOL] A structure in calcareous rocks consisting of concentrically laminated masses of calcium carbonate
and calcium-magnesium carbonate which are believed to be
of calcareous algal origin; these structures are irregular to
columnar and hemispheroidal in shape, and range from 1
millimeter to many meters in thickness. Also known as
callenia.

stromatolith [GEOL] **1.** A complex sill-like igneous intrusion
interfingered with sedimentary strata. **2.** *See* stromatite.

Stromatoporoidea [PALEON] An extinct order of fossil colonial organisms thought to belong to the class Hydrozoa; the
skeleton is a coenosteum.

stromatosis   *See* interstitial endometrosis.

Strombacéa [PALEON] An extinct superfamily of gastropod
mollusks in the order Prosobranchia.

Strombidae [INV ZOO] A family of gastropod mollusks comprising tropical conchs.

strombolian [GEOL] A type of volcanic eruption characterized by fire fountains of lava from a central crater.

stromeyerite [MIN] CuAgS A metallic-gray orthorhombic
mineral with a blue tarnish composed of silver copper sulfide
occurring in compact masses.

strong acid [CHEM] An acid with a high degree of dissociation in solution, for example, mineral acids, such as hydrochloric acid, HCl, sulfuric acid, H_2SO_4, or nitric acid, HNO_3.

strong base [CHEM] A base with a high degree of dissociation in solution, for example, sodium hydroxide, $NaOH$,
potassium hydroxide, KOH.

strong breeze [METEOROL] In the Beaufort wind scale, a
wind whose speed is from 22 to 27 knots (25 to 31 miles per
hour or 41 to 50 kilometers per hour).

strong fix [NAV] A fix determined from horizontal sextant
angles between objects well placed.

strong gale [METEOROL] In the Beaufort wind scale, a wind
whose speed is from 41 to 47 knots (47 to 54 miles per hour
or 76 to 87 kilometers per hour).

strong interaction [PARTIC PHYS] One of the fundamental
interactions of elementary particles, primarily responsible for
nuclear forces and other interactions among hadrons.

strongly damped collision   *See* quasi-fission.

strong point [ORD] A strongly fortified and heavily armed
point in a defense system, usually supported by auxiliary
armed positions.

strong topology [MATH] The topology on a normed space
obtained from the given norm; the basic open neighborhoods
of a vector x are sets consisting of all those vectors y where the
norm of $x - y$ is less than some number.

strongyle [INV ZOO] A monaxonic spicule rounded at each
end.

Strongyloidea [INV ZOO] The hookworms, an order or superfamily of roundworms which, as adults, are endoparasites of
most vertebrates, including man.

strongyloidiasis [MED] An infestation of man with one of the
roundworms of the genus *Strongyloides.*

strongylote [INV ZOO] Rounded at one end, referring to
sponge spicules.

strontia   *See* strontium oxide.

strontianite [MINERAL] $SrCO_3$ A pale-green, white, gray, or
yellowish mineral of the aragonite group having orthorhombic symmetry and occurring in veins or as masses; hardness is
3.5 on Mohs scale, and specific gravity is 3.76.

strontium [CHEM] A metallic element in group IIA, symbol
Sr, with atomic number 38, atomic weight 87.62; flammable,
soft, pale-yellow solid; soluble in alcohol and acids, decomposes in water; melts at 770°C, boils at 1380°C; chemistry is
similar to that of calcium; used as electron-tube getter.

strontium-90 [NUC PHYS] A poisonous, radioactive isotope
of strontium; 28-year half life with β radiation; derived from
reactor-fuel fission products; used in thickness gages, medical
treatment, phosphor activation, and atomic batteries.

strontium acetate [ORG CHEM] $Sr(C_2H_3O_2)_2 \cdot \frac{1}{2}H_2O$
White, water-soluble crystals, loses water at 150°C; used for
catalysts, as a chemical intermediate, and in medicine.

strontium bromide [INORG CHEM] $SrBr_2 \cdot 6H_2O$ A white,
hygroscopic powder soluble in water and alcohol; loses water
at 180°C, melts at 643°C; used in medicine and as an
analytical reagent.

strontium carbonate [INORG CHEM] $SrCO_3$ A white powder slightly soluble in water, decomposes at 1340°C; used to
make TV-tube glass, strontium salts, and ceramic ferrites, and
in pyrotechnics.

strontium chlorate [INORG CHEM] $Sr(ClO_3)_2$ Shock-sensi-
tive, highly combustible, white, water-soluble crystals that
decompose at 120°C; used in pyrotechnics and tracer bullets.

strontium chloride [INORG CHEM] SrCl₂ Water- and alcohol-soluble white crystals, melts at 872°C; used in medicine and pyrotechnics and to make strontium salts.

strontium dioxide *See* strontium peroxide.

strontium hydrate *See* strontium hydroxide.

strontium hydroxide [INORG CHEM] Sr(OH)₂ Colorless deliquescent crystals that absorb carbon dioxide from air, soluble in hot water and acids, melts at 375°C; used by the sugar industry, in lubricants and soaps, and as a plastic stabilizer. Also known as strontium hydrate.

strontium iodide [INORG CHEM] SrI₂ Air-yellowing, white crystals that decompose in moist air, melts at 515°C; used in medicine and as a chemicals intermediate.

strontium monosulfide *See* strontium sulfide.

strontium nitrate [INORG CHEM] Sr(NO₃)₂ A white, water-soluble powder melting at 570°C; used in pyrotechnics, signals and flares, medicine, and matches, and as a chemicals intermediate.

strontium oxide [INORG CHEM] SrO A grayish powder, melts at 2430°C, becomes the hydroxide in water; used in medicine, pyrotechnics, pigments, greases, soaps, and as a chemicals intermediate. Also known as strontia.

strontium peroxide [INORG CHEM] SrO₂ A strongly oxidizing, fire-hazardous, white, alcohol-soluble powder that decomposes in hot water; used in medicine, bleaching, and fireworks. Also known as strontium dioxide.

strontium sulfate [INORG CHEM] SrSO₄ White crystals insoluble in alcohol, slightly soluble in water and concentrated acids, melts at 1605°C; used in paper manufacture, pyrotechnics, ceramics, and glass.

strontium sulfide [INORG CHEM] SrS A gray powder with a hydrogen sulfide aroma in moist air, slightly soluble in water, soluble (with decomposition) in acids, melts above 2000°C; used in depilatories and luminous paints and as a chemicals intermediate. Also known as strontium monosulfide.

strontium titanate [INORG CHEM] SrTiO₃ A solid material, insoluble in water and melting at 2060°C; used in electronics and electrical insulation.

strontium unit [NUCLEO] A unit of concentration of strontium-90 in a medium relative to the concentration of calcium, equal to 10^{-12} curie of strontium per gram of calcium. Abbreviated SU. Also known as sunshine unit (deprecated usage).

strophanthin [PHARM] A glycoside or mixture of glycosides extracted from the plant *Strophanthus kombe;* used as a cardioactive drug in the treatment of various heart ailments.

Strophanthus [BOT] A genus of woody climbers of the dogbane family (Apocynaceae), natives of tropical Asia and Africa; the source of arrow poisons and of strophanthin.

strophiole [BOT] A crestlike excrescence around the hilum in certain seeds.

strophoid [MATH] The curve traced in the plane by a point P moving along a varying line L passing through a fixed point, where the distance of P to L's intersection with the y-axis always is equal to the y-intercept value.

Strophomenida [PALEON] A large diverse order of articulate brachiopods which first appeared in Lower Ordovician times and became extinct in the Late Triassic.

Strophomenidina [PALEON] A suborder of extinct, articulate brachiopods in the order Strophomenida characterized by a concavo-convex shell, the pseudodeltidium and socket plates disposed subparallel to the hinge.

Strouhal number [MECH] A dimensionless number used in studying the vibrations of a body past which a fluid is flowing; it is equal to a characteristic dimension of the body times the frequency of vibrations divided by the fluid velocity relative to the body; for a taut wire perpendicular to the fluid flow, with the characteristic dimension taken as the diameter of the wire, it has a value between 0.185 and 0.2 Symbolized S_r. Also known as reduced frequency.

Strowger system [COMMUN] An automatic telephone switching system that uses successive step-by-step selector switches actuated by current pulses produced by rotation of a telephone dial. Also known as step-by-step system.

struck capacity [MIN ENG] The volume of water a mine car, tram, hoppet, or wagon would hold if the conveyance were of watertight construction.

struck joint [CIV ENG] A mortar joint in brickwork formed by pressing the trowel in at the lower edge, so that a recess is formed at the bottom of the joint; suitable only for interior work.

structural adhesive [MATER] An adhesive capable of bearing loads of considerable magnitude; a structural adhesive will not fail when a bonded joint prepared from the thickness of metal, or other material typical for that industry, is stressed to its yield point.

structural analysis [ENG] The determination of stresses and strains in a given structure. [PETR] *See* structural petrology.

structural bench [GEOL] A bench typifying the resistant edge of a terrace that is being reduced by erosion. Also known as rock bench.

structural bulkhead [NAV ARCH] A bulkhead designed to function as a strength member of the ship's structure.

structural clay tile [MATER] Hollow burned-clay masonry unit with parallel cells used as facing tile, load-bearing tile, partition tile, fireproofing tile, furring tile, floor tile, and header tile.

structural connection [CIV ENG] A means of joining the individual members of a structure to form a complete assembly.

structural contour map [GEOL] A map representation of a subsurface stratigraphic unit; depicts the configuration of a rock surface by means of elevation contour lines.

structural deflections [MECH] The deformations or movements of a structure and its flexural members from their original positions.

structural drawings [GRAPHICS] The design and working drawings for structures such as buildings, bridges, dams, tanks, and highways.

structural drill [MECH ENG] A highly mobile diamond- or rotary-drill rig complete with hydraulically controlled derrick mounted on a truck, designed primarily for rapidly drilling holes to determine the structure in subsurface strata or for use as a shallow, slim-hole producer or seismograph drill.

structural drilling [ENG] Drilling done specifically to obtain detailed information delineating the location of folds, domes, faults, and other subsurface structural features indiscernible by studying strata exposed at the surface.

structural engineering [CIV ENG] A branch of civil engineering dealing with the design of structures such as buildings, dams, and bridges.

structural engineering system solver *See* STRESS.

structural fabric *See* fabric.

structural formula [CHEM] A system of notation used for organic compounds in which the exact structure, if it is known, is given in schematic representation.

structural gene *See* cistron.

structural geology [GEOL] A branch of geology concerned with the form, arrangement, and internal structure of the rocks.

structural high [GEOL] Any of various structural features such as a crest, culmination, anticline, or dome.

structural information [ADP] Information specifying the number of independently variable features or degrees of freedom of a pattern.

structural low [GEOL] Any of various structural features such as a basin, a syncline, a saddle, or a sag.

structural petrology [PETR] The study of the internal structure of a rock to determine its deformational history. Also known as fabric analysis; microtectonics; petrofabric analysis; petrofabrics; petrogeometry; petromorphology; structural analysis.

structural riveting [ENG] Riveting structural members by using punched holes.

structural shape [MET] A piece of metal of a standard design used in construction.

structural steel [MET] Steel used in engineering structures, usually manufactured by either open-hearth or the electric furnace process.

structural terrace [GEOL] A terracelike landform developed where generally steeply inclined and otherwise uniformly dipping strata locally flatten.

structural tile [MATER] A hollow clay product which may be load-bearing or non-load-bearing; used for facing, flooring, or partitions.

STROPHANTHUS

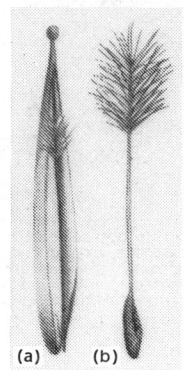

Strophanthus kombe seed (a) in folicle, (b) out of folicle.

structural trap [GEOL] Containment in a reservoir bed of oil or gas due to flexure or fracture of the bed.

structural valley [GEOL] A valley whose form and origin is attributable to the underlying geologic structure.

structural weight *See* construction weight.

structure [AERO ENG] The construction or makeup of an airplane, spacecraft, or missile, including that of the fuselage, wings, empennage, nacelles, and landing gear, but not that of the power plant, furnishings, or equipment. [CIV ENG] Something, as a bridge or a building, that is built or constructed and designed to sustain a load. [GEOL] **1.** An assemblage of rocks upon which erosive agents have been or are acting. **2.** The sum total of the structural features of an area. [MINERAL] The form taken by a mineral, such as tabular or fibrous. [PETR] A macroscopic feature of a rock mass or rock unit, best seen in an outcrop. [SCI TECH] The arrangement and interrelation of the parts of an object.

structure cell *See* unit cell.

structure contour [GEOL] A contour that portrays a structural surface, such as a fault. Also known as subsurface contour.

structure-contour map [GEOL] A map that uses structure contour lines to portray subsurface configuration. Also known as structure map.

structure factor [SOLID STATE] A factor which determines the amplitude of the beam reflected from a given atomic plane in the diffraction of an x-ray beam by a crystal, and is equal to the sum of the atomic scattering factors of the atoms in a unit cell, each multiplied by an appropriate phase factor.

structure map *See* structure-contour map.

structure number [DES ENG] A number, generally from 0 to 15, indicating the spacing of abrasive grains in a grinding wheel relative to their grit size.

structure of a system [ADP] Refers to the nature of the chain of command, the origin and type of data collected, the form and destination of results, and the procedures used to control operations.

structure section [GEOL] A vertical section showing the observed or inferred geologic structure on a vertical surface or plane.

structure type [CRYSTAL] The structural arrangement of a crystal, regardless of the atomic elements present; it corresponds to the crystal's space group.

struma lymphomatosa [MED] A type of chronic thyroiditis involving diffuse enlargement of the gland with retrogressive epithelial changes and lymphoid hyperplasia. Also known as Hashimoto's disease; Hashimoto's struma; lymphadenoid goiter.

struma ovarii [MED] An ovarian teratoma composed chiefly of thyroid tissue.

strut [AERO ENG] A bar supporting the wing or landing gear of an airplane. [CIV ENG] A long structural member of timber or metal, or a bar designed to resist pressure in the direction of its length. [ENG] **1.** A brace or supporting piece. **2.** A diagonal brace between two legs of a drill tripod or derrick. [MIN ENG] A vertical-compression member in a structure or in an underground timber set. [NAV ARCH] A bracket outside the hull of a ship, supporting the propeller shaft. Also known as propeller strut.

Struthionidae [VERT ZOO] The single family of the avian order Struthioniformes.

Struthioniformes [VERT ZOO] A monofamilial order of ratite birds containing the single living species of ostrich (*Struthio camelus*).

struvite [MINERAL] $Mg(NH_4)PO_4\cdot6H_2O$ A colorless to yellow or pale-brown mineral consisting of a hydrous ammonium magnesium phosphate, and occurring in orthorhombic crystals; hardness is 2 on Mohs scale, and specific gravity is 1.7.

strychnine [ORG CHEM] $C_{21}H_{22}O_2N_2$ An alkaloid obtained primarily from the plant nux vomica, formerly used for therapeutic stimulation of the central nervous system.

strychninization [MED] The condition resulting from large doses of strychnine.

Stuart factor [BIOCHEM] A procoagulant in normal plasma but deficient in the blood of patients with a hereditary bleeding disorder; may be closely related to prothrombin since both are formed in the liver by action of vitamin K. Also known as factor X; Stuart-Power factor.

Stuart-Power factor *See* Stuart factor.

Stuart windmill *See* Fales-Stuart windmill.

stub [ELECTROMAG] **1.** A short section of transmission line, open or shorted at the far end, connected in parallel with a transmission line to match the impedance of the line to that of an antenna or transmitter. **2.** A solid projection one-quarter-wavelength long, used as an insulating support in a waveguide or cavity.

stub angle [ELECTROMAG] Right-angle elbow for a coaxial radio-frequency transmission line which has the inner conductor supported by a quarter-wave stub.

stub axle [MECH ENG] An axle carrying only one wheel.

stubborn disease [PL PATH] A virus disease of citrus trees characterized by short internodes resulting in stiff brushy growth and chlorotic leaves.

stub cable [ELEC] Short branch off a principal cable; the end is often sealed until it is used at a later date; pairs in the stub are referred to as stubbed out pairs.

stub entry [MIN ENG] A short, narrow entry turned from another entry and driven into the solid coal, but not connected with other mine workings.

stub matching [ELECTROMAG] Use of a stub to match a transmission line to an antenna or load; matching depends on the spacing between the two wires of the stub, the position of the shorting bar, and the point at which the transmission line is connected to the stub.

stub mortise [ENG] A mortise which passes through only part of a timber.

Stubs gage [DES ENG] A number system for denoting the thickness of steel wire and drills.

stub-supported coaxial [ELECTROMAG] Coaxial whose inner conductor is supported by means of short-circuited coaxial stubs.

stub-supported line [ELECTROMAG] A transmission line that is supported by short-circuited quarter-wave sections of coaxial line; a stub exactly a quarter-wavelength long acts as an insulator because it has infinite reactance.

stub switch [ENG] A pair of short switch rails, held only at or near one end and free to move at the other end; used in mining and to some extent on narrow-gage industrial tramways.

stub tenon [ENG] A tenon that fits into a stub mortise.

stub tuner [ELECTROMAG] Stub which is terminated by movable short-circuiting means and used for matching impedance in the line to which it is joined as a branch.

stucco [MATER] A smooth plasterlike material applied to the outside wall or other exterior surface of a building or structure.

stud [BUILD] One of the vertical members in the walls of a framed building to which wallboards, lathing, or paneling is nailed or fastened. [DES ENG] **1.** A rivet, boss, or nail with a large, ornamental head. **2.** A short rod or bolt threaded at both ends without a head.

Student's distribution [STAT] The probability distribution used to test the hypothesis that a random sample of n observations comes from a normal population with a given mean.

Student's t-statistic [STAT] A one-sample test statistic computed by $T = \sqrt{n}(\bar{X} - \mu_H)/S$, where $\bar{X}$ is the mean of a collection of n observations, S is the square root of the mean square deviation, and μ_H is the hypothesized mean.

Student's t-test [STAT] A test in a one-sample problem which uses Student's t-statistic.

studio [COMMUN] A room in which television or radio programs are produced.

stud link chain [NAV ARCH] Chain in which each link has a stud at its midlength perpendicular to the major axis to maintain the shape of the link.

stud wall [BUILD] A wall formed with timbers; studs are usually spaced 12–16 inches (30–41 centimeters) on center.

stud welding [MET] Arc-welding using the heat of an electric arc produced between a metal stud and another part, and then bringing the parts together under pressure.

stuffed mineral [MINERAL] A mineral having extra ions of a foreign element within its larger interstices.

stuffing [ENG] A method of sealing the mechanical joint between two metal surfaces; packing (stuffing) material is

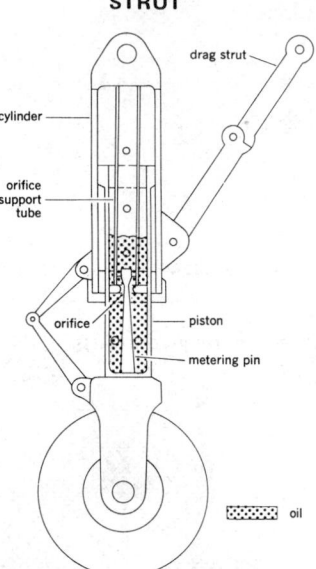

STRUT

Diagram of oleopneumatic shock strut. As airplane sinks toward ground, piston forces oil through orifice, causing force which changes path of airplane.

inserted within the seal area container (the stuffing or packing box), and compressed to a liquid-proof seal by a threaded packing ring follower. Also known as packing.

stuffing box [ENG] A packed, pressure-tight joint for a rod that moves through a hole, to reduce or eliminate fluid leakage.

stuffing nut [ENG] A nut for adjusting a stuffing box.

stull [MIN ENG] A platform laid on timbers, braced across a working from side to side, to support workers or to carry ore or waste.

stull piece [MIN ENG] **1.** A piece of timber placed slanting over the back of a level to prevent rock falling into the level from the stopes above. **2.** Timbers bracing the platform of a stull.

stull stoping [MIN ENG] Stull timbers placed between the foot and hanging walls, which constitute the only artificial support provided during the excavation of a stope.

stump [MIN ENG] A small pillar of coal left between the gangway or airway and the breasts to protect these passages; any small pillar.

stunt [PL PATH] Any of several plant diseases marked by reduction in size of the plant.

stunt box [ELEC] A device to control the nonprinting functions of a teletypewriter terminal.

stupp [MIN ENG] A black residue from distilled mercury ore, consisting of soot, hydrocarbons, mercury and mercury compounds, and ore dust.

sturgeon [VERT ZOO] Any of 10 species of large bottom-living fish which comprise the family Acipenseridae; the body has five rows of bony plates, and the snout is elongate with four barbels on its lower surface.

Sturm-Liouville problem [MATH] The general problem of solving a given linear differential equation of order $2n$ together with $2n$-boundary conditions. Also known as eigenvalue problem.

Sturm-Liouville system [MATH] A given differential equation together with its boundary conditions having Sturm-Liouville problem form.

Sturm sequence [MATH] For a polynomial $p(x)$, this is the sequence of functions $f_0(x), f_1(x), \ldots,$ where $f_0(x) = p(x), f_1(x) = p'(x)$, and $f_n(x)$ is the negative remainder that occurs by finding the greatest common divisor of $f_{n-2}(x)$ and $f_{n-1}(x)$ via the Euclidean algorithm.

Sturm's theorem [MATH] This gives a method to determine the number of real roots of a polynomial $p(x)$ which lie between two given values of x; the Sturm sequence of $p(x)$ provides the necessary information.

sturtite [MINERAL] A black mineral composed of hydrous silicate of iron, manganese, calcium, and magnesium; occurs in compact masses.

stutter [COMMUN] Series of undesired black and white lines sometimes produced when a facsimile signal undergoes a sharp amplitude change. [MED] A speech disorder marked by repetition of words, syllables, or sounds, or by hesitations in manner, without apparent awareness by the speaker.

Stuve chart [METEOROL] A thermodynamic diagram with atmospheric temperature as the x axis and atmospheric pressure to the power 0.286 as the y ordinate, increasing downward; named after G. Stuve. Also known as adiabatic chart; pseudoadiabatic chart.

S twist [TEXT] A left-handed yarn twist in which the spirals resemble the letter S.

sty See hordeolum.

Styginae [INV ZOO] A subfamily of butterflies in the family Lycaenidae in which the prothoracic legs in the male are nonfunctional.

Stygocaridacea [INV ZOO] An order of crustaceans in the superorder Syncarida characterized by having a furca.

Stylasterina [INV ZOO] An order of the class Hydrozoa, including several brightly colored branching or encrusting coral-like coelenterates of warm seas.

style [BOT] The portion of a pistil connecting the stigma and ovary. [ZOO] A slender elongated process on an animal.

stylet [GRAPHICS] A slender, pointed marking tool, as one used in graving. [INV ZOO] A slender, rigid, elongated appendage. [MED] **1.** A slender probe used for surgery. **2.** A thin wire inserted in a catheter to provide support or in a hollow needle to clear the passage.

styloglossus [ANAT] A muscle arising from the styloid process of the temporal bone, and inserted into the tongue.

stylohyoid [ANAT] Pertaining to the styloid process of the temporal bone and the hyoid bone.

styloid [ZOO] Resembling a style.

stylolite [GEOL] An irregular surface, generally parallel to a bedding plane, in which small toothlike projections on one side of the surface fit into cavities of complementary shape on the other surface; interpreted to result diagenetically by pressure solution.

stylomastoid [ANAT] Relating to the styloid and the mastoid processes of the temporal bone.

Stylommatophora [INV ZOO] A large order of the molluscan subclass Pulmonata characterized by having two pairs of retractile tentacles with eyes located on the tips of the large tentacles.

stylopodium [BOT] A conical or disk-shaped enlargement at the base of the style in plants of the family Umbelliferae.

stylotypite See tetrahedrite.

stylus [ENG ACOUS] The portion of a phonograph pickup that follows the modulations of a record groove and transmits the resulting mechanical motions to the transducer element of the pickup for conversion to corresponding audio-frequency signals. Also known as needle; phonograph needle; reproducing stylus. [GRAPHICS] A rather blunt metal point sometimes used in painting to make lightly ruled lines.

stylus printing See matrix printing.

styphnic acid [ORG CHEM] $C_6H(OH)_2(HO_2)_3$ An explosive, yellow, crystalline compound, melting at 179–180°C, slightly soluble in water; used in explosives as a priming agent. Also known as 2,4,6-trinitroresorcinol.

Stypocapitellidae [INV ZOO] A family of polychaete annelids belonging to the Sedentaria and consisting of a monotypic genus found in western Germany.

styramate [PHARM] $C_9H_{11}NO_3$ A compound that crystallizes from chloroform solution, and melts at 111–112°C; used in medicine as a muscle relaxant. Also known as carbamic acid β-hydroxyphenethyl ester.

styrene [ORG CHEM] $C_6H_5CH:CH_2$ A colorless, toxic liquid with a strong aromatic aroma; insoluble in water, soluble in alcohol and ether; polymerizes rapidly, can become explosive; boils at 145°C; used to make polymers and copolymers, polystyrene plastics, and rubbers. Also known as phenylethylene; styrene monomer; vinylbenzene.

styrene-acrylonitrile resin [ORG CHEM] A thermoplastic copolymer of styrene and acrylonitrile with good stiffness and resistance to scratching, chemicals, and stress. Also known as SAN.

styrene-butadiene rubber [MATER] The most common type of synthetic rubber, made by the copolymerization of styrene and butadiene monomers; used in tires, footwear, adhesives, and sealants. Also known as SBR.

styrene monomer See styrene.

styrene oxide [ORG CHEM] C_8H_8O A moderately toxic, combustible, colorless or straw-colored liquid miscible in acetone, ether, and benzene, and melts at 195°C; used as a chemical intermediate.

styrene plastic [ORG CHEM] A plastic made by the polymerization of styrene or the copolymerization of styrene with other unsaturated compounds.

styrene-rubber plastic [MATER] A plastic-rubber mixture consisting of at least 50% of a styrene plastic combined with rubber and various compounding ingredients.

styroflex cable [COMMUN] A radio-frequency cable whose protective covering is a special type of styrene tape.

styryl carbinol See cinnamic alcohol.

styrylformic acid See cinnamic acid.

SU See strontium unit.

SU₃ symmetry See unitary symmetry.

subacute bacterial endocarditis See bacterial endocarditis.

subadditive function [MATH] A function F is subadditive if $f(x + y)$ is less than or equal to $f(x) + f(y)$ for all x and y in its domain.

subaerial [GEOL] Pertaining to conditions and processes occurring beneath the atmosphere or in the open air, that is, on or adjacent to the land surface.

subage [GEOL] A subdivision of a geologic age.

STURGEON

Short-nosed sturgeon (*Acipenser brevirostrus*), a species of sturgeon found in United States coastal waters.

STYLOLITE

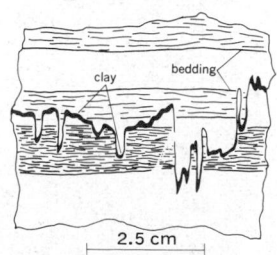

Stylolite in limestone.

subalgebra [MATH] **1.** A subset of an algebra which itself forms an algebra relative to the same operations. **2.** A subalgebra (of sets) is any algebra (of sets) contained in some given algebra.

subalkaline [GEOCHEM] Pertaining to a soil in which the pH is 8.0 to 8.5, usually in a limestone or salt-marsh region.

subalphabet [ADP] A subset of an alphabet.

subangular [SCI TECH] Somewhat angular but free from sharp edges and corners.

Subantarctic Intermediate Water [OCEANOGR] A layer of water above the deep-water layer in the South Atlantic.

subaqueous [HYD] Pertaining to conditions and processes occurring in, under, or beneath the surface of water, especially fresh water.

subaqueous mining [MIN ENG] Surface mining in which the mined material is removed from the bed of a natural body of water.

subarachnoid hemorrhage [MED] Bleeding between the pia mater and the arachnoid of the brain.

subarachnoid space [ANAT] The space between the pia mater and the arachnoid of the brain.

subarctic [GEOGR] Pertaining to regions adjacent to the Arctic Circle or having characteristics somewhat similar to those of these regions.

subarctic climate *See* taiga climate.

subarid [CLIMATOL] Pertaining to regions that are moderately or slightly arid.

subarkose [GEOL] Sandstone that is intermediate in composition between arkose and pure quartz sandstone; it contains less feldspar than arkose.

subassembly [ELECTR] Two or more components combined into a unit for convenience in assembling or servicing equipment; an intermediate-frequency strip for a receiver is an example. [ENG] A structural unit, which, though manufactured separately, was designed for incorporation with other parts in the final assembly of a finished product.

subastral point *See* substellar point.

subatmospheric heating system [MECH ENG] A system which regulates steam flow into the main throttle valve under automatic thermostatic control and maintains a fixed vacuum differential between supply and return by means of a differential controller and a vacuum pump.

subatomic particle [PHYS] A particle which is smaller than an atom, namely, an elementary particle or an atomic nucleus.

subbase for a topology [MATH] A family S of subsets of a topological space X where by taking all finite intersections of sets from S and all unions of such intersections the entire topology of open sets of X is obtained.

subbituminous coal [GEOL] Black coal intermediate in rank between lignite and bituminous coal; has more carbon and less moisture than lignite.

subboreal [ECOL] A biogeographic zone whose climatic condition approaches that of the boreal.

subbottom reflection [GEOPHYS] The return of sound energy from a discontinuity in material below the surface of the sea bottom.

subboundary structure [MET] A network of low-angle grain boundaries of less than one degree within the main crystals of a metal.

subcaliber [ORD] Smaller than the standard caliber for a gun, generally used in practice firing.

subcaliber ammunition [ORD] Ammunition used with a gun or launching tube, usually in practice firing, of a caliber smaller than the standard.

subcaliber equipment [ORD] Any item of equipment, such as small guns, adapters, tubes, and accessories, used for firing subcaliber ammunition in practice drills with larger guns.

subcaliber firing [ORD] Practice firing of subcaliber ammunition, in connection with drills in elevating, traversing, or aiming guns of larger caliber.

subcaliber gun [ORD] A gun mounted on the outside and above the tube of a larger gun; it is used in practice firing of subcaliber ammunition, in connection with aiming drills with the larger gun.

subcaliber mount [ORD] Special mount in or on the tube of a gun, upon which a gun of smaller caliber can be attached for practice firing.

subcaliber rocket [ORD] A rocket designed especially to be fired from launching tubes of larger caliber than the rocket itself.

subcapillary interstice [GEOL] An interstice in which the molecular attraction of its walls extends across the entire opening; it is smaller than a capillary interstice.

subcardinal vein [VERT ZOO] Either of a pair of longitudinal veins of the mammalian embryo or the adult of some lower vertebrates which partly replace the postcardinals in the abdominal region, ventromedial to the mesonephros.

subcarrier [COMMUN] **1.** A carrier that is applied as a modulating wave to modulate another carrier. **2.** *See* chrominance subcarrier.

subcarrier oscillator [ELECTR] **1.** The crystal oscillator that operates at the chrominance subcarrier or burst frequency of 3.579545 megahertz in a color television receiver; this oscillator, synchronized in frequency and phase with the transmitter master oscillator, furnishes the continuous subcarrier frequency required for demodulators in the receiver. **2.** An oscillator used in a telemetering system to translate variations in an electrical quantity into variations of a frequency-modulated signal at a subcarrier frequency.

subcerebral plane [ANTHRO] The plane passing through a line traversing the lower angles of the parietal bones and the juncture of the superciliary ridge and the cheek bone.

subchannel [ADP] The portion of an input/output channel associated with a specific input/output operation. [COMMUN] In a telemetry system, the route required to convey the magnitude of a single subcommutated measurand.

subclavian artery [ANAT] The proximal part of the principal artery in the arm or forelimb.

subclavian vein [ANAT] The proximal part of the principal vein in the arm or forelimb.

subclavius [ANAT] A small muscle attached to the clavicle and the first rib.

subclimax [ECOL] A community immediately preceding a climax in an ecological succession.

subclutter visibility [ELECTR] A measure of the effectiveness of moving-target indicator radar, equal to the ratio of the signal from a fixed target that can be canceled to the signal from a just visible moving target.

subcollateral [ANAT] Ventrad of the collateral sulcus of the brain.

subcommutation [COMMUN] In telemetry, commutation of additional channels with output applied to individual channels of the primary commutator.

subcomponent [DES ENG] A part of a component having characteristics of the component.

subconchoidal [GEOL] Pertaining to a fracture that is partly or vaguely conchoidal in shape.

subconscious [PSYCH] Pertaining to mental activity beyond the level of consciousness, including the preconscious and the unconscious.

subcontinent [GEOGR] **1.** A landmass such as Greenland that is large but not as large as the generally recognized continents. **2.** A large subdivision of a continent (for example, the Indian subcontinent) distinguished geologically or geomorphically from the rest of the continent.

subcontract [ENG] A contract made with a third party by one who has contracted to perform work or service for whole or part performance of that work or service.

subcontractor [ENG] A manufacturer or organization that receives a contract from a prime contractor for a portion of the work on a project.

subcritical [NUCLEO] Having an effective multiplication constant less than one, so that a self-supporting chain reaction cannot be maintained in a nuclear reactor.

subcritical assembly *See* subcritical reactor.

subcritical flow *See* subsonic flow.

subcritical mass [NUCLEO] A piece of fissionable material having an effective multiplication constant of less than one, so that it does not give rise to a self-supporting chain reaction.

subcritical reactor [NUCLEO] A reactor having an effective multiplication constant of less than one, so that a self-supporting chain reaction cannot be maintained. Also known as subcritical assembly; teaching reactor.

subcrop [GEOL] An occurrence of strata beneath the subsurface of an inclusive stratigraphic unit that succeeds an unconformity on which there is marked overstep.

subcutaneous connective tissue [HISTOL] The layer of loose connective tissue beneath the dermis.

subcycle generator [ELECTR] Frequency-reducing device used in telephone equipment which furnishes ringing power at a submultiple of the power supply frequency.

subdevice bit *See* select bit.

subdivided capacitor [ELEC] Capacitor in which several capacitors known as sections are mounted so that they may be used individually or in combination.

subdrainage [CIV ENG] Natural or artificial removal of water from beneath a lined conduit.

subdrilling [ENG] Refers to the breaking of the base in which boreholes are drilled one foot or several feet below the level of the quarry floor.

subduction [GEOL] The process by which one crustal block descends beneath another, such as the descent of the Pacific plate beneath the Andean plate along the Andean Trench.

subdural hematoma [MED] A mass of blood between the arachnoid and the dura mater.

subdural hemorrhage [MED] Bleeding between the dura mater and the arachnoid.

subdwarf star [ASTRON] An intermediate star type; luminosity is between that of main sequence stars and the white dwarf stars on the Hertzsprung-Russell diagram; spectral classes F, G, and K are most numerous.

subendothelial layer [HISTOL] The middle layer of the tunica intima of veins and of medium and larger arteries, consisting of collagenous and elastic fibers and a few fibroblasts.

suberic acid [ORG CHEM] $HOOC(CH_2)_6COOH$ A colorless, crystalline compound that melts at 143°C, and dissolves slightly in cold water; used in organic synthesis. Also known as octanedioic acid.

suberin [BIOCHEM] A fatty substance found in many plant cell walls, especially cork.

suberinite [GEOL] A variety of provitrinite composed of corky tissue.

suberization [BOT] Infiltration of plant cell walls by suberin resulting in the formation of corky tissue that is impervious to water.

suberose [BOT] Having a texture like cork due to or resembling that due to suberization.

subfeldspathic [GEOL] Referring to mature lithic wacke or arenite containing an abundance of quartz grains with less than 10% feldspar grains.

subfield [MATH] **1.** A subset of a field which itself forms a field relative to the same operations. **2.** A subfield (of sets) is any field (of sets) contained in some given field of sets.

subframe [COMMUN] In telemetry, a complete sequence of frames during which all subchannels of a specific channel are sampled once.

subgelisol [GEOL] Unfrozen ground beneath permafrost.

subgeostrophic wind [METEOROL] Any wind of lower speed than the geostrophic wind required by the existing pressure gradient.

subgiant star [ASTRON] A member of the family of stars whose luminosity is intermediate between giants and the main sequence in the Hertzsprung-Russell diagram; spectral classes G and K are most frequent.

subglacial [GEOL] Pertaining to the area in or at the bottom of, or immediately beneath, a glacier.

subgrade [CIV ENG] The soil or rock leveled off to support the foundation of a structure.

subgradient wind [METEOROL] A wind of lower speed than the gradient wind required by the existing pressure gradient and centrifugal force.

subgrain [MET] The portion of a metal crystal or grain with an orientation that differs slightly from the orientation of neighboring portions of the same crystal.

subgraph [MATH] A graph contained in a given graph which has as its vertices some subset of the vertices of the original.

subgraywacke [PETR] An argillaceous sandstone with a composition intermediate between graywacke and orthoquartzite; a clay matrix is usually present but it amounts to less than 15%.

subgroup [MATH] A subset N of a group G which is itself a group relative to the same operation.

subharmonic [PHYS] A sinusoidal quantity having a frequency that is an integral submultiple of the frequency of some other sinusoidal quantity to which it is referred. A third subharmonic would be one-third the fundamental or reference frequency.

subharmonic function [MATH] A continuous function is subharmonic in a region R of the plane if its value at any point z_0 of R is less than or equal to its integral along a circle centered at z_0.

subharmonic triggering [ELECTR] A method of frequency division which makes use of a triggered multivibrator having a period of one cycle which allows triggering only by a pulse that is an exact integral number of input pulses from the last effective trigger.

subhedral [MINERAL] **1.** Pertaining to an individual mineral crystal that is partly bounded by its own crystal faces and partly bounded by surfaces formed against preexisting crystals. **2.** Descriptive of a crystal having partially developed crystal faces.

subhumid climate [CLIMATOL] A humidity province based on its typical vegetation. Also known as grassland climate; prairie climate.

subidiomorphic *See* hypidiomorphic.

subirrigation [AGR] Natural or artificial irrigation of crops by the application of water below the ground surface.

subjacent [GEOL] Being lower than but not directly underneath.

subjacent igneous body [GEOL] An igneous intrusion without a known floor, and which presumably enlarges downward.

subjamming visibility [ORD] Relates to how well a particular radar antijam technique can see through jamming signals.

subject copy [GRAPHICS] Material that is to be transmitted for facsimile reproduction. Also known as copy.

subjective probability *See* personal probability.

subkiloton weapon [ORD] A nuclear weapon producing a yield below 1 kiloton.

sublacustrine [GEOL] Existing or formed on the bottom of a lake.

sublacustrine channel [GEOL] A channel eroded in a lake bed either before the lake existed or by a strong current in the lake.

sublethal gene *See* semilethal gene.

subleukemic [MED] Less than leukemic; usually applied to states in which the peripheral blood manifestations of leukemia are temporarily suppressed.

sublevel [ATOM PHYS] *See* subshell. [MIN ENG] An intermediate level opened a short distance below the main level; or in the caving system of mining, a 15–20-foot (4.6–6.1-meter) level below the top of the ore body, preliminary to caving the ore between it and the level above.

sublevel caving [MIN ENG] A stoping method in which relatively thin blocks of ore are caused to cave by successively undermining small panels.

sublevel drive [MIN ENG] A drive often made in a section which divides the deposit into narrower panels and zones.

sublevel stoping [MIN ENG] A mining method involving overhand, underhand, and shrinkage stoping; the characteristic feature is the use of sublevels which are worked simultaneously, the lowest on a given block being farthest advanced and the sublevels above following one another at short intervals.

sublimation [PSYCH] A defense mechanism whereby the energies of undesirable instinctual cravings and impulses are converted into socially acceptable activities. [THERMO] The process by which solids are transformed directly to the vapor state or vice versa without passing through the liquid phase.

sublimation cooling [THERMO] Cooling caused by the extraction of energy to produce sublimation.

sublimation curve [THERMO] A graph of the vapor pressure of a solid as a function of temperature.

sublimation energy [THERMO] The increase in internal energy when a unit mass, or 1 mole, of a solid is converted into a gas, at constant pressure and temperature.

sublimation nucleus [METEOROL] Any particle upon which an ice crystal may grow by the process of sublimation.

sublimation point [THERMO] The temperature at which the vapor pressure of the solid phase of a compound is equal to the total pressure of the gas phase in contact with it; analogous to the boiling point of a liquid.

sublimation pressure [THERMO] The vapor pressure of a solid.

sublimation vein [GEOL] A vein of mineral that has condensed from a vapor.

sublimatography [ANALY CHEM] A procedure of fractional sublimation in which a solid mixture is separated into bands along a condensing tube with a temperature gradient.

sublimator [CHEM] Device used for the heating of solids (usually under vacuum) to the temperature at which the solid sublimes.

sublime [THERMO] To change from the solid to the gaseous state without passing through the liquid phase.

sublimed sulfur *See* flowers of sulfur.

subliminal [PHYSIO] Below the threshold of responsiveness, consciousness, or sensation to a stimulus.

subliming rocket propellant [MATER] A propellant characterized by sublimation of the material at the heated surface.

sublingual gland [ANAT] A complex of salivary glands located in the sublingual fold on each side of the floor of the mouth.

sublittoral zone [OCEANOGR] The benthic region extending from mean low water (or 40–60 meters, according to some authorities) to a depth of about 100 fathoms (200 meters), or the edge of a continental shelf, beyond which most abundant attached plants do not grow.

sublunar point [ASTRON] The moon's geographic zenith position at any particular moment in time.

submachine gun [ORD] A type of machine gun; a short-barreled shoulder firearm using pistol-type ammunition and capable of full automatic fire.

submandibular duct [ANAT] The duct of the submandibular gland which empties into the mouth on the side of the frenulum of the tongue.

submandibular gland [ANAT] A large seromucous or mixed salivary gland located below the mandible on each side of the jaw. Also known as mandibular gland; submaxillary gland.

submarine [NAV ARCH] A ship that can operate both on the surface of the water and completely submerged. [OCEANOGR] Being or functioning in the sea.

submarine base [NAV ARCH] A base providing logistic support for submarines.

submarine bell [NAV] A bell whose signal is transmitted through water.

submarine blast [ENG] A charge of high explosives fired in boreholes drilled in the rock underwater for dislodging dangerous projections and for deepening channels.

Hydrofoil *Victoria*, with submerged foils and autopilot, designed by Gibbs and Cox, Inc.

submarine cable [ELEC] A cable designed for service under water; usually a lead-covered cable with steel armor applied between layers of jute.

submarine cave *See* submarine fan.

submarine delta *See* submarine fan.

submarine earthquake *See* seaquake.

submarine false target [ORD] A pyrotechnic item designed to be ejected from a submarine to confuse and disrupt underwater echo-ranging equipment and create a bubble wake which can be seen by aircraft and surface vessels.

submarine fan [GEOL] A shallow marine sediment that is fan- or cone-shaped and lies off the seaward opening of large rivers and submarine canyons. Also known as abyssal cave; abyssal fan; sea fan; submarine cave; submarine delta; subsea apron.

submarine geology *See* geological oceanography.

submarine isthmus [GEOL] A submarine elevation joining two land areas and separating two basins or depressions by a depth less than that of the basins.

submarine mine [MIN ENG] A mine for the extraction of minerals or ores under the sea.

submarine navigation [NAV] 1. Navigation of a submarine, whether submerged or surfaced. 2. *See* underwater navigation.

submarine oscillator [ENG ACOUS] A large, electrically operated diaphragm horn which produces a powerful sound for signaling through water.

submarine peninsula [GEOL] An elevated portion of the submarine relief resembling a peninsula.

submarine pipeline [ENG] A pipeline installed under water, resting on the bed of the waterway; frequently used for petroleum or natural gas transport across rivers, lakes, or bays.

submarine pit [GEOL] A cavity on the bottom of the sea. Also known as submarine well.

submarine plain *See* plain.

submarine relief [GEOL] Relative elevations of the ocean bed, or the representation of them on a chart.

submarine rocket *See* SUBROC.

submarine sanctuary [NAV ARCH] Any of the restricted areas established for the conduct of noncombat submarine or antisubmarine exercises; they may be either stationary or moving and are normally designated only in rear areas.

submarine sentry [ENG] A form of underwater kite towed at a predetermined constant depth in search of elevations of the bottom; the kite rises to the surface upon encountering an obstruction.

submarine sound signal [ACOUS] A sound signal transmitted through water.

submarine spring [HYD] A spring of water issuing from the bottom of the sea.

submarine station [OCEANOGR] 1. One of the places for which tide or tidal current predictions are determined by applying a correction to the predictions of a reference station. 2. A tide or tidal current station at which a short series of observations have been made; these observations are reduced by comparison with simultaneous observations at a reference station.

submarine striking forces [NAV ARCH] Submarines having guided or ballistic missile launching or guidance capabilities formed to launch offensive nuclear strikes.

submarine topography [GEOL] Configuration of a surface such as the sea bottom or of a surface of given characteristics within the water mass.

submarine trench *See* trench.

submarine trough *See* trough.

submarine valley *See* valley.

submarine weathering [GEOL] A slow alteration of the form, texture, and composition of the sea floor from chemical, thermal, and biological causes.

submarine well *See* submarine pit.

submaxillary gland *See* submandibular gland.

submerged-arc furnace [MET] An arc-heating furnace in which the arcs may be completely submerged under the charge or in the molten bath under the charge.

submerged-arc welding [MET] Arc welding with a bare metal electrode, the arc and tip of the electrode being shielded by a blanket of granular, fusible material.

submerged breakwater [OCEANOGR] A breakwater with its top below the still water level; when struck by a wave, part of the wave energy is reflected seaward and the remaining energy is largely dissipated in a breaker, transmitted shoreward as a multiple crest system, or transmitted shoreward as a simple wave system.

submerged coastal plain [GEOL] The continental shelf as the seaward extension of a coastal plain on the land. Also known as coast shelf.

submerged-combustion evaporator [ENG] A liquid-evaporation device in which heat is provided by combustion gases bubbling up through the liquid; the burner is submerged in the body of the liquid.

submerged-combustion heater [ENG] A combustion device in which fuel and combustion air are mixed and ignited below the surface of a liquid; used in heaters and evaporators where absorption of the combustion products will not be detrimental.

submerged culture [MICROBIO] A method for growing pure cultures of aerobic bacteria in which microorganisms are incubated in a liquid medium subjected to continuous, vigorous agitation.

submerged fermentation [MICROBIO] Industrial production of antibiotics, enzymes, and other substances by growing the microorganisms that produce the product in a submerged culture.

submerged-foil hydrofoil [NAV ARCH] A hydrofoil craft with foils which are completely submerged and whose angle of attack is controlled by an autopilot and sensors to maintain the height and attitude of the craft.

submerged lands [GEOL] Lands covered by water at any stage of the tide, as distinguished from tidelands which are attached to the mainland or an island and are covered or uncovered with the tide; tidelands presuppose a high-water line as the upper boundary, submerged lands do not.

submerged screw log [NAV ARCH] A type of electric log which is actuated by the flow of water past a propeller.

submerged shoreline *See* shoreline of submergence.

submerged weir [CIV ENG] A dam which, when in use, has the downstream water level at an elevation equal to or higher than the crest of the dam.

submergence [GEOL] A change in the relative levels of water and land either from a sinking of the land or a rise of the water level.

submersible pump [MECH ENG] A pump and its electric motor together in a protective housing which permits the unit to operate under water.

submetallic [OPTICS] Referring to a luster intermediate between metallic and nonmetallic, such as exhibited by the mineral chromite.

submillimeter wave [ELECTROMAG] An electromagnetic wave whose wavelength is less than 1 millimeter, corresponding to frequencies above 300 gigahertz.

subminiature tube [ELECTR] An extremely small electron tube designed for use in hearing aids and other miniaturized equipment; a typical subminiature tube is about 1½ inches (4 centimeters) long and 0.4 inch (1 centimeter) in diameter, with the pins emerging through the glass base.

submissile [ORD] One of several smaller missiles carried and released by a larger missile, especially in a warhead.

submucosa [HISTOL] The layer of fibrous connective tissue that attaches a mucous membrane to its subadjacent parts.

submucous plexus [ANAT] A visceral nerve network lying in the submucosa of the digestive tube. Also known as Meissner's plexus.

submultiple resonance [PHYS] Resonance at a frequency that is a submultiple of the frequency of the exciting impulses.

suboptimization [SYS ENG] The process of fulfilling or optimizing some chosen objective which is an integral part of a broader objective; usually the broad objective and lower-level objective are different.

Suboscines [VERT ZOO] A major division of the order Passeriformes, usually divided into the suborders Eurylaimi, Tyranni, and Memirae.

subpolar anticyclone *See* subpolar high.

subpolar glacier [HYD] A polar glacier with 10 to 20 meters of firn in the accumulation area where some melting occurs.

subpolar high [METEOROL] A high that forms over the cold continental surfaces of subpolar latitudes, principally in Northern Hemisphere winters; these highs typically migrate eastward and southward. Also known as polar anticyclone; polar high; subpolar anticyclone.

subpolar low-pressure belt [METEOROL] A belt of low pressure located, in the mean, between 50 and 70° latitude; in the Northern Hemisphere, this belt consists of the Aleutian low and the Icelandic low; in the Southern Hemisphere, it is supposed to exist around the periphery of the Antarctic continent.

subpolar westerlies *See* westerlies.

subpopulation [STAT] A subset of population. Also known as stratum.

subprogram [ADP] A part of a larger program which can be converted independently into machine language.

subrefraction [ELECTROMAG] Atmospheric refraction which is less than standard refraction.

subring [MATH] 1. A subset I of a ring R where I is also a ring relative to the operations of R. 2. A subring (of sets) is any ring (of sets) contained in some given ring (of sets).

SUBROC [ORD] A Navy missile that is fired underwater from a conventional torpedo tube, but is rocket-powered and most of its course is through the air, and it has a range of about 50 miles (80 kilometers); used against enemy submarines. Abbreviation for submarine rocket.

subroutine [ADP] 1. A body of computer instruction (and the associated constants and working-storage areas, if any) designed to be used by other routines to accomplish some particular purpose. 2. A statement in FORTRAN used to define the beginning of a closed subroutine (first definition).

subscale [MET] Oxidation occurring within a metal instead of on the surface.

subscapularis [ANAT] A muscle arising from the costal surface of the scapula and inserted on the lesser tubercle of the humerus.

subscriber line [ELEC] A telephone line between a central office and a telephone station, private branch exchange, or other end equipment. Also known as central office line; subscriber loop.

subscriber loop *See* subscriber line.

subscriber multiple [ELEC] Bank of jacks in a manual switchboard providing outgoing access to subscriber lines, and usually having more than one appearance across the face of the switchboard.

subscriber set *See* subset.

subscript [SCI TECH] A letter or symbol written below, and usually to the right, of another symbol for any of various purposes, such as to identify a particular element or elements of a set, to denote a constant value of a variable, or, in a chemical formula, to indicate the number of atoms of a particular kind in a molecule.

subscription television [COMMUN] A television system in which programs are broadcast in coded or scrambled form, for reception only by subscribers who make individual or monthly payments for use of the decoding or unscrambling devices required to obtain a clear program. Also known as pay television; toll television.

subsea apron *See* submarine fan.

subsequence [MATH] A subsequence of a given sequence is any sequence all of whose entries appear in the original sequence and in the same manner of succession.

subsequent [GEOL] Referring to a geologic feature that followed in time the development of a consequent feature of which it is a part.

subsequent drainage [HYD] Drainage by a stream developed subsequent to the system of which it is a part; drainage follows belts of weak rocks.

subsequent stream [HYD] A stream that flows in the general direction of the strike of the underlying strata and is subsequent to the formation of the consequent stream of which it is a tributary. Also known as longitudinal stream; strike stream.

subsequent valley [GEOL] A valley eroded by a stream developed subsequent to the system of which it is a part.

subsere [ECOL] A secondary community that succeeds an interrupted climax.

subset [COMMUN] A telephone or other subscriber equipment connected to a communication system, such as a modulation-demodulation device designed to make business-machine signals compatible with communication facilities. Abbreviation for subscriber set. [MATH] A subset A of a set B is a set all of whose elements are included in B.

subshell [ATOM PHYS] Electrons of an atom within the same shell (energy level) and having the same azimuthal quantum numbers. Also known as sublevel.

subsidence [METEOROL] A descending motion of air in the atmosphere, usually with the implication that the condition extends over a rather broad area. [MIN ENG] A sinking down of a part of the earth's crust due to underground excavations.

subsidence break [MIN ENG] A fracture in the rocks overlying a coal seam or mineral deposit resulting from mining operations.

subsidence inversion [METEOROL] A temperature inversion produced by the adiabatic warming of a layer of subsiding air; this inversion is enhanced by vertical mixing in the air layer below the inversion.

subsidiary conduit [CIV ENG] Terminating branch of an underground conduit run extending from a manhole or handhole to a nearby building, handhole, or pole.

subsidiary fracture *See* tension fracture.

subsidiary transport [MIN ENG] The conveying or haulage of coal or mineral from the working faces to a junction or loading point.

subsieve analysis [MET] Analysis by size distribution of metal powder particles all of which pass through a standard 44-micrometer sieve.

SUBTENSE BAR

The subtense bar used in the subtense technique of distance measurement. (*Lockwood, Kessler, and Bartlett Inc.*)

subsieve fraction [MET] The fraction of particles of a metal powder which pass through a standard 44-micrometer sieve.

subsoil [GEOL] 1. Soil underlying surface soil. 2. *See* B horizon.

subsoil ice *See* ground ice.

subsolar point [ASTRON] The sun's zenith geographic position at any particular moment in time.

subsonic [ACOUS] *See* infrasonic. [PHYS] Of, pertaining to, or dealing with speeds less than acoustic velocity, as in subsonic aerodynamics.

subsonic flight [AERO ENG] Movement of a vehicle through the atmosphere at a speed appreciably below that of sound waves; extends from zero (hovering) to a speed about 85% of sonic speed corresponding to ambient temperature.

subsonic flow [FL MECH] Flow of a fluid at a speed less than that of the speed of sound in the fluid. Also known as subcritical flow.

subsonic inlet [ENG] An entrance or orifice for the admission of fluid flowing at speeds less than the speed of sound in the fluid.

subsonic nozzle [ENG] A nozzle through which a fluid flows at speed less than the speed of sound in the fluid.

subsonic speed [FL MECH] A speed relative to surrounding fluid less than that of the speed of sound in the same fluid.

subspace [MATH] A subset of a space which, in the appropriate context, is a space in its own right.

substance [PHYS] Tangible material, occurring in macroscopic amounts.

substandard propagation [ELECTROMAG] The propagation of radio energy under conditions of substandard refraction in the atmosphere; that is, refraction by an atmosphere or section of the atmosphere in which the index of refraction decreases with height at a rate of less than 12 N-units (unit of index of refraction) per 1000 feet (304.8 meters).

substantive dye *See* direct dye.

substation [ELEC] *See* electric power substation. [ENG] An intermediate compression station to repressure a fluid being transported by pipeline over a long distance. [MIN ENG] A subsidiary station for the conversion of power to the type, usually direct current, and voltage needed for mining equipment and fed into the mine power system.

substellar point [ASTRON] The geographical position of a star; that point on the earth at which the star is in the zenith at a specified time. Also known as substar point.

substitute mode [ADP] One method of exchange buffering, in which segments of storage function alternately as buffer and as program work area.

substitution reaction [CHEM] Replacement of an atom or radical by another one in a chemical compound.

substitution solid solution [MET] A solid alloy having the atoms of the solute located at some lattice of points of the solvent.

substrate [ELECTR] The physical material on which a microcircuit is fabricated; used primarily for mechanical support and insulating purposes, as with ceramic, plastic, and glass substrates; however, semiconductor and ferrite substrates may also provide useful electrical functions. [ENG] Basic surface on which a material adheres, for example, paint or laminate.

substratosphere [METEOROL] A region of indefinite lower limit just below the stratosphere.

substratum [GEOL] Any layer underlying the true soil.

substructure [CIV ENG] The part of a structure which is below ground.

subsurface contour *See* structure contour.

subsurface current [OCEANOGR] An underwater current which is not present at the surface or whose core (region of maximum velocity) is below the surface.

subsurface flow [HYD] Interflow plus groundwater flow.

subsurface geology [GEOL] The study of geologic features beneath the land or sea-floor surface. Also known as underground geology.

subsurface tillage [AGR] A method of stirring the soil with blades that leaves stubble on or just below the surface.

subsurface waste disposal [ENG] A waste disposal method for manufacturing wastes in porous underground rock formations.

subsurface wave [ELECTROMAG] Electromagnetic wave propagated through water or land; operating frequencies for communications may be limited to approximately 35 kilohertz due to attenuation of high frequencies.

subsynchronous [ELEC] Operating at a frequency or speed that is related to a submultiple of the source frequency.

subsystem [ENG] A major part of a system which itself has the characteristics of a system, usually consisting of several components.

subtend [BOT] To lie adjacent to and below another structure, often enclosing it.

subtense bar [ENG] The horizontal bar of fixed length in the subtense technique of distance measurement method.

subtense technique [CIV ENG] A distance measuring technique in which the transit angle subtended by the subtense bar enables the computation of the transit-to-bar distance.

subterranean ice *See* ground ice.

subterranean stream [HYD] A subsurface stream that flows through a cave or a group of communicating caves.

subtilin [MICROBIO] An antibiotic substance obtained from *Bacillus subtilis*, active against gram-positive bacteria.

subtraction [MATH] The addition of one quantity with the negative of another; in a system with an additive operation this is formally the sum of one element with the additive inverse of another.

subtrahend [MATH] A quantity which is to be subtracted from another given quantity.

Subtriquetridae [INV ZOO] A family of arthropods in the suborder Porocephaloidea.

subtropic [METEOROL] An indefinite belt in each hemisphere between the tropic and temperate regions; the polar boundaries are considered to be roughly 35–40° northern and southern latitudes, but vary greatly according to continental influence, being farther poleward on the western coasts of continents and farther equatorward on the eastern coasts.

subtropical anticyclone *See* subtropical high.

Subtropical Convergence [OCEANOGR] The zone of converging currents, generally located in midlatitudes.

subtropical cyclone [METEOROL] The low-level (surface chart) manifestation of a cutoff low.

subtropical easterlies *See* tropical easterlies.

subtropical easterlies index [METEOROL] A measure of the strength of the easterly wind between the latitudes of 20° and 35°N; the index is computed from the average sea-level pressure difference between these latitudes and is expressed as the east to west component of the corresponding geostrophic wind in meters and tenths of meters per second.

subtropical forest *See* temperate rainforest.

subtropical high [METEOROL] One of the semipermanent highs of the subtropical high-pressure belt; these highs appear as centers of action on mean charts of surface pressure; they lie over oceans and are best developed in the summer season. Also known as oceanic anticyclone; oceanic high; subtropical anticyclone.

subtropical high-pressure belt [METEOROL] One of the two belts of high atmospheric pressure that are centered, in the mean, near 30°N and 30°S latitudes; these belts are formed by the subtropical highs.

subtropical westerlies *See* westerlies.

subulate [BOT] Linear, delicate, and tapering to a sharp point.

Subulitacea [PALEON] An extinct superfamily of gastropod mollusks in the order Prosobranchia which possessed a basal fold but lacked an apertural sinus.

Subuluridae [INV ZOO] The equivalent name for Heterakidae.

subumbrella [INV ZOO] The concave undersurface of the body of a jellyfish.

subvoice-grade channel [COMMUN] A channel whose bandwidth is smaller than the bandwidth of a voice-grade channel; it is usually a subchannel of a voice-grade line.

subway [CIV ENG] An underground passage.

subway-type transformer [ELEC] Transformer of submersible construction.

succession [ECOL] A gradual process brought about by the change in the number of individuals of each species of a community and by the establishment of new species populations which may gradually replace the original inhabitants.

succession of crops [AGR] **1.** Growing a crop over a long season by either repeated sowings or a single sowing of varieties of the crop that mature at different rates. **2.** Growing two or more crops in a season on the same land by planting them in succession.

successive fracture treatment [PETRO ENG] A second or third fracturing operation of an oil well in an oil reservoir to fracture a new part or zone.

succinamide [BIOCHEM] $H_2NCOCH_2CONH_2$ The amide of succinic acid.

succinate [ORG CHEM] A salt or ester of succinic acid; for example, sodium succinate, $Na_2C_4H_4O_4 \cdot 6H_2O$, the reaction product of succinic acid and sodium hydroxide.

succinic acid [ORG CHEM] $CO_2H(CH_2)_2CO_2H$ Water-soluble, colorless crystals with an acid taste; melts at 185°C; used as a chemical intermediate, in medicine, and to make perfume esters. Also known as butanedioic acid.

succinic acid dehydrogenase [BIOCHEM] An enzyme that catalyzes the dehydrogenation of succinic acid to fumaric acid in the presence of a hydrogen acceptor. Also known as succinic dehydrogenase.

succinic anhydride [ORG CHEM] $C_4H_4O_3$ Colorless or pale needles soluble in alcohol and chloroform; converts to succinic acid in water; melts at 120°C; used as a chemical and pharmaceutical intermediate and a resin hardener. Also known as butanedioic anhydride; 2,5-diketotetrahydrofurane; succinyl oxide.

succinic dehydrogenase *See* succinic acid dehydrogenase.

succinimide [ORG CHEM] $C_4H_5O_2N \cdot H_2O$ Colorless or tannish water-soluble crystals with a sweet taste; melts at 126°C; used to make plant growth stimulants and as a chemical intermediate. Also known as 2,5-diketopyrrolidine.

succinite [MINERAL] An amber-colored variety of grossularite.

succinonitrite *See* ethylene cyanide.

succinoxidase [BIOCHEM] A complex enzyme system containing succinic dehydrogenase and cytochromes that catalyzes the conversion of succinate ion and molecular oxygen to fumarate ion.

succinylcholine chloride [ORG CHEM] $[Cl(CH_3)_3N(CH_2)_2-OOCH_2]_2 \cdot 2H_2O$ Water-soluble white crystals with a bitter taste, melts at 162°C; used in medicine. Also known as choline succinate dichloride dihydrate.

succinyl oxide *See* succinic anhydride.

succinylsulfathiazole [PHARM] $C_{13}H_{13}N_3O_5S_2$ A poorly absorbed sulfonamide used as an intestinal antibacterial agent in preoperative preparation of patients for abdominal surgery, and also postoperatively to maintain a low bacterial count.

succulent [BOT] Describing a plant having juicy, fleshy tissue.

succus entericus [PHYSIO] The intestinal juice secreted by the glands of the intestinal mucous membrane; it is thin, opalescent, alkaline, and has a specific gravity of 1.011.

sucker [BOT] A shoot that develops rapidly from the lower portion of a plant, and usually at the expense of the plant. [ZOO] A disk-shaped organ in various animals for adhering to or holding onto an individual, usually of another species.

sucker rod [PETRO ENG] A connecting rod between a downhole oil-well pump and the lifting or pumping device on the surface.

sucker-rod pump [PETRO ENG] A cylinder-piston-type pump used to displace oil into the oil-well tubing string, and to the surface.

sucking louse [INV ZOO] The common name for insects of the order Anoplura, so named for the slender, tubular mouthparts.

sucrase *See* saccharase.

sucrose [ORG CHEM] $C_{12}H_{22}O_{11}$ Combustible, white crystals soluble in water, decomposes at 160 to 186°C; derived from sugarcane or sugarbeet; used as a sweetener in drinks and foods and to make syrups, preserves, and jams. Also known as saccharose; table sugar.

sucrose octoacetate [ORG CHEM] $C_{28}H_{38}O_{19}$ A bitter crystalline compound that forms needles from alcohol solution, melts at 89°C, and breaks down at 286°C or above; used as an adhesive, to impregnate and insulate paper, and in lacquers and plastics.

sucrosic *See* saccharoidal.

suction anemometer [ENG] An anemometer consisting of an inverted tube which is half-filled with water that measures the change in water level caused by the wind's force.

suction boundary layer control [AERO ENG] A technique that is used in addition to purely geometric means to control boundary layer flow; it consists of sucking away the retarded flow in the lower regions of the boundary through slots or perforations in the surface.

suction cup [ENG] A cup, often of flexible material such as rubber, in which a partial vacuum is created when it is inverted on a surface; the vacuum tends to hold the cup in place.

suction-cutter dredger [MECH ENG] A dredger in which rotary blades dislodge the material to be excavated, which is then removed by suction as in a sand-pump dredger.

suction dredge [NAV ARCH] A vessel equipped with a centrifugal pump to excavate under water.

suction head *See* suction lift.

suction lift [MECH ENG] The head, in feet, that a pump must provide on the inlet side to raise the liquid from the supply well to the level of the pump. Also known as suction head.

suction line [ENG] A pipe or tubing feeding into the inlet of a fluid impelling device (for example, pump, compressor, or blower), consequently under suction.

suction pump [MECH ENG] A pump that raises water by the force of atmospheric pressure pushing it into a partial vacuum under the valved piston, which retreats on the upstroke.

suction stroke [MECH ENG] The piston stroke that draws a fresh charge into the cylinder of a pump, compressor, or internal combustion engine.

suction wave *See* rarefaction wave.

Suctoria [INV ZOO] A small subclass of the protozoan class Ciliatea, distinguished by having tentacles which serve as mouths.

Suctorida [INV ZOO] The single order of the protozoan subclass Suctoria.

sudamen [MED] A skin disease in which sweat accumulates under the superficial horny layers of the epidermis to form small, clear, transparent vesicles.

sudatoria *See* hyperhidrosis.

sudburite [GEOL] A basic basalt composed of hypersthene, augite, and magnetite, among other minerals.

sudden commencement [GEOPHYS] Magnetic storms which start suddenly (within a few seconds) and simultaneously all over the earth.

sudden death syndrome *See* crib death.

sudden ionospheric disturbance [GEOPHYS] A complex combination of sudden changes in the condition of the ionosphere following the appearance of solar flares, and the effects of these changes. Abbreviated SID.

sudomotor [PHYSIO] Pertaining to the efferent nerves that control the activity of sweat glands.

suede [MATER] Leather with a velvet finish on the flesh side of the skin; calfskin is the commonest suede leather. Also known as napped leather.

suestada [METEOROL] Strong southeast winds occurring in winter along the coast of Argentina, Uruguay, and southern Brazil; they cause heavy seas and are accompanied by fog and rain; the counterpart of the northeast storm in North America.

suevite [GEOL] A grayish or yellowish fragmental rock associated with meteorite impact craters; resembles tuff breccia or pumiceous tuff but is of nonvolcanic origin.

suffrutescent [BOT] Of or pertaining to a stem intermediate between herbaceous and shrubby, becoming partly woody and perennial at the base.

suffruticose [BOT] Low stems which are woody, grading into herbaceous at the top.

sugar [BIOCHEM] A generic term for a class of carbohydrates usually crystalline, sweet, and water soluble; examples are glucose and fructose.

sugar alcohol [ORG CHEM] Any of the acyclic linear polyhydric alcohols; may be considered sugars in which the aldehydic group of the first carbon atom is reduced to a primary alcohol; classified according to the number of hydroxyl groups in the molecule; sorbitol (D-glucitol, sorbite) is one of

SUCKER

A sucker, characterized by fleshy lips.

SUCTION BOUNDARY LAYER CONTROL

Cross-section of airfoil with suction boundary layer control.

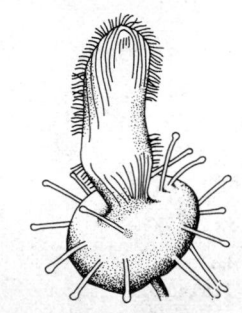

SUCTORIA

Endogenous budding in the suctorian *Podophrya*, a species which measures 10–28 micrometers.

SUGARBEET

Root with leaves of a typical sugarbeet. (USDA)

SUGARCANE

Flowering sugarcane in Florida. (USDA)

SULFENIC ACID

HO—S—R

Structural formula of sulfenic acids; R = alkyl group or aryl group such as CH₃.

the most widespread of all the naturally occurring sugar alcohols.

sugarbeet [BOT] *Beta vulgaris.* A beet characterized by a white root and cultivated for the high sugar content of the roots.

sugar berg [OCEANOGR] An iceberg of porous glacier ice.

sugarcane [BOT] *Saccharum officinarum.* A stout, perennial grass plant characterized by two-ranked leaves, and a many-jointed stalk with a terminal inflorescence in the form of a silky panicle; the source of more than 50% of the world's annual sugar production.

sugarcane gummosis *See* Cobb's disease.

sugarcane wax [MATER] Hard, tan to dark-green wax extracted from sugarcane; melts at 77°C; used in polishes, lubricants, and food wrappers.

sugarloaf sea [OCEANOGR] A sea characterized by waves that rise into sugarloaf shapes, with little wind, possibly resulting from intersecting waves.

sugar maple [BOT] *Acer saccharum.* A commercially important species of maple tree recognized by its gray furrowed bark, sharp-pointed scaly winter buds, and symmetrical oval outline of the crown.

sugar of lead *See* lead acetate.

sugar refining [FOOD ENG] The purification of sugar from sugarcane or sugarbeet by the removal of impurities (vegetable proteins and salts), followed by decolorization and crystallization of the pure sucrose.

sugar snow *See* depth hoar.

sugary *See* saccharoidal.

Suhl amplifier [SOLID STATE] A parametric microwave amplifier which utilizes the instability of certain spin waves in a ferromagnetic material subjected to intense microwave fields.

Suhl effect [ELECTR] When a strong transverse magnetic field is applied to an *n*-type semiconducting filament, holes injected into the filament are deflected to the surface, where they may recombine rapidly with electrons or be withdrawn by a probe.

Suidae [VERT ZOO] A family of paleodont artiodactyls in the superfamily Suoidae including wild and domestic pigs.

sukhovei [METEOROL] Literally dry wind; a dry, hot, dusty wind in the south Russian steppes, which blows principally from the east and frequently brings a prolonged drought and crop damage.

sulcate [ZOO] Having furrows or grooves on the surface.

sulculus [ZOO] A small sulcus.

sulcus [ZOO] A furrow or groove, especially one on the surface of the cerebrum.

sulfadiazine [PHARM] $C_{10}H_{10}O_2N_4S$ An antibacterial sulfonamide used in the treatment of a variety of infections.

sulfa drug [PHARM] Any of a family of drugs of the sulfonamide type with marked bacteriostatic properties.

sulfaguanidine [PHARM] $C_7H_{10}N_4O_2S$ An intestinal antibacterial sulfonamide proposed for treatment of dysentery and for sterilization of the colon prior to gastrointestinal tract surgery.

sulfamate [CHEM] A salt of sulfamic acid; for example, calcium sulfamate, $Ca(SO_3NH_2)_2 \cdot 4H_2O$.

sulfamerazine [PHARM] $C_{11}H_{12}N_4O_2S$ An antibacterial agent with uses similar to those of sulfadiazine, but generally used in combination with sulfadiazine and with sulfamethazine.

sulfamic acid [INORG CHEM] HSO_3NH_2 White, nonvolatile crystals slightly soluble in water and organic solvents, decomposes at 205°C; used to clean metals and ceramics, and as a plasticizer, fire retardant, chemical intermediate, and textile and paper bleach.

sulfamidyl *See* sulfanilamide.

sulfanilamide [PHARM] $C_6H_8O_2N_2S$ White crystals slightly soluble in water, soluble in alcohol and most sulfa drugs, but less effective and more toxic than its derivatives. Also known as *para*-aminobenzenesulfonamide; sulfamidyl; sulfonamide P.

sulfanilic acid [ORG CHEM] $C_6H_4NH_2 \cdot SO_3H \cdot H_2O$ Combustible, grayish-white crystals slightly soluble in water, alcohol, and ether, soluble in fuming hydrochloric acid; chars at 280–300°C; used in medicine and dyestuffs and as a chemical intermediate. Also known as *para*-aminobenzenesulfonic acid; *para*-anilinesulfonic acid.

meta-sulfanilic acid *See* metanillic acid.

sulfapyridine [PHARM] $C_{11}H_{11}N_3O_2S$ A sulfonamide formerly used for the treatment of various infections but found to be too toxic for general use; now employed only as a suppressant for dermatitis herpetiformis.

sulfatase [BIOCHEM] Any of a group of esterases that catalyze the hydrolysis of sulfuric esters.

sulfate [CHEM] **1.** A compound containing the $-SO_4$ group, as in sodium sulfate Na_2SO_4. **2.** A salt of sulfuric acid.

sulfate mineral [MINERAL] A mineral compound characterized by the sulfate radical SO_4.

sulfate paper [MATER] Paper made by the sulfate process, and which cannot be bleached as white as soda or sulfite paper; strong papers, such as kraft paper, are made from unbleached sulfate materials.

sulfate pulping [CHEM ENG] A wood-pulping process in which sodium sulfate is used in the caustic soda pulp-digestion liquor. Also known as kraft process; kraft pulping.

sulfathiazole [PHARM] $C_9H_9N_3O_2S_2$ A sulfa drug formerly widely used in the treatment of pneumococcal, staphylococcal, and urinary tract infections; it has been replaced by less toxic sulfonamides.

sulfatide lipidosis *See* metachromatic leukodystrophy.

sulfating [ELEC] The formation of lead sulfate on the plates of lead-acid storage batteries reducing the energy-storing ability of the battery and eventually causing failure.

sulfation [CHEM] The conversion of a compound into a sulfate by the oxidation of sulfur, as in sodium sulfide, Na_2S, oxidized to sodium sulfate, Na_2SO_4; or the addition of a sulfate group, as in the reaction of sodium and sulfuric acid to form Na_2SO_4.

sulfenic acid [ORG CHEM] An oxy acid of sulfur with the general formula RSOH, where R is an alkyl or aryl group such as CH_3; known as the esters and halides.

sulfenyl chloride [ORG CHEM] Any of a group of well-known organosulfur compounds with the general formula RSCl; although highly reactive compounds, they can generally be synthesized and isolated; examples are trichloromethanesulfenyl chloride and 2,4-dinitrobenzenesulfenyl chloride.

sulfhemoglobin [BIOCHEM] A greenish substance derived from hemoglobin by the action of hydrogen sulfide; it may appear in the blood following the ingestion of sulfanilamide and other substances.

sulfhydryl compound [CHEM] A compound with a $-SH$ group. Also known as a mercapto compound.

sulfidation [CHEM] The chemical insertion of a sulfur atom into a compound.

sulfide [CHEM] Any compound with one or more sulfur atoms in which the sulfur is connected directly to a carbon, metal, or other nonoxygen atom; for example, sodium sulfide, Na_2S.

sulfide dye [ORG CHEM] A dye containing sulfur and soluble in a 0.25–0.50% sodium sulfide solution, and used to dye cotton; the dyes are manufactured from aromatic polyamines or hydroxy amines; the amine group is primary, secondary, or tertiary, or may be an equivalent nitro, nitroso, or imino group; an example is the dye sulfur blue. Also known as sulfur dye.

sulfide mineral [MINERAL] A mineral compound characterized by the linkage of sulfur with a metal or semimetal.

sulfinate [ORG CHEM] **1.** A compound containing the R_2SX_2 grouping, where X is a halide. **2.** A salt of sulfinic acid having the general formula $R \cdot OH \cdot S:O$.

sulfinic acid [ORG CHEM] Any of the monobasic organic acids of sulfur with the general formula RS:O(OH); for example, ethanesulfinic acid, $C_2H_5SO_2H$.

sulfinyl bromide *See* thionyl bromide.

sulfisoxazole [PHARM] $C_{11}H_{13}N_3O_3S$ A sulfonamide of general therapeutic utility; for parenteral administration the soluble salt sulfisoxazole diethanolamine is used; for pediatric use the tasteless derivative acetyl sulfisoxazole is given.

sulfite [INORG CHEM] M_2SO_3 A salt of sulfurous acid, for example, sodium sulfite, Na_2SO_3.

sulfite paper [MATER] Paper made from sulfite pulp.

sulfite pulp [MATER] Wood chips digested with a solution of magnesium, ammonium, or calcium disulfite, with free sulfur dioxide present; used to make paper and paper products from spruce and other coniferous woods.

sulfite waste liquor [MATER] Waste reactants and other impurities from the sulfite pulping of wood; used as a foaming and emulsifying agent, in adhesives and tanning, and for road construction.

sulfo- [CHEM] Prefix for a compound with either a divalent sulfur atom, or the presence of $-SO_3H$, the sulfo group in a compound. Also spelled sulpho-.

sulfoborite [MINERAL] $Mg_6H_4(BO_3)_4(SO_4)_2 \cdot 7H_2O$ A mineral composed of hydrous acid sulfate and borate of magnesium.

sulfocarbimide *See* isothiocyanate.

sulfocyanate *See* thiocyanate.

sulfocyanic acid *See* thiocyanic acid.

sulfocyanide *See* thiocyanate.

sulfofication [GEOCHEM] Oxidation of sulfur and sulfur compounds into sulfates, occurring in soils by the agency of bacteria.

sulfohalite [MINERAL] $Na_6(SO_4)_2FCl$ A mineral composed of sulfate, chloride, and fluoride of sodium.

sulfonamide [ORG CHEM] One of a group of organosulfur compounds, RSO_2NH_2, prepared by the reaction of sulfonyl chloride and ammonia; used for sulfa drugs.

sulfonamide P *See* sulfanilamide.

sulfonate [CHEM] A sulfuric acid derivative or a sulfonic acid ester containing a SO_4^{--} group. [ORG CHEM] Any of a group of petroleum hydrocarbons derived from sulfuric-acid treatment of oils, used as synthetic detergents, emulsifying and wetting agents, and chemical intermediates.

sulfonated castor oil *See* Turkey red oil.

sulfonated oil [MATER] Mineral or vegetable oil treated with sulfuric acid to make a water-soluble (emulsifiable) form; used as lubricants, emulsifiers, defoamers, and softeners.

sulfonation [CHEM] Substitution of -SO_3H groups (from sulfuric acid) for hydrogen atoms, for example, conversion of benzene, C_6H_6, into benzenesulfonic acid, $C_6H_5SO_3H$.

sulfone [ORG CHEM] R_2SO_2 (or RSOOR) A compound formed by the oxidation of sulfides, for example, ethyl sulfone, $C_4H_{10}SO_2$, from ethyl sulfide, $C_4H_{10}S$; the use of sulfones, particularly 4,4'-sulfonyldianiline (dapsone) in the treatment of leprosy leads to apparent improvement; relapses associated with sulfone-resistant strains have been encountered.

sulfonic acid [ORG CHEM] A compound with the radical $-SO_2OH$, derived by the sulfuric acid replacement of a hydrogen atom; for example, conversion of benzene, C_6H_6, to the water-soluble benzenesulfonic acid, $C_6H_5SO_3H$, by treatment with sulfuric acid; used to make dyes and drugs.

sulfonyl [CHEM] Also known as sulfuryl. 1. A compound containing the radical $-SO_2-$. 2. A prefix denoting the presence of a sulfone group.

sulfonyl chloride *See* sulfuryl chloride.

4,4'-sulfonyldianiline [PHARM] $C_{12}H_{12}N_2O_2^{\cdot}S$ A sulfone that precipitates as crystals from alcohol; melting point 175–176°C; used in the treatment of leprosy. Also known as dapsone.

sulfoparaldehyde *See* trithioacetaldehyde.

sulfoxide [ORG CHEM] R_2SO A compound with the radical $=SO$; derived from oxidation of sulfides, the proportion of oxidant, such as hydrogen peroxide, and temperature being set to avoid excessive oxidation; an example is dimethyl sulfoxide $(CH_3)_2SO$.

sulfur [CHEM] A nonmetallic element in group VIa, symbol S, atomic number 16, atomic weight 32.064, existing in a crystalline or amorphous form and in four stable isotopes; used as a chemical intermediate and fungicide, and in rubber vulcanization. [MINERAL] A yellow orthorhombic mineral occurring in crystals, masses, or layers, and existing in several allotropic forms; the native form of the element.

sulfur-35 [NUC PHYS] Radioactive sulfur with mass number 35; radiotoxic, with 87.1-day half-life, β radiation; derived from pile irradiation; used as a tracer to study chemical reactions, engine wear, and protein metabolism.

sulfurated lime *See* calcium sulfide.

sulfuration [CHEM] The chemical act of combining an element or compound with sulfur.

sulfur bacteria [MICROBIO] Any of various bacteria having the ability to oxidize sulfur compounds.

sulfur ball [GEOL] A bubble of hot volcanic gas encased in a sulfurous mud skin that solidified on contact with air.

sulfur bichloride *See* sulfur dichloride.

sulfur bromide [INORG CHEM] S_2Br_2 A toxic, irritating, yellow liquid that reddens in air, soluble in carbon disulfide, decomposes in water, boils at 54°C. Also known as sulfur monobromide.

sulfur cement [MATER] Cement used for connecting iron parts; made of equal parts of sulfur and pitch.

sulfur chloride [INORG CHEM] S_2Cl_2 A combustible, water-soluble, oily, fuming, amber to yellow-red liquid with an irritating effect on the eyes and lungs, boils at 138°C; used to make military gas and insecticides, in rubber substitutes and cements, to purify sugar juices, and as a chemical intermediate. Also known as sulfur monochloride; sulfur subchloride.

sulfur dichloride [INORG CHEM] SCl_2 A red-brown liquid boiling (when heated rapidly) at 60°C, decomposes in water; used to make insecticides, for rubber vulcanization, and as a chemical intermediate and a solvent. Also known as sulfur bichloride.

sulfur dioxide [INORG CHEM] SO_2 A toxic, irritating, colorless gas soluble in water, alcohol, and ether, boils at -10°C; used as a chemical intermediate, in artificial ice, paper pulping, and ore refining, and as a solvent. Also known as sulfurous acid anhydride.

sulfur dome [MET] An inverted container containing a high concentration of sulfur dioxide gas, used in die casting to cover a pot of molten magnesium to prevent burning.

sulfur dye *See* sulfide dye.

sulfur hexafluoride [INORG CHEM] SF_6 A colorless gas soluble in alcohol and ether, slightly soluble in water, sublimes at -64°C; used as a dielectric in electronics.

sulfur hexameter [ENG] An instrument used to measure or to continuously monitor the amount of sulfur hexafluoride present in a waveguide or other device in which this gas is used as a dielectric.

sulfuric acid [INORG CHEM] H_2SO_4 A toxic, corrosive, strongly acid, colorless liquid that is miscible with water and dissolves most metals, and melts at 10°C; used in industry in the manufacture of chemicals, fertilizers, and explosives, and in petroleum refining. Also known as dipping acid.

sulfuric-acid alkylation [CHEM ENG] A petroleum refinery alkylation process in which three-carbon, four-carbon, and five-carbon olefins combine with isobutane in the presence of a sulfuric-acid catalyst to form high-octane, branched-chain hydrocarbons; used in motor gasoline.

sulfuric chloride *See* sulfuryl chloride.

sulfuric chlorohydrin *See* chlorosulfonic acid.

sulfur iodide *See* sulfur iodine.

sulfur iodine [INORG CHEM] I_2S_2 A gray-black brittle mass with an iodine aroma and a metallic luster, insoluble in water, soluble in carbon disulfide; used in medicine. Also known as iodine bisulfide; iodine disulfide; sulfur iodide.

sulfurized oil [MATER] Any of various mineral oils and fatty oils containing active sulfur to increase film strength and load-carrying ability; used generally as cutting fluids.

sulfur monobromide *See* sulfur bromide.

sulfur monochloride *See* sulfur chloride.

sulfur monoxide [INORG CHEM] SO A gas at ordinary temperatures; produces an orange-red deposit when cooled to temperatures of liquid air; prepared by passing an electric discharge through a mixture of sulfur vapor and sulfur dioxide at low temperature.

sulfur number [ANALY CHEM] The number of milligrams of sulfur per 100 milliliters of sample, determined by electrometric titration; used in the petroleum industry for oils.

sulfurous acid [INORG CHEM] H_2SO_3 An unstable, water-soluble, colorless liquid with a strong sulfur aroma; derived from absorption of sulfur dioxide in water; used in the synthesis of medicine and chemicals, manufacture of paper and wine, brewing, metallurgy, and ore flotation, as a bleach and analytic reagent, and to refine petroleum products.

sulfurous acid anhydride *See* sulfur dioxide.

sulfurous oxychloride *See* thionyl chloride.

sulfur oxide [INORG CHEM] An oxide of sulfur, such as sulfur dioxide, SO_2, and sulfur trioxide, SO_3.

sulfur oxychloride *See* thionyl chloride.

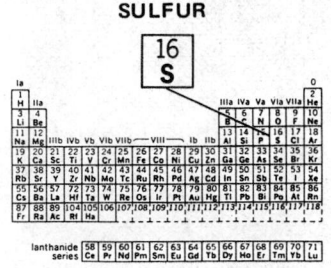

SULFUR

Periodic table of the chemical elements showing the position of sulfur.

sulfur spring [HYD] A spring containing sulfur compounds such as hydrogen sulfide.

sulfur subchloride *See* sulfur chloride.

sulfur test [ANALY CHEM] **1.** Method to determine the sulfur content of a petroleum material by combustion in a bomb. **2.** Analysis of sulfur in petroleum products by lamp combustion in which combustion of the sample is controlled by varying the flow of carbon dioxide and oxygen to the burner.

sulfur trioxide [INORG CHEM] SO_3 A toxic, irritating liquid in three forms, α, β, γ, with respective melting points of 62°C, 33°C, and 17°C; a strong oxidizing agent and fire hazard; used for sulfonation of organic chemicals.

sulfuryl *See* sulfonyl.

sulfuryl chloride [INORG CHEM] SO_2Cl_2 A colorless liquid with a pungent aroma, boils at 69°C, decomposed by hot water and alkalies; used as a chlorinating agent and solvent and for pharmaceuticals, dyestuffs, rayon, and poison gas. Also known as chlorosulfuric acid; sulfonyl chloride; sulfuric chloride.

Sulidae [VERT ZOO] A family of aquatic birds in the order Pelecaniformes including the gannets and boobies.

Sullivan angle compressor [MECH ENG] A two-stage compressor in which the low-pressure cylinder is horizontal and the high-pressure cylinder is vertical; a compact compressor driven by a belt, or directly connected to an electric motor or diesel engine.

Sullivan reaction [ORG CHEM] The formation of a red-brown color when cysteine is reacted with 1,2-naphthoquinone-4-sodium sulfate in a highly alkaline reducing medium.

sulpho- *See* sulfo-.

sulphophile element [GEOCHEM] An element occurring preferentially in an oxygen-free mineral. Also known as thiophile element.

sultriness [METEOROL] An oppressively uncomfortable state of the weather which results from the simultaneous occurrence of high temperature and high humidity, and often enhanced by calm air and cloudiness.

sulvanite [MINERAL] Cu_3VS_4 A bronze-yellow mineral composed of copper vanadium sulfide occurring in masses.

Sulzer two-cycle engine [MECH ENG] An internal combustion engine utilizing the Sulzer Co. system for the effective scavenging and charging of the two-cycle diesel engine.

sum [MATH] **1.** The addition of numbers or mathematical objects in context. **2.** The sum of an infinite series is the limit of the sequence consisting of all partial sums of the series.

sumac wax *See* Japan wax.

sumatra [METEOROL] A squall, with wind speeds occasionally exceeding 30 miles (48 kilometers) per hour, in the Malacca Strait between Malay and Sumatra during the southwest monsoon (April through November).

sumbul oil [MATER] Oil from the root of a muskroot (*Ferula sumbul*); formerly used in medicine as an antispasmodic.

sumi ink *See* india ink.

sumi reed [GRAPHICS] A reed about a foot long with a sharp point at one end, used in ink drawings.

summary plotter [ORD] A polar coordinate chart showing the tactical distribution of ships and aircraft.

summary punching [ADP] The conversion of information developed by accounting machines into holes in a punch card.

summary recorder [ADP] In computers, output equipment which records a summary of the information handled.

summation check [ADP] An error-detecting procedure involving adding together all the digits of some number and comparing this sum to a previously computed value of the same sum.

summation convention [MATH] An abbreviated notation used particularly in tensor analysis and relativity theory, in which a product of tensors is to be summed over all possible values of any index which appears twice in the expression.

summation network *See* summing network.

summation principle [METEOROL] In United States weather observing practice, the rule which governs the assignment of sky cover amount to any layer of cloud or obscuring phenomenon, and to the total sky cover; in essence, this principle states that the sky cover at any level is equal to the summation of the sky cover of the lowest layer plus the additional sky cover provided at all successively higher layers up to and including the layer in question; thus, no layer can be assigned a sky cover less than a lower layer, and no sky cover can be greater than 1.0 (10/10).

summation tone [ACOUS] Combination tone, heard under certain circumstances, whose pitch corresponds to a frequency equal to the sum of the frequencies of the two components.

summer black oil *See* tempering oil.

summer noon *See* daylight saving noon.

summer solstice [ASTRON] **1.** The sun's position on the ecliptic when it reaches its greatest northern declination. Also known as first point of Cancer. **2.** The date, about June 21, on which the sun has its greatest northern declination.

summer time *See* daylight saving time.

summerwood [BOT] The less porous, usually harder portion of an annual ring that forms in the latter part of the growing season.

summing amplifier [ELECTR] An amplifier that delivers an output voltage which is proportional to the sum of two or more input voltages or currents.

summing network [ELEC] A passive electric network whose output voltage is proportional to the sum of two or more input voltages. Also known as summation network.

summit [SCI TECH] **1.** The apex of a pyramid. **2.** The highest point of, for example, a mountain, road, trajectory, or tower.

Sumner line [NAV] **1.** A line of position established by the Sumner method. **2.** Any celestial line of position (deprecated usage).

Sumner method [NAV] The establishing of a line of position from the observation of the altitude of a celestial body by assuming two latitudes (or longitudes) and calculating the longitudes (or latitudes) through which the line of position passes; the line of position is the straight line connecting these two points (extended if necessary).

sum of states *See* partition function.

sum over states *See* partition function.

sump [ENG] A pit or tank which receives and temporarily stores drainage at the lowest point of a circulating or drainage system.

sump fuse [ENG] A fuse used for underwater blasting.

sump pump [MECH ENG] A small, single-stage vertical pump used to drain shallow pits or sumps.

sum rule [QUANT MECH] A formula which sets some quantity equal to the sum, over all the states of a system, of another quantity, usually involving the square of the magnitude of the matrix element of some operator between a given state and the state being summed over.

sun [ASTRON] The star about which the earth revolves; it is a globe of gas 1.4×10^6 kilometers in diameter, held together by its own gravity; thermonuclear reactions take place in the deep interior of the sun converting hydrogen into helium releasing energy which streams out. Also known as Sol.

sun-and-planet motion [MECH ENG] A train of two wheels moving epicyclically with a small wheel rotating a wheel on the central axis.

sunburn [MED] Skin inflammation due to overexposure to sunlight.

sun compass [NAV] A device which utilizes the shadow of a shadow pin, or gnomon, and a suitable dial to facilitate use of the sun for determination of direction.

sun crack *See* mud crack.

sun cross [METEOROL] A rare halo phenomenon in which bands of white light intersect over the sun at right angles; it appears probable that most of such observed crosses appear merely as a result of the superposition of a parhelic circle and a sun pillar.

sun crust [HYD] A type of snow crust, formed by refreezing of surface snow crystals after having been melted by the sun.

sundew [BOT] Any plant of the genus *Drosera* of the family Droseraceae; the genus comprises small, herbaceous, insectivorous plants that grow on all continents, especially Australia.

sundial [HOROL] An instrument for telling time by the sun; it is composed of a stylus that casts a shadow and a dial plate, on which hour lines are marked and upon which the shadow falls.

sun dog *See* parhelion.

SUNDEW

The sundew, *Drosera* species, a small, herbaceous, insectivorous plant. *(General Biological Supply House)*

sun drawing water [METEOROL] Popular designation for a phenomenon of the sun showing through scattered openings in a layer of clouds into a layer of turbid air that is hazy or dusty; bright bands are seen where the several beams of sunlight pass down through the subcloud layer; sailors called the phenomenon the backstays of the sun.

sunfish [VERT ZOO] Any of several species of marine and freshwater fishes in the families Centrarchidae and Molidae characterized by brilliant metallic skin coloration.

sunflower [BOT] *Helianthus annuus.* An annual plant native to the United States characterized by broad, ovate leaves growing from a single, usually long (3–20 feet or 1–6 meters) stem, and large, composite flowers with yellow petals.

sunflower oil [MATER] A combustible, pale-yellow, semi-drying oil with a pleasant scent, expressed from the seeds of the common sunflower; soluble in alcohol, ether, and carbon disulfide; consists mostly of mixed triglycerides of fatty acids; used for resins, soaps, edible oils, and margarines.

sun follower [ELECTR] A photoelectric pickup and an associated servomechanism used to maintain a sun-facing orientation, as for a space vehicle. Also known as sun seeker.

sun gear *See* central gear.

S-unit meter *See* signal-strength meter.

sunk [MIN ENG] Drilled downward, as a shaft.

sunken rock [NAV] A rock that is a potential danger to navigation; for cartographic purposes, it is a rock whose summit is below the lower limit of the zone for a rock awash.

sunlamp [ELEC] A mercury-vapor gas-discharge tube used to produce ultraviolet radiation for therapeutic or cosmetic purposes.

sun line [NAV] A line of position determined from a sextant observation of the sun.

sunlit aurora [GEOPHYS] An aurora which occurs in the part of the upper atmosphere which is in the sunlight, above the earth's shadow.

sun opal *See* fire opal.

sun pillar [METEOROL] A luminous streak of light, white or slightly reddened, extending above and below the sun, most frequently observed near sunrise or sunset; it may extend to about 20° above the sun, and generally ends in a point. Also known as light pillar.

sun-pumped laser [OPTICS] A continuous-wave laser in which pumping is achieved by concentrating the energy of the sun on the laser crystal with a parabolic mirror. Also known as solar-excited laser.

sunrise [ASTRON] The exact moment the upper limb of the sun appears above the horizon.

sun satellite *See* solar satellite.

sunscald [PL PATH] An injury to woody plants which results in local death of the plant tissues; in summer it is caused by excessive action of the sun's rays, in winter, by the great variation of temperature on the side of trees that is exposed to the sun in cold weather.

sun seeker *See* sun follower.

sun sensor *See* solar sensor.

sunset [ASTRON] The exact moment the upper limb of the sun disappears below the horizon.

sunshine [ASTRON] Direct radiation from the sun, as opposed to the shading of a location by clouds or by other obstructions.

sunshine integrator [ENG] An instrument for determining the duration of sunshine (daylight) in any locality.

sunshine recorder [ENG] An instrument designed to record the duration of sunshine without regard to intensity at a given location; sunshine recorders may be classified in two groups according to the method by which the time scale is obtained: in one group the time scale is obtained from the motion of the sun in the manner of a sun dial, in the second group the time scale is supplied by a chronograph.

sunshine unit *See* strontium unit.

sunspot [ASTRON] A dark area in the photosphere of the sun caused by a lowered surface temperature.

sunspot cycle [ASTRON] Variation of the size and number of sunspots in an 11-year cycle which is shared by all other forms of solar activity.

sunspot maximum [ASTRON] The time in the solar cycle when the number of sunspots reaches a maximum value.

sunspot number *See* relative sunspot number.

sunspot relative number *See* relative sunspot number.

sunstone [MINERAL] An aventurine feldspar containing minute flakes of hematite; usually brilliant and translucent, it emits reddish or golden billowy reflection. Also known as heliolite.

sun strobe [ELECTR] The signal display seen on a radar plan-position-indicator screen when the radar antenna is aimed at the sun; the pattern resembles that produced by continuous-wave interference, and is due to radio-frequency energy radiated by the sun.

sunstroke [MED] Heat stroke resulting from prolonged exposure to the sun, characterized by extreme pyrexia, prostration, convulsion, and coma. Also known as thermic fever.

sun's way [ASTRON] The path of the solar system through space.

sun-synchronous orbit [AERO ENG] An earth orbit of a spacecraft so that the craft is always in the same direction relative to that of the sun; as a result, the spacecraft passes over the equator at the same spots at the same times.

Suoidea [VERT ZOO] A superfamily of artiodactyls of the suborder Paleodonta which comprises the pigs and peccaries.

superacid [CHEM] 1. An excessively acid condition. 2. A solution of acetic or phosphoric acid.

superadiabatic lapse rate [METEOROL] An environmental lapse rate greater than the dry-adiabatic lapse rate, such that potential temperature decreases with height.

superaerodynamics [FL MECH] That branch of gas dynamics dealing with the flow of gases at such low density that the molecular mean free path is not negligibly small; under these conditions the gas no longer behaves as a continuous fluid.

superalloy [MET] A thermally resistant alloy for use at elevated temperatures where high stresses and oxidation are encountered.

supercalendered finish [MATER] A shiny, smooth-surface paper obtained by passing the paper between alternating fiber-filled and steel rolls with the application of steam and pressure.

supercalendering [ENG] A calendering process that uses both steam and high pressure to give calendered material, for example, paper, a high-density finish.

supercapillary interstice [GEOL] An interstice that is too large to hold water above the free water surface by surface tension; it is larger than a capillary interstice.

supercavitating propeller [NAV ARCH] A marine propeller which has special blade sections so that at sufficiently high speed the whole back of each blade becomes enveloped by a smooth sheet of cavitation.

supercentrifuge [MECH ENG] A centrifuge built to operate at faster speeds than an ordinary centrifuge.

supercharge method [ENG] A method for measuring the knock-limited power, under supercharge rich-mixture conditions, of fuels for use in spark-ignition aircraft engines.

supercharger [MECH ENG] An air pump or blower in the intake system of an internal combustion engine used to increase the weight of air charge and consequent power output from a given engine size.

supercharging [MECH ENG] A method of introducing air for combustion into the cylinder of an internal combustion engine at a pressure in excess of that which can be obtained by natural aspiration.

supercilium [ANAT] The eyebrow.

supercobalt drill [DES ENG] A drill made of 8% cobalt high-speed steel; used for drilling work-hardened stainless steels, silicon chrome, and certain chrome-nickel alloy steels.

supercompressibility factor *See* compressibility factor.

superconducting alloy [MET] An alloy capable of exhibiting superconductivity, such as an alloy of niobium and zirconium or an alloy of lead and bismuth.

superconducting circuit [CRYO] An electric circuit having elements which are in a superconducting state at least part of the time, such as a cryotron.

superconducting device *See* cryogenic device.

superconducting gyroscope *See* cryogenic gyroscope.

superconducting magnet [CRYO] An electromagnet whose coils are made of a type II superconductor with a high transition temperature and extremely high critical field, such

SUNFISH

Black crappie *(Pomoxis nigromaculatus),* a freshwater sunfish also known as the calico bass.

SUNFLOWER

Maturing sunflower on a farm in California. *(USDA)*

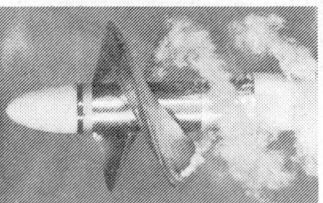

SUPERCAVITATING PROPELLER

Supercavitating propeller in 24-inch (61-centimeter) water tunnel at David Taylor Model Basin, Carderock, Maryland. *(U.S. Navy photograph)*

as niobium tin, Nb$_3$Sn; it is capable of generating magnetic fields of 100,000 oersteds and more with no steady power dissipation.

superconducting material   *See* superconductor.

superconducting memory   [CRYO] A computer memory made up of a number of cryotrons, thin-film cryotrons, superconducting thin films, or other superconducting storage devices; these operate only under cryogenic conditions and dissipate power only during the read or write operation, which permits construction of large, dense memories.

superconducting metal   [MET] A metal capable of exhibiting superconductivity.

superconducting thin film   [CRYO] A thin film of indium, tin, or other superconducting element, used as a cryogenic switching or storage device, as in a thin-film cryotron.

superconductivity   [SOLID STATE] A property of many metals, alloys, and chemical compounds at temperatures near absolute zero by virtue of which their electrical resistivity vanishes and they become strongly diamagnetic.

superconductor   [SOLID STATE] Any material capable of exhibiting superconductivity; examples include iridium, lead, mercury, niobium, tin, tantalum, vanadium, and many alloys. Also known as cryogenic conductor; superconducting material.

supercooled cloud   [METEOROL] A cloud composed of supercooled liquid waterdrops.

supercooling   [THERMO] Cooling of a substance below the temperature at which a change of state would ordinarily take place without such a change of state occurring, for example, the cooling of a liquid below its freezing point without freezing taking place; this results in a metastable state.

supercritical   [NUCLEO] Having an effective multiplication constant greater than 1, so that the rate of reaction rises rapidly in a nuclear reactor.

supercritical flow   *See* supersonic flow.

supercritical mass   [NUCLEO] A mass of nuclear-reactor fuel whose effective multiplication factor is greater than 1.

supercritical reactor   [NUCLEO] A nuclear reactor in which the effective multiplication constant is greater than 1 and consequently a reactor that is increasing its power level; if uncontrolled, a supercritical reactor will undergo a sudden and dangerous rise in power level.

SUPERCRITICAL WING

Shape of a supercritical wing.

supercritical wing   [AERO ENG] A wing developed to permit subsonic aircraft to maintain an efficient cruise speed very close to the speed of sound; the middle portion of the wing is relatively flat with substantial downward curvature near the trailing edge; in addition, the forward region of the lower surface is more curved than that of the upper surface with a substantial cusp of the rearward portion of the lower surface.

supercurrent   [SOLID STATE] In the two-fluid model of superconductivity, the current arising from motion of superconducting electrons, in contrast to the normal current.

superego   [PSYCH] The subdivision of the psyche that acts as the conscience of the unconscious; the components, derived from both the id and the ego, are associated with standards of behavior and self-criticism.

superelevation   [ORD] **1.** An added positive angle in antiaircraft gunnery that compensates for the fall of the projectile during the time of flight due to the pull of gravity. **2.** The angle the gun or launcher much be elevated above the gun-target line.

superemitron camera   *See* image iconoscope.

superexchange   [SOLID STATE] A phenomenon in which two electrons from a double negative ion (such as oxygen) in a solid go to different positive ions and couple with their spins, giving rise to a strong antiferromagnetic coupling between the positive ions, which are too far apart to have a direct exchange interaction.

superfecundation   [PHYSIO] Multiple, simultaneous fertilization by a number of sperm of many eggs released at ovulation.

superfemale   [GEN] A female with three X chromosomes and two sets of autosomes resulting in sterility and genelly early death.

superficial cleavage   [EMBRYO] Meroblastic cleavage restricted to the peripheral cytoplasm, as in the centrolecithal insect ovum.

superficial deposit   *See* surficial deposit.

superficial palmar arch   [ANAT] The arterial anastomosis formed by the ulnar artery in the palm with a branch from the radial artery. Also known as palmar arch.

superficial Rockwell hardness test   [MET] A test to determine surface hardness of thin sheet material which applies relatively light loads producing minimal penetration and damage.

superfines   [MET] The portion of a metal powder composed of particles smaller than 10 microns.

superfinishing   [MET] Fine honing of a metal surface with abrasive stones.

superfische   [GRAPHICS] A sheet of film, usually 4 by 6 inches (10 by 15 centimeters) in size, containing either negative or positive images, or frames, of printed material reduced from 40 to 100 times by photographic reduction.

superfluid   [PHYS] A collection of particles which obey Bose-Einstein statistics and are all in the lowest energy state allowed by quantum mechanics, having zero entropy and zero resistance to motion; examples are a fraction of the atoms in liquid helium II and a fraction of the pairs of electrons in a superconductor.

superfluidity   [CRYO] The frictionless flow of liquid helium at temperatures very close to absolute zero through holes as small as 10^{-7} centimeter in diameter, and for particle velocities below a few centimeters per second.

supergalaxy   [ASTRON] A hypothetical very large group of galaxies which together fill an ellipsoidal space.

supergene   [MINERAL] Referring to mineral deposits or enrichments formed by descending solutions. Also known as hypergene.

supergeostrophic wind   [METEOROL] Any wind of greater speed than the geostrophic wind required by the pressure gradient.

supergiant star   [ASTRON] A member of the family containing the intrinsically brightest stars, populating the top of the Hertzsprung-Russell diagram; supergiant stars occur at all temperatures from 30,000 to 3000K and have luminosities from 10^4 to 10^6 times that of the sun; the star Betelgeuse is an example.

superglacial   [HYD] Of or pertaining to the upper surface of a glacier or ice sheet.

supergradient wind   [METEOROL] A wind of greater speed than the gradient wind required by the existing pressure gradient and centrifugal force.

supergroup   [COMMUN] In carrier telephony, five groups (60 voice channels) multiplexed together and treated as a unit; a basic supergroup occupies the band between 312 and 552 kilohertz.

superharmonic function   [MATH] A continuous complex function f whose value at a point z_0 exceeds its average values computed by the integral of f around a circle centered at z_0.

superheated vapor   [THERMO] A vapor that has been heated above its boiling point.

superheater   [MECH ENG] A component of a steam-generating unit in which steam, after it has left the boiler drum, is heated above its saturation temperature.

superheating   [THERMO] Heating of a substance above the temperature at which a change of state would ordinarily take place without such a change of state occurring, for example, the heating of a liquid above its boiling point without boiling taking place; this results in a metastable state.

superhet   *See* superheterodyne receiver.

superheterodyne receiver   [ELECTR] A receiver in which all incoming modulated radio-frequency carrier signals are converted to a common intermediate-frequency carrier value for additional amplification and selectivity prior to demodulation, using heterodyne action; the output of the intermediate-frequency amplifier is then demodulated in the second detector to give the desired audio-frequency signal. Also known as superhet.

superhigh frequency   [COMMUN] A frequency band from 3,000 to 30,000 megahertz, corresponding to wavelengths from 1 to 10 centimeters. Abbreviated SHF.

superhighway   [CIV ENG] A broad highway, such as an expressway, freeway, turnpike, for high-speed traffic.

superimposed   [GEOL] Pertaining to layered or stratified rocks.

superimposed coding [ADP] A means of placing many keywords in a single card area, where they can be scanned simultaneously.

superimposed drainage [HYD] Drainage by superimposed streams.

superimposed fan [GEOL] An alluvial fan developed on, and having a steeper gradient than, an older fan.

superimposed glacier [GEOL] A glacier whose course is maintained despite different preexisting structures and lithologies as the glacier erodes downward.

superimposed stream [HYD] A stream, started on a new surface, that kept its course through the different preexisting lithologies and structures encountered as it eroded downward into the underlying rock. Also known as superinduced stream.

superimposed valley [GEOL] A valley eroded by or containing a superimposed stream.

superinduced stream See superimposed stream.

superinfection [VIROL] An attack on a bacterial cell by several phages due to the introduction of large numbers of viruses into the bacterial culture.

superinsulation [CHEM ENG] A multilayer insulation for cryogenic systems, comprised of many floating radiation shields in an evacuated double-wall annulus, closely spaced but thermally separated by a poor-conducting fiber.

superior [BOT] 1. Positioned above another organ or structure. 2. Referring to a calyx that is attached to the ovary. 3. Referring to an ovary that is above the insertion of the floral parts.

superior air [METEOROL] An exceptionally dry mass of air formed by subsidence and usually found aloft but occasionally reaching the earth's surface during extreme subsidence processes.

superior alveolar canals [ANAT] The alveolar canals of the maxilla.

superior conjunction [ASTRON] A conjunction when an astronomical body is opposite the earth on the other side of the sun.

superior ganglion [ANAT] 1. The upper sensory ganglion of the glossopharyngeal nerve, located in the upper part of the jugular foramen; it is inconstant. 2. The upper sensory ganglion of the vagus nerve, located in the jugular foramen.

superior mesenteric artery [ANAT] A major branch of the abdominal aorta with branches to the pancreas and intestine.

superior mirage [OPTICS] A spurious image of an object formed above the object's position by abnormal refraction conditions; opposite to an inferior mirage.

superior planet [ASTRON] Any of the planets that are farther than the earth from the sun; includes Mars, Jupiter, Saturn, Uranus, Neptune, and Pluto.

superior tide [OCEANOGR] The tide in the hemisphere in which the moon is above the horizon.

superior transit See upper transit.

superior vena cava [ANAT] The principal vein collecting blood from the head, chest wall, and upper extremities and draining into the right atrium.

superjacent roadway system See Hirschback method.

superjacent waters [OCEANOGR] The waters above the continental shelf.

superlattice [SOLID STATE] An ordered arrangement of atoms in a solid solution which forms a lattice superimposed on the normal solid solution lattice. Also known as superstructure.

supermale [GEN] A male with one X chromosome and three or more sets of autosomes, resulting in sterility and generally early death.

Supermalloy [MET] Trademark for a soft magnetic alloy having very low hysteresis loss and a maximum permeability greater than 10^6 henries per meter; it consists of 79% nickel, 16% iron, and 5% molybdenum.

supermassive star [ASTRON] A star with a mass exceeding about 50 times that of the sun.

supermature [GEOL] Pertaining to a texturally mature clastic sediment whose grains have become rounded.

Supermendur [MET] Trademark for a soft magnetic material having a nearly rectangular hysteresis loop, high saturation flux density, and low coercive force; it consists of 49% iron, 49% cobalt, and 2% vanadium.

supermode laser [OPTICS] Frequency-modulated laser, the output of which is passed through a second phase modulator driven 180° out of phase and with the same modulation index as the first modulator; all of the energy of the previously existing laser modes is compressed into a single frequency with nearly the full power of the laser concentrated in that signal.

supermultiplet [QUANT MECH] A set of quantum-mechanical states each of which has the same value of some fundamental quantum numbers and differs from the other members of the set by other quantum numbers, which take values from a range of numbers dictated by the fundamental quantum numbers.

supernatant liquor [ENG] The liquid above settled solids, as in a gravity separator.

supernova [ASTRON] A star that suddenly bursts into very great brilliance as a result of its blowing up; it is orders of magnitude brighter than a nova.

supernumerary bud See accessory bud.

supernumerary chromosome [CYTOL] A chromosome present in addition to the normal chromosome complement. Also known as accessory chromosome.

supernumerary rainbow [OPTICS] One of a set of weakly colored rainbow arcs sometimes discernible inside a primary rainbow; they are of smaller angular width, and fade toward the common center.

superphosphate [MATER] The most important phosphorus fertilizer, derived by action of sulfuric acid on phosphate rock (mostly tribasic calcium phosphate) to produce a mix of gypsum and monobasic calcium phosphate.

superposed [BOT] 1. Growing vertically over another part. 2. Of or pertaining to floral parts that are opposite each other.

superposed circuit [COMMUN] Additional channel obtained from one or more circuits, normally provided for other channels, in a way that all channels can be used simultaneously without mutual interference.

superposed stream See consequent stream.

superposition [GEOL] 1. The order in which sedimentary layers are deposited, the highest being the youngest. 2. The process by which the layering occurs. [MATH] The principle of superposition states that any given geometric figure in a euclidean space can be so moved about as not to change its size or shape. [PHYS] Addition of phenomena when the sum of two physically realizable disturbances is also physically realizable; for example, sound waves are superposable in this sense, but shock waves are not.

superposition eye [INV ZOO] A compound eye in which a given rhabdome receives light from a number of facets; visual acuity is reduced in this type of eye.

superposition integral [CONT SYS] An integral which expresses the response of a linear system to some input in terms of the impulse response or step response of the system; it may be thought of as the summation of the responses to impulses or step functions occurring at various times.

superposition principle See principle of superposition.

superposition theorem See principle of superposition.

superregeneration [ELECTR] Regeneration in which the oscillation is broken up or quenched at a frequency slightly above the upper audibility limit of the human ear by a separate oscillator circuit connected between the grid and anode of the amplifier tube, to prevent regeneration from exceeding the maximum useful amount.

superresolution [OCEANOGR] Separation of tides into components of different frequencies, without taking measurements for the full extent of the longest-period component.

Super Sabre [AERO ENG] A United States supersonic, single-engine, turbojet-powered, tactical and air superiority fighter capable of delivering either nuclear or nonnuclear bombs, rockets, and Bullpup missiles against surface targets, or cannons and Sidewinder missiles against airborne targets; it is capable of providing close support for ground forces, and it can be refueled in flight. Designated F-100.

supersaturation [METEOROL] The condition existing in a given portion of the atmosphere when the relative humidity is greater than 100%, in respect to a plane surface of pure water or pure ice. [PHYS CHEM] The condition existing in a solution when it contains more solute than is needed to cause saturation. Also known as supersolubility.

superscript [SCI TECH] A letter or symbol written above, and usually to the right, of another symbol for any of various purposes, such as to denote a power or derivative, to identify a particular element of a set, or to indicate the mass number of an isotope.

supersensitive relay [ELEC] A relay that operates on extremely small currents, generally below 250 microamperes.

supersolubility See supersaturation.

supersonic [ACOUS] See ultrasonic. [PHYS] Of, pertaining to, or dealing with speeds greater than the speed of sound.

supersonic aerodynamics [FL MECH] The study of aerodynamics of supersonic speeds.

supersonic aircraft [AERO ENG] Aircraft capable of supersonic speeds.

supersonic airfoil [AERO ENG] An airfoil designed to produce lift at supersonic speeds.

supersonic combustion ramjet See scramjet.

supersonic compressor [MECH ENG] A compressor in which a supersonic velocity is imparted to the fluid relative to the rotor blades, the stator blades, or both, producing oblique shock waves over the blades to obtain a high-pressure rise.

supersonic diffuser [MECH ENG] A diffuser designed to reduce the velocity and to increase the pressure of fluid moving at supersonic velocities.

supersonic flow [FL MECH] Flow of a fluid over a body at speeds greater than the speed of sound in the fluid, and in which the shock waves start at the surface of the body. Also known as supercritical flow.

supersonic inlet [AERO ENG] An inlet of a jet engine at which single, double, or multiple shock waves form.

supersonic nozzle See convergent-divergent nozzle.

supersonic transport [AERO ENG] A transport plane capable of flying at speeds higher than the speed of sound. Abbreviated SST.

superstandard propagation [ELECTROMAG] The propagation of radio waves under conditions of superstandard refraction in the atmosphere, that is, refraction by an atmosphere or section of the atmosphere in which the index of refraction decreases with height at a rate of greater than 12 N units (unit of index of refraction) per 1000 feet (304.8 meters).

superstructure [CIV ENG] The part of a structure that is raised on the foundation. [CRYSTAL] See superlattice. [NAV ARCH] The entire structure of a ship above the main deck.

supersystem [SCI TECH] A whole comprising systems as the major functions at the first level; conceptualized during the process of synthesis rather than analysis, it results from an awareness that a system under consideration has a relationship with one or more other systems.

supertanker [NAV ARCH] An obsolete term for an unusually large tanker, about 30,000 to 50,000 tons deadweight; such a vessel is now referred to as a very large or ultra-large crude carrier.

superturbulent flow [FL MECH] The flow of water in which the energy loss by friction is so great that Reynolds criterion for the transition of laminar to turbulent flow does not apply.

supervisor interrupt [ADP] An interruption caused by the program being executed which issues an instruction to the master control program.

supervisor mode [ADP] A method of computer operation in which the computer can execute all its own instructions, including the privileged instruction not normally allowed to the programmer, in contrast to problem mode.

supervisory computer [ADP] A minicomputer which accepts test results from satellite minicomputers, transmits new programs to the satellite minicomputers, and may further communicate with a larger computer.

supervisory control [ENG] A control panel or room showing key readings or indicators (temperature, pressure, or flow rate) from an entire operating area, allowing visual supervision and control of the overall operation.

supervisory program [ADP] A program that organizes and regulates the flow of work in a computer system, for example, it may automatically change over from one run to another and record the time of the run.

supervisory routine [ADP] A program or routine that initiates and guides the execution of several (or all) other routines

and programs; it usually forms part of (or is) the operating system.

supervisory signal [ELEC] A signal which indicates the operating condition of a circuit or a combination of circuits in a switching apparatus or other electrical equipment to an attendant.

supervisory system [ELEC] A system of control, indicating, and telemetry devices which operates between the stations of an electric power distribution system, using a single common channel to transmit signals.

supervoltage [ELEC] A voltage in the range of 500 to 2000 kilovolts, used for some x-ray tubes.

supination [ANAT] **1.** Turning the palm upward. **2.** Inversion of the foot.

supplementary angle [MATH] One angle is supplementary to another angle if their sum is 180°.

supplementary condition [QUANT MECH] In a quantized field theory, an auxiliary condition required of a state vector to make it correspond to an actual state.

supplementary group [ELEC] In wire communications, a group of trunks that directly connects local or trunk switching centers over other than a fundamental (or backbone) route.

supplied-air respirator [ENG] An atmospheric-supplying device which provides the wearer with respirable air from a source outside the contaminated area; only those with manual or motor-operated blowers are approved for immediately harmful or oxygen-deficient atmospheres.

supplies [ORD] All items necessary for the equipment, maintenance, and operation of a military command, including food, clothing, equipment, arms, ammunition, fuel, forage, materials, and machinery of all kinds.

supply control [IND ENG] The process by which an item of supply is controlled within the supply system, including requisitioning receipt, storage, stock control, shipment, disposition, identification, and accounting. [ORD] The process by which anticipated demands for supplies are balanced against the projected availability of supplies in order that qualitative acquisition or disposal action can be initiated.

supply current [GEOPHYS] The electrical current in the atmosphere which is required to balance the observed air-earth current of fair-weather regions by transporting positive charge upward or negative charge downward.

supply point [ORD] Any point where supplies are issued in detail.

supply voltage [ELEC] The voltage obtained from a power source for operation of a circuit or device.

support [MATH] The support of a real-valued function f on a topological space is the closure of the set of points where f is not zero.

support base [ENG] A place from which logistic support is provided for a group of launch complexes and their control center.

supporting artillery [ORD] Artillery which executes fire missions in support of a specific unit, usually infantry, but remains under the command of the next higher artillery commander.

supporting fire [ORD] Fire delivered by supporting units to assist or protect a unit in combat.

supporting weapon [ORD] Any weapon used to assist or protect a unit of which it is not an organic part.

suppository [PHARM] A medicated solid body of varying weight and shape, intended for introduction into different orifices of the body, as the rectum, urethra, or vagina; usually suppositories melt or are softened at body temperature, though in some instances release of medication is effected through use of a hydrophilic vehicle; typical vehicles or bases are theobroma oil (cocoa butter), glycerinated gelatin, sodium stearate, and propylene glycol monostearate.

suppressed carrier [COMMUN] A carrier that is suppressed at the transmitter; the chrominance subcarrier in a color television transmitter is an example.

suppressed-carrier modulation [COMMUN] Modulation resulting from elimination of the carrier component from an amplitude modulated wave.

suppressed-carrier transmission [COMMUN] Transmission in which the carrier component of the modulated wave is suppressed, leaving only the side bands to be transmitted.

suppressed-zero instrument [ENG] An indicating or recording instrument in which the zero position is below the lower end of the scale markings.

suppression [ADP] **1.** Removal or deletion usually of insignificant digits in a number, especially zero suppression. **2.** Optional function in either on-line or off-line printing devices that permits them to ignore certain characters or groups of characters which may be transmitted through them. [ELECTR] Elimination of any component of an emission, as a particular frequency or group of frequencies in an audio-frequency of a radio-frequency signal.

suppressor [ELEC] **1.** In general, a device used to reduce or eliminate noise or other signals that interfere with the operation of a communication system, usually at the noise source. **2.** Specifically, a resistor used in series with a spark plug or distributor of an automobile engine or other internal combustion engine to suppress spark noise that might otherwise interfere with radio reception. [ELECTR] *See* suppressor grid.

suppressor gene [GEN] A mutant gene that masks the effect of another mutant, thus causing partial or total restoration of the normal phenotype; a suppressor can be intracistronic or nonallelic.

suppressor grid [ELECTR] A grid placed between two positive electrodes in an electron tube primarily to reduce the flow of secondary electrons from one electrode to the other; it is usually used between the screen grid and the anode. Also known as suppressor.

suppressor pulse [ELECTR] Pulse used to disable an ionized flow field or beacon transponder during intervals when interference would be encountered.

suppuration [MED] Pus formation.

supracardinal veins [VERT ZOO] Paired longitudinal veins in the mammalian embryo and various adult lower vertebrates in the thoracic and abdominal regions, dorsolateral to and on the sides of the descending aorta; they replace the postcardinal and subcardinal veins.

supracrustal rocks [GEOL] Rocks that overlie basement rock.

supragelisol *See* suprapermafrost layer.

suprahyoid muscles [ANAT] The muscles attached to the upper margin of the hyoid bone.

supralateral tangent arcs [METEOROL] Two oblique luminous arcs, concave to the sun and tangent to the halo of 46° at points above the altitude of the sun.

supraliminal [PHYSIO] Above, or in excess of, a threshold.

supranuclear [ANAT] In the nervous system, central to a nucleus.

supraoccipital [ANAT] Situated above the occipital bone.

supraoptic [ANAT] Situated above the optic tract.

suprapermafrost layer [HYD] The layer of ground above permafrost; it includes the active layer and possibly occurrences of talik and perelotok. Also known as supragelisol.

suprarenal gland *See* adrenal gland.

suprascapula [ANAT] An anomalous bone sometimes found between the superior border of the scapula and the spines of the lower cervical or first thoracic vertebrae.

suprasegmental reflex [PHYSIO] A reflex employing complex multineuronal channels to integrate the body and limb musculature with fixed positions or movements of the head.

suprasternal notch [ANAT] Jugular notch of the sternum.

suprasternal space [ANAT] The triangular space above the manubrium, enclosed by the layers of the deep cervical fascia which are attached to the front and back of this bone.

supratidal sediment [GEOL] The sediment deposited immediately above the high-tide level.

supratidal zone [GEOL] Pertaining to the shore area immediately marginal to and above the high-tide level.

supravital [BIOL] Pertaining to the staining of living cells after removal from a living animal or of still living cells from a recently killed animal.

supremum *See* least upper bound.

surbase [ARCH] **1.** A molding above the base of a wall. **2.** A molding or series of moldings above the base of a pedestal.

surcharge [CIV ENG] The load supported above the level of the top of a retaining wall.

surcharged wall [CIV ENG] A retaining wall with an embankment on the top.

surf [OCEANOGR] Wave activity in the area between the shoreline and the outermost limit of breakers, that is, in the surf zone.

surface [ENG] The outer part (skin with a thickness of zero) of a body; can apply to structures, to micrometer-sized particles, or to extended-surface zeolites.

surface acoustic wave [ACOUS] A sound wave that propagates along and is bound to the surface of a solid; ordinarily it contains both compressional and shear components.

surface acoustic wave filter [ELECTR] An electric filter consisting of a piezoelectric bar with a polished surface along which surface acoustic waves can propagate, and on which are deposited metallic transducers, one of which is connected, via thermocompression-bonded leads, to the electric source, while the other drives the load.

surface-active agent [MATER] A soluble compound that reduces the surface tension of liquids, or reduces interfacial tension between two liquids or a liquid and a solid. Also known as surfactant.

surface air leakage [MIN ENG] The amount of surface air entering the fan through the casing at the top of the upcast shaft, the air-lock doors, and fan-drift walls.

surface analyzer [ENG] An instrument that measures or records irregularities in a surface by moving the stylus of a crystal pickup or similar device over the surface, amplifying the resulting voltage, and feeding the output voltage to an indicator or recorder that shows the surface irregularities magnified as much as 50,000 times.

surface area [ENG] Measurement of the extent of the area (without allowance for thickness) covered by a surface.

surface barrier [ELECTR] A potential barrier formed at a surface of a semiconductor by the trapping of carriers at the surface.

surface-barrier diode [ELECTR] A diode utilizing thin-surface layers, formed either by deposition of metal films or by surface diffusion, to serve as a rectifying junction.

surface-barrier transistor [ELECTR] A transistor in which the emitter and collector are formed on opposite sides of a semiconductor wafer, usually made of *n*-type germanium, by training two jets of electrolyte against its opposite surfaces to etch and then electroplate the surfaces.

surface boundary layer [METEOROL] That thin layer of air adjacent to the earth's surface, extending up to the so-called anemometer level (the base of the Ekman layer); within this layer the wind distribution is determined largely by the vertical temperature gradient and the nature and contours of the underlying surface, and shearing stresses are approximately constant. Also known as atmospheric boundary layer; friction layer; ground layer; surface layer.

surface carburetor [MECH ENG] A carburetor in which air is passed over the surface of gasoline to charge it with fuel.

surface chart [METEOROL] An analyzed synoptic chart of surface weather observations; essentially, a surface chart shows the distribution of sea-level pressure (therefore, the positions of highs, lows, ridges, and troughs) and the location and nature of fronts and air masses, plus the symbols of occurring weather phenomena, analysis of pressure tendency (isallobars), and indications of the movement of pressure systems and fronts. Also known as sea-level chart; sea-level-pressure chart; surface map.

surface chemistry [PHYS CHEM] The study and measurement of the forces and processes that act on the surfaces of fluids (gases and liquids) and solids, or at an interface separating two phases; for example, surface tension.

surface-coated mirror [OPTICS] Glass to which reflective coating has been applied to the front that reflects away from the glass.

surface combustion [ENG] Combustion brought about near the surface of a heated refractory material by forcing a mixture of air and combustible gases through it or through a hole in it, or having the gas impinge directly upon it; used in muffles, crucibles, and certain types of boiler furnaces.

surface condenser [MECH ENG] A heat-transfer device used to condense a vapor, usually steam under vacuum, by absorbing its latent heat in cooling fluid, ordinarily water.

surface contamination [NUCLEO] The deposition and attachment of radioactive materials to a surface.

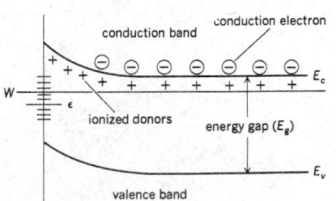

Energy diagram of a surface barrier as employed in an *n*-type semiconductor. W = Fermi level; ϵ = highest energy level of surface state filled by electrons when surface is electrically neutral; E_c = energy of bottom of conduction band; E_v = energy of top of valence band.

SURFACE-BARRIER TRANSISTOR

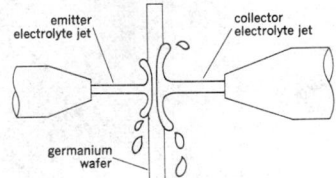

Technique for making surface-barrier transistor.

surface-controlled avalanche transistor [ELECTR] Transistor in which avalanche breakdown voltage is controlled by an external field applied through surface-insulating layers, and which permits operation at frequencies up to the 10-gigahertz range.

surface creep [GEOL] A stage of the wind erosion process in which grains of sand move each other along the surface.

surface current [OCEANOGR] 1. Water movement which extends to depths of 3–10 feet (1–3 meters) below the surface in nearshore areas, and to about 33 feet (10 meters) in deep-ocean areas. 2. Any current whose maximum velocity core is at or near the surface.

surface density [PHYS] The quantity of anything distributed over a surface per unit area of surface.

surface deposit *See* surficial deposit.

surface detention [HYD] Water in temporary storage as a thin sheet over the soil surface during the occurrence of overland flow.

surface drainage [HYD] Natural or artificial removal of excess groundwater.

surface drilling [MIN ENG] Boreholes collared at the surface of the earth, as opposed to boreholes collared in mine workings or underwater.

surface duct [GEOPHYS] Atmospheric duct for which the lower boundary is the surface of the earth.

surface-effect ship [MECH ENG] A transportation device with fixed side walls, which is supported by low-pressure, low-velocity air and operates on water only.

surface energy *See* interfacial energy.

surface finish [ENG] The surface roughness of a component after final treatment, measured by a surface profile.

surface fire [FOR] A forest fire in which only surface litter and undergrowth burn.

surface flow *See* overland flow.

surface friction [GEOPHYS] The drag or skin friction of the earth on the atmosphere, usually expressed in terms of the shearing stress of the wind on the earth's surface.

surface gage [DES ENG] 1. A scribing tool in an adjustable stand, used to mark off castings and to test the flatness of surfaces. 2. A gage for determining the distances of points on a surface from a reference plane.

surface geology [GEOL] The scientific study of the features at the surface of the earth.

surface grinder [MECH ENG] A grinding machine that produces a plane surface.

surface hardening [MET] Hardening the surface of steel by one of several processes, such as carburizing, carbonitriding, nitriding, flame or induction hardening, and surface working.

surface hoar [HYD] 1. Fernlike ice crystals formed directly on a snow surface by sublimation; a type of hoarfrost. 2. Hoarfrost that has grown primarily in two dimensions, as on a window or other smooth surface.

surface ignition [ENG] The initiation of a flame in the combustion chamber of an automobile engine by any hot surface other than the spark discharge.

surface integral [MATH] A multiple definite integral of a function $f(x,y)$ where the limits of integration determine a surface in three-dimensional space.

surface inversion [METEOROL] A temperature inversion based at the earth's surface; that is, an increase of temperature with height beginning at ground level. Also known as ground inversion.

surface irrigation [AGR] Application of water to the soil by means of pipes or furrows along the surface.

surface layer *See* surface boundary layer.

surface leakage [ELEC] The passage of current over the surface of an insulator.

surface lift [MIN ENG] In the freezing method of shaft sinking, freezing and heaving of the surface around the shaft due to the formation of ice and the variation of temperature.

surface map *See* surface chart.

surface mining [MIN ENG] Mining at or near the surface; includes placer mining, mining in open glory-hole or milling pits, mining and removing ore from opencuts by hand or with mechanical excavating and transportation equipment, and the removal of capping or overburden to uncover the ores.

surface navigation [NAV] Navigation of a craft on the surface of the earth; in particular, navigation of vessels on the surface of water.

surface noise [ELECTR] The noise component in the electric output of a phonograph pickup due to irregularities in the contact surface of the groove. Also known as needle scratch.

surface of discontinuity [METEOROL] An interface, applied to the atmosphere; for example, an atmospheric front is represented ideally by a surface of discontinuity of velocity, density, temperature, and pressure gradient.

surface of revolution [MATH] A surface realized by rotating a planar curve about some axis in its plane.

surface orientation [PHYS CHEM] Arrangement of molecules on the surface of a liquid with one part of the molecule turned toward the liquid.

surface passivation [ELECTR] A method of coating the surface of a p-type wafer for a diffused junction transistor with an oxide compound, such as silicon oxide, to prevent penetration of the impurity in undesired regions.

surface phase [GEOCHEM] A thin rock layer differing in geochemical properties from those of the volume phases on either side. Also known as volume phase.

surface-piercing hydrofoil [NAV ARCH] A hydrofoil craft which attains its stability by hydrodynamically balancing the amount of foil area above and below the water surface.

surface plate [DES ENG] A plate having a very accurate plane surface used for testing other surfaces or to provide a true surface for accurately measuring and locating testing fixtures.

surface pressure [METEOROL] The atmospheric pressure at a given location on the earth's surface; the expression is applied loosely and about equally to the more specific terms: station pressure and sea-level pressure. [PHYS] *See* film pressure.

surface reaction [CHEM] A chemical reaction carried out on a surface as on an adsorbent or solid catalyst.

surface resistivity [ELEC] The electric resistance of the surface of an insulator, measured between the opposite sides of a square on the surface; the value in ohms is independent of the size of the square and the thickness of the surface film.

surface retention *See* surface storage.

surface rights [MIN ENG] 1. Ownership of the surface land only, mineral rights being reserved. 2. Ownership of the surface land plus mineral rights. 3. The right of a mineral owner or an oil and gas lessee to use as much surface land as may be reasonably necessary for the conduct of operations under the lease.

surface rolling [MET] A cold-rolling process for hardening the surface of a metal.

surface roughness [ENG] The closely spaced unevenness of a solid surface (pits and projections) that results in friction for solid-solid movement or for fluid flow across the solid surface.

surface runoff [HYD] Runoff that moves over the soil surface to the nearest surface stream.

surface-set bit [DES ENG] A bit containing a single layer of diamonds set so that the diamonds protrude on the surface of the crown. Also known as single-layer bit.

surface sizing *See* sizing treatment.

surface soil [GEOL] The soil extending 5 to 8 inches (13 to 20 centimeters) below the surface.

surface state [SOLID STATE] An electron state in a semiconductor whose wave function is restricted to a layer near the surface.

surface storage [HYD] The part of precipitation retained temporarily at the ground surface as interception or depression storage so that it does not appear as infiltration or surface runoff either during the rainfall period or shortly thereafter. Also known as initial detention; surface retention.

surface temperature [METEOROL] Temperature of the air near the surface of the earth. [OCEANOGR] Temperature of the layer of seawater nearest the atmosphere.

surface tension [FL MECH] The force acting on the surface of a liquid, tending to minimize the area of the surface; quantitatively, the force that appears to act across a line of unit length on the surface. Also known as interfacial force; interfacial tension; surface tensity.

surface tension number [FL MECH] A dimensionless number used in studying mass transfer in packed columns equal to the square of the dynamic viscosity of a fluid times the length of the perimeter of a packing element, divided by the product of the surface area of the packing element, the surface tension, and the density of the liquid. Symbol T_s.

surface tensity *See* surface tension.

surface thermometer [ENG] A thermometer, mounted in a bucket, used to measure the temperature of the sea surface.

surface-to-air missile [ORD] A guided missile designed to be fired at an airborne target from the ground or from the deck of a surface ship; examples include Bomarc, Hawk, Nike, Talos, Tartar, Terrier, and Wizard. Abbreviated SAM.

surface-to-surface missile [ORD] A guided missile designed to be fired at a surface target from a surface position on land or water; examples include Atlas, Corporal, Dart, Jupiter, Lacross, Mace, Matador, Navaho, Pershing, Polaris, Redstone, Regulus, Sergeant, Snark, Thor, and Titan. Abbreviated SSM.

surface treating [ENG] Any method of treating a material (metal, polymer, or wood) so as to alter the surface, rendering it receptive to inks, paints, lacquers, adhesives, and various other treatments, or resistant to weather or chemical attack.

surface vibrator [MECH ENG] A vibrating device used on the surface of a pavement or flat slab to consolidate the concrete.

surface visibility [METEOROL] The visibility determined from a point on the ground, as opposed to control-tower visibility.

surface wash *See* sheet erosion.

surface water [HYD] All bodies of water on the surface of the earth. [OCEANOGR] *See* mixed layer.

surface waterproofing [ENG] Waterproofing concrete by painting a waterproofing liquid on the surface.

surface wave [COMMUN] *See* ground wave. [ELECTROMAG] A wave that can travel along an interface between two different mediums without radiation; the interface must be essentially straight in the direction of propagation; the commonest interface used is that between air and the surface of a circular wire. [FL MECH] A wave that distorts the free surface that separates two fluid phases, usually a liquid and a gas. [OCEANOGR] A progressive gravity wave in which the disturbance is of greatest amplitude at the air-water interface.

surface-wave transmission line [ELECTROMAG] A single conductor transmission line energized in such a way that a surface wave is propagated along the line with satisfactorily low attenuation.

surface weather observation [METEOROL] An evaluation of the state of the atmosphere as observed from a point at the surface of the earth, as opposed to an upper-air observation, and applied mainly to observations which are taken for the primary purpose of preparing surface synoptic charts.

surface wind [METEOROL] The wind measured at a surface observing station; customarily, it is measured at some distance above the ground itself to minimize the distorting effects of local obstacles and terrain.

surface zero *See* ground zero.

surfacing [MET] Depositing filler metal on a metal surface by welding or spraying.

surfacing mat *See* overlay.

surfactant *See* surface-active agent.

surf beat [OCEANOGR] Oscillations of water level near shore, associated with groups of high breakers.

surficial creep *See* soil creep.

surficial deposit [GEOL] Unconsolidated alluvial, residual, or glacial deposits overlying bedrock or occurring on or near the surface of the earth. Also known as superficial deposit; surface deposit.

surficial geology [GEOL] The scientific study of surficial deposits, including soils.

surf ripple [GEOL] A ripple mark formed on a sandy beach by wave-generated currents.

surf zone [OCEANOGR] The area between the landward limit of wave uprush and the farthest seaward breaker.

surge [ELEC] A momentary large increase in the current or voltage in an electric circuit. [ENG] **1.** An upheaval of fluid in a processing system, frequently causing a carryover (puking) of liquid through the vapor lines. **2.** The peak system pressure. **3.** An unstable pressure buildup in a plastic extruder leading to variable throughput and waviness of the hollow plastic tube. [FL MECH] A wave at the free surface of a liquid generated by the motion of a vertical wall, having a change in the height of the surface across the wavefront and violent eddy motion at the wavefront. [OCEANOGR] **1.** Wave

motion of low height and short period, from about ½ to 60 minutes. **2.** *See* storm surge.

surge admittance [ELEC] Reciprocal of surge impedance.

surge bunker [MIN ENG] A large-capacity storage hopper, installed near the pit bottom or at the input end of a processing plant to provide uniform bulk deliveries.

surge column [PETRO ENG] A large-sized pipe of sufficient height to provide a static head able to absorb the surging liquid discharge of the process tank to which it is connected. Also known as boot.

surge current [ELEC] A short-duration, high-amperage electric current wave that may sweep through an electrical network, as a power transmission network, when some portion of it is strongly influenced by the electrical activity of a thunderstorm.

surge drum *See* surge tank.

surge electrode current *See* fault electrode current.

surge generator [ELEC] A device for producing high-voltage pulses, usually by charging capacitors in parallel and discharging them in series.

surge impedance *See* characteristic impedance.

surge line [METEOROL] A line along which a discontinuity in the wind speed occurs.

surgery [MED] The branch of medicine that deals with conditions requiring operative procedures.

surge stress [MECH] The physical stress on process equipment or systems resulting from a sudden surge in fluid (gas or liquid) flow rate or pressure.

surge suppressor [ELECTR] A circuit that responds to the rate of change of a current or voltage to prevent a rise above a predetermined value; it may include resistors, capacitors, coils, gas tubes, and semiconducting disks. Also known as transient suppressor.

surge tank [ENG] **1.** A standpipe or storage reservoir at the downstream end of a closed aqueduct or feeder pipe, as for a water wheel, to absorb sudden rises of pressure and to furnish water quickly during a drop in pressure. Also known as surge drum. **2.** An open tank to which the top of a surge pipe is connected so as to avoid loss of water during a pressure surge. [MIN ENG] In pumping of ore pulps, a relatively small tank which maintains a steady loading of the pump.

surgical needle [MED] Any sewing needle used in a surgical operation.

surging [ENG] Motion of a ship that alternately moves forward and aft, usually when moored.

surging breaker *See* plunging breaker.

surging glacier [HYD] A glacier that alternates periodically between surges (brief periods of rapid flow) and stagnation.

surjection [MATH] A function which is onto.

suroet [METEOROL] A persistent, rain-bearing southwest wind on the west coast of France.

surprint [GRAPHICS] **1.** To superimpose an image upon a previously printed image. **2.** An image that has been surprinted.

sursassite [MINERAL] $Mn_5Al_4Si_5O_{21} \cdot 3H_2O$ A mineral composed of hydrous manganese aluminum silicate.

surveillance [ENG] Systematic observation of air, surface, or subsurface areas or volumes by visual, electronic, photographic, or other means, for intelligence or other purposes.

surveillance approach [NAV] An instrument approach conducted according to directions issued by a controller using information which appears on the surveillance radar display.

surveillance radar [NAV] Ground radar used for traffic-control purposes in the approach and landing zone; it is used to assist controllers in converting random arrivals to regular landings and in positioning such aircraft so that they may make low approaches by the use of a fixed-beam, low-approach system or by a precision radar low-approach system.

surveillance satellite [AERO ENG] A satellite whose function is to make systematic observations of the earth, usually by photographic means, for military intelligence or for other purposes.

survey [ENG] **1.** The process of determining accurately the position, extent, contour, and so on, of an area, usually for the purpose of preparing a chart. **2.** The information so obtained. [NUCLEO] Measurement of radiation in the vicinity of a nuclear reactor or other source.

SURFACE-TO-AIR MISSILE

Surface-to-air missile known as Hawk, for defense against low-flying aircraft. (*Official U.S. Army photograph*)

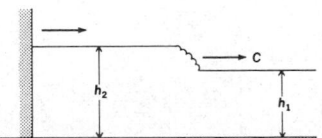

SURGE

Sketch of a surge wave (fl mech). C = speed of propagation; h_1 = height of water before passage of surge; h_2 = height of water after passage of surge.

SURVEYOR PROGRAM

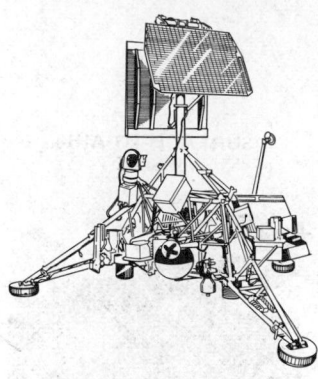

Surveyor full-scale science training model which is used in operational simulation of the instrument complement. *(NASA)*

SUSPENSION BRIDGE

Verrazano-Narrows Bridge, New York City, an example of a suspension bridge. *(Triborough Bridge and Tunnel Authority)*

survey foot [MECH] A unit of length, used by the U.S. Coast and Geodetic Survey, equal to 12/39.37 meter, or approximately 1.000002 feet.

surveying altimeter [ENG] A barometric-type instrument consisting of a pressure-sensitive element which contracts or expands in proportion to atmospheric pressure, connected through a linkage to a pointer; its dial is graduated in units of linear measurement (feet or meters) to indicate differences of elevation only.

surveying sextant *See* hydrographic sextant.

survey instrument [NUCLEO] A portable instrument used to detect and measure radiation. Also known as survey meter.

survey meter *See* survey instrument.

Surveyor program [AERO ENG] A program in which unmanned spacecraft made soft landings on the moon to take photographs, analyze samples of soil, and obtain other data that could be transmitted back to earth for guidance in planning manned landings.

surveyor's compass [ENG] An instrument used to measure horizontal angles in surveying.

surveyor's cross [ENG] An instrument for setting out right angles in surveying; consists of two bars at right angles with sights at each end.

surveyor's level [ENG] A telescope and spirit level mounted on a tripod, rotating vertically and having leveling screws for adjustment.

surveyor's measure [ENG] A system of measurement used in surveying having the engineer's, or Gunter's chain, as a unit.

survivability [ORD] The capability of a system to withstand a man-made hostile environment without suffering an abortive impairment of its ability to accomplish its designated mission.

survivable route [COMMUN] A communication cable system begun in 1960 in which the cable, main stations, amplifiers, and power feed stations are placed underground; it incorporates the latest techniques of protection against natural disasters and nuclear blasts, and avoids possible target areas.

survival curve [NUCLEO] The curve obtained by plotting the number or percentage of organisms surviving at a given time against the dose of radiation, or the number surviving at different intervals after a particular dose of radiation.

survival probability [ORD] The chance that a target will survive a given operation.

survival ratio [BIOL] The number of organisms surviving irradiation by ionizing radiation divided by the number of organisms before irradiation.

survivor curve [IND ENG] A curve showing the percentage of a group of machines or facilities surviving at a given age.

Surwell clinograph [ENG] A directional surveying instrument which records photographically the direction and magnitude of well deviations from the vertical; powered by batteries, it contains a box level gage (indicating vertical deviation), a gyroscopic compass (indicating azimuth direction) and a watch and a dial thermometer, so that a simultaneous record of amount and direction of deviation, temperature, and time can be made on 16-millimeter film.

susceptance [ELEC] The imaginary component of admittance.

susceptance standard [ELEC] Standard that introduces calibrated small values of shunt capacitance into 50-ohm coaxial transmission arrays.

susceptibility [ELEC] *See* electric susceptibility. [ELECTROMAG] *See* magnetic susceptibility. [ORD] The degree to which a device, equipment, or weapons system is open to effective attack due to one or more inherent weaknesses.

suspended ceiling [BUILD] The suspension of the furring members beneath the structural members of a ceiling.

suspended formwork [CIV ENG] Formwork suspended from supports for the floor being cast.

suspended load [GEOL] The part of the stream load that is carried for a long time in suspension. Also known as suspension load.

suspended solids *See* suspension.

suspended span [CIV ENG] A simple span supported from the free ends of cantilevers.

suspended transformation [THERMO] The cessation of change before true equilibrium is reached, or the failure of a system to change immediately after a change in conditions, such as in supercooling and other forms of metastable equilibrium.

suspended tray conveyor [MECH ENG] A vertical conveyor having pendant trays or other carriers on one or more endless chains.

suspended tubbing [MIN ENG] A permanent method of lining a circular shaft, in which the tubbing (German type) is temporarily suspended from the next wedging curb above and for which no temporary supports are required; slurry is run in behind the tubbing by means of a funnel passing through the holes provided in the segments.

suspended water *See* vadose water.

suspension [CHEM] A mixture of fine, nonsettling particles of any solid within a liquid or gas, the particles being the dispersed phase, while the suspending medium is the continuous phase. Also known as suspended solids. [ENG] A fine wire or coil spring that supports the moving element of a meter. [MIN ENG] The bolting of rock to secure fragments or sections, such as small slabs barred down after blasting blocks of rock broken by fracture or joint patterns, which may subsequently loosen and fall.

suspension bridge [CIV ENG] A long-span fixed bridge consisting of trusses carrying a roadway, suspended from two cables which pass over two towers and are anchored by backstays to a firm foundation.

suspension cable [ENG] A freely hanging cable; may carry mainly its own weight or a uniformly distributed load.

suspension current *See* turbidity current.

suspension feeder [ZOO] An animal that feeds on small particles suspended in water; particles may be minute living plants or animals, or products of excretion or decay from these or larger organisms.

suspension load *See* suspended load.

suspension roast *See* flash roast.

suspension roof [BUILD] A roof that is supported by steel cables.

suspension system [MECH ENG] A system of springs, shock absorbers, and other devices supporting the upper part of a motor vehicle on its running gear.

suspensor [BOT] A mass of cells in higher plants that pushes the embryo down into the embryo sac and into contact with the nutritive tissue. [MYCOL] A hypha which bears an apical gametangium in fungi of the Mucorales.

sussexite [MINERAL] $MnBO_2OH$ A white mineral composed of basic manganese borate occurring in fibrous veins.

sustained oscillation [CONT SYS] Continued oscillation due to insufficient attenuation in the feedback path. [PHYS] Oscillation in which forces outside the system, but controlled by the system, maintain a periodic oscillation of the system at a period or frequency that is nearly the natural period of the system.

sustained rate of fire [ORD] Actual rate of fire that a weapon can continue to deliver for an indefinite length of time without seriously overheating.

sustained yield [BIOL] In a biological resource such as timber or grain, the replacement of a harvest yield by growth or reproduction before another harvest occurs.

sustainer rocket engine [AERO ENG] A rocket engine that maintains the velocity of a rocket vehicle once it has achieved its programmed velocity by use of a booster or other engine.

sustentacular cell [HISTOL] One of the supporting cells of an epithelium as contrasted with other cells with special function, as the nonnervous cells of the olfactory epithelium or the Sertoli cells of the seminiferous tubules.

Sutherland's formula [STAT MECH] **1.** A formula which states that the absolute viscosity of a gas is proportional to $T^{3/2}/(C + T)$, where T is the absolute temperature and C is a constant for a given gas. **2.** A formula which states that the mean free path of a molecule in a gas is proportional to $1/nd\sqrt{1 + C/T}$, where n is the number of molecules per unit volume, d is the diameter of a molecule, T is the absolute temperature, and C is a constant.

Sutro weir [CIV ENG] A dam with at least one curved side and horizontal crest, so formed that the head above the crest is directly proportional to the discharge.

suture [BIOL] A distinguishable line of union between two closely united parts. [MED] A fine thread used to close a wound or surgical incision.

svabite [MINERAL] $Ca_5(AsO_4)_3F$ A colorless, yellow, rose, or reddish-brown mineral composed of fluoride-arsenate of calcium.

svanbergite [MINERAL] $SrAl_3(PO_4)(SO_4)(OH)_6$ A colorless to yellow mineral composed of basic phosphate and sulfate of strontium and aluminum; it is isomorphous with corkite, hinsdalite, and woodhouseite.

svedberg [PHYS CHEM] A unit of sedimentation coefficient, equal to 10^{-13} second.

Svedberg equation [STAT MECH] An equation which states that the amplitude of vibration of a particle which exhibits Brownian motion is proportional to its period.

sverdrup [FL MECH] A unit of volume transport equal to 1,000,000 cubic meters per second.

swab [MIN ENG] **1.** A pistonlike device provided with a rubber cap ring that is used to clean out debris inside a borehole or casing. **2.** The act of cleaning the inside of a tubular object with a swab. [PETRO ENG] In petroleum drilling, to pull the drill string so rapidly that the drill mud is sucked up and overflows the collar of the borehole, thus leaving an undesirably empty borehole.

swage bolt [DES ENG] A bolt having indentations with which it can be gripped in masonry.

swaging [MET] Tapering a rod or tube or reducing its diameter by any of several methods, such as forging, squeezing, or hammering.

swallow buoy See swallow float.

swallow float [ENG] A tubular buoy used to measure current velocities; it can be adjusted to be neutrally buoyant and to drift at a selected density level while being tracked by shipboard listening devices. Also known as neutrally buoyant float; swallow buoy.

swamp [ECOL] A waterlogged land supporting a natural vegetation predominantly of shrubs and trees.

swamp buggy [MECH ENG] A wheeled vehicle that runs on land, mud, or through shallow water; used especially in swamps.

swamper [MIN ENG] **1.** A rear brakeman in a metal mine. **2.** A laborer who assists in hauling ore and rock, coupling and uncoupling cars, throwing switches, and loading and unloading carriers.

swamp fever See infectious anemia.

swamping resistor [ELECTR] Resistor placed in the emitter lead of a transistor circuit to minimize the effects of temperature on the emitter-base junction resistance.

swamp itch See schistosome dermatitis.

swan [VERT ZOO] Any of several species of large waterfowl comprising the subfamily Anatinae; they are herbivorous stout-bodied forms with long necks and spatulate bills.

Swan See Cygnus.

Swann bands [ASTROPHYS] Particular bands seen in the visible spectra of comets; they arise from the presence of dimeric carbon (C_2) in the comet's tail.

Swanscombe man [PALEON] A partial skull recovered in Swanscombe, Kent, England, which represents an early stage of *Homo sapiens* but differing in having a vertical temporal region and a rounded occipital profile.

swarf [ENG] Chips, shavings, and other fine particles removed from the workpiece by grinding tools.

swarmer cell [MICROBIO] The daughter cell which separates from the stalked mother cell in bacteria in the genus *Caulobacter*.

Swarts reaction [ORG CHEM] The reaction of chlorinated hydrocarbons with metallic fluorides to form chlorofluorohydrocarbons, such as CCl_2F_2, which is quite inert and nontoxic.

swartzite [MINERAL] $CaMg(UO_2)(CO_3)_3 \cdot 12H_2O$ A green monoclinic mineral composed of hydrous carbonate of calcium, magnesium, and uranium.

swash [GEOL] **1.** A narrow channel or ground within a sand bank, or between a sand bank and the shore. **2.** A bar over which the sea washes. [OCEANOGR] The rush of water up onto the beach following the breaking of a wave. Also known as run-up; uprush.

swash bulkhead See swash plate.

swash mark [GEOL] A fine, wavy or arcuate line or minute ridge consisting of fine sand, seaweed, and other debris on a beach; marks the farthest advance of wave uprush. Also known as debris line; wave line; wavemark.

swash plate [NAV ARCH] A partial bulkhead in the tank of an oil tanker; used to decrease the back-and-forth surge of liquid as the ship rocks or rolls. Also known as swash bulkhead.

swash-plate pump [MECH ENG] A rotary pump in which the angle between the drive shaft and the plunger-carrying body is varied.

S wave [GEOPHYS] A seismic body wave propagated in the crust or mantle of the earth by a shearing motion of material; speed is 3.0–4.0 kilometers per second in the crust and 4.4–4.6 in the mantle. Also known as distortional wave; equivoluminal wave; rotational wave; secondary wave; shake wave; shear wave; tangential wave; transverse wave.

sway [NAV ARCH] Lateral movement of the center of gravity of a ship.

swayback [MED] Increased lumbar lordosis with compensatory increased thoracic kyphosis. [VET MED] Sinking of the back, or lordosis.

sway brace [CIV ENG] One or a pair of diagonal members designed to resist horizontal forces, such as wind.

sway frame [CIV ENG] A unit in the system of members of a bridge that provides bracing against side sway; consists of two diagonals, the verticals, the floor beam, and the bottom strut.

sweat [CHEM] Exudation of nitroglycerin from dynamite due to separation of nitroglycerin from its adsorbent. [MET] Exudate of low-melting-point constituents from a metal on solidification. [PHYSIO] The secretion of the sweat glands. Also known as perspiration. [SCI TECH] Formation of moisture beads on a surface as a result of concentration.

sweat cooling [AERO ENG] A technique for cooling combustion chambers or aerodynamically heated surfaces by forcing a coolant through a porous wall, resulting in film cooling at the interface. Also known as transpiration cooling.

sweated wax [MATER] A white, moisture-free petroleum wax with the oil removed by a sweating process, in which the unrefined wax is heated in shallow pans.

sweat gland [PHYSIO] A coiled tubular gland of the skin which secretes sweat.

sweating [CHEM ENG] Separation of paraffin oil from low-melting petroleum wax obtained from paraffin wax in a chamber (sweater) by first cooling the mixture until it is a solid cake, then warming gradually to cause partial fusion of the mixture to allow drainage of liquid from the cake. Also known as exudation.

sweating out [MET] Bringing small globules of low-melting constituents to the surface of an alloy during heat treatment (as lead out of bronze).

swedenborgite [MINERAL] $NaBe_4SbO_7$ A colorless to wine-yellow mineral composed of sodium beryllium antimony oxide.

sweep [ELECTR] **1.** The steady movement of the electron beam across the screen of a cathode-ray tube, producing a steady bright line when no signal is present; the line is straight for a linear sweep and circular for a circular sweep. **2.** The steady change in the output frequency of a signal generator from one limit of its range to the other. [MET] A profile pattern used to form molds for symmetrical articles made by sweep casting. [ORD] **1.** Swift flight of a formation of combat airplanes over enemy territory. **2.** To cover a wide area with gunfire.

sweep amplifier [ELECTR] An amplifier used with a cathode-ray tube, such as in a television receiver or cathode-ray oscilloscope, to amplify the sawtooth output voltage of the sweep oscillator, to shape the waveform for the deflection circuits of a television picture tube, or to provide balanced signals to the deflection plates.

sweepback [AERO ENG] **1.** The backward slant of a leading or trailing edge of an airfoil. **2.** The amount of this slant, expressed as the angle between a line perpendicular to the plane of symmetry and a reference line in the airfoil.

sweep circuit [ELECTR] The sweep oscillator, sweep amplifier, and any other stage used to produce the deflection voltage or current for a cathode-ray tube. Also known as scanning circuit.

sweep generator Also known as sweep oscillator. [ELECTR] **1.** An electronic circuit that generates a voltage or current, usually recurrent, as a prescribed function of time; the

SWEETGUM

Sweetgum (*Liquidambar styraciflua*); (*a*) terminal bud, (*b*) twig, and (*c*) the distinctive five-lobed leaf.

SWEET POTATO

The sweet, fleshy, tuberous roots of big-stem Jersey type sweet potatoes. (*USDA*)

SWING BRIDGE

The vertical-lift span and the swing bridge across Arthur Kill, between Staten Island and New Jersey. The vertical-lift replaced the swing, in use since 1888. (*Baltimore and Ohio Railroad Co.*)

resulting waveform is used as a time base to be applied to the deflection system of an electron-beam device, such as a cathode-ray tube. Also known as time-base generator; timing-axis oscillator. **2.** A test instrument that generates a radio-frequency voltage whose frequency varies back and forth through a given frequency range at a rapid constant rate; used to produce an input signal for circuits or devices whose frequency response is to be observed on an oscilloscope.

sweeping [NAV] **1.** The process of towing a line or object below the surface in order to determine whether an area is free from isolated submerged hazards to vessels and to determine the position of any such hazards that exist, or to determine the least depth of an area. **2.** The process of clearing an area or channel of mines or other hazards to navigation.

sweeping receivers [ELECTR] Automatically and continuously tuned receivers designed to stop and lock on when a signal is found, or to continually plot band occupancy.

sweep jamming [ELECTR] Jamming an enemy radarscope by sweeping the region of radar-beam coverage with electromagnetic waves having the same frequency as those received by the radarscope.

sweep oscillator *See* sweep generator.

sweep-out pattern [PETRO ENG] The areal pattern of water advance in a petroleum reservoir, as for a waterflood operation.

sweep rate [ELECTR] The number of times a radar radiation pattern rotates during 1 minute; sometimes expressed as the duration of one complete rotation in seconds.

sweepstakes route [ECOL] A means that allows chance migration across a sea on natural rafts, so that oceanic islands can be colonized.

sweep test [ELECTR] Test given coaxial cable with an oscilloscope to check attenuation.

sweep-through jammer [ELECTR] A jamming transmitter which is swept through a radio-frequency band in short steps to jam each frequency briefly, producing a sound like that of an aircraft engine.

sweep voltage [ELECTR] Periodically varying voltage applied to the deflection plates of a cathode-ray tube to give a beam displacement that is a function of time, frequency, or other data base.

sweet basil oil *See* basil oil.

sweet bay oil *See* volatile laurel oil.

sweet corrosion [PETRO ENG] Corrosion occurring in oil or gas wells where there is no iron-sulfide corrosion product, and there is no odor of hydrogen sulfide in the produced reservoir fluid.

sweet crudes [MATER] Crude petroleum oil containing little sulfur.

sweetening [CHEM ENG] Improvement of a petroleum-product color and odor by converting sulfur compounds into disulfides with sodium plumbite (doctor treating), or by removing them by contacting the petroleum stream with alkalies or other sweetening agents.

sweet gas [MATER] A petroleum natural gas containing no corrosive components, such as hydrogen sulfide and mercaptans.

sweetgum [BOT] *Liquidambar styraciflua.* A deciduous tree of the order Hamamelidales found in the southeastern United States, and distinguished by its five-lobed, or star-shaped, leaves, and by the corky ridges developed on the twigs.

sweet oil *See* olive oil.

sweet orange oil [MATER] A sweet, yellow, mild essential oil expressed from the peel of an orange, *Citrus aurantium;* soluble in glacial acetic acid, somewhat in alcohol; used in flavors, perfumes, and medicine. Also known as orange oil.

sweet potato [BOT] *Ipomoea batatas.* A tropical vine having variously shaped leaves, purplish flowers, and a sweet, fleshy, edible tuberous root.

sweet roast *See* dead roast.

sweet spirits of niter *See* ethyl nitrite.

swell [GEOL] **1.** The volumetric increase of soils on being removed from their compacted beds due to an increase in void ratio. **2.** A local enlargement or thickening in a vein or ore deposit. **3.** A low dome or quaquaversal anticline of considerable areal extent; long and generally symmetrical waves contribute to the mixing processes in the surface layer and

thus to its sound transmission properties. **4.** Gently rising ground, or a rounded hill above the surrounding ground or ocean floor. [MIN ENG] *See* horseback. [OCEANOGR] Ocean waves which have traveled away from their generating area; these waves are of relatively long length and period, and regular in character.

swell-and-swale topography [GEOGR] A low-relief, undulating landscape characterized by gentle slopes and rounded hills interspersed with shallow depressions.

swell diameter [AERO ENG] In a body of revolution having an ogival portion, such as a projectile, the swell diameter is in the diameter of the maximum transverse section of the geometrical ogive.

swell direction [OCEANOGR] The direction from which swell is moving.

swelled ground [GEOL] A soil or rock that expands when wetted.

swell forecast [OCEANOGR] Prediction of the frequency and height of swell waves in a remote area from the characteristics of the waves at their origin.

swelling clay [GEOL] Clay that can absorb large amounts of water, such as bentonite.

sweptback wing [AERO ENG] An airplane wing on which both the leading and trailing edges have sweepback, the trailing edge forming an acute angle with the longitudinal axis of the airplane aft of the root. Also known as swept wing.

swept wing *See* sweptback wing.

swim bladder [VERT ZOO] A gas-filled cavity found in the body cavities of most bony fishes; has various functions in different fishes, acting as a float, a lung, a hearing aid, and a sound-producing organ.

swimmeret [INV ZOO] Any of a series of paired biramous appendages under the abdomen of many crustaceans, used for swimming and egg carrying.

swimmer's conjunctivitis *See* inclusion conjunctivitis.

swimmer's itch *See* schistosome dermatitis.

swimming bird [VERT ZOO] Any bird belonging to the orders Charadriiformes and Pelacaniformes.

swimming pool conjunctivitis *See* inclusion conjunctivitis.

swimming-pool reactor *See* pool reactor.

swine [AGR] A domesticated member of the family Suidae. [VERT ZOO] Any of various species comprising the Suidae.

swine erysipelas [VET MED] An infectious bacterial disease of swine caused by *Erysipelothrix insidiosa* in which involvement of the skin is predominant.

swine influenza [VET MED] A disease of swine caused by the associated effects of a filterable virus and *Hemophilus suis,* characterized by inflammation of the upper respiratory tract.

swine plague [VET MED] Hemorrhagic septicemia of swine caused by *Pasteurella suiseptica,* characterized by pleuropneumonia.

swine pox [VET MED] A benign infection of young hogs characterized by pox lesions on the body and inner surfaces of the legs.

swinestone [PETR] Limestone containing black bituminous matter, which gives off an objectionable odor when rubbed.

swing [ELEC] Variation in frequency or amplitude of an electrical quantity. [ENG] **1.** The arc or curve described by the point of a pick or mandril when being used. **2.** Rotation of the superstructure of a power shovel on the vertical shaft in the mounting. **3.** To rotate a revolving shovel on its base.

swing-around trajectory [AERO ENG] A planetary round-trip trajectory which requires minimal propulsion at the destination planet, but instead uses the planet's gravitational field to effect the bulk of the necessary orbit change to return to earth.

swing bridge [CIV ENG] A movable bridge that pivots in a horizontal plane about a center pier.

swing cut *See* slipping cut.

swinger *See* revolver.

swing-frame grinder [MECH ENG] A grinding machine hanging by a chain so that it may swing in all directions for surface grinding heavy work.

swinging [NAV] The process of placing a craft on various headings and comparing magnetic compass readings with the corresponding magnetic directions to determine deviation; this usually follows compass adjustment or compass compensation, and is done to obtain information for making a deviation table, deviation card, or compass correction card.

swinging a claim [MIN ENG] The adjustment of the boundaries of a mining claim to more nearly conform to the strike of the vein.

swinging buoy [NAV]· A buoy placed at a favorable location to assist a vessel to adjust its compass or swing ship; the bow of the vessel is made fast to one such buoy, and the vessel is swung by means of lines to a tug or to additional buoys.

swinging choke [ELEC] An iron-core choke having a core that can be operated almost at magnetic saturation; the inductance is then a maximum for small currents, and swings to a lower value as current increases. Also known as swinging reactor.

swinging compass [NAV] An accurate portable magnetic compass used to indicate magnetic headings during aircraft magnetic compass calibration.

swinging reactor *See* swinging choke.

swinging ship [NAV] The process of determining the deviation of the ship's magnetic compass by placing a vessel or an aircraft on various headings and comparing magnetic compass readings with the corresponding but previously determined magnetic directions; this procedure usually follows compass adjustment or compass compensation, and is done to obtain information for making a deviation table; usually called swinging when referred to an aircraft compass. Also known as compass calibration.

swinging the arc [NAV] The process of rotating a sextant about the line of sight to the horizon to determine the foot of the vertical circle through a body being observed. Also known as rocking the sextant.

swinging traverse [ORD] Type of fire used against dense troop formations moving toward a machine gun position or rapidly moving targets; the traversing clamp is loosened so that a gunner makes rapid changes by exerting pressure against the pistol-grip.

swing joint [DES ENG] A pipe joint in which the parts may be rotated relative to each other.

swing pipe [ENG] A discharge pipe whose intake end can be raised or lowered on a tank.

swing shift [IND ENG] Working arrangement in a three-shift, continuously run plant with working hours changed at regular intervals; during a swing shift the morning shift becomes the afternoon shift, while the afternoon shift becomes the morning shift of the next day, with only an 8-hour break on the first day of change.

swirl error [NAV] The additional error in the reading of a magnetic compass during a turn due to friction in the compass liquid.

Swiss pattern file [DES ENG] A type of fine file used for precision filing of jewelry, instrument parts, and dies.

switch [ADP] **1.** A hardware or programmed device for indicating that one of several alternative states or conditions have been chosen, or to interchange or exchange two data items. **2.** A symbol used to indicate a branch point, or a set of instructions to condition a branch. [CIV ENG] **1.** A device for enabling a railway car to pass from one track to another. **2.** The junction of two tracks. [ELEC] A manual or mechanically actuated device for making, breaking, or changing the connections in an electric circuit. Also known as electric switch. Symbol SW.

switch angle [CIV ENG] The angle between the switch and stock rails of a railroad track, measured at the point of juncture between the gage lines.

switchback [MIN ENG] A zigzag arrangement of railroad tracks by means of which a train can reach a higher or lower level by a succession of easy grades.

switchblade knife [DES ENG] A knife in which the blade is restrained by a spring and swings open when released by a pushbutton.

switchboard [COMMUN] A manually operated apparatus at a telephone exchange, on which the various circuits from subscribers and other exchanges are terminated to enable operators to establish communications either between two subscribers on the same exchange, or between subscribers on different exchanges. Also known as telephone switchboard. [ELEC] A single large panel or assembly of panels on which are mounted switches, circuit breakers, meters, fuses, and terminals essential to the operation of electric equipment. Also known as electric switchboard.

switch function [ELECTR] A circuit having a fixed number of inputs and outputs designed such that the output information is a function of the input information, each expressed in a certain code or signal configuration or pattern.

switchgear [ELEC] The aggregate of switching devices for a power or transforming station, or for electric motor control.

switch hook [ELEC] A switch on a telephone set that operates when the receiver is placed on the hook or removed from it.

switching [ELEC] Making, breaking, or changing the connections in an electrical circuit.

switching center [COMMUN] The equipment in a relay station for automatically or semiautomatically relaying communications traffic.

switching circuit [ELEC] A constituent electric circuit of a switching or digital processing system which receives, stores, or manipulates information in coded form to accomplish the specified objectives of the system.

switching control [COMMUN] Installation in a wire system where telephone or teletypewriter switchboards are installed to interconnect circuits.

switching device [ENG] An electrical or mechanical device or mechanism, which can bring another device or circuit into an operating or nonoperating state. Also known as switching mechanism.

switching diode [ELECTR] A crystal diode that provides essentially the same function as a switch; below a specified applied voltage it has high resistance corresponding to an open switch, while above that voltage it suddenly changes to the low resistance of a closed switch.

switching gate [ELECTR] An electronic circuit in which an output having constant amplitude is registered if a particular combination of input signals exists; examples are the OR, AND, NOT, and INHIBIT circuits. Also known as logical gate.

switching key *See* key.

switching mechanism *See* switching device.

switching pad [ELECTR] Transmission-loss pad automatically cut in and out of a toll circuit for different desired operating conditions.

switching reactor [ELECTROMAG] A saturable-core reactor that has several input control windings and one or more output windings that essentially duplicate the functions of a relay.

switching system [COMMUN] An assembly of switching and control devices provided so that any station in a communications system may be connected as desired with any other station.

switching theory [ELECTR] The theory of circuits made up of ideal digital devices; included are the theory of circuits and networks for telephone switching, digital computing, digital control, and data processing.

switching-through relay [ELEC] Control relay of a line-finder selector, connector, or other stepping switch, which extends the loop of a calling telephone through to the succeeding switch in a switch train.

switching time [ELECTR] **1.** The time interval between the reference time and the last instant at which the instantaneous voltage response of a magnetic cell reaches a stated fraction of its peak value. **2.** The time interval between the reference time and the first instant at which the instantaneous integrated voltage response of a magnetic cell reaches a stated fraction of its peak value.

switching trunk [ELEC] Trunk from a long-distance office to a local exchange office used for completing a long-distance call.

switching tube [ELECTR] A gas tube used for switching high-power radio-frequency energy in the antenna circuits of radar and other pulsed radio-frequency systems; examples are ATR tube; pre-TR tube; TR tube.

switch jack [ELEC] Any of the devices that provide terminals for the control circuits of the switch.

switchman [MIN ENG] A laborer who throws switches of mine tracks in a coal mine.

switch over-travel [ELEC] That movement of a switch-operating lever which takes place after the switch has been actuated either to close or open its contacts.

switch plate [MIN ENG] An iron plate on tramroads in mines to change the direction of movement.

switch pretravel [ELEC] That movement of a switch-operating level that takes place before the switch is actuated either to close or to open its contacts.

switch register [ADP] A manual switch on the control panel by means of which a bit may be entered in a processor register.

switch room [COMMUN] Part of a central office building that houses switching mechanisms and associated apparatus.

switch train [ELEC] A series of switches in tandem.

swivel [DES ENG] A part that oscillates freely on a headed bolt or pin. [PETRO ENG] A short piece of casing having one end belled over a heavy ring, and having a large hole through both walls, the other end being threaded.

swivel block [DES ENG] A block with a swivel attached to its hook or shackle permitting it to revolve.

swivel coupling [MECH ENG] A coupling that gives complete rotary freedom to a deflecting wedge-setting assembly.

swivel gun [ORD] A gun mounted on a pedestal so that it can be turned from side to side or up and down.

swivel head [MECH ENG] The assembly of a spindle, chuck, feed nut, and feed gears on a diamond-drill machine that surrounds, rotates, and advances the drill rods and drilling stem; on a hydraulic-feed drill the feed gears are replaced by a hydraulically actuated piston assembly.

swivel hook [DES ENG] A hook with a swivel connection to its base or eye.

swivel joint [DES ENG] A joint with a packed swivel that allows one part to move relative to the other.

swivel neck *See* water swivel.

swivel pin *See* kingpin.

Swordfish *See* Dorado.

SWS *See* deep sleep.

SXAPS *See* soft x-ray appearance potential spectroscopy.

syblet [COMMUN] One of the elements of speech; others are words, syllables, and phonemes.

sycamore [BOT] **1.** Any of several species of deciduous trees of the genus *Platanus*, especially *P. occidentalis* of eastern and central North America, distinguished by simple, large, three- to five-lobed leaves and spherical fruit heads. **2.** The Eurasian maple (*Acer pseudoplatanus*).

Sycettida [INV ZOO] An order of calcareous sponges of the subclass Calcaronea in which choanocytes occur in flagellated chambers, and the spongocoel is not lined with these cells.

Sycettidae [INV ZOO] A family of sponges in the order Sycettida.

Sycidales [PALEON] A group of fossil aquatic plants assigned to the Charophyta, characterized by vertically ribbed gyrogonites.

sycon [INV ZOO] A canal system in sponges in which the flagellated layer is confined to outpocketings of the paragaster that are indirectly connected to the incurrent canals.

syconium [BOT] A fleshy fruit, as a fig, with an enlarged pulpy receptacle internally lined with minute flowers.

sycosis [MED] An inflammatory disease affecting the hair follicles, particularly of the beard, and characterized by papules, pustules, and tubercles, perforated by hairs, together with infiltration of the skin and crusting.

Sydenham's chorea *See* St. Vitus dance.

syenite [PETR] A visibly crystalline plutonic rock with granular texture composed largely of alkali feldspar, with subordinate plagioclose and mafic minerals; the intrusive equivalent of trachyte.

syenodiorite [PETR] Plutonic rock consisting of acid plagioclase, orthoclase, and a ferromagnesian mineral.

syenogabbro [PETR] Plutonic rock consisting of basic plagioclase, orthoclase, and a dark mineral such as augite.

syllabic compandor [ELECTR] A compandor in which the effective gain variations are made at speeds allowing response to the syllables of speech but not to individual cycles of the signal wave.

Syllidae [INV ZOO] A large family of polychaete annelids belonging to the Errantia; identified by their long, linear, translucent bodies with articulated cirri; size ranges from minute *Exogone* to *Trypanosyllis*, which may be 100 millimeters long.

Syllinae [INV ZOO] A subfamily of polychaete annelids of the family Syllidae.

syllogism [MATH] A statement together with a conclusion; this usually has the form "if p then q."

Sylonidae [INV ZOO] A family of parasitic crustaceans in the order Rhizocephala.

Sylopidae [INV ZOO] A family of coleopteran insects in the superfamily Meloidea in which the elytra in males are reduced to small leathery flaps while the hindwings are large and fan-shaped.

sylvanite [MINERAL] (Au,Ag)Te$_2$ A steel-gray, silver-white, or brass-yellow mineral that crystallizes in the monoclinic system and often occurs in implanted crystals. Also known as goldschmidtite; graphic tellurium; white tellurium; yellow tellurium.

sylvatic plague [VET MED] Plague occurring in rodents; may be transmitted to man. Also known as endemic rural plague.

sylvester [MIN ENG] A hand-operated device for withdrawing supports from the waste or old workings in a mine by means of a long chain which allows the device to be positioned at a safe distance from the support to be extracted.

Sylvester's theorem [MATH] If A is a matrix with distinct eigenvalues $\lambda_1, \ldots, \lambda_n$, then any analytic function $f(A)$ can be realized from the λ_i, $f(\lambda_i)$, and the matrices $A - \lambda_i I$, where I is the identity matrix.

Sylvicolidae [INV ZOO] A family of orthorrhaphous dipteran insects in the series Nematocera.

sylvine *See* sylvite.

sylvite [MINERAL] KCl A salty-tasting, white or colorless isometric mineral, occurring in cubes or crystalline masses or as a saline residue; the chief ore of potassium. Also known as leopoldite; sylvine.

sym- [ORG CHEM] A chemical prefix; denotes structure of a compound in which substituents are symmetrical with respect to a functional group or to the carbon skeleton.

symballophone [ENG] A double stethoscope for the comparison and lateralization of sounds; permits the use of the acute function of the two ears to compare intensity and varying quality of sounds arising in the body or mechanical devices.

symbiont [ECOL] A member of a symbiotic pair.

symbiosis [ECOL] **1.** An interrelationship between two different species. **2.** An interrelationship between two different organisms in which the effects of that relationship is expressed as being harmful or beneficial. Also known as consortism.

symbiotic objects [ASTRON] Stars whose spectra have characteristics of two disparate spectral classes.

symblepharon [MED] Adhesion of the eyelids to the eyeball.

symbol [CHEM] Letter or combination of letters and numbers that represent various conditions or properties of an element, for example, a normal atom, O (oxygen); with its atomic weight, O^{16}; its atomic number, $_8$O^{16}; as a molecule, O$_2$; as an ion, O^{2+}; in excited state, O*; or as an isotope, O^{18}. [SCI TECH] **1.** A design used on a diagram to represent a component or to identify specific characteristics, quantities, or objects. **2.** A sign letter or abbreviation used on a diagram or in an equation to represent a quantity or to identify an object.

symbolic address [ADP] In coding, a programmer-defined symbol that represents the location of a particular datum item, instruction, or routine. Also known as symbolic number.

symbolic age of neutrons *See* Fermi age.

symbolic coding [ADP] Instruction written in an assembly language, using symbols for operations and addresses. Also known as symbolic programming.

symbolic deck [ADP] Deck of cards punched out in programmer coding language as opposed to binary language.

symbolic language [ADP] A language which expresses addresses and operation codes of instructions in symbols convenient to humans rather than in machine language.

symbolic logic [MATH] The formal study of symbolism and its use in the foundations of mathematical logic.

symbolic number *See* symbolic address.

symbolic programming *See* symbolic coding.

symbol input [ADP] Includes all contextual symbols that may appear in a source text.

symbol sequence [ADP] A sequence of contextual symbols not interrupted by space.

symbol table [ADP] A mapping for a set of symbols to another set of symbols or numbers.

SYCAMORE

Terminal bud, leaf with seed capsule, and twig of American sycamore (*Platanus occidentalis*).

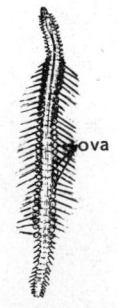

SYLLIDAE

ova

Exogone, of the Syllidae (Exogoninae), dorsal view showing cirri and ova attached to body segments.

symmetrical achromat lens [OPTICS] An older type of camera lens in which two positive achromatic meniscus lenses are symmetrically arranged about the stop.

symmetrical alternating quantity [PHYS] Alternating quantity of which all values separated by a half period have the same magnitude but opposite sign.

symmetrical avalanche rectifier [ELECTR] Avalanche rectifier that can be triggered in either direction, after which it has a low impedance in the triggered direction.

symmetrical band-pass filter [ELECTR] A band-pass filter whose attenuation as a function of frequency is symmetrical about a frequency at the center of the pass band.

symmetrical band-reject filter [ELECTR] A band-rejection filter whose attenuation as a function of frequency is symmetrical about a frequency at the center of the rejection band.

symmetrical clipper [ELECTR] A clipper in which the upper and lower limits on the amplitude of the output signal are positive and negative values of equal magnitude.

symmetrical fold [GEOL] A fold whose limbs have approximately the same angle of dip relative to the axial surface. Also known as normal fold.

symmetrical H attenuator [ELECTR] An H attenuator in which the impedance near the input terminals equals the corresponding impedance near the output terminals.

symmetrical inductive diaphragm [ELECTROMAG] A waveguide diaphragm which consists of two plates that leave a space at the center of the waveguide, and which introduces an inductance in the waveguide.

symmetrical lens [OPTICS] A lens system consisting of two parts, each of which is the mirror image of the other.

symmetrical O attenuator [ELECTR] An O attenuator in which the impedance near the input terminals equals the corresponding impedance near the output terminals.

symmetrical pi attenuator [ELECTR] A pi attenuator in which the impedance near the input terminals equals the corresponding impedance near the output terminals.

symmetrical T attenuator [ELECTR] A T attenuator in which the impedance near the input terminals equals the corresponding impedance near the output terminals.

symmetrical transducer [ELECTR] A transducer is symmetrical with respect to a specified pair of terminations when the interchange of that pair of terminations will not affect the transmission.

symmetric difference [MATH] The symmetric difference of two sets consists of all points in one or the other of the sets but not in both.

symmetric function [MATH] A function whose value is unchanged for any permutation of its variables.

symmetric group [MATH] The group consisting of all permutations of a finite set of symbols.

symmetric matrix [MATH] A matrix which equals its transpose.

symmetric relation [MATH] The property of a relation on a set that requires y to be related to x whenever x is related to y.

symmetric ripple mark [GEOL] A ripple mark whose cross-section profile is symmetric.

symmetric space [MATH] A differentiable manifold which has a differentiable multiplication operation that behaves similarly to the multiplication of a complex number and its conjugate.

symmetric top molecule [PHYS CHEM] A nonlinear molecule which has one and only one axis of threefold or higher symmetry.

Symmetrodonta [PALEON] An order of the extinct mammalian infraclass Pantotheria distinguished by the central high cusp, flanked by two smaller cusps and several low minor cusps, on the upper and lower molars.

symmetry [BIOL] The disposition of organs and other constituent parts of the body of living organisms with respect to imaginary axes. [MATH] A geometric object G has this property relative to some configuration S of its points if S determines two pieces of G which can be reflected onto each other through S. [PHYS] *See* invariance.

symmetry axis *See* axis of symmetry; rotation axis.

symmetry center *See* center of symmetry.

symmetry class *See* crystal class.

symmetry element [CRYSTAL] **1.** Some combination of rotations and reflections and translations which brings a crystal into a position that cannot be distinguished from its original position. Also known as symmetry operation; symmetry transformation. **2.** The rotational axes, mirror planes, and center of symmetry characteristic of a given crystal.

symmetry function *See* symmetry transformation.

symmetry group [MATH] A group composed of all rigid motions or similarity transformations of some geometric object onto itself.

symmetry law *See* invariance principle.

symmetry operation *See* symmetry element.

symmetry plane *See* plane of mirror symmetry.

symmetry principle [MATH] The centroid of a geometrical figure (line, area, or volume) is at a point on a line or plane of symmetry of the figure. [PHYS] *See* invariance principle.

symmetry transformation [CRYSTAL] *See* symmetry element. [MATH] A rigid motion sending a geometric object onto itself; examples are rotations and, for the case of a polygon, permutations of the vertices. Also known as symmetry function.

symmict [GEOL] Referring to a sedimentation unit that is structureless and in which coarse- and fine-grained particles are mixed more extensively in the lower part.

symmictite [PETR] An eruptive breccia that is homogenized and is made up of a mixture of country rock and intrusive rock.

symmicton *See* diamicton.

symon fault *See* horseback.

Symon's cone crusher [MIN ENG] A modified gyratory crusher used in secondary ore crushing that consists of a downward-flaring bowl within which is gyrated a conical crushing head; the main shaft is gyrated by means of a long eccentric which is driven by bevel gears.

Symon's disk crusher [MIN ENG] A mill in which the crushing is done between two cup-shaped plates that revolve on shafts set at a small angle to each other.

sympathetic detonation [ENG] Explosion caused by the transmission of a detonation wave through any medium from another explosion.

sympathetic nervous system [ANAT] The portion of the autonomic nervous system, innervating smooth muscle and glands of the body, which upon stimulation produces a functional state of preparation for flight or combat.

sympathetic ophthalmia [MED] A granulomatous inflammation of the uveal tract following ocular injury or intraocular surgery.

sympathetic vibration [PHYS] The driving of a mechanical or acoustical system at its resonant frequency by energy from an adjacent system vibrating at the same frequency.

sympathicotropic cell [HISTOL] Any of various cells possessing special affinity for the sympathetic nervous system.

sympathochromaffin cell [HISTOL] One of the precursors of sympathetic and medullary cells in the adrenal medulla.

sympatholytic [PHARM] Of or pertaining to an effect antagonistic to that of the sympathetic nervous system.

sympathomimetic [PHARM] Having the ability to produce physiologic changes similar to those caused by action of the sympathetic nervous system.

sympatric [ECOL] Of a species, occupying the same range as another species but maintaining identity by not interbreeding.

sympetalous *See* gamopetalous.

symphile [ECOL] An organism, usually a beetle, living as a guest in the nest of a social insect, such as an ant, where it is reared and bred in exchange for its exudates.

Symphyla [INV ZOO] A class of the Myriapoda comprising tiny, pale, centipedelike creatures which inhabit humus or soil.

symphysis [ANAT] An immovable articulation of bones connected by fibrocartilaginous pads.

symphysis pubis *See* pubic symphysis.

Symphyta [INV ZOO] A suborder of the Hymenoptera including the sawflies and horntails characterized by a broad base attaching the abdomen to the thorax.

symplectite *See* symplektite.

symplektite [MINERAL] An intimate intergrowth of two different minerals. Also spelled symplectite.

symplesite [MINERAL] $Fe_2(AsO_4)_3 \cdot 8H_2O$ A blue to bluish-green triclinic mineral composed of hydrous iron arsenate.

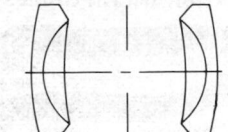

**SYMMETRICAL
ACHROMAT LENS**

Geometry of symmetrical achromat lens.

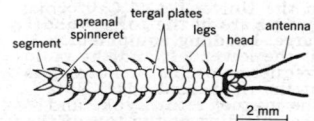

SYMPHYLA

A symphylan, *Scutigerella immaculata.* (*From R.E. Snodgrass, A Textbook of Arthropod Anatomy, Cornell University Press, 1952*)

SYNBRANCHIFORMES

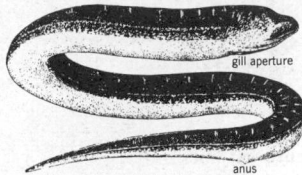

gill aperture

anus

Rice eel (*Monopterus albus*); length to 35 inches (89 centimeters). (*After M. Weber and L. F. de Beaufort, The Fishes of the Indo-Australian Archipelago, vol. 3, 1916*)

SYNCHROCYCLOTRON

View of the vacuum tank and upper magnet coil of the 184-inch (4.67-meter) synchrocyclotron at the University of California. Pumps are in the foreground; target-handling equipment is at the lower left and the radio-frequency oscillator housing is at the right-hand edge. Note the magnet return yoke and the shielding at the top of the picture. (*University of California Lawrence Berkeley Laboratory*)

SYNCHRONIZED BLOCKING OSCILLATOR

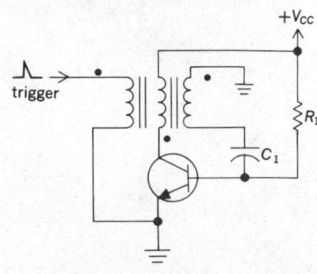

Circuit diagram of synchronized blocking oscillator. V_{CC} = collector supply voltage.

sympodium [BOT] A branching system in trees in which the main axis is comprised of successive secondary branches, each representing the dominant fork of a dichotomy.

symptom [MED] A phenomenon of physical or mental disorder or disturbance which leads to complaints on the part of the patient.

symptomatology [MED] 1. The science of symptoms. 2. In common usage, the symptoms of disease taken together as a whole.

synadelphite [MINERAL] (Mn,Mg,Ca,Pb)(AsO$_4$)(OH)$_5$ A black mineral composed of basic arsenate of manganese, often with magnesium, calcium, lead, or other metals.

Synallactidae [INV ZOO] A family of echinoderms of the order Aspidochirotida comprising mainly deep-sea forms which lack tentacle ampullae.

synandrous [BOT] Having several united stamens.

synangium [BOT] A compound sorus made up of united sporangia. [VERT ZOO] In lower vertebrates, a peripheral arterial trunk from which branches arise.

synantectic [MINERAL] Refers to a mineral that was formed by the reaction of two other minerals.

synantexis [GEOL] Deuteric alteration.

Synanthae [BOT] An equivalent name for Cyclanthales.

Synanthales [BOT] An equivalent name for Cyclanthales.

synapse [ANAT] A site where the axon of one neuron comes into contact with and influences the dendrites of another neuron or a cell body.

synapsis [CYTOL] Pairing of homologous chromosomes during the zygotene stage of meiosis.

synaptic transmission [PHYSIO] The mechanisms by which a presynaptic neuron influences the activity of an anatomically adjacent postsynaptic neuron.

synapticulum [INV ZOO] A conical or cylindrical supporting process, as those extending between septa in some corals, or connecting gill bars in *Branchiostoma*.

Synaptidae [INV ZOO] A family of large sea cucumbers of the order Apodida lacking a respiratory tree and having a reduced water-vascular system.

synaptinemal complex [CYTOL] Ribbonlike structures that extend the length of synapsing chromosomes and are believed to function in exchange pairing.

synarthrosis [ANAT] An articulation in which the connecting material (fibrous connective tissue) is continuous, immovably binding the bones.

Synbranchiformes [VERT ZOO] An order of eellike actinopterygian fishes that, unlike true eels, have the premaxillae present as distinct bones.

Synbranchii [VERT ZOO] The equivalent name for Synbranchiformes.

sync *See* synchronization.

Syncarida [INV ZOO] A superorder of crustaceans of the subclass Melacostraca lacking a carapace and oostegites and having exopodites on all thoracic limbs.

syncarp [BOT] A compound fleshy fruit.

syncarpous [BOT] Descriptive of a gynoecium having the carpels united in a compound ovary.

sync generator *See* synchronizing generator.

synchisite *See* synchysite.

synchondrosis [ANAT] A type of synarthrosis in which the bone surfaces are connected by cartilage.

synchro [ELEC] Any of several devices used for transmitting and receiving angular position or angular motion over wires, such as a synchro transmitter or synchro receiver. Also known as mag-slip (British usage); self-synchronous device; self-synchronous repeater; selsyn.

synchro- [SCI TECH] Occurring at the same time or made to occur at the same time.

synchro control transformer [ELEC] A transformer having its secondary winding on a rotor; when its three input leads are excited by angle-defining voltages, the two output leads deliver an alternating-current voltage that is proportional to the sine of the difference between the electrical input angle and the mechanical rotor angle.

synchro control transmitter [ELEC] A high-accuracy synchro transmitter, having high-impedance windings.

synchrocyclotron [NUCLEO] A circular particle accelerator for accelerating protons, deuterons, or alpha particles, in which the frequency of the accelerating voltage is modulated to maintain synchronism with the frequency of the particles which spiral outward and attain energies at which the relativistic mass increase becomes significant. Also known as frequency-modulated cyclotron; synchrophasotron.

synchro differential motor [ELEC] Motor which is electrically similar to the synchro differential generator except that a damping device is added to prevent oscillations; both its rotor and stator are connected to synchro generators, and its function is to indicate the sum or difference between the two signals transmitted by the generators.

synchro differential receiver [ELEC] A synchro receiver that subtracts one electrical angle from another and delivers the difference as a mechanical angle. Also known as differential synchro.

synchro differential transmitter [ELEC] A synchro transmitter that adds a mechanical angle to an electrical angle and delivers the sum as an electrical angle. Also known as differential synchro.

synchro generator *See* synchro transmitter.

synchromesh [MECH ENG] An automobile transmission device that minimizes clashing; acts as a friction clutch, bringing gears approximately to correct speed just before meshing.

synchro motor *See* synchro receiver.

synchronism [ELEC] Of a synchronous motor, the condition under which the motor runs at a speed which is directly related to the frequency of the power applied to the motor and is not dependent upon variables. [PHYS] Condition of two periodic quantities which have the same frequency, and whose phase difference is either constant or varies around a constant average value.

synchronization [ENG] The maintenance of one operation in step with another, as in keeping the electron beam of a television picture tube in step with the electron beam of the television camera tube at the transmitter. Also known as sync.

synchronization indicator [ENG] An indicator that presents visually the relationship between two varying quantities or moving objects.

synchronized blocking oscillator [ELECTR] A blocking oscillator which is synchronized with pulses occurring at a rate slightly faster than its own natural frequency.

synchronized shifting [MECH ENG] Changing speed gears, with the gears being brought to the same speed before the change can be made.

synchronizer [ADP] A computer storage device used to compensate for a difference in rate of flow of information or time of occurrence of events when transmitting information from one device to another. [ELECTR] The component of a radar set which generates the timing voltage for the complete set.

synchronizing generator [ELECTR] An electronic generator that supplies synchronizing pulses to television studio and transmitter equipment. Also known as sync generator; sync-signal generator.

synchronizing pulse [COMMUN] In pulse modulation, a pulse which is transmitted to synchronize the transmitter and the receiver; it is usually distinguished from signal-carrying pulses by some special characteristic.

synchronizing reactor [ELEC] Current-limiting reactor for connecting momentarily across the open contacts of a circuit-interrupting device for synchronizing purposes.

synchronizing relay [ELEC] Relay which functions when two alternating-current sources are in agreement within predetermined limits of phase angle and frequency.

synchronizing signal *See* sync signal.

synchronous [ENG] In step or in phase, as applied to two or more circuits, devices, or machines. [GEOL] Geological rock units or features formed at the same time.

synchronous bombing [ORD] Bombing done with certain bombsights, such as the Norden bombsight, in which the travel of the telescope, focused upon the target, is synchronized with the ground speed of the airplane, and the course flown is determined by manual adjustment of the bombsight, the two together determining the dropping angle and correcting for drift so that the release occurs at the right instant.

synchronous booster converter [ELEC] Synchronous converter having an alternating-current generator mounted on the same shaft and connected in series with it to adjust the voltage at the commutator of the converter.

synchronous capacitor [ELEC] A synchronous motor running without mechanical load and drawing a large leading current, like a capacitor; used to improve the power factor and voltage regulation of an alternating-current power system.

synchronous clamp circuit *See* keyed clamp circuit.

synchronous clock [HOROL] An electric clock driven by a synchronous motor, for operation from an alternating-current power system in which the frequency is accurately controlled.

synchronous computer [ADP] A digital computer designed to operate in sequential elementary steps, each step requiring a constant amount of time to complete, and being initiated by a timing pulse from a uniformly running clock.

synchronous converter [ELEC] A converter in which motor and generator windings are combined on one armature and excited by one magnetic field; normally used to change alternating to direct current. Also known as electric converter.

synchronous demodulator *See* synchronous detector.

synchronous detector [ELECTR] **1.** A detector that inserts a missing carrier signal in exact synchronism with the original carrier at the transmitter; when the input to the detector consists of two suppressed-carrier signals in phase quadrature, as in the chrominance signal of a color television receiver, the phase of the reinserted carrier can be adjusted to recover either one of the signals. Also known as synchronous demodulator. **2.** *See* cross-correlator.

synchronous gate [ELECTR] A time gate in which the output intervals are synchronized with an incoming signal.

synchronous generator [ELEC] A machine that generates an alternating voltage when its armature or field is rotated by a motor, an engine, or other means. The output frequency is exactly proportional to the speed at which the generator is driven.

synchronous growth [MICROBIO] A population of bacteria in which all cells divide at approximately the same time.

synchronous inverter *See* dynamotor.

synchronous machine [ELEC] An alternating-current machine whose average speed is proportional to the frequency of the applied or generated voltage.

synchronous motor [ELEC] A synchronous machine that transforms alternating-current electric power into mechanical power, using field magnets excited with direct current.

synchronous orbit [AERO ENG] **1.** An orbit in which a satellite makes a limited number of equatorial crossing points which are then repeated in synchronism with some defined reference (usually earth or sun). **2.** Commonly, the equatorial, circular, 24-hour case in which the satellite appears to hover over a specific point of the earth.

synchronous pluton [GEOL] Any pluton whose time of emplacement coincides with a major orogeny.

synchronous radar bombing [ORD] A kind of radar bombing in which special airborne radar equipment containing rate and steering mechanisms is used to control the direction of flight of the bombing aircraft, solve the bombing problem, and automatically drop bombs at the proper release point.

synchronous rectifier [ELECTR] A rectifier in which contacts are opened and closed at correct instants of time for rectification by a synchronous vibrator or by a commutator driven by a synchronous motor.

synchronous satellite *See* geostationary satellite.

synchronous speed [ELECTROMAG] The speed of rotation of a magnetic field in a synchronous machine; in revolutions per second, it is equal to twice the frequency of the alternating current in hertz, divided by the number of poles in the machine.

synchronous switch [ELECTR] A thyratron circuit used to control the operation of ignitrons in such applications as resistance welding.

synchronous system [COMMUN] A telecommunication system in which transmitting and receiving apparatus operate continuously at substantially the same rate, and correction devices are used, if necessary, to maintain them in a fixed relationship.

synchronous timing [MET] Regulating the welding-transformer primary current in spot, seam, or projection welding so that the following conditions prevail: the first half-cycle is initiated at the proper time in relation to the voltage to ensure a balanced current wave, each succeeding half-cycle is essentially the same as the first, and the last half-cycle is of the opposite polarity to the first.

synchronous transmission [COMMUN] Transmission of a synchronous system.

synchronous vibrator [ELECTROMAG] An electromagnetic vibrator that simultaneously converts a low direct voltage to a low alternating voltage and rectifies a high alternating voltage obtained from a power transformer to which the low alternating voltage is applied; in power packs, it eliminates the need for a rectifier tube.

synchrophasotron *See* synchrocyclotron.

synchro receiver [ELEC] A synchro that provides an angular position related to the applied angle-defining voltages; when two of its input leads are excited by an alternating-current voltage and the other three input leads are excited by the angle-defining voltages, the rotor rotates to the corresponding angular position; the torque of rotation is proportional to the sine of the difference between the mechanical and electrical angles. Also known as receiver synchro; selsyn motor; selsyn receiver; synchro motor.

synchro resolver *See* resolver.

synchroscope [ELECTR] A cathode-ray oscilloscope designed to show a short-duration pulse by using a fast sweep that is synchronized with the pulse signal to be observed. [ENG] An instrument for indicating whether two periodic quantities are synchronous; the indicator may be a rotating-pointer device or a cathode-ray oscilloscope providing a rotating pattern; the position of the rotating pointer is a measure of the instantaneous phase difference between the quantities.

synchro-shutter [ENG] A camera shutter with a circuit that flashes a light the instant the shutter opens.

synchro system [ELEC] An electric system for transmitting angular position or motion; in the simplest form it consists of a synchro transmitter connected by wires to a synchro receiver; more complex systems include synchro control transformers and synchro differential transmitters and receivers. Also known as selsyn system.

synchro transmitter [ELEC] A synchro that provides voltages related to the angular position of its rotor; when its two input leads are excited by an alternating-current voltage, the magnitudes and polarities of the voltages at the three output leads define the rotor position. Also known as selsyn generator; selsyn transmitter; synchro generator; transmitter; transmitter synchro.

synchrotron [NUCLEO] A device for accelerating electrons or protons in closed orbits in which the frequency of the accelerating voltage is varied (or held constant in the case of electrons) and the strength of the magnetic field is varied so as to keep the orbit radius constant.

synchrotron process [ELECTROMAG] The emission of electromagnetic radiation by relativistic electrons orbiting in a magnetic field.

synchrotron radiation [ELECTROMAG] Electromagnetic radiation generated by the acceleration of charged relativistic particles, usually electrons, in a magnetic field.

synchysite [MINERAL] $(Ce,La)Ca(Co_3)_2F$ A mineral composed of fluoride and carbonate of calcium, cerium, and lanthanum. Also spelled synchisite.

synclinal axis *See* trough surface.

synclinal valley [GEOL] Pertaining to a topographic valley whose sides coincide with a synclinal fold.

syncline [GEOL] A fold having stratigraphically younger rock material in its core; it is concave upward.

synclinorium [GEOL] A composite synclinal structure in a region of lesser folds.

Syncom [AERO ENG] One of a series of communication satellites placed in synchronous equatorial orbit; used for relaying television and radio communications over long distances.

syncope [MED] Swooning or fainting; temporary suspension of consciousness.

sync separator [ELECTR] A circuit that separates synchronizing pulses from the video signal in a television receiver.

SYNCHRONOUS MACHINE

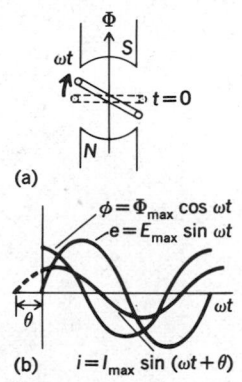

(a)

(b) $i = I_{max} \sin(\omega t + \theta)$

$\phi = \Phi_{max} \cos \omega t$
$e = E_{max} \sin \omega t$

Single-phase, two-pole synchronous machine. *(a)* Schematic diagram. Φ is direction of flux. Coil is perpendicular to pole axis at time $t = 0$. ω = angular velocity of coil; ωt = angular displacement at time t. Angular displacement is proportional to ωt. *(b)* Graphs of flux linking coil ϕ, voltage induced in coil e, and current in coil i as functions of time. θ is phase angle.

SYNCOM

Syncom communications satellite. *(NASA)*

sync signal [COMMUN] A signal transmitted after each line and field to synchronize the scanning process in a television or facsimile receiver with that of the receiver. Also known as synchronizing signal.

sync-signal generator *See* synchronizing generator.

synctial trophoblast *See* synctiotrophoblast.

syncytiotrophoblast [CYTOL] An irregular sheet or net of deeply staining cytoplasm in which nuclei are irregularly scattered. Also known as plasmoditrophoblast; synctial trophoblast.

syncytium [CYTOL] A mass of multinucleated cytoplasm without division into separate cells.

syndactyly [ANAT] The condition characterized by union of two or more digits, as in certain birds and mammals; it is a familial anomaly in man.

syndesmosis [ANAT] An articulation in which the bones are joined by collagen fibers.

syndet *See* synthetic detergent.

syndiotactic polymer [ORG CHEM] A vinyl polymer in which the side chains alternate regularly above and below the plane of the backbone.

syndrome [MED] A group of signs and symptoms which together characterize a disease. Also known as complex.

synecology [ECOL] The study of environmental relations of groups of organisms, such as communities.

Synentognathi [VERT ZOO] The equivalent name for Beloniformes.

syneresis [CHEM] Spontaneous separation of a liquid from a gel or colloidal suspension due to contraction of the gel.

synergic curve [AERO ENG] A curve plotted for the ascent of a rocket, space-air vehicle, or space vehicle, calculated to give optimum fuel economy and optimum velocity.

synergid [BOT] Either of two small cells lying in the embryo sac in seed plants adjacent to the egg cell toward the micropylar end.

synergism [ECOL] An ecological association in which the physiological processes or behavior of an individual are enhanced by the nearby presence of another organism. [MATER] An action where the total effect of two active components in a mixture is greater than the sum of their individual effects, for example, a mixture volume that is greater than the sum of the individual volumes, or in resin formulation, the use of two or more stabilizers, where the combination improves polymer stability more than expected from the additive effect of the stabilizers, a material that causes such an effect is known as a synergist.

Syngamidae [INV ZOO] A family of roundworms belonging to the Strongyloidea and including parasites of birds and mammals.

syngamy [BIOL] Sexual reproduction involving union of gametes.

syngenesis [GEOL] In place formation of unconsolidated sediments.

syngenetic [GEOL] **1.** Pertaining to a primary sedimentary structure formed contemporaneously with sediment deposition. **2.** Pertaining to a mineral deposit formed contemporaneously with the enclosing rock. Also known as ideogenous.

syngenite [MINERAL] $K_2Ca(SO_4)_2 \cdot H_2O$ A colorless or white mineral composed of hydrous potassium calcium sulfate occurring in tabular crystals.

Syngnathidae [VERT ZOO] A family of fishes in the order Gasterosteiformes including the seahorses and pipefishes.

synkinematic *See* syntectonic.

synkinesia [PHYSIO] Involuntary movement coincident with purposeful movements carried out by a distant part of the body, such as swinging the arms while walking. Also known as accessory movement; associated automatic movement.

synodic [ASTRON] Referring to conjunction of celestial bodies.

synodic month [ASTRON] A month based on the moon's phases.

synodic period [ASTRON] The time period between two successive astronomical conjunctions of the same celestial objects.

synonym [SYST] A taxonomic name that is rejected as being incorrectly applied, or incorrect in form, or not representative of a natural genetic grouping.

synoptic [METEOROL] Refers to the use of meteorological data obtained simultaneously over a wide area for the purpose of presenting a comprehensive and nearly instantaneous picture of the state of the atmosphere.

synoptic chart [METEOROL] Any chart or map on which data and analyses are presented that describe the state of the atmosphere over a large area at a given moment in time.

synoptic climatology [CLIMATOL] The study and analysis of climate in terms of synoptic weather information, principally in the form of synoptic charts; the information thus obtained gives the climate (that is, average weather) of a given locality in a given synoptic situation rather than the usual climatic parameters which represent averages over all synoptic conditions.

synoptic code [METEOROL] In general, any code by which synoptic weather observations are communicated; among the synoptic codes in use are the international synoptic code, ship synoptic code, U.S. Airways code, and RECCO code.

synoptic correlation *See* Eulerian correlation.

synoptic meteorology [METEOROL] The study and analysis of synoptic weather information.

synoptic model [METEOROL] Any model specifying a space distribution of some meteorological elements; the distribution of clouds, precipitation, wind, temperature, and pressure in the vicinity of a front is an example of a synoptic model.

synoptic oceanography [OCEANOGR] The study of the physical spatial parameters of the ocean through analysis of simultaneous observations from many stations.

synoptic report [METEOROL] An encoded and transmitted synoptic weather observation.

synoptic scale *See* cyclonic scale.

synoptic wave chart [OCEANOGR] A chart of an ocean area on which is plotted synoptic wave reports from vessels, along with computed wave heights for areas where reports are lacking; atmospheric fronts, highs, and lows are also shown; isolines of wave height and the boundaries of areas having the same dominant wave direction are drawn.

synoptic weather observation [METEOROL] A surface weather observation, made at periodic times (usually at 3- and 6-hourly intervals specified by the World Meteorological Organization), of sky cover, state of the sky, cloud height, atmospheric pressure reduced to sea level, temperature, dew point, wind speed and direction, amount of precipitation, hydrometeors and lithometeors, and special phenomena that prevail at the time of the observation or have been observed since the previous specified observation.

synorchidism [MED] Partial or complete fusion of the two testes within the abdomen or scrotum.

synorogenic [GEOL] Referring to a geologic process occurring at the same time as orogenic activity.

synostosis [ANAT] A type of synarthrosis in which the bones are continuous. [MED] Union of originally separate bones into a single bone structure.

synovia *See* synovial fluid.

synovial fluid [PHYSIO] A transparent viscid fluid secreted by synovial membranes. Also known as synovia.

synovial membrane [HISTOL] A layer of connective tissue which lines sheaths of tendons at freely moving articulations, ligamentous surfaces of articular capsules, and bursae.

synovioma [MED] Any tumor composed principally of cells similar to those covering the synovial membrane.

synovitis [MED] Inflammation of a synovial membrane.

synpelmous [VERT ZOO] Having the two main flexor tendons of the toes united beyond the branches that go to each digit, as in certain birds.

synsepalous *See* gamosepalous.

syntactic foam [MATER] A cellular polymer made by dispersing rigid, microscopic particles in a fluid polymer and then curing it.

syntax [ADP] The set of rules needed to construct valid expressions or sentences in a language.

syntaxial overgrowth [MINERAL] A crystallographically oriented overgrowth of two alternating, chemically identical substances.

syntaxis [MAP] On a map, a sheaflike pattern of mountains converging on a common center.

syntectic [GEOL] *See* syntexis. [MET] Isothermal, reversible conversion of a solid phase into two conjugate liquid phases on applying heat.

syntectonic [GEOL] Refers to a geologic process or event occurring during tectonic activity. Also known as synkinematic.

Synteliidae [INV ZOO] The sap-flow beetles, a small family of coleopteran insects in the superfamily Histeroidea.

syntenic group [GEN] The loci on the same chromosome pair, irrespective of whether or not they are known to show linkage in heredity.

Syntexidae [INV ZOO] A family of the Hymenoptera in the superfamily Siricoidea.

syntexis [GEOL] Magma made by the melting of two or more rock types and the assimilation of country rock. Also known as syntectic.

synthesis [CHEM] Any process or reaction for building up a complex compound by the union of simpler compounds or elements. [CONT SYS] *See* system design.

synthesis gas [CHEM ENG] A mixture of gases prepared as feedstock for a chemical reaction, for example, carbon monoxide and hydrogen to make hydrocarbons or organic chemicals, or hydrogen and nitrogen to make ammonia.

synthesizer [ELECTR] An electronic instrument which combines simple elements to generate more complex entities; examples are frequency synthesizer and sound synthesizer.

synthetase *See* ligase.

synthetic crude [MATER] The total liquid, multicomponent hydrocarbon mixture resulting from a process involving molecular rearrangement of charge stock, as from oil shale or synthesis gas. Also known as synthetic oil.

synthetic detergent [MATER] A liquid or solid material able to dissolve oily materials and disperse them (or emulsify them) in water. Also known as syndet.

synthetic division [MATH] A long division process for dividing a polynomial $p(x)$ by a polynomial $(x-a)$ where only the coefficients of these polynomials are used.

synthetic fiber *See* artificial fiber.

synthetic gem [MATER] A precious or semiprecious stone made by artificial processes, for example, synthetic diamonds made by extreme heat and pressure on carbon, used industrially; and synthetic rubies made by high-temperature crystallization of aluminum oxide, used in laser equipment.

synthetic graphite [MATER] Graphitic crystalline material made by the high-temperature and pressure processing of carbon.

synthetic language [ADP] A pseudocode or symbolic language; fabricated language.

synthetic lubricant [MATER] Any of a group of products, some of them based on petroleum, used as lubricants where heat, chemical resistance, and other requirements can be better met than with straight petroleum products.

synthetic mica [MATER] A fluorphlogopite mica made artificially by heating a large batch of raw material in an electric resistance furnace and allowing the mica to crystallize from the melt during controlled slow cooling.

synthetic oil *See* synthetic crude.

synthetic resin [ORG CHEM] Amorphous, organic, semisolid, or solid material derived from the polymerization of unsaturated monomers such as ethylene, butylene, propylene, and styrene.

synthetic rubber [MATER] Synthetic products whose properties are similar to those of natural rubber, including elasticity and ability to be vulcanized; usually produced by the polymerization or copolymerization of petroleum-derived olefinic or other unsaturated compounds.

synthol process [CHEM ENG] A reaction of carbon monoxide and hydrogen with an iron and sodium carbonate catalyst; produces a mixture of higher alcohols, aldehydes, ketones, higher fatty acids, and aliphatic hydrocarbons, usable as a synthetic gasoline.

syntony [ELEC] Condition in which two oscillating circuits have the same resonant frequency.

Syntrophiidina [PALEON] A suborder of extinct articulate brachiopods of the order Pentamerida characterized by a strong dorsal median fold.

syntrophism [BIOL] Mutual dependence of cells for nutritional needs, especially between strains of bacteria.

syntrophoblast [EMBRYO] The outer synctial layer of the trophoblast that forms the outermost fetal element of the placenta.

syntype [SYST] Any specimen of a series when no specimen is designated as the holotype. Also known as cotype.

synusia [ECOL] A structural unit of a community characterized by uniformity of life-form or of height.

Synxiphosura [PALEON] An extinct heterogeneous order of arthropods in the subclass Merostomata possibly representing an explosive proliferation of aberrant, terminal, and apparently blind forms.

syphilis [MED] An infectious disease caused by the spirochete *Treponema pallidum,* transmitted principally by sexual intercourse.

syphilitic meningoencephalitis *See* general paresis.

syphilophobia [PSYCH] An abnormal fear of syphilis.

syringe [MED] **1.** An apparatus commonly made of glass or plastic, fitted snugly onto a hollow needle, used to aspirate or inject fluids for diagnostic or therapeutic purposes. Also known as hypodermic syringe. **2.** A large glass barrel with a fitted rubber bulb at one end and a nozzle at the other, used primarily for irrigation purposes.

syringobulbia [MED] The presence of cavities in the medulla oblongata similar to those found in syringomyelia.

syringocystadenoma *See* syringoma.

syringocystoma *See* syringoma.

syringoma [MED] A multiple nevoid tumor of sweat glands. Also known as syringocystadenoma; syringocystoma.

syringomyelia [MED] A chronic disease characterized by the presence of cavities surrounded by gliosis near the canal of the spinal cord and often extending to the medulla.

Syringophyllidae [PALEON] A family of extinct corals in the order Tabulata.

syrinx [PALEON] A tube surrounding the pedicle in certain fossil brachiopods. [VERT ZOO] The vocal organ in birds.

Syrphidae [INV ZOO] The flower flies, a family of cyclorrhaphous dipteran insects in the series Aschiza.

syserskite [MINERAL] Mineral composed of an alloy of osmium (50–80%) and iridium (20–50%).

SYSIN [ADP] The principal input stream of an operating system. Derived from system input.

syssiderite *See* stony-iron meteorite.

Systellomatophora [INV ZOO] An order of the subclass Pulmonata in which the eyes are contractile but stalks are not retractile, the body is sluglike, oval, or lengthened, and the lung is posterior.

system [ELECTR] A combination of two or more sets generally physically separated when in operation, and such other assemblies, subassemblies, and parts necessary to perform an operational function or functions. [ENG] A combination of several pieces of equipment integrated to perform a specific function; thus a fire control system may include a tracking radar, computer, and gun. [GEOL] **1.** A major time-stratigraphic unit of worldwide significance, representing the basic unit of Phanerozoic rocks. **2.** A group of related structures, such as joints. [PHYS] A region in space or a portion of matter that has a certain amount of one or more substances, ordered in one or more phases. [SCI TECH] A method of organizing entities or terms; in particular, organizing such entities into a larger aggregate.

system analysis [CONT SYS] The use of mathematics to determine how a set of interconnected components whose individual characteristics are known will behave in response to a given input or set of inputs.

systematic distortion [ELEC] Periodic or constant distortion, such as bias or characteristic distortion; the direct opposite of fortuitous distortion.

systematic error [ENG] An error due to some known physical law by which it might be predicted; these errors produced by the same cause affect the mean in the same sense, and do not tend to balance each other but rather give a definite bias to the mean. [STAT] An error which results from some bias in the measurement process and is not due to chance, in contrast to random error.

systematic joints [GEOL] Joints occurring in patterns or sets and oriented perpendicular to the boundaries of the constituent rock unit.

SYNTROPHIIDINA

(a)   (b)

Pedicle valve of *Imbricata* in the Syntrophiidina; *(a)* exterior view, *(b)* interior view.

in the measurement process and is not due to chance, in contrast to random error.

systematic joints [GEOL] Joints occurring in patterns or sets and oriented perpendicular to the boundaries of the constituent rock unit.

systematics [BIOL] The science of animal and plant classification.

systematic sampling [MIN ENG] Extracting samples at evenly spaced periods or in fixed quantities from a unit of coal.

systematic support [MIN ENG] The regular setting of timber or steel supports at fixed intervals irrespective of the condition of the roof and sides.

system bandwidth [CONT SYS] The difference between the frequencies at which the gain of a system is $\sqrt{2}/2$ (that is, 0.707) times its peak value.

system check [ADP] A check on the overall performance of the system, usually not made by built-in computer check circuits; for example, control total, hash totals, and record counts.

system design [ADP] Determination in detail of the exact operational requirements of a system, resolution of these into file structures and input/output formats, and relation of each to management tasks and information requirements. [CONT SYS] A technique of constructing a system that performs in a specified manner, making use of available components. Also known as synthesis.

system designer [ADP] A person who prepares final system documentation, analyzes findings, and synthesizes new system design.

system effectiveness [ENG] A measure of the extent to which a system may be expected to achieve a set of specific mission requirements expressed as a function of availability, dependability, and capability.

Système International d'Unites See International System of Units.

system engineering See systems engineering.

system evaluation [ADP] A periodic evaluation of the system to assess its status in terms of original or current expectations and to chart its future direction.

system generation [ADP] A process that creates a particular and uniquely specified operating system; it combines user-specified options and parameters with manufacturer-supplied general-purpose or nonspecialized program subsections to produce an operating system (or other complex software) of the desired form and capacity.

systemic circulation [PHYSIO] The general circulation, as distinct from the pulmonary circulation.

system improvement time [ADP] The machine downtime needed for the installation and testing of new components, large or small, and machine downtime necessary for modification of existing components; this includes all programming tests following the above actions to prove the machine is operating properly.

system input See SYSIN.

system life cycle [ENG] The continuum of phases through which a system passes from conception through disposition.

system master tapes [ADP] Magnetic tapes containing programmed instructions necessary for preparing a computer prior to running programs.

system operation [ADP] The administration and operation of an automatic data-processing equipment-oriented system, including staffing, scheduling, equipment and service contract administration, equipment utilization practices, and time-sharing.

system optimization See optimization.

system reliability [ENG] The probability that a system will accurately perform its specified task under stated environmental conditions.

system response See response.

system safety [ENG] The optimum degree of safety within the constraints of operational effectiveness, time, and cost, attained through specific application of system safety engineering throughout all phases of a system.

system safety engineering [ENG] An element of systems management involving the application of scientific and engineering principles for the timely identification of hazards, and initiation of those actions necessary to prevent or control hazards within the system.

systems analysis [ENG] The analysis of an activity, procedure, method, technique, or business to determine what must be accomplished and how the necessary operations may best be accomplished.

systems ecology [ECOL] The combined approaches of systems analysis and the ecology of whole ecosystems and subsystems.

systems engineering [ENG] The design of a complex interrelation of many elements (a system) to maximize an agreed-upon measure of system performance, taking into consideration all of the elements related in any way to the system, including utilization of manpower as well as the characteristics of each of the system's components. Also known as system engineering.

systems for nuclear auxiliary power See SNAP.

systems implementation test [ENG] The test program that exercises the complete system in its actual environment to determine its capabilities and limitations; this test also demonstrates that the system is functionally operative, and is compatible with the other subsystems and supporting elements required for its operational employment.

systems test [ADP] The running of the whole system against test data, a complete simulation of the actual running system for purposes of testing out the adequacy of the system. [ENG] A test of an entire interconnected set of components for the purpose of determining proper functions and interconnections.

system study [ADP] A detailed study to determine whether, to what extent, and how automatic data-processing equipment should be used; it usually includes an analysis of the existing system and the design of the new system, including the development of system specifications which provide a basis for the selection of equipment.

system supervisor [ADP] A control program which ensures an efficient transition in running program after program and accomplishing setups and control functions.

system unit [ADP] An individual card, section of tape, or the like, which is manipulated during operation of the system; class 1 systems have one unit per document; class 2 systems have one unit per vocabulary term or concept.

systole [PHYSIO] The contraction phase of the heart cycle.

syzgy [ASTRON] **1.** One of the two points in a celestial object's orbit where it is in conjunction with or opposition to the sun. **2.** Those points in the moon's orbit where the moon, earth, and sun are in a straight line. [INV ZOO] End-to-end union of the sporonts of certain gregarine protozoans.

szaskaite See smithsonite.

Szechtman cell [CHEM ENG] An electrolytic process for manufacture of chlorine that is a variation of both the mercury cell and molten salt cell.

szmikite [MINERAL] $MrSO_4 \cdot H_2O$ A monoclinic mineral composed of hydrous manganese sulfate.

szomolnokite [MINERAL] $FeSO_4 \cdot H_2O$ A yellow or brown monoclinic mineral composed of hydrous ferrous sulfate.

t *See* troy system.

T *See* tera-; tesla.

2,4,5-T *See* 2,4,5-trichlorophenoxyacetic acid.

2,4,6-T *See* 2,4,6-trichlorophenol.

TΩ *See* teraohm.

Ta *See* tantalum.

Tabanidae [INV ZOO] The deer and horse flies, a family of orthorrhaphous dipteran insects in the series Brachycera.

tabbyite [MINERAL] A variety of solid asphalt found in the western United States; used as rubber filler and with roofing materials.

tab-card cutter [DES ENG] A device for die-cutting card stock to uniform tabulating-card size.

tabes dorsalis [MED] A form of parenchymatous neurosyphilis in which there is demyelination and sclerosis of the posterior columns of the spinal cord. Also known as locomotor ataxia.

tabetisol *See* talik.

table [ADP] A set of contiguous, related items, each uniquely identified either by its relative position in the set or by some label. [LAP] The flat face forming the top of a brilliant-cut stone. [MATH] An array or listing of computed quantities. [MECH ENG] That part of a grinding machine which directly or indirectly supports the work being ground. [MIN ENG] **1.** In placer mining, a wide, shallow sluice box designed to recover gold or other valuable material from screened gravel. **2.** A platform or plate on which coal is screened and picked.

table-driven compiler [ADP] A compiler in which the source language is described by a set of syntax rules.

tabled whelk [INV ZOO] *Neptunea tabulata*. A marine gastropod mollusk about 5 inches (13 centimeters) in length and 2 inches (5 centimeters) in diameter, found at depths of 150–200 feet (45–60 meters), off the west coast of Canada and the United States.

table flotation [MIN ENG] A flotation process in which a slurry of ore is fed to a shaking table where floatable particles become glomerules, held together by minute air bubbles and edge adhesion; the glomerules roll across the table and are discharged nearly opposite the feed end; the process is helped by jets of low-pressure air.

table iceberg *See* tabular iceberg.

table knoll [GEOGR] A knoll with a comparatively smooth, flat top.

tableland [GEOGR] A broad, elevated, nearly level, and extensive region of land that has been deeply cut at intervals by valleys or broken by escarpments. Also known as continental plateau.

table look-up [ADP] **1.** A procedure for searching identifying labels in a table so as to find the location of a desired item. **2.** By extension, a digital computer instruction which directs that the above operation be performed.

tablemount *See* guyot.

table mountain [GEOGR] A flat-topped mountain.

table reef [GEOL] A small, isolated organic reef which has a flat top and does not enclose a lagoon.

table salt *See* sodium chloride.

tablespoonful [MECH] A unit of volume used particularly in cookery, equal to 4 fluid drams or ½ fluid ounce; in the United States this is equal to approximately 14.7868 cubic centimeters, in the United Kingdom to approximately 14.2065 cubic centimeters. Abbreviated tbsp.

table sugar *See* sucrose.

tableting [ENG] A punch-and-die procedure for the compaction of powdered or granular solids; used for pharmaceuticals, food products, fireworks, vitamins, and dyes.

tabling [MIN ENG] Separation of two materials of different densities by passing a dilute suspension over a slightly inclined table having a reciprocal horizontal motion or shake with a slow forward motion and a fast return.

tabula [PALEON] A transverse septum that closes off the lower part of the polyp cavity in certain extinct corals and hydroids.

tabular [GEOL] Referring to a sedimentary particle whose length is two to three times its thickness.

tabular berg *See* tabular iceberg.

tabular crystal [CRYSTAL] A crystal that appears broad and flat due to two prominent parallel faces.

tabular iceberg [OCEANOGR] An iceberg with clifflike sides and a flat top; usually arises by detachment from an ice shelf. Also known as table iceberg; tabular berg.

tabular interpolation [MATH] Method of finding from a table the values of the dependent variable for intermediate values of the independent variable.

tabular language [ADP] A part of a program which represents the composition of a decision table required by the problem considered.

tabular spar *See* wollastonite.

Tabulata [PALEON] An extinct Paleozoic order of corals of the subclass Zoantharia characterized by an exclusively colonial mode of growth and by secretion of a calcareous exoskeleton of slender tubes.

tabulate [ADP] To order a set of data into a table form, or to print a set of data as a table, usually indicating differences and totals, or just totals.

tabulated altitude [NAV] In navigational sight reduction tables, the altitude taken directly from a table for the entering arguments.

tabulated azimuth [NAV] Azimuth taken directly from a table, before interpolation.

tabulated azimuth angle [NAV] Azimuth angle taken directly from a table, before interpolation.

tabulating card [ADP] Card into which coded holes are punched.

tabulating equipment [ADP] Machinery to punch, sense, sort, or check coded holes in tabulating cards.

tabulating system [ADP] Any group of machines which is capable of entering, converting, receiving, classifying, computing, and recording data by means of tabulating cards, and in which tabulating cards are used for storing data and for communicating with the system.

tabulator [ADP] A machine that reads information from punched cards and produces lists, tables, and totals on separate forms or continuous paper.

tabun *See* GA agent.

TAB vaccine [IMMUNOL] A vaccine containing killed typhoid bacilli and the paratyphoid organisms (*Salmonella paratyphi* A and B) most frequently involved in paratyphoid fever.

Tacan *See* tactical air navigation system.

tache noire [MED] The primary painless lesion of the tickborne typhus fevers of Africa, manifested by a raised red area with a typical black necrotic center which appears at the site of the tick bite.

Tachinidae [INV ZOO] The tachina flies, a family of bristly, grayish or black Diptera whose larvae are parasitic in caterpillars and other insects.

tachiol *See* silver fluoride.

tachometer [ENG] An instrument that measures the revolutions per minute or the angular speed of a rotating shaft.

tachycardia [MED] Excessive rapidity of the heart's action.

Tachyglossidae [VERT ZOO] A family of monotreme mammals having relatively large brains with convoluted cerebral hemispheres; comprises the echidnas or spiny anteaters.

tachylite [GEOL] A black, green, or brown volcanic glass formed from basaltic magma. Also known as basalt glass; basalt obsidian; hyalobasalt; jaspoid; sordawalite; wichtisite.

Tachyniscidae [INV ZOO] A family of myodarian cyclorrhaphous dipteran insects in the subsection Acalypteratae.

tachyon [PARTIC PHYS] A hypothetical particle that travels

TABLELAND

View of an ideal tableland: Canyon de Chelly National Monument, northeastern Arizona. (*Spence Air Photos*)

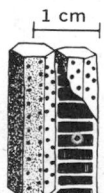

TABULA

1 cm

Part of two corallites from a specimen of the species *Favosites gothlandica* Lam in the order Tabulata, from the Corniferous Limestone (Devonian) of Woodstock, Ontario. One of the corallites is cut away to show a series of transverse tabulae. (*After H. G. Nicholson*)

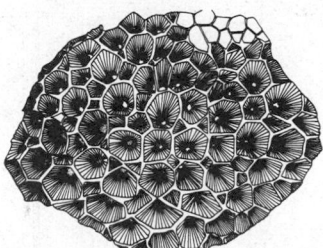

TABULATA

Specimen of *Michelinia convexa* D'Orb, a representative species of the Tabulata, seen from above, from the Carboniferous Limestone of Ontario. (*After Billings*)

faster than light, consistent with the theory of relativity.

tachyphylaxis [IMMUNOL] Rapid desensitization against doses of organ extracts or serum by the previous inoculation of small subtoxic doses of the same preparation.

tachysterol [BIOCHEM] The precursor of calciferol in the irradiation of ergosterol; an isomer of ergosterol.

tachytely [EVOL] Evolution at a rapid rate resulting in differential selection and fixation of new types.

tack [DES ENG] A small, sharp-pointed nail with a broad flat head. [MATER] Adhesive stickiness, such as occurs on the surface of a varnish or ink that has not completely dried. Also known as tackiness. [NAV] To change the course of a sailing vessel by coming about so as to take the wind from over the opposite bow (starboard or port).

tack coat [CIV ENG] A thin layer of bitumen, road tar, or emulsion laid on a road to enhance adhesion of the course above it.

tackiness *See* tack.

tackiness agent [MATER] An additive used to impart adhesive properties to otherwise nonadhesive materials, such as oils and greases.

tacking [MET] Making small, isolated tack welds.

tackle [MECH ENG] Any arrangement of ropes and pulleys to gain a mechanical advantage. [NAV ARCH] An assemblage of lines and blocks in which the line passes through more than one block.

tack range [ENG] The length of time during which an adhesive will remain in the tacky-dry condition after application to an adherent.

tack weld [MET] A joint between two pieces of metal made by welding at isolated points.

Taconian orogeny [GEOL] A process of formation of mountains in the latter part of the Ordovician period, particularly in the northern Appalachians. Also known as Taconic orogeny.

Taconic orogeny *See* Taconian orogeny.

taconite [GEOL] The siliceous iron formation from which high-grade iron ores of the Lake Superior district have been derived; consists chiefly of fine-grained silica mixed with magnetite and hematite.

tactical [ORD] Of or pertaining to the arranging, positioning, or maneuvering of forces in contact or near contact with the enemy, so as to achieve an objective or objectives in a campaign or battle.

tactical air control center [COMMUN] The principal air-operations installation (land- or ship-based) from which all aircraft and air warning functions of tactical air operations are controlled.

tactical air controller [COMMUN] The officer who is in charge of all operations of the tactical air-control center, and who is responsible to the tactical air officer for the control of all aircraft and air warning facilities within the area of responsibility.

tactical air-control system [COMMUN] The organization and equipment necessary to plan, direct, and control tactical air operations and to coordinate air operations with other services; it is composed of communications-electronics facilities which provide the means for centralized control and decentralized execution of missions.

tactical aircraft shelter [CIV ENG] A shelter to house fighter-type aircraft and to provide protection to the aircraft from attack by conventional weapons, or damage from high winds or other elemental hazards.

tactical air-direction center [AERO ENG] An air operations installation under the overall control of the tactical air-control center, from which aircraft and air warning service functions of tactical air operations in an area of responsibility are directed.

tactical air force [AERO ENG] An air force charged with carrying out tactical air operations in coordination with ground or naval forces.

tactical airlift [AERO ENG] That airlift which provides the immediate and responsive air movement and delivery of combat troops and supplies directly into objective areas through air landing, extraction, airdrop, or other delivery techniques; and the air logistic support of all theater forces, including those engaged in combat operations, to meet specific theater objectives and requirements.

tactical air navigation system [NAV] Short-range ultra-high-frequency air navigation system that provides accurate slant-range distance and bearing information; this information is presented to the pilot in two dimensions, that is, distance and bearing from a selected ground station. Also known as Tacan.

tactical air observer [AERO ENG] An officer trained as an air observer whose function is to observe from airborne aircraft and report on movement and disposition of friendly and enemy forces, on terrain and weather and hydrography, and to execute other missions as directed.

tactical air operations [AERO ENG] An air operation involving the employment of air power in coordination with ground or naval forces to gain and maintain air superiority, to prevent movement of enemy forces into and within the objective area and to seek out and destroy these forces and their supporting installations, and to join with ground or naval forces in operations within the objective area in order to assist directly in attainment of their immediate objective.

tactical air reconnaissance [AERO ENG] The use of air vehicles to obtain information concerning terrain, weather, and the disposition, composition, movement, installations, lines of communications, and electronic and communication emissions of enemy forces.

tactical air support [AERO ENG] Air operations carried out in coordination with surface forces which directly assist the land or naval battle.

tactical air transport [AERO ENG] The use of air transport in direct support of airborne assaults, carriage of air-transported forces, tactical air supply, evacuation of casualties from forward airdromes, and clandestine operations.

tactical call sign [COMMUN] Call sign which identifies a tactical communications facility.

tactical command ship [NAV ARCH] A warship designed to serve as a command ship for a fleet or force commander; it is equipped with extensive communication equipment.

tactical communications system [COMMUN] A system which provides internal communications within tactical air elements, composed of transportable and mobile equipment assigned as unit equipment to the supporting tactical unit.

tactical control radar [ENG] Antiaircraft artillery radar which has essentially the same inherent capabilities as the target acquisition radar (physically it may be the same type of set) but whose function is chiefly that of providing tactical information for the control of elements of the antiaircraft artillery defenses in battle.

tactical diameter [NAV] In marine operations, the distance gained to the right or left of the original course when a turn of 180° with a constant rudder angle has been completed.

tactical electronic warfare [ELECTR] The application of electronic warfare to tactical air operations; tactical electronic warfare encompasses the three major subdivisions of electronic warfare: electronic warfare support measures, electronic countermeasures, and electronic counter-countermeasures.

tactical fire control [ORD] The manner in which fire power is employed with regard to selection of targets, to opening, suspending, or ceasing fire, and to classes of fire.

tactical frequency [COMMUN] Radio frequency assigned to a military unit to be used in the accomplishment of a tactical mission.

tactical map [ORD] A large-scale map used for tactical and administrative purposes.

tactical missile [ORD] A missile for use in tactical operations.

tactical mobility [ORD] The capability of a unit, command, task force, or the like that enables it to be readily moved while engaged in combat.

tactical nuclear weapon [ORD] A nuclear weapon which is programmed primarily for employment against tactical targets in tactical military operations.

tactical nuclear weapon employment [ORD] The use of nuclear weapons by land, sea, or air forces against opposing forces and supporting installations or facilities, in support of operations which contribute to the accomplishment of a military mission of limited scope, or in support of the military commander's scheme of maneuver, usually limited to the area of military operations.

tactical range recorder [ENG] A sonar device in surface ships used to plot the time-range coordinates of submarines and determine firing of depth charges.

tactical reserve [ORD] A part of battalion, regiment, or similar force, held initially under the control of the commander as a maneuvering force to influence future action.

tactical target [ORD] Any physical object, person, group of persons, or position singled out for attack during the course of battle or tactical operations in order to reduce or destroy the enemy's ability to sustain combat operations.

tactical transport aircraft [AERO ENG] Aircraft designed primarily for carrying personnel or cargo over short or medium distances.

tactical unit [ORD] An organization of troops, aircraft, or ships which is intended to serve as a single unit in combat, and may include service units required for its direct support.

tactical vehicle [ORD] Any vehicle designed for field requirements in combat and tactical operations, or for training personnel for such operations.

tactic polymer [ORG CHEM] A polymer with regularity or symmetry in the structural arrangement of its molecules, as in a stereospecific polymer such as some types of polypropylene.

tactile [PHYSIO] Pertaining to the sense of touch.

tactile agnosia *See* astereognosis.

tactile hairs *See* vibrissae.

tactile receptor *See* tactoreceptor.

tactite [PETR] A rock with a complex mineralogical composition, formed by contact metamorphism and metasomatism of carbonate rocks.

tactoid [BIOCHEM] A particle that appears as a spindle-shaped body under the polarizing microscope and occurs in mosaic virus, fibrin, and myosin.

tactoreceptor [PHYSIO] A sense organ that responds to touch. Also known as tactile receptor.

tadpole [VERT ZOO] The larva of a frog or toad; at hatching it has a rounded body with a long fin-bordered tail, and the gills are external but shortly become enclosed.

tadpole shrimp [INV ZOO] Any of the phyllopod crustaceans that are members of the genus *Lepidurus*.

taele *See* frozen ground.

taenia [ANAT] A ribbon-shaped band of nerve fibers or muscle.

Taenia [INV ZOO] A genus of tapeworms which occur in humans and other mammals.

Taeniodidea [INV ZOO] An equivalent name for Cyclophyllidea.

Taeniodonta [PALEON] An order of extinct quadrupedal land mammals, known from early Cenozoic deposits in North America.

Taenioidea [INV ZOO] An equivalent name for Cyclophyllidea.

Taeniolabidoidea [PALEON] An advanced suborder of the extinct mammalian order Multituberculata having incisors that were self-sharpening in a limited way.

taeniolite [MINERAL] $KLiMg_2Si_4O_{10}F_2$ A white or colorless mica mineral.

taenite [MINERAL] A meteoritic mineral consisting of a nickel-iron alloy, with a nickel content varying from about 27 to 65%.

taenoglossate radula [INV ZOO] A long, narrow radula with seven teeth in each transverse row, found in certain pectinibranch bivalves.

taffeta [TEXT] A plain-woven, usually silk fabric that has a smooth finish and sheen on both sides.

taffrail log [ENG] A log consisting essentially of a rotator towed through the water by a braided log line attached to a distance-registering device usually secured at the taffrail, the railing at the stern. Also known as patent log.

tag [ADP] **1.** A unit of information used as a label or marker. **2.** The symbol written in the location field of an assembly-language coding form, and used to define the symbolic address of the data or instruction written on that line. [NUCLEO] *See* isotopic tracer.

Tag closed-cup tester [ANALY CHEM] A laboratory device used to determine the flash point of mobile petroleum liquids flashing below 175°F (79.4°C). Also known as Tagliabue closed tester.

tag converting unit [ADP] A device capable of reading the perforations of a price tag as input data.

tagged molecule [CHEM] A molecule having one or more atoms which are either radioactive or have a mass which differs from that of the atoms which normally make up the molecule.

Tagliabue closed tester *See* Tag closed-cup tester.

Tag-Robinson colorimeter [ENG] A laboratory device used to determine the color shades of lubricating and other oils; the color, reported as a number, is determined by varying the thickness of a column of oil until its color matches that of a standard color glass.

tagua palm [BOT] *Phytelephas macrocarpa.* A palm tree of tropical America; the endosperm of the seed is used as an ivory substitute.

Tahuian [GEOL] A local Eocene time subdivision in Australia whose identification is based on foraminiferans.

taiga [ECOL] A zone of forest vegetation encircling the Northern Hemisphere between the arctic-subarctic tundras in the north and the steppes, hardwood forests, and prairies in the south.

taiga climate [CLIMATOL] In general, a climate which produces taiga vegetation, that is, too cold for prolific tree growth but milder than the tundra climate and moist enough to promote appreciable vegetation. Also known as subarctic climate.

tail [AERO ENG] **1.** The rear part of a body, as of an aircraft or a rocket. **2.** The tail surfaces of an aircraft or rocket. [ASTRON] The part of a comet that extends from the coma in a direction opposite to the sun; it consists of dust and gas that have been blown away from the coma by the solar wind and the sun's radiation pressure. [ELECTR] **1.** A small pulse that follows the main pulse of a radar set and rises in the same direction. **2.** The trailing edge of a pulse. [VERT ZOO] **1.** The usually slender appendage that arises immediately above the anus in many vertebrates and contains the caudal vertebrae. **2.** The uropygium, and its feathers, of a bird. **3.** The caudal fin of a fish or aquatic mammal.

tail assembly *See* empennage.

tail clipping [ELECTR] Method of sharpening the trailing edge of a pulse.

tail fin [AERO ENG] A fin at the rear of a rocket or other body.

tailgate [CIV ENG] The downstream gate of a canal lock. [ENG] A hinged gate at the rear of a vehicle or railroad freight car that can be let down for convenience in loading.

tailings [ENG] The lighter particles which pass over a sieve in milling, crushing, or purifying operations. [MIN ENG] **1.** The parts, or a part, of any incoherent or fluid material separated as refuse, or separately treated as inferior in quality or value. **2.** The decomposed outcrop of a vein or bed. **3.** The refuse material resulting from processing ground ore.

tailings settling tank [MIN ENG] A vessel in which solids are removed from the tailings effluent in mineral processing plants.

tail of a comet [ASTRON] An elongated section of a comet that may appear as an extension of the comet's head or may have a structure that alone would serve to distinguish it from the comet's head; it is composed mainly of gases.

tail of a stochastic process [MATH] A tail of a stochastic process represented by $x(t_1), x(t_2), \ldots$ is the process obtained by deleting the first n terms, for some n.

tail pulley [MECH ENG] A pulley at the tail of the belt conveyor opposite the normal discharge end; may be a drive pulley or an idler pulley.

tailrace [ENG] A channel for carrying water away from a turbine, waterwheel, or other industrial application. [MIN ENG] A channel for conveying mine trailings.

tail rope [MIN ENG] **1.** The rope which passes around the return sheave in main-and-tail haulage or a scraper loader layout. **2.** The rope that is used to draw the empty cars back into a mine in a tail-rope system. **3.** A counterbalance rope attached beneath the cage when the cages are hoisted in balance. **4.** A hemp rope used for moving pumps in shafts.

tail-rope system [MIN ENG] Haulage by a hoisting engine and two separate drums in which the main rope is attached to the front end of a trip of cars, and the tail rope is attached to the rear end of the trip.

TADPOLE

Tadpole of the common frog
(*Rana temporaria*).

TADPOLE SHRIMP

Tadpole shrimp, *Lepidurus*
(Notostraca). *(From T. I. Storer
and R. L. Usinger, General
Zoology, 4th ed., McGraw-Hill,
1965)*

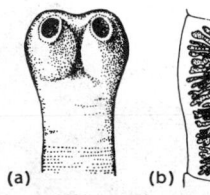

TAENIA

(a)   (b)

Morphological features of the
beef-inhabiting tapeworm *Taenia
saginata.* (a) Scolex. (b) Mature
proglottid. *(From T. I. Storer
and R. L. Usinger, General
Zoology, 3d ed., McGraw-Hill,
1957)*

TAENOGLOSSATE RADULA

Taenoglossate radula from
Lanistes, a fresh-water gastropod.
*(From R. R. Shrock and W. H.
Twenhofel, Principles of
Invertebrate Paleontology, 2d
ed., McGraw-Hill, 1953)*

tailshaft [NAV ARCH] That part of the shaft of a ship's propeller extending through the stern tube.

tail sheave [MIN ENG] The return sheave for an endless rope or the tail rope of the main-and-tail-rope system, placed at the far end of a haulageway.

tailstock [MECH ENG] A part of a lathe that holds the end of the work not being shaped, allowing it to rotate freely.

tail surface [AERO ENG] A stabilizing or control surface in the tail of an aircraft or missile.

tail track system [MIN ENG] A form of track layout for car or trip loading in which the track can be extended down the heading, turned right or left, or turned back, U-fashion, in an adjacent heading.

tail warning radar [ENG] Radar installed in the tail of an aircraft to warn the pilot that an aircraft is approaching from the rear.

tailwind [METEOROL] A wind which assists the intended progress of an exposed, moving object, for example, rendering an airborne object's ground speed greater than its airspeed; the opposite of a headwind. Also known as following wind.

taino [METEOROL] A tropical cyclone (hurricane) in parts of the Greater Antilles.

tainter gate [CIV ENG] A spillway gate whose face is a section of a cylinder; rotates about a horizontal axis on the downstream end of the gate and can be closed under its own weight. Also known as radial gate.

takedown [ADP] The actions performed at the end of an equipment operating cycle to prepare the equipment for the next setup; for example, to remove the tapes from the tape handlers at the end of a computer run is a takedown operation.

takedown time [ADP] The time required to take down a piece of equipment.

takeoff [AERO ENG] Ascent of an aircraft or rocket at any angle, as the action of a rocket vehicle departing from its launch pad or the action of an aircraft as it becomes airborne.

takeoff assist [AERO ENG] **1.** The extra thrust given to an airplane or missile during takeoff through the use of a rocket motor or other device. **2.** The device used in such a takeoff.

takeoff weight [AERO ENG] The weight of an aircraft or rocket vehicle ready for takeoff, including the weight of the vehicle, the fuel, and the payload.

take the ground [NAV] A ship takes the ground when the tide leaves it aground for want of sufficient depth of water.

takeup [MECH ENG] A tensioning device in a belt-conveyor system for taking up slack of loose parts.

takeup pulley [MECH ENG] An adjustable idler pulley to accommodate changes in the length of a conveyor belt to maintain proper belt tension.

takeup reel [ENG] The reel that accumulates magnetic tape after it is recorded or played by a tape recorder.

taku wind [METEOROL] A strong, gusty, east-northeast wind, occurring in the vicinity of Juneau, Alaska, between October and March; it sometimes attains hurricane force at the mouth of the Taku River, after which it is named.

talbot [OPTICS] A unit of luminous energy equal to the luminous energy carried by a luminous flux of 1 lumen during a period of 1 second.

Talbot's bands [OPTICS] A series of dark bands that appear in the spectrum of white light when a glass plate of the proper thickness is placed across one half of the aperture of a spectroscope on the side of the blue end of the spectrum.

Talbot's law [OPTICS] The law that apparent brightness of an object flashing at a frequency greater than about 10 hertz is equal to its actual brightness times the ratio of the exposure time to the total time.

talbutal [PHARM] $C_{11}H_{16}N_2O_3$ A crystalline compound that melts at 108–110°C, and is soluble in alcohol, chloroform, acetone, and ether; used in medicine as a short-acting hypnotic and sedative. Also known as 5-allyl-5-*sec*-butylbarbituric acid.

talc [MINERAL] $Mg_3Si_4O_{10}(OH)_2$ A whitish, greenish, or grayish hydrated magnesium silicate mineral crystallizing in the monoclinic system; it is extremely soft (hardness is 1 on Mohs scale) and has a characteristic soapy or greasy feel.

talcose rock [PETR] A rock having a soft and soapy feel, that is, resembling talc.

TALC

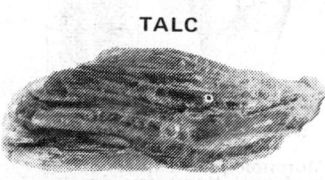

Talc from the Ural Mountains, Soviet Union. (*American Museum of Natural History*)

TALOS

Twin Talos Launcher on the USS *Chicago*. (*Official U.S. Navy photograph*)

talcosis [MED] A lung disease caused by inhalation of talc dust and characterized by chronic induration and fibrosis.

talc schist [PETR] A schist in which talc is the dominant schistose material.

talik [GEOL] A Russian term applied to permanently unfrozen ground in regions of permafrost; usually applies to a layer which lies above the permafrost but below the active layer, that is, when the permafrost table is deeper than the depth reached by winter freezing from the surface. Also known as tabetisol.

talipes [MED] Any of several foot deformities, especially of congenital origin.

talipes cavus [MED] A deformity of the foot marked by exaggeration of the longitudinal arch and by dorsal contractures of the toes.

Talitridae [INV ZOO] A family of terrestrial amphipod crustaceans in the suborder Gammaridea.

talk-back circuit *See* interphone.

talking [MIN ENG] A series of small bumps or cracking noises within the mine walls, indicating that the rock is beginning to yield to stresses.

talking battery *See* quiet battery.

talk-listen switch [ENG ACOUS] A switch provided on intercommunication units to permit using the loudspeaker as a microphone when desired.

tall oil [MATER] A yellow-black, malodorous, resinous admixture of rosin, fatty acids, sterols, high-molecular-weight alcohols, and other materials, derived from wood-pulping waste liquors; used in paint drying oils, alkyd resins, linoleum, soaps, lubricants, and greases. Also known as liquid rosin; tallol.

tallol *See* tall oil.

tallow [MATER] Animal fat with carbon chains containing 16–18 carbons, derived from cattle, sheep, and horses; used for soaps, leather dressings, candles, food, and greases, and as a chemical intermediate.

tallow-seed oil *See* stillingia oil.

tally [MIN ENG] A mark, number, or tin ticket placed by the miner on each car of coal or ore that is sent from the work place, thus facilitating a count or tally of all filled cars.

talon [VERT ZOO] A sharply hooked claw on the foot of a bird of prey.

Talos [ORD] A U.S. Navy surface-to-air guided missile having a speed of about Mach 3, a range of about 25 miles (40 kilometers), and beam-rider guidance; one version can carry nuclear warheads, and it can also be used against ships and shore bombardment targets.

Talpidae [VERT ZOO] The moles, a family of insectivoran mammals; distinguished by the forelimbs which are adapted for digging, having powerful muscles and a spadelike bony structure.

talus [ANAT] *See* astragalus. [GEOL] Also known as rubble; scree. **1.** Coarse and angular rock fragments derived from and accumulated at the base of a cliff or steep, rocky slope. **2.** The accumulated heap of such fragments.

talus creep [GEOL] The slow, downslope movement of talus.

talus glacier *See* rock glacier.

talus slope [GEOL] A steep, concave slope consisting of an accumulation of talus. Also known as debris slope.

tamarack [BOT] *Larix laricina*. A larch and a member of the pine family; it has an erect narrowly pyramidal habit, and grows in wet and moist soils in the northeastern United States, west to the Lake States and across Canada to Alaska; used for railroad ties, posts, sills, and boats. Also known as hackmarack.

Tamm-Dancoff method [QUANT MECH] A method of forming an approximate wave function of a system of interacting particles, particularly nucleons and mesons, by describing it as an algebraic sum of several possible states, the number of such states determining the order of the approximation.

tamp [ENG] To tightly pack a drilled hole with clay or other stemming material after the charge has been placed.

tamper [NUCLEO] *See* reflector. [ORD] In a weapon, any substance that resists movement for a split microsecond, used so that the active materials may build up greater pressure behind the substance.

tamping bag [ENG] A bag filled with stemming material such as sand for use in horizontal and upward sloping shotholes.

tamping bar [ENG] A piece of wood for pushing explosive cartridges or forcing the stemming into shotholes.

tamping plug [ENG] A plug of iron or wood used instead of tamping material to close up a loaded blasthole.

tampon [MED] A plug of absorbent material, such as cotton or sponge, inserted into a cavity as packing.

tan *See* tangent.

Tanaidacea [INV ZOO] An order of eumalacostracans of the crustacean superorder Peracarida; the body is linear, more or less cylindrical or dorsoventrally depressed, and the first and second thoracic segments are fused with the head, forming a carapace.

tan alt *See* shadow factor.

Tanaostigmatidae [INV ZOO] A small family of hymenopteran insects in the superfamily Chalcidoidea.

tanbark [MATER] The fibrous portion of ground oak or hemlock bark; it is burned in a mixture with other fuels to maintain combustion; also used on a running track for horses.

tandem [AERO ENG] The fore and aft configuration used in boosted missiles, long-range ballistic missiles, and satellite vehicles; stages are stacked together in series and are discarded at burnout of the propellant for each stage. [ELEC] Two-terminal pair networks are in tandem when the output terminals of one network are directly connected to the input terminals of the other network.

tandem accelerator [NUCLEO] An electrostatic accelerator in which negative hydrogen ions generated in a special ion source are accelerated as they pass from ground potential up to a high-voltage terminal, both electrons are then stripped from the negative ion by passage through a very thin foil or gas cell, and the proton is again accelerated as it passes to ground potential.

tandem central office [COMMUN] A telephone office that makes connections between local offices in an area where there is such a high density of local offices that it would be uneconomical to make direct connections between them. Also known as tandem office.

tandem compensation *See* cascade compensation.

tandem connection *See* cascade connection.

tandem-drive conveyor [MECH ENG] A conveyor having the conveyor belt in contact with two drive pulleys, both driven with the same motor.

tandem duplication [CYTOL] The occurrence of two identical sequences, one following the other, in a chromosome segment.

tandem hoisting [MIN ENG] Hoisting in a deep shaft with two skips running in one shaft; the lower skip is suspended from the tail rope of the upper skip.

tandem mill [MET] A rolling mill consisting of two or more stands in succession, synchronized so that the metal passes directly from one to another.

tandem office *See* tandem central office.

tandem propellers [NAV ARCH] Two propellers on the same shaft, rotating in the same direction.

tandem roller [MECH ENG] A steam- or gasoline-driven road roller in which the weight is divided between heavy metal rolls, of dissimilar diameter, one behind the other.

tandem switching [COMMUN] System of routing telephone calls in which calls do not travel directly between local offices, but rather through a central office.

tangeite *See* calciovolborthite.

tangelo [BOT] A tree that is hybrid between a tangerine or other mandarin and a grapefruit or shaddock; produces an edible fruit.

tangent [MATH] **1.** A line is tangent to a curve at a fixed point *P* if it is the limiting position of a line passing through *P* and a variable point on the curve *Q*, as *Q* approaches *P*. **2.** The function which is the quotient of the sine function by the cosine function. Abbreviated tan. **3.** The tangent of an angle is the ratio of its sine and cosine. Abbreviated tan.

tangent arc [METEOROL] Generic name for several types of halo arcs that form as loci tangent to other halos; the halo of 22° occasionally exhibits the horizontal and vertical tangent arcs, and the halo of 46° exhibits the infralateral tangent arcs and the supralateral tangent arcs.

tangent bending [MET] In a single piece of metal, forming a series of identical bends with parallel axes.

tangent bundle [MATH] The fiber bundle $T(M)$ associated to a differentiable manifold M which is composed of the points of M together with all their tangent vectors. Also known as tangent space.

tangent galvanometer [ENG] A galvanometer in which a small compass is mounted horizontally in the center of a large vertical coil of wire; the current through the coil is proportional to the tangent of the angle of deflection of the compass needle from its normal position parallel to the magnetic field of the earth.

tangential acceleration [MECH] The component of linear acceleration tangent to the path of a particle moving in a circular path.

tangential component [MATH] A component of a given vector acting at right angles to a given radius of a given circle.

tangential focus *See* primary focus.

tangential helical-flow turbine *See* helical-flow turbine.

tangential stress *See* shearing stress.

tangential velocity [MECH] The instantaneous linear velocity of a body moving in a circular path; its direction is tangential to the circular path at the point in question.

tangential wave *See* S wave.

tangential wave path [ELECTROMAG] In radio propagation over the earth, a path of propagation of a direct wave which is tangential to the surface of the earth; the tangential wave path is curved by atmospheric refraction.

tangent latitude error [NAV] The angle between the local meridian and the settling position or spin axis in a nonpendulous gyrocompass where damping is accomplished by offsetting the point of application of the force of a mercury ballistic.

tangent offset [ENG] In surveying, a method of plotting traverse lines; angles are laid out by linear measurement, using a constant times the natural tangent of the angle.

tangent plane to a surface [MATH] The tangent plane to a surface at a point is the plane having every line in it tangent to some curve on the surface at that point.

tangent point *See* point of tangency.

tangent screw [ENG] A screw providing tangential movement along an arc, such as the screw which provides the final angular adjustment of a marine sextant during an observation.

tangent space *See* tangent bundle.

tangent vector [MATH] A tangent vector at a point of a differentiable manifold is a vector expressing a general motion of the velocity of curves passing through the point.

tangent welding [MET] Arc welding in which two or more electrodes are in a plane parallel to the line of travel.

tangerine [BOT] Any of several trees of the species *Citrus reticulata;* the fruit is a loose-skinned mandarin with a deep-orange or scarlet rind.

tangerine oil *See* mandarin oil.

tangoreceptor [PHYSIO] A sense organ in the skin that responds to touch and pressure.

tanh *See* hyperbolic tangent.

tank [ELECTR] **1.** A unit of acoustic delay-line storage containing a set of channels, each forming a separate recirculation path. **2.** The heavy metal envelope of a large mercury-arc rectifier or other gas tube having a mercury-pool cathode. **3.** *See* tank circuit. [ENG] A large container for holding, storing, or transporting a liquid.

tankage [ENG] The contents of a storage tank. [MATER] Slaughter-house entrails and scraps used as fertilizer.

tank barge [NAV ARCH] A barge equipped with tanks that may carry any one of a great variety of liquid commodities, such as petroleum and petroleum products, chemicals, and fertilizers.

tank battery [PETRO ENG] A grouping of interconnected storage tanks situated to receive the output of one or more oil wells.

tank bottom [CHEM ENG] The liquid material in a tank below the level of the outlet pipe; often a mixture of the stored liquid with rust and other sediment.

tank car [ENG] Railroad car onto which is mounted a cylindrical, horizontal tank designed for the transport of liquids,

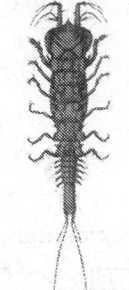

Female *Apseudes spinosus* (M. Sars), a representative species of the order Tanaidacea. *(After G. O. Sars)*

TANK BARGE

Tank barges carrying refrigerated ammonia.

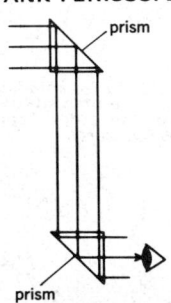

prism

prism

Diagram of a simple tank periscope with parallel reflecting surfaces.

TANTALITE

2.5 cm

Tantalite crystal in pegmatite from Raade, Norway. (*Specimen from Department of Geology, Bryn Mawr College*)

TANTALUM

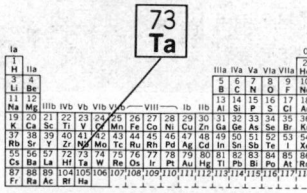

Periodic table of the chemical elements showing the position of tantalum.

chemicals, gases, meltable solids, slurries, emulsions, or fluidizable solids.

tank circuit [ELECTR] A circuit which exhibits resonance at one or more frequencies, and which is capable of storing electric energy over a band of frequencies continuously distributed about the resonant frequency, such as a coil and capacitor in parallel. Also known as electrical resonator; tank.

tank ditch *See* antitank ditch.

tank dozer [ORD] Standard tank equipped with a detachable bulldozer blade.

tanker [NAV ARCH] A steel ship in which the hull is subdivided into tanks for carrying petroleum products or other liquid cargo in bulk. Also known as tankship.

tank farm [PETRO ENG] An area in which a number of large-capacity storage tanks are located, generally used for crude oil or petroleum products.

tank gage [ENG] A device used to measure the contents of a liquid storage tank; can be manual or automatic.

tank obstacle *See* antitank obstacle.

tank periscope [OPTICS] A periscope permitting a tank occupant to observe without being exposed to bullet fire; employs a pair of plane, parallel, reflecting surfaces (either mirrors or prisms) so arranged in a mount that the path of light through the instrument forms a letter Z.

tank reactor [NUCLEO] A nuclear reactor in which the core is suspended in a closed tank, as distinct from an open-pool reactor.

tank scale [ENG] A counterweighted suspension or platform weighing mechanism for tanks, hoppers, and similar solids or liquids containers.

tankship *See* tanker.

tank switch [PETRO ENG] An automatic control of lease tanks in oil fields, including controls of lines to fill tanks and for pipeline runs; can be electrically or pneumatically actuated.

tank truck [ENG] A truck body onto which is mounted a cylindrical, horizontal tank, designed for the transport of liquids, chemicals, gases, meltable solids, slurries, emulsions, or fluidizable solids.

tannase [BIOCHEM] An enzyme that catalyzes the hydrolysis of tannic acid to gallic acid; found in cultures of *Aspergillus* and *Penicillium*.

tannic acid [ORG CHEM] 1. $C_{14}H_{10}O_9$ A yellowish powder with an astringent taste; soluble in water and alcohol, insoluble in acetone and ether; derived from nutgalls; decomposes at 210°C; used as an alcohol denaturant and a chemical intermediate, and in tanning and textiles. Also known as digallic acid; gallotannic acid; gallotannin; tannin. 2. $C_{76}H_{52}O_{46}$ Yellowish-white to light-brown amorphous powder or flakes; decomposes at 210–215°C; very soluble in alcohol and acetone; used as a mordant in dyeing, in photography, as a reagent, and in clarifying wine or beer. Also known as pentadigalloylglucose.

tannin *See* tannic acid.

tanning [ENG] A process of preserving animal hides by chemical treatment (using vegetable tannins, metallic sulfates, and sulfurized phenol compounds, or syntans) to make them immune to bacterial attack, and subsequent treatment with fats and greases to make them pliable.

tanning agent [MATER] Any one of the tannins used to treat skins and hides to preserve the hide substance and to protect it from decay.

tanning extract [MATER] Tannin-rich liquor extracted from woods and pulps; used in leather tanning.

tan rot [PL PATH] A fungus disease of strawberries caused by *Pezizella lythri* and characterized by the appearance of tan depressions on the fruit.

tantalic acid anhydride *See* tantalum oxide.

tantalic chloride *See* tantalum chloride.

tantalite [MINERAL] $(Fe,Mn)Ta_2O_6$ An iron-black mineral that crystallizes in the orthorhombic system and commonly occurs in short prismatic crystals; luster is submetallic, hardness is 6 on Mohs scale, and specific gravity is 7.95; principal ore of tantalum.

tantalum [CHEM] Metallic element in group V, symbol Ta, atomic number 73, atomic weight 180.948; black powder or steel-blue solid soluble in fused alkalies, insoluble in acids

(except hydrofluoric and fuming sulfuric); melts about 3000°C. [MET] A lustrous, platinum-gray ductile metal used in making dental and surgical tools, pen points, and electronic equipment.

tantalum capacitor [ELEC] An electrolytic capacitor that uses tantalum in the form of foil or a sintered slug as the anode, in an acid electrolyte; both weight and volume are less than for comparable aluminum electrolytic capacitors; an insulating oxide formed on the tantalum serves as the dielectric.

tantalum carbide [INORG CHEM] TaC Hard, chemical-resistant crystals melting at 3875°C; used in cutting tools and dies.

tantalum chloride [INORG CHEM] $TaCl_5$ A highly reactive, pale-yellow powder decomposing in moist air; soluble in alcohol and potassium hydroxide; melts at 221°C; used to produce tantalum and as a chemical intermediate. Also known as tantalic chloride; tantalum pentachloride.

tantalum nitride [INORG CHEM] TaN A very hard, black, water-insoluble solid, melting at 3360°C.

tantalum nitride resistor [ELECTR] A thin-film resistor consisting of tantalum nitride deposited on a substrate, such as industrial sapphire.

tantalum oxide [INORG CHEM] Ta_2O_5 Prisms insoluble in water and acids (except for hydrofluoric); melts at 1800°C; used to make tantalum, in optical glass and electronic equipment, and as a chemical intermediate. Also known as tantalic acid anhydride; tantalum pentoxide.

tantalum pentachloride *See* tantalum chloride.

tantalum pentoxide *See* tantalum oxide.

T antenna [ELECTROMAG] An antenna consisting of one or more horizontal wires, with a lead-in connection being made at the approximate center of each wire.

tanteuxenite [MINERAL] $(Y,Ce,Ca)(Ta,Nb,O)_2(O,OH)_6$ A brown or black variety of euxenite with tantalum substituting for niobium. Also known as delorenzite; eschwegeite.

tantiron [MET] An iron alloy containing silicon, carbon, manganese, phosphorus, and sulfur; used for chemical equipment where resistance to acids is needed.

Tanyderidae [INV ZOO] The primitive crane flies, a family of orthorrhaphous dipteran insects in the series Nematocera.

Tanypezidae [INV ZOO] A family of myodarian cyclorrhaphous dipteran insects in the subsection Acalypteratae.

tap [DES ENG] 1. A plug of accurate thread, form, and dimensions on which cutting edges are formed; it is screwed into a hole to cut an internal thread. 2. A threaded cone-shaped fishing tool. [ELEC] A connection made at some point other than the ends of a resistor or coil. [ENG] A small, threaded hole drilled into a pipe or process vessel; used as connection points for sampling devices, instruments, or controls. [MET] 1. A quantity of molten metal run out from a furnace at one time. 2. To remove excess slag from the floor of a pot furnace. [MIN ENG] To intersect with a borehole and withdraw or drain the contained liquid, as water from a water-bearing formation or from underground workings.

tap bolt [DES ENG] A bolt with a head that can be screwed into a hole and held in place without a nut. Also known as tap screw.

tap changer [ELEC] A device which is used to change the ratio of the input and output voltages of a transformer over any one of a definite number of steps.

tap crystal [ELECTR] Compound semiconductor that stores current when stimulated by light and then gives up energy as flashes of light when it is physically tapped.

tap density [MET] The apparent density of a volume of metal powder obtained when its receptacle is tapped or vibrated.

tap drill [MECH ENG] A drill used to make a hole of a precise size for tapping.

tape [ADP] A ribbonlike material used to store data in lengthwise sequential position. [ENG] A graduated steel ribbon used, instead of a chain, in surveying.

tape cartridge [ENG ACOUS] A cartridge that holds a length of magnetic tape in such a way that the cartridge can be slipped into a tape recorder and played without threading the tape. Also known as cartridge.

tape control unit [ADP] A device which senses which tape unit is to be accessed for read or write purpose and opens up the necessary electronic paths. Formerly known as hypertape control unit.

tape copy [COMMUN] In teletypewriter operation, message in tape form rather than page form.

tape deck [ENG ACOUS] A tape-recording mechanism that is mounted on a motor board, including the tape transport, electronics, and controls, but no power amplifier or loudspeaker.

tape drive [ADP] A tape reading or writing device consisting of a tape transport, electronics, and controls; it usually refers to magnetic tape exclusively. [ENG ACOUS] *See* tape transport.

tape editor [ADP] A routine designed to help edit, revise, and correct a routine contained on a tape.

tape-float liquid-level gage [ENG] A liquid-level measurement by a float connected by a flexible tape to a rotating member, in turn connected to an indicator mechanism.

tape gage [ENG] A box- or float-type tide gage which consists essentially of a float attached to a tape and counterpoise; the float operates in a vertical box or pipe which dampens out short-period wind waves while admitting the slower tidal movement; for the standard installation, the tape is graduated with numbers increasing toward the float and is arranged with pulleys and counterpoise to pass up and down over a fixed reading mark as the tide rises and falls.

tape grass [BOT] *Vallisnerida spiralis.* An aquatic flowering plant belonging to the family Hydrocharitaceae. Also known as eel grass.

tape label [ADP] A record appearing at the beginning or at the end of a magnetic tape to uniquely identify the tape as the one required by the system.

tape-limited [ADP] Pertaining to a computer operation in which the time required to read and write tapes exceeds the time required for computation.

tape loop [ENG ACOUS] A length of magnetic tape having the ends spliced together to form an endless loop; used in message repeater units and in some types of tape cartridges to eliminate the need for rewinding the tape.

tape mark [ADP] A special character indicating the physical end of recording on a magnetic tape.

tape player [ENG ACOUS] A machine designed only for playback of recorded magnetic tapes.

tape punch [ADP] A machine that punches code holes and feed holes in paper tape.

taper [AERO ENG] An airfoil feature in which either the thickness or the chord length or both decrease from the root to the tip. [ELEC] Continuous or gradual change in electrical properties with mechanical position such as rotation or length; for example, continuous change of cross section of a waveguide, or distribution of resistance in a potentiometer.

taper bit [DES ENG] A long, cone-shaped noncoring bit used in drilling blastholes and in wedging and reaming operations.

tape reader *See* paper-tape reader.

tape reading [ADP] A process of feeding coded tapes through a tape-to-card punch to convert the coded information into punched cards; tapes can be prepared on the typewriter tape punch or on the card-controlled tape punch; the latter is capable of punching tape that can be transmitted by telegraph.

tape recorder [ENG ACOUS] A device that records audio signals and other information on magnetic tape by selective magnetization of iron oxide particles that form a thin film on the tape; a recorder usually also includes provisions for playing back the recorded material.

tape recording [ENG ACOUS] The record made on a magnetic tape by a tape recorder.

tapered-bore [ORD] **1.** Referring to a gun with a tapered bore; the bore may be tapered throughout its length or only in the muzzle section. **2.** Pertaining to the ammunition for such a gun.

tapered core bit [DES ENG] A core bit having a conical diamond-inset crown surface tapering from a borehole size at the bit face to the next larger borehole size at its upper, shank, or reaming-shell end.

tapered joint [DES ENG] A firm, leakproof connection between two pieces of pipe having the thread formed with a slightly tapering diameter.

tapered pipeline [PETRO ENG] A changing of the pressure grade, either by change of wall thickness or material, of pipeline sections as working pressure is lessened.

tapered thread [DES ENG] A screw thread cut on the surface of a tapered part; it may be either a pine or box thread, or a V-, Acme, or square-screw thread.

tapered-tooth gear *See* beveloid gear.

tapered transmission line *See* tapered waveguide.

tapered waveguide [ELECTROMAG] A waveguide in which a physical or electrical characteristic changes continuously with distance along the axis of the waveguide. Also known as tapered transmission line.

tapered wheel [DES ENG] A flat-face grinding wheel with greater thickness at the hub than at the face.

tape-relay station [COMMUN] Component of a communications center which performs the function of receiving and forwarding messages by the tape-relay method of operation.

tape reperforator [COMMUN] A machine that automatically punches a paper tape from received signals. Also known as reperforator.

taper-in-thickness ratio [AERO ENG] A gradual change in the thickness ratio along the wing span with the chord remaining constant.

taper key [DES ENG] A rectangular machine key that is slightly tapered along its length.

taper pin [DES ENG] A small, tapered self-holding peg or nail used to connect parts together.

taper reamer [DES ENG] A reamer whose fluted portion tapers toward the front end.

taper-rolling bearing [MECH ENG] A roller bearing capable of sustaining end thrust by means of tapered rollers and coned races.

taper shank [DES ENG] A cone-shaped part on a tool that fits into a tapered sleeve on a driving member.

taper tap [DES ENG] A threaded cone-shaped tool for cutting internal screw threads.

tape search unit [ADP] Small, fully transistorized, special-purpose, digital data-processing system using a stored program to perform logical functions necessary to search a magnetic tape in off-line mode, in response to a specific request.

tape skip [ADP] A machine instruction to space forward and erase a portion of tape when a defect on the tape surface causes a write error to persist.

tape speed [ENG ACOUS] The speed at which magnetic tape moves past the recording head in a tape recorder; standard speeds are $15/16$, $1\frac{7}{8}$, $3\frac{3}{4}$, $7\frac{1}{2}$, 15, and 30 inches per second (2.38125, 4.7625, 9.525, 19.05, 38.1, and 76.2 centimeters per second); faster speeds give improved high-frequency response under given conditions.

tape station [ADP] A tape reading or writing device consisting of a tape transport, electronics, and controls; it may use either magnetic tape or paper tape.

tape-to-card [ADP] The operation, or job step, required to transfer data from magnetic or paper tape to punched card.

tape-to-card converter [ADP] A machine that converts information directly from punched or magnetic tape to cards.

tape-to-tape conversion [ADP] A routine which directs a computer to copy information from one tape to another tape of a different kind; for example, from a seven-track onto a nine-track tape.

tape transmitter [COMMUN] **1.** Code-transmitting machine actuated by previously punched paper tape; used for high-speed sending because the tape can be fed through the machine much faster than it was originally punched. **2.** Facsimile transmitter designed to transmit subject copy printed on narrow tape.

tape transport [ADP] The mechanism that physically moves a tape past a stationary head. Also known as transport. [ENG ACOUS] The mechanism of a tape recorder that holds the tape reels, drives the tape past the heads, and controls various modes of operation. Also known as tape drive.

tapetum [ANAT] **1.** A reflecting layer in the choroid coat behind the neural retina, chiefly in the eyes of nocturnal mammals. **2.** A tract of nerve fibers forming part of the roof of each lateral ventricle in the vertebrate brain. [BOT] A layer of nutritive cells surrounding the spore mother cells in

**TAPE-FLOAT
LIQUID-LEVEL GAGE**

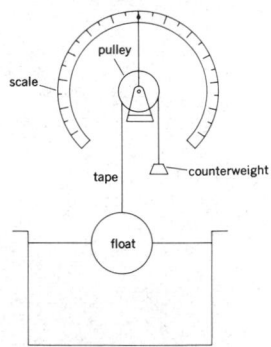

Diagram of tape-float liquid-level gage. (*D. M. Considine, ed., Process Instruments and Controls Handbook, McGraw-Hill, 1957*)

TAPE RECORDER

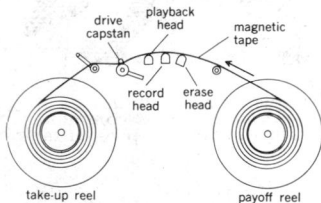

Elements of a typical magnetic tape recording and reproducing system. Capstan driven at constant angular speed ensures constant linear speed of the tape as it passes over magnetic heads. (*After H. F. Olson, Acoustical Engineering, Van Nostrand, 1957*)

TAPER KEY

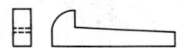

End and longitudinal views of taper key with gib head. (*From P. H. Black and O. E. Adams, Jr., Machine Design, 3d ed., 1968*)

the sporangium in higher plants; it is broken down to provide nourishment for developing spores.

tape unit [ADP] A tape reading or writing device consisting of a tape transport, electronics, controls, and possibly a cabinet; the cabinet may contain one or more magnetic tape stations.

tape verifier [ADP] A verifier for checking the accuracy of a punched paper tape by comparing it with a second manual punch of the same data; the machine stops whenever a character being punched the second time differs from that on the first tape.

tapeworm [INV ZOO] Any member of the class Cestoidea; all are vertebrate endoparasites, characterized by a ribbonlike body divided into proglottids, and the anterior end modified into a holdfast organ.

tape-wound core [ELECTROMAG] A length of ferromagnetic material in tape form, wound in such a way that each turn falls directly over the preceding turn.

taphrogenesis *See* taphrogeny.

taphrogeny [GEOL] The formation of rift or trench phenomena, characterized by block faulting and associated subsidence. Also known as taphrogenesis.

taphrogeosyncline [GEOL] A geosyncline formed as a rift basin between faults.

taping [ENG] The process of measuring distances with a surveyor's tape.

tapioca [FOOD ENG] A food, high in starch, that is made from the cassava plant.

tapioca snow *See* snow pellets.

tapiolite [MINERAL] Fe(Ta,Nb)$_2$O$_6$ A mineral that is isomorphous with mossite, occurs in pegmatites or detrital deposits; an ore of tantalum.

tapir [VERT ZOO] Any of several large odd-toed ungulates of the family Tapiridae that have a heavy, sparsely hairy body, stout legs, a prehensile muzzle, a short tail, and small eyes.

Tapiridae [VERT ZOO] The tapirs, a family of perissodactyl mammals in the superfamily Tapiroidea.

Tapiroidea [VERT ZOO] A superfamily of the mammalian order Perissodactyla in the suborder Ceratomorpha.

tapped control [ELECTR] A rheostat or potentiometer having one or more fixed taps along the resistance element, usually to provide a fixed grid bias or for automatic bass compensation.

tapped resistor [ELEC] A wire-wound fixed resistor having one or more additional terminals along its length, generally for voltage-divider applications.

tappet [MECH ENG] A lever or oscillating member moved by a cam and intended to tap or touch another part, such as a push rod or valve system.

tappet rod [MECH ENG] A rod carrying a tappet or tappets, as one for opening or closing the valves in a steam or an internal combustion engine.

tapping [MECH ENG] Forming an internal screw thread in a hole or other part by means of a tap. [MET] Opening the pouring hole of a melting furnace to remove molten metal.

tapping screw *See* self-tapping screw.

taproot [BOT] A root system in which the primary root forms a dominant central axis that penetrates vertically and rather deeply into the soil; it is generally larger in diameter than its branches.

tap screw *See* tap bolt.

tap switch [ELEC] Multicontact switch used chiefly for connecting a load to any one of a number of taps on a resistor or coil.

tar [MATER] A viscous material composed of complex, high-molecular-weight compounds derived from the distillation of petroleum or the destructive distillation of wood or coal.

tar acid [MATER] A mixture of phenols (phenols, cresols, and xylenols) found in tars and tar distillates; toxic, combustible, and soluble in alcohol and coal-tar hydrocarbons; used as a wood preservative and an insecticide for farm animals and also to make disinfectants.

tarantata [METEOROL] A strong breeze from the northwest in the Mediterranean region.

tarantula [INV ZOO] **1.** Any of various large hairy spiders of the araneid suborder Mygalomorphae. **2.** Any of the wolf spiders comprising the family Lycosidae.

Tarantula *See* Loop Nebula.

tar base [CHEM] A basic nitrogen compound found in coal tar, for example, pyridine and quinoline.

tar camphor *See* naphthalene.

Tardigrada [INV ZOO] A class of microscopic, bilaterally symmetrical invertebrates in the subphylum Malacopoda; the body consists of an anterior prostomium and five segments surrounded by a soft, nonchitinous cuticle, with four pairs of ventrolateral legs.

tar distillate [MATER] A petroleum product produced by a tar still to which is charged the tarlike bottoms from continuous crude stills, pressure stills, cracking coils, or other petroleum refinery equipment.

tare [MECH] The weight of an empty vehicle or container; subtracted from gross weight to ascertain net weight.

tare effect [FL MECH] In wind tunnel testing, the forces and moments due to support assembly and mutual interference between support assembly and model.

target [ADP] An index card or test document used to assist, reference, or calibrate equipment. [ATOM PHYS] The atom or nucleus in an atomic or nuclear reaction which is initially stationary. [ELECTR] **1.** In an x-ray tube, the anode or anticathode which emits x-rays when bombarded with electrons. **2.** In a television camera tube, the storage surface that is scanned by an electron beam to generate an output signal current corresponding to the charge-density pattern stored there. **3.** In a cathode-ray tuning indicator tube, one of the electrodes that is coated with a material that fluoresces under electron bombardment. [ENG] **1.** The sliding weight on a leveling rod used in surveying to enable the staffman to read the line of collimation. **2.** The point that a borehole or an exploratory work is intended to reach. **3.** In radar and sonar, any object capable of reflecting the transmitted beam. [ORD] **1.** A geographical area, complex, or installation planned for capture or destruction by military forces. **2.** A paper or pasteboard item of square or rectangular shape designed to be fired upon from a specified range during practice or while testing an automatic firearm such as an automatic rifle, machine gun, or submachine gun; it is used to establish a degree of accuracy; it usually consists of a series of geometric patterns of various shapes on a common background. [PHYS] An object or substance subjected to bombardment or irradiation by particles or electromagnetic radiation.

target acquisition [AERO ENG] The process of optically, manually, mechanically, or electronically orienting a tracking system in direction and range to lock on a target. [ELECTR] **1.** The first appearance of a recognizable and useful echo signal from a new target in radar and sonar. **2.** *See* acquire.

target acquisition radar [ENG] An antiaircraft artillery radar, normally of lesser range capabilities but of greater inherent accuracy than that of surveillance radar, whose normal function is to acquire aerial targets either by independent search or on direction of the surveillance radar, and to transfer these targets to tracking radars.

target analysis [ORD] Examination of potential targets to determine their military importance, their relative priority for attack, and the capabilities of available weapons for such attack.

target angle [NAV] The relative bearing of one craft from another craft, measured clockwise through 360°. [ORD] The angle at the target subtended by the observing base line.

target approach point [AERO ENG] In air transport operations, a navigational checkpoint over which the final turn-in to the drop zone–landing zone is made.

target array [ORD] A graphic representation of enemy forces, personnel, and facilities in a specific situation, accompanied by a target analysis.

target bearing [ORD] **1.** The true compass bearing of a target from a firing ship. **2.** The bearing of a target measured in the horizontal from the bow of one's own ship clockwise from 0 to 360°, or from the nose of one's own aircraft in hours of the clock.

target concentration [ORD] A grouping of geographically proximate targets.

target cross section *See* echo area.

target-designating system [ELECTR] A system for designat-

TAPIR

Brazilian tapir (*Tapirus terrestris*), the most common South American species of tapir.

TAPROOT

Taproot system of a dandelion.

ing to one instrument a target which has already been located by a second instrument; it employs electrical data transmitters and receivers which indicate on one instrument the pointing of another.

target deviation [ORD] Distance from point of impact or point of burst to the target.

target discrimination [ELECTR] The ability of a detection or guidance system to distinguish a target from its background or to discriminate between two or more targets that are close together.

target drone [AERO ENG] A pilotless aircraft controlled by radio from the ground or from a mother ship and used exclusively as a target for antiaircraft weapons.

target echo [ELECTROMAG] A radio signal reflected by an airborne or other target and received by the radar station which transmitted the original signal.

target glint *See* scintillation.

target identification [ORD] The act of determining the nature of a target, including whether it is a friend or foe.

target indicating system [ORD] A system which indicates to the tracker of an antiaircraft automatic weapon the direction of approach of a suitable target, or the approach of a new target after engagement with one target has been broken off.

target information center [ORD] An intelligence center set up afloat or ashore for assembly, evaluation, interpretation, dissemination, and coordination of target information for supporting weapons, that is, artillery, naval gunfire, and air strike.

target language [ADP] The language into which a program (or text) is to be converted.

target length [ORD] Length of a target as it appears to an observer or gunner at the moment the gun is fired.

target noise [ELECTROMAG] Statistical variations in a radar echo signal due to the presence on the target of a number of reflecting elements randomly oriented in space; target noise can cause scintillation.

target offset [ORD] Horizontal angle at the target between a line from the target to the piece and a line from the target to the observation post.

target of opportunity [ORD] **1.** A target visible to a surface or air sensor or observer, which is within range of available weapons and against which fire has not been scheduled or requested. **2.** A nuclear target observed or detected after an operation begins that has not been previously considered, analyzed, or planned for a nuclear strike.

target pattern [AERO ENG] The flight path of aircraft during the attack phase.

target program *See* object program.

target range [ORD] Area equipped for practice in shooting at targets.

target response [ORD] The effect on men, material, and equipment of blast, heat, light, and nuclear radiation resulting from the explosion of a nuclear weapon.

target routine *See* object program.

target scintillation *See* scintillation.

target seeker [ORD] **1.** A missile having a self-contained system that provides homing guidance to the target. Also known as homer. **2.** The device within such a missile that directs it to the target.

target selector [ORD] Component of both a target-designating system and a target-indicating system; it is an off-carriage observing instrument provided for the purpose of selecting an initial or new target, and it is electrically connected to the gun mount in such a manner as to slew the gun to the approximate azimuth and elevation of a selected target (when the selector is a component of a target-designating system), and to give the tracker an indication of the direction of approach of selected target (when the selector is a component of a target-indicating system).

target signal [ELECTROMAG] The radio energy returned to a radar by a target. Also known as echo signal; video signal.

target signature [ELECTR] Characteristic pattern of the target displayed by detection and classification equipment.

target spot [PL PATH] Any plant disease characterized by lesions in the form of concentric markings.

target spotter [ORD] Small, black metal disk attached to a

target in practice shooting to show the shooter exactly where the bullet has hit.

target strength [ACOUS] A measure of the reflecting power of a sonar target, which is expressed in decibels by the equation $E + 2L - S$, where E is the echo level, L is the total transmission loss, and S is the source level.

target system [ORD] **1.** All the targets situated in a particular geographic area and functionally related. **2.** A group of targets which are so related that their destruction will produce some particular effect desired by the attacker.

target timing [NAV] The timing of successive positions of a radar target, as plotted on a polar coordinate diagram, for the purpose of determining ground speed and track of a craft.

target-type flowmeter [ENG] A fluid-flow measurement device with a small circular target suspended centrally in the flow conduit; the target transmits force to a force-balance transmitter by means of a pivoted bar.

target volume [ELECTROMAG] The volume of that part of a precipitation-type radar target from which a target signal is received; if the precipitation completely fills the radar beam, the target volume is identical with the radar volume.

target vulnerability [ORD] A factor considered in target selection that relates each potential target to a standard scale in terms of the degree to which it is considered vulnerable; each target is given a scale number.

tariff [COMMUN] The rate charged by a communications common carrier for the use of a specified service or facility. [IND ENG] A government-imposed duty on imported or exported goods.

tarn [GEOGR] A landlocked pool or small lake that may occur in a marsh or swamp, or that may occupy a basin amid mountain ranges.

tarnish [MET] Discoloration of a metal surface due to the formation of a thin film of oxide, sulfide, or some other corrosion product. [MINERAL] The altered color and luster of a mineral surface; characteristic of copper-bearing minerals.

tar paper [MATER] Heavy construction paper coated or impregnated with tar.

tarpaulin [MATER] A sheet of waterproof canvas or other material; used to cover and protect construction materials and equipment, athletic fields, vehicles, or other exposed objects.

tarpon [VERT ZOO] *Megalops atlantica*. A herringlike fish of the family Elopidae weighing up to 300 pounds and reaching a length of 8 feet; it has a single soft, rayed dorsal fin, strong jaws, a bony plate under the mouth, numerous small teeth, and coarse, bony flesh covered with large scales.

tarragon [FOOD ENG] A herb prepared from the pungent leaves of the tarragon tree (*Artemisia dracunculus*).

tarragon oil *See* estragon oil.

tarring [ENG] The coating of piles for permanent underground work with prepared acid-free tar.

tarsal gland [ANAT] Any of the sebaceous glands in the tarsal plates of the eyelids. Also known as Meibomian gland.

tar sand [GEOL] A type of oil sand; a sand whose interstices are filled with asphalt that remained after the escape of the lighter fractions of crude oil.

tar seep [GEOL] Natural tar that, because of its close proximity to the ground surface, seeps from cracks in the earth or from between rocks, often forming pits or pools.

tarsia *See* intarsia.

tarsier [VERT ZOO] Any of several species of primates comprising the genus *Tarsius* of the family Tarsiidae characterized by a round skull, a flattened face, and large eyes that are separated from the temporal fossae in the orbital depression, and by adhesive pads on the expanded ends of the fingers and toes.

Tarsiidae [VERT ZOO] The tarsiers, a family of prosimian primates distinguished by incomplete postorbital closure and a greatly elongated ankle region.

Tarsonemidae [INV ZOO] A small family of phytophagous mites in the suborder Trombidiformes.

tarsus [ANAT] **1.** The instep of the foot consisting of the calcaneus, talus, cuboid, navicular, medial, intermediate, and lateral cuneiform bones. **2.** The dense connective tissues supporting an eyelid.

TARPON

The tarpon *(Megalops atlantica)*, largest of the herringlike fishes, reaching a length of 8 feet (2 meters).

TARSIER

The tarsier has adhesive pads on the expanded ends of the fingers and toes.

TARTAR

Tartar missile fired to port side of the Baltimore class heavy cruiser USS *Columbus*. (Official U.S. Navy photograph)

TAURINE

$$SO_3H$$
$$|$$
$$CH_2$$
$$|$$
$$CH_2NH_2$$

Structural formula of taurine.

TAUROCHOLIC ACID

Structural formula of taurocholic acid.

TAURUS

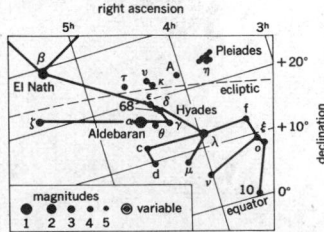

Line pattern of the constellation Taurus. The grid lines represent the coordinates of the sky. The apparent brightness, or magnitude, of the stars is shown by the sizes of the dots, which are graded by appropriate numbers as indicated.

tartar *See* dental calculi.

Tartar [ORD] A U.S. Navy surface-to-air guided missile intended primarily for use on destroyers; it is a smaller version of Terrier and has about the same range.

tartaric acid [ORG CHEM] $HOOC(CHOH)_2COOH$ Water- and alcohol-soluble colorless crystals with an acid taste, melts at 170°C; used as a chemical intermediate and a sequestrant and in tanning, effervescent beverages, baking power, ceramics, photography, textile processing, mirror silvering, and metal coloring. Also known as dihydroxysuccinic acid.

tartary buckwheat [BOT] One of three buckwheat species grown commercially; the leaves are narrower than the other two species and arrow-shaped, and the flowers are smaller with inconspicuous greenish-white sepals. Also known as duck wheat; hulless buckwheat; rye buckwheat.

tartrate [ORG CHEM] $M_2C_4H_4O_6$ A salt or ester of tartaric acid, for example, sodium tartrate, $Na_2C_4H_4O_6$.

task [ADP] A set of instructions, data, and control information capable of being executed by the central processing unit of a digital computer in order to accomplish some purpose; in a multiprogramming environment, tasks compete with one another for control of the central processing unit, but in a nonmultiprogramming environment a task is simply the current work to be done.

task fleet [ORD] A mobile command consisting of ships and aircraft necessary for the accomplishment of a specific major task which may be of a continuing nature.

Taslan [TEXT] Trade name for a bulky textured yarn manufactured by Du Pont.

tasmanite [GEOL] An impure coal, transitional between cannel coal and oil shale. Also known as combustible shale; Mersey yellow coal; white coal; yellow coal.

tassel [BOT] The male inflorescence of corn and certain other plants.

taste [PHYSIO] A chemical sense by which flavors are perceived depending on taste, tactile, and warm and cold receptors in the mouth, as well as smell receptors in the nose.

taste bud [ANAT] An end organ consisting of goblet-shaped clusters of elongate cells with microvilli on the distal end to mediate the sense of taste.

TAT *See* thematic apperception test.

T attenuator [ELEC] **1.** A resistive attenuator with three resistors forming a T network. **2.** A power-tap type of attenuator which removes part of the power from a main line through a T connection and dissipates the power, without reflection into the main line.

Tauber test [ANALY CHEM] A color test for identification of pentose sugars; the sugars produce a cherry-red color when heated with a solution of benzidine in glacial acetic acid.

tau meson [PARTIC PHYS] Former name for the K meson, especially one which decays into three pions.

Taurid meteor [ASTRON] A meteor shower occurring from about October 26 to November 16 in the Northern Hemisphere, with the maximum occurring about the first week in November; the radiant lies in the constellation Taurus.

taurine [ORG CHEM] $NH_2CH_2CH_2SO_3H$ A crystalline compound that decomposes at about 300°C; present in bile combined with cholic acid. Also known as 2-aminoethanesulfonic acid.

taurocholic acid [BIOCHEM] $C_{26}H_{45}NO_7S$ A common bile acid with a five-carbon chain; it is the product of the conjugation of taurine with cholic acid; crystallizes from an alcohol ether solution, and decomposes at about 125°C. Also known as cholaic acid; cholytaurine.

taurodont [ANAT] Of teeth, having a large pulp cavity and reduced roots.

Taurus [ASTRON] A northern constellation; right ascension 4 hours, declination 15° north; it includes the star Aldebaran, useful in navigation. Also known as Bull.

Taurus cluster [ASTRON] A cluster of stars observed in the region of the constellation Taurus; it is about 130 light-years from the sun, and 58 light years in diameter.

taut-band ammeter [ENG] A modification of the permanent-magnet movable-coil ammeter in which the jeweled bearings and control springs are replaced by a taut metallic band rigidly held at the ends; the coil is firmly attached to the band, and restoring torque is supplied by twisting of the band.

taut-line cableway [MECH ENG] A cableway whose operation is limited to the distance between two towers, usually 3000 feet (914 meters) apart, has only one carrier, and the traction cable is reeved at the carrier so that loads can be raised and lowered; the towers can be mounted on trucks or crawlers, and the machine shifted across a wide area.

tautomerism [CHEM] The reversible interconversion of structural isomers of organic chemical compounds; such interconversions usually involve transfer of a proton.

tau-value [METEOROL] The time rate of change of D-value at a fixed point defined by the relation $\tau = (\Delta_t D)/(\Delta t)$, where Δt is the change in time and $\Delta_t D$ is the change in D-value during this time interval; tau-values are expressed in terms of feet per hour; tau-value lines are drawn on 4-D charts and constitute the time dimension of these charts.

tawing [ENG] A tanning process in which alum is used as a partial tannage, supplementing or replacing chrome.

Taxales [BOT] A small order of gymnosperms in the class Pinatae; members are trees or shrubs with evergreen, often needlelike leaves, with a well-developed fleshy covering surrounding the individual seeds, which are terminal or subterminal on short shoots.

taxi channel [CIV ENG] A defined path, on a water airport, intended for the use of taxiing aircraft.

taxi-channel lights [NAV] Aeronautical ground lights arranged along a taxi channel to indicate the route to be followed by taxiing aircraft.

taxis [PHYSIO] A mechanism of orientation by means of which an animal moves in a direction related to a source of stimulation.

taxiway [CIV ENG] A specially prepared or designated path on an airport for taxiing aircraft.

Taxocrinida [PALEON] An order of flexible crinoids distributed from Ordovician to Mississippian.

Taxodonta [INV ZOO] A subclass of pelecypod mollusks in which the hinge is of the taxodontal type, that is, the dentition is a series of similar alternating teeth and sockets along the hinge margin.

taxon [SYST] A taxonomic group or entity.

taxonomic category [SYST] One of a hierarchy of levels in the biological classification of organisms; the seven major categories are kingdom, phylum, class, order, family, genus, species.

taxonomy [SYST] A study aimed at producing a hierarchical system of classification of organisms which best reflects the totality of similarities and differences.

Tayassuidae [VERT ZOO] The peccaries, a family of artiodactyl mammals in the superfamily Suoidae.

Taylor connection [ELEC] A transformer connection for converting three-phase power to two-phase power, or vice versa.

Taylor effect [FL MECH] A phenomenon in which the relative motion of a homogeneous rotating liquid tends to be the same in all planes perpendicular to the axis of rotation.

taylorite *See* bentonite.

Taylor number [FL MECH] A nondimensional number arising in problems of a rotating viscous fluid, written as $T = (f^2 h^4)/v^2$, where f is the Coriolis parameter (or, for a cylindrical system, twice the rate of rotation of the system), h is the depth of the fluid, and v is the kinematic viscosity; the square root of the Taylor number is a rotating Reynolds number, and the fourth root is proportional to the ratio of the depth h to the depth of the Ekman layer.

Taylor-Orowan dislocation *See* edge dislocation.

Taylor process [MET] A process for making extremely fine wire by stretching wire in a glass tube at elevated temperatures, or drawing wire through a bead of molten glass and then through dies.

Taylor series [MATH] The Taylor series corresponding to a function $f(x)$ at a point x_0 is the infinite series whose nth term is $(1/n!) \cdot f^{(n)}(x_0)(x-x_0)^n$, where $f^{(n)}(x)$ denotes the nth derivative of $f(x)$. [NAV ARCH] Resistance charts based upon model tests of a series of ships derived by altering the proportions of a single parent form; used to study the effects of these alterations on resistance to the ship's motion, and to predict the powering requirements for new ships.

Taylor's theorem [MATH] The theorem that a real or complex function can be realized, in a neighborhood of a point where it is differentiable, as a power series whose coefficients involve the various order derivatives evaluated at that point.

Tay-Sachs disease [MED] A form of sphingolipidosis, transmitted as an autosomal recessive, in which there is an accumulation in neuronal cells of the neuraminic fraction of gangliosides; manifested clinically within the first few months of life by hypotonia progressing to spasticity, convulsions, and visual loss accompanied by the appearance of a cherry-red spot at the macula lutea. Also known as infantile amaurotic familial idiocy.

Tb *See* terbium.

TBE *See* binding energy.

T beam [CIV ENG] A metal beam or bar with a T-shaped cross section.

T bolt [DES ENG] A bolt with a T-shaped head, made to fit into a T-shaped slot in a drill swivel head or in the bed of a machine.

TBP *See* tributyl phosphate.

tbsp *See* tablespoonful.

TBT *See* tetrabutyl titanate.

Tc *See* technetium.

TCA *See* trichloroacetic acid.

T cell [IMMUNOL] One of a heterogeneous population of thymus-derived lymphocytes which participates in the immune responses.

Tchebycheff *See* Chebyshev.

Tchornozem *See* Chernozem.

Tchuprow-Neymann allocation [STAT] A technique of stratified sampling in which the size of each strata sample is proportional to the size of the strata population and the variance of the strata.

T circulator [ELECTROMAG] A circulator in which three identical rectangular waveguides are joined asymmetrically to form a T-shaped structure, with a ferrite post or wedge at its center; power entering any waveguide emerges from only one adjacent waveguide.

T connector [ELEC] A type of electric connector that joins a through conductor to another conductor at right angles to it.

TCP *See* tricresyl phosphate.

TD *See* transmitter-distributor.

TDZ *See* touch-down zone.

Te *See* tellurium.

tea [BOT] *Thea sinensis.* A small tree of the family Theaceae having lanceolate leaves and fragrant white flowers; a caffeine beverage is made from the leaves of the plant.

teaching reactor *See* research reactor; subcritical reactor.

teakwood [MATER] The strong, durable, yellowish-brown wood obtained from the teak tree, *Tectona grandis.*

TEA laser [OPTICS] A gas laser in which a glow discharge is maintained without arc formation at atmospheric pressure (which is relatively high for a gas laser) by using a discharge which is transverse rather than parallel to the optic axis. Derived from transversely excited atmospheric pressure laser.

tear down [ENG] 1. To disassemble a drilling rig preparatory to moving it to another drill site. 2. To disassemble a machine or change the jigs and fixtures.

tear-down time [IND ENG] The downtime of a machine following a given work order which usually involves removing parts such as jigs and fixtures and which must be completely finished before setting up for the next order.

teardrop balloon [AERO ENG] A sounding balloon which, when operationally inflated, resembles an inverted teardrop; this shape was determined primarily by aerodynamic considerations of the problem of obtaining maximum stable rates of a balloon ascension.

tear fault [GEOL] A very steep to vertical fault associated with and perpendicular to the strike of an overthrust fault.

tear gas [MATER] A substance (usually liquid) which, when atomized and of a certain concentration, causes temporary but intense eye irritation and a blinding flow of tears; chloroacetophenone is a common tear gas. Also known as lacrimator.

tear gland *See* lacrimal gland.

tear strength [MECH] The force needed to initiate or to continue tearing a sheet or fabric.

teaser transformer [ELEC] Transformer, of two T-connected, single-phase units for three-phase to two-phase or two-phase to three-phase operation, which is connected between the midpoint of the main transformer and the third wire of the three-phase system.

teaspoonful [MECH] A unit of volume used particularly in cookery and pharmacy, equal to $1\frac{1}{3}$ fluid drams, or $\frac{1}{3}$ tablespoonful; in the United States this is equal to approximately 4.9289 cubic centimeters, in the United Kingdom to approximately 4.7355 cubic centimeters. Abbreviated tsp; tspn.

technetium [CHEM] A member of group VII, symbol Tc, atomic number 43; derived from uranium and plutonium fission products; chemically similar to rhenium and manganese; isotope Tc^{99} has a half-life of 2×10^5 years; used to absorb slow neutrons in reactor technology. [MET] Silver-gray metal with a high melting point, slightly magnetic.

technetron [ELECTR] High-power multichannel field-effect transistor.

technical assistance [ENG] The providing of advice, assistance, and training pertaining to the installation, operation, and maintenance of equipment.

technical atmosphere [MECH] A unit of pressure in the metric technical system equal to one kilogram-force per square centimeter. Abbreviated at.

technical characteristics [ENG] Those characteristics of equipment which pertain primarily to the engineering principles involved in producing equipment possessing desired characteristics, for example, for electronic equipment; technical characteristics include such items as circuitry, and types and arrangement of components.

technical chlorinated camphene *See* toxaphene.

technical cohesive strength [MET] Fracture stress in a notched tensile test.

technical control board [ELEC] Testing position in a switch center or relay station with provisions for testing switches and associated access lines and trunks.

technical evaluation [ENG] The study and investigation to determine the technical suitability of material, equipment, or a system.

technical information [ENG] Information, including scientific information, which relates to research, development, engineering, testing, evaluation, production, operation, use, and maintenance of equipment.

technical inspection [ENG] Inspection of equipment to determine whether it is serviceable for continued use or needs repairs.

technical intelligence [ORD] Intelligence pertaining to foreign or enemy technological developments capable of, or having, a practical application to warfare.

technical load [ELEC] Portion of a communications-electronics facility operational power load required for primary and ancillary equipment, including necessary lighting and air conditioning or ventilation required for full continuity of operation.

technical maintenance [ENG] A category of maintenance that includes the replacement of unserviceable major parts, assemblies, or subassemblies, and the precision adjustment, testing, and alignment of internal components.

technical manual [ORD] A U.S. Army publication containing detailed information on technical procedures, including instructions on the operation, handling, maintenance, and repair of equipment.

technical representative [IND ENG] A person who represents one or more manufacturers in an area and who gives technical advice on the application, installation, operation, and maintenance of their products, in addition to selling the products.

technical specifications [ENG] A detailed description of technical requirements stated in terms suitable to form the basis for the actual design-development and production processes of an item having the qualities specified in the operational characteristics.

technical white oil *See* white oil.

technology [SCI TECH] Systematic knowledge of and its application to industrial processes; closely related to engineering and science.

Tectibranchia [INV ZOO] An order of mollusks in the subclass Opisthobranchia containing the sea hares and the bubble shells; the shell may be present, rudimentary, or absent.

tectite *See* tektite.

T CONNECTOR

Drawing of a T connector.

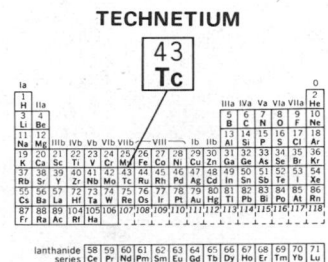

TECHNETIUM

Periodic table of the chemical elements showing the position of technetium.

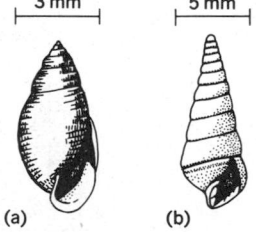

TECTIBRANCHIA

3 mm   5 mm

(a)   (b)

Shells of two genera of Tectibranchia. *(a) Acteon. (b) Pyramidella. (From A. M. Keen and J. C. Pearson, Illustrated Key to West North American Gastropod Genera, Stanford University Press, 1958)*

tectofacies [GEOL] A lithofacies that is interpreted tectonically.

tectogene [GEOL] A long, relatively narrow downward fold of sialic crust considered to be an early phase in mountain-building processes. Also known as geotectogene.

tectogenesis See orogeny.

tectonic breccia [PETR] A breccia developed from brittle rocks, formed as a result of crustal movements and produced by lateral or vertical pressure. Also known as dynamic breccia; pressure breccia.

tectonic conglomerate See crush conglomerate.

tectonic cycle [GEOL] The orogenic cycle which relates larger crustal features, such as mountain belts, to a series of stages of development. Also known as geosynclinal cycle.

tectonic framework [GEOL] The relationship in space and time of subsiding, stable, and rising tectonic elements in a sedimentary source area.

tectonic land [GEOL] Linear fold ridges and volcanic islands which existed for a short time in the interior sections of an orogenic belt during the geosynclinal phase.

tectonic lens [GEOL] An elongate, sausage-shaped body of rock formed by distortion of a continuous incompetent layer enclosed between competent layers, similar to a boudin, but genetically distinct.

tectonic map [GEOL] A map which shows the architecture of the upper portion of the earth's crust.

tectonic moraine [GEOL] An aggregation of boulders incorporated in the base of an overthrust mass.

tectonic patterns [GEOL] The arrangement of the large structural units of the earth's crust, such as mountain systems, shields or stable areas, basins, arches, and volcanic archipelagoes.

tectonic rotation [GEOL] Internal rotation of a tectonite in the direction of transport.

tectonics [CIV ENG] **1.** The science and art of construction with regard to use and design. **2.** Design relating to crustal deformations of the earth. [GEOL] A branch of geology that deals with regional structural and deformational features of the earth's crust, including the mutual relations, origin, and historical evolution of the features. Also known as geotectonics.

tectonite [PETR] A rock in which the history of its deformation is reflected in its fabric.

tectonomagnetism [GEOPHYS] Study of magnetic anomalies due to tectonic stress.

tectonometer [ENG] An apparatus, including a microammeter, used on the surface to obtain knowledge of the structure of the underlying rocks.

tectonophysicist [GEOPHYS] One who studies elastic deformation of flow and rupture of constituent materials of the earth's crust and makes deductions concerning the forces that cause these deformations.

tectonophysics [GEOPHYS] A branch of geophysics dealing with the physical processes involved in forming geological structures.

tectorial membrane [ANAT] **1.** A jellylike membrane covering the organ of Corti in the ear. **2.** A strong sheet of connective tissue running from the basilar part of the occipital bone to the dorsal surface of the bodies of the axis and third cervical vertebra.

tectosilicate [MINERAL] A structural type of silicate in which all four oxygen atoms of the silicate tetrahedra are shared with neighboring tetrahedra; tectosilicates include quartz, the feldspars, the feldspathoids, and zeolites. Also known as framework silicate.

tectosome [GEOL] A body of strata representing a tectotope.

tectum [ANAT] A rooflike structure of the body, especially the roof of the midbrain including the corpora quadrigemina.

tee [ENG] Shaped like the letter T.

tee joint [ENG] A joint in which members meet at right angles, forming a T.

teel oil See sesame oil.

teeming [MET] Pouring molten metal, usually a ferrous metal, into an ingot mold from a furnace or ladle.

teepleite [MINERAL] $Na_2BO_2Cl \cdot 2H_2O$ A mineral composed of hydrous chloride and borate of sodium.

teeth of the gale [METEOROL] An old nautical term for the direction from which the wind is blowing (upwind, windward); to sail into the teeth of the gale is to sail to windward.

teflon shakes See metal fume fever.

TEG See tetraethylene glycol; triethylene glycol.

tegmen [BIOL] An integument or covering. [BOT] The inner layer of a seed coat. [INV ZOO] A thickened forewing of Orthoptera, Coleoptera, and certain other insects.

tegmentum [ANAT] A mass of white fibers with gray matter in the cerebral peduncles of higher vertebrates. [BOT] The outer layer, or scales, of a leaf bud. [INV ZOO] The upper layer of a shell plate in Amphineura.

Tego [MATER] A thin film phenol-formaldehyde adhesive, used with thin veneers.

tehuantepecer [METEOROL] A violent squally wind from north or north-northeast in the Gulf of Tehuantepec in winter; it originates in the Gulf of Mexico, as a norther which crosses the isthmus and blows through the gap between the Mexican and Guatemalan mountains.

teichoic acid [BIOCHEM] A polymer of ribitol or glycerol phosphate with additional compounds such as glucose linked to the backbone of the polymer; found in the cell walls of some bacteria.

Teiidae [VERT ZOO] The tegus lizards, a diverse family of the suborder Sauria that is especially abundant and widespread in South America.

T-E index See temperature-efficiency index.

tektite [GEOL] A collective term applied to certain objects of natural glass of debatable origin that are widely strewn over the land and in sediments under the oceans; composition and size vary, and overall shapes resemble splash forms; most tektites are believed to be of extraterrestrial origin. Also known as obsidianite; tectite.

TEL See tetraethyllead.

telain See provitrain.

telamon [INV ZOO] A curved chitinous outgrowth of the cloacal wall in various male nematodes.

telangiectasis [MED] Localized dilation of capillaries forming dark-red, wartlike elevations varying in size from about 1 to 7 millimeters.

telautograph [COMMUN] A writing telegraph instrument in which manual movement of a pen at the transmitting position varies the current in two circuits in such a way as to cause corresponding movements of a pen at the remote receiving instrument; ordinary handwriting can thus be transmitted over wires.

TELCOMP [ADP] A computer language developed by Bolt, Beranek, and Newman, Inc., expressly as a time-sharing language; the user of the language is connected to a computer by a teletype terminal attached to a telephone line.

tele- [SCI TECH] Prefix meaning from a distance.

telecast [COMMUN] A television broadcast intended for reception by the general public, involving the transmission of the picture and sound portions of the program.

teleceptor [PHYSIO] A sense receptor which transmits information about portions of the external environment which are not necessarily in direct contact with the organism, such as the receptors of the ear, eye, and nose.

telecommunications [COMMUN] Communication over long distances.

telecommunications coordinating committee [COMMUN] Committee organized by the U.S. State Department and composed of major government departments and agencies; makes recommendations on telecommunications matters affecting international telecommunications.

teleconference [COMMUN] A conference between persons remote from one another but linked by a telecommunications system.

Telegeusidae [INV ZOO] The long-lipped beetles, a small family of colepteran insects in the superfamily Cantharoidea confined to the western United States.

telegraph alphabet See telegraph code.

telegraph bandwidth [COMMUN] The difference between the limiting frequencies of a channel used to transmit telegraph signals.

telegraph buoy [ENG] A buoy used to mark the position of a submarine telegraph cable.

telegraph cable [ELEC] A uniform conductive circuit consisting of twisted pairs of insulated wires or coaxially shielded wires or combinations of each, used to carry telegraph signals.

telegraph carrier [COMMUN] The single-frequency wave which is modulated by transmitting apparatus in carrier telegraphy.

telegraph circuit [COMMUN] The complete wire or radio circuit over which signal currents flow between transmitting and receiving apparatus in a telegraph system.

telegraph code [COMMUN] A system of symbols for transmitting telegraph messages in which each letter or other character is represented by a set of long and short electrical pulses, or by pulses of opposite polarity, or by time intervals of equal length in which a signal is present or absent. Also known as telegraph alphabet.

telegraph concentrator [ELEC] Switching arrangement by means of which a number of branch or subscriber lines or station sets may be connected to a lesser number of trunklines, operating positions, or instruments through the medium of manual or automatic switching devices to obtain more efficient use of facilities.

telegraph distributor [ELEC] Device which effectively associates one direct-current or carrier-telegraph channel in rapid succession with the elements of one or more sending or receiving devices.

telegraph emission [COMMUN] The signal transmitted by a telegraph system, classified by type of transmission, type of modulation, bandwidth, and supplementary characteristics.

telegrapher's equation [MATH] The partial differential equation $(\partial^2 f/\partial x^2) = a^2(\partial^2 f/\partial y^2) + b(\partial f/\partial y) + cf$, where a, b, and c are constants; appears in the study of atomic phenomena.

telegraph interference [COMMUN] Any undesired electrical energy that tends to interfere with the reception of telegraph signals.

telegraph receiver [ELEC] A tape reperforator, teletypewriter, or other equipment which converts telegraph signals into a pattern of holes on a tape, printed letters, or other forms of information.

telegraph repeater [ELEC] A repeater inserted at intervals in long telegraph lines to amplify weak code signals, with or without reshaping of pulses, and to retransmit them automatically over the next section of the line.

telegraph signal distortion [COMMUN] Time displacement of transitions between conditions, such as marking and spacing, with respect to their proper relative positions in perfectly timed signals; the total distortion is the algebraic sum of the bias and the characteristic and fortuitous distortions.

telegraph transmitter [ELEC] A device that controls an electric power source in order to form telegraph signals.

telegraphy [COMMUN] Communication at a distance by means of code signals consisting of current pulses sent over wires or by radio.

telemagmatic [GEOL] Pertaining to a hydrothermal mineral deposit that is distant from its magmatic source.

telemeteorograph [ENG] Any meteorological instrument, such as a radiosonde, in which the recording instrument is located at some distance from the measuring apparatus; for example, a meteorological telemeter.

telemeteorography [ENG] The science of the design, construction, and operation of various types of telemeteorographs.

telemeteorometry [METEOROL] The study of making meteorological observations at a distance.

telemeter [ENG] **1.** The complete measuring, transmitting, and receiving apparatus for indicating or recording the value of a quantity at a distance. Also known as telemetering system. **2.** To transmit the value of a measured quantity to a remote point.

telemetering [ENG] Transmitting the readings of instruments to a remote location by means of wires, radio waves, or other means. Also known as remote metering; telemetry.

telemetering antenna [ELECTROMAG] A highly directional antenna, generally mounted on a servo-controlled mount for tracking purposes, used at ground stations to receive telemetering signals from a guided missile or spacecraft.

telemetering receiver [ELECTR] A device in a telemetering system which converts electrical signals into an indication or

recording of the value of the quantity being measured at a distance.

telemetering system *See* telemeter.

telemetering transmitter [ELECTR] A device which converts the readings of instruments into electrical signals for transmission to a remote location by means of wires, radio waves, or other means.

telemetry *See* telemetering.

telencephalon [EMBRYO] The anterior subdivision of the forebrain in a vertebrate embryo; gives rise to the olfactory lobes, cerebral cortex, and corpora striata.

teleology [SCI TECH] The doctrine that explanations of phenomena are to be sought in terms of final causes, purpose, or design in nature.

teleoperator [CONT SYS] A general-purpose, remotely controlled, cybernetic, dexterous person-machine system.

Teleosauridae [PALEON] A family of Jurassic reptiles in the order Crocodilia characterized by a long snout and heavy armor.

Teleostei [VERT ZOO] An infraclass of the subclass Actinopterygii, or rayfin fishes; distinguished by paired bracing bones in the supporting skeleton of the caudal fin, a homocercal caudal fin, thin cycloid scales, and a swim bladder with a hydrostatic function.

telephone [COMMUN] A system of converting sound waves into variations in electric current that can be sent over wires and reconverted into sound waves at a distant point, used primarily for voice communication; it consists essentially of a telephone transmitter and receiver at each station, interconnecting wires, signaling devices, a central power supply, and switching facilities. Also known as telephone system. [ENG ACOUS] *See* telephone set.

telephone-answering system [COMMUN] A special type of private branch exchange system used by a telephone-answering service bureau to provide secretarial service for its customers.

telephone carrier current [ELEC] A carrier current used for telephone communication over power lines or to obtain more than one channel on a single pair of wires.

telephone central office *See* central office.

telephone channel [COMMUN] A one-way or two-way path suitable for the transmission of audio signals between two stations.

telephone circuit [ELEC] The complete circuit over which audio and signaling currents travel in a telephone system between the two telephone subscribers in communication with each other; the circuit usually consists of insulated conductors, as ground returns are now rarely used in telephony.

telephone dial [ENG] A switch operated by a finger wheel, used to make and break a pair of contacts the required number of times for setting up a telephone circuit to the party being called.

telephone emission *See* telephone signal.

telephone induction coil [ELEC] A coil used in a telephone circuit to match the impedance of the line to that of a telephone transmitter or receiver.

telephone influence factor [COMMUN] A function giving the relative interfering effect of noise induced in a telephone circuit by voltage or current in a power circuit at various frequencies. Abbreviated TIF. Also known as influence factor.

telephone line [ELEC] The conductors extending between telephone subscriber stations and central offices.

telephone loading coil *See* loading coil.

telephone modem [ELECTR] A piece of equipment that modulates and demodulates one or more separate telephone circuits, each containing one or more telephone channels; it may include multiplexing and demultiplexing circuits, individual amplifiers, and carrier-frequency sources.

telephone pickup [ELEC] A large flat coil placed under a telephone set to pick up both voices during a telephone conversation for recording purposes.

telephone plug *See* phone plug.

telephone receiver [ENG ACOUS] The portion of a telephone set that converts the audio-frequency current variations of a telephone line into sound waves, by the motion of a dia-

TELEOSAURIDAE

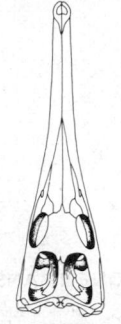

Skull of typical Teleosauridae, with long snout.

phragm activated by a magnet whose field is varied by the electrical impulses that come over the telephone wire.

telephone relay [ELEC] A relay having a multiplicity of contacts on long spring strips mounted parallel to the coil, actuated by a lever arm or other projection of the hinged armature; used chiefly for switching in telephone circuits.

telephone repeater [ELECTR] A repeater inserted at one or more intermediate points in a long telephone line to amplify telephone signals so as to maintain the required current strength.

telephone repeating coil [ELEC] A coil used in a telephone circuit for inductively coupling two sections of a line when a direct connection is undesirable.

telephone ringer [ELECTROMAG] An electromagnetic device that actuates a clapper which strikes one or more gongs to produce a ringing sound; used with a telephone set to signal a called party.

telephone set [ENG ACOUS] An assembly including a telephone transmitter, a telephone receiver, and associated switching and signaling devices. Also known as telephone.

telephone signal [COMMUN] The electrical signal transmitted by a telephone system, classified by type of transmission, type of modulation, bandwidth, and supplementary characteristics. Also known as telephone emission.

telephone switchboard *See* switchboard.

telephone system *See* telephone.

telephone theory *See* frequency theory.

telephone transmitter [ENG ACOUS] The microphone used in a telephone set to convert speech into audio-frequency electric signals.

telephony [COMMUN] The transmission of speech to a distant point by means of electric signals.

telephoto *See* facsimile.

telephotography *See* facsimile.

telephoto lens [OPTICS] A lens for photographing distant objects; it is designed in a compact manner so that the distance from the front of the lens to the film plane is less than the focal length of the lens.

telephotometer [ENG] A photometer that measures the received intensity of a distant light source.

telephotometry [OPTICS] The body of principles and techniques concerned with measuring atmospheric extinction by using various types of telephotometers.

teleprinter [ADP] Any typewriter-type device capable of being connected to a computer and of printing out a set of messages under computer control. [COMMUN] A device that responds to teletype signals and prints the corresponding characters on paper tape.

teleprinter code [COMMUN] A telegraph code used with teletypewriters or teleprinters, in which each letter is represented by a number of time intervals of equal length in which a current is present or absent; there are usually five or seven units per letter, depending on the code.

teleprinting [COMMUN] Telegraphy in which the transmitter and receiver are teleprinters or teletypewriters.

teleprocessing [ADP] **1.** The use of telecommunications equipment and systems by a computer. **2.** A computer service involving input/output at locations remote from the computer itself.

telepsychrometer [ENG] A psychrometer in which the wet- and dry-bulb thermal elements are located at a distance from the indicating elements.

Teleran *See* television and radar navigation system.

telering [ELECTR] In telephony, a frequency-selector device for the production of ringing power.

telescope [OPTICS] An optical instrument which, in order to obtain a better resolution, increases the angle under which a distant object, terrestrial or astronomical, is seen.

Telescope *See* Telescopium.

telescopic comet [ASTRON] A comet in which only the coma is observed.

telescopic derrick [ENG] A drill derrick divided into two or more sections, with the uppermost sections nesting successively into the lower sections.

telescopic loading trough [MIN ENG] A shaker conveyor trough of two sections, one nested in the other, used near the face for advancing the trough line without the necessity of adding either a standard or a short length of pan after each cut.

telescopic sight [ORD] Gunsight equipped with a telescope.

telescopic tripod [ENG] A drill or surveyor's tripod each leg of which is a series of two or more closely fitted nesting tubes, which can be locked rigidly together in an extended position to form a long leg or nested one within the other for easy transport.

telescoping valve [MECH ENG] A valve, with sliding, telescoping members, to regulate water flow in a pipe line with minimum disturbance to stream lines.

Telescopium [ASTRON] A constellation, right ascension 19 hours, declination 50° south. Also known as Telescope.

teleseism [GEOPHYS] An earthquake that is far from the recording station.

teleseismology [GEOPHYS] The aspect of seismology dealing with records made at a distance from the source of the impulse.

Telestacea [INV ZOO] An order of the subclass Alcyonaria comprised of individuals which form erect branching colonies by lateral budding from the body wall of an elongated axial polyp.

telesynd [ELECTR] Telemeter or remote-control equipment which is synchronous in both speed and position.

teletherapy [MED] Radiation treatment administered by using a source that is at a distance from the body, usually employing gamma-ray beams from radioisotope sources.

telethermal [GEOL] Pertaining to a hydrothermal mineral deposit precipitated at a shallow depth and at a mild temperature.

telethermometer [ENG] A temperature-measuring system in which the heat-sensitive element is located at a distance from the indicating element.

telethermoscope [ENG] A temperature telemeter, frequently used in a weather station to indicate the temperature at the instrument shelter located outside.

Teletype [ENG] Trademark of Teletype Corporation, used for teletypewriters.

teletypesetter [GRAPHICS] A system automatically operating a linecasting machine (Linotype or Intertype) to produce lines of type at high speed under the control of perforated tape or equivalent electrical signals.

teletypewriter [COMMUN] A special electric typewriter that produces coded electric signals corresponding to manually typed characters, and automatically types messages when fed with similarly coded signals produced by another machine. Also known as TWX machine.

teletypewriter code [COMMUN] Special code in which each code group is made up of five units, or elements, of equal length which are known as marking or spacing impulses; the five-unit start-stop code consists of five signal impulses preceded by a start impulse and followed by a stop impulse.

teletypewriter exchange service [COMMUN] A service furnished by telephone companies to 60,000 subscribers in the United States, whereby any of the subscribers can communicate directly with any other subscriber via teletypewriter. Also known as TWX service.

teletypewriter signal distortion [COMMUN] Of a start-stop teletypewriter signal, the shifting of the transition points of the signal pulses from their proper positions relative to the beginning of the start pulse; the magnitude of the distortion is expressed in percent of a perfect unit pulse length.

teletypewriter test tape [COMMUN] Perforated tape containing the identification of the transmitting station followed by repetitions of the letters RY, and a test consisting of letters and figures.

teleutospore *See* teliospore.

televise [COMMUN] To pick up a scene with a television camera and convert it into corresponding electric signals for transmission by a television station.

television [COMMUN] A system for converting a succession of visual images into corresponding electric signals and transmitting these signals by radio or over wires to distant receivers at which the signals can be used to reproduce the original images. Abbreviated TV.

TELEPHONE TRANSMITTER

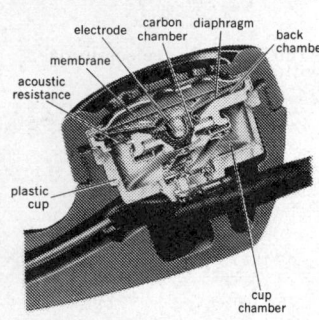

electrode, carbon chamber, diaphragm, back chamber, membrane, acoustic resistance, plastic cup, cup chamber

Cut-away view of telephone transmitter of the direct-action granular-carbon type. Sound pressure on diaphragm varies pressure of dome-shaped electrode on the carbon, causing changes in resistance of the carbon granules, and, in turn, in the magnitude of the current.

television and radar navigation system [NAV] Navigational system which employs ground-based search radar equipment along an airway to locate aircraft flying near that airway; transmits, by television, information pertaining to these aircraft and other information to the pilots of properly equipped aircraft, and provides information to the pilots appropriate for use in the landing approach. Also known as Teleran; television-radar air navigation.

television antenna [ELECTROMAG] An antenna suitable for transmitting or receiving television broadcasts; since television transmissions in the United States are horizontally polarized, the most basic type of receiving antenna is a horizontally mounted half-wave dipole.

television bandwidth [COMMUN] The difference between the limiting frequencies of a television channel; in the United States, this is 6 megahertz.

television broadcast band [COMMUN] Several groups of channels, each containing a number of 6-megahertz channels, that are available for assignment to television broadcast stations.

television broadcasting [COMMUN] Transmission of television programs by means of radio waves, for reception by the public.

television camera [ELECTR] The pickup unit used to convert a scene into corresponding electric signals; optical lenses focus the scene to be televised on the photosensitive surface of a camera tube, and the tube breaks down the visual image into small picture elements and converts the light intensity of each element in turn into a corresponding electric signal. Also known as camera.

television camera tube See camera tube.

television channel [COMMUN] A band of frequencies 6 megahertz wide in the television broadcast band, available for assignment to a television broadcast station.

television emission See television signal.

television interference [COMMUN] Interference produced in television receivers by amateur radio and other transmitters. Abbreviated TVI.

television monitor [ELECTR] **1.** A television set connected to the transmitter at a television station, used to continuously check the image picked up by a television camera and the sound picked up by the microphones. **2.** A closed-circuit television system used to provide continuous observation of such things as hazardous or remote locations, the readings of gages for process control, or microscopic or telescopic images, for greater convenience of viewing.

television network [COMMUN] An arrangement of communication channels, suitable for transmission of video and accompanying audio signals, which link together groups of television broadcasting stations or closed-circuit television users in different cities so that programs originating at one point can be fed simultaneously to all others.

television pickup station [COMMUN] A land mobile station used for the transmission of television program material and related communications from the scene of an event occurring at a remote point to a television broadcast station.

television picture tube See picture tube.

television-radar air navigation See television and radar navigation.

television receiver [ELECTR] A receiver that converts incoming television signals into the original scenes along with the associated sounds. Also known as television set.

television reconnaissance [COMMUN] Reconnaissance in which television is used to transmit a scene from the reconnoitering point to another location on the surface or in the air.

television recording [COMMUN] Permanent record of video signals that is recorded photographically, electronically, or by other means, and may be displayed through a television system or projected as a motion picture film.

television relay system See television repeater.

television repeater [ELECTR] A repeater that transmits television signals from point to point by using radio waves in free space as a medium, such transmission not being intended for direct reception by the public. Also known as television relay system.

television satellite [AERO ENG] An orbiting satellite that relays television signals between ground stations.

television scanning [COMMUN] The process of scrutinizing the brightness of each element of detail contained in the image of a scene to be transmitted by television.

television screen [ELECTR] The fluorescent screen of the picture tube in a television receiver.

television set See television receiver.

television signal [COMMUN] A general term for the aural and visual signals that are broadcast together to provide the sound and picture portions of a television program. Also known as television emission.

television station [COMMUN] The installation, assemblage of equipment, and location where radio transmissions are sent or received.

television studio [COMMUN] A complex of rooms specifically designed for the origination of live television programs; in addition to the studio room in which the program takes place, there are three support rooms: the control room, equipment room, and prop room.

television tower [ENG] A tall metal structure used as a television transmitting antenna, or used with another such structure to support a television transmitting antenna wire.

television transmitter [ELECTR] An electronic device that converts the audio and video signals of a television program into modulated radio-frequency energy that can be radiated from an antenna and received on a television receiver.

television tuner [ELECTR] A component in a television receiver that selects the desired channel and converts the frequencies received to lower frequencies within the passband of the intermediate-frequency amplifier; for very-high-frequency reception there are 12 discrete positions (channels 2–13); for ultra-high-frequency reception continuous tuning is usually employed.

telewriter [COMMUN] System in which writing movement at the transmitting end causes corresponding movement of a writing instrument at the receiving end.

Telex [COMMUN] A worldwide teleprinter exchange service providing direct send and receive teleprinter connections between subscribers. Abbreviated TEX.

telford pavement [CIV ENG] A road pavement having a firm foundation of large stones and stone fragments, and a smooth hard-rolled surface of small stones.

telinite [GEOL] A variety of provitrinite composed of plant cell-wall material.

teliospore [MYCOL] A thick-walled spore of the terminal stage of Uredinales and Ustilaginales which is a probosidium or a group of probosidia. Also known as teleutospore.

Tellerette [CHEM ENG] A type of inert packing with the appearance of a circular-wound spiral, used to create a large surface area to increase contact between falling liquid and rising vapor; used in gas-absorption operations.

Teller-Redlich rule [PHYS CHEM] For two isotopic molecules, the product of the frequency ratio values of all vibrations of a given symmetry type depends only on the geometrical structure of the molecule and the masses of the atoms, and not on the potential constants.

telltale [ENG] A marker on the outside of a tank that indicates on an exterior scale the amount of fluid inside the tank.

telltale compass [NAV] A marine magnetic compass, usually of the inverted type, frequently installed in the master's cabin.

telltale float [CIV ENG] A water-level indicator in a reservoir.

telluric acid [INORG CHEM] H_6TeO_6 Toxic white crystals, slightly soluble in cold water, soluble in hot water and alkalies; melts at 136°C; used as an analytical reagent. Also known as hydrogen tellurate.

telluric current See earth current.

telluric line [SPECT] Any of the spectral bands and lines in the spectrum of the sun and stars produced by the absorption of their light in the atmosphere of the earth.

tellurinic acid [ORG CHEM] A compound of tellurium with the general formula R_2TeOOH; an example is methanetellurinic acid, C_6H_5TeOOH.

tellurite [MINERAL] TeO_2 A white or yellowish orthorhombic mineral consisting of tellurium dioxide, and occurring in crystals; it is dimorphous with paratellurite.

tellurium [CHEM] A member of group VI, symbol Te, atomic number 52, atomic weight 127.60; dark-gray crystals, insolu-

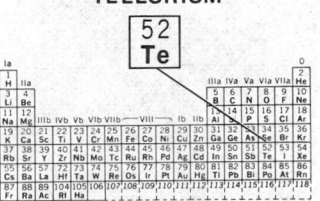

Periodic table of the chemical elements showing the position of tellurium.

ble in water, soluble in nitric and sulfuric acids and potassium hydroxide; melts at 452°C, boils at 1390°C; used in alloys (with lead or steel), glass, and ceramics.

tellurium bromide *See* tellurium dibromide.

tellurium chloride *See* tellurium dichloride.

tellurium dibromide [INORG CHEM] $TeBr_2$ Toxic, hygroscopic, green- or gray-black crystals with violet vapor, soluble in ether, decomposes in water, and melts at 210°C. Also known as tellurium bromide; tellurous bromide.

tellurium dichloride [INORG CHEM] $TeCl_2$ A toxic, amorphous, black or green-yellow powder decomposing in water, melting at 209°C. Also known as tellurium chloride; tellurous chloride.

tellurium dioxide [INORG CHEM] TeO_2 The most stable oxide of tellurium, formed when tellurium is burned in oxygen or air or by oxidation of tellurium with cold nitric acid; crystallizes as colorless, tetragonal, hexagonlike crystals that melt at 452°C.

tellurium disulfide [INORG CHEM] TeS_2 A toxic, red powder, insoluble in water and acids. Also known as tellurium sulfide.

tellurium glance *See* nagyagite.

tellurium hexafluoride [INORG CHEM] TeF_6 A colorless gas which is formed from the elements tellurium and fluorine; it is slowly hydrolyzed by water.

tellurium monoxide [INORG CHEM] TeO A black, amorphous powder, stable in cold dry air; formed by heating the mixed oxide $TeSO_3$.

tellurium sulfide *See* tellurium disulfide.

telluroketone [ORD CHEM] One of a group of compounds with the general formula R_2CTe.

telluromercaptan [ORG CHEM] One of a group of compounds with the general formula $RTeH$.

tellurous acid [INORG CHEM] H_2TeO_3 Toxic, white crystals, soluble in alkalies and acids, slightly soluble in water and alcohol; decomposes at 40°C.

tellurous bromide *See* tellurium dibromide.

tellurous chloride *See* tellurium dichloride.

teloblast [INV ZOO] A large cell that produces many smaller cells at the growing end of many embryos, especially in annelids and mollusks.

telocentric [CYTOL] Pertaining to a chromosome with a terminal centromere.

telocoel [EMBRYO] A cavity of the telencephalon.

telodendrion [ANAT] The terminal branching of an axon. Also known as telodendron.

telodendron *See* telodendrion.

telogen [PHYSIO] A quiescent phase in the cycle of hair growth when the hair is retained in the hair follicle as a dead or "club" hair.

telolecithal [CYTOL] Of an ovum, having a large, evenly dispersed volume of yolk and a small amount of cytoplasm concentrated at one pole.

telomere [CYTOL] A centromere in the terminal position on a chromosome.

telophase [CYTOL] The phase of meiosis or mitosis at which the chromosomes, having reached the poles, reorganize into interphase nuclei with the disappearance of the spindle and the reappearance of the nuclear membrane; in many organisms telophase does not occur at the end of the first meiotic division.

Telosporea [INV ZOO] A class of the protozoan subphylum Sporozoa in which the spores lack a polar capsule and develop from an oocyst.

telotaxis [BIOL] Tactic movement of an organism by the orientation of one or the other of two bilaterally symmetrical receptors toward the stimulus source.

telotroch [INV ZOO] A preanal tuft of cilia in a trochophore larva.

Telpak [COMMUN] Trademark of the Bell System for a variety of leased wide-band communication channels.

telpher [MECH ENG] An electric hoist hanging from and driven by a wheeled cab rolling on a single overhead rail or a rope.

Telsmith breaker [MECH ENG] A type of gyratory crusher, often used for primary crushing; consists of a spindle mounted in a long eccentric sleeve which rotates to impart a

gyratory motion to the crushing head, but gives a parallel stroke, that is, the axis of the spindle describes a cylinder rather than a cone, as in the suspended spindle gyratory.

telson [INV ZOO] The postabdominal segment in lobsters, amphipods, and certain other invertebrates.

Telstar satellite [AERO ENG] A spherical active repeater communications satellite, first launched on July 10, 1962.

telvar [ORG CHEM] The common name for the herbicide 3-(*para*-chlorophenyl)-1,1-dimethylurea; used as a soil sterilant.

TEM *See* triethylenemelamine.

TEMA standard [CHEM ENG] Shell-and-tube heat-exchange standard designed to supplement the ASME code for unfired pressure vessels.

TEM mode *See* transverse electromagnetic mode.

Temnocephalida [INV ZOO] A group of rhabdocoeles sometimes considered a distinct order but usually classified under the Neorhabdocoela; members are characterized by the possession of tentacles and adhesive organs.

Temnochilidae [INV ZOO] The equivalent name for Ostomidae.

Temnopleuridae [INV ZOO] A family of echinoderms in the order Temnopleuroida whose tubercles are imperforate, though usually crenulate.

Temnopleuroida [INV ZOO] An order of echinoderms in the superorder Echinacea with a camarodont lantern, smooth or sculptured test, imperforate or perforate tubercles, and bronchial slits which are usually shallow.

Temnospondyli [PALEON] An order of extinct amphibians in the subclass Labyrinthodontia having vertebrae with reduced pleurocentra and large intercentra.

TE mode *See* transverse electric mode.

temper [ENG] 1. To moisten and mix clay, plaster, or mortar to the proper consistency for use. 2. To anneal or toughen glass. [MET] 1. The hardness and strength of a rolled metal. 2. The nominal carbon content of steel. 3. To soften hardened steel or cast iron by reheating to some temperature below the eutectoid temperature. 4. An alloy added to pure tin to make the finest pewter.

tempera [MATER] 1. An opaque watercolor paint consisting of pigment ground in water and mixed with egg yolk. 2. A term used to describe a poster paint that uses glue or gum as a binder.

temperate and cold savannah [ECOL] A regional vegetation zone, very extensively represented in North America and in Eurasia at high altitudes; consists of scattered or clumped trees (very often conifers and mostly needle-leaved evergreens) and a shrub layer of varying coverage; mosses and, even more abundantly, lichens form an almost continuous carpet.

temperate and cold scrub [ECOL] Regional vegetation zone whose density and periodicity vary a good deal; requires a considerable amount of moisture in the soil, whether from mist, seasonal downpour, or snowmelt; shrubs may be evergreen or deciduous; and undergrowth of ferns and other large-leaved herbs are quite frequent, especially at subalpine level; wind shearing and very cold winters prevent tree growth. Also known as bosque; fourré; heath.

temperate belt [CLIMATOL] A belt around the earth within which the annual mean temperature is less than 20°C (68°F) and the mean temperature of the warmest month is higher than 10°C (50°F).

temperate climate [CLIMATOL] The climate of the middle latitudes; the climate between the extremes of tropical climate and polar climate.

temperate glacier [HYD] A glacier which, at the end of the melting season, is composed of firn and ice at the melting point.

temperate mixed forest [ECOL] A forest of the North Temperate Zone containing a high proportion of conifers with a few broad-leafed species.

temperate phage [VIROL] A deoxyribonucleic acid phage, the genome (DNA) of which can under certain circumstances become integrated with the genome of the host.

temperate rainforest [ECOL] A vegetation class in temperate areas of high and evenly distributed rainfall characterized by comparatively few species with large populations of each

---

TELOLECITHAL

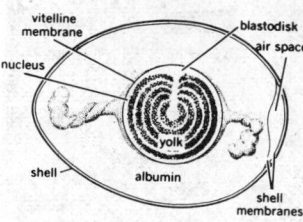

Telolecithal ovum of a hen.

TEMPERATE RAINFOREST

Temperate rainforest in Olympic National Park, Washington, showing sheath of moss on maple tree. *(Photograph by J. O. Sumner, from National Audubon Society)*

species; evergreens are somewhat short with small leaves, and there is an abundance of large tree ferns. Also known as cloud forest; laurel forest; laurisilva; moss forest; subtropical forest.

temperate rainy climate [CLIMATOL] One of the major categories in W. Kippen's climatic classification; the coldest-month mean temperature is less than 64.4°F (18°C) and greater than 26.6°F (−3°C), and the warmest-month mean temperature is more than 50°F (10°C).

temperate westerlies *See* westerlies.

temperate-westerlies index [METEOROL] A measure of the strength of the westerly wind between latitudes 35°N and 55°N; the index is computed from the average sea-level pressure difference between these latitudes and is expressed as the west to east component of geostrophic wind in meters and tenths of meters per second.

temperate woodland [ECOL] A vegetation class similar to tropical woodland in spacing, height, and stratification, but it can be either deciduous or evergreen, broad-leaved or needle-leaved. Also known as parkland; woodland.

Temperate Zone [CLIMATOL] Either of the two latitudinal zones on the earth's surface which lie between 23°27′ and 66°32′ N and S (the North Temperate Zone and South Temperate Zone, respectively).

temperature [THERMO] A property of an object which determines the direction of heat flow when the object is placed in thermal contact with another object: heat flows from a region of higher temperature to one of lower temperature; it is measured either by an empirical temperature scale, based on some convenient property of a material or instrument, or by a scale of absolute temperature, for example, the Kelvin scale.

13.0 temperature *See* annealing point.

temperature bath [THERMO] A relatively large volume of a homogeneous substance held at constant temperature, so that an object placed in thermal contact with it is maintained at the same temperature.

temperature belt [METEOROL] The belt which may be drawn on a thermograph trace or other temperature graph by connecting the daily maxima with one line and the daily minima with another.

temperature coefficient [PHYS] The rate of change of some physical quantity (such as resistance of a conductor or voltage drop across a vacuum tube) with respect to temperature.

temperature coefficient of reactivity [NUCLEO] The change in reactor reactivity (per degree of temperature) occurring when the operating temperature of the reactor changes.

temperature color scale [THERMO] The relation between an incandescent substance's temperature and the color of the light it emits.

temperature-compensated Zener diode [ELECTR] Positive-temperature-coefficient reversed-bias Zener diode (*pn* junction) connected in series with one or more negative-temperature forward-biased diodes within a single package.

temperature-compensating capacitor [ELEC] Capacitor whose capacitance varies with temperature in a known and predictable manner; used extensively in oscillator circuits to compensate for changes in the values of other parts with temperatures.

temperature compensation [ELECTR] The process of making some characteristic of a circuit or device independent of changes in ambient temperature.

temperature control [ENG] A control used to maintain the temperature of an oven, furnace, or other enclosed space within desired limits.

temperature-efficiency index [ECOL] For a given location, a measure of the long-range effectiveness of temperature (thermal efficiency) in promoting plant growth. Abbreviated T-E index. Also known as thermal-efficiency index.

temperature-efficiency ratio [ECOL] For a given location and month, a measure of thermal efficiency; it is equal to the departure, in degrees Fahrenheit, of the normal monthly temperature above 32°F (0°C) divided by 4: $(T − 32)/4$. Abbreviated T-E ratio. Also known as thermal-efficiency ratio.

temperature error [ENG] That instrument error due to non-standard temperature of the instrument.

temperature gradient [THERMO] For a given point, a vector whose direction is perpendicular to an isothermal surface at the point, and whose magnitude equals the rate of change of temperature in this direction.

temperature-humidity index [METEOROL] An index which gives a numerical value, in the general range of 70–80, reflecting outdoor atmospheric conditions of temperature and humidity as a measure of comfort (or discomfort) during the warm season of the year; equal to 15 plus 0.4 times the sum of the dry-bulb and wet-bulb temperatures in degrees Fahrenheit. Also known as comfort index; discomfort index. Abbreviated CI; DI; THI.

temperature-indicating compound [MATER] A temperature-sensitive material with a predetermined melting point; used to indicate when a predesignated temperature is reached in such processes as heat treating, welding, and molding.

temperature inversion [METEOROL] A layer in the atmosphere in which temperature increases with altitude; the principal characteristic of an inversion layer is its marked static stability, so that very little turbulent exchange can occur within it; strong wind shears often occur across inversion layers, and abrupt changes in concentrations of atmospheric particulates and atmospheric water vapor may be encountered on ascending through the inversion layer. [OCEANOGR] A layer of a large body of water in which temperature increases with depth.

temperature log [PETRO ENG] A continuous record of temperature versus depth in an oil-well borehole.

temperature province [CLIMATOL] A major division of C. W. Thornthwaite's schemes of climatic classification, determined as a function of the temperature-efficiency index or the potential evapotranspiration.

temperature resistance coefficient [ELEC] The ratio of the change of electrical resistance in a wire caused by a change in its temperature of 1°C as related to its resistance at 0°C.

temperature-salinity diagram [OCEANOGR] The plot of temperature versus salinity data of a water column; the resulting diagram identifies the water masses within the column, the column's stability, indicates the σ_T value via lines of constant σ_T printed on paper, and allows an estimate of the accuracy of the temperature and salinity measurements. Also known as T-S curve; T-S diagram; T-S relation.

temperature saturation [ELECTR] The condition in which the anode current of a thermionic vacuum tube cannot be further increased by increasing the cathode temperature at a given value of anode voltage; the effect is due to the space charge formed near the cathode. Also known as filament saturation; saturation.

temperature scale [THERMO] An assignment of numbers to temperatures in a continuous manner, such that the resulting function is single valued; it is either an empirical temperature scale, based on some convenient property of a substance or object, or it measures the absolute temperature.

temperature-sensitive mutant [GEN] A mutant gene that is functional at high (low) temperature but is inactivated by lowering (elevating) the temperature.

temperature sensor [ENG] A device designed to respond to temperature stimulation.

temperature survey [PETRO ENG] An analysis of temperature changes or differences down an oil-well borehole, based on a temperature log; used to locate the top of casing cement, lost circulation zones, or gas entry zones.

temperature transducer [ENG] A device in an automatic temperature-control system that converts the temperature into some other quantity such as mechanical movement, pressure, or electric voltage; this signal is processed in a controller, and is applied to an actuator which controls the heat of the system.

temperature wave [CRYO] A disturbance in which a variation in temperature propagates through a medium; the chief example of this is second sound. Also known as thermal wave.

temperature zone [CLIMATOL] A portion of the earth's surface defined by relatively uniform temperature characteristics, and usually bounded by selected values of some measure of temperature or temperature effect.

temper brittleness [MET] A brittle state resulting when certain low-alloy steels are slowly cooled in a range of 600–

TENDER

Polaris submarine tender USS Hunley. (Official U.S. Navy photograph)

TENDRIL

Portion of a grape plant stem with leaf and tendril.

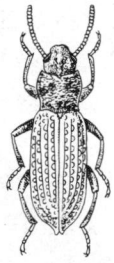

TENEBRIONIDAE

Representative species of family Tenebrionidae. (From T. I. Storer and R. L. Usinger, General Zoology, 3d ed., McGraw-Hill, 1957)

TENREC

Tenrec, showing characteristic features.

300°C or reheated in this range after quenching from 600°C.

tempering [MET] Heat treatment of hardened steels to temperatures below the transformation temperature range, usually to improve toughness.

tempering oil [MATER] A high-viscosity neutral petroleum oil, such as a steam cylinder stock, used for the drawing or tempering of steel. Also known as steam-hammer oil; summer black oil.

temper screw See stirrup.

temper time [MET] In resistance welding, that part of the postweld interval during which the current is suitable for tempering or heat treatment.

tempilstick [MATER] A crayon that, when applied to a surface, indicates when the surface temperature exceeds a given value by changing color.

template [ENG] 1. A two-dimensional representation of a machine or other equipment used for building layout design. 2. A guide or a pattern used in manufacturing items. Also spelled templet. [MOL BIO] The macromolecular model for the synthesis of another macromolecule.

templet See template.

temporal [SCI TECH] Pertaining to or limited by time.

temporal arteritis See giant-cell arteritis.

temporal bone [ANAT] The bone forming a portion of the lateral aspect of the skull and part of the base of the cranium in vertebrates.

temporale [METEOROL] A rainy wind from the southwest to west resulting from a deflection of the southeast trades of the eastern South Pacific onto the Pacific coast of Central America.

temporal-lobe epilepsy [MED] Recurrent seizures originating in lesions of the temporal lobe of the brain.

temporary base level [GEOL] Any base level, other than sea level, below which a land area temporarily cannot be reduced by erosion. Also known as local base level.

temporary geographic grid [ORD] A particular grid specified by military authorities for temporary and local use.

temporary hardness [CHEM] The portion of the total hardness of water that can be removed by boiling whereby the soluble calcium and magnesium bicarbonate are precipitated as insoluble carbonates.

temporary plankton See meroplankton.

temporary storage [ADP] The storage capacity reserved or used for retention of temporary or transient data.

Tempra [TEXT] Trade name of industrial rayon yarn made by the American Enka Corporation.

TEM wave See transverse electromagnetic wave.

tenacity [TEXT] In yarn manufacture and textile engineering, the tensile strength of a yarn or a filament for its given size.

Ten Broecke chart [THERMO] A graphical plot of heat transfer and temperature differences used to calculate the thermal efficiency of a countercurrent cool-fluid–warm-fluid heat-exchange system.

tendency [METEOROL] The local rate of change of a vector or scalar quality with time at a given point in space.

tendency chart See change chart.

tendency equation [METEOROL] An equation for the local change of pressure at any point in the atmosphere, derived by combining the equation of continuity with an integrated form of the hydrostatic equation.

tendency interval [METEOROL] The finite increment of time over which a change of the value of a meteorological element is measured in order to estimate its tendency; the most familiar example is the three-hour time interval over which local pressure differences are measured in determining pressure tendency.

tender [NAV ARCH] A naval auxiliary ship that serves as a mobile base for repair and limited resupply of other ships.

Tendipedidae [INV ZOO] The midges, a family of orthorrhaphous dipteran insects in the series Nematocera whose larvae occupy intertidal wave-swept rocks on the seacoasts.

tendon [ANAT] A white, glistening, fibrous cord which joins a muscle to some movable structure such as a bone or cartilage; tendons permit concentration of muscle force into a small area and allow the muscle to act at a distance. [CIV ENG] A steel bar or wire that is tensioned, anchored to formed

concrete, and allowed to regain its initial length to induce compressive stress in the concrete before use.

tendonitis [MED] Inflammation of a tendon.

tendon sheath [ANAT] The synovial membrane surrounding a tendon.

tendril [BOT] A stem modification in the form of a slender coiling structure capable of twining about a support to which the plant is then attached.

tenebrescence [PHYS] Darkening and bleaching under suitable irradiation; materials having this property are called scotophors; darkening may be produced by primary x-rays or cathode rays, while bleaching may be produced by heat or by photons of appropriate wavelength.

Tenebrionidae [INV ZOO] The darkling beetles, a large cosmopolitan family of coleopteran insects in the superfamily Tenebrionoidea; members are common pests of grains, dried fruits, beans, and other food products.

Tenebrionoidea [INV ZOO] A superfamily of the Coleoptera in the suborder Polyphaga.

tenggara [METEOROL] A strong, dry, hazy, east or southeast wind during the east monsoon in the Spermunde Archipelago.

teniae coli [HISTOL] The three bands comprising the longitudinal layer of the tunica muscularis of the colon: the tenia libera, tenia mesocolica, and tenia omentalis.

tennantite [MINERAL] $(Cu,Fe)_{12}As_4S_{13}$ A lead-gray mineral crystallizing in the isometric system; it is isomorphous with tetrahedrite; an important ore of copper.

tennecetin See pimaricin.

tenorite [MINERAL] CuO A triclinic mineral that occurs in small, shining, steel-gray scales, in black powder, or in black earthy masses; an ore of copper.

tenosynonitis [MED] Inflammation of a tendon and its sheath.

tenrec [VERT ZOO] Any of about 30 species of unspecialized, insectivorous mammals indigenous to Madagascar, and which have poor vision and clawed digits.

Tenrecidae [VERT ZOO] The tenrecs, a family of insectivores in the group Lipotyphla.

ten's complement [MATH] In decimal arithmetic, the unique numeral that can be added to a given N-digit numeral to form a sum equal to 10^N (that is, a one followed by N zeros).

tensile bar [ENG] A molded, cast, or machined specimen of specified cross-sectional dimensions used to determine the tensile properties of a material by use of a calibrated pull test. Also known as tensile specimen; test specimen.

tensile modulus [MECH] The tangent or secant modulus of elasticity of a material in tension.

tensile specimen See tensile bar.

tensile strength [MECH] The maximum stress a material subjected to a stretching load can withstand without tearing. Also known as hot strength.

tensile stress [MECH] Stress developed by a material bearing a tensile load.

tensile test [ENG] A test in which a specimen is subjected to increasing longitudinal pulling stress until fracture occurs.

tension [MECH] 1. The condition of a string, wire, or rod that is stretched between two points. 2. The force exerted by the stretched object on a support.

tension fault [GEOL] A fault in which crustal tension is a factor, such as a normal fault. Also known as extensional fault.

tension fracture [GEOL] A minor rock fracture developed at right angles to the direction of maximum tension. Also known as subsidiary fracture.

tension jack [MIN ENG] A type of jack with a jackscrew for wedging against the mine roof and a ratchet device for applying tension on a chain that is attached to the tail or foot section of a belt conveyor, and used to restore the proper tension to the belt.

tension joint [GEOL] A joint that is a tension fracture.

tension packer [PETRO ENG] A device to pressure-seal the annular space between an oil-well casing and tubing, held in place by tension against an upward push; a type of production packer.

tension pulley [MECH ENG] A pulley around which an endless rope passes mounted on a trolley or other movable bearing so that the slack of the rope can be readily taken up by the pull of the weights.

tension-type hanger [PETRO ENG] A type of tubing hanger for multiple-completion oil wells to allow for the varying lengths of tubing strings.

tensor [MATH] An object relative to a locally euclidean space which possesses a specified system of components for every coordinate system and which changes under a transformation of coordinates.

tensor analysis [MATH] The abstract study of mathematical objects having components which express properties similar to those of a geometric tensor; this study is fundamental to Riemannian geometry and the structure of euclidean spaces. Also known as tensor calculus.

tensor calculus See tensor analysis.

tensor contraction [MATH] For a tensor having an upper and a lower index, summation over the components in which these indexes have the same value, in order to obtain a new tensor two lower in rank.

tensor differentiation [MATH] An operation on a tensor in which a term involving a Christoffel symbol is subtracted from the ordinary derivative, to obtain another tensor of one higher rank.

tensor field [MATH] A tensor or collection of tensors defined in some open subset of a Riemann space.

tensor force [NUC PHYS] A spin-dependent force between nucleons, having the same form as the interaction between magnetic dipoles; it is introduced to account for the observed values of the magnetic dipole moment and electric quadrupole moment of the deuteron.

tensorial set [MATH] Any collection of quantities that are associated with a system of spatial coordinates and which undergo a linear transformation when this system rotates; examples are the components of a tensor and the eigenfunctions of a quantum mechanical operator.

tensor muscle [PHYSIO] A muscle that stretches a part or makes it tense.

tensor product [MATH] 1. The product of two tensors is the tensor whose components are obtained by multiplying those of the given tensors. 2. In algebra, a multiplicative operation performed between modules.

tensor quantity [MATH] A quantity mathematically represented by a tensor or possessing properties analogous to a tensor.

tensor space [MATH] A fiber bundle composed of the points of a Riemannian manifold and tensor fields.

tentacle [INV ZOO] Any of various elongate, flexible processes with tactile, prehensile, and sometimes other functions, and which are borne on the head or about the mouth of many animals.

Tentaculata [INV ZOO] A class of the phylum Ctenophora whose members are characterized by having variously modified tentacles.

tentaculocyst [INV ZOO] A sense organ located at the margin of the umbrella in some coelenterate medusoids, consisting of a modified tentacle with a cavity that often contains lithites.

tentaculozoid [INV ZOO] A slender tentacular individual of a hydrozoan colony.

tented arch [ANAT] A fingerprint pattern which possesses either an angle, an upthrust, or two of the three basic characteristics of a loop.

tenterhook [TEXT] One of the nails used to hold fabric on a frame (tenter) so that alterations in the original measurements can be corrected.

tenthmeter See angstrom.

Tenthredinidae [INV ZOO] A family of hymenopteran insects in the superfamily Tenthredinoidea including economically important species whose larvae are plant pests.

Tenthredinoidea [INV ZOO] A superfamily of Hymenoptera in the suborder Symphyla.

tentillum [INV ZOO] A contractile branch of a tentacle containing many nematocysts in certain siphonophores.

tenting [OCEANOGR] The vertical displacement upward of ice under pressure to form a flat-sided arch with a cavity beneath.

Tenuipalpidae [INV ZOO] A small family of mites in the suborder Trombidiformes.

tepary bean [BOT] *Phaseolus acutifolius* var. *latifolius.* One of the four species of beans of greatest economic importance in the United States.

tepee butte [GEOGR] A tepeelike hill or knoll, especially one of soft material capped by more resistant rock.

tepee structure [GEOL] A disharmonic sedimentary structure consisting of a fold that resembles an inverted depressed V in cross section.

tepetate See caliche.

tephigram [METEOROL] A thermodynamic diagram designed by Napier Shaw with temperature and logarithm of potential temperature as coordinates; isobars are gently curved lines and the chart is rotated so that pressure increases downward; vapor lines and saturation adiabats are curved; on this chart, energy is proportional to the area enclosed by the curve representing the process.

tephra [GEOL] Denotes all pyroclastics of a volcano.

tephrite [PETR] A group of basaltic extrusive rocks composed chiefly of calcic plagioclase, augite, and nepheline or leucite, with some sodic sanidine.

Tephritidae [INV ZOO] The fruit flies, a family of myodarian cyclorrhaphous dipteran insects in the subsection Acalypteratae.

tephrochronology [GEOL] The dating of different layers of volcanic ash for the establishment of a sequence of geologic and archeologic occurrences.

tephroite [MINERAL] Mn_2SiO_4 An olivine mineral that occurs with zinc and manganese minerals.

tequila [FOOD ENG] A distilled spirit that is Mexican in origin; obtained by the natural fermentation of the sweet sap of the agave plant and the distillation of the fermentation product.

tera- [MATH] A prefix representing 10^{12}, which is 1,000,000,-000,000 or a million million. Abbreviated T.

terahertz [PHYS] A unit of frequency, equal to 10^{12} hertz, or 1,000,000 megahertz. Abbreviated THz.

teraohm [ELEC] A unit of electrical resistance, equal to 10^{12} ohms, or 1,000,000 megohms. Abbreviated TΩ.

teraohmmeter [ENG] An ohmmeter having a teraohm range for measuring extremely high insulation resistance values.

T-E ratio See temperature-efficiency ratio.

teratocarcinoma [MED] A teratoma with carcinomatous elements.

teratogen [MED] An agent causing formation of a congenital anomaly or monstrosity.

teratogenesis [MED] The formation of a fetal monstrosity.

teratoid adenocystoma See mesonephroma.

teratology [MED] The science of fetal malformations and monstrosities.

teratoma [MED] A true neoplasm composed of bizarre and chaotically arranged tissues that are foreign embryologically as well as histologically to the area in which the tumor is found.

teratophobia [PSYCH] An abnormal fear of deformed people and pregnant women.

Teratornithidae [PALEON] An extinct family of vulturelike birds of the Pleistocene of western North America included in the order Falconiformes.

terawatt [PHYS] A unit of power, equal to 10^{12} watts, or 1,000,000 megawatts. Abbreviated TW.

terbia See terbium oxide.

terbium [CHEM] A rare-earth element, symbol Tb, in the yttrium subgroup of group III, atomic number 65, atomic weight 158.924.

terbium chloride [INORG CHEM] $TbCl_3 \cdot 6H_2O$ Water- and alcohol-soluble, hygroscopic, colorless, transparent prisms; anhydrous form melts at 588°C.

terbium nitrate [INORG CHEM] $Tb(NO_3)_3 \cdot 6H_2O$ A colorless, fire-hazardous (strong oxidant) powder, soluble in water; melts at 89°C.

terbium oxide [INORG CHEM] Tb_2O_3 A slightly hygroscopic, dark-brown powder soluble in dilute acids, absorbs carbon dioxide from air. Also known as terbia.

terbutol [ORG CHEM] The common name for the herbicide 2,6-di-*tert*-butyl-*p*-tolylmethylcarbamate; used as a selective preemergence crabgrass herbicide for turf.

TENTED ARCH

Fingerprint having tented arch pattern with upthrust. (*Federal Bureau of Investigation*)

TENTHREDINIDAE

Representative species of family Tenthredinidae. (*From T. I. Storer and R. L. Usinger, General Zoology, 3d ed., McGraw-Hill, 1957*)

TERBIUM

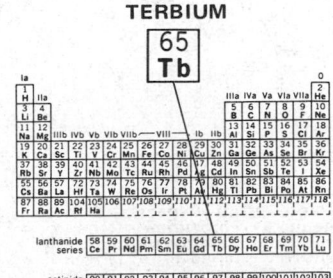

Periodic table of the chemical elements showing the position of terbium.

TEREBELLIDAE

Left-lateral view of the anterior end of *Artacamella* species in the subfamily Artacaminae of the Terebillidae.

TEREBRATELLIDINA

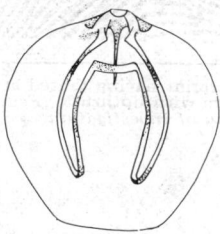

Representative species of suborder Terebratellidina, in genus *Magellania*, showing characteristic loop. *(From R. C. Moore, ed., Treatise on Invertebrate Paleontology, pt. H, Geological Society of America and University of Kansas Press, 1965)*

TEREBRATULIDINA

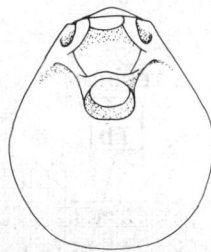

Representative species of the genus *Terebratulina* of the suborder Terebratulidina with short V-shaped loop. *(From R. C. Moore, ed., Teatise on Invertebrate Paleontology, pt. H, Geological Society of America and University of Kansas Press, 1965)*

tercentesimal thermometric scale   *See* approximate absolute temperature.

Terebellidae   [INV ZOO] A family of polychaete annelids belonging to the Sedentaria which are chiefly large, thick-bodied, tubicolous forms with the anterior end covered by a matted mass of tentacular cirri.

Terebratellidina   [PALEON] An extinct suborder of articulate brachiopods in the order Terebratulida in which the loop is long and offers substantial support to the side arms of the lophophore.

Terebratulida   [INV ZOO] An order of articulate brachiopods that has a punctate shell structure and is characterized by the possession of a loop extending anteriorly from the crural bases, providing some degree of support for the lophophore.

Terebratulidina   [INV ZOO] A suborder of articulate brachiopods in the order Terebratulida distinguished by a short V- or W-shaped loop.

Teredinidae   [INV ZOO] The pileworms or shipworms, a family of bivalve mollusks in the subclass Eulamellibranchia distinguished by having the two valves reduced to a pair of small plates at the anterior end of the animal.

terephthalic acid   [ORG CHEM] $C_6H_4(COOH)_2$ A combustible white powder, insoluble in water, soluble in alkalies, sublimes above 300°C; used to make polyester resins for fibers and films and as an analytical reagent and poultry-feed additive. Also known as benzene-*para*-dicarboxylic acid; *para*-phthalic acid; TPA.

terete   [BOT] Of a stem, cylindrical in section, but tapering at both ends.

tergite   [INV ZOO] The dorsal plate covering a somite in arthropods and certain other articulate animals.

tergum   [INV ZOO] A dorsal plate of the operculum in barnacles.

term   [SPECT] A set of $(2S+1)(2L+1)$ atomic states belonging to a definite configuration and to definite spin and orbital angular momentum quantum numbers S and L.

terminal   [ADP] A site or location at which data can leave or enter a system.   [ELEC] **1.** A screw, soldering lug, or other point to which electric connections can be made. Also known as electric terminal. **2.** The equipment at the end of a microwave relay system or other communication channel. **3.** One of the electric input or output points of a circuit or component.

terminal area   [ELECTR] The enlarged portion of conductor material surrounding a hole for a lead on a printed circuit. Also known as land; pad.

terminal bar   [CYTOL] One of the structures formed in certain epithelial cells by the combination of local modifications of contiguous surfaces and intervening intercellular substances; they become visible with the light microscope after suitable staining and appear to close the spaces between the epithelial cells of the intestine at their free surfaces.

terminal block   [COMMUN] A cluster of five captive screw terminals at which a telephone pair terminates; the center terminal is for the ground wire, and two other terminals are used for the tip and ring wires.

terminal board   [ELEC] An insulating mounting for terminal connections. Also known as terminal strip.

terminal box   [ELEC] An enclosure which includes, mounts, and protects one or more terminals or terminal boards; it may include a cover and such accessories as mounting hardware, brackets, locks, and conduit fittings.

terminal bud   [BOT] A bud that develops at the apex of a stem. Also known as apical bud.

terminal clearance capacity   [ENG] The amount of cargo or personnel that can be moved through and out of a terminal on a daily basis.

terminal control area   [NAV] A control area or a portion thereof normally situated at the confluence of air-traffic service routes in the vicinity of one or more major airfields.

terminal cutout pairs   [ELEC] Numbered, designated pairs brought out of a cable at a terminal.

terminal endocarditis   *See* verrucous endocarditis.

terminal equipment   [COMMUN] **1.** Assemblage of communications-type equipment required to transmit or receive a signal on a channel or circuit, whether it be for delivery or relay. **2.** In radio relay systems, equipment used at points where intelligence is inserted or derived, as distinct from equipment used to relay a reconstituted signal. **3.** Telephone and teletypewriter switchboards and other centrally located equipment at which wire circuits are terminated.

terminal forecast   [METEOROL] An aviation weather forecast for one or more specified air terminals.

terminal guidance   [NAV] Guidance of a craft from an arbitrary point, at which midcourse guidance ends, to the destination.

terminal hair   [ANAT] One of three types of hair in man based on hair size, time of appearance, and structural variations; the larger, coarser hair in the adult that replaces the vellus hair.

terminal leg   *See* terminal stub.

terminal moraine   [GEOL] An end moraine that extends as an arcuate or crescentic ridge across a glacial valley; marks the farthest advance of a glacier. Also known as marginal moraine.

terminal nerve   [ANAT] Either of a pair of small cranial nerves that run from the nasal area to the forebrain, present in most vertebrates; the function is not known.

terminal operations   [ENG] The reception, processing, and staging of passengers; the receipt, transit storage, and marshaling of cargo; the loading and unloading of ships or aircraft; and the manifesting and forwarding of cargo and passengers to destination.

terminal pair   [ELEC] An associated pair of accessible terminals, such as the input or output terminals of a device or network.

terminal phase   [ORD] The period of flight of a missile between the end of midcourse guidance and impact.

terminal pressure   [ENG] A pressure drop across a unit when the maximum allowable pressure drop is reached, as for a filter press.

terminal repeater   [COMMUN] **1.** Assemblage of equipment designed specifically for use at the end of a communications circuit, as contrasted with the repeater designed for an intermediate point. **2.** Two microwave terminals arranged to provide for the interconnection of separate systems, or separate sections of a system.

terminal room   [COMMUN] In telephone practice, a room associated with a central office, private branch exchange, or private exchange, which contains distributing frames, relays, and similar apparatus, except that mounted in the switchboard section.

terminal sinus   [EMBRYO] The vascular sinus bounding the area vasculosa of the blastoderm of a meroblastic ovum. Also known as marginal sinus.

terminal station   [COMMUN] Receiving equipment and associated multiplex equipment used at the ends of a radio-relay system.

terminal strip   *See* terminal board.

terminal stub   [ELEC] Piece of cable that comes with a cable terminal for splicing into the main cable. Also known as terminal leg.

terminal velocity   [FL MECH] The velocity with which a body moves relative to a fluid when the resultant force acting on it (due to friction, gravity, and so forth) is zero.   [PHYS] The maximum velocity attainable, especially by a freely falling body, under given conditions.

terminal very-high-frequency omnirange   [NAV] A low-power very-high-frequency omnidirectional radio range intended to furnish service in the local area (about 30-mile, or 48-kilometer, radius).

terminal voltage   [ELEC] The voltage at the terminals connected to the source of electricity for an electric machine.

terminated line   [ELEC] Transmission line terminated in a resistance equal to the characteristic impedance of the line, so there is no reflection and no standing waves.

terminating   [ELEC] Closing of the circuit at either end of a line or transducer by connecting some device thereto; terminating does not imply any special condition such as the elimination of reflection.

termination   [ELECTROMAG] **1.** Load connected to a transmission line or other device; to avoid wave reflections, it must match the characteristic of the line or device. **2.** In waveguide technique, the point at which energy flowing along a waveguide continues in a nonwaveguide mode of propagation.

terminator [ASTRON] The line of demarcation between the dark and light portions of the moon or planets.

Termitaphididae [INV ZOO] The termite bugs, a small family of Hemiptera in the superfamily Aradoidea.

termitarium [INV ZOO] A termites' nest.

termite [INV ZOO] A soft-bodied insect of the order Isoptera; individuals feed on cellulose and live in colonies with a caste system comprising three types of functional individuals: sterile workers and soldiers, and the reproductives. Also known as white ant.

termite shield [BUILD] A strip of metal, usually galvanized iron, bent down at the edges and placed between the foundation of a house and a timber floor, around pipes, and other places where termites can pass.

termiticole [ECOL] An organism that lives in a termites' nest.

Termitidae [INV ZOO] A large family of the order Isoptera which contains the higher termites, representing 80% of the species.

termitophile [ECOL] An organism that lives in a termites' nest in a symbiotic association with the termites.

Termopsidae [INV ZOO] A family of insects in the order Isoptera composed of damp wood-dwelling forms.

term splitting [QUANT MECH] The separation of the energies of the states in a term; in the Russell-Saunders case this is produced by the spin-orbit interaction.

ternary [SCI TECH] Consisting of three, as in a three-phase (that is, ternary) liquid system.

ternary alloy [MET] An alloy composed of three principal elements.

ternary code [COMMUN] Code in which each code element may be any one of three distinct kinds or values.

ternary compound [CHEM] A molecule consisting of three different types of atoms; for example, sulfuric acid, H_2SO_4.

ternary diagram [PETR] A triangular diagram that graphically depicts the composition of a three-component mixture or ternary system.

ternary expansion [MATH] The numerical representation of a real number relative to the base 3, the digits determined by how the given number can be written in terms of powers of 3.

ternary notation [MATH] A system of notation using the base of 3 and the characters 0, 1, and 2.

ternary pulse code modulation [COMMUN] Pulse code modulation in which each code element may be any one of three distinct kinds or values.

ternary system [CHEM] Any system with three nonreactive components; in liquid systems, the components may or may not be partially soluble.

ternate [BOT] Composed of three subdivisions, as a leaf with three leaflets.

terneplate [MATER] A sheet of iron or steel coated with a lead-tin alloy; used chiefly for roofing.

Ternifine man [PALEON] The name for a fossil human type, represented by three lower jaws and a parietal bone discovered in France and thought to be from the upper part of the middle Pleistocene.

terpane *See* menthane.

terpene [ORG CHEM] $C_{10}H_{16}$ A moderately toxic, flammable, unsaturated hydrocarbon liquid found in essential oils and plant oleoresins; used as an intermediate for camphor, menthol, and terpineol.

terpene alcohol [ORG CHEM] A generic name for an alcohol related to or derived from a terpene hydrocarbon, such as terpineol or borneol.

terpene hydrochloride [ORG CHEM] $C_{10}H_{16} \cdot HCl$ A solid, water-insoluble material melting at 125°C; used as an antiseptic. Also known as artificial camphor; dipentene hydrochloride; pinene hydrochloride; turpentine camphor.

***para*-terphenyl** [ORG CHEM] $(C_6H_5)_2C_6H_4$ A combustible, toxic liquid boiling at 405°C; crystals are used for scintillation counters; polymerized with styrene to make plastic phosphor. Also known as 1,4-diphenylbenzene.

terpineol [ORG CHEM] $C_{10}H_{17}OH$ A combustible, colorless liquid with a lilac scent, derived from pine oil, soluble in alcohol, slightly soluble in water, boils at 214–224°C; used in medicine, perfumes, soaps, and disinfectants, and as an antioxidant, a flavoring agent, and a solvent; isomeric forms are alpha-, beta- and gamma-.

terpin hydrate [ORG CHEM] $CH_3(OH)C_6H_9C(CH_3)_2OH \cdot H_2O$ Combustible, efflorescent, lustrous white prisms soluble in alcohol and ether, slightly soluble in water; melts at 116°C; used for pharmaceuticals and to make terpineol. Also known as dipentene glycol.

terpinolene [ORG CHEM] $C_{10}H_{16}$ A flammable, water-white liquid insoluble in water, soluble in alcohol, ether, and glycols, boils at 184°C; used as a solvent and as a chemical intermediate for resins and essential oils.

terpinyl acetate [ORG CHEM] $C_{10}H_{17}OOCCH_3$ A combustible, colorless, liquid slightly soluble in water and glycerol, soluble in water, boils at 220°C; used in perfumes and flavors.

terpolymer [ORG CHEM] A polymer that contains three distinct monomers; for example, acrylonitrile-butadiene-styrene terpolymer, ABS.

terra alba [MATER] Pure white uncalcined gypsum, used as a filler in paper and paints and as a nutrient in growing yeast.

terrace [BUILD] **1.** A flat roof. **2.** A colonnaded promenade. **3.** An open platform extending from a building, usually at ground level. [GEOL] **1.** A horizontal or gently sloping embankment of earth along the contours of a slope to reduce erosion, control runoff, or conserve moisture. **2.** A narrow coastal strip sloping gently toward the water. **3.** A long, narrow, nearly level surface bounded by a steeper descending slope on one side and by a steeper ascending slope on the other side. **4.** A benchlike structure bordering an undersea feature.

terraced pool [GEOGR] A shallow, rimmed pool on the surface of a reef.

terracette [GEOL] A small steplike form developed on the surface of a slumped soil mass along a steep grassy incline.

terracing *See* contour plowing.

terra-cotta [MATER] A brownish-orange clay used in the production of high-quality earthenware, vases, and statuettes, and for tile floors and roofing.

terrain-avoidance radar [NAV] Airborne radar which provides a display of terrain ahead of a low-flying aircraft to permit horizontal avoidance of obstacles.

terrain clearance indicator *See* absolute altimeter.

terrain echoes *See* ground clutter.

terrain-following radar [NAV] Airborne radar which provides a display of terrain ahead of a low-flying aircraft to permit manual control, or signals for automatic control, to maintain constant altitude above the ground.

terrain profile recorder *See* airborne profile recorder.

terrain sensing [ENG] The gathering and recording of information about terrain surfaces without actual contact with the object or area being investigated; in particular, the use of photography, radar, and infrared sensing in airplanes and artificial satellites.

terra japonica *See* gambir.

terral levante [METEOROL] A land breeze of Spain and Brazil, sometimes a northwest squall of foehn character.

Terramycin [MICROBIO] Trade name for the antibiotic oxytetracycline.

terrapin [VERT ZOO] Any of several North American tortoises in the family Testudinidae, especially the diamondback terrapin.

terra rossa [GEOL] A reddish-brown soil overlying limestone bedrock.

terrazzo [MATER] A mosaic flooring surface made by embedding marble or granite chips in mortar, allowing the mortar to harden, and then grinding and polishing the surface.

terrestrial [SCI TECH] Of or pertaining to the earth.

terrestrial coordinates *See* geographical coordinates.

terrestrial electricity [GEOPHYS] Electric phenomena and properties of the earth; used in a broad sense to include atmospheric electricity. Also known as geoelectricity.

terrestrial environment [GEOGR] The earth's land area, including its man-made and natural surface and subsurface features, and its interfaces and interactions with the atmosphere and the oceans.

terrestrial equator *See* astronomical equator.

terrestrial frozen water [HYD] Seasonally or perennially frozen waters of the earth, exclusive of the atmosphere.

terrestrial gravitation [GEOPHYS] The effect of gravitational attraction of the earth.

TERMITE

Winged female termite of the reproductive caste.

TERRAPIN

Diamondback terrapin (*Malaclemys terrapin*).

TERRIER

USS *England*, a frigate of the Leahy class, fires a Terrier. *(Official U.S. Navy photograph)*

TERTIARY

PRECAMBRIAN		
CAMBRIAN	PALEOZOIC	
ORDOVICIAN		
SILURIAN		
DEVONIAN		
Mississippian	CARBON-IFEROUS	
Pennsylvanian		
PERMIAN		
TRIASSIC	MESOZOIC	
JURASSIC		
CRETACEOUS		
TERTIARY	CENOZOIC	
QUATERNARY		

Chart showing the position of the Tertiary period in relation to the other periods and the eras of geologic time.

TESLA COIL

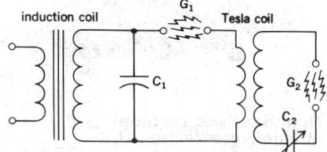

Circuit diagram of Tesla coil. G_1 and G_2 are spark gaps, C_1 is capacitor, C_2 is variable capacitor. The coils and C_1 act as a resonant circuit to produce high-frequency oscillation. Current in secondary (right-hand winding) of Tesla coil is both high-voltage and high-frequency.

terrestrial guidance *See* terrestrial-reference guidance.

terrestrial magnetism *See* geomagnetism.

terrestrial meridian *See* astronomical meridian.

terrestrial planet [ASTRON] One of the four small planets near the sun (Earth, Mercury, Venus, and Mars).

terrestrial radiation [GEOPHYS] Electromagnetic radiation originating from the earth and its atmosphere at wavelengths determined by their temperature. Also known as earth radiation; eradiation.

terrestrial-reference flight [NAV] A flight with guidance by means of some terrestrial phenomena such as the magnetic or gravitational field.

terrestrial-reference guidance [NAV] Long-range missile guidance in which the control system of the missile reacts to magnetic, gravitational, or other properties of the earth. Also known as terrestrial guidance.

terrestrial refraction [OPTICS] Any refraction phenomenon observed in the light originating from a source lying within the earth's atmosphere; this is applied only to refraction caused by inhomogeneities of the atmosphere itself, not, for example, to that caused by ice crystals suspended in the atmosphere.

terrestrial scintillation [OPTICS] A generic term for scintillation phenomena observed in light that reaches the eye from sources lying within the earth's atmosphere. Also known as atmospheric boil; atmospheric shimmer; optical haze.

terrestrial sediment [GEOL] A sedimentary deposit on land above tidal reach.

terrestrial triangle [NAV] A triangle on the surface of the earth, especially the navigational triangle.

Terrier [ORD] A U.S. Navy surface-to-air guided missile using beam-rider guidance and having a range of about 10 miles (16 kilometers).

terrigenous sediment [GEOL] Shallow marine sedimentary deposits composed of eroded terrestrial material.

territoriality [ZOO] A pattern of behavior in which one or more animals occupy and defend a definite area or territory.

tert- [ORG CHEM] $(R_1R_2R_3C-)$ Abbreviation for tertiary; trisubstituted methyl radical with the central carbon attached to three other carbons.

tertian malaria *See* vivax malaria.

Tertiary [GEOL] The older major subdivision (period) of the Cenozoic era, extending from the end of the Cretaceous to the beginning of the Quaternary, from 70,000,000 to 2,000,000 years ago.

tertiary alcohol [ORG CHEM] A trisubstituted alcohol in which the hydroxyl group is attached to a carbon that is joined to three carbons; for example, *tert*-butyl alcohol.

tertiary amine [ORG CHEM] R_3N A trisubstituted amine in which the hydroxyl group is attached to a carbon that is joined to three carbons; for example, trimethylamine, $(CH_3)_3N$.

tertiary circulation [METEOROL] The generally small, localized atmospheric circulations, represented by such phenomena as local winds, thunderstorms, and tornadoes.

tertiary creep [MET] Creep strain occurring at an accelerating rate leading to fracture.

tertiary crushing [MIN ENG] 1. The preliminary breaking down of run-of-mine ore and sometimes coal. 2. The third stage in crushing, following primary and secondary crushing.

tertiary grinding [MIN ENG] The third-stage grinding in a ball mill when a particularly fine grinding of ore is needed.

tertiary mercuric phosphate *See* mercuric phosphate.

tertiary pyroelectricity [SOLID STATE] The polarization due to temperature and gradients and corresponding nonuniform stresses and strains when the crystal is heated nonuniformly; found in pyroelectric and nonpyroelectric crystals, that is, crystals which have no polar directions. Also known as false pyroelectricity.

tertiary sodium phosphate *See* trisodium phosphate.

tertiary structure [BIOCHEM] The characteristic three-dimensional folding of the polypeptide chains in a protein molecule.

tertiary winding *See* stabilized winding.

Tesaar lens [OPTICS] An anastigmatic lens made up of a negative lens at the aperture stop with two positive lenses, one in front and the other in back; the last positive lens is a cemented doublet.

teschenite [PETR] A granular hypabyssal rock composed principally of calcic plagioclase, augite, and sometimes hornblende, with some brotite and analcime.

tesla [ELECTROMAG] The System International unit of magnetic flux density, equal to one weber per square meter. Abbreviated T.

Tesla coil [ELECTROMAG] An air-core transformer used with a spark gap and capacitor to produce a high voltage at a high frequency.

tessara [MATER] A small rectangular piece of ceramic tile, stone, or other material used in a mosaic design.

Tessaratomidae [INV ZOO] A family of large tropical Hemiptera in the superfamily Pentatomoidea.

tessellate [BOT] Marked by a pattern of small squares resembling a tiled pavement.

test [IND ENG] A procedure in which the performance of a product is measured under various conditions. [INV ZOO] A hard external covering or shell that is calcareous, siliceous, chitinous, fibrous, or membranous. [PETRO ENG] A procedure for the analysis of current, potential and ultimate product flow, and pressure-decline properties of various types of petroleum reservoirs.

testa [BOT] A seed coat. Also known as episperm.

Testacellidae [INV ZOO] A family of pulmonate gastropods that includes some species of slugs.

test ammunition [ORD] 1. In a general sense, any ammunition used, or intended to be used, for test purposes. 2. Specifically, ammunition prepared for testing firearms.

test-bed [AERO ENG] A base, mount, or frame within or upon which a piece of equipment, especially an engine, is secured for testing.

testboard [ELEC] Switchboard equipped with testing apparatus, arranged so that connections can be made from it to telephone lines or central-office equipment for testing purposes.

test chamber [AERO ENG] The test section of a wind tunnel. [ENG] A place, section, or room having special characteristics where a person or object is subjected to experimental procedures, as an altitude chamber.

test clip [ELEC] A spring clip used at the end of an insulated wire lead to make a temporary connection quickly for test purposes.

test data [ADP] A set of data developed specifically to test the adequacy of a computer run or system; the data may be actual data that has been taken from previous operations, or artificial data created for this purpose.

test firing [AERO ENG] The firing of a rocket engine, either live or static, with the purpose of making controlled observations of the engine or of an engine component.

test flight [AERO ENG] A flight to make controlled observations of the operation or performance of an aircraft or a rocket, or of a component of an aircraft or rocket.

test function [MATH] An infinitely differentiable function of several real variables used in studying solutions of partial differential equations.

test hole [MIN ENG] A drill hole or shallow excavation to assess an ore body or to obtain rock samples to determine their structural and physical characteristics.

testicular hormone [BIOCHEM] Any of various hormones secreted by the testes.

testing level [ELEC] Value of power used for reference represented by 0.001 watt working in 600 ohms.

testis [ANAT] One of a pair of male reproductive glands in vertebrates; after sexual maturity, the source of sperm and hormones.

test jack [ELEC] 1. Appearance of a circuit or circuit element in jacks for testing purposes. 2. In recent practice, a jack multipled with the switchboard operating jack.

test lead [ELEC] A flexible insulated lead, usually with a test prod at one end, used for making tests, connecting instruments to a circuit temporarily, or making other temporary connections.

test of hypothesis [STAT] A rule for rejecting or accepting a hypothesis concerning a population which is based upon a given sample of data.

test of significance [STAT] A test of a hypothetical population property against a sample property where an acceptance interval is used as the rule for rejection.

test oscillator *See* signal generator.

testosterone [BIOCHEM] $C_{19}H_{28}O_2$ The principal androgenic hormone released by the human testis; may be synthesized from cholesterol and certain other sterols.

test pattern [COMMUN] A chart having various combinations of lines, squares, circles, and graduated shading, transmitted from time to time by a television station to check definition, linearity, and contrast for the complete system from camera to receiver. Also known as resolution chart.

test pile [CIV ENG] A pile equipped with a platform on which a load of sand or pig iron is placed in order to determine the load a pile can support (usually twice the working load) without settling.

test pit [CIV ENG] An open excavation used to obtain soil samples in foundation studies.

test point [ELEC] A terminal or plug-in connector provided in a circuit to facilitate monitoring, calibration, or troubleshooting.

test prod [ELEC] A metal point attached to an insulating handle and connected to a test lead for convenience in making a temporary connection to a terminal while tests are being made. Also known as prod.

test program *See* check routine.

test reactor [NUCLEO] A nuclear reactor designed to test the behavior of materials and components under the neutron and gamma fluxes and temperature conditions of an operating reactor.

test routine *See* check routine.

test section [AERO ENG] The section of a wind tunnel where objects are tested to determine their aerodynamic characteristics.

test set [ELECTR] A combination of instruments needed for servicing a particular type of electronic equipment.

test specimen *See* tensile bar.

test stand [AERO ENG] A stationary platform or table, together with any testing apparatus attached thereto, for testing or proving engines and instruments.

Testudinata [VERT ZOO] The equivalent name for Chelonia.

Testudinellidae [INV ZOO] A family of free-swimming rotifers in the suborder Flosculariacea.

Testudinidae [VERT ZOO] A family of tortoises in the suborder Cryptodira; there are about 30 species found on all continents except Australia.

test well [PETRO ENG] A well to determine the presence of petroleum oil and its potential commercial value in terms of abundance and accessibility.

tetanolysin [BIOCHEM] A hemolysin produced by *Clostridium tetani*.

tetanophobia [PSYCH] An abnormal fear of tetanus.

tetanospasmin [BIOCHEM] A neurotoxin elaborated by the bacterium *Clostridium tetani* and which is responsible for the manifestations of tetanus.

tetanus [MED] An infectious disease of humans and animals caused by the toxin of *Clostridium tetani* and characterized by convulsive tonic contractions of voluntary muscles; infection commonly follows dirt contamination of deep wounds or other injured tissue. Also known as lockjaw.

tetanus antitoxin [IMMUNOL] A serum containing antibodies that neutralize tetanus toxin.

tetanus toxoid [IMMUNOL] Detoxified tetanus toxin used to produce active immunity against tetanus.

tetany [MED] A state of increased neuromuscular irritability caused by a decrease of serum calcium, manifested by intermittent numbness and cramps or twitchings of the extremities, laryngospasm, bizarre behavior, loss of consciousness, and convulsions.

tetany of the newborn [MED] A temporary increase of neuromuscular irritability during the first two months of life, especially in infants who are premature, are delivered by cesarean section, or receive an exchange transfusion, or in twins.

Tethinidae [INV ZOO] A family of myodarian cyclorrhaphous dipteran insects in the subsection Acalypteratae.

Tethys [ASTRON] A satellite of the planet Saturn having a diameter of about 1300 kilometers. [GEOL] **1.** A sea which existed for extensive periods of geologic time between the northern and southern continents of the Eastern Hemisphere. **2.** A composite geosyncline from which many structures of the present Alpine-Himalayan orogenic belt were formed.

Tetrabranchia [INV ZOO] A subclass of primitive mollusks of the class Cephalopoda; *Nautilus* is the only living form and is characterized by having four gills.

tetrabutyl titanate [ORG CHEM] $Ti(OC_4H_9)_4$ A combustible, colorless to yellowish liquid soluble in many solvents, boils at 312°C, decomposes in water; used in paints, surface coatings, and heat-resistant paints. Also known as butyl titanate; TBT.

tetracaine hydrochloride [ORG CHEM] $C_{15}H_{24}O_2N_2 \cdot HCl$ Bitter tasting, water-soluble crystals melting at 148°C; used as a local anesthetic. Also known as amethocaine hydrochloride; decicaine; pontocaine.

tetracene *See* naphthacene.

Tetracentraceae [BOT] A family of dicotyledonous trees in the order Trochodendrales distinguished by possession of a perianth, four stamens, palmately veined leaves, and secretory idioblasts.

tetrachlorobenzene [ORG CHEM] $C_6H_2Cl_4$ Water-insoluble, combustible white crystals that appear in two forms: 1,2,3,4-tetrachlorobenzene which melts at 47°C and is used in chemical synthesis and in dielectric fluids; and 1,2,4,5-tetrachlorobenzene which melts at 138°C and is used to make herbicides, defoliants, and electrical insulation.

***sym*-tetrachlorodifluoroethane** [ORG CHEM] CCl_2FCCl_2F A white, toxic liquid with a camphor aroma, soluble in alcohol, insoluble in water, boils at 93°C; used for metal degreasing.

***sym*-tetrachloroethane** [ORG CHEM] $CHCl_2CHCl_2$ A colorless, corrosive, toxic liquid with a chloroform scent, soluble in alcohol and ether, slightly soluble in water, boils at 147°C; used as a solvent, metal cleaner, paint remover, and weed killer. Also known as acetylene tetrachloride.

tetrachloroethylene *See* perchloroethylene.

tetrachloromethane *See* carbon tetrachloride.

tetrachlorophenol [ORG CHEM] C_6HCl_4OH Either of two toxic compounds: 2,3,4,6-tetrachlorophenol comprises brown flakes, soluble in common solvents, melting at 70°C, and is used as a fungicide; 2,4,5,6-tetrachlorophenol is a brown solid, insoluble in water, soluble in sodium hydroxide, has a phenol scent, melts at about 50°C, and is used as a fungicide and for wood preservatives.

tetrachloroquinone *See* chloranil.

Tetracorallia [PALEON] The equivalent name for Rugosa.

tetracosane [ORG CHEM] $C_{24}H_{50}$ Combustible crystals insoluble in water, soluble in alcohol, melts at 52°C; used as a chemical intermediate.

Tetractinomorpha [INV ZOO] A heterogeneous subclass of Porifera in the class Demospongiae.

tetracycline [MICROBIO] **1.** Any of a group of broad-spectrum antibiotics produced biosynthetically by fermentation with a strain of *Streptomyces aureofaciens* and certain other species or chemically by hydrogenolysis of chlortetracycline. **2.** $(C_{22}H_{24}O_8N_2)$ A broad-spectrum antibiotic belonging to the tetracycline group of antibiotics; useful because of broad antimicrobial action, with low toxicity, in the therapy of infections caused by gram-positive and gram-negative bacteria as well as rickettsiae and large viruses such as psittacosis-lymphogranuloma viruses.

tetrad [CYTOL] A group of four chromatids lying parallel to each other as a result of the longitudinal division of each of a pair of homologous chromosomes during the pachytene and later stages of the prophase of meiosis.

tetradactylous [VERT ZOO] Having four digits on a limb.

tetrad analysis [GEN] A method of genetic analysis possible in fungi, algae, bryophytes, and orchids in which the four products of an individual cell which has gone through meiosis are recovered as a group; it provides more direct and complete information regarding segregation and recombination mechanisms than is possible to obtain from meiotic products collected at random.

***n*-tetradecane** [ORG CHEM] $C_{14}H_{30}$ A combustible, colorless, water-insoluble liquid boiling at 254°C; used as a solvent and distillation chaser and in organic synthesis.

TETRACYCLINE

	R_1	R_2	R_3
Tetracycline	H	H	CH_3
Chlortetracycline	Cl	H	CH_3
Oxytetracycline	H	OH	CH_3
6-Demethyl tetracycline	H	H	H
7-Chloro-6-demethyl tetracycline	Cl	H	H

Chemical structural relationships of the tetracyclines (def. 1).

TETRAHEDRITE

Tetrahedrite crystals from Bobes, Romania. (*Specimen from Department of Geology, Bryn Mawr College*)

TETRAODONTIFORMES

Gray triggerfish (*Balistes capriscus*), a representative species of the order Tetraodontiformes. (*After G. B. Goode, Fishery Industries of the United States, sect. 1, 1884*)

TETRAPHIDALES

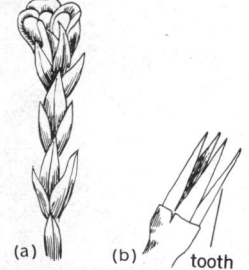

Tetraphis pellucida, a representative species of the order Tetraphidales.
(*a*) Gemmiferous branch.
(*b*) Enlarged peristome.
(*From W. H. Welch, Mosses of Indiana, Ind. Dep. Conserv., 1957*)

TETRASPORALES

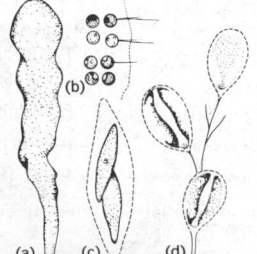

Members of Tetrasporales.
(*a*) *Tetraspora*, habit of a gelatinous thallus;
(*b*) arrangement of cells with pseudocilia. (*c*) *Elakatothrix*, simple colony. (*d*) *Chlorangium*, an attached, dendroid colony.

1-tetradecanol *See* myristyl alcohol.

1-tetradecene [ORG CHEM] $CH_2:CH(CH_2)_{11}CH_3$ A combustible, colorless, water-insoluble liquid boiling at 256°C; used as a solvent for perfumes and flavors and in medicine. Also known as α-tetradecylene.

α-tetradecylene *See* 1-tetradecene.

tetradic [MATH] An operator that transforms one dyadic into another.

tetrad of Fallot *See* tetralogy of Fallot.

tetradymite [MINERAL] Bi_2Te_2S A pale steel-gray mineral that usually occurs in foliated masses in auriferous veins; has metallic luster, hardness of 1.5–2 on Mohs scale, and specific gravity of 7.2–7.6.

tetraethylene glycol [ORG CHEM] $HO(C_2H_4O)_3C_2H_4OH$ A combustible, hygroscopic, colorless, water-soluble liquid, boils at 327°C; used as a nitrocellulose solvent and plasticizer and in lacquers and coatings. Abbreviated TEG.

tetraethyllead [ORG CHEM] $Pb(C_2H_5)_4$ A highly toxic lead compound that, when added in small proportions to gasoline, increases the fuel's antiknock quality. Also known as TEL.

tetraethylpyrophosphate [ORG CHEM] $C_8H_{20}O_7P_2$ A hygroscopic corrosive liquid miscible with although decomposed by water, and miscible with many organic solvents; inhibits the enzyme acetylcholinesterase; used as an insecticide in place of nicotine sulfate.

tetrafluoroethylene [ORG CHEM] $F_2C:CF_2$ A flammable, colorless, heavy gas, insoluble in water, boils at 78°C; used as a monomer to make polytetrafluoroethylene polymers, for example, Teflon. Also known as perfluoroethylene; TFE.

tetrafluoromethane *See* carbon tetrafluoride.

tetragonal lattice [CRYSTAL] A crystal lattice in which the axes of a unit cell are perpendicular, and two of them are equal in length to each other, but not to the third axis.

tetragonal trisoctahedron *See* trapezohedron.

tetragonal tristetrahedron *See* deltohedron.

tetrahedral symmetry [PHYS] Having the same rotation symmetries as a regular tetrahedron.

tetrahedrite [MINERAL] $(Cu,Fe,Zn,Ag)_{12}Sb_4S_{13}$ A grayish-black mineral crystallizing in the isometric system as tetrahedrons and occurring in massive or granular form; luster is metallic, hardness is 3.5–4 on Mohs scale, and specific gravity is 4.6–5.1; an important ore of copper. Also known as fahlore; gray copper ore; panabase; stylotypite.

tetrahedron [CRYSTAL] An isometric crystal form in cubic crystals, in the shape of a four-faced polyhedron, each face of which is a triangle. [MATH] A four-sided polyhedron.

tetrahexahedron [CRYSTAL] A form of regular crystal system with four triangular isosceles faces on each side of a cube; there are altogether 24 congruent faces.

tetrahydrofuran *See* butylene oxide.

tetrahydrofurfuryl alcohol [ORG CHEM] $C_4H_7OCH_2OH$ A hygroscopic, colorless liquid, miscible with water, boils at 178°C; used as a solvent for resins, in leather dyes, and in nylon. Also known as tetrahydrofurfuryl carbinol.

tetrahydrofurfuryl carbinol *See* tetrahydrofurfuryl alcohol.

tetrahydro-*para*-methoxyquinoline *See* thalline.

tetrahydronaphthalene [ORG CHEM] $C_{10}H_{12}$ A colorless, oily liquid that boils at 206°C, and is miscible with organic solvents; used as an intermediate in chemical synthesis and as a solvent.

tetrahydro-1,4-oxazine *See* morpholine.

tetrahydro-*para*-quinanisol *See* thalline.

4,6,3′,4′-tetrahydroxyaurone *See* aureusidin.

tetrahydroxy butane *See* erythritol.

3′,4′,5,7-tetrahydroxyflavanone *See* eriodictyol.

Tetrahymenina [INV ZOO] A suborder of ciliated protozoans in the order Hymenostomatida.

tetraiodofluorescein [ORG CHEM] $C_{20}H_8O_5I_4$ A yellow, water-insoluble, crystalline compound; used as a dye. Also known as pyrosin.

tetralite *See* tetryl.

tetralogy of Fallot [MED] A congenital abnormality of the heart consisting of pulmonary stenosis, defect of the interventricular septum, hypertrophy of the right ventricle, and overriding or dextroposition of the aorta. Also known as tetrad of Fallot.

Tetralophodontinae [PALEON] An extinct subfamily of proboscidean mammals in the family Gomphotheriidae.

tetramer [ORG CHEM] A polymer that results from the union of four identical monomers; for example, the tetramer C_8H_8 forms from union of four molecules of C_2H_2.

tetramerous [BIOL] Characterized by or having four parts.

tetramethyldiene *See* cyclobutadiene.

tetramethylene *See* cyclobutane.

Δ^{1,3}-tetramethylene *See* cyclobutadiene.

tetramethyllead [ORG CHEM] $Pb(CH_3)_4$ An organic compound of lead that, when added in small amounts to motor gasoline, increases the antiknock quality of the fuel; not widely used.

tetramethylolmethane *See* pentaerythritol.

tetramethylurea [ORG CHEM] $C_5H_{12}N_2O$ A liquid that boils at 176.5°C, and is miscible in water and organic solvents; used as a reagent and solvent.

tetranitromethane [ORG CHEM] $C(NO_2)_4$ A powerful oxidant; toxic, colorless liquid with a pungent aroma, insoluble in water, soluble in alcohol and ether, boils at 126°C; used in rocket fuels and as an analytical reagent.

Tetranychidae [INV ZOO] The spider mites, a family of phytophagous trombidiform mites.

Tetraodontiformes [VERT ZOO] An order of specialized teleost fishes that includes the trigger fishes, puffers, trunkfishes, and ocean sunfishes.

Tetraonidae [VERT ZOO] The ptarmigans and grouse, a family of upland game birds in the order Galliformes characterized by rounded tails and wings and feathered nostrils.

Tetraphidaceae [BOT] The single family of the plant order Tetraphidales.

Tetraphidales [BOT] A monofamilial order of mosses distinguished by scalelike protonema and the peristomes of four rigid, nonsegmented teeth.

tetraphosphorus trisulfide *See* phosphorus sesquisulfide.

Tetraphyllidea [INV ZOO] An order of small tapeworms of the subclass Cestoda characterized by the variation in the structure of the scolex; all species are intestinal parasites of elasmobranch fishes.

tetraploidy [CYTOL] The occurrence of related forms possessing in the somatic cells chromosome numbers four times the haploid number.

tetrapod [VERT ZOO] A four-footed animal.

Tetrapoda [VERT ZOO] The superclass of the subphylum Vertebrata whose members typically possess four limbs; includes all forms above fishes.

tetrapotassium pyrophosphate *See* potassium pyrophosphate.

tetrapropylene *See* dodecane.

Tetrarhynchoidea [INV ZOO] The equivalent name for Trypanorhyncha.

tetrasodium pyrophosphate *See* sodium pyrophosphate.

Tetrasporales [BOT] A heterogeneous and artificial assemblage of colonial fresh-water and marine algae in the division Chlorophyta.

tetraspore [BOT] One of the haploid asexual spores of the red algae formed in groups of four.

tetraterpene [ORG CHEM] A class of terpene compounds that contain isoprene units; best known are the carotenoid pigments from plants and animals, such as lycopene, the red coloring matter in tomatoes.

tetratohedral crystal [CRYSTAL] A crystal which has one quarter of the maximum number of faces allowed by the crystal system to which the crystal belongs.

tetraxon [INV ZOO] A type of sponge spicule with four axes.

tetrazene [ORG CHEM] $H_2NC(NH)_3N_2C(NH)_3NO$ An explosive, colorless to yellowish solid practically insoluble in water and alcohol; used as an explosive initiator and in detonators.

Tetrigidae [INV ZOO] The grouse locusts or pygmy grasshoppers in the order Orthoptera in which the front wings are reduced to small scalelike structures.

tetrode [ELECTR] A four-electrode electron tube containing an anode, a cathode, a control electrode, and one additional electrode that is ordinarily a grid.

tetrode junction transistor *See* double-base junction transistor.

tetrode thyratron [ELECTR] A thyratron with two control electrodes. Also known as gas tetrode.

tetrode transistor [ELECTR] A four-electrode transistor, such as a tetrode point-contact transistor or double-base junction transistor.

tetrodotoxin [BIOCHEM] $C_{11}H_{17}N_3O_8$ A toxin that blocks the action potential in the nerve impulse.

tetrol *See* furan.

tetrose [BIOCHEM] Any of a group of monosaccharides that have a four-carbon chain; an example is erythrose, $CH_2OH \cdot (CHOH)_2 \cdot CHO$.

tetryl [ORG CHEM] $(NO_2)_3C_6H_2N(NO_2)CH_3$ A yellow, water-insoluble, crystalline explosive material melting at 130°C; used in explosives and ammunition. Also known as *N*-methyl-*N*,2,4,6-tetranitroaniline; methylpicrylnitramine; tetralite.

tetrytol [MATER] A high-explosive mixture of tetryl and trinitrotoluene (TNT) in any of several proportions which permit melt loading.

Tettigoniidae [INV ZOO] A family of insects in the order Orthoptera which have long antennae, hindlegs fitted for jumping, and an elongate, vertically flattened ovipositor; consists of the longhorn or green grasshopper.

Teuthidae [VERT ZOO] The rabbitfishes, a family of perciform fishes in the suborder Acanthuroidei.

Teuthoidea [INV ZOO] An order of the molluscan subclass Coleoidea in which the rostrum is not developed, the proostracum is represented by the elongated pen or gladus, and ten arms are present.

TE wave *See* transverse electric wave.

Tex [TEXT] A unit of fiber fineness assessed by the weight in grams of 1000 meters of yarn; the lower the number the finer the yarn.

TEX *See* Telex.

Texas fever [VET MED] A tick-borne infectious disease of cattle caused by the parasite *Babesia annulatus* which invades erythrocytes; characterized by fever, hemoglobinuria, and splenomegaly.

Texas tower [ENG] A radar tower built in the sea offshore, to serve as part of an early-warning radar network; it resembles offshore oil derricks in the Gulf of Mexico.

textile [MATER] A material made of natural or man-made fibers and used for the manufacture of items such as clothing and furniture fittings.

textile microbiology [MICROBIO] That branch of microbiology concerned with textile materials; deals with microorganisms that are harmful either to the fibers or to the consumer, and microorganisms that are useful, as in the retting process.

textile oil [MATER] A specially compounded oil used to condition raw textile fibers, yarn, and fabric for manufacturing and finishing operations.

textile printing [TEXT] The specialized dyeing of restricted areas on fabrics; involves the preparation of printing paste, the development and fixing of color, and the addition of a thickening agent to confine paste applications to desired areas on the goods.

textile softener [MATER] A chemical that attaches molecularly to textile fibers with the polar (charged) end of the cation oriented toward the fiber and the fatty tail exposed to give a feeling of softness to the fabric.

Textulariina [INV ZOO] A suborder of foraminiferan protozoans characterized by an agglutinated wall.

texture [GEOL] The physical nature of the soil according to composition and particle size. [PETR] The physical appearance or character of a rock; applied to the megascopic or microscopic surface features of a homogeneous rock or mineral aggregate, such as grain size, shape, and arrangement.

TFE *See* tetrafluoroethylene.

th *See* thermie.

Th *See* thorium.

thalamus [ANAT] Either one of two masses of gray matter located on the sides of the third ventricle and forming part of the lateral wall of that cavity.

thalassemia [MED] A hereditary form of hemolytic anemia resulting from a defective synthesis of hemoglobin: thalassemia major is the homozygous state accompanied by clinical illness, and thalassemia minor is the heterozygous state and may not have evident clinical manifestations. Also known as Mediterranean anemia.

thalassic [OCEANOGR] Of or pertaining to the smaller seas.

Thalassinidea [INV ZOO] The mud shrimps, a group of thin-shelled, burrowing decapod crustaceans belonging to the Macrura; individuals have large chelate or subchelate first pereiopods, and no chelae on the third pereiopods.

thalassocratic [GEOL] **1.** Pertaining to a thalassocraton. **2.** Referring to a period of high sea level in the geologic past.

thalassocraton [GEOL] A craton that is part of the oceanic crust.

thalassophile element [GEOCHEM] An element that is relatively more abundant in sea water than in normal continental waters, such as sodium and chlorine.

thalassophobia [PSYCH] An abnormal fear of the sea.

Thalattosauria [PALEON] A suborder of extinct reptiles in the order Eosuchia from the Middle Triassic.

Thaliacea [INV ZOO] A small class of pelagic Tunicata in which oral and atrial apertures occur at opposite ends of the body.

thalidomide [PHARM] $C_{13}H_{10}N_2O_4$ A drug used as a sedative and hypnotic; may produce teratogenic effects when administered during pregnancy.

thalline [ORG CHEM] $C_9H_6N(OCH_3)H_4$ Colorless rhomboids soluble in water and melting at 40°C. Also known as tetrahydro-*para*-methoxyquinoline; tetrahydro-*para*-quinanisol.

thallium [CHEM] A metallic element in group III, symbol Tl, atomic number 81, atomic weight 204.37; insoluble in water, soluble in nitric and sulfuric acids, melts at 302°C, boils at 1457°C. [MET] Bluish-white metal with tinlike malleability, but a little softer; used in alloys.

thallium acetate [ORG CHEM] $TlOCOCH_3$ Toxic, white, deliquescent crystals, soluble in water and alcohol, melts at 131°C; used as an ore-flotation solvent and in medicine.

thallium-beam clock [HOROL] A device similar to a cesium-beam clock, but using a beam of thallium atoms instead of cesium; advantages over cesium are reduced sensitivity to magnetic fields and a higher frequency; the beam, however, is much harder to detect and deflect.

thallium bromide [INORG CHEM] $TlBr$ A toxic, yellowish powder soluble in alcohol, slightly soluble in water, melts at 460°C; used in infrared radiation transmitters and detectors. Also known as thallous bromide.

thallium carbonate [INORG CHEM] Tl_2CO_3 Toxic, shiny, colorless needles soluble in water, insoluble in alcohol, melts at 272°C; used as an analytical reagent and in artificial gems. Also known as thallous carbonate.

thallium chloride [INORG CHEM] $TlCl$ A white, toxic, light-sensitive powder, slightly soluble in water, insoluble in alcohol, melts at 430°C; used as a chlorination catalyst and in medicine and suntan lamps. Also known as thallous chloride.

thallium hydroxide [INORG CHEM] $TlOH \cdot H_2O$ Toxic yellow, water- and alcohol-soluble needles, decomposes at 139°C; used as an analytical reagent. Also known as thallous hydroxide.

thallium iodide [INORG CHEM] TlI A toxic, yellow powder, insoluble in alcohol, slightly soluble in water, melts at 440°C; used in infrared radiation transmitters and in medicine. Also known as thallous iodide.

thallium monoxide [INORG CHEM] Tl_2O A black, toxic, water- and alcohol-soluble powder, melts at 300°C; used as an analytical reagent and in artificial gems and optical glass. Also known as thallium oxide; thallous oxide.

thallium nitrate [INORG CHEM] $TlNO_3$ Colorless, toxic, fire-hazardous crystals soluble in hot water, insoluble in alcohol, melts at 206°C, decomposes at 450°C; used as an analytical reagent and in pyrotechnics. Also known as thallous nitrate.

thallium oxide *See* thallium monoxide.

thallium sulfate [INORG CHEM] Tl_2SO_4 Toxic, water-soluble, colorless crystals melting at 632°C; used as an analytical reagent and in medicine, rodenticides, and pesticides. Also known as thallous sulfate.

thallium sulfide [INORG CHEM] Tl_2S Lustrous, toxic, blue-

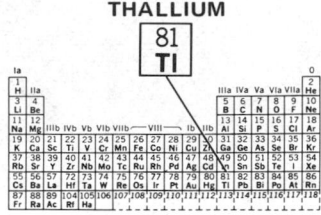

THALLIUM

Periodic table of the chemical elements showing the position of thallium.

black crystals insoluble in water, alcohol, and ether, soluble in mineral acids, melts at 448°C; used in infrared-sensitive devices. Also known as thallous sulfide.

Thallobionta [BOT] One of the two subkingdoms of plants, characterized by the absence of specialized tissues or organs and multicellular sex organs.

thallofide cell [ELECTR] A photoconductive cell in which the active light-sensitive material is thallium oxysulfide in a vacuum; it has maximum response at the red end of the visible spectrum and in the near infrared.

Thallophyta [BOT] The equivalent name for Thallobionta.

thallospore [BOT] A spore that develops by budding of hyphal cells.

thallotoxicosis [MED] Poisoning due to ingestion of thallium or its derivatives.

thallous bromide *See* thallium bromide.

thallous carbonate *See* thallium carbonate.

thallous chloride *See* thallium chloride.

thallous hydroxide *See* thallium hydroxide.

thallous iodide *See* thallium iodide.

thallous nitrate *See* thallium nitrate.

thallous oxide *See* thallium monoxide.

thallous sulfate *See* thallium sulfate.

thallous sulfide *See* thallium sulfide.

thallus [BOT] A plant body that is not differentiated into special tissue systems or organs and may vary from a single cell to a complex, branching multicellular structure.

thalweg [GEOGR] The middle of the principal navigable waterway which serves as a boundary between two states. [GEOL] **1.** A line connecting the lowest points along a stream bed or a valley. Also known as valley line. **2.** A line crossing all contour lines on a land surface perpendicularly. [HYD] Water seeping through the ground below the surface in the same direction as a surface stream course.

thanatology [MED] The study of the phenomenon of somatic death.

Thanetian [GEOL] A European stage of geologic time; uppermost Paleocene, above Montian, below Ypresian of Eocene.

Thaumaleidae [INV ZOO] A family of orthorrhaphous dipteran insects in the series Nematocera.

Thaumastellidae [INV ZOO] A monospecific family of the Hemiptera assigned to the Pentatomorpha found only in Ethiopia.

Thaumastocoridae [INV ZOO] The single family of the hemipteran superfamily Thaumastocoroidea.

Thaumastocoroidea [INV ZOO] A monofamilial superfamily of the Hemiptera in the subdivision Geocorisae which occurs in Australia and the New World tropics.

Thaumatoxenidae [INV ZOO] A family of cyclorrhaphous dipteran insects in the series Aschiza.

thaw [CLIMATOL] A warm spell during which ice and snow melt, as a January thaw.

thaw house [ENG] A small building that is designed for thawing frozen dynamite and which is capacious enough for a supply of thawed dynamite for a day's work.

thawing [ENG] Warming dynamite, to reduce risk of premature explosion. [MIN ENG] Working permanently frozen ground by pumping water at a temperature of from 50 to 60°F (10 to 15.5°C) through pipes down into the frozen gravel.

thawing index [CLIMATOL] The number of degree days, above and below 32°F, between the lowest and highest points on the cumulative degree-days time curve for one thawing season.

thawing season [CLIMATOL] The period of time between the lowest point and the succeeding highest point on the time curve of cumulative degree days above and below 32°F.

thaw pipe [MIN ENG] A string of pipe drilled into a string of drill rods that is frozen in a borehole in permafrost, through which water is circulated to thaw the ice and free the drill rods.

Theaceae [BOT] A family of dicotyledonous erect trees or shrubs in the order Theales characterized by alternate, exstipulate leaves, usually five petals, and mostly numerous stamens.

Theales [BOT] An order of dicotyledonous mostly woody plants in the subclass Dilleniidae with simple or occasionally compound leaves, petals usually separate, numerous stamens, and the calyx arranged in a tight spiral.

theater of operations [ORD] Portion of a theater of war necessary for military operations, either offensive or defensive, pursuant to an assigned mission, and for the administration incident to such military operation.

theater of war [ORD] That area of land, sea, and air which is, or may become, involved directly in the operation of war.

theater stock level [ORD] Quantity of supplies authorized by the U.S. Army to be maintained in a theater of operations as stock on hand ready for issue.

theater television [ELECTR] A large projection-type television receiver used in theaters, generally for closed-circuit showing of important sport events.

theca [ANAT] The sheath of dura mater covering the spinal cord. [BOT] **1.** A moss capsule. **2.** A pollen sac. [HISTOL] The layer of stroma surrounding a Graafian follicle. [INV ZOO] The test of a testate protozoan or a rotifer.

theca folliculi [HISTOL] The capsule surrounding a developing or mature Graafian follicle; consists of two layers, theca interna and theca externa.

Thecanephria [INV ZOO] An order of the phylum Brachiata containing a group of elongate, tube-dwelling tentaculate, deep-sea animals of bizarre structure.

thecate [BIOL] Having a theca.

Thecideidina [PALEON] An extinct suborder of articulate brachiopods doubtfully included in the order Terebratulida.

Thecodontia [PALEON] An order of archosaurian reptiles, confined to the Triassic and distinguished by the absence of a supratemporal bone, parietal foramen, and palatal teeth, and by the presence of an antorbital fenestra.

Thelastomidae [INV ZOO] A family of nematode worms in the superfamily Oxyuroidea.

thematic apperception test [PSYCH] A projective psychological test using a set of pictures suggesting life situations from which the subject constructs a story; designed to reveal to the trained interpreter some of the dominant drives, emotions, sentiments, complexes, and conflicts of personality. Abbreviated TAT.

thenar [ANAT] The ball of the thumb.

thenyl [ORG CHEM] $C_4H_3SCH_2-$ An organic radical based on methylthiophene; thus thenyl alcohol is also known as thiophenemethanol.

theobroma oil *See* cocoa butter.

theobromine [ORG CHEM] $C_7H_8N_4O_2$ A toxic alkaloid found in cocoa, chocolate products, tea, and cola nuts; closely related to caffeine. Also known as 3,7-dimethylxanthine.

theodolite [OPTICS] An optical instrument used in surveying which consists of a sighting telescope mounted so that it is free to rotate around horizontal and vertical axes, and graduated scales so that the angles of rotation may be measured; the telescope is usually fitted with a right-angle prism so that the observer continues to look horizontally into the eyepiece, whatever the variation of the elevation angle; in meteorology, it is used principally to observe the motion of a pilot balloon.

theophobia [PSYCH] Abnormal fear of a deity or of divine punishment.

Theophrastaceae [BOT] A family of tropical and subtropical dicotyledonous woody plants in the order Primulales characterized by flowers having staminodes alternate with the corolla lobes.

theophylline [ORG CHEM] $C_7H_8N_4O_2 \cdot H_2O$ Alkaloid from tea leaves; bitter-tasting white crystals slightly soluble in water and alcohol, melts at 272°C; used in medicine.

theorem [MATH] A proven mathematical statement.

theorem of corresponding states [STAT MECH] A theorem stating that two substances which have the same reduced temperature and the same reduced pressure have the same reduced volume.

theoretical cutoff frequency [ELEC] Of an electric structure, a frequency at which, disregarding the effects of dissipation, the attenuation constant changes from zero to a positive value or vice versa.

theoretical physics [PHYS] The description of natural phenomena in mathematical form.

theoretical plate [CHEM ENG] A distillation column plate or tray that produces perfect distillation (that is, produces the same difference in composition as that existing between a liquid mixture and the vapor in equilibrium with it); the packed-column equivalent of a theoretical plate is the HETP, or height (of packing) equivalent to a theoretical plate.

theory [MATH] The collection of theorems and principles associated with some mathematical object or concept. [SCI TECH] An attempt to explain a certain class of phenomena by deducing them as necessary consequences of other phenomena regarded as more primitive and less in need of explanation.

theory of antecedent conflicts [PSYCH] A theory that the effects of an emotionally disturbing event in adult life may be greatly multiplied through the conditioning or sensitizing effects of the vicissitudes of early life.

theory of equations [MATH] The study of polynomial equations from the viewpoint of solution methods, relations among roots, and connections between coefficients and roots.

theory of games See game theory.

theralite [PETR] A dark-colored, visibly crystalline rock composed chiefly of pyroxene with smaller amounts of calcic plagioclase and nepheline.

therapeutic abortion [MED] Abortion performed when pregnancy jeopardizes the health or life of the mother.

Therapsida [PALEON] An order of mammallike reptiles of the subclass Synapsida which first appeared in mid-Permian times and persisted until the end of the Triassic.

Therberg system [IND ENG] A system of categorizing hand movements that is used in the standard motion-and-time analysis technique.

therblig See elemental motion.

therblig chart [IND ENG] An operation chart with the suboperations divided into basic motions, all designated with appropriate symbols.

Therevidae [INV ZOO] The stiletto flies, a family of orthorrhaphous dipteran insects in the series Brachycera.

Theria [VERT ZOO] A subclass of the class Mammalia including all living mammals except the monotremes.

Theridiidae [INV ZOO] The comb-footed spiders, a family of the suborder Dipneumonomorphae.

therm [THERMO] A unit of heat energy, equal to 10^5 international table British thermal units, or approximately 1.055×10^8 joules.

thermal [METEOROL] A relatively small-scale, rising current of air produced when the atmosphere is heated enough locally by the earth's surface to produce absolute instability in its lower layers. [THERMO] Of or concerning heat.

thermal agitation [SOLID STATE] Random movements of the free electrons in a conductor, producing noise signals that may become noticeable when they occur at the input of a high-gain amplifier. Also known as thermal effect.

thermal ammeter See hot-wire ammeter.

thermal analysis [ANALY CHEM] Any analysis of physical or thermodynamic properties of materials in which heat (or its removal) is directly involved; for example, boiling, freezing, solidification-point determinations, heat of fusion and heat of vaporization measurements, distillation, calorimetry, and differential thermal, thermogravimetric, thermometric, and thermometric titration analyses. Also known as thermoanalysis. [MET] Determining transformations in a metal by observing the temperature-time relationship during uniform cooling or heating; phase transformations are indicated by irregularities in a smooth curve.

thermal-arrest calorimeter [ENG] A vacuum device for measurement of heats of fusion; a sample is frozen under vacuum and allowed to melt as the calorimeter warms to room temperature.

thermal aureole See aureole.

thermal barrier [AERO ENG] A limit to the speed of airplanes and rockets in the atmosphere imposed by heat from friction between the aircraft and the air, which weakens and eventually melts the surface of the aircraft. Also known as heat barrier.

thermal battery [ELEC] 1. A combination of thermal cells. Also known as fused-electrolyte battery; heat-activated battery. 2. A voltage source consisting of a number of bimetallic junctions connected to produce a voltage when heated by a flame.

thermal belt [ECOL] Any one of several possible horizontal belts of a vegetation type found in mountainous terrain, resulting primarily from vertical temperature variation. Also known as thermal zone.

thermal black [CHEM] A type of carbon black made by a thermal process using natural gas; used in the rubber industry.

thermal bond [NUCLEO] A thermally conductive bond, providing maximum transfer of heat, as between nuclear-reactor fuel and its protective cladding.

thermal breeder reactor [NUCLEO] A breeder reactor in which the fission chain reaction is sustained by thermal neutrons.

thermal bulb [ENG] A device for measurement of temperature; the liquid in a bulb expands with increasing temperature, pressuring a spiral Bourdon-type tube element and causing it to deform (unwind) in direct relation to the temperature in the bulb.

thermal burn [MED] Tissue reaction to or injury resulting from application of heat.

thermal capacitance [THERMO] The ratio of the entropy added to a body to the resulting rise in temperature.

thermal capacity See heat capacity.

thermal cell [ELEC] A reserve cell that is activated by applying heat to melt a solidified electrolyte.

thermal charge See entropy.

thermal climate [CLIMATOL] Climate as defined by temperature, and divided regionally into temperature zones.

thermal column [NUCLEO] A column of moderating material extending through the shield into the reflector of a nuclear reactor, used to provide a source of thermal neutrons for research purposes.

thermal compressor [MECH ENG] A steam-jet ejector designed to compress steam at pressures above atmospheric.

thermal conductance [THERMO] The amount of heat transmitted by a material divided by the difference in temperature of the surfaces of the material. Also known as conductance.

thermal conductimetry [THERMO] Measurement of thermal conductivities.

thermal conductivity [THERMO] The heat flow across a surface per unit area per unit time, divided by the negative of the rate of change of temperature with distance in a direction perpendicular to the surface. Also known as heat conductivity.

thermal conductivity cell See katharometer.

thermal conductivity gage [ENG] A pressure measurement device for high-vacuum systems; an electrically heated wire is exposed to the gas under pressure, the thermal conductivity of which changes with changes in the system pressure.

thermal convection [METEOROL] Atmospheric currents, predominantly vertical, arising from the release of gravitational visibility; commonly produced by solar heating of the ground; the cause of convective (cumulus) clouds. Also known as free convection; gravitational convection. [THERMO] See heat convection.

thermal converter [ELECTR] A device that converts heat energy directly into electric energy by using the Seebeck effect; it is composed of at least two dissimilar materials, one junction of which is in contact with a heat source and the other junction of which is in contact with a heat sink. Also known as thermocouple converter; thermoelectric generator; thermoelectric power generator; thermoelement.

thermal coulomb [THERMO] A unit of entropy equal to 1 joule per kelvin.

thermal cracking [CHEM ENG] A petroleum refining process that decomposes, rearranges, or combines hydrocarbon molecules by the application of heat, without the aid of catalysts.

thermal cross section [NUCLEO] The cross section of a thermal neutron.

thermal cutout [ELEC] A heat-sensitive switch that automatically opens the circuit of an electric motor or other device when the operating temperature exceeds a safe value.

thermal cutting [MET] A group of processes to sever metals by melting or by chemical reaction of oxygen with the metal at elevated temperatures.

thermal degradation [CHEM] Molecular deterioration of materials (usually organics) because of overheat; can be avoided

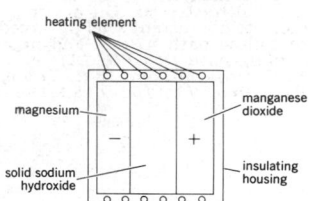

THERMAL CELL

Schematic of thermal cell. Cell is activated when heat from heating element melts sodium hydroxide.

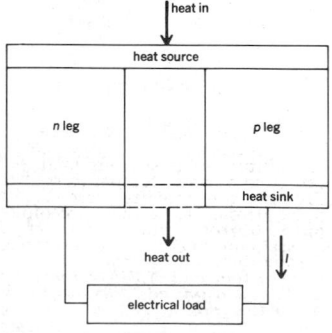

THERMAL CONVERTER

Simple thermal converter. One leg is n-type semiconductor material, the other is p-type material.

by low-temperature or vacuum processing, as for foods and pharmaceuticals.

thermal detector   *See* bolometer.

thermal diffusion   [PHYS CHEM] A phenomenon in which a temperature gradient in a mixture of fluids gives rise to a flow of one constituent relative to the mixture as a whole. Also known as thermodiffusion.

thermal diffusivity   *See* diffusivity.

thermal effect   *See* thermal agitation.

thermal efficiency   [CHEM ENG] In a tube-and-shell heat-exchange system, the ratio of the actual temperature range of the tube-side fluid (inlet versus outlet temperature) to the maximum possible temperature range.   [THERMO] *See* efficiency.

thermal-efficiency index   *See* temperature-efficiency index.

thermal-efficiency ratio   *See* temperature-efficiency ratio.

thermal electromotive force   [PHYS] An electromotive force arising from a difference in temperature at two points along a circuit, as in the Seebeck effect.

thermal emissivity   *See* emissivity.

thermal energy   [NUCLEO] Energy which is characteristic for thermal neutrons at room temperature, about 0.025 electron volt.

thermal equator   *See* heat equator.

thermal equilibrium   [THERMO] Property of a system all parts of which have attained a uniform temperature which is the same as that of the system's surroundings.

thermal excitation   [ATOM PHYS] The process in which atoms or molecules acquire internal energy in collisions with other particles.

thermal expansion   [PHYS] The dimensional changes exhibited by solids, liquids, and gases for changes in temperature while pressure is held constant.

thermal expansion coefficient   [PHYS] The fractional change in length or volume of a material for a unit change in temperature.

thermal exposure   [ORD] The total normal component of thermal radiation striking a given surface throughout the course of a detonation.

thermal farad   [THERMO] A unit of thermal capacitance equal to the thermal capacitance of a body for which an increase in entropy of 1 joule per kelvin results in a temperature rise of 1 kelvin.

thermal flasher   [ELEC] An electric device that opens and closes a circuit automatically at regular intervals because of alternate heating and cooling of a bimetallic strip that is heated by a resistance element in series with the circuit being controlled.

thermal flux   *See* heat flux.

thermal gasoline   [MATER] Gasoline produced in petroleum refineries by thermal processes, for example, thermal cracking and thermal reforming.

thermal gradient   [GEOPHYS] The rate of temperature increase with depth below the surface of the earth.

thermal henry   [THERMO] A unit of thermal inductance equal to the product of a temperature difference of 1 kelvin and a time of 1 second divided by a rate of flow of entropy of 1 watt per kelvin.

thermal high   [METEOROL] A high resulting from the cooling of air by a cold underlying surface, and remaining relatively stationary over the cold surface.

thermal horsepower   [ELEC] Electrical motor horsepower as determined by current readings from a thermal-type ammeter; will be higher than load horsepower determined from kilowatt-input methods. Also known as true motor load.

thermal hysteresis   [THERMO] A phenomenon sometimes observed in the behavior of a temperature-dependent property of a body; it is said to occur if the behavior of such a property is different when the body is heated through a given temperature range from when it is cooled through the same temperature range.

thermal imagery   [ELECTR] Imagery produced by measuring and recording electronically the thermal radiation of objects.

thermal inductance   [THERMO] The product of temperature difference and time divided by entropy flow.

thermal instability   [FL MECH] The instability resulting in free convection in a fluid heated at a boundary.

thermal ionization   *See* Saha ionization.

thermal jet   [METEOROL] A region in the atmosphere where isotherms or thickness lines are closely packed; therefore, a region of very strong thermal wind.

thermal-liquid system   [CHEM ENG] A system with a special liquid that acts as a heat sink or heat source (for example, steam, hot water, mercury, Dowtherm, molten salts, or mineral oils); used for process heating and cooling.

thermal low   [METEOROL] An area of low atmospheric pressure due to high temperatures caused by intensive heating at the earth's surface; common to the continental subtropics in summer, thermal lows remain stationary over the area that produces them, their cyclonic circulation is generally weak and diffuse, and they are nonfrontal. Also known as heat low.

thermal magnon   [SOLID STATE] A magnon with a relatively short wavelength, on the order of 10^{-6} centimeter.

thermal metamorphism   [PETR] Metamorphism that results from temperature-controlled and induced chemical reconstitution of preexisting rocks, with little influence of pressure. Also known as thermometamorphism.

thermal microphone   [ENG ACOUS] Microphone depending for its action on the variation in the resistance of an electrically heated conductor that is being alternately increased and decreased in temperature by sound waves.

thermal neutron   [NUCLEO] One of a collection of neutrons whose energy distribution is identical with or similar to the Maxwellian distribution in the material in which they are found; the average kinetic energy of such neutrons at room temperature is about 0.025 electron volt. Also known as slow neutron.

thermal noise   [ELECTR] Electric noise produced by thermal agitation of electrons in conductors and semiconductors. Also known as Johnson noise; resistance noise.

thermal noise generator   [ELECTR] A generator that uses the inherent thermal agitation of an electron tube to provide a calibrated noise source.

thermal ohm   [THERMO] A unit of thermal resistance equal to the thermal resistance for which a temperature difference of 1 kelvin produces a flow of entropy of 1 watt per kelvin. Also known as fourier.

thermal photography   *See* thermography.

thermal polymerization   [CHEM ENG] A thermal, petroleum refining process used to convert light hydrocarbon gases into liquid fuels; paraffinic hydrocarbons are cracked to produce olefinic material which is concurrently polymerized by heat and pressure to form liquids, the product being known as polymer gasoline.

thermal potential difference   [THERMO] The difference between the thermodynamic temperatures of two points.

thermal process   [CHEM ENG] Any process that utilizes heat, without the aid of a catalyst, to accomplish chemical change; for example, thermal cracking, thermal reforming, or thermal polymerization.

thermal pulse   [NUCLEO] The radiant power versus time pulse from a nuclear weapon detonation.

thermal pulse method   [SOLID STATE] A method of measuring properties of insulating and conducting crystals, in which a heat pulse of known duration is measured after propagating through a crystal; the pulse can be generated by directing a laser pulse at an absorbing film evaporated onto one face of the crystal, and detected by a thin-film circuit on the other face.

thermal quality of snow   *See* quality of snow.

thermal radiation   *See* heat radiation.

thermal reactor   [CHEM ENG] A device, system, or vessel in which chemical reactions take place because of heat (no catalysis); for example, thermal cracking, thermal reforming, or thermal polymerization.   [NUCLEO] A nuclear reactor in which fission is induced primarily by neutrons of such low energy that they are in substantial thermal equilibrium with the material of the core.

thermal reforming   [CHEM ENG] A petroleum refining process using heat (but no catalyst) to effect molecular rearrangement of a low-octane naphtha to form high-octane motor gasoline.

thermal regenerative cell   [ELEC] Fuel-cell system in which the reactants are regenerated continuously from the products formed during the cell reaction.

THERMAL HYSTERESIS

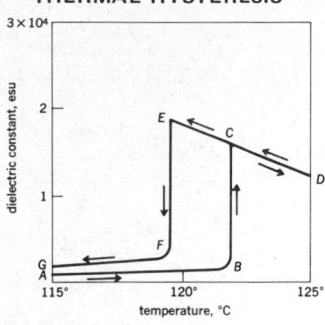

Plot of dielectric constant versus temperature for a single crystal of barium titanate, showing thermal hysteresis. On heating, dielectric constant was observed to follow path *ABCD*, and on cooling path *DCEFG*. (*After M. E. Drougard and D. R. Young, Phys. Rev., 95:1152–1153, 1954*)

THERMAL PULSE METHOD

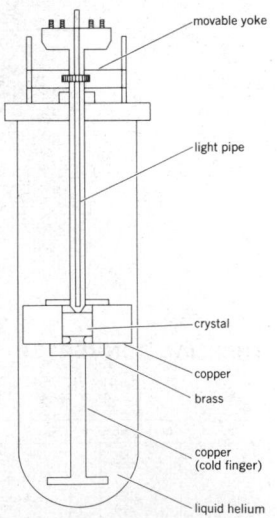

Apparatus for measuring heat pulses at various temperatures, showing components. Laser pulse propagates down light pipe onto absorbing film on face of crystal, and heat pulse propagates through crystal. (*From R. J. von Gutfeld and A. H. Nethercot, in W. P. Mason, ed., Physical Acoustics, vol. 5, 1969*)

thermal relay [ELEC] A relay operated by the heat produced by current flow.

thermal relief [ENG] A valve or other device that is preset to open when pressure becomes excessive due to increased temperature of the system.

thermal resistance [ELECTR] *See* effective thermal resistance. [THERMO] A measure of a body's ability to prevent heat from flowing through it, equal to the difference between the temperatures of opposite faces of the body divided by the rate of heat flow. Also known as heat resistance.

thermal resistivity [THERMO] The reciprocal of the thermal conductivity.

thermal resistor [ELEC] A resistor designed so its resistance varies in a known manner with changes in ambient temperature.

thermal Rossby number [FL MECH] The nondimensional ratio of the inertial force due to the thermal wind and the Coriolis force in the flow of a fluid which is heated from below.

thermal runaway [ELECTR] A condition that may occur in a power transistor when collector current increases collector junction temperature, reducing collector resistance and allowing a greater current to flow, which, in turn, increases the heating effect.

thermal scattering [SOLID STATE] Scattering of electrons, neutrons, or x-rays passing through a solid due to thermal motion of the atoms in the crystal lattice.

thermal shield [NUCLEO] A high-density heat-conducting portion of a shield placed close to the reflector of a nuclear reactor to absorb thermal neutrons, gamma rays, beta rays, and x-rays, whose absorption in the outer shield would generate excessive heat.

thermal spring [HYD] A spring whose water temperature is higher than the local mean annual temperature of the atmosphere.

thermal steering [METEOROL] The steering of an atmospheric disturbance in the direction of the thermal wind in its vicinity; equivalent to steering along thickness lines; for this purpose the thermal wind is usually taken from the earth's surface to a level in the middle troposphere.

thermal stratification [HYD] Horizontal layers of differing densities produced in a lake by temperature changes at different depths.

thermal stress [MECH] Mechanical stress induced in a body when some or all of its parts are not free to expand or contract in response to changes in temperature.

thermal stress cracking [MECH] Crazing or cracking of materials (plastics or metals) by overexposure to elevated temperatures and sudden temperature changes or large temperature differentials.

thermal structure [PETR] A distinct structural pattern, such as a dome or anticline, defined by the arrangement of metamorphic zones of increasing grade.

thermal switch [ELEC] A temperature-controlled switch. Also known as thermoswitch.

thermal telephone receiver [ENG ACOUS] A thermophone used as a telephone receiver.

thermal tide [METEOROL] A variation in atmospheric pressure due to the diurnal differential heating of the atmosphere by the sun; so-called in analogy to the conventional gravitational tide.

thermal titration *See* thermometric titration.

thermal transducer [ENG] Any device which converts energy from some form other than heat energy into heat energy; an example is the absorbing film used in the thermal pulse method.

thermal tuning [ELEC] The process of changing the operating frequency of a system by using controlled thermal expansion to alter the geometry of the system.

thermal utilization factor [NUCLEO] The probability that a thermal neutron which is absorbed is absorbed usefully, as in a fissionable material.

thermal value [THERMO] Heat produced by combustion, usually expressed in calories per gram or British thermal units per pound.

thermal volt *See* kelvin.

thermal vorticity [METEOROL] The vorticity of a thermal wind.

thermal vorticity advection [METEOROL] The advection or transport of the thermal vorticity by the thermal wind, in analogy to the advection of the vorticity by the wind.

thermal wattmeter [ENG] A wattmeter in which thermocouples are used to measure the heating produced when a current is passed through a resistance.

thermal wave [CRYO] *See* temperature wave. [SOLID STATE] A sound wave in a solid which has a short wavelength.

thermal wind [METEOROL] The mean wind-shear vector in geostrophic balance with the gradient of mean temperature of a layer bounded by two isobaric surfaces.

thermal wind equation [METEOROL] An equation for the vertical variation of the geostrophic wind in hydrostatic equilibrium which may be written $-(\partial \mathbf{V}/\partial p) = (R/pf)\mathbf{k} \times \nabla_p T$, where $\mathbf{V}$ is the vector geostrophic wind, p the pressure (used here as the vertical coordinate), R the gas constant for air, f the Coriolis parameter, $\mathbf{k}$ a vertically directed unit vector, and ∇_p the isobaric del operator.

thermal x-rays [ELECTROMAG] The electromagnetic radiation, mainly in the soft (low-energy) x-ray region.

thermal zone *See* thermal belt.

thermic boring [ENG] Boring holes into concrete by means of a high temperature, produced by a steel lance packed with steel wool which is ignited and kept burning by oxyacetylene or other gas.

thermic fever *See* sunstroke.

thermie [THERMO] A unit of heat energy equal to the heat energy needed to raise 1 tonne of water from 14.5°C to 15.5°C at a constant pressure of 1 standard atmosphere; equal to 10^6 fifteen-degrees calories or $(4.1855 \pm 0.0005) \times 10^6$ joules. Abbreviated th.

thermion [ELECTR] A charged particle, either negative or positive, emitted by a heated body, as by the hot cathode of a thermionic tube.

thermionic [ELECTR] Pertaining to the emission of electrons as a result of heat.

thermionic cathode *See* hot cathode.

thermionic converter [ELECTR] A device in which heat energy is directly converted to electric energy; it has two electrodes, one of which is raised to a sufficiently high temperature to become a thermionic electron emitter, while the other, serving as an electron collector, is operated at a significantly lower temperature. Also known as thermionic generator; thermionic power generator; thermoelectric engine.

thermionic current [ELECTR] Current due to directed movements of thermions, such as the flow of emitted electrons from the cathode to the plate in a thermionic vacuum tube.

thermionic detector [ELECTR] A detector using a hot-cathode tube.

thermionic diode [ELECTR] A diode electron tube having a heated cathode.

thermionic emission [ELECTR] 1. The outflow of electrons into vacuum from a heated electric conductor. Also known as Edison effect; Richardson effect. 2. More broadly, the liberation of electrons or ions from a substance as a result of heat.

thermionic fuel cell [ELECTR] A thermionic converter in which the space between the electrodes is filled with cesium or other gas, which lowers the work functions of the electrodes, and creates an ionized atmosphere, controlling the electron space charge.

thermionic generator *See* thermionic converter.

thermionic power generator *See* thermionic converter.

thermionics [ELECTR] The study and applications of thermionic emission.

thermionic triode [ELECTR] A three-electrode thermionic tube, containing an anode, a cathode, and a control electrode.

thermionic tube [ELECTR] An electron tube that relies upon thermally emitted electrons from a heated cathode for tube current. Also known as hot-cathode tube.

thermionic work function [ELECTR] Energy required to transfer an electron from the fermi energy in a given metal through the surface to the vacuum just outside the metal.

thermistor [ELECTR] A resistive circuit component, having a

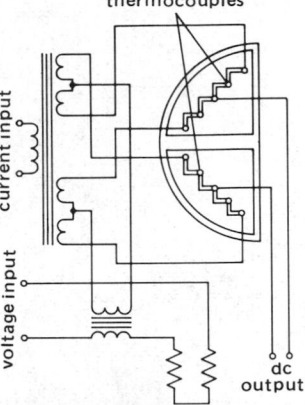

THERMAL WATTMETER

Circuit diagram of thermal wattmeter used in commercial power measurements. Thermocouples are placed in the semicylindrical structure, which is divided into two compartments designed to limit cooling by convection.

high negative temperature coefficient of resistance, so that its resistance decreases as the temperature increases; it is a stable, compact, and rugged two-terminal ceramiclike semiconductor bead, rod, or disk.

thermite [MATER] A fire-hazardous mixture of ferric oxide and powdered aluminum; upon ignition by a magnesium ribbon, it reaches a temperature of 4000°F (2200°C), sufficient to soften steel; used for industrial purposes or as an incendiary bomb.

thermit process [MET] An exothermic reaction when heating finely divided aluminum on a metal oxide causing reduction of the oxide.

thermit welding [MET] Welding with molten iron which is obtained by igniting aluminum and an iron oxide in a crucible, whereby the aluminum floats to the top of the molten metal and is poured off.

thermoammeter [ENG] An ammeter that is actuated by the voltage generated in a thermocouple through which is sent the current to be measured; used chiefly for measuring radio-frequency currents. Also known as electrothermal ammeter; thermocouple ammeter.

thermoanalysis *See* thermal analysis.

thermobalance [ANALY CHEM] An analytical balance modified for thermogravimetric analysis, the measurement of weight changes associated with the transformations of matter when heated.

thermochemical calorie *See* calorie.

thermochemistry [PHYS CHEM] The measurement, interpretation, and analysis of heat changes accompanying chemical reactions and changes in state.

thermocline [GEOPHYS] 1. A temperature gradient as in a layer of sea water, in which the temperature decrease with depth is greater than that of the overlying and underlying water. Also known as metalimnion. 2. A layer in a thermally stratified body of water in which such a gradient occurs.

thermocoagulation [MED] Destruction of tissue by means of electrocautery or a high-frequency current.

thermocompression bonding [ENG] Use of a combination of heat and pressure to make connections, as when attaching beads to integrated-circuit chips; examples include wedge bonding and ball bonding.

thermocompression evaporator [MECH ENG] A system to reduce the energy requirements for evaporation by compressing the vapor from a single-effect evaporator so that the vapor can be used as the heating medium in the same evaporator.

thermocouple [ENG] A device consisting basically of two dissimilar conductors joined together at their ends; the thermoelectric voltage developed between the two junctions is proportional to the temperature difference between the junctions, so the device can be used to measure the temperature of one of the junctions when the other is held at a fixed, known temperature, or to convert radiant energy into electric energy.

thermocouple ammeter *See* thermoammeter.

thermocouple converter *See* thermal converter.

thermocouple pyrometer *See* thermoelectric pyrometer.

thermocouple vacuum gage [ENG] A vacuum gage that depends for its operation on the thermal conduction of the gas present; pressure is measured as a function of the voltage of a thermocouple whose measuring junction is in thermal contact with a heater that carries a constant current; ordinarily, used over a pressure range of 10^{-1} to 10^{-3} millimeter of mercury.

thermocyclogenesis [METEOROL] A theory of cyclogenesis by G. Stüve, in which the disturbance is initiated in the stratosphere and is reflected in the development of a disturbance in the lower troposphere.

thermodiffusion *See* thermal diffusion.

thermoduric bacteria [MICROBIO] Bacteria which survive pasteurization, but do not grow at temperatures used in a pasteurizing process.

thermodynamic cycle [THERMO] A procedure or arrangement in which some material goes through a cyclic process and one form of energy, such as heat at an elevated temperature from combustion of a fuel, is in part converted to another form, such as mechanical energy of a shaft, the remainder being rejected to a lower temperature sink. Also known as heat cycle.

thermodynamic equilibrium [THERMO] Property of a system which is in mechanical, chemical, and thermal equilibrium.

thermodynamic function of state [THERMO] Any of the quantities defining the thermodynamic state of a substance in thermodynamic equilibrium; for a perfect gas, the pressure, temperature, and density are the fundamental thermodynamic variables, any two of which are, by the equation of state, sufficient to specify the state. Also known as state parameter; state variable; thermodynamic variable.

thermodynamic potential [THERMO] One of several extensive quantities which are determined by the instantaneous state of a thermodynamic system, independent of its previous history, and which are at a minimum when the system is in thermodynamic equilibrium under specified conditions.

thermodynamic potential at constant volume *See* free energy.

thermodynamic principles [THERMO] Laws governing the conversion of energy from one form to another.

thermodynamic probability [THERMO] Under specified conditions, the number of equally likely states in which a substance may exist; the thermodynamic probability Ω is related to the entropy S by $S = k\ln\Omega$, where k is Boltzmann's constant.

thermodynamic process [THERMO] A change of any property of an aggregation of matter and energy, accompanied by thermal effects.

thermodynamic property [THERMO] A quantity which is either an attribute of an entire system or is a function of position which is continuous and does not vary rapidly over microscopic distances, except possibly for abrupt changes at boundaries between phases of the system; examples are temperature, pressure, volume, concentration, surface tension, and viscosity. Also known as macroscopic property.

thermodynamics [PHYS] The branch of physics which seeks to derive, from a few basic postulates, relationships between properties of matter, especially those which are affected by changes in temperature, and a description of the conversion of energy from one form to another.

thermodynamic system [THERMO] A part of the physical world as described by its thermodynamic properties.

thermodynamic temperature scale [THERMO] Any temperature scale in which the ratio of the temperatures of two reservoirs is equal to the ratio of the amount of heat absorbed from one of them by a heat engine operating in a Carnot cycle to the amount of heat rejected by this engine to the other reservoir; the Kelvin scale and the Rankine scale are examples of this type.

thermodynamic variable *See* thermodynamic function of state.

thermoelasticity [PHYS] Dependence of the stress distribution of an elastic solid on its thermal state, or of its thermal conductivity on the stress distribution.

thermoelectric converter [ELECTR] A converter that changes solar or other heat energy to electric energy; used as a power source on spacecraft.

thermoelectric cooler [ENG] An electronic heat pump based on the Peltier effect, involving the absorption of heat when current is sent through a junction of two dissimilar metals; it can be mounted within the housing of a device to prevent overheating or to maintain a constant temperature.

thermoelectric cooling [ENG] Cooling of a chamber based on the Peltier effect; an electric current is sent through a thermocouple whose cold junction is thermally coupled to the cooled chamber, while the hot junction dissipates heat to the surroundings. Also known as thermoelectric refrigeration.

thermoelectric effect *See* thermoelectricity.

thermoelectric engine *See* thermionic converter.

thermoelectric generator *See* thermal converter.

thermoelectric heating [ENG] Heating based on the Peltier effect, involving a device which is in principle the same as that used in thermoelectric cooling except that the current is reversed.

thermoelectricity [PHYS] The direct conversion of heat into electrical energy, or the reverse; it encompasses the Seebeck, Peltier, and Thomson effects but, by convention, excludes other electrothermal phenomena, such as thermionic emission. Also known as thermoelectric effect.

THERMOCOUPLE

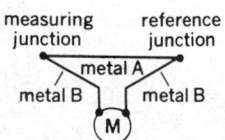

Thermocouple circuit for measuring metal temperatures. Voltmeter M measures voltage developed between the junctions.

thermoelectric junction *See* thermojunction.

thermoelectric laws [ENG] Basic relationships used in the design and application of thermocouples for temperature measurement; for example, the law of the homogeneous circuit, the law of intermediate metals, and the law of successive or intermediate temperatures.

thermoelectric material [ELECTR] A material that can be used to convert thermal energy into electric energy or provide refrigeration directly from electric energy; good thermoelectric materials include lead telluride, germanium telluride, bismuth telluride, and cesium sulfide.

thermoelectric nuclear battery [NUCLEO] A low-voltage battery in which a heat source, consisting of a radioactive isotope such as polonium-210, is hermetically sealed in a strong, dense capsule, and a series of thermocouples are alternately connected thermally, but not electrically, to the heat source and the outer surface of the capsule.

thermoelectric power generator *See* thermal converter.

thermoelectric properties [PHYS] Properties of materials associated with thermoelectricity, namely, the electromotive force generated in the Seebeck effect, the heat generated or absorbed in the Peltier and Thomson effects, and the influence of magnetic fields upon these quantities.

thermoelectric pyrometer [ENG] An instrument which uses one or more thermocouples to measure high temperatures, usually in the range between 800 and 2400°F (425 and 1315°C). Also known as thermocouple pyrometer.

thermoelectric refrigeration *See* thermoelectric cooling.

thermoelectric series [MET] A series of metals arranged in order of their thermoelectric voltage-generating ratings with respect to some reference metal, such as lead.

thermoelectric solar cell [ELECTR] A solar cell in which the sun's energy is first converted into heat by a sheet of metal, and the heat is converted into electricity by a semiconductor material sandwiched between the first metal sheet and a metal collector sheet.

thermoelectric thermometer [ENG] A type of electrical thermometer consisting of two thermocouples which are series-connected with a potentiometer and a constant-temperature bath; one couple, called the reference junction, is placed in a constant-temperature bath, while the other is used as the measuring junction.

thermoelectromotive force [ELEC] Voltage developed due to differences in temperature between parts of a circuit containing two or more different metals.

thermoelectron [ELECTR] An electron liberated by heat, as from a heated filament. Also known as negative thermion.

thermoelement *See* thermal converter.

Thermofor catalytic cracking [CHEM ENG] A proprietary, petroleum refining process with a continuously moving bed of catalyst; used to crack heavy petroleum fractions into lighter, gasoline-range liquids, and the product is known as crackate or cat crackate.

Thermofor catalytic reforming [CHEM ENG] A proprietary, petroleum refining process with a continuously moving bed of catalyst; used to rearrange low-octane naphtha molecules to produce high-octane motor gasoline, and the product is known as reformate.

Thermofor continuous percolation [CHEM ENG] A proprietary petroleum refinery clay-treating adsorbent process, used to stabilize and decolorize lubricants and waxes.

thermoforming [ENG] Forming of thermoplastic sheet by heating it and then pulling it down onto a mold surface to shape it.

thermogalvanic corrosion [MET] Corrosion associated with the passage of an electric current in which the anode and cathode are at different temperatures, the anode usually being the colder of the two.

thermogalvanometer [ENG] Instrument for measuring small high-frequency currents by their heating effect, generally consisting of a direct-current galvanometer connected to a thermocouple that is heated by a filament carrying the current to be measured.

thermograd probe [ENG] An instrument that makes a record of temperature versus depth as it is lowered to the ocean floor, and measures heat flow through the ocean floor.

thermogram [ENG] The recording made by a thermograph.

thermograph [ENG] An instrument that senses, measures, and records the temperature of the atmosphere. Also known as recording thermometer. [OPTICS] A far-infrared image-forming device that provides a thermal photograph by scanning a far-infrared image of an object or scene.

thermograph correction card [ENG] A table for quick and accurate correction of the reading of a thermograph to that of the more accurate dry-bulb thermometer at the same time and place.

thermography [ENG] A method of measuring surface temperature by using luminescent materials; the two main types are contact thermography and projection thermography. [GRAPHICS] **1.** A photocopying process in which the original copy is placed in contact with a transparent sheet and is exposed to infrared rays; heat from carbon or a metallic compound in the text ink then causes a chemical change in a substance laminated between the transparent sheet of paper and a white waxy back. **2.** Photography that uses radiation in the long-wavelength far-infrared region, emitted by objects at temperatures ranging from −170°F (−112°C) to over 300°F (149°C). Also known as thermal photography.

thermogravimetric analysis [ANALY CHEM] Chemical analysis by the measurement of weight changes of a system or compound as a function of increasing temperature.

thermogravitational column [CHEM ENG] A device in which thermal diffusion results from the countercurrent flow of hot and cold material, thus increasing the separation of materials in a solution by the formation of a concentration gradient (difference). Also known as Clusius-Dickel column.

thermohaline [OCEANOGR] Pertaining to the joint activity of salinity and temperature in the oceans.

thermohaline convection [OCEANOGR] Vertical water movement observed when sea water, due to conditions of decreasing temperature or increasing salinity, becomes heavier than the water beneath it.

thermointegrator [ENG] An apparatus, used in studying soil temperatures, for measuring the total supply of heat during a given period; it consists of a long nickel coil (inserted into the soil by an attached rod) forming a 100-ohm resistance thermometer and a 6-volt battery, the current used being recorded on a galvanometer; a mercury thermometer can be used.

thermoisopleth [CLIMATOL] An isopleth of temperature; specifically, a line on a climatic graph showing the variation of temperature in relation to two coordinates.

thermojet [AERO ENG] Air-duct-type engine in which air is scooped up from the surrounding atmosphere, compressed, heated by combustion, and then expanded and discharged at high velocity.

thermojunction [ELECTR] One of the surfaces of contact between the two conductors of a thermocouple. Also known as thermoelectric junction.

thermojunction battery [ELEC] Nuclear-type battery which converts heat into electrical energy directly by the thermoelectric or Seebeck effect.

thermokarst topography [GEOL] An irregular land surface formed in a permafrost region by melting ground ice.

thermokinetic analysis [ANALY CHEM] A type of enthalpimetric analysis which uses kinetic titrimetry; involves rapid and continuous automatic delivery of a suitable titrant, under judiciously controlled experimental conditions with temperature measurement; the end points obtained are converted by mathematical procedures into valid stoichiometric equivalence points and used for determining reaction rate constants.

thermoluminescence [ATOM PHYS] **1.** Broadly, any luminescence appearing in a material due to application of heat. **2.** Specifically, the luminescence appearing as the temperature of a material is steadily increased; it is usually caused by a process in which electrons receiving increasing amounts of thermal energy escape from a center in a solid where they have been trapped and go over to a luminescent center, giving it energy and causing it to luminesce.

thermomagnetic effect [PHYS] An electrical or thermal phenomenon occurring when a conductor or semiconductor is placed simultaneously in a temperature gradient and a magnetic field; examples are the Ettingshausen-Nernst effect and the Righi-Leduc effect.

thermometal [MET] A bimetallic strip which, on temperature

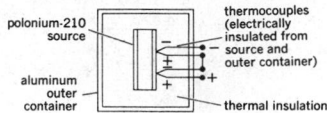

**THERMOELECTRIC
NUCLEAR BATTERY**

Diagram of thermoelectric nuclear battery.

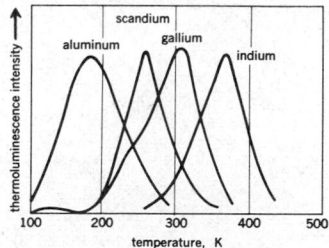

THERMOLUMINESCENCE

Plot of thermoluminescence intensity as a function of temperature in several zinc sulfide phosphors, each of which contains traces of copper, giving rise to luminescent centers, and different trivalent ions (as indicated), giving rise to traps. Curves reach a maximum and then decrease as traps are emptied.

change, deflects because of differences in the coefficients of expansion of the two bonded metals.

thermometamorphism *See* thermal metamorphism.

thermometer [ENG] An instrument that measures temperature.

thermometer anemometer [ENG] An anemometer consisting of two thermometers, one with an electric heating element connected to the bulb; the heated bulb cools in an airstream, and the difference in temperature as registered by the heated and unheated thermometers can be translated into air velocity by a conversion chart.

thermometer bird [VERT ZOO] The name applied to the brush turkey, native to Australia, because it lays its eggs in holes in mounds of earth and vegetation, with the heat from the decaying vegetation serving to incubate the eggs.

thermometer-bulb liquid-level meter [ENG] Detection of liquid level by temperature measurement changes using an immersed bulb-type thermometer.

thermometer frame [ENG] A frame designed to hold two or more reversing thermometers; such a frame is often attached directly to a Nansen bottle.

thermometer screen *See* instrument shelter.

thermometer shelter *See* instrument shelter.

thermometer support [ENG] A device used to hold liquid-in-glass maximum and minimum thermometers in the proper recording position inside an instrument shelter, and to permit them to be read and reset.

thermometric analysis [PHYS CHEM] A method for determination of the transformations a substance undergoes while being heated or cooled at an essentially constant rate, for example, freezing-point determinations.

thermometric conductivity *See* diffusivity.

thermometric depth [OCEANOGR] The ocean depth, in meters, deduced from the difference between the paired protected and unprotected reversing thermometer readings; the unprotected reversing thermometer indicates higher temperature due to pressure effects on the instrument.

thermometric titration [ANALY CHEM] A titration in an adiabatic system, yielding a plot of temperature versus volume of titrant; used for neutralization, precipitation, redox, organic condensation, and complex-formation reactions. Also known as calorimetric titration; enthalpy titration; thermal titration.

thermometry [THERMO] The science and technology of measuring temperature, and the establishment of standards of temperature measurement.

thermonuclear [NUCLEO] Referring to any process in which a very high temperature is used to bring about the fusion of light nuclei, with the accompanying liberation of energy.

thermonuclear bomb *See* fusion bomb.

thermonuclear device [NUCLEO] A fusion bomb used for peaceful purposes, tests, or experiments.

thermonuclear reaction [NUC PHYS] A nuclear fusion reaction which occurs between various nuclei of light elements when they are constituents of a gas at very high temperature.

thermonuclear rocket [NUCLEO] A type of thermal engine utilizing nuclear fusion to heat a working fluid: deuterium, tritium, and lithium are possible fuels.

thermonuclear weapon [NUCLEO] A fusion bomb in the form of a packaged unit ready for transportation or use by military forces.

thermoperiodicity [BOT] The totality of responses of a plant to appropriately fluctuating temperatures.

thermophile [BIOL] An organism that thrives at high temperatures.

thermophobia [PSYCH] An abnormal fear of heat.

thermophone [ENG ACOUS] An electroacoustic transducer in which sound waves having an accurately known strength are produced by the expansion and contraction of the air adjacent to a strip of conducting material, whose temperature varies in response to a current input that is the sum of a steady current and a sinusoidal current; used chiefly for calibrating microphones.

thermopile [ENG] An array of thermocouples connected either in series to give higher voltage output or in parallel to give higher current output, used for measuring temperature or radiant energy or for converting radiant energy into electric power.

thermopile generator [ELEC] An electricity source powered by the heating of an electrical resistor that can be connected to a thermopile to generate small amounts of electric current.

thermoplastic insulation [MATER] Electrical insulation made of a thermoplastic material.

thermoplastic recording [ELECTR] A recording process in which a modulated electron beam deposits charges on a thermoplastic film, and application of heat by radio-frequency heating electrodes softens the film enough to produce deformation that is proportional to the density of the stored electrostatic charges; an optical system is used for playback.

thermoplastic resin [MATER] A material with a linear macromolecular structure that will repeatedly soften when heated and harden when cooled; for example, styrene, acrylics, cellulosics, polyethylenes, vinyls, nylons, and fluorocarbons.

thermopower [ELEC] A measure of the temperature-induced voltage in a conductor.

thermoreception [PHYSIO] The process by which environmental temperature affects specialized sense organs (thermoreceptors).

thermoreceptor [PHYSIO] A sense receptor that responds to stimulation by heat and cold.

thermoregulation [PHYSIO] A mechanism by which mammals and birds attempt to balance heat gain and heat loss in order to maintain a constant body temperature when exposed to variations in cooling power of the external medium.

thermoregulator [ENG] A high-accuracy or high-sensitivity thermostat; one type consists of a mercury-in-glass thermometer with sealed-in electrodes, in which the rising and falling column of mercury makes and breaks an electric circuit.

thermorelay *See* thermostat.

thermoremanent remagnetization [GEOPHYS] The permanent magnetization of igneous rocks, acquired at the time of cooling from the molten state.

Thermosbaenacea [INV ZOO] An order of small crustaceans in the superorder Pancarida.

Thermosbaenidae [INV ZOO] A family of the crustacean order Thermosbaenacea.

thermoscreen *See* instrument shelter.

thermosetting resin [MATER] A plastic that solidifies when first heated under pressure, and which cannot be remelted or remolded without destroying its original characteristics; examples are epoxies, malamines, phenolics, and ureas.

thermosiphon [MECH ENG] A closed system of tubes connected to a water-cooled engine which permit natural circulation and cooling of the liquid by utilizing the difference in density of the hot and cool portions.

thermosiphon reboiler [CHEM ENG] A liquid reheater (as for distillation-column bottoms) in which natural circulation of the boiling liquid is obtained by maintaining a sufficient liquid head.

thermosphere [METEOROL] The atmospheric shell extending from the top of the mesosphere to outer space; it is a region of more or less steadily increasing temperature with height, starting at 70 or 80 kilometers; the thermosphere includes, therefore, the exosphere and most or all of the ionosphere.

thermostat [ENG] An instrument which measures changes in temperature and directly or indirectly controls sources of heating and cooling to maintain a desired temperature. Also known as thermorelay.

thermostatic switch [ELEC] A temperature-operated switch that receives its operating energy by thermal conduction or convection from the device being controlled or operated.

thermosteric anomaly [OCEANOGR] Component of the specific volume anomaly for a parcel of sea water at 1-atmosphere pressure due to its temperature being other than the standard temperature of 0°C.

thermoswitch *See* thermal switch.

thermotaxis [BIOL] Orientation movement of a motile organism in response to the stimulus of a temperature gradient.

thermotherapy [MED] The treatment of disease by heat of any kind; involves the local or general application of heat to the body.

thermotropic model [METEOROL] A model atmosphere used in numerical forecasting, in which the parameters are the height of one constant-pressure surface (usually 500 millibars) and one temperature (usually the mean temperature between 100 and 500 millibars).

THERMOSTAT

setting knob and scale

bimetal sensing element

double throw mercury switch

thermometer and scale

adjustable heat anticipator and scale

Typical thermostat uses bimetal sensing element to control room temperature by switching a heating or cooling device on and off. Heat anticipator switches device off prematurely to prevent excessive temperature swings. (*Honeywell Inc.*)

thermovac evaporator [FOOD ENG] A type of vacuum evaporator used to remove moisture from foods for the purpose of preservation; temperature and velocities are minutely, accurately, and automatically controlled in all stages of the process.

thermovoltmeter [ENG] A voltmeter in which a current from the voltage source is passed through a resistor and a fine vacuum-enclosed platinum heater wire; a thermocouple, attached to the midpoint of the heater, generates a voltage of a few millivolts, and this voltage is measured by a direct-current millivoltmeter.

therophyte [ECOL] An annual plant whose seed is the only overwintering structure.

Theropoda [PALEON] A suborder of carnivorous bipedal saurischian reptiles which first appeared in the Upper Triassic and culminated in the uppermost Cretaceous.

Theropsida [PALEON] An order of extinct mammallike reptiles in the subclass Synapsida.

Thesium [BOT] The hymenium of the apothecium in lichens.

thesocyte [INV ZOO] An amebocyte in Porifera containing ergastic cytoplasmic inclusions.

theta antigen [IMMUNOL] A cell membrane constituent which distinguishes T cells from other lymphocytes.

theta functions [MATH] Complex functions used in the study of Riemann surfaces and of elliptic functions and elliptic integrals; they are:

$$\theta_1(z) = 2\sum_{n=0}^{\infty} (-1)^n q^{(n+\frac{1}{2})^2} \sin(2n+1)z$$

$$\theta_2(z) = 2\sum_{n=0}^{\infty} q^{(n+\frac{1}{2})^2} \cos(2n+1)z$$

$$\theta_3(z) = 1 + 2\sum_{n=1}^{\infty} q^{n^2} \cos 2z$$

$$\theta_4(z) = 1 + 2\sum_{n=1}^{\infty} (-1)^n q^{n^2} \cos 2nz$$

where $q = \exp \pi i \tau$, and τ is a constant complex number with positive imaginary part.

thetagram [THERMO] A thermodynamic diagram with coordinates of pressure and temperature, both on a linear scale.

theta polarization [ELECTROMAG] State of a wave in which the E vector is tangential to the meridian lines of some given spherical frame of reference.

theta rhythm [PSYCH] A brain rhythm having a frequency of about 4–7 hertz, and somewhat greater voltage than the alpha rhythm; thought to originate in the hippocampus.

Thévenin generator [ELEC] The voltage generator in the equivalent circuit of Thévenin's theorem.

Thévenin's theorem [ELEC] A valuable theorem in network problems which allows calculation of the performance of a device from its terminal properties only: the theorem states that at any given frequency the current flowing in any impedance, connected to two terminals of a linear bilateral network containing generators of the same frequency, is equal to the current flowing in the same impedance when it is connected to a voltage generator whose generated voltage is the voltage at the terminals in question with the impedance removed, and whose series impedance is the impedance of the network looking back from the terminals into the network with all generators replaced by their internal impedances. Also known as Helmholtz's theorem.

thiacetic acid *See* thioacetic acid.

thiamine [BIOCHEM] $C_{12}H_{17}ClN_4OS$ A member of the vitamin B complex that occurs in many natural sources, frequently in the form of cocarboxylase. Also known as aneurine; vitamin B_1.

thiamine hydrochloride [ORG CHEM] $C_{12}H_{17} \cdot ON_4S \cdot HCl$ White, hygroscopic crystals soluble in water, insoluble in ether, with a yeasty aroma and a salty, nutlike taste, decomposes at 247°C; the form in which thiamine is generally employed.

thiamine pyrophosphate [BIOCHEM] The coenzyme or prosthetic component of carboxylase; catalyzes decarboxylation of various α-keto acids. Also known as cocarboxylase.

Thiaridae [INV ZOO] A family of freshwater gastropod mollusks in the order Pectinibranchia.

thiazole [ORG CHEM] C_3H_3NS A colorless to yellowish liquid with a pyridinelike aroma, slightly soluble in water, soluble in alcohol and ether; used as an intermediate for fungicides, dyes, and rubber accelerators.

thiazole dye [ORG CHEM] One of a family of dyes in which the chromophore groups are $=C=N-$, $-S-C=$, and used mainly for cotton; an example is primuline.

thick-bedded [GEOL] Pertaining to a sedimentary bed that ranges in thickness from 60 to 120 centimeters (2–4 feet).

thickened fuel *See* gelatinized gasoline.

thickened oil [MATER] Any oil to which a thickening agent has been added to increase viscosity or produce thixotropic properties; grease is a thickened oil.

thickener [ENG] A nonfilter device for the removal of liquid from a liquid-solids slurry to give a dewatered (thickened) solids product; can be by gravity settling or centrifugation.

thickening [CHEM ENG] The concentration of the solids in a suspension in order to recover a fraction with a higher concentration of solids than in the original suspension. [MIN ENG] Concentrating dilute slime pulp into a pulp containing a smaller percentage of moisture by rejecting the free liquid.

thicket *See* tropical scrub.

thick-film capacitor [ELEC] A capacitor in a thick-film circuit, made by successive screen-printing and firing processes.

thick-film circuit [ELECTR] A microcircuit in which passive components, of a ceramic-metal composition, are formed on a ceramic substrate by successive screen-printing and firing processes, and discrete active elements are attached separately.

thick-film resistor [ELEC] Fixed resistor whose resistance element is a film well over one-thousandth of an inch thick.

thickhead *See* bluetongue.

thickness [METEOROL] The vertical depth, measured in geometric or geopotential units, of a layer in the atmosphere bounded by surfaces of two different values of the same physical quantity, usually constant-pressure surfaces.

thickness chart [METEOROL] A type of synoptic chart showing the thickness of a certain physically defined layer in the atmosphere; it almost always refers to an isobaric thickness chart, that is, a chart of vertical distance between two constant-pressure surfaces.

thickness gage [ENG] A gage for measuring the thickness of a sheet of material, the thickness of an object, or the thickness of a coating; examples include penetration-type and backscattering radioactive thickness gages and ultrasonic thickness gages.

thickness line [METEOROL] A line drawn through all geographic points at which the thickness of a given atmospheric layer is the same. Also known as relative contour; relative isohypse.

thickness pattern [METEOROL] The general geometric distribution of thickness lines on a thickness chart. Also known as relative hypsography; relative topography.

thickness ratio [AERO ENG] The ratio of the maximum thickness of an airfoil section to the length of its chord.

thick-skinned structure [GEOL] Any large-scale structure, such as a fold or fault, believed to have originated as a result of basement movement beneath overlying rocks.

thick-tailed bushbaby [VERT ZOO] *Galago crassicaudatus*. A primate animal in the family Lorisidae; one of the six species of bushbaby, the thick-tailed bushbaby is more aggressive and solitary than the other species, and grows to over 1 foot (30 centimeters) in length with an equally long tail. Also known as great galago.

thick-thin chart *See* isentropic thickness chart.

thief [PETRO ENG] In the petroleum industry, a device that permits the taking of samples from a predetermined location in the liquid body to be sampled.

Thiele coordinates [CHEM ENG] A graphical method for calculating the solvent-free composition of two components being separated by solvent extraction.

Thiele-Geddes method [CHEM ENG] A method for the prediction of the product distribution from a multicomponent distillation system.

Thiele melting point apparatus [ANALY CHEM] A stirred, specially shaped test-tube device used for the determination of the melting point of a crystalline chemical.

Thiessen polygon method [METEOROL] A method of assigning areal significance to point rainfall values: perpendicular bisectors are constructed to the lines joining each measuring station with those immediately surrounding it; the bisectors

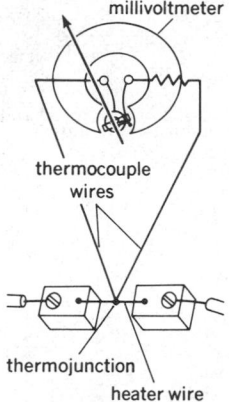

THERMOVOLTMETER

Elementary thermovoltmeter circuit. *(General Electric Co.)*

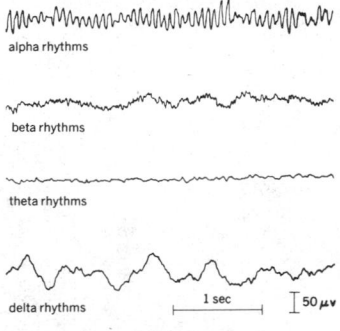

THETA RHYTHM

Electroencephalogram of brain rhythms taken from scalp leads on human subject.

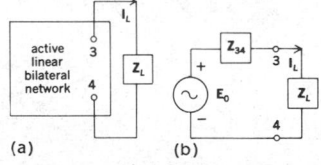

THEVENIN'S THEOREM

The theorem states that circuit to left of terminals 3 and 4 in *(a)* may be replaced by that in *(b)*. Impedance Z_L and current I_L are same in both cases. E_0 is voltage of voltage generator. Z_{34} is series impedance.

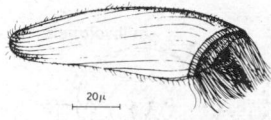

THIGMOTRICHIDA

Drawing of *Boveria*, an example of a thigmotrichid, showing ciliature.

form a series of polygons, each polygon containing one station; the value of precipitation measured at a station is assigned to the whole area covered by the enclosing polygon.

thigh [ANAT] The upper part of the leg, from the pelvis to the knee.

thigh circumference [ANTHRO] The measurement around the thigh of the left leg midway between the crotch and the knee when the subject is in a standing position.

thigmotaxis *See* stereotaxis.

Thigmotrichida [INV ZOO] An order of ciliated protozoans in the subclass Holotrichia.

thigmotropism *See* stereotropism.

thill *See* underclay.

thimerosal [ORG CHEM] $C_9H_9HgNaO_2S$ Sodium ethylmercurithiosalicylate, an organomercurial antiseptic used topically, and also as a preservative of certain biological products.

thin [METEOROL] In aviation weather observations, the description of a sky cover that is predominantly transparent.

thin-bedded [GEOL] Pertaining to a sedimentary bed that ranges in thickness from 5 to 60 centimeters (2 inches to 2 feet).

thin film [ELECTR] A film a few molecules thick deposited on a glass, ceramic, or semiconductor substrate to form a capacitor, resistor, coil, cryotron, or other circuit component.

thin-film capacitor [ELEC] A capacitor that can be constructed by evaporation of conductor and dielectric films in sequence on a substrate; silicon monoxide is generally used as the dielectric.

thin-film circuit [ELECTR] A circuit in which the passive components and conductors are produced as films on a substrate by evaporation or sputtering; active components may be similarly produced or mounted separately.

thin-film cryotron [ELECTR] A cryotron in which the transition from superconducting to normal resistivity of a thin film of tin or indium, serving as a gate, is controlled by current in a film of lead that crosses and is insulated from the gate.

thin-film ferrite coil [ELECTROMAG] An inductor made by depositing a thin flat spiral of gold or other conducting metal on a ferrite substrate.

thin-film integrated circuit [ELECTR] An integrated circuit consisting entirely of thin films deposited in a patterned relationship on a substrate.

thin-film material [ELECTR] A material that can be deposited as a thin film in a desired pattern by a variety of chemical, mechanical, or high-vacuum evaporation techniques.

thin-film memory *See* thin-film storage.

thin-film resistor [ELEC] A fixed resistor whose resistance element is a metal, alloy, carbon, or other film having a thickness of about one-millionth inch.

thin-film semiconductor [ELECTR] Semiconductor produced by the deposition of an appropriate single-crystal layer on a suitable insulator.

thin-film solar cell [ELECTR] A solar cell in which a thin film of gallium arsenide, cadmium sulfide, or other semiconductor material is evaporated on a thin, flexible metal or plastic substrate; the rather low efficiency (about 2%) is compensated by the flexibility and light weight, making these cells attractive as power sources for spacecraft.

thin-film storage [ADP] A high-speed storage device that is fabricated by depositing layers, one molecule thick, of various materials which, after etching, provide microscopic circuits which can move and store data in small amounts of time. Also known as thin-film memory.

thin-film transducer [SOLID STATE] A film a few molecules thick, usually consisting of cadmium sulfide, evaporated on a crystal substrate, used to convert microwave radiation into hypersonic sound waves in the crystal.

thin-film transistor [ELECTR] A field-effect transistor constructed entirely by thin-film techniques, for use in thin-film circuits.

thin-layer chromatography [ANALY CHEM] Chromatographing on thin layers of adsorbents rather than in columns; adsorbent can be alumina, silica gel, silicates, charcoals, or cellulose.

thin lens [OPTICS] A lens whose thickness is small enough to be neglected in calculations of such quantities as object distance, image distance, and magnification.

thin magnetic film [ELECTR] A data storage device consisting of a thin magnetic film of Permalloy deposited by vacuum evaporation or electrochemical deposition.

thinner [MATER] A liquid used to thin paint, varnish, cement, or other material to a desired consistency.

Thinocoridae [VERT ZOO] The seed snipes, family of South American birds in the order Charadriiformes.

thin-out [GEOL] Gradual thinning of a stratum, vein, or other body of rock until the upper and lower surfaces meet and the rock disappears.

thin-plate orifice [ENG] A thin-metal orifice sheet used in fluid-flow measurement in fluid conduits by means of differential pressure drop across the orifice.

thin section [GEOL] A piece of rock or mineral specifically prepared to study its optical properties; the sample is ground to 0.03-millimeter thickness then polished and placed between two microscope slides. Also known as section.

thin-skinned structure [GEOL] Any large-scale structure, such as a fold or fault, confined to and originating within a thin layer of rocks above a surface of décollement.

thio- [CHEM] A chemical prefix derived from the Greek *theion*, meaning sulfur; indicates the replacement of an oxygen in an acid radical by sulfur with a negative valence of 2, S^{-2}.

thioacetic acid [ORG CHEM] CH_3COSH A toxic, clear-yellow liquid with an unpleasant aroma, soluble in water, alcohol, and ether, boils at 82°C; used as an analytical reagent and a lachrymator. Also known as ethanethiolic acid; thiacetic acid.

thioaldehyde [ORG CHEM] An organic compound that contains the −CHS radical and has the suffix thial; for example, ethanethial, CH_3CHS.

Thiobacillus [MICROBIO] A genus of rod-shaped autotrophic bacteria, using carbon dioxide as the sole source of carbon and deriving respiratory energy from the oxidation of sulfide, sulfur, thiosulfate, polythionates, and in some cases thiocyanate.

Thiobacteriaceae [MICROBIO] A family of nonfilamentous, gram-negative bacteria of the suborder Pseudomonadineae which oxidize hydrogen sulfide, free sulfur, and inorganic sulfur compounds to sulfuric acid.

thiobarbiturate [PHARM] A derivative of thiobarbituric acid that differs from the barbiturates in the replacement of one oxygen atom by sulfur but resembles the barbiturates in its effects.

thiobarbituric acid [ORG CHEM] $C_6H_4N_2O_2S$ Malonyl thiourea, the parent compound of the thiobarbiturates; represents barbituric acid in which the oxygen atom of the urea component has been replaced by sulfur.

thiocarbamazine [PHARM] $C_{21}H_{17}AsN_2O_5S_2$ A white crystalline powder, freely soluble in dilute alkali; used in medicine as an amebicide. Also known as (*para*-ureidophenylarsylenedithio)di-*o*-benzoic acid; thiocarbamisin.

thiocarbamisin *See* thiocarbamazine.

thiocarbarsone [PHARM] $C_{11}H_{13}AsN_2O_5S_2$ A white crystalline powder, freely soluble in dilute alkali; used in medicine as an amebicide. Also known as (*para*-ureidophenylarsylenedithio)diacetic acid.

thiocyanate [INORG CHEM] A salt of thiocyanic acid that contains the −SCN radical; for example, sodium thiocyanate, NaSCN. Also known as rhodanate; rhodanide; sulfocyanate; sulfocyanide; thiocyanide.

thiocyanic acid [INORG CHEM] HSC:N A colorless, watersoluble liquid decomposing at 200°C; used to inhibit paper deterioration due to the action of light, and (in the form of organic esters) as an insecticide. Also known as rhodanic acid; sulfocyanic acid.

thiocyanide *See* thiocyanate.

thiocyanogen [INORG CHEM] NCSSCN White, light-unstable rhombic crystals melting at −2°C.

thiodiethylene glycol *See* thiodiglycol.

thiodiglycol [ORG CHEM] $(CH_2CH_2OH)_2S$ A combustible, colorless, syrupy liquid soluble in water, alcohol, acetone, and chloroform, boils at 283°C; used as a chemical intermediate, textile-dyeing solvent, and antioxidant. Also known as dihyroxyethylsulfide; thiodiethylene glycol.

thiodiglycolic acid [ORG CHEM] HOOCCH₂SCH₂COOH Combustible, colorless, water- and alcohol-soluble crystals melting at 128°C; used as an analytical reagent.

thioether [ORG CHEM] RSR A general formula for colorless, volatile organic compounds obtained from alkyl halides and alkali sulfides; the R groups can be the same, or different as in methylthioethane (CH₃SC₂H₅).

thioethyl alcohol See ethyl mercaptan.

thiofuran See thiophene.

thioglycolic acid [ORG CHEM] HSCH₂COOH A liquid with a strong unpleasant odor; used as a reagent for metals such as iron, molybdenum, silver, and tin, and in bacteriology. Also known as mercaptoacetic acid.

thioindigo [MATER] A group of sulfur dyes made by treating the appropriate organic compound with sodium sulfide; colors are fast to washing and light.

Thiokol [MATER] Trademark for any of a series of polysulfide rubbers noted for their resistance to oils, gasoline, and solvents.

thiol See mercaptan.

thiolactic acid [ORG CHEM] CH₃CH(SH)COOH An oil with a disagreeable odor; used in toiletry preparation. Also known as 2-mercaptopropionic acid; 2-thiolpropionic acid.

2-thiolpropionic acid See thiolactic acid.

thionic acid [INORG CHEM] H₂SₓO₆, where x varies from 2 to 6. [ORG CHEM] An organic acid with the radical −CSOH.

thionyl bromide [ORG CHEM] SOBr₂ A red liquid boiling at 68°C (40 mm Hg). Also known as sulfinyl bromide.

thionyl chloride [INORG CHEM] SOCl₂ A toxic, yellowish to red liquid with a pungent aroma, soluble in benzene, decomposes in water and at 140°C; boils at 79°C; used as a chemical intermediate and catalyst. Also known as sulfur oxychloride; sulfurous oxychloride.

thiopental sodium [ORG CHEM] C₁₁H₁₇O₂N₂NaS Yellow, water-soluble crystals with a characteristic aroma; used in medicine as a short-acting anesthetic. Also known as thiopentone sodium.

thiopentone sodium See thiopental sodium.

thiophene [ORG CHEM] C₄H₄S A toxic, flammable, highly reactive, colorless liquid, insoluble in water, soluble in alcohol and ether, boils at 84°C; used as a chemical intermediate and to make condensation copolymers. Also known as thiofuran.

thiophenol [ORG CHEM] C₆H₅SH A toxic, fire-hazardous, water-white liquid with a disagreeable aroma, insoluble in water, soluble in alcohol and ether, boils at 168°C; used to make pharmaceuticals. Also known as phenyl mercaptan.

thiophenylamine See phenothiazine.

thiophosphoric anhydride See phosphorus pentasulfide.

Thiorhodaceae [MICROBIO] A family of bacteria in the suborder Rhodobacteriineae composed of the purple, red, orange, and brown sulfur bacteria; nearly all are strict anaerobes, oxidize hydrogen sulfide, and store sulfur globules internally.

thiosemicarbazide [ORG CHEM] NH₂CSNHNH₂ A white, water- and alcohol-soluble powder melting at 182°C; used as an analytical reagent and in photography and rodenticides. Also known as aminothiourea.

thiosemicarbazone [PHARM] A class of chemical compounds used in treating tuberculosis; the most prominent member of the group is para-acetamidobenzaldehyde thiosemicarbazone.

thiosinamine See allylthiourea.

thiospinel [MINERAL] Any mineral with the spinel structure having the general formula AR₂S₄.

thiostreptone [MICROBIO] A polypeptide antibiotic produced by a species of Streptomyces that crystallizes from a chloroform methanol solution; used in veterinary medicine.

thiosulfate [INORG CHEM] M₂S₂O₃ A salt of thiosulfuric acid and a base; for example, reaction of sodium hydroxide and thiosulfuric acid to produce sodium thiosulfate.

thiosulfonic acid [ORG CHEM] Name for a group of oxy acids of sulfur, with the general formula RS₂O₂H; they are known as esters and salts.

thiosulfuric acid [INORG CHEM] H₂S₂O₃ An unstable acid that decomposes readily to form sulfur and sulfurous acid.

thiouracil [PHARM] C₄H₄N₂OS An antithyroid drug that acts by interfering with thyroxine synthesis.

thiourea [PHARM] CSN₂H₄ Thiocarbamide, an antithyroid drug used in the treatment of hyperthyroidism.

2-thioxo-4-thiazolidinone See rhodanine.

third-generation computer [ADP] One of the general purpose digital computers introduced in the late 1960s; it is characterized by integrated circuits and has logical organization and software which permit the computer to handle many programs at the same time, allow one to add or remove units from the computer, permit some or all input/output operations to occur at sites remote from the main processor, and allow conversational programming techniques.

third harmonic [PHYS] A sine-wave component having three times the fundamental frequency of a complex signal.

third law of motion See Newton's third law.

third law of thermodynamics [THERMO] The entropy of all perfect crystalline solids is zero at absolute zero temperature.

third-order climatological station [CLIMATOL] As defined by the World Meteorological Organization, a station, other than a precipitation station, at which the observations are of the same kind as those at a second-order climatological station, but are not so comprehensive, are made once a day only, and are made at other than the specified hours.

third-order reaction [PHYS CHEM] A chemical reaction in which the rate of reaction is determined by the concentration of three reactants.

third rail [CIV ENG] The electrified metal rail which carries current to the motor of an electric locomotive or other railway car.

thirling See holing.

thirst [PHYSIO] A sensation, as of dryness in the mouth and throat, resulting from water deprivation.

thirteen temperature See annealing point.

thirty-day forecast [METEOROL] A weather forecast for a period of 30 days; as issued by the U.S. Weather Bureau, the forecast concerns expected departures of temperature and precipitation from normal.

thirty-two nucleus [METEOROL] An unidentified type of freezing nucleus which first becomes active when a supercooled cloud is cooled to about −32°C.

thistle [BOT] Any of the various prickly plants comprising the family Compositae.

thiuram [ORG CHEM] A chemical compound containing a R₂NCS radical; occurs mainly in disulfide compounds; the most common monosulfide compound is tetramethylthiuram monosulfide.

thixotropic clay [GEOL] A clay that weakens when disturbed and increases in strength upon standing.

thixotropy [PHYS CHEM] Property of certain gels which liquefy when subjected to vibratory forces, such as ultrasonic waves or even simple shaking, and then solidify again when left standing.

Thlipsuridae [PALEON] A Paleozoic family of ostracod crustaceans in the suborder Platycopa.

tholeiite [PETR] 1. A group of basalts composed principally of plagioclase, pyroxene, and iron oxide minerals as phenocrysts in a glassy groundmass. 2. Any rock in the group.

Thoma cavitation coefficient [MECH ENG] The equation for measuring cavitation in a hydraulic turbine installation, relating vapor pressure, barometric pressure, runner setting, tail water, and head.

Thomas converter [MET] A basic Bessemer converter; that is, one in which air is forced upward through holes in the bottom of the steel container which has a basic lining, usually dolomite, and which employs a basic slag.

Thomas cyclotron [NUCLEO] A circular particle accelerator which operates like an ordinary cyclotron but employs a magnetic field that is variable in azimuth in such a way that cyclotron resonance at a fixed orbital frequency and radial and axial focusing can be maintained simultaneously.

Thomas-Fermi atom model [ATOM PHYS] A method of approximating the electrostatic potential and the electron density in an atom in its ground state, in which these two quantities are related by the Poisson equation on the one hand, and on the other hand by a semiclassical formula for the density of quantum states in phase space.

Thomas meter [ENG] An instrument used to determine the

THIOSULFONIC ACID

O
‖
H—S—S—R
‖
O

Structural formula of thiosulfonic acid.

THOR

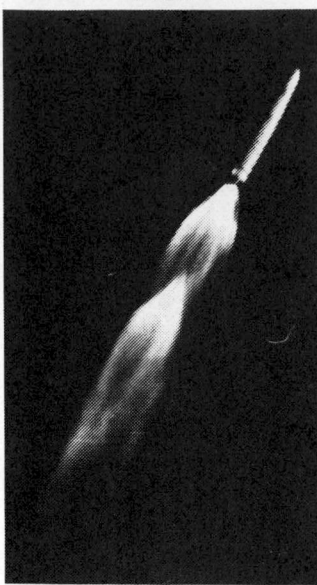

Thor missile photographed at 35 miles (56 kilometers) using 24-inch (61-centimeter) diameter, 500-inch (1270-centimeter) focal-length tracking telescope, driven by electrohydraulic servo system. *(U.S. Air Force)*

THORIANITE

1.3 cm

Thorianite crystals from Balangoda, Ceylon. *(Specimens from Department of Geology, Bryn Mawr College)*

THORIUM

Periodic table of the chemical elements showing the position of thorium.

rate of flow of a gas by measuring the rise in the gas temperature produced by a known amount of heat.

Thomas precession [RELAT] The precession of a vector in an accelerated system, relative to an observer for whom the system has a given velocity and acceleration, when this vector appears to be constant to an observer attached to the system; this precession is the kinematical basis of one type of spin-orbit coupling.

Thomas-Reiche-Kuhn sum rule *See* f-sum rule.

Thompson submachine gun [ORD] Caliber-.45, air-cooled automatic weapon that can be carried and operated by one person.

Thomson-Berthelot principle [PHYS CHEM] The assumption that the heat released in a chemical reaction is directly related to the chemical affinity, and that, in the absence of the application of external energy, that chemical reaction which releases the greatest heat is favored over others; the principle is in general incorrect, but applies in certain special cases.

Thomson bridge *See* Kelvin bridge.

Thomson coefficient [PHYS] The ratio of the voltage existing between two points on a metallic conductor to the difference in temperature of those points.

Thomson cross section [ELECTROMAG] The total scattering cross section for Thomson scattering, equal to $(8/3)\pi(e^2/mc^2)^2$, where e and m are the charge (in electrostatic units) and mass of the scattering particle, and c is the speed of light.

Thomson effect [PHYS] A thermoelectric effect in which heat flows into or out of a homogeneous conductor when an electric current flows between two points in the conductor at different temperatures, the direction of heat flow depending upon whether the current flows from colder to warmer metal or from warmer to colder.

Thomson formula [ELECTROMAG] **1.** The formula for the intensity of scattered electromagnetic radiation in Thomson scattering as a function of the scattering angle ϕ; the intensity is proportional to $1 + \cos^2 \phi$. **2.** A formula for the period of oscillation of a current when a capacitor is discharged.

thomsonite [MINERAL] $NaCa_2Al_5Si_5O_{20} \cdot 6H_2O$ Snow-white zeolite mineral forming orthorhombic crystals and occurring in masses of radiating crystals; hardness is 5–5.5 on Mohs scale.

Thomson parabolas [ELECTROMAG] A pattern of parabolas which appear on a photographic plate exposed to a beam of ions of an element which has passed through electric and magnetic fields applied in the same direction normal to the path of the ions; each parabola corresponds to a different charge-to-mass ratio, and thus to a different isotope.

Thomson relations [PHYS] Equations in the study of thermoelectricity, relating the Peltier coefficient and the Thomson coefficient to the Seebeck voltage; they are derived by thermodynamics. Also known as Kelvin relations.

Thomson scattering [ELECTROMAG] Scattering of electromagnetic radiation by free (or very loosely bound) charged particles, computed according to a classical nonrelativistic theory: energy is taken away from the primary radiation as the charged particles accelerated by the transverse electric field of the radiation, radiate in all directions.

Thomson voltage [PHYS] The voltage that exists between two points that are at different temperatures in a conductor.

Thor [ORD] A U.S. Air Force surface-to-surface intermediate-range ballistic missile using inertial guidance and having a range of about 2000 miles (3200 kilometers); it can carry nuclear warheads, and can be adapted for launching a satellite.

Thoracica [INV ZOO] An order of the subclass Cirripedia; individuals are permanently attached in the adult stage, the mantle is usually protected by calcareous plates, and six pairs of biramous thoracic appendages are present.

thoracic cavity *See* thorax.

thoracic duct [ANAT] The common lymph trunk beginning in the crura of the diaphragm at the level of the last thoracic vertebra, passing upward, and emptying into the left subclavian vein at its junction with the left internal jugular vein.

thoracic vertebrae [ANAT] The vertebrae associated with the chest and ribs in vertebrates; there are 12 in humans.

thoracoabdominal breathing [PHYSIO] The process of air breathing in reptiles, birds, and mammals that depends upon

aspiration or sucking inspiration, and involves trunk musculature to supply pulmonary ventilation.

Thoracostomopsidae [INV ZOO] A family of marine nematodes in the superfamily Enoploidea, which have the stomatal armature modified to form a hollow tube.

Thoraeus filter [NUCLEO] A primary radiological filter of tin, combined with a secondary filter of copper to absorb the characteristic radiation of the tin and a third filter of aluminum to absorb the characteristic radiation of the copper; in the range of 200 to 400 kilovolts, such a filter hardens x-rays more efficiently than the usual combination of copper and aluminum.

thorax [ANAT] The chest; the cavity of the mammalian body between the neck and the diaphragm, containing the heart, lungs, and mediastinal structures. Also known as thoracic cavity. [INV ZOO] The middle of three principal divisions of the body of certain classes of arthropods.

Thorazine [PHARM] Trademark for chlorpromazine, a tranquilizer and antiemetic drug, used as the hydrochloride salt.

thoria *See* thorium dioxide.

thorianite [MINERAL] ThO_2 A radioactive mineral that crystallizes in the isometric system, occurs in worn cubic crystals, is brownish black to reddish brown in color, and has resinous luster; hardness is 7 on the Mohs scale, and specific gravity is 9.7–9.8.

thoriated emitter *See* thoriated tungsten filament.

thoriated tungsten filament [ELECTR] A vacuum-tube filament consisting of tungsten mixed with a small quantity of thorium oxide to give improved electron emission. Also known as thoriated emitter.

Thorictidae [INV ZOO] The ant blood beetles, a family of coleopteran insects in the superfamily Dermestoidea.

thorite [MINERAL] $ThSiO_4$ A brownish-yellow to brownish-black and black radioactive mineral that is tetragonal in crystallization; hardness is about 4.5 on Mohs scale, and specific gravity is 4.3–5.4.

thorium [CHEM] An element of the actinium series, symbol Th, atomic number 90, atomic weight 232; soft, radioactive, insoluble in water and alkalies, soluble in acids, melts at 1750°C, boils at 4500°C. [MET] A heavy malleable metal that changes from silvery-white to dark gray or black in air; potential source of nuclear energy; used in manufacture of sunlamps.

thorium anhydride *See* thorium dioxide.

thorium carbide [INORG CHEM] ThC_2 A yellow solid melting at above 2630°C, decomposes in water; used in nuclear fuel.

thorium chloride [INORG CHEM] $ThCl_4$ Hygroscopic, toxic colorless crystal needles soluble in alcohol, melts at 820°C, decomposes at 928°C; used in incandescent lighting. Also known as thorium tetrachloride.

thorium dioxide [INORG CHEM] ThO_2 A heavy, white powder soluble in sulfuric acid, insoluble in water, melts at 3300°C; used in medicine, ceramics, flame spraying, and electrodes. Also known as thoria; thorium anhydride; thorium oxide.

thorium fluoride [INORG CHEM] ThF_4 A white, toxic powder, melts at 1111°C; used to make thorium metal and magnesium-thorium alloys and in high-temperature ceramics.

thorium nitrate [INORG CHEM] $Th(NO_3)_4 \cdot 4H_2O$ Explosive white crystals soluble in water and alcohol, strong oxidizer; the anhydrous form decomposes at 500°C; used in medicine and as an analytical reagent.

thorium oxalate [ORG CHEM] $Th(C_2O_4)_2 \cdot 2H_2O$ A white, toxic powder soluble in alkalies and ammonium oxalate, insoluble in water and most acids, decomposes to thorium dioxide, ThO_2, above 300–400°C; used in ceramics.

thorium oxide *See* thorium dioxide.

thorium reactor [NUCLEO] A nuclear reactor in which thorium surrounds the central enriched uranium core to give breeder operation.

thorium series [NUCLEO] The series of nuclides resulting from the decay of thorium-232.

thorium sulfate [INORG CHEM] $Th(SO_4)_2 \cdot 8H_2O$ A white powder soluble in ice water, loses water at 42° and 400°C. Also known as normal thorium sulfate.

thorium tetrachloride *See* thorium chloride.

thorn [BOT] A short, sharp, rigid, leafless branch on a plant. [ZOO] Any of various sharp spinose structures on an animal.

thornback [VERT ZOO] *Raja clavata.* A ray found in European waters and characterized by spines on its back.

thornbush [ECOL] A vegetation class that is dominated by tall succulents and profusely branching smooth-barked deciduous hardwoods which vary in density from mesquite bush in the Caribbean to the open spurge thicket in Central Africa; the climate is that of a warm desert, except for a rather short intense rainy season. Also known as Dorngeholz; Dorngestrauch; dornveld; savane armée; savane épineuse; thorn scrub.

thorn forest [ECOL] A type of forest formation, mostly tropical and subtropical, intermediate between desert and steppe; dominated by small trees and shrubs, many armed with thorns and spines; leaves are absent, succulent, or deciduous during long dry periods, which may also be cool; an example is the caatinga of northeastern Brazil.

thorn scrub *See* thornbush.

thorogummite [MINERAL] A silicate mineral and chemical variant of thorium silicate, with similar properties; isostructural with thorite and zircon; it is deficient in silica and contains small amounts of OH in substitution for oxygen.

thoron [NUC PHYS] The common name for one of the gaseous radioactive members of the thorium series; it is the isotope of radon with mass number 220. Symbolized Tn.

thoroughfare [CIV ENG] **1.** An important, unobstructed public street or highway. **2.** A street going through from one street to another. **3.** An inland waterway for passage of ships usually not between two bodies of water.

Thorpe reaction [ORG CHEM] The reaction by which, in presence of lithium amides, α,ω-dinitriles undergo base-catalyzed condensation to cyclic iminonitriles, which can be hydrolyzed and decarboxylated to cyclic ketones.

thortveitite [MINERAL] $(Sc,Y)_2Si_2O_7$ A grayish-green mineral occurring in orthorhombic crystals; a source of scandium.

thou *See* mil.

thousandth mass unit [PHYS] A unit of energy equal to the energy equivalent of a mass of 10^{-3} atomic mass unit according to the Einstein mass-energy relation, that is, to the product of 10^{-3} atomic mass unit and the square of the speed of light; equal to approximately 1.49176×10^{-13} joule.

thread [DES ENG] A continuous helical rib, as on a screw or pipe. [GEOL] An extremely small vein, even thinner than a stringer. [MIN ENG] A more or less straight line of stall faces, having no cuttings, loose ends, fast ends, or steps. [TEXT] A continuous strand formed by spinning and twisting together short strands of textile fibers.

thread blight [PL PATH] A fungus disease of a number of tropical and semitropical woody plants, including cocoa and tea, caused by species of *Pellicularia* and *Marasmius* which form filamentous mycelia on the surface of twigs and leaves.

thread contour [DES ENG] The shape of thread design as observed in a cross section along the major axis, for example, square or round.

thread count [TEXT] An index of the compactness of a fabric determined by counting the number of warp yarns and filling yarns in 1 square inch of fabric. Also known as cloth count.

thread cutter [MECH ENG] A tool used to cut screw threads on a pipe, screw, or bolt.

threadfin [VERT ZOO] Common name for one of the fishes in the family Polynemidae.

threading die [MECH ENG] A die which may be solid, adjustable, or spring adjustable, or a self-opening die head, used to produce an external thread on a part.

threading machine [MECH ENG] A tool used to cut or form threads inside or outside a cylinder or cone.

thread-lace scoria [GEOL] Scoria whose vesicle walls have collapsed and are represented only by a network of threads.

thread protector [ENG] A short-threaded ring to screw onto a pipe or into a coupling to protect the threads while the pipe is being handled or transported. Also known as pipe-thread protector.

thread rating [ENG] The maximum internal working pressure allowable for threaded pipe or tubing joints; important

for pressure systems, chemical processes, and oil-well systems.

thread waste [TEXT] The hard, thready waste left on bobbins or collected during operations such as spinning, twisting, and weaving.

three-address code [ADP] In computers, a multiple-address code which includes three addresses, usually two addresses from which data are taken and one address where the result is entered; location of the next instruction is not specified, and instructions are taken from storage in preassigned order.

three-address instruction [ADP] In computers, an instruction which includes an operation and specifies the location of three registers.

three-arm protractor [NAV] An instrument consisting essentially of a circle graduated in degrees, to which is attached one fixed arm and two arms pivoted at the center and provided with clamps so that they can be set at any angle to the fixed arm, within the limits of the instrument; used for finding a ship's position when the angles between three fixed and known points are measured. Also known as station pointer.

three-body problem [MECH] The problem of predicting the motions of three objects obeying Newton's laws of motion and attracting each other according to Newton's law of gravitation.

three-day fever *See* phlebotomus fever.

three-decibel coupler [ELECTROMAG] Junction of two waveguides having a common H wall; the two guides are coupled together by H-type aperture coupling; the coupling is such that 50% of the power from either channel will be fed into the other. Also known as Riblet coupler; short-slot coupler.

three-decision problem [STAT] A problem in which a choice must be made among three possible courses of action.

three-dimensional [SCI TECH] Giving the illusion of depth, in three dimensions.

three-dimensional display system [ELECTR] A radar display which shows range, azimuth, and elevation; for instance, a G display.

three-dimensional flow [FL MECH] Any fluid flow which is not a two-dimensional flow.

three-index symbols *See* Christoffel symbols.

three-jaw chuck [DES ENG] A drill chuck having three serrated-face movable jaws that can grip and hold fast an inserted drill rod.

three-j number [QUANT MECH] A coefficient used in coupling eigenfunctions of two commuting angular momenta to form eigenfunctions of the total angular momentum; closely related to the Clebsch-Gordan coefficients. Also known as Wigner 3-*j* symbol.

three-junction transistor [ELECTR] A *pnpn* transistor having three junctions and four regions of alternating conductivity; the emitter connection may be made to the *p* region at the left, the base connection to the adjacent *n* region, and the collector connection to the *n* region at the right, while the remaining *p* region is allowed to float.

three-layer diode [ELECTR] A junction diode with three conductivity regions.

three-level laser [OPTICS] A laser involving three energy levels, one of which is the ground state; laser action usually occurs between the intermediate and ground states.

three-level maser [PHYS] A solid-state maser in which three energy levels are used; successful operation has been obtained with crystals of gadolinium ethyl sulfate and crystals of potassium chromecyanide at the temperature of liquid helium.

threeling *See* trilling.

three-phase circuit [ELEC] A circuit energized by alternating-current voltages that differ in phase by one-third of a cycle or 120°.

three-phase current [ELEC] Current delivered through three wires, with each wire serving as the return for the other two and with the three current components differing in phase successively by one-third cycle, or 120 electrical degrees.

three-phase four-wire system [ELEC] System of alternating-current supply comprising four conductors, three of which are connected as in a three-phase, three-wire system, the fourth being connected to the neutral point of the supply, which may be grounded.

three-phase magnetic amplifier [ELECTR] A magnetic amplifier whose input is the sum of three alternating-current voltages that differ in phase by 120°.

three-phase rectifier [ELEC] A rectifier supplied by three alternating-current voltages that differ in phase by one-third of a cycle or 120°.

three-phase seven-wire system [ELEC] System of alternating-current supply from groups of three single-phase transformers connected in Y to obtain a three-phase, four-wire grounded neutral system of higher voltage for power, the neutral wire being common to both systems.

three-phase system [PHYS] Any physical system in which three distinct phases coexist; phases can be liquid, solid, vapor (gas), or three mutually insoluble liquids, or any combination thereof.

three-phase three-wire system [ELEC] System of alternating-current supply comprising three conductors between successive pairs of which are maintained alternating differences of potential successively displaced in phase by one-third cycle.

three-piece set [MIN ENG] A set of timber consisting of a cap and its two supportive posts.

three-plus-one address [ADP] An instruction format containing an operation code, three operand address parts, and a control address.

three-ply [SCI TECH] Consisting of three distinct strands, layers, or veneers.

three-point bending [MET] Bending a piece of metal by placing the specimen on two supports and then applying a load on it between the supported ends.

three-point method [GEOL] A method used to determine the dip and strike of a structural surface from three points of varying elevation along the surface.

three-point problem [ENG] The problem of locating the horizontal position of a point of observation from the two observed horizontal angles subtended by three known sides of a triangle.

three-pulse cascaded canceler [ELECTR] A moving-target indicator technique in which two "two-pulse cancelers" are cascaded together; this improves the velocity response.

three-quarters hard [MET] A temper designation for various nonferrous metals, such as aluminum, copper, and magnesium alloys, expressing degree of hardness achieved by mechanical working.

three-shift cyclic mining [MIN ENG] A system of cyclic mining on a longwall conveyor face, with coal cutting on one shift, hand filling and conveying on the next, and ripping, packing, and advancement of the face conveyor on the third shift.

three-way switch [ELEC] An electric switch with three terminals used to control a circuit from two different points.

three-wire generator [ELEC] Electric generator with a balance coil connected across the armature, the midpoint of the coil providing the potential of the neutral wire in a three-wire system.

three-wire system [ELEC] System of electric supply comprising three conductors, one of which (known as the neutral wire) is maintained at a potential midway between the potential of the other two (referred to as the outer conductors); part of the load may be connected directly between the outer conductors, the remainder being divided as evenly as possible into two parts, each of which is connected between the neutral and one outer conductor; there are thus two distinct supply voltages, one being twice the other.

threonine [BIOCHEM] $CH_3CHOHCH(NH_2)COOH$ A crystalline α-amino acid considered essential for normal growth of animals; it is biosynthesized from aspartic acid and is a precursor of isoleucine in microorganisms.

thresher [AGR] A machine that separates grain or seeds from straw. Also known as threshing machine.

thresher shark [VERT ZOO] Common name for fishes in the family Alopiidae; pelagic predacious sharks of generally wide distribution that have an extremely long, whiplike tail with which they thrash the water, destroying schools of small fishes.

threshing machine See thresher.

threshold [BUILD] A piece of stone, wood, or metal that lies under an outside door. [ELECTR] In a modulation system, the smallest value of carrier-to-noise ratio at the input to the demodulator for all values above which a small percentage change in the input carrier-to-noise ratio produces a substantially equal or smaller percentage change in the output signal-to-noise ratio. [ENG] The least value of a current, voltage, or other quantity that produces the minimum detectable response in an instrument or system. [GEOL] See riegel. [MATH] A logic operator such that, if $P, Q, R, S, \ldots$ are statements, then the threshold will be true if at least N statements are true, false otherwise. [PHYS] The minimum level of some input quantity needed for some process to take place, such as a threshold energy for a reaction, or the minimum level of pumping at which a laser can go into self-excited oscillation. [PHYSIO] The minimum level of a stimulus that will evoke a response in an irritable tissue.

threshold contrast [OPTICS] The smallest contrast of luminance (or brightness) that is perceptible to the human eye under specified conditions of adaptation luminance and target visual angle. Also known as contrast threshold; liminal contrast.

threshold depth See sill depth.

threshold detector [NUC PHYS] An element or isotope in which radioactivity is induced only by the capture of neutrons having energies in excess of a certain characteristic threshold value; used to determine the neutron spectrum from a nuclear explosion.

threshold dose [NUCLEO] The minimum radiation dose that will produce a detectable specified effect.

threshold frequency [ELECTR] The frequency of incident radiant energy below which there is no photoemissive effect.

threshold illuminance [OPTICS] The lowest value of illuminance which the eye is capable of detecting under specified conditions of background luminance and degree of dark adaptation of the eye. Also known as flux density threshold.

threshold lights [NAV] Lights so placed as to indicate the longitudinal limits of that portion of a runway, channel, or landing path usable for landing.

threshold limit value [MED] The average concentration of toxic gas to which the normal person can be exposed without injury for 8 hours per day, 5 days per week for an unlimited period; differs slightly from maximum allowable concentration in that threshold limit value is an average concentration.

threshold of audibility [PHYSIO] The minimum effective sound pressure of a specified signal that is capable of evoking an auditory sensation in a specified fraction of the trials; the threshold may be expressed in decibels relative to 0.0002 microbar or 1 microbar. Also known as threshold of detectability; threshold of hearing.

threshold of detectability See threshold of audibility.

threshold of hearing See threshold of audibility.

threshold of reaction [PHYS] The minimum energy, for an incident particle or photon, below which a particular reaction does not occur.

threshold signal [ELECTROMAG] A received radio signal (or radar echo) whose power is just above the noise level of the receiver. Also known as minimum detectable signal.

threshold speed [ENG] The minimum speed of current at which a particular current meter will measure at its rated reliability.

threshold switch [ELECTR] A voltage-sensitive alternating-current switch made from a semiconductor material deposited on a metal substrate; when the alternating-current voltage acting on the switch is increased above the threshold value, the number of free carriers present in the semiconductor material increases suddenly, and the switch changes from a high resistance of about 10 megohms to a low resistance of less than 1 ohm; in other versions of this switch, the threshold voltage is controlled by heat, pressure, light, or moisture.

threshold treatment [CHEM ENG] The process of stopping a precipitation-type reaction at the threshold of precipitate formation; used in water-treatment reactions.

threshold value [CONT SYS] The minimum input that produces a corrective action in an automatic control system.

threshold velocity [GEOPHYS] The minimum velocity at which wind or water begins to move particles of soil, sand, or other material at a given place under specified conditions.

Threskiornithidae [VERT ZOO] The ibises, a family of long-legged birds in the order Ciconiiformes.

thrip [INV ZOO] A small, slender-bodied phytophagous insect of the order Thysanoptera with suctorial mouthparts, a stout proboscis, a vestigial right mandible, and a fully developed left mandible, while wings may be present or absent.

Thripidae [INV ZOO] A large family of thrips, order Thysanoptera, which includes the most common species.

throat [ANAT] The region of the vertebrate body that includes the pharynx, the larynx, and related structures. [BOT] The upper, spreading part of the tube of a gamopetalous calyx or corolla. [DES ENG] The narrowest portion of a constricted duct, as in a diffuser or a venturi tube; specifically, a nozzle throat. [ENG] The smaller end of a horn or tapered waveguide.

throatable [DES ENG] Of a nozzle, designed to allow a change in the velocity of the exhaust stream by changing the size and shape of the throat of the nozzle.

throat balls See niter balls.

throat depth [MET] The distance from the center line of the electrodes or platens of a resistance welding machine to the nearest point of interference for flat work.

throat microphone [ENG ACOUS] A contact microphone that is strapped to the throat of a speaker and reacts directly to throat vibrations rather than to the sound waves they produce.

throat of fillet weld [MET] The thinnest part of a fillet weld, or the shortest distance from the root of a fillet weld to its face.

throat velocity See critical velocity.

thrombin [BIOCHEM] An enzyme elaborated from prothrombin in shed blood which induces clotting by converting fibrinogen to fibrin.

thrombinogen See prothrombin.

thromboangitis obliterans [MED] Thrombosis with organization and a variable degree of associated inflammation in the arteries and veins of the extremities, occasionally of the viscera, progressing to fibrosis about these structures and associated nerves, and complicated by ischemic changes in the parts supplied. Also known as Buerger's disease.

thrombocyte [HISTOL] One of the minute protoplasmic disks found in vertebrate blood; thought to be fragments of megakaryocytes. Also known as blood platelet; platelet.

thrombocythemia See thrombocytosis.

thrombocytopenia [MED] A condition characterized by a decrease in the absolute number of thrombocytes in the circulation.

thrombocytopenic purpura [MED] Hemorrhages in the skin, mucous membranes, and elsewhere associated with a decreased number of thrombocytes per unit volume of blood.

thrombocytosis [MED] A condition characterized by an increase in the absolute number of thrombocytes in the circulation. Also known as piastrenemia; thrombocythemia.

thromboembolectomy [MED] Surgical removal of an embolus that stems from a dislodged thrombus or part of a thrombus.

thromboembolism [MED] An embolism resulting from a dislodged thrombus or part of a thrombus.

thrombokinase [BIOCHEM] A proteolytic enzyme in blood plasma that, together with thromboplastin, calcium, and factor V, converts prothrombin to thrombin.

thrombophlebitis [MED] Inflammation of a vein associated with thrombosis.

thromboplastin [BIOCHEM] Any of a group of lipid and protein complexes in blood that accelerate the conversion of prothrombin to thrombin. Also known as factor III; plasma thromboplastin component (PTC).

thromboplastinogen See antihemophilic factor.

thrombosis [MED] Formation of a thrombus.

thrombotic thrombocytopenic purpura [MED] Thrombi in blood vessels associated with deposits of hyaline substances in the walls and with thrombocytopenia. Also known as Moschcowitz's disease.

thrombus [MED] A blood clot occurring on the wall of a blood vessel where the endothelium is damaged.

Throscidae [INV ZOO] The false metallic wood-boring beetles, a cosmopolitan family of the Coleoptera in the superfamily Elateroidea.

throttled flow [FL MECH] Flow which is forced to pass through a restricted area, where the velocity must increase.

throttle valve [MECH ENG] A choking device to regulate flow of a liquid, for example, in a pipeline, to an engine or turbine, from a pump or compressor.

throttling [AERO ENG] The varying of the thrust of a rocket engine during powered flight. [CONT SYS] Control by means of intermediate steps between full on and full off. [THERMO] An adiabatic, irreversible process in which a gas expands by passing from one chamber to another chamber which is at a lower pressure than the first chamber.

throttling bar [ORD] A wedge-shaped bar attached to the interior wall of a recoil cylinder of a gun; a rectangular notch in the recoil piston moves over the bar, forming an aperture through which the recoil oil must flow; the aperture decreases during recoil until at the end of recoil it is completely closed, thus stopping the flow of liquid and bringing the recoiling tube to rest.

throttling calorimeter [ENG] An instrument utilizing the principle of constant enthalpy expansion for the measurement of the moisture content of steam; steam drawn from a steampipe through sampling nozzles enters the calorimeter through a throttling orifice and moves into a well-insulated expansion chamber in which its temperature is measured. Also known as steam calorimeter.

throttling groove [ORD] A groove of varying width or depth cut on the interior wall of some recoil cylinders of a gun to control the passage of oil, hence controlling the recoil resistance.

throttling rod [ORD] A tapered rod attached to a recoil cylinder of a gun; the piston rod is hollow to receive the throttling rod during recoil, and oil is forced through the orifice around the throttling rod; as recoil proceeds, the larger section of the throttling rod comes into the throttling orifice until, at the end of recoil, the throttling rod nearly seals the throttling orifice, thus bringing the recoiling tube to rest.

throttling valve [ORD] Part of a variable recoil system of a gun, it is a spring-loaded valve through which recoil oil must flow during recoil; as the gun is elevated, the spring pressure is increased by means of a control arm; the valve thus offers greater resistance to the flow of recoil oil and the length of recoil is reduced; in a similar manner, depressing the gun decreases the spring pressure of the valve, with less resistance to oil flow and hence greater length of recoil.

through arch [CIV ENG] An arch bridge from which the roadway is suspended as distinct from one which carries the roadway on top.

through glacier [HYD] A two-ended glacier, consisting of two valley glaciers in a depression, flowing in opposite directions.

throughput [ADP] The productivity of a data-processing system, as expressed in computing work per minute or hour. [CHEM ENG] The volume of feedstock charged to a process equipment unit during a specified time. [COMMUN] A measure of the effective rate of transmission of data by a communications system. [MIN ENG] The quantity of ore or other material passed through a mill or a section of a mill in a given time or at a given rate.

through repeater [ELECTR] Microwave repeater that is not equipped to provide for connections to any local facilities other than the service channel.

through street [CIV ENG] A street at which all cross traffic is required to stop before crossing or entering. Also known as throughway.

through valley [GEOL] **1.** A depression eroded across a divide by glacier ice or meltwater streams. **2.** A valley excavated by a through glacier.

throughway See expressway; through street.

through weld [MET] A long weld made through the unbroken surface of one member to the other member in a lap or tree joint.

throw [GEOL] The vertical component of dip separation on a fault, or generally the amount of vertical displacement on any fault.

throwing power [MET] The ability of an electroplating solution to deposit metal uniformly on an irregularly shaped cathode.

THRIP

Flower thrip (*Frankliniella tritici*), a common and widely distributed species of thrip, about 1 millimeter long, showing two pairs of long featherlike wings.

THULIUM

**69
Tm**

Periodic table of the chemical elements showing the position of thulium.

THUMBSCREW

Drawing of thumbscrew showing flattened head. *(Reynolds Metals Co.)*

THUNDERSTORM

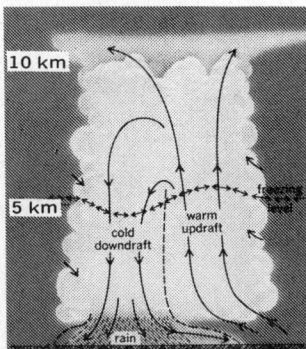

10 km

5 km

freezing level

cold downdraft

warm updraft

rain

cold diverging downdraft   squall edge

Diagram showing simplified circulation in a vertical section through a mature thunderstorm. *(After G. A. Suckstorff, H. R. Byers, and R. R. Braham, Jr.)*

throwout [MECH ENG] In automotive vehicles, the mechanism or assemblage of mechanisms by which the driven and driving plates of a clutch are separated.

throw-out spiral *See* lead-out groove.

thrush [MED] A form of candidiasis due to infection by *Candida albicans* and characterized by small whitish spots on the tip and sides of the tongue and the mucous membranes of the buccal cavity. Also known as mycotic stomatitis; parasitic stomatitis. [VET MED] A disease of the frog of a horse's foot accompanied by a fetid discharge.

thrust [AERO ENG] **1.** The pushing or pulling force developed by an aircraft engine or a rocket engine. **2.** Specifically, in rocketry, thrust is $F = mv$ where m is propellant mass flow and v is exhaust velocity relative to the vehicle. Also known as momentum thrust. [GEOL] Overriding movement of one crystal unit over another. [MECH] The force exerted in any direction by a fluid engine or a rocket engine. [MECH ENG] The weight or pressure applied to a bit to make it cut. [MIN ENG] **1.** A crushing of coal pillars caused by excess weight of the superincumbent rocks, the floor being harder than the roof. **2.** The ruins of the fallen roof, after pillars and stalls have been removed.

thrust augmentation [AERO ENG] The increasing of the thrust of an engine or power plant, especially of a jet engine and usually for a short period of time, over the thrust normally developed.

thrust augmenter [AERO ENG] Any contrivance used for thrust augmentation, as a venturi used in a rocket.

thrust axis [AERO ENG] A line or axis through an aircraft or a rocket, along which the thrust acts; an axis through the longitudinal center of a jet or rocket engine, along which the thrust of the engine acts. Also known as axis of thrust; center of thrust.

thrust block *See* thrust nappe.

thrust coefficient *See* nozzle thrust coefficient.

thruster [AERO ENG] A control jet employed in spacecraft; an example would be one utilizing hydrogen peroxide.

thrust fault [GEOL] A low-angle (less than a 45° dip) fault along which the hanging wall has moved up relative to the footwall. Also known as reverse fault; reverse slip fault; thrust slip fault.

thrust horsepower [AERO ENG] **1.** The force-velocity equivalent of the thrust developed by a jet or rocket engine. **2.** The thrust of an engine-propeller combination expressed in horsepower; it differs from the shaft horsepower of the engine by the amount the propeller efficiency varies from 100%. [NAV ARCH] The product of the speed of advance of a marine propeller through the water, in feet per second, and the thrust delivered by the propeller, in pounds, divided by 550.

thrust load [MECH ENG] A load or pressure parallel to or in the direction of the shaft of a vehicle.

thrust meter [ENG] An instrument for measuring static thrust, especially of a jet engine or rocket.

thrust nappe [GEOL] The body of rock that makes up the hanging wall of a thrust fault. Also known as thrust block; thrust plate; thrust sheet; thrust slice.

Thrustor [MECH ENG] Trademark for a hydraulic device for applying a controllable force, as to a brake.

thrust output [AERO ENG] The net thrust delivered by a jet engine, rocket engine, or rocket motor.

thrust plate *See* thrust nappe.

thrust-pound [AERO ENG] A unit of measurement for the thrust produced by a jet engine or rocket, as in an expression such as 5000 thrust-pounds.

thrust power [AERO ENG] The power usefully expended on thrust, equal to the thrust (or net thrust) times airspeed.

thrust reverser [AERO ENG] A device or apparatus for reversing thrust, especially of a jet engine.

thrust section [AERO ENG] A section in a rocket vehicle that houses or incorporates the combustion chamber or chambers and nozzles.

thrust sheet *See* thrust nappe.

thrust slice *See* thrust nappe.

thrust slip fault *See* thrust fault.

thrust terminator [AERO ENG] A device for ending the thrust in a rocket engine, either through propellant cutoff (in the case of a liquid) or through diverting the flow of gases from the nozzle.

thrust-weight ratio [AERO ENG] A quantity used to evaluate engine performance, obtained by dividing the thrust output by the engine weight less fuel.

thrust yoke [MECH ENG] The part connecting the piston rods of the feed mechanism on a hydraulically driven diamond-drill swivel head to the thrust block, which forms the connecting link between the yoke and the drive rod, by means of which link the longitudinal movements of the feed mechanism are transmitted to the swivel-head drive rod. Also known as back end.

thucolite [GEOL] Concentrations of carbonaceous matter in ancient sedimentary rocks.

thuja oil [MATER] An essential oil from white cedar leaves; pale-yellow, combustible oil with a camphor aroma, soluble in alcohol, ether, chloroform, carbon disulfide, and fixed oils; used in medicine, perfumery, and flavorings. Also known as arbor vitae oil.

Thule group [ASTRON] An accumulation of asteroids whose sidereal period of revolution is in the ratio ⅓ with that of Jupiter.

thulia *See* thulium oxide.

thulite [MINERAL] A pink, rose-red, or purplish-red variety of epidote that contains manganese; used as an ornamental stone.

thulium [CHEM] A rare-earth element, symbol Tm, group IIIB, of the lanthanide group, atomic number 69, atomic weight 168.934; reacts slowly with water, soluble in dilute acids; melts at 1550°C, boils at 1727°C; the dust is a fire hazard; used as x-ray source and to make ferrites.

thulium-170 [NUC PHYS] The radioactive isotope of thulium, with mass number 170; used as a portable x-ray source.

thulium chloride [INORG CHEM] $TmCl_3 \cdot 7H_2O$ Green, deliquescent crystals soluble in water and alcohol, melts at 824°C.

thulium oxalate [ORG CHEM] $Tm_2(C_2O_4)_3 \cdot 6H_2O$ A toxic, greenish-white solid, soluble in aqueous alkali oxalates, loses one water at 50°C; used for analytical separation of thulium from common metals.

thulium oxide [INORG CHEM] Tm_2O_3 A white, slightly hygroscopic powder that absorbs water and carbon dioxide from the air, and is slowly soluble in strong acids; used to make thulium metal. Also known as thulia.

thumbscrew [DES ENG] A screw with a head flattened in the same axis as the shaft so that it can be gripped and turned by the thumb and forefinger.

thump [ENG ACOUS] Low-frequency transient disturbance in a system or transducer characterized audibly by the vocal imitation of the word.

Thunburg technique [BIOCHEM] A technique used to study oxidation of a substrate occurring by dehydrogenation reactions; methylene blue, a reversibly oxidizable indicator, substitutes for molecular oxygen as the ultimate hydrogen acceptor (oxidant), becoming reduced to the colorless leuco form.

thunder [GEOPHYS] The sound emitted by rapidly expanding gases along the channel of a lightning discharge.

thunderbolt [GEOPHYS] In mythology, a lightning flash accompanied by a material bolt or dart and which causes great damage; it is still used as a popular term for a single lightning discharge accompanied by thunder.

Thunderchief [AERO ENG] A United States supersonic, single-engine, turbojet-powered tactical fighter capable of delivering nuclear weapons as well as nonnuclear bombs and rockets; an all-weather attack fighter, it is also capable of close support for ground forces, and its range can be extended by in-flight refueling; it is equipped with the Sidewinder missile. Designated F-105.

thundercloud [METEOROL] A convenient and often used term for the cloud mass of a thunderstorm, that is, a cumulonimbus.

thundersquall *See* rainsquall.

thunderstorm [METEOROL] A convective storm accompanied by lightning and thunder and rain, rarely snow showers but often hail, and gusty squall winds at the onset of precipitation; the characteristic cloud is the cumulonimbus.

thunderstorm cell [METEOROL] The convection cell of a cumulonimbus cloud.

thunderstorm charge separation [GEOPHYS] **1.** The process by which the large electric field found within thunderclouds is generated. **2.** The processes by which particles bearing opposite electrical charges are given the charges and are transported to different regions of the active cloud.

thunderstorm day [METEOROL] An observational day during which thunder is heard at the station; precipitation need not occur.

Thunderstreak [AERO ENG] A United States one-man fighter also used for reconnaissance; it has a range of over 2000 miles (3200 kilometers) and a speed of over 600 miles (970 kilometers) per hour; the bomb load is 6000 pounds (2700 kilograms) of conventional or nuclear bombs, incendiary gel, or rockets, and there are six .50-caliber machine guns. Designated F-84.

Thunnidae [VERT ZOO] The tunas, a family of perciform fishes; there are no scales on the posterior part of the body, and those on the anterior are fused to form an armored covering, the body is streamlined, and the tail is crescent-shaped.

Thurniaceae [BOT] A small family of monocotyledonous plants in the order Juncales distinguished by an inflorescence of one or more dense heads, vascular bundles of the leaf in vertical pairs, and silica bodies in the leaf epidermis.

Thylacinidae [VERT ZOO] A family of Australian carnivorous marsupials in the superfamily Dasyuroidea.

Thylacoleonidae [PALEON] An extinct family of carnivorous marsupials in the superfamily Phalangeroidea.

Thylox process [CHEM ENG] The conversion of coke-oven-gas, hydrogen sulfide, into elemental sulfur by absorption in a solution of arsenious oxide and soda ash in water, then air-blowing to liberate the sulfur which can be skimmed off the solution.

thyme [BOT] A perennial mint plant of the genus *Thymus*; pungent aromatic herb is made from the leaves.

thyme camphor *See* thymol.

Thymelaeceae [BOT] A family of dicotyledonous woody plants in the order Myrtales characterized by a superior ovary with a solitary ovule, and petals, if present, are scalelike.

thyme oil [MATER] An essential oil found in the flowers of the thymes *Thymus vulgaris* or *T. zygis*, a colorless to reddish-brown liquid with a sharp taste and pleasant aroma, soluble in alcohol, slightly soluble in water; used in medicine, perfumery, cosmetics, flavoring, and soap.

thymic aplasia [MED] Congenital absence of the thymus and of the parathyroids with deficient cellular immunity. Also known as Di George's syndrome.

thymic corpuscle [HISTOL] A characteristic, rounded, acidophil body in the medulla of the thymus; composed of hyalinized cells concentrically arranged about a core which is occasionally calcified. Also known as Hassal's body.

thymidine [BIOCHEM] $C_{10}H_{14}N_2O_5$ A nucleoside derived from deoxyribonucleic acid; essential growth factor for certain microorganisms in mediums lacking vitamin B_{12} and folic acid.

thymidylic acid [BIOCHEM] The phosphate ester of thymidine; nucleotide of thymine.

thymine [BIOCHEM] $C_5H_6N_2O_2$ A pyrimidine component of nucleic acid, first isolated from the thymus.

thymocyte [HISTOL] A lymphocyte formed in the thymus.

thymol [ORG CHEM] $C_{10}H_{14}O$ A naturally occurring crystalline phenol obtained from thyme or thyme oil, melting at 515°C; used to kill parasites in herbaria, to preserve anatomical specimens, and in medicine as a topical antifungal agent. Also known as isopropyl-*meta*-cresol; thyme camphor.

thymol blue [ORG CHEM] $C_6H_4SO_2OC[C_6H_2(CH_3)(OH)-CH(CH_3)_2]_2$ Brown-green crystals soluble in alcohol and dilute alkalies, insoluble in water, decomposes at 223°C; used as acid-base pH indicator. Also known as thymolsulfonphthalein.

thymol iodide [ORG CHEM] $[C_6H_2(CH_3)(OI)(C_3H_7)]_2$ A red-brown, light-sensitive powder with an aromatic aroma, soluble in ether and chloroform, insoluble in water; used in medicine and as a feed additive.

thymolphthalein [ORG CHEM] $C_6H_4COOC[C_6H_2(CH_3)(OH)-$ $CH(CH_3)_2]_2$ A white powder insoluble in water, soluble in alcohol and acetone, melts at 245°C; used in medicine and as an acid-base titration indicator.

thymolsulfonphthalein *See* thymol blue.

thymoma [MED] A usually benign primary tumor of the thymus composed principally of lymphocytic and epithelial cells in varying proportions.

thymopharyngeal duct [EMBRYO] The third pharyngo-branchial duct; it may elongate between the pharynx and thymus.

thymus gland [ANAT] A lymphoid organ in the neck or upper thorax of all vertebrates; it is most prominent in early life and is essential for normal development of the circulating pool of lymphocytes.

thyratron [ELECTR] A hot-cathode gas tube in which one or more control electrodes initiate but do not limit the anode current except under certain operating conditions. Also known as hot-cathode gas-filled tube.

thyratron gate [ELECTR] In computers, an AND gate consisting of a multielement gas-filled tube in which conduction is initiated by the coincident application of two or more signals; conduction may continue after one or more of the initiating signals are removed.

thyratron inverter [ELECTR] An inverter circuit that uses thyratrons to convert direct-current power to alternating-current power.

thyrector [ELECTR] Silicon diode that acts as an insulator up to its rated voltage, and as a conductor above rated voltage; used for alternating-current surge voltage protection.

Thyrididae [INV ZOO] The window-winged moths, a small tropical family of lepidopteran insects in the suborder Heteroneura.

thyristor [ELECTR] A transistor having a thyratronlike characteristic; as collector current is increased to a critical value, the alpha of the unit rises above unity to give high-speed triggering action.

thyrocalcitonin *See* calcitonin.

thyroglobulin [BIOCHEM] An iodinated protein found as the storage form of the iodinated hormones in the thyroid follicular lumen and epithelial cells.

thyroglossal cyst [MED] A cyst formed from the remnants of the thyroglossal duct.

thyroglossal duct [EMBRYO] A narrow temporary channel connecting the anlage of the thyroid with the surface of the tongue.

thyroid [PHARM] Dried and powdered thyroid gland which contains about 0.2% iodine in combination, especially as thyroxine, and is used therapeutically in the treatment of thyroid deficiencies.

thyroid cartilage [ANAT] The largest of the laryngeal cartilages in humans and most other mammals, located anterior to the cricoid; in man, forms the Adam's apple.

thyroidectomy [MED] Surgical removal of the thyroid gland.

thyroid gland [ANAT] An endocrine gland found in all vertebrates that produces, stores, and secretes the thyroid hormones.

thyroid hormone [BIOCHEM] Commonly, thyroxine or triiodothyronine, or both; a metabolically active compound formed and stored in the thyroid gland which functions to regulate the rate of metabolism.

thyroiditis [MED] Inflammation of the thyroid gland.

thyroid-stimulating hormone *See* thyrotropic hormone.

thyroprotein [BIOCHEM] A protein secreted in the thyroid gland, such as thyroxine.

Thyropteridae [VERT ZOO] The New-World disk-winged bats, a family of the Chiroptera found in Central and South America, characterized by a stalked sucking disk and a well-developed claw on the thumb.

thyrotoxic myopathy [MED] A chronic disease associated with hyperthyroidism resulting in muscular atrophy.

thyrotoxicosis *See* hyperthyroidism.

thyrotropic hormone [BIOCHEM] A hormone produced by the adenohypophysis which regulates thyroid gland function. Also known as thyroid-stimulating hormone (TSH).

thyrotropin [BIOCHEM] A thyroid-stimulating hormone produced by the adenohypophysis.

thyroxine [BIOCHEM] $C_{15}H_{11}I_4NO_4$ The active physiologic

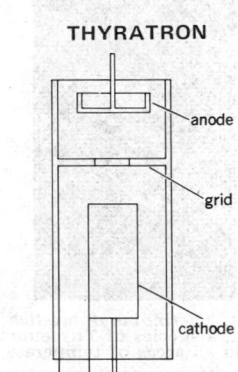

THYRATRON

Diagram of construction of a negative-grid thyratron.

principle of the thyroid gland; used in the form of the sodium salt for replacement therapy in states of hypothyroidism or absent thyroid function.

thyrse [BOT] An inflorescence with a racemose primary axis and cymose secondary and later axes.

Thysanidae [INV ZOO] A family of hymenopteran insects in the superfamily Chalcidoidea.

Thysanoptera [INV ZOO] The thrips, an order of small, slender insects having exopterygote development, sucking mouthparts, and exceptionally narrow wings with few or no veins and bordered by long hairs.

Thysanura [INV ZOO] The silverfish, machilids, and allies, an order of primarily wingless insects with soft, fusiform bodies.

THz See terahertz.

Ti See titanium.

tibia [ANAT] The larger of the two leg bones, articulating with the femur, fibula, and talus.

tibialis [ANAT] **1.** A muscle of the leg arising from the proximal end of the tibia and inserted into the first cuneiform and first metatarsal bones. **2.** A deep muscle of the leg arising proximally from the tibia and fibula and inserted into the navicular and first cuneiform bones.

tic douloureux See trigeminal neuralgia.

tick [ADP] A time interval equal to 1/60 second, used primarily in discussing computer operations. [COMMUN] A pulse broadcast at 1-second intervals by standard frequency and time broadcasting stations to indicate the exact time. [INV ZOO] Any arachnid comprising Ixodoidea; a bloodsucking parasite and important vector of various infectious diseases of humans and lower animals.

tick-bite paralysis [VET MED] A flaccid paralysis in animals, and occasionally in humans, caused by a feeding tick attached to the body.

tick-borne typhus fever of Africa [MED] Any of several infections caused by *Rickettsia conori,* transmitted by ixodid ticks, and occurring in Africa and adjacent areas; includes boutonneuse fever, Marseilles fever, Kenya tick typhus fever, and South African tick-bite fever.

ticket converting [ADP] The process of changing prepunched ticket stubs, 2.7 inches (6.9 centimeters) wide by 1 inch (2.5 centimeters) deep, into punched cards; the ticket is made up of a basic section and one or more stubs that are numerically prepunched and printed with identical information.

tick fever See Rocky Mountain spotted fever.

tickle [PHYSIO] A tingling sensation of the skin or a mucous membrane following light, tactile stimulation.

tickler coil [ELECTR] Small coil connected in series with the plate circuit of an electron tube and inductively coupled to a grid-circuit coil to establish feedback or regeneration in a radio circuit; used chiefly in regenerative detector circuits.

tick typhus See Rocky Mountain spotted fever.

tic polonga [VERT ZOO] *Vipera russellii.* A member of the Viperidae; one of the most deadly and most common snakes in India; it may reach a length of 5 feet (1.5 meters), is nocturnal in its habits, and pursues rodents into houses. Also known as Russell's viper.

tidal bore See bore.

tidal channel [OCEANOGR] A major channel followed by tidal currents, extending from the ocean into a tidal marsh or tidal flat.

tidal component See partial tide.

tidal constants [OCEANOGR] Tidal relations that remain essentially constant for any particular locality.

tidal constituent See partial tide.

tidal correction [GEOPHYS] A correction made in gravity observations to remove the effect of the earth's tides.

tidal current [OCEANOGR] The alternating horizontal movement of water associated with the rise and fall of the tide caused by the astronomical tide-producing forces.

tidal current chart [OCEANOGR] A chart showing by arrows and numbers the average direction and speed of tidal currents at a particular part of the current cycle.

tidal current tables [OCEANOGR] Tables issued annually which give daily predictions of the times of slack water and the times and velocities of the strength of flood and ebb currents for a number of reference stations, together with differences and constants for obtaining predictions at subordinate stations.

tidal cycle See tide cycle.

tidal datum [OCEANOGR] A level of the sea, defined by some phase of the tide, from which water depths and heights of tide are reckoned. Also known as tidal datum plane.

tidal datum plane See tidal datum.

tidal day [OCEANOGR] The interval between two consecutive high waters of the tide at a given place, averaging 24 hours and 51 minutes.

tidal delta [GEOL] A sand bar or shoal formed in the entrance of an inlet by the action of reversing tidal currents.

tidal difference [OCEANOGR] The difference in time or height of a high or low water at a subordinate station and at a reference station for which predictions are given in the tide tables; the difference applied to the prediction at the reference station gives the corresponding time or height for the subordinate station.

tidal energy [OCEANOGR] The energy in a tide flowing from a basin into an open sea.

tidal excursion [OCEANOGR] The net horizontal distance over which a water particle moves during one tidal cycle of flood and ebb; the distances traversed during ebb and flood are rarely equal in nature, since there is usually a layered circulation in an estuary, with a net surface flow in one direction compensated by an opposite flow at depth.

tidal flat [GEOL] A marshy, sandy, or muddy nearly horizontal coastal flatland which is alternately covered and exposed as the tide rises and falls.

tidal frequency [OCEANOGR] The rate of travel, in degrees per day, of a component of a tide, the component being created by a particular juxtaposition of forces in the sun-earth-moon system.

tidal friction [OCEANOGR] The frictional effect of the tidal wave particularly in shallow waters that lengthens the tidal epoch and tends to slow the rotational velocity of the earth, thus increasing very slowly the length of the day.

tidal glacier See tidewater glacier.

tidal harbor [OCEANOGR] A harbor affected by the tides, in distinction to a harbor in which the water level is maintained by caissons or gates.

tidal inlet [GEOL] A natural inlet maintained by tidal currents.

tidalite [GEOL] Any sediment transported and deposited by tidal currents.

tidal lights [NAV] Lights at the entrance of a harbor to indicate tide and tidal current conditions within the harbor.

tidal lock See entrance lock.

tidal marsh [GEOGR] Any marsh whose surface is covered and uncovered by tidal flow.

tidal platform ice foot [OCEANOGR] An ice foot between high and low water levels, produced by the the rise and fall of the tide.

tidal pool [OCEANOGR] An accumulation of sea water remaining in a depression on a beach or reef after the tide recedes.

tidal potential [OCEANOGR] Tidal forces expressed as components of a vector field.

tidal prism [OCEANOGR] The difference between the mean high-water volume and the mean low-water volume of an estuary.

tidal quay [CIV ENG] A quay in an open harbor or basin with sufficient depth to enable ships lying alongside to remain afloat at any state of the tide.

tidal range See tide range.

tidal scour [GEOL] Sea-floor erosion caused by strong tidal currents, resulting in removal of inshore sediments and formation of deep holes and channels. Also known as scour.

tidal stand See stand.

tidal water [OCEANOGR] Any water whose level changes periodically due to tidal action.

tidal wave [OCEANOGR] **1.** Any unusually high and generally destructive sea wave or water level along a shore. **2.** See tide wave.

tidal wind [METEOROL] A very light breeze which occurs in calm weather in inlets where the tide sets strongly; it blows onshore with rising tide and offshore with ebbing tide.

tide [OCEANOGR] The periodic rising and falling of the oceans resulting from lunar and solar tide-producing forces acting upon the rotating earth.

THYSANURA

Firebrat, *Thermobia domestica* Packard, a species of Thysanura found in all areas of temperate climate.

tide amplitude [OCEANOGR] One-half of the difference in height between consecutive high water and low water; half the tide range.

tide-bound [NAV] Referring to a vessel unable to proceed because of insufficient depth of water due to tidal action.

tide bulge *See* tide wave.

tide crack [OCEANOGR] A crack in sea ice, parallel to the shore, caused by the vertical movement of the water due to tides; several such cracks often appear as a family.

tide curve [OCEANOGR] Any graphic representation of the rise and fall of the tide; time is generally represented by the abscissas, and the height of the tide by the ordinates; for normal tides the curve so produced approximates a sine curve.

tide cycle [OCEANOGR] A period which includes a complete set of tide conditions or characteristics, such as a tidal day or a lunar month. Also known as tidal cycle.

tide gage [ENG] A device for measuring the height of a tide; may be observed visually or may consist of an elaborate recording instrument.

tide gate [CIV ENG] **1.** A restricted passage through which water runs with great speed due to tidal action. **2.** An opening through which water may flow freely when the tide sets in one direction, but which closes automatically and prevents the water from flowing in the other direction when the direction of flow is reversed.

tidehead [OCEANOGR] The inland limit of water affected by a tide.

tide hole [OCEANOGR] A hole made in ice to observe the height of the tide.

tide indicator [ENG] That part of a tide gage which indicates the height of tide at any time; the indicator may be in the immediate vicinity of the tidal water or at some distance from it.

tideland [GEOGR] Land which is under water at high tide and uncovered at low tide.

tide lock *See* entrance lock.

tide machine [ENG] An instrument that computes, sometimes for years in advance, the times and heights of high and low waters at a reference station by mechanically summing the harmonic constituents of which the tide is composed.

tidemark [OCEANOGR] **1.** A high-water mark left by tidal water. **2.** The highest point reached by a high tide.

tide notes [OCEANOGR] Notes included on nautical charts which give information on the mean range or the diurnal range of the tide, mean tide level, and extreme low water at key places on the chart.

tide pole [ENG] A graduated spar used for measuring the rise and fall of the tide. Also known as tide staff.

tide prediction [OCEANOGR] The mathematical process by which the times and heights of the tide are determined in advance from the harmonic constituents at a place.

tide-producing force [GEOPHYS] The slight local difference between the gravitational attraction of two astronomical bodies and the centrifugal force that holds them apart.

tide race [OCEANOGR] A strong tidal current or a channel in which such a current flows.

tide range [OCEANOGR] The difference in height between consecutive high and low waters. Also known as tidal range.

tide rips *See* rips.

tide-rode [NAV] Referring to a ship riding at anchor and heading into the tidal current.

tide signal [NAV] A visual signal displayed at the entrance of a harbor to indicate tidal conditions within the harbor.

tide staff *See* tide pole.

tide station [OCEANOGR] A place where observations of the tides are obtained.

tide table [OCEANOGR] A table giving daily predictions, usually a year in advance, of the times and heights of the tide for a number of reference stations.

tidewater [OCEANOGR] **1.** A body of water, such as a river, affected by tides. **2.** Water inundating land at flood tide.

tidewater glacier [HYD] A glacier that descends into the sea and usually has a terminal ice cliff. Also known as tidal glacier.

tide wave [OCEANOGR] A long-period wave associated with the tide-producing forces of the moon and sun, and identified with the rising and falling of the tide. Also known as tidal wave; tide bulge.

tideway [OCEANOGR] A channel through which a tidal current runs.

tidi-sound *See* time-division sound

tie [CIV ENG] One of the transverse supports to which railroad rails are fastened to keep them to line, gage, and grade. [ELEC] **1.** Electrical connection or strap. **2.** *See* tie wire. [ENG] A beam, post, rod, or angle to hold two pieces together; a tension member in a construction. [MIN ENG] A support for the roof in coal mines.

tieback [MIN ENG] **1.** A beam serving a purpose similar to that of a fend-off beam, but fixed at the opposite side of the shaft or inclined road. **2.** The wire ropes or stay rods that are sometimes used on the side of the tower opposite the hoisting engine, either in place of or to reinforce the engine braces.

tie bar [CIV ENG] **1.** A bar used as a tie rod. **2.** A rod connecting two switch rails on a railway to hold them to gage. [GEOL] *See* tombolo.

tie cable [ELEC] **1.** Cable between two distributing frames or distributing points. **2.** Cable between two private branch exchanges. **3.** Cable between a private branch exchange switchboard and main office. **4.** Cable connecting two other cables.

tied arch [CIV ENG] An arch having the horizontal reaction component provided by a tie between the skewbacks of the arch ends.

tied concrete column [CIV ENG] A concrete column reinforced with longitudinal bars and horizontal ties.

Tiedenmann's body [INV ZOO] One of the small glands opening into the ring vessel in many echinoderms in which amebocytes are produced.

tie-down diagram [ENG] A drawing indicating the prescribed method of securing a particular item of cargo within a specific type of vehicle.

tie-down point [ENG] An attachment point provided on or within a vehicle.

tie-down point pattern [ENG] The pattern of tie-down points within a vehicle.

tied rank [STAT] If two distinct observations have the same value, thus being given the same rank, they are said to be tied; this presents difficulties in the Wilcoxon two-sample test, the sign test, and the Fisher-Irwin test.

tie line [COMMUN] **1.** A leased communication channel or circuit. **2.** *See* data link. [PHYS CHEM] A line on a phase diagram joining the two points which represent the composition of systems in equilibrium. Also known as conode.

tie plate [CIV ENG] A metal plate between a rail and a tie to hold the rail in place and reduce wear on the tie.

tie point [ELEC] Insulated terminal to which two or more wires may be connected.

tier array [ELECTROMAG] Array of antenna elements, one above the other.

tier building [CIV ENG] A multistory skeleton frame building.

tie rod [CIV ENG] A structural member used as a brace to take tensile loads. [ENG] A round or square iron rod passing through or over a furnace and connected with buckstays to assist in binding the furnace together. [MECH ENG] A rod used as a mechanical or structural support between elements of a machine. [MIN ENG] Vertical rods mounted in overlying horizontal shaft timbers.

tie trunk [ELEC] Telephone line or channel directly connecting two private branch exchanges.

Tietze extension theorem [MATH] A topological space X is normal if and only if every continuous function of a closed subset to [0,1] can be extended to all of X.

tie wire [ELEC] A short piece of wire used to tie an open-line wire to an insulator. Also known as tie.

TIF *See* telephone influence factor.

tiger [VERT ZOO] *Felis tigris.* An Asiatic carnivorous mammal in the family Felidae characterized by a tawny coat with transverse black stripes and white underparts.

Tiger [AERO ENG] A United States single-engine, single-seat, supersonic jet fighter designed for operating from aircraft carriers for the interception and destruction of enemy aircraft, and the support of troops ashore; armament consists of Sidewinders, cannons, and rocket packs. Designated F-11.

tiger beetle [INV ZOO] The common name for any of the

bright-colored beetles in the family Cicindelidae; there are about 1300 species distributed all over the world.

tiger salamander [VERT ZOO] *Ambystoma tigrinum.* A salamander in the family Ambystomatidae, found in a variety of subspecific forms from Canada to Mexico and over most of the United States; lives in arid and humid regions, and is the only salamander in much of the Great Plains and Rocky Mountains.

tiger's-eye [MINERAL] A yellowish-brown crystalline variety of quartz; a translucent, fibrous, broadly chatoyant gemstone that may be dyed other colors.

tiger shark *See* sand shark.

tight [ENG] **1.** Unbroken, crack-free, and solid rock in which a naked hole will stand without caving. **2.** A borehole made impermeable to water by cementation or casing. [MECH ENG] Inadequate clearance or the barest minimum of clearance between working parts.

tight binding approximation [SOLID STATE] A method of calculating energy states and wave functions of electrons in a solid in which the wave function is assumed to be a sum of pure atomic wave functions centered about each of the atoms in the lattice, each multiplied by a phase factor; it is suitable for deep-lying energy levels.

tight coupling *See* close coupling.

tight fit [DES ENG] A fit between mating parts with slight negative allowance, requiring light to moderate force to assemble.

tight fold *See* closed fold.

TIG welding *See* tungsten–inert gas welding.

tile [MATER] **1.** A piece of fired clay, stone, concrete, or other material used ornamentally to cover roofs, floors, or walls. **2.** A hollow building unit made of burned clay or other material.

till [GEOL] Unsorted and unstratified drift consisting of a heterogeneous mixture of clay, sand, gravel, and boulders which is deposited by and underneath a glacier. Also known as boulder clay; glacial till; ice-laid drift.

till billow [GEOL] An undulating mass of glacial drift that is disposed in an irregular pattern with regard to the direction of movement of the ice.

tiller [NAV ARCH] A lever attached to the rudder of a boat or ship and used to turn the rudder from side to side, usually turned by hand in a boat and by mechanical devices in a ship.

Tilletiaceae [MYCOL] A family of fungi in the order Ustilaginales in which basidiospores form at the tip of the apibasidium.

tilleyite [MINERAL] $Ca_5(Si_2O_7)(CO_3)_2$ A white mineral consisting of a carbonate and silicate of calcium.

tillite [PETR] A sedimentary rock formed by lithification of till, especially pre-Pleistocene till.

Tillodontia [PALEON] An order of extinct quadrupedal land mammals known from early Cenozoic deposits in the Northern Hemisphere and distinguished by large, rodentlike incisors, blunt-cuspid cheek teeth, and five clawed toes.

tilloid [GEOL] A nonglacial till-like deposit. [PETR] A rock of uncertain origin which resembles tillite.

till plain [GEOL] An extensive, relatively flat area overlying a till.

till sheet [GEOL] A sheet, layer, or bed of till.

tilt [AERO ENG] The inclination of an aircraft, winged missile, or the like from the horizontal, measured by reference to the lateral axis or to the longitudinal axis. [ELECTROMAG] **1.** Angle which an antenna forms with the horizontal. **2.** In radar, the angle between the axis of radiation in the vertical plane and a reference axis which is normally the horizontal. [METEOROL] The inclination to the vertical of a significant feature of the circulation (or pressure) pattern or of the field of temperature or moisture; for example, troughs in the westerlies usually display a westward tilt with altitude in the lower and middle troposphere.

tilt angle [ELECTROMAG] The angle between the axis of radiation of a radar beam in the vertical plane and a reference axis (normally the horizontal).

tilt block [GEOL] A tilted fault block.

tilted iceberg [OCEANOGR] A tabular iceberg that has become unbalanced, so that the flat, level top is inclined.

tilted interface [GEOL] Oil-water interface in which water moves in a generally linear direction under an oil accumulation which is, for instance, in an anticline.

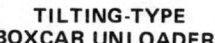
TILTING-TYPE BOXCAR UNLOADER

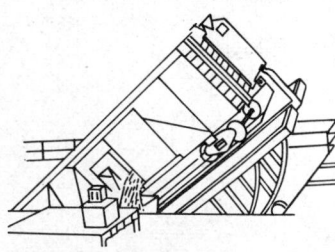

Drawing of a tilting-type unloader. (*Link-Belt Co.*)

tilt error [NAV] The error caused by the propagation of signals over the tilted ionosphere reflecting layer or, in systems requiring reception over two or more widely separated locations, by the different heights of the ionosphere reflecting layer for the various transmissions.

tilting dozer [MECH ENG] A bulldozer whose blade can be pivoted on a horizontal center pin to cut low on either side.

tilting idlers [MECH ENG] An arrangement of idler rollers in which the top set is mounted on vertical arms which pivot on spindles set low down on the frame of the roller stool.

tilting mixer [MECH ENG] A small-batch mixer consisting of a rotating drum which can be tilted to discharge the contents; used for concrete or mortar.

tilting-type boxcar unloader [CIV ENG] A mechanism that is used to unload material such as grain from a boxcar; the car, with its door open, is held by end clamps on the specialized piece of track and tilted 15% from the vertical and then tilted endwise 40% to the horizontal to discharge the material at one end of the car, and 40% in the opposite direction to discharge the material from the opposite end.

tiltmeter [ENG] An instrument used to measure small changes in the tilt of the earth's surface, usually in relation to a liquid-level surface or to the rest position of a pendulum.

tilt mold [MET] A mold that rotates from a horizontal to a vertical position during filling to reduce agitation and risk of dross entrapment.

timber [MATER] Wood used for building, carpentry, or joinery.

timbered stope [MIN ENG] A stope made of square-set timbering or any of its variations.

timbering [MIN ENG] The timber structure used for supporting the faces of an excavation during the progress of construction.

timbering machine [MIN ENG] An electrically driven machine to raise and hold timber in place while the supporting posts are being set, the posts having been cut to desired length previously by the machine's power-driven saw.

timberline [ECOL] The elevation or latitudinal limits for arboreal growth. Also known as tree line.

timber mat [MIN ENG] Broken timber forming the roof of an ore deposit that is being extracted by a caving method, such as top slicing.

timber packer *See* pack builder.

timber puller [MIN ENG] A machine used to remove the timber supports in a mine.

timber trolley [MIN ENG] A carriage consisting of a timber or steel base, mounted on wheels, with U-shaped arms.

timber truck [MIN ENG] Any truck or car used for hauling timber inside of a mine.

timbre [ACOUS] That attribute of auditory sensation in terms of which a listener can judge that two sounds similarly presented and having the same loudness and pitch are dissimilar. Also known as musical quality.

time [PHYS] **1.** The dimension of the physical universe which, at a given place, orders the sequence of events. **2.** A designated instant in this sequence, as the time of day. Also known as epoch.

time-and-altitude azimuth [NAV] In celestial navigation, the azimuth derived by a computation in which meridian angle, declination, and altitude are parameters, the values of which are either known or assumed.

time and material contract [IND ENG] A contract providing for the procurement of supplies or services on the basis of direct labor hours at specified fixed hourly rates (which rates include direct and indirect labor, overhead, and profit), and material at cost.

time and motion study [IND ENG] Observation, analysis, and measurement of the steps in the performance of a job to determine a standard time for each performance. Also known as time-motion study.

time assignment speech interpolation [COMMUN] Modulation technique based on the fact that speech is never a continuous stream of information, but consists of a large number of short signals; therefore, the period between the speech signals is used for transmitting other data.

time azimuth [NAV] In celestial navigation, the azimuth derived by a calculation in which the meridian angle, the polar distance (or declination), and the latitude are parameters, the magnitudes of which are either known or assumed.

time base [ELECTR] A device which moves the fluorescent spot rhythmically across the screen of the cathode-ray tube.

time-base generator *See* sweep generator.

time-change component [ENG] A component which because of design limitations or safety is specified to be rebuilt or overhauled after a specified period of operation (for example, an engine or propeller of an airplane).

time-code generator [ELECTR] A crystal-controlled pulse generator that produces a train of pulses with various predetermined widths and spacings, from which the time of day and sometimes also day of year can be determined; used in telemetry and other data-acquisition systems to provide the precise time of each event.

time constant [PHYS] **1.** The time required for a physical quantity to rise from zero to $1 - 1/e$ (that is, 63.2%) of its final steady value when it varies with time t as $1 - e^{-kt}$. **2.** The time required for a physical quantity to fall to $1/e$ (that is, 36.8%) of its initial value when it varies with time t as e^{-kt}. **3.** Generally, the time required for an instrument to indicate a given percentage of the final reading resulting from an input signal. Also known as lag coefficient.

time-controlled system *See* clock control system.

time-correlation [GEOL] A correlation of age or mutual time relations between stratigraphic units in separated areas.

time-current characteristics [ELEC] Of a fuse, the relation between the root-mean-square alternating current or direct current and the time for the fuse to perform the whole or some specified part of its interrupting function.

time curve *See* time front.

time delay [PHYS] The time required for a signal to travel between two points in a circuit or for a wave to travel between two points in space.

time-delay circuit [ELECTR] A circuit in which the output signal is delayed by a specified time interval with respect to the input signal. Also known as delay circuit.

time-delay relay [ELEC] A relay in which there is an appreciable interval of time between energizing or deenergizing of the coil and movement of the armature, such as a slow-acting relay and a slow-release relay.

time-derived channel [COMMUN] Any of the channels which result from time-division multiplexing of a channel.

time diagram [ASTRON] A diagram in which the celestial equator appears as a circle, and celestial meridians and hour circles as radial lines; used to facilitate solution of time problems and other problems involving arcs of the celestial equator or angles at the pole, by indicating relations between various quantities involved; conventionally, the relationships are given as viewed from a point over the South Pole, in a westward direction or counterclockwise. Also known as diagram on the plane of the celestial equator; diagram on the plane of the equinoctial.

time dilation effect *See* slowing of clocks.

time-distance graph [AERO ENG] A graph used to determine the ground distance for air-route legs of a specified time interval; time-distance relationships are often simplified by considering air, wind, and ground distances for flight legs of 1-hour duration.

time-division data links [COMMUN] Radio communications which use time-division techniques for channel separation.

time-division multiplexing [ADP] The interleaving of bits or characters in time to compensate for the slowness of input devices as compared to data transmission lines. [COMMUN] A process for transmitting two or more signals over a common path by using successive time intervals for different signals. Also known as time multiplexing.

time-division sound [COMMUN] A method under development, employing time-division multiplexing, which enables the audio signal of a television broadcast to be transmitted over the same channel as the video signal. Also known as tidi-sound.

time-division switching system [ELECTR] A type of electronic switching system in which input signals on lines and trunks are sampled periodically, and each active input is associated with the desired output for a specific phase of the period.

timed signal service [COMMUN] Radio communications service for the transmission of time signals of stated high precision, intended for general reception.

time fire [ORD] Fire in which fuses are set to act after a fixed time interval and before impact.

time formula [IND ENG] A formula to determine the standard time of an operation as a function of one or more variables in the operation.

time front [AERO ENG] A locus of points representing the maximum ground distances from a departure point that can be covered by an aircraft in a prescribed time interval. Also known as hour-out line; time curve.

time fuse [ENG] A fuse which contains a graduated time element to regulate the time interval after which the fuse will function.

time gate [ELECTR] A circuit that gives an output only during chosen time intervals.

time-height section [ELECTR] A facsimile trace of a vertically directed radar; specifically, a cloud-detection radar.

time-interval measurement [HOROL] A process that consists either in calculating the duration between two known epochs, or in counting the repetitions of a recurring phenomenon from an arbitrary starting point, as with an electronic digital-reading counter, which counts the cycles of an oscillator.

time-interval radiosonde *See* pulse-time-modulated radiosonde.

time-invariant system [CONT SYS] A system in which all quantities governing the system's behavior remain constant with time, so that the system's response to a given input does not depend on the time it is applied.

time lag [ORD] The amount by which the time of fall of a bomb, released under given conditions, would exceed that of an ideal bomb released under identical conditions. [PHYS] The time between a cause and a resultant effect, as between occurrence of a primary ionizing event and its count by a counter.

timelike vector [RELAT] A four vector in Minkowski space whose space component has a magnitude which is less than the magnitude of its time component multiplied by the speed of light.

time line [GEOL] **1.** A line that indicates equal geologic age in a correlation diagram. **2.** A rock unit represented by a time line.

time measurement [HOROL] A process that consists in counting the repetitions of any recurring phenomenon and, if the interval between successive recurrences is sensible, in subdividing it.

time meridian [ASTRON] Any meridian used as a reference for reckoning time, particularly a zone or standard meridian.

time modulation [COMMUN] Modulation in which the time of occurrence of a definite portion of a waveform is varied in accordance with a modulating signal.

time-motion study *See* time and motion study.

time-multiplexing *See* multiprogramming; time-division multiplexing.

time of delivery [COMMUN] The time at which the addressee or responsible relay agency receipts for a message.

time of flight [MECH] Elapsed time in seconds from the instant a projectile or other missile leaves a gun or launcher until the instant it strikes or bursts. [PHYS] The elapsed time from the instant a particle leaves a source to the instant it reaches a detector.

time-of-flight mass spectrometer [SPECT] A mass spectrometer in which all the positive ions of the material being analyzed are ejected into the drift region of the spectrometer tube with essentially the same energies, and spread out in accordance with their masses as they reach the cathode of a magnetic electron multiplier at the other end of the tube.

time of origin [COMMUN] The time at which a message is released for transmission.

time of receipt [COMMUN] The time at which a receiving station completes reception of a message.

time of set [MATER] The time required for freshly mixed concrete to stiffen (initial set, about 1 hour) or to attain a minimum specified hardness (final set, about 10 hours); actual times vary with the type of cement used.

time on target [ORD] **1.** A method of firing on a target in which various artillery units time their fire so that all

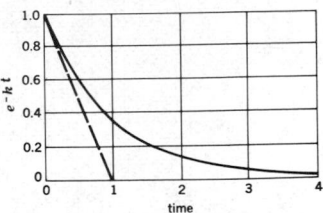

TIME CONSTANT

Graph of exponentially decreasing variable e^{-kt} (def. 2) as a function of time in time constants. Dotted line indicates variable that continues to decrease at same rate that exponential variable decreases at time $t = 0$.

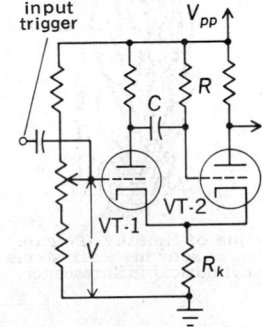

TIME-DELAY CIRCUIT

Circuit diagram of a monostable multivibrator used as a time-delay circuit. A small input trigger pulse gives rise to output signals at the plates of vacuum tubes VT-1 and VT-2 whose duration is proportional to resistance R when product of R and capacitance C is held fixed. R_k = resistance of cathode lead; V_{pp} = plate supply voltage.

projectiles reach the target simultaneously. **2.** A measure of the ability of a fire-control system or gunner to keep a weapon aimed at a moving target or to keep a weapon mounted on a moving vehicle aimed at a target; a useful evaluation of stabilizers.

time over target [ORD] The time at which an aircraft or formation of aircraft arrives over a designated point for the purpose of conducting an air mission on a target.

time-phase [PHYS] Two disturbances are in time phase if they reach corresponding peak values at the same instants of time, though not necessarily at the same points in space.

time-pulse distributor [ELECTR] A device or circuit for allocating timing pulses or clock pulses to one or more conducting paths or control lines in specified sequence.

time quadrature [PHYS] **1.** Differing by a time interval corresponding to one-fourth the time of one cycle of the frequency in question. **2.** An integration over time.

time quenching [MET] Interrupted quenching in which the time in the quenching medium is controlled.

timer [ELECTR] A circuit used in radar and in electronic navigation systems to start pulse transmission and synchronize it with other actions, such as the start of a cathode-ray sweep. [ENG] **1.** A device for automatically starting or stopping a machine or other device. **2.** See interval timer. [MECH ENG] A device that controls timing of the ignition spark of an internal combustion engine at the correct time.

time reversal [PHYS] The replacement of the time coordinate t by its negative $-t$ in the equations of motion of a dynamical system; the time reversal operator, a symmetry operator for a quantum-mechanical system, contains also the complex conjugation operator and a matrix operating on the spin coordinate.

time-rock unit See time-stratigraphic unit.

time separation [AERO ENG] The time interval between adjacent aircraft flying approximately the same path.

time series [STAT] A statistical process analogous to the taking of data at intervals of time.

time series analysis [MATH] The general study of mathematical systems or processes analogous to that of data taken at time intervals.

time-share [ADP] To perform several independent processes almost simultaneously by interleaving the operations of the processes on a single high-speed processor.

time-shared amplifier [ELECTR] An amplifier used with a synchronous switch to amplify signals from different sources one after another.

time-sharing [ADP] The simultaneous utilization of a computer system from multiple terminals.

time sight [NAV] A common method for determining longitude by celestial observations; these observations are reduced by solving the navigational triangle for meridian angle and require known or assumed values for altitude, latitude, and declination; the meridian angle is converted to local hour angle and compared with Greenwich hour angle, and the latter is read from astronomical tables.

time signal [COMMUN] An accurate signal which is broadcast by radio and marks a specified time or time interval, used for setting timepieces and for determining their errors; in particular, a radio signal broadcast at accurately known times each day on a number of different frequencies by WWV and other stations.

time standard [HOROL] A recurring phenomenon, used as a reference for establishing a unit of time; the presently accepted standard is the second, defined to be 9,192,631,770 transitions between two specified hyperfine levels of the atom of cesium-133. [IND ENG] See standard time.

time-stratigraphic facies [GEOL] A stratigraphic facies based on the amount of geologic time during which deposition and nondeposition of sediment occurred.

time-stratigraphic unit [GEOL] A stratigraphic unit based on geologic age or time of origin. Also known as chronolith; chronolithologic unit; chronostratic unit; chronostratigraphic unit; time-rock unit.

time study [IND ENG] A work measurement technique, generally using a stopwatch or other timing device, to record the actual elapsed time for performance of a task, adjusted for any observed variance from normal effort or pace, unavoidable or machine delays, rest periods, and personal needs.

time switch [ENG] A clock-controlled switch used to open or close a circuit at one or more predetermined times.

time tick [COMMUN] A radio-broadcast time signal consisting of one or more short audible sounds or beats; in particular, one which is generated by an accurately controlled pulsed radio signal.

time-transgressive unit [GEOL] A laterally continuous, lithologically homogeneous rock unit whose age differs regionally.

time-varying system [CONT SYS] A system in which certain quantities governing the system's behavior change with time, so that the system will respond differently to the same input at different times.

time zone [ASTRON] To avoid the inconvenience of the continuous change of mean solar time with longitude, the earth is divided into 24 time zones, each about 15° wide and centered on standard longitudes, 0°, 15°, 30°, and so on; within each zone the time kept is the mean solar time of the standard meridian.

timing [ORD] Adjustment of a small-arms weapon so that it will perform each function at a predetermined point in the cycle of operation.

timing-axis oscillator See sweep generator.

timing belt [MECH ENG] A positive drive belt that has axial cogs molded on the underside of the belt which fit into grooves on the pulley; prevents slip, and makes accurate timing possible; combines the advantages of belt drives with those of chains and gears. Also known as positive drive belt.

timing belt pulley [MECH ENG] A pulley that is similar to an uncrowned flat-belt pulley, except that the grooves for the belt's teeth are cut in the pulley's face parallel to the axis.

timing gears [MECH ENG] The gear train of reciprocating engine mechanisms for relating camshaft speed to crankshaft speed.

timing motor [ELEC] A motor which operates from an alternating-current power system synchronously with the alternating-current frequency, used in timing and clock mechanisms. Also known as clock motor.

timing relay [ELEC] Form of auxiliary relay used to introduce a definite time delay in the performance of a function.

timing signal [ELECTR] Any signal recorded simultaneously with data on magnetic tape for use in identifying the exact time of each recorded event.

Timken film strength [ENG] A test used on a gear lubricant to determine the amount of pressure the film of oil can withstand before rupturing.

Timken wear test [ENG] A test used on a gear lubricant to determine its abrasive effect on gear metals.

timothy [BOT] *Phleum pratense.* A perennial hay grass of the order Cyperales characterized by moderately leafy stems and a dense cylindrical inflorescence.

tin [CHEM] Metallic element in group IV, symbol Sn, atomic number 50, atomic weight 118.69; insoluble in water, soluble in acids and hot potassium hydroxide solution; melts at 232°C, boils at 2260°C. [MET] A lustrous silver-white ductile, malleable metal used in alloys, for solder, terneplate, and tin plate.

Tinamidae [VERT ZOO] The single family of the avian order Tinamiformes.

Tinamiformes [VERT ZOO] The tinamous, an order of South and Central American birds which are superficially fowllike but have fully developed wings and are weak fliers.

tin bisulfide See stannic sulfide.

tin bromide See stannic bromide; stannous bromide.

tin bronze [MET] A tin-copper alloy.

tincalconite [MINERAL] $Na_2B_4O_7 \cdot 5H_2O$ A colorless to dull-white mineral, crystallizing in the rhombohedral system; one of the principal ores of borax and boron compounds. Also known as mohavite; octahedral borax.

tin chloride See stannic chloride; stannous chloride.

tin chromate See stannic chromate; stannous chromate.

tin crystals See stannous chloride.

tincture [MATER] A dilute solution (aqueous or aqueous alcoholic) of a drug or chemical; more dilute than fluid extracts, less volatile than spirits.

tincture of iodine [PHARM] A medicinal preparation used as an anti-infective containing 20 grams iodine and 24 grams sodium iodide in 1000 milliliters of alcohol. Also known as iodine solution; iodine tincture.

TIMOTHY

Drawing of timothy *(Phleum pratense)* showing leafy stems and cylindrical inflorescence.

TIN

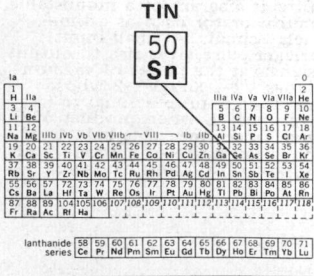

Periodic table of the chemical elements showing the position of tin.

tin dichloride   *See* stannous chloride.

tin difluoride   *See* stannous fluoride.

tin dioxide   *See* stannic oxide.

tinea   *See* ringworm.

tinea favosa   *See* favus.

Tineidae   [INV ZOO] A family of small moths in the superfamily Tineoidea distinguished by an erect, bristling vestiture on the head.

Tineoidea   [INV ZOO] A superfamily of heteroneuran Lepidoptera which includes small moths that usually have well-developed maxillary palpi.

tin fluoride   *See* stannous fluoride.

tinfoil   [MET] Foil made of tin or a tin alloy.

Tingidae   [INV ZOO] The lace bugs, the single family of the hemipteran superfamily Tingoidea.

Tingoidea   [INV ZOO] A superfamily of the Hemiptera in the subdivision Geocorisae characterized by the wings with many lacelike areolae.

tin hydride   [INORG CHEM] SnH_4 A gas boiling at $-52°C$. Also known as stannane.

tin iodide   *See* stannic iodide.

tin monosulfide   *See* stannous sulfide.

tinned wire   [MET] Copper wire that has been coated during manufacture with a layer of tin or solder to prevent corrosion and simplify soldering of connections.

tinner's rivet   [DES ENG] A special-purpose rivet that has a flat head, used in sheet metal work.

tinning   [MET] **1.** Covering or preserving with tin. **2.** A protective coating of tin.

tinnitus   [MED] A ringing, roaring, or hissing sound in one or both ears.

tin oxide   *See* stannic oxide; stannous oxide.

tin peroxide   *See* stannic oxide.

tin pest   [MET] Transformation of tin to a brittle, gray variety occurring spontaneously at temperatures below $0°C$.

tinplate   [MET] Thin sheet iron or steel coated with tin.

tin protosulfide   *See* stannous sulfide.

tin protoxide   *See* stannous oxide.

tin pyrites   *See* stannite.

tin salts   *See* stannous chloride.

tinsel cord   [ELEC] A highly flexible cord used for headphone leads and test leads, in which the conductors are strips of thin metal foil or tinsel wound around a strong but flexible central cord.

tin stone   *See* cassiterite.

tin sulfate   *See* stannous sulfate.

tin sulfide   *See* stannous sulfide.

tin sweat   [MET] Exudation of tin-rich low-melting-point material from a tin-bronze surface as a result of inverse segregation in bronze casting, or overheating of the alloy.

tin tetrabromide   *See* stannic bromide.

tin tetrachloride   *See* stannic chloride.

tin tetraiodide   *See* stannic iodide.

Tintinnida   [INV ZOO] An order of ciliated protozoans in the subclass Spirotrichia whose members are conical or trumpet-shaped pelagic forms bearing shells.

tint of passage   [OPTICS] The color produced when a plate which is colorless, but which rotates the plane of polarization of polarized light passing through it by an amount which depends on the wavelength of the light, is placed between crossed polarizers.

tintometer   [OPTICS] A device used to estimate the intensity of a colored solution by comparing it with standard solutions or colored glass slides, as with the Lovibond tintometer.

tip   [DES ENG] A piece of material secured to and differing from a cutter tooth or blade. [ELEC] The contacting part at the end of a phone plug. [ELECTR] A small protuberance on the envelope of an electron tube, resulting from the closing of the envelope after evacuation.

tipburn   [PL PATH] A disease of certain cultivated plants, such as potato and lettuce, characterized by browning of the leaf margins due to excessive loss of water.

tip clearance   [NAV ARCH] The clearance or distance between the circumference of the tip circle of a propeller and the hull of the vessel.

Tiphiidae   [INV ZOO] A family of the Hymenoptera in the superfamily Scolioidea.

tip-in   [GRAPHICS] **1.** A halftone illustration on smooth, coated paper that has been inserted on a page of a book printed on rough paper; a blank space is left on the rough page and the halftone is attached (tipped in) by a narrow coating of paste applied at the top edge. **2.** The insertion of a plate or full-page illustration on smoother or heavier stock than the rest of the book.

tip jack   [ELEC] A small single-hole jack for a single-pin contact plug. Also known as pup jack.

tipped bit   [DES ENG] A drill bit in which the cutting edge is made of especially hard material.

tipped solid cutters   [DES ENG] Cutters made of one material and having tips or cutting edges of another material bonded in place.

tipper   [MIN ENG] An apparatus for emptying coal or ore cars by turning them upside down and then righting them, with a minimum of manual labor.

tipping-bucket rain gage   [ENG] A type of recording rain gage; the precipitation collected by the receiver empties into one side of a chamber which is partitioned transversely at its center and is balanced bistably upon a horizontal axis; when a predetermined amount of water has been collected, the chamber tips, spilling out the water and placing the other half of the chamber under the receiver; each tip of the bucket is recorded on a chronograph, and the record obtained indicates the amount and rate of rainfall.

tipple   [MIN ENG] **1.** The place where the mine cars are tipped and emptied of their coal. **2.** The tracks, trestles, and screens at the entrance to a colliery, where coal is screened and loaded.

tip side   [ELEC] Conductor of a circuit which is associated with the tip of a plug or the top spring of a jack; by extension, it is common practice to designate by these terms the conductors having similar functions or arrangements in circuits where plugs or jacks may not be involved.

Tipulidae   [INV ZOO] The crane flies, a family of orthorrhaphous dipteran insects in the series Nematocera.

tire   [ENG] A continuous metal ring, or pneumatic rubber and fabric cushion, encircling and fitting the rim of a wheel.

tire iron   [DES ENG] A single metal bar having bladelike ends of various shapes to insert between the rim and the bead of a pneumatic tire to remove or replace the tire.

Tiros satellite   [AERO ENG] Television infrared observation satellite; a meteorological satellite that takes television pictures of cloud cover, using radiation sensors and cameras; it stores and transmits this information on ground command.

tirrill burner   [ENG] A modification of the bunsen burner which allows greater flexibility in the adjustment of the air-gas mixture.

Tirrill regulator   [ELEC] A device for regulating the voltage of a generator, in which the field resistance of the exciter is short-circuited temporarily when the voltage drops.

Tischenko reaction   [ORG CHEM] The formation of an ester by the condensation of two molecules of aldehyde utilizing a catalyst of aluminum alkoxides in the presence of a halide.

tissue   [HISTOL] An aggregation of cells more or less similar morphologically and functionally.

tissue culture   [CYTOL] Growth of tissue cells in artificial media.

tissue dose   [NUCLEO] The dose received by a tissue in the region of interest, expressed in roentgens for x-rays and gamma rays.

tissue paper   [MATER] Extremely lightweight paper, available in a great many colors and used in craft projects and in collage painting.

tissue roentgen   *See* rep.

Titan   [ASTRON] The largest satellite of Saturn, with a diameter estimated to be about 4200 kilometers. [ORD] A U.S. Air Force surface-to-surface intercontinental ballistic missile having a range of over 6000 miles (9700 kilometers) and a greater payload of nuclear or conventional weapons than Atlas; Titan uses inertial guidance alone or combined with radar guidance.

titanate   [INORG CHEM] A salt of titanic acid; titanates of the M_2TiO_3 type are called metatitanates, those of the M_4TiO_4 type are called orthotitanates; an example is sodium titanate, $(Na_2O)_2Ti_2O_5$.

titanellow   *See* titanium trioxide.

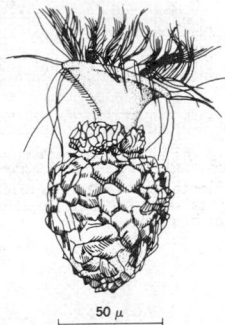

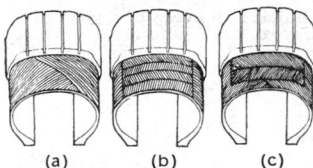

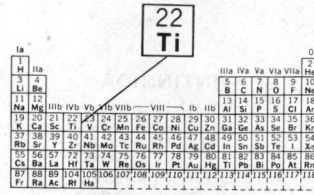

TITANIUM

Periodic table of the chemical elements showing the position of titanium.

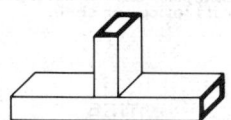

T JUNCTION

Drawing showing one type of waveguide T junction.

TOAD

A toad of the genus *Bufo*, covered with warty structures that secrete poisons for purposes of defense.

titania   *See* titanium dioxide.

Titania   [ASTRON] A satellite of Uranus, with a diameter estimated to be 1600 kilometers.

titanic acid   [INORG CHEM] H_2TiO_3 A white, water-insoluble powder; used as a dyeing mordant. Also known as metatitanic acid; titanic hydroxide.

titanic anhydride   *See* titanium dioxide.

titanic chloride   *See* titanium tetrachloride.

titanic hydroxide   *See* titanic acid.

titanic iron ore   *See* ilmenite.

titanic sulfate   *See* titanium sulfate.

titanite   *See* sphene.

titanium   [CHEM] A metallic element in group IV, symbol Ti, atomic number 22, atomic weight 47.90; ninth most abundant element in the earth's crust; insoluble in water, melts at 1660°C, boils above 3000°C.   [MET] A lustrous, silvery-gray, strong, light metal that is hard and brittle when cold, malleable when heated, and ductile when pure; used in the pure state or in alloys for aircraft and chemical-plate metals, for surgical instruments, and in cermets, and metal-ceramic brazing.

titanium boride   [INORG CHEM] TiB_2 A hard solid that resists oxidation at elevated temperatures and melts at 2980°C; used as a refractory and in alloys, high-temperature electrical conductors, and cermets.

titanium carbide   [INORG CHEM] TiC   Very hard gray crystals insoluble in water, soluble in nitric acid and aqua regia, melts at about 3140°C; used in cermets, arc-melting electrodes, and tungsten-carbide tools.

titanium chloride   *See* titanium dichloride.

titanium dichloride   [INORG CHEM] $TiCl_2$ A flammable, alcohol-soluble, black powder that decomposes in water, and in vacuum at 475°C, and burns in air. Also known as titanium chloride.

titanium dioxide   [INORG CHEM] TiO_2 A white, water-insoluble powder that melts at 1560°C, and which is produced commercially from the titanium dioxide minerals ilmenite and rutile; used in paints and cosmetics. Also known as titania; titanic anhydride; titanium oxide; titanium white.

titanium hydride   [INORG CHEM] TiH_2 A black metallic powder whose dust is an explosion hazard and which dissociates above 288°C; used in powder metallurgy, hydrogen production, foamed metals, glass solder, and refractories, and as an electronic gas getter.

titanium nitride   [INORG CHEM] TiN   Golden-brown brittle crystals melting at 2927°C; used in refractories, alloys, cermets, and semiconductors.

titanium oxalate   [ORG CHEM] $Ti_2(C_2O_4)_3 \cdot 10H_2O$ Toxic, yellow prisms soluble in water, insoluble in alcohol; used to make titanic acid and titanium metal. Also known as titanous oxalate.

titanium oxide   *See* titanium dioxide; titanium trioxide.

titanium peroxide   *See* titanium trioxide.

titanium sesquisulfate   *See* titanous sulfate.

titanium sulfate   [INORG CHEM] $Ti(SO_4)_2 \cdot 9H_2O$ Caked solid, soluble in water, toxic, highly acidic; used as a dye stripper, reducing agent, laundry chemical, and in treatment of chrome yellow colors. Also known as basic titanium sulfate; titanic sulfate; titanyl sulfate.

titanium tetrachloride   [INORG CHEM] $TiCl_4$ A colorless, toxic liquid soluble in water, fumes when exposed to moist air, boils at 136°C; used to make titanium and titanium salts, as a dye mordant and polymerization catalyst, and in smoke screens and pigments. Also known as titanic chloride.

titanium trichloride   [INORG CHEM] $TiCl_3$ Toxic, dark-violet, deliquescent crystals soluble in alcohol and some amines, decomposes in water with heat evolution, decomposes above 440°C; used as a reducing agent, chemical intermediate, polymerization catalyst, and laundry stripping agent. Also known as titanous chloride.

titanium trioxide   [INORG CHEM] TiO_3 Yellow titanium oxide used to make ivory shades in ceramics. Also known as titanellow; titanium oxide; titanium peroxide.

titanium white   *See* titanium dioxide.

Titanoideidae   [PALEON] A family of extinct land mammals in the order Pantodonta.

titanothere   [PALEON] Any member of the family Brontotheriidae.

titanous chloride   *See* titanium trichloride.

titanous oxalate   *See* titanium oxalate.

titanous sulfate   [INORG CHEM] $Ti_2(SO_4)_3$ Green crystals soluble in dilute hydrochloric and sulfuric acids, insoluble in water and alcohol; used as a textile reducing agent. Also known as titanium sesquisulfate.

titanyl sulfate   *See* titanium sulfate.

titer   [CHEM] **1.** The concentration in a solution of a dissolved substance as shown by titration. **2.** The least amount or volume needed to give a desired result in titration. **3.** The solidification point of hydrolyzed fatty acids.

Tithonian   [GEOL] Southern European equivalent of the Portlandian stage (uppermost Jurassic) of geologic time.

titrant   [ANALY CHEM] A standard solution of known concentration and composition used for analytical titrations.

titration   [ANALY CHEM] A method of analyzing the composition of a solution by adding known amounts of a standardized solution until a given reaction (color change, precipitation, or conductivity change) is produced.

tivano   [METEOROL] A night breeze blowing down the valley at Lake Como in Italy.

tjaele   *See* frozen ground.

tjenting needle   [GRAPHICS] The traditional tool for applying wax in the batik process; it is used for fine-line work and for outlining areas to be filled in by brush; the melted wax flows out from the fine needle spout.

T junction   [ELECTR] A network of waveguides with three waveguide terminals arranged in the form of a letter T; in a rectangular waveguide a symmetrical T junction is arranged by having either all three broadsides in one plane or two broadsides in one plane and the third in a perpendicular plane.

Tl   *See* thallium.

Tm   *See* thulium.

TME   *See* metric-technical unit of mass.

T method   [BOT] A budding method in which a T-shaped cut is made through the bark at the internode of the stock, the bark of the scion is separated from the xylem along the cambium and removed, and the scion is forced into the incision on the stock.

TM mode   *See* transverse magnetic mode.

TM wave   *See* transverse magnetic wave.

Tn   *See* thoron.

T network   [ELEC] A network composed of three branches, with one end of each branch connected to a common junction point, and with the three remaining ends connected to an input terminal, an output terminal, and a common input and output terminal, respectively.

TNT   *See* 2,4,6-trinitrotoluene.

TNT-ammonium nitrate explosive   [MATER] An explosive containing ammonium nitrate sensitized with trinitrotoluene; a proportion of aluminum powder or calcium silicide may be added to increase power and sensitiveness.

TNT equivalent   [NUCLEO] A measure of the energy released in the detonation of a nuclear weapon, expressed in terms of the weight of TNT that would release the same amount of energy when exploded; usually expressed in kilotons or megatons of TNT; based on the release of 10^9 calories (approximately 4.18×10^9 joules) of energy by 1 ton of TNT.

toad   [VERT ZOO] Any of several species of the amphibian order Anura, especially in the family Bufonidae; glandular structures in the skin secrete acrid, irritating substances of varying toxicity.

toadstool   [MYCOL] Any of various fleshy, poisonous or inedible fungi with a large umbrella-shaped fruiting body.

to-and-fro ropeway   *See* jig back.

Toarcian   [GEOL] A European stage of geologic time; Lower Jurassic (above Pliensbachian, below Bajocian).

tobacco   [BOT] **1.** Any plant of the genus *Nicotinia* cultivated for its leaves, which contain 1–3% of the alkaloid nicotine. **2.** The dried leaves of the plant.

tobacco jack   *See* wolframite.

tobacco mosaic   [PL PATH] Any of a complex of virus diseases of tobacco and other solanaceous plants in which the leaves are mottled with light- and dark-green patches, sometimes interspersed with yellow.

Tocco process   [MET] A patented process for local hardening of steel by applying a high-frequency current to the area for a few seconds or until heated to the desired depth, then water-spraying the surface.

tocopherol [ORG CHEM] $C_{29}H_{50}O_2$ Any of several substances having vitamin E activity that occur naturally in certain oils; alpha tocopherol is the most potent.

tocophobia [PSYCH] Abnormal fear of children.

Todidae [VERT ZOO] The todies, a family of birds in the order Coraciiformes found in the West Indies.

toe [ANAT] One of the digits on the foot of man and other vertebrates. [CIV ENG] The part of a base of a dam or retaining wall on the side opposite to the retained material. [GEOL] The leading edge of a thrust nappe. [MIN ENG] **1.** The burden of material between the bottom of the borehole and the free face. **2.** The bottom of the borehole. **3.** A spurn, or small pillar of coal. **4.** The base of a bank in an open-pit mine.

toe crack [MET] A crack in the base metal at the toe of a weld.

toe cut [ENG] In underground blasting, the cut obtained by the use of toe holes.

toe hole [ENG] A blasting hole, usually drilled horizontally or at a slight inclination into the base of a bank, bench, or slope of a quarry or open-pit mine.

toe-in [MECH ENG] The degree (usually expressed in fractions of an inch) to which the forward part of the front wheels of an automobile are closer together than the rear part, measured at hub height with the wheels in the normal "straight ahead" position of the steering gear.

toe of the weld [MET] The junction between the face of a weld and the base metal.

toe-out [MECH ENG] The outward inclination of the wheels of an automobile at the front on turns due to setting the steering arms at an angle.

Toepler-Holtz machine [ELEC] An early type of machine for continuously producing electrical charges at high voltage by electrostatic induction, superseded by the Wimhurst machine. Also known as Holtz machine.

toe-to-toe drilling [ENG] The drilling of vertical large-diameter blasting holes in quarries and opencast pits.

tofan [METEOROL] A violent spring storm common in the mountains of Indonesia.

to-from indicator [NAV] An indicator that shows whether an aircraft is flying toward or away from an omnirange station. Also known as sense indicator.

toggle [ELECTR] To switch over to an alternate state, as in a flip-flop. [MECH ENG] A form of jointed mechanism for the amplification of forces.

toggle bolt [DES ENG] A bolt having a nut with a pair of pivotal wings that close against a spring; wings open after emergence through a hole or passage in a thin or hollow wall to fasten the unit securely.

toggle press [MECH ENG] A mechanical press in which a toggle mechanism actuates the slide.

toggle switch [ELEC] A small switch that is operated by manipulation of a projecting lever that is combined with a spring to provide a snap action for opening or closing a circuit quickly. [ELECTR] An electronically operated circuit that holds either of two states until changed.

toise [GEOD] A unit of length equal to about 6.4 feet (1.95 meters); used in early geodetic surveys.

tokamak [PL PHYS] A device for confining a plasma within a toroidal chamber, which produces plasma temperatures, densities, and confinement times greater than that of any other such device; confinement is effected by a very strong externally applied toroidal field, plus a weaker poloidal field produced by a toroidally directed plasma current, and this current causes ohmic heating of the plasma.

token [ADP] A distinguishable unit in a sequence of characters.

tolazoline hydrochloride [ORG CHEM] $C_{10}H_{12}N_2 \cdot HCl$ Water-soluble white crystals, and melting at 173°C; used as a sympatholytic and vasodilator. Also known as priscol.

tolbutamide [PHARM] $C_{12}H_{18}N_2O_3S$ A hypoglycemic drug effective when administered orally.

toleragen [IMMUNOL] A substance which, in appropriate dosages, produces a state of specific immunological tolerance in humans or animals.

tolerance [DES ENG] The permissible variations in the dimensions of machine parts. [ENG] A permissible deviation

from a specified value, expressed in actual values or more often as a percentage of the nominal value. [PHARM] **1.** The ability of enduring or being less responsive to the influence of a drug or poison, particularly when acquired by continued use of the substance. **2.** The allowable deviation from a standard, as the range of variation permitted for the content of a drug in one of its dosage forms.

tolerance chart [DES ENG] A chart indicating graphically the sequence in which dimensions must be produced on a part so that the finished product will meet the prescribed tolerance limits.

tolerance dose *See* permissible dose.

tolerance limits [DES ENG] The extreme values (upper and lower) that are permitted by the tolerance.

tolerance unit [DES ENG] A unit of length used to express the degree of tolerance allowed in fitting cylinders into cylindrical holes, equal, in micrometers, to 0.45 $D^{1/3}$ + 0.001 D, where D is the cylinder diameter in millimeters.

ortho-**tolidine** [ORG CHEM] $[C_6H_3(CH_3)NH_2]_2$ Light-sensitive, combustible white to reddish crystals soluble in alcohol and ether, slightly soluble in water, melts at 130°C; used as an anlytical reagent and a curing agent for urethane resins. Also known as diaminoditolyl.

toll [COMMUN] **1.** Charge made for a connection beyond an exchange boundary. **2.** Any part of telephone plant, circuits, or services for which toll charges are made.

toll call [COMMUN] Telephone call to points beyond the area within which telephone calls are covered by a flat monthly rate or are charged for on a message unit basis.

toll center [COMMUN] A telephone central office where trunks from end offices are joined to the long-distance system, and operators are present; it is a class-4 office.

toll enrichment [NUCLEO] A proposed arrangement whereby privately owned uranium could be enriched in uranium-235 content in government facilities upon payment of a service charge by the owners.

Tollen's aldehyde test [ANALY CHEM] A test that uses an ammoniacal solution of silver oxides to test for aldehydes and ketones.

toll office [COMMUN] A telephone central office which serves mainly to terminate and interconnect toll lines and various types of trunks.

toll television *See* subscription television.

toll terminal loss [COMMUN] The part of the overall transmission loss on a toll connection that is attributable to the facilities from the toll center through the tributary office, to and including the subscriber's equipment.

α-**toluamide** *See* α-phenylacetamide.

toluene [ORG CHEM] $C_6H_5CH_3$ A colorless, aromatic liquid derived from coal tar or from the catalytic reforming of petroleum naphthas; insoluble in water, soluble in alcohol and ether, boils at 111°C; used as a chemical intermediate, for explosives, and in high-octane gasolines. Also known as methylbenzene; toluol.

para-**toluenesulfonate** *See* *para*-toluenesulfonic acid.

para-**toluenesulfonic acid** [ORG CHEM] $C_6H_4(SO_3H)(CH_3)$ Toxic, colorless, combustible crystals soluble in water, alcohol, and ether; melts at 107°C; used in dyes and as a chemical intermediate and organic catalyst. Also known as *para*-toluenesulfonate.

toluene trichloride *See* benzotrichloride.

meta-**toluic acid** [ORG CHEM] $C_6H_4CH_3COOH$ White to yellow, combustible crystals soluble in alcohol and ether, slightly soluble in water, melts at 109°C; used as a chemical intermediate and base for insect repellants. Also known as 3-methylbenzoic acid; *meta*-toluylic acid.

ortho-**toluic acid** [ORG CHEM] $C_6H_4CH_3COOH$ White, combustible crystals soluble in alcohol and chloroform, slightly soluble in water, melts at 104°C; used as a bacteriostat. Also known as 2-methylbenzoic acid; *ortho*-toluylic acid.

para-**toluic acid** [ORG CHEM] $C_6H_4CH_3COOH$ Transparent, combustible crystals soluble in alcohol and ether, slightly soluble in water, melts at 180°C; used in agricultural chemicals and as an animal feed supplement. Also known as 4-methylbenzoic acid; *para*-toluylic acid.

meta-**toluidine** [ORG CHEM] $CH_3C_6H_4NH_2$ A combustible, colorless, toxic liquid soluble in alcohol and ether, slightly

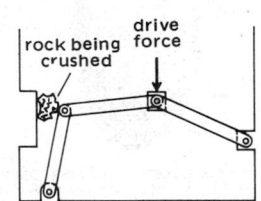

TOGGLE

Toggle mechanism used in a rock crusher; relatively small drive force causes large force to be applied to rock.

TOMATO

Drawing of tomato
(*Lycopersicon esculentum*)
showing fruit and usual
arrangement of leaves.

TONGUE

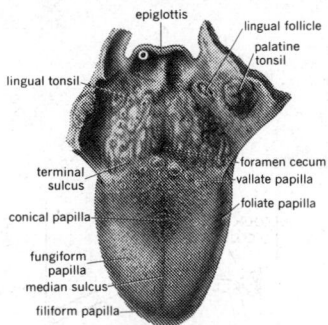

The human tongue, dorsal view.
(*From M. W. Woerdeman, Atlas
of Human Anatomy, vol. 2,
McGraw-Hill, 1950*)

soluble in water, boils at 203°C; used for dyes and as a chemical intermediate. Also known as *meta*-aminotoluene.

ortho-toluidine [ORG CHEM] $CH_3C_6H_4NH_2$ A light-green, light-sensitive, combustible, toxic liquid soluble in alcohol and ether, very slightly soluble in water, boils at 200°C; used for dyes and textile printing and as a chemical intermediate. Also known as *ortho*-aminotoluene.

para-toluidine [ORG CHEM] $CH_3C_6H_4NH_2$ Toxic, combustible, white leaflets soluble in alcohol and ether, very slightly soluble in water, boils at 200°C; used as an analytical reagent and in dyes. Also known as *para*-aminotoluene.

α-tolunitrile See benzyl cyanide.

toluol See toluene.

toluylene See stilbene.

toluylene red See neutral red.

meta-toluylic acid See *meta*-toluic acid.

ortho-toluylic acid See *ortho*-toluic acid.

para-toluylic acid See *para*-toluic acid.

tomatine [ORG CHEM] $C_{50}H_{83}NO_{21}$ A glycosidal alkaloid obtained from the leaves and stems from the tomato plant; the crude extract is known as tomatin: white, toxic crystals; used as a plant fungicide and as a precipitating agent for cholesterol.

tomato [BOT] A plant of the genus *Lycopersicon*, especially *L. esculentum*, in the family Solanaceae cultivated for its fleshy edible fruit, which is red, pink, orange, yellow, white, or green, with fleshy placentas containing many small, oval seeds with short hairs and covered with a gelatinous matrix.

tombolo [GEOL] A sand or gravel bar or spit that connects an island with another island or an island with the mainland. Also known as connecting bar; tie bar; tying bar.

tombolo cluster See complex tombolo.

tombolo series See complex tombolo.

tomentose [BOT] Covered with densely matted hairs.

tomentum [ANAT] The deep layer of the pia mater composed principally of minute blood vessels. [BIOL] Pubescence consisting of densely matted wooly hairs.

tomium [VERT ZOO] The cutting edge of a bird's beak.

Tommy gun [ORD] Trademarked name for the Thompson submachine gun; sometimes erroneously applied to other submachine guns.

tomography See sectional radiography.

Tomopteridae [INV ZOO] The glass worms, a family of pelagic polychaete annelids belonging to the group Errantia.

ton [IND ENG] A unit of volume of sea freight, equal to 40 cubic feet. Also known as freight ton; measurement ton; shipping ton. [MECH] **1.** A unit of weight in common use in the United States, equal to 2000 pounds or 907.18474 kilogram-force. Also known as just ton; net ton; short ton. **2.** A unit of mass in common use in the United Kingdom equal to 2240 pounds, or to 1016.0469088 kilogram-force. Also known as gross ton; long ton. **3.** A unit of weight in troy measure, equal to 2000 troy pounds, or to 746.4834432 kilogram-force. **4.** See tonne. [MECH ENG] A unit of refrigerating capacity, that is, of rate of heat flow, equal to the rate of extraction of latent heat when one short ton of ice of specific latent heat 144 international table British thermal units per pound is produced from water at the same temperature in 24 hours; equal to 200 British thermal units per minute, or to approximately 3516.85 watts. Also known as standard ton. [NAV ARCH] A unit of internal capacity of ships, equal to 100 cubic feet. Also known as register ton. [NUCLEO] The energy released by one metric ton of chemical high explosives calculated at the rate of 1000 calories per gram; equal to 4.18×10^9 joules; used principally in expressing the energy released by a nuclear bomb.

tonalite See quartz diorite.

tondal [MECH] A unit of force equal to the force which will impart an acceleration of 1 foot per second to a mass of 1 long ton; equal to approximately 309.6911 newtons.

tone [ACOUS] **1.** A sound oscillation capable of exciting an auditory sensation having pitch. **2.** An auditory sensation having pitch. [GRAPHICS] Each distinguishable shade variation from black to white on photographs.

tone code ranging [NAV] In navigation by satellite, a method of transmitting a high-precision time signal using the reduced peak power available to a satellite; an audio frequency is used to phase an audio oscillator in the user equipment, and a train

of short pulses is generated in synchronism with the zero crossings of the audio tone to provide high-resolution timing.

tone control [ELECTR] A control used in an audio-frequency amplifier to change the frequency response so as to secure the most pleasing proportion of bass to treble; individual bass and treble controls are provided in some amplifiers.

tone dialing See pushbutton dialing.

tone generator [ELECTR] A signal generator used to generate an audio-frequency signal suitable for signaling purposes or for testing audio-frequency equipment.

tone localizer See equisignal localizer.

tone-modulated waves [COMMUN] Waves obtained from continuous waves by amplitude-modulating them at audio frequency in a substantially periodic manner.

tone modulation [COMMUN] Type of code-signal transmission obtained by causing the radio-frequency carrier amplitude to vary at a fixed audio frequency.

tone-operated net-loss adjuster [COMMUN] System for stabilizing the net loss of a telephone circuit by a tone transmitted between conversations.

toner [GRAPHICS] The fine, black, resinous powder used in electrostatic imaging processes to make an electrostatic image readable; the toner is either deposited directly on coated paper or transferred from a charged surface to ordinary paper, then fused to the paper by heating.

tone reversal [COMMUN] Distortion of the recorder copy in facsimile which causes the various shades of black and white not to be in the proper order.

tongara [METEOROL] A hazy, southeast wind in the Macassar Strait.

tong hold [MET] The end of a forging billet that is gripped by the operator's tongs; it is removed at the end of the forging operation.

Tongrian [GEOL] A European stage of geologic time; lower Oligocene (above Ludian of Eocene, below Rupelian). Also known as Lattorfian.

tongs [DES ENG] Any of various devices for holding, handling, or lifting materials and consisting of two legs joined eccentrically by a pivot or spring.

tongue [ANAT] A muscular organ located on the floor of the mouth in man and most vertebrates which may serve various functions, such as taking and swallowing food or tasting or as a tactile organ or sometimes a prehensile organ. [GEOL] **1.** A minor rock-stratigraphic unit of limited geographic extent; it disappears laterally in one direction. **2.** A lava flow branching from a larger flow. [OCEANOGR] **1.** A protrusion of water into a region of different temperature, or salinity, or dissolved oxygen concentrating. **2.** A protrusion of one water mass into a region occupied by a different water mass.

tongue and groove [DES ENG] A joint in which a projecting rib on the edge of one board fits into a groove in the edge of another board.

tongue worm See acorn worm.

tonic convulsion See tonic postural epilepsy; tonic spasm.

tonic labyrinthine reflexes [PHYSIO] Rotation or deviation of the head causes extension of the limbs on the same side as the chin, and flexion of the opposite extremities: dorsiflexion of the head produces increased extensor tonus of the upper extremities and relaxation of the lower limbs, while ventroflexion of the head produces the reverse; seen in the young infant and patients with a lesion at the midbrain level or above.

tonic neck reflexes [PHYSIO] Reflexes in which rotation or deviation of the head causes extension of the limbs on the same side as the chin, and flexion of the opposite extremities; dorsiflexion of the head produces increased extension tonus of the upper extremities and relaxation of the lower limbs, and ventroflexion of the head, the reverse; seen normally in incomplete forms in the very young infant, and thereafter in patients with a lesion at the midbrain level or above.

tonic postural epilepsy [MED] A form of epilepsy in which seizures are characterized by a rigid posture with the arms and legs extended, hands pronated, and feet held in plantar flexion. Also known as tonic convulsion.

tonic spasm [MED] A spasm which persists for some time without relaxation. Also known as tonic convulsion.

ton-kilometer [MIN ENG] A unit of measurement equal to the weight in tons of material transported in a mine multiplied by the number of kilometers driven.

ton-mile [CIV ENG] In railroading, a standard measure of traffic, based on the rate of carriage per mile of each passenger or ton of freight.

tonnage [NAV ARCH] A measure of the size of a ship; it is usually taken to mean gross tonnage or net tonnage, but may also refer to deadweight or displacement tonnage.

tonnage deck [NAV ARCH] In vessels having three or more decks to the hull, the tonnage deck is the second deck from the keel; in all other cases, it is the upper deck of the hull.

tonne [MECH] A unit of mass in the metric system, equal to 1000 kilograms or to approximately 2204.62 pound mass. Also known as metric ton; millier; ton; tonneau.

tonneau *See* tonne.

tonofibril [CYTOL] Any of the fibrils converging on desmosomes in epithelial cells.

tonometer [MED] An electronic instrument that measures hydrostatic pressure within the eye: when placed in position, a tiny movable plate is pressed against the eye, flattening a circular section of the cornea (no eyeball anesthesia is required); a current is then sent through a small electromagnet, of such value that it will just pull the plate away from the eye; the value of the current is proportional to eye pressure; a measurement can be made in about 1 second; used in diagnosis of glaucoma. Also known as electronic tonometer.

tonoplast [BOT] The membrane surrounding a plant-cell vacuole.

Tonotron [ELECTR] Trademark for a type of direct-view storage tube.

tonsil [ANAT] **1.** Localized aggregation of diffuse and nodular lymphoid tissue found in the throat where the nasal and oral cavities open into the pharynx. **2.** *See* palatine tonsil.

tonsillectomy [MED] Surgical removal of the palatine tonsil.

tonsillitis [MED] Inflammation of the tonsils.

tonstein [GEOL] Kaolinitic bands in certain coalfields which have characteristic fossil fauna from short-lived but widespread marine invasions.

tonus [PHYSIO] The degree of muscular contraction when not undergoing shortening.

tool [ENG] Any device, instrument, or machine for the performance of an operation, for example, a hammer, saw, lathe, twist drill, drill press, grinder, planer, or screwdriver. [IND ENG] To equip a factory or industry for production by designing, making, and integrating machines, machine tools, and special dies, jigs, and instruments, so as to achieve manufacture and assembly of products on a volume basis at minimum cost.

toolbox [ENG] A box to hold tools.

tool-check system [IND ENG] A system for temporary issue of tools in which the employee is issued a number of small metal checks stamped with the same number; a check is surrendered for each tool obtained from the crib.

tool design [DES ENG] The division of mechanical design concerned with the design of tools.

tool-dresser [MECH ENG] A tool-stone-grade diamond inset in a metal shank and used to trim or form the face of a grinding wheel.

tool extractor [ENG] An implement for grasping and withdrawing drilling tools when broken, detached, or lost in a borehole.

tool mark [GEOL] Any of the wide variety of current marks, such as groove marks, prod marks, and skip marks, produced by the continuous contact or intermittent impact of solid, current-borne objects against a muddy bottom.

tool nipper [MIN ENG] A person whose duty it is to carry powder, drills, and tools to the various levels of the mine and to bring dull tools and drills to the surface.

tool steel [MET] Any of various steels capable of being hardened sufficiently so as to be a suitable material for making cutting tools.

tooth [ANAT] One of the hard bony structures supported by the jaws in mammals and by other bones of the mouth and pharynx in lower vertebrates serving principally for prehension and mastication. [DES ENG] **1.** One of the regular projections on the edge or face of a gear wheel. **2.** An angular projection on a tool or other implement, such as a rake, saw, or comb. [GRAPHICS] **1.** The coarse or abrasive quality of a

paper or a painting ground that assists in the application of charcoal, pastels, or paint. **2.** A paper texture that holds ink more readily. [INV ZOO] Any of various sharp, horny, chitinous, or calcareous processes on or about any part of an invertebrate that functions like or resembles vertebrate jaws.

toothache [MED] Pain in or about a tooth. Also known as odontalgia.

tooth decay [MED] Caries of the teeth.

tooth point [DES ENG] The chamfered cutting edge of the blade of a face mill.

tooth shell [INV ZOO] A mollusk of the class Scaphopoda characterized by the elongate, tube-shaped, or cylindrical shell which is open at both ends and slightly curved.

top [GEOL] *See* overburden. [MECH] A rigid body, one point of which is held fixed in an inertial reference frame, and which usually has an axis of symmetry passing through this point; its motion is usually studied when it is spinning rapidly about the axis of symmetry. [QUANT MECH] *See* rotator.

top and bottom process [MET] A process in which sodium sulfide is added to molten copper-nickel sulfide to form a two-layer melt, with the bulk of the nickel in the bottom layer.

topaz [MINERAL] $Al_2SiO_4(F,OH)$ A red, yellow, green, blue, or brown neosilicate mineral that crystallizes in the orthorhombic system and commonly occurs in prismatic crystals with pyramidal terminations; hardness is 8 on Mohs scale, and specific gravity is 3.4–3.6; used as a gemstone.

topaz quartz *See* citrine.

top-benching [MIN ENG] The method by which the bench is removed from above, as with a dragline.

top-blown rotary converter [MET] A rotary converter used for making nickel and steel; oxygen and other gases are fed to the furnace by a lance at the elevated end of the converter to permit formation of a metal product without oxidation.

top cager [MIN ENG] A person at the top of a mine shaft who superintends the lowering and raising of the cage, and, at most mines, the removing of loaded cars from the placing of empty cars in the cage.

top cut [MIN ENG] A machine cut made in the coal at or near the top of the working face in a mine.

top dead center [MECH ENG] The dead-center position of an engine piston and its crankshaft arm when at the top or outer end of its stroke.

top grafting [BOT] Grafting a scion of one variety of tree onto the main branch of another.

top hamper *See* rigging.

top hooker *See* lander.

tophus [MED] A localized swelling principally in cartilage and connective tissues in or adjacent to the small joints of the hands and feet; occurs specifically in gout.

top lander *See* lander.

top-loaded vertical antenna [ELECTROMAG] Vertical antenna constructed so that, because of its greater size at the top, there results modified current distribution, giving a more desirable radiation pattern in the vertical plane.

topmark [NAV] A characteristic shape secured at the top of a buoy or beacon to aid in its identification.

topocline [ECOL] A graded series of characters exhibited by a species or other closely related organisms along a geographical axis.

topogenesis *See* morphogenesis.

Topogon lens [OPTICS] A periscopic lens with supplementary thick menisci to permit the correction of aperture aberrations for a moderate aperture and a large field; one or two plane-parallel plates are sometimes added to correct distortion.

topographical latitude *See* geodetic latitude.

topographic anatomy [ANAT] The use of bony and soft tissue landmarks on the surface of the body to indicate the known location of deeper structures.

topographic climax [ECOL] A climax plant community under a uniform macroclimate over which minor topographic features such as hills, rivers, valleys, or undrained depressions exert a controlling influence.

topographic curl effect [OCEANOGR] A term in Ekman's differential equation for the effects of variable wind stress, variable depth, variable friction, and variable latitude on the deep current; tends to make the curl *G* (velocity of deep current) positive when the current flows over increasing

TOOTH SHELL

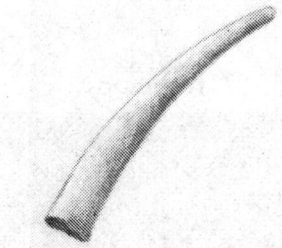

Tube-shaped tooth shell.

TOPAZ

Crystal form of topaz, from Ramona, California. (*American Museum of Natural History Specimen*)

TOPOGON LENS

Components of a Topogon lens.

depth and negative when the depth decreases in the direction of the current.

topographic infancy *See* infancy.

topographic map [MAP] A large-scale map showing relief and man-made features of a portion of a land surface distinguished by portrayal of position, relation, size, shape, and elevation of the features.

topographic maturity *See* maturity.

topographic old age *See* old age.

topographic profile *See* profile.

topographic survey [ENG] A survey that determines ground relief and location of natural and man-made features thereon.

topographic unconformity [GEOGR] A lack of harmony or conformity between two parts of a landscape or two kinds of topography.

topographic youth *See* youth.

topography [GEOGR] **1.** The general configuration of a surface, including its relief; may be a land or water-bottom surface. **2.** The natural surface features of a region, treated collectively as to form.

topological groups [MATH] Groups which also have a topology with the property that the group operation and the inverse operation determine continuous functions.

topologically closed set *See* closed set.

topological mapping *See* homeomorphism.

topological product [MATH] The topological space obtained from taking the cartesian product of topological spaces.

topological property [MATH] A property that holds true for any topological space homeomorphic to one possessing the property.

topological space [MATH] A set endowed with a topology.

topology [MATH] **1.** A collection of subsets of a set X, which includes X and the empty set, and has the property that any union or finite intersection of its members is also a member. **2.** The generalized study of properties of spaces invariant under deformations and stretchings.

topotaxis *See* tropism.

topotype [SYST] A specimen of a species not of the original type series collected at the type locality.

topped crude [MATER] A residual product remaining after the removal by distillation or other means of an appreciable quantity of the more volatile components of crude petroleum.

topping [CHEM ENG] The distillation of crude petroleum to remove the light fractions only; the unrefined distillate is called tops.

topping governor *See* limit governor.

topping lift [NAV ARCH] A rope or chain extending from the head of a boom or gaff to a mast or to the vessel's structure for the purpose of supporting the weight and permitting the boom or gaff end to be raised or lowered.

topple [MECH] In gyroscopes for marine or aeronautical use, the condition of a sudden upset gyroscope or a gyroscope platform evidenced by a sudden and rapid precession of the spin axis due to large torque disturbances such as the spin axis striking the mechanical stops. Also known as tumble.

topple axis [MECH] Of a gyroscope, the horizontal axis, perpendicular to the horizontal spin axis, around which topple occurs. Also known as tumble axis.

topset bed [GEOL] One of the nearly horizontal sedimentary layers deposited on the top surface of an advancing delta.

top shell [INV ZOO] Any of the marine snails of the family Trochidae characterized by a spiral conical shell with a flat base.

topside sounder [AERO ENG] A satellite designed to measure ion concentration in the ionosphere from above the ionosphere.

top slicing [MIN ENG] A method of stoping in which the ore is extracted by excavating a series of horizontal (sometimes inclined) timbered slices alongside each other, beginning at the top of the ore body and working progressively downward.

top slicing and cover caving [MIN ENG] A mining method that entails the working of the ore body from the top down in successive horizontal slices that may follow one another sequentially or simultaneously; the overburden or cover is caved after mining a unit.

topsoil [GEOL] **1.** Soil presumed to be fertile and used to cover areas of special planting. **2.** Surface soil, usually corresponding with the A horizon, as distinguished from subsoil.

top steam [CHEM ENG] Steam admitted near the top of a shell still to purge the still, and to prevent a vacuum from forming when pumping out the liquid contents.

topwork [BOT] A procedure employed to propagate seedless varieties of fruit and hybrids, to change the variety of fruit, and to correct pollination problems, using any of three methods: root grafting, crown grafting, and top grafting.

tor [GEOGR] An isolated, rough pinnacle or rocky peak. [MECH] *See* pascal.

torbanite [GEOL] A variety of coal that resembles a carbonaceous shale in outward appearance; it is fine-grained, black to brown, and tough. Also known as bituminite; kerosine shale.

torbernite [MINERAL] $Cu(UO_2)_2(PO_4)_2 \cdot 8\text{-}12H_2O$ A green radioactive mineral crystallizing in the tetragonal system and occurring in tabular crystals or in foliated form. Also known as chalcolite; copper uranite; cuprouranite; uran-mica.

torch [ENG] A gas burner used for brazing, cutting, or welding.

tornado [METEOROL] An intense rotary storm of small diameter, the most violent of weather phenomena; tornadoes always extend downward from the base of a convective-type cloud, generally in the vicinity of a severe thunderstorm. Also known as twister.

tornado belt [METEOROL] The district of the United States in which tornadoes are most frequent; it encompasses the great lowland areas of the central and upper Mississippi, the Ohio, and lower Missouri River valleys.

tornado cellar *See* cyclone cellar.

tornado cloud *See* tuba.

tornado echo [METEOROL] A type of radar precipitation echo which has been observed in connection with a number of tornadoes; it frequently appears, on plan-position-indicator scopes, in the form of the figure 6 in the southwest sector of the storm; this echo has not been noted with all radar-observed tornadoes.

tornadotron [ELECTR] Millimeter-wave device which generates radio-frequency power from an enclosed, orbiting electron cloud, excited by a radio-frequency field, when subjected to a strong, pulsed magnetic field.

tornaria [INV ZOO] The larva of some acorn worms (Enteropneusta) which is large and marked by complex bands of cilia.

tornote [INV ZOO] A monaxon spicule in certain Porifera having both ends terminating abruptly in points.

torn-tape relay [COMMUN] Method of receiving messages in tape form, breaking the tape, and retransmitting the message in tape form.

toroid *See* doughnut; toroidal magnetic circuit.

toroidal [SCI TECH] Shaped like a doughnut.

toroidal coil *See* toroidal magnetic circuit.

toroidal core [ELECTROMAG] The doughnut-shaped piece of magnetic material in a toroidal magnetic circuit.

toroidal magnetic circuit [ELECTROMAG] Doughnut-shaped piece of magnetic material, together with one or more coils of current-carrying wire wound about the doughnut, with the permeability of the magnetic material high enough so that the magnetic flux is almost completely confined within it. Also known as toroid; toroidal coil.

toromatic transmission [MECH ENG] A semiautomatic transmission; it contains a compound planetary gear train with a torque converter.

torose [INV ZOO] **1.** Having knobby prominences on the surface. **2.** Cylindrical with alternate swellings and contractions.

torose load cast [GEOL] One of a group of elongate load casts with alternate contractions and swellings, which may terminate down current in bulbous, teardrop, or spiral forms.

Torpedinidae [VERT ZOO] The electric rays or torpedoes, a family of batoid sharks.

torpedo [ENG] An encased explosive charge slid, lowered, or dropped into a borehole and exploded to clear the hole of obstructions or to open communications with an oil or water supply. Also known as bullet. [ORD] A missile designed to contain an explosive charge and to be launched into water, where it is self-propelling and usually directable; used against ships or other targets in the water.

torpedo air flask [ORD] A cylindrical item having various compartments for housing compressed air, fuel, water, and

TORNADO

Tornado, at Fargo, North Dakota. *(Fargo Forum photograph by C. Gebert, Grand Prize Winner, 11th Annual Graflex Contest)*

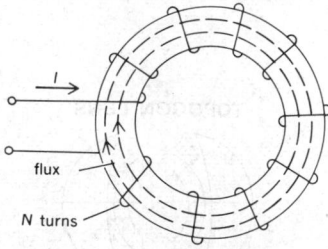

TOROIDAL MAGNETIC CIRCUIT

I
flux
N turns

Diagram of toroidal magnetic circuit. *I* represents current. *(From A. E. Fitzgerald, D. E. Higginbotham, and A. Grabel, Basic Electrical Engineering, McGraw-Hill, 1967)*

chemicals, which when combined form the propelling charge of aerial and underwater torpedoes.

torpedo boat [NAV ARCH] Small, fast vessel that is equipped with torpedo tubes, carries very light guns, and uses its inconspicuousness and speed to get within torpedo range of a target.

torpedo defense net [ORD] A net employed to close an inner harbor to torpedoes fired from seaward or to protect an individual ship at anchor or underway.

torpedo exercise head [ORD] An item designed for attachment to a torpedo main assemblage to complete a torpedo for a practice run; it may contain recording instruments.

torpedo exploder mechanism [ORD] An electrical or mechanical device designed to actuate the explosive train of a torpedo warhead by means of a physical impact or an influence signal; it may contain a disarming device.

torpedo warhead extension [ORD] A metallic cylindrical item designed to change the center of gravity of a torpedo; it may be explosive-filled.

torpor [PHYSIO] The condition in hibernating poikilotherms during winter when body temperature drops in a parallel relation to ambient environmental temperatures.

torque [MECH] **1.** For a single force, the cross product of a vector from some reference point to the point of application of the force with the force itself. Also known as moment of force; rotation moment. **2.** For several forces, the vector sum of the torques (first definition) associated with each of the forces.

torque amplifier [ADP] An analog computer device having input and output shafts and supplying work to rotate the output shaft in positional correspondence with the input shaft without imposing any significant torque on the input shaft.

torque arm [MECH ENG] In automotive vehicles, an arm to take the torque of the rear axle.

torque-coil magnetometer [ENG] A magnetometer that depends for its operation on the torque developed by a known current in a coil that can turn in the field to be measured.

torque converter [MECH ENG] A device for changing the torque speed or mechanical advantage between an input shaft and an output shaft.

torque-load characteristic [ENG] For electric motors, the armature torque developed versus the load on the motor at constant speed.

torquemeter [ENG] An instrument to measure torque.

torque motor [ELECTROMAG] A motor designed primarily to exert torque while stalled or rotating slowly.

torque reaction [MECH ENG] On a shaft-driven vehicle, the reaction between the bevel pinion with its shaft (which is supported in the rear axle housing) and the bevel ring gear (which is fastened to the differential housing) that tends to rotate the axle housing around the axle instead of rotating the axle shafts alone.

torque-speed characteristic [ELEC] For electric motors, the relationship of developed torque to armature speed.

torque-tube flowmeter [ENG] A liquid-flow measurement device in which a flexible torque tube transmits bellows motion (caused by differential pressure from the liquid flow through the pipe) to the recording pen arm.

torque-type viscometer [ENG] A device that measures liquid viscosity by the torque needed to rotate a vertical paddle submerged in the liquid; used for both Newtonian and non-Newtonian liquids and for suspensions.

torque-winding diagram [MECH ENG] A diagram showing how the winding load on a winch drum varies and is used to decide the method of balancing needed; made by plotting the turning moment in pounds per foot on the vertical axis against time, or revolutions or depth on the horizontal axis.

torque wrench [ENG] **1.** A hand or power tool used to turn a nut on a bolt that can be adjusted to deliver a predetermined amount of force to the bolt when tightening the nut. **2.** A wrench that measures torque while being turned.

torr [MECH] A unit of pressure, equal to 1/760 atmosphere; it differs from 1 millimeter of mercury by less than one part in seven million; approximately equal to 133.3224 pascals.

Torricellian barometer *See* mercury barometer.

Torricellian vacuum [FL MECH] The space enclosed above a column of mercury when a tube, closed at one end, is filled with mercury and then placed, open end downward, in a well of mercury; this space is evacuated except for mercury vapor.

Torricelli's law of efflux [FL MECH] The velocity of efflux of liquid from an orifice in a container is equal to that which would be attained by a body falling freely from rest at the free surface of the liquid to the orifice.

Torridincolidae [INV ZOO] A small family of coleopteran insects in the suborder Myxophaga found only in Africa and Brazil.

Torrid Zone [CLIMATOL] The zone of the earth's surface which lies between the Tropics of Cancer and Capricorn.

Torrox [GEOL] A suborder of the order Oxisol that is dry for some part of the year.

torsion [MECH] A twisting deformation of a solid body about an axis in which lines that were initially parallel to the axis become helices.

torsional angle [MECH] The total relative rotation of the ends of a straight cylindrical bar when subjected to a torque.

torsional modulus *See* modulus of elasticity in shear.

torsional pendulum [MECH] A device consisting of a disk or other body of large moment of inertia mounted on one end of a torsionally flexible elastic rod whose other end is held fixed; if the disk is twisted and released, it will undergo simple harmonic motion, provided the torque in the rod is proportional to the angle of twist.

torsional vibration [MECH] A periodic motion of a shaft in which the shaft is twisted about its axis first in one direction and then in the other; this motion may be superimposed on rotational or other motion.

torsion balance [ENG] An instrument, consisting essentially of a straight vertical torsion wire whose upper end is fixed while a horizontal beam is suspended from the lower end; used to measure minute gravitational, electrostatic, or magnetic forces.

torsion bar [MECH ENG] A spring flexed by twisting about its axis; found in the spring suspension of truck and passenger car wheels, in production machines where space limitations are critical, and in high-speed mechanisms where inertia forces must be minimized.

torsion damper [MECH ENG] A damper used on automobile internal combustion engines to reduce torsional vibration.

torsion fault *See* wrench fault.

torsion galvanometer [ENG] A galvanometer in which the force between the fixed and moving systems is measured by the angle through which the supporting head of the moving system must be rotated to bring the moving system back to its zero position.

torsion hygrometer [ENG] A hygrometer in which the rotation of the hygrometric element is a function of the humidity; such hygrometers are constructed by taking a substance whose length is a function of the humidity and twisting or spiraling it under tension in such a manner that a change in length will cause a further rotation of the element.

torsion of a curve [MATH] For a given space curve, the rate at which the curve turns out of its osculating plane relative to arc length; it is defined in terms of the binormals.

torsion-string galvanometer [ENG] A sensitive galvanometer in which the moving system is suspended by two parallel fibers that tend to twist around each other.

torso mountain *See* monadnoch.

torticollis [MED] A deformity of the neck resulting from contraction of the cervical muscles or fascia. Also known as wryneck.

tortoise [VERT ZOO] Any of various large terrestrial reptiles in the order Chelonia, especially the family Testudinidae.

Tortonian [GEOL] A European stage of geologic time: Miocene (above Helvetian, below Sarmatian).

Tortricidae [INV ZOO] A family of phytophagous moths in the superfamily Tortricoidea which have a stout body, lightly fringed wings, and threadlike antennae.

Tortricoidea [INV ZOO] A superfamily of small wide-winged moths in the suborder Heteroneura.

Torulopsidales [MYCOL] The equivalent name for Cryptococcales.

torulosis *See* cryptococcosis.

torus [ANAT] A rounded protuberance on a body part. [BOT] The thickened membrane closing a bordered pit. [MATH] **1.** The surface of a doughnut. **2.** The topological

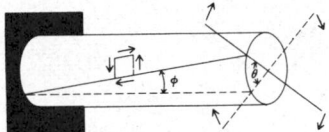

TORSION

Cylindrical bar in torsion, showing deformation of a small element of the bar; θ represents torsional angle; ϕ represents helical angle.

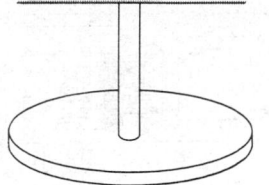

TORSIONAL PENDULUM

Diagram of a torsional pendulum.

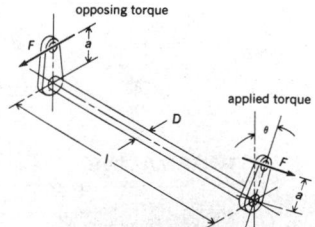

TORSION BAR

Diagram of torsion bar showing important dimensions involved in its design; θ is angle of twist, F is twisting force, a is radius arm, l is length of bar, D is diameter of bar.

TORTOISE

Desert tortoise (*Gopherus agassizi*).

space obtained by identifying the opposite sides of a rectangle. **3.** The group which is the product of two circles.

Torymidae [INV ZOO] A family of hymenopteran insects in the superfamily Chalcidoidea.

tosca [METEOROL] A southwest wind on Lake Garda in Italy.

toss bombing [ORD] A method of bombing where an aircraft flies on a line toward the target, pulls up in a vertical plane, releasing the bomb at an angle that will compensate for the effect of gravity drop on the bomb.

total binding energy *See* binding energy.

total carbon [MET] The sum of free and combined carbon in a ferrous alloy, especially steel.

total conductivity [GEOPHYS] In atmospheric electricity, the sum of the electrical conductivities of the positive and negative ions found in a given portion of the atmosphere.

total curvature [MATH] The total curvature of a surface at a point is given by the product of the principal curvatures there.

total curvature of a lens [OPTICS] The difference between the reciprocals of the radii of curvature of the two surfaces of a lens.

total cyanide [MET] Total amount of cyanide contained in an electroplating bath, including both simple and complex ions.

total differential [MATH] The total differential of a function of several variables, $f(x_1, x_2, \ldots, x_n)$, is the function given by the sum of terms $(\partial f/\partial x_i)dx_i$ as i runs from 1 to n. Also known as differential.

total displacement *See* slip.

total drift [NAV] In marine gyroscopes, the algebraic sum of the apparent and real precession or drift; the apparent drift is that caused by the gyroscope attempting to maintain its orientation in inertial space while the vehicle on which the gyroscope is mounted moves to new locations.

total eclipse [ASTRON] An eclipse that obscures the entire surface of the moon or sun.

total evaporation *See* evapotranspiration.

total harmonic distortion [ELECTR] Ratio of the power at the fundamental frequency, measured at the output of the transmission system considered, to the power of all harmonics observed at the output of the system because of its nonlinearity, when a single frequency signal of specified power is applied to the input of the system; it is expressed in decibels.

total heat *See* enthalpy.

total heat of dilution *See* heat of dilution.

total heat of solution *See* heat of solution.

total impulse [AERO ENG] The product of the thrust and the time over which the thrust is produced, expressed in pounds (force)–seconds; used especially in reference to a rocket motor or a rocket engine.

total internal reflection [OPTICS] A phenomenon in which electromagnetic radiation in a given medium which is incident on the boundary with a less-dense medium (one having a lower index of refraction) at an angle less than the critical angle is completely reflected from the boundary.

total lift [AERO ENG] The upward force produced by the gas in a balloon; it is equal to the sum of the free lift, the weight of the balloon, and the weight of auxiliary equipment carried by the balloon.

totally bounded set *See* precompact set.

totally disconnected [MATH] A topological space has this property if the largest connected subset containing any given point is only the point itself.

total porosity [GEOL] The ratio of total void space in porous oil-reservoir rock to the bulk volume of the rock itself.

total pressure [MECH] The gross load applied on a given surface. [MIN ENG] The total ventilating pressure in a mine, usually measured in the fan drift.

total radiation pyrometer [ENG] A pyrometer which focuses heat radiation emitted by a hot object on a detector (usually a thermopile or other thermal type detector), and which responds to a broad band of radiation, limited only by absorption of the focusing lens, or window and mirror.

total slip *See* net slip.

total solids [CHEM] The total content of suspended and dissolved solids in water.

total span [ANTHRO] An anthropometric determination of the distance between the tips of the middle fingers at maximum arm stretch without straining.

total specific ionization *See* specific ionization.

total variation [MATH] For a real function defined on an interval, the least upper bound of the function's variation relative to all possible partitions of the interval.

total vorticity [FL MECH] Usually, the magnitude of the vorticity vector, all components included, as opposed to the vertical (component of the) vorticity.

totipalmate [VERT ZOO] Having all four toes connected by webs, as in the Pelecaniformes.

totipotence [EMBRYO] Capacity of a blastomere to develop into a fully formed embryo.

toucan [VERT ZOO] Any of numerous fruit-eating birds, of the family Ramphastidae, noted for their large and colorful bills.

Toucan *See* Tucana.

touch [PHYSIO] The array of sensations arising from pressure sensitivity of the skin.

touch call *See* push-button dialing.

touch control [ELEC] A circuit that closes a relay when two metal areas are bridged by a finger or hand.

touch-down zone [NAV] In air operations, the first 3000 feet (914 meters) of runway beginning at the threshold. Abbreviated TDZ.

touch-down zone elevation [NAV] The highest runway centerline elevation in the touch-down zone.

touch feedback [ENG] A type of force feedback in which servos provide the manipulator fingers with a sense of resistance when an object is grasped, so that the operator does not crush the object.

Touch-Tone *See* pushbutton dialing.

toughness [MECH] A property of a material capable of absorbing energy by plastic deformation; intermediate between softness and brittleness.

tough pitch copper [MET] Copper refined in a reverberatory furnace to adjust the oxygen content to 0.2–0.5%.

tour *See* shift.

touriello [METEOROL] A south wind of foehn type descending from the Pyrenees in the Ariège valley, France; it is especially violent in February and March, when it melts the snow, flooding the rivers and sometimes causing avalanches.

tourmaline [MINERAL] $(Na,Ca)(Al,Fe,Li,Mg)_3Al_6(BO_3)_3$-$Si_6O_{18}(OH)_4$ Any of a group of cyclosilicate minerals with a complex chemical composition, vitreous to resinous luster, and variable color; crystallizes in the ditrigonal-pyramidal class of the hexagonal system, has piezoelectric properties, and is used as a gemstone.

Tournaisian [GEOL] European stage of lowermost Carboniferous time.

tourniquet [MED] An apparatus for controlling hemorrhage from, or circulation in, a limb or part of the body, where pressure can be brought upon the blood vessels by means of straps, cords, rubber tubes, or pads.

Toussaint's formula [METEOROL] A rule for the linear decrease of temperature with height in an atmosphere for which the temperature at mean sea level is 15°C, and given by the formula $t = 15 - 0.0065z$, where t is the temperature in degrees Celsius, and z is the geometric height in meters above mean sea level.

towbar [ENG] An element which connects to a vehicle that is not equipped with an integral drawbar, for the purpose of towing or moving the vehicle.

towboat [NAV ARCH] A relatively flat-bottomed vessel designed to push tows of barges on inland waterways; it has a square bow on which heavy, upright knees are fixed for the purpose of lashing the barges against the towing vessel.

towed artillery [ORD] Artillery weapons designed for movement as trailed loads behind prime movers or draft animals; some adjustment of the weapon is necessary to place it in firing position.

towed load [MECH] The weight of a carriage, trailer, or other equipment towed by a prime mover.

tower [CHEM ENG] A vertical, cylindrical vessel used in chemical and petroleum processing to increase the degree of separation of liquid mixtures by distillation or extraction. Also known as column. [ELECTROMAG] A tall metal structure used as a transmitting antenna, or used with another such structure to support a transmitting antenna wire. [ENG] A concrete, metal, or timber structure that is relatively high for its length and width, and used for various purposes, including

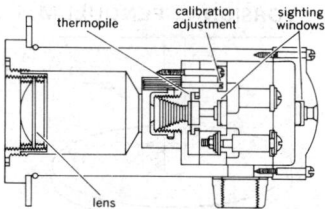

TOTAL RADIATION PYROMETER

Diagram of total radiation pyrometer. Lens focuses heat radiation onto thermopile. *(Honeywell Inc.)*

TOURMALINE

30 mm

Prismatic crystal form of tourmaline from Portland, Connecticut. *(American Museum of Natural History Specimen)*

the support of electric power transmission lines, radio and television antennas, and rockets and missiles prior to launching.

tower crane [CIV ENG] A crane mounted on top of a tower which is sometimes incorporated in the frame of a building.

tower excavator [MIN ENG] A cableway excavator designed specifically for levee work but which is used extensively in the stripping of overburden, spoil, or waste in surface mining: basically, it is a Sauerman-type excavator with towers either fixed or movable, and when the head tower is located on the spoil pile and the tail tower on the unexcavated wall, pits of almost unlimited width can be dug.

towering [METEOROL] A refraction phenomenon; a special case of looming in which the downward curvature of the light rays due to atmospheric refraction increases with elevation so that the visual image of a distant object appears to be stretched in the vertical direction.

towering cumulus [METEOROL] A descriptive term, used mostly in weather observing, for the cloud type cumulus congestus.

tower launcher [ORD] A missile launcher that is vertical (or nearly so) and high enough to give directional stability to the missile. Also known as vertical tower launcher.

tower loader [MIN ENG] A front-end loader whose bucket is lifted along tracks on a more or less vertical tower.

tower loading [ELEC] Load placed on a tower by its own weight, the weight of the wires with or without ice covering, the insulators, the wind pressure normal to the line acting both on the tower and the wires, and the pull from the wires.

tower radiator [ELECTROMAG] Metal structure used as a transmitting antenna.

towing tank *See* model basin.

town plan *See* city plan.

Townsend avalanche *See* avalanche.

Townsend characteristic [ELECTR] Current-voltage characteristic curve for a phototube at constant illumination and at voltages below that at which a glow discharge occurs.

Townsend coefficient [ELECTR] The number of ionizing collisions by an electron per centimeter of path length in the direction of the applied electric field in a radiation counter.

Townsend discharge [ELECTR] A discharge which occurs at voltages too low for it to be maintained by the electric field alone, and which must be initiated and sustained by ionization produced by other agents; it occurs at moderate pressures, above about 0.1 torr, and is free of space charges.

Townsend ionization *See* avalanche.

towrope [NAV ARCH] A hawser of either fiber or wire used for towing a vessel.

towrope horsepower *See* effective horsepower.

towrope resistance [NAV ARCH] The total resistance overcome in towing a ship or model; it equals the sum of the frictional resistance, eddy making, and wave making.

tows [TEXT] **1.** The broken, short, matted fiber that is removed during the separation of long fibers of flax, hemp, or jute. **2.** A large number of continuous rayon filaments collected in a ropelike form without a definite twist. **3.** The coarsest of linen yarns used to make crash, a coarse, plain-woven fabric.

tow target [ORD] Target for antiaircraft fire or aerial gunnery practice, towed behind an aircraft.

toxa [INV ZOO] A curved sponge spicule.

toxaphene [ORG CHEM] $C_{10}H_{10}Cl_8$ A toxic, waxy, amber solid with a mild chlorine-camphor aroma, soluble in organic solvents, melts at 65–90°C; used as an insecticide. Also known as technical chlorinated camphene.

Toxasteridae [PALEON] A family of Cretaceous echinoderms in the order Spatangoida which lacked fascioles and petals.

toxemia [MED] A condition in which the blood contains toxic substances, either of microbial origin or as by-products of abnormal protein metabolism.

toxemia of pregnancy *See* preeclampsia.

toxic [MED] Relating to a harmful effect by a poisonous substance on the human body by physical contact, ingestion, or inhalation.

toxic amaurosis [MED] Blindness following the introduction of toxic substances into the body, such as ethyl and methyl alcohol, tobacco, lead, and metabolites of uremia and diabetes.

toxic goiter *See* hyperthyroidism.

toxic hepatitis [MED] Inflammation of the liver caused by chemical agents ingested or inhaled into the body, such as chlorinated hydrocarbons and some alkaloids.

toxicity [PHARM] **1.** The quality of being toxic. **2.** The kind and amount of poison or toxin produced by a microorganism, or possessed by a chemical substance not of biological origin.

toxicology [PHARM] The study of poisons, including their nature, effects, and detection, and methods of treatment.

toxicophobia [PSYCH] Abnormal fear of being poisoned.

toxic psychosis [MED] A brain disorder due to a toxic agent such as lead or alcohol.

toxicyst [INV ZOO] A type of trichocyst in Protozoa which may, upon contact, induce paralysis or lysis of the prey.

toxin [BIOCHEM] Any of various poisonous substances produced by certain plant and animal cells, including bacterial toxins, phytotoxins, and zootoxins.

Toxodontia [PALEON] An extinct suborder of mammals representing a central stock of the order Notoungulata.

Toxoglossa [INV ZOO] A group of carnivorous marine gastropod mollusks distinguished by a highly modified radula (toxoglossate).

toxoglossate radula [INV ZOO] A radula in certain carnivorous gastropods having elongated, spearlike teeth often perforated by the ducts of large poison glands.

toxoid [IMMUNOL] Detoxified toxin, but with antigenic properties intact; toxoids of tetanus and diphtheria are used for immunization.

Toxoplasmea [INV ZOO] A class of the protozoan subphylum Sporozoa composed of small, crescent-shaped organisms that move by body flexion or gliding and are characterized by a two-layered pellicle with underlying microtubules, micropyle, paired organelles, and micronemes.

Toxoplasmida [INV ZOO] An order of the class Toxoplasmea; members are parasites of vertebrates.

toxoplasmin [BIOCHEM] The *Toxoplasma* antigen; used in a skin test to demonstrate delayed hypersensitivity to toxoplasmosis.

toxoplasmosis [MED] Infection by the protozoan *Toxoplasma gondi*, manifested clinically in severe cases by jaundice, hepatomegaly, and splenomegaly.

Toxopneustidae [INV ZOO] A family of Tertiary and extant echinoderms of the order Temnopleuroida where the branchial slits are deep and the test tends to be absent.

Toxotidae [VERT ZOO] The archerfishes, a family of small fresh-water forms in the order Perciformes.

TPA *See* terephthalic acid.

T pad [ELEC] A pad made up of resistance elements arranged as a T network (two resistors inserted in one line, with a third between their junction and the other line).

T phage [VIROL] Any of a series (T1–T7) of deoxyribonucleic acid phages which lyse strains of the gram-negative bacterium *Escherichia coli* and its relatives.

TPI test *See* Treponema pallidum immobilization test.

TPN *See* triphosphopyridine nucleotide.

TPR *See* airborne profile recorder.

Tr *See* trace.

trabecula [ANAT] A band of fibrous or muscular tissue extending from the capsule or wall into the interior of an organ.

trace [ADP] To provide a record of every step, or selected steps, executed by a computer program, and by extension, the record produced by this operation. [ELECTR] The visible path of a moving spot on the screen of a cathode-ray tube. Also known as line. [ENG] The record made by a recording device, such as a seismometer or electrocardiograph. [GEOL] The intersection of two geological surfaces. [MATH] The trace of a matrix is the sum of the entries along its principal diagonal. Designated Tr. Also known as spur. [METEOROL] A precipitation of less than 0.005 inch (0.127 millimeter). [SCI TECH] An extremely small but detectable quantity of a substance.

trace analysis [ANALY CHEM] Analysis of a very small quantity of material of a sample by such techniques as polarography or spectroscopy.

trace element [GEOCHEM] A nonessential element found in small quantities (usually less than 1.0%) in a mineral. Also known as accessory element; guest element.

trace fossil [GEOL] A trail, track, or burrow made by an

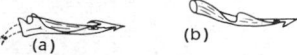
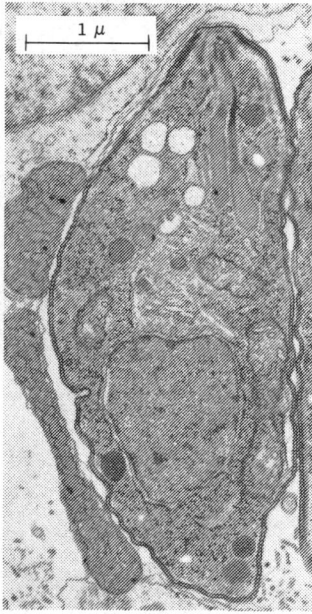

animal and found in ancient sediments such as sandstone, shale, or limestone. Also known as ichnofossil.

trace interval [ELECTR] Interval corresponding to the direction of sweep used for delineation.

tracer [CHEM] A foreign substance, usually radioactive, that is mixed with or attached to a given substance so the distribution or location of the latter can later be determined; used to trace chemical behavior of a natural element in an organism. Also known as tracer element. [ENG] A thread of contrasting color woven into the insulation of a wire for identification purposes.

tracer bullet [ORD] A bullet containing a pyrotechnic mixture to make the flight of the projectile visible by day and night.

tracer element *See* tracer.

tracer milling [MECH ENG] Cutting a duplicate of a three-dimensional form by using a mastic form to direct the tracer-controlled cutter.

tracer mixture [ORD] A pyrotechnic composition used for loading tracer bullets.

trace routine [ADP] A routine which tracks the execution of a program, step by step, to locate a program malfunction. Also known as tracing routine.

trace sensitivity [ELECTR] The ability of an oscilloscope to produce a visible trace on the scope face for a specified input voltage.

trace slip [GEOL] That component of the net slip in a fault which is parallel to the trace of an index plane on a fault plane.

trace-slip fault [GEOL] A fault whose net slip is trace slip.

trachea [ANAT] The cartilaginous and membranous tube by which air passes to and from the lungs in man and many vertebrates. [BOT] A xylem vessel resembling the trachea of vertebrates. [INV ZOO] One of the anastomosing air-conveying tubules composing the respiratory system in most insects.

tracheid [BOT] An elongate, spindle-shaped xylem cell, lacking protoplasm at maturity, and having secondary walls laid in various thicknesses and patterns over the primary wall.

Tracheophyta [BOT] A large group of plants characterized by the presence of specialized conducting tissues (xylem and phloem) in the roots, stems, and leaves.

trachoma [MED] An infectious disease of the conjunctiva and cornea caused by *Chlamydia trachomatis* producing photophobia, pain, and excessive lacrimation.

trachomatous conjunctivitis [MED] Inflammation of the conjunctiva associated with trachoma, characterized by a subepithelial cellular infiltration with a follicular distribution.

trachybasalt [PETR] An extrusive rock characterized by calcic plagioclase and sanidine, with augite, olivine, and possibly minor analcime or leucite.

Trachylina [INV ZOO] An order of moderate-sized jellyfish of the class Hydrozoa distinguished by having balancing organs and either a small polyp stage or none.

Trachymedusae [INV ZOO] A group of marine jellyfish, recognized as a separate order or as belonging to the order Trachylina whose tentacles have a solid core consisting of a single row of endodermal cells.

Trachypsammiacea [INV ZOO] An order of colonial anthozoan coelenterates characterized by a dendroid skeleton.

Trachystomata [VERT ZOO] The name given to the Meantes when the group is considered to be an order.

trachyte [PETR] The light-colored, aphanitic rock (the volcanic equivalent of syenite), composed largely of alkali feldspar with minor amounts of mafic minerals.

tracing distortion [ENG ACOUS] The nonlinear distortion introduced in the reproduction of a mechanical recording because the curve traced by the motion of the reproducing stylus is not an exact replica of the modulated groove.

tracing paper [MATER] Thin paper used both for tracing and original drawings, and of various types and surfaces.

tracing routine *See* trace routine.

track [AERO ENG] The actual line of movement of an aircraft or a rocket over the surface of the earth; it is the projection of the history of the flight path on the surface. Also known as flight track. [DES ENG] As applied to a pattern of setting diamonds in a bit crown, an arrangement of diamonds in concentric circular rows in the bit crown, with the diamonds

in a specific row following in the track cut by a preceding diamond. [ELECTR] **1.** A path for recording one channel of information on a magnetic tape, drum, or other magnetic recording medium; the location of the track is determined by the recording equipment rather than by the medium. **2.** The trace of a moving target on a plan-position-indicator radar screen or an equivalent plot. [ENG] **1.** The groove cut in a rock by a diamond inset in the crown of a bit. **2.** A pair of parallel metal rails for a railway, railroad, tramway, or for any wheeled vehicle. [MECH ENG] **1.** The slide or rack on which a diamond-drill swivel head can be moved to positions above and clear of the collar of a borehole. **2.** A crawler mechanism for earth-moving equipment. [NAV] **1.** To follow the movements of an object by keeping the reticle of an optical system or a radar beam on the object, by plotting its bearing and distance at frequent intervals, or by a combination of the two. **2.** To navigate by following the movements of a craft without regard for future positions; this is used when frequent changes of an unanticipated amount are expected in course or speed or both. **3.** A recommended route on a nautical chart, such as a North Atlantic Track. [NUCLEO] **1.** The visible path of an ionizing particle in a particle detector, such as a cloud chamber, bubble chamber, spark chamber, or nuclear photographic emulsion. **2.** *See* race track.

track angle [NAV] Track measured from 0° at the reference direction clockwise or counterclockwise through 90 or 180°; it is labeled with the reference direction as a prefix and the direction of measurement from the reference direction as a suffix; thus, track angle N44°W is 44° west of north, or 316°.

track cable [ENG] Steel wire rope, usually a locked-coil rope which supports the wheels of the carriers of a cableway.

track cable scraper [MIN ENG] A type of excavator that uses a bottomless scraper bucket which conveys its load over the ground and is operated by a two-drum hoist which controls a track cable that spans the working area and a haulage cable that leads to the front of the bucket.

track chart [NAV] A chart showing recommended, required, or established tracks, and usually indicating turning points, courses, and distances.

track crawling [NAV] Keeping a craft close to a course line by frequent small changes of heading.

Tracker [AERO ENG] A United States twin-reciprocating-engine, antisubmarine aircraft capable of operating from carriers, and designed primarily for the detection, location, and destruction of submarines. Designated S-2.

track gage [CIV ENG] The width between the rails of a railroad track; in the United States the standard gage is 4 feet 8½ inches.

track haulage [MIN ENG] Movement or transportation of excavated or mined materials in cars or trucks that run on rails.

track homing [NAV] The process of following a line of position known to pass through an objective.

track hopper [ENG] A hopper-shaped receiver mounted beside or below railroad tracks, into which railroad boxcars or bottom-dump cars are discharged; used for solid materials.

tracking [ELEC] A leakage or fault path created across the surface of an insulating material when a high-voltage current slowly but steadily forms a carbonized path. [ELECTR] The condition in which all tuned circuits in a receiver accurately follow the frequency indicated by the tuning dial over the entire tuning range. [ENG] **1.** A motion given to the major lobe of a radar or radio antenna such that some preassigned moving target in space is always within the major lobe. **2.** The process of following the movements of an object; may be accomplished by keeping the reticle of an optical system or a radar beam on the object, by plotting its bearing and distance at frequent intervals, or by a combination of techniques. [ENG ACOUS] **1.** The following of a groove by a phonograph needle. **2.** Maintaining the same ratio of loudness in the two channels of a stereophonic sound system at all settings of the ganged volume control. [NAV] Navigation which follows the movements of a craft but does not anticipate future positions.

tracking beam [ORD] The beam that is aimed directly at the target at all times in antimissile warfare; data obtained from

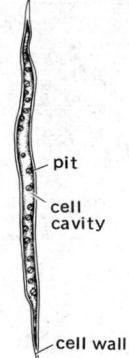

TRACHEID

Drawing of a tracheid, showing characteristic features. *(From H. J. Fuller and O. Tippo, College Botany, rev. ed., Holt, 1954)*

pit

cell cavity

cell wall

this beam are transmitted to the counterattacking guided missile over what is known as the guidance beam.

tracking error [ENG ACOUS] Deviation of the vibration axis of a phonograph pickup from tangency with a groove; true tangency is possible for only one groove when the pickup arm is pivoted; the longer the pickup arm, the less is the tracking error.

tracking filter [ELECTR] Electronic device for attenuating unwanted signals while passing desired signals, by phase-lock techniques that reduce the effective bandwidth of the circuit and eliminate amplitude variations.

tracking jitter [ENG] Minor variations in the pointing of an automatic tracking radar.

tracking network [ENG] A group of tracking stations whose operations are coordinated in tracking objects through the atmosphere or space.

tracking radar [ENG] Radar used to monitor the flight and obtain geophysical data from space probes, satellites, and high-altitude rockets.

tracking station [ENG] A radio, radar, or other station set up to track an object moving through the atmosphere or space.

tracking system [ENG] General name for apparatus, such as tracking radar, used in following and recording the position of objects in the sky.

tracking telescope [OPTICS] A long-focal-length telescope mounted to track missiles in flight precisely while collecting missile performance data.

track in range [ELECTR] To adjust the gate of a radar set so that it opens at the correct instant to accept the signal from a target of changing range from the radar.

track-laying vehicle [ORD] Vehicle which travels upon two endless tracks, one on each side of the machine; it has high mobility and can maneuver, is usually armed and frequently armored, and is intended for tactical use; tanks are one type of track-laying vehicle.

trackless mine [MIN ENG] A mine in which rubber-tired vehicles are used for haulage and transport.

trackless tunneling [MIN ENG] Tunneling by means of loaders mounted on caterpillars.

track made good [AERO ENG] The actual path of an aircraft over the surface of the earth, or its graphic representation.

track pitch [ELECTR] The physical distance between track centers.

track shifter [ENG] A machine or appliance used to shift a railway track laterally.

track telling [COMMUN] The process of communicating air surveillance and tactical data information between command and control systems and facilities within the systems.

track-while-scan [ELECTR] Electronic system used to detect a radar target, compute its velocity, and predict its future position without interfering with continuous radar scanning.

traction [GRAPHICS] A defect in a paint coating in which the film cracks and wide fissures reveal the underlying surface. [GEOL] Transport of sedimentary particles along and parallel to a bottom surface of a stream channel by rolling, sliding, dragging, pushing, or saltation. [MECH] Pulling friction of a moving body on the surface on which it moves.

tractional force [FL MECH] The force exerted on particles under flowing water by the current; it is proportional to the square of the velocity.

traction diverticulum [MED] A circumscribed sacculation, usually of the esophagus, with bulging of the full thickness of the wall; caused by the pull of adhesions arising from adjacent organs.

traction tube [ENG] A device for measuring the minimum water velocities capable of moving various sizes of sand grains; it consists of a horizontal glass tube half-filled with sand.

tractor [MECH ENG] **1.** An automotive vehicle having four wheels or a caterpillar tread used for pulling agricultural or construction implements. **2.** The front pulling section of a semitrailer. Also known as truck-tractor.

tractor drill [MECH ENG] A drill having a crawler mounting to support the feed-guide bar on an extendable arm.

tractor fuel *See* engine distillate.

tractor gate [CIV ENG] A type of outlet control gate used to

release water from a reservoir; there are two types, roller and wheel.

tractor loader [MECH ENG] A tractor equipped with a tipping bucket which can be used to dig and elevate soil and rock fragments to dump at truck height. Also known as shovel dozer; tractor shovel.

tractor shovel *See* tractor loader.

tractrix [MATH] A curve in the plane where every tangent to it has the same length.

trade air [METEOROL] The type of air of which the trade winds consist, and whose chief thermodynamic characteristic is the presence of the trade-wind inversion.

trade cumulus *See* trade-wind cumulus.

trade wind [METEOROL] The wind system, occupying most of the tropics, which blows from the subtropical highs toward the equatorial trough; a major component of the general circulation of the atmosphere; the winds are northeasterly in the Northern Hemisphere and southeasterly in the Southern Hemisphere; hence they are known as the northeast trades and southeast trades, respectively.

trade-wind cumulus [METEOROL] The characteristic cumulus cloud of the trade winds over the oceans in average, undisturbed weather conditions; the individual cloud usually exhibits a blocklike appearance since its vertical growth ends abruptly in the lower stratum of the trade-wind inversion; a group of fully grown clouds shows considerable uniformity in size and shape. Also known as trade cumulus.

trade wind desert [CLIMATOL] **1.** An area of very little rainfall and high temperature which occurs where the trade winds or their equivalent (such as the harmattan) blow over land; the best examples are the Sahara and Kalahari deserts. **2.** The arid cold-water coasts on the western shores of North and South America and Africa.

trade-wind inversion [METEOROL] A characteristic temperature inversion usually present in the trade-wind streams over the eastern portions of the tropical oceans; it is formed by broad-scale subsidence of air from high altitudes in the eastern extremities of the subtropical highs; while descending, the current meets the opposition of the low-level maritime air flowing equatorward; the inversion forms at the meeting point of these two strata which flow horizontally in the same direction.

traersu [METEOROL] A violent east wind of Lake Garda in Italy.

traffic [COMMUN] The messages transmitted and received over a communication channel. [ENG] The passage or flow of vehicles, pedestrians, ships, or planes along defined routes such as highways, sidewalks, sea lanes, or air lanes.

trafficability [CIV ENG] Capability of terrain to bear traffic, or the extent to which the terrain will permit continued movement of any or all types of traffic.

traffic-circulation map [MAP] A map showing traffic routes and the measures for traffic regulation; indicates the roads for use of certain classes of traffic, the location of traffic-control stations, and the directions in which traffic may move. Also known as circulation map.

traffic control [ENG] Control of the movement of vehicles, such as airplanes, trains, and automobiles, and the regulatory mechanisms and systems used to exert or enforce control.

traffic density [CIV ENG] The average number of vehicles that occupy 1 mile or 1 kilometer of road space, expressed in vehicles per mile or per kilometer.

traffic diagram [COMMUN] Chart or illustration used to show the movement and control of traffic over a communications system.

traffic distribution [COMMUN] Routing of communications traffic through a terminal to a switchboard or dialing center.

traffic engineering [CIV ENG] The determination of the required capacity and layout of highway and street facilities that can safely and economically serve vehicular movement between given points.

traffic flow [CIV ENG] The total number of vehicles passing a given point in a given time, expressed as vehicles per hour.

traffic flow security [COMMUN] Transmission of an uninterrupted flow of random text on a wire or radio link between two stations with no indication to an interceptor of what

TRACTOR GATE

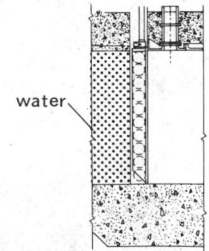

Diagram of tractor gate. *(U.S. Army Corps of Engineers and U.S. Bureau of Reclamation)*

portions of this steady stream constitute encrypted message text and what portions are merely random filler.

traffic forecast [COMMUN] Traffic level prediction on which communications system management decisions and engineering effort are based.

traffic pattern [AERO ENG] The traffic flow that is prescribed for aircraft landing at, taxiing on, and taking off from an airport; the usual components of a traffic pattern are upwind leg, crosswind leg, downwind leg, base leg, and final approach.

traffic recorder [ENG] A mechanical counter or recorder used to determine traffic movements (hourly variations and total daily volumes of traffic at a point) on an existing route; the air-impulse counter, magnetic detector, photoelectric counter, and radar detector are used.

traffic signal [CIV ENG] With the exception of traffic signs, any power-operated device for regulating, directing, or warning motorists or pedestrians.

tragacanth [MATER] The gummy exudate produced by certain Asiatic species of *Astragalus*; consists of a soluble portion containing uronic acid and arabinose, and an insoluble portion that absorbs water and swells to make a stiff opalescent mucilage.

tragion-nasal root [ANTHRO] A measure of the distance from the tragion to the deepest concavity of the nasal root.

Tragulidae [VERT ZOO] The chevrotains, a family of pecoran ruminants in the superfamily Traguloidea.

Traguloidea [VERT ZOO] A superfamily of pecoran ruminants, comprised of the most primitive forms with large canines; the chevrotain is the only extant member.

tragus [ANAT] **1.** The prominence in front of the opening of the external ear. **2.** One of the hairs in the external ear canal.

T rail [CIV ENG] A rail shaped like a T in cross section due to a wide head, web, and flanged base.

trail [ASTRON] A luminous trace left in the sky by the passage of a large meteor. [GEOL] A line of rock fragments that were picked up by glacial ice at a localized outcropping and left scattered along a fairly well-defined tract during the movement of a glacier. [GRAPHICS] One of the lines left on a photographic plate during prolonged exposure to starlight if the motion of the plate was not synchronized with the apparent motion of the sky. [ORD] **1.** In bombing, the line between the point of impact of the bomb and a point on the ground directly beneath the aircraft at the moment of impact, assuming that the aircraft stays on course after release of the bomb and maintains a constant speed. **2.** Rear part of a gun carriage which connects the piece with a limber or tractor.

trail angle [AERO ENG] The angle at an aircraft between the vertical and the line of sight to an object over which the aircraft has passed.

trailer [ELECTR] A bright streak at the right of a dark area or dark line in a television picture, or a dark area or streak at the right of a bright part; usually due to insufficient gain at low video frequencies. [MECH ENG] The section of a semitrailer that is pulled by the tractor.

trailer card [ADP] A card that contains supplemental information related to the data on the preceding cards.

trailer label [ADP] A record appearing at the end of a magnetic tape that uniquely identifies the tape as one required by the system.

trail formation [AERO ENG] Aircraft flying singly or in elements in such manner that each aircraft or element is in line behind the preceding aircraft or element. [ENG] Vehicles proceeding one behind the other at designated intervals. Also known as column formation.

trailing antenna [ELECTROMAG] An aircraft radio antenna having one end weighted and trailing free from the aircraft when in flight.

trailing edge [AERO ENG] The rear section of a multipiece airfoil, usually that portion aft of the rear spar. [ELECTR] The major portion of the decay of a pulse.

trailing-edge tab [AERO ENG] One of the devices on the aircraft elevator that reduce or eliminate hinge movements required to deflect the elevator during flight.

train [ASTRON] The bright tail of a comet or meteor. [ENG] To aim or direct a radar antenna in azimuth.

train bombing [ORD] A method of bombing in which two or more bombs are released at a predetermined interval from one aircraft as the result of a single actuation of the bomb-release mechanism.

trainer [ELECTR] A piece of equipment used for training operators of radar, sonar, and other electronic equipment by simulating signals received under operating conditions in the field.

training aid [ENG] Any item which is developed or procured primarily to assist in training and the process of learning.

training ammunition [ORD] Ammunition used for training persons in marksmanship, handling weapons, and so forth.

training time [ADP] The machine time expended in training employees in the use of the equipment, including such activities as mounting, console operation, converter operation, and printing operation, and time spent in conducting required demonstrations.

training wall [CIV ENG] A wall built along the bank of a river or estuary parallel to the direction of flow to direct and confine the flow.

train shed [CIV ENG] **1.** A structure to protect trains from weather. **2.** The part of a railroad station that covers the tracks.

trajectory [GEOPHYS] The path of a seismic wave. [MECH] The curve described by an object moving through space, as of a meteor through the atmosphere, a planet around the sun, a projectile fired from a gun, or a rocket in flight.

trajectory chart [ORD] Diagram of a side view of the paths of projectiles fired at various elevations under standard conditions; the trajectory chart is different for different guns, projectiles, and fuses.

trajectory-measuring system [ENG] A system used to provide information on the spatial position of an object at discrete time intervals throughout a portion of the trajectory or flight path.

trama [MYCOL] The loosely woven hyphal tissue between adjacent hymenia in basidiomycetes.

tramming [MIN ENG] Pushing tubs, mine cars, or trams by hand.

tramontana [METEOROL] A cold wind from the northeast or north, particularly on the west coast of Italy and northern Corsica, but also in the Balearic Islands and the Ebro valley in Catalonia.

tramp metal [MIN ENG] Unwanted metal which finds its way into the mill ore stream.

tramp metal detector [MIN ENG] A sensing device which detects presence of unwanted metal in an ore stream, and sounds an alarm or removes the metal.

tramway [MECH ENG] An overhead rail, rope, or cable on which wheeled cars run to convey a load.

Trancor [MET] A commercial magnetic alloy composed of 3.5% silicon and 96.5% iron.

tranquilizer [PHARM] **1.** Any agent that brings about a state of relief from anxiety, or peace of mind. **2.** Any agent that produces a calming or sedative effect without inducing sleep. **3.** Any drug, such as chlorpromazine, used primarily for its calming and antipsychotic effects, or such as meprobamate, used for symptomatic treatment of common psychoneuroses and as an adjunct in somatic disorders complicated by anxiety and tension.

transaction [ADP] General description of updating data relevant to any item.

transaction data [ADP] A set of data in a data-processing area in which the incidence of the data is essentially random and unpredictable; hours worked, quantities shipped, and amounts invoiced are examples from, respectively, the areas of payroll, accounts receivable, and accounts payable.

transaction file *See* detail file.

transaction tape *See* change tape.

transadmittance [ELECTR] A specific measure of transfer admittance under a given set of conditions, as in forward transadmittance, interelectrode transadmittance, short-circuit transadmittance, small-signal forward transadmittance, and transadmittance compression ratio.

transaminase [BIOCHEM] One of a group of enzymes that catalyze the transfer of the amino group of an amino acid to a keto acid to form another amino acid. Also known as aminotransferase.

transamination [CHEM] **1.** The transfer of one or more amino groups from one compound to another. **2.** The transposition of an amino group within a single compound.

transcapsidation [VIROL] Change in the capsid of PARA (particle aiding replication of adenovirus) from one type of adenovirus to another.

transceiver [ADP] A device which transmits and receives data from punch card to punch card; it is essentially a conversion device which at the sending end reads the card and transmits the data over the wire, and at the receiving end punches the data into a card. [ELECTR] A radio transmitter and receiver combined in one unit and having switching arrangements such as to permit use of one or more tubes for both transmitting and receiving. Also known as transmitter-receiver.

transceiver data link [ADP] Integrated data processing by means of punched cards, using transceivers as terminal equipment; the transmission path can be wire or radio.

transcendental element [MATH] An element of a field K is transcendental relative to a subfield F if it satisfies no polynomial whose coefficients come from F.

transcendental field extension [MATH] A field extension K of F where the elements of K not in F are all transcendental relative to F.

transcendental functions [MATH] Functions which cannot be given by any algebraic expression involving only their variables and constants.

transcendental number [MATH] An irrational number that is the root of no polynomial with rational-number coefficients.

transconductance [ELECTR] An electron-tube rating, equal to the change in plate current divided by the change in control-grid voltage that causes it, when the plate voltage and all other voltages are maintained constant. Also known as grid-anode transconductance; grid-plate transconductance; mutual conductance. Symbolized G_m; g_m. Also known as grid-plate transconductance.

transcontinental ballistic missile [ORD] A ballistic missile having a range of at least 12,500 miles (20,000 kilometers), so it can be fired from any point on the earth's surface and reach any surface target.

transcribe [ADP] To copy, with or without translating, from one external computer storage medium to another. [ELECTR] To record, as to record a radio program by means of electric transcriptions or magnetic tape for future rebroadcasting.

transcriber [ADP] The equipment used to convert information from one form to another, as for converting computer input data to the medium and language used by the computer.

transcription [ENG ACOUS] A 16-inch-diameter, 33 1/3-rpm disk recording of a complete radio program, made especially for broadcast purposes. Also known as electrical transcription.

transcrystalline [MET] Across the crystals of a metal; used of cracks in metals. Also known as intracrystalline; transgranular.

transcurrent fault [GEOL] A strike-slip fault characterized by a steeply inclined surface. Also known as transverse thrust.

transducer [ENG] Any device or element which converts an input signal into an output signal of a different form; examples include the microphone, phonograph pickup, loudspeaker, barometer, photoelectric cell, automobile horn, doorbell, and underwater sound transducer.

transducer loss [ELECTR] The ratio of the power available to a transducer from a specified source to the power that the transducer delivers to a specified load; usually expressed in decibels.

transduction [MICROBIO] Transfer of genetic material between bacterial cells by bacteriophages.

transductor *See* magnetic amplifier; saturable reactor.

transect [SCI TECH] To cut across, or to cut transversely.

transesterification [ORG CHEM] Conversion of an organic acid ester into another ester of that same acid.

transfer [ADP] *See* jump. [MIN ENG] A vertical or inclined connection between two or more levels, used as an ore pass. [NAV] **1.** The distance a vessel moves perpendicular to its initial direction in making a turn of 90° with a constant rudder angle. **2.** The distance a vessel moves perpendicular to its initial direction for turns of less than 90°.

transfer admittance [ELECTR] An admittance rating for electron tubes and other transducers or networks; it is equal to the complex alternating component of current flowing to one terminal from its external termination, divided by the complex alternating component of the voltage applied to the adjacent terminal on the cathode or reference side; all other terminals have arbitrary external terminations.

transferase [BIOCHEM] Any of various enzymes that catalyze the transfer of a chemical group from one molecule to another.

transfer car [MIN ENG] A quarry car provided with transverse tracks, on which the gang car may be conveyed to or from the saw gang.

transfer card *See* transition card.

transfer characteristic [ELECTR] **1.** Relation, usually shown by a graph, between the voltage of one electrode and the current to another electrode, with all other electrode voltages being maintained constant. **2.** Function which, multiplied by an input magnitude, will give a resulting output magnitude. **3.** Relation between the illumination on a camera tube and the corresponding output-signal current, under specified conditions of illumination.

transfer check [ADP] Check (usually automatic) on the accuracy of the transfer of a word in a computer operation.

transfer chute [ENG] A chute used at a transfer point in a conveyor system; the chute is designed with a curved base or some other feature so that the load be discharged in a centralized stream and in the same direction as the receiving conveyor.

transfer conditionally [ADP] To copy, exchange, read, record, store, transmit, or write data or to change control or jump to another location according to a certain specified rule or in accordance with a certain criterion.

transfer constant [ENG] A transducer rating, equal to one-half the natural logarithm of the complex ratio of the product of the voltage and current entering a transducer to that leaving the transducer when the latter is terminated in its image impedance; alternatively, the product may be that of force and velocity or pressure and volume velocity; the real part of the transfer constant is the image attenuation constant, and the imaginary part is the image phase constant. Also known as transfer factor.

transfer ellipse *See* transfer orbit.

transference [PSYCH] The unconscious transfer of the patient's feelings and reactions originally associated with important persons in the patient's life, usually father, mother, or siblings, toward others and in the analytic situation, toward the analyst.

transference number [PHYS CHEM] The portion of the total electrical current carried by any ion species in a fluid-state electrolyte.

transfer factor *See* transfer constant.

transfer function [CONT SYS] The mathematical relationship between the output of a control system and its input: for a linear system, it is the Laplace transform of the output divided by the Laplace transform of the input under conditions of zero initial-energy storage.

transfer impedance [ELEC] The ratio of the voltage applied at one pair of terminals of a network to the resultant current at another pair of terminals, all terminals being terminated in a specified manner.

transfer-in-channel command [ADP] A command used to direct channel control to a specified location in main storage when the next channel command word is not stored in the next location in sequence.

transfer instruction [ADP] Step in computer operation specifying the next operation to be performed, which is not necessarily the next instruction in sequence.

transfer molding [ENG] Molding of thermosetting materials in which the plastic is softened by heat and pressure in a transfer chamber, then forced at high pressure through suitable sprues, runners, and gates into a closed mold for final curing.

transfer of fire [ORD] Shifting of fire from one target to another, applying the corrections for the first target to the data for the second target.

transfer operation [ADP] An operation which moves information from one storage location or one storage medium to another (for example, read, record, copy, transmit, exchange).

transfer orbit [AERO ENG] In interplanetary travel, an elliptical trajectory tangent to the orbits of both the departure planet and the target planet. Also known as transfer ellipse.

transfer ratio [ENG] From one point to another in a transducer at a specified frequency, the complex ratio of the generalized force or velocity at the second point to the generalized force or velocity applied at the first point; the generalized force or velocity includes not only mechanical quantities, but also other analogous quantities such as acoustical and electrical; the electrical quantities are usually electromotive force and current.

transfer ribonucleic acid [MOL BIO] The smallest ribonucleic acid molecule found in cells; its structure is complementary to messenger ribonucleic acid and it functions by transferring amino acids from the free state to the polymeric form of growing polypeptide chains. Abbreviated t-RNA.

transferrin [BIOCHEM] Any of various beta globulins in blood serum which bind and transport iron to the bone marrow and storage areas.

transfer switch [ELEC] A switch for transferring one or more conductor connections from one circuit to another.

transfer test [COMMUN] Verification of transmitted information by temporary storing, retransmitting, and comparing.

transfer unit [CHEM ENG] The relationship between the overall rate coefficient (for whatever transfer operation is being calculated), column volume, and fluid volumetric flow rate in fixed-bed sorption operations.

transfinite induction [MATH] A reasoning process by which if a theorem holds true for the first element of a well-ordered set N and is true for an element n whenever it holds for all predecessors of n, then the theorem is true for all members of N.

transfinite number [MATH] Any ordinal or cardinal number.

transfluxor [ELECTROMAG] A magnetic core having two or more apertures and three or more legs for flux; used as a computer memory element, crossbar switch, channel commutator, or control element.

transform [ADP] To change the form of digital-computer information without significantly altering its meaning. [MATH] **1.** A conjugate of an element of a group. **2.** An expression, commonly used in harmonic analysis, formed from a given function f by taking an integral of $f \cdot g$, where g is a member of an orthogonal family of functions. **3.** The value of a transformation at some point.

transformation [CRYSTAL] *See* inversion. [ELEC] For two networks which are equivalent as far as conditions at the terminals are concerned, a set of equations which give the admittances or impedances of the branches of one circuit in terms of the admittances or impedances of the other. [MATH] A function, usually between vector spaces.

transformation constant *See* decay constant.

transformation group [MATH] **1.** A collection of transformations which forms a group with composition as the operation. **2.** A dynamical system or, more generally, a topological group G together with a topological space X where each g in G gives rise to a homeomorphism of X in a continuous manner with respect to the algebraic structure of G.

transformation matrix [ELECTROMAG] A two-by-two matrix which relates the amplitudes of the traveling waves on one side of a waveguide junction to those on the other.

transformation series *See* radioactive series.

transformation temperature [MET] **1.** The temperature at which a change in phase occurs in a metal during heating or cooling. **2.** The maximum or minimum temperature of a transformation temperature range.

transformation temperature ranges [MET] The ranges of temperatures within which austenite forms during heating and transforms during cooling.

transformation theory [QUANT MECH] The study of coordinate and other transformations in quantum mechanics, especially those which leave some properties of the system invariant.

transformation twin [CRYSTAL] A crystal twin developed by a growth transformation from a higher to a lower symmetry.

transformer [ELECTROMAG] An electrical component consisting of two or more multiturn coils of wire placed in close proximity to cause the magnetic field of one to link the other; used to transfer electric energy from one or more alternating-current circuits to one or more other circuits by magnetic induction.

transformer bridge [ELEC] A network consisting of a transformer and two impedances, in which the input signal is applied to the transformer primary and the output is taken between the secondary center-tap and the junction of the impedances that connect to the outer leads of the secondary.

transformer-coupled amplifier [ELECTR] Audio-frequency amplifier that uses untuned iron-core transformers to provide coupling between stages.

transformer coupling [ELEC] *See* inductive coupling. [ELECTR] Interconnection between stages of an amplifier which employs a transformer for connecting the plate circuit of one stage to the grid circuit of the following stage; a special case of inductive coupling.

transformer hybrid *See* hybrid set.

transformer load loss [ELEC] Losses in a transformer which are incident to the carrying of the load; load losses include resistance loss in the windings due to load current, stray loss due to stray fluxes in the windings, core clamps, and so on, and to circulating current, if any, in parallel windings.

transformer loss [ELEC] Ratio of the signal power that an ideal transformer of the same impedance ratio would deliver to the load impedance, to the signal power that the actual transformer delivers to the load impedance; this ratio is usually expressed in decibels.

transformer oil [MATER] A high-quality insulating oil in which windings of large power transformers are sometimes immersed to provide high dielectric strength, high insulation resistance, high flash point, freedom from moisture, and freedom from oxidization.

transformer read-only store [ADP] In computers, read-only store in which the presence or absence of mutual inductance between two circuits determines whether a binary 1 or 0 is stored.

transformer voltage ratio [ELEC] Ratio of the root-mean-square primary terminal voltage to the root-mean-square secondary terminal voltage under specified conditions of load.

transform fault [GEOL] A strike-slip fault with offset ridges characteristic of a midoceanic ridge.

transforming principle [MICROBIO] Deoxyribonucleic acid which effects transformation in bacterial cells.

transforming section [ELECTROMAG] Length of waveguide or transmission line of modified cross section, or with a metallic or dielectric insert, used for impedance transformation.

transfusion [MED] The administration of blood, or one of its components, as a part of treatment.

transgranular *See* transcrystalline.

transgression [GEOL] Geologic evidence of landward extension of the sea. Also known as invasion; marine transgression. [OCEANOGR] Extension of the sea over land areas.

transgressive deposit [GEOL] Sediment deposited during transgression of the sea or during subsidence of the land.

transgressive overlap *See* onlap.

transhybrid loss [ELEC] In a carrier telephone system, the transmission loss at a given frequency measured across a hybrid circuit joined to a given two-wire termination and balancing network.

transient [PHYS] A pulse, damped oscillation, or other temporary phenomenon occurring in a system prior to reaching a steady-state condition.

transient analyzer [ELECTR] An analyzer that generates transients in the form of a succession of equal electric surges of small amplitude and adjustable waveform, applies these transients to a circuit or device under test, and shows the resulting output waveforms on the screen of an oscilloscope.

transient distortion [ELECTR] Distortion due to inability to amplify transients linearly.

transient motion [PHYS] An oscillatory or other irregular motion occurring while a quantity is changing to a new steady-state value.

transient overshoot [PHYS] The maximum value of the overshoot of a quantity as a result of a sudden change in conditions.

transient phenomena [ELEC] Rapidly changing actions occurring in a circuit during the interval between closing of a switch and settling to a steady-state condition, or any other temporary actions occurring after some change in a circuit.

transient problem *See* initial-value problem.

transient response [PHYS] The behavior of a system following a sudden change in its input.

transient situational disturbance [PSYCH] A form of personality disorder, more or less transient, and generally an acute symptom response to a specific situation, without persistent personality disturbance.

transient suppressor *See* surge suppressor.

transistance [ELECTR] The characteristic that makes possible the control of voltages or currents so as to accomplish gain or switching action in a circuit; examples of transistance occur in transistors, diodes, and saturable reactors.

transistor [ELECTR] An active component of an electronic circuit consisting of a small block of semiconducting material to which at least three electrical contacts are made, usually two closely spaced rectifying contacts and one ohmic (nonrectifying) contact; it may be used as an amplifier, detector, or switch.

transistor amplifier [ELECTR] An amplifier in which one or more transistors provide amplification comparable to that of electron tubes.

transistor biasing [ELECTR] Maintaining a direct-current voltage between the base and some other element of a transistor.

transistor characteristics [ELECTR] The values of the impedances and gains of a transistor.

transistor chip [ELECTR] An unencapsulated transistor of very small size used in microcircuits.

transistor circuit [ELECTR] An electric circuit in which a transistor is connected.

transistor clipping circuit [ELECTR] A circuit in which a transistor is used to achieve clipping action; the bias at the input is set at such a level that output current cannot flow during a portion of the amplitude excursion of the input voltage or current waveform.

transistor gain [ELECTR] The increase in signal power produced by a transistor.

transistor input resistance [ELECTR] The resistance across the input terminals of a transistor stage. Also known as input resistance.

transistor magnetic amplifier [ELECTR] A magnetic amplifier together with a transistor preamplifier, the latter used to make the signal strong enough to change the flux in the core of the magnetic amplifier completely during a half-cycle of the power supply voltage.

transistor radio [ELECTR] A radio receiver in which transistors are used in place of electron tubes.

transistor-transistor logic [ELECTR] A logic circuit containing two transistors, for driving large output capacitances at high speed.

transit [ASTRON] **1.** A celestial body's movement across the meridian of a place. **2.** Passage of a smaller celestial body across a larger one. **3.** Passage of a satellite's shadow across the disk of its primary. [ENG] A surveying instrument with the telescope mounted so that it can measure horizontal and vertical angles. Also known as transit theodolite.

transit circle [ENG] A type of astronomical transit instrument having a micrometer eyepiece that has an extra pair of moving wires perpendicular to the vertical set to measure the zenith distance or declination of the celestial object in conjunction with readings taken from a large, accurately calibrated circle attached to the horizontal axis. Also known as meridian transit.

transit declinometer [ENG] A type of declinometer; a surveyor's transit, built to exacting specifications with respect to freedom from traces of magnetic impurities and quality of the compass needle, has a 17-power telescope for sighting on a mark and for making solar and stellar observations to determine true directions.

Transite pipe *See* asbestos-cement pipe.

transition [COMMUN] Change from one circuit condition to the other; for example, the change from mark to space or from space to mark. [MOL BIO] A mutation resulting from the substitution in deoxyribonucleic acid or ribonucleic acid of one purine or pyrimidine for another. [QUANT MECH] The

change of a quantum-mechanical system from one energy state to another. [THERMO] A change of a substance from one of the three states of matter to another.

transitional epithelium [HISTOL] A form of stratified epithelium found in the urinary bladder; cells vary between squamous, when the tissue is stretched, and columnar, when not stretched.

transitional fit [DES ENG] A fit with varying clearances due to specified tolerances on the shaft and sleeve or hole.

transitional flow [FL MECH] A flow in which the viscous and Reynolds stresses are of approximately equal magnitude; it is transitional between laminar flow and turbulent flow.

transition altitude [AERO ENG] The altitude in the vicinity of an aerodrome at or below which the vertical position of an aircraft is controlled by reference to true altitude.

transition card [ADP] In reading a deck of punched cards by a computer, a card that causes the computer to stop reading cards and begin executing a program. Also known as transfer card.

transition curve *See* easement curve.

transition element [CHEM] One of a group of metallic elements in which the members have the filling of the outermost shell to 8 electrons interrupted to bring the penultimate shell from 8 to 18 or 32 electrons; includes elements 21 through 29 (scandium through copper), 39 through 47 (yttrium through silver), 57 through 79 (lanthanum through gold), and all known elements from 89 (actinium) on. [ELECTROMAG] An element used to couple one type of transmission system to another, as for coupling a coaxial line to a waveguide.

transition factor *See* reflection factor.

transition flow [AERO ENG] A flow of fluid about an airfoil that is changing from laminar flow to turbulent flow.

transition frequency [ENG ACOUS] The frequency corresponding to the intersection of the asymptotes to the constant-amplitude and constant-velocity portions of the frequency-response curve for a disk recording; this curve is plotted with output-voltage ratio in decibels as the ordinate, and the logarithm of the frequency as the abscissa. Also known as crossover frequency; turnover frequency. [QUANT MECH] The characteristic frequency of radiation emitted or absorbed by a quantum-mechanical system as it changes from one energy state to another; equal to the energy difference between the states divided by Planck's constant.

transition lattice [MET] An unstable, intermediate configuration formed in a metal lattice during solid-state reactions such as precipitation or transformation.

transition level [NAV] The flight level below which heights are expressed in feet above mean sea level and are based on an approved station altimeter setting.

transition loss [ELEC] At a junction between a source and a load, the ratio of the available power to the power delivered to the load.

transition moment [QUANT MECH] Any type of multipole moment which determines radiative transitions between states; it consists of an integral of the product of the conjugate of the final state wave function, a multipole moment operator, and the initial state wave function.

transition point [ELECTROMAG] A point at which the constants of a circuit change in such a way as to cause reflection of a wave being propagated along the circuit. [THERMO] Either the temperature at which a substance changes from one state of aggregation to another (a first-order transition), or the temperature of culmination of a gradual change, such as the lambda point, or Curie point (a second-order transition). Also known as transition temperature.

transition probability [MATH] Conditional probability concerning a discrete Markov chain measuring change of state. [QUANT MECH] The probability per unit time that a quantum-mechanical system will make a transition from a given initial state to a given final state.

transition region [SOLID STATE] The region between two homogeneous semiconductors in which the impurity concentration changes.

transition temperature [MET] The temperature at which a fracture changes from tough to brittle in various tests, such as notched-bar impact test. [THERMO] *See* transition point.

transition time [ANALY CHEM] The time interval needed for a

Six-inch (15-centimeter) transit circle, U.S. Naval Observatory. *(Official U.S. Naval Observatory photograph)*

TRANSIT DECLINOMETER

Transit declinometer with compass needle, telescope, and microscope. *(U.S. Coast and Geodetic Survey)*

TRANSITIONAL EPITHELIUM

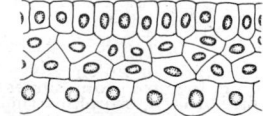

Cellular arrangement in transitional epithelium.

TRANSIT TELESCOPE

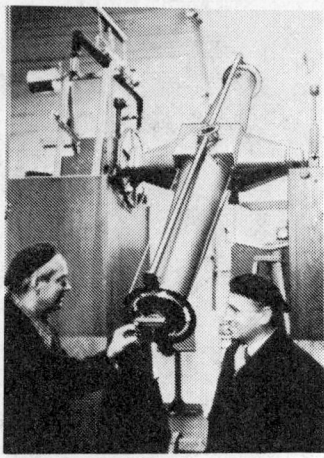

Large transit telescope, Pulkovo Observatory, Soviet Union, in use since 1838. *(Courtesy of B. L. Klock)*

TRANSKETOLASE

$$
\begin{array}{cccc}
\text{CH}_2\text{OH} & \text{CHO} & & \text{CH}_2\text{OH} \\
\text{CO} & \text{HCOH} & \text{CHO} & \text{CO} \\
\text{HOCH} & +\text{HCOH} \Longrightarrow & \text{HCOH} & +\text{HOCH} \\
\text{HCOH} & \text{HCOH} & \text{CH}_2\text{OPO}_3\text{H}_2 & \text{HCOH} \\
\text{CH}_2\text{OPO}_3\text{H}_2 & \text{CH}_2\text{OPO}_3\text{H}_2 & & \text{HCOH} \\
& & & \text{CH}_2\text{OPO}_3\text{H}_2 \\
\text{I} & \text{II} & \text{III} & \text{IV}
\end{array}
$$

Example of a reaction catalyzed by transketolase. I is xylulose-5-phosphate, II is ribulose-5-phosphate, III is glyceraldehyde-3-phosphate, and IV is sedoheptulose-7-phosphate.

TRANSLATOR

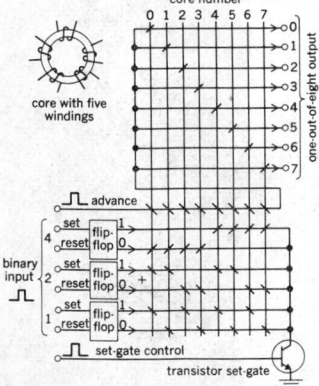

Diagram of a register circuit and translator (adp) using magnetic cores, each of which has five windings. Vertical lines represent magnetic cores. Short slanted lines represent windings and also symbolize a mirror action with respect to pulses.

TRANSLOCATION

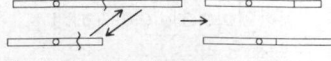

Translocation in which nonhomologous chromosomes become broken, switching their broken ends in the process of rejoining.

working (nonreference) electrode to become polarized during chronopotentiometry (time-measurement electrolysis of a sample).

transition zone [FL MECH] Those conditions of fluid flow in which the nature of the flow is changing from laminar to turbulent.

transitive relation [MATH] A relation $<$ on a set such that if $a < b$ and $b < c$, then $a < c$.

transit mix [MATER] Concrete or mortar mixed in a rotating cylinder en route to or at the construction site.

transitory target [ORD] A target that obtains only for a limited period of time, as in the case of a troop concentration which may be dissipated in a short time.

transitron [ELECTR] Thermionic-tube circuit whose action depends on the negative transconductance of the suppressor grid of a pentode with respect to the screen grid.

transitron oscillator [ELECTR] A negative-resistance oscillator in which the screen grid is more positive than the anode, and a capacitor is connected between the screen grid and the suppressor grid; the suppressor grid periodically divides the current between the screen grid and the anode, thereby producing oscillation.

transit satellite [AERO ENG] One of a system of passive, low-orbiting satellites which provide high-accuracy fixes using the Doppler technique several times a day at every point on earth, for navigation and geodesy.

transit survey [ENG] A ground surveying method in which a transit instrument is set up at a control point and oriented, and directions and distances to observed points are recorded.

transit telescope [OPTICS] A telescopic instrument adapted to the observation of the passage, or transit, of an astronomical object across the meridian of an observer; consists of a telescope mounted on a single fixed horizontal axis of rotation which has a central hollow cube (sometimes a sphere) and two conical semiaxes ending in cylindrical pivots; the objective and eyepiece halves of the instrument are also fastened to the cube of the instrument, perpendicular to the horizontal axis.

transit theodolite *See* transit.

transit time [ELECTR] The time required for an electron or other charge carrier to travel between two electrodes in an electron tube or transistor.

transketolase [BIOCHEM] An enzyme that cleaves a substrate at the position of the carbonyl carbon and transports a two-carbon fragment to an acceptor compound to form a new compound.

translate [ADP] To convert computer information from one language to another, or to convert characters from one representation set to another, and by extension, the computer instruction which directs the latter conversion to be carried out.

translating circuit *See* translator.

translating-roller [ORD] A double-thread screw by means of which a breechblock is drawn longitudinally from its position in the breech of a large-caliber gun.

translation [MATH] **1.** A function changing the coordinates of a point in a euclidean space into new coordinates relative to axes parallel to the original. **2.** A function on a group to itself given by operating on each element by some one fixed element. [MECH] The linear movement of a point in space without any rotation. [MOL BIO] The process by which the linear sequence of nucleotides in a molecule of messenger ribonucleic acid directs the specific linear sequence of amino acids, as during protein synthesis.

translational fault [GEOL] A fault in which there has been uniform movement in one direction and no rotational component of movement. Also known as translatory fault.

translation algorithm [ADP] A specific, effective, essentially computational method for obtaining a translation from one language to another.

translational motion [MECH] Motion of a rigid body in such a way that any line which is imagined rigidly attached to the body remains parallel to its original direction.

translational movement [GEOL] Movement, as of fault blocks, that is uniform, without rotation, so that parallel features maintain their orientation.

translation gliding *See* crystal gliding.

translation group [CRYSTAL] The collection of all translation operations which carry a crystal lattice into itself.

translation operation [PHYS] The process of moving an object along a straight line in such a way that any line which is fixed with respect to the object remains parallel to its original direction.

translator [ADP] A computer network or system having a number of inputs and outputs, so connected that when signals representing information expressed in a certain code are applied to the inputs, the output signals will represent the same information in a different code. Also known as translating circuit. [ELECTR] A combination television receiver and low-power television transmitter, used to pick up television signals on one frequency and retransmit them on another frequency to provide reception in areas not served directly by television stations.

translator routine [ADP] A program which accepts statements in one language and outputs them as statements in another language.

translatory fault *See* translational fault.

transliterate [ADP] To represent the characters or words of one language by corresponding characters or words of another language.

translocation [BOT] Movement of water, mineral salts, and organic substances from one part of a plant to another. [CYTOL] The transfer of a chromosome segment from its usual position to a new position in the same or in a different chromosome.

translucent attritus [GEOL] Attritus composed principally of transparent humic degradation matter. Also known as humodurite.

translucent medium [OPTICS] A medium which transmits rays of light so diffused that objects cannot be seen distinctly; examples are various forms of glass which admit considerable light but impede vision.

translucidus [METEOROL] A cloud variety occurring in a layer, patch, or extensive sheet, the greater part of which is sufficiently translucent to reveal the position of the sun, or through which higher clouds may be discerned; this variety is found in the general altocumulus, altostratus, stratocumulus, and stratus.

translunar [ASTRON] Beyond the orbit of the moon.

transmethylase [BIOCHEM] A transferase enzyme involved in catalyzing chemical reactions in which methyl groups are transferred from a substrate to a new compound.

transmethylation [BIOCHEM] A metabolic reaction in which a methyl group is transferred from one compound to another; methionine and choline are important donors of methyl groups.

transmission [ELECTR] **1.** The process of transferring a signal, message, picture, or other form of intelligence from one location to another location by means of wire lines, radio, light beams, infrared beams, or other communication systems. **2.** A message, signal, or other form of intelligence that is being transmitted. [ELECTROMAG] *See* transmittance. [MECH ENG] The gearing system by which power is transmitted from the engine to the live axle in an automobile. Also known as gearbox.

transmission anomaly [ACOUS] The ratio of the transmission loss of underwater sound at a given distance from the source to the inverse square of this distance, usually expressed in decibels.

transmission band [ELECTROMAG] Frequency range above the cutoff frequency in a waveguide, or the comparable useful frequency range for any other transmission line, system, or device.

transmission coefficient [PHYS] **1.** The value of some quantity associated with the resultant field produced by incident and reflected waves at a given point in a transmission medium divided by the corresponding quantity in the incident wave. **2.** The ratio of transmitted to incident energy flux or flux of some other quantity at a discontinuity in a transmission medium; for sound waves, it is called the sound transmission coefficient. **3.** The ratio of the transmitted flux of some quantity to the incident flux for a substance of unit thickness. [QUANT MECH] *See* penetration probability.

transmission diffraction [ANALY CHEM] A type of electron diffraction analysis in which the electron beam is transmitted through a thin film or powder whose smallest dimension is no greater than a few tenths of a micrometer.

transmission dynamometer [ENG] A device for measuring torque and power (without loss) between a propulsion power plant and the driven mechanism, for example, wheels or propellers.

transmission electron microscope [ELECTR] A type of electron microscope in which the specimen transmits an electron beam focused on it, image contrasts are formed by the scattering of electrons out of the beam, and various magnetic lenses perform functions analogous to those of ordinary lenses in a light microscope.

transmission electron radiography [ELECTR] A technique used in microradiography to obtain radiographic images of very thin specimens; the photographic plate is in close contact with the specimen, over which is placed a lead foil and then a light-tight covering; hardened x-rays shoot through the light-tight covering.

transmission facilities [COMMUN] Cable or radio facilities constituting interswitch trunks and subscriber access lines.

transmission factor [PHYS] The ratio of the flux of some quantity transmitted through a body to the incident flux.

transmission function [GEOPHYS] A mathematical formulation of relationships between infrared transmission in the atmosphere, the path length, and the concentration of absorbing gases.

transmission gain *See* gain.

transmission gate [ELECTR] A gate circuit that delivers an output waveform that is a replica of a selected input during a specific time interval which is determined by a control signal.

transmission grating [OPTICS] A diffraction grating produced on a transparent base so radiation is transmitted through the grating instead of being reflected from it.

transmission level [COMMUN] The ratio of the signal power at any point in a transmission system to the signal power at some point in the system chosen as a reference point; usually expressed in decibels.

transmission line [ELEC] A system of conductors, such as wires, waveguides, or coaxial cables, suitable for conducting electric power or signals efficiently between two or more terminals. Also known as line.

transmission-line admittance [ELEC] The complex ratio of the current flowing in a transmission line to the voltage across the line, where the current and voltage are expressed in phasor notation.

transmission-line attenuation [ELEC] The decrease in power of a transmission-line signal from one point to another, expressed as a ratio or in decibels.

transmission-line cable [ELEC] The coaxial cable, waveguide, or microstrip which forms a transmission line; a number of standard types have been designated, specified by size and materials.

transmission-line constants *See* transmission-line parameters.

transmission-line current [ELEC] The amount of electrical charge which passes a given point in a transmission line per unit time.

transmission-line efficiency [ELEC] The ratio of the power of a transmission-line signal at one end of the line to that at the other end where the signal is generated.

transmission-line impedance [ELEC] The complex ratio of the voltage across a transmission line to the current flowing in the line, where voltage and current are expressed in phasor notation.

transmission-line parameters [ELEC] The quantities which are necessary to specify the impedance per unit length of a transmission line, and the admittance per unit length between various conductors of the line. Also known as linear electrical parameters; line parameters; transmission line constants.

transmission-line power [ELEC] The amount of energy carried past a point in a transmission line per unit time.

transmission-line reflection coefficient [ELEC] The ratio of the voltage reflected from the load at the end of a transmission line to the direct voltage.

transmission-line theory [ELEC] The application of electrical and electromagnetic theory to the behavior of transmission lines.

transmission-line transducer loss [ELEC] The ratio of the

power delivered by a transmission line to a load to that produced at the generator, expressed in decibels; equal to the sum of the attenuation of the line and the mismatch loss.

transmission-line voltage [ELEC] The work that would be required to transport a unit electrical charge between two specified conductors of a transmission line at a given instant.

transmission loss [COMMUN] **1.** The ratio of the power at one point in a transmission system to the power at a point farther along the line; usually expressed in decibels. **2.** The actual power that is lost in transmitting a signal from one point to another through a medium or along a line. Also known as loss.

transmission mode *See* mode.

transmission modulation [ELECTR] Amplitude modulation of the reading-beam current in a charge storage tube as the beam passes through apertures in the storage surface; the degree of modulation is controlled by the stored charge pattern.

transmission oil [MATER] A lubricant especially compounded for automobile transmissions.

transmission plane [OPTICS] The plane of vibration of polarized light that will pass through a Nicol prism or other polarizer.

transmission primaries [COMMUN] The set of three color primaries that correspond to the three independent signals contained in the color television picture signal.

transmission range *See* night visual range.

transmission regulator [ELECTR] In electrical communications, a device that maintains substantially constant transmission levels over a system.

transmission security [COMMUN] Component of communications security which results from all measures designed to protect transmissions from unauthorized interception, traffic analysis, and imitative deception.

transmission speed [COMMUN] The number of information elements sent per unit time; usually expressed as bits, characters, bands, word groups, or records per second or per minute.

transmission time [COMMUN] Absolute time interval from transmission to reception of a signal.

transmission tower [ENG] A concrete, metal, or timber structure used to carry a transmission line.

transmissivity [ELECTROMAG] The ratio of the transmitted radiation to the radiation arriving perpendicular to the boundary between two mediums.

transmissometer [ENG] An instrument for measuring the extinction coefficient of the atmosphere and for the determination of visual range. Also known as hazemeter; transmittance meter.

transmissometry [OPTICS] The technique of determining the extinction characteristics of a medium by measuring the transmission of a light beam of known initial intensity directed into that medium.

transmit [ADP] To move data from one location to another. [COMMUN] To send a message, program, or other information to a person or place by wire, radio, or other means.

transmit-receive tube [ELECTR] A gas-filled radio-frequency switching tube used to disconnect a receiver from its antenna during the interval for pulse transmission in radar and other pulsed radio-frequency systems. Also known as TR box; TR cell (British usage); TR switch; TR tube.

transmittability [COMMUN] The ability of standard electronic and mechanical elements and automatic communications equipment to handle a code under various signal-to-noise ratios; for example, a code with a variable number of elements such as Morse presents technical problems in automatic interpretation not encountered in a fixed-length code.

transmittance [ANALY CHEM] During absorption spectroscopy, the amount of radiant energy transmitted by the solution under analysis. [ELECTROMAG] The radiant power transmitted by a body divided by the total radiant power incident upon the body. Also known as transmission.

transmittance meter *See* transmissometer.

transmittancy [ELECTROMAG] The transmittance of a solution divided by that of the pure solvent of the same thickness.

transmitted-carrier operation [COMMUN] Form of amplitude-modulated carrier transmission in which the carrier wave is transmitted.

transmitted wave *See* refracted wave.

transmitter [COMMUN] **1.** In telephony, the carbon microphone that converts sound waves into audio-frequency signals. **2.** *See* radio transmitter. [ELEC] *See* synchro transmitter.

transmitter-distributor [ELEC] In teletypewriter operations, a motor-driven device which translates teletypewriter code combinations form perforated tape into electrical impulses, and transmits these impulses to one or more receiving stations. Abbreviated TD.

transmitter noise *See* frying noise.

transmitter-receiver *See* transceiver.

transmitter synchro *See* synchro transmitter.

transmitting loop loss [COMMUN] That part of the repetition equivalent assignable to the station set, subscriber line, and battery supply circuit which is on the transmitting end.

transmitting mode [ADP] Condition of an input/output device, such as a magnetic tape when it is actually reading or writing.

transmittivity [ELECTROMAG] The internal transmittance of a piece of nondiffusing substance of unit thickness.

transmutation [NUC PHYS] A nuclear process in which one nuclide is transformed into the nuclide of a different element. Also known as nuclear transformation.

transobuoy [ENG] A free-floating or moored automatic weather station developed for the purpose of providing weather reports from the open oceans; it transmits barometric pressure, air temperature, sea-water temperature, and wind speed and direction.

transolver [ELEC] A synchro having a two-phase cylindrical rotor within a three-phase stator, for use as a transmitter or a control transformer with no degradation of accuracy or nulls.

transom [BUILD] A window above a door. [NAV ARCH] The flat, vertical aft end of a ship or boat as distinguished from a canoe-shaped or cruiser stern.

transonic [PHYS] That which occurs or is occurring within the range of speed in which flow patterns change from subsonic to supersonic (or vice versa), about Mach 0.8 to 1.2, as in transonic flight or transonic flutter.

transonic flight [AERO ENG] Flight of vehicles at speeds near the speed of sound (660 miles per hour or 1060 kilometers per hour, at 35,000 feet or 10,700 meters altitude), characterized by great increase in drag, decrease in lift at any altitude, and abrupt changes in the moments acting on the aircraft; the vehicle may shake or buffet.

transonic flow [FL MECH] Flow of a fluid over a body in the range just above and just below the acoustic velocity.

transonic range [FL MECH] The range of speeds between the speed at which one point on a body reaches supersonic speed, and the speed at which all points reach supersonic speed.

transonic speed [FL MECH] The speed of a body relative to the surrounding fluid at which the flow is in some places on the body subsonic and in other places supersonic.

transonic wind tunnel [ENG] A type of high-speed wind tunnel capable of testing the effects of airflow past an object at speeds near the speed of sound, Mach 0.7 to 1.4; sonic speed occurs where the cross section of the tunnel is at a minimum, that is, where the test object is located.

transorbital lobotomy [MED] A lobotomy performed through the roof of the orbit.

transosonde [ENG] The flight of a constant-level balloon, whose trajectory is determined by tracking with radio-direction-finding equipment; thus, it is a form of upper-air, quasi-horizontal sounding.

transparency [GRAPHICS] An image fixed on a clear base by means of a photographic, printing, chemical, or other process, especially adaptable for viewing by transmitted light.

transparent [PHYS] Permitting passage of radiation or particles.

transparent medium [OPTICS] **1.** A medium which has the property of transmitting rays of light in such a way that the human eye may see through the medium distinctly. **2.** A medium transparent to other regions of the electromagnetic spectrum, such as x-rays and microwaves.

transparent sky cover [METEOROL] In United States weather-observing practice, that portion of sky cover through which higher clouds and blue sky may be observed; opposed to opaque sky cover.

transpiration [BIOL] The passage of a gas or liquid (in the form of vapor) through the skin, a membrane, or other tissue.

transpiration cooling *See* sweat cooling.

transplantation [BIOL] **1.** The artificial removal of part of an organism and its replacement in the body of the same or of a different individual. **2.** To remove a plant from one location and replant it in another place.

transplantation antigen [IMMUNOL] An antigen in a cell which induces a histocompatibility reaction when the cell is transplanted into an organism not having that antigen.

transplanter [AGR] A special kind of equipment designed for the planting of cuttings or small plants; it transports one or more men who assist the action of the machine in placing plants in a furrow and covering them; it commonly supplies a small quantity of water to each plant.

transplutonium element [INORG CHEM] An element having an atomic number greater than that of plutonium (94).

transpolarizer [ELEC] An electrostatically controlled circuit impedance that can have about 30 discrete and reproducible impedance values: two capacitors, each having a crystalline ferroelectric dielectric with a nearly rectangular hysteresis loop, are connected in series and act as a single low impedance to an alternating-current sensing signal when both capacitors are polarized in the same direction; application of 1-microsecond pulses of appropriate polarity increases the impedance in steps.

transponder [COMMUN] A transmitter-receiver capable of accepting the challenge of an interrogator and automatically transmitting an appropriate reply.

transponder beacon *See* responder beacon.

transponder dead time [ELECTR] Time interval between the start of a pulse and the earliest instant at which a new pulse can be received or produced by a transponder.

transponder set [ELECTR] A complete electronic set which is designed to receive an interrogation signal, and which retransmits coded signals that can be interpreted by the interrogating station; it may also utilize the received signal for actuation of additional equipment such as local indicators or servo amplifiers.

transponder suppressed time delay [ELECTR] Overall fixed time delay between reception of an interrogation and transmission of a reply to this interrogation.

transport [ADP] **1.** To convey as a whole from one storage device to another in a digital computer. **2.** *See* tape transport. [ENG] Conveyance equipment such as vehicular transport, hydraulic transport, and conveyor-belt setups. [NAV ARCH] A ship designed to carry military personnel from one place to another. Also known as troop ship.

transportation [GEOL] A phase of sedimentation concerned with movement by natural agents of sediment or any loose or weathered material from one place to another.

transportation emergency [ENG] A situation which is created by a shortage of normal transportation capability and of a magnitude sufficient to frustrate movement requirements, and which requires extraordinary action by the designated authority to ensure continued movement.

transportation engineering [ENG] That branch of engineering relating to the movement of goods and people; major types of transportation are highway, water, rail, subway, air, and pipeline.

transportation lag *See* distance/velocity lag.

transportation priorities [ENG] Indicators assigned to eligible traffic which establish its movement precedence; appropriate priority systems apply to the movement of traffic by sea and air.

transportation problem [IND ENG] A programming problem that is concerned with the optimal pattern of the distribution of goods from several points of origin to several different destinations, with the specified requirements at each destination.

transport capacity [ENG] The number of persons or the tonnage (or volume) of equipment which can be carried by a vehicle under given conditions.

transport case [ENG] A moistureproof nonconductive wood, plastic, or fabric container used to transport safely small quantities of dynamite sticks to and from blasting sites.

transporter crane [MECH ENG] A long lattice girder supported by two lattice towers which may be either fixed or moved along rails laid at right angles to the girder; a crab with a hoist suspended from it travels along the girder.

transport lag *See* distance/velocity lag.

transport mean free path [NUCLEO] **1.** A path length equal to three times the diffusion coefficient of neutron flux in a nuclear reactor when Fick's law is applicable. **2.** A modification of the mean free path to take into account anisotropy of scattering and the persistence of velocities.

transport network [ENG] The complete system of the routes pertaining to all means of transport available in a particular area, made up of the network particular to each means of transport.

transport number [PHYS CHEM] The fraction of the total current carried by a given ion in an electrolyte.

transport properties [PHYS] Properties of a compound or material associated with mass or heat transport; for example, viscosity and thermal conductivity of liquids, gases, or solids.

transport vehicle [MECH ENG] Vehicle primarily intended for personnel and cargo carrying.

transpose of a matrix [MATH] The matrix obtained from the original matrix by interchanging its rows and columns.

transposition [COMMUN] Interchanging the relative positions of conductors at regular intervals along a transmission line to reduce cross talk. [MATH] A permutation of a set of symbols which exchanges exactly two while leaving all others unaffected.

transradar [COMMUN] Bandwidth compression system developed for long-range narrow-band transmission of radio signals from a radar receiver to a remote location.

transrectification [ELEC] Rectification that occurs in one circuit when an alternating voltage is applied to another circuit.

transrectification characteristic [ELECTR] Graph obtained by plotting the direct-voltage values for one electrode of a vacuum tube as abscissas against the average current values in the circuit of that electrode as ordinates, for various values of alternating voltage applied to another electrode as a parameter; the alternating voltage is held constant for each curve, and the voltages on other electrodes are maintained constant.

transrectifier [ELECTR] Device, ordinarily a vacuum tube, in which rectification occurs in one electrode circuit when an alternating voltage is applied to another electrode.

transsexual [PSYCH] An individual whose chromosomes, gonads, and body habitus mark that individual as a member of one sex, but who feels psychically to be of the other sex, with an overwhelming desire for sex reassignment through surgical and hormonal intervention.

transuranic elements [CHEM] Elements that have atomic numbers greater than 92; all are radioactive, are products of artificial nuclear changes, and are members of the actinide group. Also known as transuranium elements.

transuranium elements *See* transuranic elements.

Transvaal jade [MINERAL] A mineral that is not a true jade but a green grossularite garnet. Also known as South African jade.

transversal [MATH] **1.** A line intersecting a given family of lines. **2.** A curve orthogonal to a hypersurface. **3.** If π is a given map of a set X onto a set Y, a transversal for π is a subset T of X with the property that T contains exactly one point of $\pi^{-1}(y)$ for each $y \in Y$.

transverse bar [GEOL] A slightly submerged sand bar extending perpendicular to the shoreline.

transverse basin *See* exogeosyncline.

transverse cyclindrical orthomorphic chart *See* transverse Mercator chart.

transverse cylindrical orthomorphic projection *See* transverse Mercator projection.

transverse Doppler effect [ELECTROMAG] An aspect of the optical Doppler effect, occurring when the direction of motion of the source relative to an observer is perpendicular to the direction of the light received by the observer; the observed frequency is smaller than the source frequency by the factor $[1 - (v/c)^2]^{1/2}$, where v is the speed of the source and c is the speed of light.

transverse dune [GEOL] A sand dune having a gentle windward slope and a steep leeward slope, and elongated at right angles to the direction of prevailing winds.

transverse electric mode [ELECTROMAG] A mode in which a particular transverse electric wave is propagated in a waveguide or cavity. Abbreviated TE mode. Also known as H mode (British usage).

transverse electric wave [ELECTROMAG] An electromagnetic wave in which the electric field vector is everywhere perpendicular to the direction of propagation. Abbreviated TE wave. Also known as H wave (British usage).

transverse electromagnetic mode [ELECTROMAG] A mode in which a particular transverse electromagnetic wave is propagated in a waveguide or cavity. Abbreviated TEM mode.

transverse electromagnetic wave [ELECTROMAG] An electromagnetic wave in which both the electric and magnetic field vectors are everywhere perpendicular to the direction of propagation. Abbreviated TEM wave.

transverse equator [MAP] A meridian the plane of which is perpendicular to the axis of a transverse projection; it serves as the origin for measurement of transverse latitude.

transverse fault [GEOL] A fault whose strike is more or less perpendicular to the general structural trend of the region.

transverse fold *See* cross fold.

transverse frame [NAV ARCH] A ship frame consisting of a large number of relatively small, closely spaced, athwartship frames, reinforced in the bottom by vertical floor plates and working in conjunction with widely spaced, fore-and-aft, deep girders, such as the keel, longitudinals, and side stringers.

transverse gallery [MIN ENG] An auxiliary crosscut made in thick deposits across the ore body in order to divide it into sections along the strike.

transverse graticule [MAP] A fictitious graticule based upon a transverse projection.

transverse interference [ELEC] Interference occurring across terminals or between signal leads.

transverse joint *See* cross joint.

transverse latitude [MAP] Angular distance from a transverse equator. Also known as inverse latitude.

transversely excited atmospheric pressure laser *See* TEA laser.

transverse magnetic mode [ELECTROMAG] A mode in which a particular transverse magnetic wave is propagated in a waveguide or cavity. Abbreviated TM mode. Also known as E mode (British usage).

transverse magnetic wave [ELECTROMAG] An electromagnetic wave in which the magnetic field vector is everywhere perpendicular to the direction of propagation. Abbreviated TM wave. Also known as E wave (British usage).

transverse magnetization [ENG ACOUS] Magnetization of a magnetic recording medium in a direction perpendicular to the line of travel and parallel to the greatest cross-sectional dimension.

transverse magnetoresistance [ELECTROMAG] One of the galvanomagnetic effects, in which a magnetic field perpendicular to an electric current gives rise to an electrical potential change in the direction of the current.

transverse Mercator chart [MAP] A chart on the transverse Mercator projection. Also known as inverse cylindrical orthomorphic chart; inverse Mercator chart; transverse cylindrical orthomorphic chart.

transverse Mercator projection [MAP] A conformal map projection in which the regular Mercator projection is rotated (transversed) 90° in azimuth, the central meridian corresponding to the line which represents the equator on the regular Mercator; the characteristics as to scale are identical to those of the regular Mercator, except that the scale is dependent on distances east or west of the meridian instead of north or south of the equator. Also known as inverse cylindrical orthomorphic projection; inverse Mercator projection; transverse cylindrical orthomorphic projection.

transverse meridian [MAP] A great circle perpendicular to a transverse equator.

transverse metacenter [NAV ARCH] The point of intersection of the vertical through the center of buoyancy of a ship in the position of equilibrium with the vertical through the new center of buoyancy when the ship is slightly heeled.

transverse parallel [MAP] A circle or line parallel to a transverse equator, connecting all points of equal transverse latitude. Also known as inverse parallel.

transverse pole [MAP] One of the two points 90° from a transverse equator.

transverse recording [ELECTR] Technique for recording television signals on magnetic tape using a four-transducer rotating head.

transverse rhumb line [MAP] A line making the same oblique angle with all fictitious meridians of a transverse Mercator projection; transverse parallels and meridians may be considered special cases of the transverse rhumb line. Also known as inverse rhumb line.

transverse ripple mark [GEOL] A ripple mark formed nearly perpendicular to the direction of the current.

transverse stability [ENG] The ability of a ship or aircraft to recover an upright position after waves or wind roll it to one side.

transverse thrust *See* transcurrent fault.

transverse valley [GEOL] **1.** A valley perpendicular to the general strike of the underlying strata. **2.** A valley cutting perpendicularly across a ridge, range, or chain of mountains. Also known as cross valley.

transverse vibration [MECH] Vibration of a rod in which elements of the rod move at right angles to the axis of the rod.

transverse wave [GEOPHYS] *See* S wave. [PHYS] A wave in which the direction of the disturbance at each point of the medium is perpendicular to the wave vector and parallel to surfaces of constant phase.

transversion [MOL BIO] A mutation resulting from the substitution in deoxyribonucleic acid or ribonucleic acid of a purine for a pyrimidine or a pyrimidine for a purine.

trap [ADP] An automatic transfer of control of a computer to a known location, this transfer occurring when a specified condition is detected by hardware. [AERO ENG] That part of a rocket motor that keeps the propellant grain in place. [CIV ENG] A bend or dip in a soil drain which is always full of water, providing a water seal to prevent odors from entering the building. [ELECTR] **1.** A tuned circuit used in the radio-frequency or intermediate-frequency section of a receiver to reject undesired frequencies; traps in television receiver video circuits keep the sound signal out of the picture channel. Also known as rejector. **2.** *See* wave trap. [GEOL] *See* oil trap. [MECH ENG] A device which reduces the effect of the vapor pressure of oil or mercury on the high-vacuum side of a diffusion pump. [PETR] Any dark-colored, fine-grained, nongranitic, hypabyssal or extrusive rock. Also known as trappide; trap rock. [SOLID STATE] Any irregularity, such as a vacancy, in a semiconductor at which an electron or hole in the conduction band can be caught and trapped until released by thermal agitation. Also known as semiconductor trap.

trap address [ADP] The location at which control is transferred in case of an interrupt as soon as the current instruction is completed.

TRAPATT diode [ELECTR] A *p-n* junction diode, similar to the IMPATT diode, but characterized by the formation of a trapped space-charge plasma within the junction region; used in the generation and amplification of microwave power. Derived from trapped plasma avalanche transit time diode.

trapdoor [BUILD] A hinged, sliding, or lifting door to cover an opening in a roof, ceiling, or floor.

trapdoor fault [GEOL] A circular fault that is hinged at one end.

trapeziform [BIOL] Having the form of a trapezium.

trapezium [MATH] A quadrilateral where no sides are parallel.

Trapezium [ASTRON] Four very hot stars that appear to the eye as a single star in the Great Nebula of Orion; the star symbol is M42.

trapezohedron [CRYSTAL] An isometric crystal form of 24 faces, each face of which is an irregular four-sided figure. Also known as icositetrahedron; leucitohedron; tetragonal trisoctahedron.

trapezoid [MATH] A quadrilateral having two parallel sides.

trapezoidal excavator [MECH ENG] A digging machine which removes earth in a trapezoidal cross-section pattern for canals and ditches.

trapezoidal generator [ELECTR] Electronic stage designed to produce a trapezoidal voltage wave.

trapezoidal integration [MATH] A numerical approximation of an integral by means of the trapezoidal rule.

trapezoidal pulse [ELECTR] An electrical pulse in which the voltage rises linearly to some value, remains constant at this value for some time, and then drops linearly to the original value.

trapezoidal rule [MATH] The rule that the integral from a to b of a real function $f(x)$ is approximated by

$$\frac{b-a}{2n}\left[f(a) + \sum_{j=1}^{n-1} 2f(x_j) + f(b)\right]$$

where $x_0 = a$, $x_j = x_{j-1} + (b-a)/n$ for $j = 1, 2, \cdots, n-1$.

trapezoidal wave [ELECTR] A wave consisting of a series of trapezoidal pulses.

trap mine [ORD] Land mine designed to explode unexpectedly when enemy personnel attempt to move an object.

trapped-air process [ENG] A procedure for the blow-mold forming of closed plastic objects; the bottom pinch is conventional and, after blowing, sliding pinchers close off the top to form a sealed-air, inflated product.

trapped fuel [ENG] The fuel in an engine or fuel system that is not in the fuel tanks.

trapped plasma avalanche transit time diode *See* TRAPATT diode.

trapped radiation [GEOPHYS] Radiation from space that has become trapped in the magnetic field of the earth, as in the Van Allen belt.

trappide *See* trap.

trapping *See* guided propagation.

trapping mode [ADP] A procedure by means of which the computer, upon encountering a predetermined set of conditions, saves the program in its present status, executes a diagnostic procedure, and then resumes the processing of the program as of the moment of interruption.

trap rock *See* trap.

trash screen [CIV ENG] A screen placed in a waterway to prevent the passage of trash.

Traube's rule [PHYS CHEM] In dilute solutions, the concentration of a member of a homologous series at which a given lowering of surface tension is observed decreases threefold for each additional methylene group in a given series.

trauma [MED] An injury caused by a mechanical or physical agent. [PSYCH] A severe psychic injury.

traumatic pneumonosis [MED] The acute, noninflammatory pathologic pulmonary changes produced by a large momentary deceleration.

traumatotropism [BIOL] Orientation response of an organ of a sessile organism in response to a wound.

Trauzl test [ENG] A test to determine the relative disruptive power of explosives, in which a standard quantity of explosive (10 grams) is placed in a cavity in a lead block and exploded; the resulting volume of cavity in the block is compared with the volume produced under the same conditions by a standard explosive, usually trinitrotoluene (TNT).

traveling block [MECH ENG] The movable unit, consisting of sheaves, frame, clevis, and hook, connected to, and hoisted or lowered with, the load in a block-and-tackle system. Also known as floating block; running block.

traveling charge [ORD] A propelling charge which travels along the bore with the projectile as burning takes place. Also known as Langweiler charge.

traveling compartment [MIN ENG] The section of a mine shaft used for raising and lowering the miners.

traveling detector [ENG] Radio-frequency probe which incorporates a detector used to measure the standing-wave ratio in a slotted-line section.

traveling dune *See* wandering dune.

traveling gantry crane [ENG] A type of hoisting machine with a bridgelike structure spanning the area over which it operates and running along tracks at ground level.

traveling-grate stoker [MECH ENG] A type of furnace stoker; coal feeds by gravity into a hopper located on top of one end of a moving (traveling) grate; as the grate passes under the hopper, it carries a bed of fresh coal toward the furnace.

traveling position [ORD] Position of a weapon when ready for traveling, as opposed to firing position.

traveling road [MIN ENG] A roadway used by miners for walking to and from the face, that is, from the shaft bottom or main entry to the workings.

traveling-screen dryer [CHEM ENG] A moving screen belt on

TRANSVERSE VIBRATION

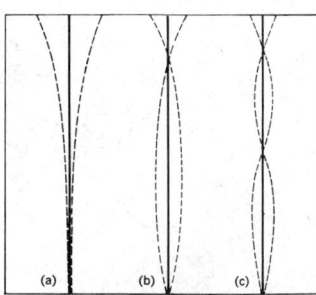

Transverse vibrations of a long circular rod, rigidly clamped at one end and free at the other. *(a)* The fundamental mode. *(b)* The first-overtone mode. *(c)* The second-overtone mode.

which damp material is conveyed through a heated drying zone. Also known as screen dryer.

traveling valve [PETRO ENG] A sucker-rod-pump (oil well) discharge valve that moves with the plunger of a stationary-barrel-type pump, and with the barrel of a traveling-barrel-type pump; contrasted with a standing valve.

traveling wave [PHYS] A wave in which energy is transported from one part of a medium to another, in contrast with a standing wave.

traveling-wave amplifier [ELECTR] An amplifier that uses one or more traveling-wave tubes to provide useful amplification of signals at frequencies of the order of thousands of megahertz.

traveling-wave antenna [ELECTROMAG] An antenna in which the current distributions are produced by waves of charges propagated in only one direction in the conductors. Also known as progressive-wave antenna.

traveling-wave magnetron [ELECTR] A traveling-wave tube in which the electrons move in crossed static electric and magnetic fields that are substantially normal to the direction of wave propagation, as in practically all modern magnetrons.

traveling-wave magnetron oscillations [ELECTR] Oscillations sustained by the interaction between the space-charge cloud of a magnetron and a traveling electromagnetic field whose phase velocity is approximately the same as the mean velocity of the cloud.

traveling-wave maser [PHYS] A ruby maser used with a comblike slow-wave structure and a number of yttrium iron garnet isolators to give L-band amplification (390 to 1550 megahertz); operation is at the temperature of liquid helium (4.2K).

traveling-wave parametric amplifier [ELECTR] Parametric amplifier which has a continuous or iterated structure incorporating nonlinear reactors and in which the signal, pump, and difference-frequency waves are propagated along the structure.

traveling-wave phototube [ELECTR] A traveling-wave tube having a photocathode and an appropriate window to admit a modulated laser beam; the modulated laser beam causes emission of a current-modulated photoelectron beam, which in turn is accelerated by an electron gun and directed into the helical slow-wave structure of the tube.

traveling-wave tube [ELECTR] An electron tube in which a stream of electrons interacts continuously or repeatedly with a guided electromagnetic wave moving substantially in synchronism with it, in such a way that there is a net transfer of energy from the stream to the wave; the tube is used as an amplifier or oscillator at frequencies in the microwave region.

travel-time curve [GEOPHYS] A plot of P-, S-, and L-wave travel times used by seismologists to locate earthquakes.

traverse [ENG] **1.** A survey consisting of a set of connecting lines of known length, meeting each other at measured angles. **2.** Movement to right or left on a pivot or mount, as of a gun, launcher, or radar antenna. [GEOL] A line of survey or sampling across a thin section of geological region. [METEOROL] A westerly wind in central France; it is moderate to strong, generally squally, humid and thundery in summer, especially on slopes facing west; it is cold in winter and spring and brings snow or hail showers. [NAV] A series of directions and distances, as those involved when a sailing vessel beats into the wind or a steam vessel zigzags.

traverse adjustment *See* balancing a survey.

traverse sailing [NAV] A method of determining the equivalent course and distance made good by a craft following a track consisting of a series of rhumb lines.

traverse table [NAV] A table giving relative values of various parts of plane right triangles, for use in solving such triangles, particularly in connection with various sailings.

traversia [METEOROL] A South American nautical term (especially Chile) for a west wind from the sea.

traversier [METEOROL] In the Mediterranean, a dangerous wind blowing directly into port.

traversing mechanism [ENG] Mechanism by which a gun or other device can be turned in a horizontal plane.

travertine [GEOL] Concretionary limestone deposited at the mouth of a hot spring.

trawl [ENG] A baglike net whose mouth is kept open by boards or by a leading diving vane or depressor at the foot of the opening and a spreader bar at the top; towed by a ship at specified depths for catching forms of marine life.

trawler [NAV ARCH] A ship designed for catching fish with a trawl.

tray elevator [MECH ENG] A device for lifting drums, barrels, or boxes; a parallel pair of vertical-mounted continuous chains turn over upper and lower drive gears, and spaced trays on the chains cradle and lift the objects to be moved.

tray tower [CHEM ENG] A vertical process tower for liquid-vapor contacting (as in distillation, absorption, stripping, evaporation, spray drying, dehumidification, humidification, flashing, rectification, dephlegmation), along the height of which is a series of trays designed to cause intimate contact between the falling liquid and the rising vapor.

TR box *See* transmit-receive tube.

TR cell *See* transmit-receive tube.

tread [CIV ENG] **1.** The horizontal part of a step in a staircase. **2.** The distance between two successive risers in a staircase. [ENG] The part of a wheel or tire that bears on the road or rail.

treater [CHEM ENG] A vessel or system for the contacting of a process stream with reagent (treating) chemicals; for example, acid treating or caustic treating.

treating [CHEM ENG] Usually, the contacting of a fluid stream (for example, water, sewage, petroleum products, or mixed gases) with chemicals to improve the fluid properties by removing, sequestering, or converting undesirable impurities.

Trebidae [INV ZOO] A family of copepod crustaceans of the order Caligoida which are external parasites on selachians.

treble [ACOUS] High audio frequencies, such as those handled by a tweeter in a sound system.

tree [BOT] A perennial woody plant at least 20 feet (6 meters) in height at maturity, having an erect stem or trunk and a well-developed crown or leaf canopy. [ELECTR] A set of connected circuit branches that includes no meshes; responds uniquely to each of the possible combinations of a number of simultaneous inputs. Also known as decoder. [MATH] A connected graph contained in a given connected graph having all the vertices of the original but without any closed circuit. [MET] A projecting treelike aggregate of crystals formed at areas of high local current density in electroplating.

tree climate [CLIMATOL] Any type of climate which supports the growth of trees, including the tropical rainy climates, temperate rainy climates, and snow-forest climates.

tree fern [BOT] The common name for plants belonging to the families Cyatheaceae and Dicksoniaceae; all are ferns that exhibit an arborescent habit.

tree frog [VERT ZOO] Any of the arboreal frogs comprising the family Hylidae characterized by expanded digital adhesive disks.

tree line *See* timberline.

trellis drainage [HYD] A drainage pattern characterized by parallel main streams and secondary tributaries intersected at right angles by tributaries. Also known as espalier drainage; grapevine drainage.

Trematoda [INV ZOO] A loose grouping of acoelomate, parasitic flatworms of the phylum Platyhelminthes; they exhibit cephalization, bilateral symmetry, and well-developed holdfast structures.

trematodiasis [MED] Infection caused by a member of the Trematoda (trematode).

Trematosauria [PALEON] A group of Triassic amphibians in the order Temnospondyli.

trembling ill *See* louping ill.

Tremellales [MYCOL] An order of basidiomycetous fungi in the subclass Heterobasidiomycetidae in which basidia have longitudinal walls.

tremie [ENG] An apparatus for placing concrete underwater, consisting of a large metal tube with a hopper at the top end and a valve arrangement at the bottom, submerged end.

tremolite [MINERAL] $Ca_2Mg_5Si_8O_{22}(OH)_2$ Magnesium-rich monoclinic calcium amphibole that forms one end member of a group of solid-solution series with iron, sodium, and aluminum; occurs in long blade-shaped or short stout prismatic crystals and also in masses or compound aggregates.

tremor [GEOPHYS] A minor earthquake. Also known as earthquake tremor; earth tremor. [MED] Involuntary, rhythmic trembling of voluntary muscles resulting from alternate contraction and relaxation of opposing muscle groups.

TRAWL

Isaacs-Kidd midwater trawl, showing depressor at foot of opening and spreader bar at top. *(Scripps Institution of Oceanography)*

trench [GEOGR] **1.** A narrow, straight, elongate, U-shaped valley between two mountain ranges. **2.** A narrow stream-eroded canyon, gulley, or depression with steep sides. [GEOL] A long, narrow, deep depression of the sea floor, with relatively steep sides. Also known as submarine trench.

trencher *See* trench excavator.

trench excavator [MECH ENG] A digging machine, usually on crawler tracks, and having either a movable wheel or a continuous chain on which buckets are mounted. Also known as bucket-ladder excavator; ditcher; trencher; trenching machine.

trench fever [MED] A louse-borne infection that is caused by *Rickettsia quintana* and is characterized by headache, chills, rash, pain in the legs and back, and often by a relapsing fever.

trenching machine *See* trench excavator.

trench mouth [MED] A common form of Vincent's angina; characterized by redness, congestion, and edema of the gums, with involvement of the entire oral cavity in severe cases.

trench sampling [MIN ENG] A slight refinement of grab sampling in which the ore material to be sampled is spread out flat and channeled in one direction with a shovel, and the material for the sample is taken at regular intervals along the channel.

trend [GEOL] The direction of an outcrop of a layer, vein, fold, or other geologic feature. Also known as direction. [STAT] The general drift, tendency, or bent of a set of statistical data as related to time or another related set of statistical data.

trennschaukel apparatus [ENG] An instrument for determining the thermal diffusion factors of gases and gas mixtures, consisting of 20 suitably interconnected tubes whose top ends are maintained at the same temperature and whose bottom ends are maintained at the same temperature, with the temperature of the top ends greater than that of the bottom ends.

Trentepohliaceae [BOT] A family of green algae belonging to the Ulotrichales having thick walls, bandlike or reticulate chloroplasts, and zoospores or isogametes produced in enlarged, specialized cells.

Trentonian [GEOL] A North American stage of geologic time; Middle Ordovician (above Wilderness, below Edenian); equivalent to the upper Mohawkian.

Treponema [MICROBIO] A genus of anaerobic spirochetes in the family Treponemataceae distinguished by uniform, somewhat angular spirals.

Treponema pallidum [MICROBIO] The spirochetal bacterium which causes syphilis.

Treponema pallidum immobilization test [IMMUNOL] A serologic test for syphilis in which suspensions of *Treponema pallidum* are immobilized in the presence of syphilitic serum and complement. Abbreviated TPI test.

Treponemataceae [MICROBIO] A family of the bacterial order Spirochaetales including the spirochetes less than 20 micrometers long and less than 5 micrometers in diameter; most species are parasitic.

treponematosis [MED] Infection caused by any species of the genus *Treponema*. Also known as treponemiasis.

treponemiasis *See* treponematosis.

Trepostomata [PALEON] An extinct order of ectoproct bryozoans in the class Stenolaemata characterized by delicate to massive colonies composed of tightly packed zooecia with solid calcareous zooecial walls.

Treroninae [VERT ZOO] The fruit pigeons, a subfamily of the avian family Columbidae distinguished by the gaudy coloration of the feathers.

trestle [ENG] **1.** A movable support usually with legs that spread diagonally. **2.** A braced structure of timber, reinforced concrete, or steel spanning a land depression to carry a road or railroad.

trestle bent [CIV ENG] A transverse frame that supports the ends of the stringers in adjoining spans of a trestle.

tretamine *See* triethylenemelamine.

Tretothoracidae [INV ZOO] A family of the Coleoptera in the superfamily Tenebrionoidea which contains a single species found in Queensland, Australia.

TRF receiver *See* tuned-radio-frequency receiver.

Triac [ELECTR] Trademark of the General Electric Company for a gate-controlled semiconductor switch designed for alternating-current power control; with phase control of the gate signal, load current can be varied over a range from 5 to 95% of full power.

triacetate [TEXT] Acetate fiber that is acetylated to a greater degree than other acetates.

triacetin [ORG CHEM] $C_3H_5(CO_2CH_3)_3$ A colorless, combustible oil with a bitter taste and a fatty aroma; found in cod liver and butter; soluble in alcohol and ether, slightly soluble in water; boils at 259°C; used in plasticizers, perfumery, cosmetics, and external medicine and as a solvent and food additive. Also known as glyceryl triacetate.

triacetyloleandomycin [MICROBIO] An antibiotic produced by *Streptomyces antibioticus* and used clinically in the treatment of pneumonia, osteomyelitis, furuncles, and carbuncles.

triad [ELECTR] A triangular group of three small phosphor dots, each emitting one of the three primary colors on the screen of a three-gun color picture tube. [NAV] *See* triplet.

triaene [INV ZOO] An elongated spicule in certain Porifera with three rays diverging from one end.

triage [MED] The process of determining which casualties (as from an accident, disaster, military battle, or explosion of nuclear weapons) need urgent treatment, which ones are well enough to go untreated, and which ones are beyond hope of benefit from treatment.

Triakidae [VERT ZOO] A family of galeoid sharks in the carcharinid line.

trial batch [ENG] A batch of concrete mixed to determine the water-cement ratio that will produce the required slump and compressive strength; from a trial batch, one can also compute the yield, cement factor, and required quantities of each material.

trial fire [ORD] Deliberate gunfire laid on a fixed point or target to determine the corrections for firing data.

trial pit [MIN ENG] A shallow hole, 2 to 3 feet (60 to 90 centimeters) in diameter, put down to test shallow minerals or to establish the nature and thickness of superficial deposits and depth to bedrock.

trial shots [ENG] The experimental shots and rounds fired in a sinking pit, tunnel, opencast, or quarry to determine the best drill-hole pattern to use.

triamcinolone [ORG CHEM] $C_{21}H_{27}FO_6$ White, toxic crystals; insoluble in water, soluble in dimethylformamide; melts at 266°C; used as an intermediate for ion-exchange resin, wetting and frothing agent, and photographic developer.

2,4,6-triamino-s-triazine *See* melamine.

triamylamine [ORG CHEM] $(C_5H_{11})_3N$ A combustible, colorless, toxic liquid; soluble in gasoline, insoluble in water; used to inhibit corrosion and in insecticides.

triamyl borate [ORG CHEM] $(C_5H_{11})_3BO_3$ A combustible, colorless liquid with an alcoholic aroma; soluble in alcohol and ether; boils at 220–280°C; used in varnishes.

triandrous [BOT] Possessing three stamens.

triangle [MATH] The figure realized by connecting three noncollinear points by line segments.

Triangle *See* Triangulum.

triangle cut [MIN ENG] A zigzag arrangement of drill holes permitting larger openings to be obtained as the drill holes can break out between the preceding row of holes.

triangle equation *See* angle equation.

triangle inequality [MATH] For real or complex numbers or vectors in a normed space x and y, the absolute value or norm of $x + y$ is less than or equal to the sum of the absolute values or norms of x and y.

triangle of velocities [NAV] The fundamental triangle associated with dead-reckoning, composed of the following vectors: heading and true airspeed, track and groundspeed, and wind speed and wind direction.

triangular facet [GEOL] A triangular-shaped steep-sloped hill or cliff formed usually by the erosion of a fault-truncated hill.

triangular ligament *See* urogenital diaphragm.

triangular matrix [MATH] A matrix where either all entries above or all entries below the principal diagonal are zero.

triangular method [MIN ENG] A method of ore reserve estimation based on the assumption that a linear relationship exists between the grade difference and the distance between all drill holes.

triangular-notch weir [CIV ENG] A measuring weir with a V-shaped notch for measuring small flows. Also known as V-notch weir.

triangular pulse [ELECTR] An electrical pulse in which the voltage rises linearly to some value, and immediately falls linearly to the original value.

triangular wave [ELECTR] A wave consisting of a series of triangular pulses.

triangulation [ENG] A surveying method for measuring a large area of land by establishing a base line from which a network of triangles is built up; in a series, each triangle has at least one side common with each adjacent triangle. [MATH] A decomposition of a topological manifold into subsets homeomorphic with a polyhedron in some euclidean space. [NAV] Determination of the position of a ship or aircraft by obtaining bearings of the moving object with reference to two fixed radio stations a known distance apart; this gives the values of one side and all angles of a triangle, from which the position can be computed.

triangulation mark [ENG] A bronze disk set in the ground to identify a point whose latitude and longitude have been determined by triangulation.

triangulation problem [MATH] The problem of whether each topological n manifold admits a piecewise linear structure.

Triangulum [ASTRON] A northern constellation, right ascension 2 hours, declination 30°N. Also known as Triangle.

Triangulum Australe [ASTRON] A southern constellation, right ascension 16 hours, declination 65°S. Also known as Southern Triangle.

Triangulum Nebula [ASTRON] A nebula that is part of a small cluster of galaxies known as the local group; the nebula is labeled M 33.

Triassic [GEOL] The first period of the Mesozoic era, lying above Permian and below Jurassic, 180–225 million years ago.

Triatominae [INV ZOO] The kissing bugs, a subfamily of hemipteran insects in the family Reduviidae, distinguished by a long, slender rostrum.

triaxial pinch [PL PHYS] A device for heating a confined plasma, in which a discharge in an annular space between two concentric cylindrical conductors forms a cylindrical sheet of plasma, and this plasma is then confined and compressed by magnetic fields produced by currents flowing in the axial direction in the discharge itself and in the two conductors.

triaxon [INV ZOO] A spicule in Porifera having three axes which cross each other at right angles.

s-triazinetriol See cyanuric acid.

triazole [ORG CHEM] A five-membered chemical ring compound with three nitrogens in the ring; for example, $C_2H_3N_3$; proposed for use as a photoconductor and for copying systems.

tribasic calcium phosphate See calcium phosphate.

tribasic zinc phosphate See zinc phosphate.

tribo- [PHYS] A prefix meaning pertaining to or resulting from friction.

triboelectric series [ELEC] A list of materials that produce an electrostatic charge when rubbed together, arranged in such an order that a material has a positive charge when rubbed with a material below it in the list, and has a negative charge when rubbed with a material above it in the list.

triboelectrification [ELEC] The production of electrostatic charges by friction.

triboluminescence [ATOM PHYS] Luminescence produced by friction between two materials.

tribromoethanol [PHARM] $C_2H_3Br_3O$ A white crystalline compound, melting at 79–82°C; used in medicine in anesthesia. Also known as tribromoethyl alcohol.

tribromoethyl alcohol See tribromoethanol.

tribromomethane See bromoform.

tributary [HYD] A stream that feeds or flows into or joins a larger stream or a lake. Also known as contributory; feeder; side stream; tributary stream.

tributary station [COMMUN] Communications terminal consisting of equipment compatible for the introduction of messages into or reception from its associated relay station.

tributary stream See tributary.

tributary waterway [HYD] Any body of water that flows into a larger body, that is, a creek in relation to a river, a river in relation to a bay, and a bay in relation to the open sea.

tributyl borate [ORG CHEM] $(C_4H_9)_3BO_3$ A combustible, water-white liquid miscible with common organic liquids; boils at 232°C; used in welding fluxes and as a chemical intermediate and textile flame-retardant. Also known as butyl borate.

tributyl citrate See butyl citrate.

tributyl phosphate [ORG CHEM] $(C_4H_9)_3PO_4$ A combustible, toxic, stable liquid; soluble in most solvents, and very slightly soluble in water; boils at 292°C; used as a heat-exchange medium, pigment-grinding assistant, antifoam agent, and solvent. Abbreviated TBP.

tributyltin acetate [ORG CHEM] $(C_4H_9)_3Sn-OOCCH_3$ An organic compound of tin, used as an antimicrobial agent in the paper, wood, plastics, leather, and textile industries.

tricalcium phosphate See calcium phosphate.

tricamera photography [GRAPHICS] Photography obtained by simultaneous exposure of three cameras systematically disposed in the air vehicle at fixed overlapping angles relative to each other in order to cover a wide field.

tricarboxylic acid cycle See Krebs cycle.

Trichechidae [VERT ZOO] The manatees, a family of nocturnal, solitary sirenian mammals in the suborder Trichechiformes.

Trichechiformes [VERT ZOO] A suborder of mammals in the order Sirenia which contains the manatees and dugongids.

trichesthesia [PHYSIO] A form of tactile sensibility in hair-covered regions of the body.

Trichiaceae [MYCOL] A family of slime molds in the order Trichiales.

Trichiales [MYCOL] An order of Myxomycetes in the subclass Myxogastromycetidae.

trichinosis [MED] Infection by the nematode *Trichinella spiralis* following ingestion of encysted larvae in raw or partially cooked pork; characterized by eosinophilia, nausea, fever, diarrhea, stiffness and painful swelling of muscles, and facial edema.

trichite [PETR] A black, straight or curved, hairlike crystallite.

Trichiuridae [VERT ZOO] The cutlass-fishes, a family of the suborder Scombroidei.

trichloroacetic acid [ORG CHEM] CCl_3COOH Toxic, deliquescent, colorless crystals with a pungent aroma; soluble in water, alcohol, and ether; boils at 198°C; used as a chemical intermediate and laboratory reagent, and in medicine, pharmacy, and herbicides. Abbreviated TCA.

trichloroacetic aldehyde See chloral.

trichlorobenzene [ORG CHEM] $C_6H_3Cl_3$ Either of two toxic compounds: 1,2,3-trichlorobenzene forms white crystals, soluble in ether, insoluble in water, boiling at 221°C, and is used as a chemical intermediate; 1,2,4-trichlorobenzene is a combustible, colorless liquid, soluble in most organic solvents and oils, insoluble in water, boiling at 213°C, and is used as a solvent and in dielectric fluids, synthetic transformer oils, lubricants, and insecticides.

trichloroethanal See chloral.

trichloroethane [ORG CHEM] $C_2H_3Cl_3$ Either of two nonflammable, irritating liquid isomeric compounds: 1,1,1-trichloroethane (CH_3CCl_3) is toxic, soluble in alcohol and ether, insoluble in water, and boils at 75°C; it is used as a solvent, aerosol propellant, and pesticide, and for metal degreasing, and is also known as methyl chloroform; 1,1,2-trichloroethane ($CHCl_2CH_2Cl$) is clear and colorless, is soluble in alcohols, ethers, esters, and ketones, insoluble in water, has a sweet aroma, and boils at 114°C; it is used as a chemical intermediate and solvent, and is also known as vinyl trichloride.

trichloroethylene [ORG CHEM] $CHCl:CCl_2$ A heavy, stable, toxic liquid with a chloroform aroma; slightly soluble in water, soluble with greases and common organic solvents; boils at 87°C; used for metal degreasing, solvent extraction, and dry cleaning and as a fumigant and chemical intermediate.

trichlorofluoromethane [ORG CHEM] CCl_3F A toxic, noncombustible, colorless liquid boiling at 24°C; used as a chemical intermediate, solvent, refrigerant, aerosol propellant, and blowing agent (plastic foams) and in fire extinguishers. Also known as fluorocarbon-11; fluorotrichloromethane.

TRIASSIC

PRECAMBRIAN		
CAMBRIAN		
ORDOVICIAN		
SILURIAN		
DEVONIAN		
Mississippian	CARBON-IFEROUS	PALEOZOIC
Pennsylvanian		
PERMIAN		
TRIASSIC		
JURASSIC		MESOZOIC
CRETACEOUS		
TERTIARY		CENOZOIC
QUATERNARY		

Chart showing position of the Triassic period in relation to the other periods and to the eras of geologic time.

TRICHOPTERA

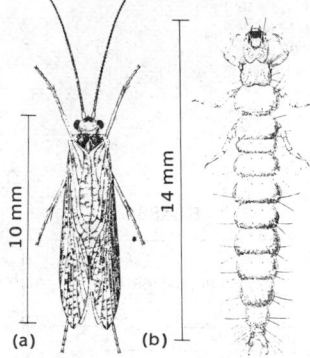

Rhyacophila, a widespread genus of Trichoptera. (a) Adult, showing well-veined wings and long antennae. (b) Free-living larva. (*Illinois Natural History Survey*)

TRICONODONT

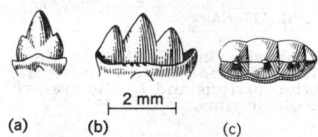

Triconodont molars, showing three main cusps in logitudinal series. (a) External view of a lower molar of *Amphilestes*, an amphilestine triconodont. (b) Internal and (c) occlusal views of a lower molar of *Priacodon*, a triconodontine triconodont. (*After G. G. Simpson, 1929*)

TRICYCLE LANDING GEAR

Typical tricycle landing gear on F8U-2 Crusader. (*LTV Aerospace Corp.*)

trichloromethane   *See* chloroform.

trichloronitromethane   *See* chloropicrin.

trichlorophenol   [ORG CHEM] $C_6H_2Cl_3OH$ Either of two toxic nonflammable compounds with a phenol aroma: 2,4,5-trichlorophenol is a gray solid, is soluble in alcohol, acetone, and ether, melts at 69°C, and is used as a fungicide and bactericide; 2,4,6-trichlorophenol forms yellow flakes, is soluble in alcohol, acetone, and ether, boils at 248°C, and is used as a fungicide, defoliant, and herbicide; it is also known as 2,4,6-T.

2,4,5-trichlorophenoxyacetic acid   [ORG CHEM] $C_6H_2Cl_3$-OCH_2CO_2H A toxic, light-tan solid; soluble in alcohol, insoluble in water; melts at 152°C; used as a defoliant, plant hormone, and herbicide. Also known as 2,4,5-T.

trichobezoar   [MED] A ball of hair or similar concretion in the stomach or intestine.

trichobothrium   [INV ZOO] An erect, bristlelike sensory hair found on certain arthropods, insects, and other invertebrates.

trichobranchiate gill   [INV ZOO] A gill with filamentous branches arranged in several series around the axis; found in some decapod crustaceans.

Trichobranchidae   [INV ZOO] A family of polychaete annelids belonging to the Sedentaria; most members are rare and live at great ocean depths.

trichocercous cercaria   [INV ZOO] A trematode larva distinguished by a spiny tail.

Trichocomaceae   [MYCOL] A small tropical family of ascomycetous fungi in the order Eurotiales with ascocarps from which a tuft of capillitial threads extrudes, releasing the ascospores after dissolution of the asci.

trichocyst   [INV ZOO] A minute structure in the cortex of certain protozoans that releases filamentous or fibrillar threads when discharged.

Trichodactylidae   [INV ZOO] A family of fresh-water crabs in the section Brachyura, found mainly in tropical regions.

trichoepithelioma   [MED] A benign tumor characterized by small, round, yellow, or flesh-colored papules, chiefly on the center of the face.

Trichogrammatidae   [INV ZOO] A family of the Hymenoptera in the superfamily Chalcidoidea whose larvae are parasitic in the eggs of other insects.

trichogyne   [BOT] A terminal portion of a procarp or archicarp which receives a spermatium.

trichome   [BOT] An appendage derived from the protoderm in plants, including hairs and scales.   [INV ZOO] A brightly colored tuft of hairs on the body of a myrmecophile that releases an aromatic substance attractive to ants.

Trichomonadida   [INV ZOO] An order of the protozoan class Zoomastigophorea which contains four families of uninucleate species.

Trichomonadidae   [INV ZOO] A family of flagellate protozoans in the order Trichomonadida.

Trichomonas   [INV ZOO] A genus of flagellate, parasitic protozoans, belonging to the Mastigophora, characterized by three to five flagella and an undulating membrane.

Trichomonas vaginalis   [INV ZOO] A species of Protozoa that causes vaginitis.

trichomoniasis   [MED] An infection caused by a species of the genus *Trichomonas*.

Trichomycetes   [MYCOL] A class of true fungi, division Fungi.

trichomycin   [MICROBIO] An antibiotic produced by *Streptomyces hachijoensis* and *S. abikoensis;* a water-soluble yellow powder that inhibits yeasts and fungi.

Trichoniscidae   [INV ZOO] A primitive family of isopod crustaceans in the suborder Oniscoidea found in damp littoral, halophilic, or riparian habitats.

trichopathophobia   [PSYCH] Extreme anxiety and fear regarding growth, color, or diseases of hair.

Trichophilopteridae   [INV ZOO] A family of lice in the order Mallophaga adapted to life upon the lemurs of Madagascar.

trichophytin   [IMMUNOL] A group antigen obtained from filtrates of *Trichophyton mentagrophytes;* used in a skin test to ascertain past or present infection with the dermatophytes.

Trichoptera   [INV ZOO] The caddis flies, an aquatic order of the class Insecta; larvae are wormlike and adults have two pairs of well-veined hairy wings, long antennae, and mouthparts capable of lapping only liquids.

Trichopterygidae   [INV ZOO] The equivalent name for Ptiliidae.

Trichostomatida   [INV ZOO] An order of ciliated protozoans in the subclass Holotrichia in which no true buccal ciliature is present but there is a vestibulum.

Trichostrongylidae   [INV ZOO] A family of parasitic roundworms belonging to the Strongyloidea; hosts are cattle, sheep, goats, swine, and cats.

trichroism   [OPTICS] Phenomenon exhibited by certain optically anisotropic transparent crystals when subjected to white light, in which a cube of the material is found to transmit a different color through each of the three pairs of parallel faces.

trichromatic theory   [OPTICS] A theory of color vision which states that three primary colors may be chosen in such a way that, combined in various proportions, they can match any color.

Trichuroidea   [INV ZOO] A group of nematodes parasitic in various vertebrates and characterized by a slender body sometimes having a thickened posterior portion.

trickle charge   [ELEC] A continuous charge of a storage battery at a low rate to maintain the battery in a fully charged condition.

trickle cooler   *See* cascade cooler.

trickle hydrodesulfurization   [CHEM ENG] A fixed-bed, petroleum refining process for desulfurization of middle distillates and gas oils; catalyst is cobalt molybdenum on alumina.

trickling filter   [CIV ENG] A bed of broken rock or other coarse aggregate onto which sewage or industrial waste is sprayed intermittently and allowed to trickle through, leaving organic matter on the surface of the rocks, where it is oxidized and removed by biological growths.

Tricladida   [INV ZOO] The planarians, an order of the Turbellaria distinguished by diverticulated intestines with a single anterior branch and two posterior branches separated by the pharynx.

triclinic crystal   [CRYSTAL] A crystal whose unit cell has axes which are not at right angles, and are unequal. Also known as anorthic crystal.

triclinic system   [CRYSTAL] The most general and least symmetric crystal system, referred to by three axes of different length which are not at right angles to one another.

tricolor picture tube   *See* color picture tube.

triconodont   [VERT ZOO] 1. A tooth with three main conical cusps. 2. Having such teeth.

Triconodonta   [PALEON] An extinct mammalian order of small flesh-eating creatures of the Mesozoic era having no angle or a pseudoangle on the lower jaw and triconodont molars.

tricosane   [ORG CHEM] $CH_3(CH_2)_{21}CH_3$ Combustible, glittering crystals; soluble in alcohol, insoluble in water; melts at 48°C; used as a chemical intermediate.

tricresyl phosphate   [ORG CHEM] $(CH_3C_6H_4O)_3PO$ A combustible, colorless liquid; insoluble in water, soluble in common solvents and vegetable oils; boils at 420°C; used as a plasticizer, plastics fire retardant, air-filter medium, and gasoline and lubricant additive. Abbreviated TCP.

Trictenotomidae   [INV ZOO] A small family of Indian and Malaysian beetles in the superfamily Tenebrionoidea.

tricuspid   [ANAT] Having three cusps or points, as a tooth.

tricuspid valve   [ANAT] A valve consisting of three flaps located between the right atrium and right ventricle of the heart.

tricycle landing gear   [AERO ENG] A landing-gear arrangement that places the nose gear well forward of the center of gravity on the fuselage and the two main gears slightly aft of the center of gravity, with sufficient distance between them to provide stability against rolling over during a yawed landing in a crosswind, or during ground maneuvers.

tricyclic dibenzopyran   *See* xanthene.

Tridacnidae   [INV ZOO] A family of bivalve mollusks in the subclass Eulamellibranchia which contains the giant clams of the tropical Pacific.

Tridactylidae   [INV ZOO] The pygmy mole crickets, a family of insects in the order Orthoptera, highly specialized for fossorial existence.

***n*-tridecane**   [ORG CHEM] $CH_3(CH_2)_{11}CH_3$ A combustible liquid; soluble in alcohol, insoluble in water; boils at 226°C;

used as a distillation chaser and chemical intermediate.

tridecanol *See* tridecyl alcohol.

tridecyl alcohol [ORG CHEM] $C_{12}H_{25}CH_2OH$ An isomer mixture; a white, combustible solid with a pleasant aroma; melts at 31°C; used in detergents and perfumery and to make synthetic lubricants. Also known as tridecanol.

trident of Newton [MATH] The curve in the plane given by the equation $xy = ax^3 + bx^2 + cx + d$, where $a \neq 0$; this cuts the x-axis in one or three points and is asymptotic to the y-axis if $d \neq 0$.

triductor [ELEC] Arrangement of iron-core transformers and capacitors used to triple a power-line frequency.

tridymite [MINERAL] SiO_2 A white or colorless crystal occurring in minute, thin, tabular crystals or scales; a high-temperature polymorph of quartz.

triethanolamine [ORG CHEM] $(HOCH_2CH_2)_3N$ A viscous, hygroscopic liquid with an ammonia aroma, soluble in chloroform, water, and alcohol, and boiling at 335°C; used in dry-cleaning soaps, cosmetics, household detergents, and textile processing, for wool scouring, and as a corrosion inhibitor.

triethylamine [ORG CHEM] $(C_2H_5)_3N$ A colorless, toxic, flammable liquid with an ammonia aroma; soluble in water and alcohol; boils at 90°C; used as a solvent, rubber-accelerator activator, corrosion inhibitor, and propellant, and in penetrating and waterproofing agents.

triethylene glycol [ORG CHEM] $HO(C_2H_4O)_3H$ A colorless, combustible, hygroscopic, water-soluble liquid; boils at 287°C; used as a chemical intermediate, solvent, bactericide, humectant, and fungicide. Abbreviated TEG.

triethylenemelamine [ORG CHEM] $NC[N(CH_2)_2]NC-[N(CH_2)_2]NC[N(CH_2)_2]$ White crystals, soluble in water, alcohol, acetone, chloroform, and methanol; polymerizes at 160°C; used in medicine and insecticides and as a chemosterilant. Abbreviated TEM. Also known as tretamine.

triethylic borate *See* ethyl borate.

trifid [BIOL] Divided into three lobes separated by narrow sinuses partway to the base.

trifluorochlorethylene resin [ORG CHEM] A fluorocarbon used as a base for polychlorotrifluoroethylene resin, marketed as Kel-F.

trifluoromethane *See* fluoroform.

trifoliate [BOT] Having three leaves or leaflets.

trifoliosis [VET MED] An acute photosensitization characterized by superficial necrosis of white or light-skinned animals feeding on certain leguminous plants.

triformal *See* sym-trioxane.

trigatron [ELECTR] Gas-filled, spark-gap switch used in line pulse modulators.

trigeminal nerve [ANAT] The fifth cranial nerve in vertebrates; either of a pair of composite nerves rising from the side of the medulla, and with three great branches: the ophthalmic, maxillary, and mandibular nerves.

trigeminal neuralgia [MED] Sudden severe pains of unknown cause along the path of one or more branches of the trigeminal nerve. Also known as tic douloureux.

trigger [ELECTR] **1.** To initiate an action, which then continues for a period of time, as by applying a pulse to a trigger circuit. **2.** The pulse used to initiate the action of a trigger circuit. **3.** *See* trigger circuit. [ORD] A metallic item, part of the firing mechanism of a firearm, designed to release a firing pin by the application of pressure by the finger.

trigger action [ELECTR] Use of a weak input pulse to initiate main current flow suddenly in a circuit or device.

trigger circuit [ELECTR] **1.** A circuit or network in which the output changes abruptly with an infinitesimal change in input at a predetermined operating point. Also known as trigger. **2.** A circuit in which an action is initiated by an input pulse, as in a radar modulator. **3.** *See* bistable multivibrator.

trigger control [ELECTR] Control of thyratrons, ignitrons, and other gas tubes in such a way that current flow may be started or stopped, but not regulated as to rate.

trigger diode [ELECTR] A symmetrical three-layer avalanche diode used in activating silicon-controlled rectifiers; it has a symmetrical switching mode, and hence fires whenever the breakover voltage is exceeded in either polarity.

triggered spark gap [ELEC] A fixed spark gap in which the discharge passes between two electrodes but is initiated by an auxiliary trigger electrode to which low-power pulses are applied at regular intervals by a pulse amplifier.

trigger electrode *See* starter.

trigger extension [ORD] A metallic item attached to a trigger in order to extend its length.

trigger housing [ORD] An item, usually of metal, designed to fit into the framework of a carbine, machine gun, pistol, or rifle; used to provide a mounting for a trigger.

triggering [ELECTR] Phenomenon observed in some high-performance magnetic amplifiers with very low leakage rectifiers; as the input current is decreased in magnitude, the amplifier remains at cutoff for some time, and the output then suddenly shoots upward.

trigger level [ELECTR] In a transponder, the minimum input to the receiver which is capable of causing a transmitter to emit a reply.

trigger motor [ORD] An electric motor on certain types of automatic weapons that operates the sear mechanism for rapid fire.

trigger pull [MECH] Resistance offered by the trigger of a rifle or other weapon; force which must be exerted to pull the trigger.

trigger pulse [ELECTR] A pulse that starts a cycle of operation. Also known as tripping pulse.

trigger shaft [ORD] The shaft which passes transversely through the breechblock and firing lock of a gun and whose arm is actuated by the firing mechanism; movement of the arm rotates the shaft and thus imparts movement to the trigger fork of the firing lock.

trigger squeeze [ORD] Method of firing a rifle or similar weapon in which the trigger is not pulled, but squeezed gradually by an independent action of the forefinger.

trigger switch [ELEC] A switch that is actuated by pulling a trigger, and is usually mounted in a pistol-grip handle.

trigistor [ELECTR] A *pnpn* device with a gating control acting as a fast-acting switch similar in nature to a thyratron.

Triglidae [VERT ZOO] The searobins, a family of perciform fishes in the suborder Cottoidei.

triglyceride [ORG CHEM] $CH_2(OOCR_1)CH(OOCR_2)-CH_2(OOCR_3)$ A naturally occurring ester of normal, fatty acids and glycerol; used in the manufacture of edible oils, fats, and monoglycerides.

Trigonalidae [INV ZOO] A small family of hymenopteran insects in the superfamily Proctotrupoidea.

trigonal lattice *See* rhombohedral lattice.

trigonal system [CRYSTAL] A crystal system which is characterized by threefold symmetry, and which is usually considered as part of the hexagonal system since the lattice may be either hexagonal or rhombohedral.

trigone [ANAT] A triangular area inside the bladder limited by the openings of the ureters and urethra. [BOT] A thickening of plant cell walls formed when three or more cells adjoin.

trigonometric functions [MATH] The real-valued functions such as $\sin(x)$, $\tan(x)$, and $\cos(x)$ obtained from studying certain ratios of the sides of a right triangle. Also known as circular functions.

trigonometric leveling [ENG] A method of determining the difference of elevation between two points, by using the principles of triangulation and trigonometric calculations.

trigonometric parallax [ASTRON] A parallax that may be determined for the nearest stars (less than 300 light-years) by a direct method utilizing trigonometry.

trigonometric polynomial [MATH] A finite series of functions of the form $a_n \cos nx + b_n \sin nx$; occasionally used synonymously with trigonometric series.

trigonometric series [MATH] An infinite series of functions with nth term of the form $a_n \cos nx + b_n \sin nx$.

trigonometry [MATH] The study of triangles and the trigonometric functions.

Trigonostylopoidea [PALEON] A suborder of Paleocene-Eocene ungulate mammals in the order Astrapotheria.

trigonous [BIOL] **1.** Having three corners. **2.** Having a triangular cross section.

trihedral [MATH] Any figure obtained from three noncoplanar lines intersecting in a common point.

1,2,4-trihydroxyanthraquinone *See* purpurin.

trihydroxycyanidine *See* cyanuric acid.

triiodomethane *See* iodoform.

TRIFOLIATE

Trifoliate leaf, with three leaflets.

TRIMERELLACEA

(a)   (b)

Internal views of Silurian trimerellacean *Trimerella*: *(a)* pedicle valve; *(b)* brachial valve. *(From C. D. Walcott, Cambrian Brachiopoda, USGS Monogr. no. 51, 1912)*

TRIMETRIC DRAWING

Trimetric drawing, in which three axes are unequally inclined to drawing surface. *(From T. E. French and C. J. Vierck, Graphic Science, McGraw-Hill, 1958)*

triisobutylene [ORG CHEM] $(C_4H_8)_3$ A mixture of isomers; combustible liquid boiling at 348–354°C; used as a chemical and resin intermediate, lubricating oil additive, and motor-fuel alkylation feedstock.

trilateration [ENG] The measurement of a series of distances between points on the surface of the earth, for the purpose of establishing relative positions of the points in surveying.

trill *See* trilling.

trilling [CRYSTAL] A cyclic crystal twin consisting of three individual crystals. Also known as threeling; trill.

Trilobita [PALEON] The trilobites, a class of extinct Cambrian-Permian arthropods characterized by an exoskeleton covering the dorsal surface, delicate biramous appendages, body segments divided by furrows on the dorsal surface, and a pygidium composed of fused segments.

Trilobitoidea [PALEON] A class of Cambrian arthropods that are closely related to the Trilobita.

Trilobitomorpha [INV ZOO] A subphylum of the Arthropoda including Trilobita.

trilocular [BIOL] Having three cavities or cells.

trim [AERO ENG] The orientation of an aircraft relative to the airstream, as indicated by the amount of control pressure required to maintain a given flight performance. [ELECTR] A small change or necessary adjustment of tuning capacitance. [NAV ARCH] **1.** The deviation of a ship from an even keel fore and aft. **2.** To add or remove water from the variable ballast tanks of a submarine to maintain neutral buoyancy.

trim by head [NAV ARCH] That condition of trim in which a vessel inclines forward so that its actual plane of flotation is not coincident with or parallel to its designed plane of flotation.

trim by stern [NAV ARCH] That condition of trim in which a vessel inclines aft so that its actual plane of flotation is not coincident with or parallel to its designed plane of flotation.

Trimenoponidae [INV ZOO] A family of lice in the order Mallophaga occurring as parasites on South American rodents.

trimer [CHEM] A condensation product of three monomer molecules; C_6H_6 is a trimer of C_2H_2.

trimercuric orthophosphate *See* mercuric phosphate.

trimercurous orthophosphate *See* mercurous phosphate.

Trimerellacea [PALEON] A superfamily of extinct inarticulate brachiopods in the order Lingulida; they have valves, usually consisting of calcium carbonate.

Trimerophytatae *See* Trimerophytopsida.

Trimerophytopsida [PALEOBOT] A group of extinct land vascular plants with leafless, dichotomously branched stems that bear terminal sporangia.

trimerous [BOT] Having parts in sets of three. [INV ZOO] In insects, having the tarsus divided or apparently divided into three segments.

***uns*-trimethylbenzene** *See* pseudocumene.

trimethyl borate [ORG CHEM] $B(OCH_3)_3$ A water-white liquid, boiling at 67–68°C; used as a solvent for resins, waxes, and oils, and as a catalyst and a reagent in analysis of paint and varnish. Also known as methyl borate.

2,2,3-trimethylbutane *See* triptane.

trimethylethylene *See* methyl butene.

2,2,4-trimethylpentane *See* isooctane.

2,4,5-trimethyl-l-propenyl benzene *See* asarone.

2,4,6-trimethylpyridine *See* 2,4,6-collidine.

trimethylvinylammonium hydroxide *See* neurine.

trimetric drawing [GRAPHICS] A form of nonperspective pictorial drawing in which the object being drawn is turned so that three mutually perpendicular edges are unequally foreshortened.

trimmer [MIN ENG] **1.** A piece of bent wire used to regulate the size of the flame of a safety lamp without removing the top of the lamp. **2.** A worker who arranges coal in the hold of a vessel (miner, ship) as the coal is discharged into it from bins. **3.** A person who cleans miners' lamps. **4.** An apparatus for trimming a pile of coal into a regular form (as a cone or prism). **5.** *See* rib holes.

trimmer capacitor [ELEC] A relatively small variable capacitor used in parallel with a larger variable or fixed capacitor to permit exact adjustment of the capacitance of the parallel combination.

trimmer conveyor [MECH ENG] A self-contained, lightweight

portable conveyor, usually of the belt type, for use in unloading and delivering bulk materials from trucks to domestic storage places, and for trimming bulk materials in bins or piles.

trimmer potentiometer [ELEC] A potentiometer which is used to provide a small-percentage adjustment and is often used with a coarse control.

trimming [MET] Removing irregular edges from a drawn part, or parting-line flash from a forging, or gates, risers, and fins from a casting.

trimorphous [BIOL] Characterized by occurring in three distinct forms, as an organ or whole organism.

trim size [GRAPHICS] The size of a map or page when the excess paper outside the margin has been trimmed off after printing.

trinickelous orthophosphate *See* nickel phosphate.

Trinidad asphalt [MATER] Natural asphaltic material found in Trinidad; contains about 47% bitumen and 28% clay, and the remainder is water.

trinitrobenzene [ORG CHEM] $C_6H_3(NO_2)_3$ A yellow crystalline compound, soluble in alcohol and ether; used as an explosive.

trinitromethane [ORG CHEM] $CH(NO_2)_3$ A crystalline compound, melting at 150°C, decomposing above 25°C; used to make explosives. Also known as nitroform.

trinitrophenol *See* picric acid.

2,4,6-trinitroresorcinol *See* styphnic acid.

2,4,6-trinitrotoluene [ORG CHEM] $CH_3C_6H_2(NO_2)_3$ Toxic, flammable, explosive, yellow crystals; soluble in alcohol and ether, insoluble in water; melts at 81°C; used as an explosive and chemical intermediate and in photographic chemicals. Also known as methyltrinitrobenzene; TNT.

trinomial [MATH] A polynomial having three terms. [SYST] A nomenclatural designation for an organism composed of three terms: genus, species, and subspecies or variety.

trinomial distribution [STAT] A multinomial distribution in which there are three distinct outcomes.

triode [ELECTR] A three-electrode electron tube containing an anode, a cathode, and a control electrode.

triode clamp [ELECTR] A keyed clamp circuit utilizing triodes, such as a circuit which contains a complementary pair of bipolar transistors.

triode clipping circuit [ELECTR] A clipping circuit that utilizes a transistor or vacuum triode.

triode laser [ELECTR] Gas laser whose light output may be modulated by signal voltages applied to an integral grid.

triode transistor [ELECTR] A transistor that has three terminals.

Trionychidae [VERT ZOO] The soft-shelled turtles, a family of reptiles in the order Chelonia.

triose [BIOCHEM] A group of monsaccharide compounds that have a three-carbon chain length.

***sym*-trioxane** [ORG CHEM] $(CH_2O)_3$ White, flammable, explosive crystals; soluble in water, alcohol, and ether; melts at 62°C; used as a chemical intermediate, disinfectant, and fuel. Also known as metaformaldehyde; triformol; trioxin.

trioxin *See* sym-trioxane.

trioxycyanidine *See* cyanuric acid.

trip [ENG] To release a lever or set free a mechanism. [MIN ENG] **1.** The line of cars hauled by mules or by motor, or run on a slope, plane, or sprag road. **2.** An automatic arrangement for dumping cars. [ORD] Part of the mechanism of some firearms, released by the action of the trigger.

tripalmitin [ORG CHEM] $C_3H_5(OOCC_{15}H_{31})_3$ A white, water-insoluble powder that melts at 65.5°C; used in the preparation of leather dressings and soaps. Also known as glyceryl tripalmitate; palmitin.

trip change [MIN ENG] The period during which the loaded cars are taken away and empties are brought back.

trip hammer [MECH ENG] A large power hammer whose head is tripped and falls by cam or lever action.

triphane *See* spodumene.

triphenylmethane dye [ORG CHEM] A family of dyes with a molecular structure derived from $(C_6H_5)_3CH$, usually by NH_2, OH, or HSO_3 substitution for one of the C_6H_5 hydrogens; includes many coal tar dyes, for example, rosaniline and fuchsin.

triphosphopyridine dinucleotide   *See* triphosphopyridine nucleotide.

triphosphopyridine nucleotide   [BIOCHEM] $C_{12}H_{28}N_7O_{17}P_3$ A grayish-white powder, soluble in methanol and in water; a coenzyme and an important component of enzymatic systems concerned with biological oxidation-reduction systems. Abbreviated TPN. Also known as codehydrogenase II; coenzyme II; triphosphopyridine dinucleotide.

triphylite   [MINERAL] $Li(Fe^{2+},Mn^{2+})PO_4$ A grayish-green or bluish-gray mineral crystallizing in the orthorhombic system; it is isomorphous with lithiophilite.

tripinnate   [BIOL] Being bipinnate and having each division pinnate.

trip lamp   [MIN ENG] A removable self-contained mine lamp, designed for marking the rear end of a train (trip) of mine cars.

triple-base propellant   [MATER] A propellant with three principal active ingredients, such as nitrocellulose, nitroglycerin, and nitroguanidine.

triple collision   [PHYS] A process in which three particles collide simultaneously.

triple-conversion receiver   [ELECTR] Communications receiver having three different intermediate frequencies to give higher adjacent-channel selectivity and greater image-frequency suppression.

triple detection   *See* double-superheterodyne reception.

triple ejection rack   [ORD] A device designated to carry up to three bombs or other munitions on a single station of an aircraft.

triple entry   [MIN ENG] A system of opening a mine by driving three parallel entries as main entries.

triple point   [PHYS CHEM] A particular temperature and pressure at which three different phases of one substance can coexist in equilibrium.

triple response   [PHYSIO] The three stages of vasomotor reaction consisting of reddening, flushing of adjacent skin, and development of wheals, when a pointed instrument is drawn heavily across the skin.

triple-stub transformer   [ELECTROMAG] Microwave transformer in which three stubs are placed a quarter-wavelength apart on a coaxial line and adjusted in length to compensate for impedance mismatch.

triple superphosphate   [MATER] A phosphatic fertilizer produced by the reaction of phosphate rock with phosphoric acid so as to give higher concentrations of calcium phosphate than for ordinary superphosphate.

triplet   [NAV] A group of three synchronous transmitting stations operating as a system to provide signals for determination of position. Also known as triad.

triplet state   [ATOM PHYS] Electronic state of an atom or molecule whose total spin angular momentum quantum number is equal to 1. [QUANT MECH] Any multiplet having three states.

triplex chain block   [MECH ENG] A geared hoist using an epicyclic train.

triplexer   [ELECTR] Dual duplexer that permits the use of two receivers simultaneously and independently in a radar system by disconnecting the receivers during the transmitted pulse.

triplex system   [COMMUN] Telegraph system in which two messages in one direction and one message in the other direction can be sent simultaneously over a single circuit.

Triploblastica   [ZOO] Animals that develop from three germ layers.

triploidy   [CYTOL] The occurrence of related forms possessing chromosome numbers three times the haploid number.

trip magnet   *See* phase magnet.

tripod   [DES ENG] An adjustable, collapsible three-legged support, as for a camera or surveying instrument.

tripod drill   [MECH ENG] A reciprocating rock drill mounted on three legs and driven by steam or compressed air; the drill steel is removed and a longer drill inserted about every 2 feet (61 centimeters).

tripoli   [GEOL] A lightweight, porous, friable, siliceous sedimentary rock that may have a white, gray, pink, red, or yellow color; used for polishing metals and stones.

tripolite   *See* diatomaceous earth.

tripotassium orthophosphate   *See* potassium phosphate.

tripper   [CIV ENG] A device activated by a passing train to work a signal or switch or to apply brakes.   [MECH ENG] A device that snubs a conveyor belt causing the load to be discharged.

tripping   [MIN ENG] **1.** The process of pulling or lowering drill-string equipment in a borehole. **2.** To open a latch or locking device, thereby allowing a door or gate to open to empty the contents of a skip or bailer.

tripping device   [ELEC] Mechanical or electromagnetic device used to bring a circuit breaker or starter to its off or open position, either when certain abnormal electrical conditions occur or when a catch is actuated manually.

tripping pulse   *See* trigger pulse.

trip rider   [MIN ENG] A rider who throws switches, gives signals, and makes couplings. Also known as rope rider.

triptane   [ORG CHEM] C_7H_{16} A hydrocarbon compound made commercially in small quantities, but having one of the highest antiknock ratings known. Also known as 2,2,3-trimethylbutane.

Tripylina   [INV ZOO] A subdivision of the protozoan order Oculosida in which the major opening (astropyle) usually contains a perforated plate.

trisaccate pollen   [BOT] A three-pored pollen grain, often having a triangular outline in cross section.

trisaccharide   [BIOCHEM] A carbohydrate which, on hydrolysis, yields three molecules of monosaccharides.

trisectrix   [MATH] The planar curve given by $x^3 + xy^2 + ay^2 - 3ax^2 = 0$ which is symmetric about the x-axis and asymptotic to the line $x = -a$; this is useful in studying the trisection of an angle problem.

trisistor   [ELECTR] Fast-switching semiconductor consisting of an alloyed junction *pnp* device in which the collector is capable of electron injection into the base; characteristics resemble those of a thyratron electron tube, and switching time is in the nanosecond range.

triskaidekaphobia   [PSYCH] Superstitious fear of the number thirteen.

trisodium orthophosphate   *See* trisodium phosphate.

trisodium phosphate   [INORG CHEM] Na_3PO_4 A water-soluble crystalline compound; used as a cleaning compound and as a water softener. Abbreviated TSP. Also known as tertiary sodium phosphate; trisodium orthophosphate.

trisomic syndrome   [MED] Any pathological condition characterized by the presence in triplicate of one of the chromosomes of a complement.

trisomy   [CYTOL] The presence in triplicate of one of the chromosomes of the complement.

trisomy 13–15   *See* D_1 trisomy.

trisomy 18 syndrome   [MED] A congenital disorder due to trisomy of all or a large part of chromosome 18, characterized by severe mental deficiency, hypertonicity with clenched hands, and anomalies of the hands, sternum, pelvis, and facies; most infants so afflicted fail to thrive. Also known as Edwards' syndrome; E trisomy.

trisomy 21 syndrome   *See* Down's syndrome.

tristearin   *See* stearin.

tristimulus colorimeter   [OPTICS] A colorimeter that measures a color stimulus in terms of tristimulus values.

tristimulus values   [OPTICS] The magnitudes of three standard stimuli needed to match a given sample of light.

trisulfide   [CHEM] A binary chemical compound that contains three sulfur atoms in its molecule, for example, iron trisulfide, Fe_2S_3.

tritanopia   [MED] A defect in a third constituent essential for color vision, as in violet blindness.

triterpene   [ORG CHEM] One of a class of compounds having molecular skeletons containing 30 carbon atoms, and theoretically composed of six isoprene units; numerous and widely distributed in nature, occurring principally in plant resins and sap; an example is ambrein.

tri-tet oscillator   [ELECTR] Crystal-controlled, electron-coupled, vacuum-tube oscillator circuit which is isolated from the output circuit through use of the screen grid electrode as the oscillator anode; used for multiband operation because it generates strong harmonics of the crystal frequency.

trithioacetaldehyde   [ORG CHEM] $(C_4H_4S_2)_3$ A colorless, water-insoluble, crystalline compound; used as a hypnotic. Also known as sulfoparaldehyde.

tritium   [NUC PHYS] The hydrogen isotope having mass num-

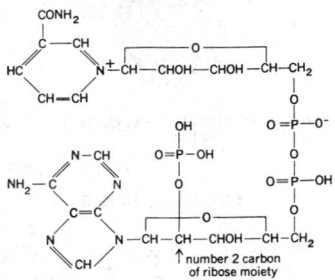

**TRIPHOSPHOPYRIDINE
NUCLEOTIDE**

Structural formula of oxidized form of triphosphopyridine nucleotide (TPN).

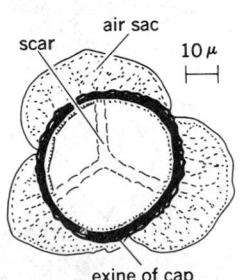

TRISACCATE POLLEN

Trisaccate pollen (young grain of *Podocarpus dacrydioides*) with scar on cap.

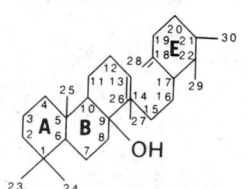

TRITERPENE

Structural formula of ambrein, a triterpene. Numbers represent positions of carbon atoms. A, B, and E identify three carbon rings used in classifying this compound as a tricyclic, tertiary alcohol.

TROCHILISCALES

1 mm

Dextrally spiraled gyrogonite of Trochiliscales.

TROCHOPHORE

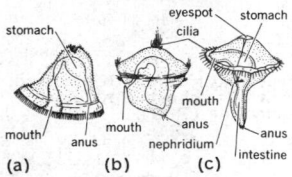

Some trochophore larvae, showing pear-shaped form, circlet of cilia, digestive tract and nephridium. (a) Bryozoan. (b) *Patella*, a mollusk. (c) *Polygardius*, an annelid. (*From T. I. Storer and R. L. Usinger, General Zoology, 3d ed., McGraw-Hill, 1957*)

TROJAN PLANET

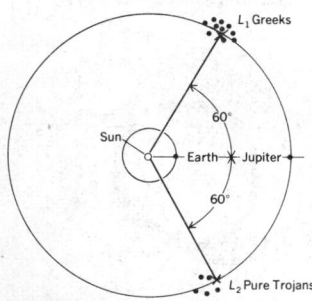

Diagram showing position of Trojan planets. Crosses indicate Lagrangian points L_1 and L_2, near which are clustered Greeks and Pure Trojans respectively.

TROMBIDIFORMES

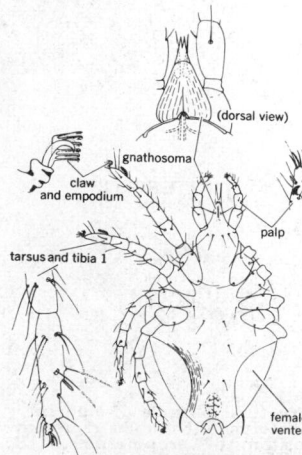

A trombidiform mite, showing characteristic features. (*Acarology Laboratory, Ohio State University*)

ber 3; it is one form of heavy hydrogen, the other being deuterium. Symbolized H³; T.

triton [NUC PHYS] The nucleus of tritium.

Triton [ASTRON] One of the two satellites of the planet Neptune with a diameter of about 4800 kilometers.

tritonal [MATER] An explosive composed of 80% trinitrotoluene (TNT) and 20% powdered aluminum; can be melt-loaded, and is used in bombs for its blast effect.

triton block [MATER] Block of pressed trinitrotoluene (TNT) used for demolition purposes.

tritonymph [INV ZOO] The third stage of development in certain acarids.

Triuridaceae [BOT] A family of monocotyledonous plants in the order Triuridales distinguished by unisexual flowers and several carpels with one seed per carpel.

Triuridales [BOT] A small order of terrestrial, mycotrophic monocots in the subclass Alismatidae without chlorophyll, and with separate carpels, trinucleate pollen, and a well-developed endosperm.

trivial name [ORG CHEM] Unsystematic nomenclature, being the name of a chemical compound derived from the names of the natural source of the compound at the time of its isolation and before anything is known about its molecular structure.

trivium [INV ZOO] The three rays opposite the madreporite in starfish.

t-RNA *See* transfer ribonucleic acid.

Trochacea [PALEON] A recent subfamily of primitive gastropod mollusks in the order Aspidobranchia.

trochal disk [INV ZOO] A flat or funnel-shaped ciliated disk at the anterior end of a rotifer that functions in locomotion and food ingestion.

trochanter [ANAT] A process on the proximal end of the femur in many vertebrates, which serves for muscle attachment and, in birds, for articulation with the ilium. [INV ZOO] The second segment of an insect leg, counting from the base.

Trochidae [INV ZOO] A family of gastropod mollusks in the order Aspidobranchia, including many of the top shells.

Trochili [VERT ZOO] A suborder of the avian order Apodiformes.

Trochilidae [VERT ZOO] The hummingbirds, a tropical New World family of the suborder Trochili with tubular tongues modified for nectar feeding; slender bills and the ability to hover are further feeding adaptations.

Trochiliscales [PALEOBOT] A group of extinct plants belonging to the Charophyta in which the gyrogonites are dextrally spiraled.

trochlea [ANAT] A pulleylike anatomical structure.

trochlear nerve [ANAT] The fourth cranial nerve; either of a pair of somatic motor nerves which innervate the superior oblique muscle of the eyeball.

trochoblast [INV ZOO] A cell bearing cilia on a trochophore.

Trochodendraceae [BOT] A family of dicotyledonous trees in the order Trochodendrales distinguished by the absence of a perianth and stipules, numerous stamens, and pinnately veined leaves.

Trochodendrales [BOT] An order of dicotyledonous trees in the subclass Hamamelidae characterized by primitively vesselless wood and unique, elongate, often branched idioblasts in the leaves.

trochoid [ANAT] *See* pivot joint. [MATH] The path in the plane obtained from a point on the radius of a circle as the circle rolls along a fixed straight line.

trochoidal mass analyzer [PHYS] A mass spectrometer in which the ion beams traverse trochoidal paths within mutually perpendicular electric and magnetic fields.

trochophore [INV ZOO] A generalized but distinct free-swimming larva found in several invertebrate groups, having a pear-shaped form with an external circlet of cilia, apical ciliary tufts, a complete functional digestive tract, and paired nephridia with excretory tubules. Also known as trochosphere.

trochosphere *See* trochophore.

trochotron *See* beam-switching tube.

trochus [INV ZOO] The inner band of cilia on a trochal disk.

troctolite [PETR] A gabbro composed principally of calcic plagioclase and olivine. Also known as forellenstein.

Troglodytidae [VERT ZOO] The wrens, a family of songbirds in the order Passeriformes.

Trogonidae [VERT ZOO] The trogons, the single, pantropical family of the avian order Trogoniformes.

Trogoniformes [VERT ZOO] An order of brightly colored, slow-moving birds characterized by a unique foot structure with the first and second toes directed backward.

troilite [MINERAL] FeS A meteorite mineral crystallizing in the hexagonal system; a variety of pyrrhotite.

Trojan planet [ASTRON] One of a group of asteroids whose periods of revolution are about equal to that of Jupiter, or about 12 years; these bodies move close to one or the other of two positions called Lagrangian points, 60° ahead of or 60° behind Jupiter; the asteroids near these positions are known as Greeks and Pure Trojans respectively.

troland [OPTICS] A unit of retinal illuminance, equal to the retinal illuminance produced by a surface whose luminance is one nit when the apparent area of the entrance pupil of the eye is 1 square millimeter. Also known as luxon; photon.

Troland and Fletcher theories [PHYSIO] Theories of hearing according to which the time nature of a sound stimulation affects the sensation of pitch.

trolley [GEOL] A basin-shaped depression in strata. Also known as lum. [MECH ENG] 1. A wheeled car running on an overhead track, rail, or ropeway. 2. An electric streetcar.

trolley locomotive [MECH ENG] A locomotive operated by electricity drawn from overhead conductors by means of a trolley pole.

trolley pole [ELEC] The pole which conducts electricity from the trolley wire to the trolley.

trolley wire [ELEC] The means by which power is conveyed to an electric trolley locomotive; it is an overhead wire which conducts power to the locomotive by the trolley pole.

Trombiculidae [INV ZOO] The chiggers, or red bugs, a family of mites in the suborder Trombidiformes whose larvae are parasites of vertebrates.

Trombidiformes [INV ZOO] The trombidiform mites, a suborder of the Acarina distinguished by the presence of a respiratory system opening at or near the base of the chelicerae.

trombone [ELECTROMAG] U-shaped, adjustable, coaxial-line matching assembly.

trommel [MIN ENG] 1. A revolving cylindrical screen used to grade coarsely crushed ore: the ore is fed into the trommel at one end, the fine material drops through the holes, and the coarse is delivered at the other end. Also known as trommel screen. 2. To separate coal into various sizes by passing it through a revolving screen.

trommel screen *See* trommel.

trona [MINERAL] $Na_2(CO_3) \cdot Na(HCO_3) \cdot 2H_2O$ A gray-white or yellowish-white mineral that crystallizes in the monoclinic system and occurs in fibrous or columnar layers or masses. Also known as urao.

Trona process [CHEM ENG] A process for separation and purification of a mixture of soda ash, anhydrous sodium sulfate, bromine, and potassium chloride from brine (Searles Lake, California).

troop ship *See* transport.

Tropaeolaceae [BOT] A family of dicotyledonous plants in the order Geraniales characterized by strongly irregular flowers, simple peltate leaves, eight stamens, and schizocarpous fruit.

tropeoline 00 [ORG CHEM] $NaSO_3C_6H_4NNC_6H_4NHC_6H_5$ An acid-base indicator with a pH range of 1.4–3.0, color change (from acid to base) red to yellow; used as a biological stain. Also known as sodium *para*-diphenylaminoazobenzene sulfonate.

Tropept [GEOL] A suborder of the order Inceptisol, characterized by moderately dark A horizons with modest additions of organic matter, B horizons with brown or reddish colors, and slightly pale C horizons; restricted to tropical regions with moderate or high rainfall.

trophallaxis [ECOL] Exchange of food between organisms, not only of the same species but between different species, especially among social insects.

trophic [BIOL] Pertaining to or functioning in nutrition.

trophic level [ECOL] Any of the feeding levels through which the passage of energy through an ecosystem proceeds; examples are photosynthetic plants, herbivorous animals, and microorganisms of decay.

trophobiosis [ECOL] A nutritional relationship associated only with certain species of ants in which alien insects supply food to the ants and are milked by the ants for their secretions.

trophoblast [EMBRYO] A layer of ectodermal epithelium covering the outer surface of the chorion and chorionic villi of many mammals.

trophocyte [INV ZOO] A nutritive cell of the ovary or testis of an insect.

trophogenic [ECOL] Originating from nutritional differences rather than resulting from genetic determinants, such as various castes of social insects.

tropholytic [ECOL] Pertaining to the deep zone in a lake where dissimilation of organic matter predominates.

trophosome [INV ZOO] The nutritional zooids of a hydroid colony.

trophotaeniae [VERT ZOO] Vascular rectal processes which establish placental relationships with the ovarian tissue in live-bearing fishes.

trophozoite [INV ZOO] A vegetative protozoan; used especially of a parasite.

trophus [INV ZOO] Masticatory apparatus in Rotifera.

tropical air [METEOROL] A type of air whose characteristics are developed over low latitudes.

tropical climate [CLIMATOL] A climate which is typical of equatorial and tropical regions, that is, one with continually high temperatures and with considerable precipitation, at least during part of the year.

tropical cyclone [METEOROL] The general term for a cyclone that originates over tropical oceans; at maturity, the tropical cyclone is one of the most intense storms of the world; winds exceeding 175 knots (324 kilometers per hour) have been measured, and the rain is torrential.

tropical disturbance [METEOROL] A cyclonic wind system of the tropics, of lesser intensity than a tropical cyclone.

tropical easterlies [METEOROL] The trade winds when shallow and exhibiting a strong vertical shear; at about 500 feet (152 meters) the easterlies give way to the upper westerlies, which are sufficiently strong and deep to govern the course of cloudiness and weather. Also known as subtropical easterlies.

tropical finish [ENG] A finish that is applied to electronic equipment to resist the high relative humidity, fungus, and insects encountered in tropical climates.

tropical front *See* intertropical front.

tropicalize [ENG] To prepare electronic equipment for use in a tropical climate by applying a coating that resists moisture and fungi.

tropical life zone [ECOL] A subdivision of the eastern division of Merriam's life zones; an example is southern Florida, where the vegetation is the broadleaf evergreen forest; typical and important plants are palms and mangroves; typical and important animals are the armadillo and alligator; typical and important crops are citrus fruits, avocado, and banana.

tropical meteorology [METEOROL] The study of the tropical atmosphere; the dividing lines, in each hemisphere, between the tropical easterlies and the mid-latitude westerlies in the middle troposphere roughly define the poleward boundaries of this region.

tropical monsoon climate [CLIMATOL] One of the tropical rainy climates; it is sufficiently warm and rainy to produce tropical rainforest vegetation, but it does exhibit the monsoon climate influences in that it has a winter dry season.

tropical month [ASTRON] The average period of the revolution of the moon about the earth with respect to the vernal equinox, a period of 27 days 7 hours 43 minutes 4.7 seconds, or approximately $27\frac{1}{3}$ days.

tropical rainforest [ECOL] A vegetation class consisting of tall, close-growing trees, their columnar trunks more or less unbranched in the lower two-thirds, and forming a spreading and frequently flat crown; occurs in areas of high temperature and high rainfall. Also known as hylaea; selva.

tropical rainforest climate [CLIMATOL] In general, the climate which produces tropical rainforest vegetation, that is, a climate of unbroken warmth, high humidity, and heavy annual precipitation. Also known as tropical wet climate.

tropical rainy climate [CLIMATOL] A major category in W. Köppen's climatic classification, characterized by a mean temperature of the coldest month of 64.4°F (18°C) or higher,

and by a mean annual precipitation, in inches, greater than 0.44 ($t - a$), where t is the mean annual temperature in degrees Fahrenheit, and a equals 32 for precipitation chiefly in winter, 19.4 for evenly distributed precipitation, and 6.8 for precipitation chiefly in summer.

tropical savanna *See* tropical woodland.

tropical savanna climate [CLIMATOL] In general, the type of climate which produces the vegetation of the tropical and subtropical savanna; thus, a climate with a winter dry season, a relatively short but heavy rainy summer season, and high year-round temperatures. Also known as savanna climate; tropical wet and dry climate.

tropical scrub [ECOL] A class of vegetation composed of low woody plants (shrubs), sometimes growing quite close together, but more often separated by large patches of bare ground, with clumps of herbs scattered throughout; an example is the Ghanaian evergreen coastal thicket. Also known as bush; brush; fourré; mallee; thicket.

tropical wet and dry climate *See* tropical savanna climate.

tropical wet climate *See* tropical rainforest climate.

tropical woodland [ECOL] A vegetation class similar to a forest but with wider spacing between trees and sparse lower strata characterized by evergreen shrubs and seasonal graminoids; the climate is warm and moist. Also known as parkland; savanna-woodland; tropical savanna.

tropical year [ASTRON] A unit of time equal to the period of one revolution of the earth about the sun measured between successive vernal equinoxes; it is 365.2422 mean solar days or 365 days 5 hours 48 minutes 46 seconds.

tropic higher-high-water interval [OCEANOGR] The lunitidal interval pertaining to the higher high waters at the time of tropic tides.

tropic high-water inequality [OCEANOGR] The average difference between the heights of the two high waters of the tidal day at the time of tropic tides.

tropic lower-low-water interval [OCEANOGR] The lunitidal interval pertaining to the lower low waters at the time of tropic tides.

tropic low-water inequality [OCEANOGR] The average difference between the heights of the two low waters of the tidal day at the time of tropic tides.

Tropic of Cancer [ASTRON] A parallel circle on the earth, latitude 23°45′ north of the Equator.

Tropic of Capricorn [ASTRON] A parallel circle on the earth, latitude 23°45′ south of the Equator.

tropics [CLIMATOL] Any portion of the earth characterized by a tropical climate.

tropic tidal currents [OCEANOGR] Tidal currents of increased diurnal inequality occurring at the time of tropic tides.

tropic tide [OCEANOGR] A tide occurring when the moon is near maximum declination; the diurnal inequality is then at a maximum.

tropic velocity [OCEANOGR] The speed of the greater flood or greater ebb at the time of tropic currents.

Tropiometridae [INV ZOO] A family of feather stars in the class Crinoidea which are bottom crawlers.

tropism [BIOL] Orientation movement of a sessile organism in response to a stimulus. Also known as topotaxis.

tropocollagen [BIOCHEM] The fundamental units of collagen fibrils.

tropomyosin [BIOCHEM] A muscle protein similar to myosin and implicated as being part of the structure of the Z bands separating sarcomeres from each other.

troponin [BIOCHEM] A protein species located at specific stations every 36.5 nanometers on the actin helix in muscle sarcomere.

tropopause [METEOROL] The boundary between the troposphere and stratosphere, usually characterized by an abrupt change of lapse rate; the change is in the direction of increased atmospheric stability from regions below to regions above the tropopause; its height varies from 15 to 20 kilometers in the tropics to about 10 kilometers in polar regions.

tropopause chart [METEOROL] A synoptic chart showing the contour lines of the tropopause and tropopause break lines.

tropopause inversion [METEOROL] The decrease in the lapse

TROPOMYOSIN

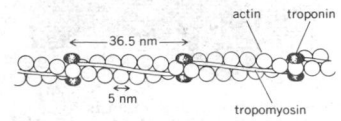

Thin filament of muscle tissue, made up primarily of protein actin, showing position of tropomyosin, and also of troponin. *(From S. Ebashi, M. Endo, and I. Ohtsuki, Control of muscle contraction, Quart. Rev. Biophys., 2:351–384, 1969)*

rate of temperature encountered at the level of the tropopause. Also known as upper inversion.

tropophytia [BOT] Plants that thrive in a climate that undergoes marked periodic changes.

troposcatter *See* tropospheric scatter.

troposphere [METEOROL] That portion of the atmosphere from the earth's surface to the tropopause, that is, the lowest 10 to 20 kilometers of the atmosphere.

tropospheric duct *See* duct.

tropospheric ducting *See* ducting.

tropospheric scatter [COMMUN] Scatter propagation of radio waves caused by irregularities in the refractive index of air in the troposphere; used for long-distance communications, with the aid of relay facilities, 300–500 kilometers apart. Also known as troposcatter.

tropospheric superrefraction [GEOPHYS] Phenomenon occurring in the troposphere whereby radio waves are bent sufficiently to be returned to the earth.

tropospheric wave [COMMUN] A radio wave that is propagated by reflection from a region of abrupt change in dielectric constant or its gradient in the troposphere.

trouble-location problem [ADP] In computers, a test problem used in a diagnostic routine.

troubleshoot [ADP] To find and correct errors and faults in a computer, usually in the hardware.

trough [GEOL] **1.** A small, straight depression formed just offshore on the bottom of a sea or lake and on the landward side of a longshore bar. **2.** Any narrow, elongate depression in the surface of the earth. **3.** An elongate depression on the sea floor that is wider and shallower than a trench. Also known as submarine trench. **4.** The line connecting the lowest points of a fold. [METEOROL] An elongated area of relatively low atmospheric pressure; the opposite of a ridge.

trough aloft *See* upper-level trough.

trough crossbedding [GEOL] A variety of crossbedding in which the lower crossbedding surfaces are smoothly curved, rather than planar.

troughed belt conveyor [MECH ENG] A belt conveyor with the conveyor belt edges elevated on the carrying run to form a trough by conforming to the shape of the troughed carrying idlers or other supporting surface.

troughed roller conveyor [MECH ENG] A roller conveyor having two rows of rolls set at an angle to form a trough over which objects are conveyed.

trough fault [GEOL] One of a set of two faults bounding a graben.

troughing idler [MECH ENG] A belt idler having two or more rolls arranged to turn up the edges of the belt so as to form the belt into a trough.

troughing rolls [MECH ENG] The rolls of a troughing idler that are so mounted on an incline as to elevate each edge of the belt into a trough.

trough plane *See* trough surface.

trough surface [GEOL] A surface or plane connecting the troughs of the bed of a syncline. Also known as synclinal axis; trough plane.

trough valley *See* U-shaped valley.

trough washer [MIN ENG] A sloping wooden trough, 1½ to 2 feet wide, 8 to 12 feet long, and 1 foot deep (about 50 by 300 by 30 centimeters), open at the tail end but closed at the head end; it is used to float adhering clay or fine stuff from the coarser portions of an ore or coal.

trout [VERT ZOO] Any of various edible fresh-water fishes in the order Salmoniformes that are generally much smaller than the salmon.

Trouton's rule [PHYS CHEM] An approximation rule for the derivation of molar heats of vaporization of normal liquids at their boiling points. [THERMO] The rule that, for a nonassociated liquid, the latent heat of vaporization in calories is equal to approximately 22 times the normal boiling point on the Kelvin scale.

trowel [DES ENG] Any of various hand tools consisting of a wide, flat or curved blade with a short wooden handle; used by gardeners, plasterers, and bricklayers.

troy ounce *See* ounce.

troy pound *See* pound.

troy system [MECH] A system of mass units used primarily to measure gold and silver; the ounce is the same as that in the apothecaries' system, being equal to 480 grains or 31.1034768 grams. Abbreviated t. Also known as troy weight.

troy weight *See* troy system.

TR switch *See* transmit-receive tube.

TR tube *See* transmit-receive tube.

Trube's correlation [PETRO ENG] An empirical correlation (based on pseudocritical properties) for compressibilities of undersaturated oil-reservoir fluids.

Trucherognathidae [PALEON] A family of conodonts in the order Conodontophorida in which the attachment scar permits the conodont to rest on the jaw ramus.

truck [MECH ENG] A self-propelled wheeled vehicle, designed primarily to transport goods and heavy equipment; it may be used to tow trailers or other mobile equipment. [MIN ENG] *See* barney.

truck crane [MECH ENG] A crane carried on the bed of a motortruck.

truck-mounted drill rig [MECH ENG] A drilling rig mounted on a lorry or caterpillar tracks.

truck-tractor *See* tractor.

true [GEOD] Related to true north. [SCI TECH] **1.** Actual, as contrasted with fictitious, such as true sun. **2.** Related to a fixed point, either on the earth or in space, in contrast with relative, which is related to a moving point.

true airspeed [AERO ENG] The actual speed of an aircraft relative to the air through which it flies, that is, the calibrated airspeed corrected for temperature, density, or compressibility.

true-airspeed indicator [AERO ENG] An instrument for measuring true airspeed. Also known as true-airspeed meter.

true-airspeed meter *See* true-airspeed indicator.

true air temperature [METEOROL] Basic air temperature corrected for heat of compression error due to high-speed motion of the thermometer through the air, as on an aircraft.

true altitude *See* corrected altitude.

true amplitude [NAV] In celestial navigation, amplitude relative to true east or true west.

true anomaly *See* anomaly.

true azimuth [NAV] Azimuth relative to true north.

true bearing [NAV] Bearing relative to true north; compass bearing corrected for magnetic deviation.

true-boiling-point analysis [CHEM ENG] A standard laboratory technique used to predict the refining qualities of crude petroleum; gives distillation cuts for gasoline, kerosine, distillate (diesel) fuel, cracking, and lube distillate stocks. Also known as true-boiling-point distillation.

true-boiling-point distillation *See* true-boiling-point analysis.

true complement *See* radix complement.

true condensing point *See* critical condensation temperature.

true convergence [GEOD] The angle at which one meridian is inclined to another on the surface of the earth.

true course [NAV] Course relative to true north.

true dip *See* dip.

true direction [NAV] Horizontal direction expressed as angular distance from true north.

true formation resistivity [GEOPHYS] Electrical resistivity of a clean (nonshaly) porous reservoir formation containing hydrocarbons and formation water; value is greater than the resistivity when there is added water incursion.

true freezing point [PHYS CHEM] The temperature at which the liquid and solid forms of a substance exist in equilibrium at a given pressure (usually 1 standard atmosphere).

true heading [NAV] Heading relative to true north.

true homing [NAV] The process of following a course such that the true bearing of the craft from the objective is maintained constant.

true horizon [OPTICS] The boundary of a horizontal plane passing through a point of vision, or in photogrammetry, the perspective center of a lens system.

true mean temperature [METEOROL] As adopted by the International Meteorological Organization, a monthly or annual mean air temperature based upon hourly observations at a given place, or on some combination of less frequent

TROPOSPHERIC SCATTER

Tropospheric scatter relay facility operated by the U.S. Air Force in Spain. High-powered ultra-high-frequency transmitters, sensitive receivers, and large parabolic antennas operating with frequency and space diversity make it possible to transmit many voice and teletype circuits simultaneously far beyond the horizon. *(Page Engineers, Inc.)*

TRUCHEROGNATHIDAE

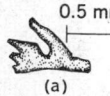
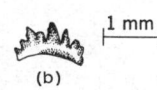

0.5 mm

1 mm

(a)   (b)

Typical examples of Trucherognathidae; (a) *Polycaulodus*, and (b) *Curtognathus*. (After illustration in R. R. Shrock and W. H. Twenhofel, *Principles of Invertebrate Paleontology*, McGraw-Hill, 2d ed., 1953)

TRUCK

Conventional truck chassis before installation of special body. *(White Motor Co.)*

observations designed to represent this mean as nearly as possible.

true-motion radar [ELECTR] A radar set which provides a true-motion radar presentation on the plan-position indicator, as opposed to the relative-motion, true-or-relative-bearing, presentation most commonly used.

true-motion radar presentation [ELECTR] A radar plan-position indicator presentation in which the center of the scope represents the same geographic position, until reset, with all moving objects, including the user's own craft, moving on the scope.

true motor load *See* thermal horsepower.

true north [NAV] **1.** The direction of the north geographical pole. **2.** The reference direction for measurement of true directions.

true porcelain *See* porcelain.

true rake [MECH ENG] The angle, measured in degrees, between a plane containing a tooth face and the axial plane through the tooth point in the direction of chip flow.

true soil *See* solum.

true solar day *See* apparent solar day.

true solar time *See* apparent solar time.

true sun [ASTRON] The actual sun as it appears in the sky.

true width [MIN ENG] The width of thickness of a vein or stratum as measured perpendicular to or normal to the dip and the strike; the true width is always the least width.

true wind [METEOROL] Wind relative to a fixed point on the earth.

true wind direction [METEOROL] The direction, with respect to true north, from which the wind is blowing.

truffle [BOT] The edible underground fruiting body of various European fungi in the family Tuberaceae, especially the genus *Tuber*.

truing [MECH ENG] Cutting a grinding wheel to make its surface run concentric with the axis.

trumpet buoy [NAV] A buoy provided with a trumpet having a distinctive tone.

trumpeter [VERT ZOO] A bird belonging to the Psophiidae, a family with three South American species; the common trumpeter (*Psophia crepitans*) is the size of a pheasant and resembles a long-legged guinea fowl.

truncate [BIOL] Abbreviated at an end, as if cut off. [MATH] **1.** To drop digits at the end of a numerical value; the number 3.14159265 is truncated to five figures in 3.1415, whereas it would be 3.1416 if rounded off to five figures. **2.** To approximate the sum of an infinite series by the sum of a finite number of its terms. **3.** To terminate an infinite sequence of successively better approximations of a quantity after a finite number of such approximations.

truncated paraboloid [ELECTROMAG] Paraboloid antenna in which a portion of the top and bottom have been cut away to broaden the main lobe in the vertical plane.

truncation [MATH] **1.** Approximating the sum of an infinite series by the sum of a finite number of its terms. **2.** *See* rounding.

truncation error [ENG] The error resulting from the analysis of a partial set of data in place of a complete or infinite set. [MATH] **1.** The computation error resulting from use of only a finite number of terms of an infinite series. **2.** The error resulting from the approximation of a derivative or differential by a finite difference.

truncus arteriosis [EMBRYO] The embryonic arterial trunk between the bulbous arteriosis and the ventral aorta in anamniotes and early stages of amniotes.

trunk [ANAT] The main mass of the human body, exclusive of the head, neck, and extremities; it is divided into thorax, abdomen, and pelvis. [BOT] The main stem of a tree. [COMMUN] **1.** A path over which information is transferred in a computer. **2.** A telephone line connecting two central offices. Also known as trunk circuit.

trunk buoy [ENG] A mooring buoy having a pendant extending through an opening in the buoy, with the ship's anchor chain or mooring line being secured to this pendant.

trunk circuit *See* trunk.

trunk exchange [COMMUN] A telephone exchange whose main function is to interconnect trunks.

trunk group [COMMUN] The collection of trunks of a given type or characteristic that connect two switching points.

trunk height [ANTHRO] A vertical measurement (taken in front) of the distance from the table top to the upper edge of the sternum, when the individual is seated as in a sitting-height measurement.

trunk roadway [MIN ENG] The main developing heading from the pit bottom and is usually driven along the strike of the coal seam.

trunk sewer [CIV ENG] A sewer receiving sewage from many tributaries serving a large territory.

trunk stream *See* main stream.

trunnion [DES ENG] **1.** Either of two opposite pivots, journals, or gudgeons, usually cylindrical and horizontal, projecting one from each side of a piece of ordnance, the cylinder of an oscillating engine, a molding flask, or a converter, and supported by bearings to provide a means of swiveling or turning. **2.** A pin or pivot usually mounted on bearings for rotating or tilting something.

truss [CIV ENG] A frame, generally of steel, timber, concrete, or a light alloy, built from members in tension and compression.

trussed beam [CIV ENG] A beam stiffened by a steel tie rod to reduce its deflection.

trussed rafter [BUILD] A triangulated beam in a trussed roof.

truth table [MATH] A table listing statements concerning an event and their respective truth values.

Tryblidiidae [PALEON] An extinct family of Paleozoic mollusks.

trypan blue [MATER] An acid disazo dye of the benzopurpurine series used as a vital stain.

Trypanorhyncha [INV ZOO] An order of tapeworms of the subclass Cestoda; all are parasites in the intestine of elasmobranch fishes.

Trypanosoma [INV ZOO] A genus of slender, polymorphic, elongate protozoans belonging to the Mastigophora, characterized by a central nucleus, posterior blepharoplast, and an undulating membrane; responsible for such infections as trypanosomiasis and Chagas' disease in man.

Trypanosomatidae [INV ZOO] A family of Protozoa, order Kinetoplastida, containing flagellated parasites which exhibit polymorphism during their life cycle.

trypanosome [INV ZOO] A flagellated protozoan of the genus *Trypanosoma*.

trypanosomiasis [MED] Any of many diseases of man and animals caused by infection with species of *Trypanosoma* and transmitted by tsetse flies and other insects.

trypsin [BIOCHEM] A proteolytic enzyme which catalyzes the hydrolysis of peptide linkages in proteins and partially hydrolyzed proteins; derived from trypsinogen by the action of enterokinase in intestinal juice.

trypsinogen [BIOCHEM] The zymogen of trypsin, secreted in the pancreatic juice. Also known as protrypsin.

tryptophan [BIOCHEM] $C_{11}H_{12}O_2N_2$ An amino acid obtained from casein, fibrin, and certain other proteins; it is a precursor of indoleacetic acid, serotonin, and nicotinic acid.

tschermakite [MINERAL] $Ca_2Mg_3(Al,Fe^{3+})_2(Al_2Si_6)O_{22}(OH,F)_2$ An amphibole mineral.

Tschudi engine [MECH ENG] A cat-and-mouse engine in which the pistons, which are sections of a torus, travel around a toroidal cylinder; motion of the pistons is controlled by two cams which bear against rollers attached to the rotors.

T-S curve *See* temperature-salinity diagram.

T-S diagram *See* temperature-salinity diagram.

T-section filter [ELEC] T network used as an electric filter.

tsetse fly [INV ZOO] Any of various South African muscoid flies of the genus *Glossina*; medically important as vectors of sleeping sickness or trypanosomiasis.

TSH *See* thyrotropic hormone.

tsi [MECH] A unit of force equal to 1 ton-force per square inch; equal to approximately 1.54444×10^7 pascals.

tsp *See* teaspoonful.

TSP *See* trisodium phosphate.

T_1 space [MATH] A topological space with the property that for any two distinct points x and y there is an open set containing one but not the other.

tspn *See* teaspoonful.

TSPP *See* sodium pyrophosphate.

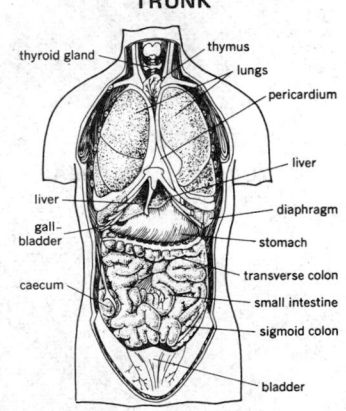

TRUNK

Human trunk, showing principal organs. (*From Franz Frohse et al., Atlas of Human Anatomy, rev. ed., Barnes and Noble, 1957*)

T square [GRAPHICS] A straightedge rule with a crosspiece at one end by which parallel lines are drawn perpendicular to the edge of the drawing board.

T-S relation *See* temperature-salinity diagram.

tsunami [OCEANOGR] A long-period sea wave produced by a seaquake or volcanic eruption; it may travel for thousands of miles. Also known as seismic sea wave.

Tsushima Current [OCEANOGR] That part of the Kuroshio Current flowing northeastward through the Korea Strait and along the Japanese coast in the Sea of Japan.

tsutsugamushi disease [MED] A rickettsial disease of man caused by *Rickettsia tsutsugamushi*, transmitted by larval mites, and characterized by headache, high fever, and a rash. Also known as scrub typhus.

T Tauri star [ASTRON] A star classification; an extrinsic variable star whose variation in luminosity may be stimulated by the associated nebulosity.

t-test [STAT] A statistical test involving means of normal populations with unknown standard deviations; small samples are used, based on a variable t equal to the difference between the mean of the sample and the mean of the population divided by a result obtained by dividing the standard deviation of the sample by the square root of the number of individuals in the sample.

tuba [METEOROL] A cloud column or inverted cloud cone, pendant from a cloud base; this supplementary feature occurs mostly with cumulus and cumulonimbus; when it reaches the earth's surface it constitutes the cloudy manifestation of an intense vortex, namely, a tornado or waterspout. Also known as pendant cloud; tornado cloud.

tubal bladder [VERT ZOO] A urine reservoir organ that is an enlargement of the mesonephric ducts in most fish; there are four types: duplex, bilobed, simplex with ureters tied, and simplex with separate ureters.

tubal ligation [MED] Surgical tying of the uterine tubes to prevent conception.

tubbing [MIN ENG] The watertight cast-iron lining of a circular shaft built up of segments with the space outside the tubbing grouted to add strength and to improve watertightness.

tube [BIOL] A narrow channel within the body of an animal or plant. [ELECTR] *See* electron tube. [ENG] A long cylindrical body with a hollow center used especially to convey fluid. [GEOL] A passage in a cave having smooth sides and an elliptical to nearly circular cross section. [ORD] The main part of a gun, the cylindrical piece of metal surrounding the bore; tube is frequently used in referring to artillery weapons, and barrel is more frequently used in referring to small arms.

tube coefficient [ELECTR] Any of the constants that describe the characteristics of a thermionic vacuum tube, such as amplification factor, mutual conductance, or alternating-current plate resistance.

tube core [AERO ENG] One type of sandwich configuration used in structural materials in aircraft; aluminum, steel, and titanium have been used for face materials with cores of wood, rubber, plastics, steel, and aluminum in the form of tubes.

tube foot [INV ZOO] One of the tentaclelike outpushings of the radial vessels of the water-vascular system in echinoderms; may be suctorial, or serve as stiltlike limbs or tentacles.

tube heating time [ELECTR] Time required for a tube to attain operating temperature.

tube mill [MECH ENG] A revolving cylinder used for fine pulverization of ore, rock, and other such materials; the material, mixed with water, is fed into the chamber from one end, and passes out the other end as slime.

tube noise [ELECTR] Noise originating in a vacuum tube, such as that due to shot effect and thermal agitation.

tube of force [ELEC] A region of space bounded by a tubular surface consisting of the lines of force which pass through a given closed curve.

tuber [BOT] The enlarged end of a rhizome in which food accumulates, as in the potato.

tuber cinereum [ANAT] An area of gray matter extending from the optic chiasma to the mammillary bodies and forming part of the floor of the third ventricle.

tubercle [BIOL] A small knoblike prominence. [MET] A mound of corrosive products on the surface of a metal that is subjected to local corrosive attack.

tubercle bacillus *See* Mycobacterium tuberculosis.

Tuberculariaceae [MYCOL] A family of fungi of the order Moniliales having short conidia that form cushion-shaped, often waxy or gelatinous aggregates (sporodochia).

tuberculate [BIOL] Having or characterized by knoblike processes.

tuberculation [MET] Corrosive attack with formation of tubercles.

tuberculin [IMMUNOL] A preparation containing tuberculoproteins derived from *Mycobacterium tuberculosis* used in the tuberculin test to determine sensitization to tubercle bacilli.

tuberculin test [IMMUNOL] A test for past or present infection with tubercle bacilli based on a delayed hypersensitivity reaction at the site where tuberculin or purified protein derivative was introduced.

tuberculosis [MED] A chronic infectious disease of humans and animals primarily involving the lungs caused by the tubercle bacillus, *Mycobacterium tuberculosis,* or by *M. bovis.* Also known as consumption; phthisis.

tube reducing [MET] Reducing the diameter and wall thickness of tubing by means of a mandrel and rolls.

tuberosity [ANAT] A large or obtuse prominence, especially as on bone for muscle attachment.

tuberous sclerosis [MED] A familial neurocutaneous syndrome characterized in its complete form by epilepsy, adenoma sebaceum, and mental deficiency, and pathologically by nodular sclerosis of the cerebral cortex. Also known as Bourneville's disease.

tube sheet [ENG] A mounting plate for elements of a larger item of equipment; for example, filter cartridges, or tubes for heat exchangers, coolers, or boilers.

tube shield [ENG] A shield designed to be placed around an electron tube.

tube socket [ENG] A socket designed to accommodate electrically and mechanically the terminals of an electron tube.

tube-still heater [CHEM ENG] A firebox containing a pipe coil through which oil for a tube still (pipe still) is pumped.

tube stock [MET] Semifinished metal tubing.

tube tester [ELECTR] A test instrument designed to measure and indicate the condition of electron tubes used in electronic equipment.

tube voltage drop [ELECTR] In a gas tube, the anode voltage during the conducting period.

tube voltmeter *See* vacuum-tube voltmeter.

Tubicola [INV ZOO] An order of sedentary polychaete annelids that surround themselves with a calcareous tube or one which is composed of agglutinated foreign particles.

tubing [ENG] Material in the form of a tube, most often seamless.

tubing hanger *See* hanger.

tubing head [PETRO ENG] A spool-type unit or housing attached to the top flange on the uppermost oil-well-casing head to support the tubing string and to seal the annulus between the tubing string and the production casing head.

tubing-head adapter flange [PETRO ENG] An intermediate flange used in oil wells to connect the top tubing-head flange to the master valve (Christmas tree) and to provide support for the tubing.

tubing pump [PETRO ENG] A type of oil-well, sucker-rod pump in which the pump barrel is attached to the tubing string, and lowered into the well bore with the tubing.

tubing safety valve *See* storm choke.

Tubulanidae [INV ZOO] A family of the order Palaeonemertini.

tubular capacitor [ELEC] A paper or electrolytic capacitor having the form of a cylinder, with leads usually projecting axially from the ends; the capacitor plates are long strips of metal foil separated by insulating strips, rolled into a compact tubular shape.

tubular gland [ANAT] A secreting structure whose secretory endpieces are tubelike or cylindrical in shape.

tubule [ANAT] A slender, elongated microscopic tube in an anatomical structure.

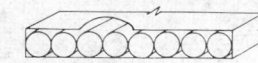

TUBE CORE

Tube core type of sandwich construction in a fuselage.

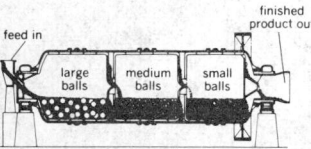

TUBE MILL

Three-compartment tube mill pulverizer, containing three sizes of balls. (*Hardinge Co., Inc.*)

TUBER

Tuber of a potato.

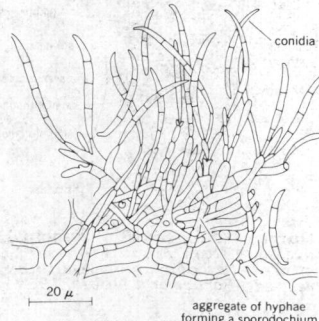

TUBERCULARIACEAE

Sporodochium of *Fusarium lini,* a representative species of Tuberculariaceae, showing sickle-shaped, multicelled conidia. (*After H. L. Bolley, 1901*)

Tubulidentata [VERT ZOO] An order of mammals which contains a single living genus, the aardvark (*Orycteropus*) of Africa.

tubuloacinous gland *See* tubuloalveolar gland.

tubuloalveolar gland [ANAT] A secreting structure having both tubular and alveolar secretory endpieces. Also known as acinotubular gland; tubuloacinous gland.

Tucana [ASTRON] A constellation in the southern hemisphere; right ascension 23 hours, declination 60° south. Also known as Toucan.

tufa [GEOL] A spongy, porous limestone formed by precipitation from evaporating spring and river waters, often onto leaves and stems of neighboring plants. Also known as calcareous sinter; calcareous tufa.

tufaceous [GEOL] Pertaining to or similar to tufa.

tuff [GEOL] Consolidated volcanic ash, composed largely of fragments (less than 4 millimeters) produced directly by volcanic eruption; much of the fragmented material represents finely comminuted crystals and rocks.

tuff ball *See* mud ball.

tuff lava *See* welded tuff.

tuft *See* mound.

tugboat [NAV ARCH] A powerful, strongly built boat with shaped hull and bow, designed to tow or push other vessels or barges in harbors, on inland waterways, and at sea.

tugger [MIN ENG] A small portable pneumatic or electric hoist mounted on a column and used in a mine.

Tukon tester [ENG] A device that uses a diamond (Knoop) indenter applying average loads of 1 to 2000 grams to determine microhardness of a metal.

tularemia [VET MED] A bacterial infection of wild rodents caused by *Pasteurella tularensis*; it may be generalized, or it may be localized in the eyes, skin, or lymph nodes, or in the respiratory tract or gastrointestinal tract; may be transmitted to humans and to some domesticated animals.

tulip [BOT] Any of various plants with showy flowers constituting the genus *Tulipa* in the family Liliaceae; characterized by coated bulbs, lanceolate leaves, and a single flower with six equal perianth segments and six stamens.

tulip poplar *See* tulip tree.

tulip tree [BOT] *Liriodendron tulipifera*. A tree belonging to the magnolia family (Magnoliaceae) distinguished by leaves which are squarish at the tip, true terminal buds, cone-shaped fruit, and large greenish-yellow and orange-colored flowers. Also known as tulip poplar.

tumble *See* topple.

tumble axis *See* topple axis.

tumble home [NAV ARCH] The curve of a boat or ship's upper side toward the centerline, causing the sides to be convex.

tumble-plating process [MET] A method of zinc-coating small metal parts by first applying zinc powder with an adhesive, then tumbling with glass beads to roll out the powder into a continuous coat.

tumbler [ENG] **1.** A device in a lock cylinder that must be moved to a particular position, as by a key, before the bolt can be thrown. **2.** A device or mechanism in which objects are tumbled.

tumbler feeder *See* drum feeder.

tumbleweed [BOT] Any of various plants that break loose from their roots in autumn and are driven by the wind in rolling masses over the ground.

tumbling [AERO ENG] An attitude situation in which the vehicle continues on its flight, but turns end over end about its center of mass. [ENG] A surface-finishing operation for small articles in which irregularities are removed or surfaces are polished by tumbling them together in a barrel, along with wooden pegs, sawdust, and polishing compounds. [MECH ENG] Loss of control in a two-frame free gyroscope, occurring when both frames of reference become coplanar.

tumbling mill [MECH ENG] A grinding and pulverizing machine consisting of a shell or drum rotating on a horizontal axis.

tumid [BIOL] Marked by swelling or inflation.

tumor [MED] Any abnormal mass of cells resulting from excessive cellular multiplication.

tumorigenic [MED] Tumor-forming.

tumuli lava [GEOL] A type of lava flow forming ovoid mounds, a few feet high and a few tens of feet long, caused by buckling up of the crust.

tuna [VERT ZOO] Any of the large, pelagic, cosmopolitan marine fishes which form the family Thunnidae including species that rank among the most valuable of food and game fish.

tunable echo box [ELECTROMAG] Echo box consisting of an adjustable cavity operating in a single mode; if calibrated, the setting of the plunger at resonance will indicate the wavelength.

tunable filter [ELECTR] An electric filter in which the frequency of the passband or rejection band can be varied by adjusting its components.

tunable laser [OPTICS] A laser whose output frequency can be varied continuously over a wide range, such as a dye laser or a spin-flip Raman laser.

tunable magnetron [ELECTR] Magnetron which can be tuned mechanically or electronically by varying its capacitance or inductance.

tuna boat *See* tuna clipper.

tuna clipper [NAV ARCH] A large craft used for tuna fishing on the Pacific coast; it is usually diesel-powered, and is equipped with refrigeration brine tanks. Also known as tuna boat.

tundra [ECOL] An area supporting some vegetation between the northern upper limit of trees and the lower limit of perennial snow on mountains, and on the fringes of the Antarctic continent and its neighboring islands.

tundra climate [CLIMATOL] The climate which produces tundra vegetation; it is too cold for the growth of trees but does not have a permanent snow-ice cover.

tune [ELECTR] To adjust for resonance at a desired frequency.

tuned amplifier [ELECTR] An amplifier in which the load is a tuned circuit; load impedance and amplifier gain then vary with frequency.

tuned-anode oscillator [ELECTR] A vacuum-tube oscillator whose frequency is determined by a tank circuit in the anode circuit, coupled to the grid to provide the required feedback. Also known as tuned-plate oscillator.

tuned-anode tuned-grid oscillator *See* tuned-grid tuned-anode oscillator.

tuned-base oscillator [ELECTR] Transistor oscillator in which the frequency-determining resonant circuit is located in the base circuit; comparable to a tuned-grid oscillator.

tuned cavity *See* cavity resonator.

tuned circuit [ELECTR] A circuit whose components can be adjusted to make the circuit responsive to a particular frequency in a tuning range. Also known as tuning circuit.

tuned-collector oscillator [ELECTR] A transistor oscillator in which the frequency-determining resonant circuit is located in the collector circuit; this is comparable to a tuned-anode electron-tube oscillator.

tuned filter [ELECTR] Filter that uses one or more tuned circuits to attenuate or pass signals at the resonant frequency.

tuned-grid oscillator [ELECTR] Oscillator whose frequency is determined by a parallel-resonant circuit in the grid coupled to the plate to provide the required feedback.

tuned-grid tuned-anode oscillator [ELECTR] A vacuum-tube oscillator whose frequency is determined by a tank circuit in the grid circuit, coupled to the anode to provide the required feedback. Also known as tuned-anode tuned-grid oscillator.

tuned-plate oscillator *See* tuned-anode oscillator.

tuned-radio-frequency receiver [ELECTR] A radio receiver consisting of a number of amplifier stages that are tuned to resonance at the carrier frequency of the desired signal by a gang capacitor; the amplified signals at the original carrier frequency are fed directly into the detector for demodulation, and the resulting audio-frequency signals are amplified by an a-f amplifier and reproduced by a loudspeaker. Abbreviated TRF receiver.

tuned-radio-frequency transformer [ELECTR] Transformer used for selective coupling in radio-frequency stages.

tuned-reed frequency meter *See* vibrating-reed frequency meter.

tuned relay [ELEC] A relay having mechanical or other resonating arrangements that limit response to currents at one particular frequency.

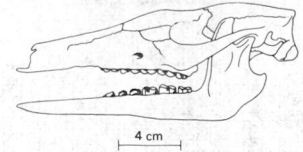

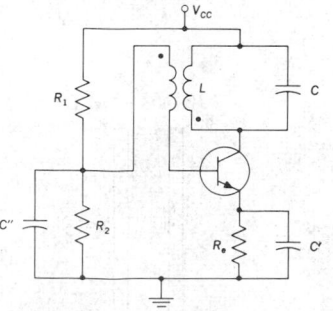

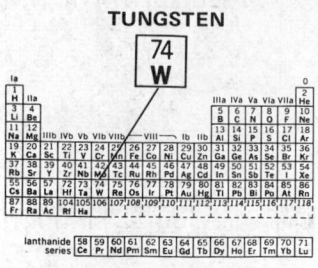

Periodic table of the chemical elements showing the position of tungsten.

TUNGSTEN–INERT GAS WELDING

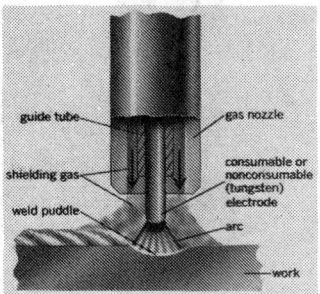

Components of tungsten–inert gas welding process.

TUNG TREE

Leaf of tung tree.

TUNING FORK

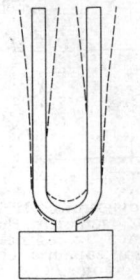

A tuning fork vibrating at its fundamental frequency.

tuned resonating cavity [ELECTROMAG] Resonating cavity half a wavelength long or some multiple of a half wavelength, used in connection with a waveguide to produce a resultant wave with the amplitude in the cavity greatly exceeding that of the wave in the waveguide.

tuned transformer [ELEC] Transformer whose associated circuit elements are adjusted as a whole to be resonant at the frequency of the alternating current supplied to the primary, thereby causing the secondary voltage to build up to higher values than would otherwise be obtained.

tuner [ELECTR] The portion of a receiver that contains circuits which can be tuned to accept the carrier frequency of the alternating current supplied to the primary, thereby causing the secondary voltage to build up to higher values than would otherwise be obtained.

tungar tube [ELECTR] A gas tube having a heated thoriated tungsten filament serving as cathode and a graphite disk serving as anode in an argon-filled bulb at a low pressure; used chiefly as a rectifier in battery chargers.

tung nut [BOT] The seed of the tung tree (*Aleurites fordii*), which is the source of tung oil.

tung oil [MATER] A yellow, combustible drying oil extracted from the seed of the tung tree; soluble in ether, chloroform, carbon disulfide, and oils; used in formulations for paints, varnishes, varnish driers, paper waterproofing, and linoleum. Also known as China wood oil.

tungstate [INORG CHEM] M_2WO_4 A salt of tungstic acid; for example, sodium tungstate, Na_2WO_4. [MINERAL] Any species of mineral containing the radical WO_4, such as wolframite.

tungsten [CHEM] Also known as wolfram. A metallic element in group VI, symbol W, atomic number 74, atomic weight 183.85; soluble in mixed nitric and hydrofluoric acids; melts at 3400°C. [MET] A hard, brittle, ductile, heavy gray-white metal used in the pure form chiefly for electrical purposes and with other substances in dentistry, pen points, x-ray-tube targets, phonograph needles, and high-speed tool metal, and as a radioactive shield.

tungsten boride [INORG CHEM] WB_2 A silvery solid; insoluble in water, soluble in aqua regia and concentrated acids; melts at 2900°C; used as a refractory for furnaces and chemical process equipment.

tungsten carbide [INORG CHEM] WC A hard, gray powder; insoluble in water; readily attacked by nitric-hydrofluoric acid mixture; melts at 2780°C; used in tools, dies, ceramics, cermets, and wear-resistant mechanical parts, and as an abrasive.

tungsten filament [ELEC] A filament used in incandescent lamps, and as an incandescent cathode in many types of electron tubes, such as thermionic vacuum tubes.

tungsten–inert gas welding [MET] Welding in which an arc plasma from a nonconsumable tungsten electrode radiates heat onto the work surface, to create a weld puddle in a protective atmosphere provided by a flow of inert shielding gas; heat must then travel by conduction from this puddle to melt the desired depth of weld. Abbreviated TIG welding.

tungsten steel [MET] Steel containing tungsten with other alloys; formerly used for cutting and forging tools but replaced by high-speed steel.

tungstic acid [INORG CHEM] H_2WO_4 A yellow powder; insoluble in water, soluble in alkalies; used as a color-resist mordant for textiles, as an ingredient in plastics, and for the manufacture of tungsten metal products. Also known as orthotungstic acid; wolframic acid.

tungstic acid anhydride *See* tungstic oxide.

tungstic anhydride *See* tungstic oxide.

tungstic oxide [INORG CHEM] WO_3 A heavy, canary-yellow powder; soluble in caustic, insoluble in water; melts at 1473°C; used in alloys, in fabric fireproofing, for ceramic pigments, and for the manufacture of tungsten metal. Also known as anhydrous wolframic acid; tungstic acid anhydride; tungstic anhydride; tungstic trioxide.

tungstic trioxide *See* tungstic oxide.

tung tree [BOT] *Aleurites fordii.* A plant of the spurge family in the order Euphorbiales, native to central and western China and grown in the southern United States.

tunica [BIOL] A membrane or layer of tissue that covers or envelops an organ or other anatomical structure.

tunica adventitia *See* adventitia.

tunica intima *See* intima.

tunica mucosa *See* mucous membrane.

Tunicata [INV ZOO] A subphylum of the Chordata characterized by restriction of the notochord to the tail and posterior body of the larva, absence of mesodermal segmentation, and secretion of an outer covering or tunic about the body.

tuning [ELECTR] The process of adjusting the inductance or the capacitance or both in a tuned circuit, for example, in a radio, television, or radar receiver or transmitter, so as to obtain optimum performance at a selected frequency.

tuning capacitor [ELEC] A variable capacitor used for tuning purposes.

tuning circuit *See* tuned circuit.

tuning core [ELECTROMAG] A ferrite core that is designed to be moved in and out of a coil or transformer to vary the inductance.

tuning fork [ENG] A U-shaped bar for hard steel, fused quartz, or other elastic material that vibrates at a definite natural frequency when struck or when set in motion by electromagnetic means; used as a frequency standard.

tuning indicator [ELECTR] A device that indicates when a radio receiver is tuned accurately to a radio station, such as a meter or a cathode-ray tuning indicator; it is connected to a circuit having a direct-current voltage that varies with the strength of the incoming carrier signal.

tuning range [ELECTR] The frequency range over which a receiver or other piece of equipment can be adjusted by means of a tuning control.

tuning screw [ELECTROMAG] A screw that is inserted into the top or bottom wall of a waveguide and adjusted as to depth of penetration inside for tuning or impedance-matching purposes.

tuning stub [ELECTROMAG] Short length of transmission line, usually shorted at its free end, connected to a transmission line for impedance-matching purposes.

tuning susceptance [ELECTR] Normalized susceptance of an anti-transmit-receive tube in its mount due to the deviation of its resonant frequency from the desired resonant frequency.

tuning wand [ELEC] Rod of insulating material having a brass plug at one end and a powered iron core at the other end; used for checking receiver alignment.

tunnel [ENG] A long, narrow, horizontal or nearly horizontal underground passage that is open to the atmosphere at both ends; used for aqueducts and sewers, carrying railroad and vehicular traffic, various underground installations, and mining.

tunnel-bearing grease [MATER] Lubricating grease for the main engine and propeller shaft (in the shaft tunnel) of ships.

tunnel blasting [ENG] A method of heavy blasting in which a heading is driven into the rock and afterward filled with explosives in large quantities, similar to a borehole, on a large scale, except that the heading is usually divided in two parts on the same level at right angles to the first heading, forming in plan a T, the ends of which are filled with explosives and the intermediate parts filled with inert material like an ordinary borehole.

tunnel borer [MECH ENG] Any boring machine for making a tunnel; often a ram armed with cutting faces operated by compressed air.

tunnel carriage [MECH ENG] A machine used for rapid tunneling, consisting of a combined drill carriage and manifold for water and air so that immediately the carriage is at the face, drilling may commence with no lost time for connecting up or waiting for drill steels; the air is supplied at pressures of 95 to 100 pounds per square inch (6.55 to 6.89 × 10⁵ newtons per square meter).

tunnel diode [ELECTR] A heavily doped junction diode that has a negative resistance at very low voltage in the forward bias direction, due to quantum-mechanical tunneling, and a short circuit in the negative bias direction. Also known as Esaki tunnel diode.

tunnel effect [QUANT MECH] The ability of a particle to pass through a region of finite extent in which the particle's potential energy is greater than its total energy; this is a quantum-mechanical phenomenon which would be impossible according to classical mechanics.

tunnel gun [ORD] A gun mounted inside an airplane fuselage and firing through an aperture.

tunnel liner [CIV ENG] Any of various materials, especially timber, concrete, and cast iron, applied to the inner surface of a vehicular or railroad tunnel. [MIN ENG] The timber, brick, concrete, or steel supports erected in a mine tunnel to maintain dimensions and safe working conditions.

tunnel rectifier [ELECT] Tunnel diode having a relatively low peak-current rating as compared with other tunnel diodes used in memory-circuit applications.

tunnel resistor [ELECTR] Resistor in which a thin layer of metal is plated across a tunneling junction, to give the combined characteristics of a tunnel diode and an ordinary resistor.

tunnel set [MIN ENG] Timbers 6 to 8 inches (15 to 20 centimeters) in diameter and of sufficient height to support the roof of the tunnel.

tunnel system [MIN ENG] A method of mining in which tunnels or drifts are extended at regular intervals from the floor of the pit into the ore body.

tunnel triode [ELECTR] Transistorlike device in which the emitter-base junction is a tunnel diode and the collector-base junction is a conventional diode.

Tupaiidae [VERT ZOO] The tree shrews, a family of mammals in the order Insectivora.

tupelo [BOT] Any of various trees belonging to the genus *Nyssa* of the sour gum family, Nyssaceae, distinguished by small, obovate, shiny leaves, a small blue-black drupaceous fruit, and branches growing at a wide angle from the axis.

Turbellaria [INV ZOO] A class of the phylum Platyhelminthes having bodies that are elongate and flat to oval or circular in cross section.

turbidimeter [OPTICS] A device that measures the loss in intensity of a light beam as it passes through a solution with particles large enough to scatter the light.

turbidimetric analysis [ANALY CHEM] A scattered-light procedure for the determination of the weight concentration of particles in cloudy, dull, or muddy solutions; uses a device that measures the loss in intensity of a light beam as it passes through the solution. Also known as turbidimetry.

turbidimetric titration [ANALY CHEM] Titration in which the end point is indicated by the developing turbidity of the titrated solution.

turbidimetry See tubidimetric analysis.

turbidite [GEOL] Any sediment or rock transported and deposited by a turbidity current, generally characterized by graded bedding, large amounts of matrix, and commonly exhibiting a Bouma sequence.

turbidity [METEOROL] Any condition of the atmosphere which reduces its transparency to radiation, especially to visible radiation. [ANALY CHEM] **1.** Measure of the clarity (using APHA or colorimetric scales) of an otherwise clear liquid. **2.** Cloudy or hazy appearance in a naturally clear liquid caused by a suspension of colloidal liquid droplets or fine solids.

turbidity coefficient [OPTICS] A factor in the absorption (light) law equation that describes the extinction of the incident light beam.

turbidity current [OCEANOGR] A highly turbid, relatively dense current carrying large quantities of clay, silt, and sand in suspension which flows down a submarine slope through less dense sea water. Also known as density current; suspension current.

turbidity factor [GEOPHYS] A measure of the atmospheric transmission of incident solar radiation; if I_0 is the flux density of the solar beam just outside the earth's atmosphere, I the flux density measured at the earth's surface with the sun at a zenith distance which implies an optical air mass m, and $I_{m,w}$ the intensity which would be observed at the earth's surface for a pure atmosphere containing 1 centimeter of precipitable water viewed through the given optical air mass, then turbidity factor θ is given by $\theta = (\ln I_0 - \ln I)/(\ln I_0 - \ln I_{m,w})$.

turbidostat [MICROBIO] A device in which a bacterial culture is maintained at a constant volume and cell density (turbidity) by adjusting the flow rate of fresh medium into the growth tube by means of a photocell and appropriate electrical connections.

turbinate [BOT] Shaped like an inverted cone. [INV ZOO] Spiral with rapidly decreasing whorls from base to apex.

turbine [MECH ENG] A fluid acceleration machine for generating rotary mechanical power from the energy in a stream of fluid.

turbine generator [ELEC] An electric generator driven by a steam, hydraulic, or gas turbine.

turbine propulsion [MECH ENG] Propulsion of a vehicle or vessel by means of a steam or gas turbine.

turbine pump See regenerative pump.

Turbinidae [INV ZOO] A family of gastropod mollusks including species of top shells.

turboblower [MECH ENG] A centrifugal or axial-flow compressor.

turbodrill [PETRO ENG] A rotary tool used in drilling oil or gas wells in which the bit is rotated by a turbine motor inside the well.

turbofan [AERO ENG] An air-breathing jet engine in which additional propulsive thrust is gained by extending a portion of the compressor or turbine blades outside the inner engine case.

turbogrid plate [CHEM ENG] A tray for distillation columns that consists of a flat grid of parallel slots extending over the entire cross-sectional area of the column; the liquid level on each tray is maintained by a dynamic balance between down-flowing liquid and up-flowing vapor.

turbojet [AERO ENG] A jet engine incorporating a turbine-driven air compressor to take in and compress the air for the combustion of fuel (or for heating by a nuclear reactor), the gases of combustion (or the heated air) being used both to rotate the turbine and to create a thrust-producing jet.

turbonada [METEOROL] A short thundersquall on the north Spanish coast, sometimes accompanied by waterspouts.

turbosphere [METEOROL] The region of the atmosphere in which turbulence frequently exists.

turbosupercharger [MECH ENG] A centrifugal air compressor, gas-turbine driven, usually used to increase induction system pressure in an internal combustion reciprocating engine.

turbulence See turbulent flow.

turbulence energy See eddy kinetic energy.

turbulent boundary layer [FL MECH] The layer in which the Reynolds stresses are much larger than the viscous stresses.

turbulent diffusion See eddy diffusion.

turbulent flow [FL MECH] Motion of fluids in which local velocities and pressures fluctuate irregularly, in a random manner. Also known as turbulence.

turbulent flux See eddy flux.

turbulent heat conduction [OCEANOGR] Conduction of heat in water by lateral and vertical eddy diffusion, with currents.

turbulent Lewis number [PHYS] A dimensionless number used in the study of combined turbulent heat and mass transfer, equal to the ratio of the eddy mass diffusivity to the eddy thermal diffusivity. Symbolized Le_T.

turbulent Prandtl number [PHYS] A dimensionless number used in the study of heat transfer in turbulent flow, equal to the ratio of the eddy viscosity to the eddy thermal diffusivity. Symbolized Pr_T.

turbulent Schmidt number [FL MECH] A dimensionless number used in the study of mass transfer in turbulent flow, equal to the ratio of the eddy viscosity to the eddy mass diffusivity. Symbolized Sc_T.

turbulent shear force [FL MECH] A shear force in a fluid which arises from turbulent flow.

Turdidae [VERT ZOO] The thrushes, a family of passeriform birds in the suborder Oscines.

turgor [BOT] Distension of a plant cell wall and membrane by the fluid contents.

turgor movement [BOT] A reversible change in the position of plant parts due to a change in turgor pressure in certain specialized cells; movement of *Mimosa* leaves when touched is an example.

turgor pressure [BOT] The actual pressure developed by the fluid content of a turgid plant cell.

Turing machine [ADP] A mathematical idealization of a

TUPELO

Branch with leaf and fruit cluster, terminal bud, and twig of black tupelo (*Nyssa sylvatica*).

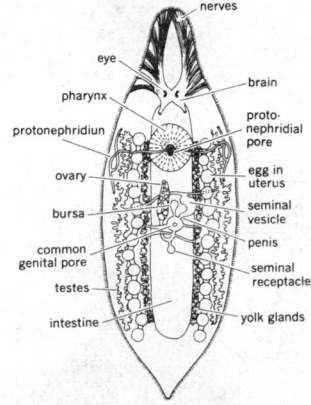

TURBELLARIA

A typical hermaphroditic turbellarian, *Mesostoma ehrenbergii wardii*, showing principal organs. (*Modified from Ruebush, 1940*)

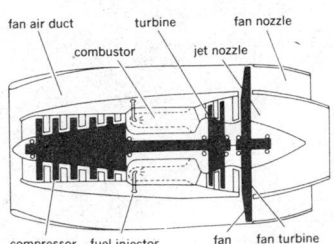

TURBOFAN

Diagram of turbofan engine.

computing automation similar in some ways to real computing machines; used by mathematicians to define the concept of computability.

turion [BOT] A scaly shoot, such as asparagus, developed from an underground bud.

turkey [VERT ZOO] Either of two species of wild birds, and any of various derived domestic breeds, in the family Meleagrididae characterized by a bare head and neck, and in the male a large pendant wattle which hangs on one side from the base of the bill.

Turkey red oil [MATER] Sulfonated castor oil, soluble in water; autoignites at 833°F (445°C); used in textiles, leather, and paper coatings, for manufacture of soaps, and as an alizarin dye assistant. Also known as soluble castor oil; sulfonated castor oil.

Turkey stone See turquoise.

Turkish geranium oil See palmarosa oil.

Turk's head rolls [MET] A group of four idler rolls, arranged in a square or rectangular pattern, through which strip metal can be drawn to form angled sections.

turmeric [BOT] Curcuma longa. An East Indian perennial of the ginger family (Zingiberaceae) with a short stem, tufted leaves, and short thick rhizomes; a spice with a pungent, bitter taste and a musky odor is derived from the rhizome. [MATER] An orange-red or reddish-brown dye obtained from the rhizome of turmeric.

turn [ELEC] One complete loop of wire. [MATH] See circle.

turnaround [CHEM ENG] In petroleum refining, the shutdown of a unit after a normal run for maintenance and repair work, then putting the unit back into operation. [ENG] The length of time between arriving at a point and departing from that point; it is used in this sense for the turnaround of vehicles, ships in ports, and aircraft.

turnaround cycle [ENG] A term used in conjunction with vehicles, ships, and aircraft, and comprising the following: loading time at home, time to and from destination, unloading and loading time at destination, unloading time at home, planned maintenance time, and, where applicable, time awaiting facilities.

turnaround system [ADP] In character recognition, a system in which the input data to be read have previously been printed by the computer with which the reader is associated; an application is invoice billing and the subsequent recording of payments.

turnaround time [ADP] The delay between submission of a job for a data-processing system and its completion.

turnbuckle [DES ENG] A sleeve with a thread at one end and a swivel at the other, or with threads of opposite hands at each end so that by turning the sleeve connected rods or wire rope will be drawn together and tightened.

Turnbull's blue [INORG CHEM] A blue pigment that precipitates from the reaction of potassium ferricyanide with a ferrous salt.

Turner's syndrome [MED] A sex aberration in humans in which the chromosome complement includes only one sex chromosome, an X.

Turnicidae [VERT ZOO] The button quails, a family of Old World birds in the order Gruiformes.

turning [MECH ENG] Shaping a member on a lathe.

turning area See enroute turning area.

turning basin [CIV ENG] An open area at the end of a canal or narrow waterway to allow boats to turn around.

turning-block linkage [MECH ENG] A variation of the sliding-block mechanical linkage in which the short link is fixed and the frame is free to rotate. Also known as the Wentworth quick-return motion.

turning buoy [NAV] A buoy marking a turn, as in a channel.

turning circle [NAV] The path approximating a circle of 360° or more described by the pivot point of the ship as it makes a turn.

turning error See northerly turning error.

turning table [ENG] In plastics molding, a rotating table or wheel carrying various molds in a multimold, single-parison blow-molding operation.

turnip [BOT] Brassica rapa or B. campestris var. rapa. An annual crucifer of Asiatic origin belonging to the family Brassiaceae in the order Capparales and grown for its foliage and edible root.

turnkey contract [ENG] A contract in which an independent agent undertakes to furnish for a fixed price all materials and labor, and to do all the work needed to complete a project.

turn-off time [ELECTR] The time that is takes a gate circuit to shut off a current.

turn of the tide See change of tide.

turn-on time [ELECTR] The time that it takes a gate circuit to allow a current to reach its full value.

turnout [ENG] **1.** A contrivance consisting of a switch, a frog, and two guardrails for passing from one track to another. **2.** The branching off of one rail track from another. **3.** A siding. [MIN ENG] To shovel coal toward the track for more convenient loading.

turnover [MOL BIO] The number of substrate molecules transformed by a single molecule of enzyme per minute, when the enzyme is operating at maximum rate.

turnover cartridge [ENG ACOUS] A phonograph pickup having two styli and a pivoted mounting that places in playing position the correct stylus for a particular record speed.

turnover frequency See transition frequency.

turnover number [BIOCHEM] The number of molecules of a substrate acted upon in a period of 1 minute by a single enzyme molecule, with the enzyme working at a maximum rate.

turnpike [CIV ENG] A toll expressway.

turnplow See moldboard.

turns ratio [ELEC] The ratio of the number of turns in a secondary winding of a transformer to the number of turns in the primary winding.

turnstile antenna [ELECTROMAG] An antenna consisting of one or more layers of crossed horizontal dipoles on a mast, usually energized so the currents in the two dipoles of a pair are equal and in quadrature; used with television, frequency modulation, and other very-high-frequency or ultra-high-frequency transmitters to obtain an essentially omnidirectional radiation pattern.

turntable [ENG ACOUS] The rotating platform on which a disk record is placed for recording or playback.

turntable rumble [ENG ACOUS] Low-frequency vibration that is mechanically transmitted to a recording or reproducing turntable and superimposed on the reproduction. Also known as rumble.

Turonian [GEOL] A European stage of geologic time: Upper or Middle Cretaceous (above Cenomanian, below Coniacian).

turpentine [MATER] An essential oil produced by steam distillation of pine woods and from gum turpentine; used as a solvent and a thinner for paints and varnishes.

turpentine camphor See terpene hydrochloride.

turquoise [MINERAL] $CuAl_6(PO_4)_4(OH)_8 \cdot 4H_2O$ A semitranslucent sky-blue, bluish-green, apple-green, or greenish-gray mineral that crystallizes in the triclinic system and occurs in veinlets or as crusts of massive, concretionary, and stalactite shapes; an important gem mineral. Also known as calaite; Turkey stone.

turret [ORD] A dome-shaped or cylindrical armored structure containing one or more guns and located on forts, warships, airplanes, and tanks.

turret coal cutter [MIN ENG] A coal cutter in which the horizontal jib can be adjusted vertically to cut at different levels in the seam; for example, an overcut.

turret ice See ropak.

turret lathe [MECH ENG] A semiautomatic lathe differing from the engine lathe in having the tailstock replaced with a multisided, indexing tool holder or turret designed to hold several tools.

turret mount [ORD] A gun mount positioned in the turret of a tank or combat vehicle; power-driven multiple gun turret mounts improve control in tracking aerial targets and also increase firepower.

turret tuner [ELECTR] A television tuner having one set of pretuned circuits for each channel, mounted on a drum that is rotated by the channel selector; rotation of the drum connects each set of tuned circuits in turn to the receiver antenna circuit, radio-frequency amplifier, and r-f oscillator.

TURNBUCKLE

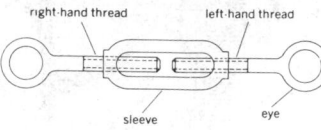

right-hand thread   left-hand thread

sleeve   eye

Turnbuckle with threads of opposite hands at each end, and with eyes for attaching rods or wire rope.

TURNIP

Turnip, showing enlarged root and foliage.

TURQUOISE

turquoise

2.5 cm

Turquoise in massive form; sample taken from Santa Fe County, New Mexico. (Specimen from Department of Geology, Bryn Mawr College)

turtle [VERT ZOO] Any of about 240 species of reptiles which constitute the order Chelonia distinguished by the two bony shells enclosing the body.

turtle oil [MATER] The oil derived from the muscles and genital glands of the giant sea turtle; melts at 25°C; used in cosmetics.

turtle stone *See* septarium.

tusche [MATER] A liquid lithographic ink that can be used with pen or brush on lithographic stones or metal plates; it is also used in the silk-screen process in a glue-resist system of pattern making; and is an extremely greasy material.

tuyere [MET] An opening in the shell and refractory lining of a furnace through which air is forced.

TV *See* television.

TV camera scanner [ADP] In optical character recognition, a device that images an input character onto a sensitive photoconductive target of a camera tube, thereby developing an electric charge pattern on the inner surface of the target; this pattern is then explored by a scanning beam which traces out a rectangular pattern with the result that a waveform is produced which represents the character's most probable identity.

TVI *See* television interference.

TW *See* terawatt.

Twaddell scale [ENG] A scale for specific gravity of solutions that is the first two digits to the right of the decimal point multiplied by two; for example, a specific gravity of 1.4202 is equal to 84.04°Tw.

tweeter [ENG ACOUS] A loudspeaker designed to handle only the higher audio frequencies, usually those well above 3000 hertz; generally used in conjunction with a crossover network and a woofer.

twilight [ASTRON] An intermediate period of illumination of the sky before sunrise and after sunset; the three forms are civil, nautical, and astronomical.

twilight arch *See* bright segment.

twilight compass [NAV] An instrument for indicating direction by use of the polarizing effect of the earth's atmosphere on sunlight. Also known as sky compass.

twilight correction [ASTRON] In the interpretation of the records of sunshine recorders, the difference between the time of sunrise and the time at which a record of sunshine first began to be made by the sunshine recorder; and conversely at sunset; this correction is added only when the horizon is clear during the period.

twilight phenomena [OPTICS] Those meteorological optical phenomena which occur during twilight, including such effects as the antitwilight arch, dark segment, bright segment, green flash, purple light, and crepuscular rays.

twilight zone [ASTRON] That zone of the earth or other planet in twilight at any time. [ELECTROMAG] Anything resembling the twilight zone of the earth, as the narrow sector on each side of the equisignal zone of a four-course radio range station, in which one signal is barely heard above the monotone on-course signal.

twill weave [TEXT] A woven pattern of diagonal or twill lines that run upward to the right or left of the fabric face.

twin [BIOL] One of two individuals born at the same time. [CRYSTAL] *See* twin crystal.

twin arithmetic units [ADP] A feature of some computers where the essential portions of the arithmetic section are virtually duplicated.

twin band [MET] A line on a polished or etched surface representing the section through crystal twins.

twin-cable ropeway [MECH ENG] An aerial ropeway which has parallel track cables with carriers running in opposite directions; both rows of carriers are pulled by the same traction rope.

twin check [ADP] Continuous check of computer operation, achieved by the duplication of equipment and automatic comparison of results.

twin crystal [CRYSTAL] A compound crystal which has one or more parts whose lattice structure is the mirror image of that in the other parts of the crystal. Also known as twin.

twine [MATER] A strong string made up of two or more strands twisted together.

twin entry [MIN ENG] A pair of parallel entries, one of which is an intake air course and the other the return air course; rooms can be worked from both entries.

twiner [BOT] A climbing stem that winds about its support, as pole beans or many tropical lianas.

twin-geared press [MECH ENG] A crank press having the drive gears attached to both ends of the crankshaft.

twinkling stars [ASTRON] Rapid fluctuations of the brightness and size of the images of stars caused by turbulence in the earth's atmosphere.

twin law [CRYSTAL] A statement relating two or more individuals of a twin to one another in terms of their crystallography (twin plane, twin axis, and so on).

twinning [CRYSTAL] The development of a twin crystal by growth, translation, or gliding.

twinning plane *See* twin plane.

twin paradox *See* clock paradox.

twin plane [CRYSTAL] The plane common to and across which the individual crystals or components of a crystal twin are symmetrically arranged or reflected. Also known as twinning plane.

Twins *See* Gemini.

twin-T filter [ELEC] An electric filter consisting of a parallel-T network with values of network elements chosen in such a way that the outputs due to each of the paths precisely cancel at a specified frequency.

twin-T network *See* parallel-T network.

twist [DES ENG] In a fiber, rope, yarn, or cord, the turns about its axis per unit length; usually expressed as TPI (turns per inch). [ELECTROMAG] A waveguide section in which there is a progressive rotation of the cross section about the longitudinal axis of the waveguide.

twist drill [DES ENG] A tool having one or more helical grooves, extending from the point to the smooth part of the shank, for ejecting cuttings and admitting a coolant.

twisted pair [ELEC] A cable composed of two small insulated conductors twisted together without a common covering.

twister [METEOROL] In the United States, a colloquial term for tornado. [SOLID STATE] A piezoelectric crystal that generates a voltage when twisted.

twisting [TEXT] In a ply yarn or cord, the process of combining two or more ends with a twist to give greater strength and smoothness, increased uniformity, or novel effects.

twist of rifling [ORD] Inclination of the spiral grooves (rifling) to the axis of the bore of a weapon; it is expressed as the number of calibers of length in which the rifling makes one complete turn.

twistor [ADP] Computer memory element consisting of a helix of a magnetic wire wound under tension at a 45° angle on a short piece of nonmagnetic wire, with a fine-wire solenoid wound over the helix.

Twitchell reagent [ORG CHEM] A catalyst for the acid hydrolysis of fats; a sulfonated addition product of naphthalene and oleic acid, that is, a naphthalenestearosulfonic acid.

two-address code [ADP] In computers, a code using two-address instructions.

two-address instruction [ADP] In computers, an instruction which includes an operation and specifies the location of two registers.

two-beam interference [PHYS] Interference between two waves.

two-body force [PHYS] A force between two particles which is not affected by the existence of other particles in the vicinity, such as a gravitational force or a Coulomb force between charged particles.

two-body problem [MECH] The problem of predicting the motions of two objects obeying Newton's laws of motion and exerting forces on each other according to some specified law such as Newton's law of gravitation, given their masses and their positions and velocities at some initial time.

two-carrier theory [SOLID STATE] A theory of the conduction properties of a material in bulk or in a rectifying barrier which takes into account the motion of both electrons and holes.

two-component neutrino theory [PARTIC PHYS] A theory according to which the neutrino and antineutrino have exactly zero rest mass, and the neutrino spin is always antiparallel to

TURTLE

Box turtle, *Terrapene carolina*, a common United States turtle. *(From J. J. Shomon, ed., Virginia Wildlife, 15(6):27, 1954)*

TWILL WEAVE

Three-shaft twill. Two warp yarns are interlaced with one filling yarn. *(From M. D. Potter and B. P. Corbman, Fiber to Fabric, 3d ed., McGraw-Hill, 1959)*

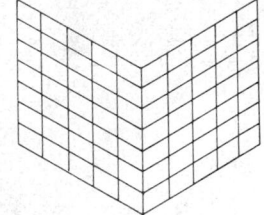

TWIN CRYSTAL

Example of twin crystal, in which the twins are mirror images of each other.

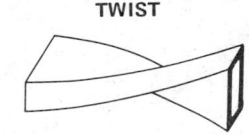

TWIST

A 90° twist for a rectangular waveguide.

its motion, while the antineutrino spin is parallel to its motion.

two-cycle engine [MECH ENG] A reciprocating internal combustion engine that requires two piston strokes or one revolution to complete a cycle.

two-decision problem [STAT] The problem of deciding, using statistical information, between two actions or decisions.

two-degrees-of-freedom gyro [MECH] A gyro whose spin axis is free to rotate about two orthogonal axes, not counting the spin axis.

two-dimensional chromatography [ANALY CHEM] A paper chromatography technique in which the sample is resolved by standard procedures (ascending, descending, or horizontal solvent movement) and then turned at right angles in a second solvent and re-resolved.

two-dimensional flow [FL MECH] Fluid flow in which all flow occurs in a set of parallel planes with no flow normal to them, and the flow is identical in each of these parallel planes.

two-fluid cell [PHYS CHEM] Cell having different electrolytes at the positive and negative electrodes.

two-hop transmission [COMMUN] Propagation of radio waves in which the waves are reflected from the ionosphere, then reflected from the ground, and then reflected from the ionosphere again before reaching the receiver.

two-layer ocean [OCEANOGR] An idealized ocean in which a layer of uniform density near the surface overlays a deep layer of uniform but distinctly higher-density water.

two-level mold [ENG] Placement of one cavity of a plastics mold above another instead of alongside it; reduces clamping force needed.

two-man concept *See* two-man rule.

two-man policy *See* two-man rule.

two-man rule [ORD] A system designed to prohibit access by an individual to nuclear weapons and certain designated components by requiring the presence at all times of at least two authorized persons, each capable of detecting incorrect or unauthorized procedures with respect to the task to be performed. Also known as two-man concept; two-man policy.

two-out-of-five code [ADP] An encoding of the decimal digits using five binary bits and having the property that every code element contains two 1s and three 0s.

two-part adhesive [MATER] A glue supplied in two parts, a resin and an accelerator, which are mixed only just before application.

two-part code [COMMUN] Randomized code consisting of an encoding section in which the plain text groups are arranged in alphabetical or other significant order accompanied by their code groups in nonalphabetical or random order, and a decoding section in which the code groups are arranged in alphabetical or numerical order and are accompanied by their meanings given in the encoding section.

two-part experiment [STAT] Experiments in which two operations or actions are performed; for example, throwing two dice, drawing two marbles from a box, throwing a die and then drawing a marble from a box.

two-person game [MATH] A game consisting of exactly two players with competing interests.

two-phase [PHYS] Having a phase difference of one quarter-cycle or 90°. Also known as quarter-phase.

two-phase alternating-current circuit [ELEC] A circuit in which there are two alternating currents on separate wires, the two currents being 90° out of phase.

two-phase current [ELEC] Current delivered through two pairs of wires or at a phase difference of one-quarter cycle (90°) between the current in the two pairs.

two-phase five-wire system [ELEC] System of alternating-current supply comprising five conductors, four of which are connected as in a two-phase four-wire system, the fifth being connected to the neutral points of each phase.

two-phase flow [CRYO] Flow of helium II, or of electrons in a superconductor, thought of as consisting of two interpenetrating, noninteracting fluids, a superfluid component which exhibits no resistance to flow and is responsible for superconducting properties, and a normal component, which behaves as does an ordinary fluid or as conduction electrons in a nonsuperconducting metal. [FL MECH] Cocurrent movement of two phases (for example, gas and liquid) through a closed conduit or duct (for example, a pipe).

two-phase four-wire system [ELEC] System of alternating-current supply comprising two pairs of conductors, between one pair of which is maintained an alternating difference of potential displaced in phase by one-quarter of a period from an alternating difference of potential of the same frequency maintained between the other pair.

two-phase three-wire system [ELEC] System of alternating-current supply comprising three conductors, between one of which (known as the common return) and each of the other two are maintained alternating difference of potential displaced in phase by one-quarter of a period with relation to each other.

two-piece set [MIN ENG] A set of timbers consisting of a cap and a single post.

two-point press [MECH ENG] A mechanical press in which the slide is actuated at two points.

two-port junction [ELECTROMAG] A waveguide junction with two openings; it can consist either of a discontinuity or obstacle in a waveguide, or of two essentially different waveguides connected together.

two-port system [CONT SYS] A system which has only one input or excitation and only one response or output.

two-position propeller [AERO ENG] An airplane propeller whose blades are limited to two angles, one for take off and climb and the other for cruising.

two-pulse canceler [ELECTR] A moving-target indicator canceler which compares the phase variation of two successive pulses received from a target; discriminates against signals with radial velocities which produce a Doppler frequency equal to a multiple of the pulse repetition frequency.

two-quadrant multiplier [ADP] Of an analog computer, a multiplier in which operation is restricted to a single sign of one input variable only.

two-range Decca [NAV] A Decca radio navigation system modified to provide circular lines of position.

two's complement [MATH] A number derived from a given n-bit number by requiring the two numbers to sum to a value of 2^n.

two-sided ideal [MATH] A two-sided ideal I is a sub-ring of a ring R where the products xy and yx are always in I for every x in R and y in I.

two-sided sampling plans [IND ENG] Any sampling plan whereby the acceptability of material is determined against upper and lower limits.

two-sided test [STAT] A test which rejects the null hypothesis when the test statistic T is either less than or equal to c or greater than or equal to d, where c and d are critical values.

two-source frequency keying [COMMUN] Keying in which the modulating wave shifts the output frequency between predetermined values derived from independent sources.

two-stage design [STAT] The design of an experiment which employs a pilot study in order to decide how to design the main experiment.

two-stage experiment [STAT] An experiment in two parts, the outcome of the first part deciding the procedure for the second.

two-stage hoisting [MIN ENG] Deep shaft hoisting with two winders, one at the surface, and the other at mid-depth in the shaft.

two-stage sampling [STAT] Sampling from a population whose members are themselves sets of objects and then sampling from the sets selected in the first sampling; for example, to first draw a sample of states and then to draw a sample of representatives to Congress from each state selected.

two-step test [MED] Repeated ascents over two 9-inch steps as a simple exercise test of cardiovascular function. Also known as Master's two-step test.

two-tone diaphone [ENG ACOUS] A diaphone producing blasts of two tones, the second tone being of a lower pitch than the first tone.

two-tone keying [COMMUN] Keying in which the modulating wave causes the carrier to be modulated with one frequency for the marking condition and modulated with a different frequency for the spacing condition.

two-tone modulation [COMMUN] In teletypewriter operation, a method of modulation in which two different carrier frequencies are employed for the two signaling conditions; the transition from one frequency to the other is abrupt, with resultant phase discontinuities.

two-valued logic [MATH] A system of logic where each statement has two possible values or states, truth or falsehood.

two-valued variable [MATH] A variable which assumes values in a set containing exactly two elements, often symbolized as 0 and 1.

two-way slab [CIV ENG] A concrete slab supported by beams along all four edges and reinforced with steel bars arranged perpendicularly.

two-wire circuit [ELEC] A metallic circuit formed by two conductors insulated from each other; in contrast with a four-wire circuit, it uses only one line or channel for transmission of electric waves in both directions.

two-wire repeater [ELECTR] Repeater that provides for transmission in both directions over a two-wire circuit; in carrier transmission, it usually operates on the principle of frequency separation for the two directions of transmission.

TWX machine *See* teletypewriter.

TWX service *See* teletypewriter exchange service.

Twyman-Green interferometer [OPTICS] An interferometer similar to the Michelson interferometer except that it is illuminated with a point source of light instead of an extended source.

Twystron [ELECTR] Very-high-power, hybrid microwave tube, combining the input section of a high-power klystron with the output section of a traveling wave tube, characterized by high operating efficiency and wide bandwidths.

Tycho [ASTRON] A crater on the near side of the moon.

Tychonic system [ASTRON] A theory of the planetary motion proposed by the astronomer Tycho Brahe in which the earth is stationary, with the sun and moon revolving about it but all the other planets revolving about the sun.

Tychonoff space *See* completely regular space.

Tychonoff theorem [MATH] A product of topological spaces is compact if and only if each individual space is compact.

Tycho's Nova [ASTRON] A supernova that appeared in the constellation Cassiopeia in 1572; the star B Cassiopeiae. Also known as Tycho's star.

Tycho's star *See* Tycho's Nova.

tyfon *See* typhon.

tying bar *See* tombolo.

Tylenchida [INV ZOO] An order of soil-dwelling or phytoparasitic nematodes in the subclass Rhabdita.

Tylenchoidea [INV ZOO] A superfamily of mainly soil and insect-associated nematodes in the order Tylenchida with a stylet for piercing live cells and sucking the juices.

Tyler screen [CHEM ENG] A screen standard for the openings in screen-type mediums based on meshes per linear inch; convertible to the U.S. Sieve Series.

Tyler Standard screen scale [ENG] A scale for classifying particles in which the particle size in micrometers is correlated with the meshes per inch of a screen.

Tylopoda [VERT ZOO] An infraorder of artiodactyls in the suborder Ruminantia that contains the camels and extinct related forms.

tylostyle [INV ZOO] A uniradiate spicule in Porifera with a point at one end and a knob at the other end.

tylote [INV ZOO] A slender sponge spicule with a knob at each end.

tympanic cavity [ANAT] The irregular, air-containing, mucous-membrane-lined space of the middle ear; contains the three auditory ossicles and communicates with the nasopharynx through the auditory tube.

tympanic membrane [ANAT] The membrane separating the external from the middle ear. Also known as eardrum; tympanum.

tympanum [ANAT] *See* tympanic membrane. [INV ZOO] A thin membrane covering an organ of hearing in insects.

Tyndall cone [OPTICS] The luminous path of a beam of light resulting from the Tyndall effect.

Tyndall effect [OPTICS] Visible scattering of light along the path of a beam of light as it passes through a system containing discontinuities, such as the surfaces of colloidal particles in a colloidal solution.

Tyndall flowers [HYDROL] Small water-filled cavities, often of basically hexagonal shape, which appear in the interior of ice masses upon which light is falling.

Tyndallization [ENG] Heat sterilization by steaming the food or medium for a few minutes at atmospheric pressure on three or four successive occasions, separated by 12- to 18-hour intervals of incubation at a temperature favorable for bacterial growth.

type [GRAPHICS] The relief or plane characters used to generate printed characters of various styles and sizes. [SYST] A specimen on which a species or subspecies is based.

type A encephalitis *See* lethargic encephalitis.

type A wave *See* continuous wave.

type A1 wave [COMMUN] An unmodulated, keyed, continuous wave.

type A2 wave [COMMUN] A modulated, keyed, continuous wave.

type A3 wave [COMMUN] A continuous wave modulated by music, speech, or other sounds.

type A4 wave [COMMUN] A superaudio frequency-modulated continuous wave, as used in facsimile systems.

type A5 wave [COMMUN] A superaudio frequency-modulated continuous wave, as used in television.

type A9 wave [COMMUN] A composite transmission and continuous wave that is not type A1, A2, A3, A4, or A5 wave.

type B wave [COMMUN] A keyed, damped wave.

Type II Cepheids *See* W Virginis stars.

type drum [ADP] A steel cylinder containing 128 to 144 lateral bands, each band containing the alphabet, the digits 0-9, and the standard set of punctuation marks such as commas and periods, and revolving at high speed; printing is achieved by a hammer facing each band and activated at the right time to cause a character to be printed on the paper flowing between hammers and drum.

typeface *See* face.

type-α leader [GEOPHYS] A stepped leader of lightning which exhibits very little branching and whose individual steps are short and so weakly luminous as to be difficult to discern.

type-β leader [GEOPHYS] A stepped leader of lightning in which the upper portion of the channel is characterized by longer and brighter steps than those found in the lower portion of the channel, a consequence of excessive branching in the upper parts under the influence of strong fields set up by heavy space charges near and around the upper end of the channel.

type-M carcinotron *See* M-type backward-wave oscillator.

type metal [MET] Any of various low-melting-point alloys, composed mainly of lead (50-90%), antimony (2-30%), and tin (2-20%), used for casting printers' type.

type-O carcinotron *See* O-type backward wave oscillator.

type I of Cori *See* von Gierke's disease.

type section [GEOL] That sequence of strata identified as the original sequence for a location or area; the standard against which other stratigraphy of parts of the area are compared. Also known as section.

type II superconductor [CRYO] A superconductor for which there are two critical magnetic fields; magnetic flux is completely excluded from the interior of the material only at field strengths below the smaller critical field, and at field strengths between the two critical fields the magnetic flux consists of flux vortices in the form of filaments embedded in the superconducting material. Also known as high field superconductor (HFS).

typewriter [GRAPHICS] A machine that produces printed copy, character by character, as the typewriter is operated; essential parts are an input keyboard, a set of raised characters, inking means, a platen, and a mechanism for advancing the position at which successive characters are imprinted.

Typhaceae [BOT] A family of monocotyledonous plants in the order Typhales characterized by an inflorescence of dense, cylindrical spikes and absence of a perianth.

Typhales [BOT] An order of marsh or aquatic monocotyledons in the subclass Commelinidae with emergent or floating stems and leaves and reduced, unisexual flowers having a single ovule in an ovary composed of a single carpel.

TYNDALL CONE

The luminous light path known as the Tyndall cone. (*H. Steeves and R. G. Babcock*)

TYROSINE

Structural formula of tyrosine.

Typhlopidae [VERT ZOO] A family of small, burrowing circumtropical snakes, suborder Serpentes, with vestigial eyes and toothless jaws.

Typhloscolecidae [INV ZOO] A family of pelagic polychaete annelids belonging to the Errantia.

typhlosole [INV ZOO] A dorsal longitudinal invagination of the intestinal wall in certain invertebrates serving to increase the absorptive surface.

typhoid fever [MED] A highly infectious, septicemic disease of humans caused by *Salmonella typhi* which enters the body by the oral route through ingestion of food or water contaminated by contact with fecal matter.

typhoid vaccine [IMMUNOL] A type of killed vaccine used for active immunity production; made from killed typhoid bacillus (*Salmonella typhi*).

typhon [ENG ACOUS] A diaphragm horn which operates under the influence of compressed air or steam. Also spelled tyfon.

Typhon [ORD] A United States surface-to-air missile of advance design for installation on carriers, cruisers, frigates, and destroyers, and for use against high-performance aircraft and short-range tactical missiles; equipped with either a nuclear or nonnuclear warhead.

typhoon [METEOROL] A severe tropical cyclone in the western Pacific.

typhoon wind *See* hurricane wind.

typhus fever [MED] Any of three louse-borne human diseases caused by *Rickettsia prowazakii* characterized by fever, stupor, headaches, and a dark-red rash.

Typotheria [PALEON] A suborder of extinct rodentlike herbivores in the order Notoungulata.

typp [TEXT] A unit of the reciprocal of line density used in the textile industry, equal to the reciprocal of the line density of a thread whose length is 1000 yards and whose length is 1 pound; equal to approximately 2015.91 meters per kilogram.

Tyranni [VERT ZOO] A suborder of suboscine Passeriformes containing birds with limited song power and having the tendon of the hind toe separate and the intrinsic muscles of the syrinx reduced to one pair.

Tyrannidae [VERT ZOO] The tyrant flycatchers, a family of passeriform birds in the suborder Tyranni confined to the Americas.

Tyrannoidea [VERT ZOO] The flycatchers, a superfamily of suboscine birds in the suborder Tyranni.

tyrocidine [MICROBIO] A peptide antibiotic produced by *Bacillus brevis*; used to control fungi, bacteria, and protozoa.

tyrosinase [BIOCHEM] An enzyme found in plants, molds, crustaceans, mollusks, and some bacteria which, in the presence of oxygen, catalyzes the oxidation of monophenols and polyphenols with the introduction of $-OH$ groups and the formation of quinones.

tyrosine [BIOCHEM] $C_9H_{11}NO_3$ A phenolic alpha amino acid found in many proteins; a precursor of the hormones epinephrine, norepinephrine, thyroxine, and triiodothyronine, and of the black pigment melanin.

tyrosinemia [MED] An inborn metabolic disorder in which there is a deficiency of the enzyme p-hydroxyphenylpyruvic acid oxidase with abnormally high blood levels of tyrosine and sometimes methionine.

tyrosinosis [MED] Excretion of excessive amounts of tyrosine and its first oxidation products in the urine.

tyrothricin [MICROBIO] A polypeptide mixture produced by *Bacillus brevis* and consisting of the antibiotic substances gramicidin and tyrocidine; effective as an antibacterial applied locally in infections due to germ-positive organisms.

Tyson's gland [ANAT] A small scent gland in the human male which secretes the smegma. Also known as preputial gland.

Tytonidae [VERT ZOO] The barn owls, a family of birds in the order Strigiformes distinguished by an unnotched sternum which is fused to large clavicles.

tyuyamunite [MINERAL] $Ca(UO_2)_2(VO_4)_2 \cdot 5-8H_2O$ A yellow orthorhombic mineral occurring in incrustations as a secondary mineral; an ore of uranium. Also known as calciocarnotite.

tyvelose [BIOCHEM] A dideoxy sugar found in bacterial lipopolysaccharides.

ubac [METEOROL] The shady (usually north) side of an Alpine mountain, characterized by a lower timberline and snow line than the sunny side.

u-band [OPTICS] The absorption band in the ultraviolet resulting from a U-center type of point lattice defect.

U-bend die [MECH ENG] A die with a square or rectangular cross section which provides two edges over which metal can be drawn.

U blades [DES ENG] Curved bulldozer blades designed to increase moving capacity of tractor equipment.

U bolt [DES ENG] A U-shaped bolt with threads at the ends of both arms to receive nuts.

UBV photometry [ASTRON] A system of three-color photometry used to obtain specific stellar magnitudes; the system is based on the comparison of stars' magnitudes with a standard sequence of about 400 stars.

U center [CRYSTAL] The color-center type of point lattice defect in ionic crystals created by the incorporation of an impurity such as hydrogen into alkali halides.

U Cephei [ASTRON] A binary star; in this double-star eclipsing system, one component has reached its Roche limit (a dynamical barrier beyond which the size of neither star can expand) while the other is distinctly smaller than this limit.

U coefficient [QUANT MECH] A coefficient that appears in the transformation between modes of coupling eigenfunctions of three angular momenta; it is equal to the product of the Racah coefficient and $[(2j_{12} + 1)(2j_{23} + 1)]^{1/2}$, where j_{12} and j_{23} are the intermediate angular momenta in the respective modes.

Uda antenna *See* Yagi-Uda antenna.

Udalf [GEOL] A suborder of the soil order Alfisol; brown soil formed in a udic moisture regime and in a mesic or warmer temperature regime.

udder [VERT ZOO] A pendulous organ consisting of several mammary glands enclosed in a single envelope; each gland has its own nipple; found in some mammals, such as the cow and goat.

Udert [GEOL] A suborder of the soil order Vertisol; formed in a humid region so that surface cracks remain open only for 2–3 months.

Udex process [CHEM ENG] A proprietary petroleum refining process used to extract aromatic hydrocarbons from a complex hydrocarbon mixture; uses a glycol-water solvent.

UDMH *See* uns-dimethylhydrazine.

Udoll [GEOL] A suborder of the Mollisol soil order; found in humid, temperate, and warm regions where maximum rainfall comes during growing season; has thick, very dark A horizons, brown B horizons, and paler C horizons.

UDP *See* uridine diphosphate.

UDPG *See* uridine diphosphoglucose.

Udult [GEOL] A suborder of the soil order Ultisol; organic-carbon content is low, argillic horizons are reddish or yellowish; formed in a udic moisture regime.

U figure *See* U index.

U format [ADP] A record format which the input/output control system treats as completely unknown and unpredictable.

U Geminorum stars [ASTRON] A class of variable stars known as dwarf novae; their light curves resemble those of novae, with range brightness variations of about 4 magnitudes; examples are U Gemini and SS Cygni. Also known as SS Cygni stars.

Ugine Sejournet process *See* Sejournet process.

UHF *See* ultra-high-frequency.

Uhlbricht sphere [OPTICS] A sphere whose inside surface has a diffusely reflecting white finish, used in an integrating sphere photometer.

U index [GEOPHYS] The difference between consecutive daily mean values of the horizontal component of the geomagnetic field. Also known as U figure.

Uintatheriidae [PALEON] The single family of the extinct mammalian order Dinocerata.

Uintatheriinae [PALEON] A subfamily of extinct herbivores in the family Uintatheriidae including all horned forms.

UJT *See* unijunction transistor.

Ulatisian [GEOL] A mammalian age in a local stage classification of the Eocene in use on the Pacific Coast based on foraminifers.

ULCC *See* ultra-large crude carrier.

ulcer [MED] Localized interruption of the continuity of an epithelial surface, with an inflamed base.

ulcerative colitis [MED] An idiopathic inflammatory disease of the mucosa and submucosa of the colon manifested clinically by pain, diarrhea, and rectal bleeding.

ulcerative endocarditis [MED] Acute bacterial endocarditis.

ulexite [MINERAL] $NaCaB_5O_9 \cdot 8H_2O$ A white mineral that crystallizes in the triclinic system and forms rounded reniform masses of extremely fine acicular crystals. Also known as boronatrocalcite; cotton ball; natronborocalcite.

ullage [ENG] The amount that a container, such as a fuel tank, lacks of being full. [NAV ARCH] The distance from the surface of the oil in a cargo tank of an oil tanker to the top of the hatch, or to the top of the inspection cover in the hatch.

ullage rocket [AERO ENG] A small rocket used in space to impart an acceleration to a tank system to ensure that the liquid propellants collect in the tank in such a manner as to flow properly into the pumps or thrust chamber.

ullmannite [MINERAL] NiSbS A steel-gray to black mineral consisting of nickel antimonide and sulfide, usually with a little arsenic, occurring massive, and having a metallic luster. Also known as nickel-antimony glance.

Ullmann reaction [ORG CHEM] A variation of the Fittig synthesis, using copper powder instead of sodium.

Ulloa's ring *See* Bouguer's halo.

Ullrich-Turner syndrome [MED] A complex of symptoms including webbing of the neck, short stature, cubitus valgus, and hypogonadism in the male. Also known as male Turner's syndrome.

Ulmaceae [BOT] A family of dicotyledonous trees in the order Urticales distinguished by alternate stipulate leaves, two styles, a pendulous ovule, and lack of a latex system.

ulmic acid *See* ulmin.

ulmin [GEOL] Alkali-soluble organic substances derived from decaying vegetable matter; occurs as amorphous brown to black gel material. Also known as carbohumin; fundamental jelly; fundamental substance; gelose; humin; humogelite; jelly; ulmic acid; vegetable jelly.

ulna [ANAT] The larger of the two bones of the forearm or forelimb in vertebrates; articulates proximally with the humerus and radius and distally with the radius.

Ulotrichaceae [BOT] A family of green algae in the suborder Ulotrichineae; contains both attached and floating filamentous species with cells having parietal, platelike or bandlike chloroplasts.

Ulotrichales [BOT] A large, artificial order of the Chlorophyta composed mostly of fresh-water, branched or unbranched filamentous species with mostly cylindrical, uninucleate cells having cellulose, but often mucilaginous walls.

Ulotrichineae [BOT] A suborder of the Ulotrichales characterized by short cylindrical cells.

ulrichite *See* uraninite.

ultimate analysis [ANALY CHEM] The determination of the percentage of elements contained in a chemical substance.

ultimate elongation [MET] The percentage of permanent deformation remaining after tensile rupture.

ultimate lines [ASTRON] Special spectral lines that can be

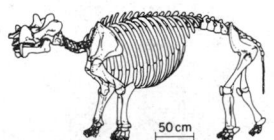
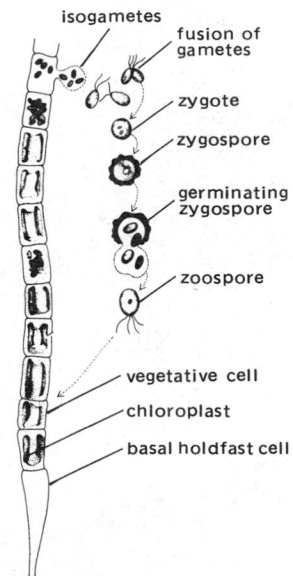

used to indicate the existence of an element in the sun or other star.

ultimate-load design [DES ENG] Design of a beam that is proportioned to carry at ultimate capacity the design load multiplied by a safety factor. Also known as limit-load design; plastic design; ultimate-strength design.

ultimate recovery [PETRO ENG] Estimated total (ultimate) recovery of hydrocarbon fluids expected from a reservoir during its productive lifetime.

ultimate-strength design *See* ultimate-load design.

ultimobranchial body [EMBRYO] One of the small, endocrine structures which originate as terminal outpocketings from each side of the embryonic vertebrate pharynx; can produce the hormone calcitonin.

Ultisol [GEOL] A soil order characterized by an argillic horizon that has a base saturation of less than 35% at a pH of 8.2.

ultra-audion circuit [ELECTR] Regenerative detector circuit in which a parallel resonant circuit is connected between the grid and the plate of a vacuum tube, and a variable capacitor is connected between the plate and cathode to control the amount of regeneration.

ultra-audion oscillator [ELECTR] Variation of the Colpitts oscillator circuit; the resonant circuit employs a section of transmission line.

ultrabasic [PETR] Of igneous rock, having a low silica content, as opposed to the higher silica contents of acidic, basic, and intermediate rocks.

ultrabasite *See* diaphorite.

ultracentrifuge [ENG] A laboratory centrifuge that develops centrifugal fields of more than 100,000 times gravity.

ultra-cold neutron [PHYS] A neutron whose energy is of the order of 10^{-7} electron volt or less, so that it is totally reflected from various materials and suitably constructed magnetic fields, regardless of the angle of incidence, and can be stored in suitably constructed bottles.

Ultrafax [COMMUN] System using radio, facsimile, and television methods to transmit printed information; characterized by the great speed with which printed material may be transmitted over distances.

ultrafiche [GRAPHICS] A sheet of film, usually 4 by 6 inches (10 by 15 centimeters) in size, containing either negative or positive images, or frames, of printed material reduced more than 100 times by photographic reduction.

ultrafilter [MATH] A filter base which has no properly subordinated filter base.

ultrafiltration [CHEM ENG] Separation of colloidal or very fine solid materials by filtration through microporous or semipermeable mediums.

Ultrafining [CHEM ENG] A proprietary fixed-bed catalytic hydrogenation process used in petroleum refineries to desulfurize and upgrade naphtha distillates.

Ultraforming [CHEM ENG] A proprietary catalytic reforming process used to increase the octane ratings of petroleum naphthas.

ultra-high-frequency [COMMUN] The band of frequencies between 300 and 3000 megahertz in the radio spectrum, corresponding to wavelengths of 10 centimeters to 1 meter. Abbreviated UHF.

ultra-high-frequency tuner [ELECTR] A tuner in a television receiver for reception of stations transmitting in the ultra-high-frequency band (channels 14–83); it usually employs continuous tuning.

ultra-high vacuum [PHYS] A vacuum in which the pressure is of the order of 10^{-10} millimeter of mercury or less.

ultra-large crude carrier [NAV ARCH] A liquid-cargo vessel over 250,000 tons. Abbreviated ULCC.

ultramafic [PETR] Referring to igneous rock composed principally of mafic minerals, such as olivine and pyroxene.

ultramarine [INORG CHEM] $Na_3Al_3Si_3S_2O_{12}$ A blue pigment; formerly made from powdered lapis lazuli. [MINERAL] *See* lazurite.

ultramarine green [INORG CHEM] An ultramarine dyestuff with a green shade.

ultramicrobalance [ENG] A differential weighing device with accuracies better than 1 microgram; used for analytical weighings in microanalysis.

ultramicroscope [OPTICS] An instrument for investigating particles of submicroscopic dimensions: it consists of a high-intensity illumination system for producing a Tyndall cone in a colloidal system, coupled with a compound microscope to examine the points of light scattered from the individual particles.

ultramicrotome [ENG] A microtome which uses a glass or diamond knife, allowing sections of cells to be cut 300 nanometers in thickness.

ultraphotic rays [ELECTROMAG] Rays outside the visible part of the spectrum, including infrared and ultraviolet rays.

ultrashort waves [COMMUN] Radio waves shorter than 10 meters in wavelength; corresponding to frequencies above 30 megahertz.

ultrasonic [ACOUS] Pertaining to signals, equipment, or phenomena involving frequencies just above the range of human hearing, hence above about 20,000 hertz. Also known as supersonic (deprecated usage).

ultrasonic atomizer [MECH ENG] An atomizer in which liquid is fed to, or caused to flow over, a surface which vibrates at an ultrasonic frequency; uniform drops may be produced at low feed rates.

ultrasonic camera [ELECTR] A device which produces a picture display of ultrasonic waves sent through a sample to be inspected or through live tissue; a piezoelectric crystal is used to convert the ultrasonic waves to voltage differences, and the voltage pattern on the crystal modulates the intensity of an electronic beam scanning the crystal; this beam in turn controls the intensity of a beam in a television tube.

ultrasonic cleaning [ENG] A method used to clean debris and swarf from surfaces by immersion in a solvent in which ultrasonic vibrations are excited.

ultrasonic coagulation [PHYS] The bonding of small particles into large aggregates by the action of ultrasonic waves.

ultrasonic communication [COMMUN] Communication accomplished through water by keying the sound output of echo-ranging sonar on ships or submarines.

ultrasonic delay line [ENG ACOUS] A delay line in which use is made of the propagation time of sound through a medium such as fused quartz, barium titanate, or mercury to obtain a time delay of a signal. Also known as ultrasonic storage cell.

ultrasonic depth finder [ENG] A direct-reading instrument which employs frequencies above the audible range to determine the depth of water; it measures the time interval between the emission of an ultrasonic signal and the return of its echo from the bottom.

ultrasonic drill [MECH ENG] A drill in which a magnetostrictive transducer is attached to a tapered cone serving as a velocity transformer; with an appropriate tool at the end of the transformer, practically any shape of hole can be drilled in hard, brittle materials such as tungsten carbide and gems.

ultrasonic drilling [MECH ENG] A vibration drilling method in which ultrasonic vibrations are generated by the compression and extension of a core of electrostrictive or magnetostrictive material in a rapidly alternating electric or magnetic field.

ultrasonic echo [ACOUS] An ultrasonic wave that has been reflected or otherwise returned with sufficient magnitude and time delay to be perceived in some manner as a wave distinct from that directly transmitted.

ultrasonic flaw detector [ENG ACOUS] An ultrasonic generator and detector used together, much as in radar, to determine the distance to a wave-reflecting internal crack or other flaw in a solid object.

ultrasonic generator [ENG ACOUS] A generator consisting of an oscillator driving an electroacoustic transducer, used to produce acoustic waves above about 20 kilohertz.

ultrasonic light modulator [OPTICS] Device containing a fluid which, by action of ultrasonic waves passing through the fluid, modulates a beam of light passed transversely through the fluid.

ultrasonic machining [MECH ENG] The removal of material by abrasive bombardment and crushing in which a flat-ended tool of soft alloy steel is made to vibrate at a frequency of about 20,000 hertz and an amplitude of 0.001–0.003 inch (0.0254–0.0762 millimeter) while a fine abrasive of silicon carbide, aluminum oxide, or boron carbide is carried by a liquid between tool and work.

ultrasonic metal inspection [MET] The application of ultrasonic vibrations to materials for detection of flaws, pits, or wall thickness.

ultrasonic microscope [OPTICS] A special type of microscope which employs ultrasonic radiation.

ultrasonic radiation [ACOUS] Ultrasonic waves propagating through a solid, liquid, or gaseous medium.

ultrasonics [ACOUS] The science of ultrasonic sound waves.

ultrasonic sealing [ENG] A method for sealing plastic film by using localized heat developed by vibratory mechanical pressure at ultrasonic frequencies.

ultrasonic soldering [MET] A method used for aluminum soldering by ultrasonically vibrating the soldering iron to disrupt the oxide film on the metal.

ultrasonic storage cell *See* ultrasonic delay line.

ultrasonic testing [ENG] A nondestructive test method that employs high-frequency mechanical vibration energy to detect and locate structural discontinuities or differences and to measure thickness of a variety of materials.

ultrasonic therapy *See* ultrasound diathermy.

ultrasonic thickness gage [ENG] A thickness gage in which the time of travel of an ultrasonic beam through a sheet of material is used as a measure of the thickness of the material.

ultrasonic transducer [ENG ACOUS] A transducer that converts alternating-current energy above 20 kilohertz to mechanical vibrations of the same frequency; it is generally either magnetostrictive or piezoelectric.

ultrasonic transmitter [ENG ACOUS] A device used to track seals, fish, and other aquatic animals: the device is fastened to the outside of the animal or fed to it, and has a loudspeaker which is made to vibrate at an ultrasonic frequency, propagating ultrasonic waves through the water to a special microphone or hydrophone.

ultrasonic wave [ACOUS] A sound wave that has a frequency above about 20,000 hertz.

ultrasonic welding [MET] A nonfusion welding process in which the atomic movement required for coalescence is stimulated by ultrasonic vibrations.

ultrasound diathermy [MED] The application of high-frequency sound waves (0.7 to 1.0 megahertz) and the conversion of this mechanical energy into heat for local thermotherapy. Also known as ultrasonic therapy.

ultraspeed welding *See* commutator-controlled welding.

ultrastrip [GRAPHICS] A piece of film, usually 1.5 inches by 7 inches (3.5 by 18 centimeters) in size, containing either negative or positive images, or frames, of printed material reduced 150 times by photographic reduction.

ultrastructure [MOL BIO] The ultimate physiochemical organization of protoplasm.

ultraviolet [PHYS] Pertaining to ultraviolet radiation. Abbreviated UV.

ultraviolet absorber [OPTICS] Any substance that absorbs ultraviolet radiant energy, then dissipates the energy in a harmless form; used in plastics and rubbers to decrease light sensitivity.

ultraviolet absorber fixative [MATER] A protective fixative that includes a material to filter ultraviolet light from the sun and from artificial light; it helps to keep colors from fading.

ultraviolet absorption [OPTICS] Absorption of specific ultraviolet radiation wavelengths by a material; for example, by a sample solution during spectroscopic analysis.

ultraviolet absorption spectrophotometry [SPECT] The study of the spectra produced by the absorption of ultraviolet radiant energy during the transformation of an electron from the ground state to an excited state as a function of the wavelength causing the transformation.

ultraviolet catastrophe [STAT MECH] The prediction of the Rayleigh-Jeans law that the energy radiated by a blackbody at extremely short wavelengths is extremely large, and the total energy radiated is infinite, whereas in reality it must be finite.

ultraviolet densitometry [SPECT] An ultraviolet-spectrophotometry technique for measurement of the colors on thin-layer chromatography absorbents following elution.

ultraviolet imagery [ELECTROMAG] That imagery produced as a result of sensing ultraviolet radiations reflected from a given target surface.

ultraviolet lamp [ELECTR] A lamp providing a high propor-

tion of ultraviolet radiation, such as various forms of mercury-vapor lamps.

ultraviolet light *See* ultraviolet radiation.

ultraviolet microscope [OPTICS] A special type of microscope which uses electromagnetic radiation in the range 180–400 nanometers; it requires reflecting optics or special quartz and crystal objectives.

ultraviolet photography [GRAPHICS] Photography in which the subject is illuminated with ultraviolet light, and either the resulting fluorescence (in the fluorescence method) or the reflected ultraviolet light (in the reflected ultraviolet method) is detected by the camera.

ultraviolet radiation [ELECTROMAG] Electromagnetic radiation in the wavelength range 4–400 nanometers; this range begins at the short-wavelength limit of visible light and overlaps the wavelengths of long x-rays (some scientists place the lower limit at higher values, up to 40 nanometers). Also known as ultraviolet light.

ultraviolet spectrometer [SPECT] A device which produces a spectrum of ultraviolet light and is provided with a calibrated scale for measurement of wavelength.

ultraviolet spectrophotometry [SPECT] Determination of the spectra of ultraviolet absorption by specific molecules in gases or liquids (for example, Cl_2, SO_2, NO_2, CS_2, ozone, mercury vapor, and various unsaturated compounds).

ultraviolet spectroscopy [SPECT] Absorption spectroscopy involving electromagnetic wavelengths in the range 4–400 nanometers.

ultraviolet spectrum [ELECTROMAG] **1.** The range of wavelengths of ultraviolet radiation, covering 4–400 nanometers. **2.** A display or graph of the intensity of ultraviolet radiation emitted or absorbed by a material as a function of wavelength or some related parameter.

ultraviolet stabilizer *See* UV stabilizer.

ultravulcanian [GEOL] A type of volcanic eruption characterized by periodic violent gaseous explosions of lithic dust and solid blocks, with little if any fiery scoria.

Ulvaceae [BOT] A large family of green algae in the order Ulvales.

Ulvales [BOT] An order of algae in the division Chlorophyta in which the thalli are macroscopic, attached tubes or sheets.

umbel [BOT] An indeterminate inflorescence with the pedicels all arising at the top of the peduncle and radiating like umbrella ribs; there are two types, simple and compound.

Umbellales [BOT] An order of dicotyledonous herbs or woody plants in the subclass Rosidae with mostly compound or conspicuously lobed or dissected leaves, well-developed schizogenous secretory canals, separate petals, and an inferior ovary.

Umbelliferae [BOT] A large family of aromatic dicotyledonous herbs in the order Umbellales; flowers have an ovary of two carpels, ripening to form a dry fruit that splits into two halves, each containing a single seed.

umbilical artery [EMBRYO] Either of a pair of arteries passing through the umbilical cord to carry impure blood from the mammalian fetus to the placenta.

umbilical connections [AERO ENG] Electrical and mechanical connections to a launch vehicle prior to lift off; the umbilical tower adjacent to the vehicle on the launch pad supports these connections which supply electrical power, control signals, data links, propellant loading, high pressure gas transfer, and air conditioning.

umbilical cord [AERO ENG] Any of the servicing electrical or fluid lines between the ground or a tower and an uprighted rocket vehicle before the launch. Also known as umbilical. [EMBRYO] The long, cylindrical structure containing the umbilical arteries and vein, and connecting the fetus with the placenta.

umbilical duct *See* vitelline duct.

umbilical hernia [MED] Herniation through the umbilical ring. Also known as annular hernia.

umbilical tower [AERO ENG] A vertical structure supporting the umbilical cords running into a rocket in launching position.

umbilical vein [EMBRYO] A vein passing through the umbilical cord and conveying purified, nutrient-rich blood from placenta to fetus.

Umbilicariaceae [BOT] The rock tripes, a family of Ascoli-

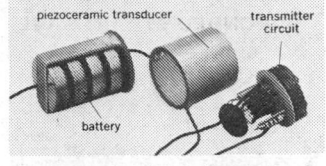

ULTRASONIC TRANSMITTER

Components of an ultrasonic fish-tag transmitter. The ceramic cylinder in the center is 0.5 inch (1.27 centimeters) in diameter. (*Sensory Systems Laboratory, Tucson, Ariz.*)

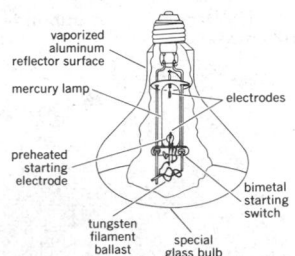

ULTRAVIOLET LAMP

Features of a sunlamp, a type of ultraviolet lamp that produces radiation in the middle ultraviolet region (280–320 nanometers) together with infrared radiation and light from the filament.

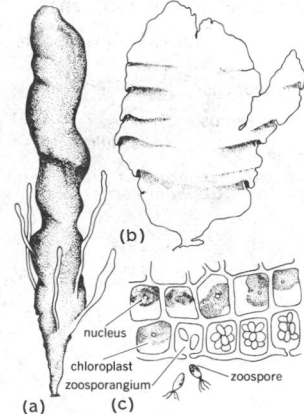

ULVALES

Ulvales. (*a*) *Enteromorpha*, a tubular form. (*b*) *Ulva*, the sea lettuce, an expanded thallus. (*c*) *Ulva*, a portion of the marginal cells of the thallus showing chloroplasts and zoospore formation.

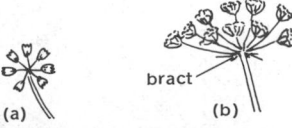

UMBEL

Two types of umbel inflorescence: (*a*) simple; (*b*) compound.

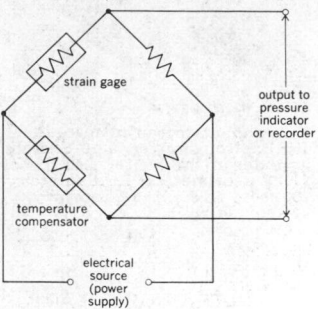

UNBONDED STRAIN GAGE

Circuit diagram of an unbonded strain gage.

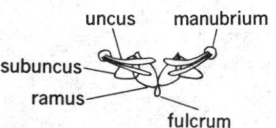

UNCINATE TROPHUS

uncus   manubrium
subuncus
ramus
fulcrum

Uncinate trophus of a predacious rotifier, *Stephanoceros*.

chenes in the order Lecanorales having a large, circular, umbilicate thallus.

umbilicus [ANAT] The navel; the round, depressed cicatrix in the median line of the abdomen, marking the site of the aperture through which passed the fetal umbilical vessels.

umbo [ANAT] A rounded elevation of the surface of the tympanic membrane. [INV ZOO] A prominence above the hinge of a bivalve mollusk shell.

umbonate [BIOL] Having or forming an umbo.

umbra [OPTICS] That portion of a shadow which is screened from light rays emanating from any part of an extended source.

umbrella antenna [ELECTROMAG] Antenna in which the wires are guyed downward in all directions from a central pole or tower to the ground, somewhat like the ribs of an open umbrella.

Umbrept [GEOL] A suborder of the Inceptisol soil order; has dark A horizon more than 10 inches (25 centimeters) thick, brown B horizons, and slightly paler C horizons; soil is strongly acid, and clay minerals are crystalline; occurs in cool or temperate climates.

Umbriel [ASTRON] One of the five satellites of the planet Uranus, with a diameter of about 400 kilometers.

Umkehr effect [OPTICS] An anomaly of the relative zenith intensities of scattered sunlight at certain wavelengths in the ultraviolet as the sun approaches the horizon; it is due to the presence of the ozone layer.

Umklapp process [SOLID STATE] The interaction of three or more waves in a solid, such as lattice waves or electron waves, in which the sum of the wave vectors is not equal to zero but, rather, is equal to a vector in the reciprocal lattice. Also known as flip-over process.

UMP *See* uridylic acid.

umpire [MIN ENG] An assay made by a third party to settle the difference in assays made by the purchaser and the seller of ore.

unaka [GEOL] A large residual mass rising above a peneplain that is less well developed than one having a monodnock.

unakite [PETR] An altered igneous rock composed principally of epidote, pink orthoclase, and quartz.

unamplified back bias [ELECTR] Degenerative voltage developed across a fast time constant circuit within an amplifier stage itself.

unarmed [ORD] The condition of a fuse (or other firing device) in which the necessary steps to put it in condition to function have not taken place; while in the unarmed state the fuse is safe to handle, store, and transport.

unary operation [MATH] An operation in which only a single operand is required to produce a unique result; some examples are negation, complementation, square root, transpose, inverse, and conjugate.

unattended operation [ADP] An operation in which components in the hardware of a communications terminal or data-processing system operate automatically, allowing handling of signals or data without human intervention.

unavailable energy [THERMO] That part of the energy which, when an irreversible process takes place, is initially in a form completely available for work and is converted to a form completely unavailable for work.

unavoidable delay [IND ENG] Any delay in a task, the occurrence of which is outside the control or responsibility of the worker.

unavoidable-delay allowance [IND ENG] An adjustment of standard time to allow for unavoidable delays in a task.

unbalanced cutter chain [MIN ENG] A cutter chain which carries more picks along the bottom line than along the top line.

unbalanced hoisting [MIN ENG] The method of hoisting in small one-compartment shafts with only one cage in operation, as opposed to balanced winding.

unbalanced line [ELEC] A transmission line in which the voltages on the two conductors are not equal with respect to ground; a coaxial line is an example.

unbalanced output [ELEC] An output in which one of the two input terminals is substantially at ground potential.

unbalanced shothole [MIN ENG] A shothole in which the explosive charge breaks down the coal at the back of the

machine cut while leaving the front portion standing or in large blocks.

unbalanced wire circuit [ELEC] A wire circuit whose two sides are inherently electrically unlike.

unbiased estimate [STAT] An estimate for a parameter θ whose expected value is θ.

unblanking pulse [ELECTR] Voltage applied to a cathode-ray tube to overcome bias and cause trace to be visible.

unbonded strain gage [ENG] A type of strain gage that consists of a grid of fine wires strung under slight tension between a stationary frame and a movable armature; pressure applied to the bellows or to the diaphragm sensing element moves the armature with respect to the frame, increasing tension in one half of the filaments and decreasing tension in the rest.

unbounded manifold [MATH] A manifold with no boundary.

unbounded set of real numbers [MATH] A set with the property that if R is any positive real number, there is a number in the set which is smaller than $-R$ or a number larger than R.

unbounded wave [PHYS] A wave which propagates through a nondissipative, homogeneous medium which is infinite in extent, without any boundaries.

unbundling [ADP] The separate pricing of software products and services from equipment charges.

uncage [ENG] To release the caging mechanism of a gyroscope, that is, the mechanism that erects the gyroscope or locks it in position.

uncertainty [SCI TECH] The estimated amount by which an observed or calculated value may depart from the true value.

uncertainty principle [QUANT MECH] The precept that the accurate measurement of an observable quantity necessarily produces uncertainties in one's knowledge of the values of other observables. Also known as Heisenberg uncertainty principle; indeterminancy principle.

uncertainty relation [QUANT MECH] The relation whereby, if one simultaneously measures values of two canonically conjugate variables, such as position and momentum, the product of the uncertainties of their measured values cannot be less than approximately Planck's constant divided by 2π. Also known as Heisenberg uncertainty relation.

uncharged [ELEC] Having no electric charge.

uncharged demolition target [ENG] A demolition target which has been prepared to receive the demolition agent, the necessary quantities of which have been calculated, packaged, and stored in a safe place.

uncinate trophus [INV ZOO] A trophus in rotifers characterized by a hooked or curved uncus.

uncompetitive enzyme inhibition [BIOCHEM] The prevention of an enzymic process as a result of the interaction of an inhibitor with the enzyme-substrate complex or a subsequent intermediate form of the enzyme, but not with the free enzyme.

unconcentrated wash *See* sheet erosion.

unconditional [ADP] Not subject to conditions external to the specific instruction.

unconditional convergence [MATH] A convergent series converges unconditionally if every series obtained by rearranging its terms also converges; equivalent to absolute convergence.

unconditional inequality [MATH] An inequality which holds true for all values of the variables involved, or which contains no variables; for example, $y + 2 > y$, or $4 > 3$. Also known as absolute inequality.

unconditional jump [ADP] A digital-computer instruction that interrupts the normal process of obtaining instructions in an ordered sequence, and specifies the address from which the next instruction must be taken. Also known as unconditional transfer.

unconditional transfer *See* unconditional jump.

unconfined explosion [ENG] Explosion occurring in the open air where the (atmospheric) pressure is constant.

unconformable [GEOL] Pertaining to strata that do not conform in position, dip, or strike to the older underlying rocks.

unconformity [GEOL] The relation between adjacent rock strata whose time of deposition was separated by a period of nondeposition or of erosion; a break in a stratigraphic sequence.

unconformity iceberg [OCEAN] An iceberg consisting of more than one kind of ice, such as blue water-formed ice and névé; such an iceberg often contains many crevasses and silt bands.

unconscious [MED] Insensible; in a state lacking conscious awareness, with reflexes abolished. [PSYCH] **1.** Pertaining to behavior or experience not controlled by the ego. **2.** The part of the mind, mental functioning, or personality not in the immediate field of awareness.

unconsolidated material [GEOL] Loosely arranged or unstratified sediment whose particles are not cemented together.

uncontrolled fragments [ORD] The segments of the casing propelled outward at detonation of an explosive device in which no design provision has been made to control the size and shape of the segments.

uncontrolled mosaic [GRAPHICS] A mosaic composed of uncorrected photographs, the details of which have been matched from print to print without ground control or other orientation; accurate measurement and direction cannot be accomplished.

unconventional warfare [ORD] A type of warfare that includes the interrelated fields of guerrilla warfare, evasion and escape, and subversion, in which operations are conducted within enemy or enemy-controlled territory by predominantly indigenous personnel, usually supported and directed in varying degrees by an external source.

uncorrecting [NAV] The process of converting true to magnetic, compass, or gyro direction, or magnetic to compass direction.

uncorrelated random variables [STAT] Two random variables whose correlation coefficient is zero.

uncountable set [MATH] An infinite set which can not be put in 1 to 1 correspondence with the set of integers; for example, the set of real numbers.

uncouple [ENG] To unscrew or disengage.

uncoupling phenomena [SPECT] Deviations of observed spectra from those predicted in a diatomic molecule as the magnitude of the angular momentum increases, caused by interactions which could be neglected at low angular momenta.

unctuous [MATER] Greasy, oily, or soapy to the touch.

undamped wave [PHYS] A continuous wave produced by oscillations having constant amplitude.

n-undecane [ORG CHEM] $CH_3(CH_2)_9CH_3$ A colorless, combustible liquid, boiling at 367°F (196°C), flash point at 149°F (65°C); used as a chemical intermediate and in petroleum research.

undecanoic acid [ORG CHEM] $CH_3(CH_2)_9COOH$ Colorless crystals, soluble in alcohol and ether, insoluble in water; melts at 29°C; used as a chemical intermediate. Also known as hendecanoic acid; n-undecylic acid.

undecyl [ORG CHEM] $C_{11}H_{23}$ The radical of undecane. Also known as hendecyl.

undecylenic acid [ORG CHEM] $CH_2 \cdot CH(CH_2)_8COOH$ A light-colored, combustible liquid with a fruity aroma; soluble in alcohol, ether, chloroform, and benzene, almost insoluble in water; used in medicine, perfumes, flavors, and plastics.

n-undecylic acid See undecanoic acid.

underbunching [ELECTR] In velocity-modulated electron streams, a condition representing less than the optimum bunching.

undercarriage [AERO ENG] The landing gear assembly for an aircraft. [ORD] Fixed or movable base on which the top carriage of a weapon moves.

undercast [METEOROL] A cloud layer of ten-tenths (1.0) coverage as viewed from an observation point above the layer; the term is used in pilot reporting of in-flight weather conditions. [MIN ENG] An air crossing in which one airway is deflected to pass under the other airway.

underchain haulage [MIN ENG] Haulage in which the chains are placed beneath the mine car at certain intervals with suitable hooks that thrust against the car axle.

underclay [GEOL] A layer of clay or other fine-grained detrital material underlying a coal bed or comprising the floor of a coal seam. Also known as coal clay; root clay; seat clay; seat earth; thill; underearth; warrant.

underclay limestone [GEOL] A thin, fresh-water limestone that is relatively free of fossils and is dense and nodular; found in underlying coal deposits.

undercliff [GEOL] A subordinate cliff or terrace formed by material which has fallen or slid from above.

undercoater [MATER] A type of solvent-thinned paint that is formulated to have good adhesion to a substrate and furnish a good base for additional coats of paint, and having a relatively small amount of pigment.

underconsolidation [GEOL] Less than normal consolidation of sedimentary material for the existing overburden.

undercooling [MET] Cooling a metal below the transformation temperature without obtaining the transformation.

undercurrent [OCEANOGR] A water current flowing beneath a surface current at a different speed or in a different direction.

undercurrent relay [ELEC] A relay designed to operate when its coil current falls below a predetermined value.

undercut [ENG] Underside recess either cut or molded into an object so as to leave a topside lip or protuberance. [MET] An unfilled groove melted into the base metal at the toe of a weld. [MIN ENG] To cut below or in the lower part of a coal bed by chipping away the coal with a pick or mining machine; cutting is usually done on the level of the floor of the mine, extending laterally the entire face and 5 or 6 feet (1.5 or 1.8 meters) into the material.

undercutting [CHEM ENG] In distillation, the technique of taking the products coming off the distillation tower at a temperature below the desired ultimate boiling point range to prevent contaminating the products with the compound that would distill just beyond the ultimate boiling point range. [GEOL] Erosion of material at the base of a steep slope, cliff, or other exposed rock.

underdamping [PHYS] Condition of a system when the amount of damping is sufficiently small so that, when the system is subjected to a single disturbance, either constant or instantaneous, one or more oscillations are executed by the system.

underdeck tonnage [NAV ARCH] The enclosed volume of a vessel below the tonnage deck, expressed in tons of 100 cubic feet (approximately 2.8317 cubic meters).

underdevelopment [GRAPHICS] Insufficient development of a photographic print; processing to a degree lower than the optimum density.

underdraft [MET] Downward curving of a metal part on leaving the rolls because of higher speed of the upper roll.

underdrain [CIV ENG] A subsurface drain with holes into which water flows when the water table reaches the drain level.

underdrive press [MECH ENG] A mechanical press having the driving mechanism located within or under the bed.

underearth See underclay.

underfeed stoker [ENG] A coal-burning system in which green coal is fed from beneath the burning fuel bed.

underfit stream [HYD] A misfit stream that appears to be too small to have eroded the valley in which it flows.

underfloor raceway [BUILD] A raceway for electric wires which runs beneath the floor.

underflow conduit [GEOL] A permeable deposit underlying a surface stream channel.

underground [ENG] Situated, done, or operating beneath the surface of the ground.

underground burst [ORD] The explosion of a nuclear weapon beneath the surface of the ground.

underground gasification See gasification.

underground geology See subsurface geology.

underground glory-hole method [MIN ENG] A mining method used in large deposits with a very strong roof: the deposit is divided by levels and on every level chutes are raised to the next level; mining starts from the mouth of the chutes in such a way as to develop a funnel-shaped excavation (mill, or glory) with slopes so steep that the broken ore falls into the chutes and thus to the cars on the lower level; a sufficiently strong pillar is left for protection at the higher level. Also known as underground milling.

underground ice See ground ice.

underground milling See underground glory-hole method.

underground stem [BOT] Any of the stems that grow underground and are often mistaken for roots; principal kinds are

UNDERFEED STOKER

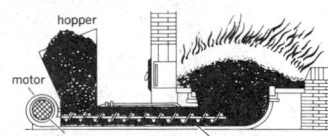

Section through an underfeed stoker.

rhizomes, tubers, corms, bulbs, and rhizomorphic droppers.

underground stream [HYD] A subsurface body of water flowing in a definite current in a distinct channel.

underhand stoping [MIN ENG] Mining downward or from upper to lower level; the stope may start below the floor of a level and be extended by successive horizontal slices, either worked sequentially or simultaneously in a series of steps; the stope may be left as an open stope or supported by stulls or pillars.

underhand work [MIN ENG] Picking or drilling downward.

underhead crack [MET] A subsurface crack in the heat-affected zone of the base metal near a weld.

underhole [MIN ENG] To mine out a portion of the bottom of a seam, by pick or powder, thus leaving the top unsupported and ready to be blown down by shots, broken down by wedges, or mined with a pick or bar.

underhung crane [MECH ENG] An overhead traveling crane in which the end trucks carry the bridge suspended below the rails.

underlap [COMMUN] 1. In facsimile transmission, the space between the recorded elemental area in one recording line and the adjacent elemental area in the next recording line, when these areas are smaller than normal; or the space between the elemental areas in the direction of the recording line. 2. The amount by which the effective height of the scanning spot falls short of the nominal width of the scanning line.

underlay shaft [MIN ENG] A shaft sunk in the footwall and following the dip of a vein. Also known as footwall shaft; underlier.

underlie [GEOL] To lie or be situated under; to occupy a lower position, or to pass beneath.

underlier *See* underlay shaft.

undermelting [HYD] The melting from below of any floating ice.

undermine [MIN ENG] To excavate the earth beneath, especially for the purpose of causing to fall; to form a mine under.

underpainting [GRAPHICS] A layer of paint that is intended to be seen through a subsequent paint layer; an underpainted section of red, for example, can be altered to a violet or purple color with the application of a thin blue glaze.

underpinning [CIV ENG] 1. Permanent supports replacing or reinforcing older ones beneath a wall or a column. 2. Braced props temporarily supporting a structure. [MIN ENG] Building up the wall of a mine shaft to join that above it.

underpunch [ADP] A second hole in a card column which is below the original hole in IBM card code.

undersaturated [PETR] Pertaining to igneous rock composed of unsaturated minerals, that is, without free silica.

undersaturated fluid [PHYS CHEM] Any fluid (liquid or gas) capable of holding additional vapor or liquid components in solution at specified conditions of pressure and temperature.

undersea mining [MIN ENG] The working of economic deposits (usually coal) situated in strata or rocks below the seabed.

undershot wheel [MECH ENG] A water wheel operated by the impact of flowing water against blades attached around the periphery of the wheel, the blades being partly or totally submerged in the moving stream of water.

undersize [ENG] That part of a crushed material (for example, ore) which passes through a screen.

underspin [MECH] Property of a projectile having insufficient rate of spin to give proper stabilization.

understressing [MET] Repeated stressing below the fatigue limit or below the final applied stress; can improve fatigue properties as a result of strain-aging effects.

underthrow distortion [COMMUN] Distortion occurring in facsimile when the maximum signal amplitude is too low.

underthrust [GEOL] A thrust fault in which the lower, active rock mass has been moved under the upper, passive rock mass.

undertow [OCEANOGR] A subsurface seaward movement by gravity flow of water carried up on a sloping beach by waves or breakers.

undervoltage protection [ELEC] An undervoltage relay which removes a motor from service when a low-voltage condition develops, so that the motor will not draw excessive

current, or which prevents a large induction or synchronous motor from starting under low-voltage conditions.

undervoltage relay [ELEC] A relay designed to operate when its coil voltage falls below a predetermined value.

underwater acoustics [ACOUS] Study of the propagation of sound waves in water, especially in the oceans, and of phenomena produced by these sound waves. Also known as hydroacoustics.

underwater antenna [ELECTROMAG] An antenna placed and used underwater.

underwater burst [ORD] The explosion of a nuclear weapon beneath the surface of the water.

underwater camera [OPTICS] A camera designed for use under the surface of the water; it is usually a conventional type enclosed in a casing to withstand water pressure, preferably with a correction lens to compensate for aberrations caused by the water.

underwater demolition [ORD] Destruction or fragmentation of underwater obstacles by use of explosive charges placed by diver personnel; primarily employed as a short-range emergency measure to accomplish a military objective with promptness and economy of material.

underwater demolition team [ORD] Naval unit organized and equipped to perform beach reconnaissance and underwater demolition in an amphibious operation.

underwater mine [ORD] A mine designed to be located underwater and exploded by means of propeller vibration, magnetic attraction, contact, or remote control.

underwater mine extender mechanism [ORD] An item designed to extend the detonator into the booster for arming an underwater mine by means of hydrostatic pressure.

underwater mine fairing [ORD] An item designed to be mounted on an underwater mine; it is shaped to reduce or equally distribute air resistance when suspended and launched from an aircraft, and it may have collapsible fins.

underwater navigation [NAV] The navigation of a submerged vessel. Also known as submarine navigation.

underwater obstacle [NAV] A natural or artificial obstacle which is located to seaward of the high-water line and wholly or partly submerged, and which acts as a barrier or obstruction to the passage of ships, landing ships, craft, vehicles, or torpedoes.

underwater ordnance [ORD] Munitions designed for use underwater; for example, torpedoes.

underwater sound [ACOUS] The production, transmission, and reception of sounds in the ocean; used for locating submarines and other submerged objects, and to determine the physical structure of the ocean and its bottom, and to study organisms found in the sea.

underwater sound projector [ENG ACOUS] A transducer used to produce sound waves in water. Also known as projector.

underwater telephone [COMMUN] A method of voice communication using underwater sound as a means of transmission; it functions similarly to a conventional telephone system except that the energy is carried by sound waves in the water rather than by electrical signals through a wire.

underwater television [COMMUN] The technique of using remotely controlled television equipment under the surface of the water.

underwater transducer [ENG ACOUS] A device used for the generation or reception of underwater sounds.

underway [NAV] 1. A ship without moorings; not secured in any way to the ground or a wharf. 2. Craft in motion, particularly the start of such motion after a standstill.

underway bottom sampler *See* underway sampler.

underway sampler [ENG] A device for collecting samples of sediment on the ocean bottom, consisting of a cup in a hollow tube; on striking the bottom, the cup scoops up a small sample which is forced into the tube which is then closed with a lid, and the device is hoisted to the surface. Also known as scoopfish; underway bottom sampler.

underwing [INV ZOO] Either of a pair of posterior wings on certain insects, as the moth.

Underwood chart [CHEM ENG] A graphical solution of mass balances for a single equilibrium stage in the calculation of a solvent-extraction operation.

Underwood distillation method [CHEM ENG] A method for

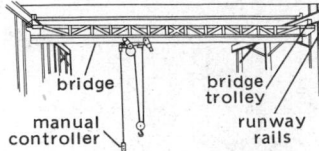

UNDERHUNG CRANE

bridge / bridge trolley / manual controller / runway rails

An underhung crane, a basic type of overhead-traveling crane.

UNDERWATER CAMERA

A picture of a photographer using underwater motion picture camera, taken at a depth of 120 feet (37 meters) using high-speed Tri-X film and camera settings of 1/100 second an *f*/4.0.

calculation of liquid separations from binary distillation systems operated at partial reflux.

undetected error rate [COMMUN] The number of bits (or other units of information) which are received but are not detected or corrected by error-control equipment, divided by the total number of bits (or other units of information) transmitted. Also known as residual error rate.

undistorted wave [COMMUN] Periodic wave in which both the attenuation and velocity of propagation are the same for all sinusoidal components, and in which no sinusoidal component is present at one point that is not present at all points.

undisturbed [ENG] Pertaining to a sample of material, as of soil, subjected to so little disturbance that it is suitable for determinations of strength, consolidation, permeability characteristics, and other properties of the material in place.

undisturbed motion [PHYS] The steady state of a system before perturbations are introduced.

undisturbed-one output [ELECTR] "One" output of a magnetic cell to which no partial-read pulses have been applied since that cell was last selected for writing.

undisturbed-zero output [ELECTR] "Zero" output of a magnetic cell to which no partial-write pulses have been applied since that cell was last selected for reading.

undulant fever *See* brucellosis.

undulating light [NAV] A continuously luminous light which alternately increases and decreases in brightness in cyclic sequence; the expression is applied primarily to aeronautical lights, the marine light equivalent being fixed and flashing light.

undulatory extinction [OPTICS] Extinction that occurs successively in adjacent areas as the microscope stage is turned. Also known as oscillatory extinction; strain shadow; wavy extinction.

undulatus *See* billow cloud.

unexploded ordnance [ORD] An object containing explosives which did not function as intended, or an object which contains some type of delay-action device.

unfavorable current [NAV] A current flowing in such a direction as to decrease the speed of a vessel.

unfavorable winds [NAV] Winds which delay the progress of a craft in a desired direction; used chiefly in connection with sailing vessels.

unfinished bolt [DES ENG] One of three degrees of finish in which standard hexagon wrench-head bolts and nuts are available; only the thread is finished.

unfired pressure vessel [CHEM ENG] A pressure vessel that is not in direct contact with a heating flame.

unfired tube [ELECTR] Condition of a TR, ATR, or pre-TR tube in which there is no radio-frequency glow discharge at either the resonant gap or resonant window.

unfreezing [GEOL] The upward movement of stones to the surface as a result of repeated freezing and thawing of the containing soil.

ungula [VERT ZOO] A nail, hoof, or claw.

ungulate [VERT ZOO] Referring to an animal that has hoofs.

ungulicutate [VERT ZOO] Having claws or nails.

unguligrade [VERT ZOO] Walking on hoofs.

uniaxial crystal [OPTICS] A doubly refracting crystal which has a single axis along which light can propagate without exhibiting double refraction.

uniaxial stress [MECH] A state of stress in which two of the three principal stresses are zero.

unicellular [BIOL] Composed of a single cell.

unicellular gland [ANAT] A gland consisting of a single cell.

unicuspid [ANAT] Having one cusp, as certain teeth.

unidentified flying object [SCI TECH] Any reported flying object which cannot be identified or explained. Abbreviated UFO. Also known as flying saucer.

unidirectional [PHYS] **1.** Flowing in only one direction, such as direct current. **2.** Radiating in only one direction.

unidirectional antenna [ELECTROMAG] An antenna that has a single well-defined direction of maximum gain.

unidirectional coupler [ELECTR] Directional coupler that samples only one direction of transmission.

unidirectional log-periodic antenna [ELECTROMAG] Broadband antenna in which the cut-out portions of a log-periodic antenna are mounted at an angle to each other, to give a

unidirectional radiation pattern in which the major radiation is in the backward direction, off the apex of the antenna; impedance is essentially constant for all frequencies, as is the radiation pattern.

unidirectional microphone [ENG ACOUS] A microphone that is responsive predominantly to sound incident from one hemisphere, without picking up sounds from the sides or rear.

unidirectional pulse-amplitude modulation [COMMUN] Modulation of pulse-amplitude type in which all pulses rise in the same direction. Also known as single-polarity pulse-amplitude modulation.

unidirectional pulses [ELECTR] Single polarity pulses which all rise in the same direction.

unidirectional transducer [ELECTR] Transducer that measures stimuli in only one direction from a reference zero or rest position. Also known as unilateral transducer.

unified field theory [RELAT] Any theory which attempts to express gravitational theory and electromagnetic theory within a single unified framework; usually, an attempt to generalize Einstein's general theory of relativity from a theory of gravitation alone to a theory of gravitation and classical electromagnetism.

unified screw thread [DES ENG] Limited to three series of threads: coarse (UNC), fine (UNF), and extra fine (UNEF); a ¼-inch (0.006 millimeter) diameter thread in the UNC series has 20 threads per inch, while in the UNF series it has 28.

uniflow engine [MECH ENG] A steam engine in which steam enters the cylinder through valves at one end and escapes through openings uncovered by the piston as it completes its stroke.

uniform bound [MATH] A number M such that $|f_n(x)| < M$ for every x and for every function in a given sequence of functions $\{f_n(x)\}$.

uniform boundedness principle [MATH] A family of pointwise bounded, real-valued continuous functions on a complete metric space X is uniformly bounded on some open subset of X.

uniform circular motion [MECH] Circular motion in which the angular velocity remains constant.

uniform continuity [MATH] A property of a function f on a set, namely: given any $\varepsilon > 0$ there is a $\delta > 0$ such that $|f(x_1) - f(x_2)| < \varepsilon$ provided $|x_1 - x_2| < \delta$ for any pair x_1, x_2 in the set.

uniform convergence [MATH] A sequence of functions $\{f_n(x)\}$ converges uniformly on E to $f(x)$ if given $\varepsilon > 0$ there is an N such that $|f_n(x) - f(x)| < \varepsilon$ for all $x \, \varepsilon E$ provided $n > N$.

uniform corrosion [MET] Corrosion which takes place uniformly over the entire exposed surfaces.

uniform distribution [STAT] The distribution of a random variable in which each value has the same probability of occurrence. Also known as rectangular distribution.

uniform field [PHYS] A field which, at the instant under consideration, has the same value at every point in the region under consideration.

uniformitarianism [GEOL] Classically, the concept that the present is the key to the past; the principle that contemporary geologic processes have occurred in the same regular manner and with essentially the same intensity throughout geologic time, and that events of the geologic past can be explained by phenomena observable today. Also known as principle of uniformity.

uniform line [ELEC] Line which has substantially identical electrical properties through its length.

uniform load [MECH] A load distributed uniformly over a portion or over the entire length of a beam; measured in pounds per foot.

uniform luminance [OPTICS] Property of a surface for which the luminous intensity of any area of the surface is proportional to the area.

uniformly most powerful test [STAT] A test which is simultaneously most powerful for all alternatives of interest in an experiment.

uniform mat [CIV ENG] A type of foundation mat, consisting of a reinforced concrete slab of constant thickness, supporting walls, and columns; it is thick, rigid, and strong.

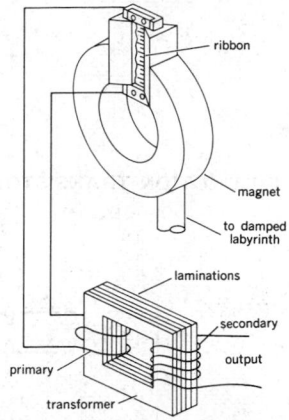

**UNIDIRECTIONAL
MICROPHONE**

Elements of a single-unit ribbon-type unidirectional microphone showing the dynamic ribbon-type transducer.

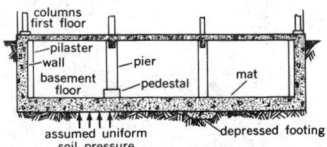

UNIFORM MAT

Foundation construction with a uniform mat.

uniform plane wave [ELECTROMAG] Plane wave in which the electric and magnetic intensities have constant amplitude over the equiphase surfaces; such a wave can only be found in free space at an infinite distance from the source.

uniform space [MATH] A topological space X whose topology is derived from a family of subsets of $X \times X$, called a uniformity; intuitively, this gives a notion of "nearness" which is uniform throughout the space.

uniform system of maritime buoyage [NAV] The international system of buoyage in accordance with fixed rules as to shape, color, lights, topmarks, position, and so forth.

uniform twist rifling [ORD] Rifling in which the degree of twist is constant from the origin of rifling to the muzzle, the path of the groove being a uniform spiral.

unijunction transistor [ELECTR] An n-type bar of semiconductor with a p-type alloy region on one side; connections are made to base contacts at either end of the bar and to the p-region. Abbreviated UJT. Formely known as double-base diode; double-base junction diode.

unilateral bearing [NAV] A bearing obtained with a radio direction finder which does not have a possible reciprocal ambiguity.

unilateral conductivity [ELECTR] Conductivity in only one direction, as in a perfect rectifier.

unilateral hermaphroditism [ZOO] A type of hermaphroditism in which there is a combination of ovatestis on one side of the body with an ovary or testis on the other side.

unilateralization [ELECTR] Use of an external feedback circuit in a high-frequency transistor amplifier to prevent undesired oscillation by canceling both the resistive and reactive changes produced in the input circuit by internal voltage feedback; with neutralization, only the reactive changes are canceled.

unilateral surface [MATH] A one-sided surface; equivalently, any nonorientable two-dimensional manifold such as the Möbius strip and the Klein bottle.

unilateral tolerance method [DES ENG] Method of dimensioning and tolerancing wherein the tolerance is taken as plus or minus from an explicitly stated dimension; the dimension represents the size or location which is nearest the critical condition (that is maximum material condition), and the tolerance is applied either in a plus or minus direction, but not in both directions, in such a way that the permissible variation in size or location is away from the critical condition.

unilateral transducer See unidirectional transducer.

unilocular [BIOL] Having a single cavity.

unimodular matrix [MATH] A unimodulus matrix with integer entries.

unimodulus matrix [MATH] A square matrix whose determinant is 1.

unimolecular reaction [PHYS CHEM] A chemical reaction involving only one molecular species as a reactant; for example, $2H_2O \rightarrow 2H_2 + O_2$, as in the electrolytic dissociation of water.

uninhibited bladder [MED] An abnormal urinary bladder that shows only a variable loss of cerebral inhibition over reflex bladder contractions, representing, of all neurogenic bladders, the least variance from normal.

union [DES ENG] A screwed or flanged pipe coupling usually in the form of a ring fitting around the outside of the joint. [MATH] A union of a given family of sets is a set consisting of those elements which are members of at least one set in the family.

union cloth [TEXT] Fabric made by using one kind of fiber for the warp and another kind for the filling.

union dye [TEXT] A special dye used on union cloth.

Unionidae [INV ZOO] The fresh-water mussels, a family of bivalve mollusks in the subclass Eulamellibranchia; the larvae, known as glochidia, are parasitic on fish.

union joint [DES ENG] A threaded assembly used for the joining of ends of lengths of installed pipe or tubing where rotation of neither length is feasible.

union of sets [MATH] A set consisting of those elements which are members of at least one set in a given family of sets.

union shop [IND ENG] An establishment in which union membership is not a requirement for original employment but becomes mandatory after a specified period of time.

Uniontown method [ENG] A road-test method to determine the knock characteristics of motor fuels.

unipath See nonshared control unit.

unipolar [ELEC] Having but one pole, polarity, or direction; when applied to amplifiers or power supplies, it means that the output can vary in only one polarity from zero and, therefore, must always contain a direct-current component.

Unipolarina [INV ZOO] A suborder of the protozoan order Myxosporida characterized by spores with one to six (never five) polar capsules located at the anterior end.

unipolar machine See homopolar generator.

unipolar transistor [ELECTR] A transistor that utilizes charge carriers of only one polarity, such as a field-effect transistor.

unipole [ELECTROMAG] A hypothetical antenna that radiates or receives signals equally well in all directions. Also known as isotropic antenna.

unipotential cathode See indirectly heated cathode.

unipotential electrostatic lens [ELECTR] An electrostatic lens in which the focusing is produced by application of a single potential difference; in its simplest form it consists of three apertures of which the outer two are at a common potential, and the central aperture is at a different, generally lower, potential.

unique factorization theorem [MATH] A positive integer may be expressed in precisely one way as a product of prime numbers.

Unisol process [CHEM ENG] A proprietary solvent-extraction process used in petroleum refineries to extract mercaptan sulfur and certain nitrogen compounds from sour gasolines or distillates.

unit [ENG] An assembly or device capable of independent operation, such as a radio receiver, cathode-ray oscilloscope, or computer subassembly that performs some inclusive operation or function. [ORD] **1.** Any military element whose structure is prescribed by competent authority, such as a table of organization and equipment; specifically, part of an organization. **2.** A standard of basic quantity into which an item of supply is divided, issued, or used. [PHYS] A quantity adopted as a standard of measurement.

unit-area acoustical ohm See rayl.

unitary air conditioner [MECH ENG] A small self-contained electrical unit enclosing a motor-driven refrigeration compressor, evaporative cooling coil, air-cooled condenser, filters, fans, and controls.

unitary decuplet [PARTIC PHYS] A collection of 10 hadrons whose isospin and hypercharge values form a symmetrical pattern, and which are related by unitary symmetry operations.

unitary group [MATH] The group of unitary transformations on a k-dimensional complex vector space. Usually denoted $U(k)$.

unitary octet [PARTIC PHYS] A collection of eight hadrons whose isospin and hypercharge values form a symmetrical pattern, and which are related by unitary symmetry operations.

unitary spin [PARTIC PHYS] A quantum number associated with SU_3 symmetry and which determines the SU_3 supermultiplet to which a particle belongs, such as singlet, octet, or decuplet.

unitary symmetry [PARTIC PHYS] An approximate symmetry law obeyed by the strong interactions of elementary particles; it may be described as the equivalence of three fundamental particles, termed quarks, out of which all hadrons could be assumed to be composed. Also known as SU_3 symmetry.

unitary transformation [MATH] A linear transformation on a complex vector space which preserves inner products and norms.

unit assembly [IND ENG] Assemblage of machine parts which constitutes a complete auxiliary part of an end item, and which performs a specific auxiliary function, and which may be removed from the parent item without itself being disassembled.

unit ball [MATH] The set of all points in euclidean n-space whose distance from the origin is at most 1.

unit binormal [MATH] A unit vector in the same direction as the binormal to a point on a surface or space curve.

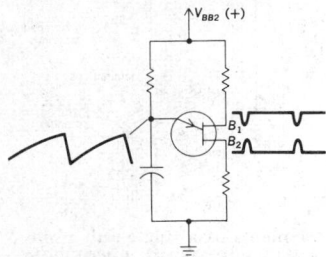
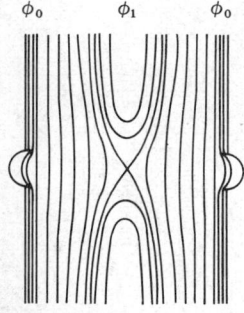

unit cell [CRYSTAL] A parallelepiped which will fill all space under the action of translations which leave the crystal lattice unchanged. Also known as structure cell. [MIN ENG] In flotation, a single cell.

unit charge *See* statcoulomb.

unit circle [MATH] The locus of points in the plane which are precisely one unit from the origin.

unit cost [IND ENG] Cost allocated to a specified unit of a product; computed as the cost over a period of time divided by the number of units produced.

unit die [MET] A die block having more than one cavity insert and allowing several different castings to be made.

United States airways code [METEOROL] A synoptic code for communicating aviation weather observations. Also known as airways code.

United States standard dry seal thread [DES ENG] A modified pipe thread used for pressure-tight connections that are to be assembled without lubricant or sealer in refrigeration pipes, automotive and aircraft fuel-line fittings, and gas and chemical shells.

unitegmic [BOT] Referring to an ovule having a single integument.

unit element [MATH] An element in a ring which acts as a multiplicative identity.

uniterm [ADP] A word, symbol, or number used as a description for retrieval of information from a collection; especially, such a description used in a coordinate indexing system.

uniterm system [ADP] An information retrieval system which uses uniterm cards; cards representing words of interest in a search are selected and compared visually; if identical members are found to appear on the uniterm card undergoing comparison, those numbers represent documents to be examined in connection with the search.

unit heater [MECH ENG] A heater consisting of a fan for circulating air over a heat-exchange surface, all enclosed in a common casing.

unit impulse *See* delta function.

unitized body [ENG] An automotive body that has the body and frame in one unit; side members are designed on the principle of a bridge truss to gain stiffness, and sheet metal of the body is stressed so that it carries some of the load.

unitized cargo [IND ENG] Grouped cargo carried aboard a ship in pallets, containers, wheeled vehicles, and barges or lighters.

unitized film [GRAPHICS] Film which is filed or used by an individual frame or a group of frames under one classification.

unitized load [IND ENG] A single item or a number of items packaged, packed, or arranged in a specified manner and capable of being handled as a unit; unitization may be accomplished by placing the item or items in a container or by banding them securely together. Also known as unit load.

unit length [COMMUN] Basic element of time used in determining code speeds in message transmission.

unit load *See* unitized load.

unit loading [ORD] The loading of troop units with their equipment and supplies in the same ships, aircraft, or land vehicles.

unit magnetic pole [ELECTROMAG] Two equal magnetic poles of the same sign have unit value when they repel each other with a force of 1 dyne if placed 1 centimeter apart in a vacuum.

unit mold [ENG] A simple plastics mold composed of a simple cavity without further mold devices; used to produce sample containers having shapes difficult to blow-mold.

unit normal [MATH] A unit vector in the direction of the principal normal to a surface or space curve.

unit of coal [MIN ENG] The quantity of coal from which the sample is taken and which the sample represents.

unit of fire [ORD] A basic load of ammunition.

unit of issue [IND ENG] In reference to special storage, the quantity of an item, such as each number, dozen, gallon, pair, pound, ream, set, or yard. [ORD] The standard or basic quantity in which an item of supply is issued.

unit operations [CHEM ENG] The basic physical operations of chemical engineering in a chemical process plant, that is, distillation, fluid transport, heat and mass transfer, evaporation, extraction, drying, crystallization, filtration, mixing, size separation, crushing and grinding, and conveying.

unit operator [MATH] The identity operator.

unitor [ADP] In computers, a device or circuit which performs a function corresponding to the Boolean operation of union.

unit power [MET] A unit describing machinability of a metal; the power needed to remove a unit volume in unit time, usually expressed as horsepower per cubic inch per minute.

unit procurement cost [IND ENG] The net basic cost paid or estimated to be paid for a unit of a particular item including, where applicable, the cost of government-furnished property and the cost of manufacturing operations performed at government-owned facilities.

unit pulse *See* baud.

unit record [ADP] Any of a collection of records, all of which have the same form and the same data elements.

unit record device [ADP] Any piece of equipment such as punch card readers, card punch, and line printers.

unit replacement [ORD] Method of repair in which a defective, worn, or damaged group of parts of a weapon or other equipment is replaced by a complete new group of parts.

unit reserves [ORD] Prescribed quantities of supplies carried by a unit as a reserve to cover emergencies.

unit sphere [MATH] The set of points in three-space (more generally *n*-space) which are precisely one unit distance from the origin.

unit strain [MECH] **1.** For tensile strain, the elongation per unit length. **2.** For compressive strain, the shortening per unit length. **3.** For shear strain, the change in angle between two lines originally perpendicular to each other.

unit strength [ORD] As applied to a friendly or enemy unit, relates to the number of personnel, amount of supplies, armament equipment and vehicles, and the total logistic capabilities.

unit stress [MECH] The load per unit of area.

unit systems [PHYS] Groups of units suitable for use in measurement of physical quantities and in the convenient statement of physical laws relating physical quantities.

unit tangent [MATH] A unit vector in the tangent plane at a point of a surface.

unit train [MIN ENG] A system for delivering coal in which a string of cars, with distinctive markings and loaded to full visible capacity, is operated without service frills or stops along the way for cars to be cut in and out.

unit vector [MATH] A vector whose length is one unit.

unity coupling [ELECTROMAG] Perfect magnetic coupling between two coils, so that all magnetic flux produced by the primary winding passes through the entire secondary winding.

unity gain bandwidth [ELECTR] Measure of the gain-frequency product of an amplifier; unity gain bandwidth is the frequency at which the open-loop gain becomes unity, based on 6 decibels per octave crossing.

unity power factor [ELEC] Power factor of 1.0, obtained when current and voltage are in phase, as in a circuit containing only resistance or in a reactive circuit at resonance.

univariant system [THERMO] A system which has only one degree of freedom according to the phase rule.

universal algebra [MATH] The study of algebraic systems such as groups, rings, modules, and fields and the examination of what families of theorems are analogous in each system.

universal donor [IMMUNOL] An individual of O blood group; can give blood to persons of all blood types.

universal gas constant *See* gas constant.

universal instrument *See* altazimuth.

universal joint [MECH ENG] A linkage that transmits rotation between two shafts whose axes are coplanar but not coinciding.

universal mill [MET] A rolling mill having both horizontal and vertical sets of rolls.

universal motor [ELEC] A motor that may be operated at approximately the same speed and output on either direct current or single-phase alternating current. Also known as ac/dc motor.

universal output transformer [ENG ACOUS] An output transformer having a number of taps on its winding, to permit its

UNITIZED BODY

Unitized construction of a body for a four-door sedan. (*American Motors Corp.*)

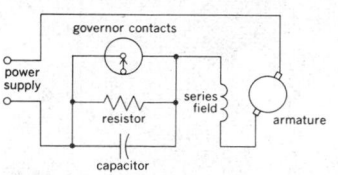

UNIVERSAL MOTOR

Universal motor circuit diagram. Centrifugal governor switches small resistor in or out, to obtain more constant speed with variations in load.

use between the audio-frequency output stage and the loud-speaker of practically any radio receiver by proper choice of connections.

universal plotting sheet [NAV] A plotting sheet on which either the latitude or longitude lines are omitted, to be drawn in by the user, making it possible to quickly construct a plotting sheet for any part of the earth's surface.

Universal Polar Stereographic Grid [MAP] A particular grid based upon the polar stereographic projection according to specifications laid down by military authorities; it may be superimposed on any map.

universal receiver *See* ac/dc receiver.

universal recipient [IMMUNOL] An individual of AB blood group; can receive a blood transfusion of all blood types, A, AB, B, or O.

universal resonance curve [ELEC] A plot of Y/Y_0 against $Q_0\delta$ for a series-resonant circuit, or of Z/Z_0 against $Q_0\delta$ for a parallel-resonant circuit, where Y and Z are the admittance and impedance of a circuit, Y_0 and Z_0 are the values of these quantities at resonance, Q_0 is the Q value of the circuit at resonance, and δ is the deviation of the frequency from resonance divided by the resonant frequency; it can be applied to all resonant circuits.

universal shunt *See* Ayrton shunt.

universal stage [OPTICS] A stage attached to the rotating stage of a polarizing microscope that has three, four, or five axes and thin sections of low-symmetry minerals to be tilted about two mutually perpendicular horizontal axes. Also known as Fedorov stage; U stage.

universal time *See* Greenwich mean time.

universal time 0 [ASTRON] The uncorrected time of the earth's rotation as measured by the transit of stars across the observer's meridian. Abbreviated UT 0.

universal time 1 [ASTRON] Universal time 0 corrected for polar motion; it is the true angular rotation. Abbreviated UT 1.

universal time 2 [ASTRON] Universal time 1 corrected for seasonal variations in the earth's rotation. Abbreviated UT 2.

universal time coordinated [ASTRON] The coordinated time kept by a uniformly running clock, approximating the measure UT 2. Abbreviated UTC.

universal transmission function [GEOPHYS] A mathematical relationship that attempts to describe quantitatively the complex infrared propagation (including absorption and reradiation) in the atmosphere.

universal transverse Mercator grid [MAP] A particular grid based upon a transverse Mercator projection, according to specifications laid down by military authorities; it may be superimposed on any map.

universal Turing machine [ADP] A Turing machine that can simulate any Turing machine.

universal wavelength function [OPTICS] One of four functions which enable one to compute easily, with reasonable accuracy, the refractive index of glass or other transparent material when this index is known for four standard wavelengths.

universe [ASTRON] The totality of astronomical things, events, relations, and energies capable of being described objectively.

univibrator *See* monostable multivibrator.

unlimited ceiling [METEOROL] A ceiling that exists when the total sky cover is less than 0.6%, or when the total transparent sky cover is 0.5% or more, or when surface-based obscuring phenomena are classed as partial obscuration (that is, they obscure 0.9% or less of the sky) and no layer aloft is reported as broken or overcast.

unloaded Q [ELECTR] The Q of a system when there is no external coupling to it.

unloader [MECH ENG] A power device for removing bulk materials from railway freight cars or highway trucks; in the case of railway cars, the car structure may aid the unloader; a transitional device between interplant transportation means and intraplant handling equipment.

unloading [CHEM ENG] **1.** The release downstream of a trapped contaminant. **2.** A filter medium failure and release of system pressure. **3.** The depressuring or emptying of a process unit.

unloading amplifer [ELECTR] Amplifier that is capable of reproducing or amplifying a given voltage signal while drawing negligible current from the voltage source.

unloading circuit [ADP] In an analog computer, a computing element or combination of computing elements capable of reproducing or amplifying a given voltage signal while drawing negligible current from the voltage source.

unloading conveyor [MECH ENG] Any of several types of portable conveyors adapted for unloading bulk materials, packages, or objects from conveyances.

unmyelinated [HISTOL] Lacking myelin, either as a normal condition or as the result of a disease.

unpack [ADP] **1.** To recover the individual data items contained in packed data. **2.** More specifically, to convert a packed decimal number into individual digits (and sometimes a sign).

unpolarized light [OPTICS] Light in which the electric vector is oriented in a random, unpredictable fashion.

unpolarized particle beam [PHYS] A beam of particles with spin in which the directions of the spins are random.

unproductive development [MIN ENG] The drifts, tunnels, and crosscuts driven in stone, preparatory to opening out production faces in a coal seam or ore body.

unprotected reversing thermometer [ENG] A reversing thermometer for sea-water temperature which is not protected against hydrostatic pressure.

unproven area [MIN ENG] An area in which it has not been established by drilling operations whether oil or gas may be found in commercial quantities.

unrelated frequencies [STAT] The long run frequency of any result in one part of an experiment is approximately equal to the long run conditional frequency of that result, given that any specified result has occurred in the other part of the experiment.

unreserved minerals [MIN ENG] Minerals which belong to the owner of the land on which or in which they are located.

unresolved pneumonia *See* organizing pneumonia.

unrestricted visibility [METEOROL] The visibility when no obstruction to vision exists in sufficient quantity to reduce the visibility to less than 7 miles (11.3 kilometers).

uns-, unsym- [ORG CHEM] A chemical prefix denoting that the substituents of an organic compound are structurally unsymmetrical with respect to the carbon skeleton, or with respect to a function group (for example, double or triple bond).

unsaturated [MINERAL] Referring to a mineral that will not form in the presence of free silica.

unsaturated compound [CHEM] Any chemical compound with more than one bond between adjacent atoms, usually carbon, and thus reactive toward the addition of other atoms at that point; for example, olefins, diolefins, and unsaturated fatty acids.

unsaturated hydrocarbon [ORG CHEM] One of a class of hydrocarbons that have at least one double or triple carbon-to-carbon bond that is not in an aromatic ring; examples are ethylene, propadiene, and acetylene.

unsaturated standard cell [ELEC] One of two types of Weston standard cells (batteries); used for voltage calibration work not requiring an accuracy greater than 0.01%.

unsaturated zone *See* zone of aeration.

unsaturation [ORG CHEM] A state in which the atomic bonds of an organic compound's chain or ring are not completely satisfied (that is, not saturated); usually applies to carbon, but can include other ring or chain atoms; unsaturation usually results in a double bond (as for olefins) or a triple bond (as for the acetylenes).

unscheduled maintenance [IND ENG] Those unpredictable maintenance requirements that had not been previously planned or programmed but require prompt attention and must be added to, integrated with, or substituted for previously scheduled workloads.

unscrambler [IND ENG] A part of a feeding and packaging line that aids in arranging cartons for the filling machines; there are rotary, straight-line, and walking-beam types.

unseen fire [ORD] Fire which is continuously aimed at the future position of an aircraft, the aim being derived from radar sources.

unsettled [METEOROL] Pertaining to fair weather which may at any time become rainy, cloudy, or stormy.

Unsin engine [MECH ENG] A type of rotary engine in which the trochoidal rotors of eccentric-rotor engines are replaced with two circular rotors, one of which has a single gear tooth upon which gas pressure acts, and the second rotor has a slot which accepts the gear tooth.

unsprung axle [MECH ENG] A rear axle in an automobile in which the housing carries the right and left rear-axle shafts and the wheels are mounted at the outer end of each shaft.

unsprung weight [MECH ENG] The weight of the various parts of a vehicle that are not carried on the springs such as wheels, axles, brakes, and so forth.

unstable [PHYS] Capable of undergoing spontaneous change, as in a radioactive nuclide or an excited nuclear system.

unstable colon *See* irritable colon.

unstable equilibrium [PHYS] An equilibrium state of a system in which any departure of the system from equilibrium gives rise to forces or tendencies moving the system further away from equilibrium; for example, mechanical equilibrium in which the potential energy is a maximum, as a sphere sitting on top of a hill.

unstable isotope *See* radioisotope.

unstable particle [PARTIC PHYS] **1.** Any elementary particle that spontaneously decays into other particles. **2.** An elementary particle that can decay through the strong interactions, as opposed to a semistable particle; it has a lifetime on the order of 10^{-23} second.

unstable wave [PHYS] A wave motion whose amplitude increases with time or whose total energy increases at the expense of its environment.

unsteady-state flow [FLUID MECH] A condition of fluid flow in which the volumetric ratios of two or more phases (liquid-gas, liquid-liquid, and so on) vary along the course of flow; can be the result of changes in temperature, pressure, or composition.

unstriated fibers *See* smooth muscle fibers.

unsurveyed area [MAP] An area on a map or chart where both relief and planimetric data are unavailable, and which is usually labeled unsurveyed; or an area on a map or chart which shows little or no charted data because accurate information is limited or not available.

untuned [ELEC] Not resonant at any of the frequencies being handled.

unwater [ENG] To remove or draw off water; to drain.

unwind [ADP] In computers, to rearrange and code a sequence of instructions to eliminate red-tape operations.

up [GRAPHICS] A flat-bed term to indicate one or more documents being in position to photograph, for example, two up, and so on.

upcast [MIN ENG] **1.** The opening through which the return air ascends and is removed from the mine; the opposite of downcast or intake. **2.** An upward current of air passing through a shaft. **3.** Material that has been thrown up, as by digging.

up-converter [ELECTR] Type of parametric amplifier which is characterized by the frequency of the output signal being greater than the frequency of the input signal.

update [ADP] In computers, to modify an instruction so that the address numbers it contains are increased by a stated amount each time the instruction is performed.

up-Doppler [ENG ACOUS] The sonar situation wherein the target is moving toward the transducer, so the frequency of the echo is greater than the frequency of the reverberations received immediately after the end of the outgoing ping; opposite of down-Doppler.

updraft carburetor [MECH ENG] For a gasoline engine, a fuel-air mixing device in which both the fuel jet and the airflow are upward.

updraft furnace [MECH ENG] A furnace in which volumes of air are supplied from below the fuel bed or supply.

upgrade [MIN ENG] **1.** To increase the commercial value of a coal or mineral product by appropriate treatment. **2.** To increase the quality rating of diamonds beyond or above the rating implied by their particular classification.

up-hole [MIN ENG] A borehole collared in an underground working place and drilled in a direction pointed above the horizontal plane of the drill-machine swivel head.

uplift pressure [CIV ENG] Pressure in an upward direction against the bottom of a structure, as a dam, a road slab, or a basement floor.

upper [GEOL] Pertaining to rocks or strata that normally overlie those of earlier formations of the same subdivision of rocks.

upper air [METEOROL] The region of the atmosphere which is above the lower troposphere; although no distinct lower limit is set, the term is generally applied to levels above that at which the pressure is 850 millibars.

upper-air chart *See* upper-level chart.

upper-air disturbance [METEOROL] A disturbance of the flow pattern in the upper air, particularly one which is more strongly developed aloft than near the ground. Also known as upper-level disturbance.

upper-air observation [METEOROL] A measurement of atmospheric conditions aloft, above the effective range of a surface weather observation. Also known as sounding; upper-air sounding.

upper-air sounding *See* upper-air observation.

upper anticyclone *See* upper-level anticyclone.

upper-arm circumference [ANTHRO] A measure of the horizontal circumference at the largest part of the biceps muscle.

upper atmosphere [METEOROL] The general term applied to the atmosphere above the troposphere.

upper atmosphere dynamics [METEOROL] Motion of the atmosphere above 500 kilometers; predominant dynamical phenomena are internal gravity waves, tides, sound waves, turbulence, and large-scale circulation.

upper band *See* upper bright band.

upper branch [GEOD] That half of a meridian or celestial meridian from pole to pole which passes through a place or its zenith.

upper bright band [METEOROL] A level of enhanced radar echo occasionally observed at a higher altitude than the bright band of the melting level; it is attributable to the growth of a layer of ice crystals in a supercooled cloud into snow pellets. Also known as radar upper band; upper band.

Upper Cambrian [GEOL] The latest epoch of the Cambrian period of geologic time, beginning approximately 510 million years ago.

Upper Carboniferous [GEOL] The European epoch of geologic time equivalent to the Pennsylvanian of North America.

upper consolute temperature *See* consolute temperature.

Upper Cretaceous [GEOL] The late epoch of the Cretaceous period of geologic time, beginning about 90 million years ago.

upper critical solution temperature *See* consolute temperature.

upper culmination *See* upper transit.

upper cyclone *See* upper-level cyclone.

Upper Devonian [GEOL] The latest epoch of the Devonian period of geologic time, beginning about 365 million years ago.

upper face height [ANTHRO] A measure of the distance from the nasion to the lower gum edge between the two central upper teeth.

upper front [METEOROL] A front which is present in the upper air but does not extend to the ground.

upper half-power frequency [ELECTR] The frequency on an amplifier response curve which is greater than the frequency for peak response and at which the output voltage is $1/\sqrt{2}$ (that is, 0.707) of its midband or other reference value.

upper high *See* upper-level anticyclone.

Upper Huronian *See* Animikean.

upper integral [MATH] The upper Riemann integral for a real-valued function $f(x)$ on an interval is computed to be the infimum of all finite sums over all partitions of the interval, the sums having terms given by $(x_i - x_{i-1})y_i$, where the x_i are from a partition, and y_i is the largest value of $f(x)$ over the interval from x_{i-1} to x_i.

upper inversion *See* tropopause inversion.

Upper Jurassic [GEOL] The latest epoch of the Jurassic period of geologic time, beginning approximately 155 million years ago.

upper-level anticyclone [METEOROL] An anticyclonic circu-

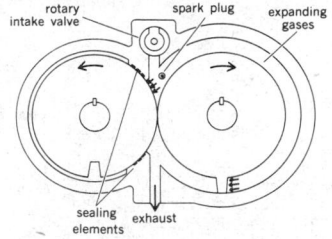

rotary intake valve — spark plug — expanding gases — sealing elements — exhaust

Section through an Unsin engine showing two circular rotors.

lation existing in the upper air; this often refers to such anticyclones only when they are much more pronounced at upper levels than at and near the earth's surface. Also known as high aloft; high-level anticyclone; upper anticyclone; upper high; upper-level high.

upper-level chart [METEOROL] A synoptic chart of meteorological conditions in the upper air, almost invariably referring to a standard constant-pressure chart. Also known as upper-air chart.

upper-level cyclone [METEOROL] A cyclonic circulation existing in the upper air, and specifically, as seen on an upper-level constant-pressure chart; often restricted to describe cyclones associated with relatively little cyclonic circulation in the lower atmosphere. Also known as high-level cyclone; low aloft; upper cyclone; upper-level low; upper low.

upper-level disturbance See upper-air disturbance.

upper-level high See upper-level anticyclone.

upper-level low See upper-level cyclone.

upper-level ridge [METEOROL] A pressure ridge existing in the upper air, especially one that is stronger aloft than near the earth's surface. Also known as high-level ridge; ridge aloft; upper ridge.

upper-level trough [METEOROL] A pressure trough existing in the upper air, but sometimes restricted to the troughs that are much more pronounced aloft than near the earth's surface. Also known as high-level trough; trough aloft; upper trough.

upper-level winds See winds aloft.

upper limb [ASTRON] That half of the outer edge of a celestial body having the greatest altitude.

upper low See upper-level cyclone.

upper mantle [GEOL] The portion of the mantle lying above a depth of about 1000 kilometers. Also known as asthenosphere; outer mantle; peridotite shell.

Upper Mississippian [GEOL] The latest epoch of the Mississippian period of geologic time.

upper mixing layer [METEOROL] The region of the upper mesosphere between about 50 and 80 kilometers (that is, immediately above the mesopeak) through which there is a rapid decrease of temperature with height and where there appears to be considerable turbulence.

Upper Ordovician [GEOL] The latest epoch of the Ordovician period of geologic time, beginning approximately 440 million years ago.

Upper Pennsylvanian [GEOL] The latest epoch of the Pennsylvanian period of geologic time.

Upper Permian [GEOL] The latest epoch of the Permian period of geologic time, beginning about 245 million years ago.

upper punch [MET] In powder metallurgy, the member of the die assembly that moves downward into the die body to transmit pressure to the metal powder in the cavity.

upper ridge See upper-level ridge.

upper semicontinuous decomposition [MATH] A partition of a topological space with the property that for every member D of the partition and for every open set U containing D there is an open set V containing D which is contained in U and is the union of members of the partition.

upper semicontinuous function [MATH] A real-valued function $f(x)$ is upper semicontinuous at a point x_0 if for any small positive ε, $f(x)$ always is less than $f(x_0) + \varepsilon$ for all x in some neighborhood of x_0.

upper side band [COMMUN] The higher of two frequencies or groups of frequencies produced by a modulation process.

Upper Silurian [GEOL] The latest epoch of the Silurian period of geologic time.

upper transit [ASTRON] The movement of a celestial body across a celestial meridian's upper branch. Also known as superior transit; upper culmination.

Upper Triassic [GEOL] The latest epoch of the Triassic period of geologic time, beginning about 200 million years ago.

upper trough See upper-level trough.

upper winds See winds aloft.

upright [CIV ENG] A vertical structural member, post, or stake.

uprush [METEOROL] The strong upward-flow air current in cumulus clouds during their stage of rapid development, often

preceding a thunderstorm. Also known as vertical jet. [OCEANOGR] See swash.

upset [ENG] To increase the diameter of a rock drill by blunting the end. [MET] A localized increase in the cross-sectional area of a metal during working, caused by the application of pressure; enables a head to be formed on fasteners such as bolts. [MIN ENG] **1.** A narrow heading connecting two levels in inclined coal. **2.** A capsized or broken skip.

upsetting test [MET] A test used to identify the role of variables in forging, which demonstrates that the force to forge is a function of the strength of the material, coefficient of friction, and ratio of the lateral to thickness dimensions of the workpiece.

upset welding [MET] Pressure butt welding in which heat is generated by resistance to the passage of current across the joint.

upslope fog [METEOROL] A type of fog formed when air flows upward over rising terrain and, consequently, is adiabatically cooled to or below its dew point.

upslope time [MET] In resistance welding, the time associated with an increase in current when slope control is used.

upstream [CHEM ENG] That portion of a process stream that has not yet entered the system or unit under consideration; for example, upstream to a refinery or to a distillation column. [HYD] Toward the source of a stream.

upstream face [CIV ENG] The side of a dam nearer the source of water.

uptake [ENG] A large pipe for exhaust gases from a boiler furnace that runs upward to a chimney or smokestack.

upthrow [GEOL] **1.** The fault side that has been thrown upward. **2.** The amount of vertical fault displacement.

up time [ADP] The time during which equipment is either producing work or is available for productive work.

Upupidae [VERT ZOO] The Old World hoopoes, a family of birds in the order Coraciiformes whose young are hatched with sparse down.

upwarp [GEOL] A broad anticline with gently sloping limbs formed as a result of differential uplift.

upwelling [OCEANOGR] The process by which water rises from a deeper to a shallower depth, usually as a result of divergence of offshore currents.

upwind [METEOROL] In the direction from which the wind is flowing.

upwind effect [METEOROL] The effect of an orographic barrier in producing orographic precipitation windward of the base of the barrier, because the airflow is forced upward before the barrier slope is actually reached.

urachus [EMBRYO] A cord or tube of epithelium connecting the apex of the urinary bladder with the allantois; its connective tissue forms the median umbilical ligament.

uracil [BIOCHEM] $C_4H_4N_2O_2$ A pyrimidine base important as a component of ribonucleic acid.

Uralean [GEOL] A stage of geologic time in Russia: uppermost Carboniferous (above Gzhelian, below Sakmarian of Permian).

uralite [MINERAL] A green variety of secondary amphibole; it is usually fibrous or acicular and is formed by alteration of pyroxene.

uralitization [GEOL] **1.** A process of replacement whereby pyroxene undergoes alteration resulting in uralite. **2.** Development of amphibole from pyroxene.

uranate [INORG CHEM] A salt of uranic acid; for example, sodium uranate, Na_2UO_4.

urania See uranium dioxide.

uranic chloride See uranium tetrachloride.

uranic oxide See uranium dioxide.

Uraniidae [INV ZOO] A tropical family of moths in the superfamily Geometroidea including some slender-bodied, brilliantly colored diurnal insects which lack a frenulum and are often mistaken for butterflies.

uranin See uranine.

uranine [ORG CHEM] $Na_2C_{20}H_{10}O_5$ A brown or orange-red hygroscopic powder soluble in water; used as a yellow dye for silk and wool, a marker in the ocean to facilitate air and sea rescues, and as an analytical reagent. Also known as sodium fluorescein; uranin; uranine yellow.

URACIL

Structural formula of uracil.

uranine yellow   *See* uranine.

uraninite   [MINERAL] UO_2 A black, brownish-black, or dark-brown radioactive mineral that is isometric in crystallization; often contains impurities such as thorium, radium, cerium, and yttrium metals, and lead; the chief ore of uranium; hardness is 5.5–6 on Mohs scale, and specific gravity of pure UO_2 is 10.9, but that of most natural material is 9.7–7.5. Also known as coracite; ulrichite.

uranium   [CHEM] A metallic element in the actinide series, symbol U, atomic number 92, atomic weight 238.03; highly toxic and radioactive; ignites spontaneously in air and reacts with nearly all nonmetals; melts at 1132°C, boils at 3818°C; used in nuclear fuel and as the source of U^{235} and plutonium. [MET] A dense, silvery, ductile, strongly electropositive metal.

uranium acetate   *See* uranyl acetate.

uranium age   [GEOL] The age of a mineral as calculated from the numbers of ionium atoms present originally, now, and when equilibrium is established with uranium.

uranium carbide   [INORG CHEM] One of the carbides of uranium, such as uranium monocarbide; used chiefly as a nuclear fuel.

uranium decay series   *See* uranium series.

uranium dioxide   [INORG CHEM] UO_2 Black, highly toxic, spontaneously flammable, radioactive crystals; insoluble in water, soluble in nitric and sulfuric acids; melts at approximately 3000°C; used to pack nuclear fuel rods and in ceramics, pigments, and photographic chemicals. Also known as urania; uranic oxide; uranium oxide.

uranium enrichment   [NUCLEO] A process carried out on uranium, in which the ratio of the abundance of the isotope uranium-235 to that of the isotope uranium-238 is increased above that found in natural uranium.

uranium hexafluoride   [INORG CHEM] UF_6 Highly toxic, radioactive, corrosive, colorless crystals; soluble in carbon tetrachloride, fluorocarbons, and liquid halogens; it reacts vigorously with alcohol, water, ether, and most metals, and it sublimes; used to separate uranium isotopes in the gaseous-diffusion process.

uranium hydride   [INORG CHEM] UH_3 A highly toxic, gray to black powder that ignites spontaneously in air, and that conducts electricity; used for making powdered uranium metal, for hydrogen-isotope separation, and as a reducing agent.

uranium-lead dating   [GEOL] A method for calculating the geologic age of a material in years based on the radioactive decay rate of uranium-238 to lead-206 and of uranium-235 to lead-207.

uranium nitrate   *See* uranyl nitrate.

uranium ocher   *See* gummite.

uranium oxide   *See* uranium dioxide; uranium trioxide.

uranium-radium series   *See* uranium series.

uranium reactor   [NUCLEO] A nuclear reactor in which the principal fuel is uranium; the uranium may be natural, with the naturally occurring ratio of 1 atom of uranium-235 to about 139 atoms of uranium-238, or may be enriched to have a higher proportion of fissile uranium-233 or uranium-235 atoms.

uranium series   [NUC PHYS] The series of nuclides resulting from the decay of uranium-238, including uranium I, II, X_1, X_2, Y, and Z, and radium A, B, C, C′, C″, D, E, E″, F, and G. Also known as uranium decay series; uranium-radium series.

uranium sulfate   *See* uranyl sulfate.

uranium tetrachloride   [INORG CHEM] UCl_4 Poisonous, radioactive, hygroscopic, dark-green crystals; soluble in alcohol and water; melts at 590°C, boils at 792°C. Also known as uranic chloride.

uranium tetrafluoride   [INORG CHEM] UF_4 Toxic, radioactive, corrosive green crystals; insoluble in water; melts at 1036°C; used in the manufacture of uranium metal. Also known as green salt.

uranium trioxide   [INORG CHEM] UO_3 A poisonous, radioactive, red to yellow powder; soluble in nitric acid, insoluble in water; decomposes when heated; used in ceramics and pigments and for uranium refining. Also known as orange oxide; uranium oxide.

uran-mica   *See* torbernite.

uranography   [ASTRON] The science of mapping stars, groups of stars, and star clusters.

uranometry   [ASTRON] The science of the measurement of the celestial sphere and celestial bodies.

uranophane   [MINERAL] $Ca(UO_2)_2Si_2O_7 \cdot 6H_2O$ A yellow or orange-yellow radioactive secondary mineral; it is dimorphous with β-uranophane. Also known as uranotile.

uranotantalite   *See* samarskite.

uranothorite   [MINERAL] A uranium-bearing variety of thorite.

uranotile   *See* uranophane.

Uranus   [ASTRON] A planet, seventh in the order of distance from the sun; it has five known satellites, and its equatorial diameter is about four times that of the earth.

uranyl acetate   [INORG CHEM] $UO_2(C_2H_3O_2)_2 \cdot 2H_2O$ Poisonous, radioactive yellow crystals, decomposed by light; soluble in cold water, decomposes in hot water; loses water of crystallization at 110°C, decomposes at 275°C; used in medicine and as an analytical reagent and bacterial oxidant. Also known as uranium acetate.

uranyl nitrate   [INORG CHEM] $UO_2(NO_3)_2 \cdot 6H_2O$ Toxic, explosive, unstable yellow crystals; soluble in water, alcohol, and ether; melts at 60°C and boils at 118°C; used in photography, in medicine, and for uranium extraction and uranium glaze. Also known as uranium nitrate; yellow salt.

uranyl salts   [INORG CHEM] Salts of UO_3 that ionize to form UO_2^{++} and that are yellow in solution; for example, uranyl chloride, UO_2Cl_2.

uranyl sulfate   [INORG CHEM] $UO_2SO_4 \cdot 3\frac{1}{2}H_2O$ and $UO_2SO_4 \cdot 3H_2O$ Poisonous, radioactive yellow crystals; soluble in water and concentrated hydrochloric acid; used as an analytical reagent. Also known as uranium sulfate.

urao   *See* trona.

urate calculi   [PATH] Kidney stones composed of uric acid salts and found particularly in people suffering from gout.

urban heat island   [METEOROL] Increased urban temperatures of 1–2°C higher for daily maxima and 1–9°C for daily minima compared to rural environs resulting from changes in moisture balance due to impermeable surfaces, decreased humidity, or alteration in heat balance.

urbanization   [CIV ENG] The state of being or becoming a community with urban characteristics.

urban renewal   [CIV ENG] Redevelopment and revitalization of a deteriorated urban community.

urban typhus   *See* murine typhus.

urceolate   [BIOL] Shaped like an urn.

urea   [BIOCHEM] $CO(NH_2)_2$ Carbamide, a product of protein metabolism; used therapeutically as a diuretic.

urea anhydride   *See* cyanamide.

urea dewaxing   [CHEM ENG] A continuous, petroleum refinery process used to produce low-pour-point oils; urea forms a filterable solid complex (adduct) with the straight-chain wax paraffins in the stock.

urea-formaldehyde resin   [ORG CHEM] A synthetic thermoset resin derived by the reaction of urea (carbamide) with formaldehyde or its polymers. Also known as urea resin.

urea nitrate   [ORG CHEM] $CO(NH_2)_2 \cdot HNO_3$ Colorless, explosive, fire-hazardous crystals; soluble in alcohol, slightly soluble in water; decomposes at 152°C; used in explosives and to make urethane.

urea peroxide   [ORG CHEM] $CO(NH_2)_2 \cdot H_2O_2$ An unstable, fire-hazardous white powder; soluble in water, alcohol, and ethylene glycol; decomposes at 75–85°C or by moisture; used as a source of water-free hydrogen peroxide, as a disinfectant, in cosmetics and pharmaceuticals, and for bleaching. Also known as carbamide peroxide.

urea resin   *See* urea-formaldehyde resin.

urease   [BIOCHEM] An enzyme that catalyzes the degradation of urea to ammonia and carbon dioxide; obtained from the seed of jack bean.

Urechinidae   [INV ZOO] A family of echinoderms in the order Holasteroida which have an ovoid test lacking a marginal fasciole.

Uredinales   [MYCOL] An order of parasitic fungi of the subclass Heterobasidiomycetidae characterized by the teleutospore, a spore with one or more cells, each of which is a

URANINITE

Uraninite crystals in pegmatite, Chester, Pennsylvania. (*Specimen from Department of Geology, Bryn Mawr College*)

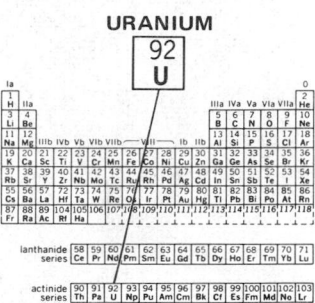

URANIUM

Periodic table of the chemical elements showing the position of uranium.

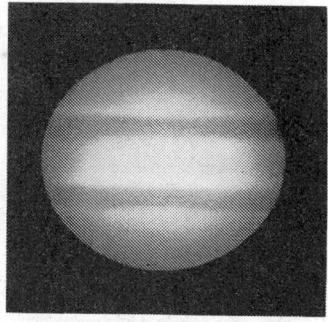

URANUS

Telescopic appearance of Uranus, when the earth is in the equatorial plane of the globe.

URIC ACID

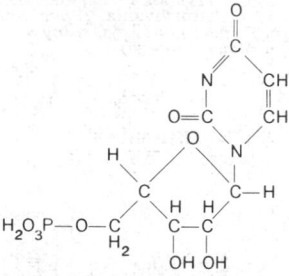

Structural formula of uric acid.

URIDYLIC ACID

Structural formula of uridylic acid.

URSA MAJOR

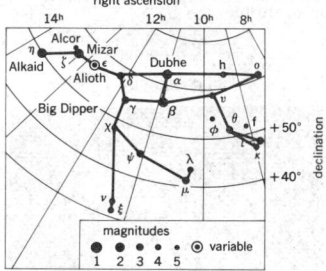

Line pattern of the constellation Ursa Major. The grid lines represent the coordinates of the sky. The apparent brightness, or magnitude, of the stars is shown by the sizes of the dots, which are graded by appropriate numbers as indicated.

URSA MINOR

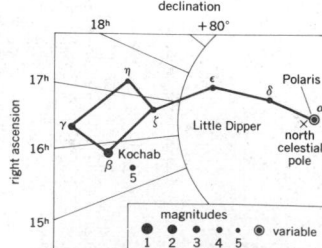

Line pattern of the constellation Ursa Minor. The grid lines represent the coordinates of the sky. The apparent brightness, or magnitude, of the stars is shown by the sizes of the dots, which are graded by appropriate numbers as indicated.

modified hypobasidium; members cause plant diseases known as rusts.

uredinium [MYCOL] The aggregation of sporebearing hyphae and urediospores of a rust fungus that forms beneath the cuticle or epidermis of a host plant.

urediospore [MYCOL] A thin-walled spore produced by rust fungi; gives rise to a vegetative mycelium which may produce more urediospores.

(para-ureidophenylarsylenedithio)diacetic acid *See* thiocarbarsone.

(para-ureidophenylarsylenedithio)di-o-benzoic acid *See* thiocarbamazine.

ureilite [GEOL] An achondritic stony meteorite consisting principally of olivine and clinobronzite, with some nickel-iron, troilite, diamond, and graphite.

uremia [MED] A condition resulting from kidney failure and characterized by azotemia, chronic acidosis, anemia, and a variety of systemic signs and symptoms.

ureter [ANAT] A long tube conveying urine from the renal pelvis to the urinary bladder or cloaca in vertebrates.

urethane [ORG CHEM] $CO(NH_2)OC_2H_5$ A combustible, toxic, colorless powder; soluble in water and alcohol; melts at 49°C; used as a solvent and chemical intermediate and in biochemical research and veterinary medicine. Also known as ethyl carbamate; ethyl urethane.

urethra [ANAT] The canal in most mammals through which urine is discharged from the urinary bladder to the outside.

urethral gland [ANAT] One of the small, branched tubular mucous glands in the mucosa lining the urethra.

urethritis [MED] Inflammation of the urethra.

ureyite [MINERAL] $NaCrSi_2O_6$ A meteoritic mineral of the pyroxene group.

uric acid [BIOCHEM] $C_5H_4N_4O_3$ A white, crystalline compound, the excretory end product in amino acid metabolism by uricotelic species. Also known as 8-hydroxyxanthine.

uricase [BIOCHEM] An enzyme present in the liver, spleen, and kidney of most mammals except man; converts uric acid to allantoin in the presence of gaseous oxygen.

uricotelism [PHYSIO] An adaptation of terrestrial reptiles and birds which effectively provides for detoxification of ammonia and also for efficient conservation of water due to a relatively low rate of glomerular filtration and active secretion of uric acid by the tubules to form a urine practically saturated with urate.

uridine [BIOCHEM] $C_9H_{12}N_2O_6$ A crystalline nucleoside composed of one molecule of uracil and one molecule of D-ribose; a component of ribonucleic acid.

uridine diphosphate [BIOCHEM] The chief transferring coenzyme in carbohydrate metabolism. Abbreviated UDP.

uridine diphosphoglucose [BIOCHEM] A compound in which α-glucopyranose is esterified, at carbon atom 1, with the terminal phosphate group of uridine-5′-pyrophosphate (that is, uridine diphosphate); occurs in animal, plant, and microbial cells; functions as a key in the transformation of glucose to other sugars. Abbreviated UDPG.

uridine monophosphate *See* uridylic acid.

uridine phosphoric acid *See* uridylic acid.

uridylic acid [BIOCHEM] $C_9H_{13}N_2O_9P$ Water- and alcohol-soluble crystals, melting at 202°C; used in biochemical research. Also known as uridine monophosphate (UMP); uridine phosphoric acid.

urinalysis [PATH] Analysis of the urine, involving chemical, physical, and microscopic tests.

urinary bladder [ANAT] A hollow organ which serves as a reservoir for urine.

urinary system [ANAT] The system which functions in the elaboration and excretion of urine in vertebrates; in man and most mammals, consists of the kidneys, ureters, urinary bladder, and urethra.

urination [PHYSIO] The discharge of urine from the bladder. Also known as micturition.

urine [PHYSIO] The fluid excreted by the kidneys.

uriniferous tubule [ANAT] One of the numerous winding tubules of the kidney. Also known as nephric tubule.

urn [BOT] The theca of a moss.

urobilin [BIOCHEM] A bile pigment produced by reduction of bilirubin by intestinal bacteria and excreted by the kidneys or removed by the liver.

urobilinogen [BIOCHEM] A chromogen, formed in feces and present in urine, from which urobilin is formed by oxidation.

urocanic acid [BIOCHEM] $C_6H_6N_2O_2$ A crystalline compound formed as an intermediate in the degradative pathway of histidine.

Urochordata [INV ZOO] The equivalent name for Tunicata.

urochrome [BIOCHEM] $C_{43}H_{51}O_{26}N$ Yellow pigment found in normal urine.

Urodela [VERT ZOO] The tailed amphibians or salamanders, an order of the class Amphibia distinguished superficially from frogs and toads by the possession of a tail, and from caecilians by the possession of limbs.

urogenital diaphragm [ANAT] The sheet of tissue stretching across the pubic arch, formed by the deep transverse perineal and the sphincter urethrae muscles. Also known as triangular ligament.

urogenital system [ANAT] The combined urinary and genital system in vertebrates, which are intimately related embryologically and anatomically. Also known as genitourinary system.

urokinase [BIOCHEM] An enzyme, present in human urine, that catalyzes the conversion of plasminogen to plasmin.

urolithiasis [MED] **1.** Condition associated with the presence of urinary calculi. **2.** Formation or presence of urinary calculi.

urology [MED] The scientific study of urine and the diseases and abnormalities of the urinary and urogenital tracts.

uronic acid [ORG CHEM] One of the compounds that are similar to sugars, except that the terminal carbon has been oxidized from the alcohol to a carboxyl group; for example, galacturonic acid and glucuronic acid.

uropepsin [BIOCHEM] The end product of the secretion of pepsinogen into the blood by gastric cells; occurs in urine.

uropod [INV ZOO] One of the flattened abdominal appendages of various crustaceans that with the telson forms the tail fan.

uroporphyrin [BIOCHEM] Any of several isomeric, metal-free porphyrins, occurring in small quantities in normal urine and feces; molecule has four acetic acid ($-CH_2COOH$) and four propionic acid ($-CH_2CH_2COOH$) groups.

Uropygi [INV ZOO] The tailed whip scorpions, an order of arachnids characterized by an elongate, flattened body which bears in front a pair of thickened, raptorial pedipalps set with sharp spines and used to hold and crush insect prey.

uropygial gland [VERT ZOO] A relatively large, compact, bilobed, secretory organ located at the base of the tail (uropygium) of most birds having a keeled sternum. Also known as oil gland; preen gland.

urostyle [VERT ZOO] An unsegmented bone representing several fused vertebrae and forming the posterior part of the vertebral column in Anura.

Urostylidae [INV ZOO] A family of hemipteran insects in the superfamily Pentatomoidea.

urotropin *See* cystamine.

Ursa Major [ASTRON] A northern constellation, right ascension 11 hours, declination 50°N; it contains a group of seven stars known as the Big Dipper.

Ursa Major cluster [ASTRON] A group of about 126 stars, including 5 stars of the constellation Ursa Major; the sun is passing through this cluster which occupies a spherical space of about 450 light-years in diameter.

Ursa Minor [ASTRON] A northern constellation, right ascension 15 hours, declination 70°N; its brightest star, Polaris, is almost at the north celestial pole; seven of the eight stars form a dipper ouline. Also known as Little Bear; Little Dipper.

Ursidae [VERT ZOO] A family of mammals in the order Carnivora including the bears and their allies.

ursids [ASTRON] A shower of meteors occurring about December 22 from a radiant in the constellation Ursa Minor.

ursolic acid [BIOCHEM] $C_{30}H_{48}O_3$ A pentacyclic terpene that crystallizes from absolute alcohol solution, found in leaves and berries of plants; used in pharmaceutical and food industries as an emulsifying agent.

urstromthal [GEOL] A large channel cut by a stream of water from melting ice, flowing along the edge of an ice sheet.

Urticaceae [BOT] A family of dicotyledonous herbs in the order Urticales characterized by a single unbranched style, a straight embryo, and the lack of milky juice (latex).

Urticales [BOT] An order of dicotyledons in the subclass Hamamelidae; woody plants or herbs with simple, usually stipulate leaves, and reduced clustered flowers that usually have a vestigial perianth.

urticaria [MED] Hives or nettle rash; a skin condition characterized by the appearance of intensely itching wheals or welts with elevated, usually white centers and a surrounding area of erythema. Also known as hives.

Urysohn lemma [MATH] If A and B are disjoint, closed sets in a normal space X, there is a real-valued function f such that $0 \leq f(x) \leq 1$ for all $x \varepsilon X$, and $f(A) = 0$ and $f(B) = 1$.

usable iron ore [MET] A steel industry term for high-grade iron ore, concentrates, or agglomerates which can be used in blast furnaces or other processing plants.

usable rate of fire [ORD] Normal rate of fire of a gun in actual use, measured in units of shots per minute; the usable rate of fire is considerably less than a gun's maximum rate of fire, which is a theoretical value based on the purely mechanical operation of a weapon.

U Sagittae [ASTRON] An eclipsing binary star in which one component has attained its Roche limit, while its mate is distinctly smaller than this limit.

use-dilution test [MICROBIO] A bioassay method for testing disinfectants for use on surfaces where a substantial reduction of bacterial contamination is not achieved by prior cleaning; the test organisms *Salmonella choleraesuis* and *Staphylococcus aureus* are deposited in stainless steel cylinders which are then exposed to the action of the test disinfectant.

user [ADP] Anyone who requires the use of services of a computing system or its products. [COMMUN] An individual, installation, or activity having access to a switching center through a local private branch exchange, or by dialing an access code.

user-to-user service [COMMUN] Method of switching that enables direct user-to-user connection that does not provide for message store-and-forward service.

U-shaped abutment [CIV ENG] A bridge abutment with wings perpendicular to the face which act as counterforts; a very stable abutment, often used for architectural effect.

U-shaped valley [GEOL] A type of valley with a broad floor and steep walls produced by glacial erosion. Also known as trough valley; U valley.

using agency [ORD] Any element of the U.S. Armed Forces having command or service functions, and requiring materiel for use in performance of its mission.

Usneaceae [BOT] The beard lichens, a family of Ascolichenes in the order Lecanorales distinguished by their conspicuous fruticose growth form.

usnic acid [BIOCHEM] $C_{18}H_{16}O_7$ Yellow crystals, insoluble in water, slightly soluble in alcohol and ether, melts about 198°C; found in lichens; used as an antibiotic. Also known as usninic acid.

usninic acid See usnic acid.

USP acid test [ANALY CHEM] A United States Pharmacopoeia test to determine the carbonizable substances present in petroleum white oils.

U.S. Survey foot [GEOD] The foot used by the U.S. Coast and Geodetic Survey in which 1 inch is equal to 2.540005 centimeters.

U stage See universal stage.

Ustalf [GEOL] A suborder of the soil order Alfisol; red or brown soil formed in a ustic moisture regime and in a mesic or warmer temperature regime.

Ustert [GEOL] A suborder of the Vertisol soil order; has a faint horizon and is dry for an appreciable period or more than one period of the year.

Ustilaginaceae [MYCOL] A family of fungi in the order Ustilaginales in which basidiospores bud from the sides of the septate epibasidium.

Ustilaginales [MYCOL] An order of the subclass Heterobasidiomycetidae comprising the smut fungi which parasitize plants and cause diseases known as smut or bunt.

Ustoll [GEOL] A suborder of the soil order Mollisol; formed in a ustic moisture regime and in a mesic or warmer temperature regime; may have a calcic, petrocalcic, or gypsic horizon.

Ustox [GEOL] A suborder of the soil order Oxisol; formed in areas that are dry for some part of the year, and has a mean annual temperature of 15°C or more.

Ustult [GEOL] A suborder of the soil order Ultisol; brownish or reddish, with low to moderate organic-carbon content; a well-drained soil of warm-temperate and tropical climates with moderate or low rainfall.

UT 0 See universal time 0.

UT 1 See universal time 1.

UT 2 See universal time 2.

utahite See jarosite.

UTC See universal time coordinated.

uterus [ANAT] The organ of gestation in mammals which receives and retains the fertilized ovum, holds the fetus during development, and becomes the principal agent of its expulsion at term.

uterus bicornis [ANAT] A uterus divided into two horns or compartments; an abnormal condition in humans but normal in many mammals, such as carnivores.

utilidor [CIV ENG] An insulated, heated conduit built below the ground surface or supported above the ground surface to protect the contained water, steam, sewage, and fire lines from freezing.

utility [ENG] One of the nonprocess (support) facilities for a manufacturing plant; usually considered as facilities for steam, cooling water, deionized water, electric power, refrigeration, compressed and instrument air, and effluent treatment.

utility routine [ADP] A program or routine of general usefulness, usually not very complicated, and applicable to many jobs or purposes.

utilization factor [ELEC] In electric power distribution, the maximum demand of a system or part of a system divided by its rated capacity.

utilization rate [AERO ENG] The amount of flying time produced in a specific period expressed in hours per period per aircraft. Also known as flying hour rate.

utricle See utriculus.

utriculus [ANAT] **1.** That part of the membranous labyrinth of the ear into which the semicircular canals open. **2.** A small, blind pouch extending from the urethra into the prostate. Also known as utricle.

U-tube heat exchanger [CHEM ENG] A heat-exchanger system consisting of a bundle of U tubes (hairpin tubes) surrounded by a shell (outer vessel); one fluid flows through the tubes, and the other fluid flows through the shell, around the tubes.

U-tube manometer [ENG] A manometer consisting of a U-shaped glass tube partly filled with a liquid of known specific gravity; when the legs of the manometer are connected to separate sources of pressure, the liquid rises in one leg and drops in the other; the difference between the levels is proportional to the difference in pressures and inversely proportional to the liquid's specific gravity. Also known as liquid-column gage.

UV See ultraviolet.

uvala [GEOGR] Broad-bottomed lowlands.

U valley See U-shaped valley.

uvarovite [MINERAL] $Ca_3Cr_2(SiO_4)_3$ The emerald-green, calcium-chromium end member of the garnet group. Also known as ouvarovite; uwarowite.

UV Ceti stars [ASTRON] A class of stars that have brief outbursts of energy over their surface areas; they may have an increase of about 1 magnitude for periods of 1 hour; the type star is UV Ceti. Also known as flare stars.

uvea [ANAT] The pigmented, vascular layer of the eye: the iris, ciliary body, and choroid.

uveitis See iridocyclochoroiditis.

uviol glass [MATER] A type of glass that is highly transparent to ultraviolet radiation.

UV stabilizer [CHEM] Any chemical compound that, admixed with a thermoplastic resin, selectively absorbs ultraviolet rays; used to prevent ultraviolet degradation of polymers. Also known as ultraviolet stabilizer.

uvula [ANAT] **1.** A fingerlike projection in the midline of the posterior border of the soft palate. **2.** A lobe of the vermiform process of the lower surface of the cerebellum.

uwarowite See uvarovite.

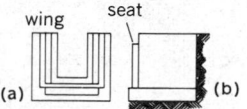

U-SHAPED ABUTMENT

U-shaped abutment design for a bridge. *(a)* Plan. *(b)* Side elevation.

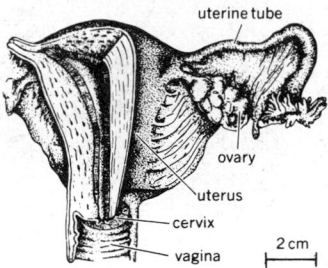

UTERUS

Human uterus and associated structures. *(From L. B. Arey, Developmental Anatomy, 7th ed., Saunders, 1965)*

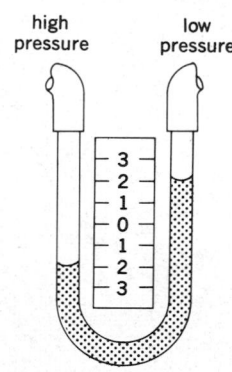

U-TUBE MANOMETER

Components of U-tube manometer for measuring liquid flow.

V *See* vanadium.

V-1 [ORD] A German robot bomb provided with wings, a horizontal stabilizer, a vertical stabilizer, a rudder, and elevators, and powered by a pulse-jet engine mounted on its back; used in World War II. Also known as buzz bomb.

V-2 [ORD] A large, German, liquid-fuel rocket developed as a ballistic missile during World War II; first launched against England on September 8, 1944, it developed 60,000 pounds (27,000 kilogram-force or 27,000 newtons) of thrust from its rocket engine. Also known as A-4.

VA *See* volt-ampere.

vac *See* millibar.

vacancy [SOLID STATE] A defect in the form of an unoccupied lattice position in a crystal.

vaccination [IMMUNOL] Inoculation of viral or bacterial organisms or antigens to produce immunity in the recipient.

vaccination encephalitis [MED] Encephalitis caused by vaccination with rabies vaccine.

vaccine [IMMUNOL] A suspension of killed or attenuated bacteria or viruses or fractions thereof, injected to produce active immunity.

vaccinia [VET MED] A contagious disease of cows which is characterized by vesicopustular lesions of the skin that are prone to appear on the teats and udder, and which is transmissible to humans by handling infected cows and by vaccination; confers immunity against smallpox. Also known as cowpox.

vacuole [CYTOL] A membrane-bound cavity within a cell; may function in digestion, storage, secretion, or excretion. [GEOL] *See* vesicle.

vacuum [PHYS] **1.** Theoretically, a space in which there is no matter. **2.** Practically, a space in which the pressure is far below normal atmospheric pressure so that the remaining gases do not affect processes being carried on in the space.

vacuum behavior [PSYCH] The carrying out of a series of model action patterns in apparent absence of any obviously appropriate behavior.

vacuum brake [MECH ENG] A form of air brake which operates by maintaining low pressure in the actuating cylinder; braking action is produced by opening one side of the cylinder to the atmosphere so that atmospheric pressure, aided in some designs by gravity, applies the brake.

vacuum breaker [ENG] A device used to relieve a vacuum formed in a water supply line to prevent backflow. Also known as backflow preventer.

vacuum capacitor [ELEC] A capacitor with separated metal plates or cylinders mounted in an evacuated glass envelope to obtain a high breakdown voltage rating.

vacuum casting [MET] Metal casting in a vacuum.

vacuum circuit breaker [ELEC] A circuit breaker in which a pair of contacts is hermetically sealed in a vacuum envelope; the contacts are separated by using a bellows to move one of them; an arc is produced by metallic vapor boiled from the electrodes, and is extinguished when the vapor particles condense on solid surfaces.

vacuum cleaner [MECH ENG] An electrically powered mechanical appliance for the dry removal of dust and loose dirt from rugs, fabrics, and other surfaces.

vacuum concrete [CIV ENG] Concrete poured into a framework that is fitted with a vacuum mat to remove water not required for setting of the cement; in this framework, concrete attains its 28-day strength in 10 days and has a 25% higher crushing strength.

vacuum condensing point [CHEM] Temperature at which the sublimate (vaporized solid) condenses in a vacuum. Abbreviated vcp.

vacuum cooling [FOOD ENG] A system of cooling fruits and vegetables prior to shipping; water is caused to evaporate from the surfaces of the food by producing a vacuum around it.

vacuum correction [PHYS] The correction to the reading of a mercury barometer required by the imperfections in the vacuum above the mercury column, due to the presence of water vapor and air; this correction is a function of both temperature and pressure.

vacuum crystallizer [CHEM ENG] Crystallizer in which a warm saturated solution is fed to a lagged, closed vessel maintained under vacuum; the solution evaporates and cools adiabatically, resulting in crystallization.

vacuum degassing [MET] A process for removing gases from a metal either by melting or heating the solid metal in a vacuum.

vacuum deposition [MET] Deposition of a thin coating of metal by condensation on a cool work surface in vacuum.

vacuum distillation [CHEM ENG] Liquid distillation under reduced (less than atmospheric) pressure; used to lower boiling temperatures and lessen the risk of thermal degradation during distillation. Also known as reduced-pressure distillation.

vacuum drying [ENG] The removal of liquid from a solid material in a vacuum system; used to lower temperatures needed for evaporation to avoid heat damage to sensitive material.

vacuum evaporator [ENG] A vacuum device used to evaporate metals and spectrographic carbon to coat (replicate) a specimen for electron spectroscopic analysis or for electron microscopy.

vacuum filter [ENG] A filter device into which a liquid-solid slurry is fed to the high-pressure side of a filter medium, with liquid pulled through to the low-pressure side of the medium and a cake of solids forming on the outside of the medium.

vacuum filtration [ENG] The separation of solids from liquids by passing the mixture through a vacuum filter.

vacuum flashing [CHEM ENG] The heating of a liquid that, upon release to a lower pressure (vacuum), undergoes considerable vaporization (flashing). Also known as flash vaporization.

vacuum forming [ENG] Plastic-sheet forming in which the sheet is clamped to a stationary frame, then heated and drawn down into a mold by vacuum.

vacuum freeze dryer [ENG] A type of indirect batch dryer used to dry materials that would be destroyed by the loss of volatile ingredients or by drying temperatures above the freezing point.

vacuum fusion [MET] A technique for determining the oxygen, hydrogen, and sometimes nitrogen content of metals; can be applied to a wide variety of metals with the exception of alkali and alkaline earth metals.

vacuum gage [ENG] A device that indicates the absolute gas pressure in a vacuum system.

vacuum heating [MECH ENG] A two-pipe steam heating system in which a vacuum pump is used to maintain a suction in the return piping, thus creating a positive return flow of air and condensate.

vacuum mat [CIV ENG] A rigid flat metal screen faced by a linen filter, the back of which is kept under partial vacuum; used to suck out surplus air and water from poured concrete to produce a dense, well-shrunk concrete.

vacuum measurement [ENG] The determination of a fluid pressure less in magnitude than the pressure of the atmosphere.

vacuum metallurgy [MET] The melting, shaping, and treating of metals and alloys under reduced pressure that ranges from subatmospheric pressure to ultra-high vacuum.

vacuum pan salt [CHEM ENG] A salt made from salt brine boiled at reduced pressure in a triple-effect evaporator.

VACUOLE

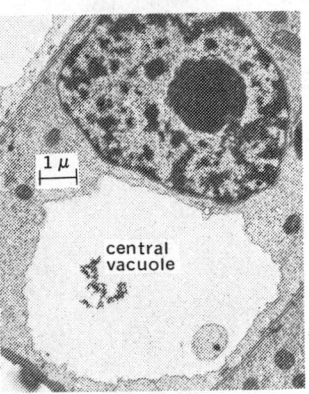

Electron micrograph of a partially mature cell from a developing root tip showing the central vacuole. The osmotic pressure in this vacuole is responsible for the turgor and provides the force to expand the cell to its mature size.

VACUUM CIRCUIT BREAKER

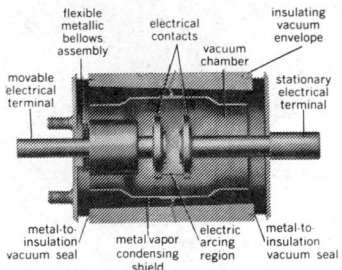

Cutaway view of vacuum circuit breaker showing component parts.

VACUUM FUSION

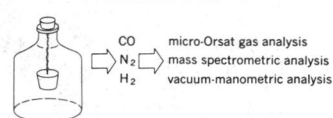

Apparatus used in the vacuum fusion method for analysis of metals. Oxygen is released from the metal as carbon monoxide by reaction of oxides or dissolved oxygen with carbon from the graphite crucible at high temperature.

vacuum phototube [ELECTR] A phototube that is evacuated to such a degree that its electrical characteristics are essentially unaffected by gaseous ionization; in a gas phototube, some gas is intentionally introduced.

vacuum plating [MET] Producing a surface film of metal on a heated surface, often in a vacuum, either by decomposition of the vapor of a compound at the work surface, or by direct reaction between the work surface and the vapor. Also known as vapor deposition.

vacuum polarization [QUANT MECH] A process in which an electromagnetic field gives rise to virtual electron-positron pairs that effectively alter the distribution of charges and currents that generated the original electromagnetic field.

vacuum printing frame [GRAPHICS] A frame employed in the graphic arts to keep materials flat or to keep them in absolute contact with one another; the frame is useful in making positives, contact prints, screened tints, and plates.

vacuum pump [MECH ENG] A compressor for exhausting air and noncondensable gases from a space that is to be maintained at subatmospheric pressure.

vacuum relay [ELEC] A sensitive relay having its contacts mounted in a highly evacuated glass housing, to permit handling radio-frequency voltages as high as 20,000 volts without flashover between contacts even though contact spacing is but a few hundredths of an inch when open.

vacuum shelf dryer [ENG] A type of indirect batch dryer which generally consists of a vacuum-tight cubical or cylindrical chamber of cast-iron or steel plate, heated supporting shelves inside the chamber, a vacuum source, and a condenser; used extensively for drying pharmaceuticals, temperature-sensitive or easily oxidizable materials, and small batches of high-cost products where any product loss must be avoided.

vacuum switch [ELEC] A switch having its contacts in an evacuated envelope to minimize sparking.

vacuum thermobalance [ANALY CHEM] An instrument used in thermogravimetry consisting of a precision balance and furnace that have been adapted for continuously measuring or recording changes in weight of a substance as a function of temperature; used in many types of physicochemical reactions where rates of reaction and energies of activation for vaporization, sublimation, and chemical reaction can be obtained.

vacuum tube [ELECTR] An electron tube evacuated to such a degree that its electrical characteristics are essentially unaffected by the presence of residual gas or vapor.

vacuum-tube amplifier [ELECTR] An amplifier employing one or more vacuum tubes to control the power obtained from a local source.

vacuum-tube circuit [ELECTR] An electric circuit in which a vacuum tube is connected.

vacuum-tube clipping circuit [ELECTR] A circuit in which a vacuum tube is used to achieve clipping action; the bias at the input is set at such a level that output current cannot flow during a portion of the amplitude excursion of the input voltage or current waveform.

vacuum-tube electrometer [ELECTR] An electrometer in which the ionization current in an ionization chamber is amplified by a special vacuum triode having an input resistance above 10,000 megohms.

vacuum-tube keying [ELECTR] Code-transmitter keying system in which a vacuum tube is connected in series with the plate supply lead of a frequency-controlling stage of the transmitter; when the key is open, the tube blocks, interrupting the plate supply to the output stage; closing the key allows the plate current to flow through the keying tube and the output tubes.

vacuum-tube modulator [ELECTR] A modulator employing a vacuum tube as a modulating element for impressing an intelligence signal on a carrier.

vacuum-tube oscillator [ELECTR] A circuit utilizing a vacuum tube to convert direct-current power into alternating-current power at a desired frequency.

vacuum-tube rectifier [ELECTR] A rectifier in which rectification is accomplished by the unidirectional passage of electrons from a heated electrode to one or more other electrodes within an evacuated space.

vacuum-tube voltmeter [ENG] Any of several types of instrument in which vacuum tubes, acting as amplifiers or rectifiers, are used in circuits for the measurement of alternating-current or direct-current voltage. Abbreviated VTVM. Also known as tube voltmeter.

vacuum-type insulation [CHEM ENG] Highly reflective double-wall structure with high vacuum between the walls; used as insulation for cryogenic systems; Dewar flasks have vacuum-type insulation.

vacuum ultraviolet radiation [ELECTROMAG] Ultraviolet radiation with a wavelength of less than 200 nanometers; absorption of radiation in this region by air and other gases requires the use of evacuated apparatus for transmission. Also known as extreme ultraviolet radiation.

vacuum ultraviolet spectroscopy [SPECT] Absorption spectroscopy involving electromagnetic wavelengths shorter than 200 nanometers; so called because the interference of the high absorption of most gases necessitates work with evacuated equipment.

vadose water [HYD] Water in the zone of aeration. Also known as kremastic water; suspended water; wandering water.

vadose zone See zone of aeration.

vaesite [MINERAL] NiS_2 An isometric mineral with pyrite structure composed of sulfide of nickel.

vagility [ECOL] The ability of organisms to disseminate.

vagina [ANAT] The canal from the vulvar opening to the cervix uteri.

vagina fibrosa tendinis [ANAT] A fibrous sheath surrounding the tendon of a muscle and usually confining the tendon in a bony groove.

vaginal protocele See rectocele.

vaginate [BIOL] Invested in a sheath.

vaginismus [MED] Painful vaginal spasm.

vaginitis [MED] 1. Inflammation of the vagina. 2. Inflammation of a tendon sheath.

vagotonine [BIOCHEM] An endocrine substance which is thought to be elaborated by cells of the pancreas and which regulates autonomic tonus.

vagus [ANAT] The tenth cranial nerve; either of a pair of sensory and motor nerves forming an important part of the parasympathetic system in vertebrates.

valais wind [METEOROL] The notable valley wind that blows along the Rhone Valley from the upper end of Lake Geneva (Valais Canton); it is sufficiently strong and regular to distort the growth of trees.

valence [BIOCHEM] The relative ability of a biological substance to react or combine. [CHEM] A positive number that characterizes the combining power of an element for other elements, as measured by the number of bonds to other atoms which one atom of the given element forms upon chemical combination; hydrogen is assigned valence 1, and the valence is the number of hydrogen atoms, or their equivalent, with which an atom of the given element combines.

valence angle See bond angle.

valence band [SOLID STATE] The highest electronic energy band in a semiconductor or insulator which can be filled with electrons.

valence bond [PHYS CHEM] The bond formed between the electrons of two or more atoms.

valence-bond method [PHYS CHEM] A method of calculating binding energies and other parameters of molecules by taking linear combinations of electronic wave functions, some of which represent covalent structures, others ionic structures; the coefficients in the linear combination are calculated by the variational method. Also known as valence-bond resonance method.

valence-bond resonance method See valence-bond method.

valence-bond theory [CHEM] A theory of the structure of chemical compounds according to which the principal requirements for the formation of a covalent bond are a pair of electrons and suitably oriented electron orbitals on each of the atoms being bonded; the geometry of the atoms in the resulting coordination polyhedron is coordinated with the orientation of the orbitals on the central atom.

valence crystal See covalent crystal.

valence electron [ATOM PHYS] An electron that belongs to the outermost shell of an atom. [SOLID STATE] See conduction electron.

VACUUM-TUBE AMPLIFIER

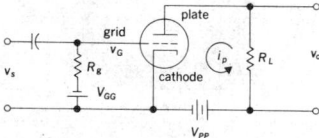

Diagram of a basic vacuum-tube amplifier. R_L = fixed load resistor; i_p = plate current; V_{GG} = bias voltage; V_{PP} = high-potential plate supply voltage; v_G = instantaneous grid voltage; v_S = signal voltage; v_O = output voltage.

valence number [CHEM] A number that is equal to the valence of an atom or ion multiplied by $+1$ or -1, depending on whether the ion is positive or negative, or equivalently on whether the atom in the molecule under consideration has lost or gained electrons from its free state.

valence shell [ATOM PHYS] The electrons that form the outermost shell of an atom.

valencianite [MINERAL] A variety of potassium feldspar from Mexico.

valeral *See* n-valeraldehyde.

n-valeraldehyde [ORG CHEM] $CH_3(CH_2)_3CHO$ A flammable liquid, soluble in ether and alcohol, slightly soluble in water; boils at $102°C$; used in flavors and as a rubber accelerator. Also known as amyl aldehyde; valeral; valeric aldehyde.

valeramide [ORG CHEM] $CH_3(CH_2)_3CONH_2$ Water-soluble, colorless crystals, melting at $127°C$. Also known as pentanamide; valeric amide.

valerianic acid *See* valeric acid.

valerian oil [MATER] A combustible, yellow to brown liquid with a penetrating aroma; soluble in alcohol, acetone, and other organic solvents; derived from the roots and rhizome of the garden heliotrope (*Valeriana officinalis*), the main components being pinene, camphene, borneol, and esters; used in medicine, flavors, and industrial odorants and to perfume tobacco.

valeric acid [ORG CHEM] $CH_3(CH_2)_3COOH$ A combustible, toxic, colorless liquid with a penetrating aroma; soluble in water, alcohol, and ether; boils at $185°C$; used to make flavors, perfumes, lubricants, plasticizers, and pharmaceuticals. Also known as n-pentanoic acid; valerianic acid.

valeric aldehyde *See* n-valeraldehyde.

valeric amide *See* valeramide.

γ-valerolactone [ORG CHEM] $C_5H_8O_2$ A combustible, mostly immiscible, colorless liquid, boiling at $205°C$; used as a dye-bath coupling agent, in brake fluids and cutting oils, and as a solvent for adhesives, lacquers, and insecticides.

valid [SYST] Describing a taxon classified on the basis of distinctive characters of accepted importance.

validation [ADP] The act of testing for compliance with a standard.

validity [MATH] Correctness; especially the degree of closeness by which iterated results approach the correct result.

validity check [ADP] Computer check of input data, based on known limits for variables in given fields.

valine [BIOCHEM] $C_5H_{11}NO_2$ An amino acid considered essential for normal growth of animals, and biosynthesized from pyruvic acid. Also known as 2-aminoisovaleric acid; α-aminoisovaleric acid; 2-amino-3-methylbutyric acid.

Valium [PHARM] Trademark for diazepam, a tranquilizer.

vallate papilla [ANAT] One of the large, flat papillae, each surrounded by a trench, in a group anterior to the sulcus terminalis of the tongue. Also known as circumvallate papilla.

vallerite [MINERAL] $CuFeS_2$ A sulfide mineral found in meteorites.

valley [GEOGR] A generally broad area of flat, low-lying land bordered by higher ground. [GEOL] A relatively shallow, wide depression of the sea floor with gentle slopes. Also known as submarine valley.

valley attenuation [ELECTR] For an electric filter with an equal ripple characteristic, the maximum attenuation occurring at a frequency between two frequencies where the attenuation reaches a minimum value.

valley bottom *See* valley floor.

valley breeze [METEOROL] A gentle wind blowing up a valley or mountain slope in the absence of cyclonic or anticyclonic winds, caused by the warming of the mountainside and valley floor by the sun.

valley fill [GEOL] Unconsolidated sedimentary deposit which fills or partly fills a valley.

valley flat [GEOL] The small plain at the bottom of a narrow valley with steep sides.

valley floor [GEOL] The broad, flat bottom of a valley. Also known as valley bottom; valley plain.

valley glacier [HYD] A glacier that flows down the walls of a mountain valley.

valley iceberg [OCEANOGR] An iceberg weathered in such a

manner that a large U-shaped slot extends through the iceberg. Also known as drydock iceberg.

valley line *See* thalweg.

valley plain *See* valley floor.

valley train [GEOL] A long, narrow body of outwash, deposited by meltwater far beyond the margin of an active glacier and extending along the floor of a valley. Also known as outwash train.

valley wind [METEOROL] A wind which ascends a mountain valley (up-valley wind) during the day; the daytime component of a mountain and valley wind system.

Valoniaceae [BOT] A family of green algae in the order Siphonocladales consisting of plants that are essentially unicellular, coenocytic vesicles, spherical or clavate, and up to 6 centimeters in diameter.

valuation [MATH] A scalar function of a field which has properties similar to those of absolute value.

value [SCI TECH] The magnitude of a quantity.

value analysis *See* value engineering.

value control *See* value engineering.

value engineering [IND ENG] The systematic application of recognized techniques which identify the function of a product or service, and provide the necessary function reliably at lowest overall cost. Also known as value analysis; value control.

value of isotope mixture [CHEM] A measure of the effort required to prepare a quantity of an isotope mixture; it is proportional to the amount of the mixture, and also depends on the composition of the mixture to be prepared and the composition of the original mixture.

value theory [SYS ENG] A concept normally associated with decision theory; it strives to evaluate relative utilities of simple and mixed parameters which can be used to describe outcomes.

Valvatacea [PALEON] A superfamily of extinct gastropod mollusks in the order Prosobranchia.

valvate [BOT] Having valvelike parts, as those which meet edge to edge or which open as if by valves.

Valvatida [INV ZOO] An order of echinoderms in the subclass Asteroidea.

Valvatina [INV ZOO] A suborder of echinoderms in the order Phanerozonida in which the upper marginals lie directly over, and not alternate with, the corresponding lower marginals.

valve [ANAT] A flat of tissue, as in the veins or between the chambers in the heart, which permits movement of fluid in one direction only. [BOT] **1.** A segment of a dehiscing capsule or legume. **2.** The lidlike portion of certain anthers. [ELECTR] *See* electron tube. [INV ZOO] **1.** One of the distinct, articulated pieces composing the shell of certain animals, such as barnacles and brachiopods. **2.** One of two shells encasing the body of a bivalve mollusk or a diatom. [MECH ENG] A device used to regulate the flow of fluids in piping systems and machinery.

valve arrester [ELEC] A type of lightning arrester which consists of a single gap or multiple gaps in series with current-limiting elements; gaps between spaced electrodes prevent flow of current through the arrester except when the voltage across them exceeds the critical gap flashover.

valve follower [MECH ENG] A linkage between the cam and the push rod of a valve train.

valve guide [MECH ENG] A channel which supports the stem of a poppet valve for maintenance of alignment.

valve head [MECH ENG] The disk part of a poppet valve that gives a tight closure on the valve seat.

valve-in-head engine *See* overhead-valve engine.

valve lifter [MECH ENG] A device for opening the valve of a cylinder as in an internal combustion engine.

valve positioner [CONT SYS] A pneumatic servomechanism which is used as a component in process control systems to improve operating characteristics of valves by reducing hysteresis. Also known as pneumatic servo.

valve seat [DES ENG] The circular metal ring on which the valve head of a poppet valve rests when closed.

valve stem [MECH ENG] The rod by means of which the disk or plug is moved to open and close a valve.

valve train [MECH ENG] The valves and valve-operating mechanism for the control of fluid flow to and from a piston-

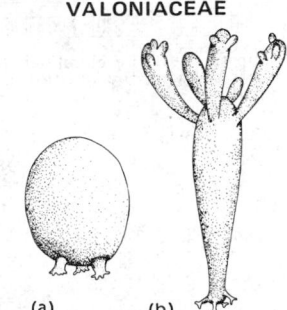

VALINE

Structural formula of valine.

VALONIACEAE

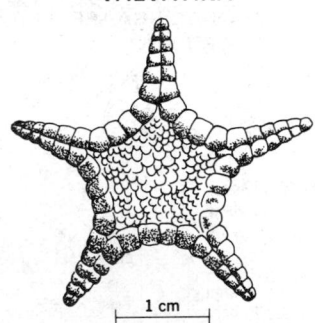

(a)   (b)

Valonia. (a) Spherical form. (b) Clavate form.

VALVATINA

|← 1 cm →|

A representative valvate starfish, *Iconaster perierctus.* (After A. G. Fisher, 1919)

VALVE ARRESTER

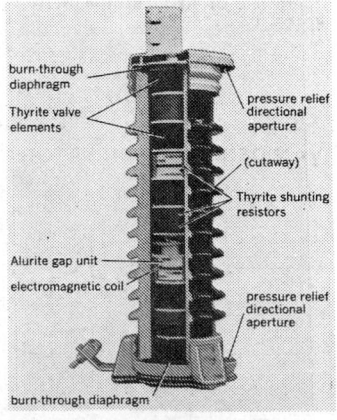

Thirty-kilovolts General Electric Alugard station lightning arrester, of valve type. (*General Electric Co.*)

VANADIUM

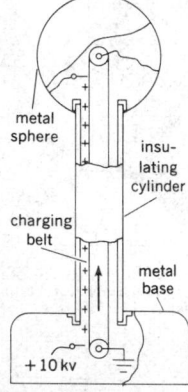

Periodic table of the chemical elements showing the position of vanadium.

VAN DE GRAAFF GENERATOR

metal sphere

insulating cylinder

charging belt

metal base

+ 10 kv

Schematic drawing of Van de Graaff generator for operation in air at atmospheric pressure.

VAN DER POL OSCILLATOR

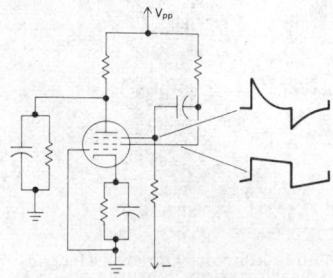

V_{pp}

Circuit diagram of Van der Pol oscillator showing waveform. V_{pp} = plate supply voltage.

cylinder machine, for example, steam, diesel, or gasoline engine.

Valvifera [INV ZOO] A suborder of isopod crustaceans distinguished by having a pair of flat, valvelike uropods which hinge laterally and fold inward beneath the rear part of the body.

valvula [BIOL] A small valve. [INV ZOO] One of the small processes forming a sheath for the ovipositor in certain insects.

valvulate [BIOL] Having valvules.

vampire [VERT ZOO] The common name for bats making up the family Desmodontidae which have teeth specialized for cutting and which subsist on a blood diet.

Vampyrellidae [INV ZOO] A family of protozoans in the order Proteomyxida including species which invade filamentous algae and sometimes higher plants.

Vampyromorpha [INV ZOO] An order of dibranchiate cephalopod mollusks represented by *Vampyroteuthis infernalis,* an inhabitant of the deeper waters of tropical and temperate seas.

van [MIN ENG] **1.** A test of the value of an ore, made by washing (vanning) a small quantity, after powdering it, on the point of a shovel. **2.** To separate, as ore from veinstone, by washing it on the point of a shovel. **3.** A shovel used in ore dressing.

vanadate [MINERAL] Any of several mineral compounds characterized by pentavalent vanadium and oxygen in the anion; an example is vanadinite.

vanadic acid [INORG CHEM] Any of various acids that do not exist in a pure state and are found in various alkali and other metal vanadates; forms are meta- (HVO_3), ortho- (H_3VO_4), and pyro- ($H_4V_2O_7$).

vanadic acid anhydride *See* vanadium pentoxide.

vanadic sulfate *See* vanadyl sulfate.

vanadic sulfide *See* vanadium sulfide.

vanadinite [MINERAL] $Pb_5(VO_4)_3Cl$ A red, yellow, or brown opatite mineral often occurring as globular masses encrusting other minerals in lead mines; an ore of vanadium and lead hardness is 2.75-3 on Mohs scale, and specific gravity is 6.66-7.10.

vanadium [CHEM] A metal in group Vb, symbol V, atomic number 23; soluble in strong acids and alkalies; melts at 1900°C, boils about 3000°C; used as a catalyst. [MET] A silvery-white, ductile metal resistant to corrosion; used in alloy steels and as an x-ray target.

vanadium carbide [INORG CHEM] VC Hard, black crystals, melting at 2800°C, boiling at 3900°C; insoluble in acids, except nitric acid; used in cutting-tool alloys and as a steel additive.

vanadium dichloride [INORG CHEM] VCl_2 Toxic, green crystals, soluble in alcohol and ether; decomposes in hot water; used as a reducing agent. Also known as vanadous chloride.

vanadium oxide [INORG CHEM] A compound of vanadium with oxygen, for example, vanadium tetroxide (V_2O_4), vanadium trioxide or sesquioxide (V_2O_3), vanadium oxide (VO), and vanadium pentoxide (V_2O_5).

vanadium oxydichloride *See* vanadyl chloride.

vanadium oxytrichloride [INORG CHEM] $VOCl_3$ A toxic, yellow liquid that dissolves or reacts with many organic substances; hydrolyzes in moisture; boils at 126°C; used as an olefin-polymerization catalyst and in organovanadium synthesis.

vanadium pentasulfide *See* vanadium sulfide.

vanadium pentoxide [INORG CHEM] V_2O_5 A toxic, yellow to red powder, soluble in alkalies and acids, slightly soluble in water; melts at 690°C; used in medicine, as a catalyst, as a ceramics coloring, for ultraviolet-resistant glass, photographic developers, textiles dyeing, and nuclear reactors. Also known as vanadic acid anhydride.

vanadium sesquioxide *See* vanadium trioxide.

vanadium steel [MET] A low-alloy steel containing 0.10-0.15% vanadium.

vanadium sulfate *See* vanadyl sulfate.

vanadium sulfide [INORG CHEM] V_2S_5 A toxic, black-green powder; insoluble in water, soluble in alkalies and acids; decomposes when heated; used to make vanadium com-

pounds. Also known as vanadic sulfide; vanadium pentasulfide.

vanadium tetrachloride [INORG CHEM] VCl_4 A toxic, red liquid; soluble in ether and absolute alcohol; boils at 154°C; used in medicine and to manufacture vanadium and organovanadium compounds.

vanadium tetraoxide [INORG CHEM] V_2O_4 A toxic blue-black powder; insoluble in water, soluble in alkalies and acids; melts at 1967°C; used as a catalyst.

vanadium trichloride [INORG CHEM] VCl_3 Toxic, deliquescent, pink crystals; soluble in ether and absolute alcohol; decomposes in water and when heated; used to prepare vanadium and organovanadium compounds.

vanadium trioxide [INORG CHEM] V_2O_3 Toxic, black crystals; soluble in alkalies and hydrofluoric acid, slightly soluble in water; melts at 1970°C; used as a catalyst. Also known as vanadium sesquioxide.

vanadous chloride *See* vanadium dichloride.

vanadyl chloride [INORG CHEM] $V_2O_2Cl_4 \cdot 5H_2O$ Toxic, deliquescent, water- and alcohol-soluble green crystals; used to mordant textiles. Also known as divanadyl tetrachloride; vanadium oxydichloride; vanadyl dichloride.

vanadyl dichloride *See* vanadyl chloride.

vanadyl sulfate [INORG CHEM] $VOSO_4 \cdot 2H_2O$ Blue, toxic, water-soluble crystals; used as a reducing agent, catalyst, glass and ceramics colorant, and mordant. Also known as vanadic sulfate; vanadium sulfate.

Van Allen radiation belt [GEOPHYS] One of the belts of intense ionizing radiation in space about the earth formed by high-energy charged particles which are trapped by the geomagnetic field.

Van Atta array [ELECTROMAG] Antenna array in which pairs of corner reflectors or other elements equidistant from the center of the array are connected together by a low-loss transmission line in such a way that the received signal is reflected back to its source in a narrow beam to give signal enhancement without amplification.

vancomycin [MICROBIO] A complex antibiotic substance produced by *Streptomyces orientalis;* useful for treatment of severe staphylococcic infections.

van Crevald–von Gierke's disease *See* von Gierke's disease.

Van Deemter rate theory [ANALY CHEM] A theory that the sample phase in gas chromatography flows continuously, not stepwise.

Van de Graaff accelerator [ELECTR] A Van de Graaff generator equipped with an evacuated tube through which charged particles may be accelerated.

Van de Graaff generator [ELECTR] A high-voltage electrostatic generator in which electrical charge is carried from ground to a high-voltage terminal by means of an insulating belt and is discharged onto a large, hollow metal electrode.

Van den Bergh reaction [PATH] A liver function test in which diazotized serum or plasma is compared with a standard solution of diazotized bilirubin.

Vandermonde determinant [MATH] The determinant of the $n \times n$ matrix whose ith row appears as $1, x_i, x_i^2, \ldots, x_i^{n-1}$ where the x_i^k appear as variables in a given polynomial equation; this provides information about the roots.

Vandermonde's theorem [MATH] Represents a binomial $(x + y)^a$, where a is an exponent involving the variables x and y, in terms of a sum of expressions $x^c y^d$ where the exponents c and d involve the variables x and y also.

Van der Pol oscillator [ELECTR] A type of relaxation oscillator which has a single pentode tube and an external circuit with a capacitance that causes the device to switch between two values of the screen voltage. [PHYS] A vibrating system governed by an equation of the form $\ddot{x} + \varepsilon(-\dot{x} + \frac{1}{3}\dot{x}^3) + x = 0$.

Van der Waals adsorption [PHYS CHEM] Adsorption in which the cohesion between gas and solid arises from Van der Waals forces.

Van der Waals attraction *See* Van der Waals force.

Van der Waals covolume [PHYS CHEM] The constant b in the Van der Waals equation, which is approximately four times the volume of an atom of the gas in question multiplied by Avogadro's number.

Van der Waals equation [PHYS CHEM] An empirical equation of state which takes into account the finite size of the molecules and the attractive forces between them: $p = [RT/(v - b)] - (a/v^2)$, where p is the pressure, v is the volume per mole, T is the absolute temperature, R is the gas constant, and a and b are constants.

Van der Waals force [PHYS CHEM] An attractive force between two atoms or nonpolar molecules, which arises because a fluctuating dipole moment in one molecule induces a dipole moment in the other, and the two dipole moments then interact. Also known as dispersion force; London dispersion force; Van der Waals attraction.

Van der Waals–London interactions [PHYS CHEM] The interaction associated with the Van der Waals force.

Van der Waals structure [CRYSTAL] The structure of a molecular crystal.

Van der Waals surface tension formula [THERMO] An empirical formula for the dependence of the surface tension on temperature: $\gamma = Kp_c^{2/3}T_c^{1/3}(1 - T/T_c)^n$, where γ is the surface tension, T is the temperature, T_c and p_c are the critical temperature and pressure, K is a constant, and n is a constant equal to approximately 1.23.

vandyke [GRAPHICS] A process used for photocopying; the material is paper-sensitized with ferric iron and silver salts and then exposed to strong light; upon processing, a negative print of white lines on a brown background results; from this a brown-line print can be made.

vane [AERO ENG] A device that projects ahead of an aircraft to sense gusts or other actions of the air so as to create impulses or signals that are transmitted to the control system to stabilize the aircraft. [MECH ENG] A flat or curved surface exposed to a flow of fluid so as to be forced to move or to rotate about an axis, to rechannel the flow, or to act as the impeller; for example, in a steam turbine, propeller fan, or hydraulic turbine. [NAV] A sight on an instrument used for observing bearings, such as on a pelorus or azimuth circle.

vane anemometer [ENG] A portable instrument used to measure low wind speeds and airspeeds in large ducts; consists of a number of vanes radiating from a common shaft and set to rotate when facing the wind.

vane-anode magnetron [ELECTR] Cavity magnetron in which the walls between adjacent cavities have parallel plane surfaces.

vane attenuator See flap attenuator.

vane feather See contour feather.

vane motor rotary actuator [MECH ENG] A type of rotary motor actuator which consists of a rotor with several spring-loaded sliding vanes in an elliptical chamber; hydraulic fluid enters the chamber and forces the vanes before it as it moves to the outlets.

vane-type instrument [ENG] A measuring instrument utilizing the force of repulsion between fixed and movable magnetized iron vanes, or the force existing between a coil and a pivoted vane-shaped piece of soft iron, to move the indicating pointer.

Vaneyellidae [INV ZOO] A family of holothurian echinoderms in the order Dactylochirotida.

vang [NAV ARCH] **1.** A rope which supports or holds steady a boom or spar. **2.** In particular, one of the two ropes which run from the top of the gaff and steady it when the sail is not set.

Vanguard satellite [AERO ENG] One of three artificial satellites launched by the United States in 1958 and 1959, using a modified Viking rocket, as a part of the International Geophysical Year program; *Vanguard 1* was the first spacecraft to use solar cells.

Vanhorniidae [INV ZOO] A monospecific family of the Hymenoptera in the superfamily Proctotrupoidea.

vanilla extract [FOOD ENG] Flavoring prepared from vanilla beans with or without the addition of sugar, dextrose, or glycerol; contains soluble matter from not less than 10 grams of vanilla beans in 100 milliliters.

vanillic aldehyde See vanillin.

vanillin [ORG CHEM] $C_8H_8O_3$ A combustible solid, soluble in water, alcohol, ether, and chloroform; melts at 82°C; used in pharmaceuticals, perfumes, and flavors, and as an analytical reagent. Also known as vanillic aldehyde.

vanishing tide [OCEANOGR] When a high water and low

water "melt" together into a period of several hours with a nearly constant water level.

vanner [MIN ENG] A machine for dressing ore; the name is given to various patented devices in which the peculiar motions of the shovel in the miner's hands in the operation of making a van are, or are supposed to be, successfully imitated. Also known as vanning machine.

vanning machine See vanner.

V antenna [ELECTROMAG] An antenna having a V-shaped arrangement of conductors fed by a balanced line at the apex; the included angle, length, and elevation of the conductors are proportioned to give the desired directivity. Also spelled vee antenna.

van't Hoff equation [PHYS CHEM] An equation for the variation with temperature T of the equilibrium constant K of a gaseous reaction in terms of the heat of reaction at constant pressure, ΔH: $d(\ln K)/dT = \Delta H/RT^2$, where R is the gas constant. Also known as van't Hoff isochore.

van't Hoff factor [PHYS] The ratio of the observed osmotic pressure of a solution to that predicted by van't Hoff's law.

van't Hoff formula [ORG CHEM] The expression that the number of stereoisomers of a sugar molecule is equal to 2^n, where n is the number of asymmetric carbon atoms.

van't Hoff isochore See van't Hoff equation.

van't Hoff isotherm [PHYS CHEM] An equation for the change in free energy during a chemical reaction in terms of the reaction, the temperature, and the concentration and number of molecules of the reactants.

van't Hoff's law [PHYS] The law that the osmotic pressure of a dissolved substance equals the gas pressure it would exert if it were an ideal gas that occupied the same volume as that of the solution.

Van Vleck equation [QUANT MECH] An equation based on quantum theory for the molar paramagnetism of a magnetically susceptible material from magnetic moment, absolute temperature, and various constants.

Van Vleck paramagnetism [QUANT MECH] The paramagnetism of a collection of atoms, ions, or molecules, as computed by quantum theory; the atoms, ions, or molecules in a magnetic field are distributed among the various allowed energy levels according to a Boltzmann distribution, and the magnetization of the system is computed by finding the average component of angular momentum parallel to the field.

vapor [THERMO] A gas at a temperature below the critical temperature, so that it can be liquefied by compression, without lowering the temperature.

vapor barrier [CIV ENG] A layer of material applied to the inner (warm) surface of a concrete wall or floor to prevent absorption and condensation of moisture.

vapor blasting [MET] Cleaning the surface of a metal with a fine abrasive suspended in water and propelled at high speed by air or steam. Also known as liquid honing; vapor honing.

vapor cycle [THERMO] A thermodynamic cycle, operating as a heat engine or a heat pump, during which the working substance is in, or passes through, the vapor state.

vapor degreasing [ENG] A type of cleaning procedure for metals to remove grease, oils, and lightly attached solids; a solvent such as trichloroethylene is boiled, and its vapors are condensed on the metal surfaces.

vapor deposition See vacuum plating.

vapor-filled thermometer [ENG] A gas- or vapor-filled temperature measurement device that moves or distorts in response to temperature-induced pressure changes from the expansion or contraction of the sealed, vapor-containing chamber.

vapor honing See vapor blasting.

vaporimeter [ENG] An instrument used to measure a substance's vapor pressure, especially that of an alcoholic liquid, in order to determine its alcohol content.

vaporization See volatilization.

vaporization cooling [ENG] Cooling by volatilization of a nonflammable liquid having a low boiling point and high dielectric strength; the liquid is flowed or sprayed on hot electronic equipment in an enclosure where it vaporizes, carrying the heat to the enclosure walls, radiators, or heat exchanger. Also known as evaporative cooling.

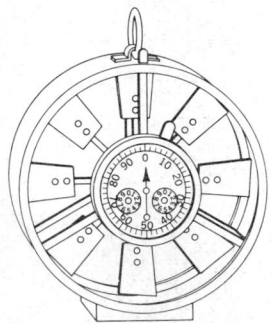

VANE ANEMOMETER

Portable revolving-vane anemometer. Counter in center indicates the number of rotations, which when timed with stopwatch serve to determine wind speed. *(From D. M. Considine, ed., Process Instruments and Controls Handbook, McGraw-Hill, 1957)*

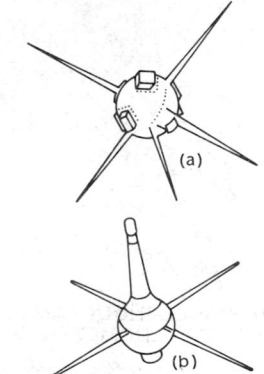

VANGUARD SATELLITE

(a)

(b)

Two spacecraft in the Vanguard program: *(a) Vanguard 1; (b) Vanguard 3.*

vaporizer [CHEM ENG] A process vessel in which a liquid is heated until it vaporizes; heat can be indirect (steam or heat-transfer fluid) or direct (hot gases or submerged combustion).

vapor lamp *See* discharge lamp.

vapor-liquid equilibrium *See* liquid-vapor equilibrium.

vapor-liquid separation [CHEM ENG] The removal of liquid droplets from a flowing stream of gas or vapor; accomplished by impingement, cyclonic action, and absorption or adsorption operations.

vapor lock [FL MECH] Interruption of the flow of fuel in a gasoline engine caused by formation of vapor or gas bubbles in the fuel-feeding system.

vapor-phase reactor [CHEM ENG] A heavy steel vessel for carrying out chemical reactions on an industrial scale where efficient control over a vapor phase is needed, for example, in an oxidation process.

vapor pressure [METEOROL] The partial pressure of water vapor in the atmosphere. [THERMO] For a liquid or solid, the pressure of the vapor in equilibrium with the liquid or solid.

vapor-pressure deficit *See* saturation deficit.

vapor-pressure osmometer [ANALY CHEM] A device for the determination of molecular weights by the decrease of vapor pressure of a solvent upon addition of a soluble sample.

vapor-pressure thermometer [ENG] A thermometer in which the vapor pressure of a homogeneous substance is measured and from which the temperature can be determined; used mostly for low-temperature measurements.

vapor rate [CHEM ENG] In distillation, the upward flow rate of vapor through a distillation column.

vapor-recovery unit [ENG] **1.** A device or system to catch vaporized materials (usually fuels or solvents) as they are vented. **2.** In petroleum refining, a process unit to which gases and vaporized gasoline from various processing operations are charged, separated, and recovered for further use.

vapor suppression [NUCLEO] A safety system that can be incorporated in the design of structures housing water-cooled nuclear reactors: the space surrounding the reactor is vented into pools of water open to the outside air; if surges of hot vapor are released from the reactor in an accident, their energy is dissipated in the pools of water; gases not condensed are scrubbed clean of radioactive particles by the bubbling. Also known as pressure suppression.

vapor trail *See* condensation trail.

vapor volume equivalent [PETRO ENG] The volume of vapor to which a specified amount of liquid would be equivalent at designated standard conditions (for example, 14.65 psia and 60°F, or 15.5°C); used in the petroleum industry to calculate the specific gravity of fluids from gas-condensate wells.

var *See* volt-ampere reactive.

VAR *See* visual-aural range.

vara [CIV ENG] A surveyors' unit of length equal to $33\frac{1}{3}$ inches (84.7 centimeters).

varactor [ELECTR] A semiconductor device characterized by a voltage-sensitive capacitance that resides in the space-charge region at the surface of a semiconductor bounded by an insulating layer. Also known as varactor diode; variable-capacitance diode; varicap; voltage-variable capacitor.

varactor diode *See* varactor.

Varanidae [VERT ZOO] The monitors, a family of reptiles in the suborder Sauria found in the hot regions of Africa, Asia, Australia, and Malaya.

vardar [METEOROL] A cold fall wind blowing from the northwest down the Vardar valley in Greece to the Gulf of Salonica; it occurs when atmospheric pressure over eastern Europe is higher than over the Aegean Sea, as is often the case in winter. Also known as vardarac.

vardarac *See* vardar.

var hour [ELEC] A unit of the integral of reactive power over time, equal to a reactive power of 1 var integrated over 1 hour; equal in magnitude to 3600 joules. Also known as reactive volt-ampere hour; volt-ampere-hour reactive.

variable [ADP] A data item, or specific area in main memory, that can assume any of a set of values. [MATH] A symbol which is used to represent some undetermined element from a given set, usually the domain of a function.

variable-area exhaust nozzle [AERO ENG] On a jet engine, an exhaust nozzle of which the exhaust exit opening can be varied in area by means of some mechanical device, permitting variation in the jet velocity.

variable-area track [ENG ACOUS] A sound track divided laterally into opaque and transparent areas; a sharp line of demarcation between these areas corresponds to the waveform of the recorded signal.

variable attenuator [ELECTR] An attenuator for reducing the strength of an alternating-current signal either continuously or in steps, without causing appreciable signal distortion, by maintaining a substantially constant impedance match.

variable-bandwidth filter [ELECTR] An electric filter whose upper and lower cutoff frequencies may be independently selected, so that almost any bandwidth may be obtained; it usually consists of several stages of *RC* filters, each separated by buffer amplifiers; tuning is accomplished by varying the resistance and capacitance values.

variable-capacitance diode *See* varactor.

variable capacitor [ELEC] A capacitor whose capacitance can be varied continuously by moving one set of metal plates with respect to another.

variable carrier modulation *See* controlled carrier modulation.

variable ceiling [METEOROL] After United States weather-observing practice, a condition in which the ceiling rapidly increases and decreases while the ceiling observation is being made; the average of the observed values is used as the reported ceiling, and it is reported only for ceilings of less than 3000 feet (914 meters).

variable connector [ADP] A flow chart symbol representing a sequence connection which is not fixed, but which can be varied by the flow-charted procedure itself; it corresponds to an assigned GO TO in a programming language such as FORTRAN.

variable costs [IND ENG] Costs which vary directly with the number of units produced; direct labor and material are examples.

variable coupling [ELEC] Inductive coupling that can be varied by moving one coil with respect to another.

variable-cycle operation [ADP] An operation that requires a variable number of regularly timed execution cycles for its completion.

variable-density sound track [ENG ACOUS] A constant-width sound track in which the average light transmission varies along the longitudinal axis in proportion to some characteristic of the applied signal.

variable-depth sonar [ENG] Sonar in which the projector and receiving transducer are mounted in a watertight pod that can be lowered below a vessel to an optimum depth for minimizing thermal effects when detecting underwater targets.

variable diode function generator [ELECTR] An improvement of a diode function generator in which fully adjustable potentiometers are used for breakpoint and slope resistances, permitting the programming of analytic, arbitrary, and empirical functions, including inflections. Abbreviated VDFG.

variable field [PHYS] Field which changes during the time under consideration.

variable flow [FL MECH] Fluid flow in which the velocity changes both with time and from point to point.

variable-focal-length lens *See* zoom lens.

variable force [MECH] A force whose direction or magnitude or both change with time.

variable geometry aircraft [AERO ENG] Aircraft with variable profile geometry, such as variable sweep wings.

variable inductance *See* variable inductor.

variable-inductance accelerometer [ENG] An accelerometer consisting of a differential transformer with three coils and a mass which passes through the coils and is suspended from springs; the center coil is excited from an external alternating-current power source, and two end coils connected in series opposition are used to produce an ac output which is proportional to the displacement of the mass.

variable inductor [ELECTROMAG] A coil whose effective inductance can be changed. Also known as variable inductance.

variable-length record [ADP] A data or file format that allows each record to be exactly as long as needed.

VARIABLE-INDUCTANCE ACCELEROMETER

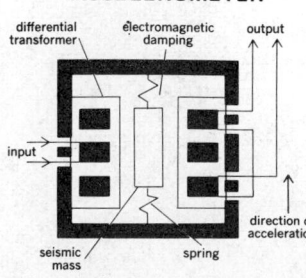

Diagram of variable-inductance accelerometer.

variable-mu tube [ELECTR] An electron tube in which the amplification factor varies in a predetermined manner with control-grid voltage; this characteristic is achieved by making the spacing of the grid wires vary regularly along the length of the grid, so that a very large negative grid bias is required to block anode current completely. Also known as remote-cutoff tube.

variable nebula [ASTRON] A nebula whose shape and brightness vary; an example is in the constellation Monoceros.

variable parameter [PHYS] A parameter which may be varied to assume any value in some range.

variable-pitch propeller [ENG] A controllable-pitch propeller whose blade angle may be adjusted to any angle between the low and high pitch limits.

variable point [ADP] A system of numeration in which the location of the decimal point is indicated by a special character at that position.

variable radio-frequency radiosonde [ENG] A radiosonde whose carrier frequency is modulated by the magnitude of the meteorological variables being sensed.

variable recoil [ORD] In recoil systems, the variation of the length of recoil according to the elevation in such manner as to prevent the gun from striking the ground when fired at high angles.

variable reduction [GRAPHICS] A characteristic of microfilming cameras; the ability to produce various sized images of a single original.

variable-reluctance microphone *See* magnetic microphone.

variable-reluctance pickup [ENG ACOUS] A phonograph pickup that depends for its operation on variations in the reluctance of a magnetic circuit due to the movements of an iron stylus assembly that is a part of the magnetic circuit. Also known as magnetic cartridge; magnetic pickup; reluctance pickup.

variable-reluctance stepper motor [ELEC] A stepper motor having a soft iron rotor with teeth or poles so positioned that they cannot simultaneously align with all the stator poles.

variable-reluctance transducer [ELECTROMAG] A transducer in which a slug of magnetic material is moved between two coils by the displacement being monitored; this changes the reluctance of the coils, thereby changing their impedance.

variable-resistance accelerometer [ENG] Any accelerometer which operates on the principle that electrical resistance of any conductor is a function of its dimensions; when the dimensions of the conductor are varied mechanically, as constant current flows through it, the voltage across it varies as a function of this mechanical excitation; examples include the strain-gage accelerometer, and an accelerometer making use of a slide-wire potentiometer.

variable resistor *See* rheostat.

variable-speed drive [MECH ENG] A mechanism transmitting motion from one shaft to another that allows the velocity ratio of the shafts to be varied continuously.

variable-speed scanning [ELECTR] Scanning method whereby the speed of deflection of the scanning beam in the cathode-ray tube of a television camera is governed by the optical density of the film being scanned.

variable star [ASTRON] A star that has a detectable change in its intensity which is often accompanied by other physical changes; changes in brightness may be a few thousandths of a magnitude to 20 magnitudes or even more.

variable-time fuse *See* proximity fuse.

variable-transconductance circuit [ELECTR] A circuit used in four-quadrant multipliers that employs a simple differential transistor pair in which one variable input to the base of one transistor controls the device's gain or transconductance, and one transistor amplifies the other's variable input, applied to the common emitter point, in proportion to the control input.

variable transformer [ELEC] An iron-core transformer having provisions for varying its output voltage over a limited range or continuously from zero to maximum output voltage, generally by means of a contact arm moving along exposed turns of the secondary winding. Also known as adjustable transformer; continuously adjustable transformer.

variable visibility [METEOROL] After United States weather observing practice, a condition in which the prevailing visibility fluctuates rapidly while the observation is being made; the average of the observed values is used as the reported visibility, and it is reported only for visibilities of less than 3 miles (4.8 kilometers).

variable word length [ADP] A phrase referring to a computer in which the number of characters addressed is not a fixed number but is varied by the data or instruction.

Variac [ELEC] Autotransformer with a toroidal winding and a rotating carbon brush, giving a continuously adjustable output voltage from zero to line voltage plus 17%.

variance [STAT] The square of the standard deviation.

variance ratio test [STAT] A technique for comparing the spreads or variabilities of two sets of figures to determine whether the two sets of figures were drawn from the same population. Also known as F test.

variant [COMMUN] **1.** One of two or more cipher or code symbols which have the same plain text equivalent. **2.** One of several plain text meanings that are represented by a single code group.

variate difference method [STAT] A technique for estimating the correlation between the random parts of two given time series.

variation [NAV] The angle between the magnetic and geographical meridians at any place, expressed in degrees and minutes east or west to indicate the direction of magnetic north from true north.

variational inequality [ASTRON] An inequality in the moon's motion, due mainly to the tangential component of the sun's attraction.

variational method [QUANT MECH] A method of calculating an upper bound on the lowest energy level of a quantum-mechanical system and an approximation for the corresponding wave function; in the integral representing the expectation value of the Hamiltonian operator, one substitutes a trial function for the true wave function, and varies parameters in the trial function to minimize the integral.

variational principle [MATH] A technique for solving boundary value problems that is applicable when the given problem can be rephrased as a minimization problem.

variation of latitude [GEOPHYS] Change of the latitude of a place on earth because of the irregular movement of the north and south poles; the movement is caused by the earth's shifting on its axis.

variation per day [GEOPHYS] The change in the value of any geophysical quantity during 1 day.

variation per hour [GEOPHYS] The change in the value of any geophysical quantity during 1 hour.

variation per minute [GEOPHYS] The change in the value of any geophysical quantity during 1 minute.

varicap *See* varactor.

varicella *See* chickenpox.

varicocele [MED] Dilatation of the veins of the pampiniform plexus of the spermatic cord, forming a soft, elastic, often uncomfortable swelling.

varicose [MED] Pertaining to blood vessels that are dilated, knotted, and tortuous.

varicose vein [ANAT] An enlarged tortuous blood vessel that occurs chiefly in the superficial veins and their tributaries in the lower extremities. Also known as varicosity.

varicosity *See* varicose vein.

variegate [BIOL] Having irregular patches of diverse colors.

variegated position effect [GEN] A phenomenon observed in some cases when a chromosome aberration causes a wild-type gene from the euchromatin to be relocated adjacent to heterochromatin; the phenotypic expression of the wild-type allele will be unstable, producing patches of phenotypically mutant tissue that differ from the surrounding wild-type tissue.

variety [SYST] A taxonomic group or category inferior in rank to a subspecies.

varifocal lens *See* zoom lens.

Varignon's theorem [MECH] The theorem that the moment of a force is the algebraic sum of the moments of its vector components acting at a common point on the line of action of the force.

varindor [ELECTROMAG] Inductor in which the inductance varies markedly with the current in the winding.

variocoupler [ELECTROMAG] In radio practice, a transformer, the self impedance of whose windings remains essentially

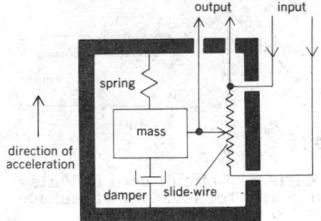

VARIABLE-RESISTANCE ACCELEROMETER

Diagram of type of variable-resistance accelerometer that makes use of slide-wire potentiometer.

constant while the mutual impedance between the windings is adjustable.

variograph [ENG] A recording variometer.

variola *See* smallpox.

variole [GEOL] A spherule the size of a pea, usually consisting of radiating plagioclase or pyroxene crystals.

variolitic [PETR] Referring to the texture of basic igneous rock composed of varioles in a finer-grained matrix.

variolosser [ELEC] Device in which loss can be controlled by a voltage or current.

variometer [ELECTROMAG] A variable inductance having two coils in series, one mounted inside the other, with provisions for rotating the inner coil in order to vary the total inductance of the unit over a wide range. [ENG] A geomagnetic device for detecting and indicating changes in one of the components of the terrestrial magnetic field vector, usually magnetic declination, the horizontal intensity component, or the vertical intensity component.

varioplex [ELEC] Telegraph switching system that establishes connections on a circuit-sharing basis between a multiplicity of telegraph transmitters in one locality and respective corresponding telegraph receivers in another locality over one or more intervening telegraph channels; maximum use of channel capacity is secured by momentarily storing the signals and allocating circuit time in rotation among the transmitters having information in storage.

Variscan orogeny [GEOL] The late Paleozoic orogenic era in Europe, extending through the Carboniferous and Permian. Also known as Hercynian orogeny.

varistor [ELECTR] A two-electrode semiconductor device having a voltage-dependent nonlinear resistance; its resistance drops as the applied voltage is increased. Also known as voltage-dependent resistor.

varix [INV ZOO] A conspicuous ridge across each whorl of certain univalves marking the ancestral position of the outer lip of the aperture. [MED] A dilated and tortuous vein, artery, or lymphatic vessel.

Varley loop test [ELEC] A method of using a Wheatstone bridge to determine the distance from the test point to a fault in a telephone or telegraph line or cable.

var measurement [ELEC] The measurement of reactive power in a circuit.

varmeter [ENG] An instrument for measuring reactive power in vars. Also known as reactive volt-ampere meter.

varnish [MATER] A transparent surface coating which is applied as a liquid and then changes to a hard solid; all varnishes are solutions of resinous materials in a solvent.

varnished cambric [TEXT] Linen or cotton fabric that has been impregnated with varnish or insulating oil and baked; used for electrical insulating purposes, especially for between-layer insulation in transformers.

varnish makers' and painters' naphtha [MATER] A petroleum naphtha that has a narrow boiling range and is used mainly as a thinner in paint and varnish. Abbreviated VM & P naphtha.

varnish paper *See* insulating paper.

varnish tree [BOT] *Rhus vernicifera.* A member of the sumac family (Anacardiaceae) cultivated in Japan; the cut bark exudes a juicy milk which darkens and thickens on exposure and is applied as a thin film to become a varnish of extreme hardness. Also known as lacquer tree.

varve [GEOL] A sedimentary bed, layer, or sequence of layers deposited in a body of still water within a year's time, and usually during a season. Also known as glacial varve.

varve clay *See* varved clay.

varved clay [GEOL] A lacustrine sediment of distinct layers consisting of varves. Also known as varve clay.

vasa vasorum [ANAT] The blood vessels supplying the walls of arteries and veins.

vascular bundle [BOT] A strandlike part of the plant vascular system containing xylem and phloem.

vascular cambium [BOT] The lateral meristem which produces secondary xylem and phloem.

vascular nevus [MED] A birthmark arising either as a developmental abnormality or as a postnatal benign neoplasm of a blood vessel.

vascular ray [BOT] A ray derived from cambium and found

in the stele of some vascular plants, often separating vascular bundles.

vascular retinopathy [MED] Pathological changes in the retina associated with diseases such as arterial hypertension, chronic nephritis, eclampsia, and advanced arteriosclerosis. Also known as retinal retinitis.

vascular tissue [BOT] The conducting tissue found in higher plants, consisting principally of xylem and phloem.

vas deferens [ANAT] The portion of the excretory duct system of the testis which runs from the epididymal duct to the ejaculatory duct. Also known as ductus deferens.

vasectomy [MED] Cutting, or removing a section from, the ductus deferens.

Vaseline [MATER] Trade name for petrolatum.

vasoconstrictor [PHYSIO] A nerve or an agent that causes blood vessel constriction.

vasodilator [PHYSIO] A nerve or an agent that causes blood vessel dilation.

vasogenic shock [MED] Failure of peripheral circulation due to vasodilation of arterioles and capillaries.

vasomotor center [PHYSIO] A large, diffuse area in the reticular formation of the lower brainstem; stimulation of different portions of this center causes either a rise in blood pressure and tachycardia (pressor area) or a fall in blood pressure and bradycardia (depressor area).

vasopressin [BIOCHEM] A peptide hormone which is elaborated by the posterior pituitary and which has a pressor effect; used medicinally as an antidiuretic. Also known as antidiuretic hormone (ADH).

vasotocin [BIOCHEM] A hormone from the neurosecretory cells of the posterior pituitary of lower vertebrates; increases permeability to water in amphibian skin and in bladder.

vat dye [MATER] One of the dyes that are easily reduced to a soluble and colorless form in which they easily impregnate fibers; subsequent oxidation produces the final color; examples are indigo and indanthrene blue.

vaterite [MINERAL] $CaCO_3$ A rare hexagonal mineral consisting of unstable calcium carbonate; it is trimorphous with calcite and aragonite.

vat printing assistant [MATER] The carrier for the dye in the printing of fabrics with vat dyes; a mixture of gums and reducing and wetting agents to assist in penetrating the fabric.

vaudaire [METEOROL] A violent south wind; a foehn of Lake Geneva in Switzerland. Also known as vauderon.

vauderon *See* vaudaire.

vault [ARCH] An arched masonry structure usually forming a ceiling or a roof. [BIOL] An anatomical structure that is arched or dome-shaped.

V band [ELECTROMAG] A radio-frequency band of 46.0 to 56.0 gigahertz. [SPECT] Absorption bands that appear in the ultraviolet part of the spectrum due to color centers produced in potassium bromide by exposure of the crystal at temperature of liquid nitrogen (81K) to intense penetrating x-rays.

V-beam radar [ELECTROMAG] A volumetric radar system that uses two fan beams to determine the distance, bearing, and height of a target: one beam is vertical and the other inclined; the beams intersect at ground level and rotate continuously about a vertical axis; the time difference between the arrivals of the echoes of the two beams is a measure of target elevation.

V belt [MECH ENG] A belt, usually endless, with a trapezoidal cross section which runs in a pulley with a V-shaped groove, with the top surface of the belt approximately flush with the top of the pulley.

V-bend die [MECH ENG] A die with a triangular cross-sectional opening to provide two edges over which bending is accomplished.

v-body *See* nucleosome.

V-bucket carrier [MECH ENG] A conveyor consisting of two strands of roller chain separated by V-shaped steel buckets; used for elevating and conveying nonabrasive materials, such as coal.

VCO *See* voltage-controlled oscillator.

V coefficient [QUANT MECH] Either of two coefficients used in the coupling of eigenfunctions of two angular momenta, differing from the Wigner 3-*j* symbol by at most a sign. Symbolized V and $\overline{V}$.

V connection *See* open-delta connection.

VARIOMETER

Variometer for measuring horizontal intensity or declination of terrestrial magnetic field, equipped with Helmholtz coil for calibration. *(U.S. Coast and Geodetic Survey)*

VARLEY LOOP TEST

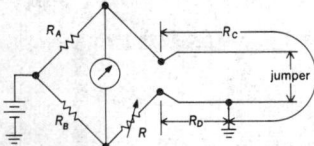

Diagram of circuit used in Varley loop test for location of leakage to ground. Resistance R is adjusted until no current flows through detector, whereupon $R_A/R_B = R_C/(R + R_D)$.

VASCULAR BUNDLE

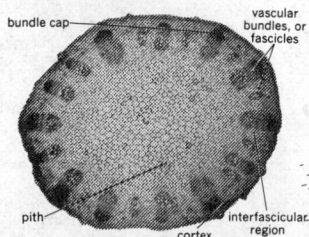

Cross section of sunflower *(Helianthus)* stem showing vascular bundles.

vcp *See* vacuum condensing point.

V cut [ENG] In mining and tunneling, a cut where the material blasted out in plan is like the letter V; usually consists of six or eight holes drilled into the face, half of which form an acute angle with the other half.

VD *See* venereal disease.

VDFG *See* variable diode function generator.

veatchite [MINERAL] $Sr_2B_{11}O_{16}(OH)_5 \cdot H_2O$ A white mineral consisting of hydrous strontium borate.

Vectian *See* Aptian.

vectograph [GRAPHICS] A picture or drawing having self-contained light polarization; at each point of such a picture, one can control the direction and magnitude of polarization, and the image can be expressed as a nonuniform vector field.

vectopluviometer [ENG] A rain gage or array of rain gages designed to measure the inclination and direction of falling rain; vectopluviometers may be constructed in the fashion of a wind vane so that the receiver always faces the wind, or they may consist of four or more receivers arranged to point in cardinal directions.

vector [MATH] An element of a vector space. [MED] An agent, such as an insect, capable of mechanically or biologically transferring a pathogen from one organism to another. [NAV] To guide a pilot, navigator, aircraft, or missile from one point to another within a given time by means of a direction communicated to the craft. [PHYS] A quantity which has both magnitude and direction, and whose components transform from one coordinate system to another in the same manner as the components of a displacement.

vector analysis [MATH] The formal study of vectors.

vectorcardiogram [PHYSIO] The part of the pathway of instantaneous vectors during one cardiac cycle. Also known as monocardiogram.

vectorcardiography [PHYSIO] A method of recording the magnitude and direction of the instantaneous cardiac vectors.

vector coupling coefficient [QUANT MECH] One of the coefficients used to express an eigenfunction of the sum of two angular momenta in terms of sums of products of eigenfunctions of the original two angular momenta. Also known as Clebsch-Gordan coefficient; Wigner coefficient.

vector current [PARTIC PHYS] A current which behaves as a vector under Lorentz transformations, rather than as an axial vector.

vectored attacks [ORD] Attacks in which a weapon carrier (air, surface, or subsurface) not holding contact on the target, is vectored to the weapon delivery point by a unit (air, surface, or subsurface) which holds contact on the target.

vector equation [MATH] An equation involving vectors.

vector field [MATH] **1.** The field of vectors arising from considering a system of differential equations on a differentiable manifold. **2.** A function whose range is in a vector space. [PHYS] A field which is characterized by a vector function.

vector function [PHYS] A function of position and time whose value at each point is a vector. Also known as vector point function.

vector gunsight [ORD] A gunsight that contains a device that computes the vector required for the bullet to follow if it is to strike its target; used especially in firing at moving targets.

vectorial structure *See* directional structure.

vector impedance meter [ENG] An instrument that not only determines the ratio between voltage and current, to give the magnitude of impedance, but also determines the phase difference between these quantities, to give the phase angle of impedance.

vector meson [PARTIC PHYS] A meson which has spin quantum number 1 and negative parity, and may be described by a vector field; examples include the ω, ρ, ϕ, and K^*-mesons.

vector model of atomic structure [ATOM PHYS] A model of atomic structure in which spin and orbital angular momenta of the electrons are represented by vectors, with special rules for their addition imposed by underlying quantum-mechanical considerations.

vector momentum *See* momentum.

vector point function *See* vector function.

vector potential [ELECTROMAG] A vector function whose curl is equal to the magnetic induction. Symbolized **A**. Also known as magnetic vector potential. [PHYS] Any vector function whose curl is equal to some solenoidal vector field.

vector power [ELEC] Vector quantity equal in magnitude to the square root of the sum of the squares of the active power and the reactive power.

vector-power factor [ELEC] Ratio of the active power to the vector power; it is the same as power factor in the case of simple sinusoidal quantities.

vector product *See* cross product.

vector resolver *See* resolver.

vector space [MATH] A system of mathematical objects which have an additive operation producing a group structure and which can be multiplied by elements from a field in a manner similar to contraction or magnification of directed line segments in euclidean space. Also known as linear space.

vector steering [AERO ENG] A steering method for rockets and spacecraft wherein one or more thrust chambers are gimbal-mounted so that the direction of the thrust force (thrust vector) may be tilted in relation to the center of gravity of the vehicle to produce a turning movement.

vector voltmeter [ENG] A two-channel high-frequency sampling voltmeter that measures phase as well as voltage of two input signals of the same frequency.

vee antenna *See* V antenna.

veer [NAV ARCH] To pay or let out, as to veer anchor chain.

veering [METEOROL] **1.** In international usage, a change in wind direction in a clockwise sense (for example, south to southwest to west) in either hemisphere of the earth. **2.** According to widespread usage among United States meteorologists, a change in wind direction in a clockwise sense in the Northern Hemisphere, counterclockwise in the Southern Hemisphere.

Vega [ASTRON] One of the brightest stars, apparent magnitude 0.1; it is a main sequence star of spectral type A0, distance is 8 parsecs, and it is 40 times brighter than the sun. Also known as α Lyrae.

Vegard's Law [MET] Linear relation between lattice parameters and composition of solid solution alloys expressed as atomic percentage.

vegetable [AGR] The edible portion of a usually herbaceous plant; customarily served with the main course of a meal. [BOT] Resembling or relating to plants.

vegetable black [MATER] Carbon made by the incomplete combustion or destructive distillation of vegetable matter, for example, wood.

vegetable diastase *See* diastase.

vegetable dye [MATER] Any colorant that is obtained from a vegetable source; for example, indigo or madder.

vegetable fat [MATER] A semisolid vegetable oil, used chiefly for food; for example, Suari fat, ucuhuba tallow, Mahuba fat, gamboge butter (gurgi, murga), Sierra Leone butter (kamga, lamy), and Mafura tallow.

vegetable glue [MATER] Mostly starch- or dextrine-based glues mixed with gums, resins, or antioxidants; tapioca paste is the most common; used on cheaper plywoods, postage stamps, envelopes, and labels.

vegetable ivory [MATER] A material from the ivory nut, a seed of the palm *Phytelephas macrocarpa*, which grows in tropical America; the nut has a white color and fine texture and is used to make buttons and similar small articles.

vegetable jelly *See* ulmin.

vegetable oil [MATER] An edible, mixed glyceride oil derived from plants (fruit, leaves, and seeds), including cottonseed, linseed, tung, and peanut; used in food oils, shortenings, soaps, and medicine, and as a paint drying oil.

vegetable parchment [MATER] A paperlike material made from a base of cotton rags or alpha cellulose called waterleaf, and containing no sizing or filling materials; used for documents and food packaging.

vegetable tanning [ENG] Leather tanning using plant extracts, such as tannic acid.

vegetable wax [MATER] A waxy substance of vegetable origin, composed of fatty acids in combination with higher alcohols (instead of glycerin, as in fats and oils); includes Japan wax, jojoba oil, candelilla, and carnauba wax.

vegetation [BOT] The total mass of plant life that occupies a given area.

vegetational plant geography [ECOL] A field of study con-

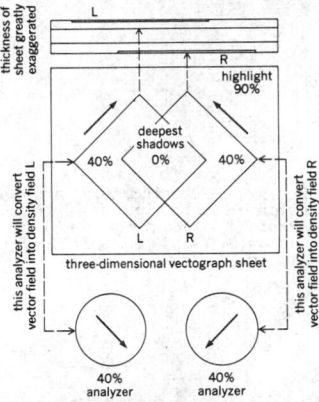

VECTOGRAPH

Diagram of superposed stereoscopic vectographs. The solid arrows show the axes of polarization; L indicates the left-eye image; R, the right-eye image. The percentages refer to light transmission.

VEIN

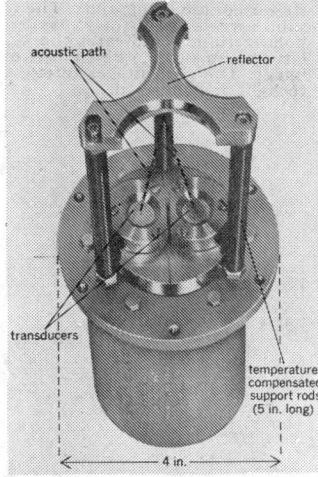

adventitia

endothelium

muscular coat

Portion of cross section through common digital vein of a human. *(From A. A. Maximow and W. Bloom, A Textbook of Histology, 6th ed., Saunders, 1953)*

VELOCIMETER

acoustic path

reflector

transducers

temperature-compensated support rods (5 in. long)

4 in.

A typical commercial velocimeter with protecting cage removed (1 inch = 2.54 centimeters). *(NUS Corp.)*

VELOCITY MICROPHONE

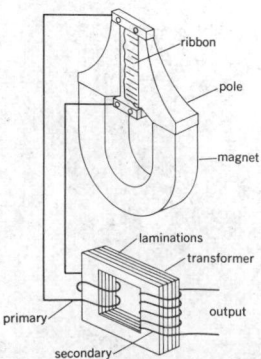

ribbon

pole

magnet

laminations

transformer

primary

output

secondary

Velocity microphone of ribbon type. Motion of ribbon in magnetic field induces voltage between two ends of ribbon.

cerned with the mapping of vegetation regions and the interpretation of these in terms of environmental or ecological influences.

vegetation and ecosystem mapping [BOT] An art and a science concerned with the drawing of maps which locate different kinds of plant cover in a geographic area.

vegetation management [ECOL] The art and practice of manipulating vegetation such as timber, forage, crops, or wild life, so as to produce a desired part or aspect of that material in higher quantity or quality.

vegetation zone [ECOL] **1.** An extensive, even transcontinental, band of physiognomically similar vegetation on the earth's surface. **2.** Plant communities assembled into regional patterns by the area's physiography, geological parent material, and history.

vegetative [BIOL] Having nutritive or growth functions, as opposed to reproductive.

vegetative propagation [BOT] Production of a new plant from a portion of another plant, such as a stem or branch.

vehicle [AERO ENG] **1.** A structure, machine, or device, such as an aircraft or rocket, designed to carry a burden through air or space. **2.** More restrictively, a rocket vehicle. [MECH ENG] A self-propelled wheeled machine that transports people or goods on or off roads; automobiles and trucks are examples.

vehicle control system [AERO ENG] A system, incorporating control surfaces or other devices, which adjusts and maintains the altitude and heading, and sometimes speed, of a vehicle in accordance with signals received from a guidance system. Also known as flight control system.

vehicle mass ratio [AERO ENG] The ratio of the final mass of a vehicle after all propellant has been used, to the initial mass.

vehicular telephony [COMMUN] The transmission of speech signals to and from mobile frequency-modulation radio stations installed in automotive vehicles; typically, each station is equipped with a transmitter and a receiver.

veil [METEOROL] A very thin cloud through which objects are visible.

Veil Nebula [ASTRON] A nebula in the constellation Cygnus; speculation is that the nebula is the remnant of a gigantic supernova which occurred 30,000 years ago.

vein [ANAT] A relatively thin-walled blood vessel that carries blood from capillaries to the heart in vertebrates. [BOT] One of the vascular bundles in a leaf. [GEOL] A mineral deposit in tabular or shell-like form filling a fracture in a host rock. [INV ZOO] **1.** One of the thick, stiff ribs providing support for the wing of an insect. **2.** A venous sinus in invertebrates.

veined gneiss [PETR] A composite gneiss with irregular layering.

veining [MET] Lines in a polished and etched metal surface marking slight imperfections in structure of an otherwise single grain.

veinite [GEOL] A genetic type of veined gneiss in which the vein material was secreted from the rock itself.

Vela [ASTRON] A southern constellation, right ascension 9 hours, declination 50°S. Also known as Sail.

velamen [BOT] The corky epidermis covering the aerial roots of an epiphytic orchid.

velarium [INV ZOO] The velum of certain scyphozoans and cubomedusans distinguished by the presence of canals lined with endoderm.

veld *See* veldt.

veldt [ECOL] Grasslands of eastern and southern Africa that are usually level and mixed with trees and shrubs. Also spelled veld.

veliger [INV ZOO] A mollusk larval stage following the trochophore, distinguished by an enlarged girdle of ciliated cells (velum).

Veliidae [INV ZOO] A family of the Hemiptera in the subdivision Amphibicorisae composed of small water striders which have short legs and a longitudinal groove between the eyes.

vellum [MATER] A high-grade paper made to rese??le genuine parchment.

vellus [ANAT] Fine body hair that is present un?? puberty.

velocimeter [ENG] An instrument for measuring the speed of sound in water; two transducers transmit acoustic pulses back and forth over a path of fixed length, each transducer immediately initiating a pulse upon receiving the previous

one; the number of pulses occurring in a unit time is measured.

Velocipedidae [INV ZOO] A tropical family of hemipteran insects in the superfamily Cimicoidea.

velocity [MECH] **1.** The time rate of change of position of a body; it is a vector quantity having direction as well as magnitude. Also known as linear velocity. **2.** The speed at which the detonating wave passes through a column of explosives, expressed in meters or feet per second.

velocity analysis [MECH] A graphical technique for the determination of the velocities of the parts of a mechanical device, especially those of a plane mechanism with rigid component links.

velocity coefficient [FL MECH] The ratio of the actual velocity of gas emerging from a nozzle to the velocity calculated under ideal conditions; it is less than 1 because of friction losses. Also known as coefficient of velocity.

velocity constant [CONT SYS] The ratio of the rate of change of the input command signal to the steady-state error, in a control system where these two quantities are proportional.

velocity curve [ASTRON] A graphical representation of the line-of-sight velocity (versus time) of a star or components of a spectroscopic binary system.

velocity discontinuity *See* seismic discontinuity.

velocity-distance relation [ASTRON] The relation wherein all the exterior galaxies are moving away from the galaxy that the sun is part of, with velocities that are greater with increasing distance of the galaxy.

velocity distribution [STAT MECH] For the molecules of a gas, a function of velocity whose value at any velocity v is proportional to the number of molecules with velocities in an infinitesimal range about v, per unit velocity range.

velocity filter [ELECTR] Storage tube device which blanks all targets that do not move more than one resolution cell in less than a predetermined number of antenna scans.

velocity fire [ORD] Preparatory fire conducted to determine the velocity for a particular combination of weapon, propellant, projectile, and fuse.

velocity-focusing mass spectrograph *See* velocity spectrograph.

velocity gradient [FL MECH] The rate of change of velocity of propagation with distance normal to the direction of flow.

velocity head [FL MECH] The square of the speed of flow of a fluid divided by twice the acceleration of gravity; it is equal to the static pressure head corresponding to a pressure equal to the kinetic energy of the fluid per unit volume.

velocity-head tachometer [ENG] A type of tachometer in which the device whose speed is to be measured drives a pump or blower, producing a fluid flow, which is converted to a pressure.

velocity level [ACOUS] A sound rating in decibels, equal to 20 times the logarithm to the base 10 of the ratio of the particle velocity of the sound to a specified reference particle velocity.

velocity microphone [ENG ACOUS] A microphone whose electric output depends on the velocity of the air particles that form a sound wave; examples are a hot-wire microphone and a ribbon microphone.

velocity-modulated oscillator [ELECTR] Oscillator which employs velocity modulation to produce radio-frequency power. Also known as klystron oscillator.

velocity modulation [ELECTR] **1.** Modulation in which a time variation in velocity is impressed on the electrons of a stream. **2.** A television system in which the intensity of the electron beam remains constant throughout a scan, and the velocity of the spot at the screen is varied to produce changes in picture brightness (not in general use).

velocity of light *See* speed of light.

velocity of sound *See* speed of sound.

velocity pickup [ELEC] A device that generates a voltage proportional to the relative velocity between two principal elements of the pickup, the two elements usually being a coil of wire and a source of magnetic field.

velocity potential [FL MECH] For a fluid flow, a scalar function whose gradient is equal to the velocity of the fluid.

velocity pressure *See* wind pressure.

velocity profile [FL MECH] A graph of the speed of a fluid flow as a function of distance perpendicular to the direction of flow.

velocity ratio [MECH ENG] The ratio of the velocity given to the effort or input of a machine to the velocity acquired by the load or output. [OCEANOGR] The ratio of the speed of tidal current at a subordinate station to the speed of the corresponding current at the reference station.

velocity resonance *See* phase resonance.

velocity servomechanism [CONT SYS] A servomechanism in which the feedback-measuring device generates a signal representing a measured value of the velocity of the output shaft. Also known as rate servomechanism.

velocity shaped canceler *See* cascaded feedback canceler.

velocity spectrograph [PHYS] A mass spectrograph in which only positive ions having a certain velocity pass through all three slits and enter a chamber where they are deflected by a magnetic field in proportion to their charge-to-mass ratio. Also known as velocity-focusing mass spectrograph.

velocity-type flowmeter [ENG] A turbine-type fluid-flow measurement device in which the fluid flow actuates the movement of a wheel or turbine-type impeller, giving a volume-time reading. Also known as current meter; rotating meter.

velour [TEXT] A fabric with a short pile on the surface.

velour paper [MATER] A paper with a velvetlike finish, produced by flocking the surface with fine bits of rayon, nylon, cotton, or wool; it is sometimes embossed in various patterns.

Velox [GRAPHICS] The trade name of one of the chloride printing papers made by Eastman Kodak; this term is sometimes used as a general reference to a photographic print.

velum [BIOL] A veil- or curtainlike membrane. [INV ZOO] A swimming organ on the larva of certain marine gastropod mollusks that develops as a contractile ciliated collar-shaped ridge. [METEOROL] An accessory cloud veil of great horizontal extent draped over or penetrated by cumuliform clouds; velum occurs with cumulus and cumulonimbus.

velvet [TEXT] A fabric with a short, thick-set pile of silk, cotton, or other fiber on a back that is closely woven and of the same or different fibers.

velveteen [TEXT] A fabric with a short pile and a cotton filling; resembles velvet.

vena cava [ANAT] One of two large veins which in air-breathing vertebrates conveys blood from the systemic circulation to the right atrium.

vena contracta [FL MECH] The contraction of a jet of liquid which comes out of an opening in a container to a cross section smaller than the opening.

venation [BOT] The system or pattern of veins in the tissues of a leaf. [INV ZOO] The arrangement of veins in an insect wing.

vendaval [METEOROL] A stormy southwest wind on the southern Mediterranean coast of Spain and in the Straits of Gibraltar; it occurs with a low advancing from the west in late autumn, winter, or early spring, and is often accompanied by thunderstorms and violent squalls.

veneer [MATER] **1.** A thin sheet of wood of uniform thickness used for facing furniture or, bonded to make plywood. **2.** A facing, as of brick or marble, on the outside of a wall.

Venera space program [AERO ENG] A series of unmanned space vehicle flights to probe the conditions near and on the planet Venus; the program was initiated by the Soviet Union.

venereal bubo *See* lymphogranuloma venereum.

venereal disease [MED] Any of several contagious diseases generally acquired during sexual intercourse; includes gonorrhea, syphilis, chancroid, granuloma inguinale, and lymphogranuloma venereum. Abbreviated VD.

venereal wart [MED] A warty growth of the penis, frequent in some parts of the world, and probably acquired during sexual intercourse.

venetian cloth [TEXT] A wool or cotton fabric with a smooth texture and a warp face.

Venetian red [INORG CHEM] A pigment with a true red hue; contains 15–40% ferric oxide and 60–80% calcium sulfate.

venite [PETR] Migmatite having mobile portions which were formed by exudation from the rock itself.

Venn diagram [MATH] A pictorial representation of set theoretic operations such as union, intersection, and complementation of sets.

venom [PHYSIO] Any of various poisonous materials secreted by certain animals, such as snakes or bees.

venous pressure [PHYSIO] Tension of the blood within the veins.

vent [ENG] **1.** A small passage made with a needle through stemming, for admitting a squib to enable the charge to be lighted. **2.** A hole, extending up through the bearing at the top of the core-barrel inner tube, which allows the water and air in the upper part of the inner tube to escape into the borehole. **3.** A small hole in the upper end of a core-barrel inner tube that allows water and air in the inner tube to escape into the annular space between the inner and outer barrels. **4.** An opening provided for the discharge of pressure or the release of pressure from tanks, vessels, reactors, processing equipment, and so on. [GEOL] The opening of a volcano on the surface of the earth. [MET] A small opening in a casting mold to allow for the escape of gases. [ZOO] The external opening of the cloaca or rectum, especially in fish, birds, and amphibians.

vent da Mùt [METEOROL] A strong, wet wind of Lake Garda in Italy.

vent des dames [METEOROL] A daily sea breeze of about 15 miles (24 kilometers) per hour from the southwest in summer on the Mediterranean coast east of the Rhone delta, extending some 20 miles (32 kilometers) inland.

vent du midi [METEOROL] A south wind in the center of the Massif Central and the southern Cevennes (France); it is warm, moist, and generally followed by a southwest wind with heavy rain.

vented baffle *See* reflex baffle.

vented battery [ELEC] A nickel-cadmium or other battery which lacks provisions for recombination of gases produced during normal operation, so that these gases must be vented to the atmosphere to avoid rupture of the cell case.

venter [ANAT] The abdomen, or other body cavity containing organs. [BOT] The thickened basal portion of an archegonium. [INV ZOO] **1.** The undersurface of an arthropod's abdomen. **2.** The outer, convex part of a curved or coiled gastropod or cephalopod shell.

ventifact [GEOL] A stone or pebble whose shape, wear, faceting, cut, or polish is the result of sandblasting. Also known as glyptolith; rillstone; wind-cut stone; wind-grooved stone; wind-polished stone; wind-scoured stone; wind-shaped stone.

ventilation [ENG] Provision for the movement, circulation, and quality control of air in an enclosed space. [METEOROL] The process of causing representative air to be in contact with the sensing elements of observing instruments; especially applied to producing a flow of air past the bulb of a wet-bulb thermometer.

ventilator [ENG] A device with an adjustable aperture for regulating the flow of fresh or stagnant air. [MECH ENG] A mechanical apparatus for producing a current of air, as a blowing or exhaust fan.

vento di sotto [METEOROL] Breezes blowing up-lake on Lake Garda in Italy.

ventral [BOT] On the lower surface of a dorsiventral plant structure, such as a leaf. [ZOO] On or belonging to the lower or anterior surface of an animal, that is, on the side opposite the back.

ventral aorta [VERT ZOO] The arterial trunk or trunks between the heart and the first aortic arch in embryos or lower vertebrates.

ventral hernia [MED] A hernia of the abdominal wall not involving the umbilical, femoral, or inguinal openings. Also known as abdominal hernia.

ventral light reflex [INV ZOO] A basic means of orientation in aquatic invertebrates, such as shrimp, which swim belly up toward the light.

ventral rib [VERT ZOO] Any of the ribs which lie in the septa dividing the trunk musculature into segments in fish. Also known as pleural rib.

ventricle [ANAT] **1.** A chamber, or one of two chambers, in the vertebrate heart which receives blood from the atrium and forces it into the arteries by contraction of the muscular wall. **2.** One of the interconnecting, fluid-filled chambers of the vertebrate brain that are continuous with the canal of the spinal cord. [ZOO] A cavity in a body part or organ.

VENTILATION

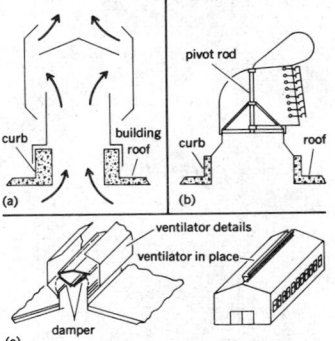

Roof exhausts for natural ventilation. (a) Cross section of round ventilator and (b) of rotating-head ventilator. (c) Continuous roof ventilator.

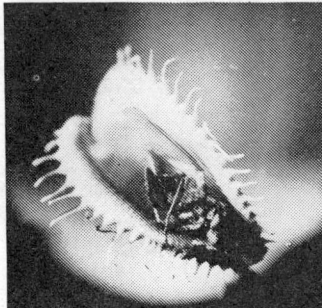

VENUS' FLYTRAP

Venus' flytrap *(Dionaea muscipula)*. The leaves capture insects.

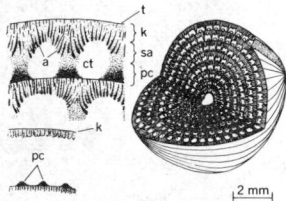

VERBEEKINIDAE

Cutaway diagram of representative species of Verbeekinidae; t = tectum, k = keriotheca, sa = septula, pc = parachromata, a = alveoli, and ct = chamberlets.

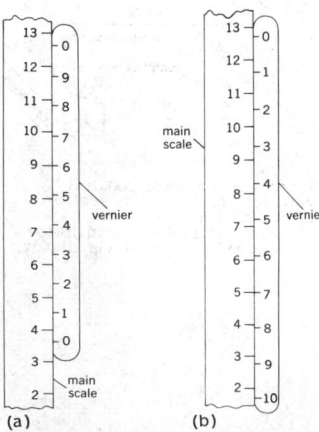

VERNIER

Two types of vernier scales. (a) Direct (reading 3.6). (b) Retrograde (reading 12.7).

ventricose [BIOL] Swollen or distended, especially on one side.

ventricular depolarization complex *See* QRS complex.

ventricular septum *See* interventricular septum.

ventriculus [ZOO] A ventricle that performs digestive functions, such as a stomach or a gizzard.

ventromedial nucleus [ANAT] A central nervous system nucleus in the hypothalamus that appears to be the satiation center; bilateral surgical damage to this nucleus results in overeating.

venturi flume [ENG] An open flume with a constricted flow which causes a drop in the hydraulic grade line; used in flow measurement.

venturi meter [ENG] An instrument for efficiently measuring fluid flow rate in a piping system; a nozzle section increases velocity and is followed by an expanding section for recovery of kinetic energy.

venturi scrubber [CHEM ENG] A gas-cleaning device in which liquid injected at the throat of a venturi is used to scrub dust and mist from the gas flowing through the venturi.

venturi tube [ENG] A constriction that is placed in a pipe and causes a drop in pressure as fluid flows through it, consisting essentially of a short straight pipe section or throat between two tapered sections; it can be used to measure fluid flow rate (a venturi meter), or to draw fuel into the main flow stream, as in a carburetor.

venule [ANAT] A small vein.

Venus [ASTRON] The planet second in distance from the sun; the linear diameter, about 12,200 kilometers, includes the top of a cloud layer; the diameter of the solid globe is about 50 kilometers less; the mass is about 0.815 (earth = 1).

Venus' flytrap [BOT] *Dionaea muscipula*. An insectivorous plant (order Sarraceniales) of North and South Carolina; the two halves of a leaf blade can swing upward and inward as though hinged, thus trapping insects between the closing halves of the leaf blade.

Venus probe [AERO ENG] A probe for exploring and reporting on conditions on or about the planet Venus, such as Pioneer and Mariner probes of the United States, and Venera probes of the Soviet Union.

veranillo [CLIMAT] The lesser dry season, made up of a few weeks of hot dry weather, that breaks up the summer rainy season on the Pacific coast of Mexico and Central America.

verano [CLIMAT] In Mexico and Central America, the main dry season, generally occurring from November through April.

veratria *See* veratrine.

veratrine [MATER] An alakaloid mixture from the seeds of sabadilla (*Schoenocaulon officinale*); which is toxic, colorless, soluble in alcohol and ether, very slightly soluble in water, and melts at about 150°C; used in medicine. Also known as veratria.

verb [ADP] In COBOL, the action indicating part of an unconditional statement.

verbal learning [PSYCH] A field of experimental psychology which studies the formation of certain verbal associations; deals with acquisition of the associations.

Verbeekinidae [PALEON] A family of extinct marine protozoans in the superfamily Fusulinacea.

Verbenaceae [BOT] A family of variously woody or herbaceous dicotyledons in the order Lamiales characterized by opposite or whorled leaves and regular or irregular flowers, usually with four or two functional stamens.

verbena oil [MATER] A volatile oil from lemon verbena leaves; contains 30% citral; used to make perfumes.

verdant zone *See* frostless zone.

Verdet constant [OPTICS] A constant of proportionality in the equation of the Faraday effect; it is equal to the angle of rotation of plane-polarized light in a magnetized substance divided by the product of the length of the light path in the substance and the strength of the magnetic field.

verdigris *See* cupric acetate.

Verel [TEXT] Trade name of a modacrylic staple fiber manufactured by Eastman Chemical Products.

verge [BUILD] The edge of a sloping roof which projects over a gable.

vergence [GEOL] The direction of overturning or of inclination of a fold.

verglas *See* glaze.

verification [ADP] The process of checking the results of one data transcription against the results of another data transcription; both transcriptions usually involve manual operations.

verification fire [ORD] Preparatory fire to test the mechanical adjustment of guns and fire-control equipment, and to measure the accuracy of corrections determined by calibration and trial fire.

verifier [ADP] A device for checking card punching semimechanically; it mimics keypunch machine operation, but reads prepunched cards without punching any new holes, and signals if the card does not agree with data entered through the verifier keyboard in some column.

verify [ADP] To determine whether an operation has been completed correctly, and in particular, to check the accuracy of keypunching by using a verifier. [COMMUN] To ensure that the meaning and phraseology of the transmitted message convey the exact intention of the originator.

Vermes [INV ZOO] An artificial taxon considered to be a phylum in some systems of classification, but variously defined as including all invertebrates except arthropods, or including all vermiform invertebrates.

vermiculite [MINERAL] $(Mg,Fe,Al)_3(Al,Si)_4O_{10}(OH)_2 \cdot 4H_2O$ A clay mineral constituent similar to chlorite and montmorillonite, and consisting of trioctahedral mica sheets separated by double water layers; sometimes used as a textural material in painting, or as an aggregate in certain plaster formulations used in sculpture.

vermiform [BIOL] Wormlike; resembling a worm.

vermiform appendix [ANAT] A small, blind sac projecting from the cecum. Also known as appendix.

Vermilingua [VERT ZOO] An infraorder of the mammalian order Edentata distinguished by lack of teeth and in having a vermiform tongue; includes the South American true anteaters.

vermilion *See* mercuric sulfide.

vermiphobia [PSYCH] An abnormal fear of worms or of infection by worms.

vermis [ANAT] The median lobe of the cerebellum.

vernadskite *See* antlerite.

vernal [GEOPHYS] Pertaining to spring.

vernal equinox [ASTRON] The sun's position on the celestial sphere about March 21; at this time the sun's path on the ecliptic crosses the celestial equator. Also known as first point of Aries; March equinox; spring equinox.

vernalization [BOT] The induction in plants of the competence or ripeness to flower by the influence of cold, that is, at temperatures below the optimal temperature for growth.

vernation [BOT] The characteristic arrangement of young leaves within the bud. Also known as prefoliation.

vernier [ENG] A short, auxiliary scale which slides along the main instrument scale to permit accurate fractional reading of the least main division of the main scale.

vernier caliper [ENG] A caliper rule with an attached vernier scale.

vernier capacitor [ELEC] Variable capacitor placed in parallel with a larger tuning capacitor to provide a finer adjustment after the larger unit has been set approximately to the desired position.

vernier dial [ENG] A tuning dial in which each complete rotation of the control knob causes only a fraction of a revolution of the main shaft, permitting fine and accurate adjustment.

vernier engine [AERO ENG] A rocket engine of small thrust used primarily to obtain a fine adjustment in the velocity and trajectory of a rocket vehicle just after the thrust cutoff of the last sustainer engine, and used secondarily to add thrust to a booster or sustainer engine. Also known as vernier rocket.

vernier rocket *See* vernier engine.

vernier sextant [NAV] A marine sextant providing a precise reading by means of a vernier used directly with the arc, and having either a clamp screw or an endless tangent screw for controlling the position of the index arm.

vernine *See* guanosine.

vernitel [ELECTR] Precision device which makes possible the transmission of data with high accuracy over standard frequency modulated-frequency modulated telemetering systems.

vernix caseosa [EMBRYO] A cheesy deposit on the surface of the fetus derived from the stratum corneum, sebaceous secretion, and remnants of the epitrichium.

Veronal [PHARM] A trade name for barbital.

verrou *See* riegel.

verruca [BIOL] A wartlike elevation on the surface of a plant or animal.

verruca peruana *See* verruca peruviana.

verruca peruviana [MED] A benign eruptive form of bartonellosis with chronic cutaneous lesions. Also known as verruca peruana.

Verrucariaceae [BOT] A family of crustose lichens in the order Pyrenulales typically found on rocks, especially in intertidal or salt-spray zones along rocky coastlines.

Verrucomorpha [INV ZOO] A suborder of the crustacean order Thoracica composed of sessile, asymmetrical barnacles.

verrucose [BIOL] Having the surface covered with wartlike protuberances.

verrucous endocarditis [MED] Small thrombotic, nonbacterial, wartlike lesions on the heart valves and endocardium, occurring frequently in systemic lupus erythematosus. Also known as terminal endocarditis.

vers *See* versed sine.

versatile anther [BOT] An anther whose attachment is near its middle, thus enabling it to swing freely.

versatile automatic test equipment [ELECTR] Computer-controlled tester, for missile electronic systems, that troubleshoots faults by deductive logic and isolates them to the plug-in module or component level.

versed cosine *See* coversed sine.

versed sine [MATH] The versed sine of A is $1 - \text{cosine } A$. Denoted vers. Also known as versine.

versine *See* versed sine.

verso [GRAPHICS] **1.** A left-hand page in a book, usually carrying an even page number. **2.** The side of a page that is to be read second.

vertebra [ANAT] One of the bones that make up the spine in vertebrates.

vertebral arch [ANAT] An arch formed by the paired pedicles and laminas of a vertebra; the posterior part of a vertebra which together with the anterior part, the body, encloses the vertebral foramen in which the spinal cord is lodged in vertebrates. Also known as neural arch.

vertebral column *See* spine.

Vertebrata [VERT ZOO] The major subphylum of the phylum Chordata including all animals with backbones, from fish to man.

vertebrate zoology [ZOO] That branch of zoology concerned with the study of members of the Vertebrata.

vertebratus [METEOROL] A cloud variety (applied mainly to the genus cirrus), the elements of which are arranged in a manner suggestive of vertebrae, ribs, or a fish skeleton.

vertex [ASTRON] **1.** The highest point that a celestial body attains. **2.** On a great circle, that point that is closest to a pole. [MATH] For a polygon or polyhedron, any of those finitely many points which together with line segments or plane pieces determine the figure or solid. [OPTICS] One of the points where the surface of a lens intersects the optical axis.

vertex power [OPTICS] The reciprocal of the back focal length of a lens.

vertical air photograph [GRAPHICS] An air photograph taken with the optical axis of the camera perpendicular to the earth's surface.

vertical anemometer [METEOROL] An instrument which records the vertical component of the wind speed.

vertical angles [MATH] The two angles produced by a pair of intersecting lines and lying on opposite sides of the point of intersection.

vertical antenna [ELECTROMAG] A vertical metal tower, rod, or suspended wire used as an antenna.

vertical axis [NAV ARCH] The vertical line near the center of gravity of a craft, perpendicular to both the longitudinal and lateral axes, around which it yaws.

vertical-axis propeller [NAV ARCH] A type of propeller wheel that consists of a circular horizontal disk set flush into a vessel's bottom, rotates about a vertical axis, and carries near its periphery a number of spadelike vertical blades.

vertical band saw [MECH ENG] A band saw whose blade operates in the vertical plane; ideal for contour cutting.

vertical blanking [ELECTR] Blanking of a television picture tube during the vertical retrace.

vertical boring mill [MECH ENG] A large type of boring machine in which a rotating workpiece is fastened to a horizontal table, which resembles a four-jaw independent chuck with extra radial T slots, and the tool has a traverse motion.

vertical broaching machine [MECH ENG] A broaching machine having the broach mounted in the vertical plane.

vertical centering control [ELECTR] The centering control provided in a television receiver or cathode-ray oscilloscope to shift the position of the entire image vertically in either direction on the screen.

vertical center keelson [NAV ARCH] The lower middle line girder, which, in conjunction with a flat plate keel on the bottom and a rider plate on top, forms the principal fore-and-aft strength member in the bottom of a ship.

vertical circle [ASTRON] A great circle of the celestial sphere, through the zenith and nadir of the celestial sphere; vertical circles are perpendicular to the horizon.

vertical compliance [ENG ACOUS] The ability of a stylus to move freely in a vertical direction while in the groove of a phonograph record.

vertical component effect *See* antenna effect.

vertical conveyor [MECH ENG] A materials-handling machine designed to move or transport bulk materials or packages upward or downward.

vertical-current recorder [ENG] An instrument which records the vertical electric current in the atmosphere.

vertical curve [CIV ENG] A curve inserted between two lengths of a road or railway which are at different slopes.

vertical danger angle [NAV] The maximum or minimum angle between the top and bottom of an object of known height, as observed from a craft, indicating the limit of safe approach to an off-lying danger.

vertical definition *See* vertical resolution.

vertical deflection oscillator [ELECTR] The oscillator that produces, under control of the vertical synchronizing signals, the sawtooth voltage waveform that is amplified to feed the vertical deflection coils on the picture tube of a television receiver. Also known as vertical oscillator.

vertical deviation [ORD] In antiaircraft artillery, the distance between the target and the point of burst in the plane normal to the line of position along a line perpendicular to the lateral deviation.

vertical differential chart [METEOROL] A synoptic chart showing the difference in value of a meteorological element between two levels in the atmosphere; a common example is the thickness chart.

vertical dip slip *See* vertical slip.

vertical drop [MECH] The drop of an object in trajectory or along a plumb line, measured vertically from its line of departure to the object.

vertical earth rate [NAV] To compensate for the effect of earth rate, the rate at which a gyroscope must be turned about its vertical axis to permit the spin axis to remain in the meridian; it is maximum at the poles, zero at the equator, and varies as the sine of the latitude.

vertical-face breakwater [CIV ENG] A breakwater whose mound of rubble does not rise above the water, but is surmounted by a vertical-face superstructure of masonry or concrete; may be built without mound rubble, provided sea bed is firm.

vertical field-strength diagram [ELECTROMAG] Representation of the field strength at a constant distance from an antenna and in a vertical plane passing through the antenna.

vertical force instrument *See* heeling adjuster.

vertical guide idlers [MECH ENG] Idler rollers about 3 inches (8 centimeters) in diameter so placed as to make contact with the edge of the belt conveyor should it run too much to one side.

vertical gyro [AERO ENG] A two-degree-of-freedom gyro with provision for maintaining its spin axis vertical; output signals

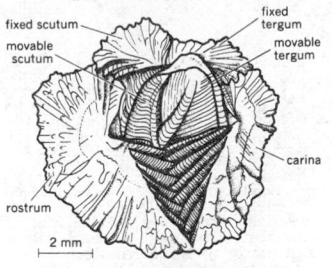

VERTICAL GYRO

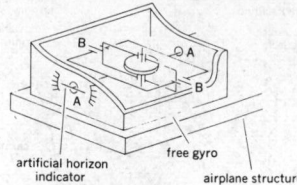

artificial horizon indicator

free gyro

airplane structure

Vertical gyro used as artificial horizon. Bearings AA and BB are made to have smallest possible friction so that axis of rotor remains vertical.

VERTICAL-LIFT GATE

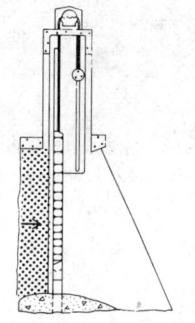

Cross-sectional diagram of vertical-lift gate. Arrow indicates direction of water pressure. (*U.S. Army Corps of Engineers and U.S. Bureau of Reclamation*)

are produced by gimbal angular displacements which correspond to components of the angular displacements of the base about two orthogonal axes; used in aircraft to measure both bank angle and pitch attitude.

vertical hold control [ELECTR] The hold control that changes the free-running period of the vertical deflection oscillator in a television receiver, so the picture remains steady in the vertical direction.

vertical illuminator [OPTICS] A microscope designed for observing surfaces of opaque substances such as metals, which has a mechanism for passing light down through the objective lens in order to illuminate the surface to be observed with a beam perpendicular to the surface.

vertical-incidence transmission [ELECTROMAG] Transmission of a radio wave vertically to the ionosphere and back.

vertical intensity [GEOPHYS] The magnetic intensity of the vertical component of the earth's magnetic field, reckoned positive if downward, negative if upward.

vertical intensity variometer [ENG] A variometer employing a large permanent magnet and equipped with very fine steel knife-edges or pivots resting on agate planes or saddles and balanced so that its magnetic axis is horizontal. Also known as Z variometer.

vertical jet *See* uprush.

vertical launch [ORD] A launch in which the missile or vehicle starts from a vertical position.

vertical lead [ORD] The vertical angle by which the gun must be moved from the line of position in order for the trajectory to pass through the target; it is the algebraic sum of the principal vertical deflection, the vertical pointing correction, and the superelevation.

vertical-lift bridge [CIV ENG] A bridge whose span is raised and lowered vertically by cables running over sheaves in towers at both ends.

vertical-lift gate [CIV ENG] A dam spillway gate of which the movable parts are raised and lowered vertically to regulate water flow.

vertical linearity control [ELECTR] A linearity control that permits narrowing or expanding the height of the image on the upper half of the screen of a television picture tube, to give linearity in the vertical direction so circular objects appear as true circles; usually mounted at the rear of the receiver.

vertically stacked loops *See* stacked loops.

vertical obstacle sonar [ENG] An active sonar used to determine heights of objects in the path of a submersible vehicle; its beam sweeps along a vertical plane, about 30° above and below the direction of the vehicle's motion. Abbreviated VOS.

vertical oscillator *See* vertical deflection oscillator.

vertical parity check *See* lateral parity check.

vertical polarization [COMMUN] Transmission of radio waves so that the electric lines of force are vertical, while the magnetic lines of force are horizontal; with this polarization, transmitting and receiving dipole antennas are placed in a vertical plane.

vertical-position welding [MET] Welding in which the weld axis is essentially vertical.

vertical recording [ENG ACOUS] A type of disk recording in which the groove modulation is perpendicular to the surface of the recording medium, so the cutting stylus moves up and down rather than from side to side during recording. Also known as hill-and-dale recording.

vertical redundancy check *See* lateral parity check.

vertical resolution [ELECTR] The number of distinct horizontal lines, alternately black and white, that can be seen in the reproduced image of a television or facsimile test pattern; it is primarily fixed by the number of horizontal lines used in scanning. Also known as vertical definition.

vertical retort process [MET] A zinc-smelting method using a vertical retort of silicon carbide brick. Also known as New Jersey retort process.

vertical retrace [ELECTR] The return of the electron beam to the top of the screen at the end of each field in television.

vertical scale [DES ENG] The ratio of the vertical dimensions of a laboratory model to those of the natural prototype; usually exaggerated in relation to the horizontal scale.

vertical separation [AERO ENG] A specified vertical distance measured in terms of space between aircraft in flight at

different altitudes or flight levels. [GEOL] The vertical component of the dip slip in a fault.

vertical separator [PETRO ENG] A gas-oil separator in the form of a vertical cylindrical tank.

vertical slip [GEOL] The vertical component of the net slip in a fault. Also known as vertical dip slip.

vertical speed indicator *See* rate-of-climb indicator.

vertical stability *See* static stability.

vertical stretching [METEOROL] A process in which ascending vertical motion of air increases with altitude, or descending motion decreases with (increasing) altitude.

vertical sweep [ELECTR] The downward movement of the scanning beam from top to bottom of the picture being televised.

vertical synchronizing pulse [ELECTR] One of the six pulses that are transmitted at the end of each field in a television system to keep the receiver in field-by-field synchronism with the transmitter. Also known as picture synchronizing pulse.

vertical tail [AERO ENG] A part of the tail assembly of an aircraft; consists of a fin (a symmetrical airfoil in line with the center line of the fuselage) fixed to the fuselage or body and a rudder which is movable by the pilot.

vertical takeoff and landing [AERO ENG] A flight technique in which an aircraft rises directly into the air and settles vertically onto the ground. Abbreviated VTOL.

vertical tower launcher *See* tower launcher.

vertical turbine pump *See* deep-well pump.

vertical turret lathe [DES ENG] Similar in principle to the horizontal turret lathe but capable of handling heavier, bulkier workpieces; it is constructed with a rotary, horizontal worktable whose diameter (30–74 inches, or 76–188 centimeters) normally designates the capacity of the machine; a crossrail mounted above the worktable carries a turret, which indexes in a vertical plane with tools that may be fed either across or downward.

vertical visibility [METEOROL] According to United States weather observing practice, the distance that an observer can see vertically into a surface-based obscuring phenomenon, such as fog, rain, or snow.

vertical vorticity [FL MECH] The vertical component of the vorticity vector.

verticillate [BOT] Whorled, in an arrangement resembling the spokes of a wheel.

vertigo [MED] The sensation that the outer world is revolving about the patient (objective vertigo) or that the patient is moving in space (subjective vertigo).

very close pack ice [OCEANOGR] Sea ice so concentrated that there is little if any open water.

very-high-frequency [COMMUN] The band of frequencies from 30 to 300 megahertz in the radio spectrum, corresponding to wavelengths of 1 meter to 10 meters. Abbreviated VHF.

very-high-frequency omnidirectional radio range [NAV] A radio navigation aid operating at very-high-frequency and supplying bearing information for the entire 360° of azimuth. Abbreviated VOR.

very-high-frequency oscillator [ELECTR] An oscillator whose frequency lies in the range from a few to several hundred megahertz; it uses distributed, rather than lumped, impedances, such as parallel wire transmission lines or coaxial cables.

very-high-frequency tuner [ELECTR] A tuner in a television receiver for reception of stations transmitting in the very-high-frequency band; it generally has 12 discrete positions corresponding to channels 2–13.

very large crude carrier [NAV ARCH] A liquid-cargo vessel in the 100,000 to 250,000 ton range. Abbreviated VLCC.

very-long-baseline interferometry [ELECTR] A method of improving angular resolution in the observation of radio sources; these are simultaneously observed by two radio telescopes which are very far apart, and the signals are recorded on magnetic tapes which are combined electronically or on a computer. Abbreviated VLBI.

very-long-range material requirements [ORD] Items required by operational and organizational concepts established for a period 10 years hence and beyond.

very-long-range radar [ELECTR] Equipment whose maximum range on a reflecting target of 1 square meter normal to the signal path exceeds 800 miles (1300 kilometers), provided line-of-sight exists between the target and the radar.

very-low-frequency [COMMUN] The band of frequencies from 3 to 30 kilohertz in the radio spectrum, corresponding to wavelengths of 10 to 100 kilometers. Abbreviated VLF.

very open pack ice [OCEANOGR] Sea ice whose concentration ranges between one-tenth and three-tenths of the sea surface.

very-short-range radar [ELECTR] Equipment whose range on a reflecting target of 1 square meter normal to the signal path is less than 50 miles (80 kilometers), provided line-of-sight exists between the target and the radar.

vesicant [PHARM] An agent that causes blistering.

vesication [MED] **1.** A blister. **2.** Formation of a blister.

vesicle [BIOL] A small, thin-walled bladderlike cavity, usually filled with fluid. [GEOL] A cavity in lava formed by entrapment of a gas bubble during solidification. Also known as air sac; bladder; saccus; vacuole; wing.

vesicular [SCI TECH] Characterized by abundant vesicles.

vesicular structure [PETR] A structure that is common in many volcanic rocks and which forms when magma is brought to or near the earth's surface; may form a structure with small cavities, or produce a pumiceous structure or a scoriaceous structure.

Vespertilionidae [VERT ZOO] The common bats, a large cosmopolitan family of the Chiroptera characterized by a long tail, extending to the edge of the uropatagium; almost all members are insect-eating.

vespertine [VERT ZOO] Active in the evening.

Vespidae [INV ZOO] A widely distributed family of Hymenoptera in the superfamily Vespoidea including hornets, yellow jackets, and potter wasps.

Vespoidea [INV ZOO] A superfamily of wasps in the suborder Apocrita.

vessel [ENG] A container or structural envelope in which materials are processed, treated, or stored; for example, pressure vessels, reactor vessels, agitator vessels, and storage vessels (tanks). [NAV ARCH] Any craft that can carry people or cargo over the surface of the water.

Vesta [ASTRON] An asteroid with a diameter of about 30 kilometers.

vestibular apparatus [ANAT] The anatomical structures concerned with the vestibular portion of the eighth cranial nerve; includes the saccule, utricle, semicircular canals, vestibular nerve, and vestibular nuclei of the ear.

vestibular membrane of Reissner See Reissner's membrane.

vestibular nerve [ANAT] A somatic sensory branch of the auditory nerve, which is distributed about the ampullae of the semicircular canals, macula sacculi, and macula utriculi.

vestibular reflexes [PHYSIO] The responses of the vestibular apparatus to strong stimulation; responses include pallor, nausea, vomiting, and postural changes.

vestibule [ANAT] **1.** The central cavity of the bony labyrinth of the ear. **2.** The parts of the membranous labyrinth within the cavity of the bony labyrinth. **3.** The space between the labia minora. **4.** See buccal cavity. [BUILD] A hall or chamber between the outer door and the interior, or rooms, of a building.

vestibule school [IND ENG] A school organized by an industrial concern to train new employees in specific tasks or prepare employees for promotion.

vestibule training [IND ENG] A procedure used in operator training in which the training location is separate from the main productive areas of the plant; includes student carrels, lecture rooms, and in many instances the same type of equipment that the trainee will use in the work station.

vestibulocerebellar [ANAT] Pertaining to the vestibular fibers and the cerebellum.

vestibulocochlear nerve See auditory nerve.

vestibulospinal tract [ANAT] A tract of nerve fibers that originates principally from the lateral vestibular nucleus and descends in the anterior funiculus of the spinal cord.

vestige [BIOL] A degenerate anatomical structure or organ that remains from one more fully developed and functional in an earlier phylogenetic form of the individual.

vestigial [BIOL] Of, being, or resembling a vestige.

vestigial side band [COMMUN] The transmitted portion of an amplitude-modulated side band that has been largely suppressed by a filter having a gradual cutoff in the neighborhood of the carrier frequency; the other side band is transmitted without much suppression. Abbreviated VSB.

vestigial-side-band filter [ELECTR] A filter that is inserted between a transmitter and its antenna to suppress part of one of the side bands.

vestigial-side-band transmission [COMMUN] A type of radio signal transmission for amplitude modulation in which the normal complete side band on one side of the carrier is transmitted, but only a part of the other side band is transmitted. Also known as asymmetrical-side-band transmission.

vesuvian See leucite; vesuvianite.

Vesuvian eruption See Vulcanian eruption.

Vesuvian garnet See leucite.

vesuvianite [MINERAL] $Ca_{10}Mg_2Al_4(SiO_4)_5(Si_2O_7)_2(OH)_4$ A brown, yellow, or green mineral found in contact-metamorphosed limestones. Also known as idocrase; vesuvian.

vetch [BOT] Any of a group of mostly annual legumes, especially of the genus *Vicia*, with weak viny stems terminating in tendrils and having compound leaves; some varieties are grown for their edible seed.

veterinary medicine [MED] The branch of medical practice which treats of the diseases and injuries of animals.

vetiver oil [MATER] A combustible, viscous essential oil from partially dried roots of *Vetiveria zizanioides* (East Indian grass), with a violet scent; soluble in fixed oils; used in perfumery. Also known as cuscus oil; vetivert.

vetivert See vetiver oil.

VF See voice frequency.

V factor [MICROBIO] Phosphopyridine nucleotide, a growth factor required by the parasitic bacteria of the genus *Haemophilus*.

V flume [MIN ENG] A V-shaped flume, supported by trestlework, and used by miners for bringing down timber and wood from the high mountains; the flume water is also used for mining purposes.

V format [ADP] A data record format in which the logical records are of variable length and each record begins with a record length indication.

VFR See visual flight rules.

VFR between layers [AERO ENG] A flight condition wherein an aircraft is operated under modified visual flight rules while in flight between two layers of clouds or obscuring phenomena, each of which constitutes a ceiling.

VFR on top [AERO ENG] A flight condition wherein an aircraft is operated under modified visual flight rules while in flight above a layer of clouds or an obscuring phenomenon sufficient to constitute a ceiling.

VFR terminal minimums [AERO ENG] A set of operational weather limits at an airport, that is, the minimum conditions of ceiling and visibility under which visual flight rules may be used.

VFR weather [METEOROL] In aviation terminology, route or terminal weather conditions which allow operation of aircraft under visual flight rules.

VGC See viscosity-gravity constant.

V guide [MECH ENG] A V-shaped groove serving to guide a wedge-shaped sliding machine element.

VHF See very-high-frequency.

VI See viscosity index.

viaduct [CIV ENG] A multiple-span bridge carrying a road or railway over a valley.

Vianaidae [INV ZOO] A small family of South American Hemiptera in the superfamily Tingordea.

vibraculum [INV ZOO] A specially modified bryozoan zooid with a bristlelike seta that sweeps debris from the surface of the colony.

vibrating capacitor [ELEC] A capacitor whose capacitance is varied in a cyclic manner to produce an alternating voltage proportional to the charge on the capacitor; used in a vibrating-reed electrometer.

vibrating conveyor See oscillating conveyor.

vibrating feeder [MECH ENG] A feeder for bulk materials (pulverized or granulated solids), which are moved by the vibration of a slightly slanted, flat vibrating surface.

vibrating grizzlies [MECH ENG] Bar grizzlies mounted on

VESUVIANITE

A crystal of vesuvianite found in Christiansand, Norway. *(Specimen from Department of Geology, Bryn Mawr College)*

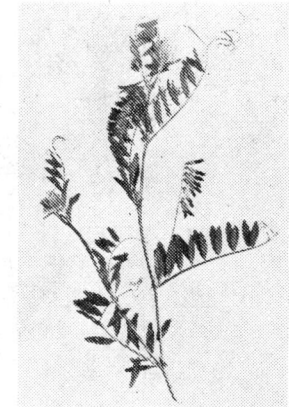

VETCH

Purple vetch *(Vicia bengalensis)*, a representative species of vetch, showing characteristic features. *(USDA)*

VIBRATING-REED FREQUENCY METER

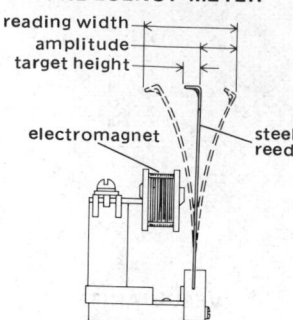

reading width
amplitude
target height

electromagnet — steel reed

Actuating system of a Frahm vibrating-reed frequency meter, consisting of electromagnet and steel reed.

VIBRATING SCREEN CLASSIFIER

Vibrating screen classifier with electric vibrator and suspension consisting of rods and springs.
(Jeffrey Mfg. Div., Dresser Industries, Inc.)

VIBROMETER

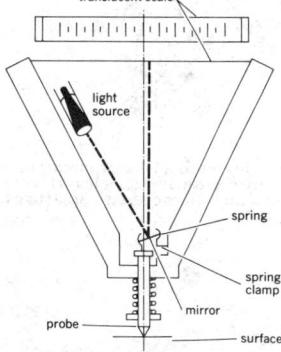

translucent scale

light source

spring

spring clamp

probe

mirror

surface

Mechanooptical vibrometer. The motion given to the probe by the vibrating surface is used to rock a mirror and thereby actuate an optical lever arm. A light beam reflected from the mirror and focused onto the scale provides an indication of the vibration amplitude.
(General Electric Co.)

eccentrics so that the entire assembly is given a forward and backward movement at a speed of some 100 strokes a minute.

vibrating needle [ENG] A magnetic needle used in compass adjustment to find the relative intensity of the horizontal components of the earth's magnetic field and the magnetic field at the compass location.

vibrating pebble mill [MECH ENG] A size-reduction device in which feed is ground by the action of vibrating, moving pebbles.

vibrating-reed electrometer [ENG] An instrument using a vibrating capacitor to measure a small charge, often in combination with an ionization chamber.

vibrating-reed frequency meter [ENG] A frequency meter consisting of steel reeds having different and known natural frequencies, all excited by an electromagnet carrying the alternating current whose frequency is to be measured. Also known as Frahm frequency meter; reed frequency meter; tuned-reed frequency meter.

vibrating-reed magnetometer [ENG] An instrument that measures magnetic fields by noting their effect on the vibration of reeds excited by an alternating magnetic field.

vibrating-reed tachometer [ENG] A tachometer consisting of a group of reeds of different lengths, each having a specific natural frequency of vibration; observation of the vibrating reed when in contact with a moving mechanical device indicates the frequency of vibration for the device.

vibrating screen [MECH ENG] A sizing screen which is vibrated by solenoid or magnetostriction, or mechanically by eccentrics or unbalanced spinning weights.

vibrating screen classifier [MECH ENG] A classifier whose screening surface is hung by rods and springs, and moves by means of electric vibrators.

vibration [MECH] A continuing periodic change in a displacement with respect to a fixed reference.

vibrational energy [PHYS CHEM] For a diatomic molecule, the difference between the energy of the molecule idealized by setting the rotational energy equal to zero, and that of a further idealized molecule which is obtained by gradually stopping the vibration of the nuclei without placing any new constraint on the motions of electrons.

vibrational spectrum [SPECT] The molecular spectrum resulting from transitions between vibrational levels of a molecule which behaves like the quantum-mechanical harmonic oscillator.

vibrational sum rule [SPECT] **1.** The rule that the sums of the band strengths of all emission bands with the same upper state is proportional to the number of molecules in the upper state, where the band strength is the emission intensity divided by the fourth power of the frequency. **2.** The sums of the band strengths of all absorption bands with the same lower state is proportional to the number of molecules in the lower state, where the band strength is the absorption intensity divided by the frequency.

vibration damping [MECH ENG] The processes and techniques used for converting the mechanical vibrational energy of solids into heat energy.

vibration drilling [MECH ENG] Drilling in which a frequency of vibration in the range of 100 to 20,000 hertz is used to fracture rock.

vibration galvanometer [ENG] An alternating-current galvanometer in which the natural oscillation frequency of the moving element is equal to the frequency of the current being measured.

vibration isolation [ENG] The isolation, in structures, of those vibrations or motions that are classified as mechanical vibration; involves the control of the supporting structure, the placement and arrangement of isolators, and control of the internal construction of the equipment to be protected.

vibration machine [MECH ENG] A device for subjecting a system to controlled and reproducible mechanical vibration. Also known as shake table.

vibration meter *See* vibrometer.

vibration pickup [ELEC] An electromechanical transducer capable of converting mechanical vibrations into electrical voltages.

vibration puddling [CIV ENG] A technique used to achieve proper consolidation of concrete; vibrating machines may be drawn vertically through the cement, or used on the surface,

or placed against the form holding the concrete in place. Also known as mechanical puddling.

vibration separation [MECH ENG] Classification or separation of grains of solids in which separation through a screen is expedited by vibration or oscillatory movement of the screening mediums.

vibrato [ACOUS] A musical embellishment that depends primarily on periodic variations of frequency which are often accompanied by variations in amplitude and waveform.

vibrator [ELEC] An electromechanical device used primarily to convert direct current to alternating current but also used as a synchronous rectifier; it contains a vibrating reed which has a set of contacts that alternately hit stationary contacts attached to the frame, reversing the direction of current flow; the reed is activated when a soft-iron slug at its tip is attracted to the pole piece of a driving coil. [MECH ENG] An instrument which produces mechanical oscillations.

vibrator power supply [ELEC] A power supply using a vibrator to produce the varying current necessary to actuate a transformer, the output of which is then rectified and filtered.

vibrator-type inverter [ELEC] A device that uses a vibrator and an associated transformer or other inductive device to change direct-current input power to alternating-current output power.

vibratory centrifuge [MECH ENG] A high-speed rotating device to remove moisture from pulverized coal or other solids.

vibratory equipment [MECH ENG] Reciprocating or oscillating devices which move, shake, dump, compact, settle, tamp, pack, screen, or feed solids or slurries in process.

vibratory hammer [MECH ENG] A type of pile hammer which uses electrically activated eccentric cams to vibrate piles into place.

vibriosis [VET MED] An infectious bacterial disease, primarily of cattle, sheep, and goats, caused by *Vibrio fetus* and characterized by abortion, retained placenta, and metritis.

vibrissae [VERT ZOO] Hairs with specialized erectile tissue; found on all mammals except man. Also known as sinus hairs; tactile hairs; whiskers.

vibroenergy separator [MECH ENG] A screen-type device for classification or separation of grains of solids by a combination of gyratory motion and auxiliary vibration caused by balls bouncing against the lower surface of the screen cloth.

vibrograph [ENG] An instrument that provides a complete oscillographic record of a mechanical vibration; in one form a moving stylus records the motion being measured on a moving paper or film.

vibrometer [ENG] An instrument designed to measure the amplitude of a vibration. Also known as vibration meter.

vibrotron [ELECTR] A triode electron tube having an anode that can be moved or vibrated by an externally applied force.

vic- [ORG CHEM] A chemical prefix indicating vicinal (neighboring or adjoining) positions on a carbon structure (ring or chain); used to identify the location of substituting groups when naming derivatives.

Vicat needle [ENG] An apparatus used to determine the setting time of cement by measuring the pressure of a special needle against the cement surface.

vicinal [ORG CHEM] Referring to neighboring or adjoining positions on a carbon structure (ring or chain).

Vickers' gun [ORD] Any of certain guns made or designed by Vickers' Sons and Maxim, Ltd., especially a type for water-cooled, belt-fed machine gun of Maxim design, having a distinctive corrugated water jacket, and widely used during and after World War I for both air and ground purposes.

Vickers hardness test *See* diamond pyramid hardness test.

Victaulic coupling [DES ENG] A development in which a groove is cut around each end of pipe instead of the usual threads; two ends of pipe are then lined up and a rubber ring is fitted around the joint; two semicircular bands, forming a sleeve, are placed around the ring and are drawn together with two bolts, which have a ridge on both edges to fit into the groove of the pipe; as the bolts are tightened, the rubber ring is compressed, making a watertight joint, while the ridges fitting in the grooves make it strong mechanically.

Victoria blue [ORG CHEM] $C_{33}H_{31}N_3 \cdot HCl$ Bronze crystals, soluble in hot water, alcohol, and ether; used as a dye for silk, wool, and cotton, as a biological stain, and to make pigment toners.

vicuna [VERT ZOO] *Lama vicugna.* A rare, wild ruminant found in the Andes mountains; the fiber of the vicuna is strong, resilient, and elastic but is the softest and most delicate of animal fibers.

video [ELECTR] **1.** Pertaining to picture signals or to the sections of a television system that carry these signals in either unmodulated or modulated form. **2.** Pertaining to the demodulated radar receiver output that is applied to a radar indicator.

video amplifier [ELECTR] A low-pass amplifier having a band-width on the order of 2–10 megahertz, used in television and radar transmission and reception; it is a modification of an RC-coupled amplifier, such that the high-frequency half-power limit is determined essentially by the load resistance, the internal transistor capacitances, and the shunt capacitance in the circuit.

Videocomp [GRAPHICS] A method developed by Radio Corporation of America for composing type electronically; instructions for writing various characters are contained in a computer data bank; the computer controls a buffer unit which supplies drive and deflection voltages to a cathode-ray tube to write and locate the character on its face, where the character is photographed.

video correlator [ELECTR] Radar circuit that enhances automatic target detection capability, provides data for digital target plotting, and gives improved immunity to noise, interference, and jamming.

video data digital processing [COMMUN] Digital processing of video signals for pictures transmitted over a television link; the computer compares each scanned line with adjacent lines and eliminates extreme changes caused by electromagnetic interference.

video discrimination [ELECTR] Radar circuit used to reduce the frequency band of the video amplifier stage in which it is used.

video frequency [COMMUN] One of the frequencies existing in the output of a television camera when an image is scanned; it may be any value from almost zero to well over 4 megahertz.

Videograph [ELECTR] Trademark of A. B. Dick Company for a high-speed cathode-ray character generator and electrostatic printer, used for printing magazine address labels under control of magnetic tape files of computer-processed addresses; the moving electron beam applies a charge on a dielectric-coated paper to form electrostatic images of characters; powder is then attracted to the image areas and fused to give readable addresses.

video integrator [ELECTR] **1.** Electric counter-countermeasures device that is used to reduce the response to nonsynchronous signals such as noise, and is useful against random pulse signals and noise. **2.** Device which uses the redundancy of repetitive signals to improve the output signal-to-noise ratio, by summing the successive video signals.

video masking [ELECTR] Method of removing chaff echoes and other extended clutter from radar displays.

video recorder [ELECTR] A magnetic tape recorder capable of storing the video signals for a television program and feeding them back later to a television transmitter or directly to a receiver.

video replay [ELECTR] Also known as video tape replay. **1.** A procedure in which the audio and video signals of a television program are recorded on magnetic tape and then the tape is run through equipment later to rebroadcast the live scene. **2.** A similar procedure in which the scene is rebroadcast almost immediately after it occurs. Also known as instant replay.

video signal [COMMUN] In television, the signal containing all of the visual information together with blanking and synchronizing pulses. [ELECTROMAG] *See* target signal.

video tape [ELECTR] A heavy-duty magnetic tape designed primarily for recording the video signals of television programs.

video tape recording [ELECTR] A method of recording television video signals on magnetic tape for later rebroadcasting of television programs. Abbreviated VTR.

video tape replay *See* video replay.

video telephone [COMMUN] A communication instrument which transmits visual images along with the attendant speech.

video transformer [ELECTR] A transformer designed to transfer, from one circuit to another, the signals containing picture information in television.

video transmitter *See* visual transmitter.

vidicon [ELECTR] A camera tube in which a charge-density pattern is formed by photoconduction and stored on a photoconductor surface that is scanned by an electron beam, usually of low-velocity electrons; used chiefly in industrial television cameras.

view camera [OPTICS] A camera that can be focused at both front and back, with adjustments for tilts, swings, shifts, and rise and fall, to control the shape of the subject in the image; it has a groundglass on the back which enables the photographer to view the image to be recorded.

viewfinder [ELECTR] An auxiliary optical or electronic device attached to a television camera so the operator can see the scene as the camera sees it. [OPTICS] A device which provides the user of a camera with the view of the subject that is focused by the lens.

viewing screen *See* screen.

viewing storage tube *See* direct-view storage tube.

viewing time [ELECTR] Time during which a storage tube is presenting a visible output corresponding to the stored information.

view printer [GRAPHICS] In microfilm technology, a reader with built-in facilities to expose and process enlargements.

vigia [NAV] **1.** A rock or shoal whose existence or position is doubtful. **2.** A warning note of a vigia condition on a chart.

Vigilante [AERO ENG] A United States supersonic, twin-engine, turbojet tactical, all-weather attack aircraft designed to operate from aircraft carriers, and capable of delivering nuclear or nonnuclear weapons; it has electronic countermeasures equipment, long-range radar, automatic pilot guidance features, inflight refueling capabilities, and a crew of pilot and bombardier. Designated A-5.

vignette [GRAPHICS] To lighten the edges of a photograph or a halftone illustration so it gradually disappears into the surrounding white of the paper, leaving no sharp edge.

vignetting [OPTICS] Reduction in intensity of illumination near the edges of an optical instrument's field of view caused by obstruction of light rays by the edge of the aperture.

vigoureux printing [TEXT] Printing worsted fibers before spinning in order to achieve a mixture of light and dark shades in the finished fabric.

Vigreaux column [ANALY CHEM] An obsolete apparatus used in laboratory fractional distillation; it is a long glass tube with indentation in its walls; a thermometer is placed at the top of the tube and a side arm is attached to a condenser.

Viking spacecraft [AERO ENG] A proposed spacecraft to be launched by the United States, to orbit Mars and land on the planet's surface.

villous adenoma [MED] A slow-growing, potentially malignant neoplasm of the mucosa of the rectum; manifested by bleeding and mucoid diarrhea.

villous placenta *See* epitheliochorial placenta.

villous tenosynovitis [MED] A chronic inflammatory reaction of a tendon sheath producing hypertrophy of the lining, with the formation of redundant folds and villi.

villus [ANAT] A fingerlike projection from the surface of a membrane.

vinasse [MATER] Residue from the fermentation of molasses or grapes; used as a fertilizer and a source of potassium salts. Also known as schlempe.

vinblastine [PHARM] $C_{46}H_{58}O_9N_4$ An alkaloid obtained from the periwinkle plant (*Vinca rosea*) and used, as the sulfate salt, as an antineoplastic drug.

Vincent's angina [MED] Vincent's infection involving tissues of the pharynx and tonsils.

Vincent's infection [MED] A noncontagious bacterial infection of the oral mucosa characterized by ulceration and formation of a gray pseudomembrane; caused by certain fusiform bacteria and spirochetes.

vincristine [PHARM] $C_{46}H_{56}O_{10}N_4$ An alkaloid extracted from the periwinkle plant (*Vinca rosea*) and used, as the sulfate salt, as an antineoplastic drug. Also known as leurocristine.

Vindobonian [GEOL] A European stage of geologic time, middle Miocene.

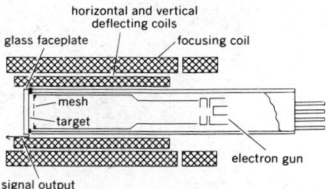
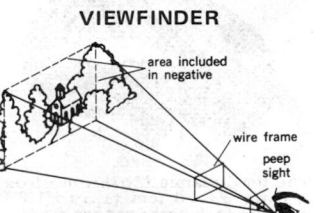

vine [BOT] A plant having a stem that is too flexible or weak to support itself.

vinegar [MATER] The product of the incomplete oxidation to acetic acid of ethyl alcohol produced by a primary fermentation of vegetable materials; contains not less than 4 grams of acetic acid per gallon; used in preparation of pickled fruits and vegetables and in salad dressing.

vinegar bacteria *See* Acetobacter.

vinegar eel [INV ZOO] *Turbatrix aceti.* A very small nematode often found in large numbers in vinegar fermentation. Also known as vinegar worm.

vinegar worm *See* vinegar eel.

vinyl acetal resin [ORG CHEM] $[CH_2CH(OC_2H_5)]_x$ A colorless, odorless, light-stable thermoplastic that is unaffected by water, gasoline, or oils; soluble in lower alcohols, benzene, and chlorinated hydrocarbons; used in lacquers, coatings, and molded objects. Also known as polyvinyl acetal resin.

vinyl acetate [ORG CHEM] $CH_3COOCH:CH_2$ A colorless, water-insoluble, flammable liquid that boils at 73°C; used as a chemical intermediate and in the production of polymers and copolymers (for example, the polyvinyl resins).

vinyl acetate resin [ORG CHEM] $(CH_2:CHOOCCH_3)_x$ An odorless thermoplastic formed by the polymerization of vinyl acetate; resists attack by water, gasoline, and oils; soluble in lower alcohols, benzene, and chlorinated hydrocarbons; used in lacquers, coatings, and molded products.

vinylacetylene [ORG CHEM] $H_2CCHCCH$ A combustible dimer of acetylene, boiling at 5°C; used for the manufacture of neoprene rubber and as a chemical intermediate.

vinyl alcohol [ORG CHEM] $CH_2:CHOH$ A flammable, unstable liquid found only in ester or polymer form. Also known as ethenol.

vinylation [CHEM] Formation of a vinyl-derived product by reaction with acetylene; for example, vinylation of alcohols gives vinyl ethers, such as vinyl ethyl ether.

vinylbenzene *See* styrene.

vinyl chloride [ORG CHEM] $CH_2:CHCl$ A flammable, explosive gas with an ethereal aroma; soluble in alcohol and ether, slightly soluble in water; boils at −14°C; an important monomer for polyvinyl chloride and its copolymers; used in organic synthesis and in adhesives. Also known as chloroethene; chloroethylene.

vinyl chloride resin [ORG CHEM] $(CH_2CHCl)_x$ A white-power polymer made by the polymerization of vinyl chloride; used to make chemical-resistant pipe (when unplasticized) or bottles and parts (when plasticized).

vinylcyanide *See* acrylonitrile.

vinyl ether [ORG CHEM] $CH_2:CHOCH:CH_2$ A colorless, light-sensitive, flammable, explosive liquid; soluble in alcohol, acetone, ether, and chloroform, slightly soluble in water; boils at 39°C; used as an anesthetic and a comonomer in polyvinyl chloride polymers. Also known as divinyl ether; divinyl oxide.

vinyl ether resin [ORG CHEM] Any of a group of vinyl ether polymers; for example, polyvinyl methyl ether, polyvinyl ethyl ether, and polyvinyl butyl ether.

vinylidene chloride [ORG CHEM] $CH_2:CCl_2$ A colorless, flammable, explosive liquid, insoluble in water; boils at 37°C; used to make polymers copolymerized with vinyl chloride or acrylonitrile (Saran).

vinylidene resin [ORG CHEM] A polymer made up of the $(-H_2CCX_2-)$ unit, with X usually a chloride, fluoride, or cyanide radical. Also known as polyvinylidene resin.

Vinylite [TEXT] Trade name for a resin made from high-molecular-weight vinyl chloride by the Union Carbide Corporation.

vinylog [ORG CHEM] Any of the organic compounds that differ from each other by a vinylene linkage ($-CH=CH-$); for example, ethyl crotonate is a vinylog of ethyl acetate and of the next higher vinylog, ethyl sorbate.

vinyl polymerization [ORG CHEM] Addition polymerization where the unsaturated monomer contains a $CH_2=C-$ group.

vinylpyridine [ORG CHEM] $C_5H_4NCH:CH_2$ A toxic, combustible liquid; soluble in water, alcohol, hydrocarbons, esters, ketones, and dilute acids; used to manufacture elastomers and pharmaceuticals.

N-vinyl-2-pyrrolidone [ORG CHEM] C_6H_9ON A colorless, toxic, combustible liquid, boiling at 148°C (100 mm Hg); used as a chemical intermediate and to make polyvinyl pyrrolidone.

vinylstyrene *See* divinylbenzene.

vinyltoluene [ORG CHEM] $CH_2:CHC_6H_4CH_3$ A colorless, flammable, moderately toxic liquid; soluble in ether and methanol, slightly soluble in water; boils at 170°C; used as a chemical intermediate and solvent. Also known as methyl styrene.

vinyl trichloride *See* 1,1,2-trichloroethane.

vinyl trichlorosilone [ORG CHEM] $CH_2CH_3SiCl_3$ A liquid that boils at 90.6°C and is soluble in organic solvents; used in silicones and adhesives.

Viocin [MICROBIO] A trademark for viomycin.

Violaceae [BOT] A family of dicotyledonous plants in the order Violales characterized by polypetalous, mostly perfect, hypogynous flowers with a single style and five stamens.

Violales [BOT] A heterogeneous order of dicotyledons in the subclass Dilleniidae distinguished by a unilocular, compound ovary and mostly parietal placentation.

violarite [MINERAL] Ni_2FeS_4 A violet-gray mineral of the linnaeite group consisting of a sulfide of nickel and iron; found in meteorites.

violet [OPTICS] The hue evoked in an average observer by monochromatic radiation having a wavelength in the approximate range from 390 to 455 nanometers; however, the same sensation can be produced in a variety of other ways.

violle [OPTICS] A unit of luminous intensity, equal to the luminous intensity of 1 square centimeter of platinum at its temperature of solidification; it is found experimentally to be equal to 20.17 candelas.

viomycin [MICROBIO] A polypeptide antibiotic or mixture of antibiotic substances produced by strains of *Streptomyces griseus* var. *purpureus* (*Streptomyces puniceus*); the sulfate salt is administered intramuscularly for treatment of tuberculosis resistant to other therapy.

viper [VERT ZOO] The common name for reptiles of the family Viperidae; thick-bodied poisonous snakes having a pair of long fangs, present on the anterior part of the upper jaw, which fold against the roof of the mouth when the jaws are closed.

Viperidae [VERT ZOO] A family of reptiles in the suborder Serpentes found in Eurasia and Africa; all species are proglyphodont.

viral encephalomyelitides [MED] A group of several encephalitis diseases caused by various viruses; includes epidemic encephalitis, equine encephalitides, and Japanese B encephalitis.

viral gastroenteritis [MED] An acute infectious gastroenteritis thought to be caused by various viruses and characterized by diarrhea, nausea, vomiting, and variable systemic symptoms.

viral hepatitis [MED] Either of two forms of hepatitis caused by a virus, serum hepatitis or infectious hepatitis.

viral pneumonia [MED] A form of pneumonia caused by a virus of various types, in which the inflammatory reaction predominates in the septa, and the alveoli contain fibrin, edema fluid, and some inflammatory cells.

virazon [METEOROL] **1.** The very strong southwesterly sea breeze experienced where the coastal chains of the Andes Mountains descend steeply to the sea; it sets in about 10 a.m. and reaches its greatest strength at about 3 p.m. **2.** A westerly sea breeze of Spain and Portugal.

viremia [MED] Presence of viral particles in the blood.

Vireonidae [VERT ZOO] The vireos, a family of New World passeriform birds in the suborder Oscines.

virga [METEOROL] Wisps or streaks of water or ice particles falling out of a cloud but evaporating before reaching the earth's surface as precipitation. Also known as fall streaks; Fallstreifen; precipitation trails.

virgate [BOT] Banded.

virgate trophus [INV ZOO] A piercing type of trophus in rotifers that is thin and slightly toothed.

virgin *See* straight-run.

Virgin *See* Virgo.

virgin neutron [NUCLEO] A neutron from any source, before it makes a collision.

virgin stock [MATER] A petroleum-derived liquid stream processed from natural (virgin) crude oil; it contains no

VIPER

The fer-de-lance (*Bothrops atrox*) is from 5 to 6 feet (1.5 to 1.8 meters) in length; ranges from southern Mexico to northern South America and is found in the West Indies.

VIRGATE TROPHUS

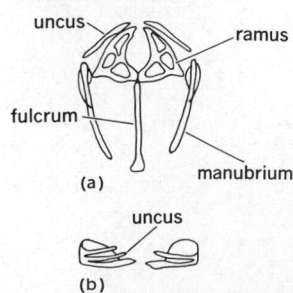

uncus

ramus

fulcrum

manubrium

(a)

uncus

(b)

Virgate trophus of *Notommata.* (a) Ventral view and (b) upper view, showing thin, slightly toothed structure. (*After Meyers*)

cracked or otherwise chemically modified material. Also known as straight-run stock.

virgin wool [TEXT] Wool used for the first time, directly after clipping from sheep.

Virgo [ASTRON] A constellation, right ascension 13 hours, declination 0°. Also known as Virgin.

Virgo A [ASTRON] A radio galaxy; it is associated with the galaxy M 87 (NGC 4486).

Virgo cluster [ASTRON] A cluster of galaxies which is the nearest to the galaxy that includes the sun; the cluster is centered in the constellation Virgo and is about 16,000,000 light-years from earth.

virgo-forcipate trophus [INV ZOO] A type of muscular chamber in rotifers containing jaws of a cuticular material intermediate between a piercing and grasping type of structure.

virial coefficients [THERMO] For a given temperature T, one of the coefficients in the expansion of P/RT in inverse powers of the molar volume, where P is the pressure and R is the gas constant.

virial of a system [STAT MECH] The average over a long period of time of $-\frac{1}{2}$ the sum over the particles in the system of the scalar product of the total force acting on the particle and its radius vector.

virial theorem See Clausius virial theorem.

viridans streptococci [MICROBIO] A group of pathogenic and saprophytic streptococci including strains not causing beta hemolysis, although many cause alpha hemolysis, and none which elaborate a C substance.

virilism [MED] See gynandry. [PSYCH] 1. Masculinity. 2. Manifestation of male behavioral patterns in the female.

virion [VIROL] The complete, mature virus particle.

Virmel engine [MECH ENG] A cat-and-mouse engine that employs vanelike pistons whose motion is controlled by a gear-and-crank system; each set of pistons stops and restarts when a chamber reaches the spark plug.

virology [MICROBIO] The study of submicroscopic organisms known as viruses.

virtual address [ADP] A symbol that can be used as a valid address part but does not necessarily designate an actual location.

virtual cathode [ELECTR] The locus of a space-charge-potential minimum such that only some of the electrons approaching it are transmitted, the remainder being reflected back to the electron-emitting cathode.

virtual entropy [THERMO] The entropy of a system, excluding that due to nuclear spin. Also known as practical entropy.

virtual gravity [METEOROL] The force of gravity on a parcel of air, reduced by centrifugal force due to the motion of the parcel relative to the earth.

virtual height [GEOPHYS] The apparent height of a layer in the ionosphere, determined from the time required for a radio pulse to travel to the layer and return, assuming that the pulse propagates at the speed of light. Also known as equivalent height.

virtual image [OPTICS] An optical image from which rays of light only appear to diverge, without actually being focused there.

virtual level [NUC PHYS] The energy of a virtual state.

virtual meridian [NAV] The meridian in which the spin axis of a gyro compass will settle as a result of speed-course-latitude error.

virtual PPI reflectoscope [ENG] A device for superimposing a virtual image of a chart on a plan-position indicator (PPI) pattern; the chart is usually prepared with white lines on a black background to the scale of the plan-position indicator range scale.

virtual pressure [METEOROL] The pressure of a parcel of moist air when it has the same density as a parcel of dry air at the same temperature.

virtual process [QUANT MECH] A process which contributes in a stage of a theoretical model but is not, by itself, physically realizable.

virtual quantum [QUANT MECH] A photon or other particle in an intermediate state which appears in matrix elements connecting initial and final states in second- and higher-order perturbation theory; energy is not conserved in the transitions to or from the intermediate state.

virtual state [NUC PHYS] An unstable state of a compound nucleus which has a lifetime many times longer than the time it takes a nucleon, with the same energy as it has in the virtual state, to cross the nucleus.

virtual temperature [METEOROL] In a system of moist air, the temperature of dry air having the same density and pressure as the moist air.

virtual work [MECH] The work done on a system during any displacement which is consistent with the constraints on the system.

virtual work principle See principle of virtual work.

virulence [MICROBIO] The disease-producing power of a microorganism; infectiousness.

virus [VIROL] A large group of infectious agents ranging from 10 to 250 nanometers in diameter, composed of a protein sheath surrounding a nucleic acid core and capable of infecting all animals, plants, and bacteria; characterized by total dependence on living cells for reproduction and by lack of independent metabolism.

virus hepatitis See infectious hepatitis.

virus interference [MICROBIO] A phenomenon which may be defined as protection of host cells against one virus, conferred as a result of prior infection with a different virus.

visbreaking See viscosity breaking.

viscera [ANAT] The organs within the cavities of the body of an organism.

visceral arch [ANAT] One of the series of mesodermal ridges covered by epithelium bounding the lateral wall of the oral and pharyngeal regions of vertebrates; embryonic in higher forms, they contribute to formation of the face and neck. [VERT ZOO] One of the first two arches of the series in gill-bearing forms.

visceral leishmaniasis [MED] A severe, generalized, and often fatal infection, caused by any of three pathogenic hemoflagellates of the genus *Leishmania*, affecting organs rich in endothelial cells; accompanied by fever, spleen and liver enlargement, anemia, leukopenia, skin pigmentation, and changes in plasma protein.

visceral peritoneum [ANAT] That portion of the peritoneum covering the organs of the abdominal cavity.

visceral pouch See pharyngeal pouch.

visceratonia [PSYCH] The behavioral type assigned to the endomorphic somatotype, manifested by a desire for assimilation and the conservation of energy through social interactions, relaxation, and love of food.

visceroptosis [MED] Prolapse of a viscus, especially the intestine; downward displacement of the intestine in the abdominal cavity. Also known as enteroptosis.

viscid [BOT] Having a sticky surface, as certain leaves.

viscoelastic fluid [FL MECH] A fluid that displays viscoelasticity.

viscoelasticity [MECH] Property of a material which is viscous but which also exhibits certain elastic properties such as the ability to store energy of deformation, and in which the application of a stress gives rise to a strain that approaches its equilibrium value slowly.

viscoelastic theory [MECH] The theory which attempts to specify the relationship between stress and strain in a material displaying viscoelasticity.

viscometer [ENG] An instrument designed to measure the viscosity of a fluid.

viscometer gage [ENG] A vacuum gage in which the gas pressure is determined from the viscosity of the gas.

viscometric analysis [FL MECH] Measurement of the flow properties of substances by viscometry.

viscometry [ENG] A branch of rheology; the study of the behavior of fluids under conditions of internal shear; the technology of measuring viscosities of fluids.

viscose process [CHEM ENG] A process for the manufacture of rayon by treating cellulose with caustic soda, and with carbon disulfide to form cellulose xanthate, which is then dissolved in a weak caustic solution to form the viscose; fibers are used as silk substitutes.

viscosity [FL MECH] Energy dissipation and generation of stresses in a fluid by the distortion of fluid elements; quantitatively, when otherwise qualified, the absolute viscosity. Also known as flow resistance; internal friction.

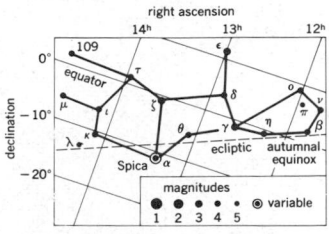

VIRGO

Line pattern of the constellation Virgo. The grid lines represent the coordinates of the sky. The apparent brightness, or magnitude, of the stars is shown by the sizes of the dots, which are graded by appropriate numbers as indicated.

VISCOMETER

Picture of a Laray viscometer; used to measure viscosity and other rheological properties of ink.

viscosity blending chart [CHEM ENG] A graphical means for estimating the viscosity at a given temperature of a blend of petroleum products.

viscosity breaking [CHEM ENG] A petroleum refinery process used to lower or break the viscosity of high-viscosity residuum by thermal cracking of molecules at relatively low temperatures. Also known as visbreaking.

viscosity coefficient [FL MECH] An empirical number used in equations of fluid mechanics to account for the effects of viscosity.

viscosity conversion table [CHEM ENG] A table or chart with which kinematic viscosity, in centistokes, can be converted to Saybolt viscosity, in seconds, at the same temperature.

viscosity curve [FL MECH] A graph showing the viscosity of a liquid or gaseous material as a function of temperature.

viscosity-gravity constant [CHEM ENG] An index of the chemical composition of crude oil; defined as the general relation between specific gravity and Saybolt Universal viscosity; the constant is low for paraffinic crude oils, high for naphthenic crude oils. Abbreviated VGC.

viscosity index [CHEM ENG] An arbitrary scale used to show the magnitude of viscosity changes in lubricating oils with changes in temperature. Abbreviated VI.

viscosity-temperature chart [CHEM ENG] A chart with which the kinematic or Saybolt viscosity of a petroleum oil at any temperature within a limited range may be ascertained, provided viscosities at two temperatures are known.

viscous damping [MECH ENG] A method of converting mechanical vibrational energy of a body into heat energy, in which a piston is attached to the body and is arranged to move through liquid or air in a cylinder or bellows that is attached to a support.

viscous dissipation function [FL MECH] A quadratic function of spatial derivatives of components of fluid velocity which gives the rate at which mechanical energy is converted into heat in a viscous fluid per unit volume. Also known as dissipation function.

viscous drag [FL MECH] That part of the rearward force on an aircraft that results from the aircraft carrying air forward with it through viscous adherence.

viscous-drag gas-density meter [ENG] A device to measure gas-mixture densities; driven impellers in sample and standard chambers create measurable turbulences (drags) against respective nonrotating impellers.

viscous fillers [MECH ENG] A packaging machine that fills viscous product into cartons; there are two basic types, straight-line and rotary plunger; the former operates intermittently on a given number of containers, while the latter fills and discharges containers continuously.

viscous flow [FL MECH] **1.** The flow of a viscous fluid. **2.** The flow of a fluid through a duct under conditions such that the mean free path is small in comparison with the smallest, transverse section of the duct.

viscous fluid [FL MECH] A fluid whose viscosity is sufficiently large to make the viscous forces a significant part of the total force field in the fluid.

viscous force [FL MECH] The force per unit volume or per unit mass arising from viscous effects in fluid flow.

viscous impingement filter [ENG] A filter made up of a relatively loosely arranged medium, such that the airstream is forced to change direction frequently as it passes through the filter medium; the medium usually consists of spun-glass fibers, metal screens, or layers of crimped expanded metal whose surfaces are coated with a tacky oil.

viscous lubrication *See* complete lubrication.

viscous magnetization *See* viscous remanent magnetization.

viscous neutral oil [MATER] The bottoms from reducing, by distillation, of a petroleum neutral oil fraction; such oils are frequently blended with bright stock to make finished oils of various viscosities.

viscous remanent magnetization [GEOPHYS] A process in which grains of magnetic minerals, which are either too small or too finely divided by undergrowths of different chemical composition to retain a permanent magnetization indefinitely, acquire a new direction of magnetization when the direction of the earth's magnetic field changes. Abbreviated VRM. Also known as viscous magnetization.

viscus [ANAT] Singular of viscera.

vise [DES ENG] A tool consisting of two jaws for holding a workpiece; opened and closed by a screw, lever, or cam mechanism.

visibility [METEOROL] In weather observing practice, the greatest distance in a given direction at which it is just possible to see and identify with the unaided eye, in the daytime, a prominent dark object against the sky at the horizon and, at nighttime, a known, preferably unfocused, moderately intense light source.

visibility factor [ELECTR] The ratio of the minimum signal input detectable by ideal instruments connected to the output of a receiver, to the minimum signal power detectable by a human operator through a display connected to the same receiver. Also known as display loss.

visibility function *See* luminosity function.

visibility meter [ENG] An instrument for making direct measurements of visual range in the atmosphere or of the physical characteristics of the atmosphere which determine the visual range. [OPTICS] A type of photometer that operates on the principle of artificially reducing the visibility of objects to threshold values (borderline of seeing and not seeing) and measuring the amount of the reduction on an appropriate scale.

visible absorption spectrophotometry [SPECT] Study of the spectra produced by the absorption of visible-light energy during the transformation of an electron from the ground state to an excited state as a function of the wavelength causing the transformation.

visible bearing [NAV] A bearing obtained by visual observation using a pelorus or azimuth circle.

visible horizon [ASTRON] That line where earth and sky appear to meet, and the projection of this line upon the celestial sphere.

visible radiation *See* light.

visible spectrophotometry [SPECT] In spectrophotometric analysis, the use of a spectrophotometer with a tungsten lamp that has an electromagnetic spectrum of 380–780 millimicrons as a light source, glass or quartz prisms or gratings in the monochromator, and a photomultiplier cell as a detector.

visible spectrum [SPECT] **1.** The range of wavelengths of visible radiation. **2.** A display or graph of the intensity of visible radiation emitted or absorbed by a material as a function of wavelength or some related parameter.

vision [PHYSIO] The sense which perceives the form, color, size, movement, and distance of objects. Also known as sight.

vision slit [ORD] A narrow opening or slit in armor to permit viewing, especially one in a tank or other armored vehicle.

visor tin [MINERAL] Twin crystals of cassiterite characterized by a notch.

visual acuity [PHYSIO] The ability to see fine details of an object; specifically, the ability to see an object whose angle subtended at the eye is 1 minute of arc.

visual aid to navigation [NAV] An object such as a tower or tripod target whose position is indicated on government-approved navigation charts and on which bearings may be taken by the naked eye or with optical assistance; the aid may or may not be illuminated.

visual angle [OPTICS] The angle which an object subtends at the nodal point of the eye of an observer.

visual-aural range [NAV] A very-high-frequency radio range that provides one course for display to the pilot on a zero-center left-right indicator and another course, at right angles to the first, in the form of aural A-N radio range signals; the A-N aural signals provide a means for differentiating between the two directions of the visual course. Abbreviated VAR.

visual binaries [ASTRON] Binary stars that to the naked eye seem to be single stars, but when viewed through the telescope, are separated into pairs. Also known as visual doubles.

visual bombing [ORD] Bombing done by sighting on an aiming point, under conditions where the aiming point is visible from the bombing aircraft.

visual bombsight [ORD] A bombsight designed for aiming a bomb when the aiming point is visible.

visual colorimetry [ANALY CHEM] A procedure for the determination of the color of an unknown solution by visual

VISCOUS DAMPING

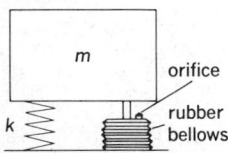

System employing viscous damping with air, used to reduce vibration of body with mass *m*, attached to spring with force constant *k*.

VISOR TIN

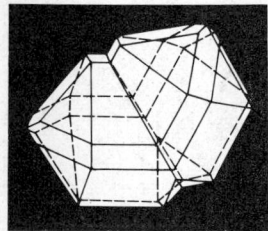

Visor tin with characteristic notch. *(From C. S. Hurlbut, Jr., Dana's Manual of Mineralogy, 17th ed., Wiley, 1959)*

comparison to color standards (solutions or color-tinted disks).

visual comparator   *See* optical comparator.

visual doubles   *See* visual binaries.

visual fire control   [ORD] Technical control of some artillery fire using an optical tracking instrument.

visual fix   [NAV] A fix established by visual observation of fixed objects.

visual flight   [AERO ENG] An aircraft flight occurring under conditions which allow navigation by visual reference to the earth's surface at a safe altitude and with sufficient horizontal visibility, and operating under visual flight rules. Also known as VFR flight.

visual flight rules   [AERO ENG] A set of regulations set down by the U.S. Civil Aeronautics Board (in Civil Air Regulations) to govern the operational control of aircraft during visual flight. Abbreviated VFR.

visual Herschel effect   *See* Herschel effect.

visual learning   [PSYCH] A type of sensory learning that is controlled by the cortical visual areas of the brain.

visual line of position   [NAV] A line of position determined by visual observation of a landmark or aid to navigation.

visual magnitude   [ASTRON] A celestial body's magnitude as seen by the eye of the observer.

visual photometer   [OPTICS] A photometer in which the luminance of two surfaces is compared by human vision; it usually utilizes the Lummer-Brodhun sightbox or some adaptation of its principles.

visual pigment   [BIOCHEM] Any of various photosensitive pigments of vertebrate and invertebrate photoreceptors.

visual projection area   [PHYSIO] The receptive center for visual images in the cortex of the brain, located in the walls and margins of the calcarine sulcus of the occipital lobe. Also known as Brodmann's area 17.

visual purple   *See* rhodopsin.

visual radio range   [NAV] Any range facility whose course is flown by visual instrumentation not associated with aural reception.

visual range   [METEOROL] The distance, under daylight conditions, at which the apparent contrast between a specified type of target and its background becomes just equal to the threshold contrast of an observer.

visual-righting reflex   [PHYSIO] A reflex mechanism whereby righting of the head and body is caused by visual stimuli. Also known as optical-righting reflex.

visual scanner   [ELECTR] Device that optically scans printed or written data and generates an analog or digital signal.

visual seizure   [MED] A form of epileptic seizure in which the patient experiences visual sensations in the form of light flashes, sometimes of varied colors.

visual storage tube   [ELECTR] Any electrostatic storage tube that also provides a visual readout.

visual telephony   [COMMUN] The transmission of picture information (television) over telephone lines.

visual transmitter   [ELECTR] Those parts of a television transmitter that act on picture signals, including parts that act on the audio signals as well. Also known as picture transmitter; video transmitter.

visual yellow   [BIOCHEM] An intermediary substance formed from rhodopsin in the retina after exposure to light; it is ultimately broken down to retinene and vitamin A.

Vitaceae   [BOT] A family of dicotyledonous plants in the order Rhamnales; mostly tendril-bearing climbers with compound or lobed leaves, as in grapes (*Vitis*).

vital capacity   [PHYSIO] The volume of air that can be forcibly expelled from the lungs after the deepest inspiration.

vitalism   [BIOL] The theory that the activities of a living organism are under the guidance of an agency which has none of the attributes of matter or energy.

Vitallium   [MET] Trademark for an alloy of cobalt (65%), chromium (30%), and molybdenum (5%); used in certain surgical appliances.

vitamin   [BIOCHEM] An organic compound present in variable, minute quantities in natural foodstuffs and essential for the normal processes of growth and maintenance of the body; vitamins do not furnish energy, but are essential for energy transformation and regulation of metabolism.

vitamin A   [BIOCHEM] $C_{20}H_{29}OH$ A pale-yellow alcohol that is soluble in fat and insoluble in water; found in liver oils and carotenoids, and produced synthetically; it is a component of visual pigments and is essential for normal growth and maintenance of epithelial tissue. Also known as antiinfective vitamin; antixerophthalmic vitamin; retinol.

vitamin B₁   *See* thiamine.

vitamin B₂   *See* riboflavin.

vitamin B₃   *See* pantothenic acid.

vitamin B₆   [BIOCHEM] A vitamin which exists as three chemically related and water-soluble forms found in food: pyridoxine, pyridoxal, and pyridoxamine; dietary requirements and physiological activities are uncertain.

vitamin B₁₂   [BIOCHEM] A group of closely related polypyrrole compounds containing trivalent cobalt; the antipernicious anemia factor, essential for normal hemopoiesis. Also known as cobalamin; cyanocobalamin; extrinsic factor.

vitamin B complex   [BIOCHEM] A group of water-soluble vitamins that include thiamine, riboflavin, nicotinic acid, pyridoxine, panthothenic acid, inositol, *p*-aminobenzoic acid, biotin, adenylic acid, folic acid, and vitamin B₁₂.

vitamin B₆ hydrochloride   *See* pyridoxine hydrochloride.

vitamin B₂ phosphate   *See* riboflavin 5′-phosphate.

vitamin C   *See* ascorbic acid.

vitamin D   [BIOCHEM] Either of two fat-soluble, sterol-like compounds, calciferol or ergocalciferol (vitamin D₂) and cholecalciferol (vitamin D₃); occurs in fish liver oils and is essential for normal calcium and phosphorus deposition in bones and teeth. Also known as antirachitic vitamin.

vitamin E   [BIOCHEM] $C_{29}H_{50}O_2$ Any of a series of eight related compounds called tocopherols, α-tocopherol having the highest biological activity; occurs in wheat germ and other oils and is believed to be needed in certain human physiological processes.

vitamin G   *See* riboflavin.

vitamin K   [BIOCHEM] Any of three yellowish oils which are fat-soluble, nonsteroid, and nonsaponifiable; it is essential for formation of prothrombin. Also known as antihemorrhagic vitamin; prothrombin factor.

vitamin P   [BIOCHEM] A substance, such as citrin or one or more of its components, believed to be concerned with maintenance of the normal state of the walls of small blood vessels.

vitamin P complex   *See* bioflavanoid.

vitavite   *See* moldavite.

vitellarium   [INV ZOO] The part of the ovary in certain rotifers and flatworms that produces nutritive cells filled with yolk. Also known as yolk larva.

vitelline artery   [EMBRYO] An artery passing from the yolk sac to the primitive aorta in young vertebrate embryos. Also known as omphalomesenteric artery.

vitelline duct   [EMBRYO] The constricted part of the yolk sac opening into the midgut region of the future ileum. Also known as omphalomesenteric duct; umbilical duct.

vitelline membrane   [CYTOL] The cytoplasmic membrane on the surface of the mammalian ovum.

vitelline vein   [EMBRYO] Any of the embryonic veins in vertebrates uniting the yolk sac and the sinus venosus; their proximal fused ends form the portal vein. Also known as omphalomesenteric vein.

vitellogenesis   [PHYSIO] The process by which yolk is formed in the ooplasm of an oocyte.

viticulture   [AGR] That division of horticulture concerned with grape growing, studies of grape varieties, methods of culture, and insect and disease control.

vitiligo   [MED] A skin disease characterized by an acquired ochromia in areas of various sizes and shapes.

vitrain   [GEOL] A brilliant black coal lithotype with vitreous luster and cubical cleavage. Also known as pure coal.

Vitreoscillaceae   [MICROBIO] A family of bacteria in the order Beggiatoales; organisms have a filamentous habit and move by gliding, but never store sulfur, and rely on organic nutrients in their metabolism.

vitreous body   *See* vitreous humor.

vitreous chamber   [ANAT] A cavity of the eye posterior to the crystalline lens and anterior to the retina, which is filled with vitreous humor.

VITAMIN A

Structural formula for vitamin A (retinol).

VITAMIN B₆

Structural formulas for the three naturally occurring forms of vitamin B₆. (a) Pyridoxine (pyridoxol). (b) Pyridoxal. (c) Pyridoxamine.

VITAMIN D

Structural formulas for the two vitamin D compounds. (a) Vitamin D₂ (calciferol). (b) Vitamin D₃ (cholecalciferol).

VIVIANITE

5 cm

Vivianite from Wannon River, Victoria, Australia. *(American Museum of Natural History specimen)*

VOCAL SAC

Toad of the genus *Bufo* giving mating call with vocal sac expanded. *(American Museum of Natural History photograph)*

vitreous enamel [MATER] A glass coating applied to a metal by covering the surface with a powdered glass frit and heating until fusion occurs. Also known as porcelain enamel.

vitreous humor [PHYSIO] The transparent gel-like substance filling the greater part of the globe of the eye, the vitreous chamber. Also known as vitreous body.

vitreous luster [OPTICS] A type of luster resembling that of glass.

vitreous silica *See* silica glass.

vitreous state [SOLID STATE] A solid state in which the atoms or molecules are not arranged in any regular order, as in a crystal, and which crystallizes only after an extremely long time. Also known as glassy state.

vitric tuff [GEOL] Tuff composed principally of volcanic glass fragments.

vitrification [GEOL] Formation of a glassy or noncrystalline material.

vitrified-clay pipe [MATER] A pipe, made of clay treated in a kiln to induce vitrification, with the surface glazed for watertightness; used for drainage.

vitrified wheel [DES ENG] A grinding wheel with a glassy or porcelanic bond.

vitrinite [GEOL] A maceral group that is rich in oxygen and composed of humic material associated with peat formation; characteristic of vitrain.

vitrinoid [GEOL] Vitrinite occurring in bituminous coking coals; characterized by a reflectance of 0.5–2.0%.

vitrophyre [PETR] Any porphyritic igneous rock whose groundmass is glassy. Also known as glass porphyry.

vittate [BOT] **1.** Having longitudinal stripes. **2.** Bearing specialized oil tubes (vittae).

viuga [METEOROL] A cold north or northeast storm of the Russian steppes, lasting about 3 days.

vivax malaria [MED] Malaria caused by *Plasmodium vivax* and characterized by typical paroxysms occurring every few days, commonly every 2 days. Also known as benign tertian malaria; tertian malaria.

Viverridae [VERT ZOO] A family of carnivorous mammals in the superfamily Feloidea composed of the civets, genets, and mongooses.

vivianite [MINERAL] $Fe_3(PO_4)_2 \cdot 8H_2O$ A colorless, blue, or green mineral in the unaltered state (darkens upon oxidation); crystallizes in the monoclinic system and occurs in earth form and as globular and encrusting fibrous masses. Also known as blue iron earth; blue ocher.

Viviparidae [INV ZOO] A family of fresh-water gastropod mollusks in the order Pectinibranchia.

viviparous [PHYSIO] Bringing forth live young.

Viyella [TEXT] Registered name of a wool and cotton blend in woven or knitted fabrics.

V jewels [DES ENG] Jewel bearings used in conjunction with a conical pivot, the bearing surface being a small radius located at the apex of a conical recess; found primarily in electric measuring instruments.

Vlasov equation [PL PHYS] A modification of the Boltzmann transport equation for the study of a plasma, in which particles interact only through the mutually induced space-charge field, and collisions are assumed to be negligible. Also known as collisionless Boltzmann equation.

VLBI *See* very-long-baseline interferometry.

VLCC *See* very large crude carrier.

VLF *See* very-low-frequency.

VM & P naphtha *See* varnish makers' and painters' naphtha.

V-notch weir *See* triangular-notch weir.

vocal cord *See* vocal fold.

vocal fold [ANAT] Either of a pair of folds of tissue covered by mucous membrane in the larynx. Also known as vocal cord.

vocal sac [VERT ZOO] An expansible pocket of skin beneath the chin or behind the jaws of certain frogs; may be inflated to a great volume and serves as a resonator.

Vochysiaceae [BOT] A family of dicotyledonous plants in the order Polygalales characterized by mostly three carpels, usually stipulate leaves, one fertile stamen, and capsular fruit.

vocoder [ELECTR] A system of electronic apparatus for synthesizing speech according to dynamic specifications derived from an analysis of that speech.

vodas [ELECTR] A voice-operated switching device used in transoceanic radiotelephone circuits to suppress echoes and singing sounds automatically; it connects a subscriber's line automatically to the transmitting station as soon as he starts speaking and simultaneously disconnects it from the receiving station, thereby permitting the use of one radio channel for both transmitting and receiving without appreciable switching delay as the parties alternately talk. Derived from voice-operated device anti-singing.

voder [ELECTR] An electronic system that uses electron tubes and filters, controlled through a keyboard, to produce voice sounds artificially. Derived from voice operation demonstrator.

vodka [FOOD ENG] A colorless and unaged alcoholic beverage distilled from rye or wheat mash or sometimes from potatoes; it is highly rectified during distillation and thus is a very pure neutral spirit without a pronounced taste.

vogad [ELECTR] An automatic gain control circuit used to maintain a constant speech output level in long-distance radiotelephony. Derived from voice-operated gain-adjusted device.

Vogel-Colson-Russell effect *See* Russell effect.

vogesite [PETR] A syenitic lamprophyre composed of phenocrysts of hornblende in a groundmass of orthoclase and hornblende.

Voges-Proskauer test [MICROBIO] One of the four tests of the IMVIC test; a qualitative test for the formation of acetyl methylcarbinol from glucose, in which solutions of α-naphthol, potassium hydroxide, and creatinine are added to an incubated culture of test organisms in a glucose broth, and a pink to rose color indicates a positive reaction.

voice call sign [COMMUN] A call sign provided primarily for voice communications.

voice channel [COMMUN] A communication channel having sufficient bandwidth to carry voice frequencies intelligibly; the minimum bandwidth for a voice channel is about 3000 hertz.

voice coder [ELECTR] Device that converts speech input into digital form prior to encipherment for secure transmission and converts the digital signals back to speech at the receiver.

voice coil [ENG ACOUS] The coil that is attached to the diaphragm of a moving-coil loudspeaker and moves through the air gap between the pole pieces due to interaction of the fixed magnetic field with that associated with the audio-frequency current flowing through the voice coil. Also known as loudspeaker voice coil; speech coil (British usage).

voice/data system [COMMUN] Integrated communications system for transmitting both voice and digital data.

voice frequency [COMMUN] An audio frequency in the range essential for transmission of speech of commercial quality, from about 300 to 3400 hertz. Abbreviated VF. Also known as speech frequency.

voice-frequency carrier telegraphy [COMMUN] Carrier telegraphy in which the carrier currents have frequencies such that the modulated currents may be transmitted over a voice-frequency telephone channel.

voice-frequency dialing [ELECTR] Method of dialing by which the direct-current pulses from the dial are transformed into voice-frequency alternating-current pulses.

voice-frequency telegraph system [COMMUN] Telegraph system permitting the use of many channels on a single circuit; a different audio frequency is used for each channel, being keyed in the conventional manner; the various audio frequencies at the receiving end are separated by suitable filter circuits and fed to their respective receiving circuits.

voice-grade channel [COMMUN] A channel whose bandwidth is large enough to transmit voice-frequency signals.

voice-operated device [ELECTR] Any of several devices in a telephone system which are brought into operation by a sound signal, or some characteristic of such a signal.

voice-operated device anti-singing *See* vodas.

voice-operated gain-adjusted device *See* vogad.

voice-operated loss control and suppressor [ELECTR] Voice-operated device which switches loss out of the transmitting branch and inserts loss in the receiving branch under control of the subscriber's speech.

voice operation demonstrator *See* voder.

voice print [ENG ACOUS] A voice spectrograph that has individually distinctive patterns of voice characteristics that can be used to identify one person's voice from other voice patterns.

void [ADP] In optical character recognition, an island of insufficiently inked paper within the area of the intended character stroke.

void channels [ENG] The open passages of a porous or packed medium through which liquid or gas can flow.

void coefficient [NUCLEO] A rate of change in the reactivity of a water reactor system resulting from a formation of steam bubbles as the power level and temperature increase.

void swelling [NUCLEO] An increase in the external dimension of solid materials after irradiation.

Voigt effect [OPTICS] Double refraction of light passing through a substance that is placed in a magnetic field perpendicular to the direction of light propagation.

Voigt notation [MECH] A notation employed in the theory of elasticity in which elastic constants and elastic moduli are labeled by replacing the pairs of letters xx, yy, zz, yz, zx, and xy by the number 1, 2, 3, 4, 5, and 6 respectively.

Voith-Schneider propulsion system [NAV ARCH] A system for ship propulsion, serving both as propeller and rudder, that consists of one or two vertical-axis rotors located underwater at the stern; rotor disks are flush with the shell plating which is approximately horizontal, and five to eight spadelike vertical-impeller blades are fitted near the periphery of the disks.

vol *See* volume.

Volan [ASTRON] A southern constellation, right ascension 8 hours, declination 70°S. Also known as Flying Fish; Pisces Volan.

volar [ANAT] Pertaining to, or on the same side as, the palm of the hand or the sole of the foot.

volatile [CHEM] Readily passing off by evaporation.

volatile component [GEOL] A component of magma whose vapor pressures are high enough to allow them to be concentrated in any gaseous phase. Also known as volatile flux.

volatile fluid [CHEM] A liquid with the tendency to become vapor at specified conditions of temperature and pressure.

volatile flux *See* volatile component.

volatile laurel oil [MATER] A bright-yellow liquid with an aromatic aroma, distilled from the leaves or berries of the laurel, *Laurus nobilis;* main components are cineole and pinene; soluble in alcohol, ether, benzene, and chloroform; used in perfumes, flavors, and medicine. Also known as sweet bay oil.

volatile memory *See* volatile storage.

volatile-oil reservoir [PETRO ENG] A type of bubble-point oil reservoir in which the temperature is high and the liquid density is low (leading to a volatilized oil situation), reducing the amount of producible liquids.

volatile storage [ADP] A storage device that must be continuously supplied with energy, or it will lose its retained data. Also known as volatile memory.

volatility [THERMO] The quality of having a low boiling point or subliming temperature at ordinary pressure or, equivalently, of having a high vapor pressure at ordinary temperatures.

volatility product [CHEM] The product of the concentrations of two or more molecules or ions that react to form a volatile substance.

volatilization [THERMO] The conversion of a chemical substance from a liquid or solid state to a gaseous or vapor state by the application of heat, by reducing pressure, or by a combination of these processes. Also known as vaporization.

volcanello *See* spatter cone.

volcanic ash [GEOL] Fine pyroclastic material; particle diameter is less than 4 millimeters.

volcanic bombs [GEOL] Pyroclastic ejecta; the lava fragments, liquid or plastic at the time of ejection, acquire rounded forms, markings, or internal structure during flight or upon landing.

volcanic breccia [PETR] A pyroclastic rock that is composed of angular volcanic fragments having a diameter larger than 2 millimeters and that may or may not have a matrix.

volcanic foam *See* pumice.

volcanic gases [GEOL] Volatile matter composed principally of about 90% water vapor, and carbon dioxide, sulfur dioxide,

hydrogen, carbon monoxide, and nitrogen, released during an eruption of a volcano.

volcanic glass [GEOL] Natural glass formed by the cooling of molten lava, or one of its liquid fractions, too rapidly to allow crystallization.

volcanicity *See* volcanism.

volcaniclastic rock [PETR] Clastic rock containing volcanic material in any proportion.

volcanic mud [GEOL] Sediment containing large quantities of ash from a volcanic eruption, mixed with water.

volcanic mudflow [GEOL] The flow of volcanic mud down the slope of a volcano.

volcanic rift zone [GEOL] A zone comprising volcanic fissures with underlying dike assemblages; occurs in Hawaii.

volcanic rock [GEOL] Finely crystalline or glassy igneous rock resulting from volcanic activity at or near the surface of the earth. Also known as extrusive rock.

volcanics [PETR] Igneous rocks that solidified after reaching or nearing the earth's surface.

volcanic vent [GEOL] Subcircular bodies composed of lava and fragmented volcanic rock; contacts are generally steep, and vents range in diameter from a few tens to thousands of feet.

volcanism [GEOL] The movement of magma and its associated gases from the interior into the crust and to the surface of the earth. Also known as volcanicity.

volcano [GEOL] A vent in the surface of the earth through which magma and associated gases and ash erupt; the structure produced by the ejected material is usually in the form of a cone.

volcanology [GEOL] The branch of geology that deals with volcanism.

vole [VERT ZOO] Any of about 79 species of rodent in the tribe Microtini of the family Cricetidae; individuals have a stout body with short legs, small ears, and a blunt nose.

Volhard titration [ANALY CHEM] Determination of the halogen content of a solution by titration with a standard thiocyanate solution.

volley [ENG] A round of holes fired at any one time. [ORD] Burst of fire, especially a salute fired by a detachment of riflemen.

volley bombing [ORD] Simultaneous or nearly simultaneous release of a number of bombs.

volley fire [ORD] Artillery fire in which each piece fires a specified number of rounds without regard to the other pieces and as fast as accuracy will permit.

volt [ELEC] The unit of potential difference or electromotive force in the meter-kilogram-second system, equal to the potential difference between two points for which 1 coulomb of electricity will do 1 joule of work in going from one point to the other.

Volta effect *See* contact potential difference.

voltage [ELEC] Potential difference or electromotive force measured in volts.

voltage amplification [ELECTR] The ratio of the magnitude of the voltage across a specified load impedance to the magnitude of the input voltage of the amplifier or other transducer feeding that load; often expressed in decibels by multiplying the common logarithm of the ratio by 20.

voltage amplifier [ELECTR] An amplifier designed primarily to build up the voltage of a signal, without supplying appreciable power.

voltage-amplitude-controlled clamp [ELECTR] A single diode clamp in which the diode functions as a clamp whenever the potential at point A rises above V_R; the diode is then in its forward-biased condition and acts as a very low resistance.

voltage coefficient [ELEC] For a resistor whose resistance varies with voltage, the ratio of the fractional change in resistance to the change in voltage.

voltage-controlled oscillator [ELECTR] An oscillator whose frequency of oscillation can be varied by changing an applied voltage. Abbreviated VCO.

voltage corrector [ELEC] Active source of regulated power placed in series with an unregulated supply to sense changes in the output voltage (or current), and to correct for these changes by automatically varying its own output in the

VOLE

Vole showing characteristic features.

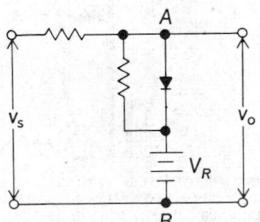

VOLTAGE-AMPLITUDE-CONTROLLED CLAMP

Circuit diagram of voltage-amplitude-controlled clamp circuit between terminals A and B. Diode functions as clamp whenever potential at A starts to rise above reference voltage V_R. Here v_S = signal voltage source; v_O = output voltage.

VOLTAGE DOUBLER

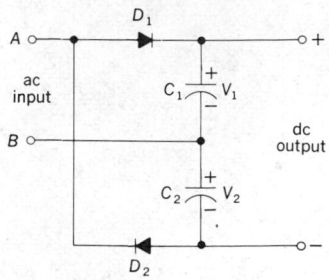

Circuit diagram of full-wave voltage doubler. When alternating-current input voltage is positive at terminal A, diode D_1 conducts, producing voltage V_1 across capacitor C_1. On other half cycle, diode D_2 conducts, producing voltage V_2 across capacitor C_2.

VOLTAGE REGULATOR

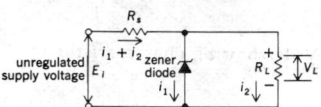

Circuit diagram of Zener-diode voltage regulator. Output load voltage V_L across load resistance R_L is maintained constant despite variation in input E_i. Current through series resistance R_s is sum of diode current i_1 and i_2.

VOLTAGE-REGULATOR TUBE

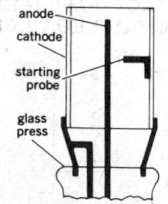

Construction of a typical voltage-regulator tube. Cathode is large cylinder; anode is thin wire.

opposite direction, thereby maintaining the total output voltage (or current) constant.

voltage-current dual [ELEC] A pair of circuits in which the elements of one circuit are replaced by their dual elements in the other circuit according to the duality principle; for example, currents are replaced by voltages, capacitances by resistances.

voltage-dependent resistor See varistor.

voltage derating [ELEC] The reduction of a voltage rating to extend the lifetime of an electric device or to permit operation at a high ambient temperature.

voltage divider [ELEC] A tapped resistor, adjustable resistor, potentiometer, or a series arrangement of two or more fixed resistors connected across a voltage source; a desired fraction of the total voltage is obtained from the intermediate tap, movable contact, or resistor junction. Also known as potential divider.

voltage doubler [ELECTR] A transformerless rectifier circuit that gives approximately double the output voltage of a conventional half-wave vacuum-tube rectifier by charging a capacitor during the normally wasted half-cycle and discharging it in series with the output voltage during the next half-cycle. Also known as doubler.

voltage drop [ELEC] The voltage developed across a component or conductor by the flow of current through the resistance or impedance of that component or conductor.

voltage feed [ELECTROMAG] Excitation of a transmitting antenna by applying voltage at a point of maximum potential (at a voltage loop or antinode).

voltage flare [ELEC] A higher than normal voltage purposely supplied to exposure lamps for a short period to produce full brilliance.

voltage gain [ELECTR] The difference between the output signal voltage level in decibels and the input signal voltage level in decibels; this value is equal to 20 times the common logarithm of the ratio of the output voltage to the input voltage.

voltage generator [ELECTR] A two-terminal circuit element in which the terminal voltage is independent of the current through the element.

voltage gradient [ELEC] The voltage per unit length along a resistor or other conductive path.

voltage level [ELEC] At any point in a transmission system, the ratio of the voltage existing at that point to an arbitrary value of voltage used as a reference.

voltage measurement [ELEC] Determination of the difference in electrostatic potential between two points.

voltage multiplier [ELEC] See instrument multiplier. [ELECTR] A rectifier circuit capable of supplying a direct-current output voltage that is two or more times the peak value of the alternating-current voltage.

voltage-multiplier circuit [ELEC] A rectifier circuit capable of supplying a direct-current output voltage that is two or more times the peak value of the alternating-current input voltage; useful for high-voltage, low-current supplies.

voltage node [ELECTROMAG] Point having zero voltage in a stationary wave system, as in an antenna or transmission line; for example, a voltage node exists at the center of a half-wave antenna.

voltage phasor [ELEC] A line whose length represents the magnitude of a sinusoidally varying voltage and whose angle with the positive x-axis represents its phase.

voltage quadrupler [ELECTR] A rectifier circuit, containing four diodes, which supplies a direct-current output voltage which is four times the peak value of the alternating-current input voltage.

voltage-range multiplier See instrument multiplier.

voltage rating [ELEC] The maximum sustained voltage that can safely be applied to an electric device without risking the possibility of electric breakdown. Also known as working voltage.

voltage ratio [ELEC] The root-mean-square primary terminal voltage of a transformer divided by the root-mean-square secondary terminal voltage under a specified load.

voltage reflection coefficient [ELECTROMAG] The ratio of the phasor representing the magnitude and phase of the electric field of the backward-traveling wave at a specified cross

section of a waveguide to the phasor representing the forward-traveling wave at the same cross section.

voltage-regulating transformer [ELECTROMAG] Saturated-core type of transformer which holds output voltage to within a few percent (5% above or below normal) with input variations up to 20% above or below normal; considerable harmonic distortion results unless extensive filters are employed.

voltage regulation [ELEC] The ratio of the difference between no-load and full-load output voltage of a device to the full-load output voltage, expressed as a percentage.

voltage regulator [ELECTR] A device that maintains the terminal voltage of a generator or other voltage source within required limits despite variations in input voltage or load. Also known as automatic voltage regulator.

voltage-regulator tube [ELECTR] A glow-discharge tube in which the tube voltage drop is approximately constant over the operating range of current; used to maintain an essentially constant direct voltage in a circuit despite changes in line voltage or load. Also known as VR tube.

voltage saturation See anode saturation.

voltage transformer [ELEC] An instrument transformer whose primary winding is connected in parallel with a circuit in which the voltage is to be measured or controlled. Also known as potential transformer.

voltage-tunable tube [ELECTR] Oscillator tube whose operating frequency can be varied by changing one or more of the electrode voltages, as in a backward-wave magnetron.

voltage-variable capacitor See varactor.

voltaic cell [ELEC] A primary cell consisting of two dissimilar metal electrodes in a solution that acts chemically on one or both of them to produce a voltage.

voltaic pile [ELEC] An early form of primary battery, consisting of a pile of alternate pairs of dissimilar metal disks, with moistened pads between pairs.

voltametry [PHYS CHEM] Any electrochemical technique in which a faradaic current passing through the electrolysis solution is measured while an appropriate potential is applied to the polarizable or indicator electrode; for example, polarography.

voltammeter [ELEC] An instrument that may be used either as a voltmeter or ammeter.

volt-ampere [ELEC] The unit of apparent power in the International System; it is equal to the apparent power in a circuit when the product of the root-mean-square value of the voltage, expressed in volts, and the root-mean-square value of the current, expressed in amperes, equals 1. Abbreviated VA.

volt-ampere hour [ELEC] A unit for expressing the integral of apparent power over time, equal to the product of 1 volt-ampere and 1 hour, or to 3600 joules.

volt-ampere-hour reactive See var hour.

volt-ampere reactive [ELEC] The unit of reactive power in the International System; it is equal to the reactive power in a circuit carrying a sinusoidal current when the product of the root-mean-square value of the voltage, expressed in volts, by the root-mean-square value of the current, expressed in amperes, and by the sine of the phase angle between the voltage and the current, equals 1. Abbreviated var. Also known as reactive volt-ampere.

Volta series See displacement series.

volt box [ELEC] A series of resistors arranged so that a desired fraction of a voltage can be measured, and the voltage thereby computed.

Volterra dislocation [SOLID STATE] A model of a dislocation which is formed in a ring of crystalline material by cutting the ring, moving the cut surfaces over each other, and then rejoining them.

Volterra equations [MATH] Given functions $f(x)$ and $K(x,y)$, these are two types of equations with unknown function y:

$$f(x) = \int_a^x K(x,t)y(t)dt$$
$$y(x) = f(x) + \lambda \int_a^x K(x,t)y(t)dt$$

voltmeter [ENG] An instrument for the measurement of potential difference between two points, in volts or in related smaller or larger units.

voltmeter-ammeter [ENG] A voltmeter and an ammeter combined in a single case but having separate terminals.

voltmeter-ammeter method [ELEC] A method of measuring resistance in which simultaneous readings of the voltmeter and ammeter are taken, and the unknown resistance is calculated from Ohm's law.

voltmeter sensitivity [ELEC] Ratio of the total resistance of the voltmeter to its full scale reading in volts, expressed in ohms per volt.

volt-ohm-milliammeter [ENG] A test instrument having a number of different ranges for measuring voltage, current, and resistance. Also known as circuit analyzer; multimeter; multiple-purpose tester.

volume [ACOUS] The intensity of a sound. [ADP] A single unit of external storage, all of which can be read or written by a single access mechanism or input/output device. [ENG ACOUS] The magnitude of a complex audio-frequency current as measured in volume units on a standard volume indicator. [MATH] A measure of the size of a body or definite region in three-dimensional space; it is equal to the least upper bound of the sum of the volumes of nonoverlapping cubes that can be fitted inside the body or region, where the volume of a cube is the cube of the length of one of its sides. Abbreviated vol.

volume compressor [ENG ACOUS] An audio-frequency circuit that limits the volume range of a radio program at the transmitter, to permit using a higher average percent modulation without risk of overmodulation; also used when making disk recordings, to permit a closer groove spacing without overcutting. Also known as automatic volume compressor.

volume control [ENG ACOUS] A potentiometer used to vary the loudness of a reproduced sound by varying the audio-frequency signal voltage at the input of the audio amplifier.

volume control system [ENG ACOUS] An electronic system that regulates the signal amplification or limits the output of a circuit, such as a volume compressor or a volume expander.

volume dose *See* integral dose.

volume expander [ENG ACOUS] An audio-frequency control circuit sometimes used to increase the volume range of a radio program or recording by making weak sounds weaker and loud sounds louder; the expander counteracts volume compression at the transmitter or recording studio. Also known as automatic volume expander.

volume flow rate [FL MECH] The volume of the fluid that passes through a given surface in a unit time.

volume indicator [ENG ACOUS] A standardized instrument for indicating the volume of a complex electric wave such as that corresponding to speech or music; the reading in volume units is equal ·to the number of decibels above a reference level which is realized when the instrument is connected across a 600-ohm resistor that is dissipating a power of 1 milliwatt at 100 hertz. Also known as volume unit meter.

volume integral [MATH] **1.** A multiple integral whose limits of integration define a solid. **2.** A definite multiple integral whose value is the volume of the solid defined by the limits of integration.

volume lifetime [SOLID STATE] Average time interval between the generation and recombination of minority carriers in a homogeneous semiconductor.

volume-limiting amplifier [ELECTR] Amplifier containing an automatic device that functions only when the input signal exceeds a predetermined level, and then reduces the gain so the output volume stays substantially constant despite further increases in input volume; the normal gain of the amplifier is restored when the input volume returns below the predetermined limiting level.

volume meter [ENG] Any flowmeter in which the actual flow is determined by the measurement of a phenomenon associated with the flow.

volume phase *See* surface phase.

volume range [ELEC] In a transmission system, the difference, expressed in decibels, between the maximum and minimum volumes that can be satisfactorily handled by the system. [ENG ACOUS] The difference, expressed in decibels, between the maximum and minimum volumes of a complex audio-frequency signal occurring over a specified period of time.

volume resistivity [ELEC] Electrical resistance between opposite faces of a 1-centimeter cube of insulating material, commonly expressed in ohm-centimeters. Also known as specific insulation resistance.

volume susceptibility [PHYS CHEM] The magnetic susceptibility of a specified volume (for example, 1 cubic centimeter) of a magnetically susceptible material.

volume table of contents [ADP] An index record near the beginning of each volume, which records the name, location, and extent of every file or data set residing on that particular volume; usually not found on magnetic tapes, but often required on all disk packs and drums. Abbreviated VTOC.

volume target [ELECTROMAG] A radar target composed of a large number of objects too close together to be resolved.

volume test [ADP] The processing of a volume of actual data to check for program malfunction.

volume transport [OCEANOGR] The volume of moving water measured between two points of reference and expressed in cubic meters per second.

volumetric analysis [ANALY CHEM] Quantitative analysis of solutions of known volume but unknown strength by adding reagents of known concentration until a reaction end point (color change or precipitation) is reached; the most common technique is by titration.

volumetric efficiency [MECH ENG] In describing an engine or gas compressor, the ratio of volume of working substance actually admitted, measured at a specified temperature and pressure, to the full piston displacement volume; for a liquid-fuel engine, such as a diesel engine, volumetric efficiency is the ratio of the volume of air drawn into a cylinder to the piston displacement.

volumetric flask [ANALY CHEM] A laboratory flask primarily intended for the preparation of definite, fixed volumes of solutions, and therefore calibrated for a single volume only.

volumetric performance [PETRO ENG] The volume production of gas and oil from a reservoir; usually expressed as gas-oil ratio.

volumetric pipet [ANALY CHEM] A graduated glass tubing used to measure quantities of a solution; the tube is open at the top and bottom, and a slight vacuum (suction) at the top pulls liquid into the calibrated section; breaking the vacuum allows liquid to leave the tube.

volumetric radar [ENG] Radar capable of producing three-dimensional position data on a multiplicity of targets.

volumetric strain [MECH] One measure of deformation; the change of volume per unit of volume.

volume unit [ENG ACOUS] A unit for expressing the audio-frequency power level of a complex electric wave, such as that corresponding to speech or music; the power level in volume units is equal to the number of decibels above a reference level of 1 milliwatt as measured with a standard volume indicator. Abbreviated VU.

volume unit meter *See* volume indicator.

volume velocity [ACOUS] The rate of flow of a medium through a specified area due to a sound wave.

voluntary muscle [PHYSIO] A muscle directly under the control of the will of the organism.

volute [DES ENG] A spiral casing for a centrifugal pump or a fan designed so that speed will be converted to pressure without shock.

volute pump [MECH ENG] A centrifugal pump housed in a spiral casing.

Volutidae [INV ZOO] A family of gastropod mollusks in the order Neogastropoda.

volutin [BIOCHEM] A basophilic substance, thought to be a nucleic acid, occurring as granules in the cytoplasm and vacuoles of algae and other microorganisms.

volva [MYCOL] A cuplike membrane surrounding the base of the stipe in certain gill fungi.

volvent nematocyst [INV ZOO] A nematocyst in the form of an unarmed, coiled tube that is closed at the end.

Volvocales [BOT] An order of one-celled or colonial green algae in the division Chlorophyta; individuals are motile with two, four, or rarely eight whiplike flagella.

Volvocida [INV ZOO] An order of the protozoan class Phyta-

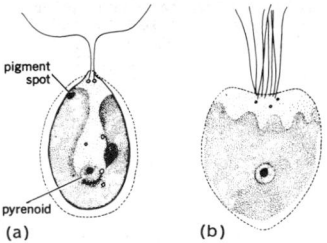
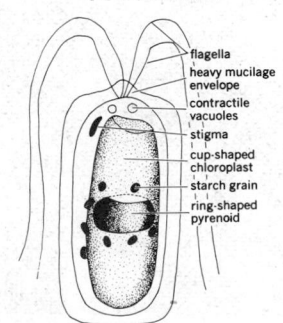

mastigophorea; individuals are grass-green or colorless, have one, two, four, or eight flagella, and thick cell walls of cellulose.

volvulus [MED] A twisting of the bowel upon itself so as to occlude the lumen and, in severe cases, compromise its circulation.

Vombatidae [VERT ZOO] A family of marsupial mammals in the order Diprotodonta in some classification systems.

vomer [ANAT] A skull bone below the ethmoid region constituting part of the nasal septum in most vertebrates.

vomeronasal cartilage [ANAT] A strip of hyaline cartilage extending from the anterior nasal spine upward and backward on either side of the septal cartilage of the nose and attached to the anterior margin of the vomer. Also known as Jacobson's cartilage.

vomiting gas [MATER] Any one of a group of toxic gases, such as adamsite, that causes coughing, sneezing, sometimes vomiting, and other effects.

von Arx current meter [ENG] A type of current-measuring device using electromagnetic induction to determine speed and, in some models, direction of deep-sea currents.

von Gierke's disease [MED] A form of glycogenosis characterized by marked diminution in or absence of hepatic glucose-6-phosphatase, resulting in hepatic glycogenosis, hypoglycemia, and acidosis. Also known as glycogen storage disease; hepatic glycogenosis; type I of Cori; van Crevald–von Gierke's disease.

von Kármán vortex street *See* Kármán vortex street.

VOR *See* very-high-frequency omnidirectional radio range.

Vorce diaphragm cell [CHEM ENG] A cylindrical cell with graphite anodes and asbestos-covered cathode, used in the electrolytic process for the manufacture of chlorine.

vorobievite *See* vorobyevite.

vorobyevite [MINERAL] A rose-red, purplish-red, or pinkish cesium-containing variety of beryl; used as a gem. Also known as morganite; rosterite; vorobievite; worobieffite.

vortac [NAV] A ground radio station consisting of a collocated very-high-frequency omnidirectional radio range (VOR) and Tacan facility; this station permits obtaining polar coordinates by the use of VOR receiver and distance-measuring equipment, or by Tacan equipment alone.

vortex [FL MECH] **1.** Any flow possessing vorticity; for example, an eddy, whirlpool, or other rotary motion. **2.** A flow with closed streamlines, such as a free vortex or line vortex. **3.** *See* vortex tube.

vortex amplifier [ENG] A fluidic device in which the supply flow is introduced at the circumference of a shallow cylindrical chamber; the vortex field developed can substantially reduce or throttle flow; used in fluidic diodes, throttles, pressure amplifiers, and a rate sensor.

vortex burner [ENG] Combustion device in which the combustion air is fed tangentially into the burner, creating a spin (vortex) to mix it with the fuel as it is injected.

vortex cage meter [ENG] In flow measurement, a type of quantity meter which exerts only a slight retardation on the flowing fluid; the elements rotate at a speed that is linear with fluid velocity; revolutions are counted either by coupling to a local mounted counter or by a proximity detector for remote transmission.

vortex distribution method [FL MECH] An analytic method used in ideal aerodynamics which ignores the thickness of the profile of the aerodynamic figure being studied.

vortex filament [FL MECH] The line of concentrated vorticity in a line vortex. Also known as vortex line.

vortex line [FL MECH] **1.** A line drawn through a fluid such that it is everywhere tangent to the vorticity. **2.** *See* vortex filament.

vortex ring [FL MECH] A line vortex in which the line of concentrated vorticity is a closed curve. Also known as collar vortex; ring vortex.

vortex shedding [FL MECH] In the flow of fluids past objects, the shedding of fluid vortices periodically downstream from the restricting object (for example, smokestacks, pipelines, or orifices).

vortex sheet [FL MECH] A surface across which there is a discontinuity in fluid velocity, such as in slippage of one layer of fluid over another; the surface may be regarded as being composed of vortex filaments.

vortex street [FL MECH] A series of vortices which are systematically shed from the downstream side of a body around which fluid is flowing rapidly. Also known as vortex trail; vortex train.

vortex thermometer [ENG] A thermometer, used in aircraft, which automatically corrects for adiabatic and frictional temperature rises by imparting a rotary motion to the air passing the thermal sensing element.

vortex trail *See* vortex street.

vortex train *See* vortex street.

vortex tube [FL MECH] A tubular surface consisting of the collection of vortex lines which pass through a small closed curve. Also known as vortex.

vorticity [FL MECH] For a fluid flow, a vector equal to the curl of the velocity of flow.

vorticity equation [FL MECH] An equation of fluid mechanics describing horizontal circulation in the motion of particles around a vertical axis: $(d/dt)(S + f) = -(S + f) \, \mathrm{div}_h c$, where $(S + f)$ is the absolute vorticity and $\mathrm{div}_h c$ is the horizontal divergence of the fluid velocity.

vorticity-transport hypothesis [FL MECH] The hypothesis that, owing to the existence of pressure fluctuations, vorticity, and not momentum, is conservative in turbulent eddy flux.

VOS *See* vertical obstacle sonar.

Voskhod program [AERO ENG] The multimanned spaceflight program of the Soviet Union which began with the flight of *Voskhod 1*, October 12, 1964.

Vostok spacecraft [AERO ENG] One of a series of manned artificial satellites launched by the Soviet Union; *Vostok 1* launched on April 12, 1961, was the first manned spacecraft.

votator [MECH ENG] Efficient heat-exchange units for chilling and mechanically working a continuous stream of emulsion; used in food industries in preparation of margarine.

vougesite [PETR] A lamprophyre having an orthoclase and hornblende groundmass in which are embedded hornblende phenocrysts.

voyage [NAV] **1.** The outward and homeward passage of a trip by sea. **2.** A trip by sea.

V particle [PARTIC PHYS] The name first used for the unstable particles whose decay is responsible for the production of characteristic V-shaped tracks observed in cloud chambers exposed to cosmic radiation; they are neutral semistable particles such as neutral K mesons or Λ hyperons.

VPR chart [NAV] A type of radar chart for use with VPR (virtual plan position indicator reflectoscope).

VRC *See* lateral parity check.

vriajem *See* friagem.

VRM *See* viscous remanent magnetization.

VR tube *See* voltage-regulator tube.

VSB *See* vestigial side band.

V-shaped depression [METEOROL] On a surface chart, a low or trough about which the isobars display a pronounced V shape, with the point of the V usually extending equatorward from the parent low.

V-shaped valley [GEOL] A valley having a cross-sectional profile in the form of the letter V, commonly produced by stream erosion. Also known as V valley.

vt fuse *See* proximity fuse.

VTOC *See* volume table of contents.

VTOL *See* vertical takeoff and landing.

VTR *See* video tape recording.

VTVM *See* vacuum-tube voltmeter.

VU *See* volume unit.

vug [PETR] A small cavity in a vein or rock usually lined with minerals differing in composition from those of the enclosing rock. Also known as bughole.

Vulcan [ASTRON] A hypothetical planet that was supposed to have an orbit within the orbit of Mercury; its existence was considered about 1859 and in the next few years, but it is generally considered by present-day astronomers to be non-existent.

Vulcanian eruption [GEOL] A volcanic eruption characterized by periodic explosive events. Also known as paroxysmal eruption; Plinian eruption; Vesuvian eruption.

vulcanization [CHEM ENG] A chemical reaction of sulfur (or

VORTEX CAGE METER

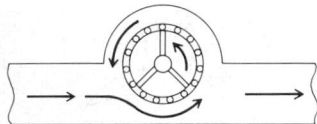

Diagram of vortex cage meter. Arrows indicate direction of fluid flow.

VOSTOK SPACECRAFT

Vostok spacecraft, a Soviet manned artificial satellite. (*NASA*)

other vulcanizing agent) with rubber or plastic to cause cross-linking of the polymer chains; it increases strength and resiliency of the polymer.

vulcanized fiber [MATER] A laminated plastic made by chemically treating layers of 100% rag-content paper to gelatinize the paper and fuse the layers into a solid mass; when dried under pressure, it forms a hard, tough material having good electrical properties along with mechanical strength and dimensional stability.

vulcan power [MATER] A high explosive composed of 30% nitroglycerin, 52.5% sodium nitrate, 10.5% charcoal, and 7% sulfur.

vulgar establishment *See* high-water full and change.

Vulpecula [ASTRON] A northern constellation, right ascension 20 hours, declination 25°N. Also known as Little Fox.

vulture [VERT ZOO] The common name for any of various birds of prey in the families Cathartidae and Accipitridae of the order Falconiformes; the head of these birds is usually naked.

vulva [ANAT] The external genital organs of women.

vulval gland [ANAT] A scent gland in the vulval tissues of the human female.

vulvovaginitis [MED] Simultaneous inflammation of the vulva and the vagina.

vuthan [METEOROL] In southern South America, an intense storm.

V valley *See* V-shaped valley.

V weapon [ORD] Either the V-1 bomb or the V-2 bomb, used by the Germans in World War II.

Vycor glass [MATER] A nearly pure silica glass formed from a soda borosilicate glass without the production problems of fused silica.

Vyrene [TEXT] Trade name of a Spandex fiber manufactured by Uniroyal, Inc.; it is a single-filament polyurethane elastomer.

wacke [PETR] Sandstone composed of a mixture of angular and unsorted or poorly sorted fragments of minerals and rocks and an abundant matrix of clay and fine silt.

Wacker process [CHEM ENG] A process for the oxidation of ethylene to acetaldehyde by oxygen in the presence of palladium chloride and cupric chloride.

wad [MINERAL] A massive, generally soft, amorphous, earthy, dark-brown or black mineral composed principally of manganese oxides with some other minerals, and formed by decomposition of manganese minerals. Also known as black ocher; bog manganese; earthy manganese. [ORD] A felt or cardboard pad used to secure the propellant in place in cartridges. Also known as wadding.

wad cutter [ORD] Bullet designed for target shooting, shaped to cut a clean hole in a paper target.

wadding See wad.

wader See shore bird.

wadi [GEOL] In the desert regions of southwestern Asia and northern Africa, a stream bed or channel, or a steep-sided ravine, gulley, or valley, which carries water only during the rainy season. Also spelled wady.

wading bird [VERT ZOO] Any of the long-legged, long-necked birds composing the order Ciconiiformes, including storks, herons, egrets, and ibises.

Wadsworth mounting [OPTICS] 1. A device in which light passes through a prism and is then reflected from a plane mirror; it has the effect of a constant-deviation prism. 2. A mounting for a diffraction grating in which the slit is placed at the principal focus of a concave mirror, so that the light falling on the grating is in a parallel beam; it greatly reduces astigmatism.

wady See wadi.

wafer [ELECTR] A thin semiconductor slice on which matrices of microcircuits can be fabricated, or which can be cut into individual dice for fabricating single transistors and diodes. [ENG] A flat element for a process unit, as in a series of stacked filter elements.

wafer lever switch [ELECTR] A lever switch in which a number of contacts are arranged on one or both sides of one or more wafers, for engaging one or more contacts on a movable wafer segment actuated by the operating lever.

wafer socket [ELECTR] An electron-tube socket consisting of one or two wafers of insulating material having holes in which are spring metal clips that grip the terminal pins of a tube.

waffle weave [TEXT] Weave pattern of fabrics that resembles the cellular honeybee comb. Also known as honeycomb weave.

wage curve [IND ENG] A graphic representation of the relationship between wage rates and point values for key jobs.

wage incentive plan [IND ENG] A wage system which provides additional pay for qualitative and quantitative performance which exceeds standard or normal levels. Also known as incentive wage system.

Wagner earth connection See Wagner ground.

Wagner ground [ELEC] A ground connection used with an alternating-current bridge to minimize stray capacitance errors when measuring high impedances; a potentiometer is connected across the bridge supply oscillator, with its movable tap grounded. Also known as Wagner earth connection.

Wagner's reagent [ANALY CHEM] An aqueous solution of iodine and potassium iodide; used for microchemical analysis of alkaloids. Also known as Wagner's solution.

Wagner's solution See Wagner's reagent.

wagon drill [MECH ENG] 1. A vertically mounted, pneumatic, percussive-type rock drill supported on a three- or four-wheeled wagon. 2. A wheel-mounted diamond drill machine.

Wahl correlation [PETRO ENG] A pressure-volume-tempera-ture (PVT) correlation used to estimate the total oil recovery from a solution-gas-drive oil reservoir; it is based on assumed PVT data, and may be in error.

Waidner-Burgess standard [OPTICS] A unit of luminous intensity equal to the luminous intensity of 1 square centimeter of a blackbody at the melting point of platinum, or to 60 candelas.

wairakite [MINERAL] $CaAl_2Si_4O_{12} \cdot 2H_2O$ A zeolite mineral that is isostructural with analcime.

waist [ENG] The center portion of a vessel or container that has a smaller cross section than the adjacent areas.

waiting line [IND ENG] A line formed by units waiting for service. Also known as queue.

waiting time See idle time.

wake [FL MECH] The region behind a body moving relative to a fluid in which the effects of the body on the fluid's motion are concentrated.

wake flow [FL MECH] Turbulent eddying flow that occurs downstream from bluff bodies.

wake gain [NAV ARCH] The increase in the effective thrust of a propeller, for a given power delivered thereto, because of the forward motion of the water dragged along behind a vessel's hull.

wake stream theory [OCEANOGR] The theory that, in a stratified ocean, a compensation current must develop on the right side of a wake stream, flowing in the same direction, and a countercurrent in the opposite direction must appear to the left.

Walden's rule [PHYS CHEM] A rule which states that the product of the viscosity and the equivalent ionic conductance at infinite dilution in electrolytic solutions is a constant, independent of the solvent; it is only approximately correct.

Waldeyer's ring [ANAT] A circular arrangement of the lymphatic tissues formed by the palatine and pharyngeal tonsils and the lymphatic follicles at the base of the tongue and behind the posterior pillars of the fauces.

wale [TEXT] 1. A rib or raised cord in a woven fabric. 2. A lengthwise row of loops in a knitted fabric.

walkie-talkie See pack unit.

walking beam [MECH ENG] A lever that oscillates on a pivot and transmits power in a manner producing a reciprocating or reversible motion; used in rock drilling and oil well pumping.

walking bird [VERT ZOO] Any bird of the order Columbiformes, including the pigeons, doves, and sandgrouse.

walking dragline [MECH ENG] A large-capacity dragline built with moving feet; disks 20 feet (6 meters) in diameter support the excavator while working.

walking machine [MECH ENG] A machine designed to carry its operator over various types of terrain; the operator sits on a platform carried on four mechanical legs, and movements of his arms control the front legs of the machine while movements of his legs control the rear legs of the machine.

walking props [MIN ENG] Self-advancing supports.

wall [ENG] A vertical structure or member forming an enclosure or defining a space. [GEOL] The side of a cave passage. [MIN ENG] 1. The side of a level or drift. 2. The country rock bounding a vein laterally. 3. The face of a longwall working or stall, commonly called coal wall.

Wallach transformation [ORG CHEM] By the use of concentrated sulfuric acid, an azoxybenzene is converted into a *para*-hydroxyazobenzene.

wall anchor [BUILD] A steel strap fastened to the end of every second or third common joist and built into the brickwork of a wall to provide lateral support. Also known as joist anchor.

wall-attachment amplifier [FL MECH] A bistable fluidic device utilizing two walls set back from the supply jet port,

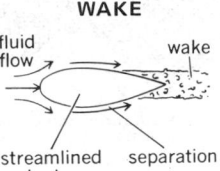
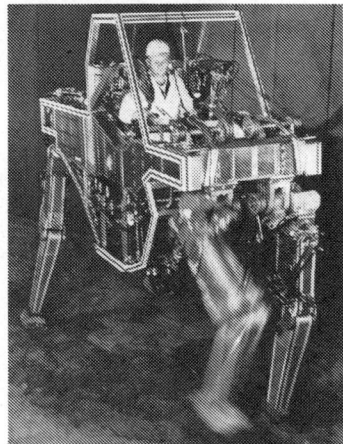

WALL CRANE

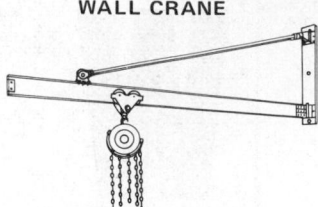

Wall crane, showing mounting on wall.

WALRUS

Walrus (*Odobenus rosmarus*), showing tusks.

WANKEL ENGINE

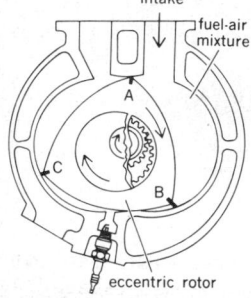

Diagram of Wankel engine showing gears on rotor and shaft. Rotor divides inner volume into three chambers, each analogous to cylinder in standard piston engine. Here, chamber AB is terminating the intake phase and commencing the compression phase, chamber BC is terminating the compression phase, and chamber CA is commencing the exhaust phase.

WARM-AIR HEATING

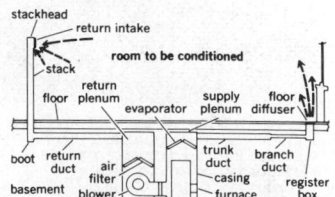

Air passage in a warm-air duct system. (*From S. Konzo, J. R. Carroll, and H. D. Bareither, Winter Air Conditioning, Industrial Press, 1958*)

control ports, and channels to define two downstream outputs. Also known as flip-flop amplifier.

wallboard [MATER] Panels of various materials for surfacing ceilings and walls, including asbestos cement sheet, plywood, gypsum plasterboard, and laminated plastics.

wall crane [MECH ENG] A jib crane mounted on a wall.

wall effect [ELECTR] The contribution to the ionization in an ionization chamber by electrons liberated from the walls.

wall energy [SOLID STATE] The energy per unit area of the boundary between two ferromagnetic domains which are oriented in different directions.

waller *See* pack builder.

Walley engine [MECH ENG] A multirotor engine employing four approximately elliptical rotors that turn in the same clockwise sense, leading to excessively high rubbing velocities.

wall friction [FL MECH] The drag created in the flow of a liquid or gas because of contact with the wall surfaces of its conductor, such as the inside surfaces of a pipe.

Wallis formulas [MATH] Formulas that determine the values of the definite integrals from 0 to $\pi/2$ of the functions $\sin^n(x)$, $\cos^n(x)$, and $\cos^m(x)\sin^n(x)$ for positive integers m and n. Also known as Wallis theorem.

Wallis product [MATH] An infinite product representation of the number $\pi/2$.

Wallis theorem *See* Wallis formulas.

wall off [ENG] To seal cracks or crevices in the wall of a borehole with cement, mud cake, compacted cuttings, or casing.

wall outlet [ELEC] An outlet mounted on a wall, from which electric power can be obtained by inserting the plug of a line cord.

wall plate [BUILD] A piece of timber laid flat along the tip of the wall; it supports the rafters.

wallplate [MIN ENG] A horizontal timber supported by posts resting on sills and extending lengthwise on each side of the tunnel; roof supports rest on the wallplates.

wall ratio [DES ENG] Ratio of the outside radius of a gun, a tube, or jacket to the inside radius; or ratio of the corresponding diameters.

wall rock [GEOL] Rock that encloses a vein.

wall rock alteration [GEOL] Alteration of wall rock adjacent to hydrothermal veins by the fluid responsible for formation of the mineral deposit.

wall-sided glacier [HYD] A glacier unconfined by a marked ravine or valley.

wall tie [BUILD] A rigid, corrosion-resistant metal tie fitted into the bed joints across the cavity of a cavity wall.

walnut [BOT] The common name for about a dozen species of deciduous trees in the genus *Juglans* characterized by pinnately compound, aromatic leaves and chambered or laminate pith; the edible nut of the tree is distinguished by a deeply furrowed or sculptured shell.

walrus [VERT ZOO] *Odobenus rosmarus*. The single species of the pinniped family Odobenidae distinguished by the upper canines in both sexes being prolonged as tusks.

Walter engine [MECH ENG] A multirotor rotary engine that uses two different-sized elliptical rotors.

wander *See* apparent wander; scintillation.

wander correction [NAV] A correction to compensate for wander error in bubble sextant readings.

wandering dune [GEOL] A sand dune that has moved as a unit in the leeward direction of the prevailing winds, and that is characterized by the lack of vegetation to anchor it. Also known as migratory dune; traveling dune.

wandering water *See* vadose water.

waning moon [ASTRON] The moon between full and new when its visible part is decreasing.

Wankel engine [MECH ENG] An eccentric-rotor-type internal combustion engine with only two primary moving parts, the rotor and the eccentric shaft; the rotor moves in one direction around the trochoidal chamber containing peripheral intake and exhaust ports.

Wannier function [SOLID STATE] The Fourier transform of a Bloch function defined for an entire band, regarded as a function of the wave vector.

want *See* nip.

Ward-Leonard speed-control system [CONT SYS] A system for controlling the speed of a direct-current motor in which the armature voltage of a separately excited direct-current motor is controlled by a motor-generator set.

warehouse [IND ENG] A building used for storing merchandise and commodities.

war game [ORD] A simulation, by whatever means, of a military operation involving two or more opposing forces, using rules, data, and procedures designed to depict an actual or assumed real-life situation.

war gas [ORD] A toxic or irritant chemical agent, regardless of its physical state, whose properties may be effectively exploited in the field of war.

warhead [ORD] An item which is designed to be mounted in or on a torpedo, guided missile, rocket, or bomb; it may contain high-explosive, nuclear, chemical, biological, or inert materials.

warhead installation [ORD] A warhead plus those additionally required items (contained in the adaptation kit) that are needed to mate the warhead with a specific carrier.

warm-air drop *See* warm pool.

warm-air heating [MECH ENG] Heating by circulating warm air; system contains a direct-fired furnace surrounded by a bonnet through which air circulates to be heated.

warm air mass [METEOROL] An air mass that is warmer than the surrounding air; an implication that the air mass is warmer than the surface over which it is moving.

warm anticyclone *See* warm high.

warm braw [METEOROL] A warm, dry, foehn wind which persists for up to 8 days during the east monsoon in the Schouten Islands off the north coast of New Guinea.

warm-core anticyclone *See* warm high.

warm-core cyclone *See* warm low.

warm-core high *See* warm high.

warm-core low *See* warm low.

warm cyclone *See* warm low.

warm drop *See* warm pool.

warm front [METEOROL] Any nonoccluded front, or portion thereof, which moves in such a way that warmer air replaces colder air.

warm high [METEOROL] At a given level in the atmosphere, any high that is warmer at its center than at its periphery. Also known as warm anticyclone; warm-core anticyclone; warm-core high.

warm low [METEOROL] At a given level in the atmosphere, any low that is warmer at its center than at its periphery; the opposite of a cold low. Also known as warm-core cyclone; warm-core low; warm cyclone.

warm pool [METEOROL] A region, or pool, of relatively warm air surrounded by colder air; the opposite of a cold pool; commonly applied to warm air of appreciable vertical extent isolated in high latitudes when a cutoff high is formed. Also known as warm-air drop; warm drop.

warm sector [METEOROL] The area of warm air, within the circulation of a wave cyclone, which lies between the cold front and warm front of a storm.

warm tongue [METEOROL] A pronounced poleward extension or protrusion of warm air.

warm-tongue steering [METEOROL] The steering influence apparently exerted upon a tropical cyclone by an upper-level warm tongue which often extends a considerable distance into regions adjacent to the cyclone.

warm wave *See* heat wave.

warning agent *See* odorant.

warning beacon *See* warning radio beacon.

warning device [ADP] A visible or audible alarm to inform the operator of a machine condition.

warning net [COMMUN] A communication system established for the purpose of disseminating warning information of enemy movement or action to all interested commands.

warning radio beacon [NAV] An auxiliary radio beacon that is located at a lightship to warn vessels of their proximity to the lightship; it is of short range, and sounds a warbling note for 1 minute immediately following the main radio beacon on the same frequency. Also known as warning beacon.

warning-receiver system [ELECTR] An electronic countermeasure system, carried on a tactical or transport aircraft, which is programmed to alert a pilot when his aircraft is being illuminated by a specific radar signal above predetermined power thresholds.

warning stage [HYD] The stage, on a fixed river gage, at which it is necessary to begin issuing warnings or river forecasts if adequate precautionary measures are to be taken before flood stage is reached.

warp [GEOL] 1. An upward or downward flexure of the earth's crust. 2. A layer of sediment deposited by water. [TEXT] Yarn extending lengthwise, under tension on a loom. Also known as end.

warpage [MECH] The action, process, or result of twisting or turning out of shape.

warp knitting [TEXT] A knitting process in which a group of yarns form rows running lengthwise by an interlocking process.

warrant *See* underclay.

Warren truss [CIV ENG] A truss having only sloping members between the top and bottom horizontal members.

war reserves [ORD] Stocks of material amassed in peacetime to meet the increase in military requirements consequent upon an outbreak of war, and intended to provide the interim support essential to sustain operations until resupply can be effected.

warringtonite *See* brochantite.

Warrior *See* Orion.

war surplus [ORD] A military article of supply or piece of equipment that has been declared surplus because it is obsolete, unserviceable, or excess to current and reserve military requirements.

wart [MED] A papillomatous growth which occurs singly or in groups on the skin surface; thought to be caused by a viral agent.

Wasatch winds [METEOROL] Strong, easterly, jet-effect winds blowing out of the mouths of the canyons of the Wasatch Mountains onto the plains of Utah.

wash [AERO ENG] The stream of air or other fluid sent backward by a jet engine or a propeller. [ENG] 1. To clean cuttings or other fragmental rock materials out of a borehole by the jetting and buoyant action of a copious flow of water or a mud-laden liquid. 2. The erosion of core or drill string equipment by the action of a rapidly flowing stream of water or mud-laden drill-circulation liquid. [FL MECH] The surge of disturbed air or other fluid resulting from the passage of something through the fluid. [GEOL] 1. An alluvial placer. 2. A piece of land washed by a sea or river. 3. Loose or eroded surface material, such as sand or gravel, accumulated, moved, and deposited by running water. [GRAPHICS] To dip negatives and prints in water after fixing to remove the soluble silver halide–fixing agent complexes. [MET] A coating applied to the face of a mold prior to casting.

washability [MIN ENG] Coal properties determining the amenability of a coal to improvement in quality by cleaning.

wash-and-strain ice foot [OCEANOGR] An ice foot formed from ice casts and slush and attached to a shelving beach, between the high and low waterlines; high waves and spray may cause it to build up above the high waterline.

washboard [GRAPHICS] An expression for the appearance of film images having narrowly spaced, diffused dark and light areas, running crosswise of the film.

washboard course [ENG] A test course for vehicles consisting of a series of waves or convolutions having arbitrary amplitude and frequency; a common type is the so-called sine-wave course.

wash boring *See* jet drilling.

washer [DES ENG] A flattened, ring-shaped device used to improve the tightness of a screw fastener.

washer thermistor [ELECTR] A thermistor in the shape of a washer, which may be as large as 0.75 inch (1.9 centimeters) in diameter and 0.50 inch (1.3 centimeters) thick; it is formed by pressing and sintering an oxide-binder mixture.

washing [ANALY CHEM] 1. In the purification of a laboratory sample, the cleaning of residual liquid impurities from precipitates by adding washing solution to the precipitates, mixing, then decanting, and repeating the operation as often as needed. 2. The removal of soluble components from a mixture of solids by using the effect of differential solubility.

[CHEM ENG] In a process operation, cleaning of a solids bed (settler) or cake (filter) with a liquid in which the solid is not soluble.

washing plant [MIN ENG] A plant where slimes are removed from relatively coarse ore by washing, tumbling, or scrubbing.

washing soda *See* sal soda.

Washita stone [MATER] A relatively porous and not very dense oilstone, used chiefly for whetstones and for sharpening coarse tools.

wash load [GEOL] The finer part of the total sediment load of a stream which is supplied from bank erosion or an external upstream source, and which can be carried in large quantities.

wash metal [MET] Molten metal used to clean out a furnace, ladle, or other container.

Washoe zephyr [METEOROL] The chinook on the Nevada side of the Sierra Nevada Mountains of northern California.

wash oil *See* absorption oil.

washout *See* horseback.

washover [GEOL] Material deposited by overwash, especially a small delta produced by storm waves and built on the landward side of a bar or barrier. Also known as storm delta; wave delta.

wash plain *See* outwash plain.

wash primer [MET] A synthetic vehicle primer containing phosphoric acid and zinc chromate; used as a corrosion-inhibiting first paint coat on metals.

wash water [CHEM ENG] Water contacted with process streams (liquid or gas), packed beds, or filter cakes to flush or dissolve out impurities.

wasp [INV ZOO] The common name for insects belonging to the order Hymenoptera; all are important as parasites or predators of injurious pests.

Wasserman test [IMMUNOL] A complement-fixation test for syphilis using sensitized lipid extracts of beef heart as antigen.

waste [ENG] 1. Rubbish from a building. 2. Dirty water from mining, industrial, and domestic use. 3. The amount of excavated material exceeding fill. [MIN ENG] 1. The barren rock in a mine. 2. The refuse from ore dressing and smelting plants. 3. The fine coal made in mining and preparing coal for market.

waste acid *See* sludge acid.

waste bank [MIN ENG] A bank made of earth excavated during the digging of a ditch and laid parallel to it.

waste disposal *See* radioactive-waste disposal.

waste filling [MIN ENG] Material used for support in heavy ground and in large stopes to prevent failure of rock walls and to minimize or control subsidence and to make it possible to extract pillars of ore in the earlier stages of mining; material used for filling includes waste rock sorted out in the stopes or mined from rock walls, milltailing, sand and gravel, smelter slag, and rock from surface open cuts or quarries.

waste-heat boiler [CHEM ENG] A heat-retrieval unit using hot by-product gas or oil from chemical processes; used to produce steam in a boiler-type system. Also known as gastube boiler.

waste lubrication [ENG] A method in which a lubricant is delivered to a bearing surface by the wicking action of cloth waste or yarn.

waste pipe [CIV ENG] A pipe to carry waste water from a basin, bath, or sink in a building.

waste raise [MIN ENG] An excavation in the mine in which barren rock and other material is broken up for use as filling at a stope.

waste rock [MIN ENG] Valueless rock that must be fractured and removed in order to gain access to or upgrade ore. Also known as muck; mullock.

waste silk [TEXT] Short silk filaments which are left on a reel after removing the long filaments; used for spun silk.

watch [COMMUN] The service performed by a qualified operator when on duty in the radio room of a vessel. Also known as radio watch. [HOROL] A small timepiece of a size convenient to be carried on the person.

watch buoy *See* station buoy.

watch error [HOROL] The amount by which watch time differs from the correct time; it is usually expressed as seconds per day and labeled fast (F) and slow (S).

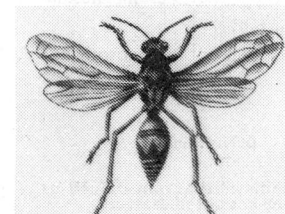

WASP

Paper wasp *(Polistes)*, a type of wasp that grows to 2.5 centimeters or more. It is common on flowers and near buildings where open-comb paper nests are constructed.

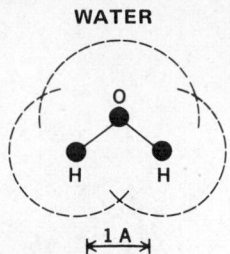

WATER

Water molecule, showing positions of atoms. Dotted circles show effective sizes of isolated atoms.

WATER-ACTIVATED BATTERY

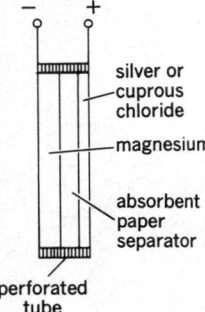

silver or cuprous chloride

magnesium

absorbent paper separator

perforated tube

Schematic diagram of cell of water-activated battery.

WATER BUG

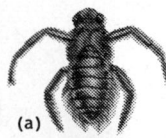

(a)

(b)

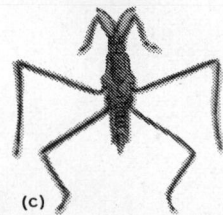

(c)

Water bugs of order Hemiptera. *(a)* Water boatman. *(b)* Giant water bug, small species. *(c)* Water strider.

watch rate [HOROL] The amount gained or lost by a watch or clock in unit time; it is usually expressed in seconds per 24 hours, to an accuracy of 0.1 second, and labeled gaining or losing. Also known as daily rate.

watch time [HOROL] The hour of the day as indicated by a watch or clock.

water [CHEM] H_2O Clear, odorless, tasteless liquid that is essential for most animal and plant life and is an excellent solvent for many substances; melting point 0°C (32°F), boiling point 100°C (212°F); the chemical compound may be termed hydrogen oxide.

water absorption tube [ANALY CHEM] A glass tube filled with a solid absorbent (calcium chloride or silica gel) to remove water from gaseous streams during or after chemical analyses.

water-activated battery [ELEC] A primary battery that contains the electrolyte but requires the addition of or immersion in water before it is usable.

water atmosphere [METEOROL] The concept of a separate atmosphere composed only of water vapor.

water ballast [NAV ARCH] Water confined to double-bottom tanks, peak tanks, or other designated compartments, for use in obtaining satisfactory draft, trim, or stability.

water ballast tank [NAV ARCH] A tank in which sea water for ballast is confined.

water-base mud [PETRO ENG] Oil-well drilling mud in which the liquid component is water, into which are mixed the thickeners and other additives.

water-base paint [MATER] Paint in which the vehicle or binder is dissolved in water or in which the vehicle or binder is dispersed as an emulsion; an example of the dispersion type is latex paint. Also known as water-thinned paint.

Water Bearer *See* Aquarius.

water-bearing strata [GEOL] Ground layers below the standing water level.

water block [PETRO ENG] The tendency of accumulated water-oil emulsion around the lower (producing) end of an oil well borehole to block the movement of formation fluids through the formation and toward the borehole.

water boiler *See* water-boiler reactor.

water-boiler reactor [NUCLEO] A homogeneous reactor that uses enriched uranium as fuel and ordinary water as moderator; the fuel is uranyl sulfate dissolved in water. Also known as water boiler.

waterborne [SCI TECH] Floating on or transported by water.

waterborne disease [MED] Disease transmitted by drinking water or by contact with potable or bathing water.

water brake [ENG] An absorption dynamometer for measuring power output of an engine shaft; the mechanical energy is converted to heat in a centrifugal pump, with a free casing where turning moment is measured.

water break [MET] A break in the continuity of the film of water on the surface of a metal withdrawn from an aqueous bath.

water budget *See* hydrologic accounting.

water bug [INV ZOO] Any insect which lives in an aquatic habitat during all phases of its life history.

water calorimeter [ENG] A calorimeter that measures radio-frequency power in terms of the rise in temperature of water in which the r-f energy is absorbed.

water clock [HOROL] An ancient device to estimate time; the operation depended upon the slow emptying of water from one graduated vessel into another, and the graduations marked the time periods.

water cloud [METEOROL] Any cloud composed entirely of liquid water drops; to be distinguished from an ice-crystal cloud and from a mixed cloud.

watercolor [MATER] A pigment ground in a solution of gum arabic, water, and plasticizer, such as glycerin; the glycerin film retards drying in the tube and prevents brittleness in the paint film.

watercolor paper [MATER] A special drawing paper with a surface texture suitable to accept watercolors; the better grades can withstand the harsh scraping that is sometimes necessary to produce highlights; for permanent painting, the paper should be 100% rag, not wood pulp.

watercolor pigment [INORG CHEM] A permanent pigment used in watercolor painting, for example, titanium oxide (white).

water conservation [ECOL] The protection, development, and efficient management of water resources for beneficial purposes.

water content [HYD] The liquid water present within a sample of snow (or soil) usually expressed in percent by weight; the water content in percent of water equivalent is 100 minus the quality of snow. Also known as free-water content; liquid-water content.

water-cooled condenser [MECH ENG] A steam condenser which is for the maintenance of vacuum, and in which water is the heat-receiving fluid.

water-cooled furnace [MECH ENG] A fuel-fired furnace containing tubes in which water is circulated to limit heat loss to the surroundings, control furnace temperature, and generate steam.

water-cooled reactor [NUCLEO] A nuclear reactor in which water is used as a primary coolant.

water-cooled tube [ELECTR] An electron tube that is cooled by circulating water through or around the anode structure.

water cooling [ELECTR] Cooling the electrodes of an electron tube by circulating water through or around them. [ENG] Cooling in which the primary coolant is water.

watercourse [HYD] **1.** A stream of water. **2.** A natural channel through which water may run or does run.

watercress [BOT] *Nasturtium officinale.* A perennial cress generally grown in flooded soil beds and used for salads and food garnishing.

water curb *See* garland.

water cycle *See* hydrologic cycle.

water demineralizing [CHEM ENG] The removal of minerals (for example, compounds of Ca, Mg, and Na) from water by chemical, ion-exchange, or distillation procedures.

water equivalent [METEOROL] The depth of water that would result from the melting of the snowpack or of a snow sample; thus, the water equivalent of a new snowfall is the same as the amount of precipitation represented by that snowfall.

water exchange [OCEANOGR] The volume and rate of water exchange between air and a body of water in a specific location, or between several bodies of water, controlled by such factors as tides, winds, river discharge, and currents.

waterfall [HYD] A perpendicular or nearly perpendicular descent of water in a stream.

waterflooding *See* flooding.

water-flow pyrheliometer [ENG] An absolute pyrheliometer, in which the radiation-sensing element is a blackened, water calorimeter; it consists of a cylinder, blackened on the interior, and surrounded by a special chamber through which water flows at a constant rate; the temperatures of the incoming and outgoing water, which are monitored continuously by thermometers, are used to compute the intensity of the radiation.

waterfowl [VERT ZOO] Aquatic birds which constitute the order Anseriformes, including the swans, ducks, geese, and screamers.

water front [GEOGR] An area partly bounded by water.

water gap [GEOL] A deep and narrow pass that cuts to the base of a mountain ridge, and through which a stream flows; the Delaware Water Gap is an example.

water garland *See* garland.

water-gas coke [MATER] Coke which is used in the manufacture of water gas, and which should have a low ash content, a softening temperature of about 2500°F (1370°C), a low sulfur content, and a size larger than 2 inches (5 centimeters).

water gas reaction [CHEM ENG] A method used to prepare carbon monoxide by passing steam over hot coke or coal at 600-1000°C.

water glass *See* sodium silicate.

water hammer [FL MECH] Pressure rise in a pipeline caused by a sudden change in the rate of flow or stoppage of flow in the line.

water heater [MECH ENG] A tank for heating and storing hot water for domestic use.

Waterhouse-Frederikson syndrome [MED] The association of bacteremia, particularly acute meningococcemia, massive skin hemorrhage, shock, and acute adrenal hemorrhage and insufficiency.

water influx [PETRO ENG] **1.** The incursion of water (natural or injected) into oil- or gas-bearing formations. **2.** One of the mechanisms of oil production in which the water movement (drive) displaces and moves the reservoir fluids toward the well borehole.

water jacket [ORD] The casing about the barrel of a water-cooled machine gun.

water knockout drum [PETRO ENG] A device for removal of water from oil well fluids (gas, or gas with oil). Also known as water knockout trap; water knockout vessel.

water knockout trap *See* water knockout drum.

water knockout vessel *See* water knockout drum.

water lane [NAV] A designated lane or strip of water marked and maintained for the takeoff and landing of seaplanes.

water leg [ENG] The vertical area of a vessel or accessory to a vessel for the collection of water. Also known as sump.

water level *See* water table.

waterline [GEOL] *See* shoreline. [HYD] *See* water table. [NAV ARCH] **1.** The intersection of the surface of the water with the side of a ship. **2.** A line painted on the hull of a ship showing the level of the water when the ship is properly trimmed.

water load [ELECTROMAG] A matched waveguide termination in which the electromagnetic energy is absorbed in water; the resulting rise in the temperature of the water is a measure of the output power.

water loss *See* evapotranspiration.

water main [CIV ENG] The water pipe in a street from which water is delivered to individual service pipes supplying domestic property.

watermark [GRAPHICS] A localized modification of the structure and opacity of a sheet of paper so that a pattern or design can be seen when the sheet is held to the light.

water mass [OCEANOGR] A body of water identified by its temperature-salinity curve or chemical composition, and normally consisting of a mixture of two or more water types.

watermelon [BOT] *Citrullus vulgaris.* An annual trailing vine with light-yellow flowers and leaves having five to seven deep lobes; the edible, oblong or roundish fruit has a smooth, hard, green rind filled with sweet, tender, juicy, pink to red tissue containing many seeds.

water meter [ENG] An instrument for measuring the amount of water passing a specified point in a piping system.

water microbiology [MICROBIO] An aspect of microbiology that deals with the normal and adventitious microflora of natural and artificial water bodies.

water mocassin [VERT ZOO] *Agkistrodon piscivorus.* A semi-aquatic venomous pit viper; skin is brownish or olive on the dorsal aspect, paler on the sides, and has indistinct black bars. Also known as cottonmouth.

water-moderated reactor [NUCLEO] A nuclear reactor in which water is the principal moderator.

Water Monster *See* Hydra.

water of hydration [CHEM] Water present in a definite amount and attached to a compound to form a hydrate; can be removed, as by heating, without altering the composition of the compound.

water opal *See* hyalite.

water opening [OCEANOGR] A break in sea ice, revealing the sea surface.

water paint [MATER] A paint in which the vehicle or binder is dissolved in water; examples are calcimine in which the vehicle is glue, and casein paints in which the vehicle is casein.

water plane [NAV ARCH] A plane coincident with or parallel to the surface of the water and limited by the line of its intersection with the vessel's hull.

water-plane area [NAV ARCH] The area of the water plane at which the ship floats.

water plane coefficient [NAV ARCH] The ratio of the area of the water plane of a ship at the surface of the water to the product of its beam and its length on the waterline.

water pollution [ECOL] Contamination of water by materials such as sewage effluent, chemicals, detergents, and fertilizer runoff.

waterpower [MECH] Power, usually electric, generated from an elevated water supply by the use of hydraulic turbines.

waterproof [ENG] Impervious to water.

waterproof grease [MATER] A viscous lubricating material that does not dissolve in water and that resists being washed out of bearings or gears; it usually has a low content of oil and metallic soaps of aluminum, barium, calcium, or strontium.

waterproofing agent [MATER] A substance used to make textiles, paper, wood, and other porous or absorbent materials impervious to penetration by water.

water-pump lubricant [MATER] A lubricating grease suitable for the types of automotive water pumps that require grease lubrication.

water purification [CIV ENG] Any of several processes in which undesirable impurities in water are removed or neutralized; for example, chlorination, filtration, primary treatment, ion exchange, and distillation.

water repellent [MATER] Chemicals used to treat textiles, leather, paper, or wood to make them resistant (but not proof) to wetting by water; includes various types of resins, aluminum of zirconium acetates, or latexes.

water requirement [HYD] The total quantity of water required to mature a specified crop under field conditions; includes applied irrigation, water precipitation, and groundwater available to the crop.

water rheostat *See* electrolytic rheostat.

water right [ENG] The right to use water for mining, agricultural, or other purposes.

water ring *See* garland.

water sample [ENG] A portion of water brought up from a depth to determine its composition.

water saturation [CHEM] **1.** A solid adsorbent that holds the maximum possible amount of water under specified conditions. **2.** A liquid solution in which additional water will cause the appearance of a second liquid phase. **3.** A gas that is at or just under its dew point because of its water content.

water scrubber [CHEM ENG] A device or system in which gases are contacted with water (either by spray or bubbling through) to wash out traces of water-soluble components of the gas stream.

water seal [ENG] A seal formed by water to prevent the passage of gas.

watershed [HYD] The drainage area of a stream.

water sky [METEOROL] The dark appearance of the underside of a cloud layer when it is over a surface of open water.

water smoke *See* steam fog.

Water Snake *See* Hydrus.

water snow [HYD] Snow that, when melted, yields a more than average amount of water; thus, any snow with a high water content.

water softening [CHEM] Removal of scale-forming calcium and magnesium ions from hard water, or replacing them by the more soluble sodium ions; can be done by chemicals or ion exchange.

waterspout [ENG] A pipe or orifice through which water is discharged or by which it is conveyed. [METEOROL] A tornado occurring over water; rarely, a lesser whirlwind over water, comparable in intensity to a dust devil over land.

water-supply engineering [CIV ENG] A branch of civil engineering concerned with the development of sources of supply, transmission, distribution, and treatment of water.

water swivel [DES ENG] A device connecting the water hose to the drill-rod string and designed to permit the drill string to be rotated in the borehole while water is pumped into it to create the circulation needed to cool the bit and remove the cuttings produced. Also known as gooseneck; swivel neck.

water table [HYD] The planar surface between the zone of saturation and the zone of aeration. Also known as free-water elevation; free-water surface; groundwater level; groundwater surface; groundwater table; level of saturation; phreatic surface; plane of saturation; saturated surface; water level; waterline.

water-thinned paint *See* water-base paint.

watertight subdivision [NAV ARCH] A part of a ship that can be sealed off so that water cannot enter it.

water tower [CIV ENG] A tower or standpipe for storing

WATER-TUBE BOILER

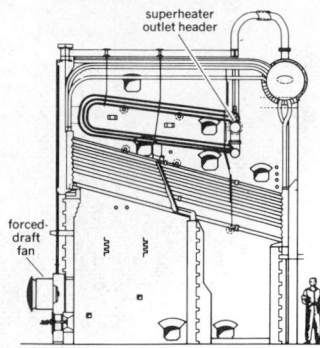

Diagram of straight-tube type of water-tube boiler.

WATT-HOUR METER

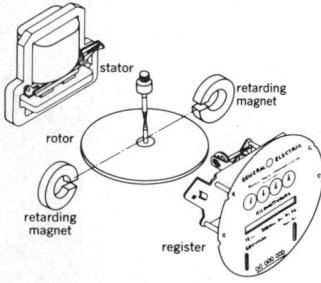

Schematic sketch showing basic elements of induction-type watt-hour meter. (General Electric Co.)

water in areas where ordinary water pressure is inadequate for distribution to consumers.

water treatment [CIV ENG] Purification of water to make it suitable for drinking or for any other use.

water-tube boiler [MECH ENG] A steam boiler in which water circulates within tubes and heat is applied from outside the tubes to generate steam.

water tunnel [AERO ENG] A device similar to a wind tunnel, but using water as the working fluid instead of air or other gas. [CIV ENG] A tunnel to transport water in a water-supply system.

water type [OCEANOGR] Ocean water of a specified temperature and salinity.

water vapor [PHYS] Water in the form of a vapor, especially when below the boiling point and diffused.

water-vapor absorption [METEOROL] The absorption of certain wavelengths of infrared radiation by atmospheric water vapor; a process of fundamental importance in the energy budget of the earth's atmosphere.

water-vapor laser [OPTICS] A laser whose active substance is water vapor, and which emits infrared radiation at wavelengths of 27.97, 47.7, 78.46, and 118.6 micrometers.

water-vascular system [INV ZOO] An internal closed system of reservoirs and ducts containing a watery fluid, found only in echinoderms.

waterway [CIV ENG] A channel for the escape or passage of water. [NAV] A navigable stream or canal.

water well [CIV ENG] A well sunk to extract water from a zone of saturation.

water-wettable [CHEM] Denoting the capability of a material to accept water, or of being hydrophilic or hydrophoric.

waterwheel [MECH ENG] A vertical wheel on a horizontal shaft that is made to revolve by the action or weight of water on or in containers attached to the rim.

water white [CHEM] A grade of color for liquids that has the appearance of clear water; for petroleum products, a plus 21 in the scale of the Saybolt chromometer.

water-white kerosine [MATER] Kerosine or refined oil from the crude still before it is treated or rerun; has the whitest (nearest to colorless) of the three standard kerosine colors, namely, water white, prime white, and standard white.

waterworks [CIV ENG] The whole system of supply and treatment utilized in acquisition and distribution of water to consumers.

water year [HYD] Any 12-month period, usually selected to begin and end during a relatively dry season, used as a basis for processing streamflow and other hydrologic data; the period from October 1 to September 30 is most widely used in the United States.

WATS See Wide Area Telephone Service.

Watson equation [PHYS CHEM] Calculation method to extend heat of vaporization data for organic compounds to within 10 or 15°C of the critical temperature; uses known latent heats of vaporization and reduced temperature data.

Watson factor See characterization factor.

Watson-Sommerfeld transformation [MATH] A procedure for transforming a series whose lth term is the product of the lth Legendre polynomial and a coefficient, a_l, having certain properties, into the sum of a contour integral of $a(l)$ and terms involving poles of $a(l)$, where $a(l)$ is a meromorphic function such that $a(l)$ equals a_l at integral values of l; used in studying rainbows, propagation of radio waves around the earth, scattering from various potentials, and scattering of elementary particles. Also known as Sommerfeld-Watson transformation.

watt [PHYS] The unit of power in the meter-kilogram-second system of units, equal to 1 joule per second. Abbreviated W.

wattage rating [ELEC] A rating expressing the maximum power that a device can safely handle continuously.

watt current See active current.

watt-hour [ELEC] A unit of energy used in electrical measurements, equal to the energy converted or consumed at a rate of 1 watt during a period of 1 hour, or to 3600 joules. Abbreviated whr.

watt-hour capacity [ELEC] Number of watt-hours which can be delivered from a storage battery under specified conditions as to temperature, rate of discharge, and final voltage.

watt-hour meter [ENG] A meter that measures and registers the integral, with respect to time, of the active power of the circuit in which it is connected; the unit of measurement is usually the kilowatt-hour.

wattle gum [MATER] A gum arabic extracted from the Australian and East African trees of the genus *Acacia*, as *A. dealbata*, and other species (called mimosa in Kenya); contains 65% tannin.

wattless current See reactive current.

wattless power See reactive power.

wattmeter [ENG] An instrument that measures electric power in watts ordinarily.

watt-second [PHYS] Amount of energy corresponding to 1 watt acting for 1 second; 1 watt-second is equal to 1 joule.

Watt's law [THERMO] A law which states that the sum of the latent heat of steam at any temperature of generation and the heat required to raise water from 0°C to that temperature is constant; it has been shown to be substantially in error.

wave [FL MECH] A disturbance which moves through or over the surface of a liquid, as of a sea. [PHYS] A disturbance which propagates from one point in a medium to other points without giving the medium as a whole any permanent displacement.

wave acoustics [ACOUS] The study of the propagation of sound based on its wave properties.

wave amplitude [PHYS] The magnitude of the greatest departure from equilibrium of the wave disturbance.

wave analyzer See harmonic analyzer.

wave angle [ELECTROMAG] The angle, either in bearing or elevation, at which a radio wave leaves a transmitting antenna or arrives at a receiving antenna.

wave antenna [ELECTROMAG] Directional antenna composed of a system of parallel, horizontal conductors, varying from a half to several wavelengths long, terminated to ground at the far end in its characteristic impedance.

wave base [HYD] The depth at which sediments are not stirred by wave action, usually about 10 meters. Also known as wave depth.

wave basin [GEOGR] A basin close to the inner entrance of a harbor in which the waves from the outer entrance are absorbed, thus reducing the size of the waves entering the inner harbor.

wave-built platform See wave-built terrace.

wave-built terrace [GEOL] A gently sloping coastal surface built up by sediment and loose material at the seaward or lakeward edge of a wave-cut platform. Also known as built terrace; wave-built platform.

wave celerity See phase velocity.

wave clutter See sea clutter.

wave converter [ELECTROMAG] Device for changing a wave of a given pattern into a wave of another pattern, for example, baffle-plate converters, grating converters, and sheath-reshaping converters for waveguides.

wave-corpuscle duality See wave-particle duality.

wave crest [PHYS] The position at which the disturbance of a progressive wave attains its maximum positive value.

wave-cut bench [GEOL] A level or nearly level narrow platform produced by wave erosion and extending outward from the base of a wave-cut cliff. Also known as beach platform; high-water platform.

wave-cut cliff [GEOL] A cliff formed by the erosive action of waves on rock.

wave-cut plain See wave-cut platform.

wave-cut platform [GEOL] A gently sloping surface which is produced by wave erosion and which extends into the sea for a considerable distance from the base of the wave-cut cliff. Also known as cut platform; erosion platform; strand flat; wave-cut plain; wave-cut terrace; wave platform.

wave-cut terrace See wave-cut platform.

wave cyclone [METEOROL] A cyclone which forms and moves along a front; the circulation about the cyclone center tends to produce a wavelike deformation of the front. Also known as wave depression.

wave delta See washover.

wave depression See wave cyclone.

wave depth See wave base.

wave disturbance [METEOROL] In synoptic meteorology, the same as wave cyclone, but usually denoting an early state in the development of a wave cyclone, or a poorly developed one.

wave duct [ELECTROMAG] 1. Waveguide, with tubular boundaries, capable of concentrating the propagation of waves within its boundaries. 2. Natural duct, formed in air by atmospheric conditions, through which waves of certain frequencies travel with more than average efficiency.

wave equation [PHYS] 1. In classical physics, a special equation governing waves that suffer no dissipative attenuation; it states that the second partial derivative with respect to time of the function characterizing the wave is equal to the square of the wave velocity times the Laplacian of this function. Also known as classical wave equation; d'Alembert's wave equation. 2. Any of several equations which relate the spatial and time dependence of a function characterizing some physical entity which can propagate as a wave, including quantum-wave equations for particles.

wave erosion *See* marine abrasion.

wave filter [ELEC] A transducer for separating waves on the basis of their frequency; it introduces relatively small insertion loss to waves in one or more frequency bands and relatively large insertion loss to waves of other frequencies.

wave forecasting [OCEANOGR] The theoretical determination of future wave characteristics based on observed or forecasted meteorological phenomena.

waveform [PHYS] The pictorial representation of the form or shape of a wave, obtained by plotting the displacement of the wave as a function of time, at a fixed point in space.

waveform-amplitude distortion *See* frequency distortion.

waveform analysis [PHYS] The determination of the amplitude and phase of the components of a complex waveform, either mathematically or by means of electronic instruments.

wavefront [PHYS] 1. A surface of constant phase. 2. The portion of a wave envelope that is between the beginning zero point and the point at which the wave reaches its crest value, as measured either in time or distance.

wavefront splitting [OPTICS] Any method of producing interference in which light from a single source is split into two parts which can then be recombined; examples include Young's two-slit experiment, the Fresnel double mirror, and the Fresnel biprism.

wave gage [ENG] A device for measuring the height and period of waves.

wave group [PHYS] A series of waves in which the wave direction, length, and height vary only slightly.

waveguide [ELECTROMAG] 1. Broadly, a device which constrains or guides the propagation of electromagnetic waves along a path defined by the physical construction of the waveguide; includes ducts, a pair of parallel wires, and a coaxial cable. Also known as guide; microwave waveguide. 2. More specifically, a metallic tube which can confine and guide the propagation of electromagnetic waves in the lengthwise direction of the tube.

waveguide assembly [ELECTROMAG] An item consisting of one or more definite lengths of straight or formed, flexible or rigid, prefabricated hollow tubing of conductive material; the tubing has a predetermined cross-section, and is designed to guide or conduct high-frequency electromagnetic energy through its interior; one or more ends are terminated.

waveguide attenuation [ELECTROMAG] The decrease from one point of a waveguide to another, in the power carried by an electromagnetic wave in the waveguide.

waveguide bend [ELECTROMAG] A section of waveguide in which the direction of the longitudinal axis is changed; an **E**-plane bend in a rectangular waveguide is bent along the narrow dimension, while an **H**-plane bend is bent along the wide dimension. Also known as waveguide elbow.

waveguide cavity [ELECTROMAG] A cavity resonator formed by enclosing a section of waveguide between a pair of waveguide windows which form shunt susceptances.

waveguide connector [ELECTROMAG] A mechanical device for electrically joining and locking together separable mating parts of a waveguide system. Also known as waveguide coupler.

waveguide coupler *See* waveguide connector.

waveguide critical dimension [ELECTROMAG] Dimension of

waveguide cross section which determines the cutoff frequency.

waveguide cutoff frequency [ELECTROMAG] Frequency limit of propagation along a waveguide for waves of a given field configuration.

waveguide discontinuity *See* discontinuity.

waveguide elbow *See* waveguide bend.

waveguide filter [ELECTROMAG] A filter made up of waveguide components, used to change the amplitude-frequency response characteristic of a waveguide system.

waveguide hybrid [ELECTROMAG] A waveguide circuit that has four arms so arranged that a signal entering through one arm will divide and emerge from the two adjacent arms, but will be unable to reach the opposite arm.

waveguide junction *See* junction.

waveguide plunger *See* piston.

waveguide probe *See* probe.

waveguide propagation [COMMUN] Long-range communications in the 10-kilohertz frequency range by the waveguide characteristics of the atmospheric duct formed by the ionospheric D layer and the surface of the earth.

waveguide resonator *See* cavity resonator.

waveguide shim [ELECTROMAG] Thin resilient metal sheet inserted between waveguide components to ensure electrical contact.

waveguide slot [ELECTROMAG] A slot in a waveguide wall, either for coupling with a coaxial cable or another waveguide, or to permit the insertion of a traveling probe for examination of standing waves.

waveguide switch [ELECTROMAG] A switch designed for mechanically positioning a waveguide section so as to couple it to one of several other sections in a waveguide system.

waveguide window *See* iris.

wave height [OCEANOGR] The height of a water-surface wave is generally taken as the height difference between the wave crest and the preceding trough. [PHYS] Twice the wave amplitude.

wave-height correction [NAV] A correction to a sextant altitude required because of the elevation of parts of the sea surface by wave action.

wave impedance [ELECTROMAG] The ratio, at every point in a specified plane of a waveguide, of the transverse component of the electric field to the transverse component of the magnetic field.

wave intensity [PHYS] The average amount of energy transported by a wave in the direction of wave propagation, per unit area per unit time.

wave interference *See* interference.

wavelength [PHYS] The distance between two points having the same phase in two consecutive cycles of a periodic wave, along a line in the direction of propagation.

wavelength shifter [ELECTR] A photofluorescent compound used with a scintillator material to increase the wavelengths of the optical photons emitted by the scintillator, thereby permitting more efficient use of the photons by the phototube or photocell.

wavelength standards [SPECT] Accurately measured lengths of waves emitted by specified light sources for the purpose of obtaining the wavelengths in other spectra by interpolating between the standards.

wave line *See* swash mark.

wavellite [MINERAL] $Al_3(PO_4)_2(OH)_3 \cdot 5H_2O$ A white to yellow, green, or black mineral crystallizing in the orthorhombic system and occurring in small hemispherical aggregates.

wavemark *See* swash mark.

wave mechanics *See* Schrödinger's wave mechanics.

wavemeter [ENG] A device for measuring the geometrical spacing between successive surfaces of equal phase in an electromagnetic wave.

wave microphone [ENG ACOUS] Any microphone whose directivity depends upon some type of wave interference, such as a line microphone or a reflector microphone.

wave motion [PHYS] The process by which a disturbance at one point is propagated to another point more remote from the source with no net transport of the material of the medium itself; examples include the motion of electromagnetic waves,

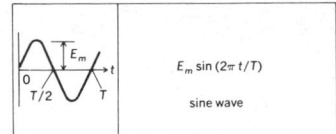
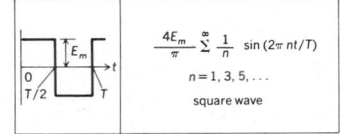
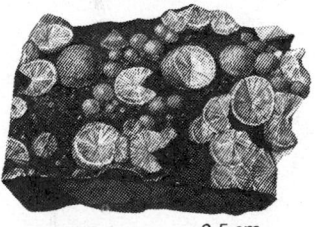
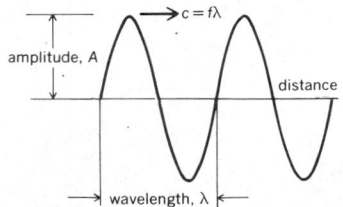

sound waves, hydrodynamic waves in liquids, and vibration waves in solids. Also known as propagation; wave propagation.

wave motor [MECH ENG] A motor that depends on the lifting power of sea waves to develop its usable energy.

wave normal [PHYS] **1.** A unit vector which is perpendicular to an equiphase surface of a wave, and has its positive direction on the same side of the surface as the direction of propagation. **2.** One of a family of curves which are everywhere perpendicular to the equiphase surfaces of a wave.

wave number [PHYS] The reciprocal of the wavelength of a wave, or sometimes 2π divided by the wavelength.

wave optics [OPTICS] The branch of optics which treats of light (or electromagnetic radiation in general) with explicit recognition of its wave nature.

wave packet [PHYS] In wave phenomena, a superposition of waves of differing lengths, so phased that the resultant amplitude is negligibly small except in a limited portion of space whose dimensions are the dimensions of the packet.

wave-particle duality [QUANT MECH] The principle that both matter and electromagnetic radiation exhibit phenomena in which they behave as waves and other phenomena in which they behave as particles, the two aspects being associated by the de Broglie relations. Also known as duality principle; wave-corpuscle duality.

wave period [PHYS] The time between the attainment of successive maxima, at a fixed point, of a quantity characterizing a wave.

wave plate [OPTICS] A plate of material which is linearly birefringent. Also known as retardation sheet.

wave platform See wave-cut platform.

wave polarization See polarization.

wave propagation See wave motion.

wave refraction [PHYS] The process by which the direction of a wave train moving in shallow water at an angle to the contours is changed.

wave ripple mark See oscillation ripple mark.

wave setdown [OCEANOGR] A decrease in the mean water level in the region in which breakers form near the seashore, caused by the presence of a pressure field.

wave setup [OCEANOGR] An increase in the mean water level shoreward of the region in which breakers form at the seashore, caused by the onshore flux of momentum against the beach.

wave shaper [ENG] Of explosives, an insert or core of inert material or of explosives having different detonation rates, used for changing the shape of the detonation wave.

wave-shaping circuit [ELECTR] An electronic circuit used to create or modify a specified time-varying electrical quantity, usually voltage or current, using combinations of electronic devices, such as vacuum tubes or transistors, and circuit elements, including resistors, capacitors, and inductors.

wave soldering See flow soldering.

wave speed See phase velocity.

wave tail [ELECTR] Part of a signal-wave envelope (in time or distance) between the steady-state value (or crest) and the end of the envelope.

wave theory of cyclones [METEOROL] A theory of cyclone development based upon the principle of wave formation on an interface between two fluids; in the atmosphere, a front is taken as such an interface.

wave theory of light [OPTICS] A theory which assumes that light is a wave motion, rather than a stream of particles.

wave tilt [ELECTROMAG] Forward inclination of a radio wave due to its proximity to ground.

wave train [PHYS] A series of waves produced by the same disturbance.

wave trap [CIV ENG] A device used to reduce the size of waves from sea or swell entering a harbor before they penetrate as far as the quayage; usually in the form of diverging breakwaters, or small projecting breakwaters situated close within the entrance. [ELECTR] A resonant circuit connected to the antenna system of a receiver to suppress signals at a particular frequency, such as that of a powerful local station that is interfering with reception of other stations. Also known as trap.

wave trough [PHYSICS] The lowest part of a wave form between successive wave crests.

wave vector [PHYS] A vector whose direction is the direction of phase propagation of a wave at each point in space, and whose magnitude is sometimes set at $2\pi/\lambda$ and sometimes at $1/\lambda$, where λ is the wavelength.

wave-vector space [SOLID STATE] The space of the wave vectors of the state functions of some system; this would be used, for example, for electron wave functions in a crystal and thermal vibrations of a lattice. Also known as k-space; reciprocal space.

wave velocity See phase velocity.

wavy extinction See undulatory extinction.

wax [MATER] Any of a group of substances resembling beeswax in appearance and character, and in general distinguished by their composition of esters and higher alcohols, and by their freedom from fatty acids.

wax-coating machine [GRAPHICS] A machine that applies a pressure-sensitive coating of wax to the backs of proofs, stats, photos, overlays, and other materials.

wax distillate [MATER] A neutral distillate from distillation of crude oil that contains a high percentage of crystallizable paraffin wax; used as a primary base for paraffin wax and neutral lubricating oils.

waxed paper [MATER] Paper that is treated or coated with wax to make it waterproof and greaseproof; used for wrapping.

wax fractionation [CHEM ENG] A continuous solvent-recovery/crystallization petroleum-refinery process for the production of waxes with low oil content from wax concentrates; for example, MEK (methyl ethyl ketone) deoiling.

wax gland See ceruminous gland.

waxing moon [ASTRON] The moon between new and full, when its visible part is increasing.

wax manufacturing [CHEM ENG] A petroleum refinery process similar to wax fractionation for the manufacture of oil-free waxes by chilling and crystallization from a solvent.

wax master See wax original.

wax original [ENG ACOUS] An original recording made on a wax surface and used to make a master. Also known as wax master.

wax stain [MATER] A semitransparent pigment mixed with beeswax, and thinned with turpentine.

wax tailings [MATER] A sticky, pitchlike substance with dark-brown color, the last volatile product distilling off an oil charge before it is coked; used as wood preservative and in manufacture of roofing paper. Also known as petroleum tailings; still wax.

waxy-electrolyte battery [ELEC] A primary battery in which the electrolyte is a waxy material, such as polyethylene glycol, in which is dissolved a small amount of a salt, such as zinc chloride; the electrodes are frequently made of zinc and manganese dioxide, and the electrolyte is melted and painted on a paper sheet to form the separator.

way point [NAV] A reference point between the point of departure and the destination, particularly a point on a course line the coordinates of which are defined in relation to an electronic aid to navigation.

ways [CIV ENG] **1.** The tracks and sliding timbers used in launching a vessel. **2.** The building slip or space upon which the sliding timbers or ways, supporting a vessel to be launched, travel. [MECH ENG] Bearing surfaces used to guide and support moving parts of machine tools; may be flat, V-shaped, or dovetailed.

W coefficient See Racah coefficient.

weak acid [CHEM] An acid that does not ionize greatly; for example, acetic acid or carbonic acid.

weak convergence [MATH] A sequence of elements $x_1, x_2, \ldots$ from a topological vector space X converges weakly if the sequence $f(x_1), f(x_2), \ldots$ converges for every continuous linear functional f on X.

weak coupling [PARTIC PHYS] The coupling of four fermion fields in the weak interaction, having a strength many orders of magnitude weaker than that of the strong or electromagnetic interactions.

weak fix [NAV] A fix determined from horizontal sextant angles between objects poorly located.

WAXY-ELECTROLYTE BATTERY

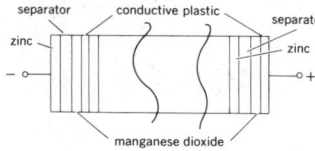

Diagram of battery stack of cells using a waxy electrolyte.

weak ground [MIN ENG] Roof and walls of underground excavations which would be in danger of collapse unless suitably supported.

weak interaction [PARTIC PHYS] One of the fundamental interactions among elementary particles, responsible for beta decay of nuclei, and for the decay of elementary particles with lifetimes greater than about 10^{-10} second, such as muons, K-mesons, and lambda hyperons; it is several orders of magnitude weaker than the strong and electromagnetic interactions, and fails to conserve strangeness or parity. Also known as beta interaction.

weak topology [MATH] A topology on a topological vector space X whose open neighborhoods around a point x are obtained from those points y of X for which every $f_i(x)$ is close to $f_i(y)$, f_i appearing in a finite list of linear functionals.

weapon [ORD] An instrument of combat, either offensive or defensive, used to destroy, injure, defeat, or threaten an enemy; for example, a gun, bayonet, bomb, or missile.

weapon delivery [ORD] The total action required to locate the target, establish the necessary release conditions, and maintain guidance to the target if required; it includes the detection recognition and the acquisition of the target, and the weapon release and weapon guidance.

weapon of mass destruction [ORD] Nuclear, bacteriological, or other weapon capable of causing widespread death or destruction.

weapon record book [ORD] Record book used to keep data on the performance, maintenance, and inspection of a gun or other weapon.

weapons system [ORD] Two or more instruments of combat operating as a single unit of striking power in military combat; specifically, a system in which two instruments of combat are required to perform a single mission.

weapon target line [ORD] An imaginary straight line from a weapon to a target.

wear [ENG] Deterioration of a surface due to material removal caused by relative motion between it and another part.

wearing course [CIV ENG] The top layer of surfacing on a road.

wear tables [ORD] Tables indicating the decrease of muzzle velocity expected as the result of firing a certain number of equivalent rounds; although tubes may vary considerably from the wear rate indicated in such tables, the tables may be used to correct calibration data between periods of calibration.

weasel [VERT ZOO] The common name for at least 12 species of small, slim carnivores which belong to the family Mustelidae and which have a reddish-brown coat with whitish underparts; species in the northern regions have white fur during the winter and are called ermine.

weather [METEOROL] 1. The state of the atmosphere, mainly with respect to its effects upon life and human activities; as distinguished from climate, weather consists of the short-term (minutes to months) variations of the atmosphere. 2. As used in the making of surface weather observations, a category of individual and combined atmospheric phenomena which must be drawn upon to describe the local atmospheric activity at the time of observation.

weather central [METEOROL] An organization which collects, collates, evaluates, and disseminates meteorological information in such a manner that it becomes a principal source of such information for a given area.

weathercocking [AERO ENG] The aerodynamic action causing alignment of the longitudinal axis of a rocket with the relative wind after launch. Also known as weather vaning.

weathercock stability See directional stability.

weather deck [NAV ARCH] 1. The uppermost deck of a ship. 2. Any deck that does not have overhead protection from the weather.

weathered crude [MATER] Crude petroleum that, owing to evaporation and other natural causes during storage and handling, has lost an appreciable quantity of its more volatile components.

weathered iceberg [OCEANOGR] An iceberg which is irregular in shape, due to an advanced stage of ablation; it may have overturned.

weathered layer [GEOPHYS] The zone of the earth which lies immediately below the surface and is characterized by low wave velocities.

weather forecast [METEOROL] A forecast of the future state of the atmosphere with specific reference to one or more associated weather elements.

weathering [GEOL] Physical disintegration and chemical decomposition of earthy and rocky materials on exposure to atmospheric agents, producing an in-place mantle of waste. Also known as clastation; demorphism.

weathering-potential index [GEOL] A measure of the susceptibility of a rock or mineral to weathering.

weather map [METEOROL] A chart portraying the state of the atmospheric circulation and weather at a particular time over a wide area; it is derived from a careful analysis of simultaneous weather observations made at many observing points in the area.

weather-map type See weather type.

weather minimum [METEOROL] The worst weather conditions under which aviation operations may be conducted under either visual or instrument flight rules; usually prescribed by directives and standing operating procedures in terms of minimum ceiling, visibility, or specific hazards to flight.

weather modification [METEOROL] The changing of natural weather phenomena by technical means; so far, only on the microscale of condensation and freezing nuclei has it been possible to exert modifying influences.

weather observation [METEOROL] An evaluation of one or more meteorological elements that describe the state of the atmosphere either at the earth's surface or aloft.

weather observation radar See weather radar.

weatherometer [ENG] A device used to subject articles and finishes to accelerated weathering conditions; for example, a rich ultraviolet source, water spray, or salt water.

weatherproof [ENG] Able to withstand exposure to weather without damage.

weather radar [ENG] Generally, any radar which is suitable or can be used for the detection of precipitation or clouds. Also known as weather observation radar.

weather shore [METEOROL] As observed from a vessel, the shore lying in the direction from which the wind is blowing.

weather side [METEOROL] The side of a ship exposed to the wind or weather.

weather signal [METEOROL] A visual signal displayed to indicate a weather forecast.

weather station [METEOROL] A place and facility for the observation, measurement, and recording and transmission of data of the variable elements of weather; one of the most effective network facilities is that of the U.S. Weather Bureau.

weather strip [BUILD] A piece of material, such as wood or rubber, applied to the joints of a window or door to stop drafts.

weather type [METEOROL] A series of generalized synoptic situations, usually presented in chart form; weather types are selected to represent typical pressure patterns, and were originally devised as a method for lengthening the effective time-range of forecasts. Also known as weather-map type.

weather vaning See weathercocking.

weave [TEXT] 1. To make cloth by interlacing strands of warp and filling threads. 2. A cloth made by weaving. 3. The pattern of a woven fabric.

web [ARCH] The portion of a ribbed vault between ribs. [CIV ENG] The vertical strip connecting the upper and lower flanges of a rail or girder. [GRAPHICS] The continuous length of paper formed when paper pulp moves through a papermaking machine; the web is then cut into sheets or wrapped onto rolls. [MATER] In a grain of propellant, the minimum thickness of the grain between any two adjacent surfaces. [MECH ENG] For twist drills and reamers, the central portion of the tool body that joins the loads. [MET] In forging, the thin section of metal remaining at the bottom of a depression or at the location of the punches. [TEXT] A fabric as it is being woven on a loom. [VERT ZOO] The membrane between digits in many birds and amphibians.

weber [ELECTROMAG] The unit of magnetic flux in the meter-kilogram-second system, equal to the magnetic flux which, linking a circuit of one turn, produces in it an electromotive

WEASEL

The common or European weasel (Mustela nivalis).

WEATHER MODIFICATION

(a)

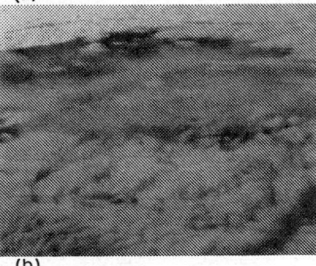

(b)

Example of cloud dissipation, a type of weather modification. (a) Three lines in a stratocumulus cloud layer 15 minutes after seeding. (b) Opening in stratocumulus layer 70 minutes after seeding. (U.S. Army ECOM, Fort Monmouth, N.J.)

force of 1 volt as it is reduced to zero at a uniform rate in 1 second.

Weber-Christian disease [MED] Febrile, relapsing, nodular nonsuppurative panniculitis.

Weber differential equation [MATH] A special case of the confluent hypergeometric equation that has as solution a confluent hypergeometric series.

Weberian apparatus [VERT ZOO] A series of bony ossicles which form a chain connecting the swim bladder with the inner ear in fishes of the superorder Ostariophysi.

Weberian ossicle [VERT ZOO] One of a chain of three or four small bones that make up the Weberian apparatus.

Weber number 1 [FL MECH] A dimensionless number used in the study of surface tension waves and bubble formation, equal to the product of the square of the velocity of the wave or the fluid velocity, the density of the fluid, and a characteristic length, divided by the surface tension. Symbolized N_{We_1}, We.

Weber number 2 [FL MECH] A dimensionless number, equal to the square root of Weber number 1. Symbolized N_{We_2}.

Weber number 3 [CHEM ENG] A dimensionless number used in interfacial area determination in distillation equipment, equal to the surface tension divided by the product of the liquid density, the acceleration of gravity, and the depth of liquid on the tray under consideration. Symbolized N_{We_3}.

Weber's law [PHYSIO] The law that the stimulus increment which can barely be detected (the just noticeable difference) is a constant fraction of the initial magnitude of the stimulus; this is only an approximate rule of thumb.

web-fed press [GRAPHICS] Printing press designed to accept paper from rolls instead of sheets; large web presses offer great speed and economy for long press runs, but high makeready and plate costs make them too costly for short runs.

web frame [NAV ARCH] A built-up member consisting of a web plate with single or double bars riveted or welded to its edges; web frames are placed several frame spaces apart with smaller frames in between.

websterite *See* aluminite.

Weddell Current [OCEANOGR] A surface current which flows in an easterly direction from the Weddell Sea outside the limit of the West Wind Drift.

Wedener-Bergeron process *See* Bergeron-Findeisen theory.

wedge [COMMUN] A convergent pattern of equally spaced black and white lines, used in a television test pattern to indicate resolution. [DES ENG] A piece of resistant material whose two major surfaces make an acute angle. [ELECTROMAG] A waveguide termination consisting of a tapered length of dissipative material introduced into the guide, such as carbon. [METEOROL] *See* ridge. [OPTICS] An optical filter in which the transmission decreases continuously or in steps from one end to the other.

wedge bit [DES ENG] A tapered-nose noncoring bit, used to ream out the borehole alongside the steel deflecting wedge in hole-deflection operations. Also known as bull-nose bit; wedge reaming bit; wedging bit.

wedge core lifter [MECH ENG] A core-gripping device consisting of a series of three or more serrated-face, tapered wedges contained in slotted and tapered recesses cut into the inner surface of a lifter case or sleeve; the case is threaded to the inner tube of a core barrel, and as the core enters the inner tube, it lifts the wedges up along the case taper; when the barrel is raised, the wedges are pulled tight, gripping the core.

wedge filter [NUCLEO] A radiation filter so constructed that its thickness or transmission characteristics vary continuously or in steps from one edge to the other; used to increase the uniformity of radiation in certain types of treatment.

wedge photometer [ENG] A photometer in which the luminous flux density of light from two sources is made equal by pushing into the beam from the brighter source a wedge of absorbing material; the wedge has a scale indicating how much it reduces the flux density, so that the luminous intensities of the sources may be compared.

wedge reaming bit *See* wedge bit.

wedge spectrograph [SPECT] A spectrograph in which the intensity of the radiation passing through the entrance slit is varied by moving an optical wedge.

wedging [ENG] **1.** A method used in quarrying to obtain

large, regular blocks of building stones; a row of holes is drilled, either by hand or by pneumatic drills, close to each other so that a longitudinal crevice is formed into which a gently sloping steel wedge is driven, and the block of stone can be detached without shattering. **2.** The act of changing the course of a borehole by using a deflecting wedge. **3.** The lodging of two or more wedge-shaped pieces of core inside a core barrel, and therefore blocking it. **4.** The material, moss, or wood used to render the shaft lining tight.

wedging bit *See* wedge bit.

weed [BOT] A plant that is useless or of low economic value, especially one growing on cultivated land to the detriment of the crop.

week [ASTRON] A time period of 7 days which has been accepted from ancient Babylon; the 7 days of the week were first given names of the seven celestial bodies: the sun, moon, and five visible planets.

weep hole [CIV ENG] A hole in a wood sill, retaining wall, or other structure to allow accumulated water to escape.

weeping spring *See* spring seepage.

weevil [INV ZOO] Any of various snout beetles whose larvae destroy crops by eating the interior of the fruit or grain, or bore through the bark into the pith of many trees.

weft *See* filling.

weft knitting [TEXT] A knitting process in which a continuous yarn is carried in crosswise rows.

Wegener's granulomatosis [MED] A rare disease of unknown causation characterized by necrotizing granulomas in the air passages, necrotizing vasculitis, and glomerulitis.

wehrlite [MINERAL] BiTe A mineral that is a native alloy of bismuth and tellurium. Also known as mirror glance. [PETR] A peridotite composed principally of olivine and clinopyroxene with accessory opaque oxides.

Weierstrass' approximation theorem [MATH] A continuous real valued function on a closed interval can be uniformly approximated by polynomials.

Weierstrass functions [MATH] Used in the calculus of variations, these determine functions satisfying the Euler-Lagrange equation and Jacobi's condition while maximizing a given definite integral.

Weierstrassian elliptic function [MATH] A function that plays a central role in the theory of elliptic functions; for z, g_2 and g_3 real or complex numbers, let y be that number such that

$$z = \int_y^\infty \frac{dt}{4t^3 - g_2t - g_3} ;$$

the Weierstrassian elliptic function of z with parameters g_2 and g_3 is $p\,(z; g_2, g_3) = y$.

Weierstrass M test [MATH] An infinite series of numbers will converge or functions will converge uniformly if each term is dominated in absolute value by a nonnegative constant M_n, where these M_n form a convergent series. Also known as Weierstrass' test for convergence.

Weierstrass' test for convergence *See* Weierstrass M test.

Weierstrass transform [MATH] This transform of a real function $f(y)$ is the function given by the integral from $-\infty$ to ∞ of $(4\pi t)^{-\frac{1}{2}}\exp[-(x-y)^2/4]\,f(y)dy$; this is used in studying the heat equation.

weighing rain gage [ENG] A type of recording rain gage, consisting of a receiver in the shape of a funnel which empties into a bucket mounted upon a weighing mechanism; the weight of the catch is recorded, on a clock-driven chart, as inches of precipitation; used at climatological stations.

weight [MECH] **1.** The gravitational force with which the earth attracts a body. **2.** By extension, the gravitational force with which a star, planet, or satellite attracts a nearby body.

weight and balance sheet [AERO ENG] A sheet which records the distribution of weight in an aircraft and shows the center of gravity of an aircraft at takeoff and landing.

weight barometer [ENG] A mercury barometer which measures atmospheric pressure by weighing the mercury in the column or the cistern.

weight density [PHYS] The weight of a body or portion of a body divided by its volume.

weighted area masks [ADP] In character recognition, a set of characters (each character residing in the character reader in

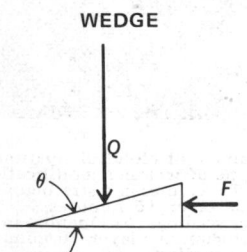

WEDGE

The shape of a wedge and a diagram of forces acting on it. F is smaller applied force, Q is larger force to be exerted, θ is angle between surfaces of wedge.

the form of weighted points) which theoretically render all input specimens unique, regardless of the size or style.

weighted average [STAT] The number obtained by adding the product of α_i times the ith number in a set of N numbers for $i = 1, 2, \ldots, N$, where α_i are numbers (weights) such that $\alpha_1 + \alpha_2 + \ldots + \alpha_N = 1$.

weight factor [STAT MECH] The number of microstates that correspond to a given macrostate.

weight function [MATH] Two real valued functions f and g are orthogonal relative to a weight function σ on an interval if the integral over the interval of $f \cdot g \cdot \sigma$ vanishes.

weighting [ENG] The artificial adjustment of measurements to account for factors that, in the normal use of the device, would otherwise be different from conditions during the measurements.

weighting network [ENG ACOUS] One of three or more circuits in a sound-level meter designed to adjust its response; the A and B weighting networks provide responses approximating the 40- and 70-phon equal loudness contours, respectively, and the C weighting network provides a flat response up to 8000 hertz.

weightlessness [MECH] A condition in which no acceleration, whether of gravity or other force, can be detected by an observer within the system in question. Also known as zero gravity.

weightlessness switch *See* zero gravity switch.

weight-loaded regulator [ENG] A pressure-regulator valve for pressure vessels or flow systems; the regulator is preloaded by counterbalancing weights to open (or close) at the upper (or lower) limit of a preset pressure range.

weight zone [ORD] A weight range having specified minimum and maximum weights; artillery projectiles of 75-millimeter caliber and larger are sometimes grouped into weight zones and marked with appropriate symbols; the selection of projectiles of a single weight zone for a specific firing problem results in improved ballistic uniformity.

Weil-Felix test [IMMUNOL] An agglutination test for various rickettsial infections based on production of nonspecific agglutinins in the blood of infected patients, and using various strains of *Proteus vulgaris* as antigen.

Weil's disease [MED] A severe form of leptospirosis characterized by jaundice, oliguria, circulatory collapse, and tendency to hemorrhage. Also known as icterohemorrhagic fever; leptospirosis icterohemorrhagia; spirochetal jaundice.

Weingarten formulas [MATH] Equations concerning the normals to a surface at a point.

weinschenkite [MINERAL] **1.** $YPO_4 \cdot 2H_2O$ A white mineral consisting of a hydrous yttrium phosphate. Also known as churchite. **2.** A dark-brown variety of hornblende high in ferric iron, aluminum, and water.

weir [CIV ENG] A dam in a waterway over which water flows, serving to regulate water level or measure flow.

weir tank [PETRO ENG] A type of oil-field storage tank with high- and low-level weir boxes and liquid-level controls for metering the liquid content of the tank.

Weissenberg method [SOLID STATE] A method of studying crystal structure by x-ray diffraction in which the crystal is rotated in a beam of x-rays, and a photographic film is moved parallel to the axis of rotation; the crystal is surrounded by a sleeve which has a slot that passes only diffraction spots from a single layer of the reciprocal lattice, permitting positive identification of each spot in the pattern.

Weiss magneton [ATOM PHYS] A unit of magnetic moment, equal to 1.853×10^{-21} erg/oersted, about one-fifth of the Bohr magneton; it is experimentally derived, the magnetic moments of certain molecules being close to integral multiples of this quantity.

Weiss molecular field [SOLID STATE] The effective magnetic field postulated in the Weiss theory of ferromagnetism, which acts on atomic magnetic moments within a domain, tending to align them, and is in turn generated by these magnetic moments.

Weiss theory [SOLID STATE] A theory of ferromagnetism based on the hypotheses that below the Curie point a ferromagnetic substance is composed of small, spontaneously magnetized regions called domains, and that each domain is spontaneously magnetized because a strong molecular magnetic field tends to align the individual atomic magnetic moments within the domain. Also known as molecular field theory.

Weisz ring oven [ANALY CHEM] A device for vaporization of solvent from filter paper, leaving the solute in a ring (circular) shape; used for qualitative analysis of very small samples.

Weizäcker-Williams method [QUANT MECH] A method of calculating the bremsstrahlung emitted when two particles, whose relative kinetic energies are much larger than their rest energies, collide; in the rest frame of one of the particles, the field of the other is equivalent to a set of virtual photons, and Compton scattering of these photons by the particle at rest is computed.

Weizsaecker's theory [ASTRON] A theory of the origin of the solar system; it hypothesizes primeval turbulent eddies which become permanent and self-gravitating; Weizsaecker does not discuss the origin of the gas clouds.

weld [MET] A union made between two metals by welding.

weldability [MET] Suitability of a metal to be welded under specified conditions.

weld bead [MET] A deposit of filler metal from a single welding pass. Also known as bead.

weldbonding [MET] A process for joining metals in which adhesive, typically an epoxy paste, is applied to the parts, which are then clamped together, spot-welded, and put into an oven (250°F, or 121°C, for 1 hour) to cure the adhesive.

weld crack [MET] A fracture in weld metal.

weld decay [MET] Intercrystalline corrosion of austenitic stainless steels near welded areas; caused by chromium carbide precipitation along grain boundaries of alloy subject to prolonged heating in the temperature range 400–850°C.

weld delay time [MET] Delay of the time current in spot, seam, or projection welding with respect to starting the forge delay timer used to synchronize pressure and heat.

welded tuff [PETR] A pyroclastic deposit hardened by the action of heat, pressure from overlying material, and hot gases. Also known as tuff lava.

welder [MET] **1.** A machine used in welding. Also known as welding machine. **2.** A person who performs a welding operation.

weld gage [ENG] A device used to check the shape and size of welds.

welding [GEOL] Consolidation of sediments by pressure; water is squeezed out and cohering particles are brought within the limits of mutual molecular attraction. [MET] Joining two metals by applying heat to melt and fuse them, with or without filler metal.

welding current [ELEC] The current that flows through a circuit while a weld is being made.

welding cycle [MET] The complete sequence of events involved in making a resistance weld.

welding electrode [MET] **1.** In arc welding, the current-carrying rod or rods used to strike an arc between rod and work. **2.** In resistance welding, the component of a machine through which current and pressure are applied to the work.

welding force *See* electrode force.

welding generator [ELEC] A generator used for supplying the welding current.

welding ground *See* work lead.

welding machine *See* welder.

welding rod [MET] Filler metal in the form of a rod or heavy wire.

welding schedule [MET] A record of all welding machine settings plus identification of the machine needed to produce a weld for a given material of a given size and finish.

welding sequence [MET] The order for welding component parts of a weldment or structure.

welding stress [MET] Residual stress resulting from localized heating and cooling during welding.

welding tip [ENG] A replaceable nozzle for a gas torch used in welding. [MET] An electrode used in spot or projection welding.

welding torch [ENG] A gas-mixing and burning tool for the welding of metal.

welding transformer [ELEC] A high-current, low-voltage power transformer used to supply current for welding.

weld interval [MET] The total heat and cool times for making one multiple-impulse weld.

WEIGHTLESSNESS

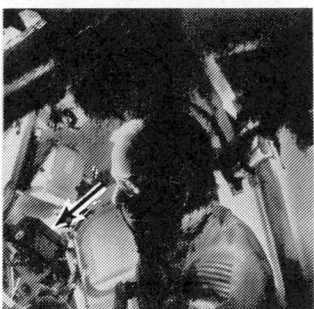

Film magazine (arrow) floats in weightless environment of space cabin. *(NASA)*

weld-interval timer [ENG] A device used to control weld interval.

weld line [MET] The junction of the weld metal and base metal, or the junction of base-metal parts when filler metal is not used.

weld mark *See* flow line.

weldment [ENG] An assembly or structure whose component parts are joined by welding.

weld metal [MET] The metal constituting the fused zone in spot, seam, or projection welding.

weld time [MET] The time that the welding current is applied to the work in single-impulse and flash welding.

Welge method [PETRO ENG] A method of calculation of the anticipated oil-recovery performance of a gas-cap-drive oil reservoir.

well [BUILD] An open shaft in a building, extending vertically through floors to accommodate stairs or an elevator. [ENG] A hole dug into the earth to reach a supply of water, oil, brine, or gas.

wellbore hydraulics [PETRO ENG] A branch of oil production engineering that deals with the motion of fluids (oil, gas, or water) in wellbore tubing or casing, or the annulus between tubing and casing.

well completion [PETRO ENG] The final sealing off of a drilled well (after drilling apparatus is removed from the borehole) with valving, safety, and flow-control devices.

well conditioning [PETRO ENG] 1. Preparation of a well for sampling procedures by control of production rate and associated pressure drawdown. 2. Removal of accumulated scale, wax, mud, and sand from the inner surfaces of a wellbore, or breakage of water blocks to increase production of oil or gas.

well core [ENG] A sample of rock penetrated in a well or other borehole obtained by use of a hollow bit that cuts a circular channel around a central column or core.

well-deck vessel [NAV ARCH] A merchant vessel having a sunken deck fitted between the forecastle and a long poop or continuous bridge house or raised quarterdeck.

well drill [MECH ENG] A drill, usually a churn drill, used to drill water wells.

wellhead [CIV ENG] The top of a well. [HYD] The place where a stream emerges from the ground.

wellhole [MIN ENG] 1. A large-diameter vertical hole used in quarries and opencast pits for taking heavy explosive charges in blasting. 2. The sump, or portion of a shaft below the place where skips are caged at the bottom of the shaft, in which water collects.

well injectivity [PETRO ENG] The ability of an injection well (water or gas) to receive injected fluid; can be negatively influenced by formation plugging, borehole scale, or liquid blocking around the lower end of the borehole.

well logging [ENG] The technique of analyzing and recording the character of a formation penetrated by a drill hole in mineral exploration and exploitation work.

well-ordered set [MATH] A linearly ordered set where every subset has a least element.

well-ordering principle [MATH] The proposition that every set can be endowed with an order so that it becomes a well-ordered set; this is equivalent to the axiom of choice.

well performance [PETRO ENG] The measurement of a well's production of oil or gas as related to the well's anticipated productive capacity, pressure drop, or flow rate.

wellpoint [CIV ENG] A component of a wellpoint system consisting of a perforated pipe about 4 feet (1.2 meters) long and about 2 inches (5 centimeters) in diameter, equipped with a ball valve, a screen, and a jetting tip.

wellpoint system [CIV ENG] A method of keeping an excavated area dry by intercepting the flow of groundwater with pipe wells located around the excavation area.

well shooting [ENG] The firing of a charge of nitroglycerin, or other high explosive, in the bottom of a well for the purpose of increasing the flow of water, oil, or gas.

well spacing [PETRO ENG] Areal location and interrelationship between producing oil or gas wells in an oil field; calculated for the maximum ultimate production from a given reservoir.

well stimulation *See* stimulation treatment.

well-type manometer [ENG] A type of double-leg, glass-tube manometer; one leg has a relatively small diameter, and the second leg is a reservoir; the level of the liquid in the reservoir does not change appreciably with change of pressure; mercury barometer is a common example.

Welwitschiales [BOT] An order of gymnosperms in the subdivision Geneticae represented by the single species *Welwitschia mirabilis* of southwestern Africa; distinguished by only two leaves and short, unbranched, cushion- or saucer-shaped woody main stem which tapers to a very long taproot.

Wenlockian [GEOL] A European stage of geologic time: Middle Silurian (above Tarannon, below Ludlovian).

Wentworth classification [GEOL] A logarithmic grade for size classification of sediment particles starting at 1 millimeter and using the ratio of $\frac{1}{2}$ in one direction (and 2 in the other), providing diameter limits to the size classes of 1, $\frac{1}{2}$, $\frac{1}{4}$, etc. and 1, 2, 4, etc.

Wentworth quick-return motion *See* turning-block linkage.

Wentworth scale [GEOL] A geometric grade scale for sedimentary particles ranging from clay particles (diameter less than $\frac{1}{250}$ millimeter) to boulders (diameters greater than 256 millimeters), in which the size classes are related to one another by a constant ratio of $\frac{1}{2}$ (4, 2, 1, $\frac{1}{2}$, etc.).

Wentzel-Kramers-Brillouin method [QUANT MECH] Method of approximating quantum-mechanical wave functions and energy levels, in which the logarithm of the wave function is expanded in powers of Planck's constant, and all except the first two terms are neglected. Also known as WKB method.

Werdnig-Hoffman disease [MED] Infantile spinal muscular atrophy.

Werfenian stage *See* Scythian stage.

Werner band [SPECT] A band in the ultraviolet spectrum of molecular hydrogen extending from 116 to 125 nanometers.

Werner complex *See* coordination compound.

wernerite *See* scapolite.

Werner's syndrome [MED] A complex of symptoms, thought to be inherited as an autosomal recessive, including premature senescence, dwarfism, cataracts, sclerodermalike changes of the skin, osteoporosis, and multiglandular dysfunction.

Wernicke's encephalopathy [MED] A disease due to thiamine deficiency, characterized by vomiting, ophthalmoplegia, ptosis, nystagmus, ataxia, weakness, dementia, and hemorrhaging of neurons around the third ventricle, cerebral aqueduct, and mammillary bodies.

Wertheim effect *See* Wiedemann effect.

west [GEOGR] The direction 90° to the left or 270° to the right of north.

West Australia Current [OCEANOGR] The complex current flowing northward along the west coast of Australia; it is strongest from November to January, and weakest and variable from May to July; it curves toward the west to join the South Equatorial Current.

westerlies [METEOROL] The dominant west-to-east motion of the atmosphere, centered over the middle latitudes of both hemispheres; at the earth's surface, the westerly belt (or westwind belt) extends, on the average, from about 35 to 65° latitude. Also known as circumpolar westerlies; middle-latitude westerlies; mid-latitude westerlies; polar westerlies; subpolar westerlies; subtropical westerlies; temperate westerlies; zonal westerlies; zonal winds.

westerly wave [METEOROL] An atmospheric wave disturbance embedded in the mid-latitude westerlies.

Western Equatorial Countercurrent [OCEANOGR] Weak, narrow bands of eastward-flowing water observed in some winter months in the western Atlantic near the equator.

Western equine encephalitis [MED] A type of equine encephalitis which occurs chiefly west of the Mississippi River; the chief vector is the culicine mosquito *Culex tarsalis*.

West Greenland Current [OCEANOGR] The current flowing northward along the west coast of Greenland into the Davis Strait; part of this current joins the Labrador Current, while the other part continues into Baffin Bay.

westing [NAV] The distance a craft makes good to the west.

West Nile fever [MED] An acute, usually mild, mosquito-borne virus disease occurring in summer, chiefly in Egypt, Israel, Africa, India, and Korea; signs are fever and lymphadenopathy, sometimes with a rash.

Weston standard cell [ELEC] A standard cell used as a highly

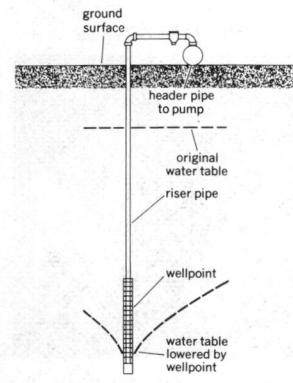
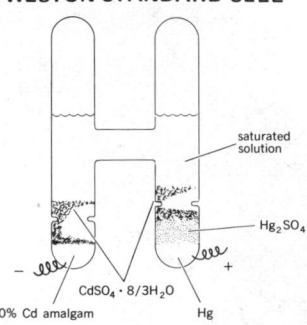

accurate voltage source for calibrating purposes; the positive electrode is mercury, the negative electrode is cadmium, and the electrolyte is a saturated cadmium sulfate solution; the Weston standard cell has a voltage of 1.018636 volts at 20°C.

Westphal balance [ENG] A direct-reading instrument for determining the densities of solids and liquids; a plummet of known mass and volume is immersed in the liquid whose density is to be measured or, alternatively, a sample of the solid whose density is to be measured is immersed in a liquid of known density, and the loss in weight is measured, using a balance with movable weights.

Westphal-Pilcz reflex   *See* pupillary reflex.

Westphal's pupillary reflex   *See* pupillary reflex.

west point [ASTRON] That point on the celestial sphere that is due west of observer; at this point the celestial equator crosses the horizon.

westward intensification [OCEANOGR] The intensification of ocean currents to the west, derived from a mathematical model that includes the effects of zonal wind stress at the sea surface and internal friction.

West Wind Drift   *See* Antarctic Circumpolar Current.

wet [PHYS] A liquid is said to wet a solid if the contact angle between the solid and the liquid, measured through the liquid, lies between 0 and 90°, and not to wet the solid if the contact angle lies between 90 and 180°.

wet adiabat   *See* saturation adiabat.

wet and dry bulb thermometer   *See* psychrometer.

wet ashing [ORG CHEM] The conversion of an organic compound into ash (decomposition) by treating the compound with nitric or sulfuric acid.

wet assay [MIN ENG] The determination of the quantity of a desired constituent in ores, metallurgical residues, and alloys by the use of the processes of solution, flotation, or other liquid means.

wet blasting [ENG] Shot firing in wet holes. [MET] Liquid honing in which an impeller wheel drives the liquid suspension.

wet-bulb depression [METEOROL] The difference in degrees between the dry-bulb temperature and the wet-bulb temperature.

wet-bulb temperature [METEOROL] **1.** Isobaric wet-bulb temperature, that is, the temperature an air parcel would have if cooled adiabatically to saturation at constant pressure by evaporation of water into it, all latent heat being supplied by the parcel. **2.** The temperature read from the wet-bulb thermometer; for practical purposes, the temperature so obtained is identified with the isobaric wet-bulb temperature.

wet-bulb thermometer [ENG] A thermometer having the bulb covered with a cloth, usually muslin or cambric, saturated with water.

wet cell [ELEC] A primary cell in which there is a substantial amount of free electrolyte in liquid form.

wet-cell caplight [MIN ENG] A rechargeable head lamp; the batteries are worn on the belt.

wet classifier [ENG] A device for the separation of solid particles in a mixture of solids and liquid into fractions, according to particle size or density by methods other than screening; operates by the difference in the settling rate between coarse and fine or heavy and light particles in a tank-confined liquid.

wet climate [CLIMATOL] A climate whose vegetation is of the rainforest type. Also known as rainforest climate.

wet collector   *See* scrubber.

wet contact [ELEC] Contact through which direct current flows.

wet cooling tower [MECH ENG] A structure in which water is cooled by atomization into a stream of air; heat is lost through evaporation. Also known as evaporative cooling tower.

wet corrosion [MET] Corrosion caused by exposure to aqueous solutions.

wet criticality [NUCLEO] Reactor criticality achieved with the coolant present.

wet drawing [MET] Drawing in which the dies and blocks are completely immersed in the lubricant.

wet drill [MECH ENG] A percussive drill with a water feed either through the machine or by means of a water swivel, to suppress the dust produced when drilling.

wet electrolytic capacitor [ELEC] An electrolytic capacitor employing a liquid electrolyte.

wet emplacement [AERO ENG] A launch emplacement that provides a deluge of water for cooling the flame bucket, the rocket engines, and other equipment during the launch of a missile.

wet engine [MECH ENG] An engine with its oil, liquid coolant (if any), and trapped fuel inside.

wet gas [MATER] Natural gas produced along with crude petroleum in oil fields or from gas-condensate fields; in addition to methane, it contains ethane, propane, butanes, and some higher hydrocarbons, such as pentane and hexane.

wet grinding [MECH ENG] **1.** The milling of materials in water or other liquid. **2.** The practice of applying a coolant to the work and the wheel to facilitate the grinding process.

wet hole [ENG] A borehole that traverses a water-bearing formation from which the flow of water is great enough to keep the hole almost full of water.

wet mill [MECH ENG] **1.** A grinder in which the solid material to be ground is mixed with liquid. **2.** A mill in which the grinding energy is developed by a fast-flowing liquid stream; for example, a jet pulverizer.

wet-reed relay [ELEC] Reed-type relay containing mercury at the relay contacts to reduce arcing and contact bounce.

wet season   *See* rainy season.

wet slip [CIV ENG] An opening between two wharves or piers where dock trials are usually conducted, and the final fitting out is done.

wet snow [METEOROL] Deposited snow that contains a great deal of liquid water.

wet spinning [TEXT] Producing synthetic and man-made filaments by extruding the chemical solution through spinnerets into a chemical bath where they coagulate.

wet stowage [ORD] Method of stowing major caliber ammunition in combat vehicles by placing it in racks surrounded by nonflammable liquid, to reduce ammunition fire hazards.

wet strength [MATER] **1.** The strength of a material saturated with water. **2.** The ability to withstand water (as for paper products) with a wet-strength additive or resin finish.

wet-strength paper [MATER] Paper with increased water resistance due to processing and interlocking of fibers, as well as impregnation with small amounts of resins (for example, melamine or urea formaldehyde). Also known as wet-strong paper.

wet-strong paper   *See* wet-strength paper.

wettability [CHEM] The ability of any solid surface to be wetted when in contact with a liquid; that is, the surface tension of the liquid is reduced so that the liquid spreads over the surface.

wet tabling [MIN ENG] A tabling process in which a pulp of two or more minerals flows across an inclined, riffled plane surface, is shaken longwise, and is water-washed crosswise.

wetted [CHEM] Pertaining to material that has accepted water or other liquid, either on its surface or within its pore structure.

wetted perimeter [GEOL] The portion of the perimeter of a steam channel cross section which is in contact with the water.

wetted surface [NAV ARCH] The surface of a ship's hull in contact with the water under specified conditions.

wetted-wall column [CHEM ENG] A vertical column that operates with the inner walls wetted by the liquid being processed; used in theoretical studies of mass transfer rates and in analytical distillations; an example is a spinning-band column.

wet-test meter [ENG] A device to measure gas flow by counting the revolutions of a shaft upon which water-sealed, gas-carrying cups of fixed capacity are mounted.

wetting [ELECTR] The coating of a contact surface with an adherent film of mercury.

wetting agent [GRAPHICS] A substance that renders a surface nonrepellent to a wetting liquid.

wetting phase [PETRO ENG] In a two-phase oil reservoir system (oil and water), one phase (water) will wet the pore surfaces of the reservoir formation, the other (oil) will not.

wet well [MECH ENG] A chamber which is used for collecting liquid, and to which the suction pipe of a pump is attached.

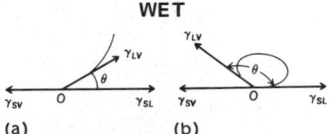

WET

Contact angle between the solid and the liquid is indicated by θ. *(a)* Liquid wets solid, *(b)* liquid does not wet solid. Arrows represent interfacial tensions γ_{SL}, γ_{SV}, and γ_{LV}, where S, L, and V refer to solid, liquid and vapor.

WHALE

Sperm whale (*Physeter catodon*), a species of whale having worldwide distribution. Head measures 20 feet (6 meters), or one-third of total body length.

WHEAT

(a) (b)   (c)   (d)   (e) (f)

Spikes of some representative varieties of wheat. (a) Wild and (b) cultivated forms of einkorn (*Triticum monococcum*); (c) wild emmer (*T. dicocciodes*); (d) emmer (*T. dicoccum*); (e) durum (*T. durum*); (f) poulard (*T. turgidum*). (Courtesy of H. Kihara)

WHEATSTONE BRIDGE

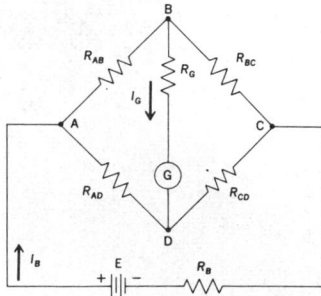

Circuit diagram of Wheatstone bridge, used to measure resistance $R_{\bar{C}D}$ in terms of known resistances R_{AB}, R_{BC}, and R_{AD}; the latter are adjusted until current I_G through detector G equals zero. R_G = internal resistance of detector. Current I_B flows through battery with open-circuit voltage E and internal resistance R_B.

wetwood [PL PATH] Wood having a water-soaked appearance because of a high water content; may be caused by bacteria or by physiological factors.

whale [VERT ZOO] A large marine mammal of the order Cetacea; the body is streamlined, the broad flat tail is used for propulsion, and the limbs are balancing structures.

Whale *See* Cetus.

whaleback dune [GEOL] A smooth, elongated mound or hill of desert sand shaped generally like a whale's back. Also known as sand levee.

whaleboat [NAV ARCH] **1.** A long, narrow rowboat that has a large sheer and both ends sharp and inclined to the perpendicular; formerly used to hunt whales. **2.** A long, narrow rowboat or motorboat that has both ends sharp and rounded in the manner of the original whaleboats, and that is equipped with buoyancy tanks; it is often carried on merchant ships and warships.

whalebone *See* baleen.

whale oil [MATER] A combustible, nontoxic, yellow-brown fixed oil obtained from whale blubber; soluble in alcohol, ether, chloroform, carbon disulfide, and benzene; used as a lubricant, illuminant, and leather dressing, and in soapmaking and fat manufacture. Also known as blubber oil.

wharf [CIV ENG] A structure of open construction built parallel to the shoreline; used by vessels to receive and discharge passengers and cargo.

Wharton's duct *See* submandibular duct.

wheat [BOT] A food grain crop of the genus *Triticum*; plants are self-pollinating; the inflorescence is a spike bearing sessile spikelets arranged alternately on a zigzag rachis.

wheat germ [BOT] The embryo of a wheat grain.

wheat germ oil [MATER] A light-yellow oil extracted from wheat germ; used as a source for vitamin E, as a dietary supplement, and in medicine.

wheat middlings *See* sharps.

Wheatstone bridge [ELEC] A four-arm bridge circuit, all arms of which are predominantly resistive; used to measure the electrical resistance of an unknown resistor by comparing it with a known standard resistance. Also known as resistance bridge; Wheatstone network.

Wheatstone network *See* Wheatstone bridge.

wheel [DES ENG] A circular frame with a hub at the center for attachment to an axle, about which it may revolve and bear a load.

wheelbarrow [ENG] A small, hand-pushed vehicle with a single wheel and axle between the front ends of two shafts that support a boxlike body and serve as handles at the rear. Also known as barrow.

wheel base [DES ENG] The distance in the direction of travel from front to rear wheels of a vehicle, measured between centers of ground contact under each wheel.

wheel-bearing lubricant [MATER] A lubricating grease with the character, structure, and consistency needed to make it suitable for use in antifriction wheel bearings.

wheeled crane [MECH ENG] A self-propelled crane that rides on a rubber-tired chassis with power for transportation provided by the same engine that is used for hoisting.

Wheeler-Feynman theory [RELAT] A relativistic action-at-a-distance theory in which it is assumed that there are enough absorbers in the universe to serve as sinks for all actions that emanate from any charged particle; radiation damping is a consequence of the theory.

wheelhouse *See* pilothouse.

wheel load capacity [CIVIL ENG] The capacity of airfield runways, taxiways, parking areas, or roadways to bear the pressures exerted by aircraft or vehicles in a gross weight static configuration.

wheel printer [ADP] A line printer that prints its characters from the rim of a wheel around which is the type for the alphabet, numerals, and other characters.

wheel sleeve [DES ENG] A flange used as an adapter on precision grinding machines where the hole in the wheel is larger than the machine arbor.

wheel static [ELECTR] Interference encountered in automobile-radio installations due to static electricity developed by friction between the tires and the street.

whelk [INV ZOO] A gastropod mollusk belonging to the order Neogastropoda; species are carnivorous but also scavenge.

whey [FOOD ENG] The watery part of milk separated from the curd in the process of making cheese.

whiffletree switch [ELECTR] In computers, a multiposition electronic switch composed of gate tubes and flip-flops, so named because its circuit diagram resembles a whiffletree.

whip antenna [ELECTROMAG] A flexible vertical rod antenna, used chiefly on vehicles. Also known as fishpole antenna.

whip grafting [BOT] A method of grafting by fitting a small tongue and notch cut in the base of the scion into corresponding cuts in the stock.

Whipple's disease [MED] A disease characterized by infiltration of the intestinal wall and lymphatics by macrophages filled with glycoprotein. Also known as intestinal lipodystrophy.

whippoorwill storm *See* frog storm.

whipworm disease [MED] A chronic, wasting diarrhea produced by heavy parasitization of the large intestine by the nematode *Trichuris trichiura*, particularly in undernourished children in the tropics.

whirlpool [OCEANOGR] Water in rapid rotary motion.

whirly [METEOROL] A small violent storm, a few yards (or meters) to 100 yards (91 meters) or more in diameter, frequent in Antarctica near the time of the equinoxes.

whiskers *See* catwhiskers; crystal whiskers; vibrissae.

whiskey [FOOD ENG] A potable alcoholic beverage made by distilling fermented grain mashes and aging the distillate in wood, usually oak; principal sources of grain are barley, wheat, rye, oats, and corn.

whistle buoy [NAV] A buoy equipped with a whistle; in the United States it is usually a conical buoy with a whistle located on its top.

whistler [GEOPHYS] An effect that occurs when a plasma disturbance, caused by a lightning discharge, travels out along lines of magnetic force of the earth's field and is reflected back to its origin from a magnetically conjugate point on the earth's surface; the disturbance may be picked up electromagnetically and converted directly to sound; the characteristic drawn-out descending pitch of the whistler is a dispersion effect due to the greater velocity of the higher-frequency components of the disturbance.

whistler wave *See* electron cyclotron wave.

whistling meteor [ELECTROMAG] Name applied to a radio meteor when a special system for detection is used in which the presence of the meteor is indicated by a rapidly changing audio-frequency radio signal.

White Alice *See* Alaska Integrated Communications Exchange.

white ant *See* termite.

white blood cell *See* leukocyte.

white body [PHYS] A hypothetical substance whose surface absorbs no electromagnetic radiation of any wavelength, that is, one which exhibits zero absorptivity for all wavelengths.

whitecap [OCEANOGR] A cloud of bubbles at the sea surface caused by a breaking wave.

white carbon black [MATER] A white silica powder made from silicon tetrachloride; used as a replacement for carbon black in rubber compounding.

white cast iron [MET] An extremely hard cast iron, rapidly cooled from the melt; contains about 3% carbon in the form of cementite and fine pearlite.

white cement [MATER] Pure white portland cement, made from raw materials with a low iron content, or by using a reducing flame to fire the clinker.

white clay *See* kaolin.

white coal *See* tasmanite.

white coat [BUILD] The finishing coat in plastering.

white compression [COMMUN] In facsimile or television the reduction in picture-signal gain at levels corresponding to light areas, with respect to the gain at the level for midrange light values; the overall effect of white compression is to reduce contrast in the highlights of the picture.

white copperas *See* zinc sulfate.

white corpuscle *See* leukocyte.

white cutch *See* gambir.

white damp [MIN ENG] In mining, carbon monoxide (CO); a gas that may be present in the afterdamp of a gas or coal-dust explosion, or in the gases given off by a mine fire; it is an important constituent of illuminating gas, supports combustion, and is very poisonous.

white diarrhea *See* pullorum disease.

white dwarf star [ASTRON] An intrinsically faint star of very small radius and high density; the mass is about 0.6 that of the sun and the average radius is about 8000 kilometers; it is one final stage of stellar evolution with thermonuclear energy sources extinct.

white feldspar *See* albite.

whitefish [VERT ZOO] Any of various food fishes in the family Salmonidae, especially of the genus *Coregonus*, characterized by an adipose dorsal fin and nearly toothless mouth.

white frost *See* hoarfrost.

white garnet *See* leucite.

whiteheart malleable iron [MET] White cast iron malleablized and decarburized by heat treatment in an oxidizing material at 900°C for 100–150 hours; decarburization produces a light-colored fracture, in contrast to blackheart malleable iron, which is not decarburized. Also known as blackheart malleable iron.

white infarct [MED] An infarct in which hemorrhage is slight, or that has been decolorized by removal of blood or its pigments.

white iron [MET] A brittle cast iron whose total carbon content is in the combined forms, and containing little or no graphite; a fresh fracture is white.

white iron ore *See* siderite.

white lead [INORG CHEM] Basic lead carbonate of variable composition, the oldest and most important lead paint pigment; also used in putty and ceramics.

white level [COMMUN] The carrier signal level corresponding to maximum picture brightness in television and facsimile.

white light [OPTICS] Any radiation producing the same color sensation as average noon sunlight.

white metal [MET] **1.** Any of several white-colored metals and their alloys of relatively low melting points, such as lead, tin, antimony, and zinc. **2.** A copper matte of about 77% copper, obtained from the smelting of sulfide copper ores.

white-metal bearing alloy *See* lead-base babbitt.

white mica *See* muscovite.

white mineral oil [MATER] A highly refined, colorless hydrocarbon oil with low volatility; used as a laxative and in medicine. Also known as liquid petrolatum; paraffinum liquidum.

white nickel *See* rammelsbergite.

whitening filter [ELECTR] An electrical filter which converts a given signal to white noise. Also known as prewhitening filter.

white noise [PHYS] Random noise that has a constant energy per unit bandwidth at every frequency in the range of interest.

white object [OPTICS] An object that reflects all wavelengths of light with substantially equal high efficiencies and with considerable diffusion.

white oil [MATER] Any of various highly refined, colorless hydrocarbon oils of low volatility and a wide range of viscosities; used for lubrication of food and textile machinery and as medicinal and mineral oils. Also known as technical white oil.

white olivine *See* forsterite.

whiteout [METEOROL] An atmospheric optical phenomenon of the polar regions in which the observer appears to be engulfed in a uniformly white glow: shadows, horizon, and clouds are not discernible; sense of depth and orientation are lost; dark objects in the field of view appear to float at an indeterminable distance. Also known as milky weather.

white phosphorus [CHEM] The element phosphorus in its allotropic form, a soft, waxy, poisonous solid melting at 44.5°C; soluble in carbon disulfide, insoluble in water and alcohol; self-igniting in air. Also known as yellow phosphorus.

white phosphorus grenade [ORD] Hand grenade or rifle grenade containing a main charge of white phosphorus and a small explosive burster charge for scattering the main charge; used for smoke and some incendiary effect.

white portland cement [MATER] Finely ground portland cement made from pure calcite limestone and white clay.

white potato *See* potato.

whiteprint [GRAPHICS] A print made by a plan copying process, produced on diazo paper or film, the emulsion containing a diazonium compound and a coupling or activating component; the process is based on sensitivity to ultraviolet light; development is by ammonia vapors, anhydrous ammonia gas, an alkaline solution which includes the coupler, or heat.

white rainbow *See* fogbow.

white schorl *See* albite.

white signal [COMMUN] Signal at any point in a facsimile system produced by the scanning of a minimum density area of the subject copy.

white squall [METEOROL] A sudden squall in tropical or subtropical waters, which lacks the usual squall cloud and whose approach is signaled only by the whiteness of a line of broken water or whitecaps.

white tellurium *See* sylvanite.

white-to-black amplitude range [COMMUN] **1.** In a facsimile system employing positive amplitude modulation, the ratio of signal voltage (or current) for picture white to the signal voltage (or current) for picture black at any point in the system. **2.** In a facsimile system employing negative amplitude modulation, the ratio of the signal voltage (or current) for picture black to the signal voltage (or current) for picture white; this ratio is often expressed in decibels.

white-to-black frequency swing [COMMUN] In a facsimile system employing frequency modulation, the numerical difference between the signal frequencies corresponding to picture white and picture black at any point in the system.

white transmission [COMMUN] **1.** In an amplitude-modulated system, that form of transmission in which the maximum transmitted power corresponds to the maximum density of the subject copy. **2.** In a frequency-modulation system, that form of transmission in which the lowest transmitted frequency corresponds to the minimum density of the subject copy.

white vitriol *See* zinc sulfate.

whitewash [MATER] A simple mixture of hydrated lime and water, used mostly for painting fences and outbuildings; common whitewash is not water-resistant and rubs off easily.

white water [OCEANOGR] Frothy water, as in whitecaps or breakers.

whiting *See* chalk.

whitleyite [GEOL] An achondritic stony meteorite consisting essentially of enstatite with fragments of black chondrite.

whitlockite [MINERAL] $Ca_9(Mg,Fe)H(PO_4)_7$ A rare mineral that forms hexagonal crystals.

Whittaker differential equation [MATH] A special form of Gauss' hypergeometric equation with solutions as special cases of the confluent hypergeometric series.

Whitworth screw thread [DES ENG] A British screw thread standardized to form and dimension.

whizzer mill *See* Jeffrey crusher.

whole-body counter [NUCLEO] A radiation counter that directly measures radioactivity in the entire human body.

whole gale [METEOROL] **1.** In storm-warning terminology, a wind of 48 to 63 knots (55 to 72 miles, or 89 to 133 kilometers, per hour). **2.** In the Beaufort wind scale, a wind whose speed is from 48 to 55 knots (55 to 63 miles, or 89 to 102 kilometers, per hour).

whole range point [AERO ENG] The point vertically below an aircraft at the moment of impact of a bomb released from that aircraft, assuming that the aircraft's velocity has remained unchanged.

whole step *See* whole tone.

whole tone [ACOUS] The interval between two sounds whose basic frequency ratio is approximately equal to the sixth root of 2. Also known as whole step.

whooping cough *See* pertussis.

whooping crane [VERT ZOO] *Grus americana.* A member of a rare North American migratory species of wading birds; the entire species forms a single population.

whorl [ANAT] A fingerprint pattern in which at least two deltas are present with a recurve in front of each. [BOT] An arrangement of several identical anatomical parts, such as petals, in a circle around the same point.

WHOOPING CRANE

Whooping cranes (*Grus americana*), a rare North American species in danger of extinction. (*Bureau of Sport Fisheries and Wildlife*)

WHORL

Plain whorl fingerprint pattern. (*Federal Bureau of Investigation*)

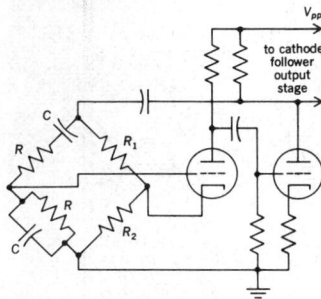

WIEN BRIDGE OSCILLATOR

Circuit diagram of Wien bridge oscillator. V_{pp} = plate supply voltage. Frequency of oscillation is $f_0 = 1/2\pi RC$.

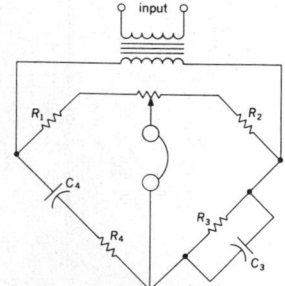

WIEN FREQUENCY BRIDGE

Circuit diagram of Wien frequency bridge. The frequency of the input can be determined from the resistances and capacitances when the latter are adjusted so that current through detector (the circles) is zero.

whr   See watt-hour.

Whytt's reflex   See pupillary reflex.

wiborgite   See rapakivi.

wichtisite   See tachylite.

wicket dam   [CIV ENG] A movable dam consisting of a number of rectangular panels of wood or iron hinged to a sill and propped vertically; the prop is hinged and can be tripped to drop the wickets flat on the sill.

wicking   [ENG] The flow of solder under the insulation of covered wire.

Widal test   [IMMUNOL] A macroscopic or microscopic agglutination test for the diagnosis of typhoid fever and other *Salmonella* infections by using killed or preserved bacteria as the antigen.

wide-angle lens   [OPTICS] An optical lens having a large angular field, generally greater than 80°.

wide area data service   [COMMUN] Automatic wide area teletypewriter data exchange service, using leased commercial lines.

Wide Area Telephone Service   [COMMUN] A special telephone service that allows a customer to call anyone in one or more of six regions into which the continental United States has been divided, on a direct dialing basis, for a flat monthly charge related to the number of regions to be called. Abbreviated WATS.

wide band   [ELECTR] Property of a tuner, amplifier, or other device that can pass a broad range of frequencies.

wide-band amplifier   [ELECTR] An amplifier that will pass a wide range of frequencies with substantially uniform amplification.

wide-band communications system   [COMMUN] Communications system which provides numerous channels of communications on a highly reliable and secure basis which are relatively invulnerable to interruption by natural phenomena or countermeasures; included are multichannel telephone cable, tropospheric scatter, and multichannel line-of-sight radio system such as microwave.

wide-band ratio   [COMMUN] Ratio of a system of the occupied frequency bandwidth to the intelligence bandwidth.

wide-band repeater   [ELECTR] Airborne system that receives a radio-frequency signal for transmission; used in reconnaissance missions when low-altitude reconnaissance aircraft require an airborne relay platform for beyond line-of-sight data transmission to a readout station.

wide-band switching   [ELECTR] Basically, four-wire circuits using correed matrices with electronic controls capable of switching wide-band facilities up to 50 kilohertz in bandwidth.

wide-band transformer   [ELEC] A transformer that can transfer electric energy from one circuit to another at any of a broad range of frequencies.

wide berth   [NAV] A vessel which keeps well away from another ship or navigational hazard is said to give the other ship or hazard a wide berth.

wide-open   [ELECTR] Refers to the untuned characteristic or lack of frequency selectivity.

width   [COMMUN] **1.** The horizontal dimension of a television or facsimile picture. **2.** The time duration of a pulse.

width coding   [COMMUN] Modifying the duration of the pulses emitted from the transponder according to a prearranged code for recognition in the display.

width control   [ELECTR] Control that adjusts the width of the pattern on the screen of a cathode-ray tube in a television receiver or oscilloscope.

Wiedemann effect   [ELECTROMAG] The twist produced in a current-carrying wire when placed in a longitudinal magnetic field. Also known as Wertheim effect.

Wiedemann-Franz law   [SOLID STATE] The law that the ratio of the thermal conductivity of a metal to its electrical conductivity is a constant, independent of the metal, times the absolute temperature. Also known as Lorentz relation.

Wiedemann's additivity law   [PHYS CHEM] The law that the mass (or specific) magnetic susceptibility of a mixture or solution of components is the sum of the proportionate (by weight fraction) susceptibilities of each component in the mixture.

Wien bridge oscillator   [ELECTR] A phase-shift feedback oscillator that uses a Wien bridge as the frequency-determining element.

Wien capacitance bridge   [ELEC] A four-arm alternating-current bridge used to measure capacitance in terms of resistance and frequency; two adjacent arms contain capacitors respectively in parallel and in series with resistors, while the other two arms are nonreactive resistors; bridge balance depends on frequency.

Wien constant   [STAT MECH] The product of the temperature and the wavelength at which the intensity of radiation from a blackbody reaches its maximum; it is equal to approximately 2898 micron-degrees.

Wien-DeSauty bridge   See DeSauty's bridge.

Wien effect   [PHYS CHEM] An increase in the conductance of an electrolyte at very high potential gradients.

Wiener experiment   [OPTICS] An experiment in which a front-faced mirror is covered with a thick photographic emulsion which is then exposed to light incident perpendicular to the surface; upon development, it is found that standing waves are set up in the emulsion whose nodes coincide with those of the electric vector, rather than those of the magnetic vector.

Wiener-Hopf equations   [MATH] Integral equations arising in the study of random walks and harmonic analysis.

Wiener-Khintchine theorem   [MATH] The theorem that determines the form of the correlation function of a given stationary stochastic process.

Wiener process   [MATH] A stochastic process with normal density at each stage, arising from the study of Brownian motion, which represents the limit of a sequence of experiments. Also known as Gaussian noise.

Wien frequency bridge   [ELEC] A modification of the Wien capacitance bridge, used to measure frequencies.

Wien inductance bridge   [ELEC] A four-arm alternating-current bridge used to measure inductance in terms of resistance and frequency; two adjacent arms contain inductors respectively in parallel and in series with resistors, while the other two arms are nonreactive resistors; bridge balance depends on frequency.

Wien-Maxwell bridge   See Maxwell bridge.

Wien's displacement law   [STAT MECH] A law for blackbody radiation which states that the wavelength at which the maximum amount of radiation occurs is a constant equal to approximately 2898 times the product of 1 micrometer and 1 kelvin. Also known as displacement law; Wien's radiation law.

Wien's distribution law   [STAT MECH] A formula for the spectral distribution of radiation from a blackbody, which is a good approximation to the Planck radiation formula at sufficiently low temperatures or wavelengths, for example, in the visible region of the spectrum below 3000K. Also known as Wien's radiation law.

Wien's radiation law   [STAT MECH] **1.** The law that the intensity of radiation emitted by a blackbody per unit wavelength, at that wavelength at which this intensity reaches a maximum, is proportional to the fifth power of the temperature. **2.** See Wien's displacement law. **3.** See Wien's distribution law.

Wierl equation   [ELECTR] A formula for the intensity of an electron beam scattered through a specified angle by diffraction from the molecules in a gas.

Wiese formula   [ENG] An empirical relationship for motor fuel antiknock values above 100 in relation to performance numbers; basis for the ASTM scale, in which octane numbers above 100 are related to increments of tetraethyllead added to isooctane.

Wiesen   See meadow.

wiggle stick   See divining rod.

Wigner coefficient   See vector coupling coefficient.

Wigner-Eckart theorem   [QUANT MECH] A theorem in the quantum theory of angular momentum which states that the matrix elements of a tensor operator can be factored into two quantities, the first of which is a vector-coupling coefficient, and the second of which contains the information about the physical properties of the particular states and operator, and is completely independent of the magnetic quantum numbers.

Wigner effect   See discomposition effect.

Wigner force   [NUCLEO] A short-range nonexchange force between nucleons, postulated to explain various phenomena.

Wigner nuclides [NUC PHYS] The most important class of mirror nuclides, comprising pairs of odd-mass-number isobars for which the atomic number and the neutron number differ by 1.

Wigner-Seitz cell [CRYSTAL] A polyhedron about an atom in a face-centered cubic structure, made by drawing planes which perpendicularly bisect the lines to the nearest neighbors; in a body-centered cubic structure, bisecting planes of lines to nearest neighbors and next-nearest neighbors are used; such polyhedra fill space.

Wigner-Seitz method [SOLID STATE] A method of approximating the band structure of a solid: Wigner-Seitz cells surrounding atoms in the solid are approximated by spheres, and band solutions of the Schrödinger equation for one electron are estimated by using the assumption that an electronic wave function is the product of a plane wave function and a function whose gradient has a vanishing radial component at the sphere's surface.

Wigner's theorem [QUANT MECH] **1.** The theorem that, if ψ is an eigenfunction of the Hamiltonian operator and R is a symmetry element of the Hamiltonian, then $R\psi$ is an eigenfunction of the Hamiltonian having the same eigenvalue as ψ. **2.** Angular momentum of the electron spin is conserved in a collision of the second kind.

Wigner supermultiplet [NUC PHYS] A set of quantum-mechanical states of a collection of nucleons which form the basis of a representation of SU(4), especially appropriate when spin and isospin dependence of the nuclear interaction may be disregarded; several combinations of spin and isospin multiplets may occur in a supermultiplet.

Wigner three-*j* symbol *See* three-*j* number.

Wijs' iodine monochloride solution [ANALY CHEM] A solution in glacial acetic acid of iodine monochloride; used to determine iodine numbers. Also known as Wijs' special solution.

Wijs' special solution *See* Wijs' iodine monochloride solution.

Wilcoxon one-sample test [STAT] A rank test for testing the hypothesis $\mu = \mu_H$ against the alternative $\mu > \mu_H$ under the assumption that observations are symmetrically distributed about μ_H; here μ_H is a given number and μ is the (unknown) mean of a random variable.

Wilcoxon paired comparison distribution [STAT] The distribution of the rank sum V_- (or V_+) of the negative differences (or positive differences) of observations in paired comparisons.

Wilcoxon paired comparison test [STAT] The test based upon the rank sum V_- (or V_+) of the negative differences (or positive differences) of observations in paired comparisons.

Wilcoxon two-sample distribution [STAT] The distribution of the Wilcoxon two-sample test statistic; it consists of the rank sums of treated subjects.

Wilcoxon two-sample test [STAT] The test based upon the rank sum of treated (or untreated) subjects.

wild boar [VERT ZOO] *Sus scrofa*. A wild hog with coarse, grizzled hair and enlarged tusks or canines on both jaws. Also known as boar.

wildcat [NAV ARCH] The drum of an anchor windlass, with projections on its rim that engage the anchor chain.

wildcat drilling [MIN ENG] The drilling of boreholes in unproved territory. Also known as cold nosing; wildcatting.

wildcatting *See* wildcat drilling.

wild cinnamon *See* bayberry.

Wild fence [ENG] A wooden enclosure about 16 feet square and 8 feet high with a precipitation gage in its center; the function of the fence is to minimize eddies around the gage, and thus ensure a catch which will be representative of the actual rainfall or snowfall.

wildfire [PL PATH] A bacterial disease of tobacco caused by *Pseudomonas tabaci* and characterized by the appearance of brown spots surrounded by yellow rings, which turn dark, rot, and fall out.

wildflysch [GEOL] A type of flysch facies that represents a stratigraphic unit with irregularly sorted boulders resulting from fragmentation, and twisted, confused beds resulting from slumping or sliding due to the influence of gravity.

wild gasoline [MATER] Unstabilized casinghead gasoline.

wildness [MET] A condition that exists when molten metal, during cooling, evolves so much gas that it becomes violently agitated, forcibly ejecting metal from its container.

wild shot [ORD] Artillery shot which is completely out of the normal pattern of dispersion; that is, shot whose impact is more than four probable errors, and more than six firing table probable errors from the center of impact.

wild snow [METEOROL] Newly deposited snow which is very fluffy and unstable; in general, it falls only during a dead calm at very low air temperatures.

Wilfley table [MIN ENG] A flat, rectangular surface that can be tilted and shaken about the long axis and has horizontal riffles for imposing restraint in removing minerals from classified sand. Also known as shaking table.

Willans line [MECH ENG] The line (nearly straight) on a graph showing steam consumption (pounds per hour) versus power output (kilowatt or horsepower) for a steam engine or turbine; frequently extended to show total fuel consumed (pounds per hour) for gas turbines, internal combustion engines, and complete power plants.

willemite [MINERAL] Zn_2SiO_4 A white, greenish-yellow, green, reddish, or brown mineral that forms rhombohedral crystals and exhibits intense bright-yellow fluorescence in ultraviolet light; a minor ore of zinc.

Williams-Hazen formula [FL MECH] In a liquid-flow system, a method for calculation of head loss due to the friction in a pipeline.

Williamsoniaceae [PALEOBOT] A family of extinct plants in the order Cycadeoidales distinguished by profuse branching.

Williamson synthesis [ORG CHEM] The synthesis of ethers utilizing an alkyl iodide and sodium alcoholate.

Williams refractometer [OPTICS] A refractometer in which light from a single slit is divided into two beams by a pentagonal prism.

Williams tube [ELECTR] A cathode-ray storage tube in which information is stored as a pattern of electric charges produced, maintained, read, and erased by suitably controlled scanning of the screen by the electron beam.

williwaw [METEOROL] A very violent squall in the Straits of Magellan; it may occur in any month but occurs most frequently in winter.

willow [BOT] A deciduous tree and shrub of the genus *Salix*, order Salicales; twigs are often yellow-green and bear alternate leaves which are characteristically long, narrow, and pointed, usually with fine teeth along the margins.

willy-willy [METEOROL] In Australia, a severe tropical cyclone.

Wilms' tumor [MED] A malignant renal tumor composed principally of mesodermal tissues. Also known as nephroblastoma.

Wilson cloud chamber [NUCLEO] A cloud chamber containing air supersaturated with water vapor by sudden expansion, in which rapidly moving nuclear particles such as alpha or beta rays produced ionization tracks by condensation of vapor on the ions produced by the rays.

Wilson electroscope [ELEC] An electroscope that has a single gold leaf which, when charged, is attracted to a grounded metal plate inclined at an angle that maximizes the instrument's sensitivity.

Wilson experiment [ELECTROMAG] An experiment that tests the validity of electromagnetic theory; a hollow cylinder of dielectric material, having layers of metal on its outer and inner cylindrical surfaces, is rotated about its axis in a magnetic field parallel to the axis; a sensitive electrometer, connected to the metal layers, indicates a charge that has the magnitude and sign predicted by theory.

Wilson's disease [MED] A hereditary disease of ceruloplasmin formation transmitted as an autosomal recessive and characterized by decreased serum ceruloplasmin and copper values, and increased excretion of copper in the urine. Also known as hepatolenticular degeneration.

Wilson's theorem [MATH] The number $(n-1)! + 1$ is divisible by n if and only if n is a prime.

wilt [PL PATH] Any of various plant diseases characterized by drooping and shriveling, following loss of turgidity.

Wimshurst machine [ELEC] An electrostatic generator consisting of two glass disks rotating in opposite directions,

WILLEMITE

troostite
franklinite
calcite
2.5 cm

Willemite from Sussex County, New Jersey. This crystal is the variety troostite, in which manganese replaces a considerable part of the zinc. (*Specimen from Department of Geology, Bryn Mawr College*)

WILLOW

Twig, terminal bud, and leaf of the Babylon weeping willow (*Salix babylonia*).

having sectors of tinfoil and collecting combs so arranged that static electricity is produced for charging Leyden jars or discharging across a gap.

winch [MECH ENG] A machine having a drum on which to coil a rope, cable, or chain for hauling, pulling, or hoisting.

winch operator *See* hoistman.

wind [ELECTR] The manner in which magnetic tape is wound onto a reel; in an A wind, the coated surface faces the hub; in a B wind, the coated surface faces away from the hub. [METEOROL] The motion of air relative to the earth's surface; usually means horizontal air motion, as distinguished from vertical motion, and air motion averaged over the response period of the particular anemometer.

windage [MECH] 1. The deflection of a bullet or other projectile due to wind. 2. The correction made for such deflection.

windage scale [ORD] Scale for adjusting a sight to allow for the effect of the wind on a bullet in flight. Also known as wind gage.

windage yaw [ORD] In aerial gunnery, the yaw produced by the relative motion of the gun and the air; the tangent of the windage yaw is the crosswind, or the component of the airspeed perpendicular to the axis of the bore, divided by the muzzle velocity (the initial velocity relative to the gun).

windbreak [ENG] Any device designed to obstruct wind flow and intended for protection against any ill effects of wind.

windburn [BOT] Injury to plant foliage, caused by strong, hot, dry winds. [MED] A superficial inflammation of the skin, analogous to sunburn, caused by exposure to wind, especially a hot dry wind, inducing a dilation of the surface blood vessels.

wind charger [ELEC] A wind-driven direct-current generator used for charging storage batteries.

wind chill [METEOROL] That part of the total cooling of a body caused by air motion.

wind-chill index [METEOROL] The cooling effect of any combination of temperature and wind, expressed as the loss of body heat in kilogram calories per hour per square meter of skin surface; it is only an approximation because of individual body variations in shape, size, and metabolic rate.

wind cone [ENG] A tapered fabric sleeve, shaped like a truncated cone and pivoted at its larger end on a standard, for the purpose of indicating wind direction; since the air enters the fixed end, the small end of the cone points away from the wind. Also known as wind sleeve; wind sock.

wind correction [ENG] Any adjustment which must be made to allow for the effect of wind; especially, the adjustments to correct for the effect on a projectile in flight, on sound received by sound ranging instruments, and on an aircraft flown by dead reckoning navigation.

wind crust [HYD] A type of snow crust, formed by the packing action of wind on previously deposited snow; wind crust may break locally but, unlike wind slab, does not constitute an avalanche hazard.

wind current [METEOROL] Generally, any of the quasi-permanent, large-scale wind systems of the atmosphere, for example, the westerlies, trade winds, equatorial easterlies, or polar easterlies.

wind-cut stone *See* ventifact.

wind deflection [MECH] Deflection caused by the influence of wind on the course of a projectile in flight.

wind direction [METEOROL] The direction from which wind blows.

wind direction indicator [ENG] A device to indicate the direction from which the wind blows; an example is a weather vane.

wind-direction shaft [METEOROL] A representational mark for wind direction on a synoptic chart, it is a straight line drawn directly upwind from the station circle; the wind arrow is completed by adding the wind-speed barbs and pennants to the outer end of the shaft.

wind divide [METEOROL] A semipermanent feature of the atmospheric circulation (usually a high-pressure ridge) on opposite sides of which the prevailing wind directions differ greatly.

wind drift [ACOUS] Shift in the apparent position of a sound source or target observed by sound apparatus; it is caused by

WINDING

Field winding (def. 1) on cylindrical rotor. *(National Electric Coil Co.)*

the effect of wind on sound waves, which changes their direction and increases or decreases sound lag. [OCEANOGR] *See* drift current.

wind-driven current *See* drift current.

wind erosion [GEOL] Detachment, transportation, and deposition of loose topsoil or sand by the action of wind.

wind factor [NAV] In air navigation, a measure of the net effect of wind on the ground speed of an aircraft; it is the magnitude of the wind vector component parallel to the heading of an aircraft, averaged over the entire flight positive if a tailwind, negative if a headwind.

wind-fire angle [ORD] The horizontal angle measured clockwise from the plane of fire to the direction from which the ballistic wind is blowing.

wind gage *See* windage scale.

wind gap [GEOL] A shallow, relatively high-level notch in the upper part of a mountain ridge, usually an abandoned water gap. Also known as air gap; wind valley.

wind-grooved stone *See* ventifact.

winding [ELEC] 1. One or more turns of wire forming a continuous coil for a transformer, relay, rotating machine, or other electric device. 2. A conductive path, usually of wire, that is inductively coupled to a magnetic storage core or cell.

winding number [MATH] The number of times a given closed curve winds in the counterclockwise direction about a designated point in the plane.

windlass *See* anchor windlass.

wind measurement [METEOROL] The determination of three parameters: the size of an air sample, its speed, and its direction of motion.

windmill [MECH ENG] Any of various mechanisms, such as a mill, pump, or electric generator, operated by the force of wind against vanes or sails radiating about a horizontal shaft.

windmill anemometer [ENG] A rotation anemometer in which the axis of rotation is horizontal; the instrument has either flat vanes (as in the air meter) or helicoidal vanes (as in the propeller anemometer); the relation between wind speed and angular rotation is almost linear.

wind noise [ACOUS] Noise caused by turbulent airflow over and around an object.

window [AERO ENG] An interval of time during which conditions are favorable for launching a spacecraft on a specific mission. [BUILD] An opening in the wall of a building or the body of a vehicle to admit light and usually to permit vision through a transparent or translucent material, usually glass. [ELECTR] A material having minimum absorption and minimum reflection of radiant energy, sealed into the vacuum envelope of a microwave or other electron tube to permit passage of the desired radiation through the envelope to the output device. [ELECTROMAG] A hole in a partition between two cavities or waveguides, used for coupling. [GEOL] A break caused by erosion of a thrust sheet or a large recumbent anticline that exposes the rocks beneath the thrust sheet. Also known as fenster. [GEOPHYS] Any range of wavelengths in the electromagnetic spectrum to which the atmosphere is transparent. [HYD] The unfrozen part of a river surrounded by river ice during the winter. [NUCLEO] 1. An aperture for the passage of particles or radiation in a nuclear reactor. 2. An energy range of relatively high transparency in the total neutron cross section of a material; such windows arise from interference between potential and resonance scattering in elements of intermediate atomic weight, and can be of importance in neutron shielding. [ORD] A confusion reflector consisting of strips of chaff, wire, or bars cut to give resonance at expected enemy radar frequencies, and dropped in clusters from aircraft or expelled from shells or rockets as a radar countermeasure.

window frost [HYD] A thin deposit of hoarfrost often found on interior surfaces of windows in winter, and frequently exhibiting beautiful fernlike patterns.

window ice [HYD] A thin deposit of ice which forms by the freezing of many tiny drops of water that have condensed on the indoors side of a cold window surface.

window rocket [ORD] A rocket filled with window that is to be expelled at a desired height.

wind-polished stone   *See* ventifact.

wind pressure   [MECH] The total force exerted upon a structure by wind. Also known as velocity pressure.

wind ripple   [METEOROL] One of a series of wavelike formations on a snow surface, an inch or so in height, at right angles to the direction of wind. Also known as snow ripple.

wind-rode   [NAV ARCH] A ship riding at anchor is said to be wind-rode when it is heading into the wind.

wind rose   [METEOROL] A diagram in which statistical information concerning direction and speed of the wind at a location may be summarized; a line segment is drawn in each of perhaps eight compass directions from a common origin; the length of a particular segment is proportional to the frequency with which winds blow from that direction; thicknesses of a segment indicate frequencies of occurrence of various classes of wind speed.

windrow   [GEOL] Any accumulation of material formed by wind or tide action.

winds aloft   [METEOROL] Generally, the wind speeds and directions at various levels in the atmosphere above the domain of surface weather observations, as determined by any method of winds-aloft observation. Also known as upper-level winds; upper winds.

winds-aloft observation   [METEOROL] The measurement and computation of wind speeds and directions at various levels above the surface of the earth.

wind scoop   [METEOROL] A saucerlike depression in the snow near obstructions such as trees, houses, and rocks, caused by the eddying action of the deflected wind.

wind-scoured stone   *See* ventifact.

wind-shaped stone   *See* ventifact.

wind shear   [METEOROL] The local variation of the wind vector or any of its components in a given direction.

wind shield   *See* rain gage shield.

windshield   [ENG] A transparent glass screen that protects the passengers and compartment of a vehicle from wind and rain.

wind-shift line   [METEOROL] A line or narrow zone along which there is an abrupt change of wind direction.

wind slab   [HYD] A type of snow crust; a patch of hard-packed snow, which is packed as it is deposited in favored spots by the wind, in contrast to wind crust.

wind sleeve   *See* wind cone.

wind sock   *See* wind cone.

wind speed   [METEOROL] The rate of motion of air.

wind star   [NAV] A method of solution for the speed and direction of the wind by observing drift on three different headings which form an approximate equilateral triangle.

windstorm   [METEOROL] A storm in which strong wind is the most prominent characteristic.

wind stress   [METEOROL] The drag or tangential force per unit area exerted on the surface of the earth by the adjacent layer of moving air.

wind tee   [ENG] A weather vane shaped like the letter T or like an airplane, situated on an airport or landing field to indicate the wind direction. Also known as landing tee.

wind triangle   [AERO ENG] A vector diagram showing the effect of the wind on the flight of an aircraft; it is composed of the wind direction and wind speed vector, the true heading and true airspeed vector, and the resultant track and ground speed vector.

wind tunnel   [ENG] A duct in which the effects of airflow past objects can be determined.

wind-tunnel balance   [ENG] A device or apparatus that measures the aerodynamic forces and moments acting upon a body tested in a wind tunnel.

wind-tunnel instrumentation   [ENG] Measuring devices used in wind-tunnel tests; in addition to conventional laboratory instruments for fluid flow, thermometry, and mechanical measurements, there are sensing devices capable of precision measurement in the small-scale environment of the test setup.

wind valley   *See* wind gap.

wind vane   [ENG] An instrument used to indicate wind direction, consisting basically of an asymmetrically shaped object mounted at its center of gravity about a vertical axis; the end which offers the greater resistance to the motion of air moves to the downwind position; the direction of the wind is determined by reference to an attached oriented compass rose.

wind velocity   [METEOROL] The speed and direction of wind.

windward   [METEOROL] In the general direction from which the wind blows.

wind wave   [OCEANOGR] A wave resulting from the action of wind on a water surface.

wine   [FOOD ENG] An alcoholic beverage made by fermentation of the juice of fruits or berries, especially grapes; classified on the basis of color, sweetness, alcoholic content, variety of grape, presence of carbon dioxide, and region where the grapes are grown.

wine lees   [FOOD ENG] Sediment or deposit that forms in the bottom of wine casks during the fermentation process; used as a source of tartaric acid and tartrates.

wing   [AERO ENG] **1.** A major airfoil. **2.** An airfoil on the side of an airplane's fuselage or cockpit, paired off by one on the other side, the two providing the principal lift for the airplane. [GEOL] *See* vesicle. [ZOO] Any of the paired appendages serving as organs of flight on many animals.

wing assembly   [AERO ENG] An aeronautical structure designed to maintain a guided missile in stable flight; it consists of all panels, sections, fastening devices, chords, spars, plumbing accessories, and electrical components necessary for a complete wing assembly.

wing axis   [AERO ENG] The locus of the aerodynamic centers of all the wing sections of an airplane.

wing dam   *See* groin.

winged headland   [GEOGR] A seacliff with two bays or spits, one on either side.

Winged Horse   *See* Pegasus.

winged missile   [ORD] A missile that has wings, distinguished from wingless missiles such as bullets, projectiles, and certain rockets.

wing gun   [ORD] A fixed gun mounted on the wing of an airplane.

wingless abutment   [CIV ENG] A straight-sided bridge abutment designed to resist pressure in back and provide a bridge seat.

wing loading   [AERO ENG] A measure of the load carried by an airplane wing per unit of wing area; commonly used units are pounds per square foot and kilograms per square meter.

wing nut   [DES ENG] An internally threaded fastener with wings to permit it to be tightened or loosened by finger pressure only. Also known as butterfly nut.

wing panel   [AERO ENG] That portion of a multipiece wing section that usually lies between the front and rear spars; it may be designed to include either the leading edge or the trailing edge as an integral part, but never both, and excludes control surfaces.

wing profile   [AERO ENG] The outline of a wing section.

wing rib   [AERO ENG] A chordwise member of the wing structure of an airplane, used to give the wing section its form and to transmit the load from the fabric to the spars.

wing section   *See* airfoil profile.

wing spot generator   [ELECTR] Electronic circuit that grows wings on the video target signal of a type G indicator; these wings are inversely proportional in size to the range.

wing structure   [AERO ENG] In an aircraft, the combination of outside fairing panels that provide the aerodynamic lifting surfaces and the inside supporting members that transmit the lifting force to the fuselage; the primary load-carrying portion of a wing is a box beam (the prime box) made up usually of two or more vertical webs, plus a major portion of the upper and lower skins of the wing, which serve as chords of the beam.

wing-tip rake   [AERO ENG] The shape of the wing when the tip edge is straight in plan but not parallel to the plane of symmetry; the amount of rake is measured by the acute angle between the straight portion of the wing tip and the plane of symmetry; the rake is positive when the trailing edge is longer than the leading edge.

Winkler titration   [ANALY CHEM] A chemical method for estimating the dissolved oxygen in seawater: manganous hydroxide is added to the sample and reacts with oxygen to produce a manganese compound which in the presence of acid

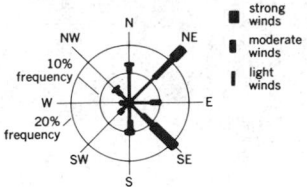
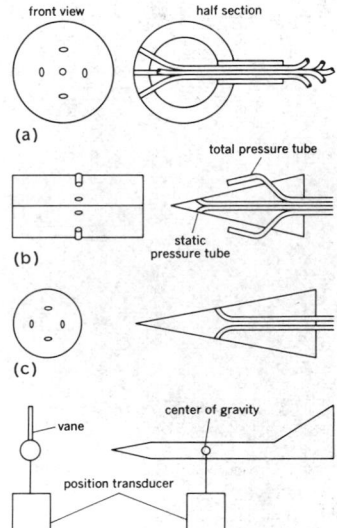
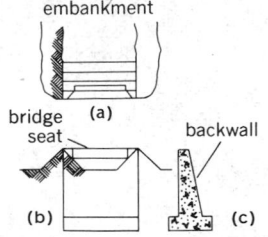

potassium iodide liberates an equivalent quantity of iodine that can be titrated with standard sodium thiosulfate.

winning [MIN ENG] **1.** A new mine opening. **2.** The portion of a coal field laid out for working. **3.** Mining.

winnowing gold [MIN ENG] Tossing up dry powdered auriferous material in air, and catching the heavier particles not blown away.

winter [ASTRON] The period from the winter solstice, about December 22, to the vernal equinox, about March 21; popularly and for most meteorological purposes, winter is taken to include December, January, and February in the Northern Hemisphere, and June, July, and August in the Southern Hemisphere.

Winteraceae [BOT] A family of dicotyledonous plants in the order Magnoliales distinguished by hypogynous flowers, exstipulate leaves, air vessels absent, and stamens usually laminar.

winter buoy [NAV] An unlighted buoy without sound signal which is maintained in certain areas during winter months when other aids are temporarily removed or extinguished.

wintergreen oil *See* methyl salicylate.

winter ice [OCEANOGR] Level sea ice more than 8 inches (20 centimeters) thick, and less than 1 year old; the stage which follows young ice.

winterization [ENG] The preparation of equipment for operation in conditions of winter weather; this applies to preparation not only for cold temperatures, but also for snow, ice, and strong winds.

winter load line [NAV ARCH] The waterline to which a vessel is allowed to load when going to sea in the wintertime.

winter marker [NAV] An unlighted marker without sound signal which is maintained in certain areas during the winter months when other aids are temporarily removed or extinguished.

winter solstice [ASTRON] **1.** The sun's position on the ecliptic (about December 22). Also known as first point of Capricorn. **2.** The date (December 22) when the greatest southern declination of the sun occurs.

winter-talus ridge [GEOL] A wall-like arcuate ridge on the floor of a cirque formed by freezing activity that dislodged boulders from a cirque wall covered with a snowbank. Also known as nivation ridge.

winze [MIN ENG] A vertical or inclined opening or excavation connecting two levels in a mine, differing from a raise only in construction; a winze is sunk underhand, and a raise is put up overhand.

wiped joint [MET] A joint wherein filler metal is applied in liquid form, and the joint is wiped mechanically to distribute the metal.

wiper [ELEC] That portion of the moving member of a selector, or other similar device, in communications practice, which makes contact with the terminals of a bank.

wiping [NAV] The process of reducing the amount of permanent magnetism in a naval vessel by placing a single coil horizontally around the vessel and moving it, while energized, up and down along the sides of the vessel.

wiping contact [ELEC] A switch or relay contact designed to move laterally with a wiping motion after it touches a mating contact. Also known as self-cleaning contact; sliding contact.

wiping effect [MET] Activation of a metal surface by mechanically rubbing or wiping to enhance the formation of a conversion coating.

wire [ELEC] A single bare or insulated metallic conductor having solid, stranded, or tinsel construction, designed to carry current in an electric circuit. Also known as electric wire. [MET] A thin, flexible, continuous length of metal, usually of circular cross section.

wire bonding [ELEC] Lead-covered tie used to connect two cable sheaths until a splice is permanently closed and covered.

wire cloth [DES ENG] Screen composed of wire crimped or woven into a pattern of squares or rectangles.

wire-cut brick [MATER] A brick cut from clay shaped by extrusion before burning; the long bar of extruded clay is cut into bricks by a set of wires 9 inches apart.

wire drag [ENG] An apparatus for surveying rocky underwater areas where normal sounding methods are insufficient to ensure the discovery of all existing submerged obstructions, small shoals, or rocks above a given depth or for determining the least depth of an area; it consists essentially of a buoyed wire towed at the desired depth by two launches.

wiredrawing [MET] Pulling a metal rod or wire through a die to reduce its cross section.

wire-fabric reinforcing [CIV ENG] Reinforcing concrete or mortar with a welded wire fabric.

wire facsimile system [COMMUN] A facsimile system in which messages are sent over wires or cables, rather than by radio.

wire fusing current [ELEC] The electric current which will cause a wire to melt.

wire gage [DES ENG] **1.** A gage for measuring the diameter of wire or thickness of sheet metal. **2.** A standard series of sizes arbitrarily indicated by numbers, to which the diameter of wire or the thickness of sheet metal is usually made, and which is used in describing the size or thickness.

wire glass [MATER] Sheet glass with woven wire mesh embedded in the center of the sheet; used in building construction for windows, doors, floors, and skylights.

wiregrating [ELECTROMAG] A series of wires placed in a waveguide that allow one or more types of waves to pass and block all others.

wire insulation [MATER] A flexible insulation used to cover an electric wire.

wire line [DES ENG] **1.** Any cable or rope made of steel wires twisted together to form the strands. **2.** A steel wire rope $\frac{5}{16}$ inch or less in diameter. [PETRO ENG] A line or cable used to lower and raise devices and gages in oil well boreholes; used for logging instruments and bottom-hole pressure gages.

wire-line coring [PETRO ENG] A method for obtaining samples of reservoir rocks during the drilling phase of oil wells.

wire-link telemetry [COMMUN] Telemetry in which electric signals are sent over transmission lines, rather than by radio. Also known as hard-wire telemetry.

wire mile [ELEC] Unit of measure of the length of two-conductor wire between two points; the length of the route multiplied by the number of circuits gives the number of wire miles.

wire nail [DES ENG] A nail made of wire and having a circular cross section.

wirephoto [COMMUN] **1.** A photograph transmitted over wires to a facsimile receiver. **2.** *See* facsimile.

wire printing *See* matrix printing.

wire recorder [ENG ACOUS] A magnetic recorder that utilizes a round stainless steel wire about 0.004 inch (0.01 centimeter) in diameter instead of magnetic tape.

wire recording [ENG ACOUS] Magnetic recording by use of a magnetized wire.

wire rod [MET] A metal rod used in wiredrawing.

wire rope [ENG] A rope formed of twisted strands of wire.

wire saw [MECH ENG] A machine employing one- or three-strand wire cable, up to 16,000 feet (4900 meters) long, running over a pulley as a belt; used in quarries to cut rock by abrasion.

wiresonde [ENG] An atmospheric sounding instrument which is supported by a captive balloon and used to obtain temperature and humidity data from the ground level to a height of a few kilometers; height is determined by means of a sensitive altimeter, or from the amount of cable released and the angle which the cable makes with the ground, and the information is telemetered to the ground through a wire cable.

wire stripper [ENG] A hand-operated tool or special machine designed to cut and remove the insulation for a predetermined distance from the end of an insulated wire, without damaging the solid or stranded wire inside.

wire tack [DES ENG] A tack made from wire stock.

wiretap [COMMUN] A secretly made and concealed connection to a telephone line, office intercommunication line, or other wiring system, for the purpose of monitoring conversations and activities in a room from a remote location without knowledge of the participants, legally or illegally.

wire telegraphy [COMMUN] Telegraphy in which messages are sent over wires or cables, rather than by radio.

wire train [ENG] An assembly that normally consists of an extruder, a crosshead and die, a means of cooling, and feed and take-up spools for the wire; used to coat wire with resin.

wireway [ENG] A trough which is lined with sheet metal and has hinged covers, designed to house electrical conductors or cables.

wire weight gage [ENG] A river gage in which a weight suspended on a wire is lowered to the water surface from a bridge or other overhead structure to measure the distance from a point of known elevation on the bridge to the water surface; the distance is usually measured by counting the number of revolutions of a drum required to lower the weight, and a counter is provided which reads the water stage directly.

wire-wound cryotron [CRYO] A cryotron that consists of a central insulated wire surrounded by a control coil; it is designed so that a relatively small current passed through the control coil produces a magnetic field which makes the gate resistive.

wire-wound potentiometer [ELEC] A potentiometer which is similar to a slide-wire potentiometer, except that the resistance wire is wound on a form and contact is made by a slider which moves along an edge from turn to turn.

wire-wound resistor [ELEC] A resistor employing as the resistance element a length of high-resistance wire or ribbon, usually Nichrome, wound on an insulating form.

wire-wound rheostat [ELEC] A rheostat in which a sliding or rolling contact moves over resistance wire that has been wound on an insulating core.

wire-wrap connection [ELEC] A solderless connection made by wrapping several turns of bare wire around a sharp-corner rectangular terminal under tension, using either a power tool or hand tool. Also known as solderless wrapped connection; wrapped connection.

wiring [ELEC] The installation and utilization of a system of wire for conduction of electricity. Also known as electric wiring. [SCI TECH] A system of wires.

wiring board *See* control panel.

wiring diagram *See* circuit diagram.

wiring harness [ELEC] An array of insulated conductors bound together by lacing cord, metal bands, or other binding, in an arrangement suitable for use only in specific equipment for which the harness was designed; it may include terminations.

Wirsung's duct [ANAT] The adult pancreatic duct in man, sheep, ganoid fish, teleost fish, and frog.

Wisconsin [GEOL] Pertaining to the fourth, and last, glacial stage of the Pleistocene epoch in North America; followed the Sangamon interglacial, beginning about 85,000 ± 15,000 years ago and ending 7000 years ago.

Wisconsin false blossom *See* false blossom.

wisper wind [METEOROL] A cold night wind, blowing out of the valley of the Wisper River in Germany during clear weather.

witches'-broom disease [PL PATH] An abnormal cluster of small branches or twigs that grow on a tree or shrub as a result of attack by fungi, viruses, dwarf mistletoes, or insect injury. Also known as hexenbesen; staghead.

witch hazel [MATER] A water extract from the dried leaves of the witch hazel shrub (*Hamamelis virginiana*); a solution of 14% alcohol with 1% witch hazel extract is commonly known as witch hazel; used as a tonic and sedative.

witch of Agnesi [MATH] The curve, symmetric about the *y*-axis and asymptotic in both directions to the *x*-axis, given by $x^2y = 4a^2(2a - y)$.

witherite [MINERAL] $BaCO_3$ A yellowish- or grayish-white mineral of the aragonite group that has orthorhombic symmetry, hardness of $3\frac{1}{4}$ on Mohs scale, and specific gravity 4.3.

withertip [PL PATH] A blighting of the terminal shoots or the tips of leaves associated with certain plant diseases, such as anthracnose of citrus plants.

Witte-Margules equation [OCEANOGR] A formula expressing the slope of the boundary layer between two water masses of different densities and velocities, taking into account the rotation of the earth. Also known as Margules equation.

Wittig ether rearrangement [ORG CHEM] The rearrangement of benzyl and alkyl ethers when reacted with a methylating agent, producing secondary and tertiary alcohols.

Witt theory [CHEM] A theory of the mechanism of dyeing stating that all colored organic compounds (called chromogens) contain certain unsaturated chromophoric groups which are responsible for the color, and if these compounds also contain certain auxochromic groups, they possess dyeing properties.

WKB method *See* Wentzel-Kramers-Brillouin method.

Wobbe index [THERMO] A measure of the amount of heat released by a gas burner with a constant orifice, equal to the gross calorific value of the gas in British thermal units per cubic foot at standard temperature and pressure divided by the square root of the specific gravity of the gas.

wobble wheel roller [MECH ENG] A roller with freely suspended pneumatic tires used in soil stabilization.

wobbulator [ELECTR] A signal generator in which a motor-driven variable capacitor is used to vary the output frequency periodically between two known limits, as required for displaying a frequency-response curve on the screen of a cathode-ray oscilloscope.

Wolbachiae [MICROBIO] A tribe of rickettsial organisms containing symbiotic forms that inhabit arthropods.

wolf [VERT ZOO] Any of several wild species of the genus *Canis* in the family Canidae which are fierce and rapacious, sometimes attacking man; includes the red wolf, gray wolf, and coyote.

Wolf 359 [ASTRON] A star of absolute magnitude 16.6; it is 7.7 light-years from the sun and is a variable flare star, which may emit bursts of light and even radio noise.

Wolfcampian [GEOL] A North American provincial series of geologic time; lowermost Permian (below Leonardian, above Virgilian of Pennsylvania).

Wolffian duct *See* mesonephric duct.

Wolf-Kishner reduction [ORG CHEM] Conversion of aldehydes and ketones to corresponding hydrocarbons by heating their semicarbazones, phenylhydrazones, and hydrazones with sodium ethoxide or by heating the carbonyl compound with excess sodium ethoxide and hydrazine sulfate.

Wolf number *See* relative sunspot number.

wolfram *See* tungsten; wolframite.

wolframic acid *See* tungstic acid.

wolframine *See* wolframite.

wolframite [MINERAL] $(Fe,Mn)WO_4$ A brownish- or grayish-black mineral occurring in short monoclinic, prismatic, bladed crystals; the most important ore of tungsten. Also known as tobacco jack; wolfram; wolframine.

Wolf-Rayet star [ASTRON] A member of a class of very hot stars (100,000–35,000K) which characteristically show broad bright emission lines in their spectra; luminosities are high, probably in the range 10^4–10^5 times that of the sun; these stars are probably very young and represent an early short-lived stage in stellar evolution.

wolfsbane *See* aconite.

Wolf-Wolfer number *See* relative sunspot number.

wollastonite [MINERAL] $CaSiO_3$ A white to gray inosilicate mineral (a pyroxenoid) that crystallizes in the triclinic system in tabular crystals and has a pearly or silky luster on the cleavages; hardness is 5–5.5 on Mohs scale, and specific gravity is 2.85. Also known as tabular spar.

Wollaston polarizing prism [OPTICS] A device for producing linearly polarized beams of light, consisting of two adjacent quartz wedges with their optic axes perpendicular to each other and to the direction of incident light.

wolverine [VERT ZOO] *Gulo gulo*. A carnivorous mammal which is the largest and most vicious member of the family Mustelidae.

wood [BOT] The hard fibrous substance that makes up the trunks and large branches of trees beneath the bark. [ECOL] A dense growth of trees, more extensive than a grove and smaller than a forest. [MATER] Lumber or timber obtained from trees.

wood alcohol *See* methyl alcohol.

wood block [GRAPHICS] A picture, design, or lettering carved on a block of wood with hand tools; the nonprinting areas of the block are carefully carved away, and only the raised areas of the design carry the ink; it differs from a wood engraving because the design is carved along the side of the block parallel with the grain, rather than perpendicular to it;

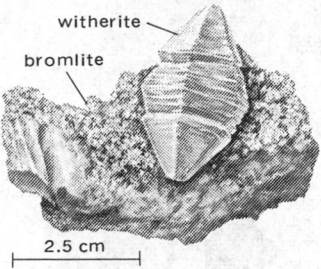

WITHERITE

witherite

bromlite

2.5 cm

Large crystal form of witherite with bromlite, from Fallowfield, Northumberland, England. (*Specimen from Department of Geology, Bryn Mawr College*)

WOLFRAMITE

4 cm

Crystal form of wolframite from Zinnwald, Czechoslovakia. (*American Museum of Natural History Specimen*)

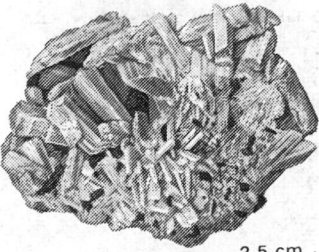

WOLLASTONITE

2.5 cm

Triclinic tabular crystals of wollastonite, embedded in limestone in this specimen, from Santa Fe, Chiapas, Mexico. (*Specimen from Department of Geology, Bryn Mawr College*)

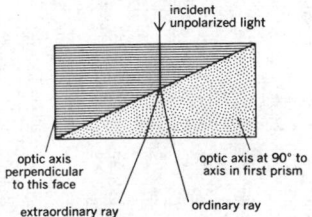

**WOLLASTON
POLARIZING PRISM**

incident
unpolarized light

optic axis
perpendicular
to this face

optic axis at 90° to
axis in first prism

extraordinary ray

ordinary ray

Diagram of Wollaston polarizing prism.

WOODPECKER

Red-headed woodpecker
(*Melanerpes erythrocephalus*),
a typical woodpecker with
clawed feet and stiff tail feathers.

WOODRUFF KEY

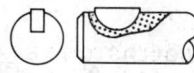

Diagrams of Woodruff key, end
and side views. The half disk
shape of the key is shown in the
side view. (*From P. H. Black and
O. E. Adams, Jr., Machine Design,
3d ed., 1968*)

it is therefore characterized by broad areas and strokes running with the grain. Also known as woodcut.

wood-block printing [GRAPHICS] Making prints from wood blocks by use of a press.

wood-carving tools [GRAPHICS] The tools normally used in wood carving; they consist of adzes, chisels, gouges, files, and rasps, all of which vary in size and shape.

wood coal *See* bituminous wood.

wood copper *See* olivenite.

woodcut *See* wood block.

Wood effect [OPTICS] Transparence of alkali metals to ultraviolet light.

wooden bomb [ORD] A concept which pictures a weapon as being completely reliable and having an infinite shelf life while at the same time requiring no special handling, storage, or surveillance.

wood engraving [GRAPHICS] An engraving carved on the edge grain of a piece of wood.

wood ether *See* dimethyl ether.

wood filler [MATER] A paste designed to fill the pores of open-grained woods such as ash, chestnut, mahogany, and oak.

wood flour [MET] Pulverized wood used in casting molds to furnish a reducing atmosphere, help overcome expansion of sand, increase flowability of the metal, and improve casting finish.

woodland *See* forest; temperate woodland.

woodpecker [VERT ZOO] A bird of the family Picidae characterized by stiff tail feathers and zygodactyl feet which enable them to cling to a tree trunk while drilling into the bark for insects.

wood physics [MATER] The area of wood science concerned with the physical and mechanical properties of wood and the factors which affect them.

wood preservative [MATER] A material used to coat wood to kill insects and fungi, but not usually classed as an insecticide; coal tar creosote and its derivatives are the most widely used wood preservatives.

wood pulp *See* pulp.

Woodruff key [DES ENG] A self-aligning machine key made by a side-milling cutter in the form of a segment of a disk.

wood screw [DES ENG] A threaded fastener with a pointed shank, a slotted or recessed head, and a sharp tapered thread of relatively coarse pitch for use only in wood.

Wood's metal [MET] A fusible alloy of the Cerro Corporation that contains 50% bismuth, 25% lead, 12.5% tin, and 12.5% cadmium, and melts at 158°F (70°C); used for automatic sprinkler plugs.

woodstave pipe [DES ENG] A pipe made of narrow strips of wood placed side by side and banded with wire, metal collars, and inserted joints, used largely for municipal water supply, outfall sewers, and mining irrigation.

woodstone *See* silicified wood.

wood sugar *See* xylose.

wood tin [MINERAL] A riniform, brownish variety of cassiterite with fibers radiating concentrically and resembling dry wood. Also known as dneprovskite.

Woodward-Hoffmann rule [ORG CHEM] A concept which can predict or explain the stereochemistry of certain types of reactions in organic chemistry; it is also described as the conservation of orbital symmetry.

woody lignite *See* bituminous wood.

woody structure [MET] A fibrous appearance in a fracture, particularly found in wrought iron and extruded aluminum alloys, usually associated with elongated inclusions or grains.

woof *See* filling.

woofer [ENG ACOUS] A large loudspeaker designed to reproduce low audio frequencies at relatively high power levels; usually used in combination with a crossover network and a high-frequency loudspeaker called a tweeter.

wool [TEXT] A textile fiber made from raw wool characterized by absorbency, resiliency, and insulation. [VERT ZOO] The soft undercoat of various animals such as sheep, angora, goat, camel, alpaca, llama, and vicuna.

wool fat *See* wool grease.

wool grease [MATER] A highly complex mixture of wax ester, alcohols, and fatty acids coating the surface of sheep

wool fibers and obtained by scouring the wool with soap or synthetic detergent; used in the manufacture of lanolin and its derivatives, for dressing leather, and in lubricating and slushing oils, soaps, and ointments. Formerly known as degras. Also known as wool fat; wool oil; wool wax.

wool-sorter's disease *See* anthrax.

wool wax *See* wool grease.

word [ADP] The fundamental unit of storage capacity for a digital computer, almost always considered to be more than eight bits in length. Also known as computer word.

word format [ADP] Arrangement of characters in a word, with each position or group of positions in the word containing certain specified data.

word length [ADP] The number of bits, digits, characters, or bytes in one word.

word mark [ADP] A nondata punctuation bit used to delimit a word in a variable-word-length computer.

word rate [ADP] In computer operations, the frequency derived from the elapsed period between the beginning of the transmission of one word and the beginning of the transmission of the next word.

words per minute [COMMUN] A measure of the speed with which messages can be transmitted by a telegraph system. Abbreviated WPM.

word time *See* minor cycle.

work [ELEC] *See* load. [MECH] The transference of energy that occurs when a force is applied to a body that is moving in such a way that the force has a component in the direction of the body's motion; it is equal to the line integral of the force over the path taken by the body.

workability [MATER] The ease with which concrete can be placed.

work angle [MET] In arc welding, the angle in a plane normal to the weld axis between the electrode and one member of the joint.

work cycle [IND ENG] A sequence of tasks, operations, and processes, or a pattern of manual motions, elements, and activities that is repeated for each unit of work.

worked-out [MIN ENG] Exhausted, referring to a coal seam or ore deposit.

worked penetration [ENG] Penetration of a sample of lubricating grease immediately after it has been brought to a specified temperature and subjected to strokes in a standard grease worker.

worker [INV ZOO] One of the neuter, usually sterile individuals making up a caste of social insects, such as ants, termites, or bees, which labor for the colony.

work function [SOLID STATE] The minimum energy needed to remove an electron from the Fermi level of a metal to infinity; usually expressed in electron volts. [THERMO] *See* free energy.

work hardening [MET] Increased hardness accompanying plastic deformation of a metal below the recrystallization temperature range.

working [COMMUN] Carrying on radio communication with a station by means of telegraphy, telephony, or facsimile for a purpose other than calling. [MIN ENG] **1.** The whole strata excavated in working a seam. **2.** Ground or rocks shifting under pressure and producing noise. [NAV] In sea ice navigation, making headway through an ice pack by boring, breaking, and slewing.

working load [ENG] The maximum load that any structural member is designed to support.

working place [MIN ENG] The place in a mine at which coal or ore is being actually mined.

working pressure [ENG] The allowable operating pressure in a pressurized vessel or conduit, usually calculated by ASME (American Society of Mechanical Engineers) or API (American Petroleum Institute) codes.

working Q *See* loaded Q.

working solution [GRAPHICS] A solution ready for use.

working space *See* working storage.

working storage [ADP] **1.** An area of main memory that is reserved by the programmer for storing temporary or intermediate values. Also known as working space. **2.** In COBOL (computer language), a section in the data division used for describing the name, structure, usage, and initial value of program variables that are neither constants nor records of input/output files.

working voltage   *See* voltage rating.

work-kinetic energy theorem   [MECH] The theorem that the change in the kinetic energy of a particle during a displacement is equal to the work done by the resultant force on the particle during this displacement.

work lead   [MET] The electrical conductor connecting the source of current to the work in arc welding. Also known as ground lead; welding ground.

Workman-Reynolds effect   [GEOPHYS] A mechanism for electric charge separation during freezing of slightly impure water; when a very dilute solution of certain salts freezes rapidly, a strong potential difference is established between the solid and liquid phases; for some salts, the ice attains negative charge, for others, positive; this mechanism has been suggested as one possible mode of thunderstorm charge separation in those portions of a thunderstorm downdraft where snow-pellet or hail particles sweep out supercooled waterdrops.

work print   [GRAPHICS] The first print of a motion picture; used for preliminary screening and editing.

work sampling   [IND ENG] A technique to measure work activity as related to delays consisting of intermittent observations of actual work and delays. Also known as activity sampling; frequency study; ratio delay study.

Workshop   *See* Sculptor.

work standardization   [IND ENG] The establishment of uniformity of working conditions, tools, equipment, technical procedures, administrative procedures, workplace arrangements, motion sequences, materials, quality requirements, and similar factors which affect the performance of work.

work task   [IND ENG] A specified amount of work, set of responsibilities, or occupation assigned to an individual or to a group.

world   [RELAT] Pertaining to Lorentz transformations and four-dimensional space-time, rather than rotations and three-dimensional space, as in world scalar, world vector, world line.

world calendar   [ASTRON] A proposed calendar in which the present 12 months are retained but the days are divided into four equal quarters; January, April, July, and October begin on Sunday and have 31 days, the other months have 30 days, so that there are 364 days with the 365th day following December 30 in no month; leap-year days would follow June 30.

world geographic reference system   [NAV] A geographic reference system used by the U.S. Air Force for aircraft position reports, target designations, and other tactical air operations. Abbreviated georef.

world rift system   [GEOL] The system of interconnected midocean ridges which is the locus of tensional splitting and magma upwelling believed responsible for sea-floor spreading.

worm   [DES ENG] A shank having at least one complete tooth (thread) around the pitch surface; the driver of a worm gear. [INV ZOO] **1.** The common name for members of the Annelida. **2.** Any of various elongated, naked, soft-bodied animals resembling an earthworm. [MET] Sweat of molten metal which exudes through the crust of solidifying metal in a casting, and is caused by gas evolution.

worm conveyor   *See* screw conveyor.

worm gear   [MECH ENG] A gear of a worm and a worm wheel; used to connect nonparallel, nonintersecting, perpendicular shafts.

wormseed oil   [MATER] An essential oil distilled from the seeds and leaf stems of the plant *Chenopodium anthelminticum*, which is grown in Maryland, and which contains the alkaloid ascoridole; used in worm treatment of animals. Also known as Baltimore oil.

worm wheel   [DES ENG] A gear wheel with curved teeth that meshes with a worm.

worobieffite   *See* vorobyevite.

worsted   [TEXT] **1.** Yarn spun from combed long-staple wool fibers. **2.** Fabric made from this yarn.

wort   [FOOD ENG] A clear infusion of malt grain extract, used by brewers for fermentation.

wound shock   *See* hypovolemic shock.

woven-screen storage   [ADP] Digital storage plane made by weaving wires coated with thin magnetic films; when currents are sent through a selected pair of wires that are at right angles in the screen, storage and readout occur at the intersection of the two wires.

wow   [ENG ACOUS] A low-frequency flutter; when caused by an off-center hole in a disk record, occurs once per revolution of the turntable.

WPM   *See* words per minute.

wrap   [GRAPHICS] In a book or magazine, an insert, usually on a different stock, which is not tipped in but wrapped around one of the signatures, and which usually consists of at least two leaves (four pages) securely stitched into the binding.

wrap-around hanger   [PETRO ENG] An oil-well tubing hanger made up of two hinged halves with a resilient sealing element between two steel mandrels.

wrap forming   *See* stretch forming.

wrapped connection   *See* wire-wrap connection.

Wratten filter   [OPTICS] A gelatin or glass filter designed to have specific light-transmission characteristics.

wrecking ball   *See* skull cracker.

wren   [VERT ZOO] Any of the various small brown singing birds in the family Troglodytidae; they are insectivorous and tend to inhabit dense, low vegetation.

wrench   [ENG] A manual or power tool with adapted or adjustable jaws or sockets either at the end or between the ends of a lever for holding or turning a bolt, pipe, or other object.

wrench fault   [GEOL] A lateral fault with a more or less vertical fault surface. Also known as basculating fault; torsion fault.

wrench-head bolt   [DES ENG] A bolt with a square or hexagonal head designed to be gripped between the jaws of a wrench.

Wright system   [PETRO ENG] A method for mining oil from partially drained sands that involves drilling a shaft through the productive strata, followed by long, slanting holes drilled radially in all directions from the shaft bottom into the oil sands.

wringing fit   [DES ENG] A fit of zero-to-negative allowance.

wrinkling   [MET] Waviness around the edges of a drawn metal product.

wrist   [ANAT] The part joining forearm and hand.

wrist breadth   [ANTHRO] A measurement from the outside projection of the distal part of the ulna to the radius at the wrist joint.

wrist thickness   [ANTHRO] The measurement transverse to the wrist breadth.

writable control storage   [ADP] A section of the control storage holding microprograms which can be loaded from a console file or under microprogramming control.

write   [ADP] **1.** To transmit data from any source onto an internal storage medium. **2.** A command directing that an output operation be performed.

write head   [ELECTR] Device that stores digital information as coded electrical pulses on a magnetic drum, disk, or tape.

writing speed   [ELECTR] Lineal scanning rate of the electron beam across the storage surface in writing information on a cathode-ray storage tube.

Wronskian   [MATH] An $n \times n$ matrix whose ith row is a list of the $(i-1)$st derivatives of a set of functions $f_1, \ldots, f_n$; ordinarily used to determine linear independence of solutions of linear homogeneous differential equations.

wrought alloy   [MET] An alloy that has been mechanically worked after casting.

wrought iron   [MET] A commercial iron consisting of slag fibers, primarily iron silicate, embedded in a ferrite matrix.

wryneck   *See* torticollis.

W stars   [ASTRON] Stars of the W spectral class; their spectra contain an abundance of highly ionized elements such as He, C, N, and O, and they are intensely hot with surface temperatures of about 50,000 to 100,000K.

Wuchereria   [INV ZOO] A genus of filarial worms parasitic in man in all worm regions of the world.

WRIGHT SYSTEM

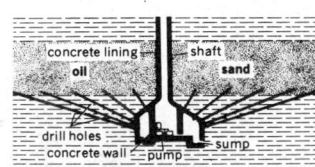

Sketch of Wright system for draining oil sands by a series of boreholes drilled from mine shaft. *(After L. C. Uren, Petroleum Production Engineering: Oil Field Exploitation, 3d ed., McGraw-Hill, 1953)*

wuchereriasis [MED] Infection with worms of the genus *Wuchereria*. Also known as Bancroft's filariasis.

wulfenite [MINERAL] $PbMoO_4$ A yellow, orange, orange-yellow, or orange-red tetragonal mineral occurring in tabular crystals or granular masses; an ore of molybdenum. Also known as yellow lead ore.

Wulff process [CHEM ENG] A chemical process to make acetylene and ethylene by cracking a hydrocarbon gas (for example, butane) with high-temperature steam in a regenerative furnace.

Wullenweber antenna [ELECTROMAG] An antenna array consisting of two concentric circles of masts, connected to be electronically steerable; used for ground-to-air communication at Strategic Air Command bases.

Würm [GEOL] **1.** A European stage of geologic time: uppermost Pleistocene (above Riss, below Holocene). **2.** Pertaining to the fourth glaciation of the Pleistocene epoch in the Alps, equivalent to the Wisconsin glaciation in North America, following the Riss-Würm interglacial.

W Ursae Majoris stars [ASTRON] Eclipsing variable stars whose brightness is continuously varying in periods of a few hours; they are composed of two close stars that have a common gaseous envelope.

Wurtz-Fittig reaction [ORG CHEM] A modified Wurtz reaction in which an aromatic halide reacts with an aklyl halide in the presence of sodium and an anhydrous solvent to form alkylated aromatic hydrocarbons.

wurtzilite [GEOL] A black, massive, sectile, infusible, asphaltic pyrobitumen derived from the metamorphosis of petroleum.

wurtzite [MINERAL] $(Zn,Fe)S$ A brownish-black hexagonal mineral consisting of zinc sulfide and occurring in hemimorphic pyramidal crystals, or in radiating needles and bundles.

Wurtz reaction [ORG CHEM] Synthesis of hydrocarbons by treating alkyl iodides in ethereal solution with sodium according to the reaction $2CH_3I + 2Na \rightarrow CH_3CH_3 + 2NaI$.

wustite [MINERAL] FeO An artificial mineral that consists of ferric oxide.

W Virginis stars [ASTRON] Periodic variable stars with periods of about 10 to 30 days; they exhibit two surges of activity from the same star so that there is a doubling of their spectral lines. Also known as Type II Cepheids.

WWV [COMMUN] The call letters of a radio station maintained by the National Bureau of Standards to provide standard radio and audio frequencies and other technical services, such as precision time signals and radio propagation disturbance warnings; the station broadcasts on 2.5, 5, 10, 15, 20, 25, 30, and 35 megahertz at various times.

WWVH [COMMUN] The National Bureau of Standards radio station at Maui, Hawaii, broadcasting services similar to those of WWV on 5, 10, and 15 megahertz.

wye [ELEC] Polyphase circuit whose phase differences are 120° and which when drawn resembles the letter Y. [ENG] A pipe branching off a straight main run at an angle of 45°. Also known as Y; yoke.

wye level *See* Y level.

Wynyardiidae [PALEON] An extinct family of herbivorous marsupial mammals in the order Diprotodonta.

xanthan gum [ORG CHEM] A high-molecular-weight (5–10 million) water-soluble natural gum; a heteropolysaccharide made up of building blocks of D-glucose, D-mannose, and D-glucuronic acid residues; produced by pure culture fermentation of glucose with *Xanthomonas campestris*.

xanthate [ORG CHEM] A water-soluble salt of xanthic acid, usually potassium or sodium; used as an ore-flotation collector.

xanthelasma [MED] Raised yellow plaques occurring around the eyelids, resulting from lipid-filled cells in the dermis.

xanthene [ORG CHEM] $CH_2(C_6H_4)_2O$ Yellowish crystals that are soluble in ether, slightly soluble in water and alcohol; melts at 100°C; used as a fungicide and chemical intermediate. Also known as tricyclic dibenzopyran.

xanthene dye [ORG CHEM] Any of a family of dyes related to the xanthenes; the chromophore groups are (C_6H_4).

xanthene ketone *See* xanthone.

Xanthidae [INV ZOO] The mud crabs, a family of decapod crustaceans in the section Brachyura.

xanthine [ORG CHEM] $C_5H_4N_4O_2$ A toxic yellow-white purine base that is found in blood and urine, and occasionally in plants; it is a powder, insoluble in water and acids, soluble in caustic soda; sublimes when heated; used in medicine and as a chemical intermediate. Also known as dioxopurine.

xanthine oxidase [BIOCHEM] A flavoprotein enzyme catalyzing the oxidation of certain purines.

xanthochroite *See* greenockite.

xanthoma [MED] A yellowish mass of lipid-filled histocytes occurring in subcutaneous tissue, often around tendons.

xanthomatosis [MED] A condition marked by the deposit of a yellowish or orange lipoid material in the reticuloendothelial cells, the skin, and the internal organs.

xanthomycin [MICROBIO] An antibiotic produced by a strain of *Streptomyces* and composed of two varieties, A and B; active in low concentrations against a number of gram-positive microorganisms.

xanthone [ORG CHEM] $CO(C_6H_4)_2O$ White needle crystals that are found in some plant pigments; insoluble in water, soluble in alcohol, chloroform, and benzene; melts at 173°C, sublimes at 350°C; used as a larvicide, as a dye intermediate, and in perfumes and pharmaceuticals. Also known as benzophenone oxide; dibenzopyrone; xanthene ketone.

xanthophore [CYTOL] A yellow chromatophore.

Xanthophyceae [BOT] A class of yellow-green to green flagellate organisms of the division Chrysophyta; zoologists classify these organisms in the order Heterochlorida.

xanthophyll [BIOCHEM] $C_{40}H_{56}O_2$ Any of a group of yellow, alcohol-soluble carotenoid pigments that are oxygen derivatives of the carotenes, and are found in certain flowers, fruits, and leaves. Also known as lutein.

xanthophyllite *See* clintonite.

xanthosiderite *See* goethite.

Xantusiidae [VERT ZOO] The night lizards, a family of reptiles in the suborder Sauria.

x axis [CRYSTAL] A reference axis in a quartz crystal. [MATH] **1.** A horizontal axis in a system of rectangular coordinates. **2.** That line on which distances to the right or left (east or west) of the reference line are marked, especially on a map, chart, or graph.

X band [COMMUN] A radio-frequency band extending from 5200 to 10,900 megahertz, corresponding to wavelengths of 5.77 to 2.75 centimeters.

X chromosome [GEN] The sex chromosome occurring in double dose in the homogametic sex and in single dose in the heterogametic sex.

X coefficient *See* nine-*j* symbol.

x-coordinate [MATH] One of the coordinates of a point in a two- or three-dimensional cartesian coordinate system, equal to the directed distance of a point from the *y*-axis in a two-dimensional system, or from the plane of the *y*- and *z*-axes in a three-dimensional system, measured along a line parallel to the *x*-axis.

x-cut [CRYSTAL] A quartz-crystal cut made in such a manner that the *x*-axis is perpendicular to the faces of the resulting slab.

Xe *See* xenon.

Xenarthra [VERT ZOO] A suborder of mammals in the order Edentata including sloths, anteaters, and related forms; posterior vertebrae have extra articular facets and vertebrae in the hip, and shoulder regions tend to be fused.

X engine [MECH ENG] An in-line engine with the cylinder banks so arranged around the crankshaft that they resemble the letter X when the engine is viewed from the end.

xenoblast [MINERAL] A mineral which has grown during metamorphism without development of its characteristic crystal faces. Also known as allotrioblast.

xenocryst [CRYSTAL] A crystal in igneous rock that resembles a phenocryst and is foreign to the enclosing body of rock. Also known as chadacryst.

xenogamy [BOT] Cross-fertilization between flowers on different plants.

***para*-xenol** *See* phenylphenol.

xenolith [PETR] An inclusion in an igneous rock which is not genetically related, such as an unmelted fragment of country rock. Also known as accidental inclusion; exogenous inclusion.

xenomorphic *See* allotriomorphic.

xenon [CHEM] An element, symbol Xe, member of the noble gas family, group 0, atomic number 54, atomic weight 131.30; colorless, boiling point −108°C (1 atm), noncombustible, nontoxic, and nonreactive; used in photographic flash lamps, luminescent tubes, and lasers, and as an anesthetic.

xenon-135 [NUC PHYS] A radioactive isotope of xenon produced in nuclear reactors; readily absorbs neutrons; half-life is 9.2 hours.

xenon arc lamp [ELEC] An arc lamp filled with xenon giving a light intensity approaching that of the carbon arc; particularly valuable in projecting motion pictures.

xenon flash lamp [ELEC] A flash tube containing xenon gas, which produces an intense peak of radiant energy at a wavelength of 566 nanometers when a high direct-current pulsed voltage is applied between electrodes at opposite ends of the tube.

xenon override [NUCLEO] In a nuclear reactor, the excess reactivity provided to compensate for the poisoning effect of xenon buildup.

xenon poisoning [NUCLEO] The accumulation in a nuclear reactor of xenon-135, formed by beta decay of iodine-135; xenon-135 has the highest cross section for thermal neutron capture of any known reactor poison.

Xenophyophorida [INV ZOO] An order of Protozoa in the subclass Granuloreticulosia; includes deep-sea forms that develop as discoid to fan-shaped branching forms which are multinucleate at maturity.

Xenopneusta [INV ZOO] A small order of wormlike animals belonging to the Echiurida.

Xenopterygii [VERT ZOO] The equivalent name for Gobiesociformes.

Xenosauridae [VERT ZOO] A family of four rare species of lizards in the suborder Sauria; composed of the Chinese lizard (*Shinisaurus crocodilurus*) and three Central American species of the genus *Xenosaurus*.

xenothermal [MINERAL] Pertaining to a mineral deposit formed at high temperature but at shallow to moderate depth.

Xenungulata [PALEON] An order of large, digitigrade, extinct, tapirlike mammals with relatively short, slender limbs

XENON

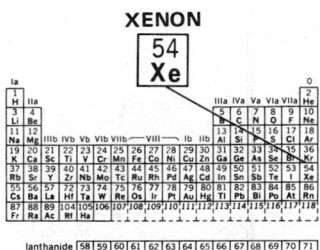

Periodic table of the chemical elements showing the position of xenon.

XENOPHYOPHORIDA

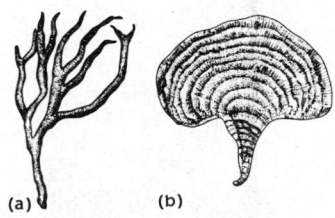

Representative fan-shaped branching forms of xenophyophorids. (a) *Stanomma dendroides* (after Schulze). (b) *Stannophyllum zonarium* (after Schulze). (Doflein and Reichenon, Lehrbuch der Protozoenkunde, 1929)

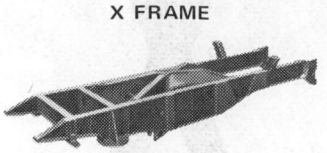

X FRAME

X frame with X-shaped member joining side rails. (*GMC Truck and Coach Division, General Motors Corp.*)

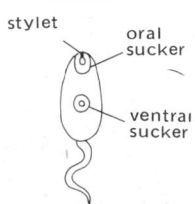

XIPHIDIO CERCARIA

stylet

oral sucker

ventral sucker

Xiphidio cercaria showing stylet in oral sucker. (*From R. M. Cable, An Illustrated Laboratory Manual of Parasitology, Burgess, 1958*)

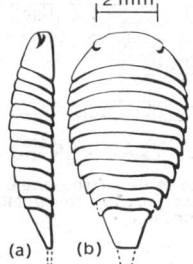

XIPHOSURIDA

├─ 2 mm ─┤

(a)   (b)

Drawing of *Paleomerus* species of the subclass Xiphosurida. (a) Side view. (b) Dorsal view.

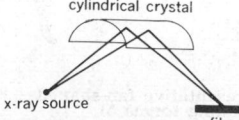

X-RAY IMAGE SPECTROGRAPHY

cylindrical crystal

x-ray source

film

Arrangement of x-ray source, cylindrical crystal, and photographic film in x-ray image spectrography.

---

and five-toed feet with broad, flat phalanges; restricted to the Paleocene deposits of Brazil and Argentina.

xenyl [ORG CHEM] The chemical radical $C_6H_5C_6H_4-$.

Xeralf [GEOL] A suborder of the soil order Alfisol; brown or red soil formed in a xeric moisture regime.

Xerert [GEOL] A suborder of the soil order Vertisol, formed in a Mediterranean climate; wide surface cracks open and close once a year.

xeric [ECOL] 1. Of or pertaining to a habitat having a low or inadequate supply of moisture. 2. Of or pertaining to an organism living in such an environment.

xeroderma pigmentosum [MED] A genodermatosis characterized by premature degenerative changes in the form of keratoses, malignant epitheliomatosis, and hyper- and hypo-pigmentation.

xerodermosteosis *See* Sjögren's syndrome.

xerography [GRAPHICS] A printing method developed by the Xerox Corporation; a negative image is formed by a resinous powder on an electrically charged plate, and this image is transferred and thermally fixed onto a paper as a positive.

Xeroll [GEOL] A suborder of the soil order Mollisol, formed in a xeric moisture regime; may have a calcic, petrocalcic, or gypsic horizon, or a duripan.

xerophthalmia [MED] Dryness and thickening of the conjunctiva, sometimes following chronic conjunctivitis, disease of the lacrimal apparatus, or vitamin A deficiency.

xerophyte [ECOL] A plant adapted to life in areas where the water supply is limited.

xeroradiography [GRAPHICS] An electrostatic image-forming process in which x-rays or gamma rays form an electrostatic image on a photoconductive insulating medium; the charged image areas attract and hold a fine powder called a toner, and then the powder image is transferred to paper and fused there by heat.

xerosere [ECOL] A temporary community in an ecological succession on dry, sterile ground such as rock, sand, or clay.

xerosis conjunctivae [MED] A condition marked by silver-gray, shiny, triangular spots on both sides of the cornea, within the region of the palpebral aperture, consisting of dried epithelium, flaky masses, and microorganisms.

xerothermal period *See* xerothermic period.

xerothermic [CLIMATOL] Characterized by dryness and heat.

xerothermic period [GEOL] A postglacial interval of a warmer, drier climate. Also known as xerothermal period.

Xerult [GEOL] A suborder of the soil order Ultisol, formed in a xeric moisture regime; brownish or reddish soil with a low to moderate organic-carbon content.

X frame [DES ENG] An automotive frame which either has side rails bent in at the center of the vehicle, making the overall form that of an X, or has an X-shaped member which joins the side rails with diagonals for added strength and resistance to torsional stresses.

xi hyperon [PARTIC PHYS] Also known as xi particle. 1. Collective name for the xi-minus and xi-zero particles, which form an isotopic-spin multiplet of quasi-stable baryons, designated Ξ, having a hypercharge of -1, a total isotopic spin of $1/2$, a spin of $1/2$, positive parity, and an average mass of approximately 1318 MeV (million electron volts). Also known as cascade hyperon; cascade particle. 2. A baryon belonging to any isotopic-spin multiplet having a hypercharge of -1 and a total isotopic spin of $1/2$; designated by $\Xi_{JP}(m)$, where m is the mass of the baryon in MeV, and J and P are its spin and parity (if known); the $\Xi_{3/2}+$ (1530) is sometimes designated Ξ^*.

xi-minus particle [PARTIC PHYS] A negatively charged xi hyperon, designated Ξ^-. Also known as cascade particle.

xi particle *See* xi hyperon.

xiphidio cercaria [INV ZOO] A digenetic trematode larva having a stylet in the oral sucker.

Xiphiidae [VERT ZOO] The swordfishes, a family of perciform fishes in the suborder Scombroidei characterized by a tremendously produced bill.

xiphisternum [ANAT] The elongated posterior portion of the sternum.

Xiphodontidae [PALEON] A family of primitive tylopod ruminants in the superfamily Anaplotherioidea from the late Eocene to the middle Oligocene of Europe.

Xiphosura [INV ZOO] The equivalent name for Xiphosurida.

Xiphosurida [INV ZOO] A subclass of primitive arthropods in the class Merostomata characterized by cephalothoracic appendages, ocelli, book lungs, a somewhat trilobed body, and freely articulating styliform telson.

Xiphydriidae [INV ZOO] A family of the Hymenoptera in the superfamily Siricoidea.

xi-zero particle [PARTIC PHYS] An uncharged xi hyperon, designated Ξ^0.

X organ [INV ZOO] A cluster of neurosecretory cells of the medulla terminalis, a portion of the brain lying in the eyestalk in stalk-eyed crustaceans.

x-parallax *See* absolute stereoscopic parallax.

X punch [ADP] In an 80-column punched card, any hole in the second row from the top. Also known as eleven punch.

x-radiation *See* x-rays.

x-ray absorption [ELECTROMAG] The taking up of energy from an x-ray beam by a medium through which the beam is passing.

x-ray analysis [PHYS] The use of x-ray radiations to detect heavy elements in the presence of lighter ones, to give critical-edge absorption to identify elemental composition, and to identify crystal structures by diffraction patterns.

x-ray astronomy [ASTRON] The study of x-rays mainly from sources outside the solar system; it includes the study of novae and supernovae in the Milky Way Galaxy, together with extragalactic radio sources.

x-ray crystallography [CRYSTAL] The study of crystal structure by x-ray diffraction techniques. Also known as roentgen diffractometry.

x-ray crystal spectrometer [SPECT] An instrument designed to produce an x-ray spectrum and measure the wavelengths of its components, by diffracting x-rays from a crystal with known lattice spacing.

x-ray diffraction [PHYS] The scattering of x-rays by matter, especially crystals, with accompanying variation in intensity due to interference effects. Also known as x-ray microdiffraction.

x-ray diffraction analysis [CRYSTAL] Analysis of the crystal structure of materials by passing x-rays through them and registering the diffraction (scattering) image of the rays.

x-ray diffractometer [ENG] An instrument used in x-ray analysis to measure the intensities of the diffracted beams at different angles.

x-ray emission *See* x-ray fluorescence.

x-ray film [GRAPHICS] A film base coated, usually on both sides, with an emulsion designed for use with x-rays.

x-ray fluorescence [ATOM PHYS] Emission by a substance of its characteristic x-ray line spectrum upon exposure to x-rays. Also known as x-ray emission.

x-ray fluorescence analysis [ANALY CHEM] A nondestructive physical method used for chemical analyses of solids and liquids; the specimen is irradiated by an intense x-ray beam and the lines in the spectrum of the resulting x-ray fluorescence are diffracted at various angles by a crystal with known lattice spacing; the elements in the specimen are identified by the wavelengths of their spectral lines, and their concentrations are determined by the intensities of these lines. Also known as x-ray fluorimetry.

x-ray fluorescent emission spectrometer [SPECT] An x-ray crystal spectrometer used to measure wavelengths of x-ray fluorescence; in order to concentrate beams of low intensity, it has bent reflecting or transmitting crystals arranged so that the theoretical curvature required can be varied with the diffraction angle of a spectrum line.

x-ray fluorimetry *See* x-ray fluorescence analysis.

x-ray generator [ELECTR] A metal from whose surface large amounts of x-rays are emitted when it is bombarded with high-velocity electrons; metals with high atomic weight are the most efficient generators.

x-ray goniometer [ENG] A scale designed to measure the angle between the incident and refracted beams in x-ray diffraction analysis.

x-ray hardness [ELECTROMAG] The penetrating ability of x-rays; it is an inverse function of the wavelength.

x-ray image spectrography [SPECT] A modification of x-ray fluorescence analysis in which x-rays irradiate a cylindrically bent crystal, and Bragg diffraction of the resulting emissions produces a slightly enlarged image with a resolution of about 50 micrometers.

x-ray irradiation [PHYS] Subjection of a material, object, or patient to x-rays.

x-ray microdiffraction *See* x-ray diffraction.

x-ray microprobe *See* microprobe.

x-ray microscope [ENG] **1.** A device in which an ultra-fine-focus x-ray tube or electron gun produces an electron beam focused to an extremely small image on a transmission-type x-ray target that serves as a vacuum seal; the magnification is by projection; specimens being examined can thus be in air, as also can the photographic film that records the magnified image. **2.** Any of several instruments which utilize x-radiation for chemical analysis and for magnification of 100–1000 diameters; it is based on contact or projection microradiography, reflection x-ray microscopy, or x-ray image spectrography.

x-ray monochromator [ENG] An instrument in which x-rays are diffracted from a crystal to produce a beam having a narrow range of wavelengths.

x-ray nebulae [ASTRON] The remnant of an ancient supernova that has been identified as a source of x-rays; an example is the Crab Nebula.

x-ray optics [ELECTROMAG] A title-by-analogy of those phases of x-ray physics in which x-rays demonstrate properties similar to those of light waves. Also known as roentgen optics.

x-ray powder diffractometer *See* powder diffraction camera.

x-ray powder method *See* powder method.

x-ray projection microscopy *See* projection microradiography.

x-rays [PHYS] A penetrating electromagnetic radiation, usually generated by accelerating electrons to high velocity and suddenly stopping them by collision with a solid body, or by inner-shell transitions of atoms with atomic number greater than 10; their wavelengths range from about 10^{-5} angstrom to 10^3 angstroms, the average wavelength used in research being about 1 angstrom. Also known as roentgen rays; x-radiation.

x-ray spectrograph [SPECT] An x-ray spectrometer equipped with photographic or other recording apparatus; one application is fluorescence analysis.

x-ray spectrometer [SPECT] An instrument for producing the x-ray spectrum of a material and measuring the wavelengths of the various components.

x-ray spectrometry [SPECT] The measure of wavelengths of x-rays by observing their diffraction by crystals of known lattice spacing. Also known as roentgen spectrometry; x-ray spectroscopy.

x-ray spectroscopy *See* x-ray spectrometry.

x-ray spectrum [SPECT] A display or graph of the intensity of x-rays, produced when electrons strike a solid object, as a function of wavelengths or some related parameter; it consists of a continuous bremsstrahlung spectrum on which are superimposed groups of sharp lines characteristic of the elements in the target.

x-ray star [ASTROPHYS] A source of x-rays from outside the solar system; examples are the point x-ray sources Scorpius X-1, Cygnus X-2, and the Crab x-ray source.

x-ray target [ELECTR] The metal body with which high-velocity electrons collide, in a vacuum tube designed to produce x-rays.

x-ray telescope [ENG] An instrument designed to detect x-rays emanating from a source outside the earth's atmosphere and to resolve the x-rays into an image; they are carried to high altitudes by balloons, rockets, or space vehicles; although several types of x-ray detector, involving gas counters, scintillation counters, and collimators, have been used, only one, making use of the phenomenon of total external reflection of x-rays from a surface at grazing incidence, is strictly an x-ray telescope.

x-ray therapy [MED] Medical treatment by controlled application of x-rays; a type of radiotherapy.

x-ray thickness gage [ENG] A thickness gage used for measuring and indicating the thickness of moving cold-rolled sheet steel during the rolling process without making contact with the sheet; an x-ray beam directed through the sheet is absorbed in proportion to the thickness of the material and its atomic number.

x-ray tube [ELECTR] A vacuum tube designed to produce x-rays by accelerating electrons to a high velocity by means of an electrostatic field, then suddenly stopping them by collision with a target.

x-ray unit *See* siegbahn.

X test [STAT] A one-sample test which rejects the hypothesis $\mu = \mu_H$ in favor of the alternative $\mu > \mu_H$ if $X - \mu_H \geq c$ where c is an appropriate critical value, X is the arithmetic mean of observations, μ_H is a given number, and μ is the (unknown) expected value of the random variable X.

XU *See* siegbahn.

X unit *See* siegbahn.

X wave *See* extraordinary wave.

x,y chromaticity diagram *See* Maxwell triangle.

XY coordinate plotter *See* coordinate plotter.

Xyelidae [INV ZOO] A family of hymenopteran insects in the superfamily Megalodontoidea.

xylem [BOT] The principal water-conducting tissue and the chief supporting tissue of higher plants; composed of tracheids, vessel members, fibers, and parenchyma.

xylene [ORG CHEM] $C_6H_4(CH_3)_2$ Any one of the family of isomeric, colorless aromatic hydrocarbon liquids, produced by the destructive distillation of coal or by the catalytic reforming of petroleum naphthenic fractions; used for high-octane and aviation gasolines, solvents, chemical intermediates, and the manufacture of polyester resins. Also known as dimethylbenzene; xylol.

***meta*-xylene** [ORG CHEM] $1,3\text{-}C_6H_4(CH_3)_2$ A flammable, toxic liquid; insoluble in water, soluble in alcohol and ether; boils at 139°C; used as an intermediate for dyes, a chemical intermediate, and a solvent, and in insecticides and aviation fuel. Also known as 1,3-dimethylbenzene.

***ortho*-xylene** [ORG CHEM] $1,2\text{-}C_6H_4(CH_3)_2$ A flammable, moderately toxic liquid; insoluble in water, soluble in alcohol and ether; boils at 144°C; used to make phthalic anhydride, vitamins, pharmaceuticals, and dyes, and in insecticides and motor fuels. Also known as 1,2-dimethylbenzene.

***para*-xylene** [ORG CHEM] $1,4\text{-}C_6H_4(CH_3)_2$ A toxic, combustible liquid; insoluble in water, soluble in alcohol and ether; boils at 139°C; used as a chemical intermediate, and to synthesize terephthalic acid, vitamins, and pharmaceuticals, and in insecticides. Also known as 1,4-dimethylbenzene.

xylenol [ORG CHEM] $(CH_3)_2C_6H_3OH$ Highly toxic, combustible crystals; slightly soluble in water, soluble in most organic solvents; melts at 20–76°C; used as a chemical intermediate, disinfectant, solvent, and fungicide, and for pharmaceuticals and dyestuffs. Also known as dimethylhydroxybenzene; dimethylphenol; hydroxydimethylbenzene.

xylidine [ORG CHEM] $(CH_3)_2C_6H_3NH_2$ A toxic, combustible liquid; soluble in alcohol and ether, slightly soluble in water; boils about 220°C; used as a chemical intermediate and to make dyes and pharmaceuticals. Also known as aminodimethylbenzene; aminoxylene.

xylite *See* xylitol.

xylitol [ORG CHEM] $CH_2OH(CHOH)_3CH_2OH$ Pentahydric alcohols derived from xylose. Also known as xylite.

Xylocopidae [INV ZOO] A family of hairy tropical bees in the superfamily Apoidea.

xyloid coal *See* bituminous wood.

xyloid lignite *See* bituminous wood.

xylol *See* xylene.

Xylomyiidae [INV ZOO] A family of orthorrhaphous dipteran insects in the series Brachycera.

xylose [BIOCHEM] $C_5H_{10}O_5$ A pentose sugar found in many woody materials; combustible, white crystals with a sweet taste; soluble in water and alcohol; melts about 148°C; used as a nonnutritive sweetener and in dyeing and tanning. Also known as wood sugar.

XY plotter *See* coordinate plotter.

XY recorder [ENG] A recorder that traces on a chart the relation of two variables, neither of which is time.

Xyridaceae [BOT] A family of terrestrial monocotyledonous plants in the order Commelinales characterized by an open leaf sheath, three stamens, and a simple racemose head for the inflorescence.

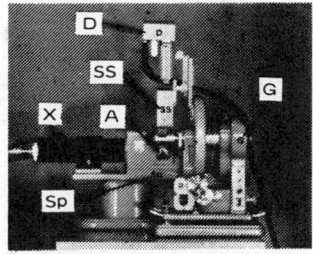

X-RAY SPECTROGRAPH

Modern x-ray spectrograph for fluorescence analysis. X is x-ray tube; Sp, specimen; A, crystal analyzer; SS, Soller (parallel) slits; D, counter tube detector; G, goniometer. *(Philips Electronic Instruments, Inc.)*

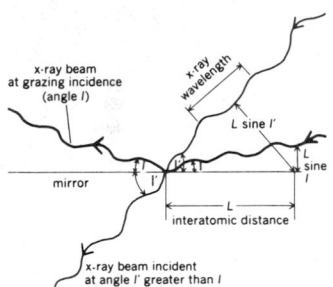

X-RAY TELESCOPE

Diagram illustrating phenomenon of total external reflection of x-rays at grazing incidence, on which x-ray telescope is based. Beam incident at angle I is totally reflected, because L sine I is much smaller than x-ray wavelength, while beam incident at angle I' greater than I is not reflected.

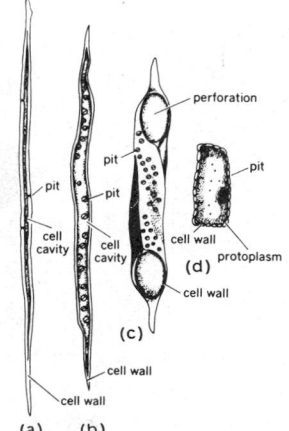

XYLEM

Xylem cell types. (*a*) Wood (xylem) fiber. (*b*) Tracheid. (*c*) Vessel member. (*d*) Xylem parenchyma cell. All can vary widely in structure. *(From H. J. Fuller and O. Tippo, College Botany, rev. ed., Holt, 1954)*

XY switching system [ELECTR] A telephone switching system consisting of a series of flat bank and wiper switches in which the wipers move in a horizontal plane, first in one direction and then in another under the control of pulses from a subscriber's dial; the switches are stacked one on top of another on frames, and are operated one after another.

Y   *See* wye; yttrium.

yacca gum   *See* acaroid resin.

yacht   [NAV ARCH] A sailing or power boat used for pleasure cruises or racing.

Yagi antenna   *See* Yagi-Uda antenna.

Yagi-Uda antenna   [ELECTROMAG] An end-fire antenna array having maximum radiation in the direction of the array line; it has one dipole connected to the transmission line and a number of equally spaced unconnected dipoles mounted parallel to the first in the same horizontal plane to serve as directors and reflectors. Also known as Uda antenna; Yagi antenna.

yag laser   *See* yttrium-aluminum-garnet laser.

yak   [VERT ZOO] *Poephagus grunniens.* A heavily built, long-haired mammal of the order Artiodactyla, with a shoulder hump; related to the bison, and resembles it in having 14 pairs of ribs.

yalca   [METEOROL] A local name for a severe snowstorm with a strong squally wind which occurs in the Andes Mountain passes of northern Peru.

yam   [BOT] An erroneous name for the Puerto Rico variety of sweet potato; the edible, starchy tuberous root of the plant.

yamase   [METEOROL] A cool, onshore, easterly wind in the Senriku district of Japan in summer.

yard   [CIV ENG] A facility for building and repairing ships. [MECH] A unit of length in common use in the United States and United Kingdom, equal to 0.9144 meter, or 3 feet. Abbreviated yd. [NAV ARCH] A long spar, tapered at the ends, attached at its middle to a mast and running athwartships, and used to support a sail.

yardage   [MECH] An amount expressed in yards. [MIN ENG] The extra compensation a miner receives in addition to the mining price for working in a narrow place or in deficient coal, usually at a certain price per yard advanced.

yardarm   [NAV ARCH] One of the ends of a yard.

yard crane   *See* crane truck.

yard maintenance   [ENG] A category of maintenance that includes the complete rebuilding of parts, subassemblies, or components.

yardstick compass   [GRAPHICS] A simple arrangement consisting of two fully adjustable clips that slide on an ordinary yardstick; one clip has a sharp point, the other a pencil holder; the compass is used to make large, accurate circles.

Yarmouth interglacial   [GEOL] The second interglacial stage of the Pleistocene epoch in North America, following the Kansan glacial stage and before the Illinoian.

yarn   [TEXT] A continuous strand of two or more plies of carded or combed fibers twisted together, or a single filament of natural or synthetic fibers used for weaving or knitting.

yarn-dyed   [TEXT] Pertaining to a fabric made of yarns dyed before weaving or knitting.

yaw   [MECH] **1.** The rotational or oscillatory movement of a ship, aircraft, rocket, or the like about a vertical axis. Also known as yawing. **2.** The amount of this movement, that is, the angle of yaw. **3.** To rotate or oscillate about a vertical axis.

yaw acceleration   [MECH] The angular acceleration of an aircraft or missile about its normal or *Z* axis.

yaw angle   *See* angle of yaw.

yaw axis   [MECH] A vertical axis through an aircraft, rocket, or similar body, about which the body yaws; it may be a body, wind, or stability axis. Also known as yawing axis.

yaw damper   [AERO ENG] A control system or device that reduces the yaw of an aircraft, guided missile, or the like.

yaw in bore   [ORD] The maximum angle between the axis of the bore of a gun and the axis of the projectile which can occur due to the clearance between the bore diameter and the bourrelets.

yaw indicator   [AERO ENG] A device that measures the angular direction of the airflow relative to the longitudinal vertical plane of the aircraft; this may be accomplished by a balanced vane or by a differential pressure sensor that aligns the detector to the airflow, and in so doing transmits the measured angle between the normal axis and the detector as the yaw angle.

yawing   *See* yaw.

yawing axis   *See* yaw axis.

yaws   [MED] An infectious tropical disease of humans caused by the spirochete *Treponema pertenue;* manifested by a primary cutaneous lesion followed by a granulomatous skin eruption.

yaw simulator   [CONT SYS] A test instrument used to derive and thereby permit study of probable aerodynamic behavior in controlled flight under specific initial conditions; certain components of the missile guidance system, such as the receiver or servo loop, are connected into the simulator circuitry; also, certain aerodynamic parameters of the specific missile must be known and set into the simulator; applicable to the yaw plane.

y-axis   [CRYSTAL] A line perpendicular to two opposite parallel faces of a quartz crystal. [MATH] **1.** A vertical axis in a system of rectangular coordinates. **2.** That line on which distances above or below (north or south) the reference line are marked, especially on a map, chart, or graph.

Yb   *See* ytterbium.

Y block   [MET] A Y-shaped test casting used to appraise low-shrinkage alloys.

Y chromosome   [GEN] The sex chromosome found only in the heterogametic sex.

Y circulator   [ELECTROMAG] Circulator in which three identical rectangular waveguides are joined to form a symmetrical Y-shaped configuration, with a ferrite post or wedge at its center; power entering any waveguide will emerge from only one adjacent waveguide.

Y connection   *See* Y network.

y-coordinate   [MATH] One of the coordinates of a point in a two- or three-dimensional coordinate system, equal to the directed distance of a point from the *x*-axis in a two dimensional system, or from the plane of the *x*- and *z*-axes in a three-dimensional coordinate system, measured along a line parallel to the *y*-axis.

Y cut   [CRYSTAL] A quartz-crystal cut such that the *y*-axis is perpendicular to the faces of the resulting slab.

yd   *See* yard.

Y-delta transformation   [ELEC] One of two electrically equivalent networks with three terminals, one being connected internally by a Y configuration and the other being connected internally by a delta transformation. Also known as delta-Y transformation; pi-T transformation.

year   [ASTRON] Any of several units of time based on the revolution of the earth about the sun; the tropical year to which the calendar is adjusted is the period required for the sun's longitude to increase 360°; it is about 365.24220 mean solar days. Abbreviated yr.

yeast   [MYCOL] A collective name for those fungi which possess, under normal conditions of growth, a vegetative body (thallus) consisting, at least in part, of simple, individual cells.

yellow   [OPTICS] The hue evoked in an average observer by monochromatic radiation having a wavelength in the approximate range from 577 to 597 nanometers; however, the same sensation can be produced in a variety of other ways.

yellow arsenic   *See* orpiment.

yellow cake   [MIN ENG] The final precipitate formed in the milling of uranium ores.

yellow coal   *See* tasmanite.

yellow copperas   *See* copiapite.

YAGI-UDA ANTENNA

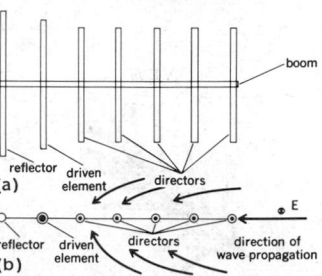

Yagi-Uda antenna. *(a)* View from above. *(b)* Side view; arrows show energy flow close to parasitic elements.

YAK

Yak *(Poephagus grunniens)* showing characteristic features.

YAM

Puerto Rico sweet potato known as the yam. *(USDA)*

YEW

Female branchlet and single leaf of Japanese yew (*Taxus cuspidata*).

YOUNGINIFORMES

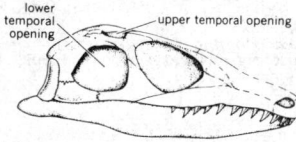

lower temporal opening / upper temporal opening

Lateral view of skull of *Youngina*, a representative of Younginiformes, Upper Permian to Lower Triassic. (*From A. S. Romer, Vertebrate Paleontology, 3d ed., University of Chicago Press, 1966*)

YOUNG'S TWO-SLIT INTERFERENCE

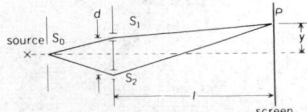

Young's two-slit interference. S_1 and S_2 are the two parallel slits, illuminated by light from slit S_0; d is distance between slits; l is distance from plane of slits to screen. Light bands on screen are located at points P for which distance $y = n\lambda l/d$, where λ = wavelength of light.

yellow dwarf [PL PATH] Any of several plant viral diseases characterized by yellowing of the foliage and stunting of the plant.

yellow fat cell [HISTOL] A large, generally spherical fat cell with a thin shell of protoplasm and a single enlarged fat droplet which appears yellowish.

yellow fever [MED] An acute, febrile, mosquito-borne viral disease characterized in severe cases by jaundice, albuminuria, and hemorrhage.

yellow-green algae [BOT] The common name for members of the class Xanthophyceae.

yellow lead ore *See* wulfenite.

yellow leaf blotch [PL PATH] A fungus disease of alfalfa caused by *Pyrenopeziza medicaginis* characterized by the appearance of yellow or orange blotches with small black dots on the foliage.

yellow mud [GEOL] Mud containing sediment having a characteristic yellow color, resulting from certain iron compounds.

yellow phosphorus *See* white phosphorus.

yellow precipitate *See* mercuric oxide.

yellow prussiate of potash *See* potassium ferrocyanide.

yellow prussiate of soda *See* sodium ferrocyanide.

yellow pyoktannin *See* auramine hydrochloride.

yellows [PL PATH] Any of various fungus diseases of plants characterized by yellowing of the leaves which later turn brown, become brittle, and die; affects cabbage, lettuce, cauliflower, peach, sugarbeet, and other plants.

yellow salt *See* uranyl nitrate.

yellow scale [MATER] The commercial name for low-grade paraffin wax.

Yellow Sea [GEOGR] An inlet of the Pacific Ocean between northeastern China and Korea.

yellow snow [HYD] Snow with a golden or yellow appearance because of the presence of pine or cypress pollen.

yellow tellurium *See* sylvanite.

yew [BOT] A genus of evergreen trees and shrubs, *Taxus*, with the fruit, an aril, containing a single seed surrounded by a scarlet, fleshy, cuplike envelope; the leaves are flat and acicular.

Y factor [PETRO ENG] An empirical relationship of bubble-point data (pressure and formation volume) used to smooth oil reservoir solution-gas/oil-ratio data for graphical presentation.

Y gun [ORD] Two-barreled, antisubmarine gun shaped like the letter Y, used to throw depth charges to either side of the stern of the vessel on which the gun is mounted.

yield [ENG] Product of a reaction or process as in chemical reactions or food processing.   [MECH] That stress in a material at which plastic deformation occurs.   [ORD] The total effective energy released in a nuclear explosion; usually expressed in terms of the equivalent tonnage of trinitrotoluene (TNT) required to produce the same energy release.

yielding arches [MIN ENG] Steel arches installed in underground openings as the ground is removed to support loads caused by changing ground movement or by faulted and fractured rock; when the ground load exceeds the design load of the arch as installed, yielding takes places in the joint of the arch, permitting the overburden to settle into a natural arch of its own and thus tending to bring all forces into equilibrium.

yielding floor [MIN ENG] A soft floor which heaves and flows into open spaces when subjected to heavy pressure from packs or pillars.

yielding prop [MIN ENG] A steel prop which is adjustable in length and incorporates a sliding or flexible joint which comes into operation when the roof pressure exceeds a set load or value.

yield-pillar system [MIN ENG] A method of roof control whereby the natural strength of the roof strata is maintained by the relief of pressure in working areas and the controlled transference of load to abutments which are clear of the workings and roadways.

yield point [MECH] The lowest stress at which strain increases without increase in stress.

yield strength [MECH] The stress at which a material exhibits a specified deviation from proportionality of stress and strain.

yield stress [MECH] The lowest stress at which extension of the tensile test piece increases without increase in load.

yig device [ELECTR] A filter, oscillator, parametric amplifier, or other device that uses an yttrium-iron-garnet crystal in combination with a variable magnetic field to achieve wide-band tuning in microwave circuits. Derived from yttrium-iron-garnet device.

yig filter [ELECTR] A filter consisting of an yttrium-iron-garnet crystal positioned in a magnetic field provided by a permanent magnet and a solenoid; tuning is achieved by varying the amount of direct current through the solenoid; the bias magnet serves to tune the filter to the center of the band, thus minimizing the solenoid power required to tune over wide bandwidths.

yig-tuned parametric amplifier [ELECTR] A parametric amplifier in which tuning is achieved by varying the amount of direct current flowing through the solenoid of a yig filter.

yig-tuned tunnel-diode oscillator [ELECTR] Microwave oscillator in which precisely controlled wide-band tuning is achieved by varying the current through a tuning solenoid that acts on a yig filter in the tunnel-diode oscillator circuit.

Y junction [ELECTROMAG] A waveguide in which the longitudinal axes of the waveguide form a Y.

ylem [ASTROPHYS] The primordial matter which according to the big bang theory existed just prior to the formation of the chemical elements.

Y level [ENG] A surveyor's level with Y-shaped rests to support the telescope. Also known as wye level.

Y ligament *See* iliofemoral ligament.

Y network [ELEC] A star network having three branches. Also known as Y connection.

yogurt [FOOD ENG] A fermented milk food made by adding cultures of *Lactobacillus acidophilus* and *Streptococcus thermophilus* to skimmed cow's milk and milk solids.

yoke [ARCH] A horizontal member forming the head of a window frame.   [DES ENG] A clamp or similar device to embrace and hold two other parts.   [ELECTROMAG] **1.** Piece of ferromagnetic material without windings, which permanently connects two or more magnet cores. **2.** *See* deflection yoke.   [ENG] **1.** A bar of wood used to join the necks of draft animals for working together. **2.** *See* wye.   [MECH ENG] A slotted crosshead used instead of a connecting rod in some steam engines.

yoked basin *See* zeugogeosyncline.

yolk [BIOCHEM] **1.** Nutritive material stored in an ovum. **2.** The yellow spherical mass of food material that makes up the central portion of the egg of a bird or reptile.

yolk larva *See* vitellarium.

yolk sac [EMBRYO] A distended extraembryonic extension, heavy-laden with yolk, through the umbilicus of the midgut of the vertebrate embryo.

Y organ [INV ZOO] Either of a pair of nonneural structures found in the anterior portion of the crustacean body; source of the molting hormone, ecdysone.

York-Scheibel column *See* Scheibel extractor.

youg [METEOROL] A hot wind during unsettled summer weather in the Mediterranean.

Young construction [OPTICS] A graphical procedure for tracing a light ray through a boundary between two media having different refractive indices.

Young-Helmholtz theory [PHYSIO] A theory of color vision according to which there are three types of color receptors that respond to short, medium, and long waves respectively; primary colors are those that stimulate most successfully the three types of receptors. Also known as Helmholtz theory.

young ice [HYD] Newly formed ice in the transitional stage of development from ice crust to winter ice.

Younginiformes [PALEON] A suborder of extinct small lizardlike reptiles in the order Eosuchia, ranging from the Middle Permian to the Lower Triassic in South Africa.

Young's modulus [MECH] The ratio of a simple tension stress applied to a material to the resulting strain parallel to the tension.

Young's two-slit interference [OPTICS] Interference of light from two parallel slits which are illuminated by light from a single slit, which in turn is illuminated by a source; the interference can be seen by letting the light fall on a screen, which then shows a series of parallel fringes.

youth [GEOL] The first stage of the cycle of erosion in which the original surface or structure is the dominant topographic feature; characterized by broad, flat-topped interstream divides, numerous swamps and shallow lakes, and progressive increase of local relief. Also known as topographic youth.

Yponomeutidae [INV ZOO] A heterogeneous family of small, often brightly colored moths in the superfamily Tineoidea; the head is usually smooth with reduced or absent ocelli.

Ypsilothuriidae [INV ZOO] A family of echinoderms in the order Dactylochirotida having 8–10 tentacles, a permanent spire on the plates of the test, and the body fusiform or U-shaped.

Y punch [ADP] In a standard punched card, a hole in the topmost row.

yr *See* year.

Y search [NAV] A sector search with one of the legs modified to provide for coverage of the interior of the sector.

Y signal *See* luminance signal.

ytterbia *See* ytterbium oxide.

ytterbium [CHEM] A rare-earth metal of the yttrium subgroup, symbol Yb, atomic number 70, atomic weight 173.04; lustrous, malleable, soluble in dilute acids and liquid ammonia, reacts slowly with water; melts at 824°C, boils at 1427°C; used in chemical research, lasers, garnet doping, and x-ray tubes.

ytterbium oxide [INORG CHEM] Yb_2O_3 A colorless compound, melts at 2346°C, dissolves in hot dilute acids; used to prepare alloys, ceramics, and special glasses. Also known as ytterbia.

yttria *See* yttrium oxide.

yttrium [CHEM] A rare-earth metal, symbol Y, atomic number 39, atomic weight 88.905; dark-gray, flammable (as powder), soluble in dilute acids and potassium hydroxide solution, and decomposes in water; melts at 1500°C, boils at 2927°C; used in alloys and nuclear technology and as a metal deoxidizer.

yttrium acetate [ORG CHEM] $Y(C_2H_3O_2)_3 \cdot 8H_2O$ Colorless, water-soluble crystals used as an analytical reagent.

yttrium-aluminum-garnet laser [OPTICS] A four-level infrared laser in which the active material is neodymium ions in an yttrium-aluminum-garnet crystal; it can provide a continuous output power of several watts. Abbreviated yag laser.

yttrium chloride [INORG CHEM] $YCl_3 \cdot 6H_2O$ Reddish, transparent, water- and alcohol-soluble prisms; decomposes at 100°C; used as an analytical reagent.

yttrium-iron-garnet device *See* yig device.

yttrium oxide [INORG CHEM] Y_2O_3 A yellowish powder, insoluble in water, soluble in dilute acids; used as television tube phosphor and microwave filters. Also known as yttria.

yttrium sulfate [INORG CHEM] $Y_2(SO_4)_3 \cdot 8H_2O$ Reddish crystals that are soluble in concentrated sulfuric acid, slightly soluble in water; decomposes at 700°C; used as an analytical reagent.

Yucatán Current [OCEANOGR] A rapid northward flowing current along the western side of the Yucatán Strait; generally loops to the north and exits as the Florida Current.

yugawaralite [MINERAL] $CaAl_2Si_6O_{16} \cdot 4H_2O$ A zeolite mineral consisting of hydrous calcium aluminum silicate.

Yukawa force [NUC PHYS] The strong, short-range force between nucleons, as calculated on the assumption that this force is due to the exchange of a particle of finite mass (Yukawa meson), just as electrostatic forces are interpreted in quantum electrodynamics as being due to the exchange of photons.

Yukawa meson [PARTIC PHYS] A particle, having a finite rest mass, whose exchange between nucleons is postulated to account for the strong, short-range forces between them; such a contributor is the pi meson.

Yukawa potential [NUC PHYS] The potential function that is associated with the Yukawa force, with the form $V(r) = -V_0(b/r) \exp(-r/b)$, where r is the distance between the nucleons and V_0 and b are constants, giving measures of the strength and range of the force respectively.

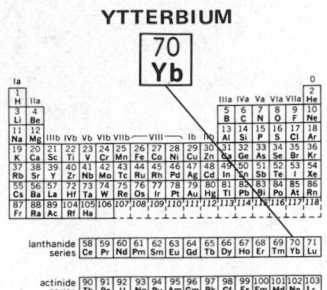

Periodic table of the chemical elements showing the position of ytterbium.

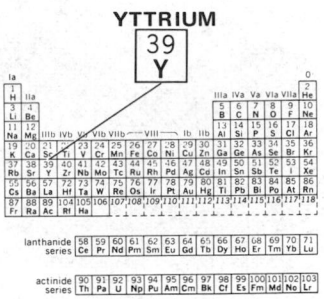

Periodic table of the chemical elements showing the position of yttrium.

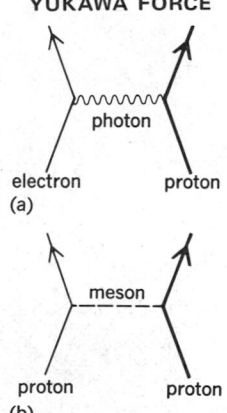

Analogy between *(a)* electromagnetic force via photon exchange and *(b)* nuclear Yukawa force via meson exchange.

Zalambdalestidae [PALEON] A family of extinct insectivorous mammals belonging to the group Proteutherea; they occur in the Late Cretaceous of Mongolia.

Zanclidae [VERT ZOO] The Moorish idols, a family of Indo-Pacific perciform fishes in the suborder Acanthuroidei.

Zantrel [TEXT] Trade name for a cellulosic rayon made by the American Enka Corporation.

Zanzibar gum [MATER] A combustible, hard, fossil-type copal, which is insoluble in most solvents and melts at about 245°C; used in varnishes.

Zapoididae [VERT ZOO] The Northern Hemisphere jumping mice, a family of the order Rodentia with long legs and large feet adapted for jumping.

zaratite [MINERAL] $Ni_3(CO_3)(OH)_4 \cdot 4H_2O$ An emerald-green mineral consisting of a hydrous basic nickel carbonate and occurring in incrustations or compact masses.

zastruga *See* sastruga.

z-axis [CRYSTAL] The optical axis of a quartz crystal, perpendicular to both the *x*- and *y*-axes. [MATH] One of the three axes in a three-dimensional cartesian coordinate system; in a rectangular coordinate system it is perpendicular to the *x*- and *y*-axes.

Z cam [ASTRON] A representative type of variable star; it is eruptive with a cycle of about 10-600 days; magnitude ranges from 2 to 6.

Z coefficient [QUANT MECH] A coefficient used in the transformation between modes of coupling eigenfunctions of three angular momenta, and especially in calculating matrix elements in beta decay and similar problems.

z-coordinate [MATH] One of the coordinates of a point in a three-dimensional coordinate system, equal to the directed distance of a point from the plane of the *x*- and *y*-axes, measured along a line parallel to the *z*-axis.

zebra [VERT ZOO] Any of three species of African mammals belonging to the family Equidae distinguished by a coat of black and white stripes.

zebu [VERT ZOO] A domestic breed of cattle, indigenous to India, belonging to the family Bovidae, distinguished by long drooping ears, a dorsal hump between the shoulders, and a dewlap under the neck; known as the Brahman in the United States.

Zechstein [GEOL] A European series of geologic time, especially in Germany: Upper Permian (above Rothliegende).

Zeeman displacement [SPECT] The separation, in wave numbers, of adjacent spectral lines in the normal Zeeman effect in a unit magnetic field, equal (in centimeter-gram-second Gaussian units) to $e/4\pi mc^2$, where e and m are the charge and mass of the electron, or to approximately 4.67×10^{-5} (centimeter)$^{-1}$(gauss)$^{-1}$.

Zeeman effect [SPECT] A splitting of spectral lines in the radiation emitted by atoms or molecules in a static magnetic field.

Zeeman energy [ATOM PHYS] The energy of interaction between an atomic or molecular magnetic moment and an applied magnetic field.

Zefkrome [TEXT] Trade name for an acrylic fiber made by the Dow Badische Company, used for double-knit fabric.

Zefran [TEXT] Trade name of the Dow Badische Company for a basic dyeable acrylic fiber.

Zeiformes [VERT ZOO] The dories, a small order of teleost fishes, distinguished by the absence of an orbitosphenoid bone, a spinous dorsal fin, and a pelvic fin with a spine and five to nine soft rays.

zein [MATER] A combustible, white to yellowish protein powder derived from corn; insoluble in water, soluble in dilute alcohol; used in inks, fibers, microencapsulation, and coatings for paper and food.

Zener breakdown [ELECTR] Nondestructive breakdown in a semiconductor, occurring when the electric field across the barrier region becomes high enough to produce a form of field emission that suddenly increases the number of carriers in this region. Also known as Zener effect.

Zener diode [ELECTR] A special type of silicon diode that acts like a rectifier until the applied voltage reaches a value known as the avalanche breakdown voltage or Zener voltage; at this point the diode becomes conducting, with the voltage drop across the diode remaining essentially constant independent of current. Also known as avalanche diode; breakdown diode; silicon reference diode.

Zener diode voltage regulator *See* diode voltage regulator.

Zener effect *See* Zener breakdown.

Zener voltage *See* breakdown voltage.

zenith [ASTRON] That point of the celestial sphere vertically overhead.

zenithal chart *See* azimuthal chart.

zenithal rain [METEOROL] In the tropics or subtropics, the rainy season which recurs annually or semiannually at about the time that the sun is most nearly overhead (at zenith).

zenith distance [ASTRON] Angular distance from the zenith; the arc of a vertical circle between the zenith and a point on the celestial sphere, measured from the zenith through 90°, for bodies above the horizon. Also known as co-altitude.

zenith telescope [OPTICS] A type of telescope that is fixed in the vertical or moves only a small amount from the vertical; it is used to get positional measurement of stars moving near the zenith.

zenodiagnosis [MED] A procedure of using a suitable arthropod to transfer an infectious agent from a patient to a susceptible laboratory animal.

Zeno's paradox [MATH] An erroneous group of paradoxes dealing with motion; the most famous one concerns two objects, one chasing the other which has a given head start, where the chasing one moves faster yet seemingly never catches the other.

Zeoidea [VERT ZOO] An equivalent name for Zeiformes.

zeolite [MINERAL] 1. A group of white or colorless, sometimes red or yellow, hydrous tectosilicate minerals characterized by an aluminosilicate tetrahedral framework, ion-exchangeable large cations, and loosely held water molecules permitting reversible dehydration. 2. Any mineral of the zeolite group, such as analcime, chabazite, natrolite, and stilbite.

zeolite catalyst [INORG CHEM] Hydrated aluminum and calcium (or sodium) silicates (for example, $CaO \cdot 2Al_2O_3 \cdot 5SiO_2$ or $Na_2O \cdot 2Al_2O_3 \cdot 5SiO_2$) made with controlled porosity; used as a catalytic cracking catalyst in petroleum refineries, or loaded with catalyst for other chemical reactions.

zeolite facies [PETR] Metamorphic rocks formed in the transitional period from diagenesis to metamorphism, at pressures of about 2000-3000 bars and temperatures of 200-300°C.

zeolitization [GEOL] Introduction of or replacement by a zeolite mineral.

Zeomorphi [VERT ZOO] An equivalent name for Zeiformes.

zeotrope [PHYS CHEM] A nonazeotropic liquid mixture which may be separated by distillation, and in which the components are miscible in all proportions (homogeneous zeotrope or homozeotrope) or not miscible in all proportions (heterogeneous zeotrope or heterozeotrope).

Zepel [TEXT] Trade name of a stain- and water-repellent fluorochemical made by DuPont.

zephyr [METEOROL] Any soft, gentle breeze.

Zepp antenna [ELECTROMAG] Horizontal antenna which is a multiple of a half-wavelength long and is fed at one end by one lead of a two-wire transmission line that is some multiple of a quarter-wavelength long.

ZEBRA

Grevy's zebra *(Equus grevyi)*, the largest species of zebra.

ZEBU

Zebu, or Brahman, a domestic breed of cattle.

ZEIFORMES

John dory *(Zenopsis ocellata)*, a representative of Zeiformes, showing characteristic features. *(After G. B. Goode, Fishery Industries of the United States, sect. 1, 1884)*

Zerewitinoff reagent [ANALY CHEM] A light-colored methyl-magnesium iodide–*n*-butyl ether solution that reacts rapidly with moisture and oxygen; used to determine water, alcohols, and amines in inert solvents.

zero [MATH] The additive identity element of an algebraic system.

zero-access instruction [ADP] An instruction consisting of an operation which does not require the designation of an address in the usual sense; for example, the instruction, "shift left 0003," has in its normal address position the amount of the shift desired.

zero-access storage [ADP] Computer storage for which waiting time is negligible.

zero adjuster [ENG] A device for adjusting the pointer position of an instrument or meter to read zero when the measured quantity is zero.

zero beat [ELEC] The condition in which a circuit is oscillating at the exact frequency of an input signal, so no beat tone is produced or heard.

zero-beat reception *See* homodyne reception.

zero bevel gears [DES ENG] A special form of bevel gear having curved teeth with a zero-degree spiral angle.

zero bias [ELECTR] The condition in which the control grid and cathode of an electron tube are at the same direct-current voltage.

zero-bias tube [ELECTR] Vacuum tube which is designed so that it may be operated as a class B amplifier without applying a negative bias to its control grid.

zero branch [SPECT] A spectral band whose Fortrat parabola lies between two other Fortrat parabolas, with its vertex almost on the wave number axis.

zero compression [ADP] Any of a number of techniques used to eliminate the storage of nonsignificant leading zeros during data processing in a computer.

zero curtain [GEOL] The layer of ground between the active layer and permafrost where the temperature remains nearly constant at 0°C.

zero defects [IND ENG] A program for improving product quality to the point of perfection, so there will be no failures due to defects in construction.

zero deflection [ORD] Adjustment of a sight exactly parallel to the axis of the bore of the gun to which it is attached.

zero error [ELECTR] Delay time occurring within the transmitter and receiver circuits of a radar system; for accurate range data, this delay time must be compensated for in the calibration of the range unit.

zero-field emission *See* field-free emission current.

zerogel [CHEM] A gel which has dried until apparently solid; sometimes it will swell or redisperse to form a sol when treated with a suitable solvent.

zero geodesic *See* null geodesic.

zero gravity *See* weightlessness.

zero-gravity switch [ELEC] A switch that closes as weightlessness or zero gravity is approached; in one version, conductive sphere of mercury encompasses two contacts at zero gravity but flattens away from the upper contact under the influence of gravity. Also known as weightlessness switch.

zero height of burst [ORD] Condition obtained when rounds fired with the same fuse setting and the same quadrant elevation result in an equal number of airbursts and bursting of the projectile at the instant of ground impact.

zero in [ORD] 1. To adjust the sight settings of a weapon by calibrated results of firings. 2. To adjust any device to another so that automatic synchronization results.

zero layer [OCEANOGR] A reference level in the ocean, at which horizontal motion is at a minimum.

zero length [AERO ENG] In rocket launchers, zero length indicates that the launcher is designed to hold the rocket in position for launching but not to give it guidance.

zero-length launcher [ORD] A launcher that holds a vehicle in position and releases the rocket simultaneously at two points so that the buildup of thrust, normally rocket thrust, is sufficient to take the missile or vehicle directly into the air without need of a takeoff run and without imposing a pitch rate release.

zero-length rocket [ORD] A rocket with a sufficient thrust to launch a vehicle directly into the air.

zero level [ENG ACOUS] Reference level used for comparing sound or signal intensities; in audio-frequency work, a power of 0.006 watt is generally used as zero level; in sound, the threshold of hearing is generally assumed as the zero level.

zero-level address [ADP] The operand contained in an instruction so structured as to make immediate use of the operand.

zero-lift angle [AERO ENG] The angle of attack of an airfoil when its lift is zero.

zero-lift chord [AERO ENG] A chord taken through the trailing edge of an airfoil in the direction of the relative wind when the airfoil is at zero-lift angle of attack.

zero method *See* null method.

zero of a function [MATH] Any point where a function assumes the value zero.

zero-order hold [CONT SYS] A device which converts a sampled output into an output which is held constant between samples at the last sampled value.

zero-order reaction [PHYS CHEM] A reaction for which reaction rate is independent of the concentrations of the reactants; for example, a photochemical reaction in which the rate is determined by the intensity of light.

zero output [ELECTR] 1. Voltage response obtained from a magnetic cell in a zero state by a reading or resetting process. 2. Integrated voltage response obtained from a magnetic cell in a zero state by a reading or resetting process; a ratio of a one output to a zero output is a one-to-zero ratio.

zero phase-sequence relay [ELEC] Relay which functions in conformance with the zero phase-sequence component of the current, voltage, or power of the circuit.

zero point [ORD] The location of the center of a burst of an atomic missile at the instant of detonation; the zero point may be in the air, or on or beneath the surface of land or water, dependent upon the type of burst, and it is thus to be distinguished from ground zero.

zero-point energy [STAT MECH] The kinetic energy retained by the molecules of a substance at a temperature of absolute zero.

zero-point entropy [STAT MECH] The entropy that a substance such as glass, which is not in thermodynamic equilibrium, retains at a temperature of absolute zero.

zero-point vibration [STAT MECH] The vibrational motion which molecules in a crystal lattice, or particles in any oscillator potential, retain at a temperature of absolute zero; it is quantum-mechanical in origin. Also known as residual vibration.

zero potential [ELEC] Expression usually applied to the potential of the earth, as a convenient reference for comparison.

zero-power reactor [NUCLEO] An experimental nuclear reactor operated at low neutron flux and at a power level so low that no forced cooling is required; fission product activity in the fuel is then sufficiently low to permit handling the fuel after use.

zero subcarrier chromaticity [COMMUN] Chromaticity, in color television, which is intended to be displayed when the subcarrier amplitude is zero.

zero-sum game [MATH] A two-person game where the sum of the payoffs to the two players is zero for each move.

zero suppression [ADP] A process of replacing leading (nonsignificant) zeros in a numeral by blanks; it is an editing operation designed to make computable numerals easily readable to the human eye.

zero time reference [ELECTR] Reference point in time from which the operations of various radar circuits are measured.

zero twist [ORD] Rifling with no twist; some designs have this condition at the origin of rifling for guns with increasing twist rifling.

zeta function *See* Riemann zeta function.

Zeta Geminorum stars [ASTRON] A subgroup of classical Cepheid variable stars whose variation of magnitude with time for one complete cycle produces a quasi-bell-shaped curve.

zeta potential [PHYS] The electrical potential that exists across the interface of all solids and liquids. Also known as electrokinetic potential.

zeugogeosyncline [GEOL] A geosyncline in a craton or stable area, within which is also an uplifted area, receiving clastic sediments. Also known as yoked basin.

zeunerite [MINERAL] $Cu(UO_2)_2(AsO_4)_2 \cdot 10-16H_2O$ A green secondary mineral of the autunite group consisting of a hydrous copper uranium arsenate; it is isomorphous with uranospinite.

zeylanite See ceylonite.

Ziegler catalyst [MATER] A special catalyst developed to produce stereospecific polymers, and derived from a transition-metal halide and a metal hydride or metal alkyl.

Ziegler process [CHEM ENG] A process for the low-pressure linear polymerization of ethylene and stereospecific polymerization of propylene; the product is a high-density polymer or elastomer.

Ziehl-Neelsen stain [MICROBIO] A procedure for acid-fast staining of tubercle bacilli with carbol fuchsin.

zigzag lightning [GEOPHYS] Ordinary lightning of a cloud-to-ground discharge that appears to have a single, but very irregular, lightning channel; viewed from the right angle, this may be observed as beaded lightning.

zigzag reflections [ELECTROMAG] From a layer of the ionosphere, high-order multiple reflections which may be of abnormal intensity; they occur in waves which travel by multihop ionosphere reflections and finally turn back toward their starting point by repeated reflections from a slightly curved or sloping portion of an ionized layer.

Zimm plot [ANALY CHEM] A graphical determination of the root-square-mean end-to-end distances of coillike polymer molecules during scattered-light photometric analyses.

zinc [CHEM] A metal of group IIb, symbol Zn, atomic number 30, atomic weight 65.37; explosive as powder; soluble in acids and alkalies, insoluble in water; strongly electropositive; melts at 419°C, boils at 907°C. [MET] A shiny, bluish-white, lustrous metal that is ductile when pure; used in alloys, metal coatings, electrical fuses, anodes, and dry cells.

zinc-65 [NUC PHYS] A radioactive isotope of zinc, which has a 250-day half-life with beta and gamma radiation; used in alloy-wear tracer studies and body metabolism studies.

zinc acetate [ORG CHEM] $Zn(C_2H_3O_2)_2 \cdot 2H_2O$ Pearly-white crystals with an astringent taste; soluble in water and alcohol; decomposes at 200°C; used to preserve wood in textile dyeing, and as an analytical reagent, a feed additive, and a polymer cross-linking agent.

zinc arsenate [INORG CHEM] $ZnHAsO_4$ A toxic white powder that is insoluble in water, soluble in alkalies; used as an insecticide. Also known as zinc orthoarsenate.

zinc arsenite [INORG CHEM] $Zn(AsO_2)_2$ A toxic white powder that is insoluble in water, soluble in alkalies; used as an insecticide and timber preservative. Also known as zinc metaarsenite.

zincate [INORG CHEM] A reaction product of zinc with an alkali metal or with ammonia; for example, sodium zincate, Na_2ZnO_2.

zinc baryta white See lithopone.

zinc borate [INORG CHEM] $3ZnO \cdot 2B_2O_3$ A white, amorphous powder that is soluble in dilute acids, slightly soluble in water; melts at 980°C; used in medicine, as a ceramics flux, as an inhibitor for mildew, and to fireproof textiles.

zinc bromide [INORG CHEM] $ZnBr_2$ Water- and alcohol-soluble, white crystals that melt at 294°C; used in medicine, manufacture of rayon, and photography, and as a radiation viewing screen.

zinc carbonate [INORG CHEM] $ZnCO_3$ White crystals that are insoluble in water, soluble in alkalies and acids; used in ceramics and ointments, and as a fireproofing agent and feed additive.

zinc chills See metal fume fever.

zinc chloride [INORG CHEM] $ZnCl_2$ Water- and alcohol-soluble, white, fire-hazardous crystals that melt at 290°C, and are irritating to the skin; used as a catalyst and in electroplating, wood preservation, textile processing, petroleum refining, medicine, and feed additives.

zinc chromate [INORG CHEM] $ZnCrO_4$ A toxic, yellow powder that is insoluble in water, soluble in acids; used as a pigment in paints (artists', automotive, primer), varnishes, linoleum, and epoxy laminates.

zinc cyanide [INORG CHEM] $Zn(CN)_2$ A toxic, white powder that is insoluble in water and alcohol, soluble in alkalies and dilute acids; melts at 800°C; used as an analytical reagent and insecticide, and in medicine and metal plating.

zinc fluoride [INORG CHEM] ZnF_2 A toxic white powder that is slightly soluble in water and melts at 872°C; used in enamels, ceramic glazes, and galvanizing.

zinc formate [ORG CHEM] $Zn(CHO_2)_2 \cdot 2H_2O$ Toxic, white crystals that are soluble in water, insoluble in alcohol; used as a catalyst, weatherproofing agent, and wood preservative.

zinc halide [INORG CHEM] A binary compound of zinc and a halogen; for example, $ZnBr_2$, $ZnCl_2$, ZnF_2, and ZnI_2.

zinc hydroxide [INORG CHEM] $Zn(OH)_2$ Colorless, water-soluble crystals that decompose at 125°C; used as a chemical intermediate and in rubber compounding and surgical dressings.

zincite [MINERAL] $(Zn,Mn)O$ A deep-red to orange-yellow brittle mineral; an ore of zinc. Also known as red oxide of zinc; red zinc ore; ruby zinc; spartalite.

zinckenite See zinkenite.

zinc metaarsenite See zinc arsenite.

zinc naphthenate [ORG CHEM] $Zn(C_6H_5COO)_2$ A combustible, viscous, acetone-soluble solid; used in paints, varnishes, and resins, and as a drier and wetting agent, insecticide, fungicide, and mildewstat.

zinc orthoarsenate See zinc arsenate.

zinc orthophosphate See zinc phosphate.

zinc oxide [INORG CHEM] ZnO A bitter-tasting, white to gray powder that is insoluble in water, soluble in alkalies and acids; melts at 1978°C; used as a pigment, mold-growth inhibitor, and dietary supplement, and in cosmetics, electronics, and color photography.

zinc phosphate [INORG CHEM] $Zn_3(PO_4)_2$ A white powder that is insoluble in water, soluble in acids and ammonium hydroxide; melts at 900°C; used in coatings for steel, aluminum, and other metals, and in dental cements and phosphors. Also known as tribasic zinc phosphate; zinc orthophosphate.

zinc phosphide [INORG CHEM] Zn_3P_2 A toxic, alcohol-insoluble, gray gritty powder that reacts violently with oxidizing agents; melts at over 420°C, decomposes in water; used as a rat poison and in medicine.

zinc selenide [INORG CHEM] $ZnSe$ A water-insoluble, moderately toxic, yellow to reddish solid that is a fire hazard when in contact with water and acids; melts above 1100°C; used as infrared optical windows.

zinc-silver-chloride primary cell [ELEC] A reserve primary cell that is activated by adding water; it can have a high capacity, up to 40 watt-hours per pound, and long life after activation.

zinc spar See smithsonite.

zinc spinel See gahnite.

zinc sulfate [INORG CHEM] $ZnSO_4 \cdot 7H_2O$ Efflorescent, water-soluble, colorless crystals with an astringent taste; used to preserve skins and wood and as a paper bleach, analytical reagent, feed additive, and fungicide. Also known as white copperas; white vitriol; zinc vitriol.

zinc sulfide [INORG CHEM] ZnS A yellowish powder that is insoluble in water, soluble in acids; exists in two crystalline forms (alpha, or wartzite, and beta, or sphalerite); beta becomes alpha at 1020°C, and sublimes at 1180°C; used as a pigment for paints and linoleum, in opaque glass, rubber, and plastics, for hydrosulfite dyeing process, as x-ray and television screen phosphor, and as a fungicide.

zinc sulfide white See lithopone.

zinc telluride [INORG CHEM] $ZnTe$ Moderately toxic, reddish crystals that melt at 1238°C and decompose in water.

zinc vitriol See zinc sulfate.

zinc white See Chinese white.

Zingiberaceae [BOT] A family of aromatic monocotyledonous plants in the order Zingiberales characterized by one functional stamen with two pollen sacs, distichously arranged leaves and bracts, and abundant oil cells.

Zingiberales [BOT] An order of monocotyledonous herbs or scarcely branched shrubs in the subclass Commelinidae characterized by pinnately veined leaves and irregular flowers that have well-differentiated sepals and petals, an inferior ovary, and either one or five functional stamens.

ZINC

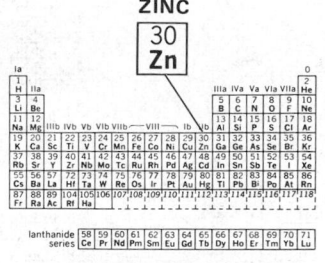

Periodic table of the chemical elements showing the position of zinc.

ZINCITE

Zincite crystals from Franklin, N.J. (Specimen from Department of Geology, Bryn Mawr College)

zinkenite [MINERAL] $Pb_6Sb_{14}S_{27}$ A steel-gray orthorhombic mineral consisting of a lead antimony sulfide and occurring in crystals and in masses; has metallic luster, hardness of 3–3.5 on Mohs scale, and specific gravity of 5.30–5.35. Also spelled zinckenite.

zinnwaldite [MINERAL] $K_2(Li,Fe,Al)_6(Si,Al)_8O_{20}(OH,F)_4$ A pale-violet, yellowish, brown, or dark-gray mica mineral; an iron-bearing variety of lepidolite; the characteristic mica of greisens.

zipper [ENG] A generic name for slide fasteners in which two sets of interlocking teeth of the same design provide sturdy and continuous closure for adjacent pieces of textile, leather, and other materials.

zipper conveyor [MECH ENG] A type of conveyor belt with zipperlike teeth that mesh to form a closed tube; used to handle fragile materials.

zircon [MINERAL] $ZrSiO_4$ A brown, green, pale-blue, red, orange, golden-yellow, grayish, or colorless neosilicate mineral occurring in tetragonal prisms; it is the chief source of zirconium; the colorless varieties provide brilliant gemstones. Also known as hyacinth; jacinth; zirconite.

zirconia See zirconium oxide.

zirconia brick [MATER] A type of brick containing zirconium oxide, used to line metallurgical furnaces.

zirconic anhydride See zirconium oxide.

zirconite See zircon.

zirconium [CHEM] A metallic element of group IVb, symbol Zr, atomic number 40, atomic weight 91.22; occurs as crystals, flammable as powder; insoluble in water, soluble in hot, concentrated acids; melts at 1850°C, boils at 4377°C. [MET] A hard, lustrous, grayish metal that is strong and ductile; used in alloys, pyrotechnics, welding fluxes, and explosives.

zirconium-95 [NUC PHYS] A radioactive isotope of zirconium; half-life of 63 days with beta and gamma radiation; used to trace petroleum-pipeline flows and in the circulation of a catalyst in a cracking plant.

zirconium boride [INORG CHEM] ZrB_2 A hard, toxic, gray powder that melts at 3000°C; used as an aerospace refractory, in cutting tools, and to protect thermocouple tubes. Also known as zirconium diboride.

zirconium carbide [INORG CHEM] ZrC Hard, gray crystals that are soluble in water, soluble in acids; as powder, it ignites spontaneously in air; melts at 3400°C, boils at 5100°C; used as an abrasive, refractory, and metal cladding, and in cermets, incandescent filaments, and cutting tools.

zirconium chloride See zirconium tetrachloride.

zirconium diboride See zirconium boride.

zirconium dioxide See zirconium oxide.

zirconium halide [INORG CHEM] A compound of zirconium with a halogen; for example, $ZrBr_2$, $ZrCl_2$, $ZrCl_3$, $ZrCl_4$, $ZrBr_2$, $ZrBr_3$, ZrF_4, and ZrI_4.

zirconium hydride [INORG CHEM] ZrH_2 A flammable, gray-black powder; used in powder metallurgy and nuclear moderators, and as a reducing agent, vacuum-tube getter, and metal-foaming agent.

zirconium hydroxide [INORG CHEM] $Zr(OH)_4$ A toxic, amorphous white powder; insoluble in water, soluble in dilute mineral acids; decomposes at 550°C; used in pigments, glass, and dyes, and to make zirconium compounds.

zirconium lamp [ELECTR] A high-intensity point-source lamp having a zirconium oxide cathode in an argon-filled bulb, used because of its low emanation of long-wavelength light and its concentrated source.

zirconium nitride [INORG CHEM] ZrN A hard, brassy powder that is soluble in concentrated acids; melts at 2930°C; used in refractories, cermets, and laboratory crucibles.

zirconium orthophosphate See zirconium phosphate.

zirconium oxide [INORG CHEM] ZrO_2 A toxic, heavy white powder that is insoluble in water, soluble in mineral acids; melts at 2700°C; used in ceramic glazes, special glasses, and medicine, and to make piezoelectric crystals. Also known as zirconia; zirconic anhydride; zirconium dioxide.

zirconium oxychloride [INORG CHEM] $ZrOCl_2 \cdot 8H_2O$ White crystals that are soluble in water, insoluble in organic solvents, and acidic in aqueous solution; used for textile dyeing and oil-field acidizing, in cosmetics and greases, and for antiperspirants and water repellents. Also known as basic zirconium chloride; zirconyl chloride.

zirconium phosphate [INORG CHEM] $ZrO(H_2PO_4)_2 \cdot 3H_2O$ A toxic, dense white powder that is insoluble in water, soluble in acids and organic solvents; decomposes on heating; used as an analytical reagent, coagulant, and radioactive-phosphor carrier. Also known as basic zirconium phosphate; zirconium orthophosphate.

zirconium tetrachloride [INORG CHEM] $ZrCl_4$ Toxic, alcohol-soluble, white lustrous crystals; sublimes above 300°C and decomposes in water; used to make pure zirconium and for water-repellent textiles and as an analytical reagent. Also known as zirconium chloride.

zirconyl chloride See zirconium oxychloride.

zitterbewegung [QUANT MECH] An oscillatory motion of an electron suggested in some interpretations of the Dirac electron theory, having a frequency greater than $4\pi mc^2/h$, where m is the electron's mass, c is the speed of light, and h is Planck's constant, or approximately 1.5×10^{21} hertz.

Z line [HISTOL] The line formed by attachment of the actin filaments between two sarcomeres.

Z-marker beacon [NAV] Transmitter equipment installed as part of a four-course radio range; it radiates vertically to indicate to aircraft when they pass directly over the range station; it is usually not keyed for identification.

Zn See zinc.

Zoantharia [INV ZOO] A subclass of the class Anthozoa; individuals are monomorphic and most have retractile, simple, tubular tentacles.

Zoanthidea [INV ZOO] An order of anthozoans in the subclass Zoantharia; these are mostly colonial, sedentary, skeletonless, anemonelike animals that live in warm, shallow waters and coral reefs.

Zoarcidae [VERT ZOO] The eelpouts, a family of actinopterygian fishes in the order Gadiformes which inhabit cold northern and far southern seas.

zobaa [METEOROL] In Egypt, a lofty whirlwind of sand resembling a pillar, moving with great velocity.

Zodiac [ASTRON] A band of the sky extending 8° on each side of the ecliptic, within which the moon and principal planets remain.

zodiacal cone See zodiacal pyramid.

zodiacal constellations [ASTRON] The constellations Aries, Taurus, Gemini, Cancer, Leo, Virgo, Libra, Scorpio, Sagittarius, Capricorn, Aquarius, and Pisces which are assigned to 12 equal portions of the zodiac.

zodiacal counterglow See gegenschein.

zodiacal light [GEOPHYS] A diffuse band of luminosity occasionally visible on the ecliptic; it is sunlight diffracted and reflected by dust particles in the solar system within and beyond the orbit of the earth.

zodiacal pyramid [GEOPHYS] The pattern formed by the zodiacal light. Also known as zodiacal cone.

zoea [INV ZOO] An early larval stage of decapod crustaceans distinguished by a relatively large cephalothorax, conspicuous eyes, and large, fringed antennae.

zoisite [MINERAL] $Ca_2Al_3Si_3O_{12}(OH)$ A white, gray, brown, green, or rose-red orthorhombic mineral of the epidote group consisting of a basic calcium aluminum silicate and occurring massive or in prismatic crystals.

Zollinger-Ellison syndrome [MED] Gastric hypersecretion and hyperacidity, fulminating intractable atypical peptic ulceration, and hyperplasia of the islet cells of the pancreas.

zonal [METEOROL] Latitudinal, easterly or westerly, opposed to meridional.

zonal centrifuge [BIOL] A centrifuge that uses a rotating chamber of large capacity in which to separate cell organelles by density-gradient centrifugation.

zonal circulation See zonal flow.

zonal flow [METEOROL] The flow of air along a latitude circle; more specifically, the latitudinal (east or west) component of existing flow. Also known as zonal circulation.

zonal harmonics [MATH] Spherical harmonics which do not depend on the azimuthal angle; they are proportional to Legendre polynomials of $\cos \theta$, where θ is the colatitude.

zonal index [METEOROL] A measure of strength of the middle-latitude westerlies, expressed as the horizontal pressure difference between 35° and 55° latitude, or as the corresponding geostrophic wind.

zonal kinetic energy [METEOROL] The kinetic energy of the

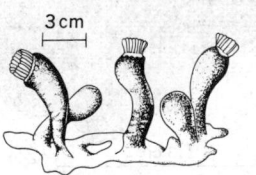

ZIRCONIUM

40
Zr

Periodic table of the chemical elements showing the position of zirconium.

ZOANTHIDEA

Colony of *Zoanthina tentaculata*, a representative species of Zoanthidea. (*After Y. Dalage*)

ZOEA

Zoea larva of crab, showing characteristic features. (*Smithsonian Institution*)

mean zonal wind, obtained by averaging the zonal component of the wind along a fixed latitude circle.

zonal soil [GEOL] In early classification systems in the United States, a soil order including soils with well-developed characteristics that reflect the influence of agents of soil genesis. Also known as mature soil.

zonal theory [GEOL] A theory of the formation of mineral deposition and sequence patterns, based on the changes in a mineral-bearing fluid as it passes upward from a magmatic source.

zonal westerlies *See* westerlies.

zonal wind [METEOROL] The wind, or wind component, along the local parallel of latitude, as distinguished from the meridional wind.

zonal winds *See* westerlies.

zonal wind-speed profile [METEOROL] A diagram in which the speed of the zonal flow is one coordinate and latitude the other.

zona pellucida [HISTOL] The thick, solid, elastic envelope of the ovum. Also known as oolemma.

zonation [ECOL] Arrangement of organisms in biogeographic zones. [GEOL] The condition of being arranged in zones.

zonda [METEOROL] A hot wind in Argentina.

Zond spacecraft [AERO ENG] One of a series of Soviet space probes which have photographed the moon and made observations in interplanetary space.

zone [ADP] **1.** One of the top three rows of a punched card, namely, the 11, 12, and zero rows. **2.** *See* storage area. [GEOGR] An area or region of latitudinal character. [GEOL] A belt, layer, band, or strip of earth material such as rock or soil. [MATH] The portion of a sphere lying between two parallel planes that intersect the sphere. [ORD] **1.** Any tactical area of importance, generally parallel to the front, such as a fortified area, a defensive position, a combat zone, or a traffic-control zone. **2.** An area in which projectiles will fall when a given propelling charge is used and the elevation is varied between the minimum and the maximum; in practice, generally limited to howitzer and mortar firings.

zone bit [ADP] A set of bits; for example, it may indicate whether the set of bits represents a numeric or alphabetic character.

zone blanking [ELECTR] Method of turning off the cathode-ray tube during part of the sweep of an antenna.

zone charge [ORD] The number of increments of propellant in a propellant charge of semifixed rounds, corresponding to the intended zone of fire; for example, zone charge five consists of five increments of propellant.

zone control [ENG] The zoning of a process or building, and the independent heating or temperature controls for each zone.

zoned decimal [ADP] A format for use with EBCDIC input and output permitting a sign overpunch in the low order position of the field; thus, + 1234 would be represented as: 1111/0001/1111/0010/1111/0011/1100/0100.

zone description [NAV] The number, with its sign, that must be added to or subtracted from the zone time to obtain the Greenwich mean time; the zone description is usually a whole number of hours.

zone fire [ORD] Artillery or mortar fire that is designed to cover an area in which a target is situated.

zone marker [NAV] Very-high-frequency radio station designed to radiate signals vertically in a cone-shaped pattern to define a zone above a radio range station.

zone melting crystallization [CHEM ENG] A method for purification of crystalline solids; the sample, packed in a narrow column, is heated so that a molten zone passes down through the sample, carrying impurities with it.

zone meridian [ASTRON] The meridian used for reckoning zone time; this is generally the nearest meridian whose longitude is exactly divisible by 15°.

zone noon [ASTRON] Twelve o'clock zone time, or the instant the mean sun is over the upper branch of the zone meridian.

zone of aeration [GEOL] A subsurface zone containing water below atmospheric pressure and air or gases at atmospheric pressure. Also known as unsaturated zone; vadose zone; zone of suspended water.

zone of avoidance [ASTRON] An irregularly shaped area in the Milky Way Galaxy in which no extragalactic nebulae are observed because of the presence of interstellar matter.

zone of cementation [GEOL] The layer of the earth's crust in which unconsolidated deposits are cemented by percolating water containing dissolved minerals from the overlying zone of weathering. Also known as belt of cementation.

zone of intersection [NAV] That part of a civil airway which overlaps and lies within any part of another civil airway.

zone of maximum precipitation [METEOROL] In a mountain region, the belt of elevation at which the annual precipitation is greatest.

zone of optimal proportion [IMMUNOL] One of three zones considered to appear when antigen and antibody are mixed; it is that zone in which there is no uncombined antigen or antibody. Also known as equivalence zone.

zone of saturation [HYD] A subsurface zone in which water fills the interstices and is under pressure greater than atmospheric pressure. Also known as phreatic zone; saturated zone.

zone of silence *See* skip zone.

zone of soil water *See* belt of soil water.

zone of suspended water *See* zone of aeration.

zone plate [OPTICS] A plate with alternate transparent and opaque rings, designed to block off every other Fresnel half-period zone; light from a point source passing through the plate produces an intense point image much like that produced by a lens.

zone-position indicator [ENG] Auxiliary radar set for indicating the general position of an object to another radar set with a narrower field.

zone punch [ADP] In a punched card, the 11 or 12 punch according to any code, the zero punch if another numeric punch is present in the same column, and sometimes the 8 and 9 punches in EBCDIC-coded cards.

zone purification *See* zone refining.

zone refining [MET] A technique to purify materials in which a narrow molten zone is moved slowly along the complete length of the specimen to bring about impurity segregation, and which depends on differences in composition of the liquid and solid in equilibrium. Also known as zone purification.

zone theory *See* stage theory.

zone time [ASTRON] The local mean time of a reference or zone meridian whose time is kept throughout a designated zone; the zone meridian is usually the nearest meridian whose longitude is exactly divisible by 15°.

zoning [CIV ENG] Designation and reservation under a master plan of land use for light and heavy industry, dwellings, offices, and other buildings; use is enforced by restrictions on types of buildings in each zone. [ELECTROMAG] The displacement of various portions of the lens or surface of a microwave reflector so the resulting phase front in the near field remains unchanged. Also known as stepping.

zonite [INV ZOO] A body segment in Diplopoda.

zonochlorite *See* pumpellyite.

zoocecidium [PL PATH] A plant gall usually caused by an insect.

zoochlorellae [BIOL] Unicellular green algae which live as symbionts in the cytoplasm of certain protozoans, sponges, and other invertebrates.

zoochory [BOT] Dispersal of plant disseminules by animals.

zooecium [INV ZOO] The exoskeleton of a feeding zooid in bryozoans.

zoogeographic region [ECOL] A major unit of the earth's surface characterized by faunal homogeneity.

zoogeography [BIOL] The science that attempts to describe and explain the distribution of animals in space and time.

zoogloea [MICROBIO] A gelatinous or mucilaginous mass characteristic of certain bacteria grown in organic-rich fluid media.

zooid [INV ZOO] A more or less independent individual of colonial animals such as bryozoans and coral.

zoology [BIOL] The science that deals with knowledge of animal life.

Zoomar lens [OPTICS] Trademark for a type of zoom lens used in a television camera in which the focal length is determined by the cameraman by means of a handle which he moves forward or backward.

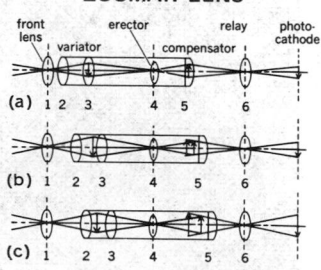

ZOOMAR LENS

Zoomar lens in three operating positions. (a) Wide-angle. (b) Medium-angle. (c) Telephoto. Lens elements 2, 3, and 5 are mounted in movable barrel connected to zoom handle. Lens elements 1, 4, and 6 are stationary, mounted in lens housing. *(From D. G. Fink, ed., Television Engineering Handbook, McGraw-Hill, 1957)*

ZOSTEROPHYLLATAE

Reconstruction of *Zosterophyllum*, a typical genus of Zosterophyllatae, showing characteristic features. *(Modified from Kräusel and Weyland, 1935)*

ZYGNEMATACEAE

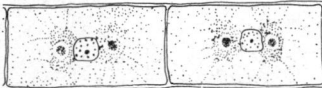

Vegetative cells of *Zygnema* species, type genus of the family Zygnemataceae, showing characteristic pair of stellate chloroplasts in each cell. *(From G. M. Smith, Cryptogamic Botany, 2d ed., McGraw-Hill, 1955)*

ZYGOPTERA

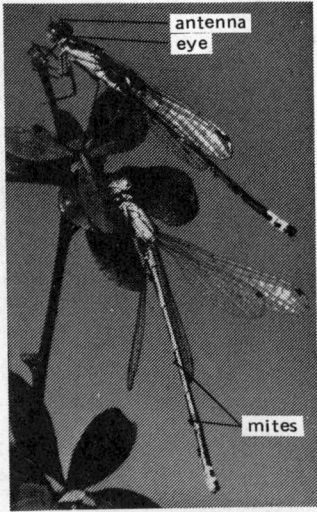

antenna
eye
mites

Two adult damsel flies of the suborder Zygoptera. Note mites attached.

Zoomastigina [INV ZOO] The equivalent name for Zoomastigophorea.

Zoomastigophorea [INV ZOO] A class of flagellate protozoans in the subphylum Sarcomastigophora; some are simple, some are specialized, and all are colorless.

zoom lens [OPTICS] A system of lenses in which two or more parts are moved with respect to each other to obtain a continuously variable focal length and hence magnification, while the image is kept in the same image plane. Also known as variable-focal-length lens; varifocal lens.

zoonoses [BIOL] Diseases which are biologically adapted to and normally found in lower animals but which under some conditions also infect man.

zoophobia [PSYCH] An abnormal fear of animals.

zooplankton [ECOL] Microscopic animals which move passively in aquatic ecosystems.

zoosphere [ECOL] The world community of animals.

zoosporangium [BOT] A spore case bearing zoospores.

zoospore [BIOL] An independently motile spore.

zooxanthellae [BIOL] Microscopic yellow-green algae which live symbiotically in certain radiolarians and marine invertebrates.

Zoraptera [INV ZOO] An order of insects, related to termites and psocids, which live in decaying wood, sheltered from light; most individuals are wingless, pale in color, and blind.

Zorn's lemma [MATH] Each linearly ordered subset of a partially ordered set contains a maximal element.

Zoroasteridae [INV ZOO] A family of deep-water asteroid echinoderms in the order Forcipulatida.

Zorotypidae [INV ZOO] The single family, containing one genus, *Zorotypus*, in the order Zoraptera.

zorsite [MINERAL] $Ca_2Al_3Si_3O_{12}(OH)$ White, gray, brown, green, or rose-red orthorhombic mineral of the epidote group; an essential constituent of saussurite.

zoster *See* herpes zoster.

Zosteraceae [BOT] A family of monocotyledonous plants in the order Najadales; the group is unique among flowering plants in that they grow submerged in shallow ocean waters near the shore.

Zosterophyllatae [PALEOBOT] *See* Zosterophyllopsida.

Zosterophyllopsida [PALEOBOT] A group of early land vascular plants ranging from the Lower to the Upper Devonian; individuals were leafless and rootless.

Zr *See* zirconium.

Z time *See* Greenwich mean time.

z-transfer function *See* pulsed transfer function.

z-transform [MATH] The z-transform of a sequence whose general term is f_n is the sum of a series whose general term is $f_n z^{-n}$, where z is a complex variable; n runs over the positive integers for a one-sided transform, over all the integers for a two-sided transform.

Z twist [TEXT] A right-handed yarn twist in which the spiral slants like the middle part of the letter Z.

Zulu time *See* Greenwich mean time.

Zuni [ORD] A United States air-to-surface unguided rocket with solid propellant; can be armed with various types of heads, including flares, fragmentation, and armor-piercing.

Zurich number *See* relative sunspot number.

Zwischengebirge *See* median mass.

zwitterion *See* dipolar ion.

Zygaenidae [INV ZOO] A diverse family of small, often brightly colored African moths in the superfamily Zygaenoidea.

Zygaenoidea [INV ZOO] A superfamily of moths in the suborder Heteroneura characterized by complete venation, rudimentary palpi, and usually a rudimentary proboscis.

Zyglo method [ENG] A procedure for visualizing incipient cracks caused by fatigue failure, in which the part is immersed in a special activated penetrating oil and viewed under black light.

Zygnemataceae [BOT] A family of filamentous plants in the order Conjugales; they are differentiated into genera by chloroplast morphology, which may be spiral, bandlike, or cushionlike.

zygodactyl [VERT ZOO] Of birds, having a toe arrangement of two in front and two behind.

zygomatic bone [ANAT] A bone of the side of the face below the eye; forms part of the zygomatic arch and part of the orbit in mammals. Also known as malar bone.

zygomorphic [BIOL] Bilaterally symmetrical.

Zygomycetes [MYCOL] A class of fungi in the division Eumycetes.

zygopophysis [ANAT] One of the articular processes of the neural arch of a vertebra.

Zygoptera [INV ZOO] The damsel flies, a suborder of insects in the order Odonata; individuals are slender, dainty creatures, often with bright-blue or orange coloring and usually with clear or transparent wings.

zygote [EMBRYO] 1. An organism produced by the union of two gametes. 2. The fertilized ovum before cleavage.

zygotene [CYTOL] The stage of meiotic prophase during which homologous chromosomes synapse; visible bodies in the nucleus are now bivalents. Also known as amphitene.

zygotic induction [VIROL] Phage induction following conjugation of a lysogenic bacterium with a nonlysogenic one.

zymase [BIOCHEM] A complex of enzymes that catalyze glycosis.

zymogen [BIOCHEM] The inactive precursor of an enzyme; liberates an active enzyme on reaction with an appropriate kinose. Also known as proenzyme.

zymogen granules [BIOCHEM] Granules of zymogen in gland cells, particularly those of the pancreatic acini and of the gastric chief cells.

zymogenic [MICROBIO] Obtaining energy by amylolitic processes.

zymophore [BIOCHEM] The active portion of an enzyme.

zymosis *See* fermentation.

zymosterol [BIOCHEM] $C_{27}H_{43}OH$ An unsaturated sterol obtained from yeast fat; yields cholesterol on hydrogenation.

Zythiaceae [MYCOL] A family of fungi of the order Sphaeropsidales which contains many plant and insect pathogens.

Appendix

Appendix

U.S. Customary System and the metric system

To date, scientists and engineers have used two major systems of units in measurement. These are commonly called the U.S. Customary System (inherited from the British Imperial System) and the metric system.

In the U.S. Customary System the units yard and pound with their divisions, such as the inch, and multiples, such as the ton, are basic. The metric system was evolved during the 18th century and has been adopted for general use by most countries. Nearly everywhere it is used for precise measurements in science. The meter and kilogram with their multiples, such as the kilometer, and fractions, such as the gram, are basic to the metric system.

In the U.S. Customary System, units of the same kind are related almost at random. For example, there are the units of length, the inch, yard, and mile. In the metric system the relationships between units of the same kind are strictly decimal (millimeter, meter, and kilometer).

However, to complicate matters in scientific writing, there is no uniformity within each of these two systems as to the choice of units for the same quantities. For example, the hour or the second, the foot or the inch, and the centimeter or the millimeter could be chosen by a scientist as the unit of measurement for the quantities time and length.

The International System, or SI

To simplify matters and to make communication more understandable, an internationally accepted system of units is coming into use. This is termed the International System of Units, which is abbreviated SI in all languages.

Fundamentally the system is metric with the base units derived from scientific formulas or natural constants. For example, the meter in the SI is defined as the length equal to 1 650 763.73 wavelengths in vacuum of the radiation corresponding to the transition between the electronic energy levels $2p_{10}$ and $5d_5$ of the krypton-86 atom. Previously, in the metric system, the meter was defined as the distance between two marks on a specific metal bar.

In a similar way the second in the SI is defined as the duration of 9 192 631 770 periods of the radiation corresponding to the transition between two hyperfine levels of the ground state of the cesium-133 atom.

Interestingly, the kilogram, the SI unit of mass, is still the mass of the kilogram kept at Sèvres, France. However, it is possible that eventually the unit will be redefined in terms of atomic mass.

Although the SI is increasing in usage by scientists and engineers, there are some units in everyday use which will probably remain, for example, minute, hour, day, degree (angle), and liter. The point should be made, however, that these terms will not be employed in a scientific context if the SI is fully adopted.

Because of their extremely common use among scientists, several units are still permitted in conjunction with SI units, for example, the electron volt, gauss, barn, and curie. In time their usage might be phased out.

One further point is that in October, 1967, the Thirteenth General Conference of Weights and Measures decided to name the SI unit of thermodynamic temperature "kelvin" (symbol K) instead of "degree Kelvin" (symbol °K). For example, the notation is 273K and not 273°K.

The base units and derived units of the SI are shown in **Table 1** and **Table 2.**

In the SI the prefixes differ from a unit in steps of 10^3. A list of prefix terms, symbols, and their factors is given in **Table 3.** Some examples of the use of these prefixes follow:

$$1000 \text{ m} = 1 \text{ kilometer} = 1 \text{ km}$$
$$1000 \text{ V} = 1 \text{ kilovolt} = 1 \text{ kV}$$
$$1\,000\,000 \ \Omega = 1 \text{ megohm} = 1 \text{ M}\Omega$$
$$0.000\,000\,001 \text{ s} = 1 \text{ nanosecond} = 1 \text{ ns}$$

Only one prefix is to be employed for a unit. For example:

$$1000 \text{ kg} = 1 \text{ Mg} \qquad \text{not } 1 \text{ kkg}$$
$$10^{-9} \text{ s} = 1 \text{ ns} \qquad \text{not } 1 \text{ m}\mu\text{s}$$
$$1\,000\,000 \text{ m} = 1 \text{ Mm} \qquad \text{not } 1 \text{ kkm}$$

Also, when a unit is raised to a power, the power applies to the whole unit including the prefix. For example:

$$\text{km}^2 = (\text{km})^2 = (1000 \text{ m})^2 = 10^6 \text{ m}^2 \qquad \text{not } 1000 \text{ m}^2$$

Table 1. Base units of the International System

Quantity	Name of unit	Unit symbol
length	meter	m
mass	kilogram	kg
time	second	s
electric current	ampere	A
temperature	kelvin	K
luminous intensity	candela	cd
amount of substance	mole	mol

Table 2. Derived units of the International System

Quantity	Name of unit	Unit symbol or abbreviation, where differing from base form	Unit expressed in terms of base or supplementary units*
area	square meter		m²
volume	cubic meter		m³
frequency	hertz	Hz	s⁻¹
density	kilogram per cubic meter		kg/m³
velocity	meter per second		m/s
angular velocity	radian per second		rad/s
acceleration	meter per second squared		m/s²
angular acceleration	radian per second squared		rad/s²
volumetric flow rate	cubic meter per second		m³/s
force	newton	N	kg·m/s²
surface tension	newton per meter, joule per square meter	N/m, J/m²	kg/s²
pressure	newton per square meter, pascal	N/m², Pa	kg/m·s²
viscosity, dynamic	newton-second per square meter, pascal-second	N·s/m², Pa·s	kg/m·s
viscosity, kinematic	meter squared per second		m²/s
work, torque, energy, quantity of heat	joule, newton-meter, watt-second	J, N·m, W·s	kg·m²/s²
power, heat flux	watt, joule per second	W, J/s	kg·m²/s³
heat flux density	watt per square meter	W/m²	kg/s³
volumetric heat release rate	watt per cubic meter	W/m³	kg/m·s³
heat transfer coefficient	watt per square meter degree	W/m²·deg	kg/s³·deg
heat capacity (specific)	joule per kilogram degree	J/kg·deg	m²/s²·deg
capacity rate	watt per degree	W/deg	kg·m²/s³·deg
thermal conductivity	watt per meter degree	W/m·deg, $\dfrac{\text{J·m}}{\text{s·m}^2\text{·deg}}$	kg·m/s³·deg
quantity of electricity	coulomb	C	A·s
electromotive force	volt	V, W/A	kg·m²/A·s³
electric field strength	volt per meter		V/m
electric resistance	ohm	Ω, V/A	kg·m²/A²·s³
electric conductivity	ampere per volt meter	A/V·m	A²·s³/kg·m³
electric capacitance	farad	F, A·s/V	A³·s⁴/kg·m²
magnetic flux	weber	Wb, V·s	kg·m²/A·s²
inductance	henry	H, V·s/A	kg·m²/A²·s²
magnetic permeability	henry per meter	H/m	kg·m/A²·s²
magnetic flux density	tesla, weber per square meter	T, Wb/m²	kg/A·s²
magnetic field strength	ampere per meter		A/m
magnetomotive force	ampere		A
luminous flux	lumen	lm	cd·sr
luminance	candela per square meter		cd/m²
illumination	lux, lumen per square meter	lx, lm/m²	cd·sr/m²

*Supplementary units are: plane angle, radian (rad), solid angle, steradian (sr).

Table 3. Prefixes for units in the International System

Prefix	Symbol	Power	Example
tera	T	10^{12}	
giga	G	10^9	
mega	M	10^6	megahertz (MHz)
kilo	k	10^3	kilometer (km)
hecto	h	10^2	
deca	da	10^1	
deci	d	10^{-1}	
centi	c	10^{-2}	
milli	m	10^{-3}	milligram (mg)
micro	μ	10^{-6}	microgram (μg)
nano	n	10^{-9}	nanosecond (ns)
pico	p	10^{-12}	picofarad (pf)
femto	f	10^{-15}	
atto	a	10^{-18}	

Some common units defined in terms of SI units are given in **Table 4** (the definitions in the fourth column are exact).

Table 4. Some common units defined in terms of SI units

Quantity	Name of unit	Unit symbol	Definition of unit
length	inch	in	2.54×10^{-2} m
mass	pound (avoirdupois)	lb	0.45359237 kg
force	kilogram-force	kgf	9.80665 N
pressure	atmosphere	atm	101325 N·m^{-2}
pressure	torr	Torr	(101325/760) N·m^{-2}
pressure	conventional millimeter of mercury*	mmHg	$13.5951 \times 980.665 \times 10^{-2}$ N·m^{-2}
energy	kilowatt-hour	kWh	3.6×10^6 J
energy	thermochemical calorie	cal	4.184 J
energy	international steam table calorie	cal$_{IT}$	4.1868 J
thermodynamic temperature (T)	degree Rankine	°R	(5/9) K
customary temperature (t)	degree Celsius	°C	$t(°C) = T(K) - 273.16$
customary temperature (t)	degree Fahrenheit	°F	$t(°F) = T(°R) - 459.68$
radioactivity	curie	Ci	3.7×10^{10} s^{-1}
energy†	electron volt	eV	eV $\approx 1.6021 \times 10^{-19}$ J
mass†	unified atomic mass unit	u	u $\approx 1.66041 \times 10^{-27}$ kg

*The conventional millimeter of mercury, symbol mmHg (not mm Hg), is the pressure exerted by a column exactly 1 mm high of a fluid of density exactly 13.5951 g·cm^{-3} in a place where the gravitational acceleration is exactly 980.665 cm·s^{-2}. The mmHg differs from the Torr by less than 2×10^{-7} Torr.
†These units defined in terms of the best available experimental values of certain physical constants may be converted to SI units. The factors for conversion of these units are subject to change in the light of new experimental measurements of the constants involved.

Conversion factors for the measurement systems

Because it will take some years for all scientists and engineers to convert to the SI, this dictionary has retained the U.S. Customary and metric systems. Conversion factors between the two systems and SI are given in **Table 5** for some prevalent units; in each of the subtables the user proceeds as follows:

To convert a quantity expressed in a unit in the left-hand column to the equivalent in a unit in the top row of a subtable, multiply the quantity by the factor common to both units.

The factors have been carried out to seven significant figures, as derived from the fundamental constants and the definitions of the units. However, this does not mean that the factors are always known to that accuracy. Numbers followed by ellipses are to be continued indefinitely with repetition of the same pattern of digits. Factors written with fewer than seven significant digits are exact values. Numbers followed by an asterisk are definitions of the relation between the two units. In "G. UNITS OF ENERGY," the electrical units are those in terms of which certification of standard cells, standard resistances, and so forth, is made by the National Bureau of Standards; unless otherwise indicated, all electrical units are absolute.

Table 5. Conversion factors for the U.S. Customary System, metric system, and International System

A. UNITS OF LENGTH

Units	cm	m	in.	ft	yd	mile
1 cm	= 1	0.01^*	0.3937008	0.03280840	0.01093613	6.213712×10^{-6}
1 m	= 100.	1	39.37008	3.280840	1.093613	6.213712×10^{-4}
1 in.	$= 2.54^*$	0.0254	1	0.08333333...	0.02777777...	1.578283×10^{-5}
1 ft	= 30.48	0.3048	$12.^*$	1	0.3333333...	1.893939×10^{-4}...
1 yd	= 91.44	0.9144	36.	$3.^*$	1	5.681818×10^{-4}...
1 mile	$= 1.609344 \times 10^5$	1.609344×10^3	6.336×10^4	$5280.^*$	1760.	1

B. UNITS OF AREA

Units	cm^2	m^2	$in.^2$	ft^2	yd^2	$mile^2$
1 cm^2	= 1	10^{-4*}	0.1550003	1.076391×10^{-3}	1.195990×10^{-4}	3.861022×10^{-11}
1 m^2	$= 10^4$	1	1550.003	10.76391	1.195990	3.861022×10^{-7}
1 $in.^2$	$= 6.4516^*$	6.4516×10^{-4}	1	6.944444×10^{-3}...	7.716049×10^{-4}	2.490977×10^{-10}
1 ft^2	= 929.0304	0.09290304	$144.^*$	1	0.1111111...	3.587007×10^{-8}
1 yd^2	= 8361.273	0.8361273	1296.	$9.^*$	1	3.228306×10^{-7}
1 $mile^2$	$= 2.589988 \times 10^{10}$	2.589988×10^6	4.014490×10^9	$2.78784 \times 10^{7*}$	3.0976×10^6	1

C. UNITS OF VOLUME

Units	cm^3	liter	$in.^3$	ft^3	qt	gal
1 cm^3	= 1	10^{-3}	0.06102374	3.531467×10^{-5}	1.056688×10^{-3}	2.641721×10^{-4}
1 liter	$= 1000.^*$	1	61.02374	0.03531467	1.056688	0.2641721
1 $in.^3$	$= 16.38706^*$	0.01638706	1	5.787037×10^{-4}	0.01731602	4.329004×10^{-3}
1 ft^3	= 28316.85	28.31685	$1728.^*$	1	2.992208	7.480520
1 qt	= 946.353	0.946353	57.75	0.0342014	1	0.25
1 gal (U.S.)	= 3785.412	3.785412	$231.^*$	0.1336806	$4.^*$	1

Table 5. Conversion factors for the U.S. Customary System, metric system, and International System (cont.)

D. UNITS OF MASS

Units	g	kg	oz	lb	metric ton	ton
1 g	= 1	10^{-3}	0.03527396	2.204623×10^{-3}	10^{-6}	1.102311×10^{-6}
1 kg	= 1000.	1	35.27396	2.204623	10^{-3}	1.102311×10^{-3}
1 oz (avdp)	= 28.34952	0.02834952	1	0.0625	2.834952×10^{-5}	$5. \times 10^{-4}$
1 lb (avdp)	= 453.5924	0.4535924	16.*	1	4.535924×10^{-4}	0.0005
1 metric ton	= 10^{6}	1000.*	35273.96	2204.623	1	1.102311
1 ton	= 907184.7	907.1847	32000.	2000.*	0.9071847	1

E. UNITS OF DENSITY

Units	g cm^{-3}	g l.$^{-1}$	oz in.$^{-3}$	lb in.$^{-3}$	lb ft^{-3}	lb gal^{-1}
1 g cm^{-3}	= 1	1000.	0.5780365	0.03612728	62.42795	8.345403
1 g l.$^{-1}$	= 10^{-3}	1	5.780365×10^{-4}	3.612728×10^{-5}	0.06242795	8.345403×10^{-3}
1 oz in.$^{-3}$	= 1.729994	1729.994	1	0.0625	108.	14.4375
1 lb in.$^{-3}$	= 27.67991	27679.91	16.	1	1728.	231.
1 lb ft^{-3}	= 0.01601847	16.01847	9.259259×10^{-3}	5.7870370×10^{-4}	1	0.1336806
1 lb gal^{-1}	= 0.1198264	119.8264	4.749536×10^{-3}	4.3290043×10^{-3}	7.480519	1

F. UNITS OF PRESSURE

Units	dyn cm^{-2}	bar	atm	kg (wt) cm^{-2}	mmHg (Torr)	in. Hg	lb (wt) in.$^{-2}$
1 dyn cm^{-2}	= 1	10^{-6}	9.869233×10^{-7}	1.019716×10^{-6}	7.500617×10^{-4}	2.952999×10^{-5}	1.450377×10^{-5}
1 bar	= 10^{6}*	1	0.9869233	1.019716	750.0617	29.52999	14.50377
1 atm	= 1013250.*	1.013250	1	1.033227	760.	29.92126	14.69595
1 kg (wt) cm^{-2}	= 980665.	0.980665	0.9678411	1	735.5592	28.95903	14.22334
1 mmHg (Torr)	= 1333.224	1.333224×10^{-3}	1.3157895×10^{-3}	1.3595099×10^{-3}	1	0.03937008	0.01933678
1 in. Hg	= 33863.88	0.03386388	0.03342105	0.03453155	25.4	1	0.4911541
1 lb (wt) in.$^{-2}$	= 68947.57	0.06894757	0.06804596	0.07030696	51.71493	2.036021	1

G. UNITS OF ENERGY

Units	g mass (energy equiv)	J	int J	cal	cal$_{IT}$	Btu$_{IT}$	kW hr	hp hr	ft-lb (wt)	cu ft-lb (wt) in.$^{-2}$	l.-atm
1 g mass (energy equiv)	= 1	8.987554×10^{13}	8.986071×10^{13}	2.148077×10^{13}	2.146640×10^{13}	8.518558×10^{10}	2.496543×10^{7}	3.347919×10^{7}	6.628880×10^{13}	4.603399×10^{11}	8.870026×10^{11}
1 J	= 1.112650×10^{-14}	1	0.999835	0.2390057	0.2388459	9.478172×10^{-4}	$2.777777... \times 10^{-7}$	3.725062	0.7375622	5.121960×10^{-3}	9.869233×10^{-3}
1 int J	= 1.112833×10^{-14}	1.000165	1	0.2390452	0.2388853	9.479735×10^{-4}	2.778236×10^{-7}	3.725676×10^{-7}	0.7376839	5.122805×10^{-3}	9.870862×10^{-3}
1 cal	= 4.655327×10^{-14}	4.184	4.183310	1	0.9993312	3.965667×10^{-3}	$1.1622222... \times 10^{-6}$	1.558562×10^{-6}	3.085960	2.143028×10^{-2}	0.04129287
1 cal$_{IT}$	= 4.658442×10^{-14}	4.1868	4.186109	1.000669	1	3.968321×10^{-3}	1.163000×10^{-6}	1.559609×10^{-6}	3.088025	2.144462×10^{-2}	0.04132050

continued

Table 5. Conversion factors for the U.S. Customary System, metric system, and International System (cont.)

G. UNITS OF ENERGY (cont.)

Units	g mass (energy equiv)	J	int J	cal	cal$_{IT}$	Btu$_{IT}$	kW hr	hp hr	ft-lb (wt)	cu ft-lb (wt) in.$^{-2}$	l.-atm
1 Btu$_{IT}$	= 1.173908 × 10^{-11}	1055.056	1054.882	252.1644	251.9958*	1	2.930711 × 10^{-4}	3.930148 × 10^{-4}	778.1693	5.403953	10.41259
1 kW hr	= 4.005539 × 10^{-8}	3600000.*	3599406.	860420.7	859845.2	3412.142	1	1.341022	2655224.	18439.06	35529.24
1 hp hr	= 2.986930 × 10^{-8}	2684519.	2684077.	641615.6	641186.5	2544.33	0.7456998	1	1980000.*	13750.	26494.15
1 ft-lb (wt)	= 1.508550 × 10^{-14}	1.355818	1.355594	0.3240483	0.3238315	1.285067 × 10^{-3}	3.766161 × 10^{-7}	5.050505... × 10^{-7}	1	6.944444... × 10^{-3}	0.01338088
1 cu ft-lb (wt) in.$^{-2}$	= 2.172313 × 10^{-12}	195.2378	195.2056	46.66295	46.63174	0.1850497	5.423272 × 10^{-5}	7.272727... × 10^{-5}	144.*	1	1.926847
1 l.-atm	= 1.127392 × 10^{-12}	101.3250	101.3083	24.21726	24.20106	0.09603757	2.814583 × 10^{-5}	3.774419 × 10^{-5}	74.73349	0.5189825	1

Greek alphabet

Upper and lower cases	Name	Upper and lower cases	Name	Upper and lower cases	Name	Upper and lower cases	Name
Aα	Alpha	Hη	Eta	Nν	Nu	Tτ	Tau
Bβ	Beta	Θθ	Theta	Ξξ	Xi	Υυ	Upsilon
Γγ	Gamma	Iι	Iota	Oo	Omicron	Φφ	Phi
Δδ	Delta	Kκ	Kappa	Ππ	Pi	Xχ	Chi
Eε	Epsilon	Λλ	Lambda	Pρ	Rho	Ψψ	Psi
Zζ	Zeta	Mμ	Mu	Σσς	Sigma	Ωω	Omega

Symbols and atomic numbers for the chemical elements*

Name	Symbol	At. no.	Name	Symbol	At. no.	Name	Symbol	At. no.	Name	Symbol	At. no.
Actinium	Ac	89	Erbium	Er	68	Mercury	Hg	80	Samarium	Sm	62
Aluminum	Al	13	Europium	Eu	63	Molybdenum	Mo	42	Scandium	Sc	21
Americium	Am	95	Fermium	Fm	100	Neodymium	Nd	60	Selenium	Se	34
Antimony	Sb	51	Fluorine	F	9	Neon	Ne	10	Silicon	Si	14
Argon	Ar	18	Francium	Fr	87	Neptunium	Np	93	Silver	Ag	47
Arsenic	As	33	Gadolinium	Gd	64	Nickel	Ni	28	Sodium	Na	11
Astatine	At	85	Gallium	Ga	31	Niobium	Nb	41	Strontium	Sr	38
Barium	Ba	56	Germanium	Ge	32	Nitrogen	N	7	Sulfur	S	16
Berkelium	Bk	97	Gold	Au	79	Nobelium	No	102	Tantalum	Ta	73
Beryllium	Be	4	Hafnium	Hf	72	Osmium	Os	76	Technetium	Tc	43
Bismuth	Bi	83	Helium	He	2	Oxygen	O	8	Tellurium	Te	52
Boron	B	5	Holmium	Ho	67	Palladium	Pd	46	Terbium	Tb	65
Bromine	Br	35	Hydrogen	H	1	Phosphorus	P	15	Thallium	Tl	81
Cadmium	Cd	48	Indium	In	49	Platinum	Pt	78	Thorium	Th	90
Calcium	Ca	20	Iodine	I	53	Plutonium	Pu	94	Thulium	Tm	69
Californium	Cf	98	Iridium	Ir	77	Polonium	Po	84	Tin	Sn	50
Carbon	C	6	Iron	Fe	26	Potassium	K	19	Titanium	Ti	22
Cerium	Ce	58	Krypton	Kr	36	Praseodymium	Pr	59	Tungsten	W	74
Cesium	Cs	55	Lanthanum	La	57	Promethium	Pm	61	Uranium	U	92
Chlorine	Cl	17	Lawrencium	Lr (Lw)	103	Protactinium	Pa	91	Vanadium	V	23
Chromium	Cr	24	Lead	Pb	82	Radium	Ra	88	Xenon	Xe	54
Cobalt	Co	27	Lithium	Li	3	Radon	Rn	86	Ytterbium	Yb	70
Copper	Cu	29	Lutetium	Lu	71	Rhenium	Re	75	Yttrium	Y	39
Curium	Cm	96	Magnesium	Mg	12	Rhodium	Rh	45	Zinc	Zn	30
Dysprosium	Dy	66	Manganese	Mn	25	Rubidium	Rb	37	Zirconium	Zr	40
Einsteinium	Es	99	Mendelevium	Md	101	Ruthenium	Ru	44			

*Elements 104 and 105 have been reported, but no official names or symbols have yet been assigned.

Mathematical signs and symbols

Symbol	Definition	Symbol	Definition	Symbol	Definition
$+$	plus (sign of addition)	$\propto$	varies as	$\oint$	line integral around a closed path
$+$	positive	∞	infinity	Σ	(sigma) summation of
$-$	minus (sign of subtraction)	$\sqrt{}$	square root of	$f(x), F(x)$	functions of x
$-$	negative	$\sqrt[3]{}$	cube root of	$\exp x = e^x$	(e = naperian log base) (abbreviation for e^x)
$\pm\,(\mp)$	plus or minus (minus or plus)	$\therefore$	therefore		
$\times$	times, by (multiplication sign)	$\parallel$	parallel to	∇	del or nabla, vector differential operator
$\cdot$	multiplied by	$()[]\{\}$	parentheses, brackets and braces; quantities enclosed by them to be taken together in multiplying, dividing, etc.	∇^2	Laplacian operator
$\div$	sign of division			$\pounds$	Laplace operational symbol
$/$	divided by			$4!$	factorial $4 = 1 \times 2 \times 3 \times 4$
$:$	ratio sign, divided by, is to	$\overline{AB}$	length of line from A to B	$\|x\|$	absolute value of x
$::$	equals, as (proportion)	π	(pi), $= 3.14159 +$	$\dot{x}$	first derivative of x with respect to time
$<$	less than	$\circ$	degrees		
$>$	greater than	$'$	minutes	$\ddot{x}$	second derivative of x with respect to time
$\ll$	much less than	$''$	seconds		
$\gg$	much greater than	$\angle$	angle	$\mathbf{A} \times \mathbf{B}$	vector product; magnitude of $\mathbf{A}$ times magnitude of $\mathbf{B}$ times sine of the angle from $\mathbf{A}$ to $\mathbf{B}$; $AB \sin \overline{AB}$
$=$	equals	dx	differential of x		
$\equiv$	identical with	Δ	(delta) difference		
$\sim$	similar to	Δx	increment of x	$\mathbf{A} \cdot \mathbf{B}$	scalar product of $\mathbf{A}$ and $\mathbf{B}$; magnitude of $\mathbf{A}$ times magnitude of $\mathbf{B}$ times cosine of the angle from $\mathbf{A}$ to $\mathbf{B}$; $AB \cos \overline{AB}$
$\approx$	approximately equals	$\partial u/\partial x$	partial derivative of u with respect to x		
$\cong$	approximately equals, congruent	$\int$	integral of		
$\leq$	equal to or less than	$\int_b^a$	integral of, between limits a and b		
$\geq$	equal to or greater than				
$\neq\,\neq$	not equal to				
$\rightarrow \doteq$	approaches				

Mathematical notation

Mathematical logic.

$p, q, P(x)$	Sentences, propositional functions, propositions
$-p, \sim p,$ non p, Np	Negation, read "not p" ($\neq$: read "not equal")
$p \vee q, p + q, Apq$	Disjunction, read "p or q," "p, q," or both
$p \wedge q, p \cdot q, p \& q, Kpq$	Conjunction, read "p and q"
$p \rightarrow q, p \supset q, p \Rightarrow q, Cpq$	Implication, read "p implies q" or "if p then q"
$p \leftrightarrow q, p \equiv q, p \Longleftrightarrow q, Epq, p$ iff q	Equivalence, read "p is equivalent to q" or "p if and only if q"
n.a.s.c.	Read "necessary and sufficient condition"
$(), [], \{\}, \cdot\cdot, \cdot$	Parentheses
$V, \forall, \Sigma$	Universal quantifier, read "for all" or "for every"
$\exists, \boldsymbol{\exists}, \Pi$	Existential quantifier, read "there is a" or "there exists"
$\vdash$	Assertion sign ($p \vdash q$: read "q follows from p"; $\vdash p$: read "p is or follows from an axiom," or "p is a tautology"

$0, 1$	Truth, falsity (values)
$=$	Identity
$\overset{\mathrm{Df}}{=}, \overset{\mathrm{df}}{=}, \underset{\mathrm{df}}{=}, \equiv$	Definitional identity
$\blacksquare$	"End of proof"; "QED"

Set theory, relations, functions.

X, Y	Sets
$x \in X$	x is a member of the set X
$x \notin X$	x is not a member of X
$A \subset X, A \subseteq X$	Set A is contained in set X
$A \not\subset X, A \not\subseteq X$	A is not contained in X
$X \cup Y, X + Y$	Union of sets X and Y
$X \cap Y, X \cdot Y$	Intersection of sets X and Y
$+, \dot{+}, \bigcirc$	Symmetric difference of sets
$\bigcup X_i, \Sigma X_i$	Union of all the sets X_i
$\bigcap X_i, \Pi X_i$	Intersection of all the sets X_i
$\emptyset, 0, \Lambda$	Null set, empty set
$X', \mathbf{C}X, CX$	Complement of the set X
$X - Y, X \backslash Y$	Difference of sets X and Y
$\hat{x}(P(x)), \{x \| P(x)\}, \{x : P(x)\}$	The set of all x with the property P

(x,y,z), $\langle x,y,z \rangle$ — Ordered set of elements x, y, and z; to be distinguished from (x,z,y), for example

$\{x,y,z\}$ — Unordered set, the set whose elements are x, y, z, and no others

$\{a_1, a_2, \ldots, a_n\}$, $\{a_i\}_{i=1,2,\ldots,n}$, $\{a_i\}_{i=1}^n$ — The set whose members are a_i, where i is any whole number from 1 to n

$\{a_1, a_2, \ldots\}$, $\{a_i\}_{i=1,2,\ldots}$, $\{a_i\}_{i=1}^\infty$ — The set whose members are a_i, where i is any positive whole number

$X \times Y$ — Cartesian product, set of all (x,y) such that $x \in X$, $y \in Y$

$\{a_i\}_{i \in I}$ — The set whose elements are a_i, where $i \in I$

xRy, $R\{x,y\}$ — Relation

$\equiv, \cong, \sim, \simeq$ — Equivalence relations, for example, congruence

$\geqq, \geq, >, \succ, \gg, \leqq, \leq, <$ — Transitive relations, for example, numerical order

$f: X \to Y$, $X \xrightarrow{f} Y$, $X \to Y$, $f \in Y^X$ — Function, mapping, transformation

$f^{-1}, f, X \xleftarrow{f^{-1}} Y$ — Inverse mapping

$g \circ f$ — Composite functions: $(g \circ f)(x) = g(f(x))$

$f(X)$ — Image of X by f

$f^{-1}(X)$ — Inverse-image set, counter image

1-1, one-one — Read "one-to-one correspondence"

$\begin{array}{ccc} X & \xrightarrow{f} & Y \\ \phi \downarrow & & \downarrow \psi \\ W & \xrightarrow{g} & Z \end{array}$ — Diagram: the diagram is commutative in case $\psi \circ f = g \ \phi$

$f|A$ — Partial mapping, restriction of function f to set A

X, card X, $|X|$ — Cardinal of the set A

$\aleph_0, d$ — Denumerable infinity

$\mathfrak{c}, c, 2^{\aleph_0}$ — Power of continuum

ω — Order type of the set of positive integers

σ- — Read "countably"

Number, numerical functions.

1.4; 1,4; 1·4 — Read "one and four-tenths"

1(1)20(10)100 — Read "from 1 to 20 in intervals of 1, and from 20 to 100 in intervals of 10"

const — Constant

$A \geqq 0$ — The number A is nonnegative, or, the matrix A is positive definite, or, the matrix A has nonnegative entries

$x|y$ — Read "x divides y"

$x \equiv y \bmod p$ — Read "x congruent to y modulo p"

$a_0 + \cfrac{1}{a_1 +} \cfrac{1}{a_2 +} \cdots,$ $a_0 + \cfrac{1|}{|a_1} + \cdots$ — Continued fractions

$[a,b]$ — Closed interval

$[a,b)$, $[a,b[$ — Half-open interval (open at the right)

(a,b), $]a,b[$ — Open interval

$[a,\infty)$, $[a,\to[$ — Interval closed at the left, infinite to the right

$(-\infty, \infty)$, $]\leftarrow,\to[$ — Set of all real numbers

$\max_{x \in X} f(x)$, $\max\{f(x)|x \in X\}$ — Maximum of $f(x)$ when x is in the set X

min — Minimum

sup, l.u.b. — Supremum, least upper bound

inf, g.l.b. — Infimum, greatest lower bound

$\lim_{x\to a} f(x) = b$, $\lim_{x=a} f(x) = b$, $f(x) \to b$ as $x \to a$ — b is the limit of $f(x)$ as x approaches a

$\lim_{x\to a-} f(x)$, $\lim_{x=a-0} f(x), f(a-)$ — Limit of $f(x)$ as x approaches a from the left

$\limsup, \overline{\lim}$ — Limit superior

$\liminf, \underline{\lim}$ — Limit inferior

l.i.m. — Limit in the mean

$z = x + iy = re^{i\theta}$, $\zeta = \xi + i\eta$, $w = u + iv = \rho e^{i\phi}$ — Complex variables

$\bar{z}, z^*$ — Complex conjugate

Re, $\Re$ — Real part

Im, $\Im$ — Imaginary part

arg — Argument

$\dfrac{\partial(u,v)}{\partial(x,y)}$, $\dfrac{D(u,v)}{D(x,y)}$ — Jacobian, functional determinant

$\displaystyle\int_E f(x)\,d\mu(x)$ — Integral (for example, Lebesgue integral) of function f over set E with respect to measure μ

$f(n) \sim \log n$ as $n \to \infty$ — $f(n)/\log n$ approaches 1 as $n \to \infty$

$f(n) = O(\log n)$ as $n \to \infty$ — $f(n)/\log n$ is bounded as $n \to \infty$

$f(n) = o(\log n)$ — $f(n)/\log n$ approaches zero

$f(x) \nearrow b, f(x) \uparrow b$ — $f(x)$ increases, approaching the limit b

$f(x) \downarrow b, f(x) \searrow b$ — $f(x)$ decreases, approaching the limit b

a.e., p.p. — Almost everywhere

ess sup — Essential supremum

$C^0, C^0(X), C(X)$ — Space of continuous functions

$C^k, C^k[a,b]$ — The class of functions having continuous kth derivative (on $[a,b]$)

C'	Same as C^1
Lip_α, $\text{Lip}\,\alpha$	Lipschitz class of functions
$L^p, L_p, L^p[a,b]$	Space of functions having integrable absolute pth power (on $[a,b]$)
L'	Same as L^1
(C,α), (C,p)	Cesàro summability

Special functions.

$[x]$	The integral part of x
$\binom{n}{k}$, ${}^nC_k$, ${}_nC_k$	Binomial coefficient $n!/k!(n-k)!$
$\left(\dfrac{n}{p}\right)$	Legendre symbol
e^x, $\exp x$	Exponential function
$\sinh x$, $\cosh x$, $\tanh x$	Hyperbolic functions
$\text{sn}\,x$, $\text{cn}\,x$, $\text{dn}\,x$	Jacobi elliptic functions
$\wp(x)$	Weierstrass elliptic function
$\Gamma(x)$	Gamma function
$J_\nu(x)$	Bessel function
$\chi_X(x)$	Characteristic function of the set X: $\chi_X(x)=1$ in case $x \in X$, otherwise $\chi_X(x)=0$
$\text{sgn}\,x$	Signum: $\text{sgn}\,0 = 0$, while $\text{sgn}\,x = x/|x|$ for $x \neq 0$
$\delta(x)$	Dirac delta function

Algebra, tensors, operators.

$+, \cdot, \times, \circ, \mathsf{T}, \tau$	Laws of composition in algebraic systems
$e, 0$	Identity, unit, neutral element (of an additive system)
$e, 1, I$	Identity, unit, neutral element (of a general algebraic system)
e, e, E, P	Idempotent
a^{-1}	Inverse of a
$\text{Hom}(M,N)$	Group of all homomorphisms of M into N
G/H	Factor group, group of cosets
$[K:k]$	Dimension of K over k
$\oplus, \dotplus$	Direct sum
$\otimes$	Tensor product, Kronecker product
$\wedge$	Exterior product, Grassmann product
$\vec{x}, \mathbf{x}, \mathfrak{x}, x$	Vector
$\vec{x}\cdot\vec{y}$, $\mathbf{x}\cdot\mathbf{y}$, $(\mathfrak{x},\mathfrak{h})$	Inner product, scalar product, dot product
$\mathbf{x}\times\mathbf{y}$, $[\mathfrak{x},\mathfrak{h}]$, $\mathbf{x}\wedge\mathbf{y}$	Outer product, vector product, cross product
$\lvert x\rvert, |x|, \|x\|, \|x\|_p$	Norm of the vector x
Ax, xA	The image of x under the transformation A
δ_{ij}	Kronecker delta: $\delta_{ii}=1$, while $\delta_{ij}=0$ for $i \neq j$
A', tA, A^t, ${}^tA$	Transpose of the matrix A
$A^*, \tilde{A}$	Adjoint, Hermitian conjugate of A
$\text{tr}\,A$, $\text{Sp}\,A$	Trace of the matrix A
$\det A$, $|A|$	Determinant of the matrix A
$\Delta^n f(x)$, $\Delta_h{}^n f$, $\Delta_h^n f(x)$	Finite differences
$[x_0,x_1]$, $[x_0,x_1,x_2]$, $\underset{x_1}{\Delta}u_{x_0}$, $[x_0,x_1]_f$	Divided differences
∇f, $\text{grad}\,f$	Read "gradient of f"
$\nabla\cdot\mathbf{v}$, $\text{div}\,\mathbf{v}$	Read "divergence of $\mathbf{v}$"
$\nabla\times\mathbf{v}$, $\text{curl}\,\mathbf{v}$, $\text{rot}\,\mathbf{v}$	Read "curl of $\mathbf{v}$"
∇^2, Δ, div grad	Laplacian
$[X,Y]$	Poisson bracket, or commutator, or Lie product
$\text{GL}(n,R)$	Full linear group of degree n over field R
$\text{O}(n,R)$	Full orthogonal group
$\text{SO}(n,R)$, $\text{O}^+(n,R)$	Special orthogonal group

Topology.

E^n	Euclidean n space
S^n	n sphere
$\rho(p,q)$, $d(p,q)$	Metric, distance (between points p and q)
$\overline{X}, X^-, \text{cl}\,X, X^c$	Closure of the set X
$\text{Fr}X$, $\text{fr}X$, ∂X, $\text{bdry}\,X$	Frontier, boundary of X
$\text{int}\,X$, $\mathring{X}$	Interior of X
T_2 space	Hausdorff space
F_σ	Union of countably many closed sets
G_δ	Intersection of countably many open sets
$\dim X$	Dimensionality, dimension of X
$\pi_1(X)$	Fundamental group of the space X
$\pi_n(X)$, $\pi_n(X,A)$	Homotopy groups
$H_n(X)$, $H_n(X,A;G)$, $H_*(X)$	Homology groups
$H^n(X)$, $H^n(X,A;G)$, $H^*(X)$	Cohomology groups

Probability and statistics.

X, Y	Random variables
$P(X \leq 2)$, $\text{Pr}\{X \leq 2\}$	Probability that $X \leq 2$
$P(X \leq 2 | Y \geq 1)$	Conditional probability
$E(X)$, $\text{E}(X)$	Expectation of X
$E(X | Y \geq 1)$	Conditional expectation
c.d.f.	Cumulative distribution function
p.d.f.	Probability density function
c.f.	Characteristic function
$\bar{x}$	Mean (especially, sample mean)
σ, s.d.	Standard deviation
σ^2, Var, var	Variance
$\mu_1, \mu_2, \mu_3, \mu_i, \mu_{ij}$	Moments of a distribution
ρ	Coefficient of correlation
$\rho_{12\cdot34}$	Partial correlation coefficient

Fundamental constants

Compiled by E. R. Cohen and B. N. Taylor under the auspices of the CODATA Task Group on Fundamental Constants. This set has been officially adopted by CODATA and is taken from J. Phys. Chem. Ref. Data, Vol. 2, No. 4, p. 663 (1973) and CODATA Bulletin No. 11 (December 1973).

Quantity	Symbol	Numerical Value *	Uncert. (ppm)	SI † ← Units →	cgs ‡
Speed of light in vacuum	c	299792458(1.2)	0.004	$m \cdot s^{-1}$	10^2 $cm \cdot s^{-1}$
Permeability of vacuum	μ_0	4π		10^{-7} $H \cdot m^{-1}$	
		$=12.5663706144$		10^{-7} $H \cdot m^{-1}$	
Permittivity of vacuum, $1/\mu_0 c^2$	ϵ_0	8.854187818(71)	0.008	10^{-12} $F \cdot m^{-1}$	
Fine-structure constant, $[\mu_0 c^2/4\pi](e^2\hbar c)$	α	7.2973506(60)	0.82	10^{-3}	10^{-3}
	α^{-1}	137.03604(11)	0.82		
Elementary charge	e	1.6021892(46)	2.9	10^{-19} C	10^{-20} emu
		4.803242(14)	2.9		10^{-10} esu
Planck constant	h	6.626176(36)	5.4	10^{-34} $J \cdot s$	10^{-27} $erg \cdot s$
	$\hbar = h/2\pi$	1.0545887(57)	5.4	10^{-34} $J \cdot s$	10^{-27} $erg \cdot s$
Avogadro constant	N_A	6.022045(31)	5.1	10^{23} mol^{-1}	10^{23} mol^{-1}
Atomic mass unit, $10^{-3} kg \cdot mol^{-1} N_A^{-1}$	u	1.6605655(86)	5.1	10^{-27} kg	10^{-24} g
Electron rest mass	m_e	9.109534(47)	5.1	10^{-31} kg	10^{-28} g
		5.4858026(21)	0.38	10^{-4} u	10^{-4} u
Proton rest mass	m_p	1.6726485(86)	5.1	10^{-27} kg	10^{-24} g
		1.007276470(11)	0.011	u	u
Ratio of proton mass to electron mass	m_p/m_e	1836.15152(70)	0.38		
Neutron rest mass	m_n	1.6749543(86)	5.1	10^{-27} kg	10^{-24} g
		1.008665012(37)	0.037	u	u
Electron charge to mass ratio	e/m_e	1.7588047(49)	2.8	10^{11} $C \cdot kg^{-1}$	10^7 $emu \cdot g^{-1}$
		5.272764(15)	2.8		10^{17} $esu \cdot g^{-1}$
Magnetic flux quantum, $[c]^{-1}(hc/2e)$	Φ_0	2.0678506(54)	2.6	10^{-15} Wb	10^{-7} $G \cdot cm^2$
	h/e	4.135701(11)	2.6	10^{-15} $J \cdot s \cdot C^{-1}$	10^{-7} $erg \cdot s \cdot emu^{-1}$
		1.3795215(36)	2.6		10^{-17} $erg \cdot s \cdot esu^{-1}$
Josephson frequency-voltage ratio	$2e/h$	4.835939(13)	2.6	10^{14} $Hz \cdot V^{-1}$	
Quantum of circulation	$h/2m_e$	3.6369455(60)	1.6	10^{-4} $J \cdot s \cdot kg^{-1}$	$erg \cdot s \cdot g^{-1}$
	h/m_e	7.273891(12)	1.6	10^{-4} $J \cdot s \cdot kg^{-1}$	$erg \cdot s \cdot g^{-1}$
Faraday constant, $N_A e$	F	9.648456(27)	2.8	10^4 $C \cdot mol^{-1}$	10^3 $emu \cdot mol^{-1}$
		2.8925342(82)	2.8		10^{14} $esu \cdot mol^{-1}$
Rydberg constant, $[\mu_0 c^2/4\pi]^2(m_e e^4/4\pi\hbar^3 c)$	R_∞	1.097373177(83)	0.075	10^7 m^{-1}	10^5 cm^{-1}
Bohr radius, $[\mu_0 c^2/4\pi]^{-1}(\hbar^2/m_e e^2)=\alpha/4\pi R_\infty$	a_0	5.2917706(44)	0.82	10^{-11} m	10^{-9} cm
Classical electron radius, $[\mu_0 c^2/4\pi](e^2/m_e c^2)=\alpha^3/4\pi R_\infty$	$r_e=\alpha\lambda_C$	2.8179380(70)	2.5	10^{-15} m	10^{-13} cm
Thomson cross section, $(8/3)\pi r_e^2$	σ_e	0.6652448(33)	4.9	10^{-28} m^2	10^{-24} cm^2
Free electron g-factor, or electron magnetic moment in Bohr magnetons	$g_e/2=\mu_e/\mu_B$	1.0011596567(35)	0.0035		
Free muon g-factor, or muon magnetic moment in units of $[c](e\hbar/2m_\mu c)$	$g_\mu/2$	1.00116616(31)	0.31		
Bohr magneton, $[c](e\hbar/2m_e c)$	μ_B	9.274078(36)	3.9	10^{-24} $J \cdot T^{-1}$	10^{-21} $erg \cdot G^{-1}$
Electron magnetic moment	μ_e	9.284832(36)	3.9	10^{-24} $J \cdot T^{-1}$	10^{-21} $erg \cdot G^{-1}$
Gyromagnetic ratio of protons in H_2O	γ'_p	2.6751301(75)	2.8	10^8 $s^{-1} \cdot T^{-1}$	10^4 $s^{-1} \cdot G^{-1}$
	$\gamma'_p/2\pi$	4.257602(12)	2.8	10^7 $Hz \cdot T^{-1}$	10^3 $Hz \cdot G^{-1}$
γ'_p corrected for diamagnetism of H_2O	γ_p	2.6751987(75)	2.8	10^8 $s^{-1} \cdot T^{-1}$	10^4 $s^{-1} \cdot G^{-1}$
	$\gamma_p/2\pi$	4.257711(12)	2.8	10^7 $Hz \cdot T^{-1}$	10^3 $Hz \cdot G^{-1}$
Magnetic moment of protons in H_2O in Bohr magnetons	μ'_p/μ_B	1.52099322(10)	0.066	10^{-3}	10^{-3}
Proton magnetic moment in Bohr magnetons	μ_p/μ_B	1.521032209(16)	0.011	10^{-3}	10^{-3}
Ratio of electron and proton magnetic moments	μ_e/μ_p	658.2106880(66)	0.010		
Proton magnetic moment	μ_p	1.4106171(55)	3.9	10^{-26} $J \cdot T^{-1}$	10^{-23} $erg \cdot G^{-1}$
Magnetic moment of protons in H_2O in nuclear magnetons	μ'_p/μ_N	2.7927740(11)	0.38		
μ'_p/μ_N corrected for diamagnetism of H_2O	μ_p/μ_N	2.7928456(11)	0.38		
Nuclear magneton, $[c](e\hbar/2m_p c)$	μ_N	5.050824(20)	3.9	10^{-27} $J \cdot T^{-1}$	10^{-24} $erg \cdot G^{-1}$
Ratio of muon and proton magnetic moments	μ_μ/μ_p	3.1833402(72)	2.3		
Muon magnetic moment	μ_μ	4.490474(18)	3.9	10^{-26} $J \cdot T^{-1}$	10^{-23} $erg \cdot G^{-1}$
Ratio of muon mass to electron mass	m_μ/m_e	206.76865(47)	2.3		

Fundamental constants (cont.)

Quantity	Symbol	Numerical Value *	Uncert. (ppm)	SI †	← Units → cgs ‡	
Muon rest mass	m_μ	1.883566(11)	5.6	10^{-28} kg	10^{-25} g	
		0.11342920(26)	2.3	u	u	
Compton wavelength of the electron, $h/m_ec = \alpha^2/2R_\infty$	λ_C	2.4263089(40)	1.6	10^{-12} m	10^{-10} cm	
	$\bar\lambda_C = \lambda_C/2\pi = \alpha a_0$	3.8615905(64)	1.6	10^{-13} m	10^{-11} cm	
Compton wavelength of the proton, h/m_pc	$\lambda_{C,p}$	1.3214099(22)	1.7	10^{-15} m	10^{-13} cm	
	$\bar\lambda_{C,p} = \lambda_{C,p}/2\pi$	2.1030892(36)	1.7	10^{-16} m	10^{-14} cm	
Compton wavelength of the neutron, h/m_nc	$\lambda_{C,n}$	1.3195909(22)	1.7	10^{-15} m	10^{-13} cm	
	$\bar\lambda_{C,n} = \lambda_{C,n}/2\pi$	2.1001941(35)	1.7	10^{-16} m	10^{-14} cm	
Molar volume of ideal gas at s.t.p.	V_m	22.41383(70)	31	10^{-3} m³·mol⁻¹	10^3 cm³·mol⁻¹	
Molar gas constant, V_mp_0/T_0 ($T_0 \equiv 273.15$ K; $p_0 \equiv 101325$ Pa≡1atm)	R	8.31441(26)	31	J·mol⁻¹·K⁻¹	10^7 erg·mol⁻¹·K⁻¹	
		8.20568(26)	31	10^{-5} m³·atm·mol⁻¹·K⁻¹	10 cm³·atm·mol⁻¹·K⁻¹	
Boltzmann constant, R/N_A	k	1.380662(44)	32	10^{-23} J·K⁻¹	10^{-16} erg·K⁻¹	
Stefan-Boltzmann constant, $\pi^2k^4/60\hbar^3c^2$	σ	5.67032(71)	125	10^{-8} W·m⁻²·K⁻⁴	10^{-5} erg·s⁻¹·cm⁻²·K⁻⁴	
First radiation constant, $2\pi hc^2$	c_1	3.741832(20)	5.4	10^{-16} W·m²	10^{-5} erg·cm²·s⁻¹	
Second radiation constant, hc/k	c_2	1.438786(45)	31	10^{-2} m·K	cm·K	
Gravitational constant	G	6.6720(41)	615	10^{-11} m³·s⁻²·kg⁻¹	10^{-8} cm³·s⁻²·g⁻¹	
Ratio, kx-unit to ångström, $\Lambda = \lambda(\text{Å})/\lambda(\text{kxu})$; $\lambda(\text{Cu}K\alpha_1) \equiv 1.537400$ kxu	Λ	1.0020772(54)	5.3			
Ratio, Å* to ångström, $\Lambda^* = \lambda(\text{Å})/\lambda(\text{Å}^*)$; $\lambda(\text{W}K\alpha_1) \equiv 0.2090100$ Å*	Λ^*	1.0000205(56)	5.6			

ENERGY CONVERSION FACTORS AND EQUIVALENTS

Quantity	Symbol	Numerical Value *	Units	Uncert. (ppm)
1 kilogram (kg·c²)		8.987551786(72)	10^{16} J	0.008
		5.609545(16)	10^{29} MeV	2.9
1 Atomic mass unit (u·c²)		1.4924418(77)	10^{-10} J	5.1
		931.5016(26)	MeV	2.8
1 Electron mass m_e·c²)		8.187241(42)	10^{-11} J	5.1
		0.5110034(14)	MeV	2.8
1 Muon mass (m_μ·c²)		1.6928648(96)	10^{-11} J	5.6
		105.65948(35)	MeV	3.3
1 Proton mass (m_p·c²)		1.5033015(77)	10^{-10} J	5.1
		938.2796(27)	MeV	2.8
1 Neutron mass (m_n·c²)		1.5053738(78)	10^{-10} J	5.1
		939.5731(27)	MeV	2.8
1 Electron volt		1.6021892(46)	10^{-19} J	2.9
			10^{-12} erg	2.9
	1 eV/h	2.4179696(63)	10^{14} Hz	2.6
	1 eV/hc	8.065479(21)	10^5 m⁻¹	2.6
			10^3 cm⁻¹	2.6
	1 eV/k	1.160450(36)	10^4 K	31
Voltage-wavelength conversion, hc		1.986478(11)	10^{-25} J·m	5.4
		1.2398520(32)	10^{-6} eV·m	2.6
			10^{-4} eV·cm	2.6
Rydberg constant	$R_\infty hc$	2.179907(12)	10^{-18} J	5.4
			10^{-11} erg	5.4
		13.605804(36)	eV	2.6
	$R_\infty c$	3.28984200(25)	10^{15} Hz	0.075
	$R_\infty hc/k$	1.578885(49)	10^5 K	31
Bohr magneton	μ_B	9.274078(36)	10^{-24} J·T⁻¹	3.9
		5.7883785(95)	10^{-5} eV·T⁻¹	1.6
	μ_B/h	1.3996123(39)	10^{10} Hz·T⁻¹	2.8
	μ_B/hc	46.68604(13)	m⁻¹·T⁻¹	2.8
			10^{-2} cm⁻¹·T⁻¹	2.8
	μ_B/k	0.671712(21)	K·T⁻¹	31
Nuclear magneton	μ_N	5.505824(20)	10^{-27} J·T⁻¹	3.9
		3.1524515(53)	10^{-8} eV·T⁻¹	1.7
	μ_N/h	7.622532(22)	10^6 Hz·T⁻¹	2.8
	μ_N/hc	2.5426030(72)	10^{-2} m⁻¹·T⁻¹	2.8
			10^{-4} cm⁻¹·T⁻¹	2.8
	μ_N/k	3.65826(12)	10^{-4} K·T⁻¹	31

* Note that the numbers in parentheses are the one standard-deviation uncertainties in the last digits of the quoted value computed on the basis of internal consistency, that the unified atomic mass scale $^{12}C \triangleq 12$ has been used throughout, that u=atomic mass unit, C=coulomb, F=farad, G=gauss, H=henry, Hz=hertz=cycle/s, J=joule, K=kelvin (degree Kelvin), Pa=pascal=N·m⁻², T=tesla (10^4 G), V=volt, Wb=weber=T·m², and W=watt. In cases where formulas for constants are given (e.g., R_∞), the relations are written as the product of two factors. The second factor, in parentheses, is the expression to be used when all quantities are expressed in cgs units, with the electron charge in electrostatic units. The first factor, in brackets, is to be included only if all quantities are expressed in SI units. We remind the reader that with the exception of those auxiliary constants which have been taken to be exact, the uncertainties of these constants are correlated, and therefore the general law of error propagation must be used in calculating additional quantities requiring two or more of these constants.

† Quantities given in u and atm are for the convenience of the reader; these units are not part of the International System of Units (SI).

‡ In order to avoid separate columns for "electromagnetic" and "electrostatic" units, both are given under the single heading "cgs Units." When using these units, the elementary charge e in the second column should be understood to be replaced by e_m or e_s, respectively.

Abbreviations for scientific and technical organizations

Abbrev.	Organization	Abbrev.	Organization	Abbrev.	Organization
AAA	American Automobile Association	ASRE	American Society of Refrigerating Engineers (see ASHRAE)	NACM	National Association of Chain Manufacturers
AAR	Association of American Railroads	ASST	American Society for Steel Treating	NASA	National Aeronautics and Space Administration
AAS	American Astronautical Society	ASTM	American Society for Testing and Materials	NBC	National Broadcasting Company
ABAI	American Boiler & Affiliated Industries	ASTME	American Society of Tool & Manufacturing Engineers	NBFU	National Board of Fire Underwriters
ACI	American Concrete Institute	AWPA	American Wood Preservation Association	NBS	National Bureau of Standards
ACM	Association for Computing Machinery	AWS	American Welding Society	NDHA	National District Heating Association
ACRMA	Air Conditioning and Refrigerating Manufacturers Association	AWWA	American Water Works Association	NEC	National Electrical Code
ACS	American Chemical Society	CAB	Civil Aeronautics Board	NEMA	National Electrical Manufacturers Association
AEC	Atomic Energy Commission (U.S.)	CAGI	Compressed Air & Gas Institute	NFPA	National Fire Protection Association
AFBMA	Anti-friction Bearings Manufacturers' Association	CME	Chartered Mechanical Engineers (IMechE)	NLGI	National Lubricating Grease Institute
AFS	American Foundrymen's Society	COESA	U.S. Committee on Extension to the Standard Atmosphere	OSW	Office of Saline Water
AGA	American Gas Association	CSA	Canadian Standards Association	OTS	Office of Technical Services, U.S. Dept. of Commerce
AGMA	American Gear Manufacturers' Association	EIA	Electronic Industries Association	PFI	Pipe Fabrication Institute
AIChE	American Institute of Chemical Engineers	EEI	Edison Electric Institute	RMA	Rubber Manufacturers Association
AIEE	American Institute of Electrical Engineers (see IEEE)	FAA	Federal Aviation Agency	SAE	Society of Automotive Engineers
AIME	American Institute of Mining Engineers	FCC	Federal Communications Commission; Federal Construction Council	SBI	Steel Boiler Institute
AIP	American Institute of Physics	FPC	Federal Power Commission	SNA&ME	Society of Naval Architects and Marine Engineers
AISE	American Iron & Steel Engineers	FSB	Federal Specifications Board	TAC	Technical Advisory Committee on Weather Design Conditions (ASHRAE)
AISI	American Iron and Steel Institute	IAeS	Institute of Aerospace Sciences	UKAEA	United Kingdom Atomic Energy Authority
AMA	Automobile Manufacturers' Association; Acoustical Materials Association	ICAO	International Civil Aviation Organization	UL	Underwriters' Laboratory
AMCA	Air Moving & Conditioning Association, Inc.	ICC	Interstate Commerce Commission	USAF	U.S. Air Force
ANC	Army-Navy Civil Aeronautics Committee	ICE	Institute of Civil Engineers	USAS or USASI	USA Standard (USA Standards Institute, successor to ASA)
ANS	American Nuclear Society	ICI	International Commission on Illumination	USCG	U.S. Coast Guard
API	American Petroleum Institute	IEEE	Institute of Electrical & Electronics Engineers (successor to AIEE)	USCS	U.S. Commercial Standard
APWA	American Public Works Association	IES	Illuminating Engineering Society	USDA	U.S. Department of Agriculture
AREA	American Railroad Engineering Association	IGT	Institute of Gas Technology	USFPL	U.S. Forest Products Laboratory
ARI	Air Conditioning and Refrigeration Institute	IMechE	Institute of Mechanical Engineers	USGS	U.S. Geologic Survey
ARS	American Rocket Society	INA	Institute of Naval Architects	USHEW	U.S. Department of Health, Education and Welfare
ASA	American Standards Association (see USASI)	IRE	Institute of Radio Engineers	USN	U.S. Navy
ASCE	American Society of Civil Engineers	ISO	International Organization for Standardization	USP	U.S. Pharmacopoeia
ASHAE	see ASHRAE	ISTM	International Society for Testing Materials	USPHS	U.S. Public Health Service
ASHRAE	American Society of Heating, Refrigerating, and Air Conditioning Engineers (formerly ASHAE and ASH&VE)	IUPAC	International Union of Pure & Applied Chemistry	VDI	Verein Deutscher Ingenieure
		MSS	Manufacturers Standardization Society of the Valve & Fittings Industry	WHO	World Health Organization
ASH&VE	see ASHRAE	NAA	National Association of Accountants	WPA	Western Pine Association
ASLE	American Society of Lubricating Engineers				
ASM	American Society for Metals	NACA	National Advisory Committee on Aeronautics (see NASA)		
ASME	American Society of Mechanical Engineers				

Semiconductor symbols and abbreviations*

A, a — Anode

B, b — Base

$\mathbf{b_{fs}}$ — Common-source small-signal forward transfer susceptance

$\mathbf{b_{is}}$ — Common-source small-signal input susceptance

$\mathbf{b_{os}}$ — Common-source small-signal output susceptance

$\mathbf{b_{rs}}$ — Common-source small-signal reverse transfer susceptance

C, c — Collector

$\mathbf{C_{cb}}$ — Collector-base interterminal capacitance

$\mathbf{C_{ce}}$ — Collector-emitter interterminal capacitance

$\mathbf{C_{ds}}$ — Drain-source capacitance

$\mathbf{C_{du}}$ — Drain-substrate capacitance

$\mathbf{C_{eb}}$ — Emitter-base interterminal capacitance

$\mathbf{C_{ibo}}$ — Common-base open-circuit input capacitance

$\mathbf{C_{ibs}}$ — Common-base short-circuit input capacitance

$\mathbf{C_{ieo}}$ — Common-emitter open-circuit input capacitance

$\mathbf{C_{ies}}$ — Common-emitter short-circuit input capacitance

$\mathbf{C_{iss}}$ — Common-source short-circuit input capacitance

$\mathbf{C_{obo}}$ — Common-base open-circuit output capacitance

$\mathbf{C_{obs}}$ — Common-base short-circuit output capacitance

$\mathbf{C_{oeo}}$ — Common-emitter open-circuit output capacitance

$\mathbf{C_{oes}}$ — Common-emitter short-circuit output capacitance

$\mathbf{C_{oss}}$ — Common-source short-circuit output capacitance

$\mathbf{C_{rbs}}$ — Common-base short-circuit reverse transfer capacitance

$\mathbf{C_{rcs}}$ — Common-collector short-circuit reverse transfer capacitance

$\mathbf{C_{res}}$ — Common-emitter short-circuit reverse transfer capacitance

$\mathbf{C_{rss}}$ — Common-source short-circuit reverse transfer capacitance

$\mathbf{C_{tc}}$ — Collector depletion-layer capacitance

$\mathbf{C_{te}}$ — Emitter depletion-layer capacitance

D, d — Drain

E, e — Emitter

η — Intrinsic standoff ratio

$\mathbf{f_{hfb}}$ — Common-base small-signal short-circuit forward current transfer ratio cutoff frequency

$\mathbf{f_{hfc}}$ — Common-collector small-signal short-circuit forward current transfer ratio cutoff frequency

$\mathbf{f_{hfe}}$ — Common-emitter small-signal short-circuit forward current transfer ratio cutoff frequency

$\mathbf{f_{max}}$ — Maximum frequency of oscillation

$\mathbf{f_T}$ — Transition frequency (frequency at which common-emitter small-signal forward current transfer ratio extrapolates to unity)

G, g — Gate

$\mathbf{g_{fs}}$ — Common-source small-signal forward transfer conductance

$\mathbf{g_{is}}$ — Common-source small-signal input conductance

$\mathbf{g_{MB}}$ — Common-base static transconductance

$\mathbf{g_{MC}}$ — Common-collector static transconductance

$\mathbf{g_{ME}}$ — Common-emitter static transconductance

$\mathbf{g_{os}}$ — Common-source small-signal output conductance

$\mathbf{G_{PB}}$ — Common-base large-signal insertion power gain

$\mathbf{G_{pb}}$ — Common-base small-signal insertion power gain

$\mathbf{G_{PC}}$ — Common-collector large-signal insertion power gain

$\mathbf{G_{pc}}$ — Common-collector small-signal insertion power gain

$\mathbf{G_{PE}}$ — Common-emitter large-signal insertion power gain

$\mathbf{G_{pe}}$ — Common-emitter small-signal insertion power gain

$\mathbf{G_{pg}}$ — Common-gate small-signal insertion power gain

$\mathbf{G_{ps}}$ — Common-source small-signal insertion power gain

$\mathbf{g_{rs}}$ — Common-source small-signal reverse transfer conductance

$\mathbf{G_{TB}}$ — Common-base large-signal transducer power gain

$\mathbf{G_{tb}}$ — Common-base small-signal transducer power gain

$\mathbf{G_{TC}}$ — Common-collector large-signal transducer power gain

$\mathbf{G_{tc}}$ — Common-collector small-signal transducer power gain

$\mathbf{G_{TE}}$ — Common-emitter large-signal transducer power gain

$\mathbf{G_{te}}$ — Common-emitter small-signal transducer power gain

$\mathbf{G_{tg}}$ — Common-gate small-signal transducer power gain

$\mathbf{G_{ts}}$ — Common-source small-signal transducer power gain

$\mathbf{h_{FB}}$ — Common-base static forward current transfer ratio

$\mathbf{h_{fb}}$ — Common-base small-signal short-circuit forward current transfer ratio

$\mathbf{h_{FC}}$ — Common-collector static forward current transfer ratio

$\mathbf{h_{fc}}$ — Common-collector small-signal short-circuit forward current transfer ratio

$\mathbf{h_{FE}}$ — Common-emitter static forward current transfer ratio

$\mathbf{h_{fe}}$ — Common-emitter small-signal short-circuit forward current transfer ratio

$\mathbf{h_{FEL}}$ — Inherent large-signal forward current transfer ratio

$\mathbf{h_{IB}}$ — Common-base static input resistance

$\mathbf{h_{ib}}$ — Common-base small-signal short-circuit input impedance

$\mathbf{h_{IC}}$ — Common-collector statis input resistance

$\mathbf{h_{ic}}$ — Common-collector small-signal short-circuit input impedance

$\mathbf{h_{IE}}$ — Common-emitter static input resistance

$\mathbf{h_{ie}}$ — Common-emitter small-signal short-circuit input impedance

$\mathbf{h_{ie(imag)}}$ — Imaginary part of common-emitter small-signal short-circuit input impedance

$\mathbf{h_{ie(real)}}$ — Real part of common-emitter small-signal short-circuit input impedance

$\mathbf{h_{ob}}$ — Common-base small-signal open-circuit output admittance

$\mathbf{h_{oc}}$ — Common-collector small-signal open-circuit output admittance

$\mathbf{h_{oe}}$ — Common-emitter small-signal open-circuit output admittance

$\mathbf{h_{oe(imag)}}$ — Imaginary part of common-emitter small-signal open-circuit output admittance

$\mathbf{h_{oe(real)}}$ — Real part of common-emitter small-signal open-circuit output admittance

$\mathbf{h_{rb}}$ — Common-base small-signal open-circuit reverse voltage transfer ratio

$\mathbf{h_{rc}}$ — Common-collector small-signal open-circuit reverse voltage transfer ratio

$\mathbf{h_{ro}}$ — Common-emitter small-signal open-circuit reverse voltage transfer ratio

$\mathbf{I_B}$ — Base-terminal dc current

$\mathbf{I_b}$ — Alternating component (rms value) of base-terminal current

$\mathbf{i_B}$ — Instantaneous total value of base-terminal current

$\mathbf{I_{BEV}}$ — Base cutoff current, dc

$\mathbf{I_{B2(mod)}}$ — Interbase modulated current

$\mathbf{I_C}$ — Collector-terminal dc current

$\mathbf{I_c}$ — Alternating component (rms value) of collector-terminal current

$\mathbf{i_C}$ — Instantaneous total value of collector-terminal current

$\mathbf{I_{CBO}}$ — Collector cutoff current (dc), emitter open

$\mathbf{I_{CEO}}$ — Collector cutoff current (dc), base open

$\mathbf{I_{CER}}$ — Collector cutoff current (dc), specified resistance between base and emitter

$\mathbf{I_{CES}}$ — Collector cutoff current (dc), base shorted to emitter

$\mathbf{I_{CEV}}$ — Collector cutoff current (dc), specified voltage between base and emitter

$\mathbf{I_{CEX}}$ — Collector cutoff current (dc), specified circuit between base and emitter

$\mathbf{I_D}$ — Drain current; dc

$\mathbf{I_{D(off)}}$ — Drain cutoff current

$\mathbf{I_{D(on)}}$ — On-state drain current

$\mathbf{I_{DSS}}$ — Zero-gate-voltage drain current

$\mathbf{I_E}$ — Emitter-terminal dc current

$\mathbf{I_e}$ — Alternating component (rms value) of emitter-terminal current

$\mathbf{i_E}$ — Instantaneous total value of emitter-terminal current

$\mathbf{I_{EBO}}$ — Emitter cutoff current (dc), collector open

$\mathbf{I_{EB20}}$ — Emitter reverse current

*Recommended by the Joint Electron Device Engineering Council of the Electronic Industries Association and the National Electrical Manufacturers Association for use in semiconductor device data sheets and specifications. From R. F. Graf, *Modern Dictionary of Electronics*, 4th ed., Indianapolis: Harold Sams, 1972.

Semiconductor symbols and abbreviations (cont.)

$I_{EC(ofs)}$ — Emitter-collector offset current

I_{ECS} — Emitter cutoff current (dc), base short-circuited to collector

$I_{EIE2(off)}$ — Emitter cutoff current

I_F — For voltage-regulator and voltage-reference diodes: dc forward current. For signal diodes and rectifier diodes: dc forward current (no alternating component)

I_f — Alternating component of forward current (rms value)

i_F — Instantaneous total forward current

$I_{F(AV)}$ — Forward current, dc (with alternating component)

I_{FM} — Maximum (peak) total forward current

$I_{F(OV)}$ — Forward current, overload

I_{FRM} — Maximum (peak) forward current, repetitive

$I_{F(RMS)}$ — Total rms forward current

I_{FSM} — Maximum (peak) forward current, surge

I_G — Gate current, dc

I_{GF} — Forward gate current

I_{GR} — Reverse gate current

I_{GSS} — Reverse gate current, drain short-circuited to source

I_{GSSF} — Forward gate current, drain short-circuited to source

I_{GSSR} — Reverse gate current, drain short-circuited to source

I_I — Inflection-point current

$Im(h_{oe})$ — Imaginary part of common-emitter small-signal short-circuit input impedance

$Im(h_{oe})$ — Imaginary part of common-emitter small-signal open-circuit output admittance

I_O — Average forward current, 180° conduction angle, 60-Hz half sine wave

I_P — Peak-point current

I_R — For voltage-regulator and voltage-reference diodes: dc reverse current. For signal diodes and rectifier diodes: dc reverse current (no alternating component)

I_r — Alternating component of reverse current (rms value)

i_R — Instantaneous total reverse current

$I_{R(AV)}$ — Reverse current, dc (with alternating component)

I_{RM} — Maximum (peak) total reverse current

I_{RRM} — Maximum (peak) reverse current, repetitive

$I_{R(RMS)}$ — Total rms reverse current

I_{RSM} — Maximum (peak) surge reverse current

I_S — Source current, dc

I_{SDS} — Zero-gate-voltage source current

$I_{S(off)}$ — Source cutoff current

I_V — Valley-point current

I_Z — Regulator current, reference current (dc)

I_{ZK} — Regulator current, reference current (dc near breakdown knee)

I_{ZM} — Regulator current, reference current (dc maximum rated current)

K, k — Cathode

L_c — Conversion loss

M — Figure of merit

NF_o — Overall noise figure

NR_o — Output noise ratio

P_{BE} — Power input (dc) to base, common emitter

p_{BE} — Instantaneous total power input to base, common emitter

P_{CB} — Power input (dc) to collector, common base

p_{CB} — Instantaneous total power input to collector, common base

P_{CE} — Power input (dc) to collector, common emitter

p_{CE} — Instantaneous total power input to collector, common emitter

P_{EB} — Power input (dc) to emitter, common base

p_{EB} — Instantaneous total power input to emitter, common base

P_F — Forward power dissipation, dc (no alternating component)

p_F — Instantaneous total forward power dissipation

$P_{F(AV)}$ — Forward power dissipation, dc (with alternating component)

P_{FM} — Maximum (peak) total forward power dissipation

P_{IB} — Common-base large-signal input power

p_{ib} — Common-base small-signal input power

P_{IC} — Common-collector large-signal input power

p_{ic} — Common collector small-signal input power

P_{IE} — Common-emitter large-signal input power

p_{ie} — Common-emitter small-signal input power

P_{OB} — Common-base large-signal output power

p_{ob} — Common-base small-signal output power

P_{OC} — Common-collector large-signal output power

p_{oc} — Common-collector small-signal output power

P_{OE} — Common-emitter large-signal output power

p_{oe} — Common-emitter small-signal output power

P_R — Reverse power dissipation, dc (no alternating component)

p_R — Instantaneous total reverse power dissipation

$P_{R(AV)}$ — Reverse power dissipation, dc (with alternating component)

P_{RM} — Maximum (peak) total reverse power dissipation

P_T — Total nonreactive power input to all terminals

p_T — Nonreactive power input, instantaneous total, to all terminals

Q_S — Stored charge

r_{BB} — Interbase resistance

$r_b{}'C_c$ — Collector-base time constant

$r_{CE(sat)}$ — Saturation resistance, collector-to-emitter

$r_{DS(on)}$ — Static drain-source on-state resistance

$r_{ds(on)}$ — Small-signal drain-source on-state resistance

$Re(h_{ie})$ — Real part of common-emitter small-signal short-circuit input impedance

$Re(h_{oe})$ — Real part of common-emitter small-signal open-circuit output admittance

$r_{ele2(on)}$ — Small-signal emitter-emitter on-state resistance

r_I — Dynamic resistance at inflection point

R_θ — Thermal resistance

$R_{\theta CA}$ — Thermal resistance, case to ambient

$R_{\theta JA}$ — Thermal resistance, junction to ambient

$R_{\theta JC}$ — Thermal resistance, junction to case

S, s — Source

T_A — Ambient temperature or free-air temperature

T_C — Case temperature

t_d — Delay time

$t_{d(off)}$ — Turn-off delay time

$t_{d(on)}$ — Turn-on delay time

t_f — Fall time

t_{fr} — Forward recovery time

T_j — Junction temperature

t_{off} — Turn-off time

t_{on} — Turn-on time

t_p — Pulse time

t_r — Rise time

t_{rr} — Reverse recovery time

t_s — Storage time

TSS — Tangential signal sensitivity

T_{stg} — Storage temperature

t_w — Pulse average time

U, u — Bulk (substrate)

V_{BB} — Base supply voltage (dc)

V_{BC} — Average or dc voltage, base to collector

v_{bc} — Instantaneous value of alternating component of base-collector voltage

V_{BE} — Average or dc voltage, base to emitter

v_{be} — Instantaneous value of alternating component of base-emitter voltage

$V_{(BR)}$ — Breakdown voltage (dc)

$v_{(BR)}$ — Breakdown voltage (instantaneous total)

$V_{(BR)CBO}$ — Collector-base breakdown voltage, emitter open

$V_{(BR)CEO}$ — Collector-emitter breakdown voltage, base open

$V_{(BR)CER}$ — Collector-emitter breakdown voltage, resistance between base and emitter

$V_{(BR)CES}$ — Collector-emitter breakdown voltage, base shorted to emitter

$V_{(BR)CEV}$ — Collector-emitter breakdown voltage, specified voltage between base and emitter

Semiconductor symbols and abbreviations (cont.)

$V_{(BR)CEX}$ — Collector-emitter breakdown voltage, specified circuit between base and emitter

$V_{(BR)EBO}$ — Emitter-base breakdown voltage, collector open

$V_{(BR)ECO}$ — Emitter-collector breakdown voltage, base open

$V_{(BR)E1E2}$ — Emitter-emitter breakdown voltage

$V_{(BR)GSS}$ — Gate-source breakdown voltage

$V_{(BR)GSSF}$ -- Forward gate-source breakdown voltage

$V_{(BR)GSSR}$ — Reverse gate-source breakdown voltage

V_{B2B1} — Interbase voltage

V_{CB} — Average or dc voltage, collector to base

v_{cb} — Instantaneous value of alternating component of collector-base voltage

$V_{CB(fl)}$ — Collector-base dc open-circuit voltage (floating potential)

V_{CBO} — Collector-base voltage, dc, emitter open

V_{CC} — Collector supply voltage (dc)

V_{CE} — Average or dc voltage, collector to emitter

v_{ce} — Instantaneous value of alternating component of collector-emitter voltage

$V_{CE(fl)}$ — Collector-emitter dc open-circuit voltage (floating potential)

V_{CEO} — Collector-emitter voltage (dc), base open

$V_{CE(ofs)}$ — Collector-emitter offset voltage

V_{CER} — Collector-emitter voltage (dc), resistance between base and emitter

V_{CES} — Collector-emitter voltage (dc), base shorted to emitter

$V_{CE(sat)}$ — Collector-emitter dc saturation voltage

V_{CEV} — Collector-emitter voltage (dc), specified voltage between base and emitter

V_{CEX} — Collector-emitter voltage (dc), specified circuit between base and emitter

V_{DD} — Drain supply voltage (dc)

V_{DG} — Drain-gate voltage

V_{DS} — Drain-source voltage

$V_{DS(on)}$ — Drain-source on-state voltage

V_{DU} — Drain-substrate voltage

V_{EB} — Average or dc voltage, emitter to base

v_{eb} — Instantaneous value of alternating component of emitter-base voltage

$V_{EB(fl)}$ — Emitter-base dc open-circuit voltage (floating potential)

V_{EBO} — Emitter-base voltage (dc), collector open

$V_{EBI(sat)}$ — Emitter saturation voltage

V_{EC} — Average or dc voltage, emitter to collector

v_{ec} — Instantaneous value of alternating component of emitter-collector voltage

$V_{EC(fl)}$ — Emitter-collector dc open-circuit voltage (floating potential)

$V_{EC(ofs)}$ — Emitter-collector offset voltage

V_{EE} — Emitter supply voltage (dc)

V_F — For voltage-regulator and voltage-reference diodes: dc forward voltage. For signal diodes and rectifier diodes: dc forward voltage (no alternating component)

V_f — Alternating component of forward voltage (rms value)

v_F — Instantaneous total forward voltage

$V_{F(AV)}$ — Forward voltage, dc (with alternating component)

V_{FM} — Maximum (peak) total forward voltage

$V_{F(RMS)}$ — Total rms forward voltage

V_{GG} — Gate supply voltage (dc)

V_{GS} — Gate-source voltage

V_{GSF} — Forward gate-source voltage

$V_{GS(off)}$ — Gate-source cutoff voltage

V_{GSR} — Reverse gate-source voltage

$V_{GS(th)}$ — Gate-source threshold voltage

V_{GU} — Gate-substrate voltage

V_I — Inflection-point voltage

V_{OBI} — Base-1 peak voltage

V_P — Peak-point voltage

V_{PP} — Projected peak-point voltage

V_R — For voltage-regulator and voltage-reference diodes: dc reverse voltage. For signal diodes and rectifier diodes: dc reverse voltage (no alternating component)

V_r — Alternating component of reverse voltage (rms value)

v_R — Instantaneous total reverse voltage

$V_{R(AV)}$ — Reverse voltage, dc (with alternating component)

V_{RM} — Maximum (peak) total reverse voltage

V_{RRM} — Repetitive peak reverse voltage

$V_{R(RMS)}$ — Total rms reverse voltage

V_{RSM} — Nonrepetitive peak reverse voltage

V_{RT} — Reach-through voltage

V_{RWM} — Working peak reverse voltage

V_{SS} — Source supply voltage (dc)

V_{SU} — Source-substrate voltage

$V_{(TO)}$ — Threshold voltage

V_V — Valley-point voltage

V_Z — Regulator voltage, reference voltage (dc)

V_{ZM} — Regulator voltage, reference voltage (dc at maximum rated current)

y_{fb} — Common base small-signal short-circuit forward transfer admittance

y_{fc} — Common-collector small-signal short-circuit forward transfer admittance

y_{fe} — Common-emitter small-signal short-circuit forward transfer admittance

y_{fs} — Common-source small-signal short-circuit forward transfer admittance

$y_{fs(imag)}$ — Common-source small-signal forward transfer susceptance

$y_{fs(real)}$ — Common-source small-signal forward transfer conductance

y_{ib} — Common-base small-signal short-circuit input admittance

y_{ic} — Common-collector small-signal short-circuit input admittance

y_{ie} — Common-emitter small-signal short-circuit input admittance

$y_{ie(imag)}$ — Imaginary part of small-signal short-circuit input admittance (common-emitter)

$y_{ie(real)}$ — Real part of small-signal short-circuit input admittance (common-emitter)

y_{is} — Common-source small-signal short-circuit input admittance

$y_{is(imag)}$ — Common-source small-signal input susceptance

$y_{is(real)}$ — Common-source small-signal input conductance

y_{ob} — Common-base small-signal short-circuit output admittance

y_{oc} — Common-collector small-signal short-circuit output admittance

y_{oe} — Common-emitter small-signal short-circuit output admittance

$y_{oe(imag)}$ — Imaginary part of small-signal short-circuit output admittance (common-emitter)

$y_{oe(real)}$ — Real part of small-signal short-circuit output admittance (common-emitter)

y_{os} — Common-source small-signal short-circuit output admittance

$y_{os(imag)}$ — Common-source small-signal output susceptance

$Y_{os(real)}$ — Common-source small-signal output conductance

y_{rb} — Common-base small-signal short-circuit reverse transfer admittance

y_{rc} — Common-collector small-signal short-circuit reverse transfer admittance

y_{re} — Common-emitter small-signal short-circuit reverse transfer admittance

y_{rs} — Common-source small-signal short-circuit reverse transfer admittance

$y_{rs(imag)}$ — Common-source small-signal reverse transfer susceptance

$y_{rs(real)}$ — Common-source small-signal reverse transfer conductance

z_{if} — Intermediate-frequency impedance

z_m — Modulator-frequency load impedance

z_{rf} — Radio-frequency impedance

$Z_{\theta JA(t)}$ — Junction-to-ambient transient thermal impedance

$Z_{\theta JC(t)}$ — Junction-to-case transient thermal impedance

$Z_{\theta(t)}$ — Transient thermal impedance

z_v — Video impedance

z_z — Regulator impedance, reference impedance (small-signal at I_z)

z_{zk} — Regulator impedance, reference impedance (small-signal at I_{ZK})

Z_{zm} — Regulator impedance, reference impedance (small-signal at I_{ZM})

Schematic electronic symbols*

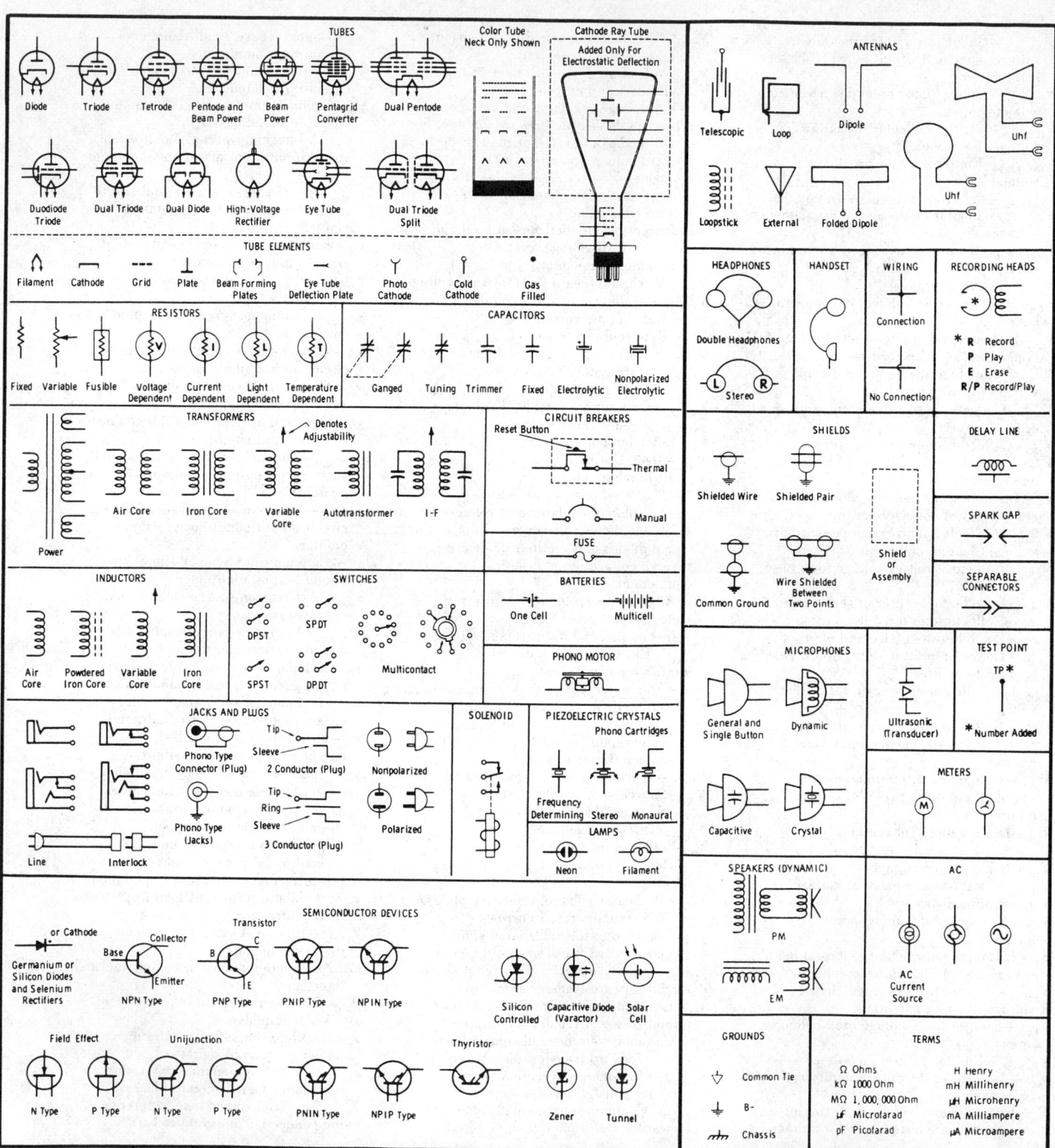

*From R. F. Graf, *Modern Dictionary of Electronics*, 4th ed., Indianapolis: Harold Sams, 1972.

Elementary particles

The quasi-stable elementary particles (mean lives longer than 10^{-20} sec)

Symbol and name	J^P (and C, if self-conjugate)	Mass, MeV	Mean life, sec	Principal decay modes and branching ratios, %
Classons				
(graviton)	$2^+(C=+)$	0	Stable	
γ(photon)	$1^-(C=-)$	0	Stable	
Leptons				
ν_e(e neutrino) $\bar{\nu}_e$(e antineutrino)	1/2	0	Stable	
e^-(electron) e^+(positron)	1/2	0.511	Stable	
ν_μ(μ neutrino) $\bar{\nu}_\mu$(μ antineutrino)	1/2	0	Stable	
μ^- μ^+ (muon)	1/2	105.7	2.20×10^{-6}	$e^-\bar{\nu}_e\nu_\mu$ 100 $e^+\nu_e\bar{\nu}_\mu$ 100
Hadrons	Hadronic quantum numbers			
Mesons				
π (pion) $\begin{bmatrix}Y=0\\I_G=1_-\end{bmatrix}0^-$	$I_3=+1:\pi^+$	139.6	2.60×10^{-8}	$\mu^+\nu_\mu$ 100
	$I_3=0:\pi^0(C=+)$	135.0	0.8×10^{-16}	$\gamma\gamma$ 99 γe^+e^- 1
	$I_3=-1:\pi^-(=\overline{\pi^+})$	139.6	2.60×10^{-8}	$\mu^-\bar{\nu}_\mu$ 100
$\left.\begin{array}{c}K\\K\end{array}\right\}$ (kaon) $\begin{bmatrix}Y=\left\{\begin{array}{c}+1\\-1\end{array}\right\}\\I=1/2\end{bmatrix}0^-$ $Y=+1$	$I_3=+1/2:K^+$	493.7	1.24×10^{-8}	$\mu^+\nu_\mu$ 64 $\pi^+\pi^0$ 21 $\pi^+\pi\pi$ 7 $\pi^0 e^+\nu_e$ 5 $\pi^0\mu^+\nu_\mu$ 3
	$I_3=-1/2:K^0$ — $K_S(CP\approx+)$	498	0.89×10^{-10}	$\pi^+\pi^-$ 69 $\pi^0\pi^0$ 31
	$I_3=+1/2:\overline{K}^0$ — $K_L(CP\approx-)$		0.52×10^{-7}	$\pi^0\pi^0\pi^0$ 21 $\pi^+\pi^-\pi^0$ 12 $\pi e\nu$ 39 $\pi\mu\nu$ 27 $\pi e\nu\gamma\approx$ 1 $\pi^+\pi$ 0.2 $\pi^0\pi^0$ 0.1
$Y=-1$	$I_3=-1/2:K^-(=\overline{K^+},\text{ see }K^+)$			
η (eta meson) $\begin{bmatrix}Y=0\\I_G=0_+\end{bmatrix}$	$0^-(C=+)$	549	$\approx3\times10^{-19}$ (Full width = 2.6 ± 0.6 keV)	$\gamma\gamma$ 38 $\pi^0\pi^0\pi^0$ 30 $\pi^+\pi^-\pi^0$ 24 $\pi^+\pi^-\gamma$ 5 $\pi°\gamma\gamma$ 3(?)
Baryons (all have distinct antiparticles)				
N (nucleon) $\begin{bmatrix}Y=+1\\I=1/2\end{bmatrix}1/2^+$	$I=+1/2:p$ (proton)	938.3	Stable	
	$I=-1/2:n$ (neutron)	939.6	0.9×10^3	$pe^-\bar{\nu}_e$ 100
Λ (lambda hyperon) $\begin{bmatrix}Y=0\\I_G=0\end{bmatrix}1/2^+$	Λ^0	1115.6	2.6×10^{-10}	$p\pi^-$ 64 $n\pi^0$ 36
Σ (sigma hyperon) $\begin{bmatrix}Y=0\\I=1\end{bmatrix}1/2^+$	$I_3=+1:\Sigma^+$	1189.4	0.80×10^{-10}	$p\pi^0$ 52 $n\pi^+$ 48
	$I_3=0:\Sigma^0$	1192.5	(3×10^{-19})	$\Lambda\gamma$ 100
	$I_3=-1:\Sigma^-$	1197.3	1.5×10^{-10}	$n\pi^-$ 100
Ξ (xi or cascade hyperon) $\begin{bmatrix}Y=-1\\I=1/2\end{bmatrix}1/2^+$	$I_3=+1/2:\Xi^0$	1315	3×10^{-10}	$\Lambda\pi^0$ 100
	$I_3=-1/2:\Xi^-$	1321	1.7×10^{-10}	$\Lambda\pi^-$ 100
Ω (omega hyperon) $\begin{bmatrix}Y=-2\\I=0\end{bmatrix}(3/2^+)$	Ω^-	1672	$\approx10^{-10}$	$\Xi\pi$? Λk ?

International graphic symbols*

Architecture

DIMENSION LINE	LONG BREAK LINE	CUTTING PLANE	SECTIONING	SINGLE OUTLET	SINGLE OUTLET, FLOOR	SINGLE OUTLET, SPECIAL PURPOSE	DUPLEX OUTLET	DUPLEX OUTLET, SPLIT WIRED
DUPLEX OUTLET, SPECIAL PURPOSE	RANGE OUTLET	THERMOSTAT	TRANSFORMER	LIGHT OUTLET, CEILING	LIGHT OUTLET, WALL	OUTLET, Surface Fluorescent Fixture	OUTLET, Bare-lamp Fluorescent Strip	FLOODLIGHT
ELECTRIC EYE (Beam Source)	FLOOR OUTLET, Public Telephone	FLOOR OUTLET, Private Telephone	FLOODLIGHT, SPECIAL	SPOTLIGHT, REFLECTOR	SPOTLIGHT, SEALED-BEAM	SPOTLIGHT, LENS	SPOTLIGHT, FRESNEL	SPOTLIGHT, PROFILE
SYSTEM DEVICES, PAGING	SYSTEM DEVICES, FIRE ALARM	SIREN	SOFTLIGHT	EFFECTS PROJECTOR	SPOTLIGHT, BIFOCAL	BATH	SHOWER STALL	
ELECTRICITY METER	SOUND SYSTEM	POLE, Electric Distribution	SHOWER HEAD	URINAL	WATER CLOSET		BIDET	
SYSTEM DEVICES, Public Telephone	SYSTEM DEVICES, Private Telephone	WH WATER HEATER	METER	R RANGE	DF DRINKING FOUNTAIN	WATER CISTERN	HB HOSE BIBB	CORK (Linoleum)
METAL	EARTH	ROCK	NATURAL RUBBLE	RUBBLE	NATURAL ASHLAR	TERRAZZO	FINISH WOOD, END GRAIN	FINISHED WOOD with GRAIN
ASHLAR, CAST or NATURAL	CAST STONE	MARBLE	SLATE	BRICK	COMMON BRICK	GLASS, Small Scale	GLASS, Large Scale	SAND, PLASTER, or CEMENT FINISH
FACE BRICK	FIRE BRICK	BRICK-COTTA	TERRA COTTA	TERRA COTTA, UNGLAZED	TERRA COTTA, GLAZED FACE	WOOD, Large Pieces	ROUGH WOOD	GLASS
TERRA COTTA, ARCHITECTURAL	TILE, Encaustic, Faience, or Ceramic	TILE, Small Scale	STONE CONCRETE	CINDER CONCRETE	CONCRETE BLOCK	GYPSUM	INSULATION, LOOSE	INSULATION, SOLID
METAL, Large Scale	SHINGLES; SIDING							

*H. Dreyfuss, *Symbol Sourcebook*, New York: McGraw-Hill, 1972.

Astronomy

SUN	NEW MOON	FIRST QUARTER MOON	FULL MOON	LAST QUARTER MOON	MERCURY	TAURUS	GEMINI	CANCER
VENUS	EARTH; GLOBULAR CLUSTER	MARS	JUPITER	SATURN	URANUS	SCORPIUS	SAGITTARIUS	CAPRICORNUS
NEPTUNE	PLUTO	STAR	COMET	GALACTIC CLUSTER	PLANETARY NEBULA	LEO	VIRGO	LIBRA; AUTUMNAL EQUINOX
GALAXY	CONJUNCTION	OPPOSITION	ASCENDING NODE	DESCENDING NODE	ARIES; VERNAL EQUINOX	AQUARIUS	PISCES	

Biology

MALE OR	FEMALE OR	NEUTER	NEUTER HERMAPHRODITE	AUTOSOMAL INHERITANCE	SEX-LINKED INHERITANCE	MATED WITH
SEX UNKNOWN or UNSPECIFIED	DEATH, MALE	MATING	CONSANGUINEOUS MATING	MATING, MALE PROGENY / DIZYGOTIC TWINS, MALE	MONOZYGOTIC TWINS, MALE / DIOECIOUS	ANNUAL
STAMINATE OR	PISTILLATE	HERMAPHRODITE (Monoclinous)	DICLINOUS	MONOECIOUS	BIENNIAL / PERENNIAL	EVERGREEN

Business

DOLLAR; ESCUDO; PESO; CRUZEIRO	CENT(S)	COLON	POUND STERLING	YEN	RUPEE	PENNY	SHILLING OR
MAGNETIC INK CHARACTERS (Numerals)	Branch Bank Identification / Amount of Check / Customer Account Number / Dash				NUMBER; POUND	PER CENT; ORDER OF	REGISTERED / COPYRIGHT

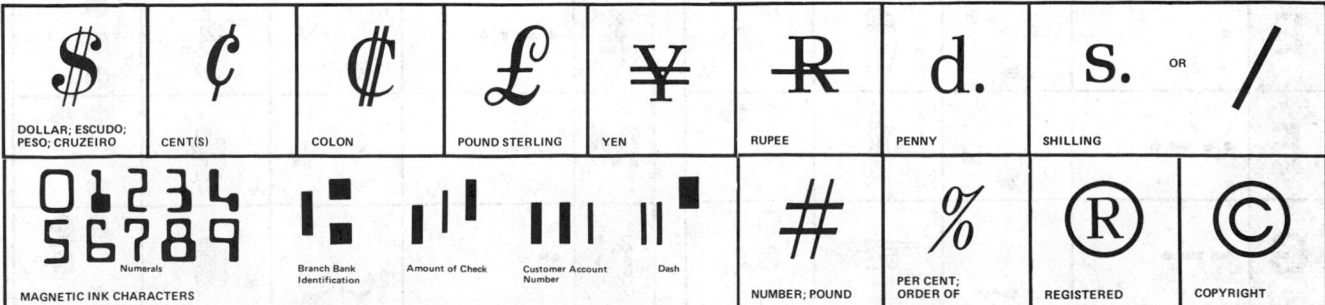

continued

Chemistry

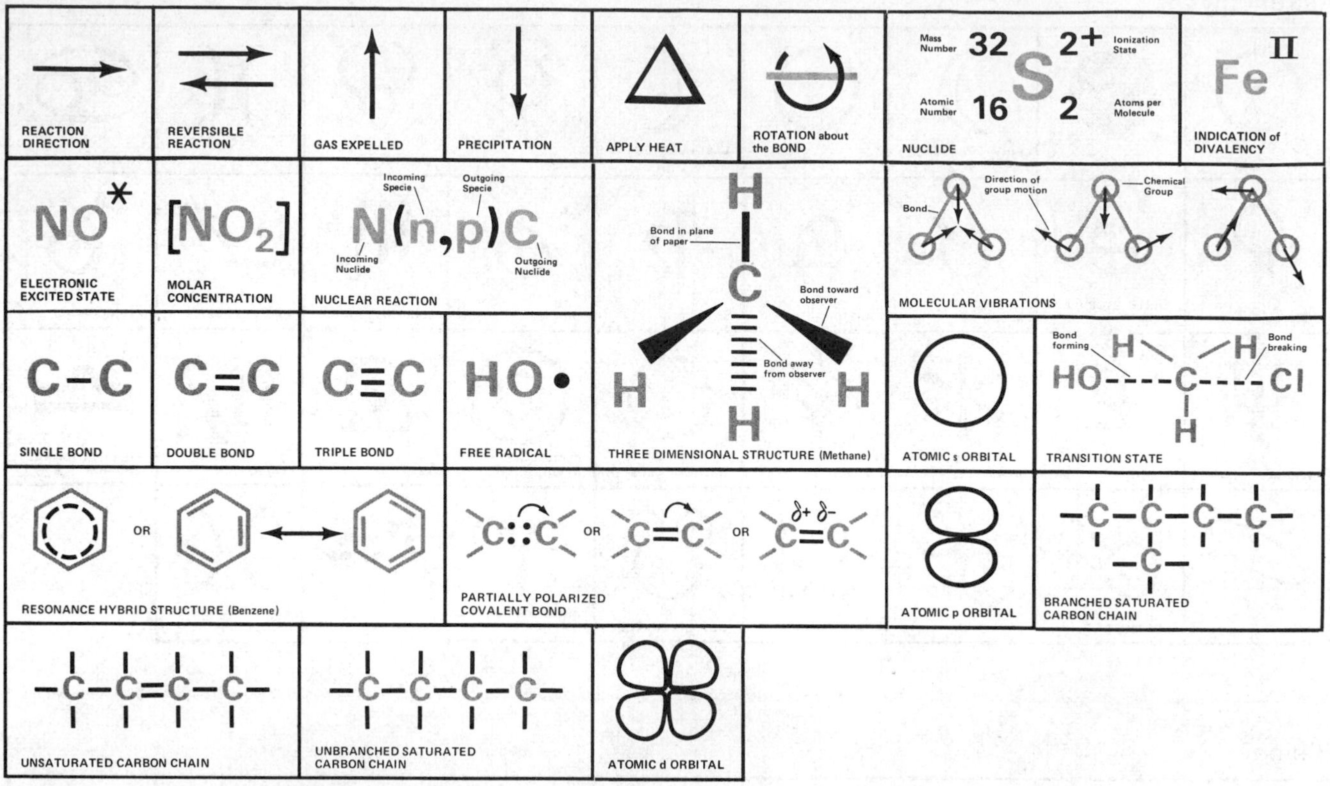

REACTION DIRECTION	REVERSIBLE REACTION	GAS EXPELLED	PRECIPITATION	APPLY HEAT	ROTATION about the BOND	NUCLIDE — Mass Number 32, Atomic Number 16, S, $2+$ Ionization State, 2 Atoms per Molecule	Fe^{II} INDICATION of DIVALENCY

NO^{*} — ELECTRONIC EXCITED STATE

$[NO_2]$ — MOLAR CONCENTRATION

$N(n,p)C$ — NUCLEAR REACTION (Incoming Specie / Incoming Nuclide / Outgoing Specie / Outgoing Nuclide)

THREE DIMENSIONAL STRUCTURE (Methane) — Bond in plane of paper, Bond toward observer, Bond away from observer

MOLECULAR VIBRATIONS — Bond, Direction of group motion, Chemical Group

$C-C$ — SINGLE BOND

$C=C$ — DOUBLE BOND

$C\equiv C$ — TRIPLE BOND

$HO\bullet$ — FREE RADICAL

ATOMIC s ORBITAL

TRANSITION STATE — Bond forming, Bond breaking, $HO----C----Cl$

RESONANCE HYBRID STRUCTURE (Benzene) ... OR ...

PARTIALLY POLARIZED COVALENT BOND — $C::C$ OR $C=C$ OR $\overset{\delta+}{C}=\overset{\delta-}{C}$

ATOMIC p ORBITAL

BRANCHED SATURATED CARBON CHAIN

UNSATURATED CARBON CHAIN

UNBRANCHED SATURATED CARBON CHAIN

ATOMIC d ORBITAL

Communications (non-graphic alphabet)

	Morse Code	Semaphore	Manual (Deaf)	Braille		Morse Code	Semaphore	Manual (Deaf)	Braille
A	· —				H	· · · ·			
B	— · · ·				I	· ·			
C	— · — ·				J	· — — —			
D	— · ·				K	— · —			
E	·				L	· — · ·			
F	· · — ·				M	— —			
G	— — ·				N	— ·			

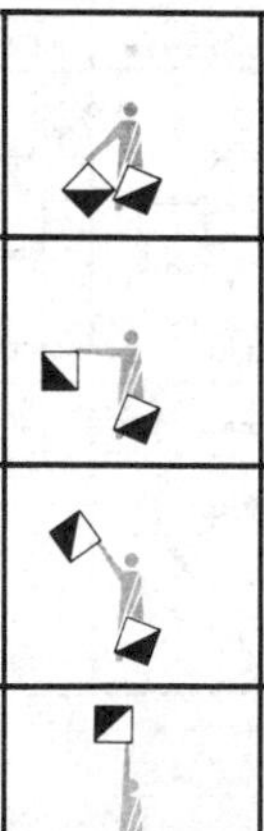

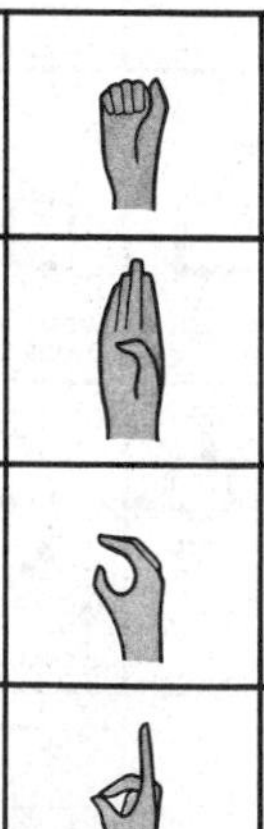

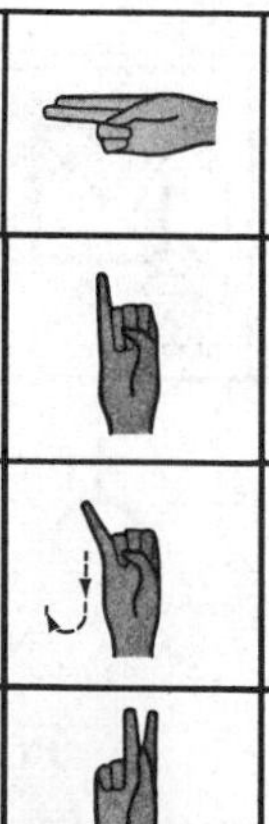

Communications (cont.)

	Morse Code	Semaphore	Manual (Deaf)	Braille		Morse Code	Semaphore	Manual (Deaf)	Braille
O	---				Z	--··			
P	·--·				1	·----			
Q	--·-				2	··---			
R	·-·				3	···--			
S	···				4	····-			
T	-				5	·····			
U	··-				6	-····			
V	···-				7	--···			
W	·--				8	---··			
X	-··-				9	----·			
Y	-·--				0	-----			

Computer

PROCESS	DECISION	PREPARATION	PREDEFINED PROCESS	MANUAL OPERATION	MANUAL INPUT	AUXILIARY OPERATION	MERGE	EXTRACT
INPUT/OUTPUT	ONLINE STORAGE	OFFLINE STORAGE	DOCUMENT	PUNCHED CARD	DECK of CARDS	COLLATE	SORT	CORE MEMORY

continued

Computer (cont.)

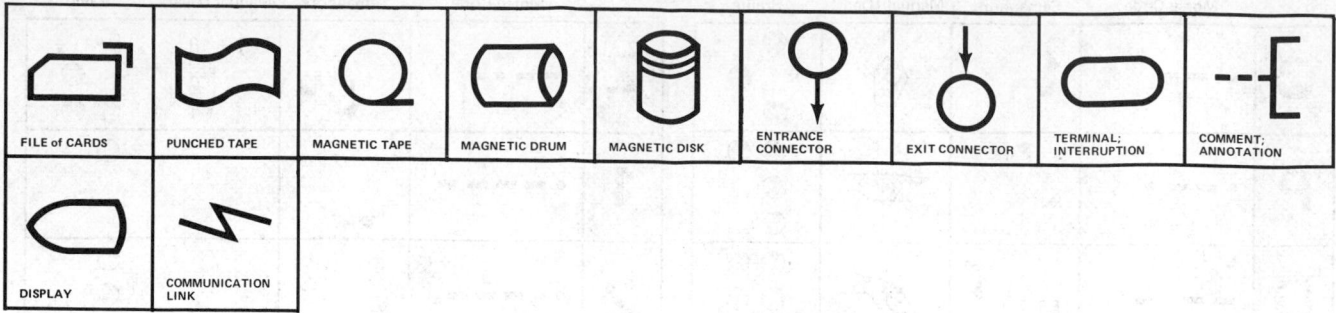

FILE of CARDS	PUNCHED TAPE	MAGNETIC TAPE	MAGNETIC DRUM	MAGNETIC DISK	ENTRANCE CONNECTOR	EXIT CONNECTOR	TERMINAL; INTERRUPTION	COMMENT; ANNOTATION
DISPLAY	COMMUNICATION LINK							

Crystal structures

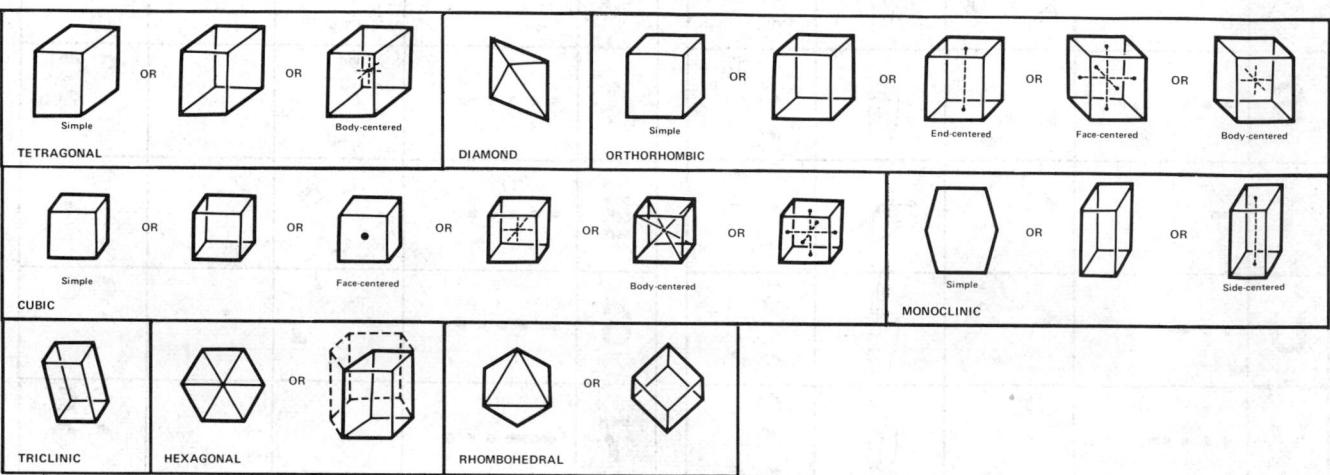

Meteorology

VISIBILITY reduced by smoke	HAZE	LIGHT FOG	HEAVY FOG; ICE FOG	DUST WHIRLS	DUST or SAND STORM	SHOWERS	HAIL	RAIN SHOWERS, moderate or heavy
TORNADO (Funnel Cloud)	TROPICAL STORM	HURRICANE	SQUALL	DRIZZLE	SLIGHT FREEZING DRIZZLE	RAIN SHOWERS, violent	SLIGHT SHOWERS of SNOW PELLETS	SLIGHT SHOWERS of HAIL
SLIGHT RAIN, INTERMITTENT	SLIGHT RAIN, CONTINUOUS	MODERATE RAIN, INTERMITTENT	MODERATE RAIN, CONTINUOUS	HEAVY RAIN, INTERMITTENT	HEAVY RAIN, CONTINUOUS	LIGHTNING	THUNDERSTORM	THUNDERSTORM moderate, with hail
PRECIPITATION during Past Hour	INCREASED Phenomenon during Past Hour	DECREASED Phenomenon during Past Hour	PRECIPITATION not REACHING GROUND	PRECIPITATION landing far from station	PRECIPITATION landing near station	THUNDERSTORM heavy, with hail	ICE PRISMS	SNOW GRAINS

Meteorology (cont.)

ICE PELLETS (Sleet)	SNOW	STARLIKE SNOW CRYSTALS	DRIFTING SNOW, slight to moderate	DRIFTING SNOW, heavy	BLOWING SNOW, slight to moderate	WIND calm	WIND approx. 1mph (1 Knot)	WIND approx. 6mph (5 Knots)
WIND approx. 12mph (10 Knots)	WIND approx. 58mph (50 Knots)	WARM FRONT, ALOFT	WARM FRONT, SURFACE	COLD FRONT, ALOFT	COLD FRONT, SURFACE	OCCLUDED FRONT, SURFACE	STATIONARY FRONT, SURFACE	CLEAR SKY
SCATTERED CLOUDS, 0.1 or less	SCATTERED CLOUDS, 0.2 or 0.3	SCATTERED CLOUDS, 0.4	SCATTERED CLOUDS, 0.5	BROKEN CLOUDS, 0.6 - 0.9	BROKEN CLOUDS, 0.6	BROKEN CLOUDS, 0.7 or 0.8	BROKEN CLOUDS, 0.9	OVERCAST
OVERCAST, COMPLETE	OVERCAST; SKY OBSCURED	STRATUS and/or FRACTOSTRATUS	FRACTOSTRATUS, Fractocumulus (Scud)	ALTOSTRATUS, thin, semi-transparent	ALTOSTRATUS, thick	STRATOCUMULUS, spreading from cumulus	STRATOCUMULUS, not from cumulus	CUMULUS, little vertical development
CUMULUS and STRATOCUMULUS	CUMULUS, considerable development	CUMULONIMBUS, clear-cut tops lacking	CUMULONIMBUS clear top	ALTOCUMULUS, thin, semi-transparent	ALTOCUMULUS, thin, patches	ALTOCUMULUS, in bands and thickening	ALTOCUMULUS, double-layered	CIRROCUMULUS
ALTOCUMULUS spreading from cumulus	ALTOCUMULUS, tufts or turrets	ALTOCUMULUS of chaotic sky	CIRRUS filaments (Mare's Tails)	CIRRUS, dense, patches, tufts	CIRRUS, dense, anvil shaped	CIRRUS, hook-shaped, thickening	CIRRUS and Cirrostratus, over 45°	CIRRUS and Cirrostratus, not 45°
CIRROSTRATUS, not increasing	CIRROSTRATUS, veil covering sky							

TAXONOMY OF BACTERIA

Bergey's Manual of Determinative Bacteriology is the standard reference work for taxonomic organization of bacteria. This dictionary observes the taxonomic organization presented in the seventh edition (R. S. Breed, E. G. D. Murray, N. R. Smith, eds.; Baltimore: Williams and Wilkins, 1957). However, the eighth edition (R. E. Buchanan and N. E. Gibbons, eds.; 1974), published while this dictionary was in press, shows a much modified organization. For comparison, the taxonomic schemes for both editions are outlined in the following pages.

This change in the microbiologists' view of identification and classification of bacteria has evolved slowly from the application of three sets of experimental-observational approaches: (1) the use of comparative cytology with the light microscope and staining methods to observe the behavior and form of that part of the nucleus that contains DNA; (2) the use of the electron microscope for the extension of comparative cytology to the level of the ultrastructure of the bacterial cell; and (3) the use of the techniques of biochemistry and biophysics to further define those features of cellular organization unique to bacteria.

The seventh edition of *Bergey's Manual* classifies bacteria in the form of a complete hierarchy, while the eighth abandons this approach and groups bacteria into 19 sections based on a few readily determined criteria, including structure, genetic data, and biochemical, nutritional, physiological, staining, and ecological characteristics. Many families, orders, and classes, have been eliminated and a large number of genera are considered to be of uncertain affiliation.

Outline from Bergey's eighth edition

The Phototrophic Bacteria
  Order I. *Rhodospirillales*
    Suborder: *Rhodospirillineae*
      Family I. *Rhodospirillaceae*
        Genus   I. *Rhodospirillum*
        Genus II. *Rhodopseudomonas*
        Genus III. *Rhodomicrobium*
      Family II. *Chromatiaceae*
        Genus   I. *Chromatium*
        Genus II. *Thiocystis*
        Genus III. *Thiosarcina*
        Genus IV. *Thiospirillum*
        Genus   V. *Thiocapsa*
        Genus VI. *Lamprocystis*
        Genus VII. *Thiodictyon*
        Genus VIII. *Thiopedia*
        Genus IX. *Amoebobacter*
        Genus   X. *Ectothiorhodospira*
    Suborder: *Chlorobiineae*
      Family III. *Chlorobiaceae*
        Genus   I. *Chlorobium*
        Genus II. *Prosthecochloris*
        Genus III. *Chloropseudomonas*
        Genus IV. *Pelodictyon*
        Genus   V. *Clathrochloris*
    Addenda
        Genus: *Chlorochromatium*
        Genus: *Cylindrogloea*
        Genus: *Chlorobacterium*
The Gliding Bacteria
  Order I. *Myxobacterales*
    Family I. *Myxococcaceae*
      Genus I. *Myxococcus*
    Family II. *Archangiaceae*
      Genus I. *Archangium*
    Family III. *Cystobacteraceae*
      Genus   I. *Cystobacter*
      Genus II. *Melittangium*
      Genus III. *Stigmatella*
    Family IV. *Polyangiaceae*
      Genus   I. *Polyangium*
      Genus II. *Nannocystis*
      Genus III. *Chondromyces*
  Order II. *Cytophagales*
    Family I. *Cytophagaceae*
      Genus   I. *Cytophaga*
      Genus II. *Flexibacter*
      Genus III. *Herpetosiphon*
      Genus IV. *Flexithrix*
      Genus   V. *Saprospira*
      Genus VI. *Sporocytophaga*
    Family II. *Beggiatoaceae*
      Genus   I. *Beggiatoa*
      Genus II. *Vitreoscilla*
      Genus III. *Thioploca*
    Family III. *Simonsiellaceae*
      Genus   I. *Simonsiella*
      Genus II. *Alysiella*
    Family IV. *Leucotrichaceae*
      Genus   I. *Leucothrix*
      Genus II. *Thiothrix*
    Incertae sedis
      Genus: *Toxothrix*
    Familiae incertae sedis
      Achromatiaceae
      Genus: *Achromatium*
      Pelonemataceae
      Genus   I. *Pelonema*
      Genus II. *Achroonema*
      Genus III. *Peloploca*

Genus IV. *Desmanthos*
The Sheathed Bacteria
      Genus: *Sphaerotilus*
      Genus: *Leptothrix*
      Genus: *Streptothrix*
      Genus: *Lieskeela*
      Genus: *Phragmidiothrix*
      Genus: *Chrenothrix*
      Genus: *Clonothrix*
Budding and/or Appendaged Bacteria
      Genus: *Hyphomicrobium*
      Genus: *Hyphomonas*
      Genus: *Pedomicrobium*
      Genus: *Caulobacter*
      Genus: *Asticcacaulis*
      Genus: *Ancalomicrobium*
      Genus: *Prosthecomicrobium*
      Genus: *Thiodendron*
      Genus: *Pasteuria*
      Genus: *Blastobacter*
      Genus: *Seliberia*
      Genus: *Gallionella*
      Genus: *Nevskia*
      Genus: *Planctomyces*
      Genus: *Metallogenium*
      Genus: *Caulococcus*
      Genus: *Kusnezonia*
The Spirochetes
  Order I. *Spirochaetales*
    Family I. *Spirochaetaceae*
      Genus   I. *Spirochaeta*
      Genus II. *Cristispira*
      Genus III. *Treponema*
      Genus IV. *Borrelia*
      Genus   V. *Leptospira*
Spiral and Curved Bacteria
    Family I. *Spirillaceae*
      Genus   I. *Spirillum*
      Genus II. *Compylobacter*
    Incertae sedis
      Genus: *Bdellovibrio*
      Genus: *Microcyclus*
      Genus: *Pelosigma*
      Genus: *Brachyarcus*
Gram-negative Aerobic Rods and Cocci
    Family   I. *Pseudomonadaceae*
      Genus   I. *Pseudomonas*
      Genus II. *Xanthomonas*
      Genus III. *Zoogloea*
      Genus IV. *Glauconobacter*
    Family II. *Azotobacteraceae*
      Genus   I. *Azotobacter*
      Genus II. *Azomonas*
      Genus III. *Beijerinckia*
      Genus IV. *Derxia*
    Family III. *Rhizobiaceae*
      Genus   I. *Rhizobium*
      Genus II. *Agrobacterium*
    Family IV. *Methylomonadaceae*
      Genus   I. *Methylomonas*
      Genus II. *Methylococcus*
    Family V. *Halobacteriaceae*
      Genus   I. *Halobacterium*
      Genus II. *Halococcus*
    Incertae sedis
      Genus: *Alcaligenes*
      Genus: *Acetobacter*
      Genus: *Brucella*
      Genus: *Bordetella*
      Genus: *Francisella*

      Genus: *Thermus*
Gram-negative Facultatively Anaerobic Rods
    Family I. *Enterobacteriaceae*
      Genus   I. *Escherichia*
      Genus   II. *Edwardsiella*
      Genus III. *Citrobacter*
      Genus   IV. *Salmonella*
      Genus   V. *Shigella*
      Genus   VI. *Klebsiella*
      Genus VII. *Enterobacter*
      Genus VIII. *Hafnia*
      Genus   IX. *Serratia*
      Genus   X. *Proteus*
      Genus   XI. *Yersinia*
      Genus XII. *Erwinia*
    Family II. *Vibrionaceae*
      Genus   I. *Vibrio*
      Genus II. *Aeromonas*
      Genus III. *Plesiomonas*
      Genus IV. *Photobacterium*
      Genus   V. *Lucibacterium*
    Incertae sedis
      Genus: *Zymomonas*
      Genus: *Chromobacterium*
      Genus: *Flavobacterium*
      Genus: *Haemophilus*
      Genus: *Pasteurella*
      Genus: *Actinobacillus*
      Genus: *Cardiobacterium*
      Genus: *Streptobacillus*
      Genus: *Calymmatobacterium*
Gram-negative Anaerobic Bacteria
    Family I. *Bacteroidaceae*
      Genus   I. *Bacteroides*
      Genus II. *Fusobacterium*
      Genus III. *Leptotrichia*
    Incertae sedis
      Genus: *Desulfovibrio*
      Genus: *Butyrivibrio*
      Genus: *Succinivibrio*
      Genus: *Succinimonas*
      Genus: *Lachnospira*
      Genus: *Selenomonas*
Gram-negative Cocci and Coccobacilli
    Family I. *Neisseriaceae*
      Genus   I. *Neisseria*
      Genus II. *Branhamella*
      Genus III. *Moraxella*
      Genus IV. *Acinetobacter*
    Incertae sedis
      Genus: *Paracoccus*
      Genus: *Lampropedia*
Gram-negative Anaerobic Cocci
    Family I. *Veillonellaceae*
      Genus   I. *Veillonella*
      Genus II. *Acidaminococcus*
      Genus III. *Megasphaera*
Gram-negative Chemolithotrophic Bacteria
    Family I. *Nitrobacteraceae*
      Genus   I. *Nitrobacter*
      Genus II. *Nitrospina*
      Genus III. *Nitrococcus*
      Genus IV. *Nitrosomonas*
      Genus   V. *Nitrospira*
      Genus VI. *Nitrosococcus*
      Genus VII. *Nitrosolobus*
Organisms Metabolizing Sulfur
      Genus 1. *Thiobacillus*
      Genus 2. *Sulfolobus*
      Genus 3. *Thiobacterium*

Genus 4. *Macromonas*
Genus 5. *Thiovulum*
Genus 6. *Thiospira*
Family II. *Siderocapsaceae*
  Genus I. *Siderocapsa*
  Genus II. *Ochrobium*
  Genus III. *Siderococcus*
Methane-producing Bacteria
  Family I. *Methanobacteriaceae*
    Genus I. *Methanobacterium*
    Genus II. *Methanosarcina*
    Genus III. *Methanococcus*
Gram-positive Cocci
  Family I. *Micrococcaceae*
    Genus I. *Micrococcus*
    Genus II. *Staphylococcus*
    Genus III. *Planococcus*
  Family II. *Streptococcaceae*
    Genus I. *Streptococcus*
    Genus II. *Leuconostoc*
    Genus III. *Pediococcus*
    Genus IV. *Aerococcus*
    Genus V. *Gemella*
  Family III. *Peptococcaceae*
    Genus I. *Peptococcus*
    Genus II. *Peptostreptococcus*
    Genus III. *Ruminococcus*
    Genus IV. *Sarcina*
Endospore-forming Rods and Cocci
  Family I. *Bacillaceae*
    Genus I. *Bacillus*
    Genus II. *Sporolactobacillus*
    Genus III. *Clostridum*
    Genus IV. *Desulfotomaculum*
    Genus V. *Sporosarcina*
  Incertae sedis
    Genus: *Oscillospira*
Gram-positive, Asporogenous Rod-shaped Bacteria
  Family I. *Lactobacillaceae*
    Genus I. *Lactobacillus*
  Incertae sedis
    Genus: *Listeria*
    Genus: *Erysipelothrix*
    Genus: *Caryophanon*
Actinomycetes and Related Organisms
  Coryneform Group of Bacteria

Genus I. *Corynebacterium*
Genus II. *Arthrobacter*
Incertae sedis
  Genus A. *Brevibacterium*
  Genus B. *Microbacterium*
  Genus III. *Cellulomonas*
  Genus IV. *Kurthia*
Family I. *Propionibacteriaceae*
  Genus I. *Propionibacterium*
  Genus II. *Eubacterium*
Order I. *Actinomycetales*
  Family I. *Actinomycetaceae*
    Genus I. *Actinomyces*
    Genus II. *Arachnia*
    Genus III. *Bifidobacterium*
    Genus IV. *Bacterionema*
    Genus V. *Rothia*
  Family II. *Mycobacteriaceae*
    Genus I. *Mycobacterium*
  Family III. *Frankiaceae*
    Genus I. *Frankia*
  Family IV. *Actinoplanaceae*
    Genus I. *Actinoplanes*
    Genus II. *Spirillospora*
    Genus III. *Streptosporangium*
    Genus IV. *Amorphosporangium*
    Genus V. *Ampullariella*
    Genus VI. *Pilemelia*
    Genus VII. *Planomonospora*
    Genus VIII. *Planobispora*
    Genus IX. *Dactylosporangium*
    Genus X. *Kitastoa*
  Family V. *Dermatophilaceae*
    Genus I. *Dermatophilus*
    Genus II. *Geodermatophilus*
  Family VI. *Nocardiaceae*
    Genus I. *Nocardia*
    Genus II. *Pseudonocardia*
  Family VII. *Streptomycetaceae*
    Genus I. *Streptomyces*
    Genus II. *Streptoverticillium*
    Genus III. *Sporichthya*
    Genus IV. *Microellobosporia*
  Family VIII. *Micromonosporaceae*
    Genus I. *Micromonospora*
    Genus II. *Thermoactinomyces*

Genus III. *Actinobifida*
Genus IV. *Thermomonospora*
Genus V. *Microbispora*
Genus VI. *Micropolyspora*
The Rickettsias
  Order I. *Rickettsiales*
    Family: *Rickettsiaceae*
      Tribe I. *Rickettsieae*
        Genus I. *Rickettsia*
        Genus II. *Rochalimaea*
        Genus III. *Coxiella*
      Tribe II. *Ehrlichieae*
        Genus IV. *Ehrlichia*
        Genus V. *Cowdria*
        Genus VI. *Neorickettsia*
      Tribe III. *Wolbachieae*
        Genus VII. *Wolbachia*
        Genus VIII. *Symbiotes*
        Genus IX. *Blattabacterium*
        Genus X. *Rickettsiella*
    Family: *Bartonellaceae*
      Genus I. *Bartonella*
      Genus II. *Grahamella*
    Family: *Anaplasmataceae*
      Genus I. *Anaplasma*
      Genus II. *Paranaplasma*
      Genus III. *Aegyptianella*
      Genus IV. *Haemobartonella*
      Genus V. *Eperythrozoon*
  Order II. *Chlamydiales*
    Family I. *Chlamydiaceae*
      Genus I. *Chlamydia*
The Mycoplasmas
  Class: *Mollicutes*
    Order I. *Mycoplasmatales*
      Family I. *Mycoplasmataceae*
        Genus I. *Mycoplasma*
      Family II. *Acholeplasmataceae*
        Genus I. *Acholeplasma*
    Incertae sedis
      Genus: *Thermoplasma*
  Incertae sedis
    Genus: *Spiroplasma*

Outline from Bergey's seventh edition

Division I. *Protophyta*
  Class I. *Schizophyceae*
  Class II. *Schizomycetes*
    Order I. *Pseudomonadales*
      Suborder I. *Rhodobacteriineae*
        Family I. *Thiorhodaceae*
          Genus I. *Thiosarcina*
          Genus II. *Thiopedia*
          Genus III. *Thiocapsa*
          Genus IV. *Thiodictyon*
          Genus V. *Thiothece*
          Genus VI. *Thiocystis*
          Genus VII. *Lamprocystis*
          Genus VIII. *Amoebobacter*
          Genus IX. *Thiopolycoccus*
          Genus X. *Thiospirillum*
          Genus XI. *Rhabdomonas*
          Genus XII. *Rhodothece*
          Genus XIII. *Chromatium*
        Family II. *Athiorhodaceae*
          Genus I. *Rhodopseudomonas*
          Genus II. *Rhodospirillum*
        Family III. *Chlorobacteriaceae*
          Genus I. *Chlorobium*

Genus II. *Pelodictyon*
Genus III. *Clathrochloris*
Genus IV. *Chlorobacterium*
Genus V. *Chlorochromatium*
Genus VI. *Cylindrogloea*
Suborder II. *Pseudomonadineae*
  Family I. *Nitrobacteraceae*
    Genus I. *Nitrosomonas*
    Genus II. *Nitrosococcus*
    Genus III. *Nitrosospira*
    Genus IV. *Nitrosocystis*
    Genus V. *Nitrosogloea*
    Genus VI. *Nitrobacter*
    Genus VII. *Nitrocystis*
  Family II. *Methanomonadaceae*
    Genus I. *Methanomonas*
    Genus II. *Hydrogenomonas*
    Genus III. *Carboxydomonas*
  Family III. *Thiobacteriaceae*
    Genus I. *Thiobacterium*
    Genus II. *Macromonas*
    Genus III. *Thiovulum*
    Genus IV. *Thiospira*
    Genus V. *Thiobacillus*

Family IV. *Pseudomonadaceae*
  Genus I. *Pseudomonas*
  Genus II. *Xanthomonas*
  Genus III. *Acetobacter*
  Genus IV. *Aeromonas*
  Genus V. *Photobacterium*
  Genus VI. *Azotomonas*
  Genus VII. *Zymomonas*
  Genus VIII. *Protaminobacter*
  Genus IX. *Alginomonas*
  Genus X. *Mycoplana*
  Genus XI. *Zoogloea*
  Genus XII. *Halobacterium*
Family V. *Caulobacteraceae*
  Genus I. *Caulobacter*
  Genus II. *Gallionella*
  Genus III. *Siderophacus*
  Genus IV. *Nevskia*
Family VI. *Siderocapsaceae*
  Genus I. *Siderocapsa*
  Genus II. *Siderosphaera*
  Genus III. *Sideronema*
  Genus IV. *Ferribacterium*
  Genus V. *Sideromonas*

Genus VI. *Naumanniella*
Genus VII. *Ochrobium*
Genus VIII. *Siderococcus*
Genus IX. *Siderobacter*
Genus X. *Ferrobacillus*
Family VII. *Spirillaceae*
Genus I. *Vibrio*
Genus II. *Desulfovibrio*
Genus III. *Methanobacterium*
Genus IV. *Cellvibrio*
Genus V. *Cellfalcicula*
Genus VI. *Microcyclus*
Genus VII. *Spirillum*
Genus VIII. *Paraspirillum*
Genus IX. *Selenomonas*
Genus X. *Myconostoc*
Order II. *Chlamydobacteriales*
Family I. *Chlamydobacteriaceae*
Genus I. *Sphaerotilus*
Genus II. *Leptothrix*
Genus III. *Toxothrix*
Family II. *Peloplocaceae*
Genus I. *Peloploca*
Genus II. *Pelonema*
Family III. *Crenotrichaceae*
Genus I. *Crenothrix*
Genus II. *Phragmidiothrix*
Genus III. *Clonothrix*
Order III. *Hyphomicrobiales*
Family I. *Hyphomicrobiaceae*
Genus I. *Hyphomicrobium*
Genus II. *Rhodomicrobium*
Family II. *Pasteuriaceae*
Genus I. *Pasteuria*
Genus II. *Blastocaulis*
Order IV. *Eubacteriales*
Family I. *Azotobacteraceae*
Genus I. *Azotobacter*
Family II. *Rhizobiaceae*
Genus I. *Rhizobium*
Genus II. *Agrobacterium*
Genus III. *Chromobacterium*
Family III. *Achromobacteraceae*
Genus I. *Alcaligenes*
Genus II. *Achromobacter* ·
Genus III. *Flavobacterium*
Genus IV. *Agarbacterium*
Genus V. *Beneckea*
Family IV. *Enterobacteriaceae*
Tribe I. *Escherichieae*
Genus I. *Escherichia*
Genus II. *Aerobacter*
Genus III. *Klebsiella*
Genus IV. *Paracolobactrum*
Genus V. *Alginobacter*
Tribe II. *Erwinieae*
Genus VI. *Erwinia*
Tribe III. *Serratieae*
Genus VII. *Serratia*
Tribe IV. *Proteeae*
Genus VIII. *Proteus*
Tribe V. *Salmonelleae*
Genus IX. *Salmonella*
Genus X. *Shigella*
Family V. *Brucellaceae*
Genus I. *Pasturella*
Genus II. *Bordetella*
Genus III. *Brucella*
Genus IV. *Haemophilus*
Genus V. *Actinobacillus*
Genus VI. *Calymmato-
     bacterium*
Genus VII. *Morazella*
Genus VIII. *Noguchia*
Family VI. *Bacteroidaceae*
Genus I. *Bacteroides*

Genus II. *Fusobacterium*
Genus III. *Dialister*
Genus IV. *Sphaerophorus*
Genus V. *Streptobacillus*
Family VII. *Micrococcaceae*
Genus I. *Micrococcus*
Genus II. *Staphylococcus*
Genus III. *Gaffkya*
Genus IV. *Sarcina*
Subgenus I. *Zymosarcina*
Subgenus II. *Methanosarcina*
Subgenus III. *Sarcinococcus*
Subgenus IV. *Urosarcina*
Genus V. *Methanococcus*
Genus VI. *Peptococcus*
Family VIII. *Neisseriaceae*
Genus I. *Neisseria*
Genus II. *Veillonella*
Family IX. *Brevibacteriaceae*
Genus I. *Brevibacterium*
Genus II. *Kurthia*
Family X. *Lactobacillaceae*
Tribe I. *Streptococceae*
Genus I. *Diplococcus*
Genus II. *Streptococcus*
Genus III. *Pediococcus*
Genus IV. *Leuconostoc*
Genus V. *Peptostreptococcus*
Tribe II. *Lactobacilleae*
Genus I. *Lactobacillus*
Subgenus I. *Lactobacillus*
Subgenus II. *Saccharobacillus*
Genus II. *Eubacterium*
Genus III. *Catenabacterium*
Genus IV. *Ramibacterium*
Genus V. *Cillobacterium*
Family XI. *Propionibacteriaceae*
Genus I. *Propionibacterium*
Genus II. *Butyribacterium*
Genus III. *Zymobacterium*
Family XII. *Corynebacteriaceae*
Genus I. *Corynebacterium*
Genus II. *Listeria*
Genus III. *Erysipelothrix*
Genus IV. *Microbacterium*
Genus V. *Cellulomonas*
Genus VI. *Arthrobacter*
Family XIII. *Bacillaceae*
Genus I. *Bacillus*
Genus II. *Clostridium*
Order V. *Caryophanales*
Family I. *Caryophanaceae*
Genus I. *Caryophanon*
Genus II. *Lineola*
Genus III. *Simonsiella*
Family II. *Oscillospiraceae*
Genus I. *Oscillospira*
Family III. *Arthromitaceae*
Genus I. *Arthromitus*
Genus II. *Coleomitus*
Order VI. *Actinomycetales*
Family I. *Mycobacteriaceae*
Genus I. *Mycobacterium*
Genus II. *Mycococcus*
Family II. *Actinomycetaceae*
Genus I. *Nocardia*
Genus II. *Actinomyces*
Family III. *Streptomycetaceae*
Genus I. *Streptomyces*
Genus II. *Micromonospora*
Genus III. *Thermoactinomyces*
Family IV. *Actinoplanaceae*
Genus I. *Actinoplanes*
Genus II. *Streptosporangium*
Order VII. *Beggiatoales*
Family I. *Beggiatoaceae*

Genus I. *Beggiatoa*
Genus II. *Thiospirillopsis*
Genus III. *Thioploca*
Genus IV. *Thiothrix*
Family II. *Vitreoscillaceae*
Genus I. *Vitreoscilla*
Genus II. *Bactoscilla*
Genus III. *Microscilla*
Family III. *Leucotrichaceae*
Genus I. *Leucothrix*
Family IV. *Achromatiaceae*
Genus I. *Achromatium*
Order VIII. *Myxobacterales*
Family I. *Cytophagaceae*
Genus I. *Cytophaga*
Family II. *Archangiaceae*
Genus I. *Archangium*
Genus II. *Stelangium*
Family III. *Sorangiaceae*
Genus I. *Sorangium*
Family IV. *Polyangiaceae*
Genus I. *Polyangium*
Genus II. *Synangium*
Genus III. *Podangium*
Genus IV. *Chondromyces*
Family V. *Myxococcaceae*
Genus I. *Myxococcus*
Genus II. *Chondrococcus*
Genus III. *Angiococcus*
Genus IV. *Sporocytophaga*
Order IX. *Spirochaetales*
Family I. *Spirochaetaceae*
Genus I. *Spirochaeta*
Genus II. *Saprospira*
Genus III. *Cristispira*
Family II. *Treponemataceae*
Genus I. *Borrelia*
Genus II. *Treponema*
Genus III. *Leptospira*
Order X. *Mycoplasmatales*
Family I. *Mycoplasmataceae*
Genus I. *Mycoplasma*
Addendum to Class II: *Schizomycetes*
Class III. *Microtatobiotes*
Order I. *Rickettsiales*
Family I. *Rickettsiaceae*
Tribe I. *Rickettsieae*
Genus I. *Rickettsia*
Subgenus A. *Rickettsia*
Subgenus B. *Zinssera*
Subgenus C. *Dermacentroxenus*
Subgenus D. *Rochalimaea*
Genus II. *Coxiella*
Tribe II. *Ehrlichieae*
Genus III. *Ehrlichia*
Genus IV. *Cowdria*
Genus V. *Neorickettsia*
Tribe III. *Wolbachieae*
Genus VI. *Wolbachia*
Genus VII. *Symbiotes*
Genus VIII. *Rickettsiella*
Family II. *Chlamydiaceae*
Genus I. *Chlamydia*
Genus II. *Colesiota*
Genus III. *Ricolesia*
Genus IV. *Colettsia*
Genus V. *Miyagawanella*
Family III. *Bartonellaceae*
Genus I. *Bartonella*
Genus II. *Grahamella*
Genus III. *Haemobartonella*
Genus IV. *Eperythrozoon*
Family IV. *Anaplasmataceae*
Genus I. *Anaplasma*
Order II. *Virales*